Y0-BXH-234

EPA/NIH Mass Spectral Data Base

Volume 1
Molecular Weights 30-186

S.R. Heller

Office of Planning and Management
Environmental Protection Agency
Washington, D.C. 20460

and

G.W.A. Milne

National Heart, Lung, and Blood Institute
National Institutes of Health
Bethesda, Maryland 20014

U.S. DEPARTMENT OF COMMERCE, Juanita M. Kreps, Secretary

Dr. Sidney Harman, Under Secretary

Jordan J. Baruch, Assistant Secretary for Science and Technology

NATIONAL BUREAU OF STANDARDS, Ernest Ambler, Director

Issued December 1978

Library of Congress Cataloging in Publication Data

Heller, S.R.
EPA/NIH mass spectral data base.

(Nat. stand. ref. data ser. ; NSRDS-NBS 63)
"National Standard Reference Data System."
Supt. of Docs. no.: C 13.48:63.
CONTENTS: v. 1. Molecular weights 30-186. —v. 2. Molecular weights 186-273. — v. 3. Molecular weights 273-381. [etc.]
1. Organic compounds—Spectra—Indexes. 2. Mass spectrometry—Indexes. 3. United States. Environmental Protection Agency—Indexes. 4. United States. National Institutes of Health—Indexes. I. Milne, George W.A., 1937– joint author. II. Title. III. Series: United States. National Bureau of Standards. National Standard Reference Data Series; NSRDS-NBS 63.
QC100.U573 no. 63 [QC462.85] 602'.1s [547'.308'7]
78-606175

NSRDS-NBS 63

Nat. Stand. Ref. Data Ser., Nat. Bur. Stand.(U.S.), 63, Vol. 3, 1001 pages (Dec. 1978)
CODEN: NSRDAP

U.S. GOVERNMENT PRINTING OFFICE
WASHINGTON: 1978

For sale by the Superintendent of Documents, U.S. Governmnent Printing Office, Washington, D.C. 20402
(Sold only in sets.)
Stock No. 003-003-01987-9

Foreword

The National Standard Reference Data System provides access to the quantitative data of physical science, critically evaluated and compiled for convenience and readily accessible through a variety of distribution channels. The System was established in 1963 by action of the President's Office of Science and Technology and the Federal Council for Science and Technology, and responsibility to administer it was assigned to the National Bureau of Standards.

NSRDS receives advice and planning assistance from a Review Committee of the National Research Council of the National Academy of Sciences-National Academy of Engineering. A number of Advisory Panels, each concerned with a single technical area, meet regularly to examine major portions of the program, assign relative priorities, and identify specific key problems in need of further attention. For selected specific topics, the Advisory Panels sponsor subpanels which make detailed studies of users' needs, the present state of knowledge, and existing data resources as a basis for recommending one or more data compilation activities. This assembly of advisory services contributes greatly to the guidance of NSRDS activities.

The System now includes a complex of data centers and other activities in academic institutions and other laboratories. Components of the NSRDS produce compilations of critically evaluated data, reviews of the state of quantitative knowledge in specialized areas, and computations of useful functions derived from standard reference data. The centers and projects also establish criteria for evaluation and compilation of data and recommend improvements in experimental techniques. They are normally associated with research in the relevant field.

The technical scope of NSRDS is indicated by the categories of projects active or being planned: nuclear properties, atomic and molecular properties, solid state properties, thermodynamic and transport properties, chemical kinetics, and colloid and surface properties.

Reliable data on the properties of matter and materials are a major foundation of scientific and technical progress. Such important activities as basic scientific research, industrial quality control, development of new materials for building and other technologies, measuring and correcting environmental pollution depend on quality reference data. In NSRDS, the Bureau's responsibility to support American science, industry, and commerce is vitally fulfilled.

E. Ambler.

ERNEST AMBLER, *Director*

Preface

In carrying out its mandate to provide reliable reference data needed by the U.S. technical community, the Office of Standard Reference Data works closely with other Agencies of the Federal Government which are concerned with data bases in support of their mission responsibilities. The present publication is an example of this collaboration. The collection of mass spectral data contained herein represents a major effort of the Environmental Protection Agency and the National Institutes of Health. The National Bureau of Standards is participating in this effort by helping to assure adequate quality control of the data and by providing suitable dissemination mechanisms.

This is a continuing program; both an expansion of the size of the mass spectral data base and an enhancement of the quality of the data are planned. Contributions of new or improved spectra, as well as general comments on the format and utility of the data, are strongly encouraged.

DAVID R. LIDE, JR., *Chief*
Office of Standard Reference Data
National Bureau of Standards

Contents

EPA/NIH Mass Spectral Data Base

S. R. Heller

Environmental Protection Agency, Washington, D.C. 20460

and

G. W. A. Milne

National Institutes of Health, Bethesda, Maryland 20014

This publication presents a collection of 25,556 verified mass spectra of individual substances compiled from the EPA/NIH mass spectral file. The spectra are given in bar graph format over the full mass range. Each spectrum is accompanied by a Chemical Abstracts Index substance name, molecular formula, molecular weight, structural formula, and Chemical Abstracts Service Registry Number.

Key words: Analytical data; mass spectra; organic substances; verified spectra.

1. Introduction

During the last seven years, two agencies of the United States Government, the Environmental Protection Agency (EPA) and the National Institutes of Health (NIH), have developed a mass spectrometry data base together with computer software to search this data base. The Mass Spectrometry Data Centre (MSDC), supported by the United Kingdom Department of Industry, has collaborated in this project. A major objective of the project has been to provide a means of rapid identification of unknown chemical substances by matching their mass spectra against a comprehensive library of verified spectra.

The result of this international cooperative effort has been a machine-readable data base of mass spectra of 25,556 different compounds. This EPA/NIH Mass Spectral Data Base is available in two ways: through lease of a magnetic tape from the National Bureau of Standards (NBS) [1][1] and through interactive access via an international time-sharing computer network [2]. The magnetic tape contains the same data listed herein. They include the spectrum, and indexes listing name, formula, molecular weight, and the CAS Registry Number.

While many organizations and laboratories have been well served by one or the other of these mechanisms for the dissemination of mass spectral data, neither of the approaches was suitable for those lacking access to computers or computer terminals. Hence, at the suggestion of EPA's Industrial Environmental Research Laboratory (IERL) in Research Triangle Park, North Carolina, we undertook to prepare this book of mass spectra in bar-graph format derived from the EPA/NIH Mass Spectral Data Base.

This publication, which has been produced by the Photocomposition System APS-IV [3] of the Chemical Abstracts Service (CAS) under contract to EPA (Contract 68-01-2731), differs in a number of significant ways from earlier published collections of mass spectral data such as the Eight Peak Index of Mass Spectra [4], the Registry of Mass Spectral Data [5], and the Compilation of Mass Spectral Data [6]. The 25,556 mass spectra herein are of different substances; there are no duplicate spectra. Neither are there any spectra of substances labelled with less-abundant isotopes (e.g., deuterium, oxygen-18).

The current data base was derived from an initial file of about 48,000 mass spectra. This larger file was processed by CAS, which identified the CAS Registry Number for every substance. These CAS Registry Numbers are unique numerical identifiers assigned to chemical substances in the CAS Chemical Registry System [7]. The Chemical Registry System identifies a chemical substance on the basis of an unambiguous, computer-language description of its molecular structure. Currently, there are some 4 million chemical substances in the master file of the Registry System which was begun in 1965. Approximately 8,000 of these substances were first entered into the Registry System during the processing of the EPA/NIH Mass Spectral Data Base, i.e., these substances had not previously been reported in the open literature.

Once the Registry Numbers for all substances in the file of 48,000 spectra were identified, all the duplicate entries in the file were found using these Registry Numbers. Finally, the best spectrum from each group of duplicate spectra was selected and the resultant data base is presented in this book. The Quality Index of a mass spectrum is calculated using an algorithm devised by McLafferty [8] which detects common errors in mass spectra. This index takes account of 9 different parameters including such items as ionization conditions, impurity peaks, etc.

All the 25,556 substances whose spectra are given in this book are identified by their respective CAS Registry Numbers and the name under which the substances appear in the CAS 8th (1967-1971) or 9th (1972-1976) Collective Indexes—the "8CI" or "9CI" name. Synonyms from the CAS Registry System, as well as the Collective Index names, are used in the indexes. For further information regarding the Collective Index names, the reader is referred to the CAS Index Guide [9].

2. Presentation of Spectra

Spectra are presented as shown in figure 1. The conventional bar-graph format is used in "boxes" of width 150 mass units.

[1] Figures in brackets indicate literature references.

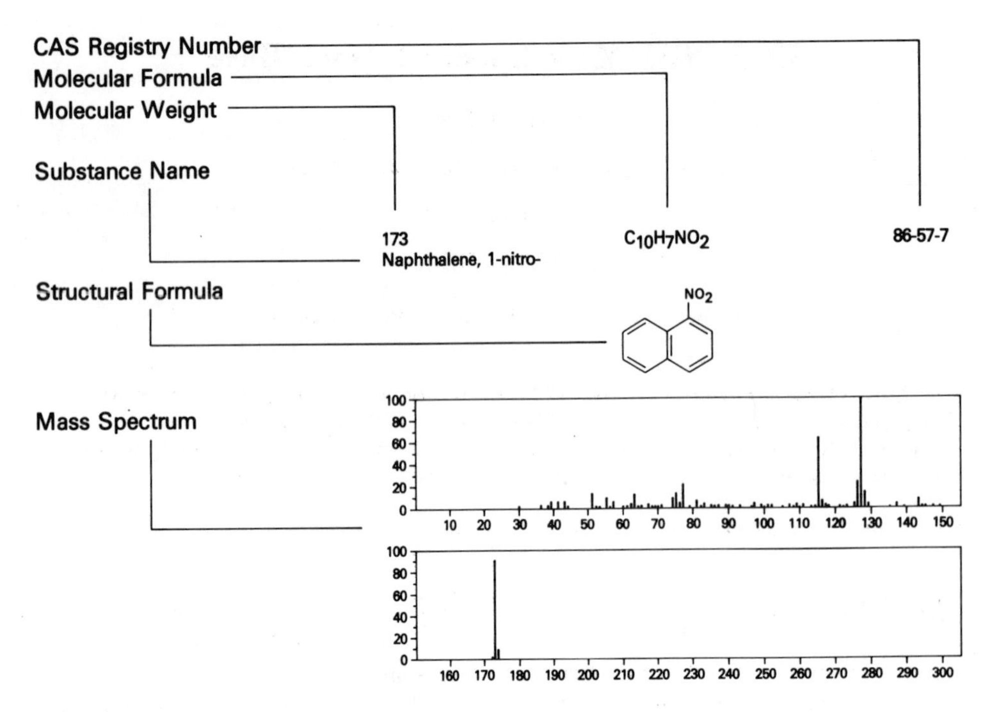

FIGURE 1. Typical mass spectrum.

Additional boxes are provided only if there are peaks in a broader mass range. Each spectrum is also accompanied by the molecular formula of the substance, the molecular weight, and the structure diagram. The molecular formulas are ordered by the Hill convention [10], and were derived by computer from the CAS-supplied connection table record. Molecular weights were derived by computer from the same records, using the integral atomic masses C=12, H=1, O=16, and so on. The structure diagrams, a unique feature of this book, are also machine derived from the connection table record for each substance [11].

The spectra reported in this book do not contain any metastable or multiple-charge ions, and all *m/e* and intensity data are rounded off to the nearest integer. Relative intensities between 0.1% and 2.0% are arbitrarily adjusted to 2% for the sake of clarity. It should be noted that the relative intensities of the peaks are sensitive to instrumental conditions and can only be taken as a rough guide. Spectra are presented in order of the molecular weights of the corresponding substances. In cases where substances have the same (integral) molecular weight, those spectra are further ordered according to molecular formula and, if necessary, further ordered by CAS Registry Number, in increasing order of the first set of digits.

The mass spectra themselves have been checked before entry into the file, and many further corrections have been made as a result of the registration processing by the CAS. In addition, a large number of changes have been made as a result of comments from the many users of the computerized Mass Spectral Search System and users of the tape copies of the data base supplied by NBS.

The four indexes provided in the final volume of this book may be used to find the spectrum of a specific substance, as opposed to finding the substance that might give a specific mass spectrum. The first of these gives the substance name in alphabetical order, the second is arranged by formula, the third is ordered by increasing molecular weight, and the fourth by CAS Registry Number. It is planned to produce further volumes of this compilation as the data base increases in size. When this is done, completely new indexes will be produced and may be used to replace the older versions. It is for this reason that the indexes are in a separate volume and paper bound rather than cloth bound. It is expected that this will facilitate additions and revisions to the data.

It is hoped that this book and the subsequent additions to it will be of value to the mass spectrometry community. Your comments, criticisms, and suggestions are most welcome and will certainly lead to further improvements in future volumes.

3. Acknowledgments

It is a pleasure to acknowledge, first, the large number of mass spectroscopists, too many by far to be identified individually, who labored to produce the mass spectra presented in these volumes.

The impetus to publish this book came from Joe McSorley and Gene Tucker of EPA-IERL, Research Triangle Park, North Carolina, who were responsible for providing the necessary funding to develop the software used in the production of the book, and also defrayed some of the production costs. The remainder of the production costs were provided by EPA-MIDSD, and in this connection, we thank Willis Greenstreet, Morris Yaguda, and Mike Springer for their support. The printing of the books was carried out by the Government Printing Office and the cost of printing was borne by NBS. Further EPA financial and technical support was provided by William Budde and John McGuire of EPA's Office of Research and Development. Computer support for the management of the data base has been provided by NIH-NHLBI. The master mass spectra file was assembled and is being maintained by Alvin Fein, Gary Marquart, and David Martinsen of Fein-Marquart Associates, Inc.

We are also very grateful to the many people responsible for technical improvements in the data base. Included among these are: Klaus Biemann, Alice Bridy, Henry Fales, Rachelle Heller, David Maxwell, Andrew McCormick, and Fred McLafferty.

4. References

[1] For further information regarding the leasing of this magnetic tape, please contact:
Office of Standard Reference Data
National Bureau of Standards
A537 Administration Building
Washington, D.C. 20234

[2] For further information regarding access to the Mass Spectral Search System, please contact:
Dr. H. J. Bernstein
Department of Chemistry
Brookhaven National Laboratory
Upton, New York 11973

[3] Blake, J. E., Farmer, N., and Haines, R. C., J. Chem. Inf. Comput. Sci. **17**, 223 (1977).

[4] Eight Peak Index of Mass Spectra, 2nd Edition, Her Majesty's Stationery Office, London (1975).

[5] Cornu, A., and Massot, R., Compilation of Mass Spectral Data, 2nd Edition, Heyden, London (1975).

[6] Stenhagen, E., Abrahamsson, S., and McLafferty, F. W., Registry of Mass Spectral Data, 1st Edition, Wiley, New York (1974).

[7] CA Volume 76-85 Cumulative Index Guide (1972-1976), Chemical Abstracts Service.

[8] Speck, D. D., Venkataraghavan, R., and McLafferty, F. W., J. Org. Mass Spectrom. **13**, 209 (1978).

[9] CAS Index Guide, Chem. Abs. **76**, (1972).

[10] Hill, E. A., J. Amer. Chem. Soc. **22**, 478 (1900).

[11] Dittmar, P. G., et al., J. Chem. Inf. Comput. Sci. **17**, 186 (1977).

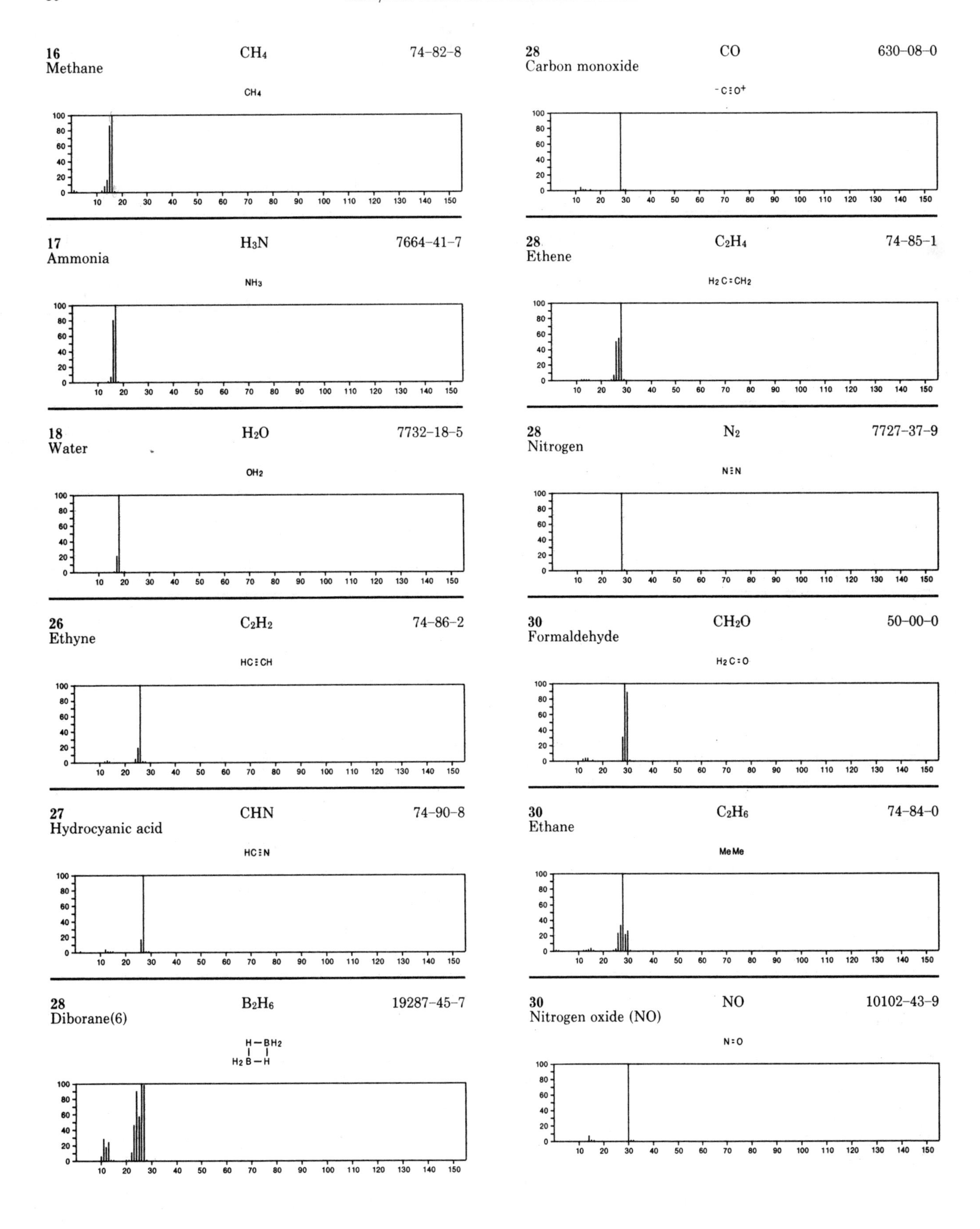
16
Methane
CH_4
74-82-8
CH4
17
Ammonia
H_3N
7664-41-7
NH3
18
Water
H_2O
7732-18-5
OH2
26
Ethyne
C_2H_2
74-86-2
HC≡CH
27
Hydrocyanic acid
CHN
74-90-8
HC≡N
28
Diborane(6)
B_2H_6
19287-45-7
28
Carbon monoxide
CO
630-08-0
28
Ethene
C_2H_4
74-85-1
H2C=CH2
28
Nitrogen
N_2
7727-37-9
N≡N
30
Formaldehyde
CH_2O
50-00-0
H2C=O
30
Ethane
C_2H_6
74-84-0
MeMe
30
Nitrogen oxide (NO)
NO
10102-43-9
N=O

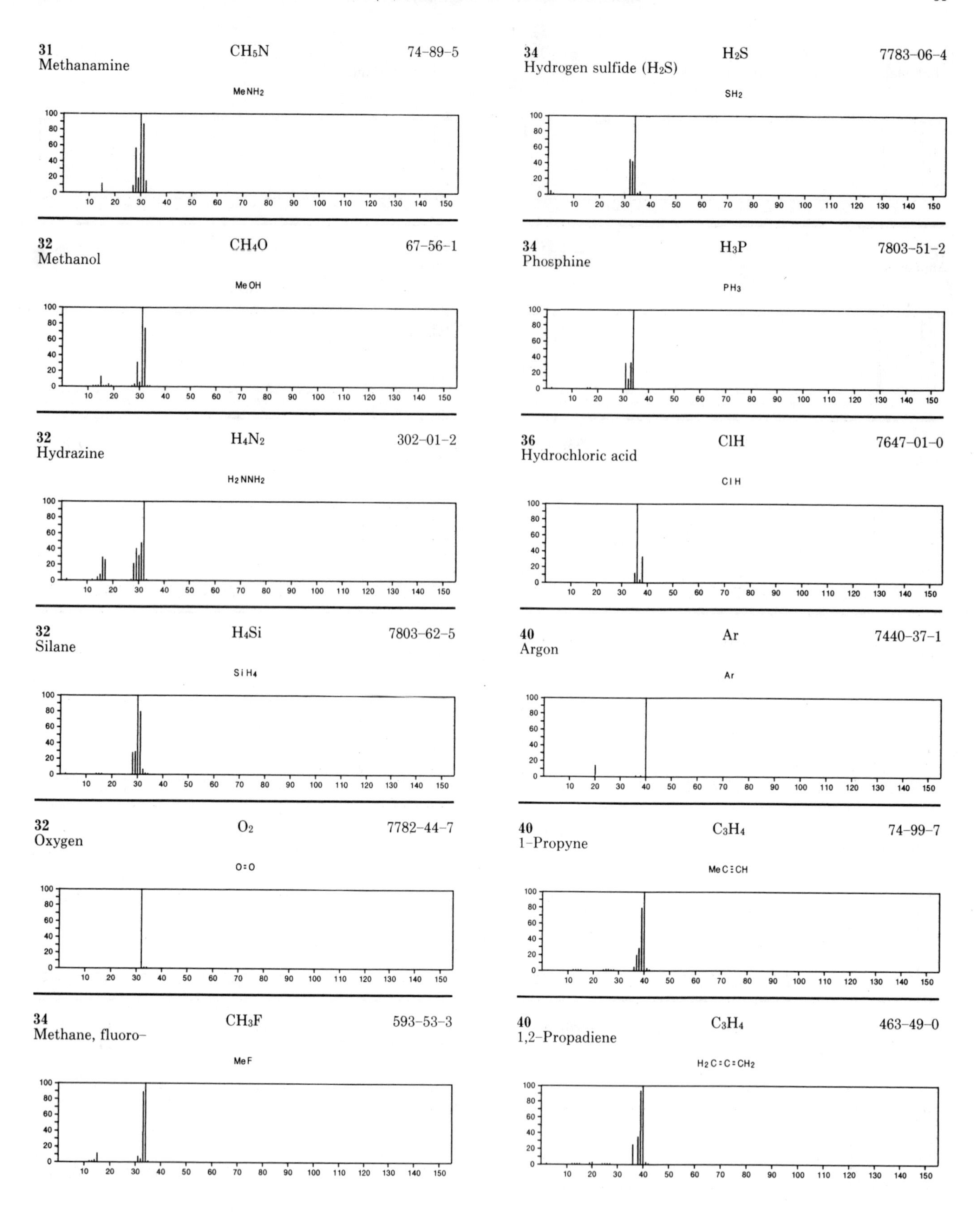
31
Methanamine
CH5N
74-89-5
MeNH2
34
Hydrogen sulfide (H2S)
H2S
7783-06-4
SH2
32
Methanol
CH4O
67-56-1
MeOH
34
Phosphine
H3P
7803-51-2
PH3
32
Hydrazine
H4N2
302-01-2
H2NNH2
36
Hydrochloric acid
ClH
7647-01-0
ClH
32
Silane
H4Si
7803-62-5
SiH4
40
Argon
Ar
7440-37-1
Ar
32
Oxygen
O2
7782-44-7
O=O
40
1-Propyne
C3H4
74-99-7
MeC≡CH
34
Methane, fluoro-
CH3F
593-53-3
MeF
40
1,2-Propadiene
C3H4
463-49-0
H2C=C=CH2

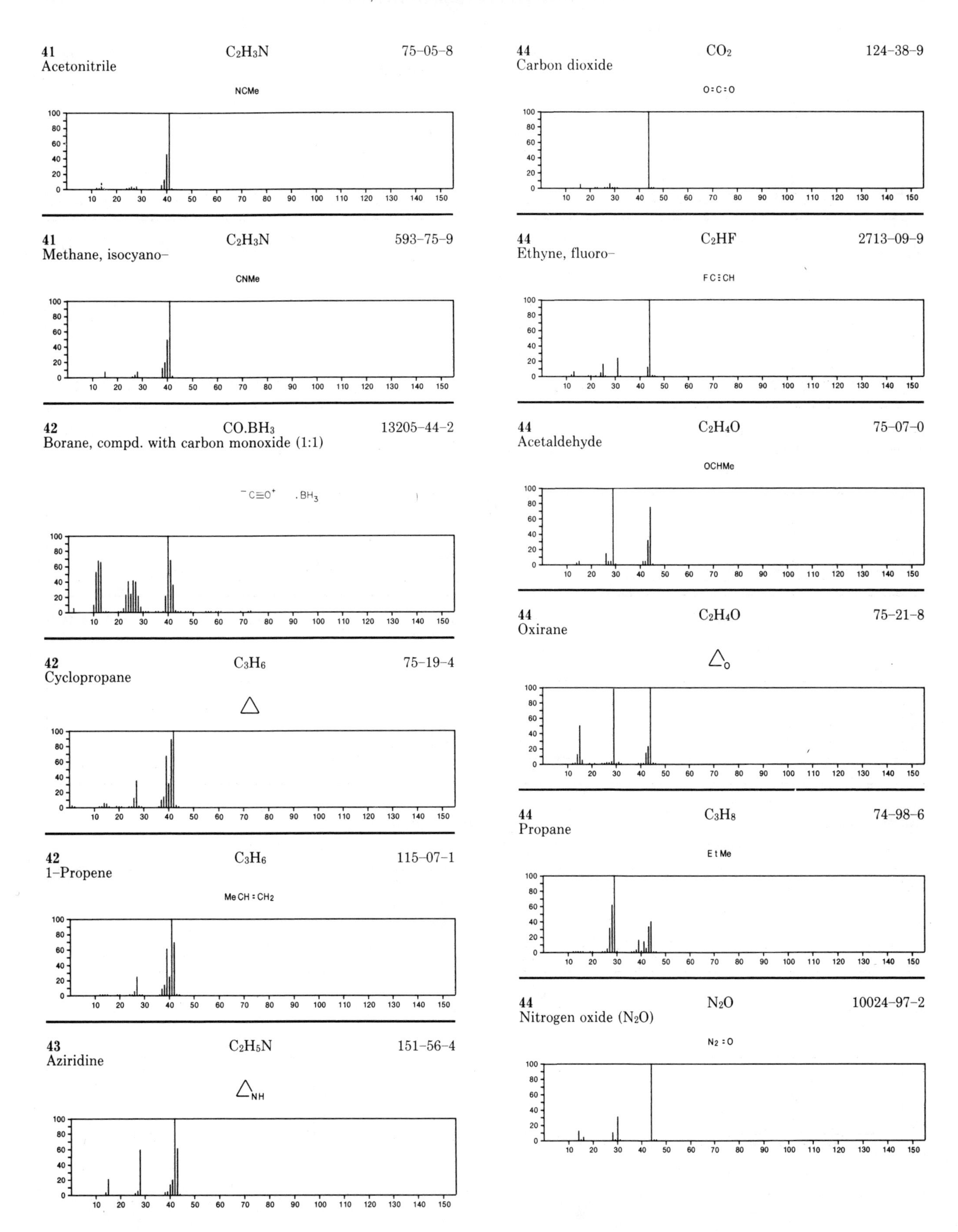
41
Acetonitrile
C2H3N
75-05-8
NCMe
41
Methane, isocyano-
C2H3N
593-75-9
CNMe
42
Borane, compd. with carbon monoxide (1:1)
CO.BH3
13205-44-2
-C≡O+ .BH3
42
Cyclopropane
C3H6
75-19-4
42
1-Propene
C3H6
115-07-1
Me CH = CH2
43
Aziridine
C2H5N
151-56-4
NH
44
Carbon dioxide
CO2
124-38-9
O:C:O
44
Ethyne, fluoro-
C2HF
2713-09-9
FC⋮CH
44
Acetaldehyde
C2H4O
75-07-0
OCHMe
44
Oxirane
C2H4O
75-21-8
O
44
Propane
C3H8
74-98-6
Et Me
44
Nitrogen oxide (N2O)
N2O
10024-97-2
N2 : O
100
80
60
40
20
0
10 20 30 40 50 60 70 80 90 100 110 120 130 140 150

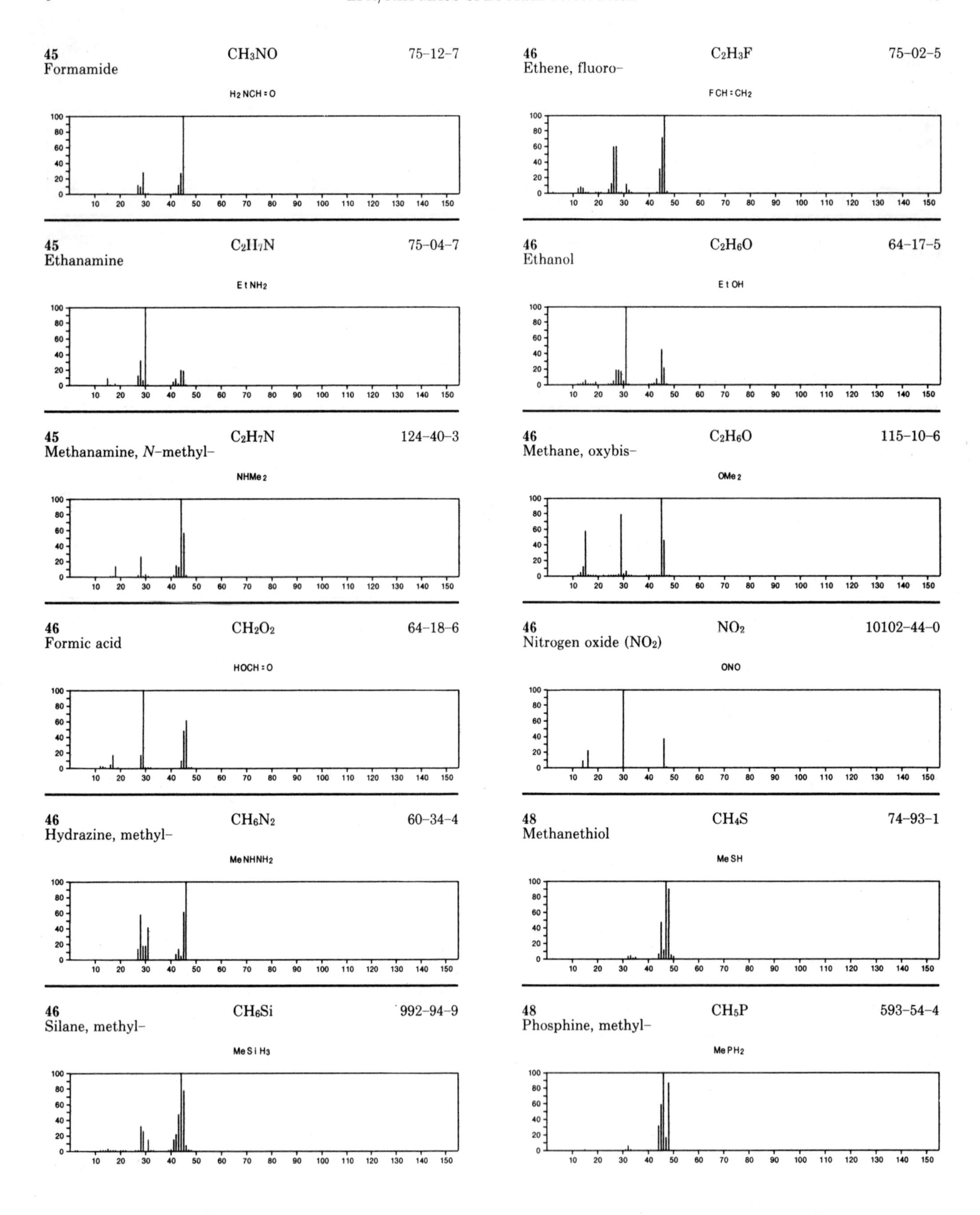
45
Formamide
CH3NO
75-12-7
H2NCH=O
45
Ethanamine
C2H7N
75-04-7
EtNH2
45
Methanamine, N-methyl-
C2H7N
124-40-3
NHMe2
46
Formic acid
CH2O2
64-18-6
HOCH=O
46
Hydrazine, methyl-
CH6N2
60-34-4
MeNHNH2
46
Silane, methyl-
CH6Si
992-94-9
MeSiH3
46
Ethene, fluoro-
C2H3F
75-02-5
FCH=CH2
46
Ethanol
C2H6O
64-17-5
EtOH
46
Methane, oxybis-
C2H6O
115-10-6
OMe2
46
Nitrogen oxide (NO2)
NO2
10102-44-0
ONO
48
Methanethiol
CH4S
74-93-1
MeSH
48
Phosphine, methyl-
CH5P
593-54-4
MePH2

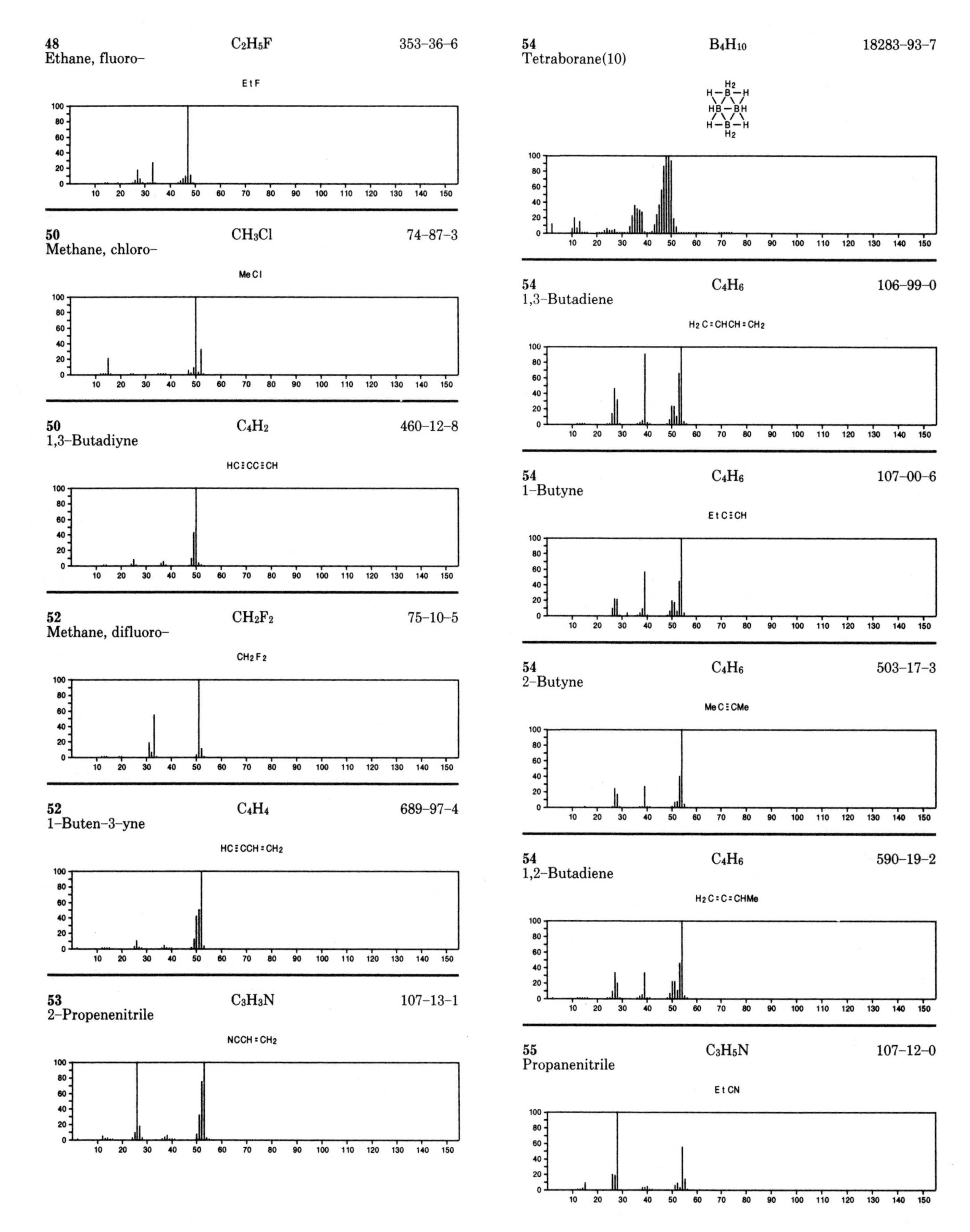
48
C2H5F
353-36-6
Ethane, fluoro-
EtF
50
CH3Cl
74-87-3
Methane, chloro-
MeCl
50
C4H2
460-12-8
1,3-Butadiyne
HC≡CC≡CH
52
CH2F2
75-10-5
Methane, difluoro-
CH2F2
52
C4H4
689-97-4
1-Buten-3-yne
HC≡CCH=CH2
53
C3H3N
107-13-1
2-Propenenitrile
NCCH=CH2
54
B4H10
18283-93-7
Tetraborane(10)
54
C4H6
106-99-0
1,3-Butadiene
H2C=CHCH=CH2
54
C4H6
107-00-6
1-Butyne
EtC≡CH
54
C4H6
503-17-3
2-Butyne
MeC≡CMe
54
C4H6
590-19-2
1,2-Butadiene
H2C=C=CHMe
55
C3H5N
107-12-0
Propanenitrile
EtCN

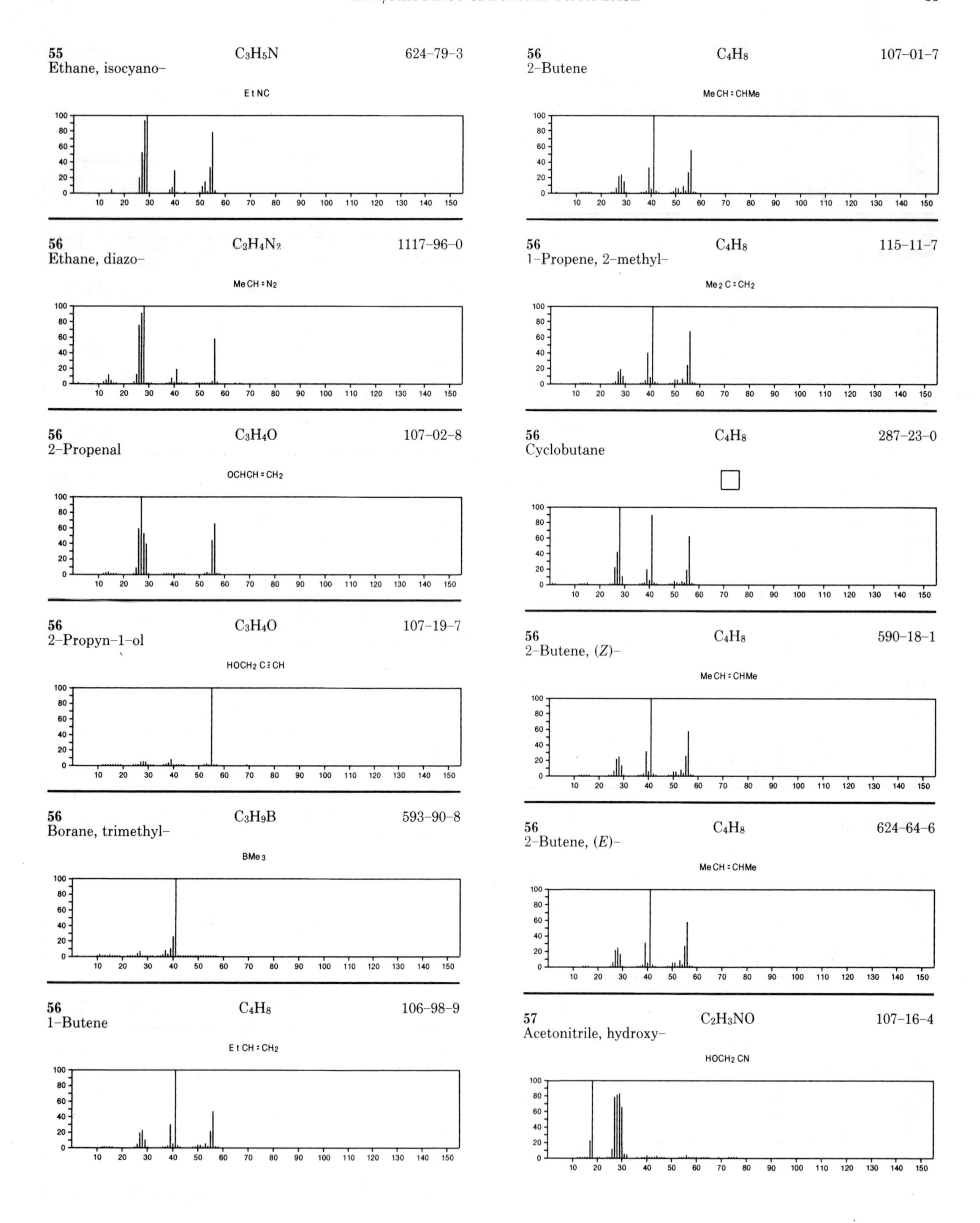
55
Ethane, isocyano–
C3H5N
624–79–3
Et NC
56
Ethane, diazo–
C2H4N2
1117–96–0
Me CH = N2
56
2–Propenal
C3H4O
107–02–8
OCHCH = CH2
56
2–Propyn–1–ol
C3H4O
107–19–7
HOCH2 C ≡ CH
56
Borane, trimethyl–
C3H9B
593–90–8
BMe3
56
1–Butene
C4H8
106–98–9
Et CH = CH2
56
2–Butene
C4H8
107–01–7
Me CH = CHMe
56
1–Propene, 2–methyl–
C4H8
115–11–7
Me2 C = CH2
56
Cyclobutane
C4H8
287–23–0
56
2–Butene, (Z)–
C4H8
590–18–1
Me CH = CHMe
56
2–Butene, (E)–
C4H8
624–64–6
Me CH = CHMe
57
Acetonitrile, hydroxy–
C2H3NO
107–16–4
HOCH2 CN

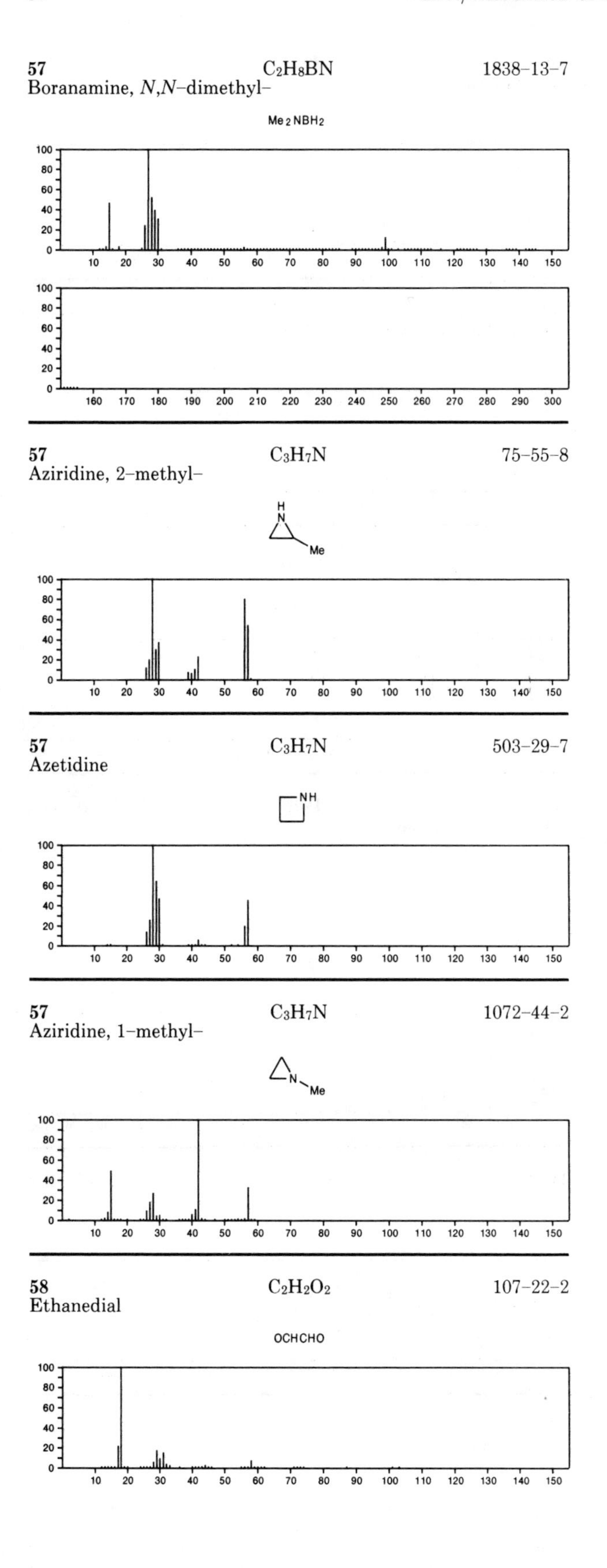

57 C_2H_8BN 1838-13-7
Boranamine, *N*,*N*-dimethyl-

57 C_3H_7N 75-55-8
Aziridine, 2-methyl-

57 C_3H_7N 503-29-7
Azetidine

57 C_3H_7N 1072-44-2
Aziridine, 1-methyl-

58 $C_2H_2O_2$ 107-22-2
Ethanedial

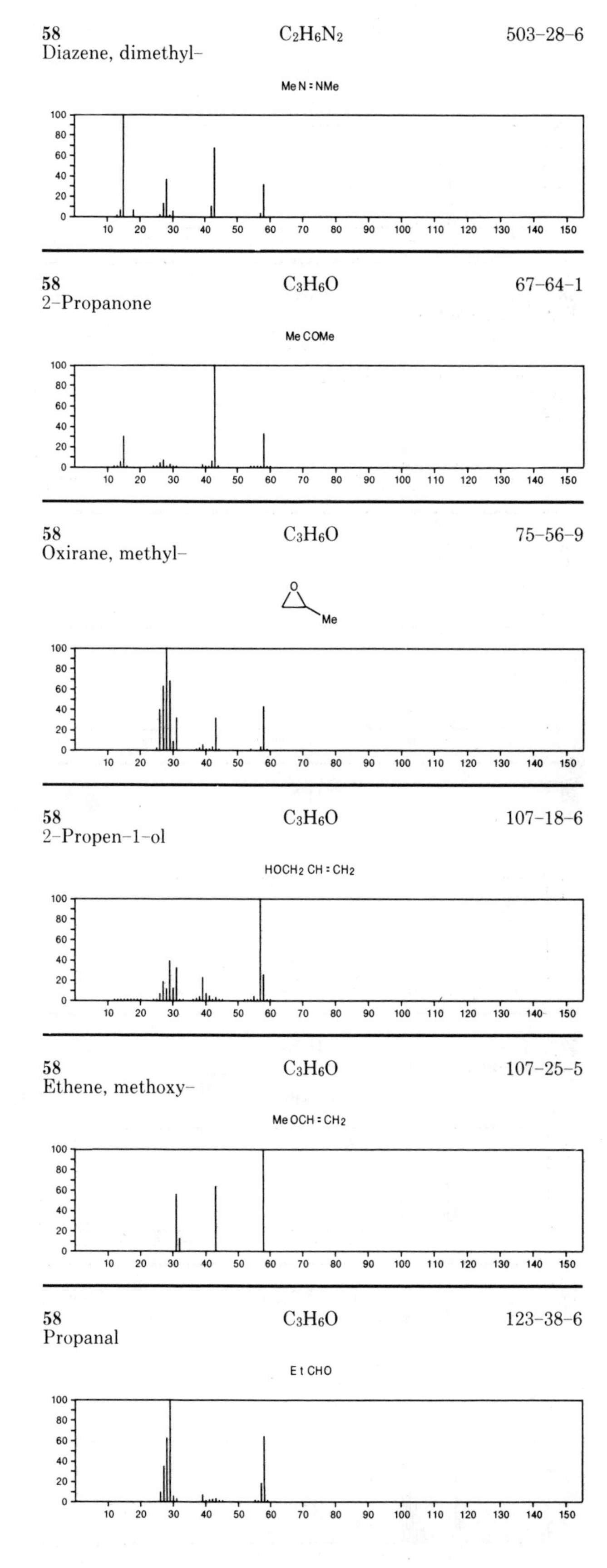

58 $C_2H_6N_2$ 503-28-6
Diazene, dimethyl-

58 C_3H_6O 67-64-1
2-Propanone

58 C_3H_6O 75-56-9
Oxirane, methyl-

58 C_3H_6O 107-18-6
2-Propen-1-ol

58 C_3H_6O 107-25-5
Ethene, methoxy-

58 C_3H_6O 123-38-6
Propanal

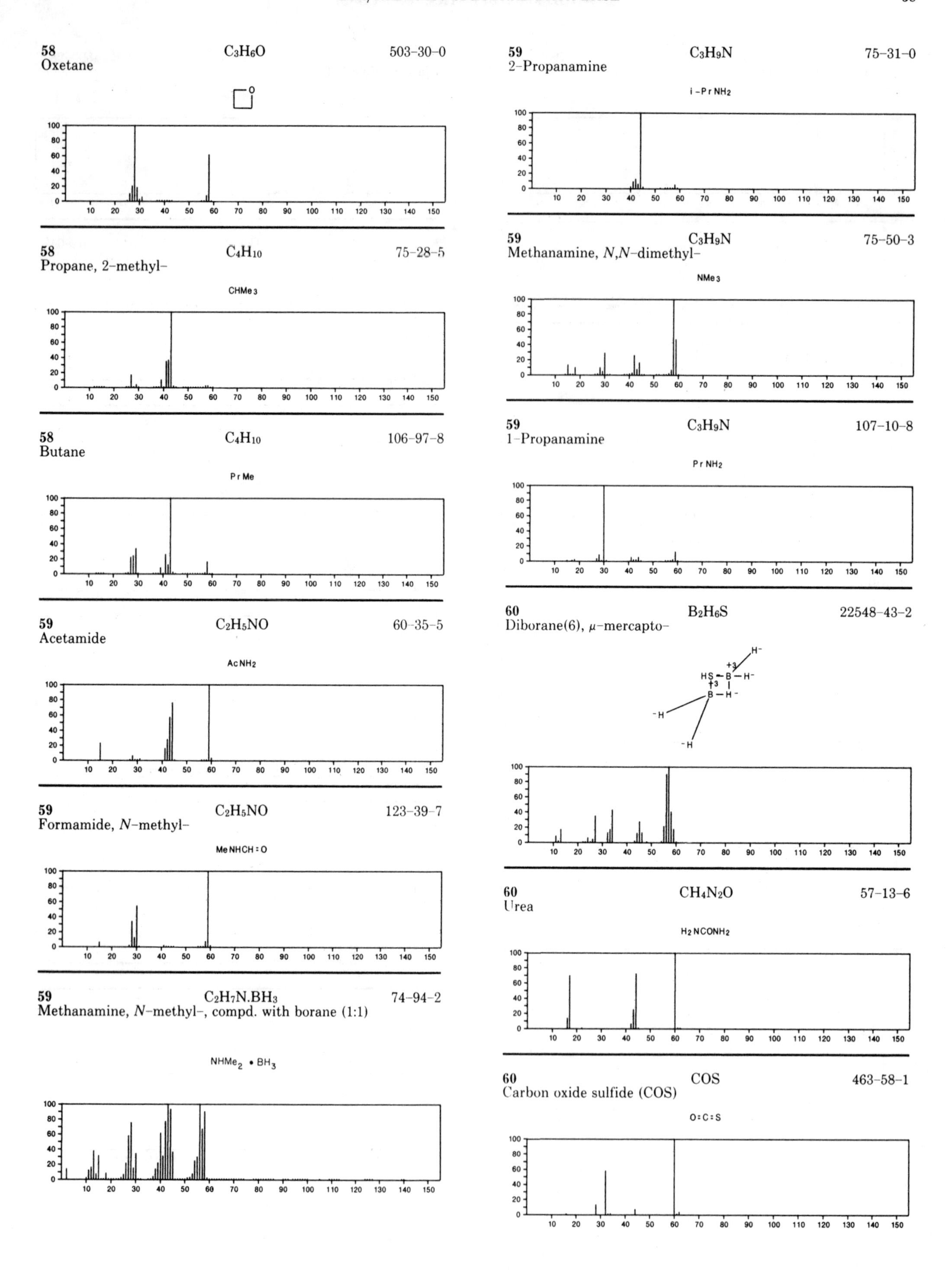

58 Oxetane C_3H_6O 503-30-0
58 Propane, 2-methyl- C_4H_{10} 75-28-5
CHMe3
58 Butane C_4H_{10} 106-97-8
PrMe
59 Acetamide C_2H_5NO 60-35-5
AcNH2
59 Formamide, N-methyl- C_2H_5NO 123-39-7
MeNHCH=O
59 Methanamine, N-methyl-, compd. with borane (1:1) $C_2H_7N.BH_3$ 74-94-2
NHMe2 • BH3
59 2-Propanamine C_3H_9N 75-31-0
i-PrNH2
59 Methanamine, N,N-dimethyl- C_3H_9N 75-50-3
NMe3
59 1-Propanamine C_3H_9N 107-10-8
PrNH2
60 Diborane(6), μ-mercapto- B_2H_6S 22548-43-2
60 Urea CH_4N_2O 57-13-6
H2NCONH2
60 Carbon oxide sulfide (COS) COS 463-58-1
O=C=S

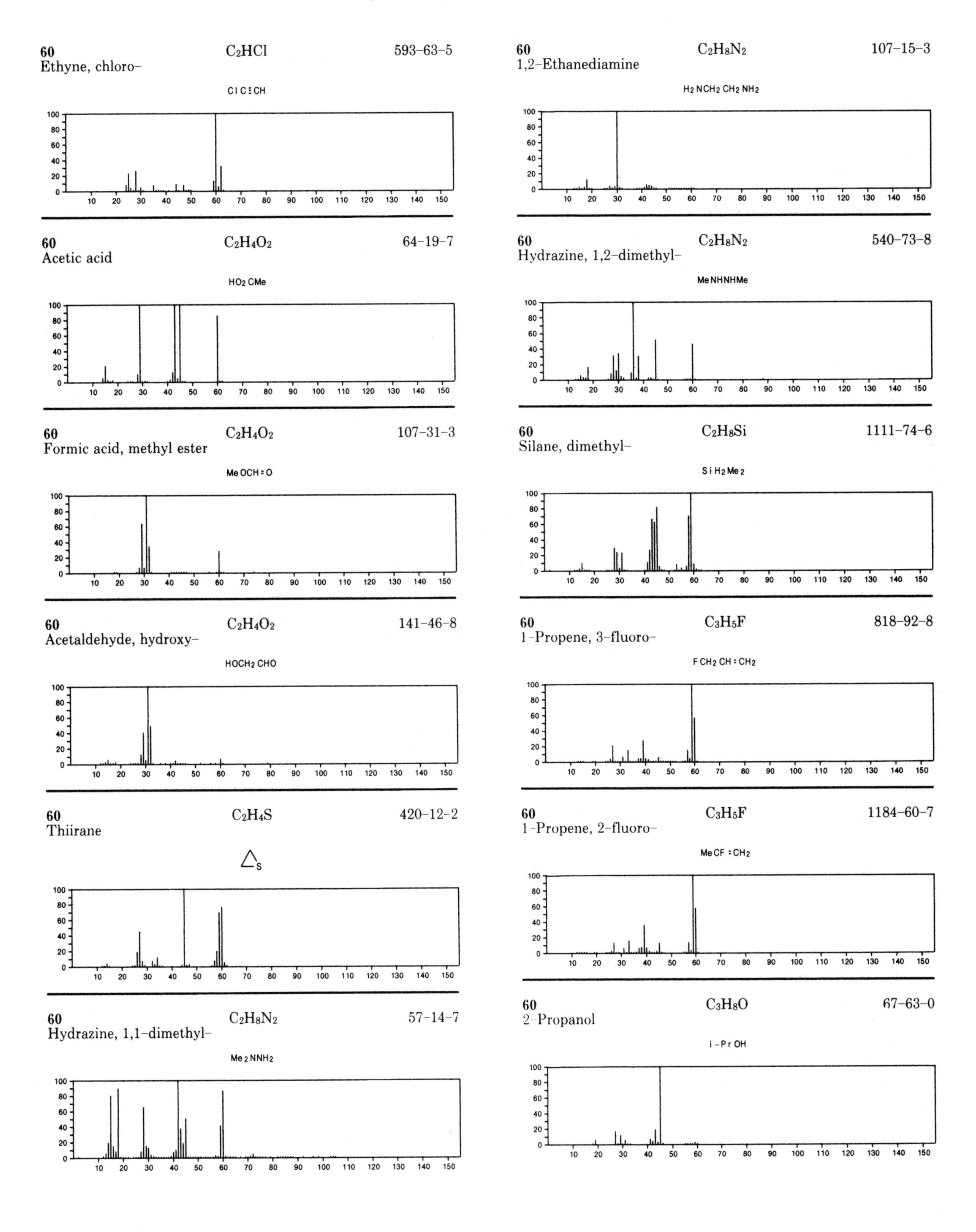

60 C2HCl 593-63-5
Ethyne, chloro-
ClC⋮CH
60 C2H4O2 64-19-7
Acetic acid
HO2CMe
60 C2H4O2 107-31-3
Formic acid, methyl ester
MeOCH=O
60 C2H4O2 141-46-8
Acetaldehyde, hydroxy-
HOCH2CHO
60 C2H4S 420-12-2
Thiirane
S
60 C2H8N2 57-14-7
Hydrazine, 1,1-dimethyl-
Me2NNH2
60 C2H8N2 107-15-3
1,2-Ethanediamine
H2NCH2CH2NH2
60 C2H8N2 540-73-8
Hydrazine, 1,2-dimethyl-
MeNHNHMe
60 C2H8Si 1111-74-6
Silane, dimethyl-
SiH2Me2
60 C3H5F 818-92-8
1-Propene, 3-fluoro-
FCH2CH=CH2
60 C3H5F 1184-60-7
1-Propene, 2-fluoro-
MeCF=CH2
60 C3H8O 67-63-0
2-Propanol
i-PrOH

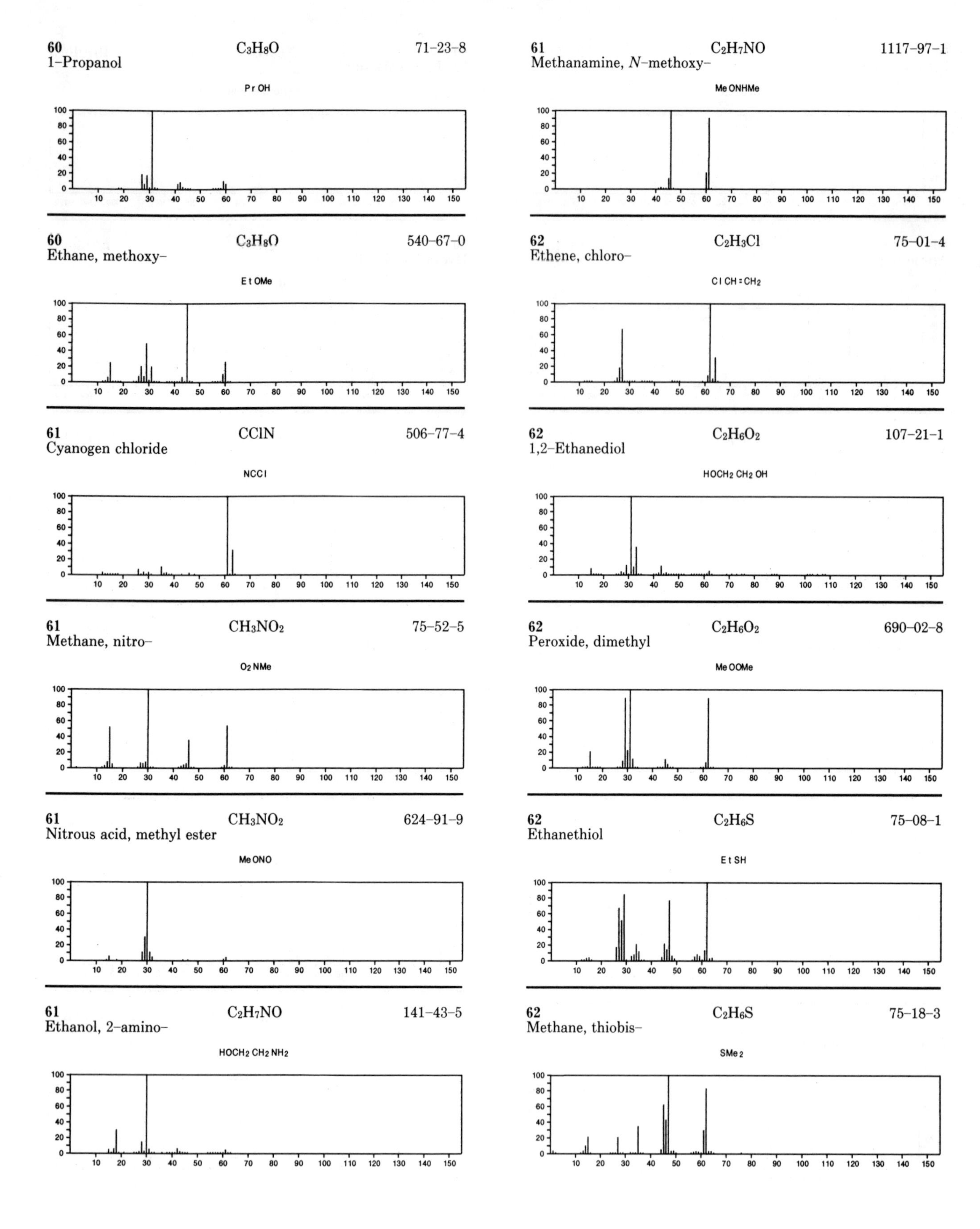

60 1-Propanol C_3H_8O 71-23-8

PrOH

60 Ethane, methoxy- C_3H_8O 540-67-0

EtOMe

61 Cyanogen chloride CClN 506-77-4

NCCl

61 Methane, nitro- CH_3NO_2 75-52-5

O_2NMe

61 Nitrous acid, methyl ester CH_3NO_2 624-91-9

MeONO

61 Ethanol, 2-amino- C_2H_7NO 141-43-5

$HOCH_2CH_2NH_2$

61 Methanamine, *N*-methoxy- C_2H_7NO 1117-97-1

MeONHMe

62 Ethene, chloro- C_2H_3Cl 75-01-4

$ClCH=CH_2$

62 1,2-Ethanediol $C_2H_6O_2$ 107-21-1

$HOCH_2CH_2OH$

62 Peroxide, dimethyl $C_2H_6O_2$ 690-02-8

MeOOMe

62 Ethanethiol C_2H_6S 75-08-1

EtSH

62 Methane, thiobis- C_2H_6S 75-18-3

SMe_2

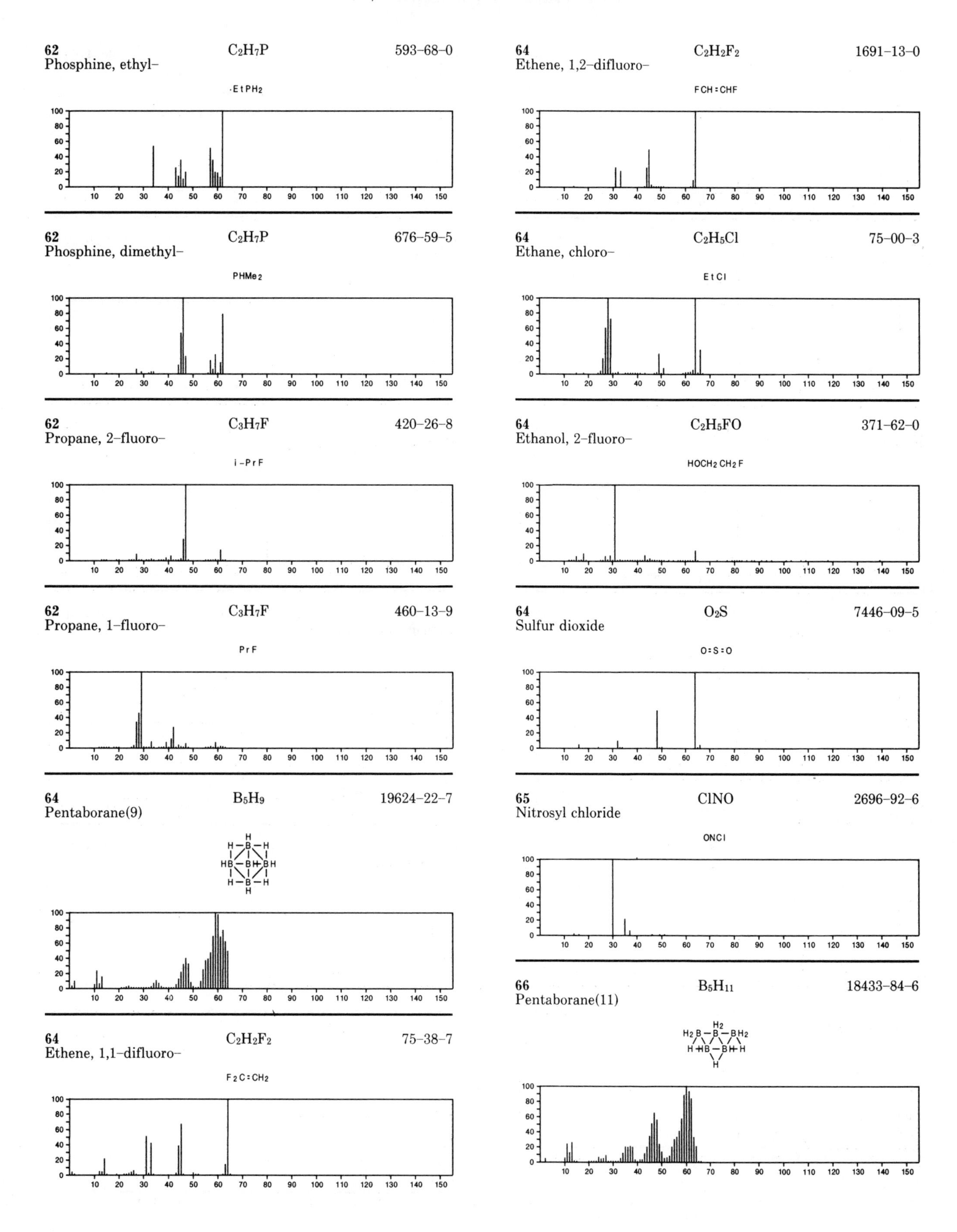

62 Phosphine, ethyl- C2H7P 593-68-0
EtPH2
62 Phosphine, dimethyl- C2H7P 676-59-5
PHMe2
62 Propane, 2-fluoro- C3H7F 420-26-8
i-PrF
62 Propane, 1-fluoro- C3H7F 460-13-9
PrF
64 Pentaborane(9) B5H9 19624-22-7
64 Ethene, 1,1-difluoro- C2H2F2 75-38-7
F2C:CH2
64 Ethene, 1,2-difluoro- C2H2F2 1691-13-0
FCH:CHF
64 Ethane, chloro- C2H5Cl 75-00-3
EtCl
64 Ethanol, 2-fluoro- C2H5FO 371-62-0
HOCH2CH2F
64 Sulfur dioxide O2S 7446-09-5
O:S:O
65 Nitrosyl chloride ClNO 2696-92-6
ONCl
66 Pentaborane(11) B5H11 18433-84-6

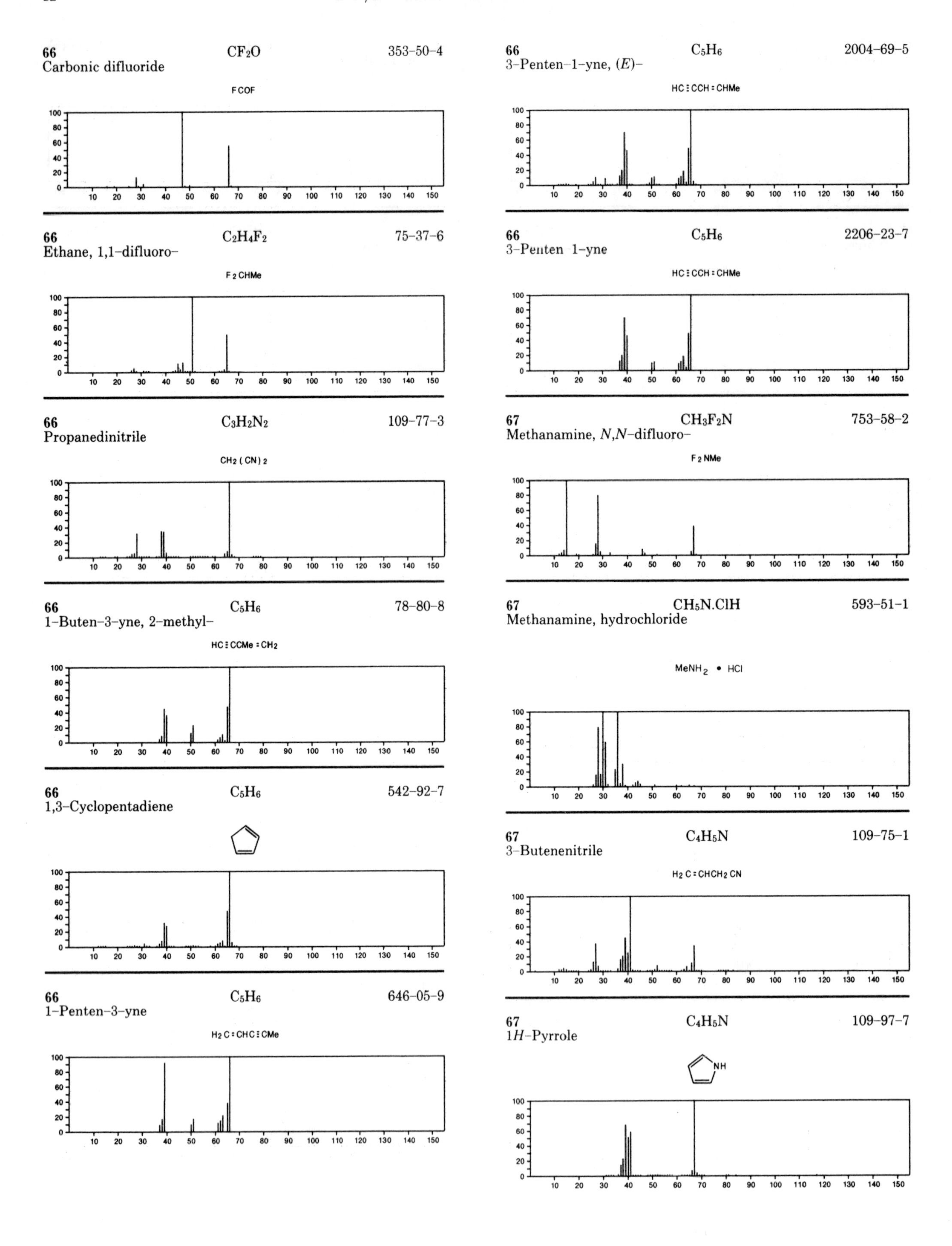
66 CF2O 353-50-4
Carbonic difluoride
FCOF
66 C2H4F2 75-37-6
Ethane, 1,1-difluoro-
F2CHMe
66 C3H2N2 109-77-3
Propanedinitrile
CH2(CN)2
66 C5H6 78-80-8
1-Buten-3-yne, 2-methyl-
HC≡CCMe=CH2
66 C5H6 542-92-7
1,3-Cyclopentadiene
66 C5H6 646-05-9
1-Penten-3-yne
H2C=CHC≡CMe
66 C5H6 2004-69-5
3-Penten-1-yne, (E)-
HC≡CCH=CHMe
66 C5H6 2206-23-7
3-Penten-1-yne
HC≡CCH=CHMe
67 CH3F2N 753-58-2
Methanamine, N,N-difluoro-
F2NMe
67 CH5N.ClH 593-51-1
Methanamine, hydrochloride
MeNH2 • HCl
67 C4H5N 109-75-1
3-Butenenitrile
H2C=CHCH2CN
67 C4H5N 109-97-7
1H-Pyrrole
NH

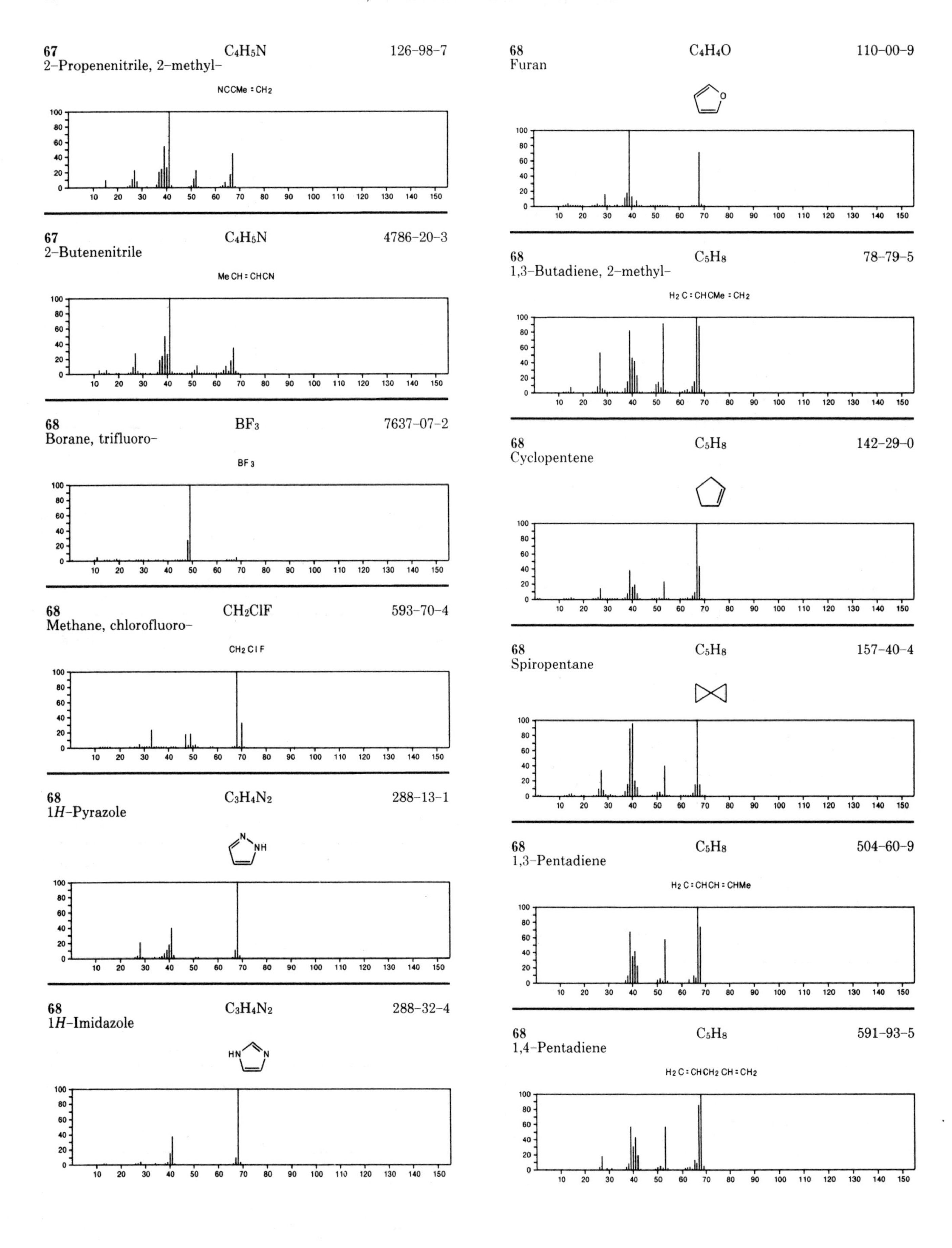
67 C_4H_5N 126-98-7
2-Propenenitrile, 2-methyl-
NCCMe = CH2
67 C_4H_5N 4786-20-3
2-Butenenitrile
Me CH = CHCN
68 BF_3 7637-07-2
Borane, trifluoro-
BF3
68 CH_2ClF 593-70-4
Methane, chlorofluoro-
CH2 Cl F
68 $C_3H_4N_2$ 288-13-1
1*H*-Pyrazole
68 $C_3H_4N_2$ 288-32-4
1*H*-Imidazole
68 C_4H_4O 110-00-9
Furan
68 C_5H_8 78-79-5
1,3-Butadiene, 2-methyl-
H2 C = CHCMe = CH2
68 C_5H_8 142-29-0
Cyclopentene
68 C_5H_8 157-40-4
Spiropentane
68 C_5H_8 504-60-9
1,3-Pentadiene
H2 C = CHCH = CHMe
68 C_5H_8 591-93-5
1,4-Pentadiene
H2 C = CHCH2 CH = CH2

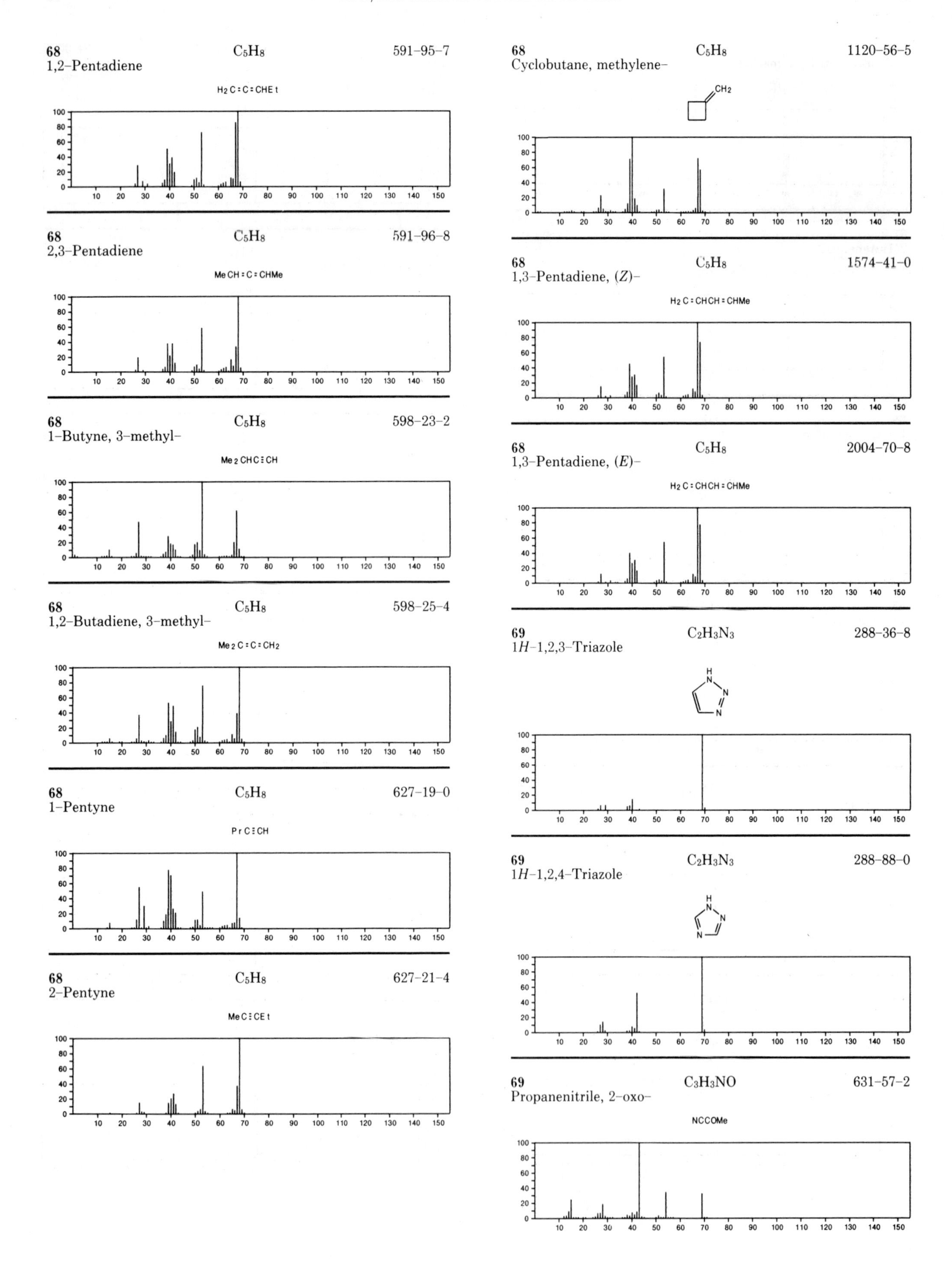
68 C5H8 591-95-7
1,2-Pentadiene
H2C=C=CHEt
68 C5H8 591-96-8
2,3-Pentadiene
MeCH=C=CHMe
68 C5H8 598-23-2
1-Butyne, 3-methyl-
Me2CHC≡CH
68 C5H8 598-25-4
1,2-Butadiene, 3-methyl-
Me2C=C=CH2
68 C5H8 627-19-0
1-Pentyne
PrC≡CH
68 C5H8 627-21-4
2-Pentyne
MeC≡CEt
68 C5H8 1120-56-5
Cyclobutane, methylene-
CH2
68 C5H8 1574-41-0
1,3-Pentadiene, (Z)-
H2C=CHCH=CHMe
68 C5H8 2004-70-8
1,3-Pentadiene, (E)-
H2C=CHCH=CHMe
69 C2H3N3 288-36-8
1H-1,2,3-Triazole
H
N
N
N
69 C2H3N3 288-88-0
1H-1,2,4-Triazole
H
N
N
N
69 C3H3NO 631-57-2
Propanenitrile, 2-oxo-
NCCOMe
100
80
60
40
20
0
10 20 30 40 50 60 70 80 90 100 110 120 130 140 150

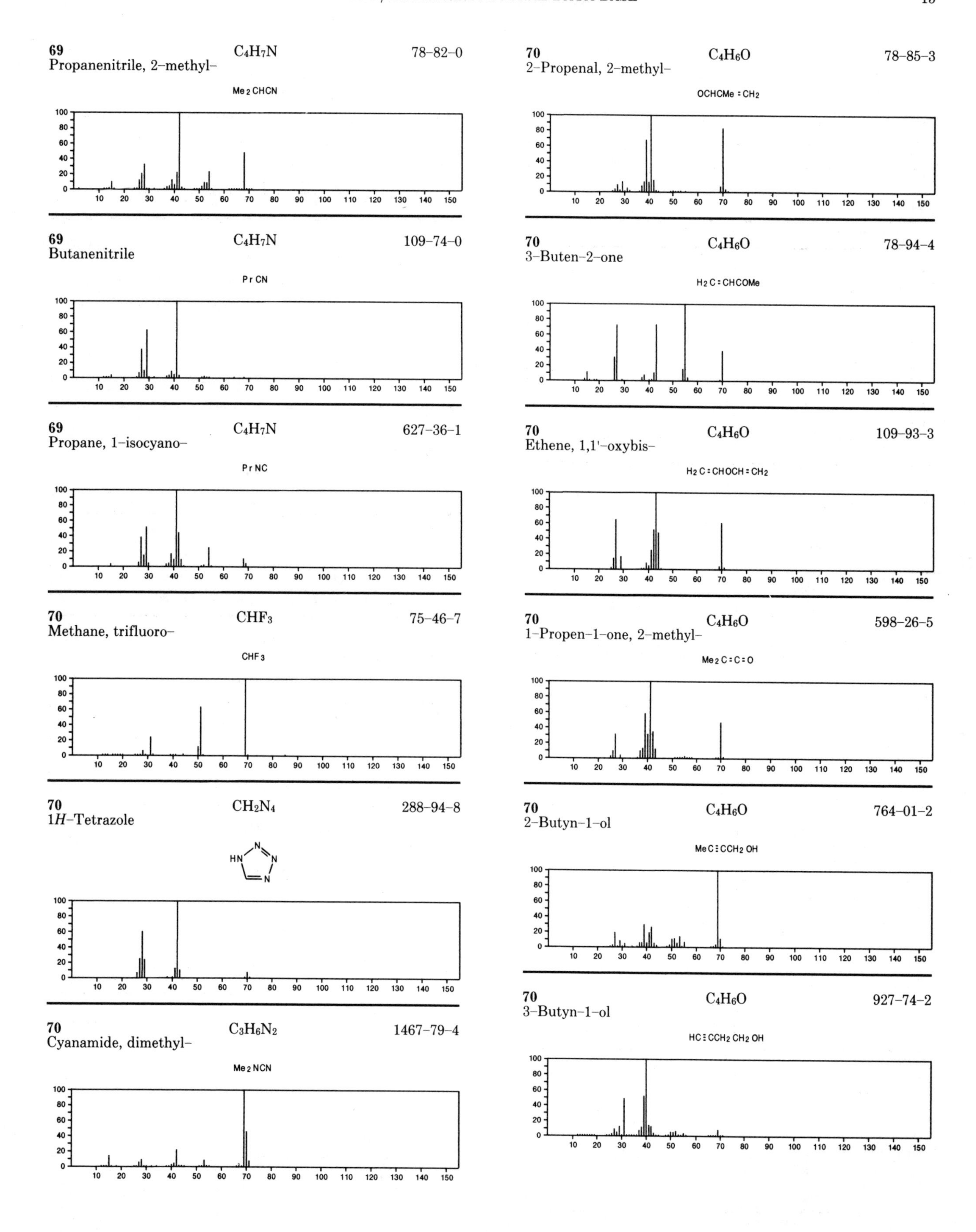
69
Propanenitrile, 2-methyl-
C4H7N
78-82-0
Me2CHCN
69
Butanenitrile
C4H7N
109-74-0
PrCN
69
Propane, 1-isocyano-
C4H7N
627-36-1
PrNC
70
Methane, trifluoro-
CHF3
75-46-7
CHF3
70
1H-Tetrazole
CH2N4
288-94-8
HN
N
N
N
70
Cyanamide, dimethyl-
C3H6N2
1467-79-4
Me2NCN
70
2-Propenal, 2-methyl-
C4H6O
78-85-3
OCHCMe=CH2
70
3-Buten-2-one
C4H6O
78-94-4
H2C=CHCOMe
70
Ethene, 1,1'-oxybis-
C4H6O
109-93-3
H2C=CHOCH=CH2
70
1-Propen-1-one, 2-methyl-
C4H6O
598-26-5
Me2C=C=O
70
2-Butyn-1-ol
C4H6O
764-01-2
MeC≡CCH2OH
70
3-Butyn-1-ol
C4H6O
927-74-2
HC≡CCH2CH2OH

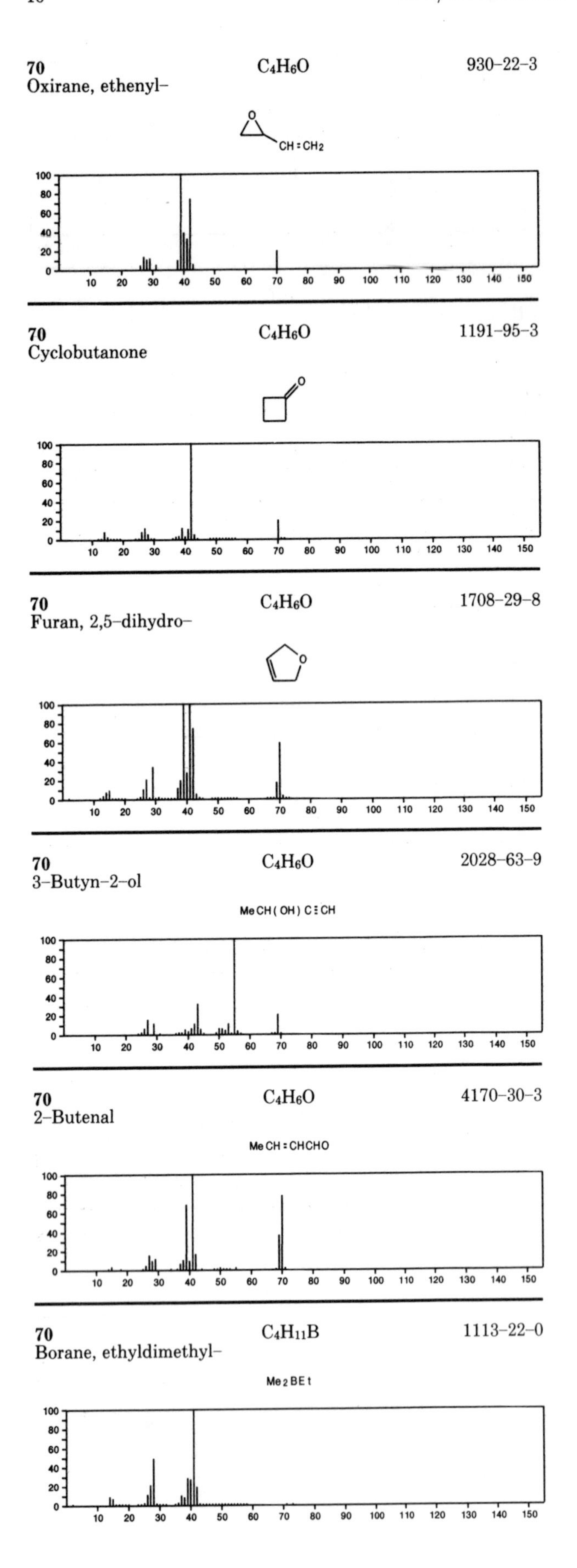

70 Oxirane, ethenyl– C_4H_6O 930–22–3

70 Cyclobutanone C_4H_6O 1191–95–3

70 Furan, 2,5–dihydro– C_4H_6O 1708–29–8

70 3–Butyn–2–ol C_4H_6O 2028–63–9

MeCH(OH)C≡CH

70 2–Butenal C_4H_6O 4170–30–3

MeCH=CHCHO

70 Borane, ethyldimethyl– $C_4H_{11}B$ 1113–22–0

Me2BEt

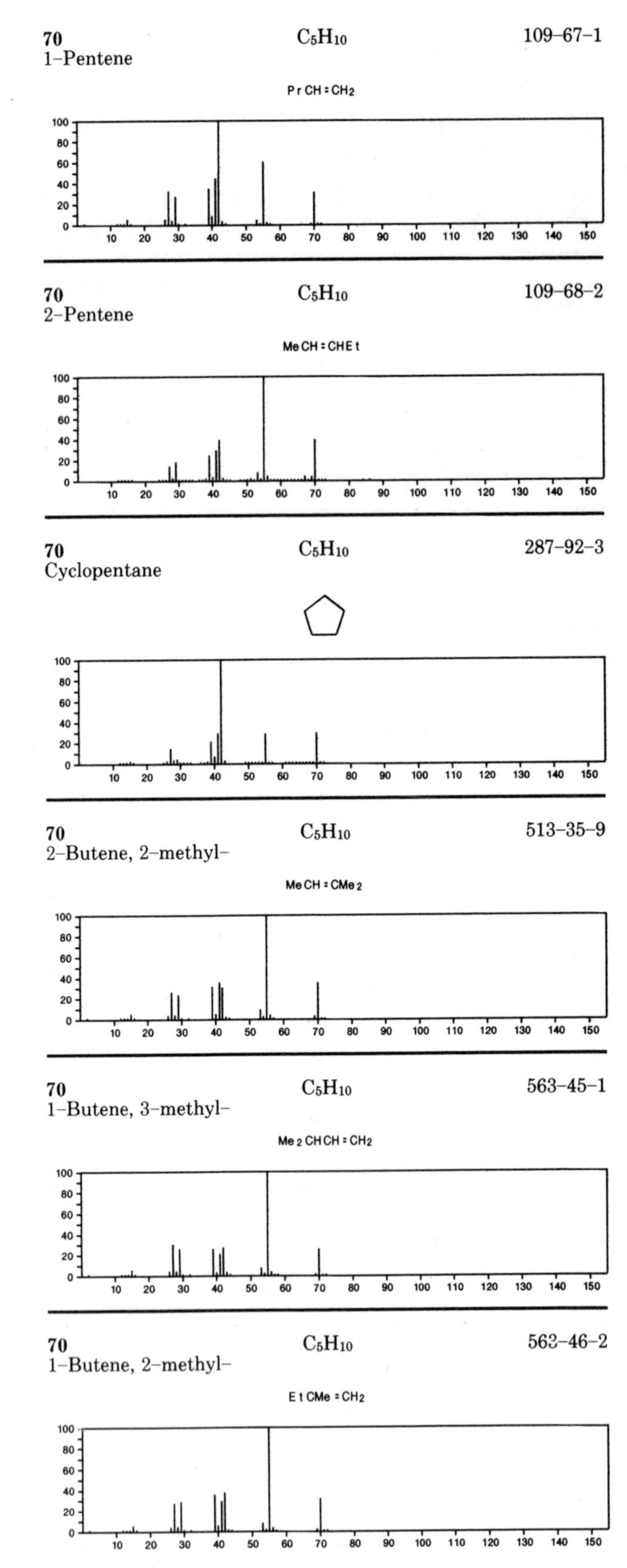

70 1–Pentene C_5H_{10} 109–67–1

PrCH=CH2

70 2–Pentene C_5H_{10} 109–68–2

MeCH=CHEt

70 Cyclopentane C_5H_{10} 287–92–3

70 2–Butene, 2–methyl– C_5H_{10} 513–35–9

MeCH=CMe2

70 1–Butene, 3–methyl– C_5H_{10} 563–45–1

Me2CHCH=CH2

70 1–Butene, 2–methyl– C_5H_{10} 563–46–2

EtCMe=CH2

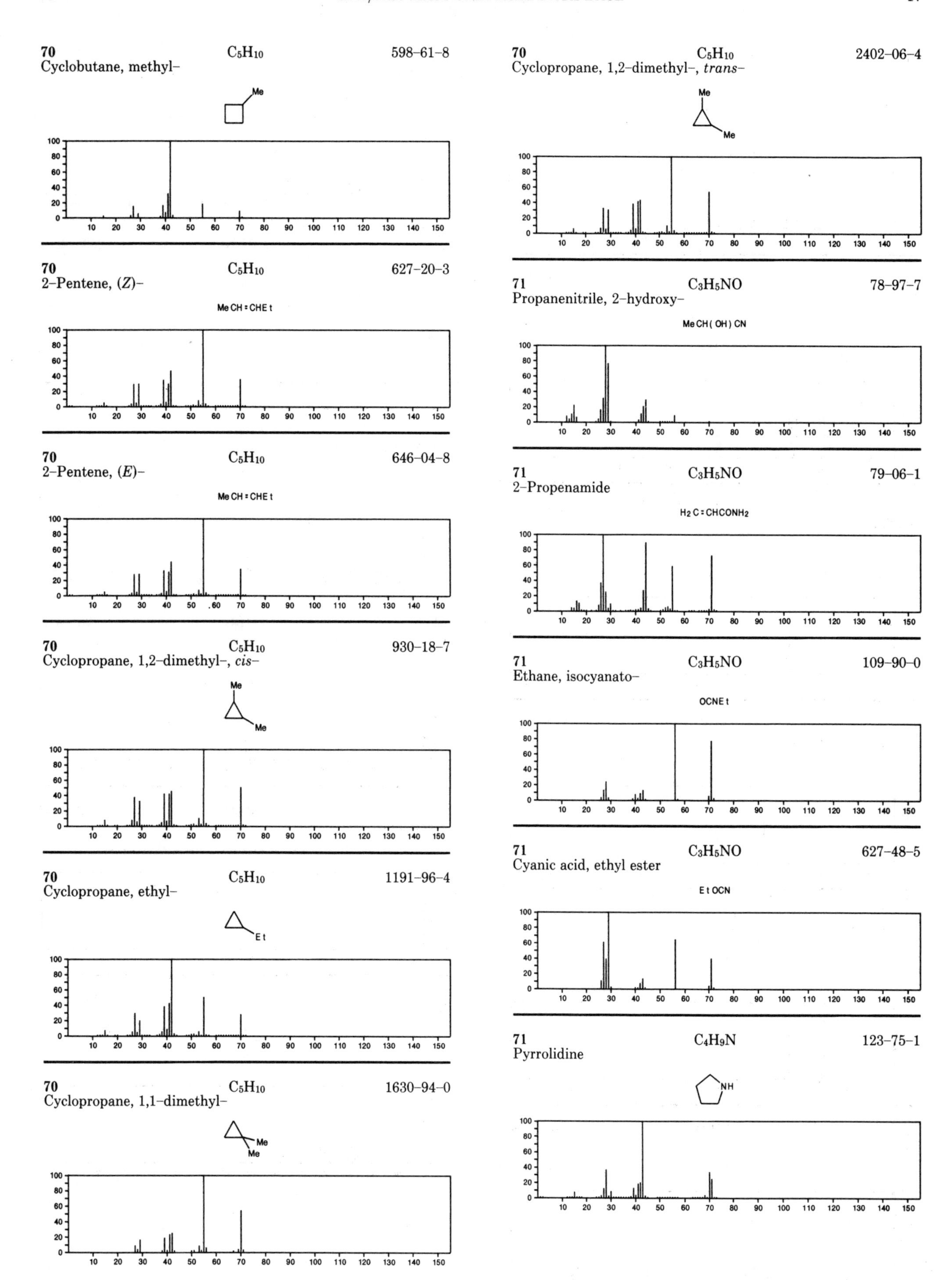
70 C_5H_{10} 598-61-8
Cyclobutane, methyl-
Me
70 C_5H_{10} 627-20-3
2-Pentene, (*Z*)-
MeCH=CHEt
70 C_5H_{10} 646-04-8
2-Pentene, (*E*)-
MeCH=CHEt
70 C_5H_{10} 930-18-7
Cyclopropane, 1,2-dimethyl-, *cis*-
Me
Me
70 C_5H_{10} 1191-96-4
Cyclopropane, ethyl-
Et
70 C_5H_{10} 1630-94-0
Cyclopropane, 1,1-dimethyl-
Me
Me
70 C_5H_{10} 2402-06-4
Cyclopropane, 1,2-dimethyl-, *trans*-
Me
Me
71 C_3H_5NO 78-97-7
Propanenitrile, 2-hydroxy-
MeCH(OH)CN
71 C_3H_5NO 79-06-1
2-Propenamide
H_2C=CHCONH$_2$
71 C_3H_5NO 109-90-0
Ethane, isocyanato-
OCNEt
71 C_3H_5NO 627-48-5
Cyanic acid, ethyl ester
EtOCN
71 C_4H_9N 123-75-1
Pyrrolidine
NH

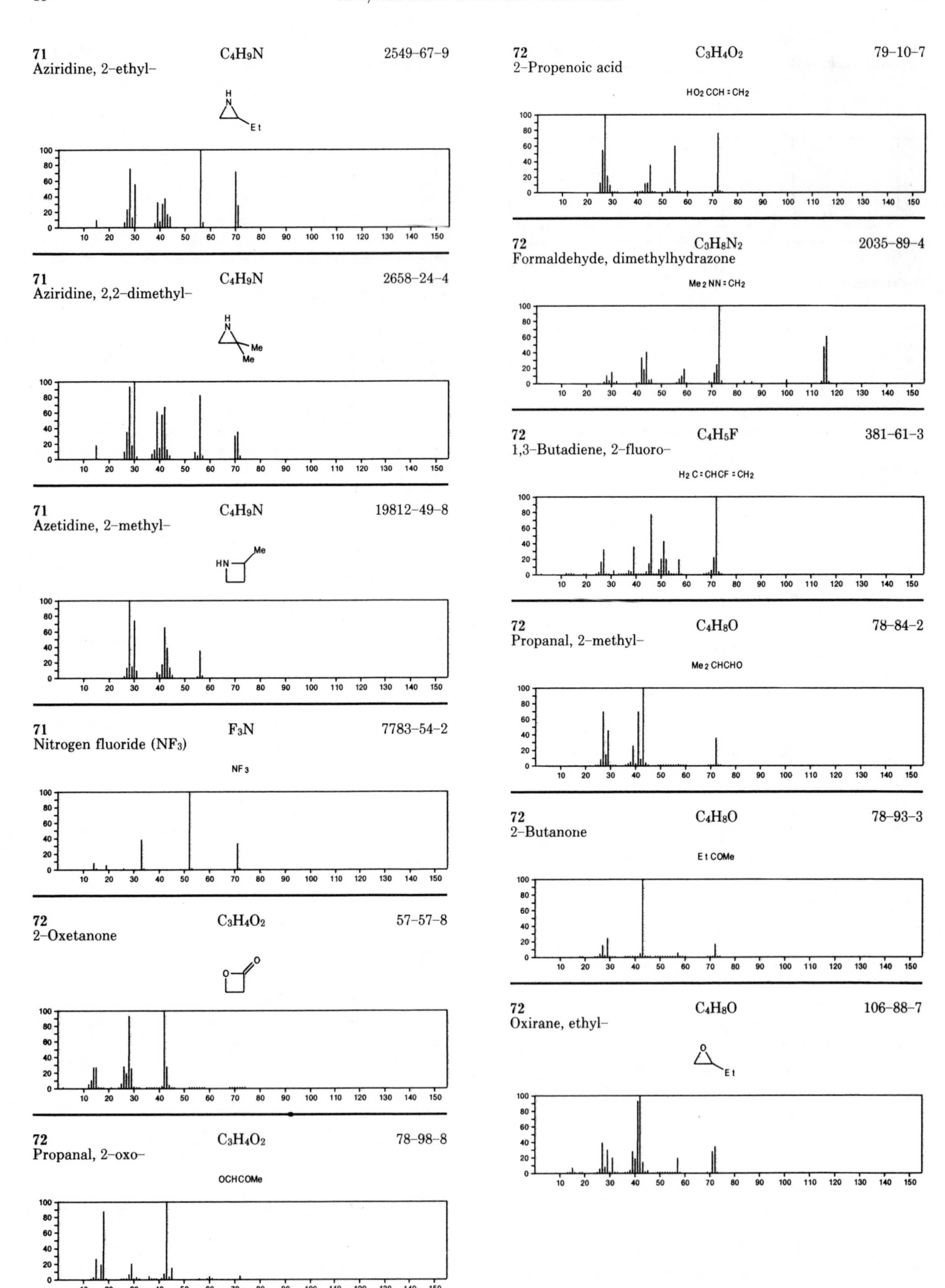
71 C4H9N 2549-67-9
Aziridine, 2-ethyl-
71 C4H9N 2658-24-4
Aziridine, 2,2-dimethyl-
71 C4H9N 19812-49-8
Azetidine, 2-methyl-
71 F3N 7783-54-2
Nitrogen fluoride (NF3)
NF3
72 C3H4O2 57-57-8
2-Oxetanone
72 C3H4O2 78-98-8
Propanal, 2-oxo-
OCHCOMe
72 C3H4O2 79-10-7
2-Propenoic acid
HO2CCH=CH2
72 C3H8N2 2035-89-4
Formaldehyde, dimethylhydrazone
Me2NN=CH2
72 C4H5F 381-61-3
1,3-Butadiene, 2-fluoro-
H2C=CHCF=CH2
72 C4H8O 78-84-2
Propanal, 2-methyl-
Me2CHCHO
72 C4H8O 78-93-3
2-Butanone
EtCOMe
72 C4H8O 106-88-7
Oxirane, ethyl-

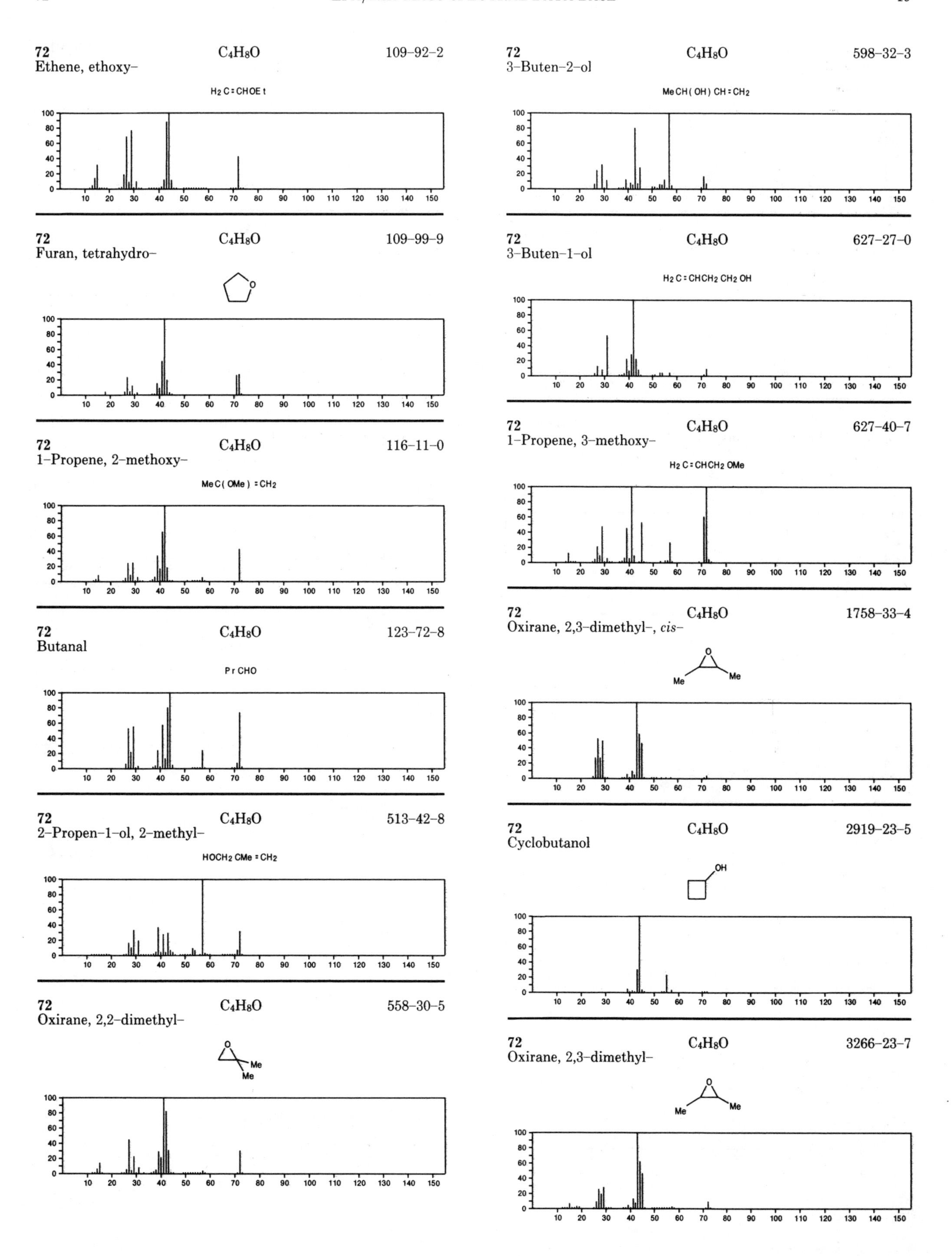

72 C4H8O 109-92-2
Ethene, ethoxy-
H2C=CHOEt
72 C4H8O 598-32-3
3-Buten-2-ol
MeCH(OH)CH=CH2
72 C4H8O 109-99-9
Furan, tetrahydro-
72 C4H8O 627-27-0
3-Buten-1-ol
H2C=CHCH2CH2OH
72 C4H8O 116-11-0
1-Propene, 2-methoxy-
MeC(OMe)=CH2
72 C4H8O 627-40-7
1-Propene, 3-methoxy-
H2C=CHCH2OMe
72 C4H8O 123-72-8
Butanal
PrCHO
72 C4H8O 1758-33-4
Oxirane, 2,3-dimethyl-, cis-
72 C4H8O 513-42-8
2-Propen-1-ol, 2-methyl-
HOCH2CMe=CH2
72 C4H8O 2919-23-5
Cyclobutanol
72 C4H8O 558-30-5
Oxirane, 2,2-dimethyl-
72 C4H8O 3266-23-7
Oxirane, 2,3-dimethyl-

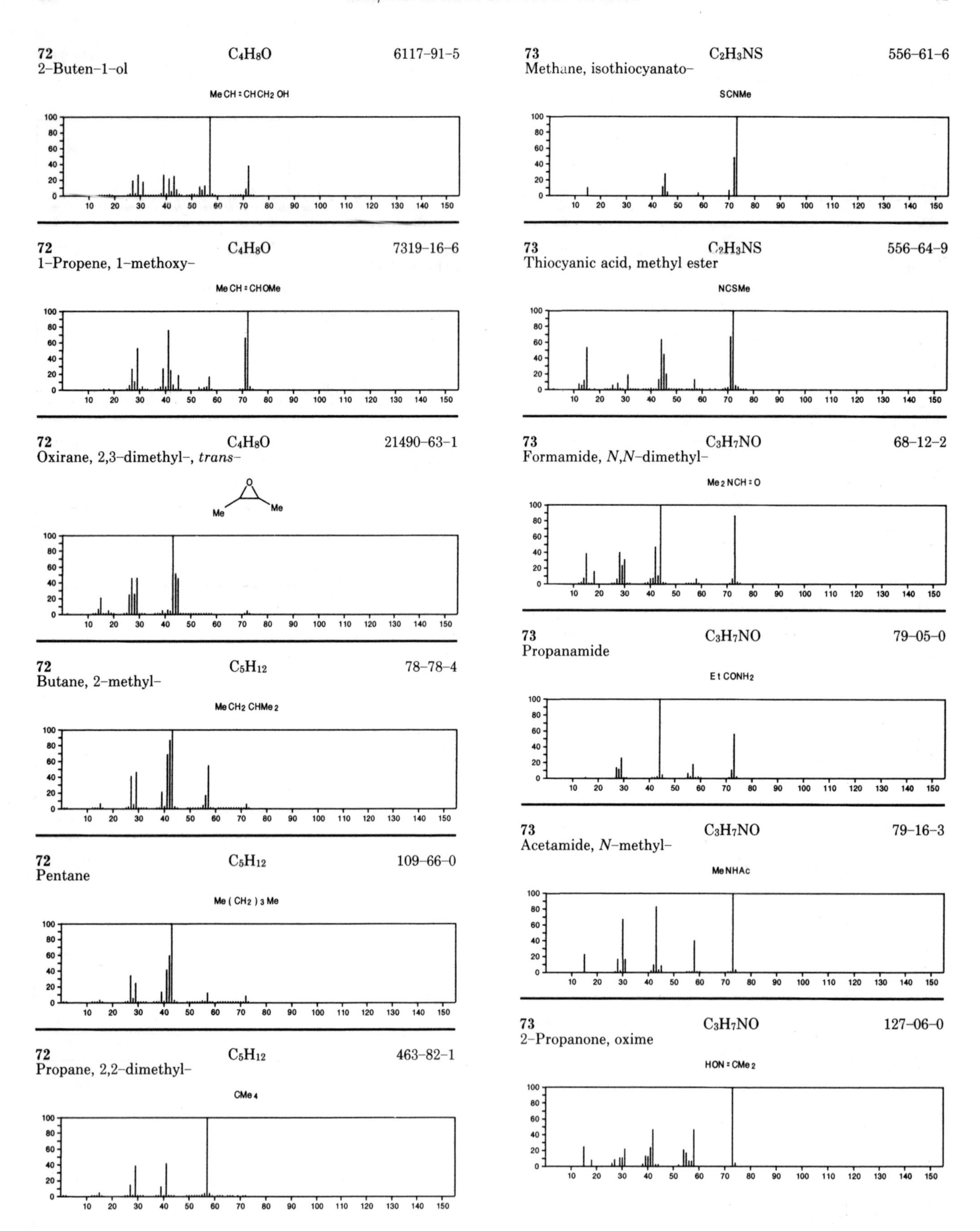
72 C_4H_8O 6117-91-5
2-Buten-1-ol
MeCH=CHCH2OH
72 C_4H_8O 7319-16-6
1-Propene, 1-methoxy-
MeCH=CHOMe
72 C_4H_8O 21490-63-1
Oxirane, 2,3-dimethyl-, *trans-*
O
Me
Me
72 C_5H_{12} 78-78-4
Butane, 2-methyl-
MeCH2CHMe2
72 C_5H_{12} 109-66-0
Pentane
Me(CH2)3Me
72 C_5H_{12} 463-82-1
Propane, 2,2-dimethyl-
CMe4
73 C_2H_3NS 556-61-6
Methane, isothiocyanato-
SCNMe
73 C_2H_3NS 556-64-9
Thiocyanic acid, methyl ester
NCSMe
73 C_3H_7NO 68-12-2
Formamide, *N,N*-dimethyl-
Me2NCH=O
73 C_3H_7NO 79-05-0
Propanamide
EtCONH2
73 C_3H_7NO 79-16-3
Acetamide, *N*-methyl-
MeNHAc
73 C_3H_7NO 127-06-0
2-Propanone, oxime
HON=CMe2
100
80
60
40
20
0
10 20 30 40 50 60 70 80 90 100 110 120 130 140 150

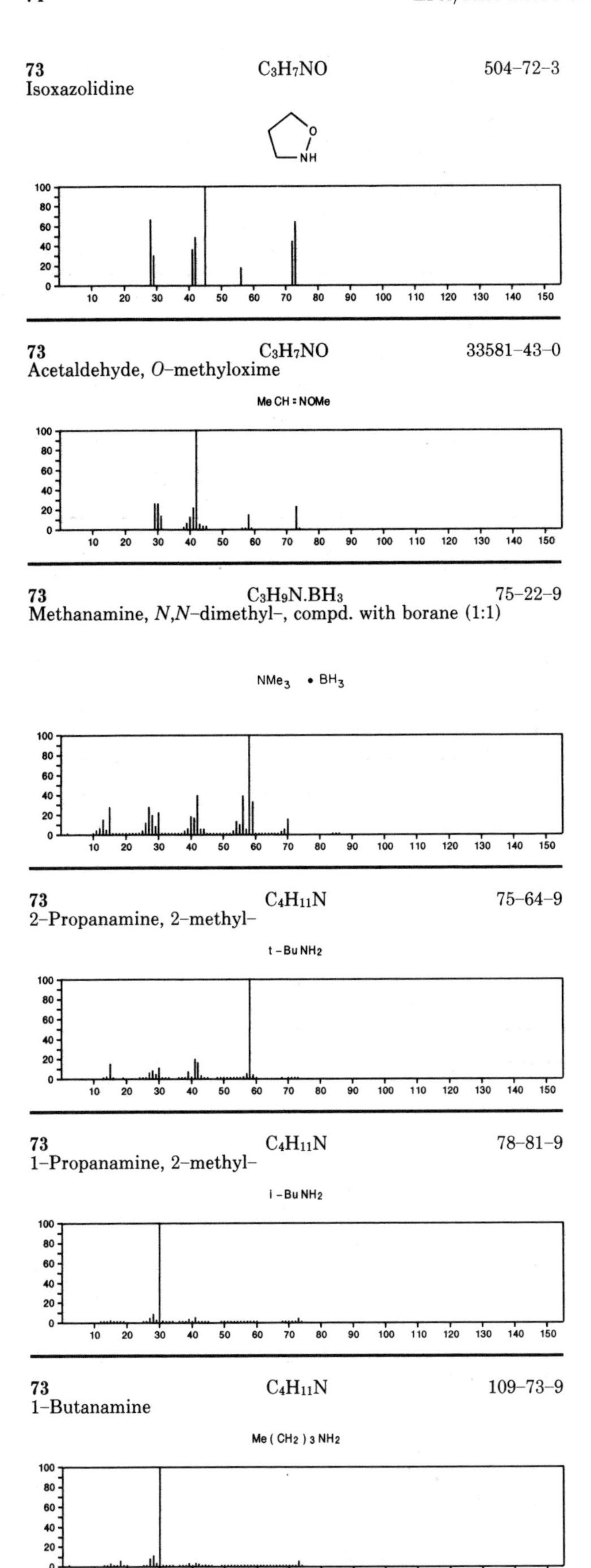

73 C_3H_7NO 504-72-3
Isoxazolidine

73 C_3H_7NO 33581-43-0
Acetaldehyde, *O*-methyloxime
MeCH=NOMe

73 $C_3H_9N.BH_3$ 75-22-9
Methanamine, *N,N*-dimethyl-, compd. with borane (1:1)
$NMe_3 \cdot BH_3$

73 $C_4H_{11}N$ 75-64-9
2-Propanamine, 2-methyl-
$t\text{-}BuNH_2$

73 $C_4H_{11}N$ 78-81-9
1-Propanamine, 2-methyl-
$i\text{-}BuNH_2$

73 $C_4H_{11}N$ 109-73-9
1-Butanamine
$Me(CH_2)_3NH_2$

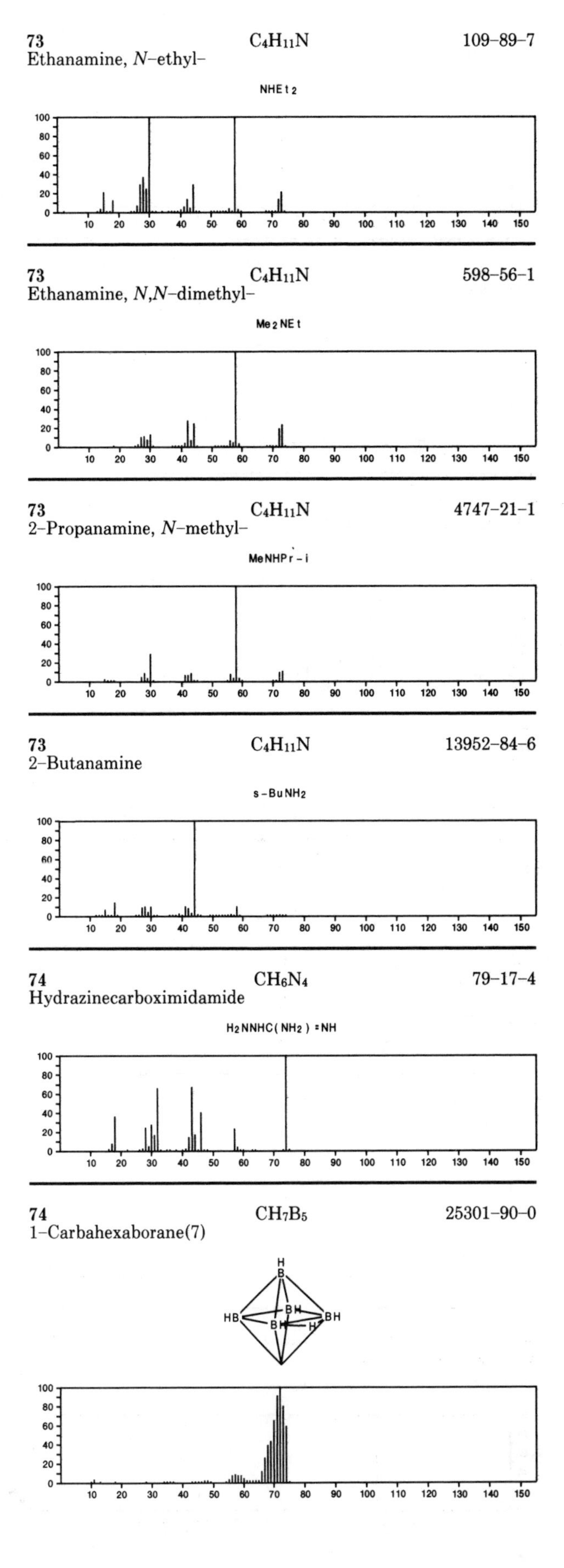

73 $C_4H_{11}N$ 109-89-7
Ethanamine, *N*-ethyl-
$NHEt_2$

73 $C_4H_{11}N$ 598-56-1
Ethanamine, *N,N*-dimethyl-
Me_2NEt

73 $C_4H_{11}N$ 4747-21-1
2-Propanamine, *N*-methyl-
MeNHPr-i

73 $C_4H_{11}N$ 13952-84-6
2-Butanamine
$s\text{-}BuNH_2$

74 CH_6N_4 79-17-4
Hydrazinecarboximidamide
$H_2NNHC(NH_2)=NH$

74 CH_7B_5 25301-90-0
1-Carbahexaborane(7)

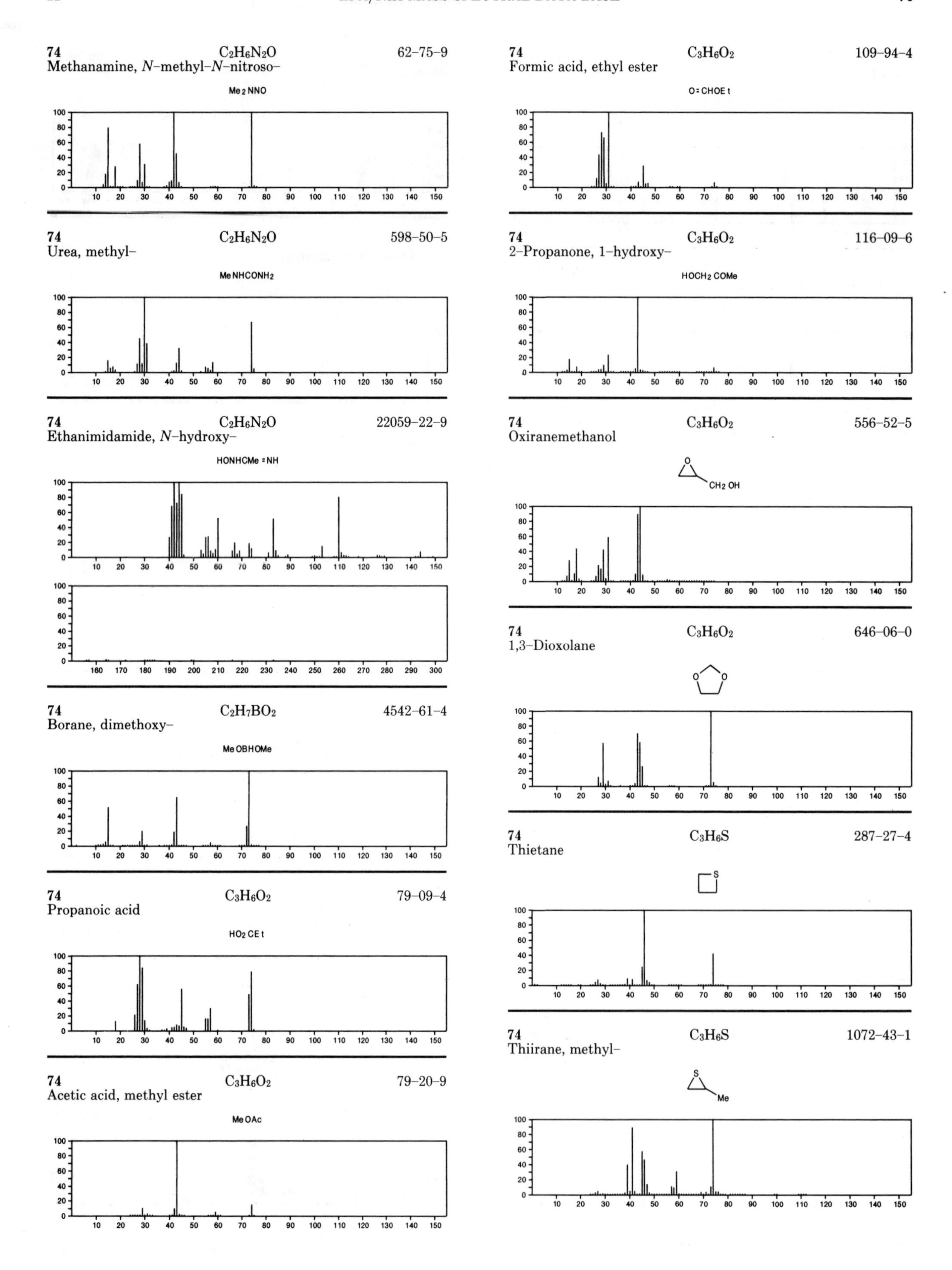
74 C2H6N2O 62-75-9
Methanamine, N-methyl-N-nitroso-
Me2NNO
74 C2H6N2O 598-50-5
Urea, methyl-
MeNHCONH2
74 C2H6N2O 22059-22-9
Ethanimidamide, N-hydroxy-
HONHCMe=NH
74 C2H7BO2 4542-61-4
Borane, dimethoxy-
MeOBHOMe
74 C3H6O2 79-09-4
Propanoic acid
HO2CEt
74 C3H6O2 79-20-9
Acetic acid, methyl ester
MeOAc
74 C3H6O2 109-94-4
Formic acid, ethyl ester
O=CHOEt
74 C3H6O2 116-09-6
2-Propanone, 1-hydroxy-
HOCH2COMe
74 C3H6O2 556-52-5
Oxiranemethanol
CH2OH
74 C3H6O2 646-06-0
1,3-Dioxolane
74 C3H6S 287-27-4
Thietane
74 C3H6S 1072-43-1
Thiirane, methyl-
Me

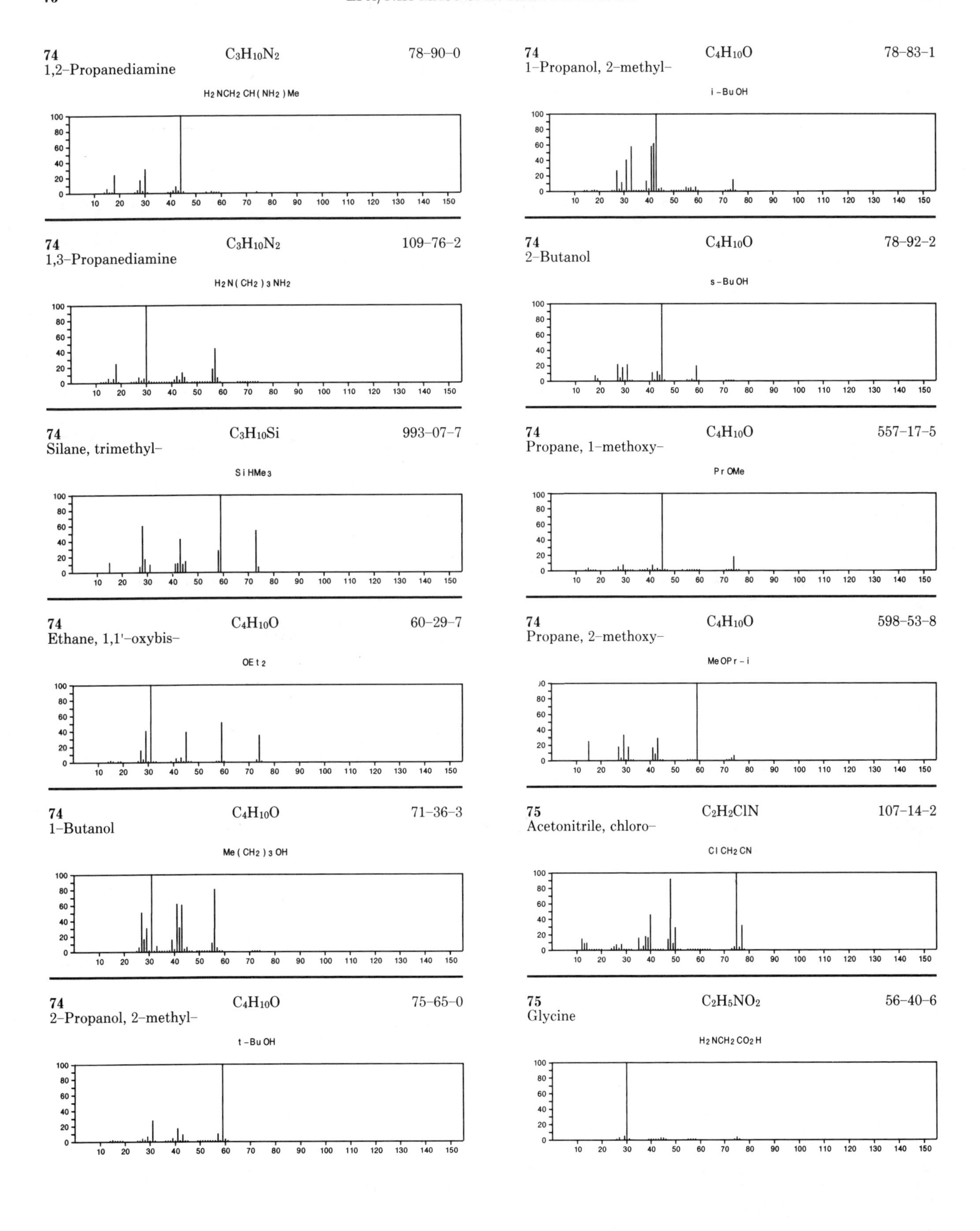
74 C3H10N2 78-90-0
1,2-Propanediamine
H2NCH2CH(NH2)Me
74 C3H10N2 109-76-2
1,3-Propanediamine
H2N(CH2)3NH2
74 C3H10Si 993-07-7
Silane, trimethyl-
SiHMe3
74 C4H10O 60-29-7
Ethane, 1,1'-oxybis-
OEt2
74 C4H10O 71-36-3
1-Butanol
Me(CH2)3OH
74 C4H10O 75-65-0
2-Propanol, 2-methyl-
t-BuOH
74 C4H10O 78-83-1
1-Propanol, 2-methyl-
i-BuOH
74 C4H10O 78-92-2
2-Butanol
s-BuOH
74 C4H10O 557-17-5
Propane, 1-methoxy-
PrOMe
74 C4H10O 598-53-8
Propane, 2-methoxy-
MeOPr-i
75 C2H2ClN 107-14-2
Acetonitrile, chloro-
ClCH2CN
75 C2H5NO2 56-40-6
Glycine
H2NCH2CO2H
100
80
60
40
20
0
10 20 30 40 50 60 70 80 90 100 110 120 130 140 150

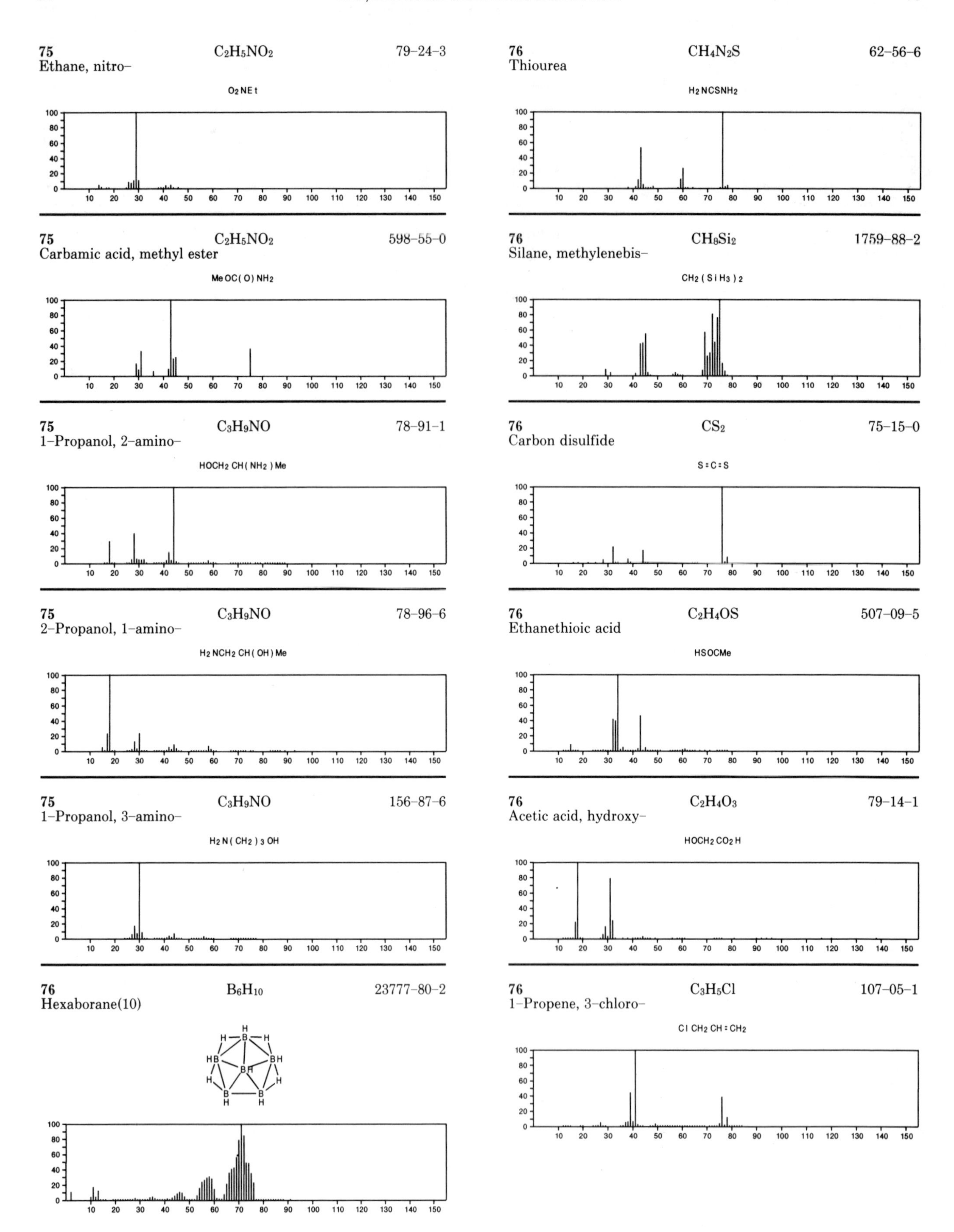
75 C2H5NO2 79-24-3
Ethane, nitro-
O2NEt
75 C2H5NO2 598-55-0
Carbamic acid, methyl ester
MeOC(O)NH2
75 C3H9NO 78-91-1
1-Propanol, 2-amino-
HOCH2CH(NH2)Me
75 C3H9NO 78-96-6
2-Propanol, 1-amino-
H2NCH2CH(OH)Me
75 C3H9NO 156-87-6
1-Propanol, 3-amino-
H2N(CH2)3OH
76 B6H10 23777-80-2
Hexaborane(10)
76 CH4N2S 62-56-6
Thiourea
H2NCSNH2
76 CH8Si2 1759-88-2
Silane, methylenebis-
CH2(SiH3)2
76 CS2 75-15-0
Carbon disulfide
S=C=S
76 C2H4OS 507-09-5
Ethanethioic acid
HSOCMe
76 C2H4O3 79-14-1
Acetic acid, hydroxy-
HOCH2CO2H
76 C3H5Cl 107-05-1
1-Propene, 3-chloro-
ClCH2CH=CH2

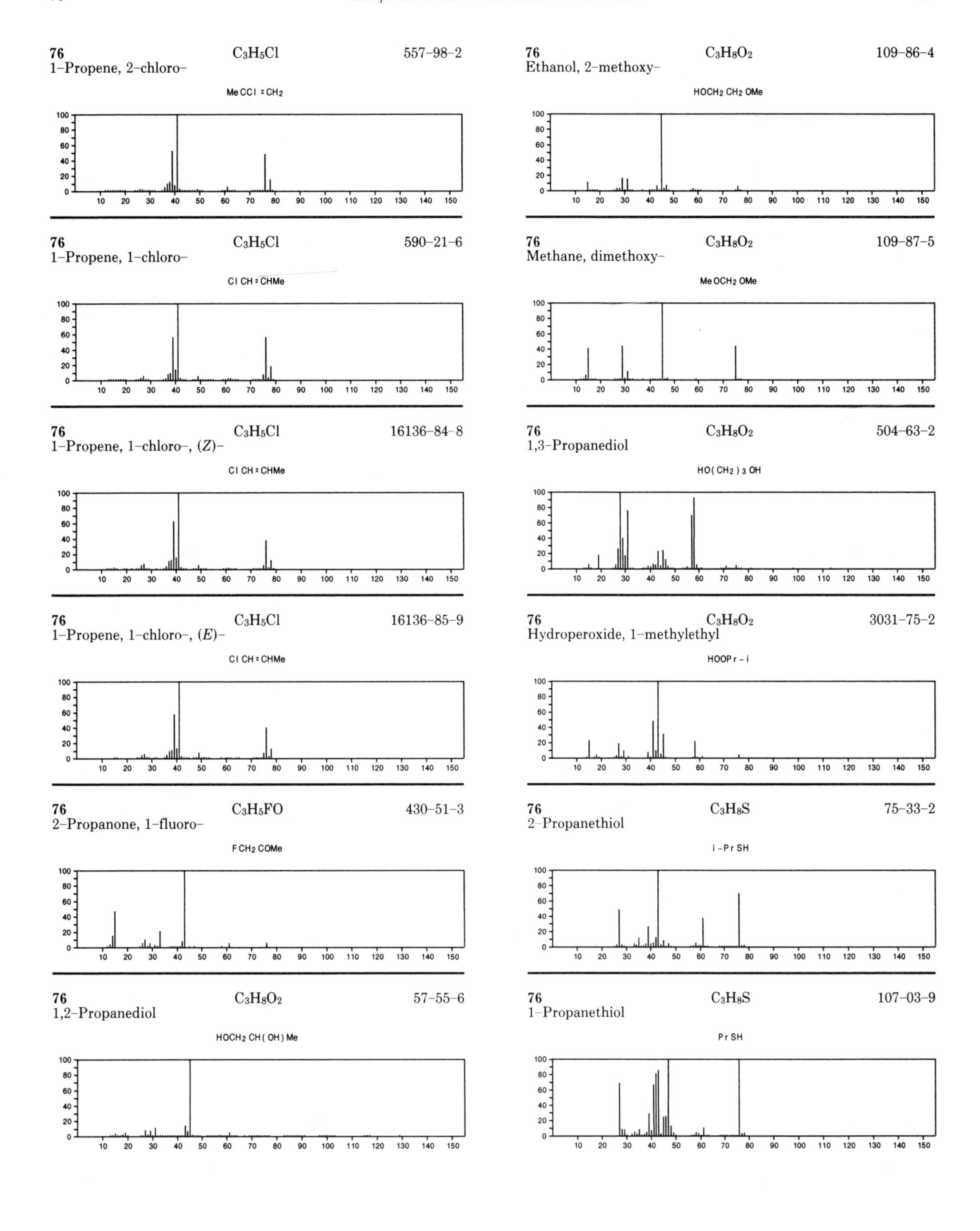

76 1-Propene, 2-chloro- C_3H_5Cl 557-98-2
MeCCl=CH2

76 1-Propene, 1-chloro- C_3H_5Cl 590-21-6
ClCH=CHMe

76 1-Propene, 1-chloro-, (*Z*)- C_3H_5Cl 16136-84-8
ClCH=CHMe

76 1-Propene, 1-chloro-, (*E*)- C_3H_5Cl 16136-85-9
ClCH=CHMe

76 2-Propanone, 1-fluoro- C_3H_5FO 430-51-3
FCH2COMe

76 1,2-Propanediol $C_3H_8O_2$ 57-55-6
HOCH2CH(OH)Me

76 Ethanol, 2-methoxy- $C_3H_8O_2$ 109-86-4
HOCH2CH2OMe

76 Methane, dimethoxy- $C_3H_8O_2$ 109-87-5
MeOCH2OMe

76 1,3-Propanediol $C_3H_8O_2$ 504-63-2
HO(CH2)3OH

76 Hydroperoxide, 1-methylethyl $C_3H_8O_2$ 3031-75-2
HOOPr-i

76 2-Propanethiol C_3H_8S 75-33-2
i-PrSH

76 1-Propanethiol C_3H_8S 107-03-9
PrSH

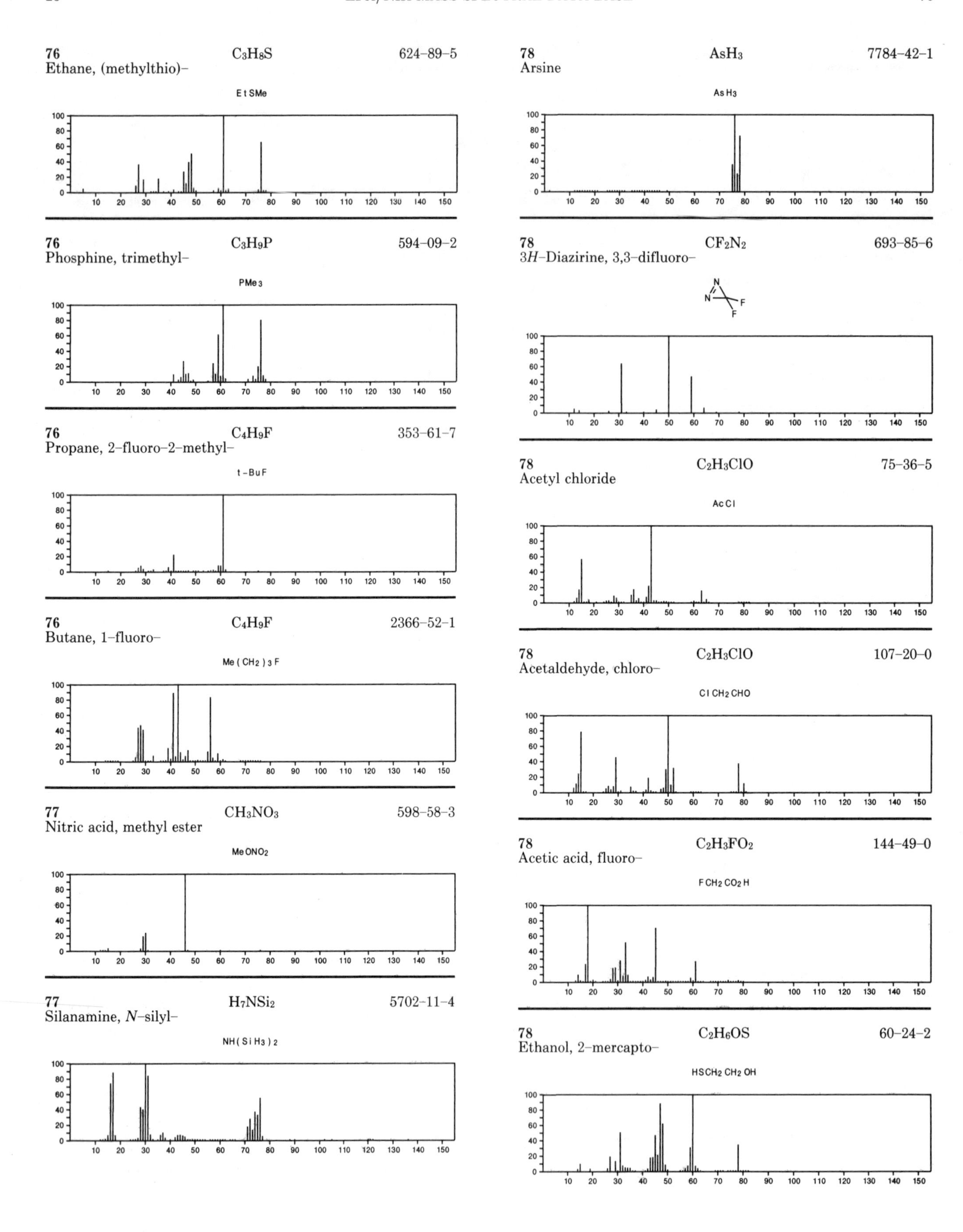
76 C3H8S 624-89-5
Ethane, (methylthio)-
EtSMe
76 C3H9P 594-09-2
Phosphine, trimethyl-
PMe3
76 C4H9F 353-61-7
Propane, 2-fluoro-2-methyl-
t-BuF
76 C4H9F 2366-52-1
Butane, 1-fluoro-
Me(CH2)3F
77 CH3NO3 598-58-3
Nitric acid, methyl ester
MeONO2
77 H7NSi2 5702-11-4
Silanamine, N-silyl-
NH(SiH3)2
78 AsH3 7784-42-1
Arsine
AsH3
78 CF2N2 693-85-6
3H-Diazirine, 3,3-difluoro-
78 C2H3ClO 75-36-5
Acetyl chloride
AcCl
78 C2H3ClO 107-20-0
Acetaldehyde, chloro-
ClCH2CHO
78 C2H3FO2 144-49-0
Acetic acid, fluoro-
FCH2CO2H
78 C2H6OS 60-24-2
Ethanol, 2-mercapto-
HSCH2CH2OH

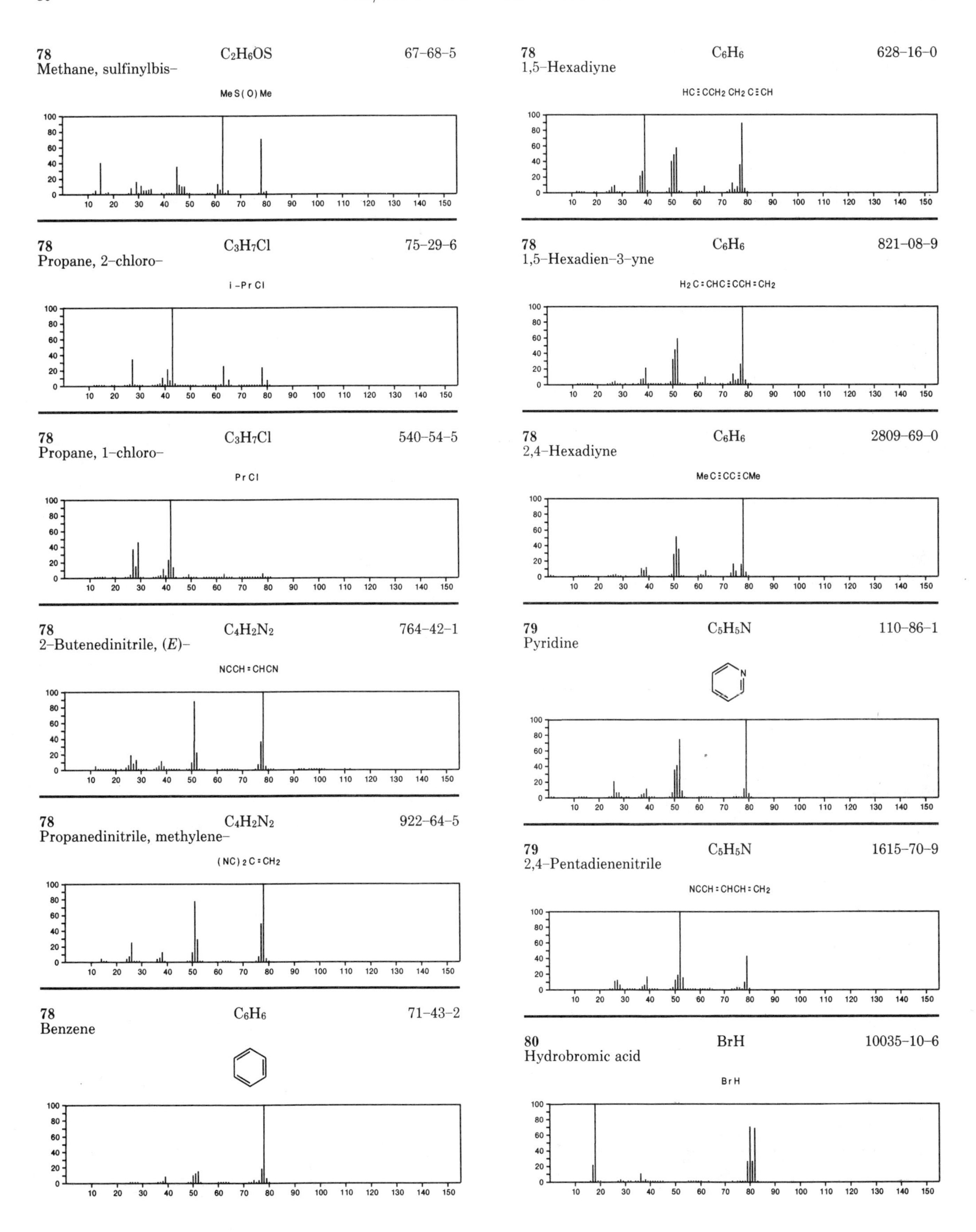
78
C_2H_6OS
67-68-5
Methane, sulfinylbis-
Me S(O) Me
78
C_3H_7Cl
75-29-6
Propane, 2-chloro-
i-Pr Cl
78
C_3H_7Cl
540-54-5
Propane, 1-chloro-
Pr Cl
78
$C_4H_2N_2$
764-42-1
2-Butenedinitrile, (E)-
NCCH=CHCN
78
$C_4H_2N_2$
922-64-5
Propanedinitrile, methylene-
(NC)2C=CH2
78
C_6H_6
71-43-2
Benzene
78
C_6H_6
628-16-0
1,5-Hexadiyne
HC⋮CCH2CH2C⋮CH
78
C_6H_6
821-08-9
1,5-Hexadien-3-yne
H2C=CHC⋮CCH=CH2
78
C_6H_6
2809-69-0
2,4-Hexadiyne
MeC⋮CC⋮CMe
79
C_5H_5N
110-86-1
Pyridine
N
79
C_5H_5N
1615-70-9
2,4-Pentadienenitrile
NCCH=CHCH=CH2
80
BrH
10035-10-6
Hydrobromic acid
BrH
100
80
60
40
20
0
10
20
30
40
50
60
70
80
90
100
110
120
130
140
150

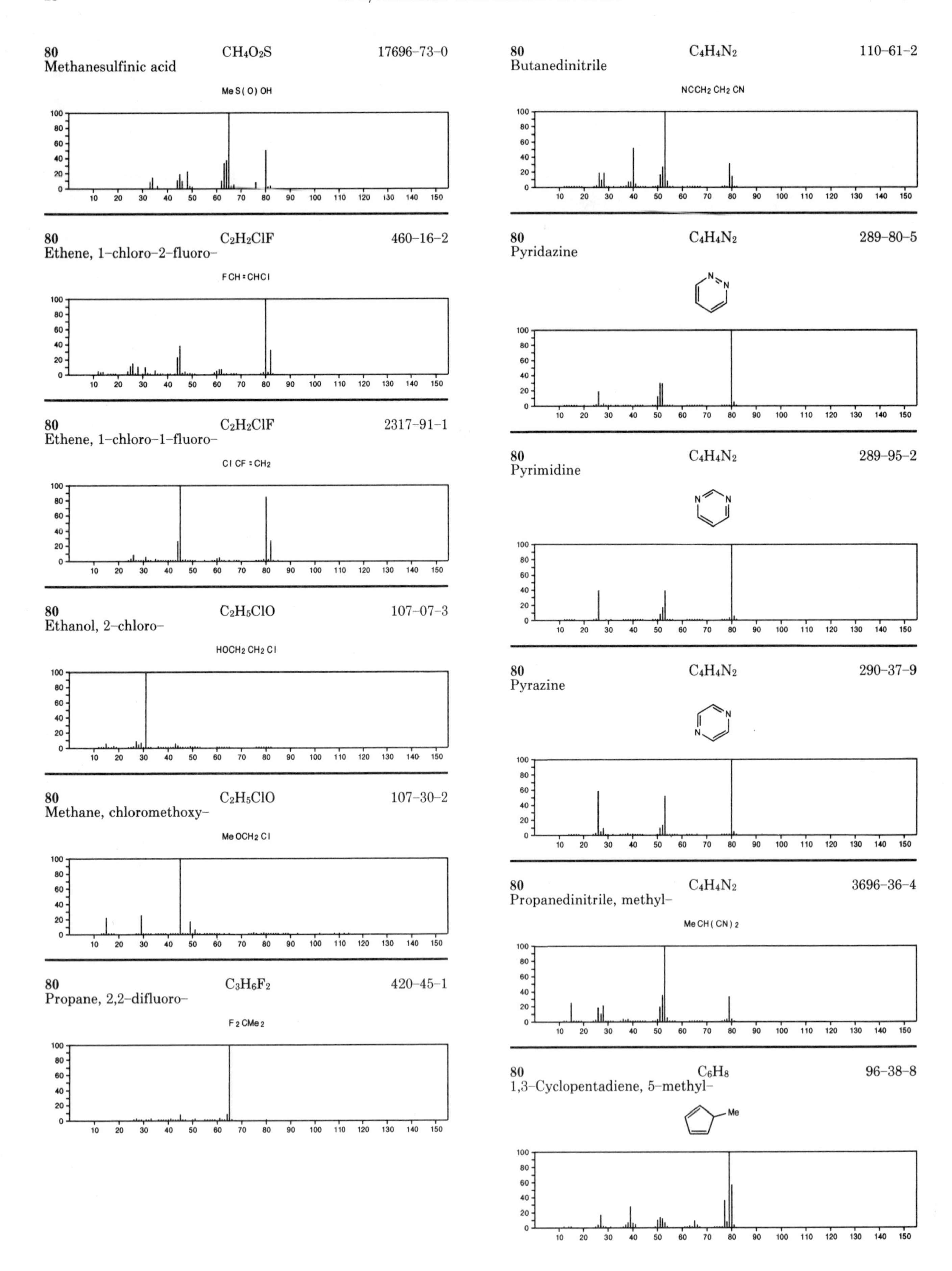
80 CH4O2S 17696-73-0
Methanesulfinic acid
Me S (O) OH
80 C2H2ClF 460-16-2
Ethene, 1-chloro-2-fluoro-
F CH = CHCl
80 C2H2ClF 2317-91-1
Ethene, 1-chloro-1-fluoro-
Cl CF = CH2
80 C2H5ClO 107-07-3
Ethanol, 2-chloro-
HOCH2 CH2 Cl
80 C2H5ClO 107-30-2
Methane, chloromethoxy-
Me OCH2 Cl
80 C3H6F2 420-45-1
Propane, 2,2-difluoro-
F2 CMe2
80 C4H4N2 110-61-2
Butanedinitrile
NCCH2 CH2 CN
80 C4H4N2 289-80-5
Pyridazine
80 C4H4N2 289-95-2
Pyrimidine
80 C4H4N2 290-37-9
Pyrazine
80 C4H4N2 3696-36-4
Propanedinitrile, methyl-
Me CH (CN) 2
80 C6H8 96-38-8
1,3-Cyclopentadiene, 5-methyl-
Me

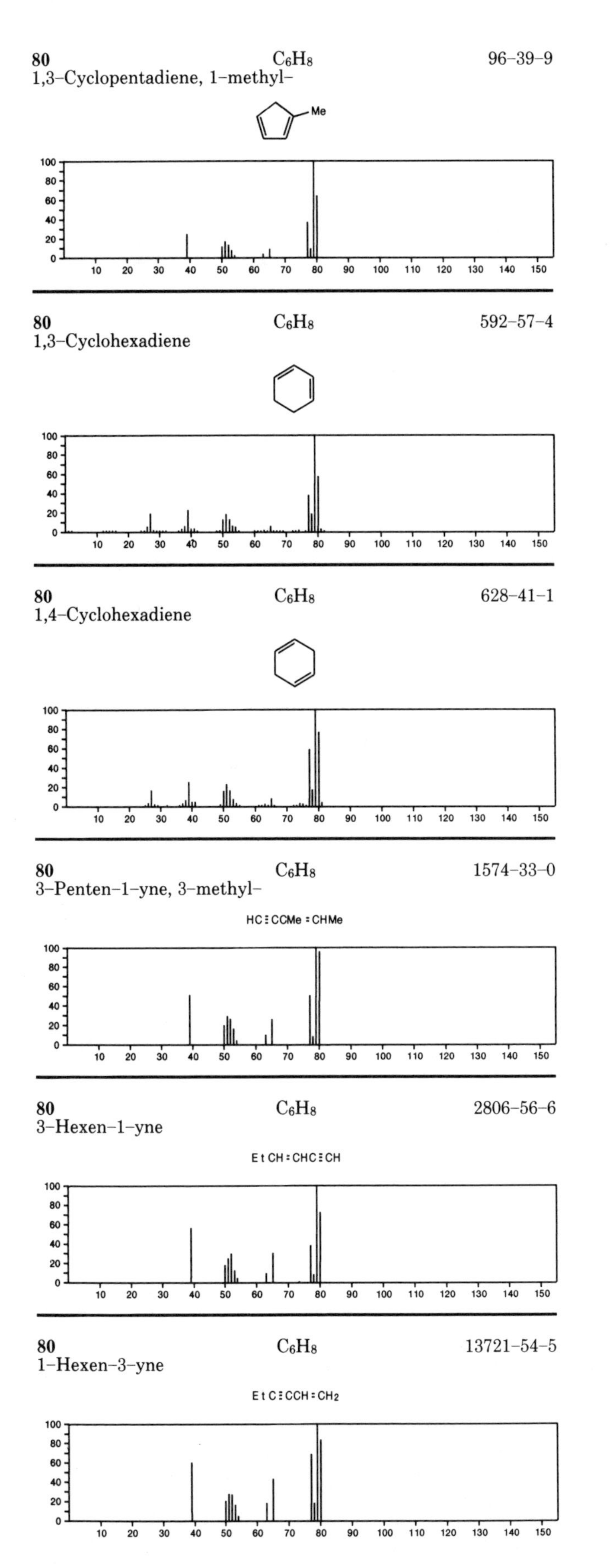

80 C_6H_8 96-39-9
1,3-Cyclopentadiene, 1-methyl-

80 C_6H_8 592-57-4
1,3-Cyclohexadiene

80 C_6H_8 628-41-1
1,4-Cyclohexadiene

80 C_6H_8 1574-33-0
3-Penten-1-yne, 3-methyl-

HC≡CCMe=CHMe

80 C_6H_8 2806-56-6
3-Hexen-1-yne

EtCH=CHC≡CH

80 C_6H_8 13721-54-5
1-Hexen-3-yne

EtC≡CCH=CH2

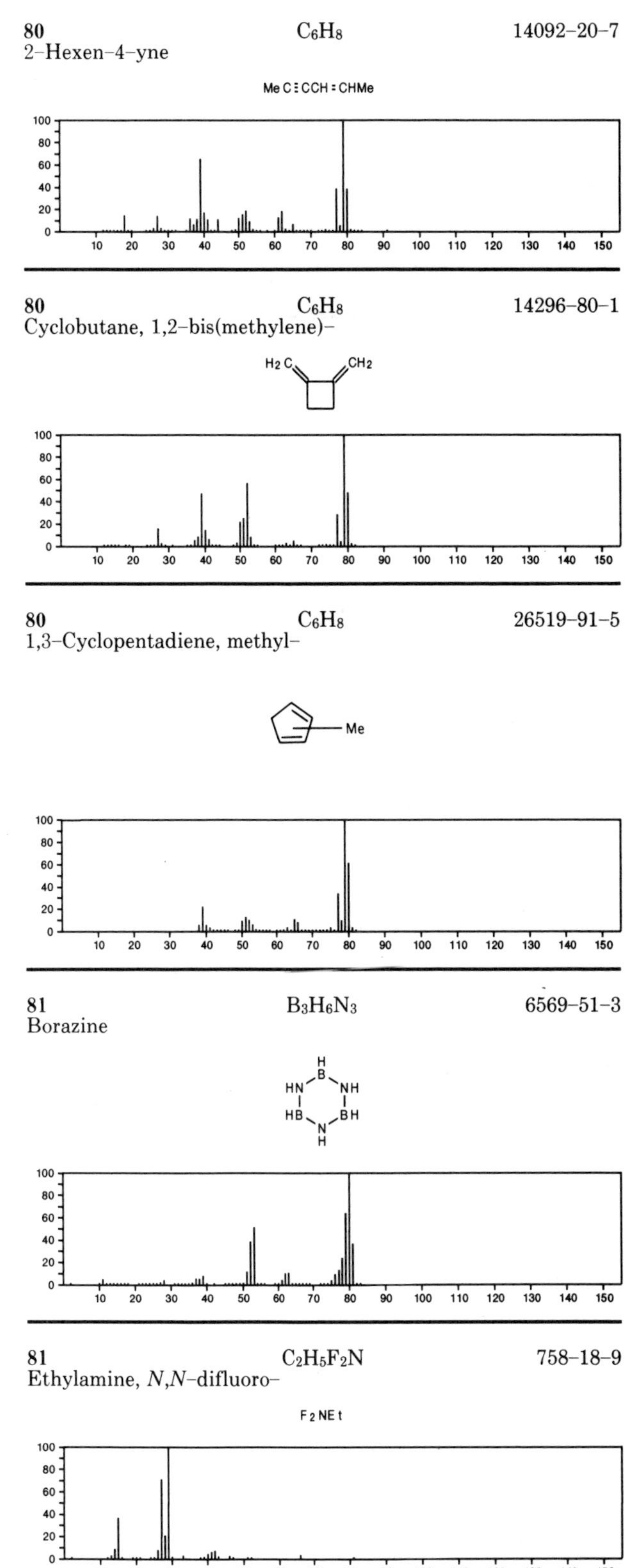

80 C_6H_8 14092-20-7
2-Hexen-4-yne

MeC≡CCH=CHMe

80 C_6H_8 14296-80-1
Cyclobutane, 1,2-bis(methylene)-

80 C_6H_8 26519-91-5
1,3-Cyclopentadiene, methyl-

81 $B_3H_6N_3$ 6569-51-3
Borazine

81 $C_2H_5F_2N$ 758-18-9
Ethylamine, *N,N*-difluoro-

F2NEt

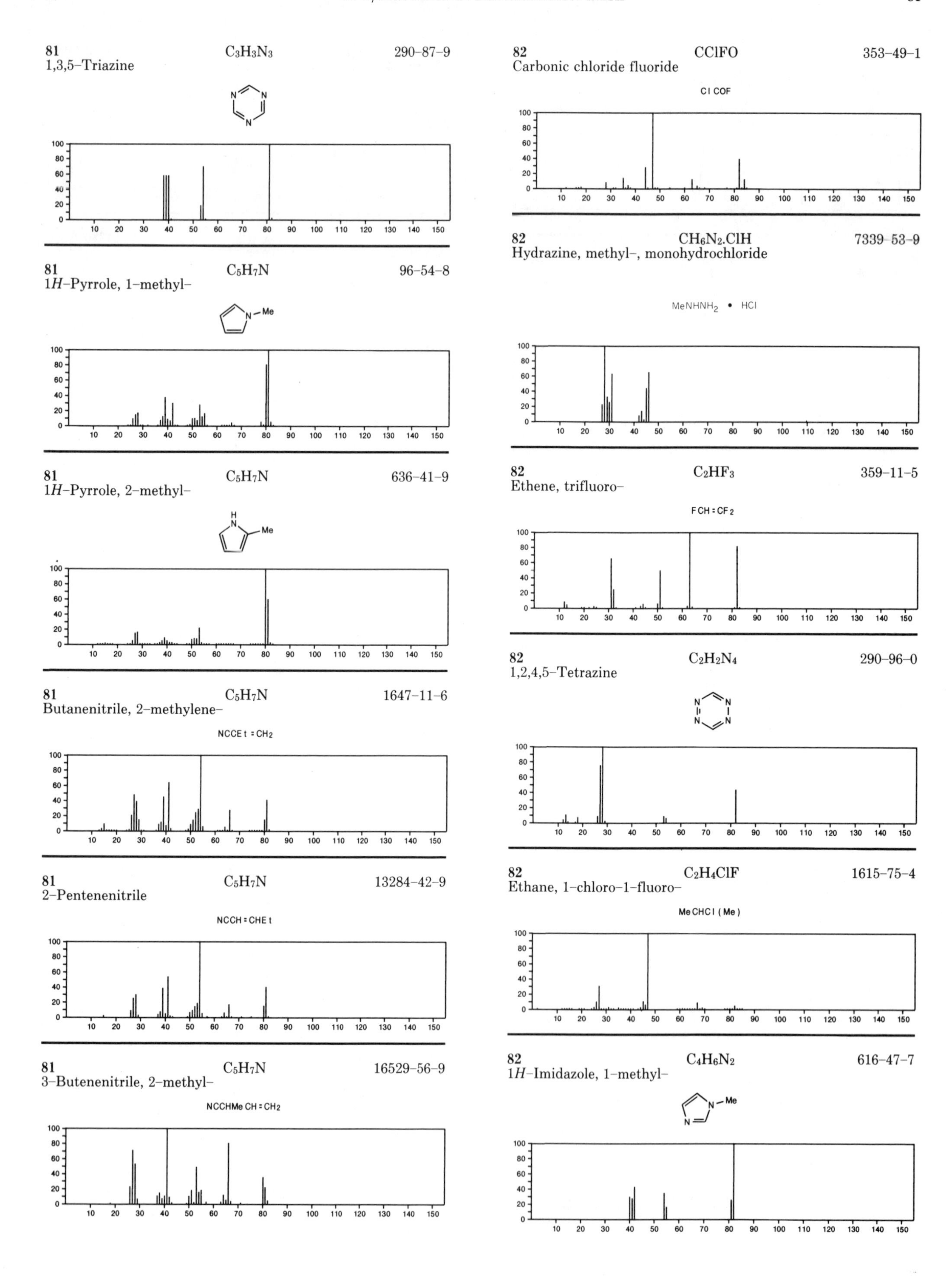
81 C3H3N3 290-87-9
1,3,5-Triazine
81 C5H7N 96-54-8
1H-Pyrrole, 1-methyl-
N-Me
81 C5H7N 636-41-9
1H-Pyrrole, 2-methyl-
Me
81 C5H7N 1647-11-6
Butanenitrile, 2-methylene-
NCCEt = CH2
81 C5H7N 13284-42-9
2-Pentenenitrile
NCCH = CHEt
81 C5H7N 16529-56-9
3-Butenenitrile, 2-methyl-
NCCHMe CH = CH2
82 CClFO 353-49-1
Carbonic chloride fluoride
Cl COF
82 CH6N2.ClH 7339-53-9
Hydrazine, methyl-, monohydrochloride
MeNHNH2 • HCl
82 C2HF3 359-11-5
Ethene, trifluoro-
FCH = CF2
82 C2H2N4 290-96-0
1,2,4,5-Tetrazine
82 C2H4ClF 1615-75-4
Ethane, 1-chloro-1-fluoro-
MeCHCl (Me)
82 C4H6N2 616-47-7
1H-Imidazole, 1-methyl-
N-Me

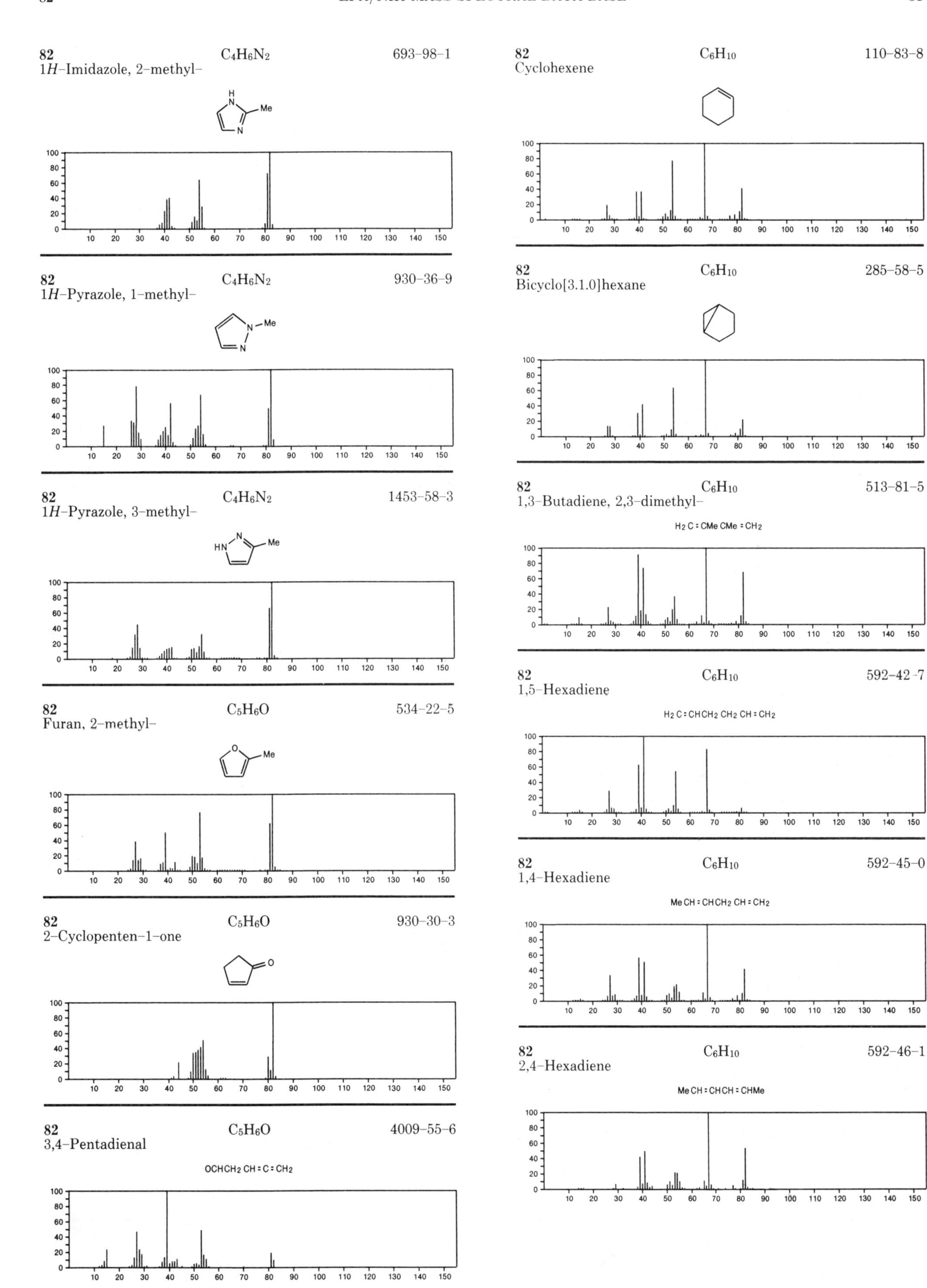

82 $C_4H_6N_2$ 693-98-1
1*H*-Imidazole, 2-methyl-

82 $C_4H_6N_2$ 930-36-9
1*H*-Pyrazole, 1-methyl-

82 $C_4H_6N_2$ 1453-58-3
1*H*-Pyrazole, 3-methyl-

82 C_5H_6O 534-22-5
Furan, 2-methyl-

82 C_5H_6O 930-30-3
2-Cyclopenten-1-one

82 C_5H_6O 4009-55-6
3,4-Pentadienal

OCHCH2 CH = C = CH2

82 C_6H_{10} 110-83-8
Cyclohexene

82 C_6H_{10} 285-58-5
Bicyclo[3.1.0]hexane

82 C_6H_{10} 513-81-5
1,3-Butadiene, 2,3-dimethyl-

H2 C = CMe CMe = CH2

82 C_6H_{10} 592-42-7
1,5-Hexadiene

H2 C = CHCH2 CH2 CH = CH2

82 C_6H_{10} 592-45-0
1,4-Hexadiene

Me CH = CHCH2 CH = CH2

82 C_6H_{10} 592-46-1
2,4-Hexadiene

Me CH = CHCH = CHMe

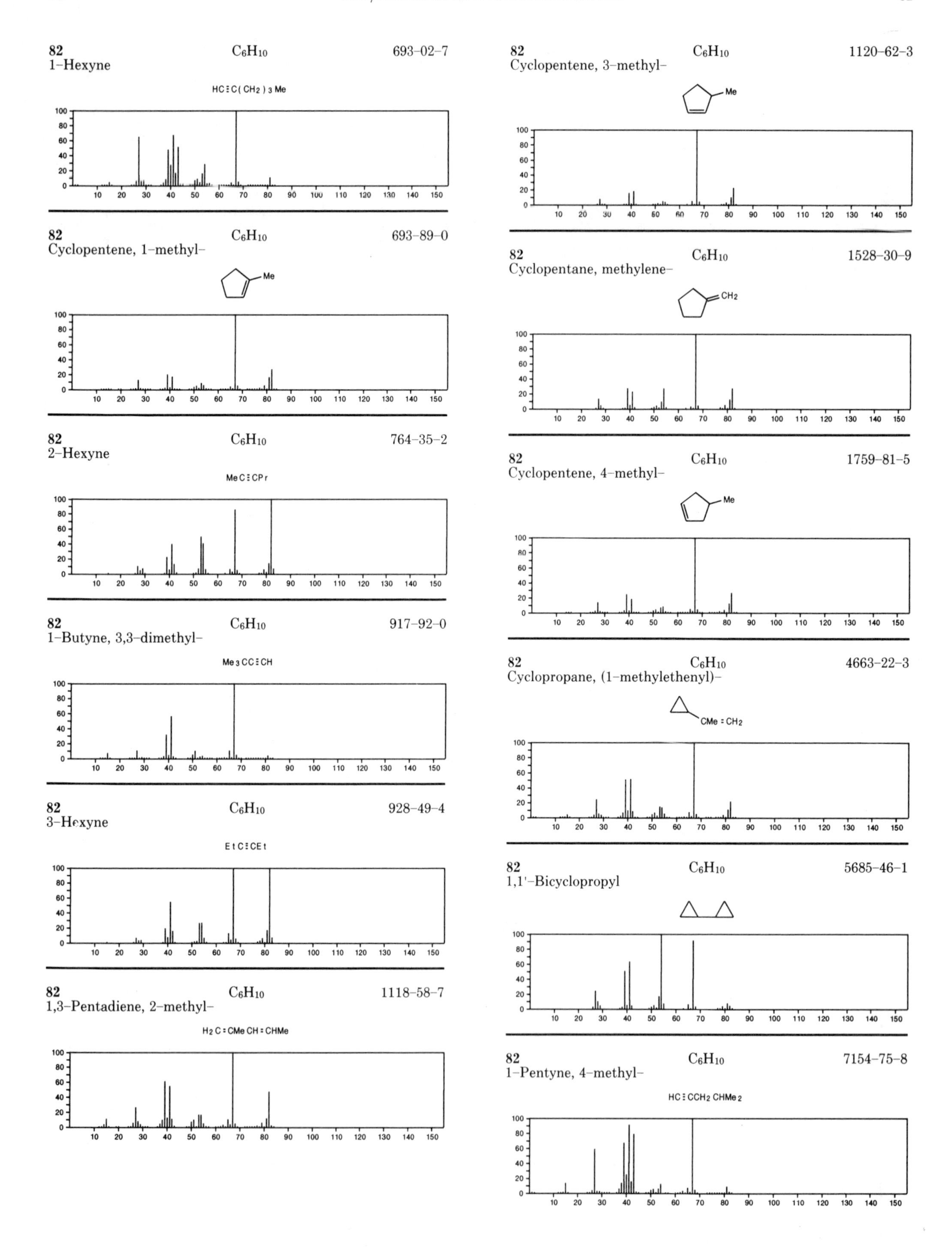
82
1-Hexyne
C_6H_{10}
693-02-7
HC≡C(CH2)3Me
82
Cyclopentene, 1-methyl-
C_6H_{10}
693-89-0
Me
82
2-Hexyne
C_6H_{10}
764-35-2
MeC≡CPr
82
1-Butyne, 3,3-dimethyl-
C_6H_{10}
917-92-0
Me3CC≡CH
82
3-Hexyne
C_6H_{10}
928-49-4
EtC≡CEt
82
1,3-Pentadiene, 2-methyl-
C_6H_{10}
1118-58-7
H2C=CMeCH=CHMe
82
Cyclopentene, 3-methyl-
C_6H_{10}
1120-62-3
Me
82
Cyclopentane, methylene-
C_6H_{10}
1528-30-9
CH2
82
Cyclopentene, 4-methyl-
C_6H_{10}
1759-81-5
Me
82
Cyclopropane, (1-methylethenyl)-
C_6H_{10}
4663-22-3
CMe=CH2
82
1,1'-Bicyclopropyl
C_6H_{10}
5685-46-1
82
1-Pentyne, 4-methyl-
C_6H_{10}
7154-75-8
HC≡CCH2CHMe2

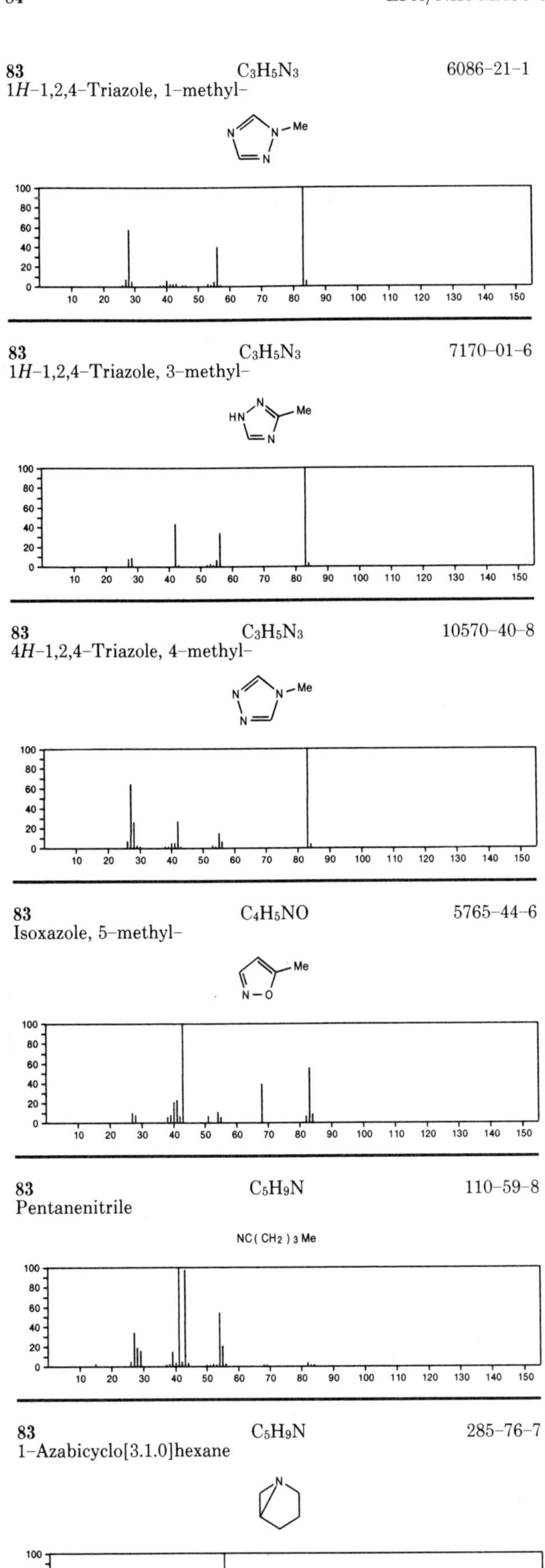

83 $C_3H_5N_3$ 6086-21-1
1*H*-1,2,4-Triazole, 1-methyl-

83 $C_3H_5N_3$ 7170-01-6
1*H*-1,2,4-Triazole, 3-methyl-

83 $C_3H_5N_3$ 10570-40-8
4*H*-1,2,4-Triazole, 4-methyl-

83 C_4H_5NO 5765-44-6
Isoxazole, 5-methyl-

83 C_5H_9N 110-59-8
Pentanenitrile
NC(CH2) 3 Me

83 C_5H_9N 285-76-7
1-Azabicyclo[3.1.0]hexane

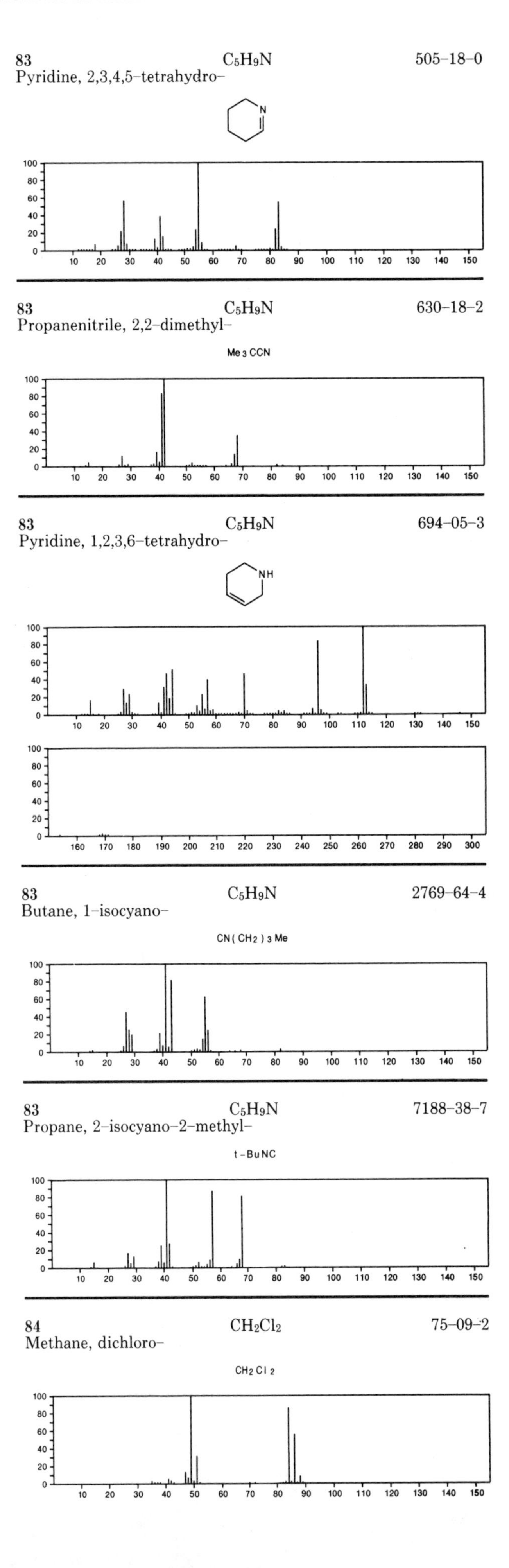

83 C_5H_9N 505-18-0
Pyridine, 2,3,4,5-tetrahydro-

83 C_5H_9N 630-18-2
Propanenitrile, 2,2-dimethyl-
Me 3 CCN

83 C_5H_9N 694-05-3
Pyridine, 1,2,3,6-tetrahydro-

83 C_5H_9N 2769-64-4
Butane, 1-isocyano-
CN(CH2) 3 Me

83 C_5H_9N 7188-38-7
Propane, 2-isocyano-2-methyl-
t-BuNC

84 CH_2Cl_2 75-09-2
Methane, dichloro-
CH2 Cl 2

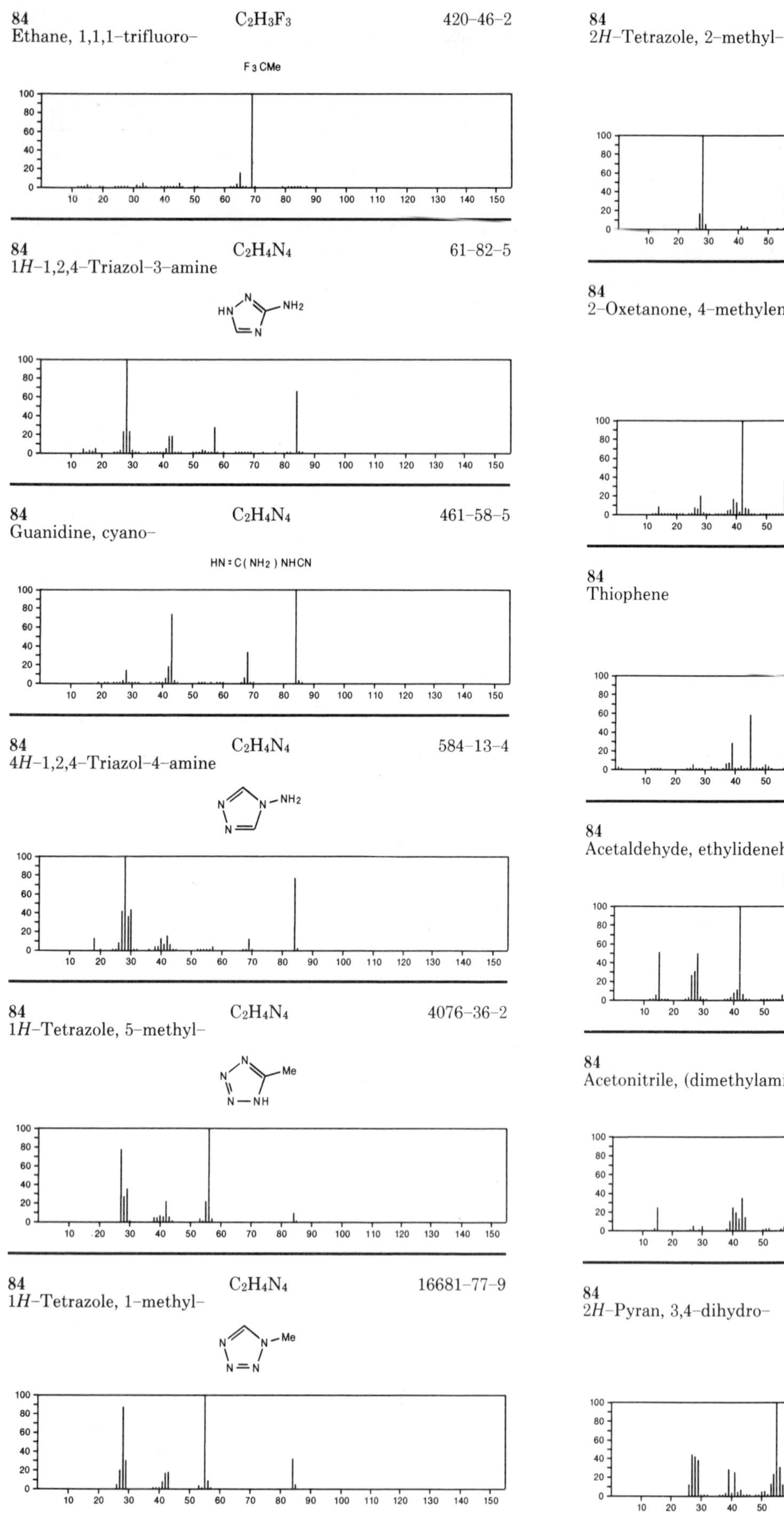

84 $C_2H_3F_3$ 420–46–2
Ethane, 1,1,1–trifluoro–

84 $C_2H_4N_4$ 61–82–5
1*H*–1,2,4–Triazol–3–amine

84 $C_2H_4N_4$ 461–58–5
Guanidine, cyano–

84 $C_2H_4N_4$ 584–13–4
4*H*–1,2,4–Triazol–4–amine

84 $C_2H_4N_4$ 4076–36–2
1*H*–Tetrazole, 5–methyl–

84 $C_2H_4N_4$ 16681–77–9
1*H*–Tetrazole, 1–methyl–

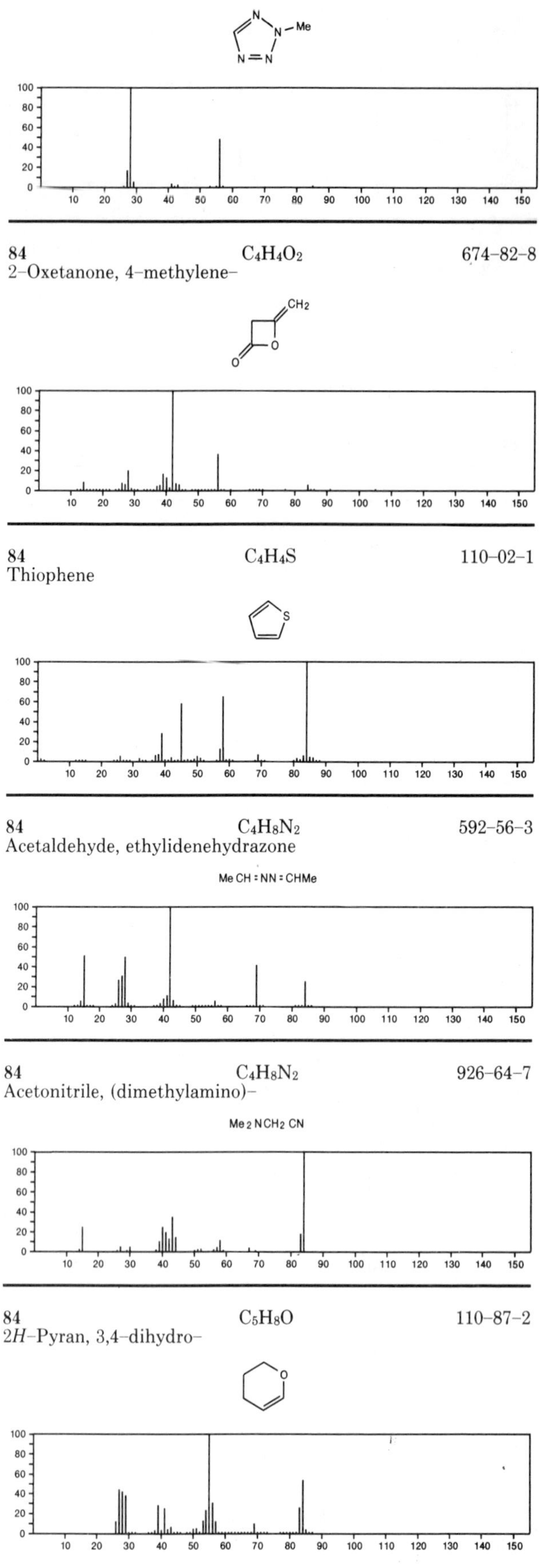

84 $C_2H_4N_4$ 16681–78–0
2*H*–Tetrazole, 2–methyl–

84 $C_4H_4O_2$ 674–82–8
2–Oxetanone, 4–methylene–

84 C_4H_4S 110–02–1
Thiophene

84 $C_4H_8N_2$ 592–56–3
Acetaldehyde, ethylidenehydrazone

84 $C_4H_8N_2$ 926–64–7
Acetonitrile, (dimethylamino)–

84 C_5H_8O 110–87–2
2*H*–Pyran, 3,4–dihydro–

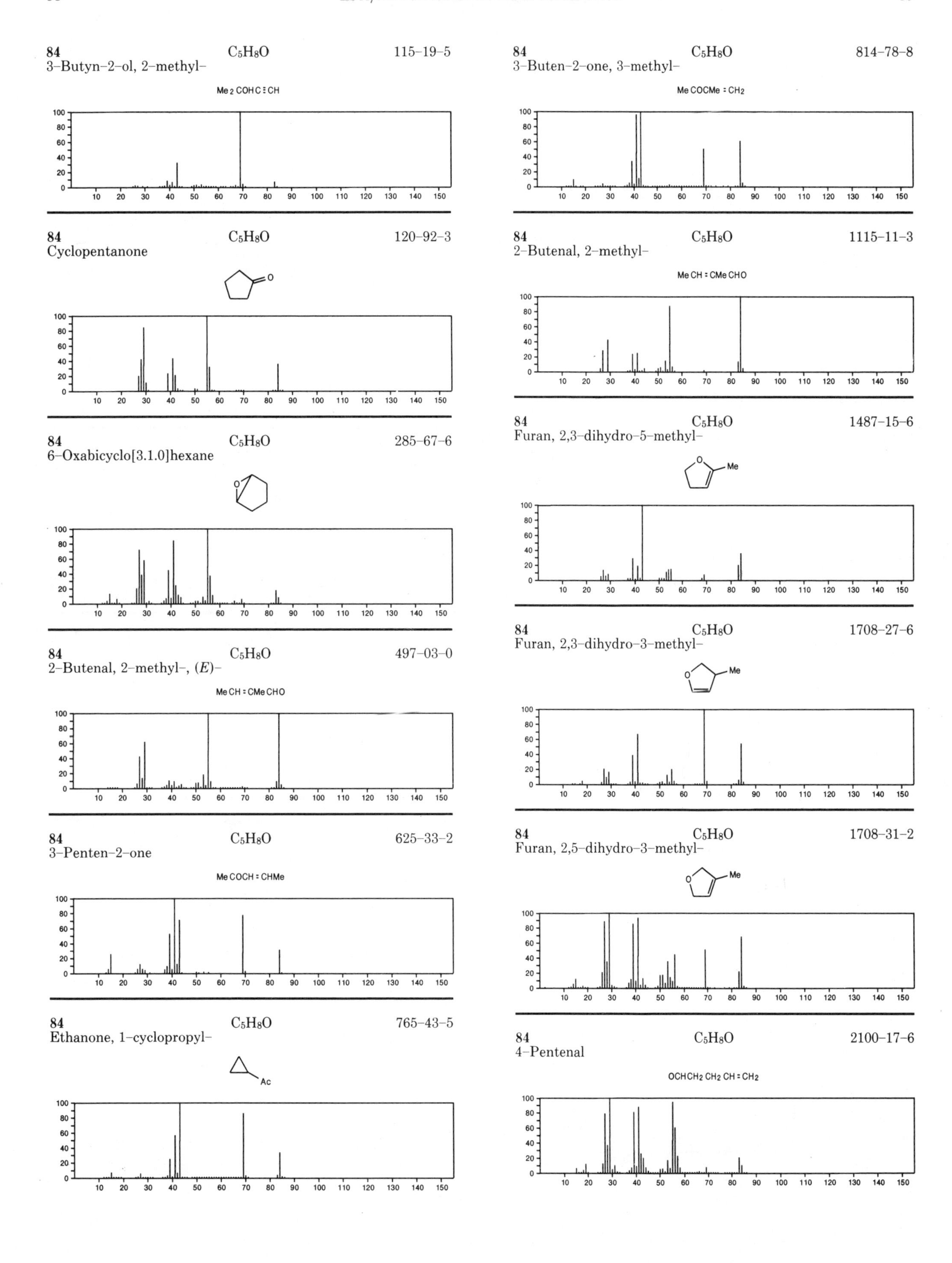
84 C_5H_8O 115-19-5
3-Butyn-2-ol, 2-methyl-
Me2COHC≡CH
84 C_5H_8O 120-92-3
Cyclopentanone
84 C_5H_8O 285-67-6
6-Oxabicyclo[3.1.0]hexane
84 C_5H_8O 497-03-0
2-Butenal, 2-methyl-, (*E*)-
MeCH=CMeCHO
84 C_5H_8O 625-33-2
3-Penten-2-one
MeCOCH=CHMe
84 C_5H_8O 765-43-5
Ethanone, 1-cyclopropyl-
Ac
84 C_5H_8O 814-78-8
3-Buten-2-one, 3-methyl-
MeCOCMe=CH2
84 C_5H_8O 1115-11-3
2-Butenal, 2-methyl-
MeCH=CMeCHO
84 C_5H_8O 1487-15-6
Furan, 2,3-dihydro-5-methyl-
Me
84 C_5H_8O 1708-27-6
Furan, 2,3-dihydro-3-methyl-
Me
84 C_5H_8O 1708-31-2
Furan, 2,5-dihydro-3-methyl-
Me
84 C_5H_8O 2100-17-6
4-Pentenal
OCHCH2CH2CH=CH2

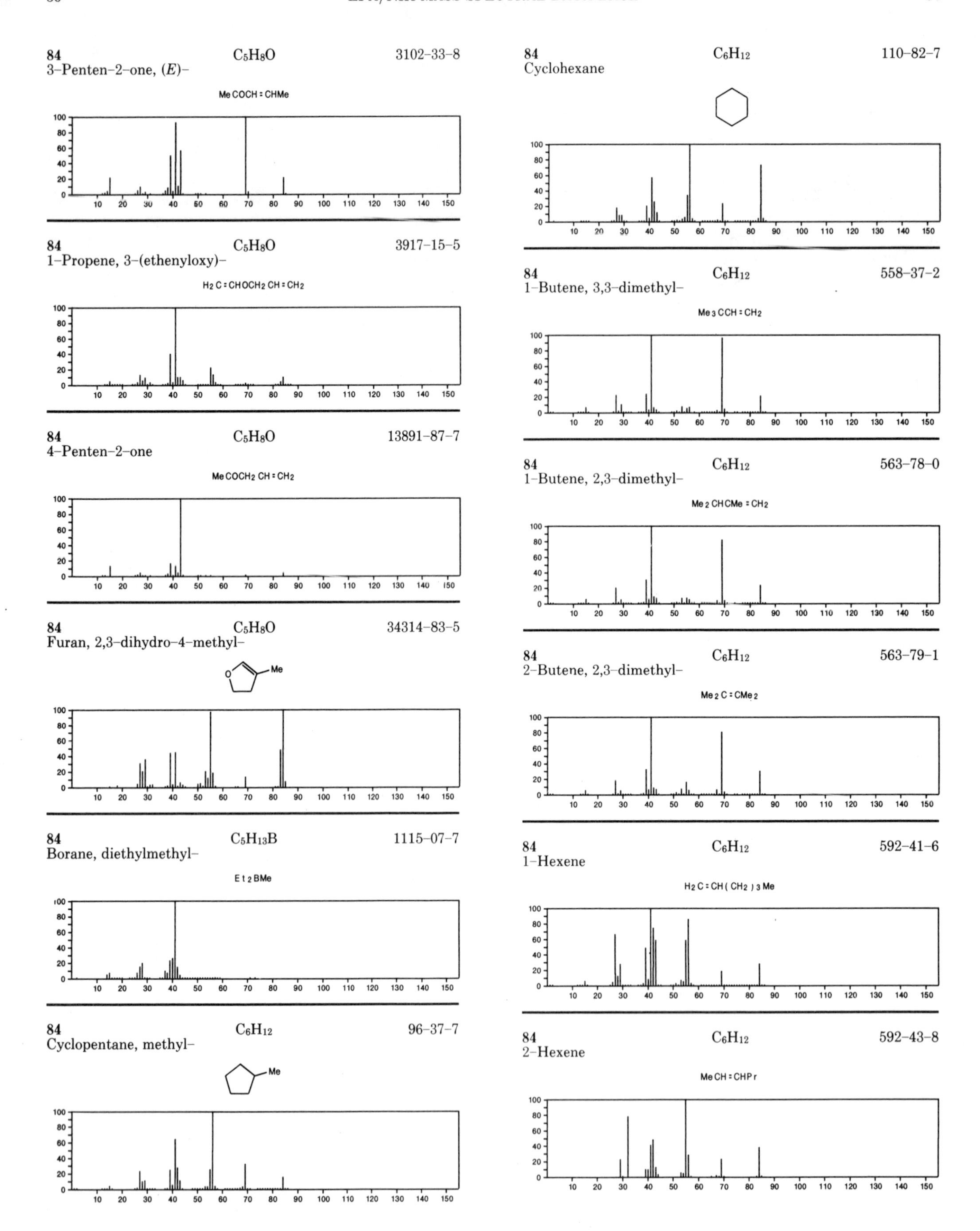
84
C_5H_8O
3102-33-8
3-Penten-2-one, (E)-
MeCOCH=CHMe
84
C_5H_8O
3917-15-5
1-Propene, 3-(ethenyloxy)-
H2C=CHOCH2CH=CH2
84
C_5H_8O
13891-87-7
4-Penten-2-one
MeCOCH2CH=CH2
84
C_5H_8O
34314-83-5
Furan, 2,3-dihydro-4-methyl-
84
$C_5H_{13}B$
1115-07-7
Borane, diethylmethyl-
Et2BMe
84
C_6H_{12}
96-37-7
Cyclopentane, methyl-
84
C_6H_{12}
110-82-7
Cyclohexane
84
C_6H_{12}
558-37-2
1-Butene, 3,3-dimethyl-
Me3CCH=CH2
84
C_6H_{12}
563-78-0
1-Butene, 2,3-dimethyl-
Me2CHCMe=CH2
84
C_6H_{12}
563-79-1
2-Butene, 2,3-dimethyl-
Me2C=CMe2
84
C_6H_{12}
592-41-6
1-Hexene
H2C=CH(CH2)3Me
84
C_6H_{12}
592-43-8
2-Hexene
MeCH=CHPr

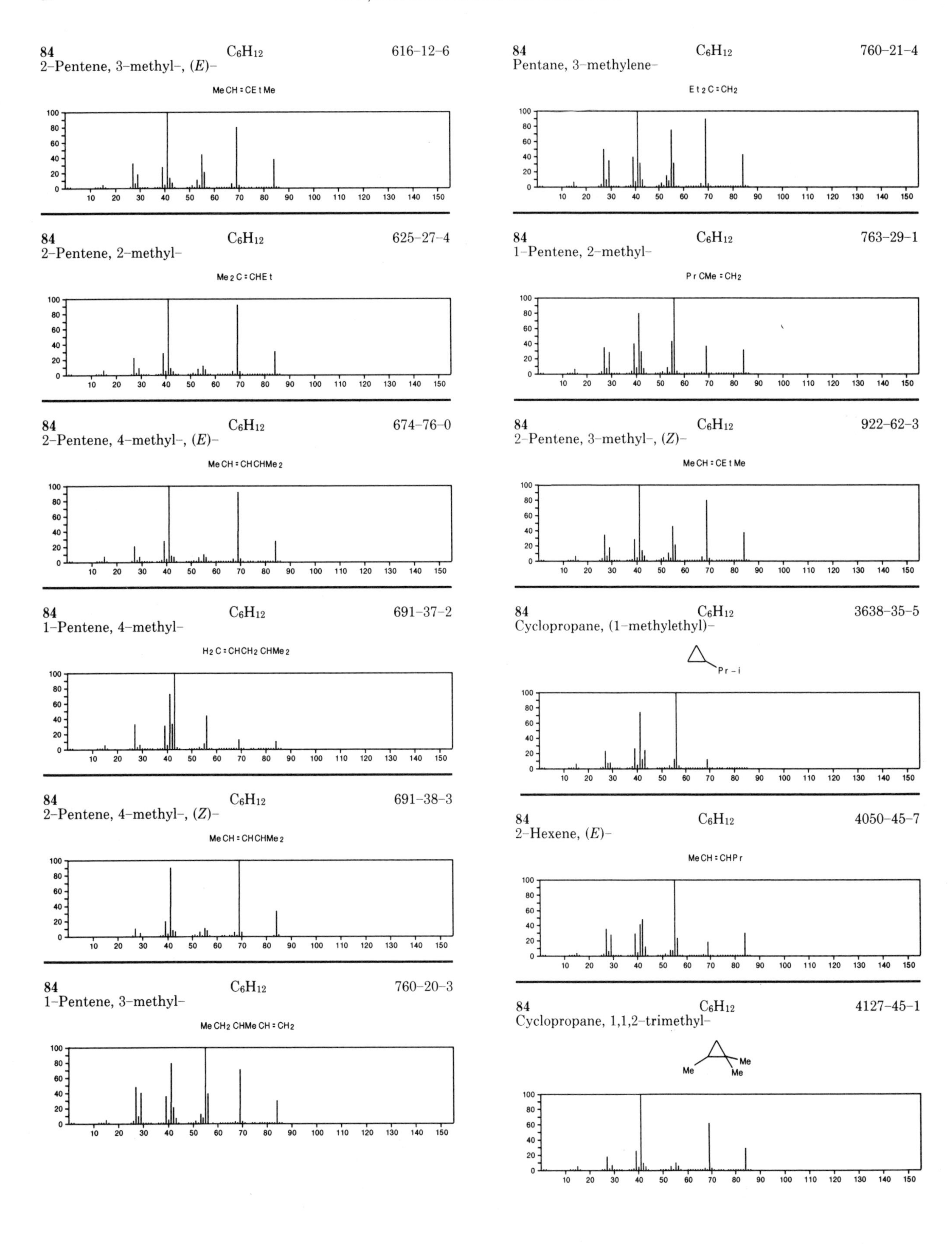
84 C6H12 616-12-6
2-Pentene, 3-methyl-, (E)-
Me CH = CE t Me
84 C6H12 625-27-4
2-Pentene, 2-methyl-
Me 2 C = CHE t
84 C6H12 674-76-0
2-Pentene, 4-methyl-, (E)-
Me CH = CHCHMe 2
84 C6H12 691-37-2
1-Pentene, 4-methyl-
H2 C = CHCH2 CHMe 2
84 C6H12 691-38-3
2-Pentene, 4-methyl-, (Z)-
Me CH = CHCHMe 2
84 C6H12 760-20-3
1-Pentene, 3-methyl-
Me CH2 CHMe CH = CH2
84 C6H12 760-21-4
Pentane, 3-methylene-
E t 2 C = CH2
84 C6H12 763-29-1
1-Pentene, 2-methyl-
P r CMe = CH2
84 C6H12 922-62-3
2-Pentene, 3-methyl-, (Z)-
Me CH = CE t Me
84 C6H12 3638-35-5
Cyclopropane, (1-methylethyl)-
P r - i
84 C6H12 4050-45-7
2-Hexene, (E)-
Me CH = CHP r
84 C6H12 4127-45-1
Cyclopropane, 1,1,2-trimethyl-
Me
Me
Me

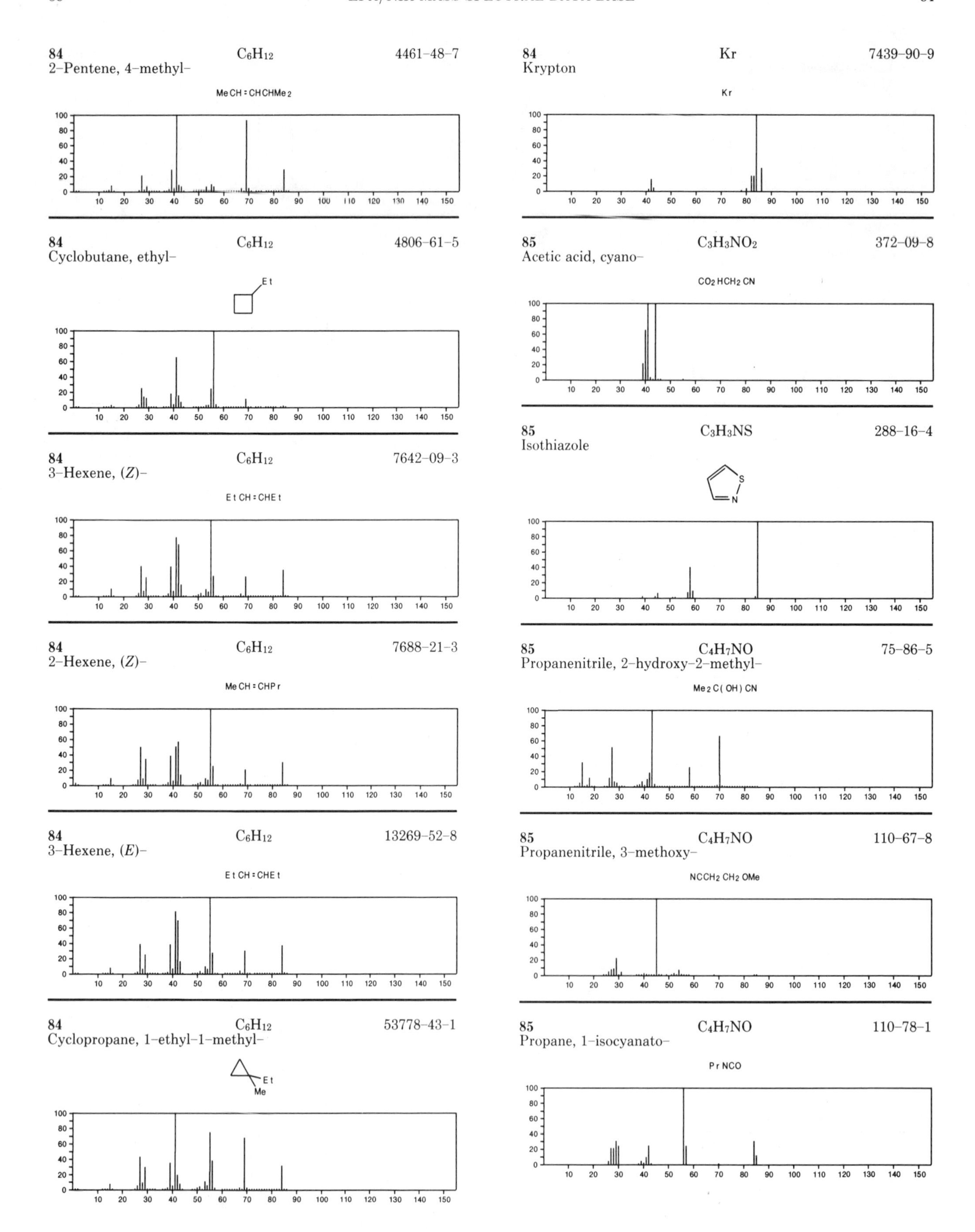
84 C6H12 4461-48-7
2-Pentene, 4-methyl-
MeCH=CHCHMe2
84 Cyclobutane, ethyl- C6H12 4806-61-5
Et
84 C6H12 7642-09-3
3-Hexene, (Z)-
EtCH=CHEt
84 C6H12 7688-21-3
2-Hexene, (Z)-
MeCH=CHPr
84 C6H12 13269-52-8
3-Hexene, (E)-
EtCH=CHEt
84 C6H12 53778-43-1
Cyclopropane, 1-ethyl-1-methyl-
Et
Me
84 Kr 7439-90-9
Krypton
Kr
85 C3H3NO2 372-09-8
Acetic acid, cyano-
CO2HCH2CN
85 C3H3NS 288-16-4
Isothiazole
S
N
85 C4H7NO 75-86-5
Propanenitrile, 2-hydroxy-2-methyl-
Me2C(OH)CN
85 C4H7NO 110-67-8
Propanenitrile, 3-methoxy-
NCCH2CH2OMe
85 C4H7NO 110-78-1
Propane, 1-isocyanato-
PrNCO
100
80
60
40
20
0
10 20 30 40 50 60 70 80 90 100 110 120 130 140 150

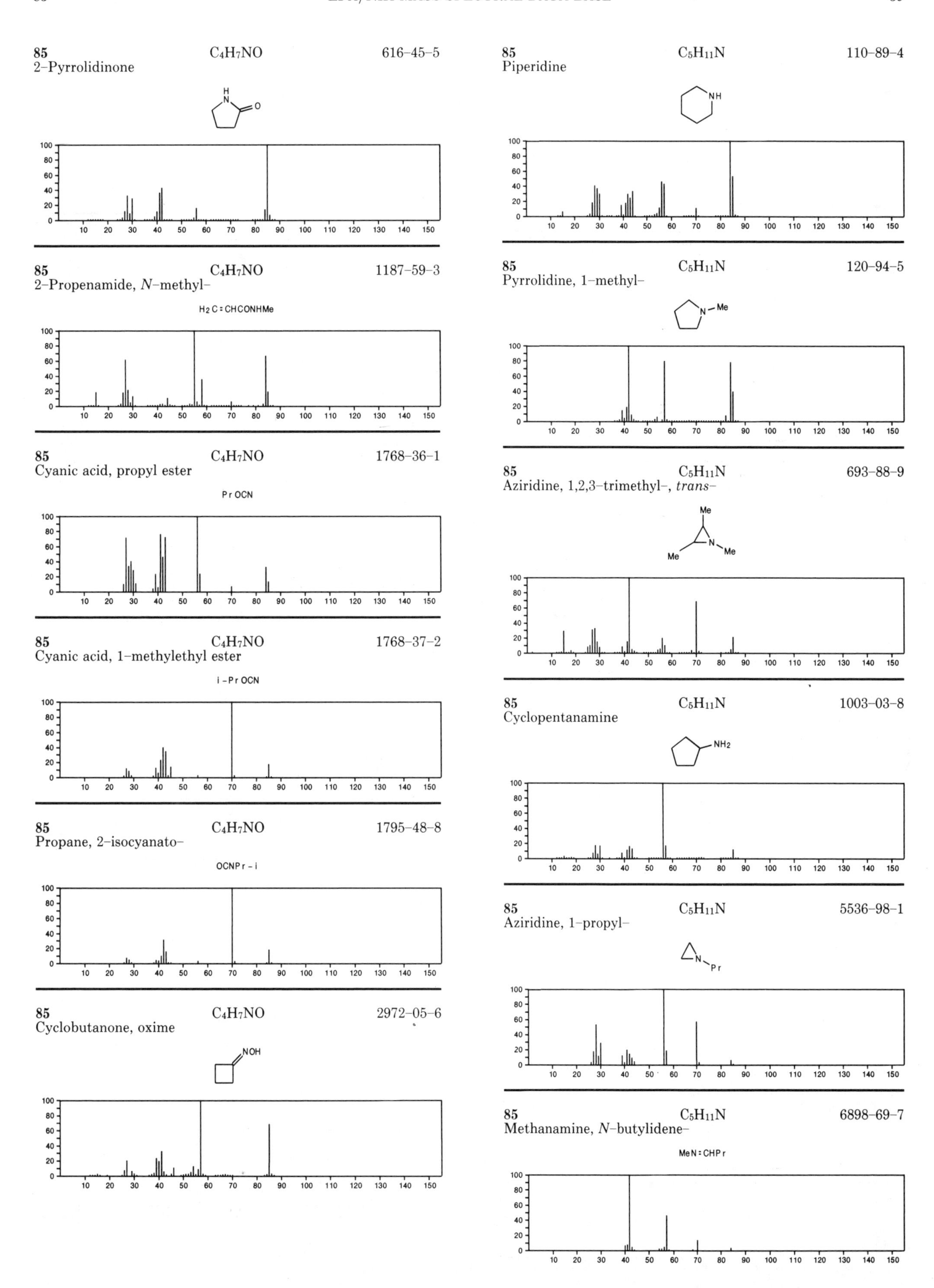
85
2-Pyrrolidinone
C4H7NO
616-45-5
85
2-Propenamide, N-methyl-
C4H7NO
1187-59-3
H2C=CHCONHMe
85
Cyanic acid, propyl ester
C4H7NO
1768-36-1
PrOCN
85
Cyanic acid, 1-methylethyl ester
C4H7NO
1768-37-2
i-PrOCN
85
Propane, 2-isocyanato-
C4H7NO
1795-48-8
OCNPr-i
85
Cyclobutanone, oxime
C4H7NO
2972-05-6
NOH
85
Piperidine
C5H11N
110-89-4
NH
85
Pyrrolidine, 1-methyl-
C5H11N
120-94-5
N-Me
85
Aziridine, 1,2,3-trimethyl-, trans-
C5H11N
693-88-9
Me
N
85
Cyclopentanamine
C5H11N
1003-03-8
NH2
85
Aziridine, 1-propyl-
C5H11N
5536-98-1
N
Pr
85
Methanamine, N-butylidene-
C5H11N
6898-69-7
MeN=CHPr

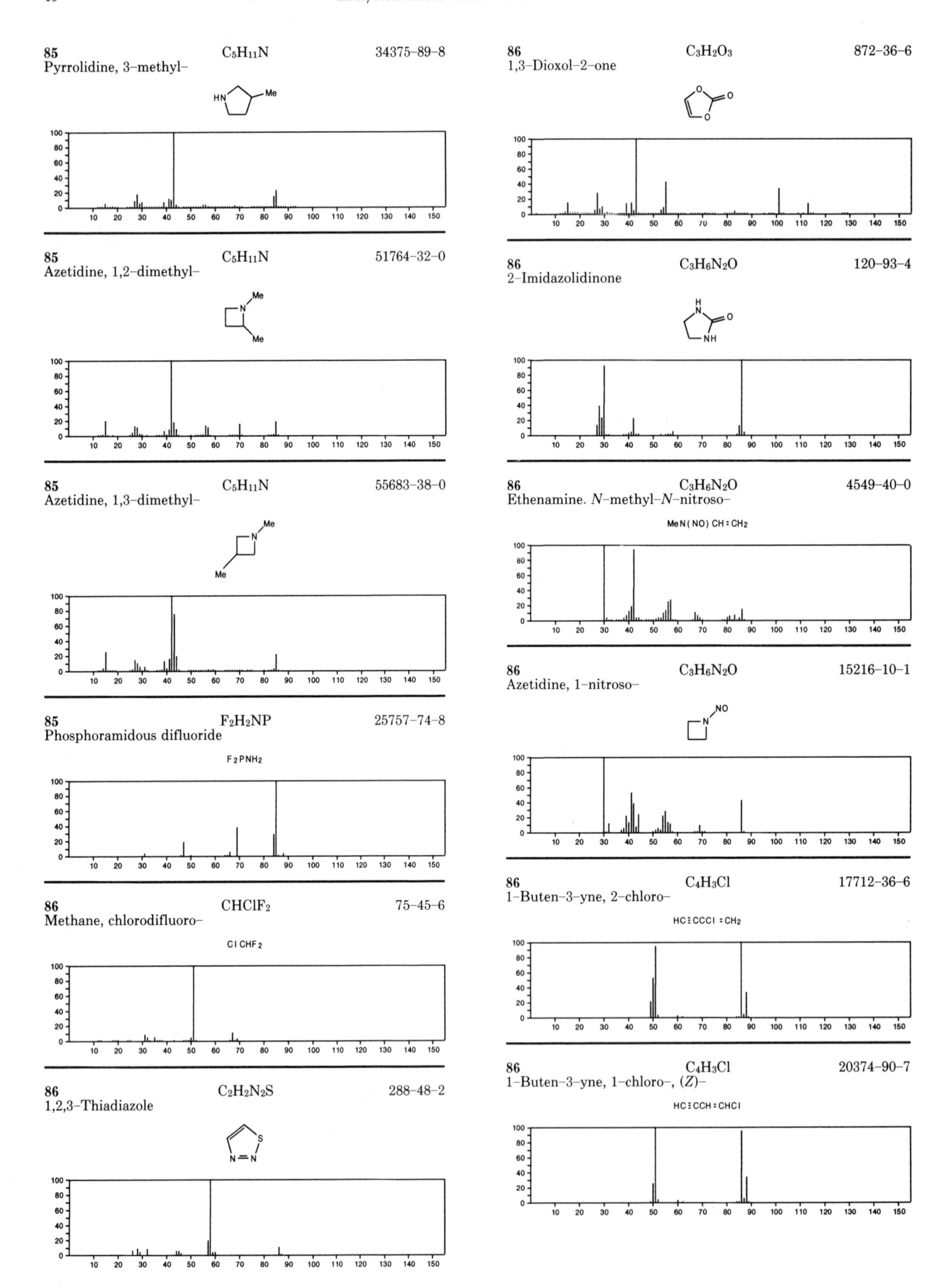
85 $C_5H_{11}N$ 34375-89-8
Pyrrolidine, 3-methyl-
85 $C_5H_{11}N$ 51764-32-0
Azetidine, 1,2-dimethyl-
85 $C_5H_{11}N$ 55683-38-0
Azetidine, 1,3-dimethyl-
85 F_2H_2NP 25757-74-8
Phosphoramidous difluoride
F2PNH2
86 $CHClF_2$ 75-45-6
Methane, chlorodifluoro-
ClCHF2
86 $C_2H_2N_2S$ 288-48-2
1,2,3-Thiadiazole
86 $C_3H_2O_3$ 872-36-6
1,3-Dioxol-2-one
86 $C_3H_6N_2O$ 120-93-4
2-Imidazolidinone
86 $C_3H_6N_2O$ 4549-40-0
Ethenamine. N-methyl-N-nitroso-
MeN(NO)CH=CH2
86 $C_3H_6N_2O$ 15216-10-1
Azetidine, 1-nitroso-
86 C_4H_3Cl 17712-36-6
1-Buten-3-yne, 2-chloro-
HC≡CCCl=CH2
86 C_4H_3Cl 20374-90-7
1-Buten-3-yne, 1-chloro-, (Z)-
HC≡CCH=CHCl

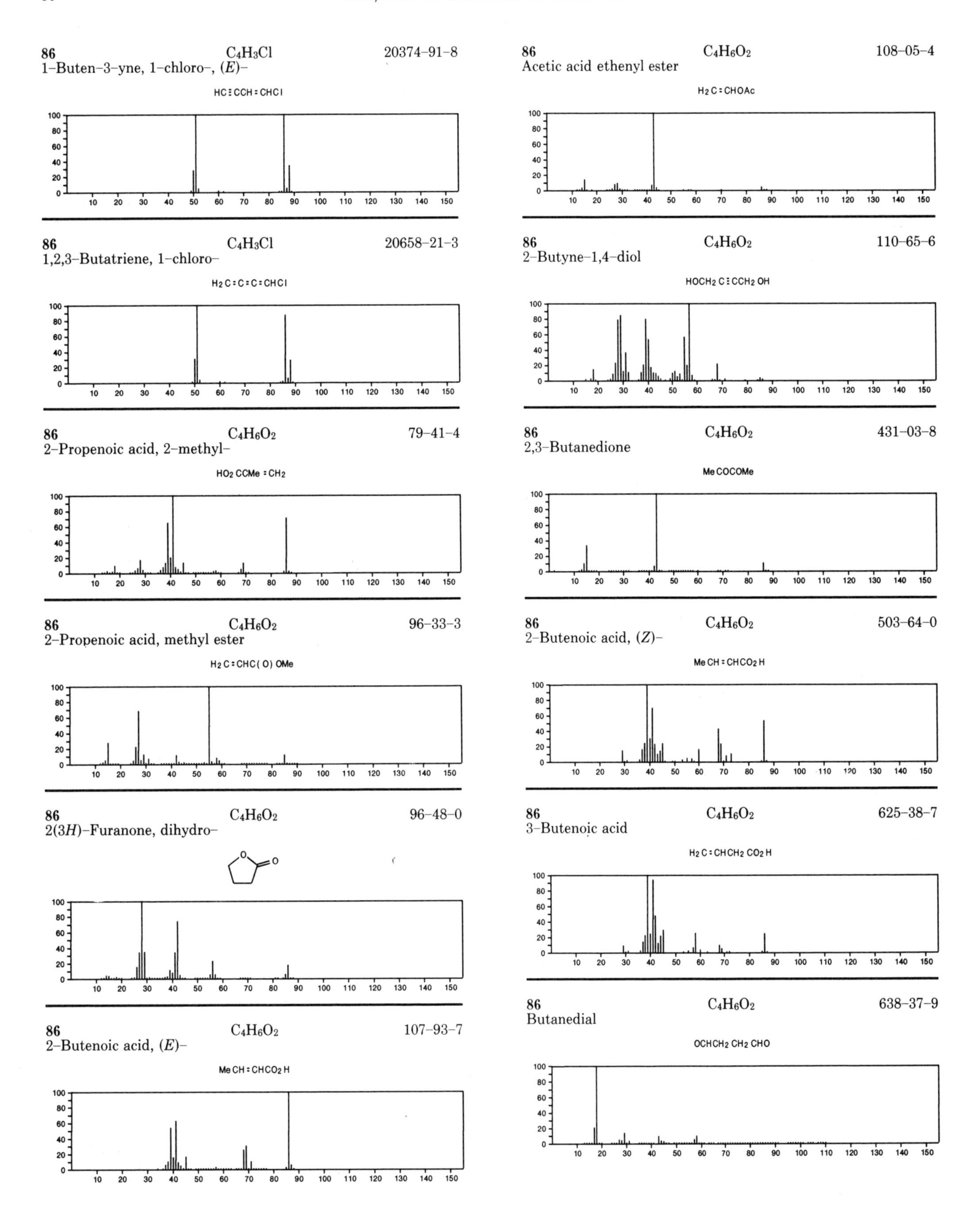
86 C4H3Cl 20374-91-8
1-Buten-3-yne, 1-chloro-, (E)-
HC≡CCH=CHCl
86 C4H3Cl 20658-21-3
1,2,3-Butatriene, 1-chloro-
H2C=C=C=CHCl
86 C4H6O2 79-41-4
2-Propenoic acid, 2-methyl-
HO2CCMe=CH2
86 C4H6O2 96-33-3
2-Propenoic acid, methyl ester
H2C=CHC(O)OMe
86 C4H6O2 96-48-0
2(3H)-Furanone, dihydro-
86 C4H6O2 107-93-7
2-Butenoic acid, (E)-
MeCH=CHCO2H
86 C4H6O2 108-05-4
Acetic acid ethenyl ester
H2C=CHOAc
86 C4H6O2 110-65-6
2-Butyne-1,4-diol
HOCH2C≡CCH2OH
86 C4H6O2 431-03-8
2,3-Butanedione
MeCOCOMe
86 C4H6O2 503-64-0
2-Butenoic acid, (Z)-
MeCH=CHCO2H
86 C4H6O2 625-38-7
3-Butenoic acid
H2C=CHCH2CO2H
86 C4H6O2 638-37-9
Butanedial
OCHCH2CH2CHO

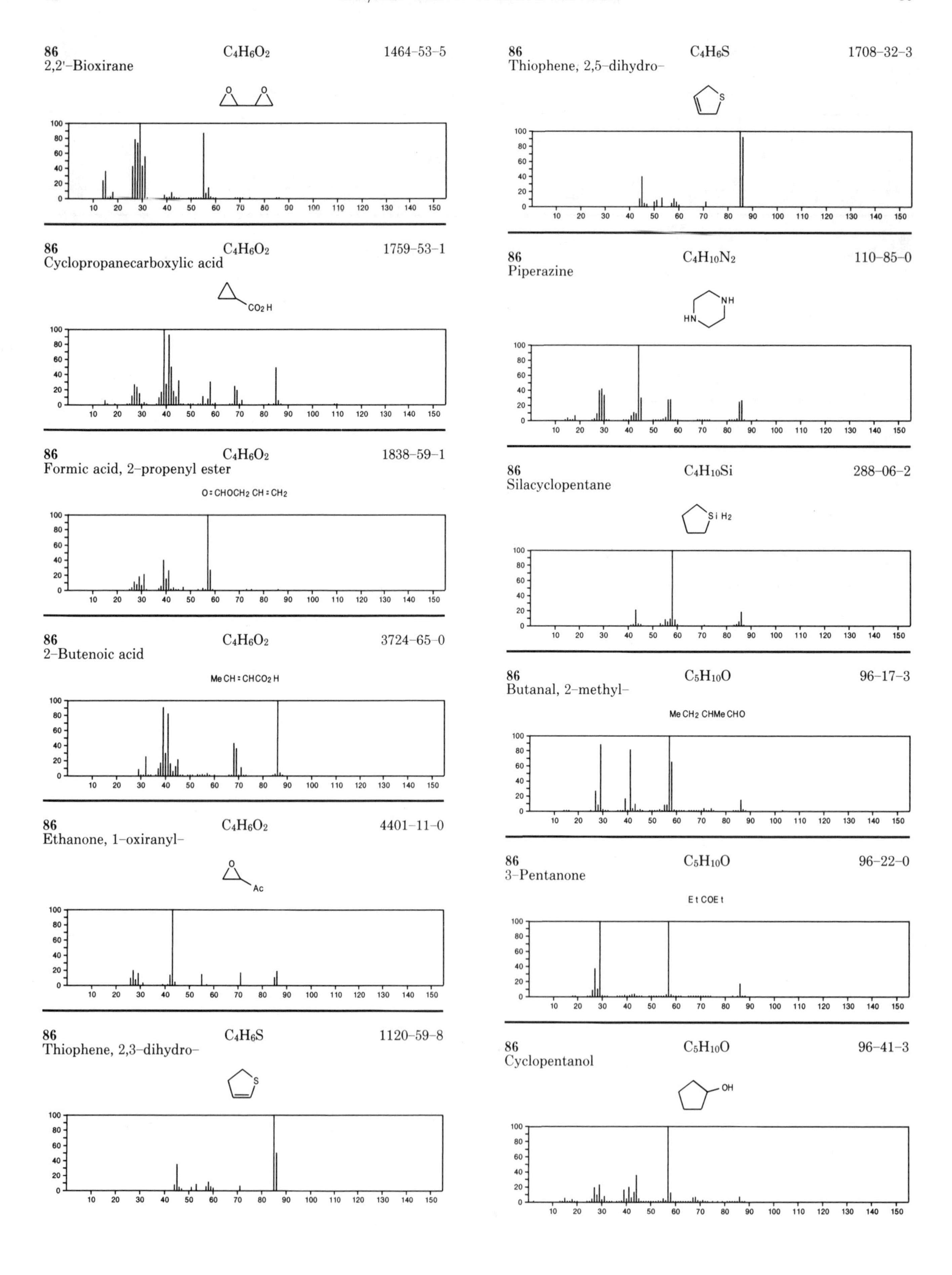

86 $C_4H_6O_2$ 1464-53-5
2,2'-Bioxirane
86 $C_4H_6O_2$ 1759-53-1
Cyclopropanecarboxylic acid
CO_2H
86 $C_4H_6O_2$ 1838-59-1
Formic acid, 2-propenyl ester
O=CHOCH2 CH=CH2
86 $C_4H_6O_2$ 3724-65-0
2-Butenoic acid
MeCH=CHCO2H
86 $C_4H_6O_2$ 4401-11-0
Ethanone, 1-oxiranyl-
Ac
86 C_4H_6S 1120-59-8
Thiophene, 2,3-dihydro-
86 C_4H_6S 1708-32-3
Thiophene, 2,5-dihydro-
86 $C_4H_{10}N_2$ 110-85-0
Piperazine
NH
HN
86 $C_4H_{10}Si$ 288-06-2
Silacyclopentane
SiH2
86 $C_5H_{10}O$ 96-17-3
Butanal, 2-methyl-
MeCH2 CHMe CHO
86 $C_5H_{10}O$ 96-22-0
3-Pentanone
EtCOEt
86 $C_5H_{10}O$ 96-41-3
Cyclopentanol
OH

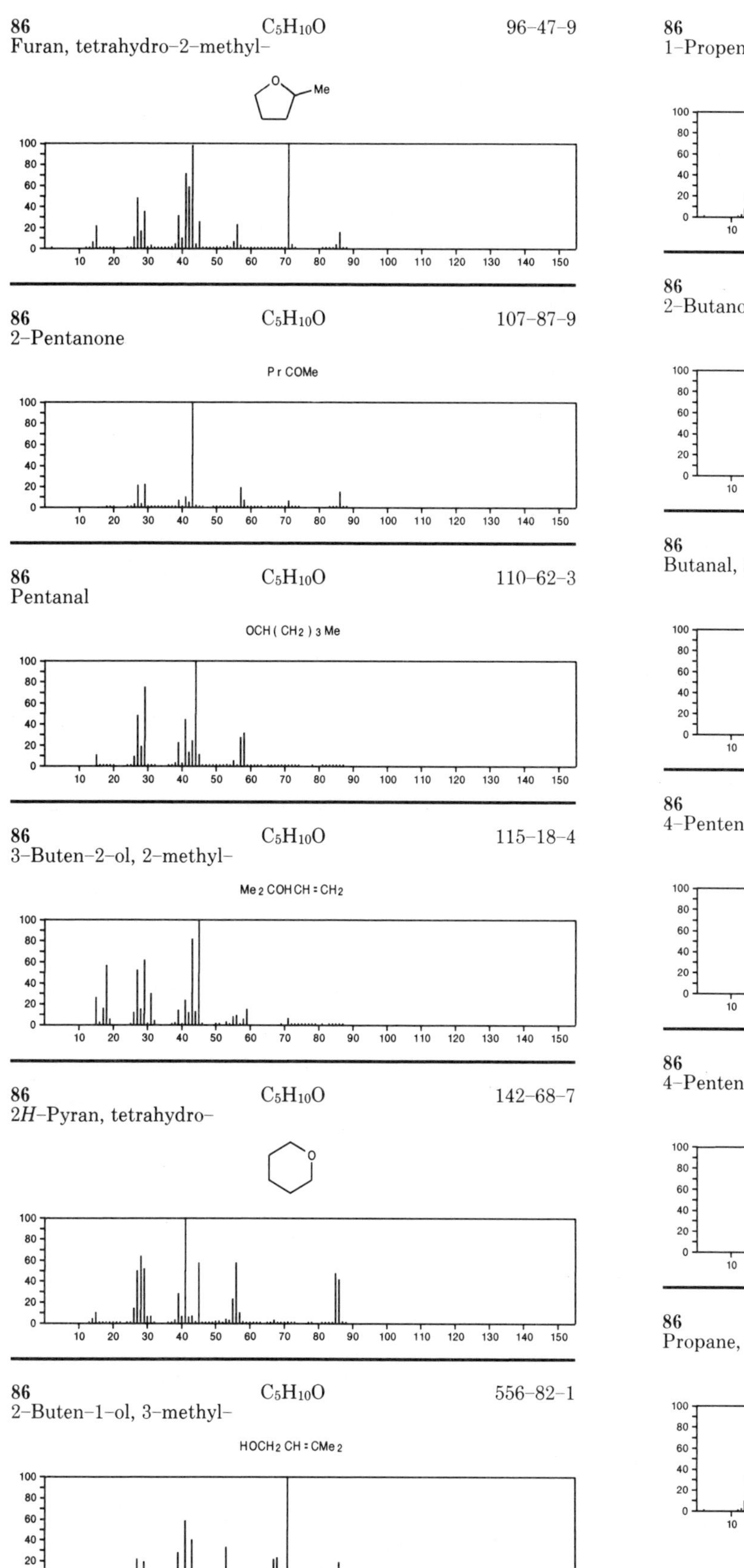
86 C5H10O 96-47-9
Furan, tetrahydro-2-methyl-
Me
86 C5H10O 107-87-9
2-Pentanone
Pr COMe
86 C5H10O 110-62-3
Pentanal
OCH (CH2) 3 Me
86 C5H10O 115-18-4
3-Buten-2-ol, 2-methyl-
Me2 COHCH = CH2
86 C5H10O 142-68-7
2H-Pyran, tetrahydro-
O
86 C5H10O 556-82-1
2-Buten-1-ol, 3-methyl-
HOCH2 CH = CMe2

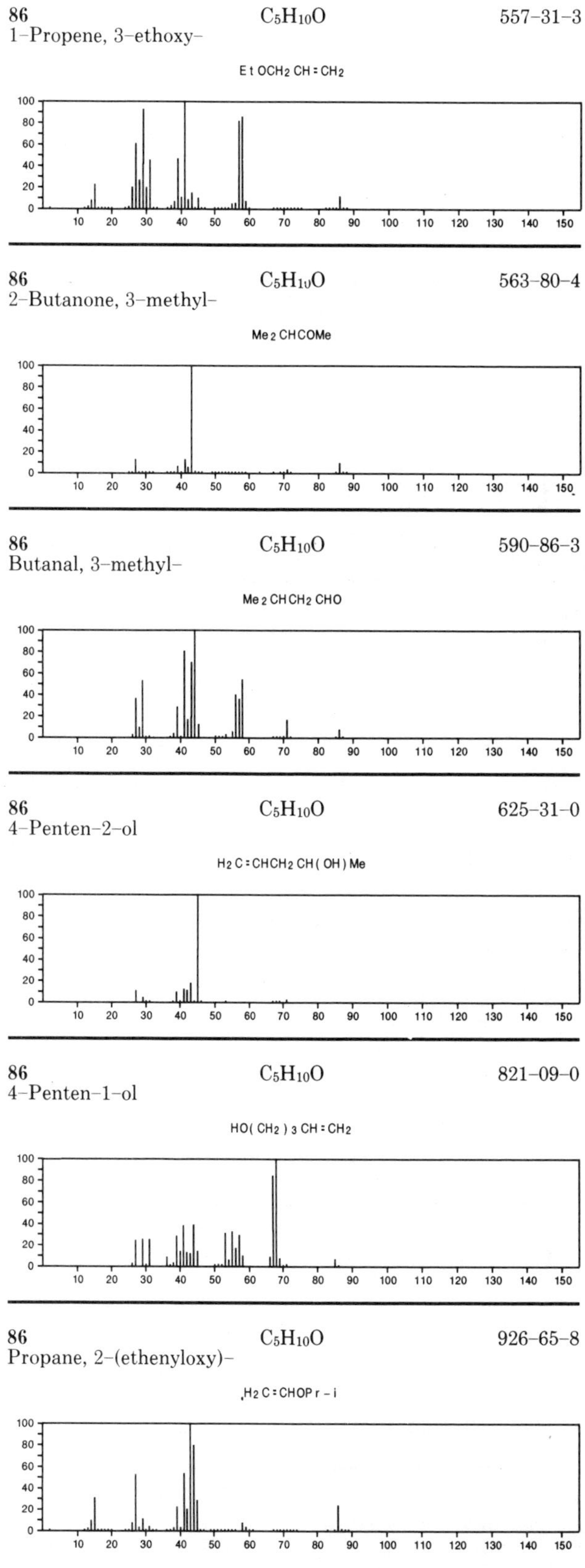
86 C5H10O 557-31-3
1-Propene, 3-ethoxy-
E t OCH2 CH = CH2
86 C5H10O 563-80-4
2-Butanone, 3-methyl-
Me2 CHCOMe
86 C5H10O 590-86-3
Butanal, 3-methyl-
Me2 CHCH2 CHO
86 C5H10O 625-31-0
4-Penten-2-ol
H2 C = CHCH2 CH (OH) Me
86 C5H10O 821-09-0
4-Penten-1-ol
HO(CH2) 3 CH = CH2
86 C5H10O 926-65-8
Propane, 2-(ethenyloxy)-
H2 C = CHOPr - i

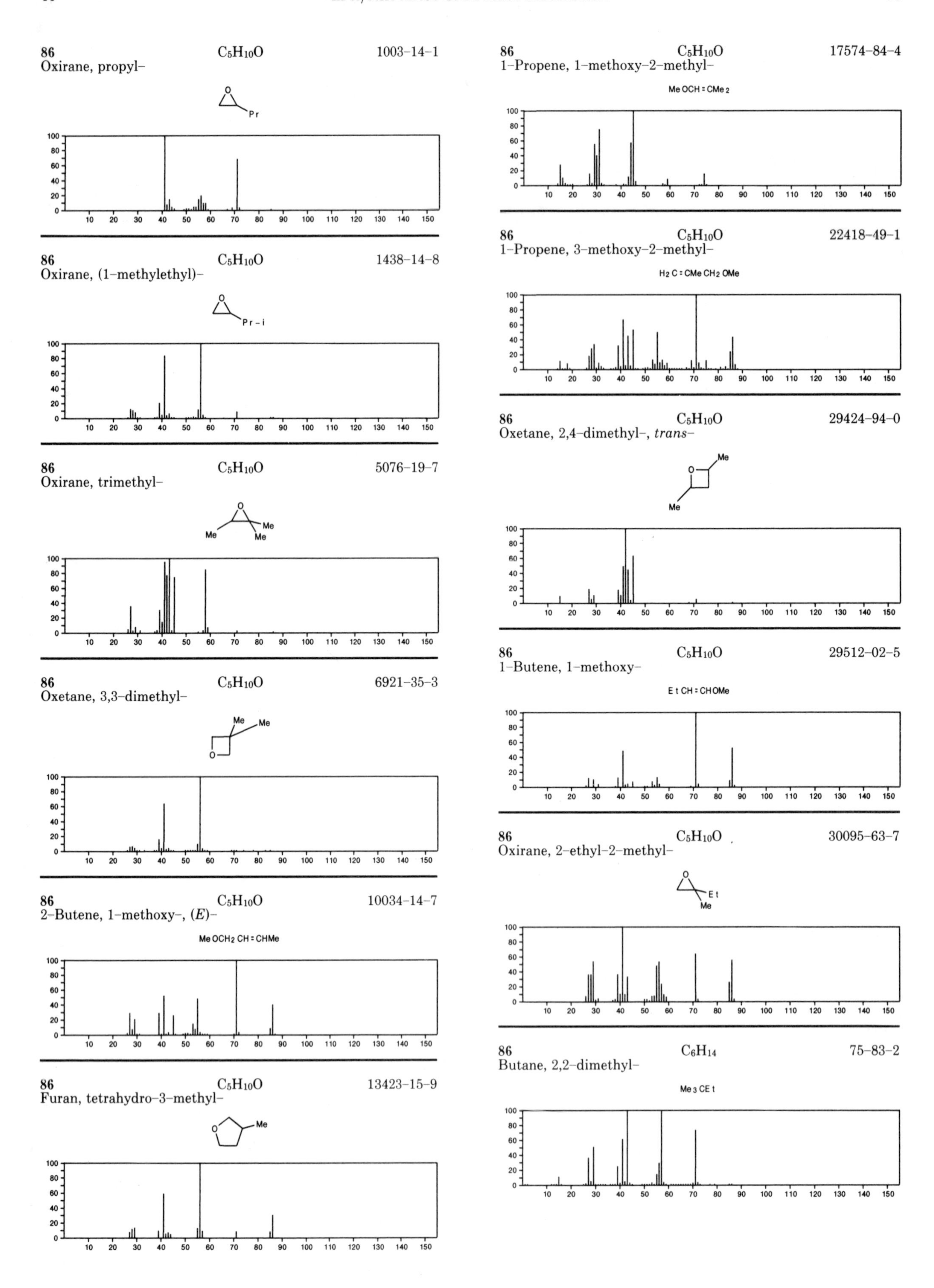
86 C5H10O 1003-14-1
Oxirane, propyl-
Pr
86 C5H10O 1438-14-8
Oxirane, (1-methylethyl)-
Pr-i
86 C5H10O 5076-19-7
Oxirane, trimethyl-
Me Me Me
86 C5H10O 6921-35-3
Oxetane, 3,3-dimethyl-
Me Me
86 C5H10O 10034-14-7
2-Butene, 1-methoxy-, (E)-
Me OCH2 CH = CHMe
86 C5H10O 13423-15-9
Furan, tetrahydro-3-methyl-
Me
86 C5H10O 17574-84-4
1-Propene, 1-methoxy-2-methyl-
Me OCH = CMe 2
86 C5H10O 22418-49-1
1-Propene, 3-methoxy-2-methyl-
H2 C = CMe CH2 OMe
86 C5H10O 29424-94-0
Oxetane, 2,4-dimethyl-, trans-
Me Me
86 C5H10O 29512-02-5
1-Butene, 1-methoxy-
Et CH = CHOMe
86 C5H10O 30095-63-7
Oxirane, 2-ethyl-2-methyl-
Et Me
86 C6H14 75-83-2
Butane, 2,2-dimethyl-
Me 3 CEt

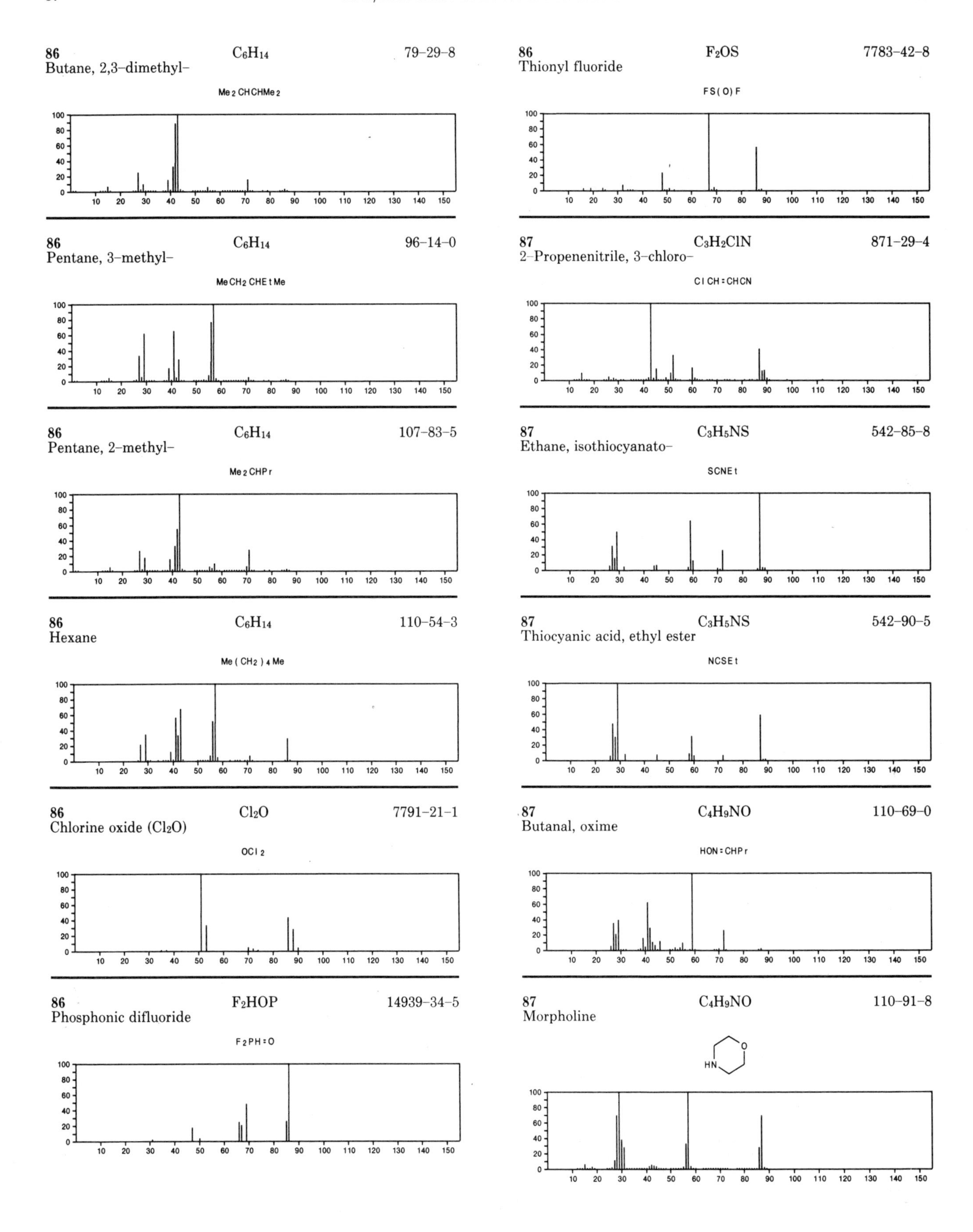
86 Butane, 2,3-dimethyl- C6H14 79-29-8
Me2CHCHMe2
86 Pentane, 3-methyl- C6H14 96-14-0
MeCH2CHEtMe
86 Pentane, 2-methyl- C6H14 107-83-5
Me2CHPr
86 Hexane C6H14 110-54-3
Me(CH2)4Me
86 Chlorine oxide (Cl2O) Cl2O 7791-21-1
OCl2
86 Phosphonic difluoride F2HOP 14939-34-5
F2PH=O
86 Thionyl fluoride F2OS 7783-42-8
FS(O)F
87 2-Propenenitrile, 3-chloro- C3H2ClN 871-29-4
ClCH=CHCN
87 Ethane, isothiocyanato- C3H5NS 542-85-8
SCNEt
87 Thiocyanic acid, ethyl ester C3H5NS 542-90-5
NCSEt
87 Butanal, oxime C4H9NO 110-69-0
HON=CHPr
87 Morpholine C4H9NO 110-91-8
HN
O

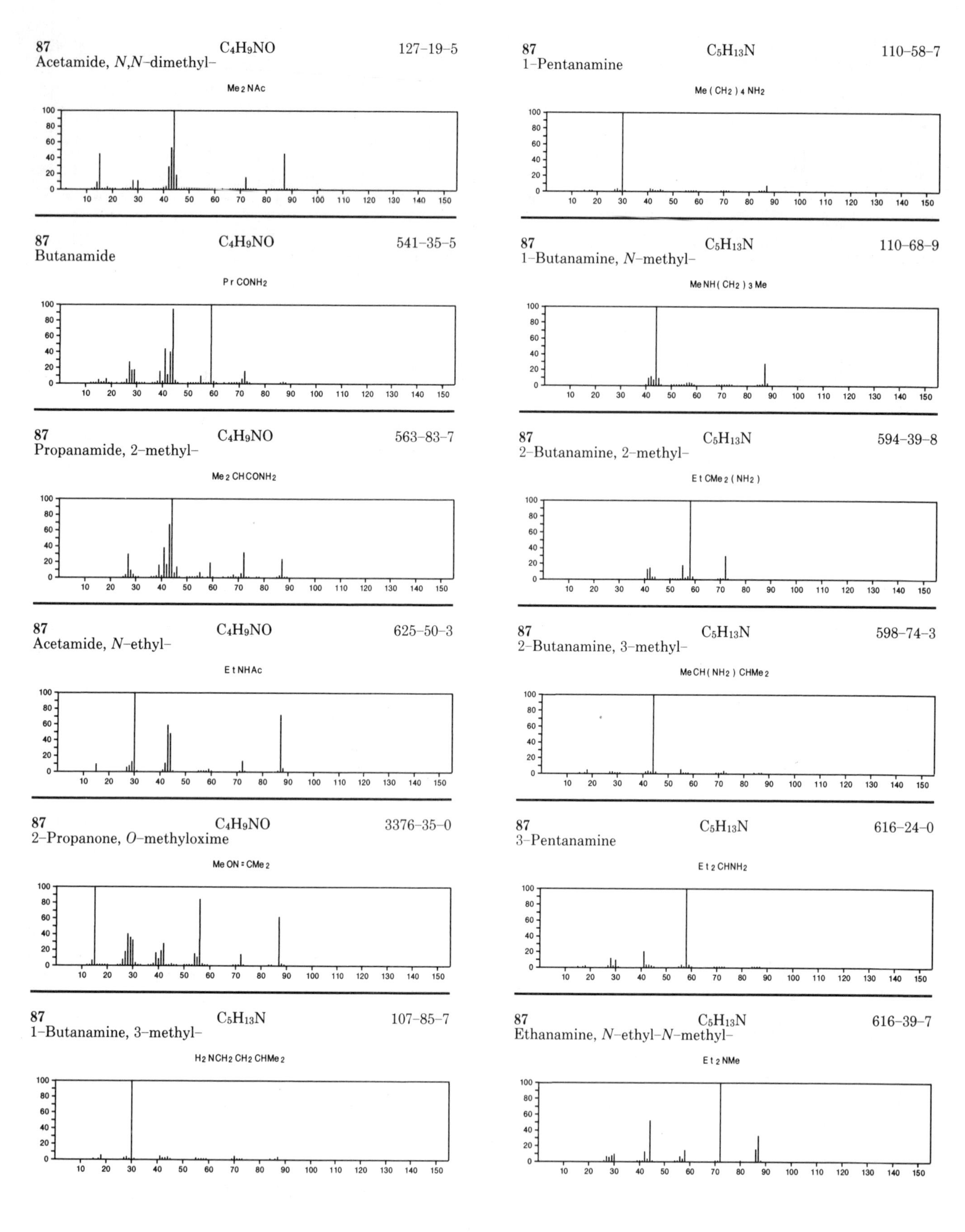
87
C4H9NO
127-19-5
Acetamide, N,N-dimethyl-
Me2NAc
87
C4H9NO
541-35-5
Butanamide
PrCONH2
87
C4H9NO
563-83-7
Propanamide, 2-methyl-
Me2CHCONH2
87
C4H9NO
625-50-3
Acetamide, N-ethyl-
EtNHAc
87
C4H9NO
3376-35-0
2-Propanone, O-methyloxime
MeON=CMe2
87
C5H13N
107-85-7
1-Butanamine, 3-methyl-
H2NCH2CH2CHMe2
87
C5H13N
110-58-7
1-Pentanamine
Me(CH2)4NH2
87
C5H13N
110-68-9
1-Butanamine, N-methyl-
MeNH(CH2)3Me
87
C5H13N
594-39-8
2-Butanamine, 2-methyl-
EtCMe2(NH2)
87
C5H13N
598-74-3
2-Butanamine, 3-methyl-
MeCH(NH2)CHMe2
87
C5H13N
616-24-0
3-Pentanamine
Et2CHNH2
87
C5H13N
616-39-7
Ethanamine, N-ethyl-N-methyl-
Et2NMe

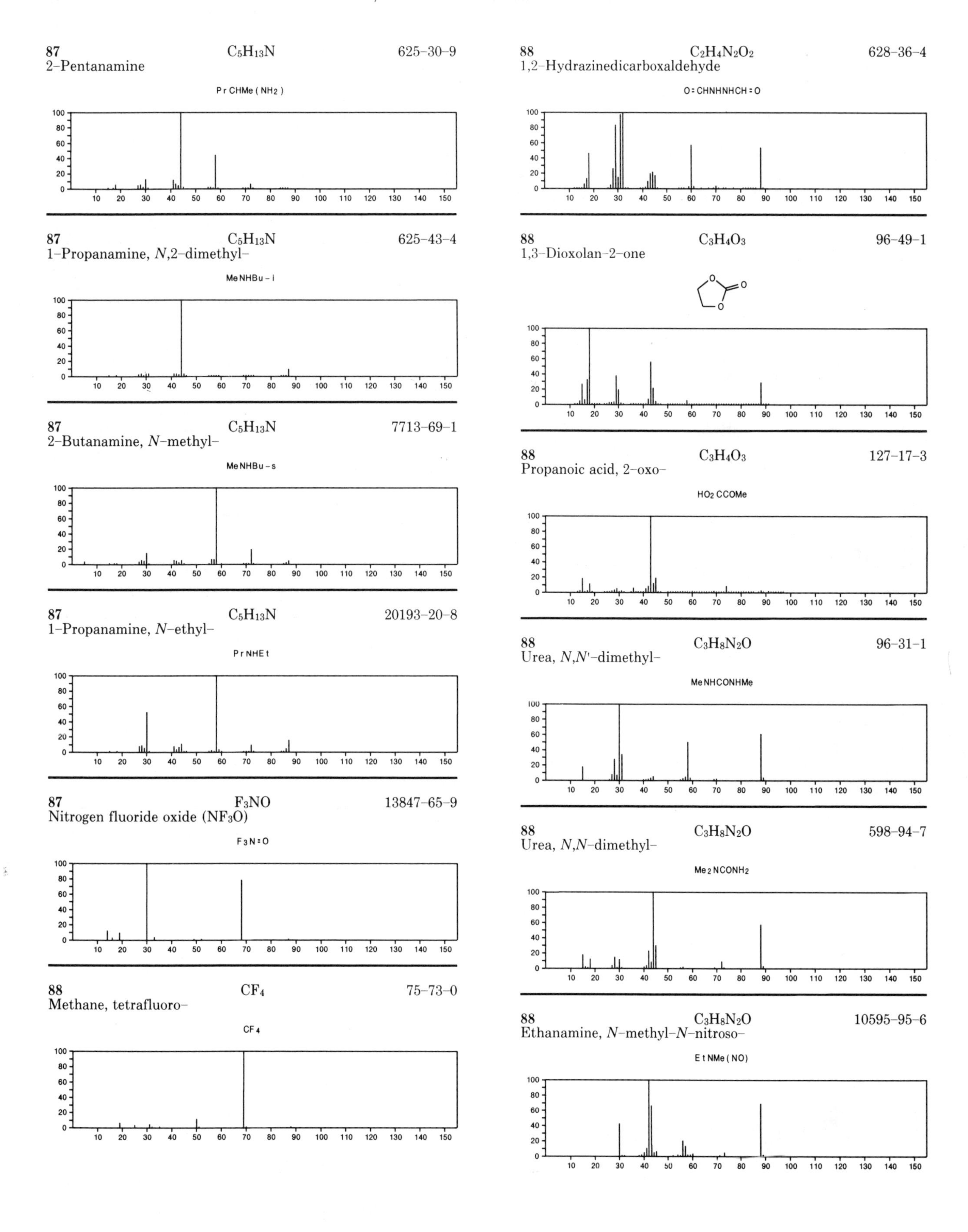
87 C5H13N 625-30-9
2-Pentanamine
PrCHMe(NH2)
87 C5H13N 625-43-4
1-Propanamine, N,2-dimethyl-
MeNHBu-i
87 C5H13N 7713-69-1
2-Butanamine, N-methyl-
MeNHBu-s
87 C5H13N 20193-20-8
1-Propanamine, N-ethyl-
PrNHEt
87 F3NO 13847-65-9
Nitrogen fluoride oxide (NF3O)
F3N=O
88 CF4 75-73-0
Methane, tetrafluoro-
CF4
88 C2H4N2O2 628-36-4
1,2-Hydrazinedicarboxaldehyde
O=CHNHNHCH=O
88 C3H4O3 96-49-1
1,3-Dioxolan-2-one
88 C3H4O3 127-17-3
Propanoic acid, 2-oxo-
HO2CCOMe
88 C3H8N2O 96-31-1
Urea, N,N'-dimethyl-
MeNHCONHMe
88 C3H8N2O 598-94-7
Urea, N,N-dimethyl-
Me2NCONH2
88 C3H8N2O 10595-95-6
Ethanamine, N-methyl-N-nitroso-
EtNMe(NO)
100
80
60
40
20
0
10 20 30 40 50 60 70 80 90 100 110 120 130 140 150

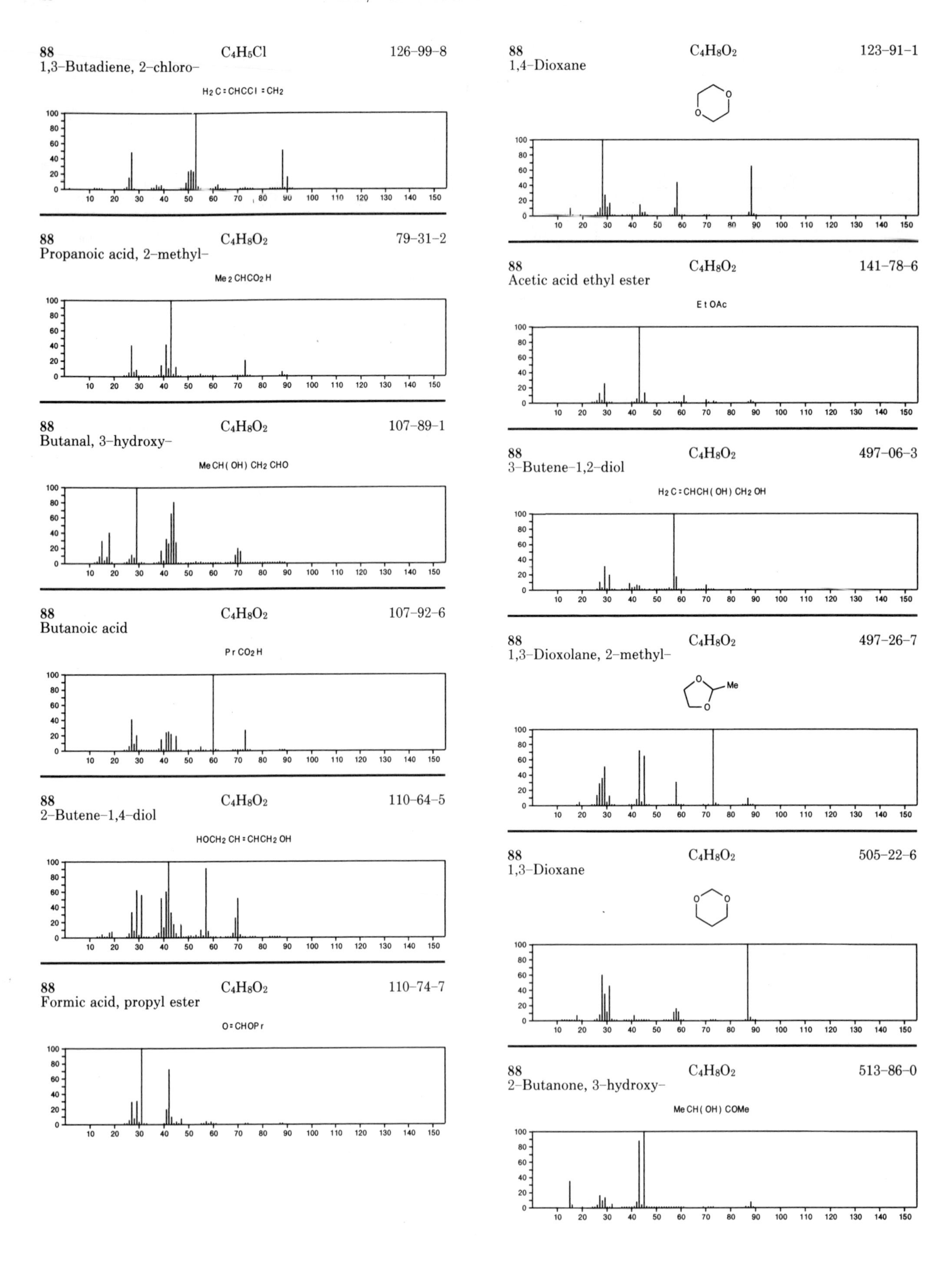
88 C_4H_5Cl 126-99-8
1,3-Butadiene, 2-chloro-
$H_2C=CHCCl=CH_2$
88 $C_4H_8O_2$ 79-31-2
Propanoic acid, 2-methyl-
Me_2CHCO_2H
88 $C_4H_8O_2$ 107-89-1
Butanal, 3-hydroxy-
$MeCH(OH)CH_2CHO$
88 $C_4H_8O_2$ 107-92-6
Butanoic acid
$PrCO_2H$
88 $C_4H_8O_2$ 110-64-5
2-Butene-1,4-diol
$HOCH_2CH=CHCH_2OH$
88 $C_4H_8O_2$ 110-74-7
Formic acid, propyl ester
O=CHOPr
88 $C_4H_8O_2$ 123-91-1
1,4-Dioxane
88 $C_4H_8O_2$ 141-78-6
Acetic acid ethyl ester
EtOAc
88 $C_4H_8O_2$ 497-06-3
3-Butene-1,2-diol
$H_2C=CHCH(OH)CH_2OH$
88 $C_4H_8O_2$ 497-26-7
1,3-Dioxolane, 2-methyl-
Me
88 $C_4H_8O_2$ 505-22-6
1,3-Dioxane
88 $C_4H_8O_2$ 513-86-0
2-Butanone, 3-hydroxy-
MeCH(OH)COMe

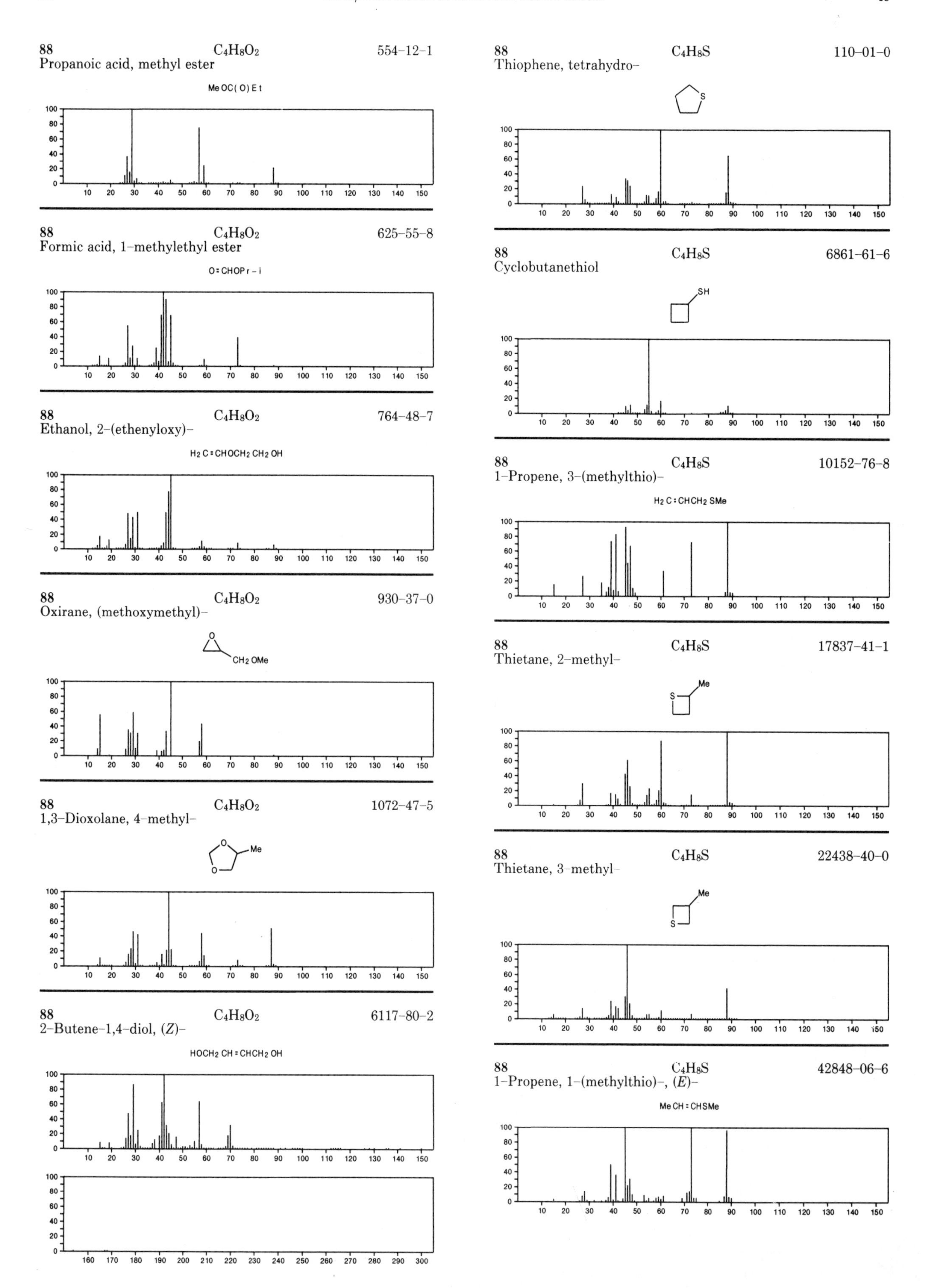
88 C4H8O2 554-12-1
Propanoic acid, methyl ester
MeOC(O)Et
88 C4H8O2 625-55-8
Formic acid, 1-methylethyl ester
O=CHOPr-i
88 C4H8O2 764-48-7
Ethanol, 2-(ethenyloxy)-
H2C=CHOCH2CH2OH
88 C4H8O2 930-37-0
Oxirane, (methoxymethyl)-
CH2OMe
88 C4H8O2 1072-47-5
1,3-Dioxolane, 4-methyl-
88 C4H8O2 6117-80-2
2-Butene-1,4-diol, (Z)-
HOCH2CH=CHCH2OH
88 C4H8S 110-01-0
Thiophene, tetrahydro-
88 C4H8S 6861-61-6
Cyclobutanethiol
SH
88 C4H8S 10152-76-8
1-Propene, 3-(methylthio)-
H2C=CHCH2SMe
88 C4H8S 17837-41-1
Thietane, 2-methyl-
88 C4H8S 22438-40-0
Thietane, 3-methyl-
88 C4H8S 42848-06-6
1-Propene, 1-(methylthio)-, (E)-
MeCH=CHSMe

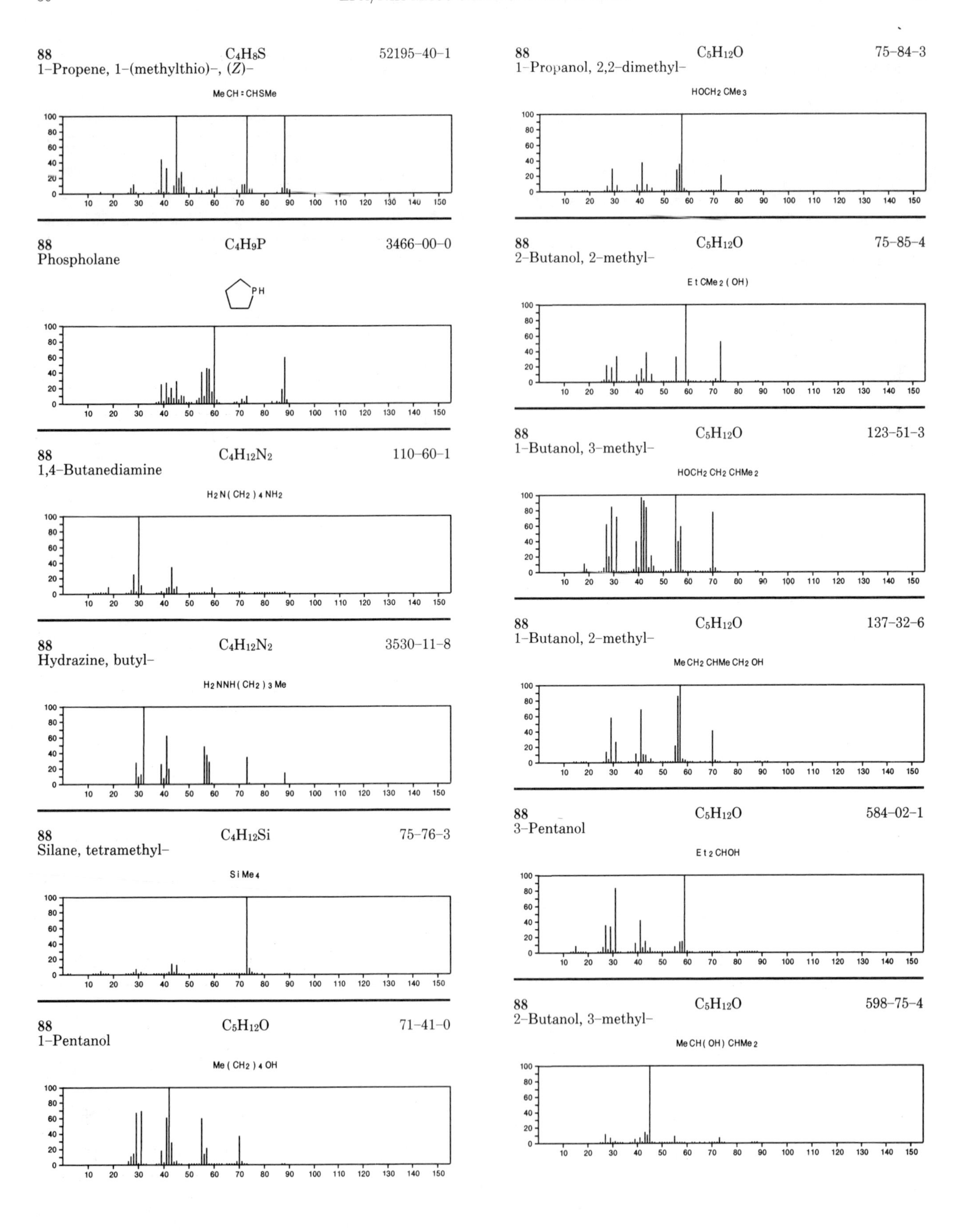

88 C4H8S 52195-40-1
1-Propene, 1-(methylthio)-, (Z)-
MeCH=CHSMe
88 C4H9P 3466-00-0
Phospholane
PH
88 C4H12N2 110-60-1
1,4-Butanediamine
H2N(CH2)4NH2
88 C4H12N2 3530-11-8
Hydrazine, butyl-
H2NNH(CH2)3Me
88 C4H12Si 75-76-3
Silane, tetramethyl-
SiMe4
88 C5H12O 71-41-0
1-Pentanol
Me(CH2)4OH
88 C5H12O 75-84-3
1-Propanol, 2,2-dimethyl-
HOCH2CMe3
88 C5H12O 75-85-4
2-Butanol, 2-methyl-
EtCMe2(OH)
88 C5H12O 123-51-3
1-Butanol, 3-methyl-
HOCH2CH2CHMe2
88 C5H12O 137-32-6
1-Butanol, 2-methyl-
MeCH2CHMeCH2OH
88 C5H12O 584-02-1
3-Pentanol
Et2CHOH
88 C5H12O 598-75-4
2-Butanol, 3-methyl-
MeCH(OH)CHMe2

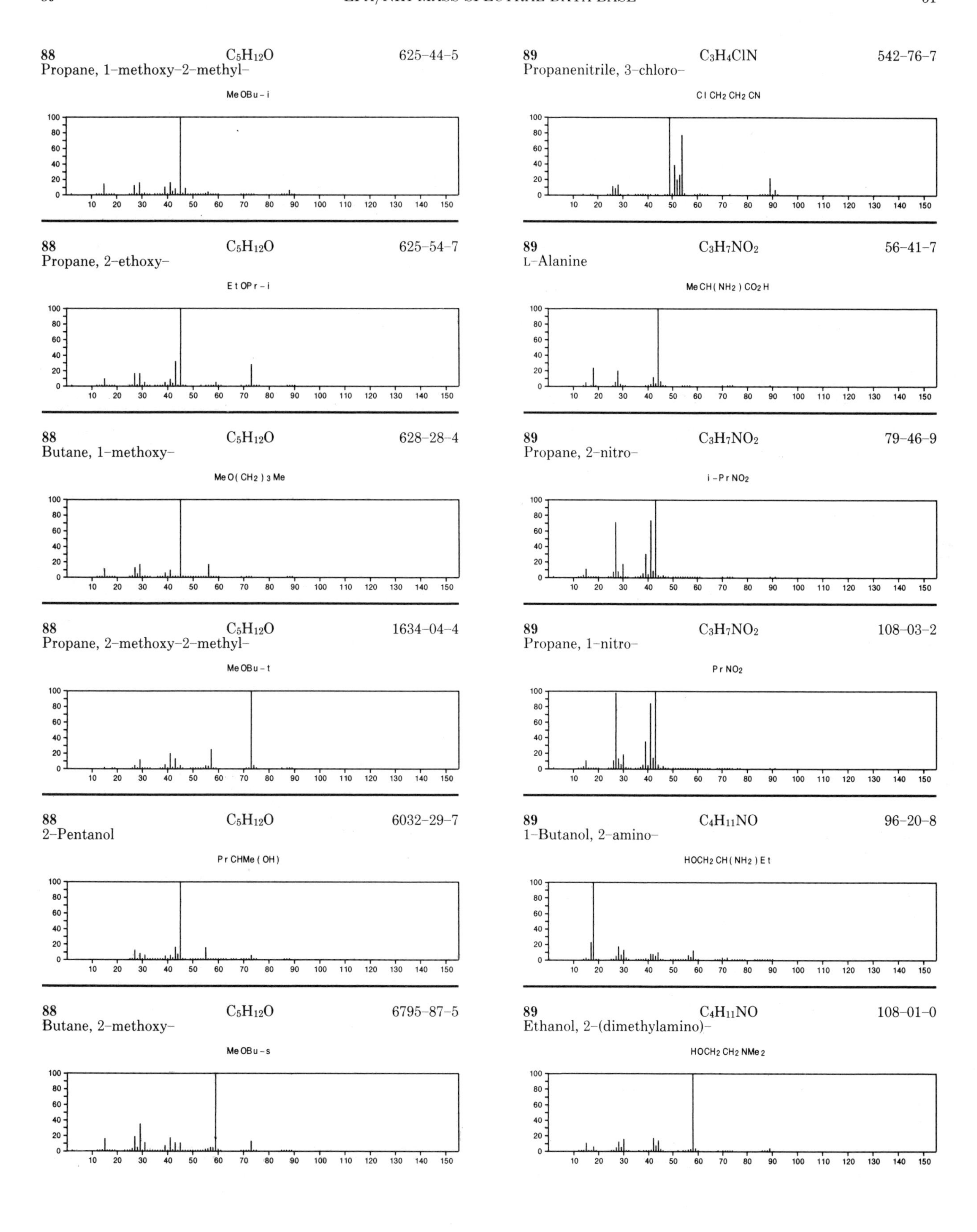

88 $C_5H_{12}O$ 625-44-5
Propane, 1-methoxy-2-methyl-
MeOBu-i

88 $C_5H_{12}O$ 625-54-7
Propane, 2-ethoxy-
EtOPr-i

88 $C_5H_{12}O$ 628-28-4
Butane, 1-methoxy-
$MeO(CH_2)_3Me$

88 $C_5H_{12}O$ 1634-04-4
Propane, 2-methoxy-2-methyl-
MeOBu-t

88 $C_5H_{12}O$ 6032-29-7
2-Pentanol
PrCHMe(OH)

88 $C_5H_{12}O$ 6795-87-5
Butane, 2-methoxy-
MeOBu-s

89 C_3H_4ClN 542-76-7
Propanenitrile, 3-chloro-
$ClCH_2CH_2CN$

89 $C_3H_7NO_2$ 56-41-7
L-Alanine
$MeCH(NH_2)CO_2H$

89 $C_3H_7NO_2$ 79-46-9
Propane, 2-nitro-
$i\text{-}PrNO_2$

89 $C_3H_7NO_2$ 108-03-2
Propane, 1-nitro-
$PrNO_2$

89 $C_4H_{11}NO$ 96-20-8
1-Butanol, 2-amino-
$HOCH_2CH(NH_2)Et$

89 $C_4H_{11}NO$ 108-01-0
Ethanol, 2-(dimethylamino)-
$HOCH_2CH_2NMe_2$

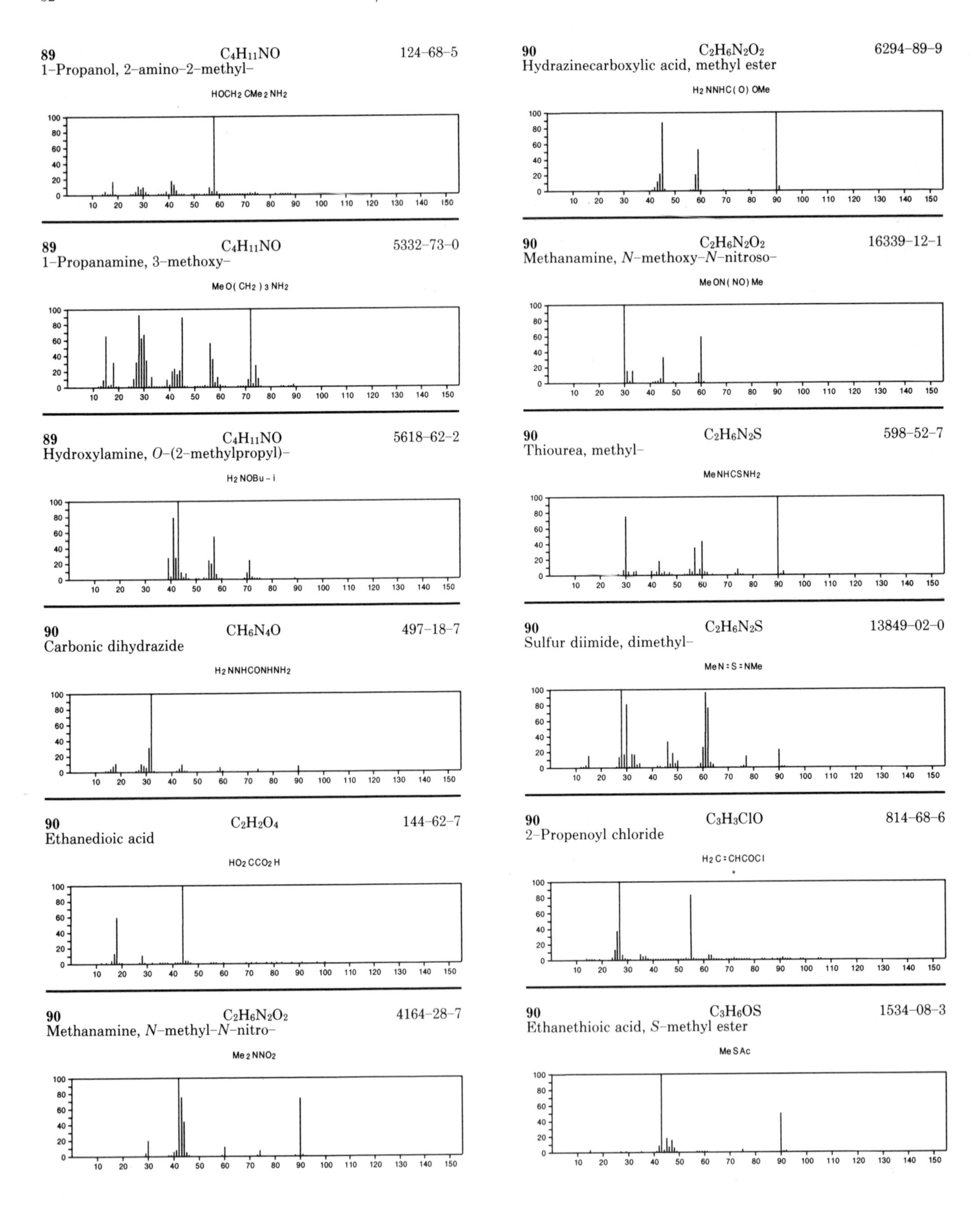
89 C4H11NO 124-68-5
1-Propanol, 2-amino-2-methyl-
HOCH2 CMe2 NH2
89 C4H11NO 5332-73-0
1-Propanamine, 3-methoxy-
MeO(CH2)3 NH2
89 C4H11NO 5618-62-2
Hydroxylamine, O-(2-methylpropyl)-
H2 NOBu-i
90 CH6N4O 497-18-7
Carbonic dihydrazide
H2 NNHCONHNH2
90 C2H2O4 144-62-7
Ethanedioic acid
HO2 CCO2 H
90 C2H6N2O2 4164-28-7
Methanamine, N-methyl-N-nitro-
Me2 NNO2
90 C2H6N2O2 6294-89-9
Hydrazinecarboxylic acid, methyl ester
H2 NNHC(O)OMe
90 C2H6N2O2 16339-12-1
Methanamine, N-methoxy-N-nitroso-
MeON(NO)Me
90 C2H6N2S 598-52-7
Thiourea, methyl-
MeNHCSNH2
90 C2H6N2S 13849-02-0
Sulfur diimide, dimethyl-
MeN=S=NMe
90 C3H3ClO 814-68-6
2-Propenoyl chloride
H2C=CHCOCl
90 C3H6OS 1534-08-3
Ethanethioic acid, S-methyl ester
MeSAc

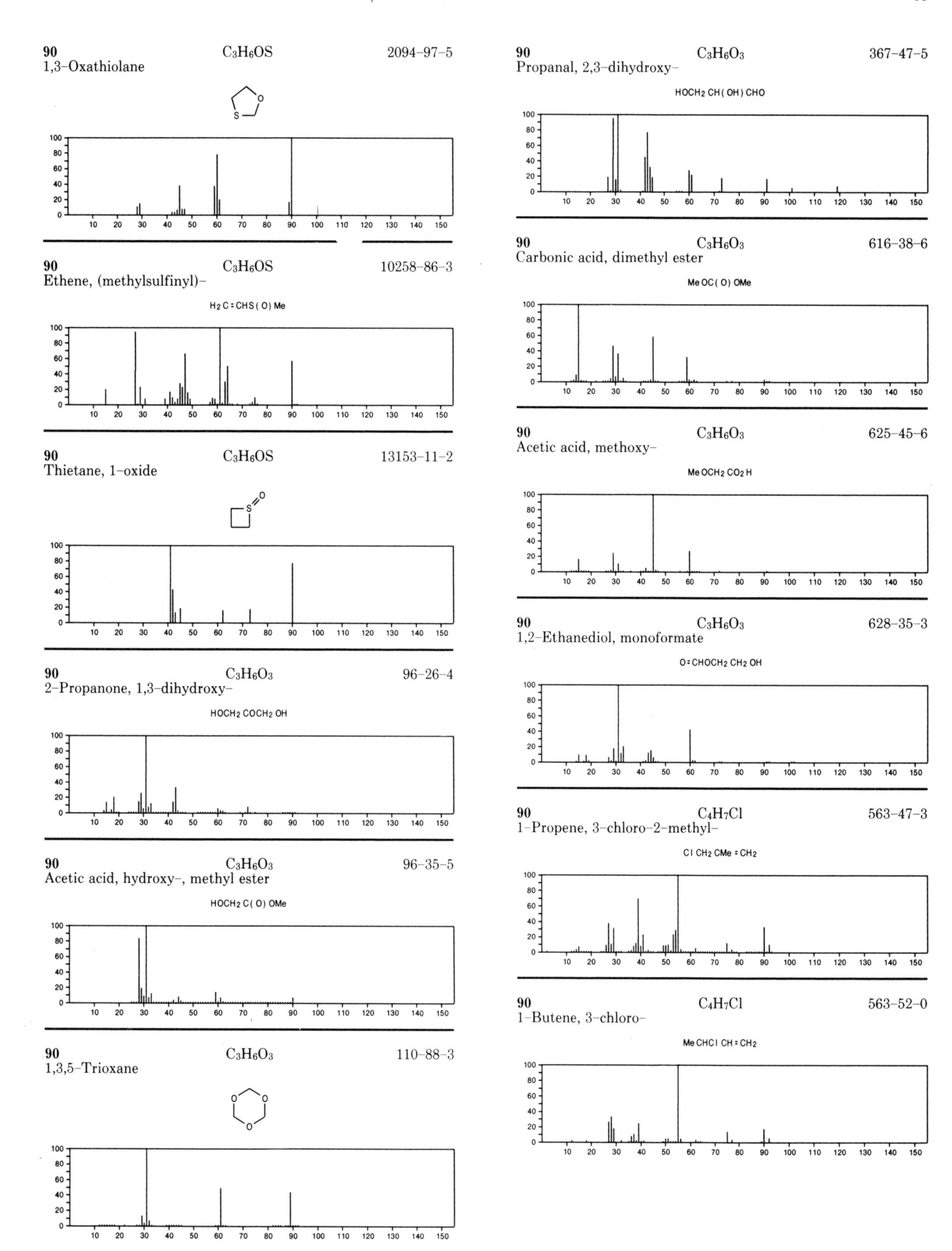
90 C3H6OS 2094-97-5
1,3-Oxathiolane
90 C3H6OS 10258-86-3
Ethene, (methylsulfinyl)-
H2C=CHS(O)Me
90 C3H6OS 13153-11-2
Thietane, 1-oxide
90 C3H6O3 96-26-4
2-Propanone, 1,3-dihydroxy-
HOCH2COCH2OH
90 C3H6O3 96-35-5
Acetic acid, hydroxy-, methyl ester
HOCH2C(O)OMe
90 C3H6O3 110-88-3
1,3,5-Trioxane
90 C3H6O3 367-47-5
Propanal, 2,3-dihydroxy-
HOCH2CH(OH)CHO
90 C3H6O3 616-38-6
Carbonic acid, dimethyl ester
MeOC(O)OMe
90 C3H6O3 625-45-6
Acetic acid, methoxy-
MeOCH2CO2H
90 C3H6O3 628-35-3
1,2-Ethanediol, monoformate
O=CHOCH2CH2OH
90 C4H7Cl 563-47-3
1-Propene, 3-chloro-2-methyl-
ClCH2CMe=CH2
90 C4H7Cl 563-52-0
1-Butene, 3-chloro-
MeCHClCH=CH2

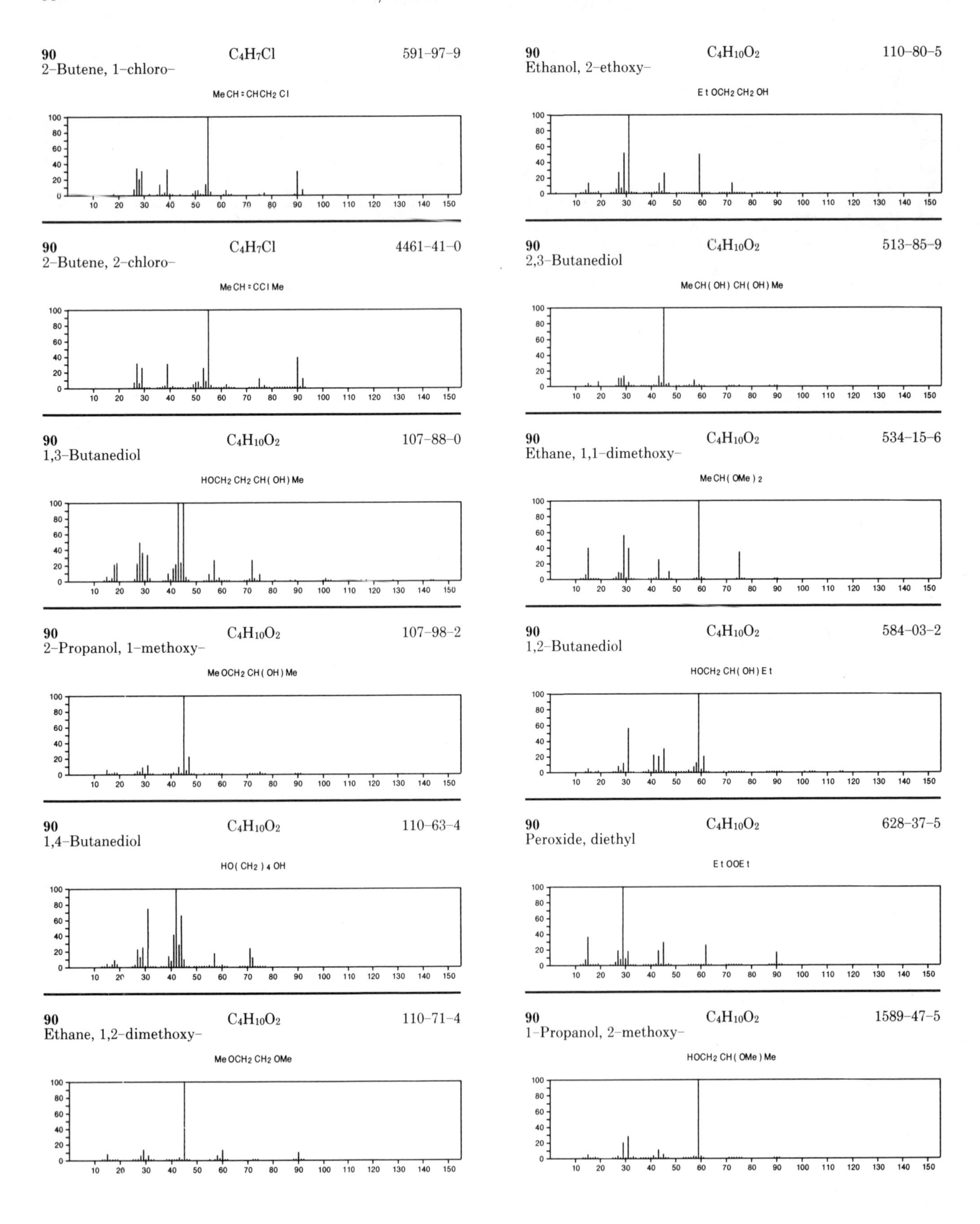
90
C_4H_7Cl
591–97–9
2–Butene, 1–chloro–
MeCH=CHCH2Cl
90
C_4H_7Cl
4461–41–0
2–Butene, 2–chloro–
MeCH=CClMe
90
$C_4H_{10}O_2$
107–88–0
1,3–Butanediol
HOCH2CH2CH(OH)Me
90
$C_4H_{10}O_2$
107–98–2
2–Propanol, 1–methoxy–
MeOCH2CH(OH)Me
90
$C_4H_{10}O_2$
110–63–4
1,4–Butanediol
HO(CH2)4OH
90
$C_4H_{10}O_2$
110–71–4
Ethane, 1,2–dimethoxy–
MeOCH2CH2OMe
90
$C_4H_{10}O_2$
110–80–5
Ethanol, 2–ethoxy–
EtOCH2CH2OH
90
$C_4H_{10}O_2$
513–85–9
2,3–Butanediol
MeCH(OH)CH(OH)Me
90
$C_4H_{10}O_2$
534–15–6
Ethane, 1,1–dimethoxy–
MeCH(OMe)2
90
$C_4H_{10}O_2$
584–03–2
1,2–Butanediol
HOCH2CH(OH)Et
90
$C_4H_{10}O_2$
628–37–5
Peroxide, diethyl
EtOOEt
90
$C_4H_{10}O_2$
1589–47–5
1–Propanol, 2–methoxy–
HOCH2CH(OMe)Me

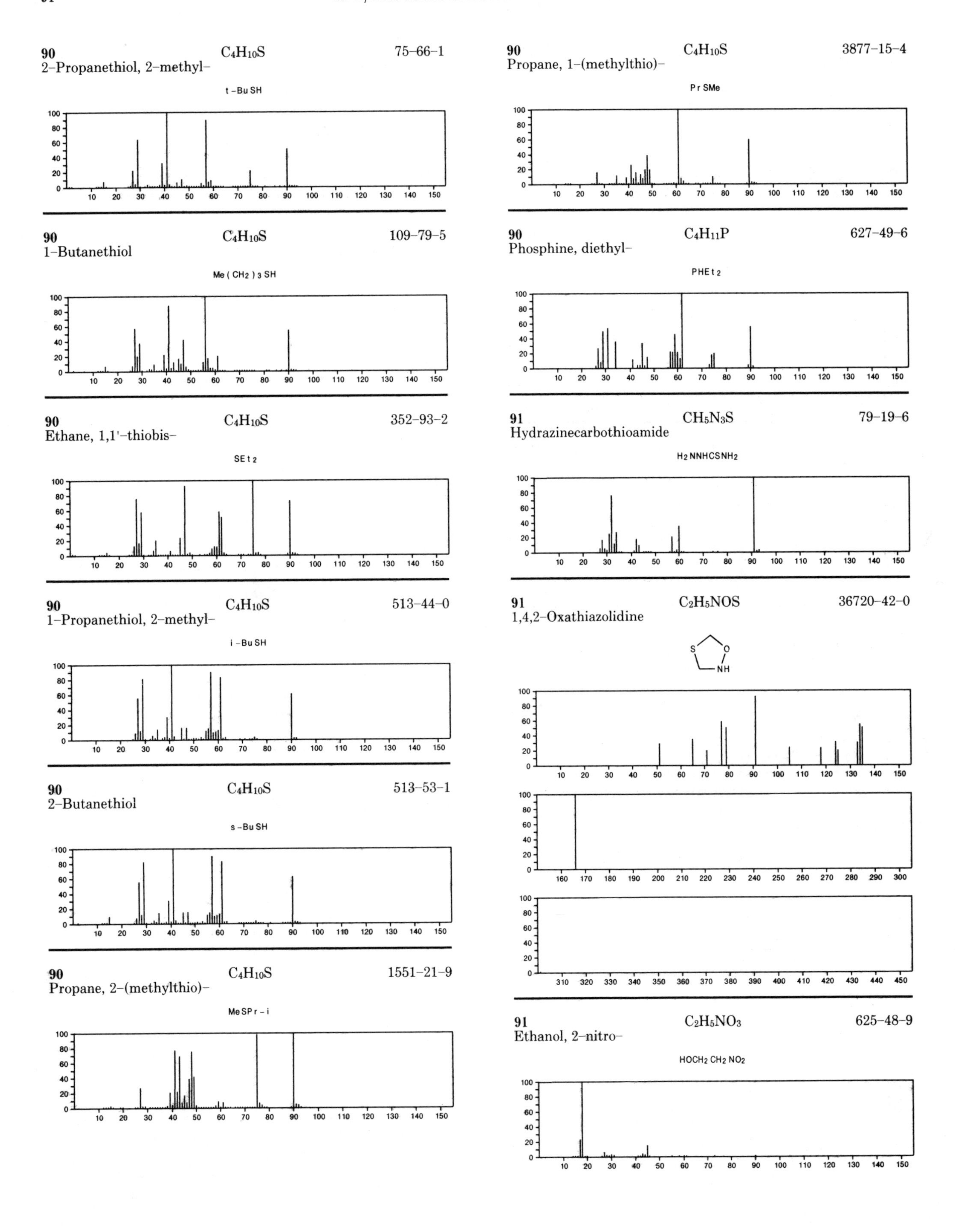
90
C4H10S
75-66-1
2-Propanethiol, 2-methyl-
t-Bu SH
90
C4H10S
3877-15-4
Propane, 1-(methylthio)-
Pr SMe
90
C4H10S
109-79-5
1-Butanethiol
Me (CH2) 3 SH
90
C4H11P
627-49-6
Phosphine, diethyl-
PHEt2
90
C4H10S
352-93-2
Ethane, 1,1'-thiobis-
SEt2
91
CH5N3S
79-19-6
Hydrazinecarbothioamide
H2NNHCSNH2
90
C4H10S
513-44-0
1-Propanethiol, 2-methyl-
i-Bu SH
91
C2H5NOS
36720-42-0
1,4,2-Oxathiazolidine
S
O
NH
90
C4H10S
513-53-1
2-Butanethiol
s-Bu SH
90
C4H10S
1551-21-9
Propane, 2-(methylthio)-
MeSPr-i
91
C2H5NO3
625-48-9
Ethanol, 2-nitro-
HOCH2 CH2 NO2

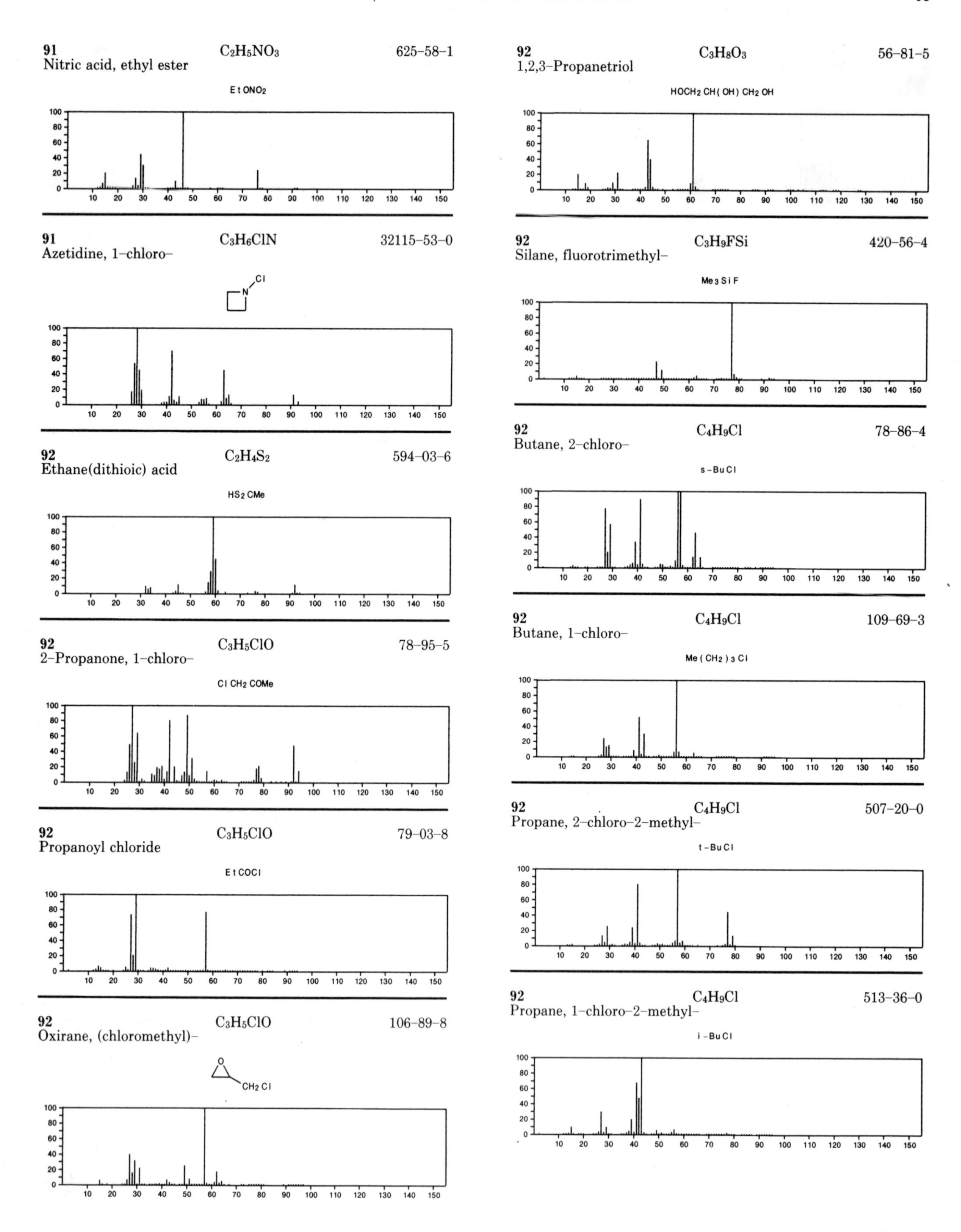
91 Nitric acid, ethyl ester C2H5NO3 625-58-1
EtONO2
91 Azetidine, 1-chloro- C3H6ClN 32115-53-0
92 Ethane(dithioic) acid C2H4S2 594-03-6
HS2CMe
92 2-Propanone, 1-chloro- C3H5ClO 78-95-5
ClCH2COMe
92 Propanoyl chloride C3H5ClO 79-03-8
EtCOCl
92 Oxirane, (chloromethyl)- C3H5ClO 106-89-8
CH2Cl
92 1,2,3-Propanetriol C3H8O3 56-81-5
HOCH2CH(OH)CH2OH
92 Silane, fluorotrimethyl- C3H9FSi 420-56-4
Me3SiF
92 Butane, 2-chloro- C4H9Cl 78-86-4
s-BuCl
92 Butane, 1-chloro- C4H9Cl 109-69-3
Me(CH2)3Cl
92 Propane, 2-chloro-2-methyl- C4H9Cl 507-20-0
t-BuCl
92 Propane, 1-chloro-2-methyl- C4H9Cl 513-36-0
i-BuCl

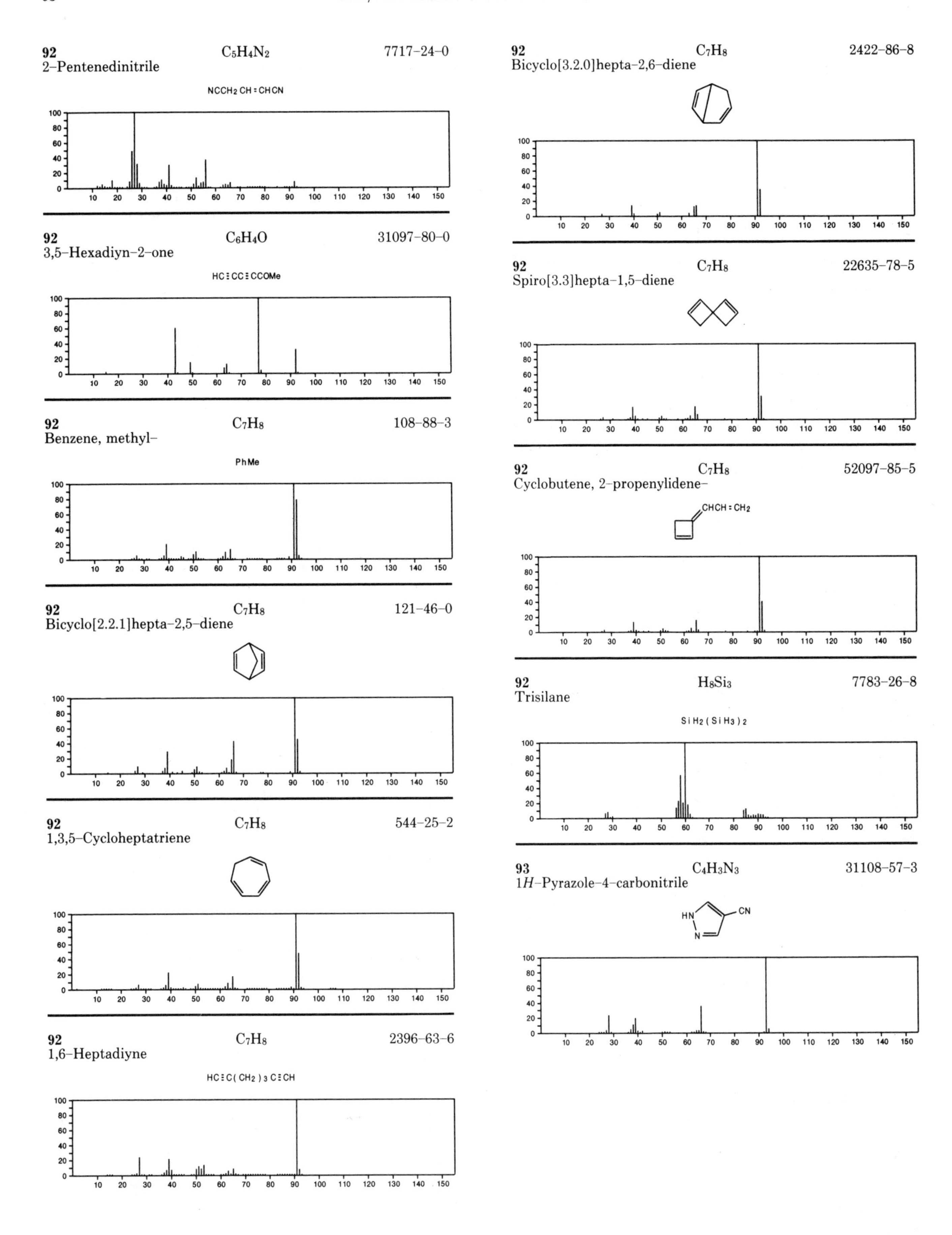

92 $C_5H_4N_2$ 7717-24-0
2-Pentenedinitrile

92 C_6H_4O 31097-80-0
3,5-Hexadiyn-2-one

92 C_7H_8 108-88-3
Benzene, methyl-

92 C_7H_8 121-46-0
Bicyclo[2.2.1]hepta-2,5-diene

92 C_7H_8 544-25-2
1,3,5-Cycloheptatriene

92 C_7H_8 2396-63-6
1,6-Heptadiyne

92 C_7H_8 2422-86-8
Bicyclo[3.2.0]hepta-2,6-diene

92 C_7H_8 22635-78-5
Spiro[3.3]hepta-1,5-diene

92 C_7H_8 52097-85-5
Cyclobutene, 2-propenylidene-

92 H_8Si_3 7783-26-8
Trisilane

93 $C_4H_3N_3$ 31108-57-3
1*H*-Pyrazole-4-carbonitrile

93 $C_5H_5N.BH_3$ 110–51–0 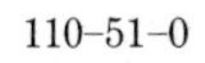

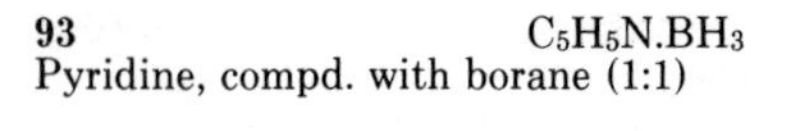
Pyridine, compd. with borane (1:1)

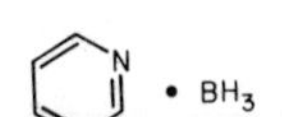

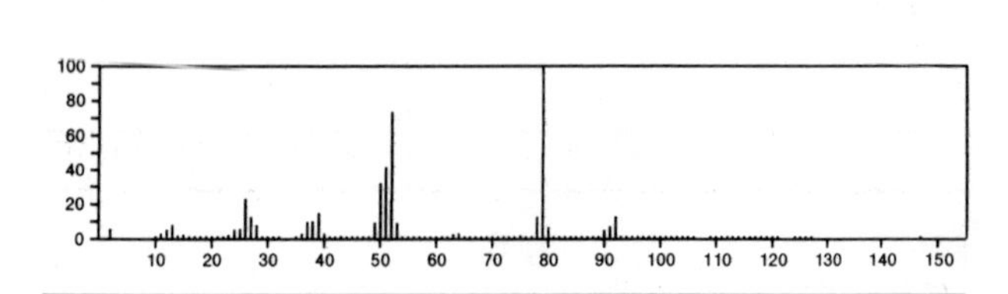

93 C_6H_7N 62–53–3
Benzenamine

PhNH2

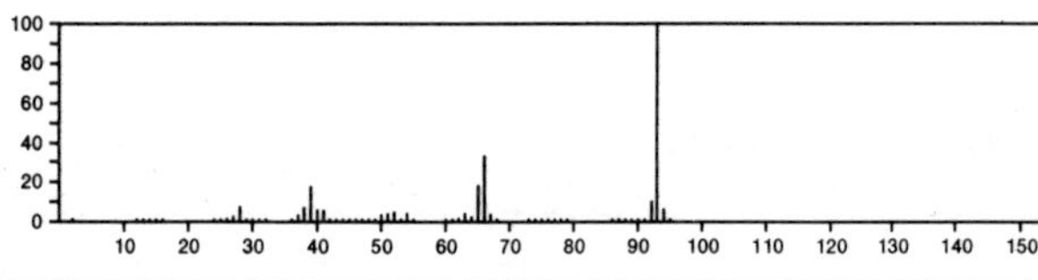

93 C_6H_7N 108–89–4
Pyridine, 4–methyl–

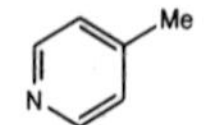

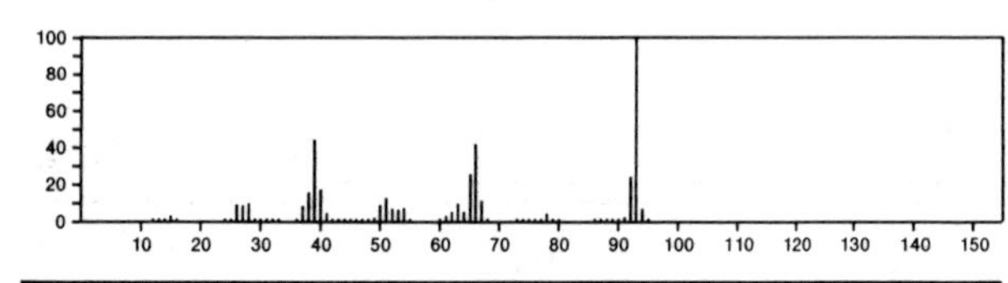

93 C_6H_7N 108–99–6
Pyridine, 3–methyl–

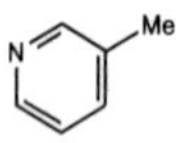

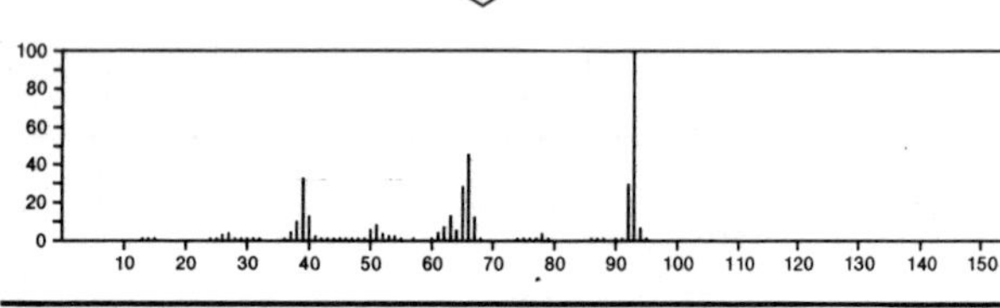

93 C_6H_7N 109–06–8
Pyridine, 2–methyl–

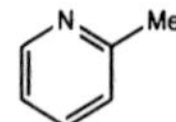

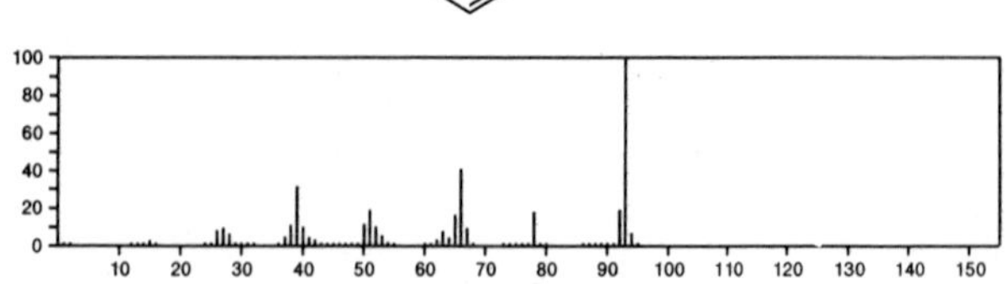

93 C_6H_7N 28769–50–8 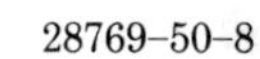
4–Pentenenitrile, 2–methylene–

H2 C = C(CN) CH2 CH = CH2

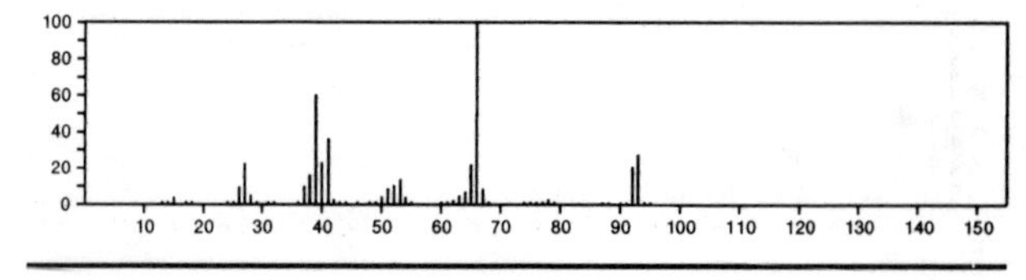

94  CH_3Br 74–83–9
Methane, bromo–

Br Me

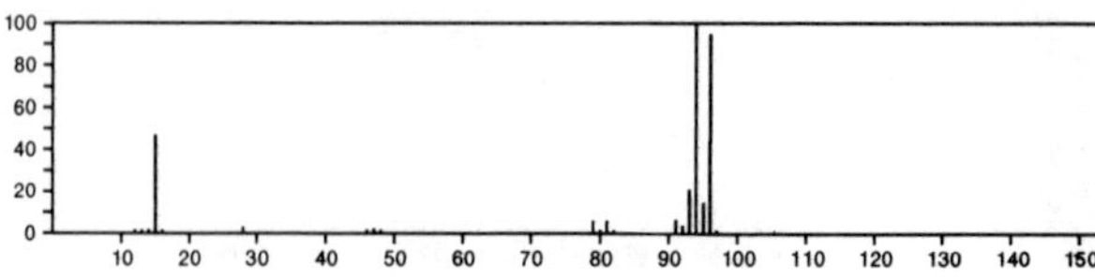

94 C_2Cl_2 7572–29–4
Ethyne, dichloro–

Cl C≡CCl

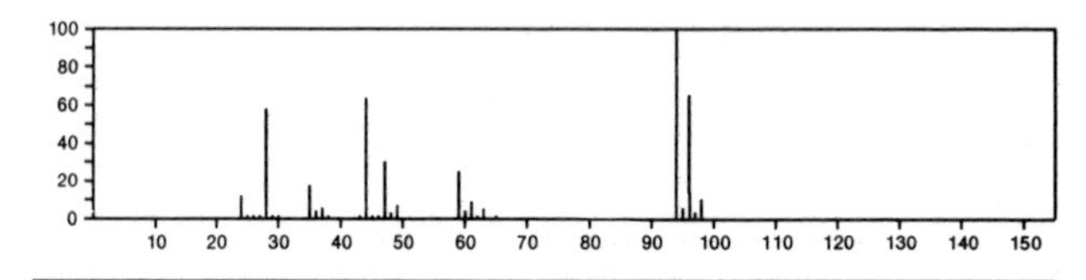

94 $C_2H_3ClO_2$ 79–11–8
Acetic acid, chloro–

Cl CH2 CO2 H

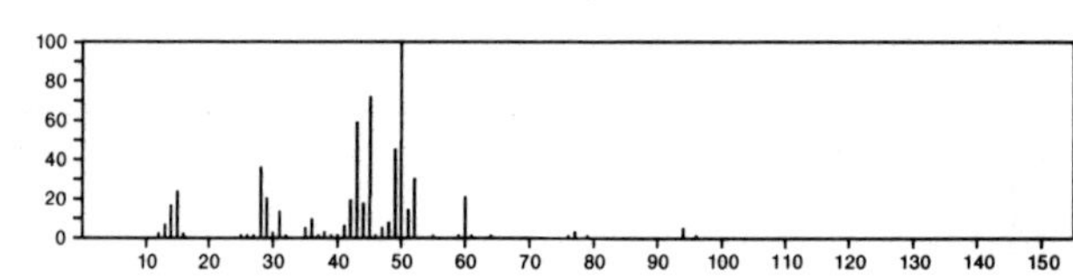

94 $C_2H_3ClO_2$ 79–22–1
Carbonochloridic acid, methyl ester

Me OC(O) Cl

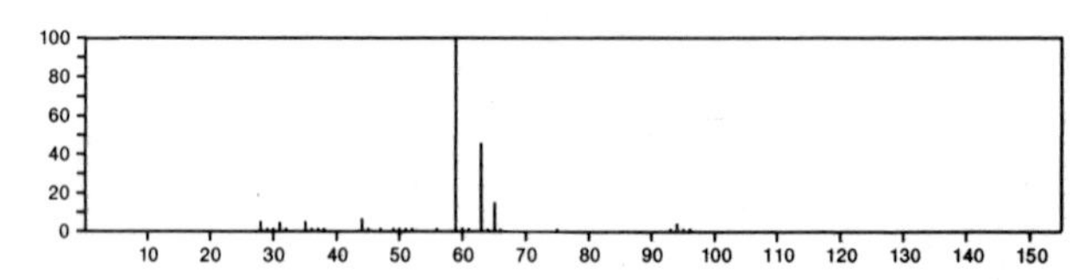

94 $C_2H_6O_2S$ 67–71–0
Methane, sulfonylbis–

Me SO2 Me

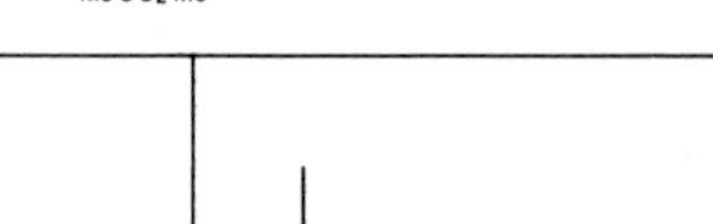

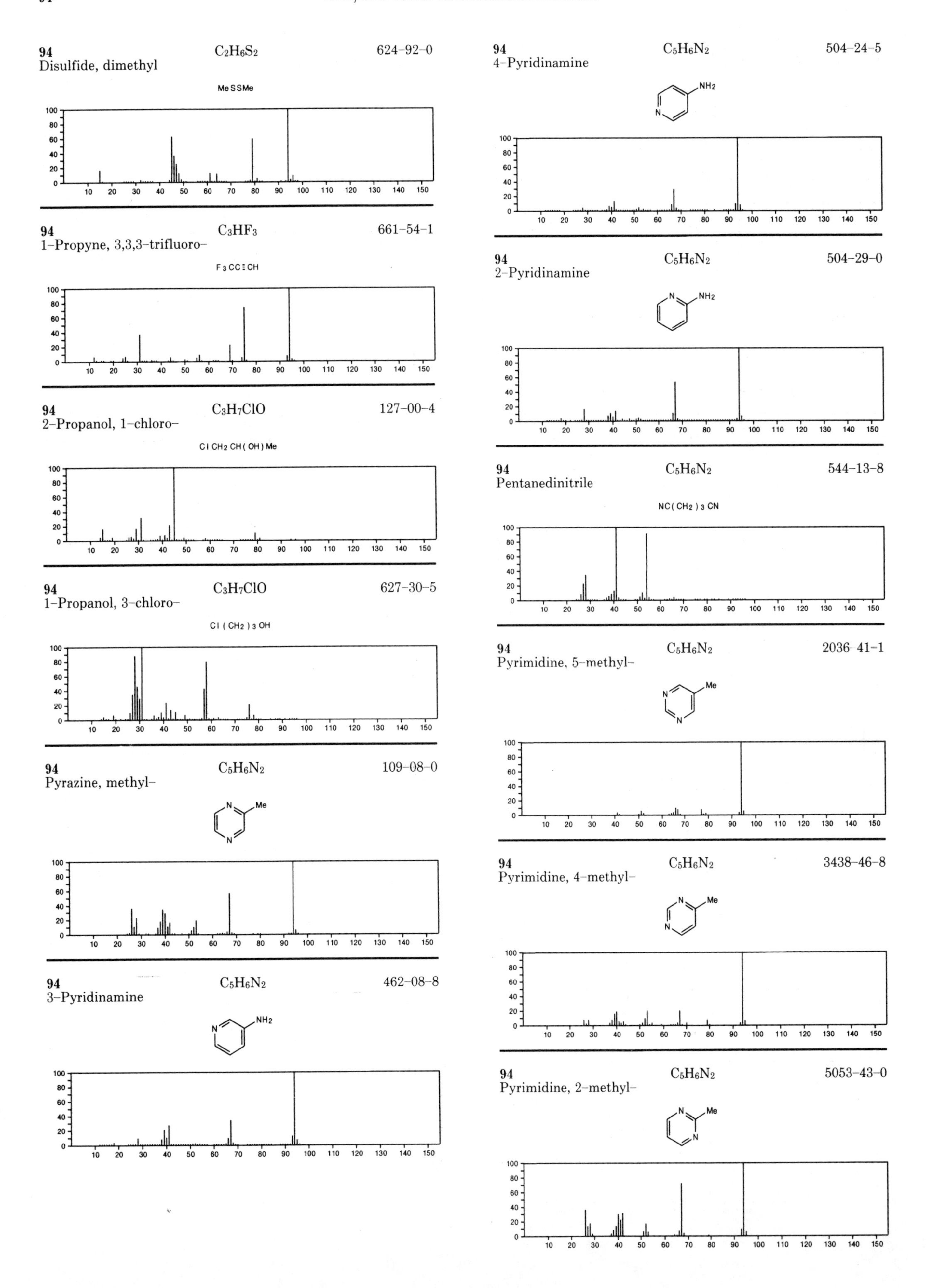

94 Disulfide, dimethyl C2H6S2 624-92-0
MeSSMe
94 1-Propyne, 3,3,3-trifluoro- C3HF3 661-54-1
F3CC≡CH
94 2-Propanol, 1-chloro- C3H7ClO 127-00-4
ClCH2CH(OH)Me
94 1-Propanol, 3-chloro- C3H7ClO 627-30-5
Cl(CH2)3OH
94 Pyrazine, methyl- C5H6N2 109-08-0
N
Me
N
94 3-Pyridinamine C5H6N2 462-08-8
N
NH2
94 4-Pyridinamine C5H6N2 504-24-5
NH2
N
94 2-Pyridinamine C5H6N2 504-29-0
N
NH2
94 Pentanedinitrile C5H6N2 544-13-8
NC(CH2)3CN
94 Pyrimidine, 5-methyl- C5H6N2 2036-41-1
N
Me
N
94 Pyrimidine, 4-methyl- C5H6N2 3438-46-8
N
Me
N
94 Pyrimidine, 2-methyl- C5H6N2 5053-43-0
N
Me
N

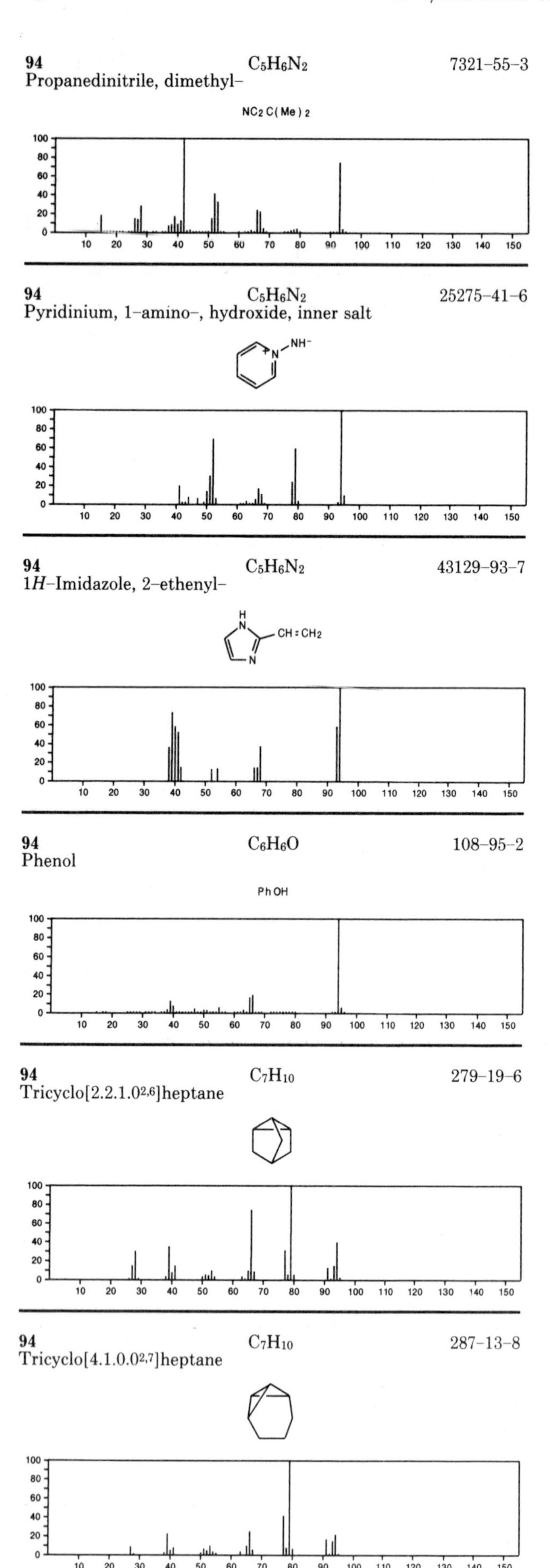

94 $C_5H_6N_2$ 7321-55-3
Propanedinitrile, dimethyl-

94 $C_5H_6N_2$ 25275-41-6
Pyridinium, 1-amino-, hydroxide, inner salt

94 $C_5H_6N_2$ 43129-93-7
1*H*-Imidazole, 2-ethenyl-

94 C_6H_6O 108-95-2
Phenol

94 C_7H_{10} 279-19-6
Tricyclo[2.2.1.0^{2,6}]heptane

94 C_7H_{10} 287-13-8
Tricyclo[4.1.0.0^{2,7}]heptane

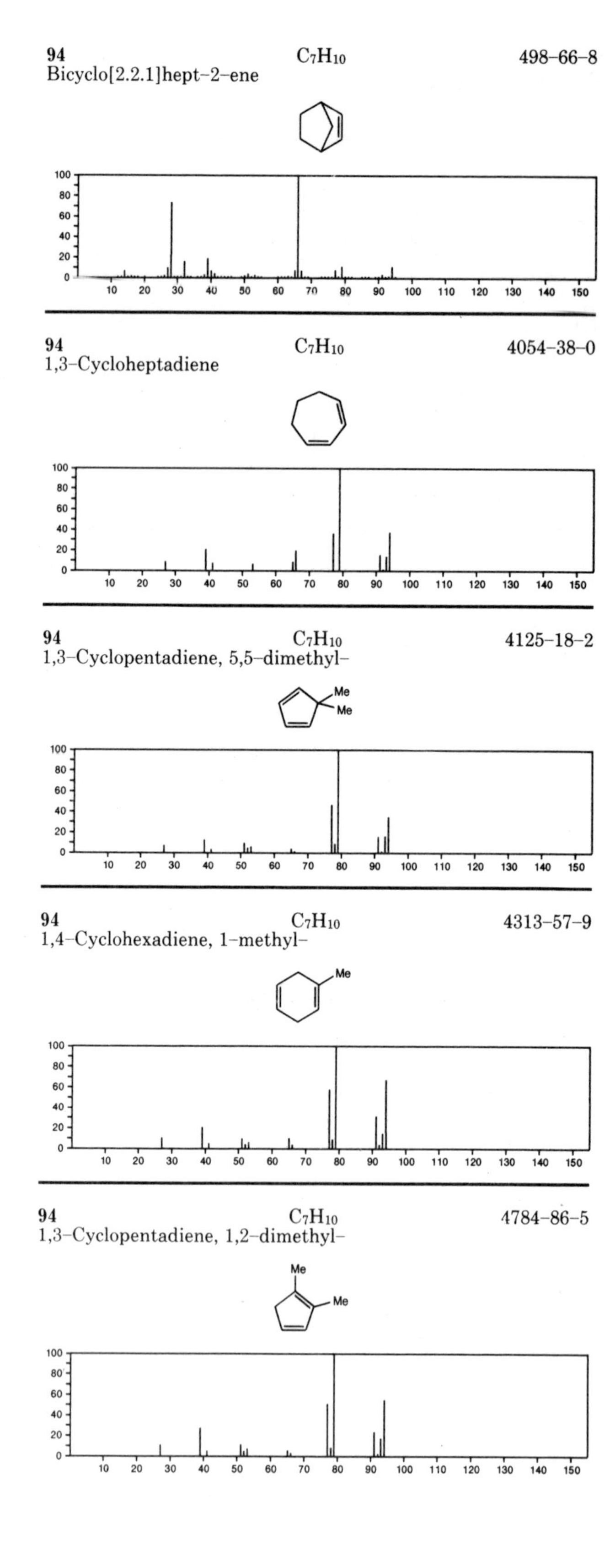

94 C_7H_{10} 498-66-8
Bicyclo[2.2.1]hept-2-ene

94 C_7H_{10} 4054-38-0
1,3-Cycloheptadiene

94 C_7H_{10} 4125-18-2
1,3-Cyclopentadiene, 5,5-dimethyl-

94 C_7H_{10} 4313-57-9
1,4-Cyclohexadiene, 1-methyl-

94 C_7H_{10} 4784-86-5
1,3-Cyclopentadiene, 1,2-dimethyl-

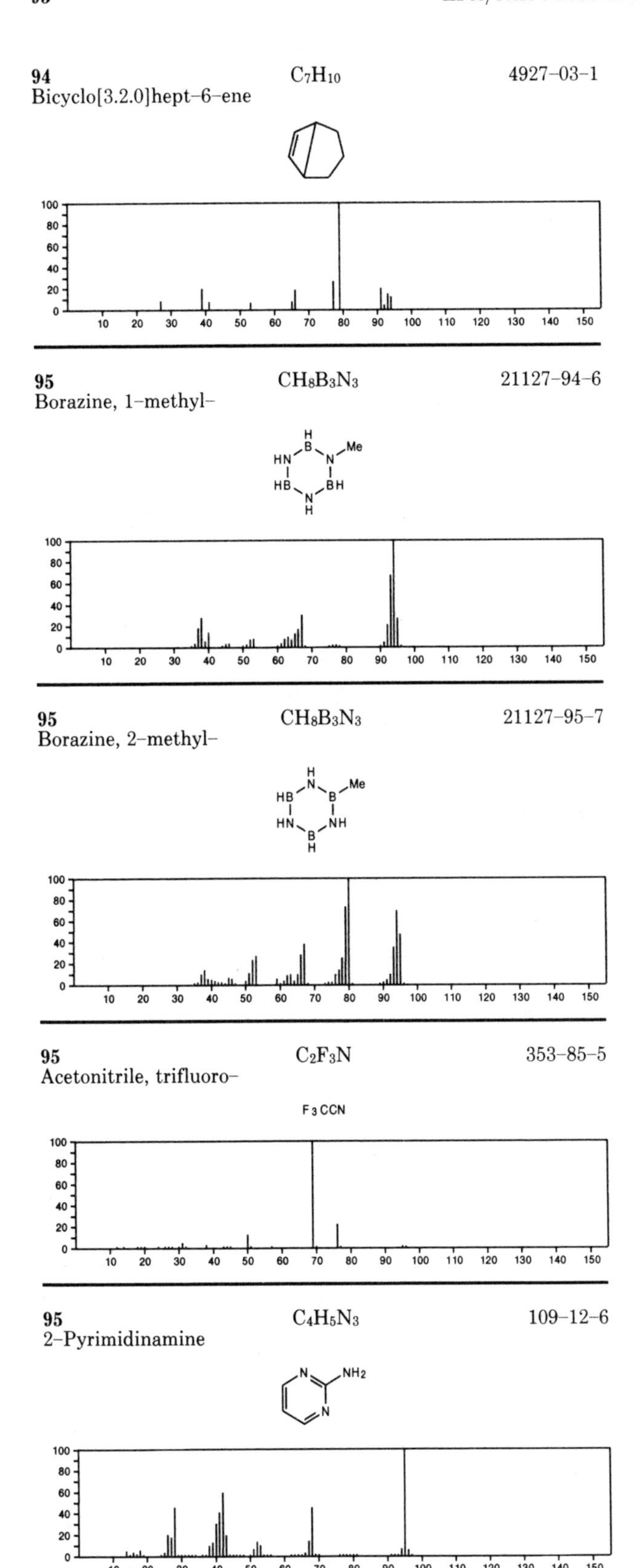

94 C_7H_{10} 4927-03-1
Bicyclo[3.2.0]hept-6-ene

95 $CH_8B_3N_3$ 21127-94-6
Borazine, 1-methyl-

95 $CH_8B_3N_3$ 21127-95-7
Borazine, 2-methyl-

95 C_2F_3N 353-85-5
Acetonitrile, trifluoro-

F3CCN

95 $C_4H_5N_3$ 109-12-6
2-Pyrimidinamine

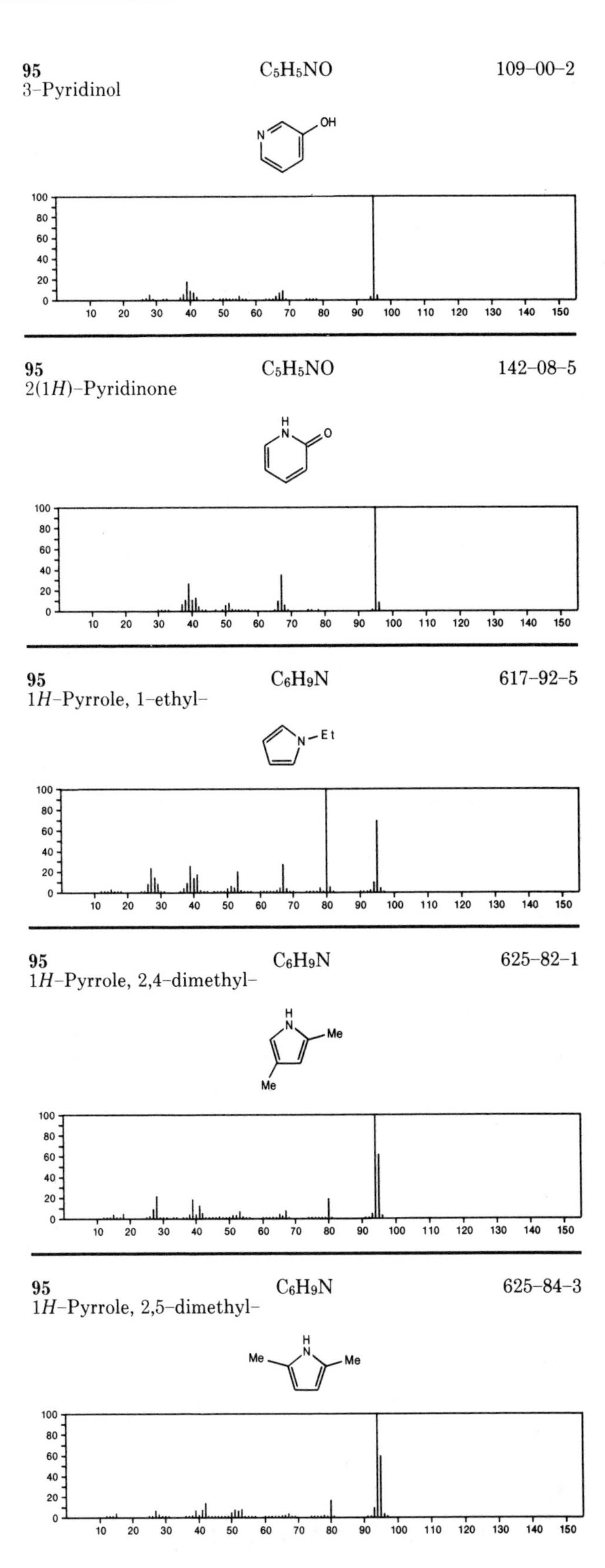

95 C_5H_5NO 109-00-2
3-Pyridinol

95 C_5H_5NO 142-08-5
2(1*H*)-Pyridinone

95 C_6H_9N 617-92-5
1*H*-Pyrrole, 1-ethyl-

95 C_6H_9N 625-82-1
1*H*-Pyrrole, 2,4-dimethyl-

95 C_6H_9N 625-84-3
1*H*-Pyrrole, 2,5-dimethyl-

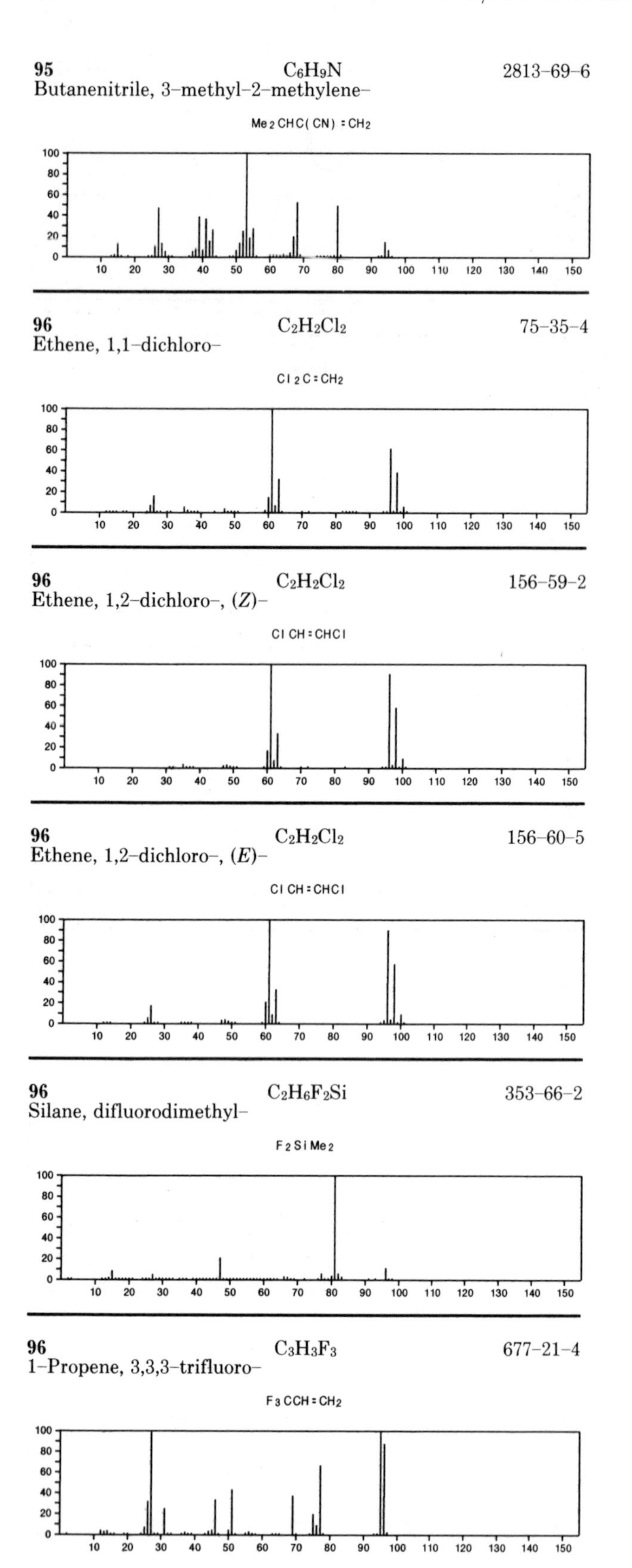
95 C_6H_9N 2813-69-6
Butanenitrile, 3-methyl-2-methylene-
Me2CHC(CN)=CH2
96 $C_2H_2Cl_2$ 75-35-4
Ethene, 1,1-dichloro-
Cl2C=CH2
96 $C_2H_2Cl_2$ 156-59-2
Ethene, 1,2-dichloro-, (Z)-
ClCH=CHCl
96 $C_2H_2Cl_2$ 156-60-5
Ethene, 1,2-dichloro-, (E)-
ClCH=CHCl
96 $C_2H_6F_2Si$ 353-66-2
Silane, difluorodimethyl-
F2SiMe2
96 $C_3H_3F_3$ 677-21-4
1-Propene, 3,3,3-trifluoro-
F3CCH=CH2

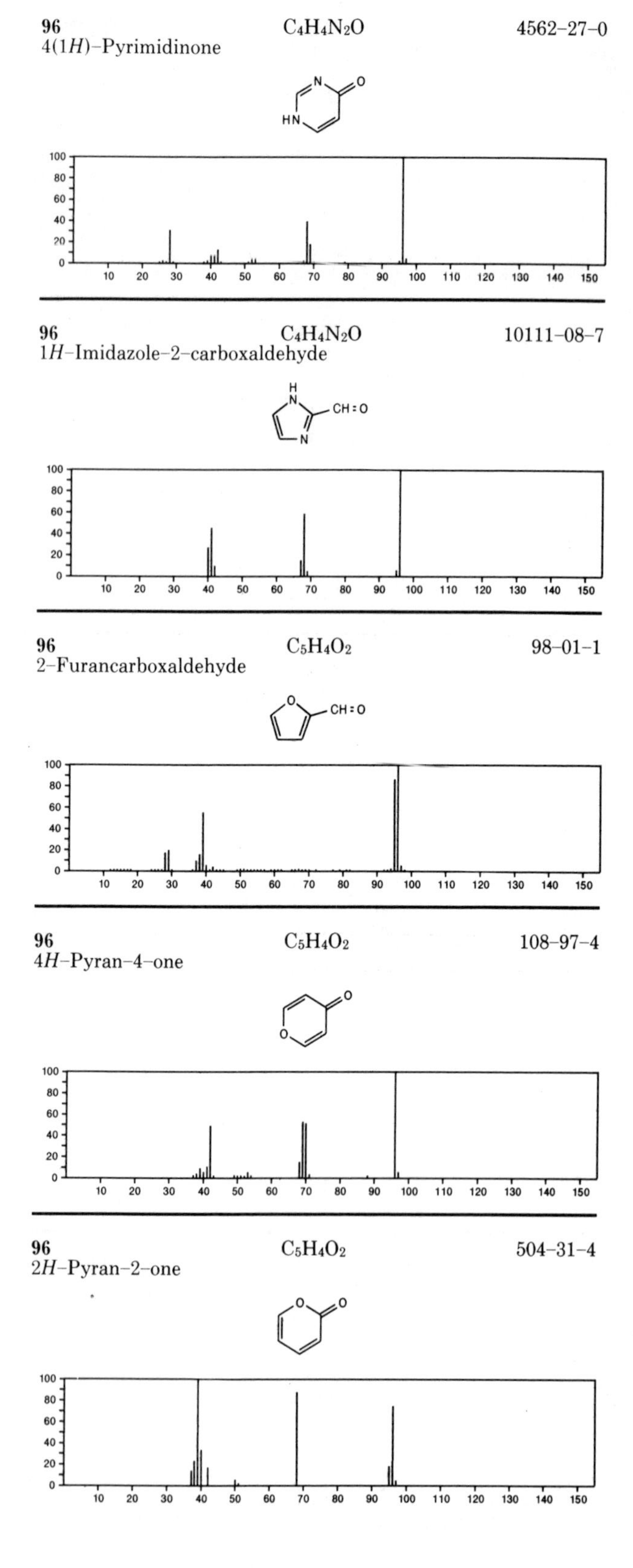
96 $C_4H_4N_2O$ 4562-27-0
4(1H)-Pyrimidinone
96 $C_4H_4N_2O$ 10111-08-7
1H-Imidazole-2-carboxaldehyde
CH=O
96 $C_5H_4O_2$ 98-01-1
2-Furancarboxaldehyde
CH=O
96 $C_5H_4O_2$ 108-97-4
4H-Pyran-4-one
96 $C_5H_4O_2$ 504-31-4
2H-Pyran-2-one

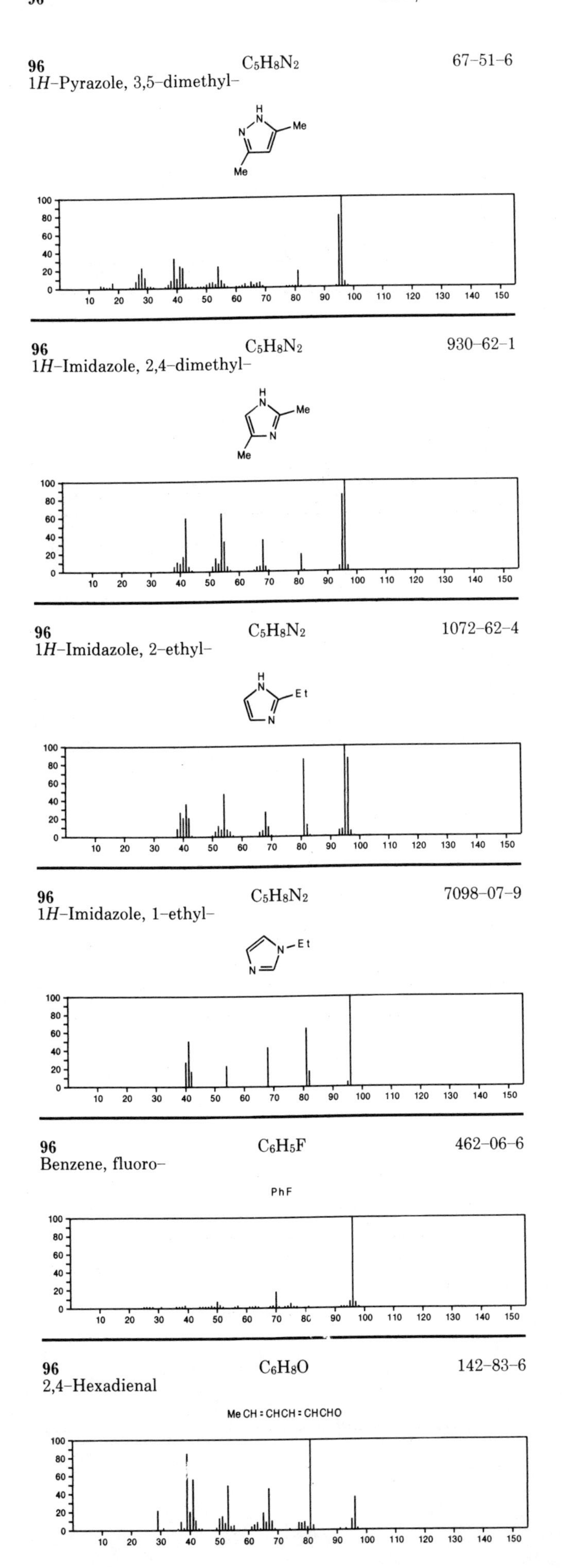

96 $C_5H_8N_2$ 67–51–6
1*H*–Pyrazole, 3,5–dimethyl–

96 $C_5H_8N_2$ 930–62–1
1*H*–Imidazole, 2,4–dimethyl–

96 $C_5H_8N_2$ 1072–62–4
1*H*–Imidazole, 2–ethyl–

96 $C_5H_8N_2$ 7098–07–9
1*H*–Imidazole, 1–ethyl–

96 C_6H_5F 462–06–6
Benzene, fluoro–

PhF

96 C_6H_8O 142–83–6
2,4–Hexadienal

MeCH:CHCH:CHCHO

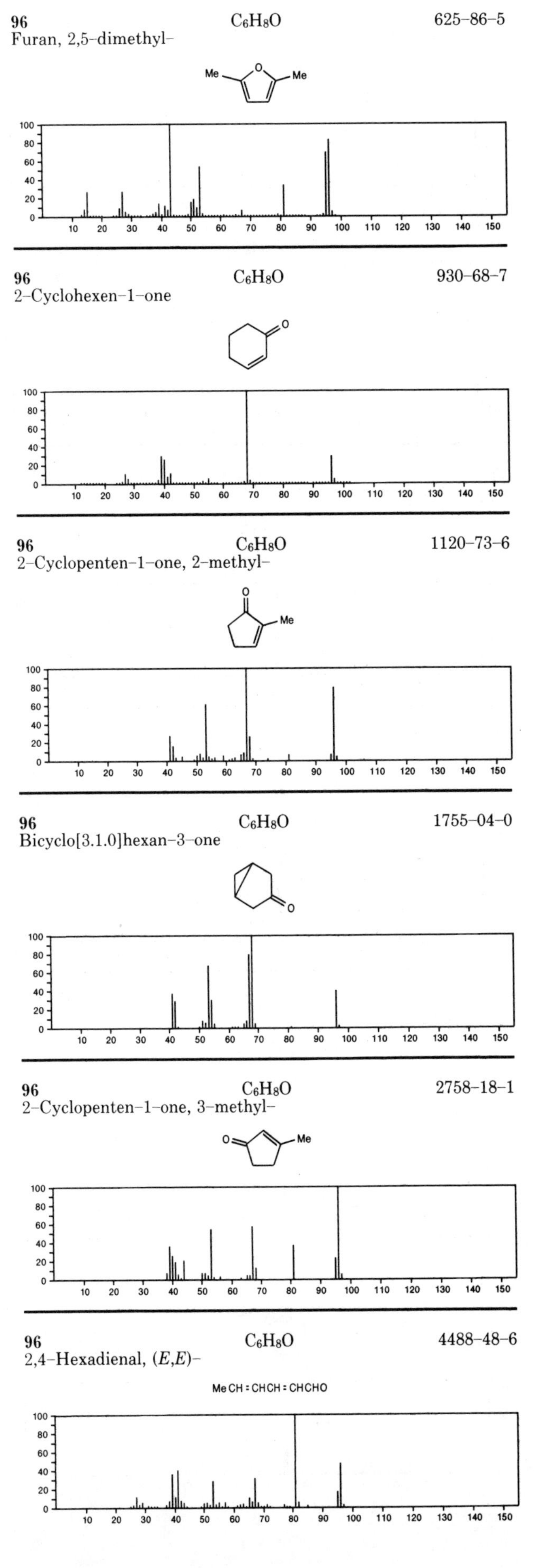

96 C_6H_8O 625–86–5
Furan, 2,5–dimethyl–

96 C_6H_8O 930–68–7
2–Cyclohexen–1–one

96 C_6H_8O 1120–73–6
2–Cyclopenten–1–one, 2–methyl–

96 C_6H_8O 1755–04–0
Bicyclo[3.1.0]hexan–3–one

96 C_6H_8O 2758–18–1
2–Cyclopenten–1–one, 3–methyl–

96 C_6H_8O 4488–48–6
2,4–Hexadienal, (*E*,*E*)–

MeCH:CHCH:CHCHO

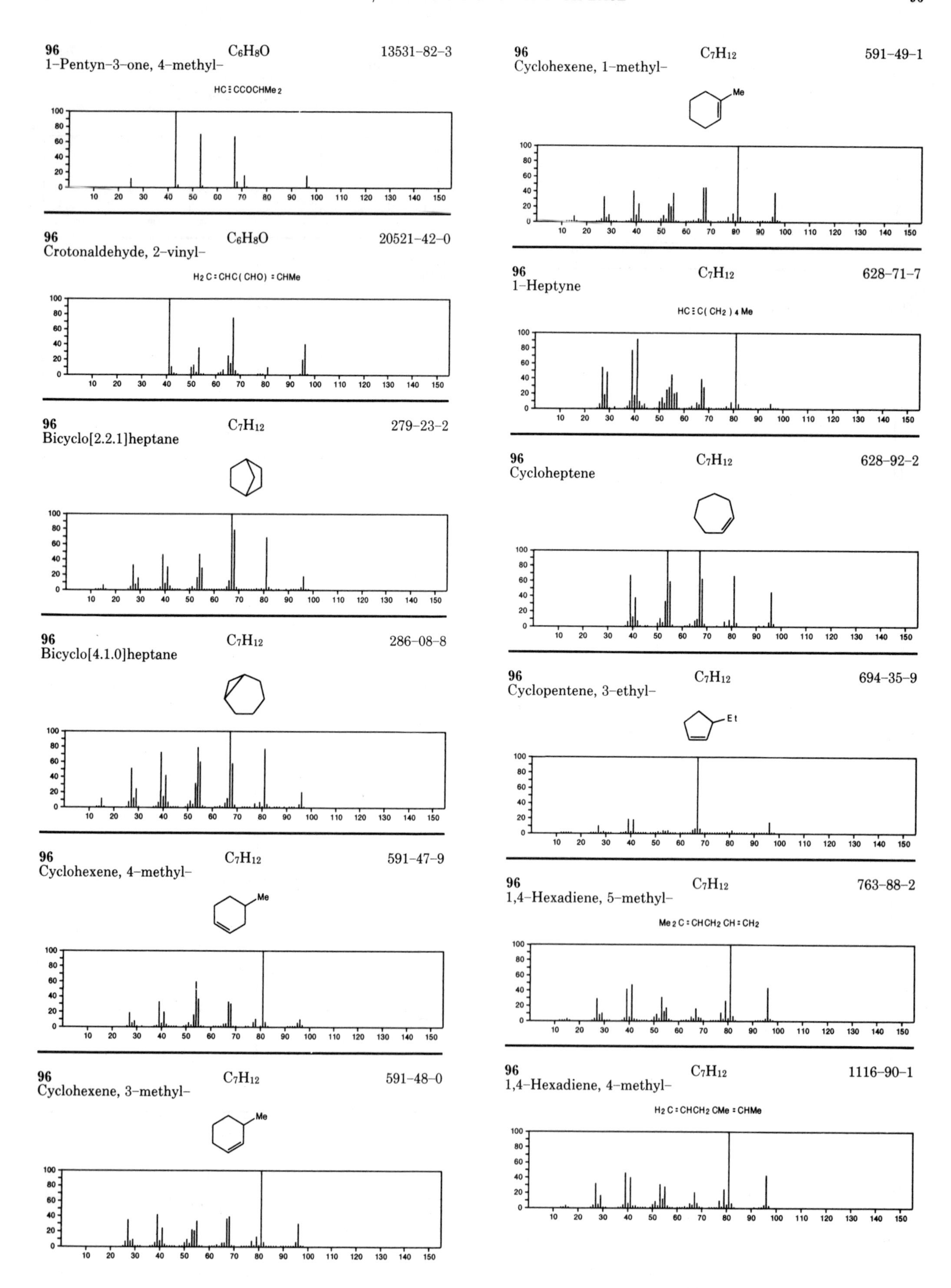
96 C6H8O 13531-82-3
1-Pentyn-3-one, 4-methyl-
HC:CCOCHMe2
96 C6H8O 20521-42-0
Crotonaldehyde, 2-vinyl-
H2C:CHC(CHO):CHMe
96 C7H12 279-23-2
Bicyclo[2.2.1]heptane
96 C7H12 286-08-8
Bicyclo[4.1.0]heptane
96 C7H12 591-47-9
Cyclohexene, 4-methyl-
Me
96 C7H12 591-48-0
Cyclohexene, 3-methyl-
Me
96 C7H12 591-49-1
Cyclohexene, 1-methyl-
Me
96 C7H12 628-71-7
1-Heptyne
HC:C(CH2)4Me
96 C7H12 628-92-2
Cycloheptene
96 C7H12 694-35-9
Cyclopentene, 3-ethyl-
Et
96 C7H12 763-88-2
1,4-Hexadiene, 5-methyl-
Me2C:CHCH2CH:CH2
96 C7H12 1116-90-1
1,4-Hexadiene, 4-methyl-
H2C:CHCH2CMe:CHMe

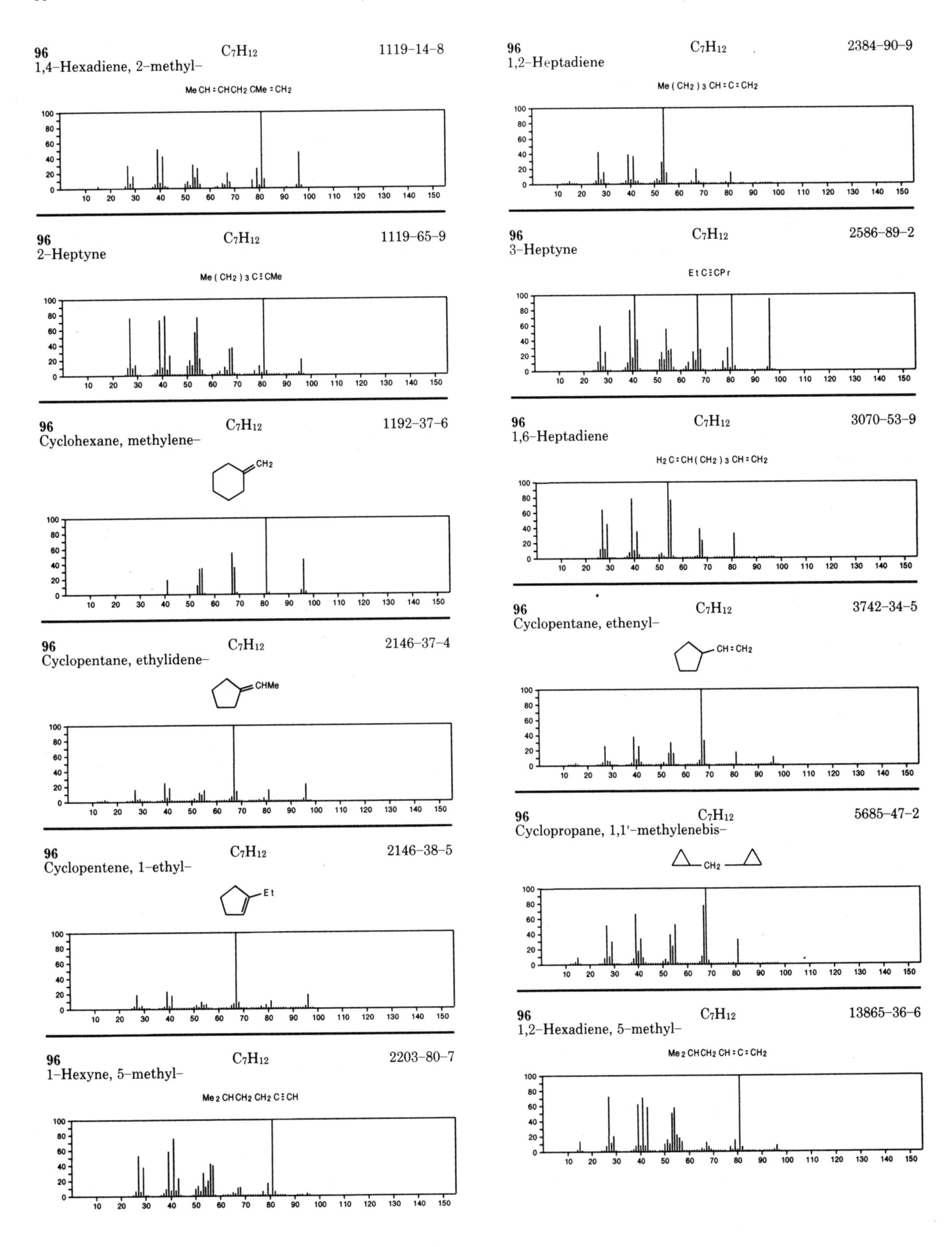
96
1,4-Hexadiene, 2-methyl-
C_7H_{12}
1119-14-8
Me CH = CHCH2 CMe = CH2
96
2-Heptyne
C_7H_{12}
1119-65-9
Me (CH2) 3 C ≡ CMe
96
Cyclohexane, methylene-
C_7H_{12}
1192-37-6
CH2
96
Cyclopentane, ethylidene-
C_7H_{12}
2146-37-4
CHMe
96
Cyclopentene, 1-ethyl-
C_7H_{12}
2146-38-5
Et
96
1-Hexyne, 5-methyl-
C_7H_{12}
2203-80-7
Me 2 CHCH2 CH2 C ≡ CH
96
1,2-Heptadiene
C_7H_{12}
2384-90-9
Me (CH2) 3 CH = C = CH2
96
3-Heptyne
C_7H_{12}
2586-89-2
Et C ≡ CPr
96
1,6-Heptadiene
C_7H_{12}
3070-53-9
H2 C = CH (CH2) 3 CH = CH2
96
Cyclopentane, ethenyl-
C_7H_{12}
3742-34-5
CH = CH2
96
Cyclopropane, 1,1'-methylenebis-
C_7H_{12}
5685-47-2
CH2
96
1,2-Hexadiene, 5-methyl-
C_7H_{12}
13865-36-6
Me 2 CHCH2 CH = C = CH2

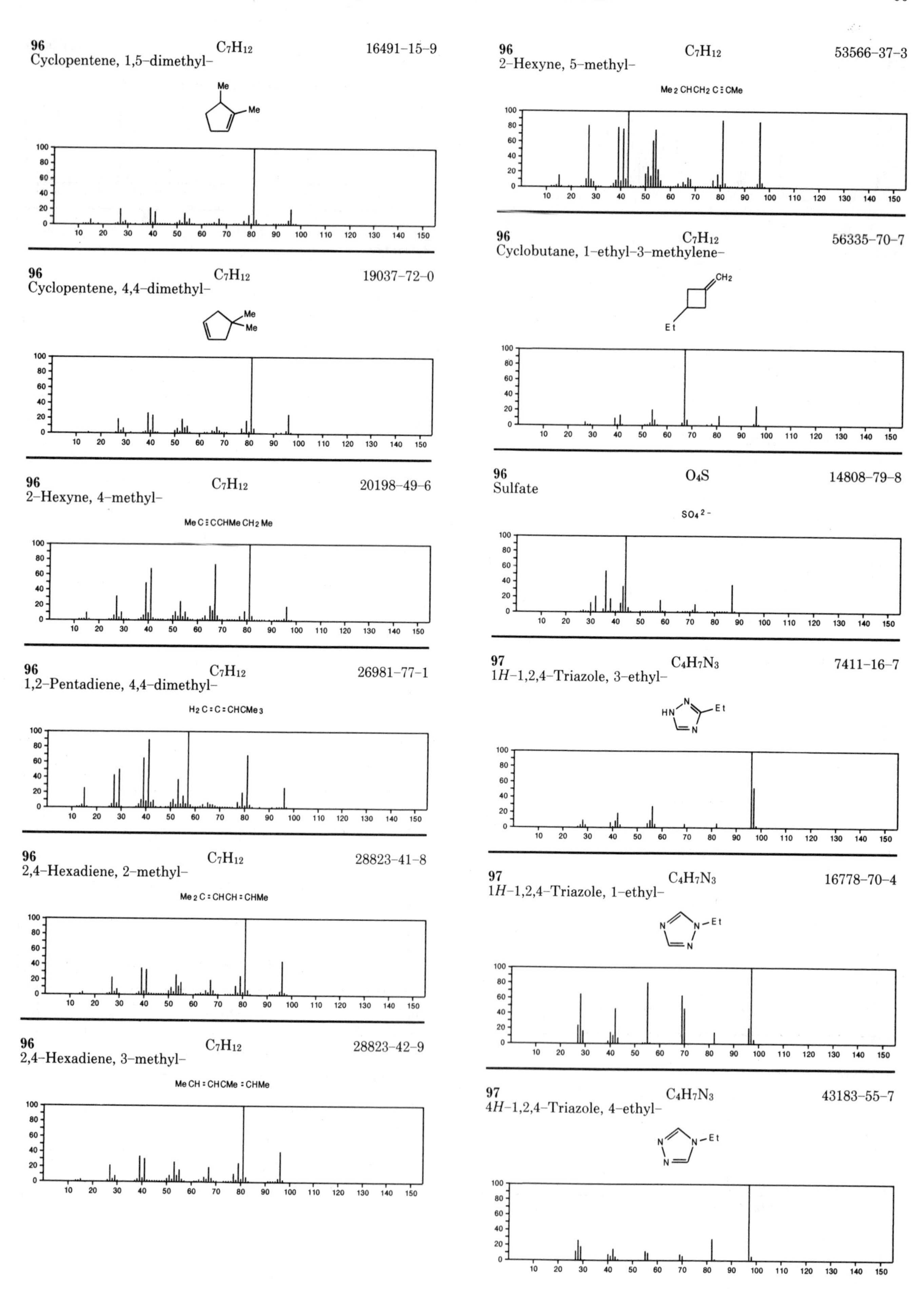
96
Cyclopentene, 1,5-dimethyl-
C7H12
16491-15-9
Me
Me
96
Cyclopentene, 4,4-dimethyl-
C7H12
19037-72-0
Me
Me
96
2-Hexyne, 4-methyl-
C7H12
20198-49-6
MeC≡CCHMeCH2Me
96
1,2-Pentadiene, 4,4-dimethyl-
C7H12
26981-77-1
H2C=C=CHCMe3
96
2,4-Hexadiene, 2-methyl-
C7H12
28823-41-8
Me2C=CHCH=CHMe
96
2,4-Hexadiene, 3-methyl-
C7H12
28823-42-9
MeCH=CHCMe=CHMe
96
2-Hexyne, 5-methyl-
C7H12
53566-37-3
Me2CHCH2C≡CMe
96
Cyclobutane, 1-ethyl-3-methylene-
C7H12
56335-70-7
CH2
Et
96
Sulfate
O4S
14808-79-8
SO4 2-
97
1H-1,2,4-Triazole, 3-ethyl-
C4H7N3
7411-16-7
HN
N
Et
97
1H-1,2,4-Triazole, 1-ethyl-
C4H7N3
16778-70-4
N-Et
97
4H-1,2,4-Triazole, 4-ethyl-
C4H7N3
43183-55-7
N-Et

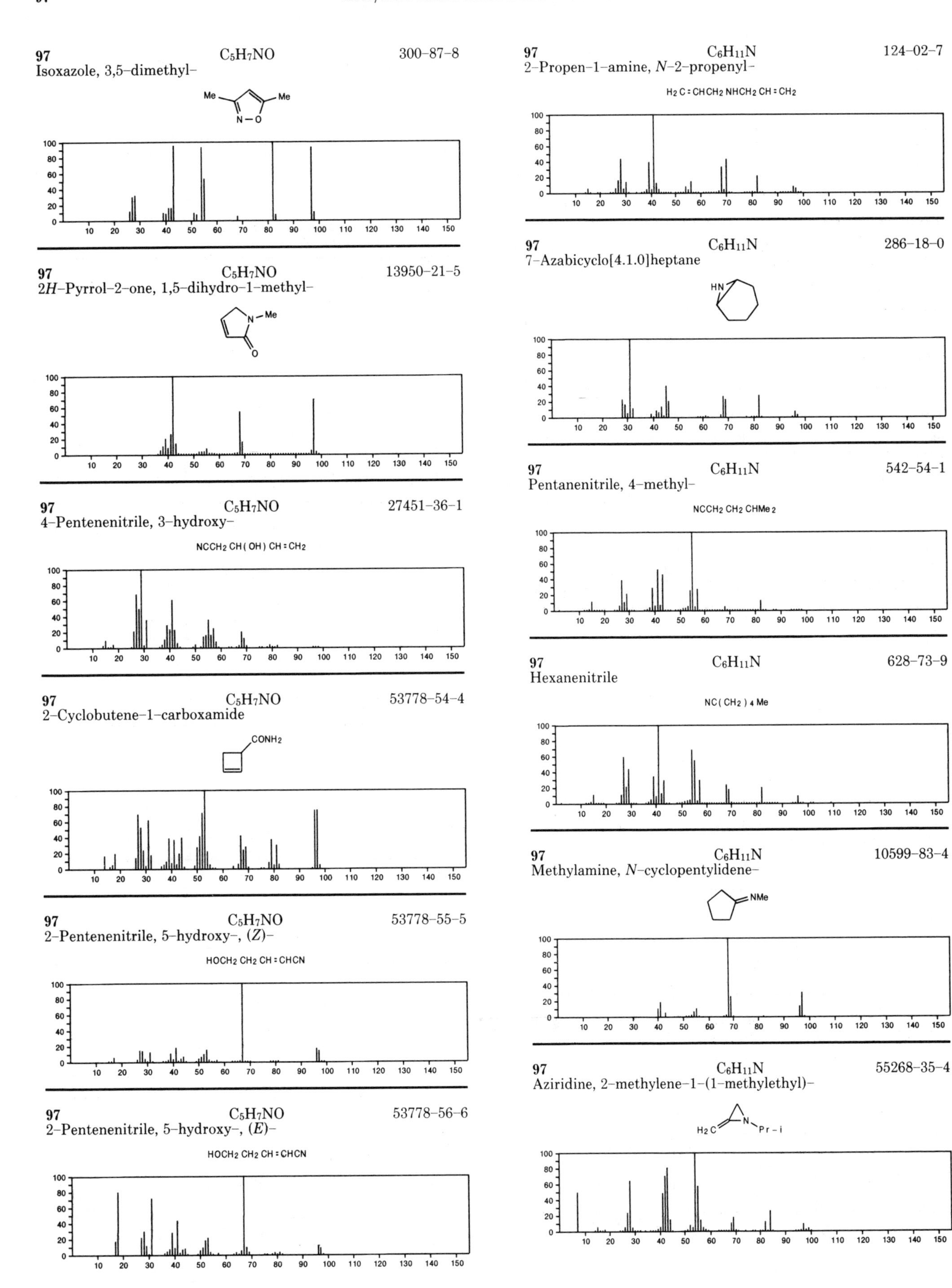
97
C_5H_7NO
300-87-8
Isoxazole, 3,5-dimethyl-
Me
Me
N-O
97
C_5H_7NO
13950-21-5
2H-Pyrrol-2-one, 1,5-dihydro-1-methyl-
N-Me
O
97
C_5H_7NO
27451-36-1
4-Pentenenitrile, 3-hydroxy-
NCCH2 CH(OH) CH:CH2
97
C_5H_7NO
53778-54-4
2-Cyclobutene-1-carboxamide
CONH2
97
C_5H_7NO
53778-55-5
2-Pentenenitrile, 5-hydroxy-, (Z)-
HOCH2 CH2 CH:CHCN
97
C_5H_7NO
53778-56-6
2-Pentenenitrile, 5-hydroxy-, (E)-
HOCH2 CH2 CH:CHCN
97
$C_6H_{11}N$
124-02-7
2-Propen-1-amine, N-2-propenyl-
H2C:CHCH2 NHCH2 CH:CH2
97
$C_6H_{11}N$
286-18-0
7-Azabicyclo[4.1.0]heptane
HN
97
$C_6H_{11}N$
542-54-1
Pentanenitrile, 4-methyl-
NCCH2 CH2 CHMe2
97
$C_6H_{11}N$
628-73-9
Hexanenitrile
NC(CH2)4 Me
97
$C_6H_{11}N$
10599-83-4
Methylamine, N-cyclopentylidene-
NMe
97
$C_6H_{11}N$
55268-35-4
Aziridine, 2-methylene-1-(1-methylethyl)-
H2C
N
Pr-i

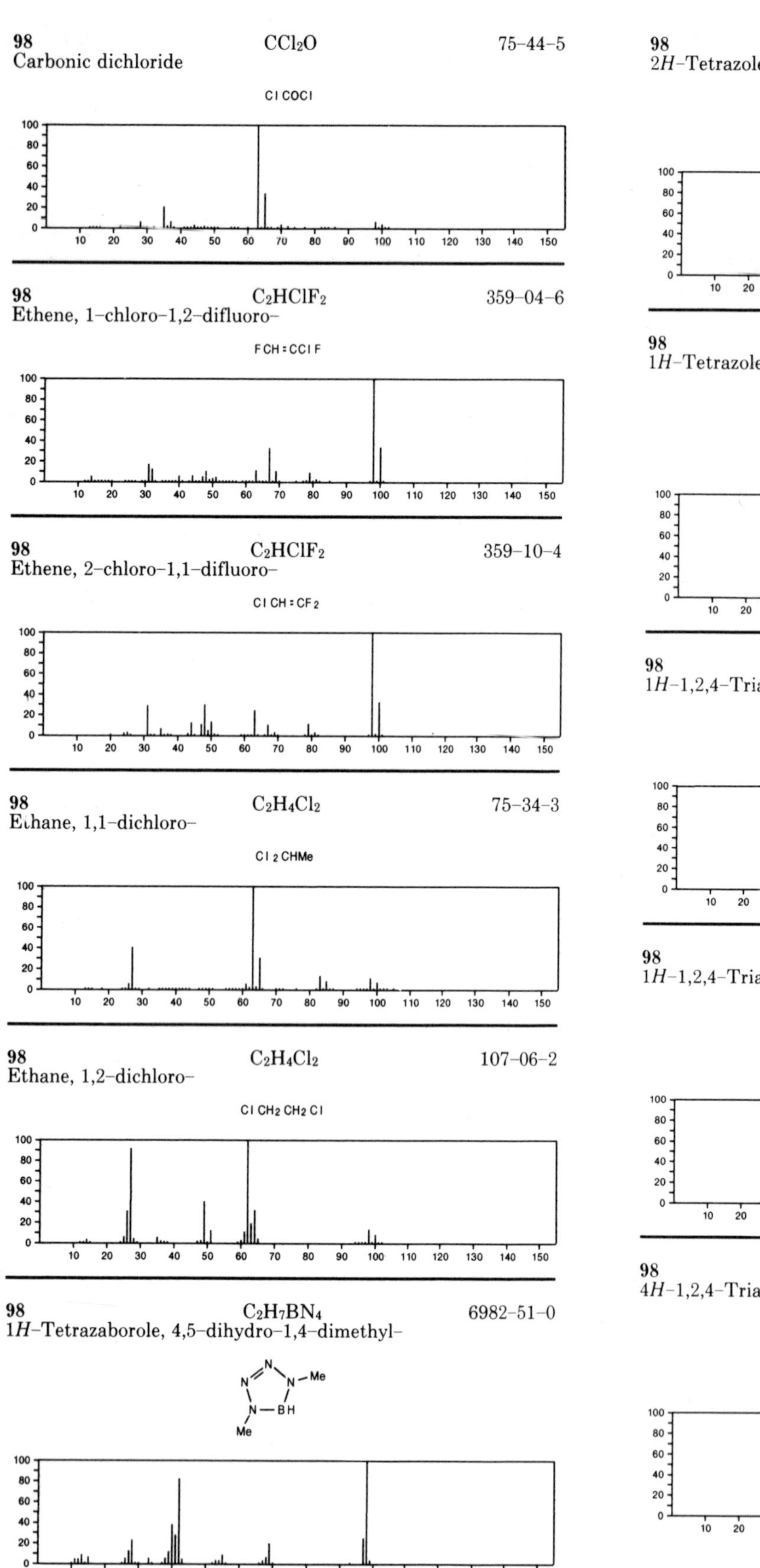

98 CCl_2O 75-44-5
Carbonic dichloride
Cl COCl

98 C_2HClF_2 359-04-6
Ethene, 1-chloro-1,2-difluoro-
FCH=CClF

98 C_2HClF_2 359-10-4
Ethene, 2-chloro-1,1-difluoro-
Cl CH=CF2

98 $C_2H_4Cl_2$ 75-34-3
Ethane, 1,1-dichloro-
Cl2CHMe

98 $C_2H_4Cl_2$ 107-06-2
Ethane, 1,2-dichloro-
Cl CH2 CH2 Cl

98 $C_2H_7BN_4$ 6982-51-0
1*H*-Tetrazaborole, 4,5-dihydro-1,4-dimethyl-

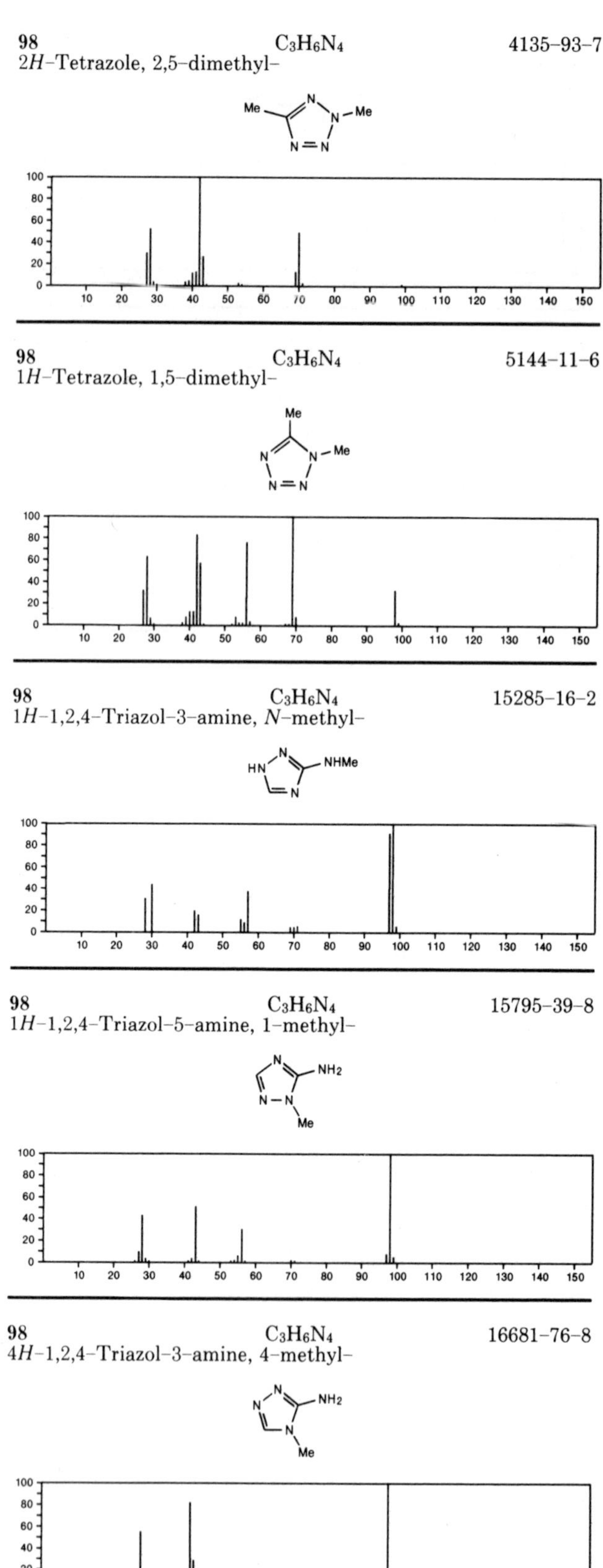

98 $C_3H_6N_4$ 4135-93-7
2*H*-Tetrazole, 2,5-dimethyl-

98 $C_3H_6N_4$ 5144-11-6
1*H*-Tetrazole, 1,5-dimethyl-

98 $C_3H_6N_4$ 15285-16-2
1*H*-1,2,4-Triazol-3-amine, *N*-methyl-

98 $C_3H_6N_4$ 15795-39-8
1*H*-1,2,4-Triazol-5-amine, 1-methyl-

98 $C_3H_6N_4$ 16681-76-8
4*H*-1,2,4-Triazol-3-amine, 4-methyl-

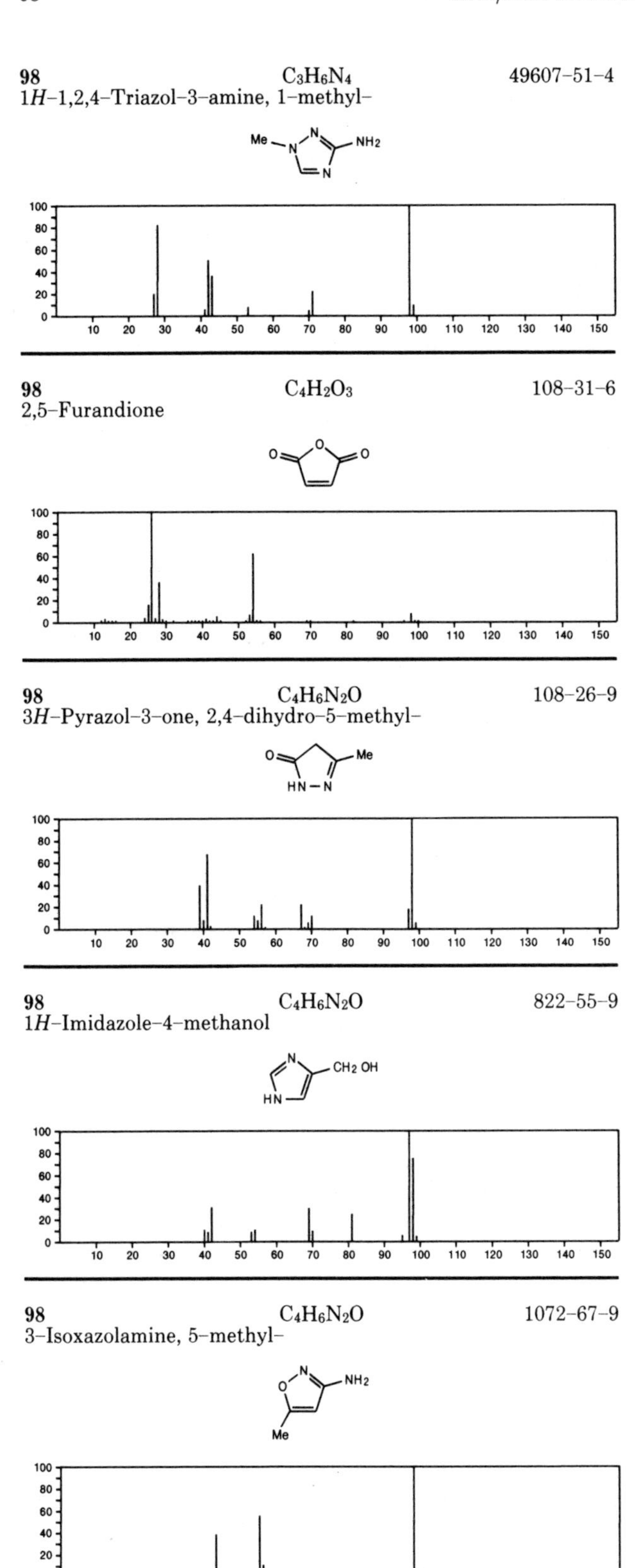

98 $C_3H_6N_4$ 49607–51–4
1*H*–1,2,4–Triazol–3–amine, 1–methyl–

98 $C_4H_2O_3$ 108–31–6
2,5–Furandione

98 $C_4H_6N_2O$ 108–26–9
3*H*–Pyrazol–3–one, 2,4–dihydro–5–methyl–

98 $C_4H_6N_2O$ 822–55–9
1*H*–Imidazole–4–methanol

98 $C_4H_6N_2O$ 1072–67–9
3–Isoxazolamine, 5–methyl–

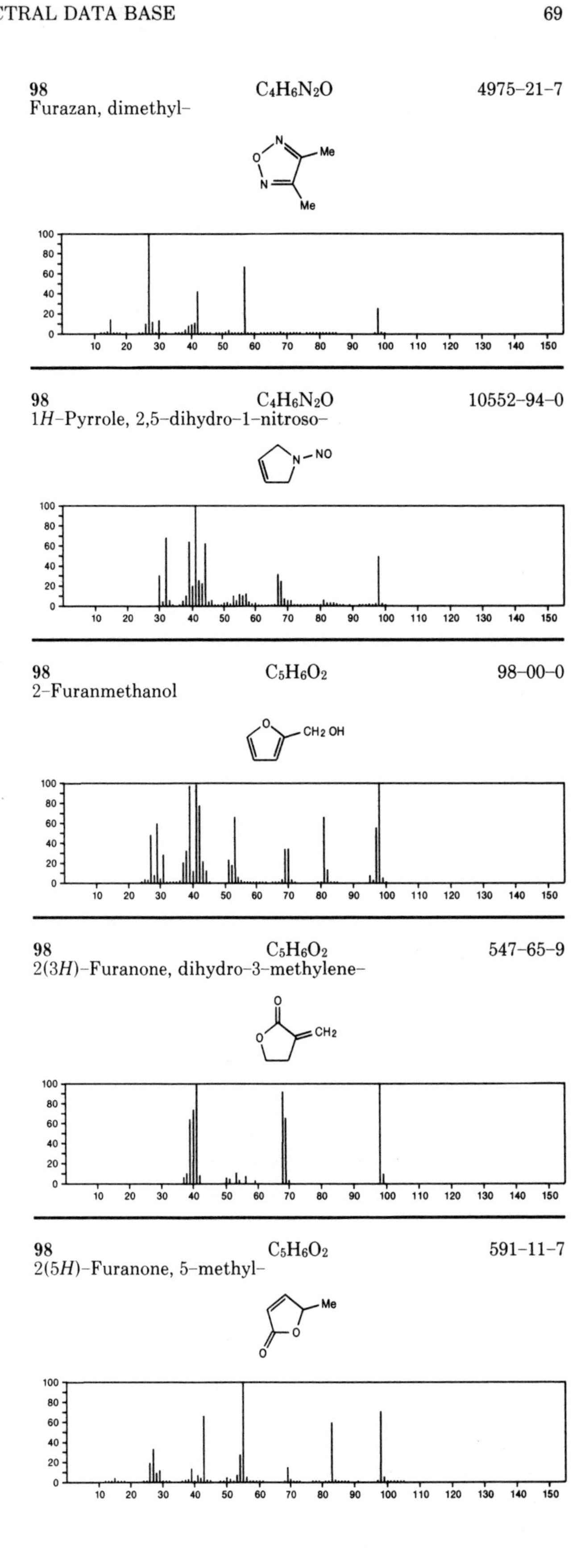

98 $C_4H_6N_2O$ 4975–21–7
Furazan, dimethyl–

98 $C_4H_6N_2O$ 10552–94–0
1*H*–Pyrrole, 2,5–dihydro–1–nitroso–

98 $C_5H_6O_2$ 98–00–0
2–Furanmethanol

98 $C_5H_6O_2$ 547–65–9
2(3*H*)–Furanone, dihydro–3–methylene–

98 $C_5H_6O_2$ 591–11–7
2(5*H*)–Furanone, 5–methyl–

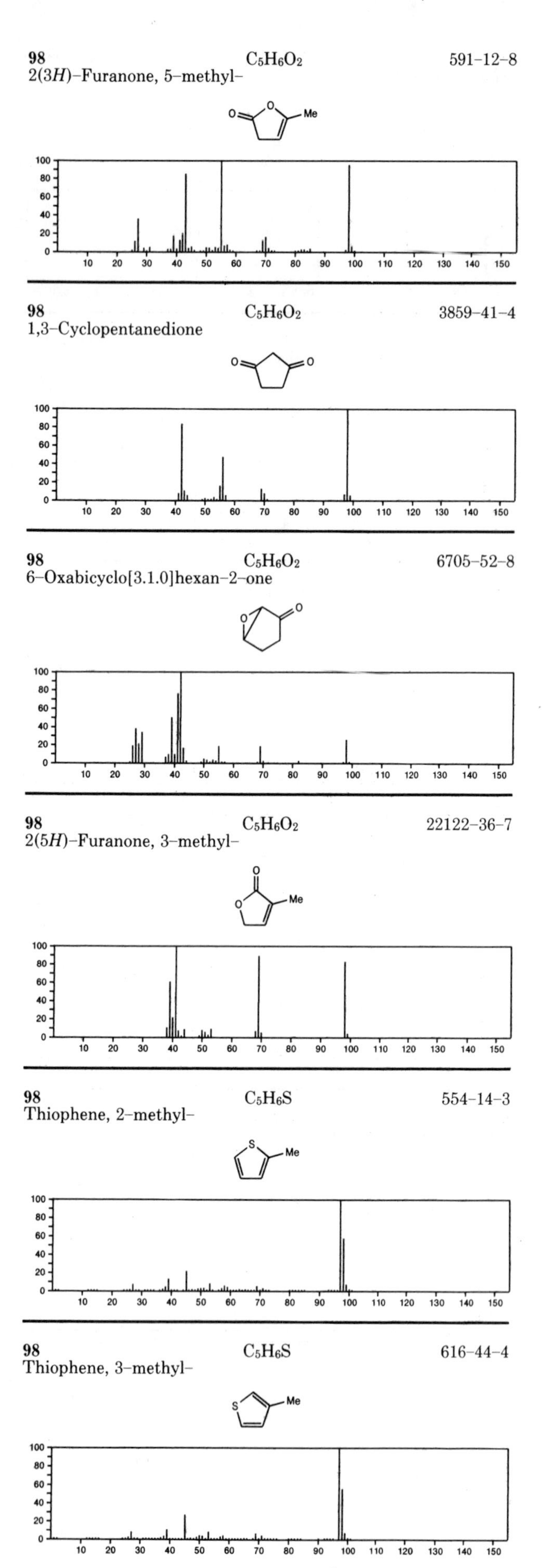
98 $C_5H_6O_2$ 591-12-8
2(3*H*)-Furanone, 5-methyl-
98 $C_5H_6O_2$ 3859-41-4
1,3-Cyclopentanedione
98 $C_5H_6O_2$ 6705-52-8
6-Oxabicyclo[3.1.0]hexan-2-one
98 $C_5H_6O_2$ 22122-36-7
2(5*H*)-Furanone, 3-methyl-
98 C_5H_6S 554-14-3
Thiophene, 2-methyl-
98 C_5H_6S 616-44-4
Thiophene, 3-methyl-

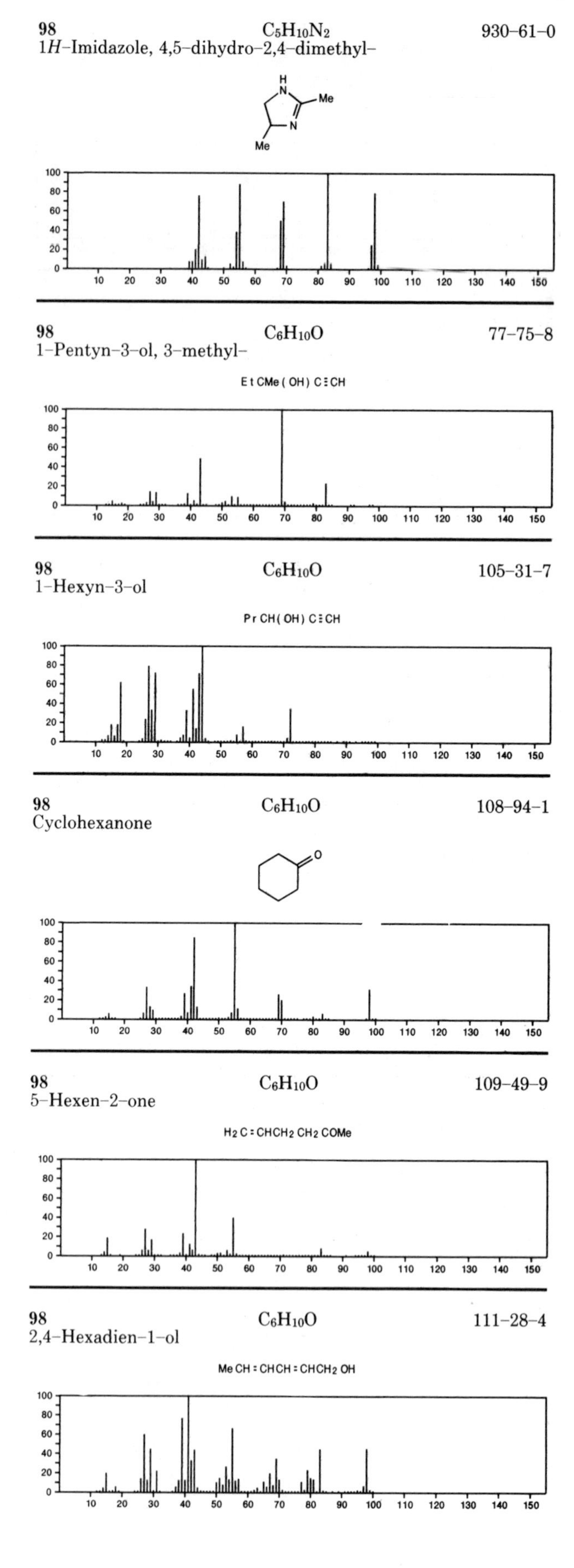
98 $C_5H_{10}N_2$ 930-61-0
1*H*-Imidazole, 4,5-dihydro-2,4-dimethyl-
98 $C_6H_{10}O$ 77-75-8
1-Pentyn-3-ol, 3-methyl-
EtCMe(OH)C≡CH
98 $C_6H_{10}O$ 105-31-7
1-Hexyn-3-ol
PrCH(OH)C≡CH
98 $C_6H_{10}O$ 108-94-1
Cyclohexanone
98 $C_6H_{10}O$ 109-49-9
5-Hexen-2-one
$H_2C=CHCH_2CH_2COMe$
98 $C_6H_{10}O$ 111-28-4
2,4-Hexadien-1-ol
$MeCH=CHCH=CHCH_2OH$

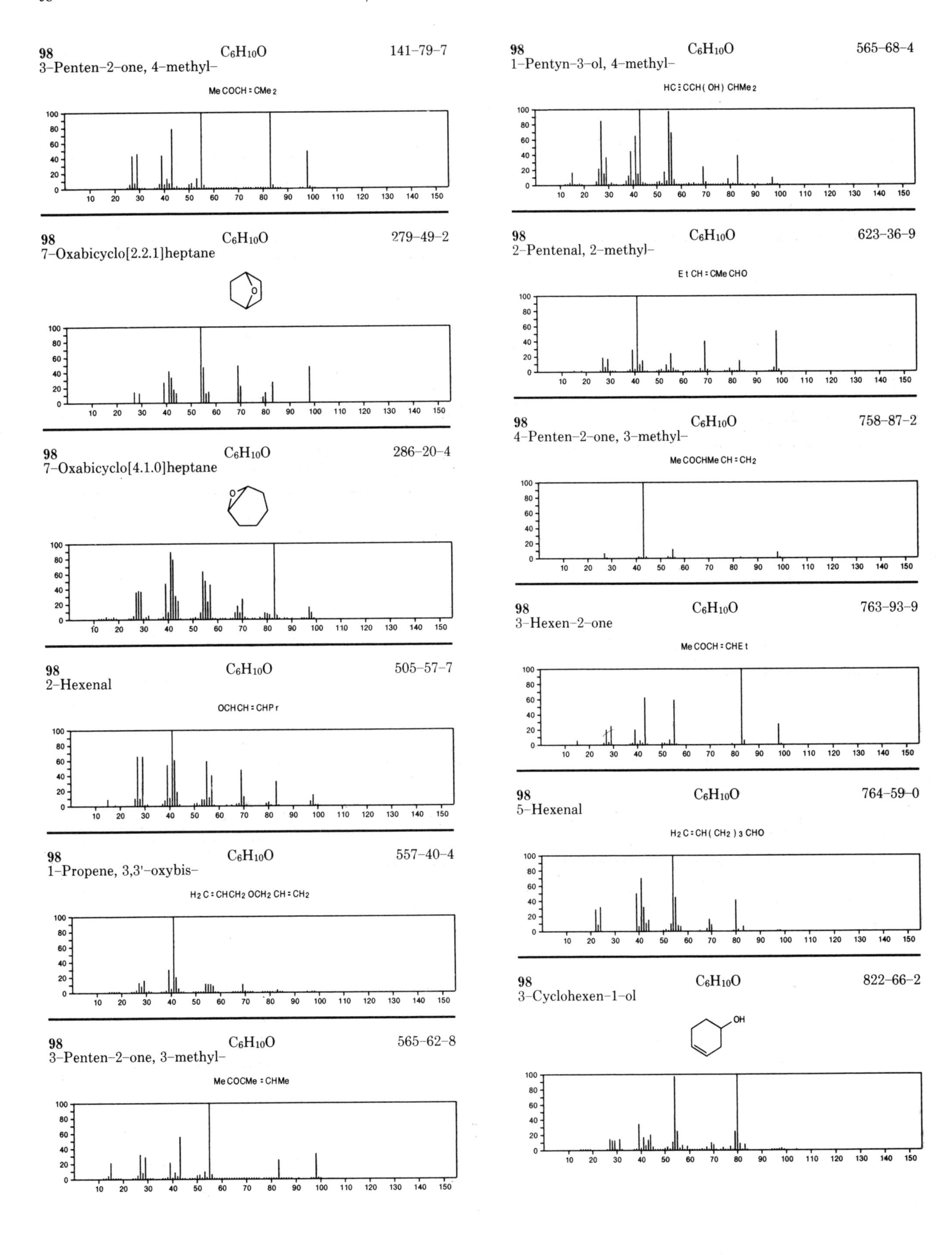
98 C6H10O 141-79-7
3-Penten-2-one, 4-methyl-
Me COCH = CMe 2
98 C6H10O 279-49-2
7-Oxabicyclo[2.2.1]heptane
98 C6H10O 286-20-4
7-Oxabicyclo[4.1.0]heptane
98 C6H10O 505-57-7
2-Hexenal
OCHCH = CHPr
98 C6H10O 557-40-4
1-Propene, 3,3'-oxybis-
H2 C = CHCH2 OCH2 CH = CH2
98 C6H10O 565-62-8
3-Penten-2-one, 3-methyl-
Me COCMe = CHMe
98 C6H10O 565-68-4
1-Pentyn-3-ol, 4-methyl-
HC ≡ CCH(OH) CHMe 2
98 C6H10O 623-36-9
2-Pentenal, 2-methyl-
Et CH = CMe CHO
98 C6H10O 758-87-2
4-Penten-2-one, 3-methyl-
Me COCHMe CH = CH2
98 C6H10O 763-93-9
3-Hexen-2-one
Me COCH = CHEt
98 C6H10O 764-59-0
5-Hexenal
H2 C = CH(CH2) 3 CHO
98 C6H10O 822-66-2
3-Cyclohexen-1-ol
OH

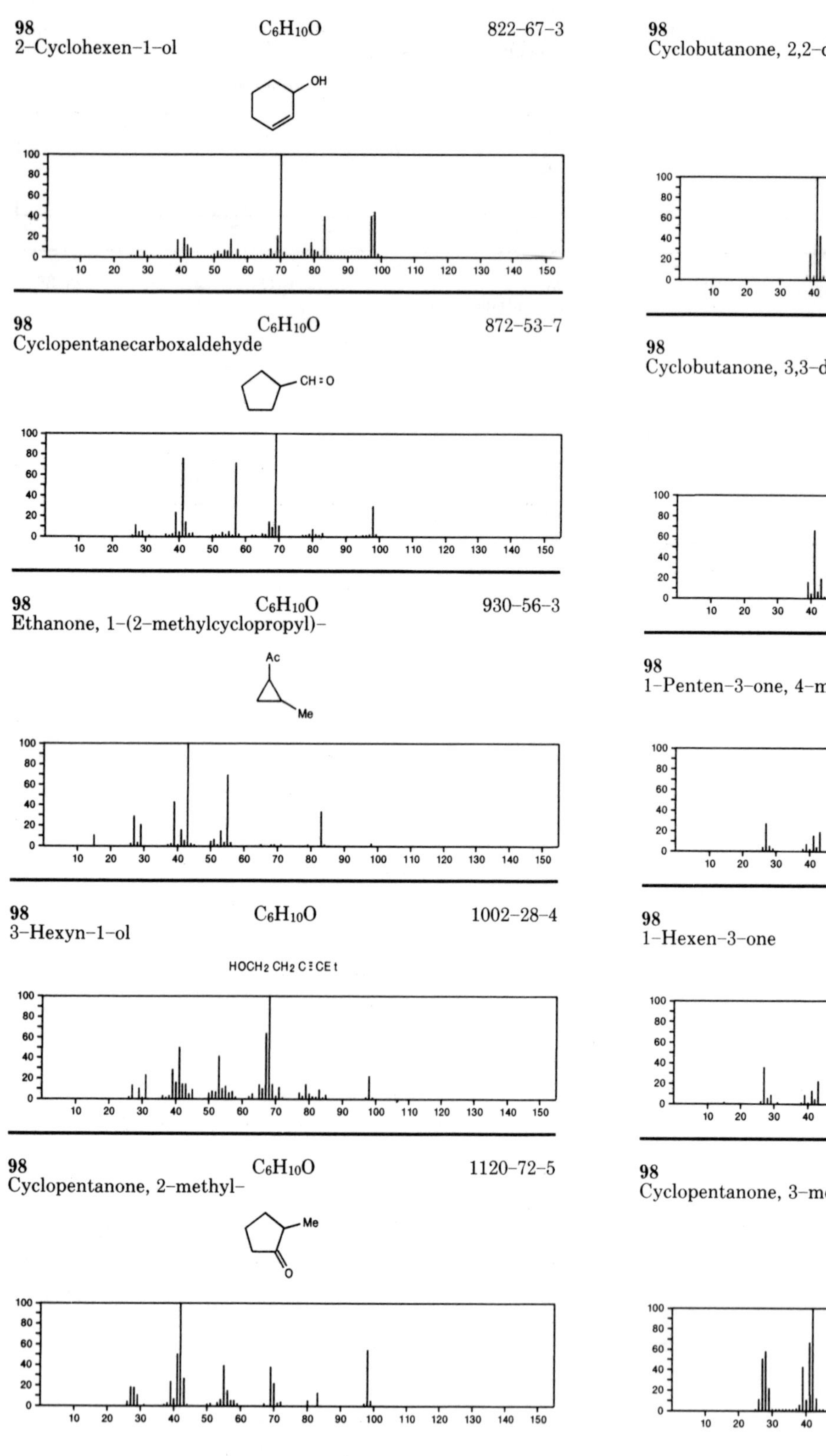

98 2-Cyclohexen-1-ol — $C_6H_{10}O$ — 822-67-3

98 Cyclopentanecarboxaldehyde — $C_6H_{10}O$ — 872-53-7

98 Ethanone, 1-(2-methylcyclopropyl)- — $C_6H_{10}O$ — 930-56-3

98 3-Hexyn-1-ol — $C_6H_{10}O$ — 1002-28-4

HOCH2 CH2 C≡CEt

98 Cyclopentanone, 2-methyl- — $C_6H_{10}O$ — 1120-72-5

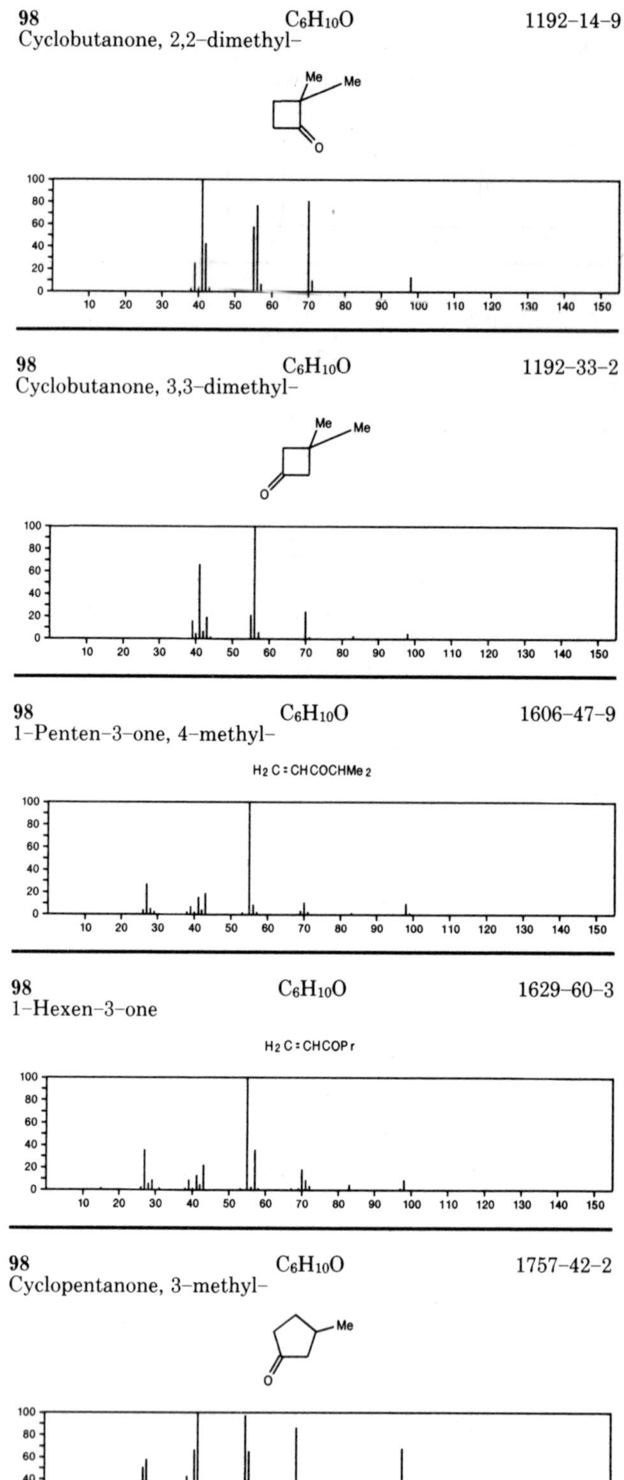

98 Cyclobutanone, 2,2-dimethyl- — $C_6H_{10}O$ — 1192-14-9

98 Cyclobutanone, 3,3-dimethyl- — $C_6H_{10}O$ — 1192-33-2

98 1-Penten-3-one, 4-methyl- — $C_6H_{10}O$ — 1606-47-9

H2C=CHCOCHMe2

98 1-Hexen-3-one — $C_6H_{10}O$ — 1629-60-3

H2C=CHCOPr

98 Cyclopentanone, 3-methyl- — $C_6H_{10}O$ — 1757-42-2

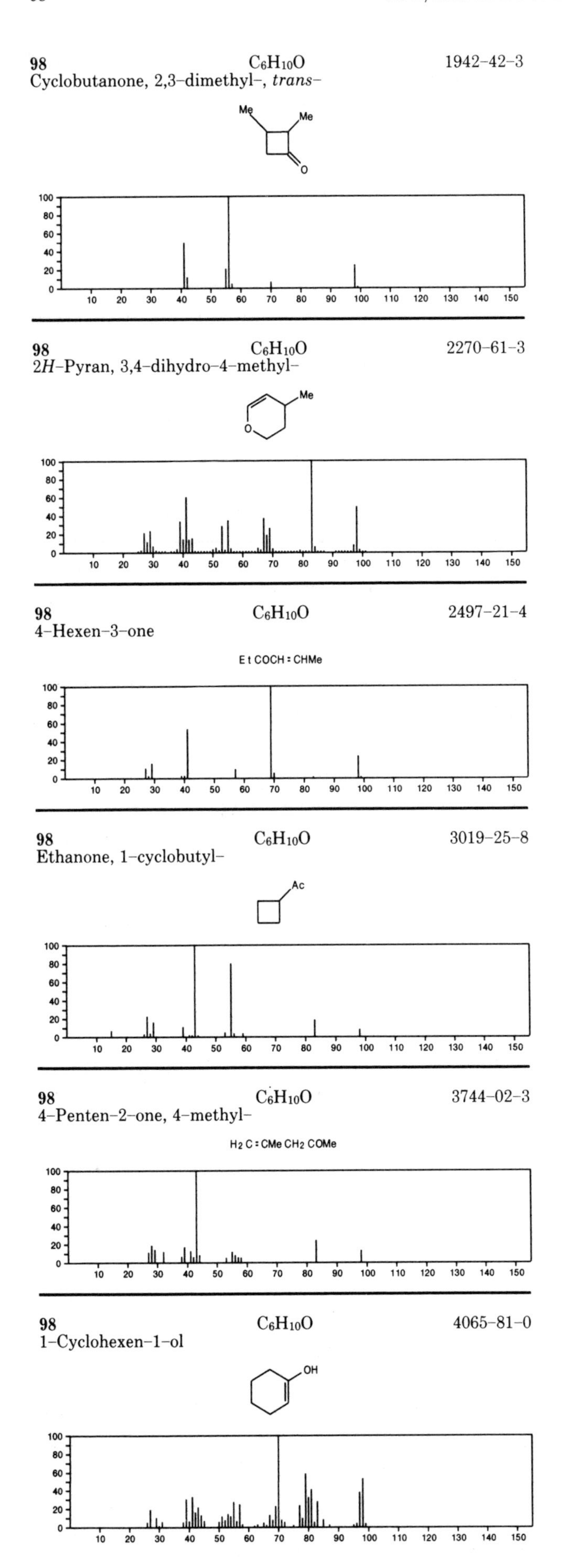

98 $C_6H_{10}O$ 1942-42-3
Cyclobutanone, 2,3-dimethyl-, *trans-*

98 $C_6H_{10}O$ 2270-61-3
2H-Pyran, 3,4-dihydro-4-methyl-

98 $C_6H_{10}O$ 2497-21-4
4-Hexen-3-one
EtCOCH=CHMe

98 $C_6H_{10}O$ 3019-25-8
Ethanone, 1-cyclobutyl-

98 $C_6H_{10}O$ 3744-02-3
4-Penten-2-one, 4-methyl-
H_2C=CMeCH_2COMe

98 $C_6H_{10}O$ 4065-81-0
1-Cyclohexen-1-ol

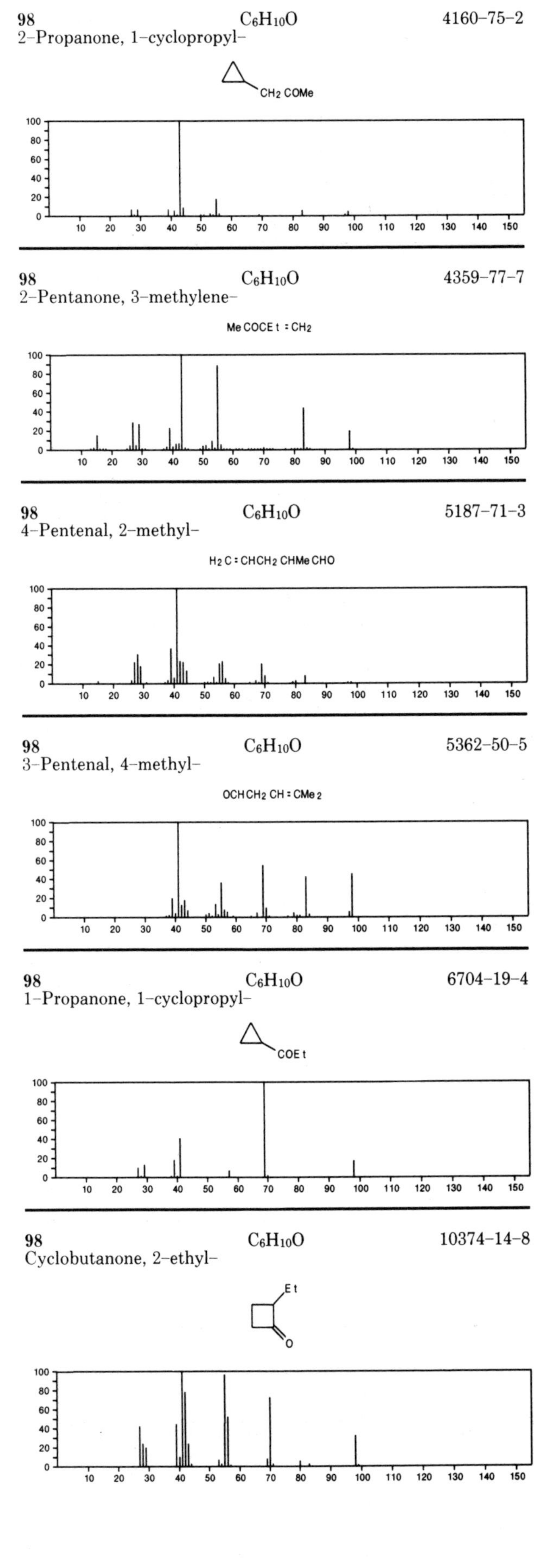

98 $C_6H_{10}O$ 4160-75-2
2-Propanone, 1-cyclopropyl-

98 $C_6H_{10}O$ 4359-77-7
2-Pentanone, 3-methylene-
MeCOCEt=CH_2

98 $C_6H_{10}O$ 5187-71-3
4-Pentenal, 2-methyl-
H_2C=CHCH_2CHMeCHO

98 $C_6H_{10}O$ 5362-50-5
3-Pentenal, 4-methyl-
OCHCH_2CH=CMe_2

98 $C_6H_{10}O$ 6704-19-4
1-Propanone, 1-cyclopropyl-

98 $C_6H_{10}O$ 10374-14-8
Cyclobutanone, 2-ethyl-

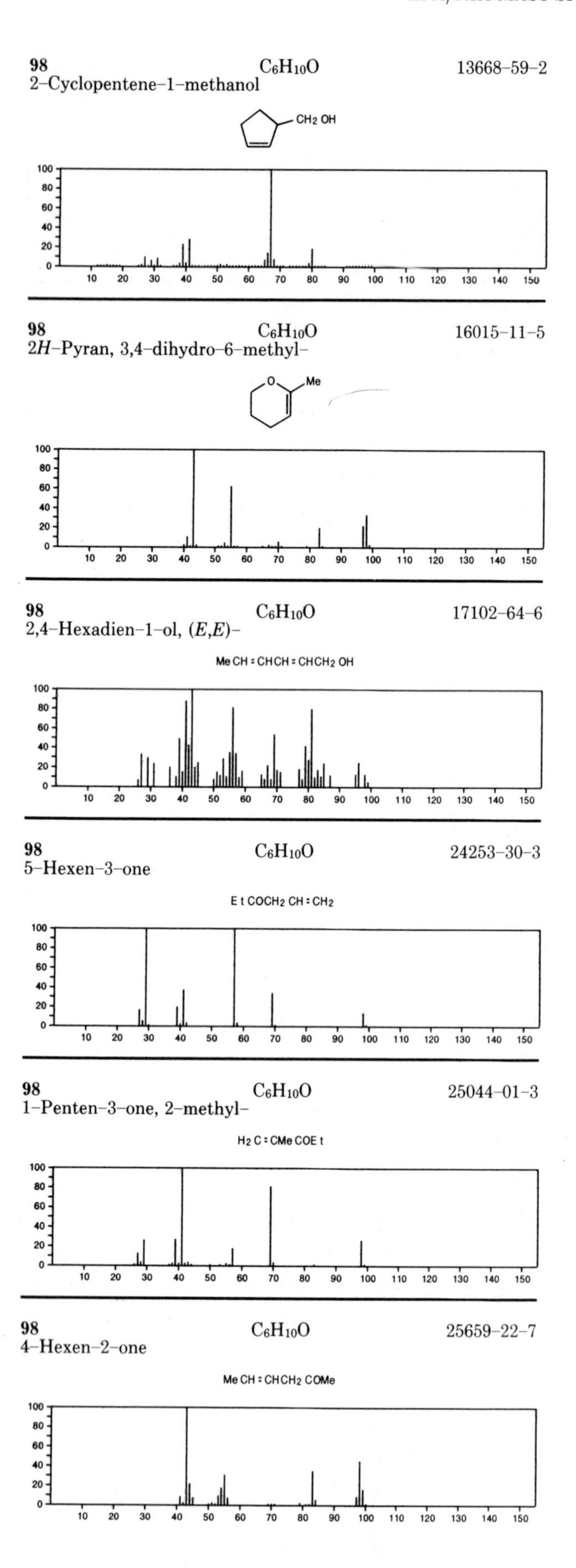

98 $C_6H_{10}O$ 13668–59–2
2–Cyclopentene–1–methanol

98 $C_6H_{10}O$ 16015–11–5
2*H*–Pyran, 3,4–dihydro–6–methyl–

98 $C_6H_{10}O$ 17102–64–6
2,4–Hexadien–1–ol, (*E*,*E*)–

Me CH : CHCH : CHCH2 OH

98 $C_6H_{10}O$ 24253–30–3
5–Hexen–3–one

E t COCH2 CH : CH2

98 $C_6H_{10}O$ 25044–01–3
1–Penten–3–one, 2–methyl–

H2 C : CMe COE t

98 $C_6H_{10}O$ 25659–22–7
4–Hexen–2–one

Me CH : CHCH2 COMe

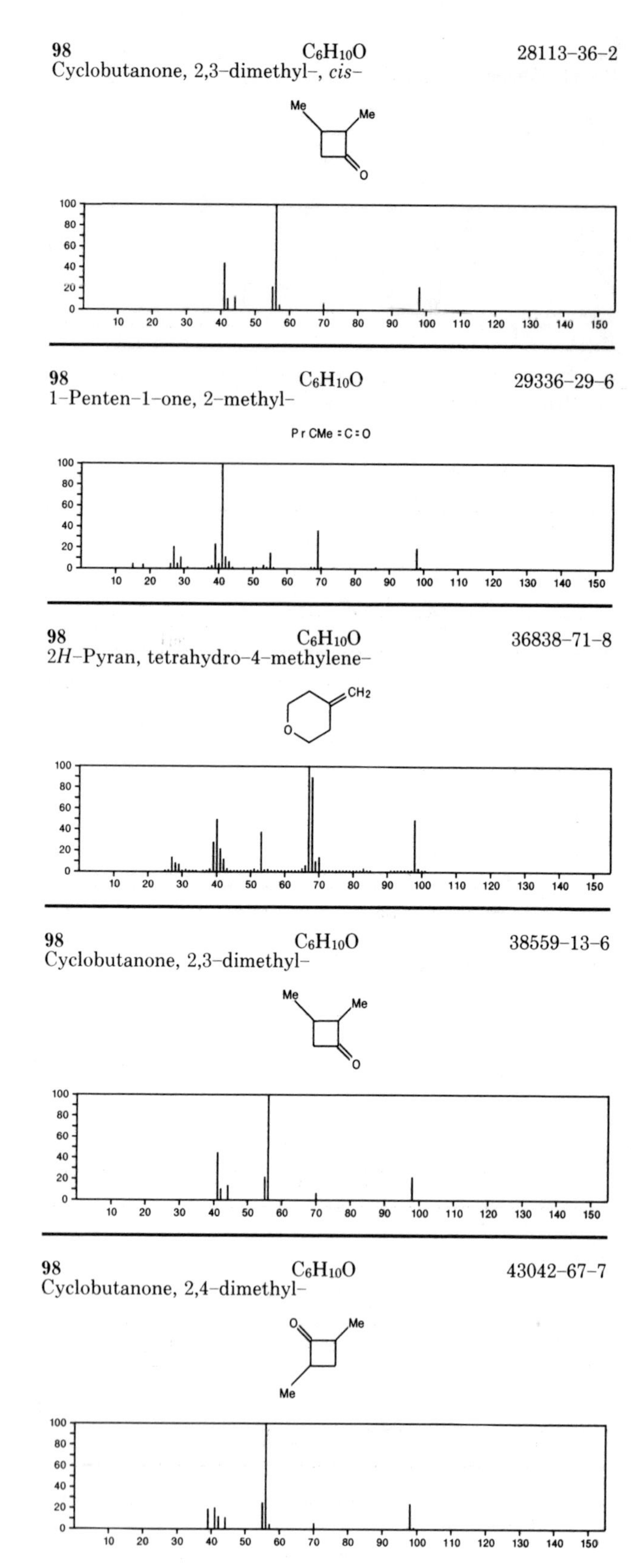

98 $C_6H_{10}O$ 28113–36–2
Cyclobutanone, 2,3–dimethyl–, *cis*–

98 $C_6H_{10}O$ 29336–29–6
1–Penten–1–one, 2–methyl–

P r CMe : C : O

98 $C_6H_{10}O$ 36838–71–8
2*H*–Pyran, tetrahydro–4–methylene–

98 $C_6H_{10}O$ 38559–13–6
Cyclobutanone, 2,3–dimethyl–

98 $C_6H_{10}O$ 43042–67–7
Cyclobutanone, 2,4–dimethyl–

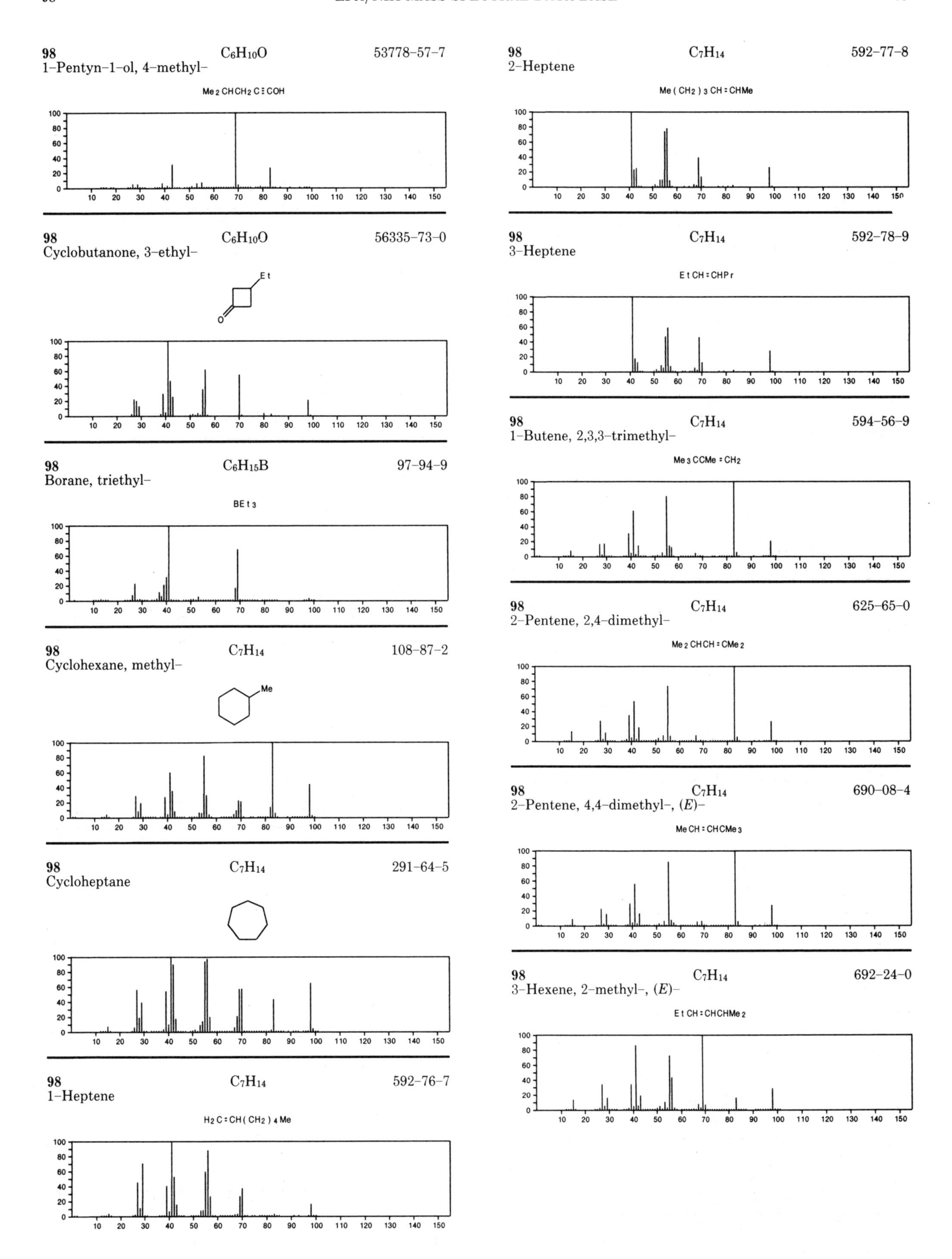
98 C6H10O 53778-57-7
1-Pentyn-1-ol, 4-methyl-
Me2CHCH2C≡COH
98 C6H10O 56335-73-0
Cyclobutanone, 3-ethyl-
Et
O
98 C6H15B 97-94-9
Borane, triethyl-
BEt3
98 C7H14 108-87-2
Cyclohexane, methyl-
Me
98 C7H14 291-64-5
Cycloheptane
98 C7H14 592-76-7
1-Heptene
H2C=CH(CH2)4Me
98 C7H14 592-77-8
2-Heptene
Me(CH2)3CH=CHMe
98 C7H14 592-78-9
3-Heptene
EtCH=CHPr
98 C7H14 594-56-9
1-Butene, 2,3,3-trimethyl-
Me3CCMe=CH2
98 C7H14 625-65-0
2-Pentene, 2,4-dimethyl-
Me2CHCH=CMe2
98 C7H14 690-08-4
2-Pentene, 4,4-dimethyl-, (E)-
MeCH=CHCMe3
98 C7H14 692-24-0
3-Hexene, 2-methyl-, (E)-
EtCH=CHCHMe2

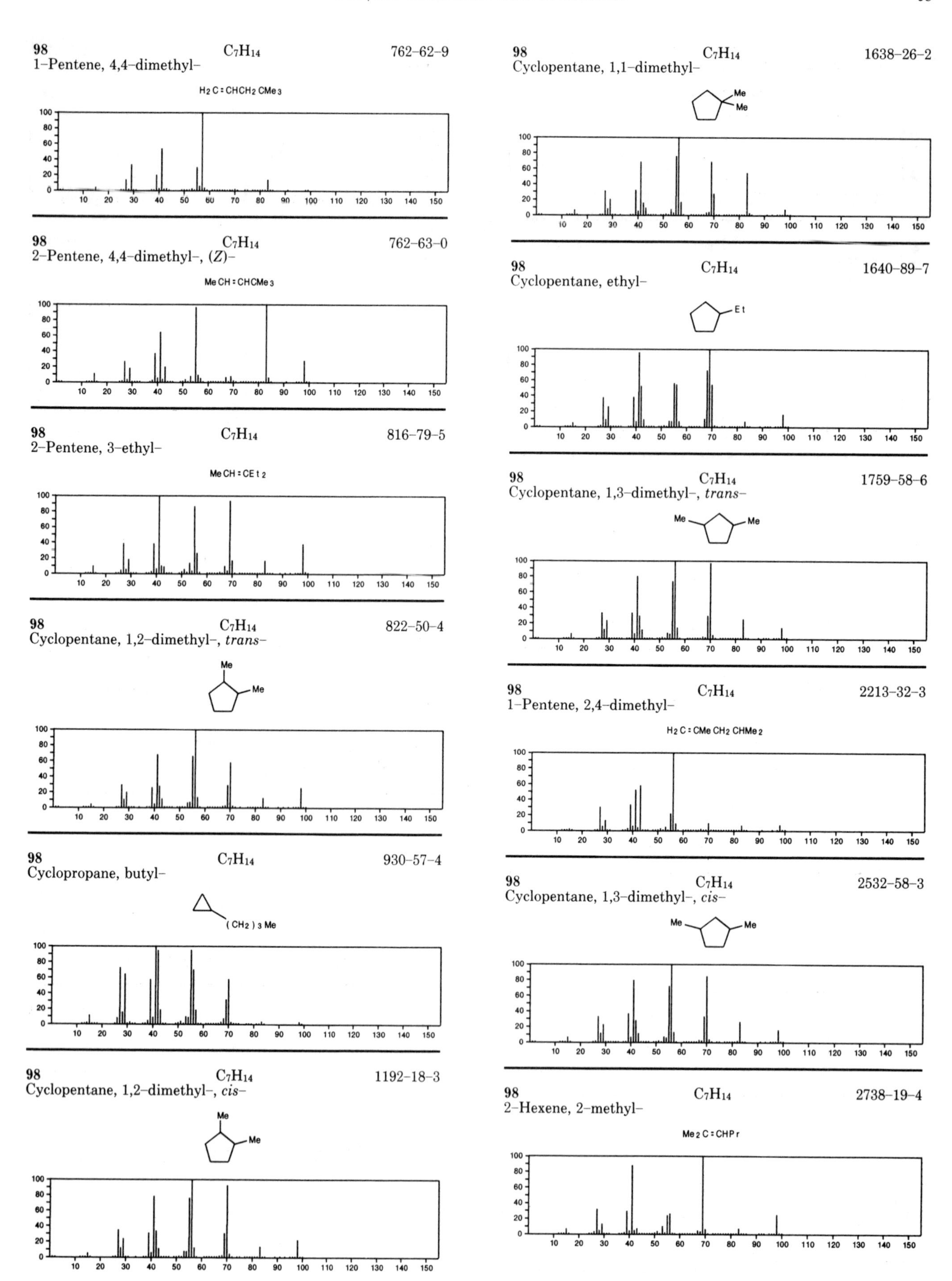
98 C7H14 762-62-9
1-Pentene, 4,4-dimethyl-
H2C=CHCH2CMe3
98 C7H14 762-63-0
2-Pentene, 4,4-dimethyl-, (Z)-
MeCH=CHCMe3
98 C7H14 816-79-5
2-Pentene, 3-ethyl-
MeCH=CEt2
98 C7H14 822-50-4
Cyclopentane, 1,2-dimethyl-, trans-
98 C7H14 930-57-4
Cyclopropane, butyl-
(CH2)3Me
98 C7H14 1192-18-3
Cyclopentane, 1,2-dimethyl-, cis-
98 C7H14 1638-26-2
Cyclopentane, 1,1-dimethyl-
98 C7H14 1640-89-7
Cyclopentane, ethyl-
Et
98 C7H14 1759-58-6
Cyclopentane, 1,3-dimethyl-, trans-
98 C7H14 2213-32-3
1-Pentene, 2,4-dimethyl-
H2C=CMeCH2CHMe2
98 C7H14 2532-58-3
Cyclopentane, 1,3-dimethyl-, cis-
98 C7H14 2738-19-4
2-Hexene, 2-methyl-
Me2C=CHPr

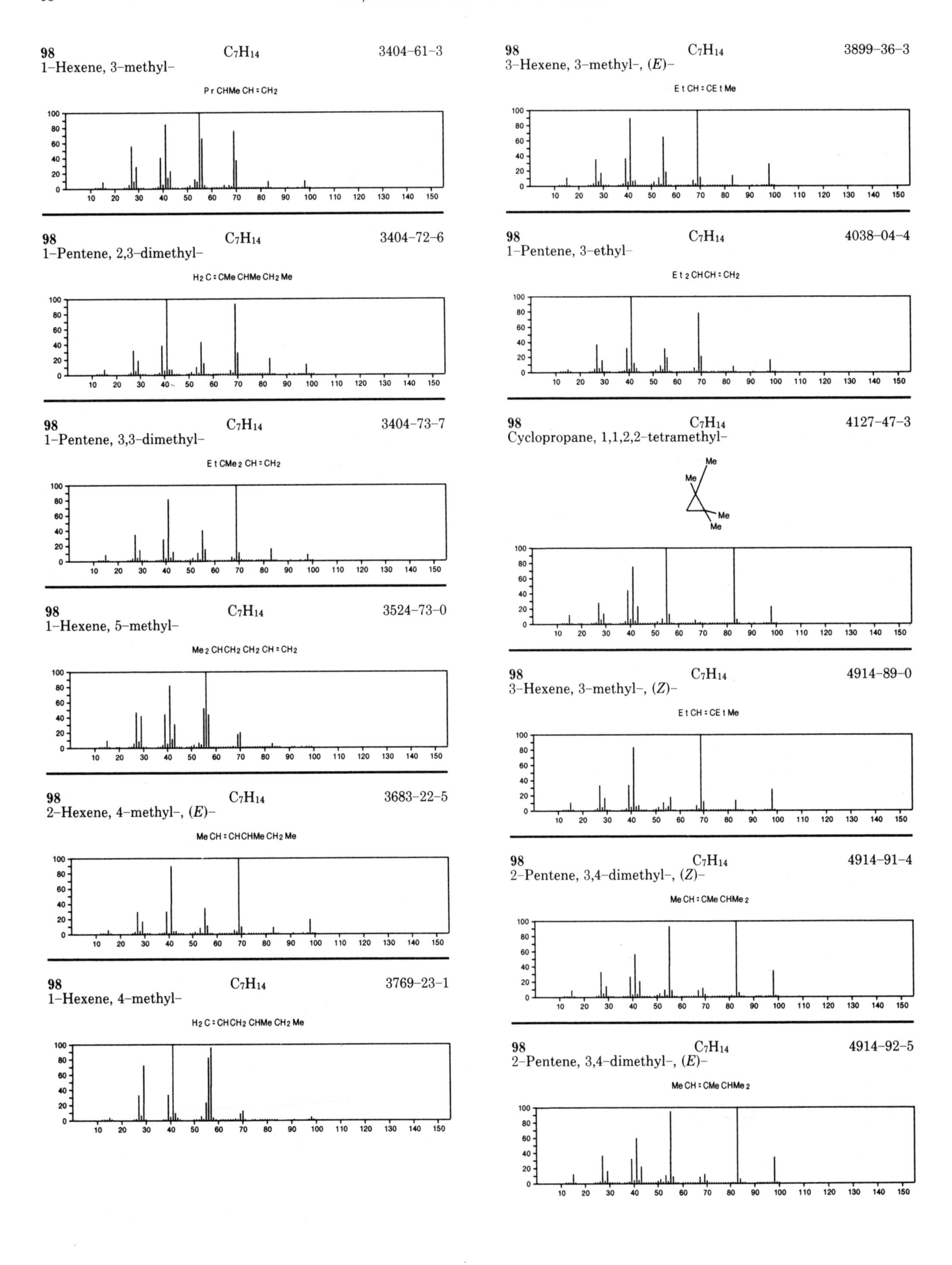
98 C7H14 3404-61-3
1-Hexene, 3-methyl-
PrCHMeCH=CH2
98 C7H14 3899-36-3
3-Hexene, 3-methyl-, (E)-
EtCH=CEtMe
98 C7H14 3404-72-6
1-Pentene, 2,3-dimethyl-
H2C=CMeCHMeCH2Me
98 C7H14 4038-04-4
1-Pentene, 3-ethyl-
Et2CHCH=CH2
98 C7H14 3404-73-7
1-Pentene, 3,3-dimethyl-
EtCMe2CH=CH2
98 C7H14 4127-47-3
Cyclopropane, 1,1,2,2-tetramethyl-
Me
Me
Me
Me
98 C7H14 3524-73-0
1-Hexene, 5-methyl-
Me2CHCH2CH2CH=CH2
98 C7H14 4914-89-0
3-Hexene, 3-methyl-, (Z)-
EtCH=CEtMe
98 C7H14 3683-22-5
2-Hexene, 4-methyl-, (E)-
MeCH=CHCHMeCH2Me
98 C7H14 4914-91-4
2-Pentene, 3,4-dimethyl-, (Z)-
MeCH=CMeCHMe2
98 C7H14 3769-23-1
1-Hexene, 4-methyl-
H2C=CHCH2CHMeCH2Me
98 C7H14 4914-92-5
2-Pentene, 3,4-dimethyl-, (E)-
MeCH=CMeCHMe2

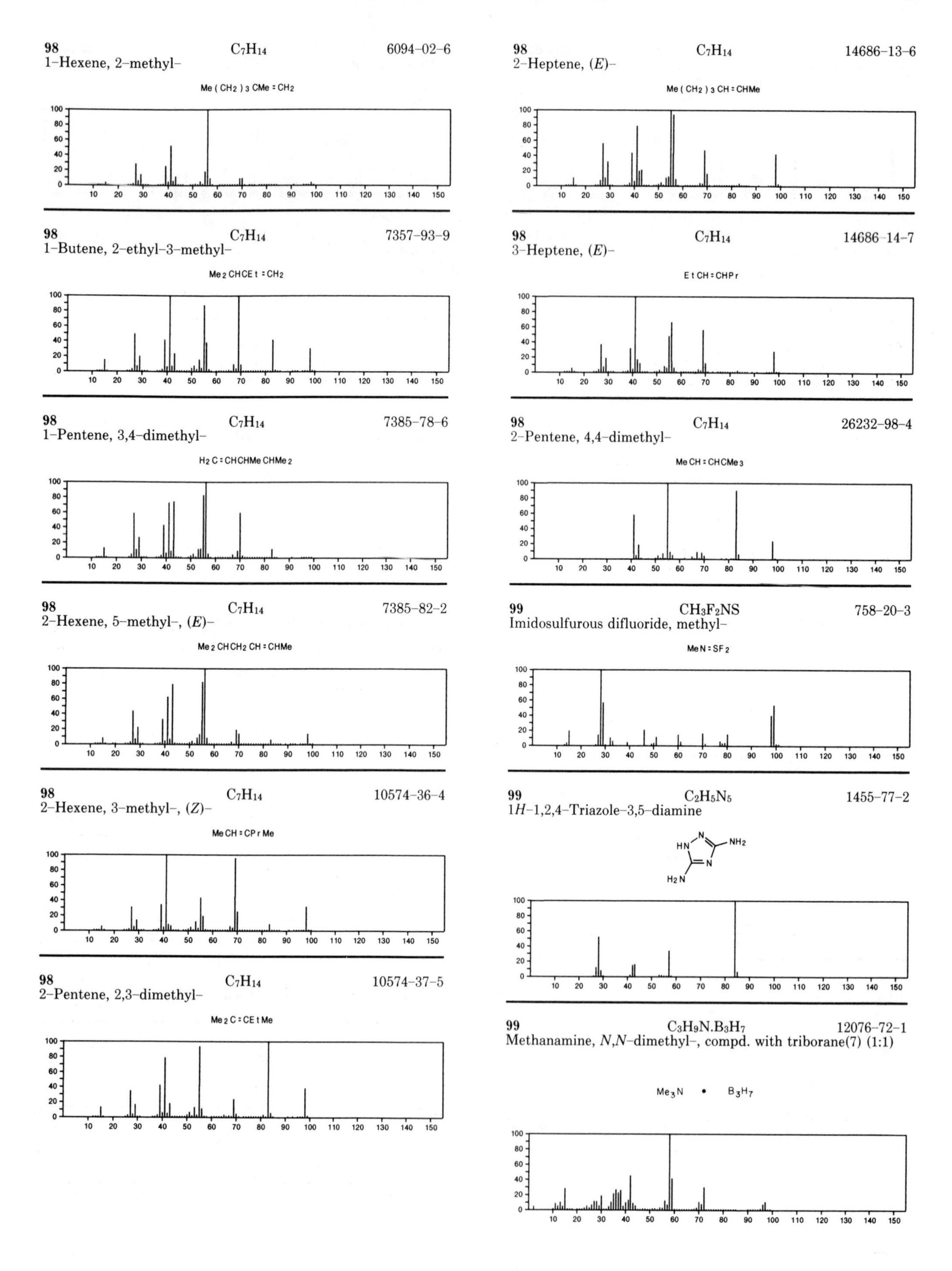

98 C_7H_{14} 6094-02-6
1-Hexene, 2-methyl-
Me (CH2) 3 CMe = CH2

98 C_7H_{14} 7357-93-9
1-Butene, 2-ethyl-3-methyl-
Me 2 CHCE t = CH2

98 C_7H_{14} 7385-78-6
1-Pentene, 3,4-dimethyl-
H2 C = CHCHMe CHMe 2

98 C_7H_{14} 7385-82-2
2-Hexene, 5-methyl-, (*E*)-
Me 2 CHCH2 CH = CHMe

98 C_7H_{14} 10574-36-4
2-Hexene, 3-methyl-, (*Z*)-
Me CH = CP r Me

98 C_7H_{14} 10574-37-5
2-Pentene, 2,3-dimethyl-
Me 2 C = CE t Me

98 C_7H_{14} 14686-13-6
2-Heptene, (*E*)-
Me (CH2) 3 CH = CHMe

98 C_7H_{14} 14686-14-7
3-Heptene, (*E*)-
E t CH = CHP r

98 C_7H_{14} 26232-98-4
2-Pentene, 4,4-dimethyl-
Me CH = CHCMe 3

99 CH_3F_2NS 758-20-3
Imidosulfurous difluoride, methyl-
Me N = SF 2

99 $C_2H_5N_5$ 1455-77-2
1*H*-1,2,4-Triazole-3,5-diamine

99 $C_3H_9N.B_3H_7$ 12076-72-1
Methanamine, *N*,*N*-dimethyl-, compd. with triborane(7) (1:1)
Me3N • B3H7

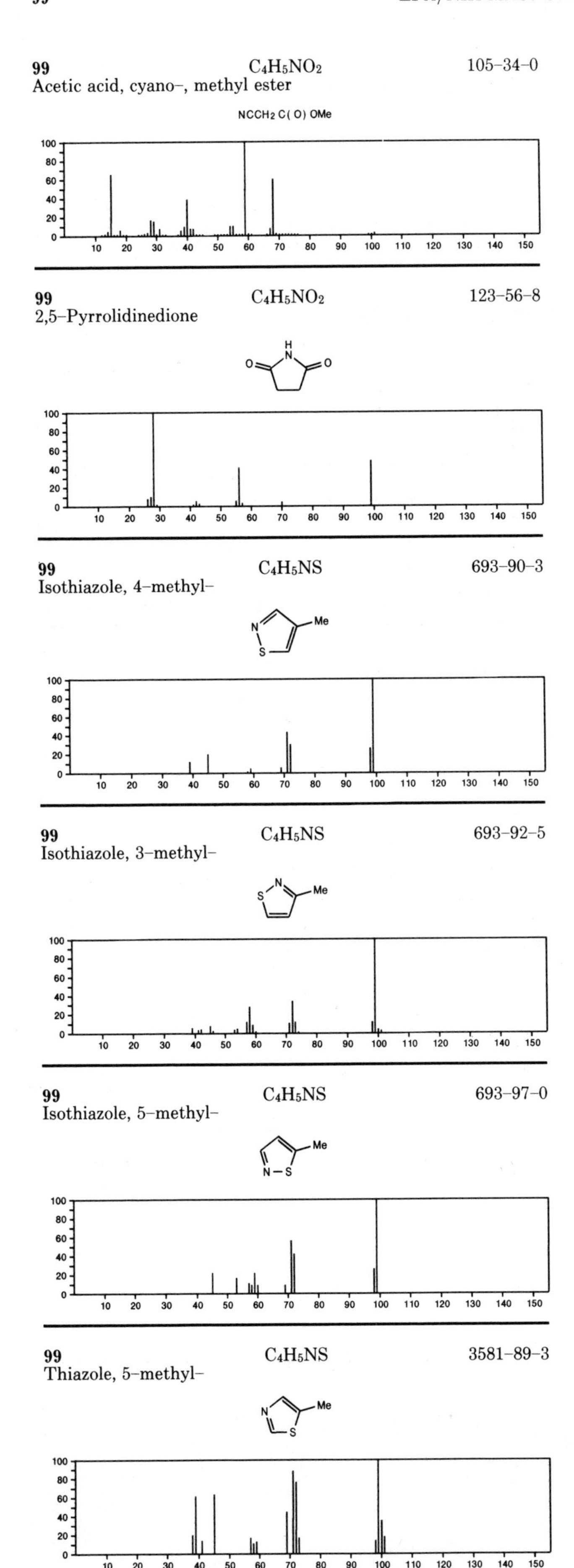

99 $C_4H_5NO_2$ 105-34-0
Acetic acid, cyano-, methyl ester

99 $C_4H_5NO_2$ 123-56-8
2,5-Pyrrolidinedione

99 C_4H_5NS 693-90-3
Isothiazole, 4-methyl-

99 C_4H_5NS 693-92-5
Isothiazole, 3-methyl-

99 C_4H_5NS 693-97-0
Isothiazole, 5-methyl-

99 C_4H_5NS 3581-89-3
Thiazole, 5-methyl-

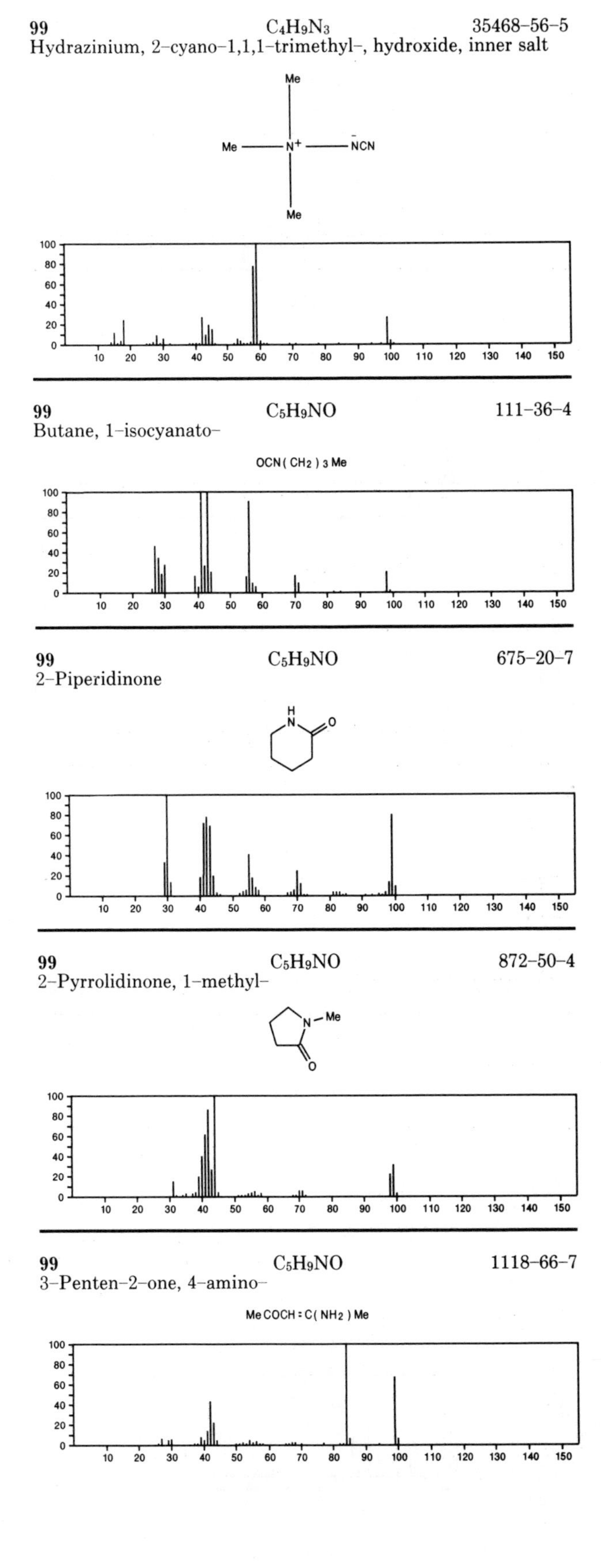

99 $C_4H_9N_3$ 35468-56-5
Hydrazinium, 2-cyano-1,1,1-trimethyl-, hydroxide, inner salt

99 C_5H_9NO 111-36-4
Butane, 1-isocyanato-

99 C_5H_9NO 675-20-7
2-Piperidinone

99 C_5H_9NO 872-50-4
2-Pyrrolidinone, 1-methyl-

99 C_5H_9NO 1118-66-7
3-Penten-2-one, 4-amino-

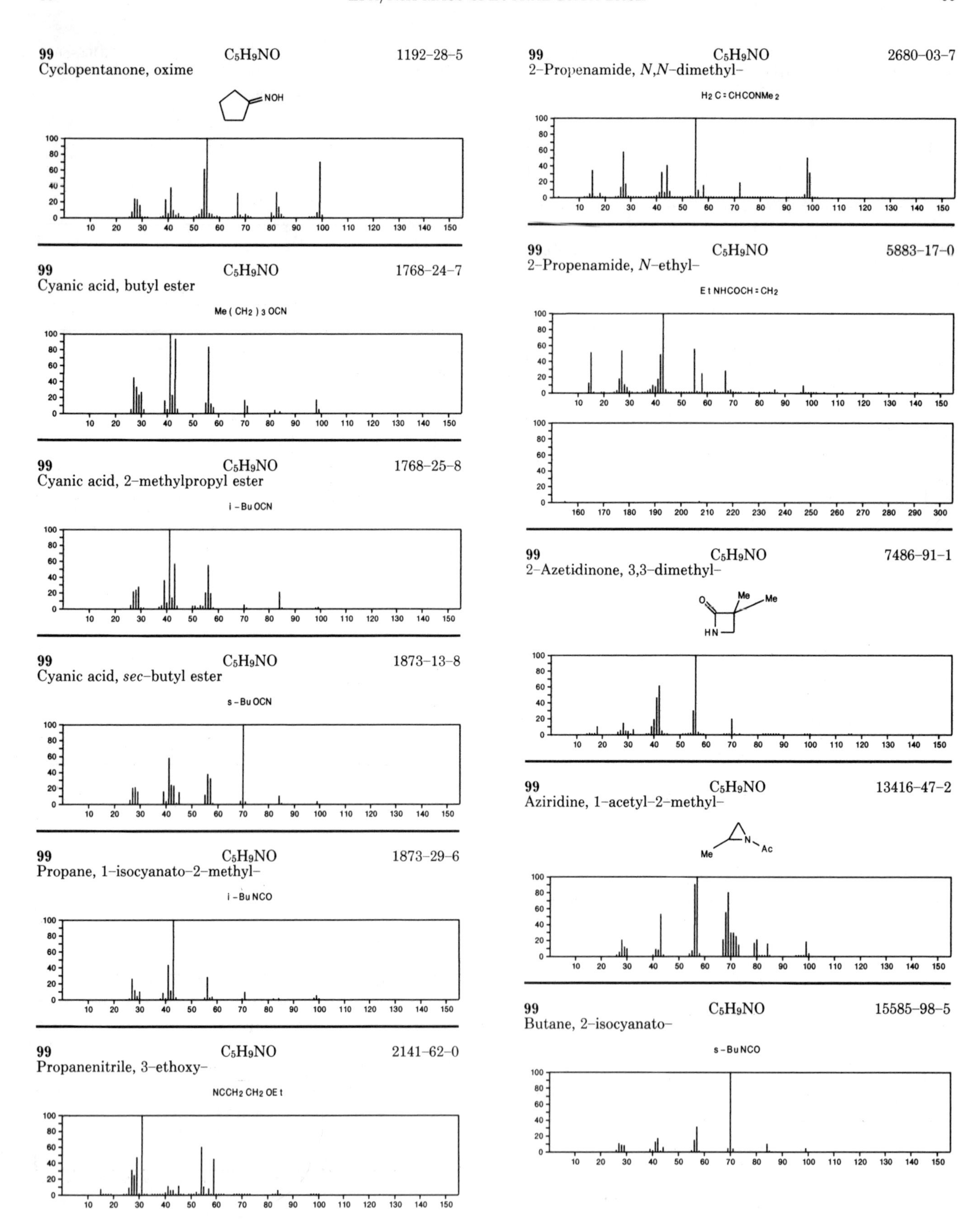
99 C5H9NO 1192-28-5
Cyclopentanone, oxime
NOH
99 C5H9NO 1768-24-7
Cyanic acid, butyl ester
Me (CH2) 3 OCN
99 C5H9NO 1768-25-8
Cyanic acid, 2-methylpropyl ester
i - Bu OCN
99 C5H9NO 1873-13-8
Cyanic acid, sec-butyl ester
s - Bu OCN
99 C5H9NO 1873-29-6
Propane, 1-isocyanato-2-methyl-
i - Bu NCO
99 C5H9NO 2141-62-0
Propanenitrile, 3-ethoxy-
NCCH2 CH2 OE t
99 C5H9NO 2680-03-7
2-Propenamide, N,N-dimethyl-
H2 C = CHCONMe 2
99 C5H9NO 5883-17-0
2-Propenamide, N-ethyl-
E t NHCOCH = CH2
99 C5H9NO 7486-91-1
2-Azetidinone, 3,3-dimethyl-
O
Me
Me
HN
99 C5H9NO 13416-47-2
Aziridine, 1-acetyl-2-methyl-
Me
N
Ac
99 C5H9NO 15585-98-5
Butane, 2-isocyanato-
s - Bu NCO

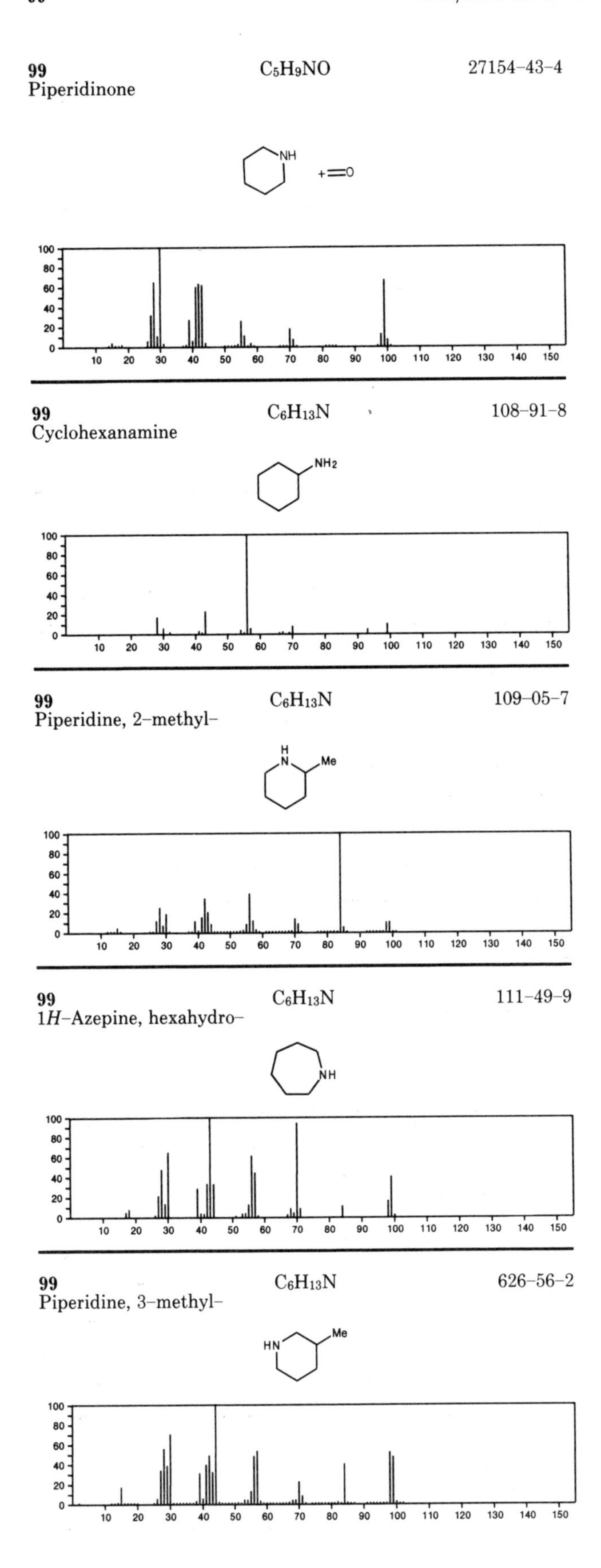
99 C5H9NO 27154-43-4
Piperidinone
NH
+=O
99 C6H13N 108-91-8
Cyclohexanamine
NH2
99 C6H13N 109-05-7
Piperidine, 2-methyl-
H
N
Me
99 C6H13N 111-49-9
1H-Azepine, hexahydro-
NH
99 C6H13N 626-56-2
Piperidine, 3-methyl-
HN
Me

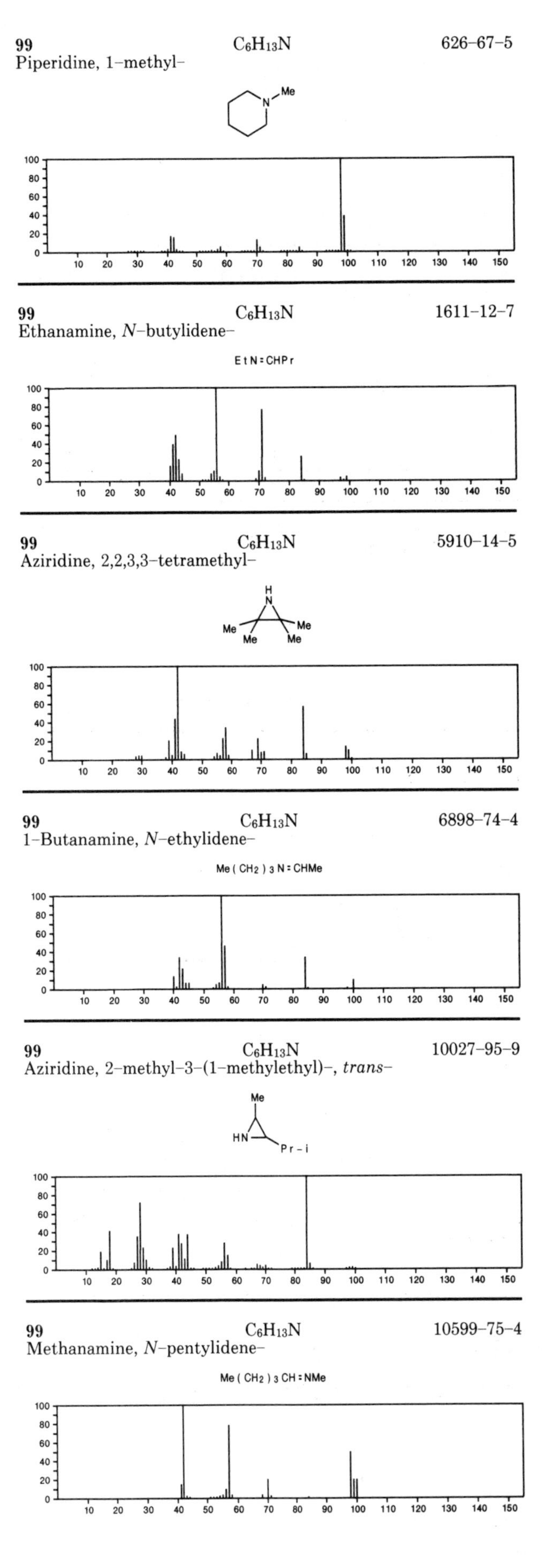
99 C6H13N 626-67-5
Piperidine, 1-methyl-
N
Me
99 C6H13N 1611-12-7
Ethanamine, N-butylidene-
EtN=CHPr
99 C6H13N 5910-14-5
Aziridine, 2,2,3,3-tetramethyl-
H
N
Me Me Me Me
99 C6H13N 6898-74-4
1-Butanamine, N-ethylidene-
Me(CH2)3N=CHMe
99 C6H13N 10027-95-9
Aziridine, 2-methyl-3-(1-methylethyl)-, trans-
Me
HN
Pr-i
99 C6H13N 10599-75-4
Methanamine, N-pentylidene-
Me(CH2)3CH=NMe

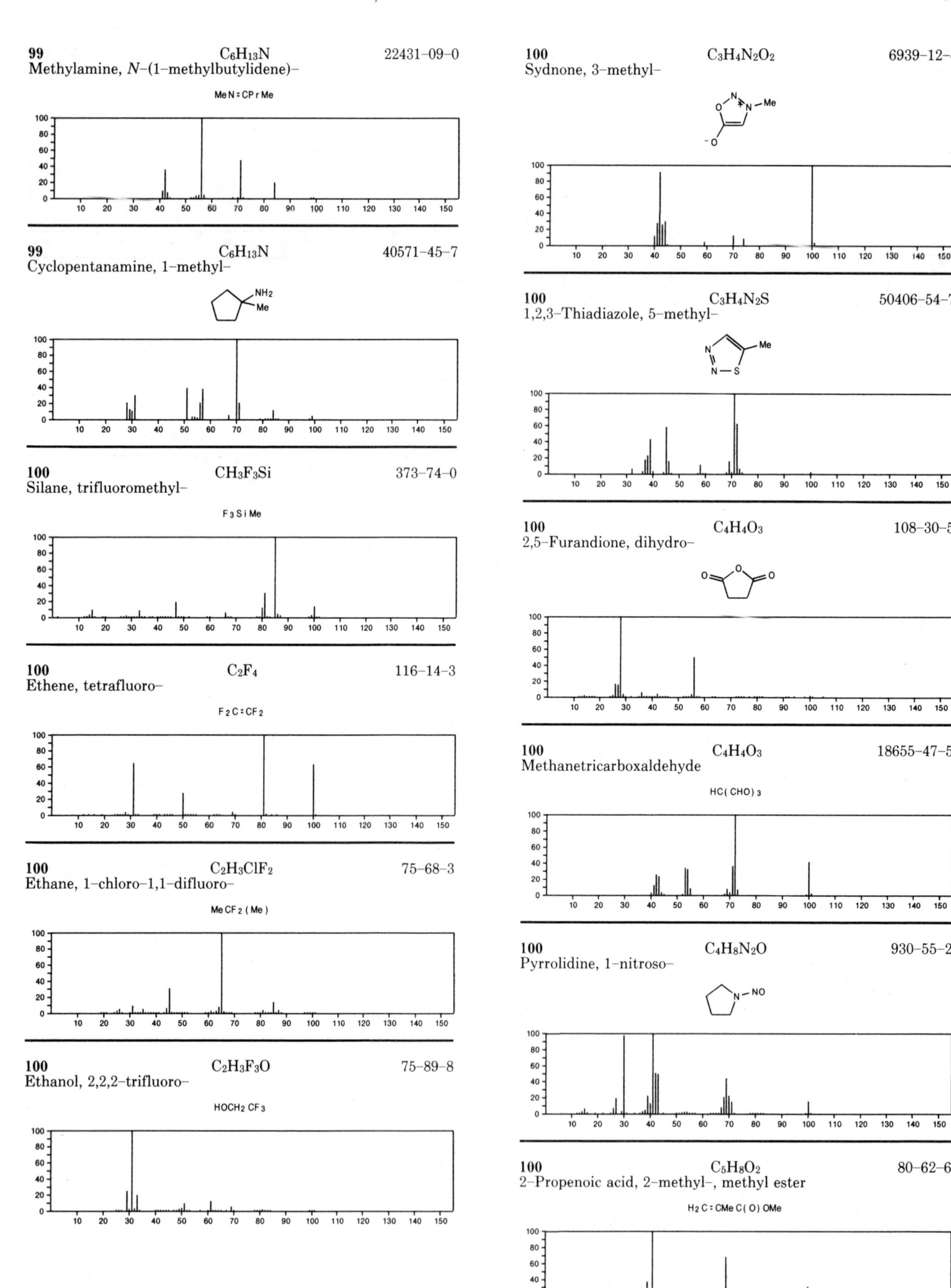

99 $C_6H_{13}N$ 22431–09–0
Methylamine, *N*–(1–methylbutylidene)–
MeN:CPrMe

99 $C_6H_{13}N$ 40571–45–7
Cyclopentanamine, 1–methyl–

100 CH_3F_3Si 373–74–0
Silane, trifluoromethyl–
F3SiMe

100 C_2F_4 116–14–3
Ethene, tetrafluoro–
F2C:CF2

100 $C_2H_3ClF_2$ 75–68–3
Ethane, 1–chloro–1,1–difluoro–
MeCF2(Me)

100 $C_2H_3F_3O$ 75–89–8
Ethanol, 2,2,2–trifluoro–
HOCH2CF3

100 $C_3H_4N_2O_2$ 6939–12–4
Sydnone, 3–methyl–

100 $C_3H_4N_2S$ 50406–54–7
1,2,3–Thiadiazole, 5–methyl–

100 $C_4H_4O_3$ 108–30–5
2,5–Furandione, dihydro–

100 $C_4H_4O_3$ 18655–47–5
Methanetricarboxaldehyde
HC(CHO)3

100 $C_4H_8N_2O$ 930–55–2
Pyrrolidine, 1–nitroso–

100 $C_5H_8O_2$ 80–62–6
2–Propenoic acid, 2–methyl–, methyl ester
H2C:CMeC(O)OMe

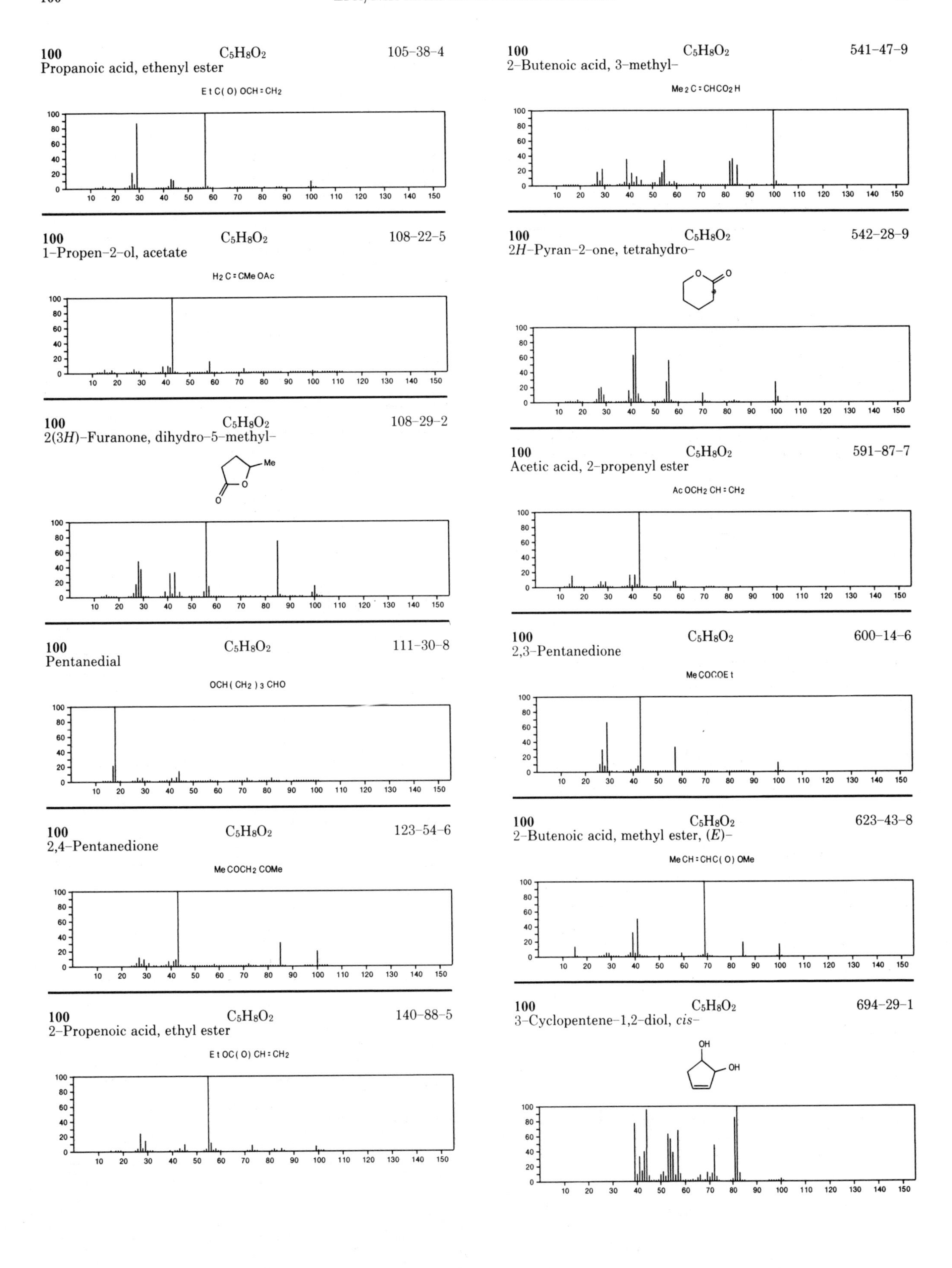

100 $C_5H_8O_2$ 105-38-4
Propanoic acid, ethenyl ester
$EtC(O)OCH{=}CH_2$

100 $C_5H_8O_2$ 108-22-5
1-Propen-2-ol, acetate
$H_2C{=}CMeOAc$

100 $C_5H_8O_2$ 108-29-2
2(3*H*)-Furanone, dihydro-5-methyl-

100 $C_5H_8O_2$ 111-30-8
Pentanedial
$OCH(CH_2)_3CHO$

100 $C_5H_8O_2$ 123-54-6
2,4-Pentanedione
$MeCOCH_2COMe$

100 $C_5H_8O_2$ 140-88-5
2-Propenoic acid, ethyl ester
$EtOC(O)CH{=}CH_2$

100 $C_5H_8O_2$ 541-47-9
2-Butenoic acid, 3-methyl-
$Me_2C{=}CHCO_2H$

100 $C_5H_8O_2$ 542-28-9
2*H*-Pyran-2-one, tetrahydro-

100 $C_5H_8O_2$ 591-87-7
Acetic acid, 2-propenyl ester
$AcOCH_2CH{=}CH_2$

100 $C_5H_8O_2$ 600-14-6
2,3-Pentanedione
$MeCOCOEt$

100 $C_5H_8O_2$ 623-43-8
2-Butenoic acid, methyl ester, (*E*)-
$MeCH{=}CHC(O)OMe$

100 $C_5H_8O_2$ 694-29-1
3-Cyclopentene-1,2-diol, *cis*-

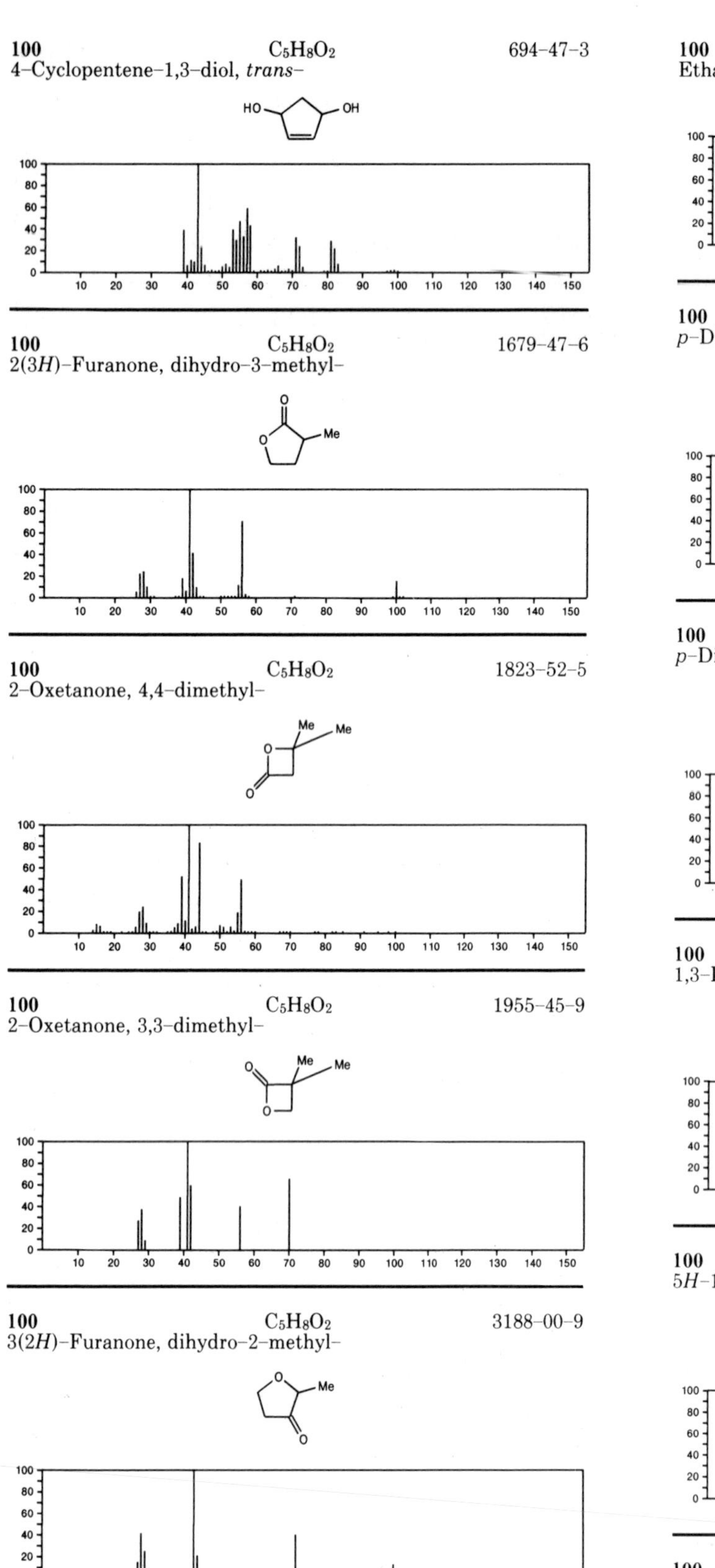

100 $C_5H_8O_2$ 694–47–3
4–Cyclopentene–1,3–diol, *trans*–

100 $C_5H_8O_2$ 1679–47–6
2(3*H*)–Furanone, dihydro–3–methyl–

100 $C_5H_8O_2$ 1823–52–5
2–Oxetanone, 4,4–dimethyl–

100 $C_5H_8O_2$ 1955–45–9
2–Oxetanone, 3,3–dimethyl–

100 $C_5H_8O_2$ 3188–00–9
3(2*H*)–Furanone, dihydro–2–methyl–

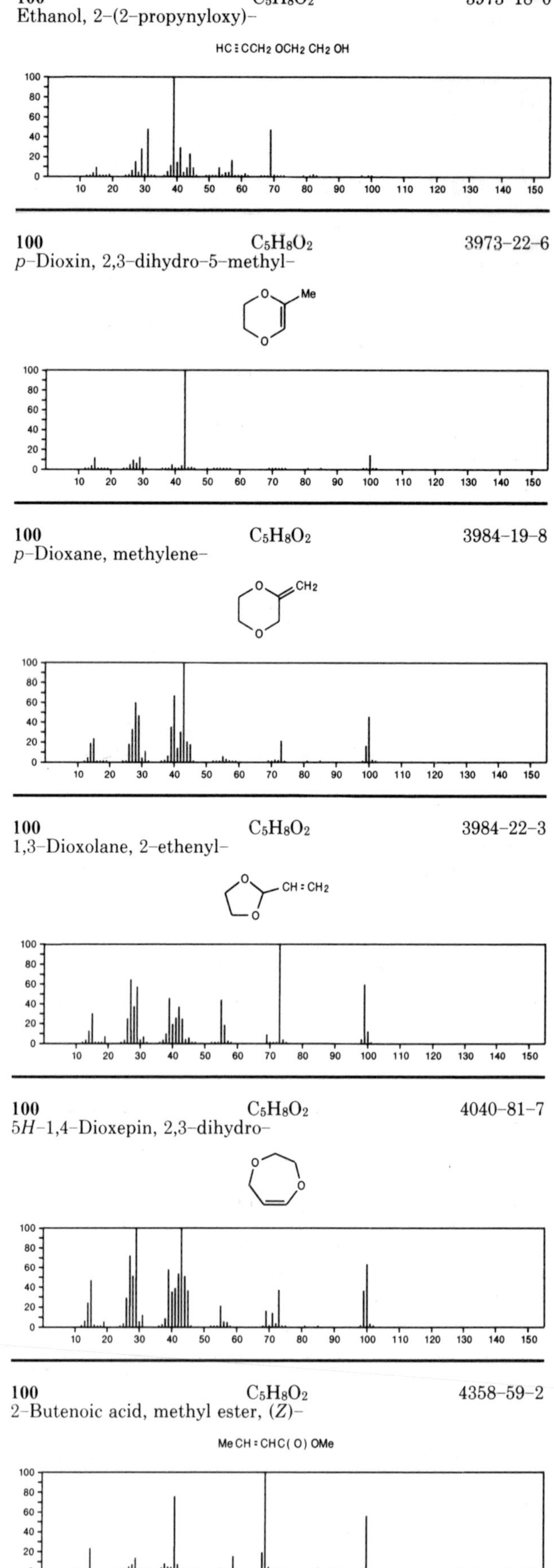

100 $C_5H_8O_2$ 3973–18–0
Ethanol, 2–(2–propynyloxy)–

100 $C_5H_8O_2$ 3973–22–6
p–Dioxin, 2,3–dihydro–5–methyl–

100 $C_5H_8O_2$ 3984–19–8
p–Dioxane, methylene–

100 $C_5H_8O_2$ 3984–22–3
1,3–Dioxolane, 2–ethenyl–

100 $C_5H_8O_2$ 4040–81–7
5*H*–1,4–Dioxepin, 2,3–dihydro–

100 $C_5H_8O_2$ 4358–59–2
2–Butenoic acid, methyl ester, (*Z*)–

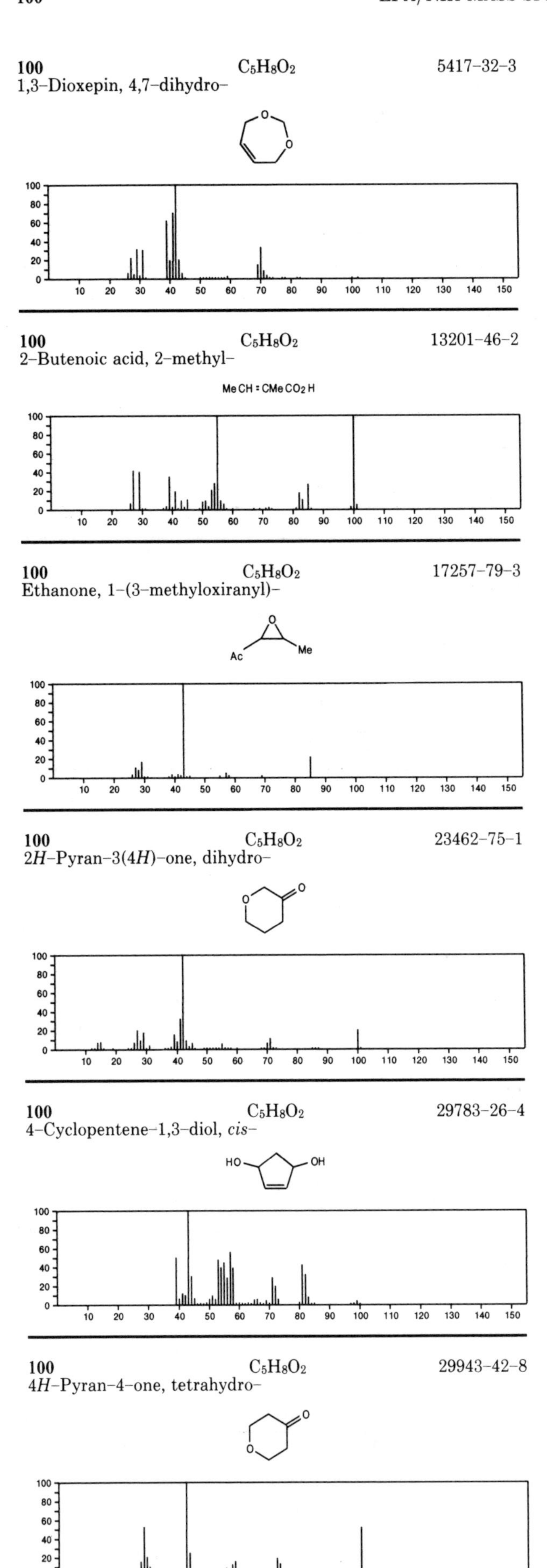

100 $C_5H_8O_2$ 5417–32–3
1,3–Dioxepin, 4,7–dihydro–

100 $C_5H_8O_2$ 13201–46–2
2–Butenoic acid, 2–methyl–

MeCH = CMeCO$_2$H

100 $C_5H_8O_2$ 17257–79–3
Ethanone, 1–(3–methyloxiranyl)–

100 $C_5H_8O_2$ 23462–75–1
2*H*–Pyran–3(4*H*)–one, dihydro–

100 $C_5H_8O_2$ 29783–26–4
4–Cyclopentene–1,3–diol, *cis*–

100 $C_5H_8O_2$ 29943–42–8
4*H*–Pyran–4–one, tetrahydro–

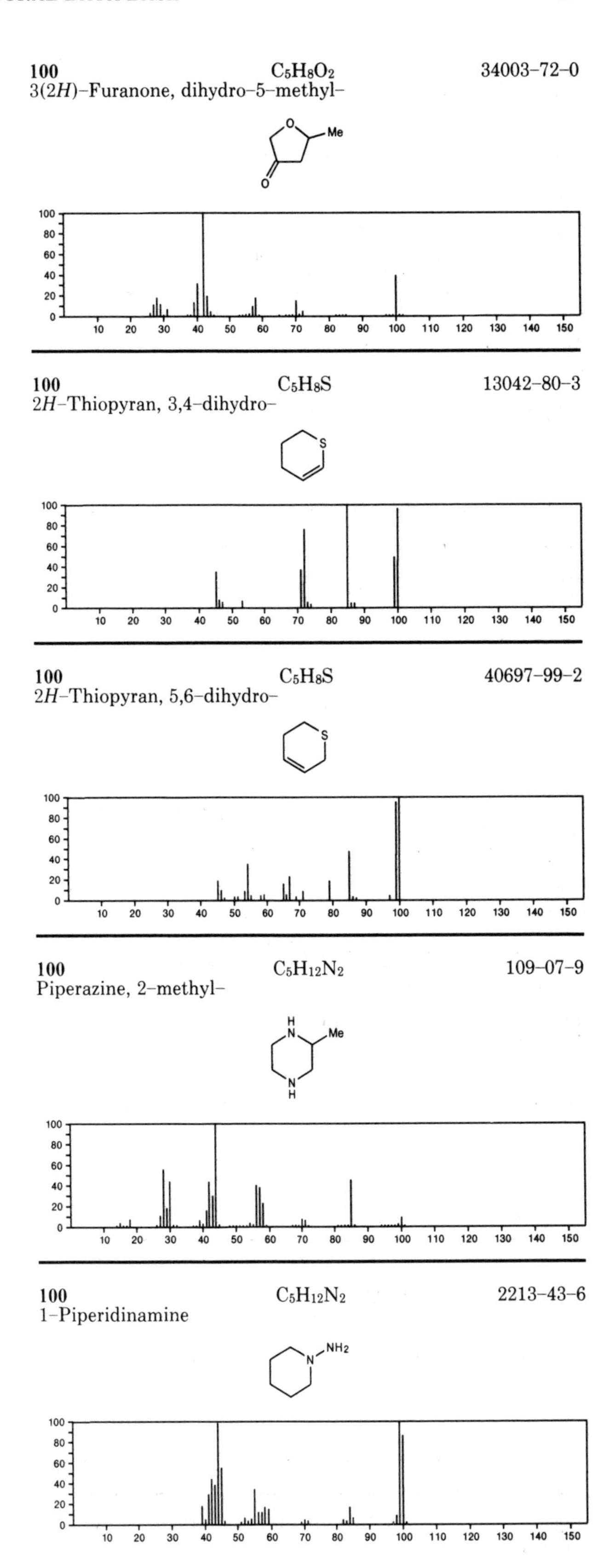

100 $C_5H_8O_2$ 34003–72–0
3(2*H*)–Furanone, dihydro–5–methyl–

100 C_5H_8S 13042–80–3
2*H*–Thiopyran, 3,4–dihydro–

100 C_5H_8S 40697–99–2
2*H*–Thiopyran, 5,6–dihydro–

100 $C_5H_{12}N_2$ 109–07–9
Piperazine, 2–methyl–

100 $C_5H_{12}N_2$ 2213–43–6
1–Piperidinamine

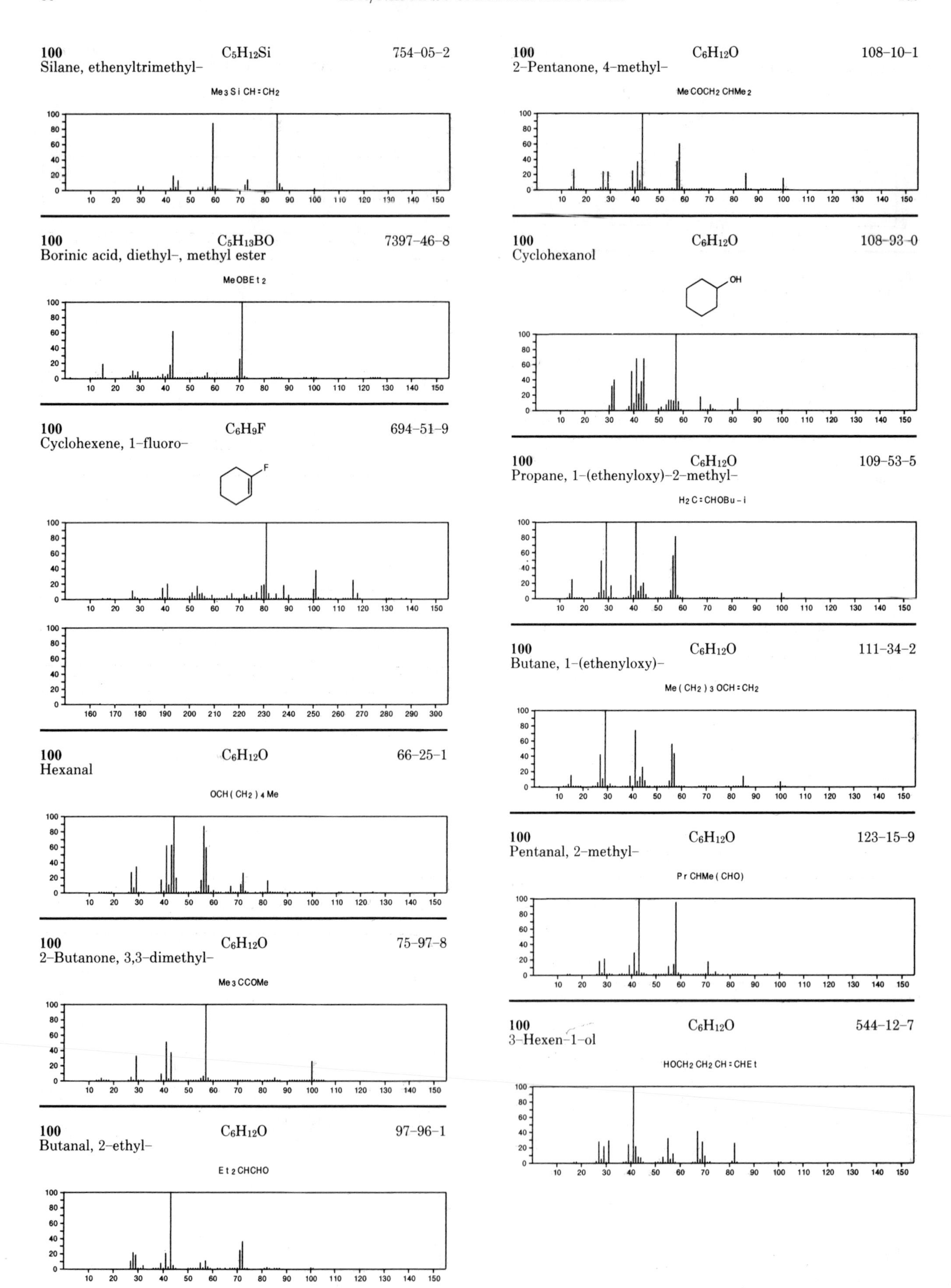
100 $C_5H_{12}Si$ 754-05-2
Silane, ethenyltrimethyl-
Me3SiCH=CH2
100 $C_5H_{13}BO$ 7397-46-8
Borinic acid, diethyl-, methyl ester
MeOBEt2
100 C_6H_9F 694-51-9
Cyclohexene, 1-fluoro-
F
100 $C_6H_{12}O$ 66-25-1
Hexanal
OCH(CH2)4Me
100 $C_6H_{12}O$ 75-97-8
2-Butanone, 3,3-dimethyl-
Me3CCOMe
100 $C_6H_{12}O$ 97-96-1
Butanal, 2-ethyl-
Et2CHCHO
100 $C_6H_{12}O$ 108-10-1
2-Pentanone, 4-methyl-
MeCOCH2CHMe2
100 $C_6H_{12}O$ 108-93-0
Cyclohexanol
OH
100 $C_6H_{12}O$ 109-53-5
Propane, 1-(ethenyloxy)-2-methyl-
H2C=CHOBu-i
100 $C_6H_{12}O$ 111-34-2
Butane, 1-(ethenyloxy)-
Me(CH2)3OCH=CH2
100 $C_6H_{12}O$ 123-15-9
Pentanal, 2-methyl-
PrCHMe(CHO)
100 $C_6H_{12}O$ 544-12-7
3-Hexen-1-ol
HOCH2CH2CH=CHEt

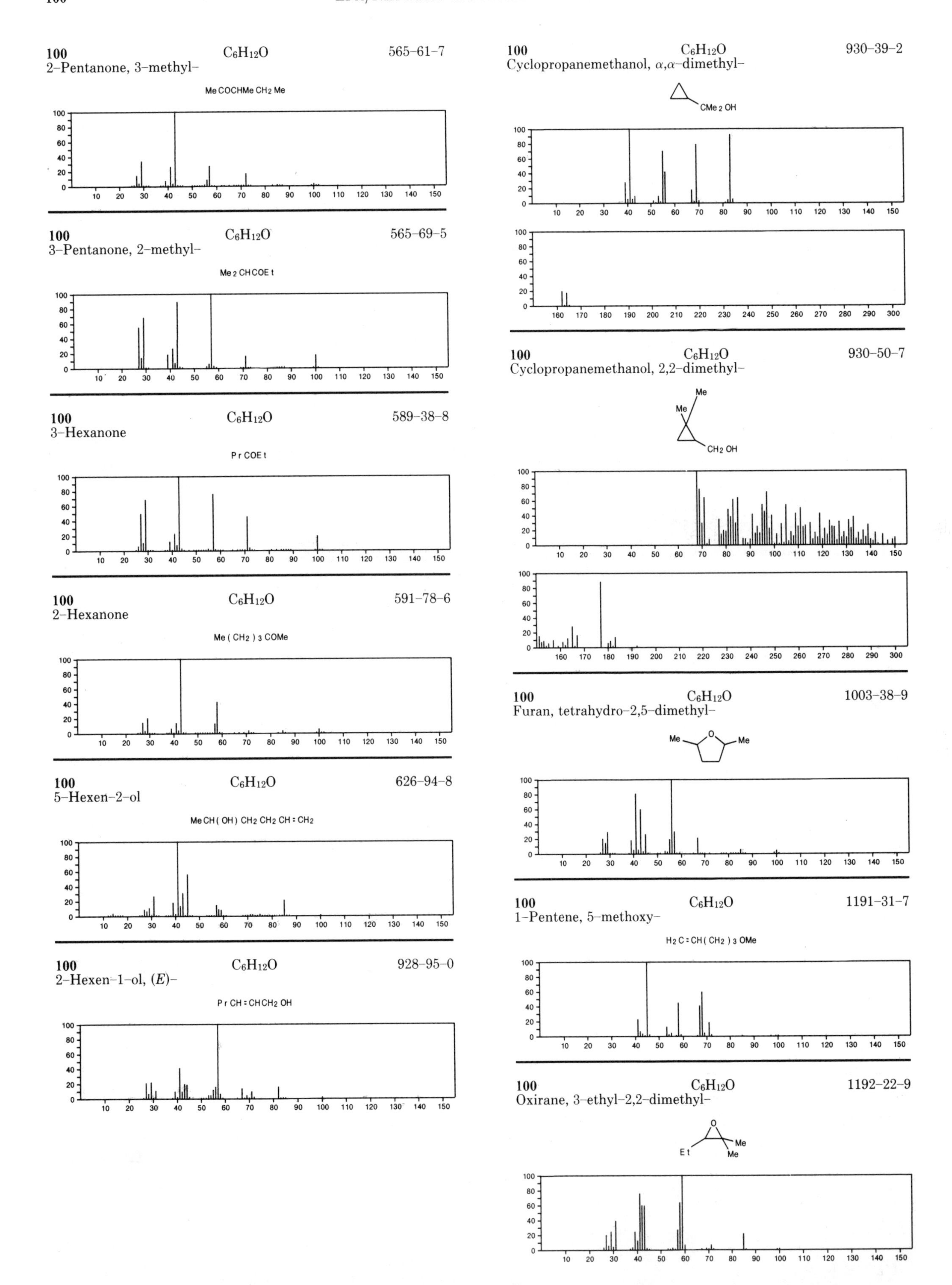
100 C6H12O 565-61-7
2-Pentanone, 3-methyl-
Me COCHMe CH2 Me
100 C6H12O 565-69-5
3-Pentanone, 2-methyl-
Me 2 CHCOE t
100 C6H12O 589-38-8
3-Hexanone
P r COE t
100 C6H12O 591-78-6
2-Hexanone
Me (CH2) 3 COMe
100 C6H12O 626-94-8
5-Hexen-2-ol
Me CH (OH) CH2 CH2 CH = CH2
100 C6H12O 928-95-0
2-Hexen-1-ol, (E)-
P r CH = CHCH2 OH
100 C6H12O 930-39-2
Cyclopropanemethanol, α,α-dimethyl-
CMe 2 OH
100 C6H12O 930-50-7
Cyclopropanemethanol, 2,2-dimethyl-
Me
Me
CH2 OH
100 C6H12O 1003-38-9
Furan, tetrahydro-2,5-dimethyl-
Me
O
Me
100 C6H12O 1191-31-7
1-Pentene, 5-methoxy-
H2 C = CH (CH2) 3 OMe
100 C6H12O 1192-22-9
Oxirane, 3-ethyl-2,2-dimethyl-
O
Me
E t
Me

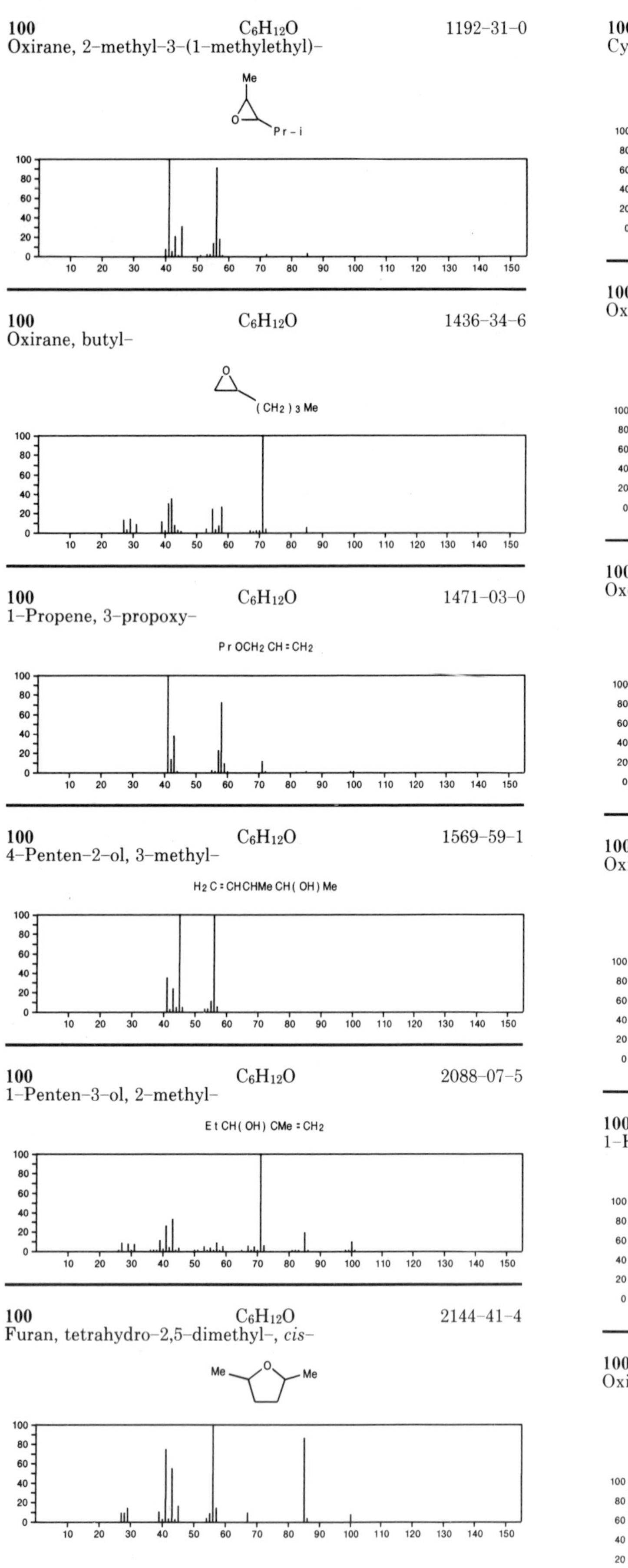

100 $C_6H_{12}O$ 1192-31-0
Oxirane, 2-methyl-3-(1-methylethyl)-

100 $C_6H_{12}O$ 1436-34-6
Oxirane, butyl-

100 $C_6H_{12}O$ 1471-03-0
1-Propene, 3-propoxy-

PrOCH2CH=CH2

100 $C_6H_{12}O$ 1569-59-1
4-Penten-2-ol, 3-methyl-

H2C=CHCHMeCH(OH)Me

100 $C_6H_{12}O$ 2088-07-5
1-Penten-3-ol, 2-methyl-

EtCH(OH)CMe=CH2

100 $C_6H_{12}O$ 2144-41-4
Furan, tetrahydro-2,5-dimethyl-, *cis*-

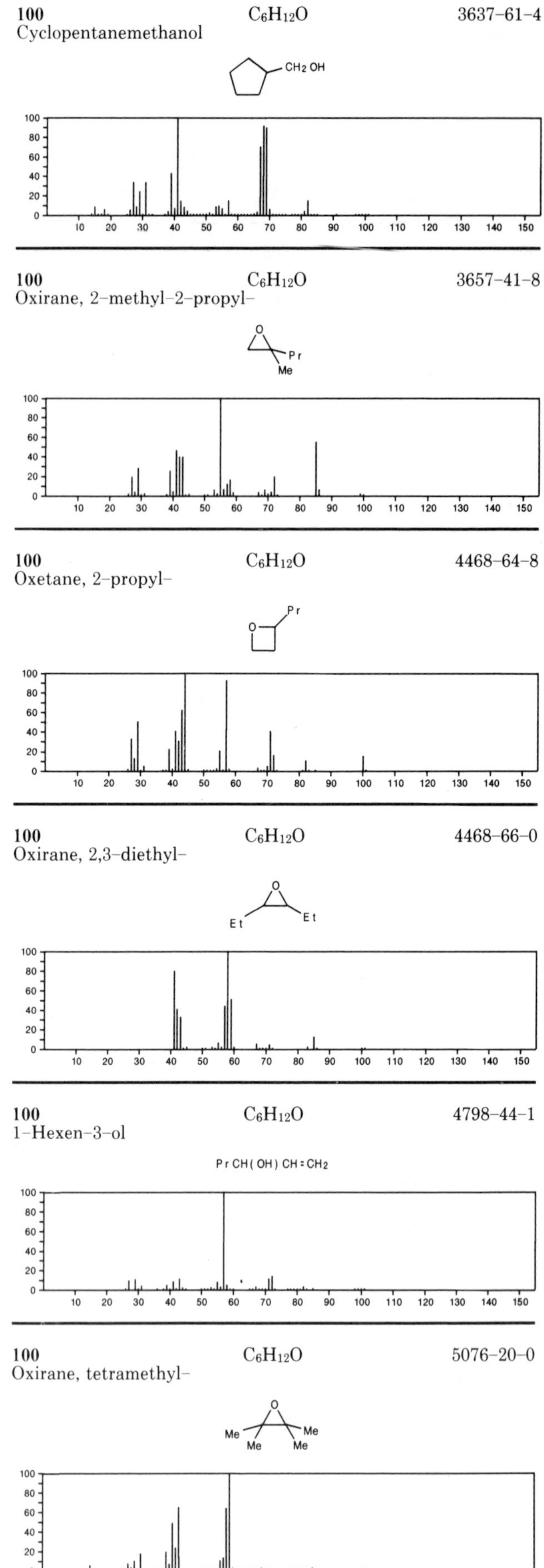

100 $C_6H_{12}O$ 3637-61-4
Cyclopentanemethanol

100 $C_6H_{12}O$ 3657-41-8
Oxirane, 2-methyl-2-propyl-

100 $C_6H_{12}O$ 4468-64-8
Oxetane, 2-propyl-

100 $C_6H_{12}O$ 4468-66-0
Oxirane, 2,3-diethyl-

100 $C_6H_{12}O$ 4798-44-1
1-Hexen-3-ol

PrCH(OH)CH=CH2

100 $C_6H_{12}O$ 5076-20-0
Oxirane, tetramethyl-

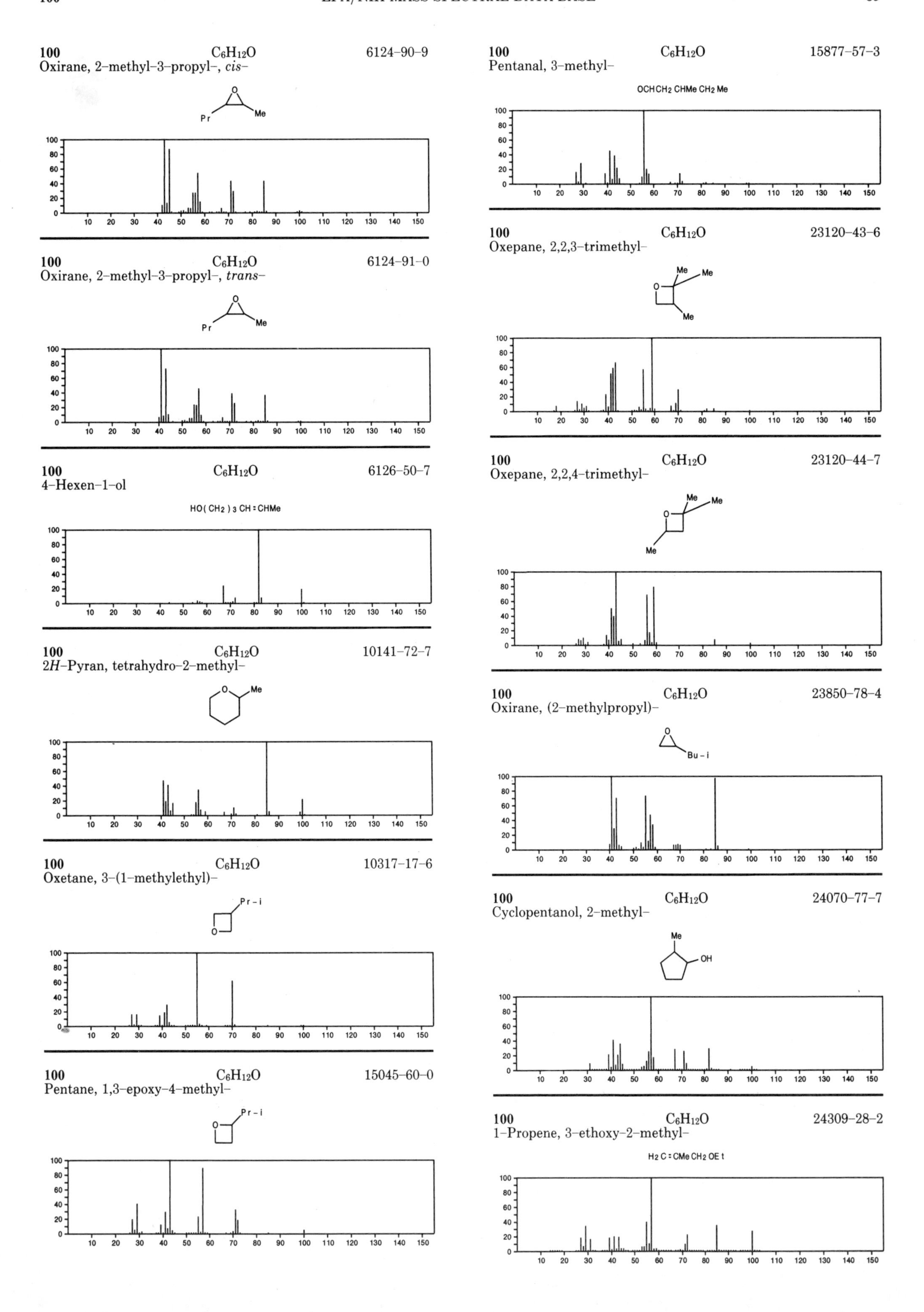
100 C6H12O 6124-90-9
Oxirane, 2-methyl-3-propyl-, cis-
100 C6H12O 6124-91-0
Oxirane, 2-methyl-3-propyl-, trans-
100 C6H12O 6126-50-7
4-Hexen-1-ol
HO(CH2) 3 CH = CHMe
100 C6H12O 10141-72-7
2H-Pyran, tetrahydro-2-methyl-
100 C6H12O 10317-17-6
Oxetane, 3-(1-methylethyl)-
100 C6H12O 15045-60-0
Pentane, 1,3-epoxy-4-methyl-
100 C6H12O 15877-57-3
Pentanal, 3-methyl-
OCHCH2 CHMe CH2 Me
100 C6H12O 23120-43-6
Oxepane, 2,2,3-trimethyl-
100 C6H12O 23120-44-7
Oxepane, 2,2,4-trimethyl-
100 C6H12O 23850-78-4
Oxirane, (2-methylpropyl)-
100 C6H12O 24070-77-7
Cyclopentanol, 2-methyl-
100 C6H12O 24309-28-2
1-Propene, 3-ethoxy-2-methyl-
H2 C = CMe CH2 OE t

100 $C_6H_{12}O$ 25144-04-1
Cyclopentanol, 2-methyl-, *trans*-

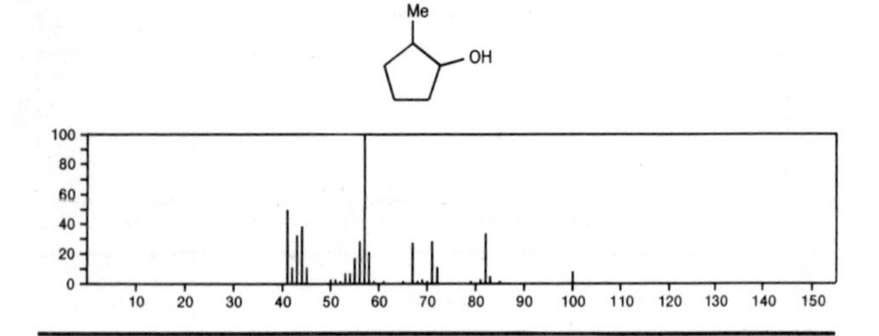

100 $C_6H_{12}O$ 25144-05-2
Cyclopentanol, 2-methyl-, *cis*-

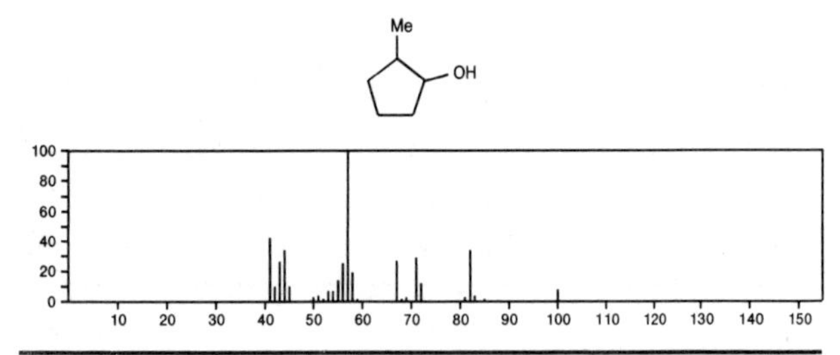

100 $C_6H_{12}O$ 26093-63-0
2*H*-Pyran, tetrahydro-3-methyl-

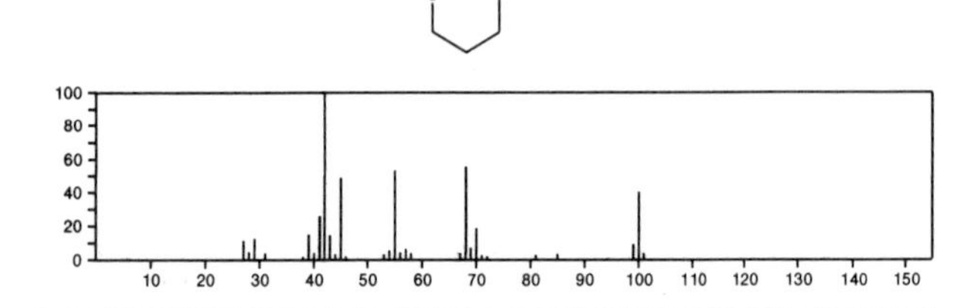

100 $C_6H_{12}O$ 27953-97-5
Furan, tetrahydro-3,4-dimethyl-, *cis*-

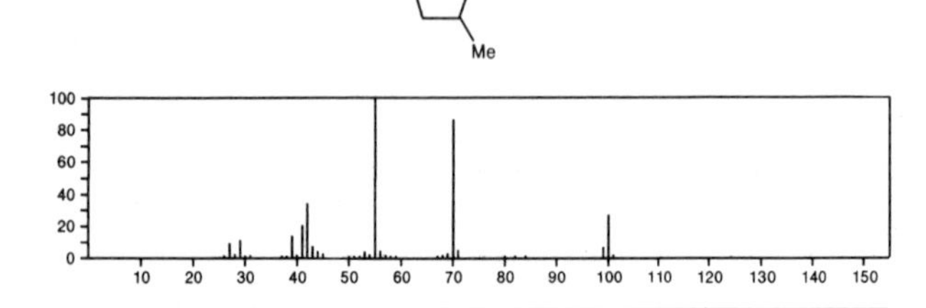

100 $C_6H_{12}O$ 32347-12-9
Oxetane, 2,3,4-trimethyl-, $(2\alpha,3\alpha,4\beta)$-

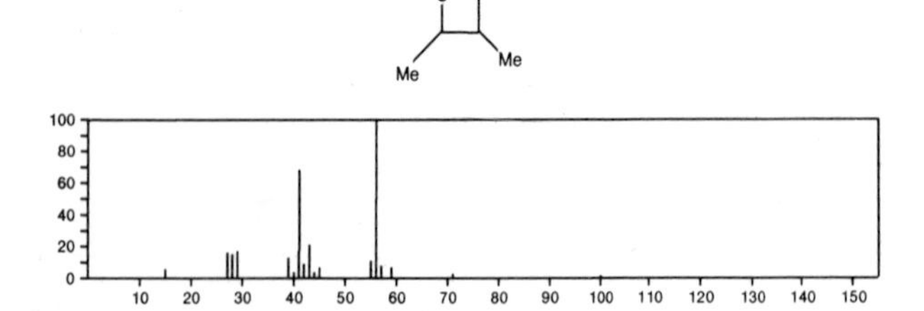

100 $C_6H_{12}O$ 35301-43-0
Cyclobutanol, 2-ethyl-

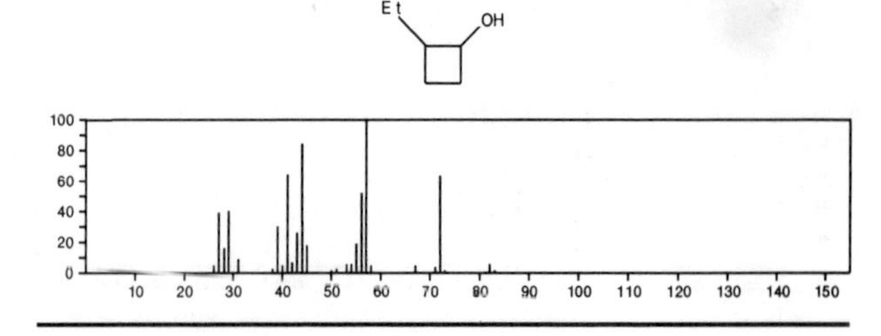

100 $C_6H_{12}O$ 38484-59-2
Furan, tetrahydro-2,5-dimethyl-, *trans*-(±)-

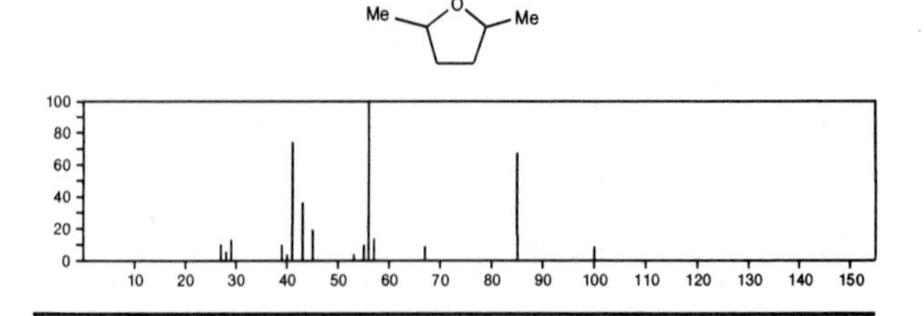

100 $C_6H_{12}O$ 39168-01-9
Furan, tetrahydro-2,4-dimethyl-, *cis*-

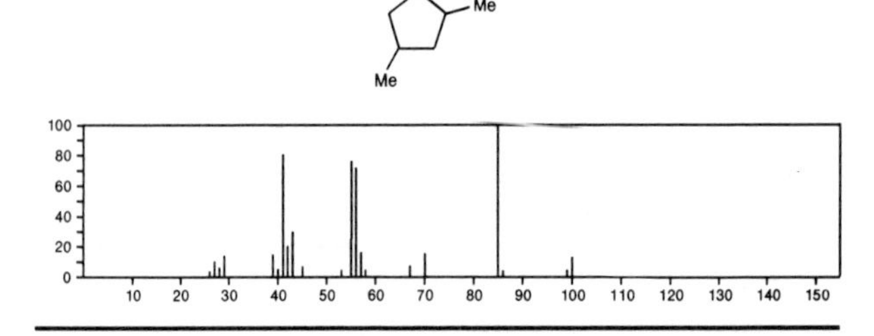

100 $C_6H_{12}O$ 39168-02-0
Furan, tetrahydro-2,4-dimethyl-, *trans*-

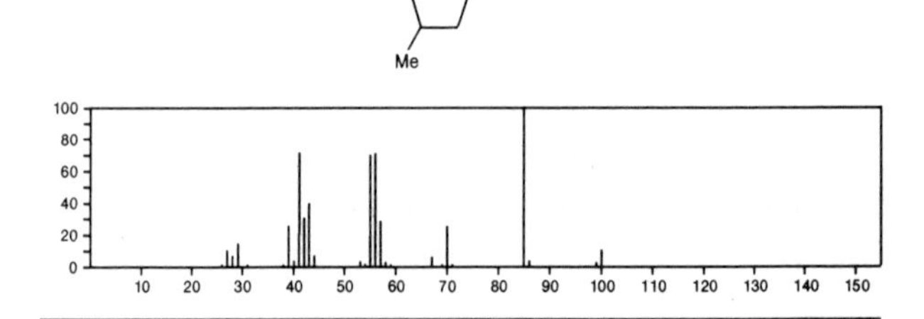

100 $C_6H_{12}O$ 53778-61-3
Oxetane, 2,3,4-trimethyl-

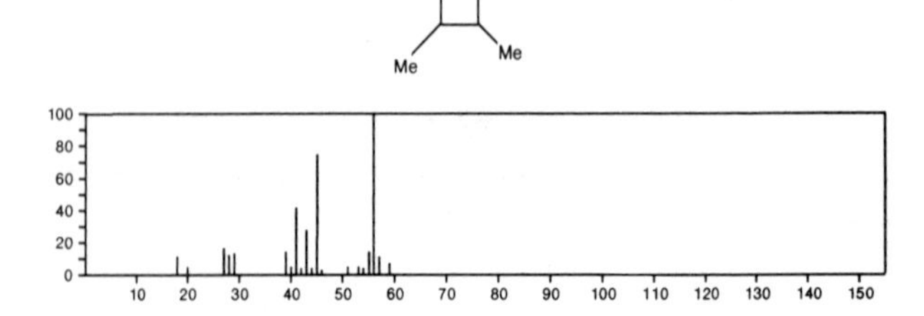

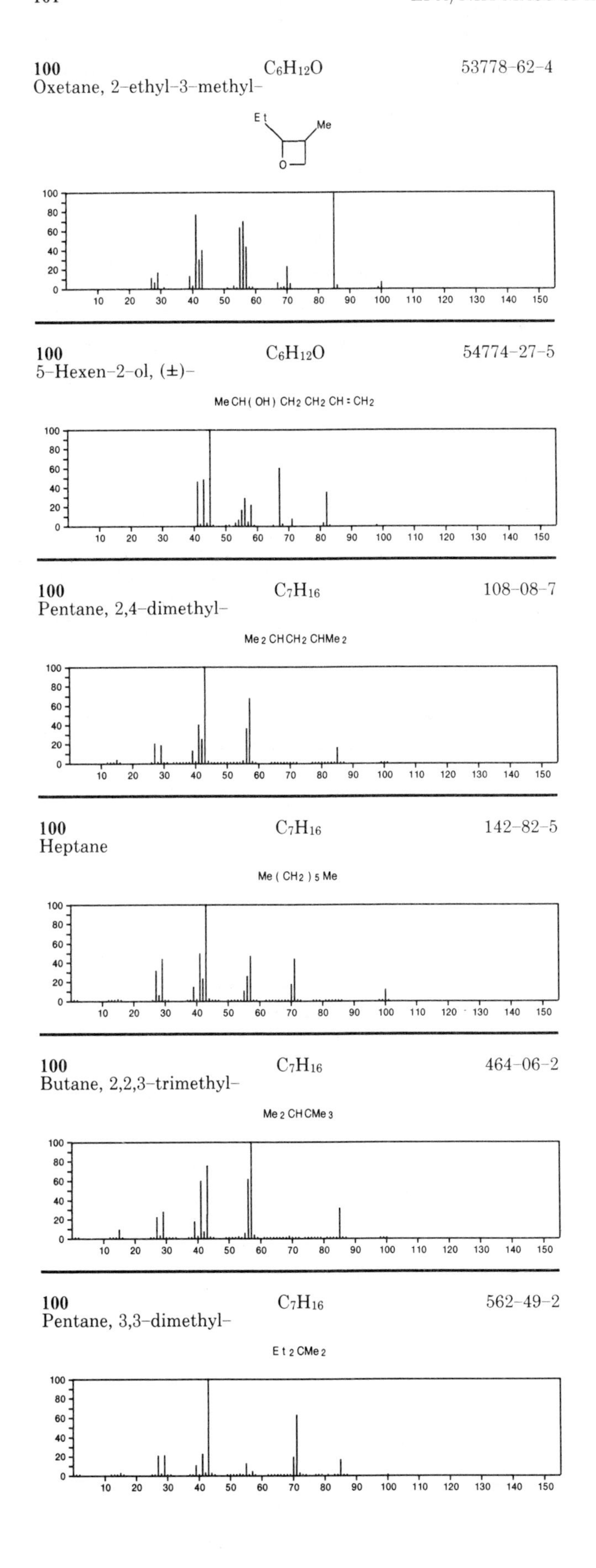

100 $C_6H_{12}O$ 53778-62-4
Oxetane, 2-ethyl-3-methyl-

100 $C_6H_{12}O$ 54774-27-5
5-Hexen-2-ol, (±)-
Me CH(OH) CH2 CH2 CH = CH2

100 C_7H_{16} 108-08-7
Pentane, 2,4-dimethyl-
Me 2 CHCH2 CHMe 2

100 C_7H_{16} 142-82-5
Heptane
Me (CH2) 5 Me

100 C_7H_{16} 464-06-2
Butane, 2,2,3-trimethyl-
Me 2 CHCMe 3

100 C_7H_{16} 562-49-2
Pentane, 3,3-dimethyl-
E t 2 CMe 2

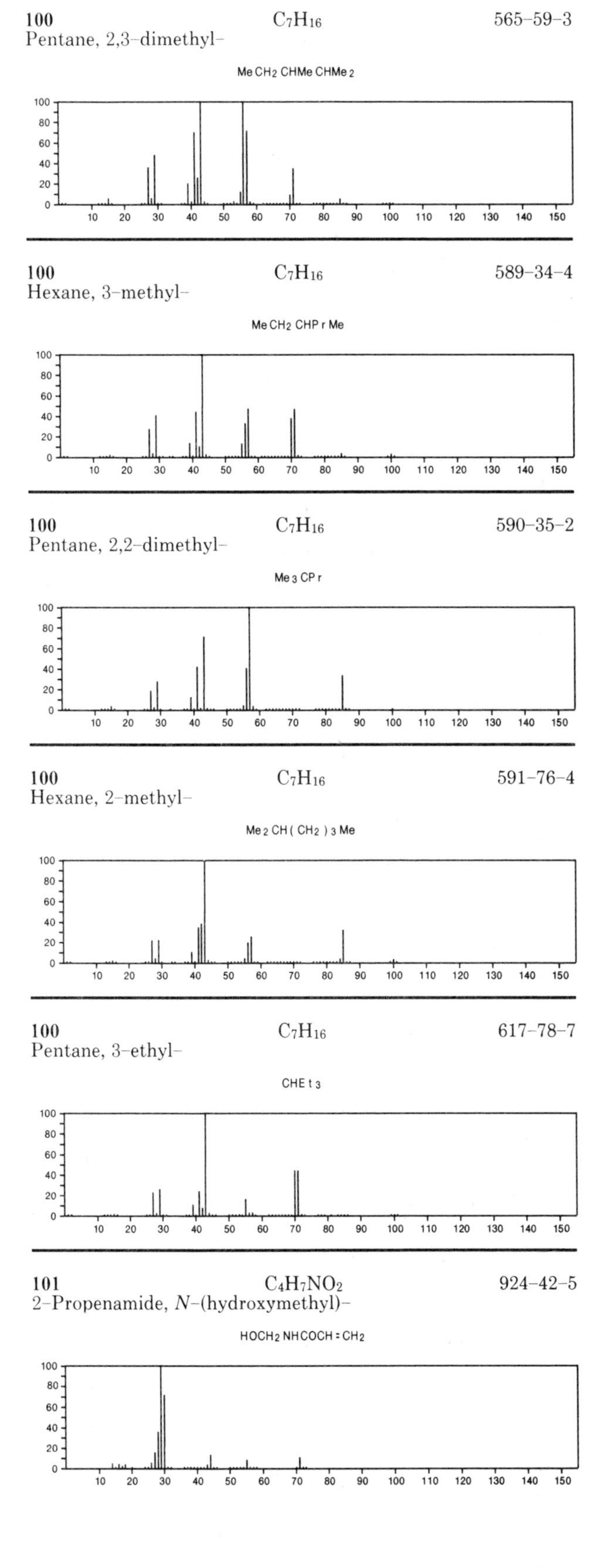

100 C_7H_{16} 565-59-3
Pentane, 2,3-dimethyl-
Me CH2 CHMe CHMe 2

100 C_7H_{16} 589-34-4
Hexane, 3-methyl-
Me CH2 CHP r Me

100 C_7H_{16} 590-35-2
Pentane, 2,2-dimethyl-
Me 3 CP r

100 C_7H_{16} 591-76-4
Hexane, 2-methyl-
Me 2 CH (CH2) 3 Me

100 C_7H_{16} 617-78-7
Pentane, 3-ethyl-
CHE t 3

101 $C_4H_7NO_2$ 924-42-5
2-Propenamide, *N*-(hydroxymethyl)-
HOCH2 NHCOCH = CH2

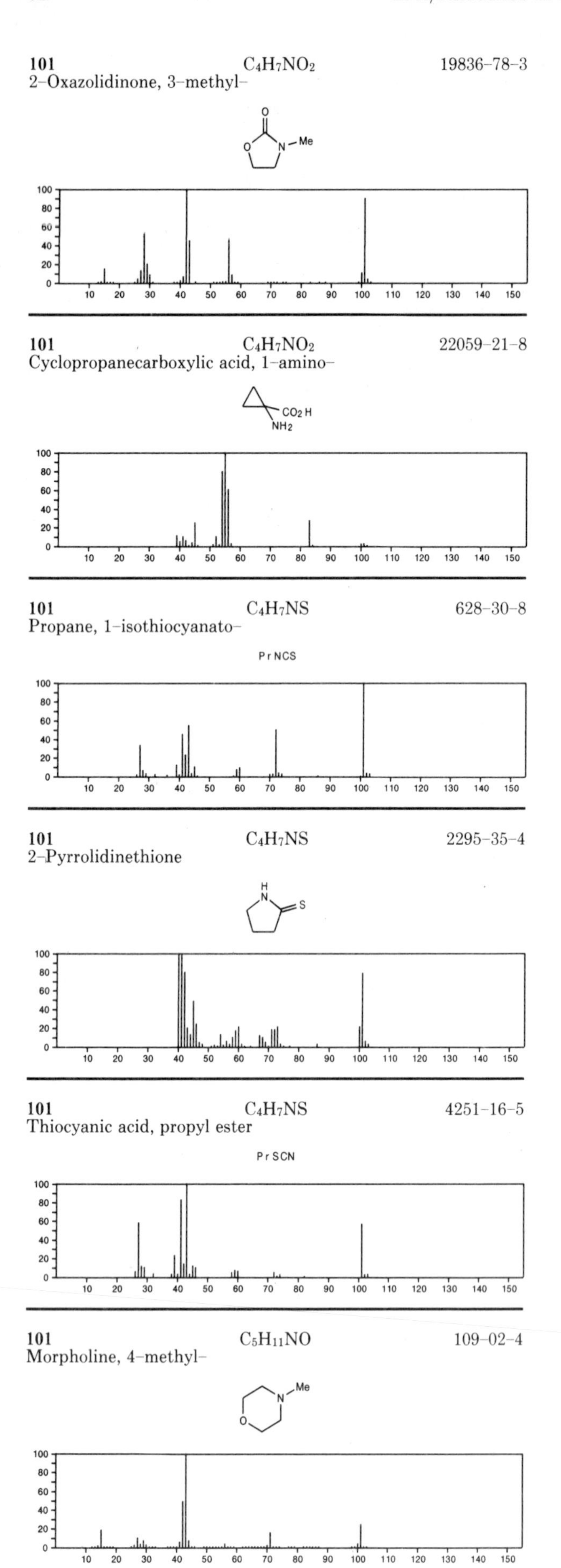

101 $C_4H_7NO_2$ 19836-78-3
2-Oxazolidinone, 3-methyl-

101 $C_4H_7NO_2$ 22059-21-8
Cyclopropanecarboxylic acid, 1-amino-

101 C_4H_7NS 628-30-8
Propane, 1-isothiocyanato-

PrNCS

101 C_4H_7NS 2295-35-4
2-Pyrrolidinethione

101 C_4H_7NS 4251-16-5
Thiocyanic acid, propyl ester

PrSCN

101 $C_5H_{11}NO$ 109-02-4
Morpholine, 4-methyl-

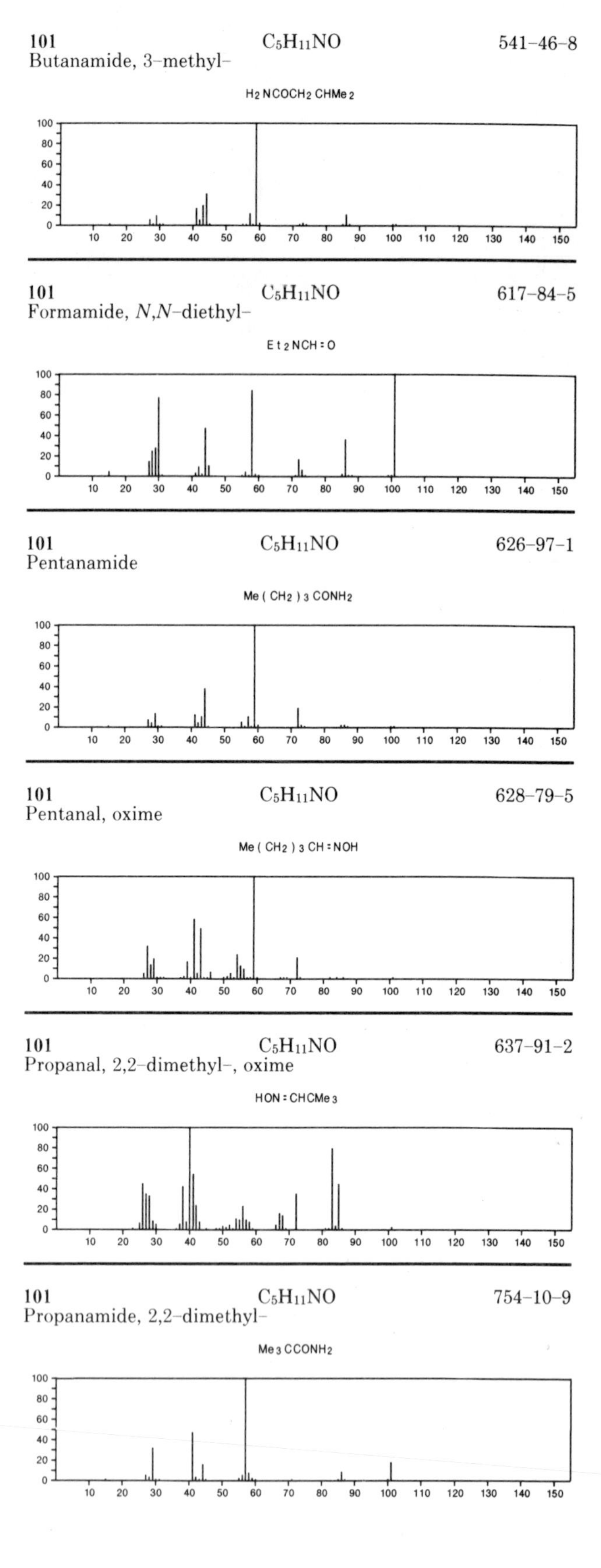

101 $C_5H_{11}NO$ 541-46-8
Butanamide, 3-methyl-

$H_2NCOCH_2CHMe_2$

101 $C_5H_{11}NO$ 617-84-5
Formamide, *N,N*-diethyl-

$Et_2NCH{:}O$

101 $C_5H_{11}NO$ 626-97-1
Pentanamide

$Me(CH_2)_3CONH_2$

101 $C_5H_{11}NO$ 628-79-5
Pentanal, oxime

$Me(CH_2)_3CH{:}NOH$

101 $C_5H_{11}NO$ 637-91-2
Propanal, 2,2-dimethyl-, oxime

$HON{:}CHCMe_3$

101 $C_5H_{11}NO$ 754-10-9
Propanamide, 2,2-dimethyl-

Me_3CCONH_2

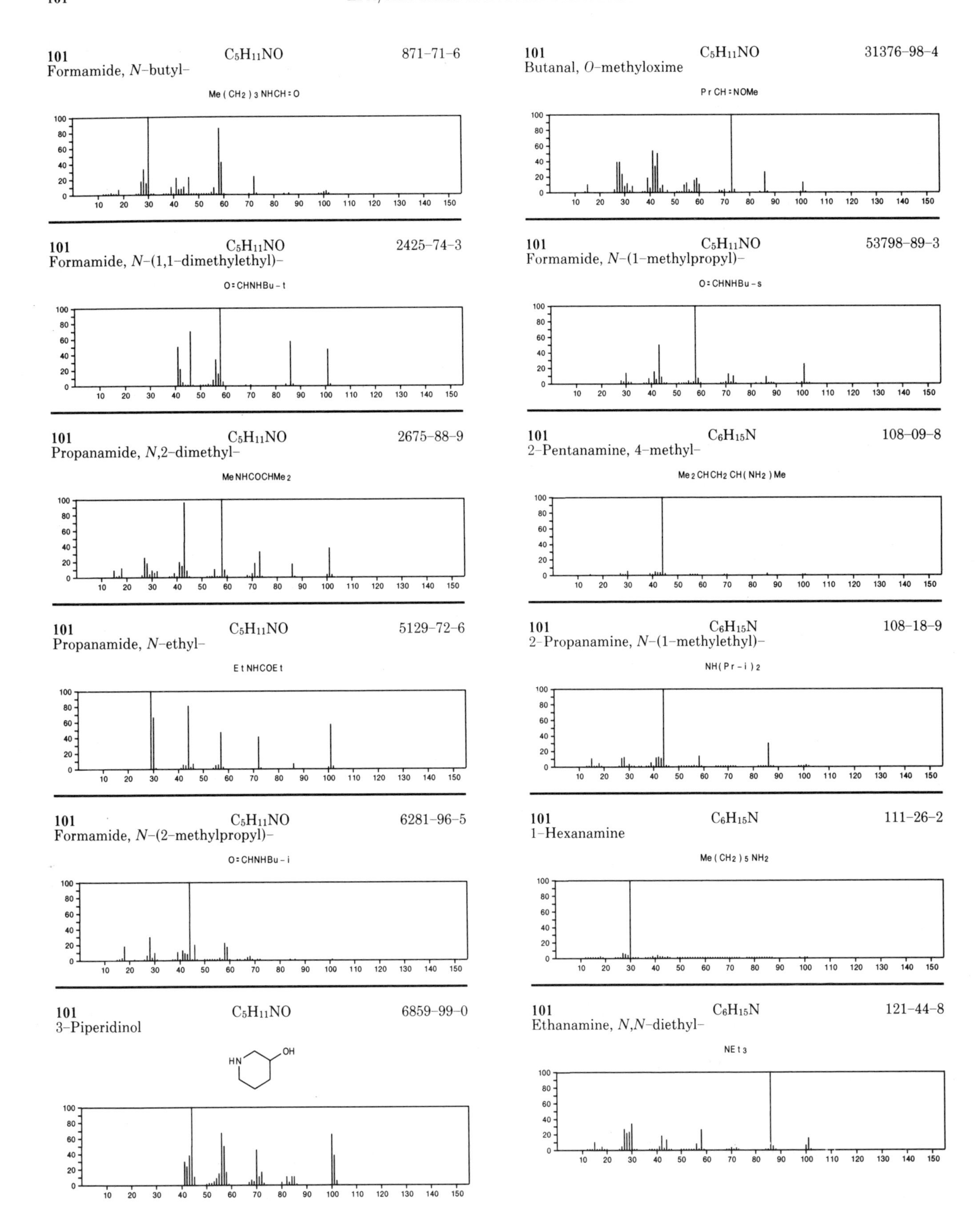
101 C5H11NO 871-71-6
Formamide, N-butyl-
Me(CH2)3NHCH=O
101 C5H11NO 2425-74-3
Formamide, N-(1,1-dimethylethyl)-
O=CHNHBu-t
101 C5H11NO 2675-88-9
Propanamide, N,2-dimethyl-
MeNHCOCHMe2
101 C5H11NO 5129-72-6
Propanamide, N-ethyl-
EtNHCOEt
101 C5H11NO 6281-96-5
Formamide, N-(2-methylpropyl)-
O=CHNHBu-i
101 C5H11NO 6859-99-0
3-Piperidinol
HN
OH
101 C5H11NO 31376-98-4
Butanal, O-methyloxime
PrCH=NOMe
101 C5H11NO 53798-89-3
Formamide, N-(1-methylpropyl)-
O=CHNHBu-s
101 C6H15N 108-09-8
2-Pentanamine, 4-methyl-
Me2CHCH2CH(NH2)Me
101 C6H15N 108-18-9
2-Propanamine, N-(1-methylethyl)-
NH(Pr-i)2
101 C6H15N 111-26-2
1-Hexanamine
Me(CH2)5NH2
101 C6H15N 121-44-8
Ethanamine, N,N-diethyl-
NEt3

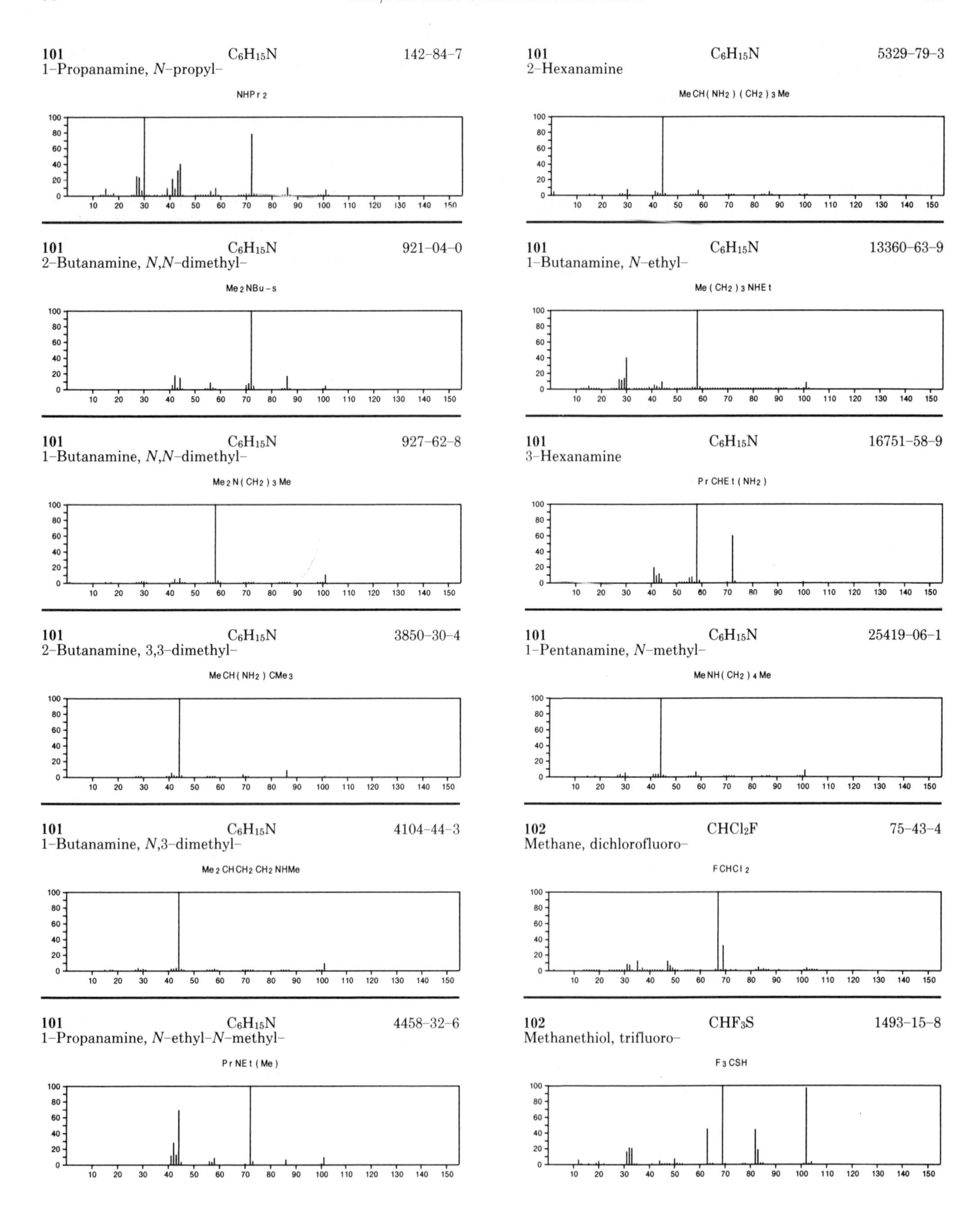

101 $C_6H_{15}N$ 142-84-7
1-Propanamine, *N*-propyl-
NHPr2
101 $C_6H_{15}N$ 5329-79-3
2-Hexanamine
MeCH(NH2)(CH2)3Me
101 $C_6H_{15}N$ 921-04-0
2-Butanamine, *N*,*N*-dimethyl-
Me2NBu-s
101 $C_6H_{15}N$ 13360-63-9
1-Butanamine, *N*-ethyl-
Me(CH2)3NHEt
101 $C_6H_{15}N$ 927-62-8
1-Butanamine, *N*,*N*-dimethyl-
Me2N(CH2)3Me
101 $C_6H_{15}N$ 16751-58-9
3-Hexanamine
PrCHEt(NH2)
101 $C_6H_{15}N$ 3850-30-4
2-Butanamine, 3,3-dimethyl-
MeCH(NH2)CMe3
101 $C_6H_{15}N$ 25419-06-1
1-Pentanamine, *N*-methyl-
MeNH(CH2)4Me
101 $C_6H_{15}N$ 4104-44-3
1-Butanamine, *N*,3-dimethyl-
Me2CHCH2CH2NHMe
102 $CHCl_2F$ 75-43-4
Methane, dichlorofluoro-
FCHCl2
101 $C_6H_{15}N$ 4458-32-6
1-Propanamine, *N*-ethyl-*N*-methyl-
PrNEt(Me)
102 CHF_3S 1493-15-8
Methanethiol, trifluoro-
F3CSH

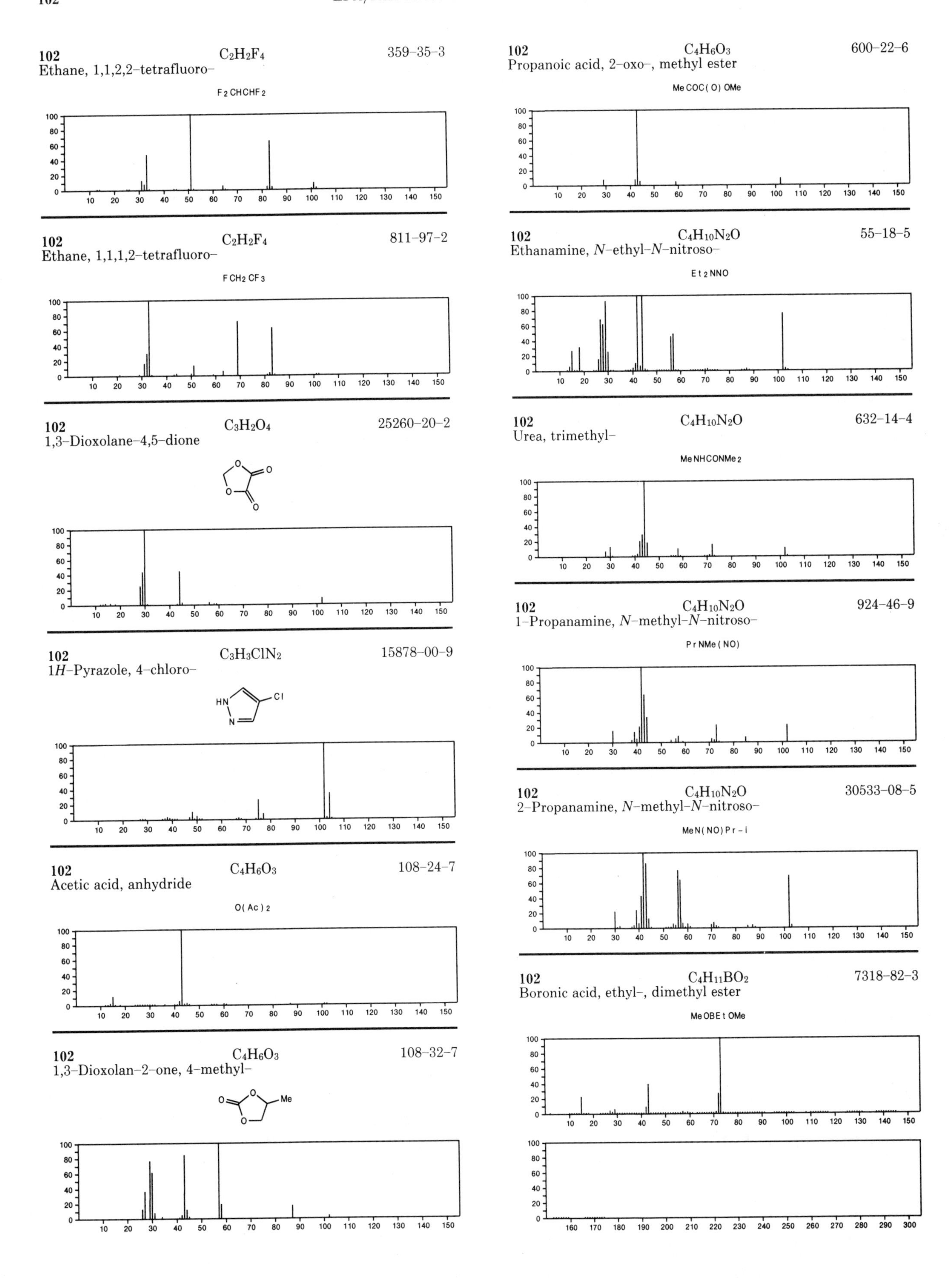

102 $C_2H_2F_4$ 359-35-3
Ethane, 1,1,2,2-tetrafluoro-
F_2CHCHF_2

102 $C_2H_2F_4$ 811-97-2
Ethane, 1,1,1,2-tetrafluoro-
FCH_2CF_3

102 $C_3H_2O_4$ 25260-20-2
1,3-Dioxolane-4,5-dione

102 $C_3H_3ClN_2$ 15878-00-9
1*H*-Pyrazole, 4-chloro-

102 $C_4H_6O_3$ 108-24-7
Acetic acid, anhydride
$O(Ac)_2$

102 $C_4H_6O_3$ 108-32-7
1,3-Dioxolan-2-one, 4-methyl-

102 $C_4H_6O_3$ 600-22-6
Propanoic acid, 2-oxo-, methyl ester
MeCOC(O)OMe

102 $C_4H_{10}N_2O$ 55-18-5
Ethanamine, *N*-ethyl-*N*-nitroso-
Et_2NNO

102 $C_4H_{10}N_2O$ 632-14-4
Urea, trimethyl-
$MeNHCONMe_2$

102 $C_4H_{10}N_2O$ 924-46-9
1-Propanamine, *N*-methyl-*N*-nitroso-
PrNMe(NO)

102 $C_4H_{10}N_2O$ 30533-08-5
2-Propanamine, *N*-methyl-*N*-nitroso-
MeN(NO)Pr-i

102 $C_4H_{11}BO_2$ 7318-82-3
Boronic acid, ethyl-, dimethyl ester
MeOBEtOMe

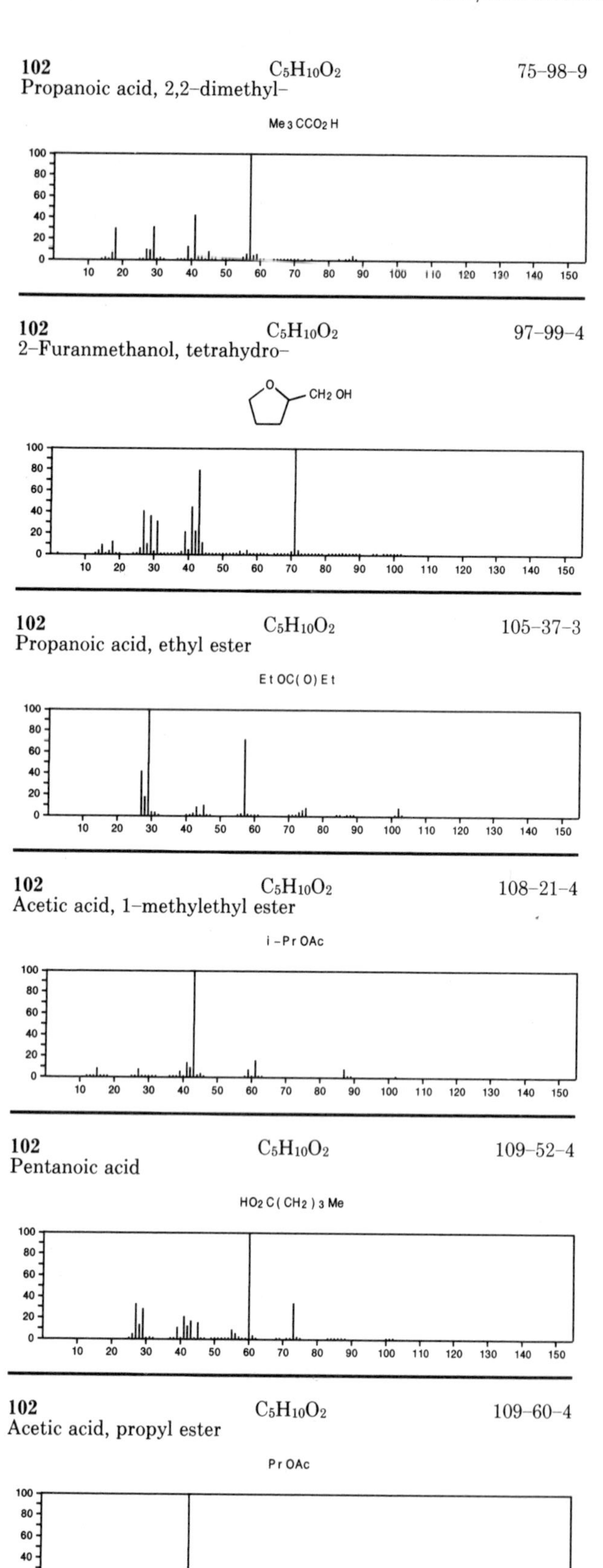

102 $C_5H_{10}O_2$ 75-98-9
Propanoic acid, 2,2-dimethyl-

Me_3CCO_2H

102 $C_5H_{10}O_2$ 97-99-4
2-Furanmethanol, tetrahydro-

CH_2OH

102 $C_5H_{10}O_2$ 105-37-3
Propanoic acid, ethyl ester

EtOC(O)Et

102 $C_5H_{10}O_2$ 108-21-4
Acetic acid, 1-methylethyl ester

i-PrOAc

102 $C_5H_{10}O_2$ 109-52-4
Pentanoic acid

$HO_2C(CH_2)_3Me$

102 $C_5H_{10}O_2$ 109-60-4
Acetic acid, propyl ester

PrOAc

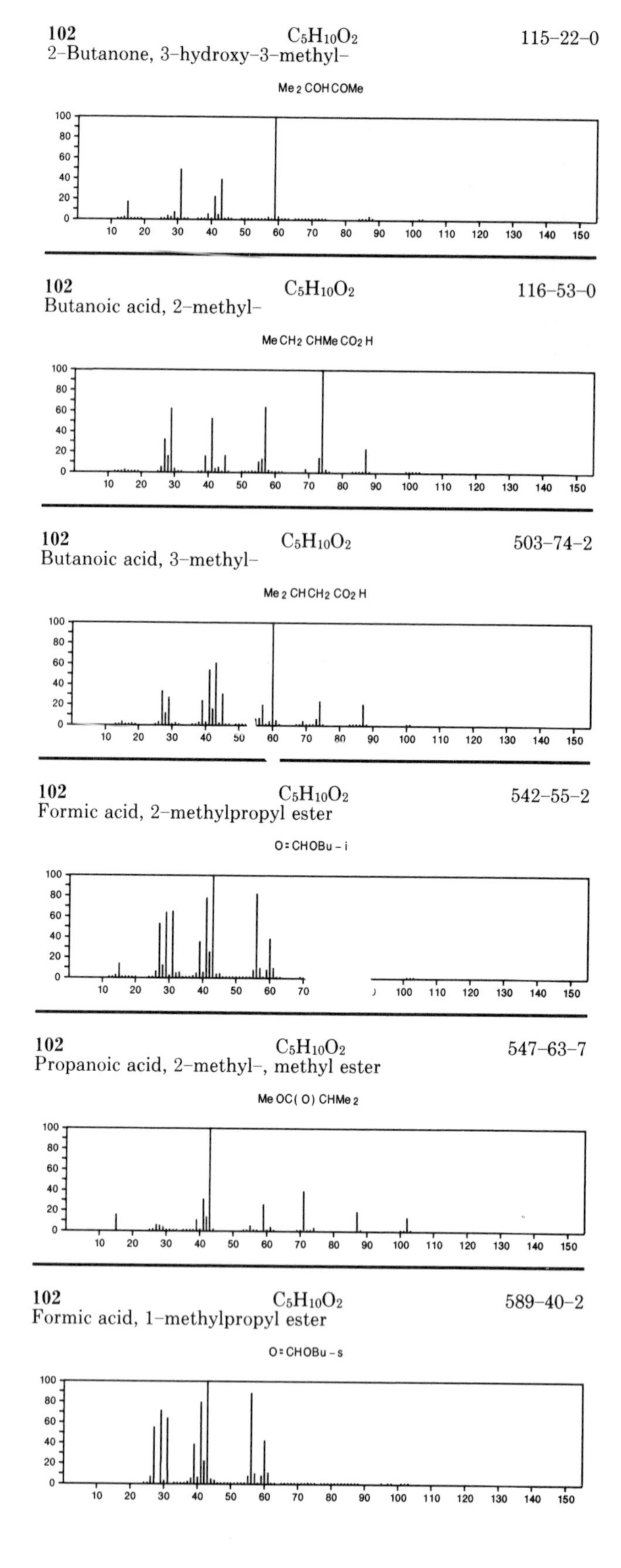

102 $C_5H_{10}O_2$ 115-22-0
2-Butanone, 3-hydroxy-3-methyl-

$Me_2COHCOMe$

102 $C_5H_{10}O_2$ 116-53-0
Butanoic acid, 2-methyl-

$MeCH_2CHMeCO_2H$

102 $C_5H_{10}O_2$ 503-74-2
Butanoic acid, 3-methyl-

$Me_2CHCH_2CO_2H$

102 $C_5H_{10}O_2$ 542-55-2
Formic acid, 2-methylpropyl ester

O=CHOBu-i

102 $C_5H_{10}O_2$ 547-63-7
Propanoic acid, 2-methyl-, methyl ester

$MeOC(O)CHMe_2$

102 $C_5H_{10}O_2$ 589-40-2
Formic acid, 1-methylpropyl ester

O=CHOBu-s

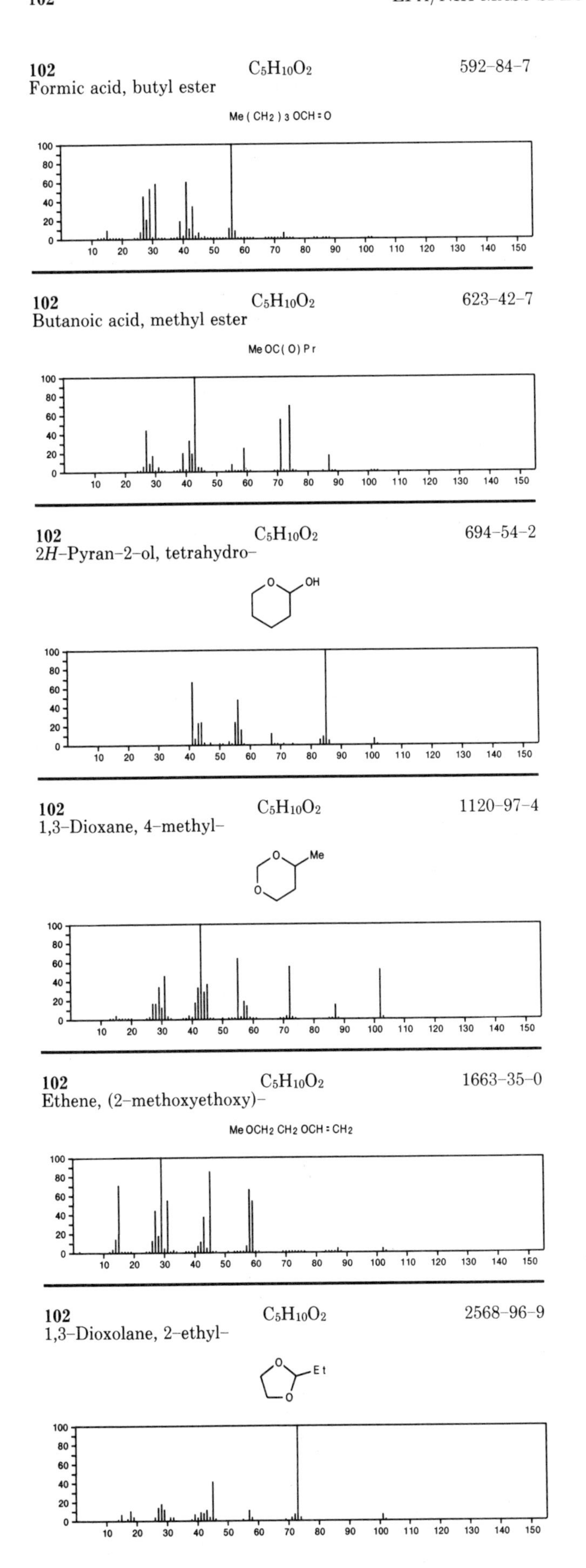
102 $C_5H_{10}O_2$ 592-84-7
Formic acid, butyl ester
Me (CH2) 3 OCH = O
102 $C_5H_{10}O_2$ 623-42-7
Butanoic acid, methyl ester
Me OC(O) Pr
102 $C_5H_{10}O_2$ 694-54-2
2*H*-Pyran-2-ol, tetrahydro-
OH
102 $C_5H_{10}O_2$ 1120-97-4
1,3-Dioxane, 4-methyl-
Me
102 $C_5H_{10}O_2$ 1663-35-0
Ethene, (2-methoxyethoxy)-
Me OCH2 CH2 OCH = CH2
102 $C_5H_{10}O_2$ 2568-96-9
1,3-Dioxolane, 2-ethyl-
Et

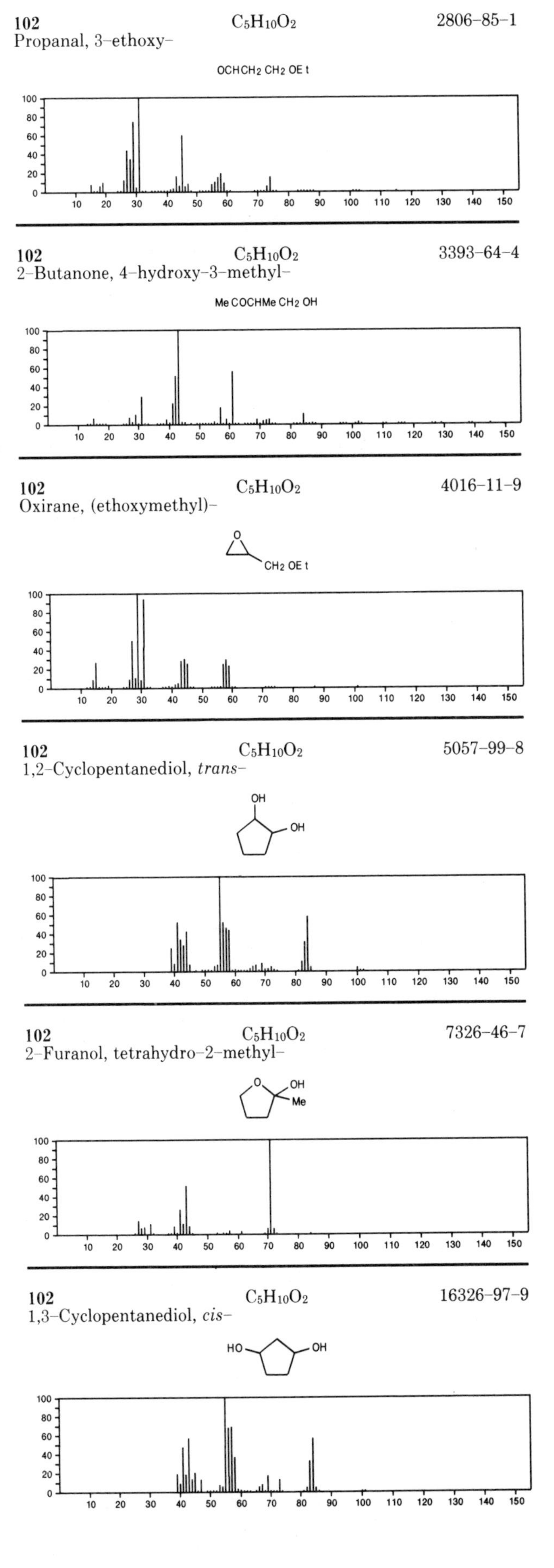
102 $C_5H_{10}O_2$ 2806-85-1
Propanal, 3-ethoxy-
OCHCH2 CH2 OEt
102 $C_5H_{10}O_2$ 3393-64-4
2-Butanone, 4-hydroxy-3-methyl-
Me COCHMe CH2 OH
102 $C_5H_{10}O_2$ 4016-11-9
Oxirane, (ethoxymethyl)-
CH2 OEt
102 $C_5H_{10}O_2$ 5057-99-8
1,2-Cyclopentanediol, *trans*-
OH
OH
102 $C_5H_{10}O_2$ 7326-46-7
2-Furanol, tetrahydro-2-methyl-
OH
Me
102 $C_5H_{10}O_2$ 16326-97-9
1,3-Cyclopentanediol, *cis*-
HO
OH

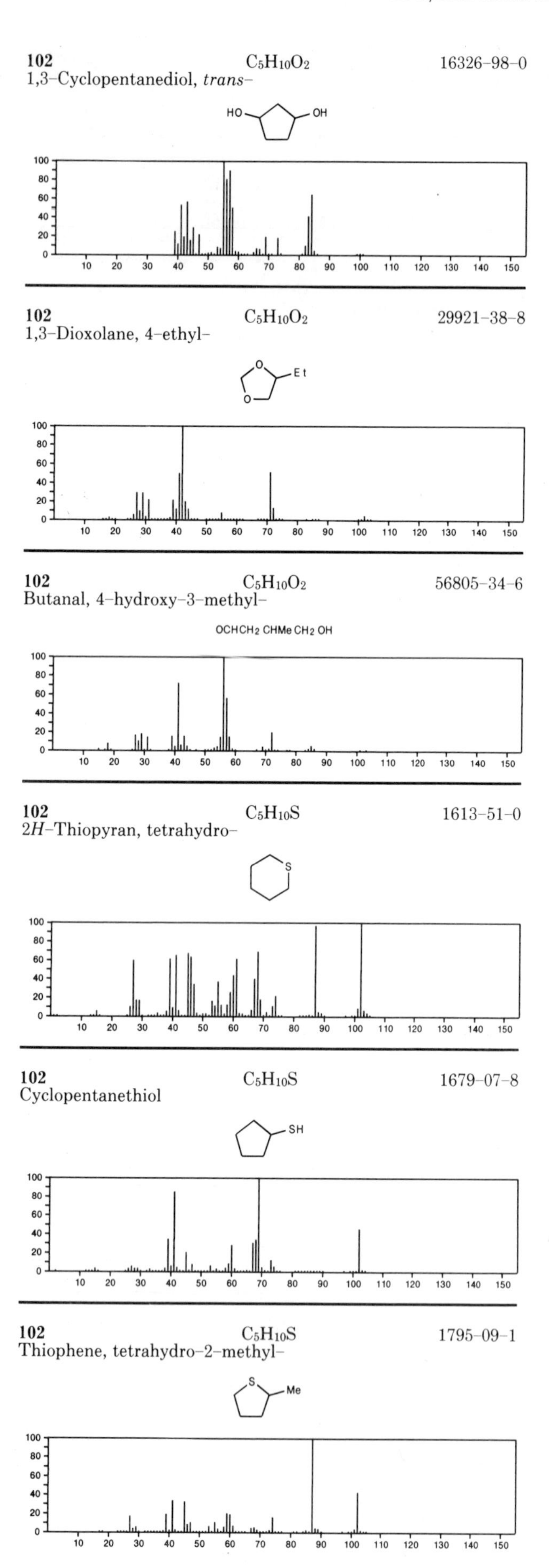

102 $C_5H_{10}O_2$ 16326-98-0
1,3-Cyclopentanediol, *trans*-

102 $C_5H_{10}O_2$ 29921-38-8
1,3-Dioxolane, 4-ethyl-

102 $C_5H_{10}O_2$ 56805-34-6
Butanal, 4-hydroxy-3-methyl-
OCHCH2 CHMe CH2 OH

102 $C_5H_{10}S$ 1613-51-0
2H-Thiopyran, tetrahydro-

102 $C_5H_{10}S$ 1679-07-8
Cyclopentanethiol

102 $C_5H_{10}S$ 1795-09-1
Thiophene, tetrahydro-2-methyl-

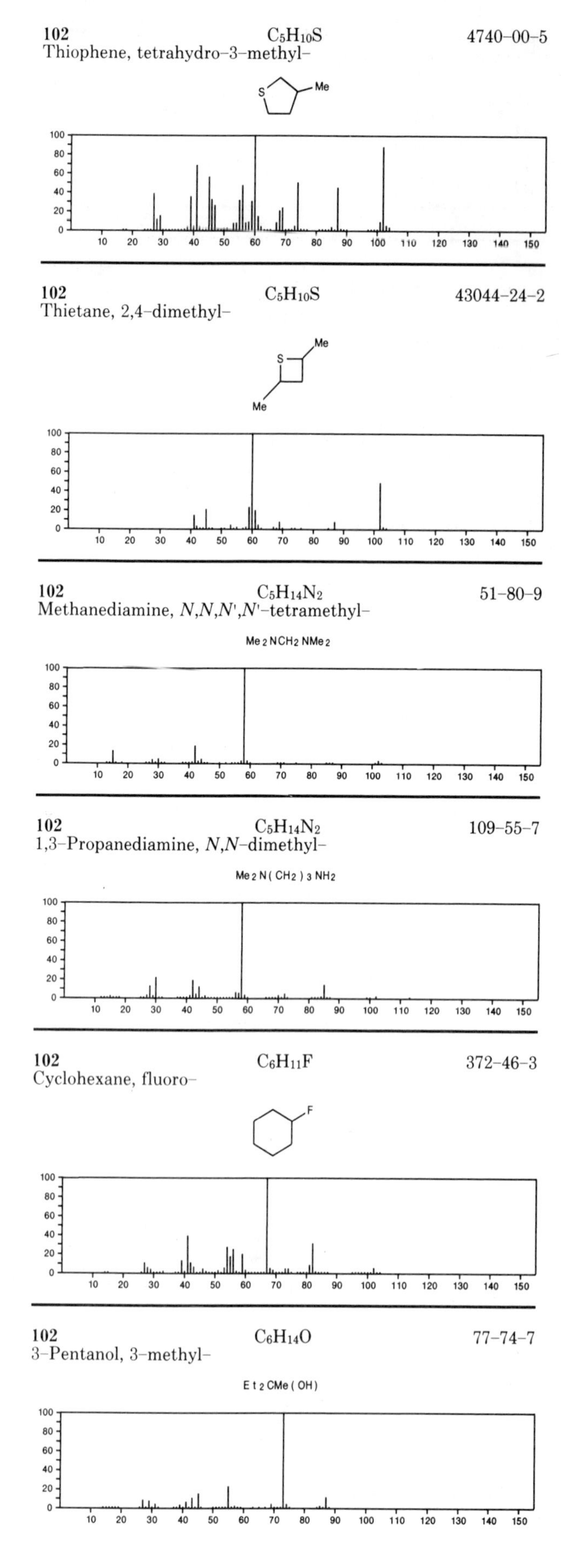

102 $C_5H_{10}S$ 4740-00-5
Thiophene, tetrahydro-3-methyl-

102 $C_5H_{10}S$ 43044-24-2
Thietane, 2,4-dimethyl-

102 $C_5H_{14}N_2$ 51-80-9
Methanediamine, *N,N,N',N'*-tetramethyl-
Me2NCH2NMe2

102 $C_5H_{14}N_2$ 109-55-7
1,3-Propanediamine, *N,N*-dimethyl-
Me2N(CH2)3NH2

102 $C_6H_{11}F$ 372-46-3
Cyclohexane, fluoro-

102 $C_6H_{14}O$ 77-74-7
3-Pentanol, 3-methyl-
Et2CMe(OH)

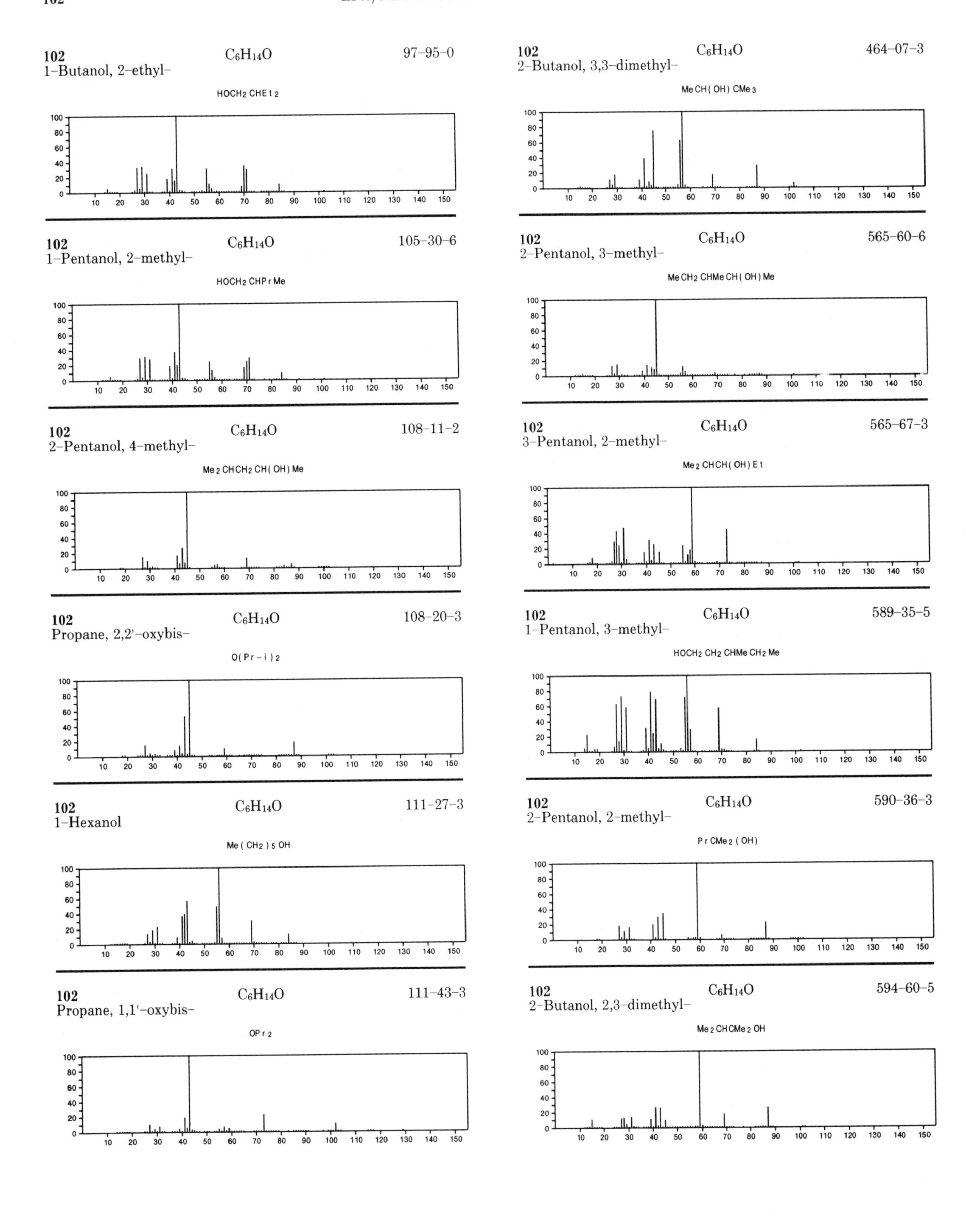
102
C6H14O
97-95-0
1-Butanol, 2-ethyl-
HOCH2 CHEt2
102
C6H14O
105-30-6
1-Pentanol, 2-methyl-
HOCH2 CHPrMe
102
C6H14O
108-11-2
2-Pentanol, 4-methyl-
Me2 CHCH2 CH(OH)Me
102
C6H14O
108-20-3
Propane, 2,2'-oxybis-
O(Pr-i)2
102
C6H14O
111-27-3
1-Hexanol
Me(CH2)5 OH
102
C6H14O
111-43-3
Propane, 1,1'-oxybis-
OPr2
102
C6H14O
464-07-3
2-Butanol, 3,3-dimethyl-
MeCH(OH)CMe3
102
C6H14O
565-60-6
2-Pentanol, 3-methyl-
MeCH2 CHMeCH(OH)Me
102
C6H14O
565-67-3
3-Pentanol, 2-methyl-
Me2 CHCH(OH)Et
102
C6H14O
589-35-5
1-Pentanol, 3-methyl-
HOCH2 CH2 CHMeCH2 Me
102
C6H14O
590-36-3
2-Pentanol, 2-methyl-
PrCMe2(OH)
102
C6H14O
594-60-5
2-Butanol, 2,3-dimethyl-
Me2 CHCMe2 OH

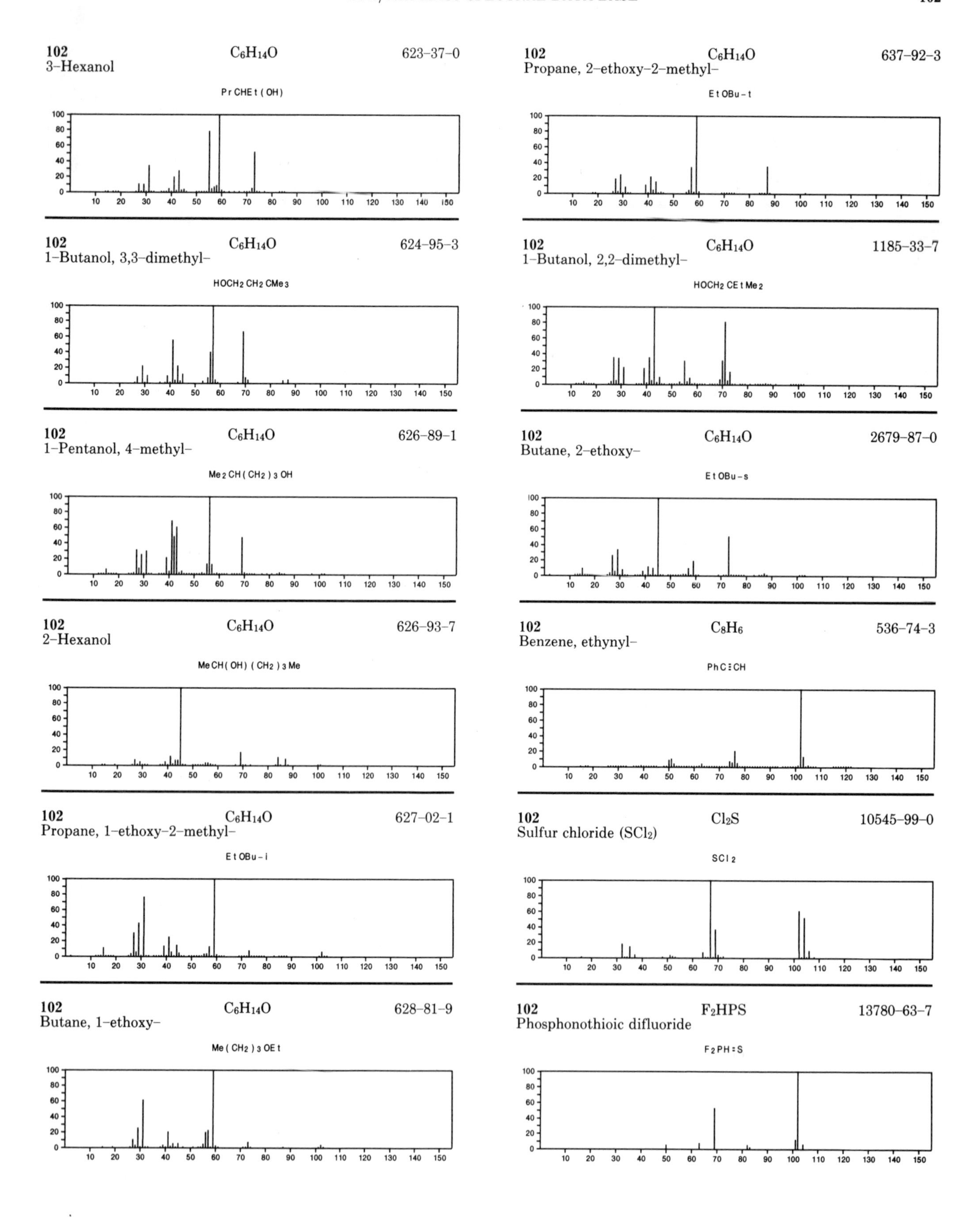
102 C6H14O 623-37-0
3-Hexanol
PrCHEt(OH)
102 C6H14O 637-92-3
Propane, 2-ethoxy-2-methyl-
EtOBu-t
102 C6H14O 624-95-3
1-Butanol, 3,3-dimethyl-
HOCH2CH2CMe3
102 C6H14O 1185-33-7
1-Butanol, 2,2-dimethyl-
HOCH2CEtMe2
102 C6H14O 626-89-1
1-Pentanol, 4-methyl-
Me2CH(CH2)3OH
102 C6H14O 2679-87-0
Butane, 2-ethoxy-
EtOBu-s
102 C6H14O 626-93-7
2-Hexanol
MeCH(OH)(CH2)3Me
102 C8H6 536-74-3
Benzene, ethynyl-
PhC≡CH
102 C6H14O 627-02-1
Propane, 1-ethoxy-2-methyl-
EtOBu-i
102 Cl2S 10545-99-0
Sulfur chloride (SCl2)
SCl2
102 C6H14O 628-81-9
Butane, 1-ethoxy-
Me(CH2)3OEt
102 F2HPS 13780-63-7
Phosphonothioic difluoride
F2PH=S

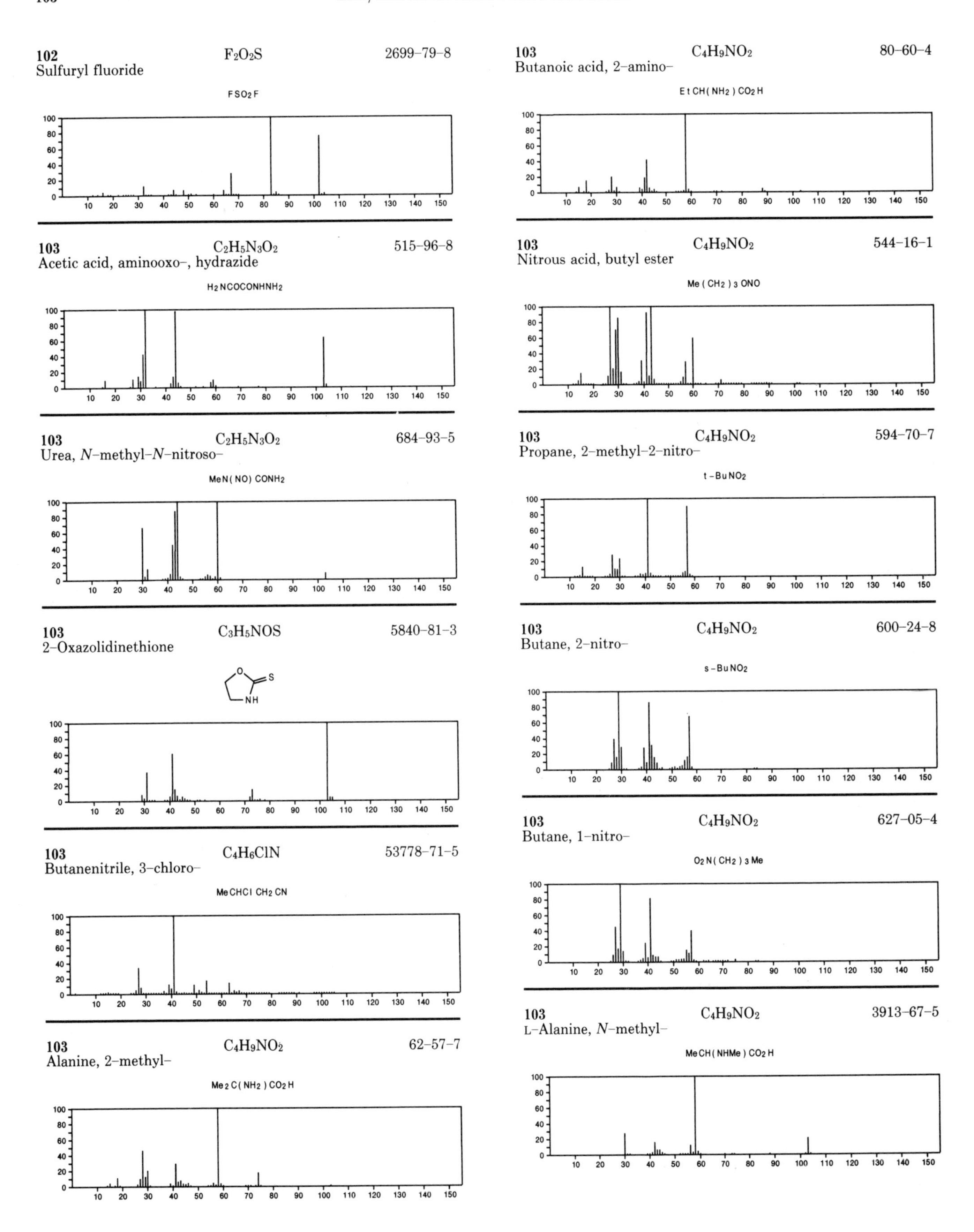
102 F_2O_2S 2699–79–8
Sulfuryl fluoride
FSO_2F
103 $C_2H_5N_3O_2$ 515–96–8
Acetic acid, aminooxo–, hydrazide
$H_2NCOCONHNH_2$
103 $C_2H_5N_3O_2$ 684–93–5
Urea, N–methyl–N–nitroso–
$MeN(NO)CONH_2$
103 C_3H_5NOS 5840–81–3
2–Oxazolidinethione
103 C_4H_6ClN 53778–71–5
Butanenitrile, 3–chloro–
$MeCHClCH_2CN$
103 $C_4H_9NO_2$ 62–57–7
Alanine, 2–methyl–
$Me_2C(NH_2)CO_2H$
103 $C_4H_9NO_2$ 80–60–4
Butanoic acid, 2–amino–
$EtCH(NH_2)CO_2H$
103 $C_4H_9NO_2$ 544–16–1
Nitrous acid, butyl ester
$Me(CH_2)_3ONO$
103 $C_4H_9NO_2$ 594–70–7
Propane, 2–methyl–2–nitro–
$t-BuNO_2$
103 $C_4H_9NO_2$ 600–24–8
Butane, 2–nitro–
$s-BuNO_2$
103 $C_4H_9NO_2$ 627–05–4
Butane, 1–nitro–
$O_2N(CH_2)_3Me$
103 $C_4H_9NO_2$ 3913–67–5
L–Alanine, N–methyl–
$MeCH(NHMe)CO_2H$

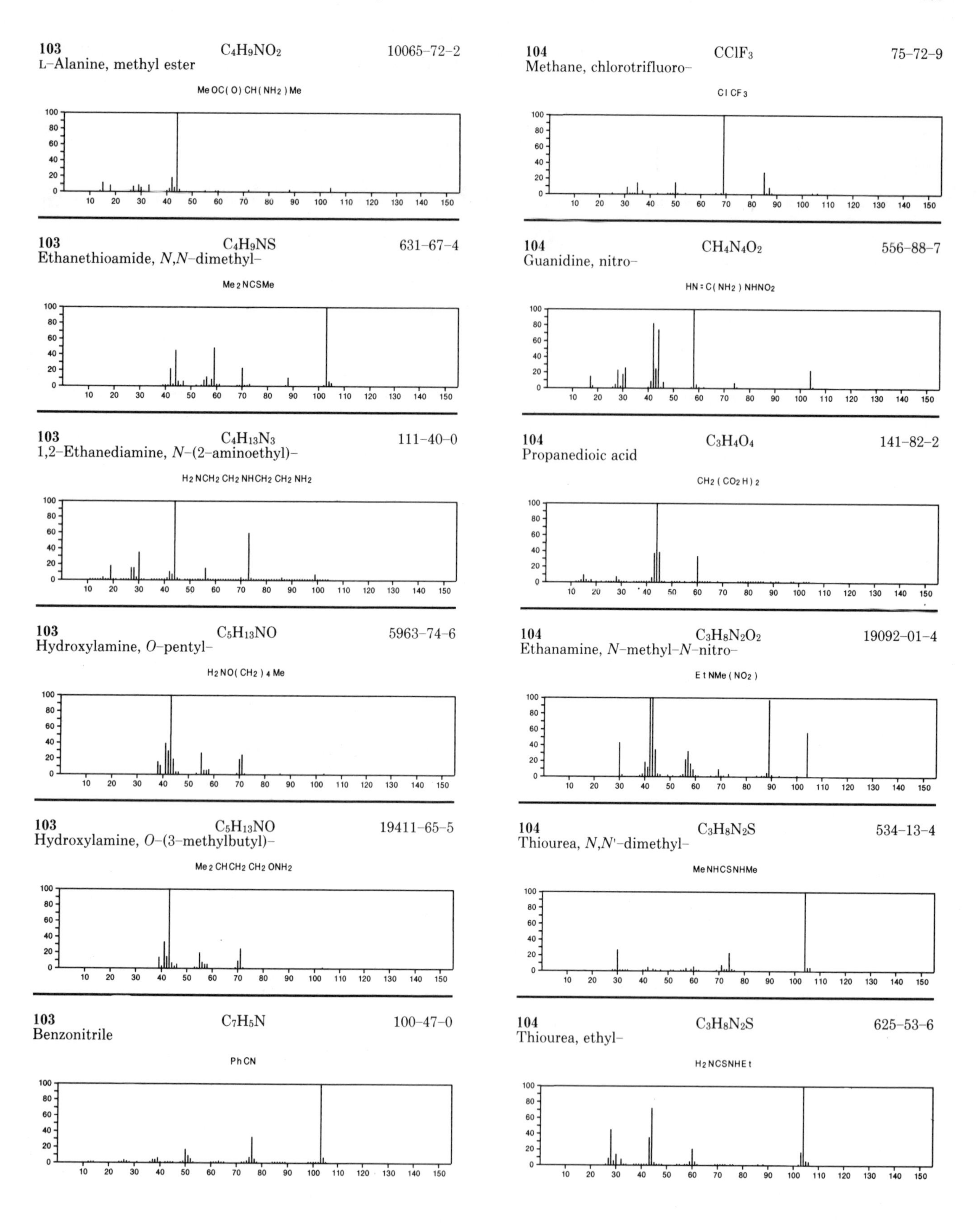

103 C4H9NO2 10065-72-2
L-Alanine, methyl ester
MeOC(O)CH(NH2)Me
104 CClF3 75-72-9
Methane, chlorotrifluoro-
ClCF3
103 C4H9NS 631-67-4
Ethanethioamide, N,N-dimethyl-
Me2NCSMe
104 CH4N4O2 556-88-7
Guanidine, nitro-
HN=C(NH2)NHNO2
103 C4H13N3 111-40-0
1,2-Ethanediamine, N-(2-aminoethyl)-
H2NCH2CH2NHCH2CH2NH2
104 C3H4O4 141-82-2
Propanedioic acid
CH2(CO2H)2
103 C5H13NO 5963-74-6
Hydroxylamine, O-pentyl-
H2NO(CH2)4Me
104 C3H8N2O2 19092-01-4
Ethanamine, N-methyl-N-nitro-
EtNMe(NO2)
103 C5H13NO 19411-65-5
Hydroxylamine, O-(3-methylbutyl)-
Me2CHCH2CH2ONH2
104 C3H8N2S 534-13-4
Thiourea, N,N'-dimethyl-
MeNHCSNHMe
103 C7H5N 100-47-0
Benzonitrile
PhCN
104 C3H8N2S 625-53-6
Thiourea, ethyl-
H2NCSNHEt

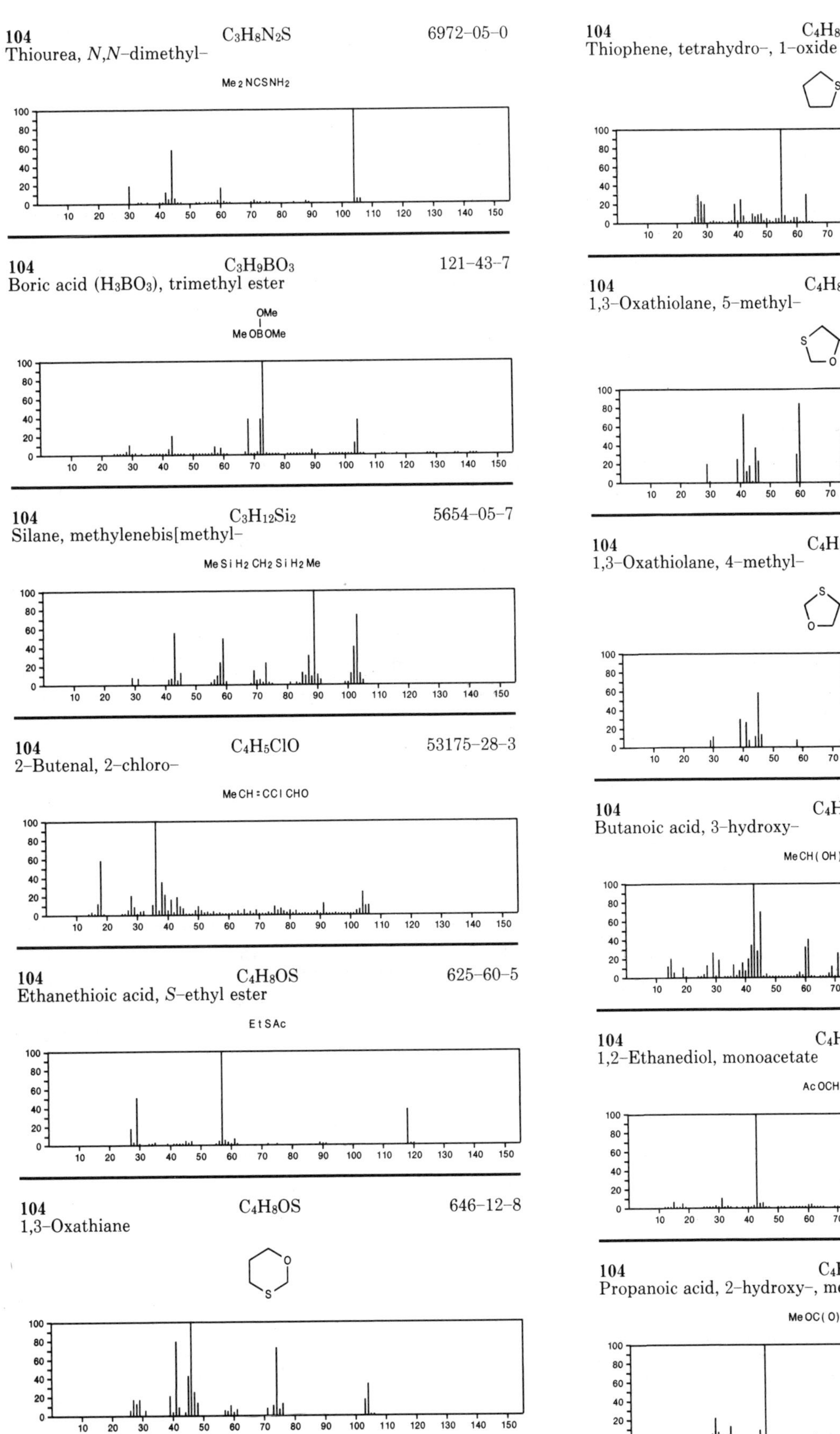

104 $C_3H_8N_2S$ 6972-05-0
Thiourea, *N,N*-dimethyl-

Me_2NCSNH_2

104 $C_3H_9BO_3$ 121-43-7
Boric acid (H_3BO_3), trimethyl ester

MeOB(OMe)OMe

104 $C_3H_{12}Si_2$ 5654-05-7
Silane, methylenebis[methyl-

$MeSiH_2CH_2SiH_2Me$

104 C_4H_5ClO 53175-28-3
2-Butenal, 2-chloro-

MeCH=CClCHO

104 C_4H_8OS 625-60-5
Ethanethioic acid, *S*-ethyl ester

EtSAc

104 C_4H_8OS 646-12-8
1,3-Oxathiane

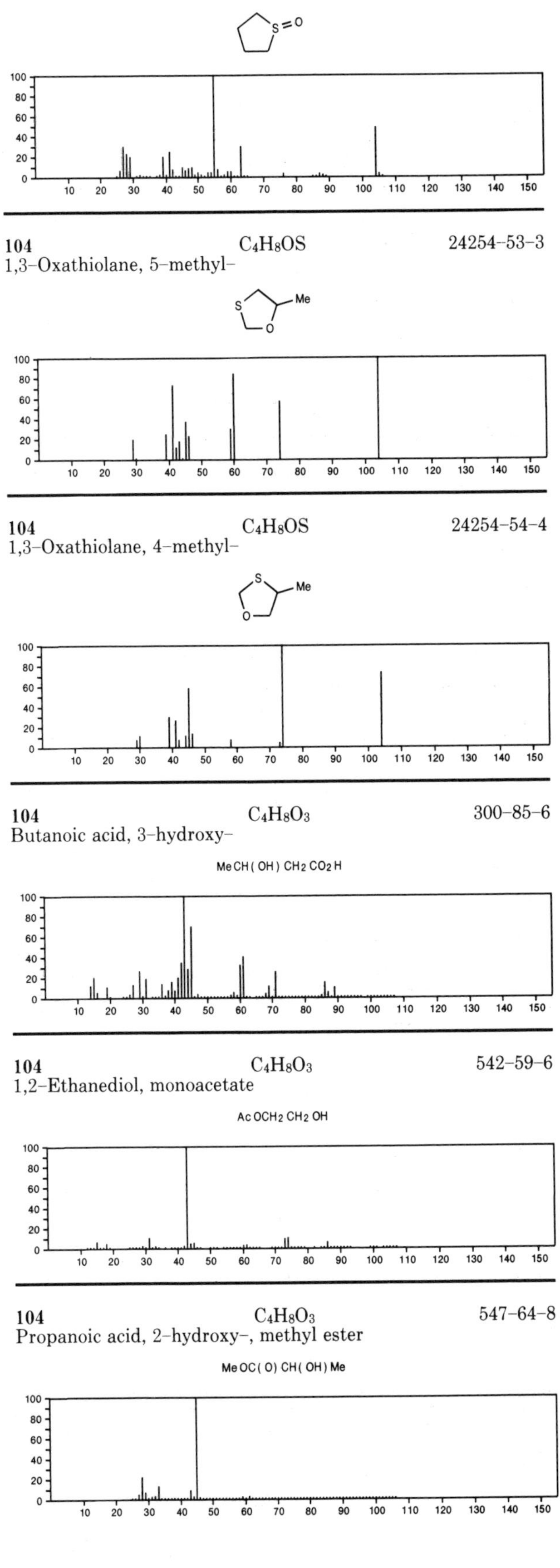

104 C_4H_8OS 1600-44-8
Thiophene, tetrahydro-, 1-oxide

104 C_4H_8OS 24254-53-3
1,3-Oxathiolane, 5-methyl-

104 C_4H_8OS 24254-54-4
1,3-Oxathiolane, 4-methyl-

104 $C_4H_8O_3$ 300-85-6
Butanoic acid, 3-hydroxy-

$MeCH(OH)CH_2CO_2H$

104 $C_4H_8O_3$ 542-59-6
1,2-Ethanediol, monoacetate

$AcOCH_2CH_2OH$

104 $C_4H_8O_3$ 547-64-8
Propanoic acid, 2-hydroxy-, methyl ester

MeOC(O)CH(OH)Me

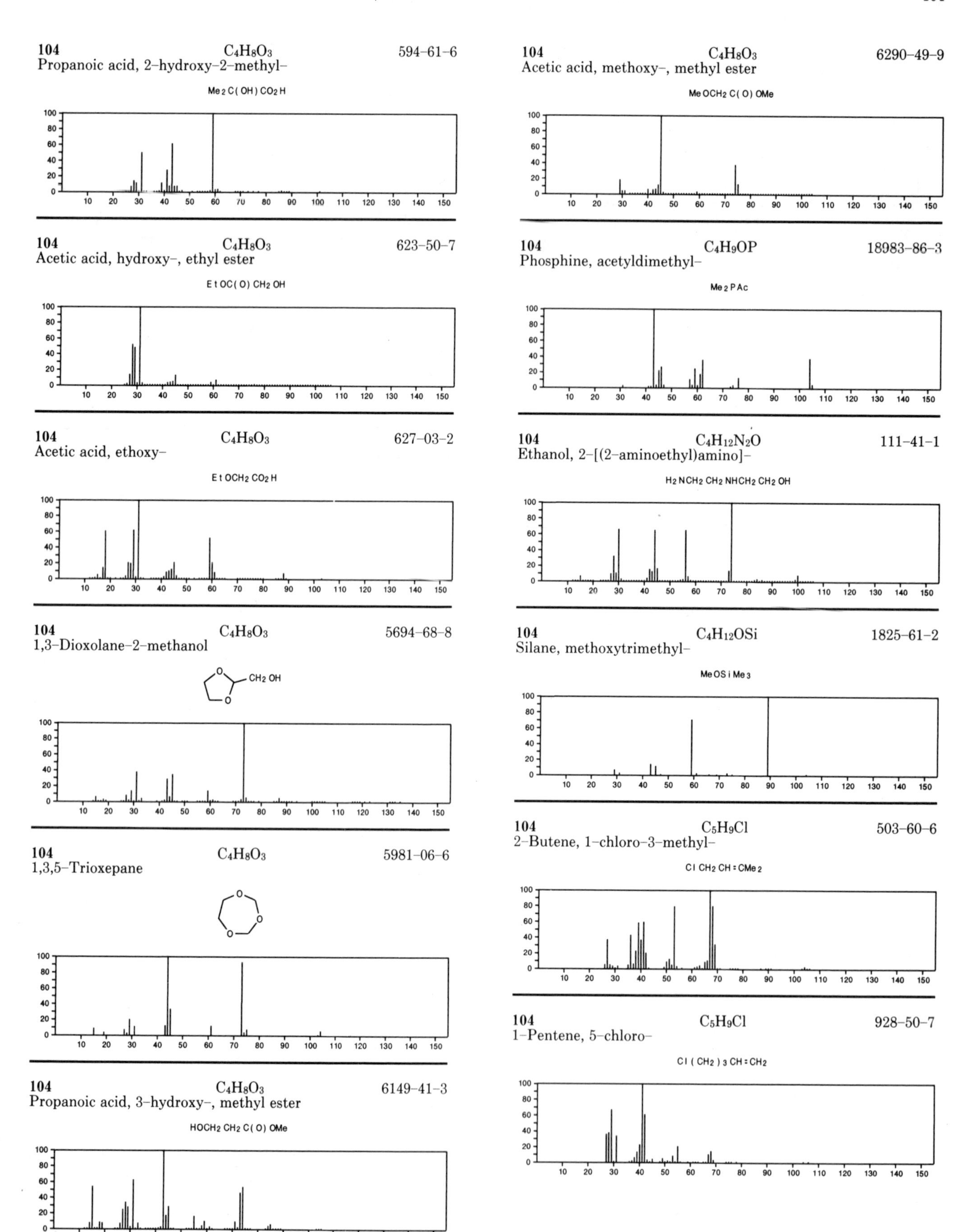
104 C4H8O3 594-61-6
Propanoic acid, 2-hydroxy-2-methyl-
Me2C(OH)CO2H
104 C4H8O3 623-50-7
Acetic acid, hydroxy-, ethyl ester
EtOC(O)CH2OH
104 C4H8O3 627-03-2
Acetic acid, ethoxy-
EtOCH2CO2H
104 C4H8O3 5694-68-8
1,3-Dioxolane-2-methanol
CH2OH
104 C4H8O3 5981-06-6
1,3,5-Trioxepane
104 C4H8O3 6149-41-3
Propanoic acid, 3-hydroxy-, methyl ester
HOCH2CH2C(O)OMe
104 C4H8O3 6290-49-9
Acetic acid, methoxy-, methyl ester
MeOCH2C(O)OMe
104 C4H9OP 18983-86-3
Phosphine, acetyldimethyl-
Me2PAc
104 C4H12N2O 111-41-1
Ethanol, 2-[(2-aminoethyl)amino]-
H2NCH2CH2NHCH2CH2OH
104 C4H12OSi 1825-61-2
Silane, methoxytrimethyl-
MeOSiMe3
104 C5H9Cl 503-60-6
2-Butene, 1-chloro-3-methyl-
ClCH2CH=CMe2
104 C5H9Cl 928-50-7
1-Pentene, 5-chloro-
Cl(CH2)3CH=CH2

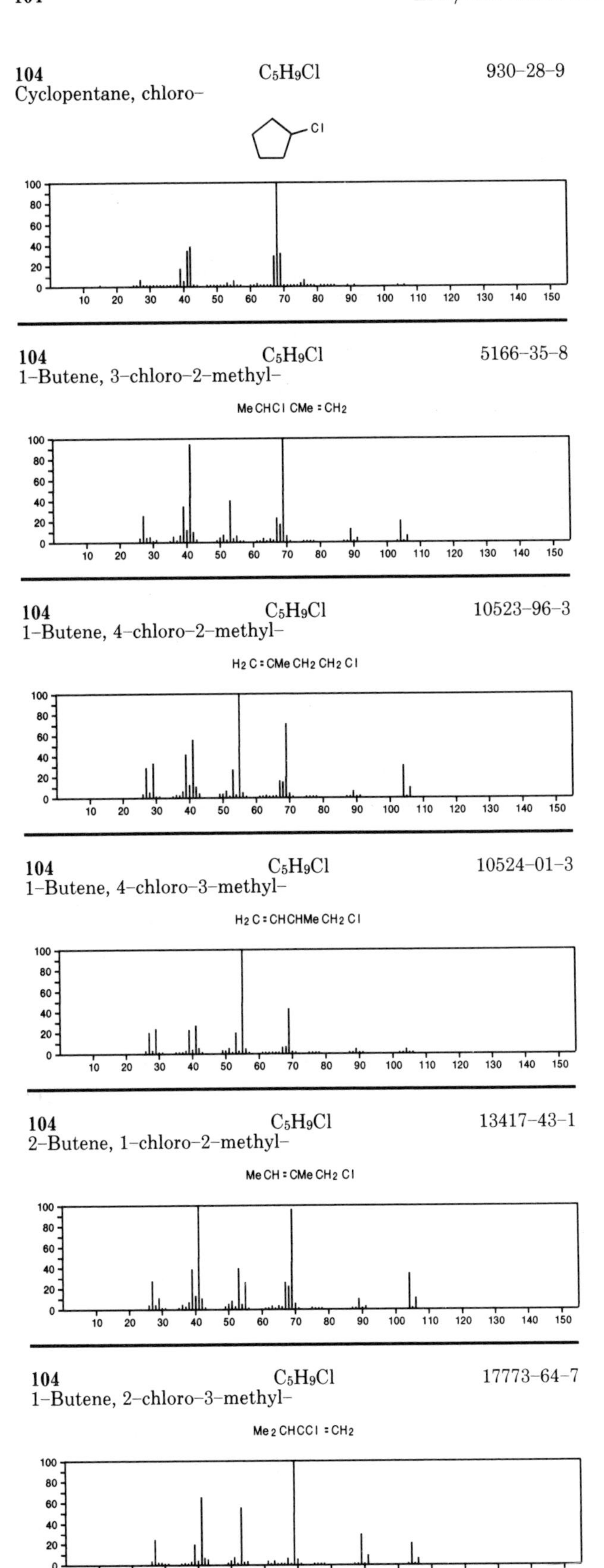

104 C_5H_9Cl 930-28-9
Cyclopentane, chloro-

104 C_5H_9Cl 5166-35-8
1-Butene, 3-chloro-2-methyl-
MeCHCl CMe = CH_2

104 C_5H_9Cl 10523-96-3
1-Butene, 4-chloro-2-methyl-
H_2C = CMe CH_2 CH_2 Cl

104 C_5H_9Cl 10524-01-3
1-Butene, 4-chloro-3-methyl-
H_2C = CHCHMe CH_2 Cl

104 C_5H_9Cl 13417-43-1
2-Butene, 1-chloro-2-methyl-
MeCH = CMe CH_2 Cl

104 C_5H_9Cl 17773-64-7
1-Butene, 2-chloro-3-methyl-
Me_2 CHCCl = CH_2

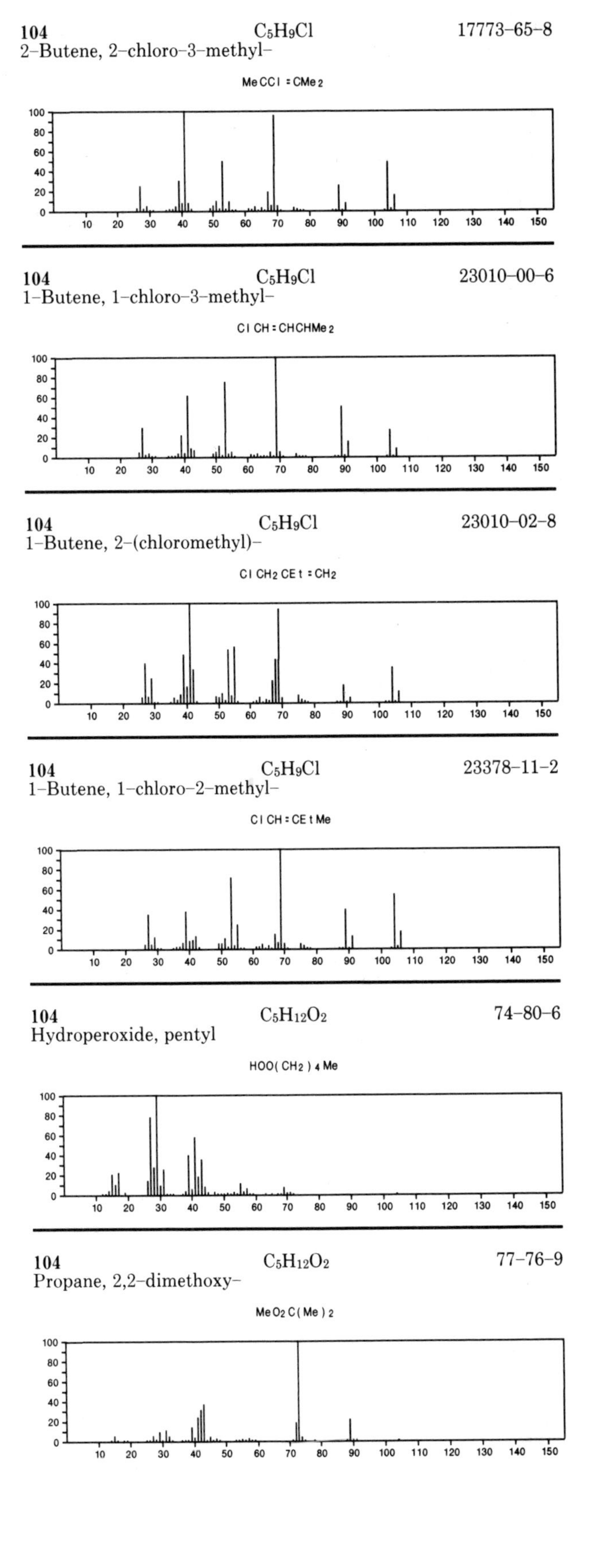

104 C_5H_9Cl 17773-65-8
2-Butene, 2-chloro-3-methyl-
MeCCl = CMe_2

104 C_5H_9Cl 23010-00-6
1-Butene, 1-chloro-3-methyl-
Cl CH = $CHCHMe_2$

104 C_5H_9Cl 23010-02-8
1-Butene, 2-(chloromethyl)-
Cl CH_2 CEt = CH_2

104 C_5H_9Cl 23378-11-2
1-Butene, 1-chloro-2-methyl-
Cl CH = CEt Me

104 $C_5H_{12}O_2$ 74-80-6
Hydroperoxide, pentyl
HOO(CH_2)$_4$ Me

104 $C_5H_{12}O_2$ 77-76-9
Propane, 2,2-dimethoxy-
$MeO_2C(Me)_2$

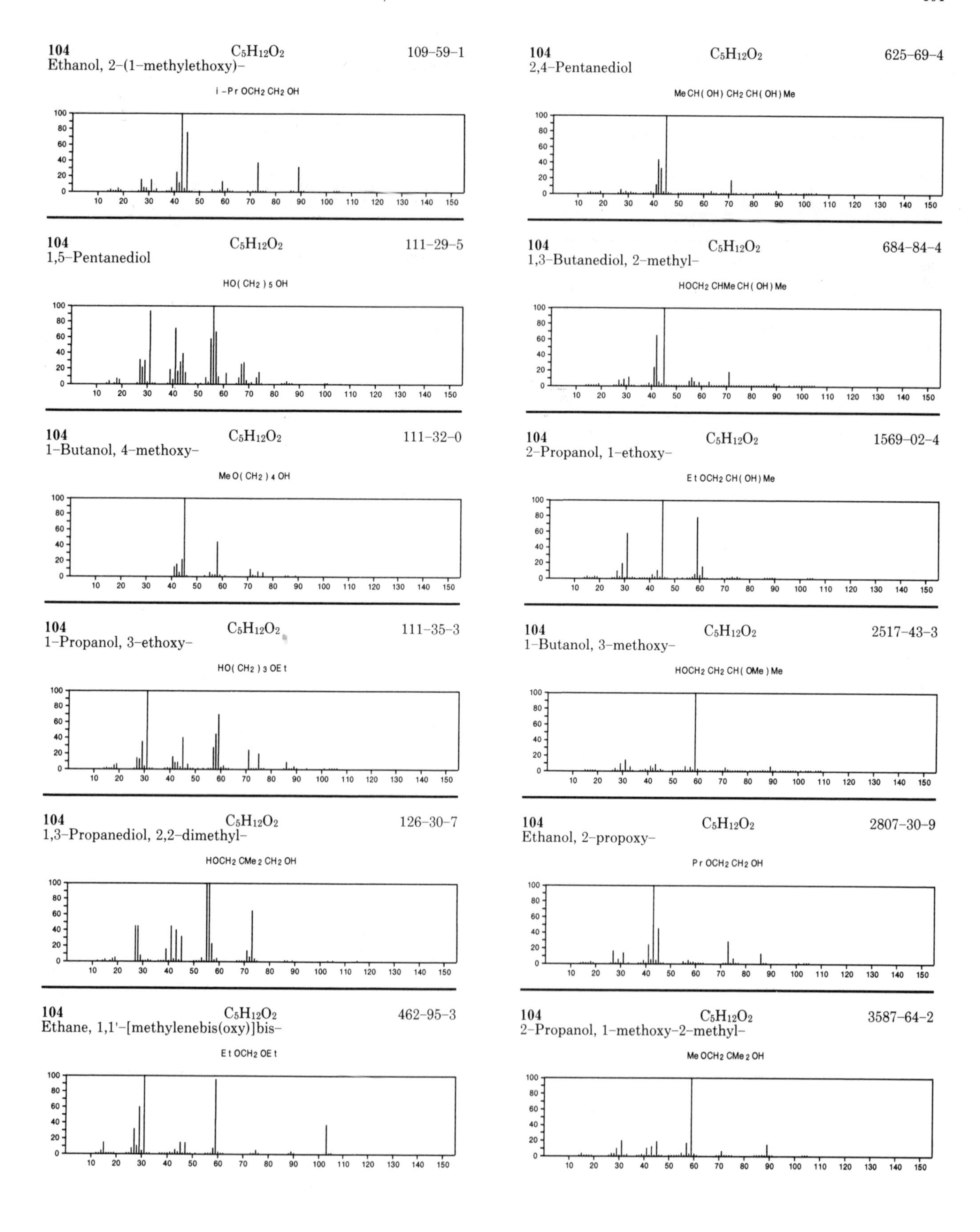
104 C5H12O2 109-59-1
Ethanol, 2-(1-methylethoxy)-
i-Pr OCH2 CH2 OH
104 C5H12O2 625-69-4
2,4-Pentanediol
Me CH(OH) CH2 CH(OH) Me
104 C5H12O2 111-29-5
1,5-Pentanediol
HO(CH2)5 OH
104 C5H12O2 684-84-4
1,3-Butanediol, 2-methyl-
HOCH2 CHMe CH(OH) Me
104 C5H12O2 111-32-0
1-Butanol, 4-methoxy-
Me O(CH2)4 OH
104 C5H12O2 1569-02-4
2-Propanol, 1-ethoxy-
Et OCH2 CH(OH) Me
104 C5H12O2 111-35-3
1-Propanol, 3-ethoxy-
HO(CH2)3 OEt
104 C5H12O2 2517-43-3
1-Butanol, 3-methoxy-
HOCH2 CH2 CH(OMe) Me
104 C5H12O2 126-30-7
1,3-Propanediol, 2,2-dimethyl-
HOCH2 CMe2 CH2 OH
104 C5H12O2 2807-30-9
Ethanol, 2-propoxy-
Pr OCH2 CH2 OH
104 C5H12O2 462-95-3
Ethane, 1,1'-[methylenebis(oxy)]bis-
Et OCH2 OEt
104 C5H12O2 3587-64-2
2-Propanol, 1-methoxy-2-methyl-
Me OCH2 CMe2 OH

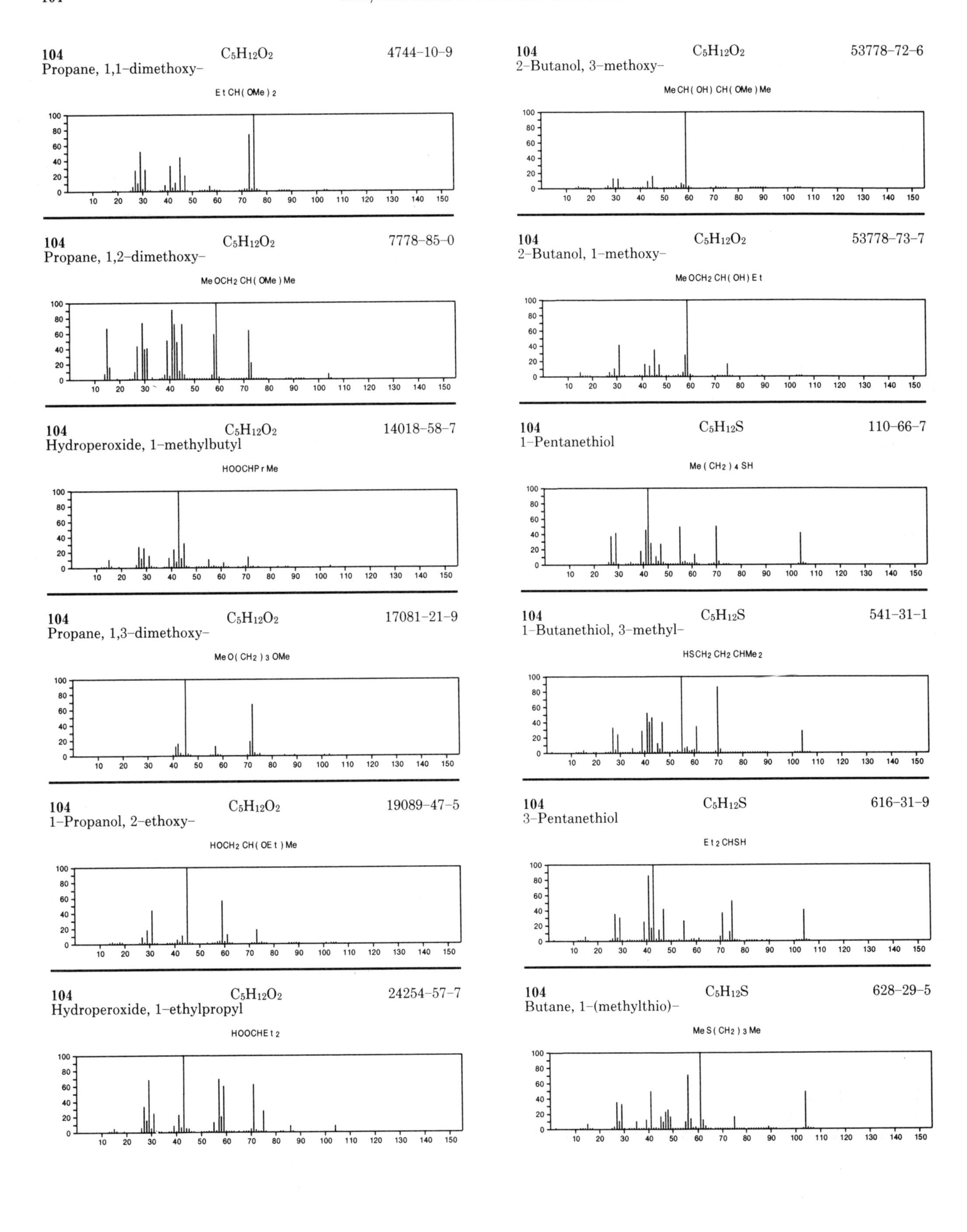
104 $C_5H_{12}O_2$ 4744-10-9
Propane, 1,1-dimethoxy-
Et CH(OMe)2
104 $C_5H_{12}O_2$ 7778-85-0
Propane, 1,2-dimethoxy-
MeOCH2 CH(OMe)Me
104 $C_5H_{12}O_2$ 14018-58-7
Hydroperoxide, 1-methylbutyl
HOOCHPrMe
104 $C_5H_{12}O_2$ 17081-21-9
Propane, 1,3-dimethoxy-
MeO(CH2)3 OMe
104 $C_5H_{12}O_2$ 19089-47-5
1-Propanol, 2-ethoxy-
HOCH2 CH(OEt)Me
104 $C_5H_{12}O_2$ 24254-57-7
Hydroperoxide, 1-ethylpropyl
HOOCHEt2
104 $C_5H_{12}O_2$ 53778-72-6
2-Butanol, 3-methoxy-
MeCH(OH)CH(OMe)Me
104 $C_5H_{12}O_2$ 53778-73-7
2-Butanol, 1-methoxy-
MeOCH2 CH(OH)Et
104 $C_5H_{12}S$ 110-66-7
1-Pentanethiol
Me(CH2)4 SH
104 $C_5H_{12}S$ 541-31-1
1-Butanethiol, 3-methyl-
HSCH2 CH2 CHMe2
104 $C_5H_{12}S$ 616-31-9
3-Pentanethiol
Et2 CHSH
104 $C_5H_{12}S$ 628-29-5
Butane, 1-(methylthio)-
MeS(CH2)3 Me

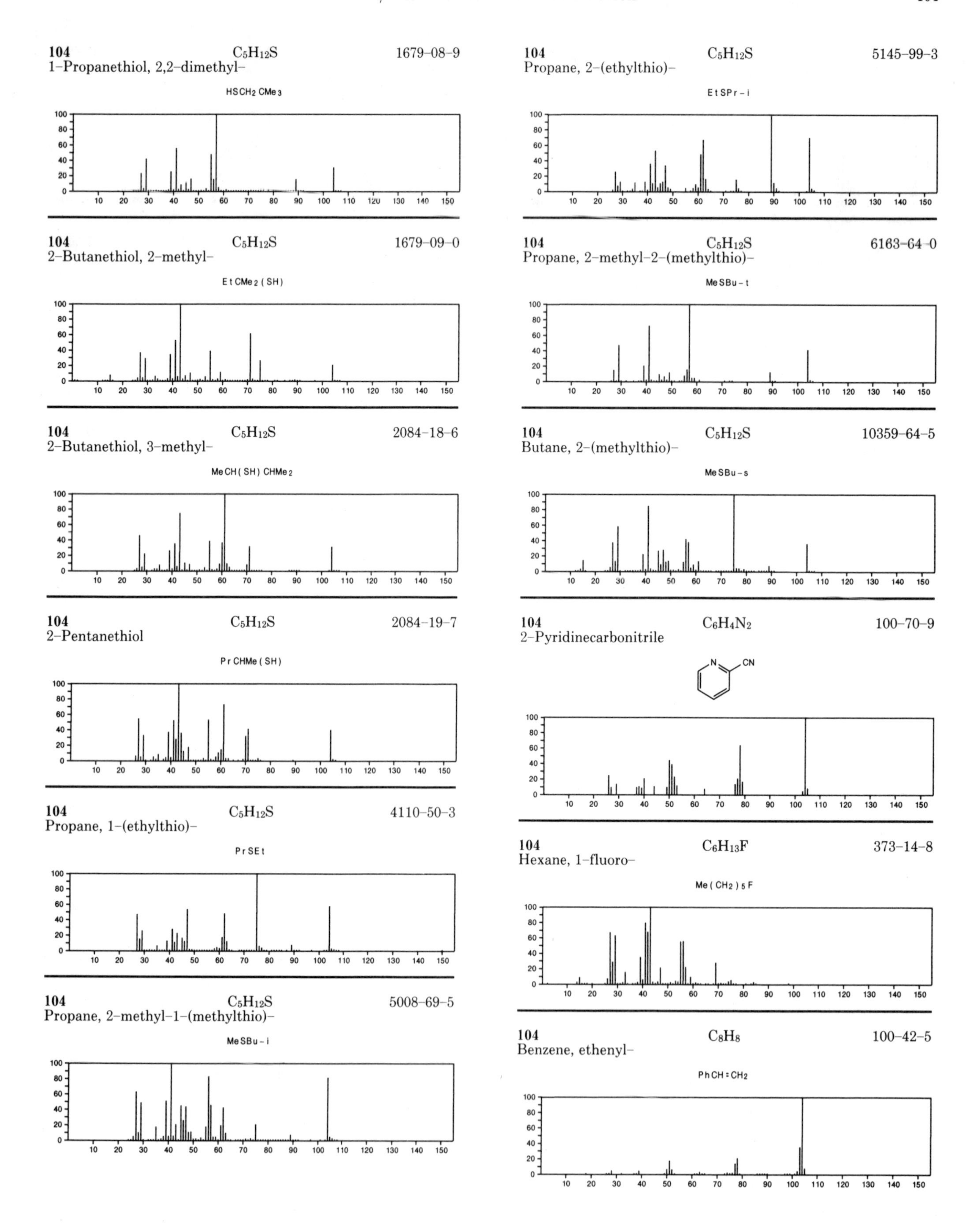
104 C5H12S 1679-08-9
1-Propanethiol, 2,2-dimethyl-
HSCH2 CMe3
104 C5H12S 1679-09-0
2-Butanethiol, 2-methyl-
Et CMe2 (SH)
104 C5H12S 2084-18-6
2-Butanethiol, 3-methyl-
Me CH(SH) CHMe2
104 C5H12S 2084-19-7
2-Pentanethiol
Pr CHMe (SH)
104 C5H12S 4110-50-3
Propane, 1-(ethylthio)-
Pr SEt
104 C5H12S 5008-69-5
Propane, 2-methyl-1-(methylthio)-
Me SBu - i
104 C5H12S 5145-99-3
Propane, 2-(ethylthio)-
Et SPr - i
104 C5H12S 6163-64-0
Propane, 2-methyl-2-(methylthio)-
Me SBu - t
104 C5H12S 10359-64-5
Butane, 2-(methylthio)-
Me SBu - s
104 C6H4N2 100-70-9
2-Pyridinecarbonitrile
N
CN
104 C6H13F 373-14-8
Hexane, 1-fluoro-
Me (CH2) 5 F
104 C8H8 100-42-5
Benzene, ethenyl-
Ph CH = CH2

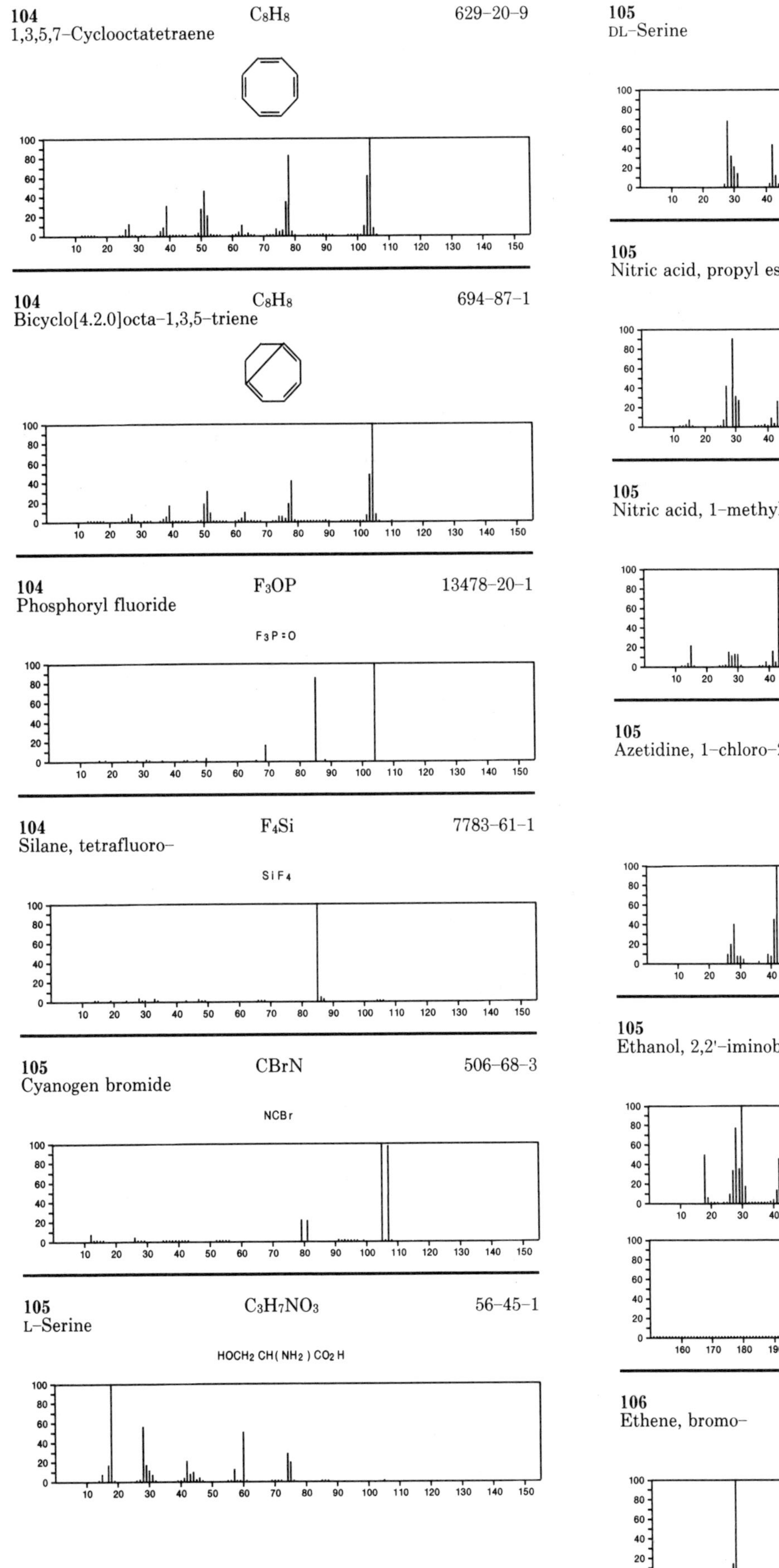

104 C_8H_8 629-20-9
1,3,5,7-Cyclooctatetraene

104 C_8H_8 694-87-1
Bicyclo[4.2.0]octa-1,3,5-triene

104 F_3OP 13478-20-1
Phosphoryl fluoride
F3P=O

104 F_4Si 7783-61-1
Silane, tetrafluoro-
SiF4

105 CBrN 506-68-3
Cyanogen bromide
NCBr

105 $C_3H_7NO_3$ 56-45-1
L-Serine
HOCH2CH(NH2)CO2H

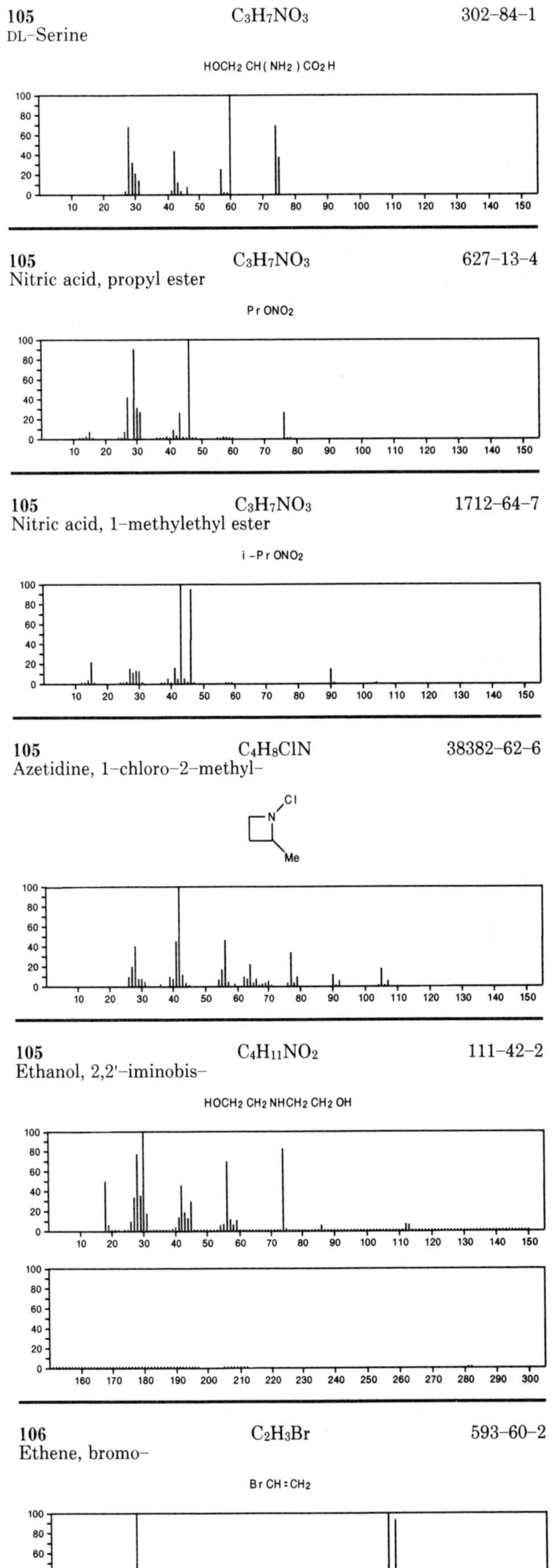

105 $C_3H_7NO_3$ 302-84-1
DL-Serine
HOCH2CH(NH2)CO2H

105 $C_3H_7NO_3$ 627-13-4
Nitric acid, propyl ester
PrONO2

105 $C_3H_7NO_3$ 1712-64-7
Nitric acid, 1-methylethyl ester
i-PrONO2

105 C_4H_8ClN 38382-62-6
Azetidine, 1-chloro-2-methyl-

105 $C_4H_{11}NO_2$ 111-42-2
Ethanol, 2,2'-iminobis-
HOCH2CH2NHCH2CH2OH

106 C_2H_3Br 593-60-2
Ethene, bromo-
BrCH=CH2

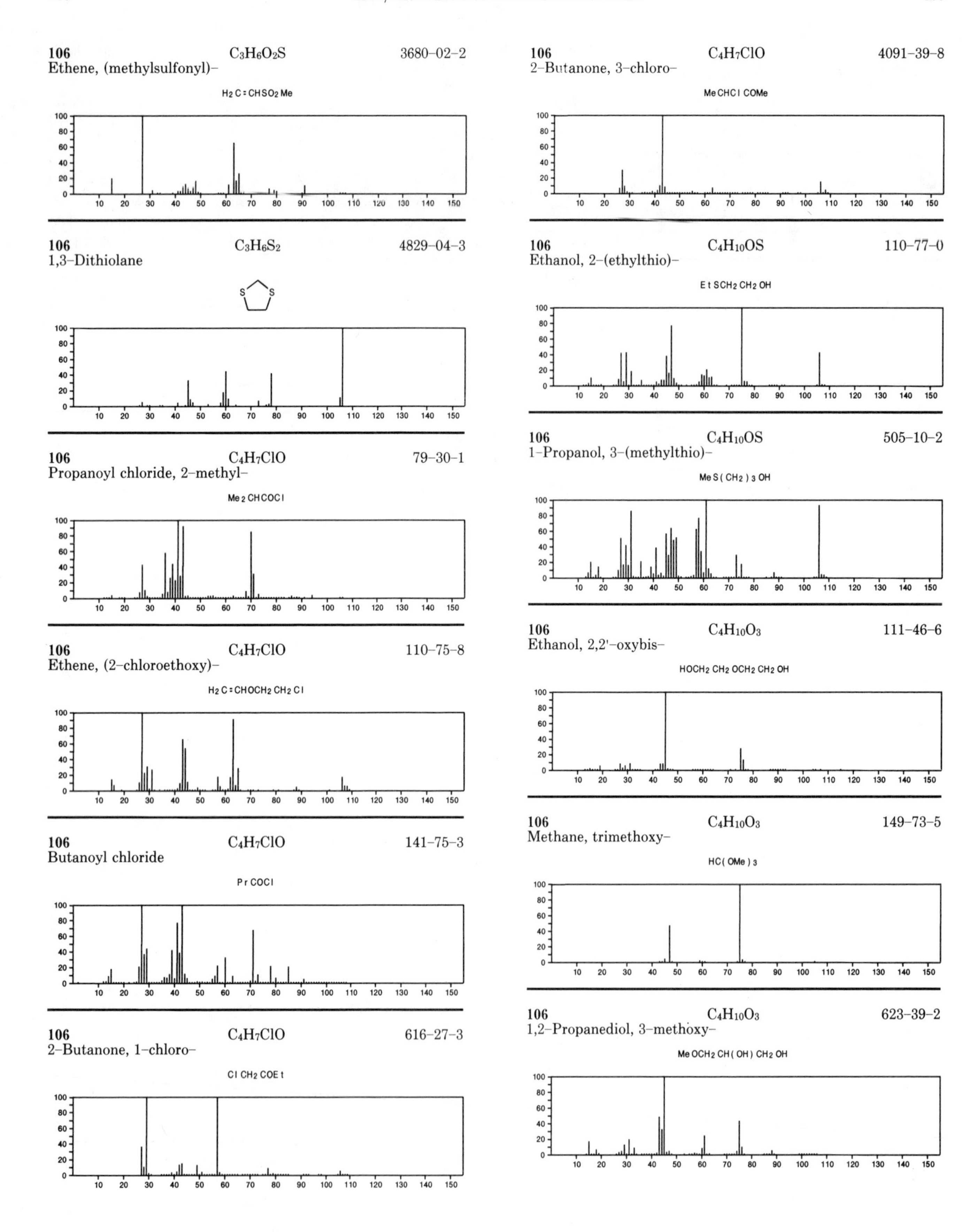

106 $C_3H_6O_2S$ 3680-02-2
Ethene, (methylsulfonyl)-
$H_2C{=}CHSO_2Me$

106 $C_3H_6S_2$ 4829-04-3
1,3-Dithiolane

106 C_4H_7ClO 79-30-1
Propanoyl chloride, 2-methyl-
$Me_2CHCOCl$

106 C_4H_7ClO 110-75-8
Ethene, (2-chloroethoxy)-
$H_2C{=}CHOCH_2CH_2Cl$

106 C_4H_7ClO 141-75-3
Butanoyl chloride
PrCOCl

106 C_4H_7ClO 616-27-3
2-Butanone, 1-chloro-
$ClCH_2COEt$

106 C_4H_7ClO 4091-39-8
2-Butanone, 3-chloro-
MeCHClCOMe

106 $C_4H_{10}OS$ 110-77-0
Ethanol, 2-(ethylthio)-
$EtSCH_2CH_2OH$

106 $C_4H_{10}OS$ 505-10-2
1-Propanol, 3-(methylthio)-
$MeS(CH_2)_3OH$

106 $C_4H_{10}O_3$ 111-46-6
Ethanol, 2,2'-oxybis-
$HOCH_2CH_2OCH_2CH_2OH$

106 $C_4H_{10}O_3$ 149-73-5
Methane, trimethoxy-
$HC(OMe)_3$

106 $C_4H_{10}O_3$ 623-39-2
1,2-Propanediol, 3-methoxy-
$MeOCH_2CH(OH)CH_2OH$

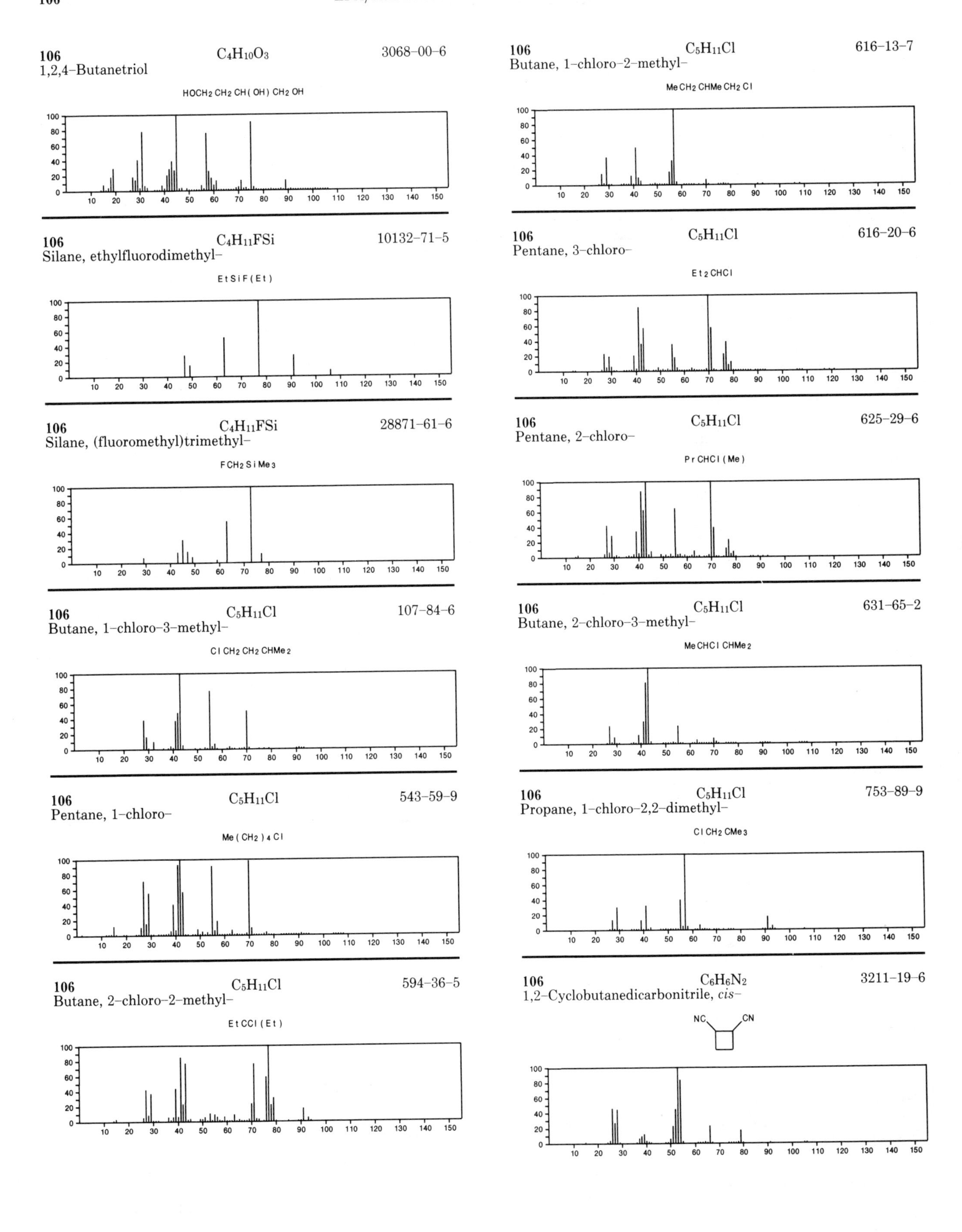
106 C4H10O3 3068–00–6
1,2,4–Butanetriol
HOCH2 CH2 CH(OH) CH2 OH
106 C4H11FSi 10132–71–5
Silane, ethylfluorodimethyl–
Et Si F (Et)
106 C4H11FSi 28871–61–6
Silane, (fluoromethyl)trimethyl–
F CH2 Si Me3
106 C5H11Cl 107–84–6
Butane, 1–chloro–3–methyl–
Cl CH2 CH2 CHMe2
106 C5H11Cl 543–59–9
Pentane, 1–chloro–
Me (CH2)4 Cl
106 C5H11Cl 594–36–5
Butane, 2–chloro–2–methyl–
Et CCl (Et)
106 C5H11Cl 616–13–7
Butane, 1–chloro–2–methyl–
Me CH2 CHMe CH2 Cl
106 C5H11Cl 616–20–6
Pentane, 3–chloro–
Et2 CHCl
106 C5H11Cl 625–29–6
Pentane, 2–chloro–
Pr CHCl (Me)
106 C5H11Cl 631–65–2
Butane, 2–chloro–3–methyl–
Me CHCl CHMe2
106 C5H11Cl 753–89–9
Propane, 1–chloro–2,2–dimethyl–
Cl CH2 CMe3
106 C6H6N2 3211–19–6
1,2–Cyclobutanedicarbonitrile, cis–
NC
CN

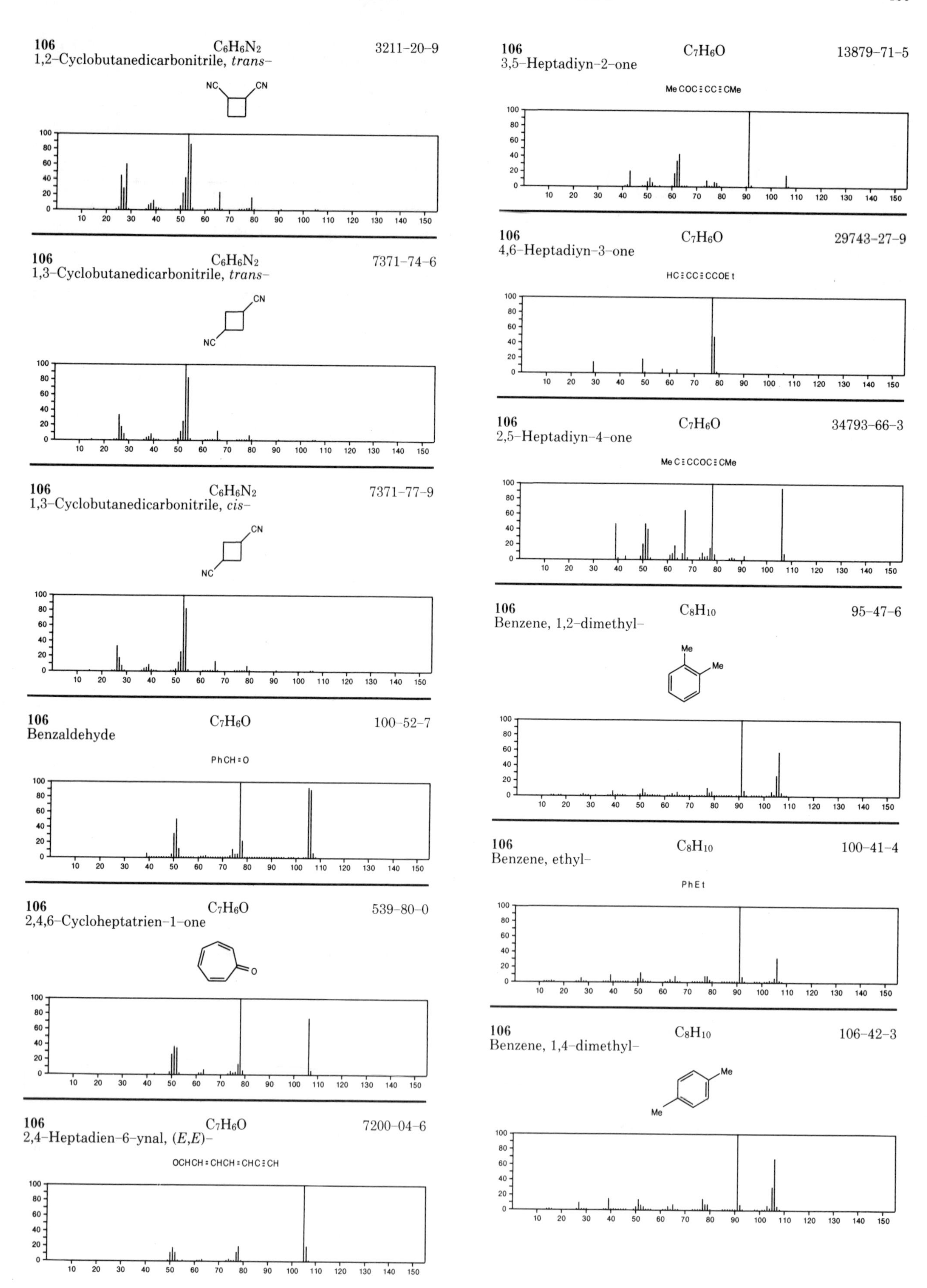
106 $C_6H_6N_2$ 3211-20-9
1,2-Cyclobutanedicarbonitrile, *trans*-
NC CN
106 $C_6H_6N_2$ 7371-74-6
1,3-Cyclobutanedicarbonitrile, *trans*-
CN
NC
106 $C_6H_6N_2$ 7371-77-9
1,3-Cyclobutanedicarbonitrile, *cis*-
CN
NC
106 C_7H_6O 100-52-7
Benzaldehyde
PhCH=O
106 C_7H_6O 539-80-0
2,4,6-Cycloheptatrien-1-one
O
106 C_7H_6O 7200-04-6
2,4-Heptadien-6-ynal, (*E,E*)-
OCHCH=CHCH=CHC≡CH
106 C_7H_6O 13879-71-5
3,5-Heptadiyn-2-one
MeCOC≡CC≡CMe
106 C_7H_6O 29743-27-9
4,6-Heptadiyn-3-one
HC≡CC≡CCOEt
106 C_7H_6O 34793-66-3
2,5-Heptadiyn-4-one
MeC≡CCOC≡CMe
106 C_8H_{10} 95-47-6
Benzene, 1,2-dimethyl-
Me
Me
106 C_8H_{10} 100-41-4
Benzene, ethyl-
PhEt
106 C_8H_{10} 106-42-3
Benzene, 1,4-dimethyl-
Me
Me

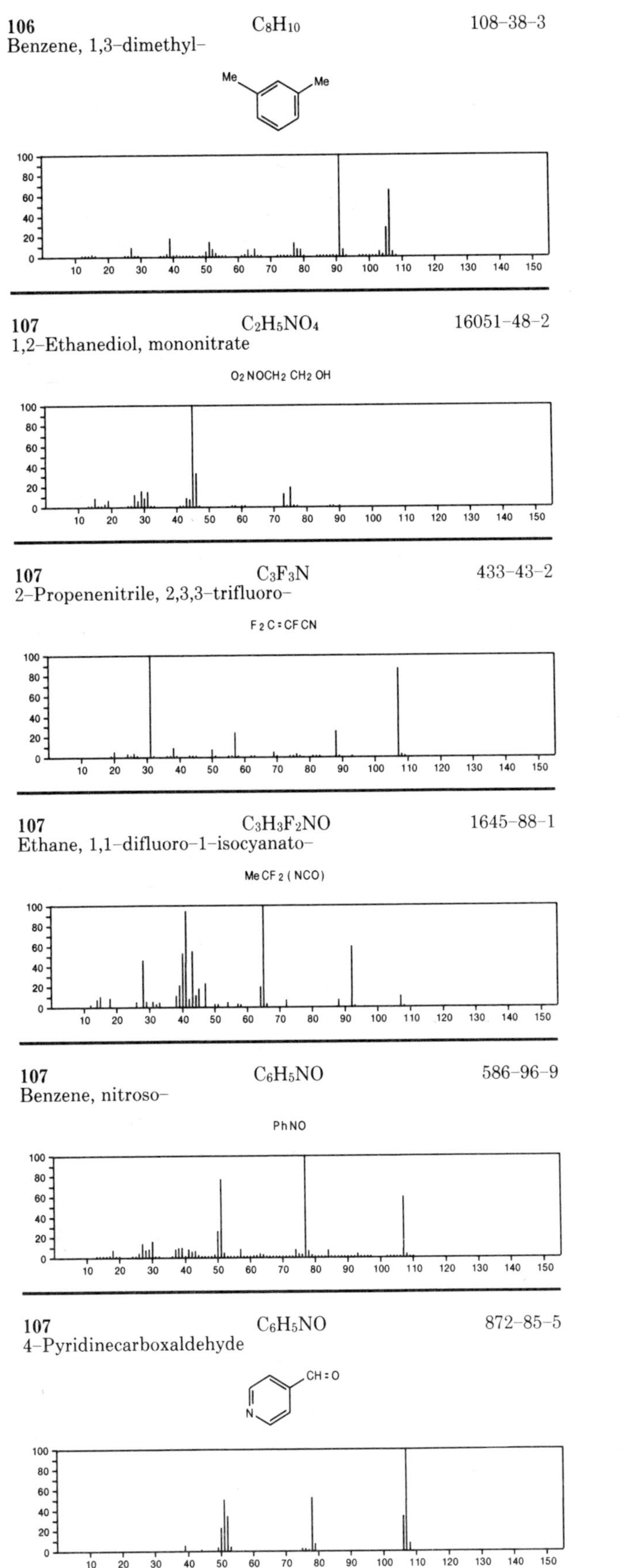

106 C_8H_{10} 108-38-3
Benzene, 1,3-dimethyl-

107 $C_2H_5NO_4$ 16051-48-2
1,2-Ethanediol, mononitrate
$O_2NOCH_2CH_2OH$

107 C_3F_3N 433-43-2
2-Propenenitrile, 2,3,3-trifluoro-
$F_2C=CFCN$

107 $C_3H_3F_2NO$ 1645-88-1
Ethane, 1,1-difluoro-1-isocyanato-
$MeCF_2(NCO)$

107 C_6H_5NO 586-96-9
Benzene, nitroso-
PhNO

107 C_6H_5NO 872-85-5
4-Pyridinecarboxaldehyde

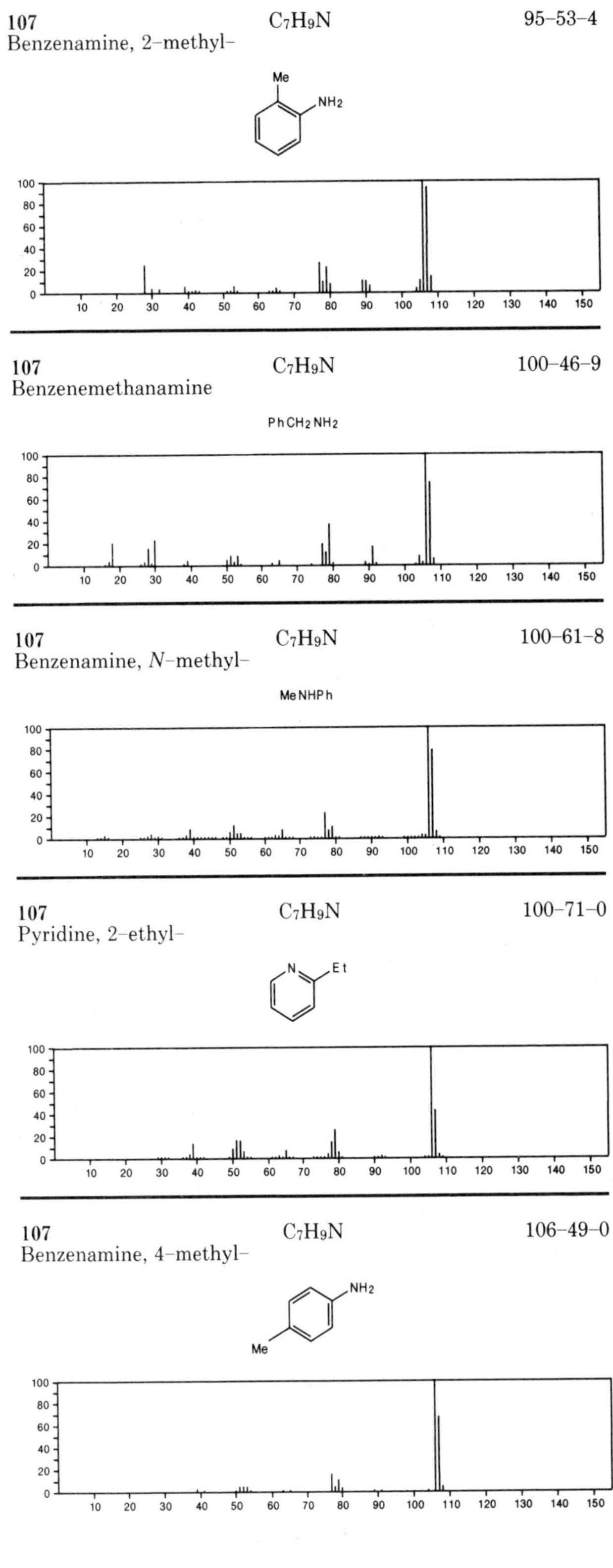

107 C_7H_9N 95-53-4
Benzenamine, 2-methyl-

107 C_7H_9N 100-46-9
Benzenemethanamine
$PhCH_2NH_2$

107 C_7H_9N 100-61-8
Benzenamine, *N*-methyl-
MeNHPh

107 C_7H_9N 100-71-0
Pyridine, 2-ethyl-

107 C_7H_9N 106-49-0
Benzenamine, 4-methyl-

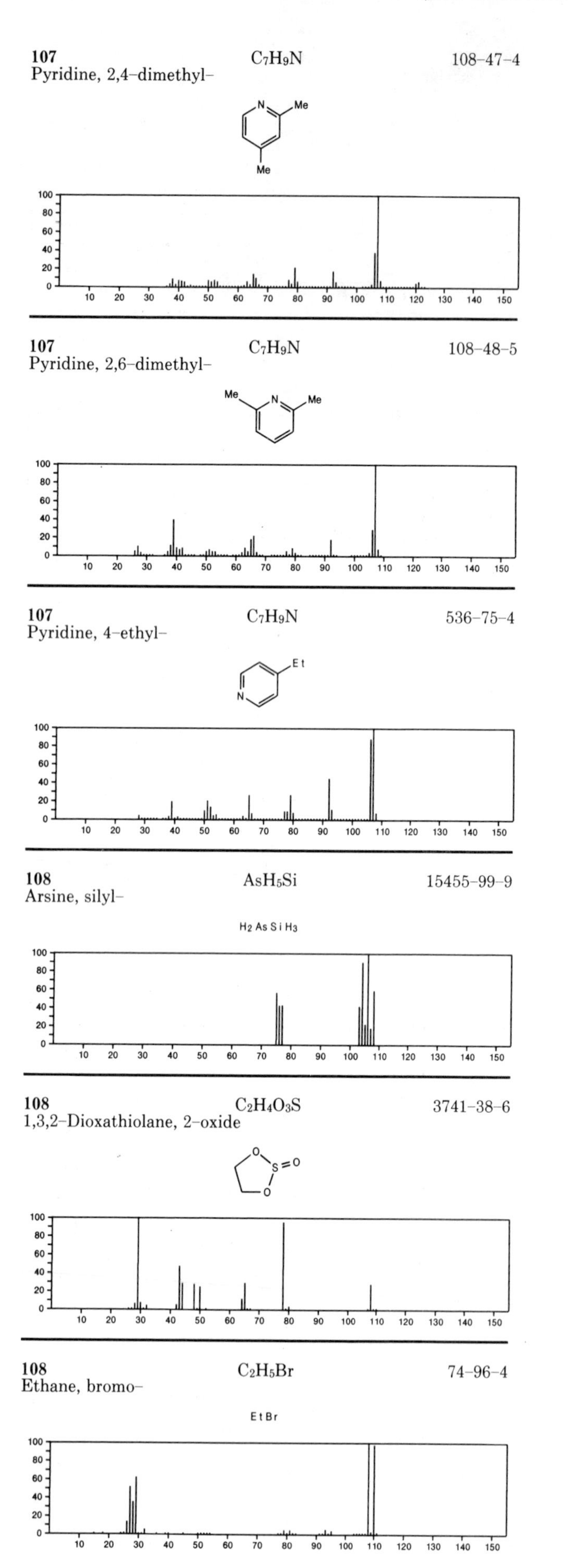
107 C_7H_9N 108-47-4
Pyridine, 2,4-dimethyl-
107 C_7H_9N 108-48-5
Pyridine, 2,6-dimethyl-
107 C_7H_9N 536-75-4
Pyridine, 4-ethyl-
108 AsH_5Si 15455-99-9
Arsine, silyl-
H2 As Si H3
108 $C_2H_4O_3S$ 3741-38-6
1,3,2-Dioxathiolane, 2-oxide
108 C_2H_5Br 74-96-4
Ethane, bromo-
EtBr

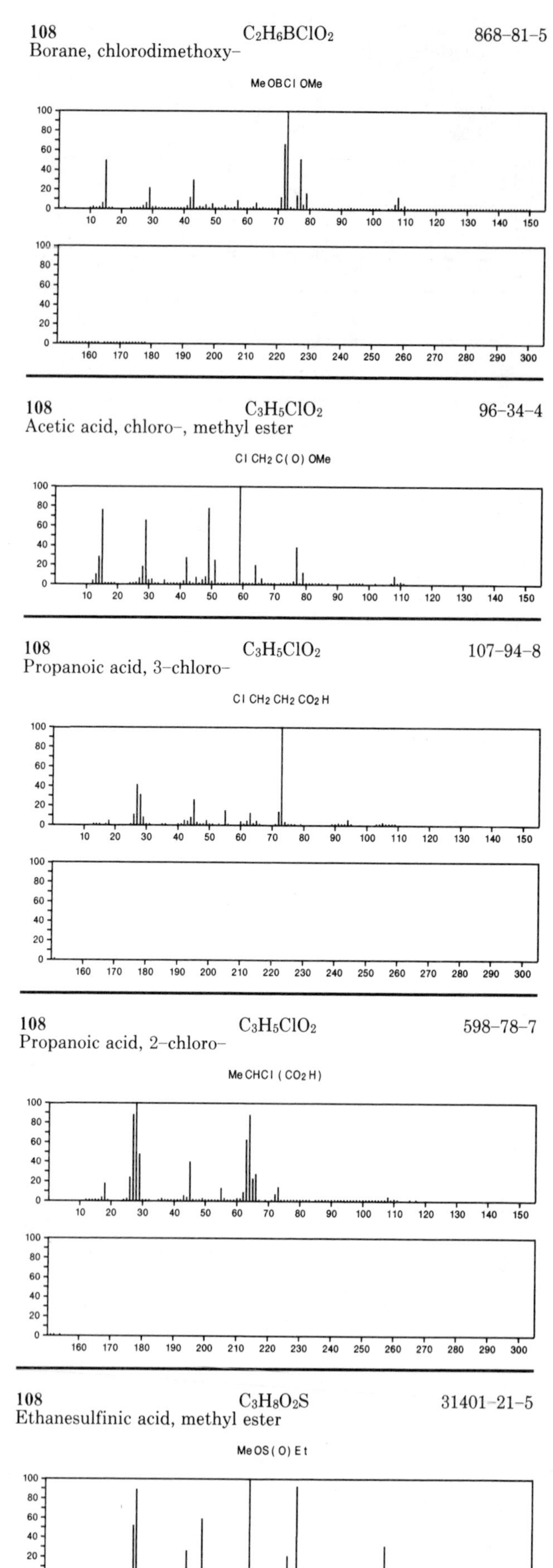
108 $C_2H_6BClO_2$ 868-81-5
Borane, chlorodimethoxy-
MeOBCl OMe
108 $C_3H_5ClO_2$ 96-34-4
Acetic acid, chloro-, methyl ester
Cl CH2 C(O) OMe
108 $C_3H_5ClO_2$ 107-94-8
Propanoic acid, 3-chloro-
Cl CH2 CH2 CO2 H
108 $C_3H_5ClO_2$ 598-78-7
Propanoic acid, 2-chloro-
MeCHCl (CO2H)
108 $C_3H_8O_2S$ 31401-21-5
Ethanesulfinic acid, methyl ester
MeOS(O)Et

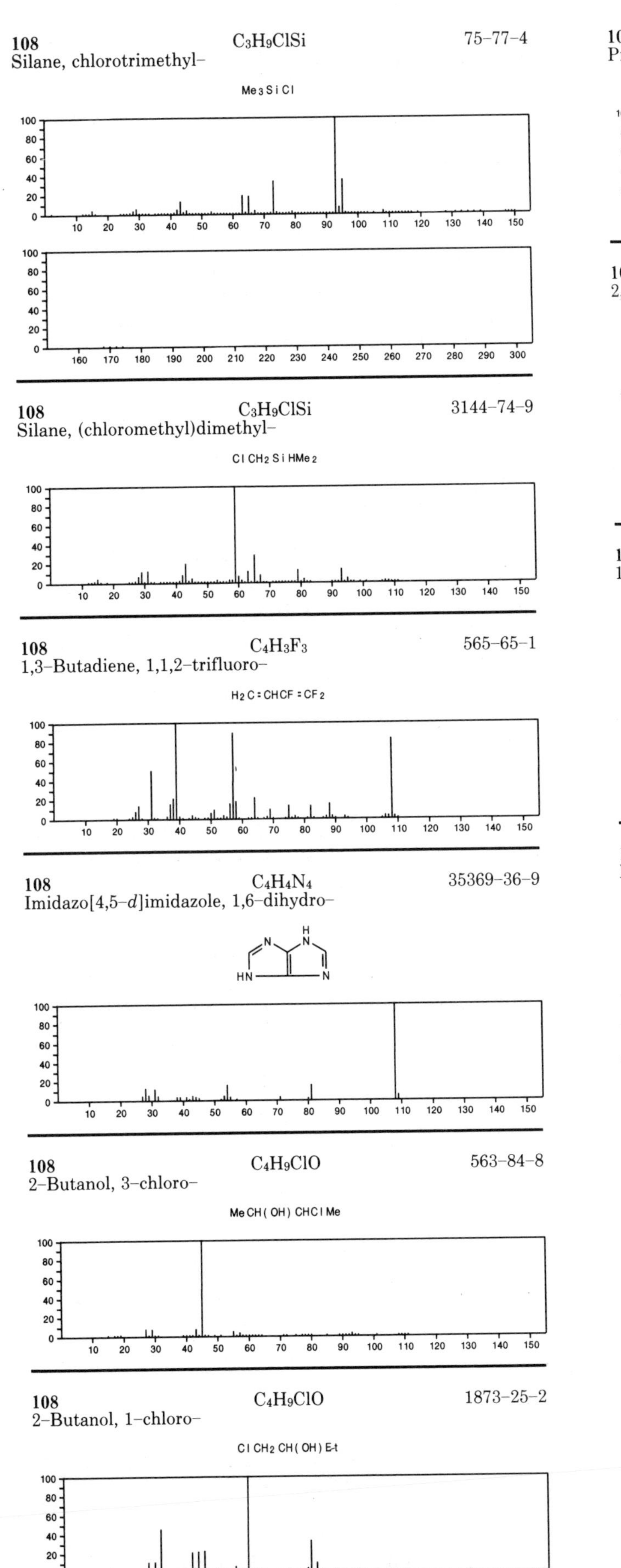

108 C_3H_9ClSi 75–77–4
Silane, chlorotrimethyl–
Me3SiCl
108 C_3H_9ClSi 3144–74–9
Silane, (chloromethyl)dimethyl–
ClCH2SiHMe2
108 $C_4H_3F_3$ 565–65–1
1,3–Butadiene, 1,1,2–trifluoro–
H2C=CHCF=CF2
108 $C_4H_4N_4$ 35369–36–9
Imidazo[4,5–d]imidazole, 1,6–dihydro–
108 C_4H_9ClO 563–84–8
2–Butanol, 3–chloro–
MeCH(OH)CHClMe
108 C_4H_9ClO 1873–25–2
2–Butanol, 1–chloro–
ClCH2CH(OH)Et

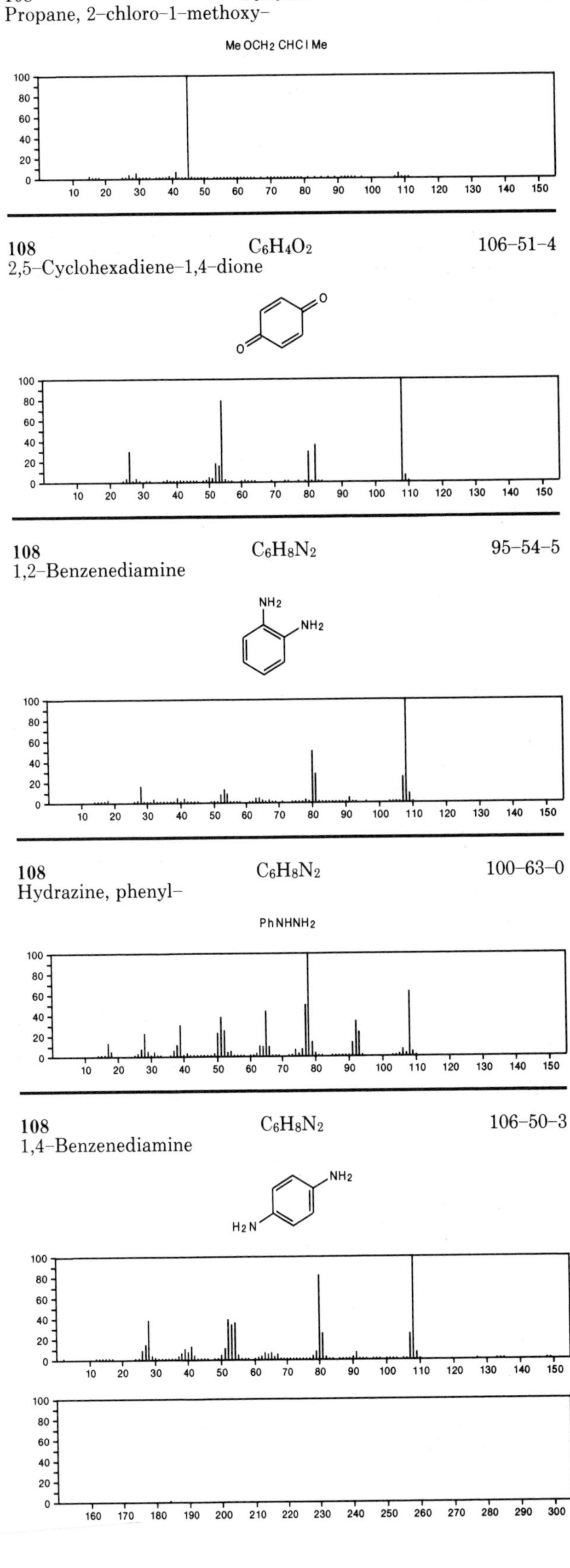

108 C_4H_9ClO 5390–71–6
Propane, 2–chloro–1–methoxy–
MeOCH2CHClMe
108 $C_6H_4O_2$ 106–51–4
2,5–Cyclohexadiene–1,4–dione
108 $C_6H_8N_2$ 95–54–5
1,2–Benzenediamine
108 $C_6H_8N_2$ 100–63–0
Hydrazine, phenyl–
PhNHNH2
108 $C_6H_8N_2$ 106–50–3
1,4–Benzenediamine

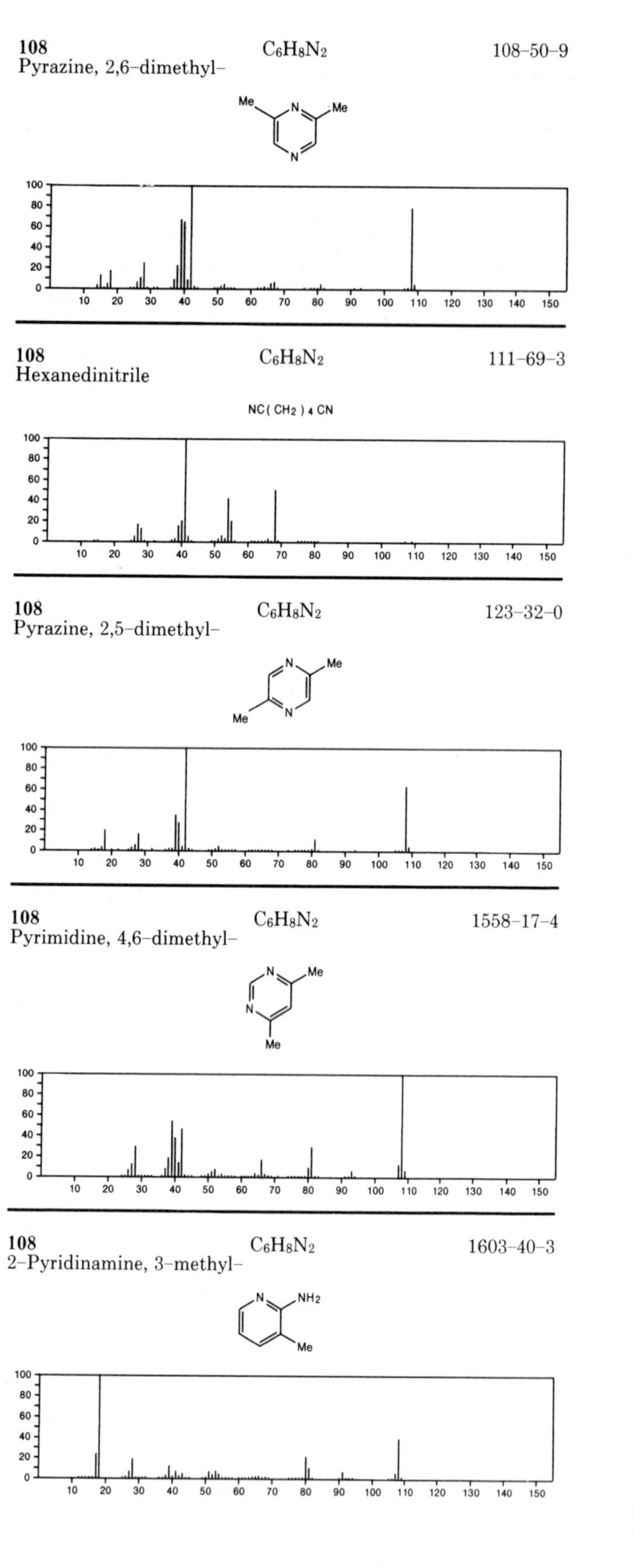

108 $C_6H_8N_2$ 108-50-9
Pyrazine, 2,6-dimethyl-

108 $C_6H_8N_2$ 111-69-3
Hexanedinitrile

NC(CH2)4CN

108 $C_6H_8N_2$ 123-32-0
Pyrazine, 2,5-dimethyl-

108 $C_6H_8N_2$ 1558-17-4
Pyrimidine, 4,6-dimethyl-

108 $C_6H_8N_2$ 1603-40-3
2-Pyridinamine, 3-methyl-

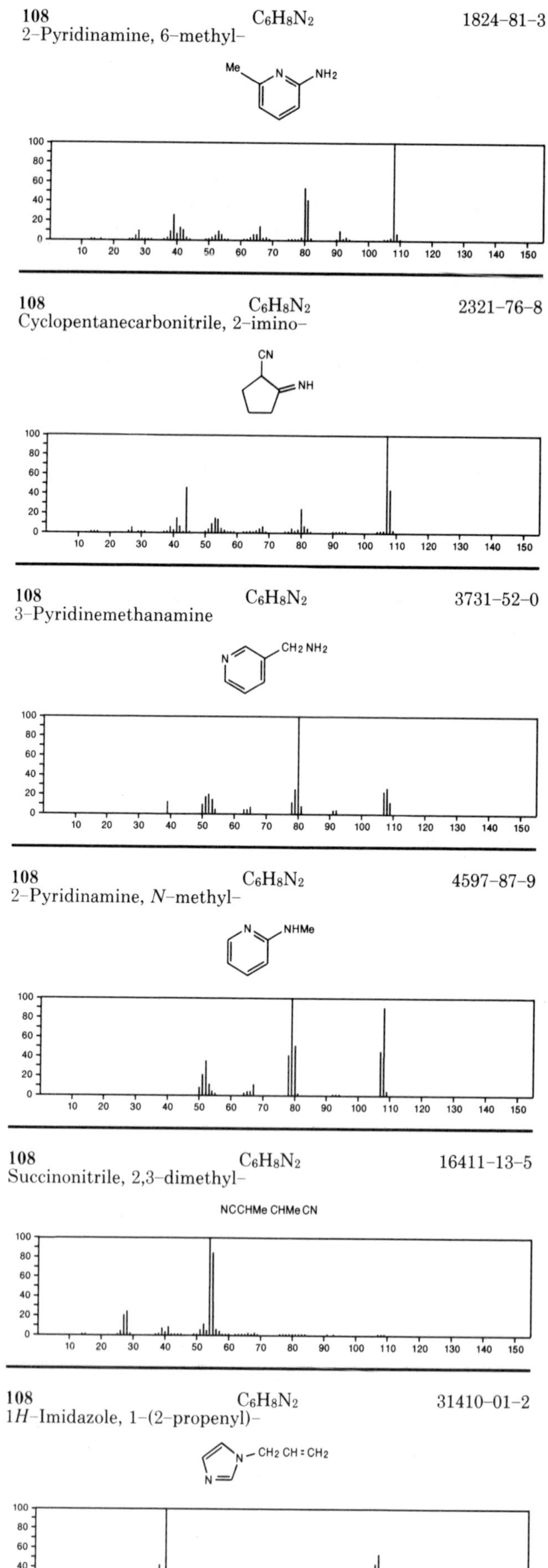

108 $C_6H_8N_2$ 1824-81-3
2-Pyridinamine, 6-methyl-

108 $C_6H_8N_2$ 2321-76-8
Cyclopentanecarbonitrile, 2-imino-

108 $C_6H_8N_2$ 3731-52-0
3-Pyridinemethanamine

108 $C_6H_8N_2$ 4597-87-9
2-Pyridinamine, *N*-methyl-

108 $C_6H_8N_2$ 16411-13-5
Succinonitrile, 2,3-dimethyl-

NCCHMe CHMe CN

108 $C_6H_8N_2$ 31410-01-2
1*H*-Imidazole, 1-(2-propenyl)-

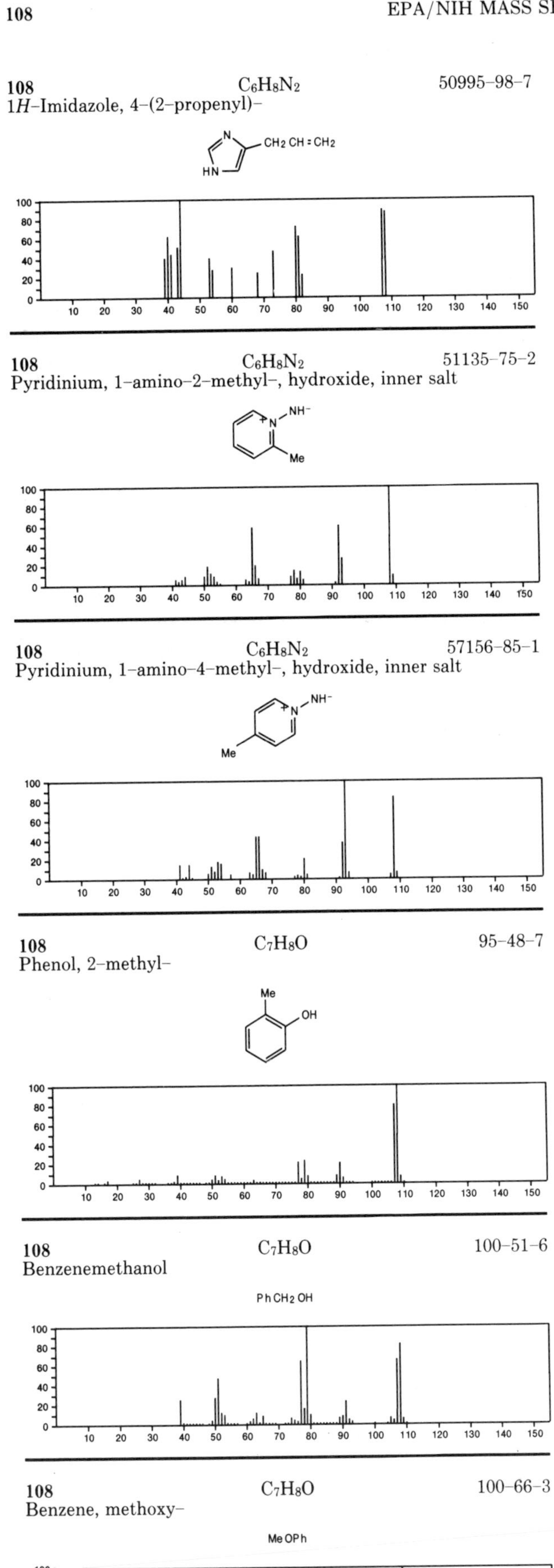
108 $C_6H_8N_2$ 50995-98-7
1*H*-Imidazole, 4-(2-propenyl)-
108 $C_6H_8N_2$ 51135-75-2
Pyridinium, 1-amino-2-methyl-, hydroxide, inner salt
108 $C_6H_8N_2$ 57156-85-1
Pyridinium, 1-amino-4-methyl-, hydroxide, inner salt
108 C_7H_8O 95-48-7
Phenol, 2-methyl-
108 C_7H_8O 100-51-6
Benzenemethanol
PhCH2OH
108 C_7H_8O 100-66-3
Benzene, methoxy-
MeOPh

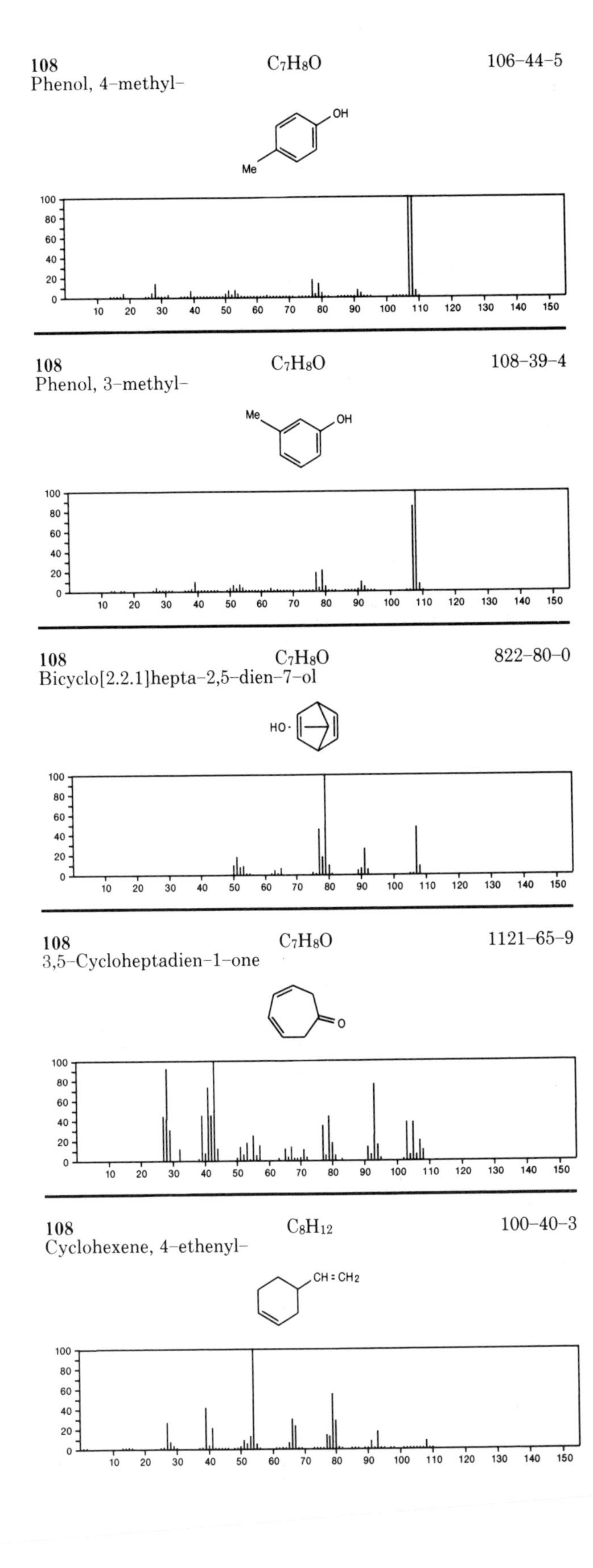
108 C_7H_8O 106-44-5
Phenol, 4-methyl-
108 C_7H_8O 108-39-4
Phenol, 3-methyl-
108 C_7H_8O 822-80-0
Bicyclo[2.2.1]hepta-2,5-dien-7-ol
108 C_7H_8O 1121-65-9
3,5-Cycloheptadien-1-one
108 C_8H_{12} 100-40-3
Cyclohexene, 4-ethenyl-

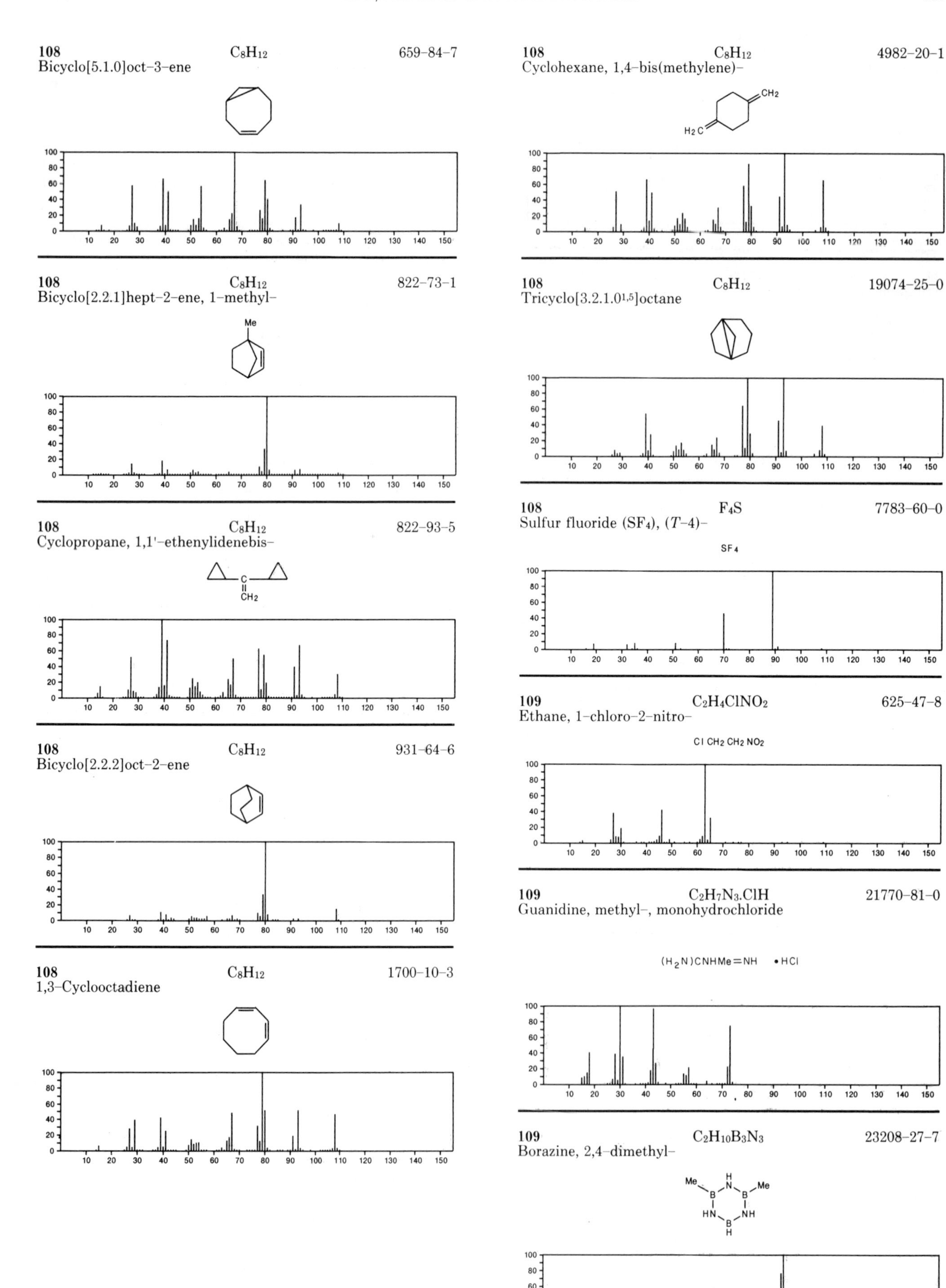

108 C_8H_{12} 659-84-7
Bicyclo[5.1.0]oct-3-ene
108 C_8H_{12} 822-73-1
Bicyclo[2.2.1]hept-2-ene, 1-methyl-
Me
108 C_8H_{12} 822-93-5
Cyclopropane, 1,1'-ethenylidenebis-
C
CH2
108 C_8H_{12} 931-64-6
Bicyclo[2.2.2]oct-2-ene
108 C_8H_{12} 1700-10-3
1,3-Cyclooctadiene
108 C_8H_{12} 4982-20-1
Cyclohexane, 1,4-bis(methylene)-
CH2
H2C
108 C_8H_{12} 19074-25-0
Tricyclo[3.2.1.0[1,5]]octane
108 F_4S 7783-60-0
Sulfur fluoride (SF_4), (*T*-4)-
SF4
109 $C_2H_4ClNO_2$ 625-47-8
Ethane, 1-chloro-2-nitro-
Cl CH2 CH2 NO2
109 $C_2H_7N_3.ClH$ 21770-81-0
Guanidine, methyl-, monohydrochloride
(H2N)CNHMe=NH •HCl
109 $C_2H_{10}B_3N_3$ 23208-27-7
Borazine, 2,4-dimethyl-
Me B N H B Me HN B NH H

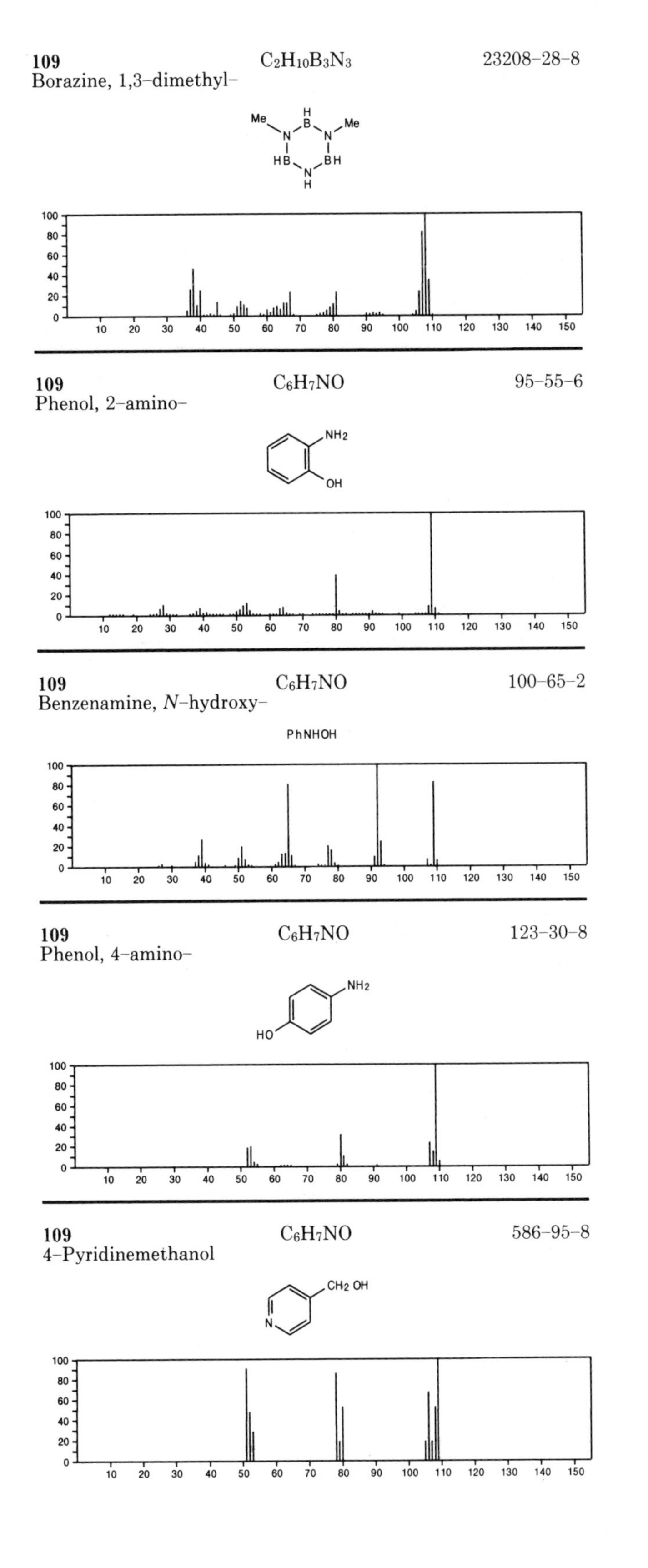

109 $C_2H_{10}B_3N_3$ 23208-28-8
Borazine, 1,3-dimethyl-
109 C_6H_7NO 95-55-6
Phenol, 2-amino-
109 C_6H_7NO 100-65-2
Benzenamine, *N*-hydroxy-
PhNHOH
109 C_6H_7NO 123-30-8
Phenol, 4-amino-
109 C_6H_7NO 586-95-8
4-Pyridinemethanol

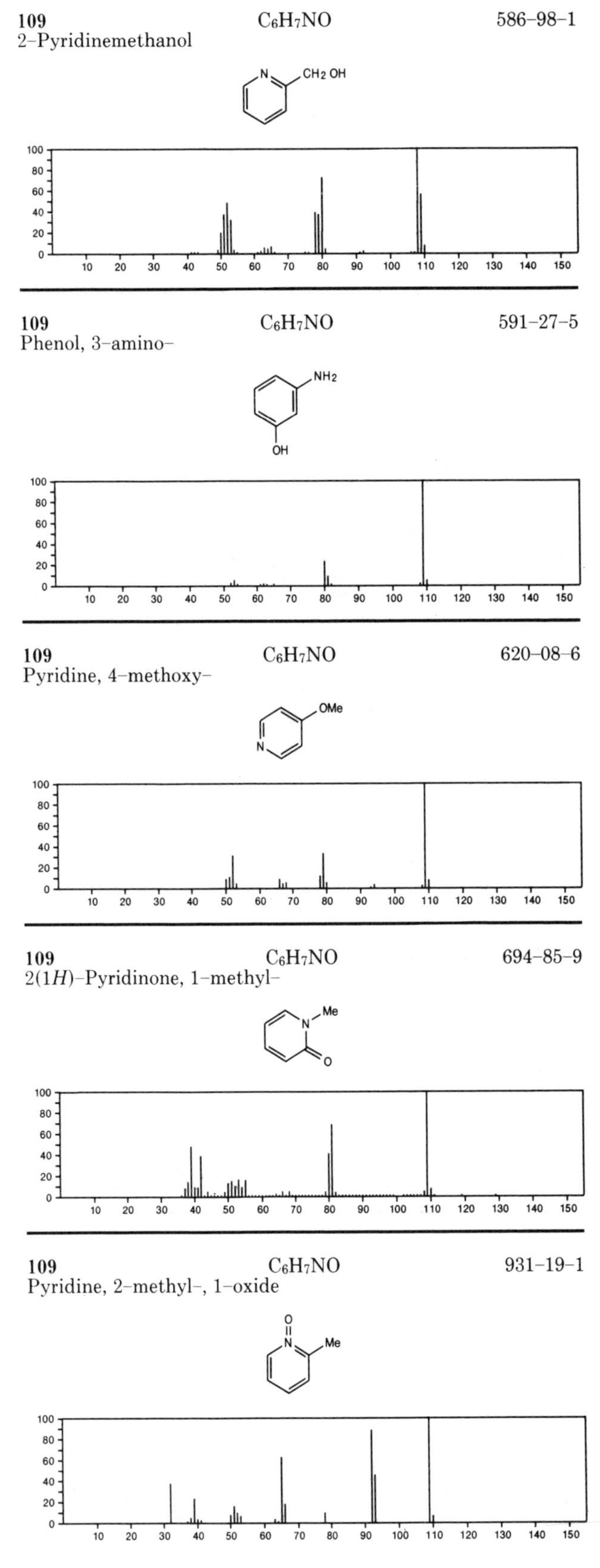

109 C_6H_7NO 586-98-1
2-Pyridinemethanol
109 C_6H_7NO 591-27-5
Phenol, 3-amino-
109 C_6H_7NO 620-08-6
Pyridine, 4-methoxy-
109 C_6H_7NO 694-85-9
2(1*H*)-Pyridinone, 1-methyl-
109 C_6H_7NO 931-19-1
Pyridine, 2-methyl-, 1-oxide

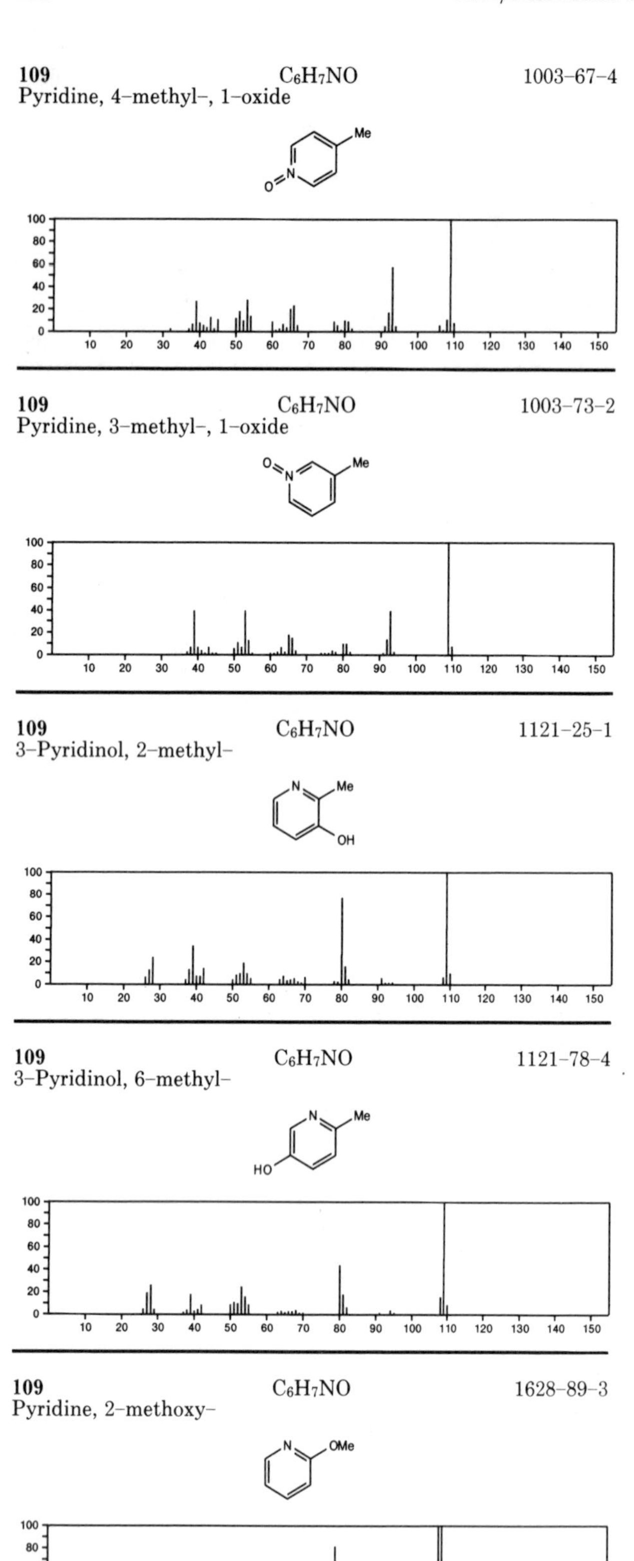

109 C_6H_7NO 1003-67-4
Pyridine, 4-methyl-, 1-oxide

109 C_6H_7NO 1003-73-2
Pyridine, 3-methyl-, 1-oxide

109 C_6H_7NO 1121-25-1
3-Pyridinol, 2-methyl-

109 C_6H_7NO 1121-78-4
3-Pyridinol, 6-methyl-

109 C_6H_7NO 1628-89-3
Pyridine, 2-methoxy-

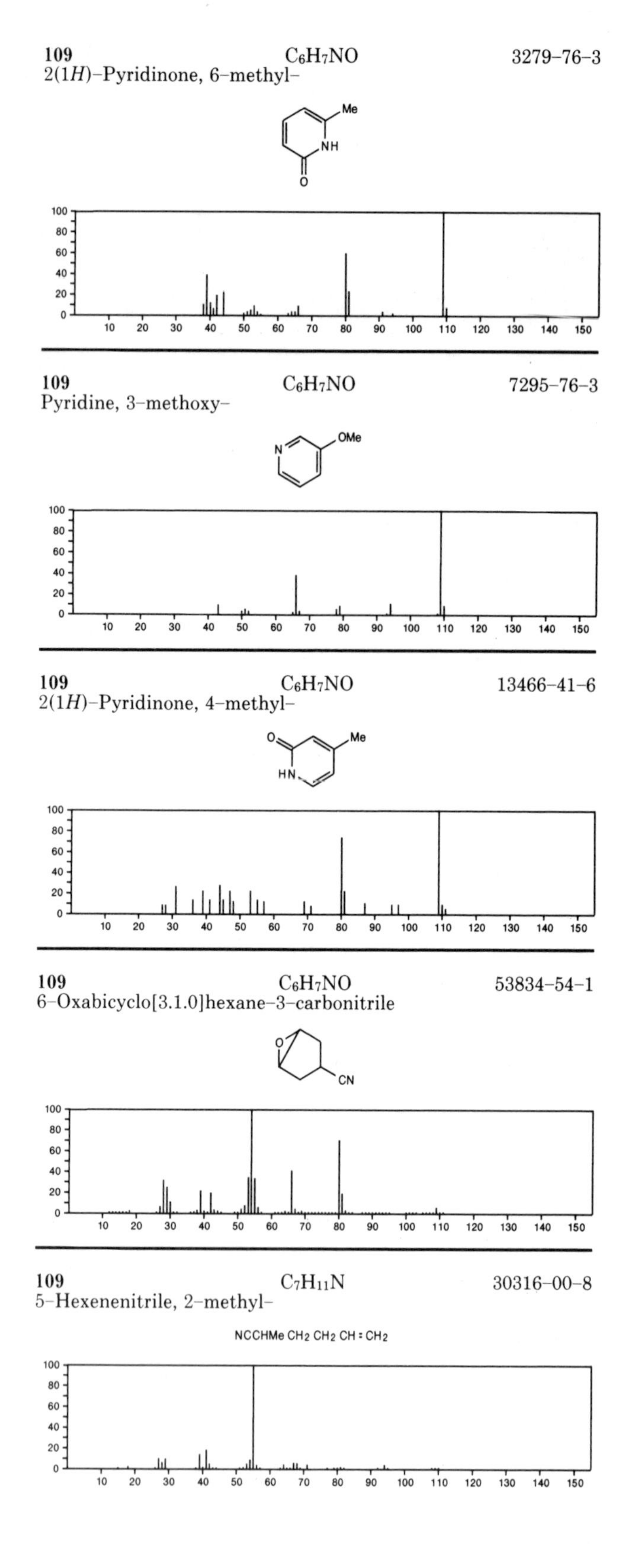

109 C_6H_7NO 3279-76-3
2(1*H*)-Pyridinone, 6-methyl-

109 C_6H_7NO 7295-76-3
Pyridine, 3-methoxy-

109 C_6H_7NO 13466-41-6
2(1*H*)-Pyridinone, 4-methyl-

109 C_6H_7NO 53834-54-1
6-Oxabicyclo[3.1.0]hexane-3-carbonitrile

109 $C_7H_{11}N$ 30316-00-8
5-Hexenenitrile, 2-methyl-

NCCHMe CH2 CH2 CH = CH2

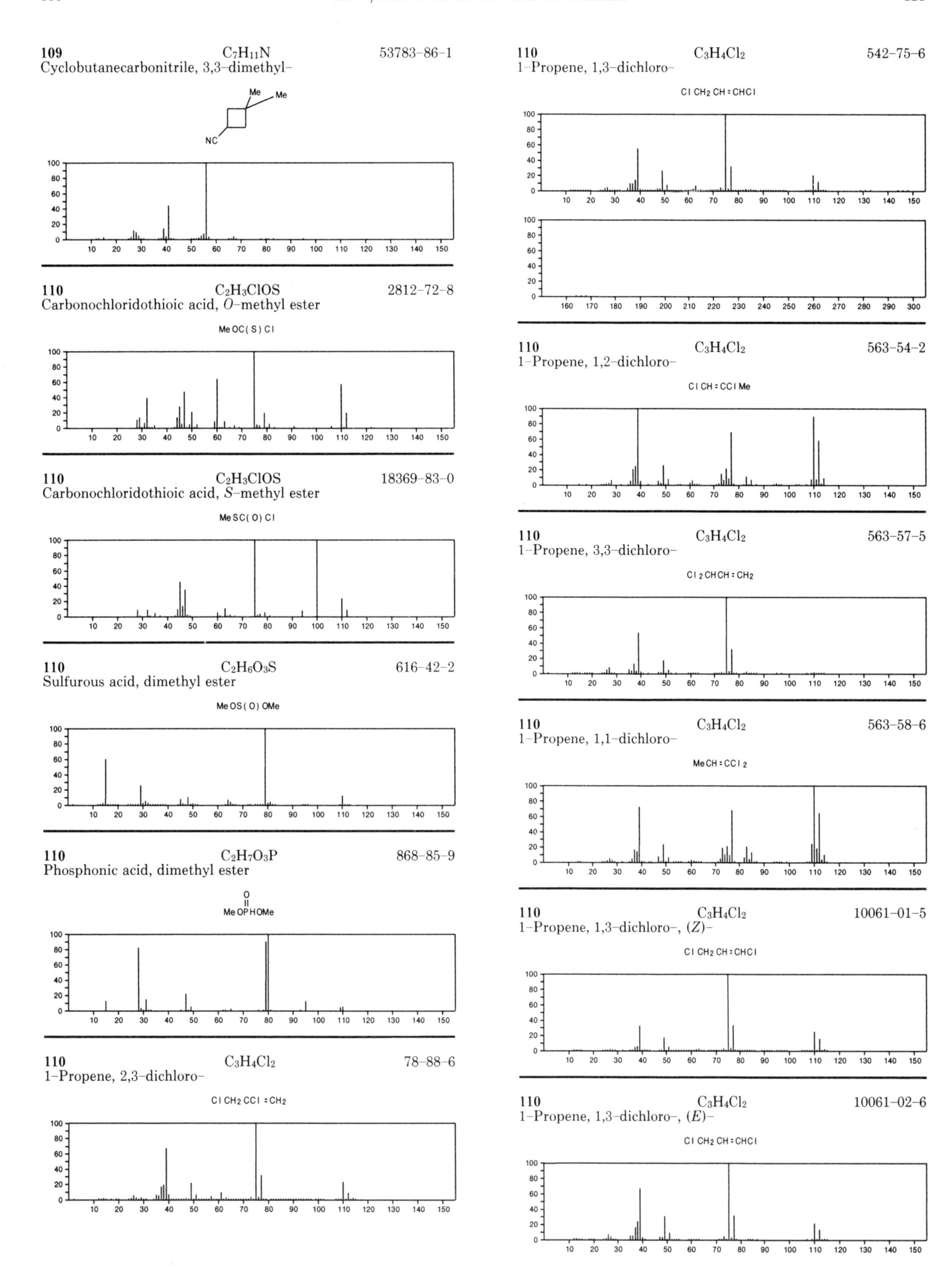
109 C7H11N 53783-86-1
Cyclobutanecarbonitrile, 3,3-dimethyl-
Me Me NC
110 C2H3ClOS 2812-72-8
Carbonochloridothioic acid, O-methyl ester
MeOC(S)Cl
110 C2H3ClOS 18369-83-0
Carbonochloridothioic acid, S-methyl ester
MeSC(O)Cl
110 C2H6O3S 616-42-2
Sulfurous acid, dimethyl ester
MeOS(O)OMe
110 C2H7O3P 868-85-9
Phosphonic acid, dimethyl ester
O
MeOPHOMe
110 C3H4Cl2 78-88-6
1-Propene, 2,3-dichloro-
ClCH2CCl=CH2
110 C3H4Cl2 542-75-6
1-Propene, 1,3-dichloro-
ClCH2CH=CHCl
110 C3H4Cl2 563-54-2
1-Propene, 1,2-dichloro-
ClCH=CClMe
110 C3H4Cl2 563-57-5
1-Propene, 3,3-dichloro-
Cl2CHCH=CH2
110 C3H4Cl2 563-58-6
1-Propene, 1,1-dichloro-
MeCH=CCl2
110 C3H4Cl2 10061-01-5
1-Propene, 1,3-dichloro-, (Z)-
ClCH2CH=CHCl
110 C3H4Cl2 10061-02-6
1-Propene, 1,3-dichloro-, (E)-
ClCH2CH=CHCl

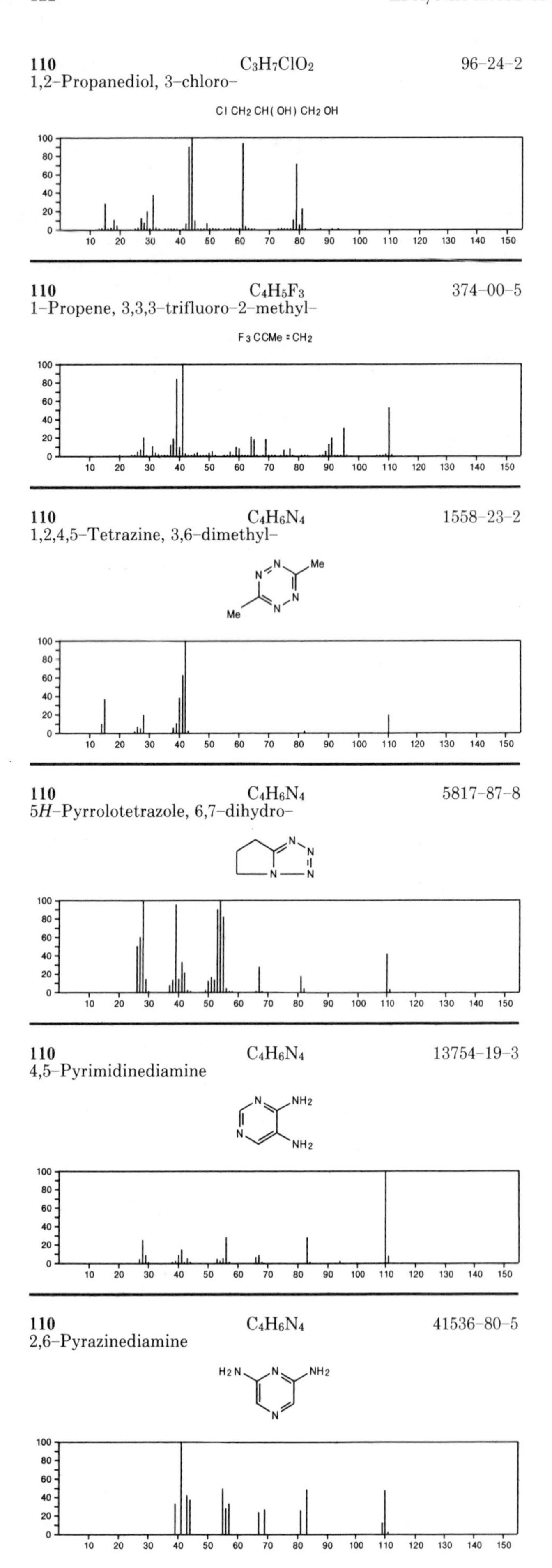

110 $C_3H_7ClO_2$ 96-24-2
1,2-Propanediol, 3-chloro-

Cl CH2 CH(OH) CH2 OH

110 $C_4H_5F_3$ 374-00-5
1-Propene, 3,3,3-trifluoro-2-methyl-

F3 CCMe = CH2

110 $C_4H_6N_4$ 1558-23-2
1,2,4,5-Tetrazine, 3,6-dimethyl-

110 $C_4H_6N_4$ 5817-87-8
5*H*-Pyrrolotetrazole, 6,7-dihydro-

110 $C_4H_6N_4$ 13754-19-3
4,5-Pyrimidinediamine

110 $C_4H_6N_4$ 41536-80-5
2,6-Pyrazinediamine

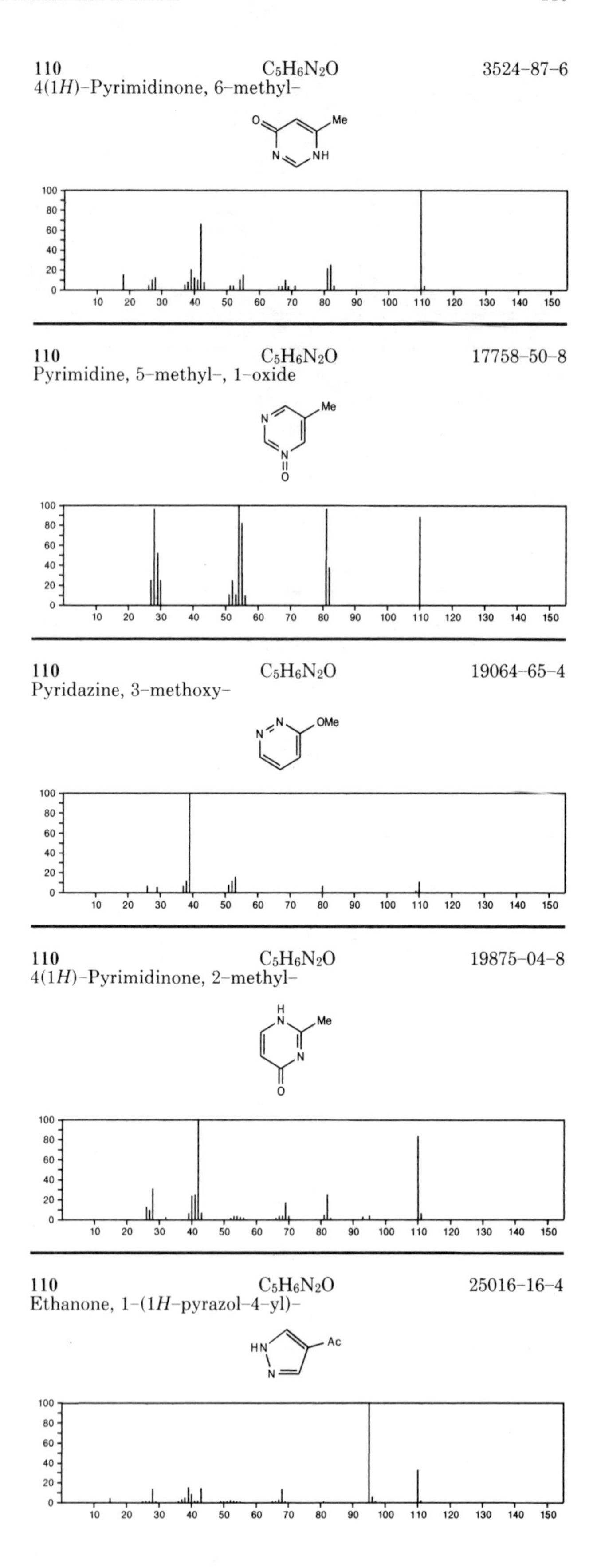

110 $C_5H_6N_2O$ 3524-87-6
4(1*H*)-Pyrimidinone, 6-methyl-

110 $C_5H_6N_2O$ 17758-50-8
Pyrimidine, 5-methyl-, 1-oxide

110 $C_5H_6N_2O$ 19064-65-4
Pyridazine, 3-methoxy-

110 $C_5H_6N_2O$ 19875-04-8
4(1*H*)-Pyrimidinone, 2-methyl-

110 $C_5H_6N_2O$ 25016-16-4
Ethanone, 1-(1*H*-pyrazol-4-yl)-

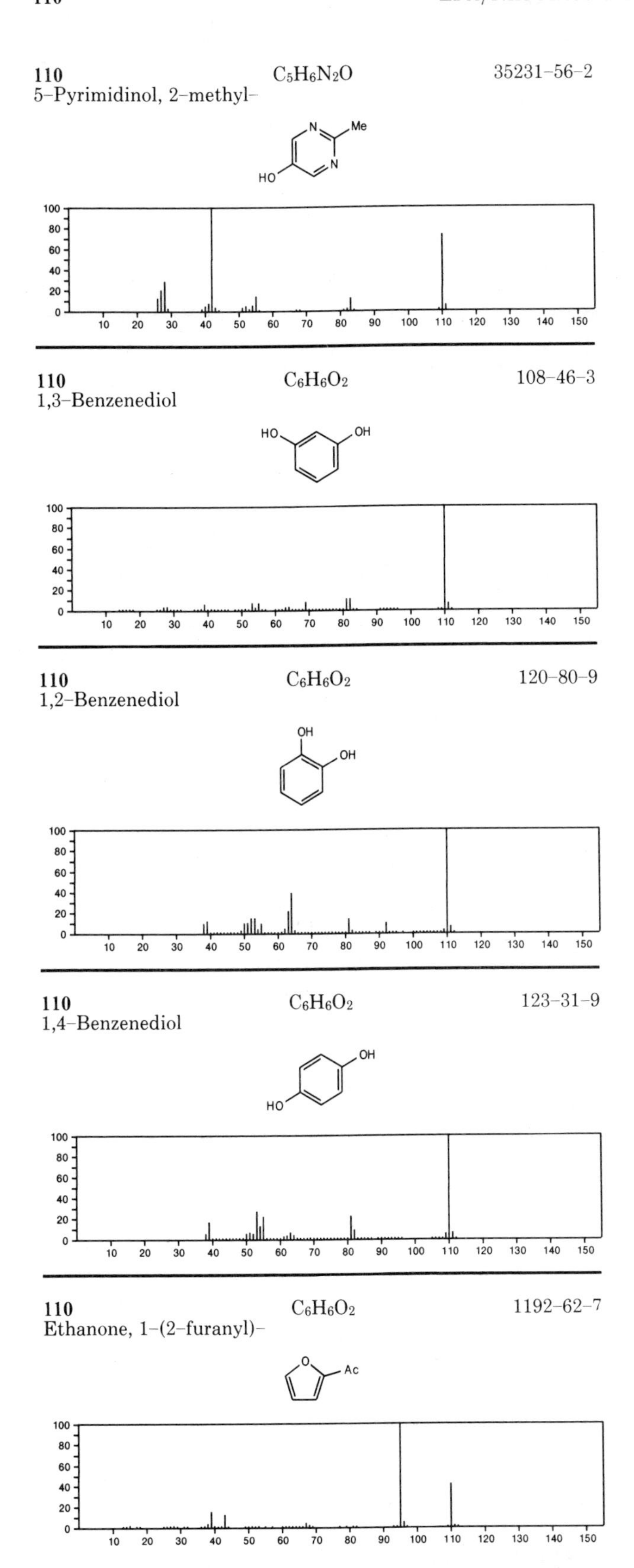

110 $C_5H_6N_2O$ 35231-56-2
5-Pyrimidinol, 2-methyl-

110 $C_6H_6O_2$ 108-46-3
1,3-Benzenediol

110 $C_6H_6O_2$ 120-80-9
1,2-Benzenediol

110 $C_6H_6O_2$ 123-31-9
1,4-Benzenediol

110 $C_6H_6O_2$ 1192-62-7
Ethanone, 1-(2-furanyl)-

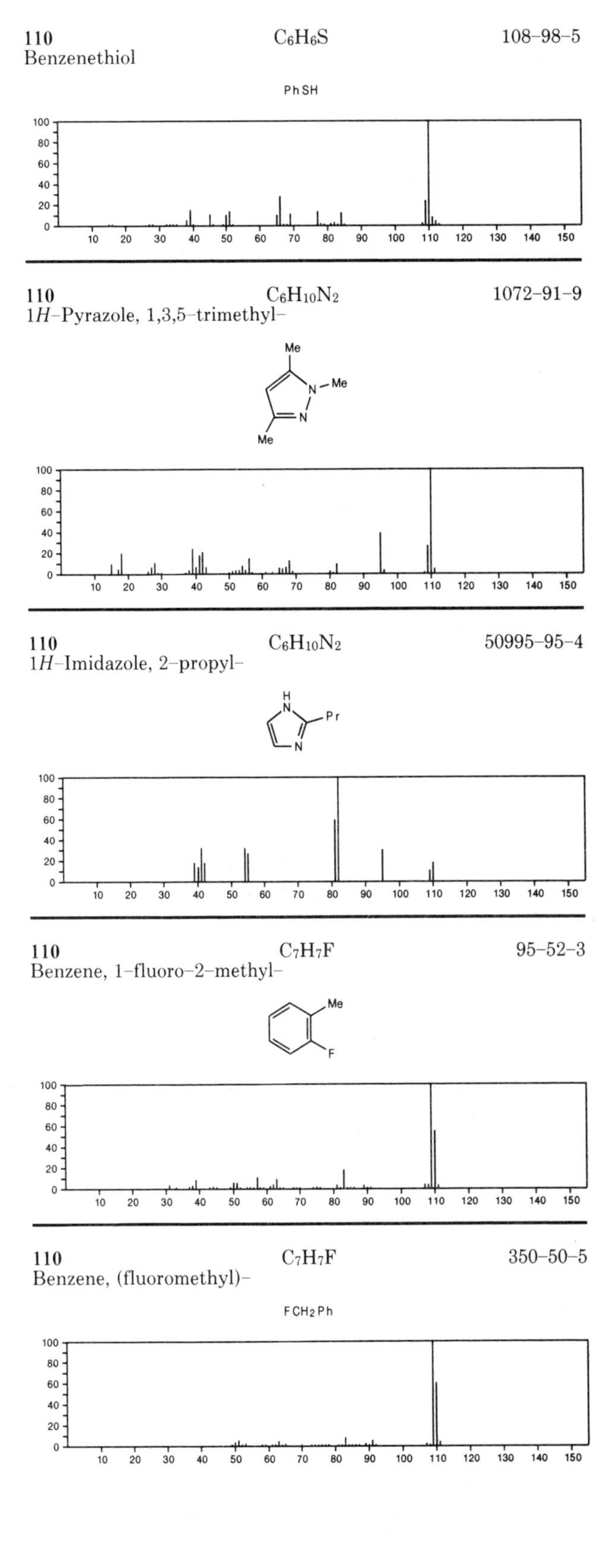

110 C_6H_6S 108-98-5
Benzenethiol

110 $C_6H_{10}N_2$ 1072-91-9
1*H*-Pyrazole, 1,3,5-trimethyl-

110 $C_6H_{10}N_2$ 50995-95-4
1*H*-Imidazole, 2-propyl-

110 C_7H_7F 95-52-3
Benzene, 1-fluoro-2-methyl-

110 C_7H_7F 350-50-5
Benzene, (fluoromethyl)-

110 C_7H_7F 352-32-9
Benzene, 1-fluoro-4-methyl-

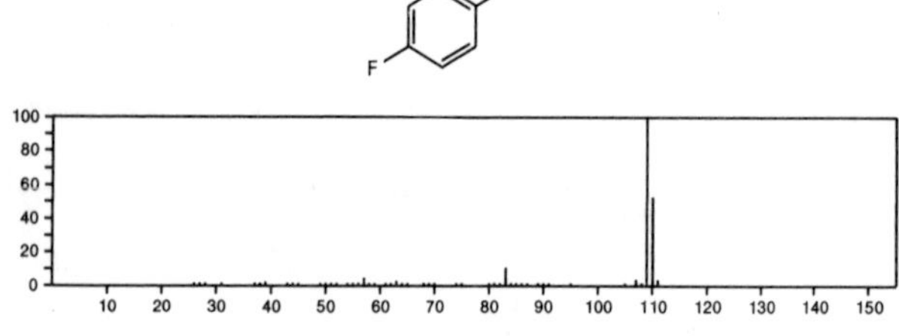

110 C_7H_7F 352-70-5
Benzene, 1-fluoro-3-methyl-

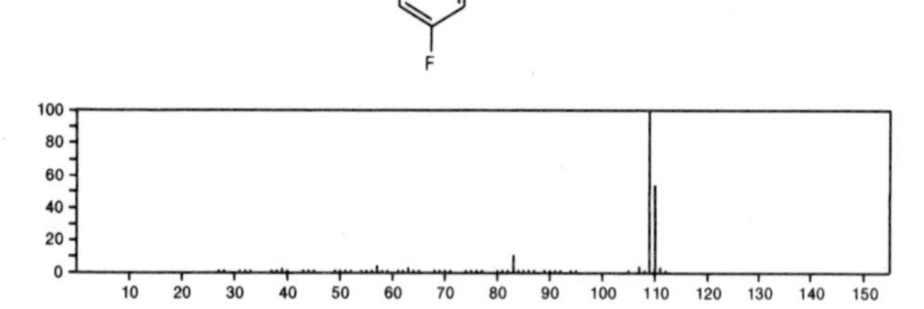

110 $C_7H_{10}O$ 100-50-5
3-Cyclohexene-1-carboxaldehyde

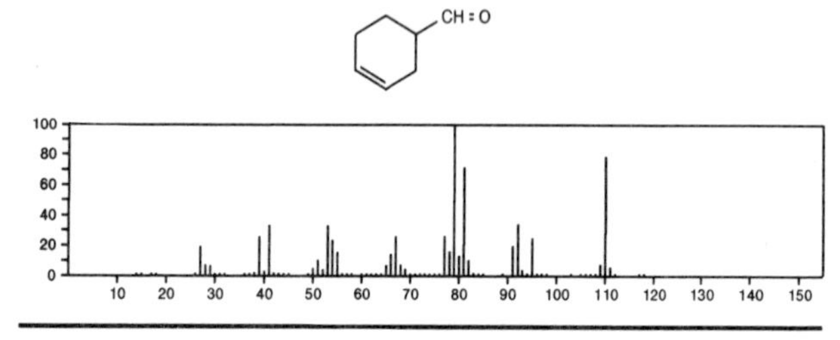

110 $C_7H_{10}O$ 497-38-1
Bicyclo[2.2.1]heptan-2-one

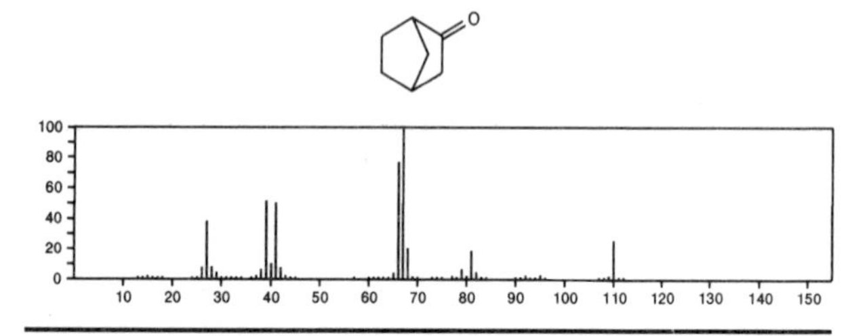

110 $C_7H_{10}O$ 695-04-5
Tricyclo[2.2.1.0^{2,6}]heptan-3-ol

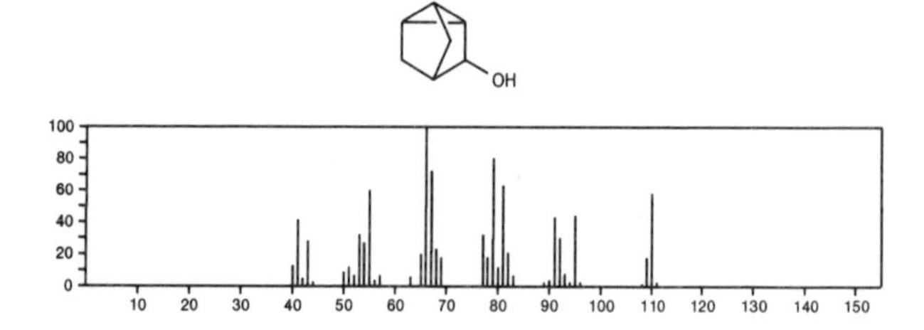

110 $C_7H_{10}O$ 1121-37-5
Methanone, dicyclopropyl-

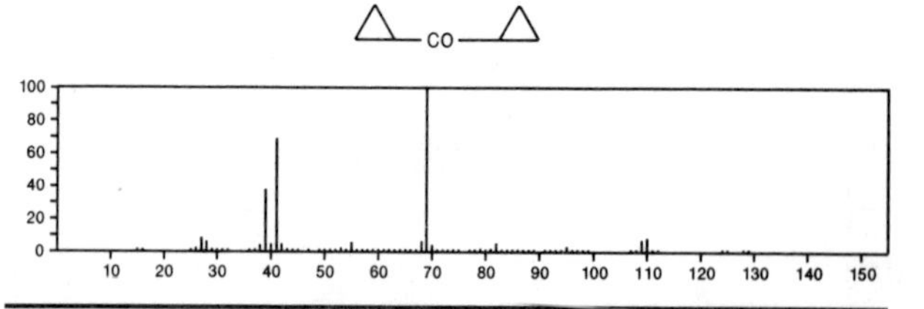

110 $C_7H_{10}O$ 1121-64-8
3-Cyclohepten-1-one

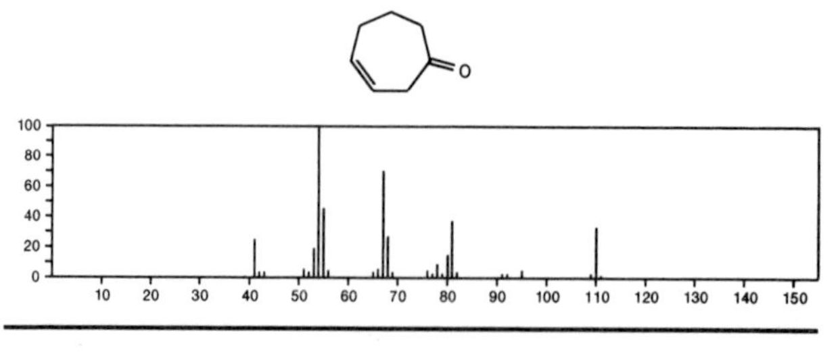

110 $C_7H_{10}O$ 1121-66-0
2-Cyclohepten-1-one

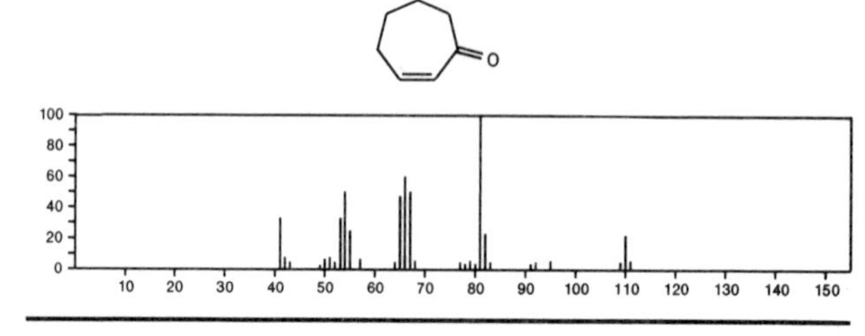

110 $C_7H_{10}O$ 1193-18-6
2-Cyclohexen-1-one, 3-methyl-

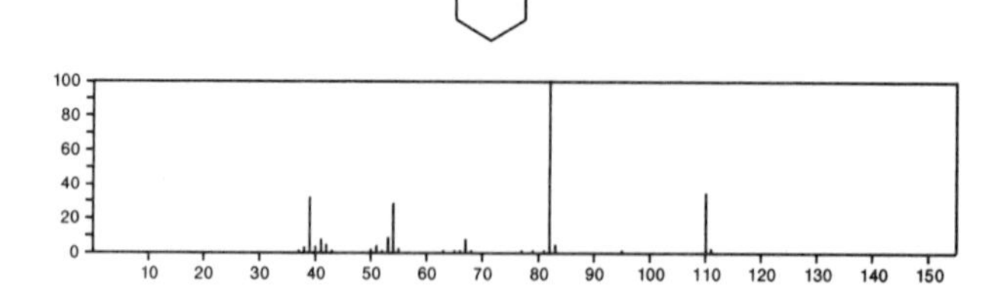

110 $C_7H_{10}O$ 3146-39-2
3-Oxatricyclo[3.2.1.0^{2,4}]octane, (1α,2β,4β,5α)-

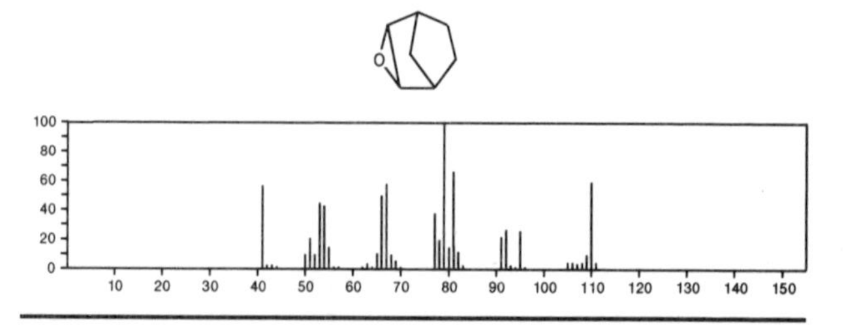

110 $C_7H_{10}O$ 4058-51-9
3,4-Pentadienal, 2,2-dimethyl-

$H_2C=C=CHCMe_2CHO$

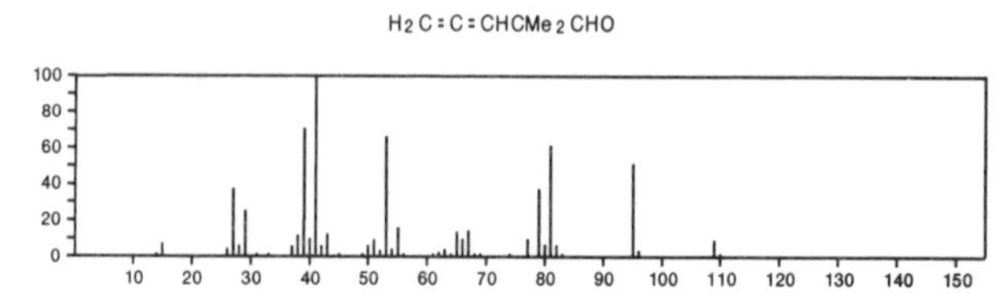

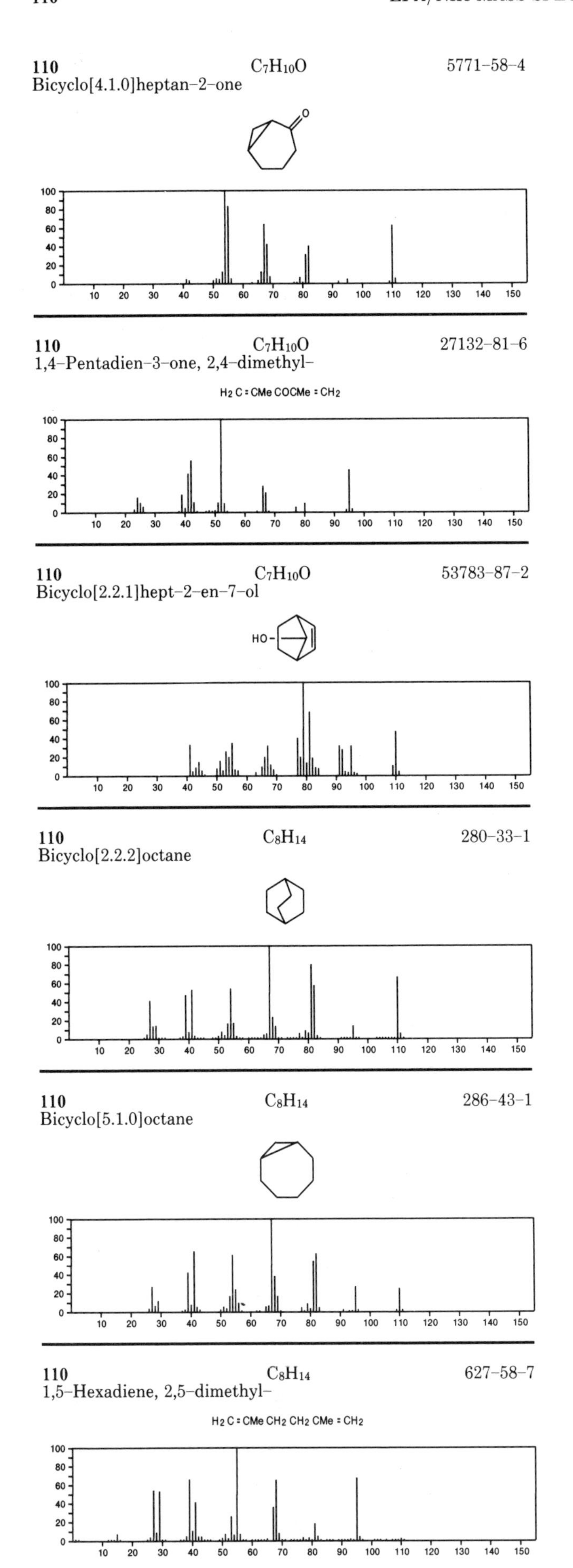
110 C7H10O 5771-58-4
Bicyclo[4.1.0]heptan-2-one
110 C7H10O 27132-81-6
1,4-Pentadien-3-one, 2,4-dimethyl-
H2C=CMeCOCMe=CH2
110 C7H10O 53783-87-2
Bicyclo[2.2.1]hept-2-en-7-ol
HO-
110 C8H14 280-33-1
Bicyclo[2.2.2]octane
110 C8H14 286-43-1
Bicyclo[5.1.0]octane
110 C8H14 627-58-7
1,5-Hexadiene, 2,5-dimethyl-
H2C=CMeCH2CH2CMe=CH2

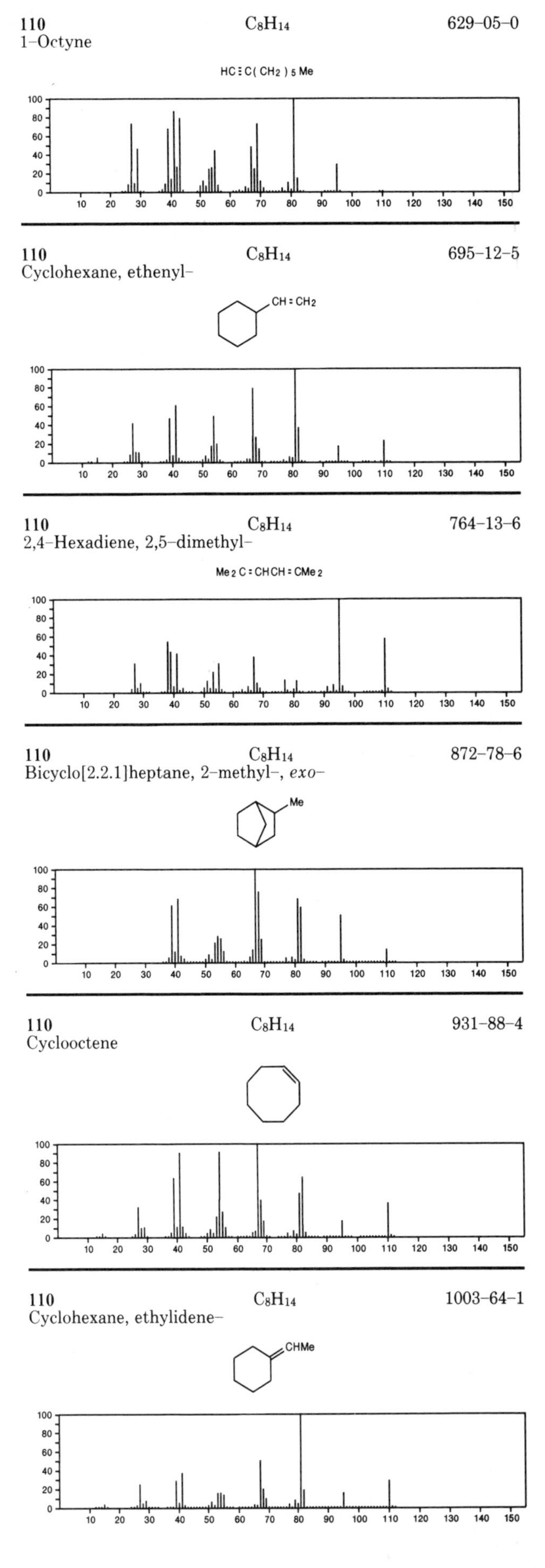
110 C8H14 629-05-0
1-Octyne
HC≡C(CH2)5Me
110 C8H14 695-12-5
Cyclohexane, ethenyl-
CH=CH2
110 C8H14 764-13-6
2,4-Hexadiene, 2,5-dimethyl-
Me2C=CHCH=CMe2
110 C8H14 872-78-6
Bicyclo[2.2.1]heptane, 2-methyl-, exo-
Me
110 C8H14 931-88-4
Cyclooctene
110 C8H14 1003-64-1
Cyclohexane, ethylidene-
CHMe

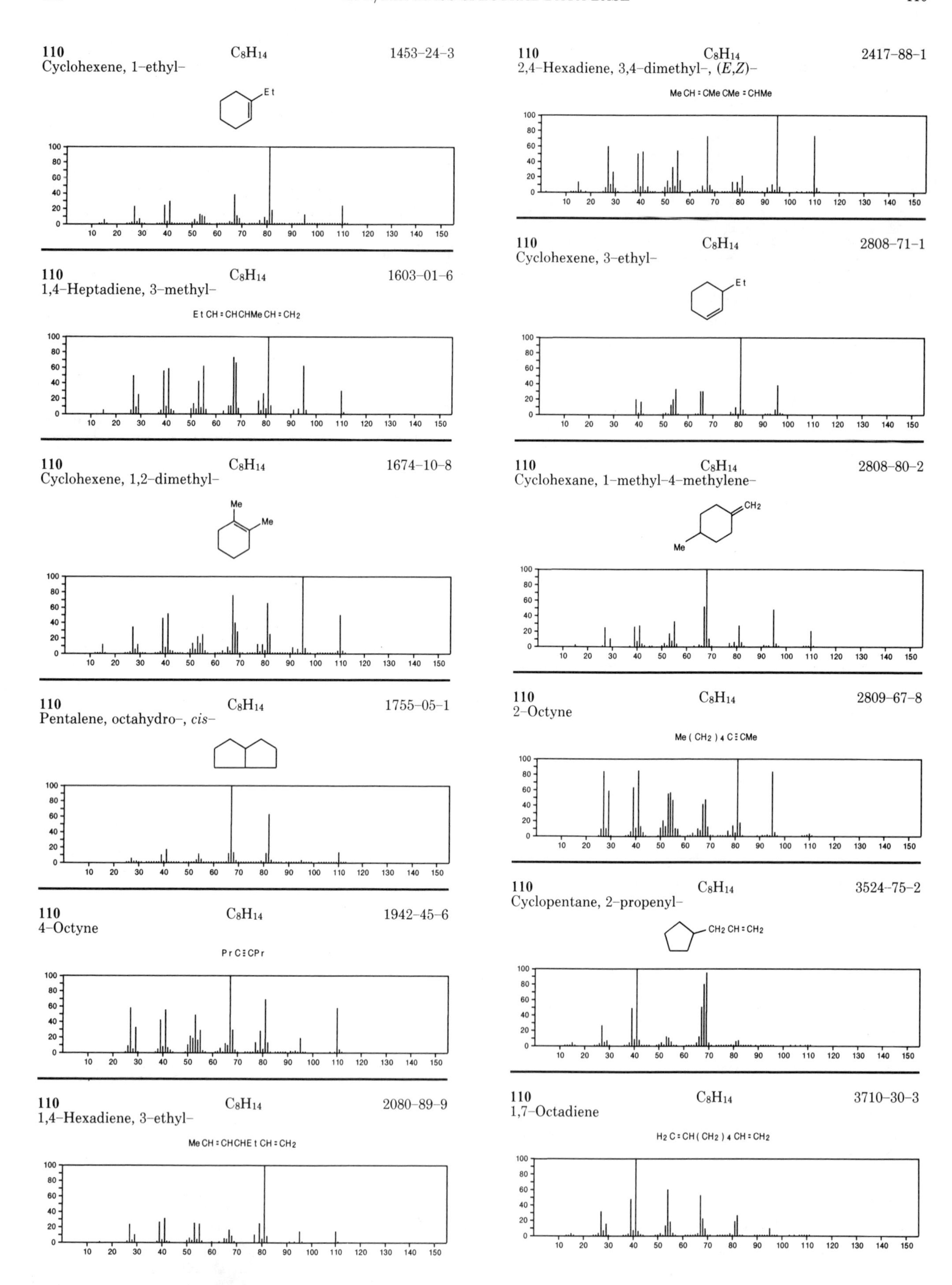
110 C_8H_{14} 1453-24-3
Cyclohexene, 1-ethyl-
Et
110 C_8H_{14} 1603-01-6
1,4-Heptadiene, 3-methyl-
Et CH = CHCHMe CH = CH2
110 C_8H_{14} 1674-10-8
Cyclohexene, 1,2-dimethyl-
Me
Me
110 C_8H_{14} 1755-05-1
Pentalene, octahydro-, *cis*-
110 C_8H_{14} 1942-45-6
4-Octyne
Pr C≡CPr
110 C_8H_{14} 2080-89-9
1,4-Hexadiene, 3-ethyl-
Me CH = CHCHE t CH = CH2
110 C_8H_{14} 2417-88-1
2,4-Hexadiene, 3,4-dimethyl-, (*E*,*Z*)-
Me CH = CMe CMe = CHMe
110 C_8H_{14} 2808-71-1
Cyclohexene, 3-ethyl-
Et
110 C_8H_{14} 2808-80-2
Cyclohexane, 1-methyl-4-methylene-
CH2
Me
110 C_8H_{14} 2809-67-8
2-Octyne
Me (CH2) 4 C≡CMe
110 C_8H_{14} 3524-75-2
Cyclopentane, 2-propenyl-
CH2 CH = CH2
110 C_8H_{14} 3710-30-3
1,7-Octadiene
H2 C = CH (CH2) 4 CH = CH2

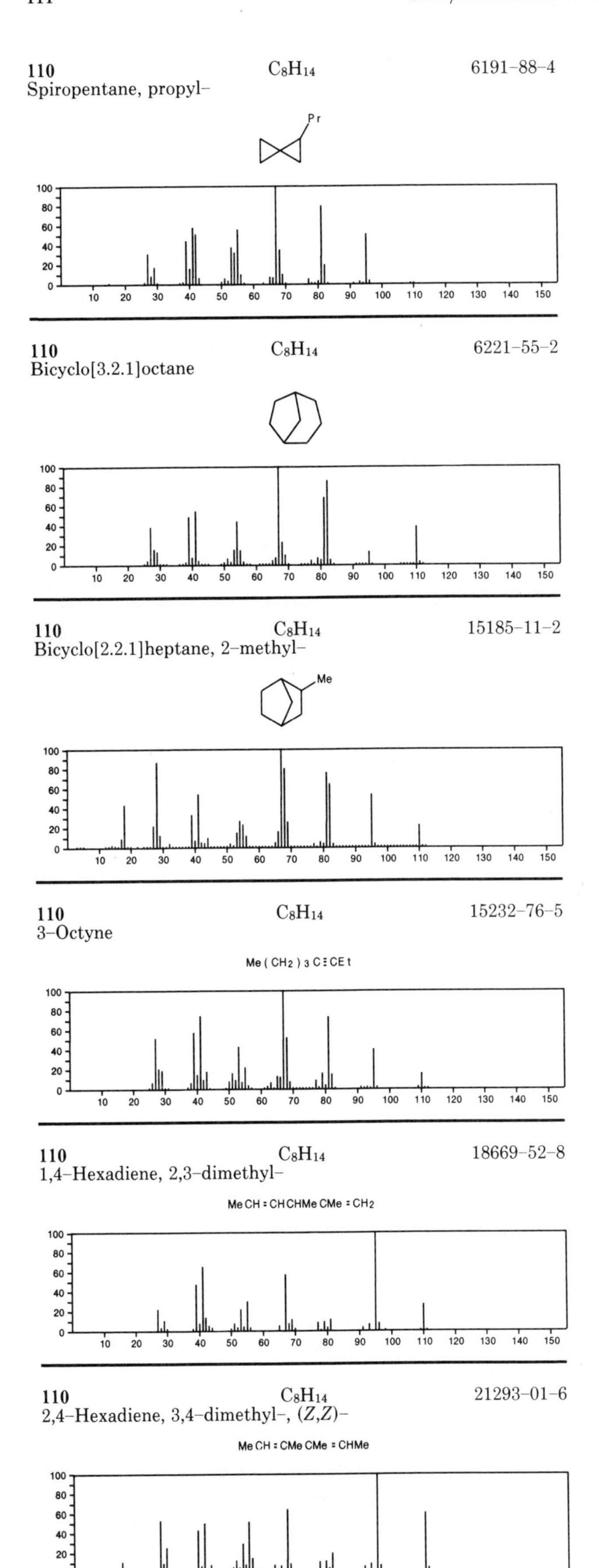
110 C8H14 6191-88-4
Spiropentane, propyl-
Pr
110 C8H14 6221-55-2
Bicyclo[3.2.1]octane
110 C8H14 15185-11-2
Bicyclo[2.2.1]heptane, 2-methyl-
Me
110 C8H14 15232-76-5
3-Octyne
Me(CH2)3C≡CEt
110 C8H14 18669-52-8
1,4-Hexadiene, 2,3-dimethyl-
MeCH=CHCHMeCMe=CH2
110 C8H14 21293-01-6
2,4-Hexadiene, 3,4-dimethyl-, (Z,Z)-
MeCH=CMeCMe=CHMe

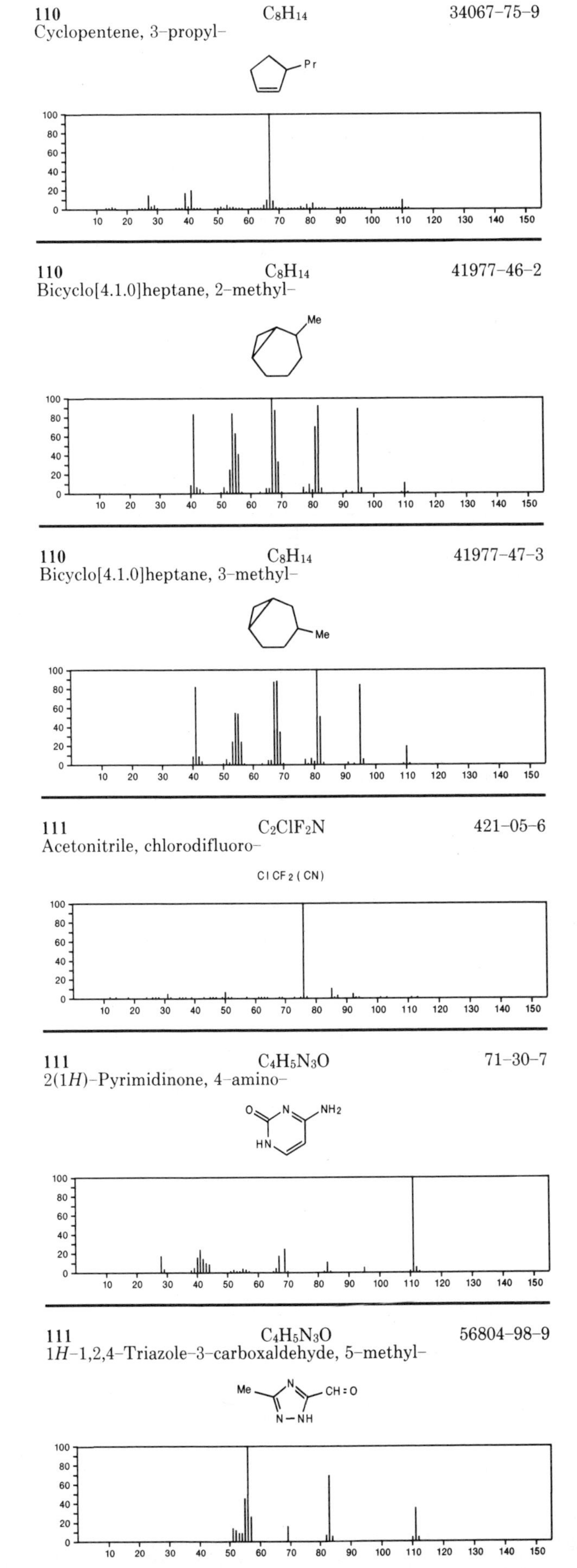
110 C8H14 34067-75-9
Cyclopentene, 3-propyl-
Pr
110 C8H14 41977-46-2
Bicyclo[4.1.0]heptane, 2-methyl-
Me
110 C8H14 41977-47-3
Bicyclo[4.1.0]heptane, 3-methyl-
Me
111 C2ClF2N 421-05-6
Acetonitrile, chlorodifluoro-
ClCF2(CN)
111 C4H5N3O 71-30-7
2(1H)-Pyrimidinone, 4-amino-
O N NH2 HN
111 C4H5N3O 56804-98-9
1H-1,2,4-Triazole-3-carboxaldehyde, 5-methyl-
Me N CH=O N-NH

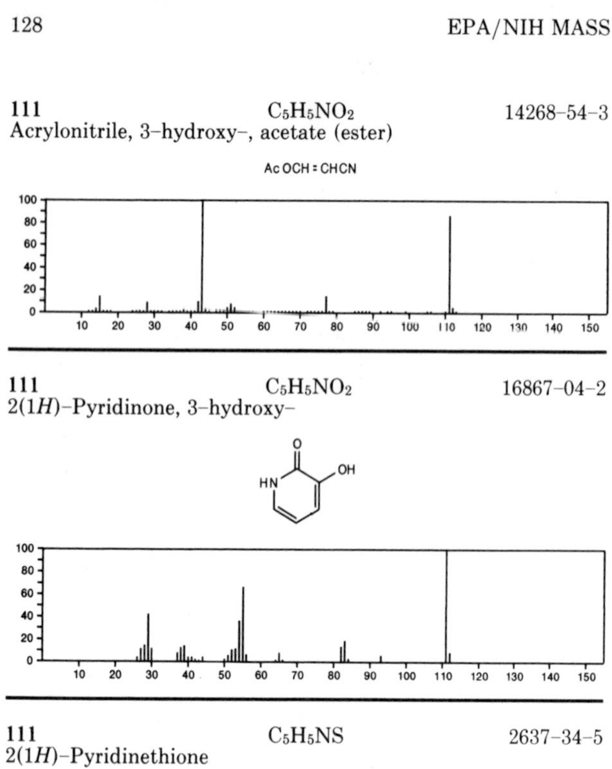

111 $C_5H_5NO_2$ 14268-54-3
Acrylonitrile, 3-hydroxy-, acetate (ester)

111 $C_5H_5NO_2$ 16867-04-2
2(1*H*)-Pyridinone, 3-hydroxy-

111 C_5H_5NS 2637-34-5
2(1*H*)-Pyridinethione

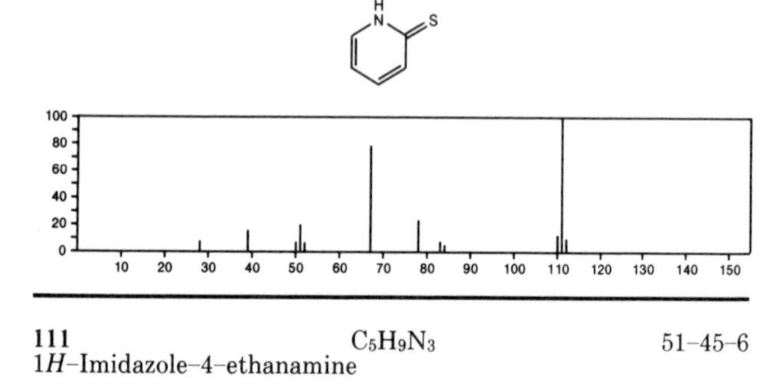

111 $C_5H_9N_3$ 51-45-6
1*H*-Imidazole-4-ethanamine

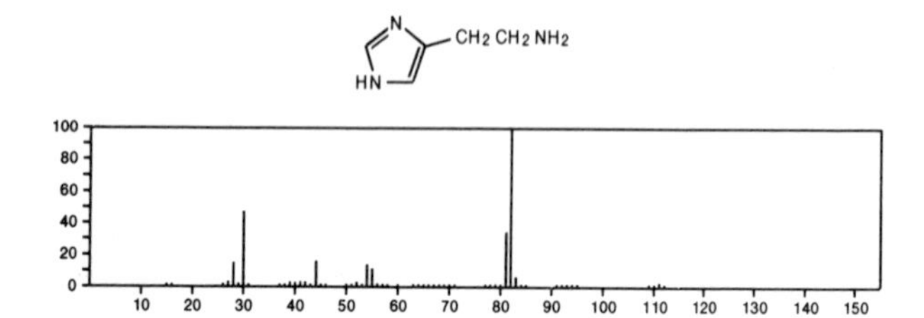

111 $C_5H_9N_3$ 19932-60-6
1*H*-1,2,4-Triazole, 3-propyl-

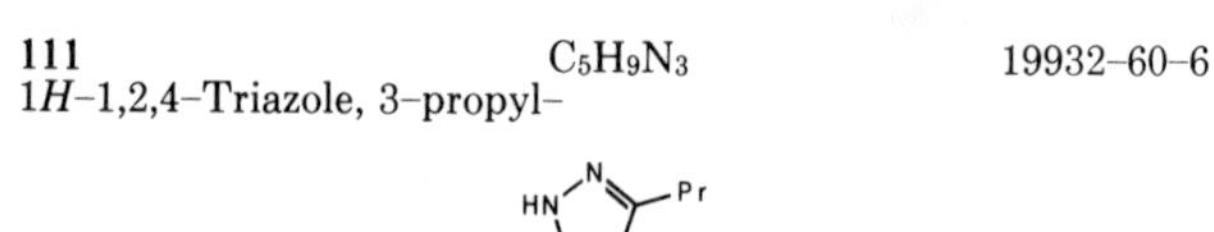

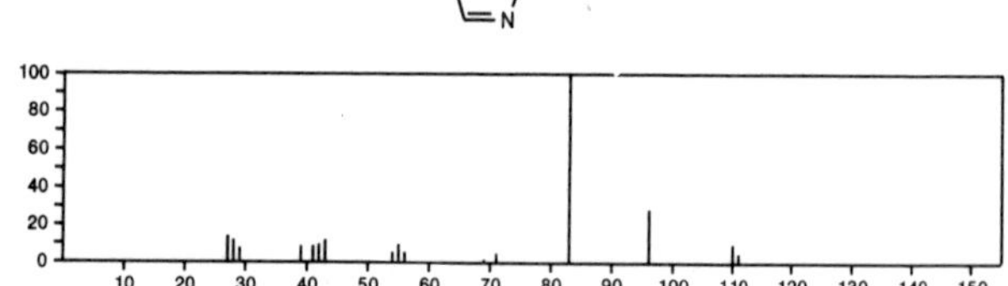

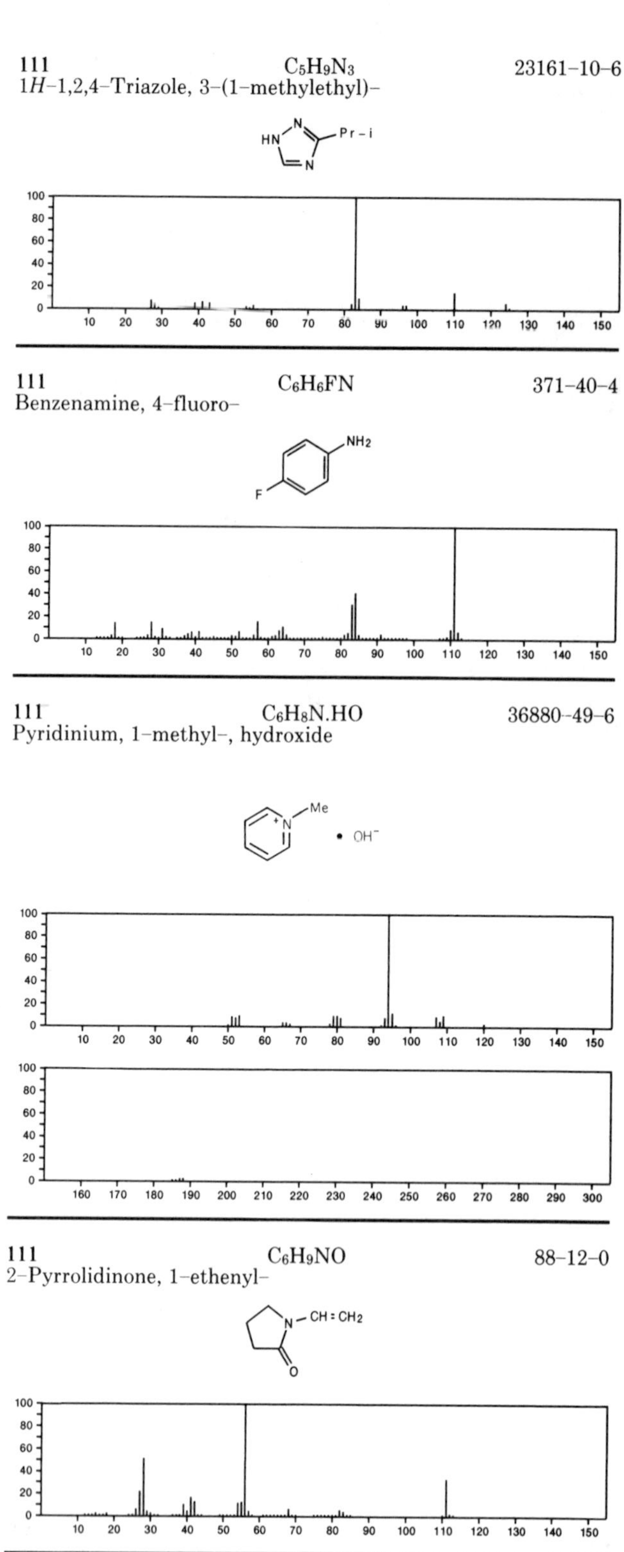

111 $C_5H_9N_3$ 23161-10-6
1*H*-1,2,4-Triazole, 3-(1-methylethyl)-

111 C_6H_6FN 371-40-4
Benzenamine, 4-fluoro-

111 $C_6H_8N.HO$ 36880-49-6
Pyridinium, 1-methyl-, hydroxide

111 C_6H_9NO 88-12-0
2-Pyrrolidinone, 1-ethenyl-

111 C_6H_9NO 2228-79-7
2*H*-Azepin-2-one, 1,5,6,7-tetrahydro-

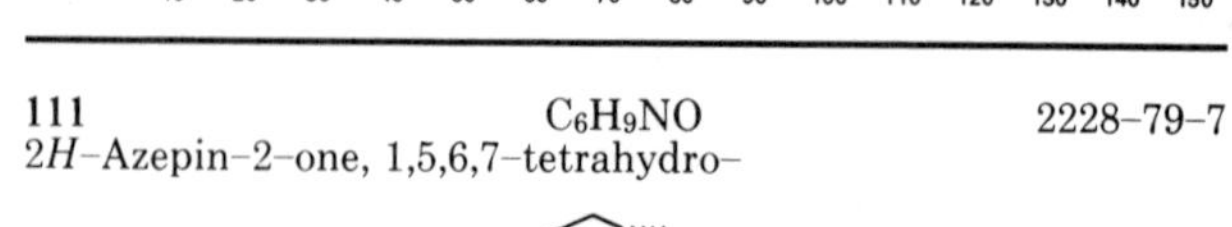

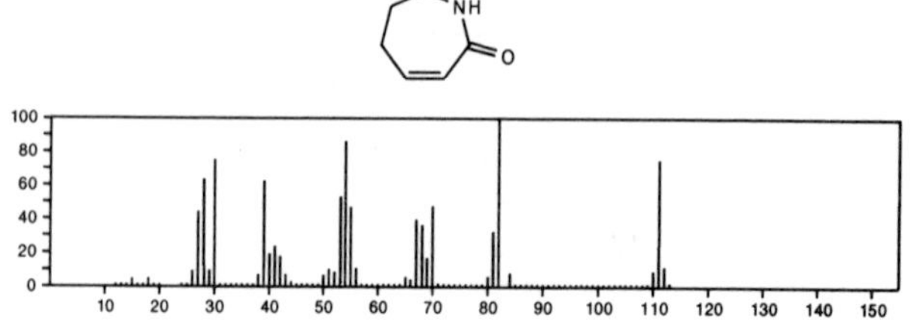

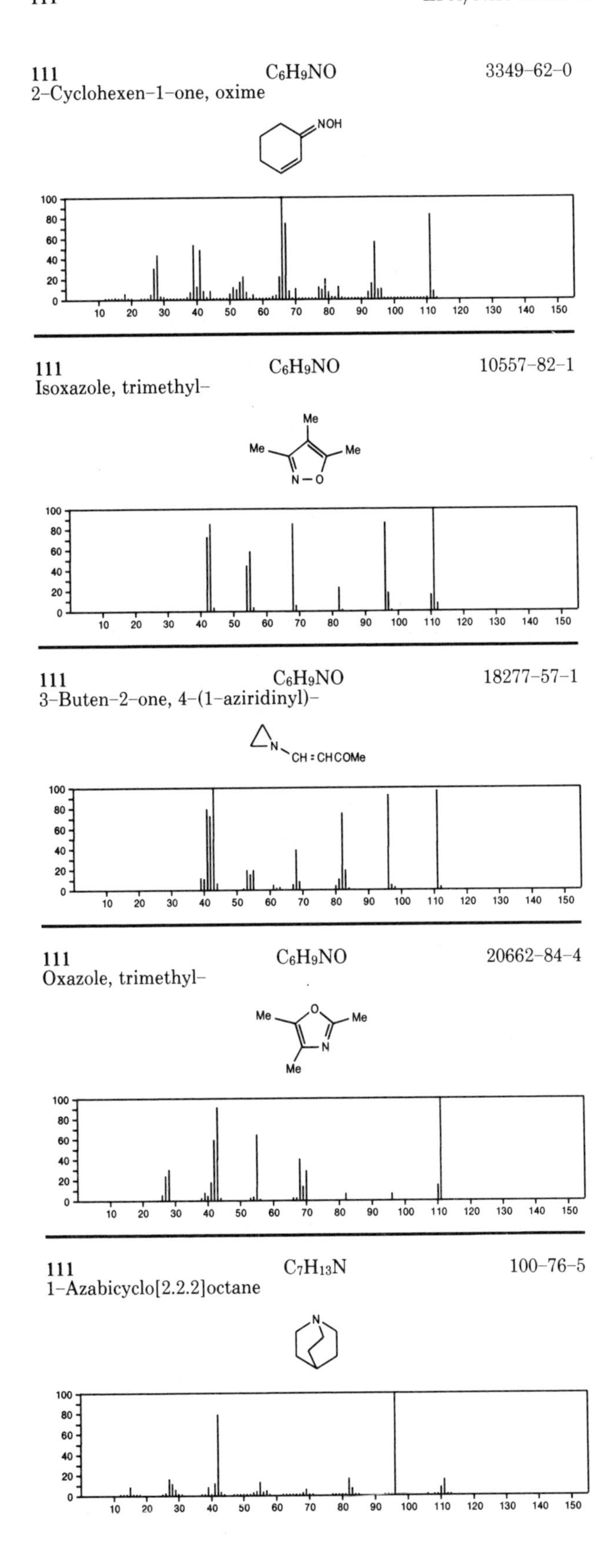

111 C_6H_9NO 3349–62–0
2–Cyclohexen–1–one, oxime

111 C_6H_9NO 10557–82–1
Isoxazole, trimethyl–

111 C_6H_9NO 18277–57–1
3–Buten–2–one, 4–(1–aziridinyl)–

111 C_6H_9NO 20662–84–4
Oxazole, trimethyl–

111 $C_7H_{13}N$ 100–76–5
1–Azabicyclo[2.2.2]octane

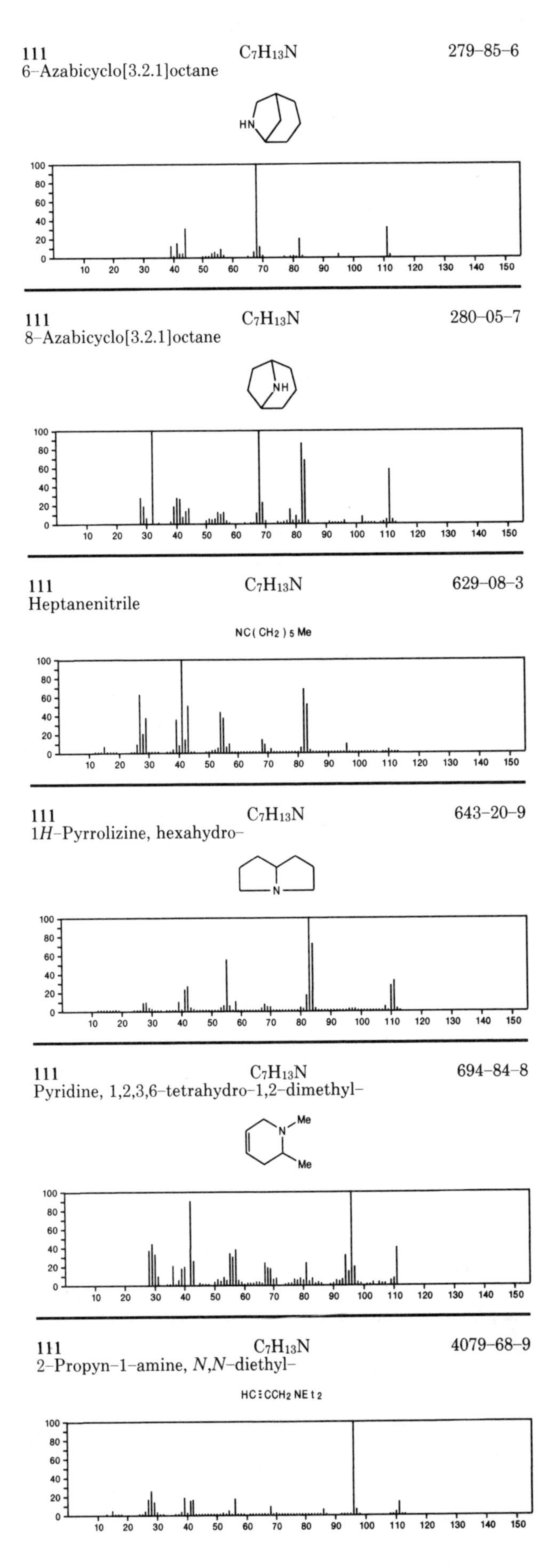

111 $C_7H_{13}N$ 279–85–6
6–Azabicyclo[3.2.1]octane

111 $C_7H_{13}N$ 280–05–7
8–Azabicyclo[3.2.1]octane

111 $C_7H_{13}N$ 629–08–3
Heptanenitrile

111 $C_7H_{13}N$ 643–20–9
1*H*–Pyrrolizine, hexahydro–

111 $C_7H_{13}N$ 694–84–8
Pyridine, 1,2,3,6–tetrahydro–1,2–dimethyl–

111 $C_7H_{13}N$ 4079–68–9
2–Propyn–1–amine, *N*,*N*–diethyl–

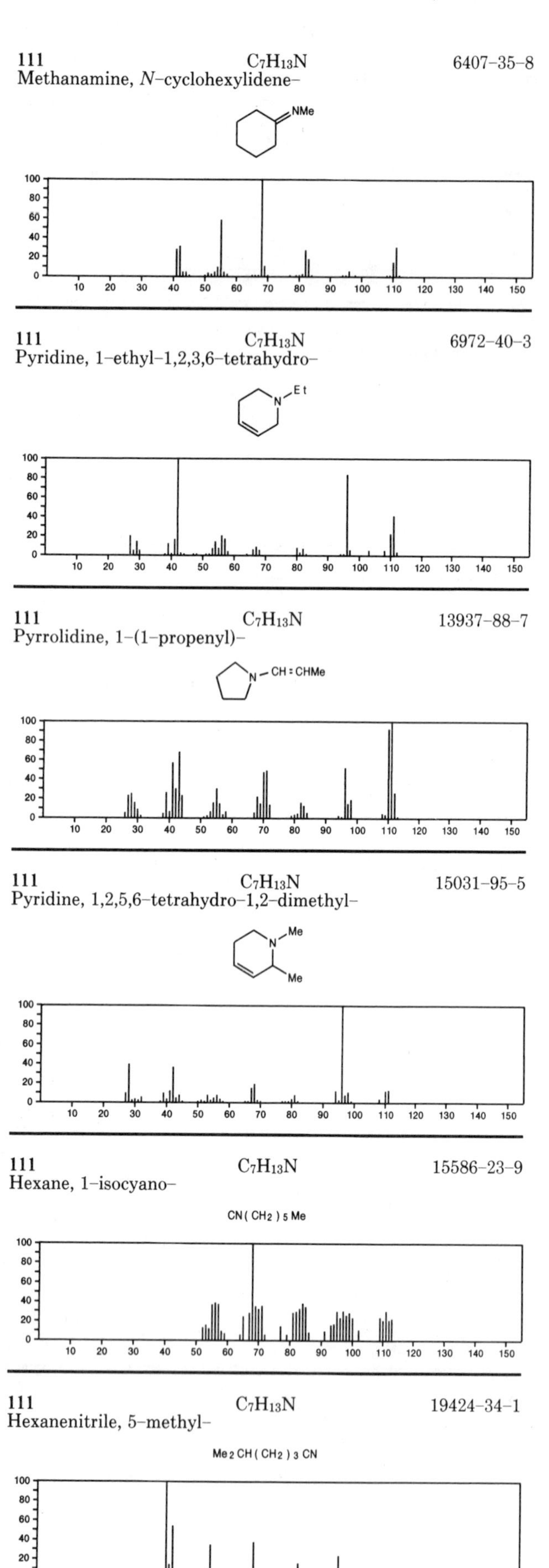

111 $C_7H_{13}N$ 6407-35-8
Methanamine, *N*-cyclohexylidene-
NMe
111 $C_7H_{13}N$ 6972-40-3
Pyridine, 1-ethyl-1,2,3,6-tetrahydro-
N Et
111 $C_7H_{13}N$ 13937-88-7
Pyrrolidine, 1-(1-propenyl)-
N CH:CHMe
111 $C_7H_{13}N$ 15031-95-5
Pyridine, 1,2,5,6-tetrahydro-1,2-dimethyl-
N Me
Me
111 $C_7H_{13}N$ 15586-23-9
Hexane, 1-isocyano-
CN(CH2)5Me
111 $C_7H_{13}N$ 19424-34-1
Hexanenitrile, 5-methyl-
Me2CH(CH2)3CN

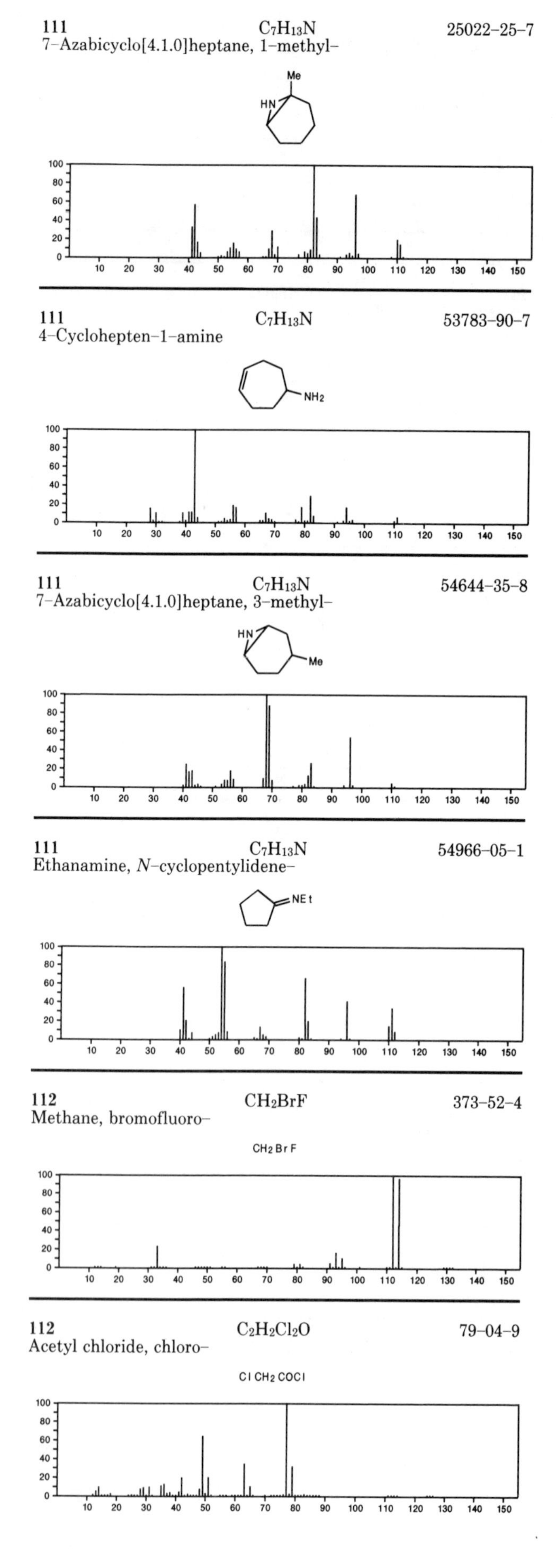

111 $C_7H_{13}N$ 25022-25-7
7-Azabicyclo[4.1.0]heptane, 1-methyl-
Me
HN
111 $C_7H_{13}N$ 53783-90-7
4-Cyclohepten-1-amine
NH2
111 $C_7H_{13}N$ 54644-35-8
7-Azabicyclo[4.1.0]heptane, 3-methyl-
HN
Me
111 $C_7H_{13}N$ 54966-05-1
Ethanamine, *N*-cyclopentylidene-
NEt
112 CH_2BrF 373-52-4
Methane, bromofluoro-
CH2BrF
112 $C_2H_2Cl_2O$ 79-04-9
Acetyl chloride, chloro-
ClCH2COCl

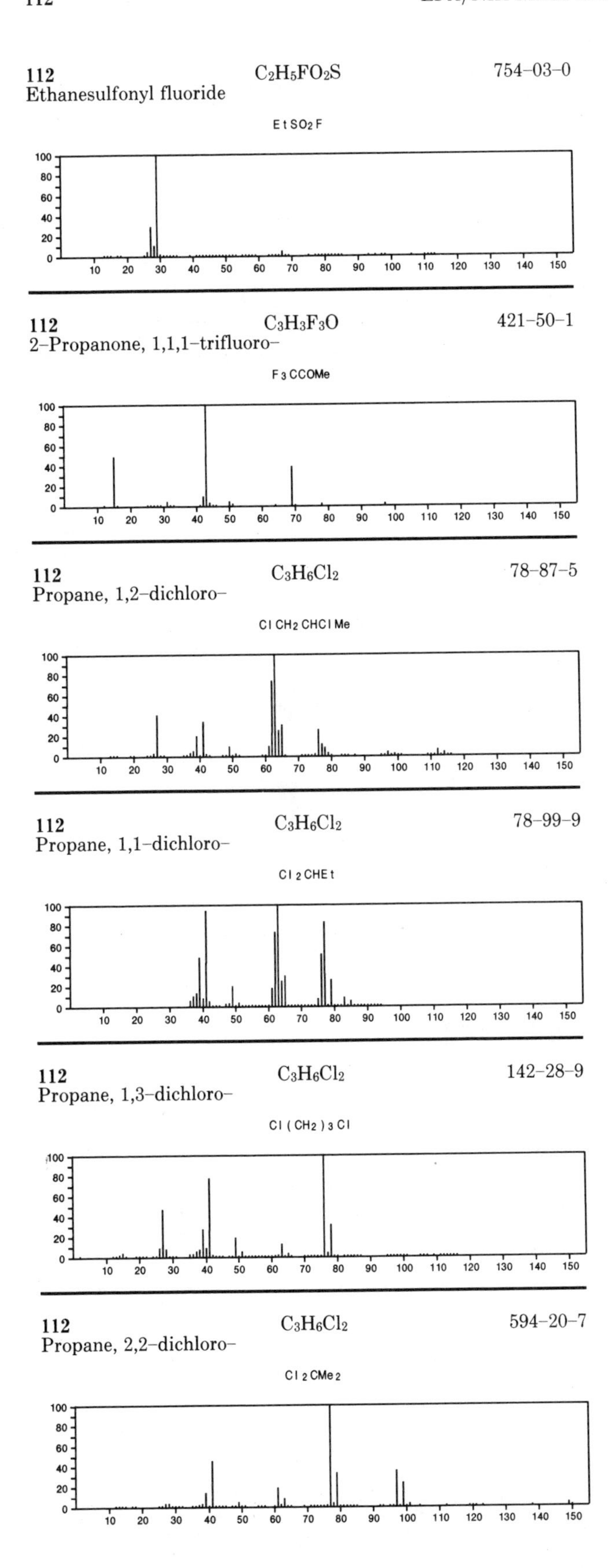
112 C2H5FO2S 754-03-0
Ethanesulfonyl fluoride
EtSO2F
112 C3H3F3O 421-50-1
2-Propanone, 1,1,1-trifluoro-
F3CCOMe
112 C3H6Cl2 78-87-5
Propane, 1,2-dichloro-
ClCH2CHClMe
112 C3H6Cl2 78-99-9
Propane, 1,1-dichloro-
Cl2CHEt
112 C3H6Cl2 142-28-9
Propane, 1,3-dichloro-
Cl(CH2)3Cl
112 C3H6Cl2 594-20-7
Propane, 2,2-dichloro-
Cl2CMe2

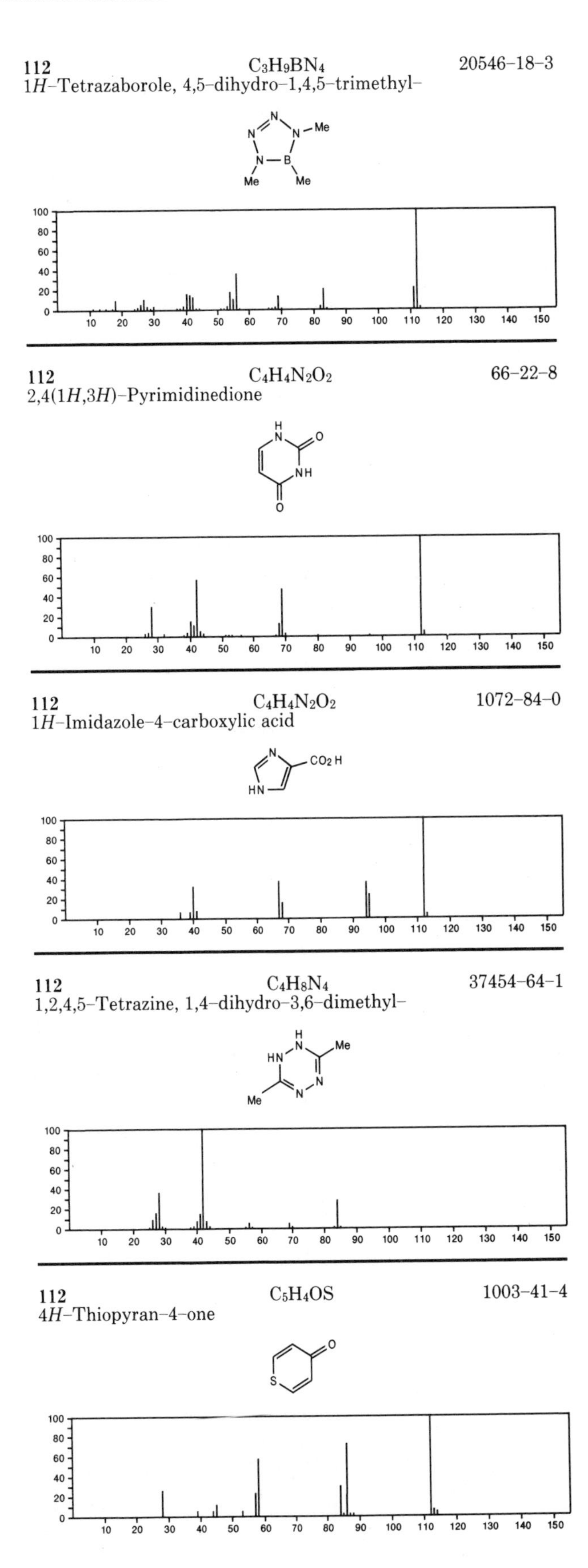
112 C3H9BN4 20546-18-3
1H-Tetrazaborole, 4,5-dihydro-1,4,5-trimethyl-
112 C4H4N2O2 66-22-8
2,4(1H,3H)-Pyrimidinedione
112 C4H4N2O2 1072-84-0
1H-Imidazole-4-carboxylic acid
112 C4H8N4 37454-64-1
1,2,4,5-Tetrazine, 1,4-dihydro-3,6-dimethyl-
112 C5H4OS 1003-41-4
4H-Thiopyran-4-one

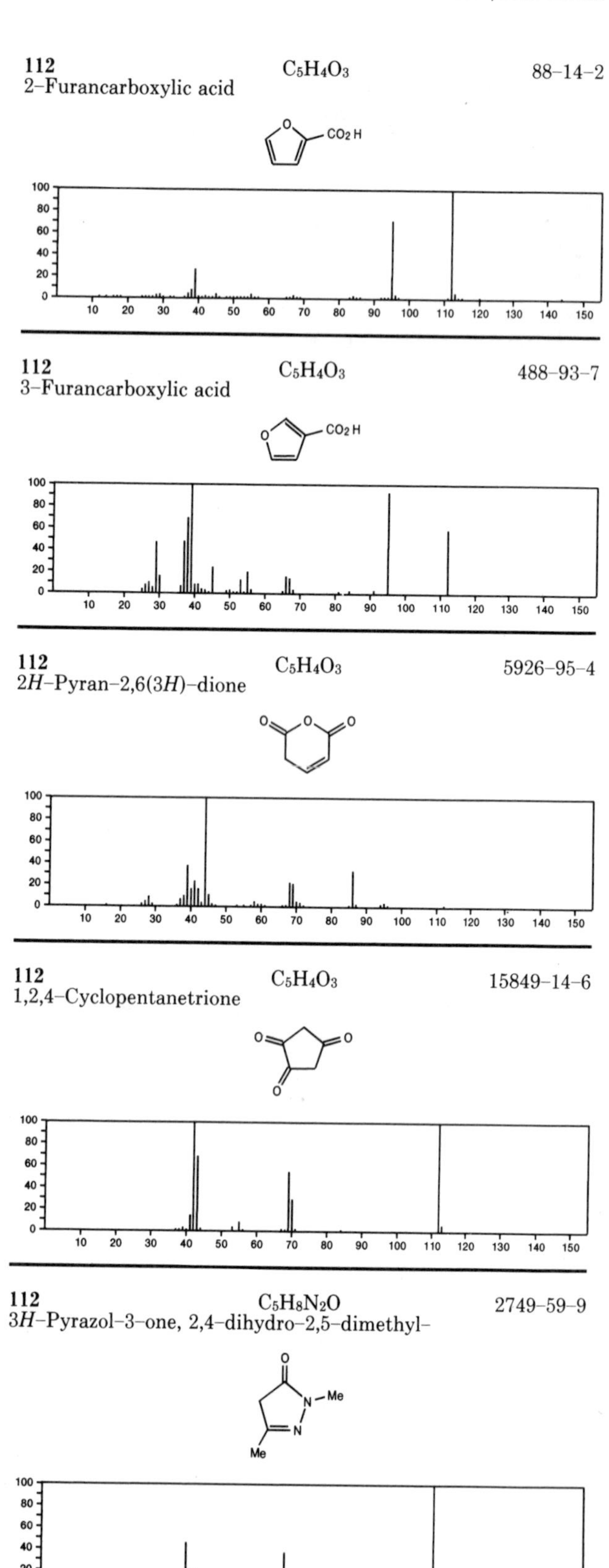

112 $C_5H_4O_3$ 88-14-2
2-Furancarboxylic acid

112 $C_5H_4O_3$ 488-93-7
3-Furancarboxylic acid

112 $C_5H_4O_3$ 5926-95-4
2*H*-Pyran-2,6(3*H*)-dione

112 $C_5H_4O_3$ 15849-14-6
1,2,4-Cyclopentanetrione

112 $C_5H_8N_2O$ 2749-59-9
3*H*-Pyrazol-3-one, 2,4-dihydro-2,5-dimethyl-

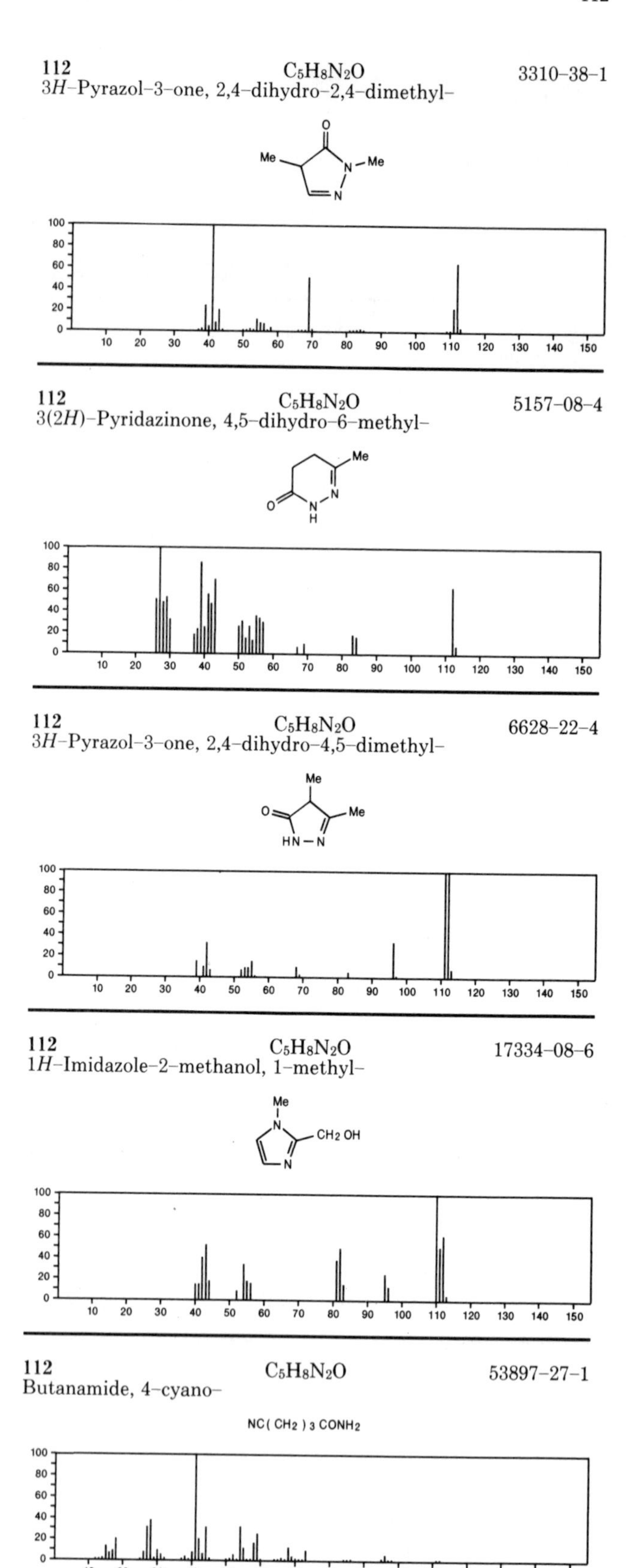

112 $C_5H_8N_2O$ 3310-38-1
3*H*-Pyrazol-3-one, 2,4-dihydro-2,4-dimethyl-

112 $C_5H_8N_2O$ 5157-08-4
3(2*H*)-Pyridazinone, 4,5-dihydro-6-methyl-

112 $C_5H_8N_2O$ 6628-22-4
3*H*-Pyrazol-3-one, 2,4-dihydro-4,5-dimethyl-

112 $C_5H_8N_2O$ 17334-08-6
1*H*-Imidazole-2-methanol, 1-methyl-

112 $C_5H_8N_2O$ 53897-27-1
Butanamide, 4-cyano-

NC(CH2)3CONH2

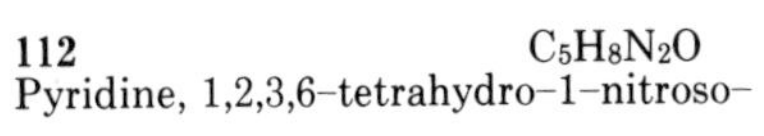

112 $C_5H_8N_2O$ 55556-92-8
Pyridine, 1,2,3,6-tetrahydro-1-nitroso-

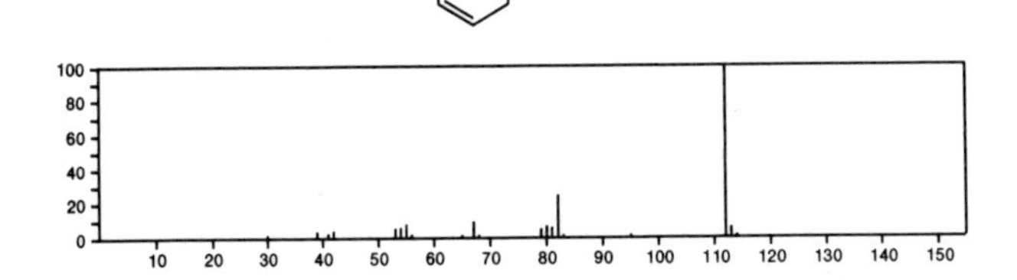

112 C_6H_5Cl 108-90-7
Benzene, chloro-

PhCl

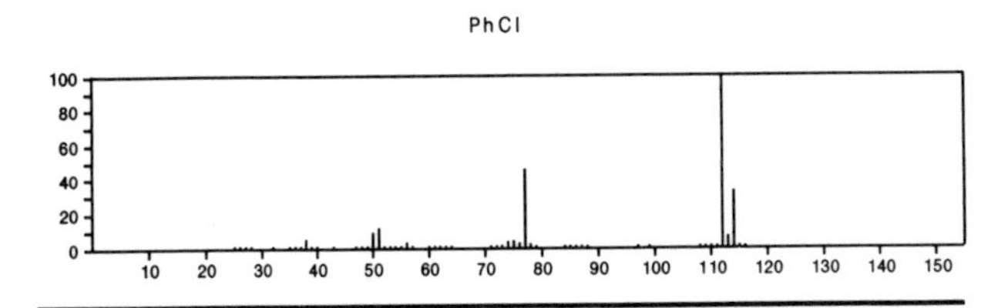

112 C_6H_5FO 367-12-4
Phenol, 2-fluoro-

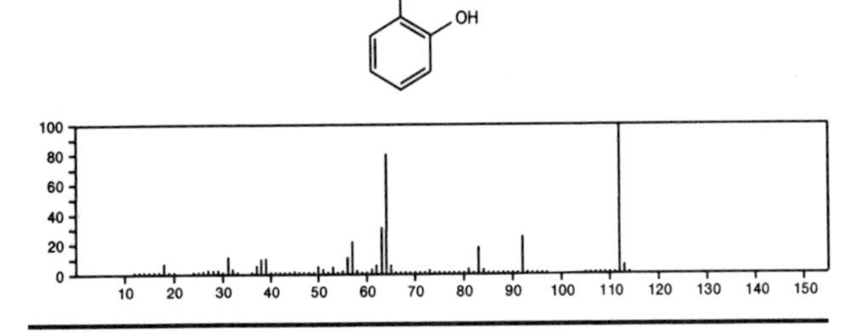

112 C_6H_5FO 371-41-5
Phenol, 4-fluoro-

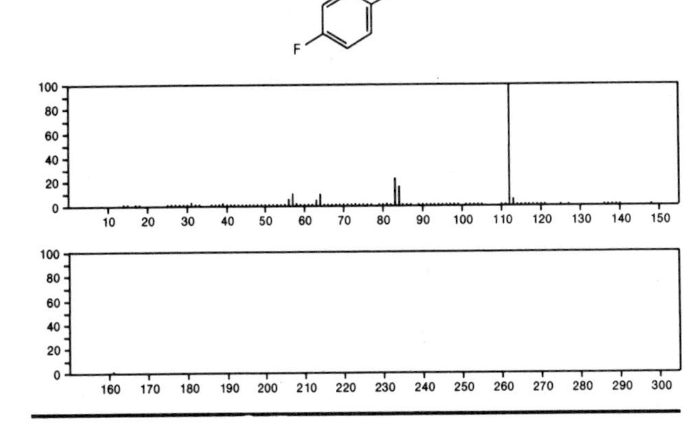

112 $C_6H_8O_2$ 80-71-7
2-Cyclopenten-1-one, 2-hydroxy-3-methyl-

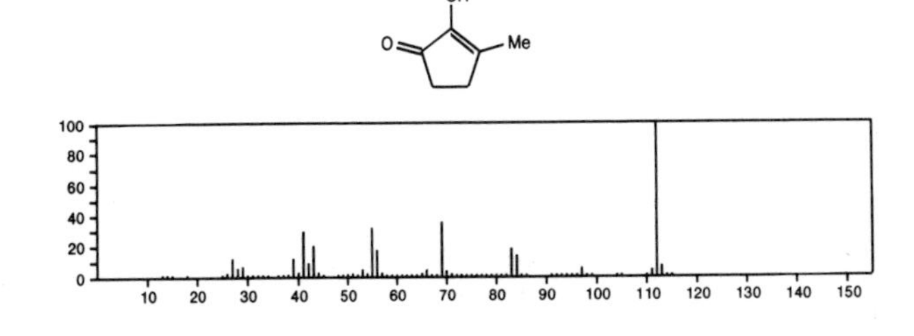

112 $C_6H_8O_2$ 100-73-2
2*H*-Pyran-2-carboxaldehyde, 3,4-dihydro-

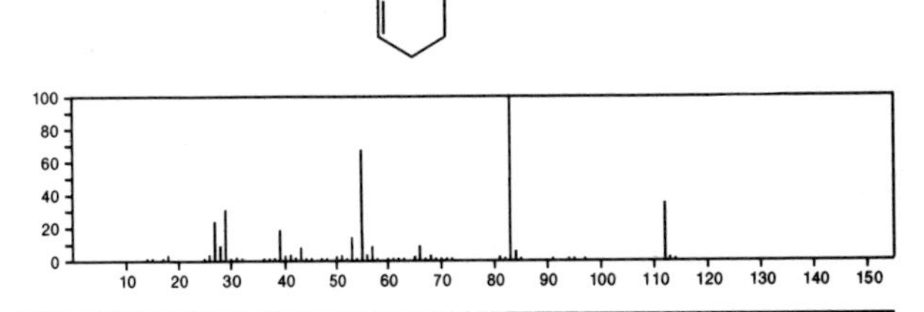

112 $C_6H_8O_2$ 110-44-1
2,4-Hexadienoic acid, (*E*,*E*)-

MeCH=CHCH=CHCO2H

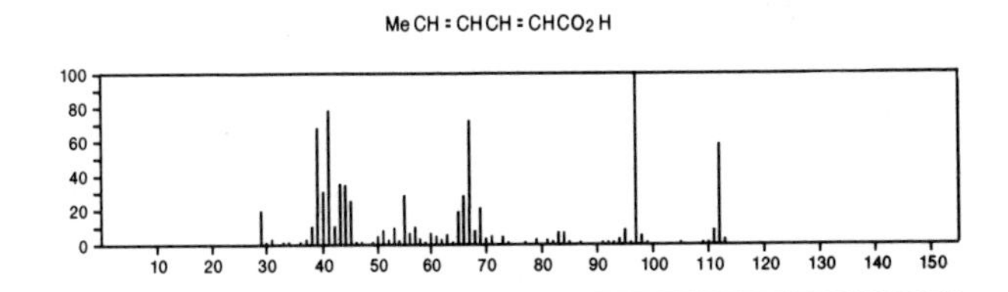

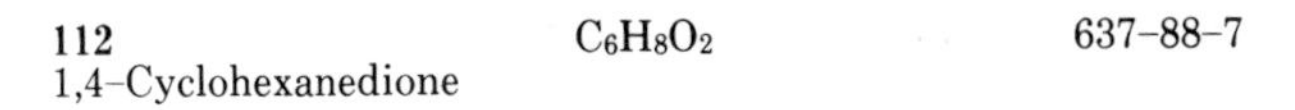

112 $C_6H_8O_2$ 637-88-7
1,4-Cyclohexanedione

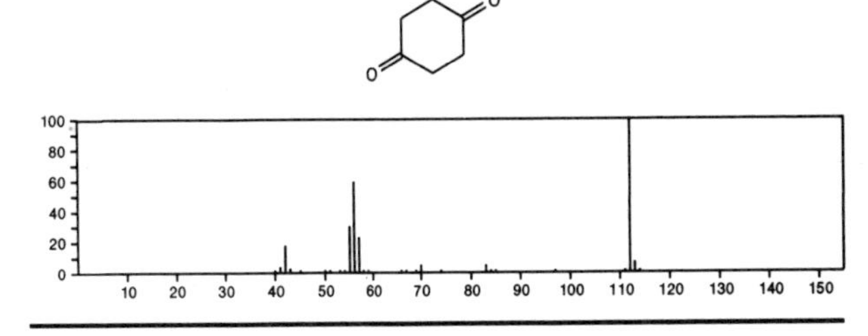

112 $C_6H_8O_2$ 765-69-5
1,3-Cyclopentanedione, 2-methyl-

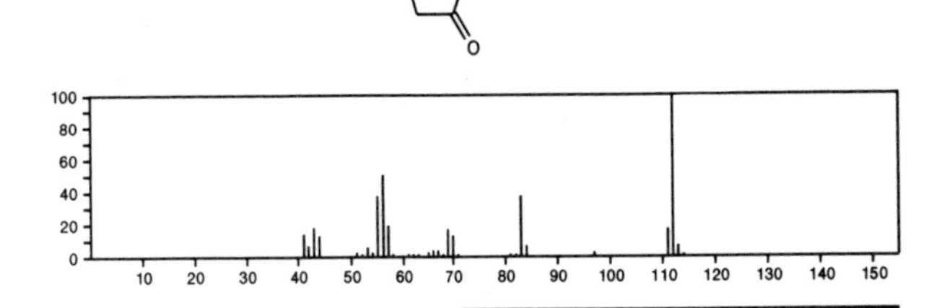

112 $C_6H_8O_2$ 765-87-7
1,2-Cyclohexanedione

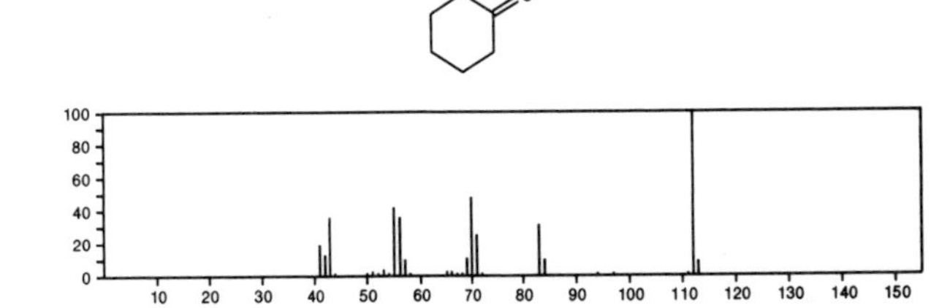

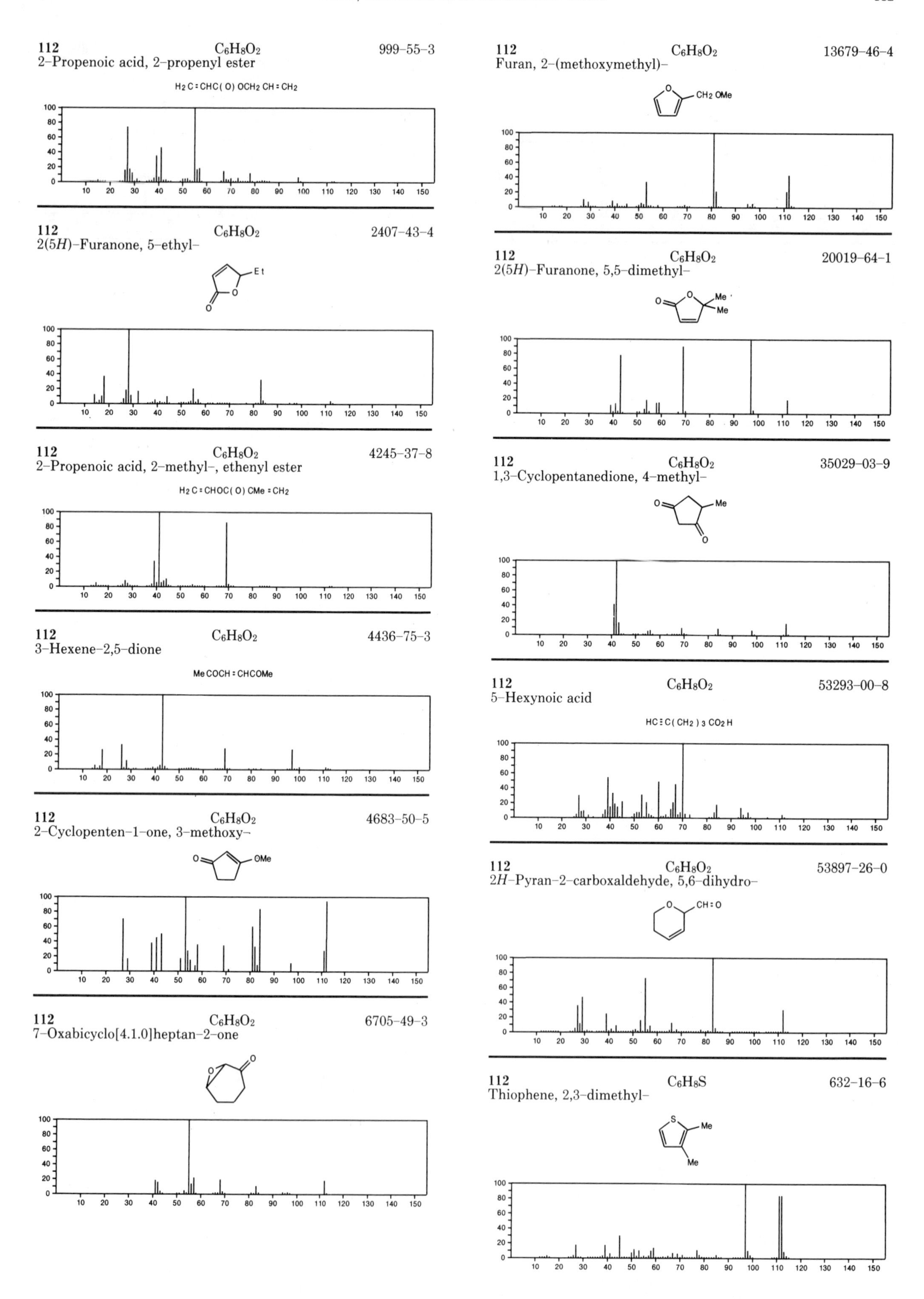
112 C6H8O2 999-55-3
2-Propenoic acid, 2-propenyl ester
H2C=CHC(O)OCH2CH=CH2
112 C6H8O2 2407-43-4
2(5H)-Furanone, 5-ethyl-
Et
O
112 C6H8O2 4245-37-8
2-Propenoic acid, 2-methyl-, ethenyl ester
H2C=CHOC(O)CMe=CH2
112 C6H8O2 4436-75-3
3-Hexene-2,5-dione
MeCOCH=CHCOMe
112 C6H8O2 4683-50-5
2-Cyclopenten-1-one, 3-methoxy-
O
OMe
112 C6H8O2 6705-49-3
7-Oxabicyclo[4.1.0]heptan-2-one
O
112 C6H8O2 13679-46-4
Furan, 2-(methoxymethyl)-
CH2OMe
112 C6H8O2 20019-64-1
2(5H)-Furanone, 5,5-dimethyl-
Me
112 C6H8O2 35029-03-9
1,3-Cyclopentanedione, 4-methyl-
Me
112 C6H8O2 53293-00-8
5-Hexynoic acid
HC≡C(CH2)3CO2H
112 C6H8O2 53897-26-0
2H-Pyran-2-carboxaldehyde, 5,6-dihydro-
CH=O
112 C6H8S 632-16-6
Thiophene, 2,3-dimethyl-
S
Me
100
80
60
40
20
0
10 20 30 40 50 60 70 80 90 100 110 120 130 140 150

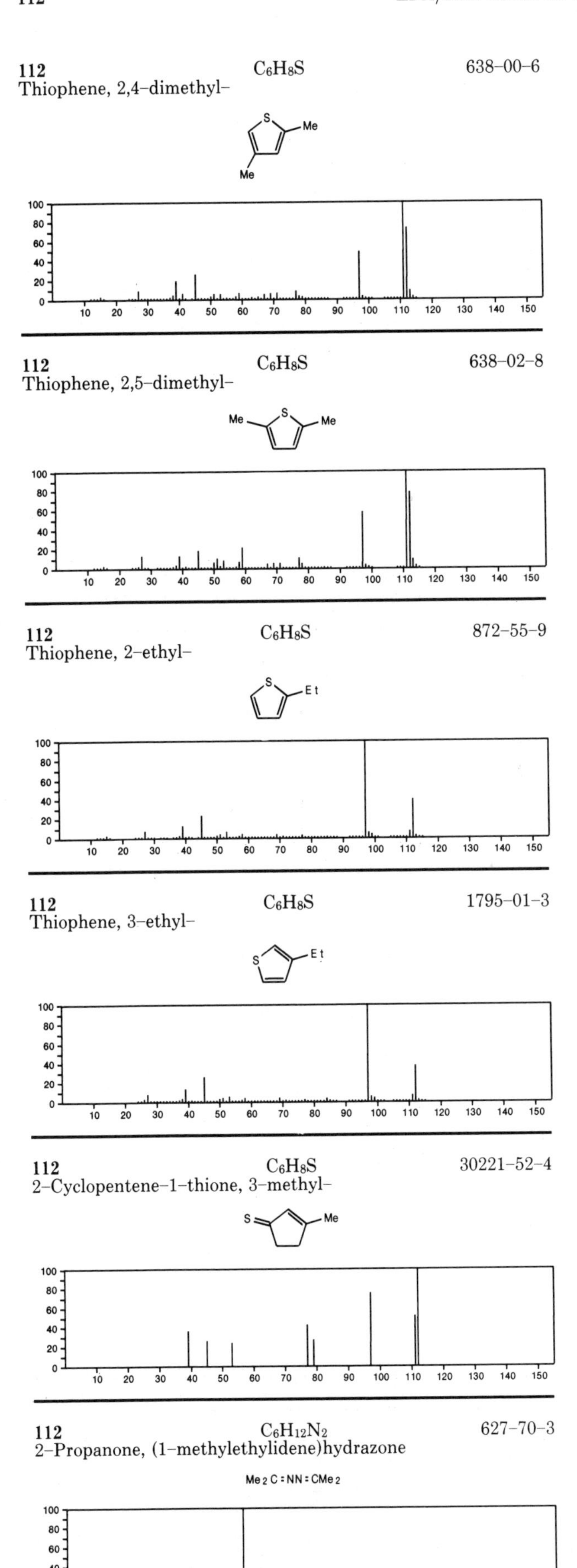

112 C_6H_8S 638-00-6
Thiophene, 2,4-dimethyl-

112 C_6H_8S 638-02-8
Thiophene, 2,5-dimethyl-

112 C_6H_8S 872-55-9
Thiophene, 2-ethyl-

112 C_6H_8S 1795-01-3
Thiophene, 3-ethyl-

112 C_6H_8S 30221-52-4
2-Cyclopentene-1-thione, 3-methyl-

112 $C_6H_{12}N_2$ 627-70-3
2-Propanone, (1-methylethylidene)hydrazone

$Me_2C{:}NN{:}CMe_2$

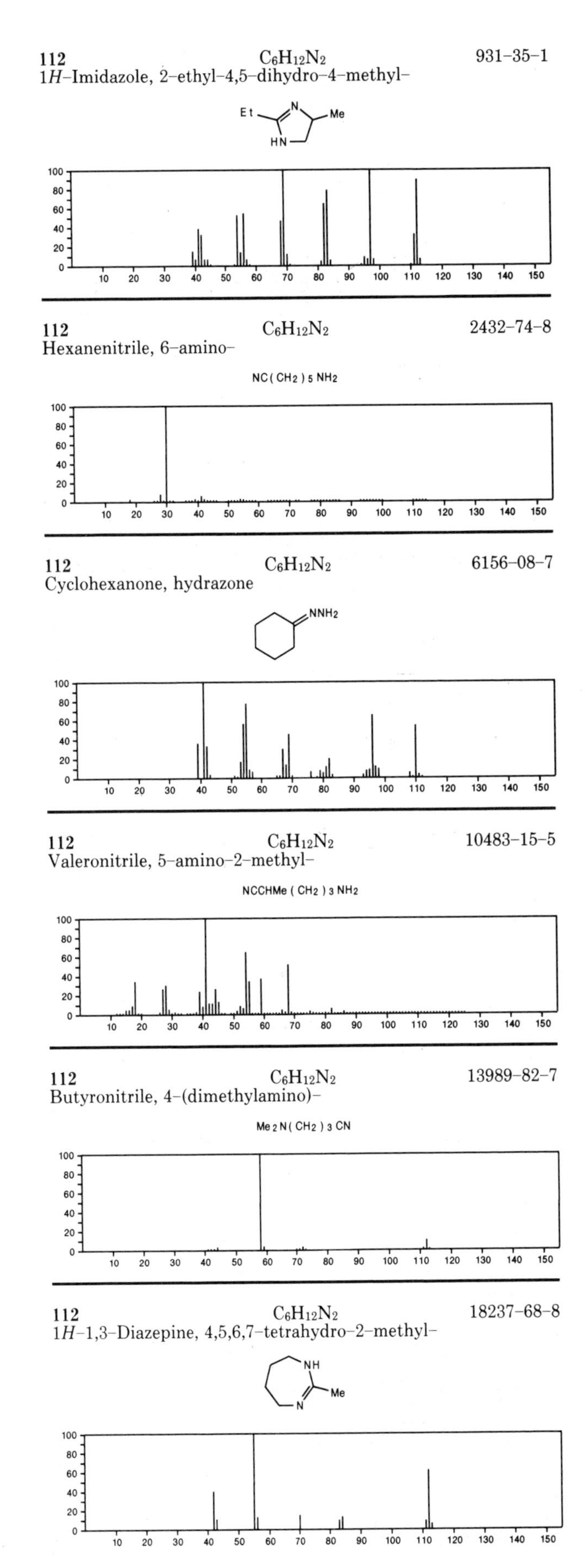

112 $C_6H_{12}N_2$ 931-35-1
1*H*-Imidazole, 2-ethyl-4,5-dihydro-4-methyl-

112 $C_6H_{12}N_2$ 2432-74-8
Hexanenitrile, 6-amino-

$NC(CH_2)_5NH_2$

112 $C_6H_{12}N_2$ 6156-08-7
Cyclohexanone, hydrazone

112 $C_6H_{12}N_2$ 10483-15-5
Valeronitrile, 5-amino-2-methyl-

$NCCHMe(CH_2)_3NH_2$

112 $C_6H_{12}N_2$ 13989-82-7
Butyronitrile, 4-(dimethylamino)-

$Me_2N(CH_2)_3CN$

112 $C_6H_{12}N_2$ 18237-68-8
1*H*-1,3-Diazepine, 4,5,6,7-tetrahydro-2-methyl-

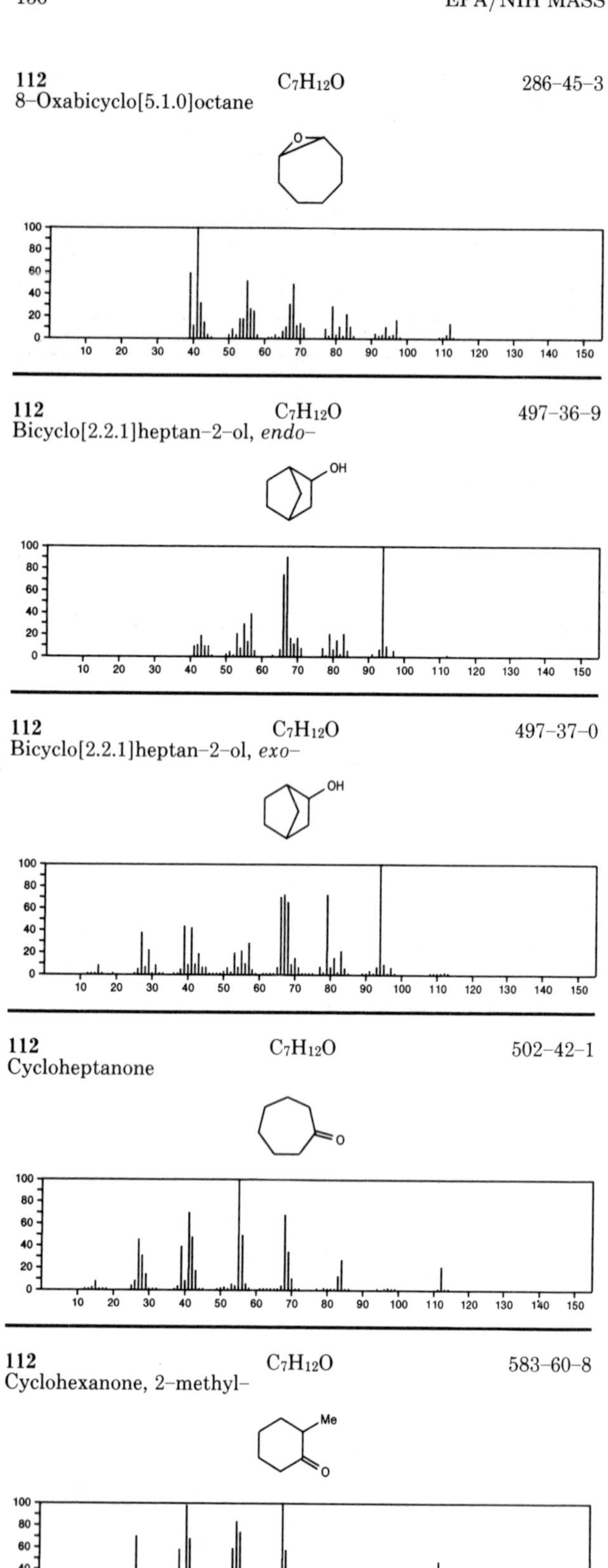
112 C7H12O 286-45-3
8-Oxabicyclo[5.1.0]octane
112 C7H12O 497-36-9
Bicyclo[2.2.1]heptan-2-ol, endo-
OH
112 C7H12O 497-37-0
Bicyclo[2.2.1]heptan-2-ol, exo-
OH
112 C7H12O 502-42-1
Cycloheptanone
O
112 C7H12O 583-60-8
Cyclohexanone, 2-methyl-
Me
O

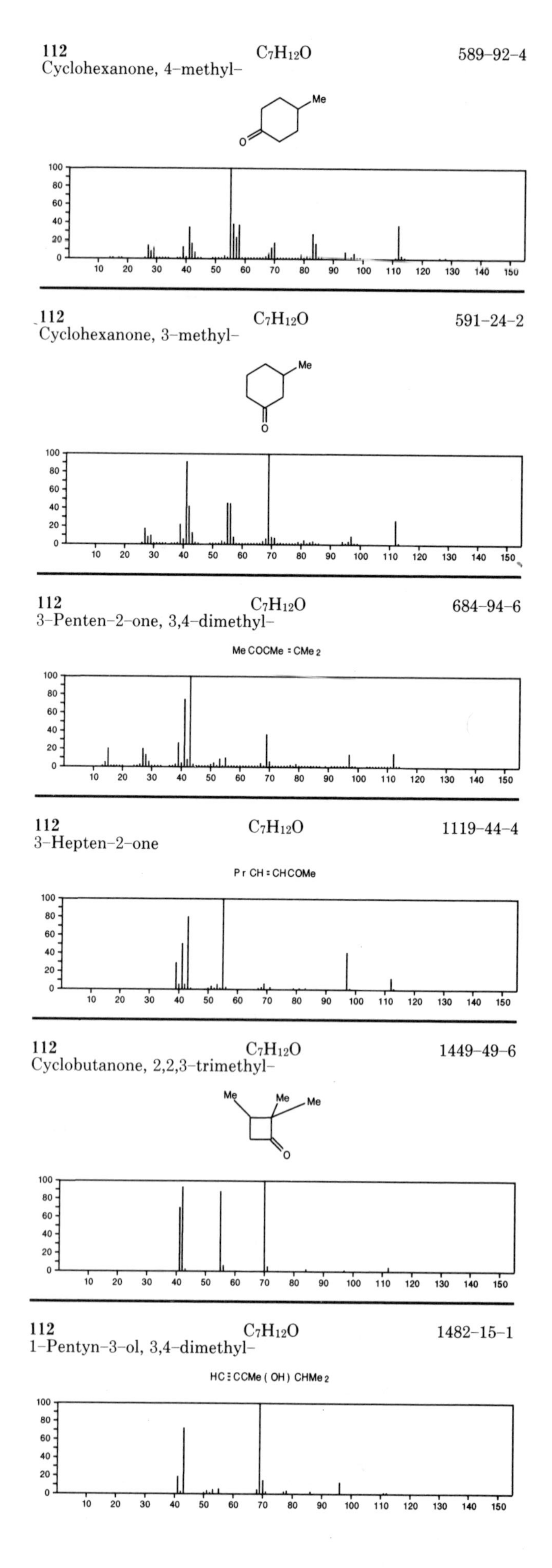
112 C7H12O 589-92-4
Cyclohexanone, 4-methyl-
Me
O
112 C7H12O 591-24-2
Cyclohexanone, 3-methyl-
Me
O
112 C7H12O 684-94-6
3-Penten-2-one, 3,4-dimethyl-
Me COCMe = CMe 2
112 C7H12O 1119-44-4
3-Hepten-2-one
Pr CH = CHCOMe
112 C7H12O 1449-49-6
Cyclobutanone, 2,2,3-trimethyl-
Me Me Me
O
112 C7H12O 1482-15-1
1-Pentyn-3-ol, 3,4-dimethyl-
HC ≡ CCMe (OH) CHMe 2

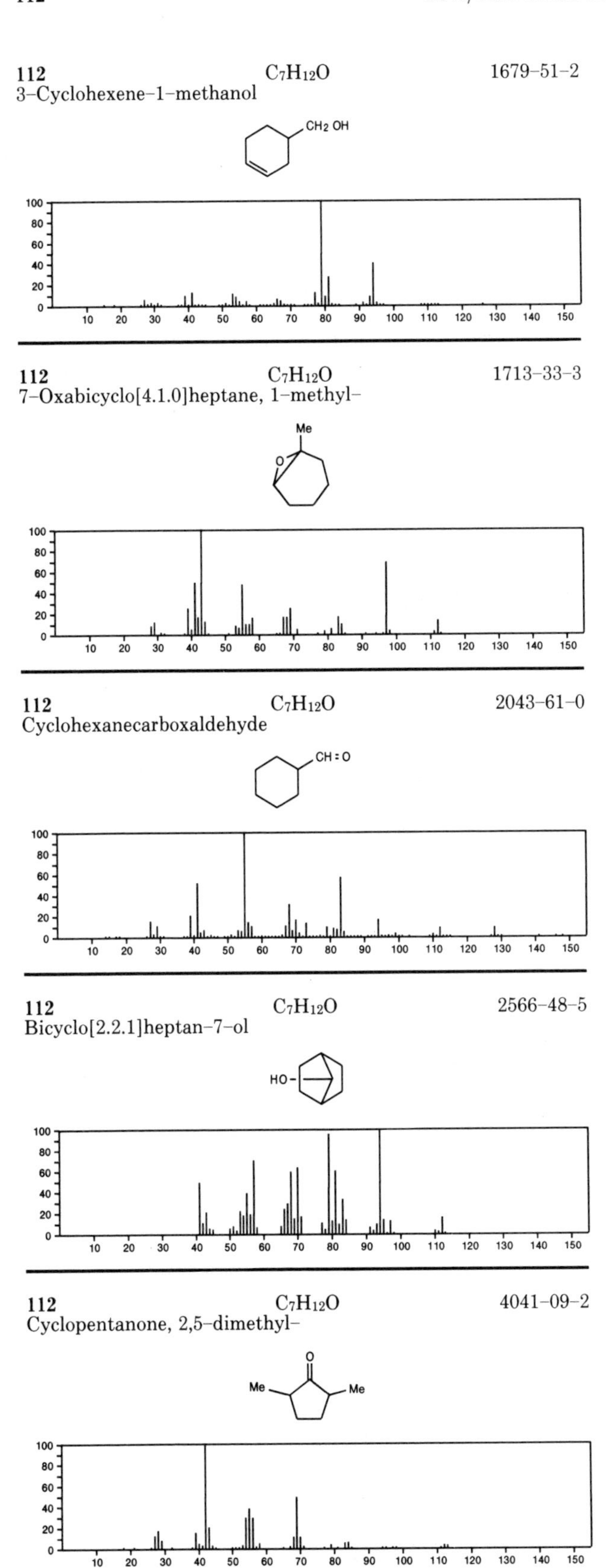

112 $C_7H_{12}O$ 1679-51-2
3-Cyclohexene-1-methanol

112 $C_7H_{12}O$ 1713-33-3
7-Oxabicyclo[4.1.0]heptane, 1-methyl-

112 $C_7H_{12}O$ 2043-61-0
Cyclohexanecarboxaldehyde

112 $C_7H_{12}O$ 2566-48-5
Bicyclo[2.2.1]heptan-7-ol

112 $C_7H_{12}O$ 4041-09-2
Cyclopentanone, 2,5-dimethyl-

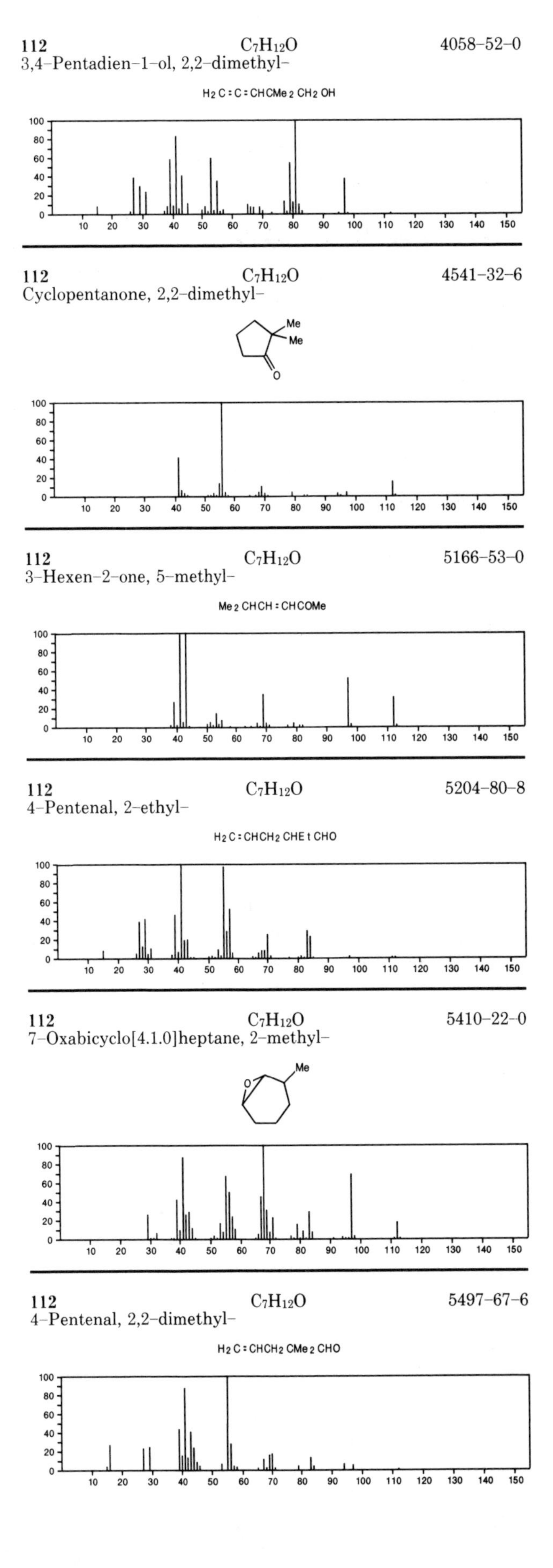

112 $C_7H_{12}O$ 4058-52-0
3,4-Pentadien-1-ol, 2,2-dimethyl-

$H_2C:C:CHCMe_2CH_2OH$

112 $C_7H_{12}O$ 4541-32-6
Cyclopentanone, 2,2-dimethyl-

112 $C_7H_{12}O$ 5166-53-0
3-Hexen-2-one, 5-methyl-

$Me_2CHCH:CHCOMe$

112 $C_7H_{12}O$ 5204-80-8
4-Pentenal, 2-ethyl-

$H_2C:CHCH_2CHEtCHO$

112 $C_7H_{12}O$ 5410-22-0
7-Oxabicyclo[4.1.0]heptane, 2-methyl-

112 $C_7H_{12}O$ 5497-67-6
4-Pentenal, 2,2-dimethyl-

$H_2C:CHCH_2CMe_2CHO$

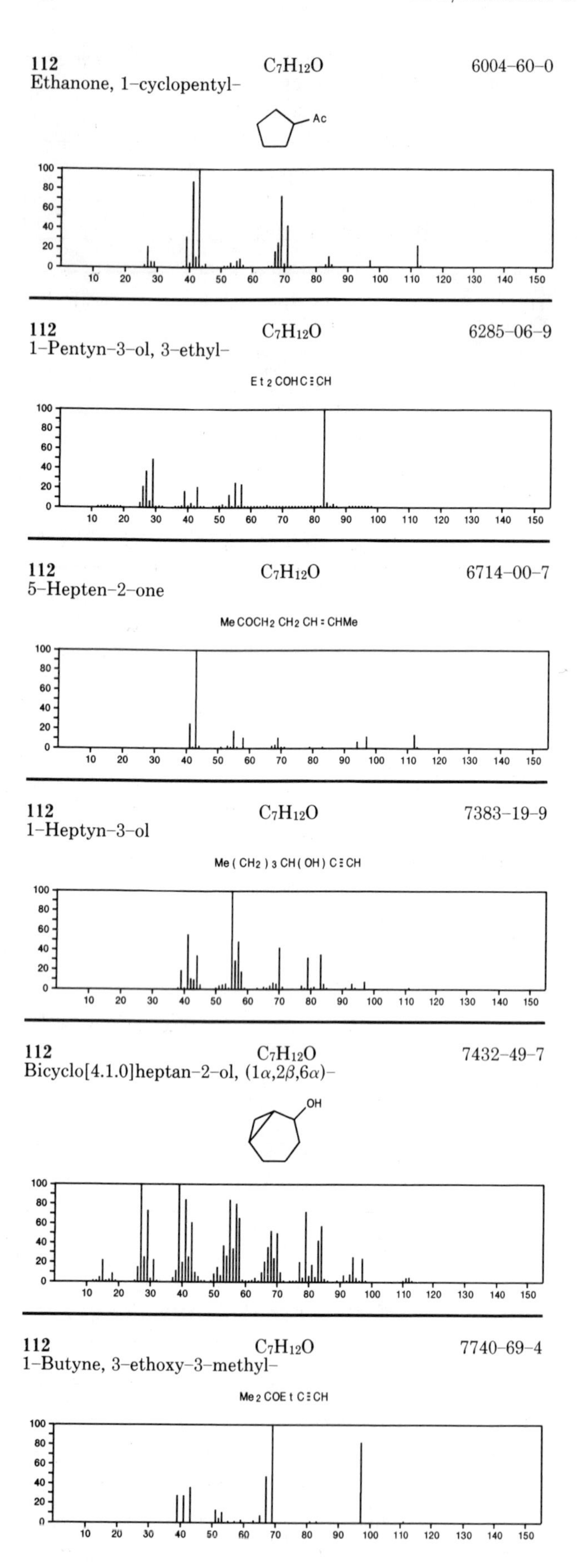

112 C7H12O 6004-60-0
Ethanone, 1-cyclopentyl-
Ac
112 C7H12O 6285-06-9
1-Pentyn-3-ol, 3-ethyl-
Et2COHC≡CH
112 C7H12O 6714-00-7
5-Hepten-2-one
MeCOCH2CH2CH=CHMe
112 C7H12O 7383-19-9
1-Heptyn-3-ol
Me(CH2)3CH(OH)C≡CH
112 C7H12O 7432-49-7
Bicyclo[4.1.0]heptan-2-ol, (1α,2β,6α)-
OH
112 C7H12O 7740-69-4
1-Butyne, 3-ethoxy-3-methyl-
Me2COEtC≡CH

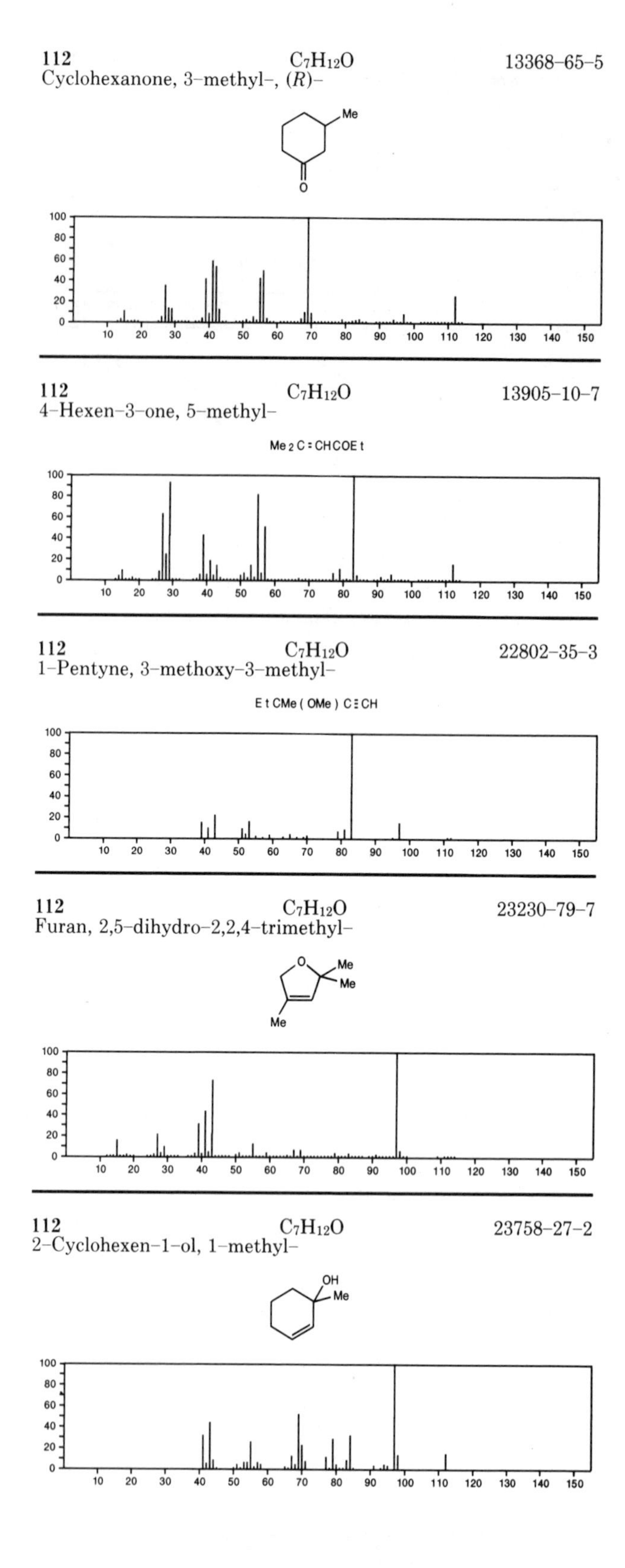

112 C7H12O 13368-65-5
Cyclohexanone, 3-methyl-, (R)-
Me
O
112 C7H12O 13905-10-7
4-Hexen-3-one, 5-methyl-
Me2C=CHCOEt
112 C7H12O 22802-35-3
1-Pentyne, 3-methoxy-3-methyl-
EtCMe(OMe)C≡CH
112 C7H12O 23230-79-7
Furan, 2,5-dihydro-2,2,4-trimethyl-
O
Me
Me
Me
112 C7H12O 23758-27-2
2-Cyclohexen-1-ol, 1-methyl-
OH
Me

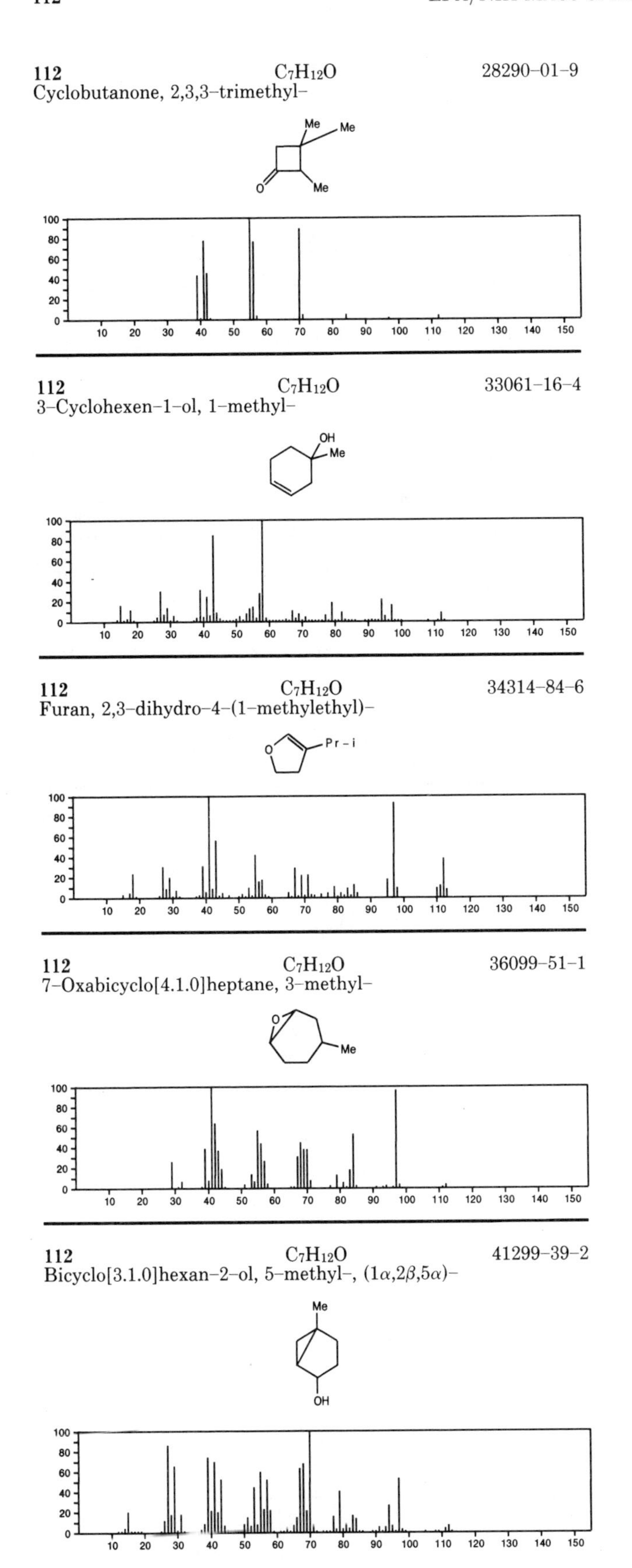
112 $C_7H_{12}O$ 28290-01-9
Cyclobutanone, 2,3,3-trimethyl-
112 $C_7H_{12}O$ 33061-16-4
3-Cyclohexen-1-ol, 1-methyl-
112 $C_7H_{12}O$ 34314-84-6
Furan, 2,3-dihydro-4-(1-methylethyl)-
112 $C_7H_{12}O$ 36099-51-1
7-Oxabicyclo[4.1.0]heptane, 3-methyl-
112 $C_7H_{12}O$ 41299-39-2
Bicyclo[3.1.0]hexan-2-ol, 5-methyl-, (1α,2β,5α)-

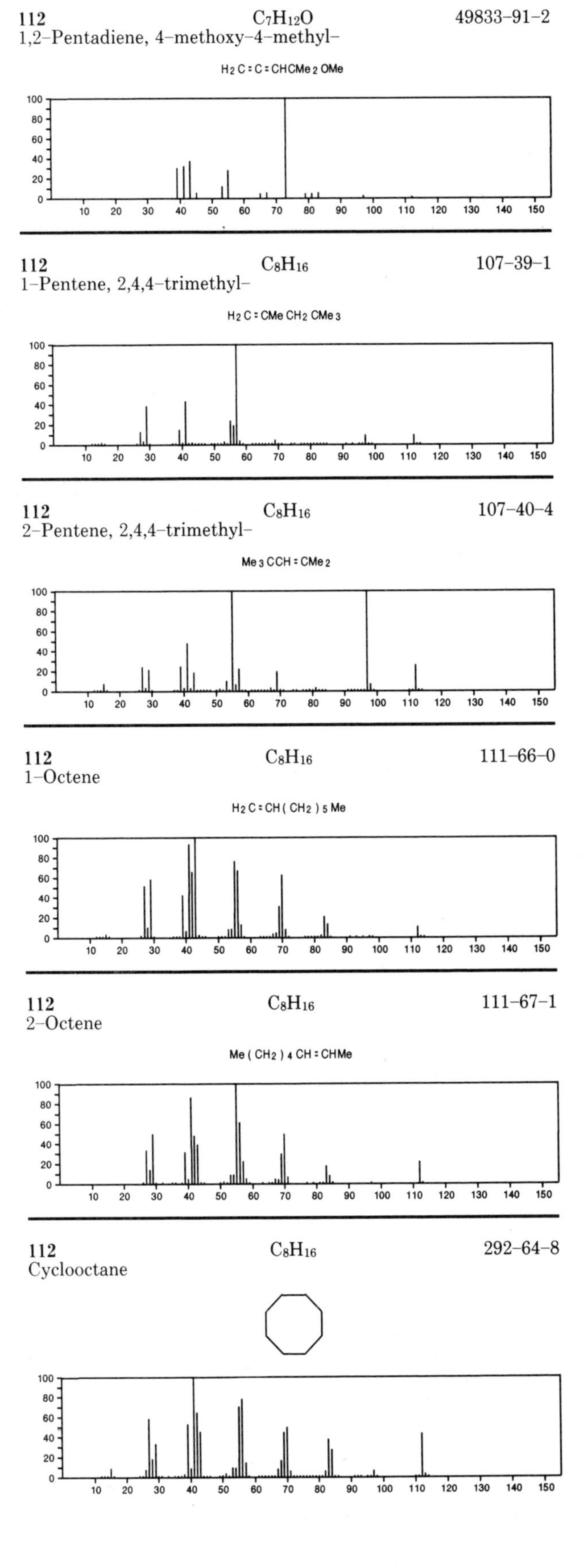
112 $C_7H_{12}O$ 49833-91-2
1,2-Pentadiene, 4-methoxy-4-methyl-
H2C=C=CHCMe2OMe
112 C_8H_{16} 107-39-1
1-Pentene, 2,4,4-trimethyl-
H2C=CMeCH2CMe3
112 C_8H_{16} 107-40-4
2-Pentene, 2,4,4-trimethyl-
Me3CCH=CMe2
112 C_8H_{16} 111-66-0
1-Octene
H2C=CH(CH2)5Me
112 C_8H_{16} 111-67-1
2-Octene
Me(CH2)4CH=CHMe
112 C_8H_{16} 292-64-8
Cyclooctane

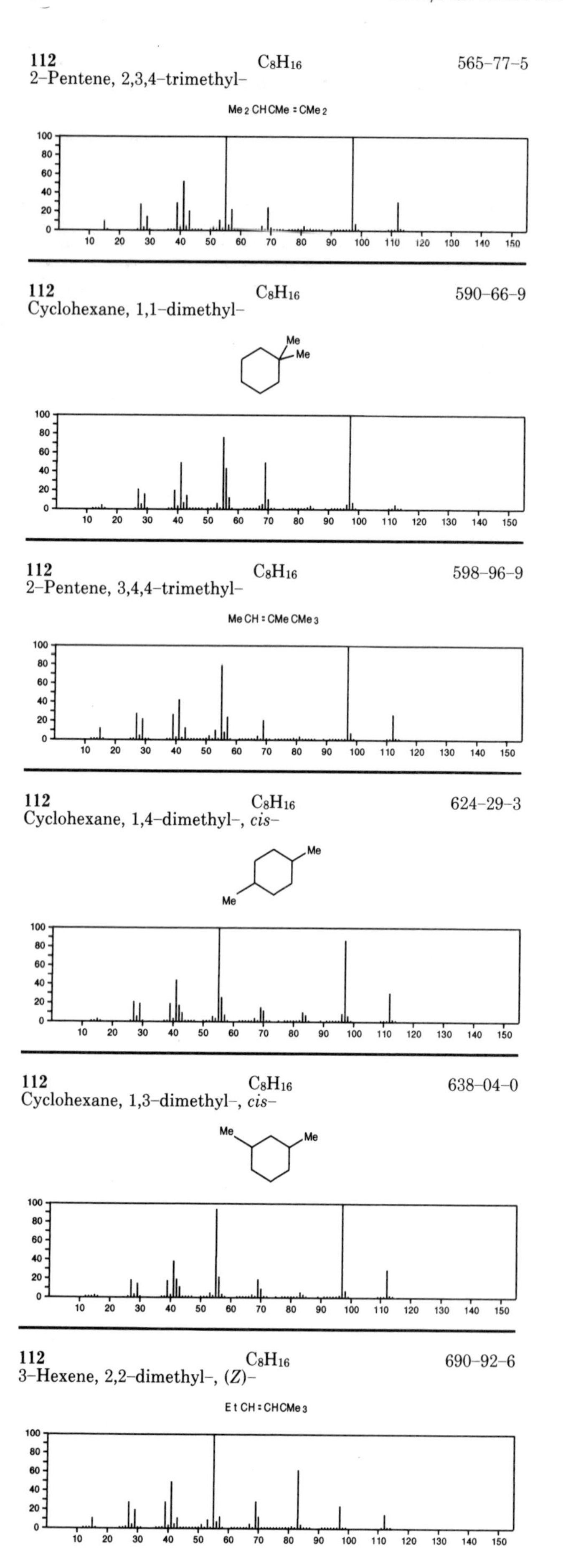

112 C_8H_{16} 565–77–5
2–Pentene, 2,3,4–trimethyl–
Me2CHCMe=CMe2

112 C_8H_{16} 590–66–9
Cyclohexane, 1,1–dimethyl–

112 C_8H_{16} 598–96–9
2–Pentene, 3,4,4–trimethyl–
MeCH=CMeCMe3

112 C_8H_{16} 624–29–3
Cyclohexane, 1,4–dimethyl–, *cis*–

112 C_8H_{16} 638–04–0
Cyclohexane, 1,3–dimethyl–, *cis*–

112 C_8H_{16} 690–92–6
3–Hexene, 2,2–dimethyl–, (*Z*)–
EtCH=CHCMe3

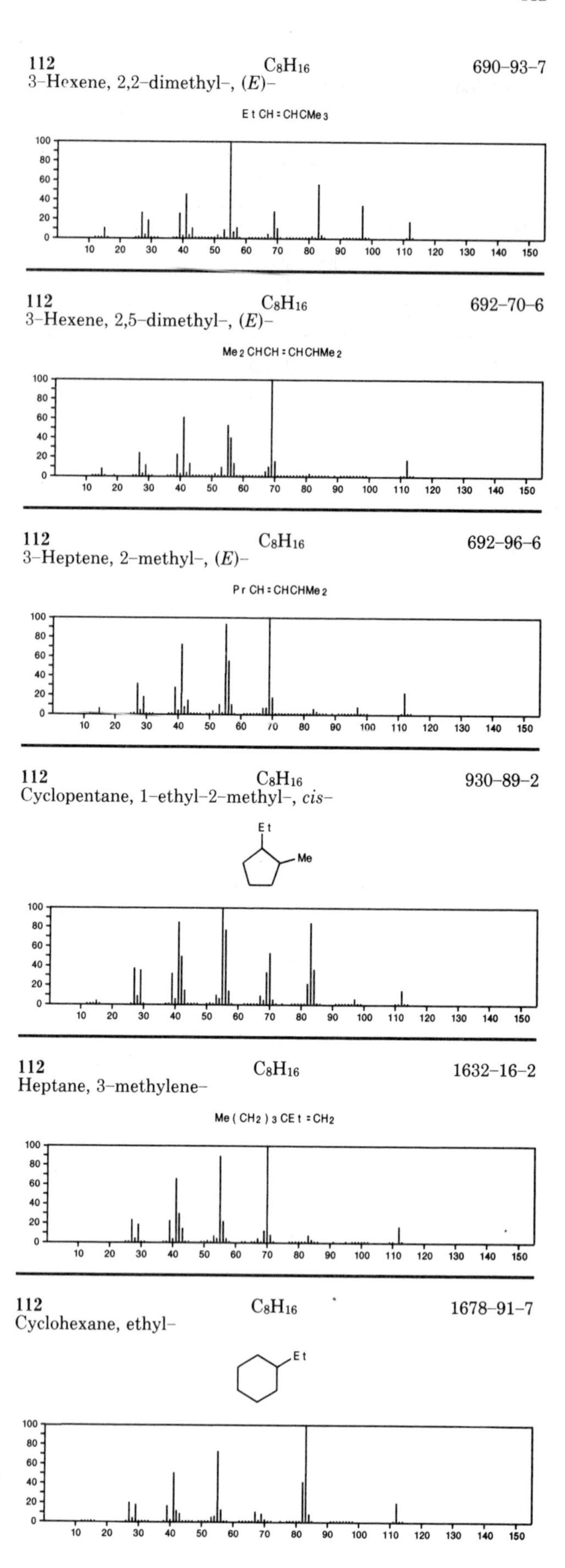

112 C_8H_{16} 690–93–7
3–Hexene, 2,2–dimethyl–, (*E*)–
EtCH=CHCMe3

112 C_8H_{16} 692–70–6
3–Hexene, 2,5–dimethyl–, (*E*)–
Me2CHCH=CHCHMe2

112 C_8H_{16} 692–96–6
3–Heptene, 2–methyl–, (*E*)–
PrCH=CHCHMe2

112 C_8H_{16} 930–89–2
Cyclopentane, 1–ethyl–2–methyl–, *cis*–

112 C_8H_{16} 1632–16–2
Heptane, 3–methylene–
Me(CH2)3CEt=CH2

112 C_8H_{16} 1678–91–7
Cyclohexane, ethyl–

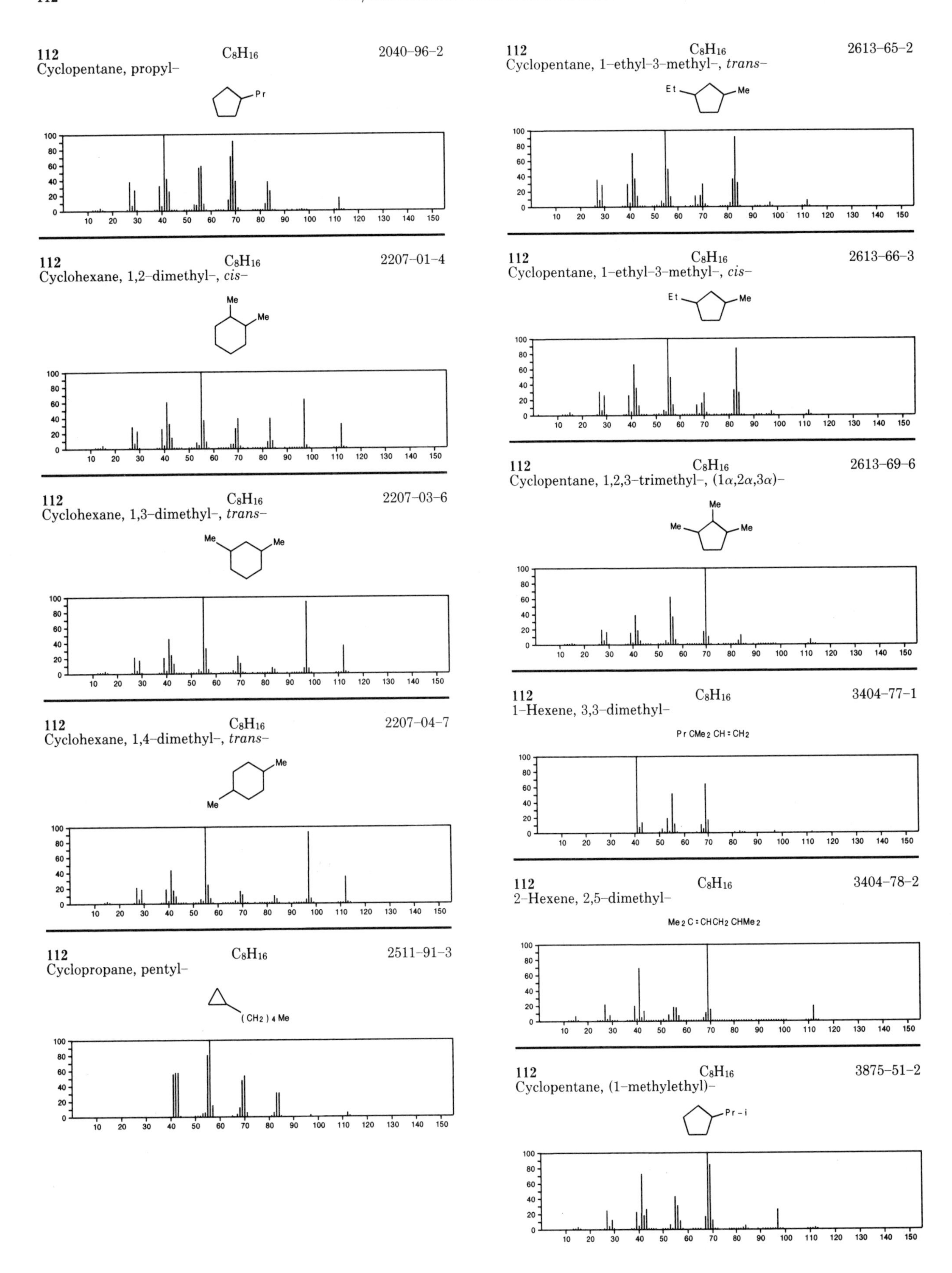

112 C8H16 2040-96-2
Cyclopentane, propyl-
Pr
112 C8H16 2207-01-4
Cyclohexane, 1,2-dimethyl-, cis-
Me
Me
112 C8H16 2207-03-6
Cyclohexane, 1,3-dimethyl-, trans-
Me
Me
112 C8H16 2207-04-7
Cyclohexane, 1,4-dimethyl-, trans-
Me
Me
112 C8H16 2511-91-3
Cyclopropane, pentyl-
(CH2)4Me
112 C8H16 2613-65-2
Cyclopentane, 1-ethyl-3-methyl-, trans-
Et
Me
112 C8H16 2613-66-3
Cyclopentane, 1-ethyl-3-methyl-, cis-
Et
Me
112 C8H16 2613-69-6
Cyclopentane, 1,2,3-trimethyl-, (1α,2α,3α)-
Me
Me
Me
112 C8H16 3404-77-1
1-Hexene, 3,3-dimethyl-
PrCMe2CH=CH2
112 C8H16 3404-78-2
2-Hexene, 2,5-dimethyl-
Me2C=CHCH2CHMe2
112 C8H16 3875-51-2
Cyclopentane, (1-methylethyl)-
Pr-i

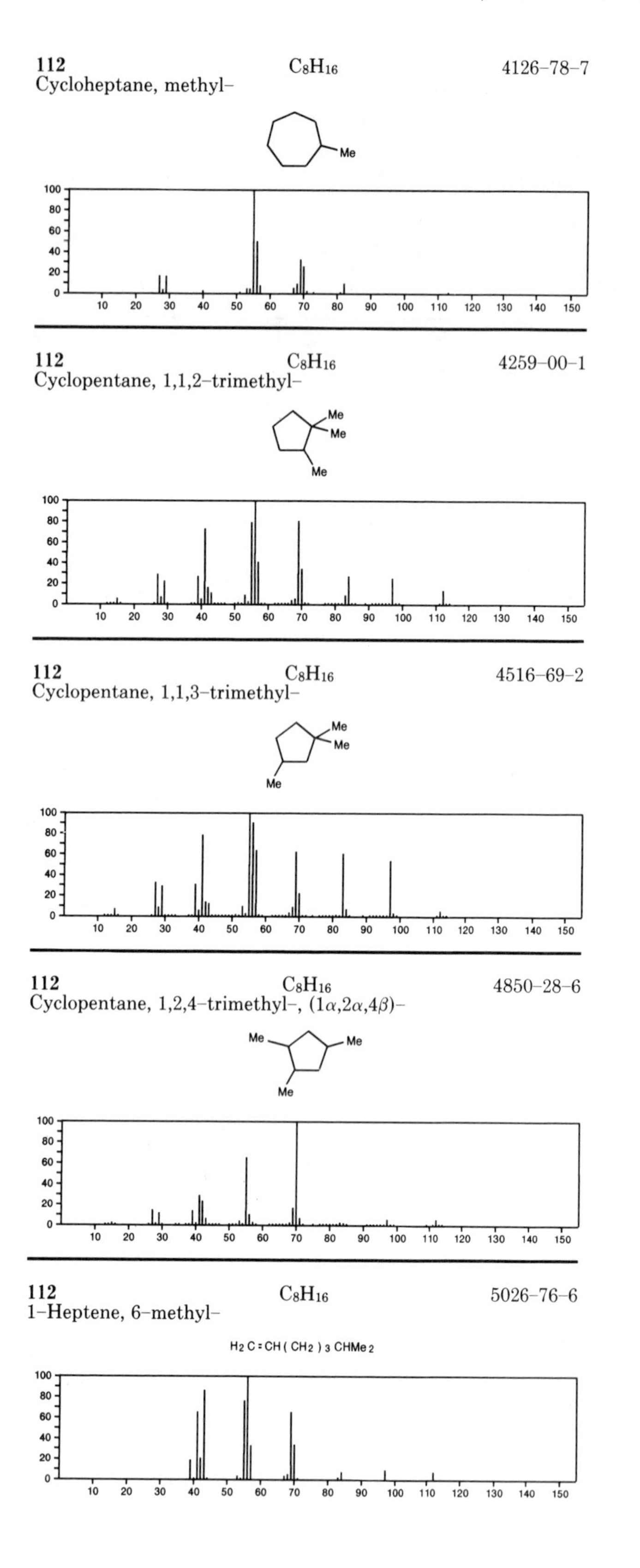
112 C8H16 4126–78–7
Cycloheptane, methyl–
Me
112 C8H16 4259–00–1
Cyclopentane, 1,1,2–trimethyl–
Me
Me
Me
112 C8H16 4516–69–2
Cyclopentane, 1,1,3–trimethyl–
Me
Me
Me
112 C8H16 4850–28–6
Cyclopentane, 1,2,4–trimethyl–, (1α,2α,4β)–
Me
Me
Me
112 C8H16 5026–76–6
1–Heptene, 6–methyl–
H2C=CH(CH2)3CHMe2

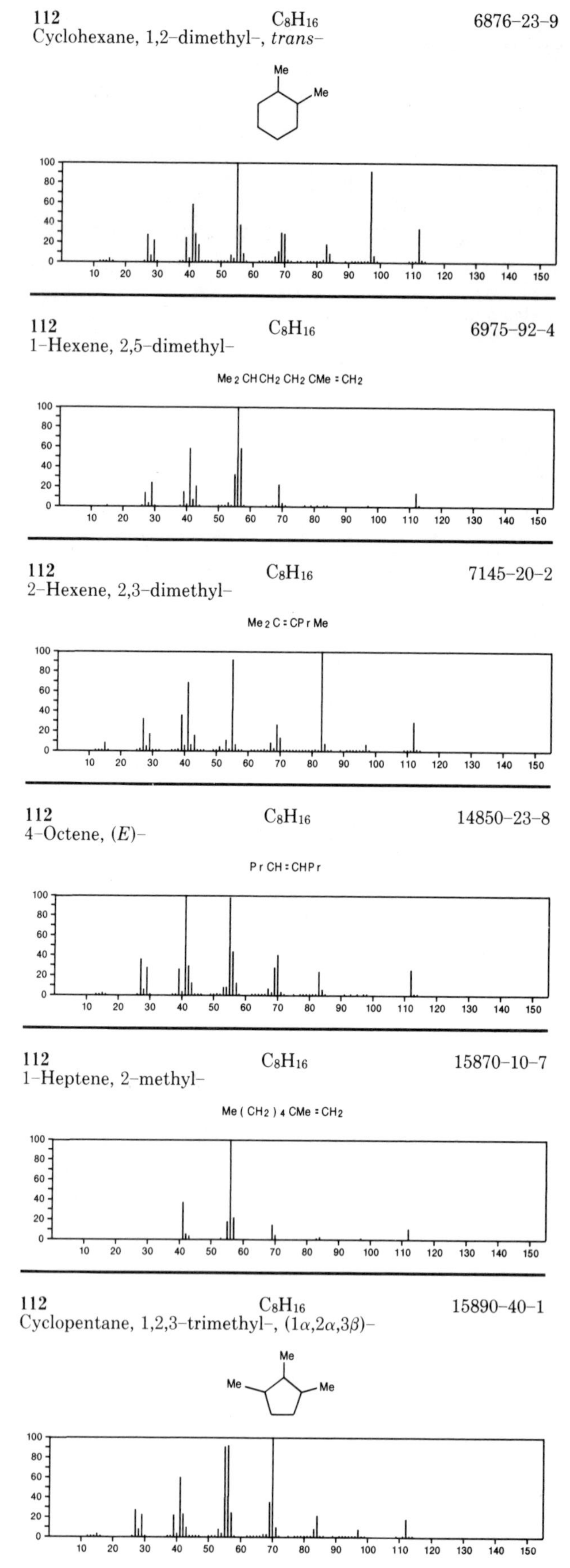
112 C8H16 6876–23–9
Cyclohexane, 1,2–dimethyl–, trans–
Me
Me
112 C8H16 6975–92–4
1–Hexene, 2,5–dimethyl–
Me2CHCH2CH2CMe=CH2
112 C8H16 7145–20–2
2–Hexene, 2,3–dimethyl–
Me2C=CPrMe
112 C8H16 14850–23–8
4–Octene, (E)–
PrCH=CHPr
112 C8H16 15870–10–7
1–Heptene, 2–methyl–
Me(CH2)4CMe=CH2
112 C8H16 15890–40–1
Cyclopentane, 1,2,3–trimethyl–, (1α,2α,3β)–
Me
Me
Me

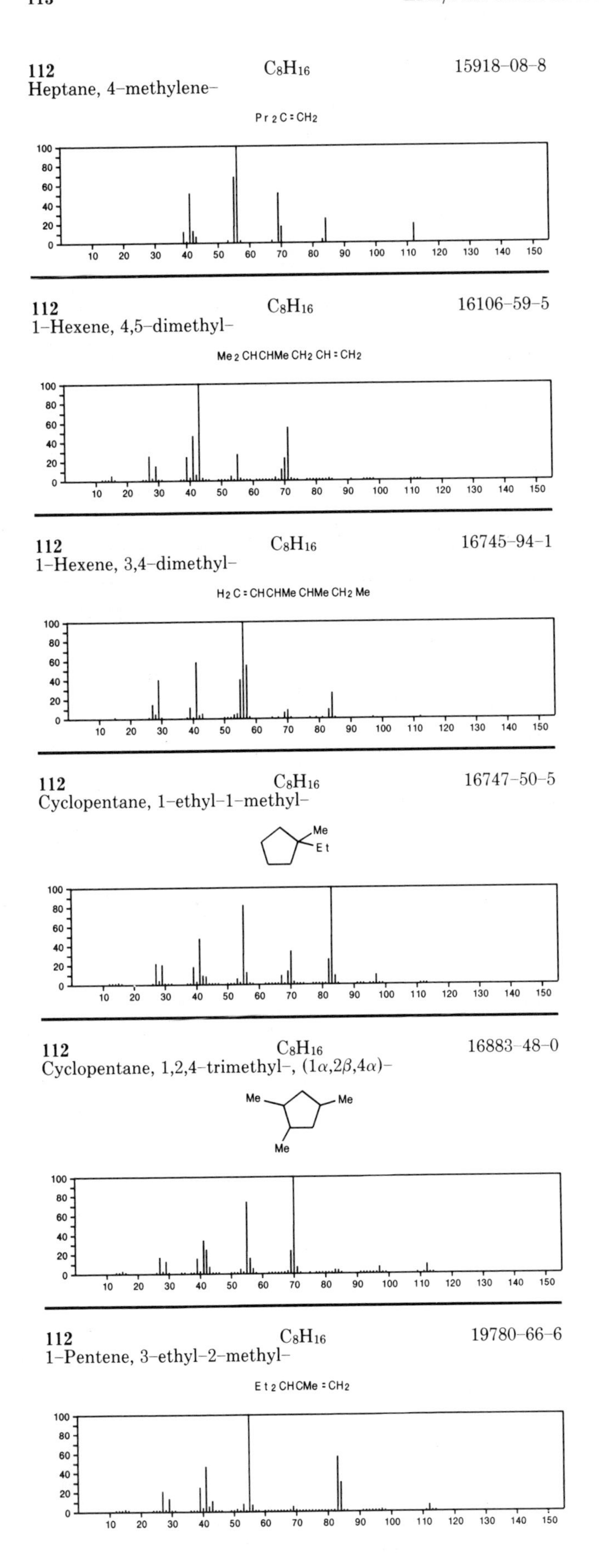

112 C_8H_{16} 15918-08-8
Heptane, 4-methylene-
$Pr_2C=CH_2$
112 C_8H_{16} 16106-59-5
1-Hexene, 4,5-dimethyl-
$Me_2CHCHMeCH_2CH=CH_2$
112 C_8H_{16} 16745-94-1
1-Hexene, 3,4-dimethyl-
$H_2C=CHCHMeCHMeCH_2Me$
112 C_8H_{16} 16747-50-5
Cyclopentane, 1-ethyl-1-methyl-
112 C_8H_{16} 16883-48-0
Cyclopentane, 1,2,4-trimethyl-, $(1\alpha,2\beta,4\alpha)$-
112 C_8H_{16} 19780-66-6
1-Pentene, 3-ethyl-2-methyl-
$Et_2CHCMe=CH_2$

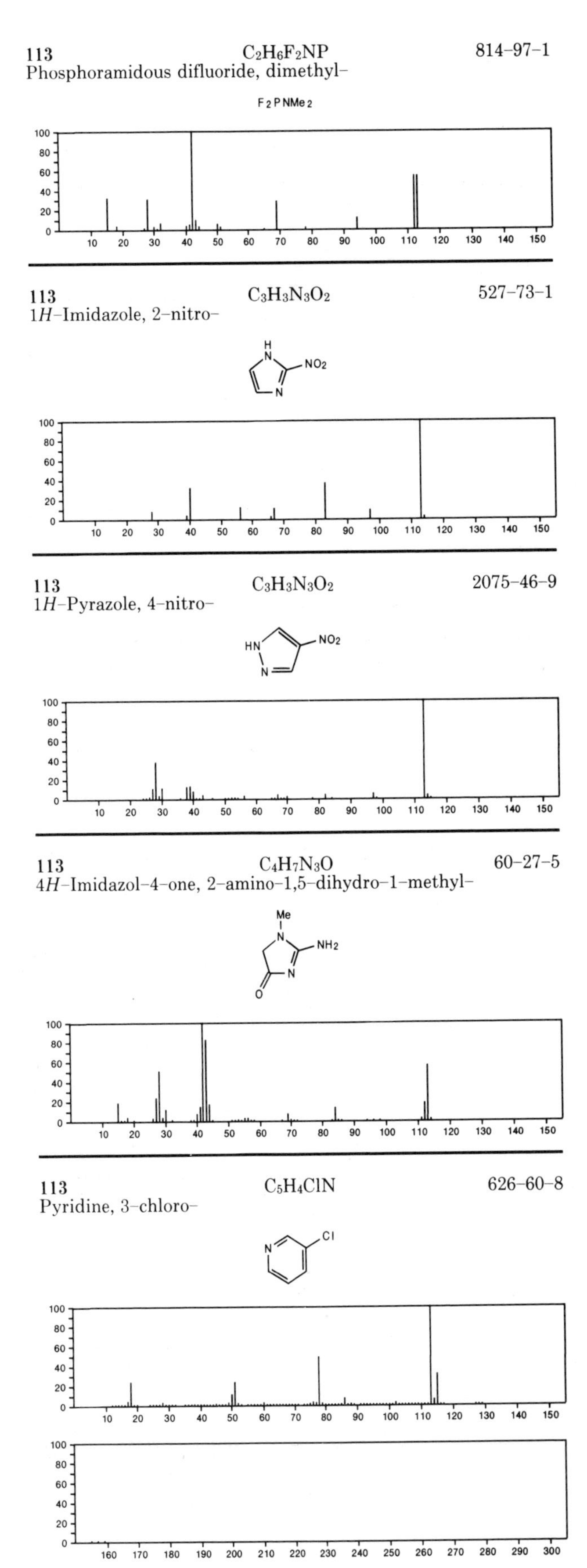

113 $C_2H_6F_2NP$ 814-97-1
Phosphoramidous difluoride, dimethyl-
F_2PNMe_2
113 $C_3H_3N_3O_2$ 527-73-1
1*H*-Imidazole, 2-nitro-
113 $C_3H_3N_3O_2$ 2075-46-9
1*H*-Pyrazole, 4-nitro-
113 $C_4H_7N_3O$ 60-27-5
4*H*-Imidazol-4-one, 2-amino-1,5-dihydro-1-methyl-
113 C_5H_4ClN 626-60-8
Pyridine, 3-chloro-

113 $C_5H_7NO_2$ 930–83–6
3(2*H*)–Isoxazolone, 4,5–dimethyl–

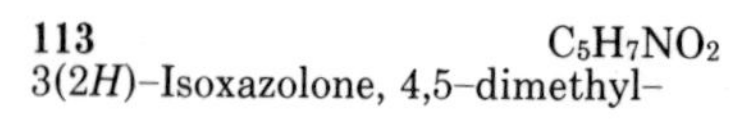

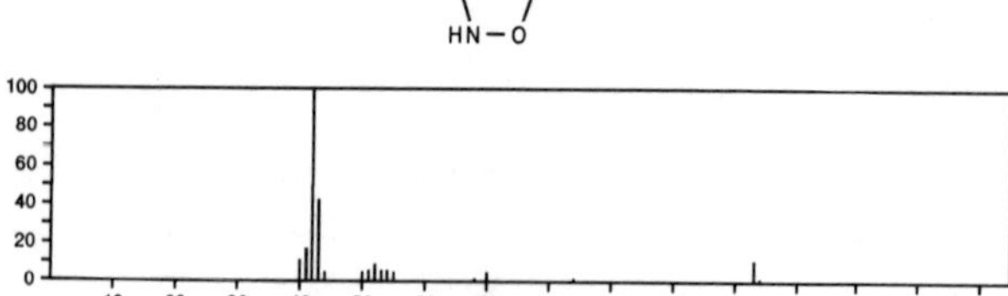

113 $C_5H_7NO_2$ 1121–07–9
2,5–Pyrrolidinedione, 1–methyl–

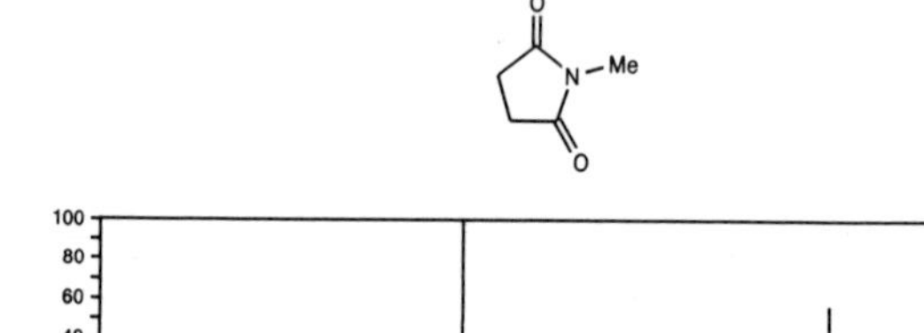

113 $C_5H_7NO_2$ 4107–62–4
Propanoic acid, 3–cyano–, methyl ester

NCCH2 CH2 C(O) OMe

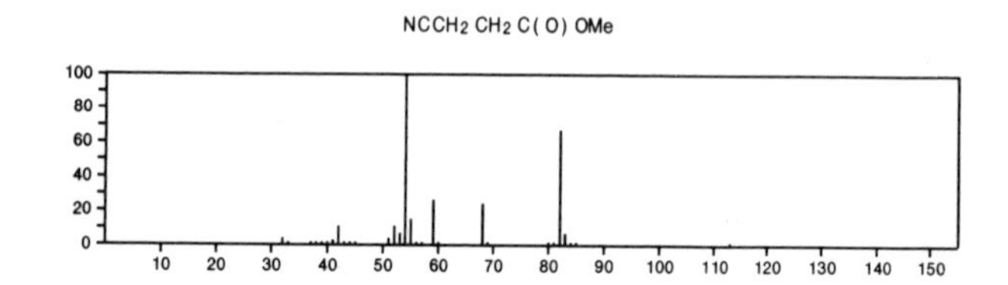

113 $C_5H_7NO_2$ 4271–26–5
2–Oxazolidinone, 3–ethenyl–

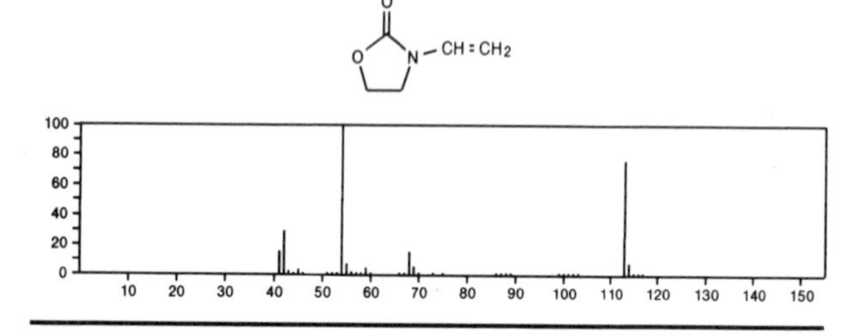

113 C_5H_7NS 3581–91–7
Thiazole, 4,5–dimethyl–

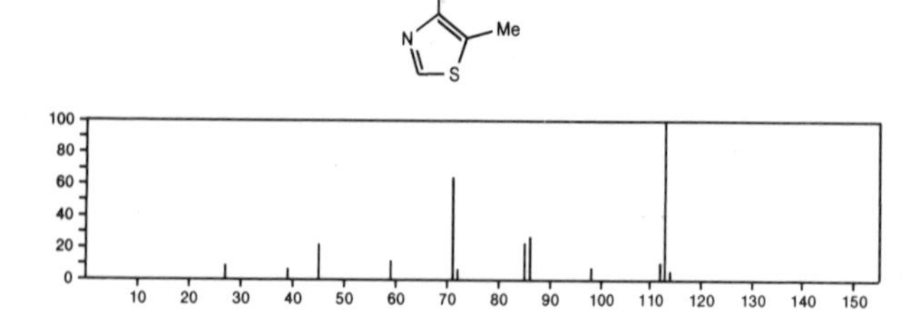

113 C_5H_7NS 4175–66–0
Thiazole, 2,5–dimethyl–

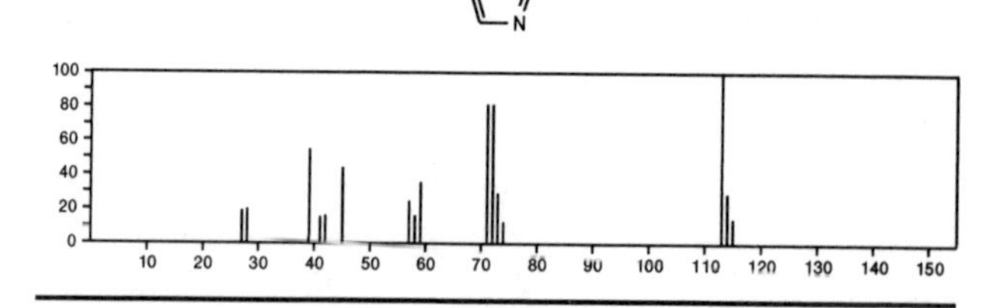

113 C_5H_7NS 15679–09–1
Thiazole, 2–ethyl–

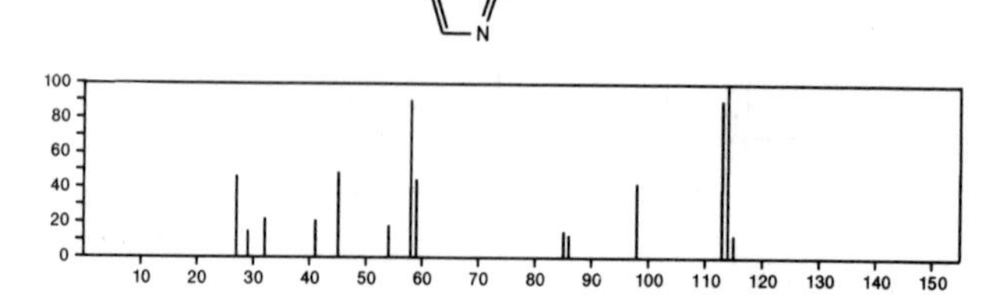

113 C_5H_7NS 17626–73–2
Thiazole, 5–ethyl–

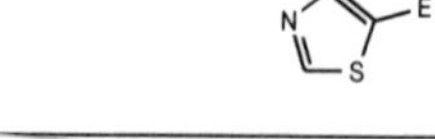

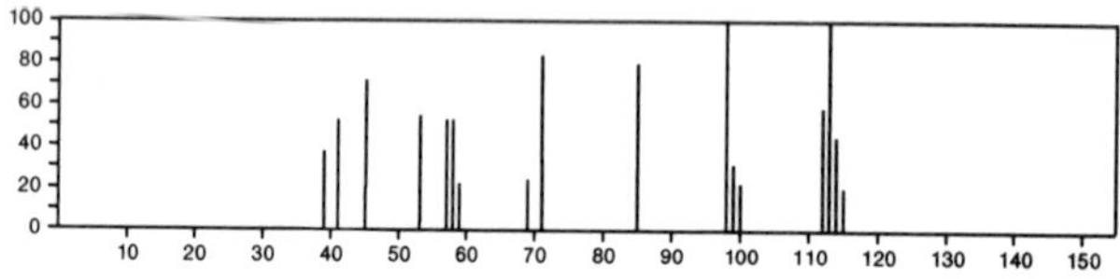

113 C_5H_7NS 24260–24–0
Isothiazole, 3,5–dimethyl–

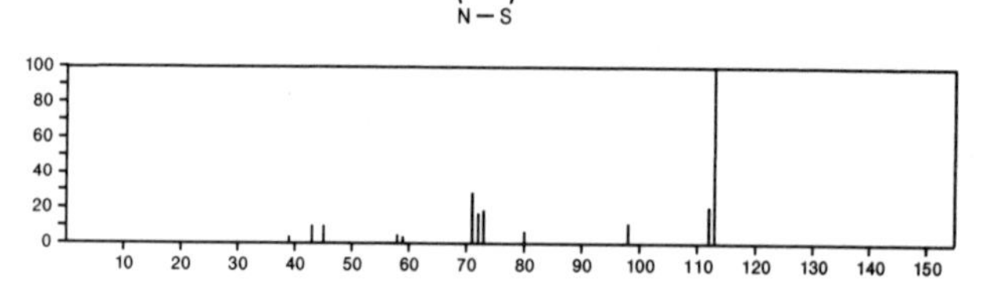

113 C_5H_7NS 27330–46–7
Isothiazole, 3,4–dimethyl–

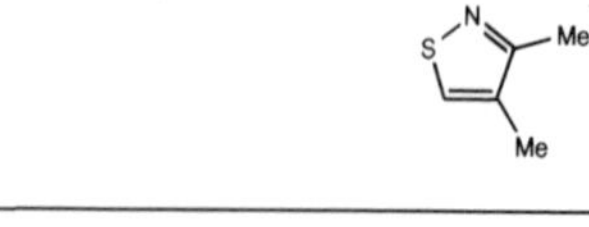

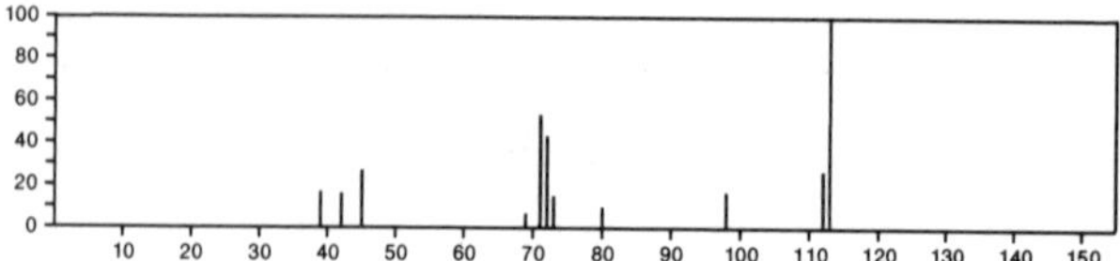

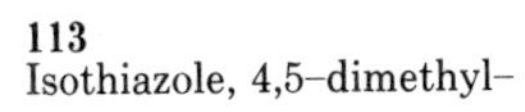

113 C_5H_7NS 27330-47-8
Isothiazole, 4,5-dimethyl-

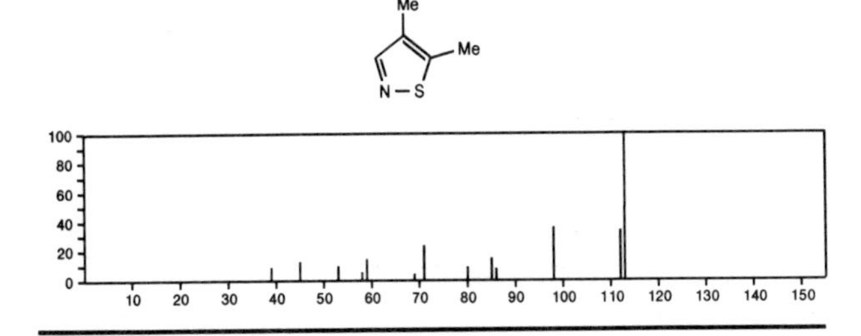

113 $C_6H_8N.F$ 36880-52-1
Pyridinium, 1-methyl-, fluoride

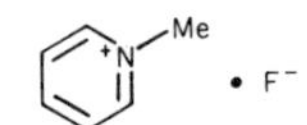

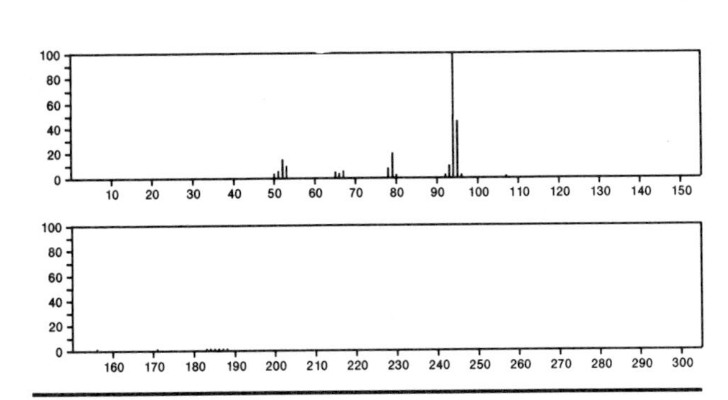

113 $C_6H_{11}NO$ 100-64-1
Cyclohexanone, oxime

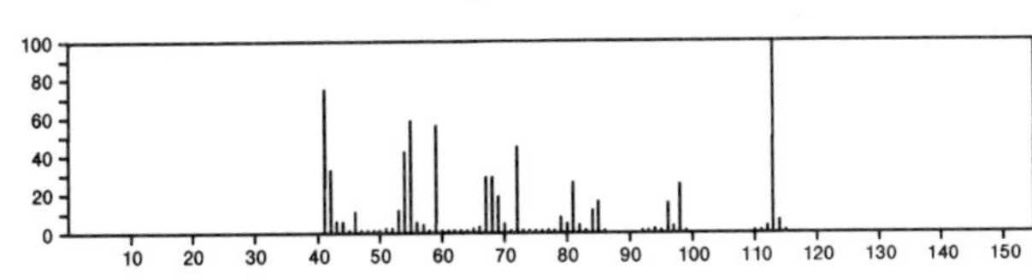

113 $C_6H_{11}NO$ 105-60-2
2*H*-Azepin-2-one, hexahydro-

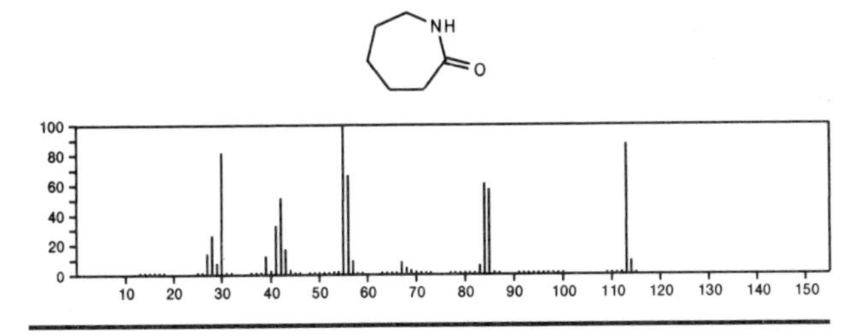

113 $C_6H_{11}NO$ 931-20-4
2-Piperidinone, 1-methyl-

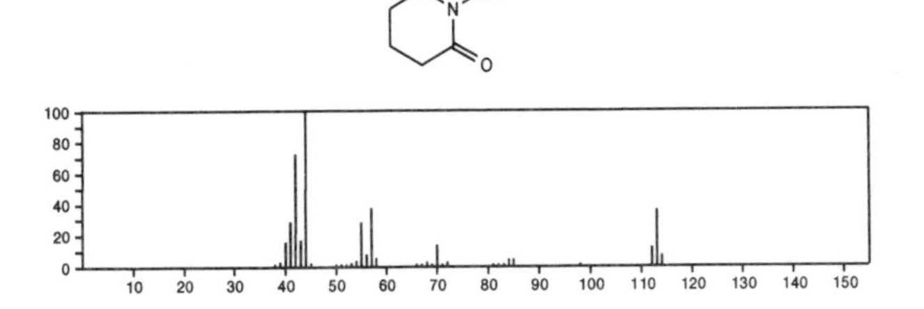

113 $C_6H_{11}NO$ 1190-91-6
3-Buten-2-one, 4-(dimethylamino)-

$Me_2NCH=CHCOMe$

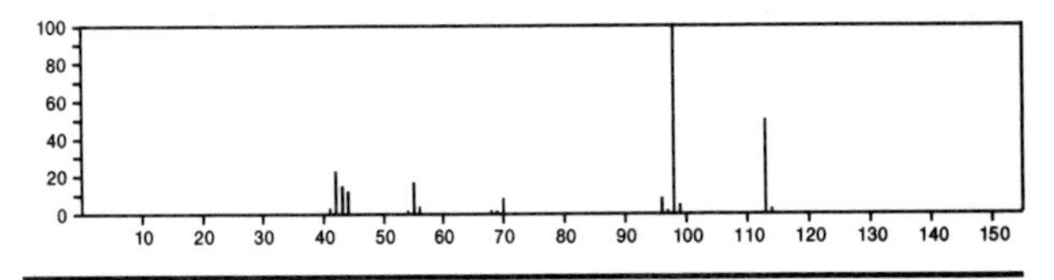

113 $C_6H_{11}NO$ 1445-73-4
4-Piperidinone, 1-methyl-

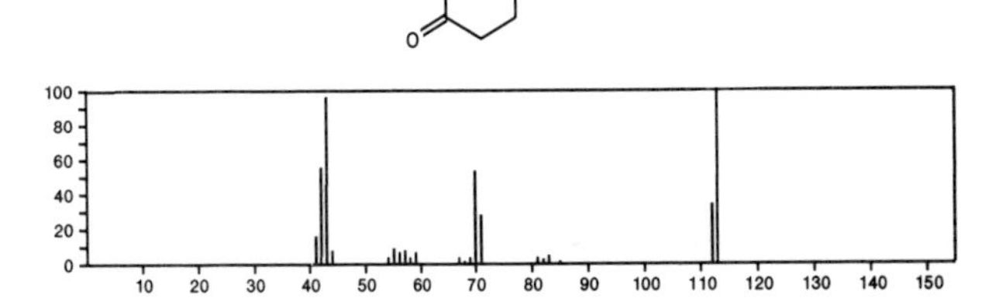

113 $C_6H_{11}NO$ 1459-44-5
Cyanic acid, 2,2-dimethylpropyl ester

$NCOCH_2CMe_3$

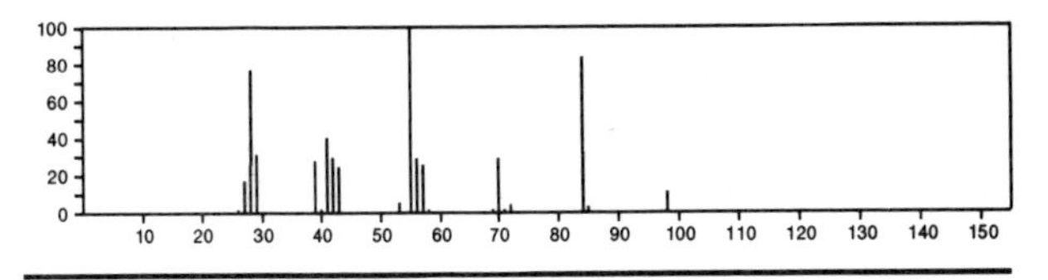

113 $C_6H_{11}NO$ 2591-86-8
1-Piperidinecarboxaldehyde

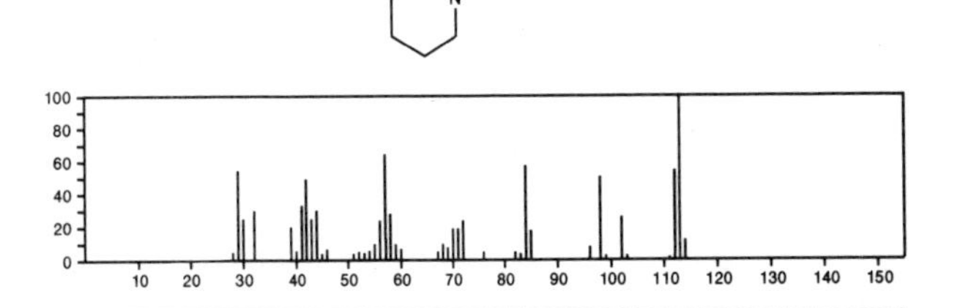

113 $C_6H_{11}NO$ 3376-37-2
Cyclopentanone, *O*-methyloxime

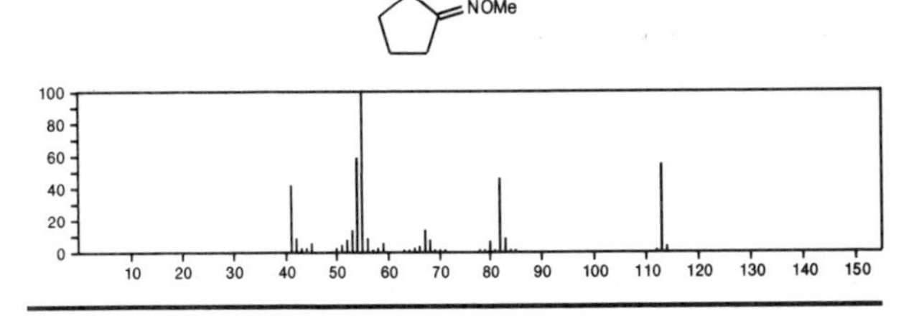

113 $C_6H_{11}NO$ 4030-18-6
Pyrrolidine, 1-acetyl-

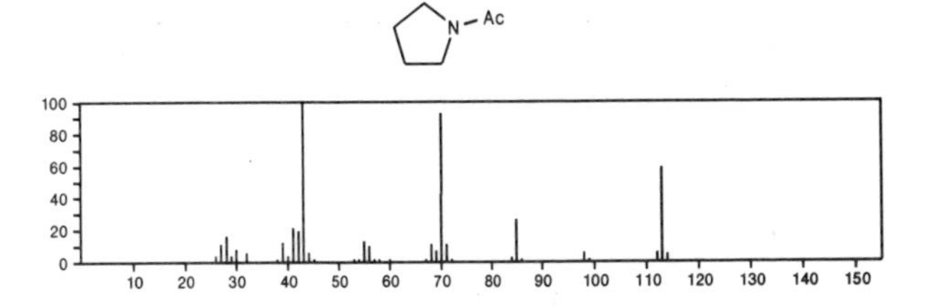

113 $C_6H_{11}NO$ 4775-98-8
2-Piperidinone, 6-methyl-

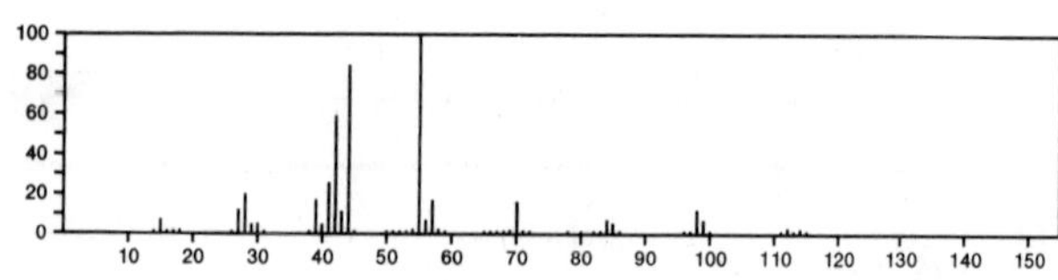

113 $C_6H_{11}NO$ 5519-50-6
3-Piperidinone, 1-methyl-

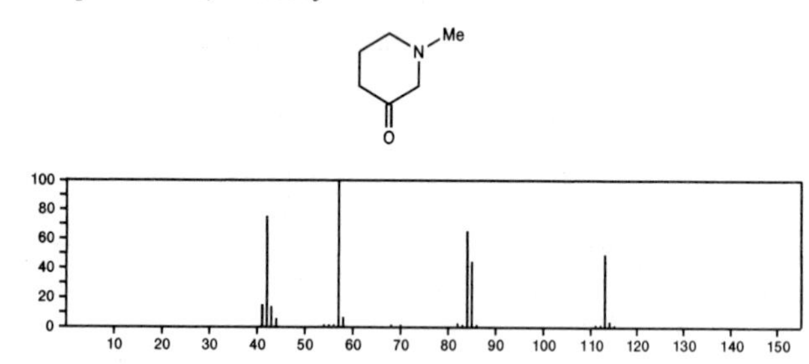

113 $C_6H_{11}NO$ 14092-14-9
3-Penten-2-one, 4-(methylamino)-

MeCOCH = C(NHMe)Me

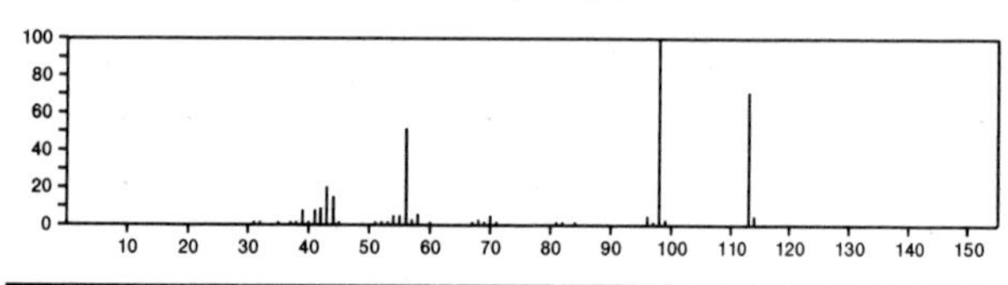

113 $C_6H_{11}NO$ 50837-77-9
Azetidine, 1-acetyl-2-methyl-

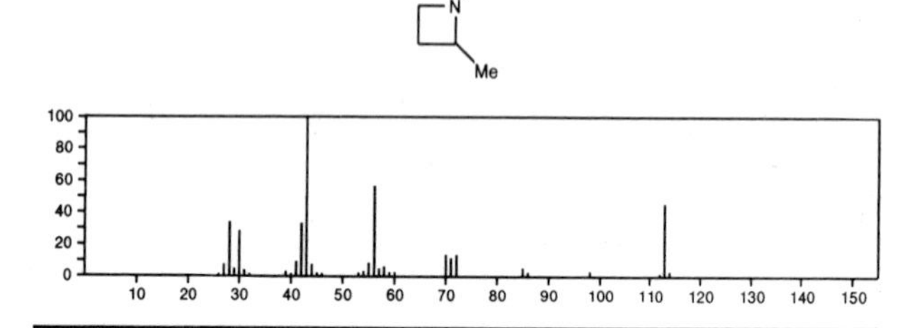

113 $C_7H_{15}N$ 100-60-7
Cyclohexanamine, *N*-methyl-

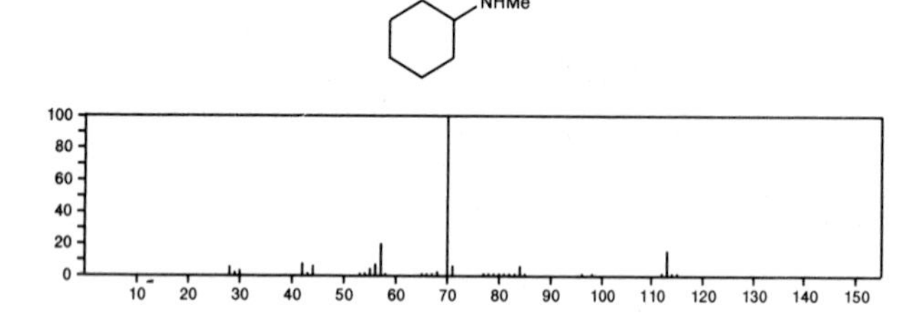

113 $C_7H_{15}N$ 504-03-0
Piperidine, 2,6-dimethyl-

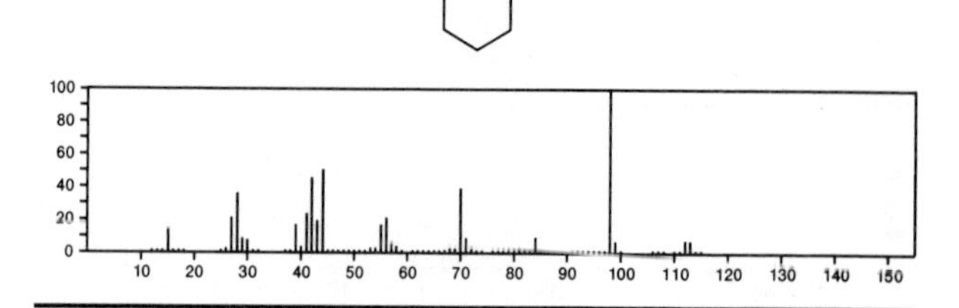

113 $C_7H_{15}N$ 671-36-3
Piperidine, 1,2-dimethyl-

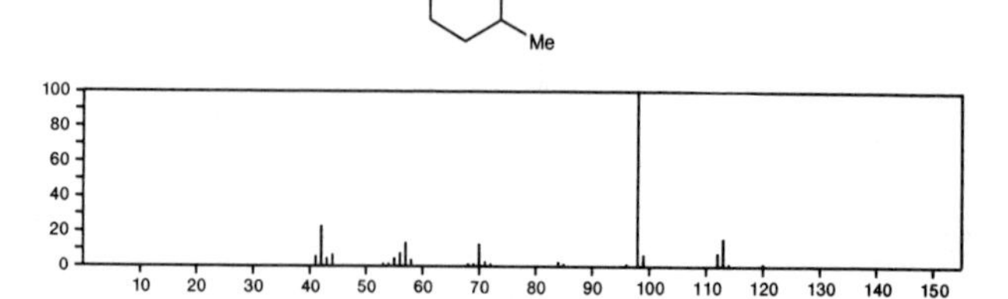

113 $C_7H_{15}N$ 766-09-6
Piperidine, 1-ethyl-

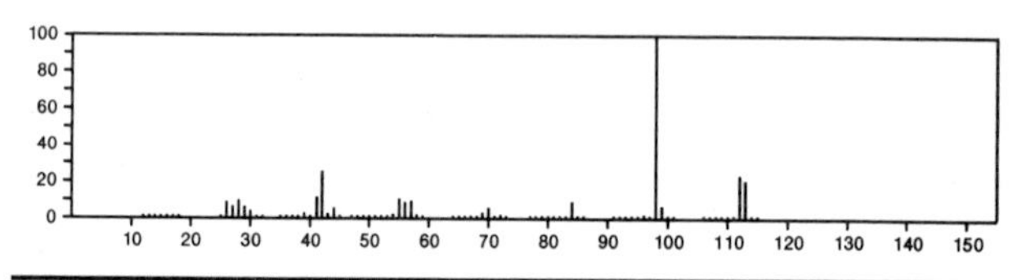

113 $C_7H_{15}N$ 931-30-6
Aziridine, 2-isopropyl-1,3-dimethyl-, *trans*-

113 $C_7H_{15}N$ 1121-92-2
Azocine, octahydro-

113 $C_7H_{15}N$ 1484-80-6
Piperidine, 2-ethyl-

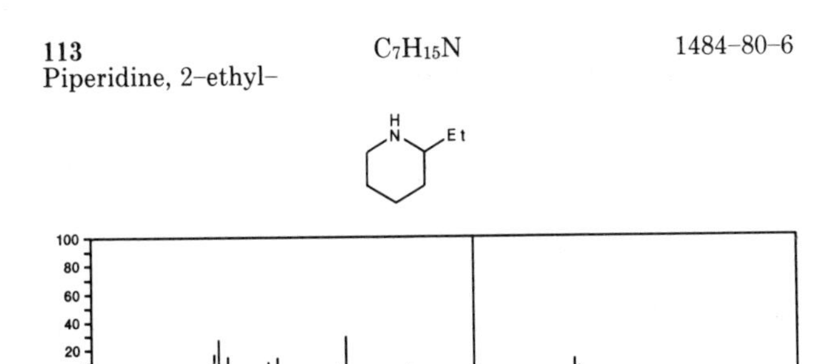

113 $C_7H_{15}N$ 10599-76-5
Ethanamine, *N*-pentylidene-

Me(CH2)3CH=NEt

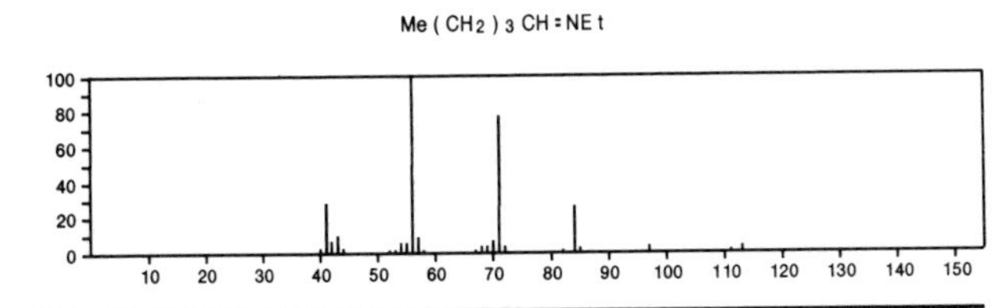

113 $C_7H_{15}N$ 26158-82-7
Pyrrolidine, 2-ethyl-1-methyl-

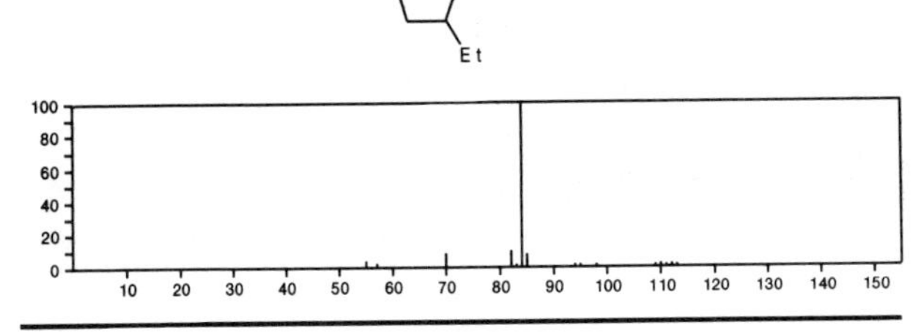

113 $C_7H_{15}N$ 45592-46-9
Cyclopentanamine, *N*-ethyl-

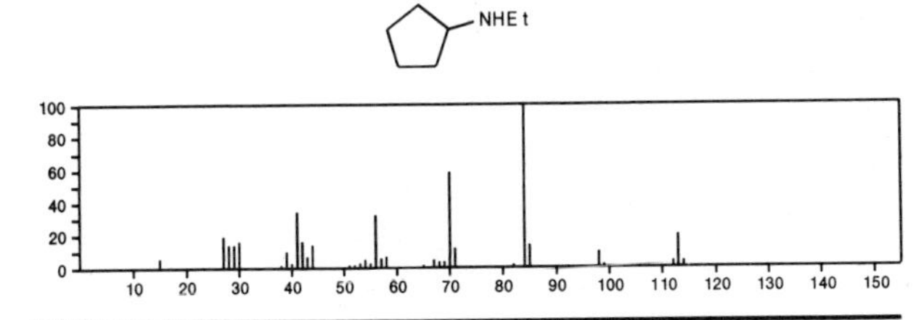

113 $C_7H_{15}N$ 55669-75-5
Aziridine, 2,3-dimethyl-1-(1-methylethyl)-, *cis*-

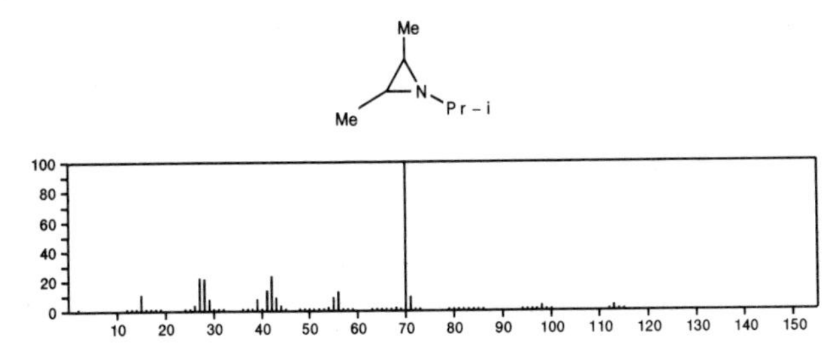

113 $C_7H_{15}N$ 55669-76-6
Aziridine, 2-(1,1-dimethylethyl)-3-methyl-, *trans*-

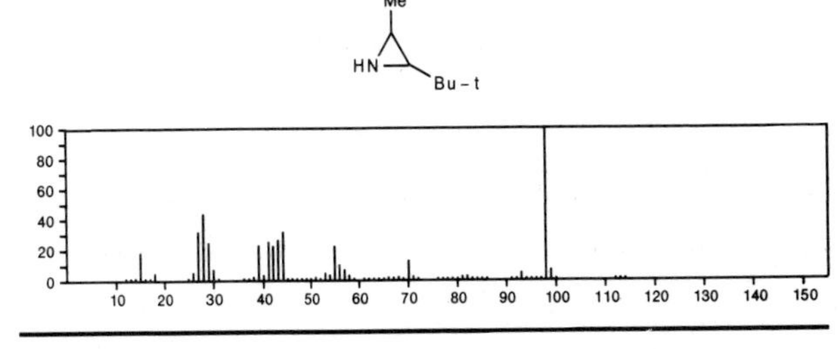

113 $C_7H_{15}N$ 55683-32-4
Azetidine, 2-methyl-1-(1-methylethyl)-

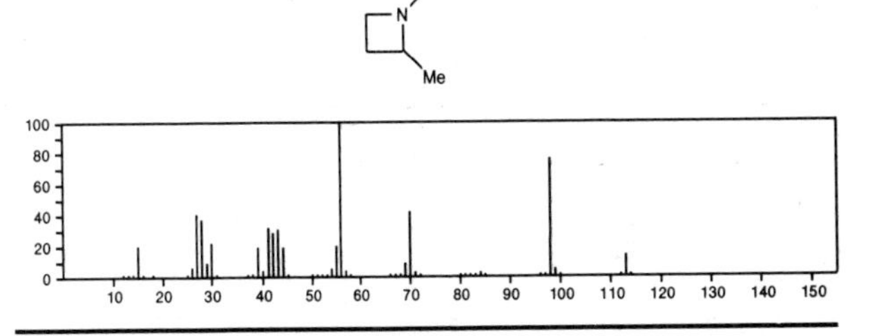

113 $C_7H_{15}N$ 55683-33-5
Azetidine, 3-methyl-1-(1-methylethyl)-

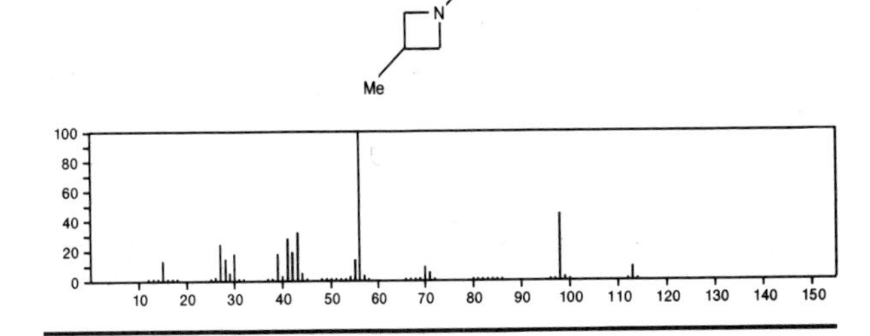

114 C_2HClF_2O 811-96-1
Acetaldehyde, chlorodifluoro-

ClCF2(CHO)

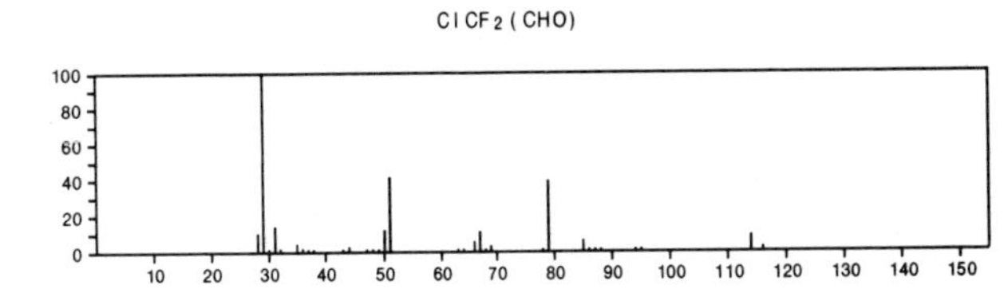

114 $C_2HF_3O_2$ 76-05-1
Acetic acid, trifluoro-

F3CCO2H

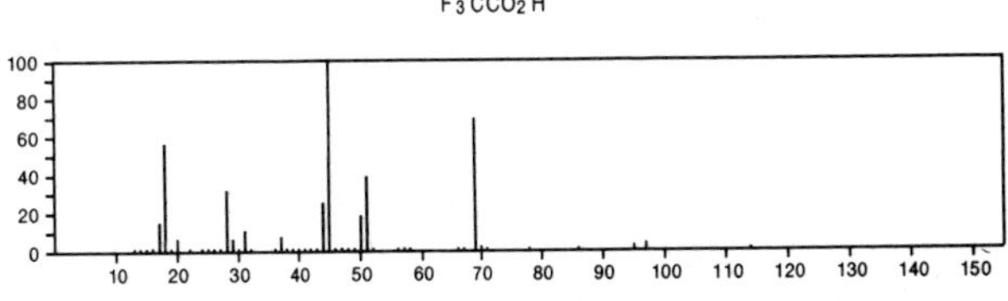

114 $C_2H_2N_4O_2$ 24807-55-4
1*H*-1,2,4-Triazole, 3-nitro-

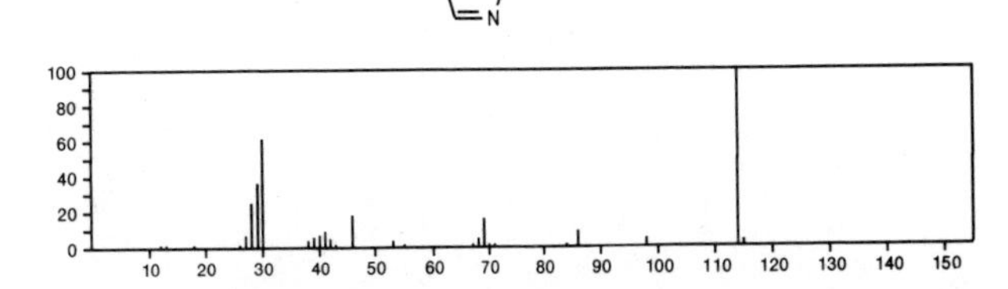

114 $C_2H_4Cl_2O$ 542-88-1
Methane, oxybis[chloro-

114 $C_2H_4Cl_2O$ 598-38-9
Ethanol, 2,2-dichloro-

114 $C_3H_5ClF_2$ 420-99-5
Propane, 1-chloro-2,2-difluoro-

114 $C_4H_2O_4$ 2892-51-5
3-Cyclobutene-1,2-dione, 3,4-dihydroxy-

114 $C_4H_6N_2O_2$ 616-03-5
2,4-Imidazolidinedione, 5-methyl-

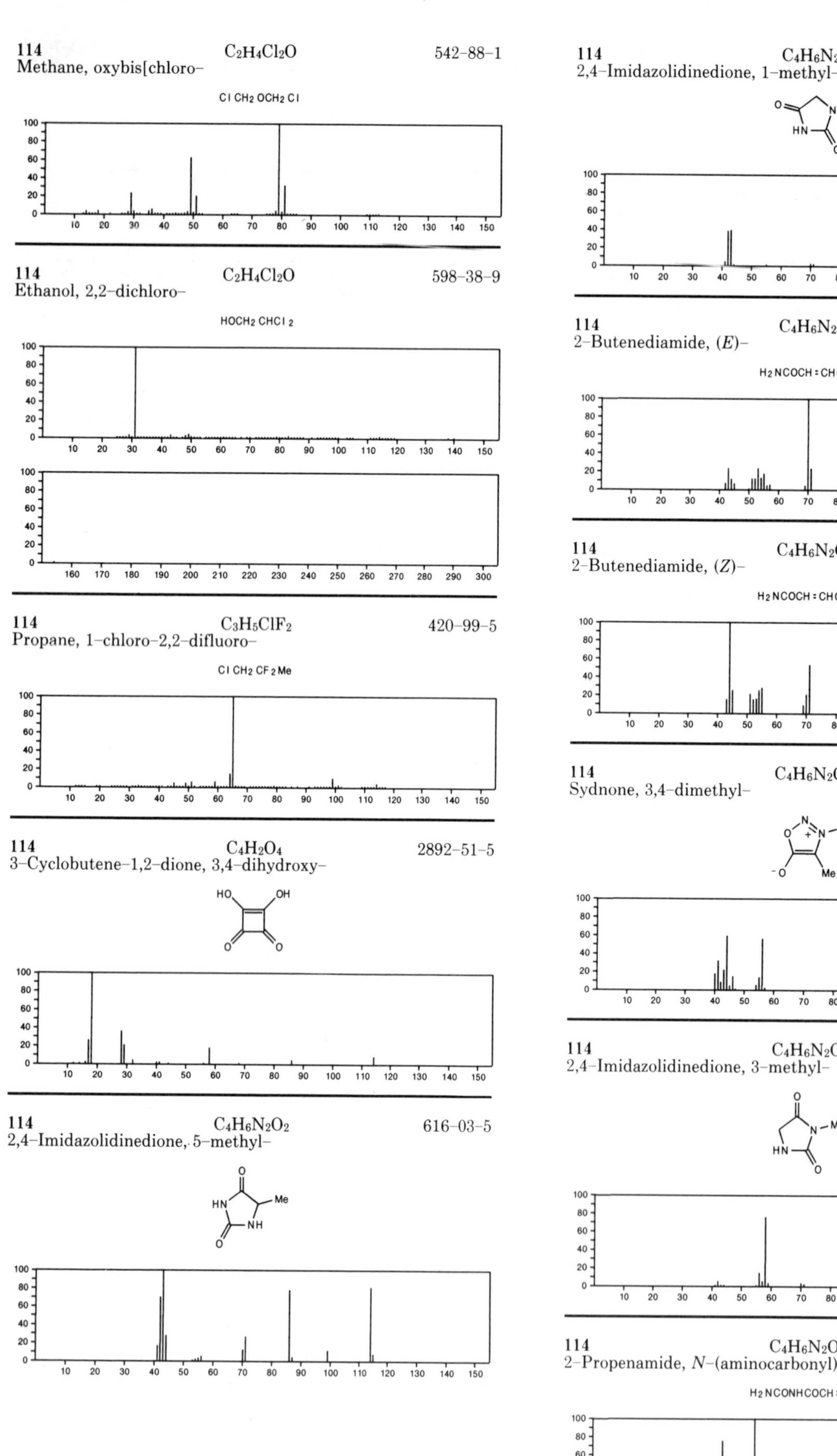

114 $C_4H_6N_2O_2$ 616-04-6
2,4-Imidazolidinedione, 1-methyl-

114 $C_4H_6N_2O_2$ 627-64-5
2-Butenediamide, (*E*)-

114 $C_4H_6N_2O_2$ 928-01-8
2-Butenediamide, (*Z*)-

114 $C_4H_6N_2O_2$ 4007-18-5
Sydnone, 3,4-dimethyl-

114 $C_4H_6N_2O_2$ 6843-45-4
2,4-Imidazolidinedione, 3-methyl-

114 $C_4H_6N_2O_2$ 20602-79-3
2-Propenamide, *N*-(aminocarbonyl)-

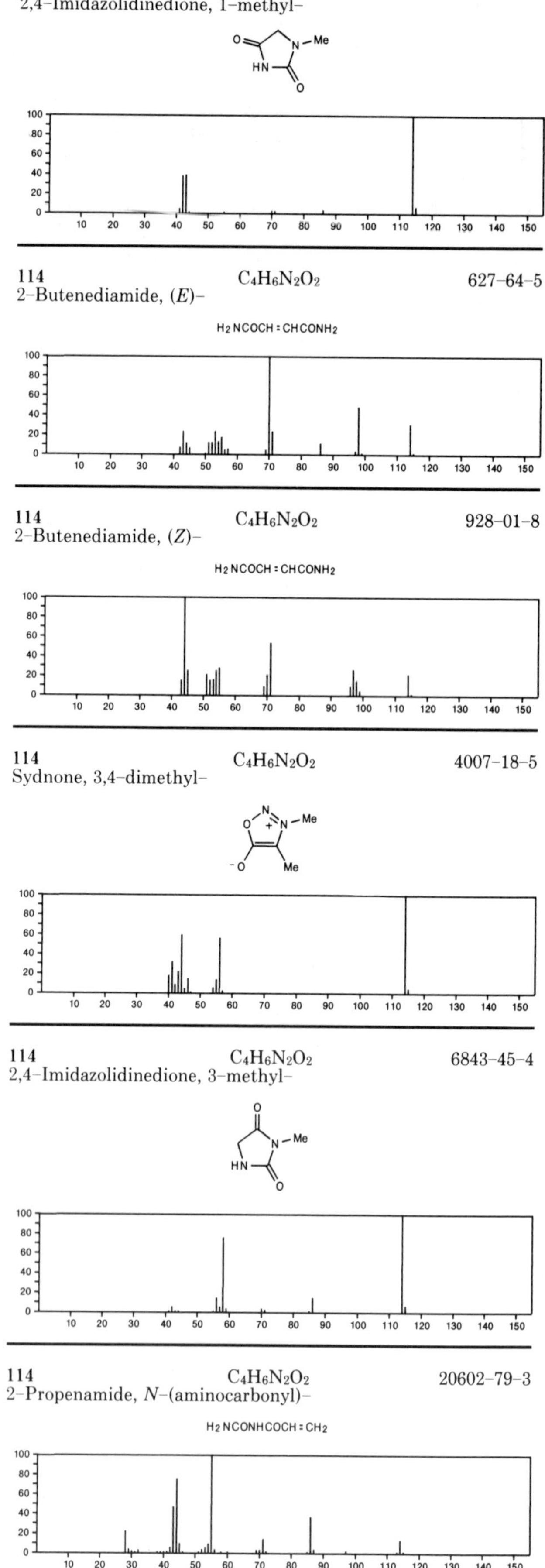

114 $C_4H_6N_2S$ 16703-47-2
Carbonocyanidothioic amide, dimethyl-

Me_2NCSCN

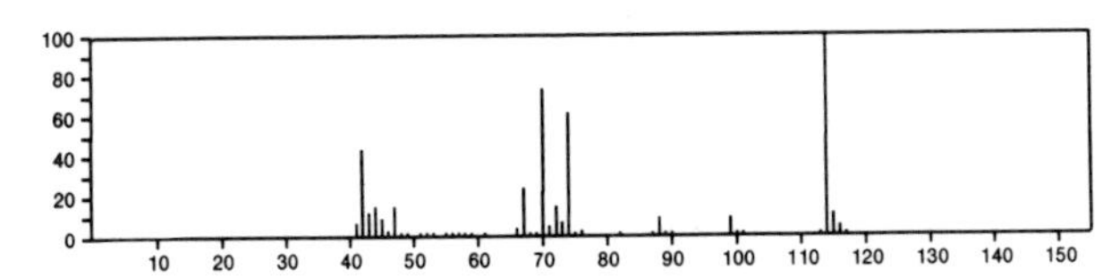

114 $C_4H_{10}N_4$ 37454-59-4
1,2,4,5-Tetrazine, 1,2,3,4-tetrahydro-3,6-dimethyl-

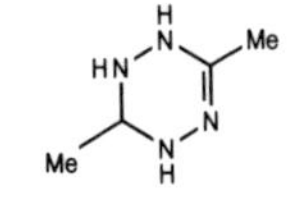

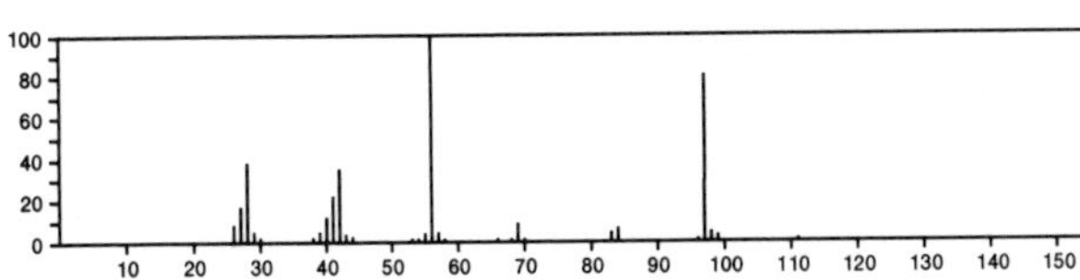

114 $C_4H_{10}N_4$ 56051-77-5
1,2,4,5-Tetrazine, 1,2,3,6-tetrahydro-3,6-dimethyl-

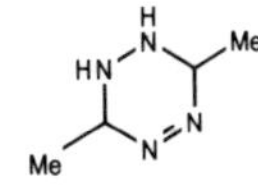

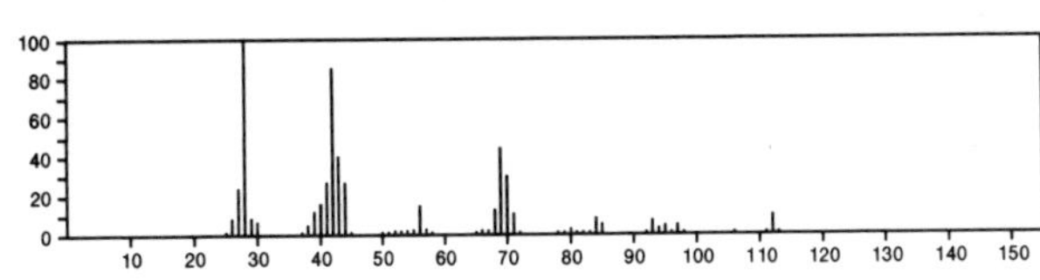

114 $C_4H_{11}B_5$ 21687-53-6
2,4-Dicarbaheptaborane(7), 2,4-dimethyl-

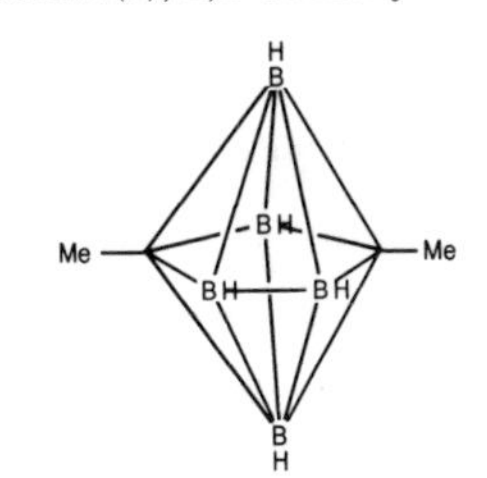

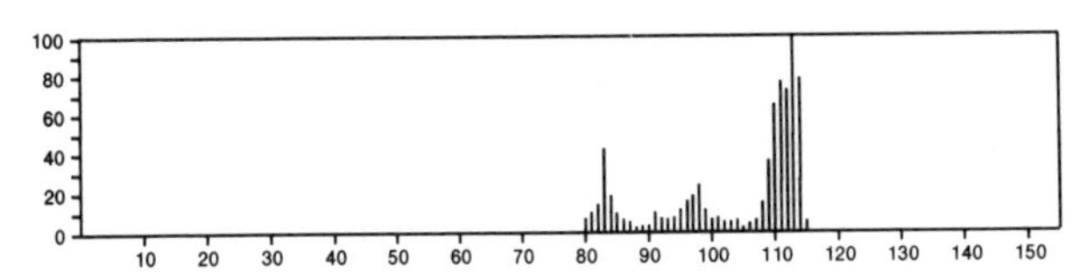

114 $C_4H_{11}B_5$ 31566-10-6
2,3-Dicarbaheptaborane(7), 2,3-dimethyl-

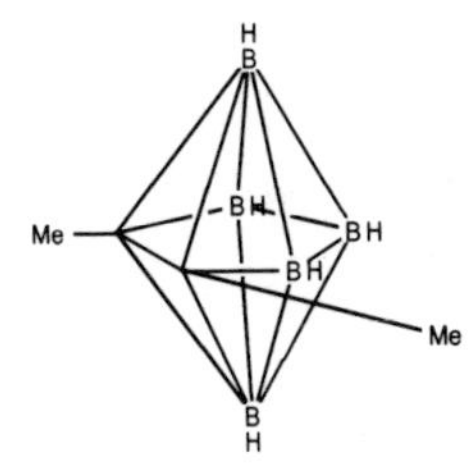

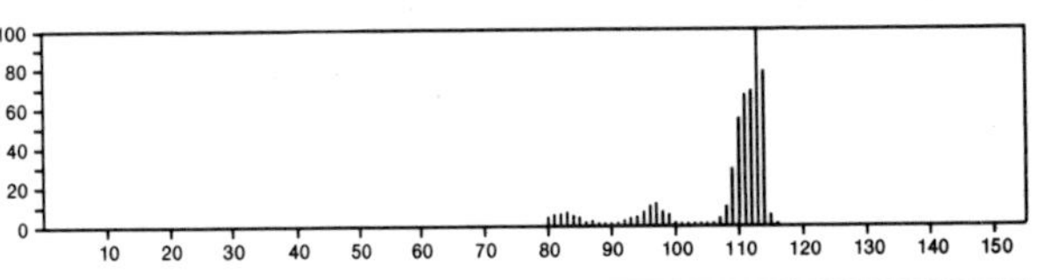

114 C_5H_6OS 98-02-2
2-Furanmethanethiol

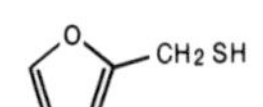

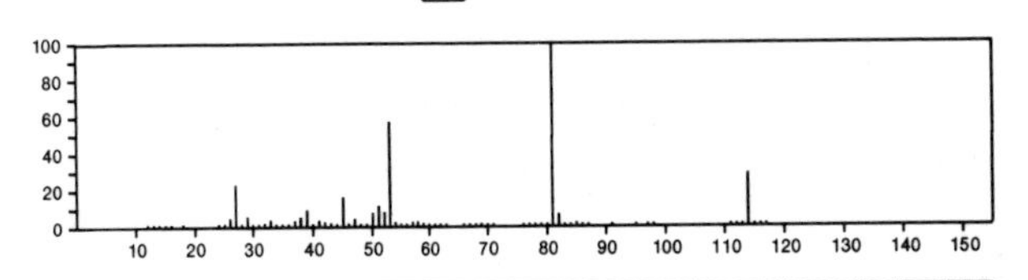

114 $C_5H_6O_3$ 108-55-4
2H-Pyran-2,6(*3H*)-dione, dihydro-

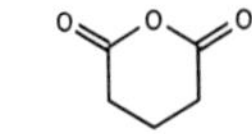

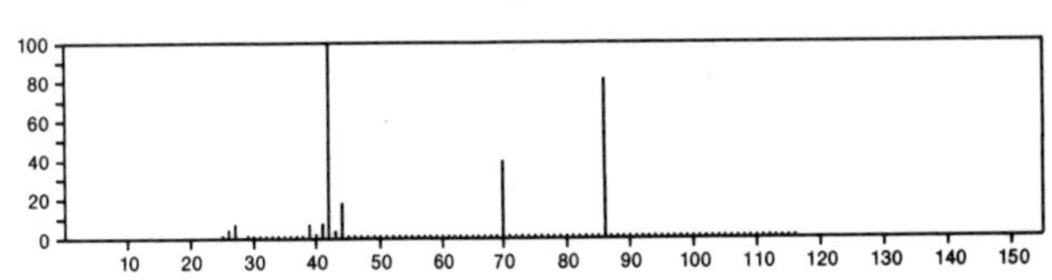

114 $C_5H_6O_3$ 4100-80-5
2,5-Furandione, dihydro-3-methyl-

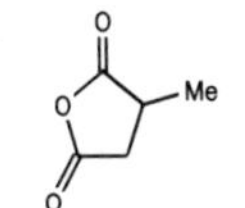

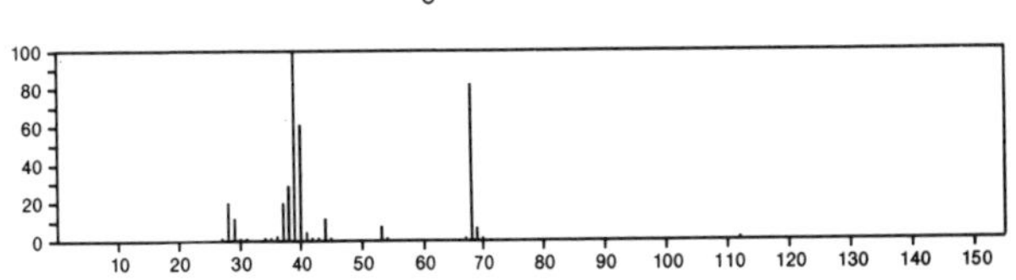

114 $C_5H_{10}N_2O$ 100-75-4
Piperidine, 1-nitroso-

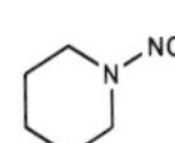

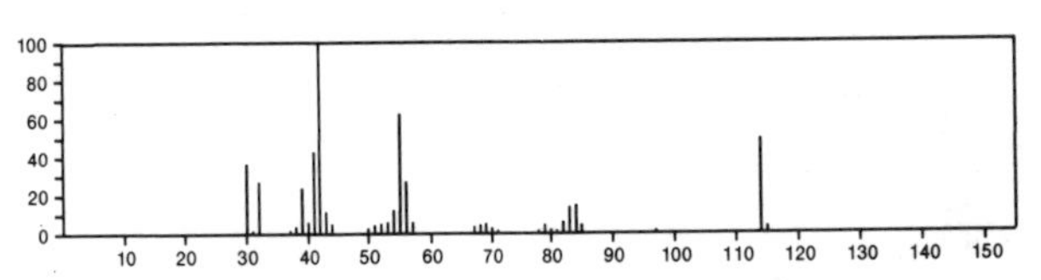

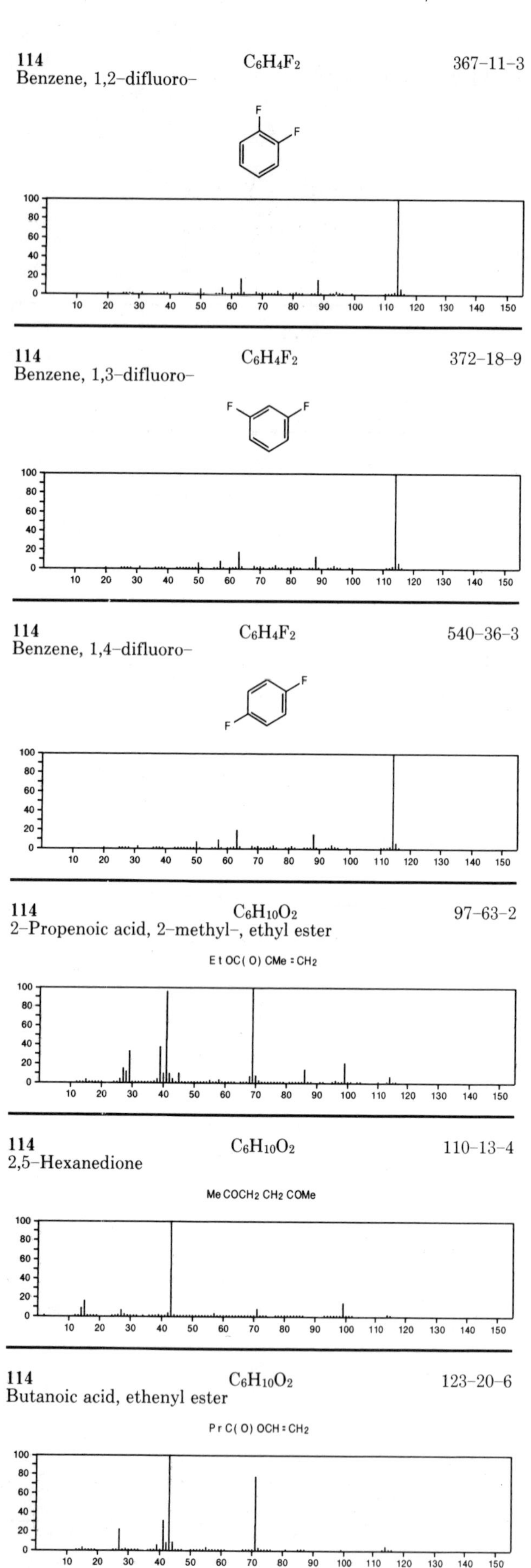
114
Benzene, 1,2-difluoro-
C6H4F2
367-11-3
114
Benzene, 1,3-difluoro-
C6H4F2
372-18-9
114
Benzene, 1,4-difluoro-
C6H4F2
540-36-3
114
2-Propenoic acid, 2-methyl-, ethyl ester
C6H10O2
97-63-2
Et OC(O) CMe = CH2
114
2,5-Hexanedione
C6H10O2
110-13-4
Me COCH2 CH2 COMe
114
Butanoic acid, ethenyl ester
C6H10O2
123-20-6
Pr C(O) OCH = CH2

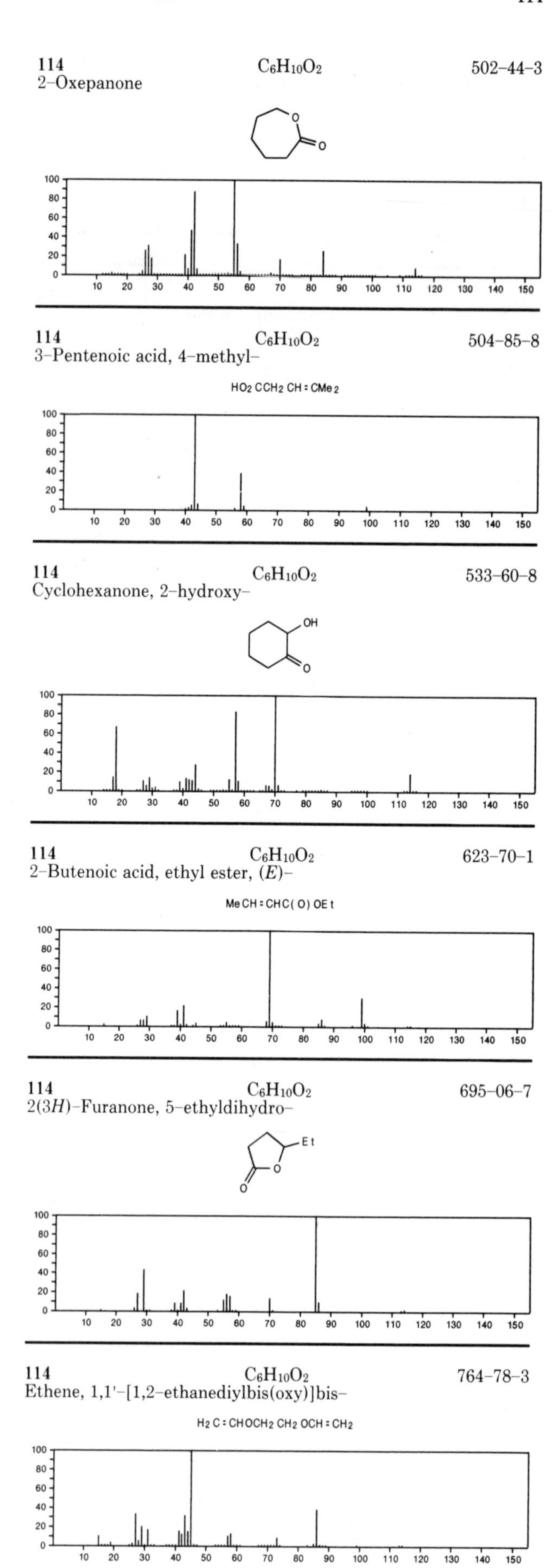
114
2-Oxepanone
C6H10O2
502-44-3
114
3-Pentenoic acid, 4-methyl-
C6H10O2
504-85-8
HO2 CCH2 CH = CMe2
114
Cyclohexanone, 2-hydroxy-
C6H10O2
533-60-8
114
2-Butenoic acid, ethyl ester, (E)-
C6H10O2
623-70-1
Me CH = CHC(O) OEt
114
2(3H)-Furanone, 5-ethyldihydro-
C6H10O2
695-06-7
114
Ethene, 1,1'-[1,2-ethanediylbis(oxy)]bis-
C6H10O2
764-78-3
H2 C = CHOCH2 CH2 OCH = CH2

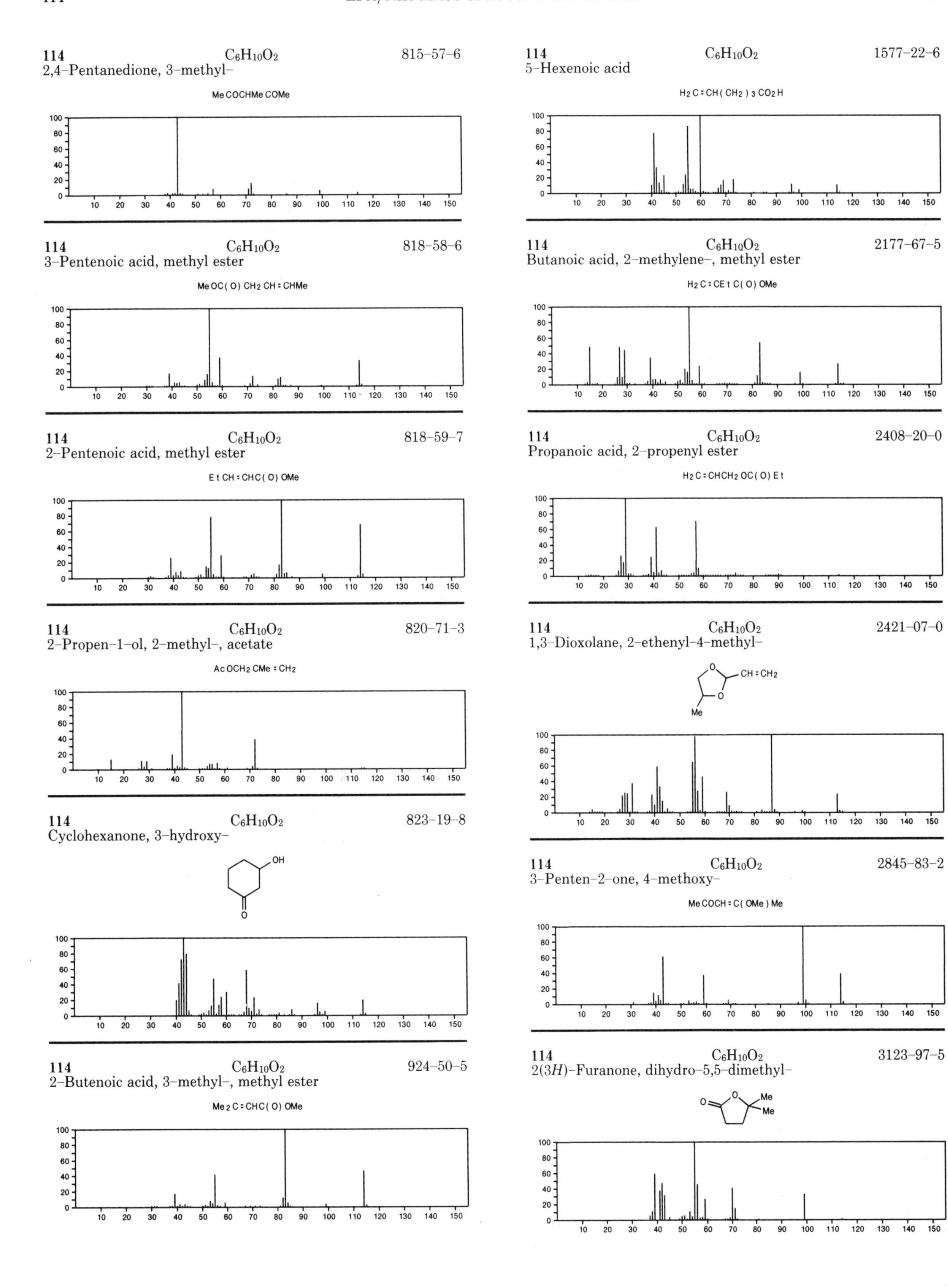
114 $C_6H_{10}O_2$ 815-57-6
2,4-Pentanedione, 3-methyl-
Me COCHMe COMe
114 $C_6H_{10}O_2$ 818-58-6
3-Pentenoic acid, methyl ester
Me OC(O) CH2 CH = CHMe
114 $C_6H_{10}O_2$ 818-59-7
2-Pentenoic acid, methyl ester
E t CH = CHC(O) OMe
114 $C_6H_{10}O_2$ 820-71-3
2-Propen-1-ol, 2-methyl-, acetate
Ac OCH2 CMe = CH2
114 $C_6H_{10}O_2$ 823-19-8
Cyclohexanone, 3-hydroxy-
OH
O
114 $C_6H_{10}O_2$ 924-50-5
2-Butenoic acid, 3-methyl-, methyl ester
Me 2 C = CHC(O) OMe
114 $C_6H_{10}O_2$ 1577-22-6
5-Hexenoic acid
H2 C = CH(CH2) 3 CO2 H
114 $C_6H_{10}O_2$ 2177-67-5
Butanoic acid, 2-methylene-, methyl ester
H2 C = CE t C(O) OMe
114 $C_6H_{10}O_2$ 2408-20-0
Propanoic acid, 2-propenyl ester
H2 C = CHCH2 OC(O) E t
114 $C_6H_{10}O_2$ 2421-07-0
1,3-Dioxolane, 2-ethenyl-4-methyl-
O
CH = CH2
O
Me
114 $C_6H_{10}O_2$ 2845-83-2
3-Penten-2-one, 4-methoxy-
Me COCH = C(OMe) Me
114 $C_6H_{10}O_2$ 3123-97-5
2(3*H*)-Furanone, dihydro-5,5-dimethyl-
O
O
Me
Me

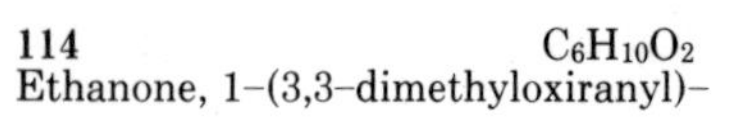

114 $C_6H_{10}O_2$ 4478-63-1
Ethanone, 1-(3,3-dimethyloxiranyl)-

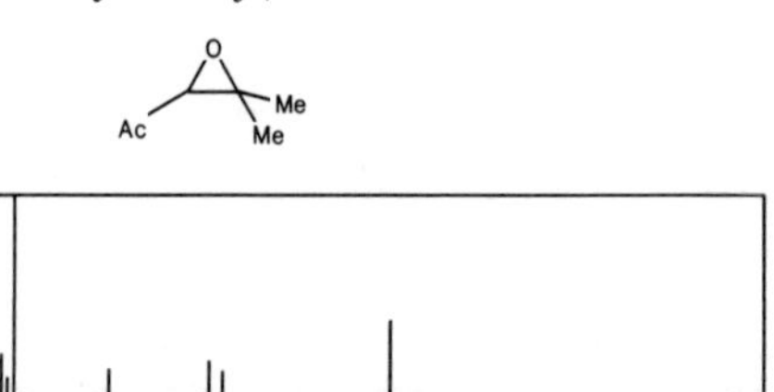

114 $C_6H_{10}O_2$ 4528-26-1
1,3-Dioxolane, 2-(1-propenyl)-

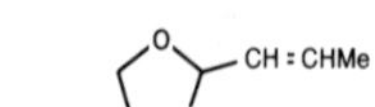

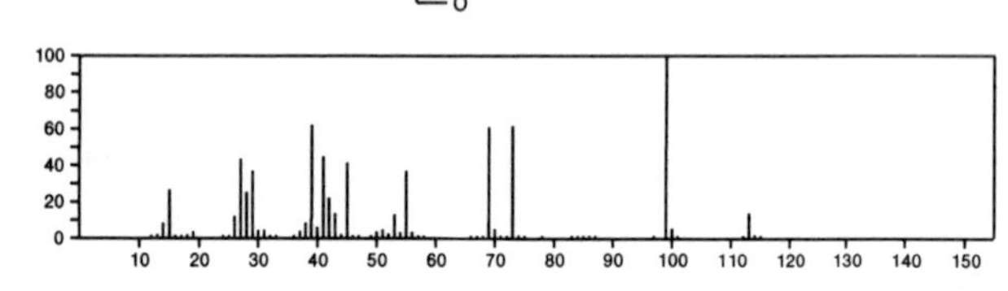

114 $C_6H_{10}O_2$ 5145-01-7
2(3*H*)-Furanone, dihydro-3,5-dimethyl-

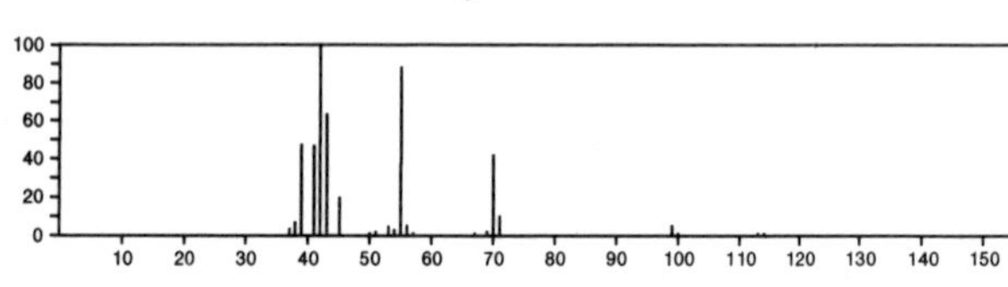

114 $C_6H_{10}O_2$ 6971-63-7
2(3*H*)-Furanone, dihydro-4,5-dimethyl-

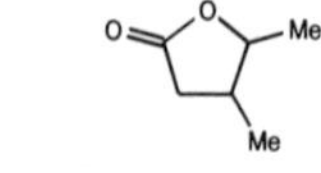

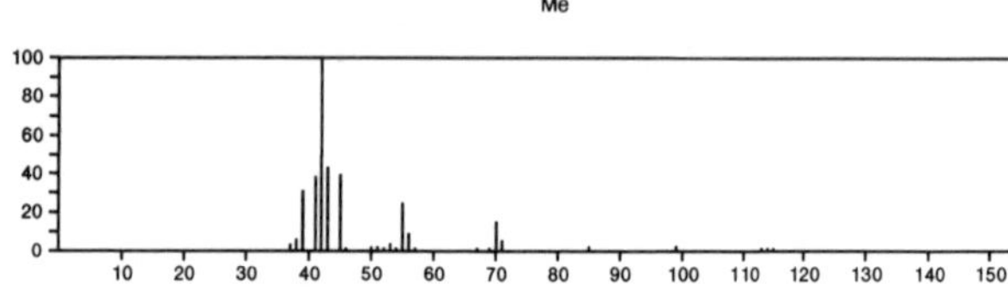

114 $C_6H_{10}O_2$ 7493-58-5
2,3-Pentanedione, 4-methyl-

MeCOCOCHMe2

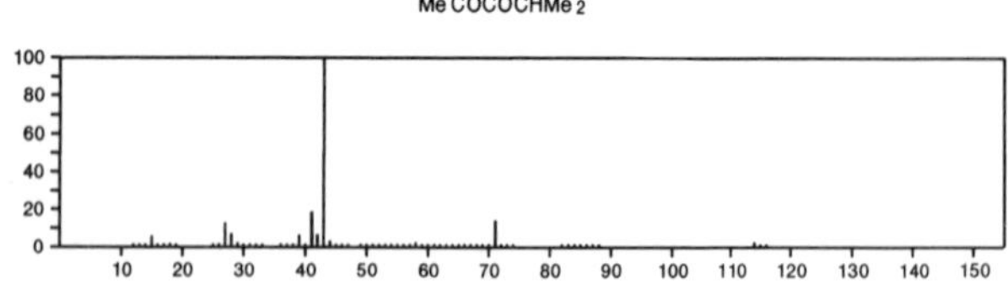

114 $C_6H_{10}O_2$ 10276-09-2
3-Butenoic acid, 2,2-dimethyl-

HO2CCMe2CH=CH2

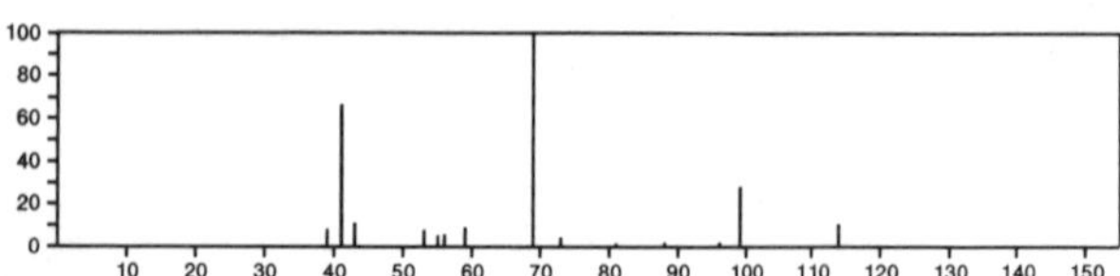

114 $C_6H_{10}O_2$ 13482-22-9
Cyclohexanone, 4-hydroxy-

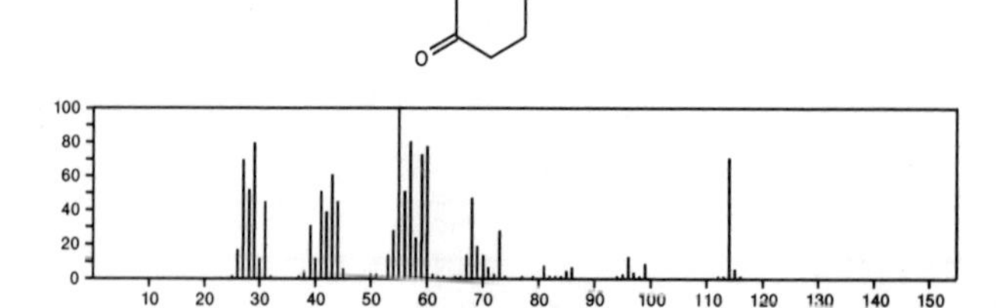

114 $C_6H_{10}O_2$ 13861-97-7
2(3*H*)-Furanone, dihydro-4,4-dimethyl-

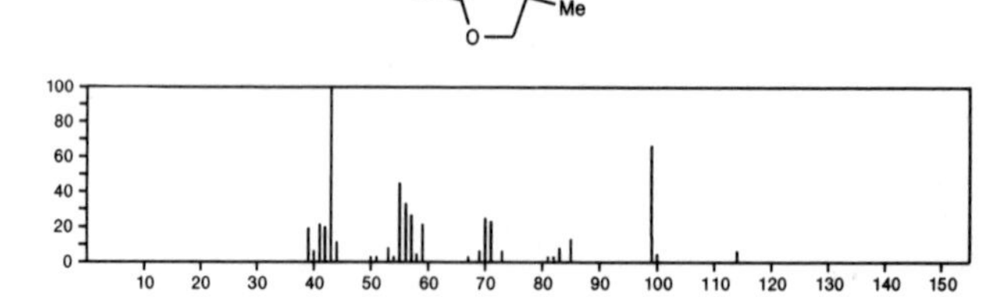

114 $C_6H_{10}O_2$ 17257-81-7
2-Hexanone, 3,4-epoxy-

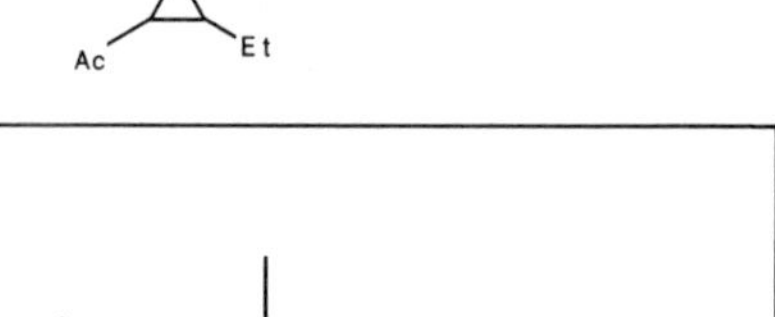

114 $C_6H_{10}O_2$ 17257-82-8
2-Butanone, 3,4-epoxy-3-ethyl-

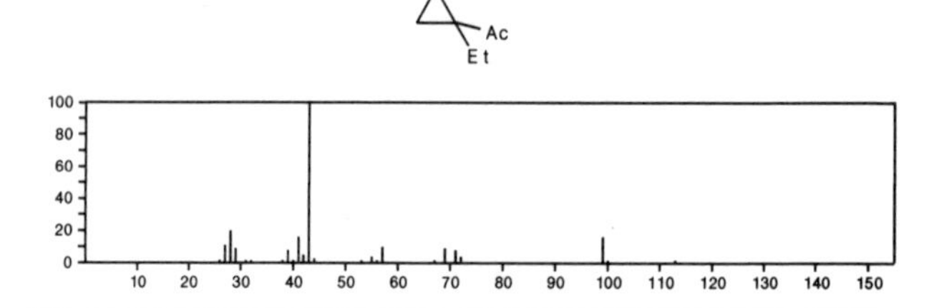

114 $C_6H_{10}O_2$ 18668-75-2
Ethanol, 2-[(1-methyl-2-propynyl)oxy]-

HC≡CCHMeOCH2CH2OH

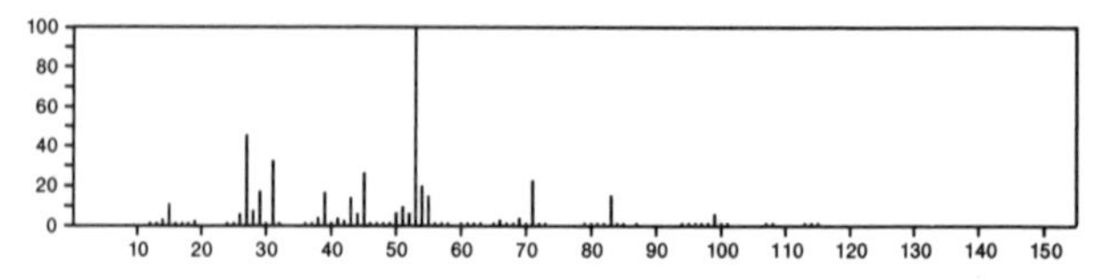

114 $C_6H_{10}O_2$ 25465-18-3
1,4-Dioxin, 2,3-dihydro-5,6-dimethyl-

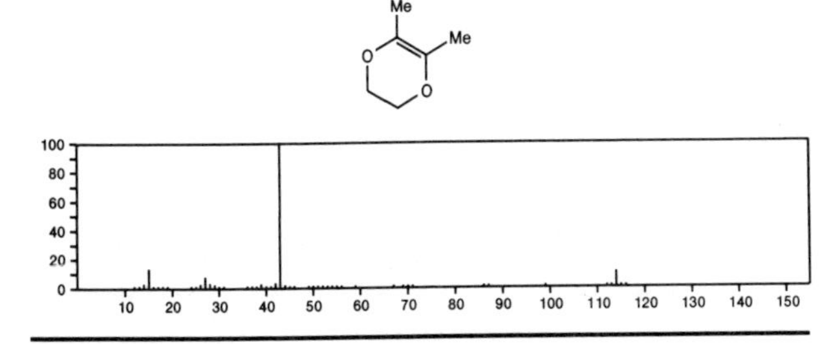

114 $C_6H_{10}O_2$ 26924-35-6
1,3-Dioxolane, 2-ethenyl-2-methyl-

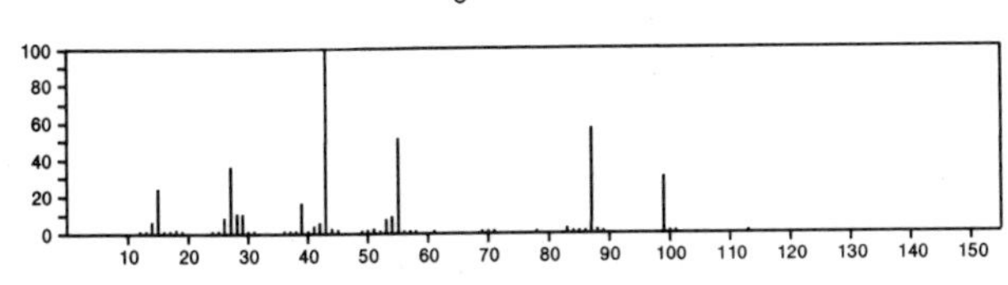

114 $C_6H_{10}O_2$ 28125-74-8
1,4-Dioxane, 2-methyl-3-methylene-

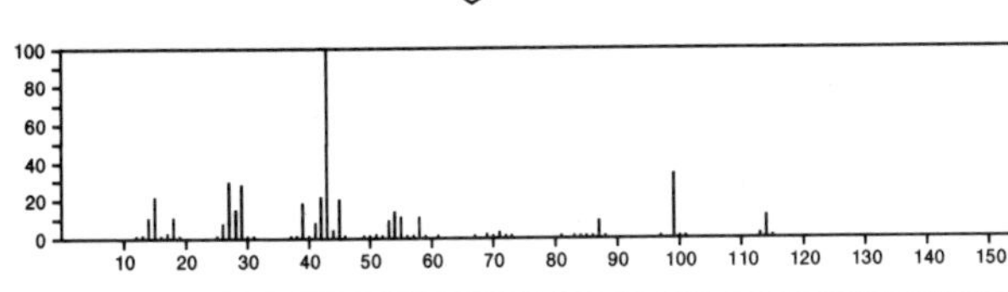

114 $C_6H_{10}O_2$ 35194-36-6
4-Hexenoic acid

MeCH=CHCH2CH2CO2H

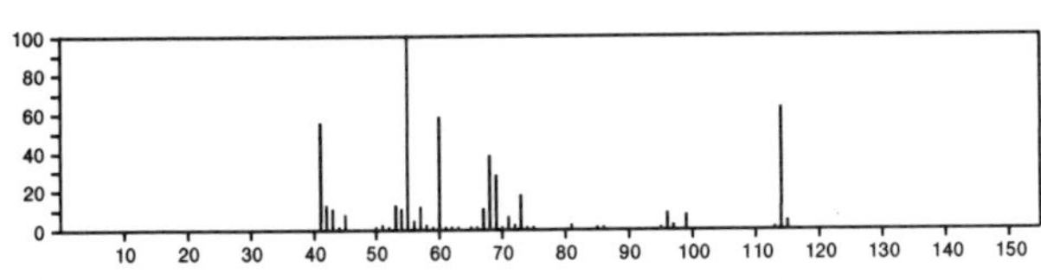

114 $C_6H_{10}O_2$ 38644-91-6
Ethanol, 2-(2-butynyloxy)-

HOCH2CH2OCH2C≡CMe

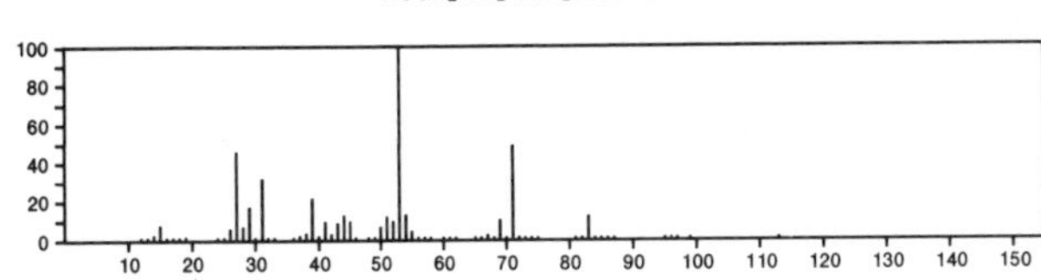

114 $C_6H_{10}O_2$ 38653-35-9
5*H*-1,4-Dioxepin, 2,3-dihydro-7-methyl-

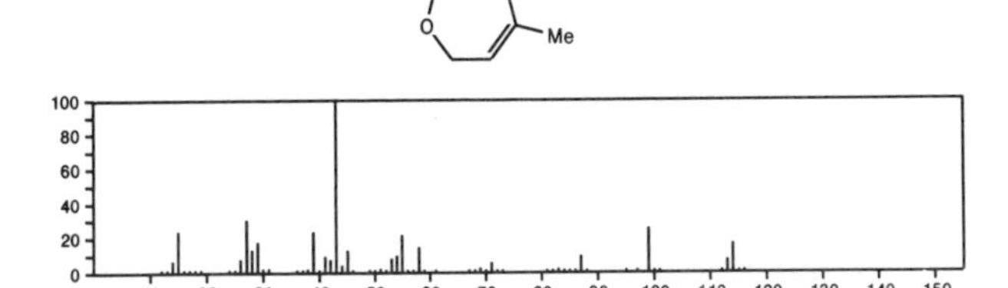

114 $C_6H_{10}O_2$ 38653-36-0
5*H*-1,4-Dioxepin, 2,3-dihydro-5-methyl-

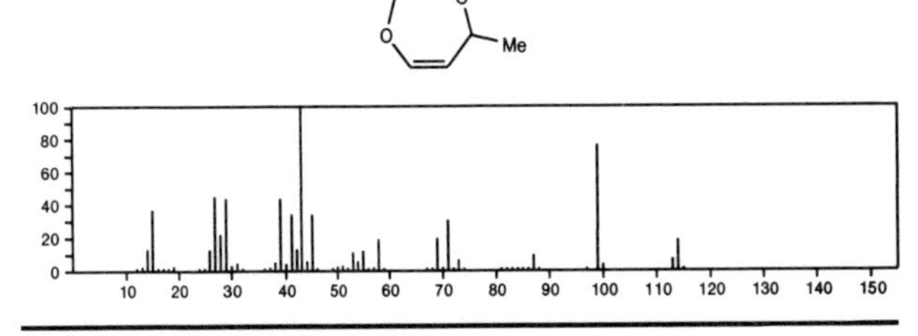

114 $C_6H_{10}O_2$ 38653-49-5
1,3-Dioxolane, 2-(2-propenyl)-

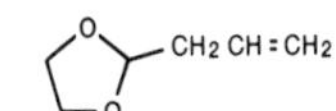

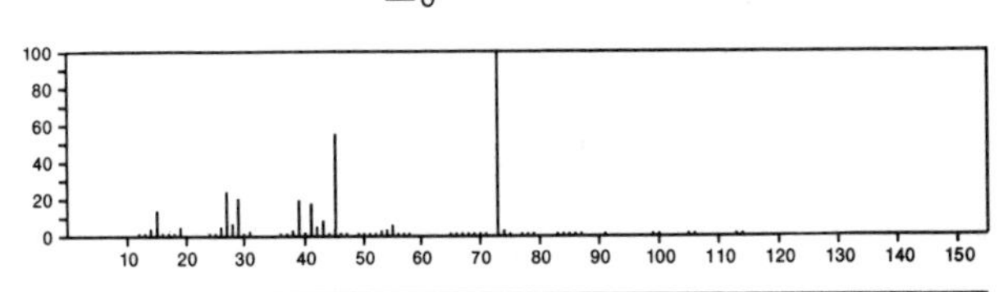

114 $C_6H_{10}O_2$ 38653-50-8
Ethanol, 2-(2,3-butadienyloxy)-

HOCH2CH2OCH2CH=C=CH2

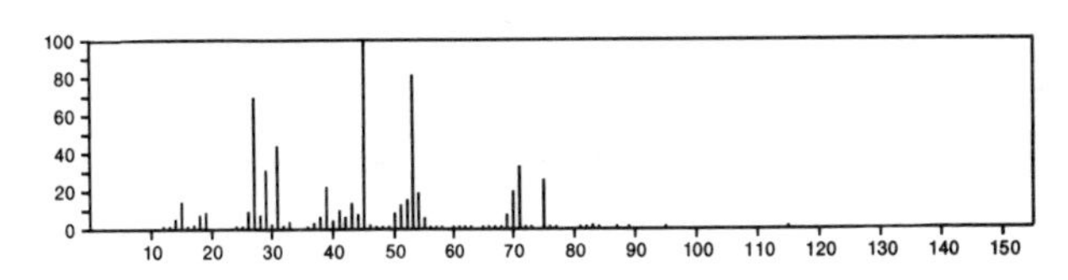

114 $C_6H_{10}O_2$ 38653-51-9
Ethanol, 2-[(1-methylene-2-propenyl)oxy]-

CH2
||
HOCH2CH2OCCH=CH2

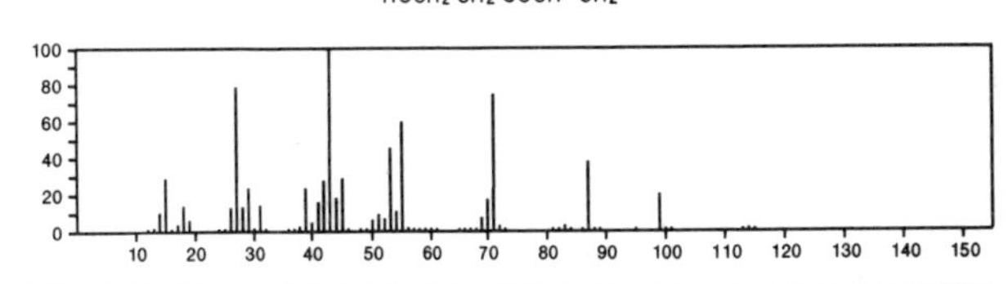

114 $C_6H_{10}O_2$ 43152-89-2
2*H*-Pyran-3(4*H*)-one, dihydro-6-methyl-

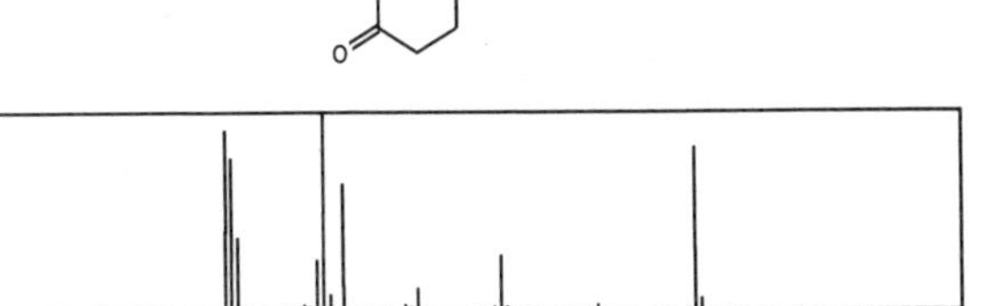

114 $C_6H_{10}O_2$ 50521-50-1
1,4-Butanediol, 2,3-bis(methylene)-

CH2
||CH2
|||
HOCH2CCCH2OH

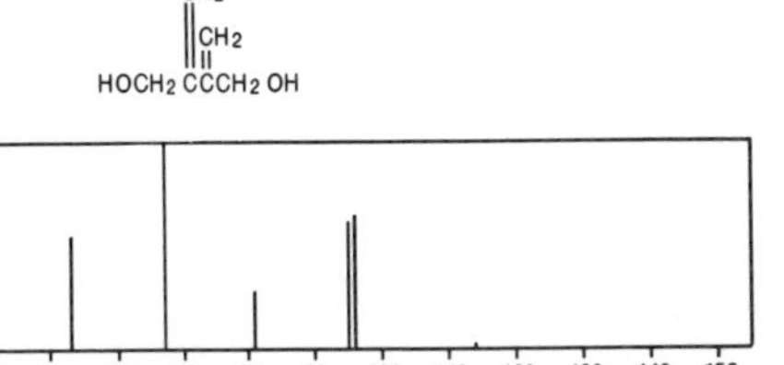

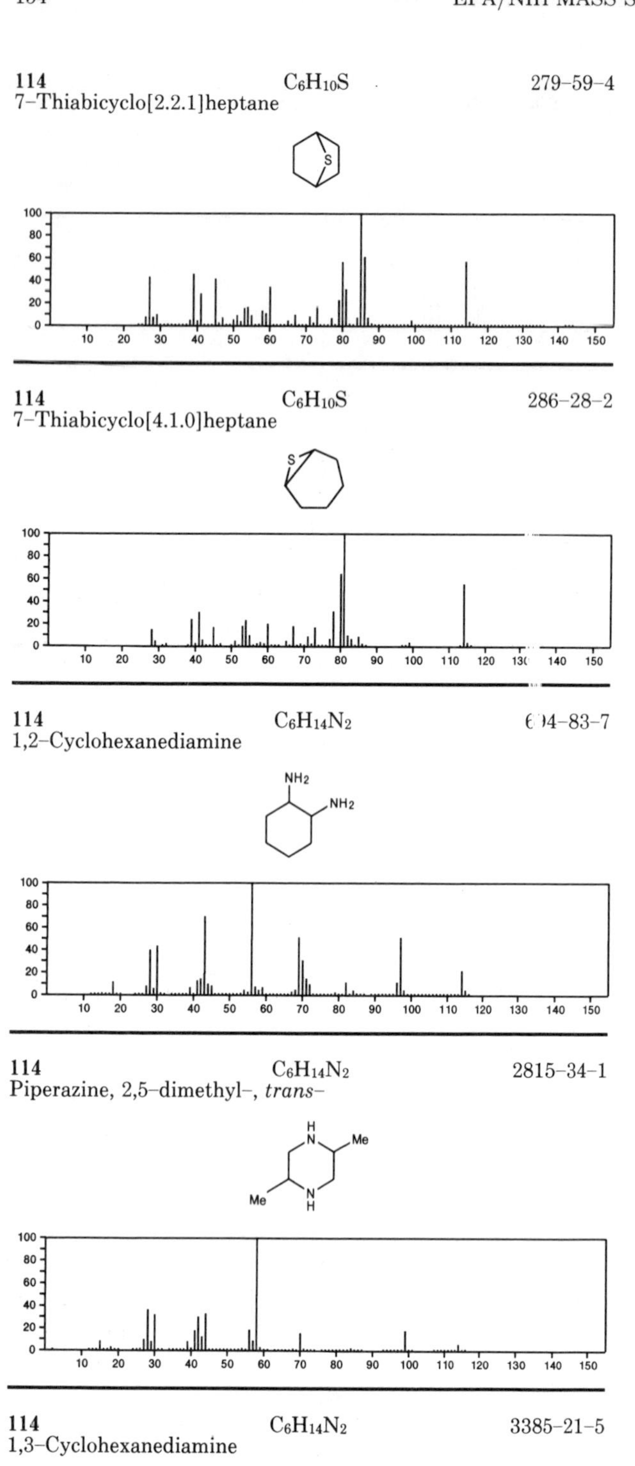

114 $C_6H_{10}S$ 279-59-4
7-Thiabicyclo[2.2.1]heptane

114 $C_6H_{10}S$ 286-28-2
7-Thiabicyclo[4.1.0]heptane

114 $C_6H_{14}N_2$ 694-83-7
1,2-Cyclohexanediamine

114 $C_6H_{14}N_2$ 2815-34-1
Piperazine, 2,5-dimethyl-, *trans*-

114 $C_6H_{14}N_2$ 3385-21-5
1,3-Cyclohexanediamine

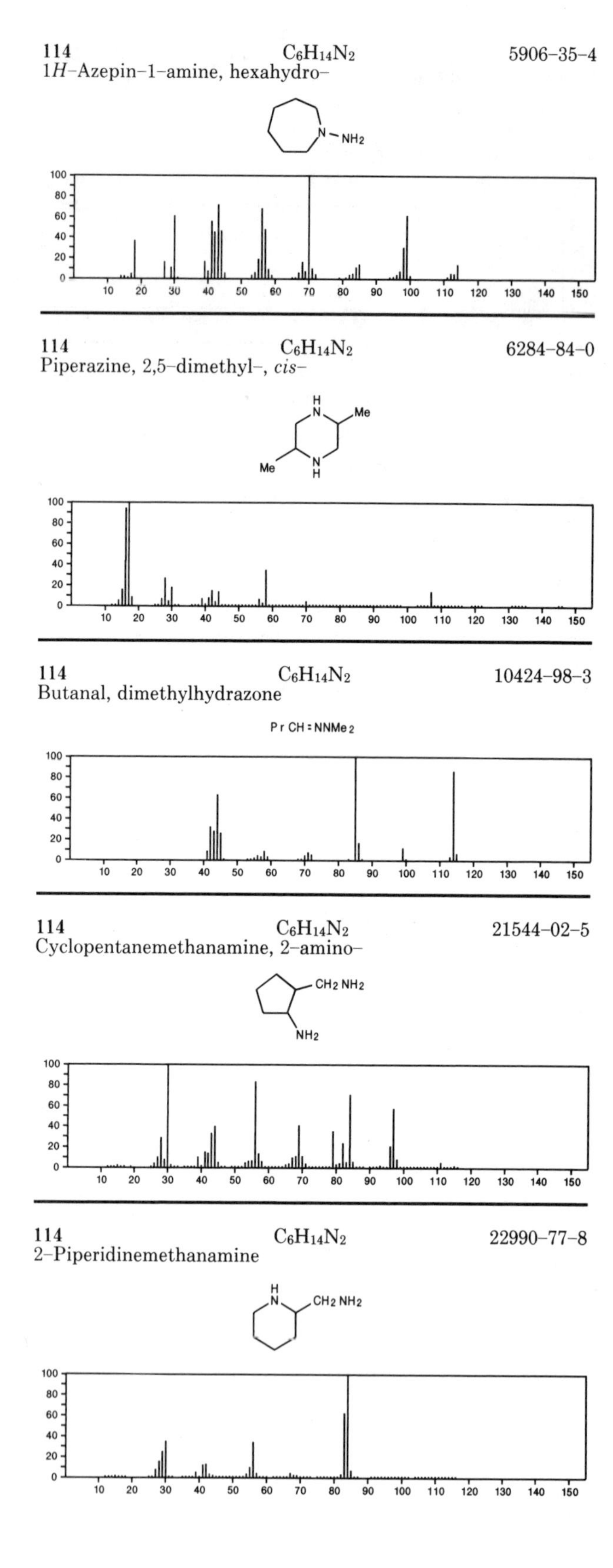

114 $C_6H_{14}N_2$ 5906-35-4
1*H*-Azepin-1-amine, hexahydro-

114 $C_6H_{14}N_2$ 6284-84-0
Piperazine, 2,5-dimethyl-, *cis*-

114 $C_6H_{14}N_2$ 10424-98-3
Butanal, dimethylhydrazone

Pr CH = NNMe2

114 $C_6H_{14}N_2$ 21544-02-5
Cyclopentanemethanamine, 2-amino-

114 $C_6H_{14}N_2$ 22990-77-8
2-Piperidinemethanamine

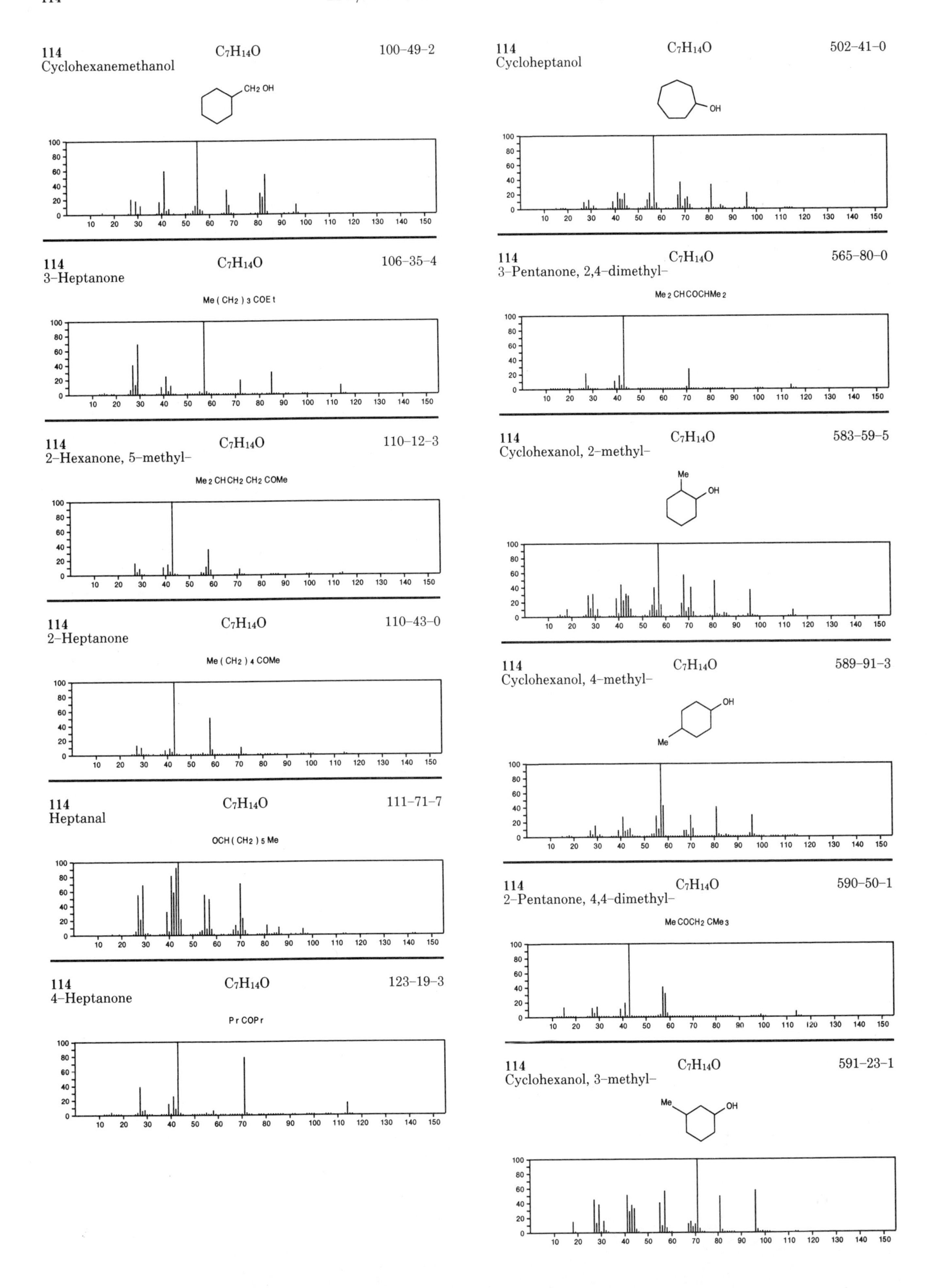
114 Cyclohexanemethanol $C_7H_{14}O$ 100-49-2
CH2 OH
114 3-Heptanone $C_7H_{14}O$ 106-35-4
Me (CH2) 3 COE t
114 2-Hexanone, 5-methyl- $C_7H_{14}O$ 110-12-3
Me 2 CH CH2 CH2 COMe
114 2-Heptanone $C_7H_{14}O$ 110-43-0
Me (CH2) 4 COMe
114 Heptanal $C_7H_{14}O$ 111-71-7
OCH (CH2) 5 Me
114 4-Heptanone $C_7H_{14}O$ 123-19-3
P r COP r
114 Cycloheptanol $C_7H_{14}O$ 502-41-0
OH
114 3-Pentanone, 2,4-dimethyl- $C_7H_{14}O$ 565-80-0
Me 2 CHCOCHMe 2
114 Cyclohexanol, 2-methyl- $C_7H_{14}O$ 583-59-5
Me
OH
114 Cyclohexanol, 4-methyl- $C_7H_{14}O$ 589-91-3
OH
Me
114 2-Pentanone, 4,4-dimethyl- $C_7H_{14}O$ 590-50-1
Me COCH2 CMe 3
114 Cyclohexanol, 3-methyl- $C_7H_{14}O$ 591-23-1
Me
OH

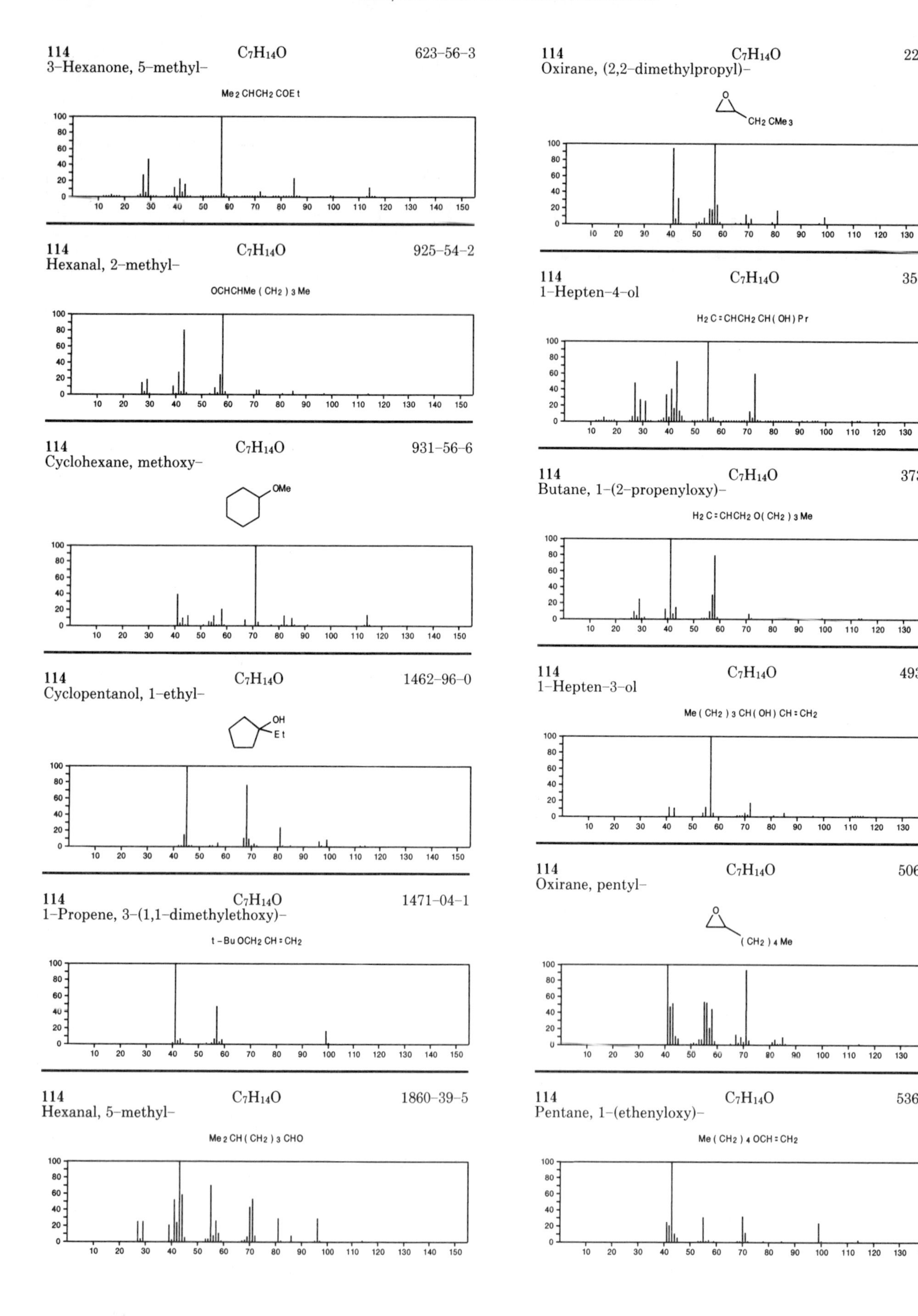
114 C7H14O 623-56-3
3-Hexanone, 5-methyl-
Me2CHCH2COEt
114 C7H14O 925-54-2
Hexanal, 2-methyl-
OCHCHMe(CH2)3Me
114 C7H14O 931-56-6
Cyclohexane, methoxy-
OMe
114 C7H14O 1462-96-0
Cyclopentanol, 1-ethyl-
OH
Et
114 C7H14O 1471-04-1
1-Propene, 3-(1,1-dimethylethoxy)-
t-BuOCH2CH=CH2
114 C7H14O 1860-39-5
Hexanal, 5-methyl-
Me2CH(CH2)3CHO
114 C7H14O 2245-29-6
Oxirane, (2,2-dimethylpropyl)-
O
CH2CMe3
114 C7H14O 3521-91-3
1-Hepten-4-ol
H2C=CHCH2CH(OH)Pr
114 C7H14O 3739-64-8
Butane, 1-(2-propenyloxy)-
H2C=CHCH2O(CH2)3Me
114 C7H14O 4938-52-7
1-Hepten-3-ol
Me(CH2)3CH(OH)CH=CH2
114 C7H14O 5063-65-0
Oxirane, pentyl-
O
(CH2)4Me
114 C7H14O 5363-63-3
Pentane, 1-(ethenyloxy)-
Me(CH2)4OCH=CH2

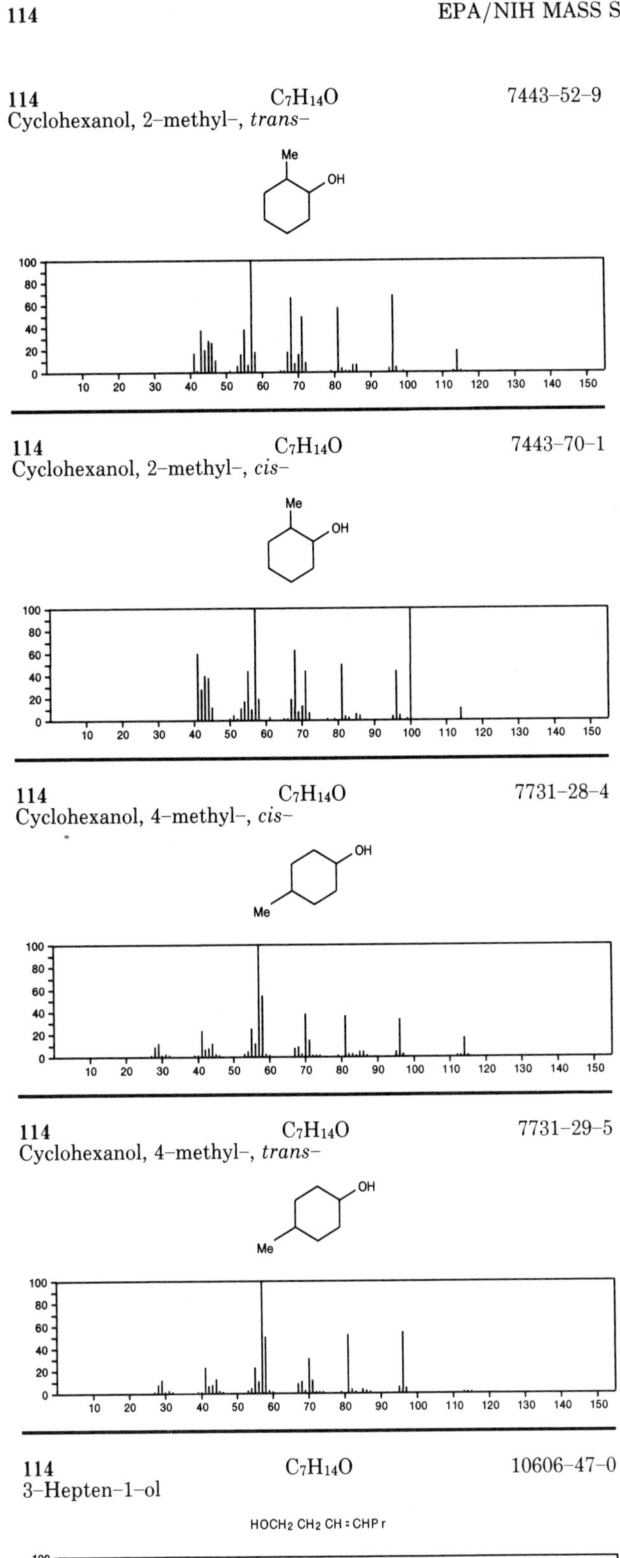

114 $C_7H_{14}O$ 7443-52-9
Cyclohexanol, 2-methyl-, *trans-*

114 $C_7H_{14}O$ 7443-70-1
Cyclohexanol, 2-methyl-, *cis-*

114 $C_7H_{14}O$ 7731-28-4
Cyclohexanol, 4-methyl-, *cis-*

114 $C_7H_{14}O$ 7731-29-5
Cyclohexanol, 4-methyl-, *trans-*

114 $C_7H_{14}O$ 10606-47-0
3-Hepten-1-ol

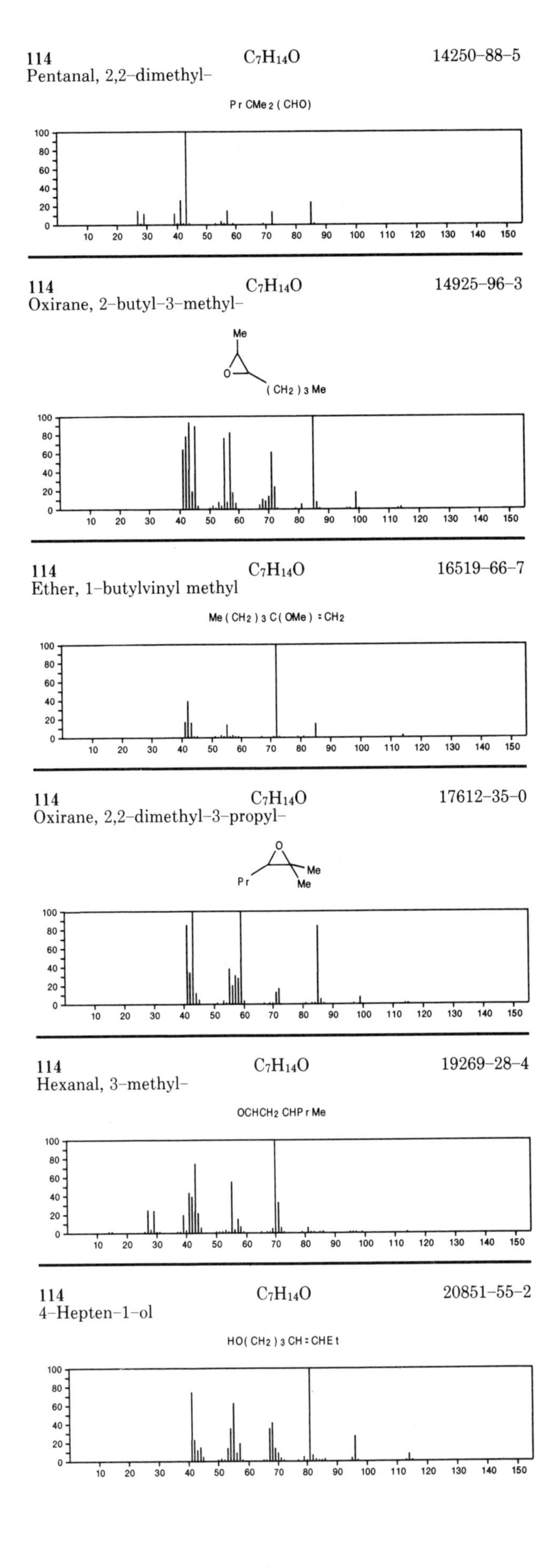

114 $C_7H_{14}O$ 14250-88-5
Pentanal, 2,2-dimethyl-

114 $C_7H_{14}O$ 14925-96-3
Oxirane, 2-butyl-3-methyl-

114 $C_7H_{14}O$ 16519-66-7
Ether, 1-butylvinyl methyl

114 $C_7H_{14}O$ 17612-35-0
Oxirane, 2,2-dimethyl-3-propyl-

114 $C_7H_{14}O$ 19269-28-4
Hexanal, 3-methyl-

114 $C_7H_{14}O$ 20851-55-2
4-Hepten-1-ol

114 $C_7H_{14}O$ 25639-42-3
Cyclohexanol, methyl-

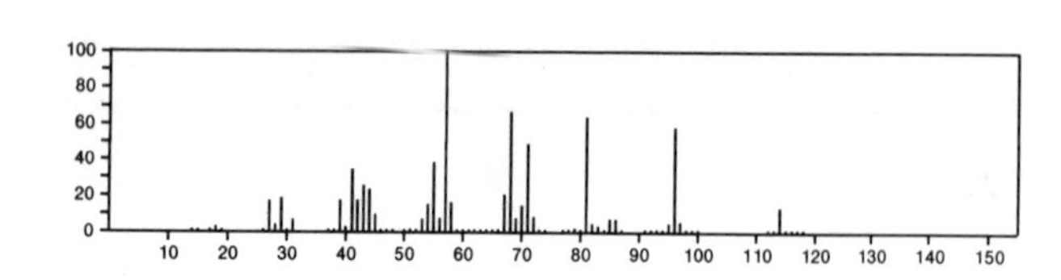

114 $C_7H_{14}O$ 27944-79-2
Pentanal, 2,4-dimethyl-

$OCHCHMe\,CH_2\,CHMe_2$

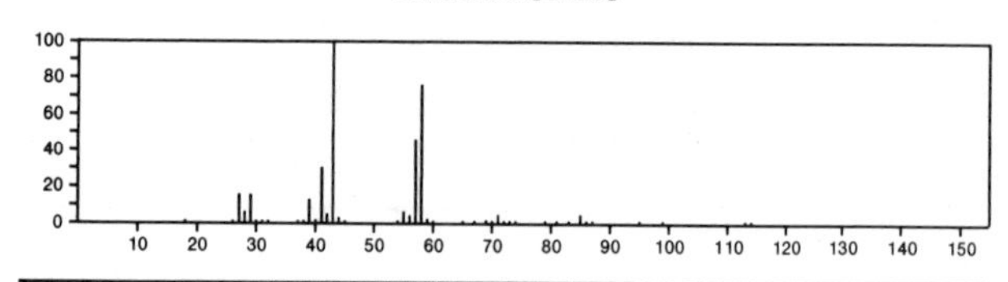

114 $C_7H_{14}O$ 29281-39-8
Butane, 2-(ethenyloxy)-2-methyl-

$H_2C{=}CHOCEtMe_2$

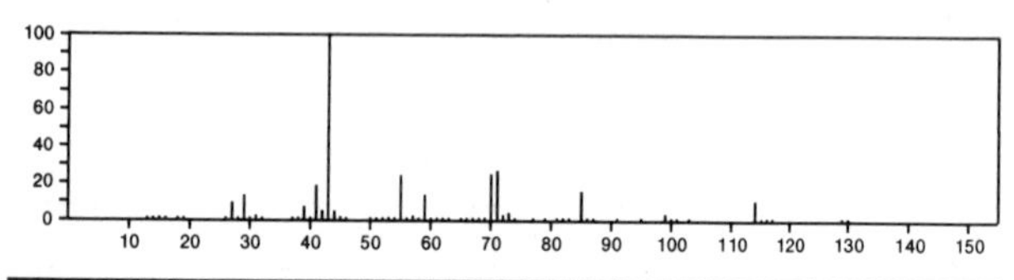

114 $C_7H_{14}O$ 30801-96-8
2-Hexen-1-ol, 3-methyl-, (*E*)-

$HOCH_2\,CH{=}CPrMe$

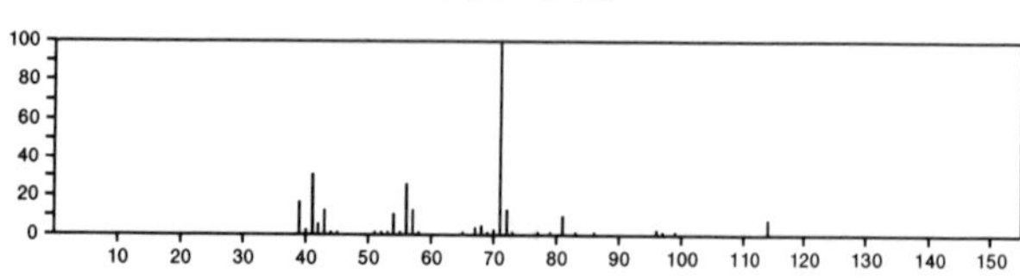

114 $C_7H_{14}O$ 32749-94-3
Pentanal, 2,3-dimethyl-

$MeCH_2\,CHMe\,CHMe\,CHO$

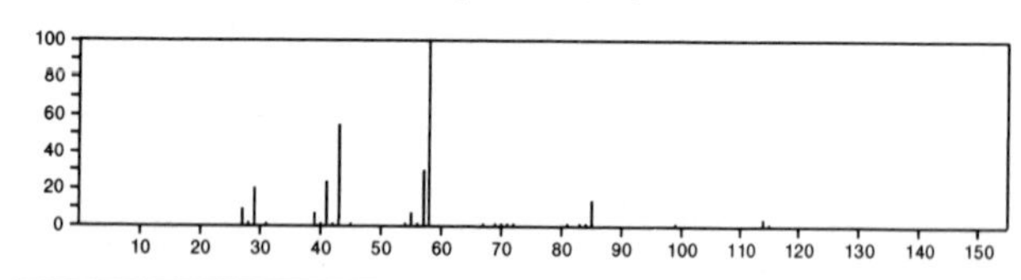

114 $C_7H_{14}O$ 34061-75-1
Ether, 3-butenyl propyl

$H_2C{=}CHCH_2\,CH_2\,OPr$

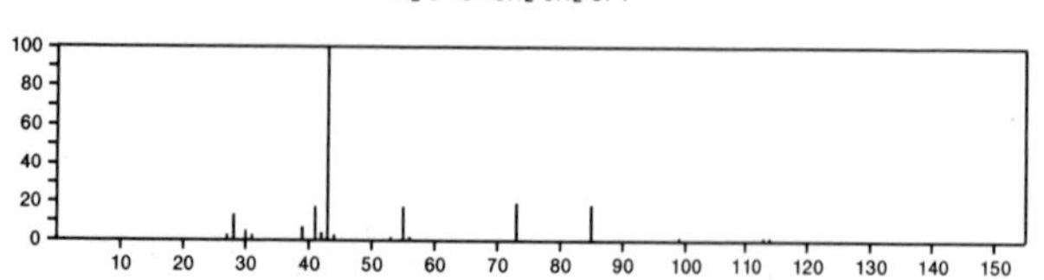

114 $C_7H_{14}O$ 39782-38-2
Butane, 1-(ethenyloxy)-3-methyl-

$H_2C{=}CHOCH_2\,CH_2\,CHMe_2$

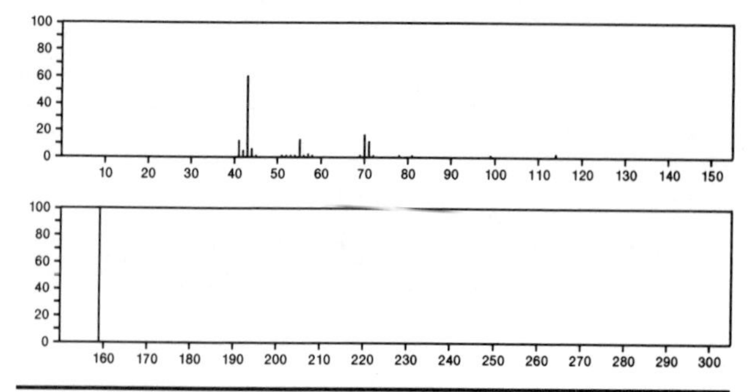

114 $C_7H_{14}O$ 41065-97-8
Hexanal, 4-methyl-

$OCHCH_2\,CH_2\,CHMe\,CH_2\,Me$

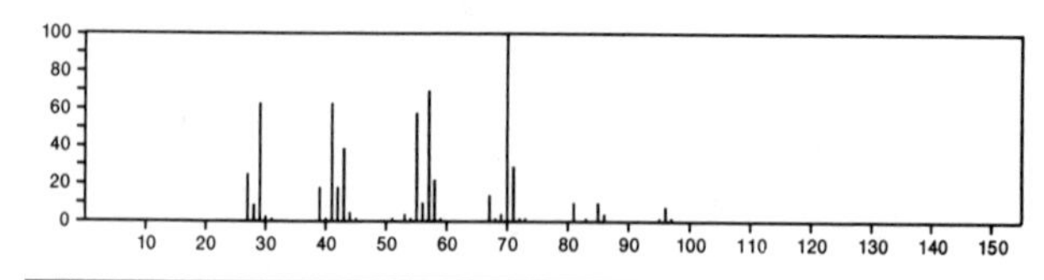

114 $C_7H_{14}O$ 42328-43-8
Oxirane, 2-methyl-2-(1-methylpropyl)-

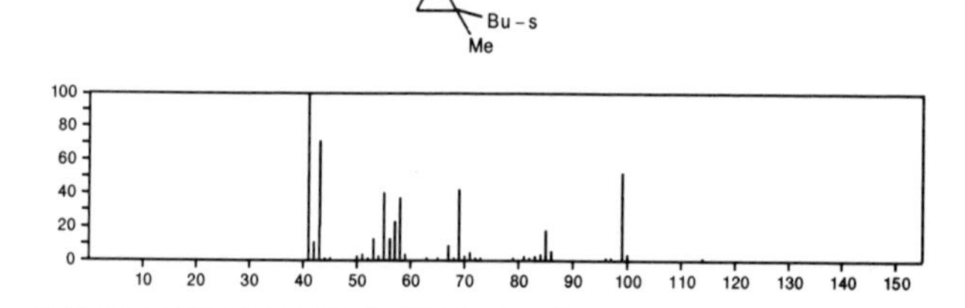

114 $C_7H_{14}O$ 52829-98-8
Cyclopentanemethanol, α-methyl-

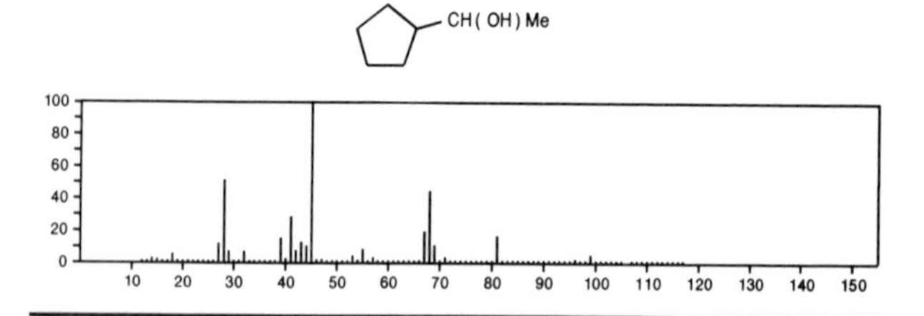

114 $C_7H_{14}O$ 53229-39-3
Oxirane, (1-methylbutyl)-

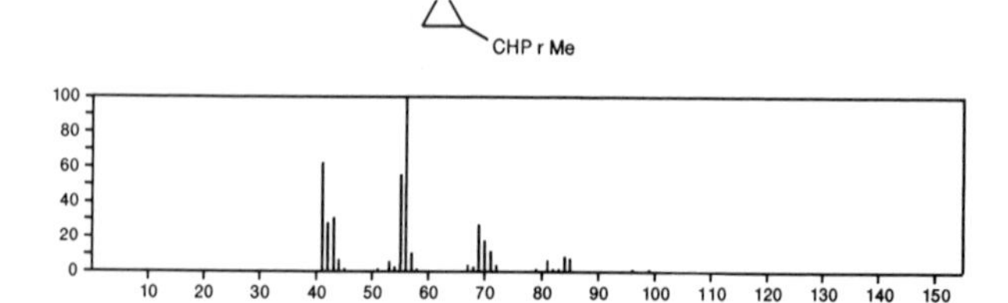

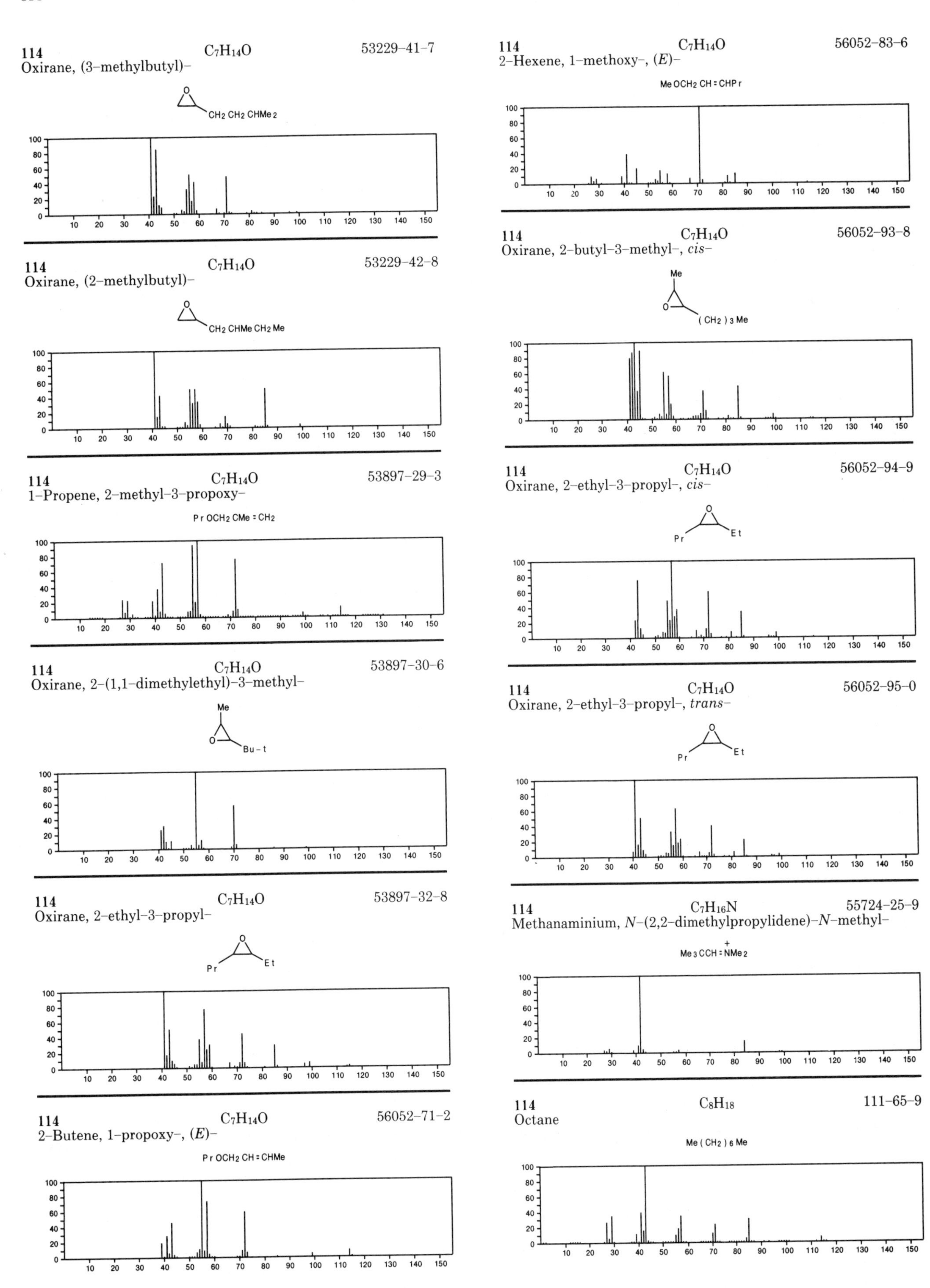

114 $C_7H_{14}O$ 53229-41-7
Oxirane, (3-methylbutyl)-
CH2 CH2 CHMe 2
114 $C_7H_{14}O$ 53229-42-8
Oxirane, (2-methylbutyl)-
CH2 CHMe CH2 Me
114 $C_7H_{14}O$ 53897-29-3
1-Propene, 2-methyl-3-propoxy-
Pr OCH2 CMe = CH2
114 $C_7H_{14}O$ 53897-30-6
Oxirane, 2-(1,1-dimethylethyl)-3-methyl-
Me
Bu-t
114 $C_7H_{14}O$ 53897-32-8
Oxirane, 2-ethyl-3-propyl-
Pr
Et
114 $C_7H_{14}O$ 56052-71-2
2-Butene, 1-propoxy-, (E)-
Pr OCH2 CH = CHMe
114 $C_7H_{14}O$ 56052-83-6
2-Hexene, 1-methoxy-, (E)-
Me OCH2 CH = CHPr
114 $C_7H_{14}O$ 56052-93-8
Oxirane, 2-butyl-3-methyl-, cis-
Me
(CH2)3 Me
114 $C_7H_{14}O$ 56052-94-9
Oxirane, 2-ethyl-3-propyl-, cis-
Pr
Et
114 $C_7H_{14}O$ 56052-95-0
Oxirane, 2-ethyl-3-propyl-, trans-
Pr
Et
114 $C_7H_{16}N$ 55724-25-9
Methanaminium, N-(2,2-dimethylpropylidene)-N-methyl-
Me3 CCH = NMe2 +
114 C_8H_{18} 111-65-9
Octane
Me (CH2)6 Me

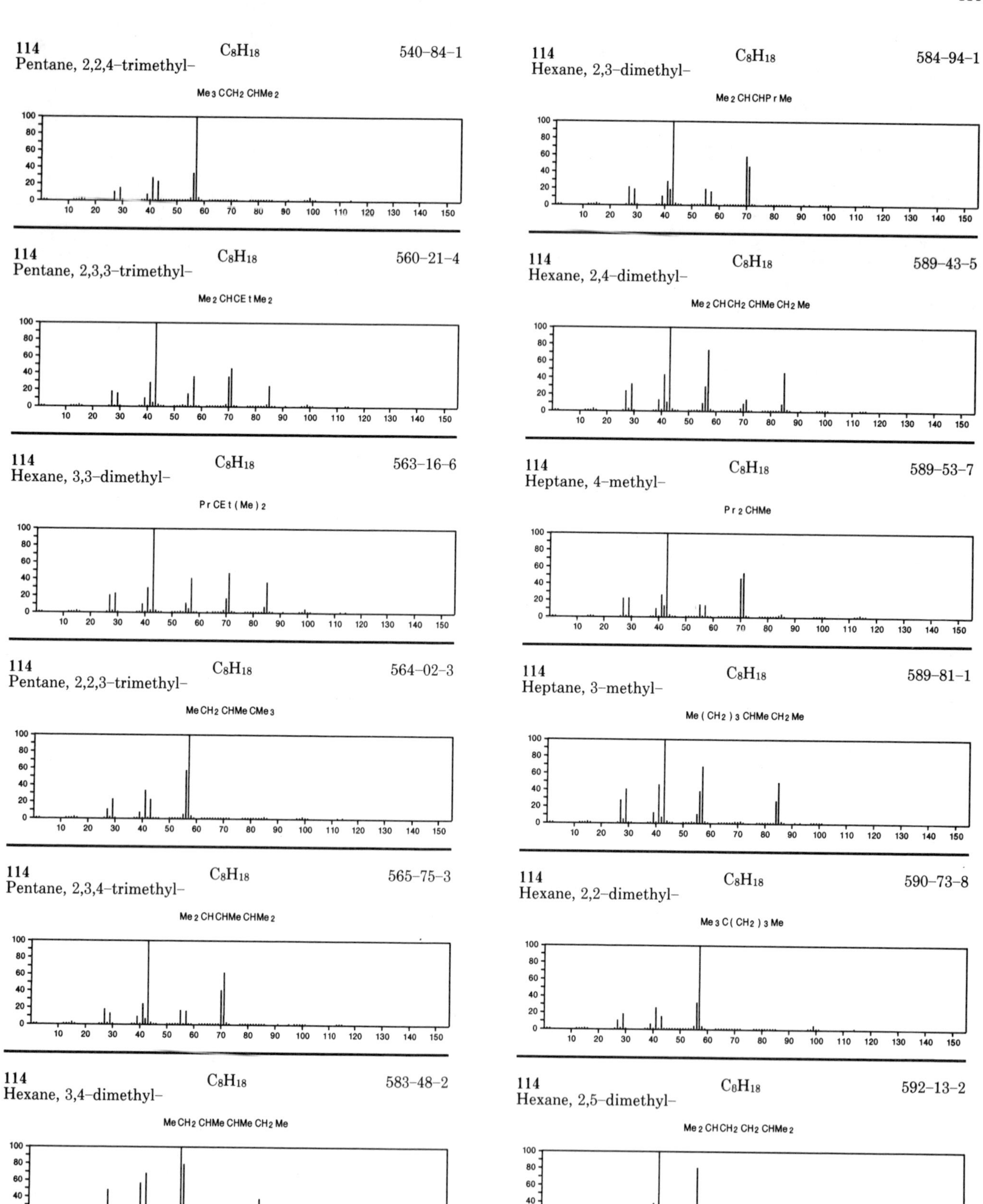
114 Pentane, 2,2,4-trimethyl- C8H18 540-84-1
Me3CCH2CHMe2
114 Hexane, 2,3-dimethyl- C8H18 584-94-1
Me2CHCHPrMe
114 Pentane, 2,3,3-trimethyl- C8H18 560-21-4
Me2CHCEtMe2
114 Hexane, 2,4-dimethyl- C8H18 589-43-5
Me2CHCH2CHMeCH2Me
114 Hexane, 3,3-dimethyl- C8H18 563-16-6
PrCEt(Me)2
114 Heptane, 4-methyl- C8H18 589-53-7
Pr2CHMe
114 Pentane, 2,2,3-trimethyl- C8H18 564-02-3
MeCH2CHMeCMe3
114 Heptane, 3-methyl- C8H18 589-81-1
Me(CH2)3CHMeCH2Me
114 Pentane, 2,3,4-trimethyl- C8H18 565-75-3
Me2CHCHMeCHMe2
114 Hexane, 2,2-dimethyl- C8H18 590-73-8
Me3C(CH2)3Me
114 Hexane, 3,4-dimethyl- C8H18 583-48-2
MeCH2CHMeCHMeCH2Me
114 Hexane, 2,5-dimethyl- C8H18 592-13-2
Me2CHCH2CH2CHMe2

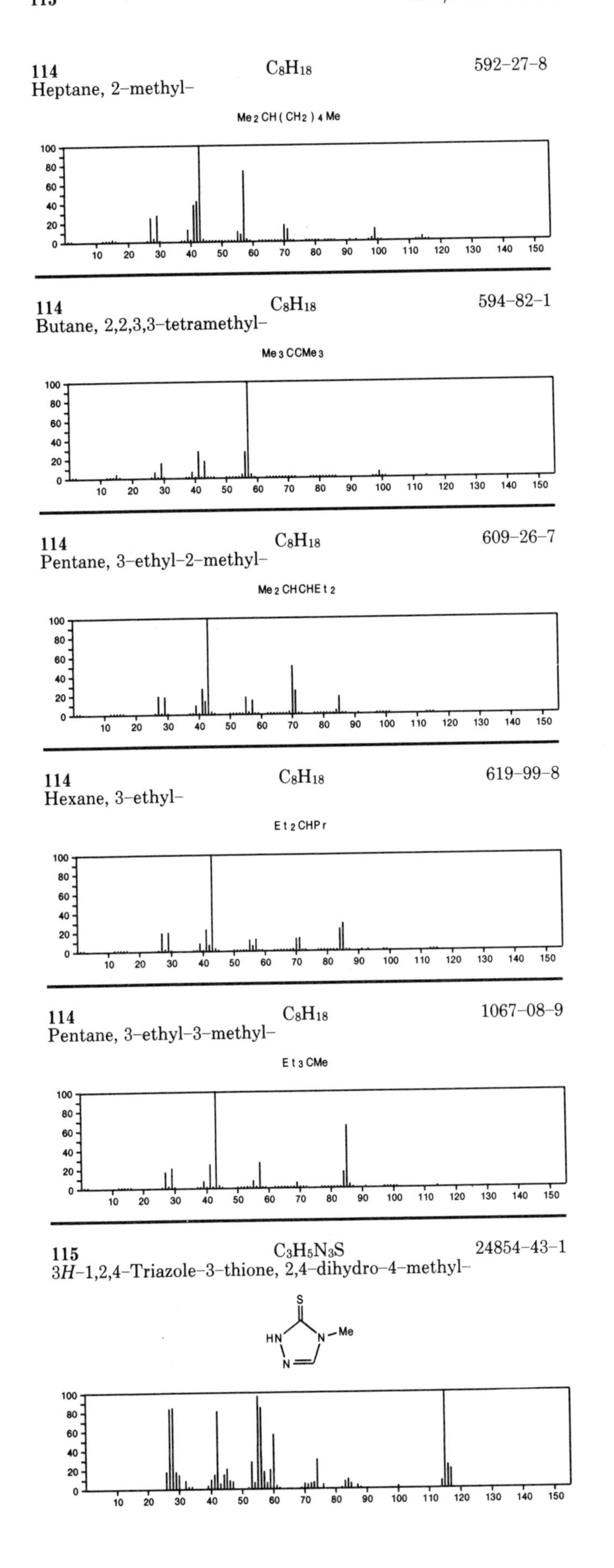
114 C8H18 592-27-8
Heptane, 2-methyl-
Me2CH(CH2)4Me
114 C8H18 594-82-1
Butane, 2,2,3,3-tetramethyl-
Me3CCMe3
114 C8H18 609-26-7
Pentane, 3-ethyl-2-methyl-
Me2CHCHEt2
114 C8H18 619-99-8
Hexane, 3-ethyl-
Et2CHPr
114 C8H18 1067-08-9
Pentane, 3-ethyl-3-methyl-
Et3CMe
115 C3H5N3S 24854-43-1
3H-1,2,4-Triazole-3-thione, 2,4-dihydro-4-methyl-

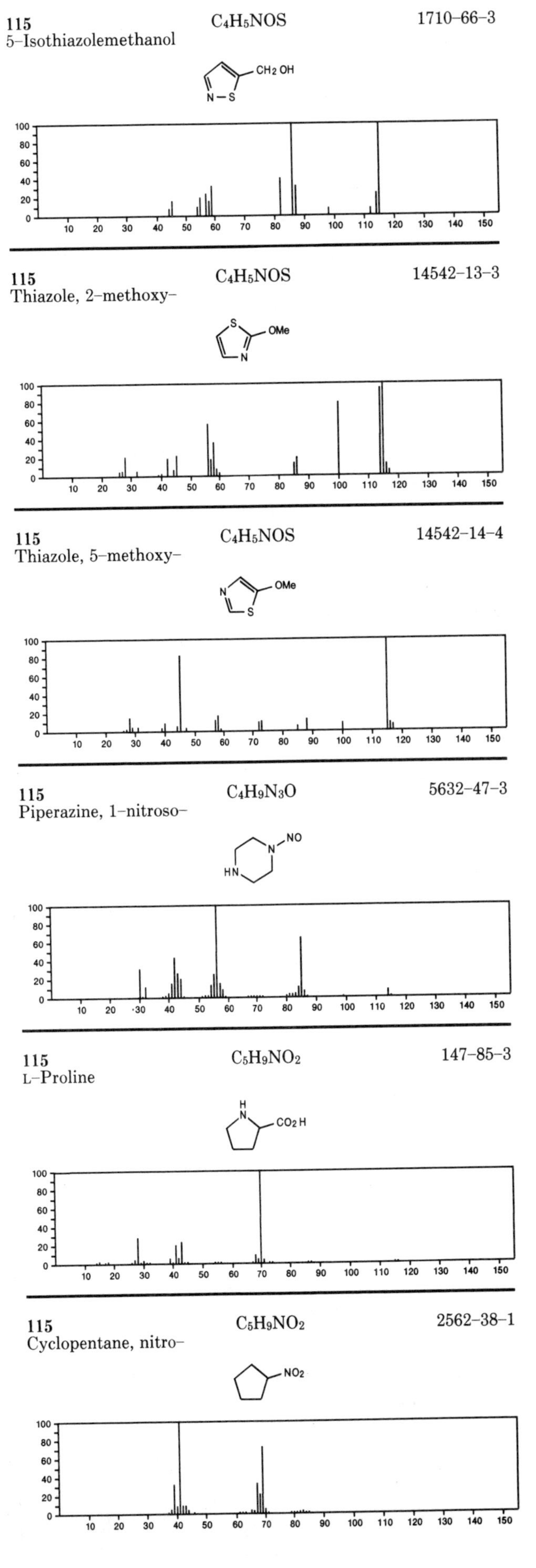
115 C4H5NOS 1710-66-3
5-Isothiazolemethanol
115 C4H5NOS 14542-13-3
Thiazole, 2-methoxy-
115 C4H5NOS 14542-14-4
Thiazole, 5-methoxy-
115 C4H9N3O 5632-47-3
Piperazine, 1-nitroso-
115 C5H9NO2 147-85-3
L-Proline
115 C5H9NO2 2562-38-1
Cyclopentane, nitro-

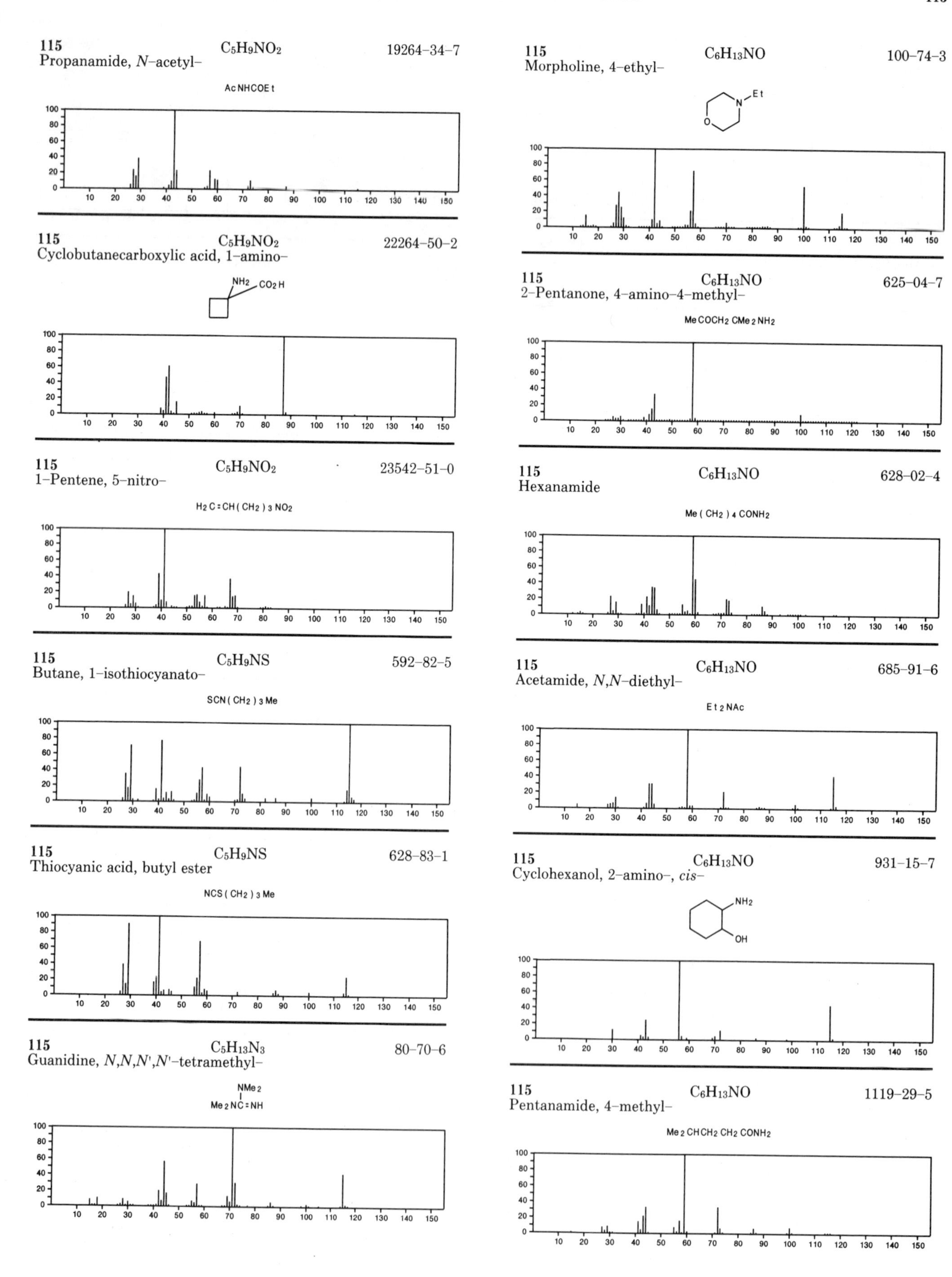
115
Propanamide, N-acetyl-
C5H9NO2
19264-34-7
AcNHCOEt
115
Cyclobutanecarboxylic acid, 1-amino-
C5H9NO2
22264-50-2
NH2 CO2H
115
1-Pentene, 5-nitro-
C5H9NO2
23542-51-0
H2C=CH(CH2)3NO2
115
Butane, 1-isothiocyanato-
C5H9NS
592-82-5
SCN(CH2)3Me
115
Thiocyanic acid, butyl ester
C5H9NS
628-83-1
NCS(CH2)3Me
115
Guanidine, N,N,N',N'-tetramethyl-
C5H13N3
80-70-6
NMe2
Me2NC=NH
115
Morpholine, 4-ethyl-
C6H13NO
100-74-3
Et
115
2-Pentanone, 4-amino-4-methyl-
C6H13NO
625-04-7
MeCOCH2CMe2NH2
115
Hexanamide
C6H13NO
628-02-4
Me(CH2)4CONH2
115
Acetamide, N,N-diethyl-
C6H13NO
685-91-6
Et2NAc
115
Cyclohexanol, 2-amino-, cis-
C6H13NO
931-15-7
NH2
OH
115
Pentanamide, 4-methyl-
C6H13NO
1119-29-5
Me2CHCH2CH2CONH2

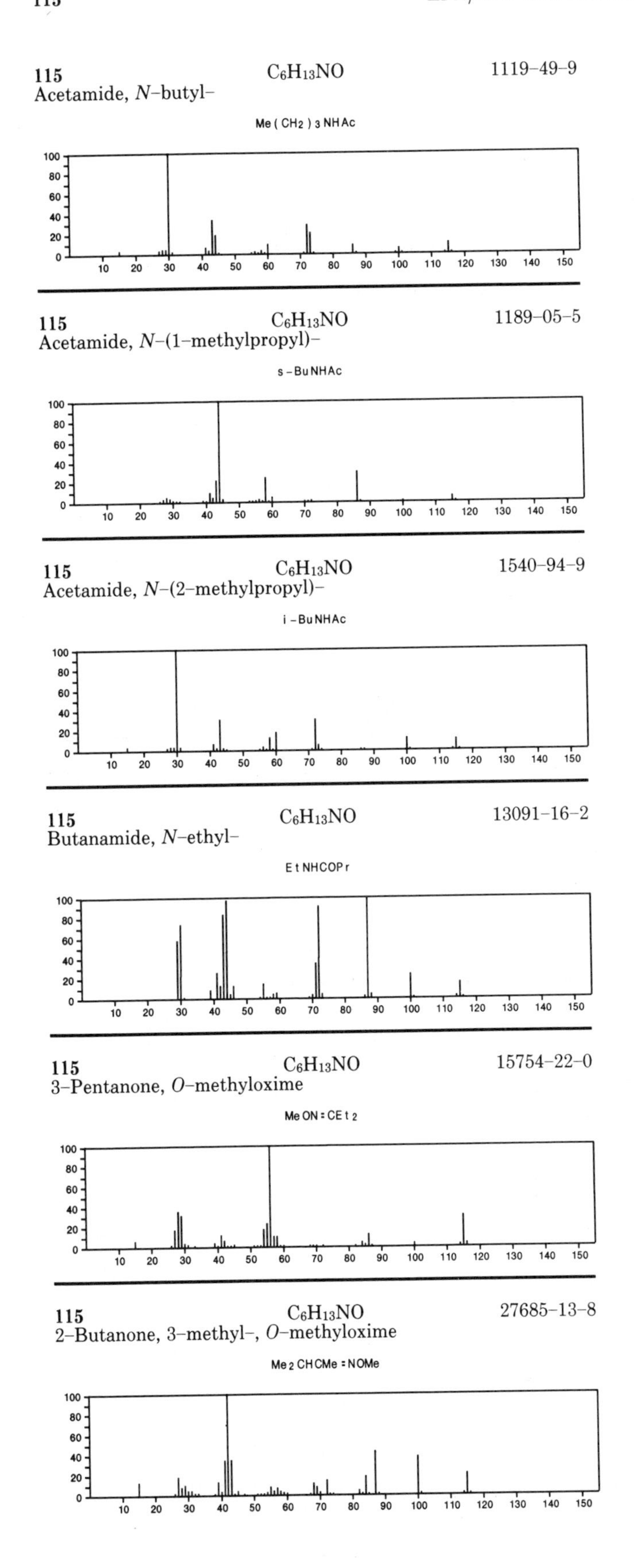
115 $C_6H_{13}NO$ 1119-49-9
Acetamide, *N*-butyl-
Me (CH2) 3 NHAc
115 $C_6H_{13}NO$ 1189-05-5
Acetamide, *N*-(1-methylpropyl)-
s - Bu NHAc
115 $C_6H_{13}NO$ 1540-94-9
Acetamide, *N*-(2-methylpropyl)-
i - Bu NHAc
115 $C_6H_{13}NO$ 13091-16-2
Butanamide, *N*-ethyl-
Et NHCOPr
115 $C_6H_{13}NO$ 15754-22-0
3-Pentanone, *O*-methyloxime
Me ON = CEt 2
115 $C_6H_{13}NO$ 27685-13-8
2-Butanone, 3-methyl-, *O*-methyloxime
Me 2 CHCMe = NOMe

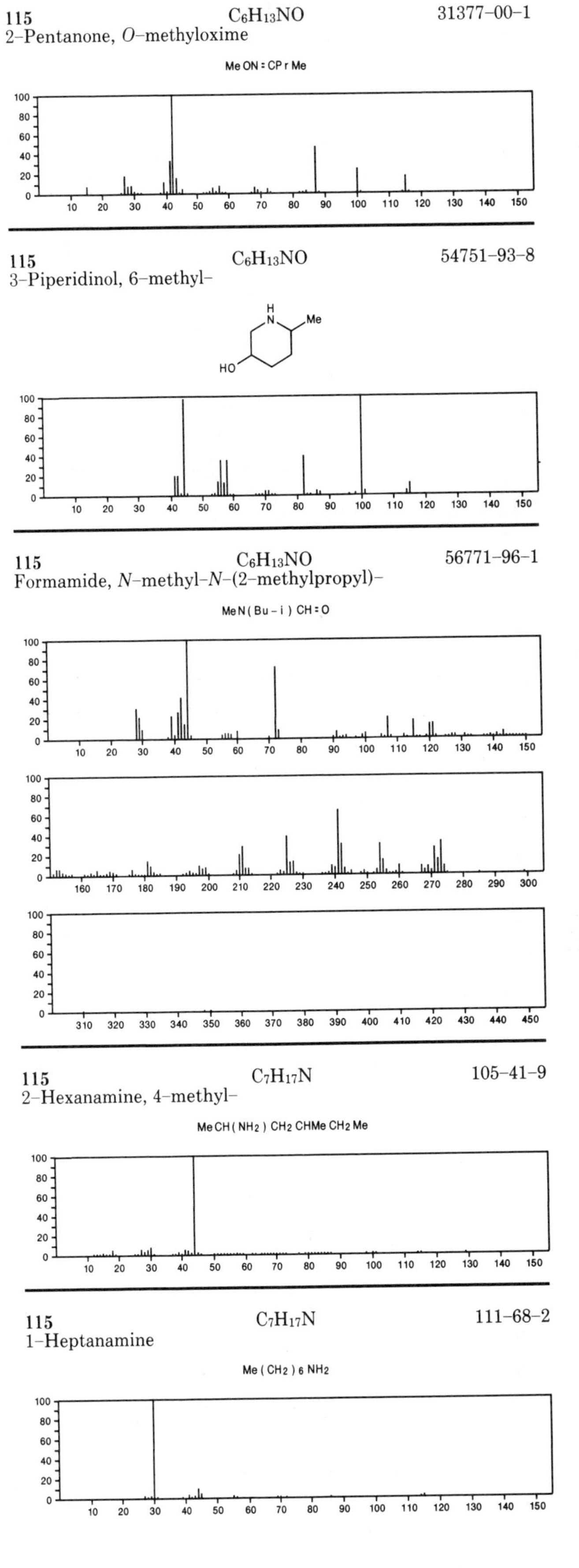
115 $C_6H_{13}NO$ 31377-00-1
2-Pentanone, *O*-methyloxime
Me ON = CPr Me
115 $C_6H_{13}NO$ 54751-93-8
3-Piperidinol, 6-methyl-
H
N
Me
HO
115 $C_6H_{13}NO$ 56771-96-1
Formamide, *N*-methyl-*N*-(2-methylpropyl)-
Me N (Bu - i) CH = O
115 $C_7H_{17}N$ 105-41-9
2-Hexanamine, 4-methyl-
Me CH (NH2) CH2 CHMe CH2 Me
115 $C_7H_{17}N$ 111-68-2
1-Heptanamine
Me (CH2) 6 NH2

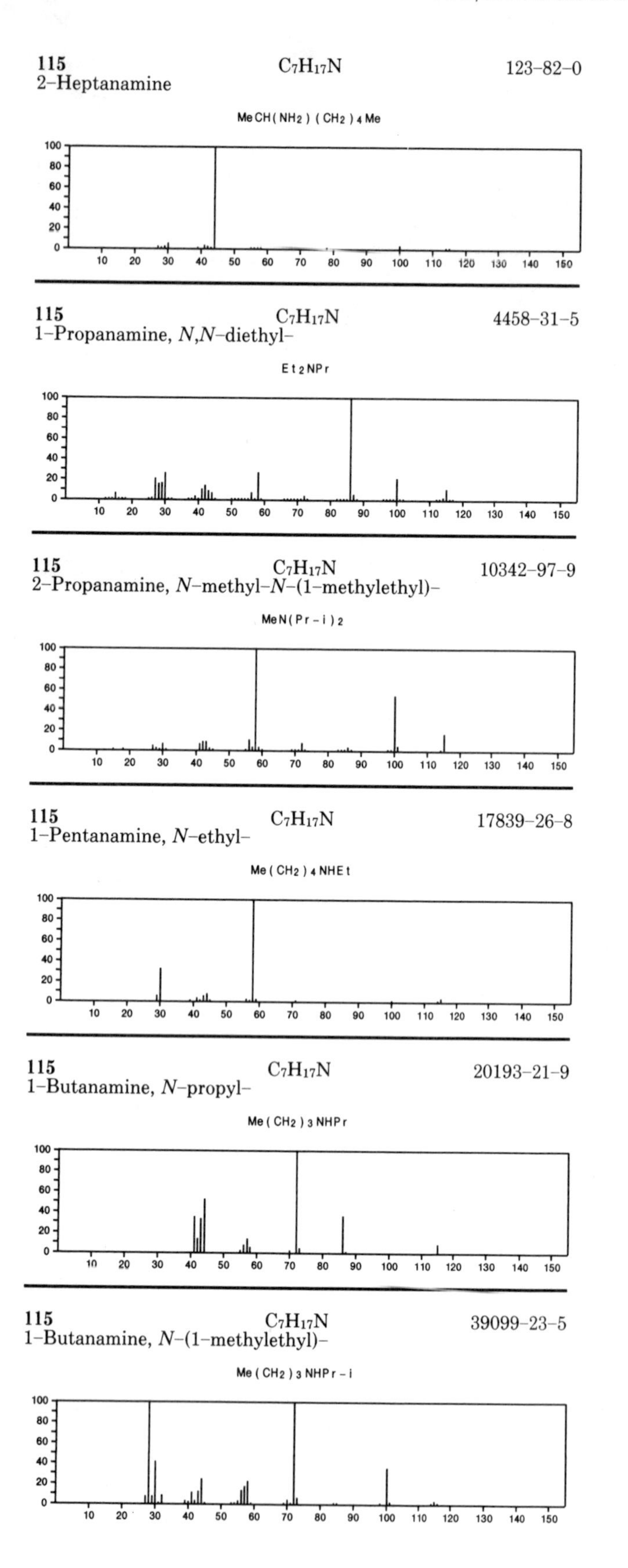

115 $C_7H_{17}N$ 123–82–0
2–Heptanamine
MeCH(NH2)(CH2)4Me

115 $C_7H_{17}N$ 4458–31–5
1–Propanamine, *N*,*N*–diethyl–
Et2NPr

115 $C_7H_{17}N$ 10342–97–9
2–Propanamine, *N*–methyl–*N*–(1–methylethyl)–
MeN(Pr-i)2

115 $C_7H_{17}N$ 17839–26–8
1–Pentanamine, *N*–ethyl–
Me(CH2)4NHEt

115 $C_7H_{17}N$ 20193–21–9
1–Butanamine, *N*–propyl–
Me(CH2)3NHPr

115 $C_7H_{17}N$ 39099–23–5
1–Butanamine, *N*–(1–methylethyl)–
Me(CH2)3NHPr-i

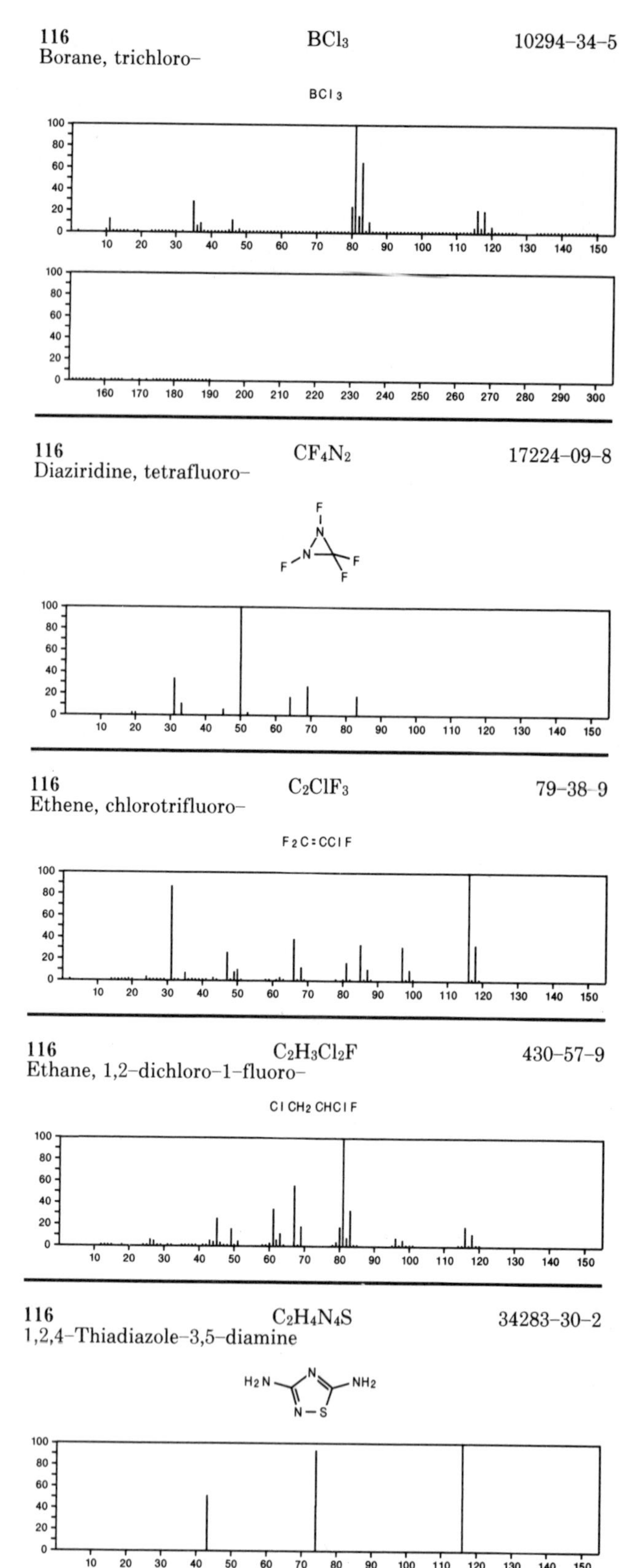

116 BCl_3 10294–34–5
Borane, trichloro–
BCl3

116 CF_4N_2 17224–09–8
Diaziridine, tetrafluoro–

116 C_2ClF_3 79–38–9
Ethene, chlorotrifluoro–
F2C=CClF

116 $C_2H_3Cl_2F$ 430–57–9
Ethane, 1,2–dichloro–1–fluoro–
ClCH2CHClF

116 $C_2H_4N_4S$ 34283–30–2
1,2,4–Thiadiazole–3,5–diamine

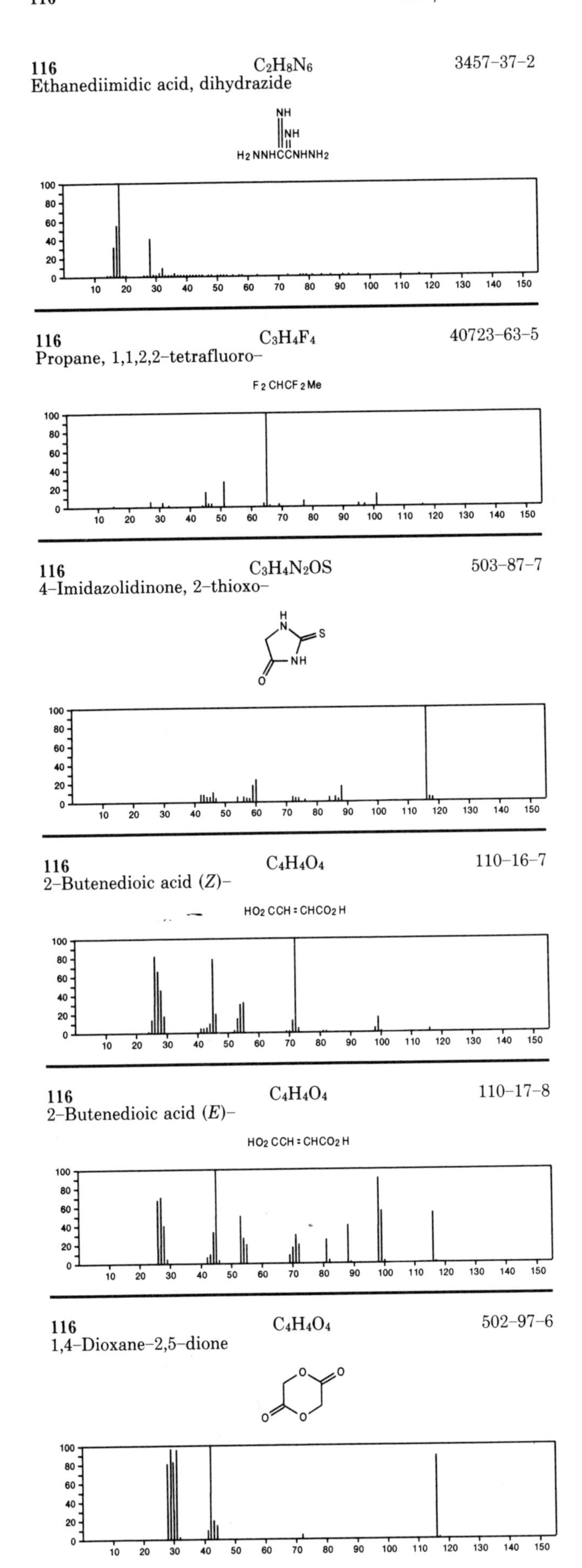

116 $C_2H_8N_6$ 3457-37-2
Ethanediimidic acid, dihydrazide

116 $C_3H_4F_4$ 40723-63-5
Propane, 1,1,2,2-tetrafluoro-

116 $C_3H_4N_2OS$ 503-87-7
4-Imidazolidinone, 2-thioxo-

116 $C_4H_4O_4$ 110-16-7
2-Butenedioic acid (*Z*)-

116 $C_4H_4O_4$ 110-17-8
2-Butenedioic acid (*E*)-

116 $C_4H_4O_4$ 502-97-6
1,4-Dioxane-2,5-dione

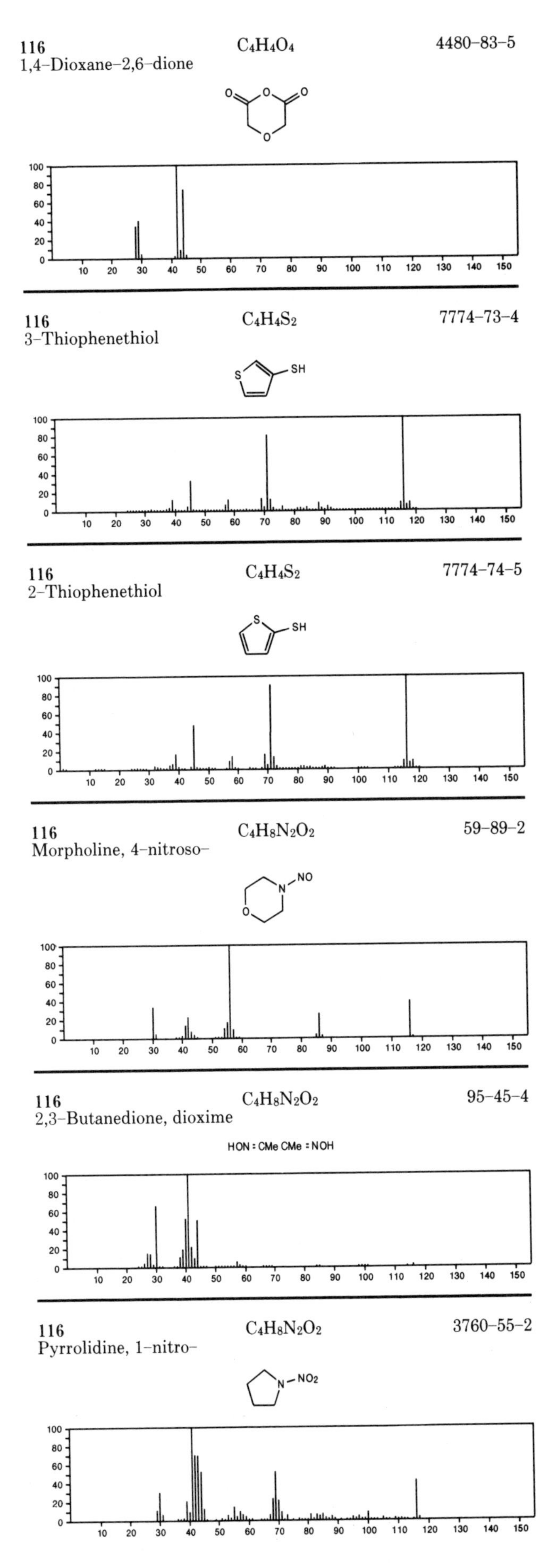

116 $C_4H_4O_4$ 4480-83-5
1,4-Dioxane-2,6-dione

116 $C_4H_4S_2$ 7774-73-4
3-Thiophenethiol

116 $C_4H_4S_2$ 7774-74-5
2-Thiophenethiol

116 $C_4H_8N_2O_2$ 59-89-2
Morpholine, 4-nitroso-

116 $C_4H_8N_2O_2$ 95-45-4
2,3-Butanedione, dioxime

116 $C_4H_8N_2O_2$ 3760-55-2
Pyrrolidine, 1-nitro-

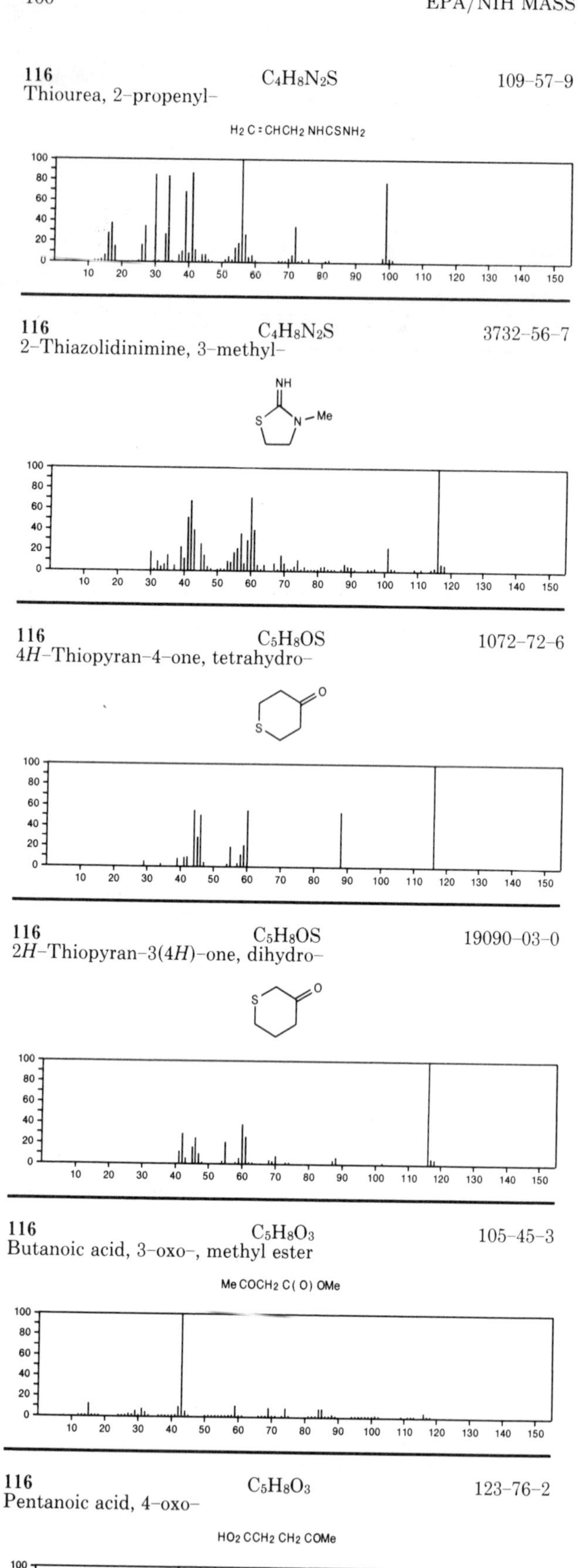

116 Thiourea, 2-propenyl- $C_4H_8N_2S$ 109-57-9

$H_2C=CHCH_2NHCSNH_2$

116 2-Thiazolidinimine, 3-methyl- $C_4H_8N_2S$ 3732-56-7

116 4*H*-Thiopyran-4-one, tetrahydro- C_5H_8OS 1072-72-6

116 2*H*-Thiopyran-3(4*H*)-one, dihydro- C_5H_8OS 19090-03-0

116 Butanoic acid, 3-oxo-, methyl ester $C_5H_8O_3$ 105-45-3

$MeCOCH_2C(O)OMe$

116 Pentanoic acid, 4-oxo- $C_5H_8O_3$ 123-76-2

$HO_2CCH_2CH_2COMe$

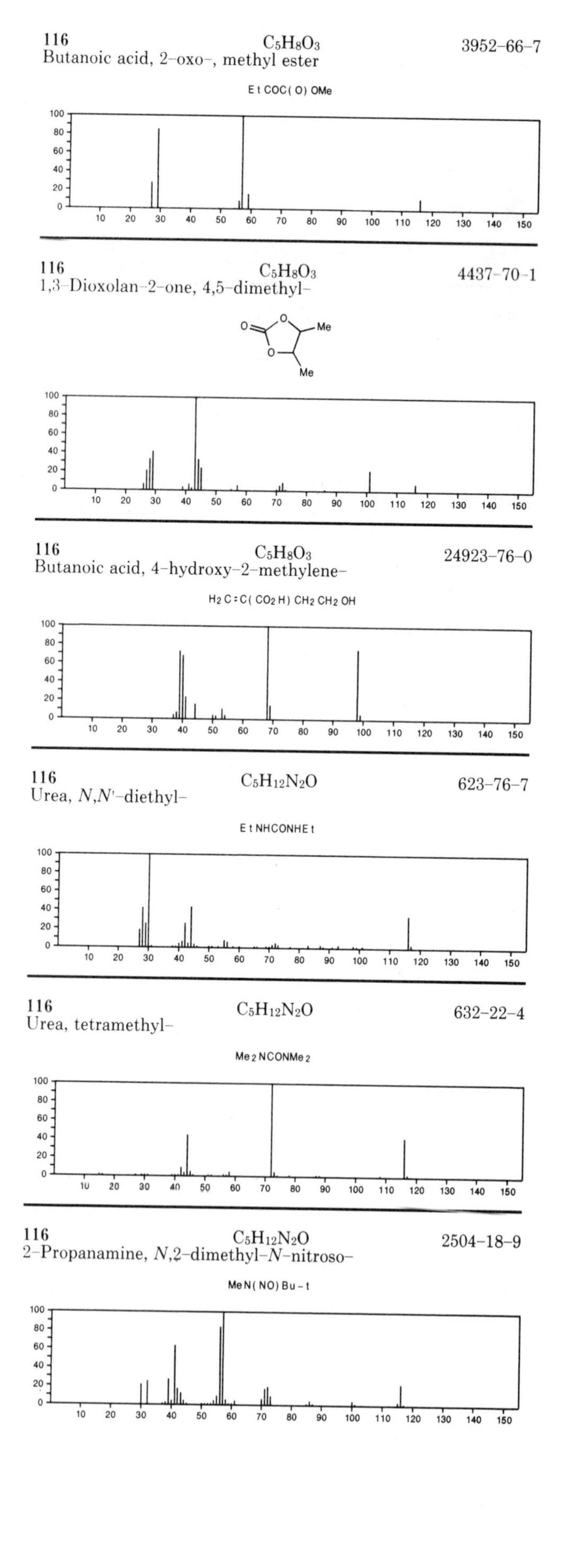

116 Butanoic acid, 2-oxo-, methyl ester $C_5H_8O_3$ 3952-66-7

EtCOC(O)OMe

116 1,3-Dioxolan-2-one, 4,5-dimethyl- $C_5H_8O_3$ 4437-70-1

116 Butanoic acid, 4-hydroxy-2-methylene- $C_5H_8O_3$ 24923-76-0

$H_2C=C(CO_2H)CH_2CH_2OH$

116 Urea, *N*,*N'*-diethyl- $C_5H_{12}N_2O$ 623-76-7

EtNHCONHEt

116 Urea, tetramethyl- $C_5H_{12}N_2O$ 632-22-4

$Me_2NCONMe_2$

116 2-Propanamine, *N*,2-dimethyl-*N*-nitroso- $C_5H_{12}N_2O$ 2504-18-9

MeN(NO)Bu-t

116 $C_5H_{12}N_2O$ 7068-83-9
1-Butanamine, *N*-methyl-*N*-nitroso-

116 $C_5H_{12}N_2O$ 13848-79-8
Hydrazinium, 2-acetyl-1,1,1-trimethyl-, hydroxide, inner salt

116 $C_5H_{12}N_2O$ 16339-04-1
2-Propanamine, *N*-ethyl-*N*-nitroso-

116 $C_5H_{12}N_2O$ 25413-61-0
1-Propanamine, *N*-ethyl-*N*-nitroso-

116 $C_5H_{12}N_2O$ 34419-76-6
1-Propanamine, *N*,2-dimethyl-*N*-nitroso-

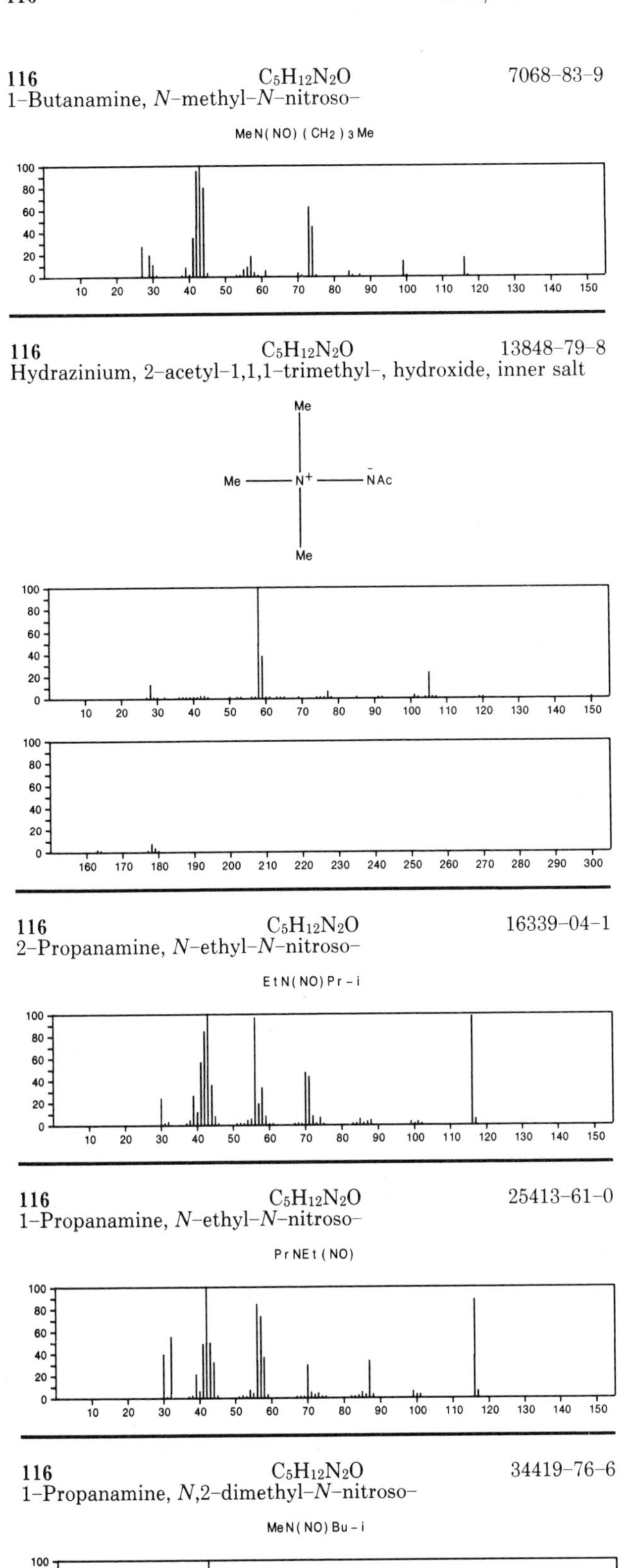

116 $C_5H_{12}N_2O$ 35606-37-2
2-Butanamine, *N*-methyl-*N*-nitroso-

116 C_6H_9Cl 930-65-4
Cyclohexene, 4-chloro-

116 C_6H_9Cl 930-66-5
Cyclohexene, 1-chloro-

116 C_6H_9Cl 2441-97-6
Cyclohexene, 3-chloro-

116 C_6H_9Cl 28077-73-8
2-Hexyne, 6-chloro-

116 C_6H_9Cl 28374-86-9
1,5-Hexadiene, 3-chloro-

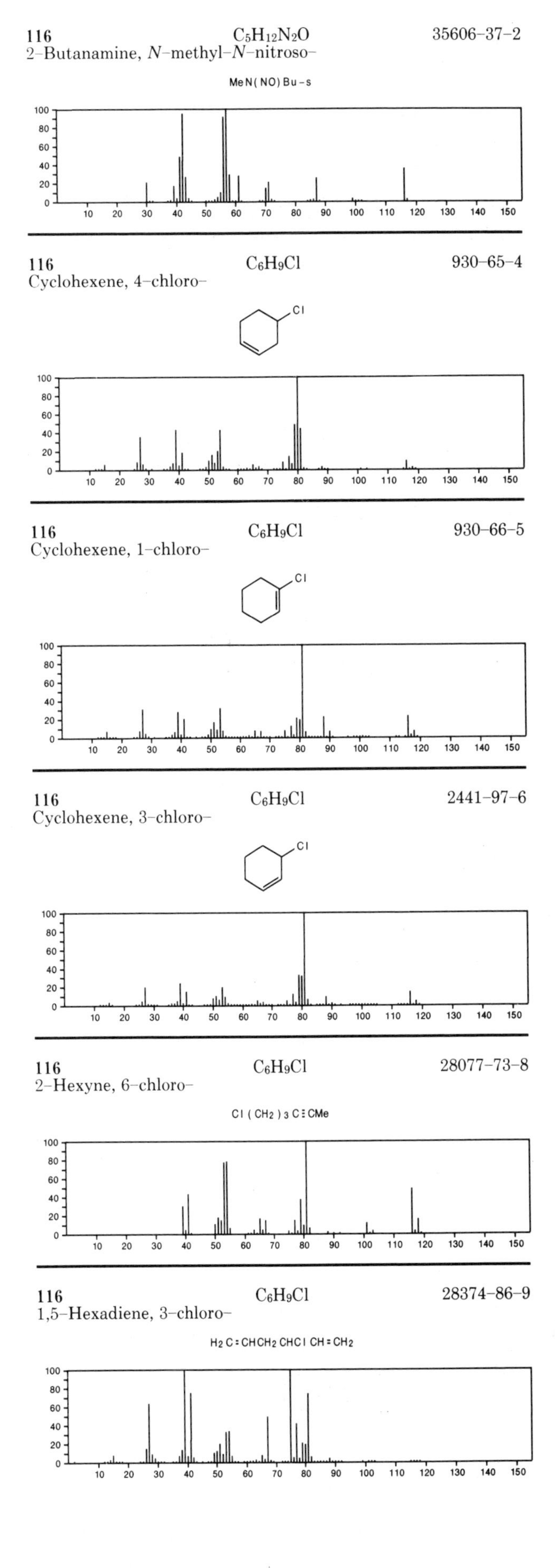

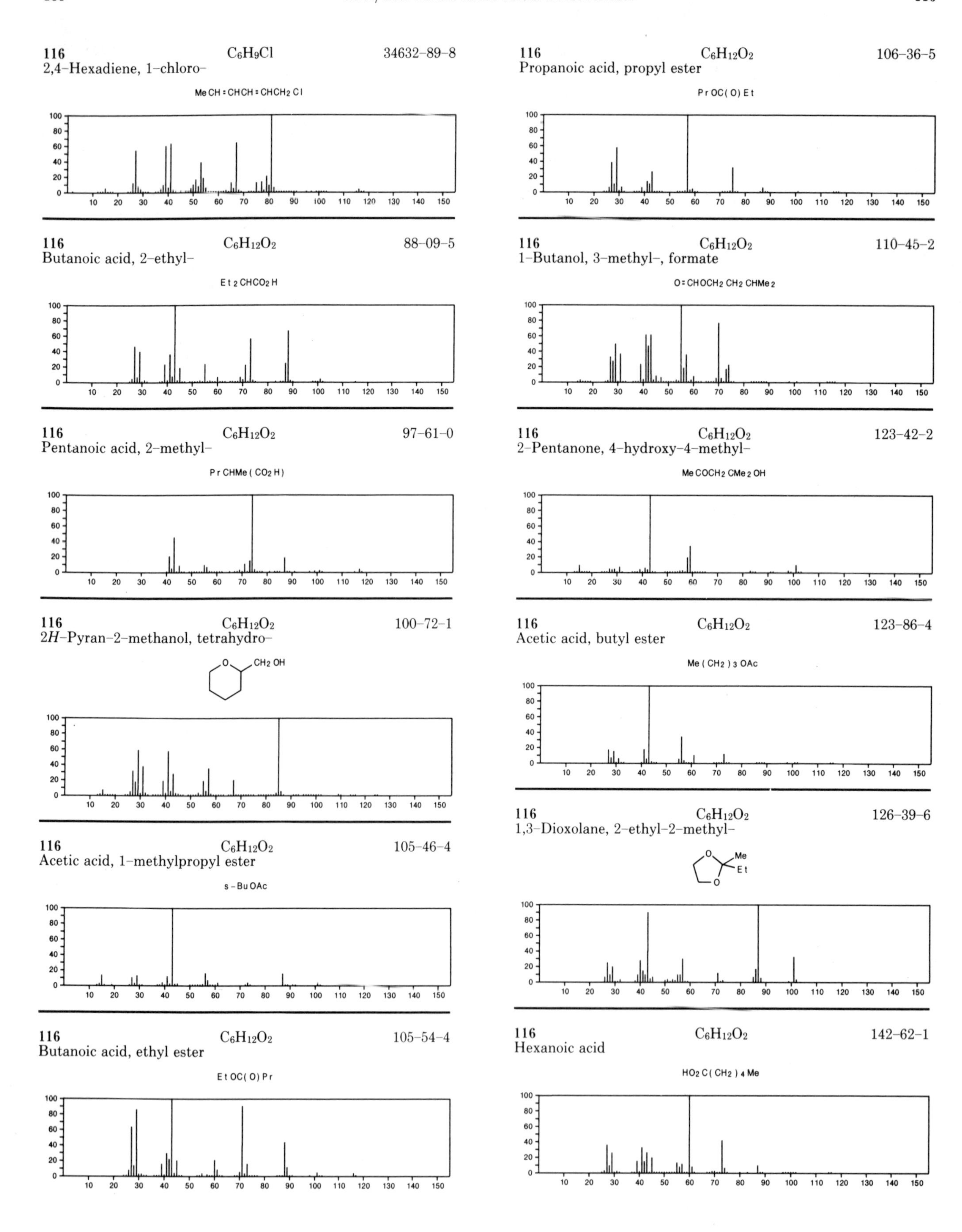
116 C6H9Cl 34632-89-8
2,4-Hexadiene, 1-chloro-
MeCH=CHCH=CHCH2Cl
116 C6H12O2 106-36-5
Propanoic acid, propyl ester
PrOC(O)Et
116 C6H12O2 88-09-5
Butanoic acid, 2-ethyl-
Et2CHCO2H
116 C6H12O2 110-45-2
1-Butanol, 3-methyl-, formate
O=CHOCH2CH2CHMe2
116 C6H12O2 97-61-0
Pentanoic acid, 2-methyl-
PrCHMe(CO2H)
116 C6H12O2 123-42-2
2-Pentanone, 4-hydroxy-4-methyl-
MeCOCH2CMe2OH
116 C6H12O2 100-72-1
2H-Pyran-2-methanol, tetrahydro-
O
CH2 OH
116 C6H12O2 123-86-4
Acetic acid, butyl ester
Me(CH2)3OAc
116 C6H12O2 105-46-4
Acetic acid, 1-methylpropyl ester
s-BuOAc
116 C6H12O2 126-39-6
1,3-Dioxolane, 2-ethyl-2-methyl-
O
Me
Et
O
116 C6H12O2 105-54-4
Butanoic acid, ethyl ester
EtOC(O)Pr
116 C6H12O2 142-62-1
Hexanoic acid
HO2C(CH2)4Me

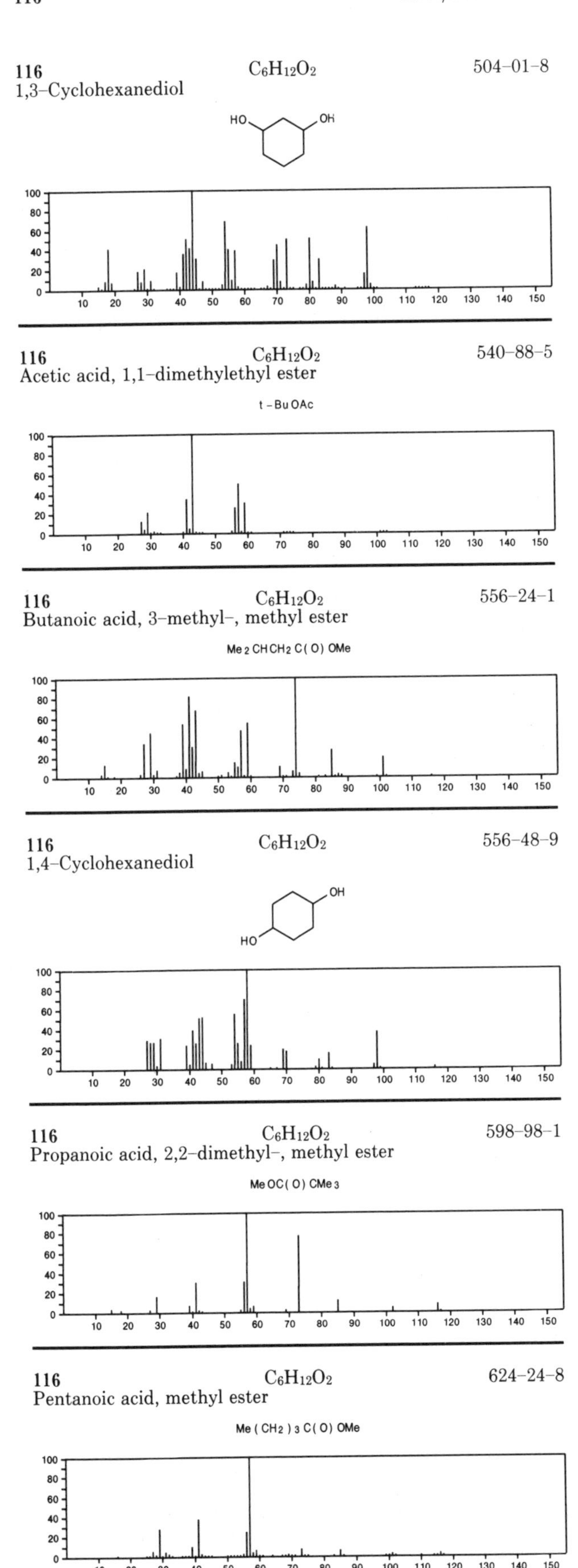

116 $C_6H_{12}O_2$ 504-01-8
1,3-Cyclohexanediol
116 $C_6H_{12}O_2$ 540-88-5
Acetic acid, 1,1-dimethylethyl ester
t-BuOAc
116 $C_6H_{12}O_2$ 556-24-1
Butanoic acid, 3-methyl-, methyl ester
Me2CHCH2C(O)OMe
116 $C_6H_{12}O_2$ 556-48-9
1,4-Cyclohexanediol
116 $C_6H_{12}O_2$ 598-98-1
Propanoic acid, 2,2-dimethyl-, methyl ester
MeOC(O)CMe3
116 $C_6H_{12}O_2$ 624-24-8
Pentanoic acid, methyl ester
Me(CH2)3C(O)OMe

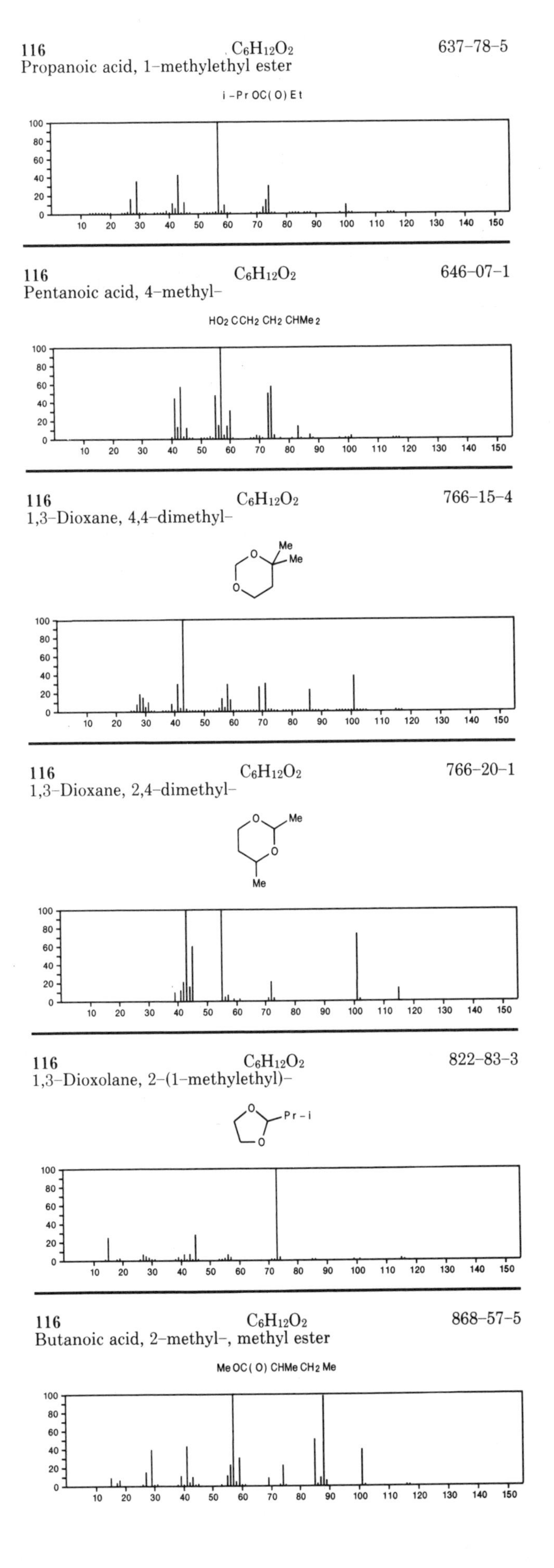

116 $C_6H_{12}O_2$ 637-78-5
Propanoic acid, 1-methylethyl ester
i-PrOC(O)Et
116 $C_6H_{12}O_2$ 646-07-1
Pentanoic acid, 4-methyl-
HO2CCH2CH2CHMe2
116 $C_6H_{12}O_2$ 766-15-4
1,3-Dioxane, 4,4-dimethyl-
116 $C_6H_{12}O_2$ 766-20-1
1,3-Dioxane, 2,4-dimethyl-
116 $C_6H_{12}O_2$ 822-83-3
1,3-Dioxolane, 2-(1-methylethyl)-
116 $C_6H_{12}O_2$ 868-57-5
Butanoic acid, 2-methyl-, methyl ester
MeOC(O)CHMeCH2Me

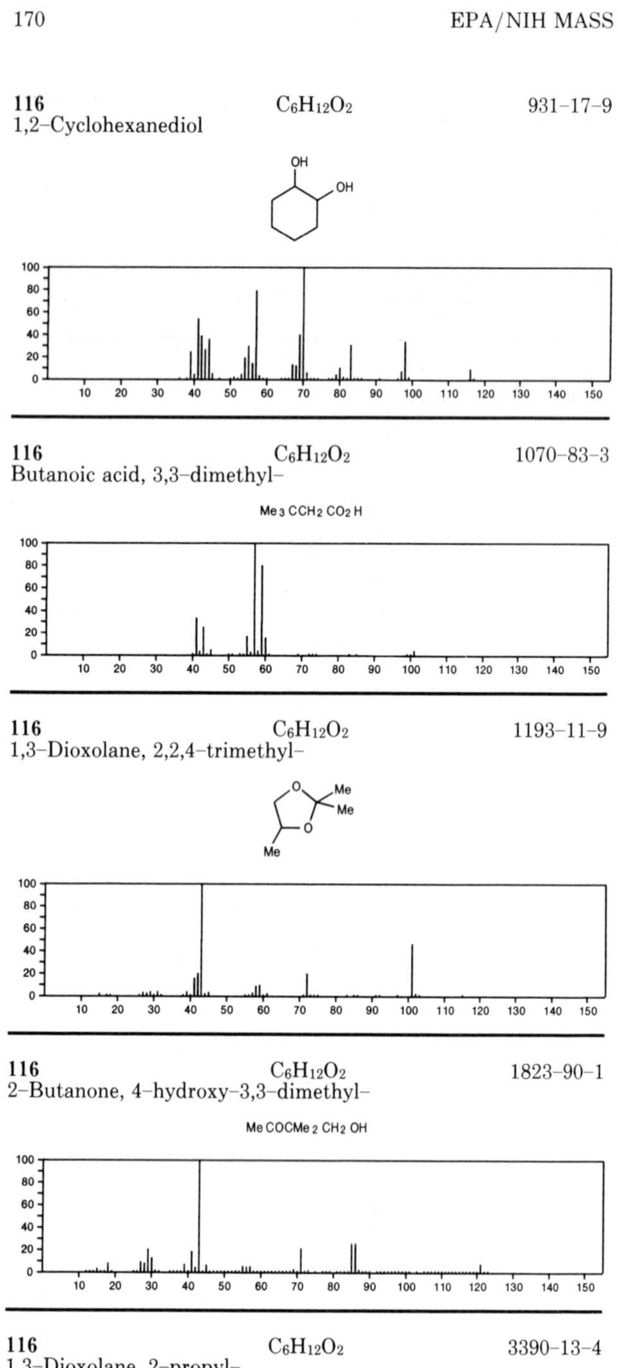

116 $C_6H_{12}O_2$ 931-17-9
1,2-Cyclohexanediol

116 $C_6H_{12}O_2$ 1070-83-3
Butanoic acid, 3,3-dimethyl-

Me3 CCH2 CO2 H

116 $C_6H_{12}O_2$ 1193-11-9
1,3-Dioxolane, 2,2,4-trimethyl-

116 $C_6H_{12}O_2$ 1823-90-1
2-Butanone, 4-hydroxy-3,3-dimethyl-

Me COCMe 2 CH2 OH

116 $C_6H_{12}O_2$ 3390-13-4
1,3-Dioxolane, 2-propyl-

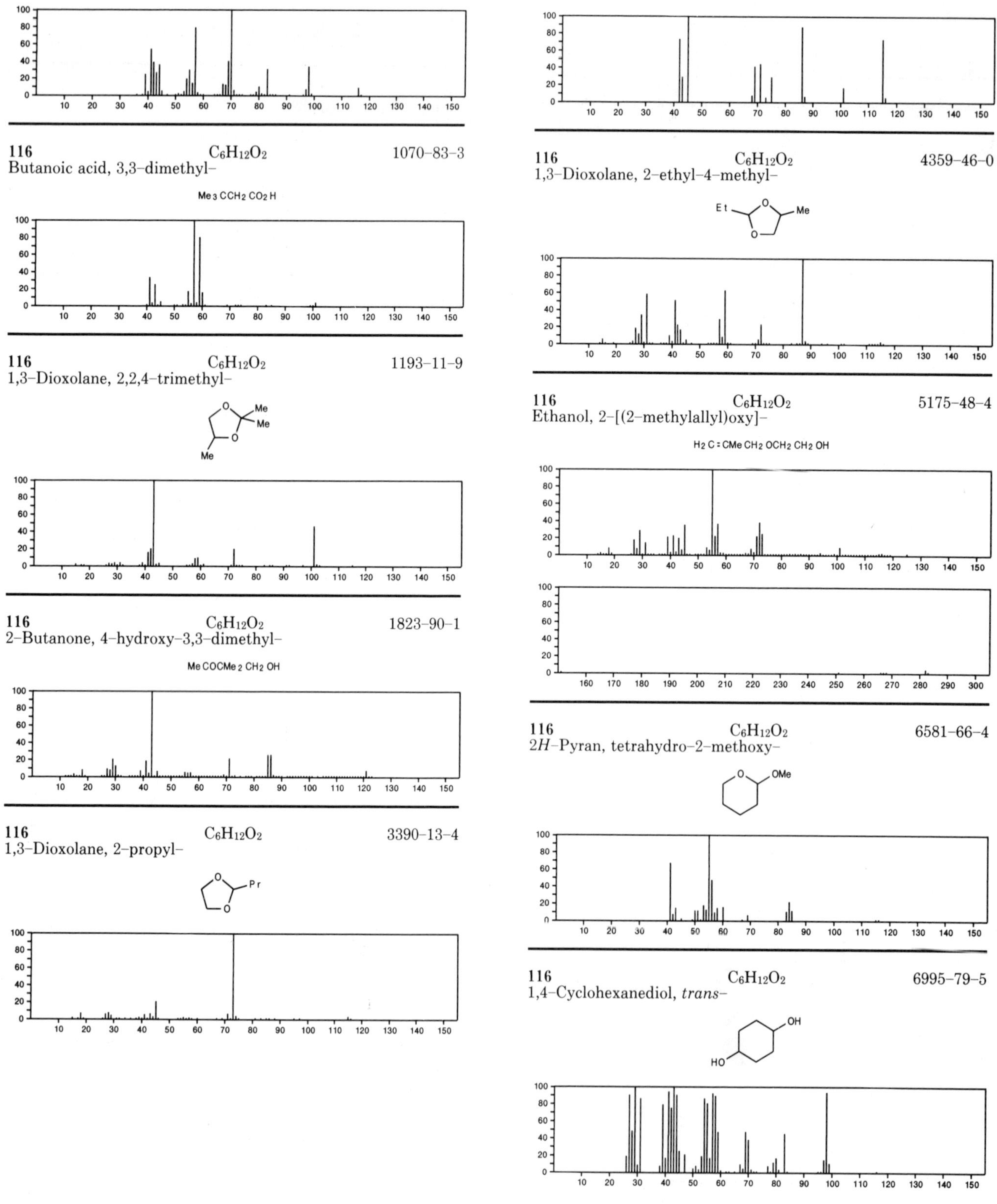

116 $C_6H_{12}O_2$ 3390-18-9
1,3-Dioxane, 4,6-dimethyl-, *cis*-

116 $C_6H_{12}O_2$ 4359-46-0
1,3-Dioxolane, 2-ethyl-4-methyl-

116 $C_6H_{12}O_2$ 5175-48-4
Ethanol, 2-[(2-methylallyl)oxy]-

H2 C = CMe CH2 OCH2 CH2 OH

116 $C_6H_{12}O_2$ 6581-66-4
2H-Pyran, tetrahydro-2-methoxy-

116 $C_6H_{12}O_2$ 6995-79-5
1,4-Cyclohexanediol, *trans*-

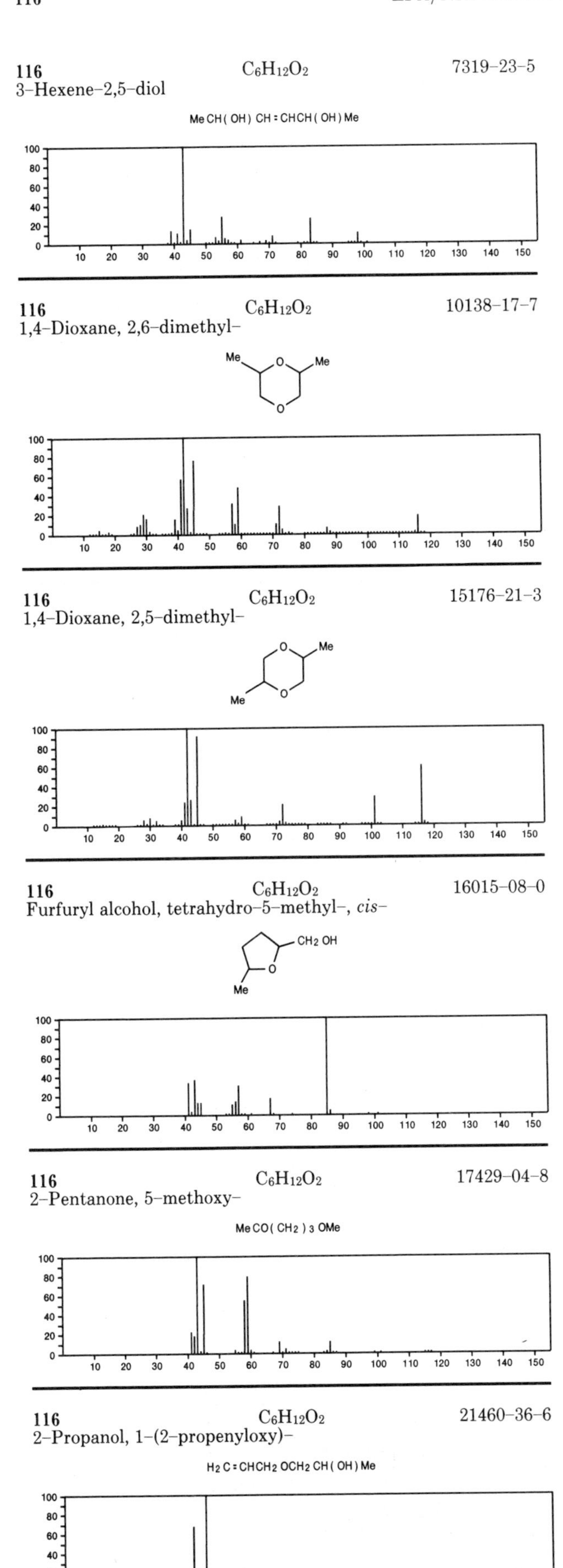
116 C6H12O2 7319-23-5
3-Hexene-2,5-diol
MeCH(OH)CH:CHCH(OH)Me
116 C6H12O2 10138-17-7
1,4-Dioxane, 2,6-dimethyl-
116 C6H12O2 15176-21-3
1,4-Dioxane, 2,5-dimethyl-
116 C6H12O2 16015-08-0
Furfuryl alcohol, tetrahydro-5-methyl-, cis-
CH2OH
116 C6H12O2 17429-04-8
2-Pentanone, 5-methoxy-
MeCO(CH2)3OMe
116 C6H12O2 21460-36-6
2-Propanol, 1-(2-propenyloxy)-
H2C:CHCH2OCH2CH(OH)Me

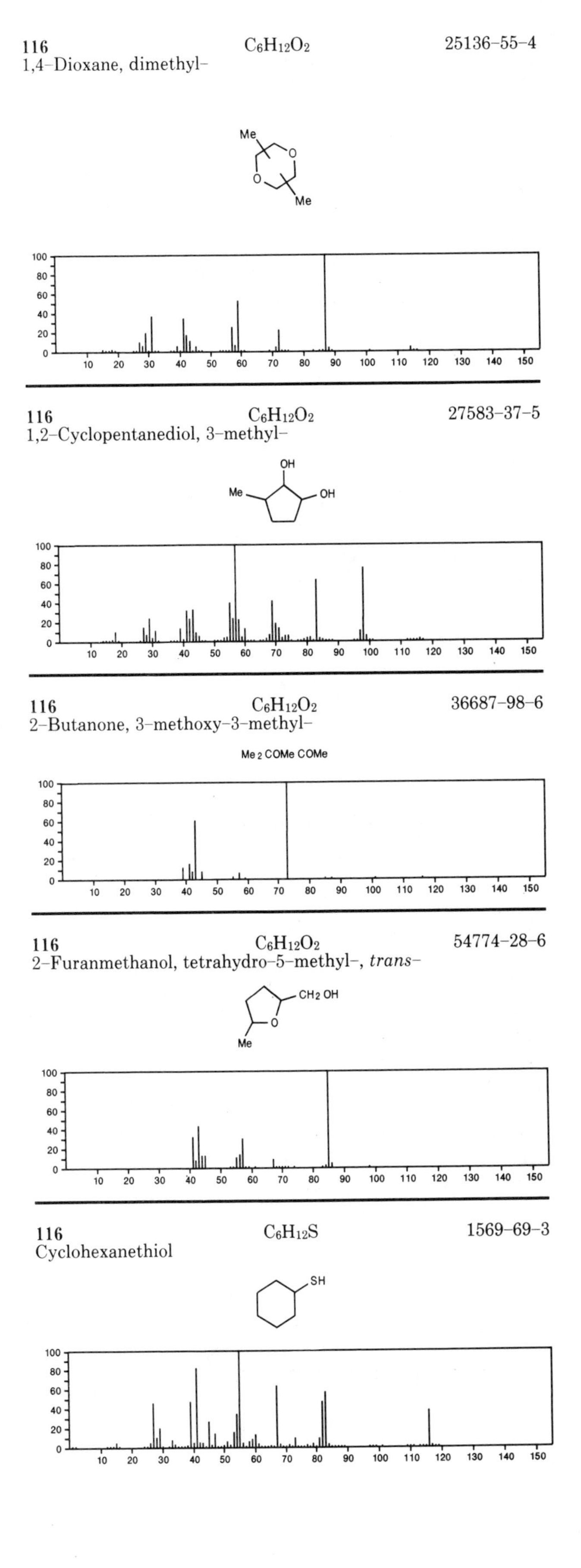
116 C6H12O2 25136-55-4
1,4-Dioxane, dimethyl-
116 C6H12O2 27583-37-5
1,2-Cyclopentanediol, 3-methyl-
116 C6H12O2 36687-98-6
2-Butanone, 3-methoxy-3-methyl-
Me2COMeCOMe
116 C6H12O2 54774-28-6
2-Furanmethanol, tetrahydro-5-methyl-, trans-
CH2OH
116 C6H12S 1569-69-3
Cyclohexanethiol
SH

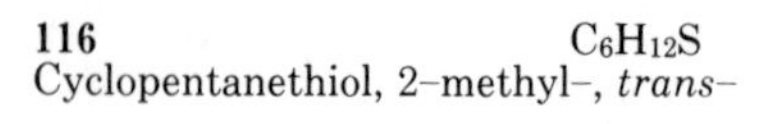

116 $C_6H_{12}S$ 1638-93-3
Cyclopentanethiol, 2-methyl-, *trans-*

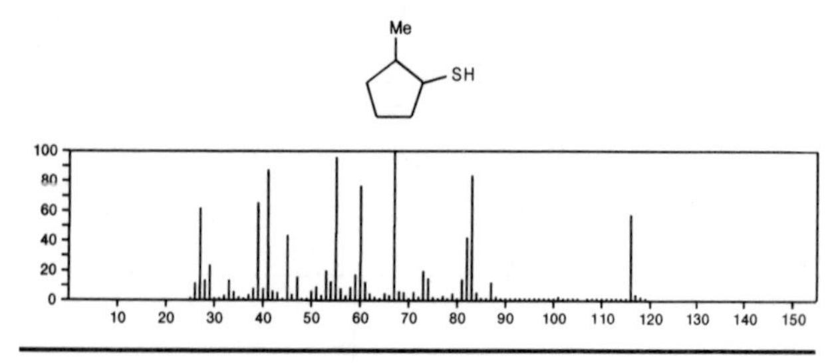

116 $C_6H_{12}S$ 1638-94-4
Cyclopentanethiol, 2-methyl-, *cis-*

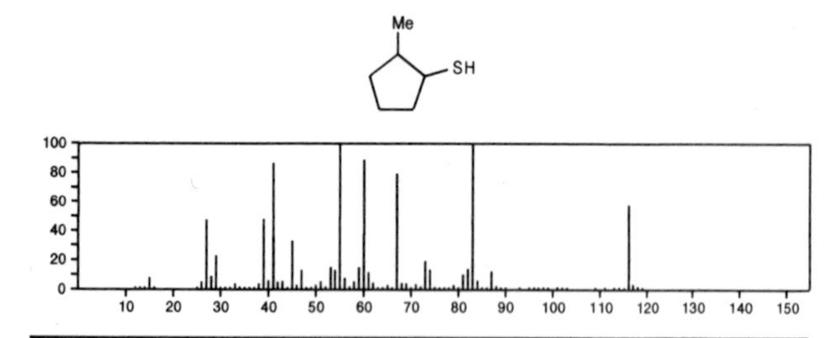

116 $C_6H_{12}S$ 1638-95-5
Cyclopentanethiol, 1-methyl-

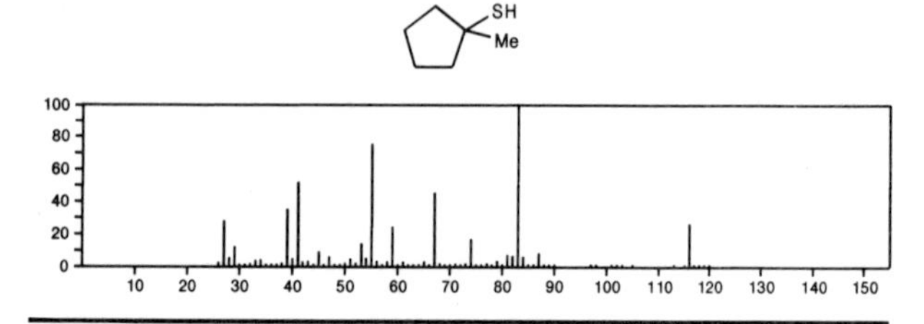

116 $C_6H_{12}S$ 4753-80-4
Thiepane

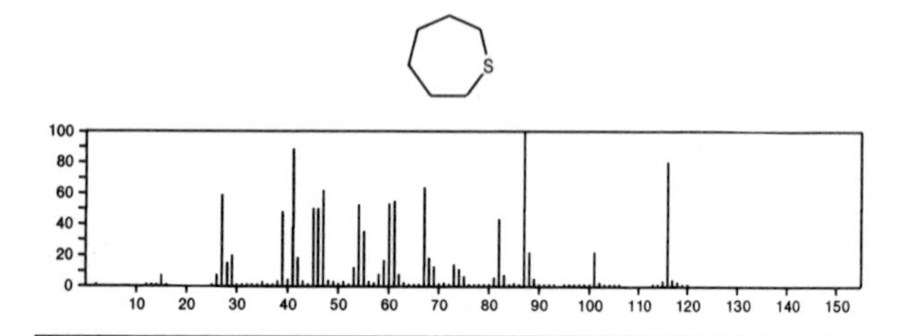

116 $C_6H_{12}S$ 5161-13-7
Thiophene, tetrahydro-2,5-dimethyl-, *cis-*

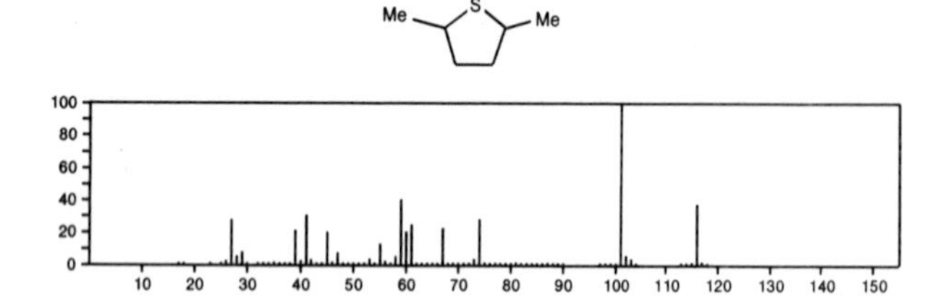

116 $C_6H_{12}S$ 5161-14-8
Thiophene, tetrahydro-2,5-dimethyl-, *trans-*

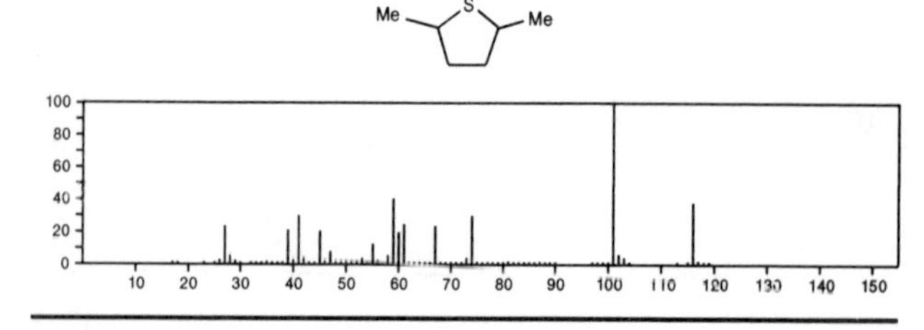

116 $C_6H_{12}S$ 5161-16-0
2*H*-Thiopyran, tetrahydro-2-methyl-

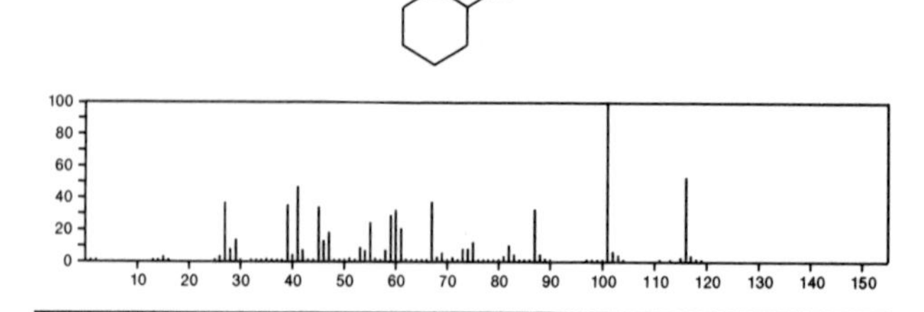

116 $C_6H_{12}S$ 5161-17-1
2*H*-Thiopyran, tetrahydro-4-methyl-

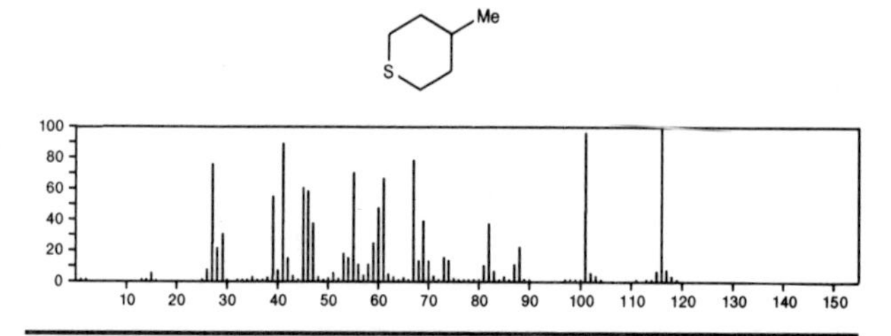

116 $C_6H_{12}S$ 5258-50-4
2*H*-Thiopyran, tetrahydro-3-methyl-

116 $C_6H_{12}S$ 7133-36-0
Cyclopentane, (methylthio)-

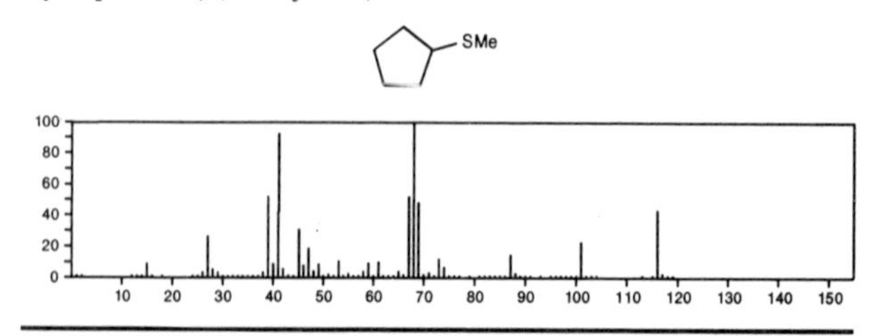

116 $C_6H_{12}S$ 50996-72-0
1-Propene, 3-[(1-methylethyl)thio]-

i-Pr SCH2 CH=CH2

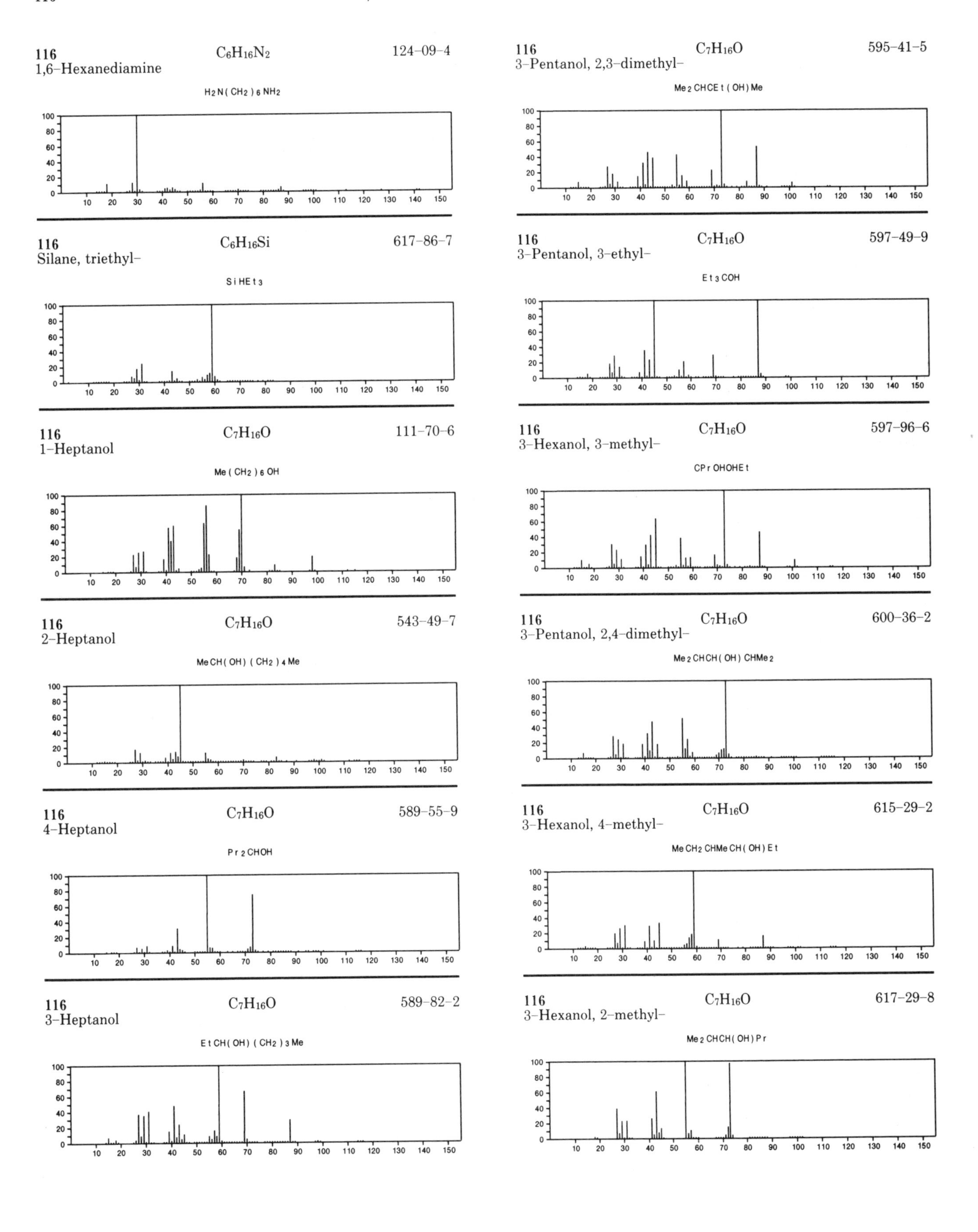
116 1,6-Hexanediamine C6H16N2 124-09-4
H2N(CH2)6NH2
116 Silane, triethyl- C6H16Si 617-86-7
SiHEt3
116 1-Heptanol C7H16O 111-70-6
Me(CH2)6OH
116 2-Heptanol C7H16O 543-49-7
MeCH(OH)(CH2)4Me
116 4-Heptanol C7H16O 589-55-9
Pr2CHOH
116 3-Heptanol C7H16O 589-82-2
EtCH(OH)(CH2)3Me
116 3-Pentanol, 2,3-dimethyl- C7H16O 595-41-5
Me2CHCEt(OH)Me
116 3-Pentanol, 3-ethyl- C7H16O 597-49-9
Et3COH
116 3-Hexanol, 3-methyl- C7H16O 597-96-6
CPrOHOHEt
116 3-Pentanol, 2,4-dimethyl- C7H16O 600-36-2
Me2CHCH(OH)CHMe2
116 3-Hexanol, 4-methyl- C7H16O 615-29-2
MeCH2CHMeCH(OH)Et
116 3-Hexanol, 2-methyl- C7H16O 617-29-8
Me2CHCH(OH)Pr

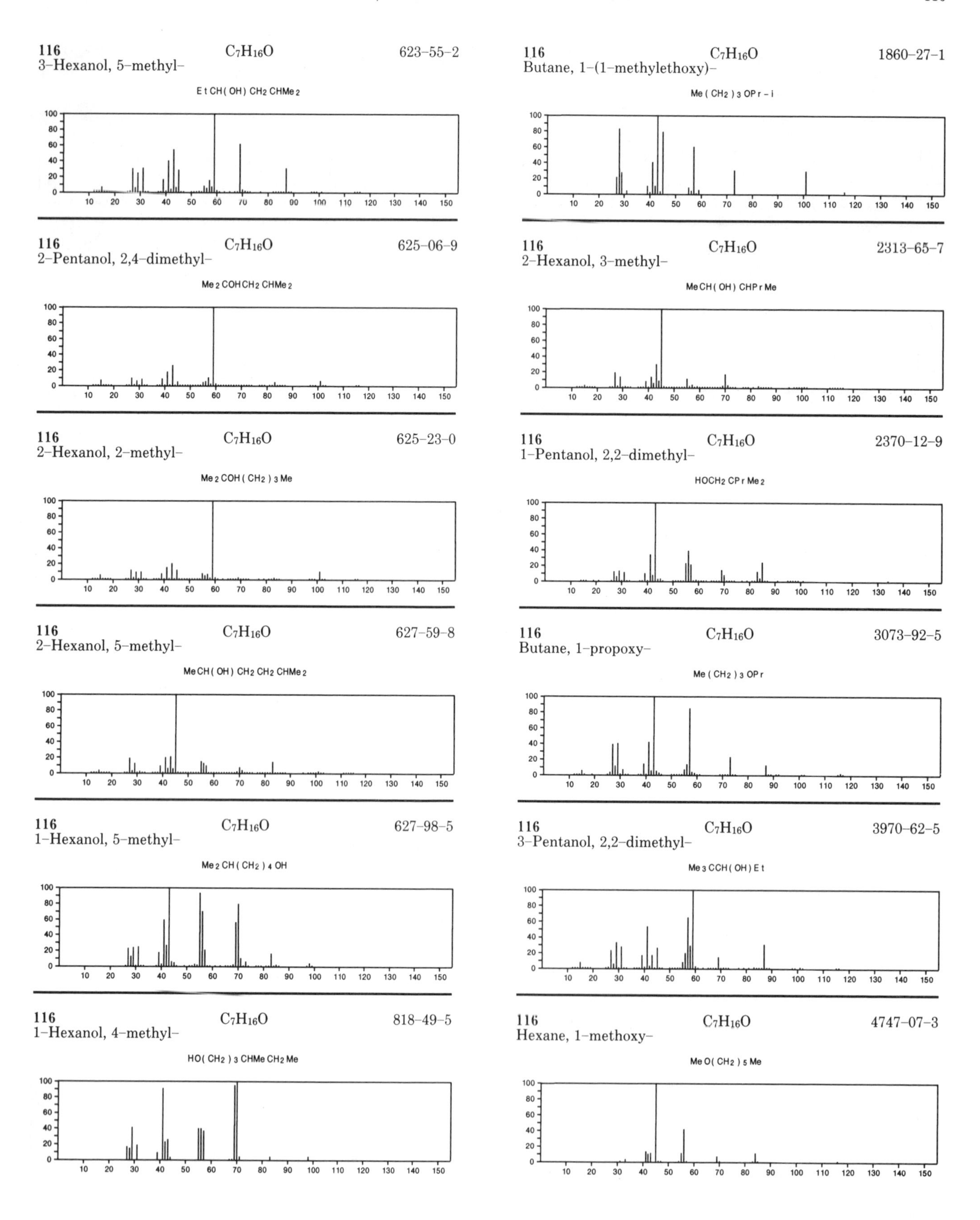
116 3-Hexanol, 5-methyl- $C_7H_{16}O$ 623-55-2
$EtCH(OH)CH_2CHMe_2$
116 Butane, 1-(1-methylethoxy)- $C_7H_{16}O$ 1860-27-1
$Me(CH_2)_3OPr-i$
116 2-Pentanol, 2,4-dimethyl- $C_7H_{16}O$ 625-06-9
$Me_2COHCH_2CHMe_2$
116 2-Hexanol, 3-methyl- $C_7H_{16}O$ 2313-65-7
$MeCH(OH)CHPrMe$
116 2-Hexanol, 2-methyl- $C_7H_{16}O$ 625-23-0
$Me_2COH(CH_2)_3Me$
116 1-Pentanol, 2,2-dimethyl- $C_7H_{16}O$ 2370-12-9
$HOCH_2CPrMe_2$
116 2-Hexanol, 5-methyl- $C_7H_{16}O$ 627-59-8
$MeCH(OH)CH_2CH_2CHMe_2$
116 Butane, 1-propoxy- $C_7H_{16}O$ 3073-92-5
$Me(CH_2)_3OPr$
116 1-Hexanol, 5-methyl- $C_7H_{16}O$ 627-98-5
$Me_2CH(CH_2)_4OH$
116 3-Pentanol, 2,2-dimethyl- $C_7H_{16}O$ 3970-62-5
$Me_3CCH(OH)Et$
116 1-Hexanol, 4-methyl- $C_7H_{16}O$ 818-49-5
$HO(CH_2)_3CHMeCH_2Me$
116 Hexane, 1-methoxy- $C_7H_{16}O$ 4747-07-3
$MeO(CH_2)_5Me$

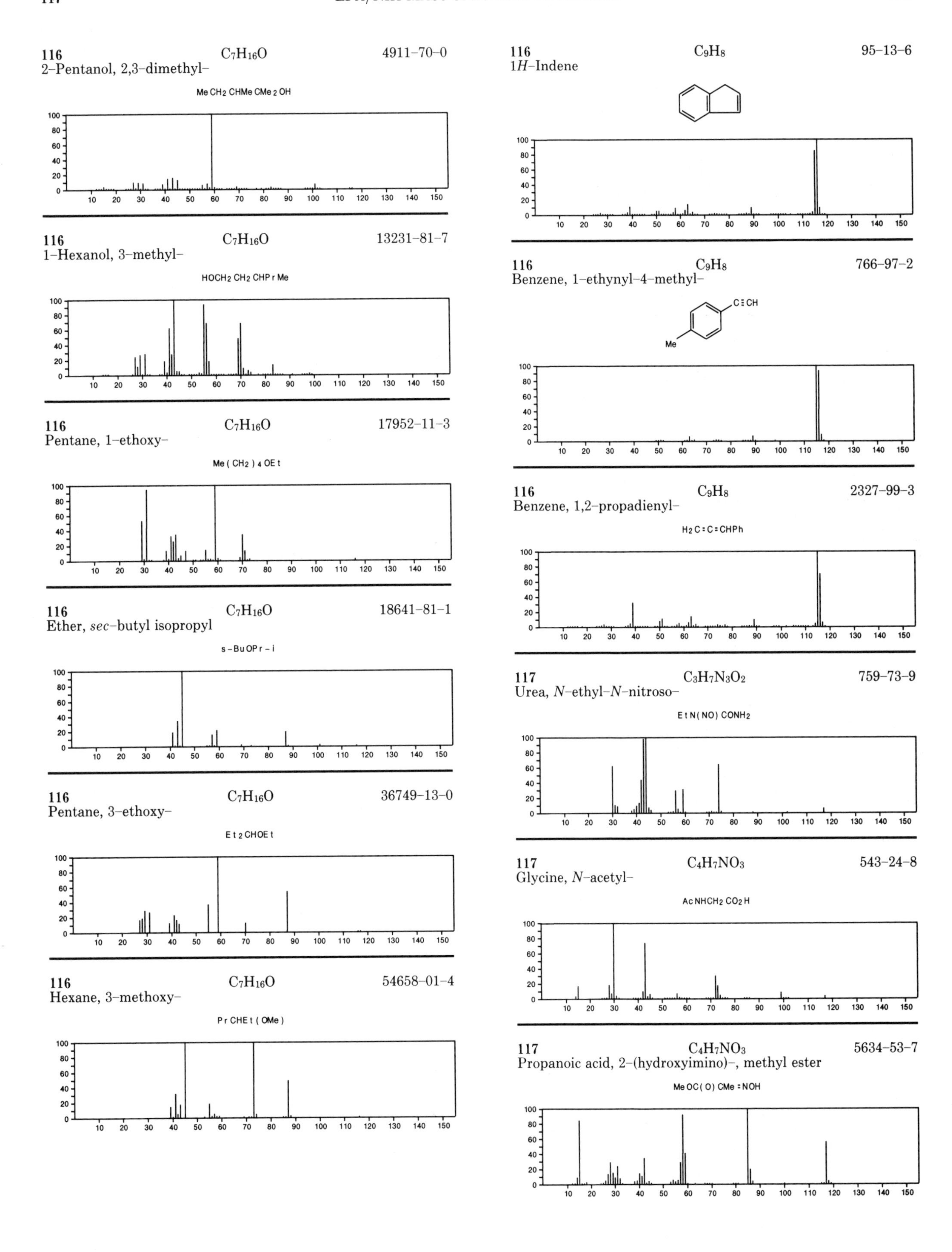
116 $C_7H_{16}O$ 4911-70-0
2-Pentanol, 2,3-dimethyl-
Me CH_2 CHMe CMe_2 OH
116 $C_7H_{16}O$ 13231-81-7
1-Hexanol, 3-methyl-
$HOCH_2$ CH_2 CHPr Me
116 $C_7H_{16}O$ 17952-11-3
Pentane, 1-ethoxy-
Me (CH_2) 4 OEt
116 $C_7H_{16}O$ 18641-81-1
Ether, *sec*-butyl isopropyl
s-BuOPr-i
116 $C_7H_{16}O$ 36749-13-0
Pentane, 3-ethoxy-
Et_2 CHOEt
116 $C_7H_{16}O$ 54658-01-4
Hexane, 3-methoxy-
PrCHEt (OMe)
116 C_9H_8 95-13-6
1*H*-Indene
116 C_9H_8 766-97-2
Benzene, 1-ethynyl-4-methyl-
C≡CH
Me
116 C_9H_8 2327-99-3
Benzene, 1,2-propadienyl-
H_2C=C=CHPh
117 $C_3H_7N_3O_2$ 759-73-9
Urea, *N*-ethyl-*N*-nitroso-
EtN(NO) $CONH_2$
117 $C_4H_7NO_3$ 543-24-8
Glycine, *N*-acetyl-
AcNHCH_2 CO_2H
117 $C_4H_7NO_3$ 5634-53-7
Propanoic acid, 2-(hydroxyimino)-, methyl ester
MeOC(O) CMe =NOH

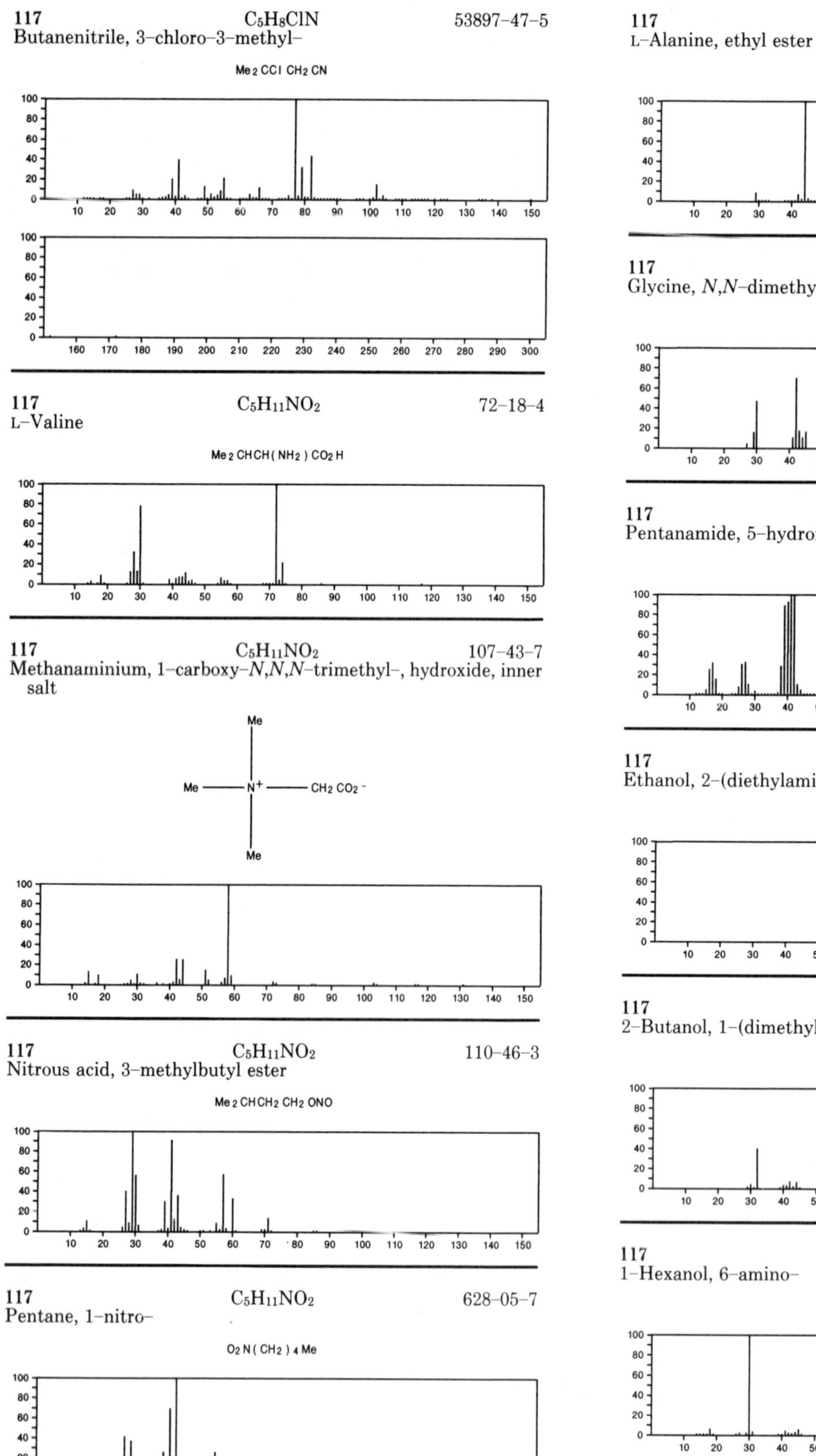

117 C_5H_8ClN 53897-47-5
Butanenitrile, 3-chloro-3-methyl-

Me2CCl CH2CN

117 $C_5H_{11}NO_2$ 72-18-4
L-Valine

Me2CHCH(NH2)CO2H

117 $C_5H_{11}NO_2$ 107-43-7
Methanaminium, 1-carboxy-*N,N,N*-trimethyl-, hydroxide, inner salt

Me3N+CH2CO2−

117 $C_5H_{11}NO_2$ 110-46-3
Nitrous acid, 3-methylbutyl ester

Me2CHCH2CH2ONO

117 $C_5H_{11}NO_2$ 628-05-7
Pentane, 1-nitro-

O2N(CH2)4Me

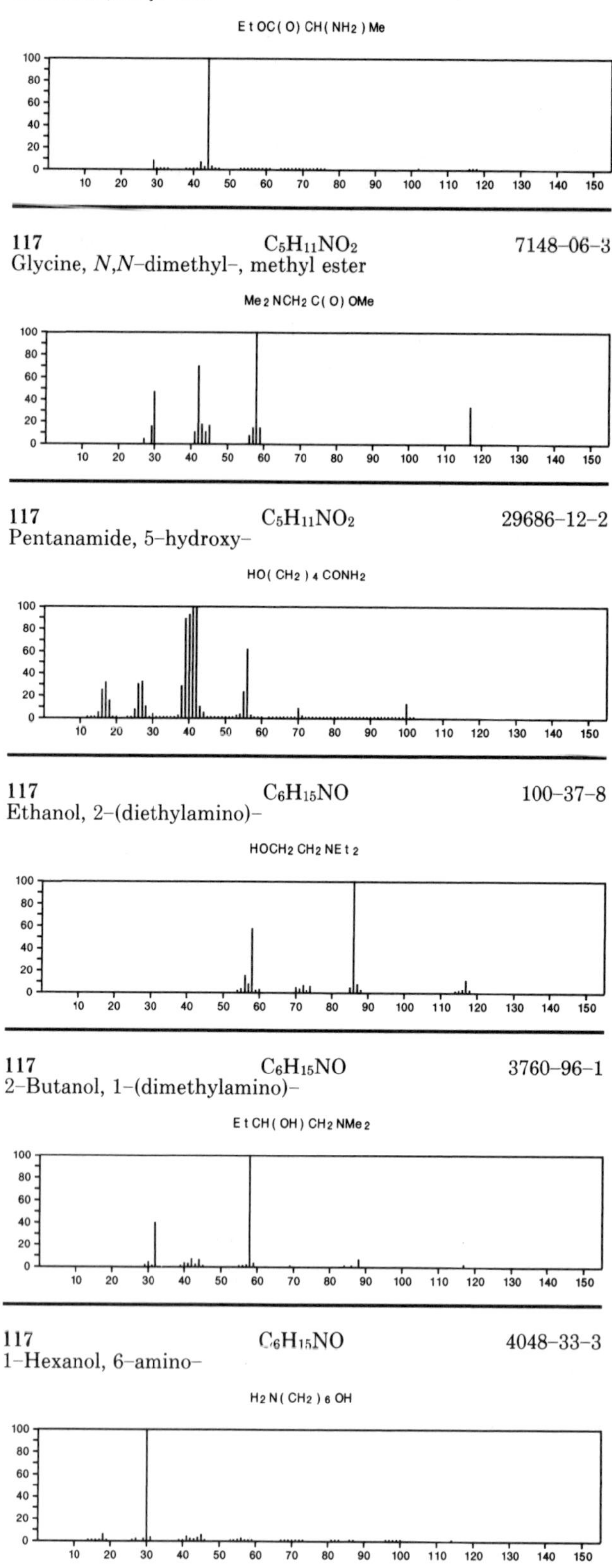

117 $C_5H_{11}NO_2$ 3082-75-5
L-Alanine, ethyl ester

EtOC(O)CH(NH2)Me

117 $C_5H_{11}NO_2$ 7148-06-3
Glycine, *N,N*-dimethyl-, methyl ester

Me2NCH2C(O)OMe

117 $C_5H_{11}NO_2$ 29686-12-2
Pentanamide, 5-hydroxy-

HO(CH2)4CONH2

117 $C_6H_{15}NO$ 100-37-8
Ethanol, 2-(diethylamino)-

HOCH2CH2NEt2

117 $C_6H_{15}NO$ 3760-96-1
2-Butanol, 1-(dimethylamino)-

EtCH(OH)CH2NMe2

117 $C_6H_{15}NO$ 4048-33-3
1-Hexanol, 6-amino-

H2N(CH2)6OH

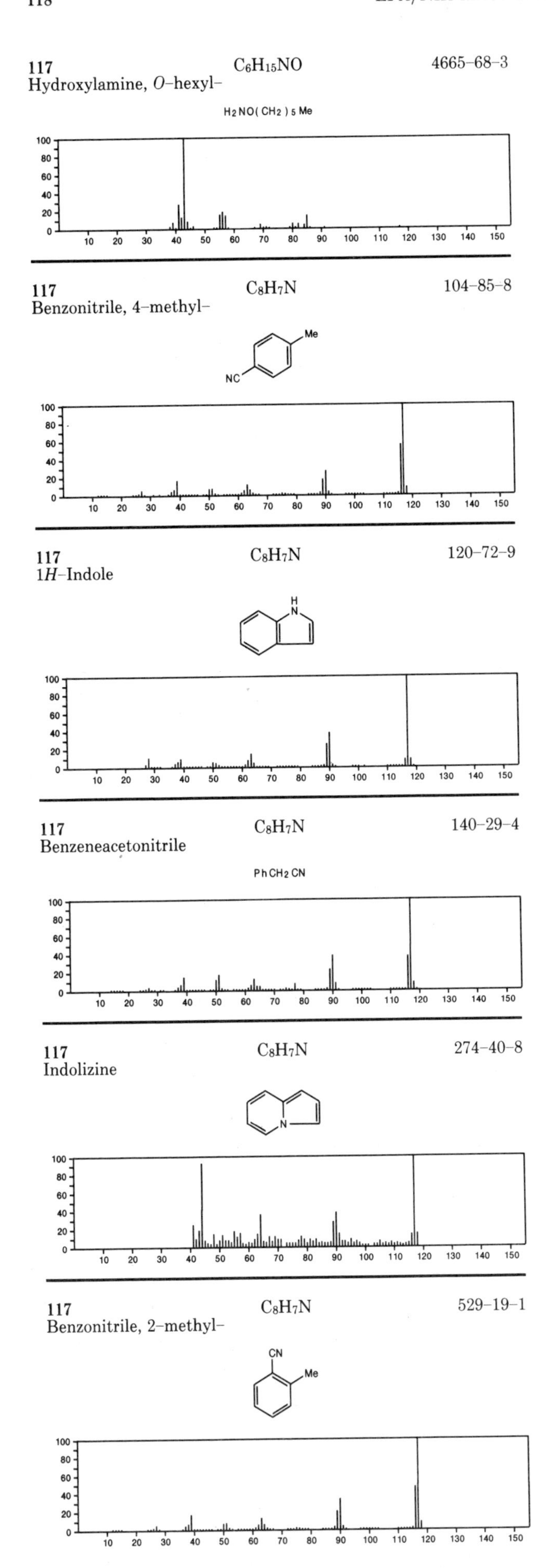

117 $C_6H_{15}NO$ 4665-68-3
Hydroxylamine, *O*-hexyl-

117 C_8H_7N 104-85-8
Benzonitrile, 4-methyl-

117 C_8H_7N 120-72-9
1*H*-Indole

117 C_8H_7N 140-29-4
Benzeneacetonitrile

117 C_8H_7N 274-40-8
Indolizine

117 C_8H_7N 529-19-1
Benzonitrile, 2-methyl-

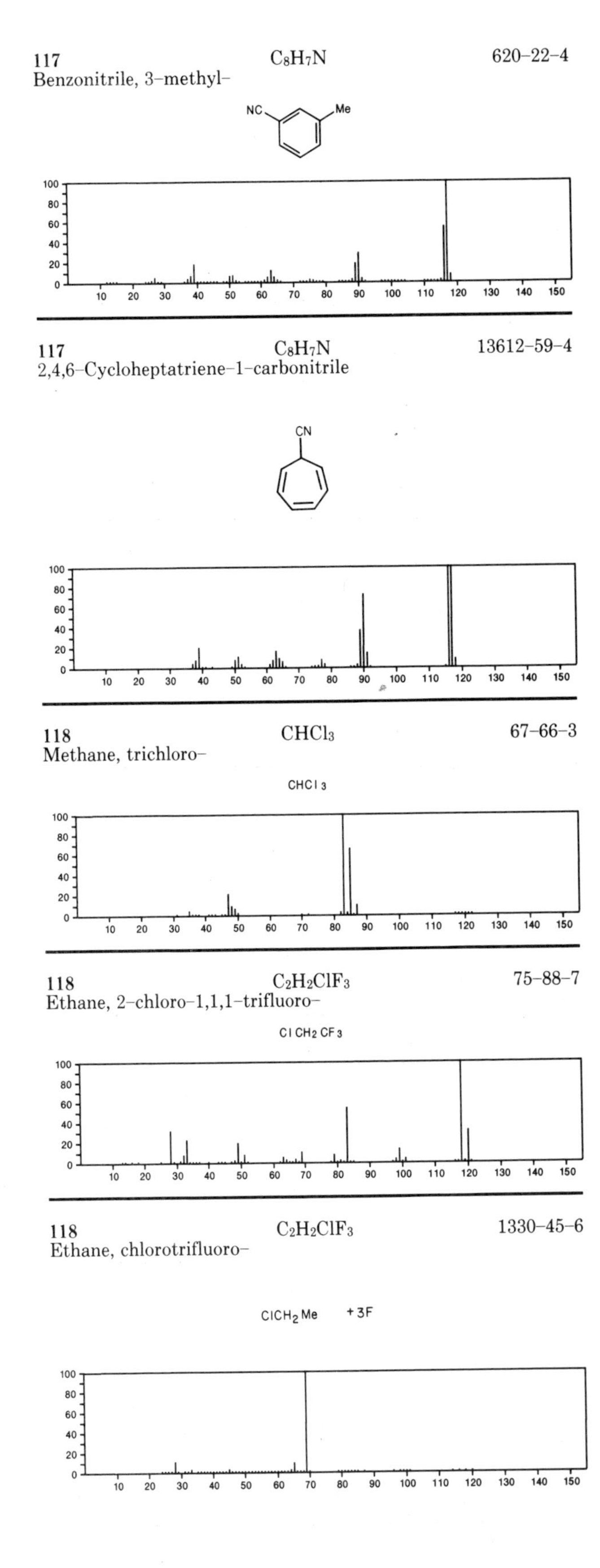

117 C_8H_7N 620-22-4
Benzonitrile, 3-methyl-

117 C_8H_7N 13612-59-4
2,4,6-Cycloheptatriene-1-carbonitrile

118 $CHCl_3$ 67-66-3
Methane, trichloro-

118 $C_2H_2ClF_3$ 75-88-7
Ethane, 2-chloro-1,1,1-trifluoro-

118 $C_2H_2ClF_3$ 1330-45-6
Ethane, chlorotrifluoro-

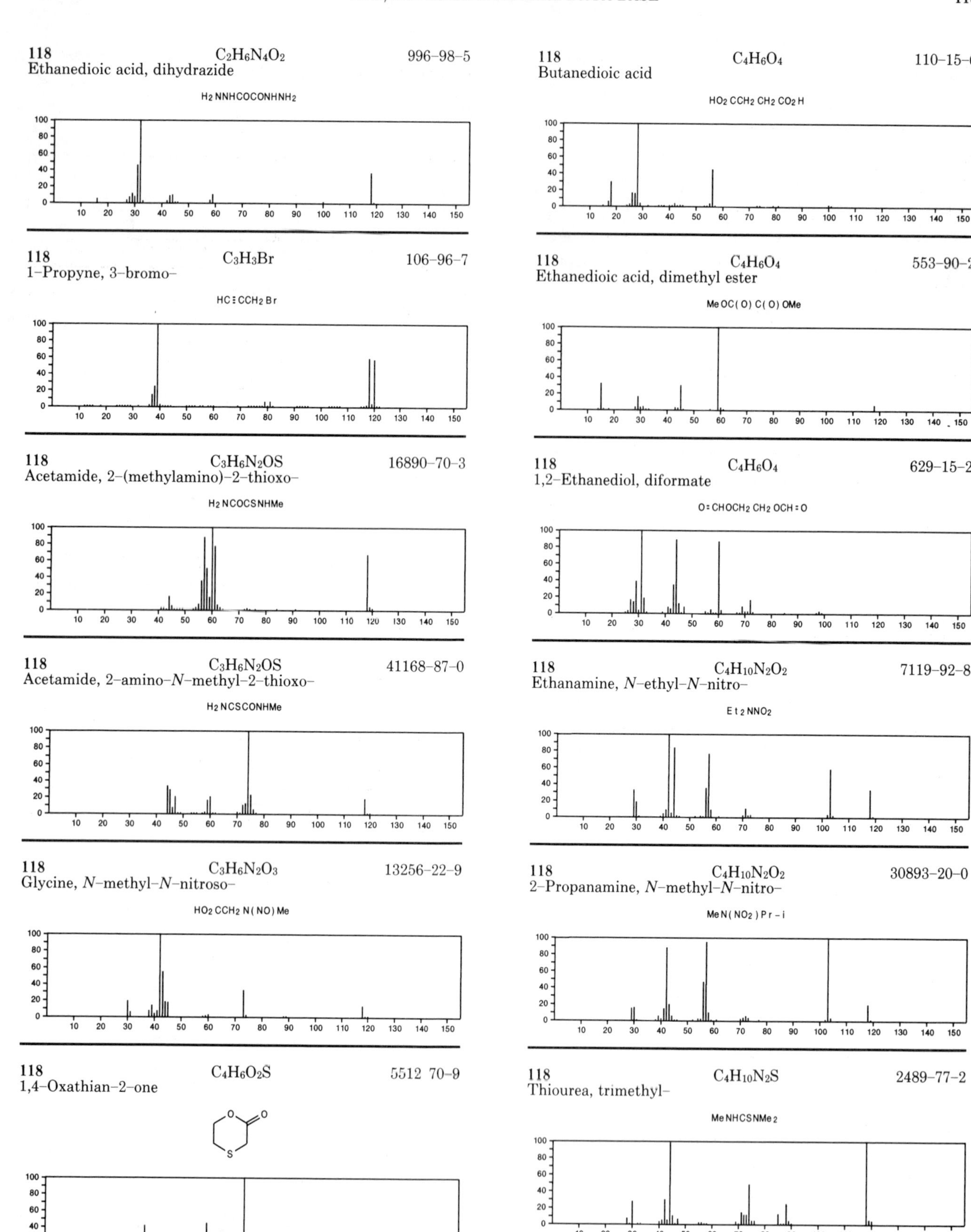
118 C2H6N4O2 996-98-5
Ethanedioic acid, dihydrazide
H2NNHCOCONHNH2
118 C4H6O4 110-15-6
Butanedioic acid
HO2CCH2CH2CO2H
118 C3H3Br 106-96-7
1-Propyne, 3-bromo-
HC≡CCH2Br
118 C4H6O4 553-90-2
Ethanedioic acid, dimethyl ester
MeOC(O)C(O)OMe
118 C3H6N2OS 16890-70-3
Acetamide, 2-(methylamino)-2-thioxo-
H2NCOCSNHMe
118 C4H6O4 629-15-2
1,2-Ethanediol, diformate
O=CHOCH2CH2OCH=O
118 C3H6N2OS 41168-87-0
Acetamide, 2-amino-N-methyl-2-thioxo-
H2NCSCONHMe
118 C4H10N2O2 7119-92-8
Ethanamine, N-ethyl-N-nitro-
Et2NNO2
118 C3H6N2O3 13256-22-9
Glycine, N-methyl-N-nitroso-
HO2CCH2N(NO)Me
118 C4H10N2O2 30893-20-0
2-Propanamine, N-methyl-N-nitro-
MeN(NO2)Pr-i
118 C4H6O2S 5512 70-9
1,4-Oxathian-2-one
118 C4H10N2S 2489-77-2
Thiourea, trimethyl-
MeNHCSNMe2

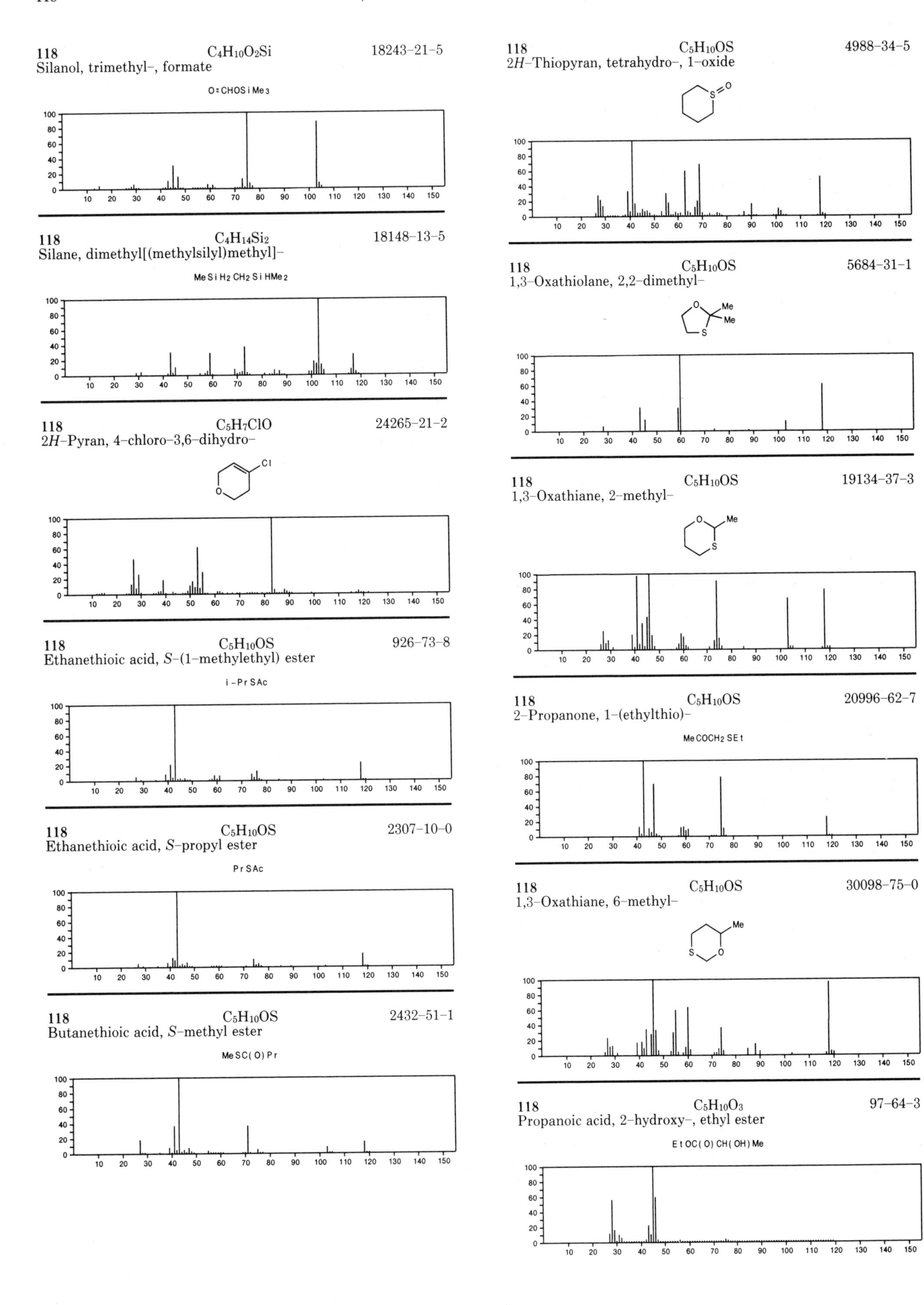
118 C4H10O2Si 18243-21-5
Silanol, trimethyl-, formate
O=CHOSiMe3
118 C4H14Si2 18148-13-5
Silane, dimethyl[(methylsilyl)methyl]-
MeSiH2CH2SiHMe2
118 C5H7ClO 24265-21-2
2H-Pyran, 4-chloro-3,6-dihydro-
Cl
O
118 C5H10OS 926-73-8
Ethanethioic acid, S-(1-methylethyl) ester
i-PrSAc
118 C5H10OS 2307-10-0
Ethanethioic acid, S-propyl ester
PrSAc
118 C5H10OS 2432-51-1
Butanethioic acid, S-methyl ester
MeSC(O)Pr
118 C5H10OS 4988-34-5
2H-Thiopyran, tetrahydro-, 1-oxide
S=O
118 C5H10OS 5684-31-1
1,3-Oxathiolane, 2,2-dimethyl-
O
Me
Me
S
118 C5H10OS 19134-37-3
1,3-Oxathiane, 2-methyl-
O
Me
S
118 C5H10OS 20996-62-7
2-Propanone, 1-(ethylthio)-
MeCOCH2SEt
118 C5H10OS 30098-75-0
1,3-Oxathiane, 6-methyl-
Me
S
O
118 C5H10O3 97-64-3
Propanoic acid, 2-hydroxy-, ethyl ester
EtOC(O)CH(OH)Me

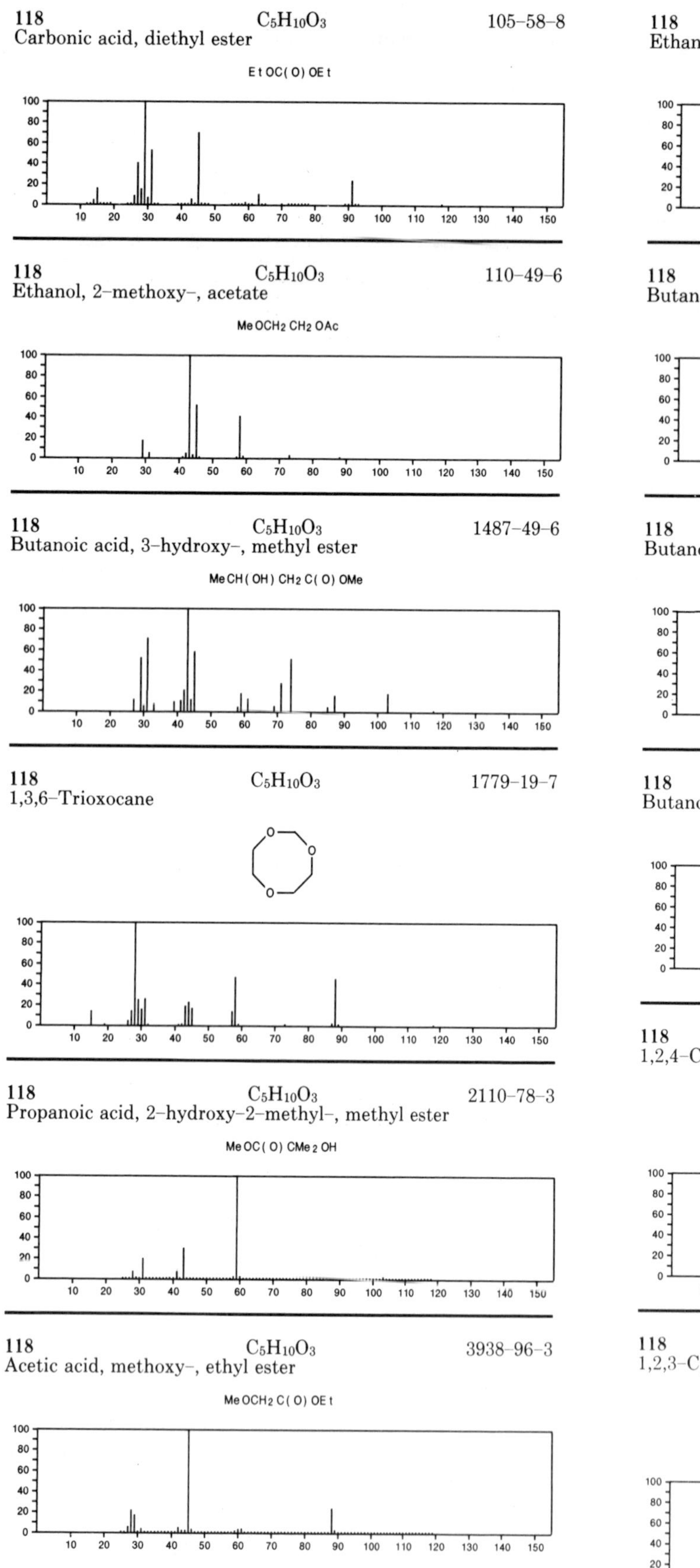

118 $C_5H_{10}O_3$ 105-58-8
Carbonic acid, diethyl ester
EtOC(O)OEt

118 $C_5H_{10}O_3$ 110-49-6
Ethanol, 2-methoxy-, acetate
MeOCH2CH2OAc

118 $C_5H_{10}O_3$ 1487-49-6
Butanoic acid, 3-hydroxy-, methyl ester
MeCH(OH)CH2C(O)OMe

118 $C_5H_{10}O_3$ 1779-19-7
1,3,6-Trioxocane

118 $C_5H_{10}O_3$ 2110-78-3
Propanoic acid, 2-hydroxy-2-methyl-, methyl ester
MeOC(O)CMe2OH

118 $C_5H_{10}O_3$ 3938-96-3
Acetic acid, methoxy-, ethyl ester
MeOCH2C(O)OEt

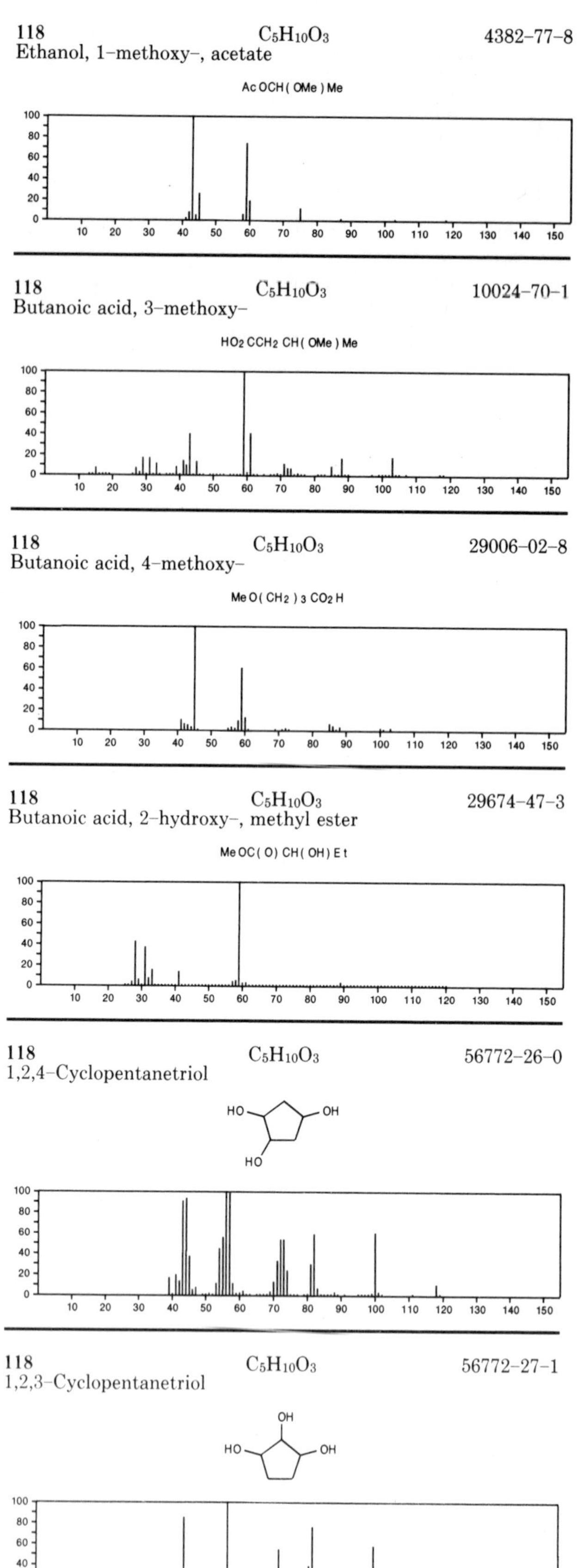

118 $C_5H_{10}O_3$ 4382-77-8
Ethanol, 1-methoxy-, acetate
AcOCH(OMe)Me

118 $C_5H_{10}O_3$ 10024-70-1
Butanoic acid, 3-methoxy-
HO2CCH2CH(OMe)Me

118 $C_5H_{10}O_3$ 29006-02-8
Butanoic acid, 4-methoxy-
MeO(CH2)3CO2H

118 $C_5H_{10}O_3$ 29674-47-3
Butanoic acid, 2-hydroxy-, methyl ester
MeOC(O)CH(OH)Et

118 $C_5H_{10}O_3$ 56772-26-0
1,2,4-Cyclopentanetriol

118 $C_5H_{10}O_3$ 56772-27-1
1,2,3-Cyclopentanetriol

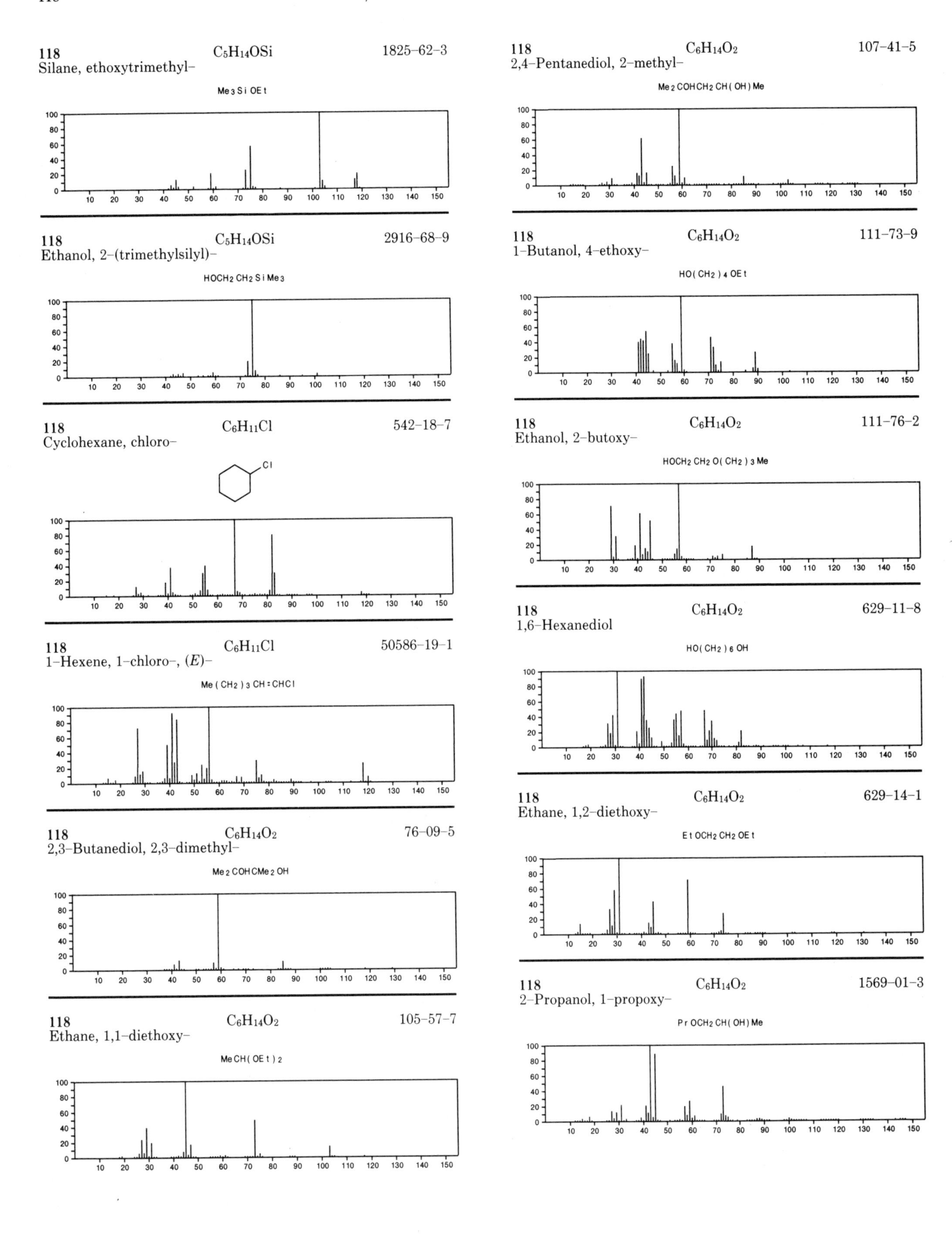
118 C5H14OSi 1825-62-3
Silane, ethoxytrimethyl-
Me3SiOEt
118 C5H14OSi 2916-68-9
Ethanol, 2-(trimethylsilyl)-
HOCH2CH2SiMe3
118 C6H11Cl 542-18-7
Cyclohexane, chloro-
Cl
118 C6H11Cl 50586-19-1
1-Hexene, 1-chloro-, (E)-
Me(CH2)3CH=CHCl
118 C6H14O2 76-09-5
2,3-Butanediol, 2,3-dimethyl-
Me2COHCMe2OH
118 C6H14O2 105-57-7
Ethane, 1,1-diethoxy-
MeCH(OEt)2
118 C6H14O2 107-41-5
2,4-Pentanediol, 2-methyl-
Me2COHCH2CH(OH)Me
118 C6H14O2 111-73-9
1-Butanol, 4-ethoxy-
HO(CH2)4OEt
118 C6H14O2 111-76-2
Ethanol, 2-butoxy-
HOCH2CH2O(CH2)3Me
118 C6H14O2 629-11-8
1,6-Hexanediol
HO(CH2)6OH
118 C6H14O2 629-14-1
Ethane, 1,2-diethoxy-
EtOCH2CH2OEt
118 C6H14O2 1569-01-3
2-Propanol, 1-propoxy-
PrOCH2CH(OH)Me

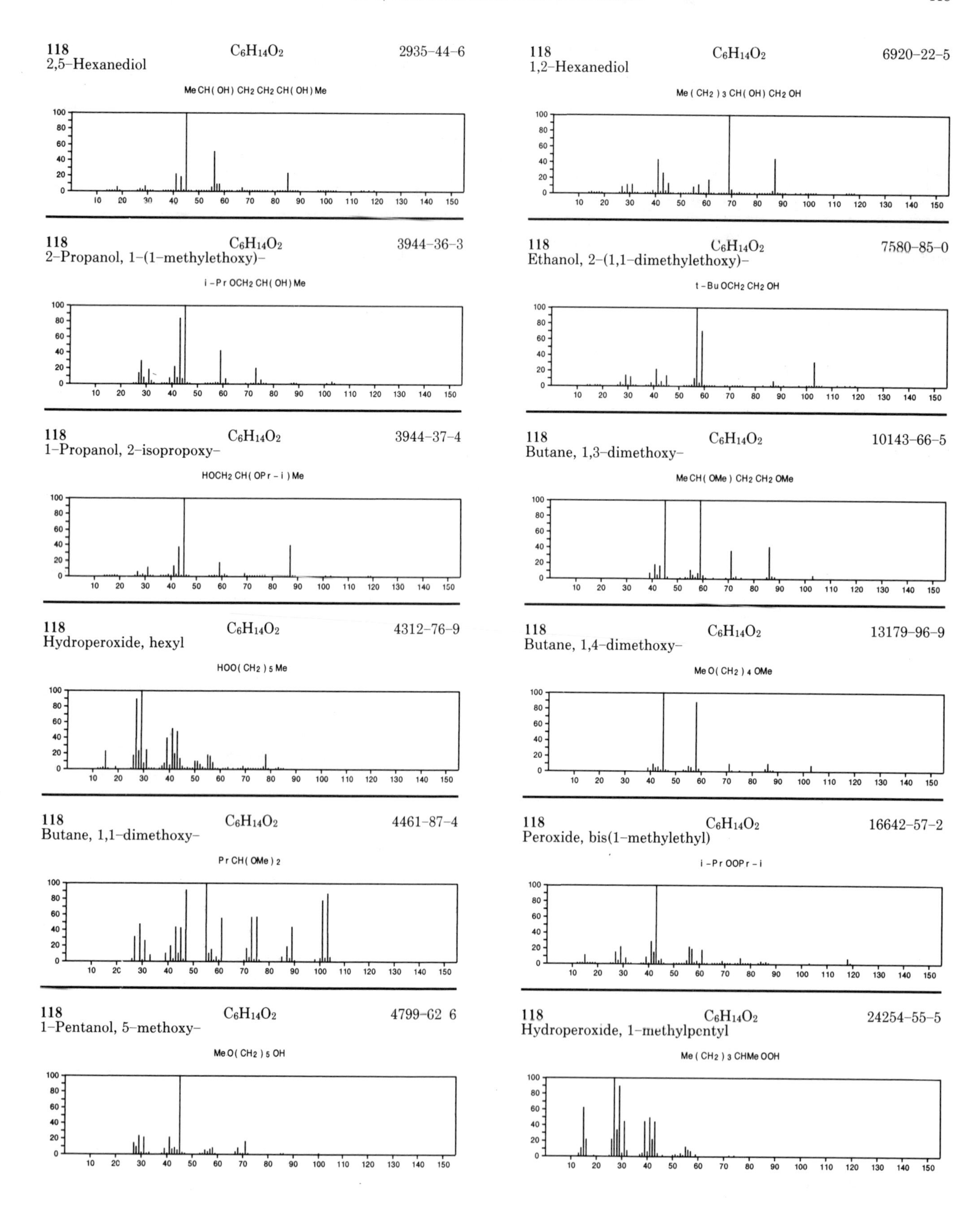
118
C6H14O2
2935-44-6
2,5-Hexanediol
Me CH(OH) CH2 CH2 CH(OH) Me
118
C6H14O2
6920-22-5
1,2-Hexanediol
Me (CH2) 3 CH(OH) CH2 OH
118
C6H14O2
3944-36-3
2-Propanol, 1-(1-methylethoxy)-
i - Pr OCH2 CH(OH) Me
118
C6H14O2
7580-85-0
Ethanol, 2-(1,1-dimethylethoxy)-
t - Bu OCH2 CH2 OH
118
C6H14O2
3944-37-4
1-Propanol, 2-isopropoxy-
HOCH2 CH(OPr - i) Me
118
C6H14O2
10143-66-5
Butane, 1,3-dimethoxy-
Me CH(OMe) CH2 CH2 OMe
118
C6H14O2
4312-76-9
Hydroperoxide, hexyl
HOO(CH2) 5 Me
118
C6H14O2
13179-96-9
Butane, 1,4-dimethoxy-
Me O(CH2) 4 OMe
118
C6H14O2
4461-87-4
Butane, 1,1-dimethoxy-
Pr CH(OMe) 2
118
C6H14O2
16642-57-2
Peroxide, bis(1-methylethyl)
i - Pr OOPr - i
118
C6H14O2
4799-62 6
1-Pentanol, 5-methoxy-
Me O(CH2) 5 OH
118
C6H14O2
24254-55-5
Hydroperoxide, 1-methylpentyl
Me (CH2) 3 CHMe OOH

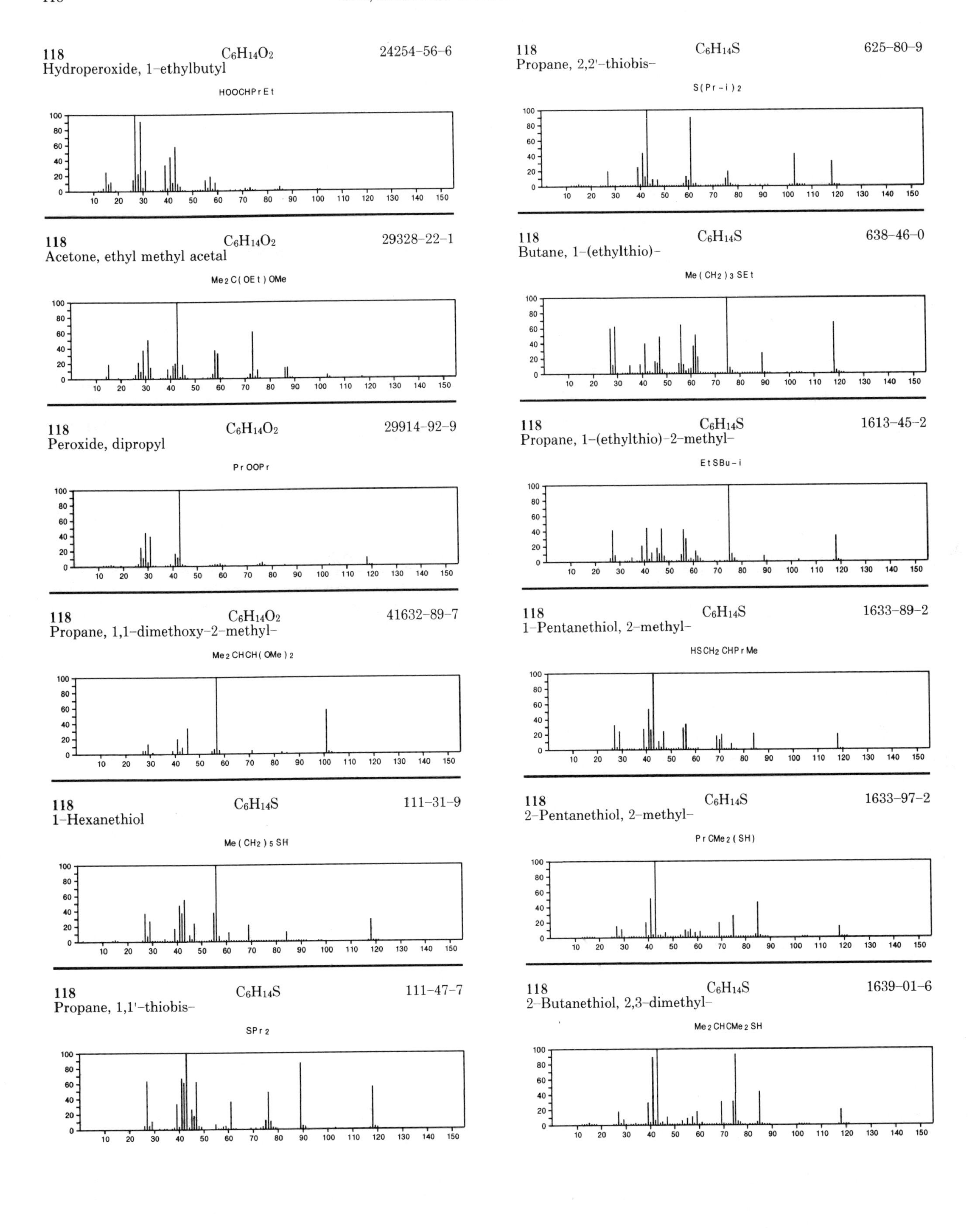

118 C6H14O2 24254-56-6
Hydroperoxide, 1-ethylbutyl
HOOCHPrEt
118 C6H14O2 29328-22-1
Acetone, ethyl methyl acetal
Me2C(OEt)OMe
118 C6H14O2 29914-92-9
Peroxide, dipropyl
PrOOPr
118 C6H14O2 41632-89-7
Propane, 1,1-dimethoxy-2-methyl-
Me2CHCH(OMe)2
118 C6H14S 111-31-9
1-Hexanethiol
Me(CH2)5SH
118 C6H14S 111-47-7
Propane, 1,1'-thiobis-
SPr2
118 C6H14S 625-80-9
Propane, 2,2'-thiobis-
S(Pr-i)2
118 C6H14S 638-46-0
Butane, 1-(ethylthio)-
Me(CH2)3SEt
118 C6H14S 1613-45-2
Propane, 1-(ethylthio)-2-methyl-
EtSBu-i
118 C6H14S 1633-89-2
1-Pentanethiol, 2-methyl-
HSCH2CHPrMe
118 C6H14S 1633-97-2
2-Pentanethiol, 2-methyl-
PrCMe2(SH)
118 C6H14S 1639-01-6
2-Butanethiol, 2,3-dimethyl-
Me2CHCMe2SH

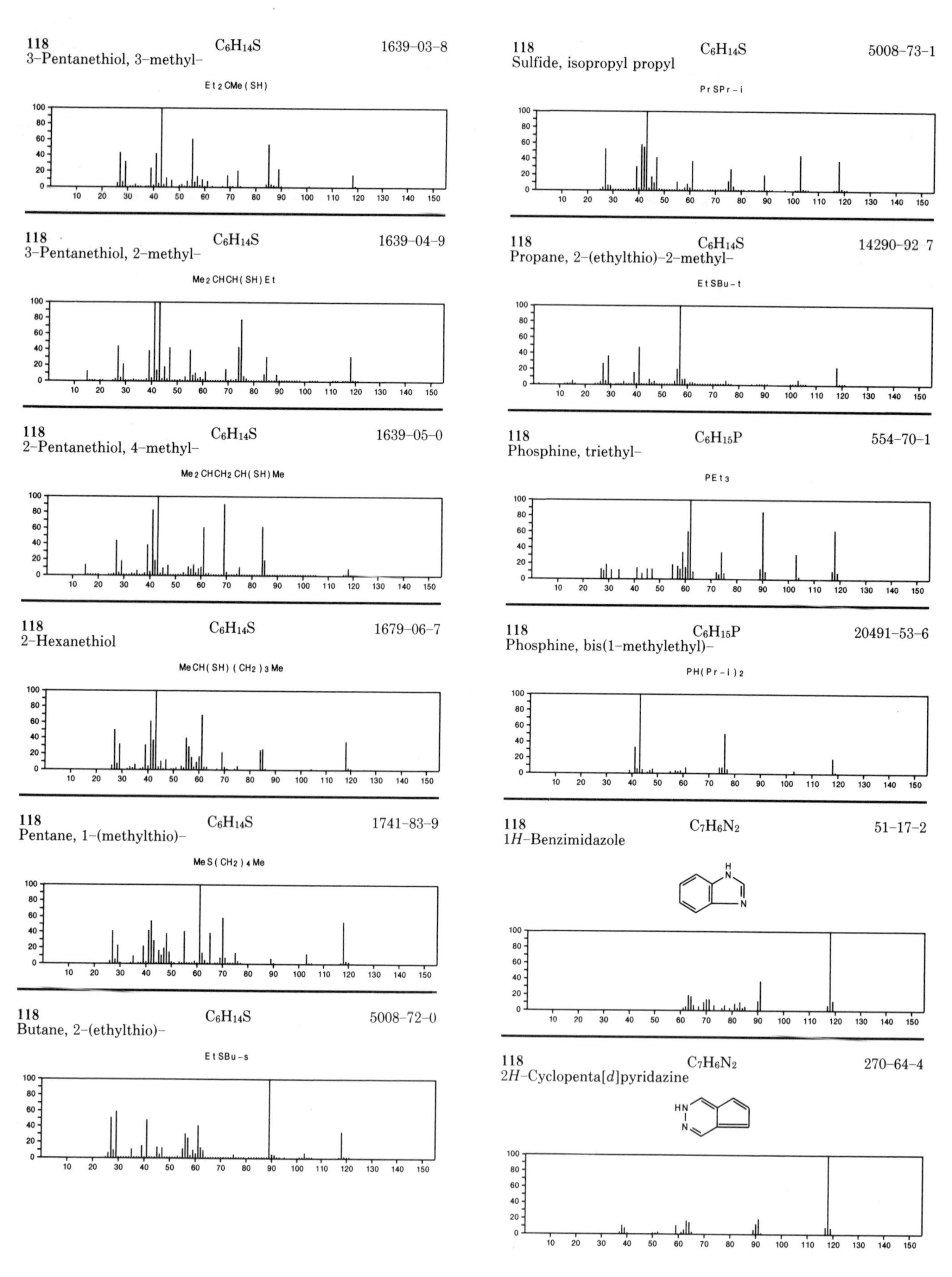

118 C6H14S 1639-03-8
3-Pentanethiol, 3-methyl-
Et2CMe(SH)
118 C6H14S 1639-04-9
3-Pentanethiol, 2-methyl-
Me2CHCH(SH)Et
118 C6H14S 1639-05-0
2-Pentanethiol, 4-methyl-
Me2CHCH2CH(SH)Me
118 C6H14S 1679-06-7
2-Hexanethiol
MeCH(SH)(CH2)3Me
118 C6H14S 1741-83-9
Pentane, 1-(methylthio)-
MeS(CH2)4Me
118 C6H14S 5008-72-0
Butane, 2-(ethylthio)-
EtSBu-s
118 C6H14S 5008-73-1
Sulfide, isopropyl propyl
PrSPr-i
118 C6H14S 14290-92-7
Propane, 2-(ethylthio)-2-methyl-
EtSBu-t
118 C6H15P 554-70-1
Phosphine, triethyl-
PEt3
118 C6H15P 20491-53-6
Phosphine, bis(1-methylethyl)-
PH(Pr-i)2
118 C7H6N2 51-17-2
1H-Benzimidazole
118 C7H6N2 270-64-4
2H-Cyclopenta[d]pyridazine

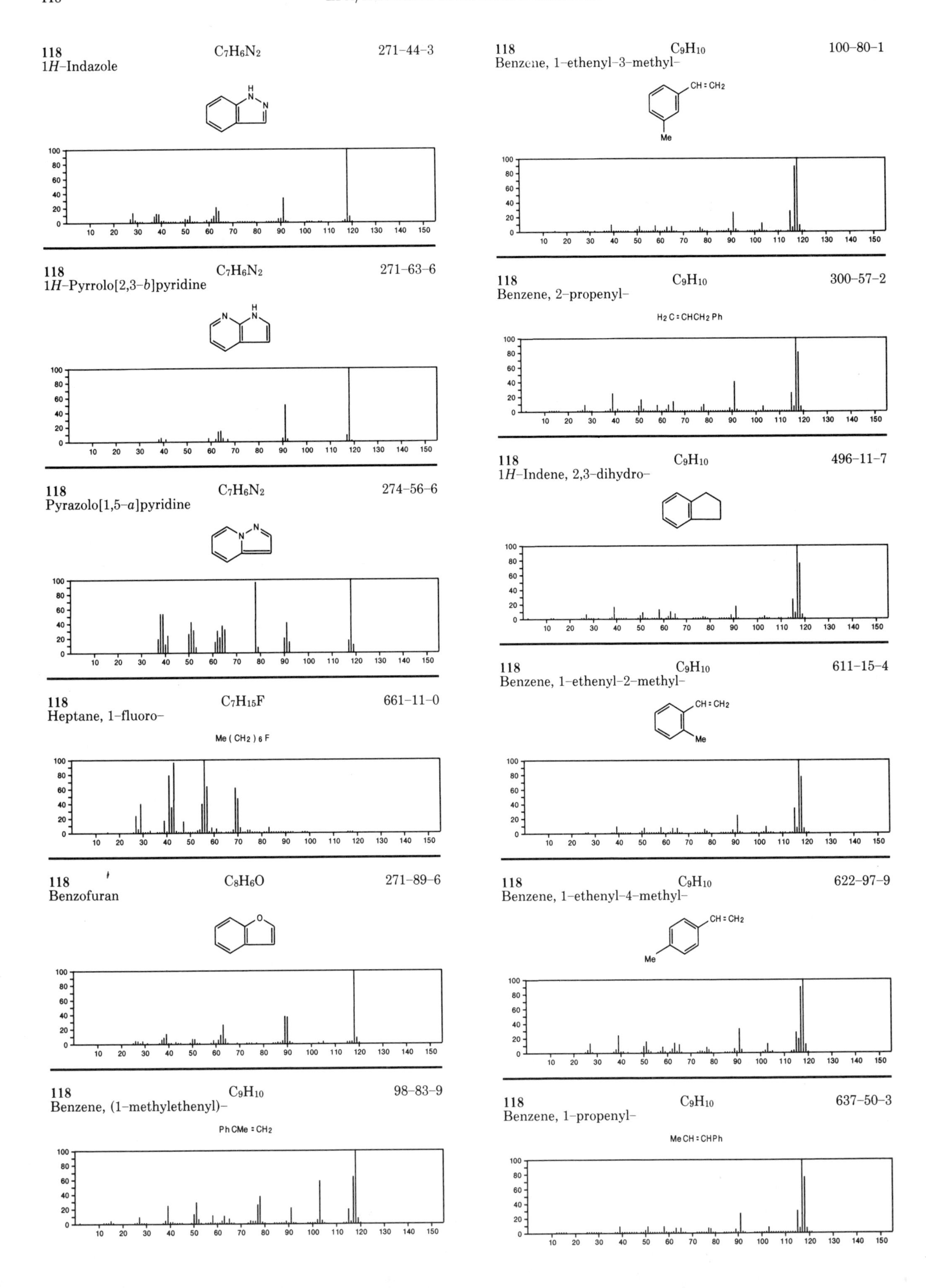
118 C7H6N2 271-44-3
1H-Indazole
118 C7H6N2 271-63-6
1H-Pyrrolo[2,3-b]pyridine
118 C7H6N2 274-56-6
Pyrazolo[1,5-a]pyridine
118 C7H15F 661-11-0
Heptane, 1-fluoro-
Me(CH2)6F
118 C8H6O 271-89-6
Benzofuran
118 C9H10 98-83-9
Benzene, (1-methylethenyl)-
PhCMe=CH2
118 C9H10 100-80-1
Benzene, 1-ethenyl-3-methyl-
CH=CH2
Me
118 C9H10 300-57-2
Benzene, 2-propenyl-
H2C=CHCH2Ph
118 C9H10 496-11-7
1H-Indene, 2,3-dihydro-
118 C9H10 611-15-4
Benzene, 1-ethenyl-2-methyl-
CH=CH2
Me
118 C9H10 622-97-9
Benzene, 1-ethenyl-4-methyl-
CH=CH2
Me
118 C9H10 637-50-3
Benzene, 1-propenyl-
MeCH=CHPh

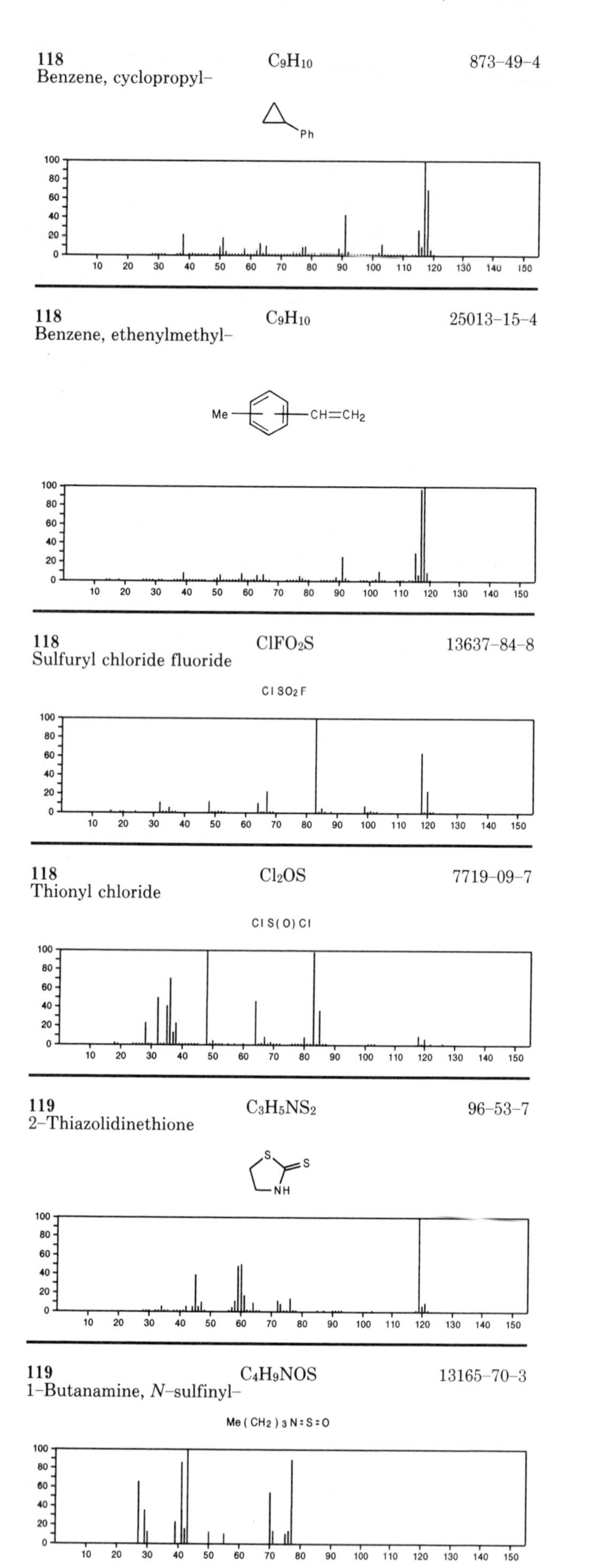

118 C_9H_{10} 873–49–4
Benzene, cyclopropyl–

118 C_9H_{10} 25013–15–4
Benzene, ethenylmethyl–

118 $ClFO_2S$ 13637–84–8
Sulfuryl chloride fluoride

Cl SO2 F

118 Cl_2OS 7719–09–7
Thionyl chloride

Cl S(O) Cl

119 $C_3H_5NS_2$ 96–53–7
2–Thiazolidinethione

119 C_4H_9NOS 13165–70–3
1–Butanamine, *N*–sulfinyl–

Me(CH2)3N=S=O

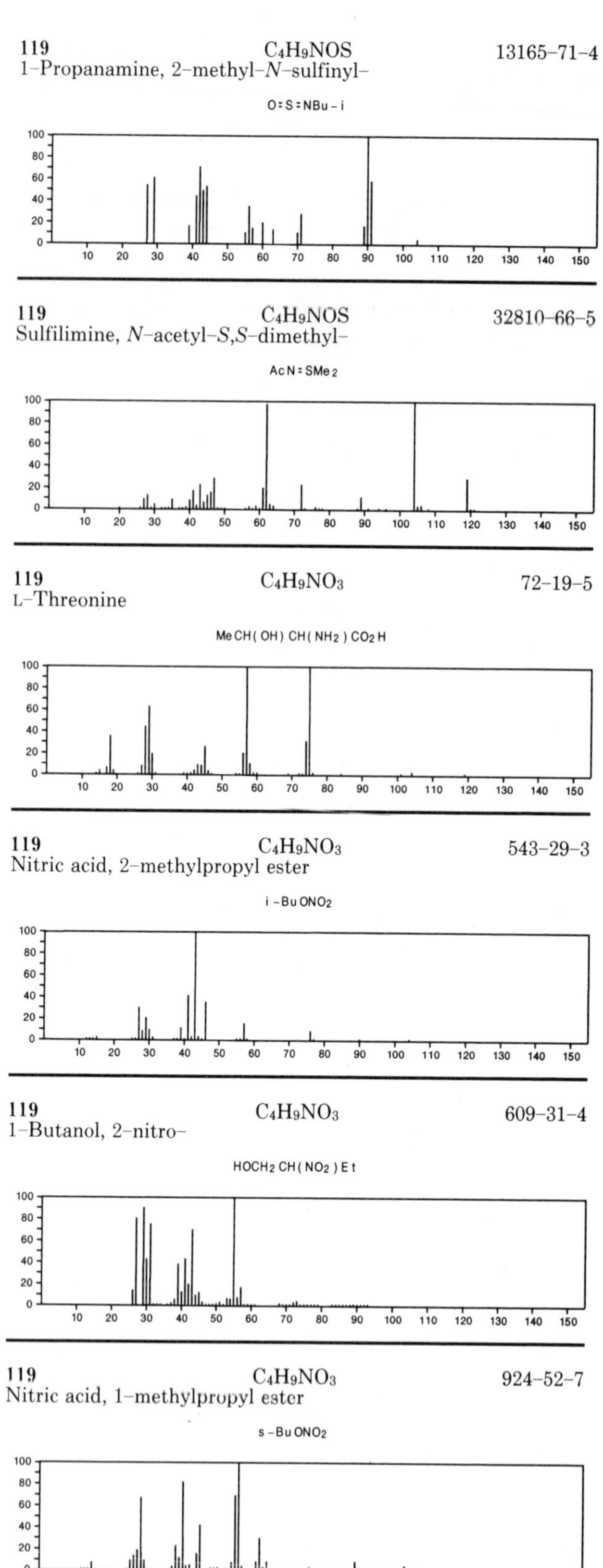

119 C_4H_9NOS 13165–71–4
1–Propanamine, 2–methyl–*N*–sulfinyl–

O=S=NBu-i

119 C_4H_9NOS 32810–66–5
Sulfilimine, *N*–acetyl–*S*,*S*–dimethyl–

AcN=SMe2

119 $C_4H_9NO_3$ 72–19–5
L–Threonine

MeCH(OH)CH(NH2)CO2H

119 $C_4H_9NO_3$ 543–29–3
Nitric acid, 2–methylpropyl ester

i-BuONO2

119 $C_4H_9NO_3$ 609–31–4
1–Butanol, 2–nitro–

HOCH2CH(NO2)Et

119 $C_4H_9NO_3$ 924–52–7
Nitric acid, 1–methylpropyl ester

s-BuONO2

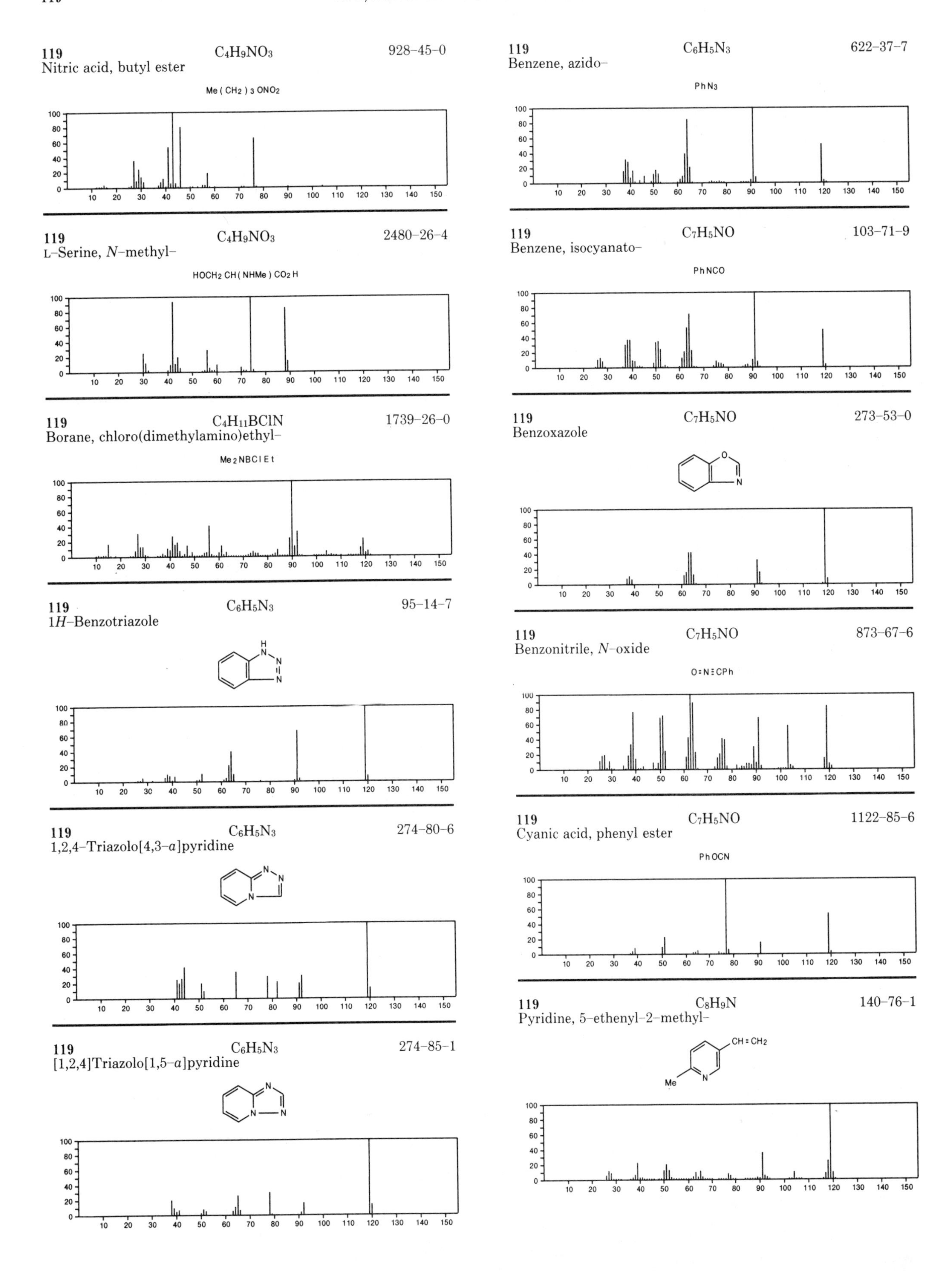
119 C4H9NO3 928-45-0
Nitric acid, butyl ester
Me(CH2)3ONO2
119 C4H9NO3 2480-26-4
L-Serine, N-methyl-
HOCH2CH(NHMe)CO2H
119 C4H11BClN 1739-26-0
Borane, chloro(dimethylamino)ethyl-
Me2NBClEt
119 C6H5N3 95-14-7
1H-Benzotriazole
119 C6H5N3 274-80-6
1,2,4-Triazolo[4,3-a]pyridine
119 C6H5N3 274-85-1
[1,2,4]Triazolo[1,5-a]pyridine
119 C6H5N3 622-37-7
Benzene, azido-
PhN3
119 C7H5NO 103-71-9
Benzene, isocyanato-
PhNCO
119 C7H5NO 273-53-0
Benzoxazole
119 C7H5NO 873-67-6
Benzonitrile, N-oxide
O=N≡CPh
119 C7H5NO 1122-85-6
Cyanic acid, phenyl ester
PhOCN
119 C8H9N 140-76-1
Pyridine, 5-ethenyl-2-methyl-
CH=CH2
Me

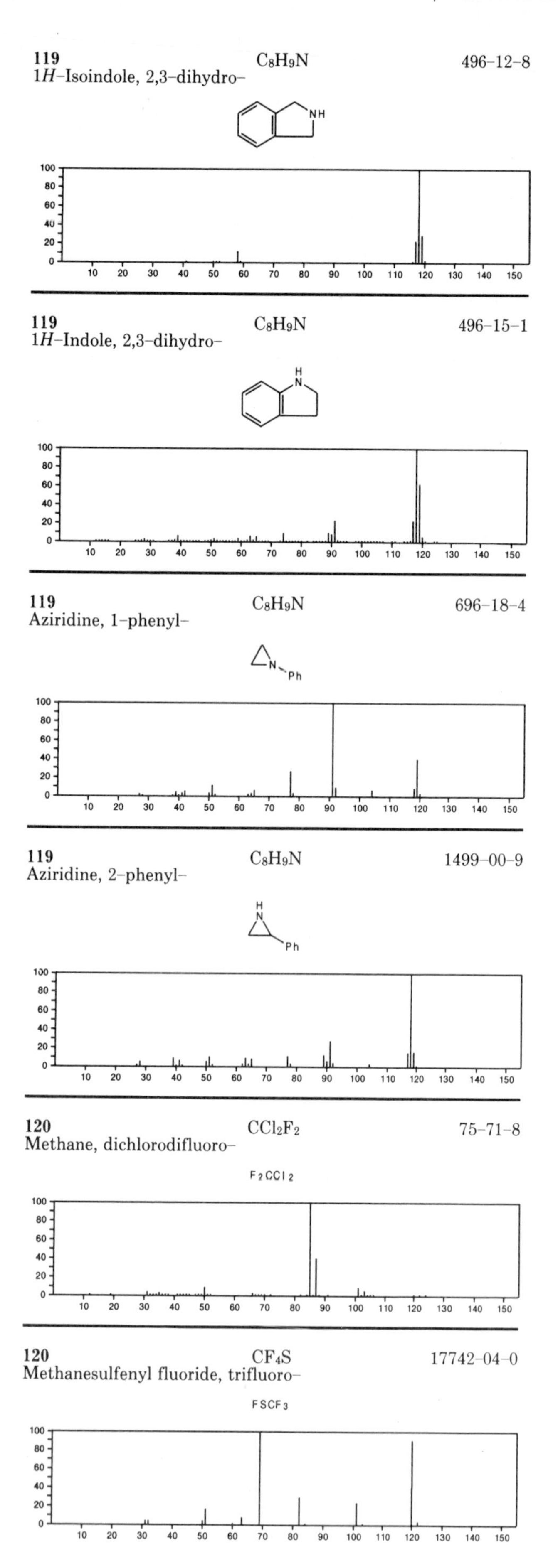
119 C_8H_9N 496-12-8
1H-Isoindole, 2,3-dihydro-
NH
119 C_8H_9N 496-15-1
1H-Indole, 2,3-dihydro-
H
N
119 C_8H_9N 696-18-4
Aziridine, 1-phenyl-
N
Ph
119 C_8H_9N 1499-00-9
Aziridine, 2-phenyl-
H
N
Ph
120 CCl_2F_2 75-71-8
Methane, dichlorodifluoro-
F_2CCl_2
120 CF_4S 17742-04-0
Methanesulfenyl fluoride, trifluoro-
$FSCF_3$

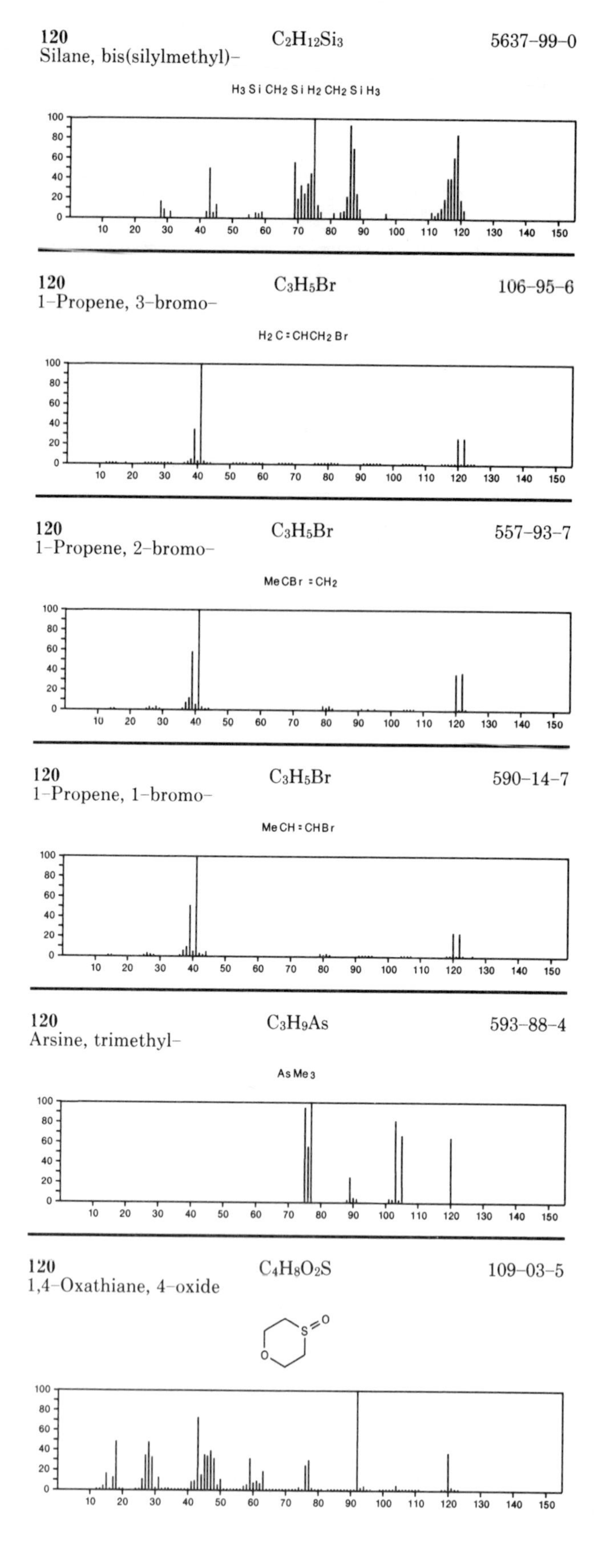
120 $C_2H_{12}Si_3$ 5637-99-0
Silane, bis(silylmethyl)-
$H_3SiCH_2SiH_2CH_2SiH_3$
120 C_3H_5Br 106-95-6
1-Propene, 3-bromo-
$H_2C=CHCH_2Br$
120 C_3H_5Br 557-93-7
1-Propene, 2-bromo-
$MeCBr=CH_2$
120 C_3H_5Br 590-14-7
1-Propene, 1-bromo-
$MeCH=CHBr$
120 C_3H_9As 593-88-4
Arsine, trimethyl-
$AsMe_3$
120 $C_4H_8O_2S$ 109-03-5
1,4-Oxathiane, 4-oxide
S=O
O

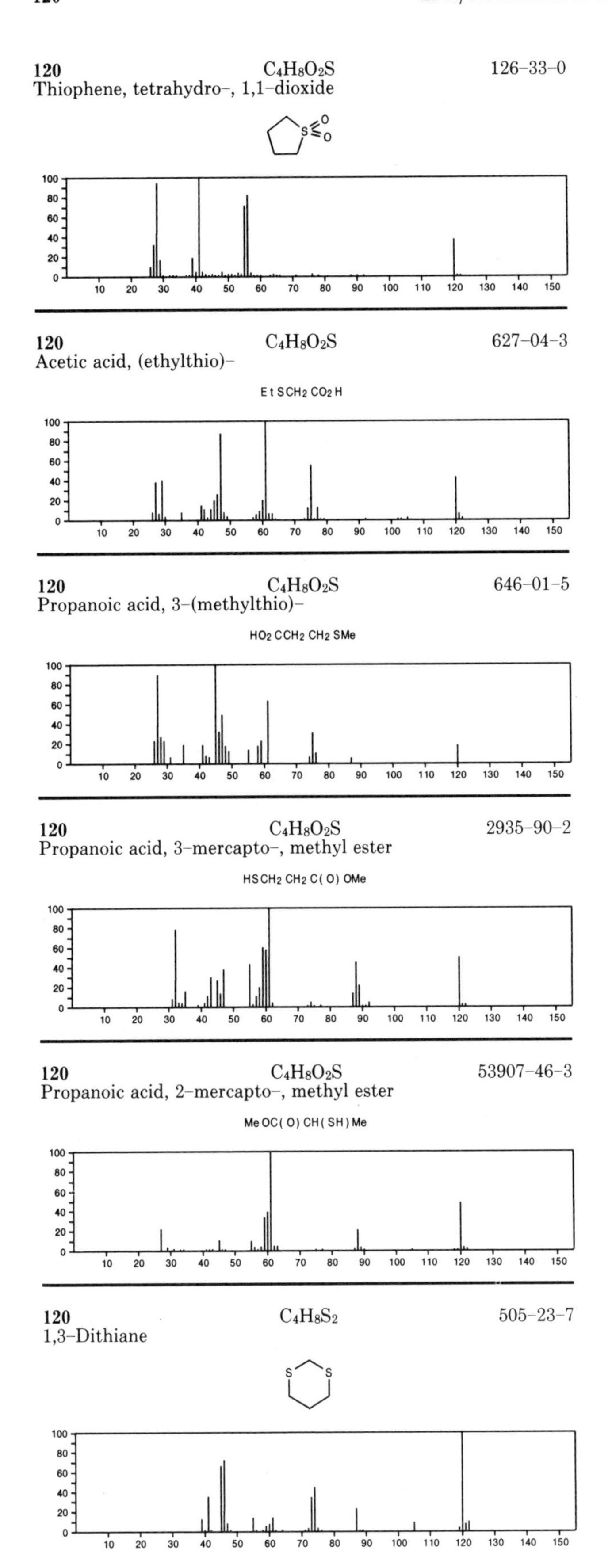
120 C4H8O2S 126-33-0
Thiophene, tetrahydro-, 1,1-dioxide
120 C4H8O2S 627-04-3
Acetic acid, (ethylthio)-
EtSCH2CO2H
120 C4H8O2S 646-01-5
Propanoic acid, 3-(methylthio)-
HO2CCH2CH2SMe
120 C4H8O2S 2935-90-2
Propanoic acid, 3-mercapto-, methyl ester
HSCH2CH2C(O)OMe
120 C4H8O2S 53907-46-3
Propanoic acid, 2-mercapto-, methyl ester
MeOC(O)CH(SH)Me
120 C4H8S2 505-23-7
1,3-Dithiane

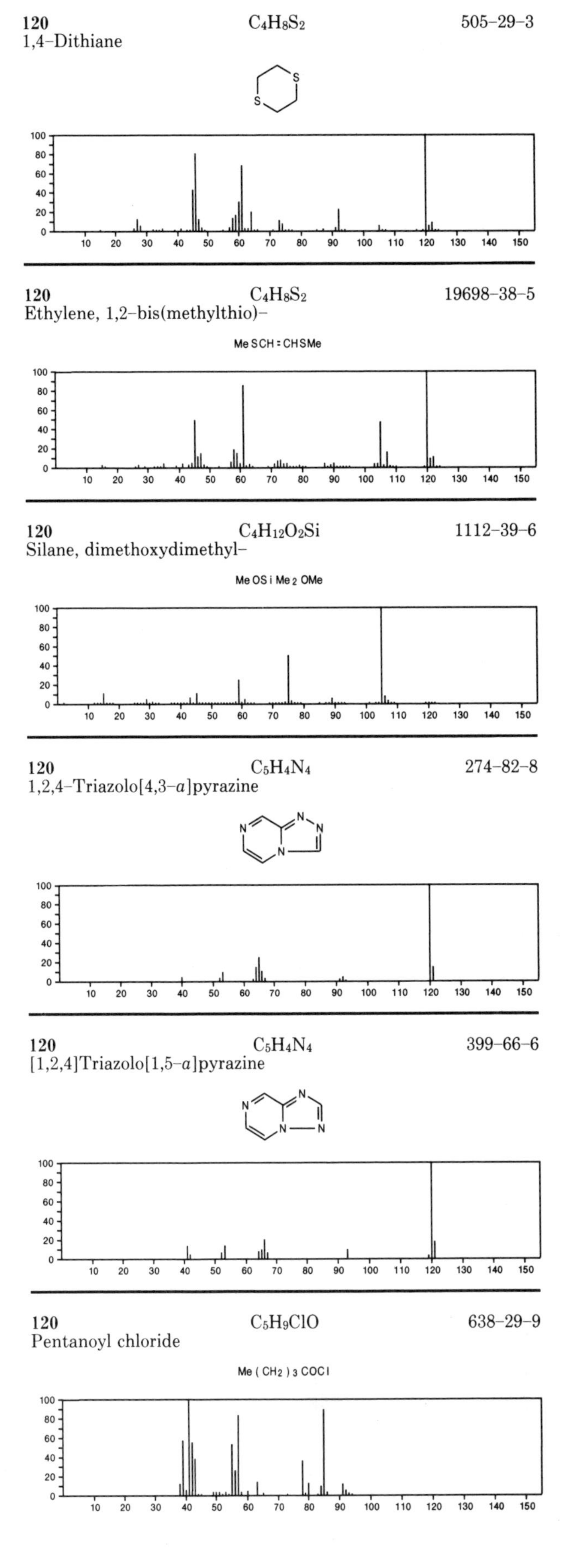
120 C4H8S2 505-29-3
1,4-Dithiane
120 C4H8S2 19698-38-5
Ethylene, 1,2-bis(methylthio)-
MeSCH=CHSMe
120 C4H12O2Si 1112-39-6
Silane, dimethoxydimethyl-
MeOSiMe2OMe
120 C5H4N4 274-82-8
1,2,4-Triazolo[4,3-a]pyrazine
120 C5H4N4 399-66-6
[1,2,4]Triazolo[1,5-a]pyrazine
120 C5H9ClO 638-29-9
Pentanoyl chloride
Me(CH2)3COCl

120 C_5H_9ClO 3282-30-2
Propanoyl chloride, 2,2-dimethyl-

Me_3CCOCl

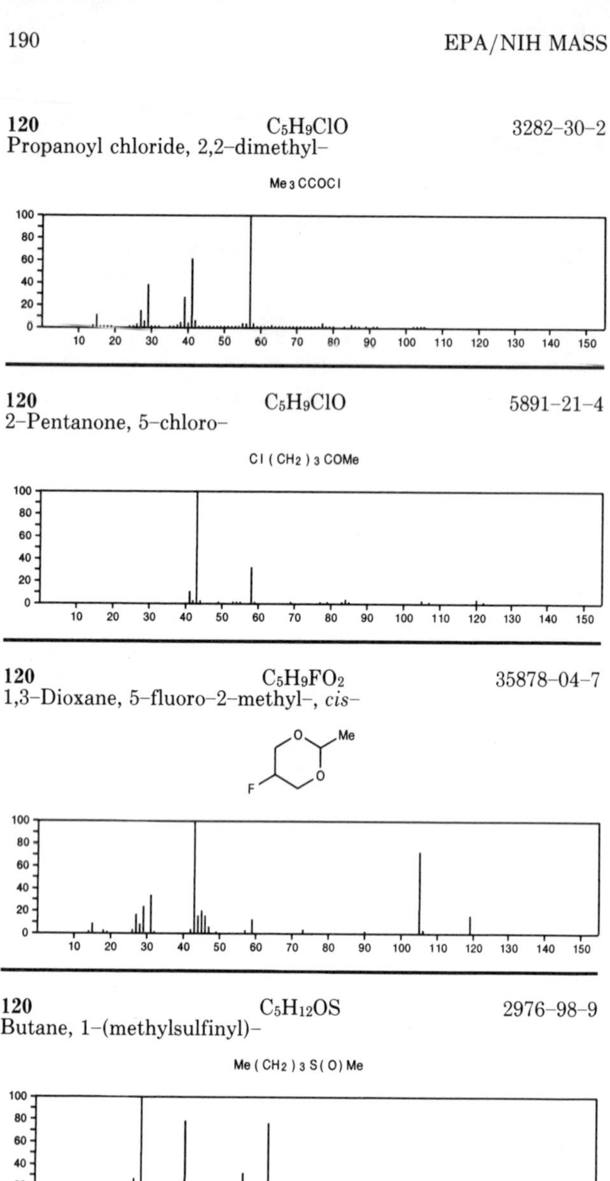

120 C_5H_9ClO 5891-21-4
2-Pentanone, 5-chloro-

$Cl(CH_2)_3COMe$

120 $C_5H_9FO_2$ 35878-04-7
1,3-Dioxane, 5-fluoro-2-methyl-, *cis*-

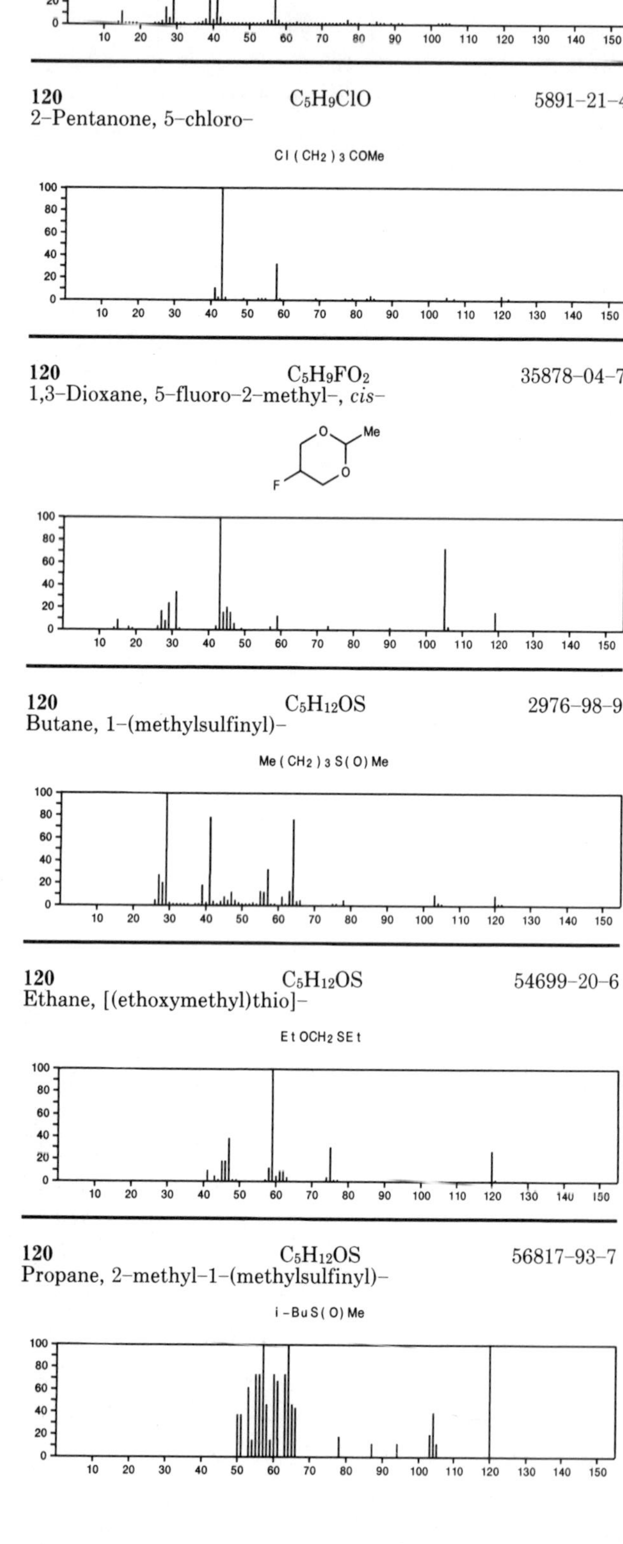

120 $C_5H_{12}OS$ 2976-98-9
Butane, 1-(methylsulfinyl)-

$Me(CH_2)_3S(O)Me$

120 $C_5H_{12}OS$ 54699-20-6
Ethane, [(ethoxymethyl)thio]-

$EtOCH_2SEt$

120 $C_5H_{12}OS$ 56817-93-7
Propane, 2-methyl-1-(methylsulfinyl)-

$i\text{-}BuS(O)Me$

120 $C_5H_{12}O_3$ 111-77-3
Ethanol, 2-(2-methoxyethoxy)-

$HOCH_2CH_2OCH_2CH_2OMe$

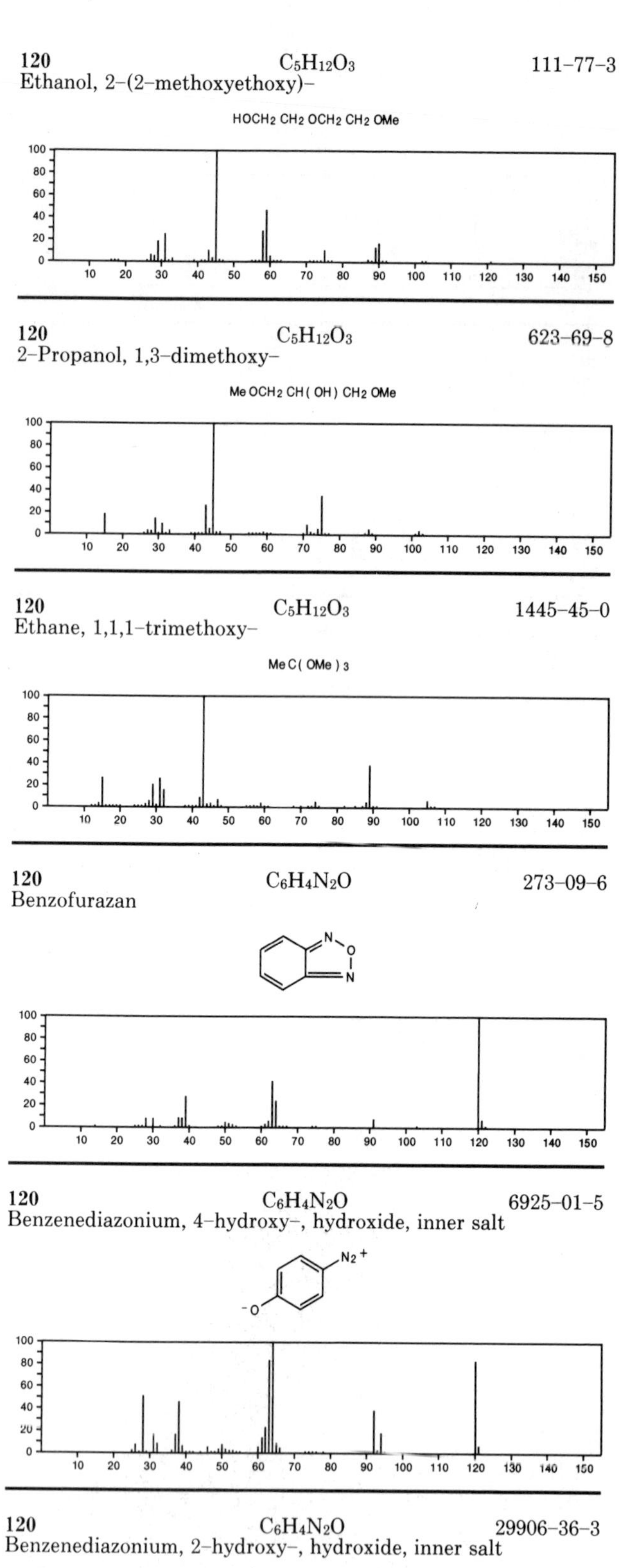

120 $C_5H_{12}O_3$ 623-69-8
2-Propanol, 1,3-dimethoxy-

$MeOCH_2CH(OH)CH_2OMe$

120 $C_5H_{12}O_3$ 1445-45-0
Ethane, 1,1,1-trimethoxy-

$MeC(OMe)_3$

120 $C_6H_4N_2O$ 273-09-6
Benzofurazan

120 $C_6H_4N_2O$ 6925-01-5
Benzenediazonium, 4-hydroxy-, hydroxide, inner salt

120 $C_6H_4N_2O$ 29906-36-3
Benzenediazonium, 2-hydroxy-, hydroxide, inner salt

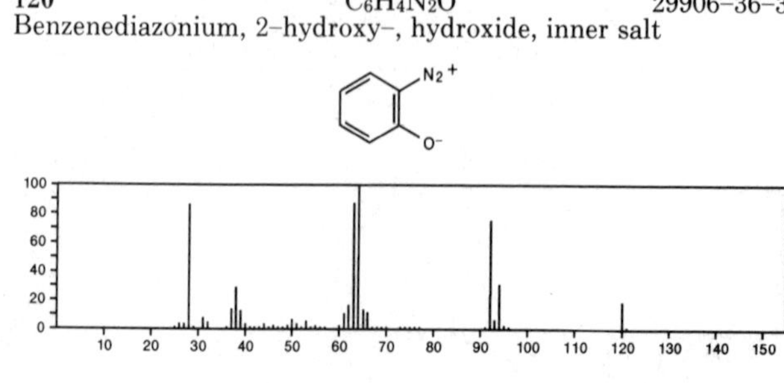

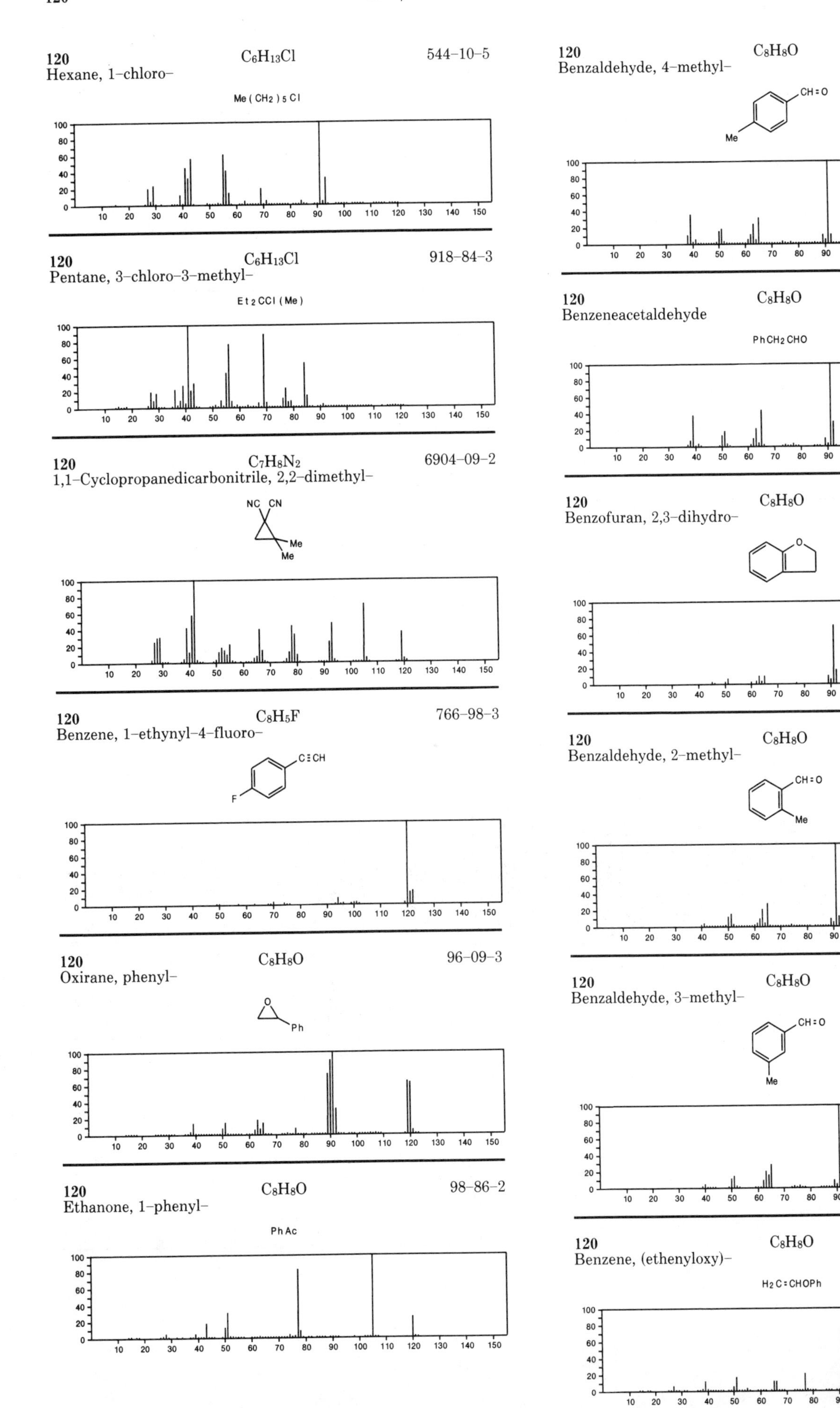

120 $C_6H_{13}Cl$ 544–10–5
Hexane, 1–chloro–
Me(CH2)5Cl

120 $C_6H_{13}Cl$ 918–84–3
Pentane, 3–chloro–3–methyl–
Et2CCl(Me)

120 $C_7H_8N_2$ 6904–09–2
1,1–Cyclopropanedicarbonitrile, 2,2–dimethyl–

120 C_8H_5F 766–98–3
Benzene, 1–ethynyl–4–fluoro–

120 C_8H_8O 96–09–3
Oxirane, phenyl–

120 C_8H_8O 98–86–2
Ethanone, 1–phenyl–
PhAc

120 C_8H_8O 104–87–0
Benzaldehyde, 4–methyl–

120 C_8H_8O 122–78–1
Benzeneacetaldehyde
PhCH2CHO

120 C_8H_8O 496–16–2
Benzofuran, 2,3–dihydro–

120 C_8H_8O 529–20–4
Benzaldehyde, 2–methyl–

120 C_8H_8O 620–23–5
Benzaldehyde, 3–methyl–

120 C_8H_8O 766–94–9
Benzene, (ethenyloxy)–
H2C=CHOPh

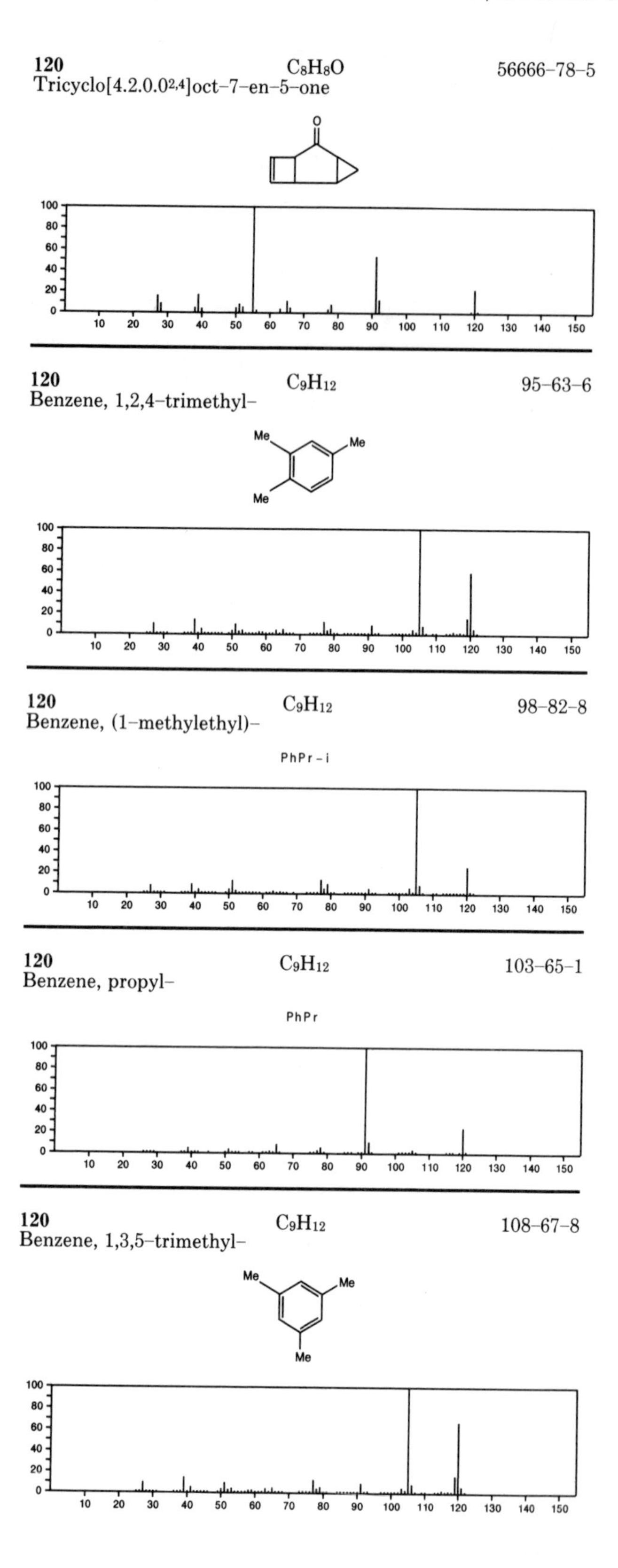

120 C_8H_8O 56666-78-5
Tricyclo[4.2.0.0^{2,4}]oct-7-en-5-one

120 C_9H_{12} 95-63-6
Benzene, 1,2,4-trimethyl-

120 C_9H_{12} 98-82-8
Benzene, (1-methylethyl)-

120 C_9H_{12} 103-65-1
Benzene, propyl-

120 C_9H_{12} 108-67-8
Benzene, 1,3,5-trimethyl-

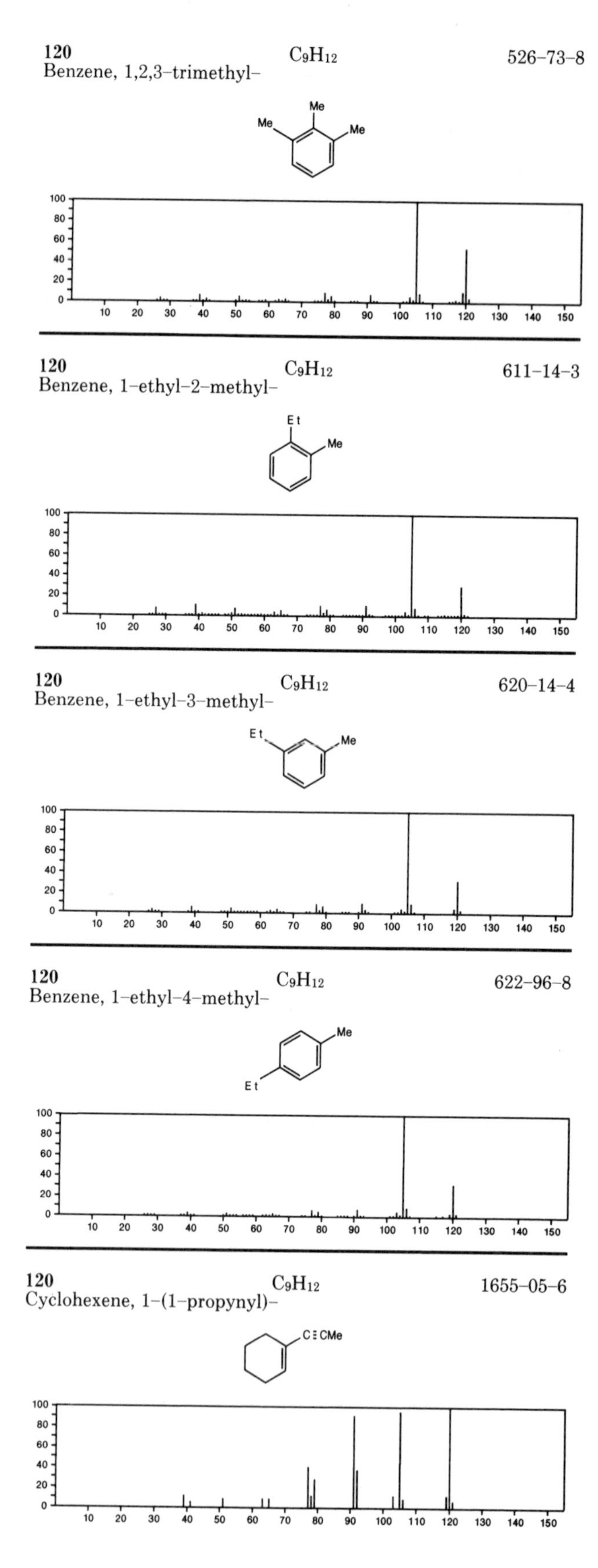

120 C_9H_{12} 526-73-8
Benzene, 1,2,3-trimethyl-

120 C_9H_{12} 611-14-3
Benzene, 1-ethyl-2-methyl-

120 C_9H_{12} 620-14-4
Benzene, 1-ethyl-3-methyl-

120 C_9H_{12} 622-96-8
Benzene, 1-ethyl-4-methyl-

120 C_9H_{12} 1655-05-6
Cyclohexene, 1-(1-propynyl)-

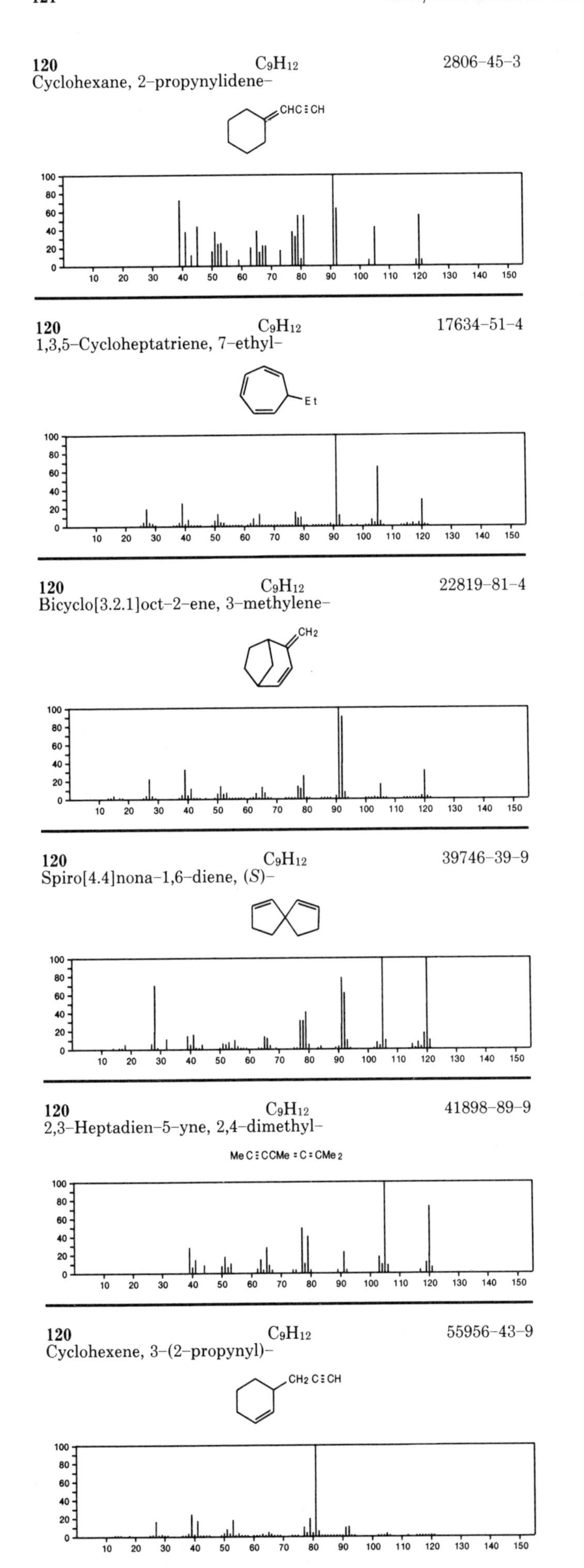

120 C_9H_{12} 2806–45–3
Cyclohexane, 2–propynylidene–

120 C_9H_{12} 17634–51–4
1,3,5–Cycloheptatriene, 7–ethyl–

120 C_9H_{12} 22819–81–4
Bicyclo[3.2.1]oct–2–ene, 3–methylene–

120 C_9H_{12} 39746–39–9
Spiro[4.4]nona–1,6–diene, (*S*)–

120 C_9H_{12} 41898–89–9
2,3–Heptadien–5–yne, 2,4–dimethyl–

120 C_9H_{12} 55956–43–9
Cyclohexene, 3–(2–propynyl)–

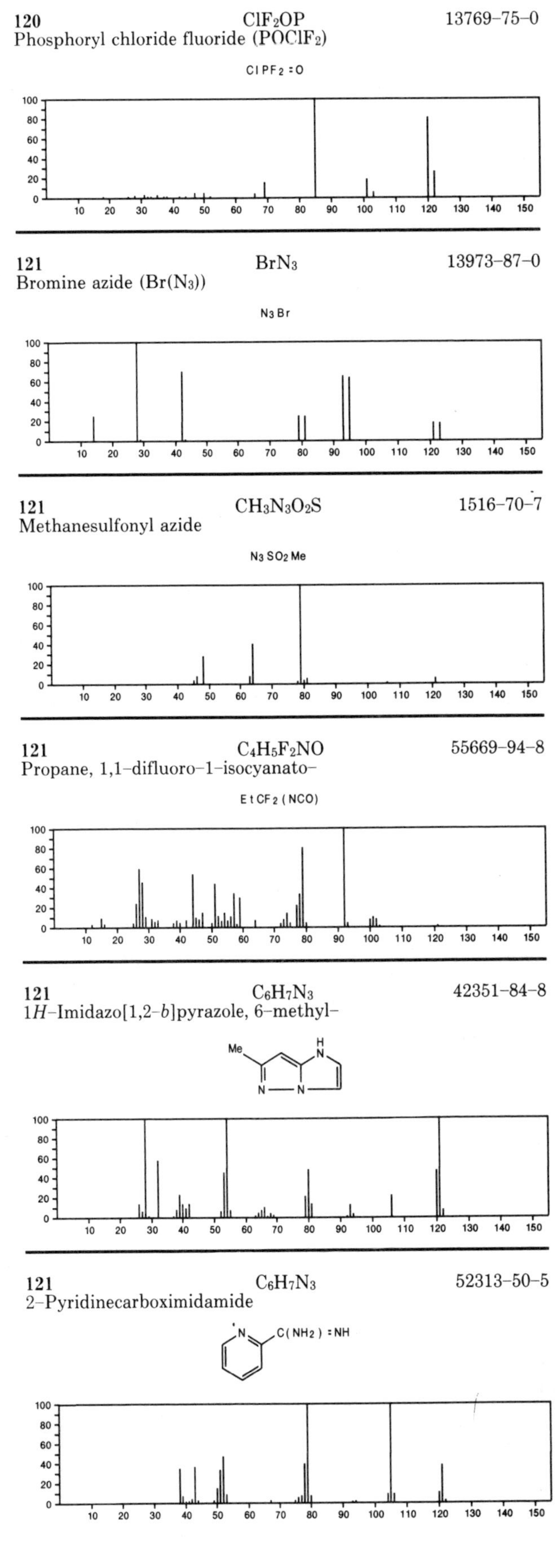

120 ClF_2OP 13769–75–0
Phosphoryl chloride fluoride ($POClF_2$)

121 BrN_3 13973–87–0
Bromine azide ($Br(N_3)$)

121 $CH_3N_3O_2S$ 1516–70–7
Methanesulfonyl azide

121 $C_4H_5F_2NO$ 55669–94–8
Propane, 1,1–difluoro–1–isocyanato–

121 $C_6H_7N_3$ 42351–84–8
1*H*–Imidazo[1,2–*b*]pyrazole, 6–methyl–

121 $C_6H_7N_3$ 52313–50–5
2–Pyridinecarboximidamide

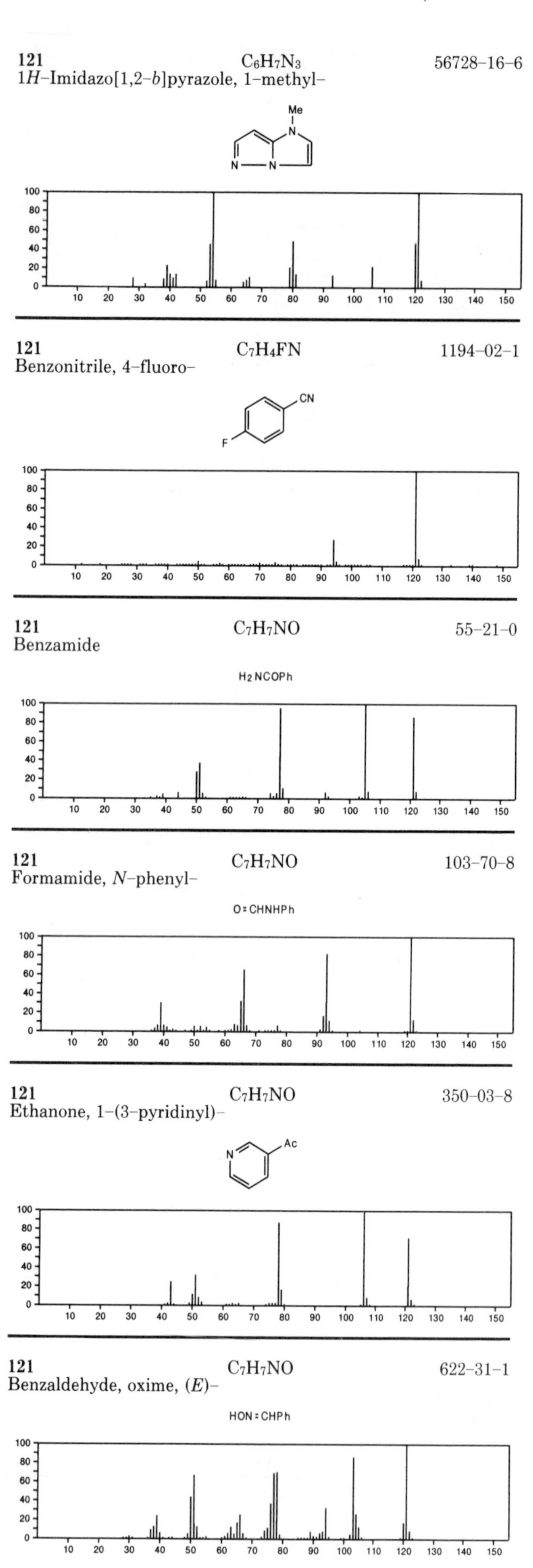

121 $C_6H_7N_3$ 56728-16-6
1*H*-Imidazo[1,2-*b*]pyrazole, 1-methyl-

121 C_7H_4FN 1194-02-1
Benzonitrile, 4-fluoro-

121 C_7H_7NO 55-21-0
Benzamide

H2NCOPh

121 C_7H_7NO 103-70-8
Formamide, *N*-phenyl-

O=CHNHPh

121 C_7H_7NO 350-03-8
Ethanone, 1-(3-pyridinyl)-

121 C_7H_7NO 622-31-1
Benzaldehyde, oxime, (*E*)-

HON=CHPh

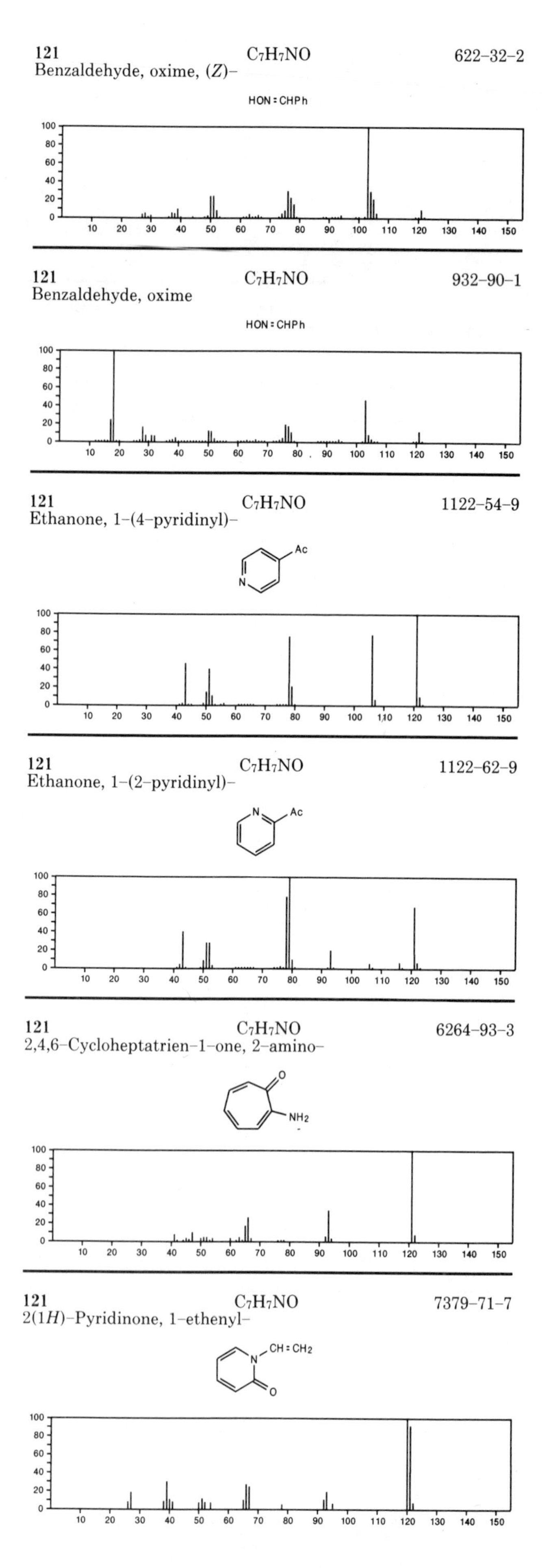

121 C_7H_7NO 622-32-2
Benzaldehyde, oxime, (*Z*)-

HON=CHPh

121 C_7H_7NO 932-90-1
Benzaldehyde, oxime

HON=CHPh

121 C_7H_7NO 1122-54-9
Ethanone, 1-(4-pyridinyl)-

121 C_7H_7NO 1122-62-9
Ethanone, 1-(2-pyridinyl)-

121 C_7H_7NO 6264-93-3
2,4,6-Cycloheptatrien-1-one, 2-amino-

121 C_7H_7NO 7379-71-7
2(1*H*)-Pyridinone, 1-ethenyl-

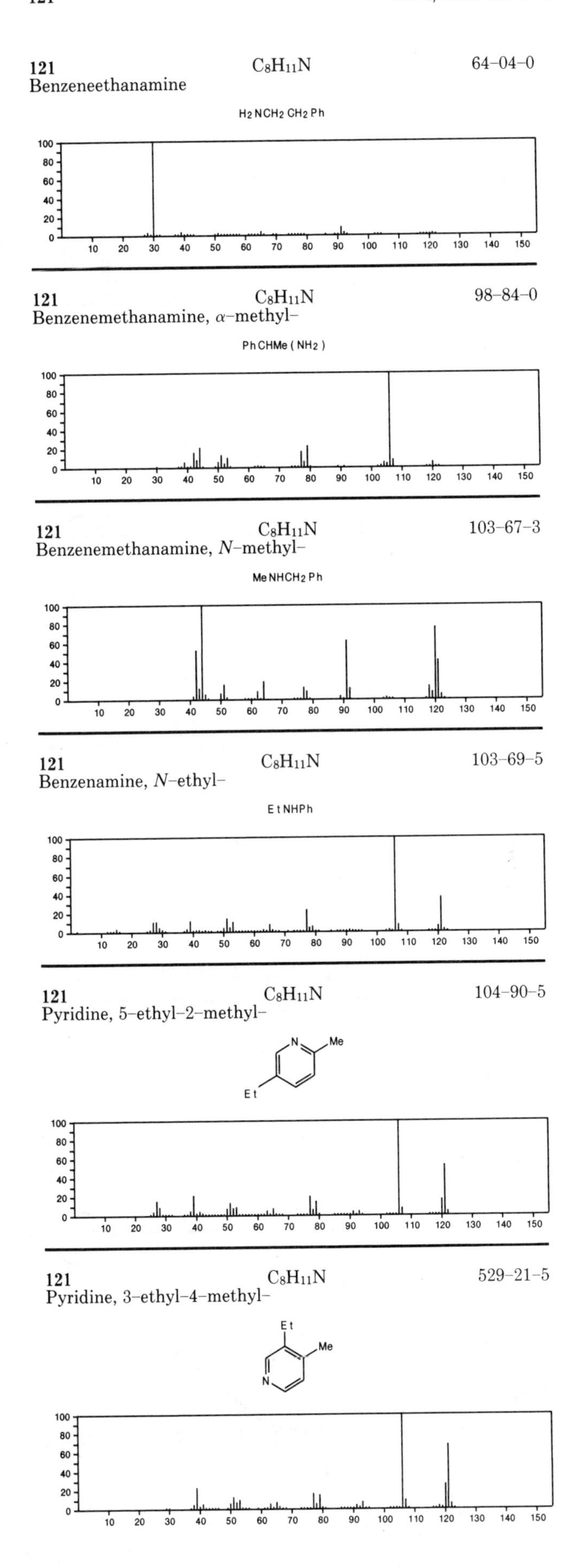

121 $C_8H_{11}N$ 64–04–0
Benzeneethanamine

$H_2NCH_2CH_2Ph$

121 $C_8H_{11}N$ 98–84–0
Benzenemethanamine, α–methyl–

PhCHMe(NH_2)

121 $C_8H_{11}N$ 103–67–3
Benzenemethanamine, *N*–methyl–

MeNHCH$_2$Ph

121 $C_8H_{11}N$ 103–69–5
Benzenamine, *N*–ethyl–

EtNHPh

121 $C_8H_{11}N$ 104–90–5
Pyridine, 5–ethyl–2–methyl–

121 $C_8H_{11}N$ 529–21–5
Pyridine, 3–ethyl–4–methyl–

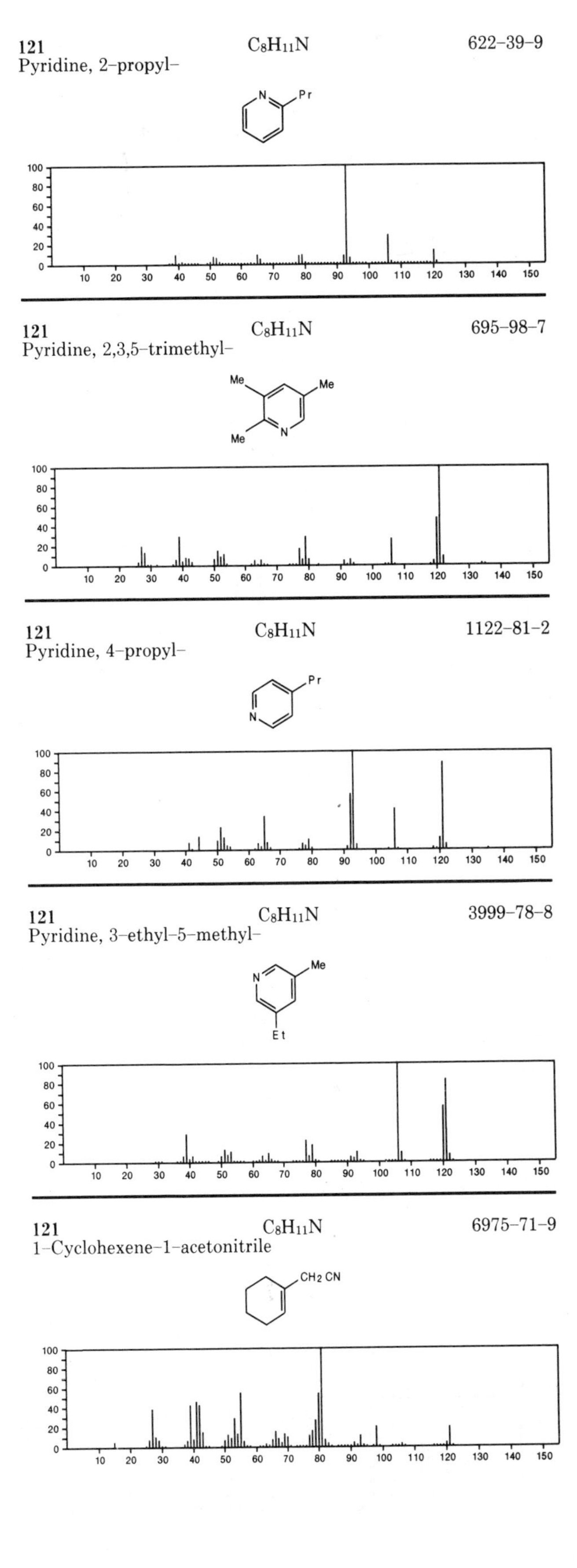

121 $C_8H_{11}N$ 622–39–9
Pyridine, 2–propyl–

121 $C_8H_{11}N$ 695–98–7
Pyridine, 2,3,5–trimethyl–

121 $C_8H_{11}N$ 1122–81–2
Pyridine, 4–propyl–

121 $C_8H_{11}N$ 3999–78–8
Pyridine, 3–ethyl–5–methyl–

121 $C_8H_{11}N$ 6975–71–9
1–Cyclohexene–1–acetonitrile

CH_2CN

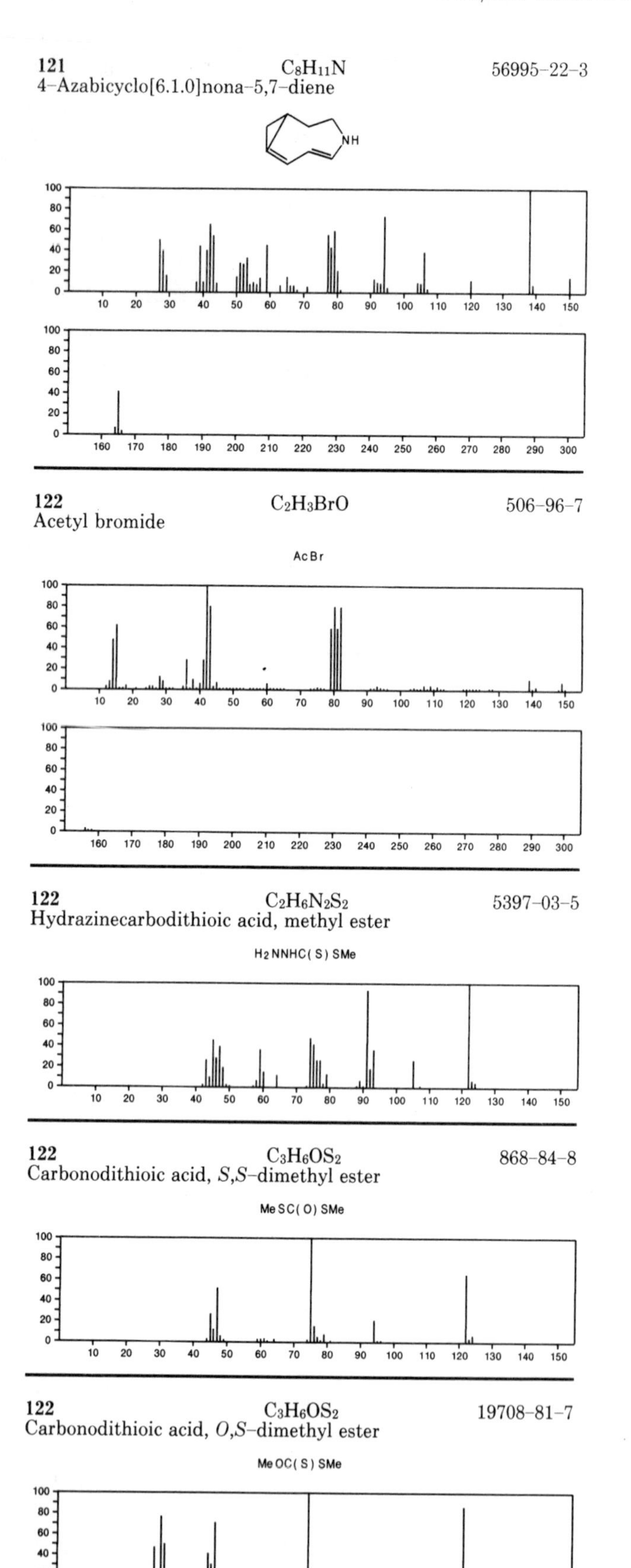

121 $C_8H_{11}N$ 56995-22-3
4-Azabicyclo[6.1.0]nona-5,7-diene

122 C_2H_3BrO 506-96-7
Acetyl bromide
AcBr

122 $C_2H_6N_2S_2$ 5397-03-5
Hydrazinecarbodithioic acid, methyl ester
H2NNHC(S)SMe

122 $C_3H_6OS_2$ 868-84-8
Carbonodithioic acid, *S,S*-dimethyl ester
MeSC(O)SMe

122 $C_3H_6OS_2$ 19708-81-7
Carbonodithioic acid, *O,S*-dimethyl ester
MeOC(S)SMe

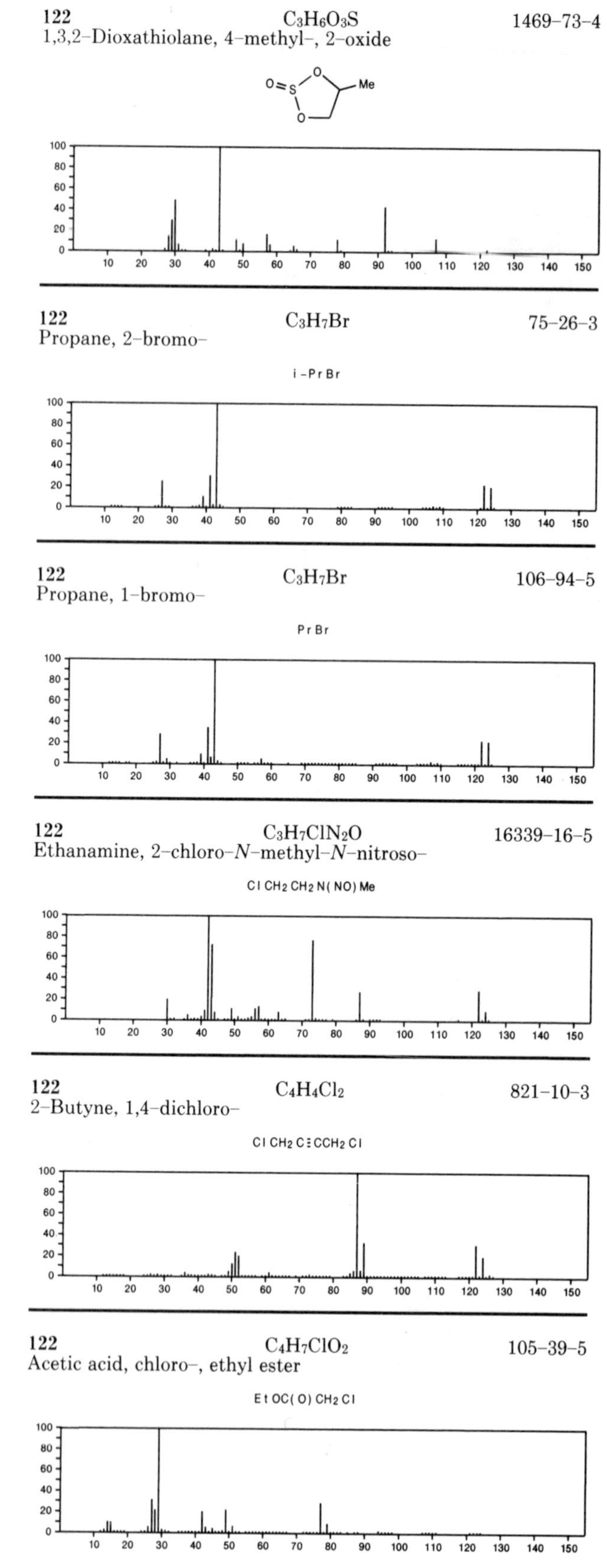

122 $C_3H_6O_3S$ 1469-73-4
1,3,2-Dioxathiolane, 4-methyl-, 2-oxide

122 C_3H_7Br 75-26-3
Propane, 2-bromo-
i-PrBr

122 C_3H_7Br 106-94-5
Propane, 1-bromo-
PrBr

122 $C_3H_7ClN_2O$ 16339-16-5
Ethanamine, 2-chloro-*N*-methyl-*N*-nitroso-
ClCH2CH2N(NO)Me

122 $C_4H_4Cl_2$ 821-10-3
2-Butyne, 1,4-dichloro-
ClCH2C≡CCH2Cl

122 $C_4H_7ClO_2$ 105-39-5
Acetic acid, chloro-, ethyl ester
EtOC(O)CH2Cl

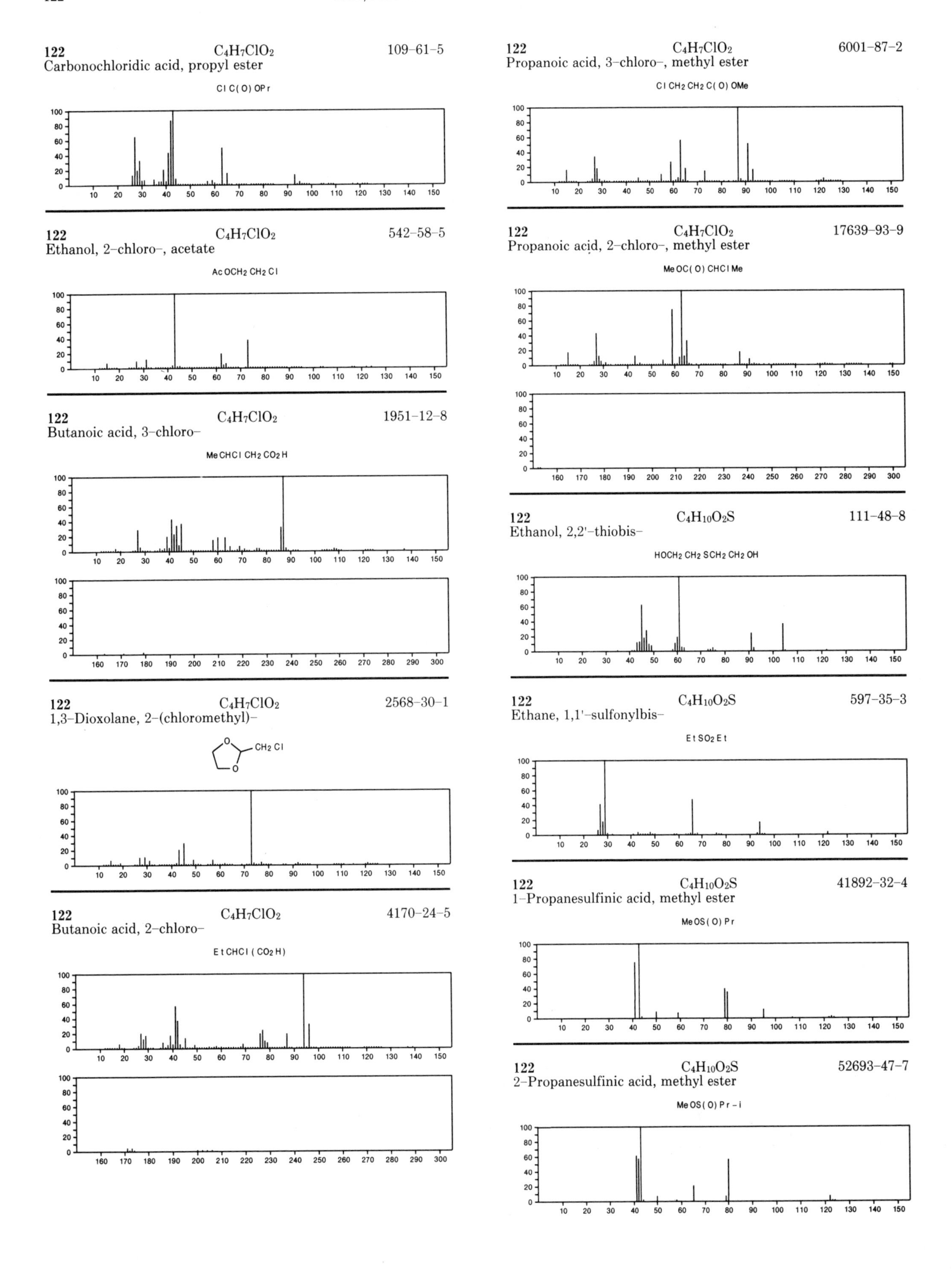

122 $C_4H_7ClO_2$ 109-61-5
Carbonochloridic acid, propyl ester
Cl C(O) OPr
122 $C_4H_7ClO_2$ 542-58-5
Ethanol, 2-chloro-, acetate
AcOCH2 CH2 Cl
122 $C_4H_7ClO_2$ 1951-12-8
Butanoic acid, 3-chloro-
MeCHCl CH2 CO2H
122 $C_4H_7ClO_2$ 2568-30-1
1,3-Dioxolane, 2-(chloromethyl)-
CH2 Cl
122 $C_4H_7ClO_2$ 4170-24-5
Butanoic acid, 2-chloro-
EtCHCl (CO2H)
122 $C_4H_7ClO_2$ 6001-87-2
Propanoic acid, 3-chloro-, methyl ester
Cl CH2 CH2 C(O) OMe
122 $C_4H_7ClO_2$ 17639-93-9
Propanoic acid, 2-chloro-, methyl ester
MeOC(O) CHCl Me
122 $C_4H_{10}O_2S$ 111-48-8
Ethanol, 2,2'-thiobis-
HOCH2 CH2 SCH2 CH2 OH
122 $C_4H_{10}O_2S$ 597-35-3
Ethane, 1,1'-sulfonylbis-
Et SO2 Et
122 $C_4H_{10}O_2S$ 41892-32-4
1-Propanesulfinic acid, methyl ester
MeOS(O) Pr
122 $C_4H_{10}O_2S$ 52693-47-7
2-Propanesulfinic acid, methyl ester
MeOS(O) Pr-i

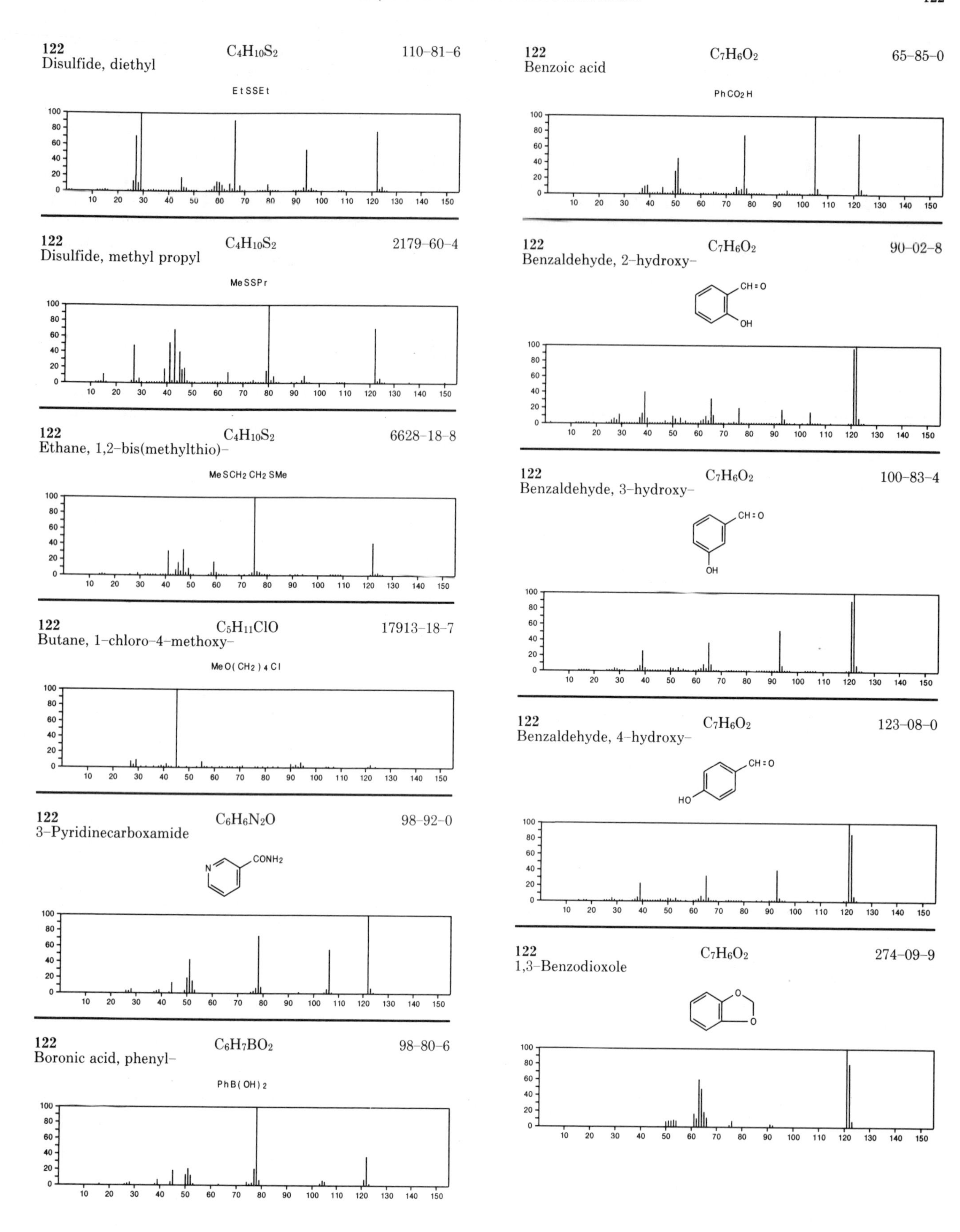
122
Disulfide, diethyl
$C_4H_{10}S_2$
110-81-6
EtSSEt
122
Disulfide, methyl propyl
$C_4H_{10}S_2$
2179-60-4
MeSSPr
122
Ethane, 1,2-bis(methylthio)-
$C_4H_{10}S_2$
6628-18-8
MeSCH2CH2SMe
122
Butane, 1-chloro-4-methoxy-
$C_5H_{11}ClO$
17913-18-7
MeO(CH2)4Cl
122
3-Pyridinecarboxamide
$C_6H_6N_2O$
98-92-0
CONH2
N
122
Boronic acid, phenyl-
$C_6H_7BO_2$
98-80-6
PhB(OH)2
122
Benzoic acid
$C_7H_6O_2$
65-85-0
PhCO2H
122
Benzaldehyde, 2-hydroxy-
$C_7H_6O_2$
90-02-8
CH=O
OH
122
Benzaldehyde, 3-hydroxy-
$C_7H_6O_2$
100-83-4
CH=O
OH
122
Benzaldehyde, 4-hydroxy-
$C_7H_6O_2$
123-08-0
CH=O
HO
122
1,3-Benzodioxole
$C_7H_6O_2$
274-09-9
O
O

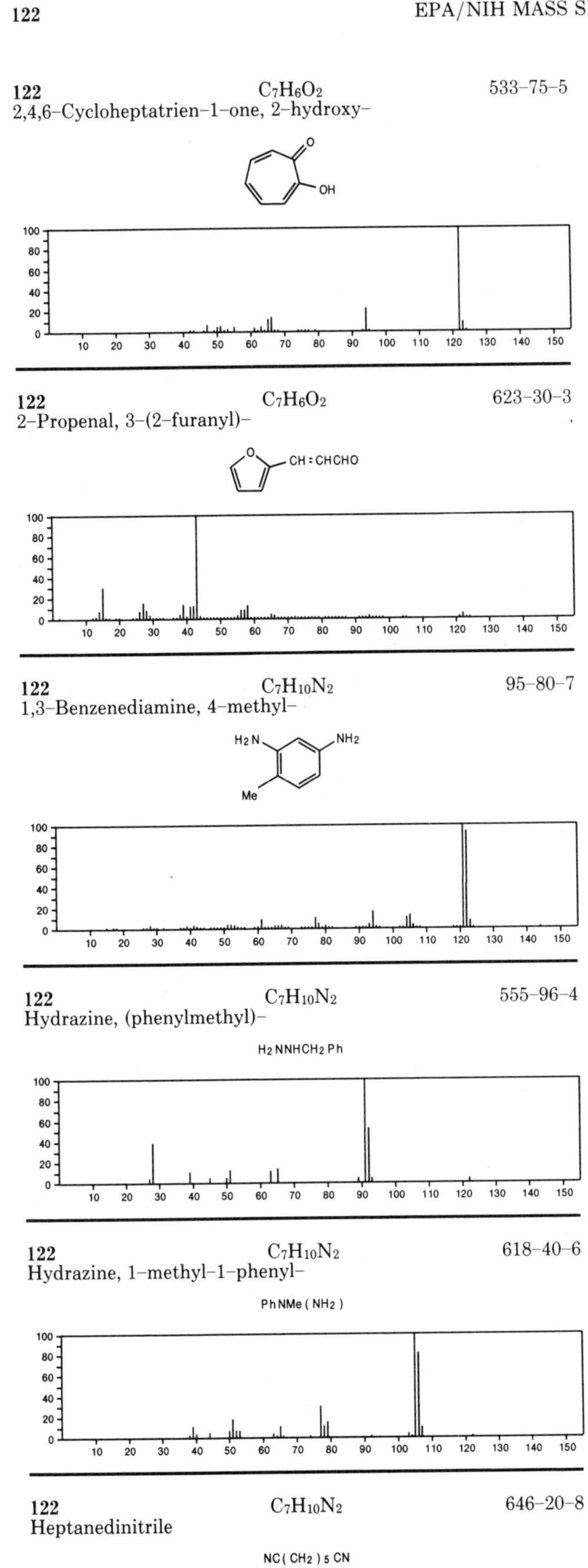

122 $C_7H_6O_2$ 533-75-5
2,4,6-Cycloheptatrien-1-one, 2-hydroxy-

122 $C_7H_6O_2$ 623-30-3
2-Propenal, 3-(2-furanyl)-

122 $C_7H_{10}N_2$ 95-80-7
1,3-Benzenediamine, 4-methyl-

122 $C_7H_{10}N_2$ 555-96-4
Hydrazine, (phenylmethyl)-

H2NNHCH2Ph

122 $C_7H_{10}N_2$ 618-40-6
Hydrazine, 1-methyl-1-phenyl-

PhNMe(NH2)

122 $C_7H_{10}N_2$ 646-20-8
Heptanedinitrile

NC(CH2)5CN

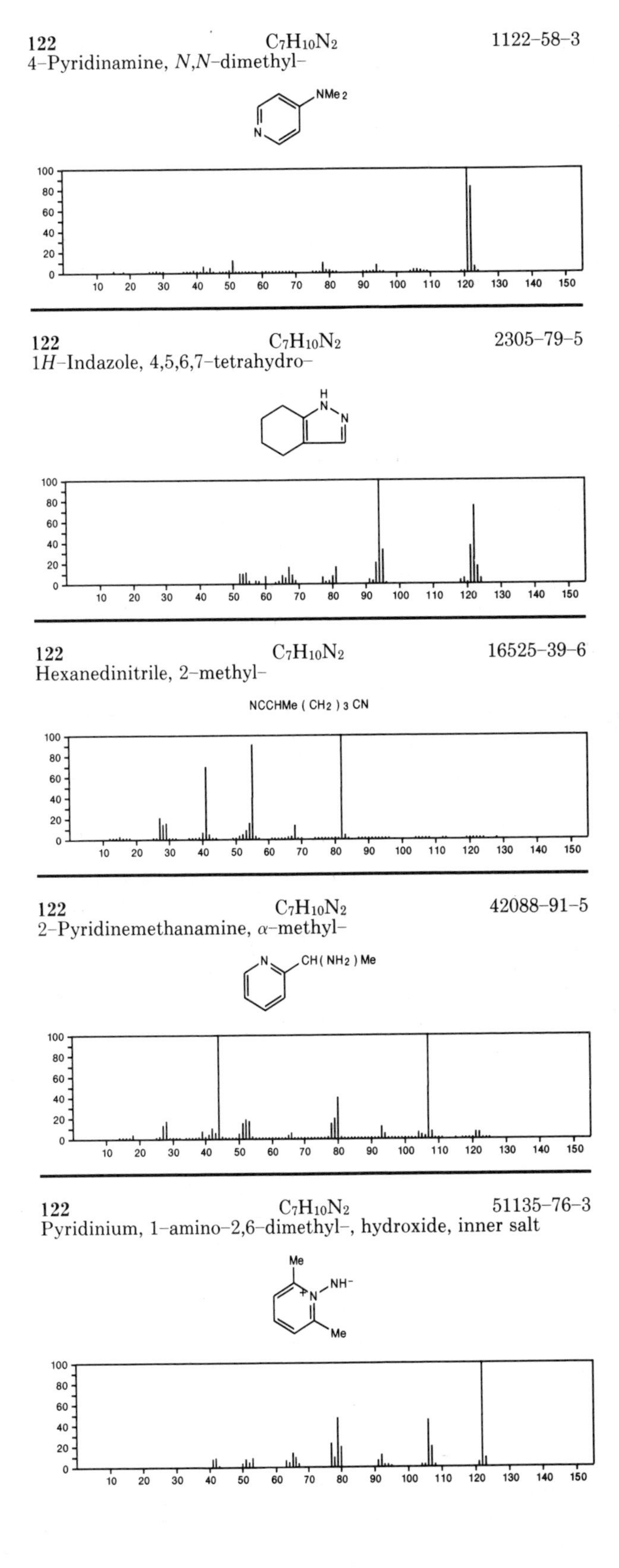

122 $C_7H_{10}N_2$ 1122-58-3
4-Pyridinamine, *N,N*-dimethyl-

122 $C_7H_{10}N_2$ 2305-79-5
1*H*-Indazole, 4,5,6,7-tetrahydro-

122 $C_7H_{10}N_2$ 16525-39-6
Hexanedinitrile, 2-methyl-

NCCHMe(CH2)3CN

122 $C_7H_{10}N_2$ 42088-91-5
2-Pyridinemethanamine, α-methyl-

122 $C_7H_{10}N_2$ 51135-76-3
Pyridinium, 1-amino-2,6-dimethyl-, hydroxide, inner salt

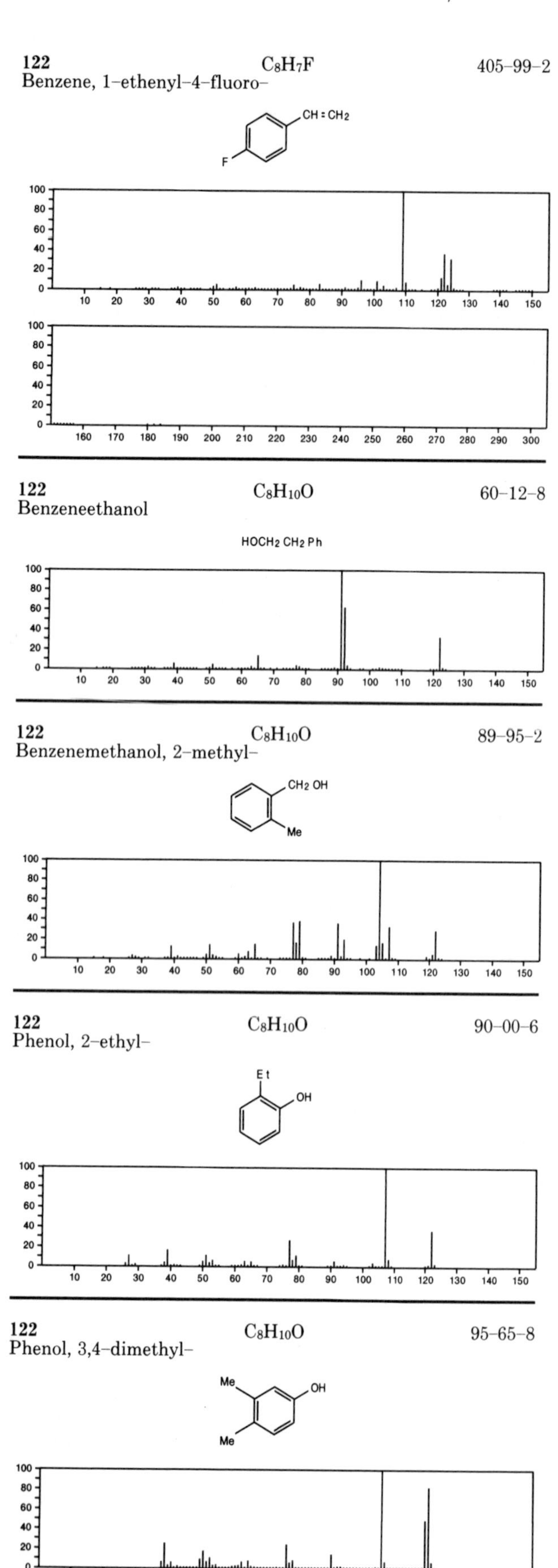

122 C_8H_7F 405-99-2
Benzene, 1-ethenyl-4-fluoro-
CH = CH2
F
122 $C_8H_{10}O$ 60-12-8
Benzeneethanol
HOCH2 CH2 Ph
122 $C_8H_{10}O$ 89-95-2
Benzenemethanol, 2-methyl-
CH2 OH
Me
122 $C_8H_{10}O$ 90-00-6
Phenol, 2-ethyl-
Et
OH
122 $C_8H_{10}O$ 95-65-8
Phenol, 3,4-dimethyl-
Me
OH
Me

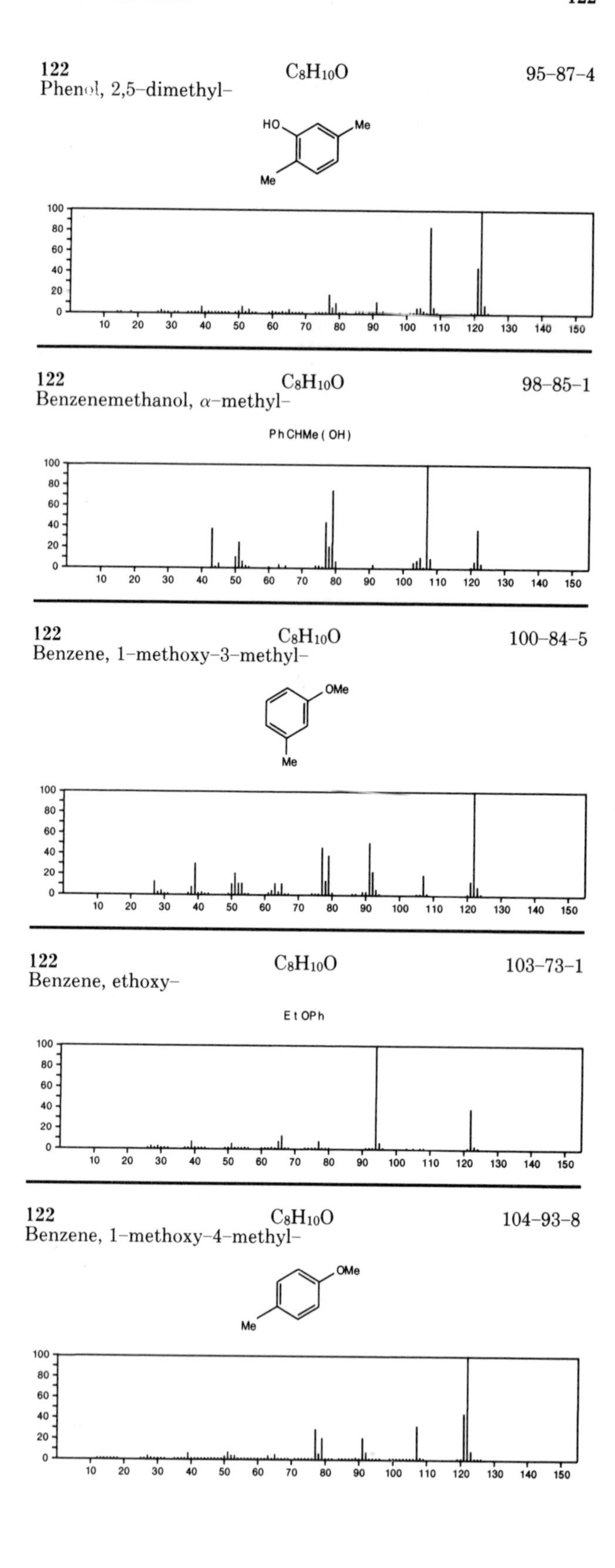

122 $C_8H_{10}O$ 95-87-4
Phenol, 2,5-dimethyl-
HO
Me
Me
122 $C_8H_{10}O$ 98-85-1
Benzenemethanol, α-methyl-
Ph CHMe (OH)
122 $C_8H_{10}O$ 100-84-5
Benzene, 1-methoxy-3-methyl-
OMe
Me
122 $C_8H_{10}O$ 103-73-1
Benzene, ethoxy-
Et OPh
122 $C_8H_{10}O$ 104-93-8
Benzene, 1-methoxy-4-methyl-
OMe
Me

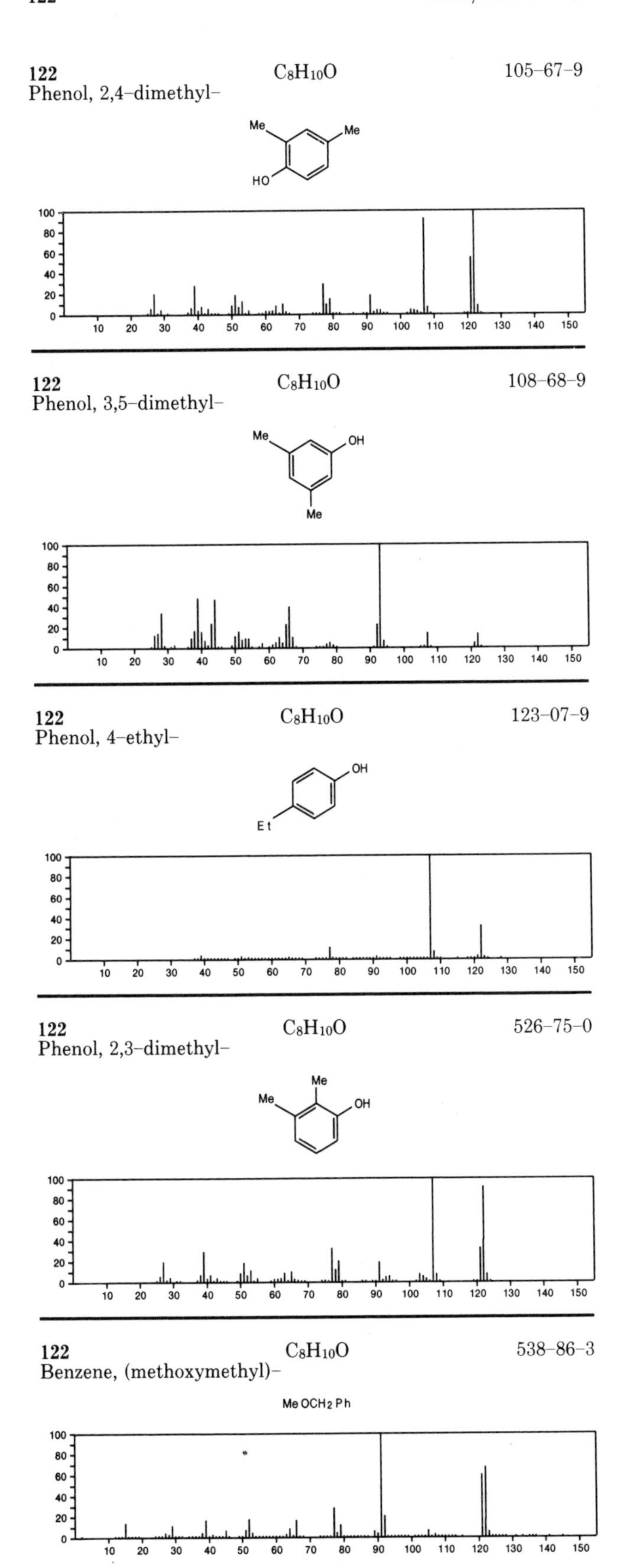

122 $C_8H_{10}O$ 105-67-9
Phenol, 2,4-dimethyl-

122 $C_8H_{10}O$ 108-68-9
Phenol, 3,5-dimethyl-

122 $C_8H_{10}O$ 123-07-9
Phenol, 4-ethyl-

122 $C_8H_{10}O$ 526-75-0
Phenol, 2,3-dimethyl-

122 $C_8H_{10}O$ 538-86-3
Benzene, (methoxymethyl)-

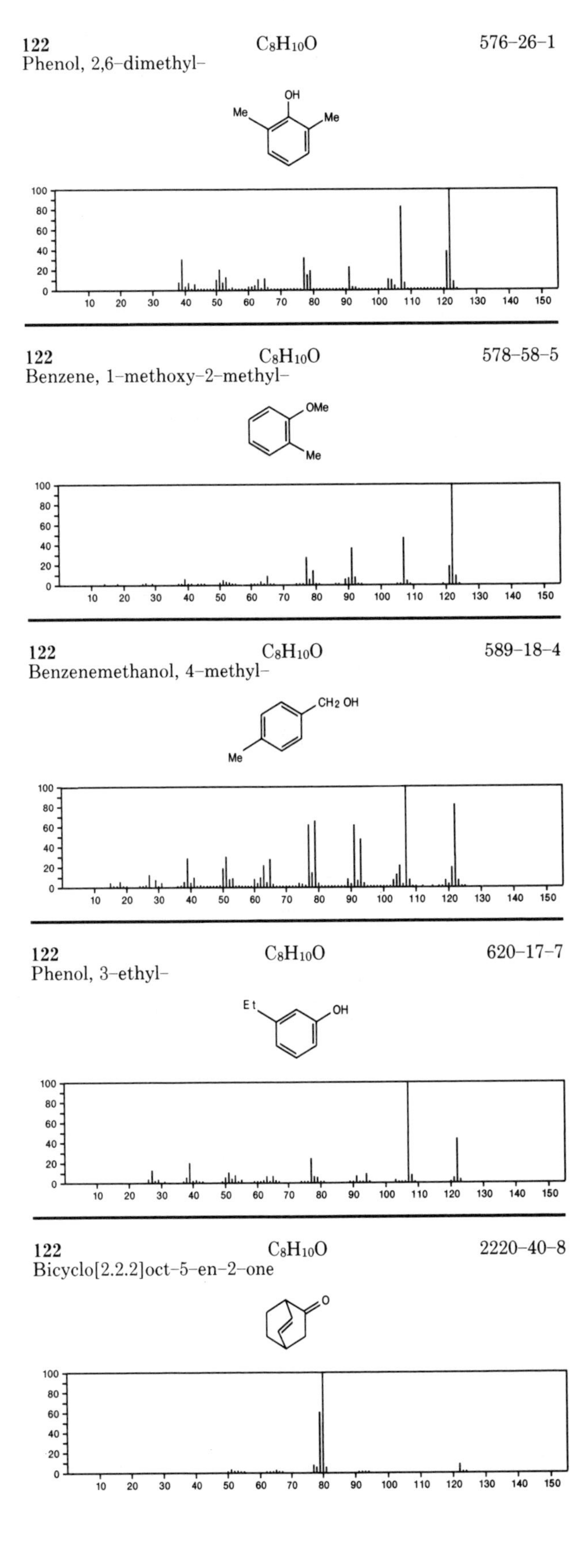

122 $C_8H_{10}O$ 576-26-1
Phenol, 2,6-dimethyl-

122 $C_8H_{10}O$ 578-58-5
Benzene, 1-methoxy-2-methyl-

122 $C_8H_{10}O$ 589-18-4
Benzenemethanol, 4-methyl-

122 $C_8H_{10}O$ 620-17-7
Phenol, 3-ethyl-

122 $C_8H_{10}O$ 2220-40-8
Bicyclo[2.2.2]oct-5-en-2-one

122 $C_8H_{10}O$ 19004-82-1
Tricyclo[4.2.0.0^{2,4}]octan-5-one, (1α,2β,4β,6a)-

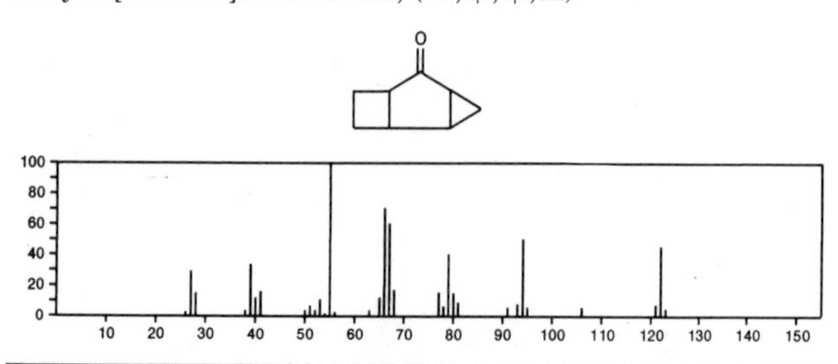

122 $C_8H_{10}O$ 19093-14-2
Tricyclo[4.2.0.0^{2,4}]octan-5-one

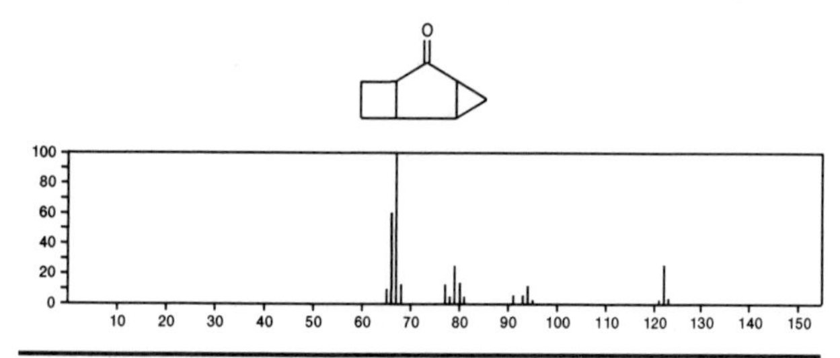

122 C_9H_{14} 529-16-8
Bicyclo[2.2.1]hept-2-ene, 2,3-dimethyl-

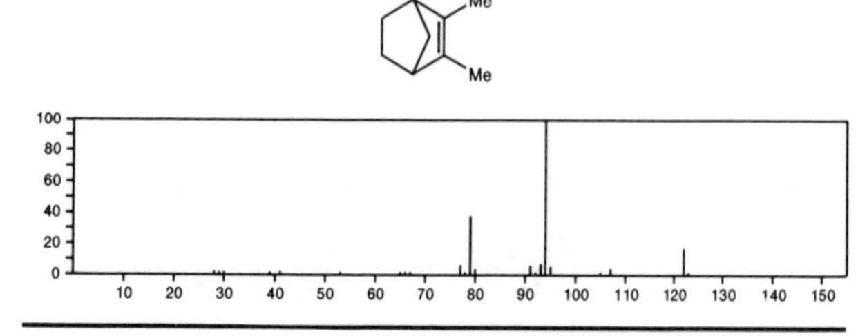

122 C_9H_{14} 5664-17-5
Cyclohexane, 1,2-propadienyl-

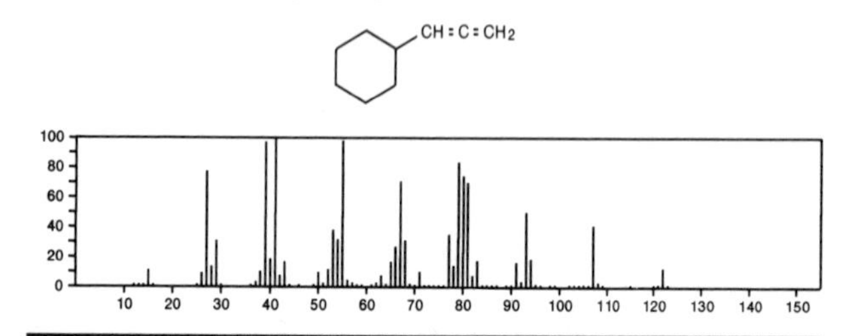

122 C_9H_{14} 22704-00-3
Cyclobutane, 1,2-diethenyl-3-methyl-

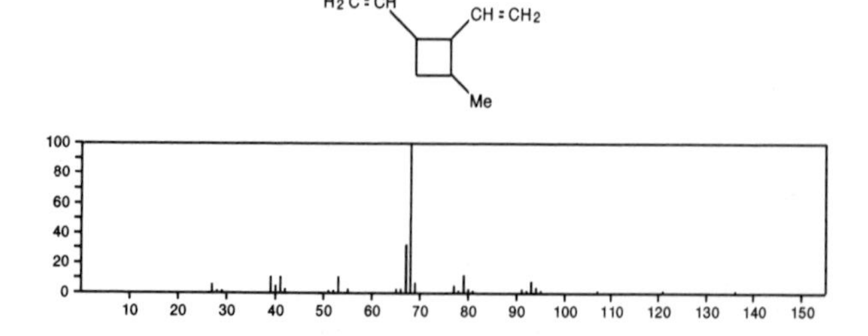

122 C_9H_{14} 24524-58-1
Bicyclo[3.1.0]hexane, 6-isopropylidene-

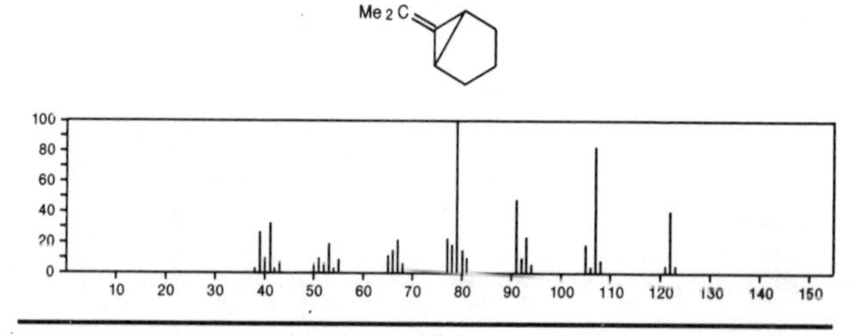

122 CrF_2O_2 7788-96-7
Chromium, difluorodioxo-

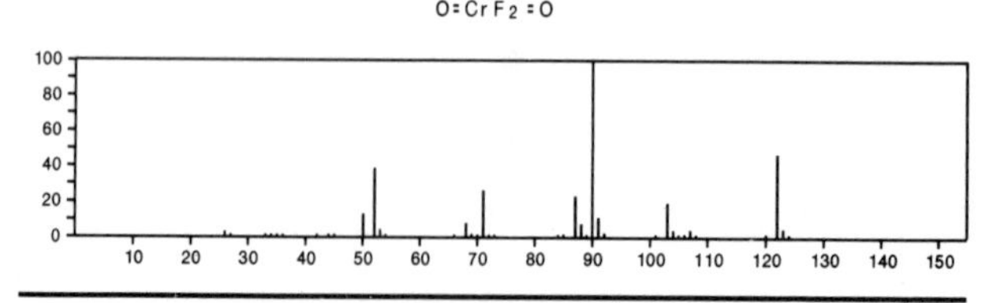

123 $C_3H_9NO_2S$ 918-05-8
Methanesulfonamide, *N*,*N*-dimethyl-

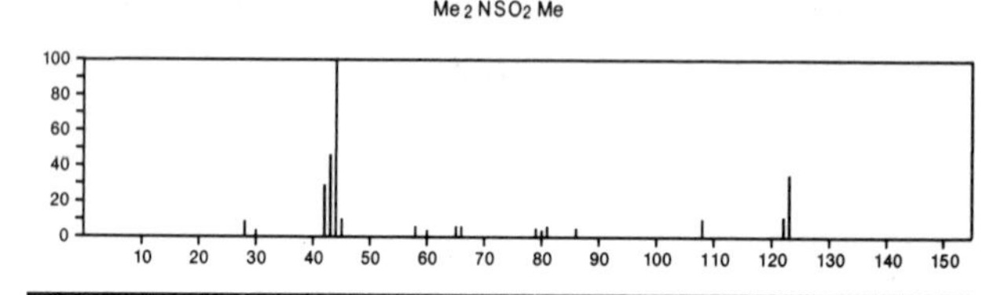

123 $C_3H_{12}B_3N_3$ 1004-35-9
Borazine, 1,3,5-trimethyl-

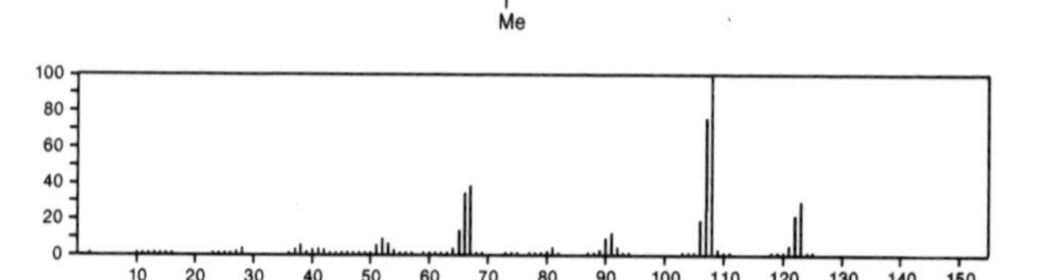

123 $C_3H_{12}B_3N_3$ 5314-85-2
Borazine, 2,4,6-trimethyl-

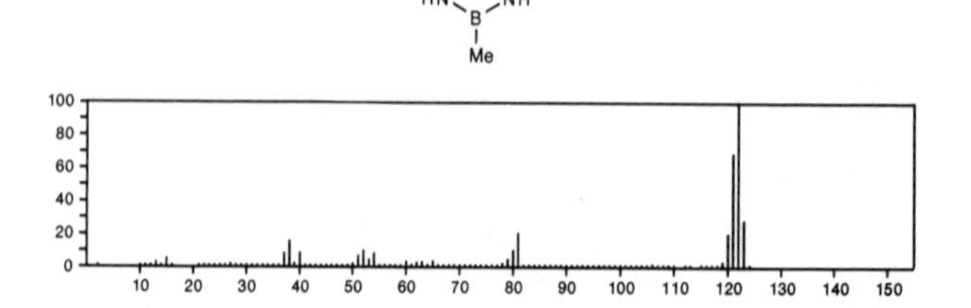

123 $C_4H_4F_3N$ 690-95-9
Butyronitrile, 4,4,4-trifluoro-

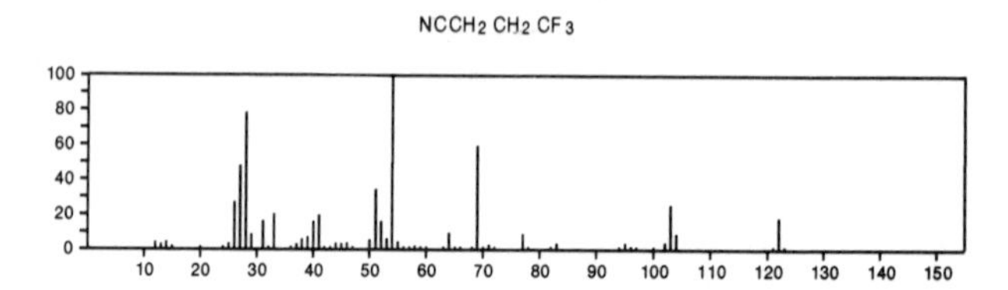

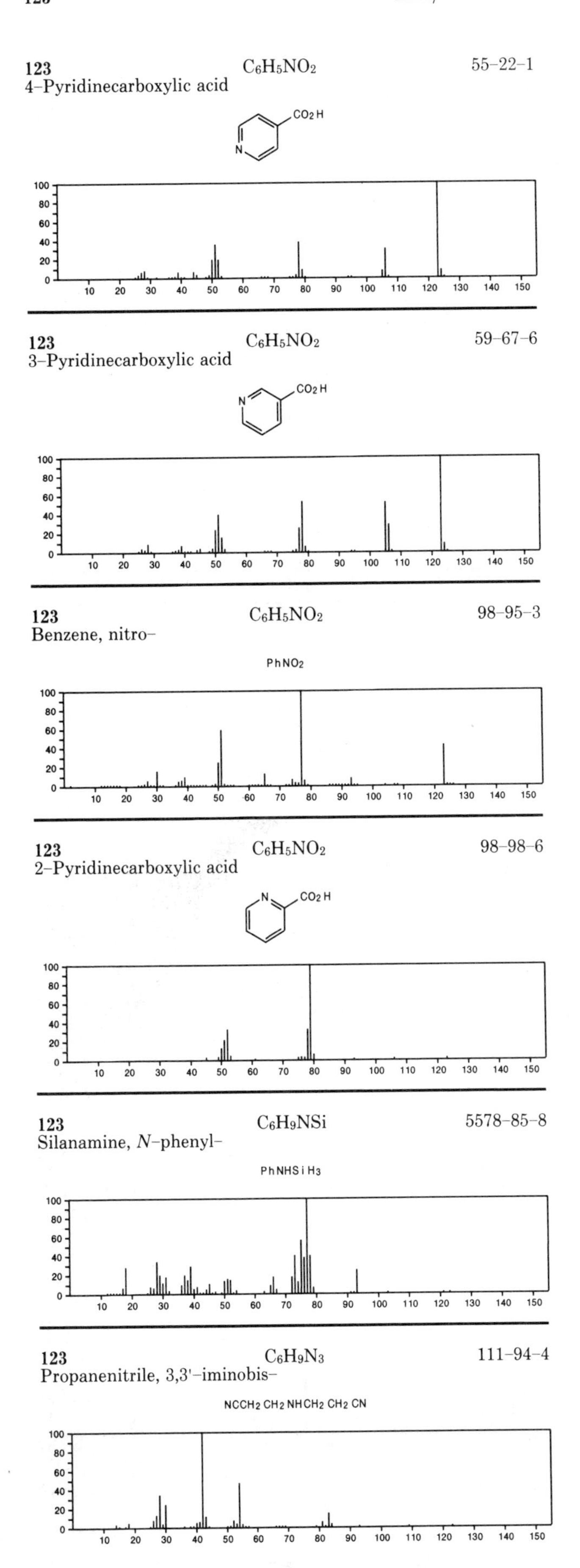

123 $C_6H_5NO_2$ 55-22-1
4-Pyridinecarboxylic acid

123 $C_6H_5NO_2$ 59-67-6
3-Pyridinecarboxylic acid

123 $C_6H_5NO_2$ 98-95-3
Benzene, nitro-
PhNO2

123 $C_6H_5NO_2$ 98-98-6
2-Pyridinecarboxylic acid

123 C_6H_9NSi 5578-85-8
Silanamine, *N*-phenyl-
PhNHSiH3

123 $C_6H_9N_3$ 111-94-4
Propanenitrile, 3,3'-iminobis-
NCCH2CH2NHCH2CH2CN

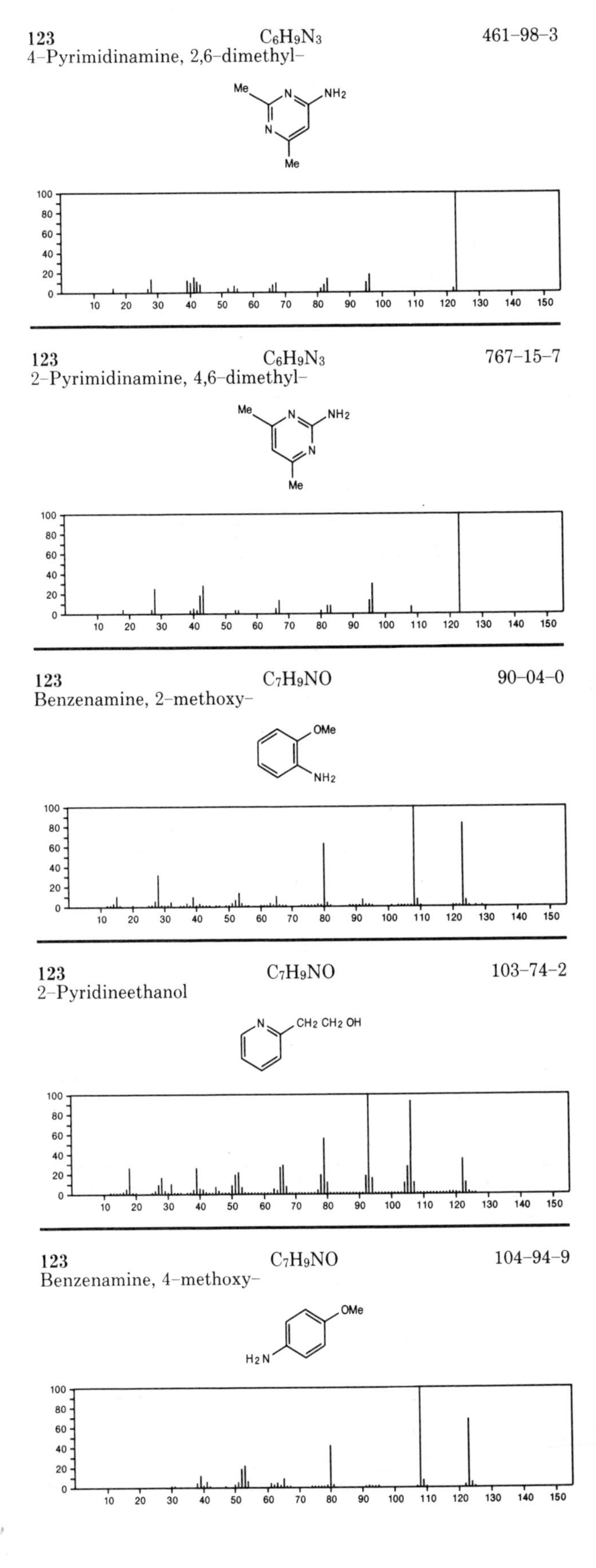

123 $C_6H_9N_3$ 461-98-3
4-Pyrimidinamine, 2,6-dimethyl-

123 $C_6H_9N_3$ 767-15-7
2-Pyrimidinamine, 4,6-dimethyl-

123 C_7H_9NO 90-04-0
Benzenamine, 2-methoxy-

123 C_7H_9NO 103-74-2
2-Pyridineethanol

123 C_7H_9NO 104-94-9
Benzenamine, 4-methoxy-

123 C_7H_9NO 536-90-3
Benzenamine, 3-methoxy-

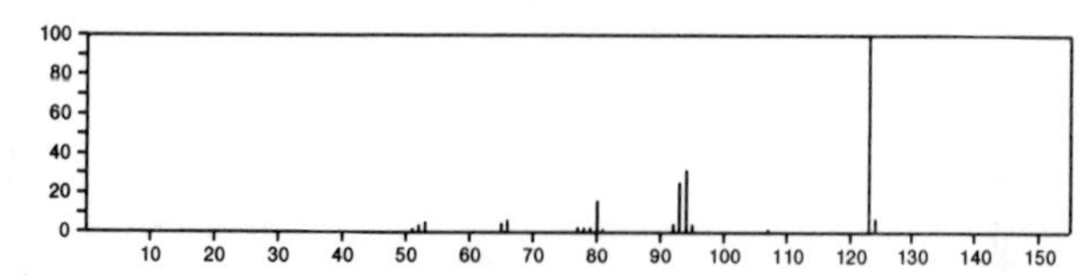

123 C_7H_9NO 622-33-3
Hydroxylamine, *O*-(phenylmethyl)-

H_2NOCH_2Ph

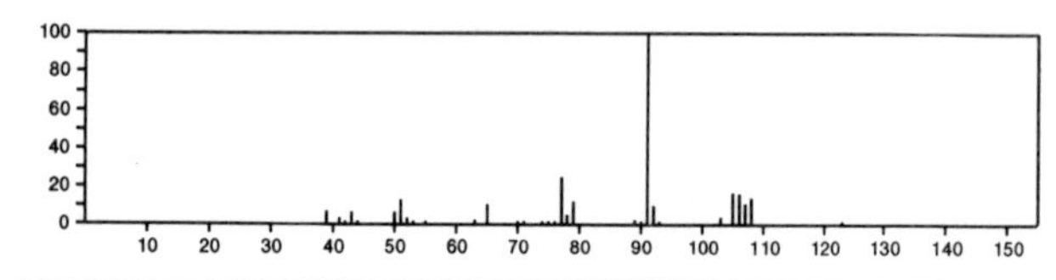

123 C_7H_9NO 623-04-1
Benzenemethanol, 4-amino-

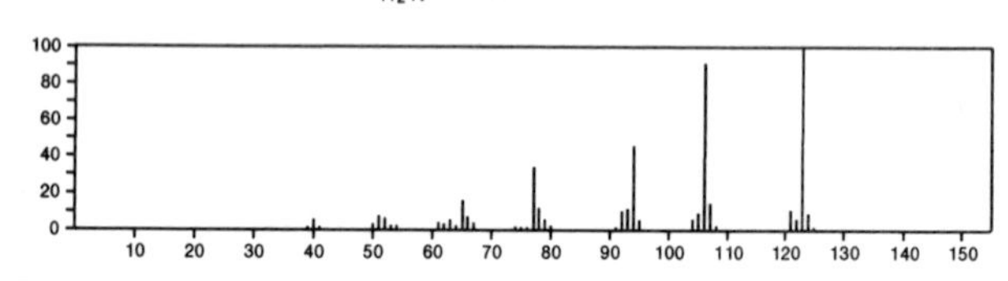

123 C_7H_9NO 1877-77-6
Benzenemethanol, 3-amino-

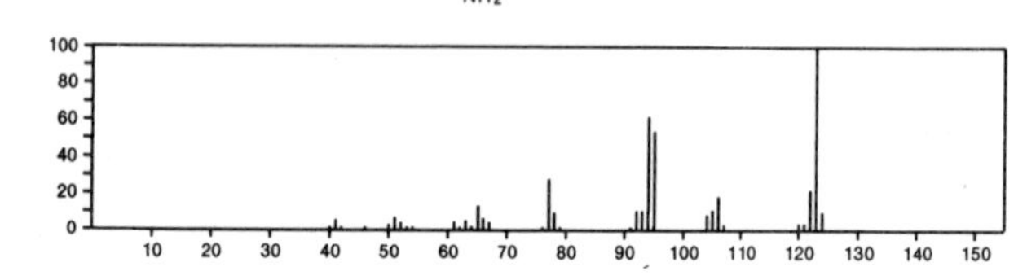

123 C_7H_9NO 2836-00-2
Phenol, 3-amino-4-methyl-

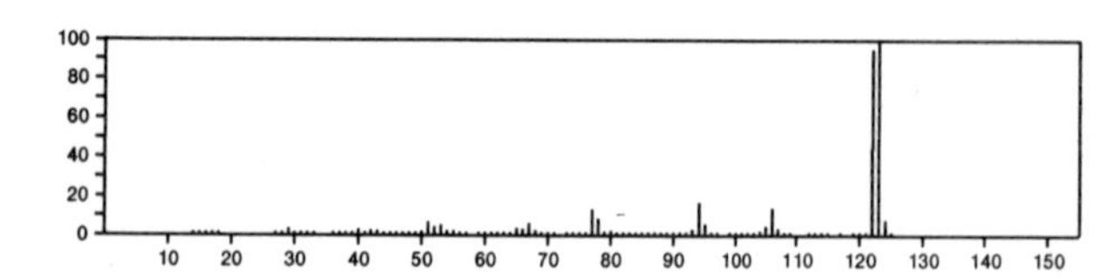

123 C_7H_9NO 4833-24-3
Pyridine, 2-ethyl-, 1-oxide

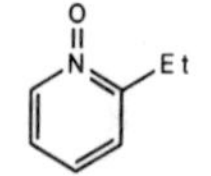

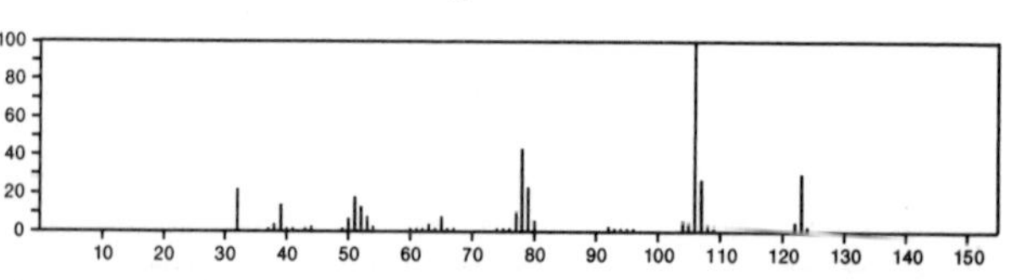

123 C_7H_9NO 5344-90-1
Benzenemethanol, 2-amino-

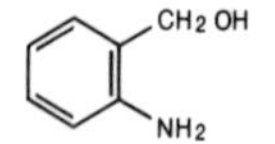

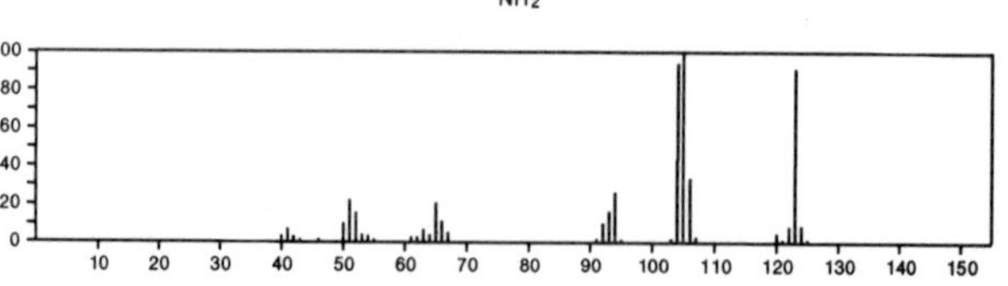

123 C_7H_9NO 6456-92-4
2(1*H*)-Pyridinone, 1,3-dimethyl-

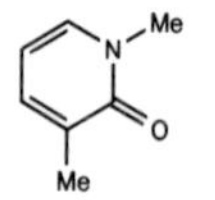

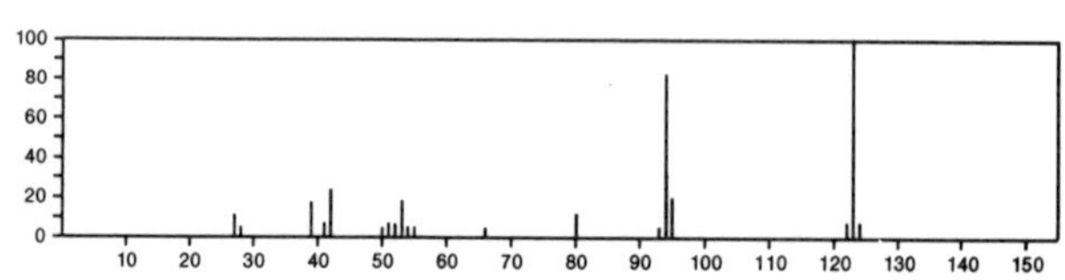

123 C_7H_9NO 6456-93-5
2(1*H*)-Pyridinone, 1,5-dimethyl-

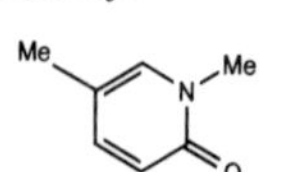

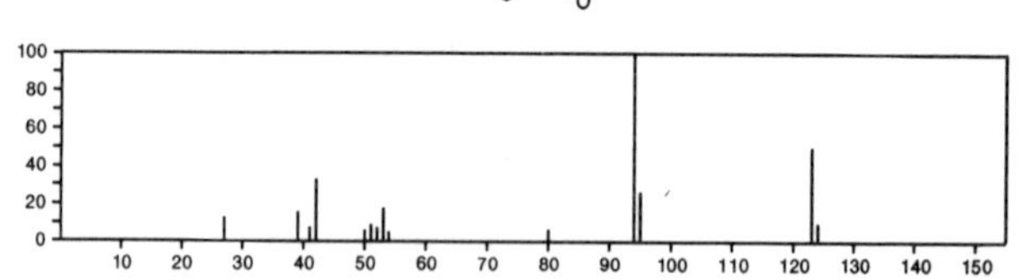

123 C_7H_9NO 13337-79-6
2(1*H*)-Pyridinone, 1-ethyl-

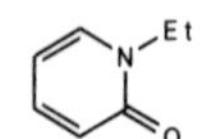

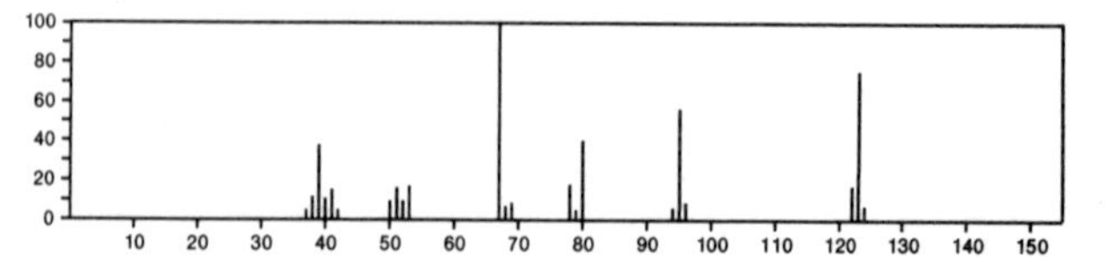

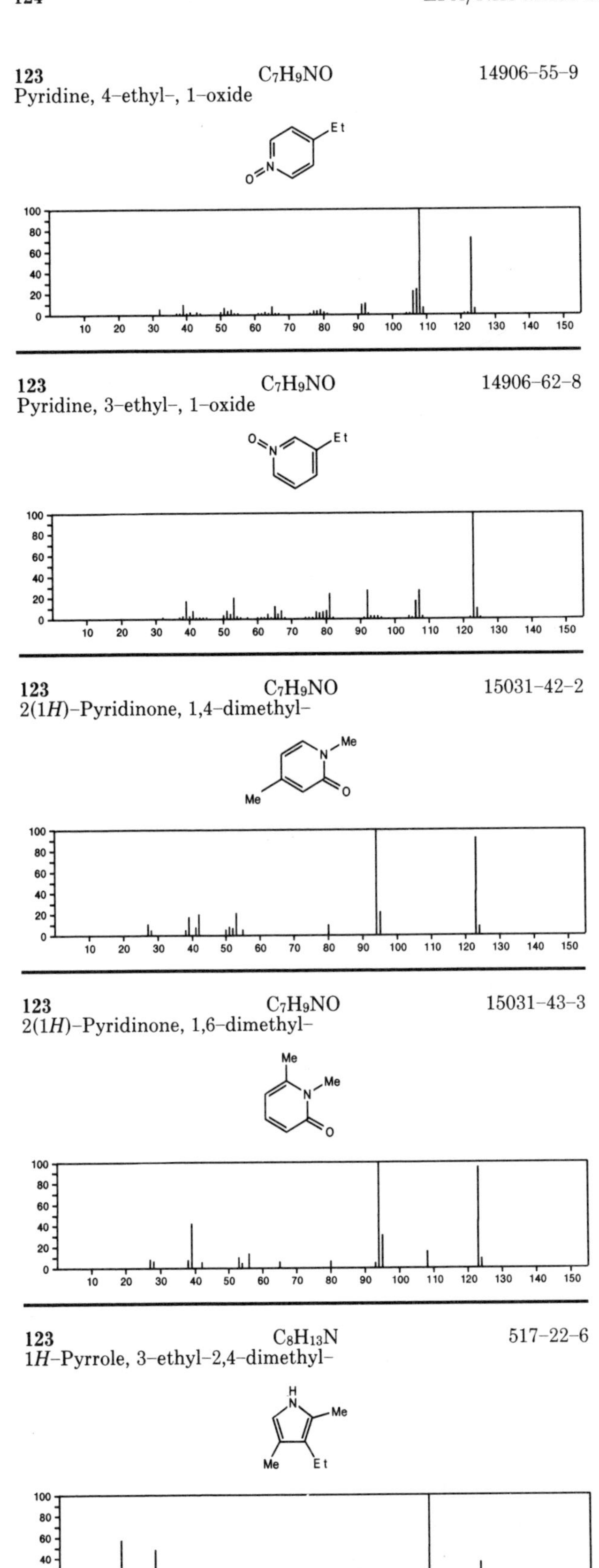

123 C_7H_9NO 14906–55–9
Pyridine, 4–ethyl–, 1–oxide

123 C_7H_9NO 14906–62–8
Pyridine, 3–ethyl–, 1–oxide

123 C_7H_9NO 15031–42–2
2(1*H*)–Pyridinone, 1,4–dimethyl–

123 C_7H_9NO 15031–43–3
2(1*H*)–Pyridinone, 1,6–dimethyl–

123 $C_8H_{13}N$ 517–22–6
1*H*–Pyrrole, 3–ethyl–2,4–dimethyl–

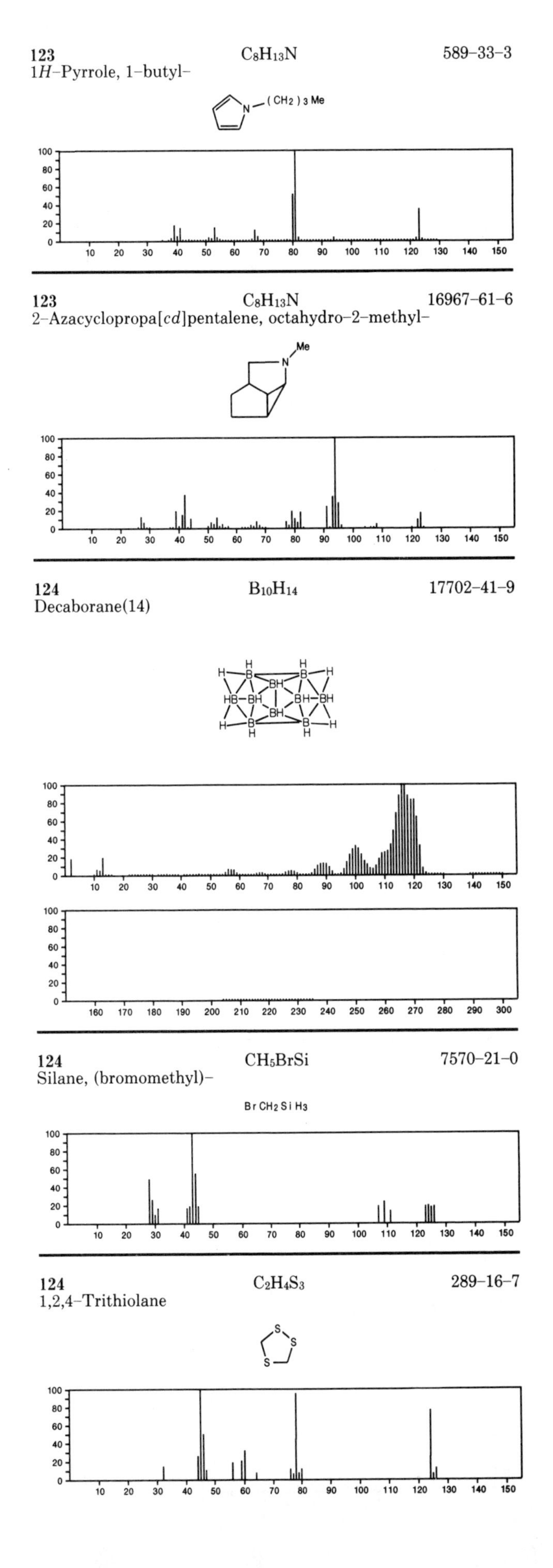

123 $C_8H_{13}N$ 589–33–3
1*H*–Pyrrole, 1–butyl–

123 $C_8H_{13}N$ 16967–61–6
2–Azacyclopropa[*cd*]pentalene, octahydro–2–methyl–

124 $B_{10}H_{14}$ 17702–41–9
Decaborane(14)

124 CH_5BrSi 7570–21–0
Silane, (bromomethyl)–

124 $C_2H_4S_3$ 289–16–7
1,2,4–Trithiolane

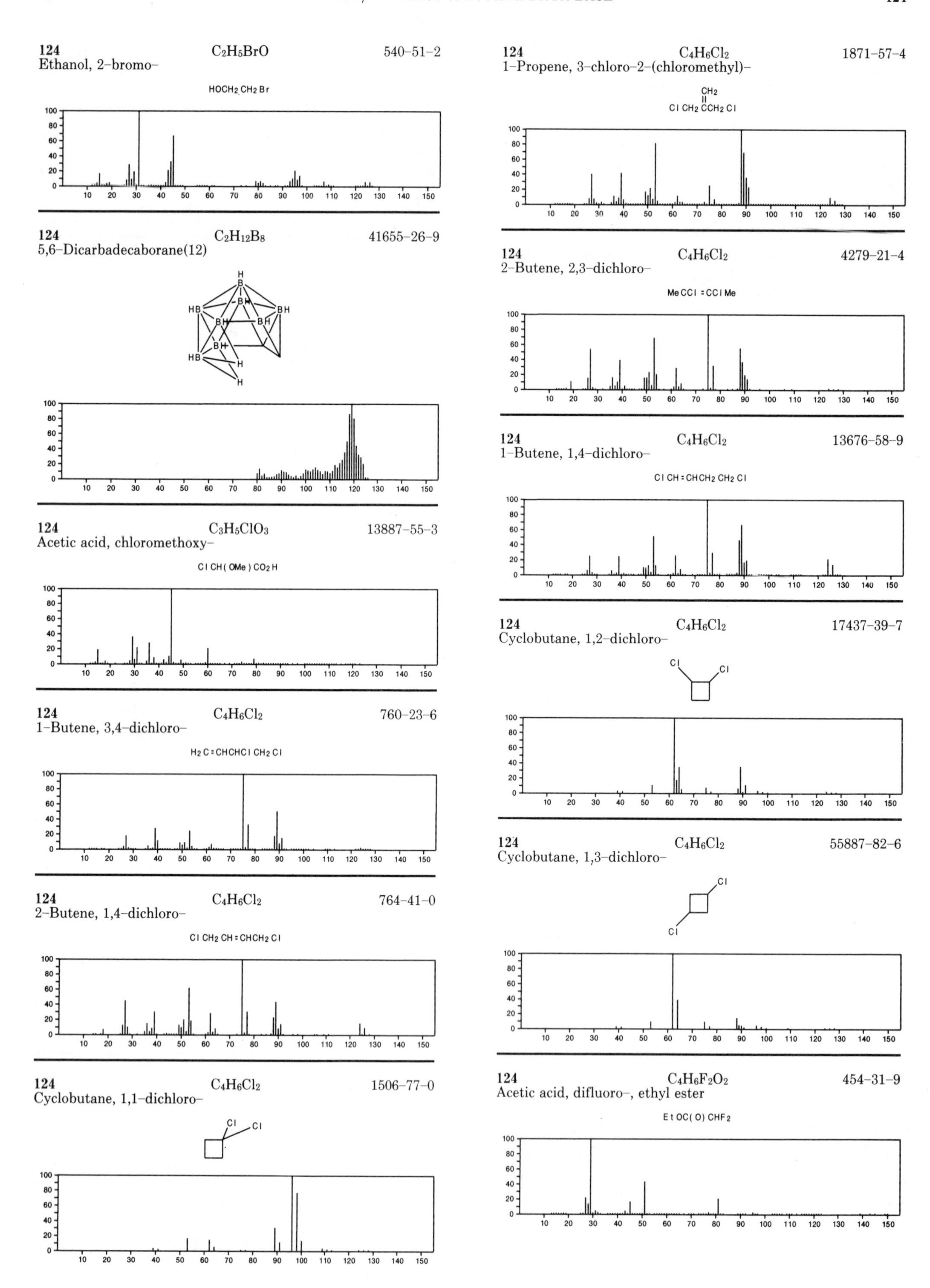
124 C2H5BrO 540-51-2
Ethanol, 2-bromo-
HOCH2 CH2 Br
124 C2H12B8 41655-26-9
5,6-Dicarbadecaborane(12)
124 C3H5ClO3 13887-55-3
Acetic acid, chloromethoxy-
Cl CH(OMe) CO2 H
124 C4H6Cl2 760-23-6
1-Butene, 3,4-dichloro-
H2 C=CHCHCl CH2 Cl
124 C4H6Cl2 764-41-0
2-Butene, 1,4-dichloro-
Cl CH2 CH=CHCH2 Cl
124 C4H6Cl2 1506-77-0
Cyclobutane, 1,1-dichloro-
124 C4H6Cl2 1871-57-4
1-Propene, 3-chloro-2-(chloromethyl)-
CH2
Cl CH2 CCH2 Cl
124 C4H6Cl2 4279-21-4
2-Butene, 2,3-dichloro-
MeCCl =CClMe
124 C4H6Cl2 13676-58-9
1-Butene, 1,4-dichloro-
Cl CH=CHCH2 CH2 Cl
124 C4H6Cl2 17437-39-7
Cyclobutane, 1,2-dichloro-
124 C4H6Cl2 55887-82-6
Cyclobutane, 1,3-dichloro-
124 C4H6F2O2 454-31-9
Acetic acid, difluoro-, ethyl ester
Et OC(O) CHF2

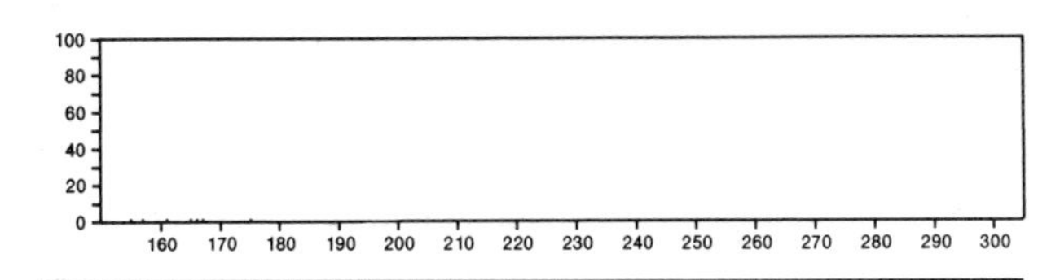

124 $C_4H_6F_2O_2$ 36301-44-7
1,3-Dioxane, 5,5-difluoro-

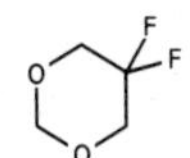

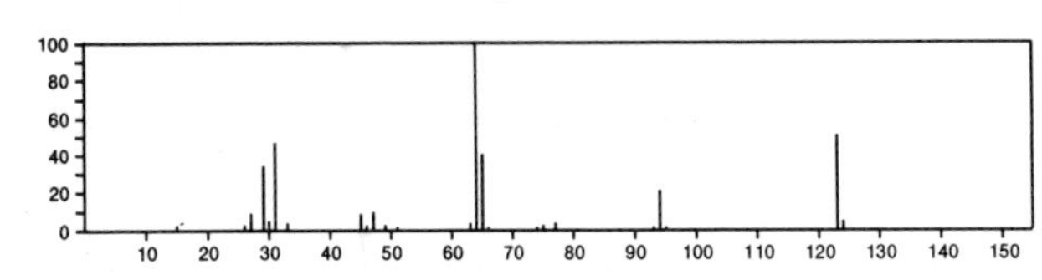

124 $C_4H_9ClO_2$ 97-97-2
Ethane, 2-chloro-1,1-dimethoxy-

Cl CH2 CH(OMe)2

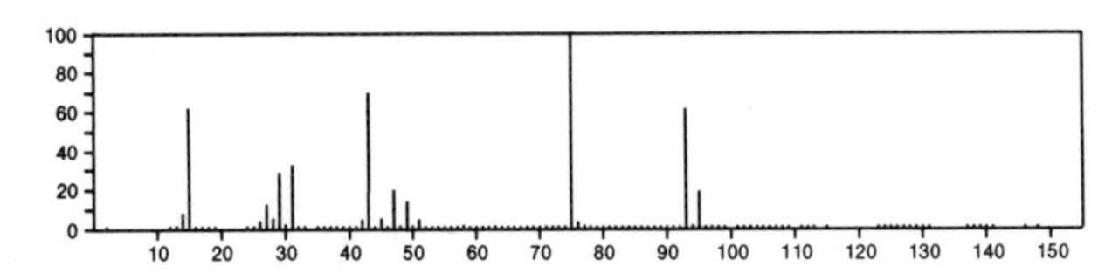

124 $C_4H_9ClO_2$ 628-89-7
Ethanol, 2-(2-chloroethoxy)-

Cl CH2 CH2 OCH2 CH2 OH

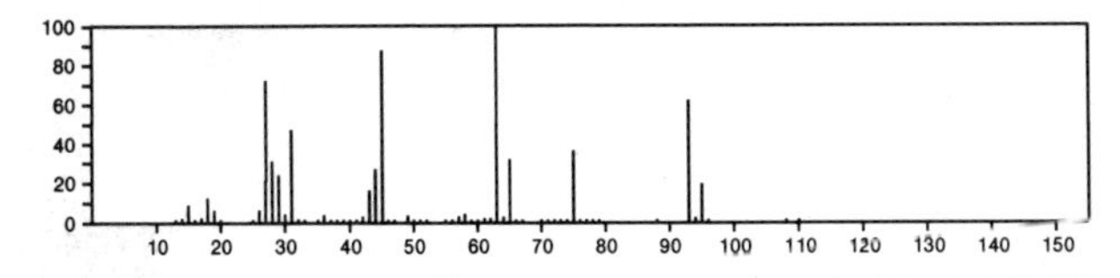

124 $C_4H_{12}N_2.ClH$ 7400-27-3
Hydrazine, (1,1-dimethylethyl)-, monohydrochloride

H2NNHBu-t • HCl

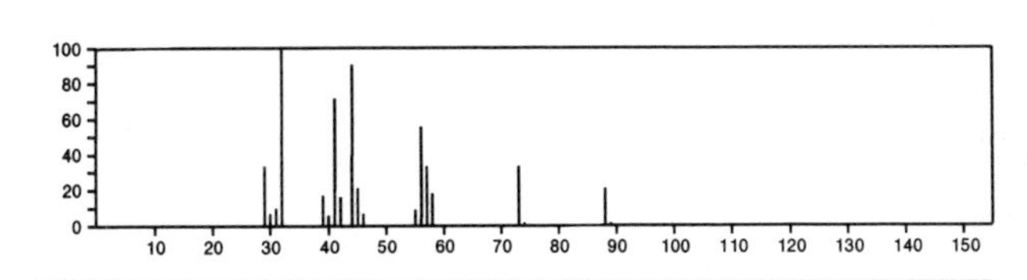

124 $C_5H_8N_4$ 7465-48-7
Tetrazolo[1,5-*a*]pyridine, 5,6,7,8-tetrahydro-

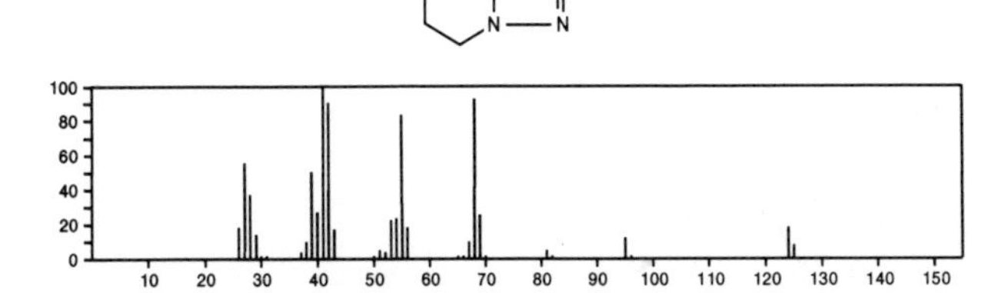

124 $C_5H_8N_4$ 22715-28-2
Pyrimidine, 4,5-diamino-6-methyl-

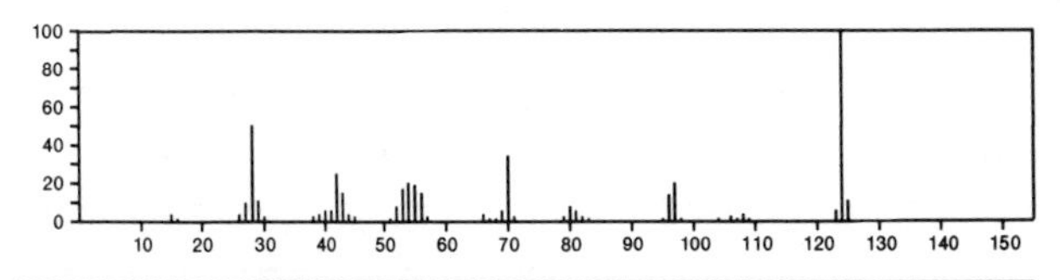

124 $C_6H_8N_2O$ 1656-48-0
Propanenitrile, 3,3'-oxybis-

NCCH2 CH2 OCH2 CH2 CN

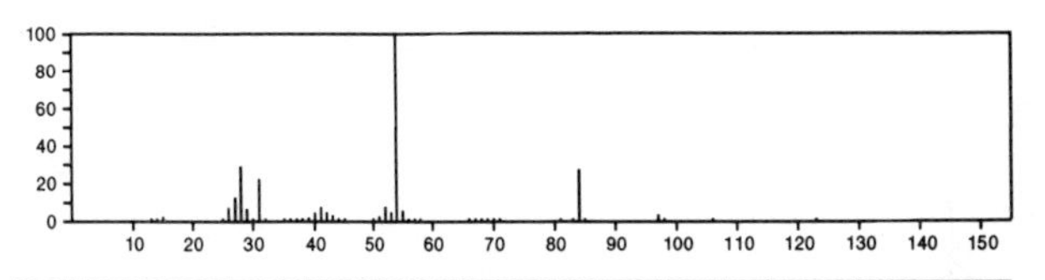

124 $C_6H_8N_2O$ 2258-73-3
Benzofurazan, 4,5,6,7-tetrahydro-

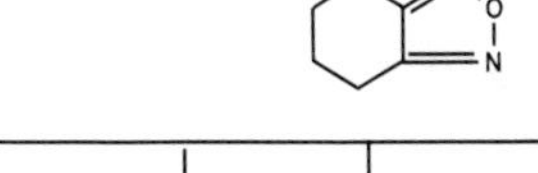

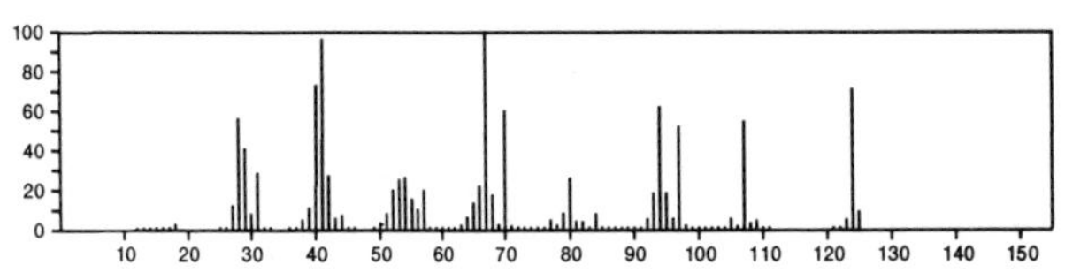

124 $C_6H_8N_2O$ 2524-90-5
Ethanone, 1-(4-methyl-1*H*-imidazol-2-yl)-

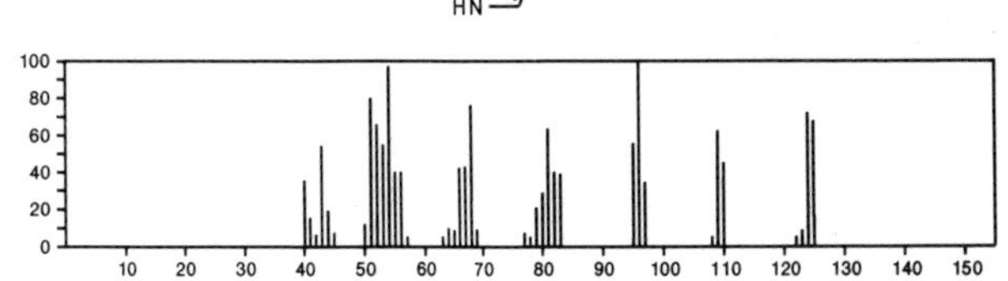

124 $C_6H_8N_2O$ 2847-30-5
Pyrazine, 2-methoxy-3-methyl-

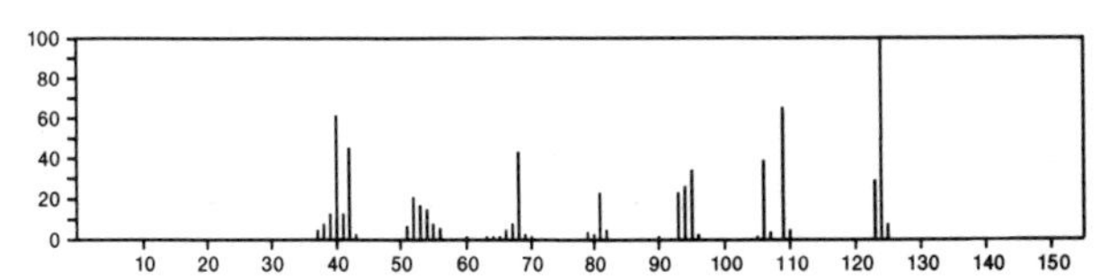

124 $C_6H_8N_2O$ 2882-21-5
Pyrazine, 2-methoxy-6-methyl-

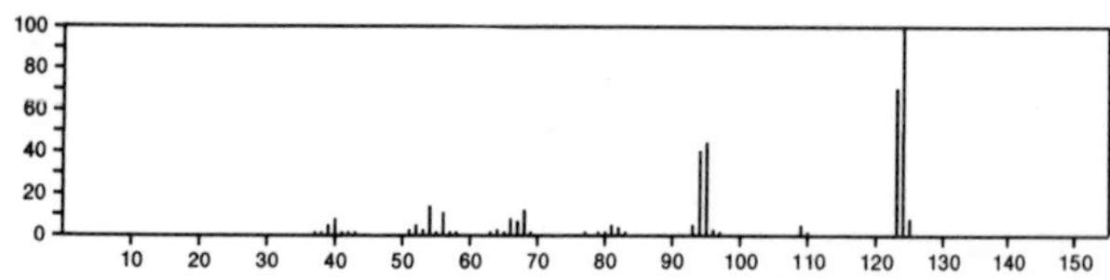

124 $C_6H_8N_2O$ 6622-92-0
4(1*H*)-Pyrimidinone, 2,6-dimethyl-

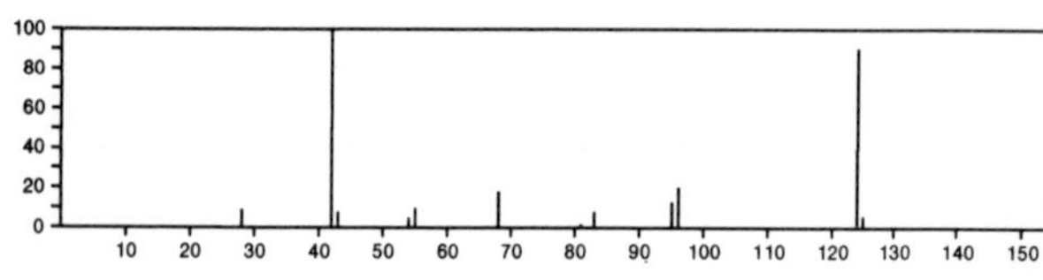

124 $C_6H_8N_2O$ 7314-65-0
Pyrimidine, 4-methoxy-2-methyl-

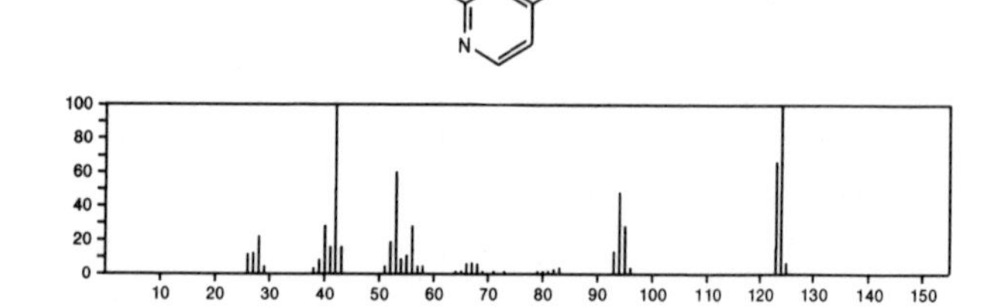

124 $C_6H_8N_2O$ 17758-38-2
4(3*H*)-Pyrimidinone, 2,3-dimethyl-

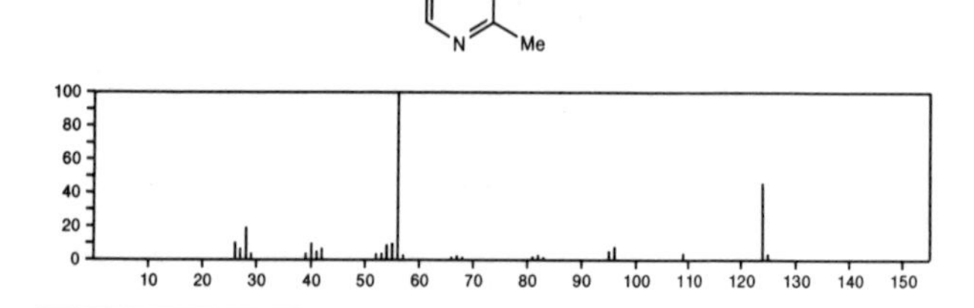

124 C_7H_5FO 455-32-3
Benzoyl fluoride

FCOPh

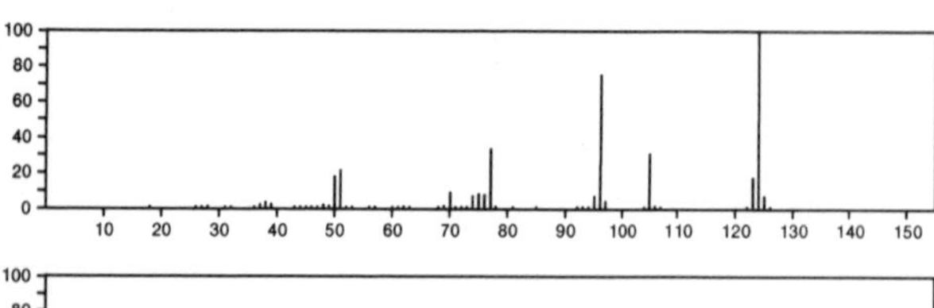

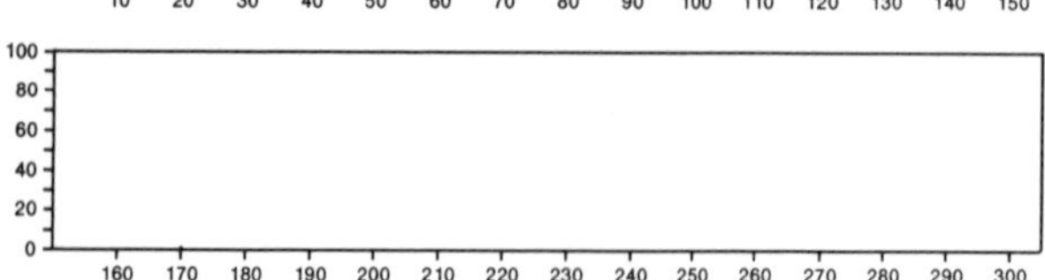

124 $C_7H_8O_2$ 90-01-7
Benzenemethanol, 2-hydroxy-

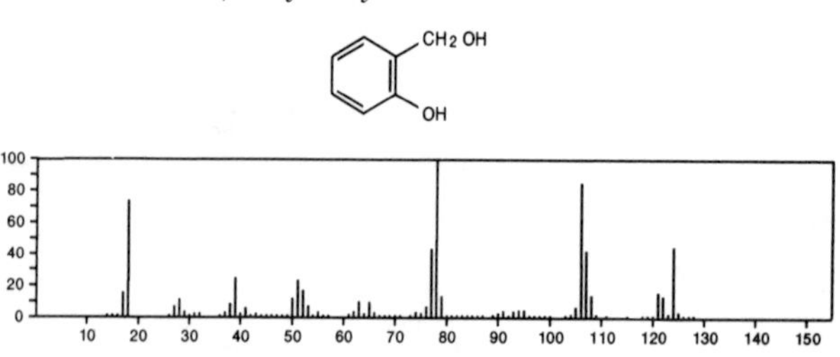

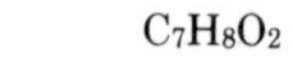

124 $C_7H_8O_2$ 90-05-1
Phenol, 2-methoxy-

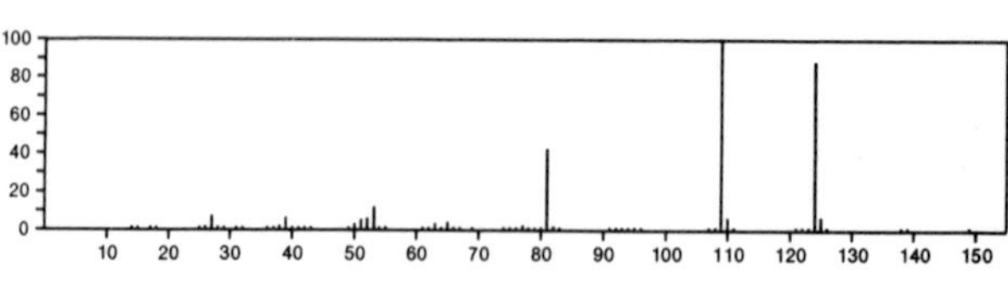

124 $C_7H_8O_2$ 150-19-6
Phenol, 3-methoxy-

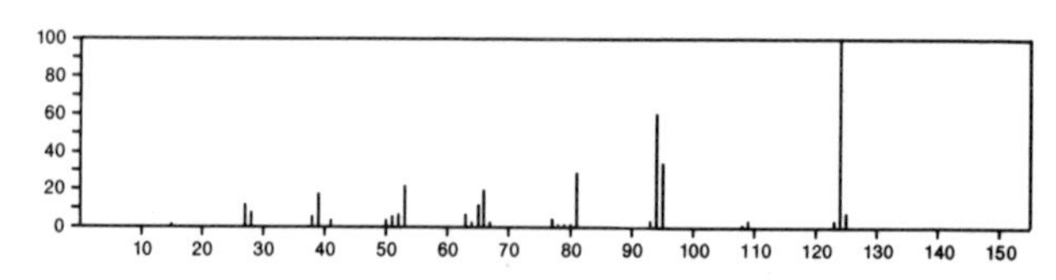

124 $C_7H_8O_2$ 150-76-5
Phenol, 4-methoxy-

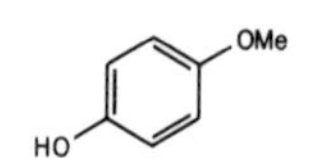

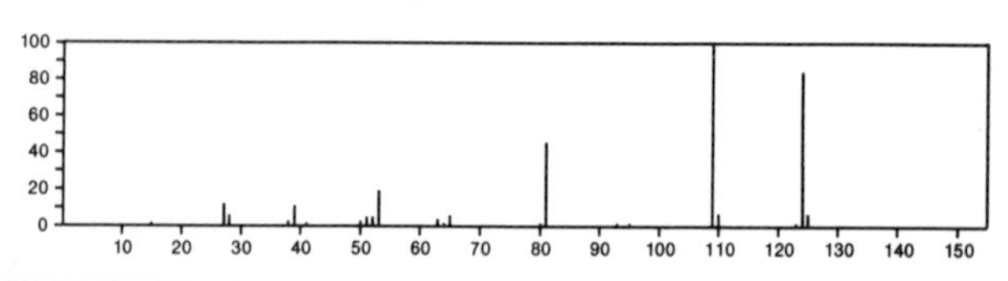

124 $C_7H_8O_2$ 488-17-5
1,2-Benzenediol, 3-methyl-

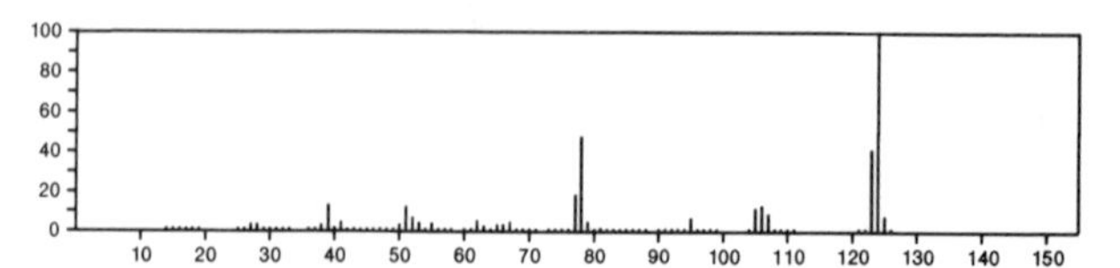

124 $C_7H_8O_2$ 504–15–4
1,3–Benzenediol, 5–methyl–

Me OH OH

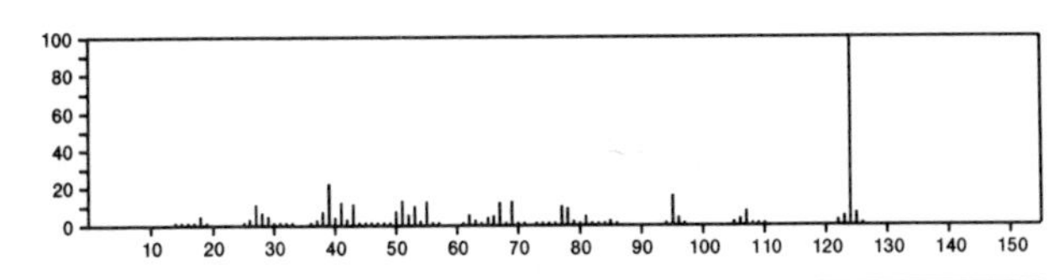

124 $C_7H_8O_2$ 608–25–3
1,3–Benzenediol, 2–methyl–

Me HO OH

100 80 60 40 20 0
10 20 30 40 50 60 70 80 90 100 110 120 130 140 150

124 $C_7H_8O_2$ 620–24–6
Benzenemethanol, 3–hydroxy–

CH2 OH OH

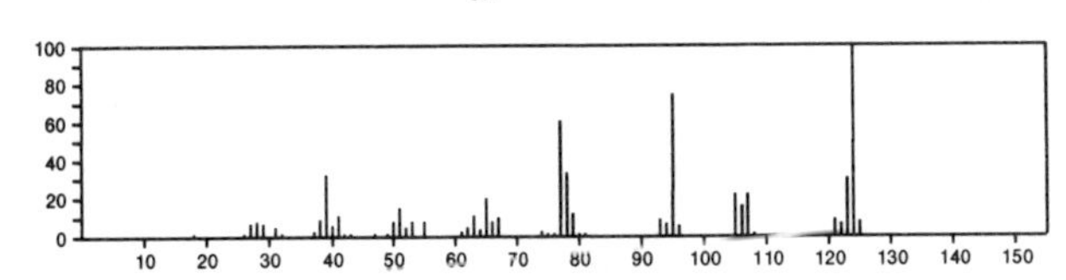

124 $C_7H_8O_2$ 623–05–2
Benzenemethanol, 4–hydroxy–

CH2 OH HO

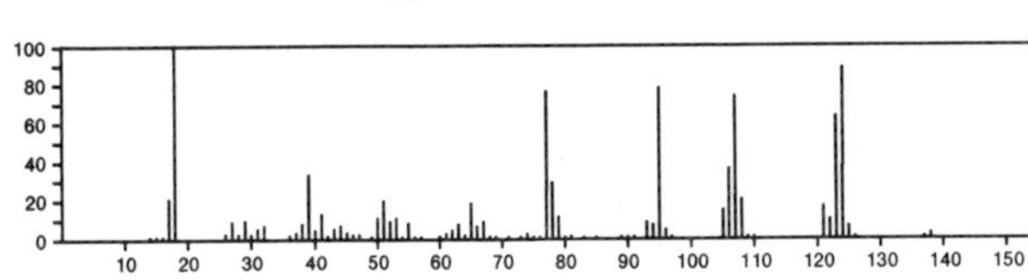

124 $C_7H_8O_2$ 675–09–2
2*H*–Pyran–2–one, 4,6–dimethyl–

Me Me O O

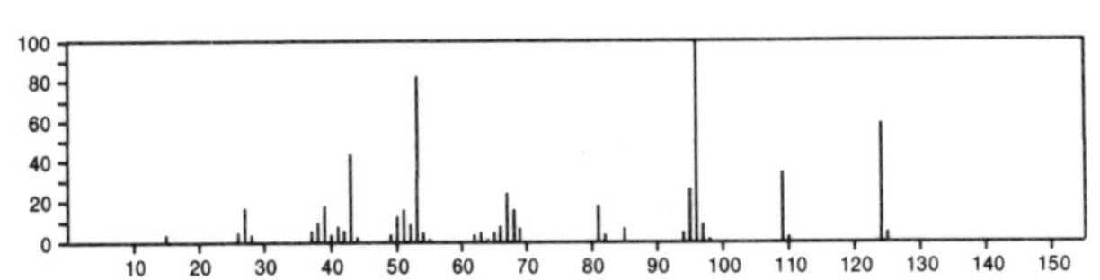

124 $C_7H_8O_2$ 1004–36–0
4*H*–Pyran–4–one, 2,6–dimethyl–

Me O Me O

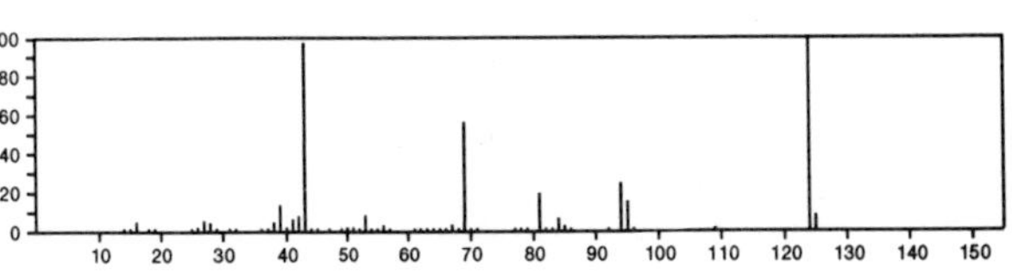

124 $C_7H_8O_2$ 3194–15–8
1–Propanone, 1–(2–furanyl)–

O COEt

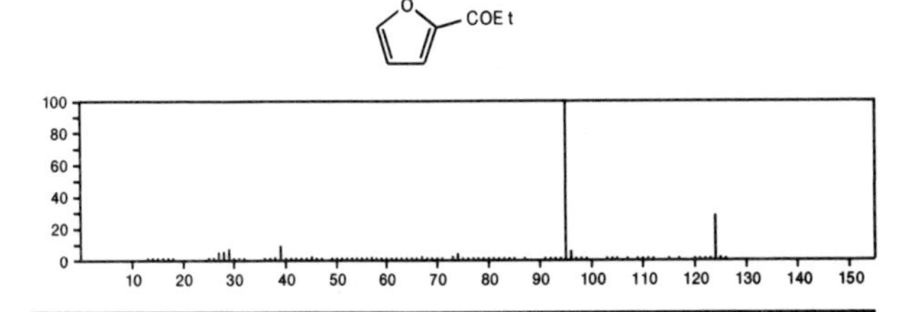

124 $C_7H_8O_2$ 19083–61–5
4*H*–Pyran–4–one, 3,5–dimethyl–

O Me O Me

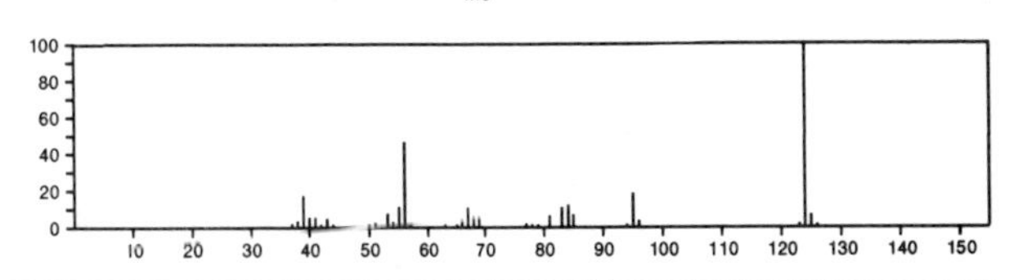

124 C_7H_8S 100–53–8
Benzenemethanethiol

PhCH2SH

100 80 60 40 20 0
10 20 30 40 50 60 70 80 90 100 110 120 130 140 150

124 C_7H_8S 100–68–5
Benzene, (methylthio)–

MeSPh

100 80 60 40 20 0
10 20 30 40 50 60 70 80 90 100 110 120 130 140 150

124 $C_7H_{12}N_2$ 17629-26-4
1*H*-Pyrazole, 1-ethyl-3,5-dimethyl-

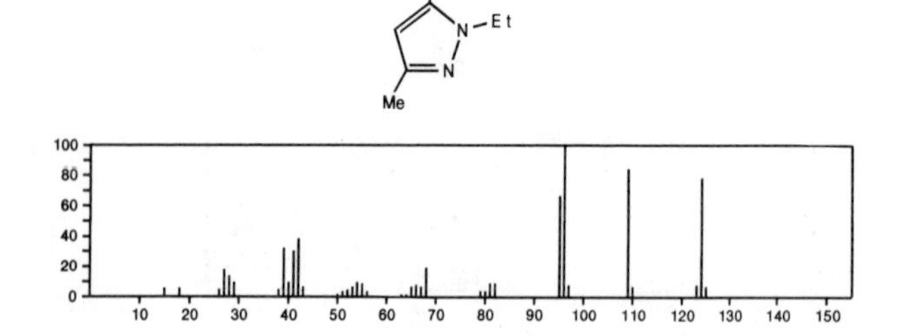

124 C_8H_9F 459-47-2
Benzene, 1-ethyl-4-fluoro-

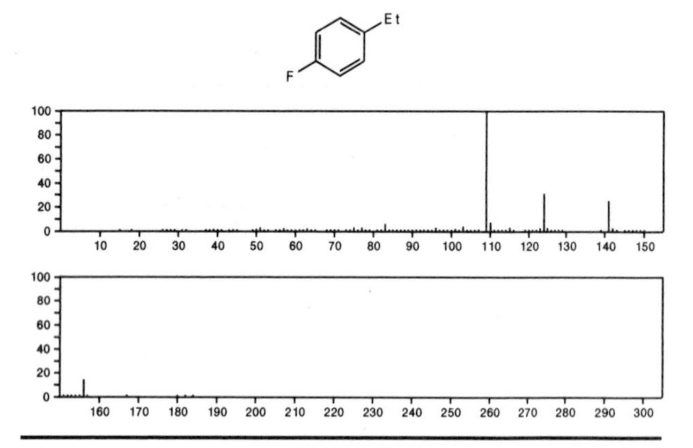

124 $C_8H_{12}O$ 78-27-3
Cyclohexanol, 1-ethynyl-

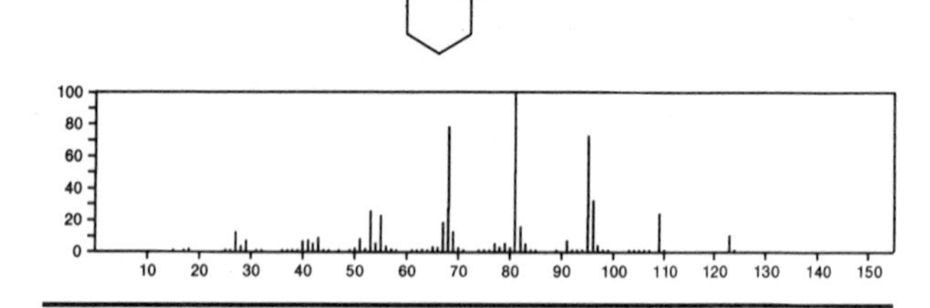

124 $C_8H_{12}O$ 278-84-2
3-Oxatricyclo[3.2.2.0^{2,4}]nonane

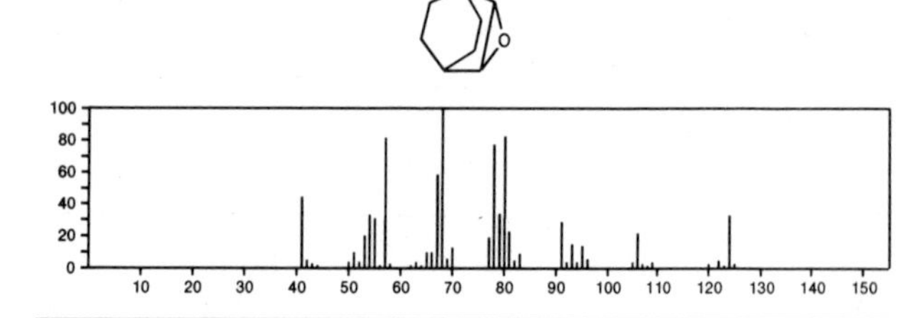

124 $C_8H_{12}O$ 637-90-1
9-Oxabicyclo[6.1.0]non-4-ene

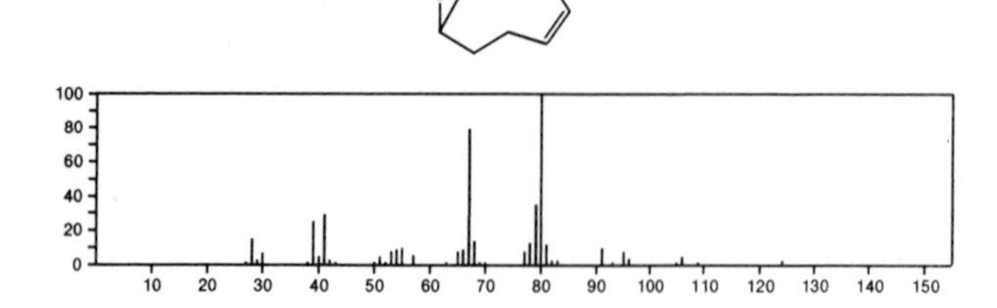

124 $C_8H_{12}O$ 931-96-4
3-Cyclohexene-1-carboxaldehyde, 1-methyl-

124 $C_8H_{12}O$ 932-66-1
Ethanone, 1-(1-cyclohexen-1-yl)-

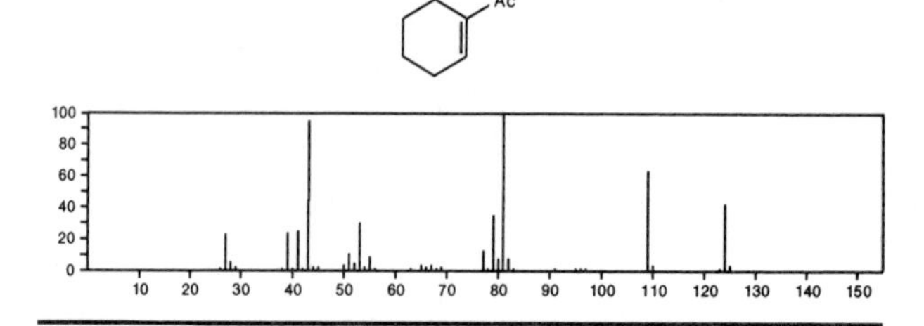

124 $C_8H_{12}O$ 1073-13-8
2-Cyclohexen-1-one, 4,4-dimethyl-

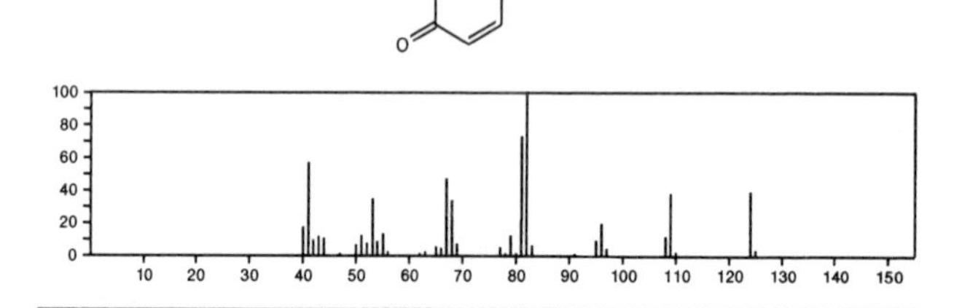

124 $C_8H_{12}O$ 1123-09-7
2-Cyclohexen-1-one, 3,5-dimethyl-

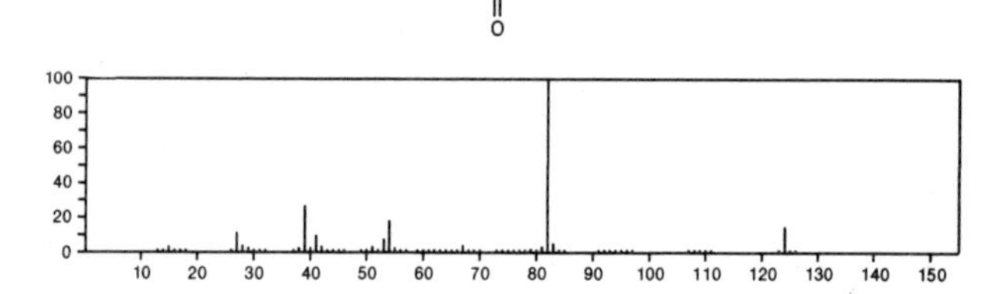

124 $C_8H_{12}O$ 1740-63-2
Cyclohexanone, 3-ethenyl-

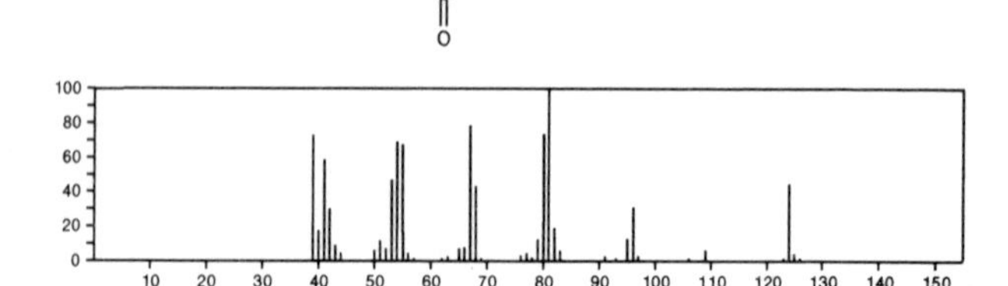

124 $C_8H_{12}O$ 1767-84-6
Ethanone, 1-(2-methyl-2-cyclopenten-1-yl)-

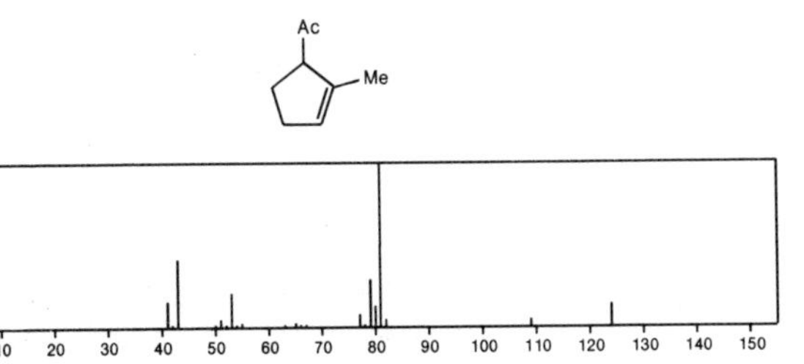

124 $C_8H_{12}O$ 2716-23-6
Bicyclo[2.2.2]octanone

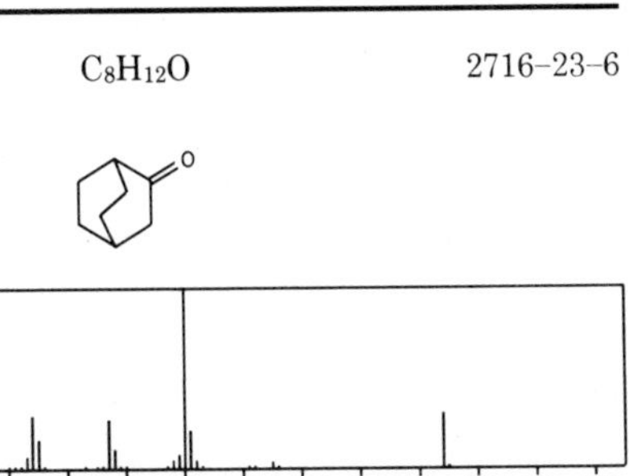

124 $C_8H_{12}O$ 3168-90-9
Ethanone, 1-(2-methyl-1-cyclopenten-1-yl)-

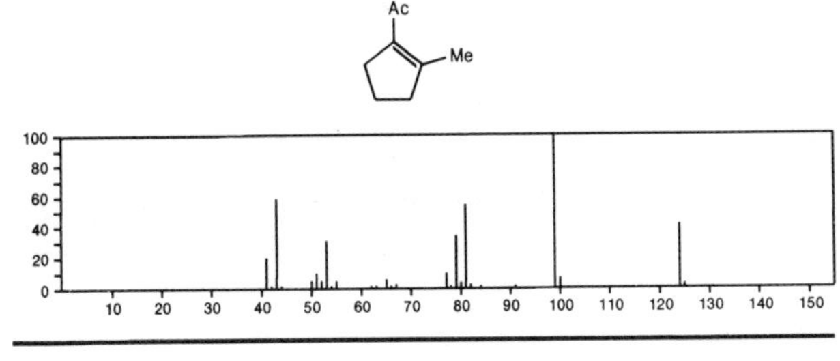

124 $C_8H_{12}O$ 5019-82-9
Bicyclo[3.2.1]octan-2-one

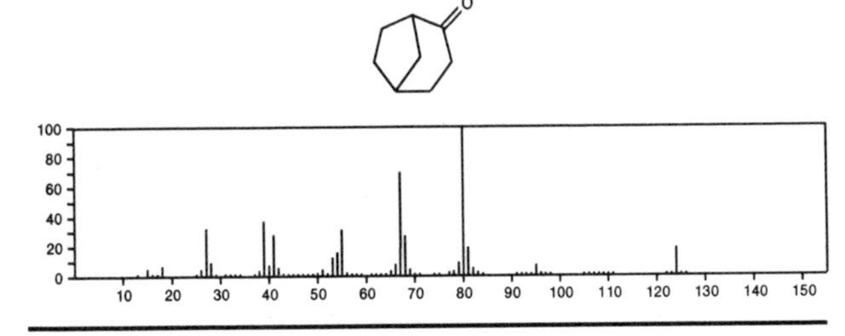

124 $C_8H_{12}O$ 5715-25-3
2-Cyclohexen-1-one, 4,5-dimethyl-

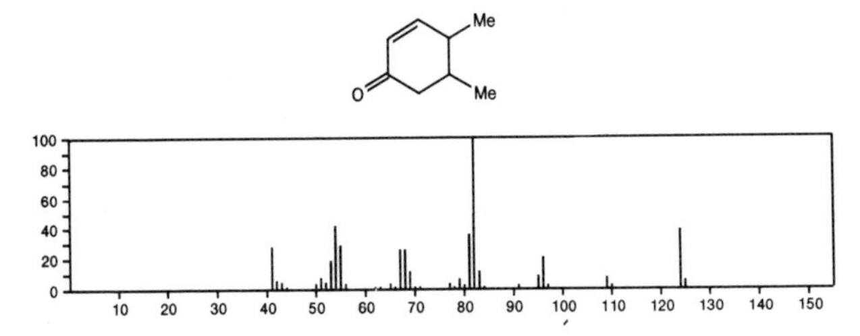

124 $C_8H_{12}O$ 7353-76-6
Ethanone, 1-(3-cyclohexen-1-yl)-

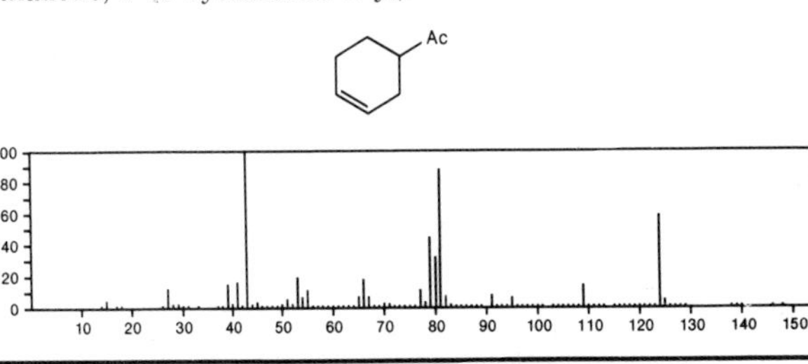

124 $C_8H_{12}O$ 14252-05-2
Bicyclo[3.2.1]octan-3-one

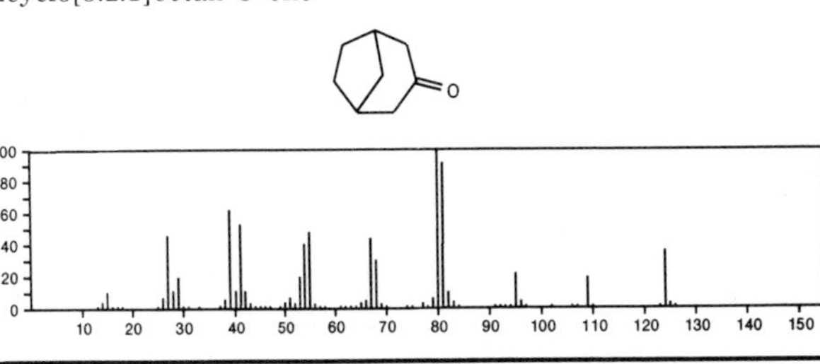

124 $C_8H_{12}O$ 14845-40-0
Bicyclo[4.1.0]heptan-2-one, 1-methyl-

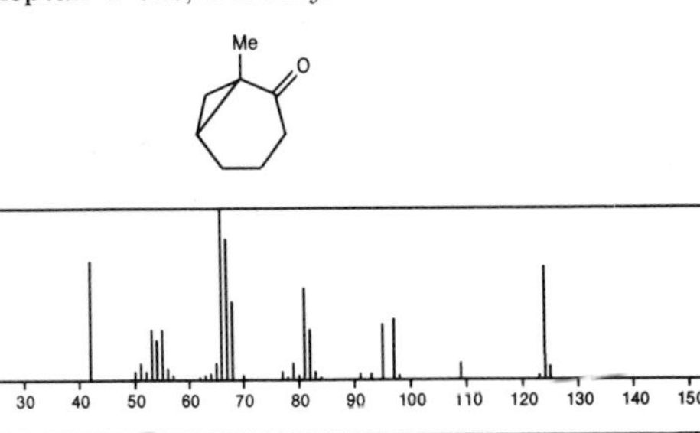

124 $C_8H_{12}O$ 14845-41-1
Bicyclo[4.1.0]heptan-2-one, 6-methyl-

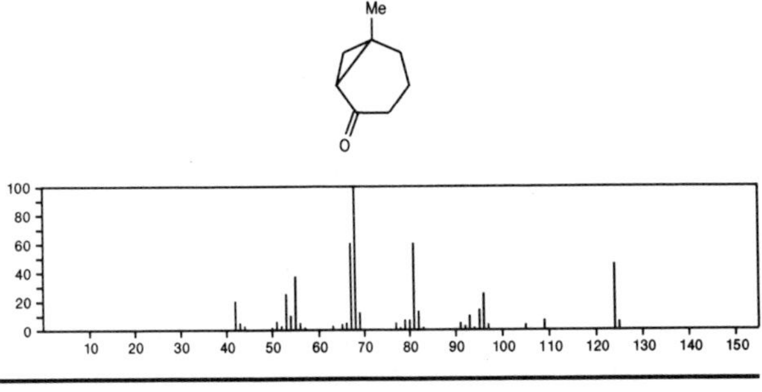

124 $C_8H_{12}O$ 16647-04-4
3,5-Heptadien-2-one, 6-methyl-, (*E*)-

MeCOCH=CHCH=CMe2

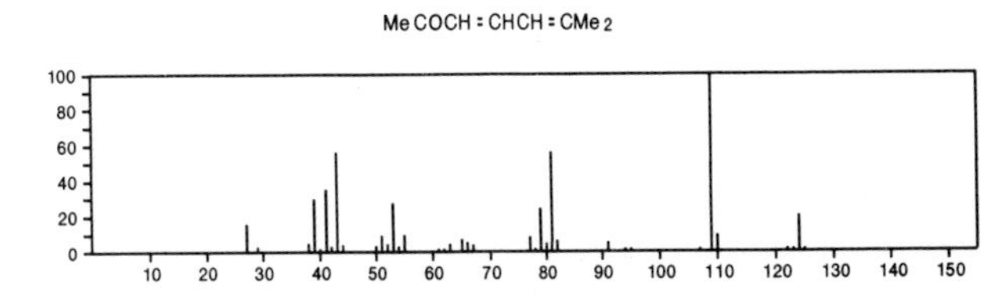

124 $C_8H_{12}O$ 24156-95-4
2-Cyclopenten-1-one, 3,5,5-trimethyl-

---

124 $C_8H_{12}O$ 25172-06-9
3,7-Octadien-2-one, (*E*)-

$H_2C=CHCH_2CH_2CH=CHCOMe$

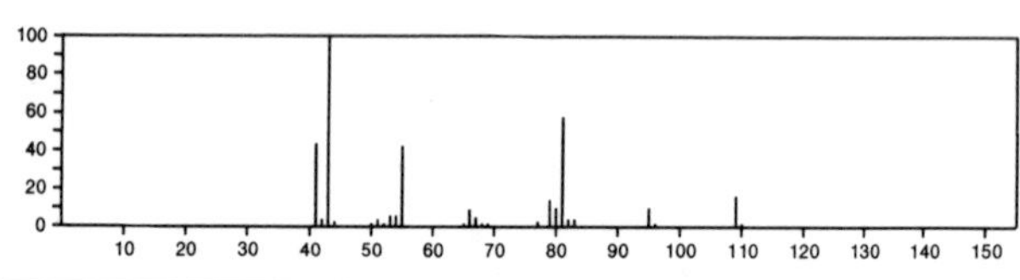

---

124 $C_8H_{12}O$ 28790-86-5
2-Cyclopenten-1-one, 2,3,4-trimethyl-

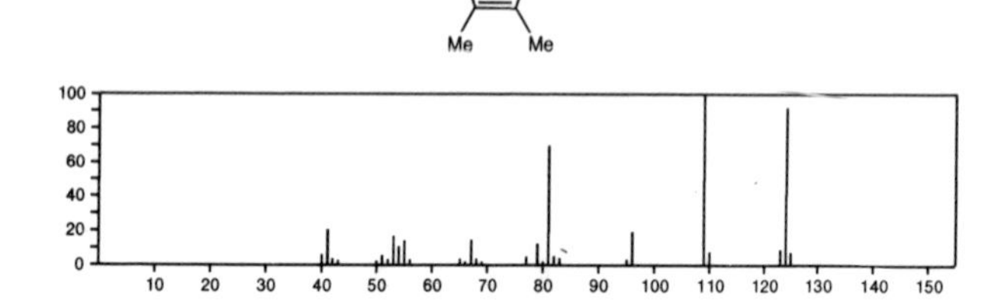

---

124 $C_8H_{12}O$ 29750-22-9
2-Norcaranone, 3-methyl-, stereoisomer

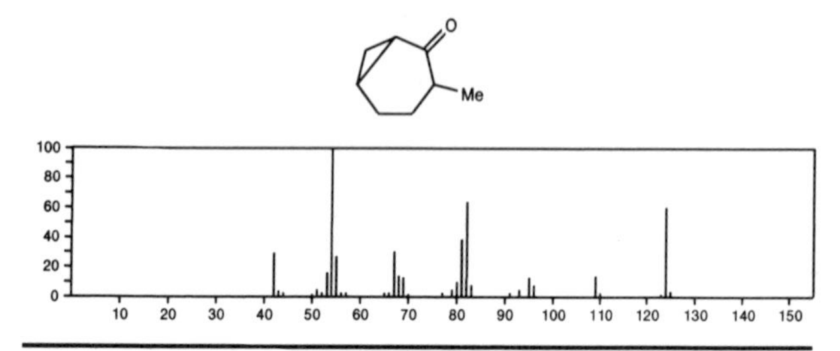

---

124 $C_8H_{12}O$ 30434-65-2
2-Cyclopenten-1-one, 3,4,4-trimethyl-

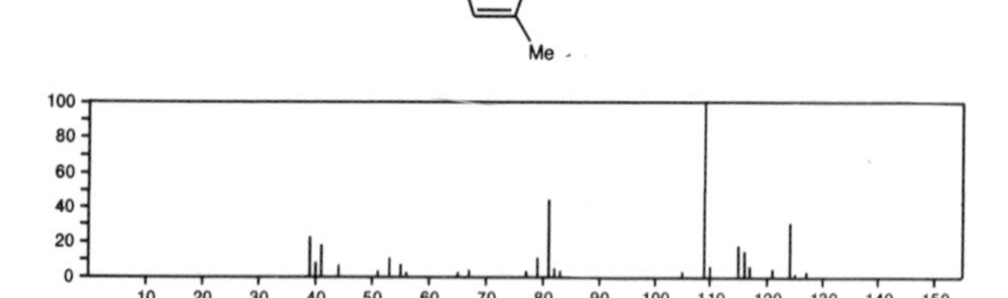

---

124 $C_8H_{12}O$ 51756-18-4
5-Hexen-2-one, 5-methyl-3-methylene-

$H_2C=CMeCH_2C(=CH_2)COMe$

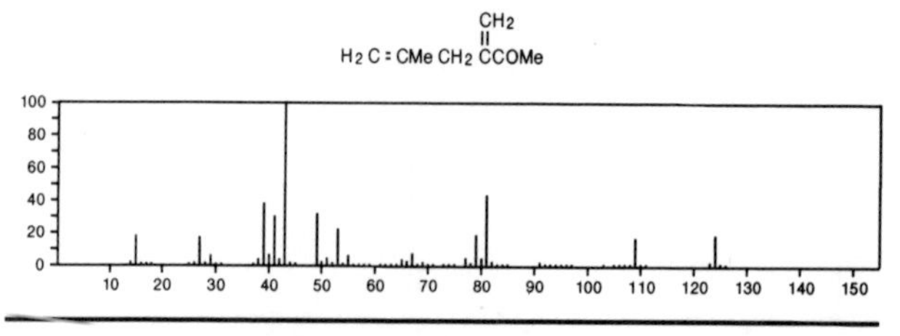

---

124 $C_8H_{12}O$ 55320-40-6
Bicyclo[2.2.2]oct-5-en-2-ol

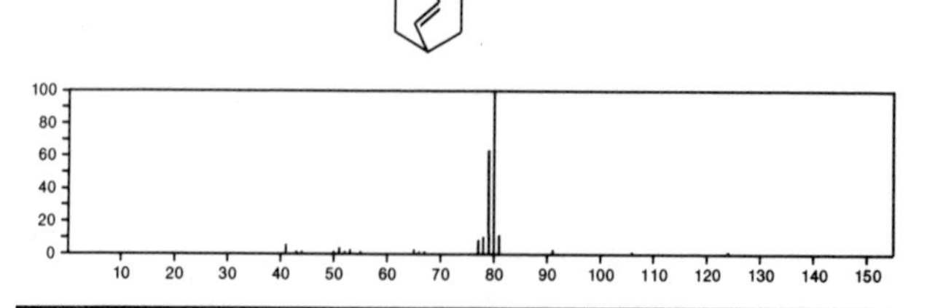

---

124 $C_8H_{12}O$ 55449-70-2
Pentaleno[1,2-*b*]oxirene, octahydro-, (1aα,1bβ,4aα,5aα)-

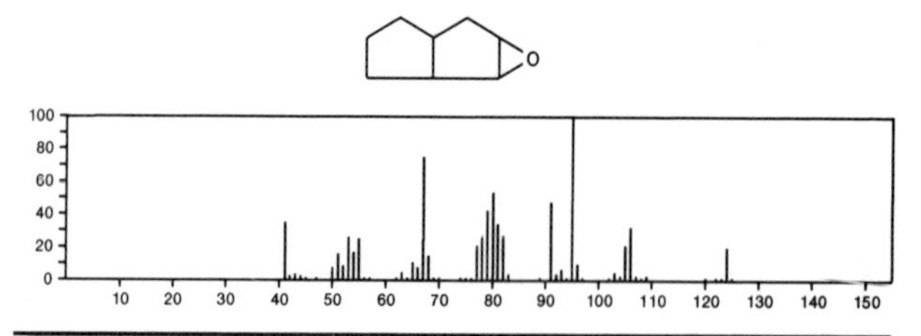

---

124 $C_8H_{12}O$ 55449-71-3
Pentaleno[1,2-*b*]oxirene, octahydro-, (1aα,1bα,4aβ,5aα)-

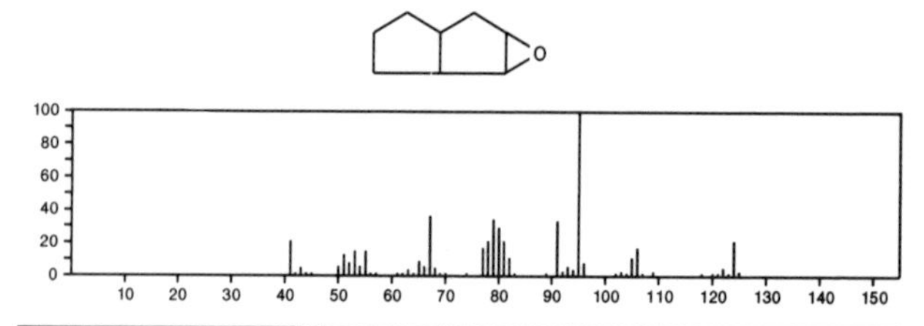

---

124 $C_8H_{12}O$ 55683-21-1
2-Cyclopenten-1-one, 3,4,5-trimethyl-

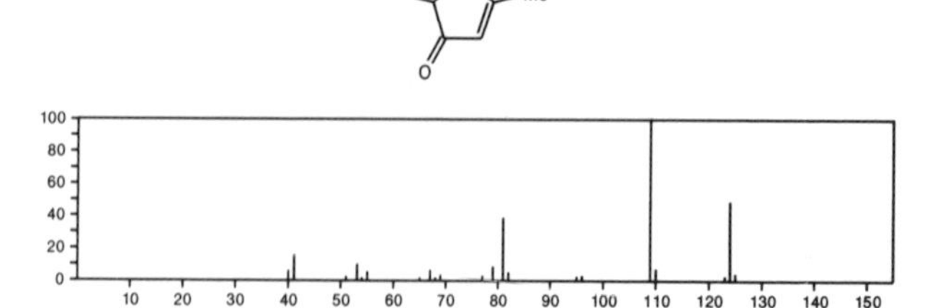

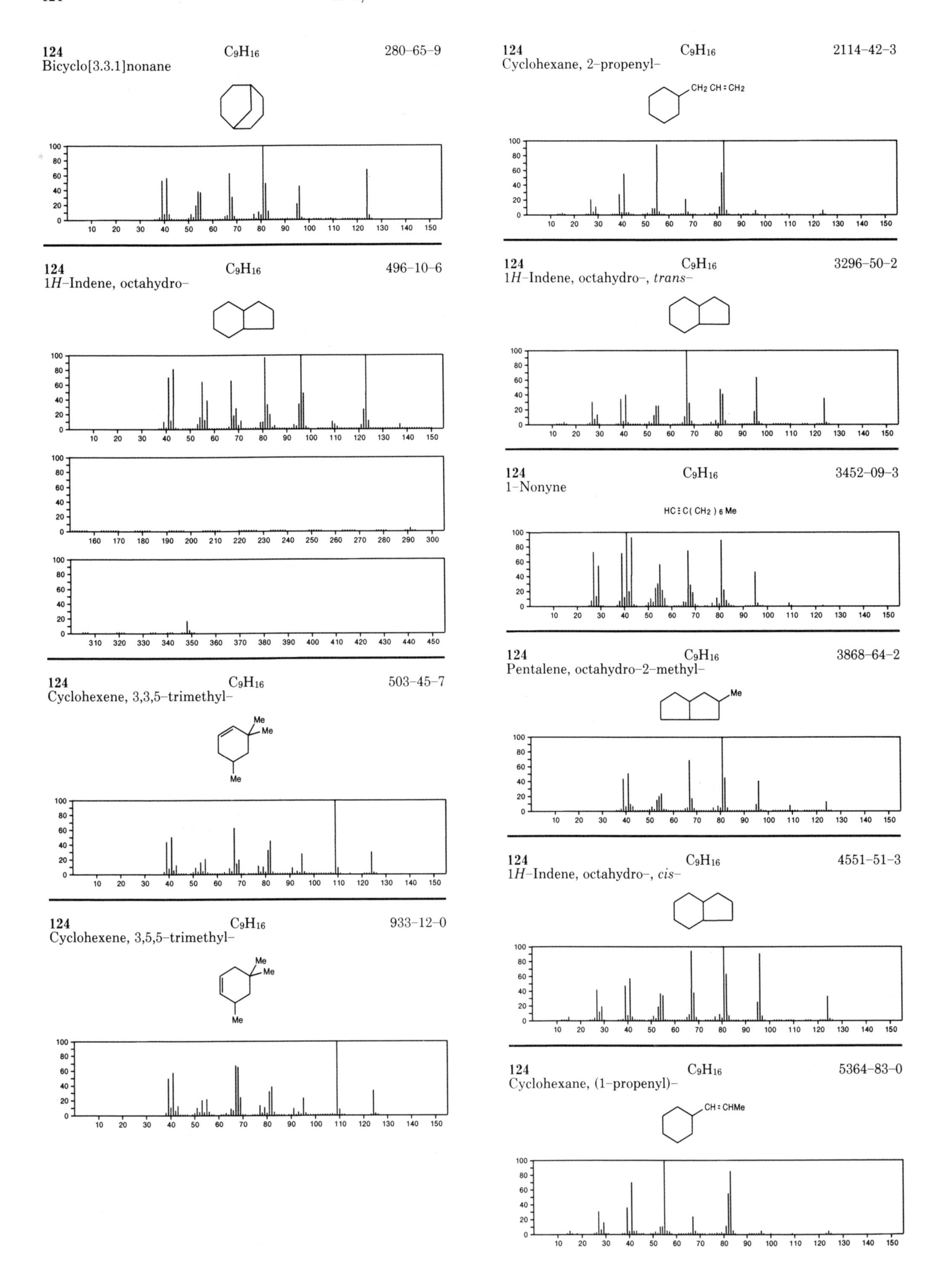
124 C_9H_{16} 280-65-9
Bicyclo[3.3.1]nonane
124 C_9H_{16} 496-10-6
1H-Indene, octahydro-
124 C_9H_{16} 503-45-7
Cyclohexene, 3,3,5-trimethyl-
Me
Me
Me
124 C_9H_{16} 933-12-0
Cyclohexene, 3,5,5-trimethyl-
Me
Me
Me
124 C_9H_{16} 2114-42-3
Cyclohexane, 2-propenyl-
CH2 CH = CH2
124 C_9H_{16} 3296-50-2
1H-Indene, octahydro-, trans-
124 C_9H_{16} 3452-09-3
1-Nonyne
HC≡C(CH2)6Me
124 C_9H_{16} 3868-64-2
Pentalene, octahydro-2-methyl-
Me
124 C_9H_{16} 4551-51-3
1H-Indene, octahydro-, cis-
124 C_9H_{16} 5364-83-0
Cyclohexane, (1-propenyl)-
CH = CHMe

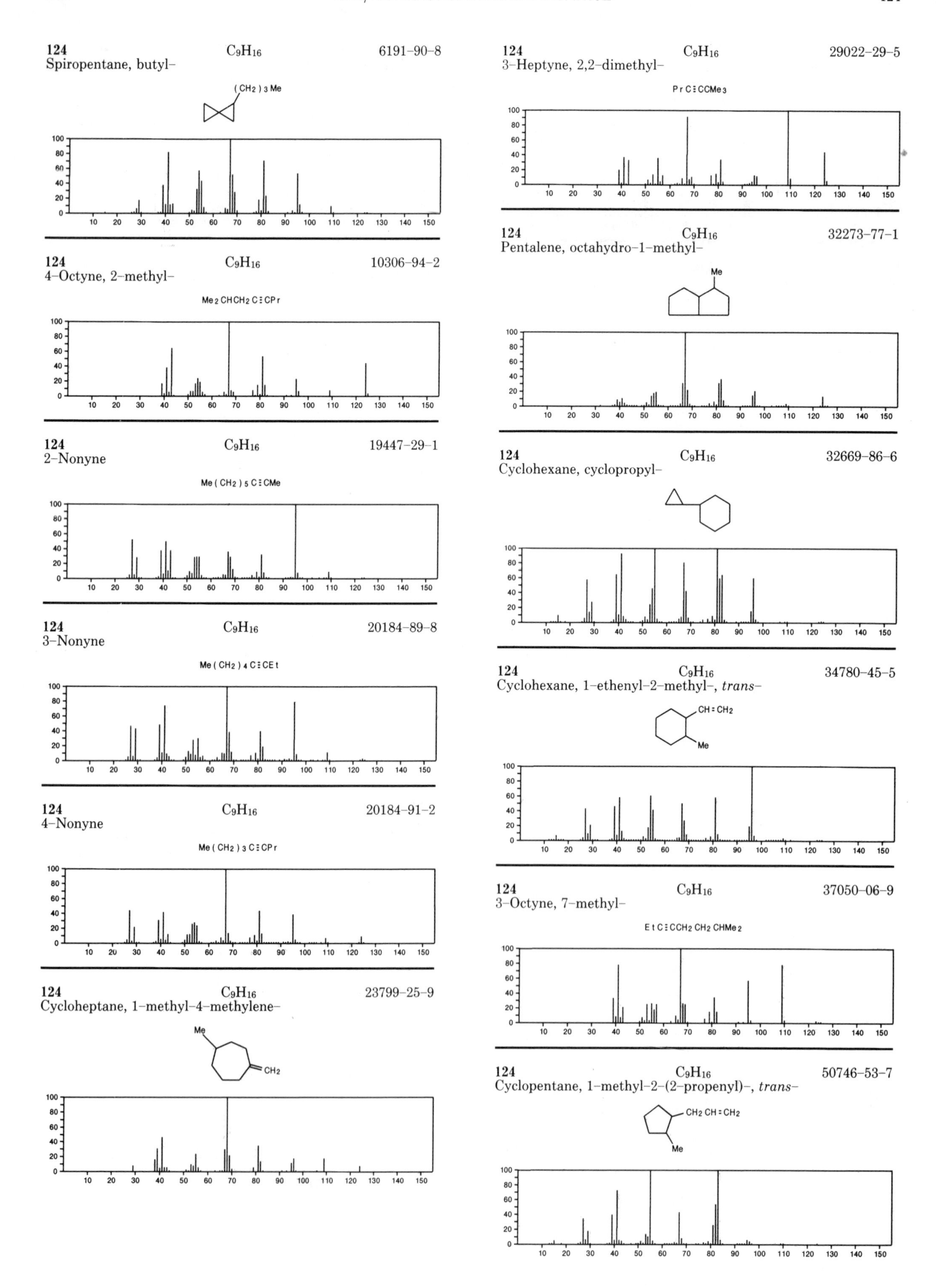
124 C9H16 6191-90-8
Spiropentane, butyl-
(CH2)3Me
124 C9H16 10306-94-2
4-Octyne, 2-methyl-
Me2CHCH2C≡CPr
124 C9H16 19447-29-1
2-Nonyne
Me(CH2)5C≡CMe
124 C9H16 20184-89-8
3-Nonyne
Me(CH2)4C≡CEt
124 C9H16 20184-91-2
4-Nonyne
Me(CH2)3C≡CPr
124 C9H16 23799-25-9
Cycloheptane, 1-methyl-4-methylene-
Me
CH2
124 C9H16 29022-29-5
3-Heptyne, 2,2-dimethyl-
PrC≡CCMe3
124 C9H16 32273-77-1
Pentalene, octahydro-1-methyl-
Me
124 C9H16 32669-86-6
Cyclohexane, cyclopropyl-
124 C9H16 34780-45-5
Cyclohexane, 1-ethenyl-2-methyl-, trans-
CH=CH2
Me
124 C9H16 37050-06-9
3-Octyne, 7-methyl-
EtC≡CCH2CH2CHMe2
124 C9H16 50746-53-7
Cyclopentane, 1-methyl-2-(2-propenyl)-, trans-
CH2CH=CH2
Me

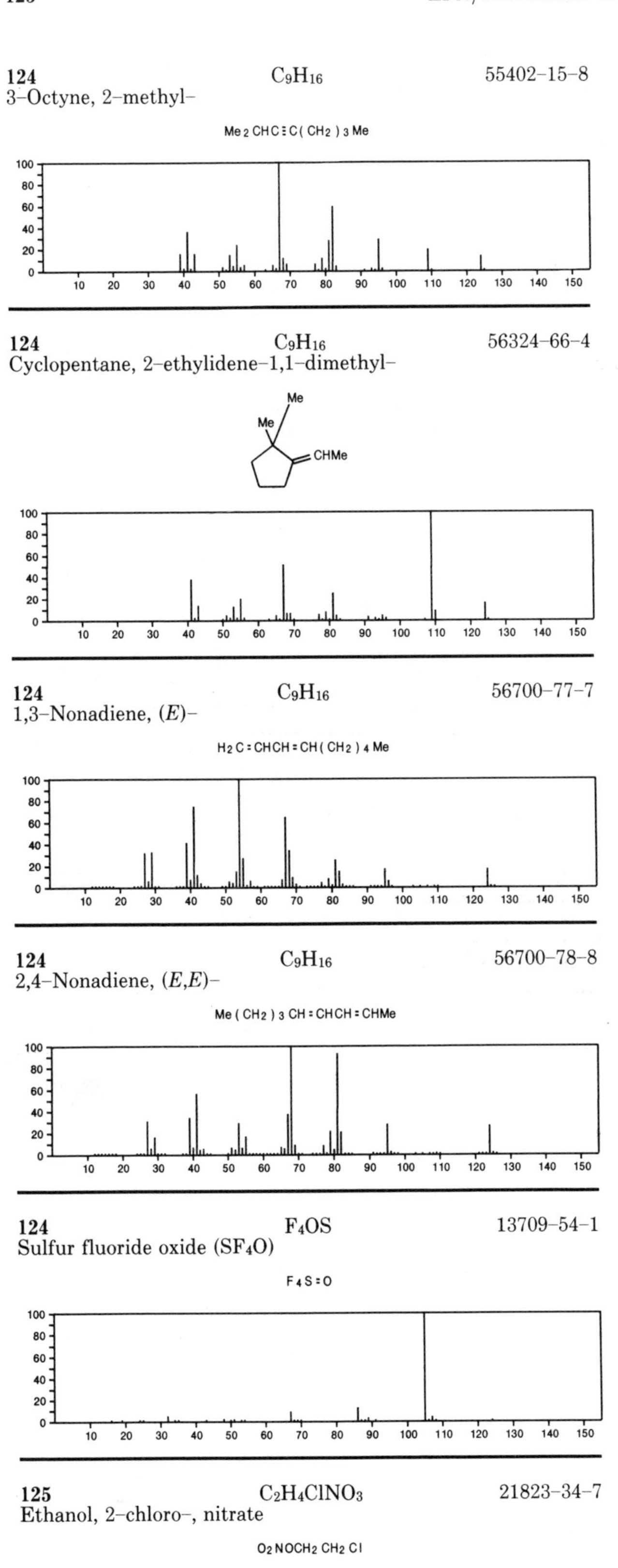

124 C_9H_{16} 55402–15–8
3–Octyne, 2–methyl–
Me2 CHC≡C(CH2)3 Me
124 C_9H_{16} 56324–66–4
Cyclopentane, 2–ethylidene–1,1–dimethyl–
124 C_9H_{16} 56700–77–7
1,3–Nonadiene, (E)–
H2C=CHCH=CH(CH2)4 Me
124 C_9H_{16} 56700–78–8
2,4–Nonadiene, (E,E)–
Me(CH2)3 CH=CHCH=CHMe
124 F_4OS 13709–54–1
Sulfur fluoride oxide (SF_4O)
F4S=O
125 $C_2H_4ClNO_3$ 21823–34–7
Ethanol, 2–chloro–, nitrate
O2NOCH2 CH2 Cl

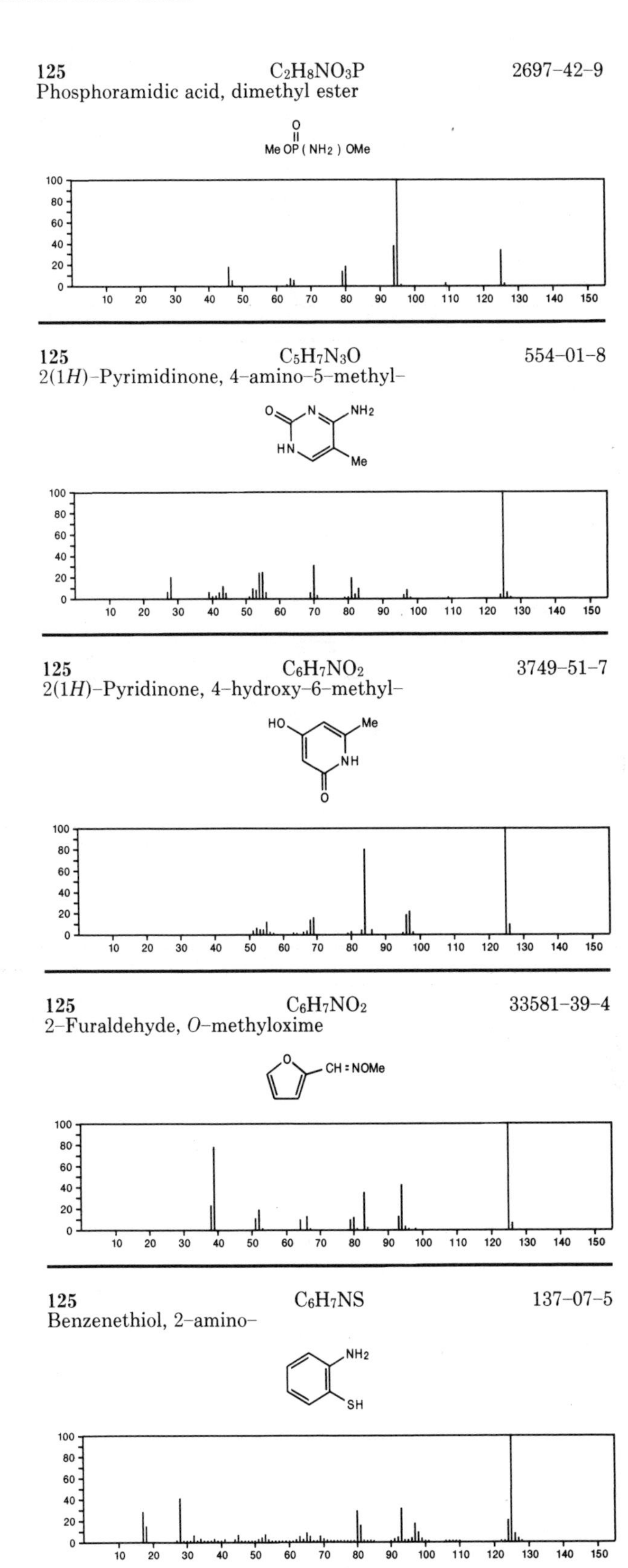

125 $C_2H_8NO_3P$ 2697–42–9
Phosphoramidic acid, dimethyl ester
MeOP(NH2)OMe
125 $C_5H_7N_3O$ 554–01–8
2(1H)–Pyrimidinone, 4–amino–5–methyl–
125 $C_6H_7NO_2$ 3749–51–7
2(1H)–Pyridinone, 4–hydroxy–6–methyl–
125 $C_6H_7NO_2$ 33581–39–4
2–Furaldehyde, O–methyloxime
CH=NOMe
125 C_6H_7NS 137–07–5
Benzenethiol, 2–amino–

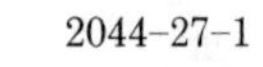

125 C_6H_7NS 2044-27-1
2(1*H*)-Pyridinethione, 1-methyl-

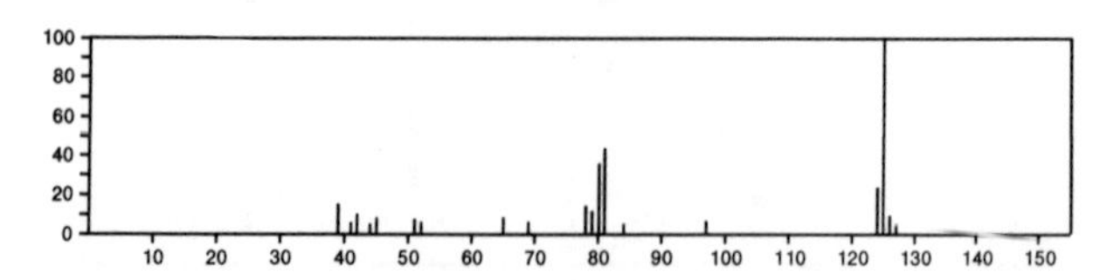

125 C_6H_7NS 18368-57-5
2(1*H*)-Pyridinethione, 6-methyl-

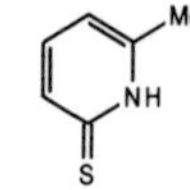

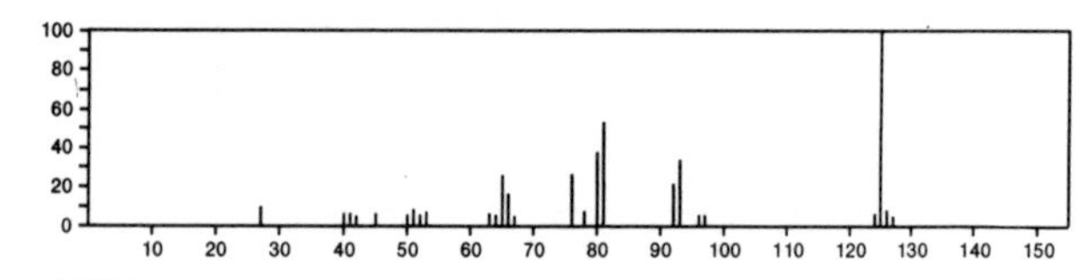

125 C_6H_7NS 18368-65-5
2(1*H*)-Pyridinethione, 4-methyl-

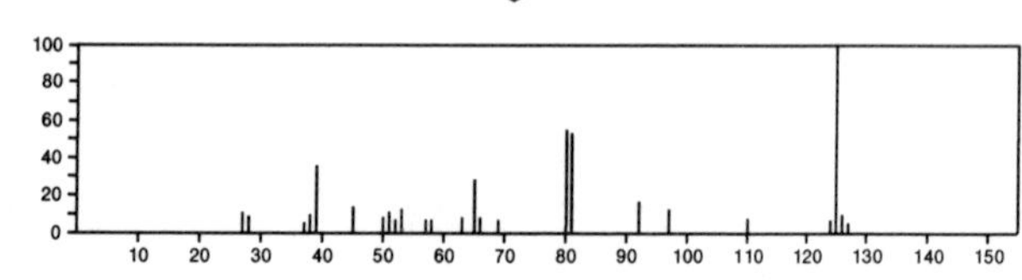

125 C_6H_7NS 18438-38-5
Pyridine, 2-(methylthio)-

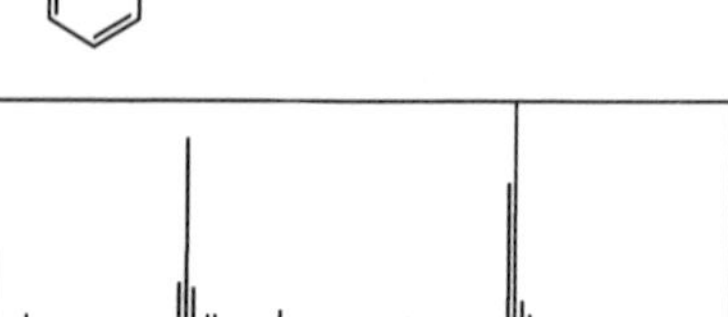

125 C_6H_7NS 18794-33-7
Pyridine, 3-(methylthio)-

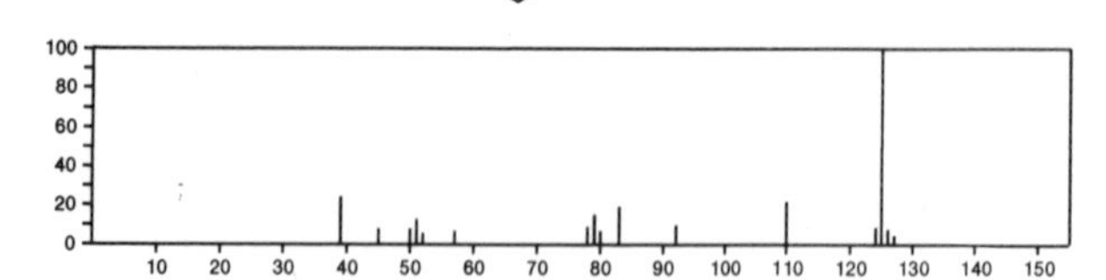

125 C_6H_7NS 36880-58-7
Pyridinium, 3-mercapto-1-methyl-, hydroxide, inner salt

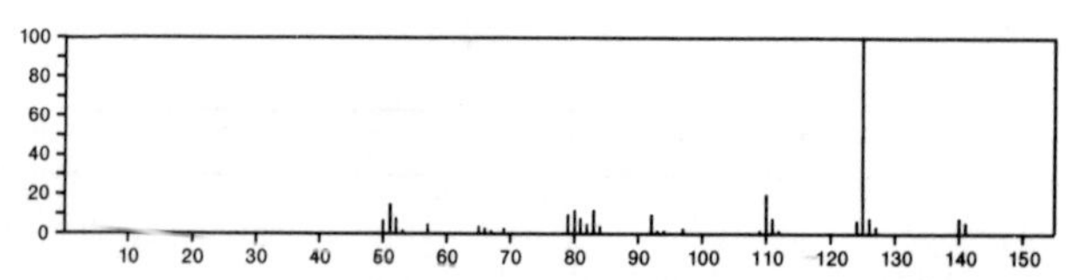

125 $C_6H_{11}N_3$ 501-75-7
1*H*-Imidazole-4-ethanamine, 1-methyl-

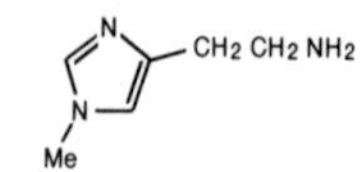

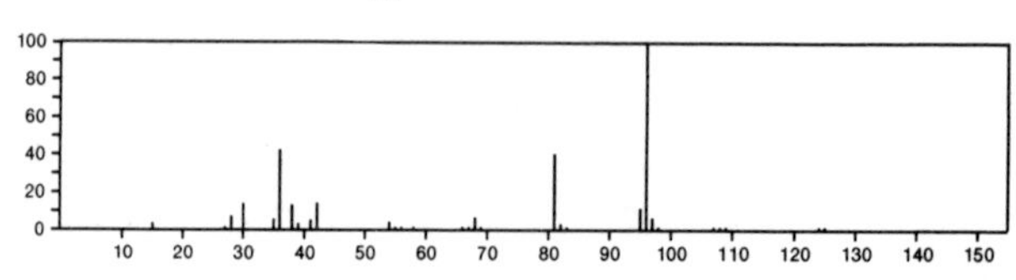

125 $C_6H_{11}N_3$ 644-42-8
1*H*-Imidazole-5-ethanamine, 1-methyl-

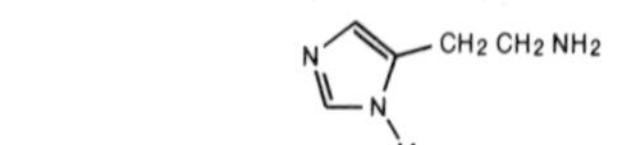

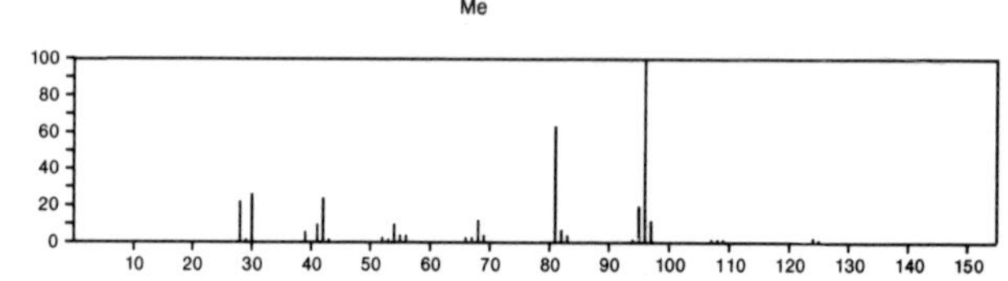

125 $C_6H_{11}N_3$ 673-50-7
1*H*-Imidazole-4-ethanamine, *N*-methyl-

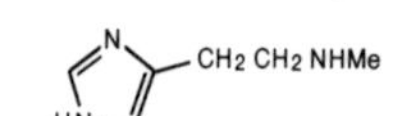

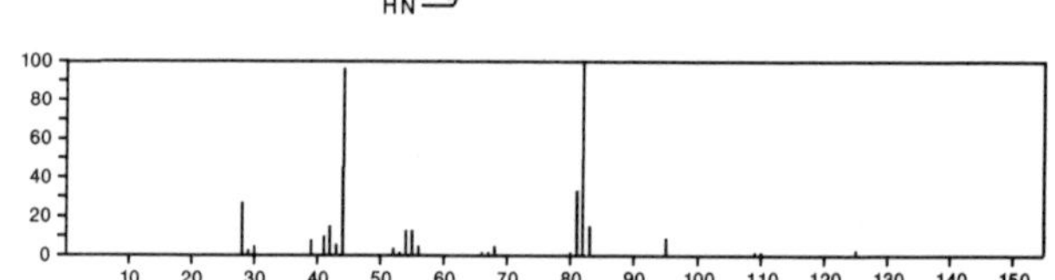

125 $C_6H_{11}N_3$ 6086-22-2
1*H*-1,2,4-Triazole, 1-butyl-

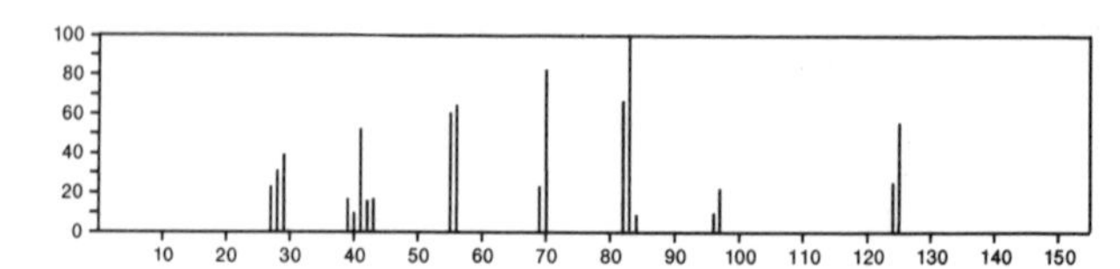

125 $C_6H_{11}N_3$ 6986-90-9
1*H*-Imidazole-4-ethanamine, α-methyl-

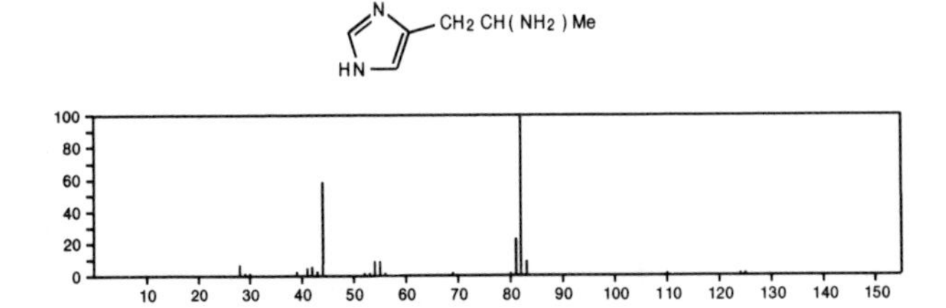

125 $C_6H_{11}N_3$ 19573-22-9
Cyclohexane, azido-

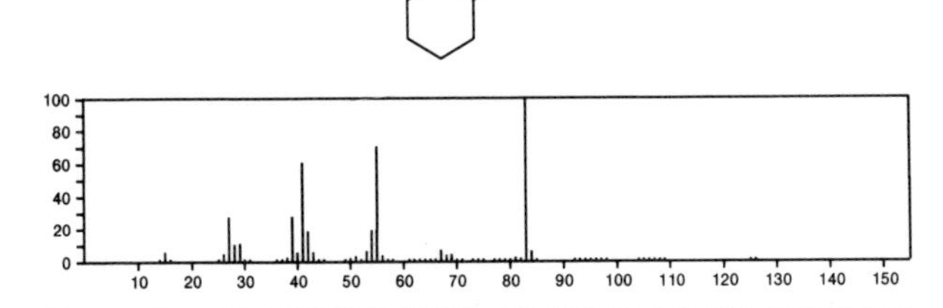

125 $C_6H_{11}N_3$ 24160-42-7
1*H*-Imidazole-4-ethanamine, β-methyl-

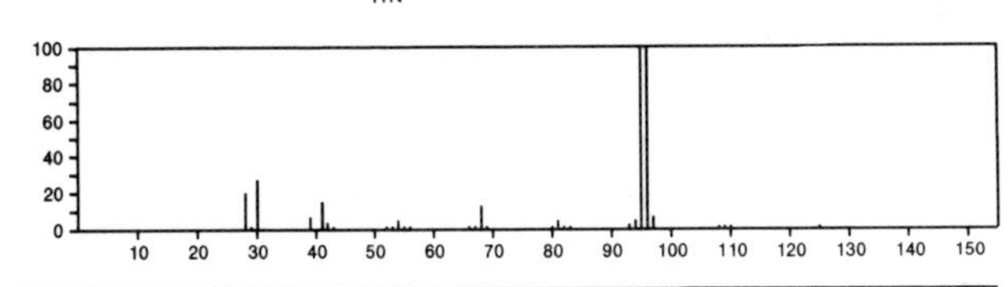

125 $C_6H_{11}N_3$ 34392-54-6
1*H*-Imidazole-4-ethanamine, 2-methyl-

125 $C_6H_{11}N_3$ 36507-31-0
1*H*-Imidazole-4-ethanamine, 5-methyl-

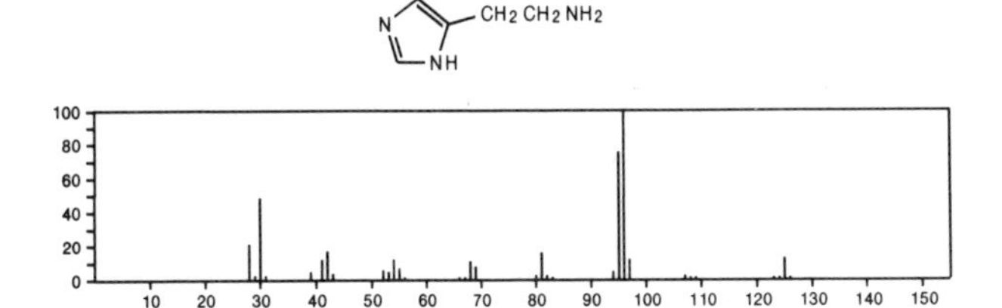

125 $C_7H_{10}N.HO$ 36880-53-2
Pyridinium, 1-ethyl-, hydroxide

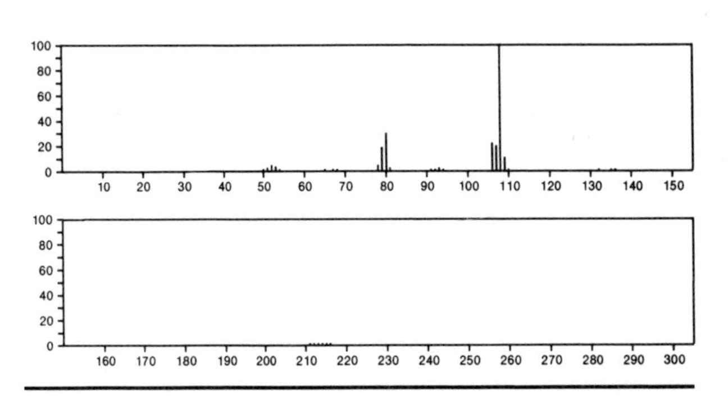

125 $C_7H_{11}NO$ 3731-38-2
1-Azabicyclo[2.2.2]octan-3-one

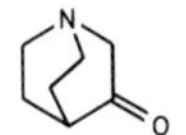

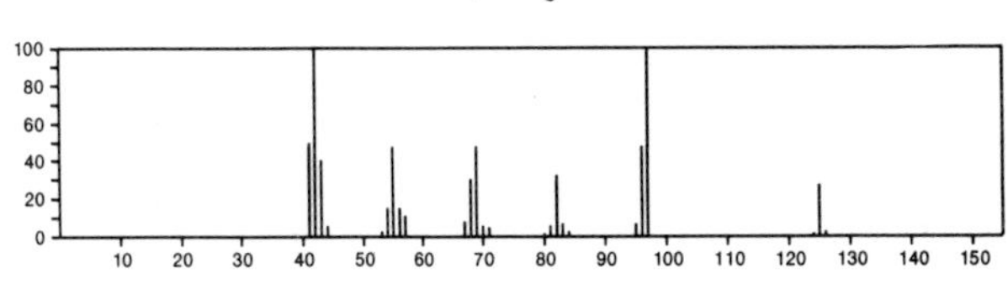

125 $C_7H_{11}NO$ 19615-27-1
Pyridine, 1-acetyl-1,2,3,4-tetrahydro-

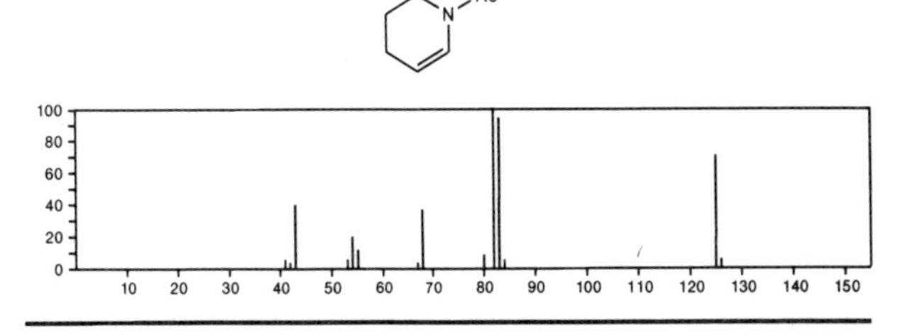

125 $C_8H_{15}N$ 283-24-9
3-Azabicyclo[3.2.2]nonane

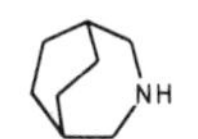

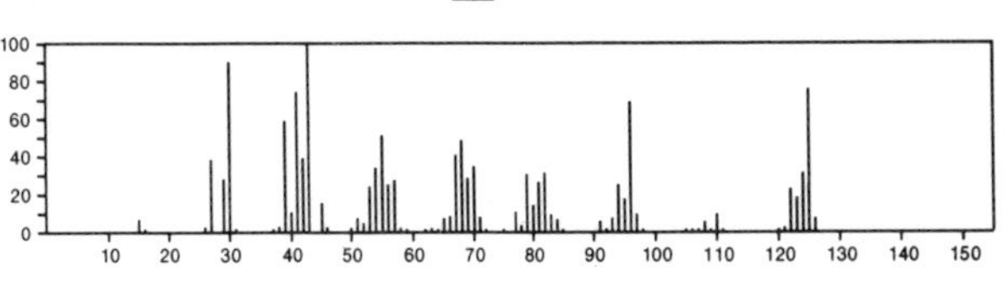

125 $C_8H_{15}N$ 529-17-9
8-Azabicyclo[3.2.1]octane, 8-methyl-

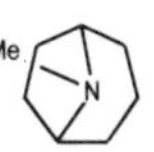

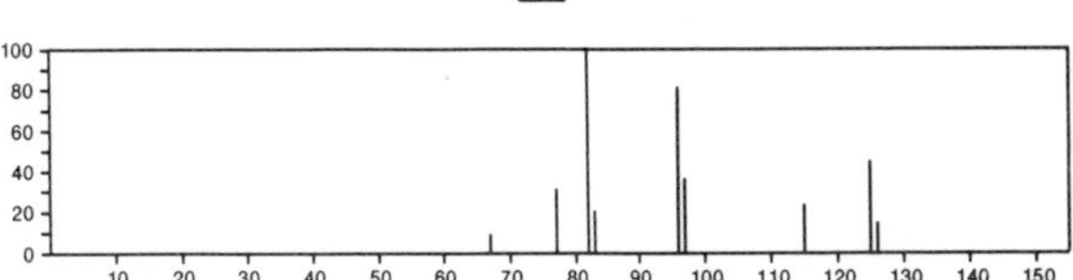

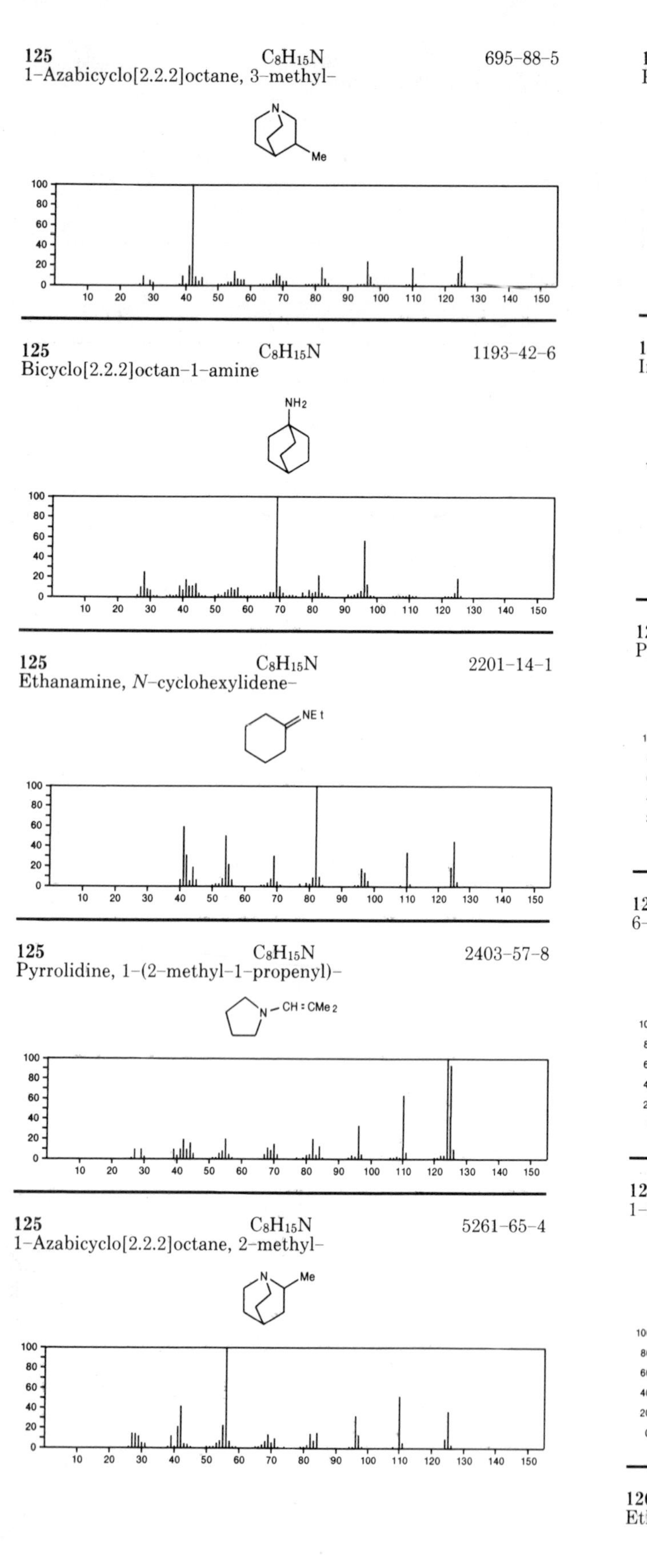

125 $C_8H_{15}N$ 695-88-5
1-Azabicyclo[2.2.2]octane, 3-methyl-

125 $C_8H_{15}N$ 1193-42-6
Bicyclo[2.2.2]octan-1-amine

125 $C_8H_{15}N$ 2201-14-1
Ethanamine, *N*-cyclohexylidene-

125 $C_8H_{15}N$ 2403-57-8
Pyrrolidine, 1-(2-methyl-1-propenyl)-

125 $C_8H_{15}N$ 5261-65-4
1-Azabicyclo[2.2.2]octane, 2-methyl-

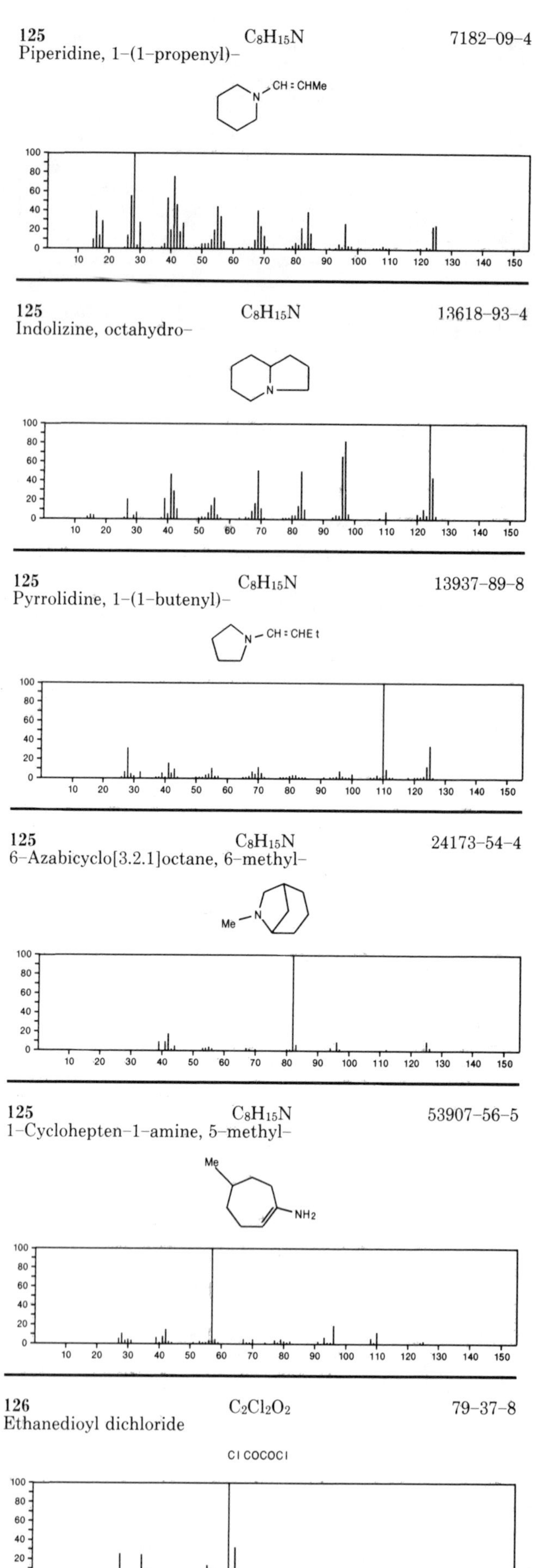

125 $C_8H_{15}N$ 7182-09-4
Piperidine, 1-(1-propenyl)-

125 $C_8H_{15}N$ 13618-93-4
Indolizine, octahydro-

125 $C_8H_{15}N$ 13937-89-8
Pyrrolidine, 1-(1-butenyl)-

125 $C_8H_{15}N$ 24173-54-4
6-Azabicyclo[3.2.1]octane, 6-methyl-

125 $C_8H_{15}N$ 53907-56-5
1-Cyclohepten-1-amine, 5-methyl-

126 $C_2Cl_2O_2$ 79-37-8
Ethanedioyl dichloride

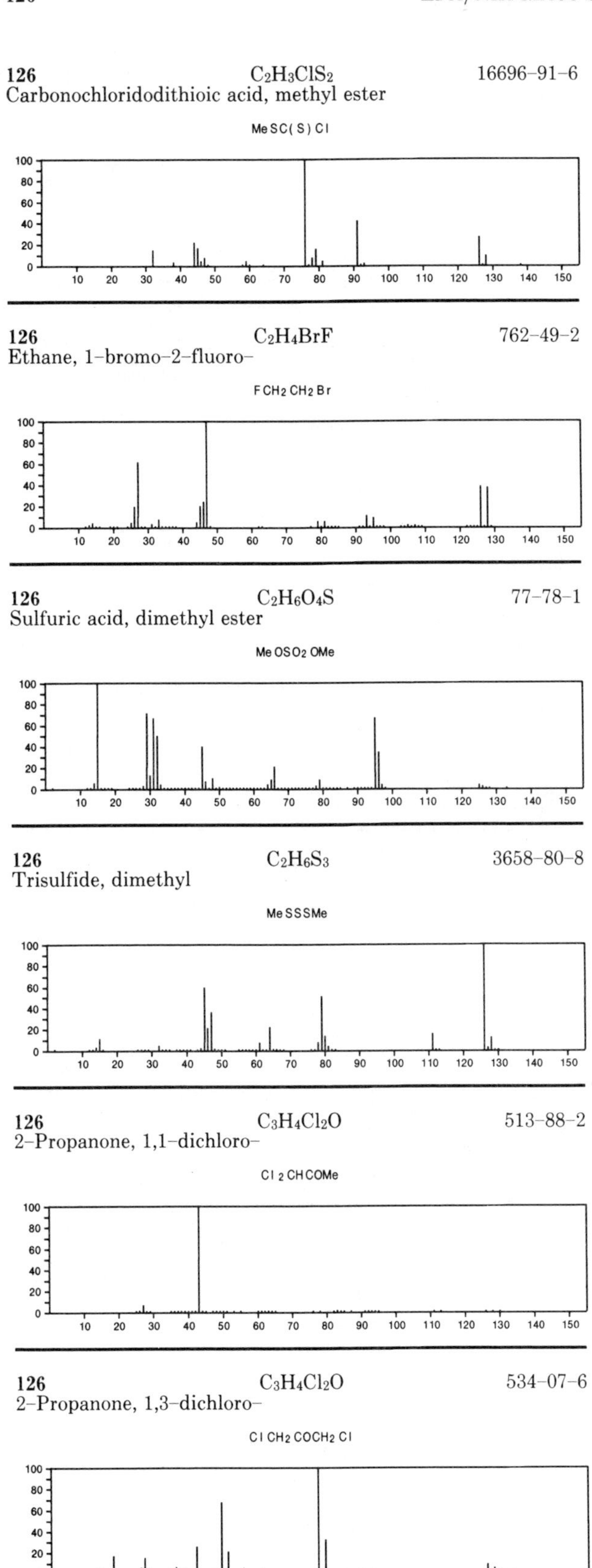
126 $C_2H_3ClS_2$ 16696-91-6
Carbonochloridodithioic acid, methyl ester
MeSC(S)Cl
126 C_2H_4BrF 762-49-2
Ethane, 1-bromo-2-fluoro-
FCH_2CH_2Br
126 $C_2H_6O_4S$ 77-78-1
Sulfuric acid, dimethyl ester
$MeOSO_2OMe$
126 $C_2H_6S_3$ 3658-80-8
Trisulfide, dimethyl
MeSSSMe
126 $C_3H_4Cl_2O$ 513-88-2
2-Propanone, 1,1-dichloro-
$Cl_2CHCOMe$
126 $C_3H_4Cl_2O$ 534-07-6
2-Propanone, 1,3-dichloro-
$ClCH_2COCH_2Cl$

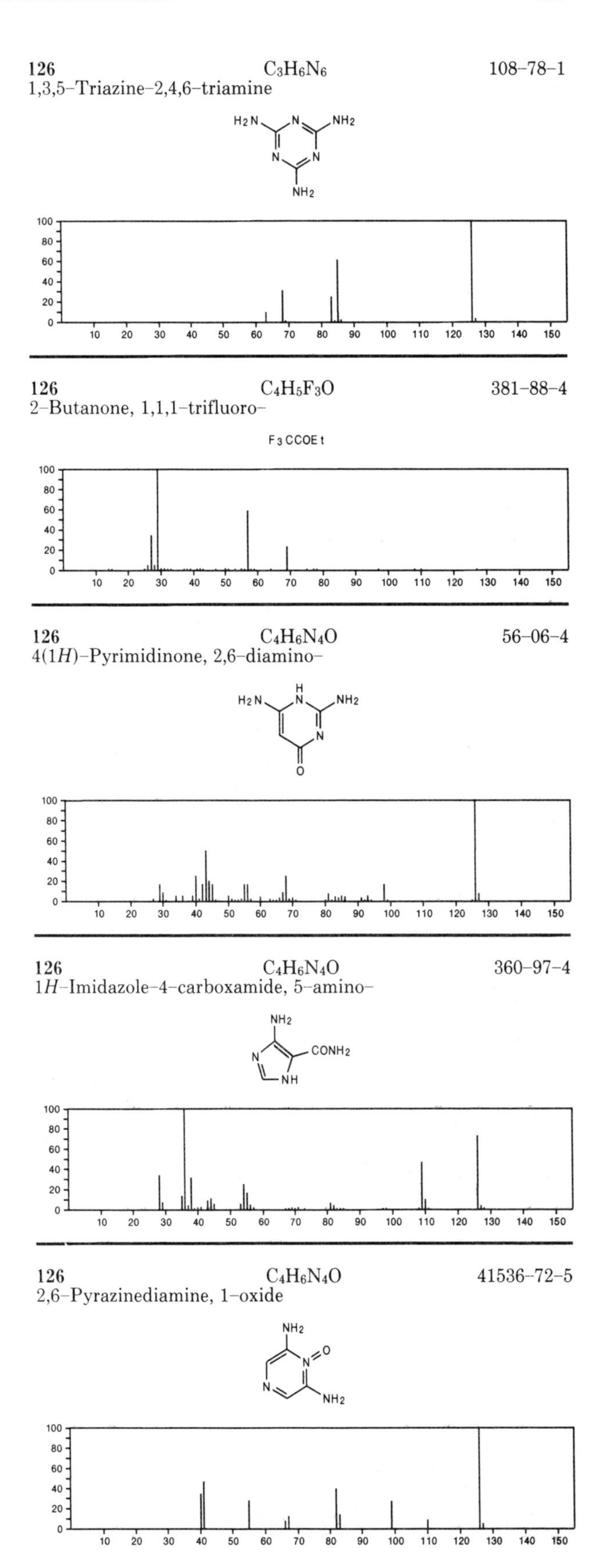
126 $C_3H_6N_6$ 108-78-1
1,3,5-Triazine-2,4,6-triamine
126 $C_4H_5F_3O$ 381-88-4
2-Butanone, 1,1,1-trifluoro-
F_3CCOEt
126 $C_4H_6N_4O$ 56-06-4
4(1*H*)-Pyrimidinone, 2,6-diamino-
126 $C_4H_6N_4O$ 360-97-4
1*H*-Imidazole-4-carboxamide, 5-amino-
126 $C_4H_6N_4O$ 41536-72-5
2,6-Pyrazinediamine, 1-oxide

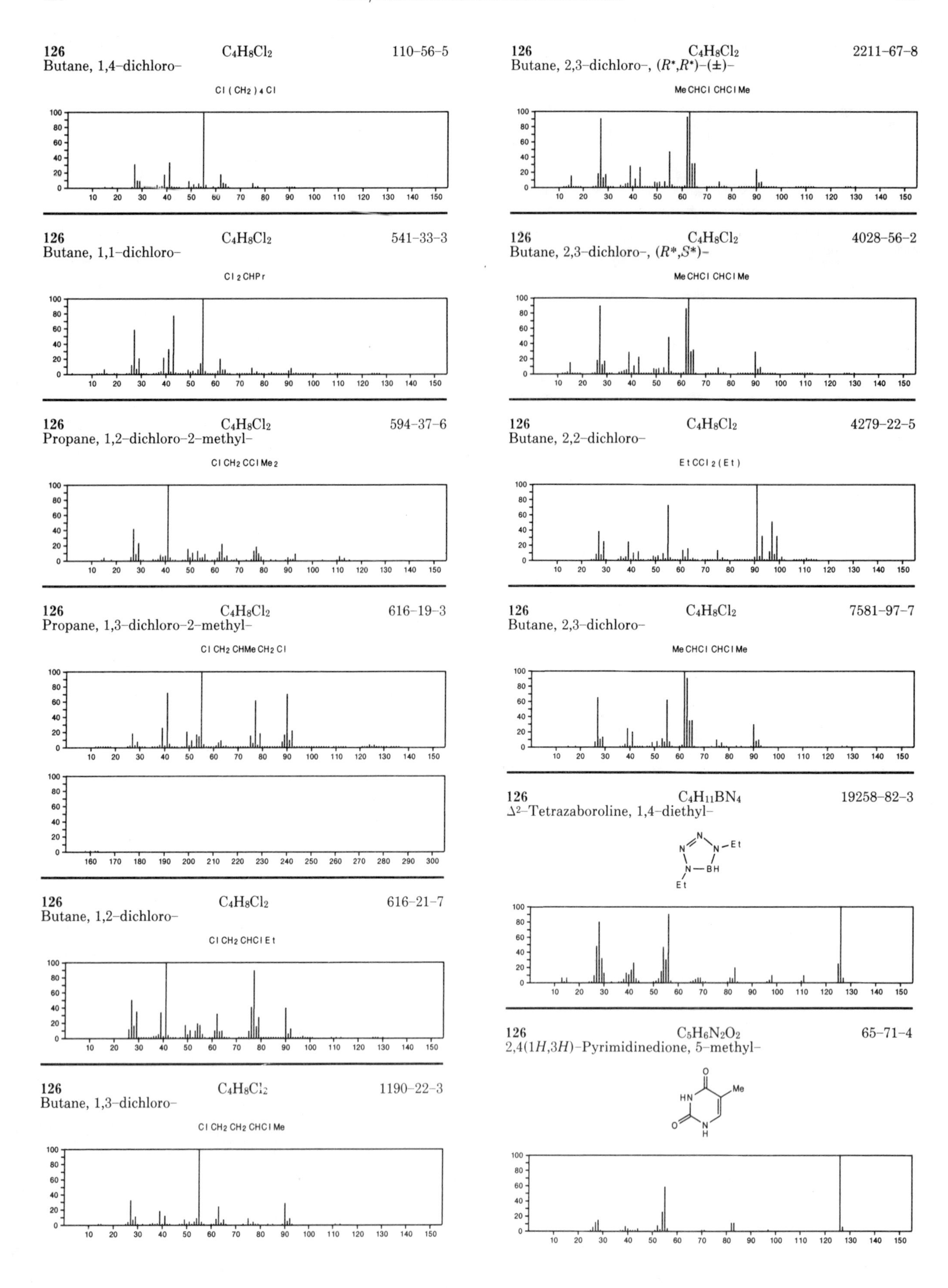
126 $C_4H_8Cl_2$ 110-56-5
Butane, 1,4-dichloro-
Cl(CH2)4Cl
126 $C_4H_8Cl_2$ 541-33-3
Butane, 1,1-dichloro-
Cl2CHPr
126 $C_4H_8Cl_2$ 594-37-6
Propane, 1,2-dichloro-2-methyl-
ClCH2CClMe2
126 $C_4H_8Cl_2$ 616-19-3
Propane, 1,3-dichloro-2-methyl-
ClCH2CHMeCH2Cl
126 $C_4H_8Cl_2$ 616-21-7
Butane, 1,2-dichloro-
ClCH2CHClEt
126 $C_4H_8Cl_2$ 1190-22-3
Butane, 1,3-dichloro-
ClCH2CH2CHClMe
126 $C_4H_8Cl_2$ 2211-67-8
Butane, 2,3-dichloro-, (R^*,R^*)-(±)-
MeCHClCHClMe
126 $C_4H_8Cl_2$ 4028-56-2
Butane, 2,3-dichloro-, (R^*,S^*)-
MeCHClCHClMe
126 $C_4H_8Cl_2$ 4279-22-5
Butane, 2,2-dichloro-
EtCCl2(Et)
126 $C_4H_8Cl_2$ 7581-97-7
Butane, 2,3-dichloro-
MeCHClCHClMe
126 $C_4H_{11}BN_4$ 19258-82-3
Δ2-Tetrazaboroline, 1,4-diethyl-
N
N-Et
BH
Et
126 $C_5H_6N_2O_2$ 65-71-4
2,4(1H,3H)-Pyrimidinedione, 5-methyl-
O
HN
Me
N
H

126 $C_5H_6N_2O_2$ 626–48–2

2,4(1*H*,3*H*)–Pyrimidinedione, 6–methyl–

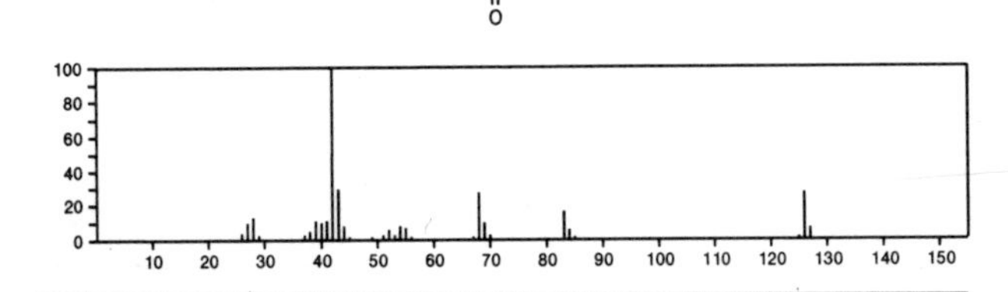

126 $C_5H_6N_2O_2$ 5754–18–7

3,6–Pyridazinedione, 1,2–dihydro–4–methyl–

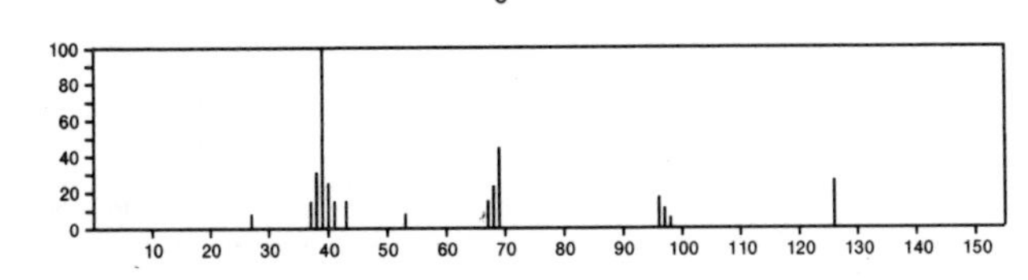

126 $C_5H_6N_2O_2$ 24614–14–0

4(1*H*)–Pyrimidinone, 5–hydroxy–2–methyl–

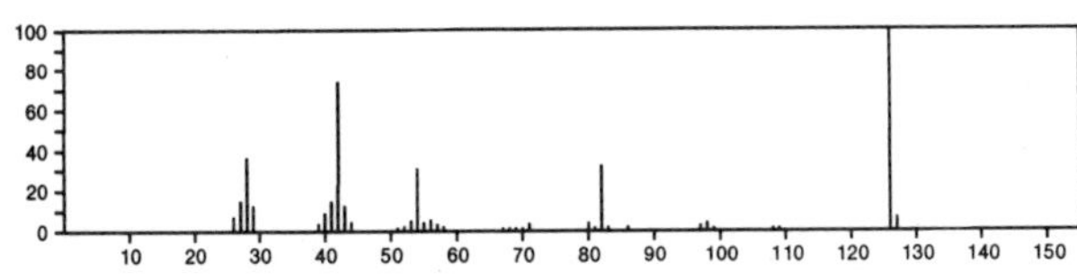

126 $C_5H_6N_2O_2$ 29397–21–5

2,4–Pentanedione, 3–diazo–

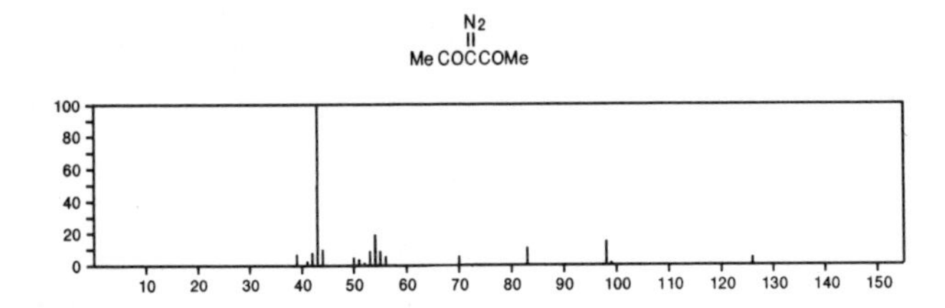

126 $C_5H_6N_2O_2$ 40704–11–8

1*H*–Pyrazole–4–carboxylic acid, 3–methyl–

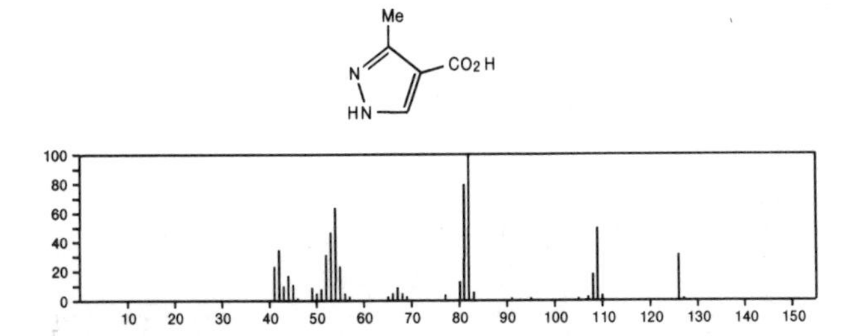

126 $C_5H_6N_2O_2$ 53907–67–8

Formamide, *N*–(3–methyl–5–isoxazolyl)–

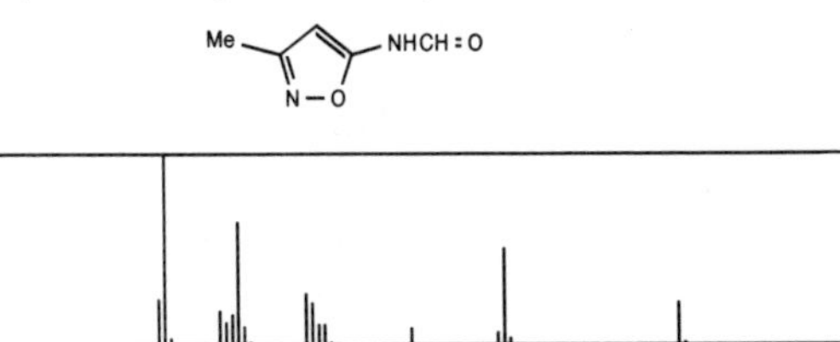

126 $C_5H_6N_2S$ 33643–86–6

4(1*H*)–Pyrimidinethione, 2–methyl–

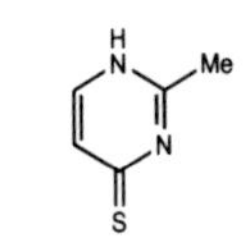

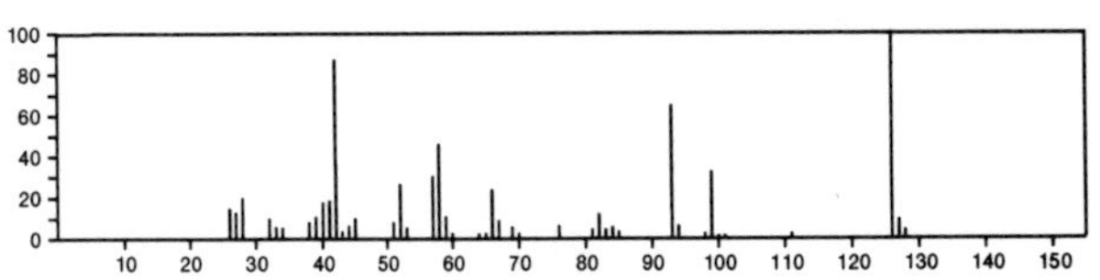

126 $C_6H_6O_3$ 87–66–1

1,2,3–Benzenetriol

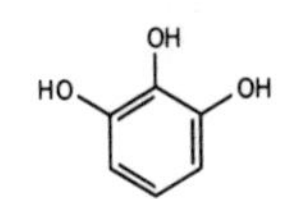

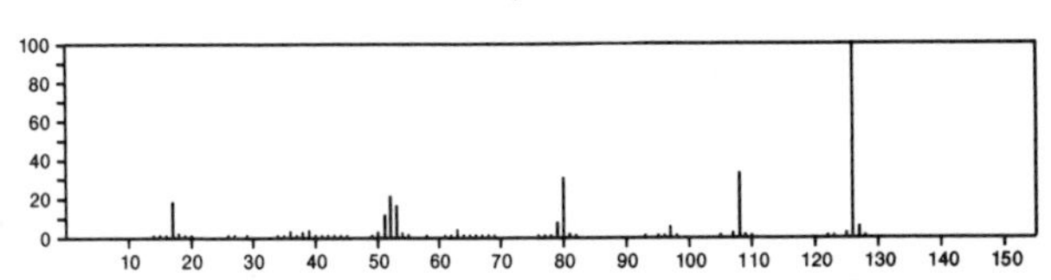

126 $C_6H_6O_3$ 118–71–8

4*H*–Pyran–4–one, 3–hydroxy–2–methyl–

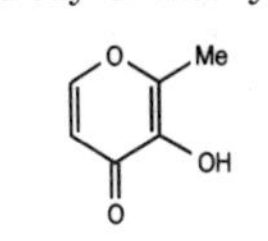

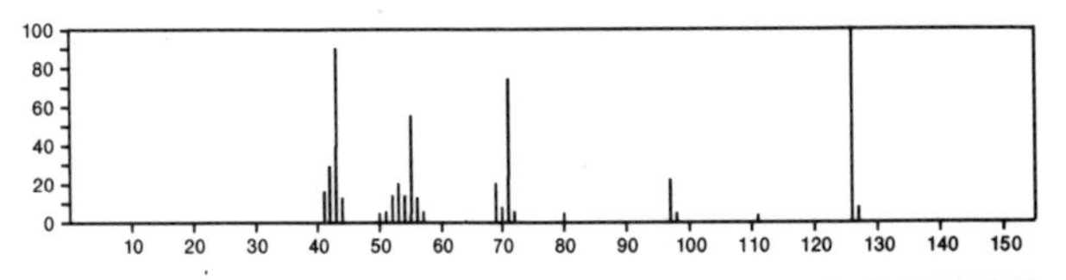

126 $C_6H_6O_3$ 611–13–2

2–Furancarboxylic acid, methyl ester

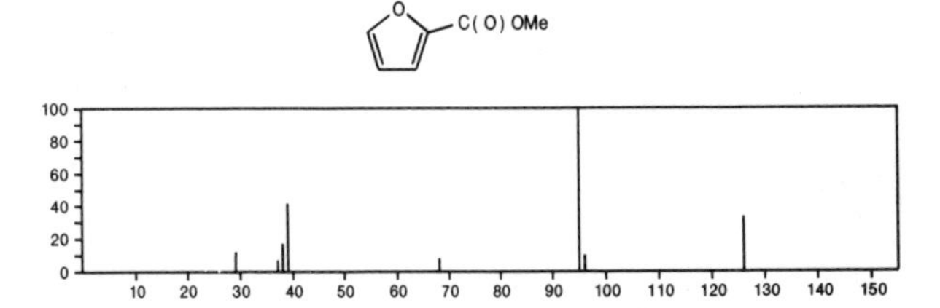

126 $C_6H_6O_3$ 644-46-2
4*H*-Pyran-4-one, 5-hydroxy-2-methyl-

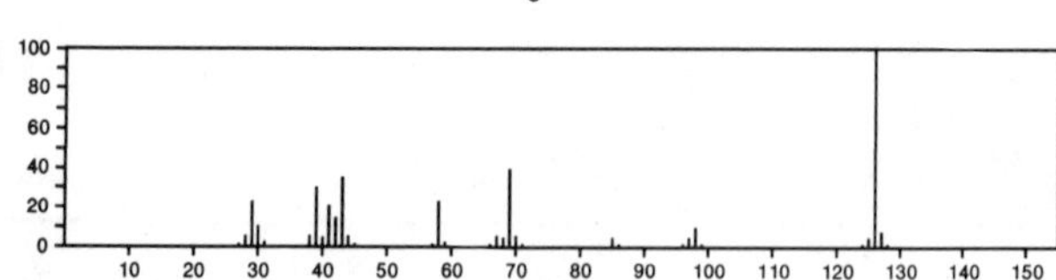

126 $C_6H_6O_3$ 675-10-5
2*H*-Pyran-2-one, 4-hydroxy-6-methyl-

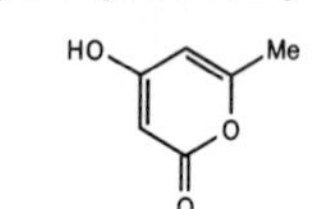

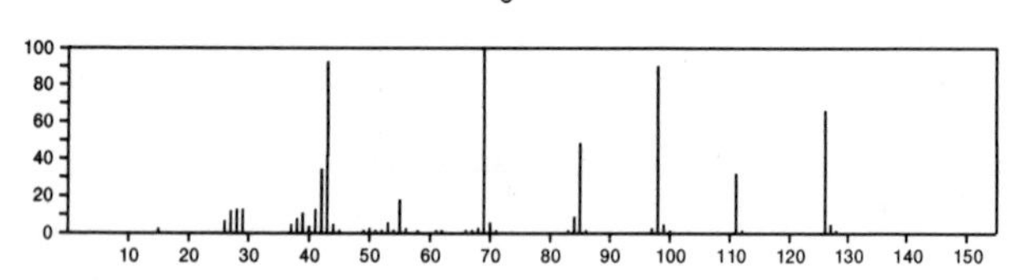

126 $C_6H_6O_3$ 3420-59-5
Ethanone, 1-(3-hydroxy-2-furanyl)-

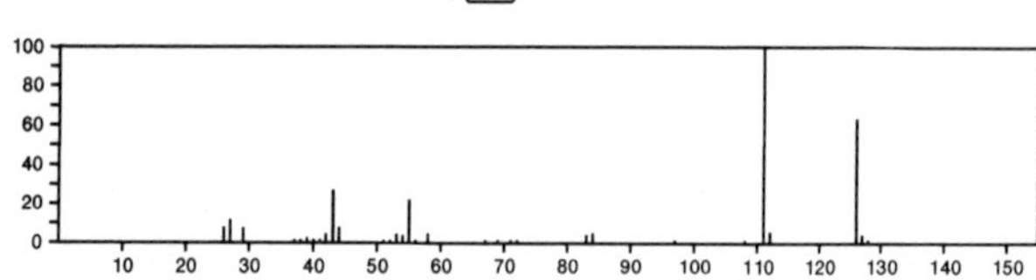

126 $C_6H_6O_3$ 4505-54-8
1,2,4-Cyclopentanetrione, 3-methyl-

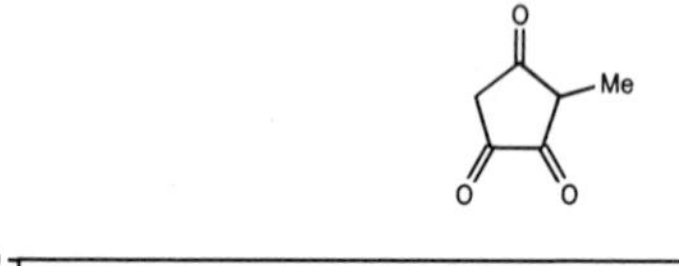

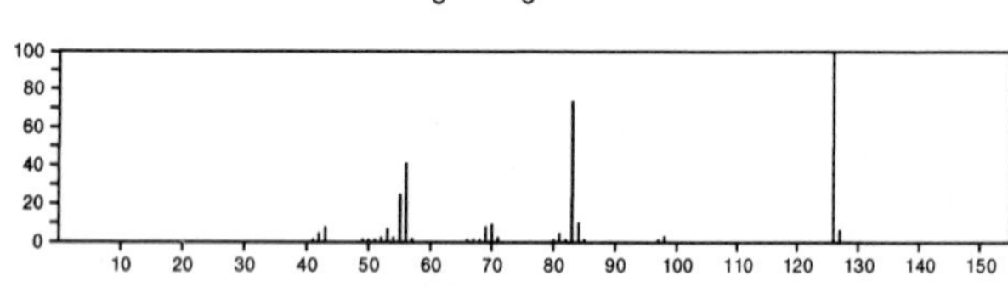

126 $C_6H_6O_3$ 13129-23-2
3-Furancarboxylic acid, methyl ester

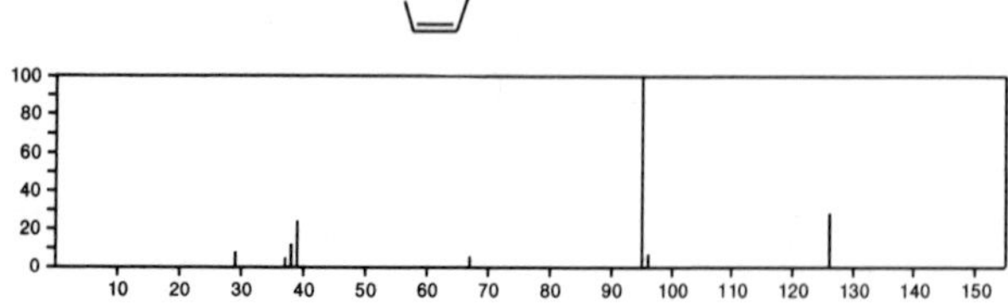

126 $C_6H_{10}N_2O$ 1632-26-4
1,2-Diazabicyclo[2.2.2]octan-3-one

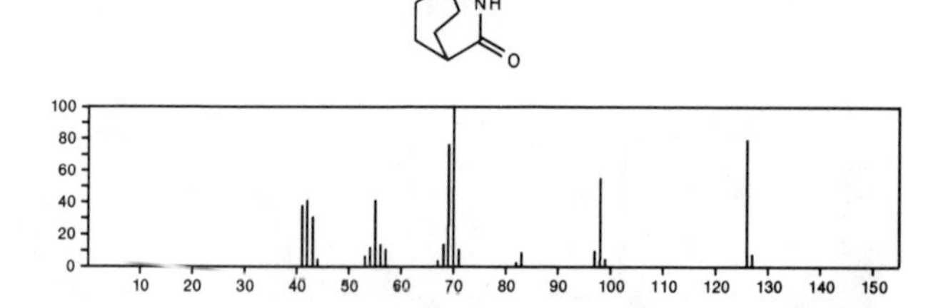

126 $C_6H_{10}N_2O$ 3201-20-5
3*H*-Pyrazol-3-one, 2,4-dihydro-4,4,5-trimethyl-

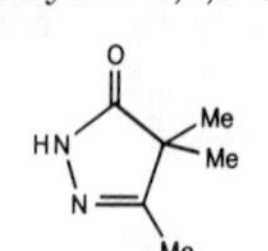

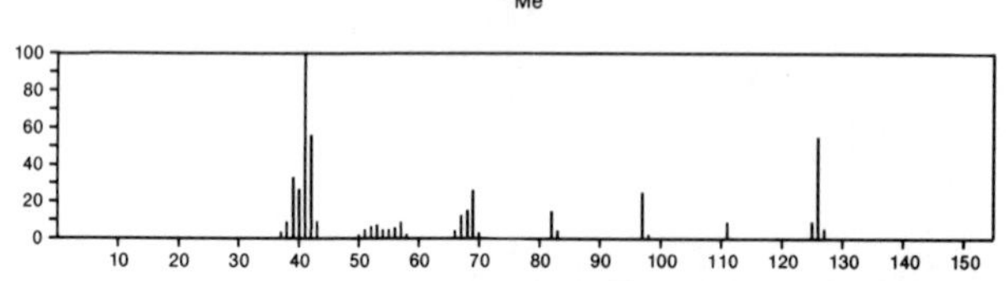

126 $C_6H_{10}N_2O$ 3201-21-6
1*H*-Pyrazole, 3-ethoxy-5-methyl-

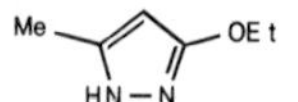

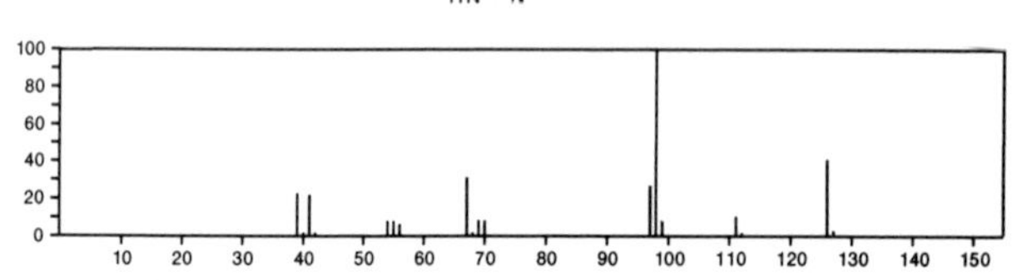

126 $C_6H_{10}N_2O$ 3201-26-1
3*H*-Pyrazol-3-one, 1,2-dihydro-1,2,5-trimethyl-

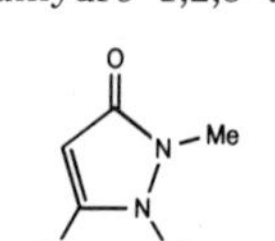

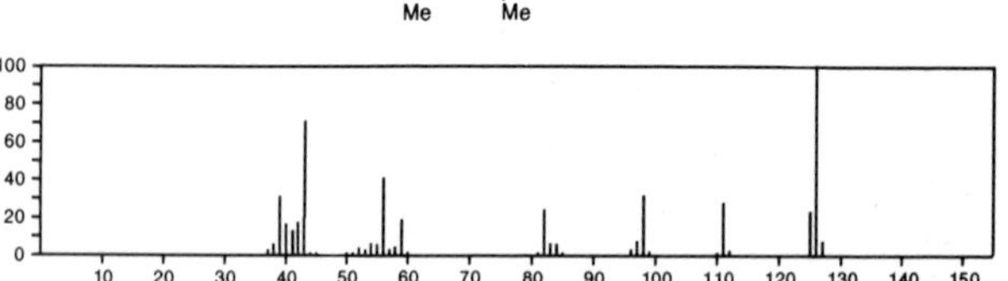

126 $C_6H_{10}N_2O$ 17826-82-3
3*H*-Pyrazol-3-one, 2,4-dihydro-2,4,5-trimethyl-

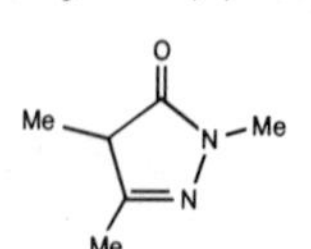

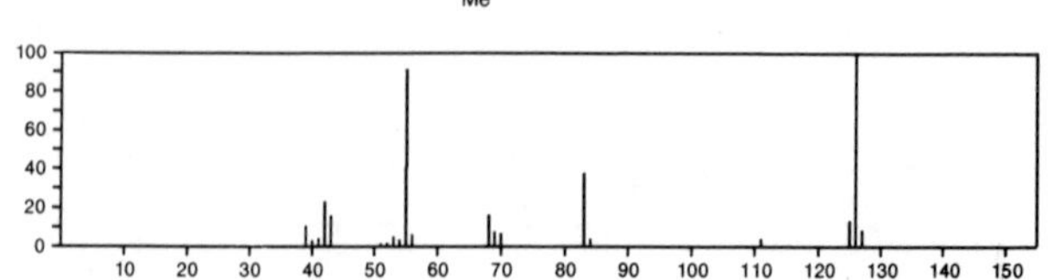

126 $C_6H_{10}N_2O$ 53091-80-8
1*H*-Pyrazole, 5-methoxy-1,3-dimethyl-

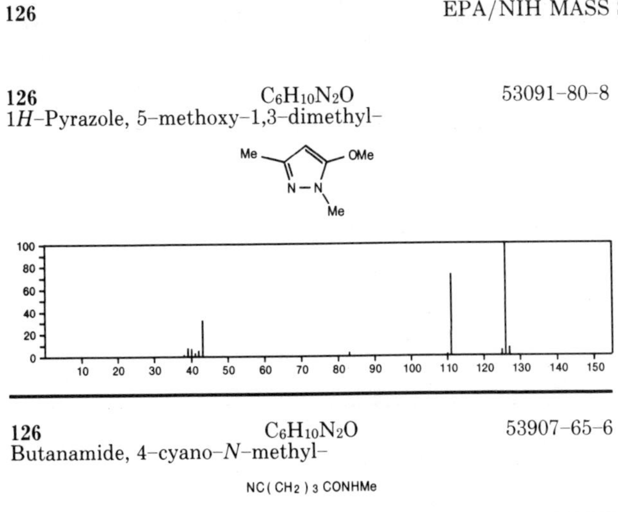

126 $C_6H_{10}N_2O$ 53907-65-6
Butanamide, 4-cyano-*N*-methyl-

NC(CH2)3CONHMe

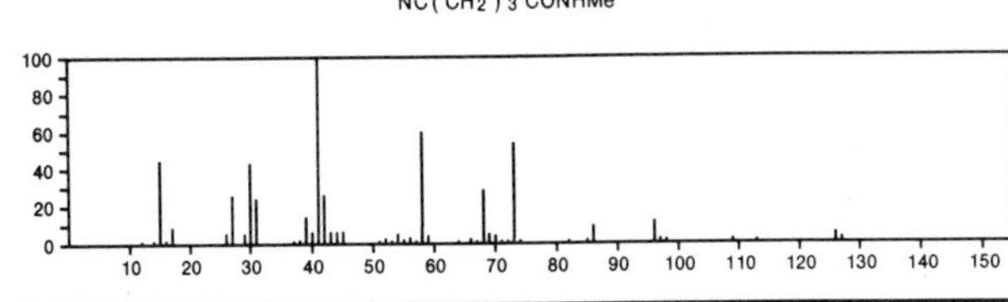

126 $C_6H_{11}BO_2$ 10534-18-6
Boric acid (HBO_2), cyclohexyl ester

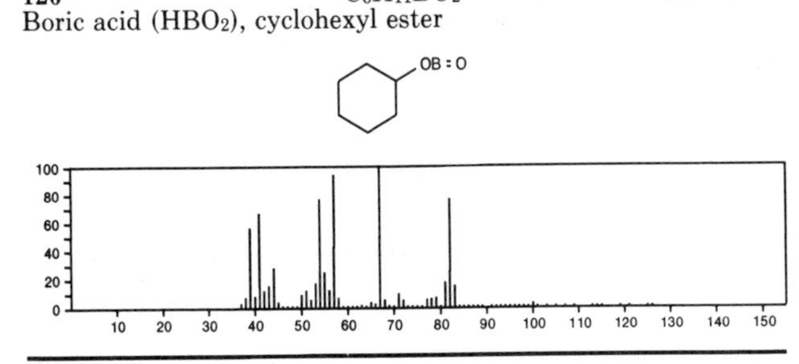

126 C_7H_7Cl 95-49-8
Benzene, 1-chloro-2-methyl-

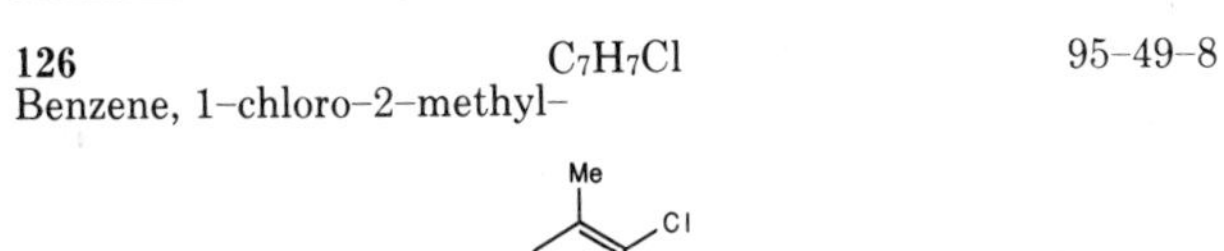

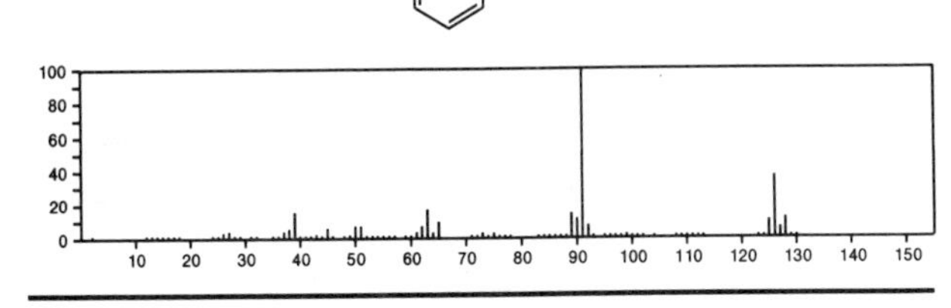

126 C_7H_7Cl 100-44-7
Benzene, (chloromethyl)-

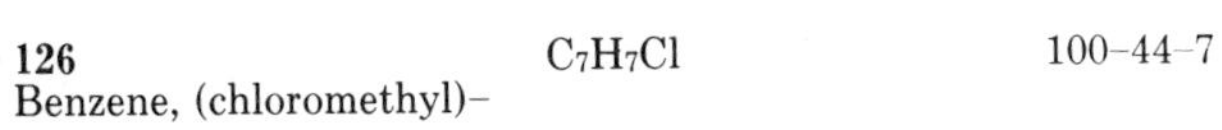

ClCH2Ph

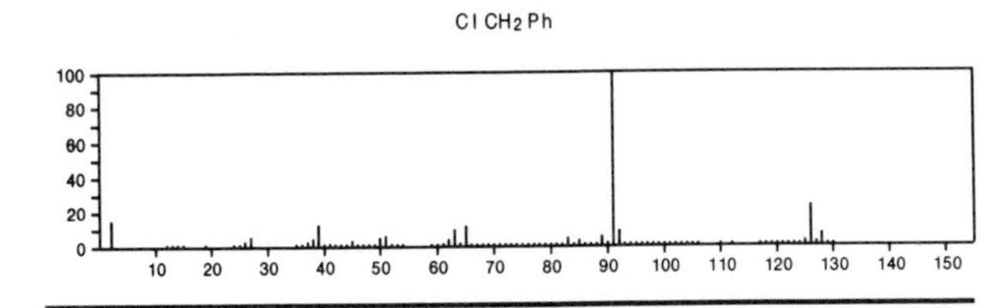

126 C_7H_7Cl 106-43-4
Benzene, 1-chloro-4-methyl-

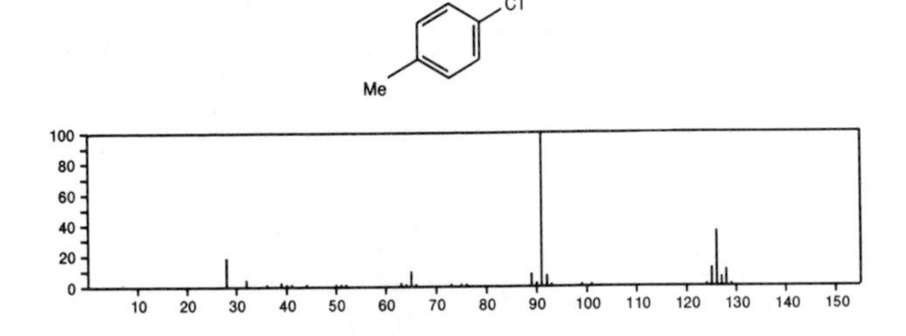

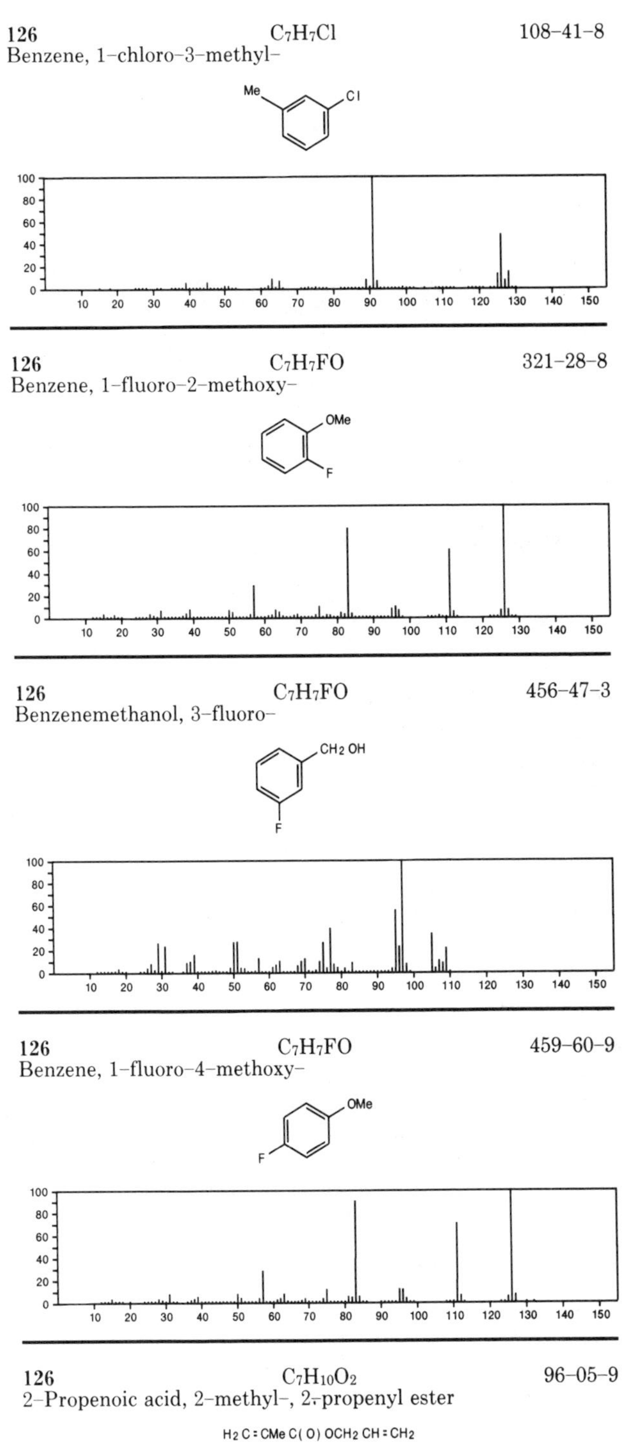

126 C_7H_7Cl 108-41-8
Benzene, 1-chloro-3-methyl-

126 C_7H_7FO 321-28-8
Benzene, 1-fluoro-2-methoxy-

126 C_7H_7FO 456-47-3
Benzenemethanol, 3-fluoro-

126 C_7H_7FO 459-60-9
Benzene, 1-fluoro-4-methoxy-

126 $C_7H_{10}O_2$ 96-05-9
2-Propenoic acid, 2-methyl-, 2-propenyl ester

H2C=CMeC(O)OCH2CH=CH2

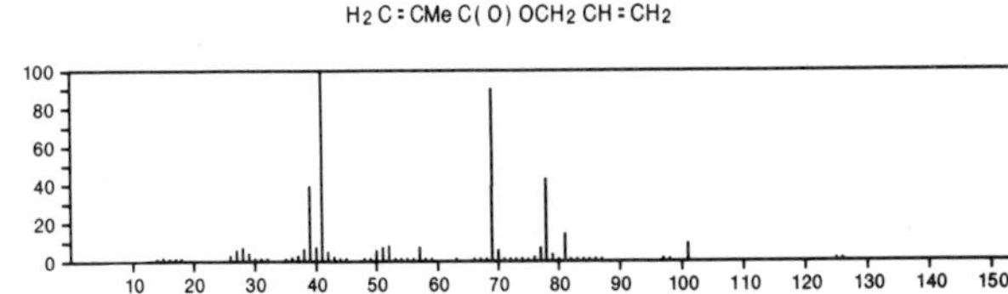

126 $C_7H_{10}O_2$ 636-82-8
1-Cyclohexene-1-carboxylic acid

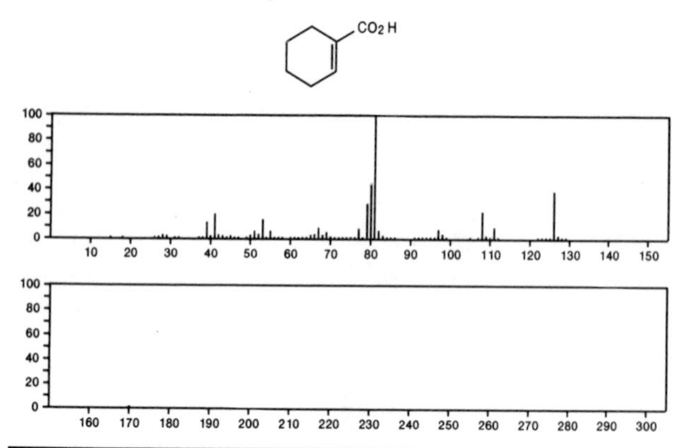

126 $C_7H_{10}O_2$ 823-36-9
1,3-Cyclopentanedione, 2-ethyl-

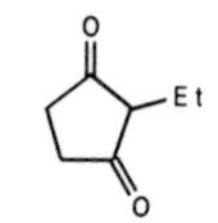

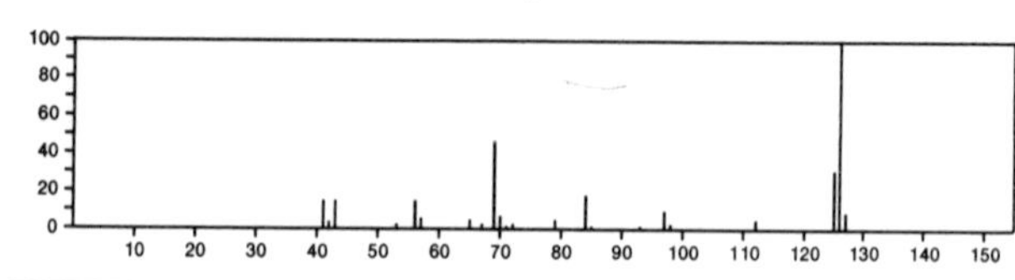

126 $C_7H_{10}O_2$ 1073-11-6
2(3*H*)-Furanone, 5-ethenyldihydro-5-methyl-

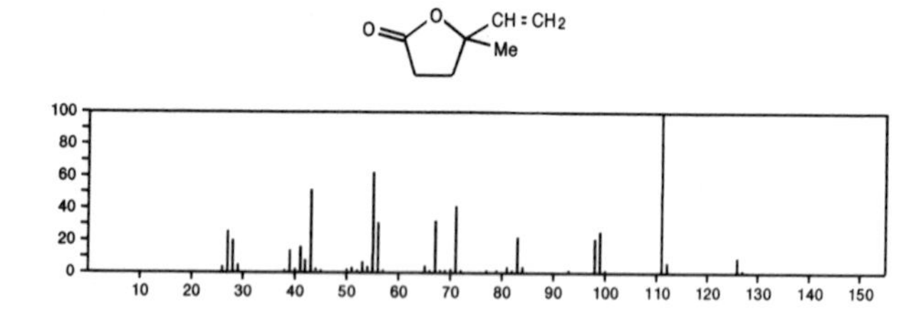

126 $C_7H_{10}O_2$ 1515-80-6
2,4-Hexadienoic acid, methyl ester

MeCH=CHCH=CHC(O)OMe

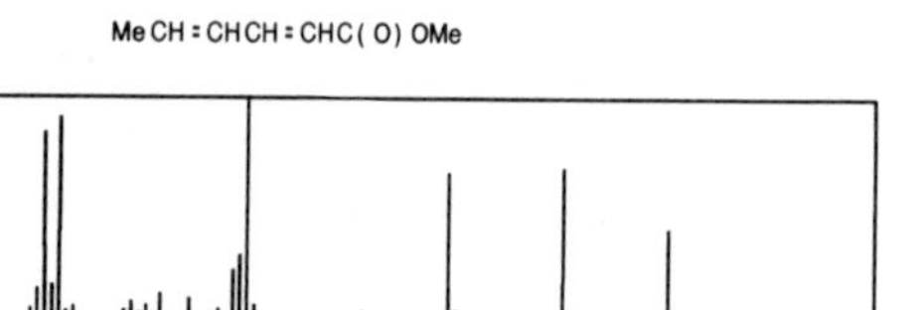

126 $C_7H_{10}O_2$ 3883-58-7
1,3-Cyclopentanedione, 2,2-dimethyl-

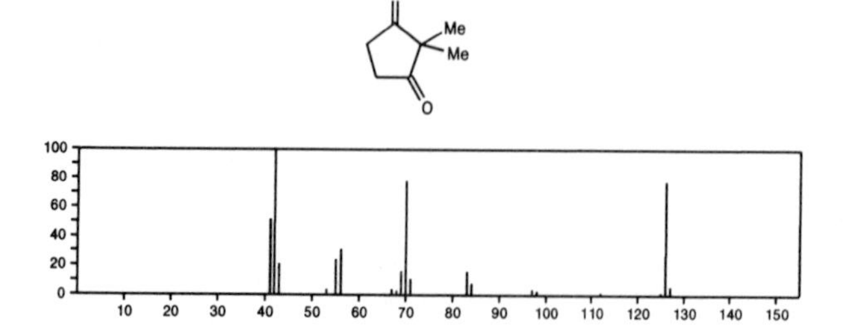

126 $C_7H_{10}O_2$ 4683-51-6
1,3-Cyclopentanedione, 4,4-dimethyl-

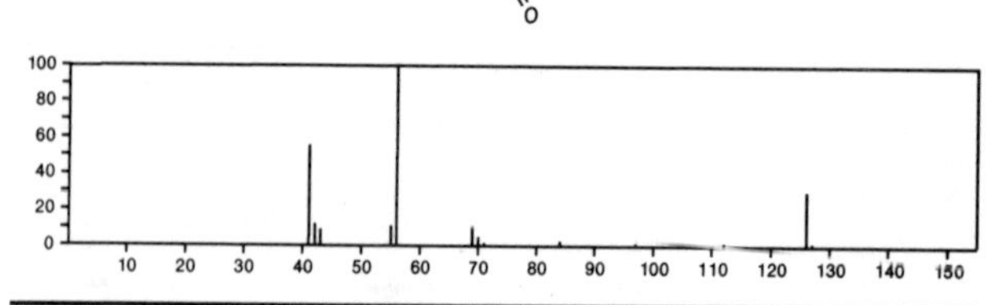

126 $C_7H_{10}O_2$ 4771-80-6
3-Cyclohexene-1-carboxylic acid

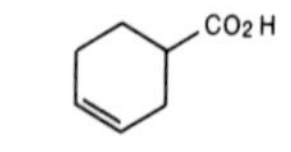

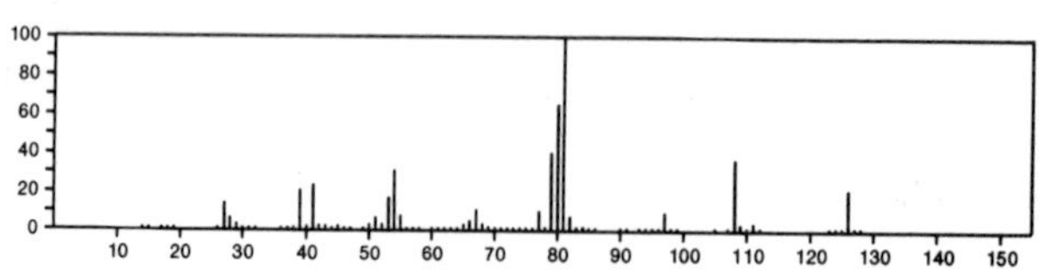

126 $C_7H_{10}O_2$ 7180-60-1
2-Cyclopenten-1-one, 3-methoxy-5-methyl-

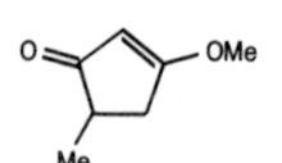

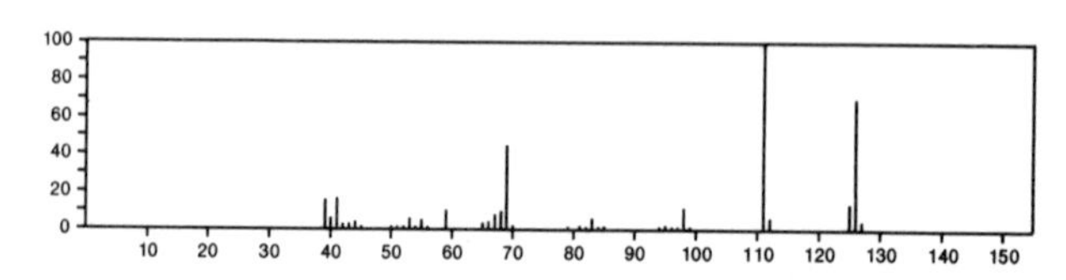

126 $C_7H_{10}O_2$ 7180-61-2
2-Cyclopenten-1-one, 3-methoxy-4-methyl-

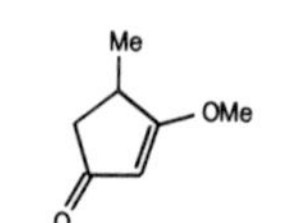

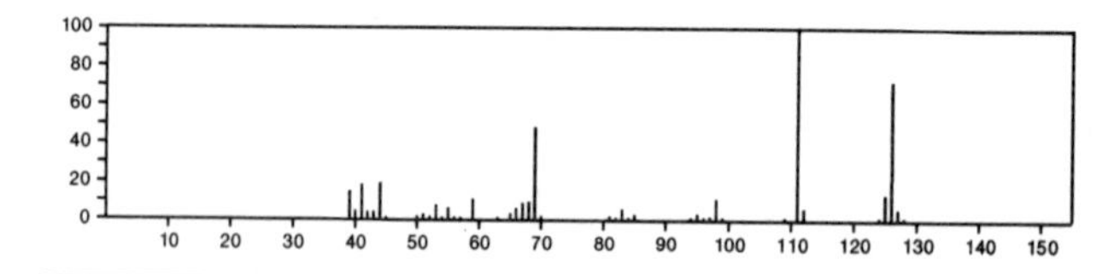

126 $C_7H_{10}O_2$ 17714-49-7
5,8-Dioxaspiro[3.4]octane, 1-methylene-

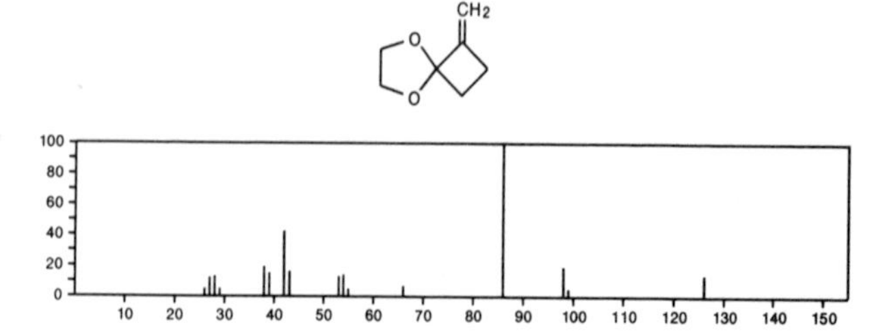

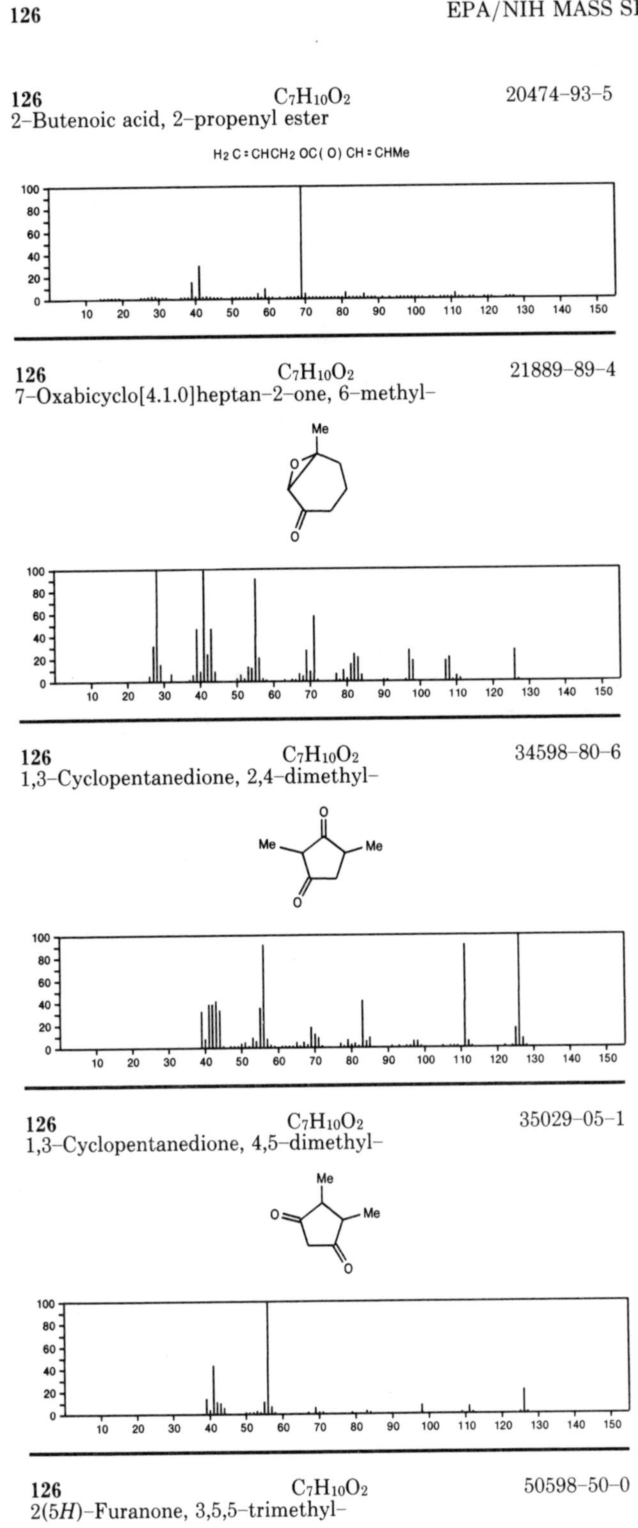

126 $C_7H_{10}O_2$ 20474-93-5
2-Butenoic acid, 2-propenyl ester

$H_2C=CHCH_2OC(O)CH=CHMe$

126 $C_7H_{10}O_2$ 21889-89-4
7-Oxabicyclo[4.1.0]heptan-2-one, 6-methyl-

126 $C_7H_{10}O_2$ 34598-80-6
1,3-Cyclopentanedione, 2,4-dimethyl-

126 $C_7H_{10}O_2$ 35029-05-1
1,3-Cyclopentanedione, 4,5-dimethyl-

126 $C_7H_{10}O_2$ 50598-50-0
2(*5H*)-Furanone, 3,5,5-trimethyl-

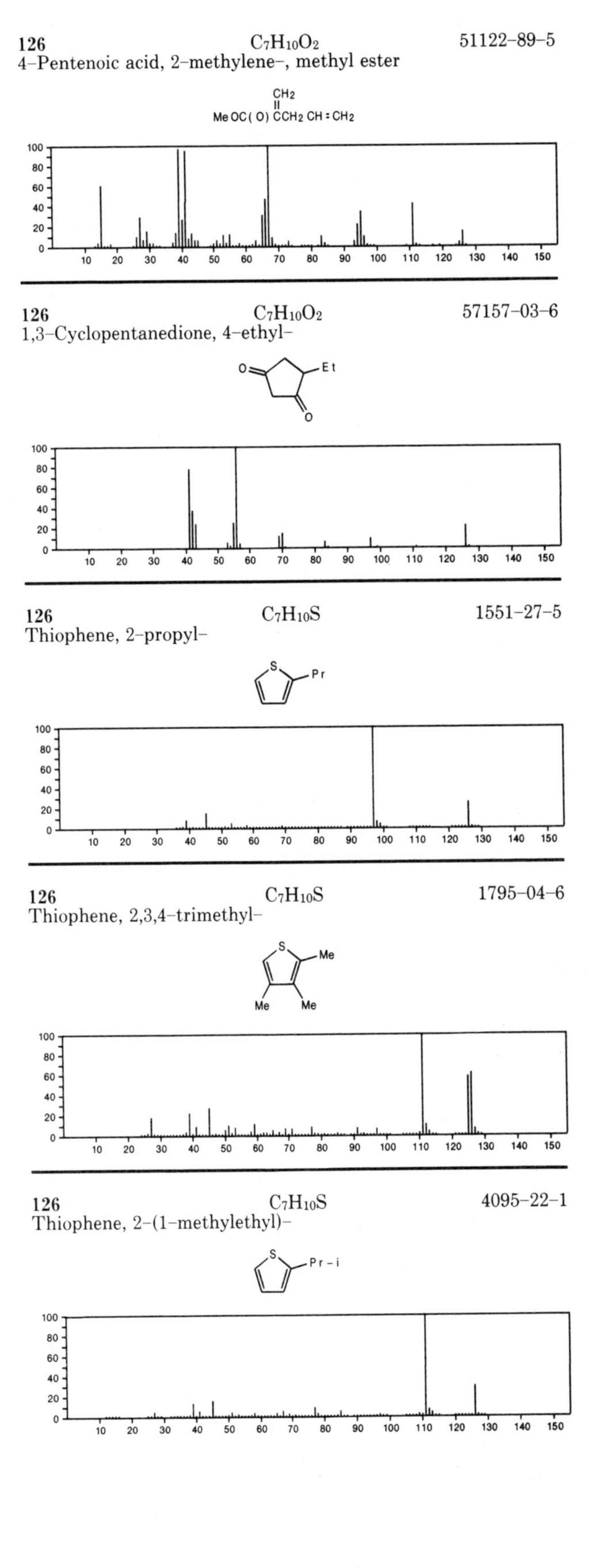

126 $C_7H_{10}O_2$ 51122-89-5
4-Pentenoic acid, 2-methylene-, methyl ester

$MeOC(O)C(=CH_2)CH_2CH=CH_2$

126 $C_7H_{10}O_2$ 57157-03-6
1,3-Cyclopentanedione, 4-ethyl-

126 $C_7H_{10}S$ 1551-27-5
Thiophene, 2-propyl-

126 $C_7H_{10}S$ 1795-04-6
Thiophene, 2,3,4-trimethyl-

126 $C_7H_{10}S$ 4095-22-1
Thiophene, 2-(1-methylethyl)-

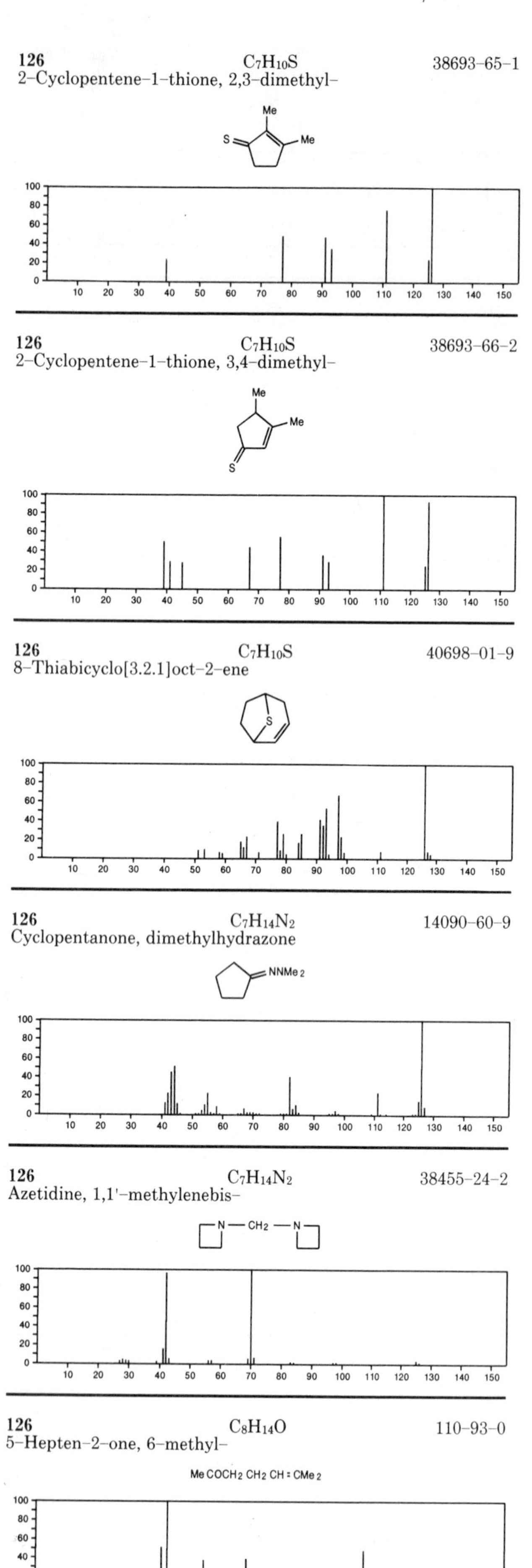

126 $C_7H_{10}S$ 38693-65-1
2-Cyclopentene-1-thione, 2,3-dimethyl-

126 $C_7H_{10}S$ 38693-66-2
2-Cyclopentene-1-thione, 3,4-dimethyl-

126 $C_7H_{10}S$ 40698-01-9
8-Thiabicyclo[3.2.1]oct-2-ene

126 $C_7H_{14}N_2$ 14090-60-9
Cyclopentanone, dimethylhydrazone

126 $C_7H_{14}N_2$ 38455-24-2
Azetidine, 1,1'-methylenebis-

126 $C_8H_{14}O$ 110-93-0
5-Hepten-2-one, 6-methyl-

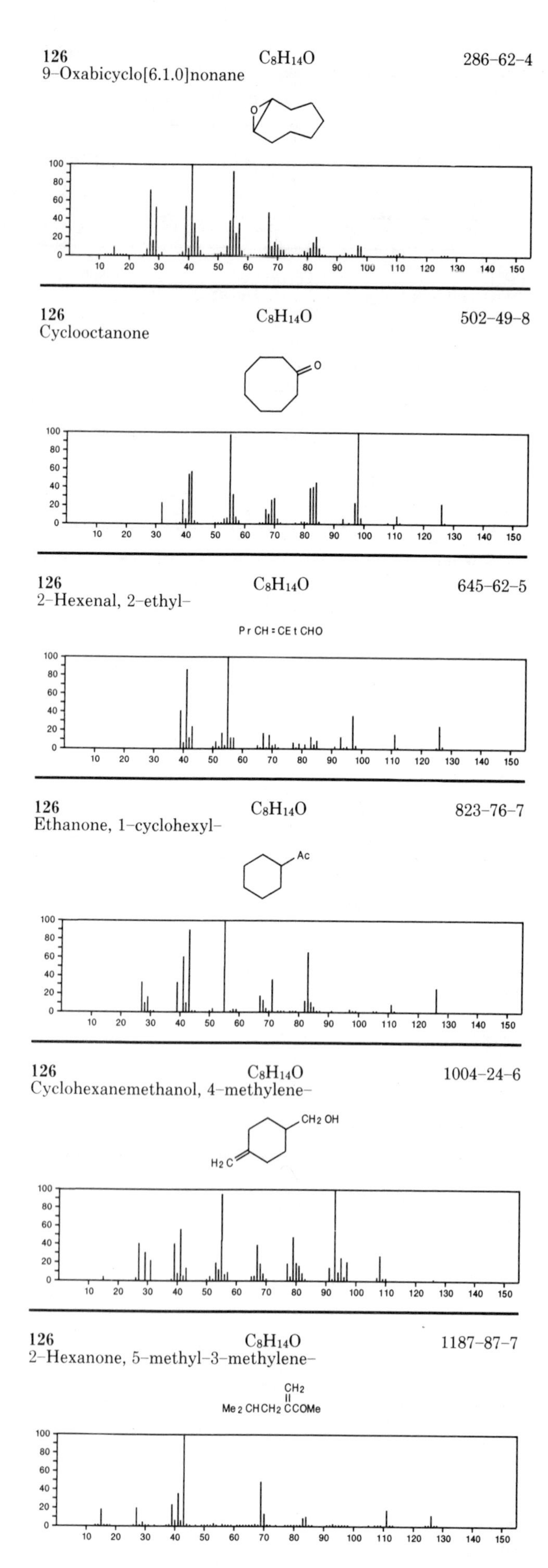

126 $C_8H_{14}O$ 286-62-4
9-Oxabicyclo[6.1.0]nonane

126 $C_8H_{14}O$ 502-49-8
Cyclooctanone

126 $C_8H_{14}O$ 645-62-5
2-Hexenal, 2-ethyl-

126 $C_8H_{14}O$ 823-76-7
Ethanone, 1-cyclohexyl-

126 $C_8H_{14}O$ 1004-24-6
Cyclohexanemethanol, 4-methylene-

126 $C_8H_{14}O$ 1187-87-7
2-Hexanone, 5-methyl-3-methylene-

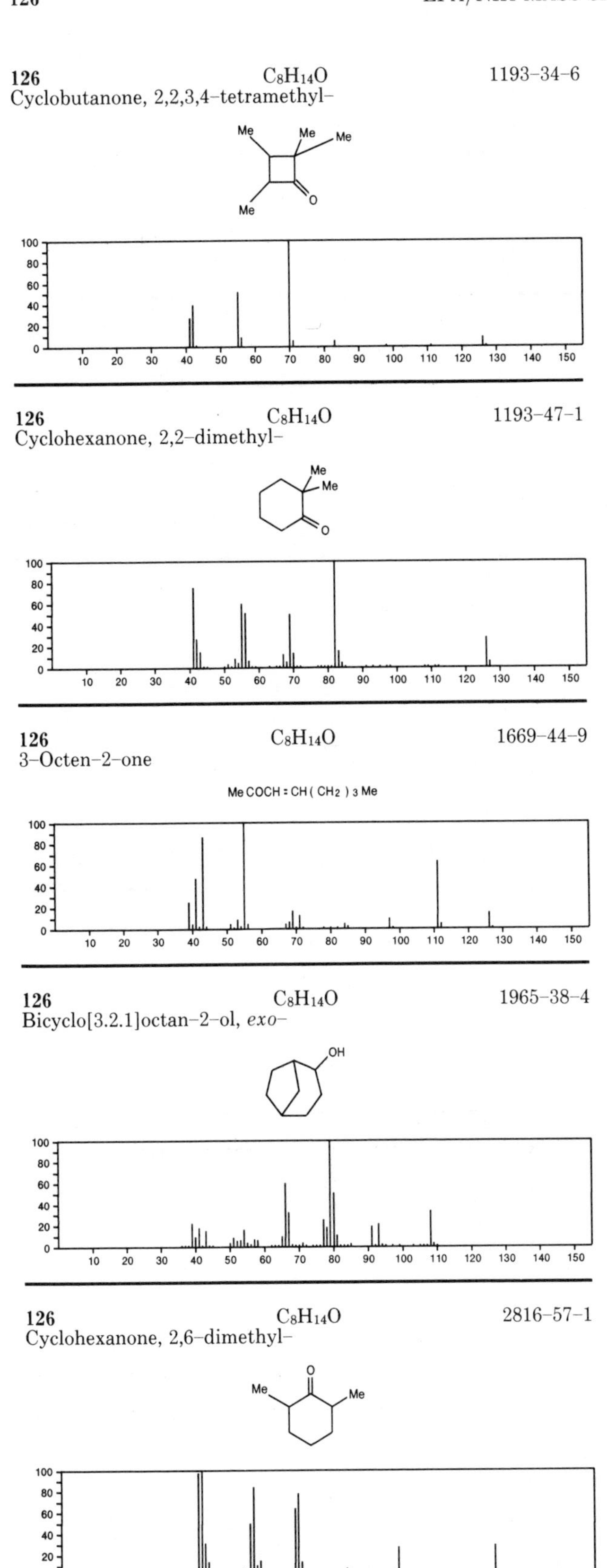

126 $C_8H_{14}O$ 1193–34–6
Cyclobutanone, 2,2,3,4–tetramethyl–

126 $C_8H_{14}O$ 1193–47–1
Cyclohexanone, 2,2–dimethyl–

126 $C_8H_{14}O$ 1669–44–9
3–Octen–2–one

MeCOCH=CH(CH2)3Me

126 $C_8H_{14}O$ 1965–38–4
Bicyclo[3.2.1]octan–2–ol, *exo*–

126 $C_8H_{14}O$ 2816–57–1
Cyclohexanone, 2,6–dimethyl–

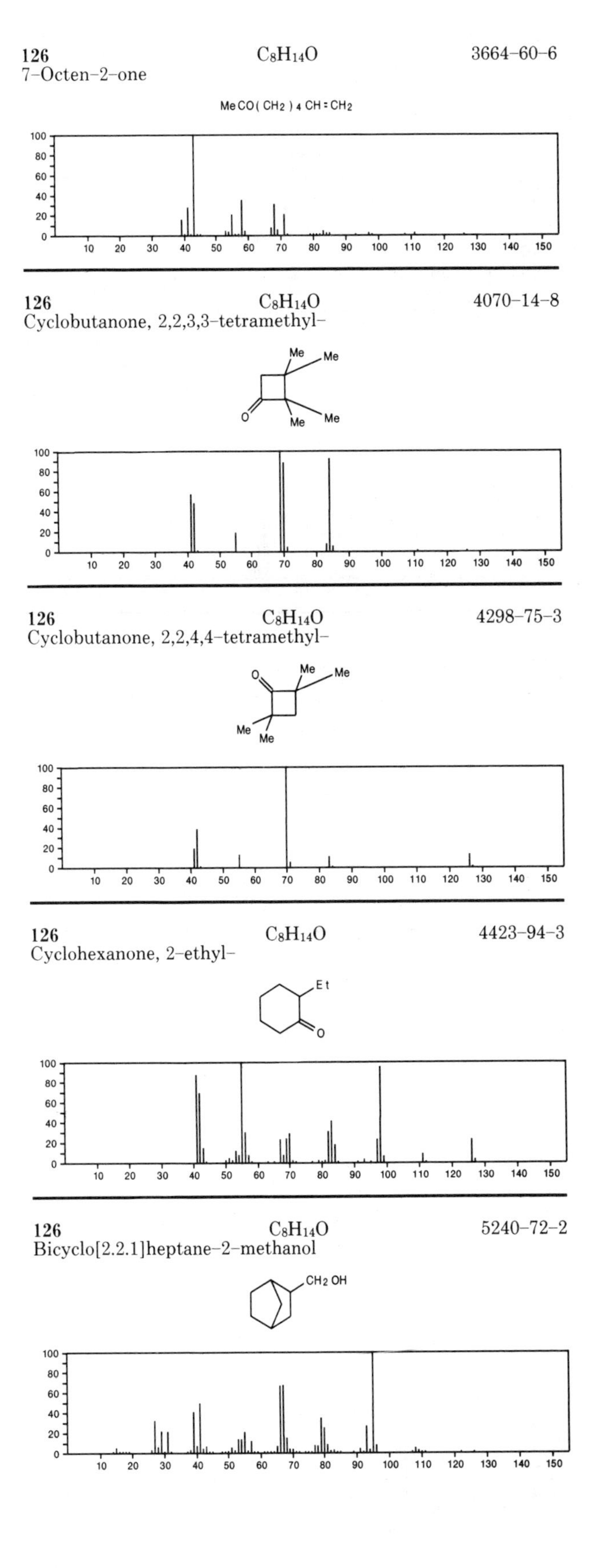

126 $C_8H_{14}O$ 3664–60–6
7–Octen–2–one

MeCO(CH2)4CH=CH2

126 $C_8H_{14}O$ 4070–14–8
Cyclobutanone, 2,2,3,3–tetramethyl–

126 $C_8H_{14}O$ 4298–75–3
Cyclobutanone, 2,2,4,4–tetramethyl–

126 $C_8H_{14}O$ 4423–94–3
Cyclohexanone, 2–ethyl–

126 $C_8H_{14}O$ 5240–72–2
Bicyclo[2.2.1]heptane–2–methanol

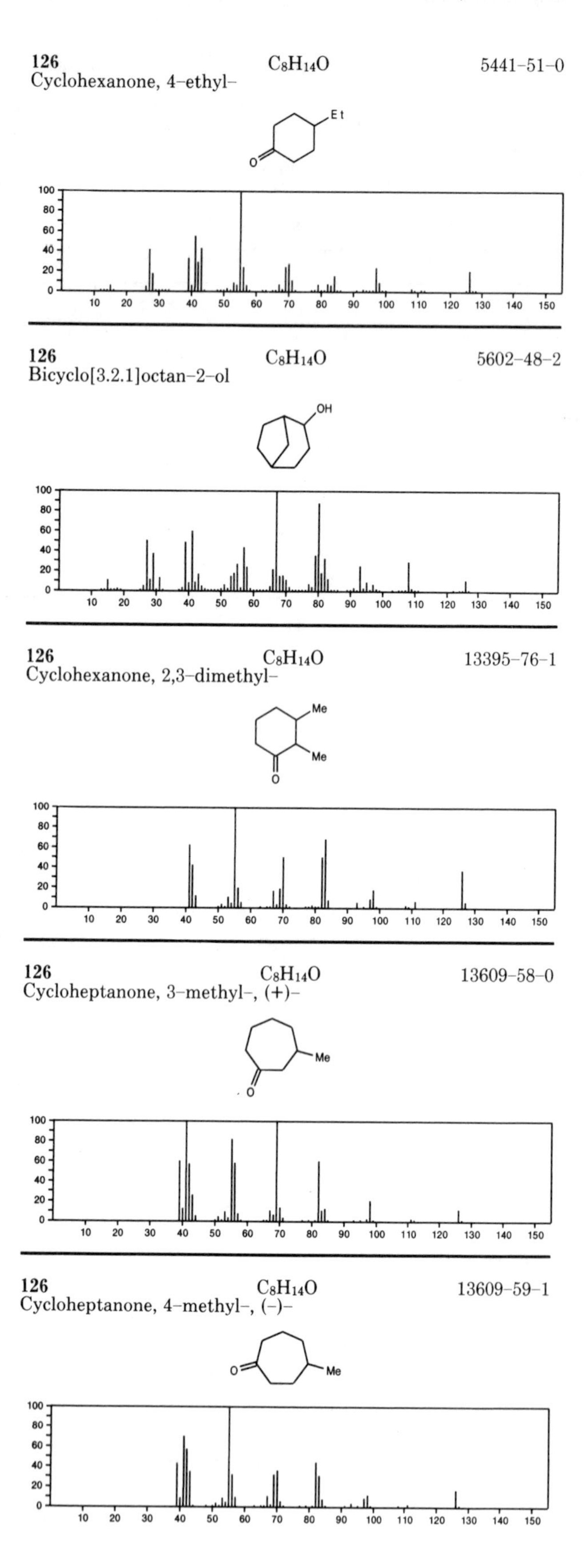
126
$C_8H_{14}O$
5441-51-0
Cyclohexanone, 4-ethyl-
Et
O
126
$C_8H_{14}O$
5602-48-2
Bicyclo[3.2.1]octan-2-ol
OH
126
$C_8H_{14}O$
13395-76-1
Cyclohexanone, 2,3-dimethyl-
Me
Me
O
126
$C_8H_{14}O$
13609-58-0
Cycloheptanone, 3-methyl-, (+)-
Me
O
126
$C_8H_{14}O$
13609-59-1
Cycloheptanone, 4-methyl-, (-)-
O
Me

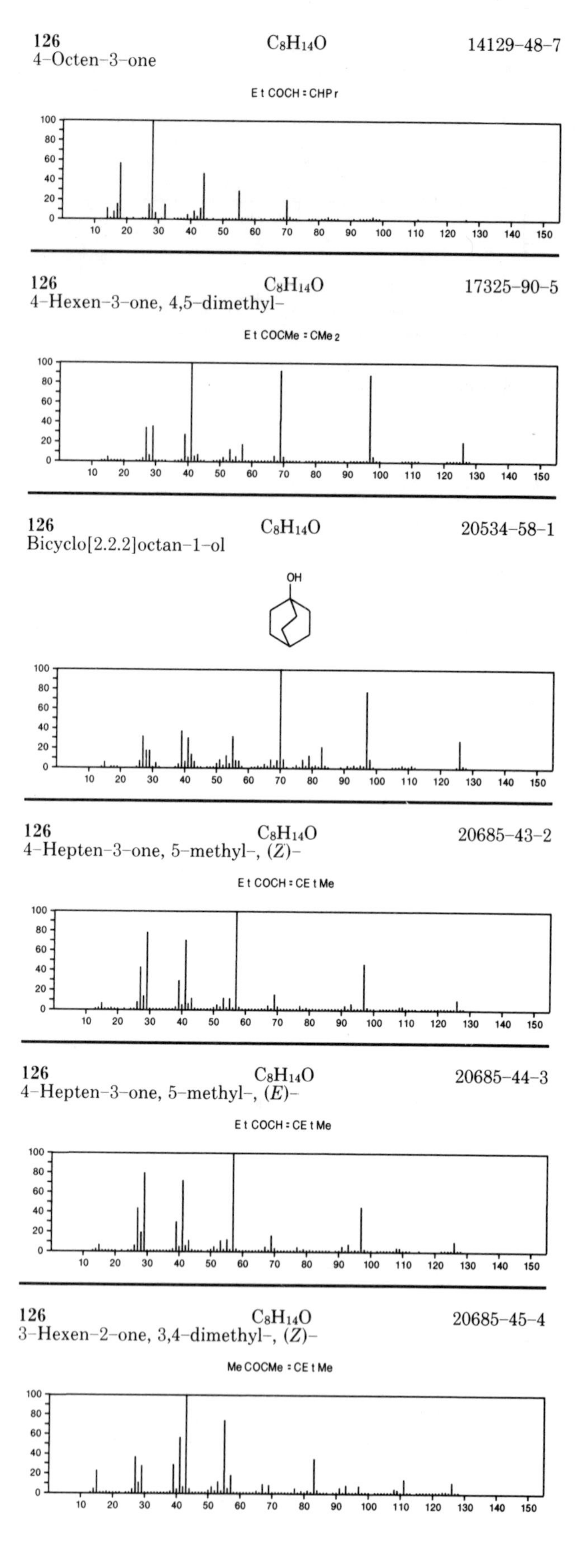
126
$C_8H_{14}O$
14129-48-7
4-Octen-3-one
Et COCH : CHPr
126
$C_8H_{14}O$
17325-90-5
4-Hexen-3-one, 4,5-dimethyl-
Et COCMe : CMe2
126
$C_8H_{14}O$
20534-58-1
Bicyclo[2.2.2]octan-1-ol
OH
126
$C_8H_{14}O$
20685-43-2
4-Hepten-3-one, 5-methyl-, (Z)-
Et COCH : CEt Me
126
$C_8H_{14}O$
20685-44-3
4-Hepten-3-one, 5-methyl-, (E)-
Et COCH : CEt Me
126
$C_8H_{14}O$
20685-45-4
3-Hexen-2-one, 3,4-dimethyl-, (Z)-
Me COCMe : CEt Me

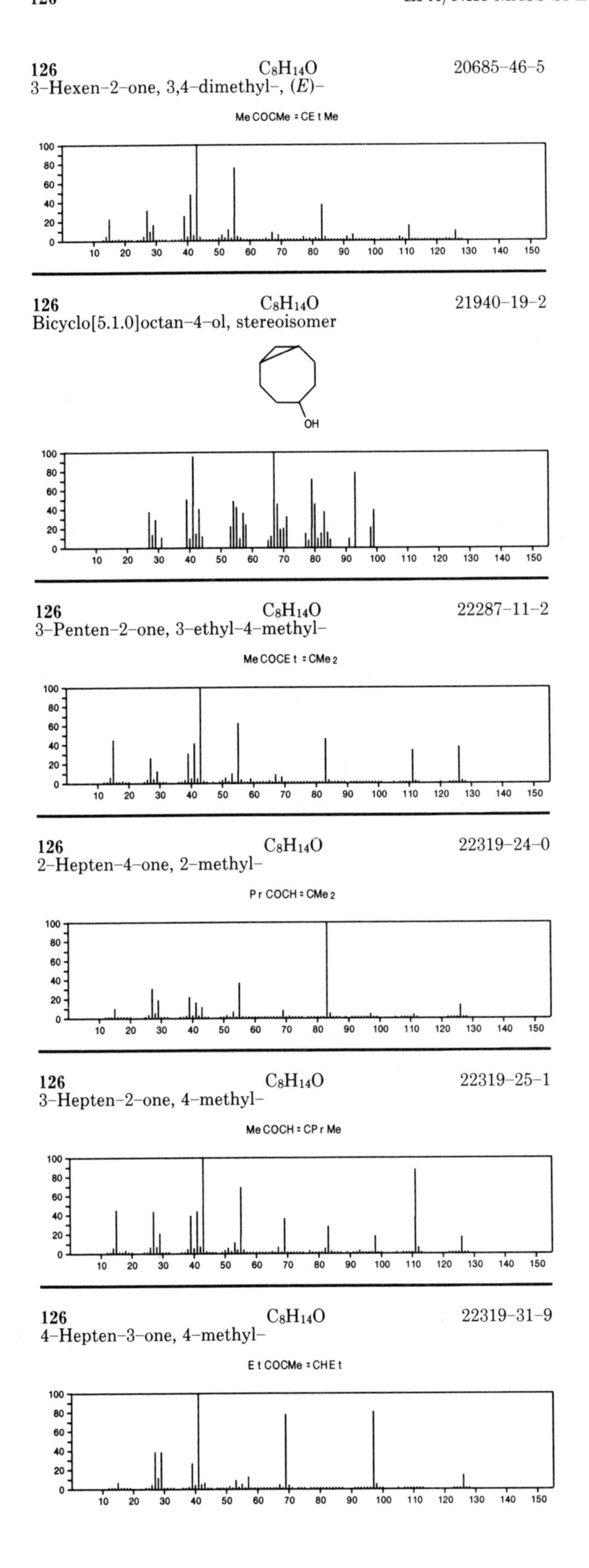
126 $C_8H_{14}O$ 20685-46-5
3-Hexen-2-one, 3,4-dimethyl-, (*E*)-
Me COCMe = CEt Me
126 $C_8H_{14}O$ 21940-19-2
Bicyclo[5.1.0]octan-4-ol, stereoisomer
OH
126 $C_8H_{14}O$ 22287-11-2
3-Penten-2-one, 3-ethyl-4-methyl-
Me COCEt = CMe 2
126 $C_8H_{14}O$ 22319-24-0
2-Hepten-4-one, 2-methyl-
Pr COCH = CMe 2
126 $C_8H_{14}O$ 22319-25-1
3-Hepten-2-one, 4-methyl-
Me COCH = CPr Me
126 $C_8H_{14}O$ 22319-31-9
4-Hepten-3-one, 4-methyl-
Et COCMe = CHEt

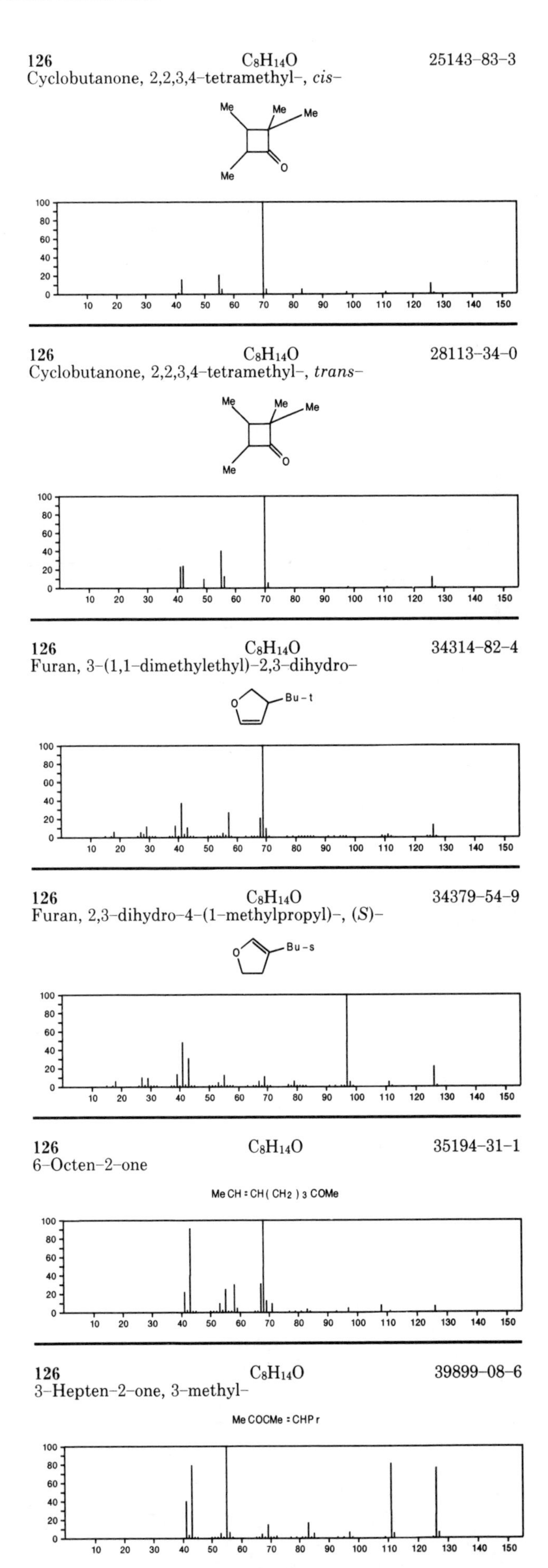
126 $C_8H_{14}O$ 25143-83-3
Cyclobutanone, 2,2,3,4-tetramethyl-, *cis*-
Me Me Me Me O
126 $C_8H_{14}O$ 28113-34-0
Cyclobutanone, 2,2,3,4-tetramethyl-, *trans*-
Me Me Me Me O
126 $C_8H_{14}O$ 34314-82-4
Furan, 3-(1,1-dimethylethyl)-2,3-dihydro-
O Bu-t
126 $C_8H_{14}O$ 34379-54-9
Furan, 2,3-dihydro-4-(1-methylpropyl)-, (*S*)-
O Bu-s
126 $C_8H_{14}O$ 35194-31-1
6-Octen-2-one
Me CH = CH(CH2) 3 COMe
126 $C_8H_{14}O$ 39899-08-6
3-Hepten-2-one, 3-methyl-
Me COCMe = CHPr

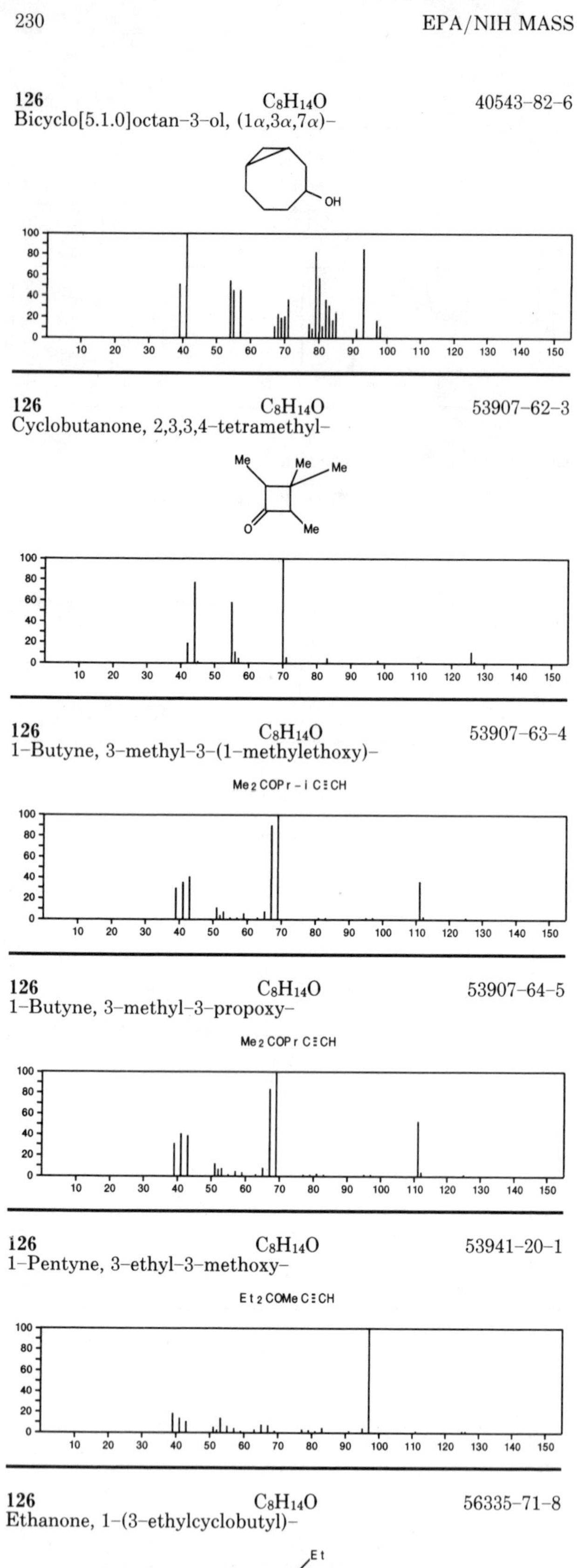

126 $C_8H_{14}O$ 40543-82-6
Bicyclo[5.1.0]octan-3-ol, (1α,3α,7α)-

126 $C_8H_{14}O$ 53907-62-3
Cyclobutanone, 2,3,3,4-tetramethyl-

126 $C_8H_{14}O$ 53907-63-4
1-Butyne, 3-methyl-3-(1-methylethoxy)-

126 $C_8H_{14}O$ 53907-64-5
1-Butyne, 3-methyl-3-propoxy-

126 $C_8H_{14}O$ 53941-20-1
1-Pentyne, 3-ethyl-3-methoxy-

126 $C_8H_{14}O$ 56335-71-8
Ethanone, 1-(3-ethylcyclobutyl)-

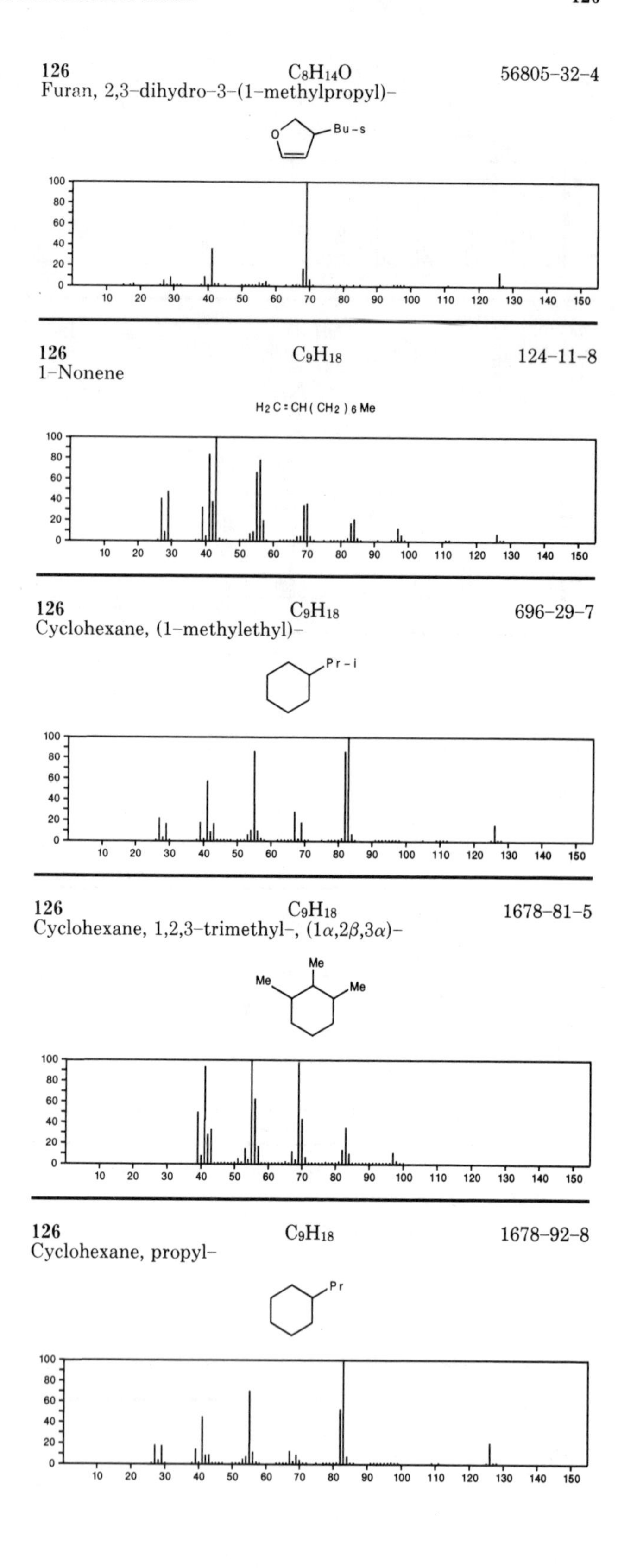

126 $C_8H_{14}O$ 56805-32-4
Furan, 2,3-dihydro-3-(1-methylpropyl)-

126 C_9H_{18} 124-11-8
1-Nonene

126 C_9H_{18} 696-29-7
Cyclohexane, (1-methylethyl)-

126 C_9H_{18} 1678-81-5
Cyclohexane, 1,2,3-trimethyl-, (1α,2β,3α)-

126 C_9H_{18} 1678-92-8
Cyclohexane, propyl-

126 C_9H_{18} 1678–97–3
Cyclohexane, 1,2,3–trimethyl–

126 C_9H_{18} 1795–26–2
Cyclohexane, 1,3,5–trimethyl–, (1α,3α,5β)–

126 C_9H_{18} 1795–27–3
Cyclohexane, 1,3,5–trimethyl–, (1α,3α,5α)–

126 C_9H_{18} 1839–63–0
Cyclohexane, 1,3,5–trimethyl–

126 C_9H_{18} 2040–95–1
Cyclopentane, butyl–

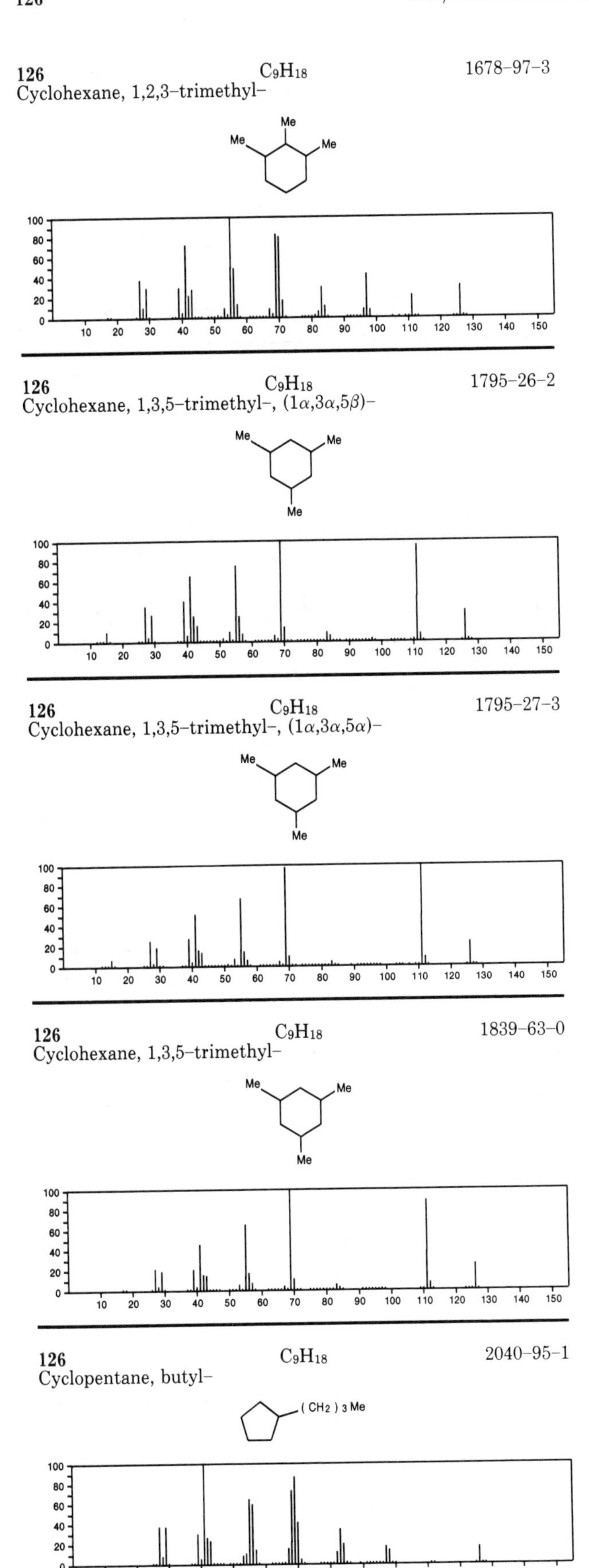

126 C_9H_{18} 2198–23–4
4–Nonene

126 C_9H_{18} 2234–75–5
Cyclohexane, 1,2,4–trimethyl–

126 C_9H_{18} 2738–18–3
3–Heptene, 2,6–dimethyl–

126 C_9H_{18} 3073–66–3
Cyclohexane, 1,1,3–trimethyl–

126 C_9H_{18} 3074–78–0
1–Heptene, 2,6–dimethyl–

126 C_9H_{18} 3788–32–7
Cyclopentane, (2–methylpropyl)–

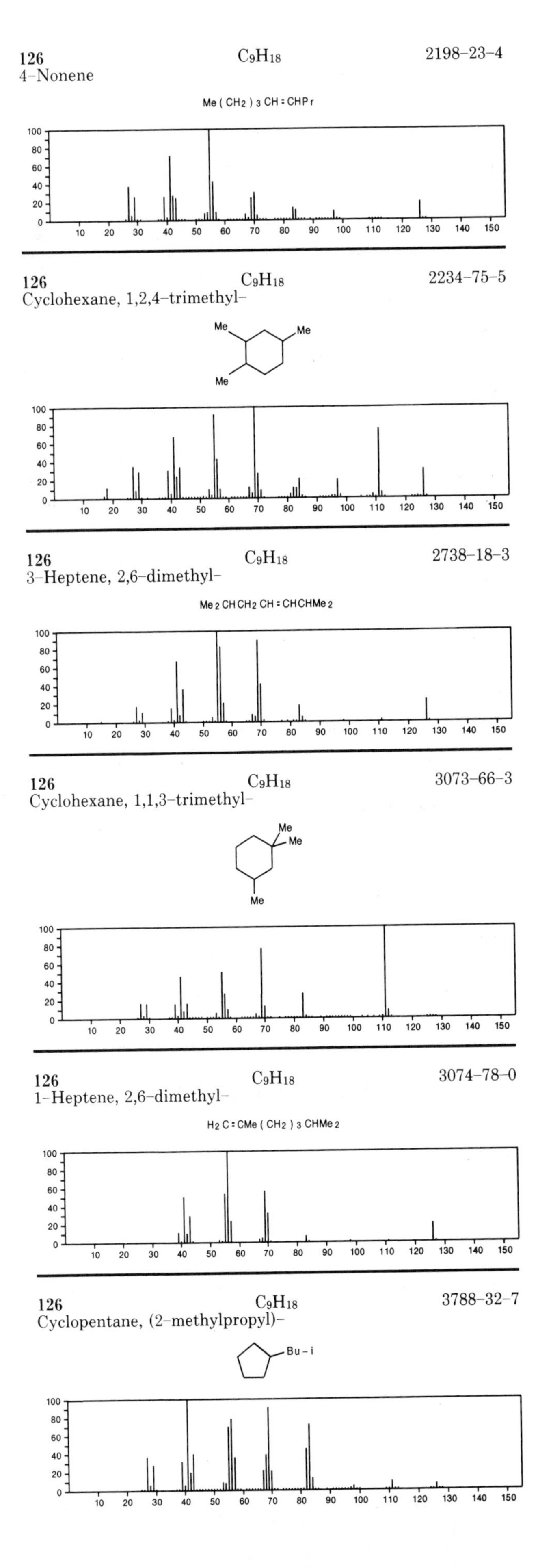

126 C_9H_{18} 4316-65-8
1-Hexene, 3,5,5-trimethyl-

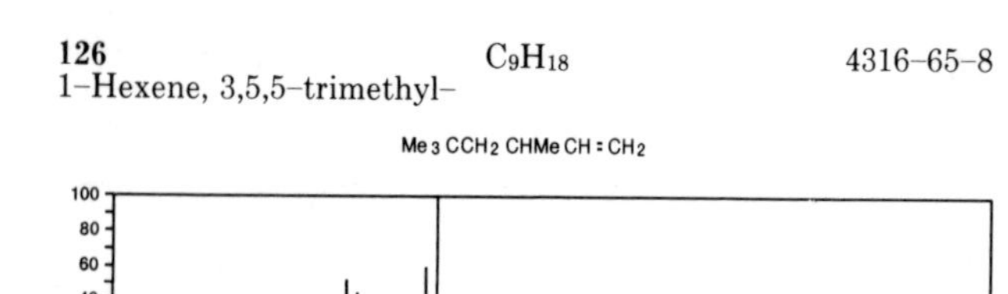

126 C_9H_{18} 4588-18-5
1-Octene, 2-methyl-

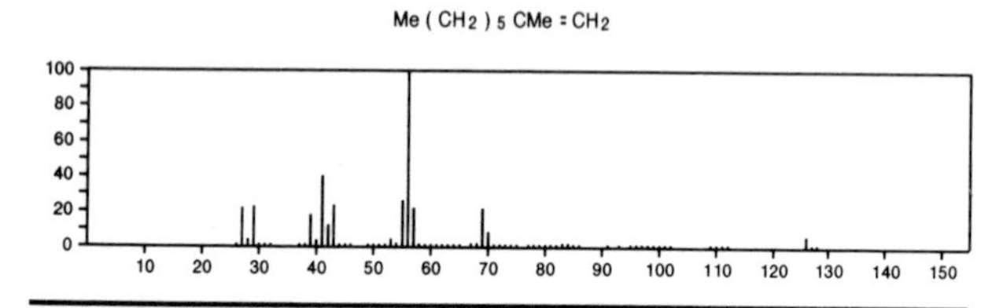

126 C_9H_{18} 4923-77-7
Cyclohexane, 1-ethyl-2-methyl-, *cis*-

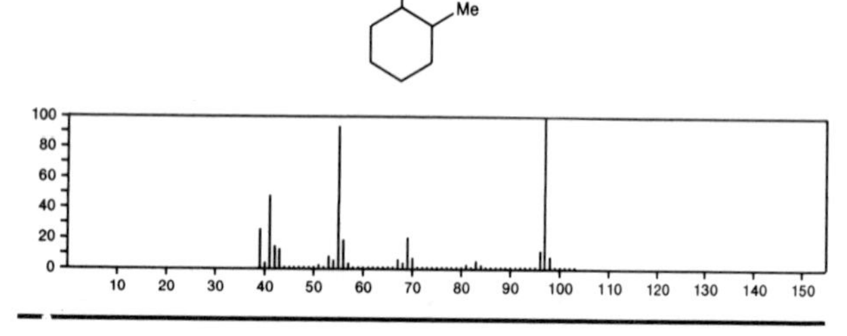

126 C_9H_{18} 4923-78-8
Cyclohexane, 1-ethyl-2-methyl-, *trans*-

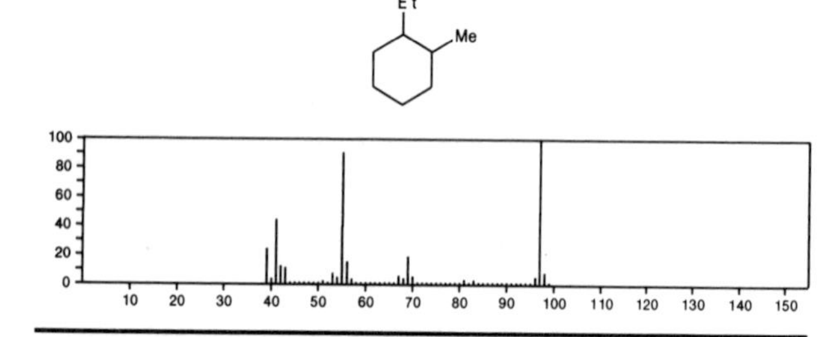

126 C_9H_{18} 4926-78-7
Cyclohexane, 1-ethyl-4-methyl-, *cis*-

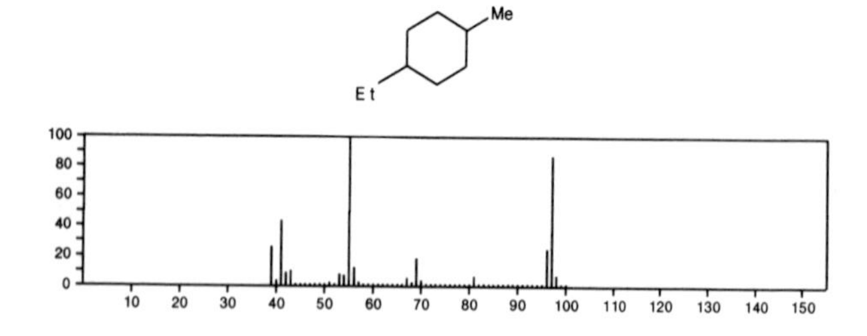

126 C_9H_{18} 4926-90-3
Cyclohexane, 1-ethyl-1-methyl-

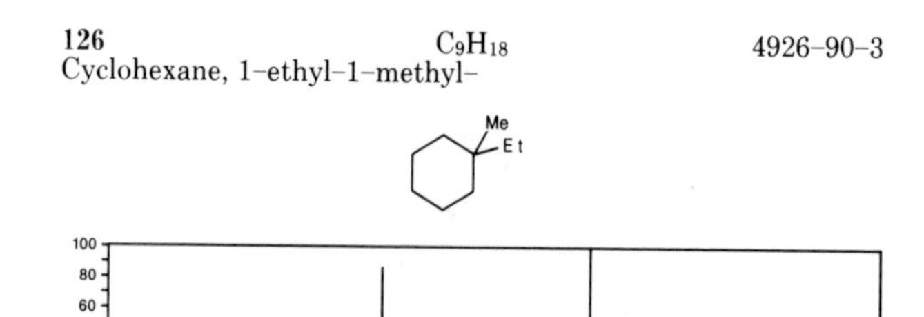

126 C_9H_{18} 6236-88-0
Cyclohexane, 1-ethyl-4-methyl-, *trans*-

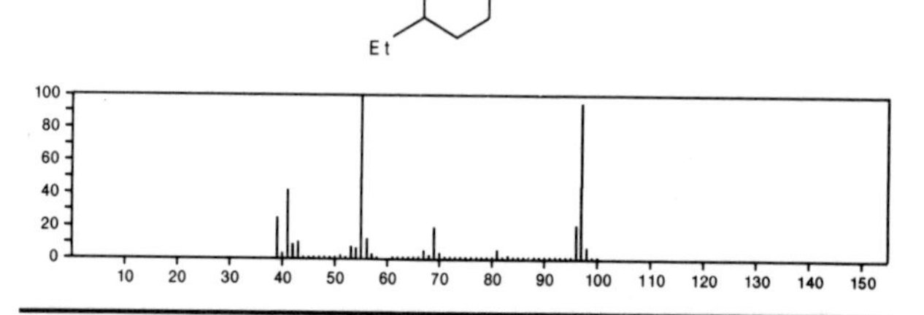

126 C_9H_{18} 7094-26-0
Cyclohexane, 1,1,2-trimethyl-

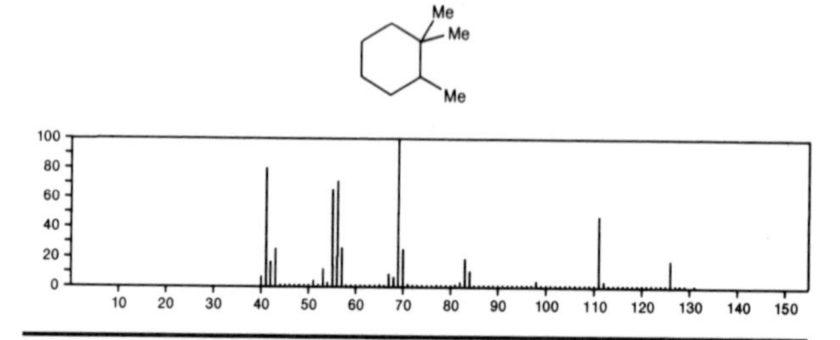

126 C_9H_{18} 7667-55-2
Cyclohexane, 1,2,3-trimethyl-, $(1\alpha,2\alpha,3\beta)$-

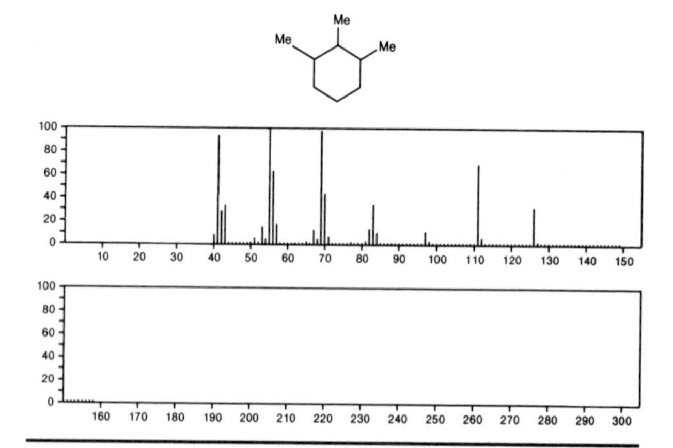

126 C_9H_{18} 7667-60-9
Cyclohexane, 1,2,4-trimethyl-, $(1\alpha,2\beta,4\beta)$-

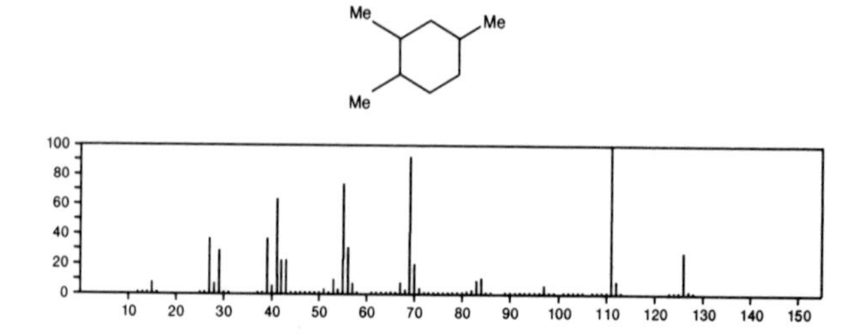

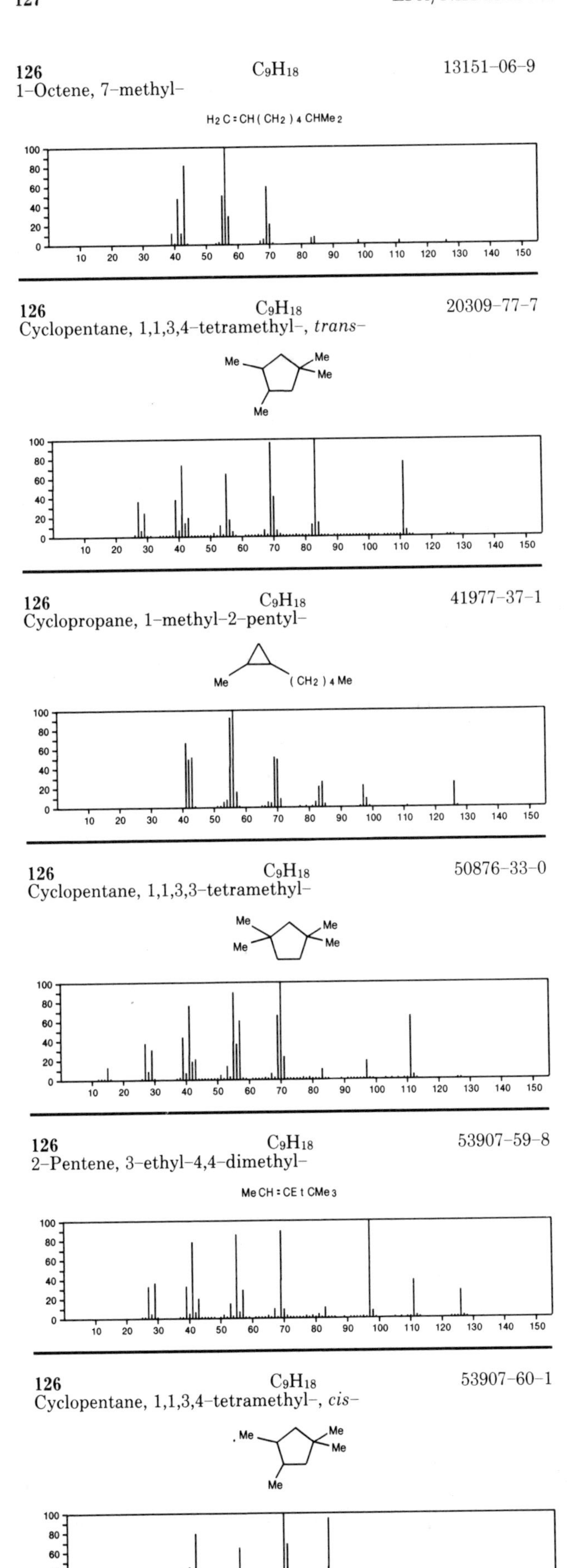

126 C_9H_{18} 13151-06-9
1-Octene, 7-methyl-
$H_2C{:}CH(CH_2)_4CHMe_2$

126 C_9H_{18} 20309-77-7
Cyclopentane, 1,1,3,4-tetramethyl-, *trans*-

126 C_9H_{18} 41977-37-1
Cyclopropane, 1-methyl-2-pentyl-

126 C_9H_{18} 50876-33-0
Cyclopentane, 1,1,3,3-tetramethyl-

126 C_9H_{18} 53907-59-8
2-Pentene, 3-ethyl-4,4-dimethyl-
$MeCH{:}CEtCMe_3$

126 C_9H_{18} 53907-60-1
Cyclopentane, 1,1,3,4-tetramethyl-, *cis*-

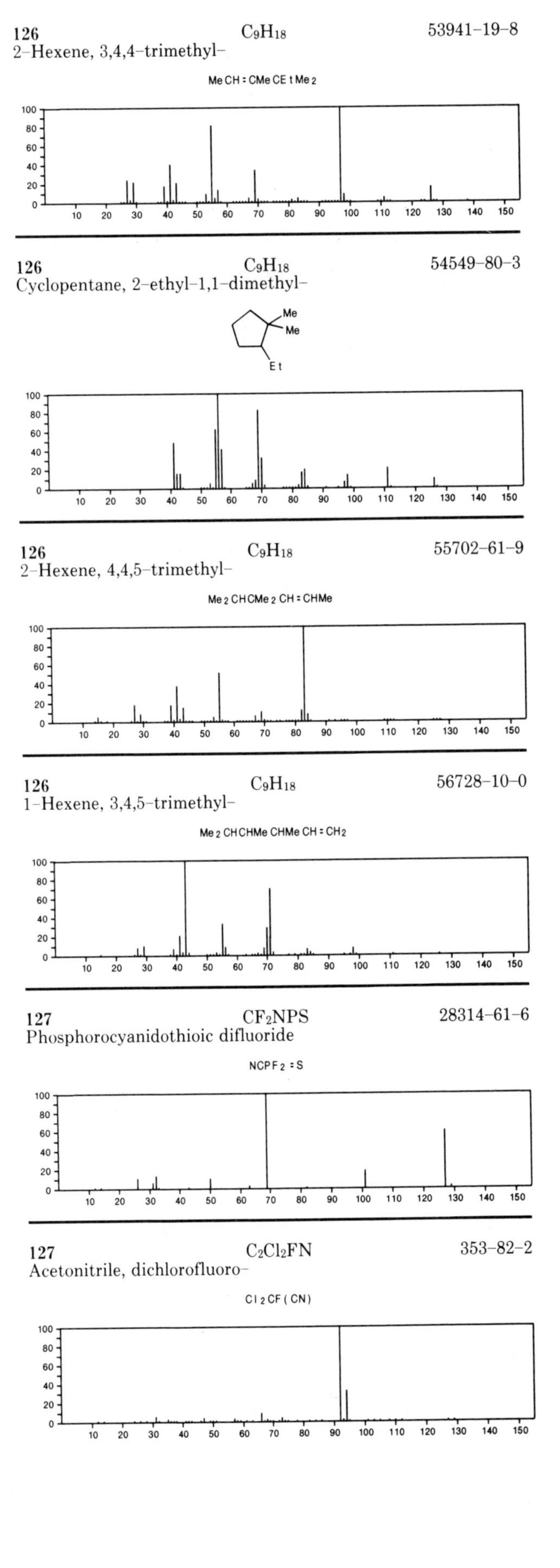

126 C_9H_{18} 53941-19-8
2-Hexene, 3,4,4-trimethyl-
$MeCH{:}CMeCEtMe_2$

126 C_9H_{18} 54549-80-3
Cyclopentane, 2-ethyl-1,1-dimethyl-

126 C_9H_{18} 55702-61-9
2-Hexene, 4,4,5-trimethyl-
$Me_2CHCMe_2CH{:}CHMe$

126 C_9H_{18} 56728-10-0
1-Hexene, 3,4,5-trimethyl-
$Me_2CHCHMeCHMeCH{:}CH_2$

127 CF_2NPS 28314-61-6
Phosphorocyanidothioic difluoride
$NCPF_2{:}S$

127 C_2Cl_2FN 353-82-2
Acetonitrile, dichlorofluoro-
$Cl_2CF(CN)$

127 $C_3H_4F_3NO$ 815-06-5
Acetamide, 2,2,2-trifluoro-*N*-methyl-

MeNHCOCF₃

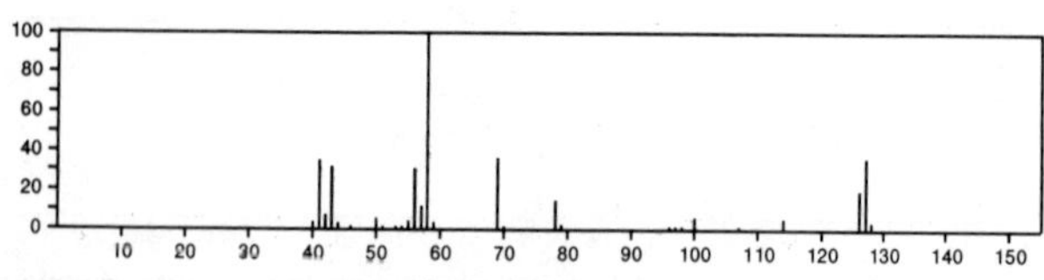

127 $C_3H_5N_5O$ 645-92-1
1,3,5-Triazin-2(1*H*)-one, 4,6-diamino-

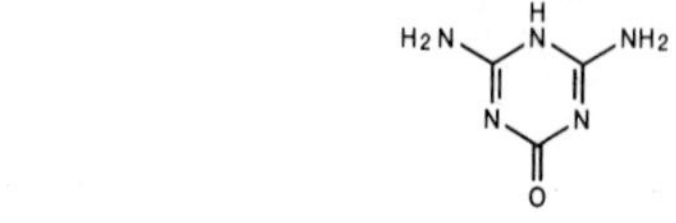

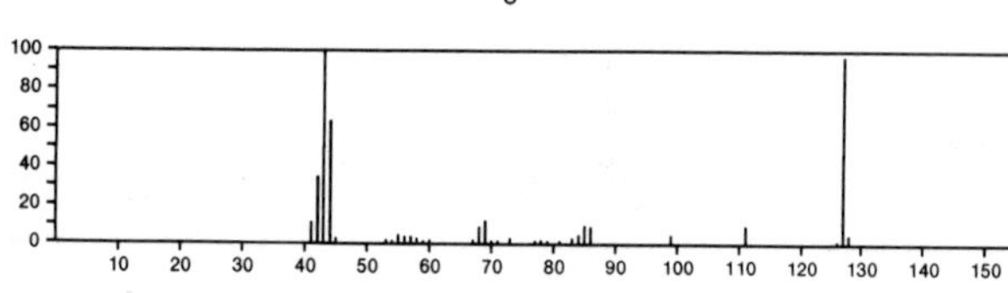

127 $C_4H_5N_3O_2$ 932-52-5
2,4(1*H*,3*H*)-Pyrimidinedione, 5-amino-

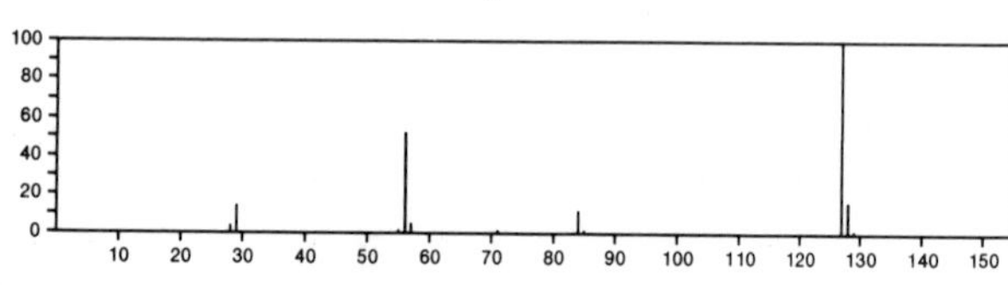

127 $C_4H_5N_3S$ 333-49-3
2(1*H*)-Pyrimidinethione, 4-amino-

127 C_5H_5NOS 23003-22-7
2(1*H*)-Pyridinethione, 3-hydroxy-

127 $C_5H_5NO_3$ 5063-96-7
1*H*-Pyrrole-2,5-dione, 1-(hydroxymethyl)-

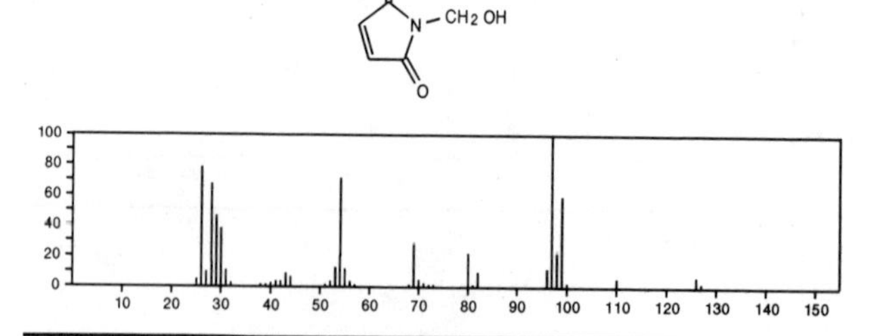

127 C_6H_6ClN 95-51-2
Benzenamine, 2-chloro-

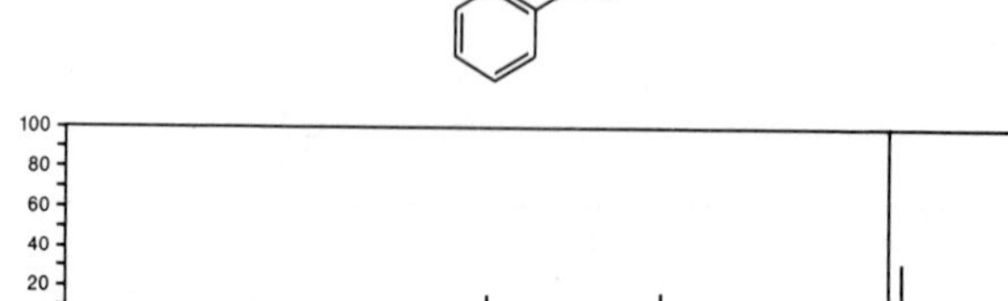

127 C_6H_6ClN 106-47-8
Benzenamine, 4-chloro-

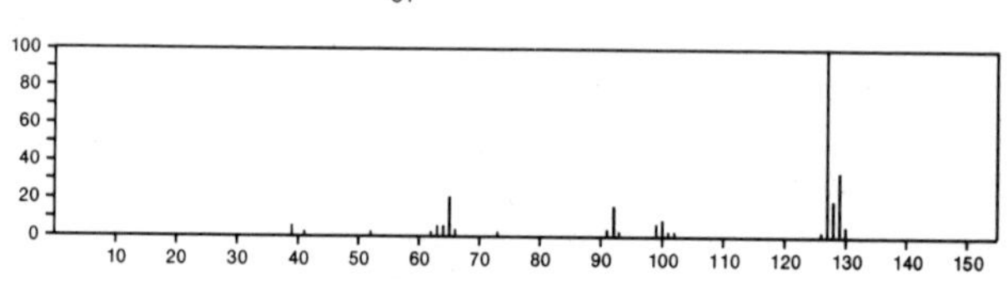

127 C_6H_6ClN 108-42-9
Benzenamine, 3-chloro-

127 $C_6H_9NO_2$ 2314-78-5
2,5-Pyrrolidinedione, 1-ethyl-

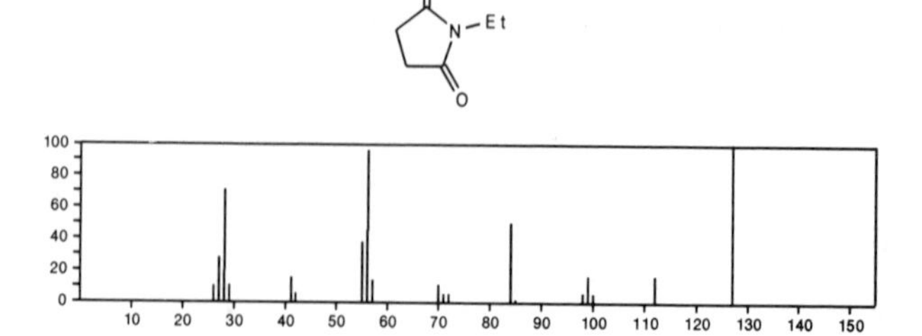

127 $C_6H_9NO_2$ 2407-60-5
2-Propenoic acid, 3-(1-aziridinyl)-, methyl ester

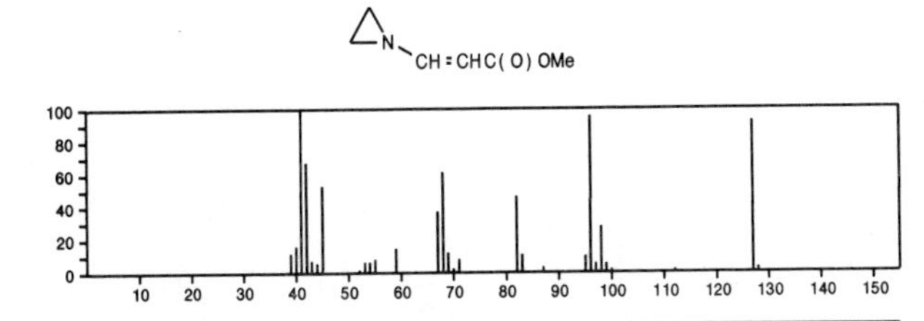

127 $C_6H_9NO_2$ 3395-98-0
2-Oxazolidinone, 3-ethenyl-5-methyl-

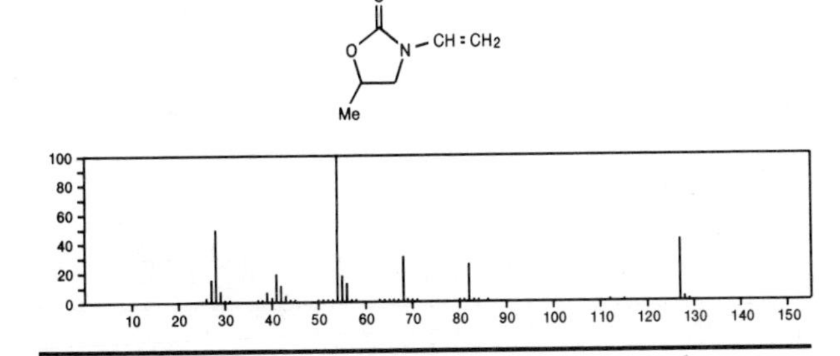

127 $C_6H_9NO_2$ 19788-36-4
4-Isoxazolemethanol, 3,5-dimethyl-

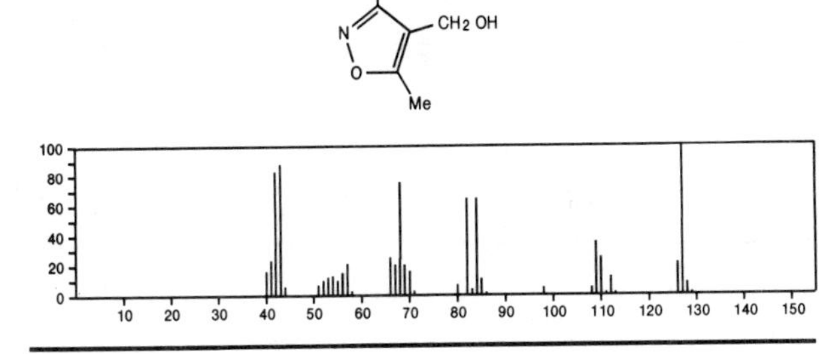

127 $C_6H_9NO_2$ 53692-87-8
Butanoic acid, 2-cyano-, methyl ester

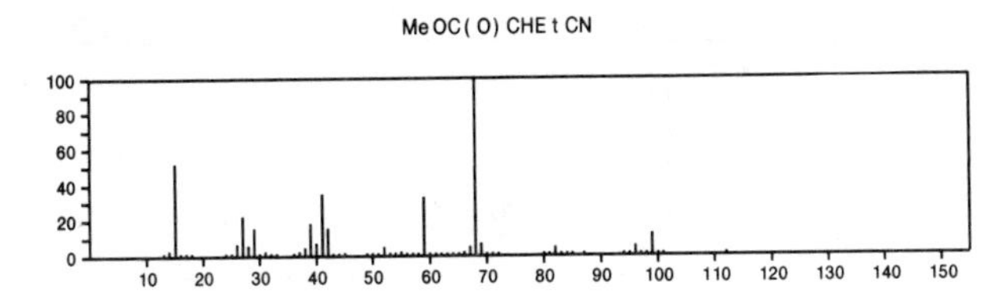

127 C_6H_9NS 13623-11-5
Thiazole, 2,4,5-trimethyl-

Me S Me N Me

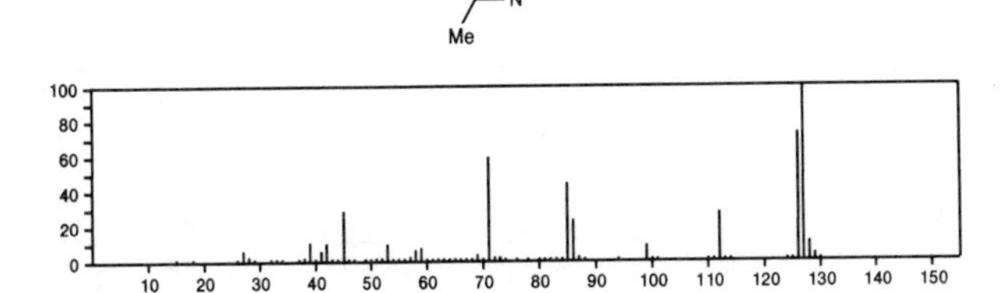

127 C_6H_9NS 15679-12-6
Thiazole, 2-ethyl-4-methyl-

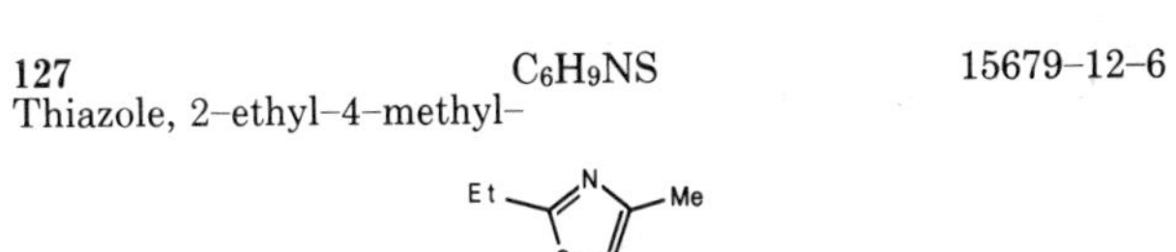

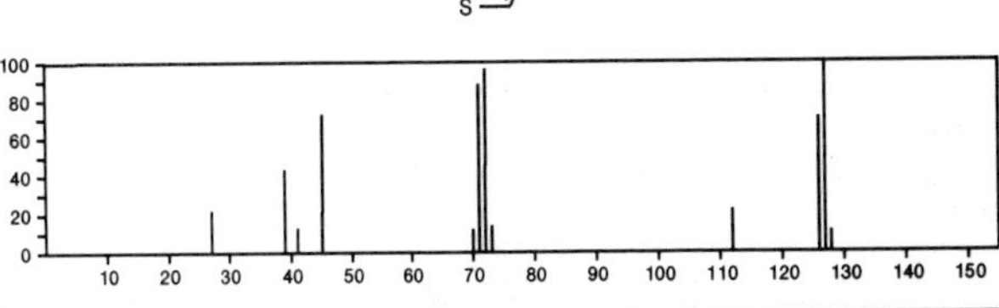

127 C_6H_9NS 17626-75-4
Thiazole, 2-propyl-

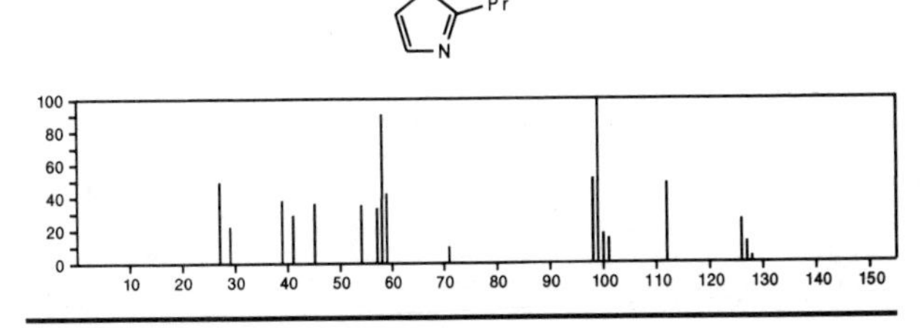

127 C_6H_9NS 19961-53-6
Thiazole, 2-ethyl-5-methyl-

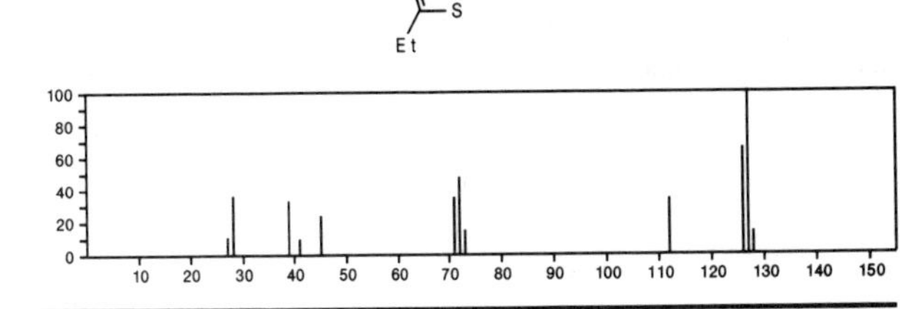

127 C_6H_9NS 31883-01-9
Thiazole, 5-ethyl-4-methyl-

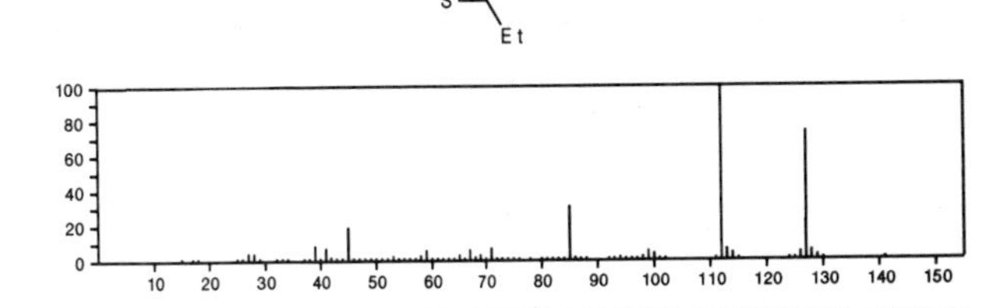

127 C_6H_9NS 32272-48-3
Thiazole, 4-ethyl-2-methyl-

S Me N Et

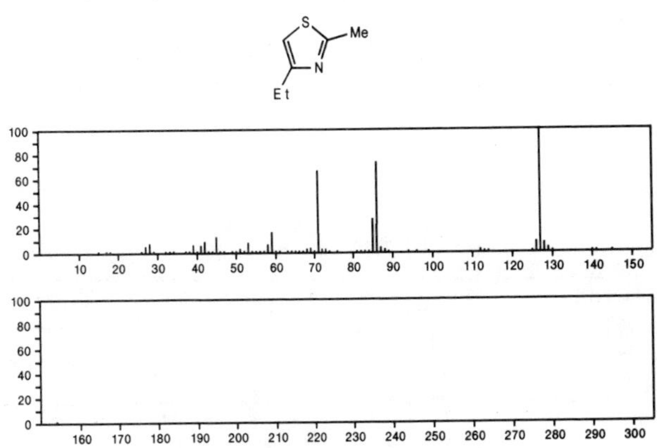

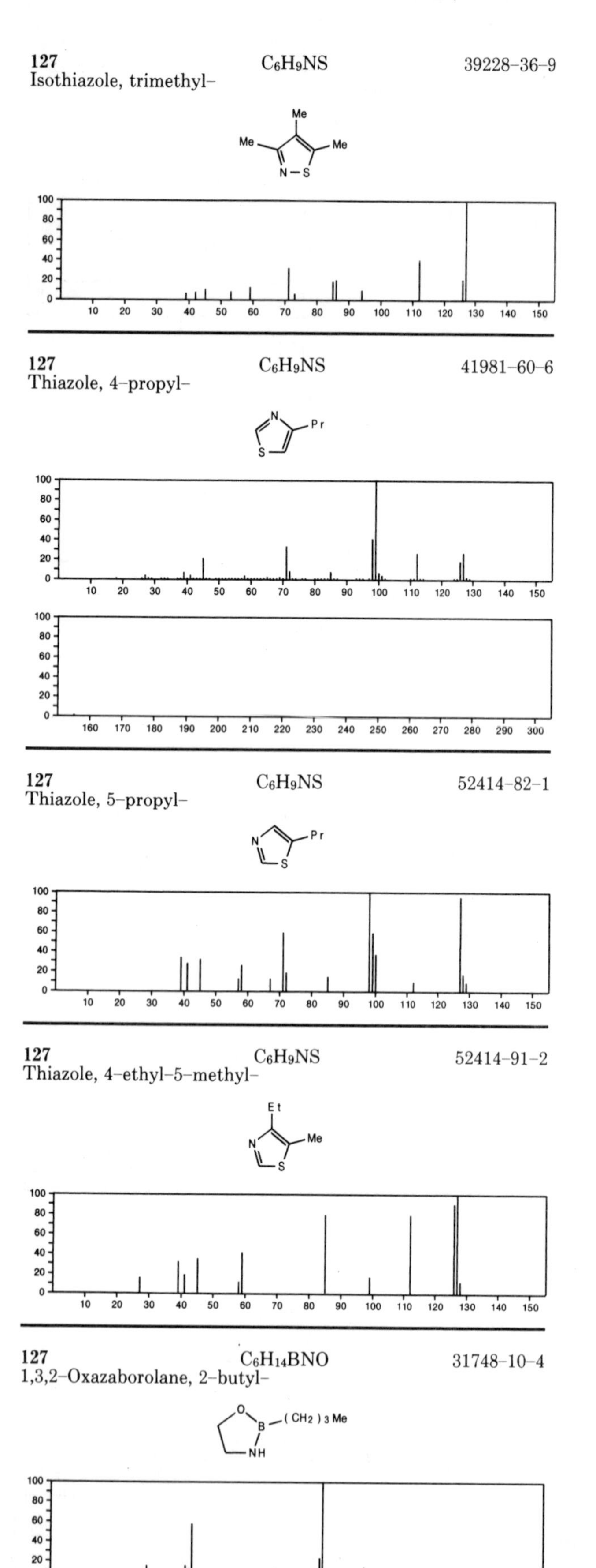

127 C_6H_9NS 39228–36–9
Isothiazole, trimethyl–

127 C_6H_9NS 41981–60–6
Thiazole, 4–propyl–

127 C_6H_9NS 52414–82–1
Thiazole, 5–propyl–

127 C_6H_9NS 52414–91–2
Thiazole, 4–ethyl–5–methyl–

127 $C_6H_{14}BNO$ 31748–10–4
1,3,2–Oxazaborolane, 2–butyl–

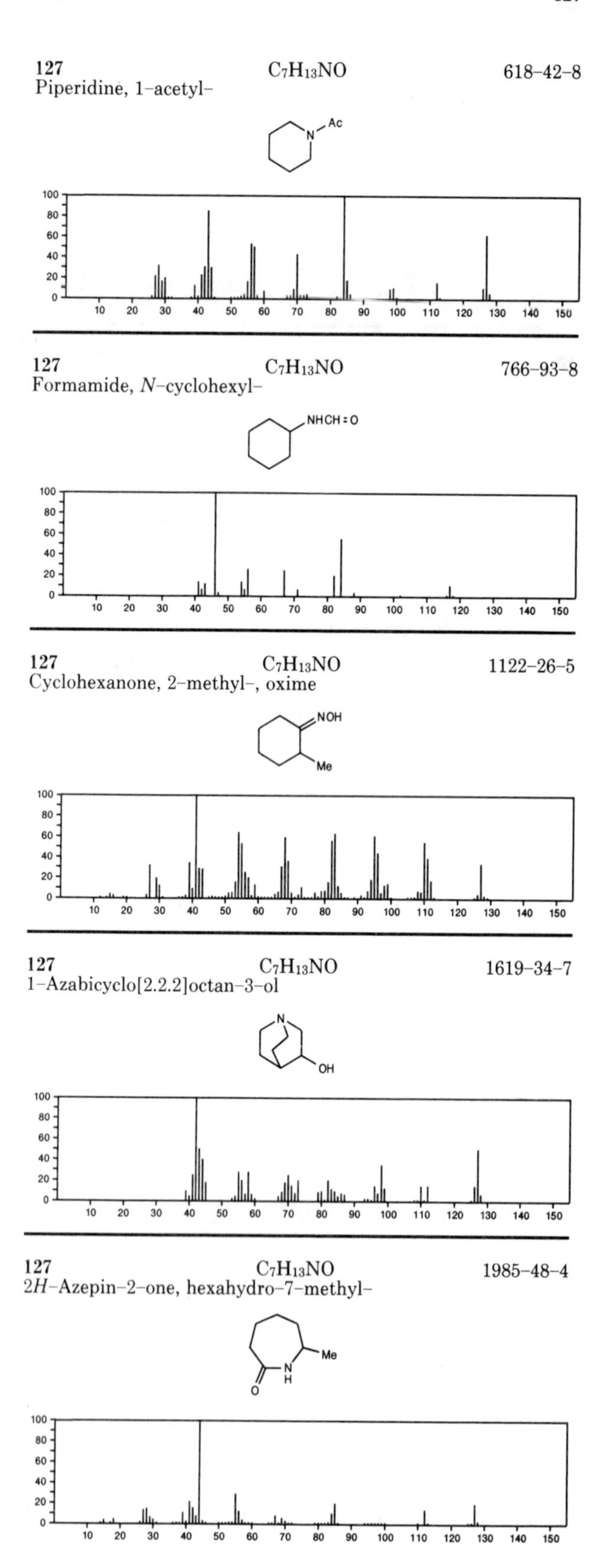

127 $C_7H_{13}NO$ 618–42–8
Piperidine, 1–acetyl–

127 $C_7H_{13}NO$ 766–93–8
Formamide, *N*–cyclohexyl–

127 $C_7H_{13}NO$ 1122–26–5
Cyclohexanone, 2–methyl–, oxime

127 $C_7H_{13}NO$ 1619–34–7
1–Azabicyclo[2.2.2]octan–3–ol

127 $C_7H_{13}NO$ 1985–48–4
2*H*–Azepin–2–one, hexahydro–7–methyl–

127 $C_7H_{13}NO$ 2073-32-7
2*H*-Azepin-2-one, hexahydro-3-methyl-

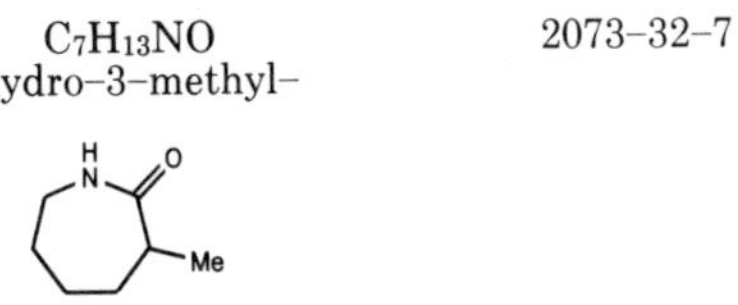

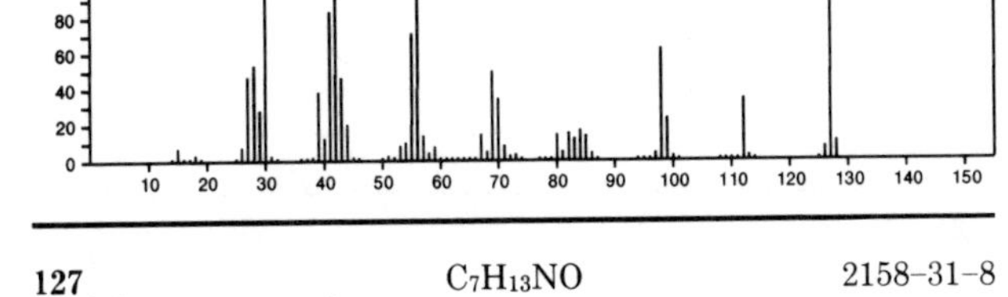

127 $C_7H_{13}NO$ 2158-31-8
Cycloheptanone, oxime

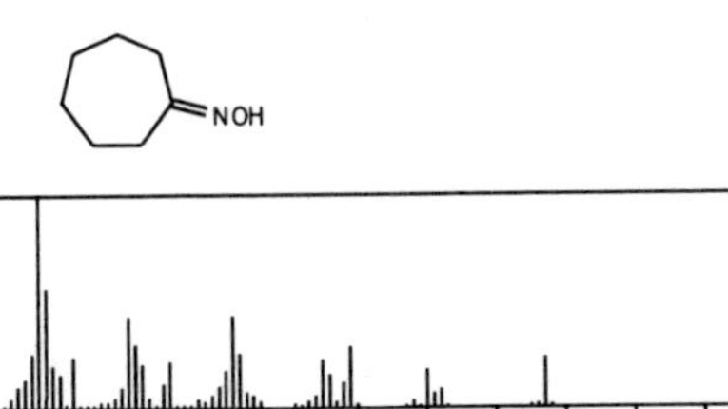

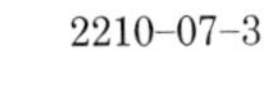

127 $C_7H_{13}NO$ 2210-07-3
2*H*-Azepin-2-one, hexahydro-5-methyl-

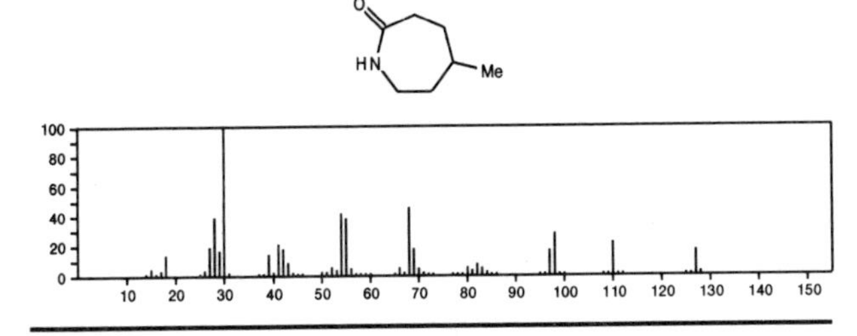

127 $C_7H_{13}NO$ 2556-73-2
2*H*-Azepin-2-one, hexahydro-1-methyl-

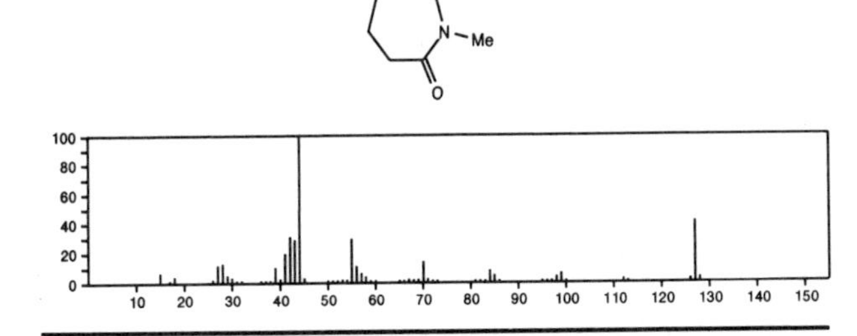

127 $C_7H_{13}NO$ 3433-62-3
3-Penten-2-one, 4-(dimethylamino)-

$Me_2NCMe=CHCOMe$

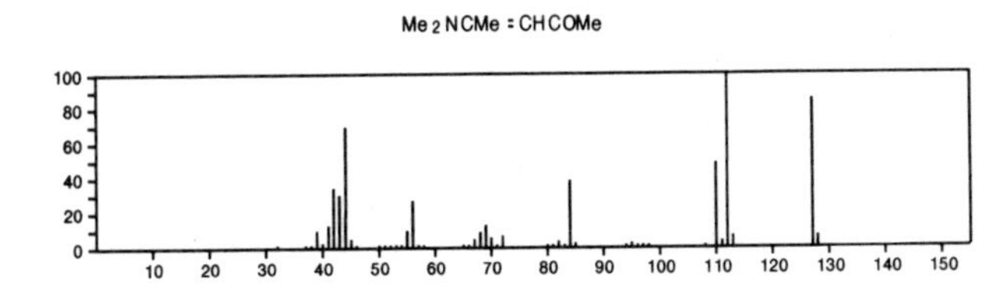

127 $C_7H_{13}NO$ 3470-99-3
2-Pyrrolidinone, 1-propyl-

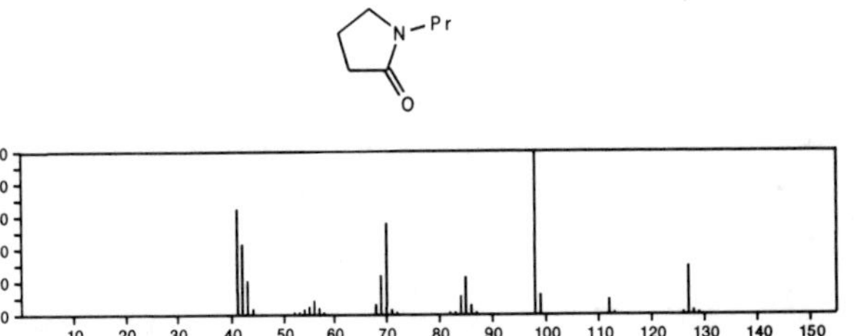

127 $C_7H_{13}NO$ 3623-05-0
2*H*-Azepin-2-one, hexahydro-4-methyl-

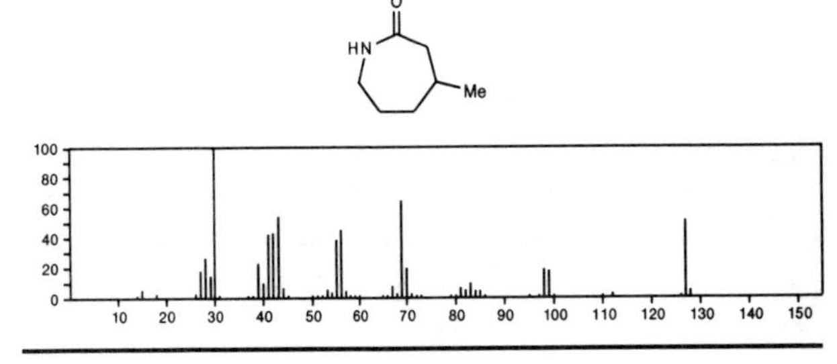

127 $C_7H_{13}NO$ 4553-05-3
Pyrrolidine, 1-(1-oxopropyl)-

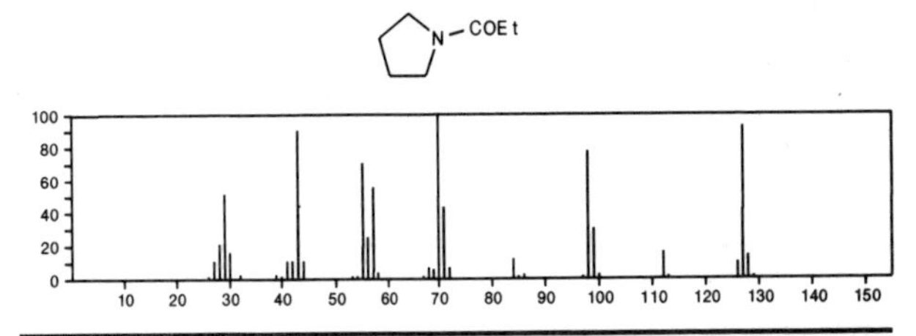

127 $C_7H_{13}NO$ 4701-95-5
Cyclohexanone, 3-methyl-, oxime

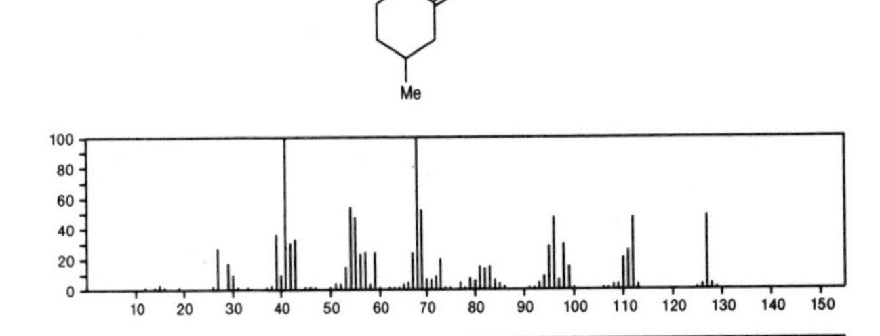

127 $C_7H_{13}NO$ 4994-13-2
Cyclohexanone, 4-methyl-, oxime

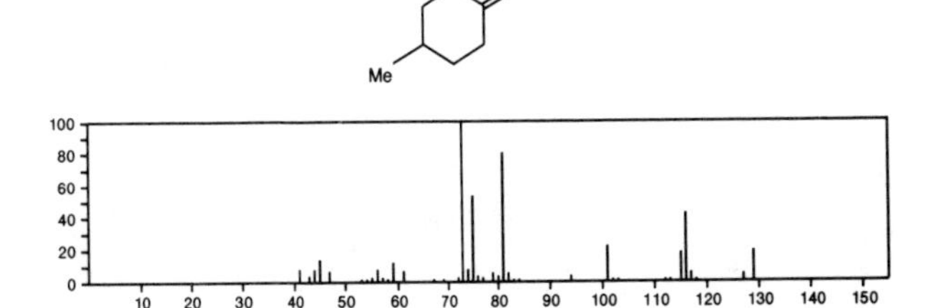

127 $C_7H_{13}NO$ 6142–55–8
2*H*–Azepin–2–one, hexahydro–6–methyl–

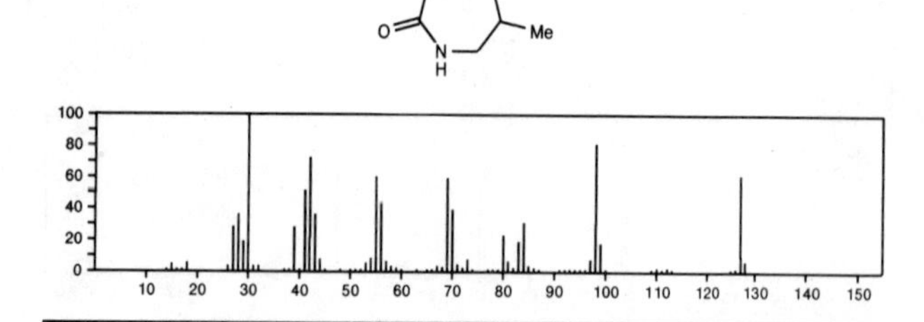

127 $C_7H_{13}NO$ 6959–71–3
Propanenitrile, 3–butoxy–

$NCCH_2CH_2O(CH_2)_3Me$

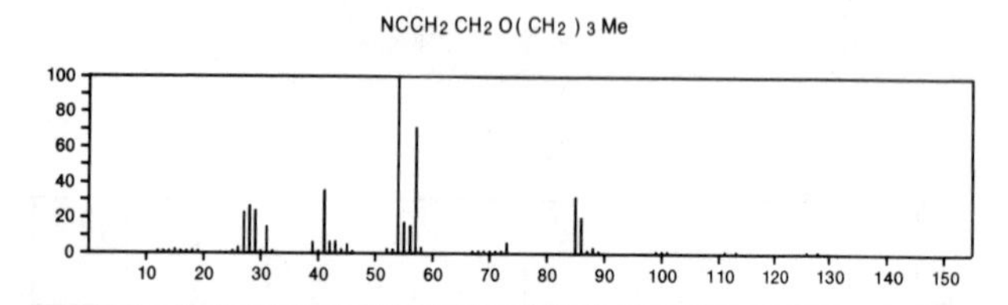

127 $C_7H_{13}NO$ 13858–85–0
Cyclohexanone, *O*–methyloxime

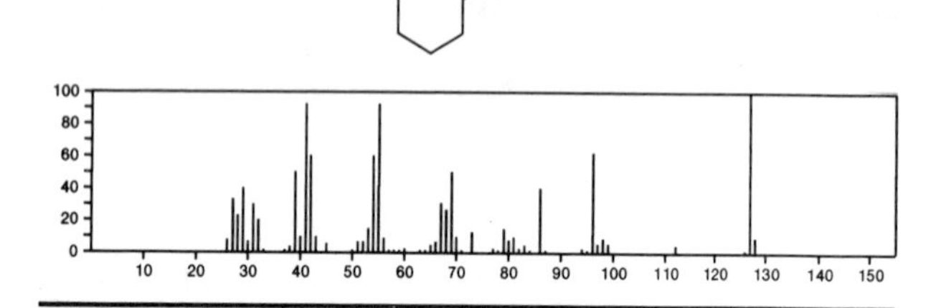

127 $C_7H_{13}NO$ 25291–41–2
Acetamide, *N*–cyclopentyl–

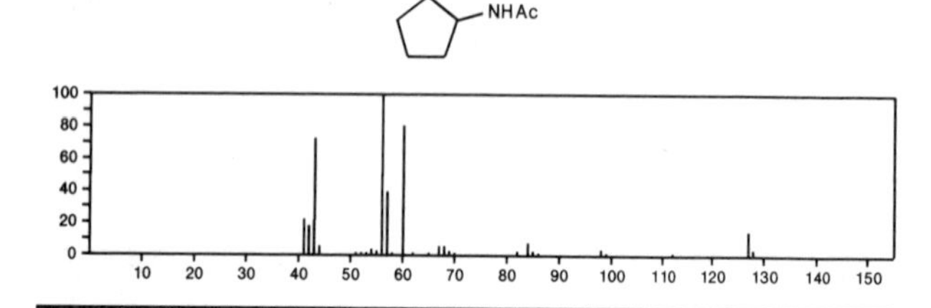

127 $C_7H_{13}NO$ 39209–03–5
3–Penten–2–one, 3–methyl–, *O*–methyloxime

$MeCH{=}CMeCMe{=}NOMe$

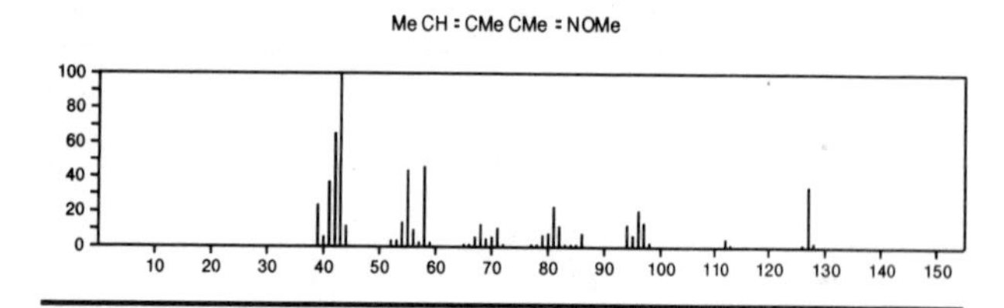

127 $C_7H_{13}NO$ 39209–04–6
4–Hexen–3–one, *O*–methyloxime

$MeCH{=}CHCEt{=}NOMe$

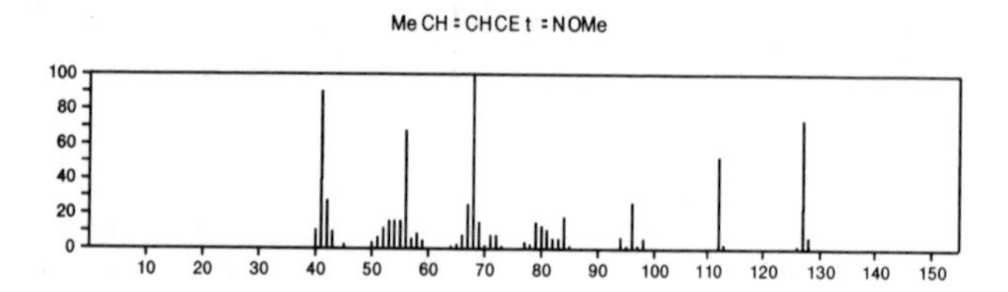

127 $C_7H_{13}NO$ 43152–92–7
3–Piperidinone, 1,6–dimethyl–

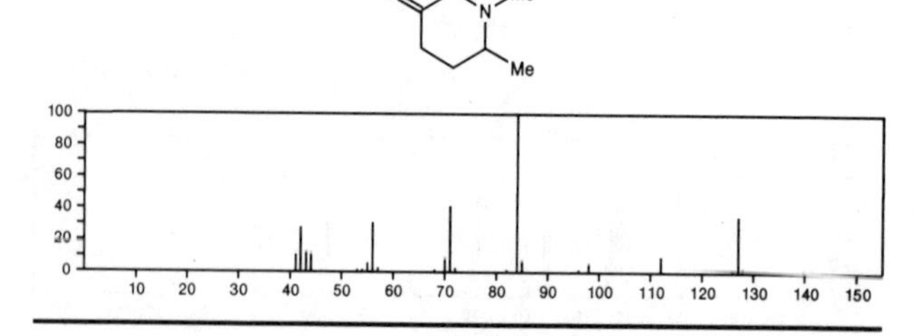

127 $C_7H_{13}NO$ 43152–93–8
3–Piperidinone, 1–ethyl–

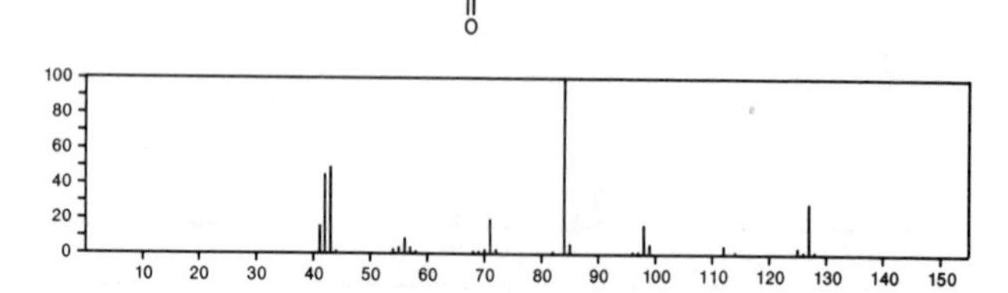

127 $C_7H_{13}NO$ 56335–97–8
5–Hexen–2–one, *O*–methyloxime

$H_2C{=}CHCH_2CH_2CMe{=}NOMe$

127 $C_7H_{13}NO$ 56335–98–9
4–Penten–2–one, 3–methyl–, *O*–methyloxime

$H_2C{=}CHCHMeCMe{=}NOMe$

127 $C_7H_{13}NO$ 56335–99–0
4–Hexen–2–one, *O*–methyloxime

$MeCH{=}CHCH_2CMe{=}NOMe$

127 $C_7H_{13}NO$ 56336–10–8
1–Propanone, 1–cyclopropyl–, *O*–methyloxime

$CEt{=}NOMe$

127 $C_7H_{13}NO$ 56336-11-9
3-Penten-2-one, 4-methyl-, *O*-methyloxime

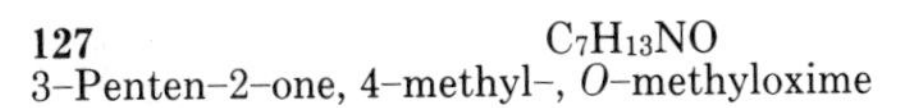

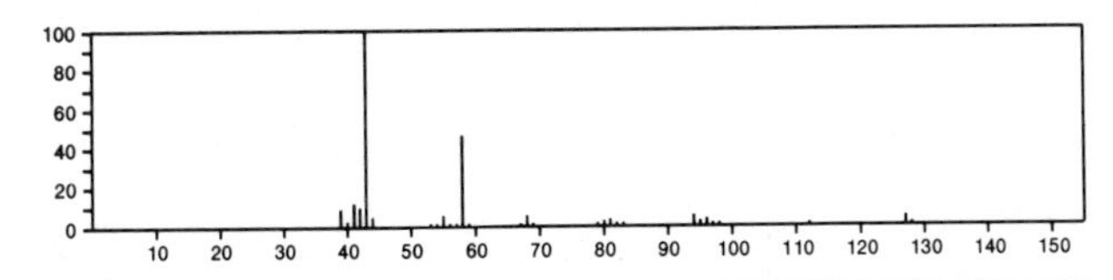

127 $C_8H_{17}N$ 98-94-2
Cyclohexanamine, *N*,*N*-dimethyl-

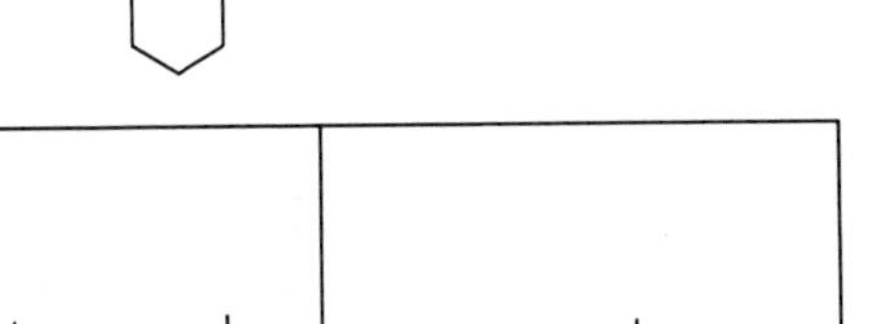

127 $C_8H_{17}N$ 766-52-9
Piperidine, 1-ethyl-2-methyl-

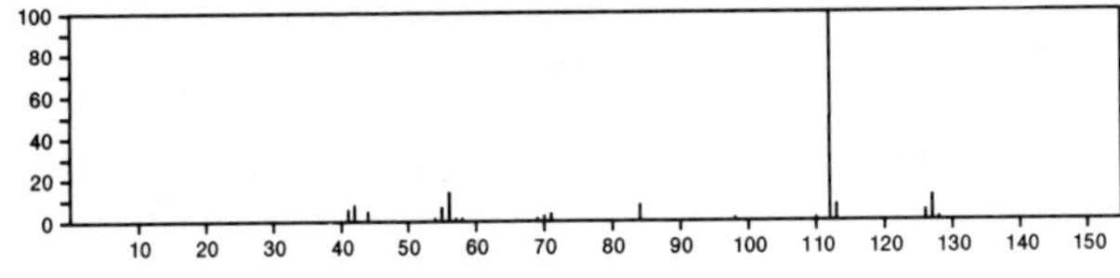

127 $C_8H_{17}N$ 2439-13-6
Piperidine, 1,2,6-trimethyl-, *cis*-

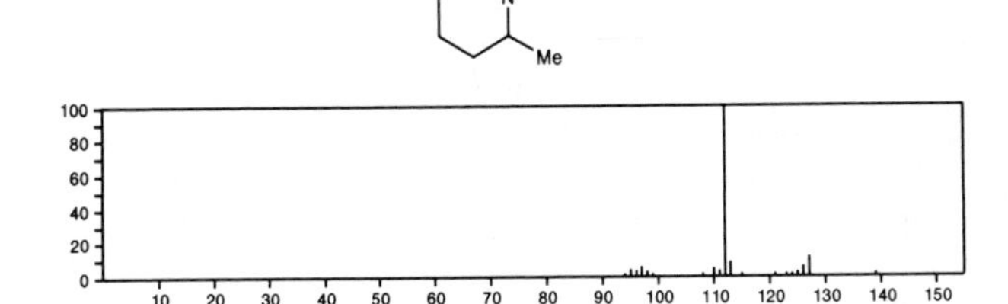

127 $C_8H_{17}N$ 3203-94-9
Aziridine, 2-*tert*-butyl-1,3-dimethyl-, *trans*-

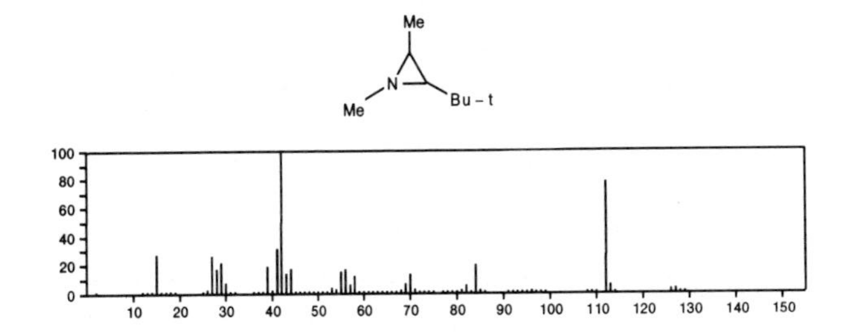

127 $C_8H_{17}N$ 4853-56-9
1-Butanamine, *N*-butylidene-

Me (CH2) 3 N = CHPr

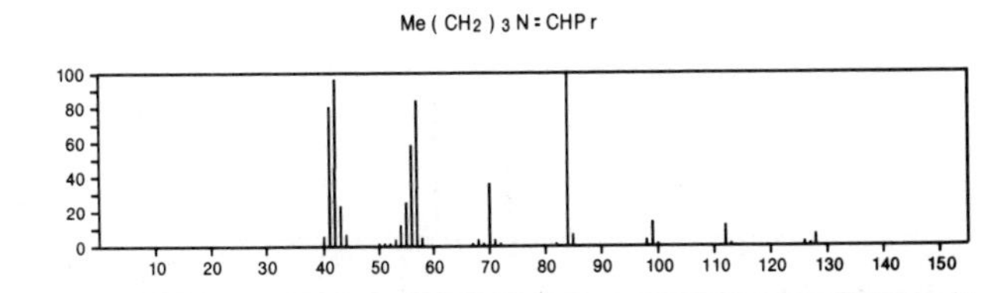

127 $C_8H_{17}N$ 5459-93-8
Cyclohexanamine, *N*-ethyl-

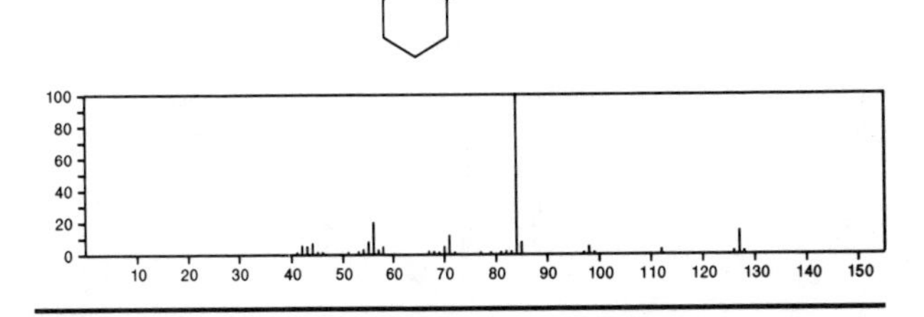

127 $C_8H_{17}N$ 5661-71-2
1*H*-Azonine, octahydro-

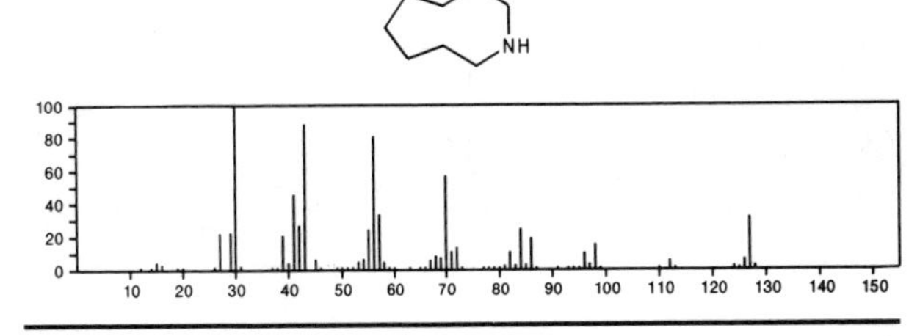

127 $C_8H_{17}N$ 6125-02-6
Aziridine, 1-(1,1-dimethylethyl)-2,3-dimethyl-, *trans*-

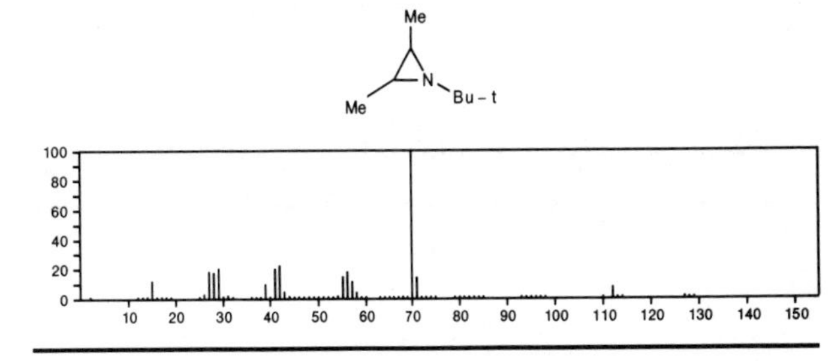

127 $C_8H_{17}N$ 6898-71-1
Methylamine, *N*-heptylidene-

Me (CH2) 5 CH = NMe

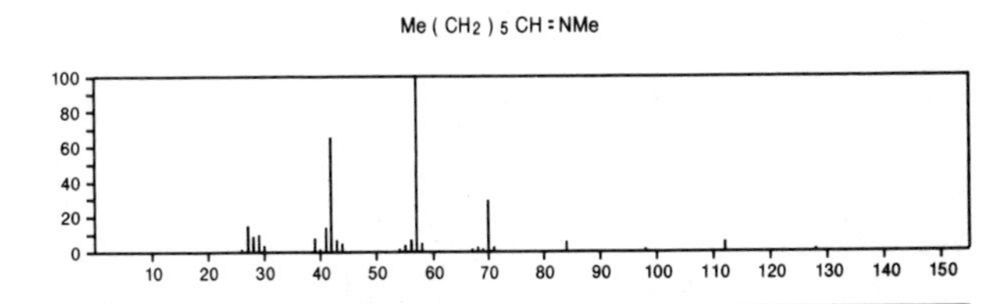

127 $C_8H_{17}N$ 6898-82-4
1-Propanamine, 2-methyl-*N*-(2-methylpropylidene)-

Me2 CHCH = NBu - i

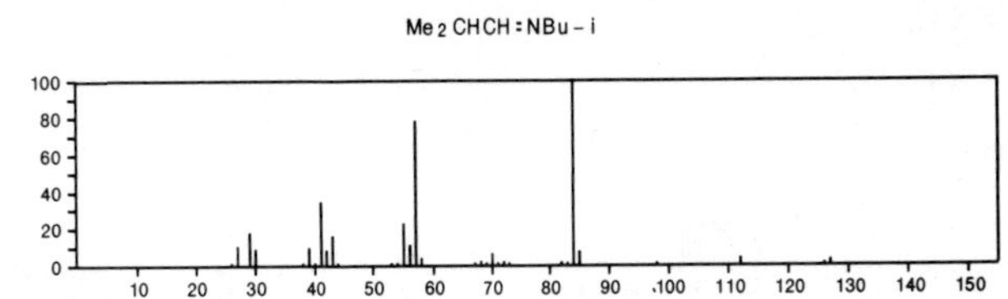

127 $C_8H_{17}N$ 10599–78–7
Methylamine, *N*–(1–propylbutylidene)–

MeN:CPr2

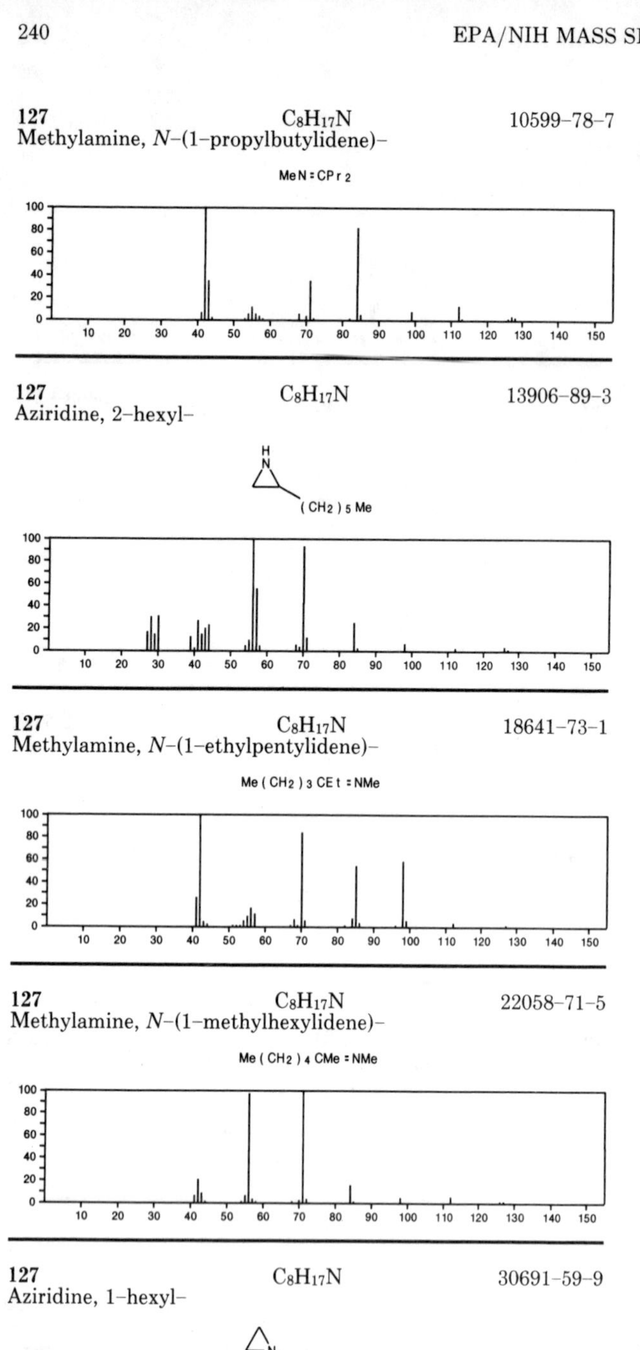

127 $C_8H_{17}N$ 13906–89–3
Aziridine, 2–hexyl–

(CH2)5Me

127 $C_8H_{17}N$ 18641–73–1
Methylamine, *N*–(1–ethylpentylidene)–

Me(CH2)3CEt:NMe

127 $C_8H_{17}N$ 22058–71–5
Methylamine, *N*–(1–methylhexylidene)–

Me(CH2)4CMe:NMe

127 $C_8H_{17}N$ 30691–59–9
Aziridine, 1–hexyl–

(CH2)5Me

127 $C_8H_{17}N$ 55702–65–3
Azetidine, 1–(1,1–dimethylethyl)–3–methyl–

Bu-t
Me

127 $C_8H_{17}N$ 55702–73–3
Aziridine, 1–ethyl–2–methyl–3–(1–methylethyl)–, *trans*–

Me
Et
Pr-i

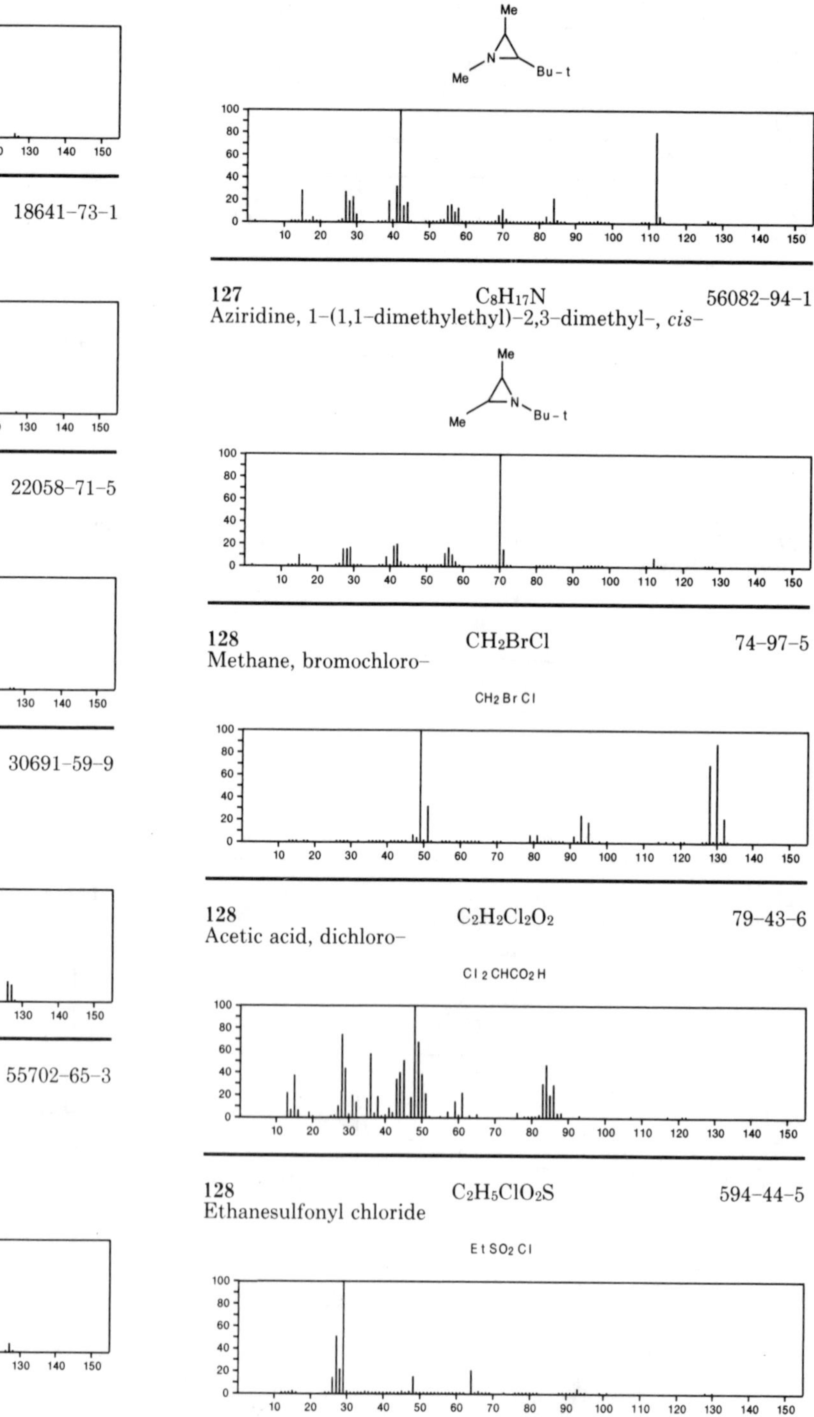

127 $C_8H_{17}N$ 55712–58–8
Aziridine, 2–(1,1–dimethylethyl)–1,3–dimethyl–

Me
Me
Bu-t

127 $C_8H_{17}N$ 56082–94–1
Aziridine, 1–(1,1–dimethylethyl)–2,3–dimethyl–, *cis*–

Me
Me
Bu-t

128 CH_2BrCl 74–97–5
Methane, bromochloro–

CH2BrCl

128 $C_2H_2Cl_2O_2$ 79–43–6
Acetic acid, dichloro–

Cl2CHCO2H

128 $C_2H_5ClO_2S$ 594–44–5
Ethanesulfonyl chloride

EtSO2Cl

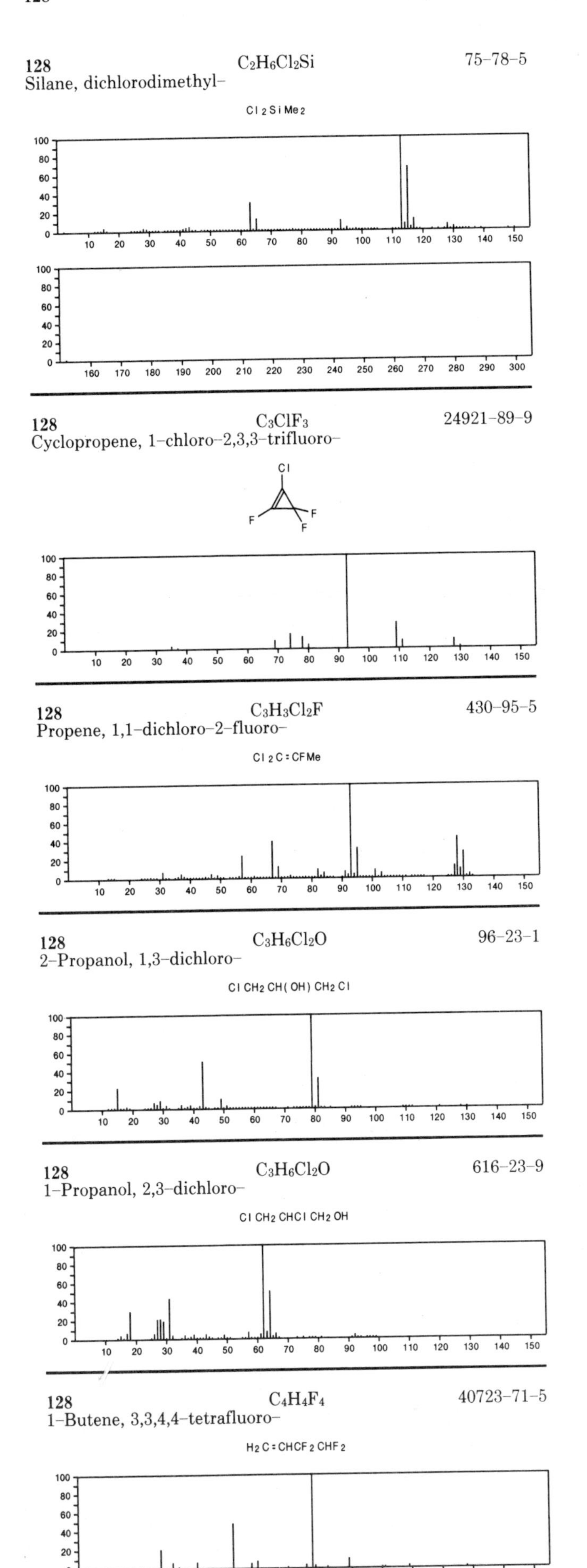

128 $C_2H_6Cl_2Si$ 75–78–5
Silane, dichlorodimethyl–
Cl_2SiMe_2

128 C_3ClF_3 24921–89–9
Cyclopropene, 1–chloro–2,3,3–trifluoro–

128 $C_3H_3Cl_2F$ 430–95–5
Propene, 1,1–dichloro–2–fluoro–
$Cl_2C{:}CFMe$

128 $C_3H_6Cl_2O$ 96–23–1
2–Propanol, 1,3–dichloro–
$ClCH_2CH(OH)CH_2Cl$

128 $C_3H_6Cl_2O$ 616–23–9
1–Propanol, 2,3–dichloro–
$ClCH_2CHClCH_2OH$

128 $C_4H_4F_4$ 40723–71–5
1–Butene, 3,3,4,4–tetrafluoro–
$H_2C{:}CHCF_2CHF_2$

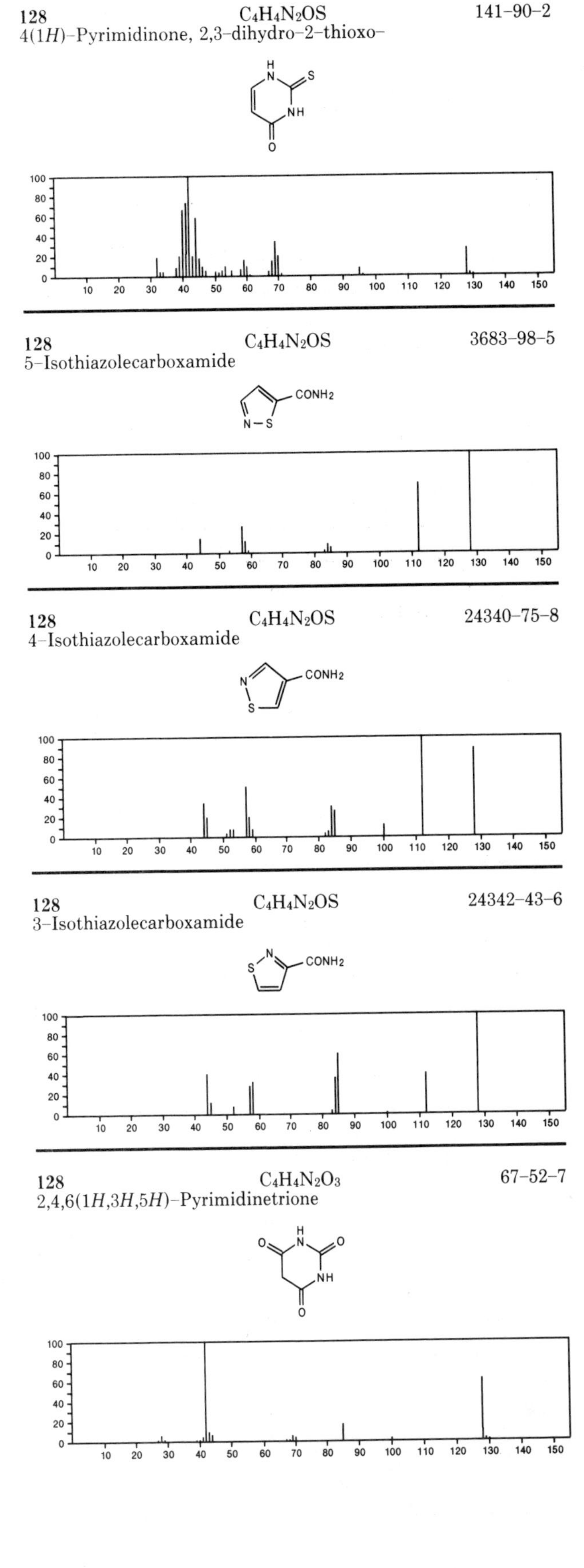

128 $C_4H_4N_2OS$ 141–90–2
4(1*H*)–Pyrimidinone, 2,3–dihydro–2–thioxo–

128 $C_4H_4N_2OS$ 3683–98–5
5–Isothiazolecarboxamide

128 $C_4H_4N_2OS$ 24340–75–8
4–Isothiazolecarboxamide

128 $C_4H_4N_2OS$ 24342–43–6
3–Isothiazolecarboxamide

128 $C_4H_4N_2O_3$ 67–52–7
2,4,6(1*H*,3*H*,5*H*)–Pyrimidinetrione

128 $C_5H_5ClN_2$ 4994-86-9
Pyrimidine, 4-chloro-2-methyl-

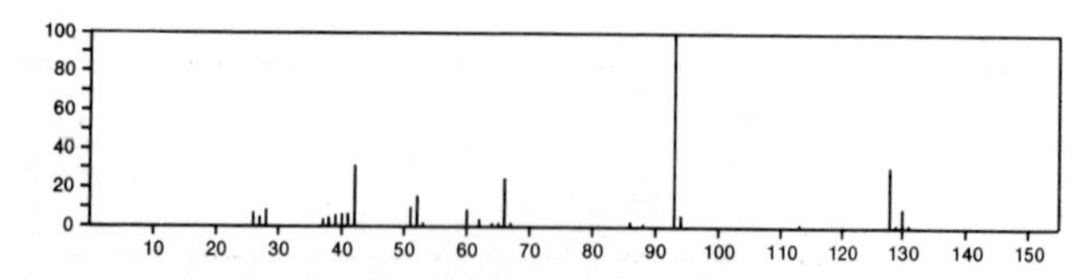

128 $C_5H_8N_2O_2$ 3786-29-6
Butanediamide, 2-methylene-

$H_2NCOCH_2C(=CH_2)CONH_2$

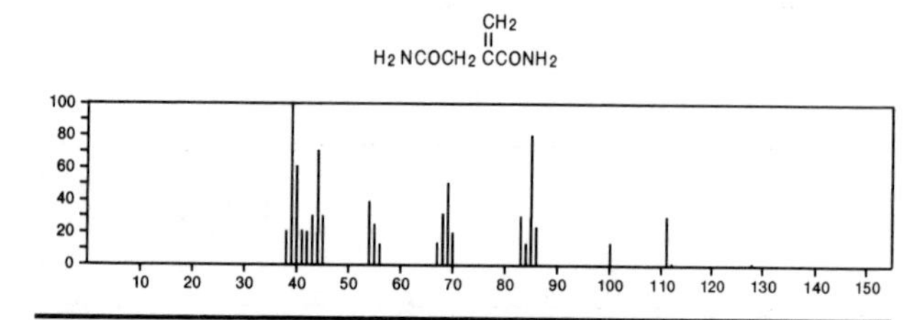

128 $C_5H_8N_2O_2$ 6939-17-9
Sydnone, 3-(1-methylethyl)-

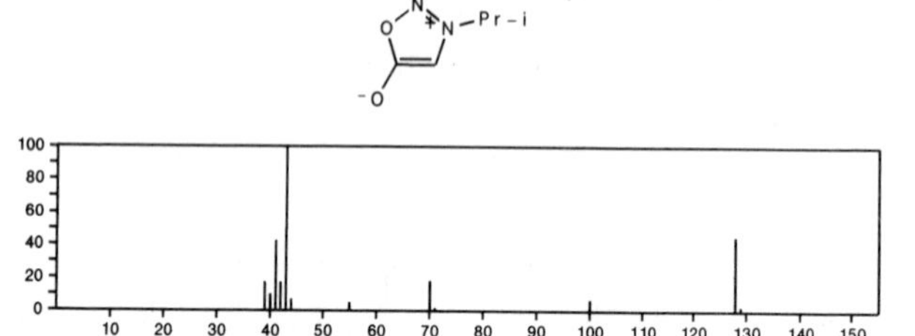

128 $C_5H_8N_2O_2$ 41138-17-4
2-Butenediamide, 2-methyl-, (*Z*)-

$H_2NCOCH=CMeCONH_2$

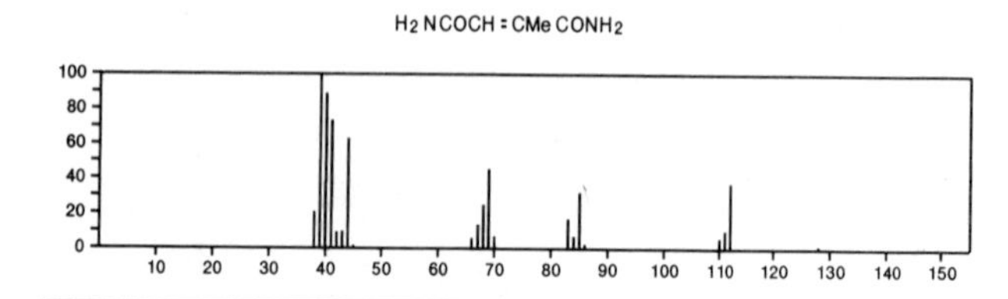

128 $C_5H_8N_2O_2$ 41138-18-5
2-Butenediamide, 2-methyl-, (*E*)-

$H_2NCOCH=CMeCONH_2$

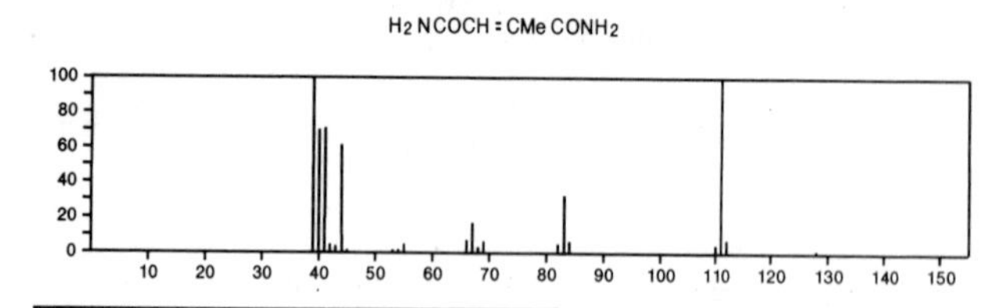

128 $C_5H_8N_2O_2$ 55556-91-7
4-Piperidinone, 1-nitroso-

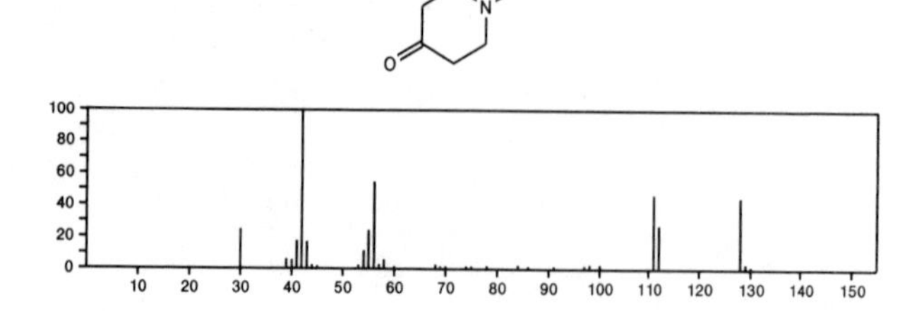

128 C_6H_5ClO 95-57-8
Phenol, 2-chloro-

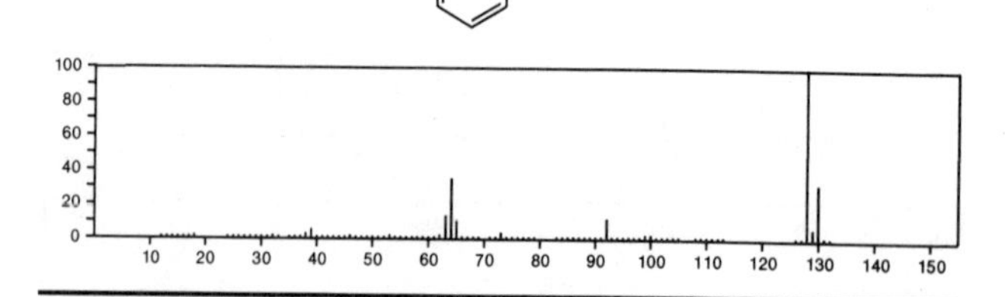

128 C_6H_5ClO 106-48-9
Phenol, 4-chloro-

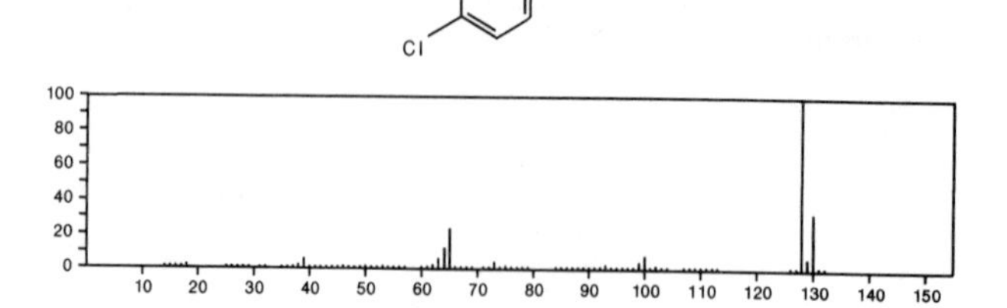

128 C_6H_5ClO 108-43-0
Phenol, 3-chloro-

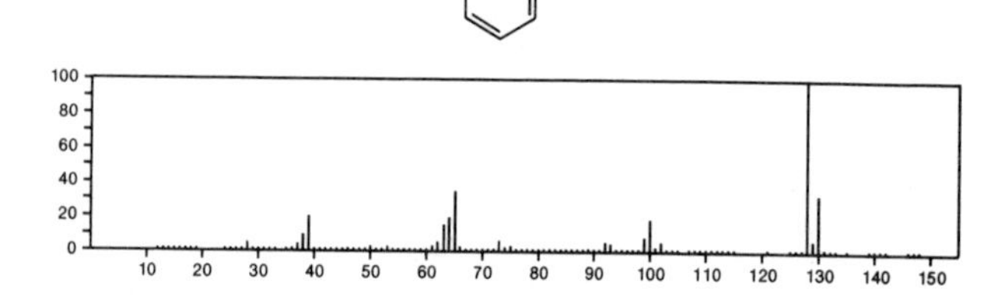

128 $C_6H_5FO_2$ 363-52-0
1,2-Benzenediol, 3-fluoro-

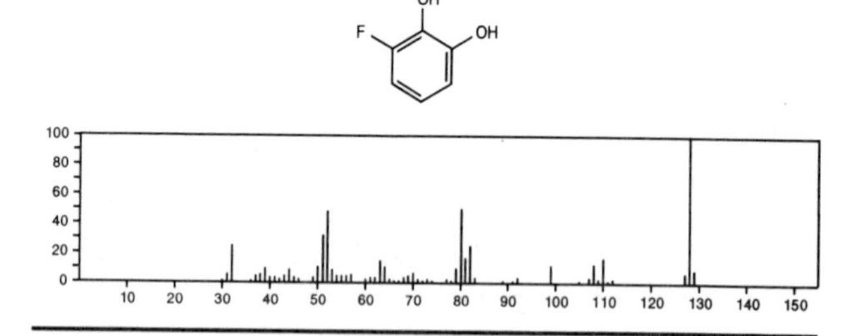

128 C_6H_8OS 31053-55-1
Thiophene, 2-methoxy-5-methyl-

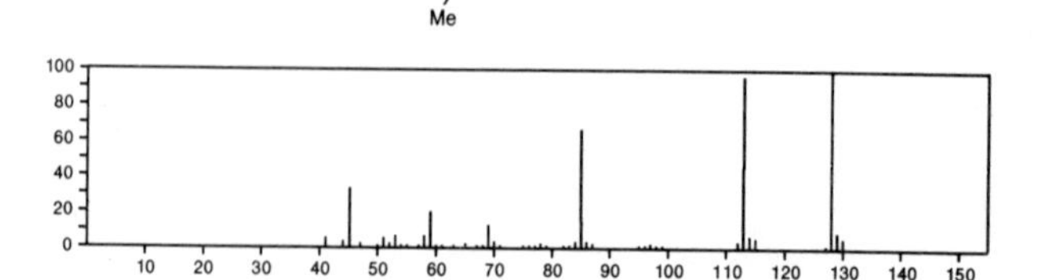

128 $C_6H_8O_3$ 517-23-7
2(3*H*)-Furanone, 3-acetyldihydro-

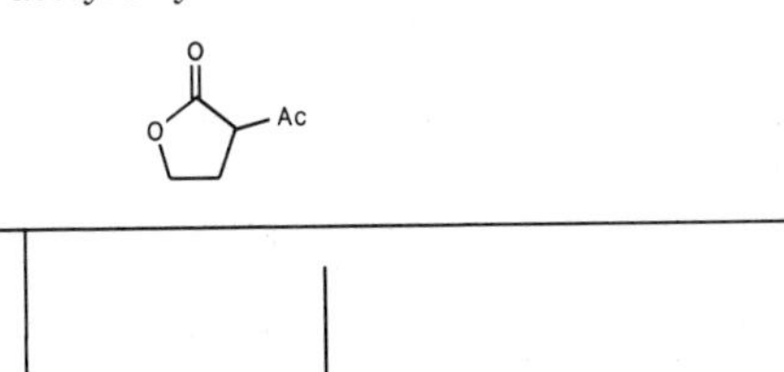

128 $C_6H_8O_3$ 4800-04-8
1,3-Cyclopentanedione, 4-hydroxy-2-methyl-

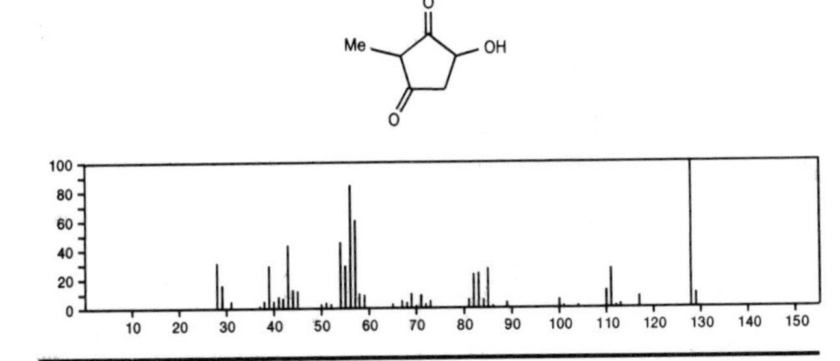

128 $C_6H_8O_3$ 55683-37-9
Propanoic acid, 3-(2-propynyloxy)-

$HO_2CCH_2CH_2OCH_2C\equiv CH$

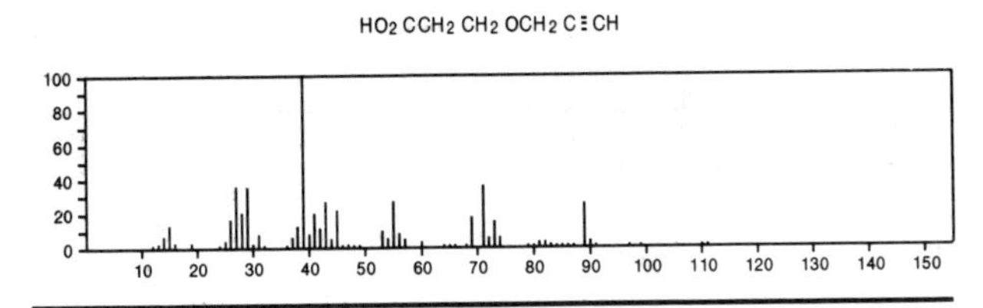

128 $C_6H_8O_3$ 57156-98-6
1,3-Cyclopentanedione, 4-hydroxy-5-methyl-

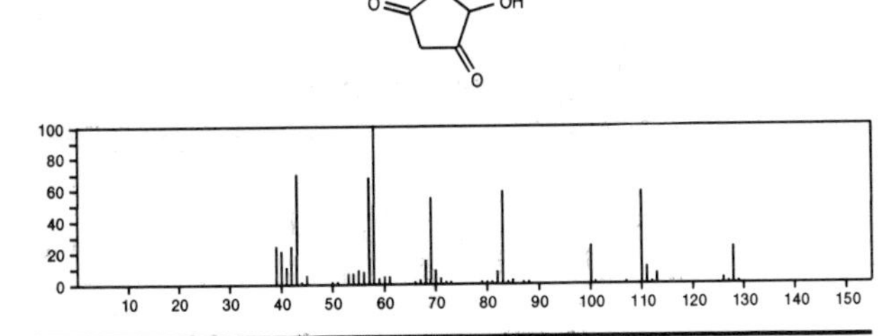

128 $C_6H_{12}N_2O$ 932-83-2
1*H*-Azepine, hexahydro-1-nitroso-

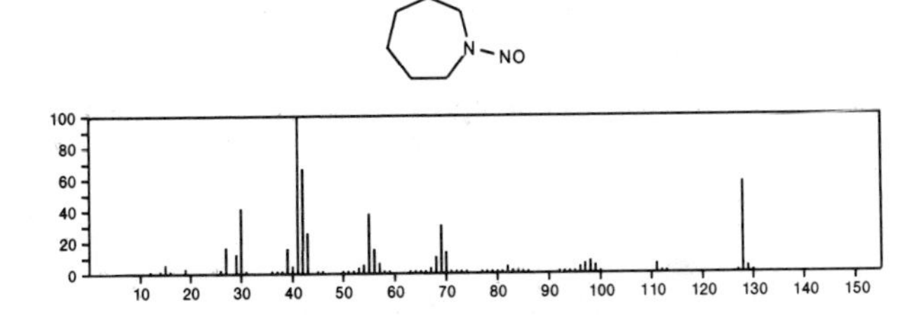

128 $C_6H_{12}N_2O$ 7247-89-4
Piperidine, 2-methyl-1-nitroso-

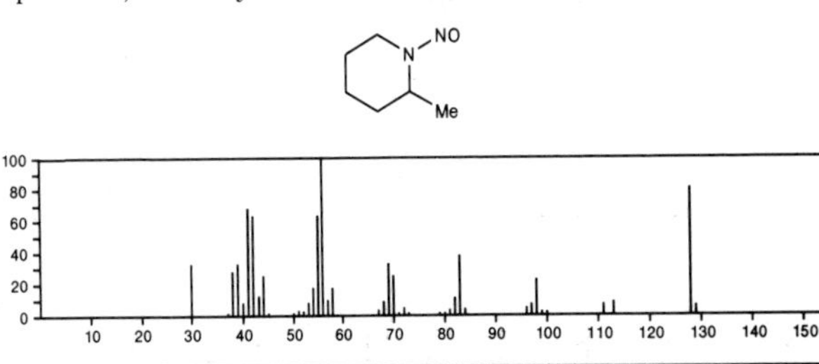

128 $C_6H_{12}N_2O$ 13603-07-1
Piperidine, 3-methyl-1-nitroso-

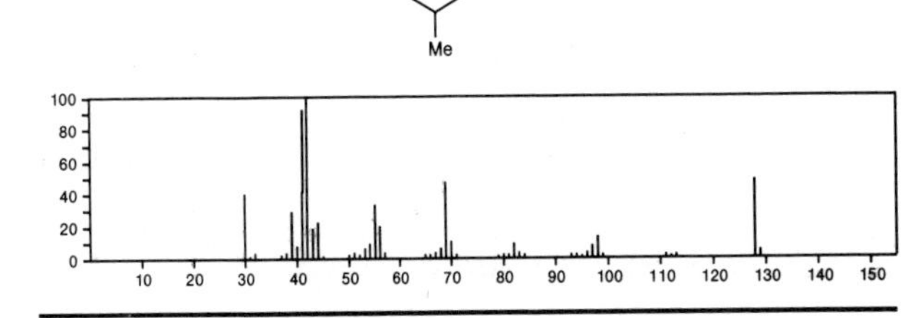

128 $C_6H_{12}N_2O$ 15104-03-7
Piperidine, 4-methyl-1-nitroso-

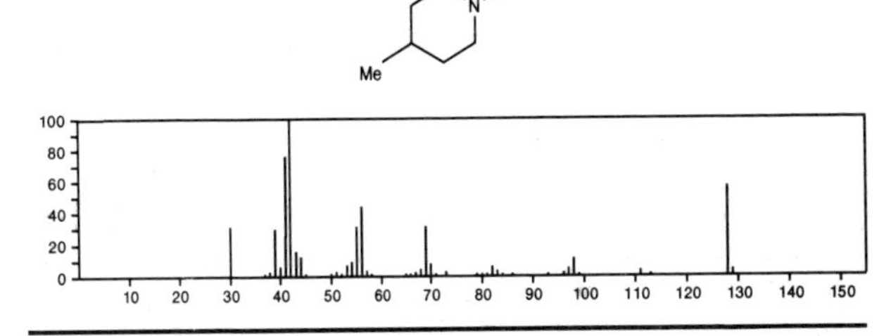

128 $C_6H_{12}N_2O$ 49582-51-6
2-Propenal, 3-(dimethylamino)-3-(methylamino)-

$OCHCH=C(NHMe)NMe_2$

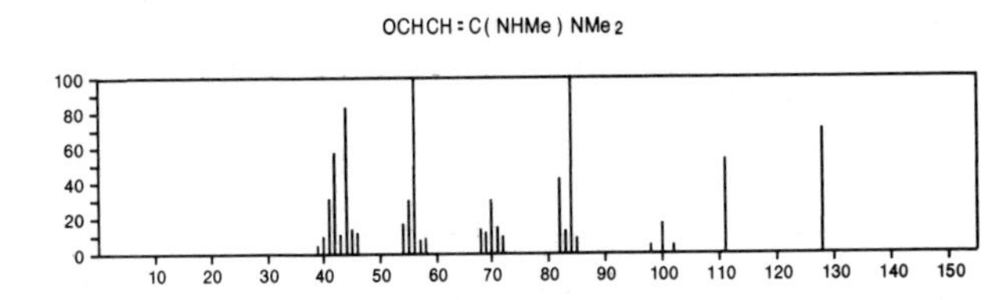

128 $C_6H_{12}N_2O$ 49582-62-9
2-Propenal, 3-(dimethylamino)-2-(methylamino)-

$OCHC(NHMe)=CHNMe_2$

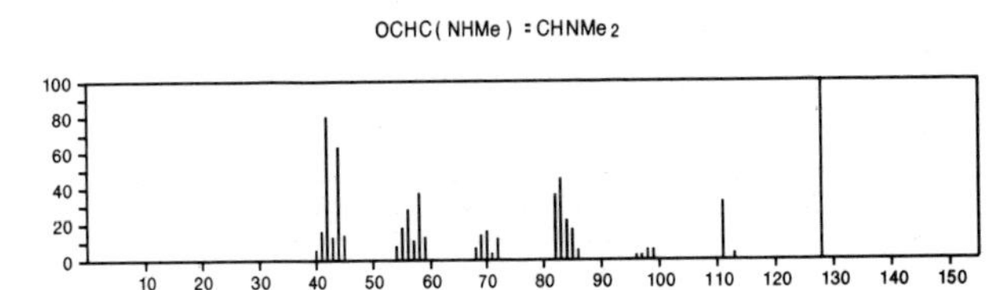

128 $C_6H_{12}N_2O$ 55556-86-0
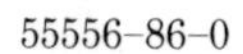
Pyrrolidine, 2,5-dimethyl-1-nitroso-
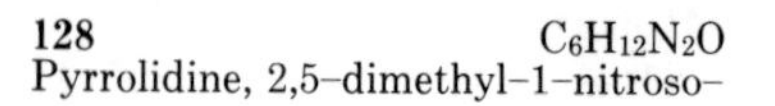

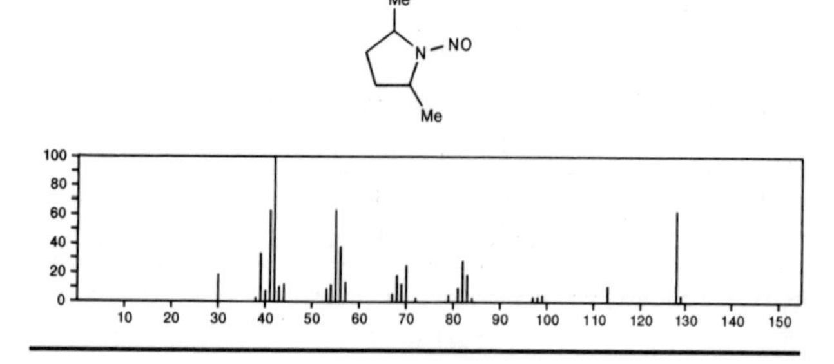

128 $C_7H_{12}O_2$ 98-89-5
Cyclohexanecarboxylic acid
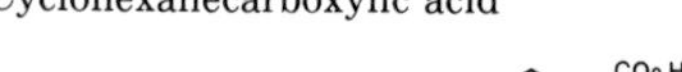

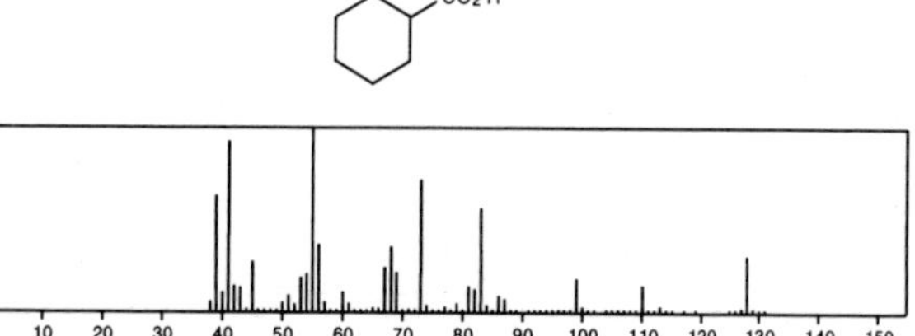

128 $C_7H_{12}O_2$ 141-32-2
2-Propenoic acid, butyl ester

Me (CH_2) 3 OC(O) CH = CH_2
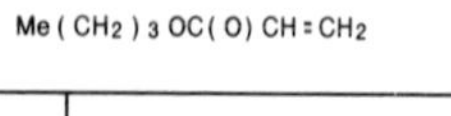

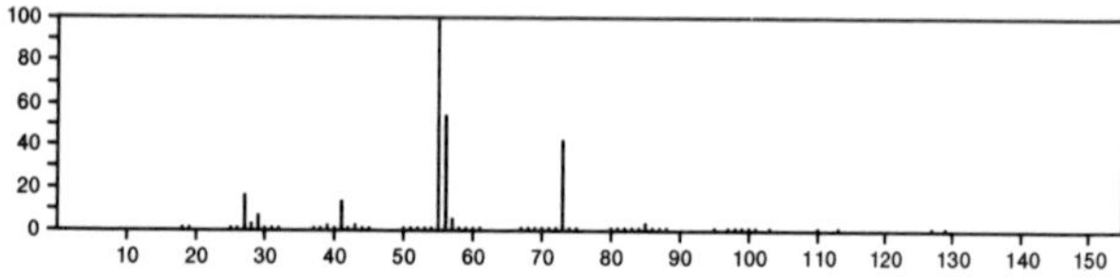

128 $C_7H_{12}O_2$ 638-10-8
2-Butenoic acid, 3-methyl-, ethyl ester

Me_2C = CHC(O) OEt

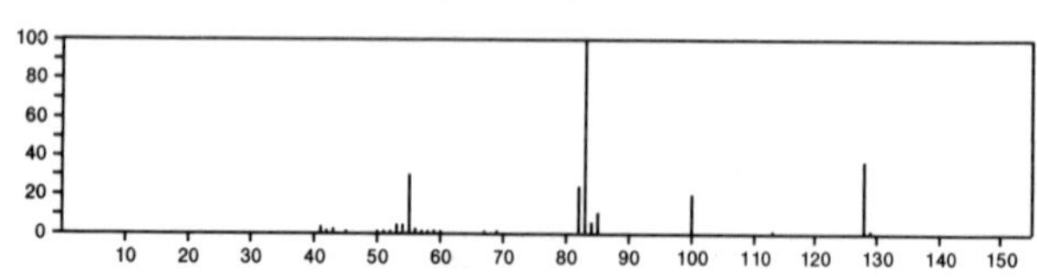

128 $C_7H_{12}O_2$ 1540-34-7
2,4-Pentanedione, 3-ethyl-

MeCOCHEtCOMe

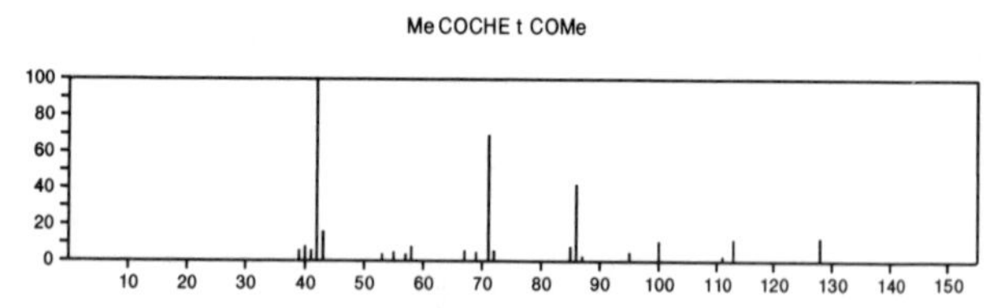

128 $C_7H_{12}O_2$ 2396-77-2
2-Hexenoic acid, methyl ester

PrCH = CHC(O) OMe

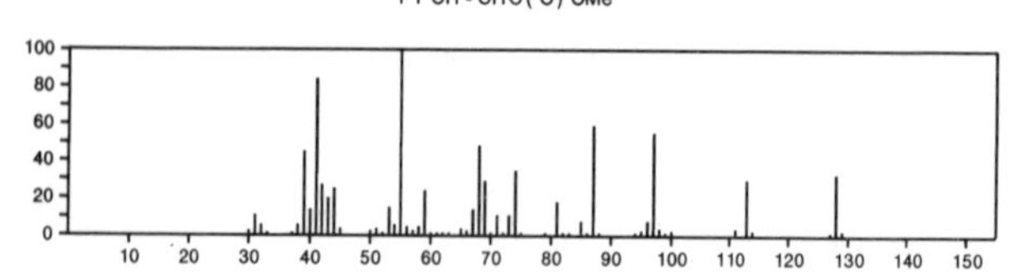

128 $C_7H_{12}O_2$ 2396-78-3
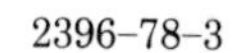
3-Hexenoic acid, methyl ester

MeOC(O) CH_2 CH = CHEt

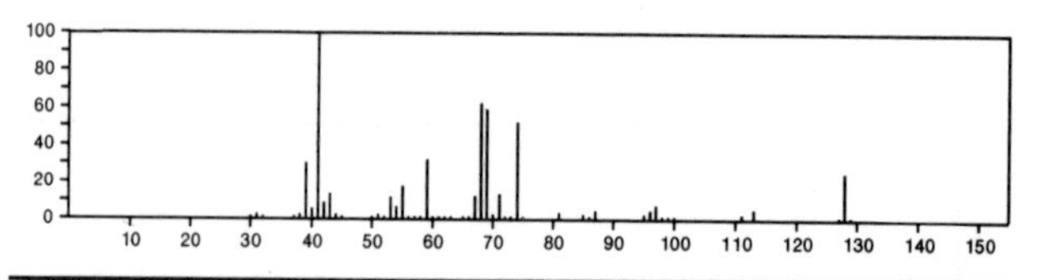

128 $C_7H_{12}O_2$ 2396-80-7
5-Hexenoic acid, methyl ester

H_2C = CH (CH_2) 3 C(O) OMe

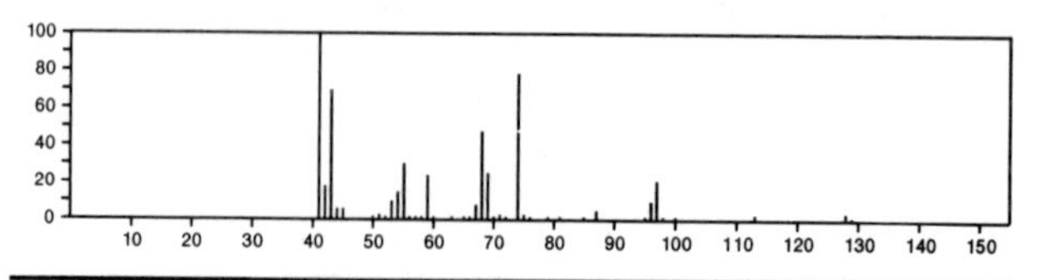

128 $C_7H_{12}O_2$ 2610-95-9
2*H*-Pyran-2-one, tetrahydro-6,6-dimethyl-

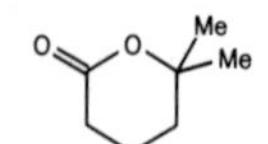

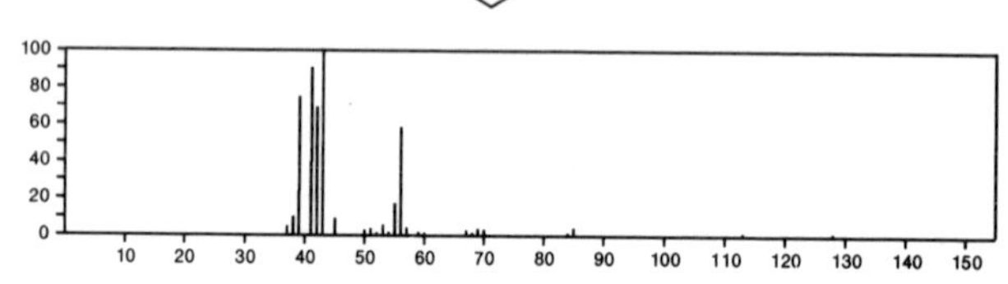

128 $C_7H_{12}O_2$ 2865-82-9
2(3*H*)-Furanone, 5-ethyldihydro-5-methyl-

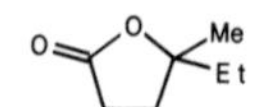

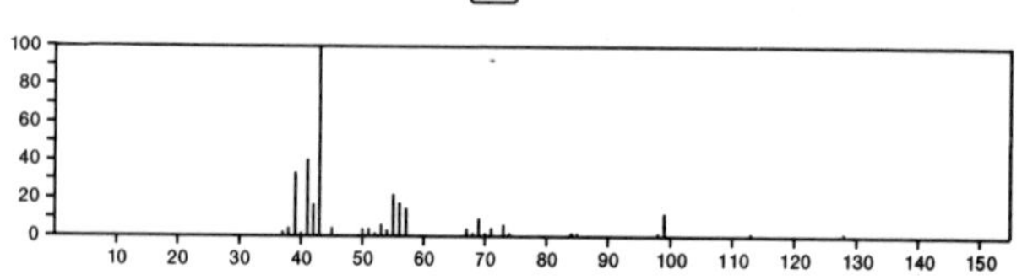

128 $C_7H_{12}O_2$ 2998-08-5
2-Propenoic acid, 1-methylpropyl ester

s-BuOC(O) CH = CH_2

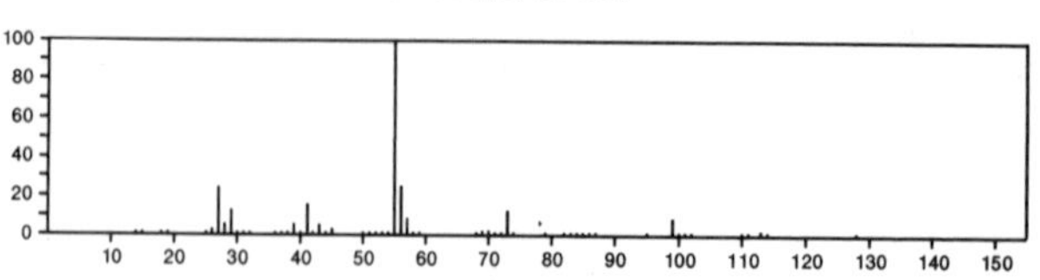

128 $C_7H_{12}O_2$ 3070-67-5
Butanoic acid, 3-methyl-2-methylene-, methyl ester

CH_2
||
MeOC(O) CCHMe 2

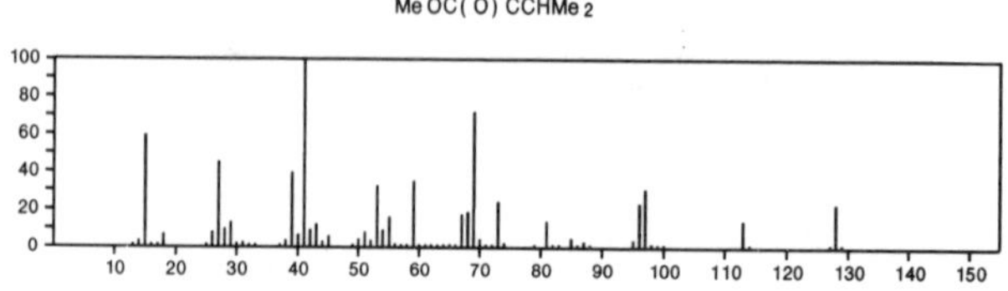

128 $C_7H_{12}O_2$ 3290-57-1
2*H*-Pyran-2-one, tetrahydro-3,5-dimethyl-

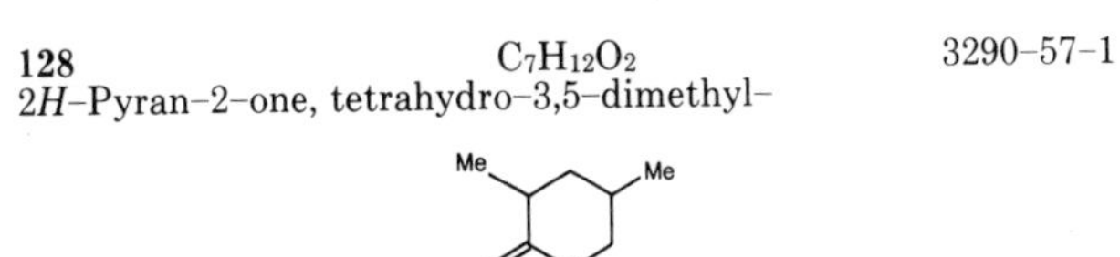

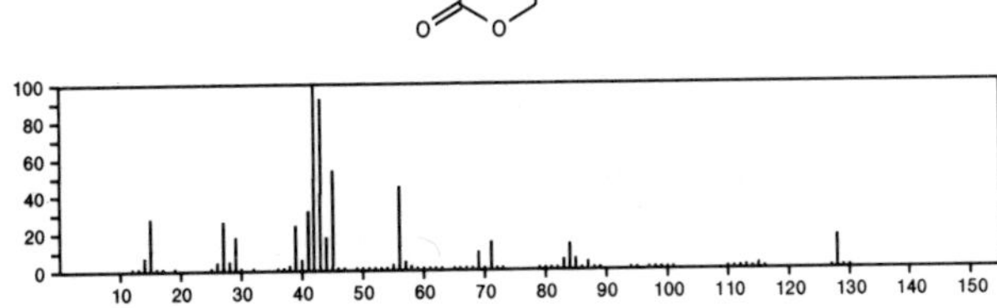

128 $C_7H_{12}O_2$ 3720-22-7
2*H*-Pyran-2-one, tetrahydro-3,6-dimethyl-

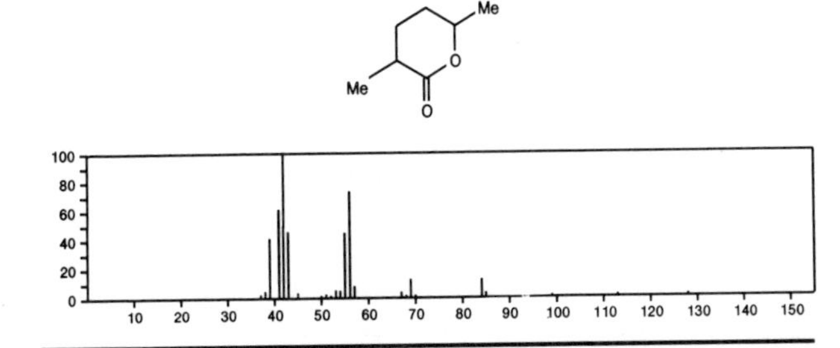

128 $C_7H_{12}O_2$ 3973-21-5
2-Propanol, 1-[(1-methyl-2-propynyl)oxy]-

MeCH(OH)CH2OCHMeC≡CH

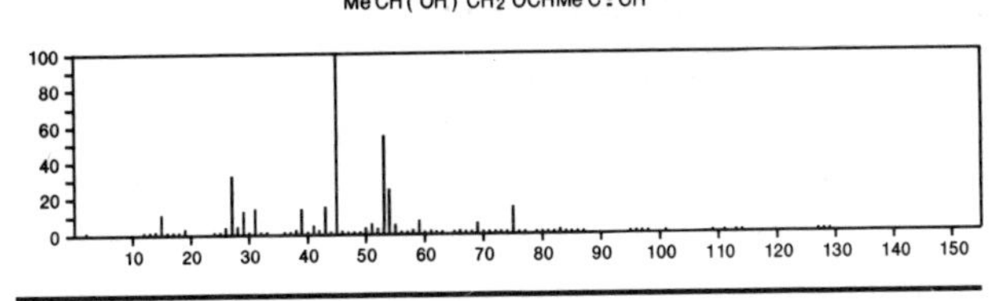

128 $C_7H_{12}O_2$ 3973-27-1
p-Dioxin, 2,3-dihydro-2,5,6-trimethyl-

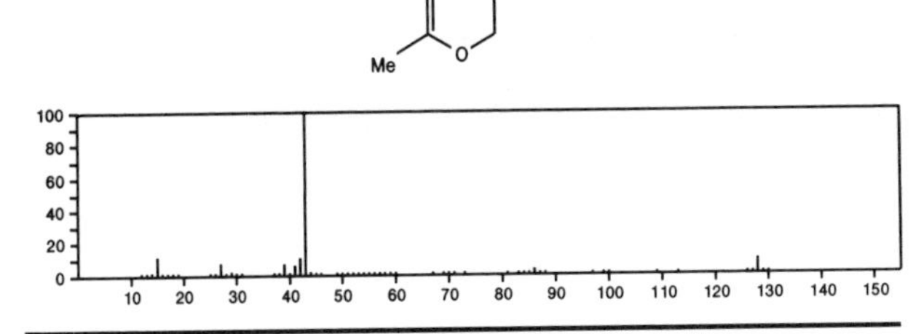

128 $C_7H_{12}O_2$ 3984-21-2
p-Dioxane, 2,5-dimethyl-3-methylene-

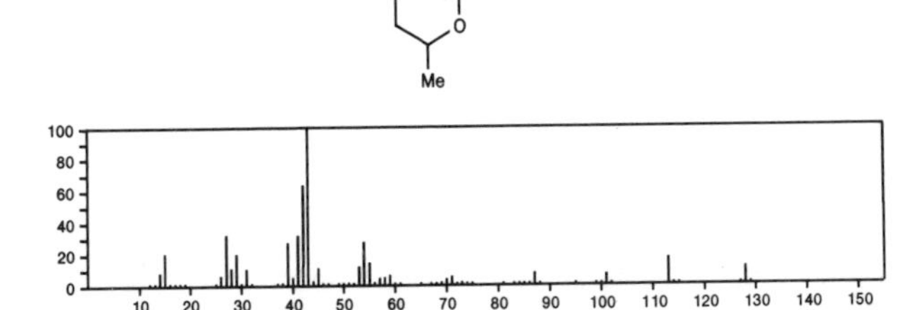

128 $C_7H_{12}O_2$ 4168-01-8
3-Butenoic acid, 2,2,3-trimethyl-

HO2CCMe2CMe=CH2

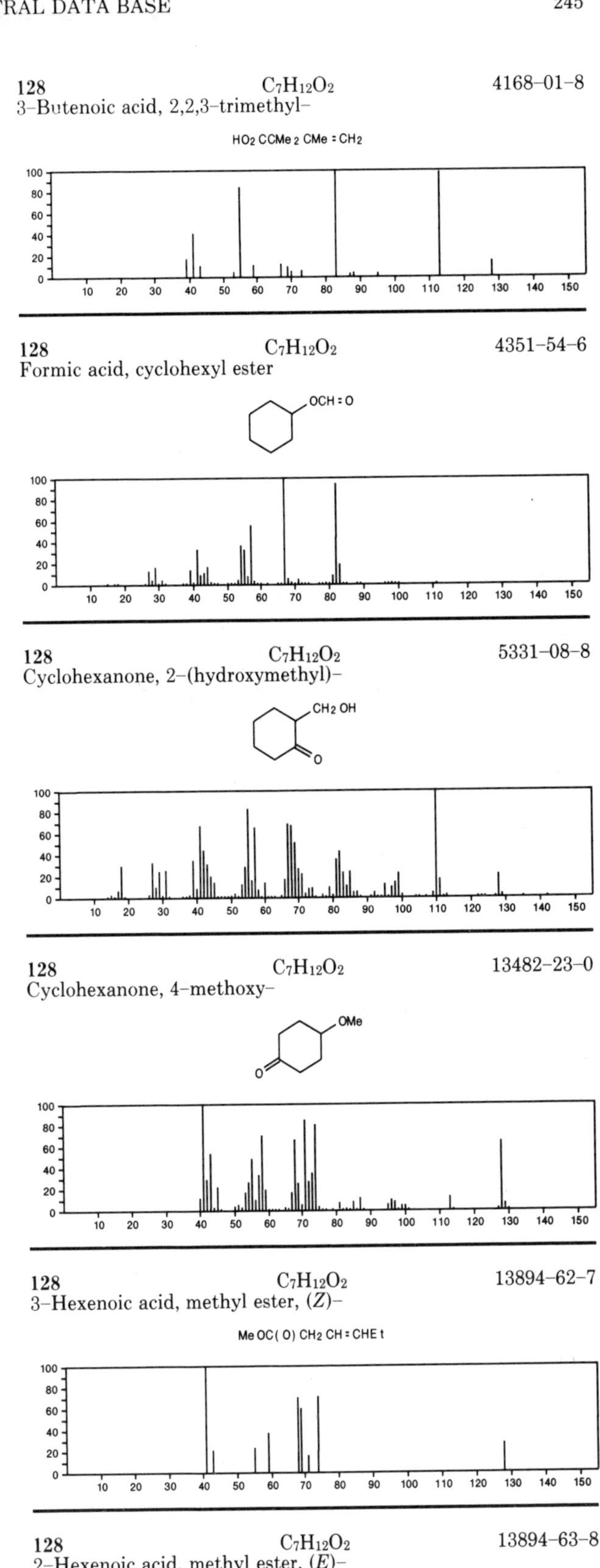

128 $C_7H_{12}O_2$ 4351-54-6
Formic acid, cyclohexyl ester

OCH=O

128 $C_7H_{12}O_2$ 5331-08-8
Cyclohexanone, 2-(hydroxymethyl)-

CH2OH

128 $C_7H_{12}O_2$ 13482-23-0
Cyclohexanone, 4-methoxy-

OMe

128 $C_7H_{12}O_2$ 13894-62-7
3-Hexenoic acid, methyl ester, (*Z*)-

MeOC(O)CH2CH=CHEt

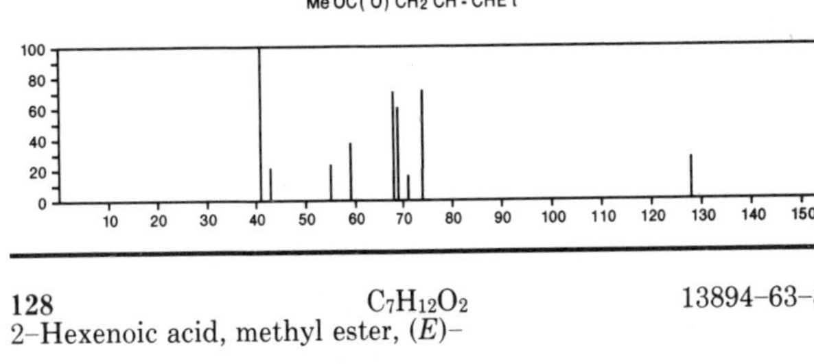

128 $C_7H_{12}O_2$ 13894-63-8
2-Hexenoic acid, methyl ester, (*E*)-

PrCH=CHC(O)OMe

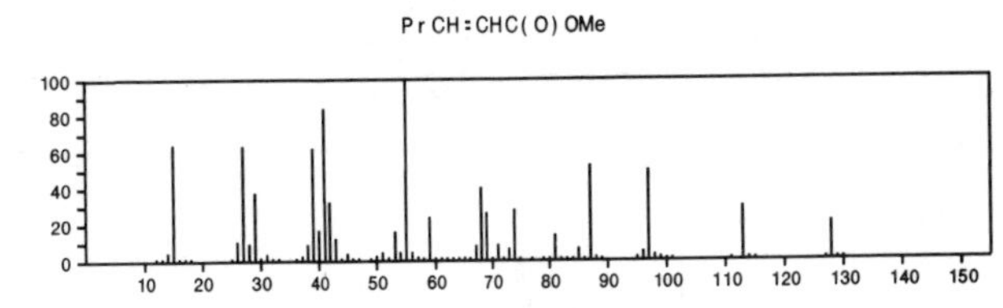

128 $C_7H_{12}O_2$ 14132-44-6
Cyclobutanecarboxylic acid, 2-methyl-, methyl ester

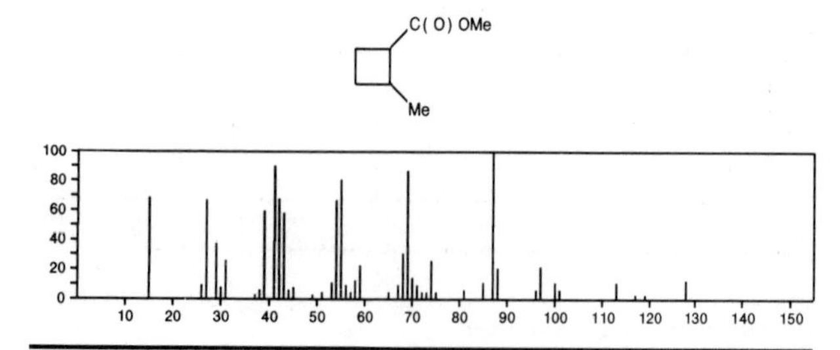

128 $C_7H_{12}O_2$ 15120-99-7
Ethanone, 1-(trimethyloxiranyl)-

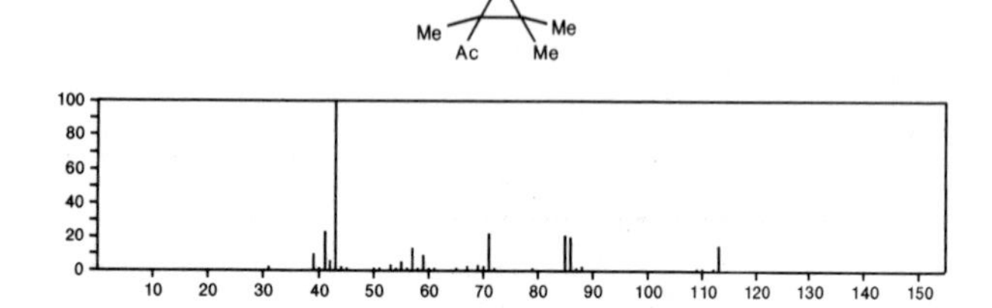

128 $C_7H_{12}O_2$ 16642-52-7
3-Pentenoic acid, 2,2-dimethyl-

MeCH=CHCMe2CO2H

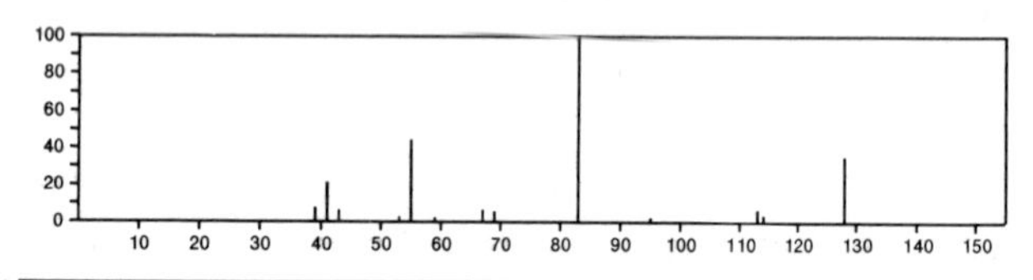

128 $C_7H_{12}O_2$ 17429-00-4
Cyclohexanone, 3-methoxy-

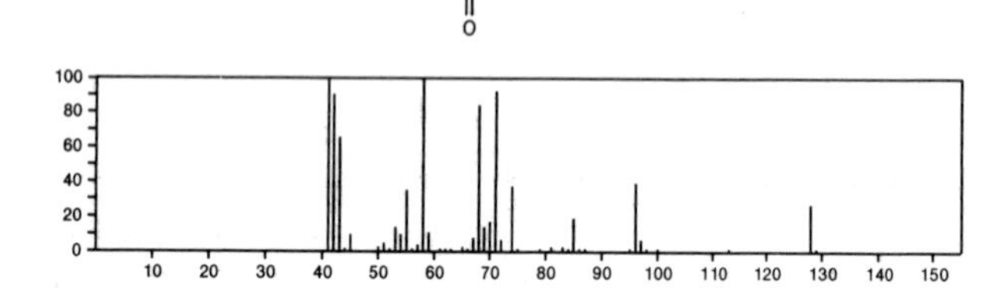

128 $C_7H_{12}O_2$ 18060-77-0
2-Butenoic acid, 1-methylethyl ester

MeCH=CHC(O)OPr-i

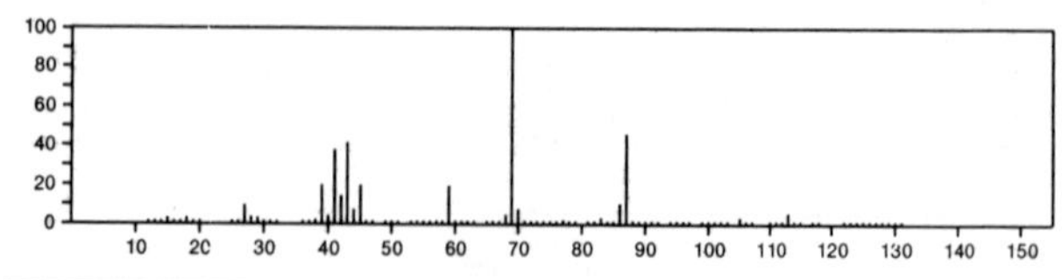

128 $C_7H_{12}O_2$ 41654-12-0
3-Pentenoic acid, 3-methyl-, methyl ester, (*E*)-

MeCH=CMeCH2C(O)OMe

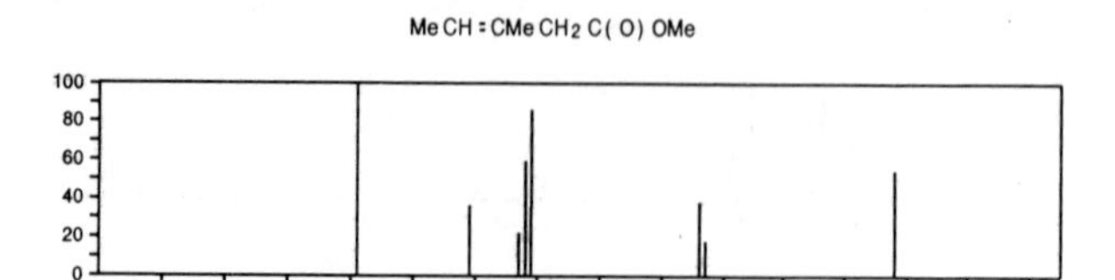

128 $C_7H_{12}O_2$ 50652-78-3
2-Pentenoic acid, 4-methyl-, methyl ester

Me2CHCH=CHC(O)OMe

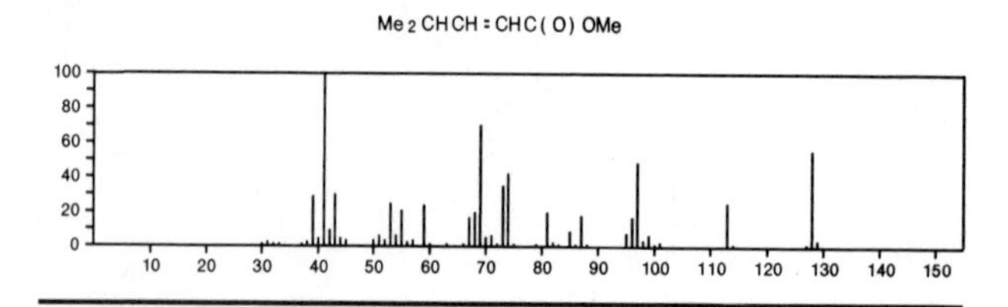

128 $C_7H_{12}O_2$ 50652-79-4
2-Pentenoic acid, 3-methyl-, methyl ester

EtCMe=CHC(O)OMe

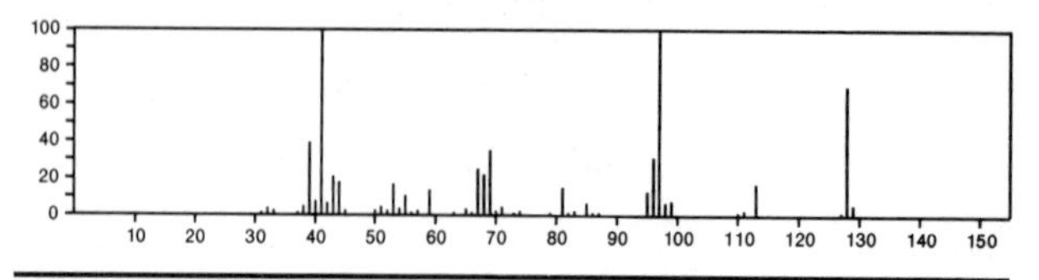

128 $C_7H_{12}O_2$ 51945-98-3
1,5-Heptadiene-3,4-diol

H2C=CHCH(OH)CH(OH)CH=CHMe

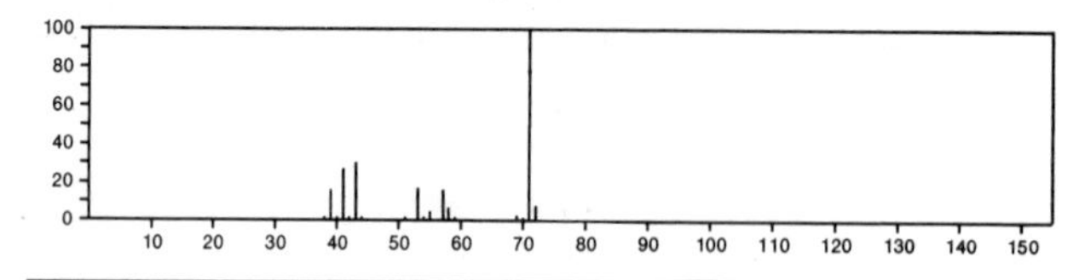

128 $C_7H_{12}O_2$ 55670-09-2
1-Butanol, 2-methylene-, acetate

H2C=CEtCH2OAc

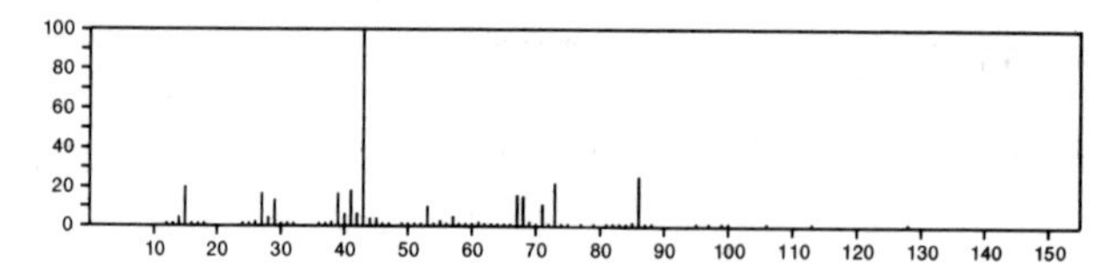

128 $C_7H_{12}O_2$ 55683-34-6
1,3-Dioxolane, 2-ethenyl-2,4-dimethyl-, *trans*-

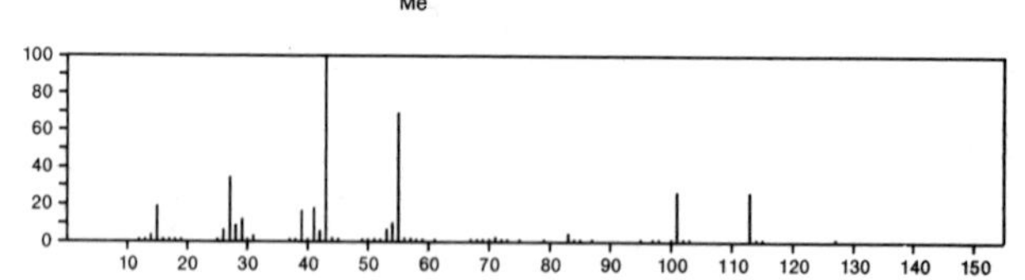

128 $C_7H_{12}O_2$ 55683-35-7
5*H*-1,4-Dioxepin, 2,3-dihydro-2,5-dimethyl-

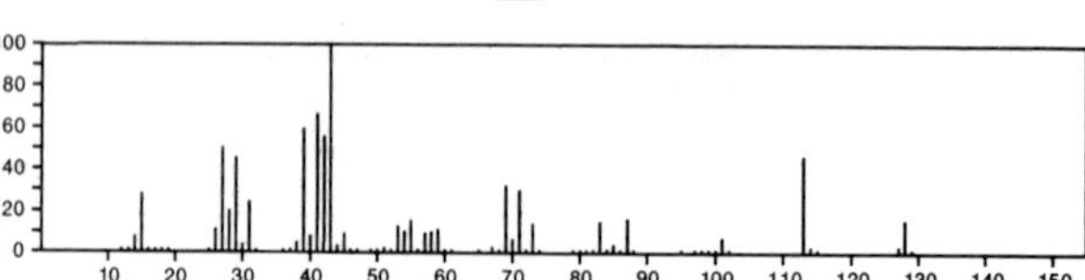

128 $C_7H_{12}O_2$ 55956-45-1
2-Pentanone, 1-methoxy-3-methylene-

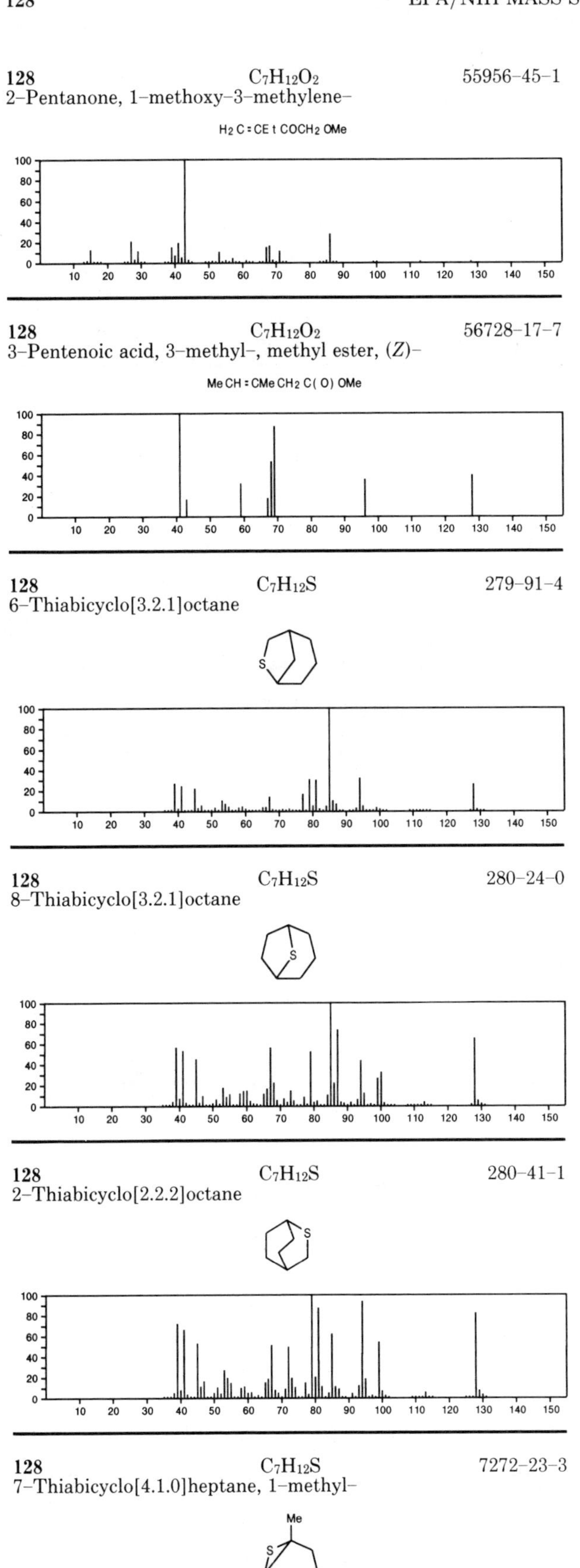

128 $C_7H_{12}O_2$ 56728-17-7
3-Pentenoic acid, 3-methyl-, methyl ester, (*Z*)-

128 $C_7H_{12}S$ 279-91-4
6-Thiabicyclo[3.2.1]octane

128 $C_7H_{12}S$ 280-24-0
8-Thiabicyclo[3.2.1]octane

128 $C_7H_{12}S$ 280-41-1
2-Thiabicyclo[2.2.2]octane

128 $C_7H_{12}S$ 7272-23-3
7-Thiabicyclo[4.1.0]heptane, 1-methyl-

128 $C_7H_{12}S$ 53907-80-5
1*H*-Cyclopenta[*c*]thiophene, hexahydro-, *cis*-

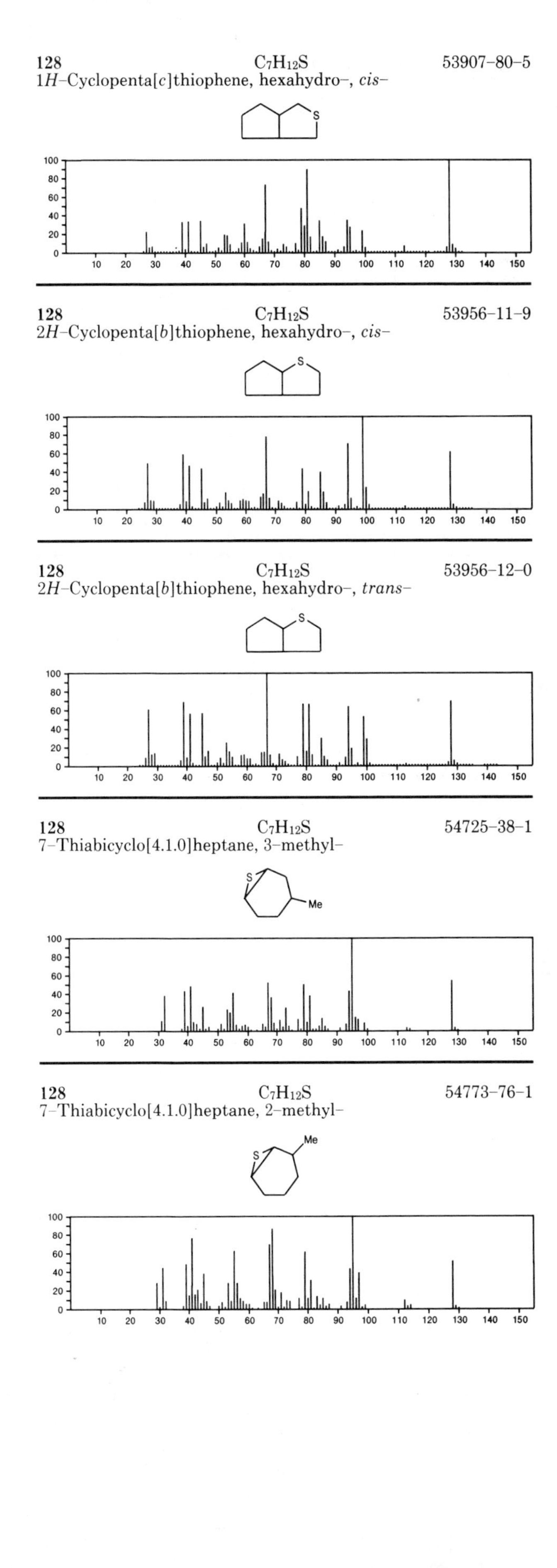

128 $C_7H_{12}S$ 53956-11-9
2*H*-Cyclopenta[*b*]thiophene, hexahydro-, *cis*-

128 $C_7H_{12}S$ 53956-12-0
2*H*-Cyclopenta[*b*]thiophene, hexahydro-, *trans*-

128 $C_7H_{12}S$ 54725-38-1
7-Thiabicyclo[4.1.0]heptane, 3-methyl-

128 $C_7H_{12}S$ 54773-76-1
7-Thiabicyclo[4.1.0]heptane, 2-methyl-

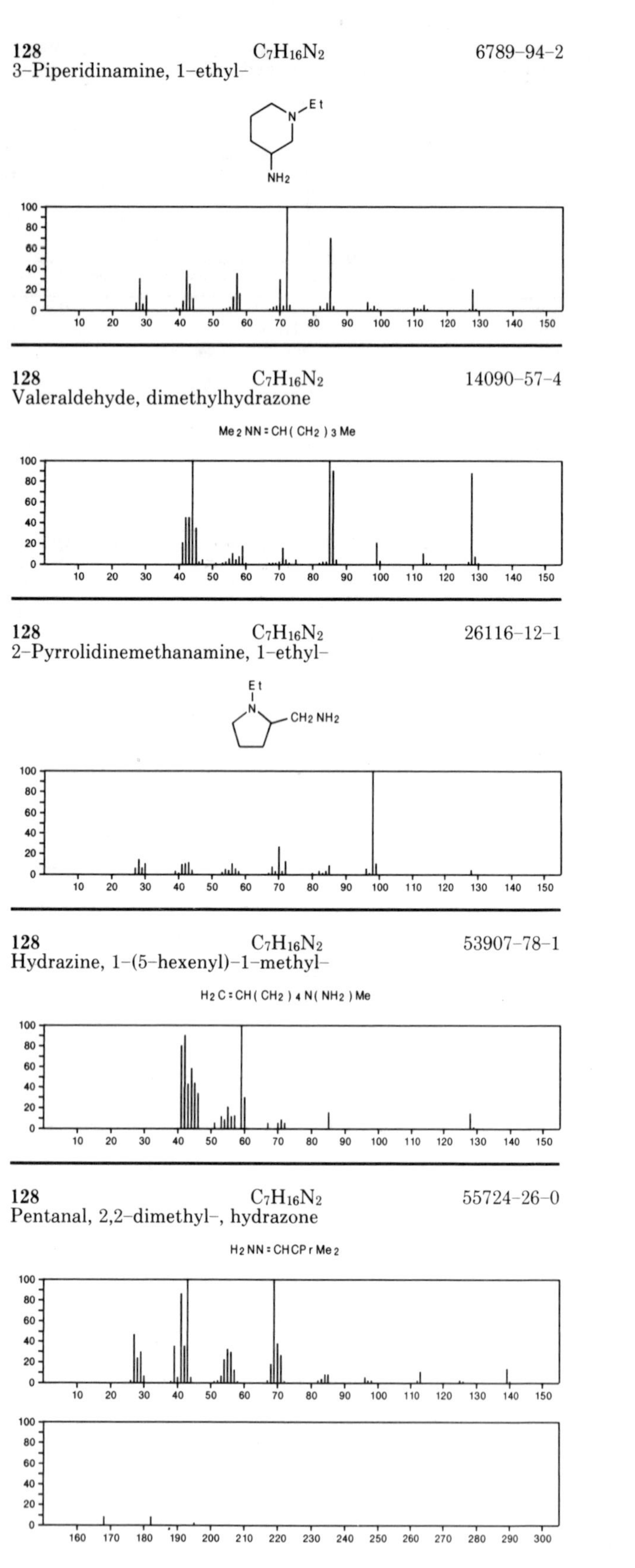

128 $C_7H_{16}N_2$ 6789–94–2
3–Piperidinamine, 1–ethyl–

128 $C_7H_{16}N_2$ 14090–57–4
Valeraldehyde, dimethylhydrazone

128 $C_7H_{16}N_2$ 26116–12–1
2–Pyrrolidinemethanamine, 1–ethyl–

128 $C_7H_{16}N_2$ 53907–78–1
Hydrazine, 1–(5–hexenyl)–1–methyl–

128 $C_7H_{16}N_2$ 55724–26–0
Pentanal, 2,2–dimethyl–, hydrazone

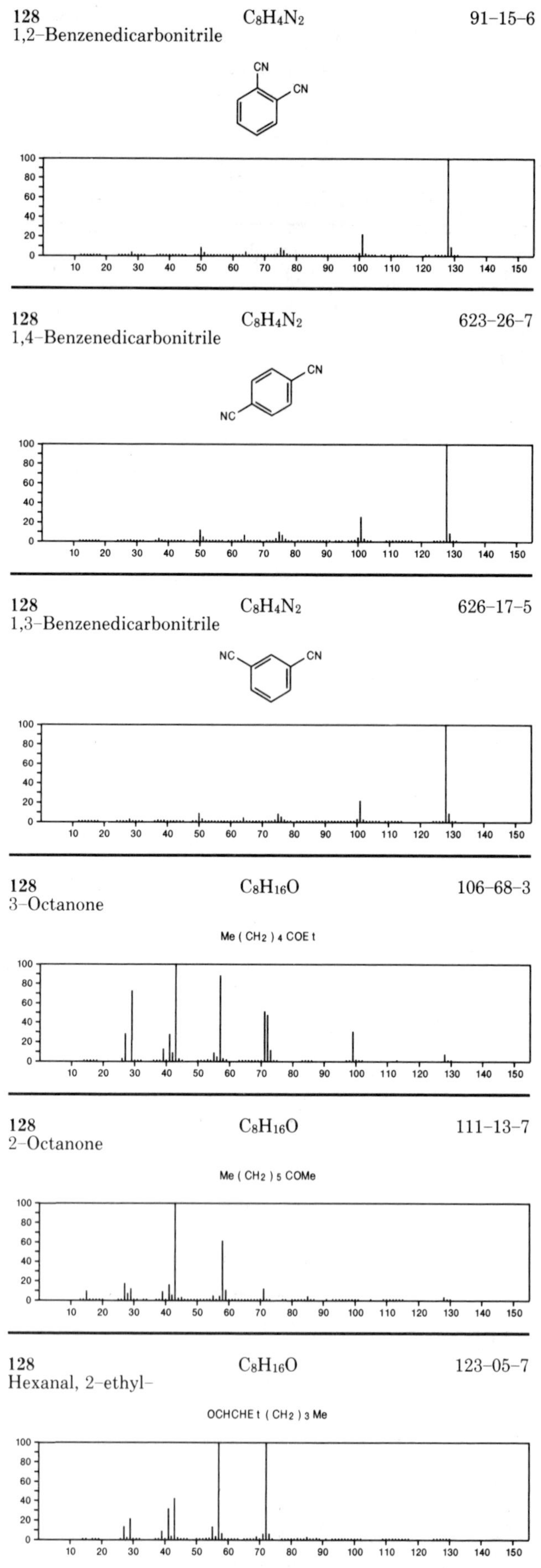

128 $C_8H_4N_2$ 91–15–6
1,2–Benzenedicarbonitrile

128 $C_8H_4N_2$ 623–26–7
1,4–Benzenedicarbonitrile

128 $C_8H_4N_2$ 626–17–5
1,3–Benzenedicarbonitrile

128 $C_8H_{16}O$ 106–68–3
3–Octanone

128 $C_8H_{16}O$ 111–13–7
2–Octanone

128 $C_8H_{16}O$ 123–05–7
Hexanal, 2–ethyl–

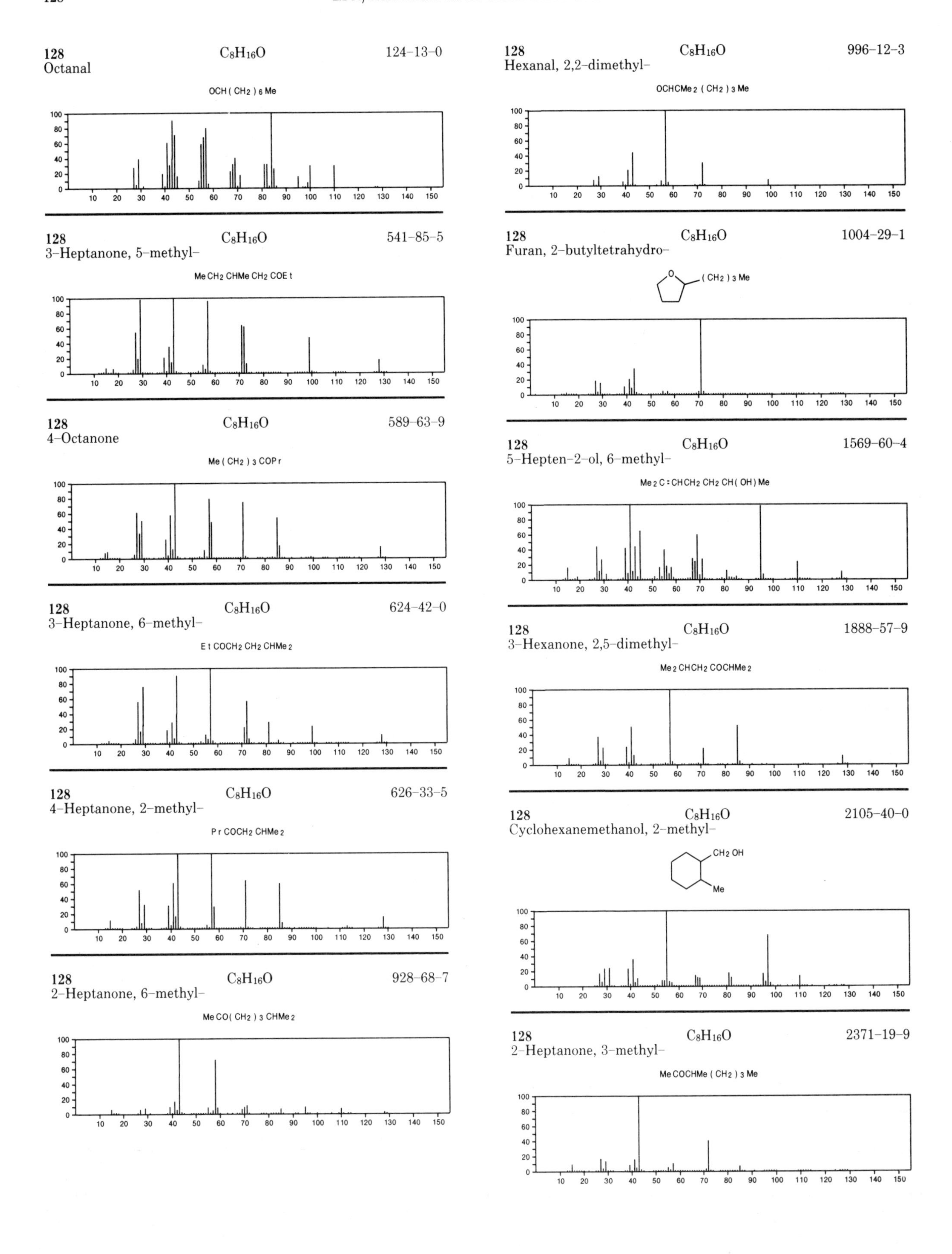
128 C8H16O 124-13-0
Octanal
OCH(CH2)6Me
128 C8H16O 541-85-5
3-Heptanone, 5-methyl-
MeCH2CHMeCH2COEt
128 C8H16O 589-63-9
4-Octanone
Me(CH2)3COPr
128 C8H16O 624-42-0
3-Heptanone, 6-methyl-
EtCOCH2CH2CHMe2
128 C8H16O 626-33-5
4-Heptanone, 2-methyl-
PrCOCH2CHMe2
128 C8H16O 928-68-7
2-Heptanone, 6-methyl-
MeCO(CH2)3CHMe2
128 C8H16O 996-12-3
Hexanal, 2,2-dimethyl-
OCHCMe2(CH2)3Me
128 C8H16O 1004-29-1
Furan, 2-butyltetrahydro-
(CH2)3Me
128 C8H16O 1569-60-4
5-Hepten-2-ol, 6-methyl-
Me2C:CHCH2CH2CH(OH)Me
128 C8H16O 1888-57-9
3-Hexanone, 2,5-dimethyl-
Me2CHCH2COCHMe2
128 C8H16O 2105-40-0
Cyclohexanemethanol, 2-methyl-
CH2OH
Me
128 C8H16O 2371-19-9
2-Heptanone, 3-methyl-
MeCOCHMe(CH2)3Me

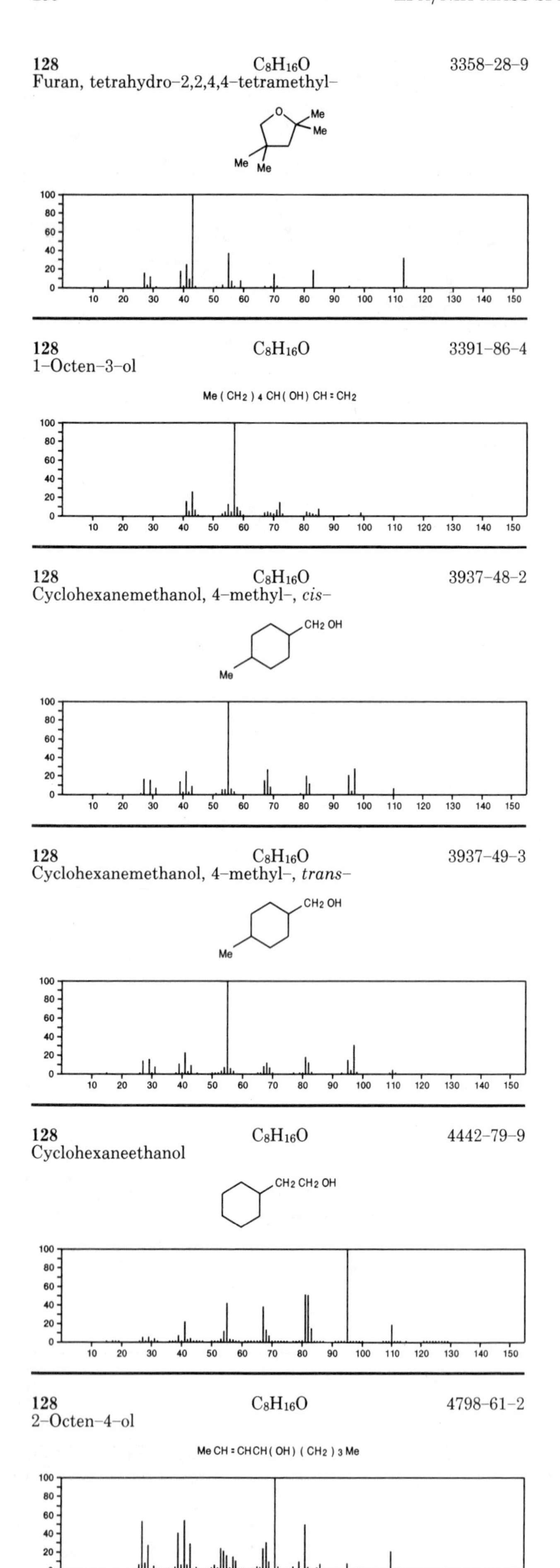
128 C8H16O 3358-28-9
Furan, tetrahydro-2,2,4,4-tetramethyl-
128 C8H16O 3391-86-4
1-Octen-3-ol
Me(CH2)4CH(OH)CH=CH2
128 C8H16O 3937-48-2
Cyclohexanemethanol, 4-methyl-, cis-
128 C8H16O 3937-49-3
Cyclohexanemethanol, 4-methyl-, trans-
128 C8H16O 4442-79-9
Cyclohexaneethanol
128 C8H16O 4798-61-2
2-Octen-4-ol
MeCH=CHCH(OH)(CH2)3Me

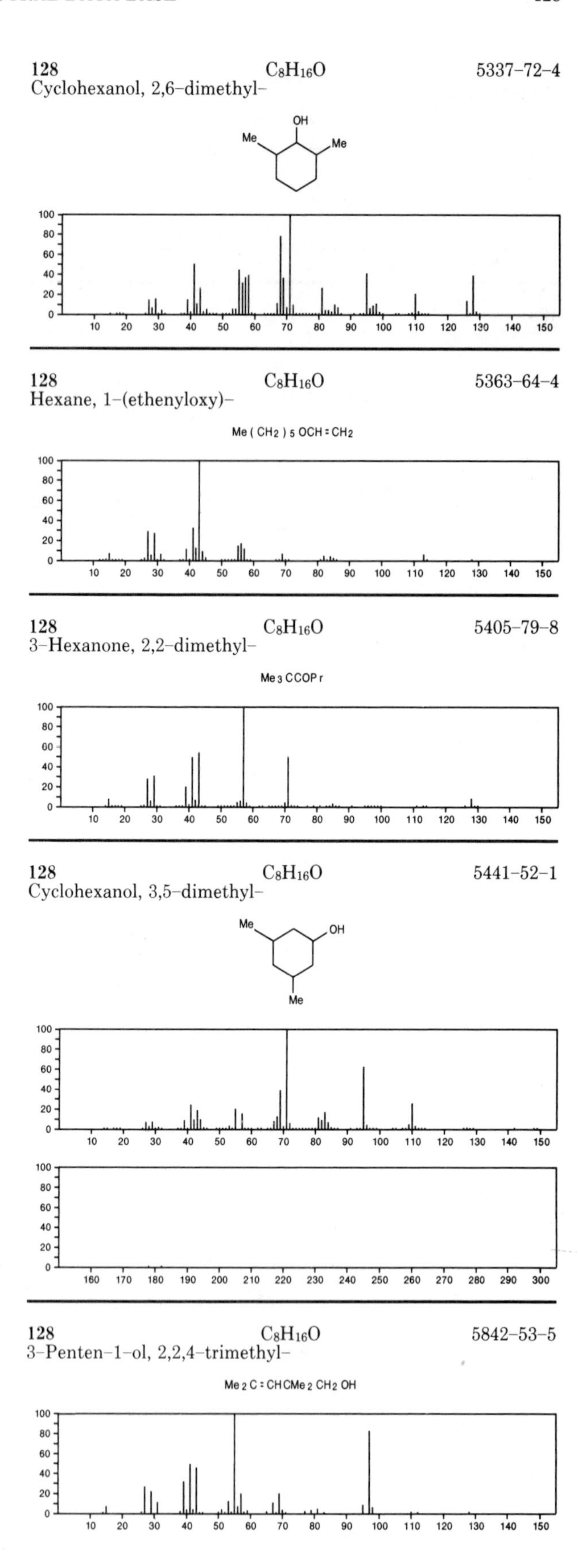
128 C8H16O 5337-72-4
Cyclohexanol, 2,6-dimethyl-
128 C8H16O 5363-64-4
Hexane, 1-(ethenyloxy)-
Me(CH2)5OCH=CH2
128 C8H16O 5405-79-8
3-Hexanone, 2,2-dimethyl-
Me3CCOPr
128 C8H16O 5441-52-1
Cyclohexanol, 3,5-dimethyl-
128 C8H16O 5842-53-5
3-Penten-1-ol, 2,2,4-trimethyl-
Me2C=CHCMe2CH2OH

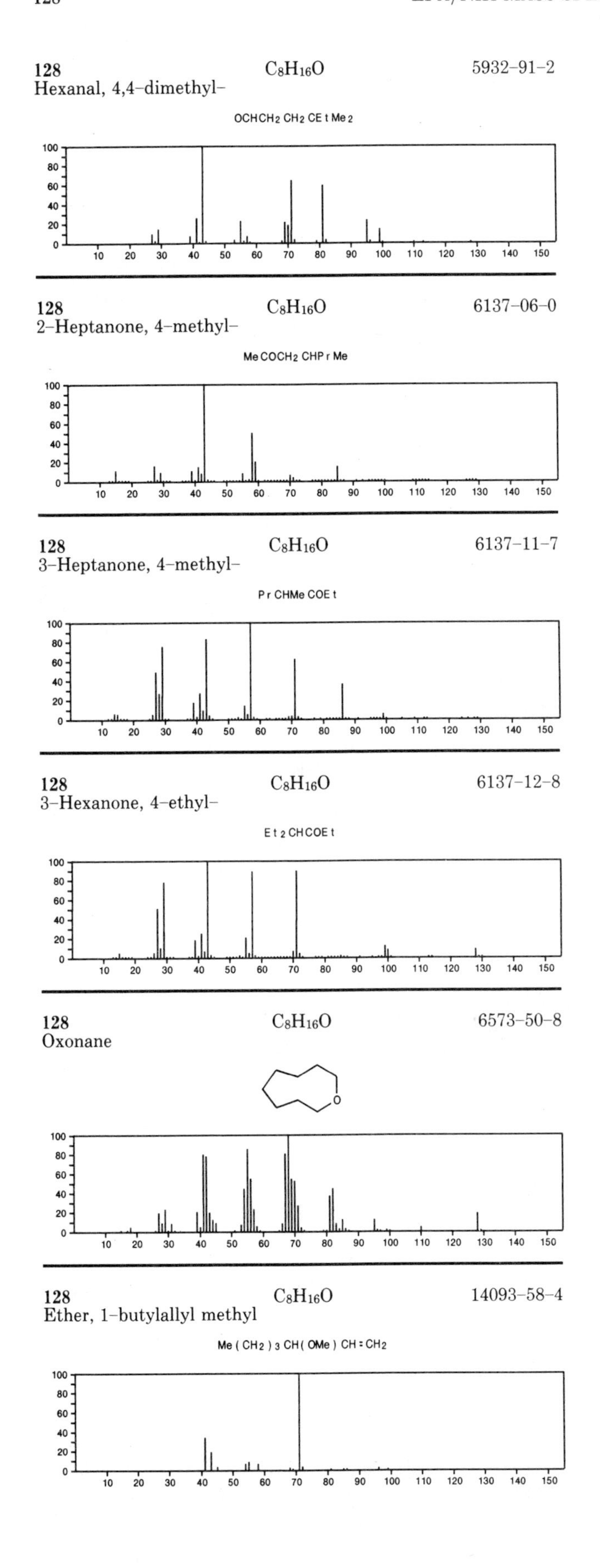

128 $C_8H_{16}O$ 5932-91-2
Hexanal, 4,4-dimethyl-

128 $C_8H_{16}O$ 6137-06-0
2-Heptanone, 4-methyl-

128 $C_8H_{16}O$ 6137-11-7
3-Heptanone, 4-methyl-

128 $C_8H_{16}O$ 6137-12-8
3-Hexanone, 4-ethyl-

128 $C_8H_{16}O$ 6573-50-8
Oxonane

128 $C_8H_{16}O$ 14093-58-4
Ether, 1-butylallyl methyl

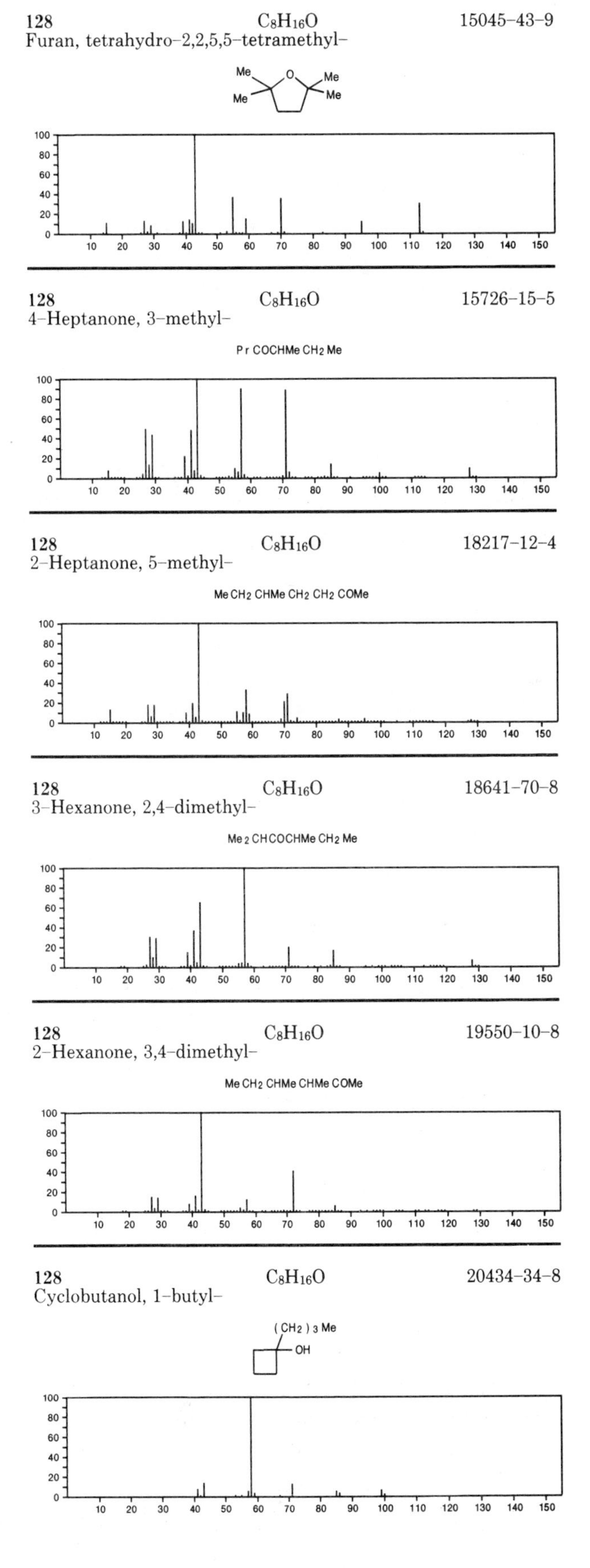

128 $C_8H_{16}O$ 15045-43-9
Furan, tetrahydro-2,2,5,5-tetramethyl-

128 $C_8H_{16}O$ 15726-15-5
4-Heptanone, 3-methyl-

128 $C_8H_{16}O$ 18217-12-4
2-Heptanone, 5-methyl-

128 $C_8H_{16}O$ 18641-70-8
3-Hexanone, 2,4-dimethyl-

128 $C_8H_{16}O$ 19550-10-8
2-Hexanone, 3,4-dimethyl-

128 $C_8H_{16}O$ 20434-34-8
Cyclobutanol, 1-butyl-

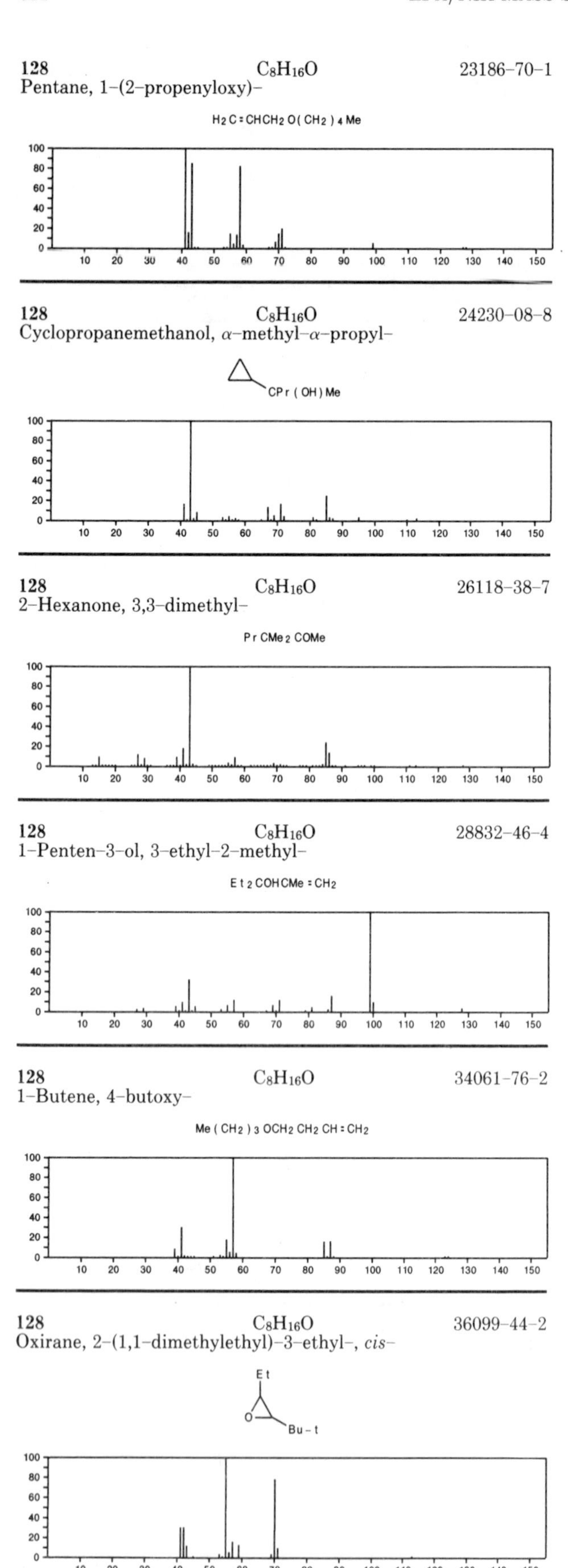
128 $C_8H_{16}O$ 23186-70-1
Pentane, 1-(2-propenyloxy)-
H2C=CHCH2O(CH2)4Me
128 $C_8H_{16}O$ 24230-08-8
Cyclopropanemethanol, α-methyl-α-propyl-
CPr(OH)Me
128 $C_8H_{16}O$ 26118-38-7
2-Hexanone, 3,3-dimethyl-
PrCMe2COMe
128 $C_8H_{16}O$ 28832-46-4
1-Penten-3-ol, 3-ethyl-2-methyl-
Et2COHCMe=CH2
128 $C_8H_{16}O$ 34061-76-2
1-Butene, 4-butoxy-
Me(CH2)3OCH2CH2CH=CH2
128 $C_8H_{16}O$ 36099-44-2
Oxirane, 2-(1,1-dimethylethyl)-3-ethyl-, *cis*-
Et
O
Bu-t

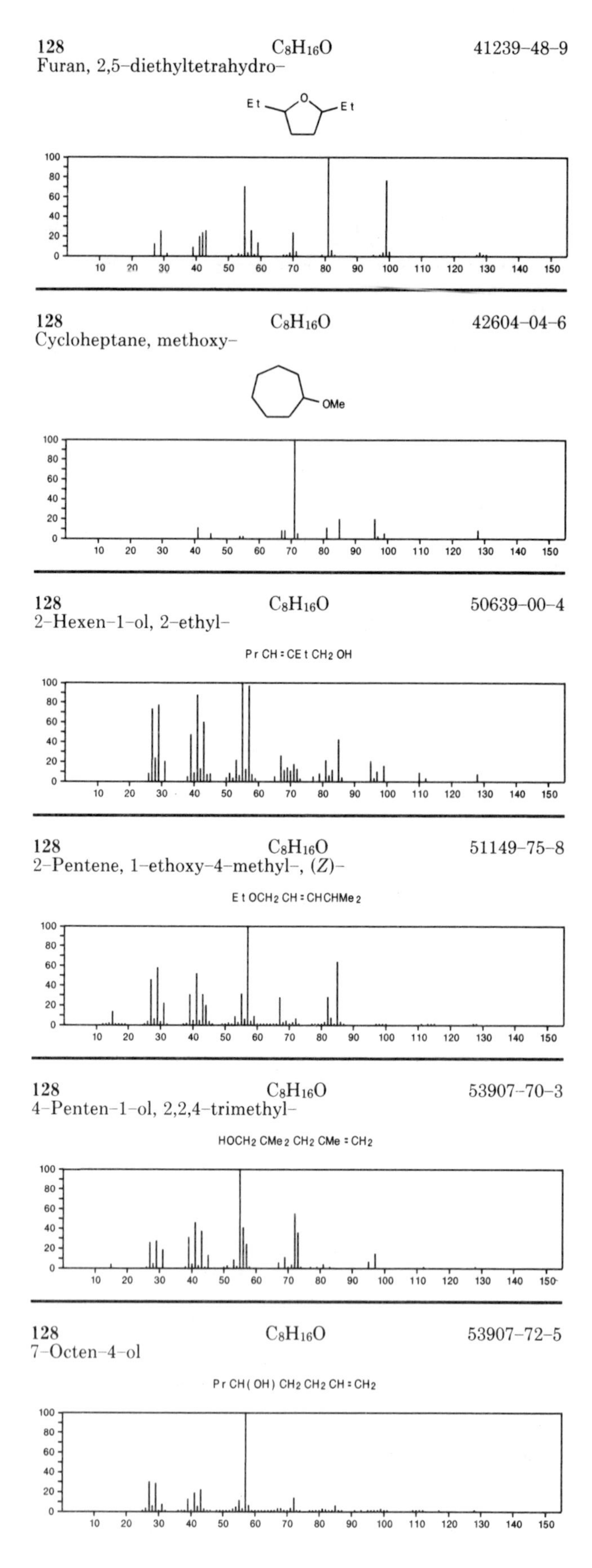
128 $C_8H_{16}O$ 41239-48-9
Furan, 2,5-diethyltetrahydro-
Et
O
Et
128 $C_8H_{16}O$ 42604-04-6
Cycloheptane, methoxy-
OMe
128 $C_8H_{16}O$ 50639-00-4
2-Hexen-1-ol, 2-ethyl-
PrCH=CEtCH2OH
128 $C_8H_{16}O$ 51149-75-8
2-Pentene, 1-ethoxy-4-methyl-, (*Z*)-
EtOCH2CH=CHCHMe2
128 $C_8H_{16}O$ 53907-70-3
4-Penten-1-ol, 2,2,4-trimethyl-
HOCH2CMe2CH2CMe=CH2
128 $C_8H_{16}O$ 53907-72-5
7-Octen-4-ol
PrCH(OH)CH2CH2CH=CH2

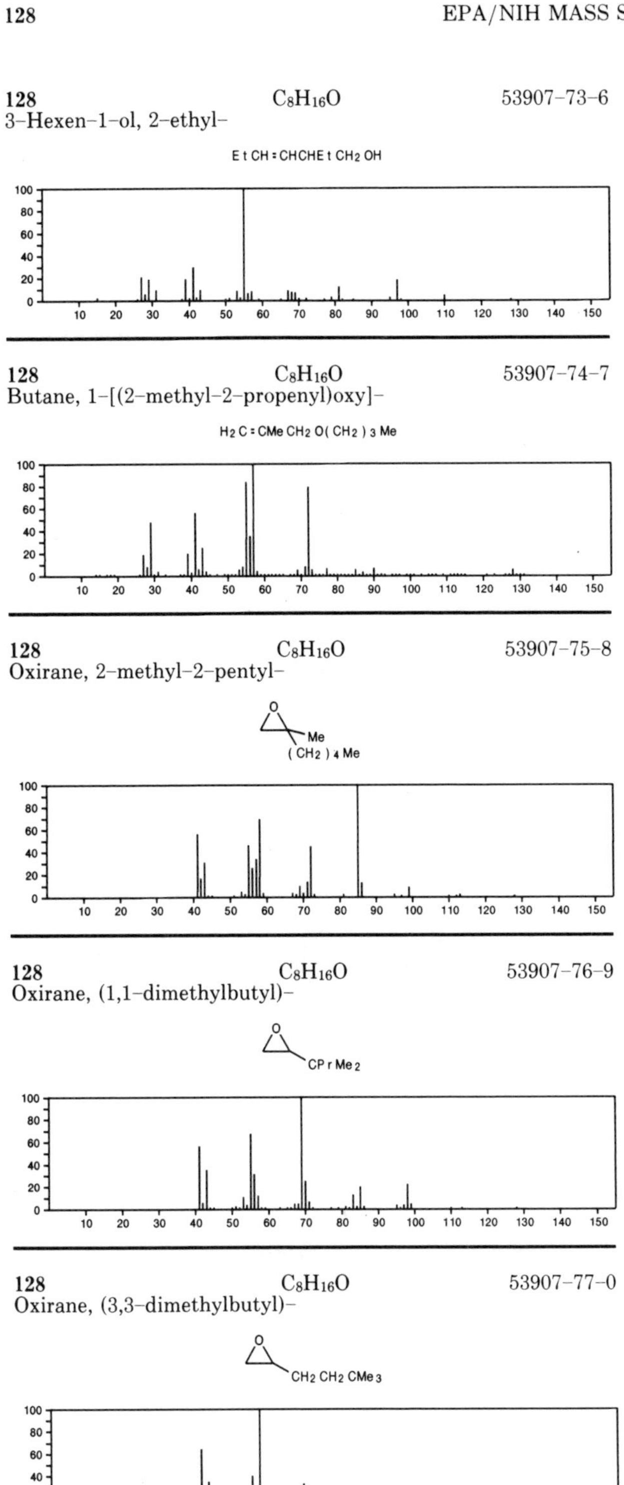

128 $C_8H_{16}O$ 53907-73-6
3-Hexen-1-ol, 2-ethyl-

EtCH=CHCHEtCH2OH

128 $C_8H_{16}O$ 53907-74-7
Butane, 1-[(2-methyl-2-propenyl)oxy]-

H2C=CMeCH2O(CH2)3Me

128 $C_8H_{16}O$ 53907-75-8
Oxirane, 2-methyl-2-pentyl-

128 $C_8H_{16}O$ 53907-76-9
Oxirane, (1,1-dimethylbutyl)-

128 $C_8H_{16}O$ 53907-77-0
Oxirane, (3,3-dimethylbutyl)-

128 $C_8H_{16}O$ 54644-32-5
Oxirane, 2,3-bis(1-methylethyl)-, *trans*-

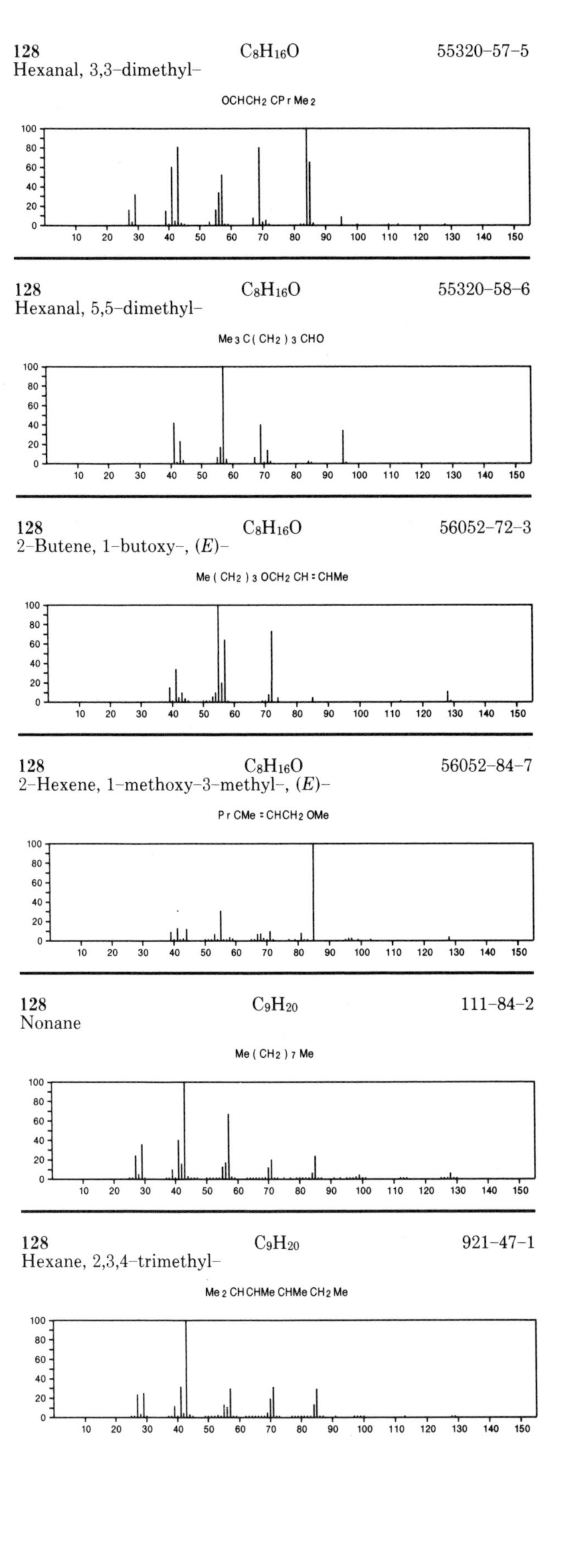

128 $C_8H_{16}O$ 55320-57-5
Hexanal, 3,3-dimethyl-

OCHCH2CPrMe2

128 $C_8H_{16}O$ 55320-58-6
Hexanal, 5,5-dimethyl-

Me3C(CH2)3CHO

128 $C_8H_{16}O$ 56052-72-3
2-Butene, 1-butoxy-, (*E*)-

Me(CH2)3OCH2CH=CHMe

128 $C_8H_{16}O$ 56052-84-7
2-Hexene, 1-methoxy-3-methyl-, (*E*)-

PrCMe=CHCH2OMe

128 C_9H_{20} 111-84-2
Nonane

Me(CH2)7Me

128 C_9H_{20} 921-47-1
Hexane, 2,3,4-trimethyl-

Me2CHCHMeCHMeCH2Me

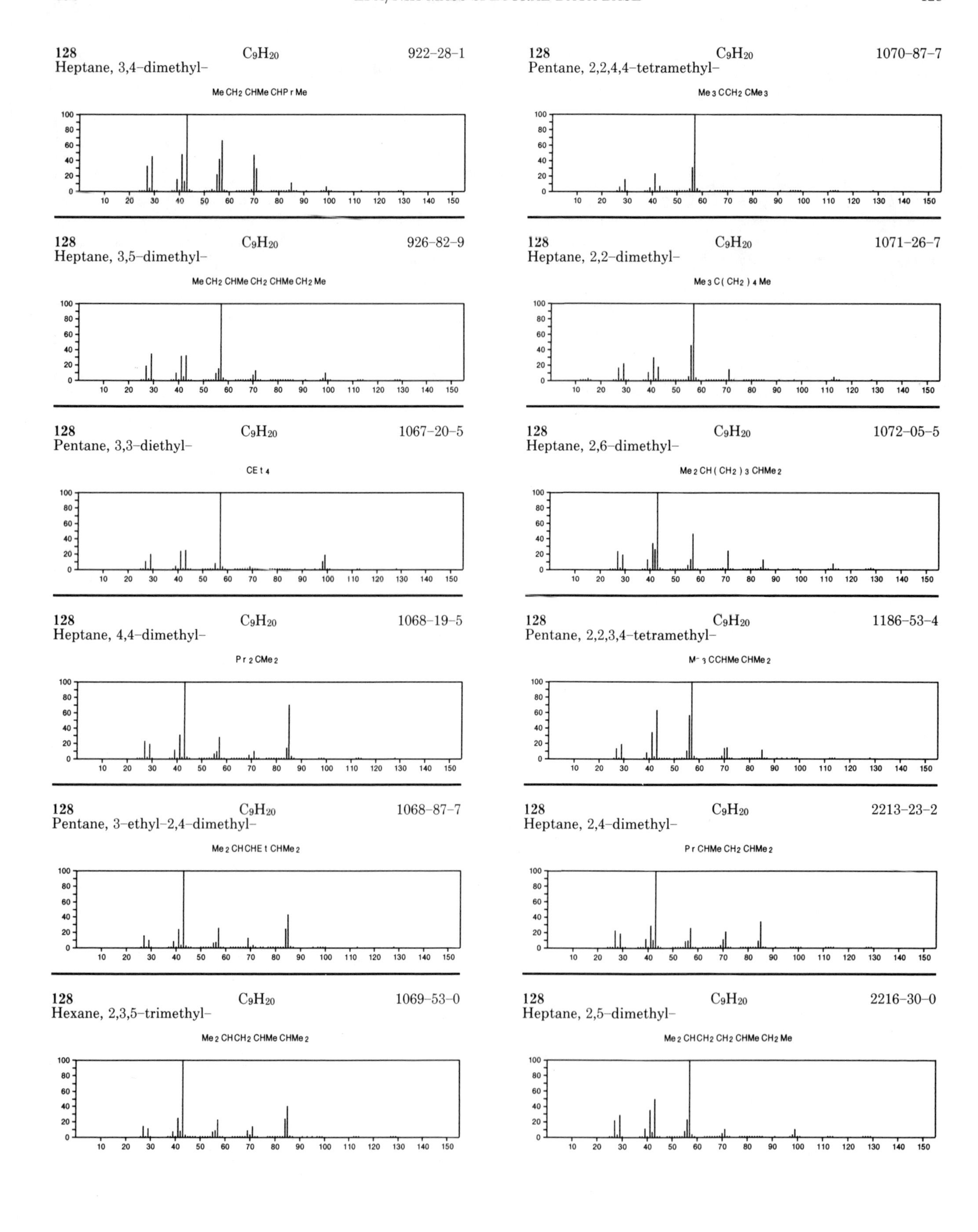
128
C9H20
922-28-1
Heptane, 3,4-dimethyl-
Me CH2 CHMe CHP r Me
128
C9H20
1070-87-7
Pentane, 2,2,4,4-tetramethyl-
Me 3 CCH2 CMe 3
128
C9H20
926-82-9
Heptane, 3,5-dimethyl-
Me CH2 CHMe CH2 CHMe CH2 Me
128
C9H20
1071-26-7
Heptane, 2,2-dimethyl-
Me 3 C(CH2) 4 Me
128
C9H20
1067-20-5
Pentane, 3,3-diethyl-
CE t 4
128
C9H20
1072-05-5
Heptane, 2,6-dimethyl-
Me 2 CH (CH2) 3 CHMe 2
128
C9H20
1068-19-5
Heptane, 4,4-dimethyl-
P r 2 CMe 2
128
C9H20
1186-53-4
Pentane, 2,2,3,4-tetramethyl-
M- 3 CCHMe CHMe 2
128
C9H20
1068-87-7
Pentane, 3-ethyl-2,4-dimethyl-
Me 2 CHCHE t CHMe 2
128
C9H20
2213-23-2
Heptane, 2,4-dimethyl-
P r CHMe CH2 CHMe 2
128
C9H20
1069-53-0
Hexane, 2,3,5-trimethyl-
Me 2 CHCH2 CHMe CHMe 2
128
C9H20
2216-30-0
Heptane, 2,5-dimethyl-
Me 2 CHCH2 CH2 CHMe CH2 Me

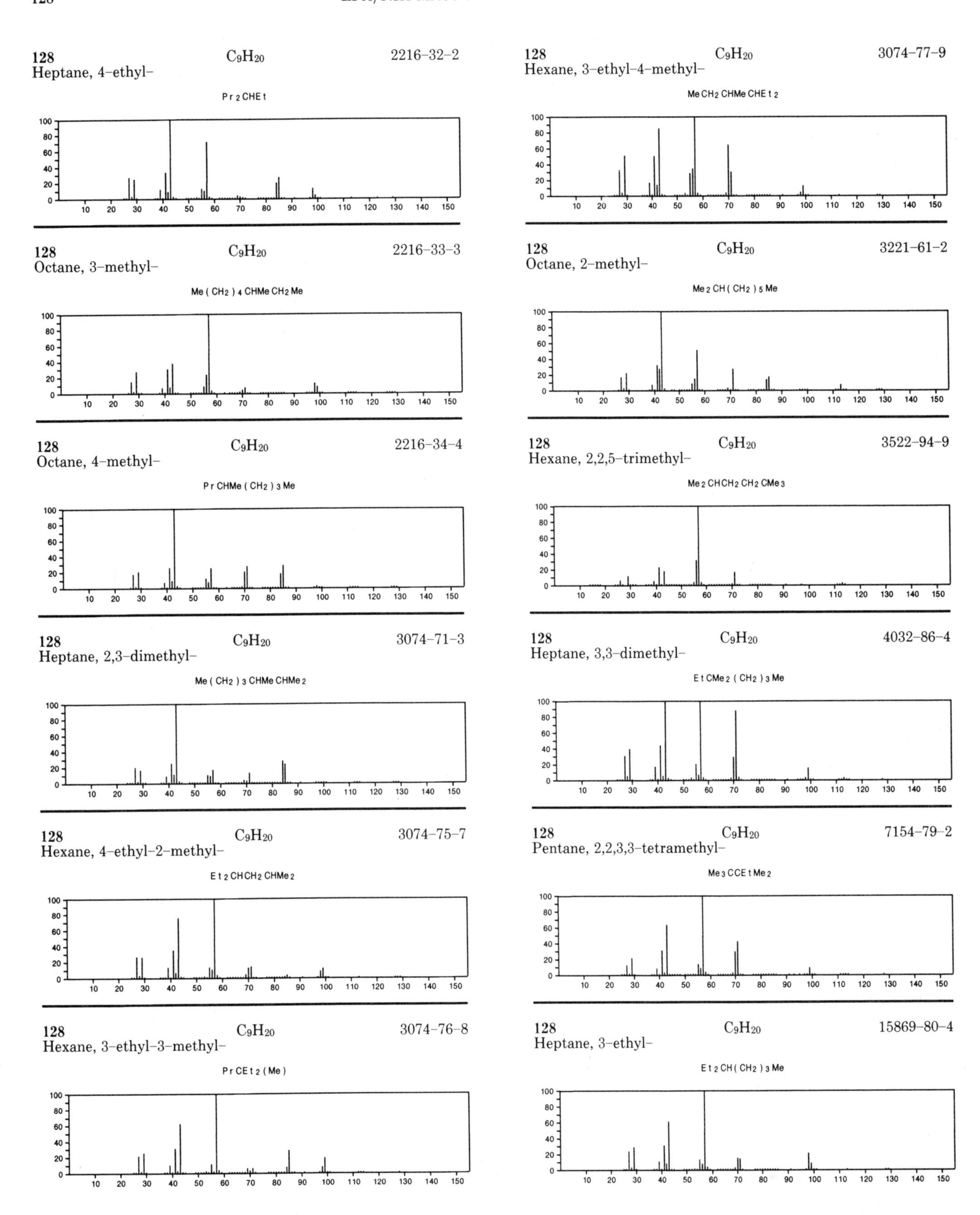
128
Heptane, 4-ethyl-
C_9H_{20}
2216-32-2
Pr 2 CHEt
128
Hexane, 3-ethyl-4-methyl-
C_9H_{20}
3074-77-9
Me CH2 CHMe CHEt 2
128
Octane, 3-methyl-
C_9H_{20}
2216-33-3
Me (CH2) 4 CHMe CH2 Me
128
Octane, 2-methyl-
C_9H_{20}
3221-61-2
Me 2 CH (CH2) 5 Me
128
Octane, 4-methyl-
C_9H_{20}
2216-34-4
Pr CHMe (CH2) 3 Me
128
Hexane, 2,2,5-trimethyl-
C_9H_{20}
3522-94-9
Me 2 CHCH2 CH2 CMe 3
128
Heptane, 2,3-dimethyl-
C_9H_{20}
3074-71-3
Me (CH2) 3 CHMe CHMe 2
128
Heptane, 3,3-dimethyl-
C_9H_{20}
4032-86-4
Et CMe 2 (CH2) 3 Me
128
Hexane, 4-ethyl-2-methyl-
C_9H_{20}
3074-75-7
Et 2 CHCH2 CHMe 2
128
Pentane, 2,2,3,3-tetramethyl-
C_9H_{20}
7154-79-2
Me 3 CCEt Me 2
128
Hexane, 3-ethyl-3-methyl-
C_9H_{20}
3074-76-8
Pr CEt 2 (Me)
128
Heptane, 3-ethyl-
C_9H_{20}
15869-80-4
Et 2 CH (CH2) 3 Me

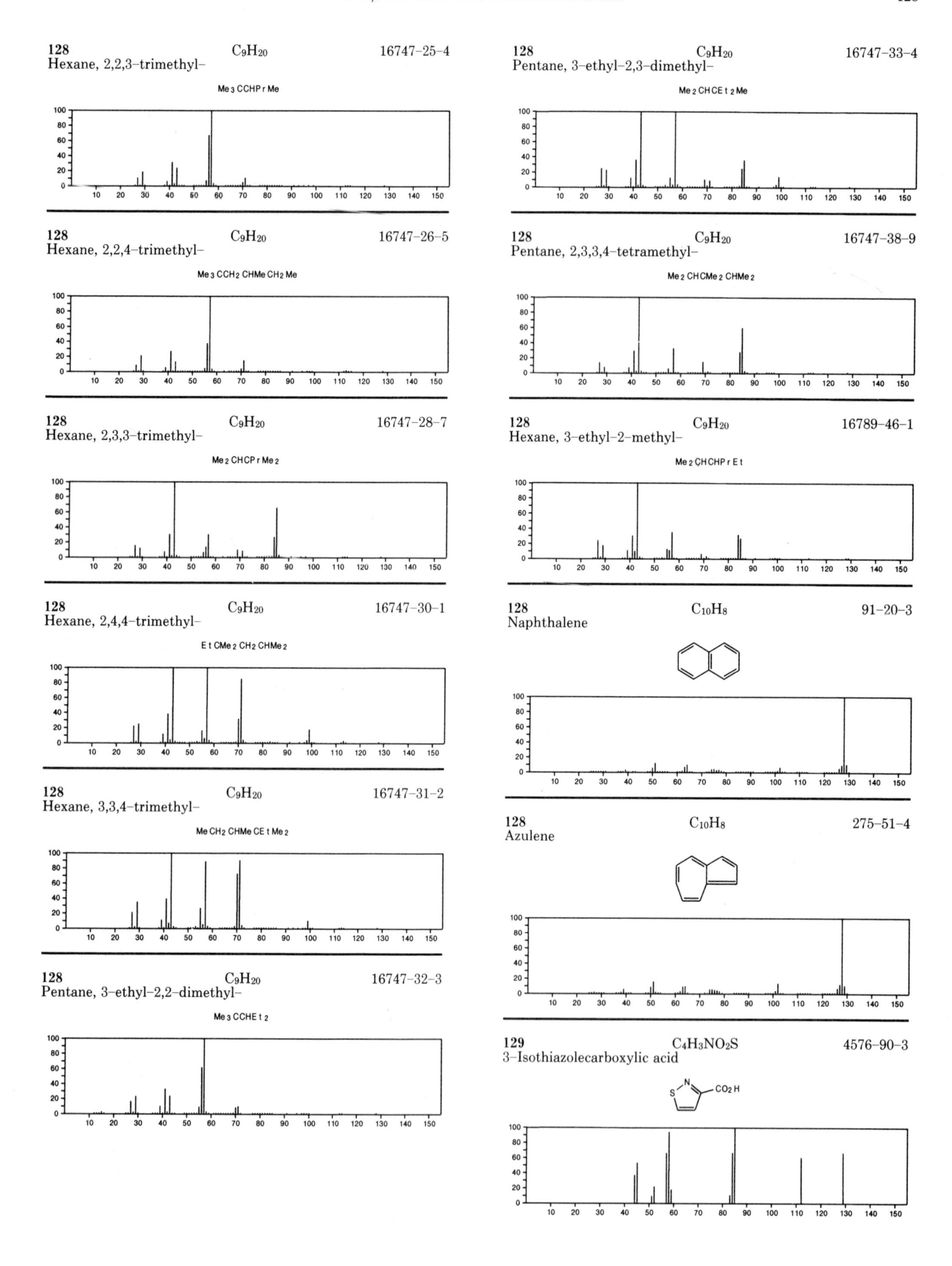

128 C9H20 16747-25-4
Hexane, 2,2,3-trimethyl-
Me3CCHPrMe
128 C9H20 16747-26-5
Hexane, 2,2,4-trimethyl-
Me3CCH2CHMeCH2Me
128 C9H20 16747-28-7
Hexane, 2,3,3-trimethyl-
Me2CHCPrMe2
128 C9H20 16747-30-1
Hexane, 2,4,4-trimethyl-
EtCMe2CH2CHMe2
128 C9H20 16747-31-2
Hexane, 3,3,4-trimethyl-
MeCH2CHMeCEtMe2
128 C9H20 16747-32-3
Pentane, 3-ethyl-2,2-dimethyl-
Me3CCHEt2
128 C9H20 16747-33-4
Pentane, 3-ethyl-2,3-dimethyl-
Me2CHCEt2Me
128 C9H20 16747-38-9
Pentane, 2,3,3,4-tetramethyl-
Me2CHCMe2CHMe2
128 C9H20 16789-46-1
Hexane, 3-ethyl-2-methyl-
Me2CHCHPrEt
128 C10H8 91-20-3
Naphthalene
128 C10H8 275-51-4
Azulene
129 C4H3NO2S 4576-90-3
3-Isothiazolecarboxylic acid
S
N
CO2H

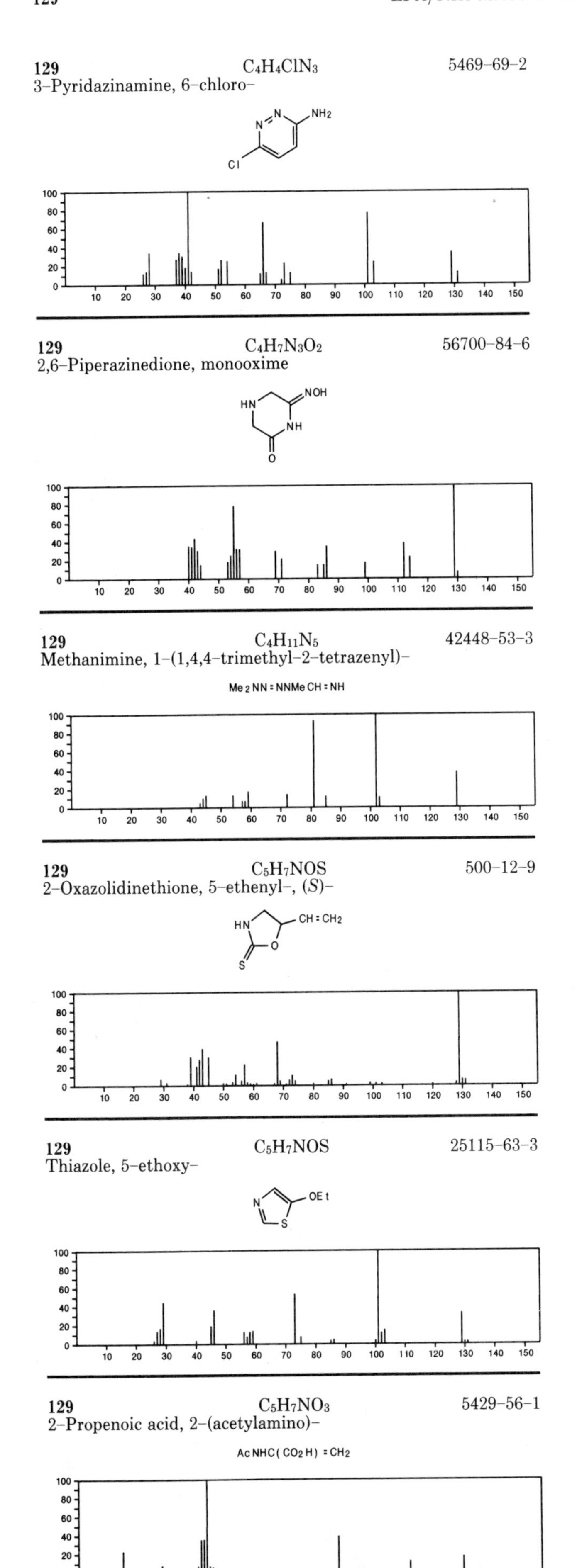

129 $C_4H_4ClN_3$ 5469–69–2
3–Pyridazinamine, 6–chloro–

129 $C_4H_7N_3O_2$ 56700–84–6
2,6–Piperazinedione, monooxime

129 $C_4H_{11}N_5$ 42448–53–3
Methanimine, 1–(1,4,4–trimethyl–2–tetrazenyl)–

$Me_2NN=NNMeCH=NH$

129 C_5H_7NOS 500–12–9
2–Oxazolidinethione, 5–ethenyl–, (*S*)–

129 C_5H_7NOS 25115–63–3
Thiazole, 5–ethoxy–

129 $C_5H_7NO_3$ 5429–56–1
2–Propenoic acid, 2–(acetylamino)–

$AcNHC(CO_2H)=CH_2$

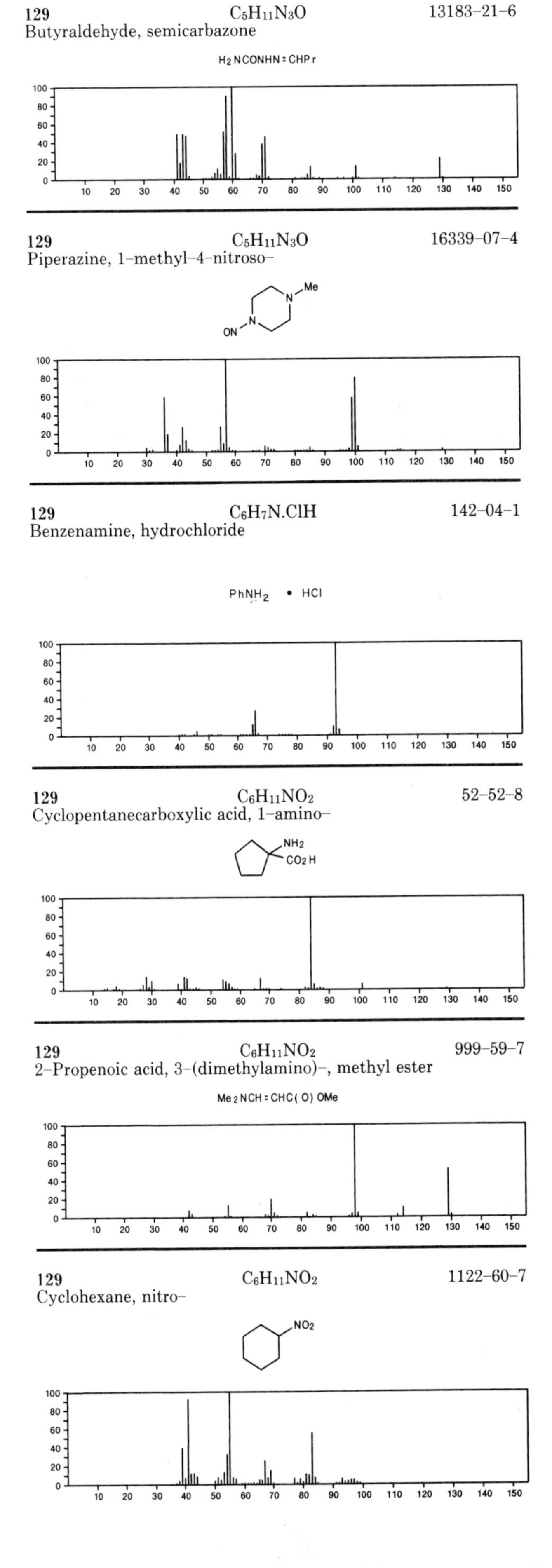

129 $C_5H_{11}N_3O$ 13183–21–6
Butyraldehyde, semicarbazone

$H_2NCONHN=CHPr$

129 $C_5H_{11}N_3O$ 16339–07–4
Piperazine, 1–methyl–4–nitroso–

129 $C_6H_7N.ClH$ 142–04–1
Benzenamine, hydrochloride

$PhNH_2 \cdot HCl$

129 $C_6H_{11}NO_2$ 52–52–8
Cyclopentanecarboxylic acid, 1–amino–

129 $C_6H_{11}NO_2$ 999–59–7
2–Propenoic acid, 3–(dimethylamino)–, methyl ester

$Me_2NCH=CHC(O)OMe$

129 $C_6H_{11}NO_2$ 1122–60–7
Cyclohexane, nitro–

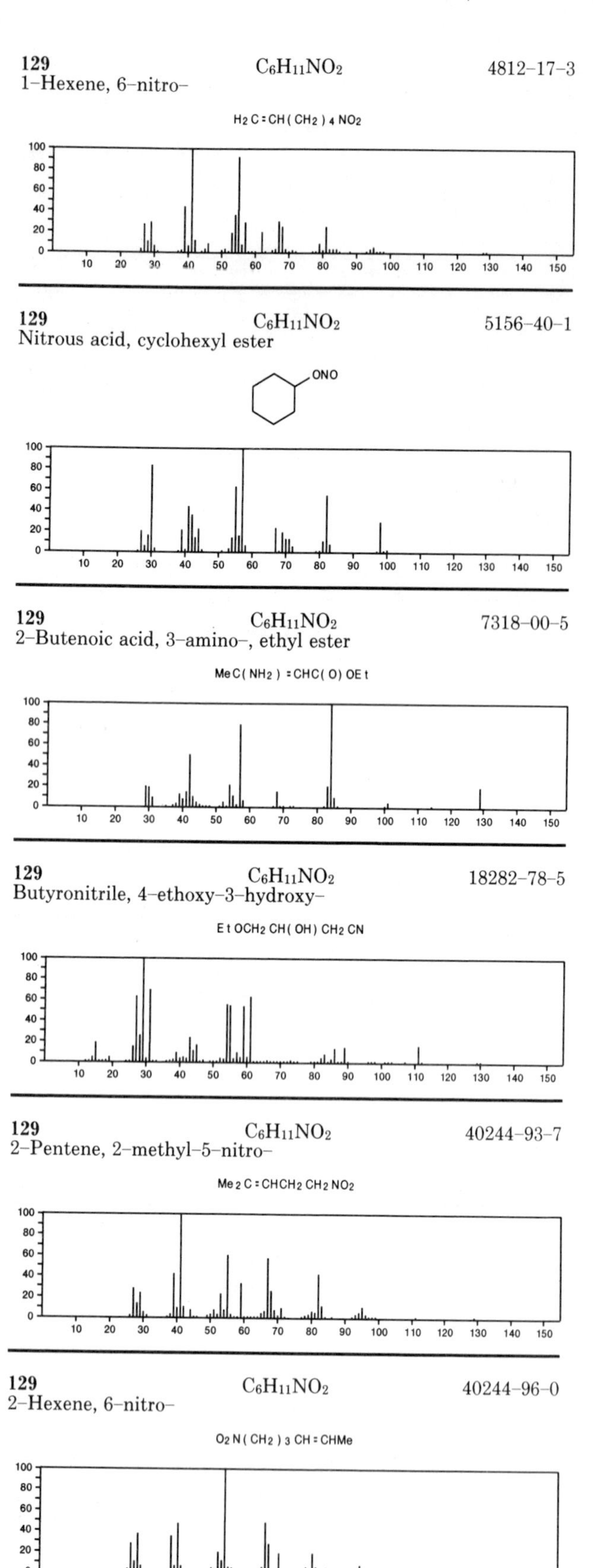

129 $C_6H_{11}NO_2$ 4812–17–3
1–Hexene, 6–nitro–

129 $C_6H_{11}NO_2$ 5156–40–1
Nitrous acid, cyclohexyl ester

129 $C_6H_{11}NO_2$ 7318–00–5
2–Butenoic acid, 3–amino–, ethyl ester

129 $C_6H_{11}NO_2$ 18282–78–5
Butyronitrile, 4–ethoxy–3–hydroxy–

129 $C_6H_{11}NO_2$ 40244–93–7
2–Pentene, 2–methyl–5–nitro–

129 $C_6H_{11}NO_2$ 40244–96–0
2–Hexene, 6–nitro–

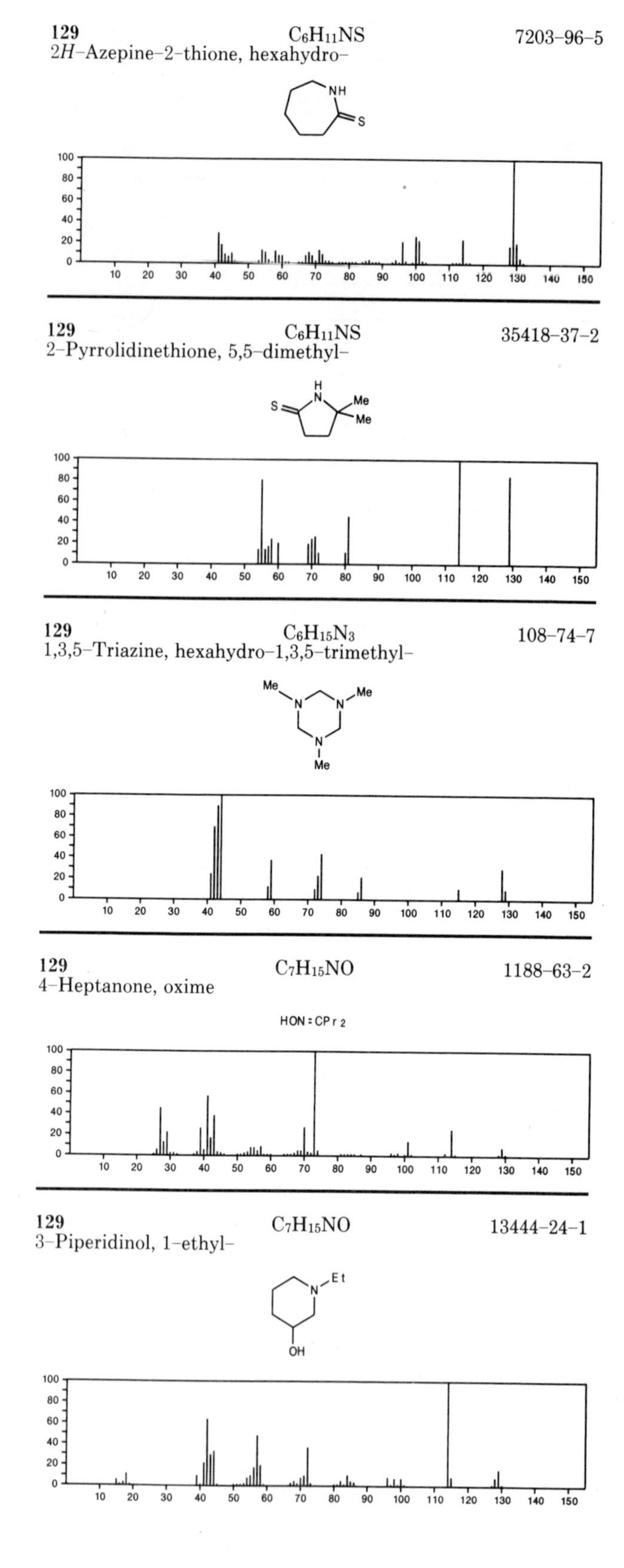

129 $C_6H_{11}NS$ 7203–96–5
2*H*–Azepine–2–thione, hexahydro–

129 $C_6H_{11}NS$ 35418–37–2
2–Pyrrolidinethione, 5,5–dimethyl–

129 $C_6H_{15}N_3$ 108–74–7
1,3,5–Triazine, hexahydro–1,3,5–trimethyl–

129 $C_7H_{15}NO$ 1188–63–2
4–Heptanone, oxime

129 $C_7H_{15}NO$ 13444–24–1
3–Piperidinol, 1–ethyl–

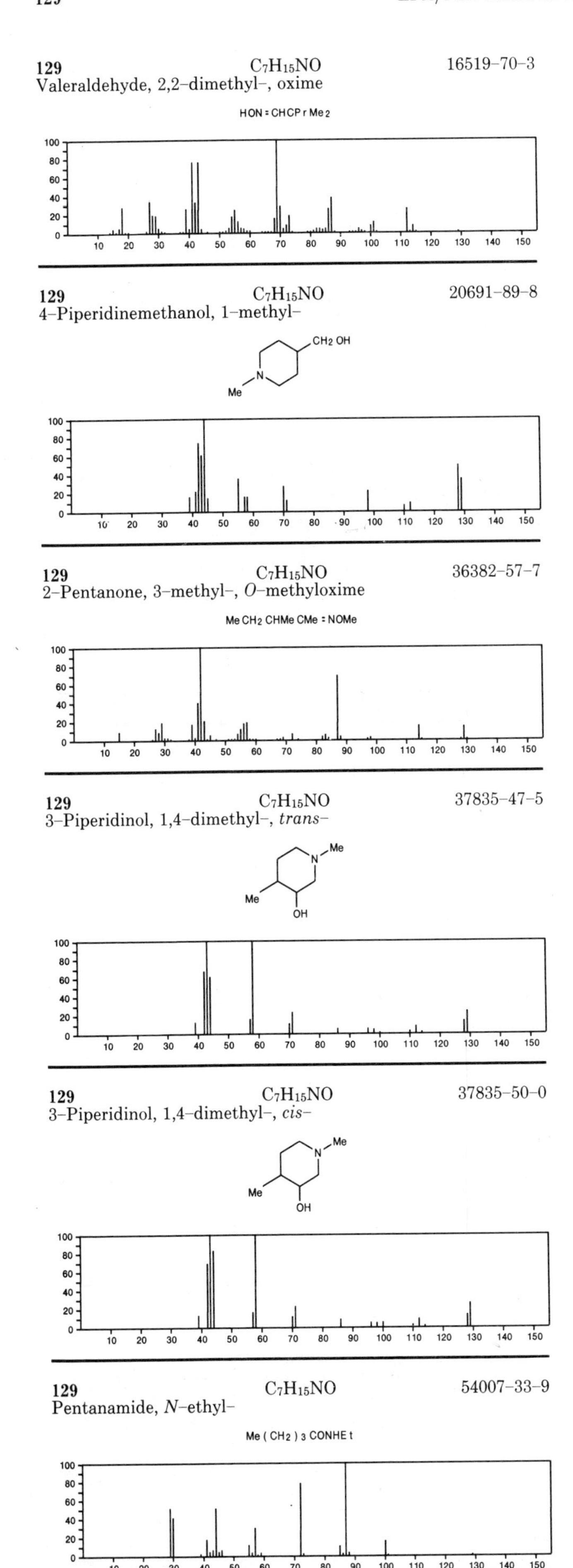

129 $C_7H_{15}NO$ 16519-70-3
Valeraldehyde, 2,2-dimethyl-, oxime

HON=CHCPrMe2

129 $C_7H_{15}NO$ 20691-89-8
4-Piperidinemethanol, 1-methyl-

129 $C_7H_{15}NO$ 36382-57-7
2-Pentanone, 3-methyl-, *O*-methyloxime

MeCH2CHMeCMe=NOMe

129 $C_7H_{15}NO$ 37835-47-5
3-Piperidinol, 1,4-dimethyl-, *trans*-

129 $C_7H_{15}NO$ 37835-50-0
3-Piperidinol, 1,4-dimethyl-, *cis*-

129 $C_7H_{15}NO$ 54007-33-9
Pentanamide, *N*-ethyl-

Me(CH2)3CONHEt

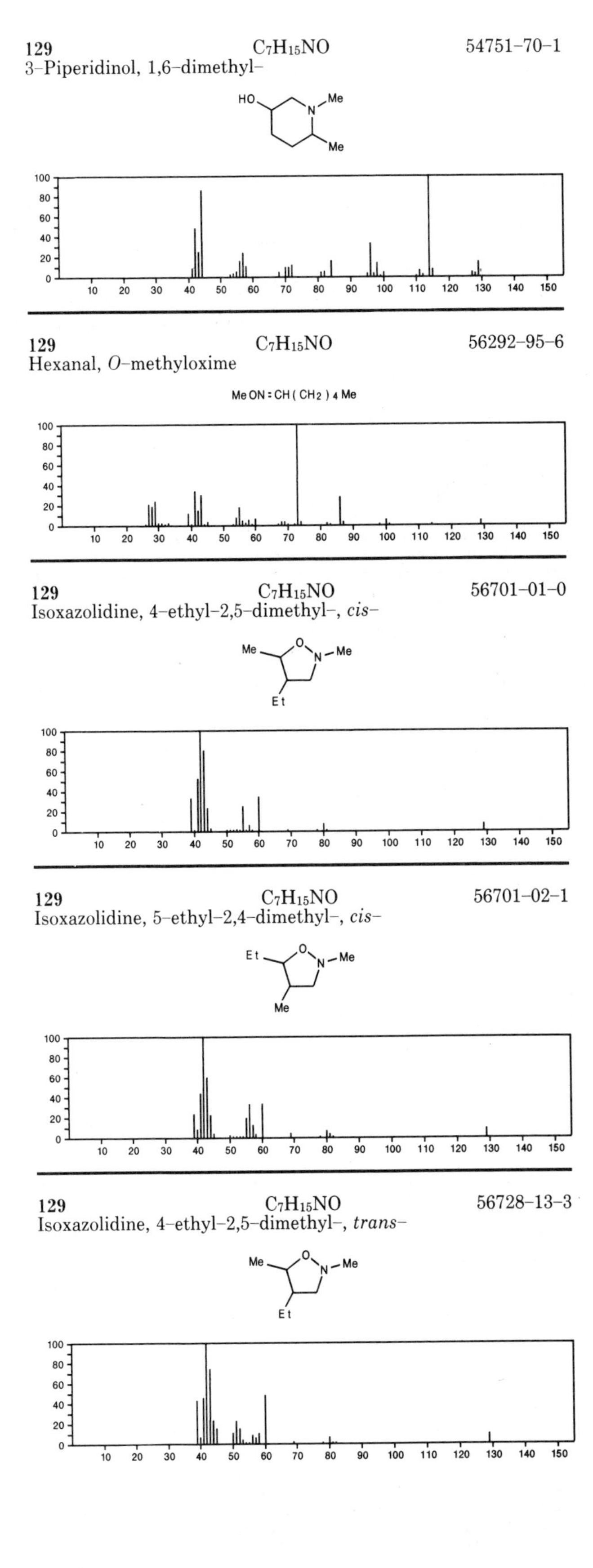

129 $C_7H_{15}NO$ 54751-70-1
3-Piperidinol, 1,6-dimethyl-

129 $C_7H_{15}NO$ 56292-95-6
Hexanal, *O*-methyloxime

MeON=CH(CH2)4Me

129 $C_7H_{15}NO$ 56701-01-0
Isoxazolidine, 4-ethyl-2,5-dimethyl-, *cis*-

129 $C_7H_{15}NO$ 56701-02-1
Isoxazolidine, 5-ethyl-2,4-dimethyl-, *cis*-

129 $C_7H_{15}NO$ 56728-13-3
Isoxazolidine, 4-ethyl-2,5-dimethyl-, *trans*-

129 $C_7H_{15}NO$ 56728-14-4
Isoxazolidine, 5-ethyl-2,4-dimethyl-, *trans*-

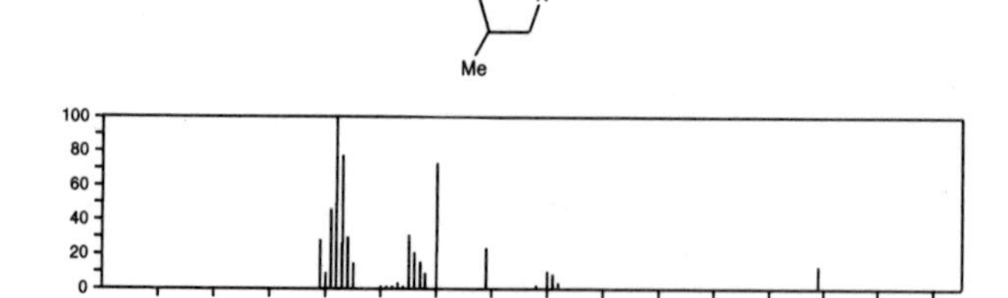

129 $C_8H_{19}N$ 104-75-6
1-Hexanamine, 2-ethyl-

$Me(CH_2)_3CHEtCH_2NH_2$

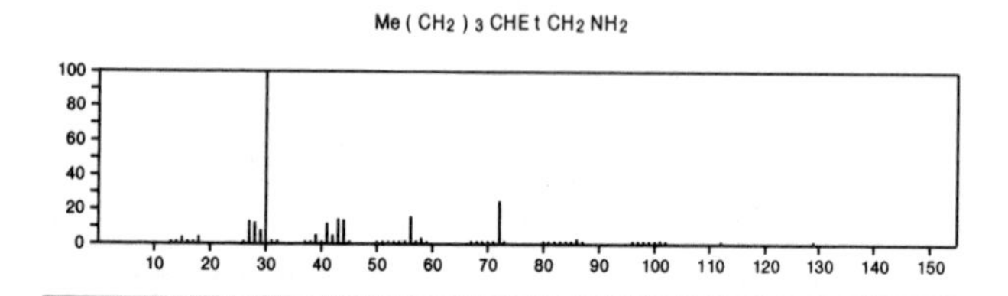

129 $C_8H_{19}N$ 107-45-9
2-Pentanamine, 2,4,4-trimethyl-

$Me_3CCH_2CMe_2NH_2$

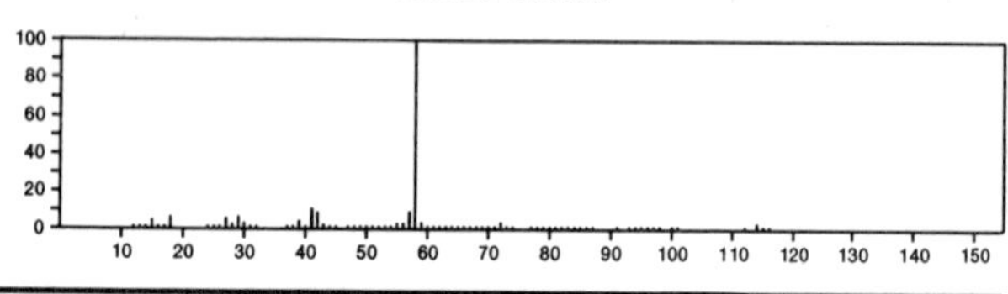

129 $C_8H_{19}N$ 110-96-3
1-Propanamine, 2-methyl-*N*-(2-methylpropyl)-

NH(Bu-i)2

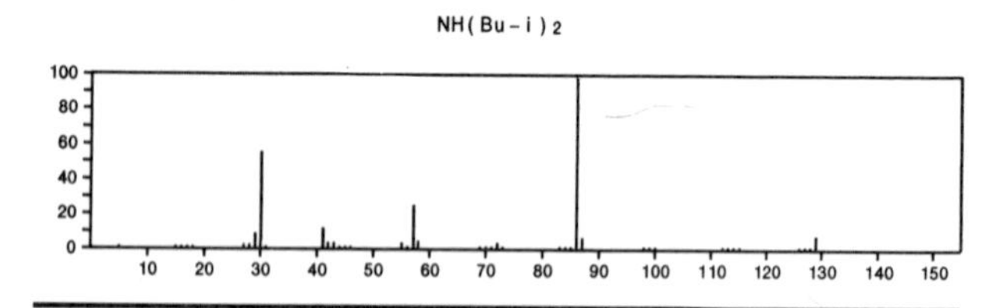

129 $C_8H_{19}N$ 111-86-4
1-Octanamine

$Me(CH_2)_7NH_2$

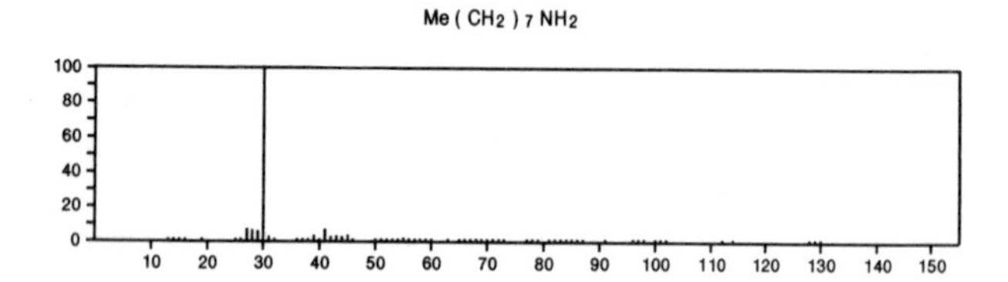

129 $C_8H_{19}N$ 111-92-2
1-Butanamine, *N*-butyl-

$Me(CH_2)_3NH(CH_2)_3Me$

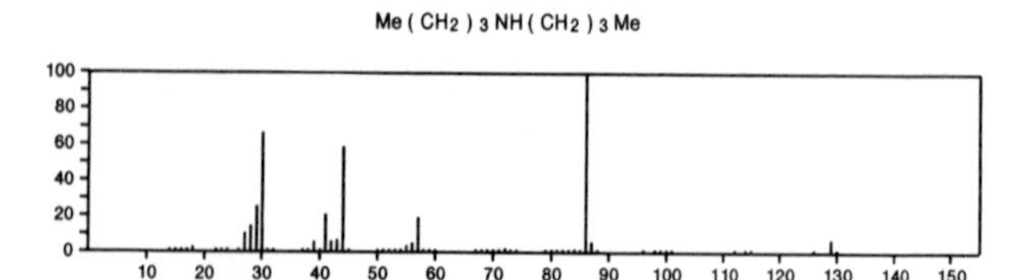

129 $C_8H_{19}N$ 543-82-8
2-Heptanamine, 6-methyl-

$Me_2CH(CH_2)_3CH(NH_2)Me$

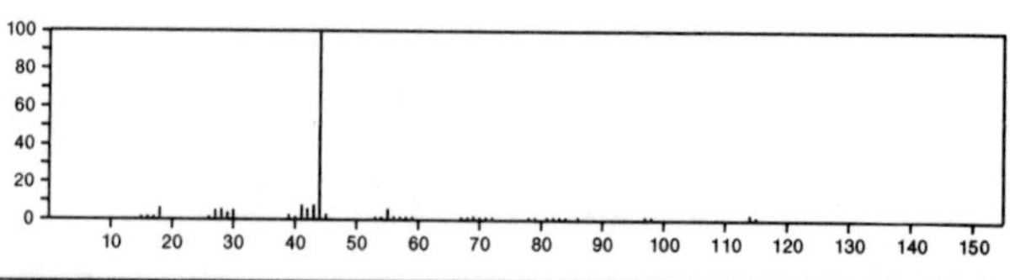

129 $C_8H_{19}N$ 626-23-3
2-Butanamine, *N*-(1-methylpropyl)-

NH(Bu-s)2

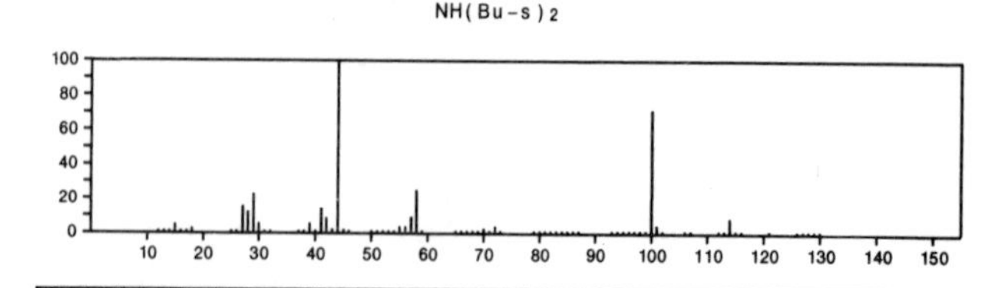

129 $C_8H_{19}N$ 693-16-3
2-Octanamine

$MeCH(NH_2)(CH_2)_5Me$

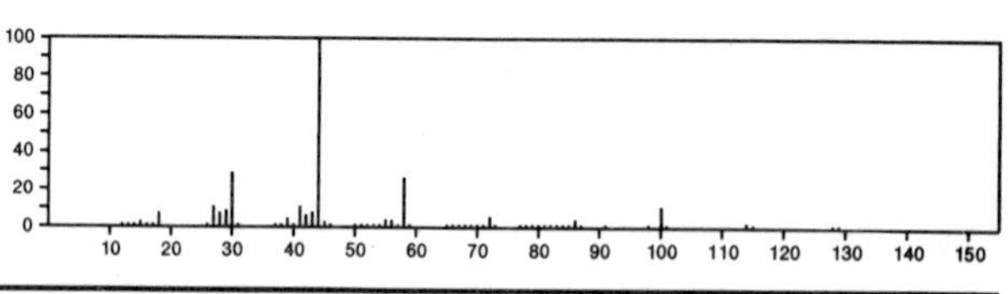

129 $C_8H_{19}N$ 4444-68-2
1-Butanamine, *N*,*N*-diethyl-

$Et_2N(CH_2)_3Me$

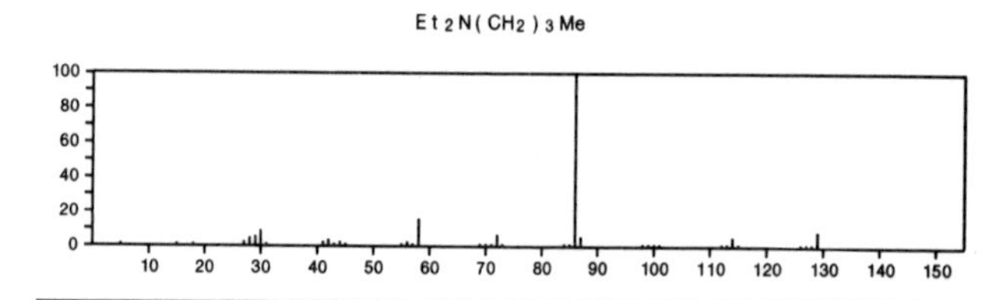

129 $C_8H_{19}N$ 5756-45-6
1-Butanamine, *N*-methyl-*N*-(1-methylethyl)-

$MeN(Pr\text{-}i)(CH_2)_3Me$

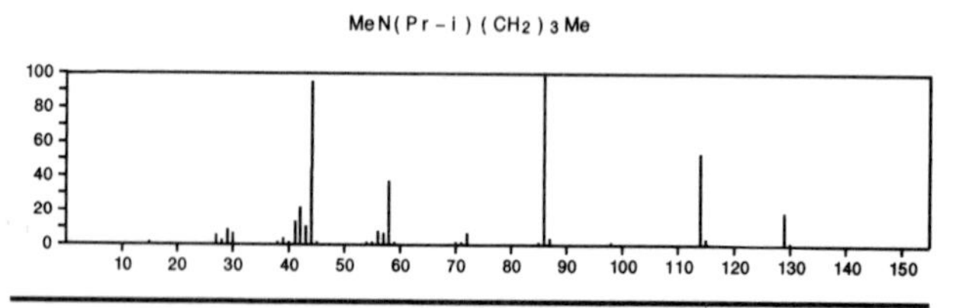

129 $C_8H_{19}N$ 7087-68-5
2-Propanamine, *N*-ethyl-*N*-(1-methylethyl)-

$EtN(Pr\text{-}i)_2$

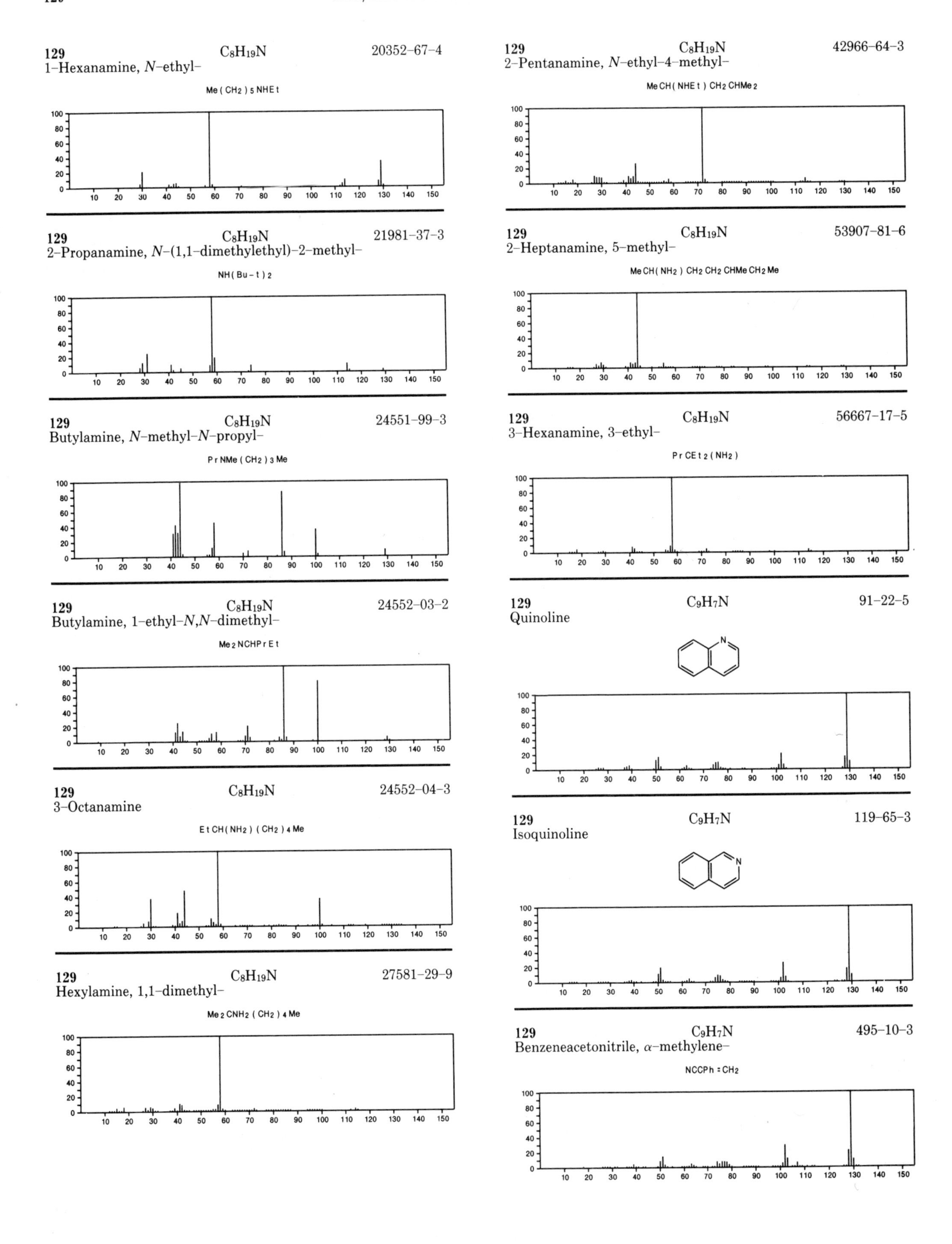
129 C8H19N 20352-67-4
1-Hexanamine, N-ethyl-
Me (CH2) 5 NHEt
129 C8H19N 21981-37-3
2-Propanamine, N-(1,1-dimethylethyl)-2-methyl-
NH(Bu-t) 2
129 C8H19N 24551-99-3
Butylamine, N-methyl-N-propyl-
PrNMe (CH2) 3 Me
129 C8H19N 24552-03-2
Butylamine, 1-ethyl-N,N-dimethyl-
Me2NCHPrEt
129 C8H19N 24552-04-3
3-Octanamine
EtCH(NH2) (CH2) 4 Me
129 C8H19N 27581-29-9
Hexylamine, 1,1-dimethyl-
Me2CNH2 (CH2) 4 Me
129 C8H19N 42966-64-3
2-Pentanamine, N-ethyl-4-methyl-
MeCH(NHEt) CH2CHMe2
129 C8H19N 53907-81-6
2-Heptanamine, 5-methyl-
MeCH(NH2) CH2CH2CHMeCH2Me
129 C8H19N 56667-17-5
3-Hexanamine, 3-ethyl-
PrCEt2 (NH2)
129 C9H7N 91-22-5
Quinoline
N
129 C9H7N 119-65-3
Isoquinoline
N
129 C9H7N 495-10-3
Benzeneacetonitrile, α-methylene-
NCCPh = CH2

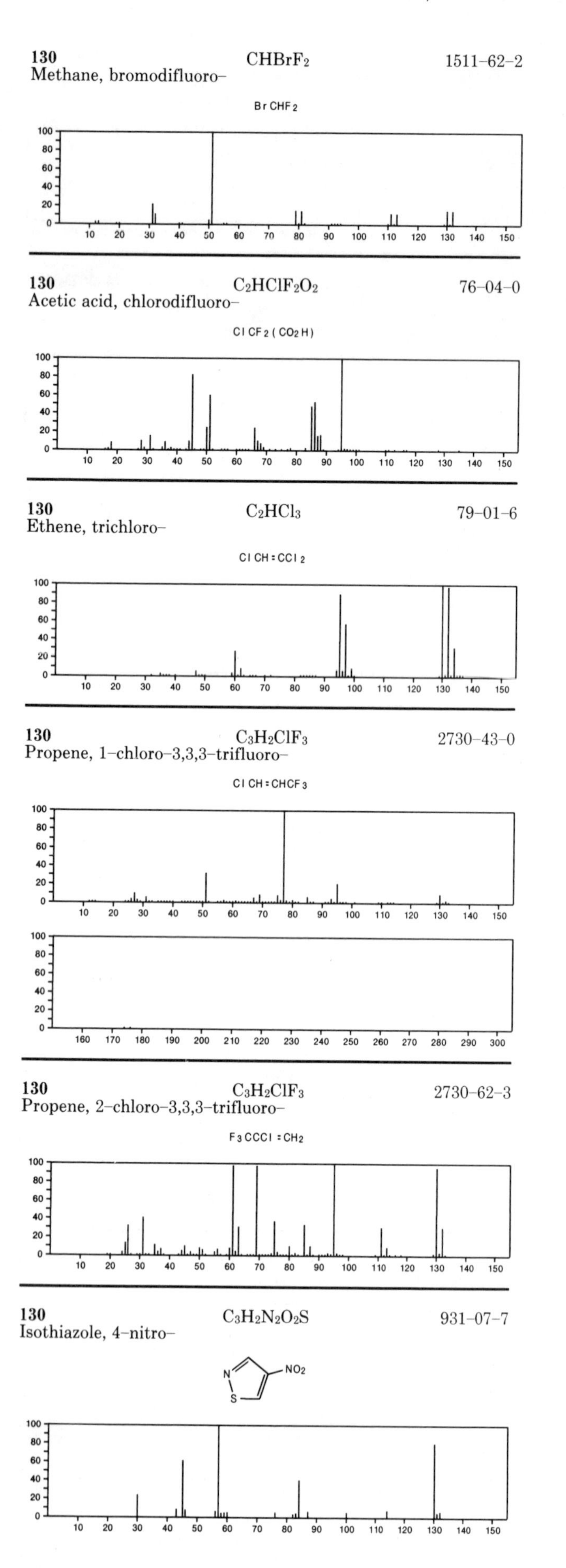
130
CHBrF2
1511-62-2
Methane, bromodifluoro-
Br CHF2
130
C2HClF2O2
76-04-0
Acetic acid, chlorodifluoro-
Cl CF2 (CO2 H)
130
C2HCl3
79-01-6
Ethene, trichloro-
Cl CH = CCl 2
130
C3H2ClF3
2730-43-0
Propene, 1-chloro-3,3,3-trifluoro-
Cl CH = CHCF3
130
C3H2ClF3
2730-62-3
Propene, 2-chloro-3,3,3-trifluoro-
F3 CCCl = CH2
130
C3H2N2O2S
931-07-7
Isothiazole, 4-nitro-
NO2

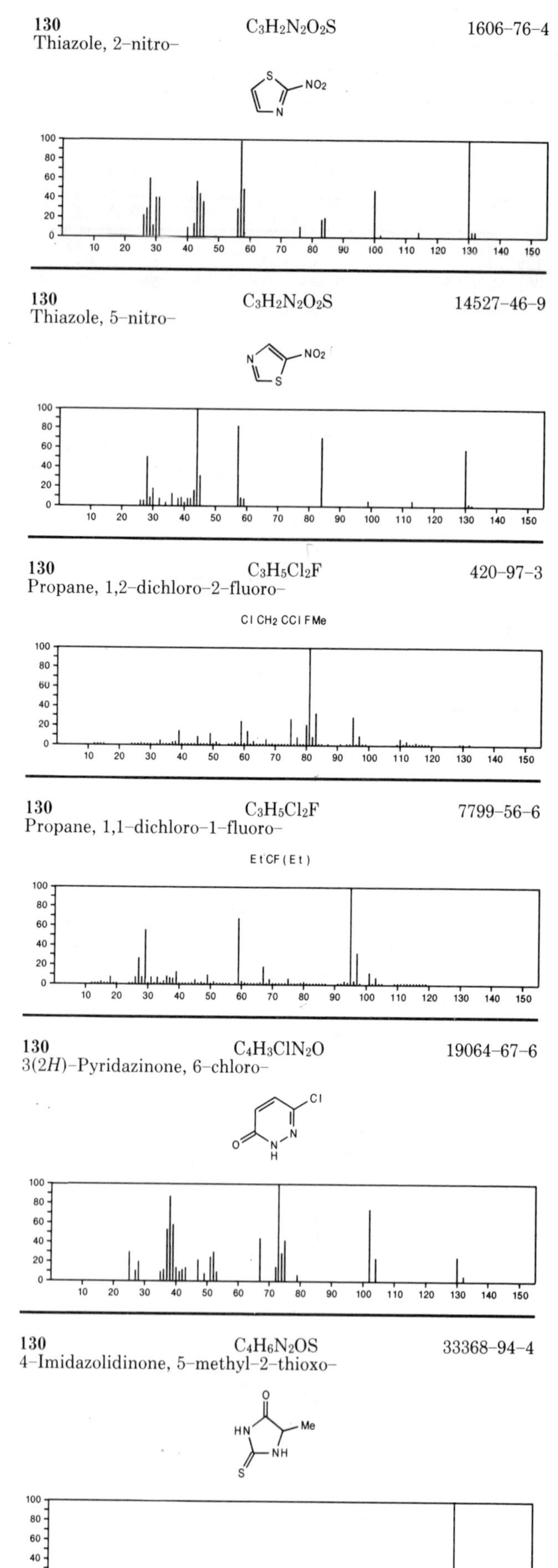
130
C3H2N2O2S
1606-76-4
Thiazole, 2-nitro-
NO2
130
C3H2N2O2S
14527-46-9
Thiazole, 5-nitro-
NO2
130
C3H5Cl2F
420-97-3
Propane, 1,2-dichloro-2-fluoro-
Cl CH2 CCl FMe
130
C3H5Cl2F
7799-56-6
Propane, 1,1-dichloro-1-fluoro-
EtCF (Et)
130
C4H3ClN2O
19064-67-6
3(2H)-Pyridazinone, 6-chloro-
130
C4H6N2OS
33368-94-4
4-Imidazolidinone, 5-methyl-2-thioxo-
Me

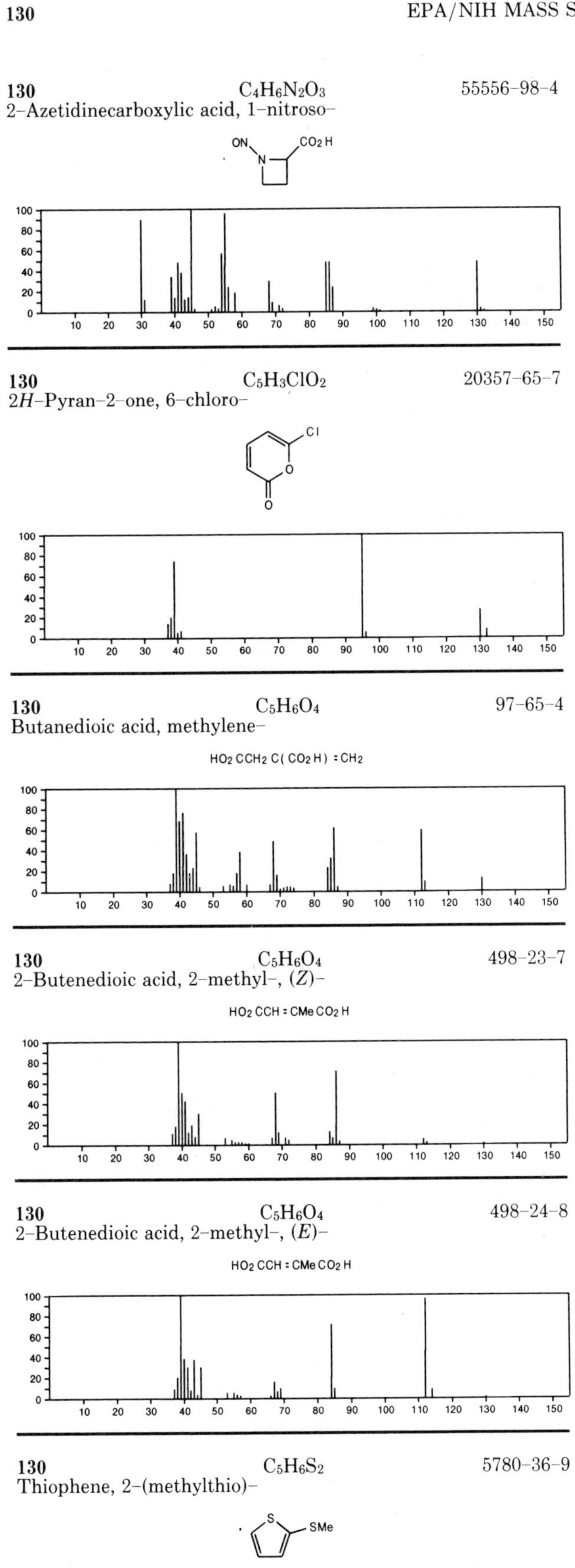

130 $C_4H_6N_2O_3$ 55556-98-4
2-Azetidinecarboxylic acid, 1-nitroso-

130 $C_5H_3ClO_2$ 20357-65-7
2H-Pyran-2-one, 6-chloro-

130 $C_5H_6O_4$ 97-65-4
Butanedioic acid, methylene-

$HO_2CCH_2C(CO_2H)=CH_2$

130 $C_5H_6O_4$ 498-23-7
2-Butenedioic acid, 2-methyl-, (*Z*)-

$HO_2CCH=CMeCO_2H$

130 $C_5H_6O_4$ 498-24-8
2-Butenedioic acid, 2-methyl-, (*E*)-

$HO_2CCH=CMeCO_2H$

130 $C_5H_6S_2$ 5780-36-9
Thiophene, 2-(methylthio)-

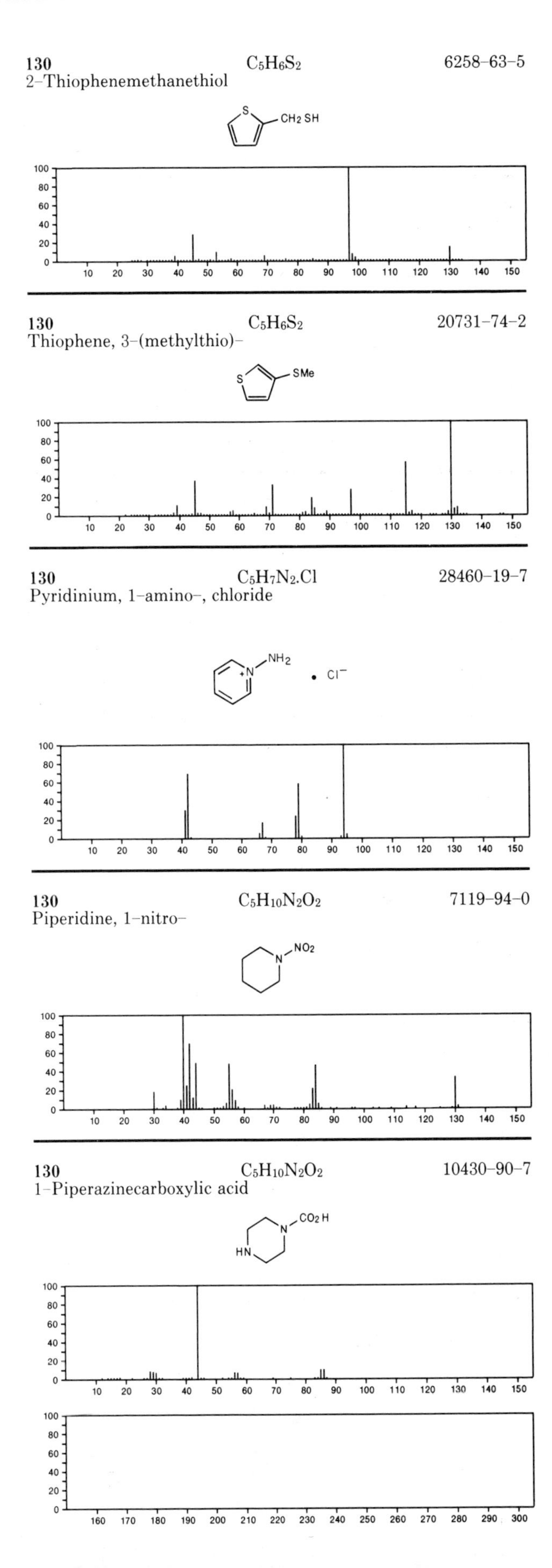

130 $C_5H_6S_2$ 6258-63-5
2-Thiophenemethanethiol

130 $C_5H_6S_2$ 20731-74-2
Thiophene, 3-(methylthio)-

130 $C_5H_7N_2.Cl$ 28460-19-7
Pyridinium, 1-amino-, chloride

130 $C_5H_{10}N_2O_2$ 7119-94-0
Piperidine, 1-nitro-

130 $C_5H_{10}N_2O_2$ 10430-90-7
1-Piperazinecarboxylic acid

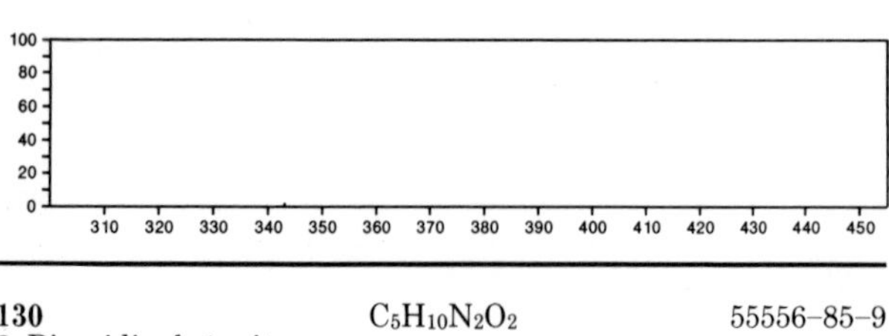

130 $C_5H_{10}N_2O_2$ 55556-85-9
3-Piperidinol, 1-nitroso-

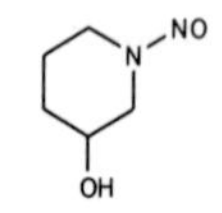

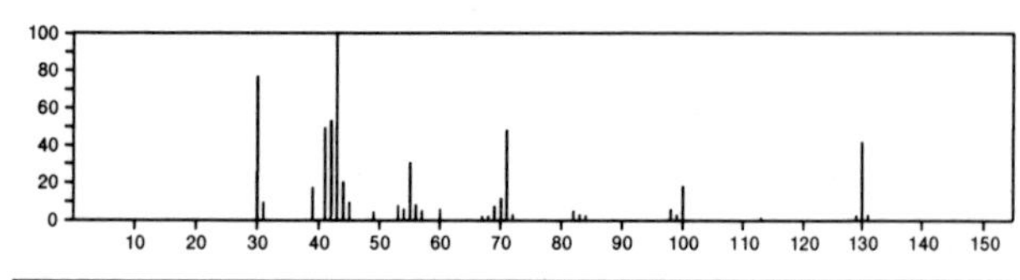

130 $C_5H_{10}N_2O_2$ 55556-93-9
4-Piperidinol, 1-nitroso-

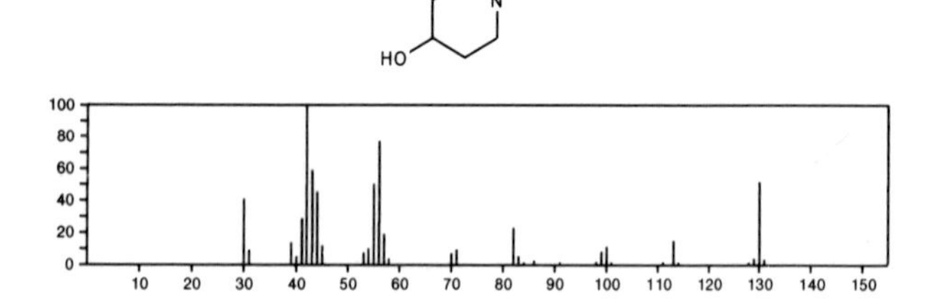

130 C_6H_4ClF 348-51-6
Benzene, 1-chloro-2-fluoro-

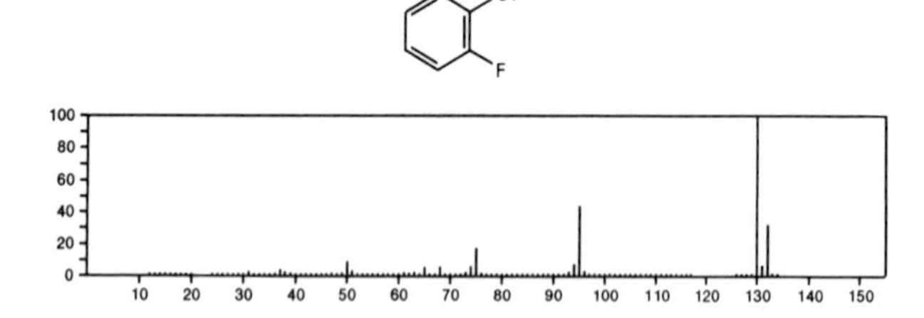

130 C_6H_4ClF 352-33-0
Benzene, 1-chloro-4-fluoro-

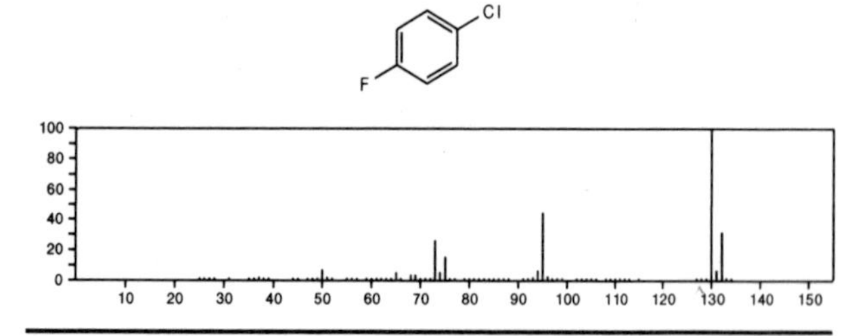

130 C_6H_4ClF 625-98-9
Benzene, 1-chloro-3-fluoro-

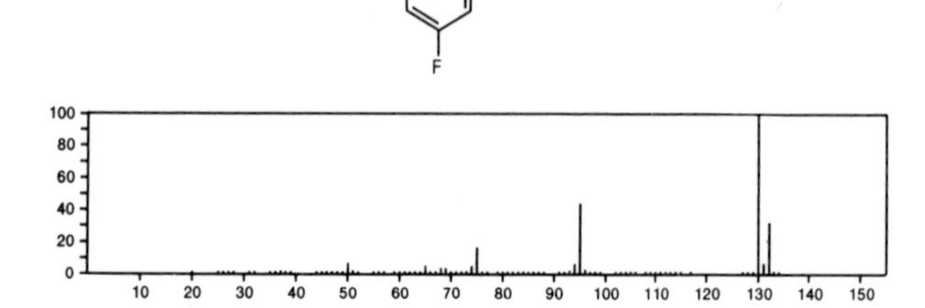

130 C_6H_4ClF 55256-17-2
Benzene, chlorofluoro-

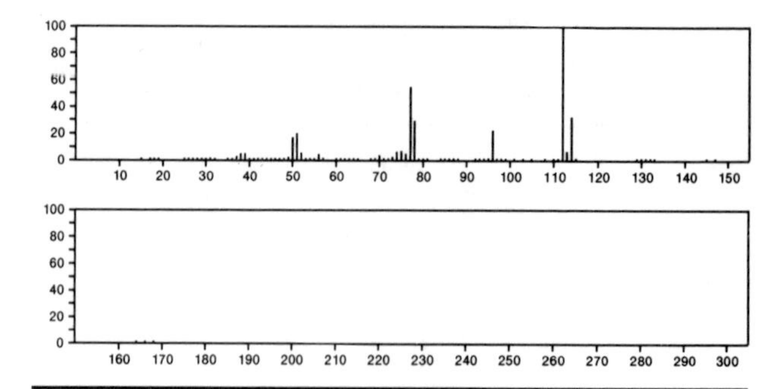

130 $C_6H_{10}OS$ 6683-25-6
7-Thiabicyclo[2.2.1]heptane, 7-oxide

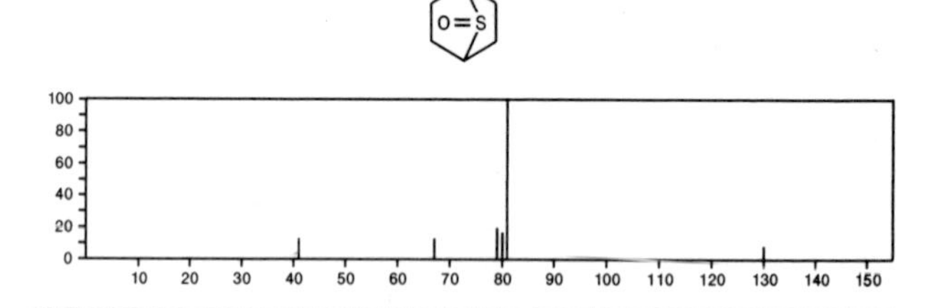

130 $C_6H_{10}OS$ 43152-90-5
2H-Thiopyran-3(4*H*)-one, dihydro-6-methyl-

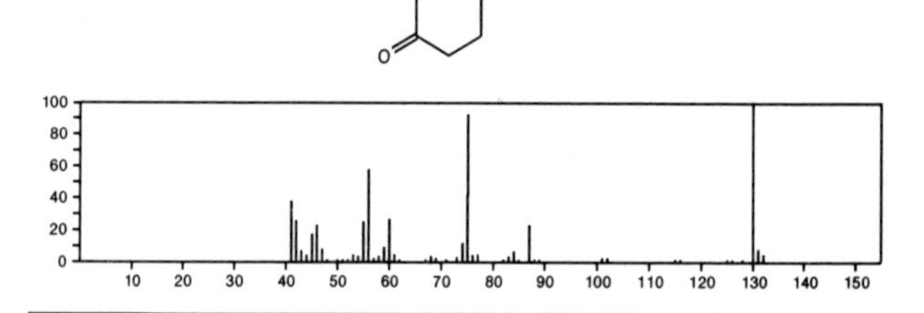

130 $C_6H_{10}O_3$ 123-62-6
Propanoic acid, anhydride

EtC(O)OC(O)Et

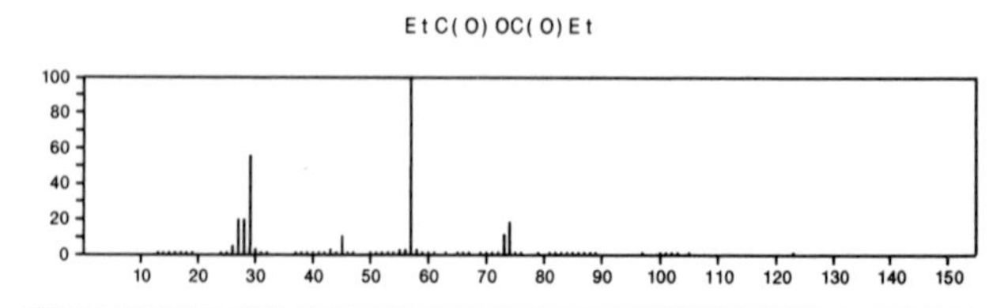

130 $C_6H_{10}O_3$ 141-31-1
Hexanedial, 2-hydroxy-

OCHCH(OH)(CH2)3CHO

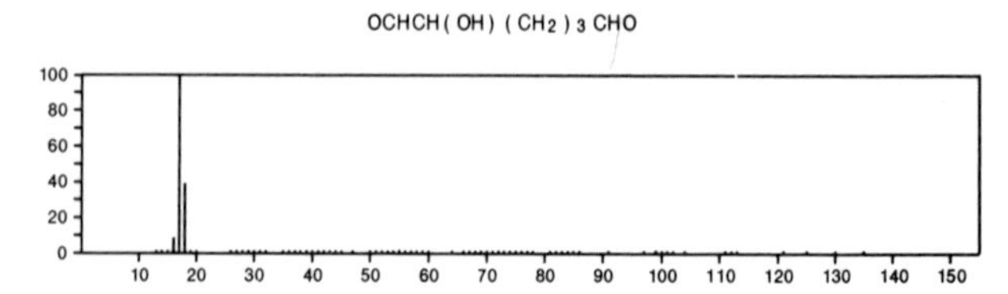

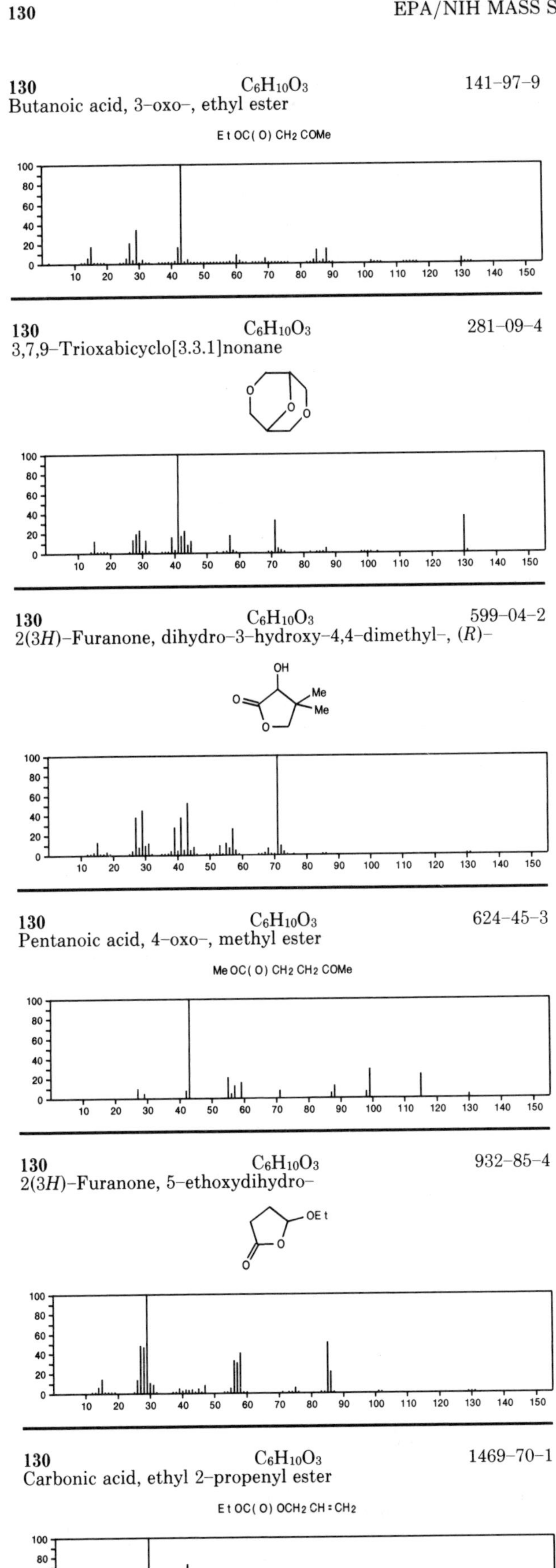
130 $C_6H_{10}O_3$ 141-97-9
Butanoic acid, 3-oxo-, ethyl ester
Et OC(O) CH2 COMe
130 $C_6H_{10}O_3$ 281-09-4
3,7,9-Trioxabicyclo[3.3.1]nonane
130 $C_6H_{10}O_3$ 599-04-2
2(3*H*)-Furanone, dihydro-3-hydroxy-4,4-dimethyl-, (*R*)-
OH
Me
Me
130 $C_6H_{10}O_3$ 624-45-3
Pentanoic acid, 4-oxo-, methyl ester
Me OC(O) CH2 CH2 COMe
130 $C_6H_{10}O_3$ 932-85-4
2(3*H*)-Furanone, 5-ethoxydihydro-
OEt
130 $C_6H_{10}O_3$ 1469-70-1
Carbonic acid, ethyl 2-propenyl ester
Et OC(O) OCH2 CH=CH2

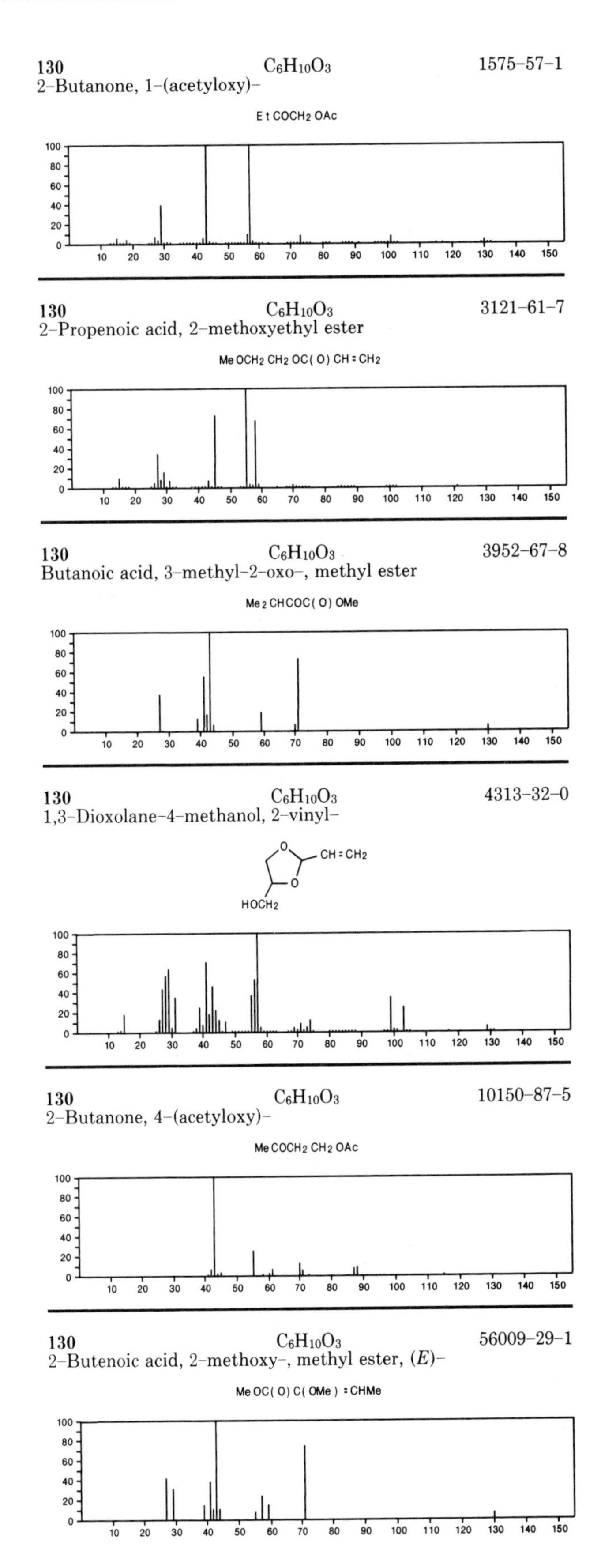
130 $C_6H_{10}O_3$ 1575-57-1
2-Butanone, 1-(acetyloxy)-
Et COCH2 OAc
130 $C_6H_{10}O_3$ 3121-61-7
2-Propenoic acid, 2-methoxyethyl ester
Me OCH2 CH2 OC(O) CH=CH2
130 $C_6H_{10}O_3$ 3952-67-8
Butanoic acid, 3-methyl-2-oxo-, methyl ester
Me2 CHCOC(O) OMe
130 $C_6H_{10}O_3$ 4313-32-0
1,3-Dioxolane-4-methanol, 2-vinyl-
CH=CH2
HOCH2
130 $C_6H_{10}O_3$ 10150-87-5
2-Butanone, 4-(acetyloxy)-
Me COCH2 CH2 OAc
130 $C_6H_{10}O_3$ 56009-29-1
2-Butenoic acid, 2-methoxy-, methyl ester, (*E*)-
Me OC(O) C(OMe)=CHMe

130 $C_6H_{10}O_3$ 56009-30-4
2-Butenoic acid, 2-methoxy-, methyl ester, (*Z*)-

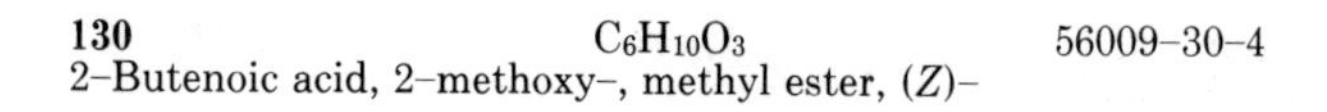

MeOC(O)C(OMe)=CHMe

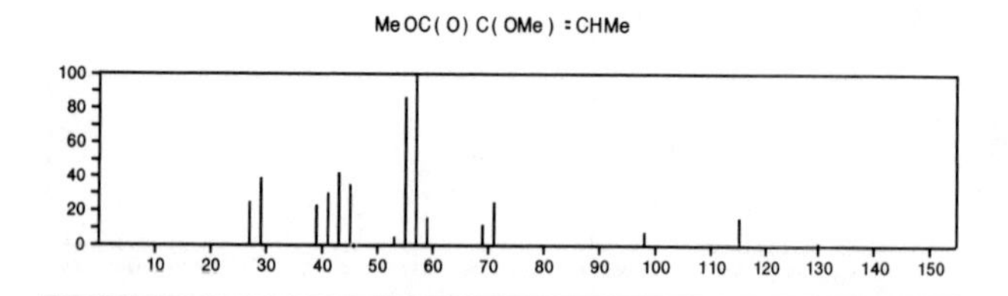

130 $C_6H_{11}OP$ 16327-48-3
4-Phosphorinanone, 1-methyl-

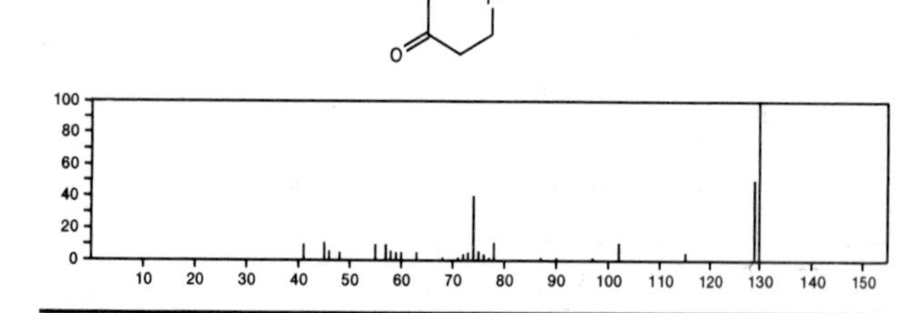

130 $C_6H_{14}N_2O$ 601-77-4
2-Propanamine, *N*-(1-methylethyl)-*N*-nitroso-

i-Pr2NNO

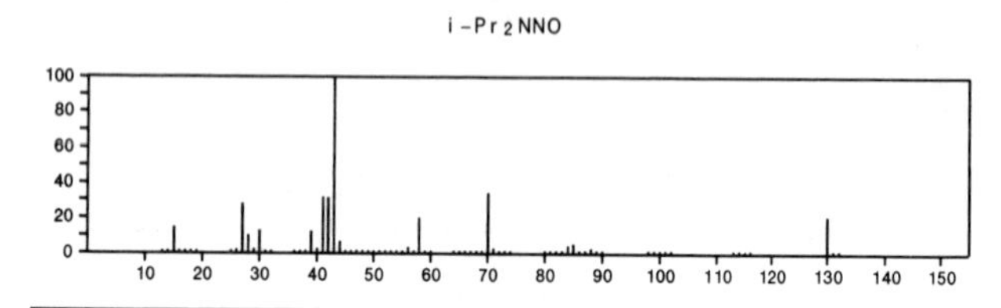

130 $C_6H_{14}N_2O$ 621-64-7
1-Propanamine, *N*-nitroso-*N*-propyl-

Pr2NNO

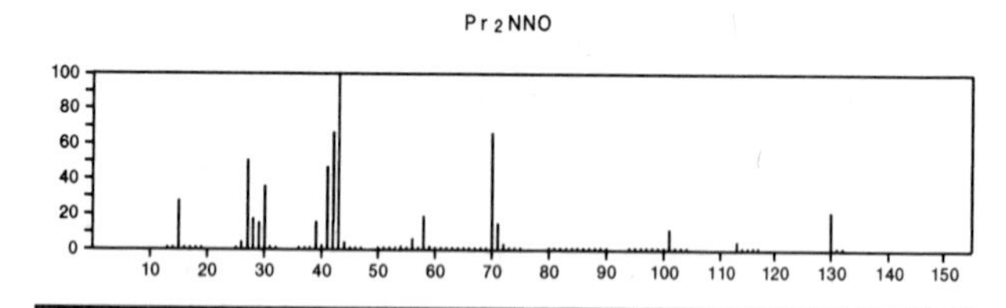

130 $C_6H_{14}N_2O$ 4549-44-4
1-Butanamine, *N*-ethyl-*N*-nitroso-

EtN(NO)(CH2)3Me

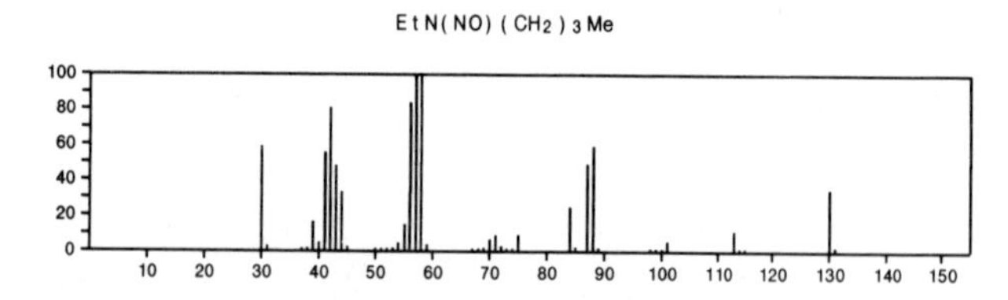

130 $C_6H_{14}N_2O$ 13256-07-0
1-Pentanamine, *N*-methyl-*N*-nitroso-

MeN(NO)(CH2)4Me

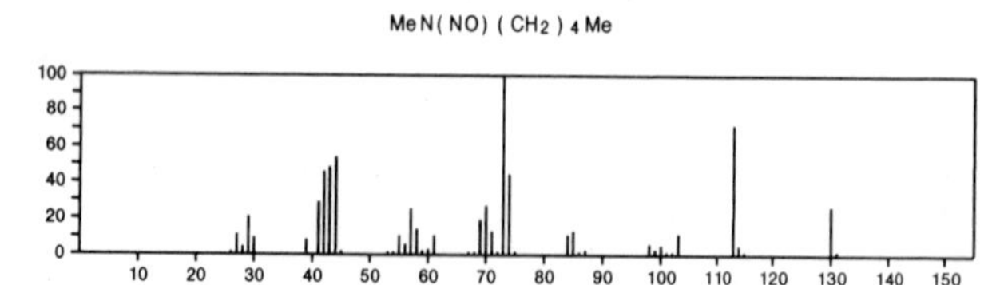

130 $C_6H_{14}N_2O$ 28023-82-7
1-Propanamine, *N*-(1-methylethyl)-2-nitroso-

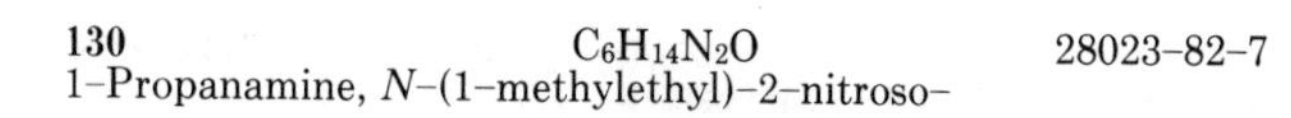

PrN(NO)Pr-i

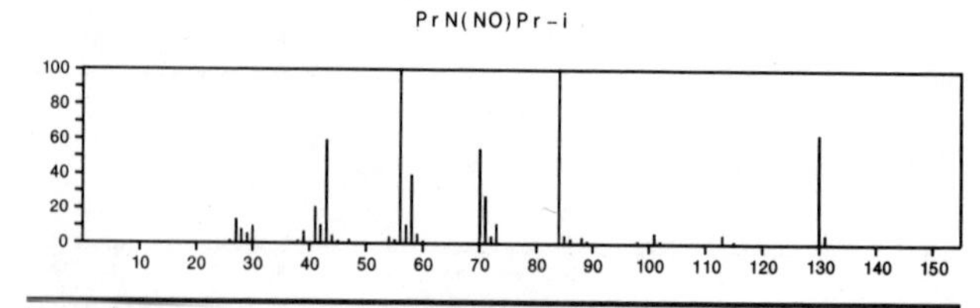

130 $C_6H_{15}BO_2$ 53907-92-9
Boronic acid, ethyl-, diethyl ester

EtOBEtOEt

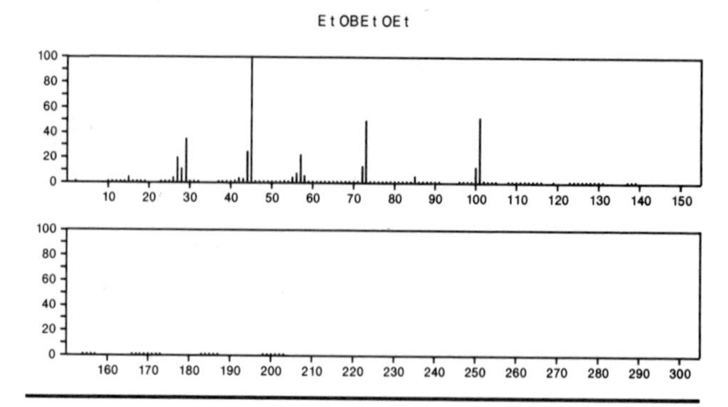

130 $C_7H_{11}Cl$ 765-91-3
Bicyclo[2.2.1]heptane, 2-chloro-, *exo*-

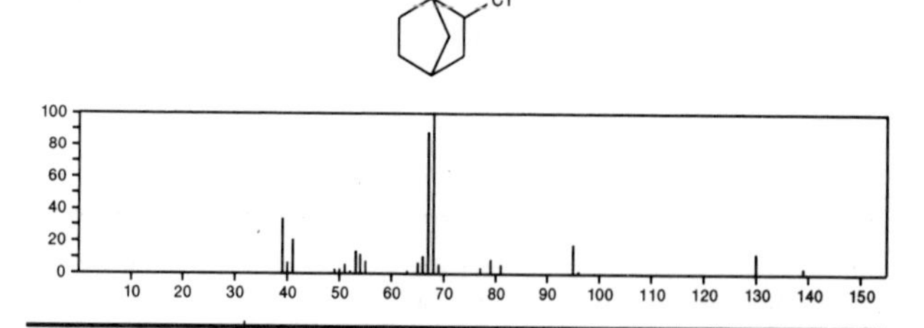

130 $C_7H_{11}Cl$ 13294-30-9
Cycloheptene, 1-chloro-

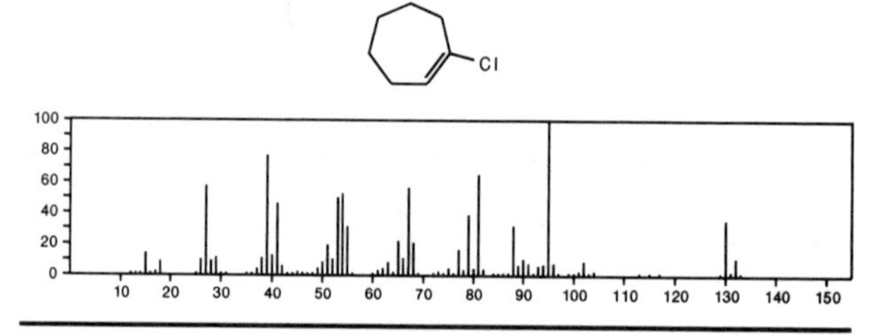

130 $C_7H_{11}Cl$ 16642-49-2
Cyclohexene, 1-chloro-2-methyl-

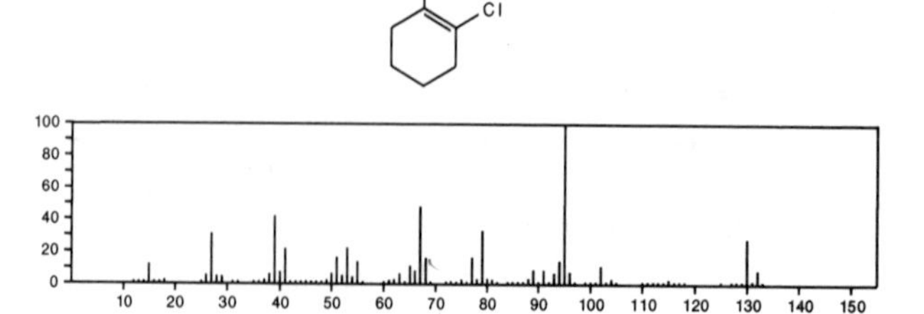

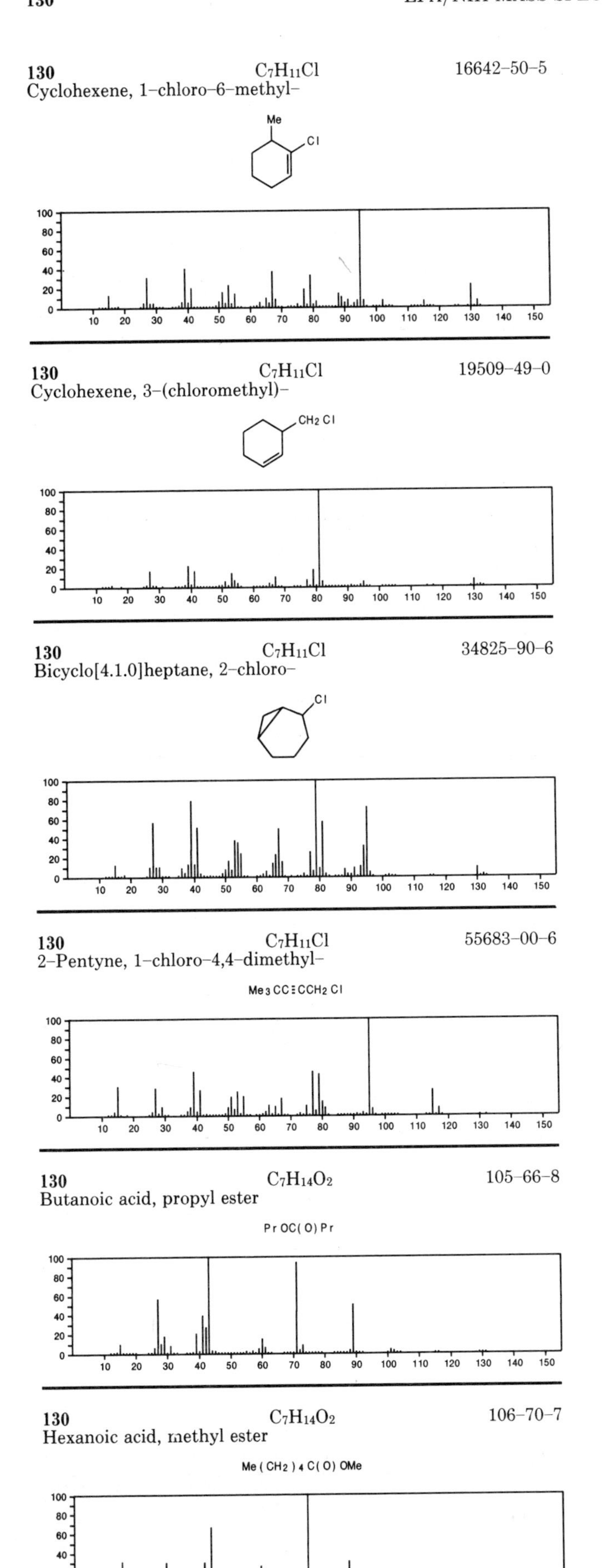
130 C7H11Cl 16642-50-5
Cyclohexene, 1-chloro-6-methyl-
Me
Cl
130 C7H11Cl 19509-49-0
Cyclohexene, 3-(chloromethyl)-
CH2 Cl
130 C7H11Cl 34825-90-6
Bicyclo[4.1.0]heptane, 2-chloro-
Cl
130 C7H11Cl 55683-00-6
2-Pentyne, 1-chloro-4,4-dimethyl-
Me3 CC≡CCH2 Cl
130 C7H14O2 105-66-8
Butanoic acid, propyl ester
Pr OC(O) Pr
130 C7H14O2 106-70-7
Hexanoic acid, methyl ester
Me (CH2)4 C(O) OMe

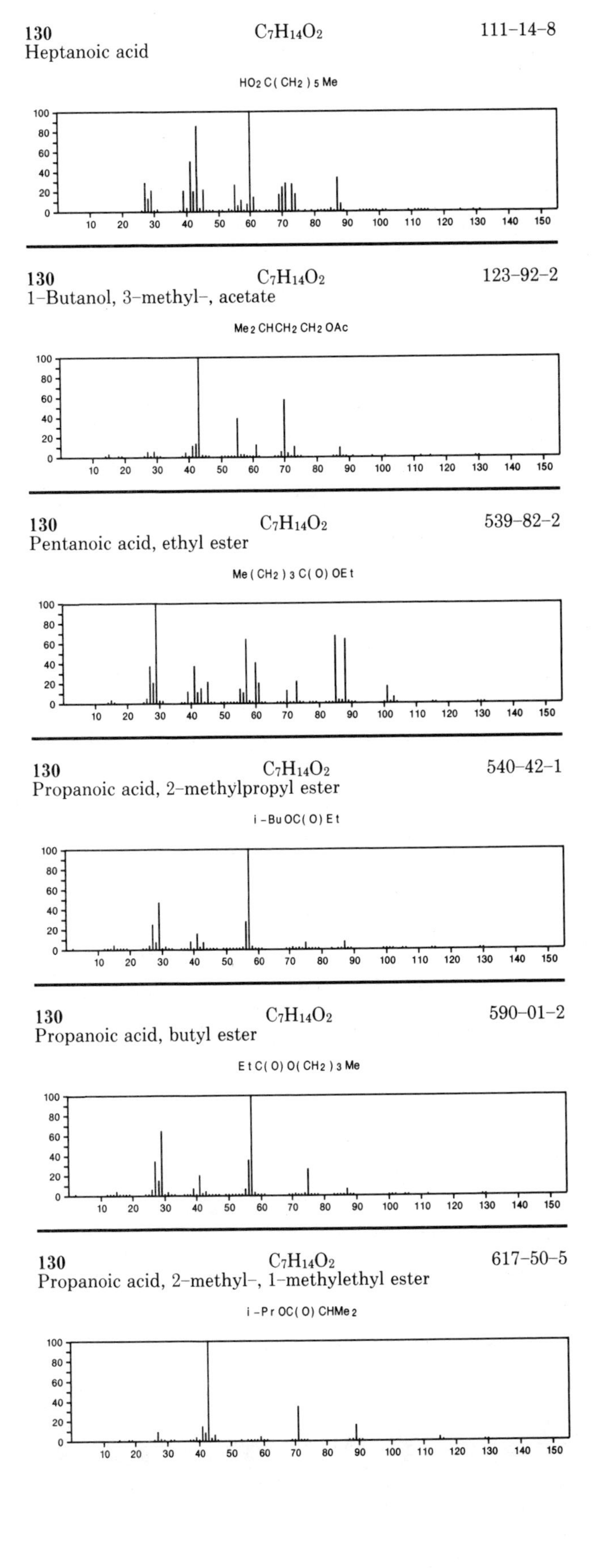
130 C7H14O2 111-14-8
Heptanoic acid
HO2 C(CH2)5 Me
130 C7H14O2 123-92-2
1-Butanol, 3-methyl-, acetate
Me2 CHCH2 CH2 OAc
130 C7H14O2 539-82-2
Pentanoic acid, ethyl ester
Me (CH2)3 C(O) OEt
130 C7H14O2 540-42-1
Propanoic acid, 2-methylpropyl ester
i-BuOC(O)Et
130 C7H14O2 590-01-2
Propanoic acid, butyl ester
EtC(O)O(CH2)3 Me
130 C7H14O2 617-50-5
Propanoic acid, 2-methyl-, 1-methylethyl ester
i-Pr OC(O) CHMe2

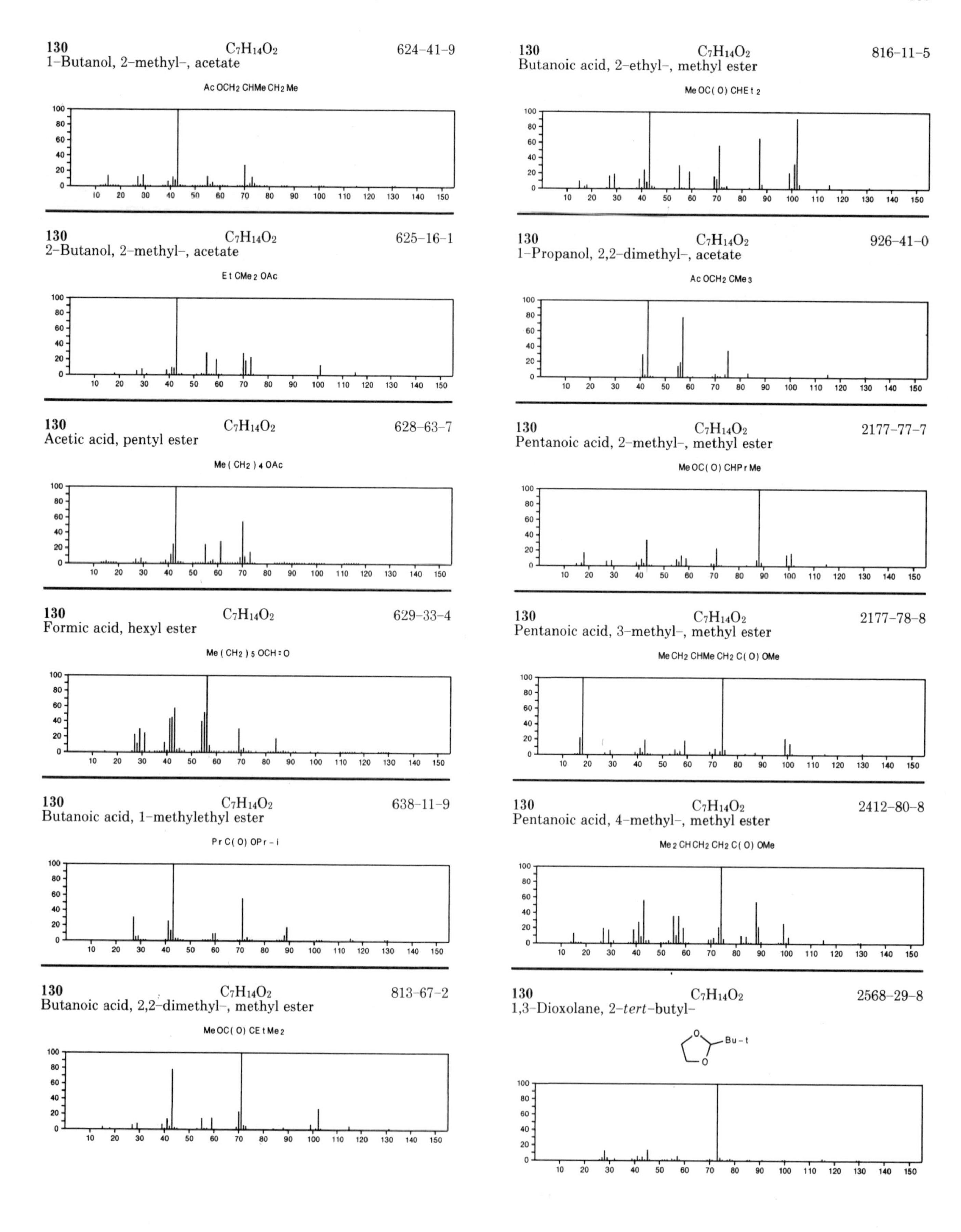
130 C7H14O2 624-41-9
1-Butanol, 2-methyl-, acetate
Ac OCH2 CHMe CH2 Me
130 C7H14O2 816-11-5
Butanoic acid, 2-ethyl-, methyl ester
Me OC(O) CHE t 2
130 C7H14O2 625-16-1
2-Butanol, 2-methyl-, acetate
E t CMe 2 OAc
130 C7H14O2 926-41-0
1-Propanol, 2,2-dimethyl-, acetate
Ac OCH2 CMe 3
130 C7H14O2 628-63-7
Acetic acid, pentyl ester
Me (CH2) 4 OAc
130 C7H14O2 2177-77-7
Pentanoic acid, 2-methyl-, methyl ester
Me OC(O) CHPr Me
130 C7H14O2 629-33-4
Formic acid, hexyl ester
Me (CH2) 5 OCH = O
130 C7H14O2 2177-78-8
Pentanoic acid, 3-methyl-, methyl ester
Me CH2 CHMe CH2 C(O) OMe
130 C7H14O2 638-11-9
Butanoic acid, 1-methylethyl ester
Pr C(O) OPr - i
130 C7H14O2 2412-80-8
Pentanoic acid, 4-methyl-, methyl ester
Me 2 CHCH2 CH2 C(O) OMe
130 C7H14O2 813-67-2
Butanoic acid, 2,2-dimethyl-, methyl ester
Me OC(O) CE t Me 2
130 C7H14O2 2568-29-8
1,3-Dioxolane, 2-tert-butyl-
Bu - t

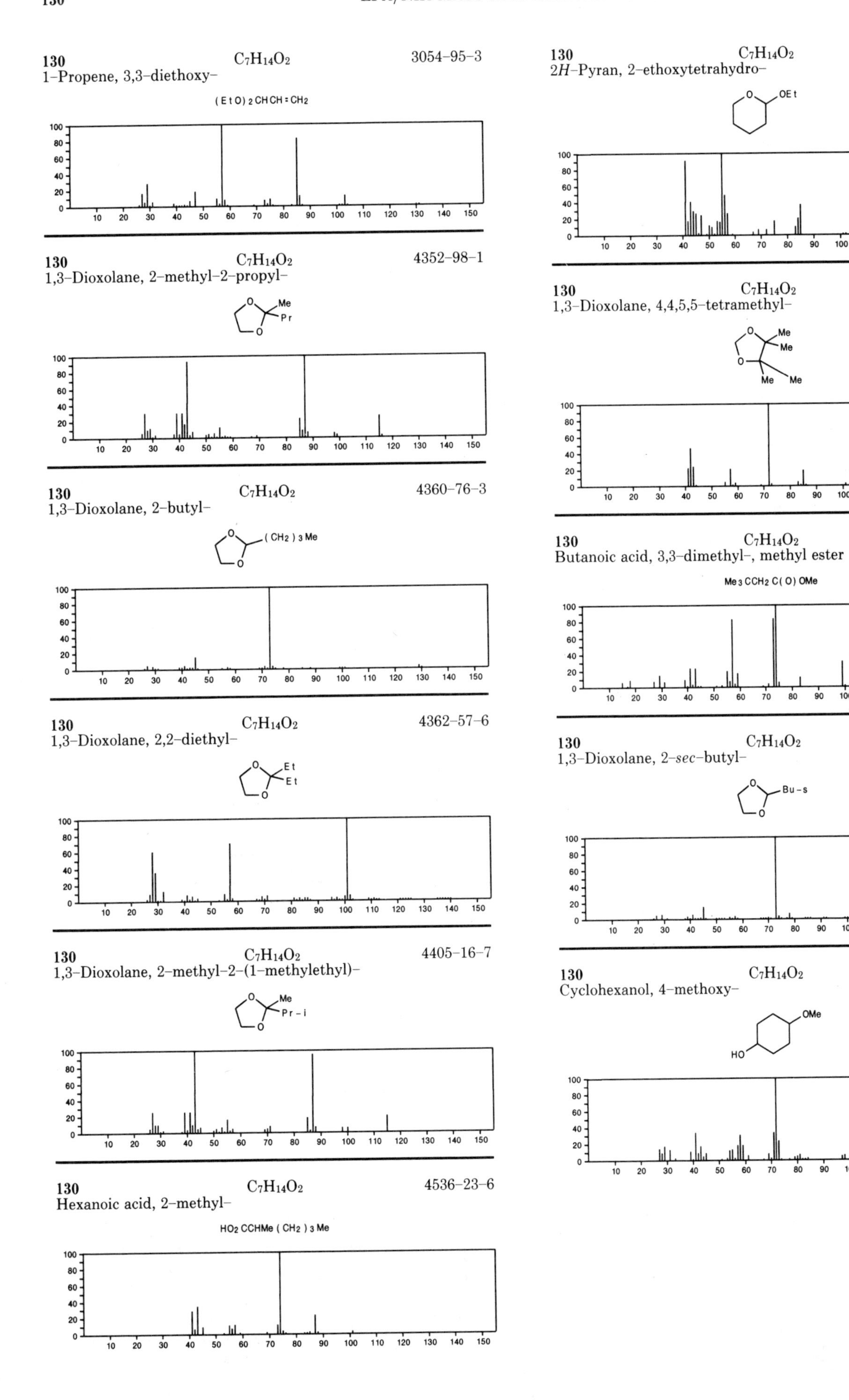

130 $C_7H_{14}O_2$ 3054–95–3
1–Propene, 3,3–diethoxy–
(EtO)2CHCH=CH2

130 $C_7H_{14}O_2$ 4352–98–1
1,3–Dioxolane, 2–methyl–2–propyl–

130 $C_7H_{14}O_2$ 4360–76–3
1,3–Dioxolane, 2–butyl–

130 $C_7H_{14}O_2$ 4362–57–6
1,3–Dioxolane, 2,2–diethyl–

130 $C_7H_{14}O_2$ 4405–16–7
1,3–Dioxolane, 2–methyl–2–(1–methylethyl)–

130 $C_7H_{14}O_2$ 4536–23–6
Hexanoic acid, 2–methyl–
HO2CCHMe(CH2)3Me

130 $C_7H_{14}O_2$ 4819–83–4
2*H*–Pyran, 2–ethoxytetrahydro–

130 $C_7H_{14}O_2$ 5660–63–9
1,3–Dioxolane, 4,4,5,5–tetramethyl–

130 $C_7H_{14}O_2$ 10250–48–3
Butanoic acid, 3,3–dimethyl–, methyl ester
Me3CCH2C(O)OMe

130 $C_7H_{14}O_2$ 14447–25–7
1,3–Dioxolane, 2–*sec*–butyl–

130 $C_7H_{14}O_2$ 18068–06–9
Cyclohexanol, 4–methoxy–

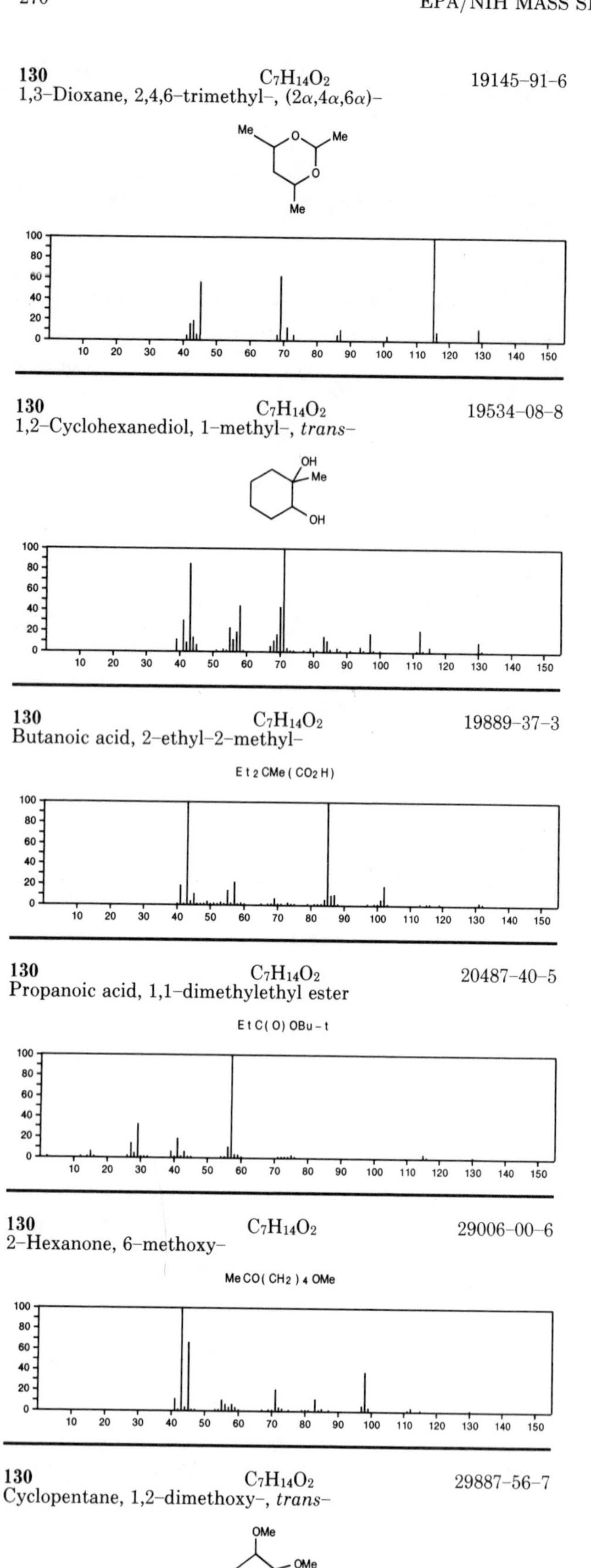
130 C7H14O2 19145-91-6
1,3-Dioxane, 2,4,6-trimethyl-, (2α,4α,6α)-
Me
O
Me
O
Me
130 C7H14O2 19534-08-8
1,2-Cyclohexanediol, 1-methyl-, trans-
OH
Me
OH
130 C7H14O2 19889-37-3
Butanoic acid, 2-ethyl-2-methyl-
Et2CMe(CO2H)
130 C7H14O2 20487-40-5
Propanoic acid, 1,1-dimethylethyl ester
EtC(O)OBu-t
130 C7H14O2 29006-00-6
2-Hexanone, 6-methoxy-
MeCO(CH2)4OMe
130 C7H14O2 29887-56-7
Cyclopentane, 1,2-dimethoxy-, trans-
OMe
OMe

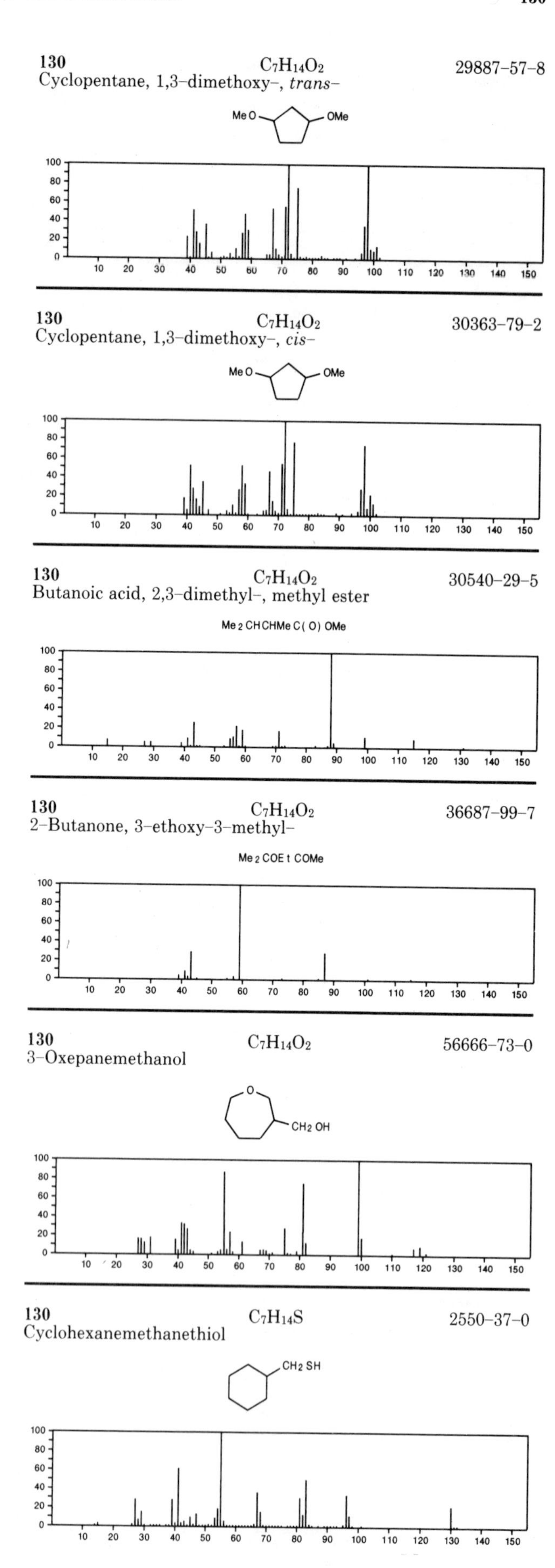
130 C7H14O2 29887-57-8
Cyclopentane, 1,3-dimethoxy-, trans-
MeO
OMe
130 C7H14O2 30363-79-2
Cyclopentane, 1,3-dimethoxy-, cis-
MeO
OMe
130 C7H14O2 30540-29-5
Butanoic acid, 2,3-dimethyl-, methyl ester
Me2CHCHMeC(O)OMe
130 C7H14O2 36687-99-7
2-Butanone, 3-ethoxy-3-methyl-
Me2COEtCOMe
130 C7H14O2 56666-73-0
3-Oxepanemethanol
O
CH2OH
130 C7H14S 2550-37-0
Cyclohexanemethanethiol
CH2SH

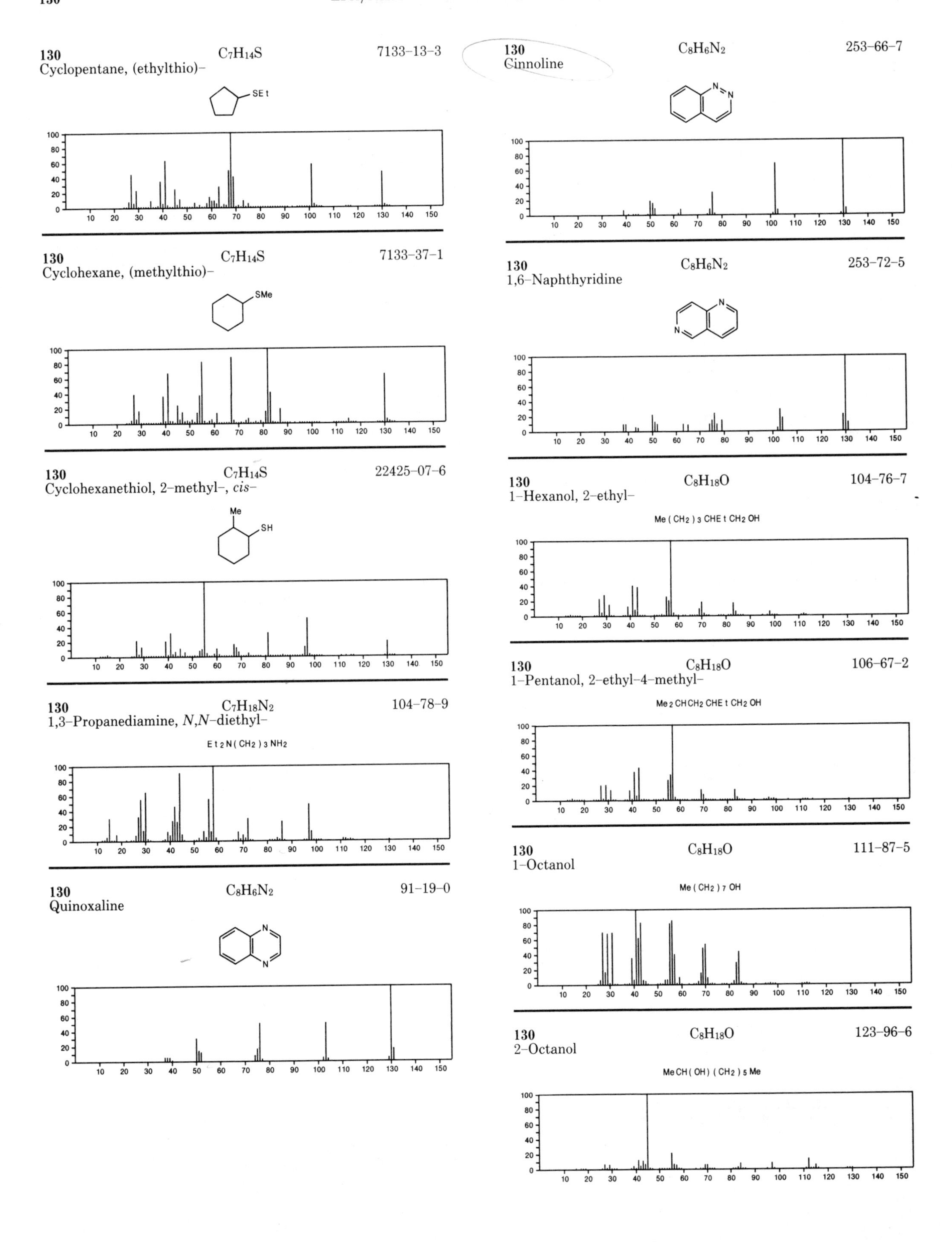
130
Cyclopentane, (ethylthio)–
C7H14S
7133–13–3
SEt
130
Cyclohexane, (methylthio)–
C7H14S
7133–37–1
SMe
130
Cyclohexanethiol, 2–methyl–, cis–
C7H14S
22425–07–6
Me
SH
130
1,3–Propanediamine, N,N–diethyl–
C7H18N2
104–78–9
Et2N(CH2)3NH2
130
Quinoxaline
C8H6N2
91–19–0
N
130
Cinnoline
C8H6N2
253–66–7
N
130
1,6–Naphthyridine
C8H6N2
253–72–5
N
130
1–Hexanol, 2–ethyl–
C8H18O
104–76–7
Me(CH2)3CHEtCH2OH
130
1–Pentanol, 2–ethyl–4–methyl–
C8H18O
106–67–2
Me2CHCH2CHEtCH2OH
130
1–Octanol
C8H18O
111–87–5
Me(CH2)7OH
130
2–Octanol
C8H18O
123–96–6
MeCH(OH)(CH2)5Me
0
20
40
60
80
100
10 20 30 40 50 60 70 80 90 100 110 120 130 140 150

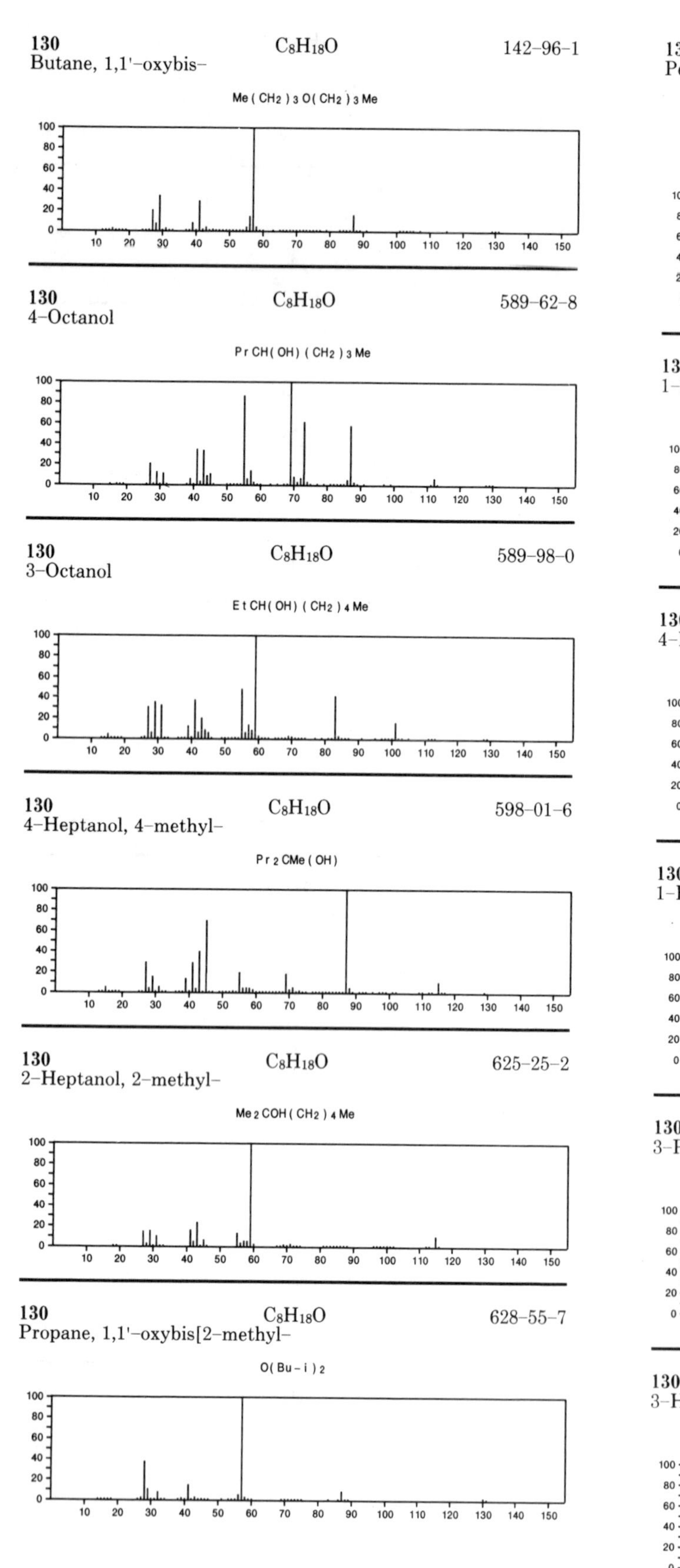
130 Butane, 1,1'-oxybis- C8H18O 142-96-1
Me (CH2) 3 O(CH2) 3 Me
130 4-Octanol C8H18O 589-62-8
Pr CH(OH) (CH2) 3 Me
130 3-Octanol C8H18O 589-98-0
Et CH(OH) (CH2) 4 Me
130 4-Heptanol, 4-methyl- C8H18O 598-01-6
Pr 2 CMe (OH)
130 2-Heptanol, 2-methyl- C8H18O 625-25-2
Me 2 COH (CH2) 4 Me
130 Propane, 1,1'-oxybis[2-methyl- C8H18O 628-55-7
O(Bu - i) 2

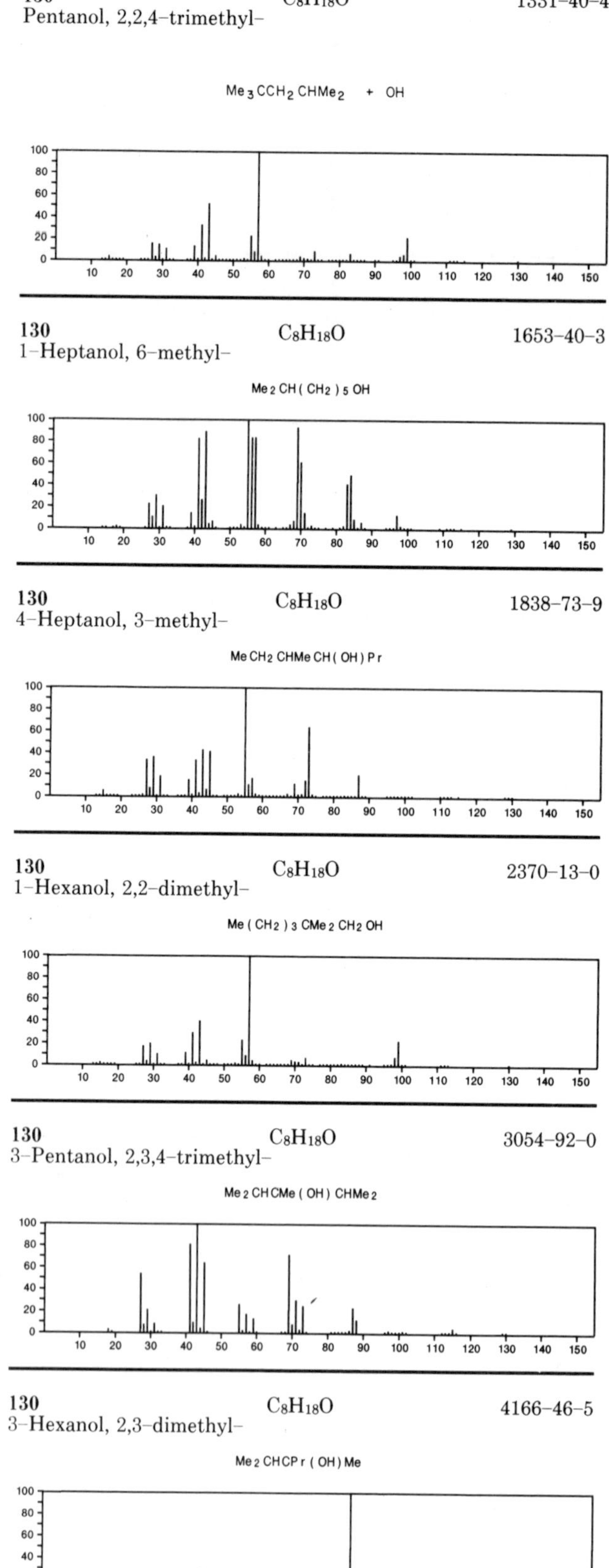
130 Pentanol, 2,2,4-trimethyl- C8H18O 1331-40-4
Me 3 CCH2 CHMe 2 + OH
130 1-Heptanol, 6-methyl- C8H18O 1653-40-3
Me 2 CH (CH2) 5 OH
130 4-Heptanol, 3-methyl- C8H18O 1838-73-9
Me CH2 CHMe CH(OH) Pr
130 1-Hexanol, 2,2-dimethyl- C8H18O 2370-13-0
Me (CH2) 3 CMe 2 CH2 OH
130 3-Pentanol, 2,3,4-trimethyl- C8H18O 3054-92-0
Me 2 CHCMe (OH) CHMe 2
130 3-Hexanol, 2,3-dimethyl- C8H18O 4166-46-5
Me 2 CHCPr (OH) Me

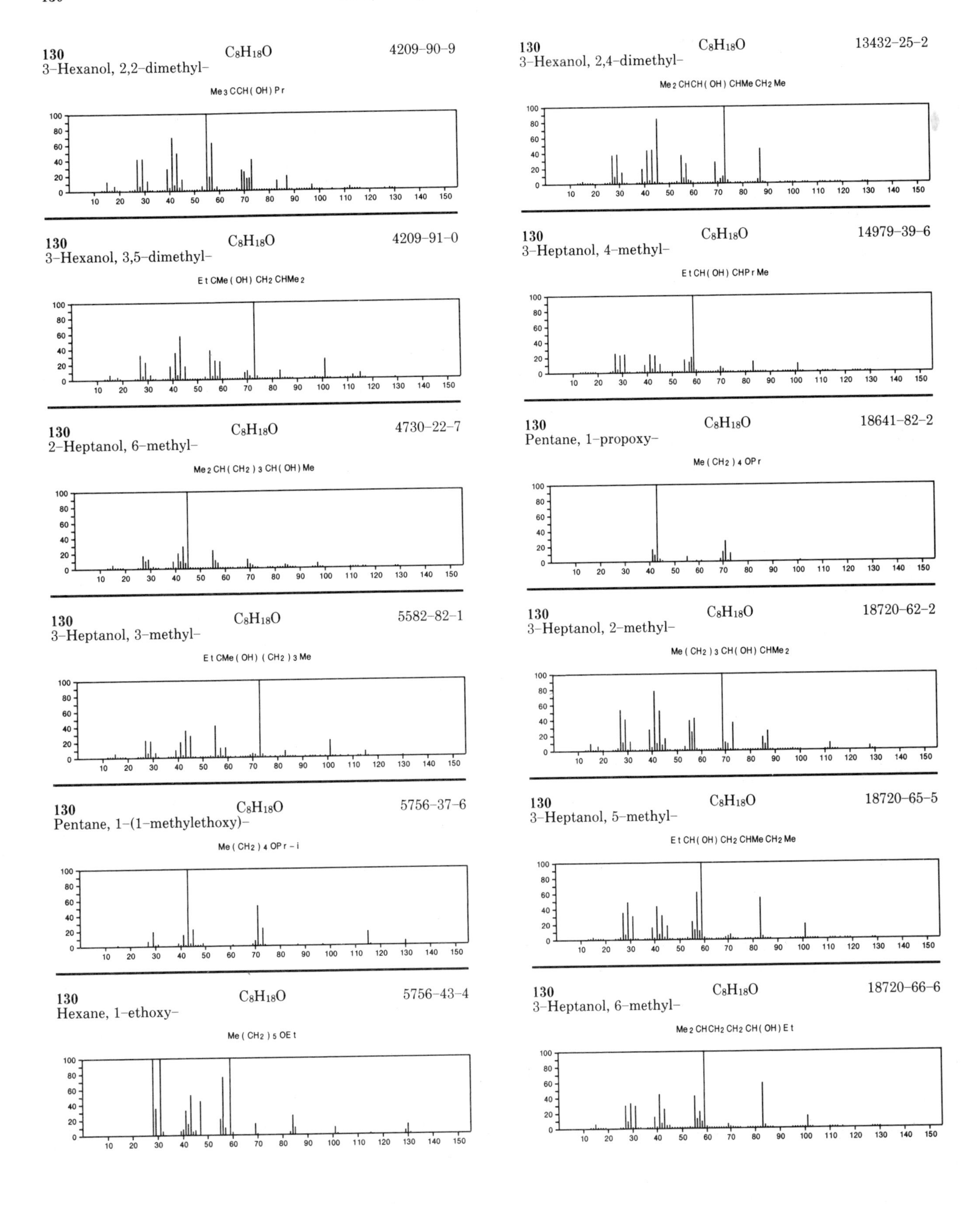
130 C8H18O 4209-90-9
3-Hexanol, 2,2-dimethyl-
Me3CCH(OH)Pr
130 C8H18O 13432-25-2
3-Hexanol, 2,4-dimethyl-
Me2CHCH(OH)CHMeCH2Me
130 C8H18O 4209-91-0
3-Hexanol, 3,5-dimethyl-
EtCMe(OH)CH2CHMe2
130 C8H18O 14979-39-6
3-Heptanol, 4-methyl-
EtCH(OH)CHPrMe
130 C8H18O 4730-22-7
2-Heptanol, 6-methyl-
Me2CH(CH2)3CH(OH)Me
130 C8H18O 18641-82-2
Pentane, 1-propoxy-
Me(CH2)4OPr
130 C8H18O 5582-82-1
3-Heptanol, 3-methyl-
EtCMe(OH)(CH2)3Me
130 C8H18O 18720-62-2
3-Heptanol, 2-methyl-
Me(CH2)3CH(OH)CHMe2
130 C8H18O 5756-37-6
Pentane, 1-(1-methylethoxy)-
Me(CH2)4OPr-i
130 C8H18O 18720-65-5
3-Heptanol, 5-methyl-
EtCH(OH)CH2CHMeCH2Me
130 C8H18O 5756-43-4
Hexane, 1-ethoxy-
Me(CH2)5OEt
130 C8H18O 18720-66-6
3-Heptanol, 6-methyl-
Me2CHCH2CH2CH(OH)Et

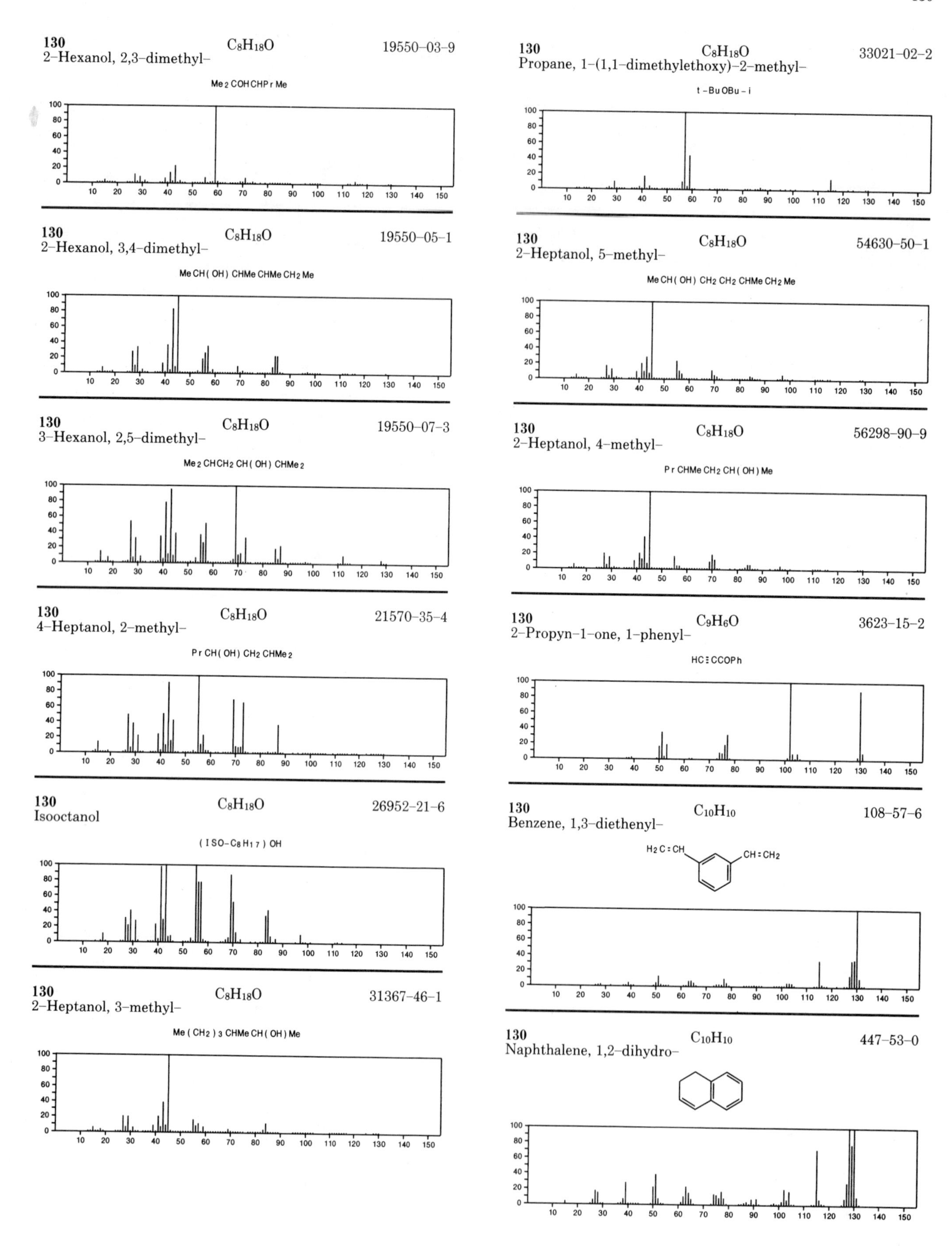

130
2-Hexanol, 2,3-dimethyl-
$C_8H_{18}O$
19550-03-9
Me2 COHCHPr Me
130
2-Hexanol, 3,4-dimethyl-
$C_8H_{18}O$
19550-05-1
Me CH(OH) CHMe CHMe CH2 Me
130
3-Hexanol, 2,5-dimethyl-
$C_8H_{18}O$
19550-07-3
Me2 CHCH2 CH(OH) CHMe2
130
4-Heptanol, 2-methyl-
$C_8H_{18}O$
21570-35-4
Pr CH(OH) CH2 CHMe2
130
Isooctanol
$C_8H_{18}O$
26952-21-6
(ISO-C8 H17) OH
130
2-Heptanol, 3-methyl-
$C_8H_{18}O$
31367-46-1
Me (CH2)3 CHMe CH(OH) Me
130
Propane, 1-(1,1-dimethylethoxy)-2-methyl-
$C_8H_{18}O$
33021-02-2
t-Bu OBu-i
130
2-Heptanol, 5-methyl-
$C_8H_{18}O$
54630-50-1
Me CH(OH) CH2 CH2 CHMe CH2 Me
130
2-Heptanol, 4-methyl-
$C_8H_{18}O$
56298-90-9
Pr CHMe CH2 CH(OH) Me
130
2-Propyn-1-one, 1-phenyl-
C_9H_6O
3623-15-2
HC≡CCOPh
130
Benzene, 1,3-diethenyl-
$C_{10}H_{10}$
108-57-6
H2C=CH
CH=CH2
130
Naphthalene, 1,2-dihydro-
$C_{10}H_{10}$
447-53-0

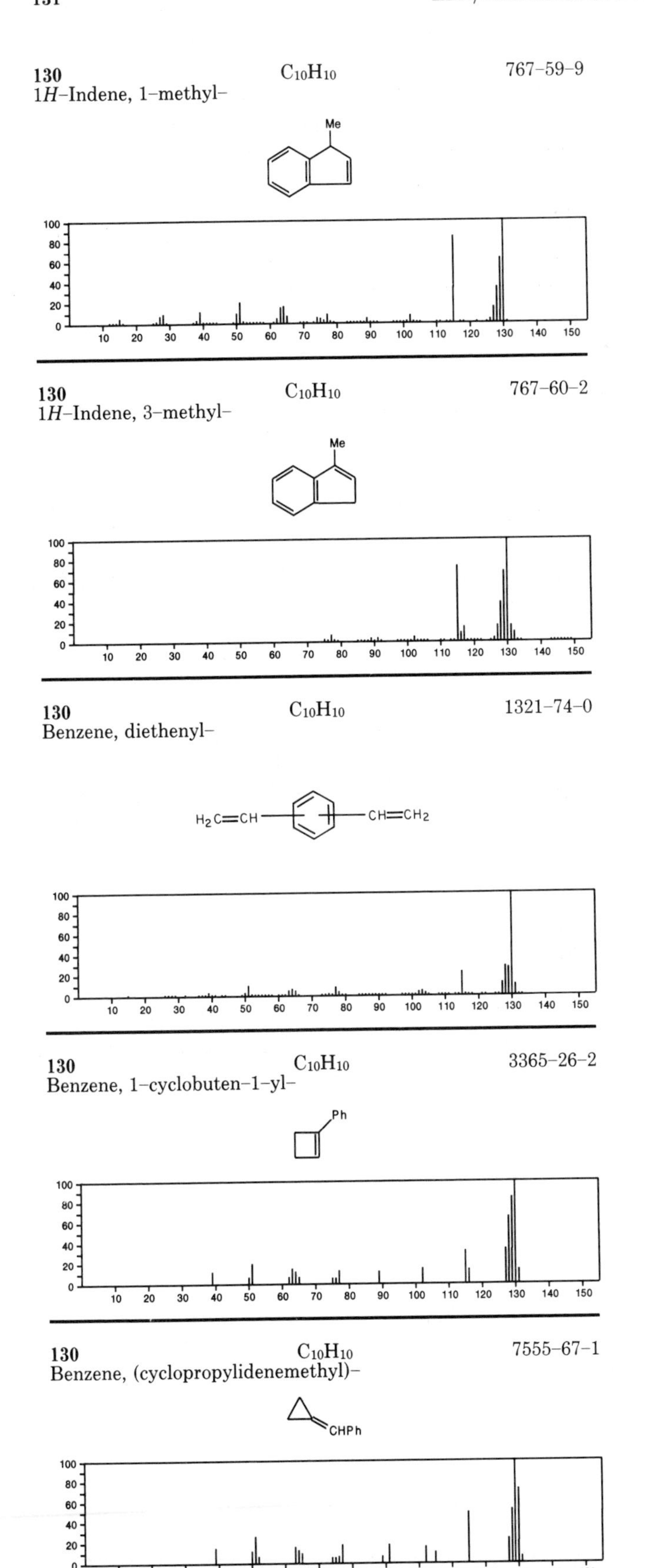
130 $C_{10}H_{10}$ 767-59-9
1*H*-Indene, 1-methyl-
Me
130 $C_{10}H_{10}$ 767-60-2
1*H*-Indene, 3-methyl-
Me
130 $C_{10}H_{10}$ 1321-74-0
Benzene, diethenyl-
H2C=CH
CH=CH2
130 $C_{10}H_{10}$ 3365-26-2
Benzene, 1-cyclobuten-1-yl-
Ph
130 $C_{10}H_{10}$ 7555-67-1
Benzene, (cyclopropylidenemethyl)-
CHPh

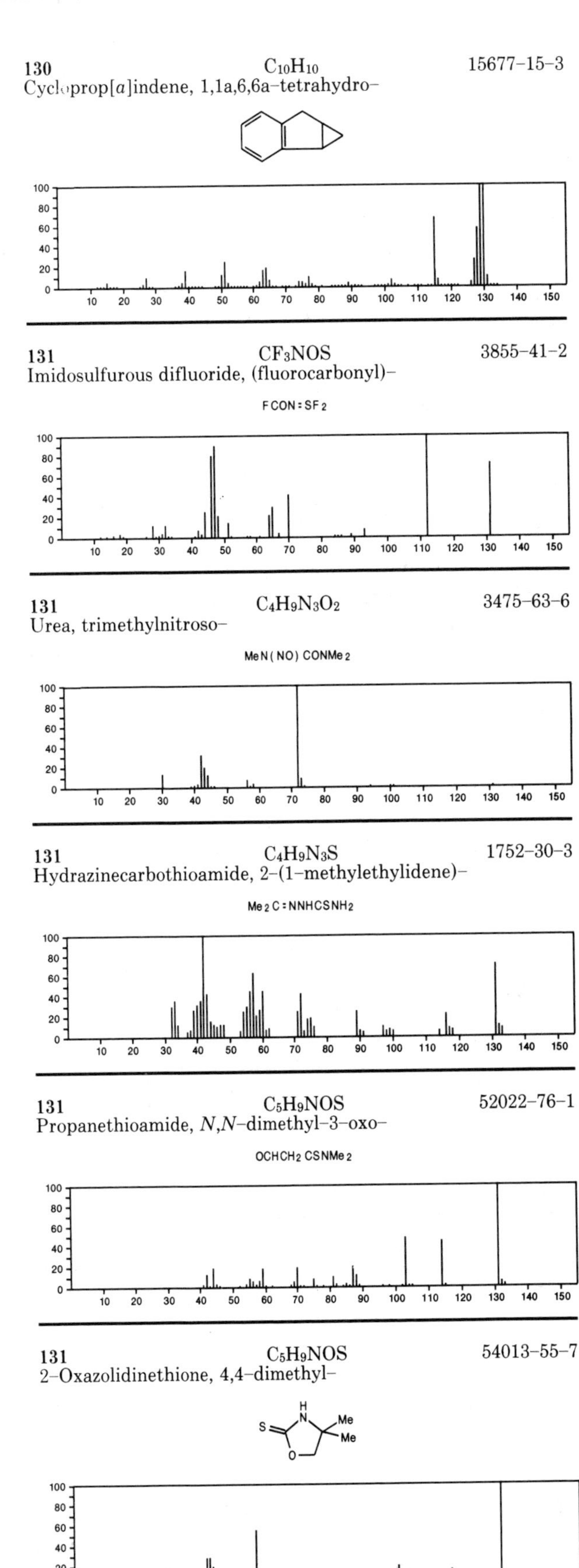
130 $C_{10}H_{10}$ 15677-15-3
Cycloprop[*a*]indene, 1,1a,6,6a-tetrahydro-
131 CF_3NOS 3855-41-2
Imidosulfurous difluoride, (fluorocarbonyl)-
FCON:SF2
131 $C_4H_9N_3O_2$ 3475-63-6
Urea, trimethylnitroso-
MeN(NO)CONMe2
131 $C_4H_9N_3S$ 1752-30-3
Hydrazinecarbothioamide, 2-(1-methylethylidene)-
Me2C:NNHCSNH2
131 C_5H_9NOS 52022-76-1
Propanethioamide, *N*,*N*-dimethyl-3-oxo-
OCHCH2CSNMe2
131 C_5H_9NOS 54013-55-7
2-Oxazolidinethione, 4,4-dimethyl-
H
N
S
O
Me
Me

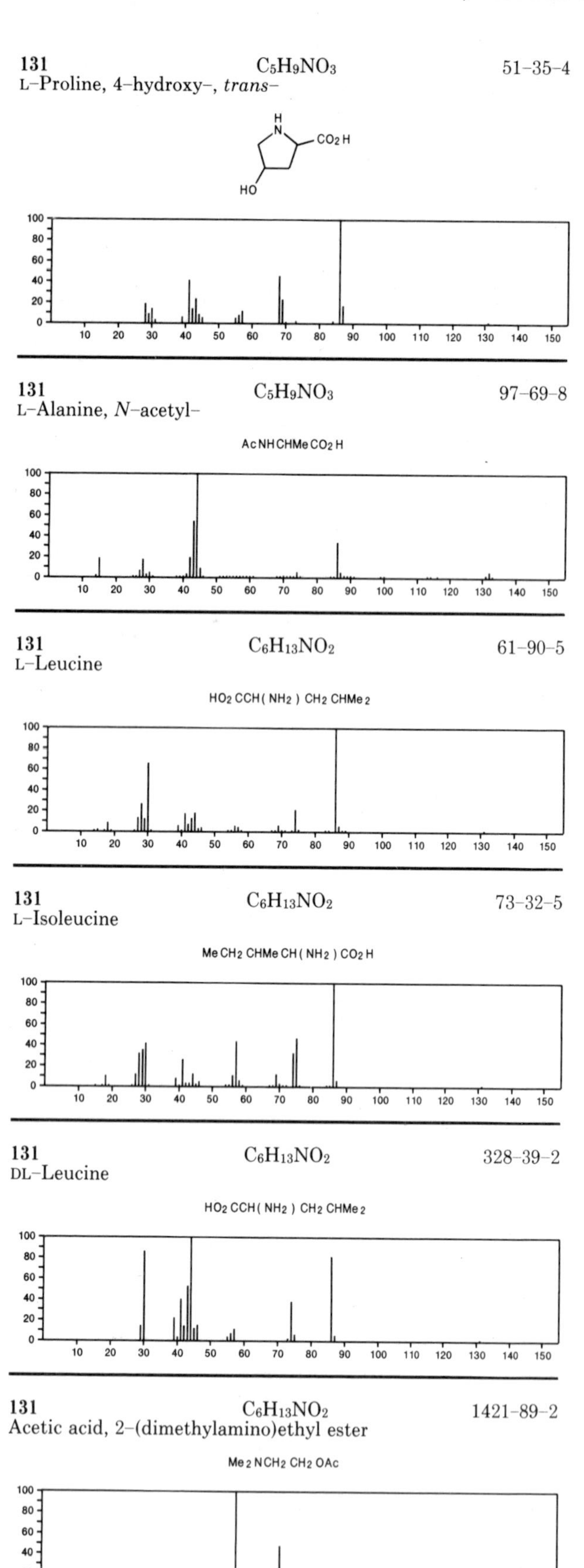
131 C5H9NO3 51-35-4
L-Proline, 4-hydroxy-, trans-
CO2H
HO
131 C5H9NO3 97-69-8
L-Alanine, N-acetyl-
AcNHCHMeCO2H
131 C6H13NO2 61-90-5
L-Leucine
HO2CCH(NH2)CH2CHMe2
131 C6H13NO2 73-32-5
L-Isoleucine
MeCH2CHMeCH(NH2)CO2H
131 C6H13NO2 328-39-2
DL-Leucine
HO2CCH(NH2)CH2CHMe2
131 C6H13NO2 1421-89-2
Acetic acid, 2-(dimethylamino)ethyl ester
Me2NCH2CH2OAc

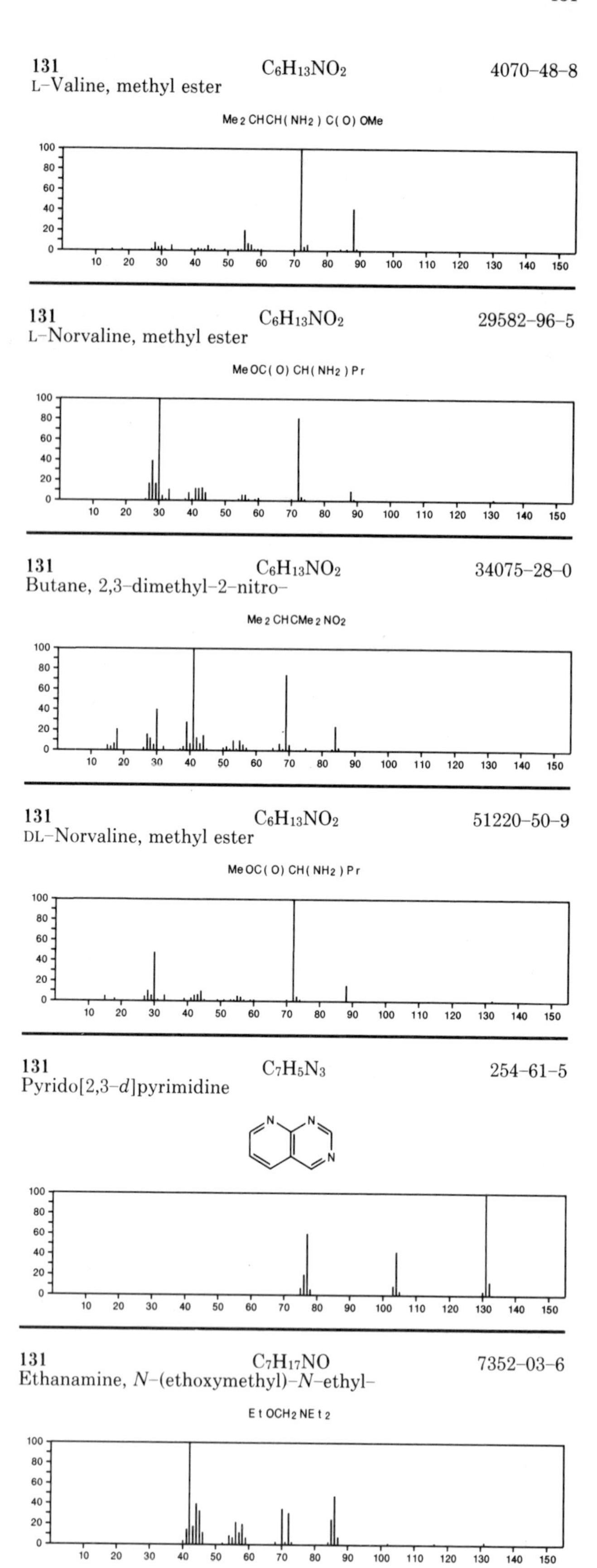
131 C6H13NO2 4070-48-8
L-Valine, methyl ester
Me2CHCH(NH2)C(O)OMe
131 C6H13NO2 29582-96-5
L-Norvaline, methyl ester
MeOC(O)CH(NH2)Pr
131 C6H13NO2 34075-28-0
Butane, 2,3-dimethyl-2-nitro-
Me2CHCMe2NO2
131 C6H13NO2 51220-50-9
DL-Norvaline, methyl ester
MeOC(O)CH(NH2)Pr
131 C7H5N3 254-61-5
Pyrido[2,3-d]pyrimidine
131 C7H17NO 7352-03-6
Ethanamine, N-(ethoxymethyl)-N-ethyl-
EtOCH2NEt2

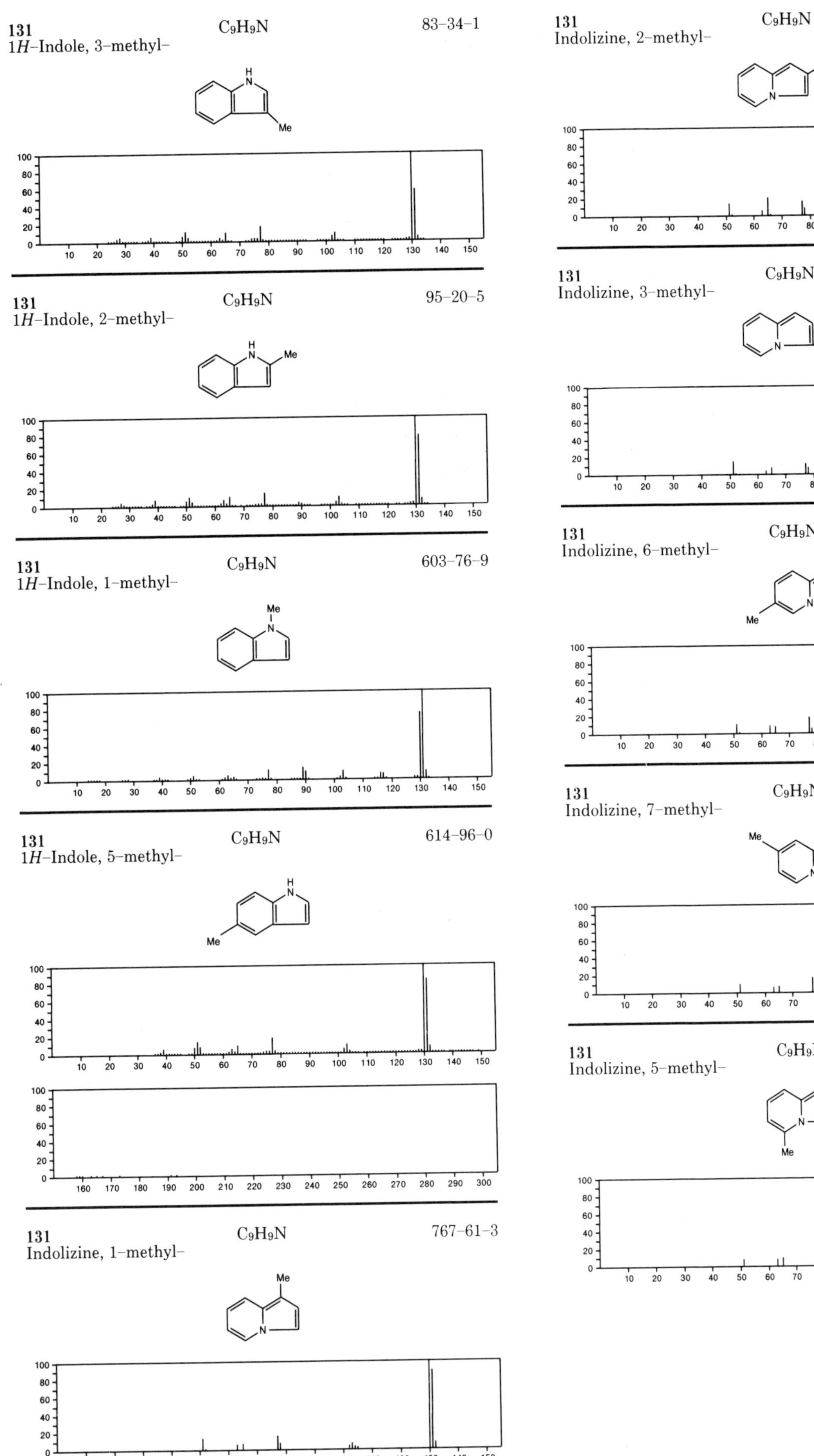

131 C_9H_9N 83-34-1
1H-Indole, 3-methyl-

131 C_9H_9N 95-20-5
1H-Indole, 2-methyl-

131 C_9H_9N 603-76-9
1H-Indole, 1-methyl-

131 C_9H_9N 614-96-0
1H-Indole, 5-methyl-

131 C_9H_9N 767-61-3
Indolizine, 1-methyl-

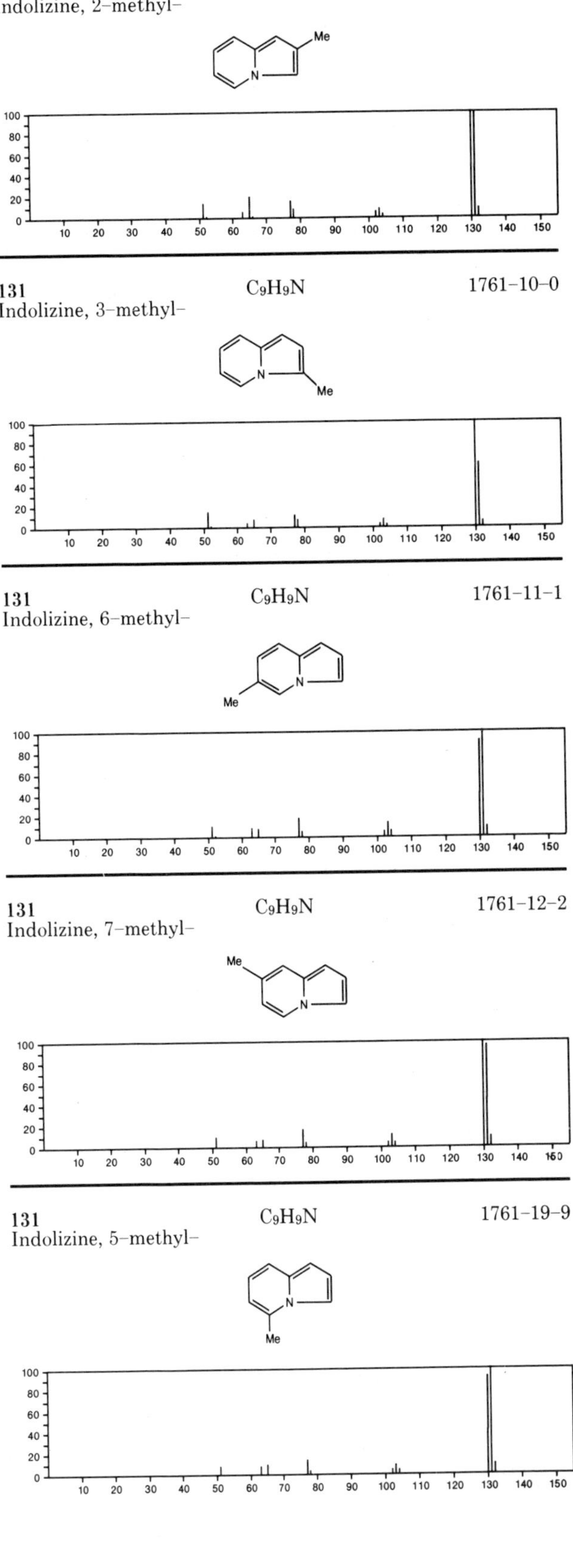

131 C_9H_9N 768-18-3
Indolizine, 2-methyl-

131 C_9H_9N 1761-10-0
Indolizine, 3-methyl-

131 C_9H_9N 1761-11-1
Indolizine, 6-methyl-

131 C_9H_9N 1761-12-2
Indolizine, 7-methyl-

131 C_9H_9N 1761-19-9
Indolizine, 5-methyl-

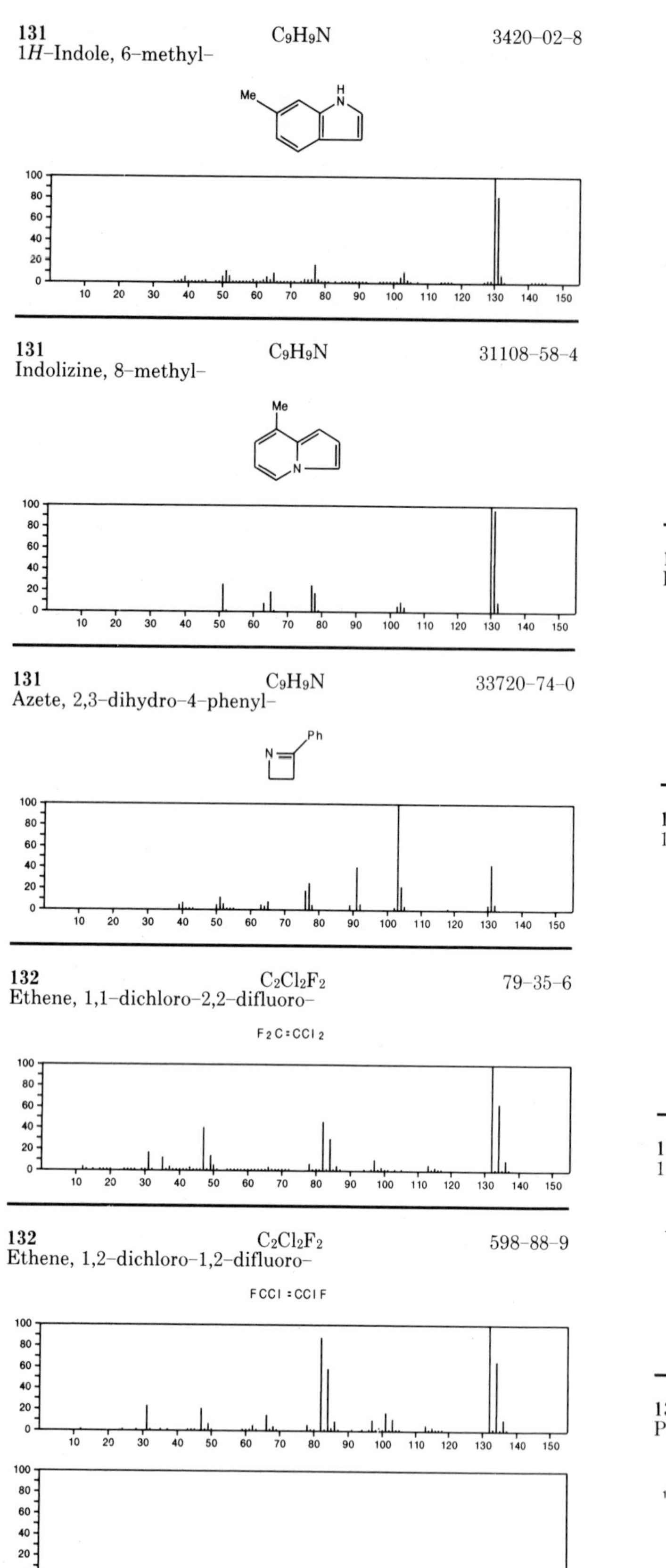
131
1H-Indole, 6-methyl-
C9H9N
3420-02-8
131
Indolizine, 8-methyl-
C9H9N
31108-58-4
131
Azete, 2,3-dihydro-4-phenyl-
C9H9N
33720-74-0
132
Ethene, 1,1-dichloro-2,2-difluoro-
C2Cl2F2
79-35-6
F2C=CCl2
132
Ethene, 1,2-dichloro-1,2-difluoro-
C2Cl2F2
598-88-9
FCCl=CClF

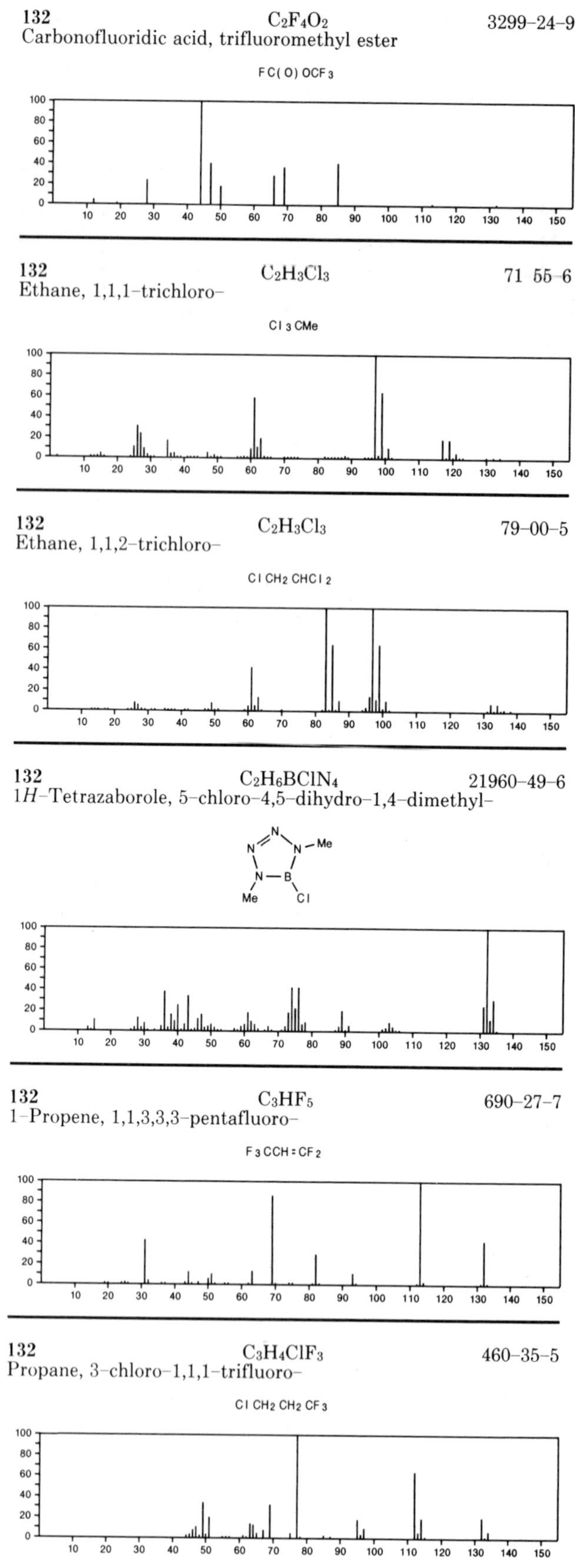
132
Carbonofluoridic acid, trifluoromethyl ester
C2F4O2
3299-24-9
FC(O)OCF3
132
Ethane, 1,1,1-trichloro-
C2H3Cl3
71-55-6
Cl3CMe
132
Ethane, 1,1,2-trichloro-
C2H3Cl3
79-00-5
ClCH2CHCl2
132
1H-Tetrazaborole, 5-chloro-4,5-dihydro-1,4-dimethyl-
C2H6BClN4
21960-49-6
132
1-Propene, 1,1,3,3,3-pentafluoro-
C3HF5
690-27-7
F3CCH=CF2
132
Propane, 3-chloro-1,1,1-trifluoro-
C3H4ClF3
460-35-5
ClCH2CH2CF3

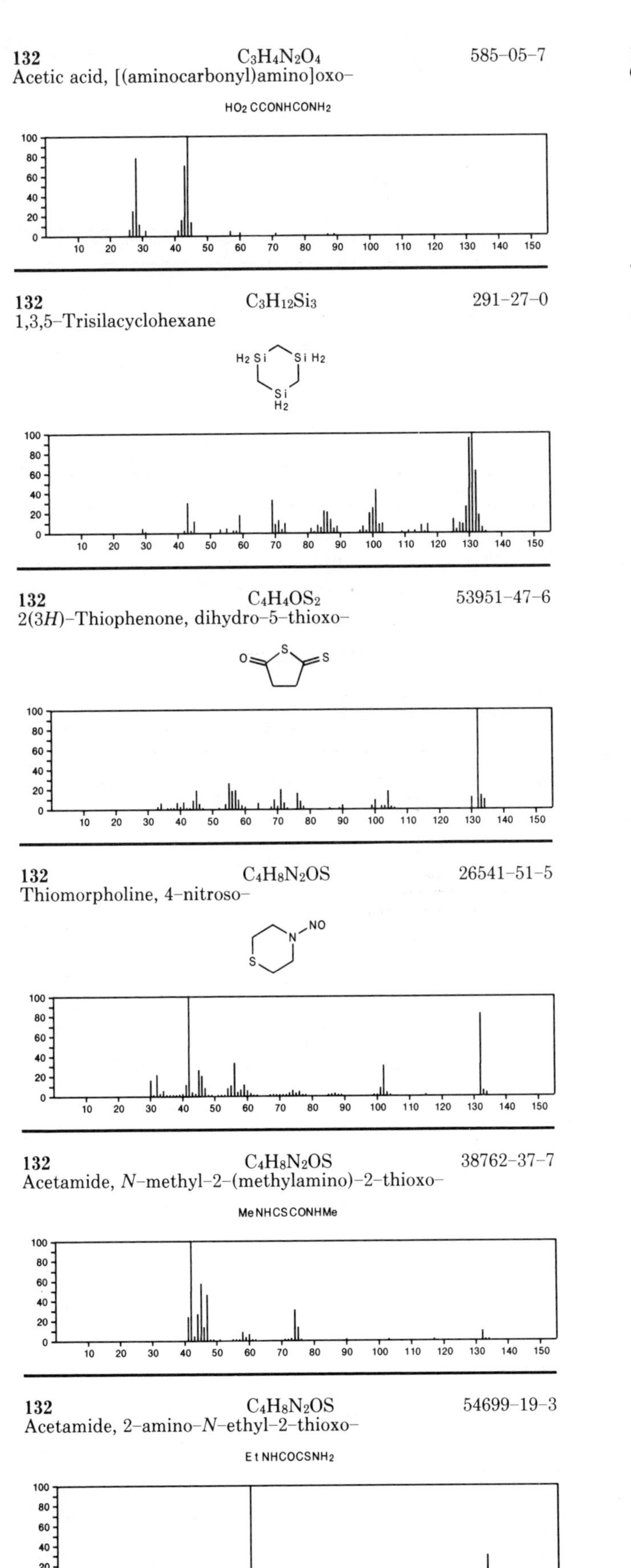

132 $C_3H_4N_2O_4$ 585–05–7
Acetic acid, [(aminocarbonyl)amino]oxo–
$HO_2CCONHCONH_2$

132 $C_3H_{12}Si_3$ 291–27–0
1,3,5–Trisilacyclohexane

132 $C_4H_4OS_2$ 53951–47–6
2(3*H*)–Thiophenenone, dihydro–5–thioxo–

132 $C_4H_8N_2OS$ 26541–51–5
Thiomorpholine, 4–nitroso–

132 $C_4H_8N_2OS$ 38762–37–7
Acetamide, *N*–methyl–2–(methylamino)–2–thioxo–
MeNHCSCONHMe

132 $C_4H_8N_2OS$ 54699–19–3
Acetamide, 2–amino–*N*–ethyl–2–thioxo–
EtNHCOCSNH$_2$

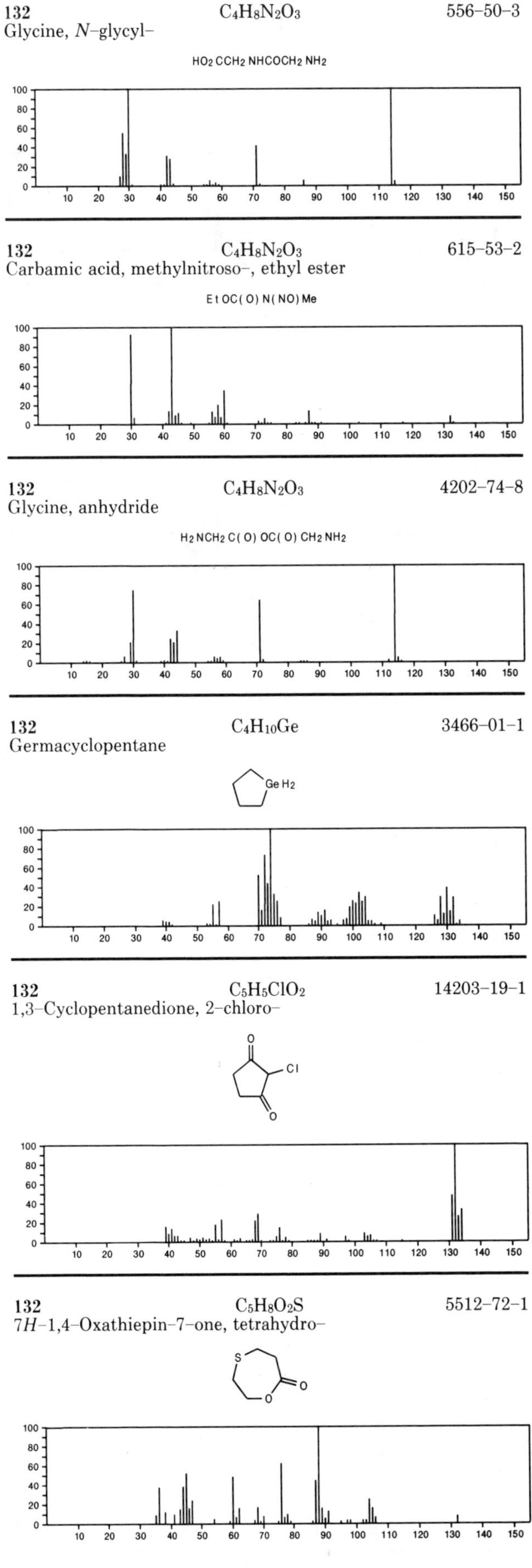

132 $C_4H_8N_2O_3$ 556–50–3
Glycine, *N*–glycyl–
$HO_2CCH_2NHCOCH_2NH_2$

132 $C_4H_8N_2O_3$ 615–53–2
Carbamic acid, methylnitroso–, ethyl ester
EtOC(O)N(NO)Me

132 $C_4H_8N_2O_3$ 4202–74–8
Glycine, anhydride
$H_2NCH_2C(O)OC(O)CH_2NH_2$

132 $C_4H_{10}Ge$ 3466–01–1
Germacyclopentane

132 $C_5H_5ClO_2$ 14203–19–1
1,3–Cyclopentanedione, 2–chloro–

132 $C_5H_8O_2S$ 5512–72–1
7*H*–1,4–Oxathiepin–7–one, tetrahydro–

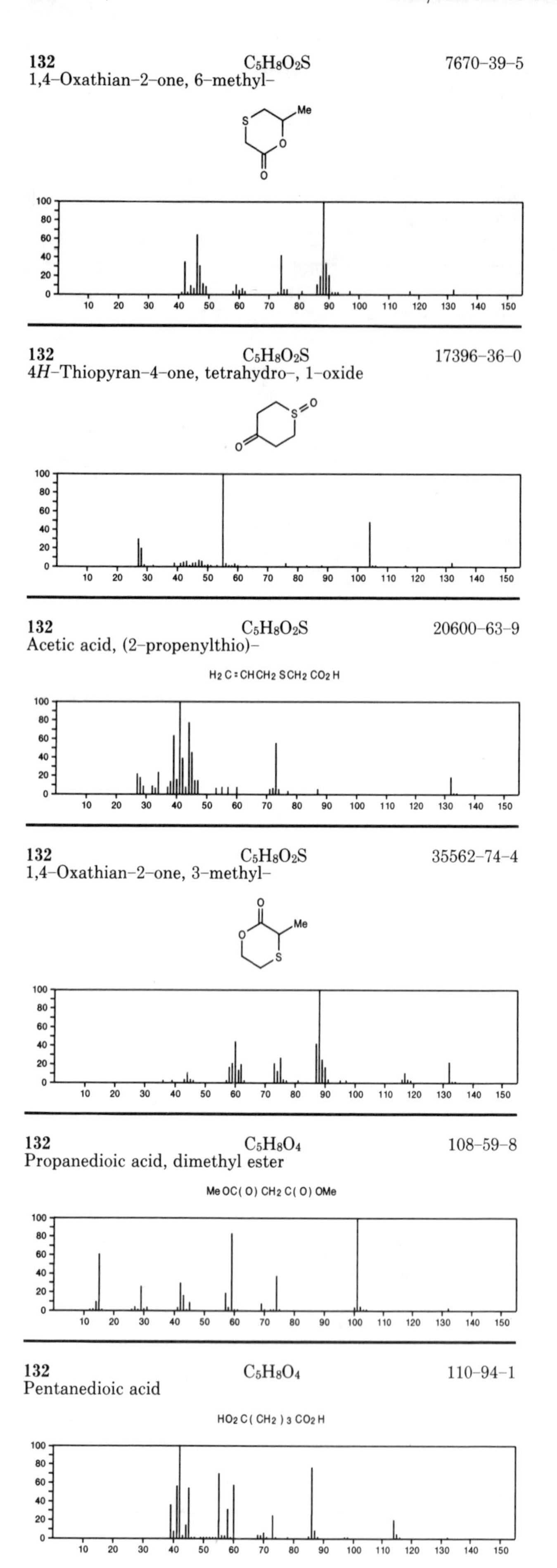

132 $C_5H_8O_2S$ 7670-39-5
1,4-Oxathian-2-one, 6-methyl-

132 $C_5H_8O_2S$ 17396-36-0
4*H*-Thiopyran-4-one, tetrahydro-, 1-oxide

132 $C_5H_8O_2S$ 20600-63-9
Acetic acid, (2-propenylthio)-
$H_2C{=}CHCH_2SCH_2CO_2H$

132 $C_5H_8O_2S$ 35562-74-4
1,4-Oxathian-2-one, 3-methyl-

132 $C_5H_8O_4$ 108-59-8
Propanedioic acid, dimethyl ester
MeOC(O)CH$_2$C(O)OMe

132 $C_5H_8O_4$ 110-94-1
Pentanedioic acid
$HO_2C(CH_2)_3CO_2H$

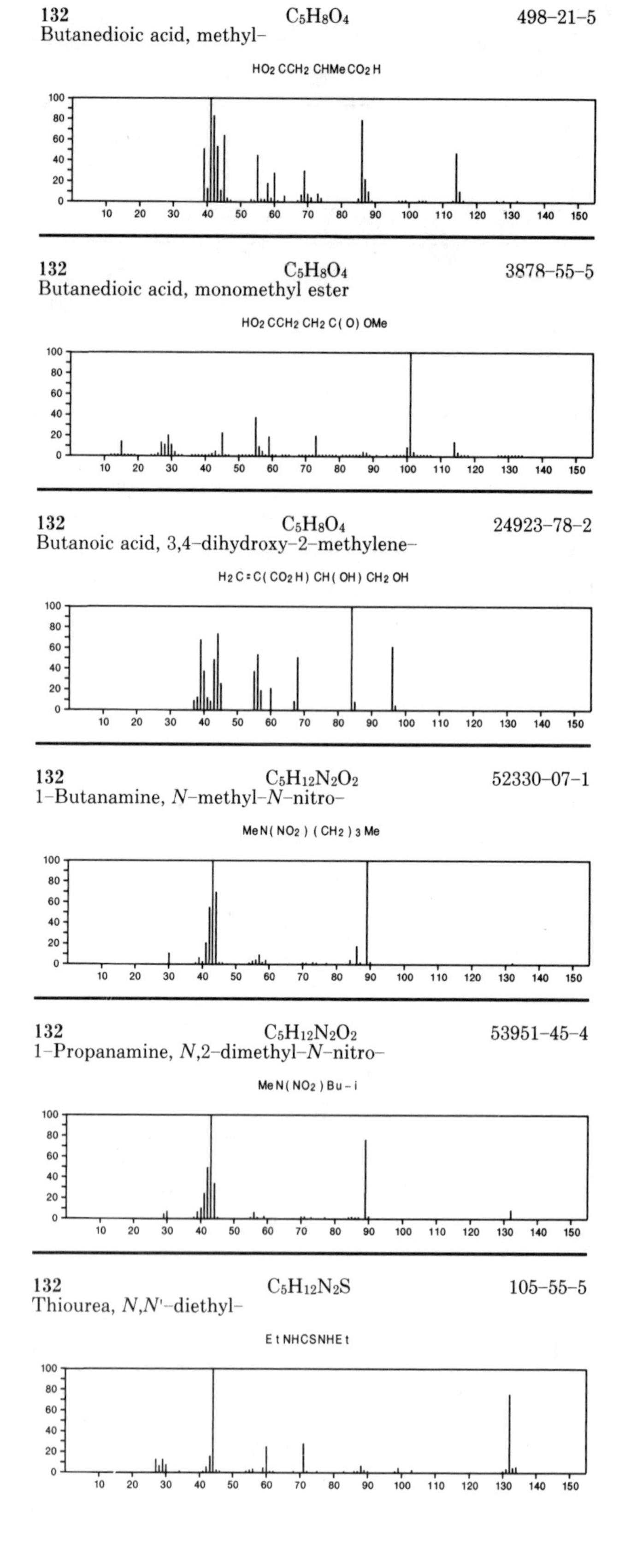

132 $C_5H_8O_4$ 498-21-5
Butanedioic acid, methyl-
$HO_2CCH_2CHMeCO_2H$

132 $C_5H_8O_4$ 3878-55-5
Butanedioic acid, monomethyl ester
$HO_2CCH_2CH_2C(O)OMe$

132 $C_5H_8O_4$ 24923-78-2
Butanoic acid, 3,4-dihydroxy-2-methylene-
$H_2C{=}C(CO_2H)CH(OH)CH_2OH$

132 $C_5H_{12}N_2O_2$ 52330-07-1
1-Butanamine, *N*-methyl-*N*-nitro-
$MeN(NO_2)(CH_2)_3Me$

132 $C_5H_{12}N_2O_2$ 53951-45-4
1-Propanamine, *N*,2-dimethyl-*N*-nitro-
MeN(NO$_2$)Bu-i

132 $C_5H_{12}N_2S$ 105-55-5
Thiourea, *N*,*N'*-diethyl-
EtNHCSNHEt

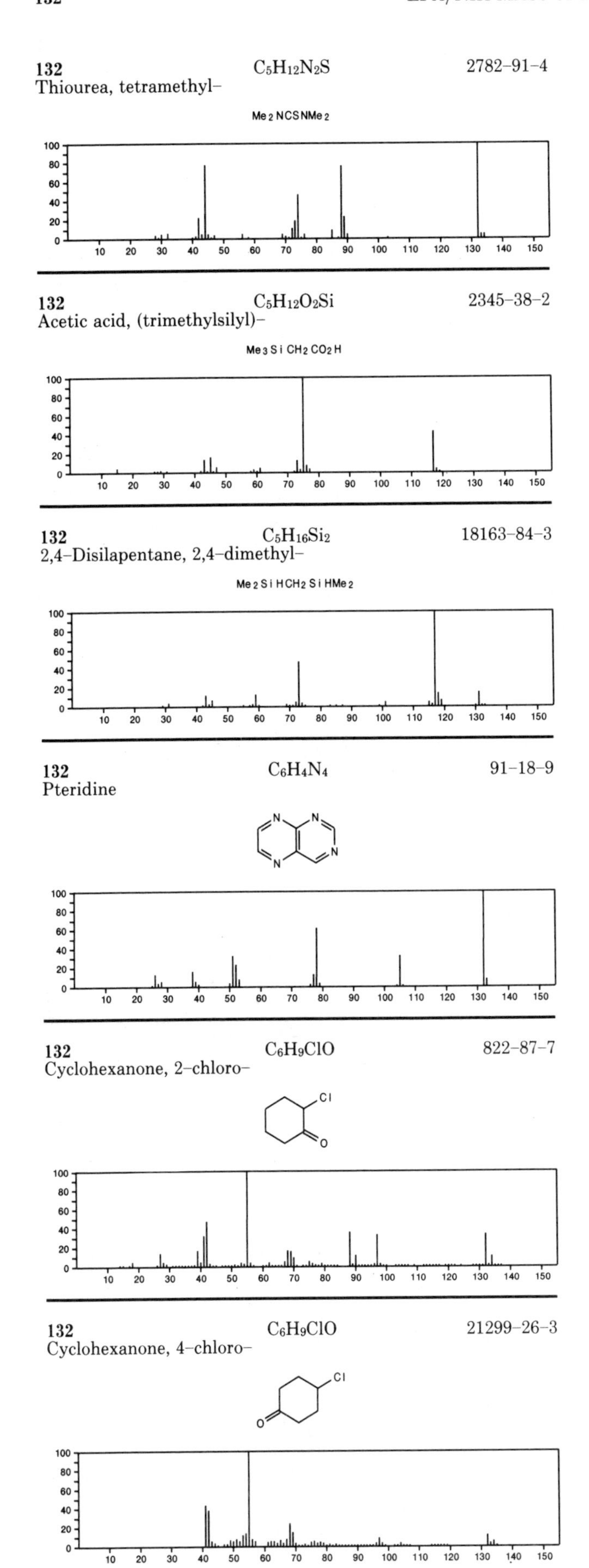

132 $C_5H_{12}N_2S$ 2782-91-4
Thiourea, tetramethyl-

$Me_2NCSNMe_2$

132 $C_5H_{12}O_2Si$ 2345-38-2
Acetic acid, (trimethylsilyl)-

$Me_3SiCH_2CO_2H$

132 $C_5H_{16}Si_2$ 18163-84-3
2,4-Disilapentane, 2,4-dimethyl-

$Me_2SiHCH_2SiHMe_2$

132 $C_6H_4N_4$ 91-18-9
Pteridine

132 C_6H_9ClO 822-87-7
Cyclohexanone, 2-chloro-

132 C_6H_9ClO 21299-26-3
Cyclohexanone, 4-chloro-

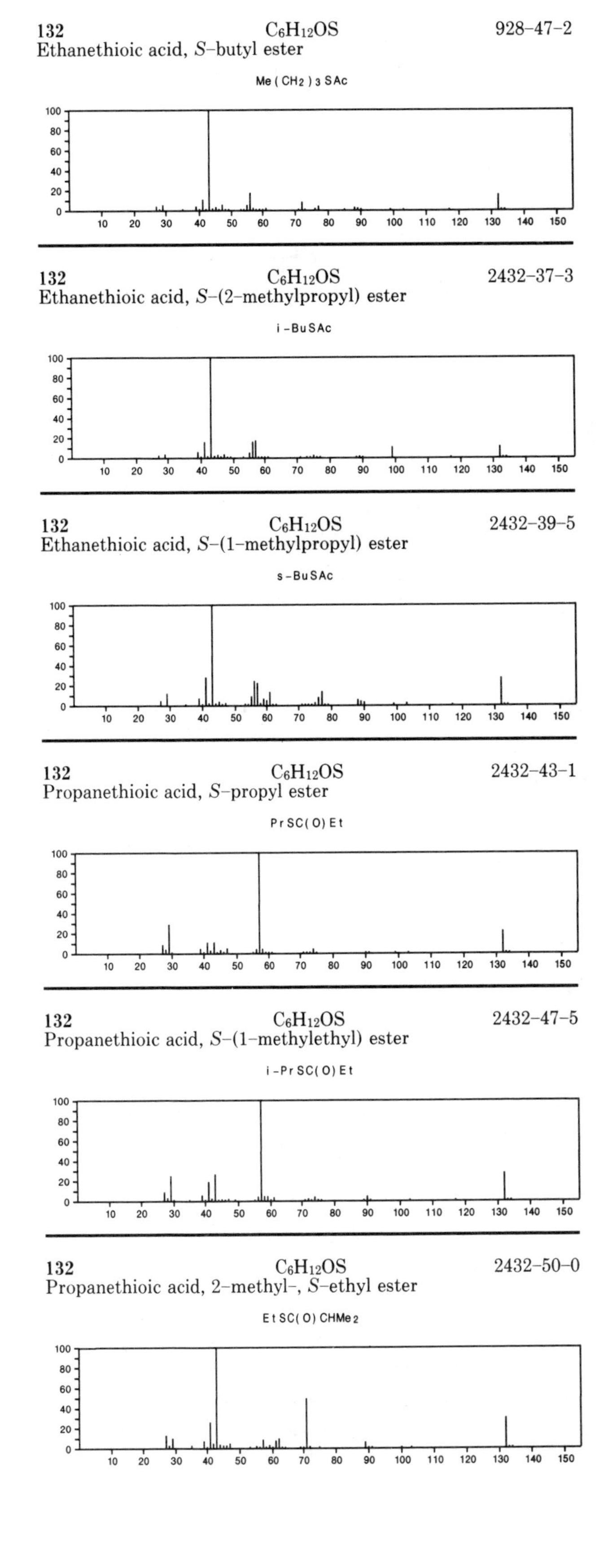

132 $C_6H_{12}OS$ 928-47-2
Ethanethioic acid, *S*-butyl ester

$Me(CH_2)_3SAc$

132 $C_6H_{12}OS$ 2432-37-3
Ethanethioic acid, *S*-(2-methylpropyl) ester

i-BuSAc

132 $C_6H_{12}OS$ 2432-39-5
Ethanethioic acid, *S*-(1-methylpropyl) ester

s-BuSAc

132 $C_6H_{12}OS$ 2432-43-1
Propanethioic acid, *S*-propyl ester

PrSC(O)Et

132 $C_6H_{12}OS$ 2432-47-5
Propanethioic acid, *S*-(1-methylethyl) ester

i-PrSC(O)Et

132 $C_6H_{12}OS$ 2432-50-0
Propanethioic acid, 2-methyl-, *S*-ethyl ester

EtSC(O)CHMe$_2$

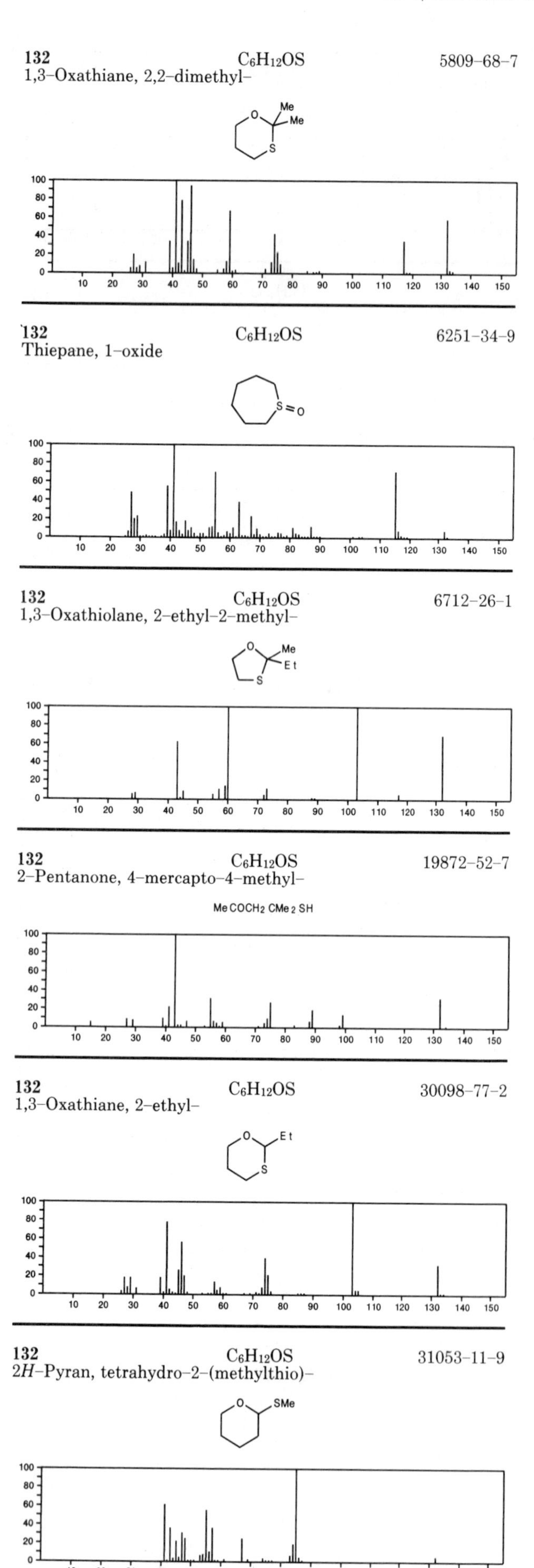

132 $C_6H_{12}OS$ 5809-68-7
1,3-Oxathiane, 2,2-dimethyl-

132 $C_6H_{12}OS$ 6251-34-9
Thiepane, 1-oxide

132 $C_6H_{12}OS$ 6712-26-1
1,3-Oxathiolane, 2-ethyl-2-methyl-

132 $C_6H_{12}OS$ 19872-52-7
2-Pentanone, 4-mercapto-4-methyl-

Me COCH2 CMe2 SH

132 $C_6H_{12}OS$ 30098-77-2
1,3-Oxathiane, 2-ethyl-

132 $C_6H_{12}OS$ 31053-11-9
2*H*-Pyran, tetrahydro-2-(methylthio)-

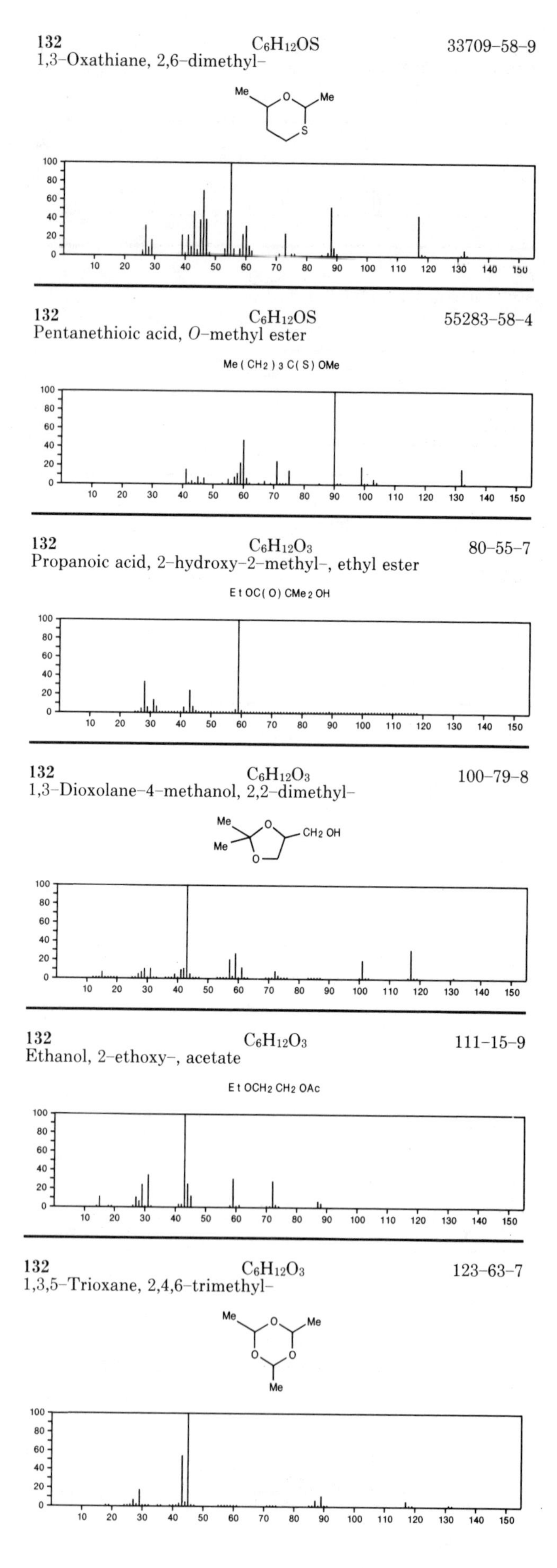

132 $C_6H_{12}OS$ 33709-58-9
1,3-Oxathiane, 2,6-dimethyl-

132 $C_6H_{12}OS$ 55283-58-4
Pentanethioic acid, *O*-methyl ester

Me (CH2) 3 C(S) OMe

132 $C_6H_{12}O_3$ 80-55-7
Propanoic acid, 2-hydroxy-2-methyl-, ethyl ester

Et OC(O) CMe2 OH

132 $C_6H_{12}O_3$ 100-79-8
1,3-Dioxolane-4-methanol, 2,2-dimethyl-

132 $C_6H_{12}O_3$ 111-15-9
Ethanol, 2-ethoxy-, acetate

Et OCH2 CH2 OAc

132 $C_6H_{12}O_3$ 123-63-7
1,3,5-Trioxane, 2,4,6-trimethyl-

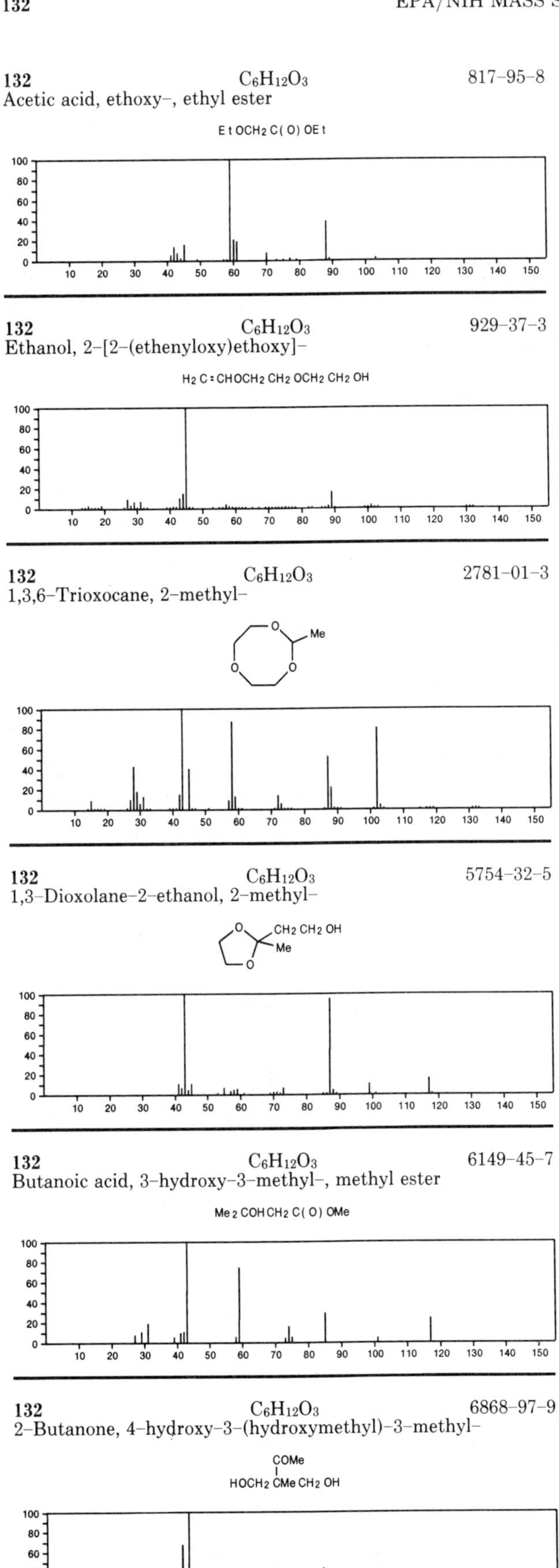

132 $C_6H_{12}O_3$ 817-95-8
Acetic acid, ethoxy-, ethyl ester

EtOCH2C(O)OEt

132 $C_6H_{12}O_3$ 929-37-3
Ethanol, 2-[2-(ethenyloxy)ethoxy]-

H2C=CHOCH2CH2OCH2CH2OH

132 $C_6H_{12}O_3$ 2781-01-3
1,3,6-Trioxocane, 2-methyl-

132 $C_6H_{12}O_3$ 5754-32-5
1,3-Dioxolane-2-ethanol, 2-methyl-

132 $C_6H_{12}O_3$ 6149-45-7
Butanoic acid, 3-hydroxy-3-methyl-, methyl ester

Me2COHCH2C(O)OMe

132 $C_6H_{12}O_3$ 6868-97-9
2-Butanone, 4-hydroxy-3-(hydroxymethyl)-3-methyl-

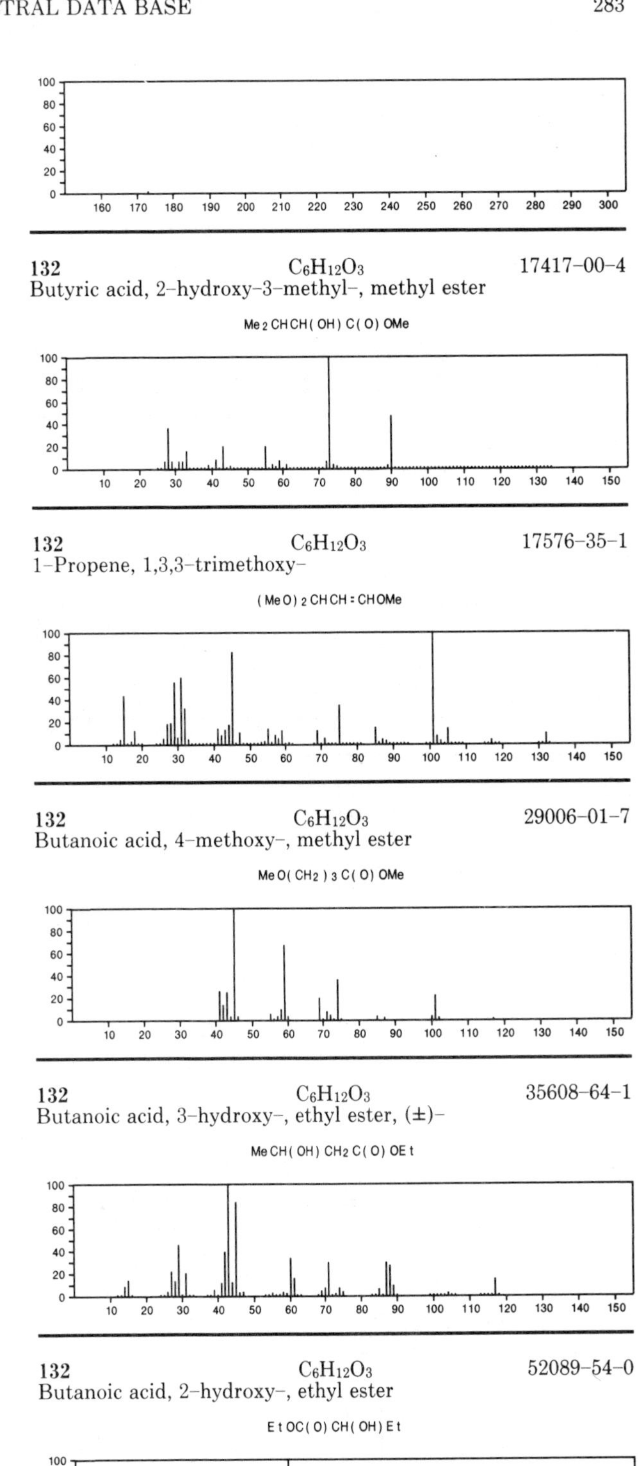

132 $C_6H_{12}O_3$ 17417-00-4
Butyric acid, 2-hydroxy-3-methyl-, methyl ester

Me2CHCH(OH)C(O)OMe

132 $C_6H_{12}O_3$ 17576-35-1
1-Propene, 1,3,3-trimethoxy-

(MeO)2CHCH=CHOMe

132 $C_6H_{12}O_3$ 29006-01-7
Butanoic acid, 4-methoxy-, methyl ester

MeO(CH2)3C(O)OMe

132 $C_6H_{12}O_3$ 35608-64-1
Butanoic acid, 3-hydroxy-, ethyl ester, (±)-

MeCH(OH)CH2C(O)OEt

132 $C_6H_{12}O_3$ 52089-54-0
Butanoic acid, 2-hydroxy-, ethyl ester

EtOC(O)CH(OH)Et

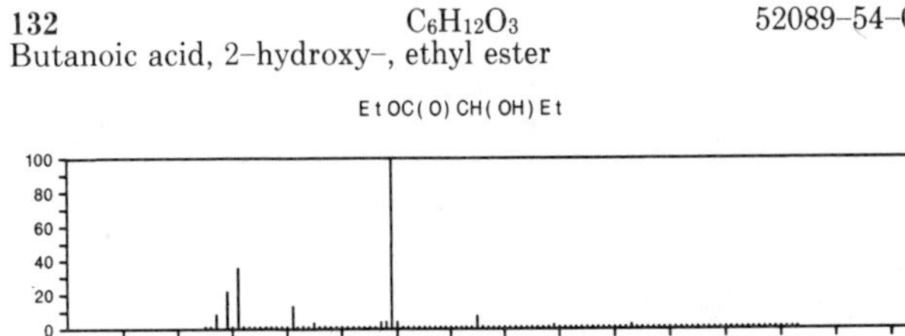

132 $C_6H_{12}O_3$ 53951-42-1

2H-Pyran-3,4-diol, tetrahydro-2-methyl-

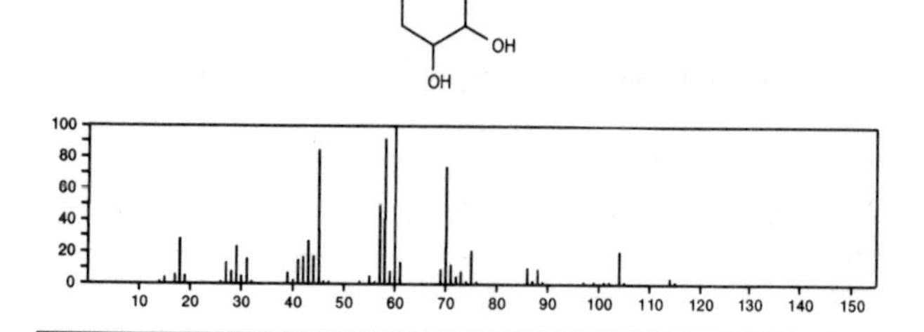

132 $C_6H_{12}O_3$ 53951-43-2

1,3-Dioxolane-2-methanol, 2,4-dimethyl-

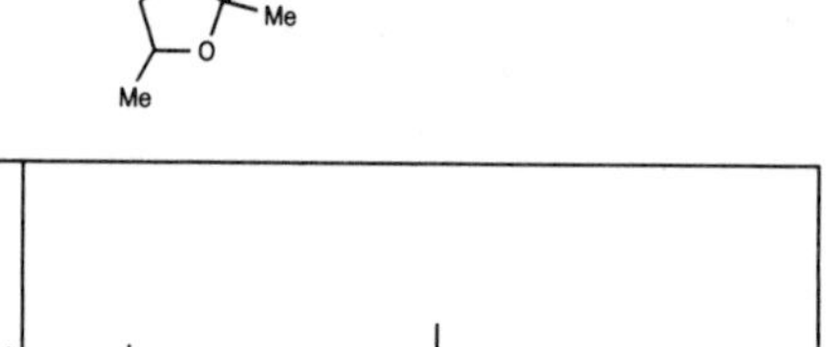

132 $C_6H_{12}O_3$ 53951-44-3

1,3-Dioxolane-4-methanol, 2-ethyl-

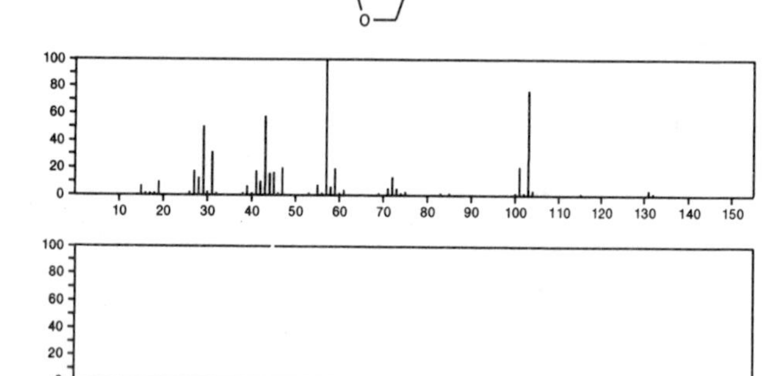

132 $C_6H_{12}O_3$ 56009-31-5

Pentanoic acid, 3-hydroxy-, methyl ester

EtCH(OH)CH2C(O)OMe

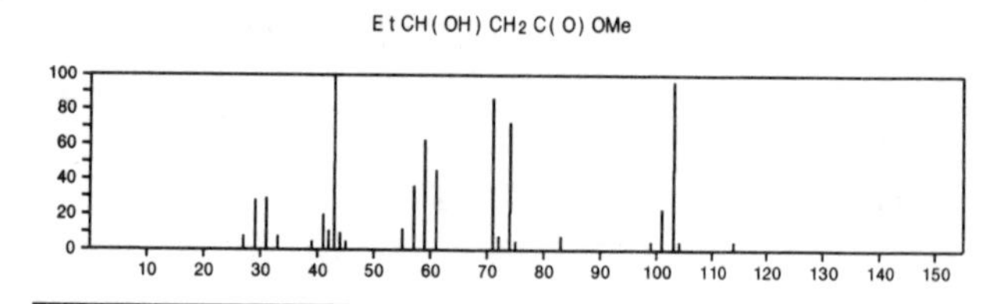

132 $C_6H_{16}OSi$ 1825-63-4

Silane, trimethylpropoxy-

Me3SiOPr

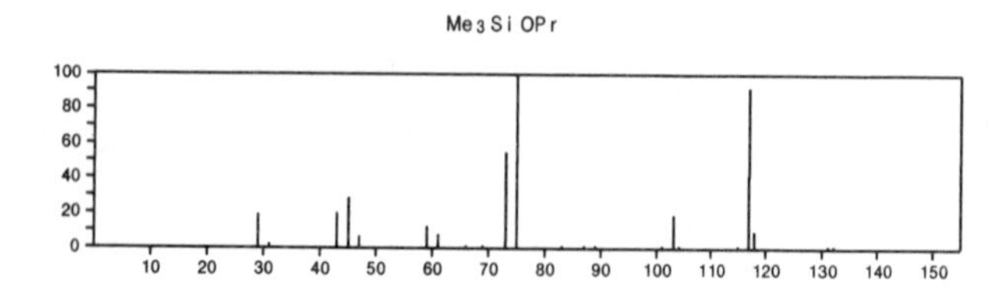

132 $C_6H_{16}OSi$ 1825-64-5

Silane, trimethyl(1-methylethoxy)-

Me3SiOPr-i

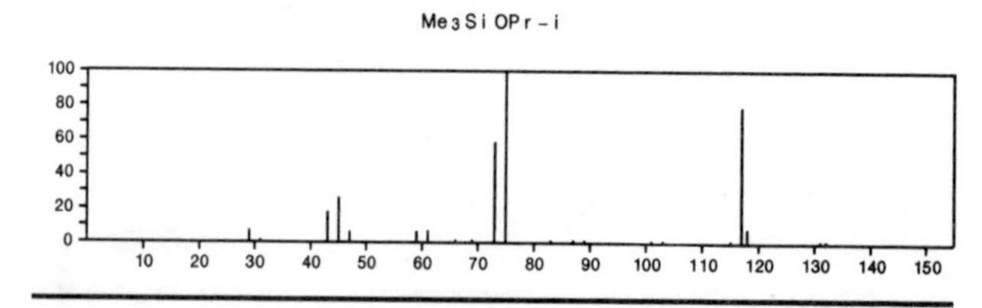

132 $C_6H_{16}OSi$ 18173-63-2

Silane, (2-methoxyethyl)trimethyl-

Me3SiCH2CH2OMe

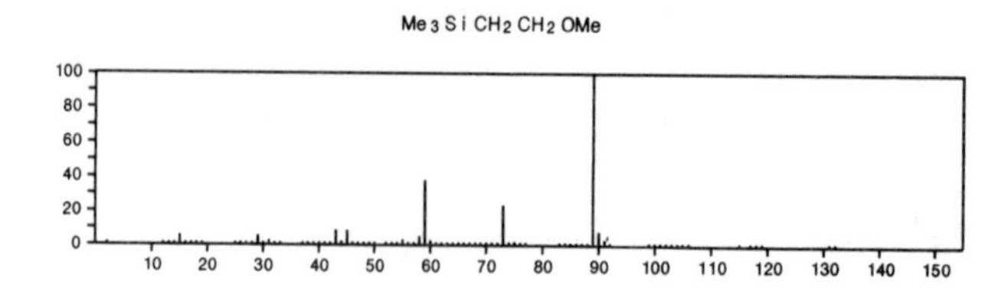

132 $C_7H_{13}Cl$ 55638-53-4

2-Heptene, 1-chloro-, (*Z*)-

ClCH2CH=CH(CH2)3Me

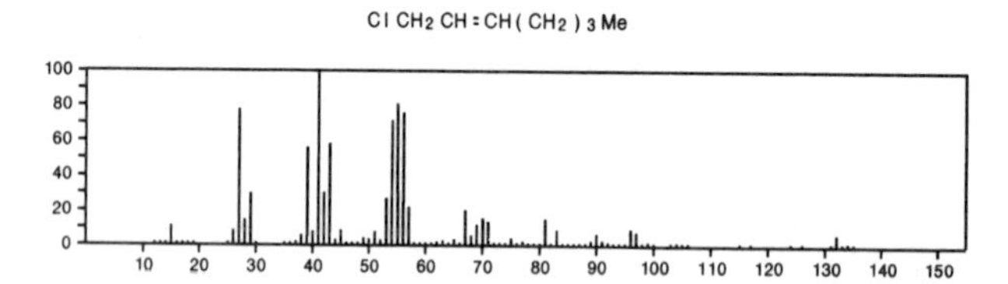

132 $C_7H_{13}Cl$ 55682-98-9

1-Heptene, 3-chloro-

Me(CH2)3CHClCH=CH2

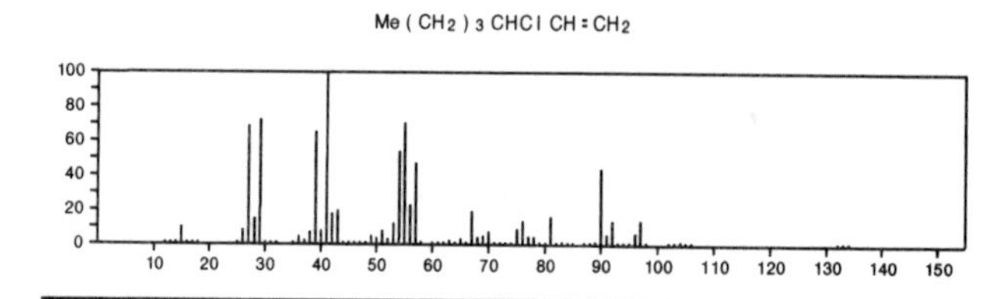

132 $C_7H_{16}O_2$ 78-26-2

1,3-Propanediol, 2-methyl-2-propyl-

HOCH2CPrMeCH2OH

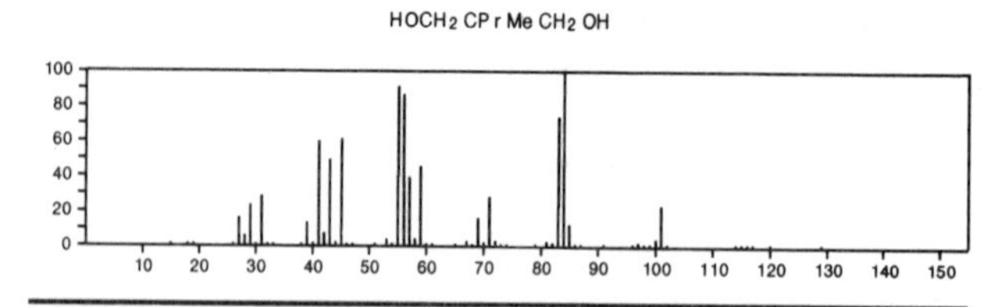

132 $C_7H_{16}O_2$ 111-89-7

Pentane, 1,5-dimethoxy-

MeO(CH2)5OMe

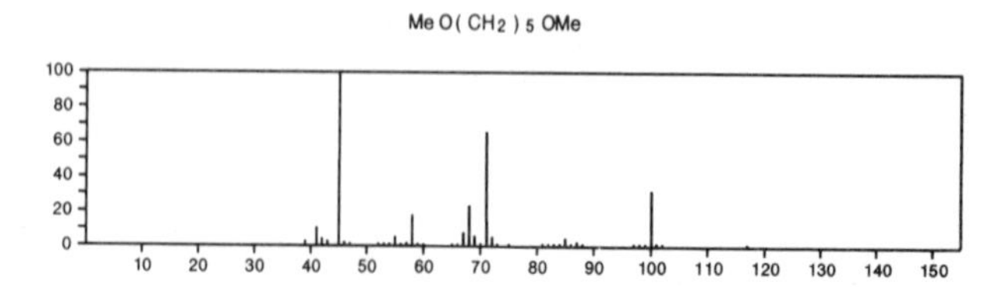

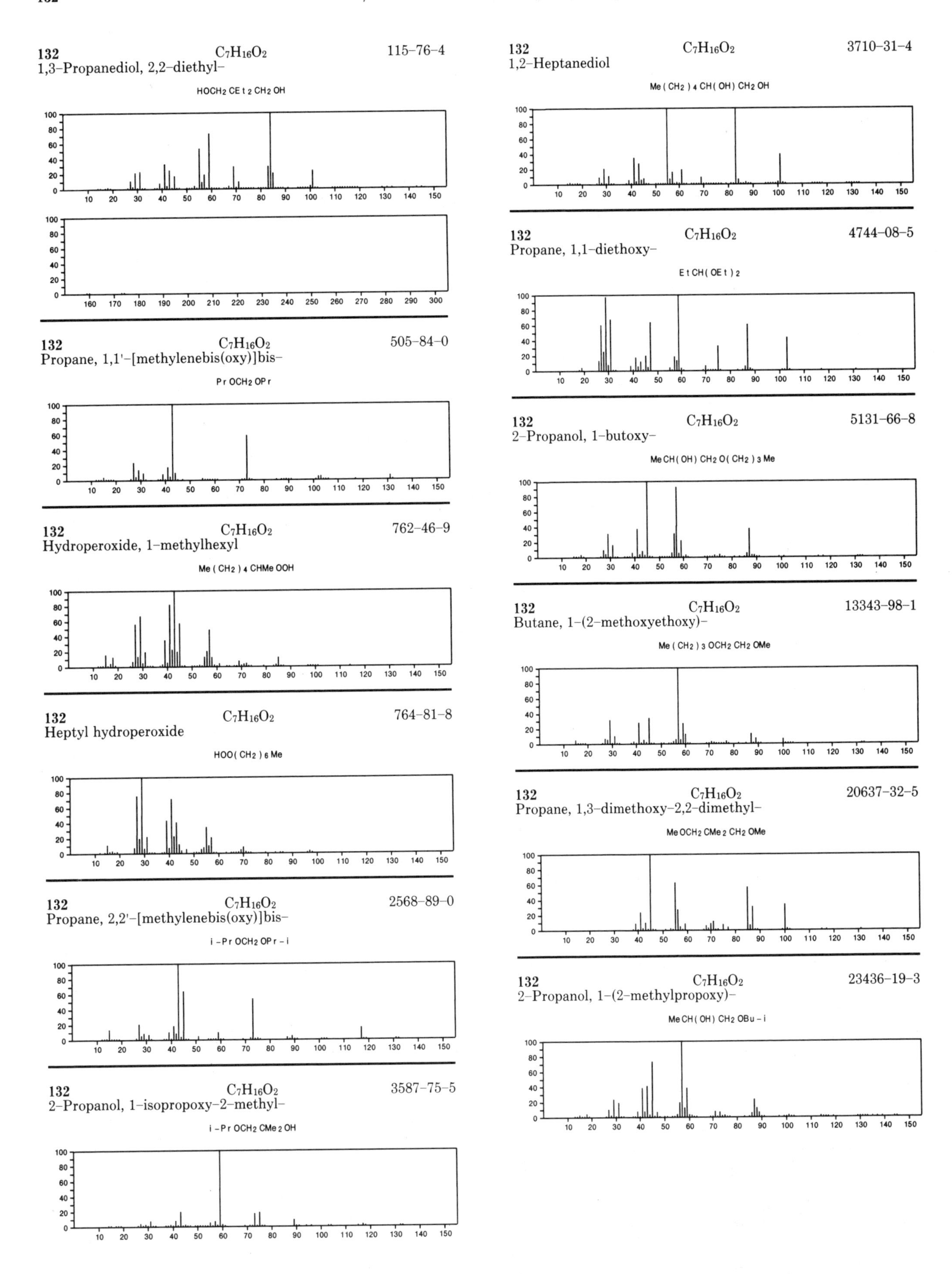
132 C7H16O2 115-76-4
1,3-Propanediol, 2,2-diethyl-
HOCH2 CEt2 CH2 OH
132 C7H16O2 505-84-0
Propane, 1,1'-[methylenebis(oxy)]bis-
Pr OCH2 OPr
132 C7H16O2 762-46-9
Hydroperoxide, 1-methylhexyl
Me(CH2)4 CHMe OOH
132 C7H16O2 764-81-8
Heptyl hydroperoxide
HOO(CH2)6 Me
132 C7H16O2 2568-89-0
Propane, 2,2'-[methylenebis(oxy)]bis-
i-Pr OCH2 OPr-i
132 C7H16O2 3587-75-5
2-Propanol, 1-isopropoxy-2-methyl-
i-Pr OCH2 CMe2 OH
132 C7H16O2 3710-31-4
1,2-Heptanediol
Me(CH2)4 CH(OH) CH2 OH
132 C7H16O2 4744-08-5
Propane, 1,1-diethoxy-
Et CH(OEt)2
132 C7H16O2 5131-66-8
2-Propanol, 1-butoxy-
Me CH(OH) CH2 O(CH2)3 Me
132 C7H16O2 13343-98-1
Butane, 1-(2-methoxyethoxy)-
Me(CH2)3 OCH2 CH2 OMe
132 C7H16O2 20637-32-5
Propane, 1,3-dimethoxy-2,2-dimethyl-
Me OCH2 CMe2 CH2 OMe
132 C7H16O2 23436-19-3
2-Propanol, 1-(2-methylpropoxy)-
Me CH(OH) CH2 OBu-i

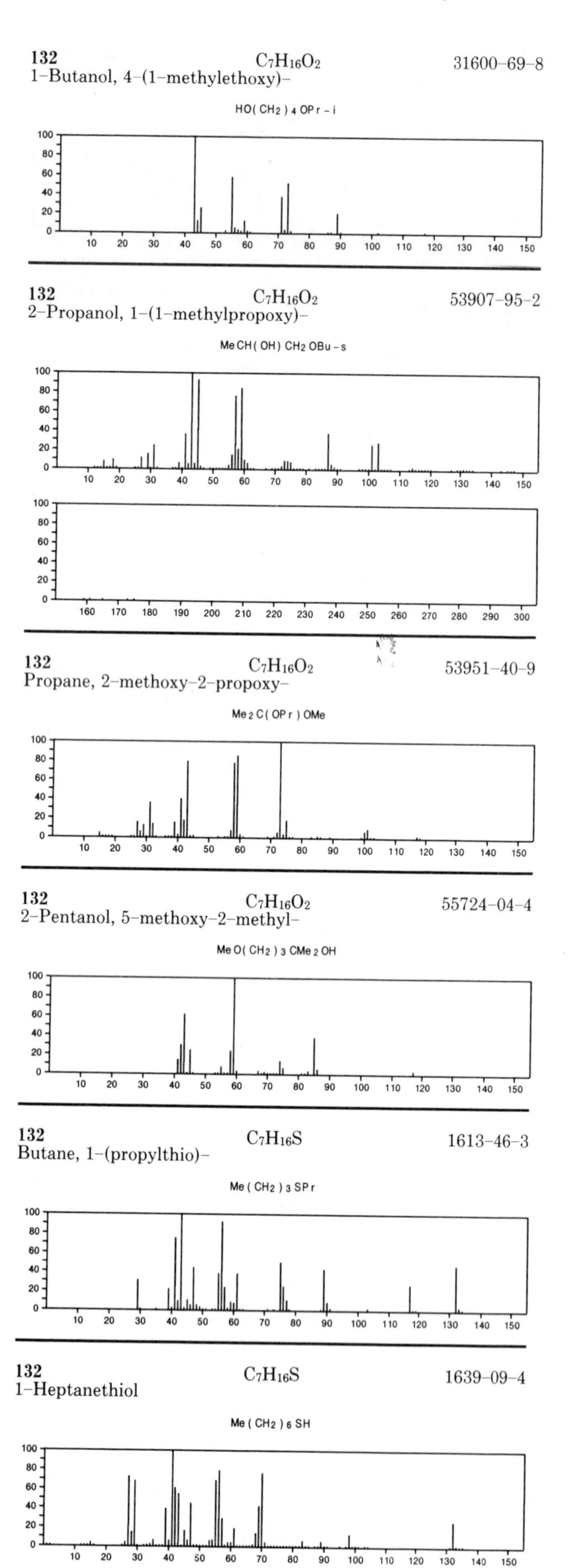
132
$C_7H_{16}O_2$
31600-69-8
1-Butanol, 4-(1-methylethoxy)-
HO(CH2) 4 OPr - i
132
$C_7H_{16}O_2$
53907-95-2
2-Propanol, 1-(1-methylpropoxy)-
Me CH(OH) CH2 OBu - s
132
$C_7H_{16}O_2$
53951-40-9
Propane, 2-methoxy-2-propoxy-
Me 2 C(OPr) OMe
132
$C_7H_{16}O_2$
55724-04-4
2-Pentanol, 5-methoxy-2-methyl-
Me O(CH2) 3 CMe 2 OH
132
$C_7H_{16}S$
1613-46-3
Butane, 1-(propylthio)-
Me (CH2) 3 SPr
132
$C_7H_{16}S$
1639-09-4
1-Heptanethiol
Me (CH2) 6 SH

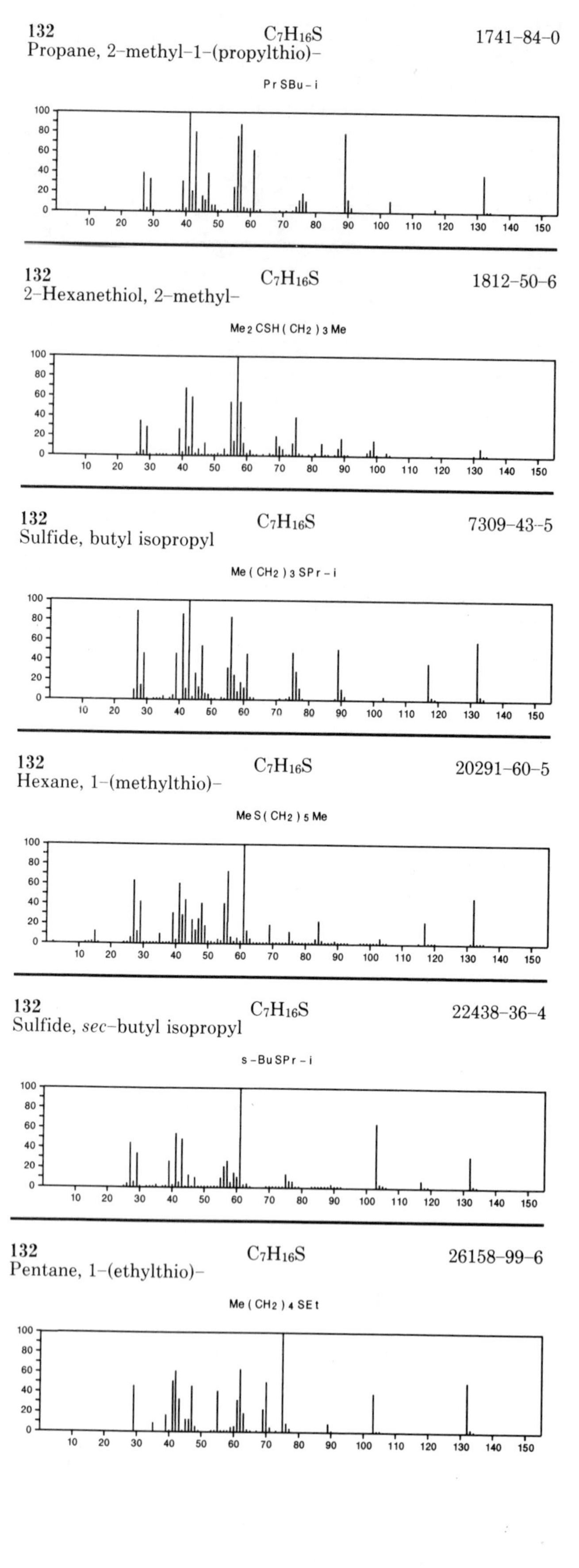
132
$C_7H_{16}S$
1741-84-0
Propane, 2-methyl-1-(propylthio)-
Pr SBu - i
132
$C_7H_{16}S$
1812-50-6
2-Hexanethiol, 2-methyl-
Me 2 CSH (CH2) 3 Me
132
$C_7H_{16}S$
7309-43-5
Sulfide, butyl isopropyl
Me (CH2) 3 SPr - i
132
$C_7H_{16}S$
20291-60-5
Hexane, 1-(methylthio)-
Me S(CH2) 5 Me
132
$C_7H_{16}S$
22438-36-4
Sulfide, *sec*-butyl isopropyl
s -Bu SPr - i
132
$C_7H_{16}S$
26158-99-6
Pentane, 1-(ethylthio)-
Me (CH2) 4 SEt

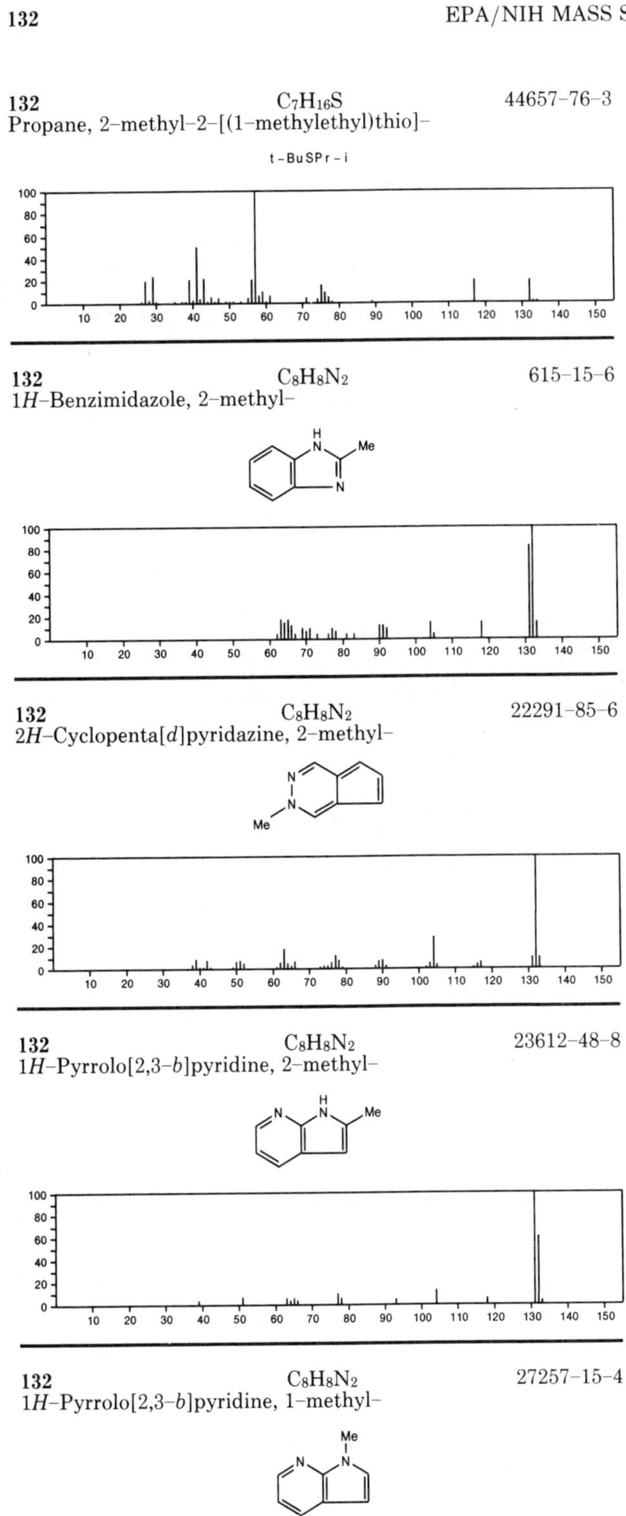

132 $C_7H_{16}S$ 44657-76-3
Propane, 2-methyl-2-[(1-methylethyl)thio]-
t-BuSPr-i

132 $C_8H_8N_2$ 615-15-6
1*H*-Benzimidazole, 2-methyl-

132 $C_8H_8N_2$ 22291-85-6
2*H*-Cyclopenta[*d*]pyridazine, 2-methyl-

132 $C_8H_8N_2$ 23612-48-8
1*H*-Pyrrolo[2,3-*b*]pyridine, 2-methyl-

132 $C_8H_8N_2$ 27257-15-4
1*H*-Pyrrolo[2,3-*b*]pyridine, 1-methyl-

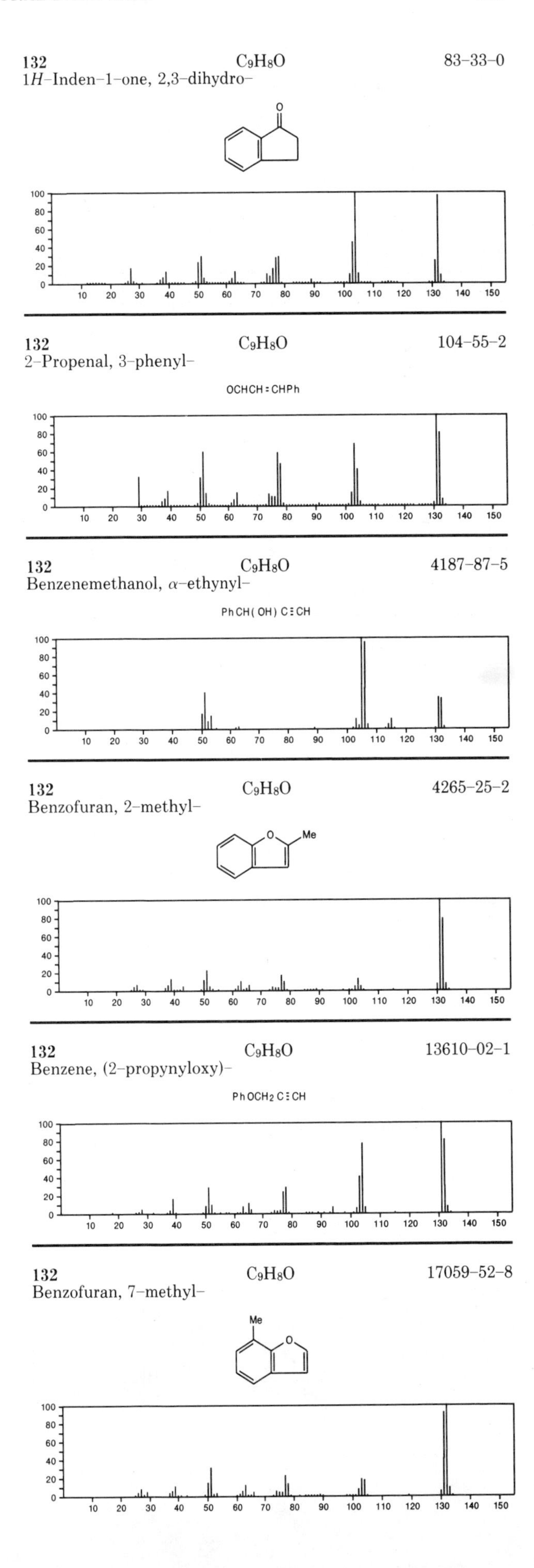

132 C_9H_8O 83-33-0
1*H*-Inden-1-one, 2,3-dihydro-

132 C_9H_8O 104-55-2
2-Propenal, 3-phenyl-
OCHCH=CHPh

132 C_9H_8O 4187-87-5
Benzenemethanol, α-ethynyl-
PhCH(OH)C≡CH

132 C_9H_8O 4265-25-2
Benzofuran, 2-methyl-

132 C_9H_8O 13610-02-1
Benzene, (2-propynyloxy)-
PhOCH₂C≡CH

132 C_9H_8O 17059-52-8
Benzofuran, 7-methyl-

132 C_9H_8O 30844-12-3
1,3,5,7-Cyclooctatetraene-1-carboxaldehyde

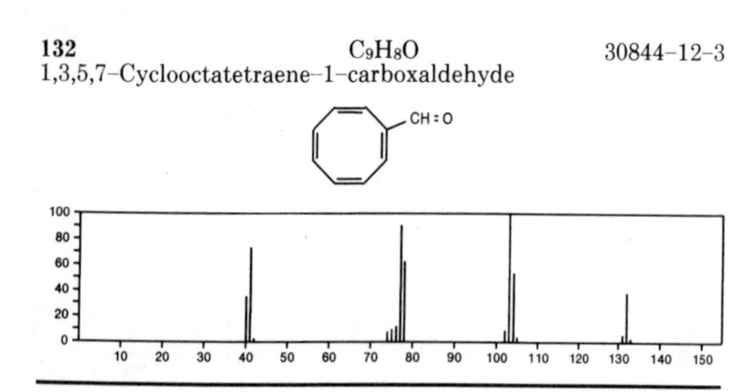

132 C_9H_8O 43145-54-6
Benzaldehyde, ethenyl-

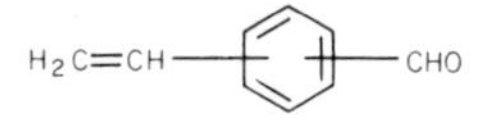

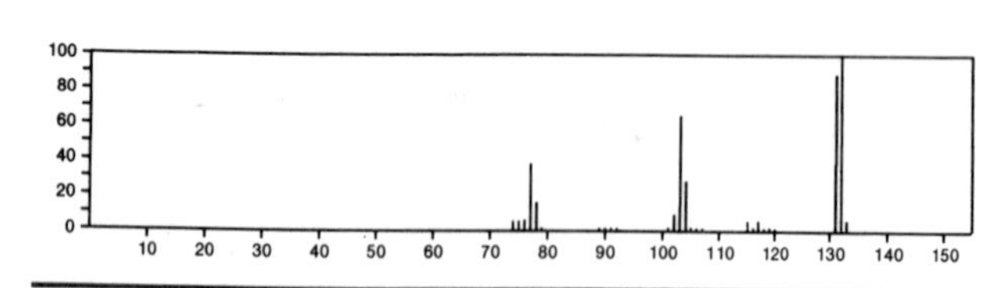

132 $C_{10}H_{12}$ 77-73-6
4,7-Methano-1*H*-indene, 3a,4,7,7a-tetrahydro-

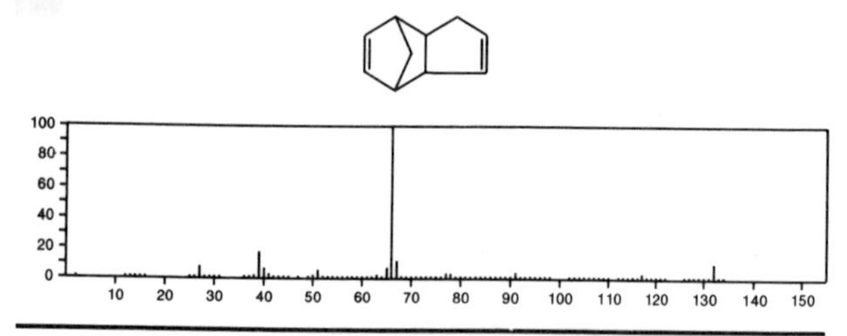

132 $C_{10}H_{12}$ 119-64-2
Naphthalene, 1,2,3,4-tetrahydro-

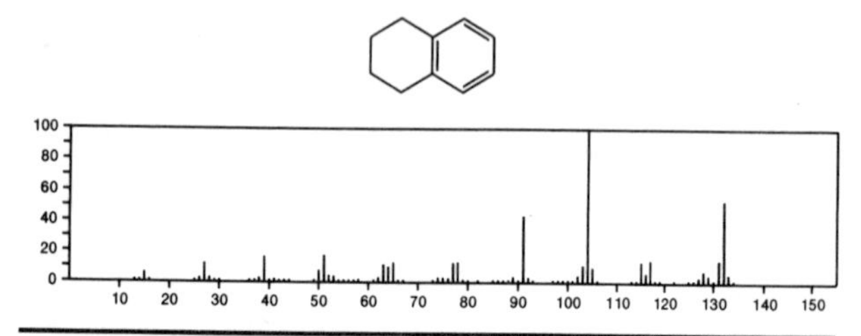

132 $C_{10}H_{12}$ 767-58-8
1*H*-Indene, 2,3-dihydro-1-methyl-

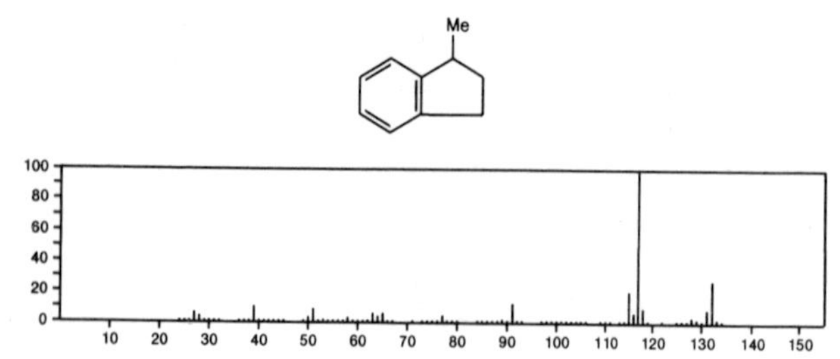

132 $C_{10}H_{12}$ 768-49-0
Benzene, (2-methyl-1-propenyl)-

Me2C=CHPh

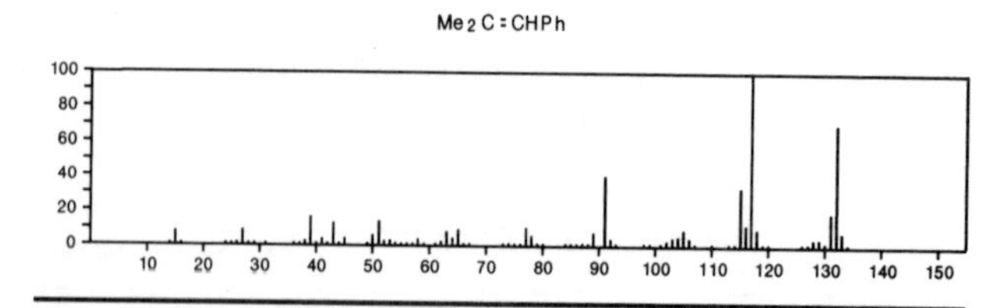

132 $C_{10}H_{12}$ 824-22-6
1*H*-Indene, 2,3-dihydro-4-methyl-

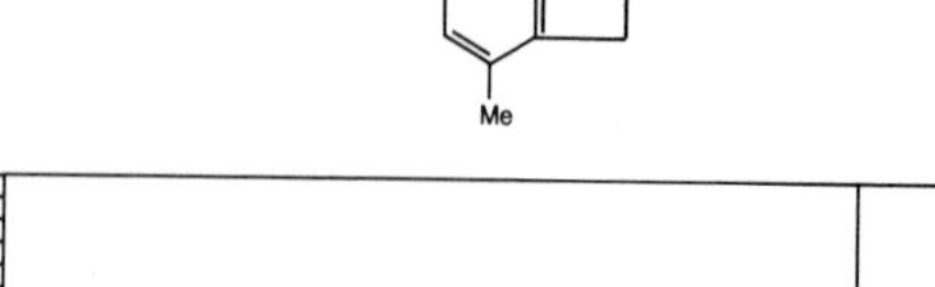

132 $C_{10}H_{12}$ 824-63-5
1*H*-Indene, 2,3-dihydro-2-methyl-

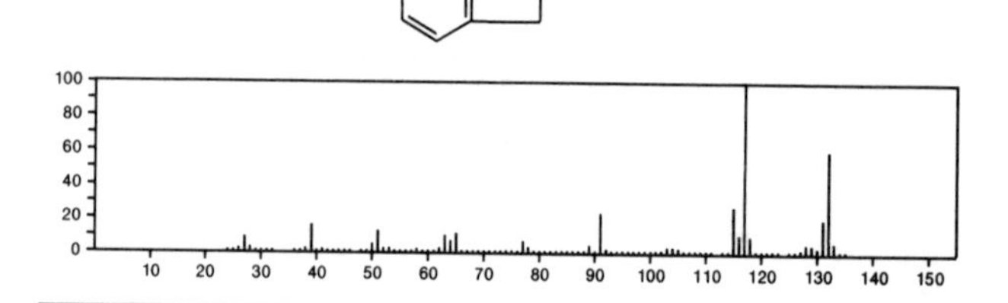

132 $C_{10}H_{12}$ 874-35-1
1*H*-Indene, 2,3-dihydro-5-methyl-

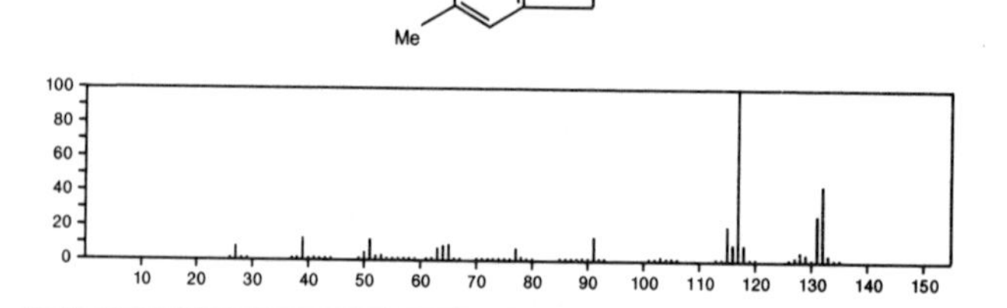

132 $C_{10}H_{12}$ 1195-32-0
Benzene, 1-methyl-4-(1-methylethenyl)-

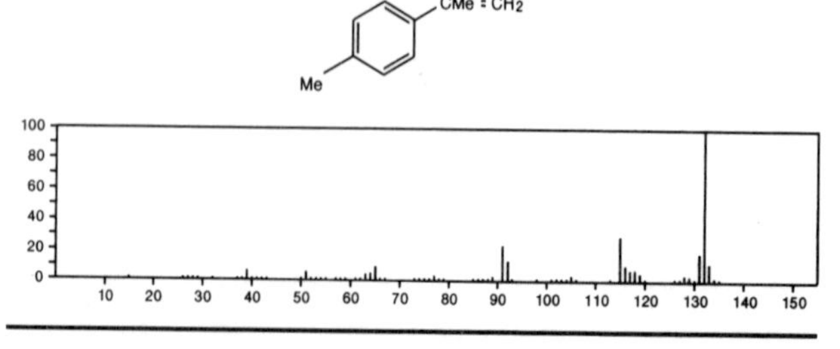

132 $C_{10}H_{12}$ 1560-06-1
Benzene, 2-butenyl-

PhCH2CH=CHMe

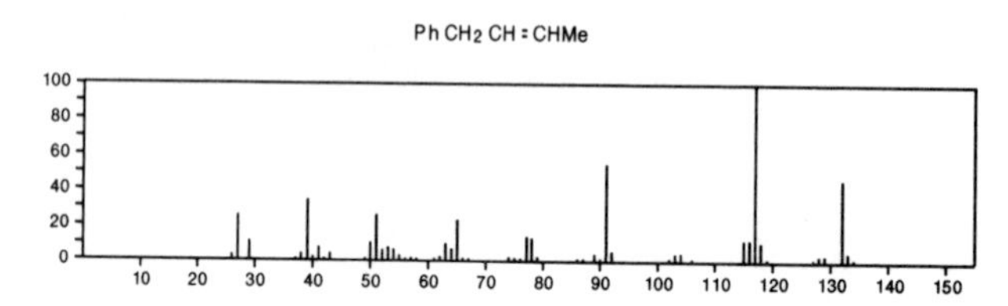

132 $C_{10}H_{12}$ 1587-04-8
Benzene, 1-methyl-2-(2-propenyl)-

132 $C_{10}H_{12}$ 2039-90-9
Benzene, 2-ethenyl-1,3-dimethyl-

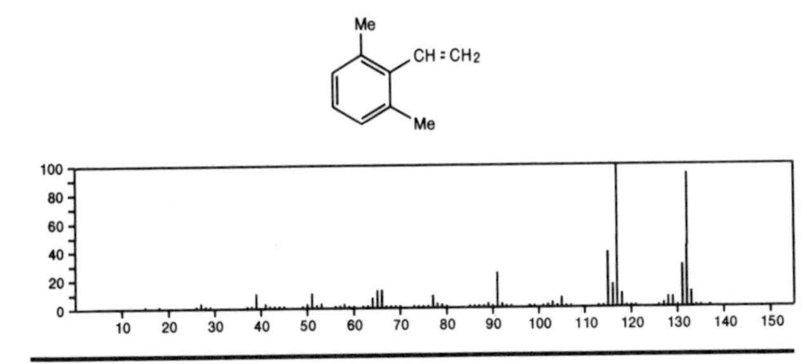

132 $C_{10}H_{12}$ 3290-53-7
Benzene, (2-methyl-2-propenyl)-

$H_2C=CMeCH_2Ph$

132 $C_{10}H_{12}$ 3454-07-7
Benzene, 1-ethenyl-4-ethyl-

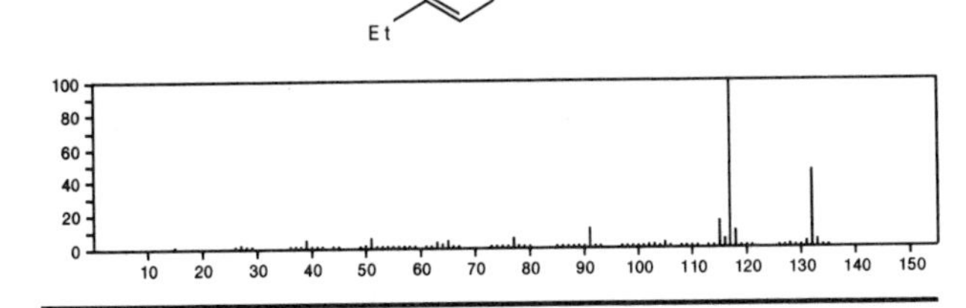

132 $C_{10}H_{12}$ 4392-30-7
Benzene, cyclobutyl-

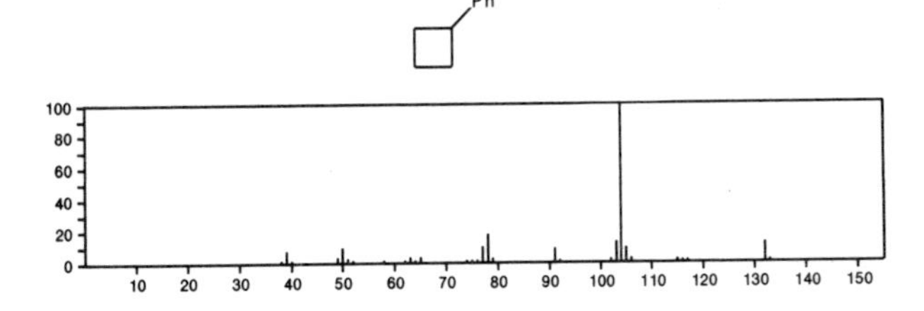

132 $C_{10}H_{12}$ 5379-20-4
Benzene, 1-ethenyl-3,5-dimethyl-

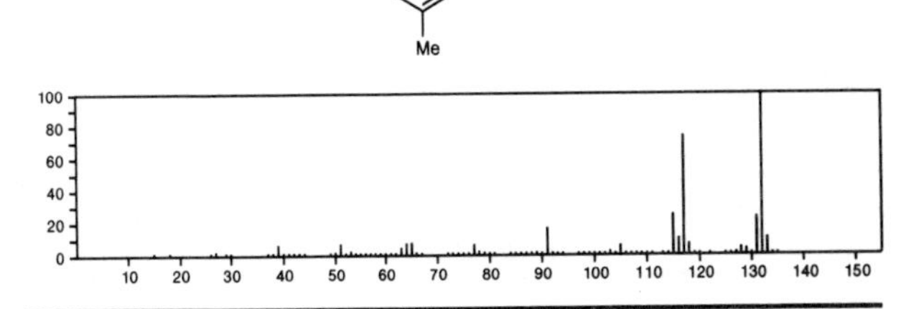

132 $C_{10}H_{12}$ 7525-62-4
Benzene, 1-ethenyl-3-ethyl-

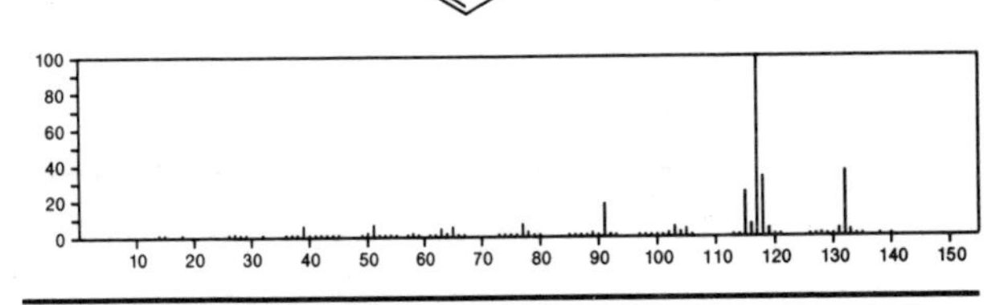

132 $C_{10}H_{12}$ 26444-18-8
Benzene, methyl(1-methylethenyl)-

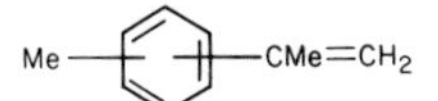

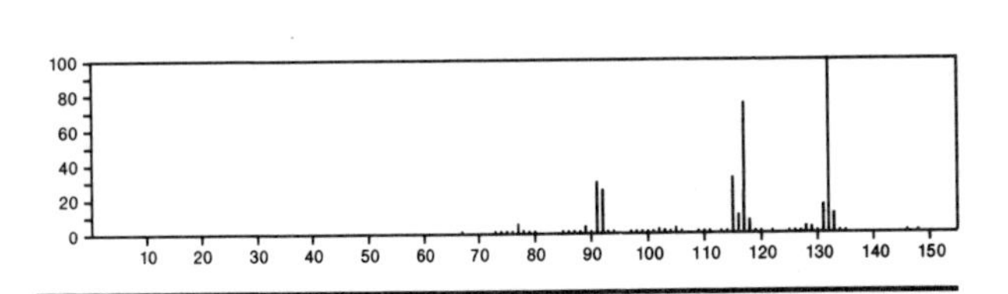

132 $C_{10}H_{12}$ 27576-03-0
Benzene, ethenyl-, dimethyl deriv.

$PhCH=CH_2$ + 2Me

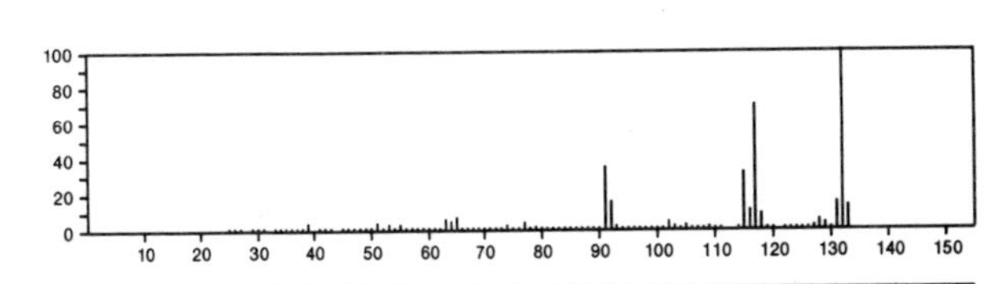

132 $C_{10}H_{12}$ 27831-13-6
Benzene, 4-ethenyl-1,2-dimethyl-

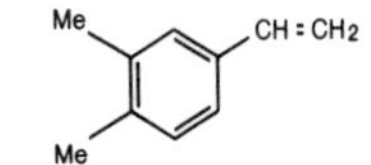

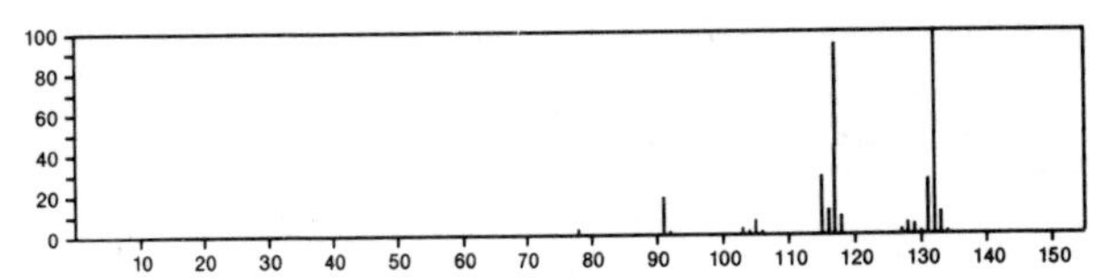

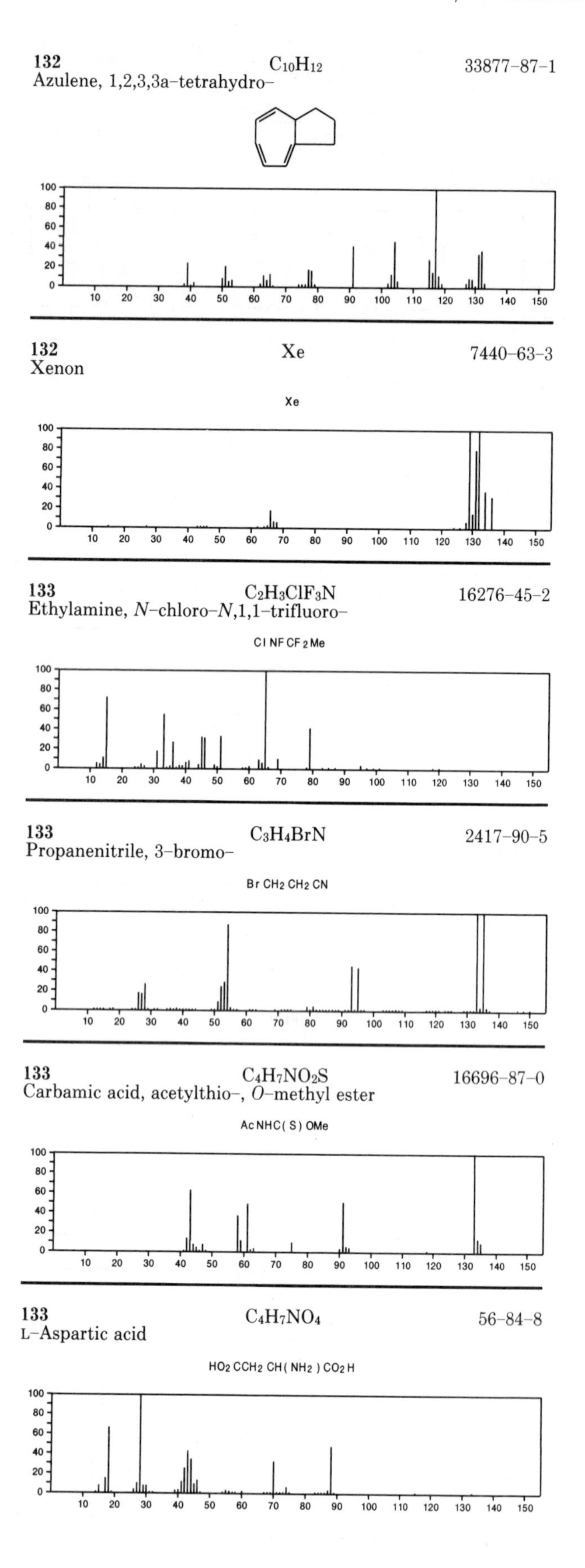

132 $C_{10}H_{12}$ 33877-87-1
Azulene, 1,2,3,3a-tetrahydro-

132 Xe 7440-63-3
Xenon

Xe

133 $C_2H_3ClF_3N$ 16276-45-2
Ethylamine, *N*-chloro-*N*,1,1-trifluoro-

Cl NF CF$_2$Me

133 C_3H_4BrN 2417-90-5
Propanenitrile, 3-bromo-

Br CH$_2$ CH$_2$ CN

133 $C_4H_7NO_2S$ 16696-87-0
Carbamic acid, acetylthio-, *O*-methyl ester

AcNHC(S)OMe

133 $C_4H_7NO_4$ 56-84-8
L-Aspartic acid

HO$_2$CCH$_2$ CH(NH$_2$)CO$_2$H

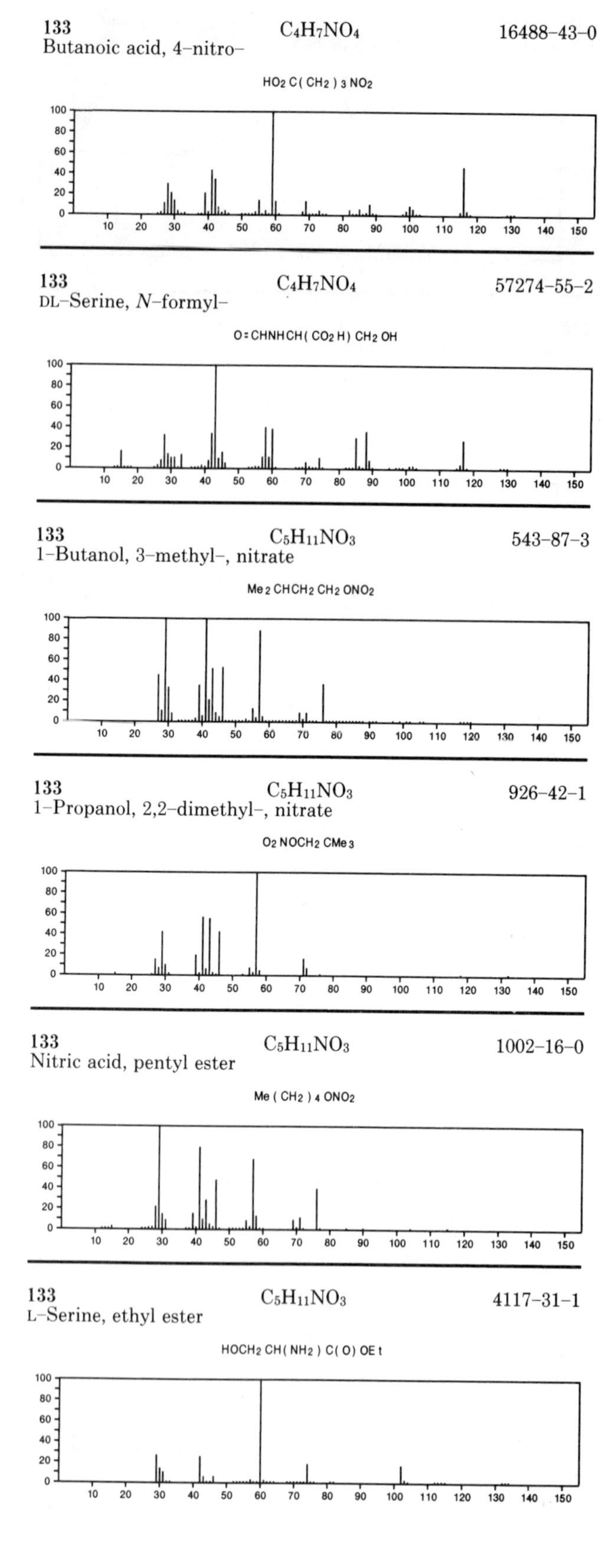

133 $C_4H_7NO_4$ 16488-43-0
Butanoic acid, 4-nitro-

HO$_2$C(CH$_2$)$_3$NO$_2$

133 $C_4H_7NO_4$ 57274-55-2
DL-Serine, *N*-formyl-

O=CHNHCH(CO$_2$H)CH$_2$OH

133 $C_5H_{11}NO_3$ 543-87-3
1-Butanol, 3-methyl-, nitrate

Me$_2$CHCH$_2$CH$_2$ONO$_2$

133 $C_5H_{11}NO_3$ 926-42-1
1-Propanol, 2,2-dimethyl-, nitrate

O$_2$NOCH$_2$CMe$_3$

133 $C_5H_{11}NO_3$ 1002-16-0
Nitric acid, pentyl ester

Me(CH$_2$)$_4$ONO$_2$

133 $C_5H_{11}NO_3$ 4117-31-1
L-Serine, ethyl ester

HOCH$_2$CH(NH$_2$)C(O)OEt

133 $C_6H_{15}NO_2$ 16684–49–4
Ethanol, 2–(diethylamino)–, *N*–oxide

133 $C_7H_7N_3$ 768–19–4
[1,2,4]Triazolo[1,5–*a*]pyridine, 2–methyl–

133 $C_7H_7N_3$ 1004–65–5
1,2,4–Triazolo[4,3–*a*]pyridine, 3–methyl–

133 $C_7H_7N_3$ 2101–86–2
Benzene, 1–azido–4–methyl–

133 $C_7H_7N_3$ 4113–72–8
Benzene, 1–azido–3–methyl–

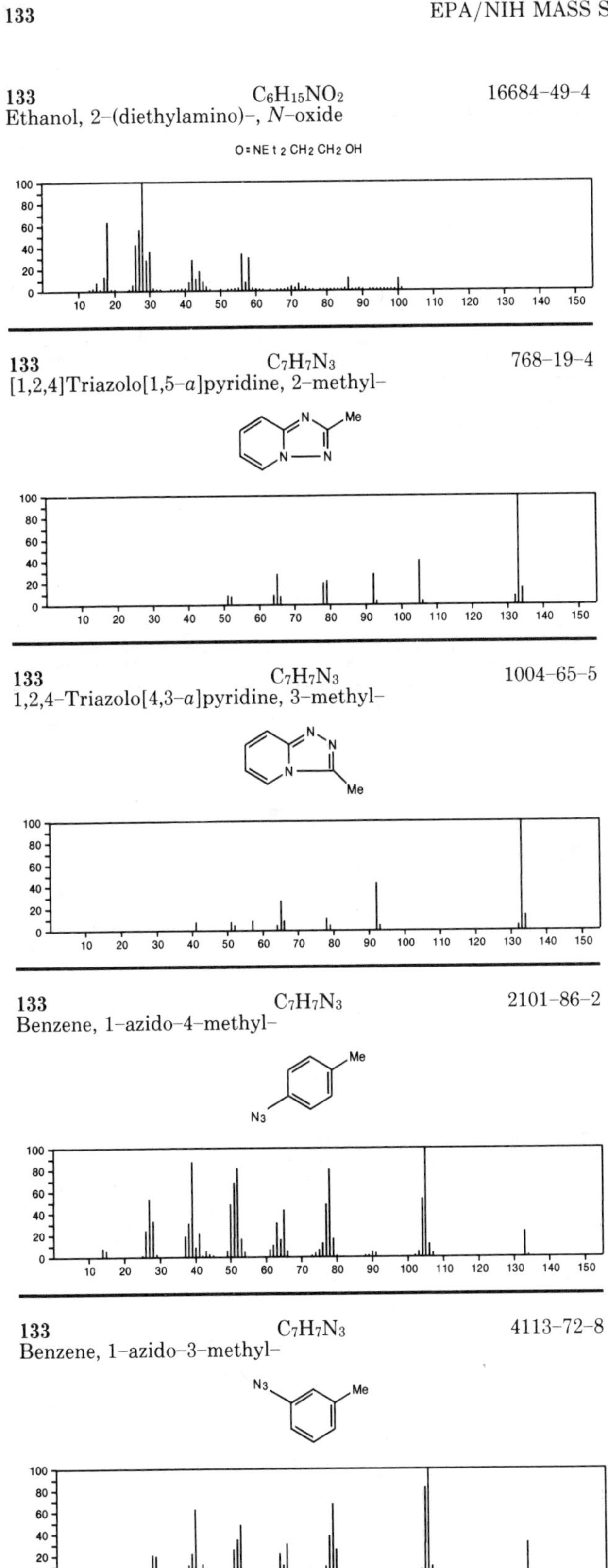

133 $C_7H_7N_3$ 4919–09–9
s–Triazolo[4,3–*a*]pyridine, 6–methyl–

133 $C_7H_7N_3$ 4919–10–2
s–Triazolo[4,3–*a*]pyridine, 7–methyl–

133 $C_7H_7N_3$ 13351–73–0
1*H*–Benzotriazole, 1–methyl–

133 $C_7H_7N_3$ 16584–00–2
2*H*–Benzotriazole, 2–methyl–

133 $C_7H_7N_3$ 31656–92–5
Benzene, 1–azido–2–methyl–

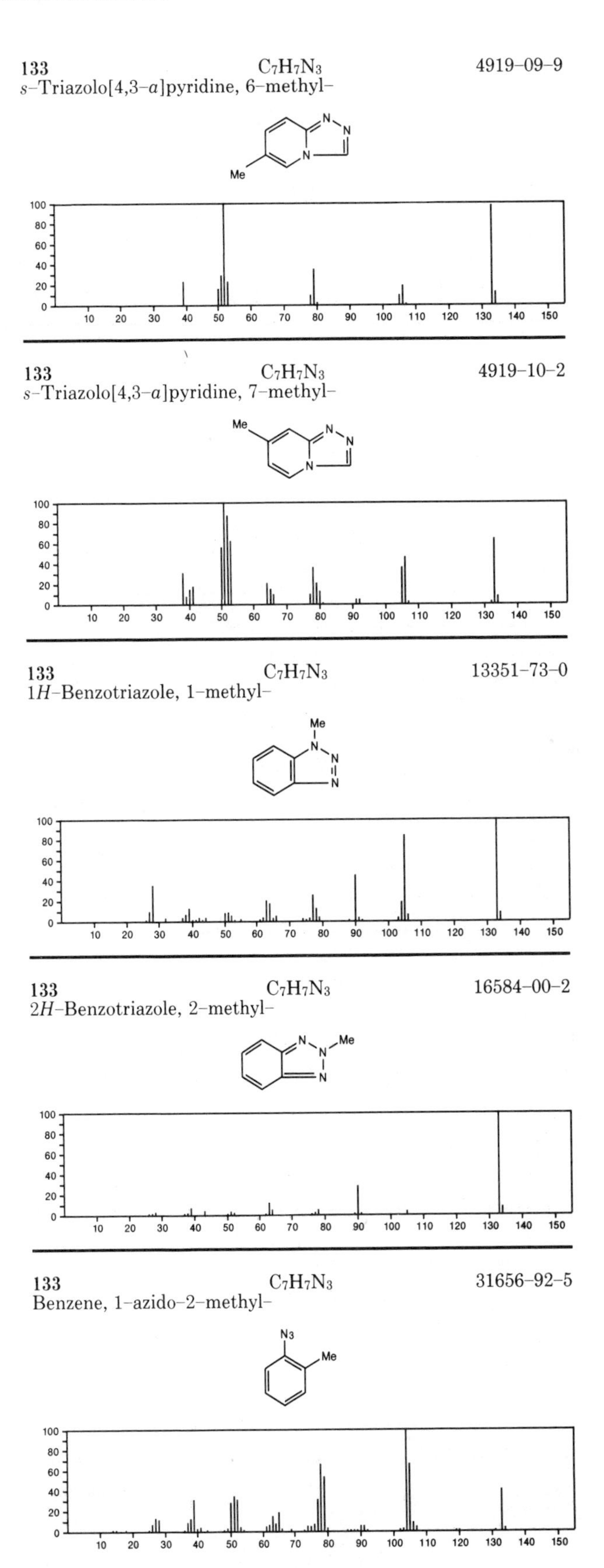

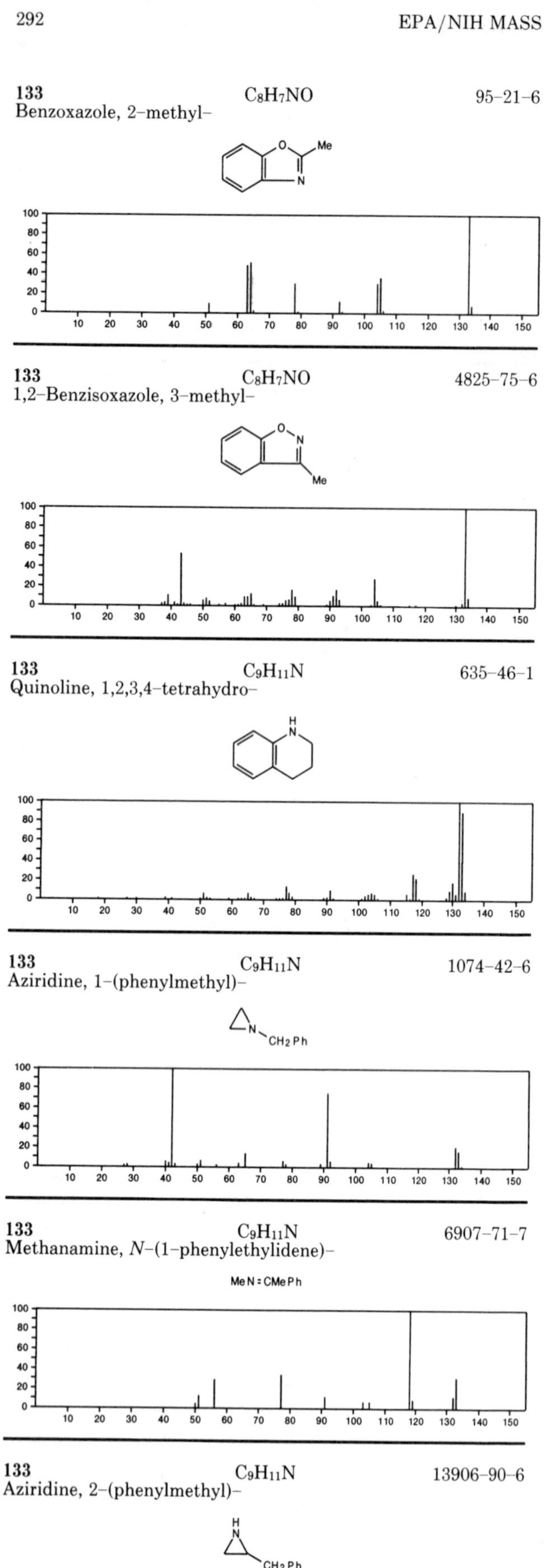

133 C_8H_7NO 95–21–6
Benzoxazole, 2–methyl–

133 C_8H_7NO 4825–75–6
1,2–Benzisoxazole, 3–methyl–

133 $C_9H_{11}N$ 635–46–1
Quinoline, 1,2,3,4–tetrahydro–

133 $C_9H_{11}N$ 1074–42–6
Aziridine, 1–(phenylmethyl)–

133 $C_9H_{11}N$ 6907–71–7
Methanamine, *N*–(1–phenylethylidene)–

MeN=CMePh

133 $C_9H_{11}N$ 13906–90–6
Aziridine, 2–(phenylmethyl)–

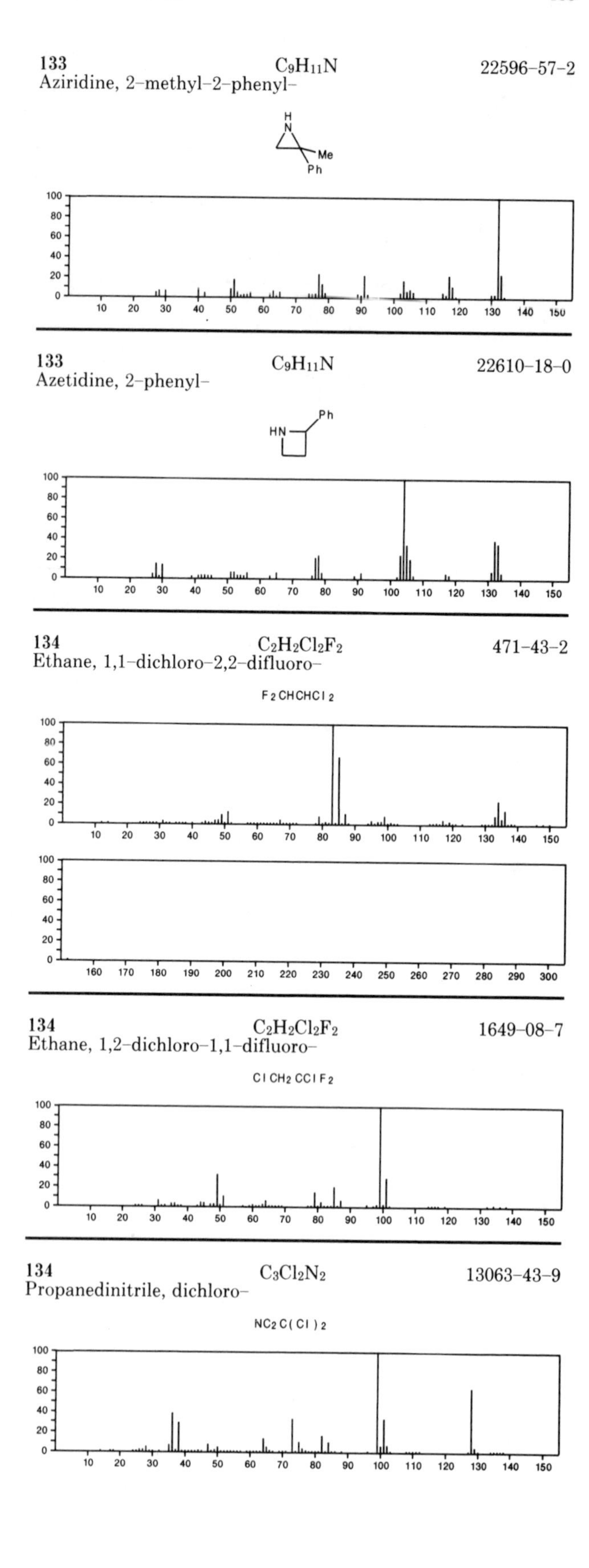

133 $C_9H_{11}N$ 22596–57–2
Aziridine, 2–methyl–2–phenyl–

133 $C_9H_{11}N$ 22610–18–0
Azetidine, 2–phenyl–

134 $C_2H_2Cl_2F_2$ 471–43–2
Ethane, 1,1–dichloro–2,2–difluoro–

$F_2CHCHCl_2$

134 $C_2H_2Cl_2F_2$ 1649–08–7
Ethane, 1,2–dichloro–1,1–difluoro–

$ClCH_2CClF_2$

134 $C_3Cl_2N_2$ 13063–43–9
Propanedinitrile, dichloro–

$NC_2C(Cl)_2$

134 $C_3H_3F_5$ 679-86-7
Propane, 1,1,2,2,3-pentafluoro-

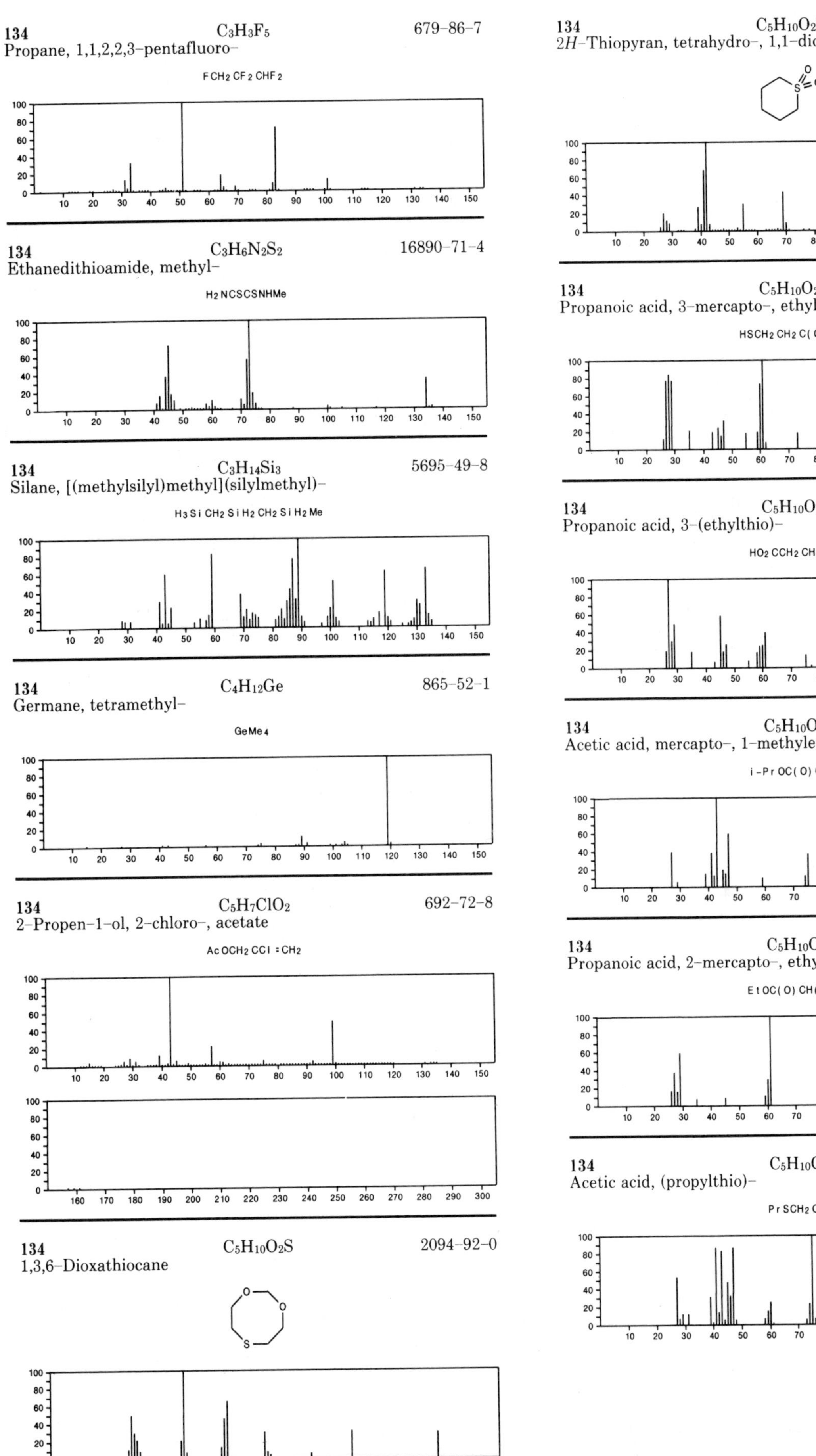

134 $C_3H_6N_2S_2$ 16890-71-4
Ethanedithioamide, methyl-

134 $C_3H_{14}Si_3$ 5695-49-8
Silane, [(methylsilyl)methyl](silylmethyl)-

134 $C_4H_{12}Ge$ 865-52-1
Germane, tetramethyl-

134 $C_5H_7ClO_2$ 692-72-8
2-Propen-1-ol, 2-chloro-, acetate

134 $C_5H_{10}O_2S$ 2094-92-0
1,3,6-Dioxathiocane

134 $C_5H_{10}O_2S$ 4988-33-4
2*H*-Thiopyran, tetrahydro-, 1,1-dioxide

134 $C_5H_{10}O_2S$ 5466-06-8
Propanoic acid, 3-mercapto-, ethyl ester

HSCH2CH2C(O)OEt

134 $C_5H_{10}O_2S$ 7244-82-8
Propanoic acid, 3-(ethylthio)-

HO2CCH2CH2SEt

134 $C_5H_{10}O_2S$ 7383-61-1
Acetic acid, mercapto-, 1-methylethyl ester

i-PrOC(O)CH2SH

134 $C_5H_{10}O_2S$ 19788-49-9
Propanoic acid, 2-mercapto-, ethyl ester

EtOC(O)CH(SH)Me

134 $C_5H_{10}O_2S$ 20600-60-6
Acetic acid, (propylthio)-

PrSCH2CO2H

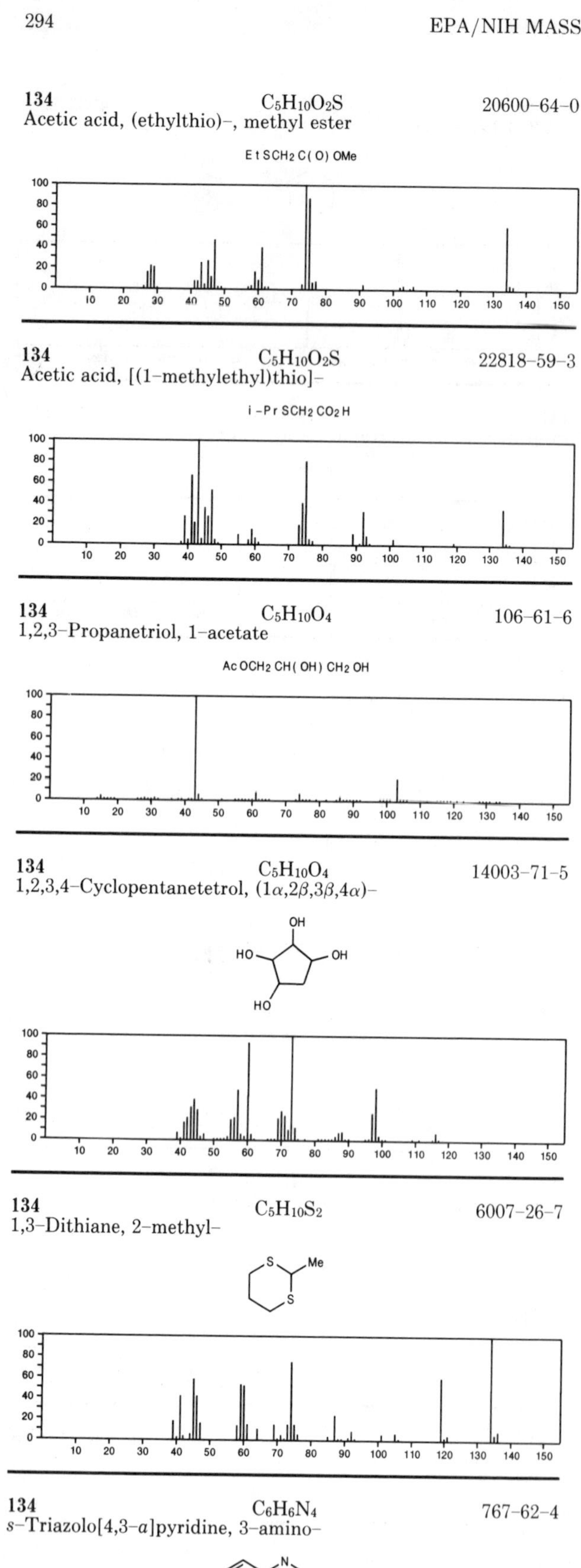

134 $C_5H_{10}O_2S$ 20600-64-0
Acetic acid, (ethylthio)-, methyl ester

EtSCH2C(O)OMe

134 $C_5H_{10}O_2S$ 22818-59-3
Acetic acid, [(1-methylethyl)thio]-

i-PrSCH2CO2H

134 $C_5H_{10}O_4$ 106-61-6
1,2,3-Propanetriol, 1-acetate

AcOCH2CH(OH)CH2OH

134 $C_5H_{10}O_4$ 14003-71-5
1,2,3,4-Cyclopentanetetrol, $(1\alpha,2\beta,3\beta,4\alpha)$-

134 $C_5H_{10}S_2$ 6007-26-7
1,3-Dithiane, 2-methyl-

134 $C_6H_6N_4$ 767-62-4
s-Triazolo[4,3-*a*]pyridine, 3-amino-

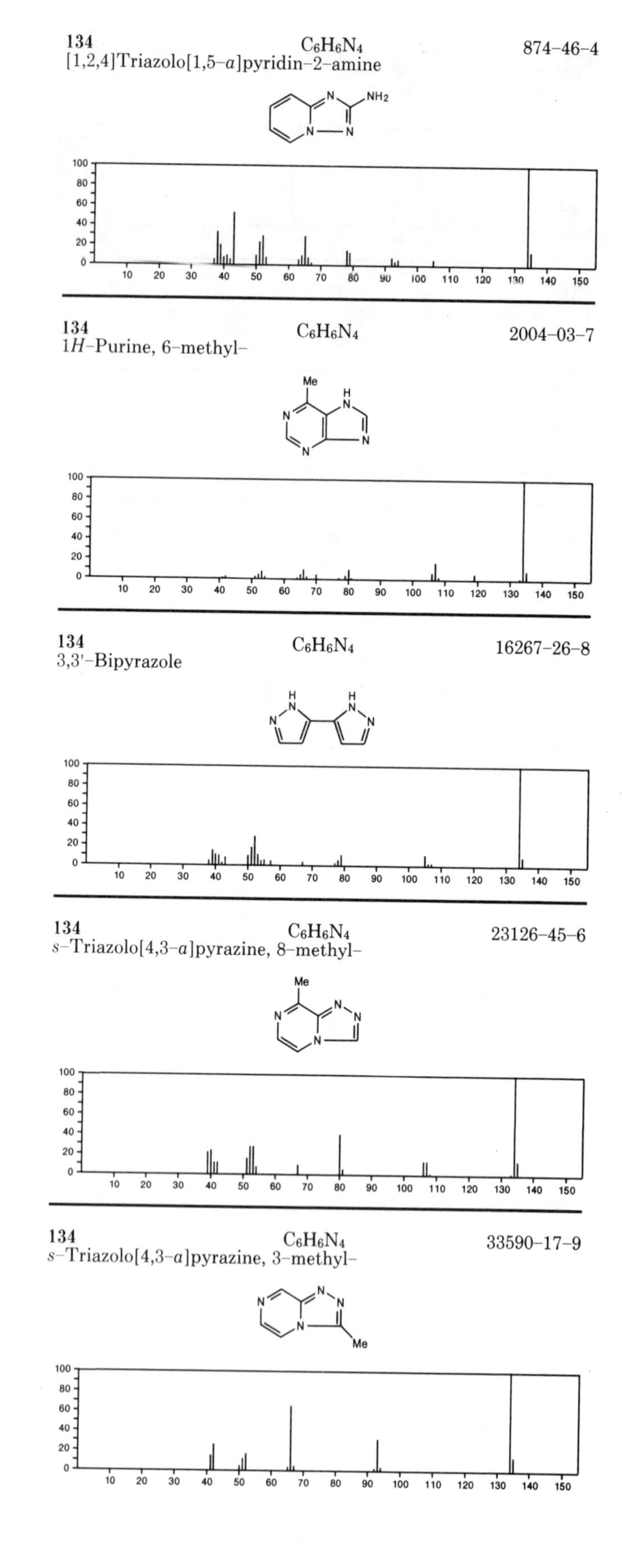

134 $C_6H_6N_4$ 874-46-4
[1,2,4]Triazolo[1,5-*a*]pyridin-2-amine

134 $C_6H_6N_4$ 2004-03-7
1*H*-Purine, 6-methyl-

134 $C_6H_6N_4$ 16267-26-8
3,3'-Bipyrazole

134 $C_6H_6N_4$ 23126-45-6
s-Triazolo[4,3-*a*]pyrazine, 8-methyl-

134 $C_6H_6N_4$ 33590-17-9
s-Triazolo[4,3-*a*]pyrazine, 3-methyl-

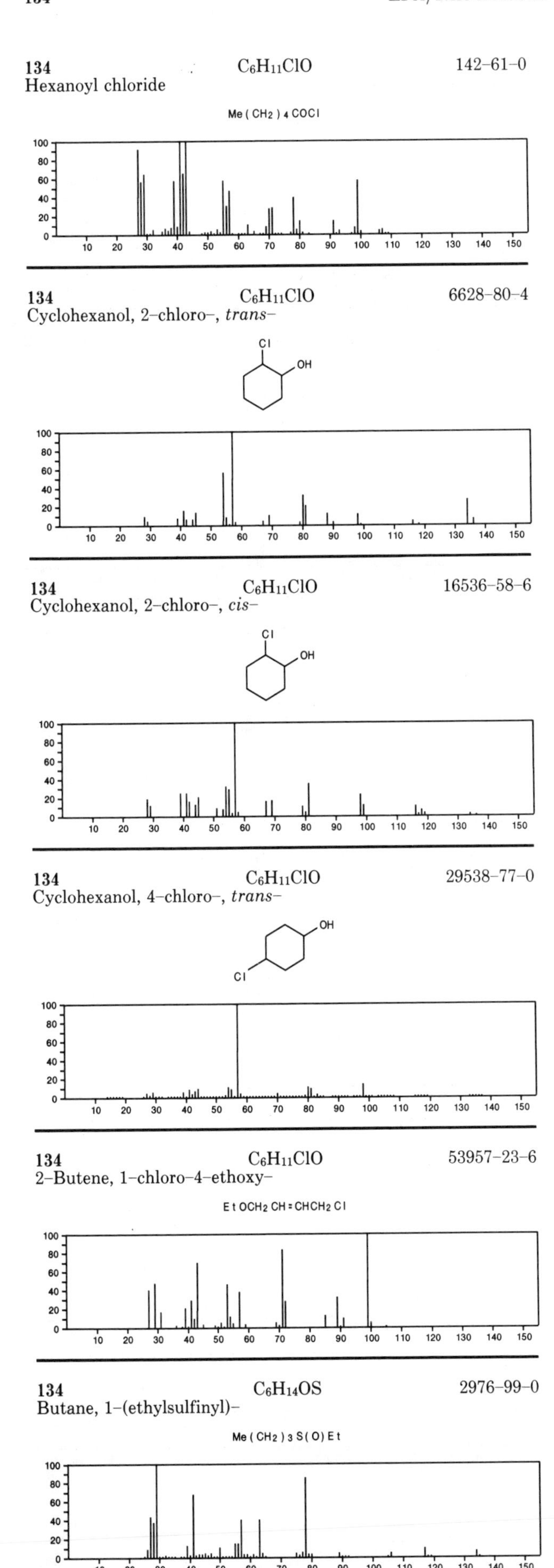

134 $C_6H_{11}ClO$ 142-61-0
Hexanoyl chloride
Me(CH2)4COCl

134 $C_6H_{11}ClO$ 6628-80-4
Cyclohexanol, 2-chloro-, *trans-*

134 $C_6H_{11}ClO$ 16536-58-6
Cyclohexanol, 2-chloro-, *cis-*

134 $C_6H_{11}ClO$ 29538-77-0
Cyclohexanol, 4-chloro-, *trans-*

134 $C_6H_{11}ClO$ 53957-23-6
2-Butene, 1-chloro-4-ethoxy-
EtOCH2CH=CHCH2Cl

134 $C_6H_{14}OS$ 2976-99-0
Butane, 1-(ethylsulfinyl)-
Me(CH2)3S(O)Et

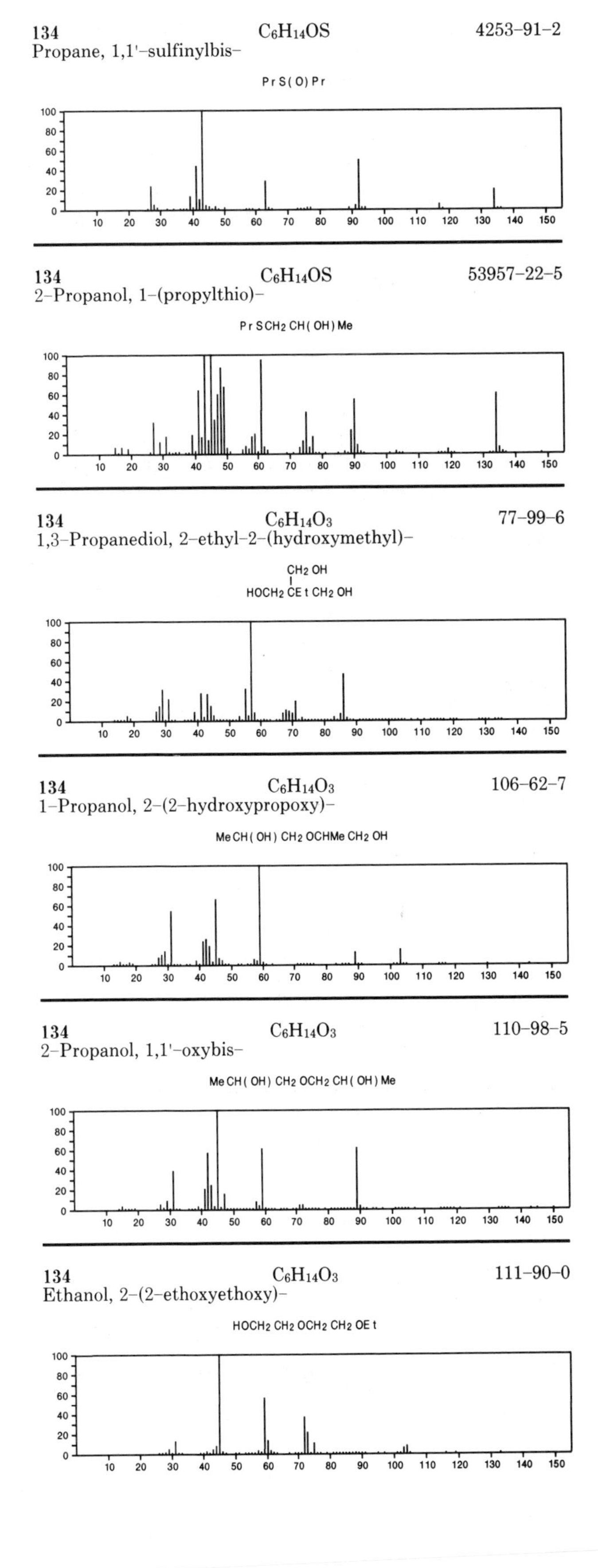

134 $C_6H_{14}OS$ 4253-91-2
Propane, 1,1'-sulfinylbis-
PrS(O)Pr

134 $C_6H_{14}OS$ 53957-22-5
2-Propanol, 1-(propylthio)-
PrSCH2CH(OH)Me

134 $C_6H_{14}O_3$ 77-99-6
1,3-Propanediol, 2-ethyl-2-(hydroxymethyl)-
HOCH2CEt(CH2OH)CH2OH

134 $C_6H_{14}O_3$ 106-62-7
1-Propanol, 2-(2-hydroxypropoxy)-
MeCH(OH)CH2OCHMeCH2OH

134 $C_6H_{14}O_3$ 110-98-5
2-Propanol, 1,1'-oxybis-
MeCH(OH)CH2OCH2CH(OH)Me

134 $C_6H_{14}O_3$ 111-90-0
Ethanol, 2-(2-ethoxyethoxy)-
HOCH2CH2OCH2CH2OEt

134 $C_6H_{14}O_3$ 111-96-6
Ethane, 1,1'-oxybis[2-methoxy-

Me OCH2 CH2 OCH2 CH2 OMe

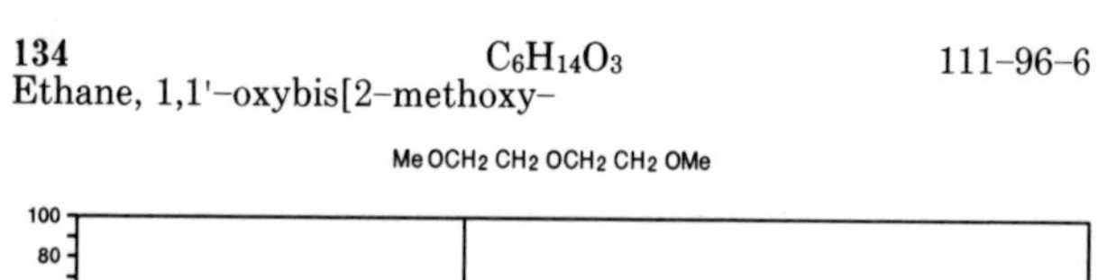

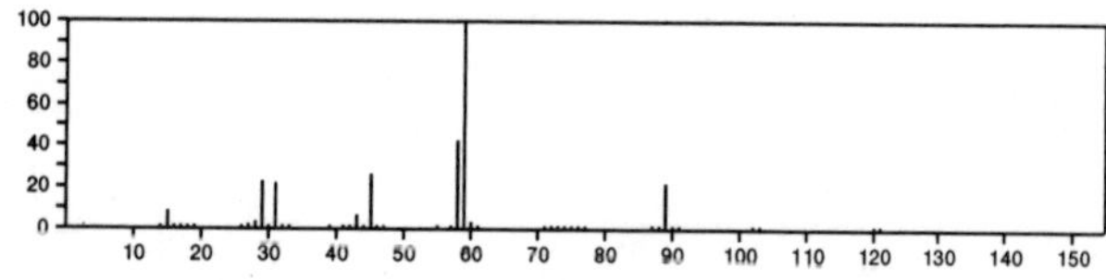

134 $C_6H_{14}O_3$ 24823-81-2
Propane, 1,1,1-trimethoxy-

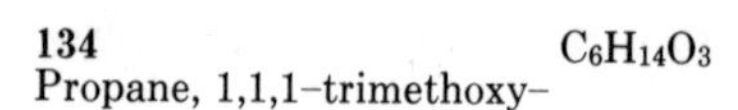

Et C(OMe) 3

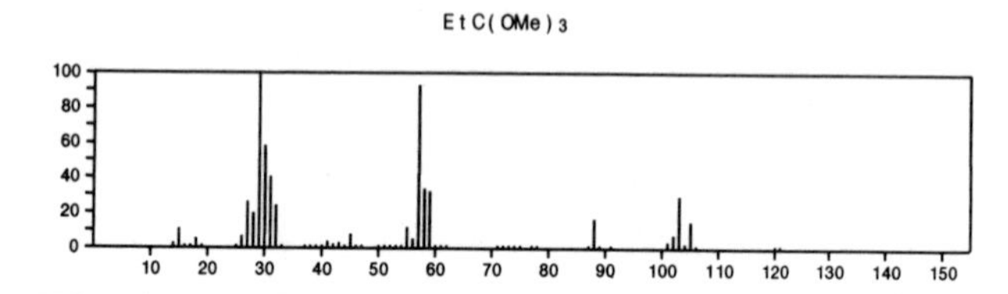

134 $C_7H_6N_2O$ 767-98-6
Nicotinonitrile, 1,4-dihydro-1-methyl-4-oxo-

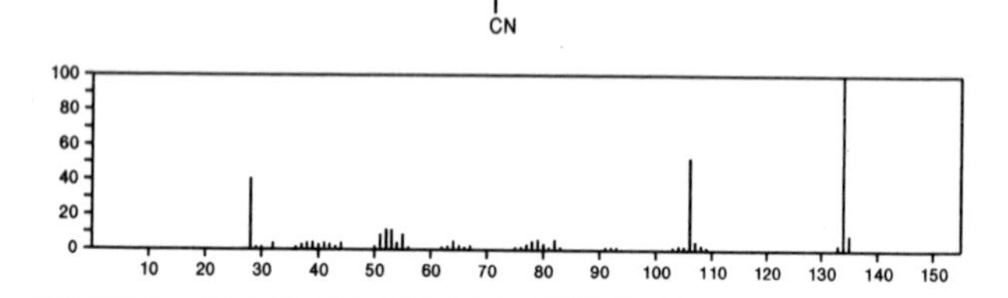

134 $C_7H_6N_2O$ 3999-06-2
Imidazo[1,2-*a*]pyridin-2(3*H*)-one

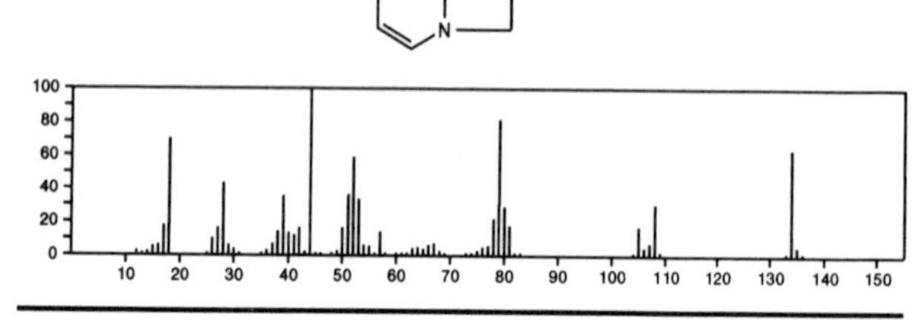

134 $C_7H_6N_2O$ 4241-27-4
3-Pyridinecarbonitrile, 1,2-dihydro-6-methyl-2-oxo-

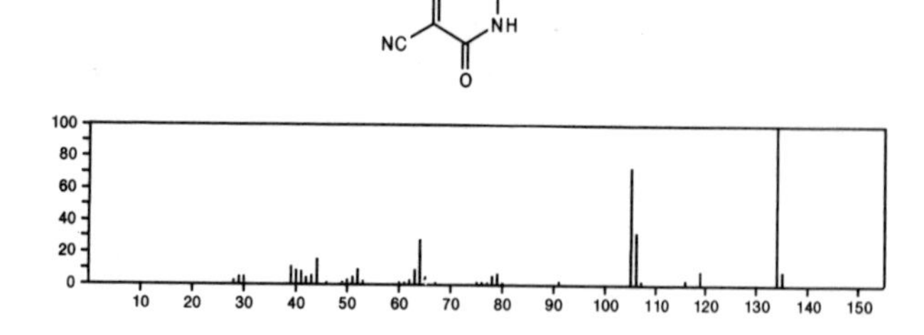

134 $C_7H_6N_2O$ 4570-41-6
2-Benzoxazolamine

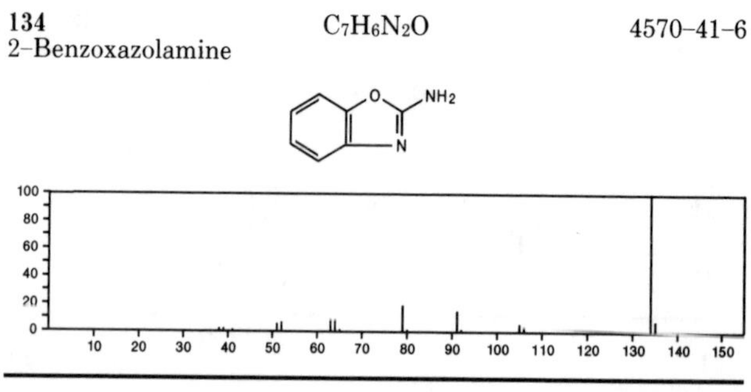

134 $C_7H_6N_2O$ 7364-25-2
3*H*-Indazol-3-one, 1,2-dihydro-

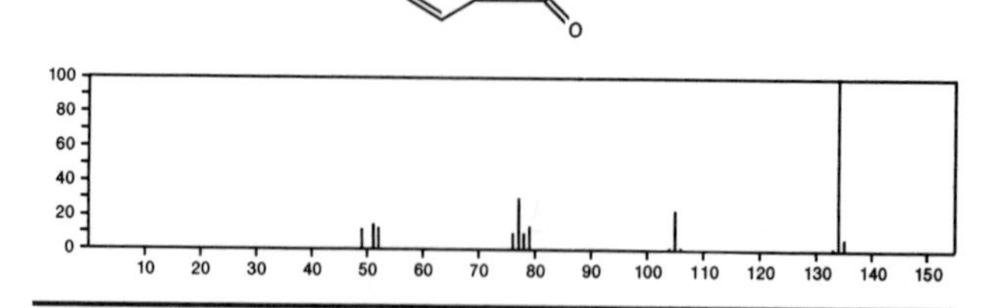

134 $C_7H_6N_2O$ 18916-43-3
1*H*-Benzimidazole, 3-oxide

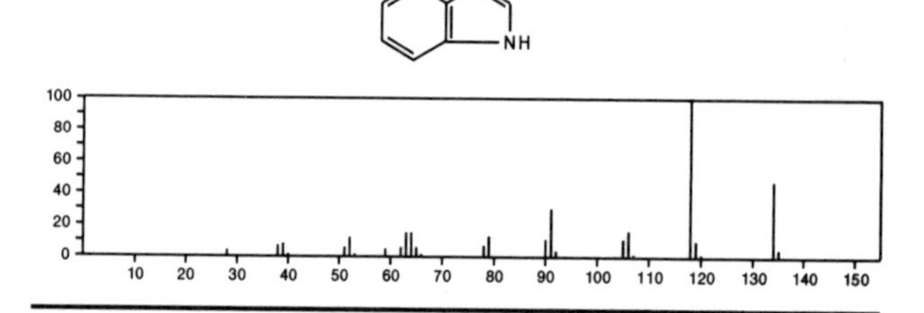

134 $C_7H_{15}Cl$ 629-06-1
Heptane, 1-chloro-

Me (CH2) 6 Cl

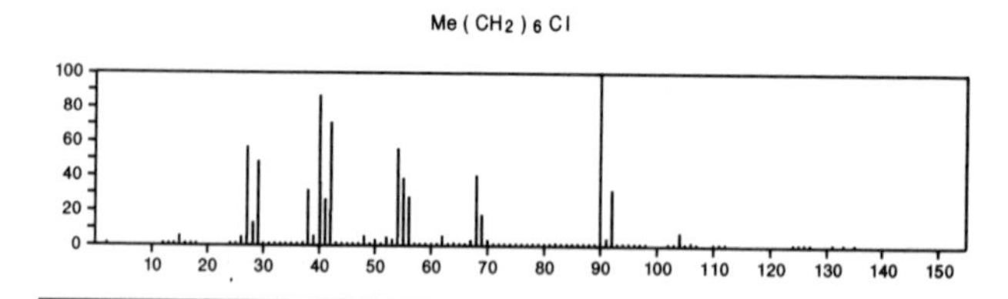

134 $C_7H_{15}Cl$ 1001-89-4
Heptane, 2-chloro-

MeCHCl (CH2) 4 Me

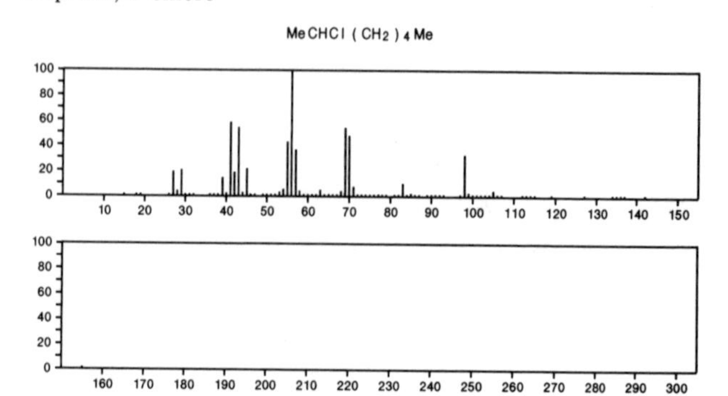

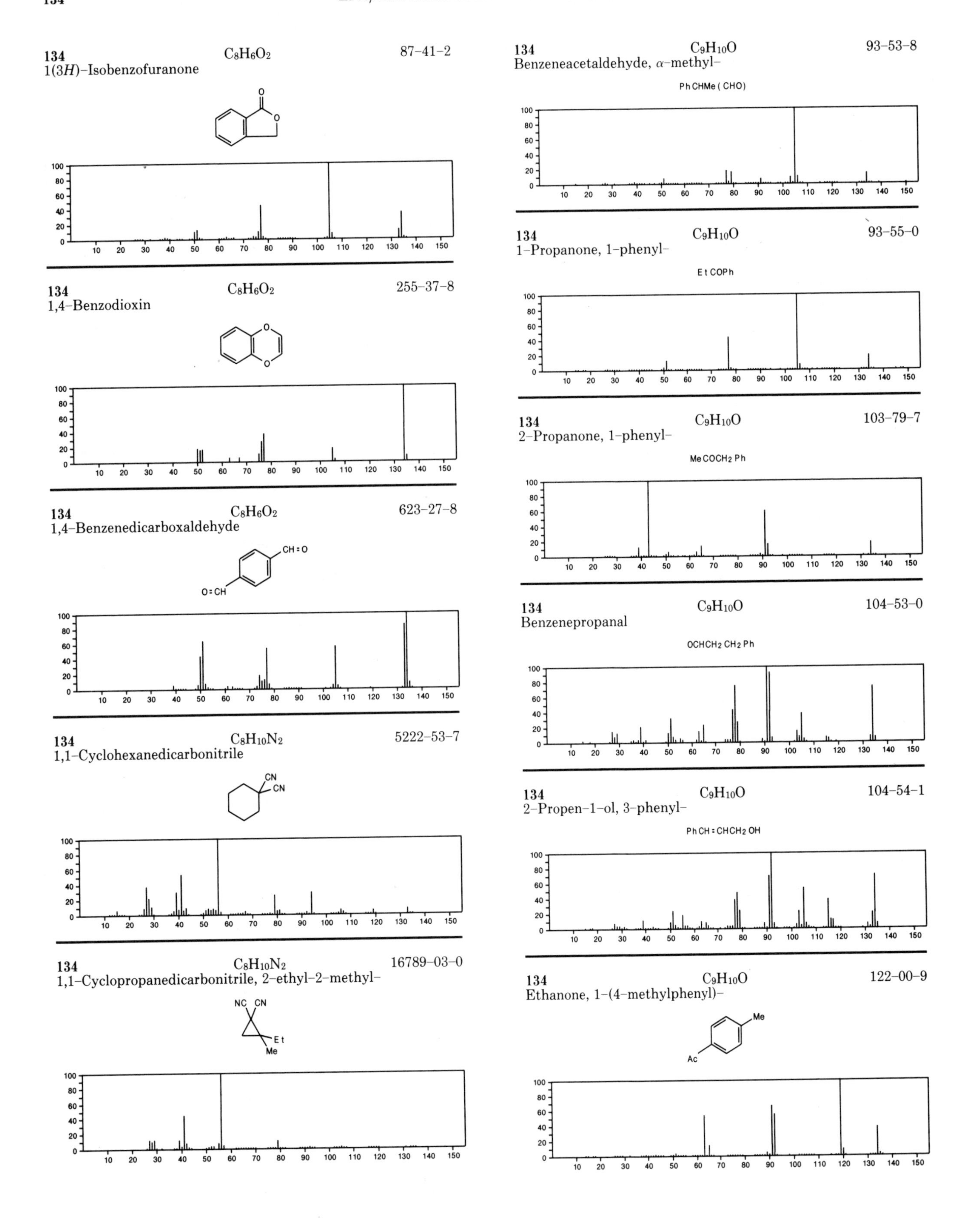

134 $C_8H_6O_2$ 87-41-2
1(3*H*)-Isobenzofuranone

134 $C_8H_6O_2$ 255-37-8
1,4-Benzodioxin

134 $C_8H_6O_2$ 623-27-8
1,4-Benzenedicarboxaldehyde

134 $C_8H_{10}N_2$ 5222-53-7
1,1-Cyclohexanedicarbonitrile

134 $C_8H_{10}N_2$ 16789-03-0
1,1-Cyclopropanedicarbonitrile, 2-ethyl-2-methyl-

134 $C_9H_{10}O$ 93-53-8
Benzeneacetaldehyde, α-methyl-

PhCHMe(CHO)

134 $C_9H_{10}O$ 93-55-0
1-Propanone, 1-phenyl-

EtCOPh

134 $C_9H_{10}O$ 103-79-7
2-Propanone, 1-phenyl-

MeCOCH2Ph

134 $C_9H_{10}O$ 104-53-0
Benzenepropanal

OCHCH2CH2Ph

134 $C_9H_{10}O$ 104-54-1
2-Propen-1-ol, 3-phenyl-

PhCH=CHCH2OH

134 $C_9H_{10}O$ 122-00-9
Ethanone, 1-(4-methylphenyl)-

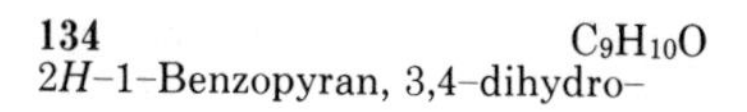

134 $C_9H_{10}O$ 493-08-3
2*H*-1-Benzopyran, 3,4-dihydro-

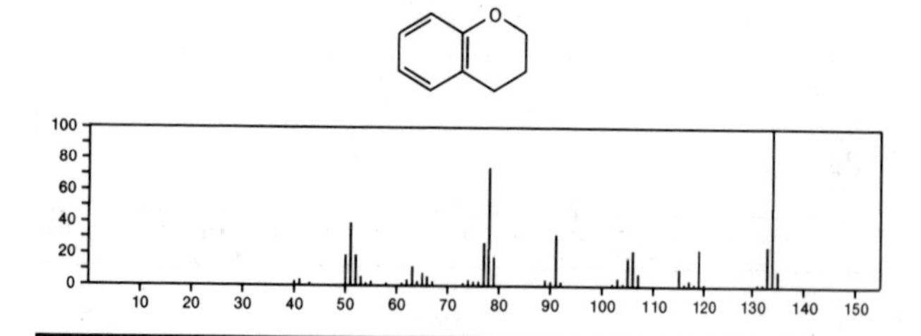

134 $C_9H_{10}O$ 577-16-2
Ethanone, 1-(2-methylphenyl)-

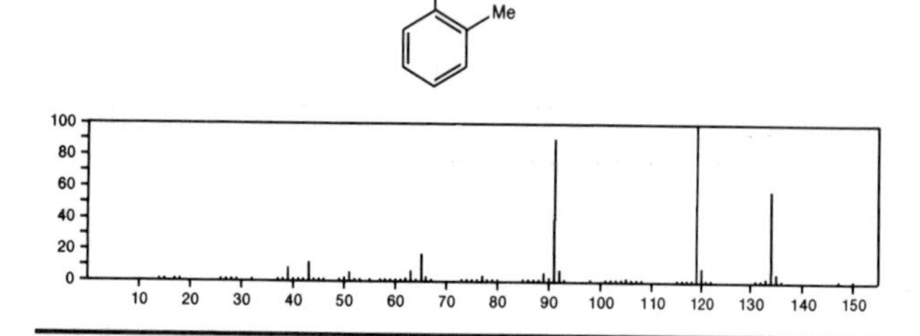

134 $C_9H_{10}O$ 1745-81-9
Phenol, 2-(2-propenyl)-

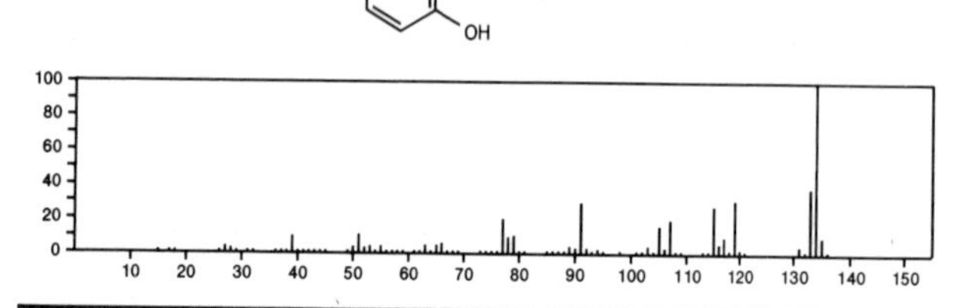

134 $C_9H_{10}O$ 1746-11-8
Benzofuran, 2,3-dihydro-2-methyl-

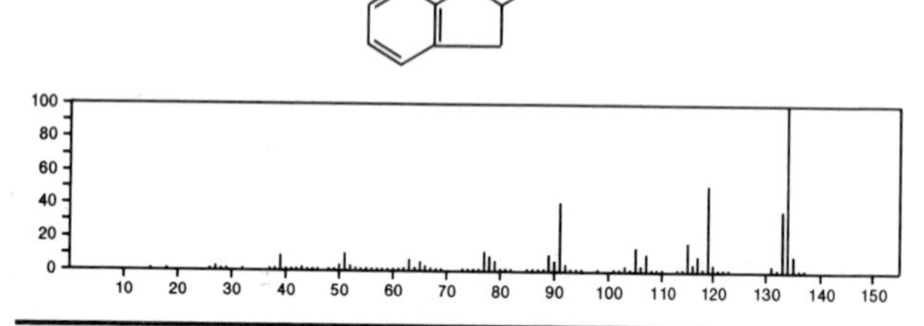

134 $C_9H_{10}O$ 1746-13-0
Benzene, (2-propenyloxy)-

PhOCH2 CH=CH2

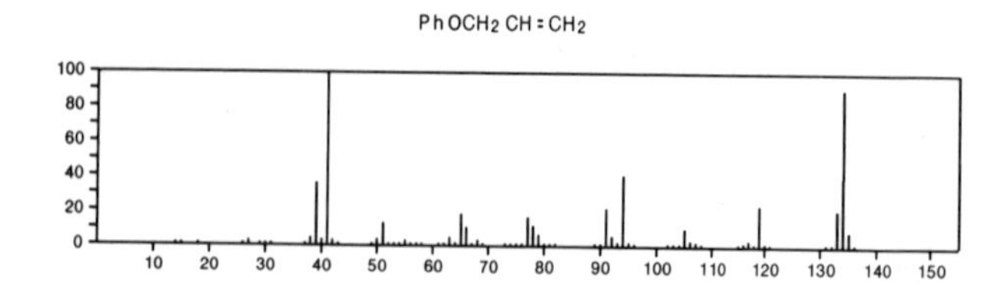

134 $C_9H_{10}O$ 2085-88-3
Oxirane, 2-methyl-2-phenyl-

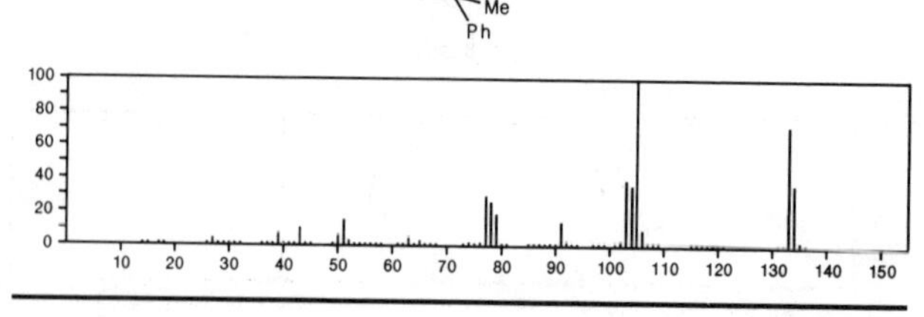

134 $C_9H_{10}O$ 4254-29-9
1*H*-Inden-2-ol, 2,3-dihydro-

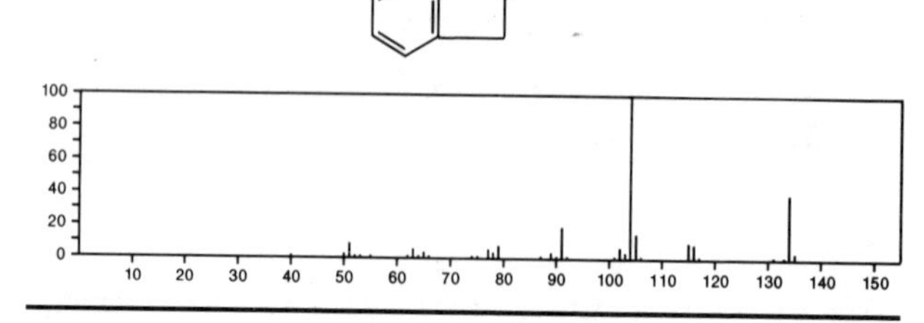

134 $C_9H_{10}O$ 4436-22-0
Oxirane, 2-methyl-3-phenyl-

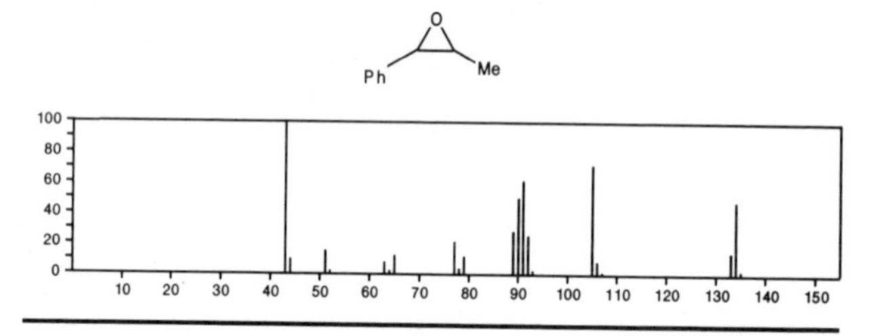

134 $C_9H_{10}O$ 4747-13-1
Benzene, (1-methoxyethenyl)-

PhC(OMe)=CH2

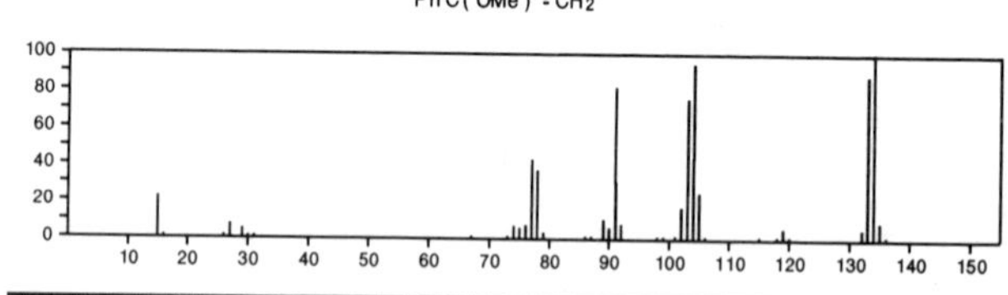

134 $C_9H_{10}O$ 5779-94-2
Benzaldehyde, 2,5-dimethyl-

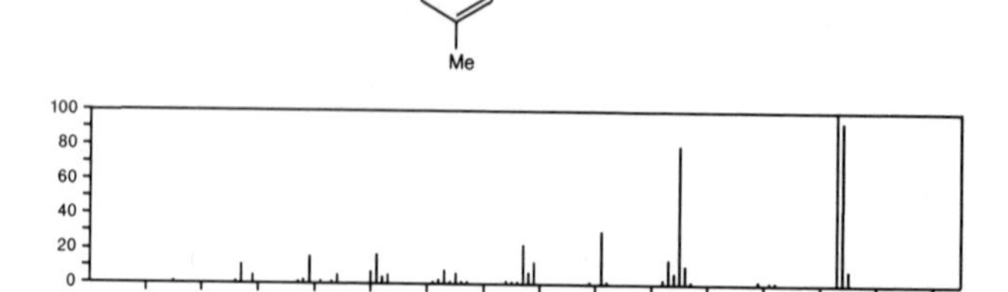

134 $C_9H_{10}O$ 5973-71-7
Benzaldehyde, 3,4-dimethyl-

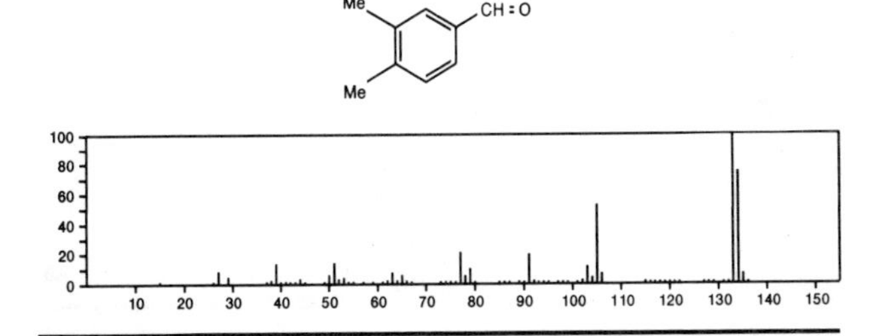

134 $C_9H_{10}O$ 6351-10-6
1*H*-Inden-1-ol, 2,3-dihydro-

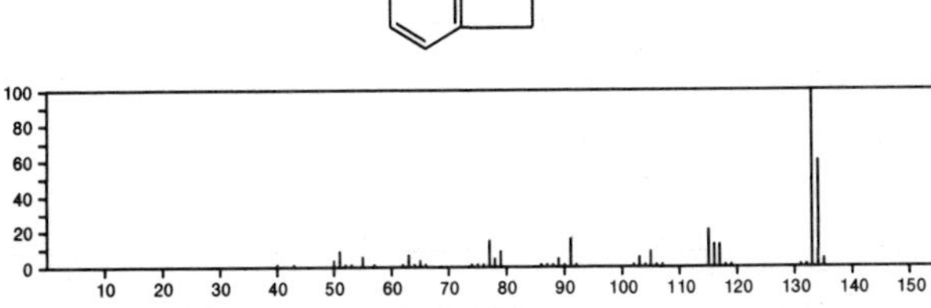

134 $C_9H_{10}O$ 14371-19-8
Benzene, (2-methoxyethenyl)-, (*Z*)-

PhCH=CHOMe

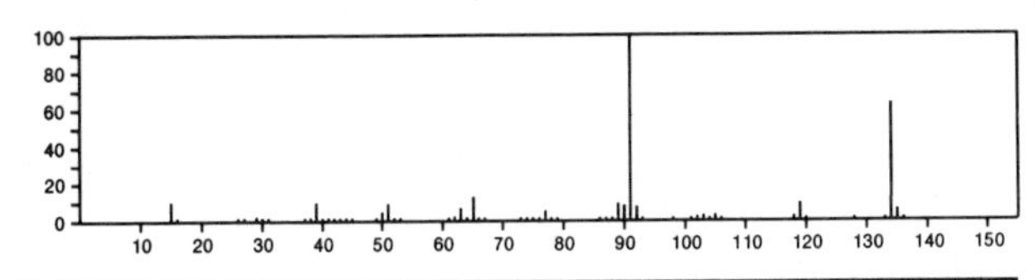

134 $C_9H_{10}O$ 15764-16-6
Benzaldehyde, 2,4-dimethyl-

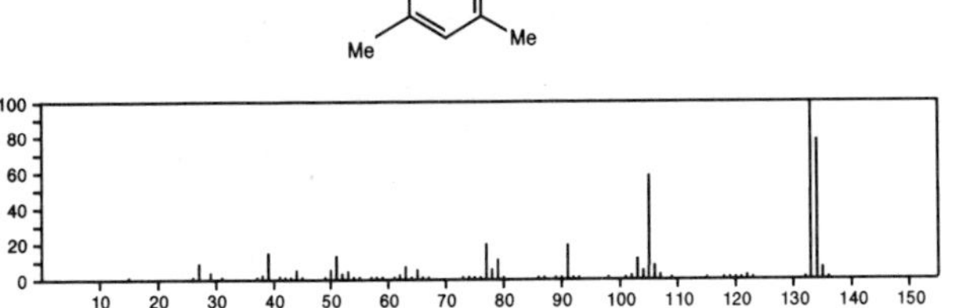

134 $C_9H_{10}O$ 26444-19-9
Ethanone, 1-(methylphenyl)-

Me—⟨phenyl⟩—Ac

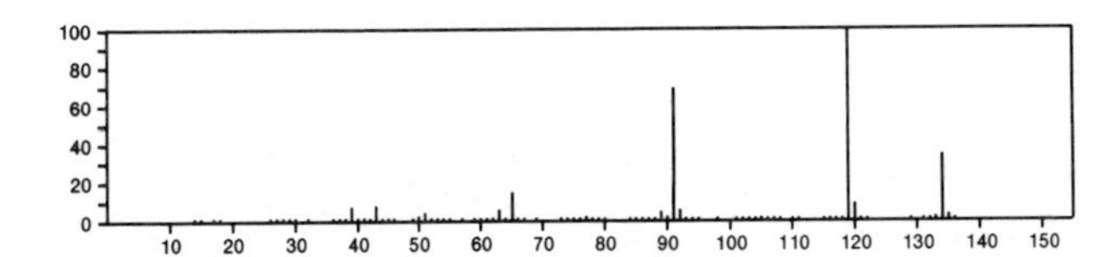

134 $C_9H_{10}O$ 29743-33-7
4,6-Octadiyn-3-one, 2-methyl-

$Me_2CHCOC{\equiv}CC{\equiv}CMe$

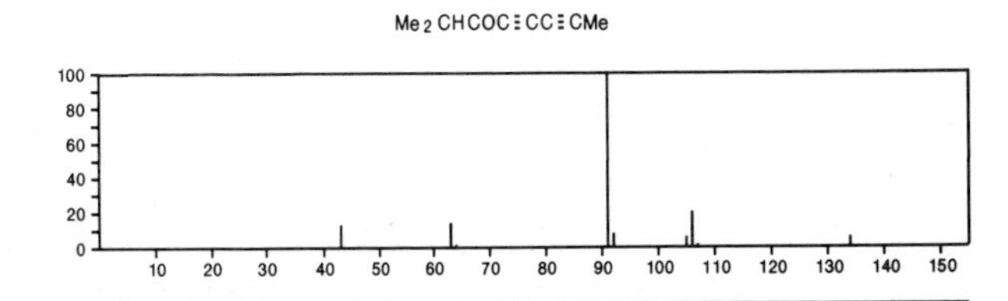

134 $C_9H_{10}O$ 30584-69-1
Benzenemethanol, *ar*-ethenyl-

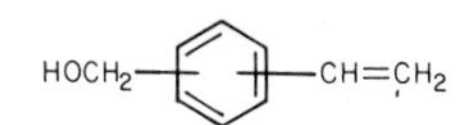

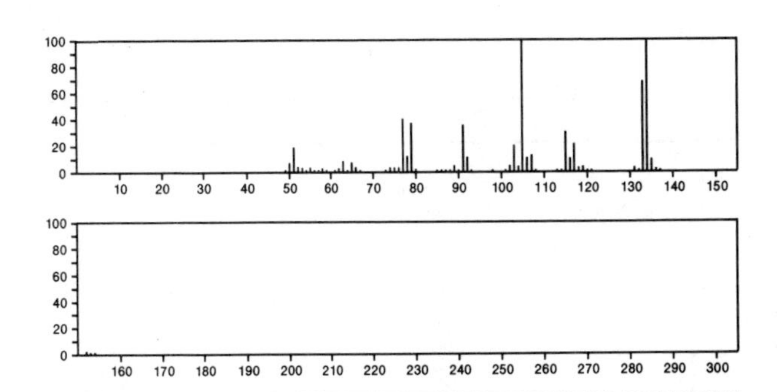

134 $C_9H_{10}O$ 53951-50-1
Benzaldehyde, ethyl-

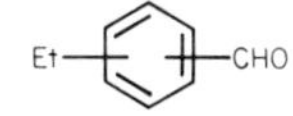

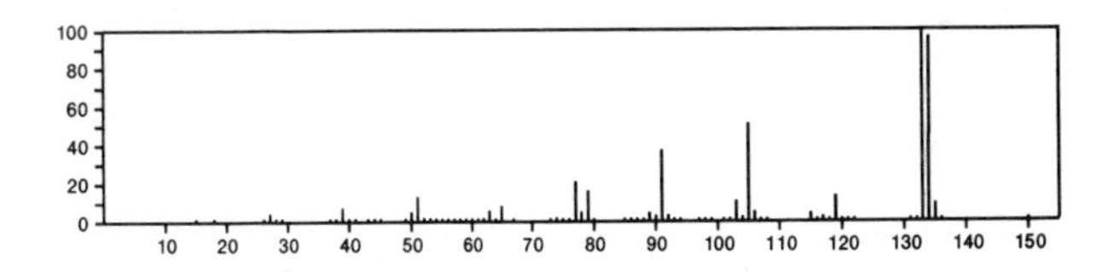

134 $C_9H_{10}O$ 56666-74-1
Bicyclo[6.1.0]nona-5,8-dien-4-one

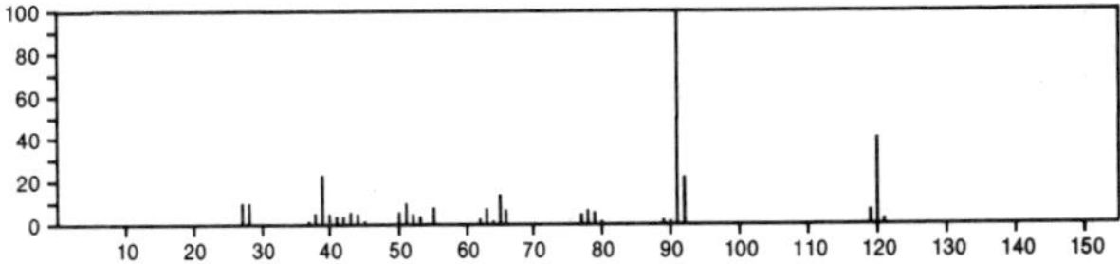

134 $C_{10}H_{14}$ 95-93-2
Benzene, 1,2,4,5-tetramethyl-

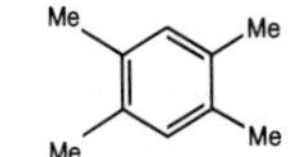

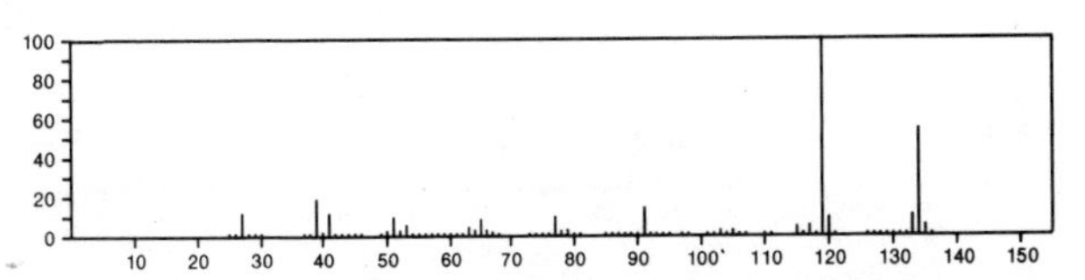

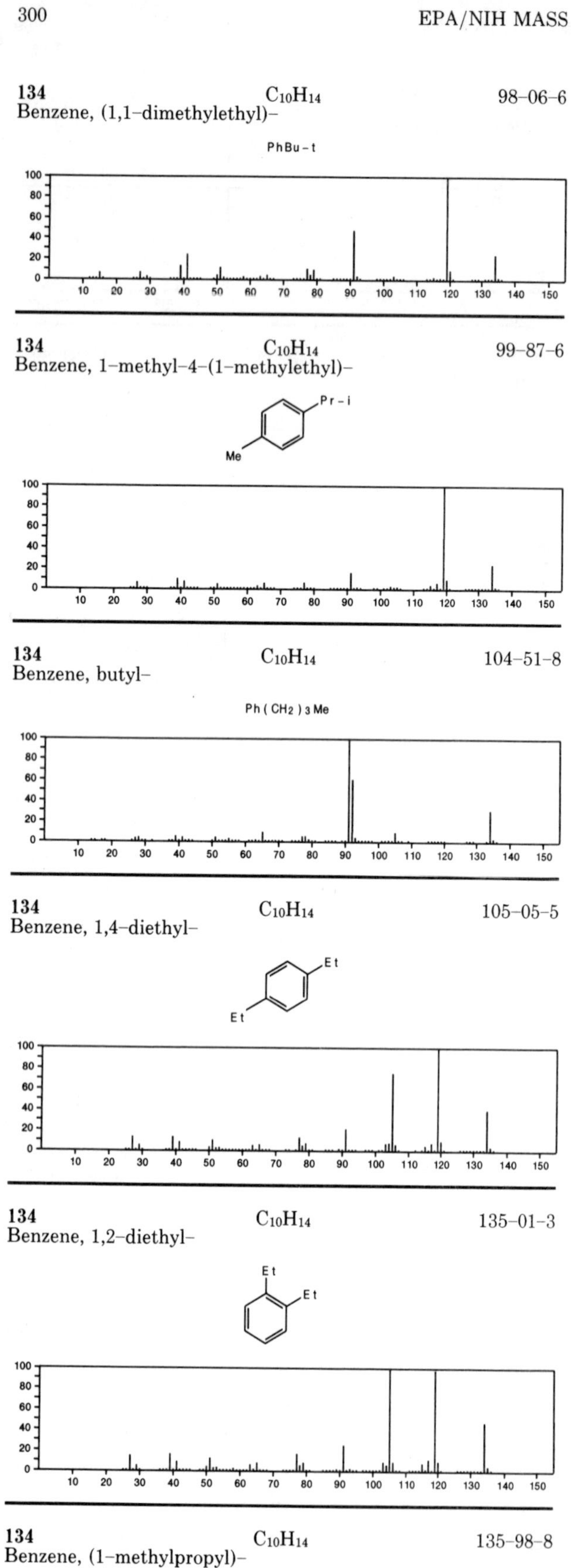

134 $C_{10}H_{14}$ 98-06-6
Benzene, (1,1-dimethylethyl)-

134 $C_{10}H_{14}$ 99-87-6
Benzene, 1-methyl-4-(1-methylethyl)-

134 $C_{10}H_{14}$ 104-51-8
Benzene, butyl-

134 $C_{10}H_{14}$ 105-05-5
Benzene, 1,4-diethyl-

134 $C_{10}H_{14}$ 135-01-3
Benzene, 1,2-diethyl-

134 $C_{10}H_{14}$ 135-98-8
Benzene, (1-methylpropyl)-

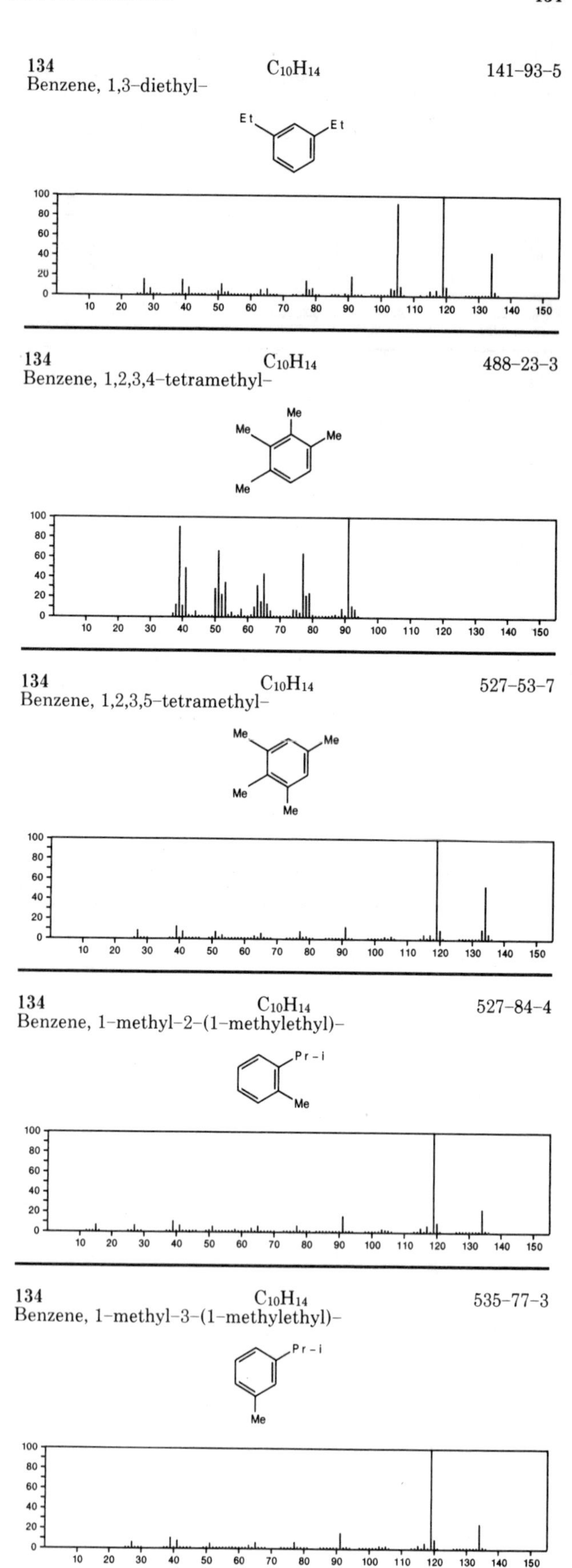

134 $C_{10}H_{14}$ 141-93-5
Benzene, 1,3-diethyl-

134 $C_{10}H_{14}$ 488-23-3
Benzene, 1,2,3,4-tetramethyl-

134 $C_{10}H_{14}$ 527-53-7
Benzene, 1,2,3,5-tetramethyl-

134 $C_{10}H_{14}$ 527-84-4
Benzene, 1-methyl-2-(1-methylethyl)-

134 $C_{10}H_{14}$ 535-77-3
Benzene, 1-methyl-3-(1-methylethyl)-

134 $C_{10}H_{14}$ 538-93-2
Benzene, (2-methylpropyl)-

PhBu-i

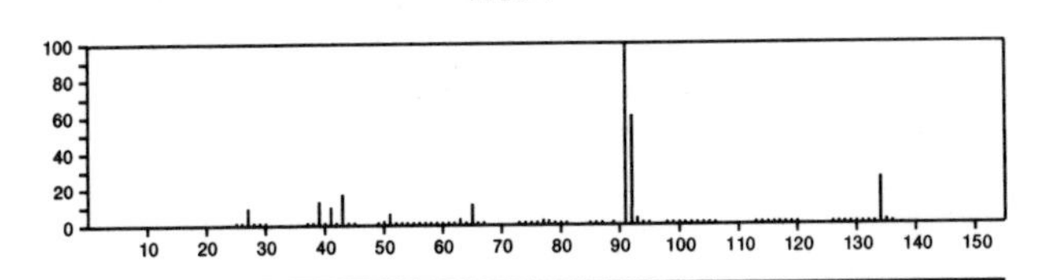

134 $C_{10}H_{14}$ 874-41-9
Benzene, 1-ethyl-2,4-dimethyl-

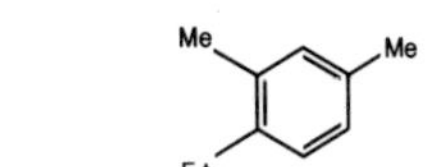

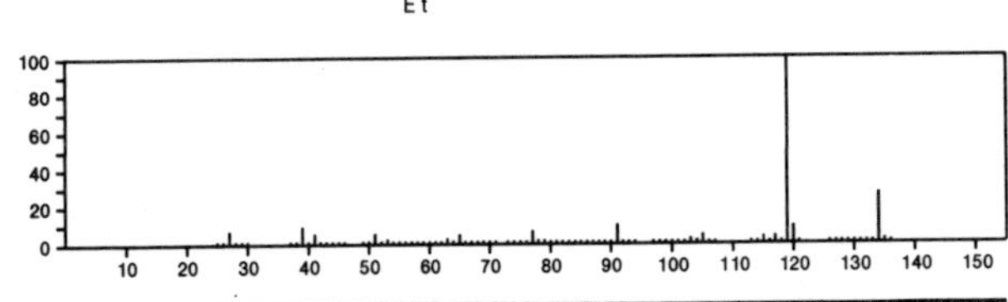

134 $C_{10}H_{14}$ 933-98-2
Benzene, 1-ethyl-2,3-dimethyl-

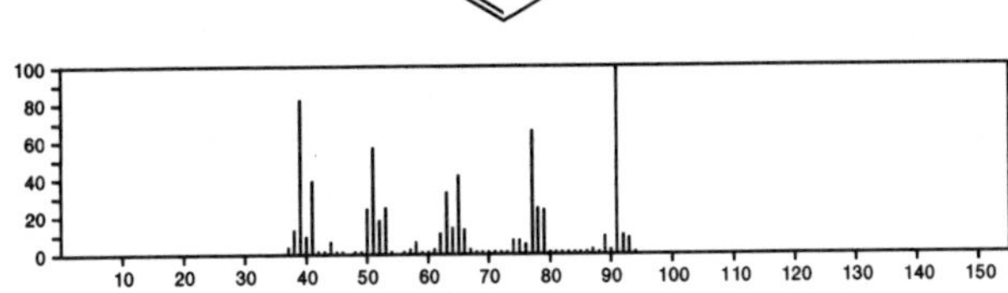

134 $C_{10}H_{14}$ 934-74-7
Benzene, 1-ethyl-3,5-dimethyl-

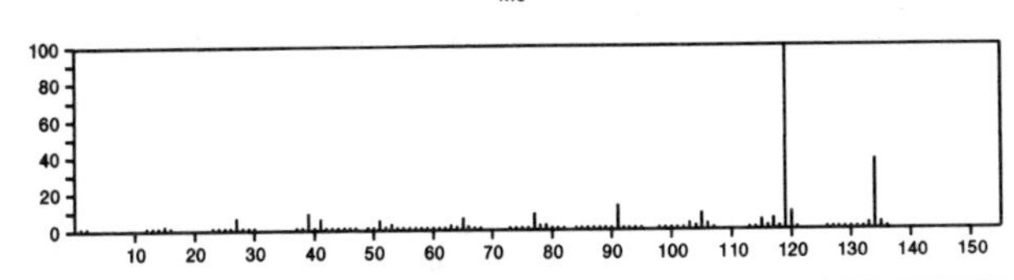

134 $C_{10}H_{14}$ 934-80-5
Benzene, 4-ethyl-1,2-dimethyl-

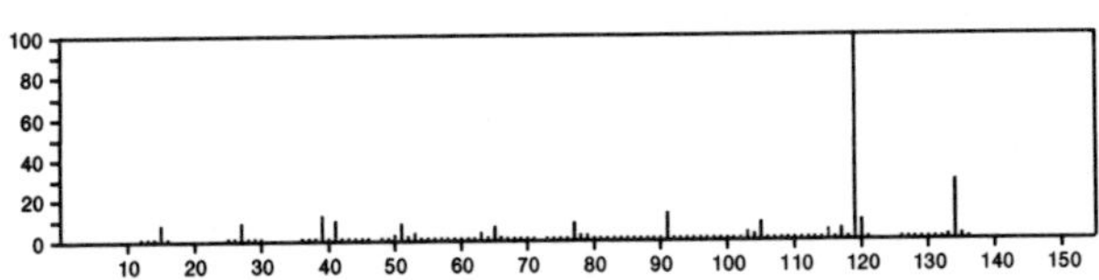

134 $C_{10}H_{14}$ 1074-17-5
Benzene, 1-methyl-2-propyl-

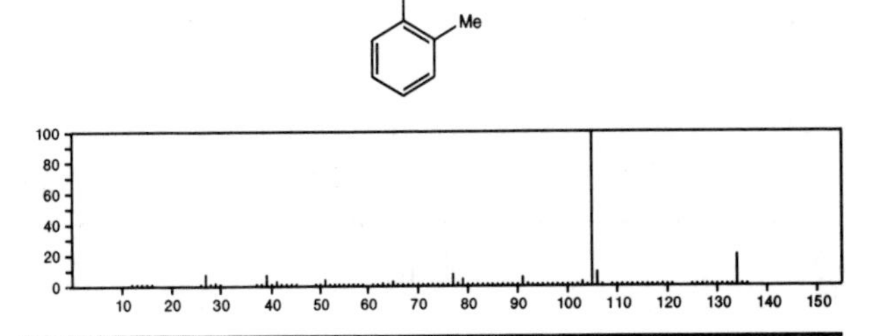

134 $C_{10}H_{14}$ 1074-43-7
Benzene, 1-methyl-3-propyl-

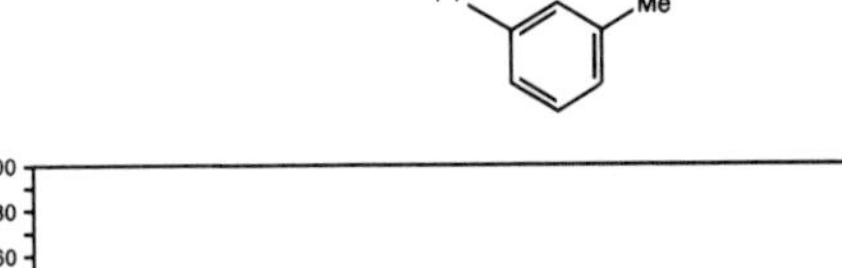

134 $C_{10}H_{14}$ 1074-55-1
Benzene, 1-methyl-4-propyl-

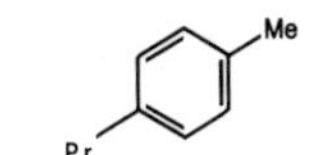

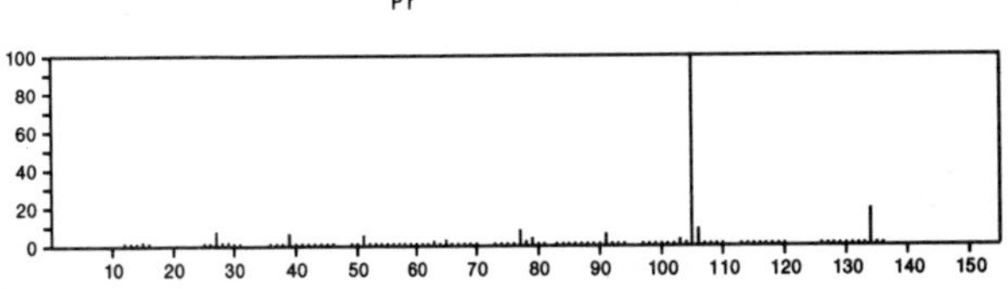

134 $C_{10}H_{14}$ 1758-88-9
Benzene, 2-ethyl-1,4-dimethyl-

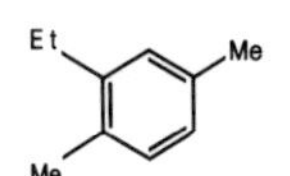

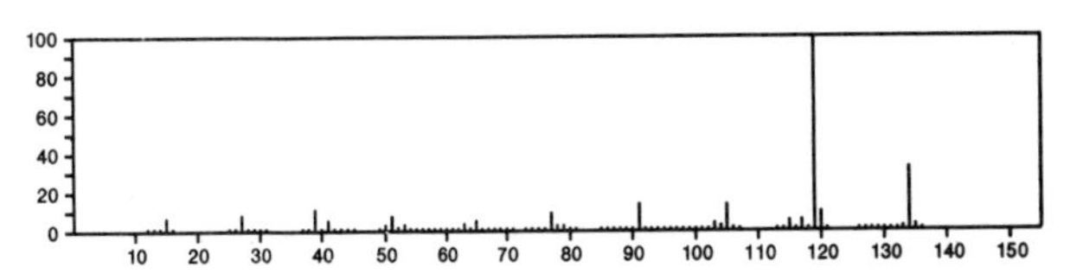

134 $C_{10}H_{14}$ 2870-04-4
Benzene, 2-ethyl-1,3-dimethyl-

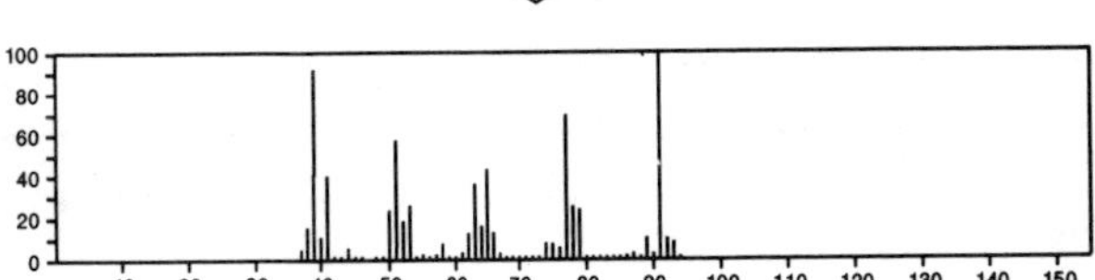

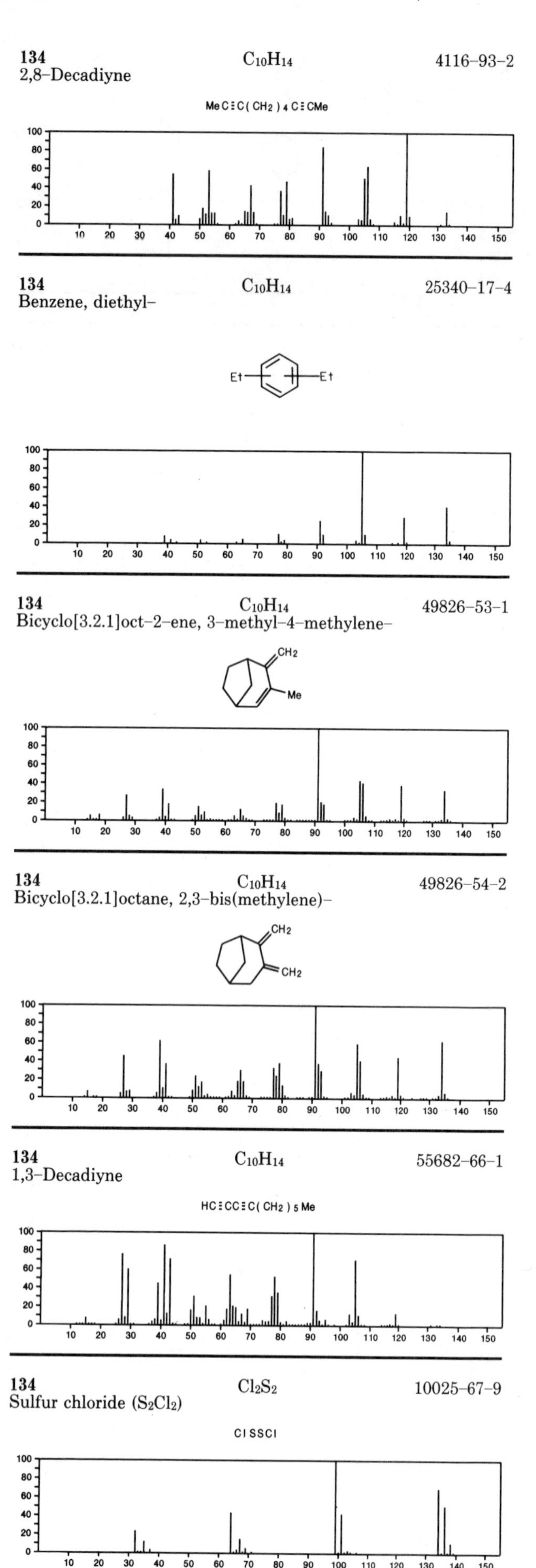
134 $C_{10}H_{14}$ 4116-93-2
2,8-Decadiyne
MeC≡C(CH2)4C≡CMe
134 $C_{10}H_{14}$ 25340-17-4
Benzene, diethyl-
134 $C_{10}H_{14}$ 49826-53-1
Bicyclo[3.2.1]oct-2-ene, 3-methyl-4-methylene-
134 $C_{10}H_{14}$ 49826-54-2
Bicyclo[3.2.1]octane, 2,3-bis(methylene)-
134 $C_{10}H_{14}$ 55682-66-1
1,3-Decadiyne
HC≡CC≡C(CH2)5Me
134 Cl_2S_2 10025-67-9
Sulfur chloride (S_2Cl_2)
ClSSCl

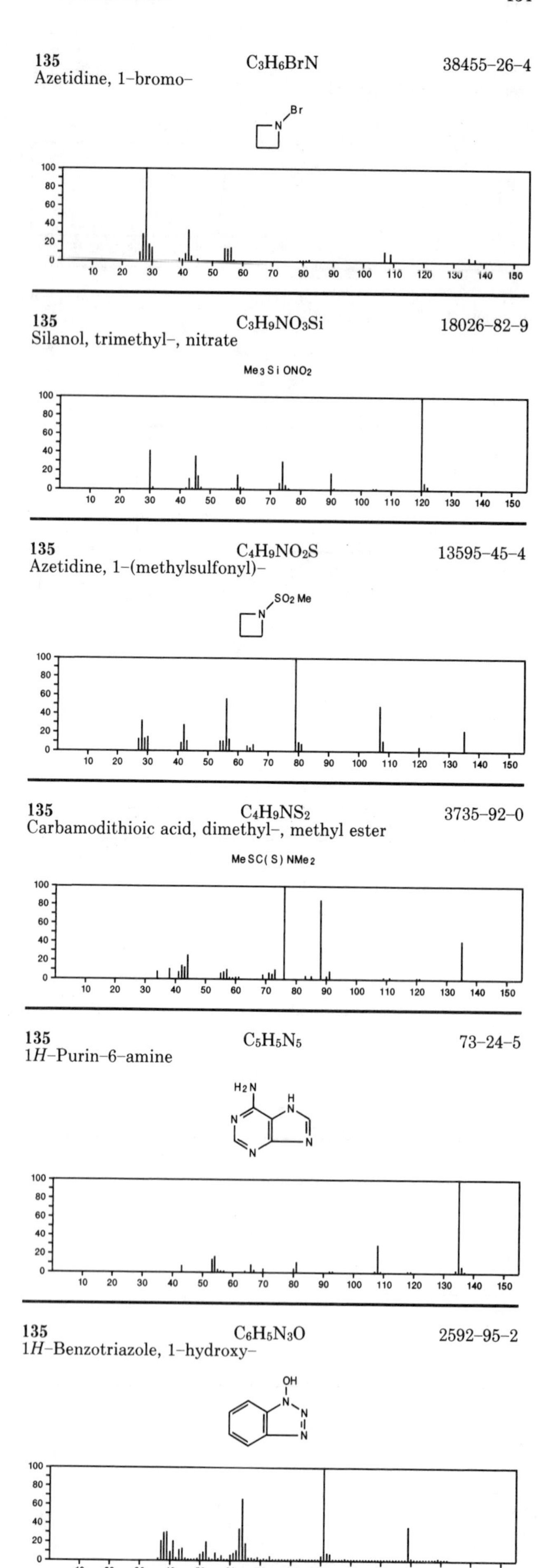
135 C_3H_6BrN 38455-26-4
Azetidine, 1-bromo-
135 $C_3H_9NO_3Si$ 18026-82-9
Silanol, trimethyl-, nitrate
Me3SiONO2
135 $C_4H_9NO_2S$ 13595-45-4
Azetidine, 1-(methylsulfonyl)-
135 $C_4H_9NS_2$ 3735-92-0
Carbamodithioic acid, dimethyl-, methyl ester
MeSC(S)NMe2
135 $C_5H_5N_5$ 73-24-5
1H-Purin-6-amine
135 $C_6H_5N_3O$ 2592-95-2
1H-Benzotriazole, 1-hydroxy-

135 $C_6H_5N_3O$ 6969-71-7
1,2,4-Triazolo[4,3-*a*]pyridin-3(2*H*)-one

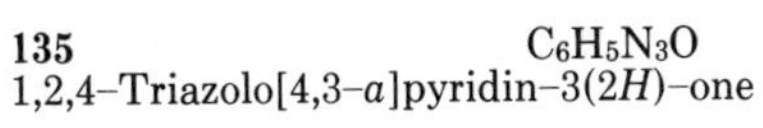

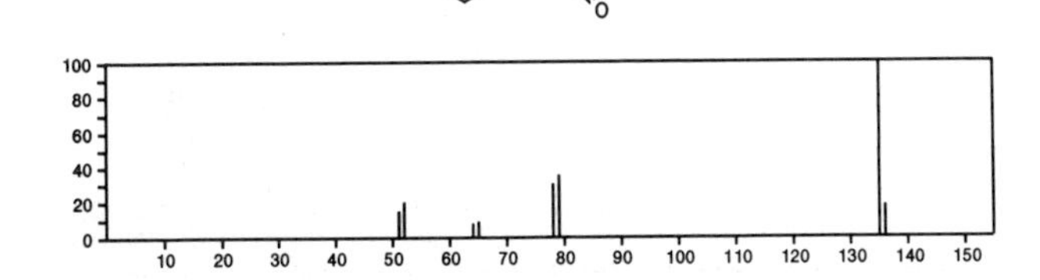

135 $C_6H_5N_3O$ 16328-62-4
2*H*-Imidazo[4,5-*b*]pyridin-2-one, 1,3-dihydro-

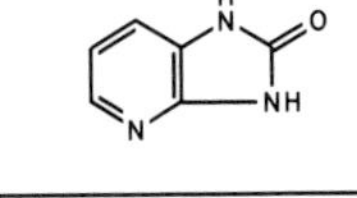
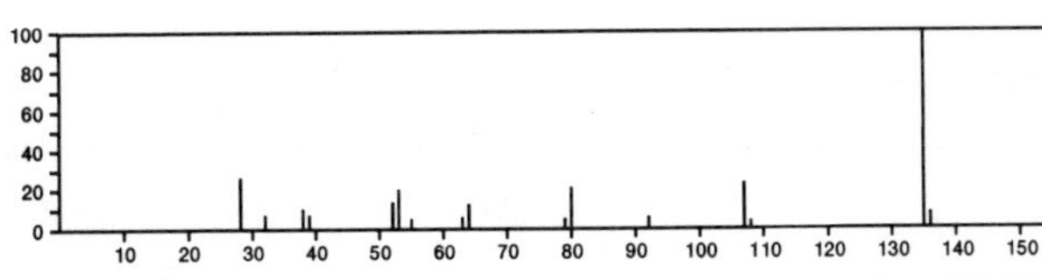

135 $C_6H_5N_3O$ 53975-70-5
3*H*-Pyrazolo[3,4-*c*]pyridin-3-one, 1,2-dihydro-

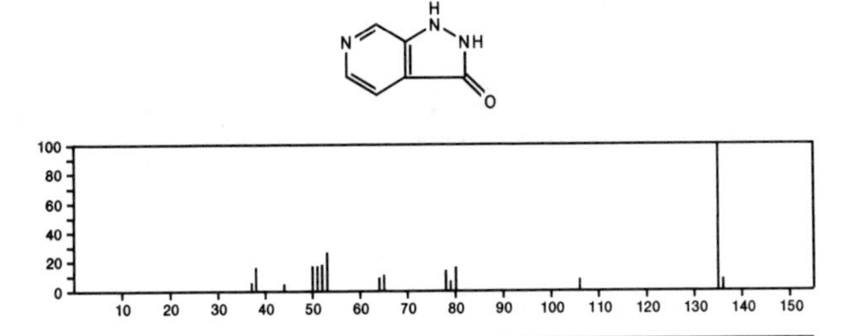

135 $C_7H_5NO_2$ 59-49-4
2(3*H*)-Benzoxazolone

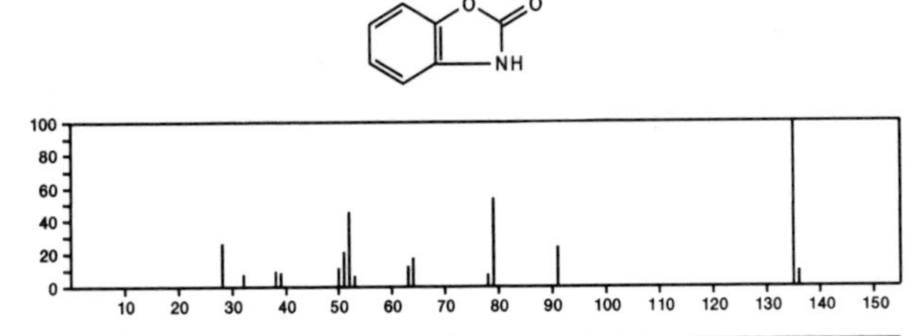

135 $C_7H_5NO_2$ 29809-25-4
Benzaldehyde, *o*-nitroso-

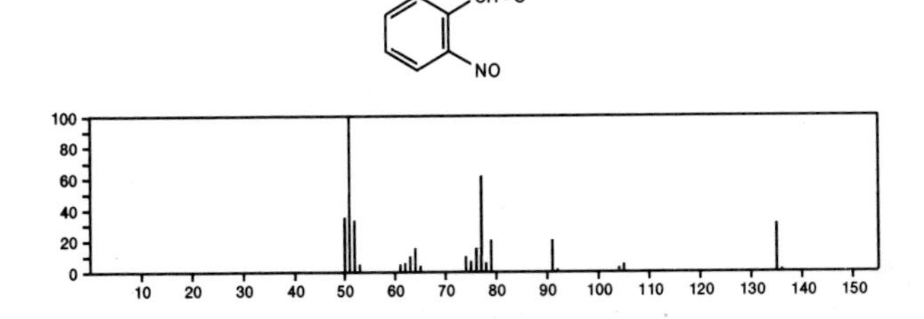

135 $C_7H_5NO_2$ 31499-90-8
2,1-Benzisoxazol-3(1*H*)-one

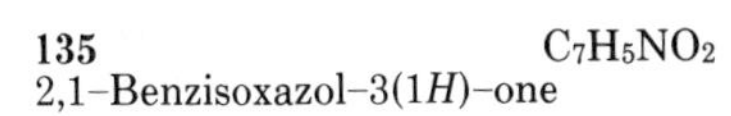
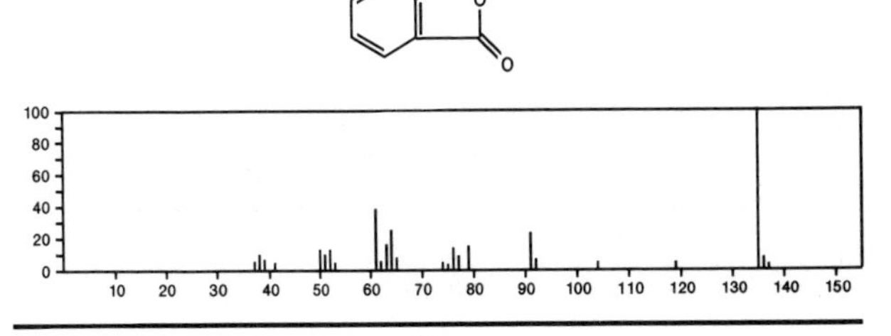

135 C_7H_5NS 95-16-9
Benzothiazole

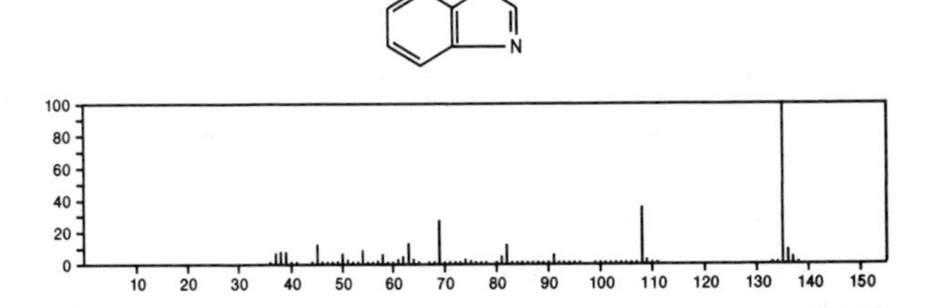

135 C_7H_5NS 272-12-8
Thieno[2,3-*c*]pyridine

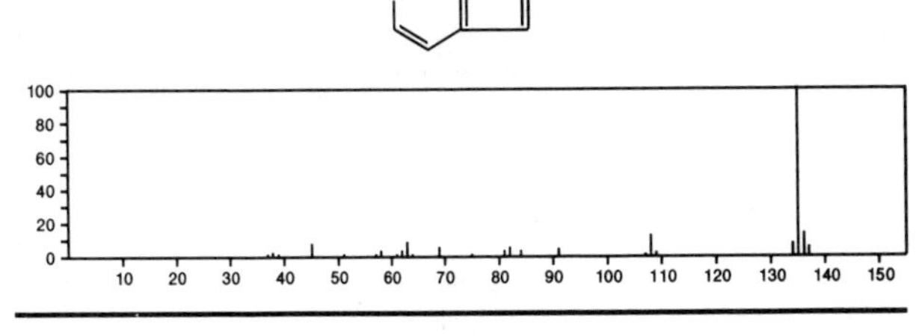

135 C_7H_5NS 272-14-0
Thieno[3,2-*c*]pyridine

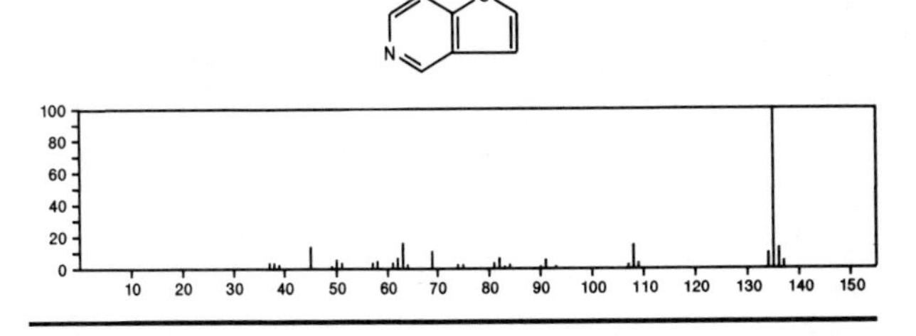

135 C_7H_5NS 272-16-2
1,2-Benzisothiazole

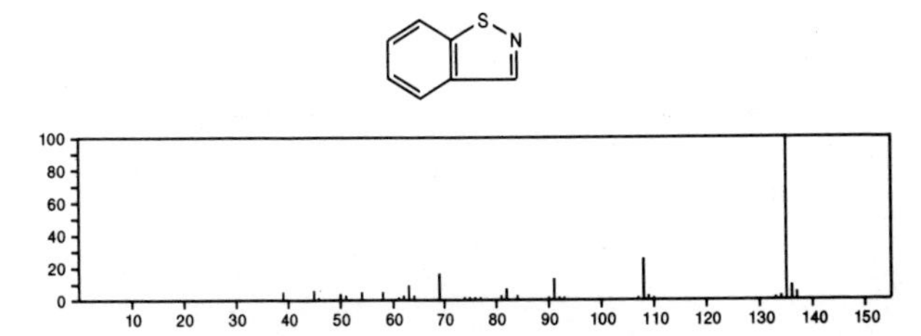

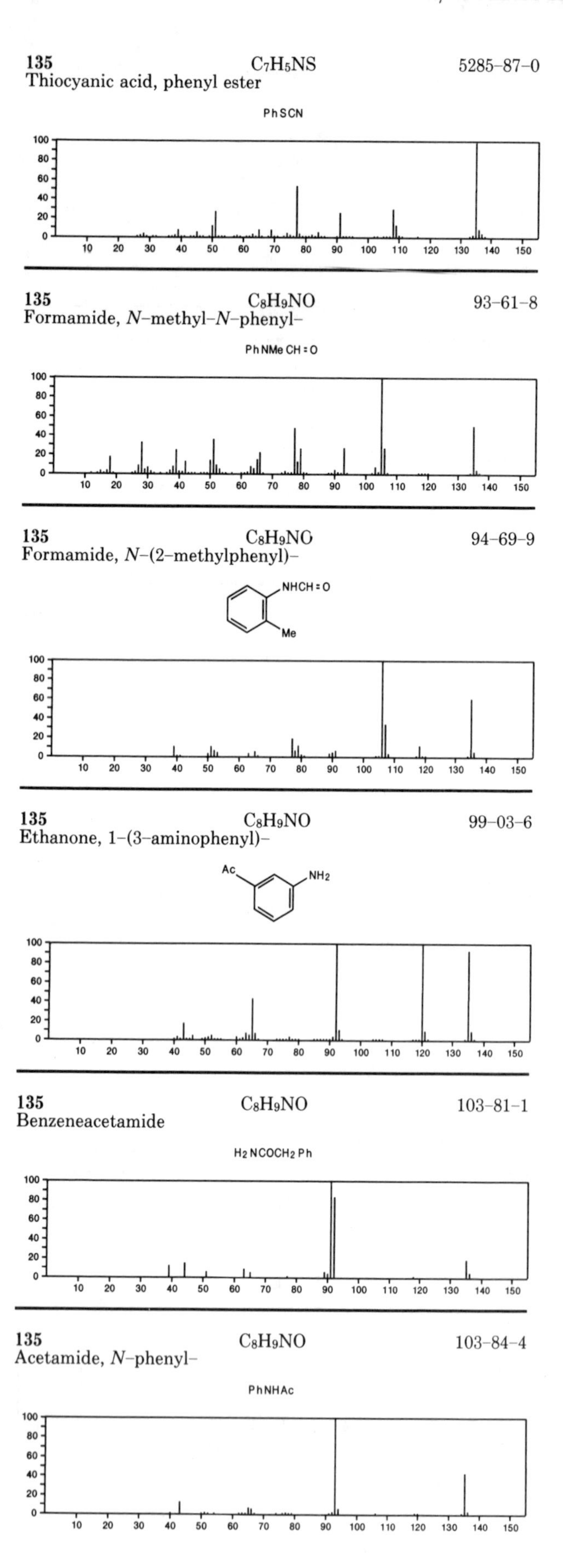

135 C7H5NS 5285-87-0
Thiocyanic acid, phenyl ester
PhSCN
135 C8H9NO 93-61-8
Formamide, N-methyl-N-phenyl-
PhNMe CH=O
135 C8H9NO 94-69-9
Formamide, N-(2-methylphenyl)-
NHCH=O
Me
135 C8H9NO 99-03-6
Ethanone, 1-(3-aminophenyl)-
Ac
NH2
135 C8H9NO 103-81-1
Benzeneacetamide
H2NCOCH2Ph
135 C8H9NO 103-84-4
Acetamide, N-phenyl-
PhNHAc

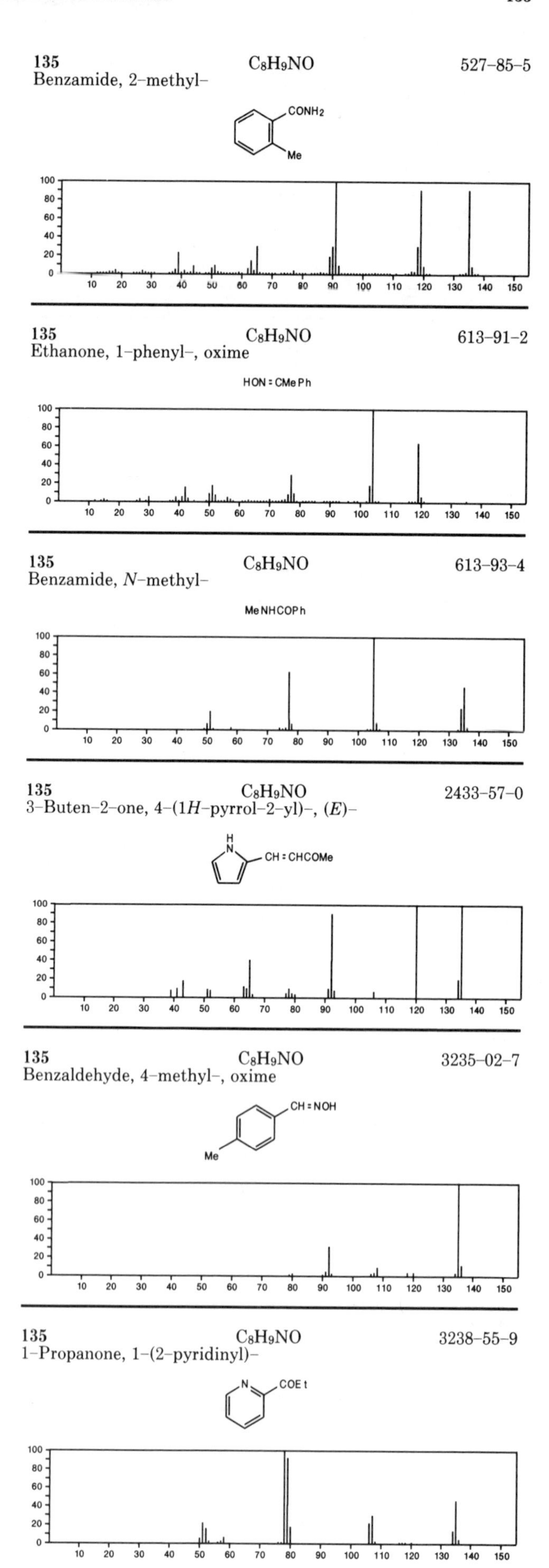

135 C8H9NO 527-85-5
Benzamide, 2-methyl-
CONH2
Me
135 C8H9NO 613-91-2
Ethanone, 1-phenyl-, oxime
HON=CMePh
135 C8H9NO 613-93-4
Benzamide, N-methyl-
MeNHCOPh
135 C8H9NO 2433-57-0
3-Buten-2-one, 4-(1H-pyrrol-2-yl)-, (E)-
CH=CHCOMe
135 C8H9NO 3235-02-7
Benzaldehyde, 4-methyl-, oxime
CH=NOH
Me
135 C8H9NO 3238-55-9
1-Propanone, 1-(2-pyridinyl)-
COEt

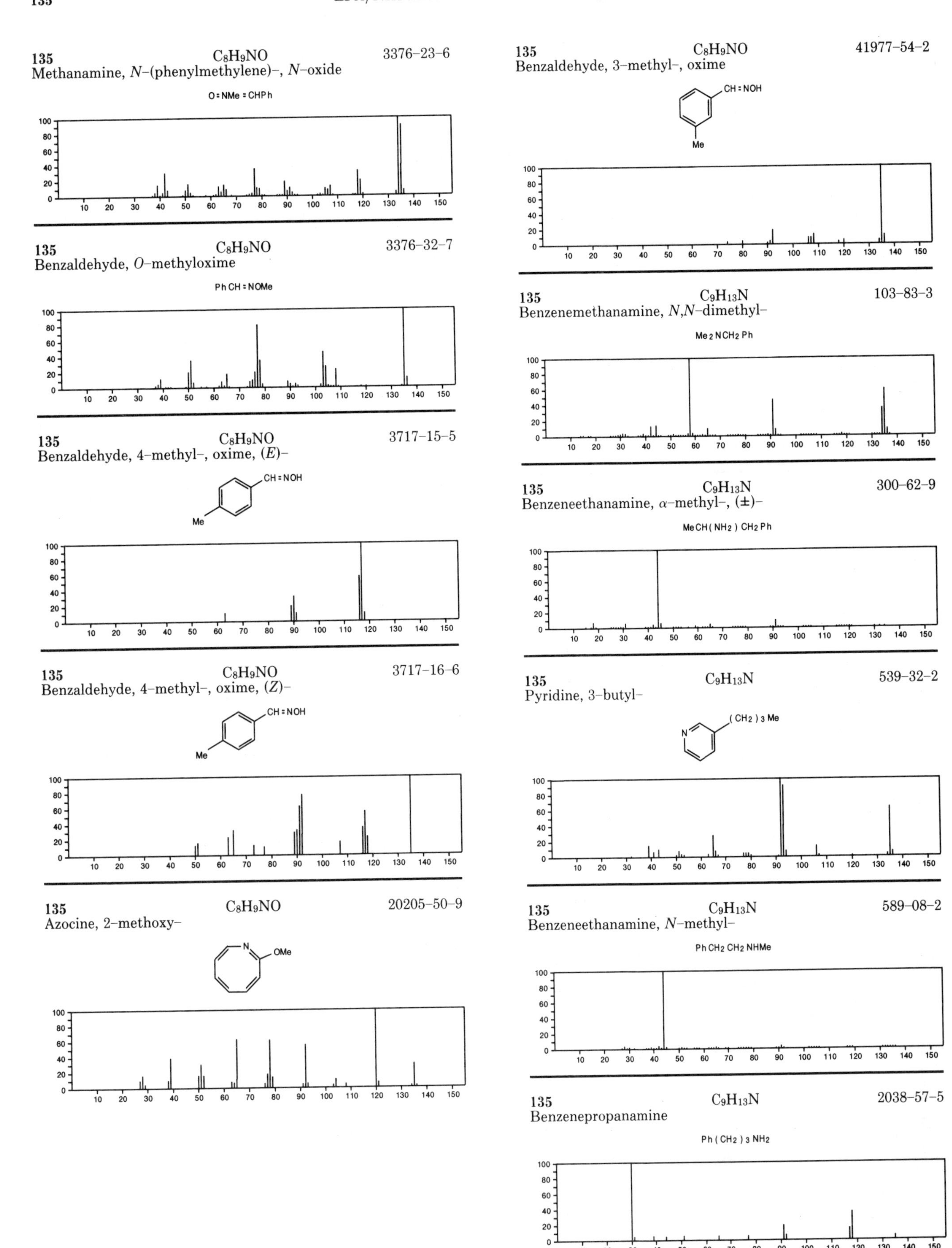
135 C8H9NO 3376-23-6
Methanamine, N-(phenylmethylene)-, N-oxide
O=NMe=CHPh
135 C8H9NO 3376-32-7
Benzaldehyde, O-methyloxime
PhCH=NOMe
135 C8H9NO 3717-15-5
Benzaldehyde, 4-methyl-, oxime, (E)-
CH=NOH
Me
135 C8H9NO 3717-16-6
Benzaldehyde, 4-methyl-, oxime, (Z)-
CH=NOH
Me
135 C8H9NO 20205-50-9
Azocine, 2-methoxy-
N
OMe
135 C8H9NO 41977-54-2
Benzaldehyde, 3-methyl-, oxime
CH=NOH
Me
135 C9H13N 103-83-3
Benzenemethanamine, N,N-dimethyl-
Me2NCH2Ph
135 C9H13N 300-62-9
Benzeneethanamine, α-methyl-, (±)-
MeCH(NH2)CH2Ph
135 C9H13N 539-32-2
Pyridine, 3-butyl-
(CH2)3Me
N
135 C9H13N 589-08-2
Benzeneethanamine, N-methyl-
PhCH2CH2NHMe
135 C9H13N 2038-57-5
Benzenepropanamine
Ph(CH2)3NH2

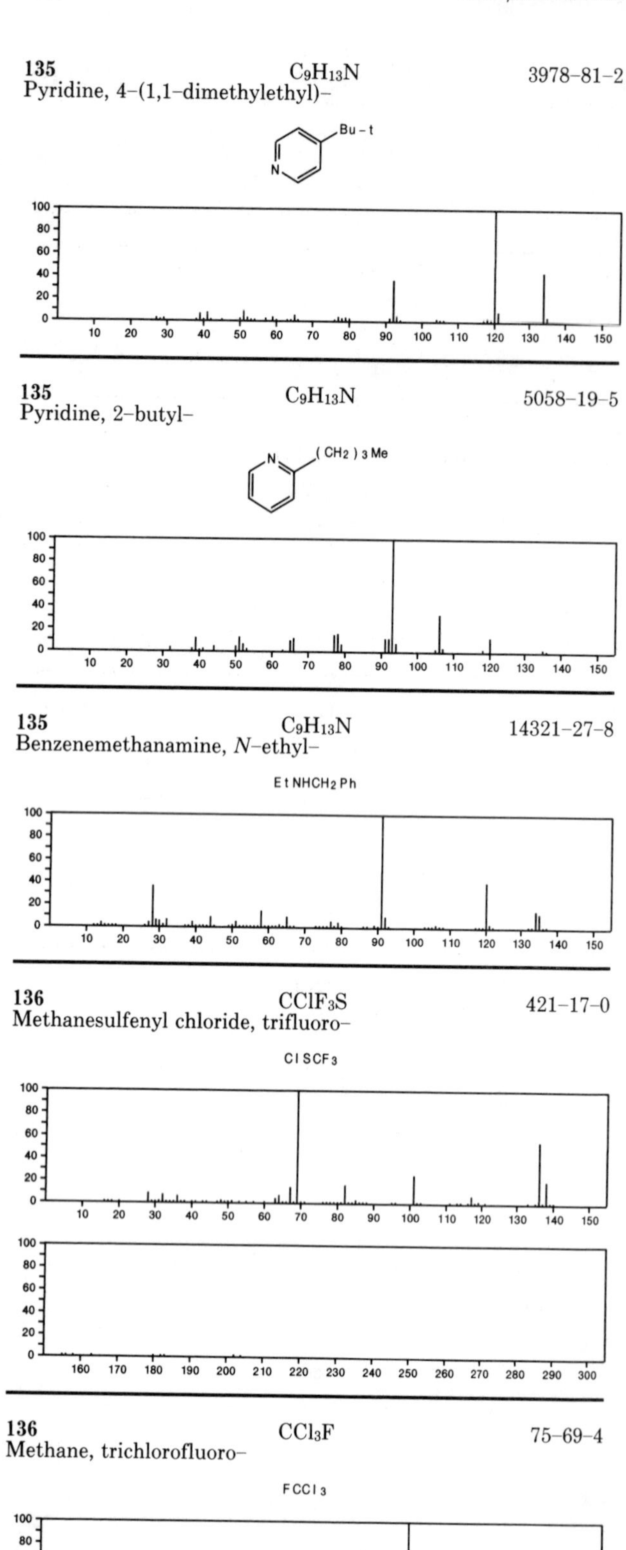

135 $C_9H_{13}N$ 3978-81-2
Pyridine, 4-(1,1-dimethylethyl)-

135 $C_9H_{13}N$ 5058-19-5
Pyridine, 2-butyl-

135 $C_9H_{13}N$ 14321-27-8
Benzenemethanamine, *N*-ethyl-

136 $CClF_3S$ 421-17-0
Methanesulfenyl chloride, trifluoro-

136 CCl_3F 75-69-4
Methane, trichlorofluoro-

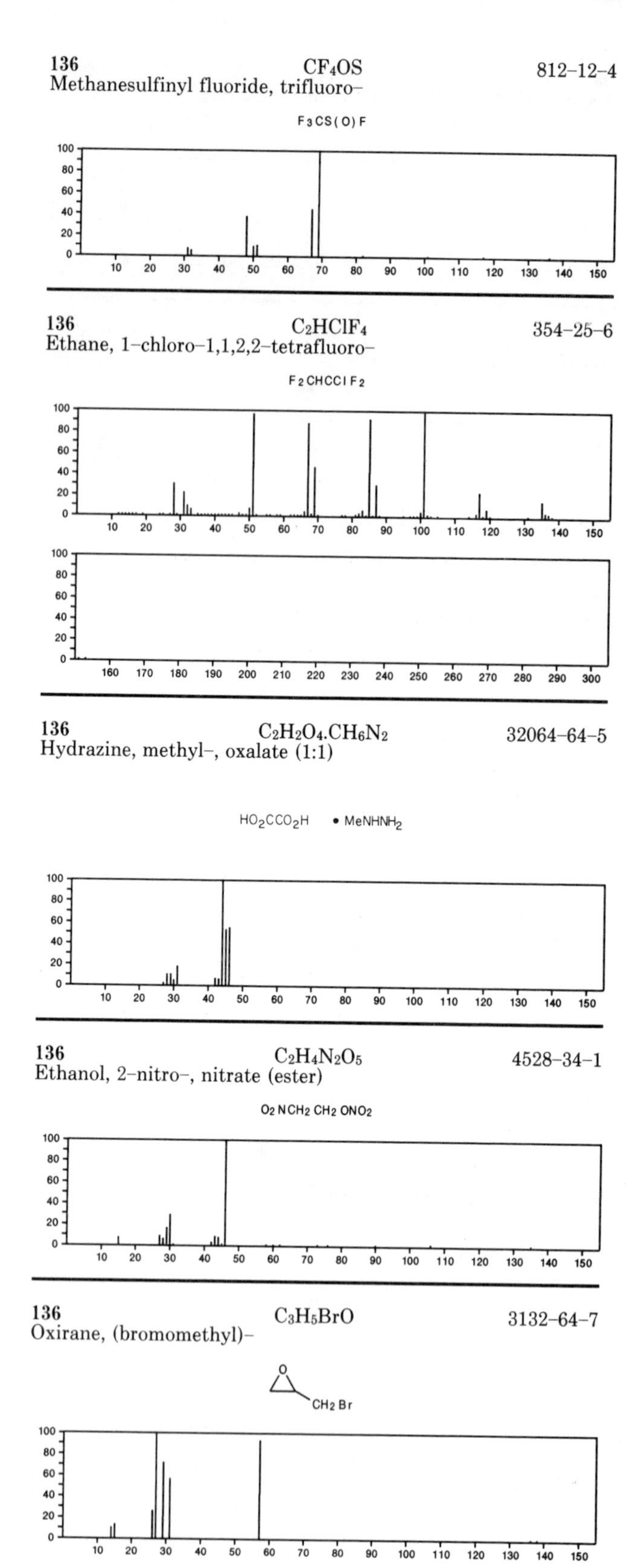

136 CF_4OS 812-12-4
Methanesulfinyl fluoride, trifluoro-

136 C_2HClF_4 354-25-6
Ethane, 1-chloro-1,1,2,2-tetrafluoro-

136 $C_2H_2O_4.CH_6N_2$ 32064-64-5
Hydrazine, methyl-, oxalate (1:1)

136 $C_2H_4N_2O_5$ 4528-34-1
Ethanol, 2-nitro-, nitrate (ester)

136 C_3H_5BrO 3132-64-7
Oxirane, (bromomethyl)-

136 $C_4H_8O_3S$ 4440–90–8
1,3,2-Dioxathiolane, 4,5-dimethyl-, 2-oxide

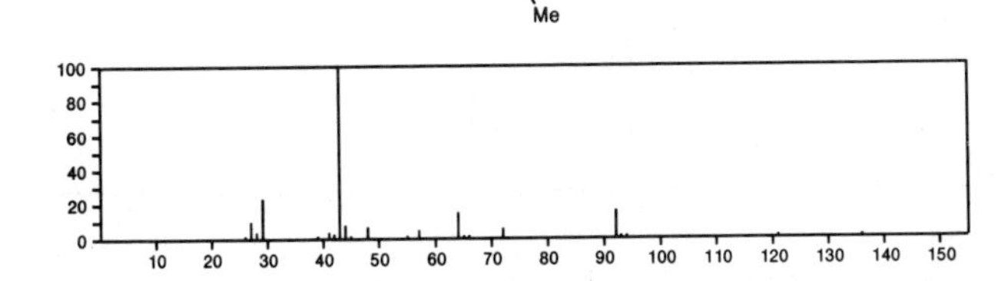

136 C_4H_8Se 3465–98–3
Selenophene, tetrahydro-

Se

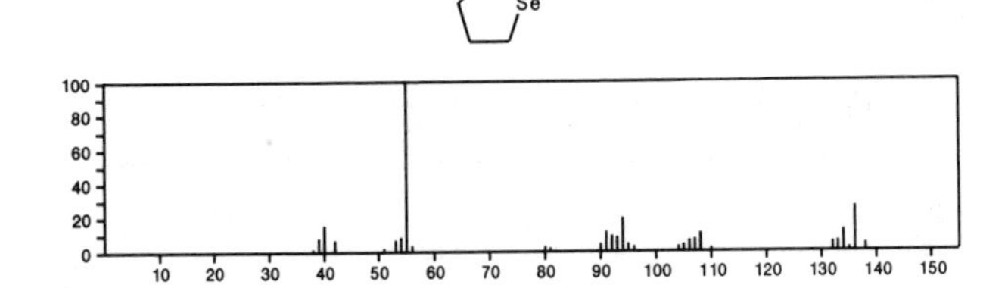

136 C_4H_9Br 78–76–2
Butane, 2-bromo-

s-BuBr

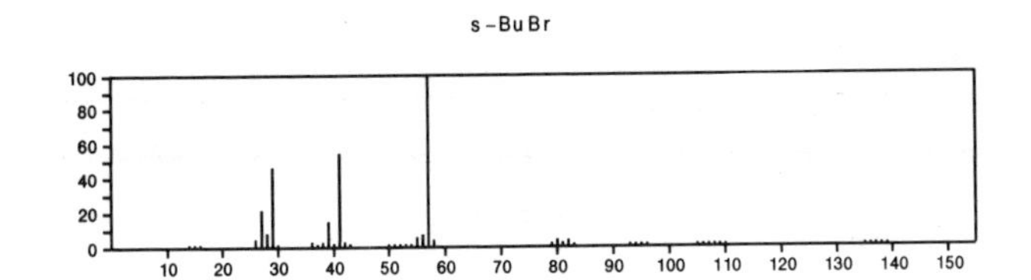

136 C_4H_9Br 78–77–3
Propane, 1-bromo-2-methyl-

i-BuBr

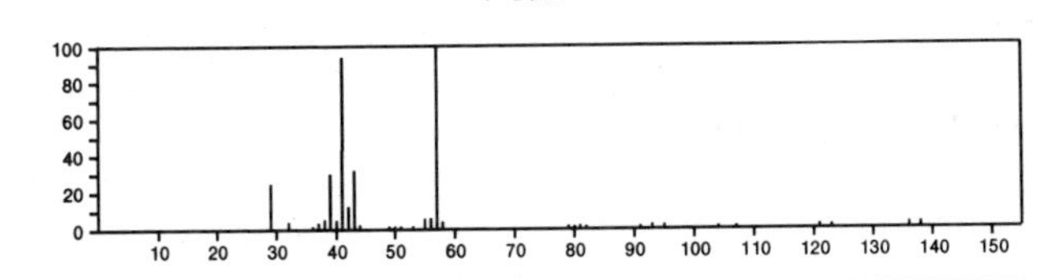

136 C_4H_9Br 109–65–9
Butane, 1-bromo-

Br$(CH_2)_3$Me

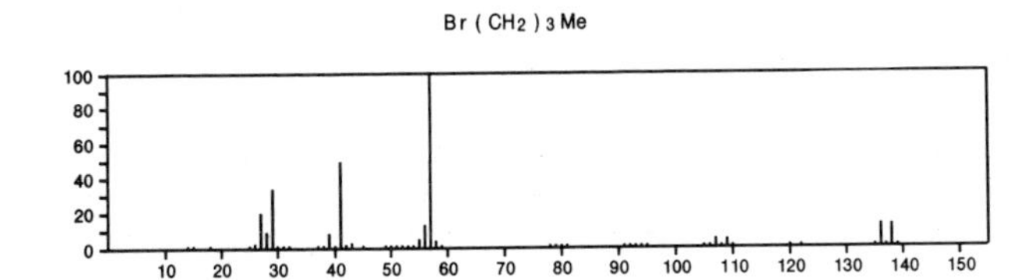

136 C_4H_9Br 507–19–7
Propane, 2-bromo-2-methyl-

t-BuBr

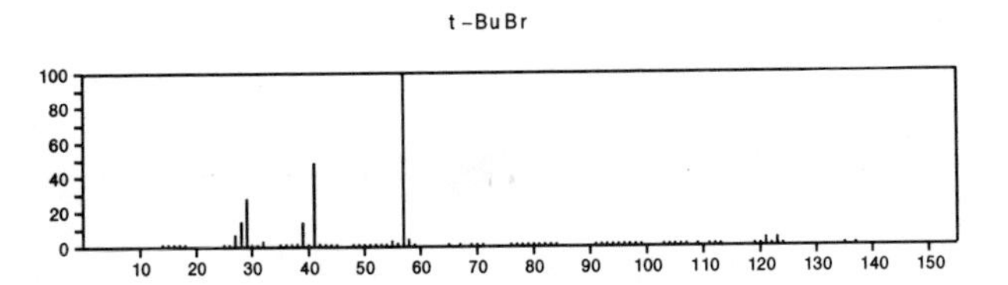

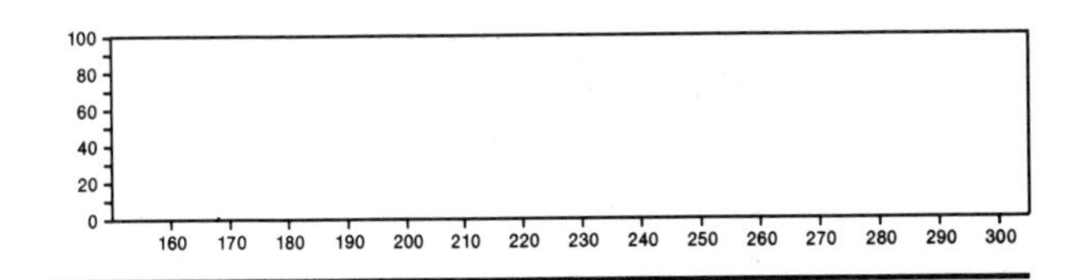

136 $C_5H_4N_4O$ 68–94–0
6*H*-Purin-6-one, 1,7-dihydro-

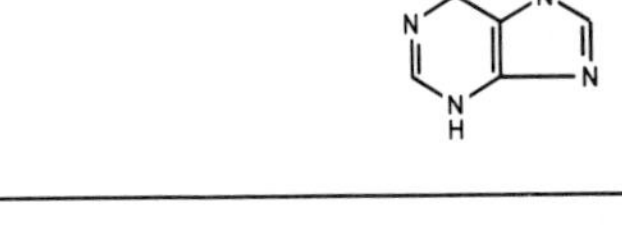

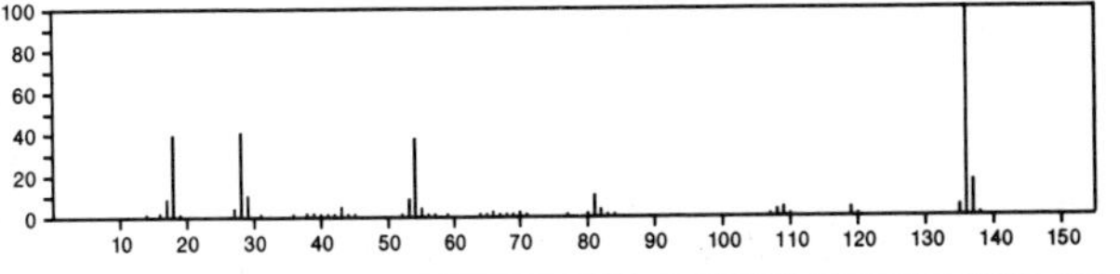

136 $C_5H_4N_4O$ 315–30–0
4*H*-Pyrazolo[3,4-*d*]pyrimidin-4-one, 1,5-dihydro-

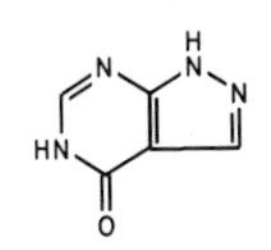

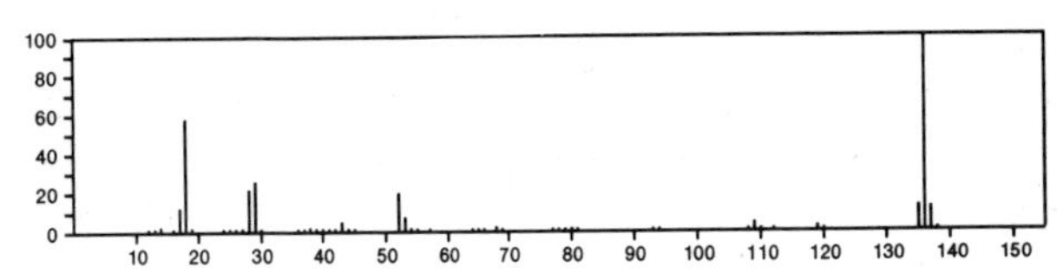

136 $C_5H_4N_4O$ 16328–63–5
2*H*-Imidazo[4,5-*b*]pyrazin-2-one

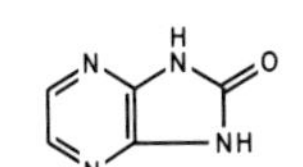

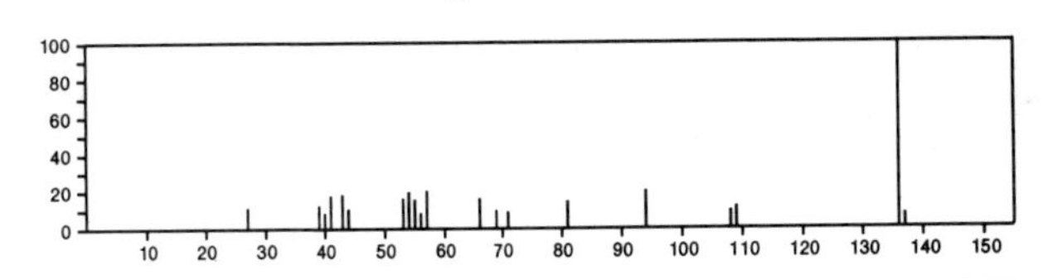

136 $C_5H_6Cl_2$ 694–33–7
Cyclopropane, 1,1-dichloro-2-ethenyl-

Cl
Cl
$CH=CH_2$

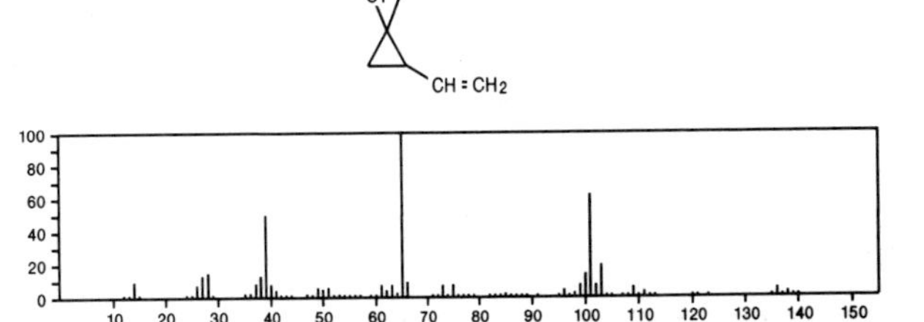

136 $C_5H_6F_2O_2$ 1597–40–6
2-Propen-1-ol, 3,3-difluoro-, acetate

$F_2C=CHCH_2OAc$

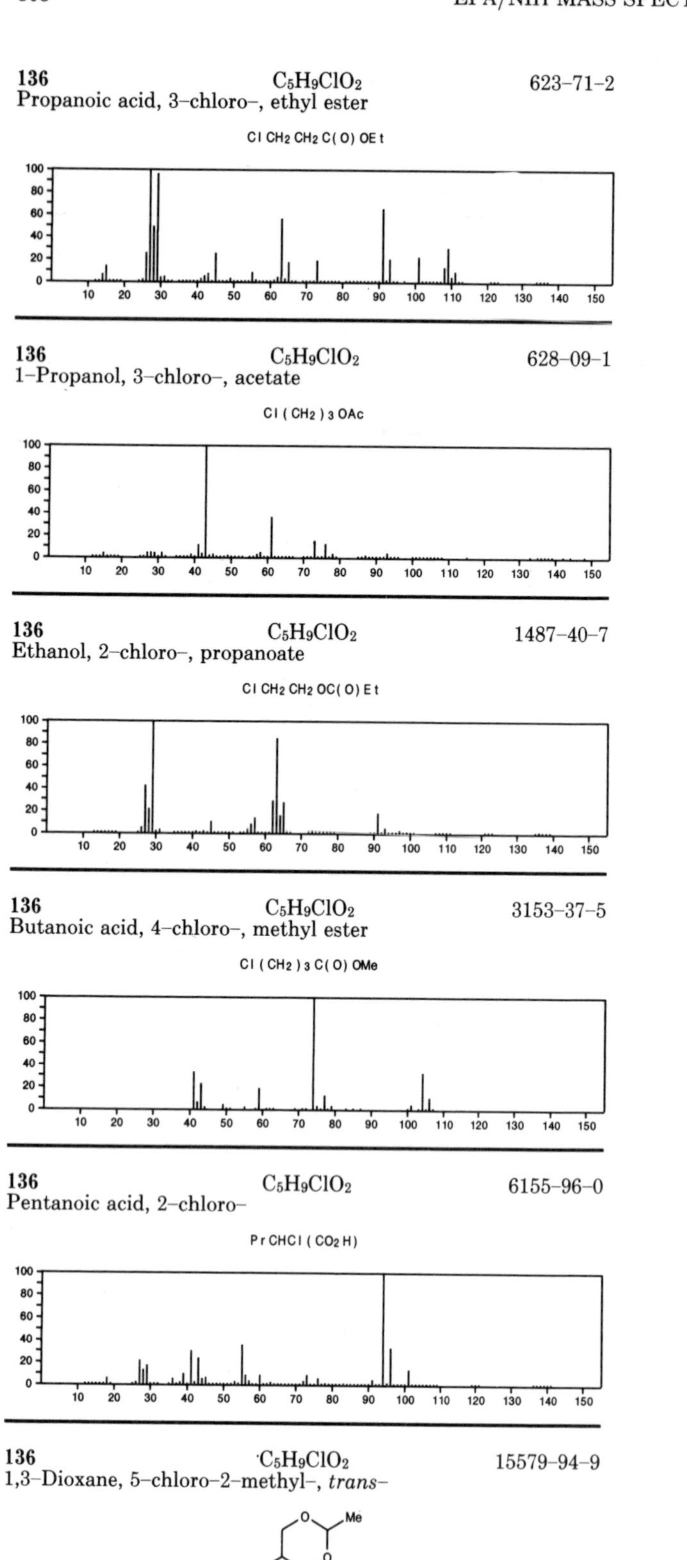
136 $C_5H_9ClO_2$ 623–71–2
Propanoic acid, 3–chloro–, ethyl ester
Cl CH2 CH2 C(O) OEt
136 $C_5H_9ClO_2$ 628–09–1
1–Propanol, 3–chloro–, acetate
Cl (CH2)3 OAc
136 $C_5H_9ClO_2$ 1487–40–7
Ethanol, 2–chloro–, propanoate
Cl CH2 CH2 OC(O) Et
136 $C_5H_9ClO_2$ 3153–37–5
Butanoic acid, 4–chloro–, methyl ester
Cl (CH2)3 C(O) OMe
136 $C_5H_9ClO_2$ 6155–96–0
Pentanoic acid, 2–chloro–
Pr CHCl (CO2H)
136 $C_5H_9ClO_2$ 15579–94–9
1,3–Dioxane, 5–chloro–2–methyl–, trans–

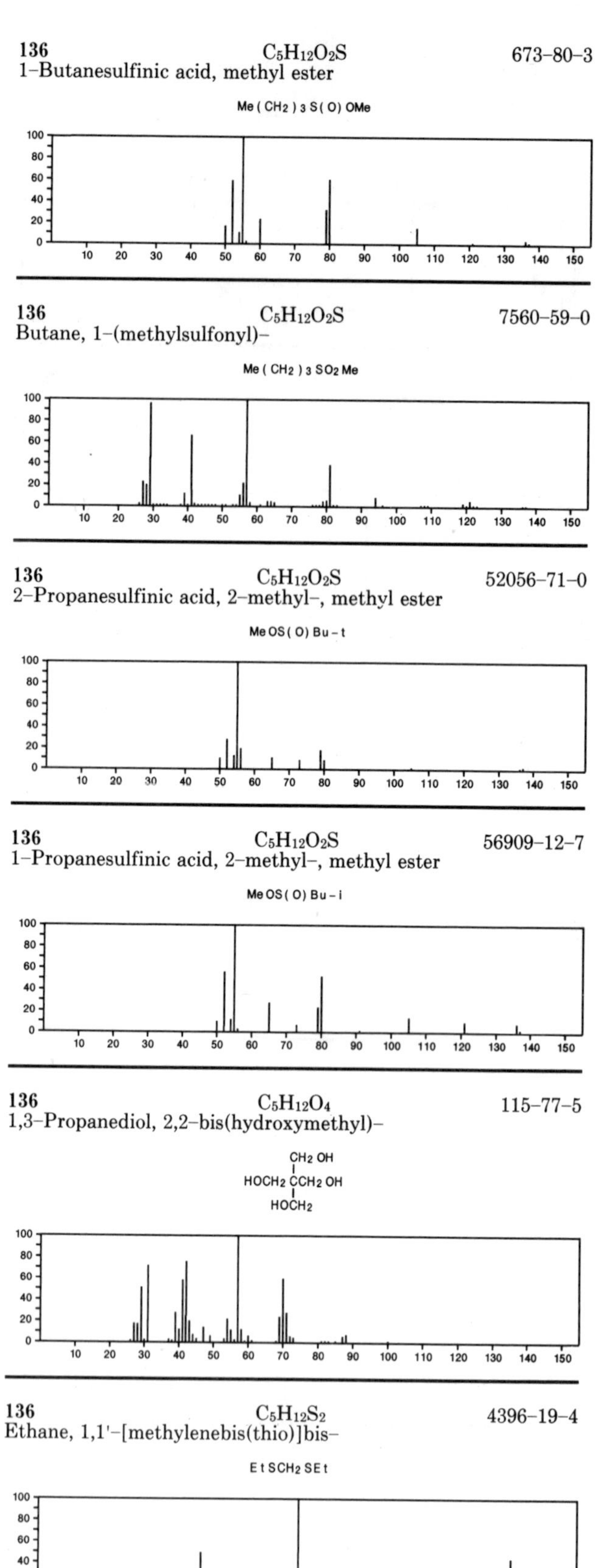
136 $C_5H_{12}O_2S$ 673–80–3
1–Butanesulfinic acid, methyl ester
Me (CH2)3 S(O) OMe
136 $C_5H_{12}O_2S$ 7560–59–0
Butane, 1–(methylsulfonyl)–
Me (CH2)3 SO2 Me
136 $C_5H_{12}O_2S$ 52056–71–0
2–Propanesulfinic acid, 2–methyl–, methyl ester
Me OS(O) Bu–t
136 $C_5H_{12}O_2S$ 56909–12–7
1–Propanesulfinic acid, 2–methyl–, methyl ester
Me OS(O) Bu–i
136 $C_5H_{12}O_4$ 115–77–5
1,3–Propanediol, 2,2–bis(hydroxymethyl)–
136 $C_5H_{12}S_2$ 4396–19–4
Ethane, 1,1'–[methylenebis(thio)]bis–
Et SCH2 SEt

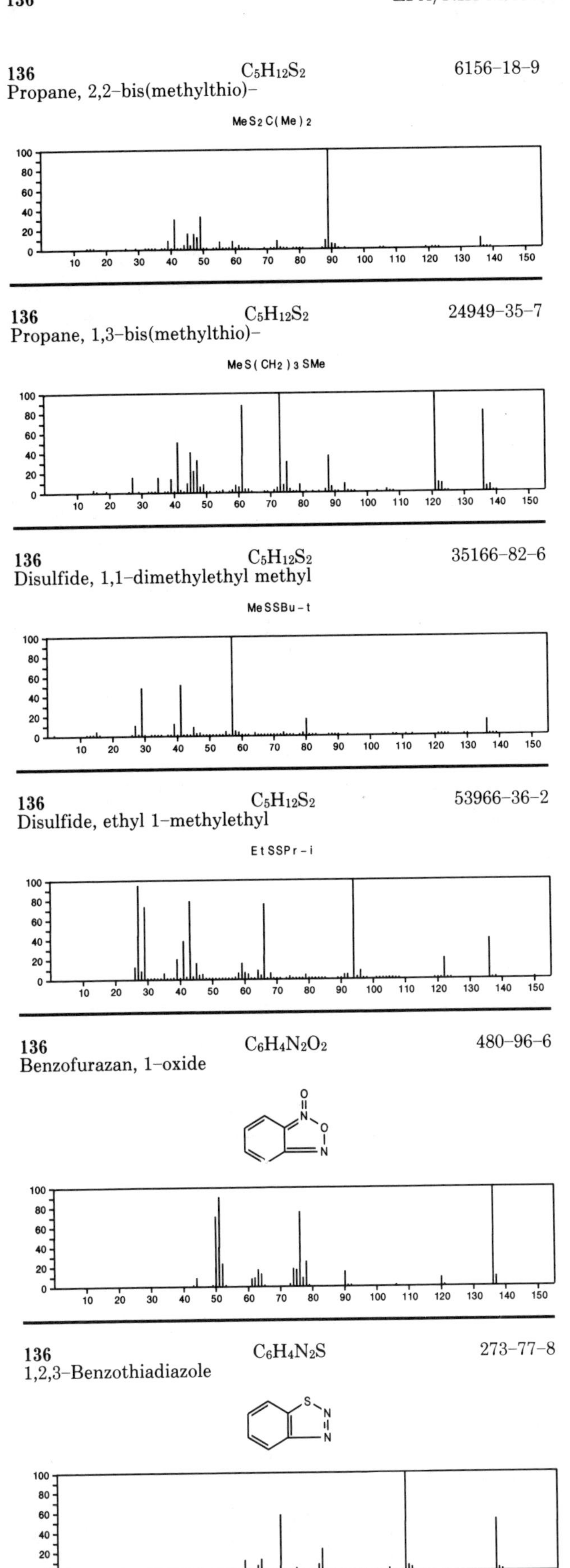

136 $C_5H_{12}S_2$ 6156-18-9
Propane, 2,2-bis(methylthio)-

136 $C_5H_{12}S_2$ 24949-35-7
Propane, 1,3-bis(methylthio)-

136 $C_5H_{12}S_2$ 35166-82-6
Disulfide, 1,1-dimethylethyl methyl

136 $C_5H_{12}S_2$ 53966-36-2
Disulfide, ethyl 1-methylethyl

136 $C_6H_4N_2O_2$ 480-96-6
Benzofurazan, 1-oxide

136 $C_6H_4N_2S$ 273-77-8
1,2,3-Benzothiadiazole

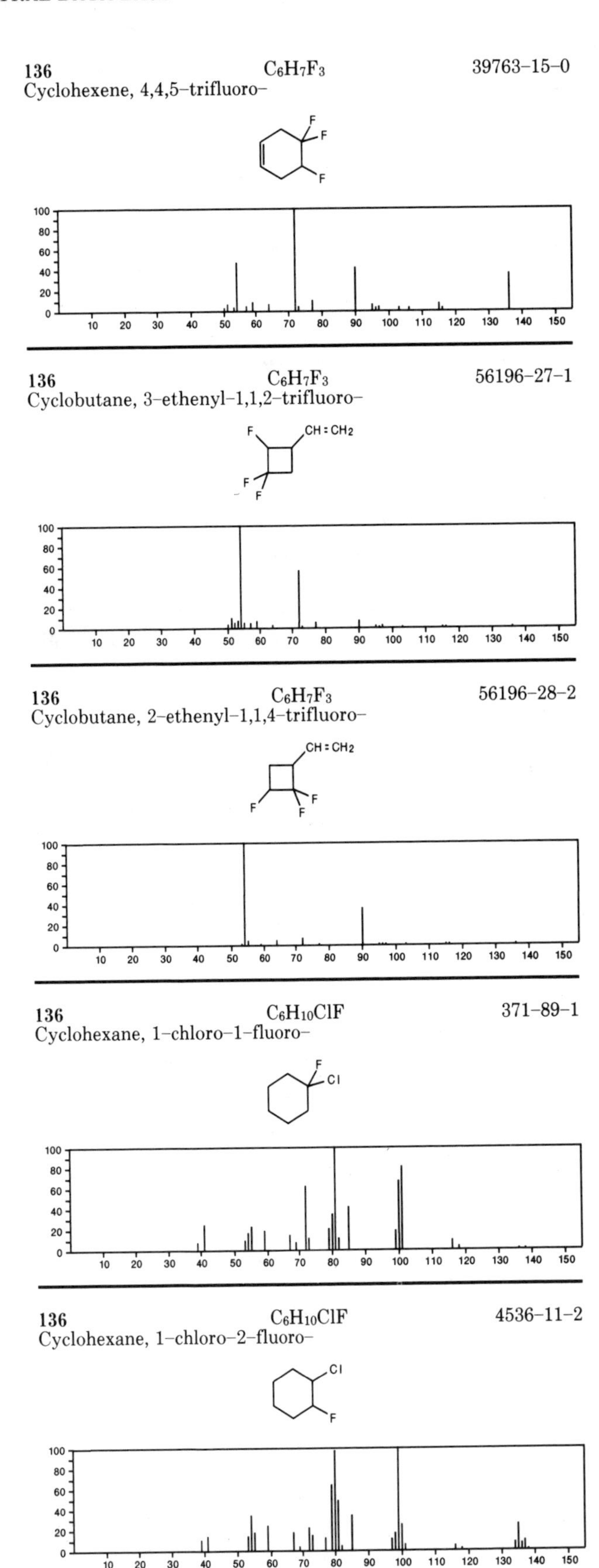

136 $C_6H_7F_3$ 39763-15-0
Cyclohexene, 4,4,5-trifluoro-

136 $C_6H_7F_3$ 56196-27-1
Cyclobutane, 3-ethenyl-1,1,2-trifluoro-

136 $C_6H_7F_3$ 56196-28-2
Cyclobutane, 2-ethenyl-1,1,4-trifluoro-

136 $C_6H_{10}ClF$ 371-89-1
Cyclohexane, 1-chloro-1-fluoro-

136 $C_6H_{10}ClF$ 4536-11-2
Cyclohexane, 1-chloro-2-fluoro-

136 $C_6H_{10}ClF$ 55887–79–1
Cyclohexane, 1–chloro–3–fluoro–

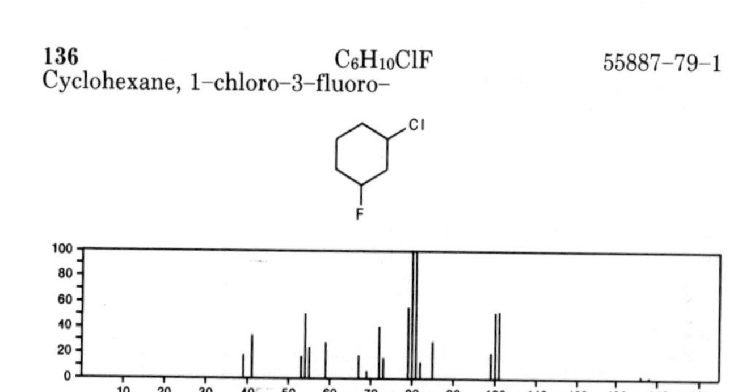

136 $C_6H_{10}ClF$ 55887–80–4
Cyclohexane, 1–chloro–4–fluoro–

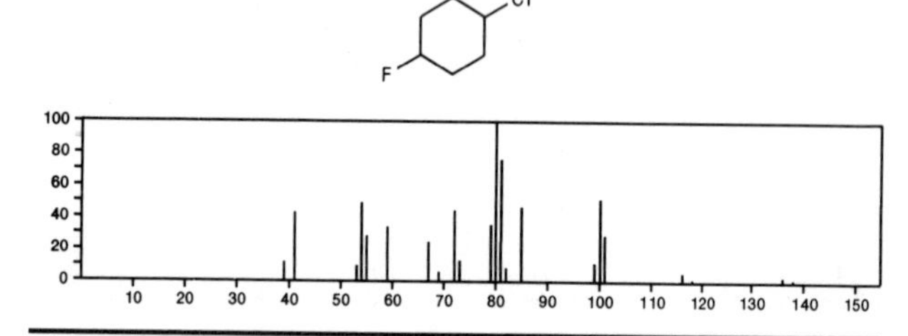

136 $C_7H_4O_3$ 2171–74–6
1,3–Benzodioxol–2–one

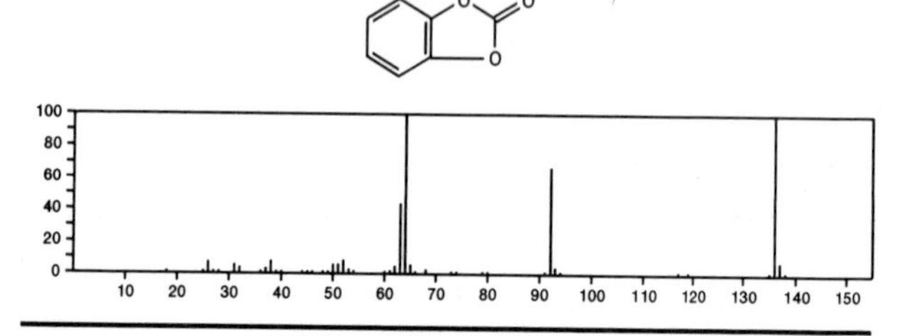

136 $C_7H_8N_2O$ 88–68–6
Benzamide, 2–amino–

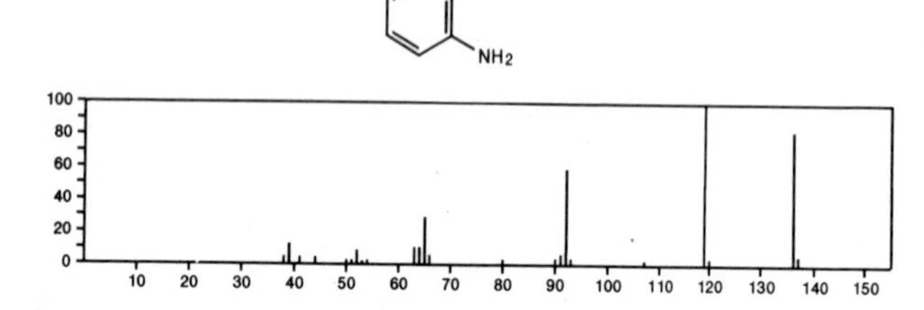

136 $C_7H_8N_2O$ 114–33–0
3–Pyridinecarboxamide, *N*–methyl–

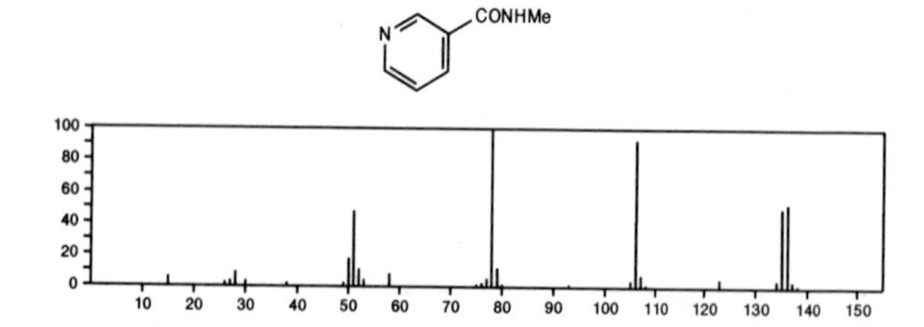

136 $C_7H_8N_2O$ 613–94–5
Benzoic acid, hydrazide

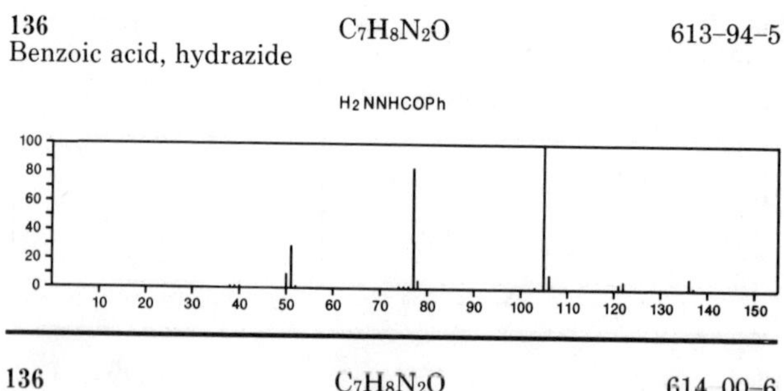

136 $C_7H_8N_2O$ 614–00–6
Benzenamine, *N*–methyl–*N*–nitroso–

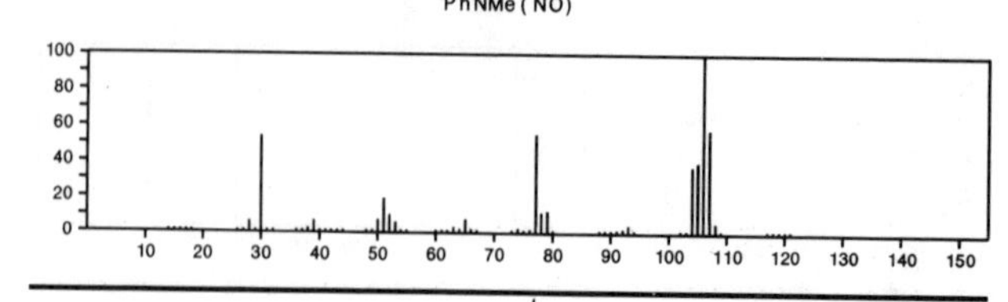

136 $C_7H_8N_2O$ 1468–29–7
Pyridinium, 1–(acetylamino)–, hydroxide, inner salt

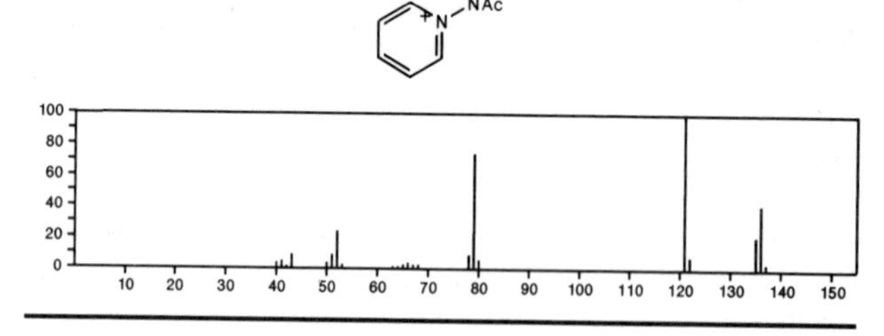

136 $C_7H_8N_2O$ 2835–66–7
Benzaldehyde, 3–amino–, oxime

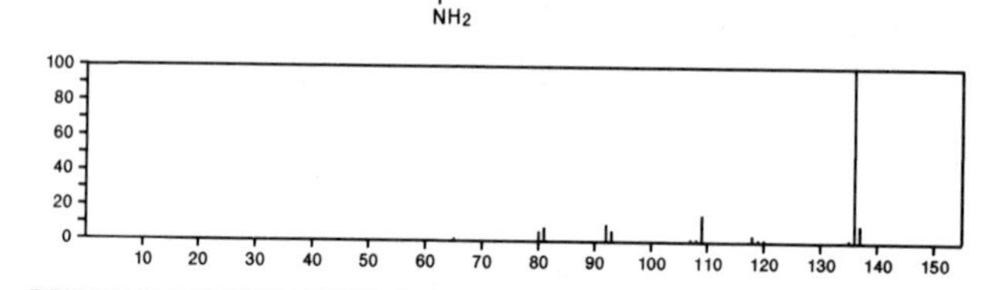

136 $C_7H_8N_2O$ 2835–68–9
Benzamide, 4–amino–

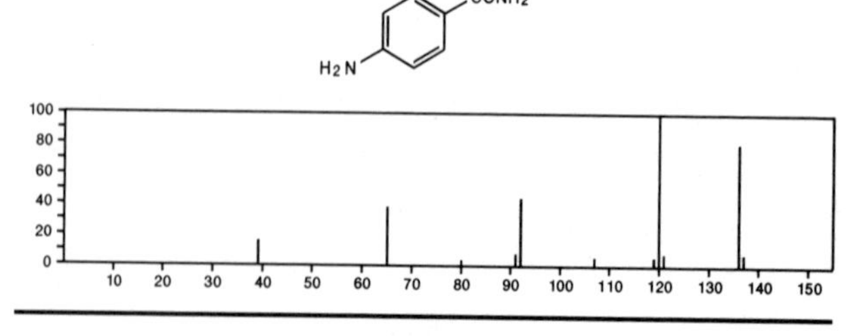

136 $C_7H_8N_2O$ 3419–18–9
Benzaldehyde, 4–amino–, oxime

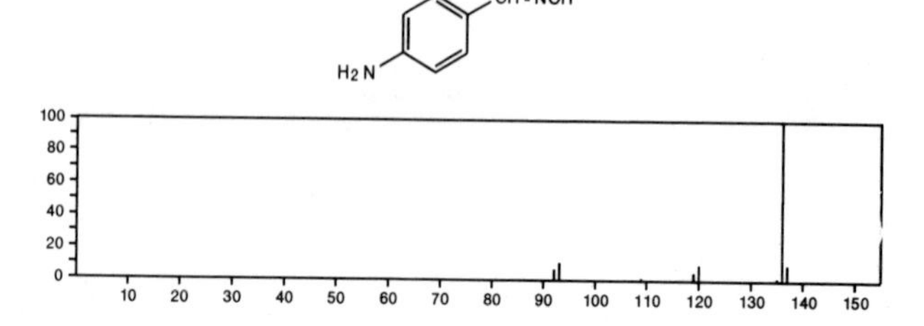

136 $C_7H_8N_2O$ 3544-24-9
Benzamide, 3-amino-

136 $C_7H_8N_2O$ 4406-68-2
Diazene, methylphenyl-, 2-oxide

MeN=NPh=O

136 $C_7H_8N_2O$ 35150-71-1
Diazene, methylphenyl-, 2-oxide, (*Z*)-

MeN=NPh=O

136 $C_7H_8N_2O$ 35150-73-3
Diazene, methylphenyl-, 2-oxide, (*E*)-

MeN=NPh=O

136 $C_7H_8N_2O$ 35150-74-4
Diazene, methylphenyl-, 1-oxide, (*Z*)-

O=NMe=NPh

136 $C_7H_8N_2O$ 35150-75-5
Diazene, methylphenyl-, 1-oxide, (*E*)-

O=NMe=NPh

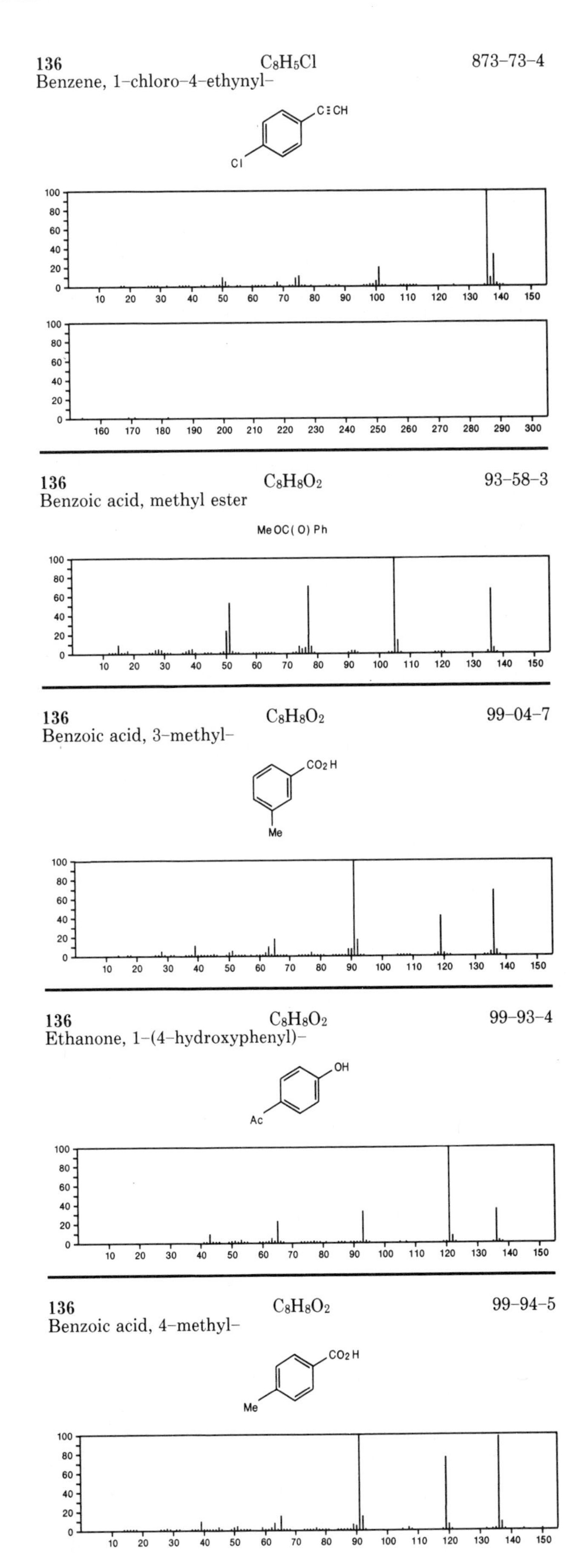

136 C_8H_5Cl 873-73-4
Benzene, 1-chloro-4-ethynyl-

136 $C_8H_8O_2$ 93-58-3
Benzoic acid, methyl ester

MeOC(O)Ph

136 $C_8H_8O_2$ 99-04-7
Benzoic acid, 3-methyl-

136 $C_8H_8O_2$ 99-93-4
Ethanone, 1-(4-hydroxyphenyl)-

136 $C_8H_8O_2$ 99-94-5
Benzoic acid, 4-methyl-

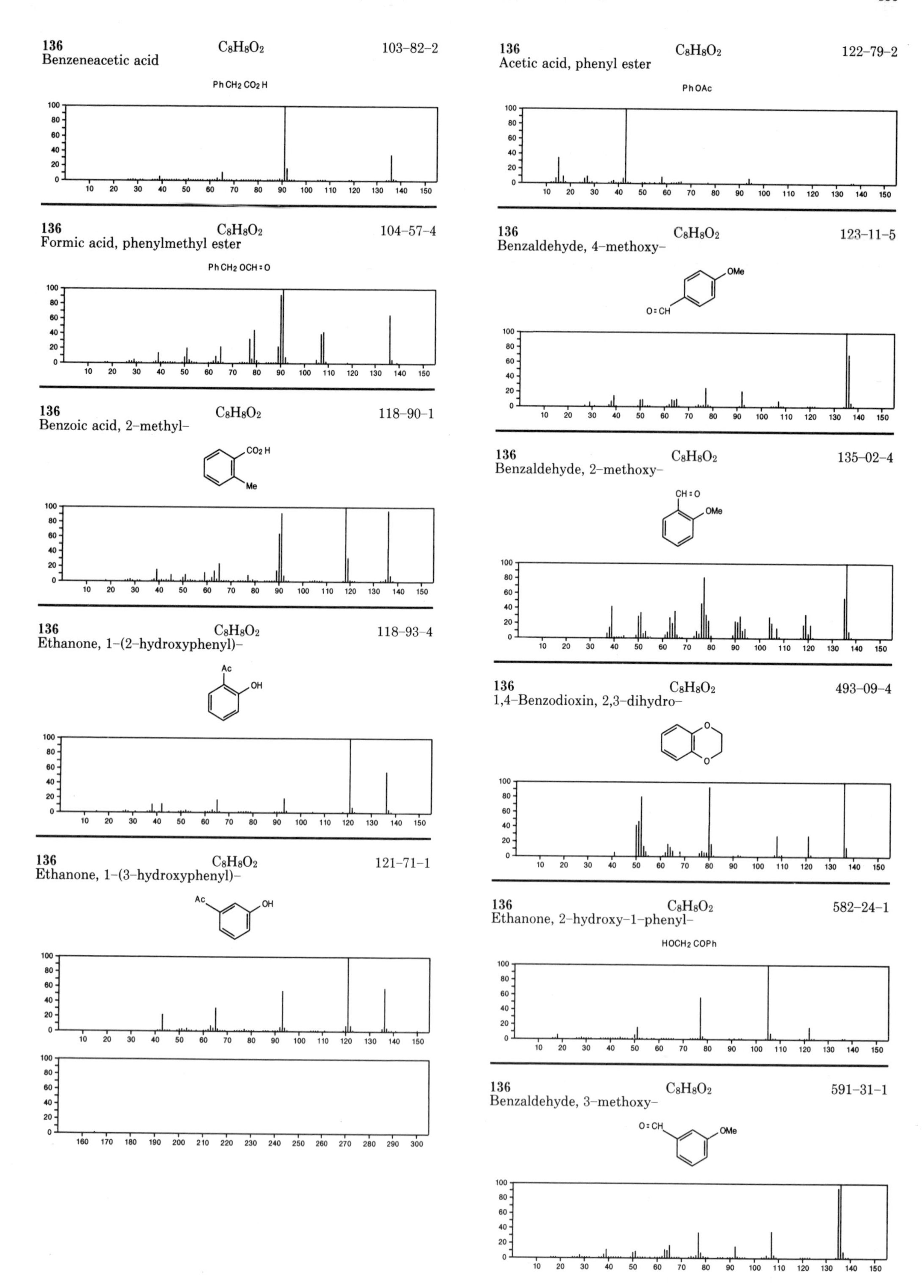
136 $C_8H_8O_2$ 103-82-2
Benzeneacetic acid
PhCH2CO2H
136 $C_8H_8O_2$ 104-57-4
Formic acid, phenylmethyl ester
PhCH2OCH=O
136 $C_8H_8O_2$ 118-90-1
Benzoic acid, 2-methyl-
CO2H
Me
136 $C_8H_8O_2$ 118-93-4
Ethanone, 1-(2-hydroxyphenyl)-
Ac
OH
136 $C_8H_8O_2$ 121-71-1
Ethanone, 1-(3-hydroxyphenyl)-
Ac
OH
136 $C_8H_8O_2$ 122-79-2
Acetic acid, phenyl ester
PhOAc
136 $C_8H_8O_2$ 123-11-5
Benzaldehyde, 4-methoxy-
OMe
O=CH
136 $C_8H_8O_2$ 135-02-4
Benzaldehyde, 2-methoxy-
CH=O
OMe
136 $C_8H_8O_2$ 493-09-4
1,4-Benzodioxin, 2,3-dihydro-
O
O
136 $C_8H_8O_2$ 582-24-1
Ethanone, 2-hydroxy-1-phenyl-
HOCH2COPh
136 $C_8H_8O_2$ 591-31-1
Benzaldehyde, 3-methoxy-
O=CH
OMe

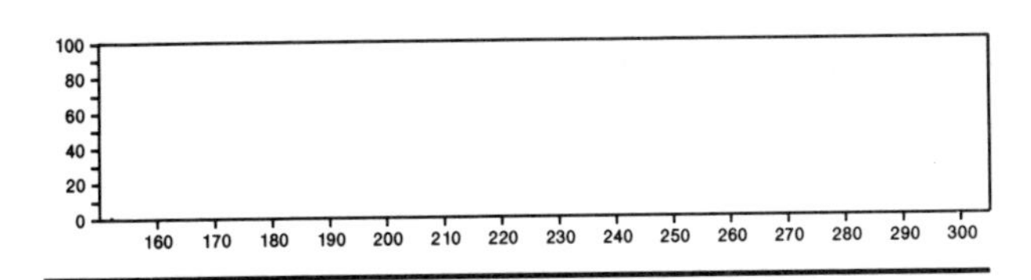

136 $C_8H_8O_2$ 4754-26-1
2,5-Cyclohexadiene-1,4-dione, 2-ethyl-

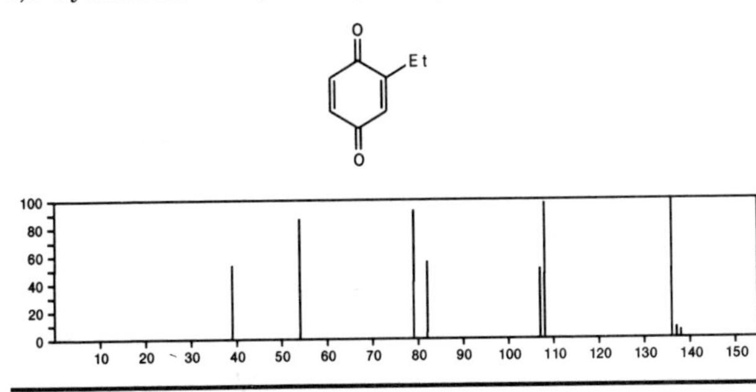

136 $C_8H_8O_2$ 17660-74-1
Bicyclo[2.2.2]oct-7-ene-2,5-dione

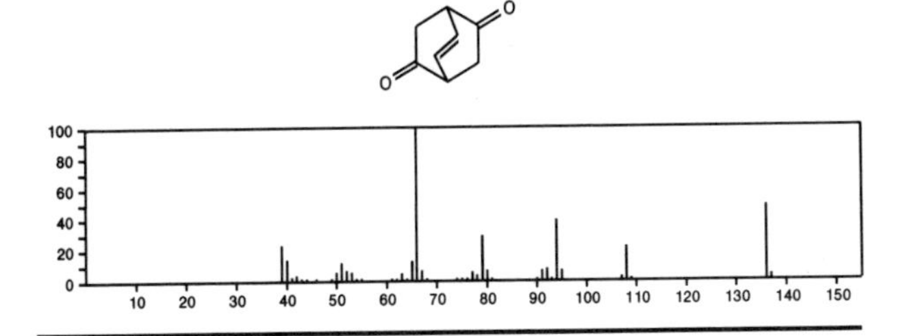

136 $C_8H_8O_2$ 28000-13-7
7-Oxabicyclo[4.2.1]nona-2,4-dien-8-one

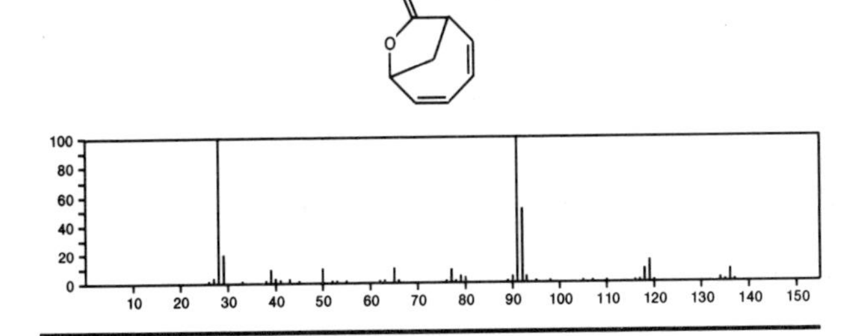

136 C_8H_8S 2471-92-3
Benzo[*c*]thiophene, 1,3-dihydro-

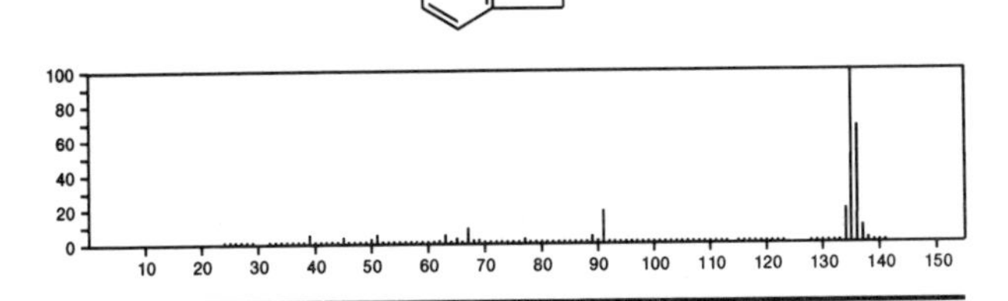

136 C_8H_8S 4565-32-6
Benzo[*b*]thiophene, 2,3-dihydro-

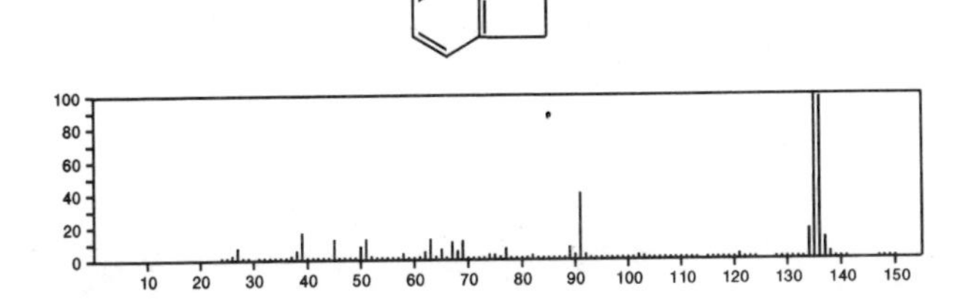

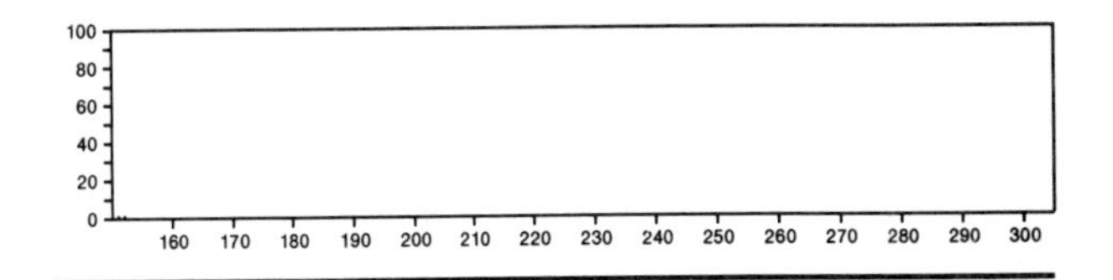

136 $C_8H_{12}N_2$ 629-40-3
Octanedinitrile

$NC(CH_2)_6CN$

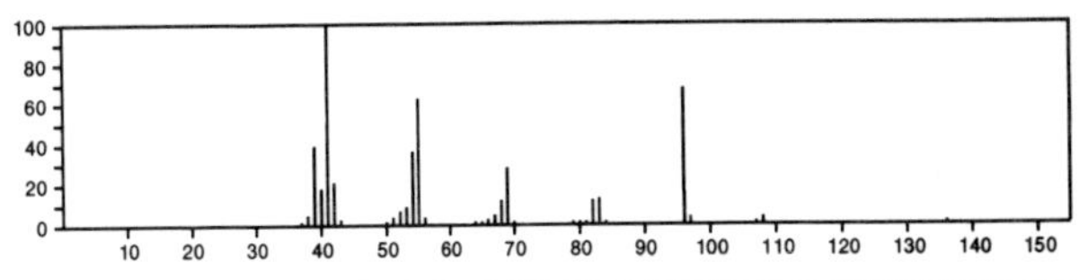

136 $C_8H_{12}N_2$ 1124-11-4
Pyrazine, tetramethyl-

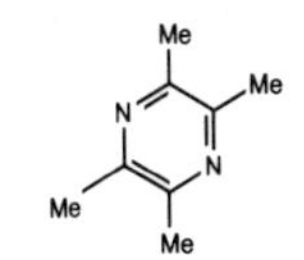

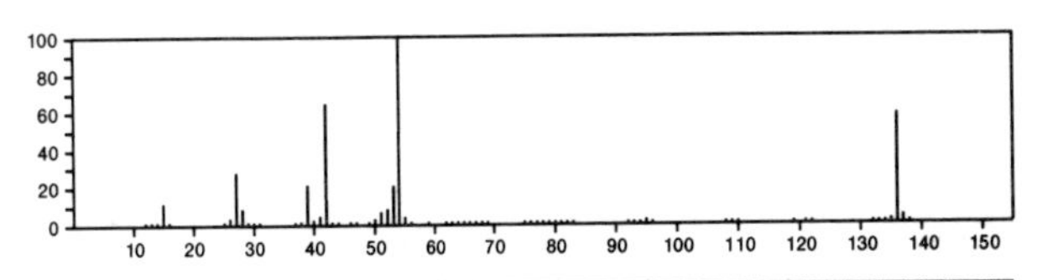

136 $C_8H_{12}N_2$ 1664-40-0
1,2-Ethanediamine, *N*-phenyl-

$PhNHCH_2CH_2NH_2$

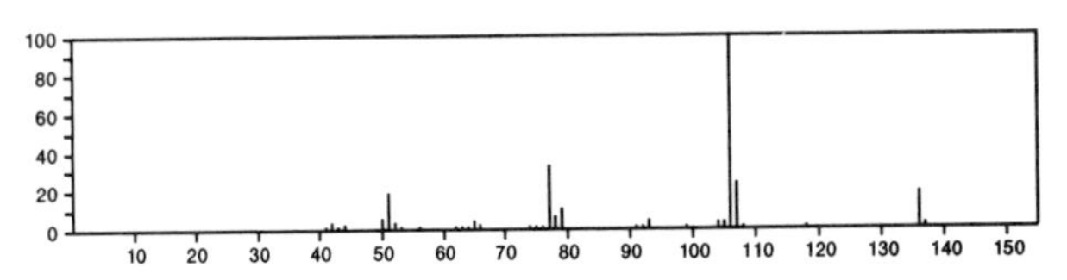

136 $C_8H_{12}N_2$ 32286-94-5
1*H*-Indazole, 4,5,6,7-tetrahydro-7-methyl-

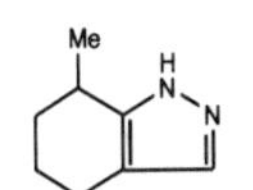

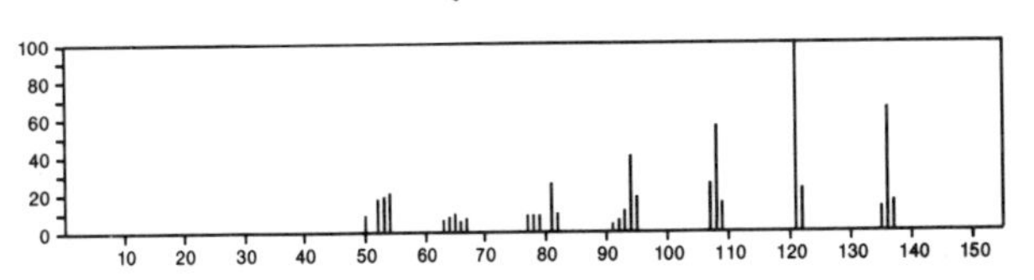

136 $C_9H_{12}O$ 88-69-7
Phenol, 2-(1-methylethyl)-

Pr-i

OH

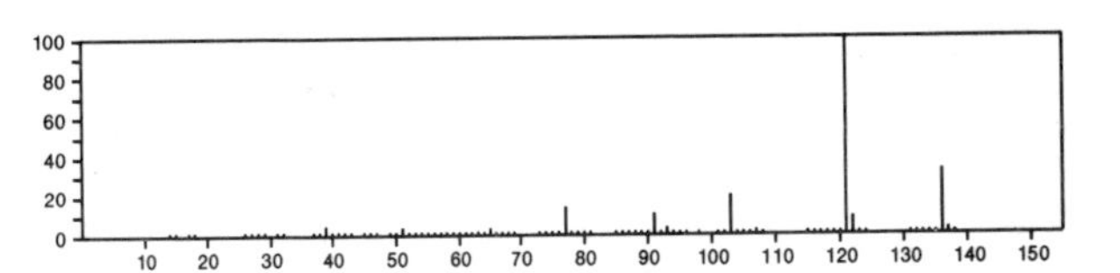

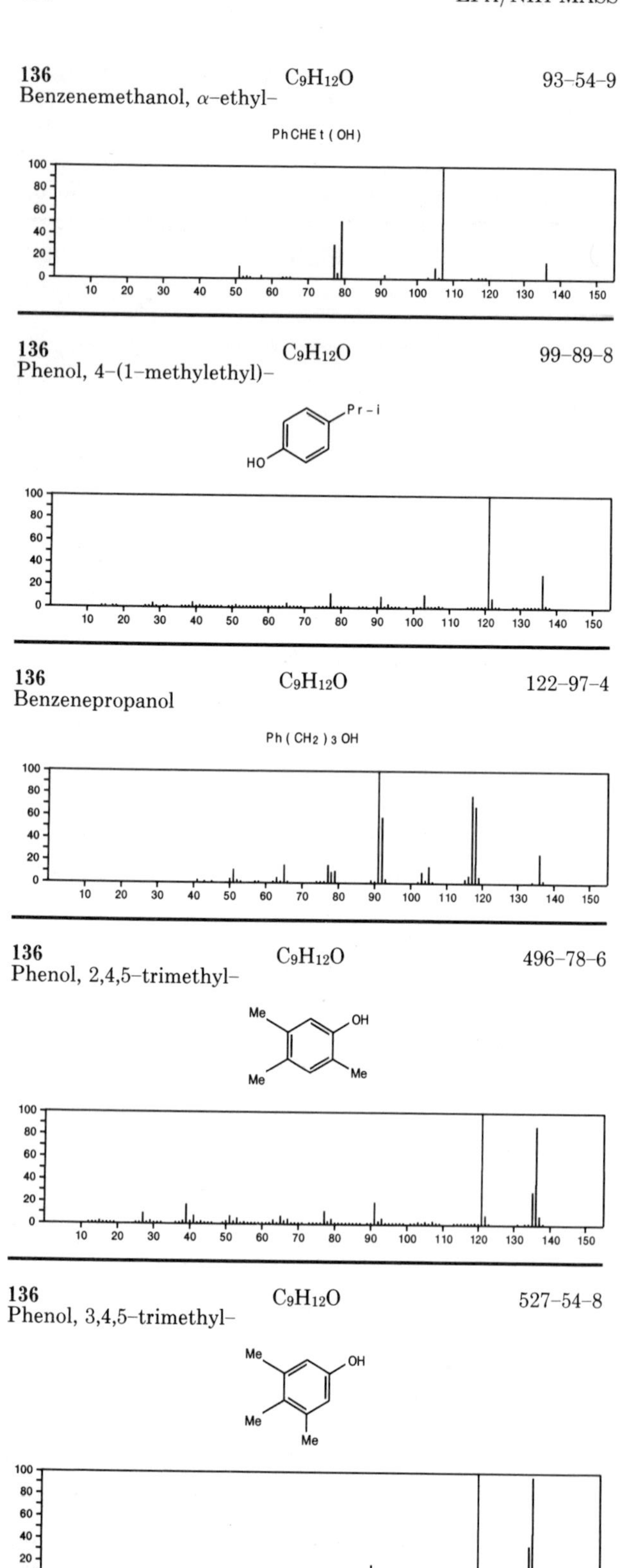
136
$C_9H_{12}O$
93–54–9
Benzenemethanol, α–ethyl–
PhCHEt(OH)
136
$C_9H_{12}O$
99–89–8
Phenol, 4–(1–methylethyl)–
Pr-i
HO
136
$C_9H_{12}O$
122–97–4
Benzenepropanol
Ph(CH2)3OH
136
$C_9H_{12}O$
496–78–6
Phenol, 2,4,5–trimethyl–
Me
OH
Me
Me
136
$C_9H_{12}O$
527–54–8
Phenol, 3,4,5–trimethyl–
Me
OH
Me
Me

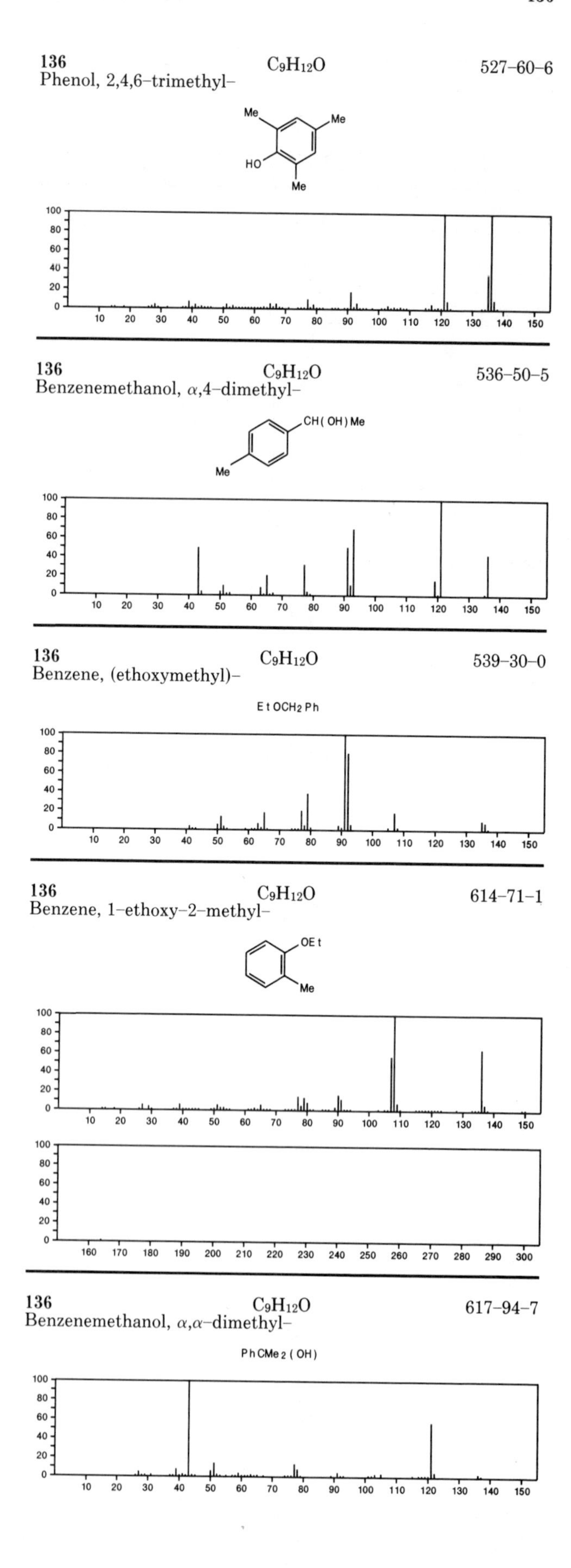
136
$C_9H_{12}O$
527–60–6
Phenol, 2,4,6–trimethyl–
Me
Me
HO
Me
136
$C_9H_{12}O$
536–50–5
Benzenemethanol, α,4–dimethyl–
CH(OH)Me
Me
136
$C_9H_{12}O$
539–30–0
Benzene, (ethoxymethyl)–
EtOCH2Ph
136
$C_9H_{12}O$
614–71–1
Benzene, 1–ethoxy–2–methyl–
OEt
Me
136
$C_9H_{12}O$
617–94–7
Benzenemethanol, α,α–dimethyl–
PhCMe2(OH)

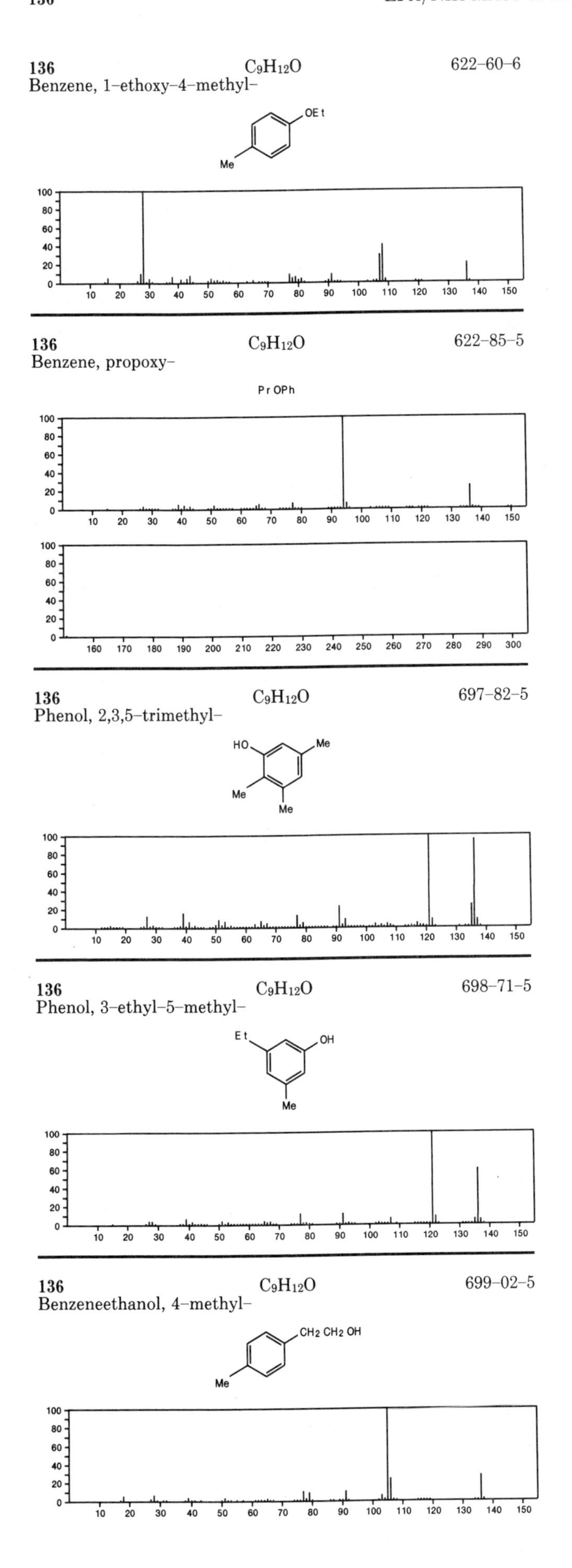

136 $C_9H_{12}O$ 622–60–6
Benzene, 1–ethoxy–4–methyl–

136 $C_9H_{12}O$ 622–85–5
Benzene, propoxy–

136 $C_9H_{12}O$ 697–82–5
Phenol, 2,3,5–trimethyl–

136 $C_9H_{12}O$ 698–71–5
Phenol, 3–ethyl–5–methyl–

136 $C_9H_{12}O$ 699–02–5
Benzeneethanol, 4–methyl–

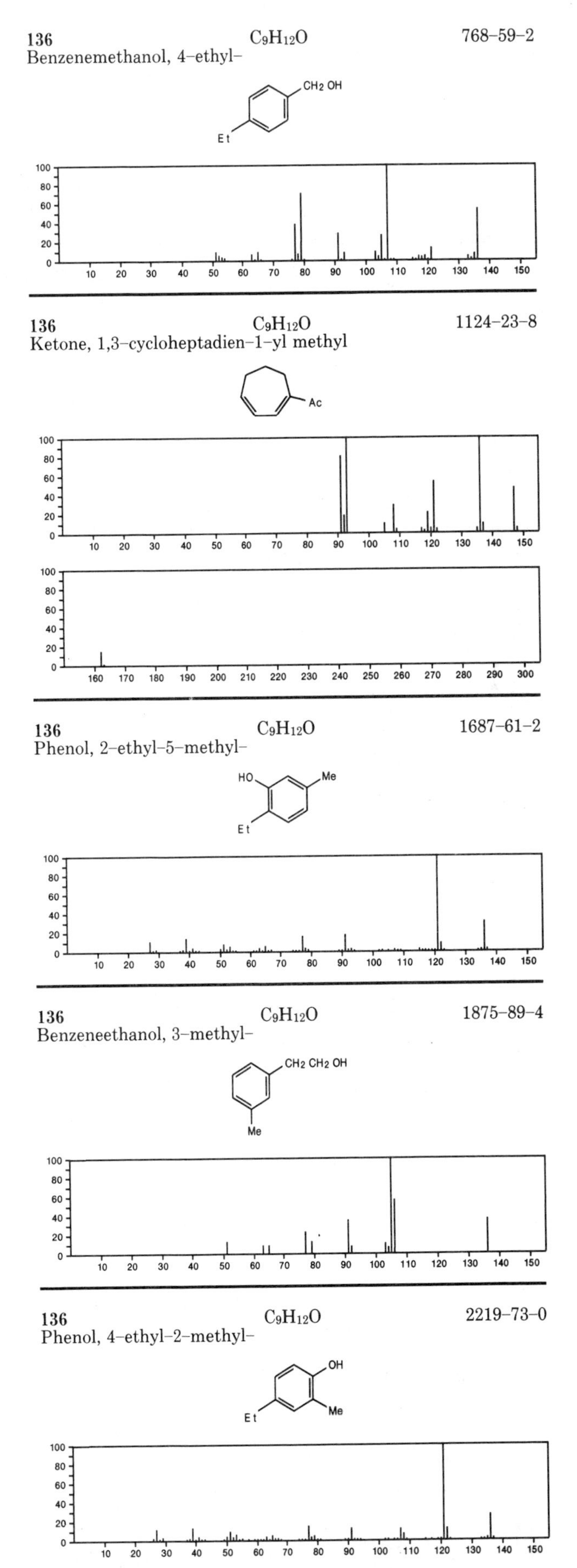

136 $C_9H_{12}O$ 768–59–2
Benzenemethanol, 4–ethyl–

136 $C_9H_{12}O$ 1124–23–8
Ketone, 1,3–cycloheptadien–1–yl methyl

136 $C_9H_{12}O$ 1687–61–2
Phenol, 2–ethyl–5–methyl–

136 $C_9H_{12}O$ 1875–89–4
Benzeneethanol, 3–methyl–

136 $C_9H_{12}O$ 2219–73–0
Phenol, 4–ethyl–2–methyl–

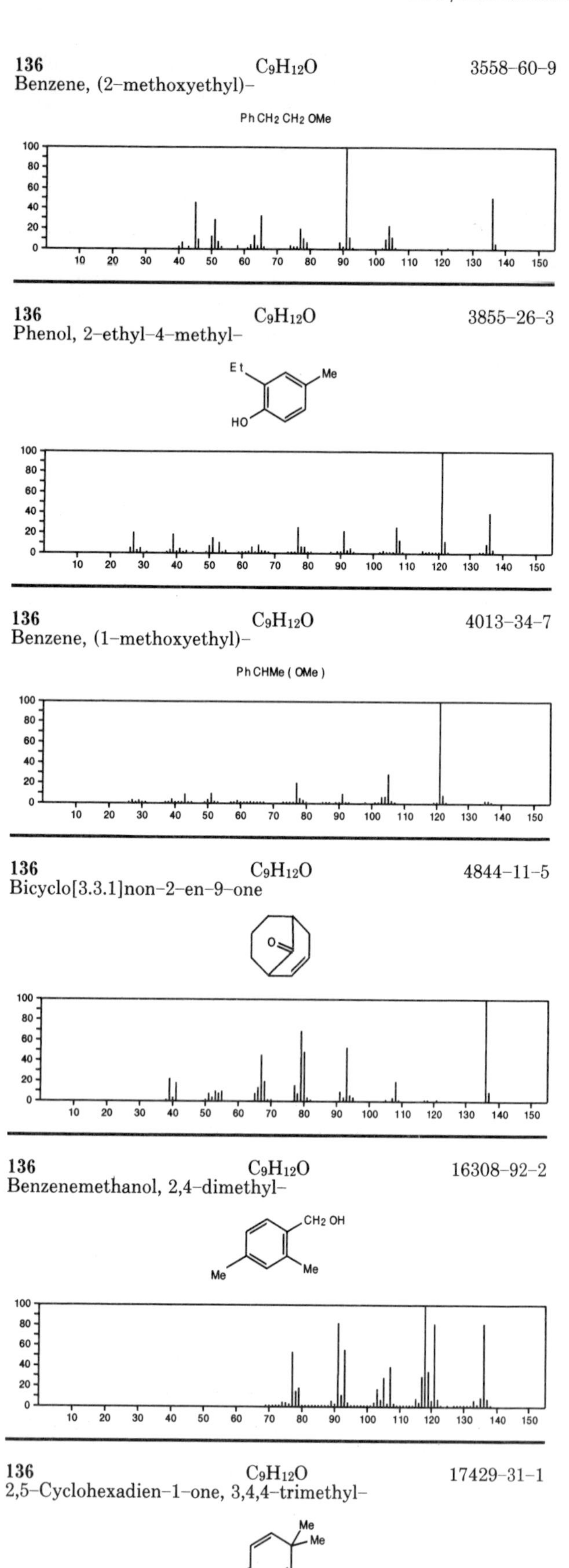

136 $C_9H_{12}O$ 3558-60-9
Benzene, (2-methoxyethyl)-
PhCH2CH2OMe
136 $C_9H_{12}O$ 3855-26-3
Phenol, 2-ethyl-4-methyl-
Et
Me
HO
136 $C_9H_{12}O$ 4013-34-7
Benzene, (1-methoxyethyl)-
PhCHMe(OMe)
136 $C_9H_{12}O$ 4844-11-5
Bicyclo[3.3.1]non-2-en-9-one
136 $C_9H_{12}O$ 16308-92-2
Benzenemethanol, 2,4-dimethyl-
CH2OH
Me
Me
136 $C_9H_{12}O$ 17429-31-1
2,5-Cyclohexadien-1-one, 3,4,4-trimethyl-
Me
Me
Me

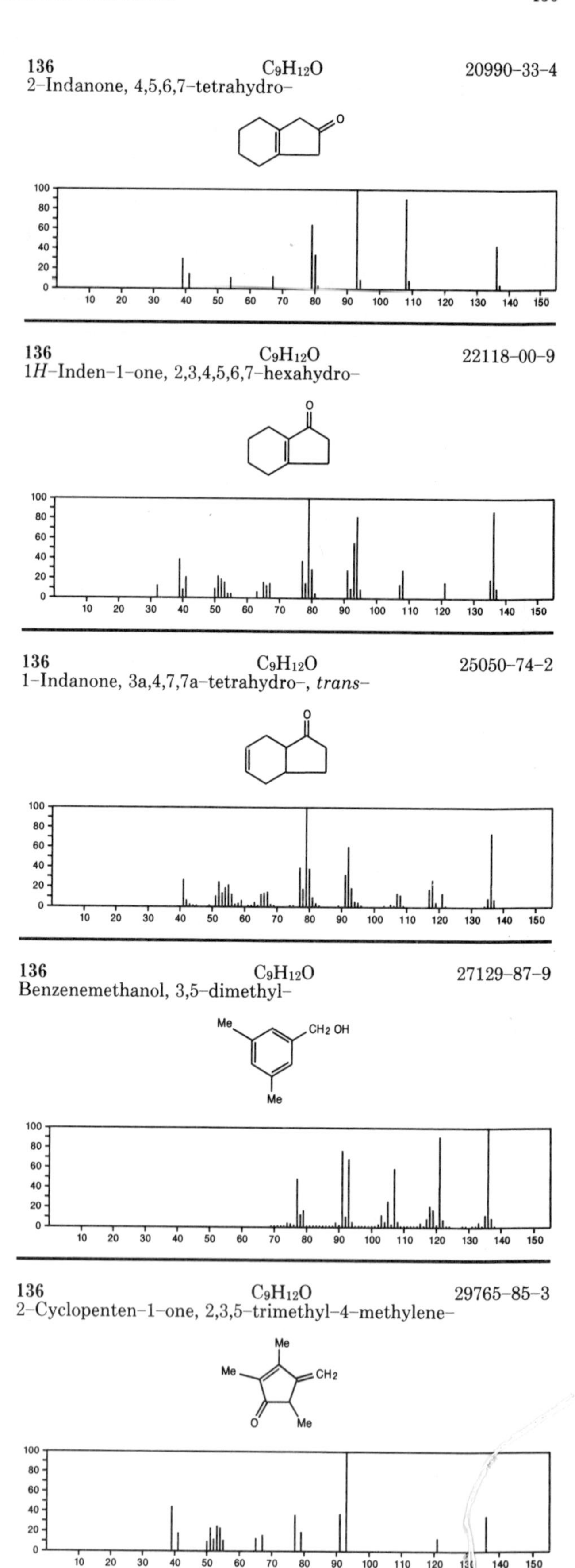

136 $C_9H_{12}O$ 20990-33-4
2-Indanone, 4,5,6,7-tetrahydro-
136 $C_9H_{12}O$ 22118-00-9
1H-Inden-1-one, 2,3,4,5,6,7-hexahydro-
136 $C_9H_{12}O$ 25050-74-2
1-Indanone, 3a,4,7,7a-tetrahydro-, trans-
136 $C_9H_{12}O$ 27129-87-9
Benzenemethanol, 3,5-dimethyl-
Me
CH2OH
Me
136 $C_9H_{12}O$ 29765-85-3
2-Cyclopenten-1-one, 2,3,5-trimethyl-4-methylene-
Me
Me
CH2
Me

136 $C_9H_{12}O$ 37778-99-7
Benzeneethanol, *β*-methyl-, (*S*)-

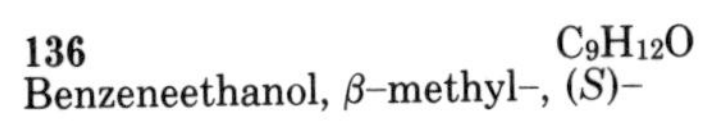

HOCH2 CHMePh

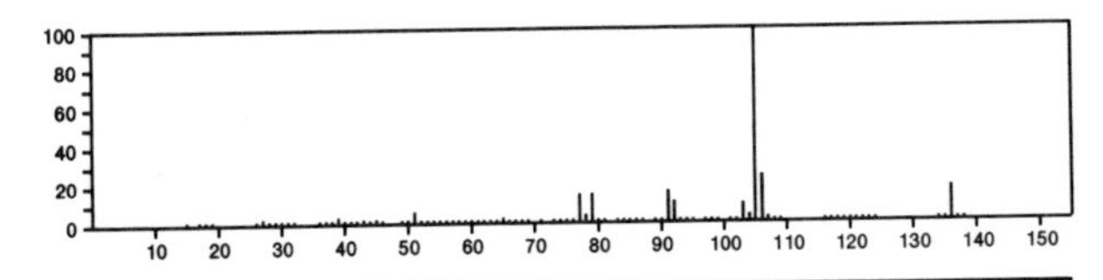

136 $C_9H_{12}O$ 43142-42-3
4,6-Nonadien-8-yn-3-ol, (*E,E*)-

EtCH(OH)CH=CHCH=CHC≡CH

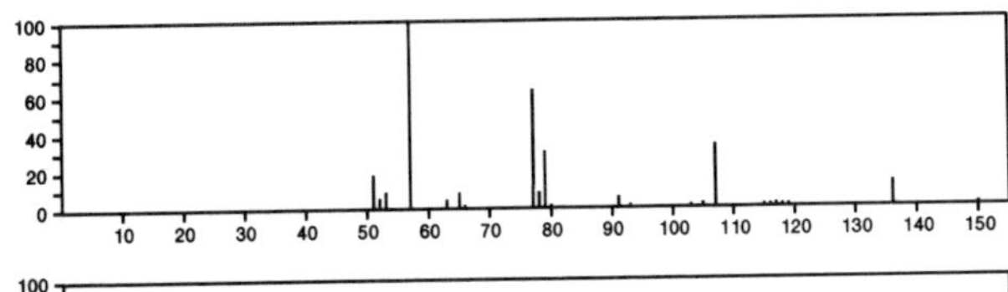

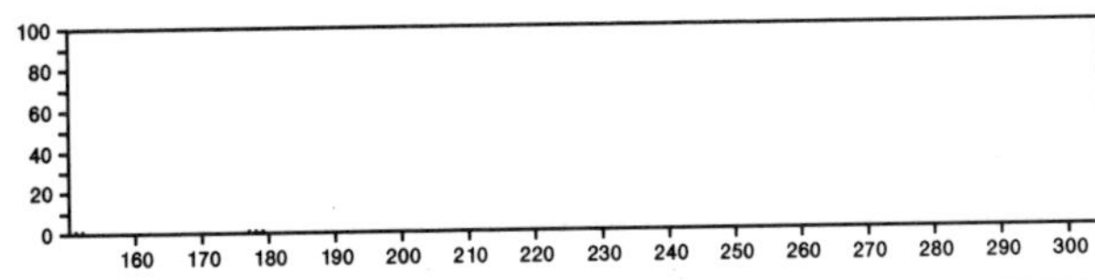

136 $C_9H_{12}O$ 43142-43-4
3,5-Nonadien-7-yn-2-ol, (*E,E*)-

MeC≡CCH=CHCH=CHCH(OH)Me

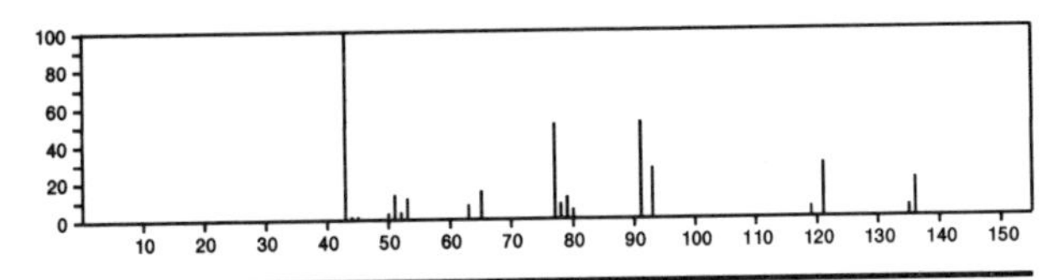

136 $C_9H_{12}O$ 43212-86-8
2,4-Nonadien-6-yn-1-ol, (*E,E*)-

EtC≡CCH=CHCH=CHCH2OH

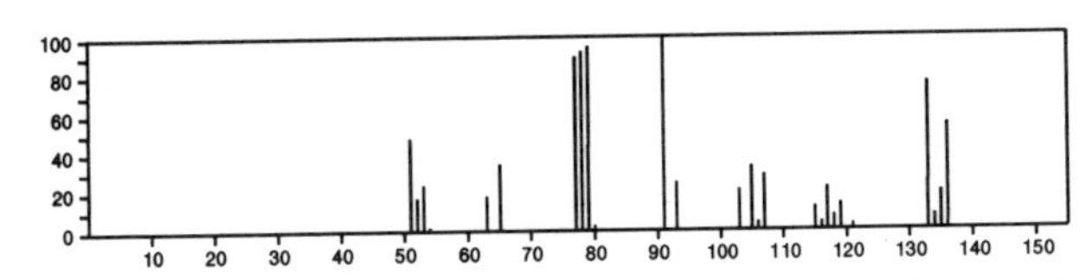

136 $C_9H_{12}O$ 53921-54-3
1*H*-Inden-1-one, 2,3,3a,4,7,7a-hexahydro-, *cis*-

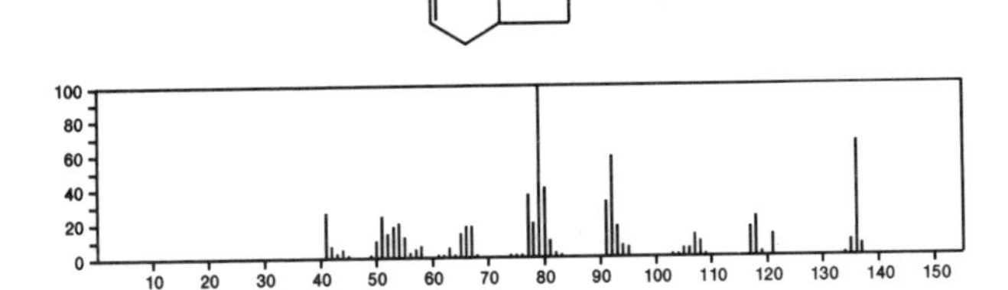

136 $C_9H_{12}O$ 53957-33-8
Benzenemethanol, 2,5-dimethyl-

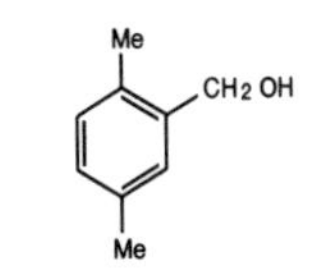

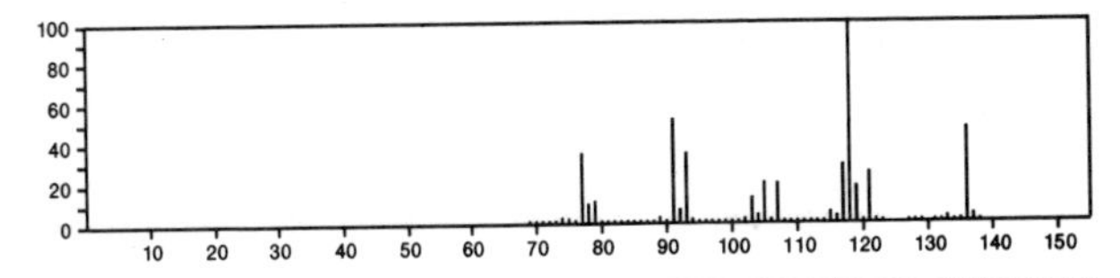

136 $C_9H_{12}O$ 53957-34-9
Benzenemethanol, *ar*-ethyl-

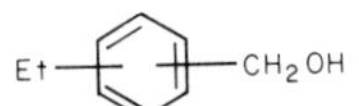

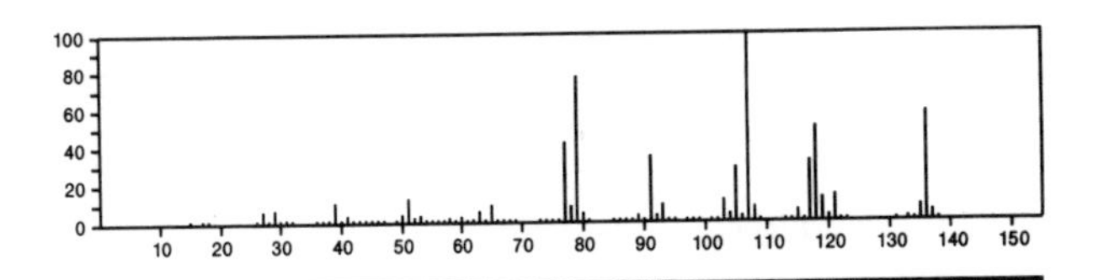

136 $C_9H_{12}O$ 56771-49-4
Spiro[oxirane-2,1'(2'*H*)-pentalene], 3',4',5',6'-tetrahydro-

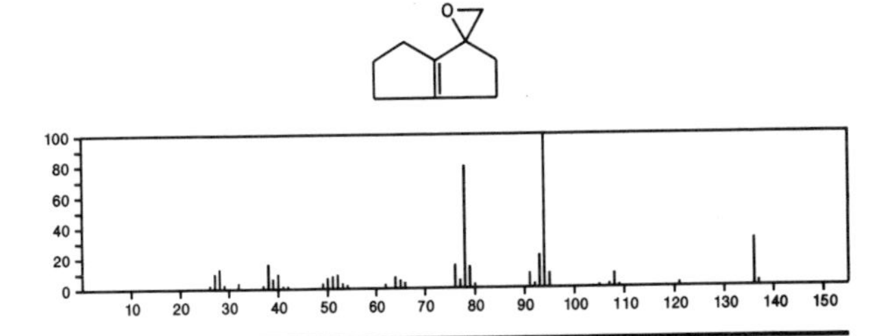

136 $C_{10}H_{16}$ 79-92-5
Bicyclo[2.2.1]heptane, 2,2-dimethyl-3-methylene-

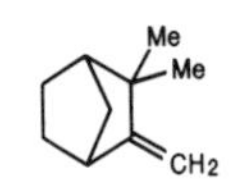

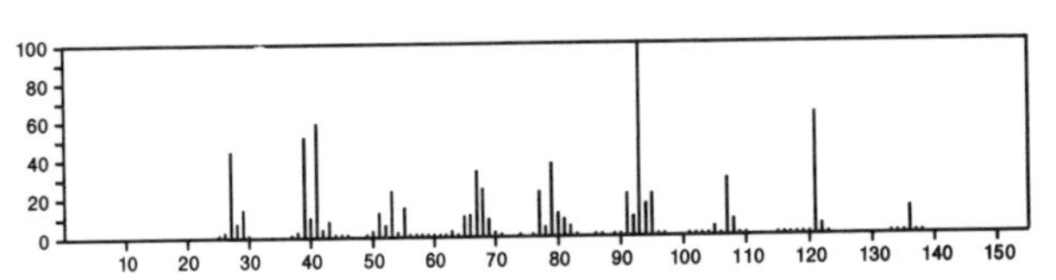

136 $C_{10}H_{16}$ 80-56-8
Bicyclo[3.1.1]hept-2-ene, 2,6,6-trimethyl-

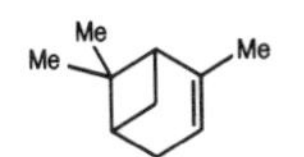

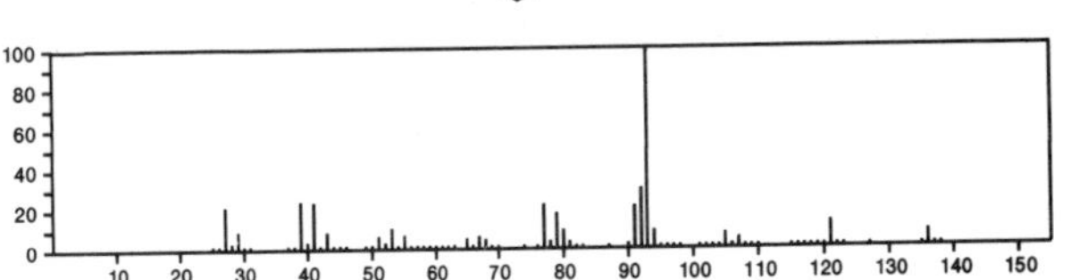

136 $C_{10}H_{16}$ 99-83-2
1,3-Cyclohexadiene, 2-methyl-5-(1-methylethyl)-

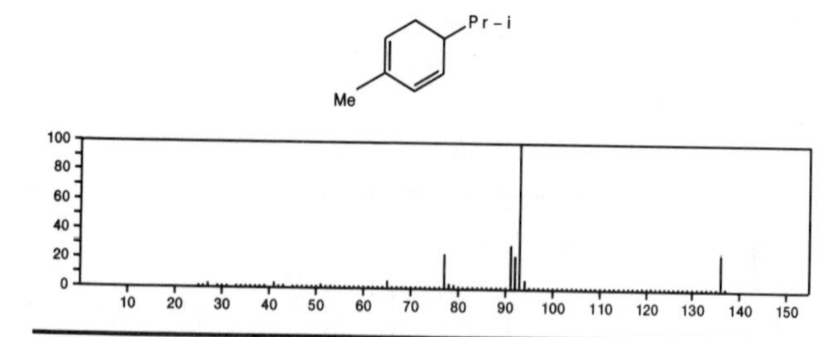

136 $C_{10}H_{16}$ 99-85-4
1,4-Cyclohexadiene, 1-methyl-4-(1-methylethyl)-

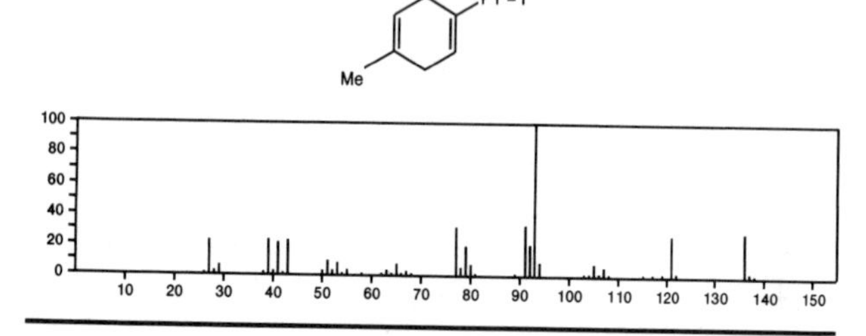

136 $C_{10}H_{16}$ 99-86-5
1,3-Cyclohexadiene, 1-methyl-4-(1-methylethyl)-

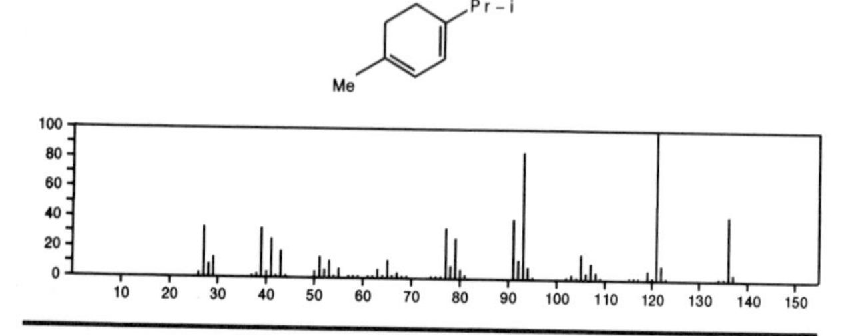

136 $C_{10}H_{16}$ 123-35-3
1,6-Octadiene, 7-methyl-3-methylene-

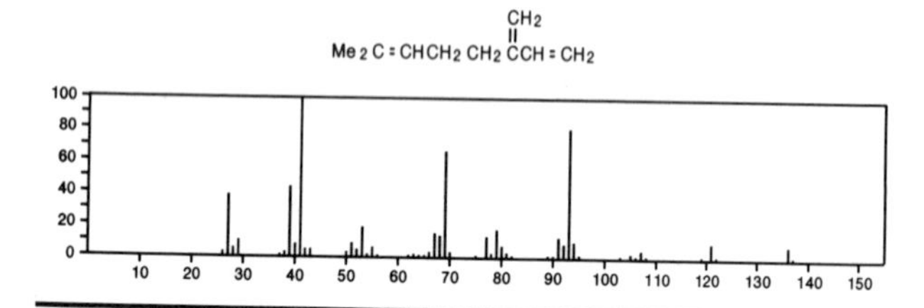

136 $C_{10}H_{16}$ 127-91-3
Bicyclo[3.1.1]heptane, 6,6-dimethyl-2-methylene-

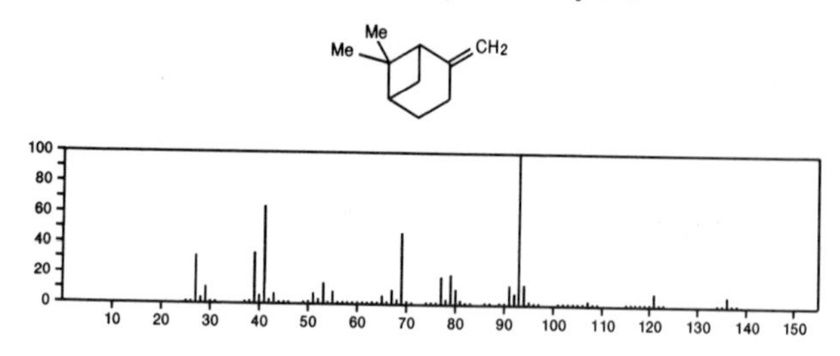

136 $C_{10}H_{16}$ 138-86-3
Cyclohexene, 1-methyl-4-(1-methylethenyl)-

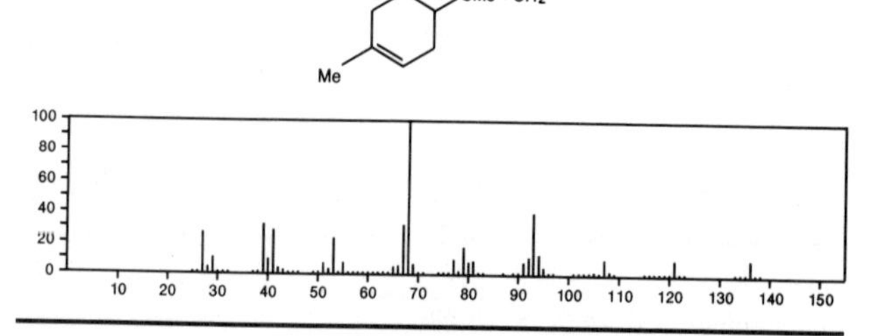

136 $C_{10}H_{16}$ 281-23-2
Tricyclo[3.3.1.1^{3,7}]decane

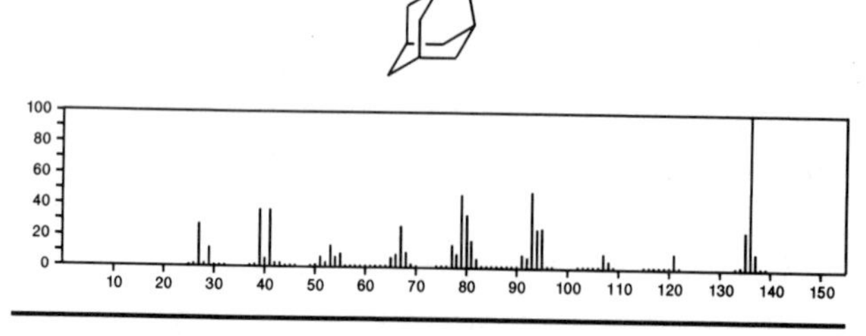

136 $C_{10}H_{16}$ 471-84-1
Bicyclo[2.2.1]heptane, 7,7-dimethyl-2-methylene-

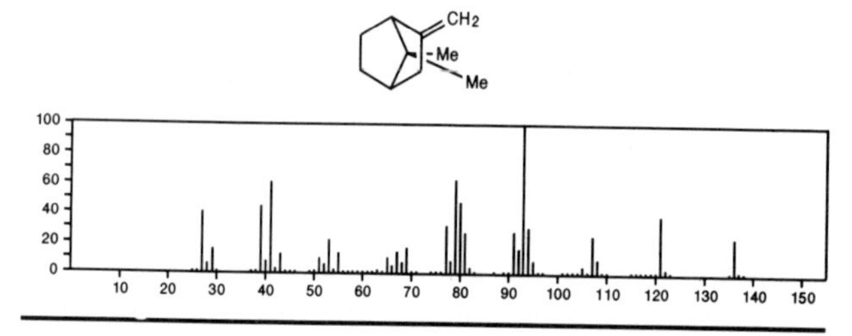

136 $C_{10}H_{16}$ 488-97-1
Tricyclo[2.2.1.0^{2,6}]heptane, 1,3,3-trimethyl-

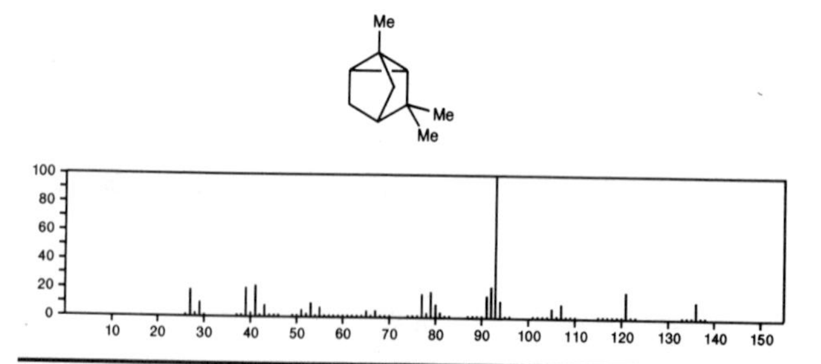

136 $C_{10}H_{16}$ 499-03-6
Cyclohexene, 1-methyl-3-(1-methylethenyl)-, (±)-

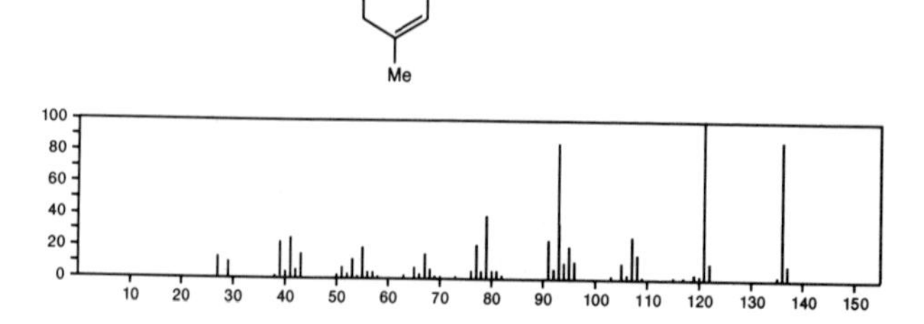

136 $C_{10}H_{16}$ 499-97-8
Cyclohexane, 1-methylene-4-(1-methylethenyl)-

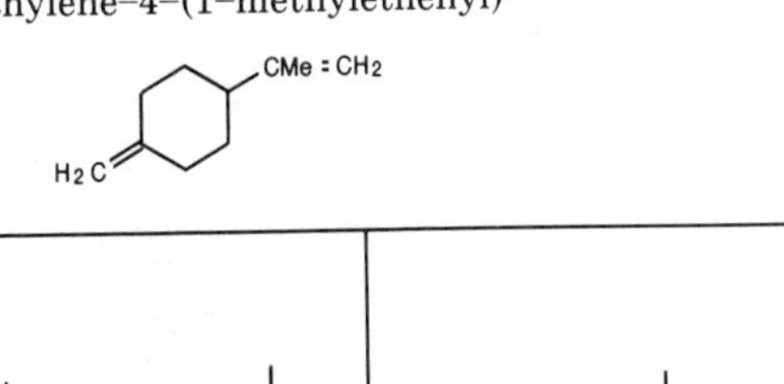

136 $C_{10}H_{16}$ 502-99-8
1,3,7-Octatriene, 3,7-dimethyl-

H2C = CMe CH2 CH2 CH = CMe CH = CH2

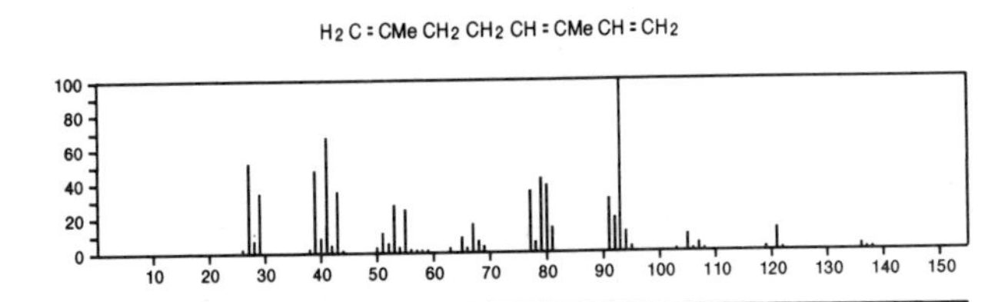

136 $C_{10}H_{16}$ 508-32-7
Tricyclo[2.2.1.0^{2,6}]heptane, 1,7,7-trimethyl-

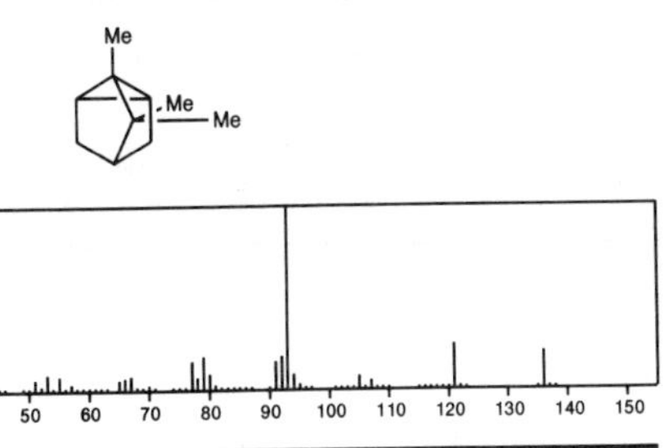

136 $C_{10}H_{16}$ 514-94-3
1,3-Cyclohexadiene, 1,5,5,6-tetramethyl-

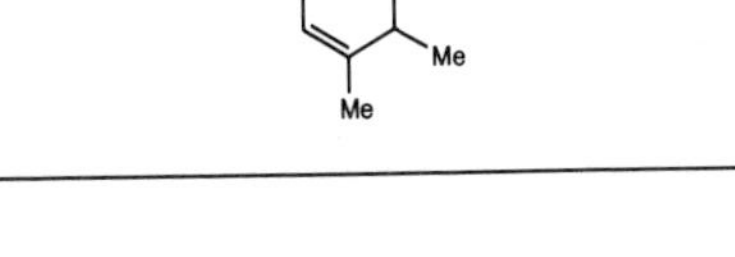

136 $C_{10}H_{16}$ 514-96-5
1,3-Cyclohexadiene, 1,2,6,6-tetramethyl-

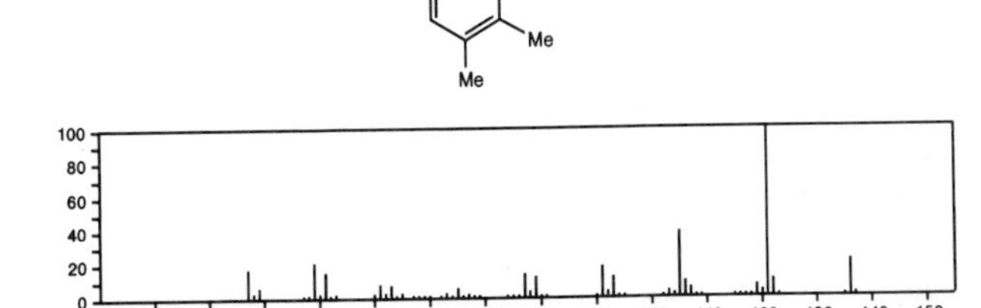

136 $C_{10}H_{16}$ 554-61-0
Bicyclo[4.1.0]hept-2-ene, 3,7,7-trimethyl-

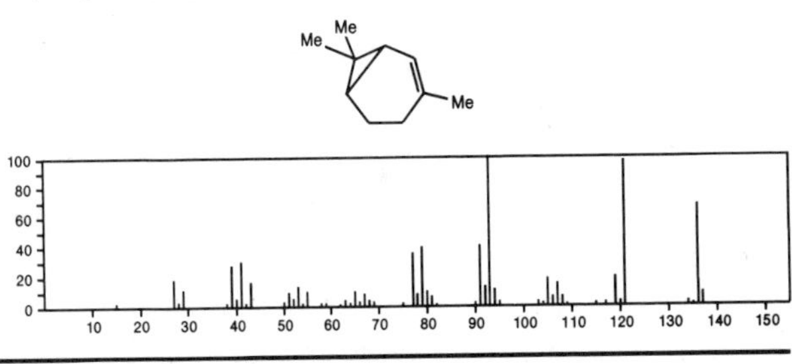

136 $C_{10}H_{16}$ 555-10-2
Cyclohexene, 3-methylene-6-(1-methylethyl)-

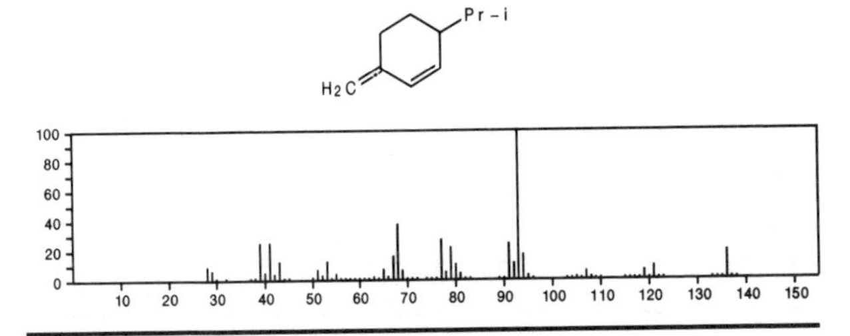

136 $C_{10}H_{16}$ 586-62-9
Cyclohexene, 1-methyl-4-(1-methylethylidene)-

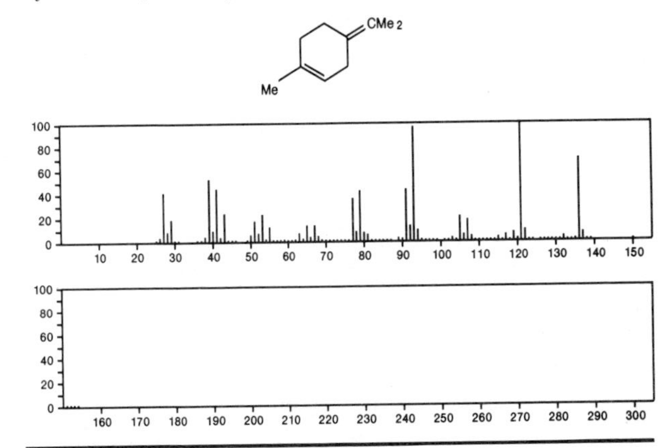

136 $C_{10}H_{16}$ 673-84-7
2,4,6-Octatriene, 2,6-dimethyl-

MeCH = CMe CH = CHCH = CMe2

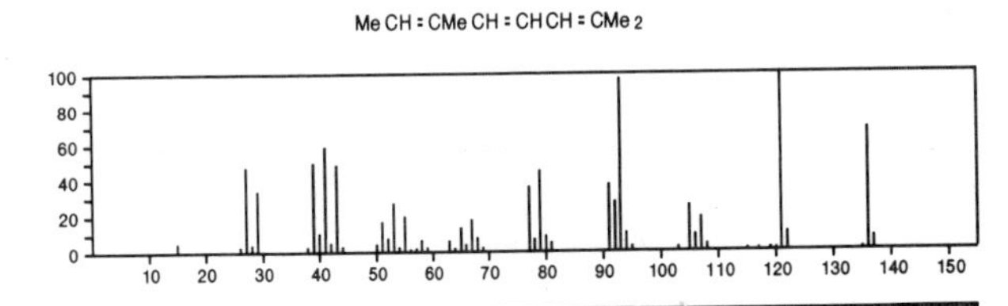

136 $C_{10}H_{16}$ 1461-27-4
Cyclohexene, 1-methyl-5-(1-methylethenyl)-, (*R*)-

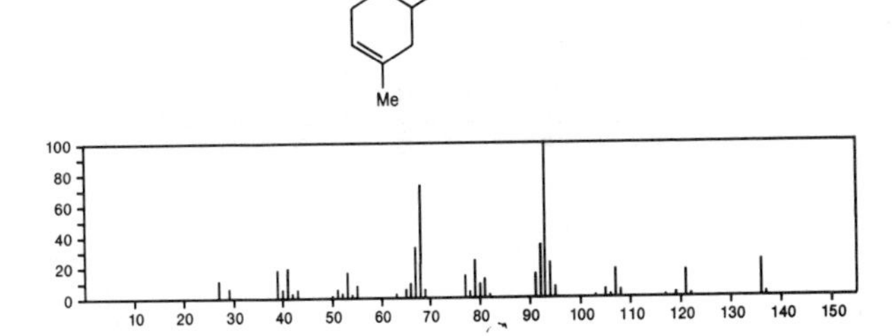

136 $C_{10}H_{16}$ 3387-41-5
Bicyclo[3.1.0]hexane, 4-methylene-1-(1-methylethyl)-

Pr-i
CH2

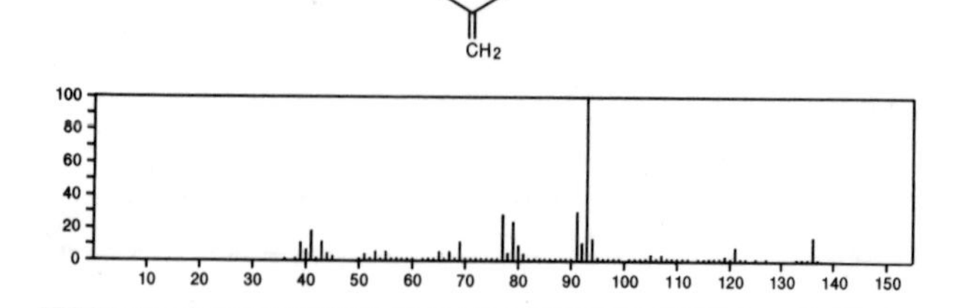

136 $C_{10}H_{16}$ 3779-61-1
1,3,6-Octatriene, 3,7-dimethyl-, (*E*)-

H2C=CHCMe=CHCH2CH=CMe2

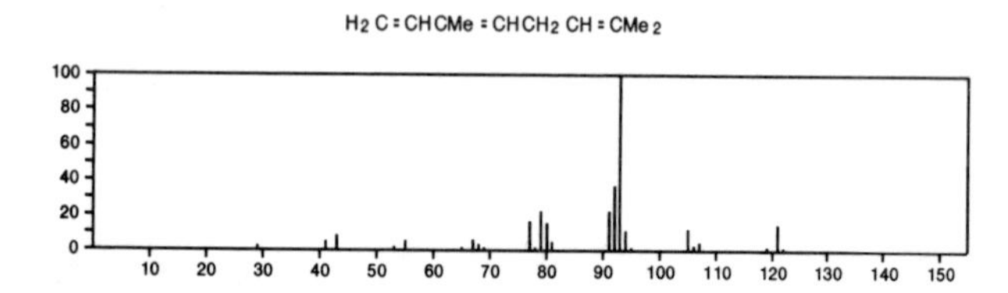

136 $C_{10}H_{16}$ 5113-87-1
Cyclohexene, 3-methyl-6-(1-methylethenyl)-, (3*R*-*trans*)-

CMe=CH2
Me

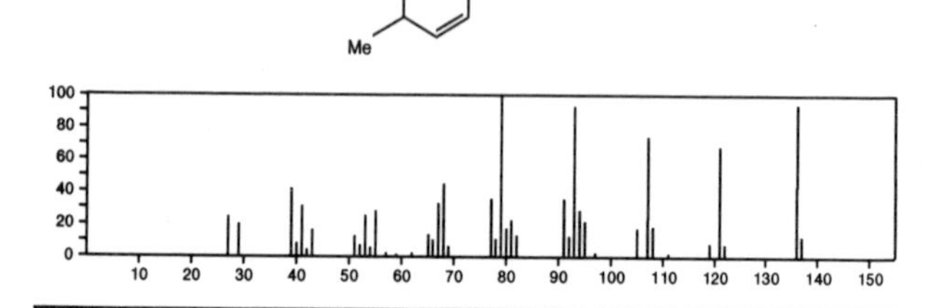

136 $C_{10}H_{16}$ 5208-50-4
4-Carene, (1*S*,3*S*,6*R*)-(-)-

Me Me
Me

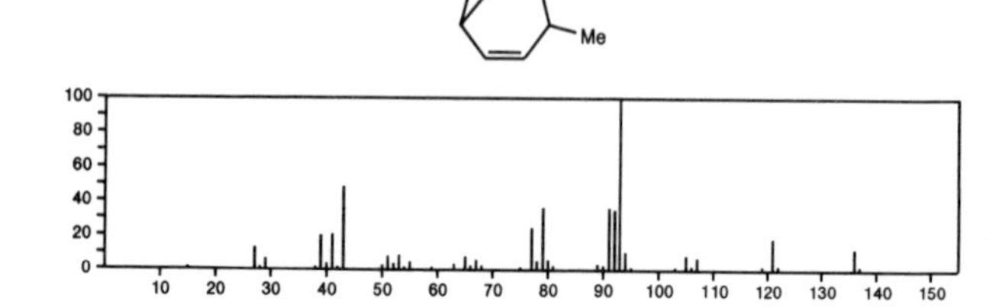

136 $C_{10}H_{16}$ 5208-51-5
m-Mentha-4,8-diene, (1*S*,3*S*)-(+)-

Me CMe=CH2

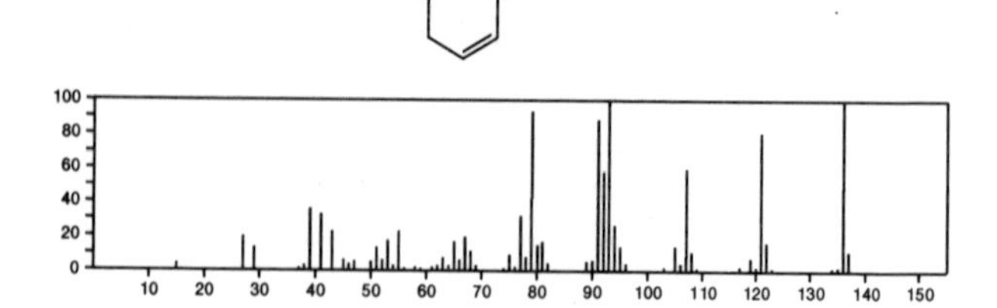

136 $C_{10}H_{16}$ 5989-27-5
Cyclohexene, 1-methyl-4-(1-methylethenyl)-, (*R*)-

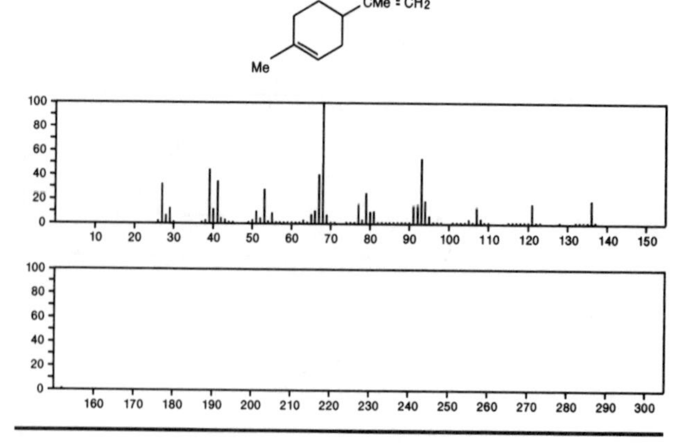

136 $C_{10}H_{16}$ 5989-54-8
Cyclohexene, 1-methyl-4-(1-methylethenyl)-, (*S*)-

CMe=CH2
Me

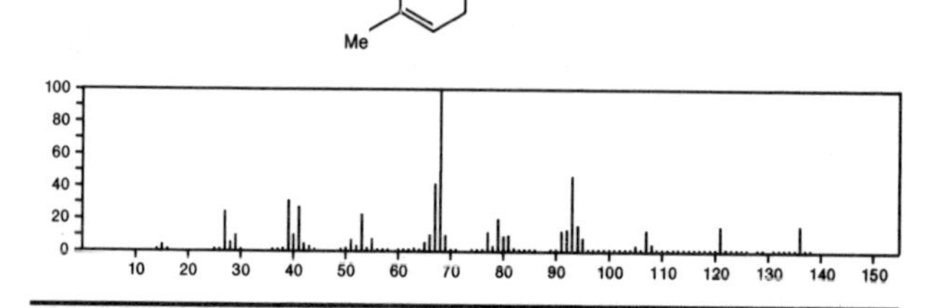

136 $C_{10}H_{16}$ 13466-78-9
Bicyclo[4.1.0]hept-3-ene, 3,7,7-trimethyl-

Me Me
Me

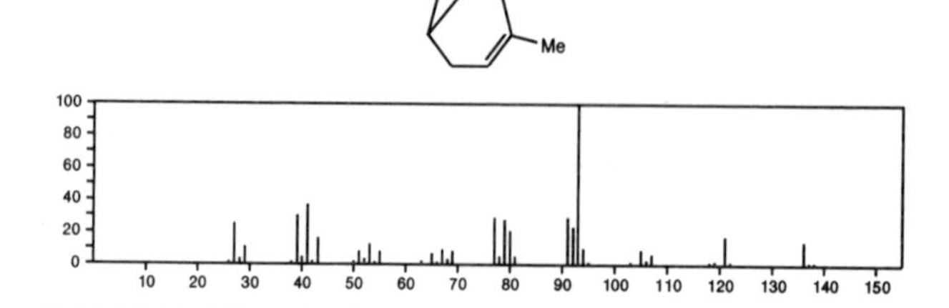

136 $C_{10}H_{16}$ 13837-95-1
Cyclohexane, 1-methylene-3-(1-methylethenyl)-, (*R*)-

CMe=CH2
CH2

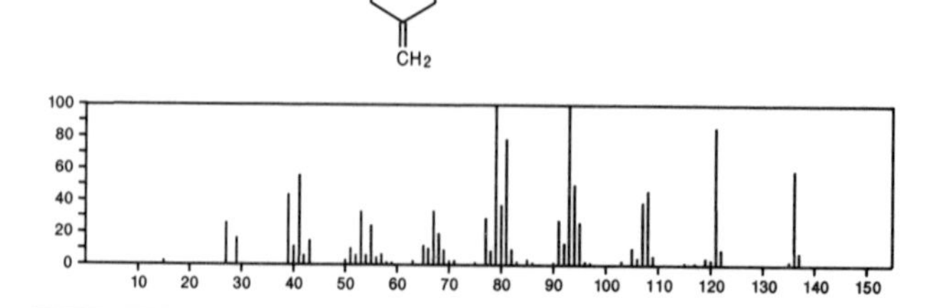

136 $C_{10}H_{16}$ 13877-91-3
1,3,6-Octatriene, 3,7-dimethyl-

H2C=CHCMe=CHCH2CH=CMe2

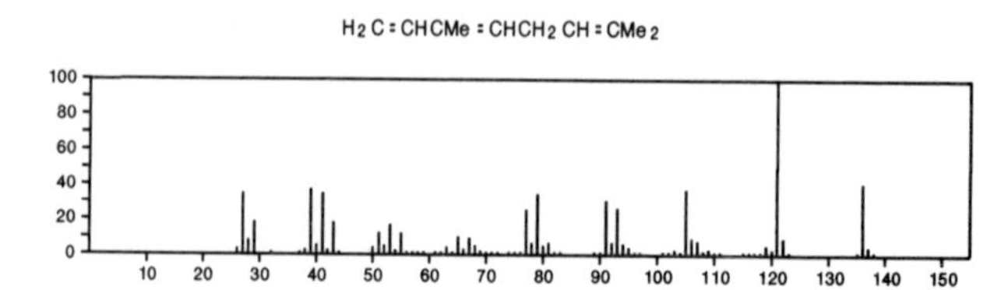

136 $C_{10}H_{16}$ 14803-30-6
Cyclopropane, trimethyl(2-methyl-1-propenylidene)-

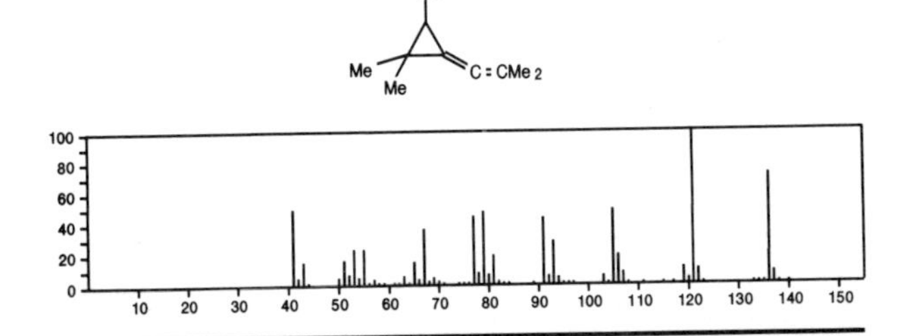

136 $C_{10}H_{16}$ 15402-94-5
Cycloheptene, 5-ethylidene-1-methyl-

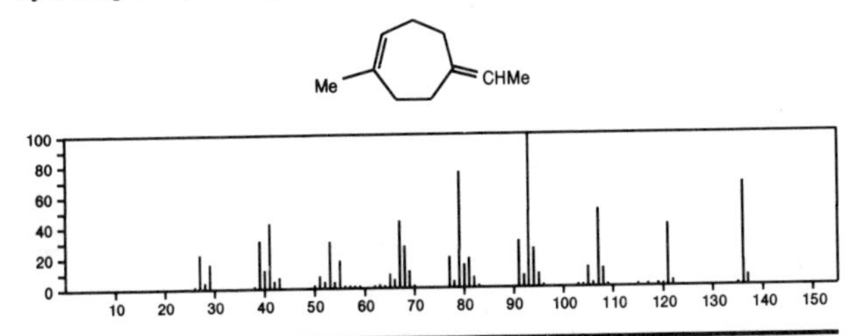

136 $C_{10}H_{16}$ 18172-67-3
Bicyclo[3.1.1]heptane, 6,6-dimethyl-2-methylene-, (1*S*)-

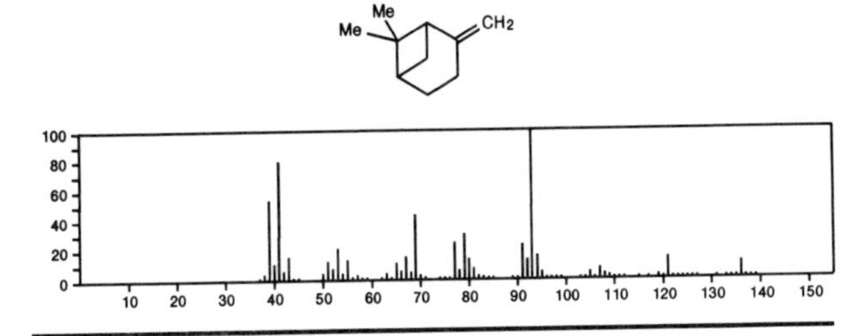

136 $C_{10}H_{16}$ 19026-94-9
2,5-Methano-1*H*-indene, octahydro-

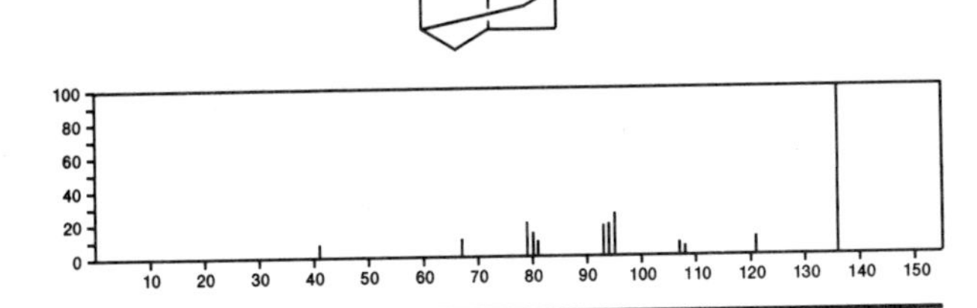

136 $C_{10}H_{16}$ 22769-00-2
Cyclobutane, 1,2-dipropenyl-

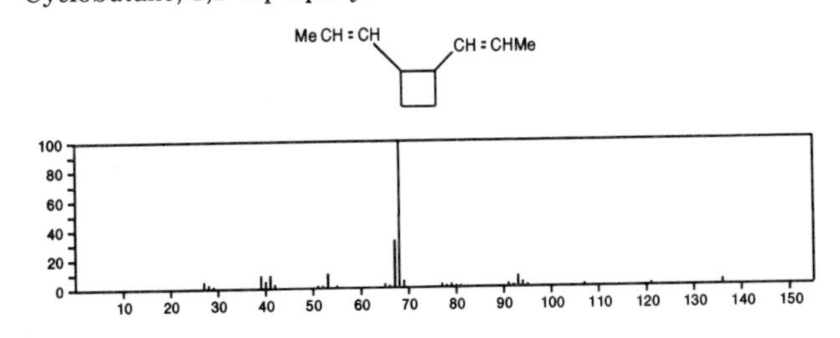

136 $C_{10}H_{16}$ 24524-57-0
Bicyclo[3.1.0]hexane, 6-isopropylidene-1-methyl-

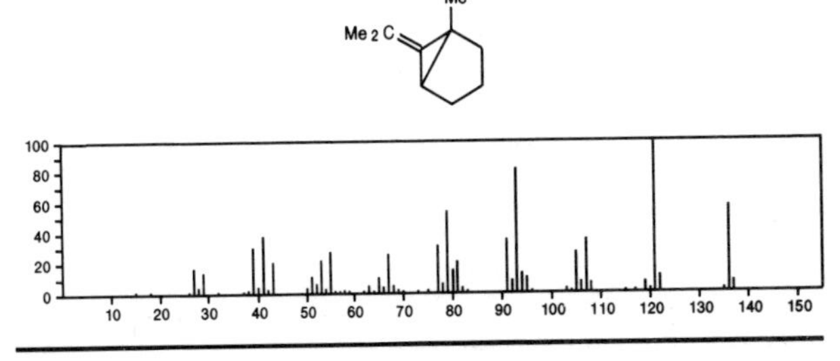

136 $C_{10}H_{16}$ 28935-76-4
3-Octen-5-yne, 2,7-dimethyl-, (*Z*)-

$Me_2CHC≡CCH=CHCHMe_2$

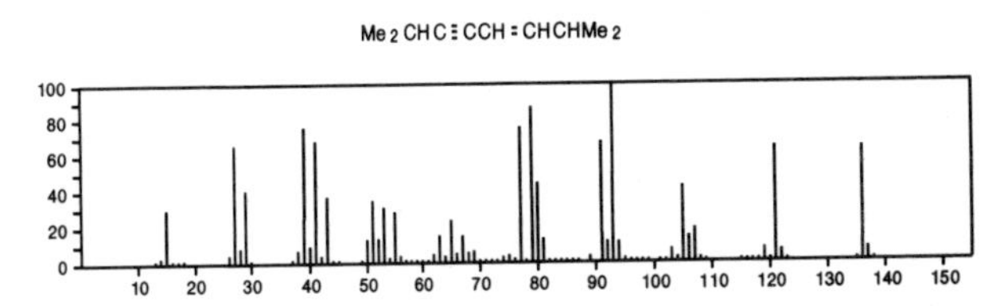

136 $C_{10}H_{16}$ 29714-87-2
Octane, 2,6-dimethyl-, hexadehydro deriv.

$Me_2CH(CH_2)_3CHMeCH_2Me$ − 6H

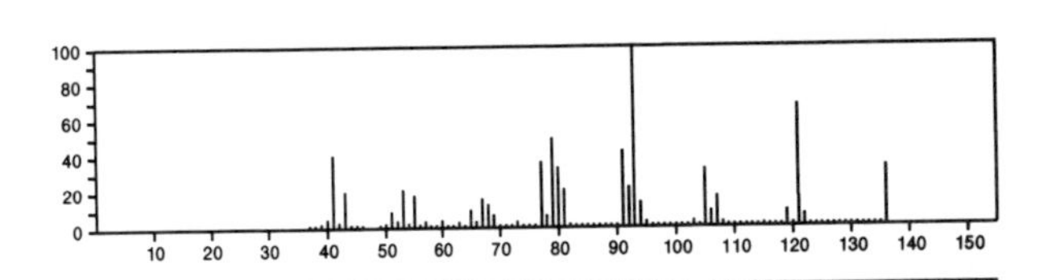

136 $C_{10}H_{16}$ 53282-47-6
Bicyclo[4.1.0]heptane, 7-(1-methylethylidene)-

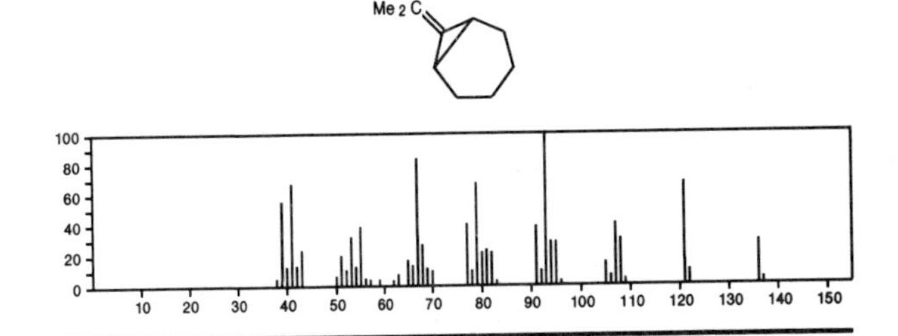

136 $C_{10}H_{16}$ 55956-33-7
3-Octen-5-yne, 2,7-dimethyl-, (*E*)-

$Me_2CHC≡CCH=CHCHMe_2$

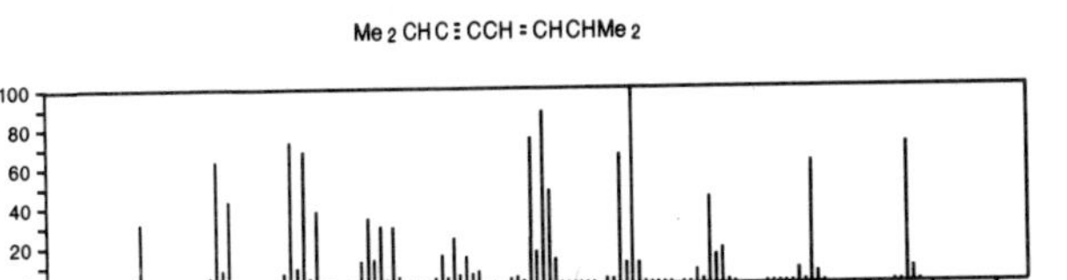

136 $C_{10}H_{16}$ 56701-52-1
Bicyclo[3.1.0]hexane, 1-methyl-6-(1-methylethylidene)-

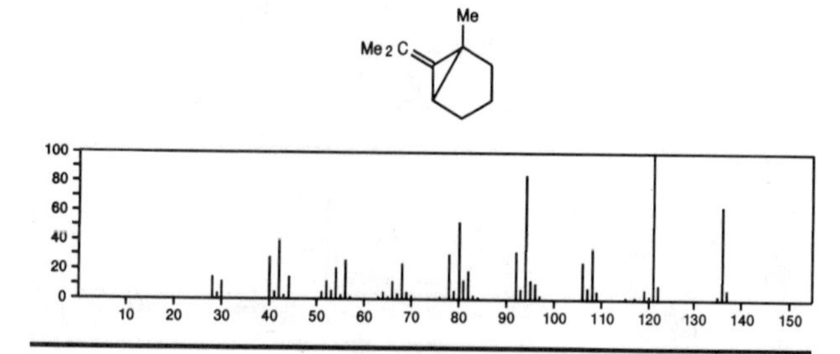

136 $C_{10}H_{16}$ 56710-83-9
Cyclopentane, 2-methyl-1-methylene-3-(1-methylethenyl)-

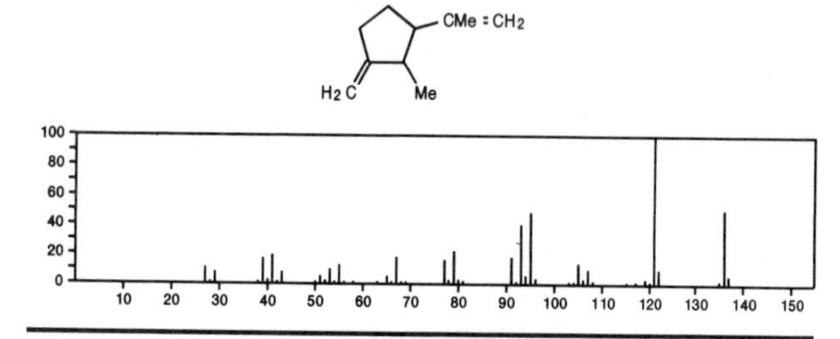

136 $C_{10}H_{16}$ 56816-08-1
Cyclohexene, 5-methyl-3-(1-methylethenyl)-, *trans*-(-)-

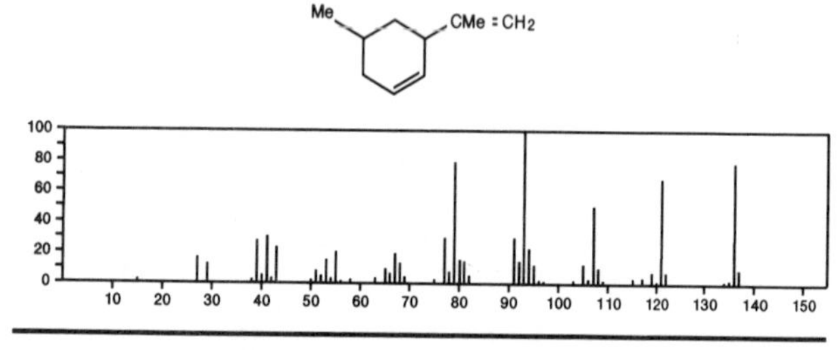

136 $C_{10}H_{16}$ 57396-75-5
2,4,6-Octatriene, 3,4-dimethyl-

MeCH : CHCH : CMeCMe : CHMe

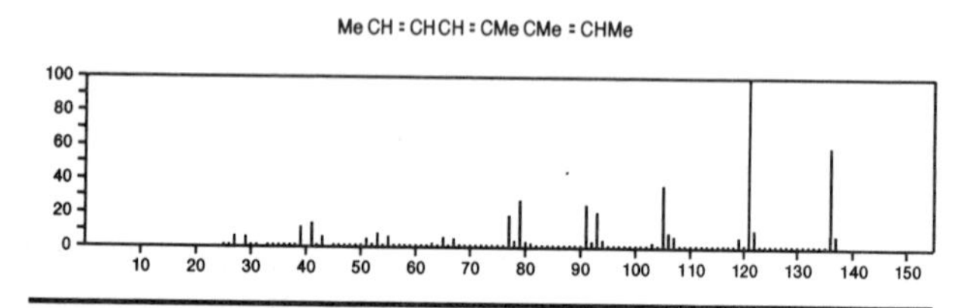

137 $C_3H_7NO_5$ 624-43-1
1,2,3-Propanetriol, 1-nitrate

O2NOCH2CH(OH)CH2OH

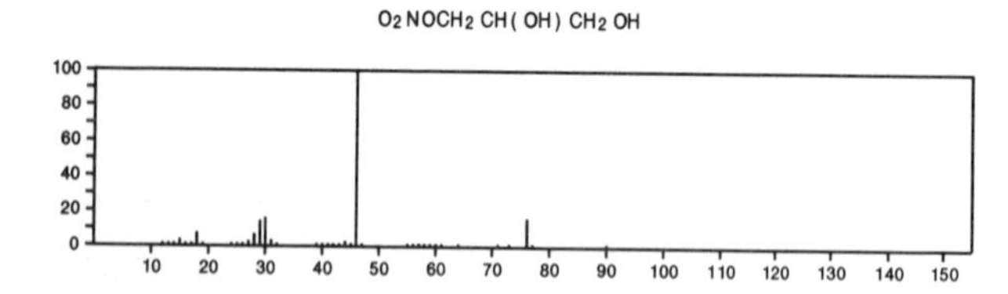

137 $C_5H_3N_3S$ 273-86-9
Thiazolo[5,4-*d*]pyrimidine

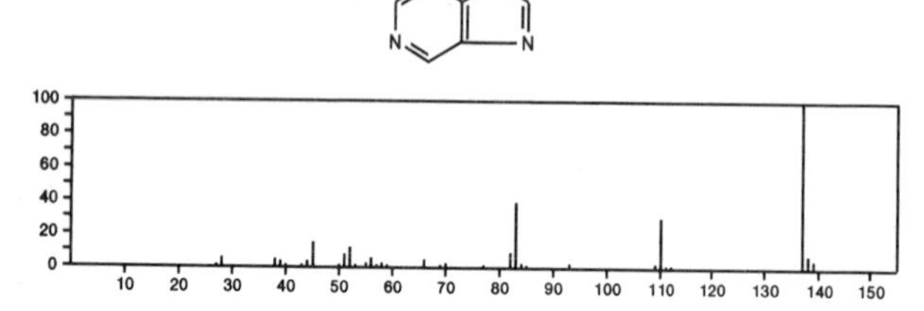

137 $C_5H_6F_3N$ 2214-85-9
Butyronitrile, 4,4,4-trifluoro-2-methyl-

F3CCH2CHMeCN

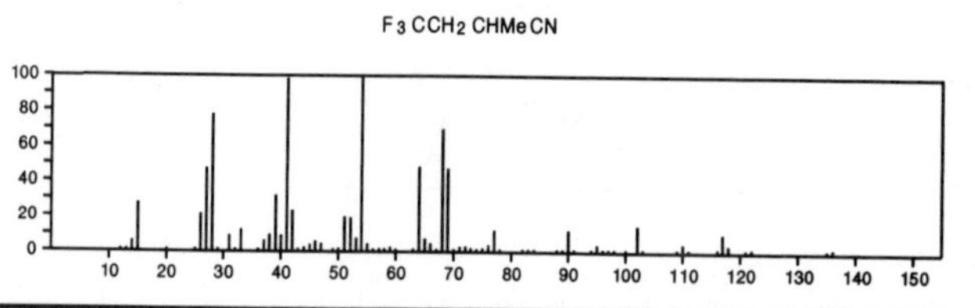

137 $C_6H_7N_3O$ 54-85-3
4-Pyridinecarboxylic acid, hydrazide

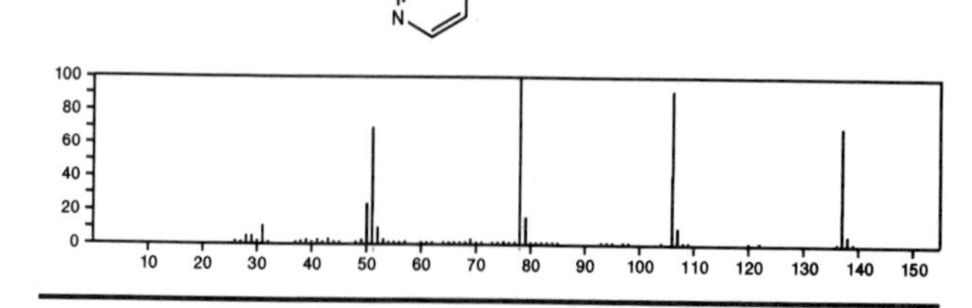

137 $C_6H_7N_3O$ 553-53-7
3-Pyridinecarboxylic acid, hydrazide

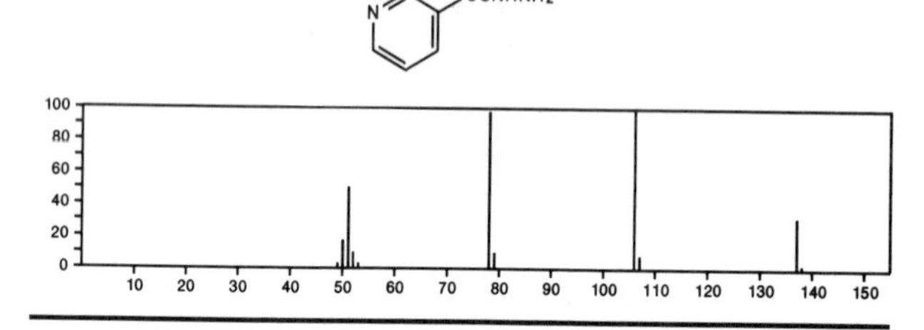

137 $C_6H_7N_3O$ 31378-80-0
Pyridinium, 1-ureido-, hydroxide, inner salt

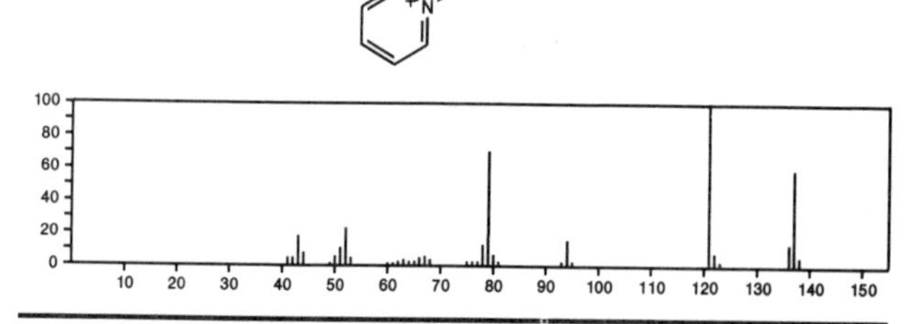

137 C_7H_4ClN 623-03-0
Benzonitrile, 4-chloro-

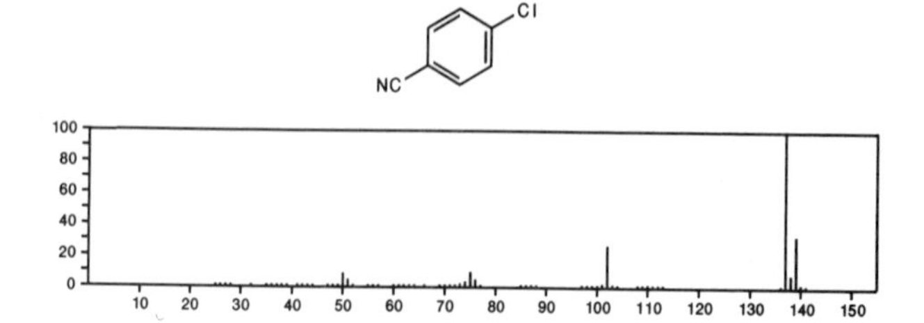

137 $C_7H_7NO_2$ 65-45-2

Benzamide, 2-hydroxy-

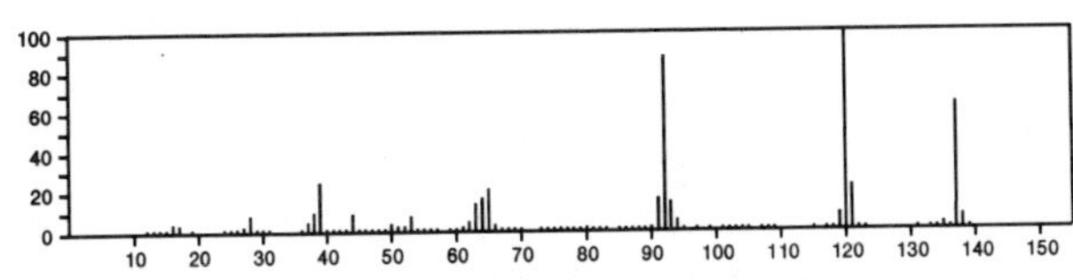

137 $C_7H_7NO_2$ 99-99-0

Benzene, 1-methyl-4-nitro-

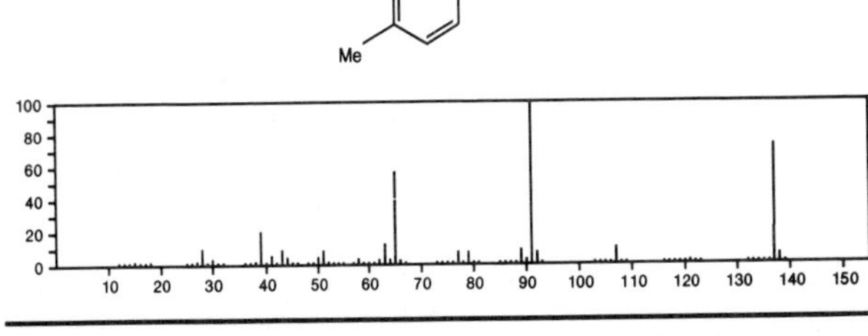

137 $C_7H_7NO_2$ 88-72-2

Benzene, 1-methyl-2-nitro-

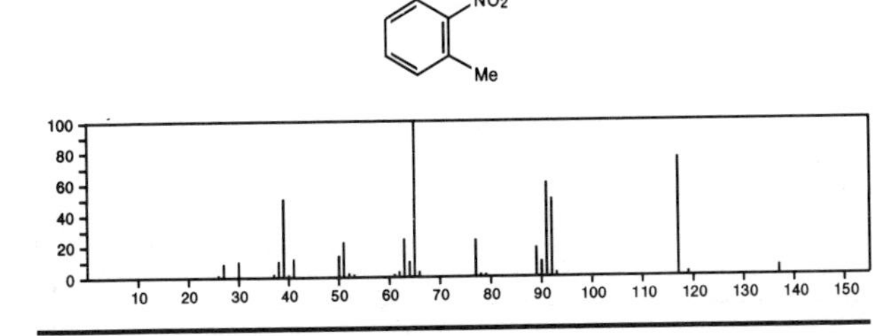

137 $C_7H_7NO_2$ 118-92-3

Benzoic acid, 2-amino-

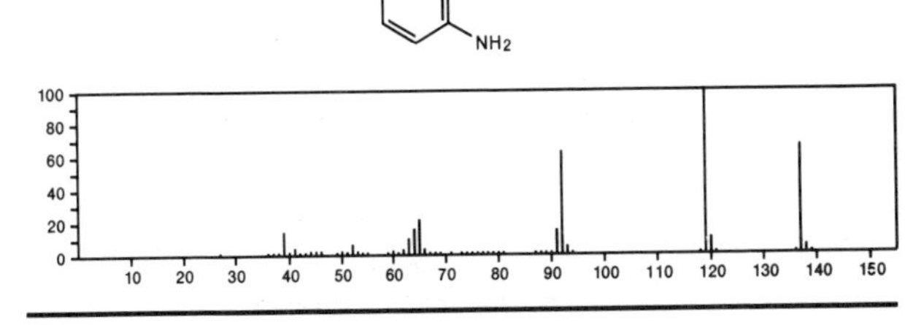

137 $C_7H_7NO_2$ 93-60-7

3-Pyridinecarboxylic acid, methyl ester

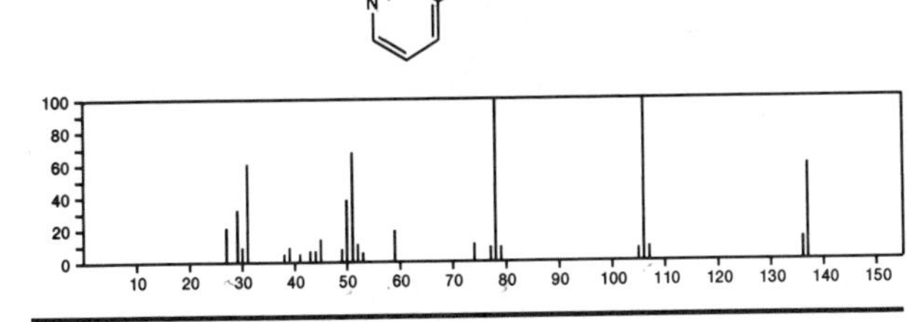

137 $C_7H_7NO_2$ 150-13-0

Benzoic acid, 4-amino-

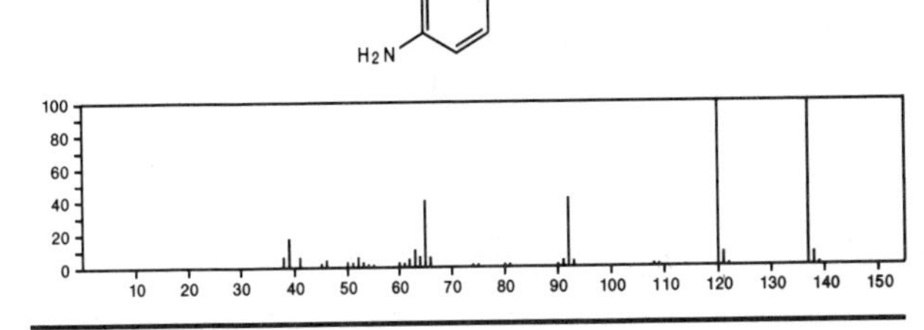

137 $C_7H_7NO_2$ 94-67-7

Benzaldehyde, 2-hydroxy-, oxime

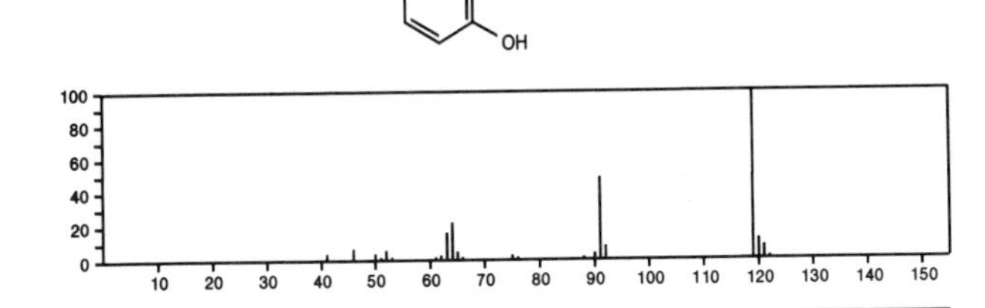

137 $C_7H_7NO_2$ 445-30-7

Pyridinium, 2-carboxy-1-methyl-, hydroxide, inner salt

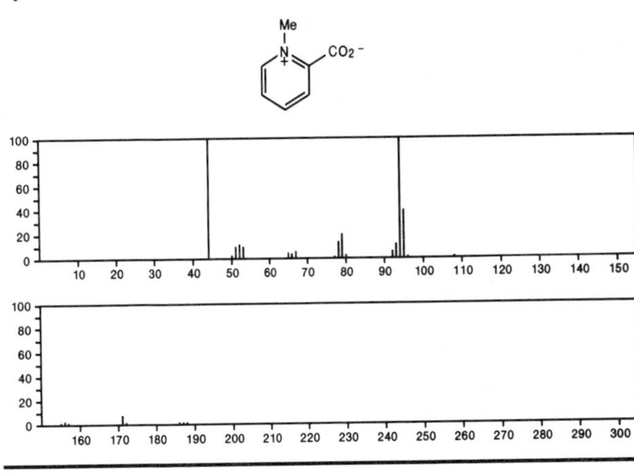

137 $C_7H_7NO_2$ 99-05-8

Benzoic acid, 3-amino-

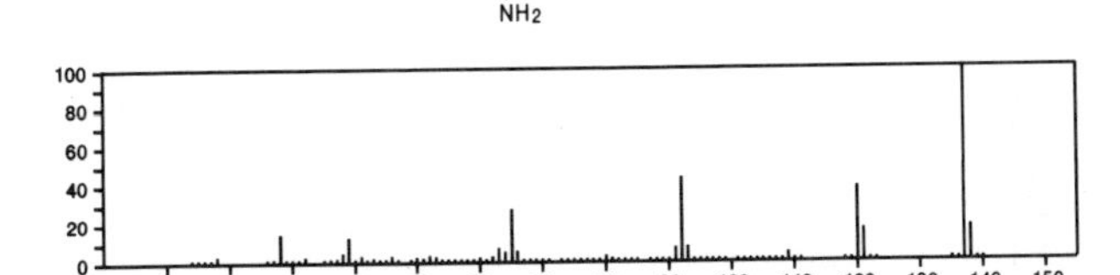

137 $C_7H_7NO_2$ 501-81-5

3-Pyridineacetic acid

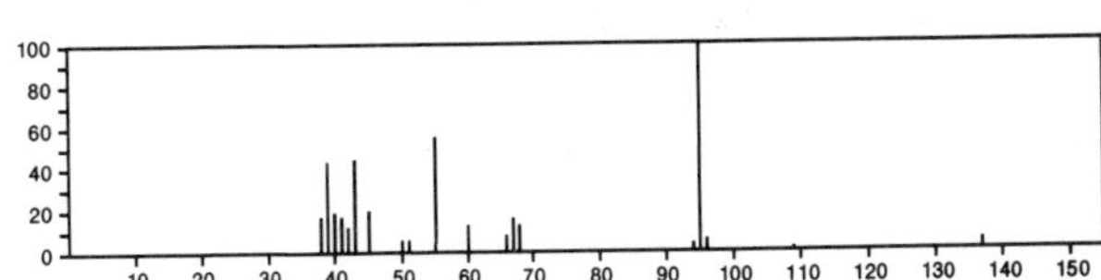

137 $C_7H_7NO_2$ 535–83–1
Pyridinium, 3–carboxy–1–methyl–, hydroxide, inner salt

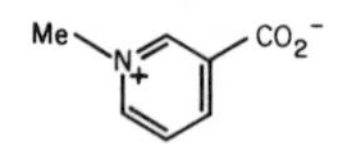

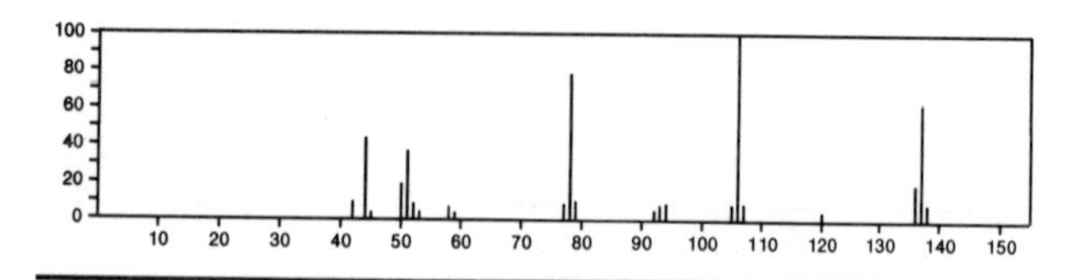

137 $C_7H_7NO_2$ 622–42–4
Benzene, (nitromethyl)–

PhCH2NO2

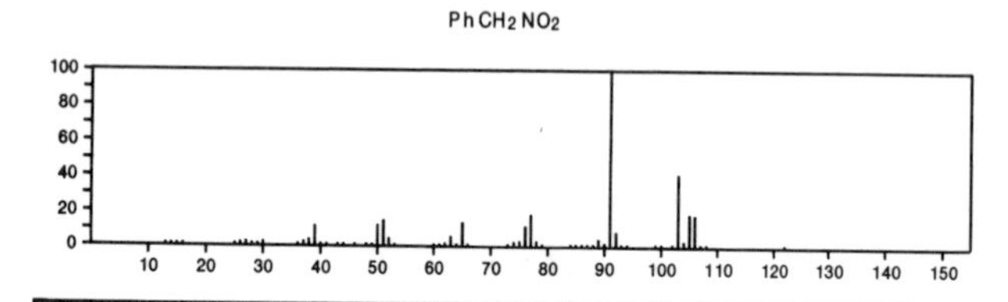

137 $C_7H_7NO_2$ 699–06–9
Benzaldehyde, 4–hydroxy–, oxime

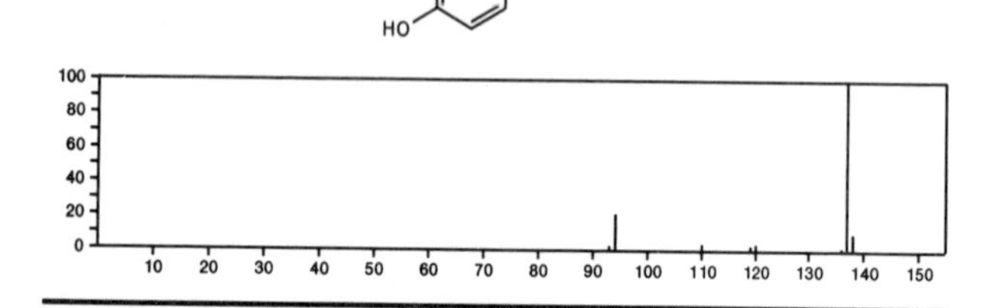

137 $C_7H_7NO_2$ 934–60–1
2–Pyridinecarboxylic acid, 6–methyl–

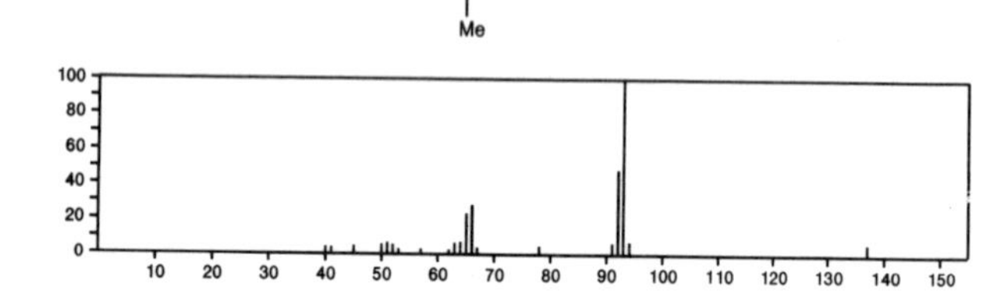

137 $C_7H_7NO_2$ 2459–07–6
2–Pyridinecarboxylic acid, methyl ester

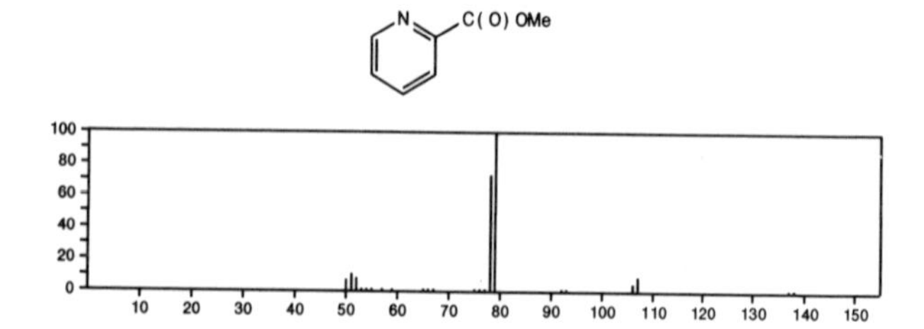

137 $C_7H_7NO_2$ 2459–09–8
4–Pyridinecarboxylic acid, methyl ester

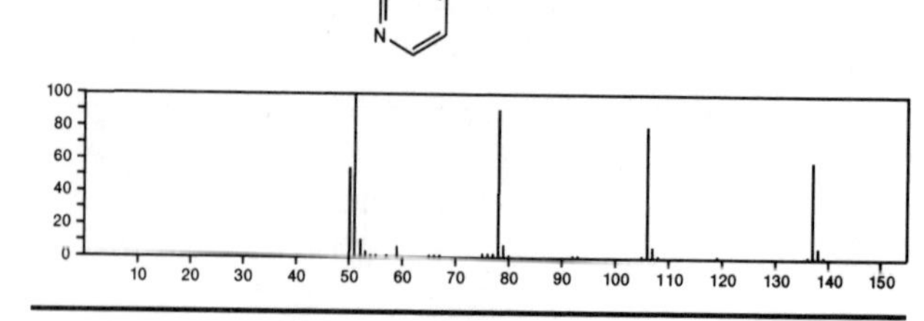

137 $C_7H_7NO_2$ 14210–20–9
4–Pyridinol, acetate (ester)

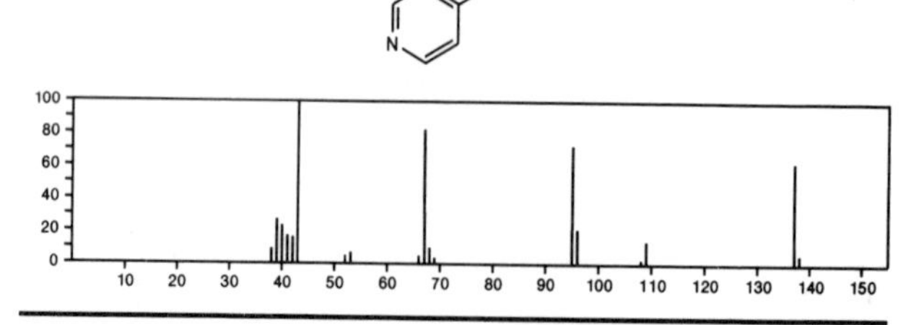

137 $C_7H_7NO_2$ 21494–57–5
1*H*–Pyrrole–2,5–dione, 3–ethenyl–4–methyl–

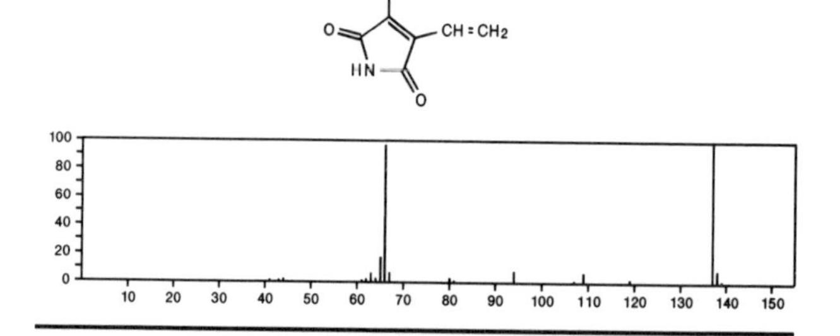

137 $C_7H_7NO_2$ 22241–18–5
Benzaldehyde, 3–hydroxy–, oxime

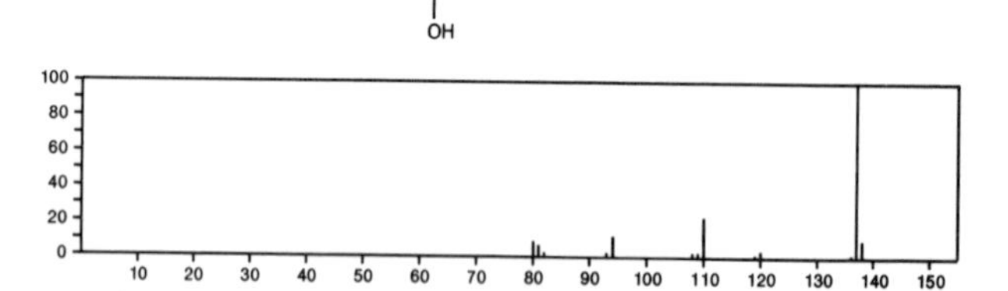

137 $C_7H_7NO_2$ 24608–93–3
Pyridinium, 1–(carboxymethyl)–, hydroxide, inner salt

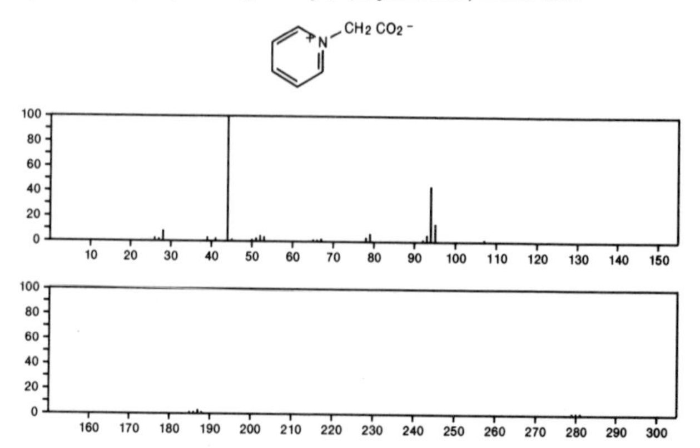

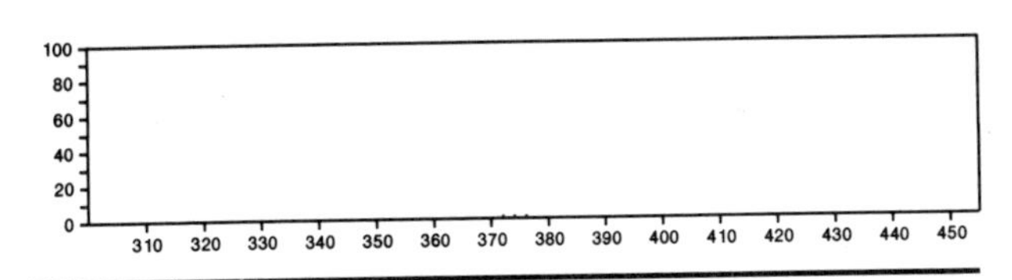

137 $C_7H_7NO_2$ 34600–55–0
1*H*–Pyrrole–2–carboxylic acid, 1–ethenyl–

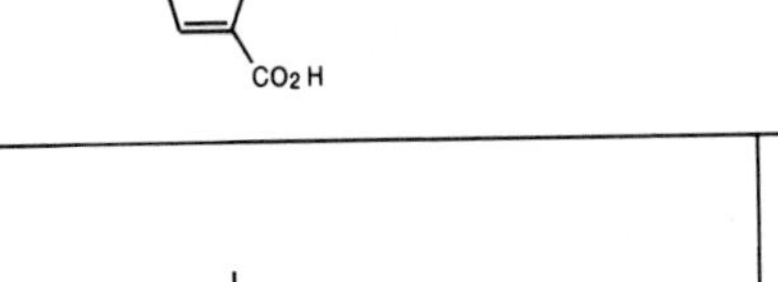

137 C_7H_7NS 637–51–4
Methanethioamide, *N*–phenyl–

S=CHNHPh

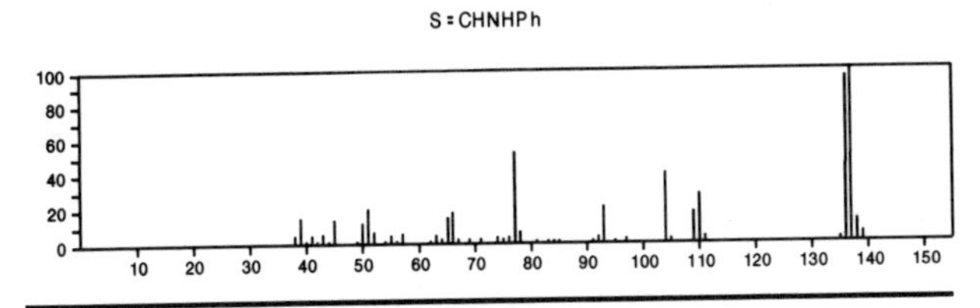

137 C_7H_7NS 2227–79–4
Benzenecarbothioamide

H_2NCSPh

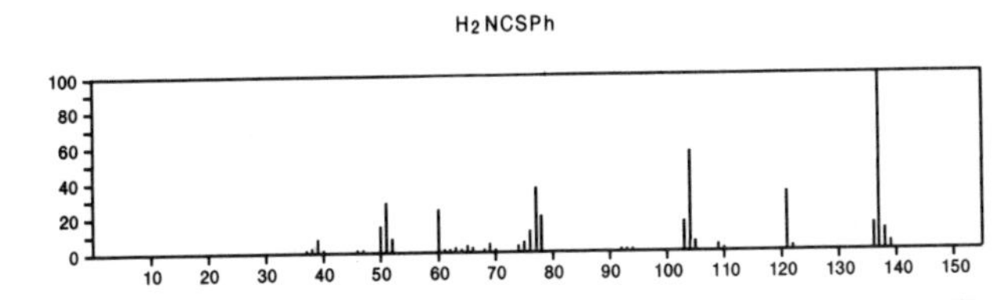

137 C_7H_7NS 19006–83–8
2(1*H*)–Pyridinethione, 1–ethenyl–

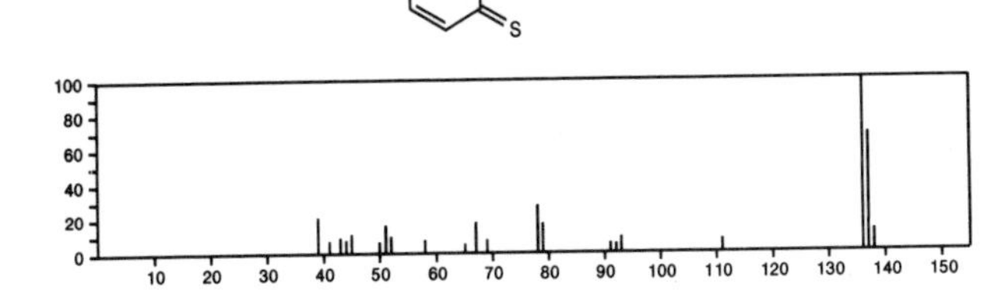

137 $C_8H_{11}NO$ 51–67–2
Phenol, 4–(2–aminoethyl)–

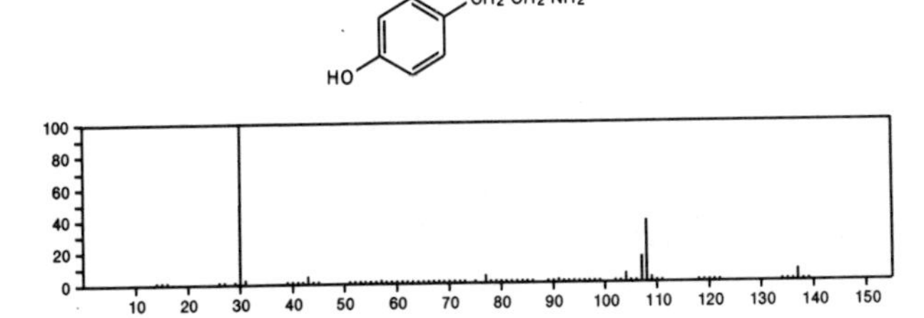

137 $C_8H_{11}NO$ 99–07–0
Phenol, 3–(dimethylamino)–

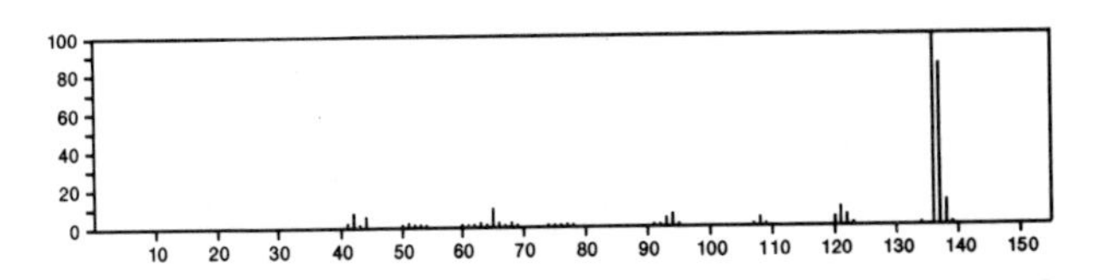

137 $C_8H_{11}NO$ 2386–25–6
Ethanone, 1–(2,4–dimethyl–1*H*–pyrrol–3–yl)–

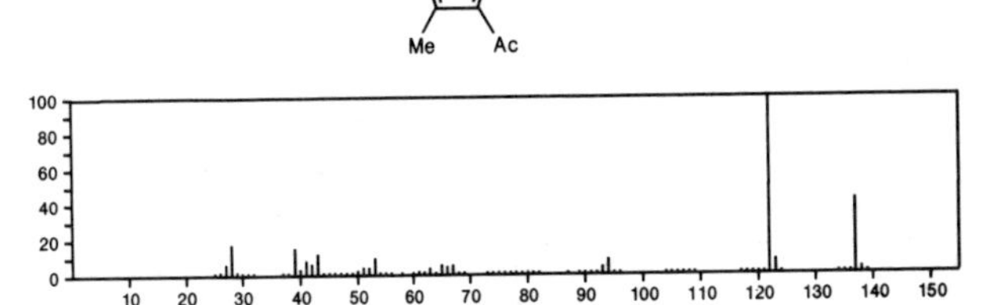

137 $C_8H_{11}NO$ 2393–23–9
Benzenemethanamine, 4–methoxy–

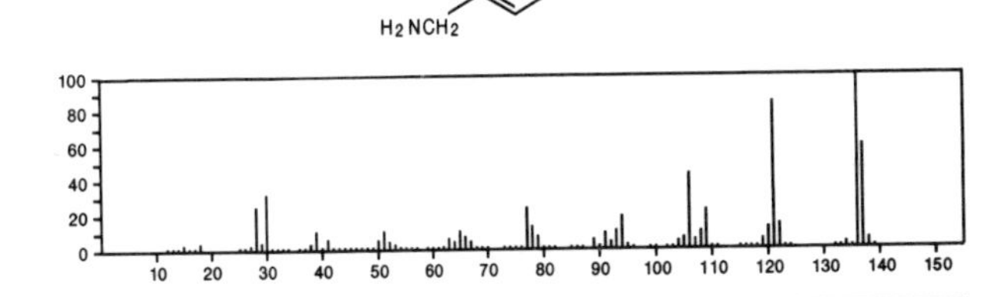

137 $C_8H_{11}NO$ 4438–38–4
8–Azabicyclo[3.2.1]oct–6–en–3–one, 8–methyl–

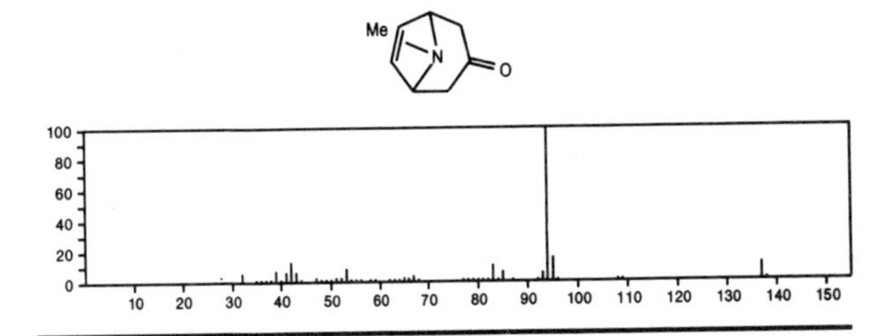

137 $C_8H_{11}NO$ 5961–59–1
Benzenamine, 4–methoxy–*N*–methyl–

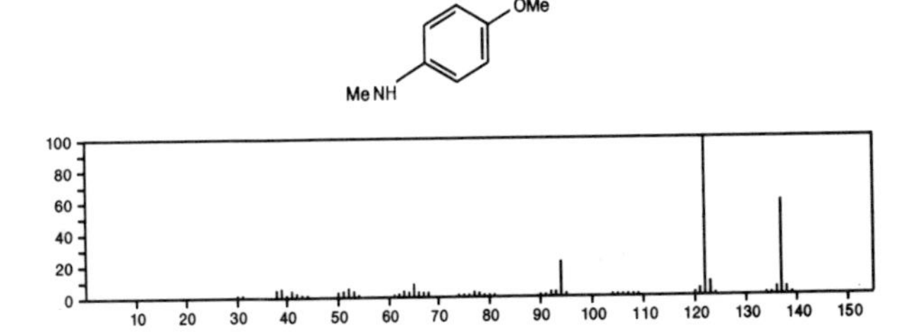

137 $C_8H_{11}NO$ 7568–93–6
Benzenemethanol, α–(aminomethyl)–

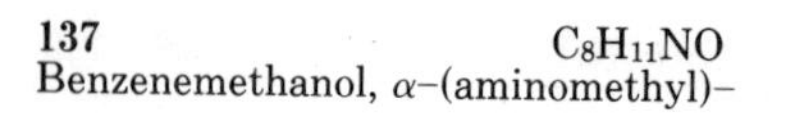

$H_2NCH_2CH(OH)Ph$

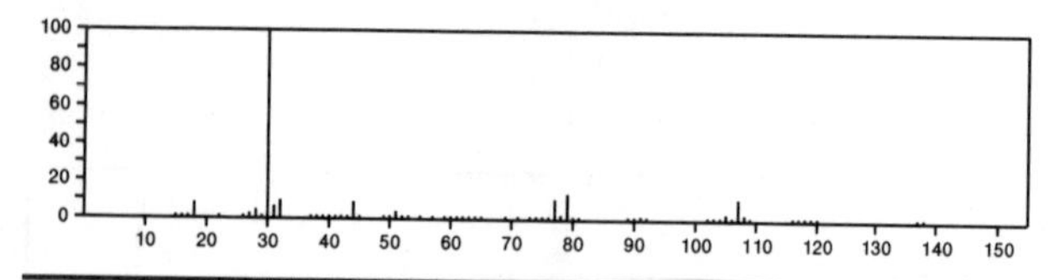

137 $C_8H_{11}NO$ 15031–89–7
2(1*H*)–Pyridinone, 1,4,6–trimethyl–

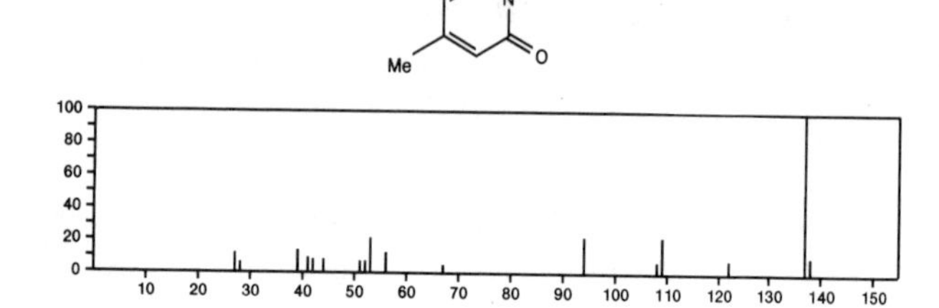

137 $C_8H_{11}NO$ 19006–62–3
2(1*H*)–Pyridone, 1–ethyl–4–methyl–

137 $C_8H_{11}NO$ 19006–63–4
2(1*H*)–Pyridinone, 1–propyl–

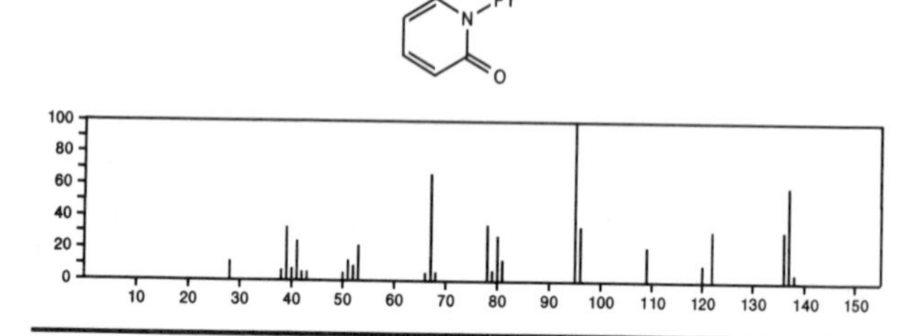

137 $C_8H_{11}NO$ 19038–36–9
2(1*H*)–Pyridinone, 1–ethyl–6–methyl–

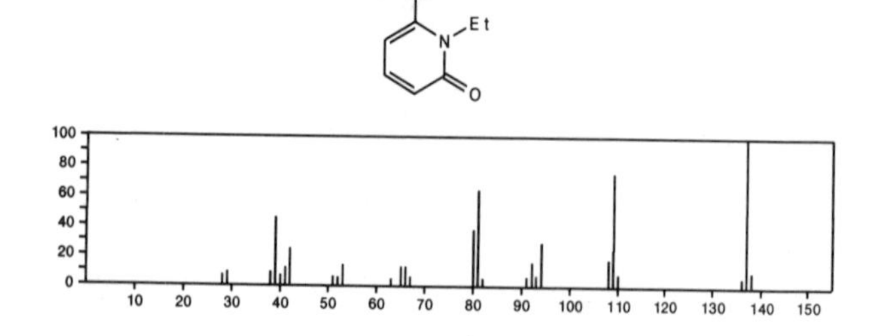

137 $C_8H_{11}NO$ 29055–08–1
Benzenemethanol, 2–(methylamino)–

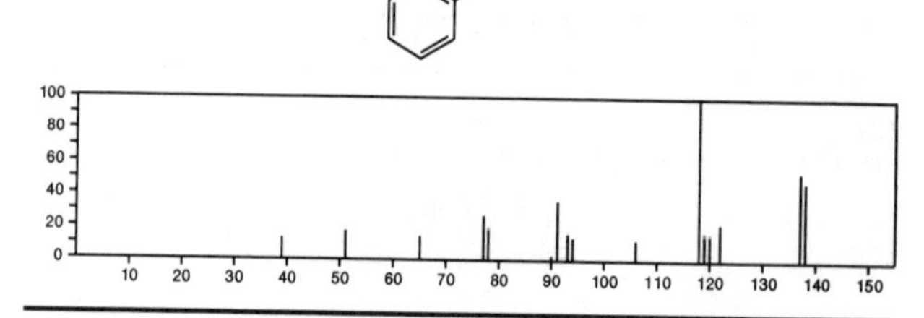

137 $C_8H_{11}NO$ 42185–27–3
Cyclohexaneacetonitrile, 2–oxo–

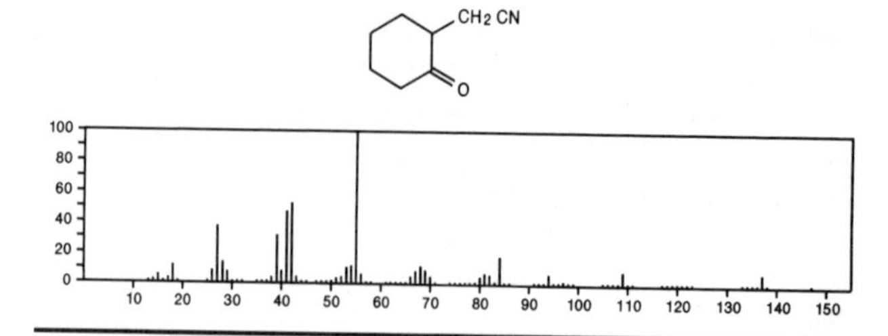

137 $C_9H_{15}N$ 102–70–5
2–Propen–1–amine, *N*,*N*–di–2–propenyl–

$H_2C=CHCH_2N(CH_2CH=CH_2)CH_2CH=CH_2$

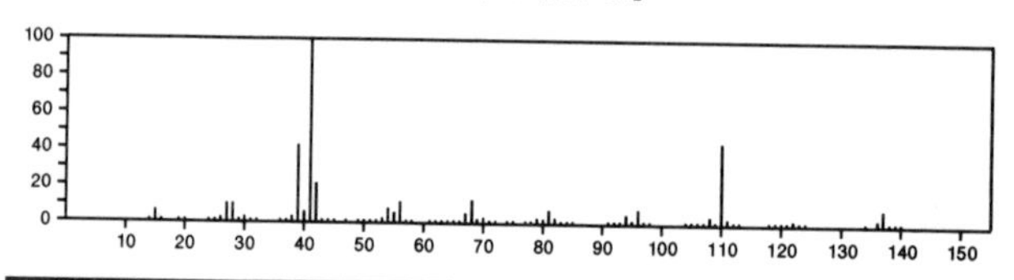

137 $C_9H_{15}N$ 699–22–9
1*H*–Pyrrole, 1–pentyl–

137 $C_9H_{15}N$ 5633–86–3
Octanenitrile, 2–methylene–

$Me(CH_2)_5C(CN)=CH_2$

137 $C_9H_{15}N$ 7148–07–4
Pyrrolidine, 1–(1–cyclopenten–1–yl)–

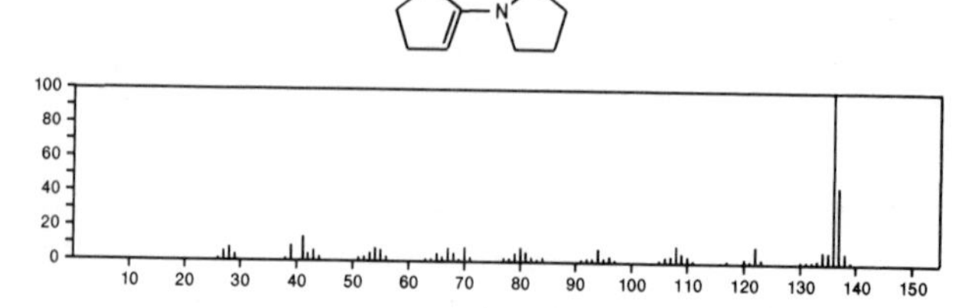

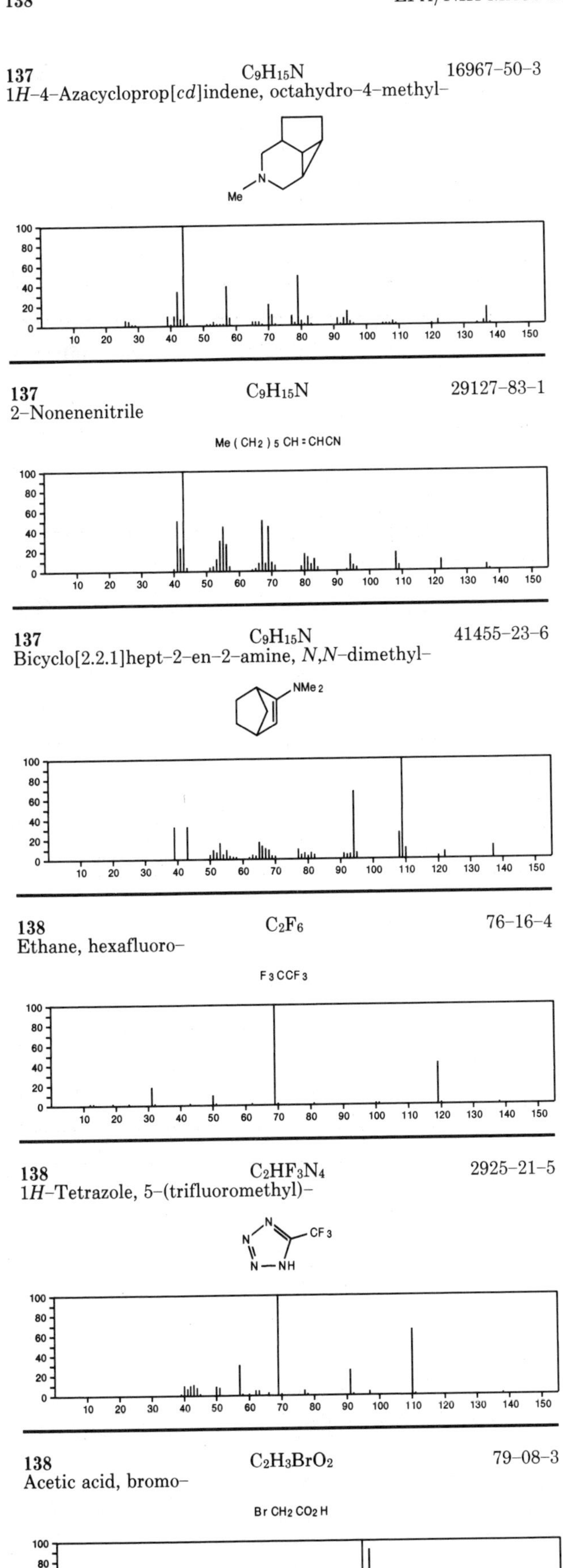
137 $C_9H_{15}N$ 16967-50-3
1H-4-Azacycloprop[cd]indene, octahydro-4-methyl-
Me
N
137 $C_9H_{15}N$ 29127-83-1
2-Nonenenitrile
Me(CH2)5CH=CHCN
137 $C_9H_{15}N$ 41455-23-6
Bicyclo[2.2.1]hept-2-en-2-amine, N,N-dimethyl-
NMe2
138 C_2F_6 76-16-4
Ethane, hexafluoro-
F3CCF3
138 $C_2HF_3N_4$ 2925-21-5
1H-Tetrazole, 5-(trifluoromethyl)-
N
N
N—NH
CF3
138 $C_2H_3BrO_2$ 79-08-3
Acetic acid, bromo-
BrCH2CO2H

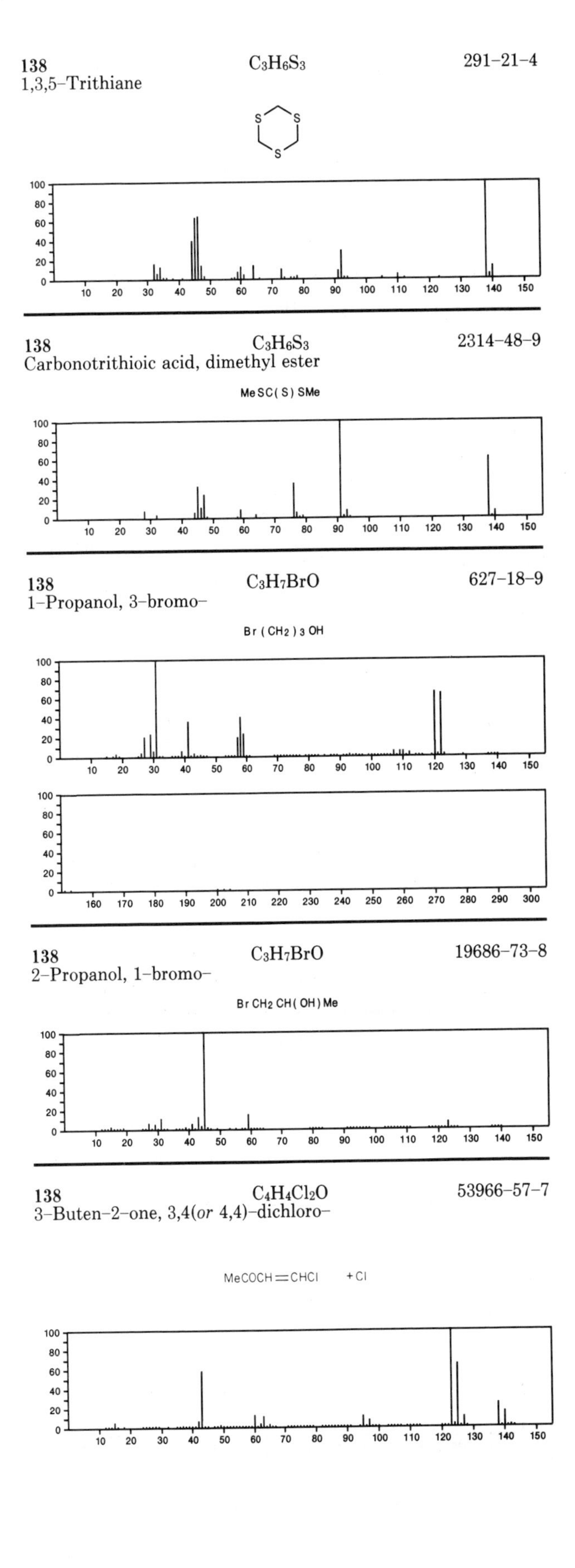
138 $C_3H_6S_3$ 291-21-4
1,3,5-Trithiane
S
S
S
138 $C_3H_6S_3$ 2314-48-9
Carbonotrithioic acid, dimethyl ester
MeSC(S)SMe
138 C_3H_7BrO 627-18-9
1-Propanol, 3-bromo-
Br(CH2)3OH
138 C_3H_7BrO 19686-73-8
2-Propanol, 1-bromo-
BrCH2CH(OH)Me
138 $C_4H_4Cl_2O$ 53966-57-7
3-Buten-2-one, 3,4(or 4,4)-dichloro-
MeCOCH=CHCl +Cl

138 $C_4H_{10}O_3S$ 623–81–4
Sulfurous acid, diethyl ester

EtOS(O)OEt

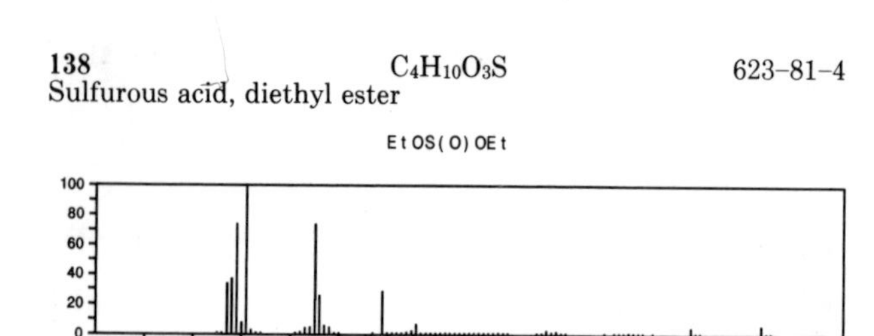

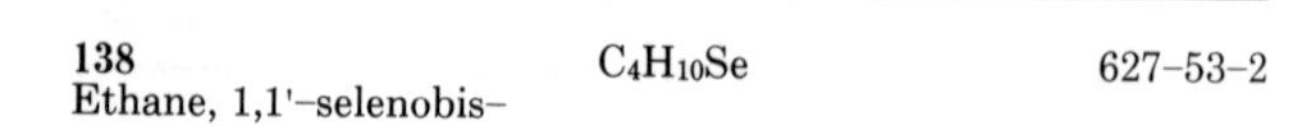

138 $C_4H_{10}Se$ 627–53–2
Ethane, 1,1'–selenobis–

SeEt2

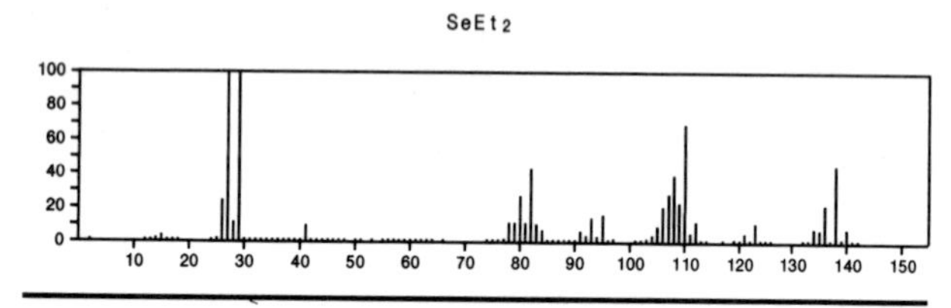

138 $C_4H_{11}O_3P$ 762–04–9
Phosphonic acid, diethyl ester

EtOP(O)HOEt

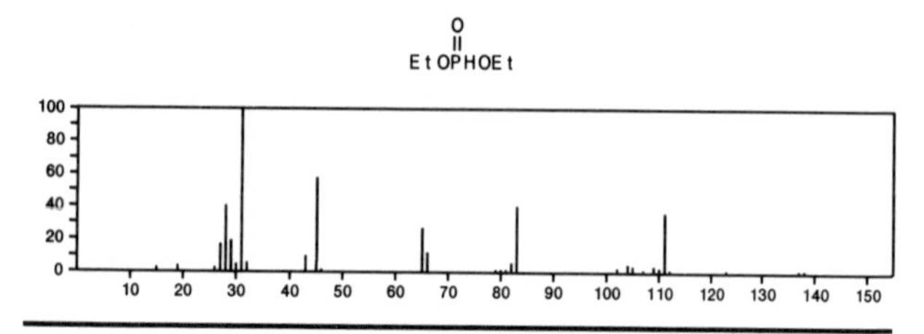

138 $C_4H_{12}FN_2P$ 1735–82–6
Phosphorodiamidous fluoride, tetramethyl–

Me2NPFNMe2

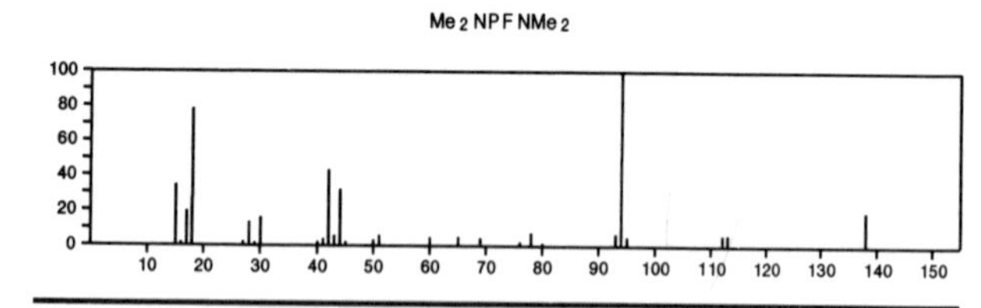

138 $C_5H_8Cl_2$ 694–16–6
Cyclopropane, 1,1–dichloro–2,2–dimethyl–

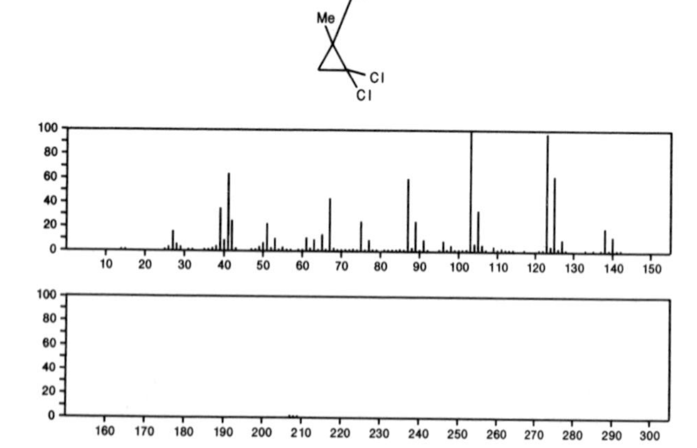

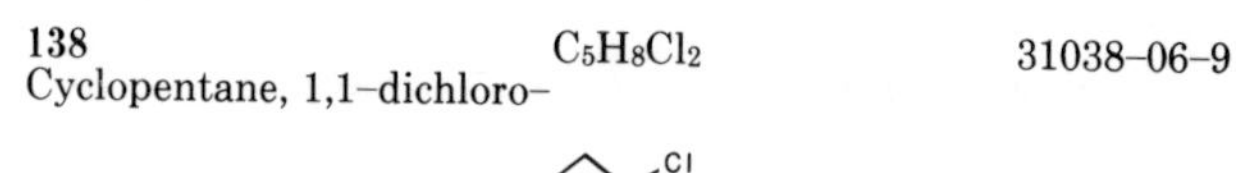

138 $C_5H_8Cl_2$ 31038–06–9
Cyclopentane, 1,1–dichloro–

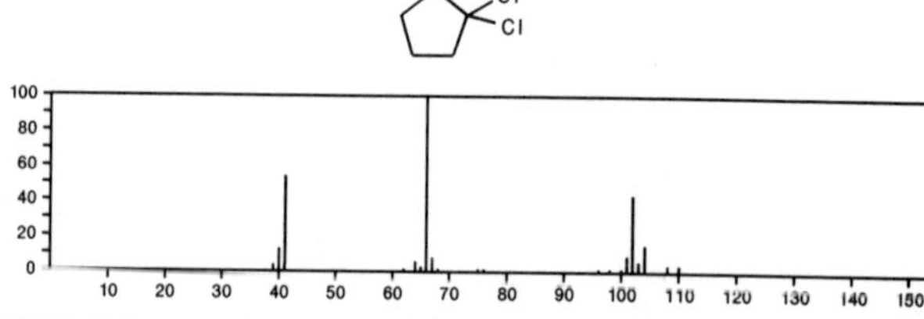

138 $C_5H_{11}ClO_2$ 4151–98–8
2–Propanol, 1–chloro–3–ethoxy–

EtOCH2CH(OH)CH2Cl

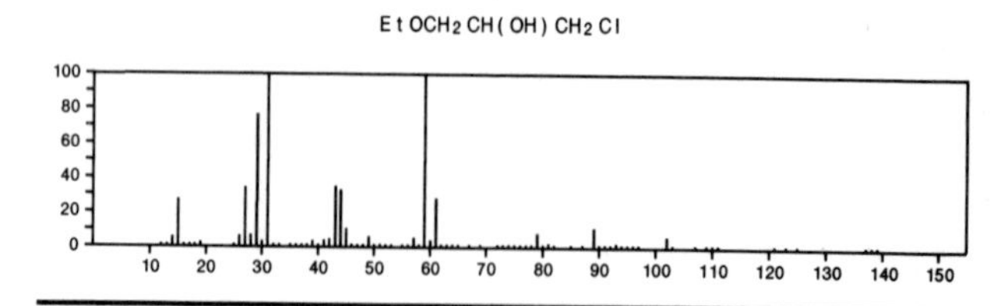

138 $C_6H_6N_2O_2$ 88–74–4
Benzenamine, 2–nitro–

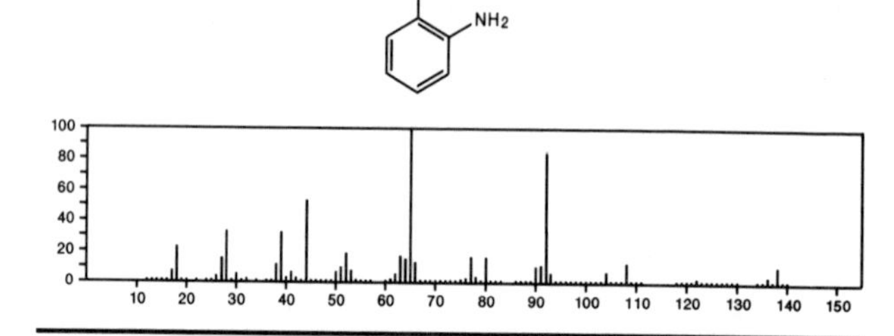

138 $C_6H_6N_2O_2$ 99–09–2
Benzenamine, 3–nitro–

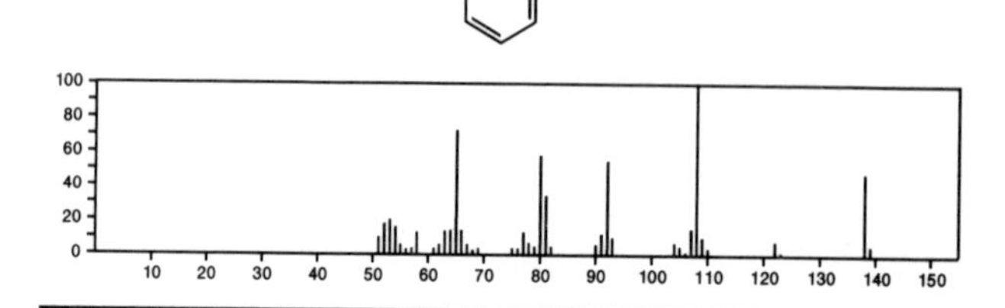

138 $C_6H_6N_2O_2$ 100–01–6
Benzenamine, 4–nitro–

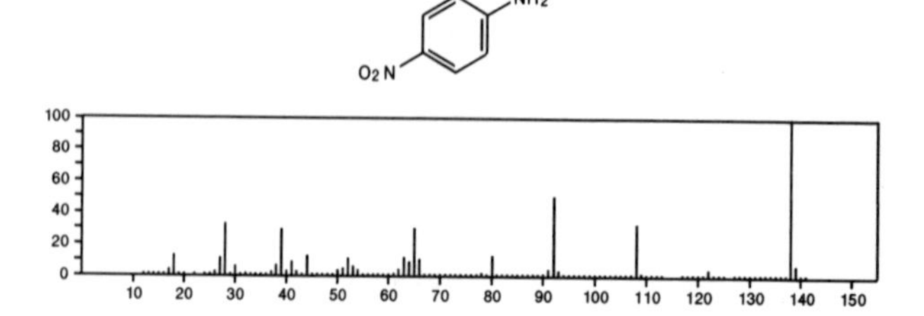

138 $C_6H_6N_2O_2$ 104–98–3
2–Propenoic acid, 3–(1*H*–imidazol–4–yl)–

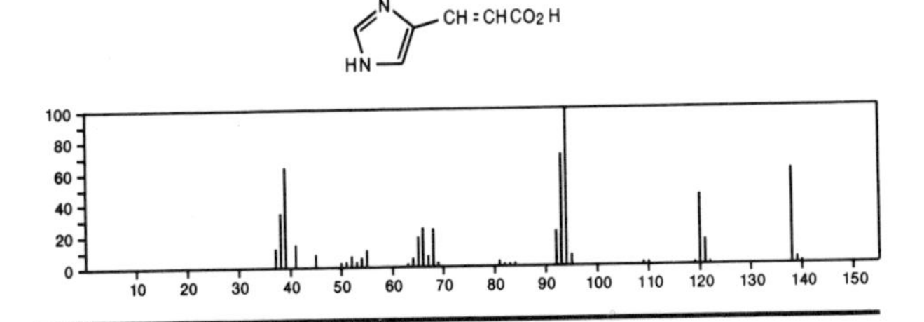

138 $C_6H_6N_2O_2$ 1986–81–8
3–Pyridinecarboxamide, 1–oxide

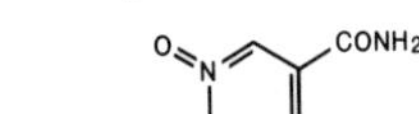

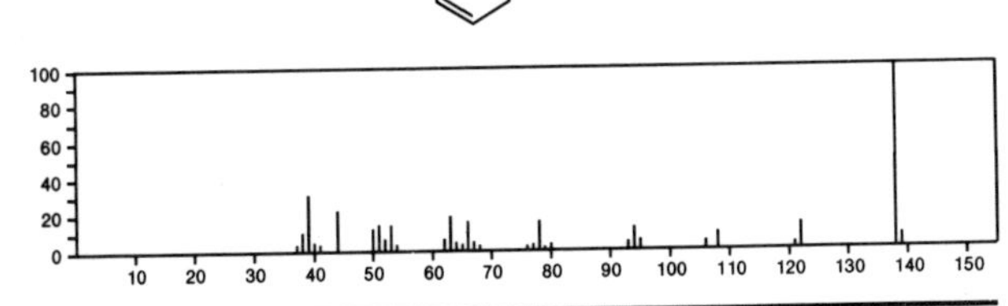

138 $C_6H_6N_2O_2$ 3670–59–5
3–Pyridinecarboxamide, 1,6–dihydro–6–oxo–

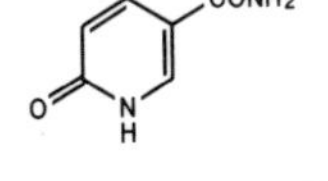

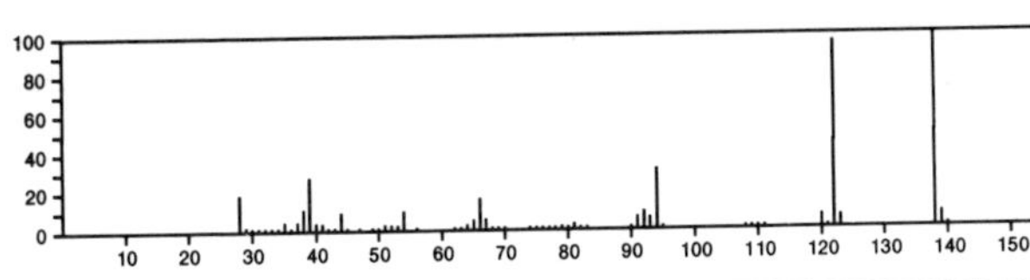

138 $C_6H_6N_2O_2$ 4427–22–9
4–Pyridinecarboxamide, *N*–hydroxy–

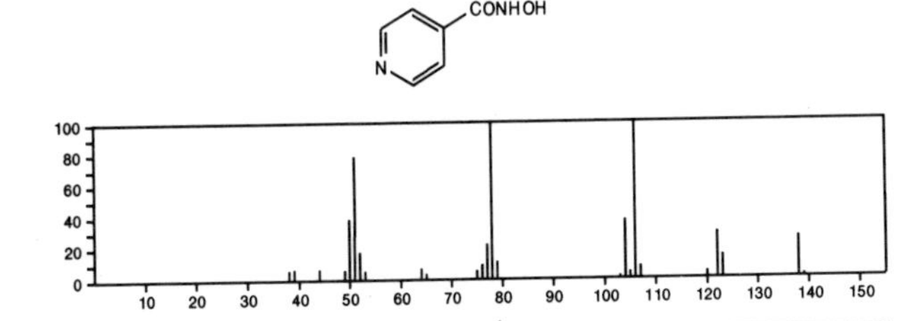

138 $C_6H_6N_2O_2$ 5345–47–1
3–Pyridinecarboxylic acid, 2–amino–

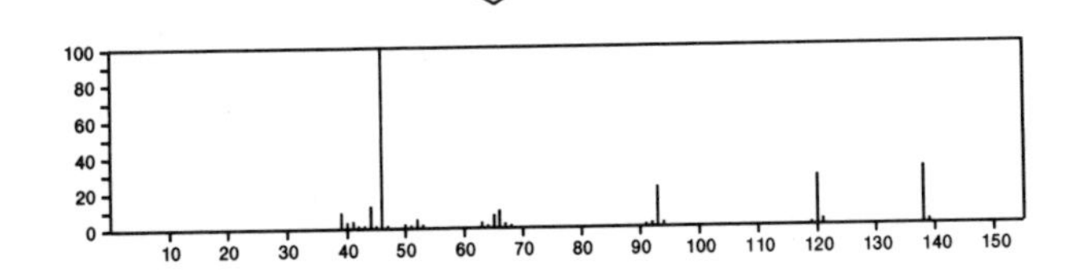

138 $C_6H_6N_2O_2$ 7790–01–4
Propanedinitrile, (acetyloxy)methyl–

AcOCMe(CN)2

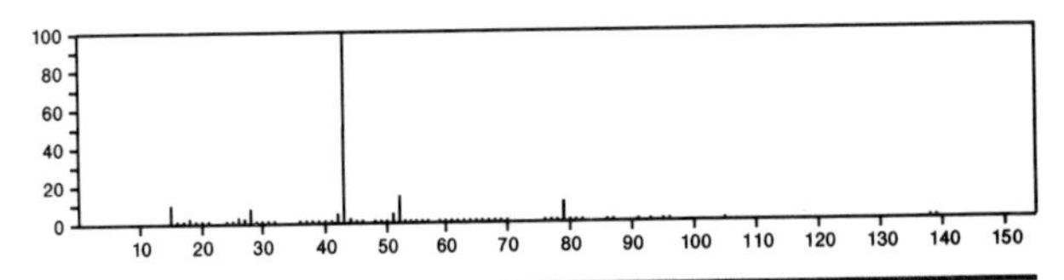

138 $C_6H_6N_2O_2$ 23628–31–1
2–Pyridinecarboxylic acid, 6–amino–

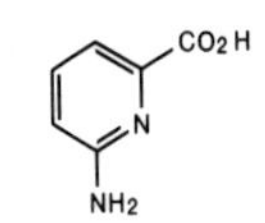

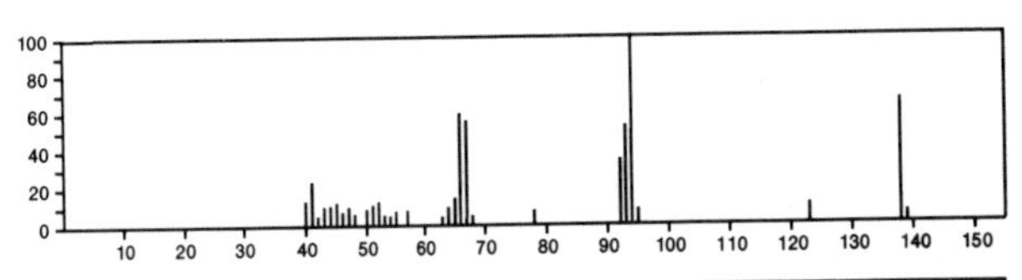

138 $C_6H_6N_2S$ 2196–13–6
4–Pyridinecarbothioamide

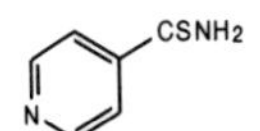

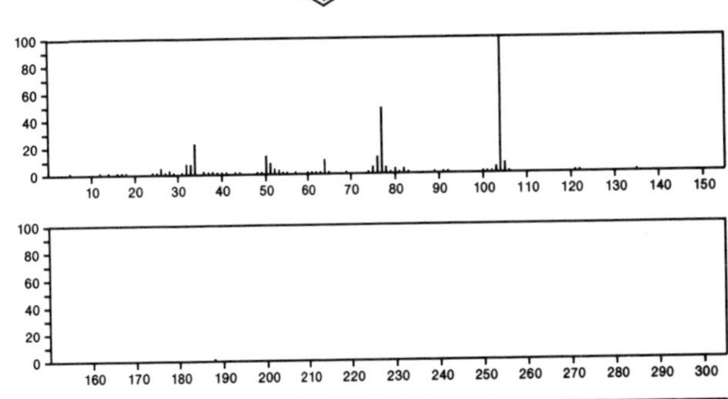

138 $C_6H_{10}N_4$ 54–95–5
5*H*–Tetrazolo[1,5–*a*]azepine, 6,7,8,9–tetrahydro–

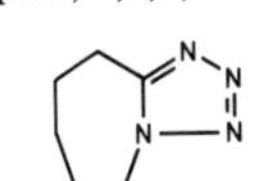

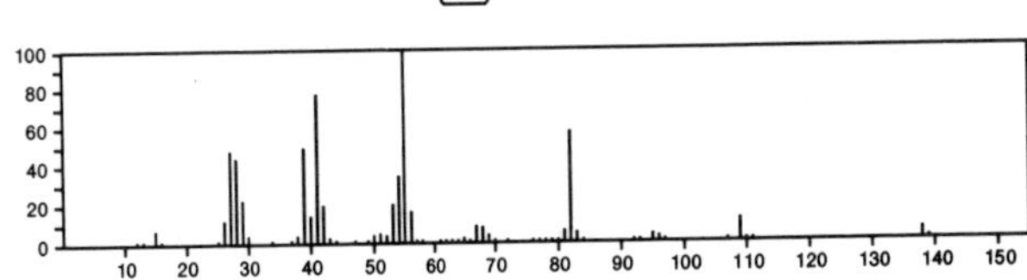

138 $C_6H_{10}N_4$ 13717–91–4
1,2,4,5–Tetrazine, 3,6–diethyl–

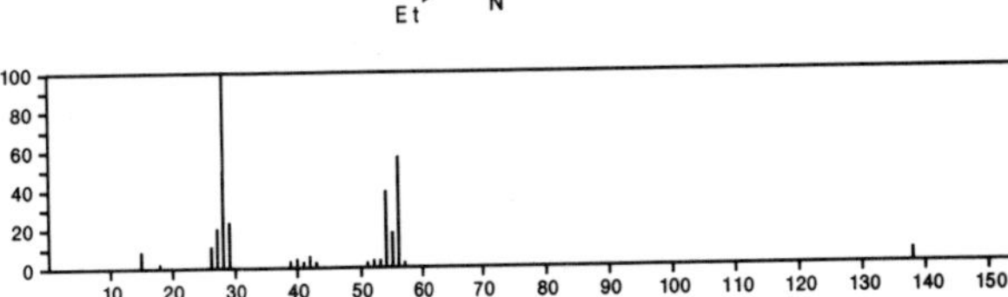

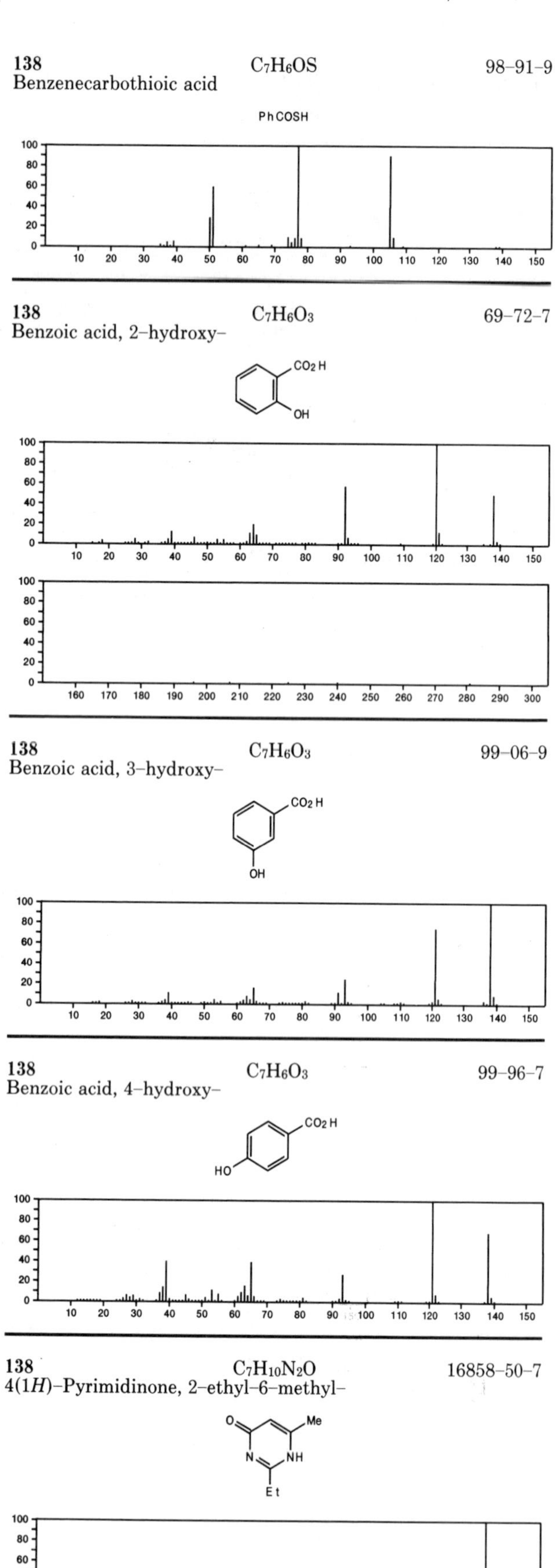

138 C_7H_6OS 98-91-9
Benzenecarbothioic acid

138 $C_7H_6O_3$ 69-72-7
Benzoic acid, 2-hydroxy-

138 $C_7H_6O_3$ 99-06-9
Benzoic acid, 3-hydroxy-

138 $C_7H_6O_3$ 99-96-7
Benzoic acid, 4-hydroxy-

138 $C_7H_{10}N_2O$ 16858-50-7
4(1*H*)-Pyrimidinone, 2-ethyl-6-methyl-

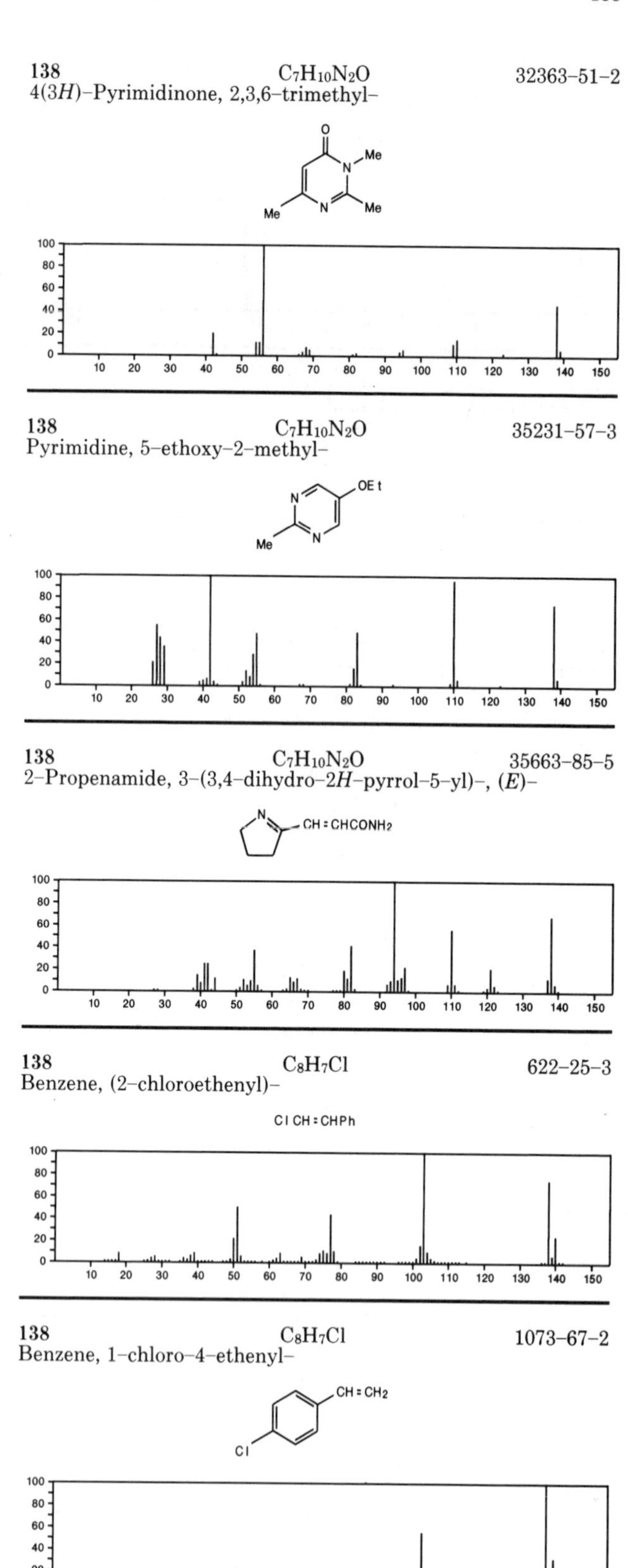

138 $C_7H_{10}N_2O$ 32363-51-2
4(3*H*)-Pyrimidinone, 2,3,6-trimethyl-

138 $C_7H_{10}N_2O$ 35231-57-3
Pyrimidine, 5-ethoxy-2-methyl-

138 $C_7H_{10}N_2O$ 35663-85-5
2-Propenamide, 3-(3,4-dihydro-2*H*-pyrrol-5-yl)-, (*E*)-

138 C_8H_7Cl 622-25-3
Benzene, (2-chloroethenyl)-

138 C_8H_7Cl 1073-67-2
Benzene, 1-chloro-4-ethenyl-

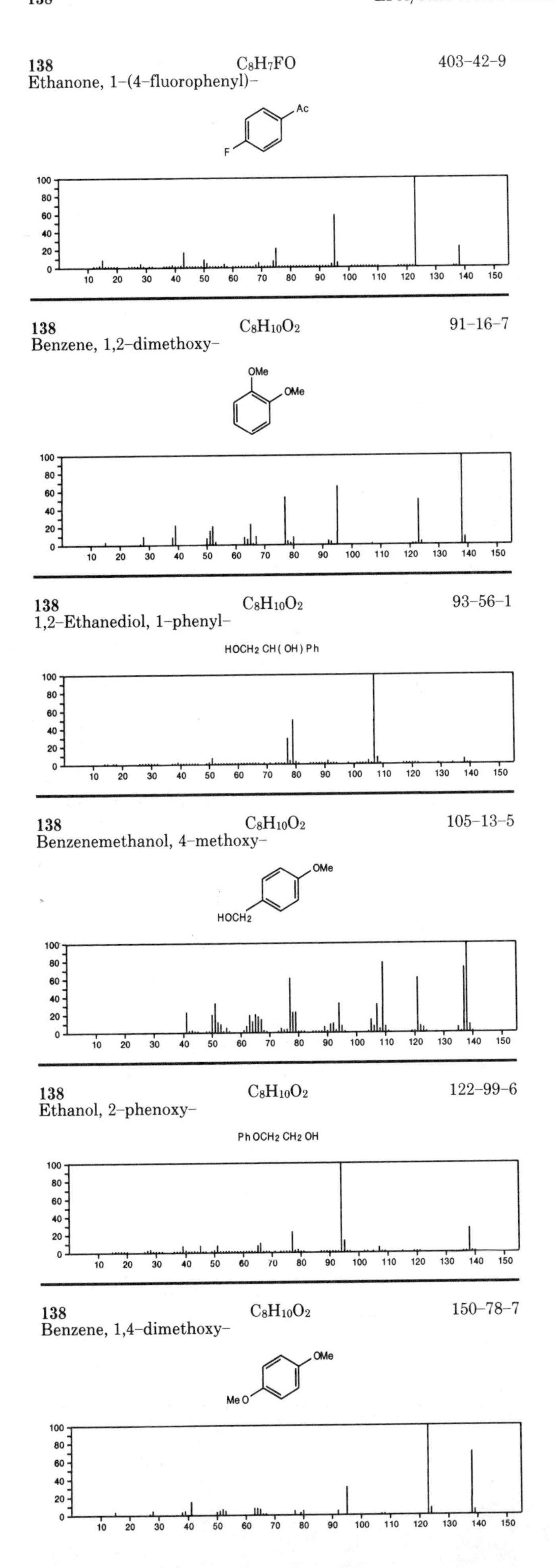
138 C8H7FO 403-42-9
Ethanone, 1-(4-fluorophenyl)-
Ac
F
138 C8H10O2 91-16-7
Benzene, 1,2-dimethoxy-
OMe
OMe
138 C8H10O2 93-56-1
1,2-Ethanediol, 1-phenyl-
HOCH2 CH(OH) Ph
138 C8H10O2 105-13-5
Benzenemethanol, 4-methoxy-
OMe
HOCH2
138 C8H10O2 122-99-6
Ethanol, 2-phenoxy-
Ph OCH2 CH2 OH
138 C8H10O2 150-78-7
Benzene, 1,4-dimethoxy-
OMe
MeO

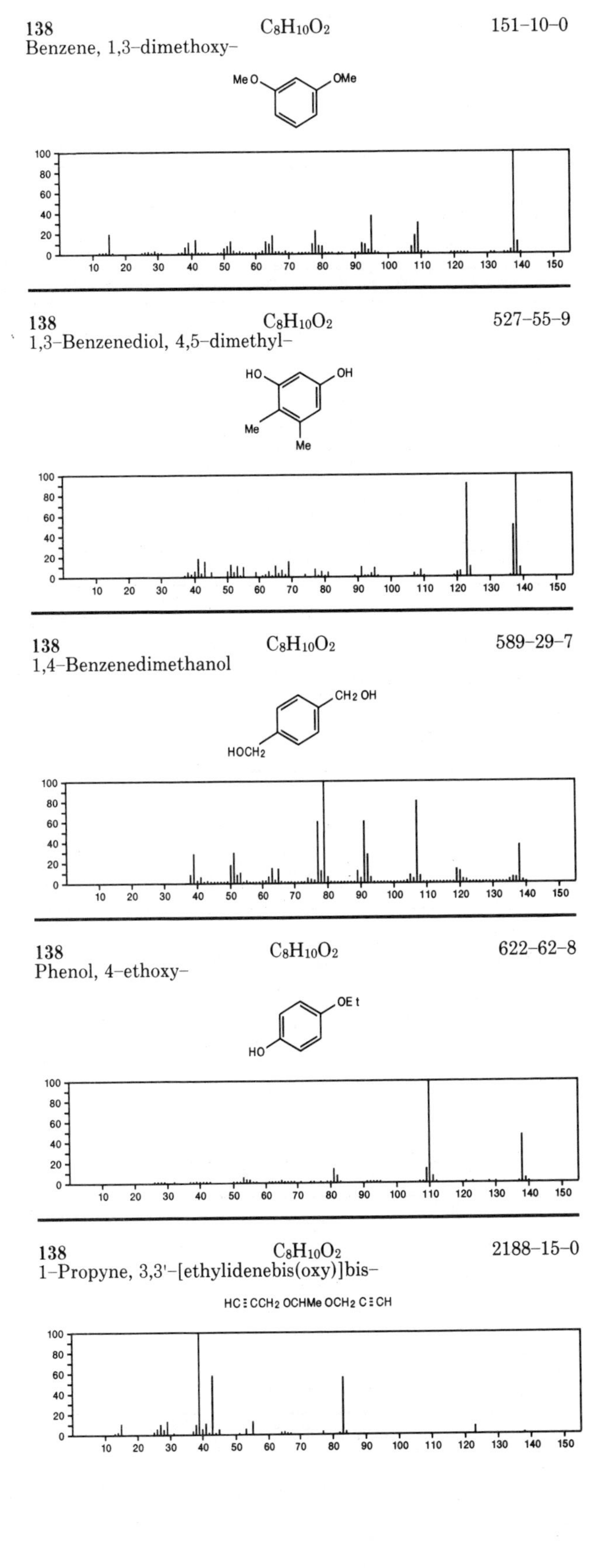
138 C8H10O2 151-10-0
Benzene, 1,3-dimethoxy-
MeO
OMe
138 C8H10O2 527-55-9
1,3-Benzenediol, 4,5-dimethyl-
HO
OH
Me
Me
138 C8H10O2 589-29-7
1,4-Benzenedimethanol
CH2 OH
HOCH2
138 C8H10O2 622-62-8
Phenol, 4-ethoxy-
OEt
HO
138 C8H10O2 2188-15-0
1-Propyne, 3,3'-[ethylidenebis(oxy)]bis-
HC≡CCH2 OCHMe OCH2 C≡CH

138 $C_8H_{10}O_2$ 2896-60-8
1,3-Benzenediol, 4-ethyl-

HO OH Et

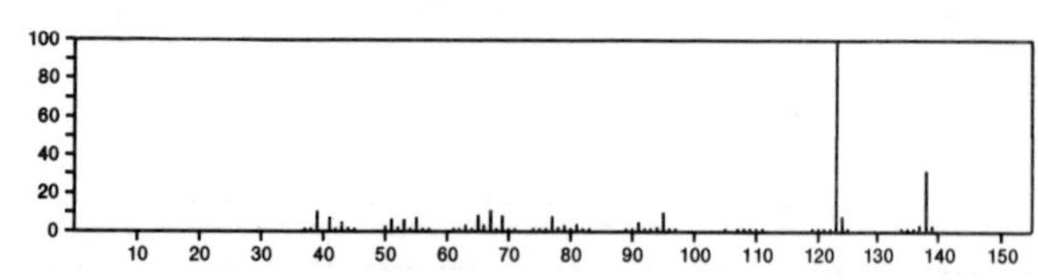

138 $C_8H_{10}O_2$ 4208-63-3
2-Butanone, 1-(2-furanyl)-

CH_2COEt

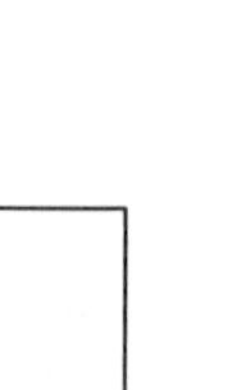

138 $C_8H_{10}O_2$ 13398-94-2
Benzeneethanol, 3-hydroxy-

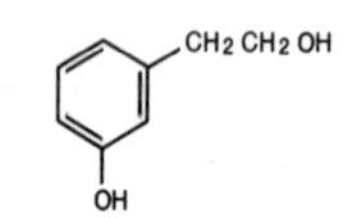

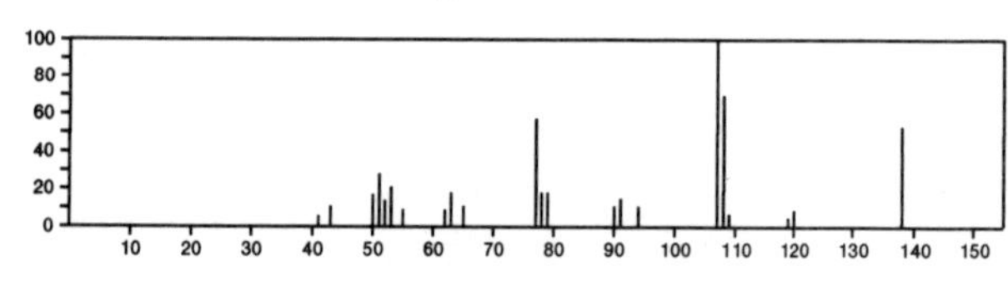

138 $C_8H_{10}O_2$ 50267-08-8
3,10-Dioxatricyclo[4.3.1.0^{2,4}]dec-7-ene, (1α,2α,4α,6α)-

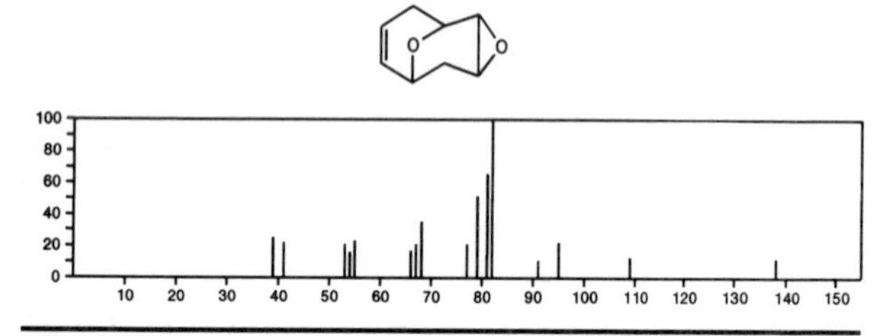

138 $C_8H_{10}S$ 622-38-8
Benzene, (ethylthio)-

EtSPh

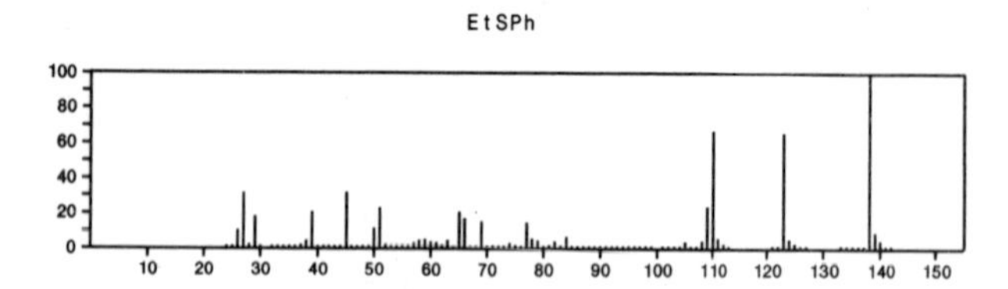

138 $C_8H_{10}S$ 623-13-2
Benzene, 1-methyl-4-(methylthio)-

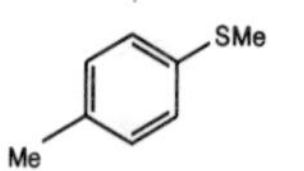

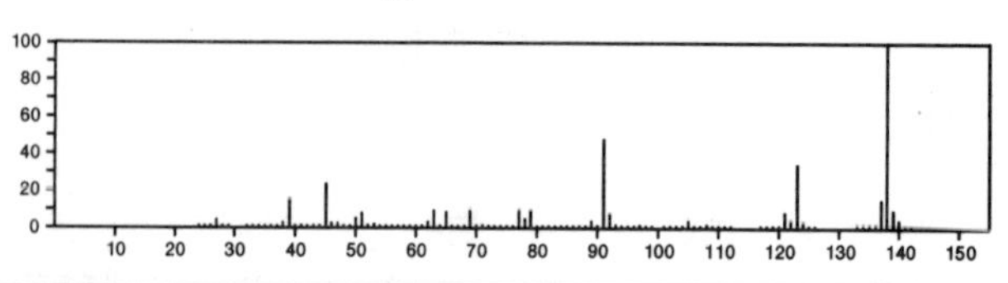

138 $C_8H_{10}S$ 766-92-7
Benzene, [(methylthio)methyl]-

MeSCH2Ph

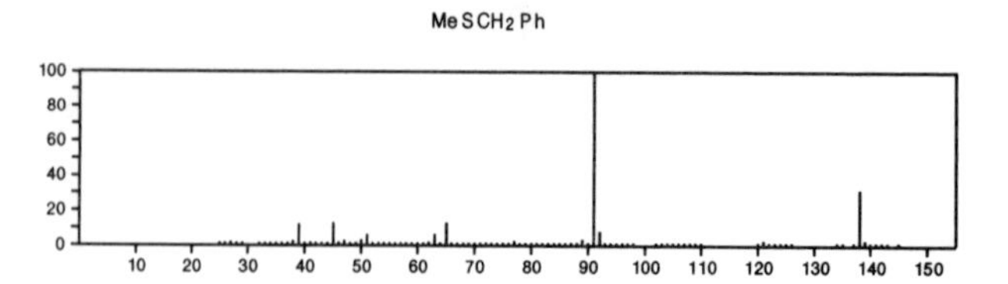

138 $C_8H_{10}S$ 4886-77-5
Benzene, 1-methyl-3-(methylthio)-

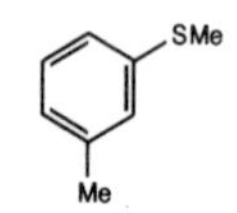

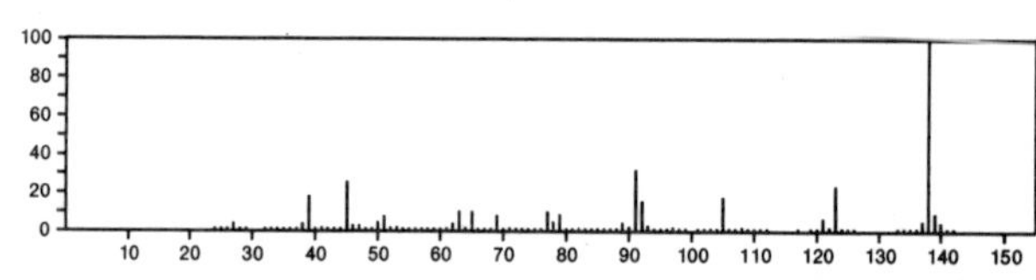

138 $C_8H_{10}S$ 14092-00-3
Benzene, 1-methyl-2-(methylthio)-

SMe Me

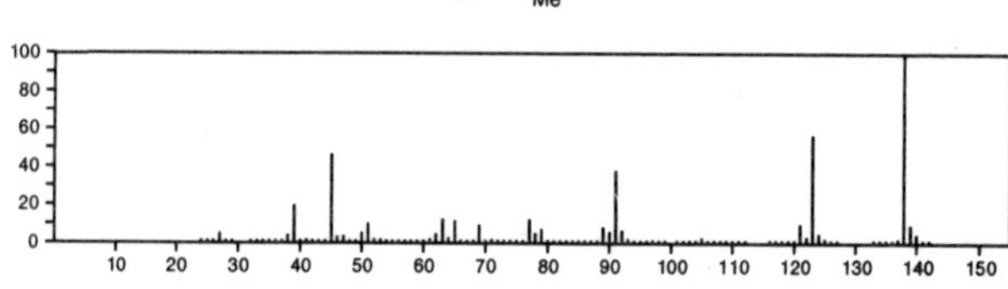

138 $C_8H_{10}S$ 52008-15-8
Thiophene, 2-(2-butenyl)-, (*E*)-

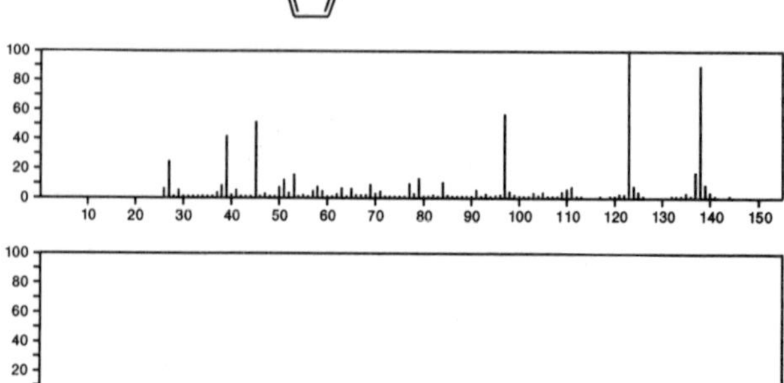

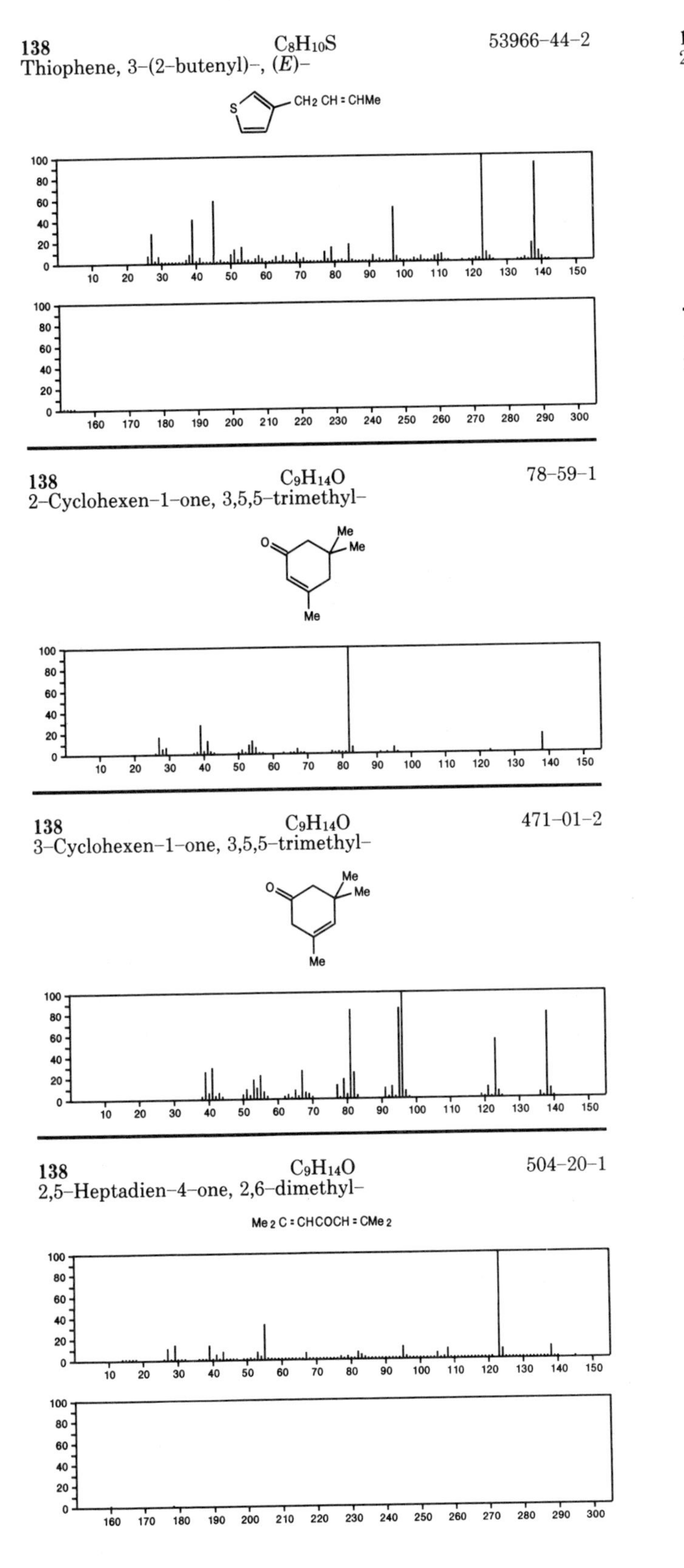

138 $C_8H_{10}S$ 53966-44-2
Thiophene, 3-(2-butenyl)-, (*E*)-

138 $C_9H_{14}O$ 78-59-1
2-Cyclohexen-1-one, 3,5,5-trimethyl-

138 $C_9H_{14}O$ 471-01-2
3-Cyclohexen-1-one, 3,5,5-trimethyl-

138 $C_9H_{14}O$ 504-20-1
2,5-Heptadien-4-one, 2,6-dimethyl-

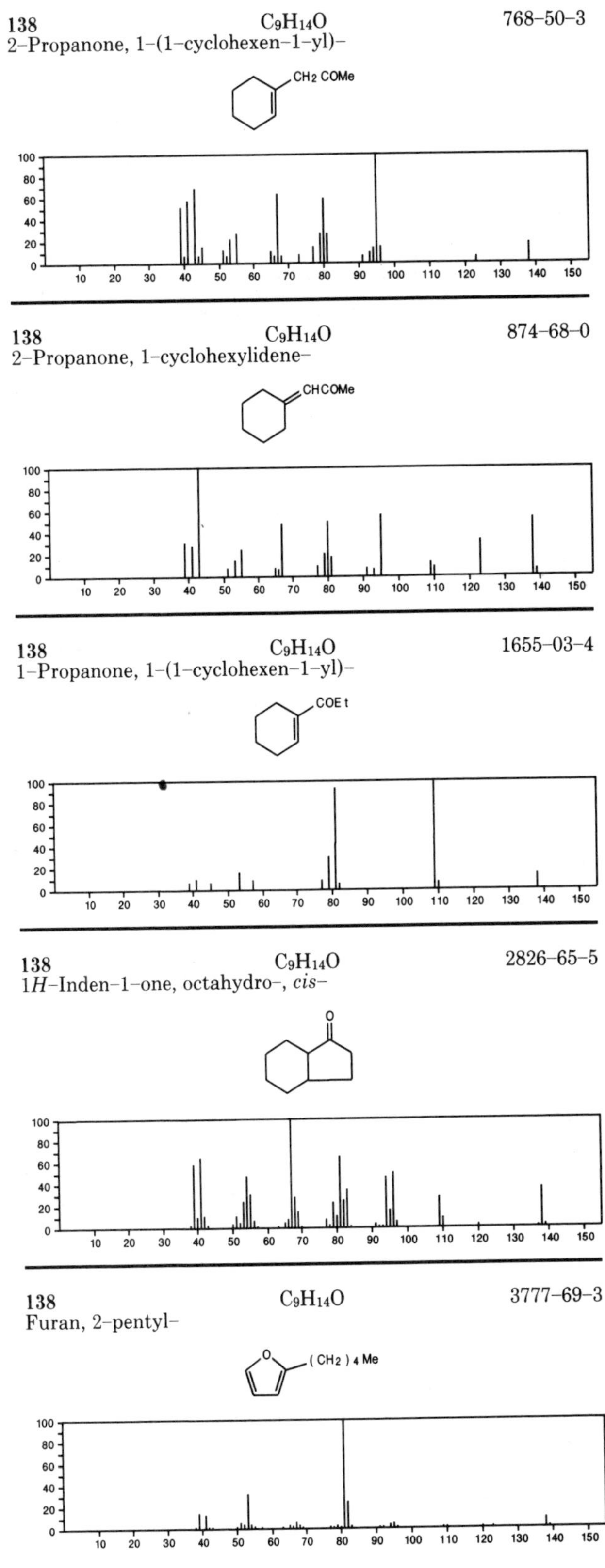

138 $C_9H_{14}O$ 768-50-3
2-Propanone, 1-(1-cyclohexen-1-yl)-

138 $C_9H_{14}O$ 874-68-0
2-Propanone, 1-cyclohexylidene-

138 $C_9H_{14}O$ 1655-03-4
1-Propanone, 1-(1-cyclohexen-1-yl)-

138 $C_9H_{14}O$ 2826-65-5
1*H*-Inden-1-one, octahydro-, *cis*-

138 $C_9H_{14}O$ 3777-69-3
Furan, 2-pentyl-

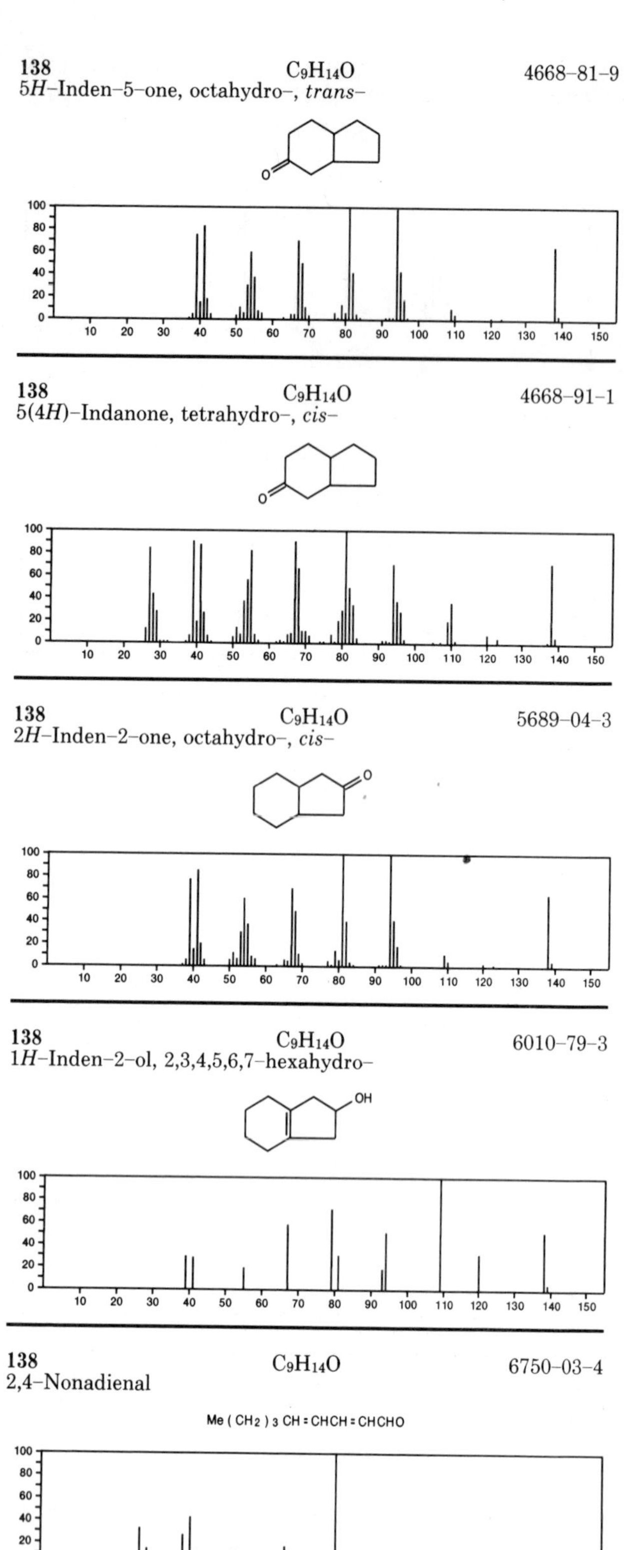

138 $C_9H_{14}O$ 4668-81-9
5*H*-Inden-5-one, octahydro-, *trans*-

138 $C_9H_{14}O$ 4668-91-1
5(4*H*)-Indanone, tetrahydro-, *cis*-

138 $C_9H_{14}O$ 5689-04-3
2*H*-Inden-2-one, octahydro-, *cis*-

138 $C_9H_{14}O$ 6010-79-3
1*H*-Inden-2-ol, 2,3,4,5,6,7-hexahydro-

138 $C_9H_{14}O$ 6750-03-4
2,4-Nonadienal

Me(CH2)3CH=CHCH=CHCHO

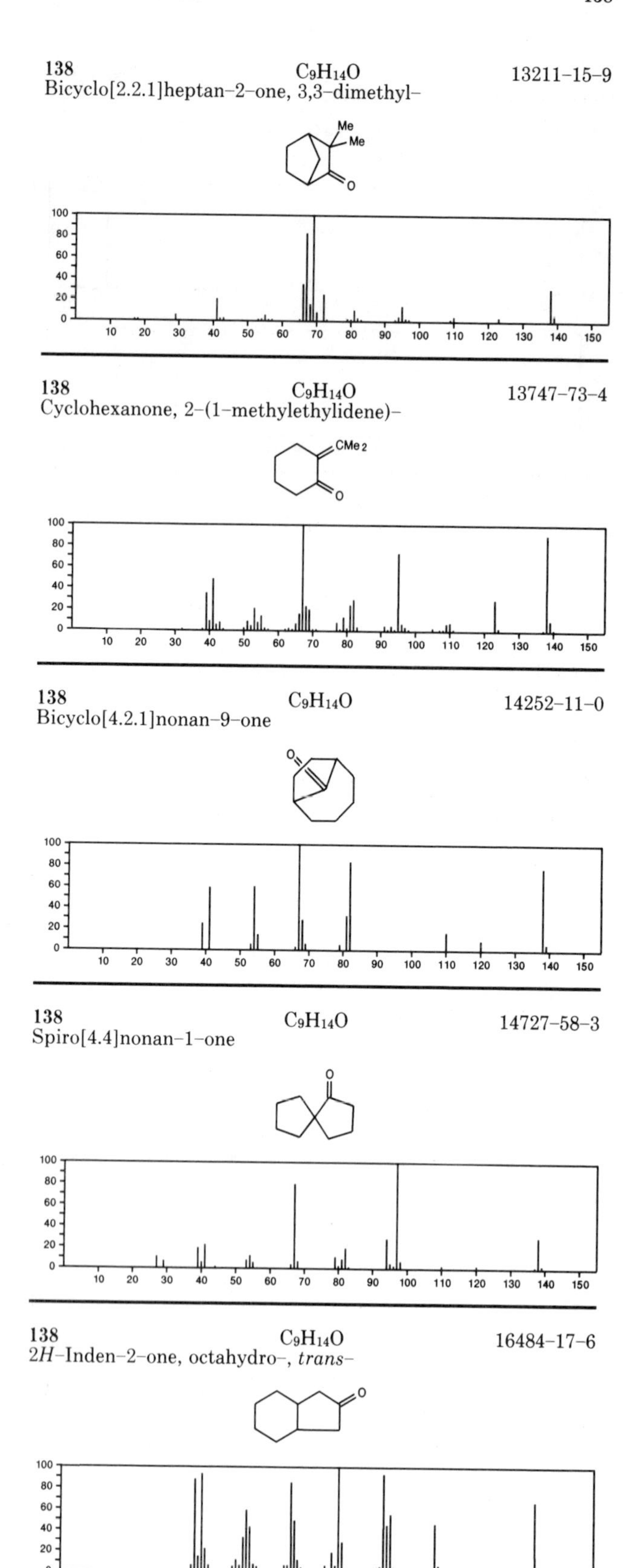

138 $C_9H_{14}O$ 13211-15-9
Bicyclo[2.2.1]heptan-2-one, 3,3-dimethyl-

138 $C_9H_{14}O$ 13747-73-4
Cyclohexanone, 2-(1-methylethylidene)-

138 $C_9H_{14}O$ 14252-11-0
Bicyclo[4.2.1]nonan-9-one

138 $C_9H_{14}O$ 14727-58-3
Spiro[4.4]nonan-1-one

138 $C_9H_{14}O$ 16484-17-6
2*H*-Inden-2-one, octahydro-, *trans*-

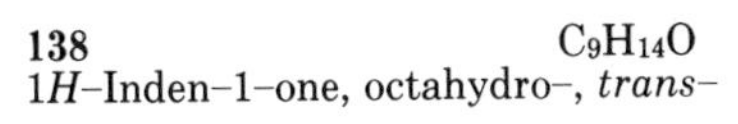

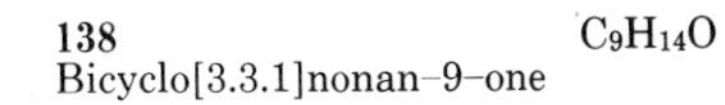

138 $C_9H_{14}O$ 16783-22-5
1*H*-Inden-1-one, octahydro-, *trans*-

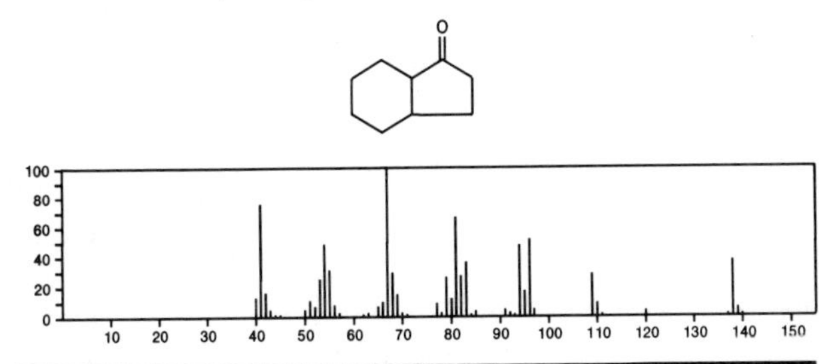

138 $C_9H_{14}O$ 17299-41-1
2-Cyclohexen-1-one, 3,4,4-trimethyl-

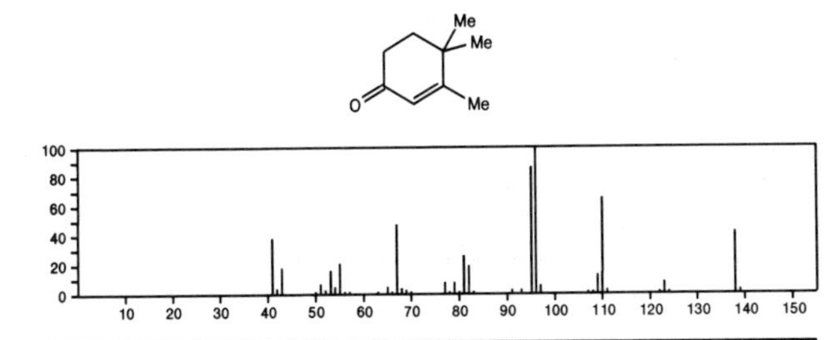

138 $C_9H_{14}O$ 17429-29-7
2-Cyclohexen-1-one, 4,4,5-trimethyl-

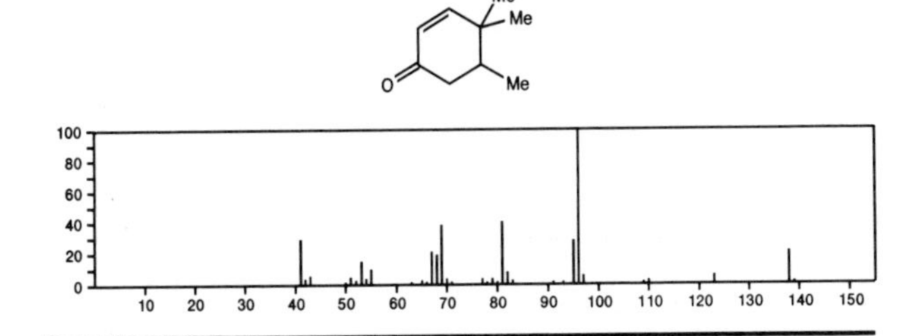

138 $C_9H_{14}O$ 17429-32-2
2-Cyclohexen-1-one, 4-ethyl-4-methyl-

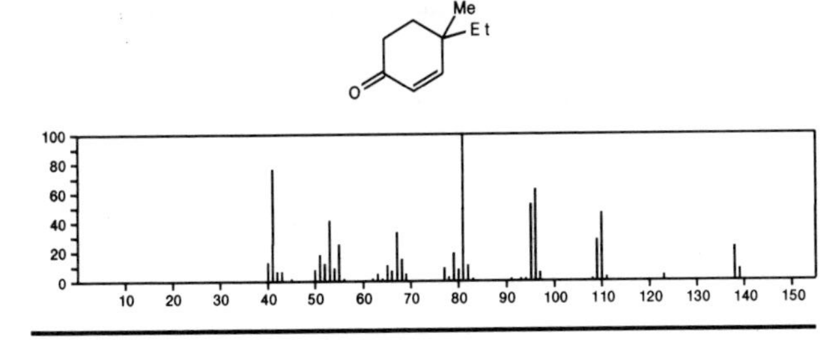

138 $C_9H_{14}O$ 17587-33-6
2,6-Nonadienal, (*E,E*)-

$OCHCH=CHCH_2CH_2CH=CHEt$

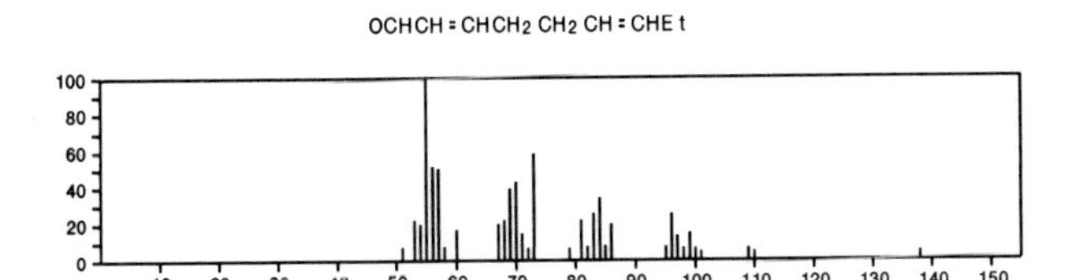

138 $C_9H_{14}O$ 17931-55-4
Bicyclo[3.3.1]nonan-9-one

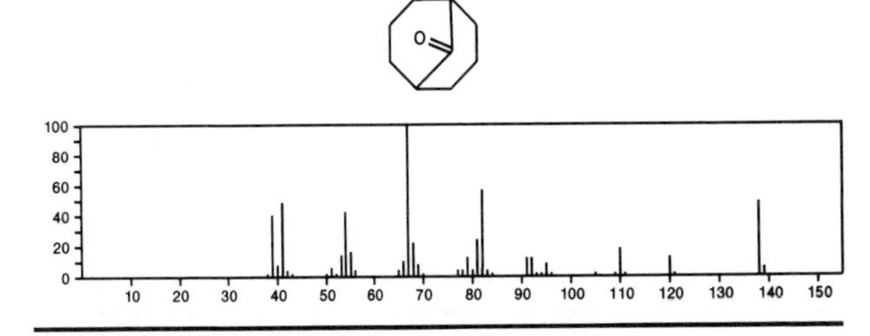

138 $C_9H_{14}O$ 19877-78-2
Bicyclo[3.3.1]non-2-en-9-ol, *syn*-

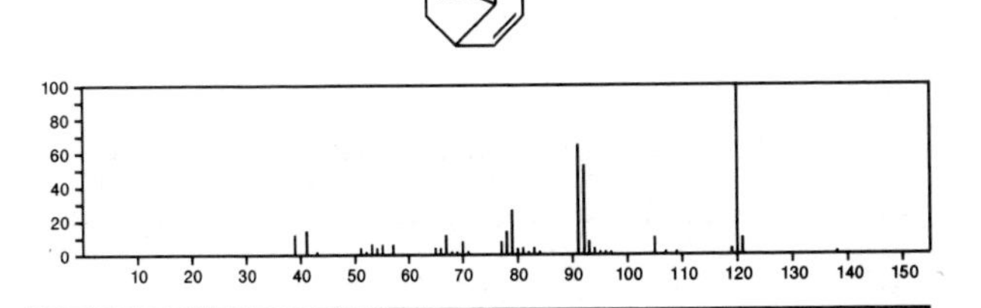

138 $C_9H_{14}O$ 24844-19-7
Bicyclo[3.3.1]non-2-en-9-ol, *anti*-

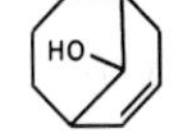

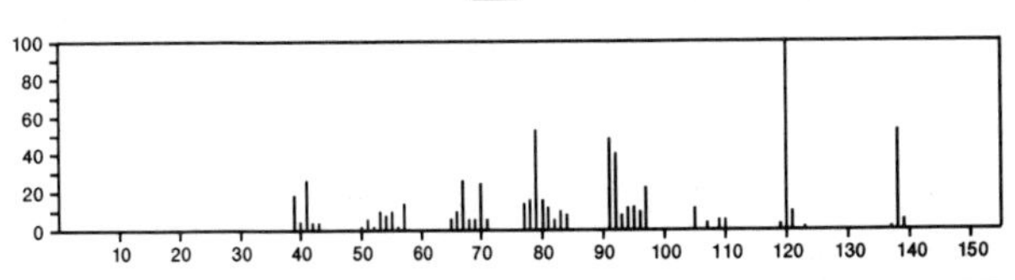

138 $C_9H_{14}O$ 26051-25-2
Bicyclo[2.2.2]octanone, 3-methyl-

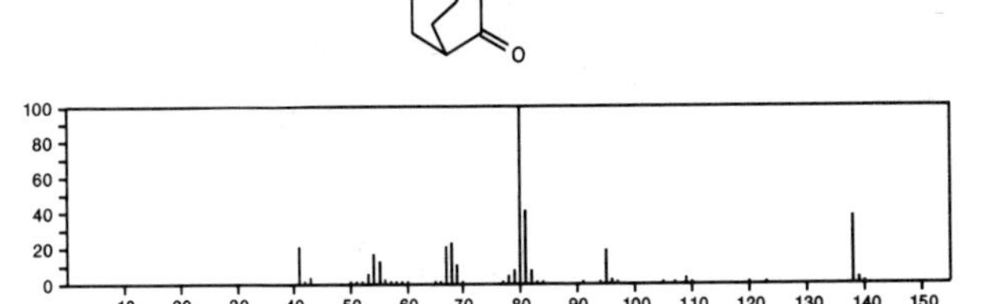

138 $C_9H_{14}O$ 29927-85-3
1*H*-Inden-1-one, octahydro-

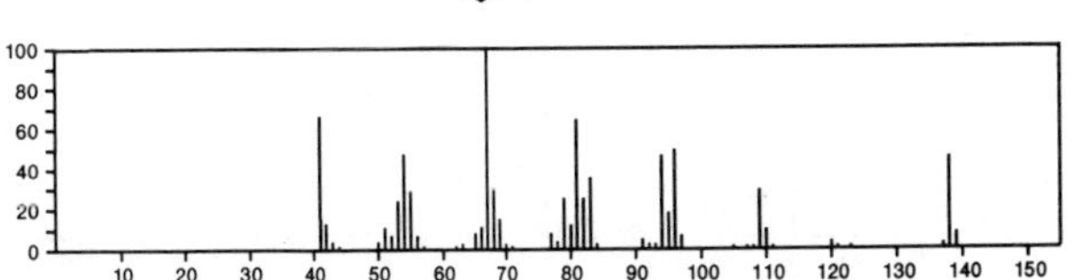

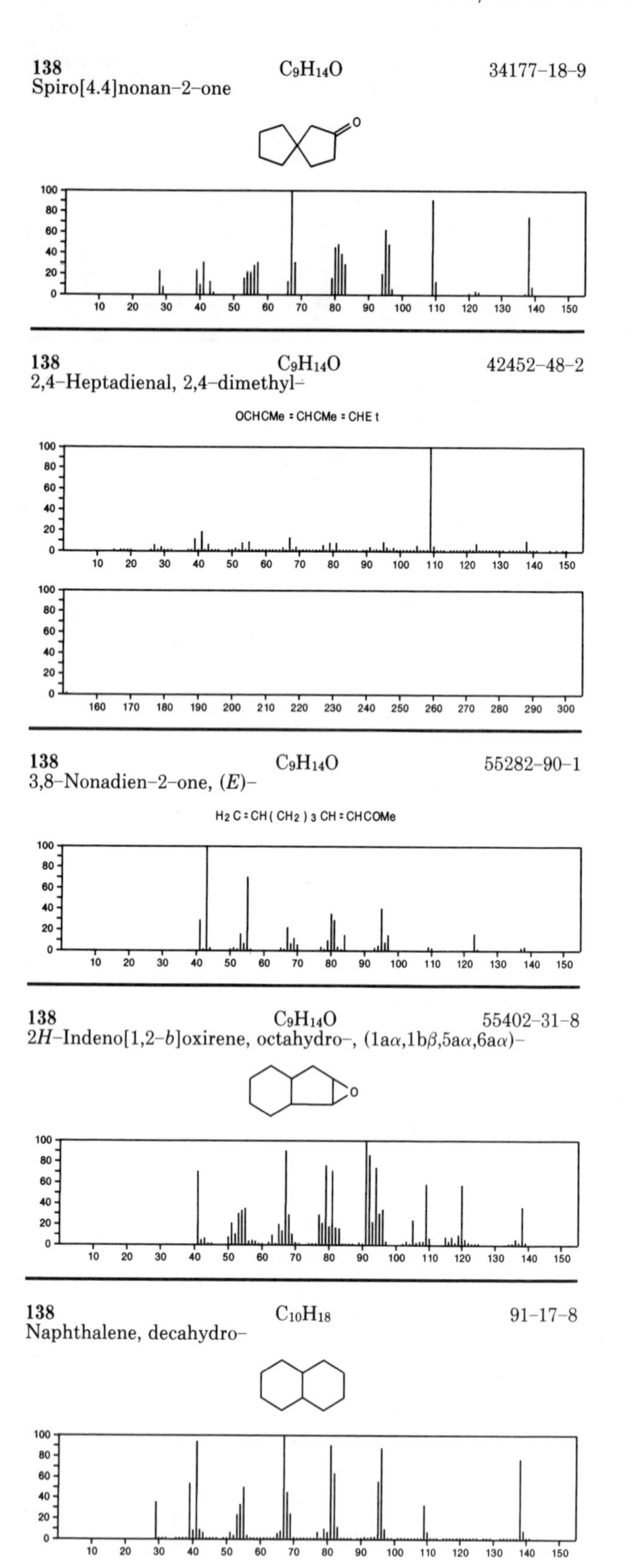

138 $C_9H_{14}O$ 34177-18-9
Spiro[4.4]nonan-2-one

138 $C_9H_{14}O$ 42452-48-2
2,4-Heptadienal, 2,4-dimethyl-

138 $C_9H_{14}O$ 55282-90-1
3,8-Nonadien-2-one, (*E*)-

138 $C_9H_{14}O$ 55402-31-8
2*H*-Indeno[1,2-*b*]oxirene, octahydro-, (1aα,1bβ,5aα,6aα)-

138 $C_{10}H_{18}$ 91-17-8
Naphthalene, decahydro-

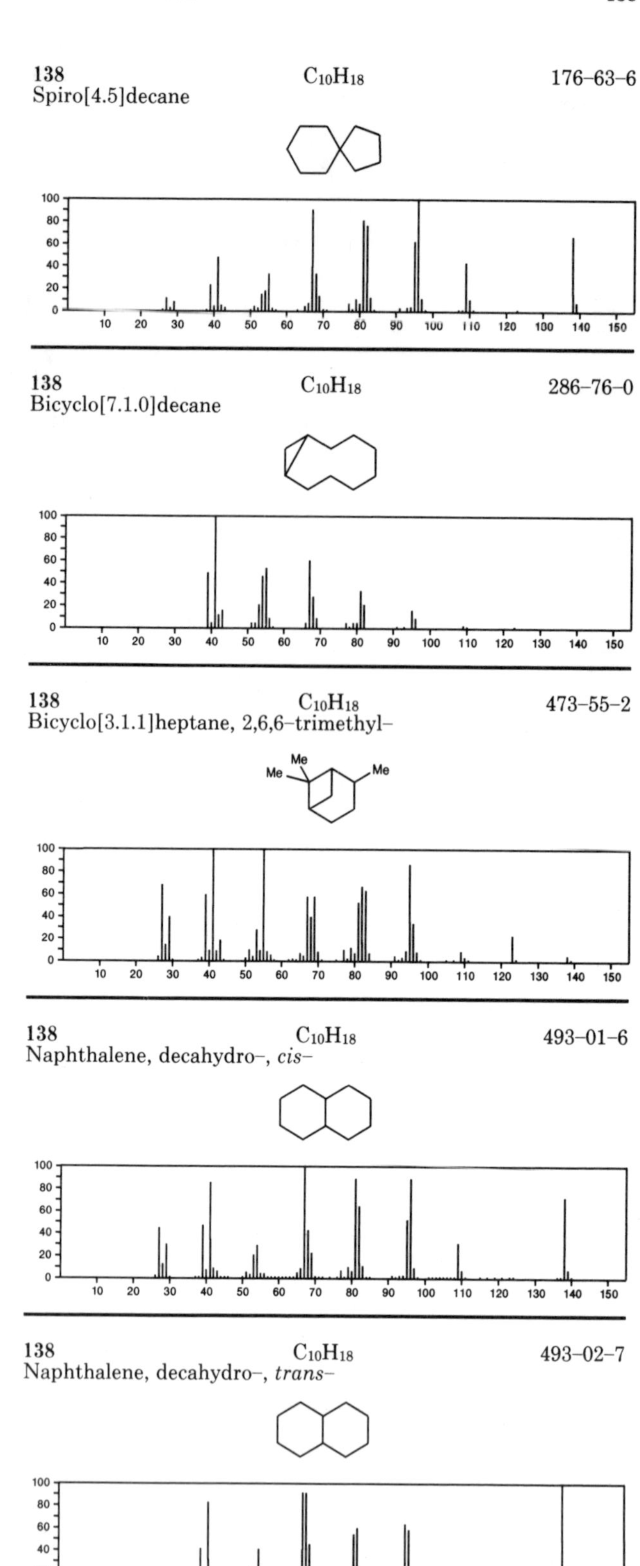

138 $C_{10}H_{18}$ 176-63-6
Spiro[4.5]decane

138 $C_{10}H_{18}$ 286-76-0
Bicyclo[7.1.0]decane

138 $C_{10}H_{18}$ 473-55-2
Bicyclo[3.1.1]heptane, 2,6,6-trimethyl-

138 $C_{10}H_{18}$ 493-01-6
Naphthalene, decahydro-, *cis*-

138 $C_{10}H_{18}$ 493-02-7
Naphthalene, decahydro-, *trans*-

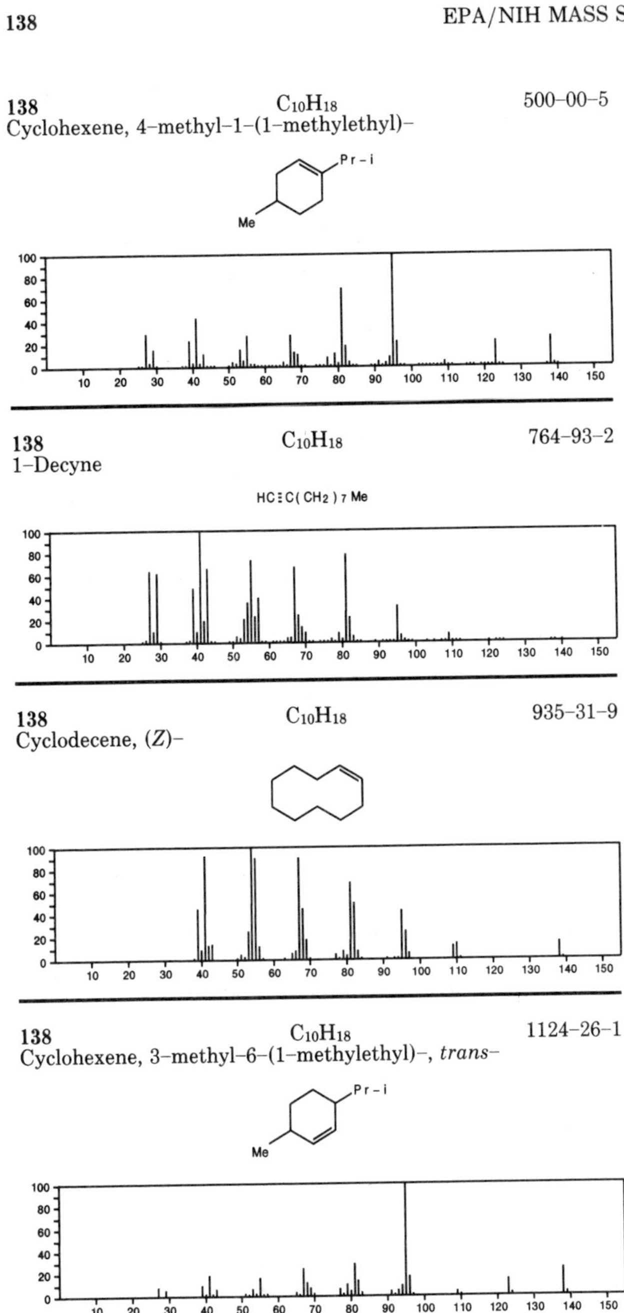

138 $C_{10}H_{18}$ 500-00-5
Cyclohexene, 4-methyl-1-(1-methylethyl)-

138 $C_{10}H_{18}$ 764-93-2
1-Decyne

138 $C_{10}H_{18}$ 935-31-9
Cyclodecene, (*Z*)-

138 $C_{10}H_{18}$ 1124-26-1
Cyclohexene, 3-methyl-6-(1-methylethyl)-, *trans*-

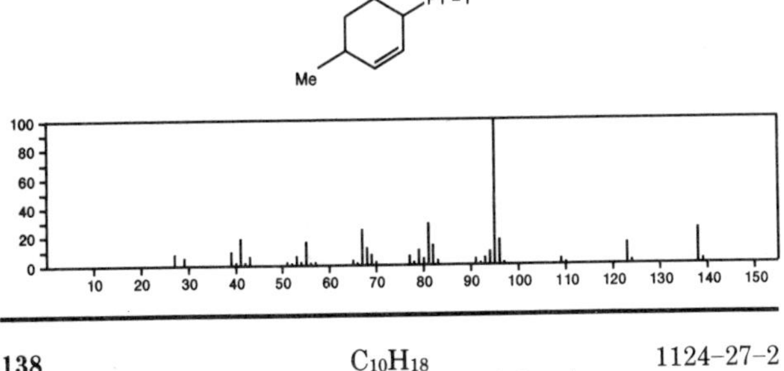

138 $C_{10}H_{18}$ 1124-27-2
Cyclohexane, 1-methyl-4-(1-methylethylidene)-

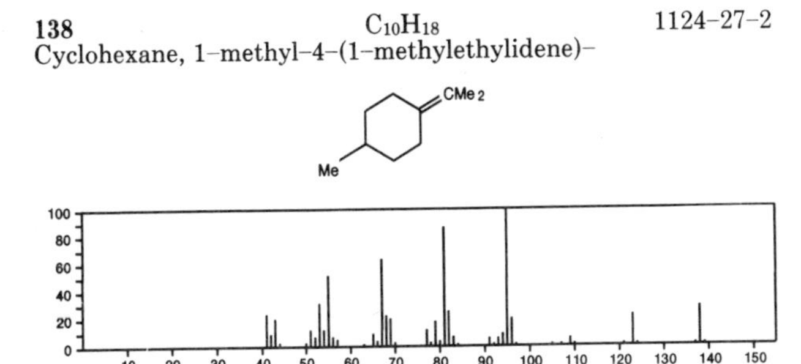

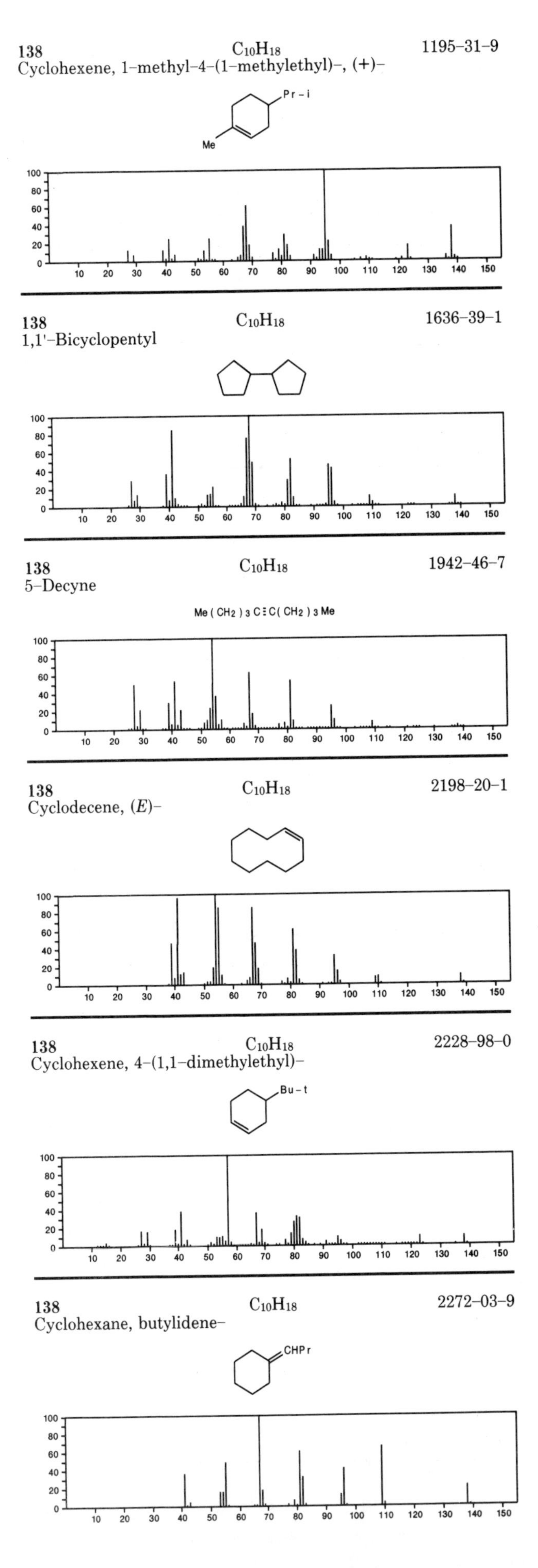

138 $C_{10}H_{18}$ 1195-31-9
Cyclohexene, 1-methyl-4-(1-methylethyl)-, (+)-

138 $C_{10}H_{18}$ 1636-39-1
1,1'-Bicyclopentyl

138 $C_{10}H_{18}$ 1942-46-7
5-Decyne

138 $C_{10}H_{18}$ 2198-20-1
Cyclodecene, (*E*)-

138 $C_{10}H_{18}$ 2228-98-0
Cyclohexene, 4-(1,1-dimethylethyl)-

138 $C_{10}H_{18}$ 2272-03-9
Cyclohexane, butylidene-

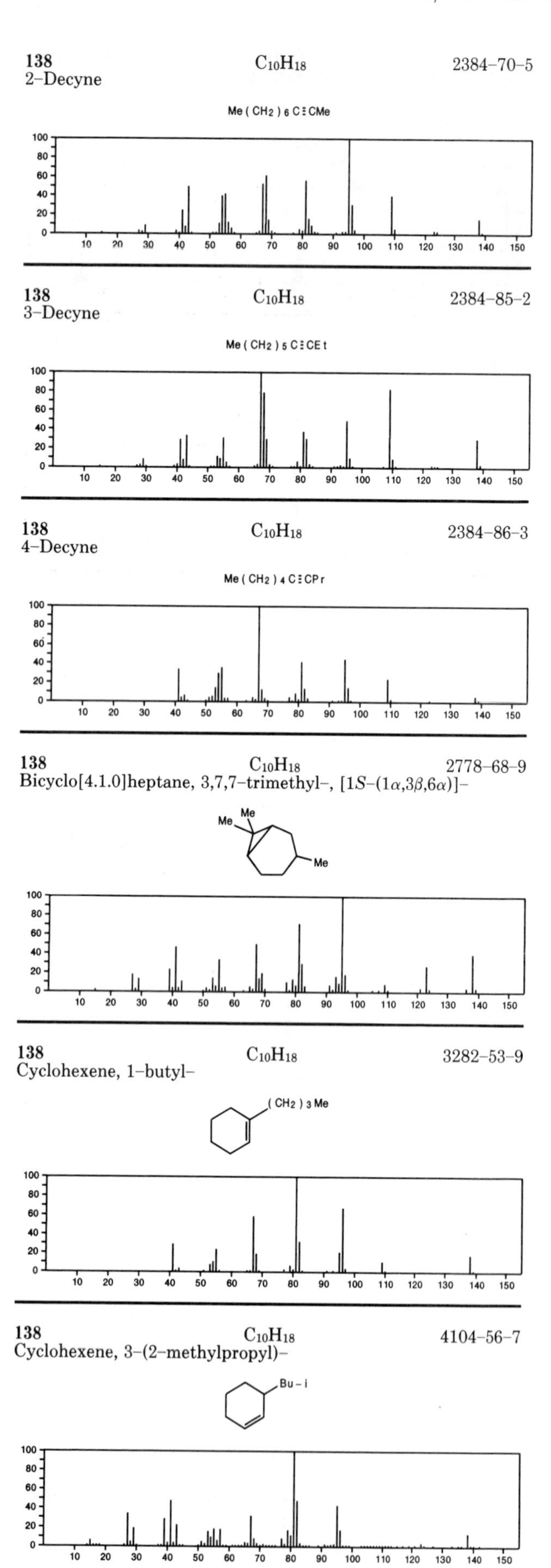

138 $C_{10}H_{18}$ 2384–70–5
2–Decyne

Me(CH2)6C≡CMe

138 $C_{10}H_{18}$ 2384–85–2
3–Decyne

Me(CH2)5C≡CEt

138 $C_{10}H_{18}$ 2384–86–3
4–Decyne

Me(CH2)4C≡CPr

138 $C_{10}H_{18}$ 2778–68–9
Bicyclo[4.1.0]heptane, 3,7,7–trimethyl–, [1S–(1α,3β,6α)]–

138 $C_{10}H_{18}$ 3282–53–9
Cyclohexene, 1–butyl–

138 $C_{10}H_{18}$ 4104–56–7
Cyclohexene, 3–(2–methylpropyl)–

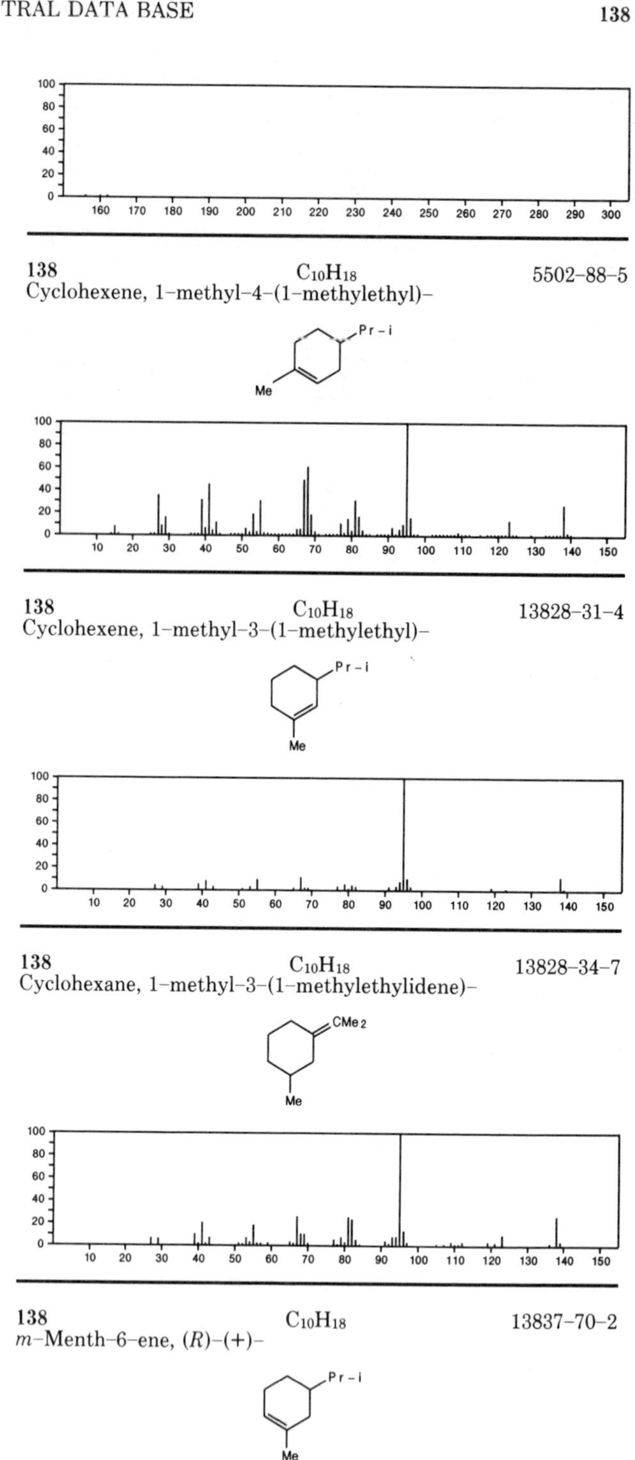

138 $C_{10}H_{18}$ 5502–88–5
Cyclohexene, 1–methyl–4–(1–methylethyl)–

138 $C_{10}H_{18}$ 13828–31–4
Cyclohexene, 1–methyl–3–(1–methylethyl)–

138 $C_{10}H_{18}$ 13828–34–7
Cyclohexane, 1–methyl–3–(1–methylethylidene)–

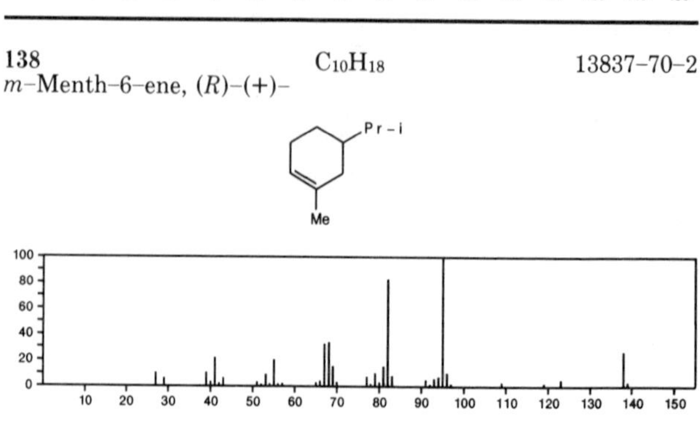

138 $C_{10}H_{18}$ 13837–70–2
m–Menth–6–ene, (R)–(+)–

138 $C_{10}H_{18}$ 13837–71–3
m–Menth–1(7)–ene, (*R*)–(–)–

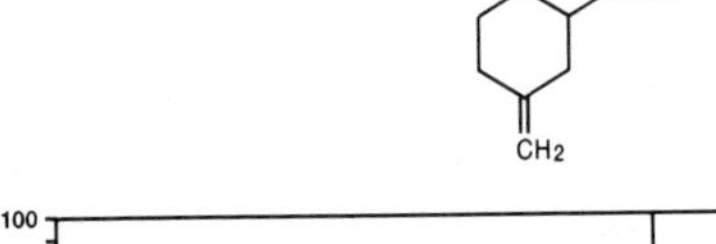

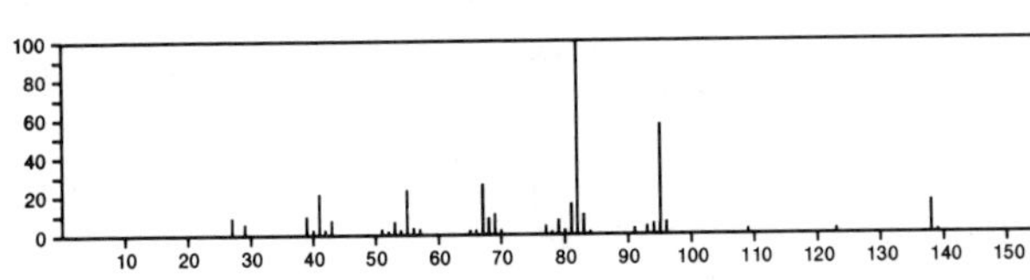

138 $C_{10}H_{18}$ 18968–23–5
Bicyclo[4.1.0]heptane, 3,7,7–trimethyl–, (1α,3α,6α)–

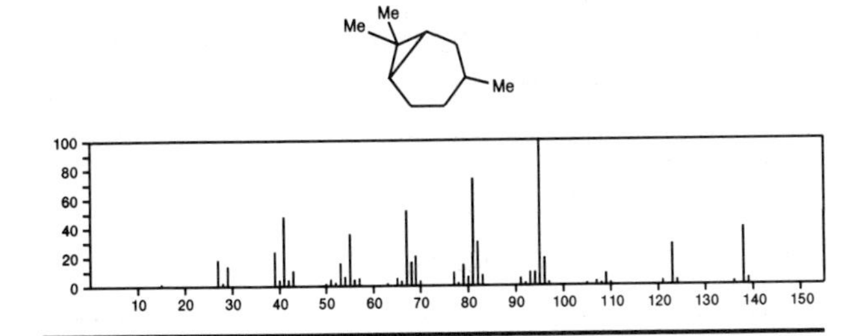

138 $C_{10}H_{18}$ 19482–57–6
3–Octyne, 2,2–dimethyl–

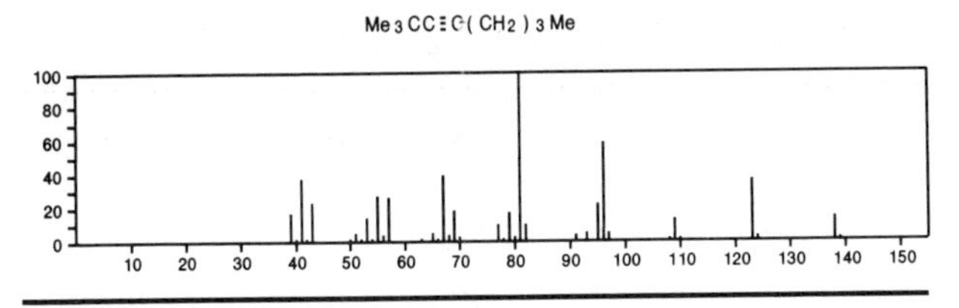

138 $C_{10}H_{18}$ 24399–15–3
Cyclohexane, 1–methyl–3–(1–methylethenyl)–, *cis*–

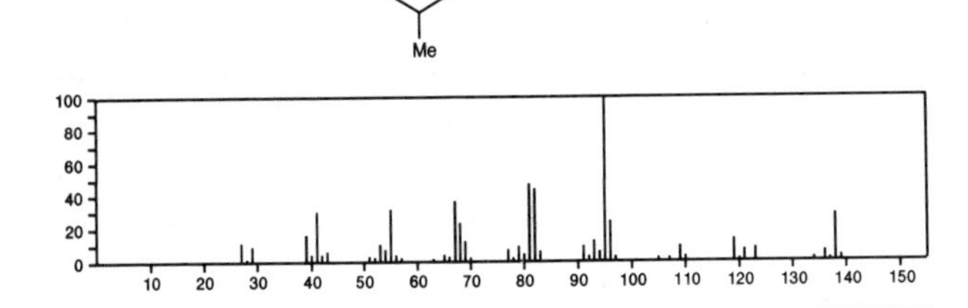

138 $C_{10}H_{18}$ 24519–04–8
Cyclopropane, tetramethylpropylidene–

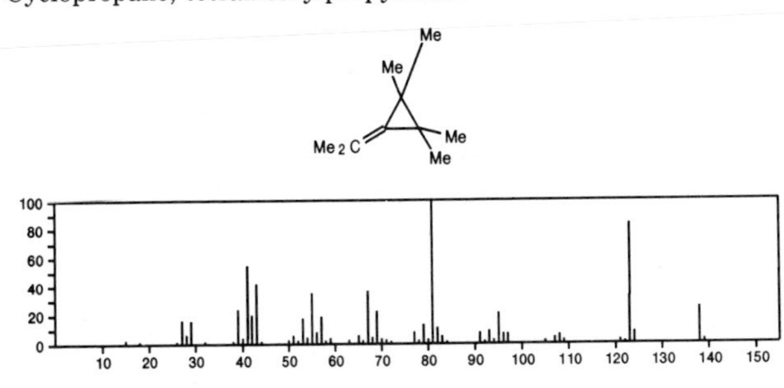

138 $C_{10}H_{18}$ 24524–51–4
Propane, 2–(2–isopropylidene–3–methylcyclopropyl)–, *trans*–

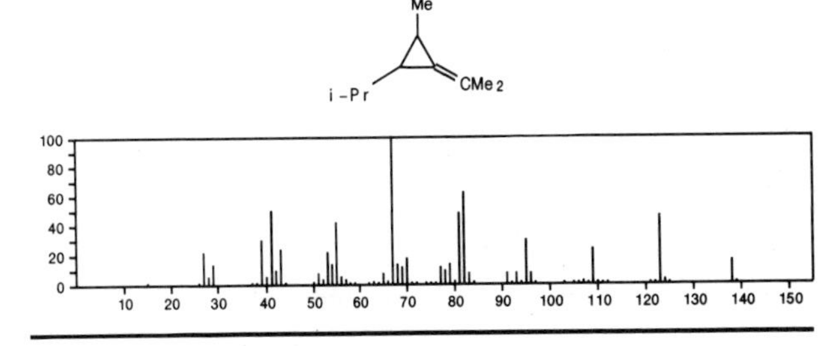

138 $C_{10}H_{18}$ 24524–52–5
Propane, 2–(2–isopropylidene–3–methylcyclopropyl)–, *cis*–

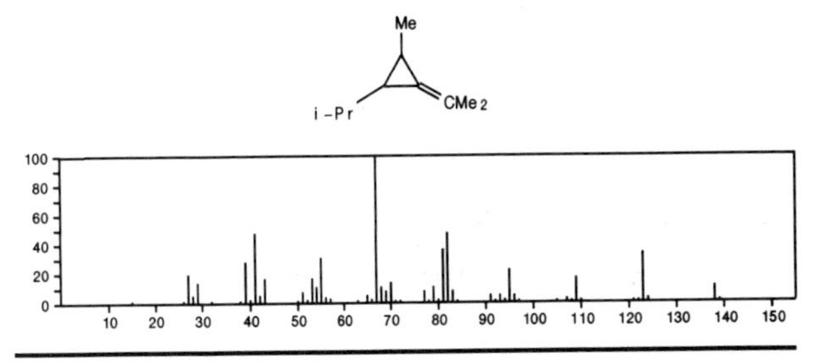

138 $C_{10}H_{18}$ 28588–55–8
Pentalene, octahydro–2,5–dimethyl–

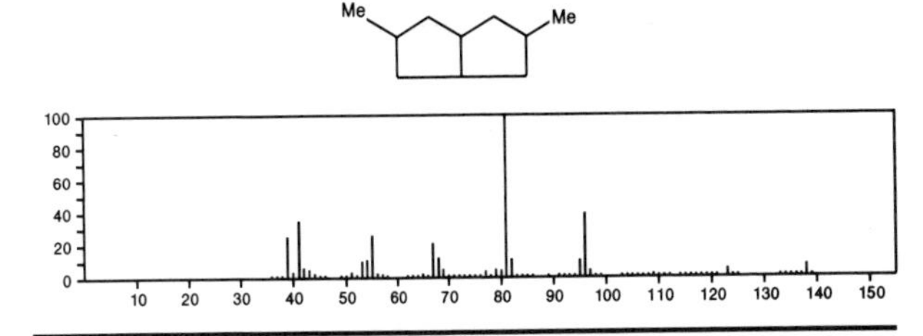

138 $C_{10}H_{18}$ 28980–73–6
3,5–Octadiene, 2,7–dimethyl–, (*Z*,*Z*)–

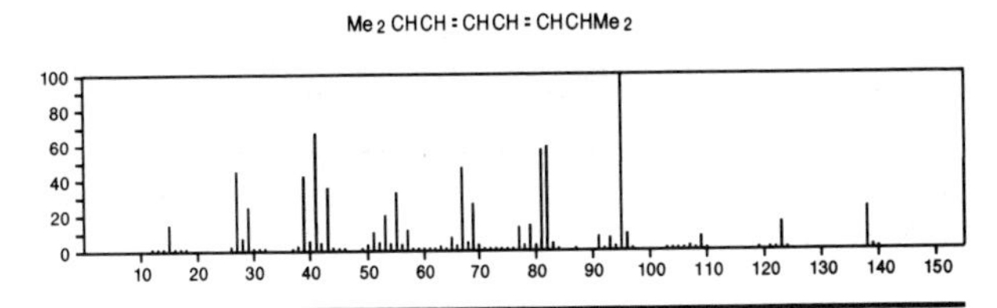

138 $C_{10}H_{18}$ 51504–54–2
1,4–Hexadiene, 2,3,4,5–tetramethyl–

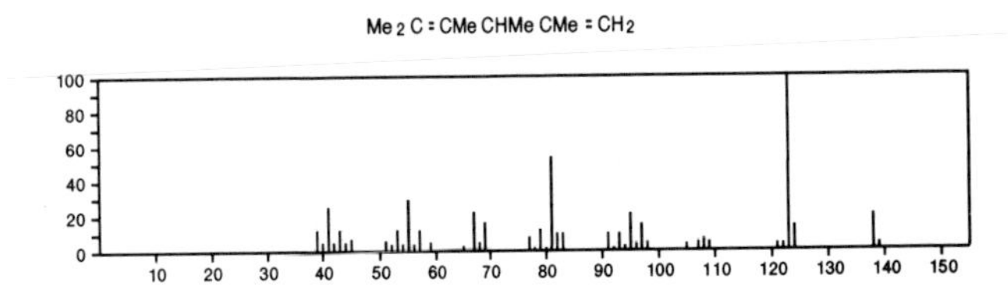

138 $C_{10}H_{18}$ 54764-57-7
Cyclopropane, 1,1,2-trimethyl-3-(2-methyl-1-propenyl)-

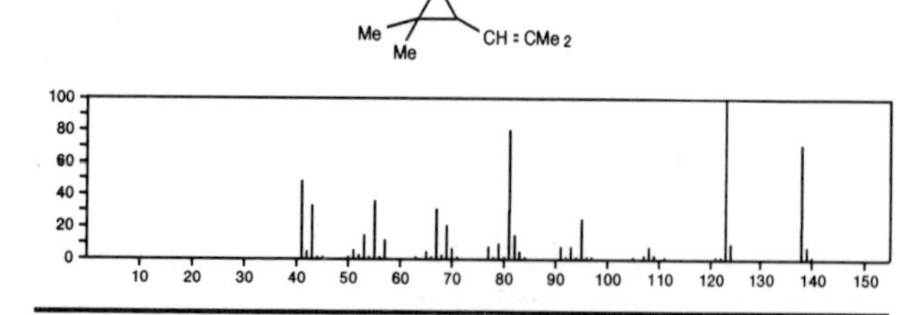

138 $C_{10}H_{18}$ 55682-64-9
3,5-Octadiene, 2,7-dimethyl-, (*E,Z*)-

$Me_2CHCH=CHCH=CHCHMe_2$

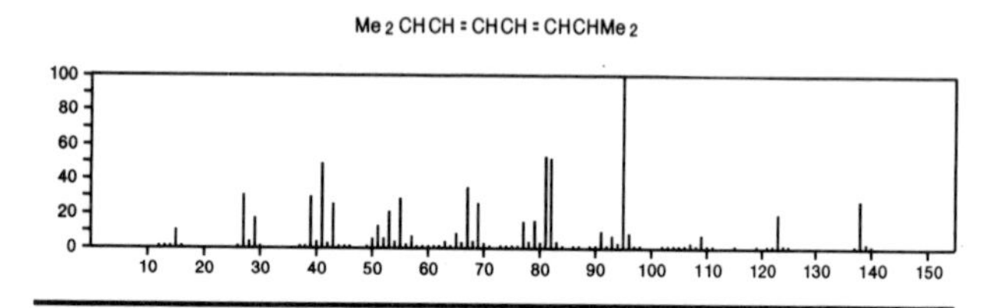

138 $C_{10}H_{18}$ 55682-65-0
4,6-Decadiene

PrCH=CHCH=CHPr

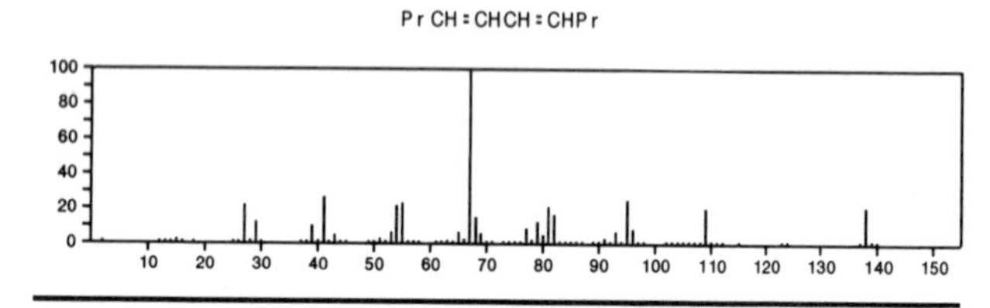

138 $C_{10}H_{18}$ 55956-32-6
4,5-Nonadiene, 2-methyl-

$PrCH=C=CHCH_2CHMe_2$

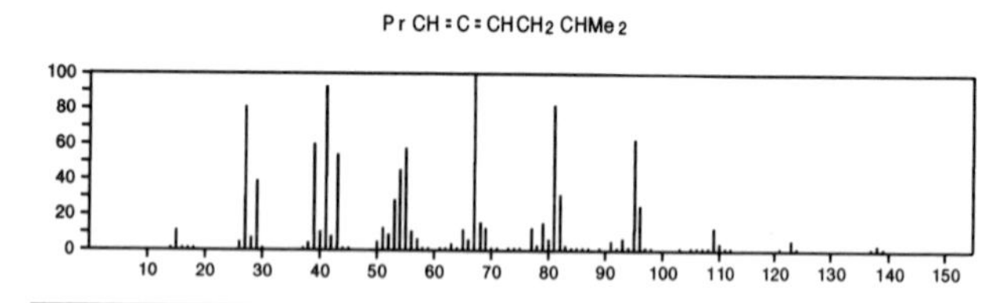

139 $C_5H_5N_3O_2$ 15862-54-1
3-Pyridinamine, *N*-nitro-

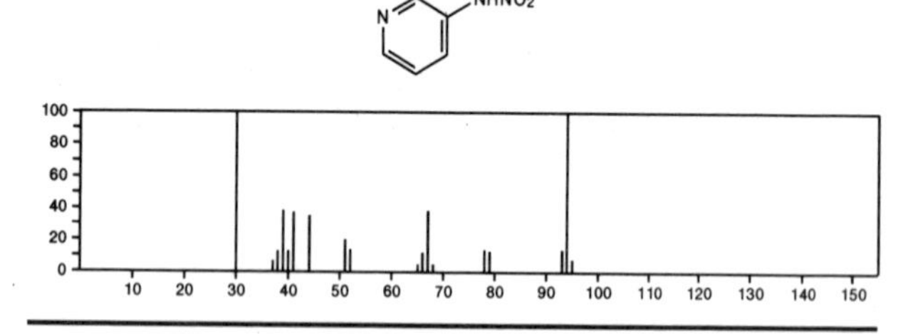

139 $C_5H_5N_3O_2$ 26482-54-2
2-Pyridinamine, *N*-nitro-

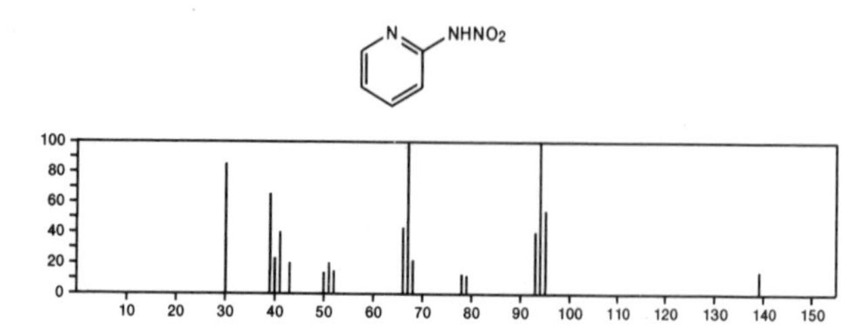

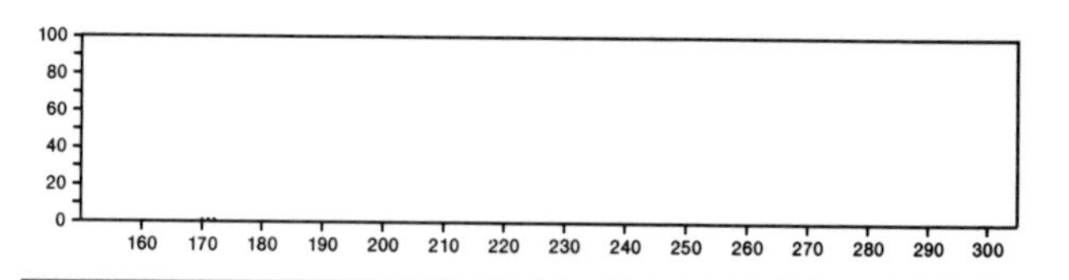

139 $C_5H_5N_3O_2$ 26482-55-3
4-Pyridinamine, *N*-nitro-

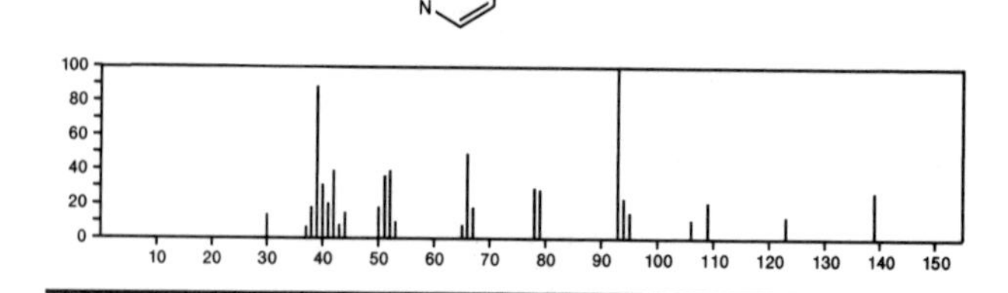

139 $C_6H_5NO_3$ 88-75-5
Phenol, 2-nitro-

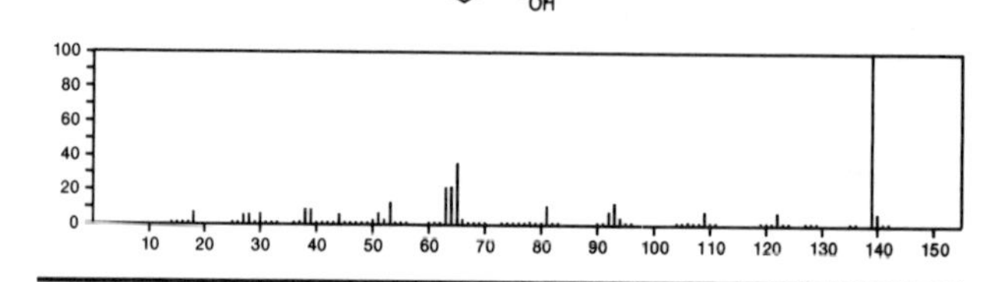

139 $C_6H_5NO_3$ 100-02-7
Phenol, 4-nitro-

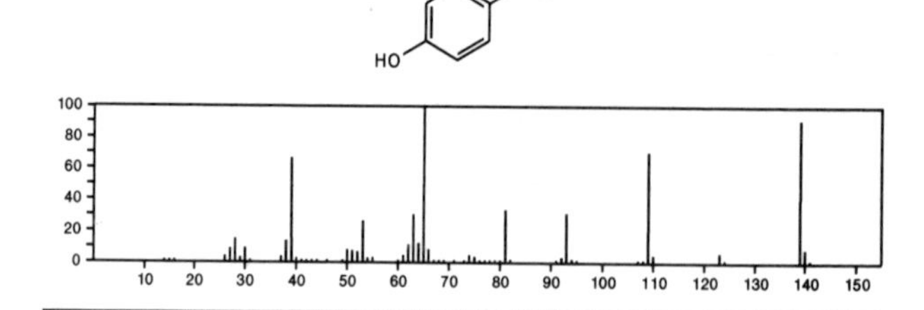

139 $C_6H_5NO_3$ 554-84-7
Phenol, 3-nitro-

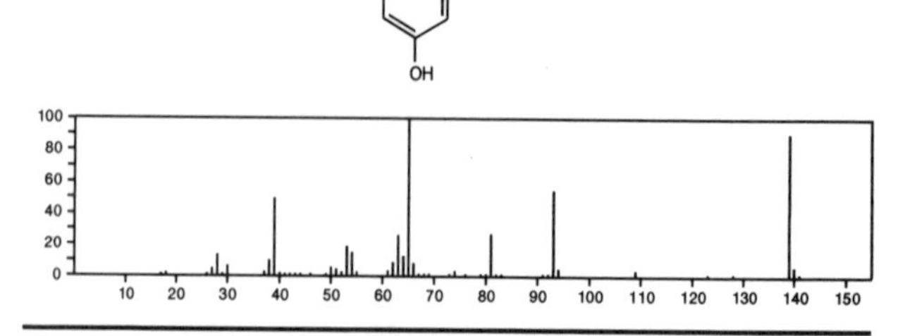

139 $C_6H_5NO_3$ 5006-66-6
3-Pyridinecarboxylic acid, 1,6-dihydro-6-oxo-

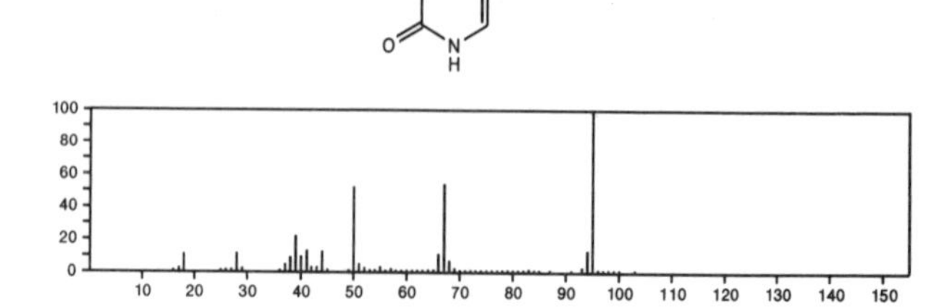

139 $C_6H_9N_3O$ 7749-47-5
2-Pyrimidinamine, 4-methoxy-6-methyl-

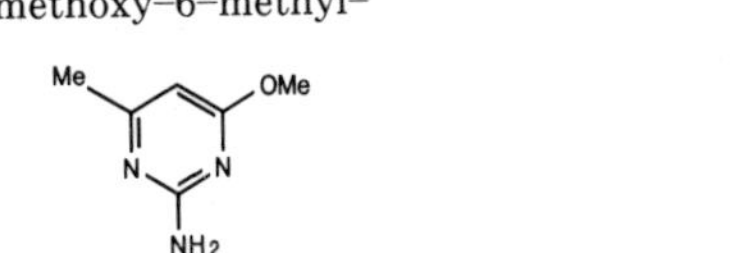

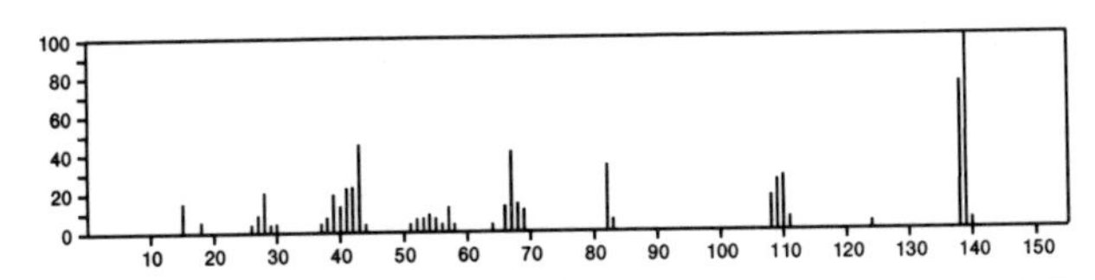

139 $C_7H_9NO_2$ 6052-75-1
2(1*H*)-Pyridinone, 4-hydroxy-1,6-dimethyl-

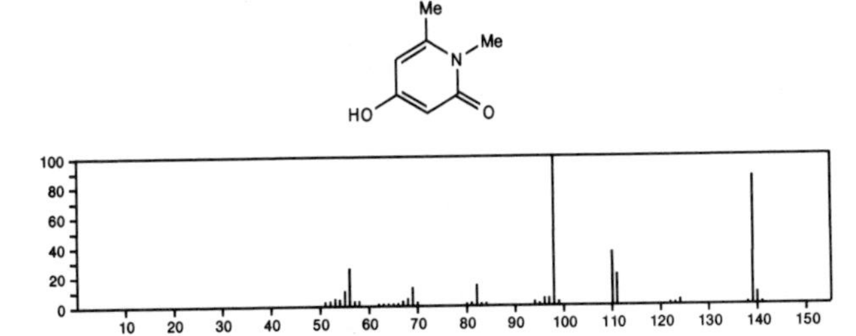

139 $C_7H_9NO_2$ 50618-94-5
2-Furanacetamide, *N*-methyl-

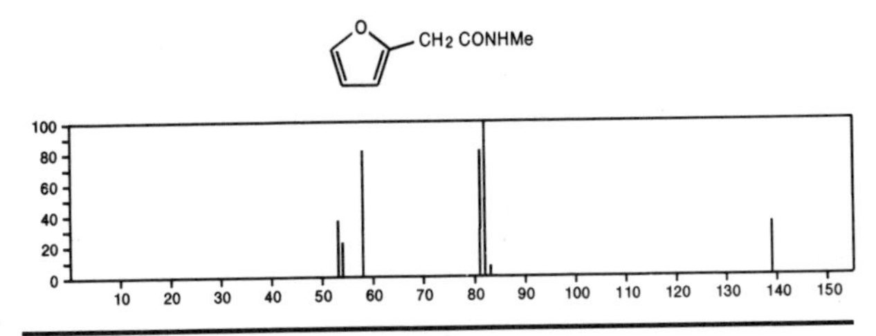

139 $C_7H_9NO_2$ 53603-10-4
4-Pyridinol, 5-methoxy-2-methyl-

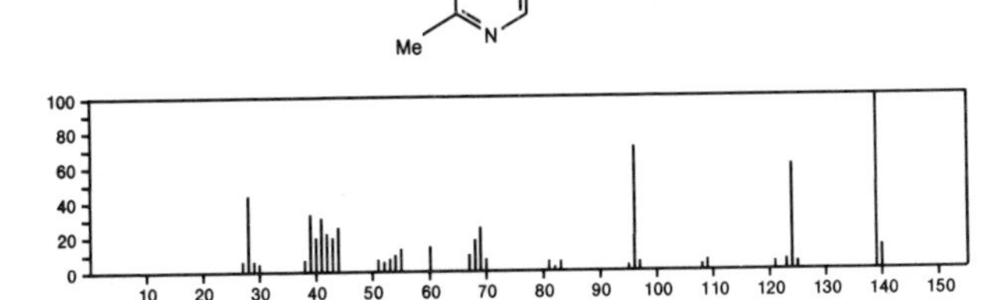

139 $C_7H_9NO_2$ 53603-11-5
4-Pyridinol, 3-methoxy-2-methyl-

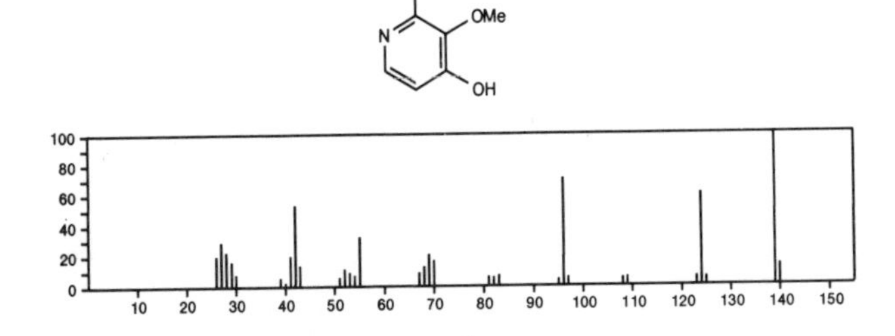

139 $C_7H_9NO_2$ 53603-12-6
4-Pyridinol, 2-methoxy-6-methyl-

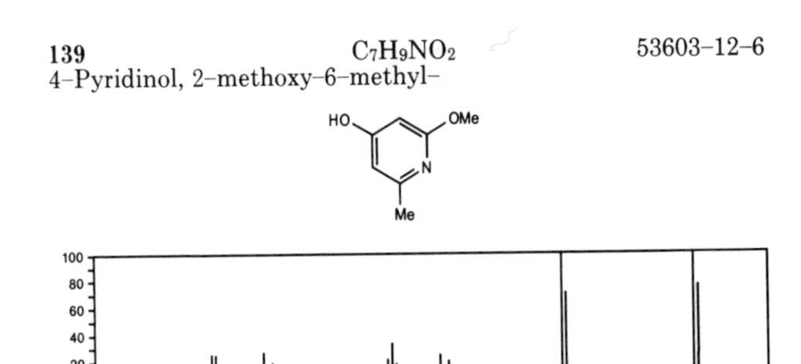

139 C_7H_9NS 19006-66-7
2(1*H*)-Pyridinethione, 1,3-dimethyl-

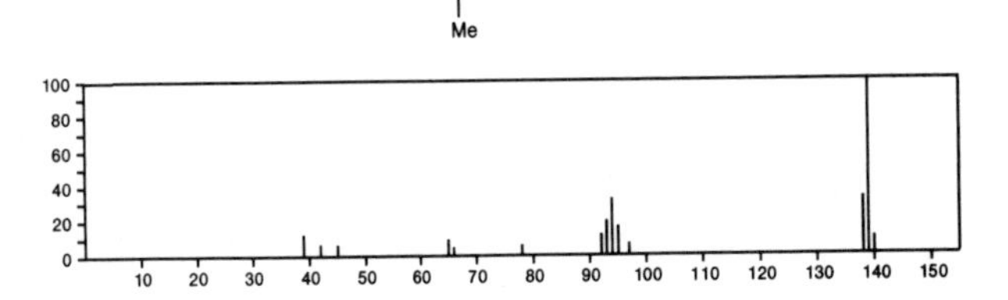

139 C_7H_9NS 19006-67-8
2(1*H*)-Pyridinethione, 1,4-dimethyl-

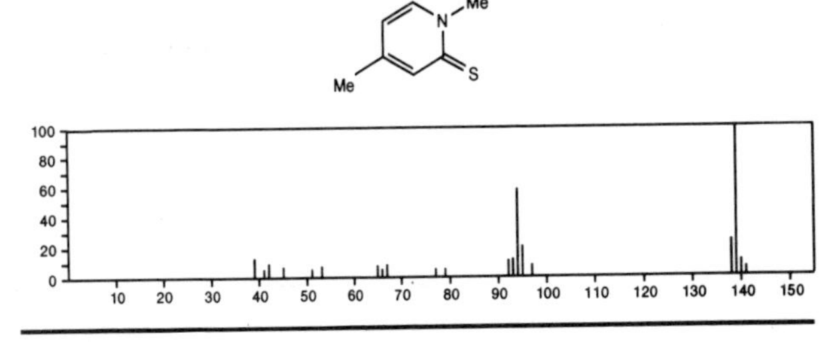

139 C_7H_9NS 19006-68-9
2(1*H*)-Pyridinethione, 1,5-dimethyl-

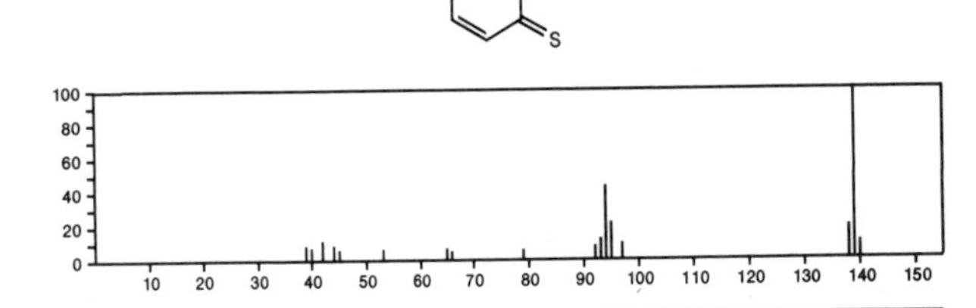

139 C_7H_9NS 19006-69-0
2(1*H*)-Pyridinethione, 1,6-dimethyl-

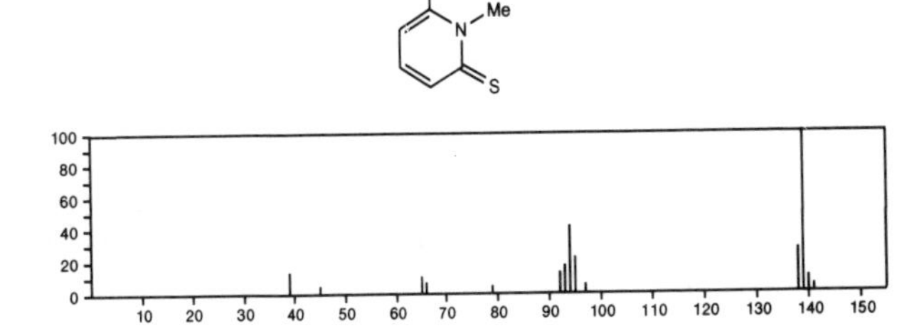

139 C_7H_9NS 19006–71–4

2(1*H*)–Pyridinethione, 1–ethyl–

139 C_7H_9NS 19006–76–9

Pyridine, 2–(ethylthio)–

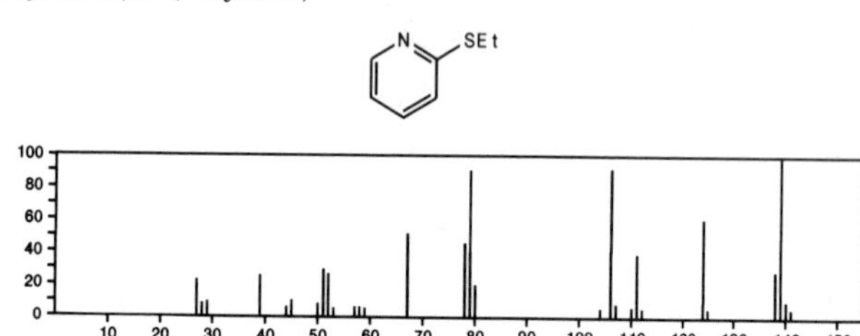

139 C_7H_9NS 19006–77–0

4–Picoline, 2–(methylthio)–

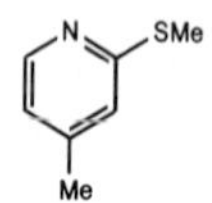

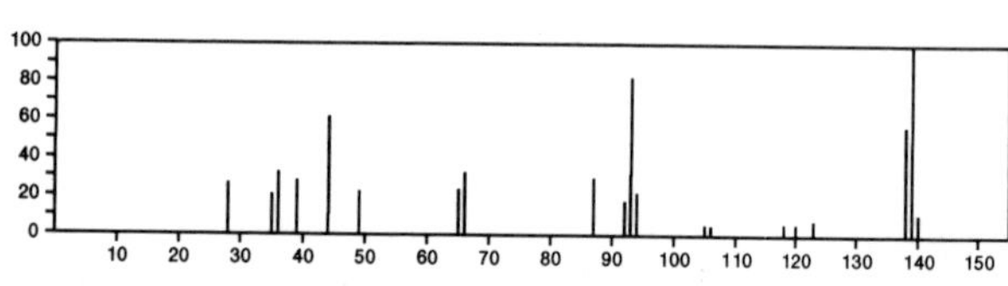

139 C_7H_9NS 20329–37–7

2–Picoline, 6–(methylthio)–

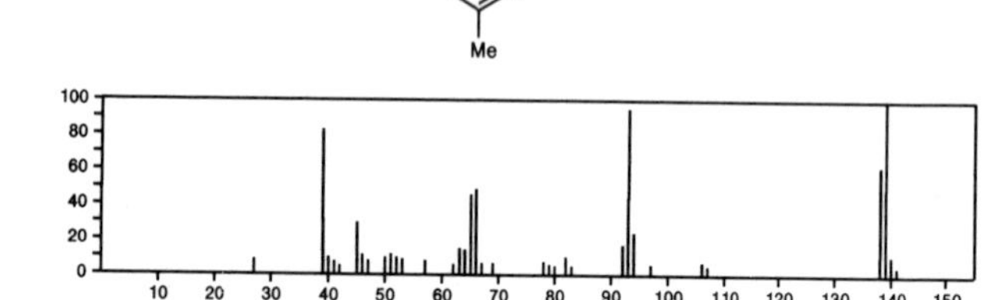

139 $C_7H_{13}N_3$ 673–46–1

1*H*–Imidazole–4–ethanamine, *N*,*N*–dimethyl–

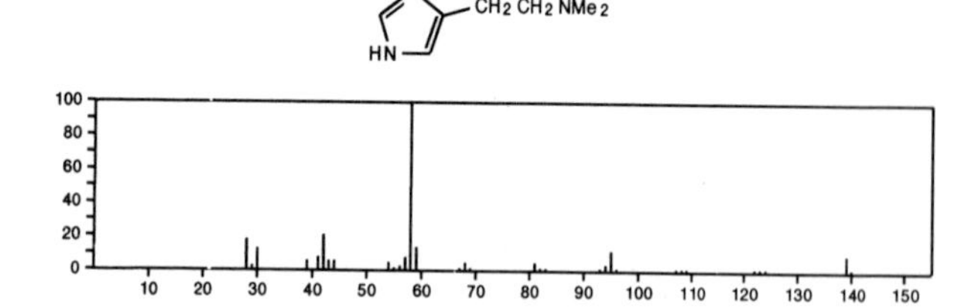

139 $C_7H_{13}N_3$ 21150–01–6

1*H*–Imidazole–4–ethanamine, β,β–dimethyl–

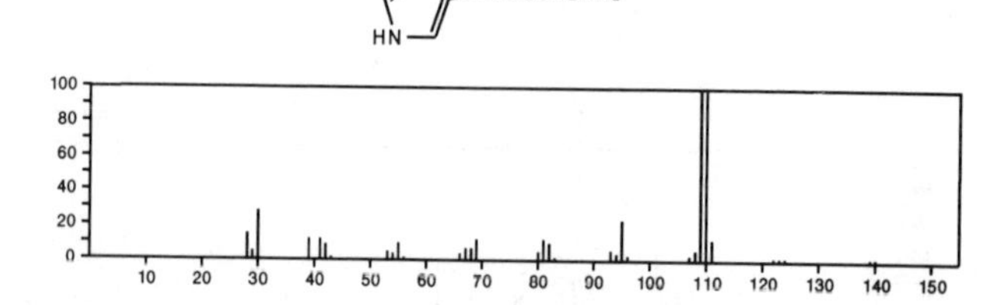

139 $C_7H_{13}N_3$ 40515–30–8

1*H*–1,2,4–Triazole, 3–(2,2–dimethylpropyl)–

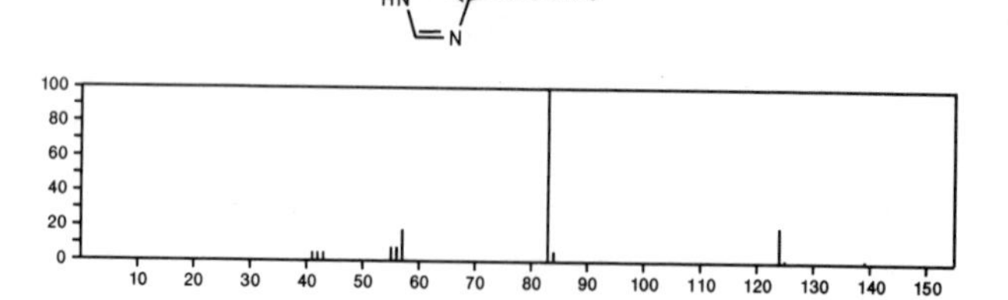

139 $C_7H_{13}N_3$ 53966–45–3

1*H*–Imidazole–4–ethanamine, 1,5–dimethyl–

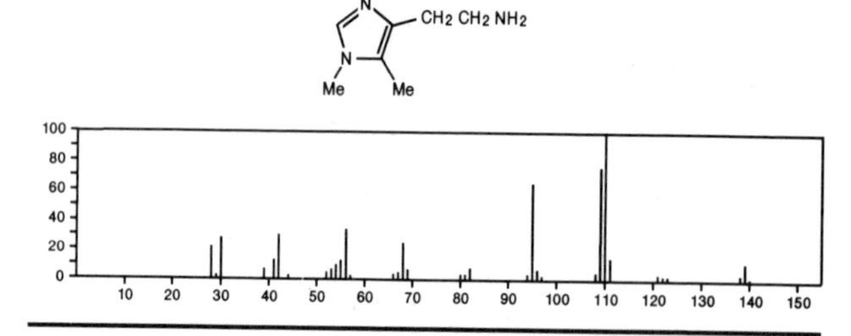

139 $C_7H_{13}N_3$ 53966–46–4

1*H*–Imidazole–4–ethanamine, *N*,5–dimethyl–

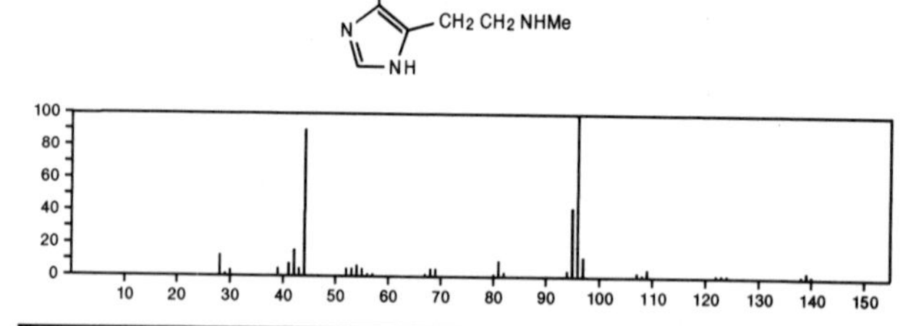

139 $C_8H_{13}NO$ 532–24–1

8–Azabicyclo[3.2.1]octan–3–one, 8–methyl–

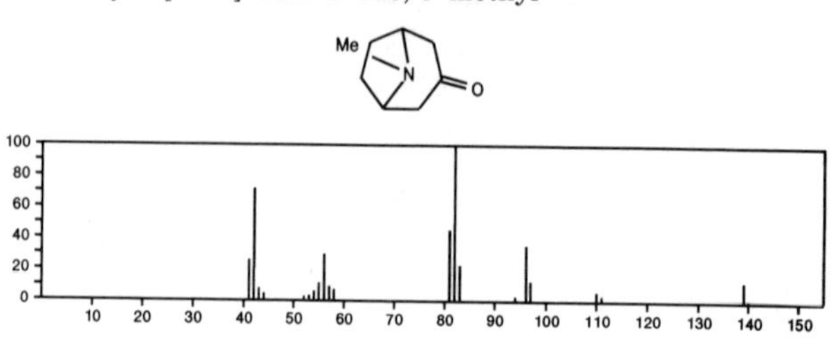

139 $C_8H_{13}NO$ 19352-93-3
Morpholine, 4-(1,3-butadienyl)-

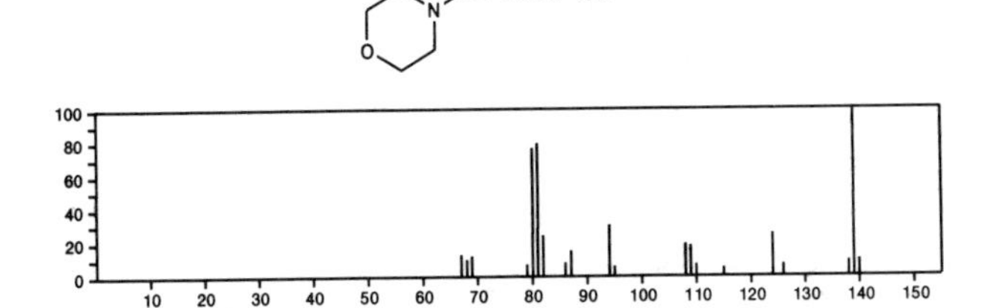

139 $C_8H_{13}NO$ 20513-09-1
8-Azabicyclo[3.2.1]oct-6-en-3-ol, 8-methyl-, *endo*-

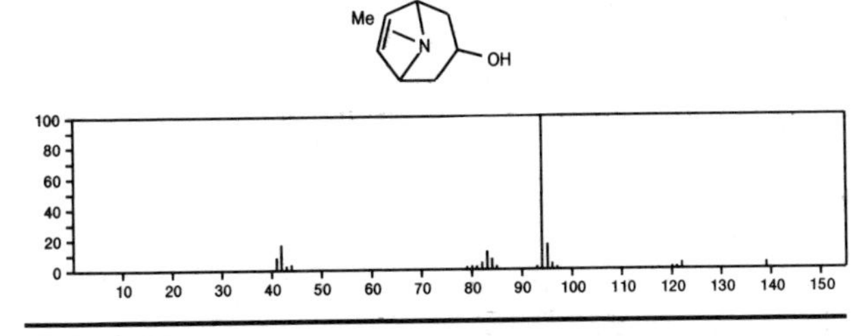

139 $C_8H_{13}NO$ 26625-33-2
6-Azabicyclo[3.2.1]octan-3-one, 6-methyl-

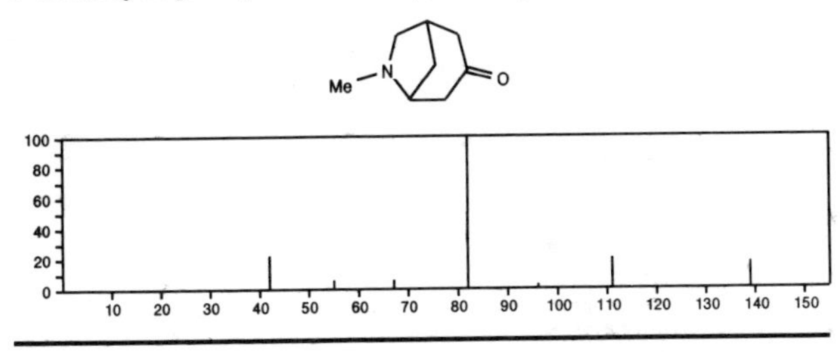

139 $C_8H_{13}NO$ 32810-62-1
6-Azabicyclo[3.2.1]octan-4-one, 6-methyl-

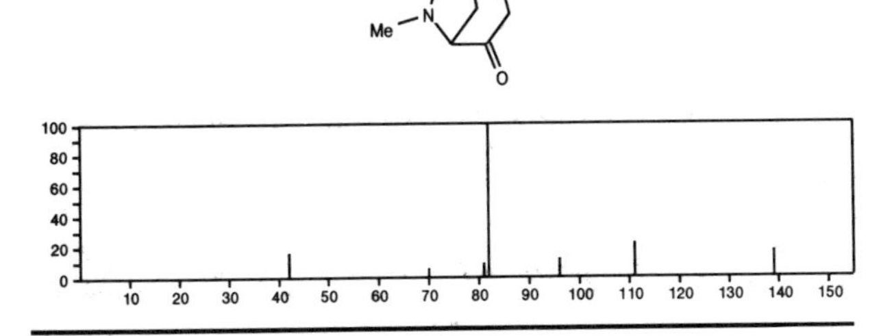

139 $C_8H_{13}NO$ 38225-15-9
2,5-Methano-2*H*-furo[3,2-*b*]pyrrole, hexahydro-4-methyl-

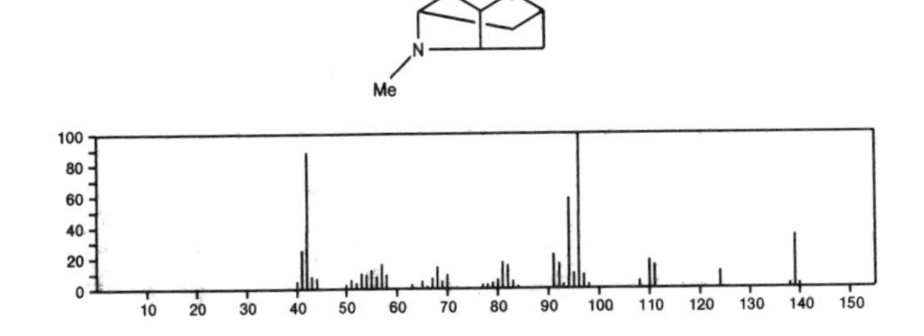

139 $C_8H_{13}NO$ 50838-17-0
3-Buten-2-one, 4-(2,2-dimethyl-1-aziridinyl)-

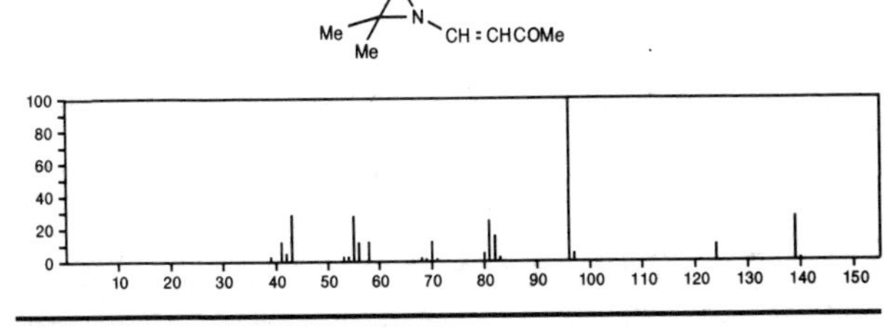

139 $C_8H_{13}NO$ 54725-49-4
8-Azabicyclo[3.2.1]oct-6-en-3-ol, 8-methyl-

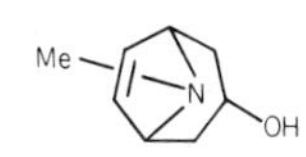

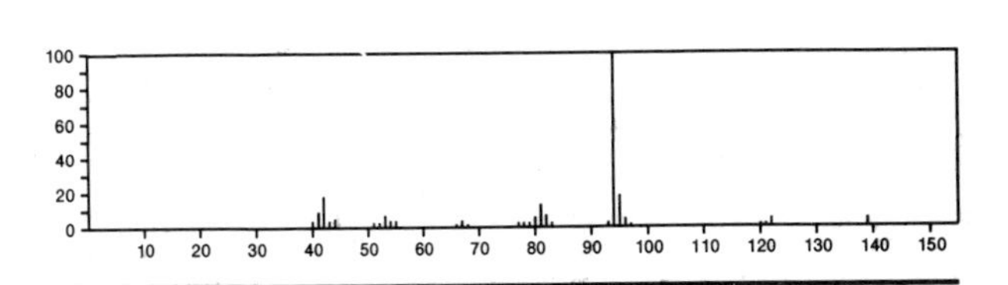

139 $C_8H_{13}NO$ 56336-07-3
2-Cyclohexen-1-one, 3-methyl-, *O*-methyloxime

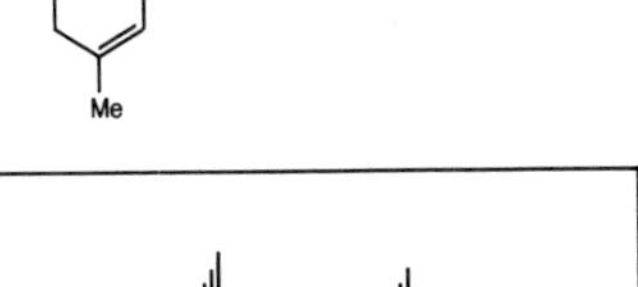

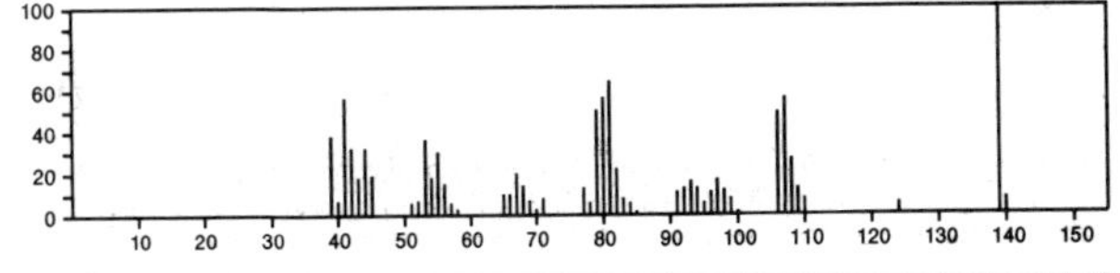

139 $C_8H_{13}NO$ 56771-95-0
8-Azabicyclo[3.2.1]octane-8-carboxaldehyde

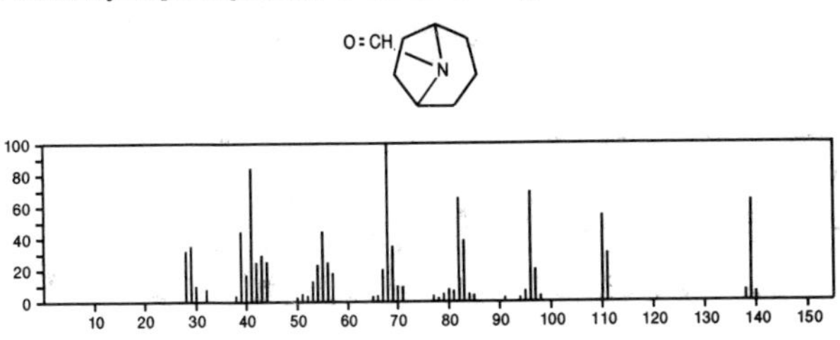

139 $C_8H_{18}BN$ 42843-12-9
Boranamine, 1-ethyl-*N*,*N*-dimethyl-1-(1-methyl-2-propenyl)-

$Me_2NBEt\ CHMe\ CH=CH_2$

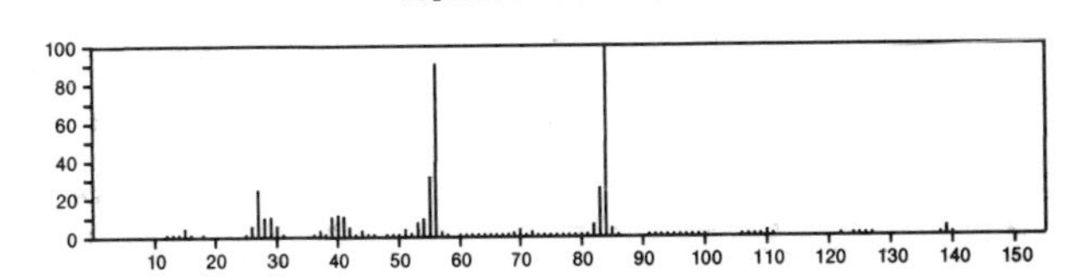

139 $C_9H_{17}N$ 673–33–6
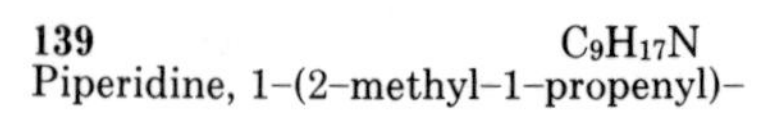
Piperidine, 1–(2–methyl–1–propenyl)–

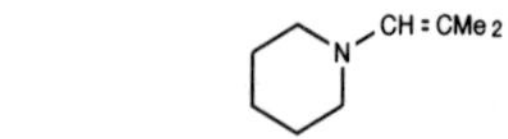

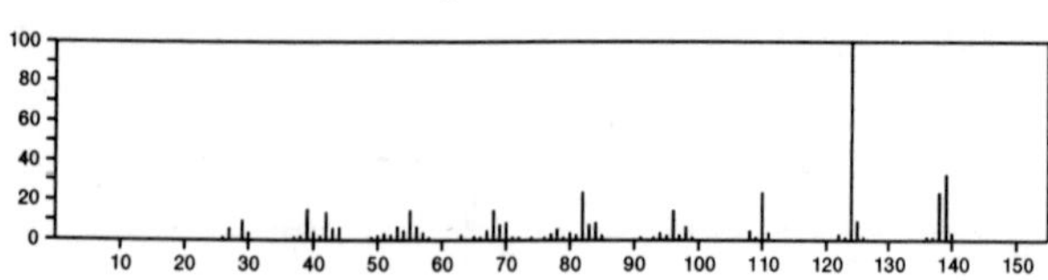

139 $C_9H_{17}N$ 697–75–6

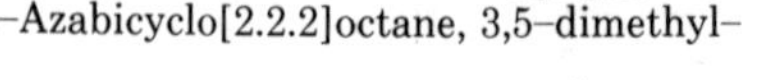
1–Azabicyclo[2.2.2]octane, 3,5–dimethyl–

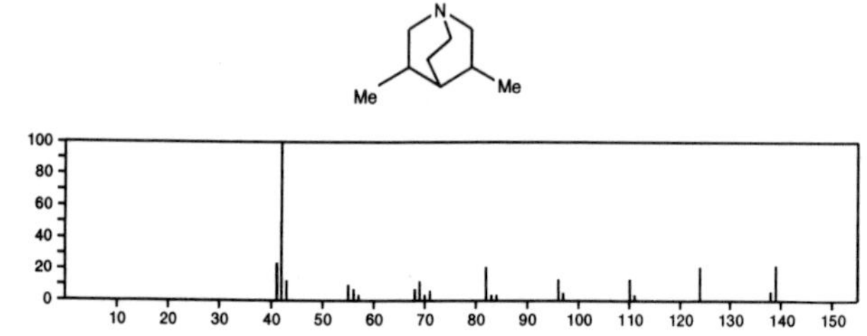

139 $C_9H_{17}N$ 767–92–0
Quinoline, decahydro–, *trans*–

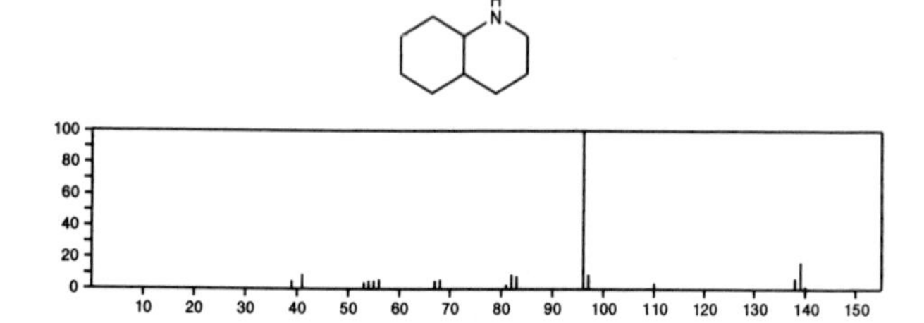

139 $C_9H_{17}N$ 2051–28–7
Quinoline, decahydro–

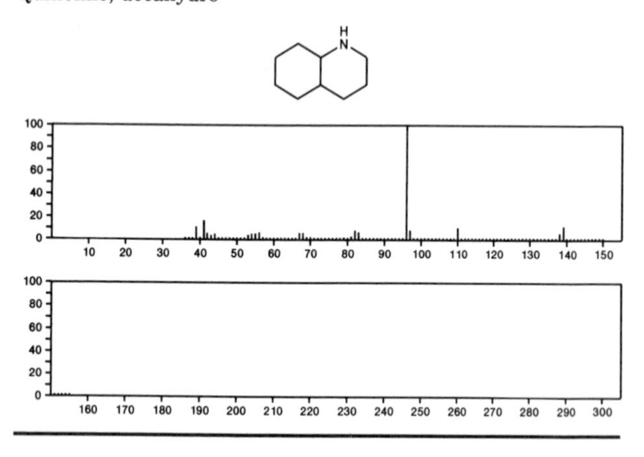

139 $C_9H_{17}N$ 6323–87–1
2–Propynylamine, *N,N*–diisopropyl–

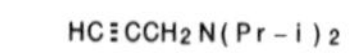

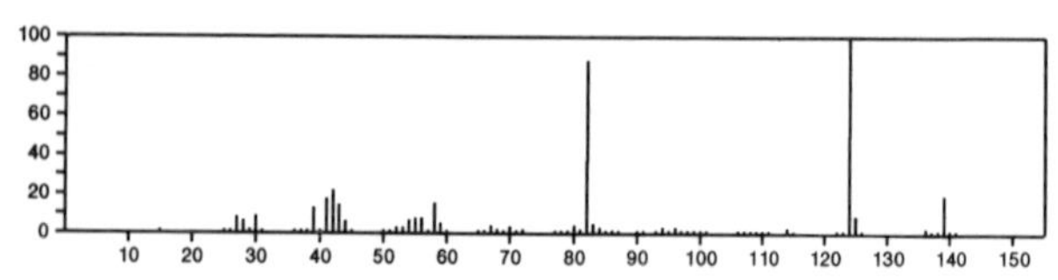

139 $C_9H_{17}N$ 7182–10–7
Piperidine, 1–(1–butenyl)–

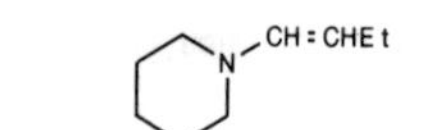

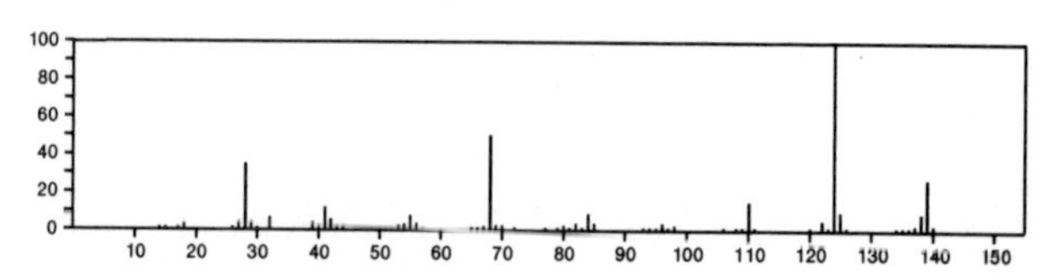

139 $C_9H_{17}N$ 10343–99–4
Quinoline, decahydro–, *cis*–

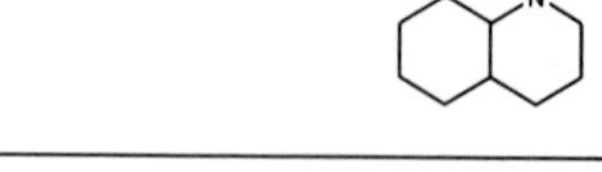

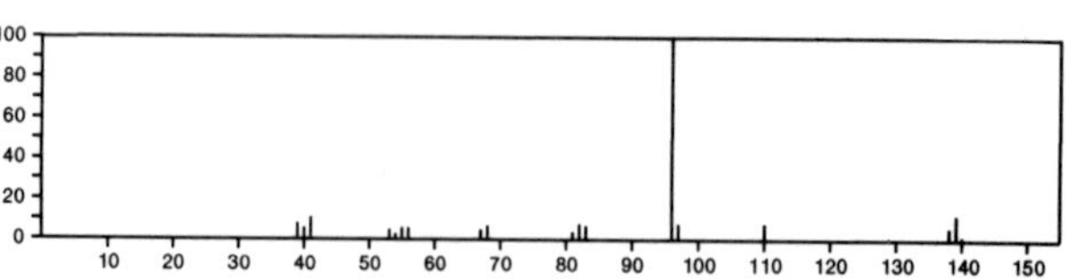

139 $C_9H_{17}N$ 13218–09–2
1–Azabicyclo[2.2.2]octane, 2,6–dimethyl–, (1α,2α,4α,6α)–

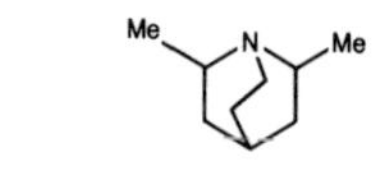

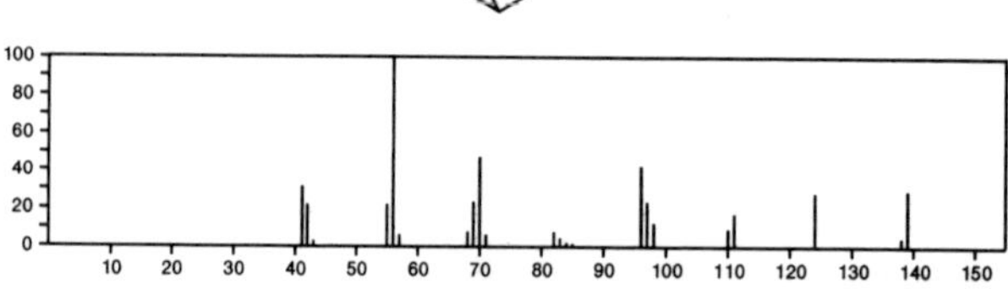

139 $C_9H_{17}N$ 13218–10–5
1–Azabicyclo[2.2.2]octane, 2,6–dimethyl–, (1α,2β,4α,6β)–

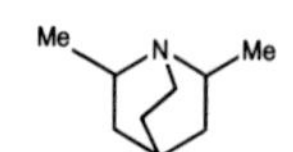

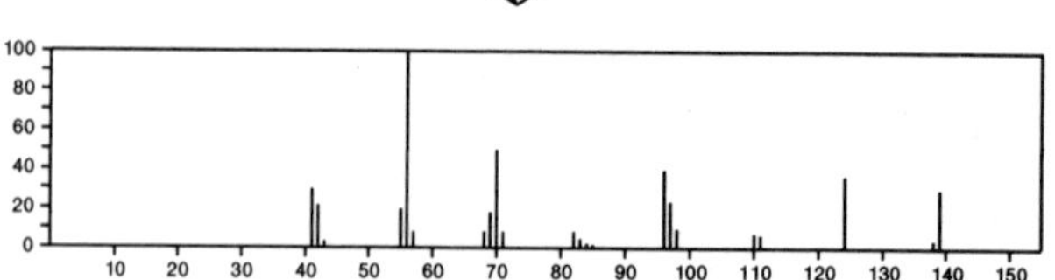

139 $C_9H_{17}N$ 13937–90–1
Pyrrolidine, 1–(1–pentenyl)–

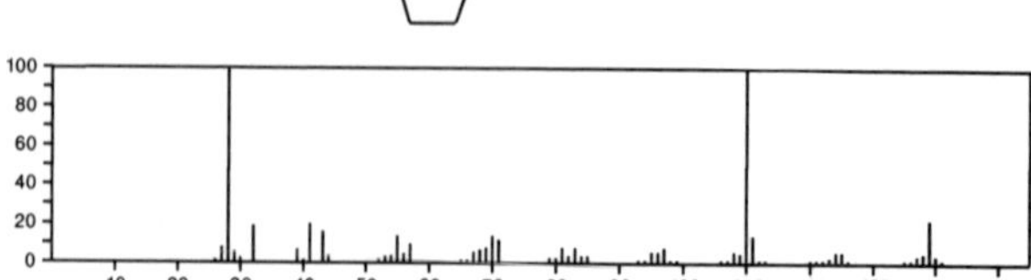

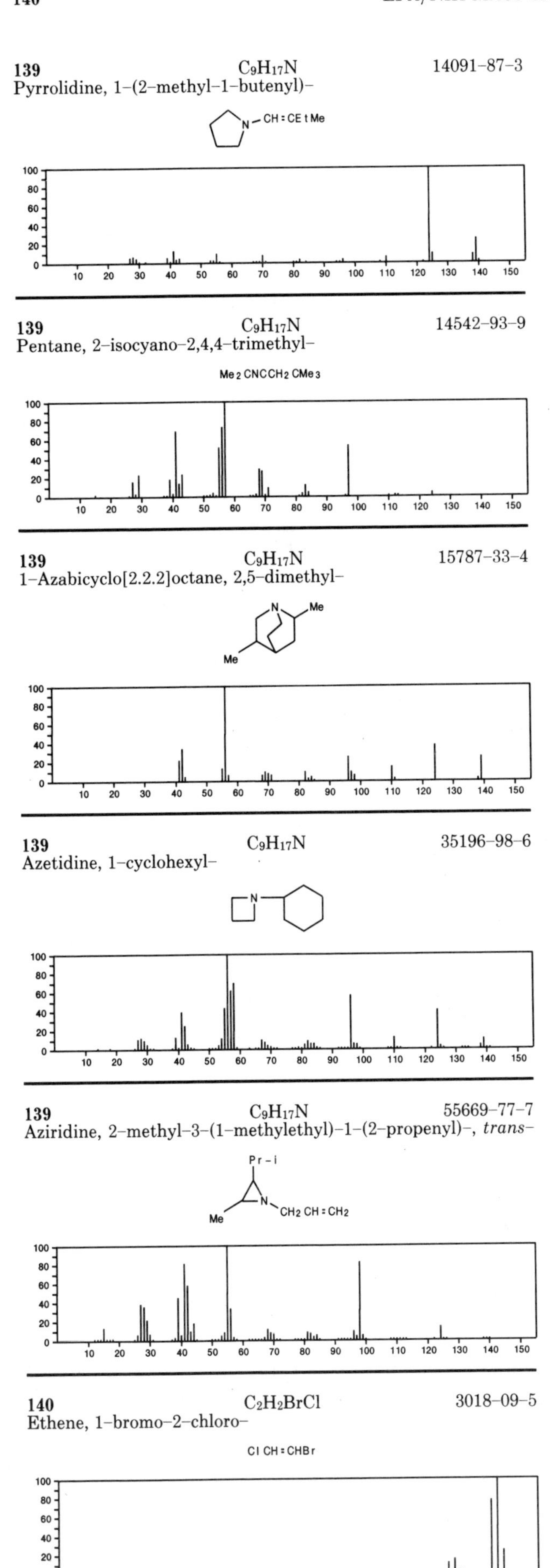
139 $C_9H_{17}N$ 14091-87-3
Pyrrolidine, 1-(2-methyl-1-butenyl)-
CH=CEtMe
139 $C_9H_{17}N$ 14542-93-9
Pentane, 2-isocyano-2,4,4-trimethyl-
$Me_2CNCCH_2CMe_3$
139 $C_9H_{17}N$ 15787-33-4
1-Azabicyclo[2.2.2]octane, 2,5-dimethyl-
N
Me
Me
139 $C_9H_{17}N$ 35196-98-6
Azetidine, 1-cyclohexyl-
N
139 $C_9H_{17}N$ 55669-77-7
Aziridine, 2-methyl-3-(1-methylethyl)-1-(2-propenyl)-, *trans*-
Pr-i
N
$CH_2CH=CH_2$
Me
140 C_2H_2BrCl 3018-09-5
Ethene, 1-bromo-2-chloro-
ClCH=CHBr
100
80
60
40
20
0
10 20 30 40 50 60 70 80 90 100 110 120 130 140 150

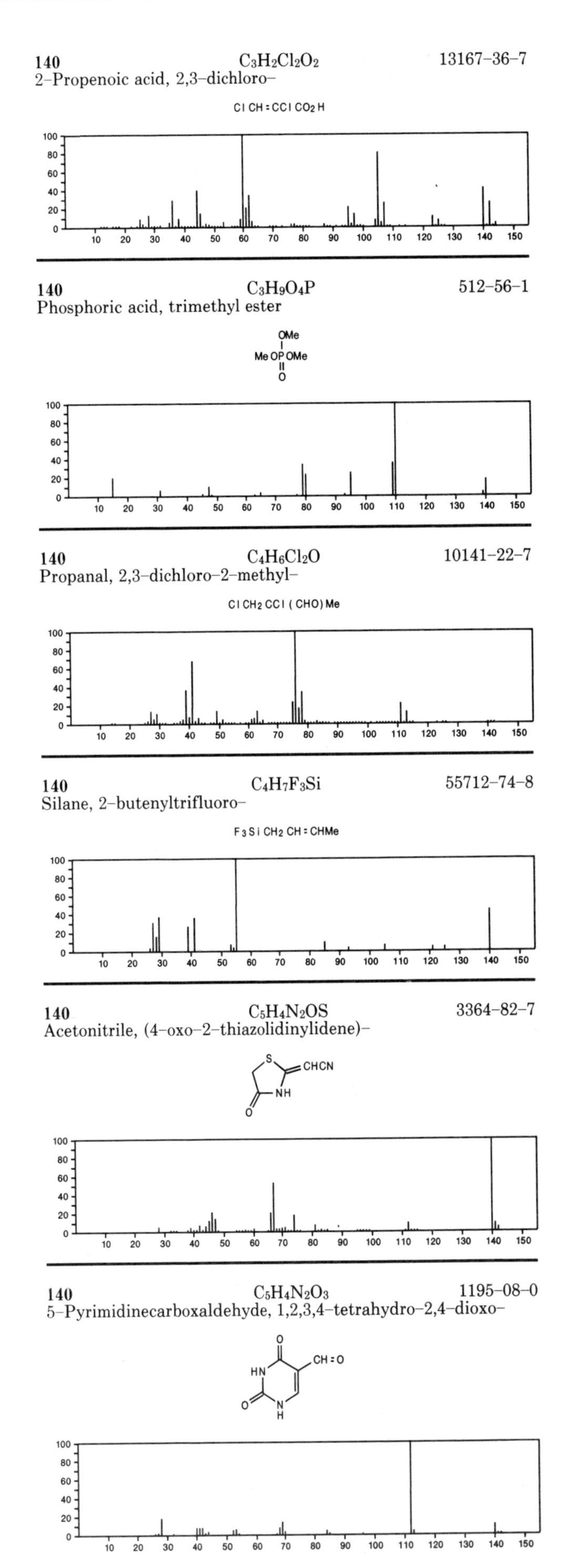
140 $C_3H_2Cl_2O_2$ 13167-36-7
2-Propenoic acid, 2,3-dichloro-
$ClCH=CClCO_2H$
140 $C_3H_9O_4P$ 512-56-1
Phosphoric acid, trimethyl ester
OMe
MeOPOMe
O
140 $C_4H_6Cl_2O$ 10141-22-7
Propanal, 2,3-dichloro-2-methyl-
$ClCH_2CCl(CHO)Me$
140 $C_4H_7F_3Si$ 55712-74-8
Silane, 2-butenyltrifluoro-
$F_3SiCH_2CH=CHMe$
140 $C_5H_4N_2OS$ 3364-82-7
Acetonitrile, (4-oxo-2-thiazolidinylidene)-
S
CHCN
NH
O
140 $C_5H_4N_2O_3$ 1195-08-0
5-Pyrimidinecarboxaldehyde, 1,2,3,4-tetrahydro-2,4-dioxo-
O
HN
CH=O
O
N
H
100
80
60
40
20
0
10 20 30 40 50 60 70 80 90 100 110 120 130 140 150

140 $C_5H_8N_4O$ 54004–20–5

4(1*H*)–Pyrimidinone, 2–amino–6–(methylamino)–

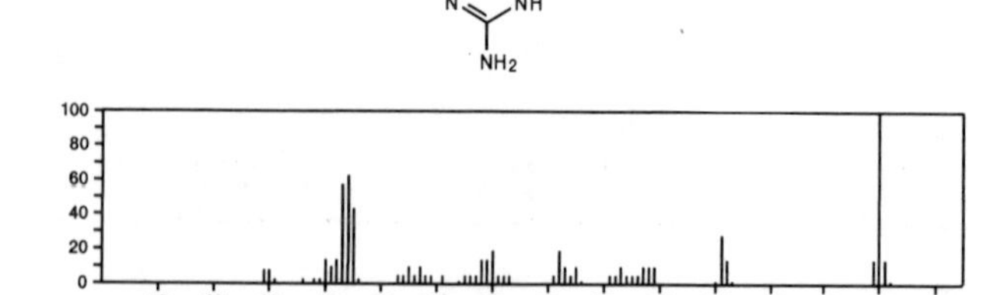

140 $C_5H_{10}Cl_2$ 507–45–9

Butane, 2,3–dichloro–2–methyl–

$MeCHClCClMe_2$

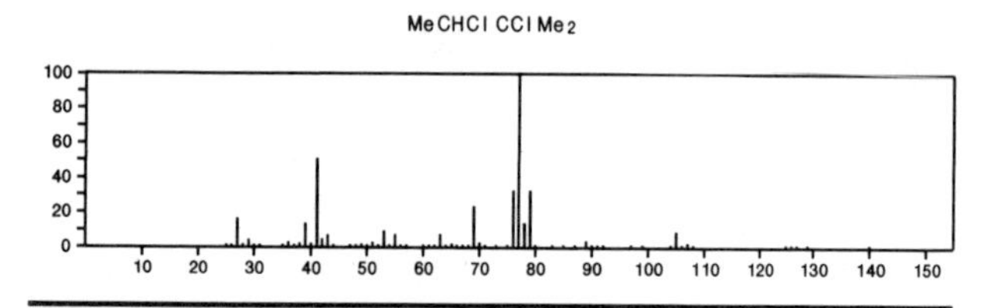

140 $C_5H_{10}Cl_2$ 623–34–7

Butane, 1,4–dichloro–2–methyl–

$ClCH_2CH_2CHMeCH_2Cl$

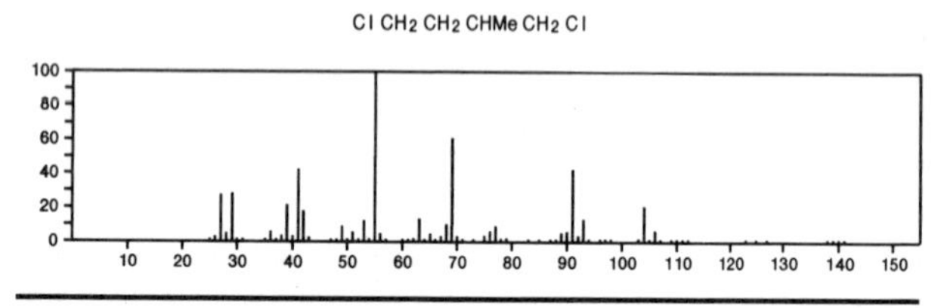

140 $C_5H_{10}Cl_2$ 625–66–1

Butane, 1,1–dichloro–3–methyl–

$Me_2CHCH_2CHCl_2$

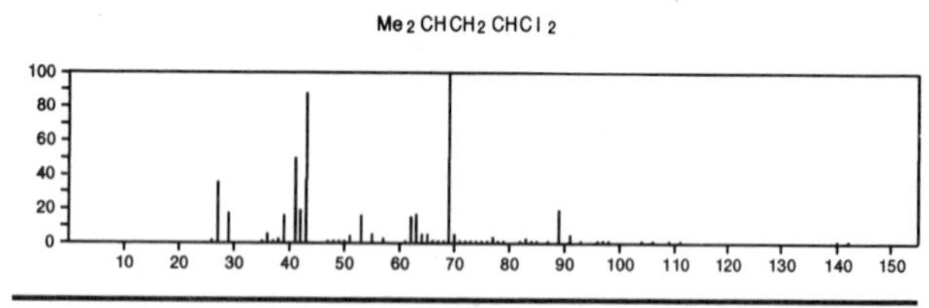

140 $C_5H_{10}Cl_2$ 625–67–2

Pentane, 2,4–dichloro–

$MeCHClCH_2CHClMe$

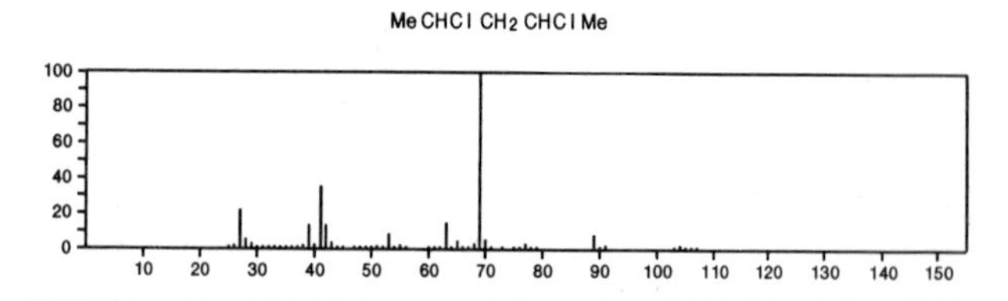

140 $C_5H_{10}Cl_2$ 626–92–6

Pentane, 1,4–dichloro–

$MeCHCl(CH_2)_3Cl$

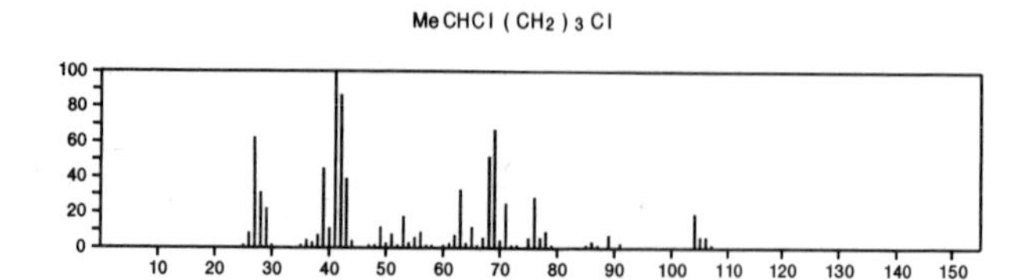

140 $C_5H_{10}Cl_2$ 628–76–2

Pentane, 1,5–dichloro–

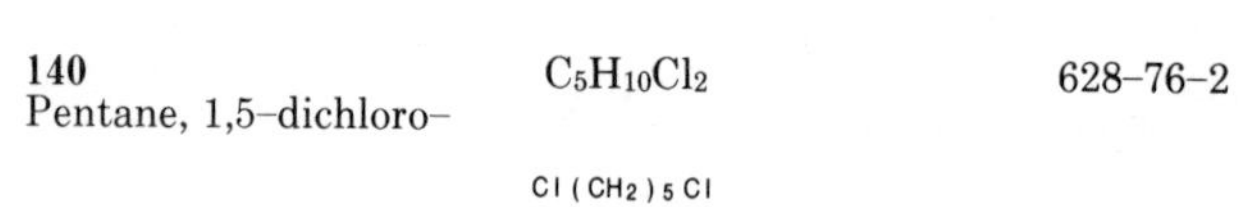

$Cl(CH_2)_5Cl$

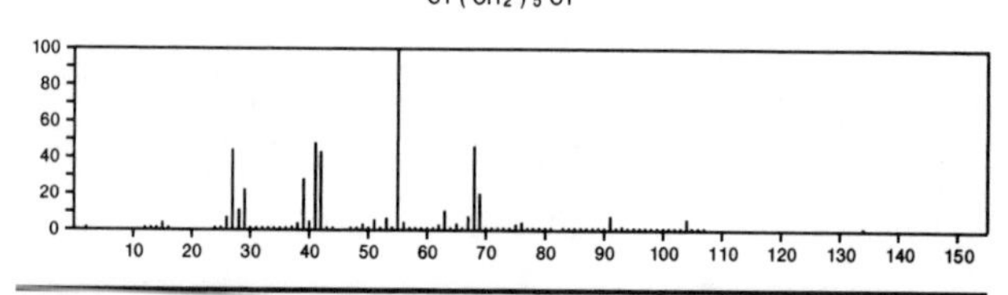

140 $C_5H_{10}Cl_2$ 1674–33–5

Pentane, 1,2–dichloro–

$ClCH_2CHClPr$

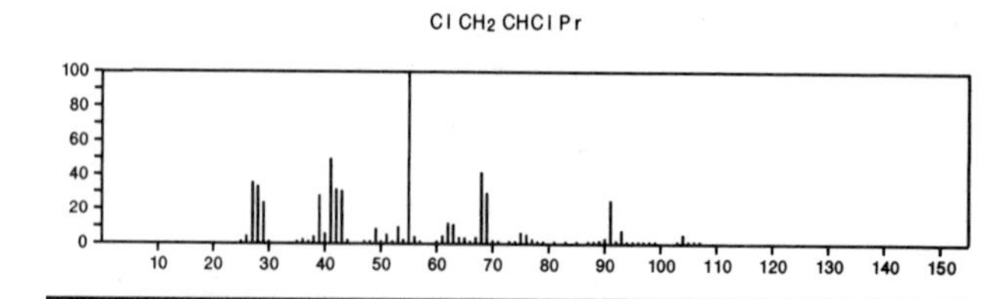

140 $C_5H_{10}Cl_2$ 17773–66–9

Butane, 2,2–dichloro–3–methyl–

Me_2CHCCl_2Me

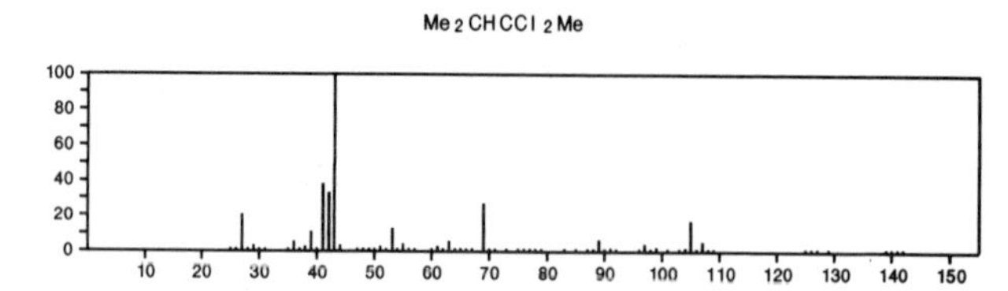

140 $C_5H_{10}Cl_2$ 23010–04–0

Butane, 1,2–dichloro–2–methyl–

$ClCH_2CClEtMe$

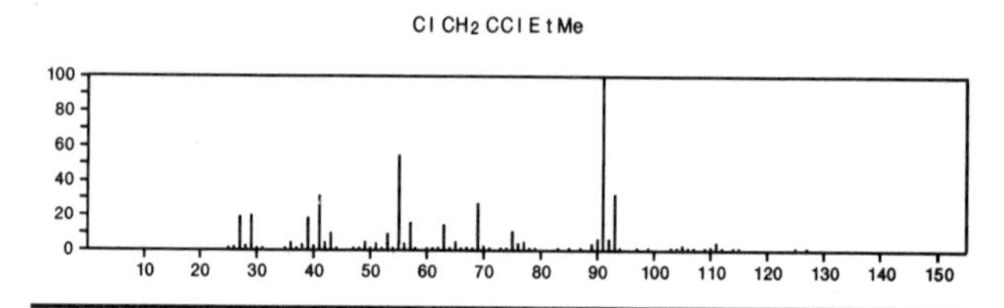

140 $C_5H_{10}Cl_2$ 23010–07–3

Butane, 1,3–dichloro–2–methyl–

$ClCH_2CHMeCHClMe$

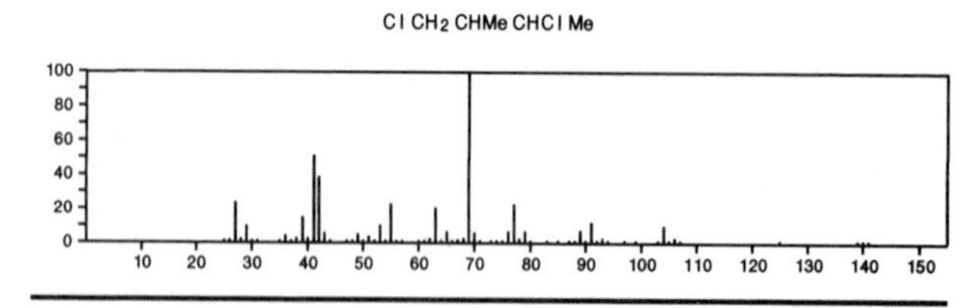

140 $C_5H_{10}Cl_2$ 34887–14–4

Pentane, 2,2–dichloro–

$PrCCl_2(Pr)$

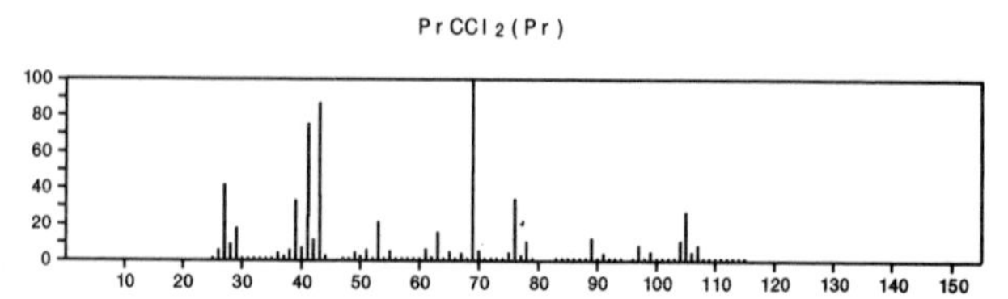

140 $C_6H_4O_4$ 500-05-0

2*H*-Pyran-5-carboxylic acid, 2-oxo-

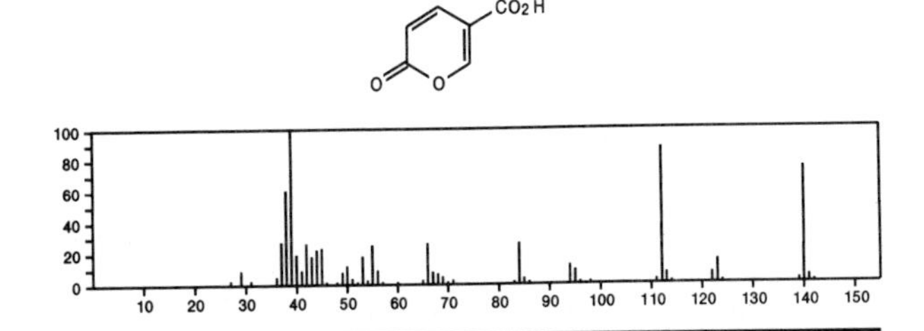

140 $C_6H_4O_4$ 35069-70-6

2,5-Cyclohexadiene-1,4-dione, 2,6-dihydroxy-

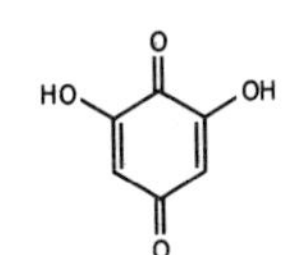

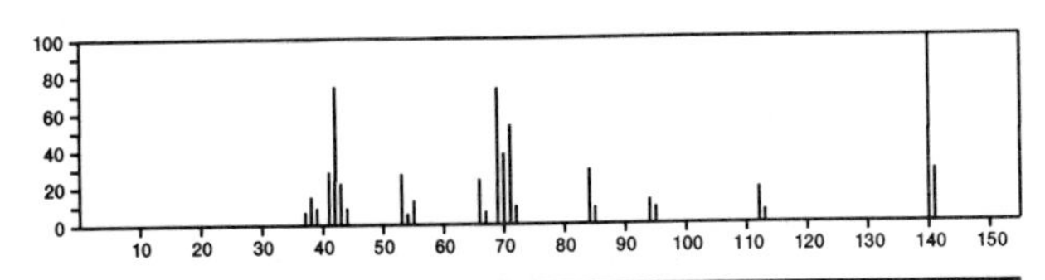

140 $C_6H_8N_2O_2$ 874-14-6

2,4(1*H*,3*H*)-Pyrimidinedione, 1,3-dimethyl-

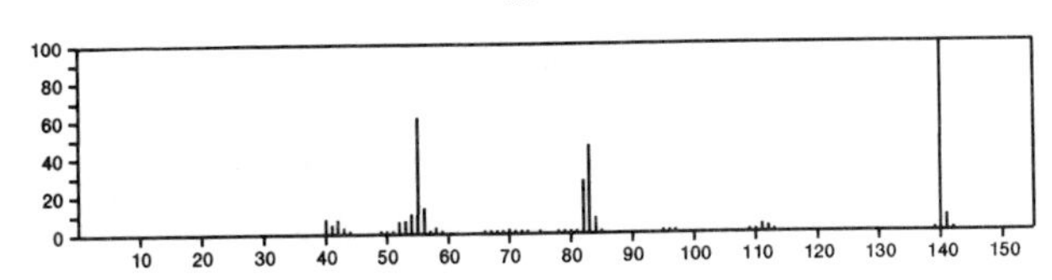

140 $C_6H_8N_2O_2$ 5203-92-9

2-Pyrazolin-5-one, 1-acetyl-3-methyl-

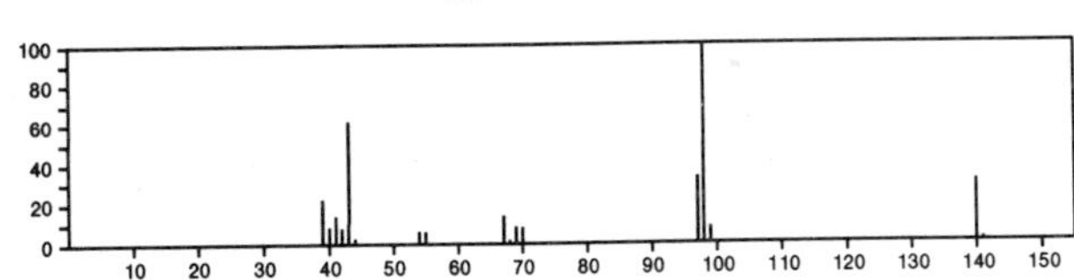

140 $C_6H_8N_2O_2$ 13223-74-0

Acetamide, *N*-(5-methyl-3-isoxazolyl)-

140 $C_6H_8N_2O_2$ 19674-60-3

2,4(1*H*,3*H*)-Pyrimidinedione, 3,6-dimethyl-

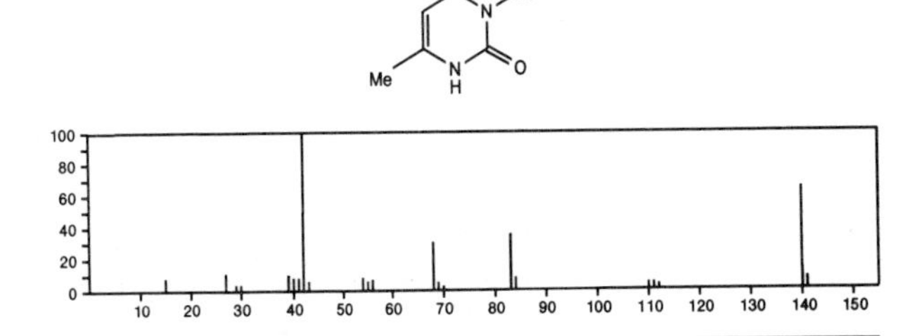

140 $C_6H_8N_2S$ 2882-20-4

Pyrazine, 2-methyl-3-(methylthio)-

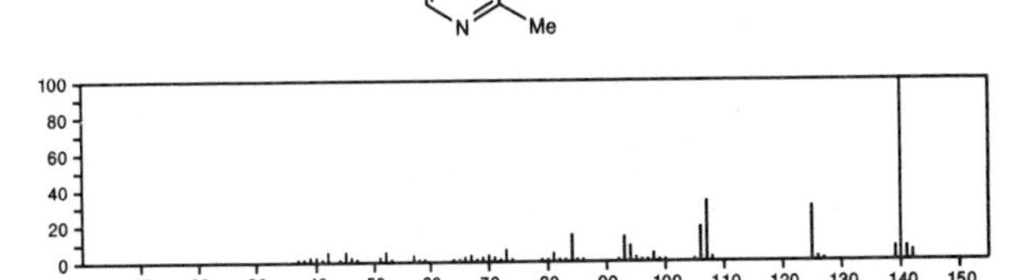

140 $C_6H_8N_2S$ 2884-13-1

Pyrazine, 2-methyl-6-(methylthio)-

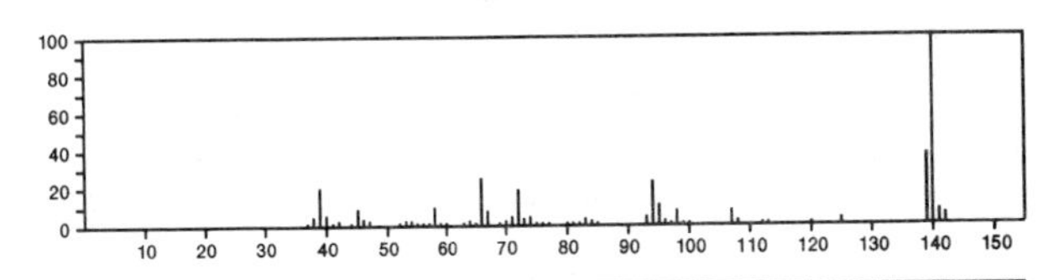

140 $C_6H_8N_2S$ 33779-33-8

Pyrimidine, 2-methyl-4-(methylthio)-

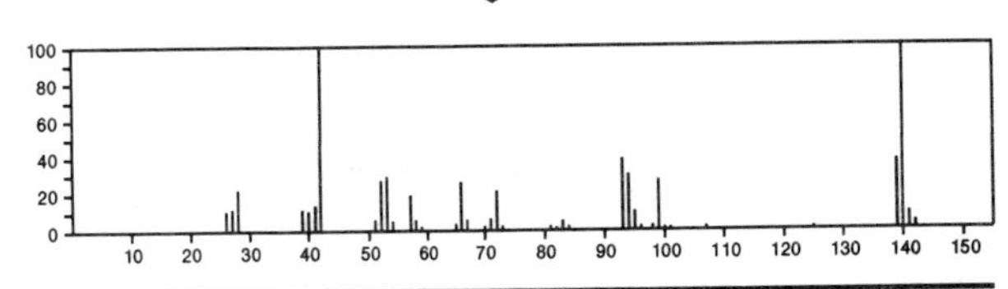

140 $C_6H_{12}N_2Si$ 18156-74-6

1*H*-Imidazole, 1-(trimethylsilyl)-

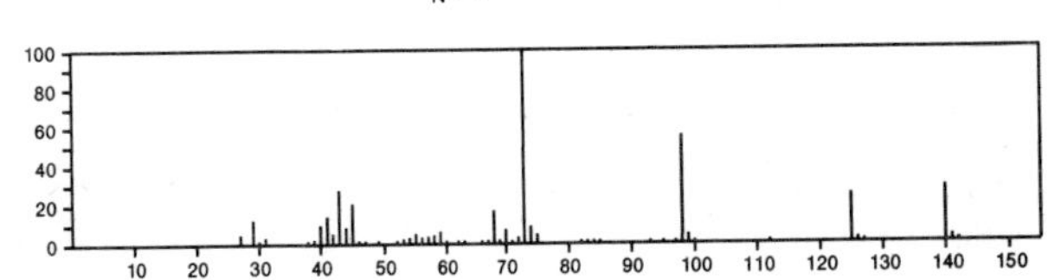

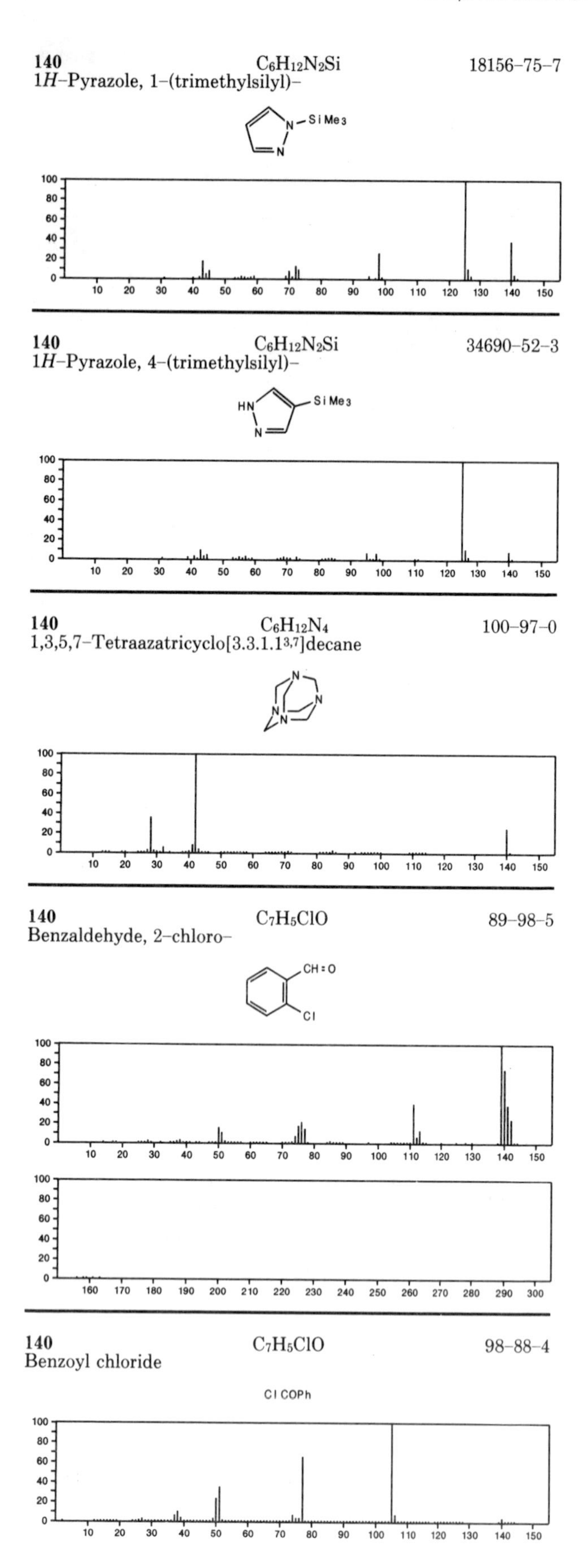

140 $C_6H_{12}N_2Si$ 18156-75-7
1*H*-Pyrazole, 1-(trimethylsilyl)-

140 $C_6H_{12}N_2Si$ 34690-52-3
1*H*-Pyrazole, 4-(trimethylsilyl)-

140 $C_6H_{12}N_4$ 100-97-0
1,3,5,7-Tetraazatricyclo[3.3.1.1^{3,7}]decane

140 C_7H_5ClO 89-98-5
Benzaldehyde, 2-chloro-

140 C_7H_5ClO 98-88-4
Benzoyl chloride

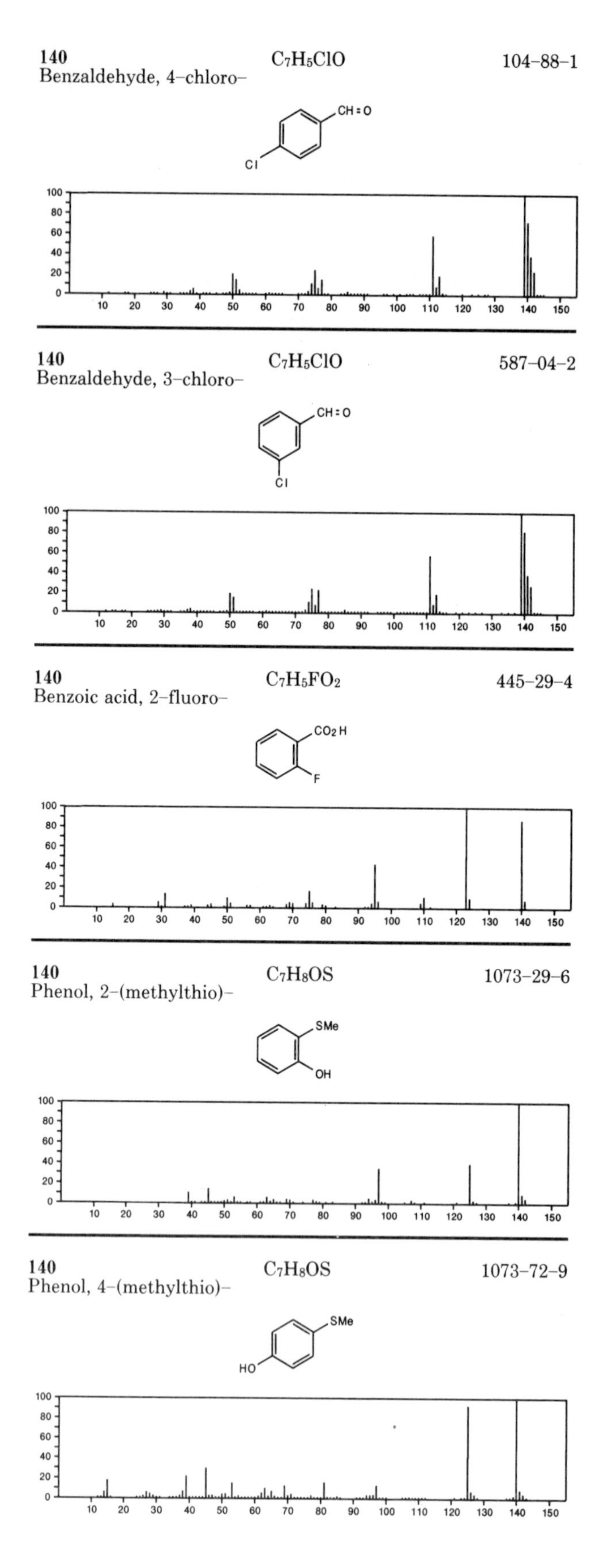

140 C_7H_5ClO 104-88-1
Benzaldehyde, 4-chloro-

140 C_7H_5ClO 587-04-2
Benzaldehyde, 3-chloro-

140 $C_7H_5FO_2$ 445-29-4
Benzoic acid, 2-fluoro-

140 C_7H_8OS 1073-29-6
Phenol, 2-(methylthio)-

140 C_7H_8OS 1073-72-9
Phenol, 4-(methylthio)-

140 $C_7H_8O_3$ 623-17-6
2-Furanmethanol, acetate

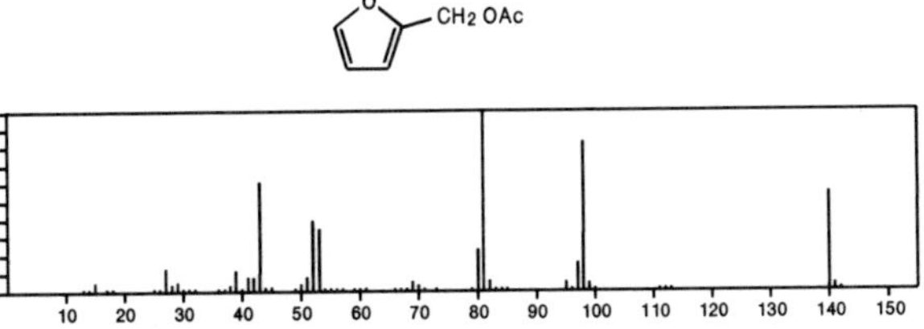

140 $C_7H_8O_3$ 672-89-9
2*H*-Pyran-2-one, 4-methoxy-6-methyl-

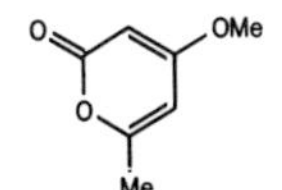

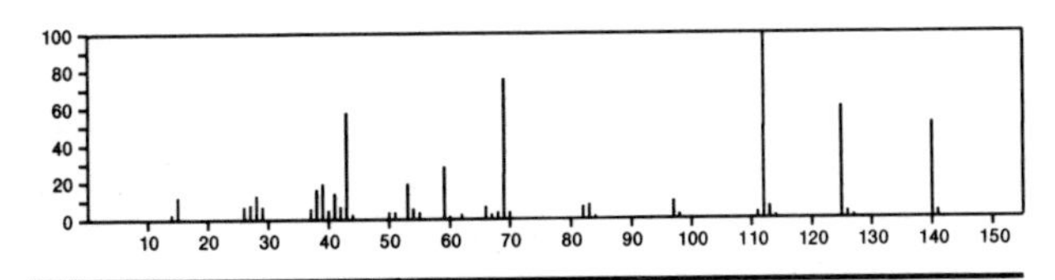

140 $C_7H_8O_3$ 2298-99-9
4*H*-Pyran-4-one, 3-hydroxy-2,6-dimethyl-

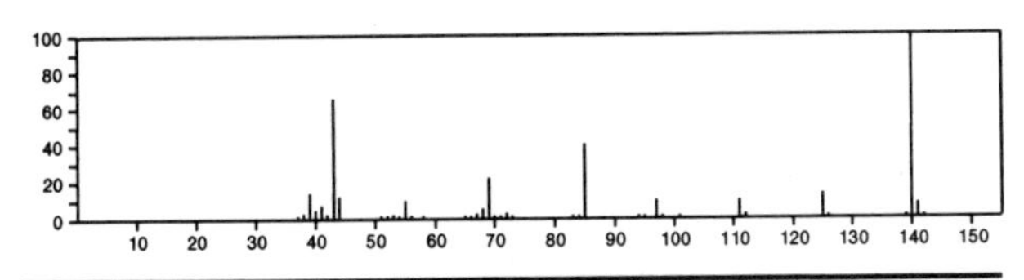

140 $C_7H_8O_3$ 3552-33-8
2,5-Furandione, 3-ethyl-4-methyl-

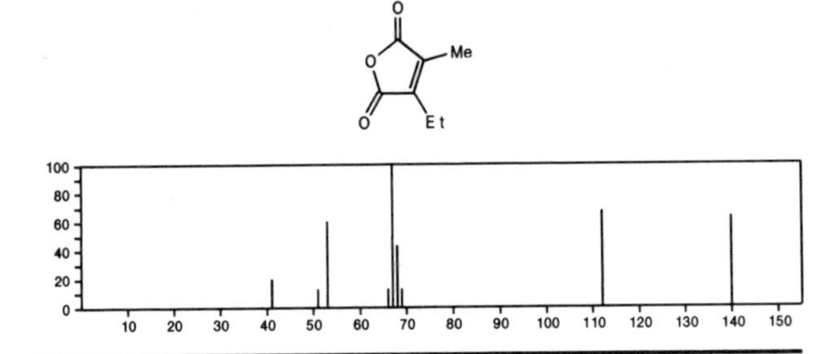

140 $C_7H_8O_3$ 3859-39-0
1,3-Cyclopentanedione, 2-acetyl-

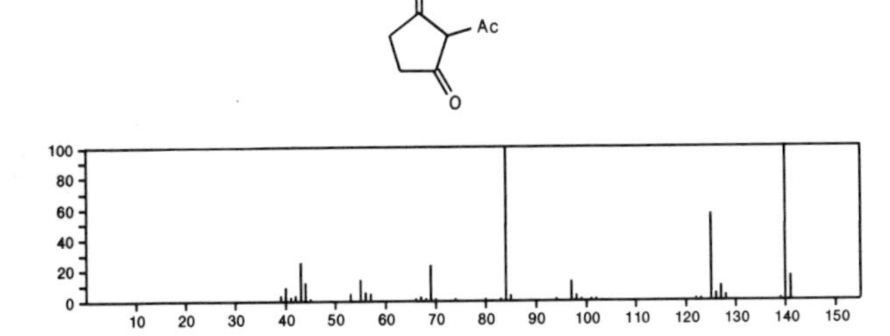

140 $C_7H_8O_3$ 4225-42-7
4*H*-Pyran-4-one, 2-methoxy-6-methyl-

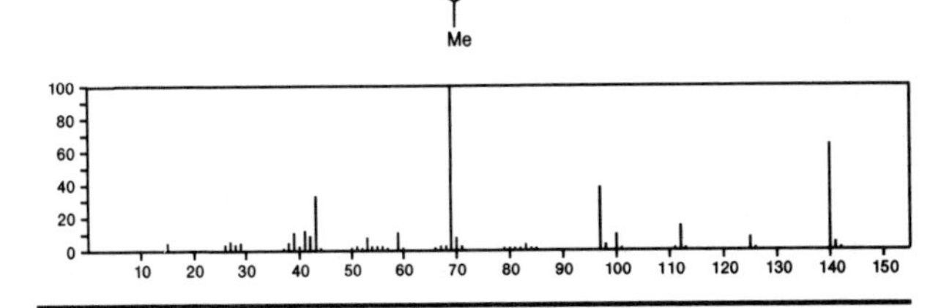

140 $C_7H_8O_3$ 4505-53-7
1,2,4-Cyclopentanetrione, 3-ethyl-

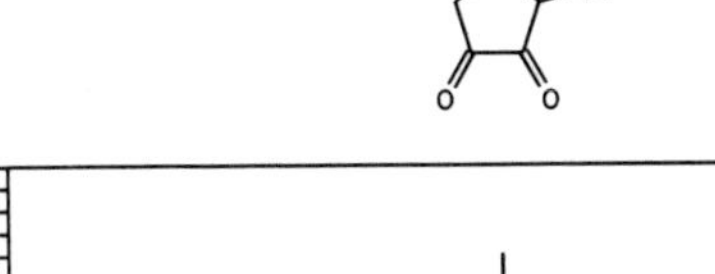

140 $C_7H_8O_3$ 4940-11-8
4*H*-Pyran-4-one, 2-ethyl-3-hydroxy-

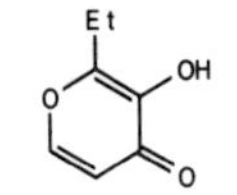

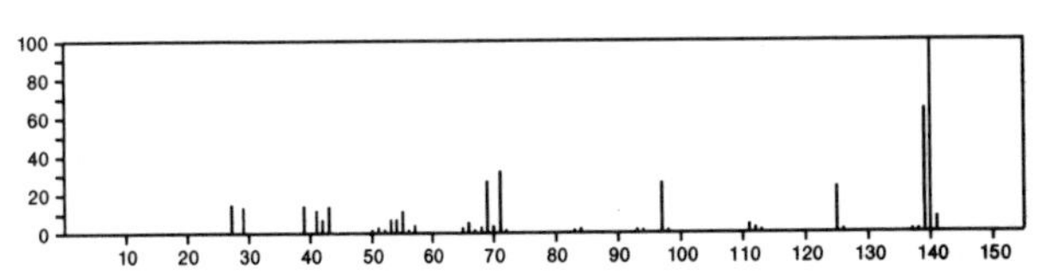

140 $C_7H_8O_3$ 5192-62-1
2*H*-Pyran-2-one, 4-hydroxy-3,6-dimethyl-

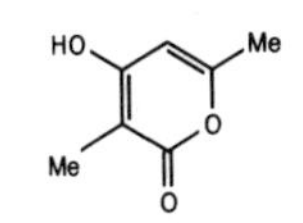

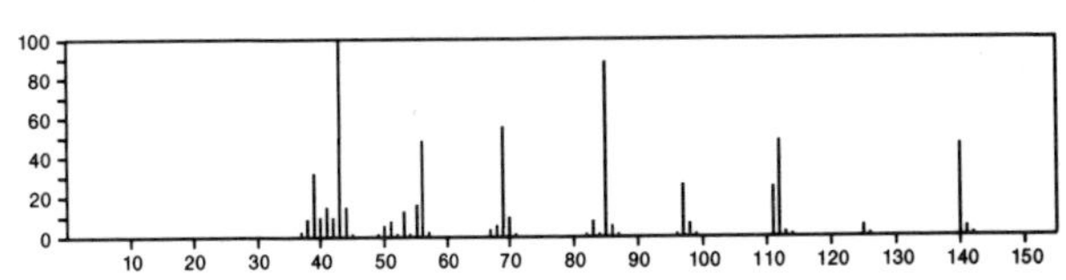

140 $C_7H_8O_3$ 7180-62-3
4-Cyclopentene-1,3-dione, 4-methoxy-5-methyl-

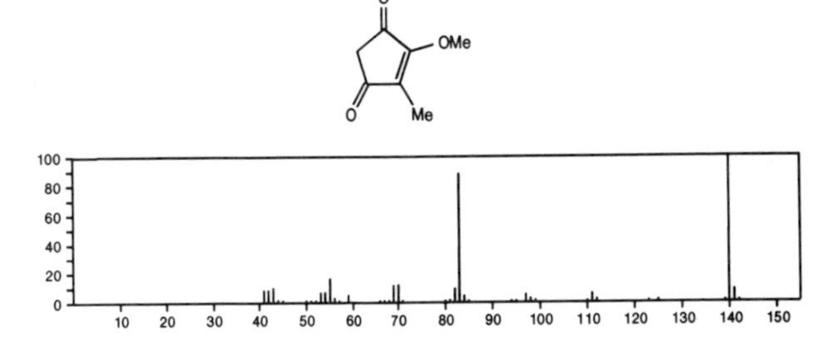

140 $C_7H_8O_3$ 17530–56–2
1,2,4–Cyclopentanetrione, 3,3–dimethyl–

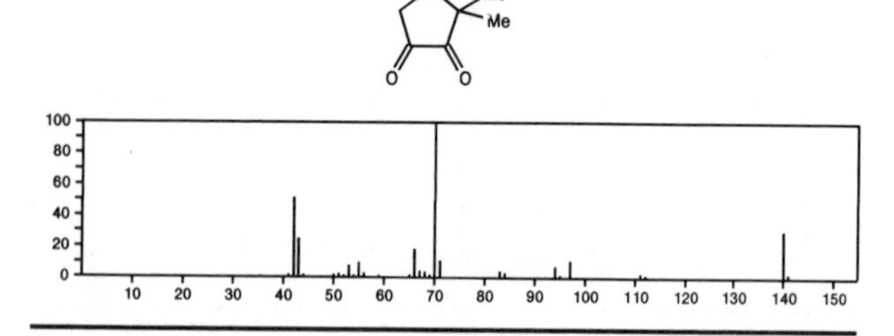

140 $C_7H_8O_3$ 54966–49–3
2–Pentynoic acid, 4–oxo–, ethyl ester

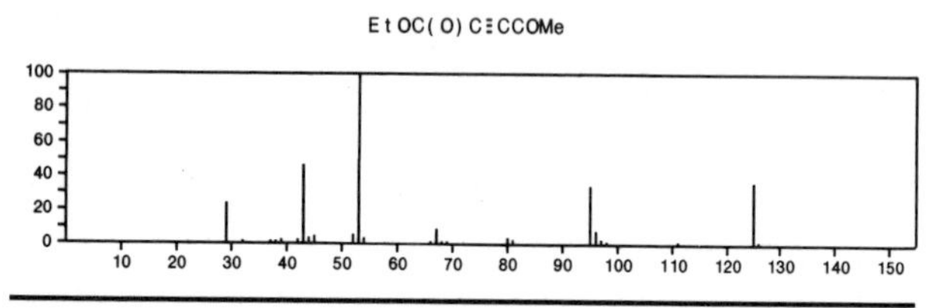

140 $C_7H_8O_3$ 56666–81–0
2–Oxabicyclo[3.1.0]hex–3–ene–4–carboxylic acid, methyl ester

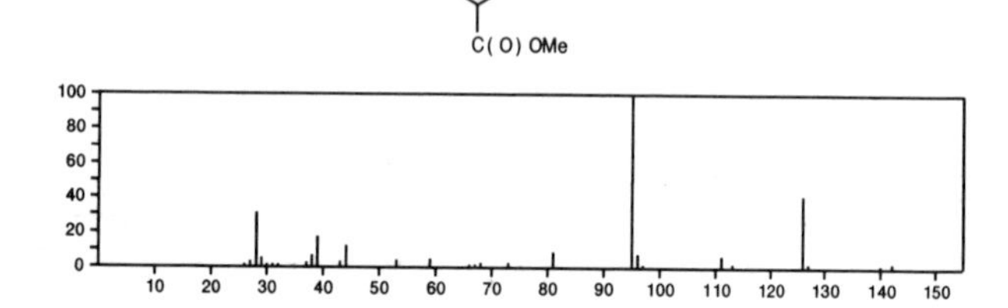

140 $C_7H_{12}N_2O$ 1801–72–5
Urea, *N,N'*–di–2–propenyl–

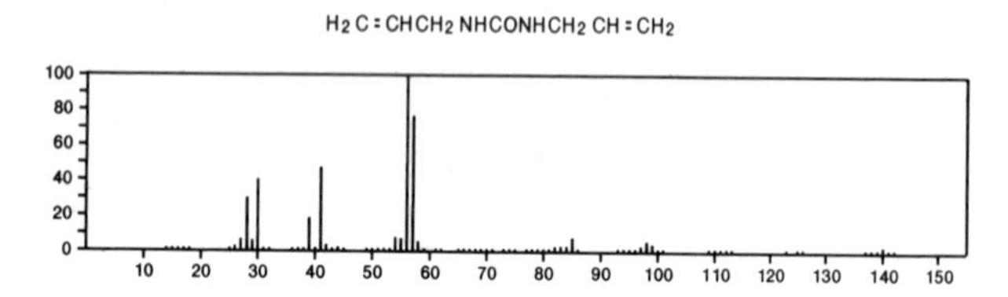

140 $C_7H_{12}N_2O$ 3201–25–0
3*H*–Pyrazol–3–one, 2,4–dihydro–2,4,4,5–tetramethyl–

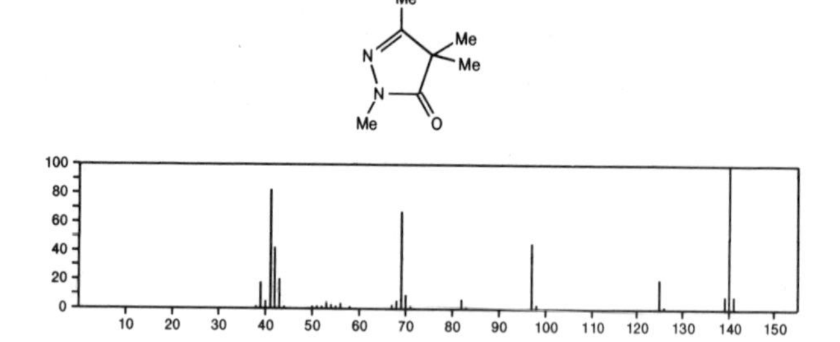

140 $C_7H_{12}N_2O$ 49582–42–5
2–Propenal, 3–(1–aziridinyl)–3–(dimethylamino)–

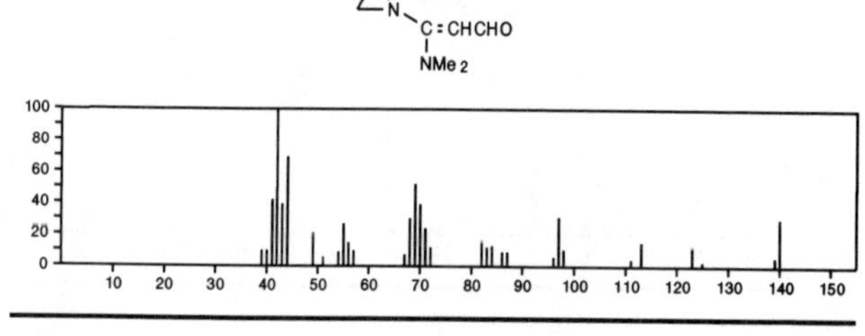

140 $C_8H_6F_2$ 405–42–5
Benzene, (2,2–difluoroethenyl)–

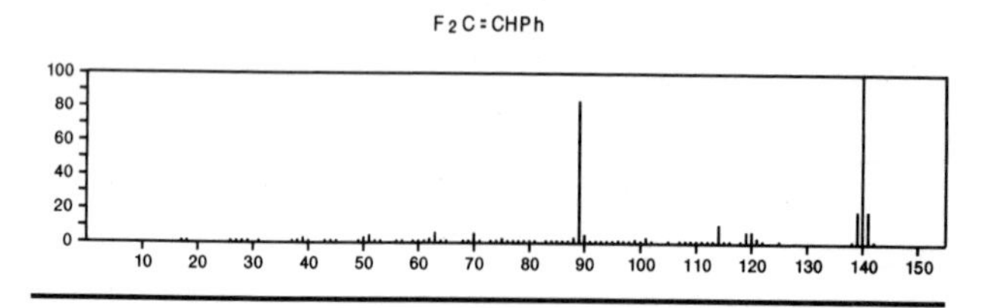

140 C_8H_9Cl 89–96–3
Benzene, 1–chloro–2–ethyl–

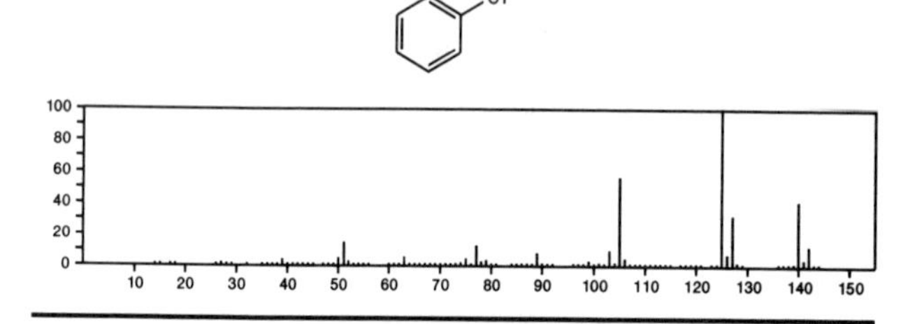

140 C_8H_9Cl 95–66–9
Benzene, 1–chloro–2,4–dimethyl–

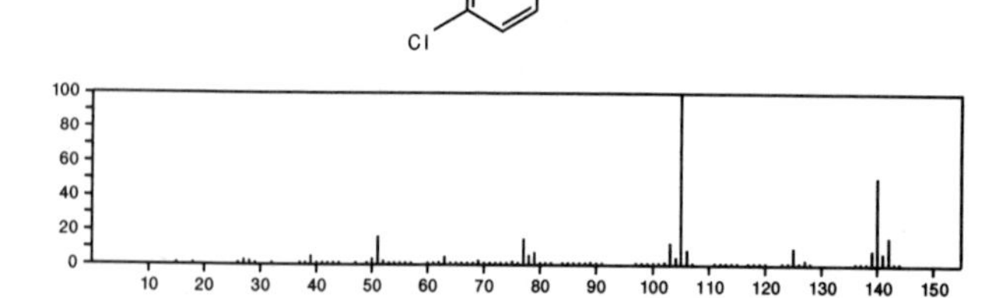

140 C_8H_9Cl 95–72–7
Benzene, 2–chloro–1,4–dimethyl–

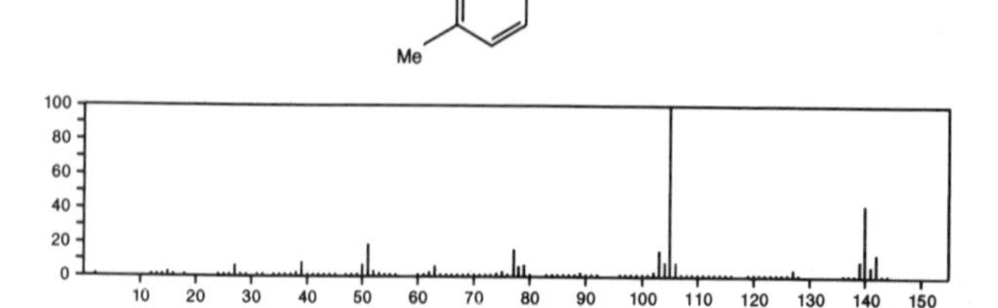

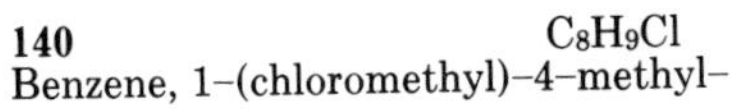

140 C_8H_9Cl 104–82–5
Benzene, 1–(chloromethyl)–4–methyl–

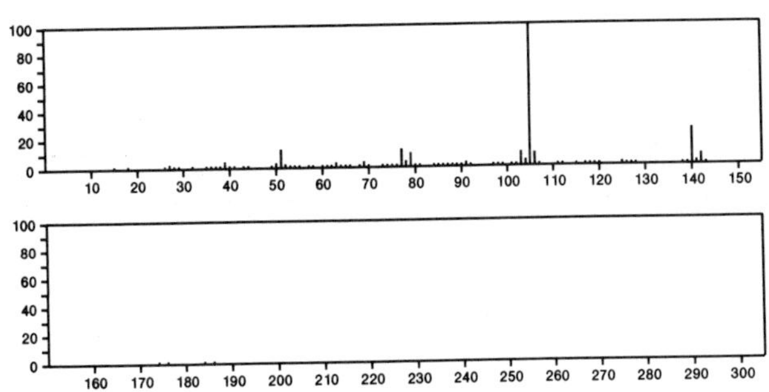

140 C_8H_9Cl 552–45–4
Benzene, 1–(chloromethyl)–2–methyl–

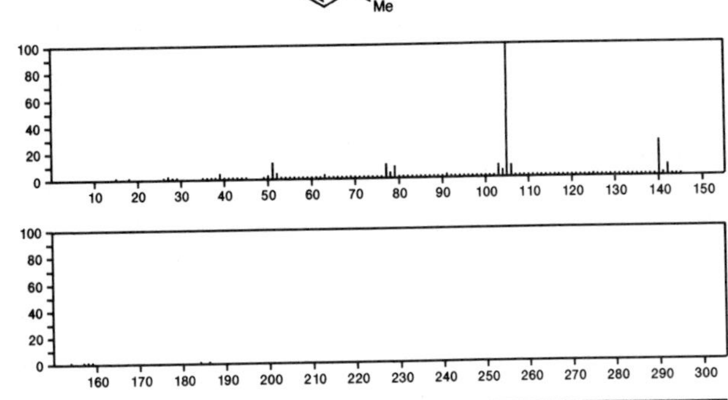

140 C_8H_9Cl 615–60–1
Benzene, 4–chloro–1,2–dimethyl–

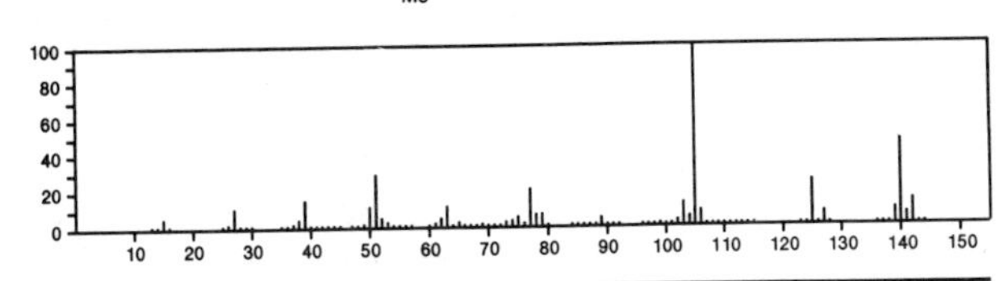

140 C_8H_9Cl 620–16–6
Benzene, 1–chloro–3–ethyl–

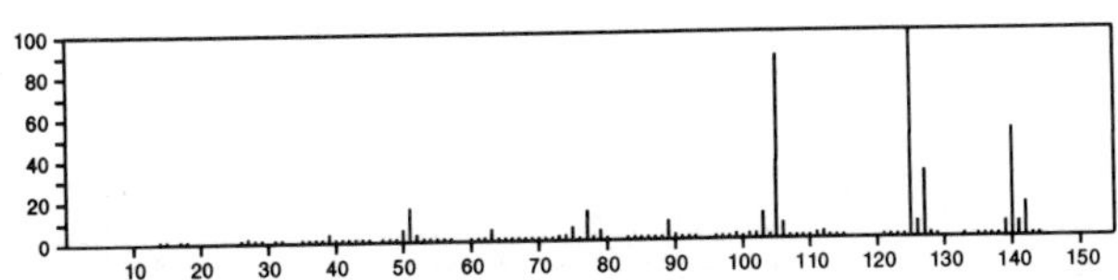

140 C_8H_9Cl 620–19–9
Benzene, 1–(chloromethyl)–3–methyl–

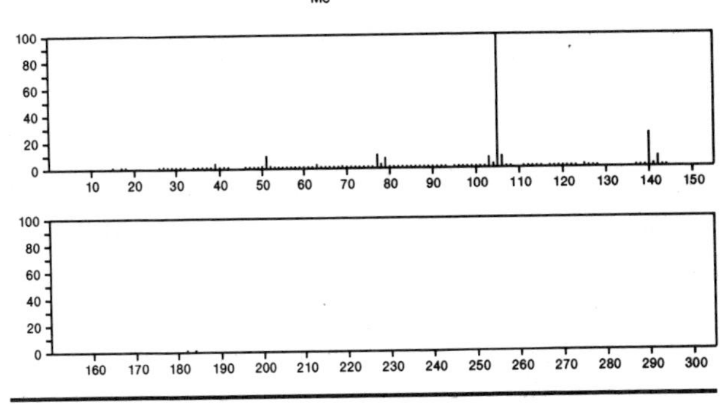

140 C_8H_9Cl 622–24–2
Benzene, (2–chloroethyl)–

Cl CH2 CH2 Ph

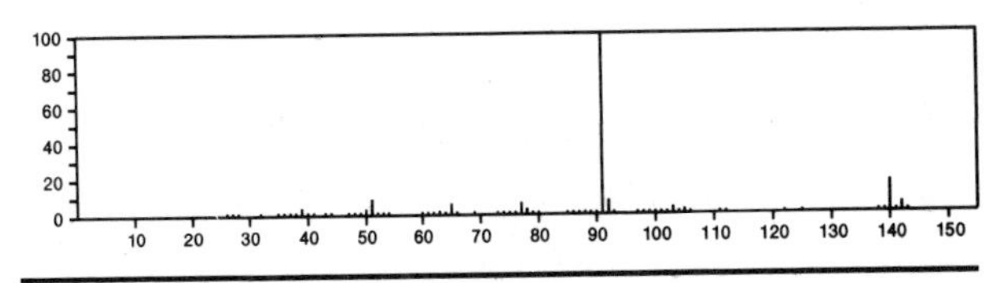

140 C_8H_9Cl 622–98–0
Benzene, 1–chloro–4–ethyl–

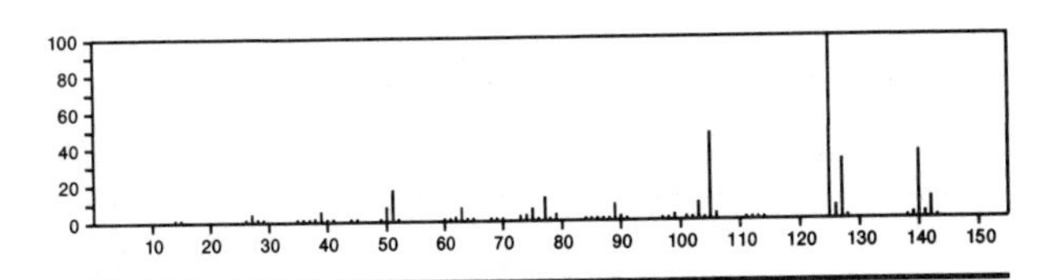

140 C_8H_9Cl 672–65–1
Benzene, (1–chloroethyl)–

PhCHCl (Ph)

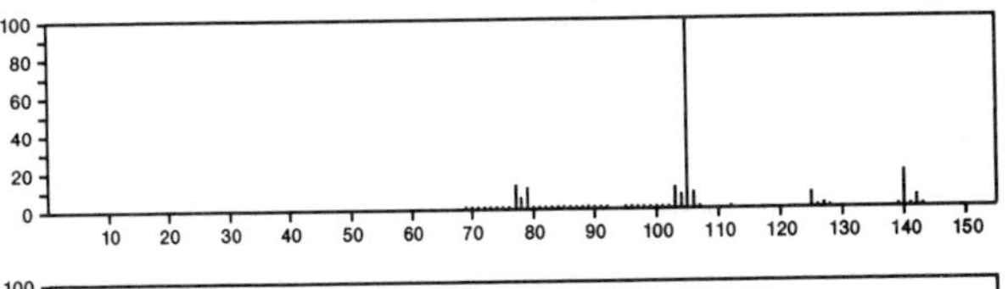

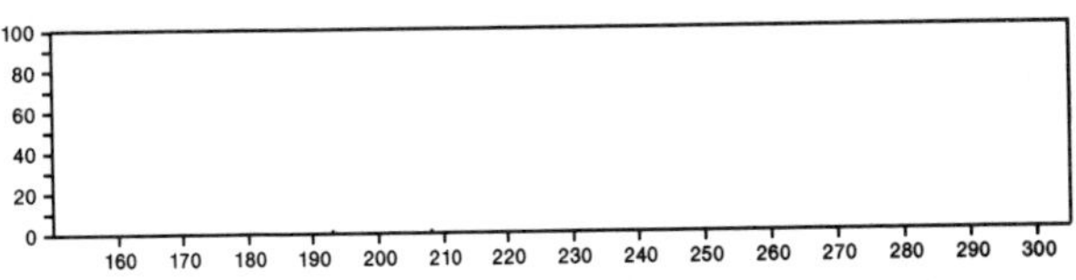

140 C_8H_9Cl 6781–98–2
Benzene, 2–chloro–1,3–dimethyl–

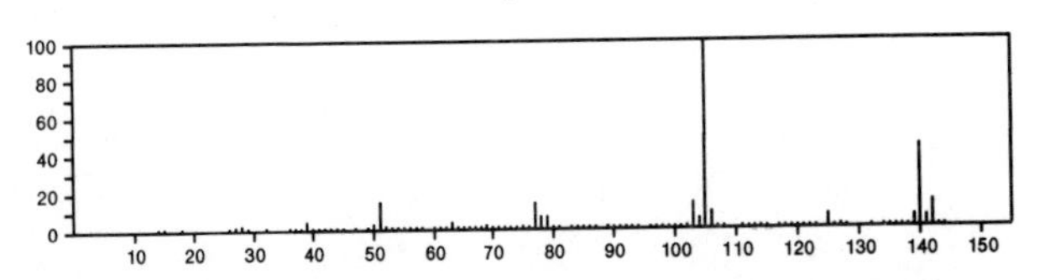

140 C_8H_9Cl 26445-11-4
o-Xylene, chloro-

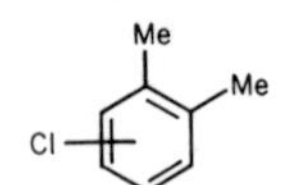

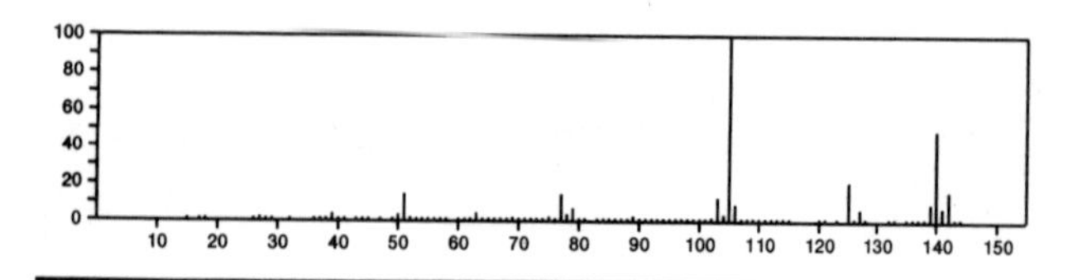

140 C_8H_9FO 451-80-9
Benzene, 1-ethoxy-2-fluoro-

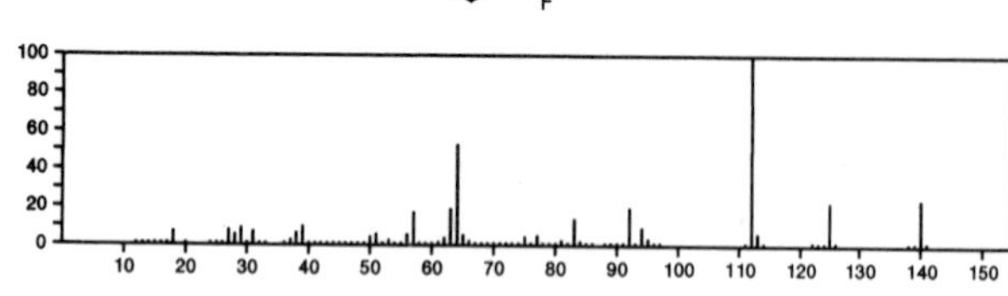

140 C_8H_9FO 459-26-7
Benzene, 1-ethoxy-4-fluoro-

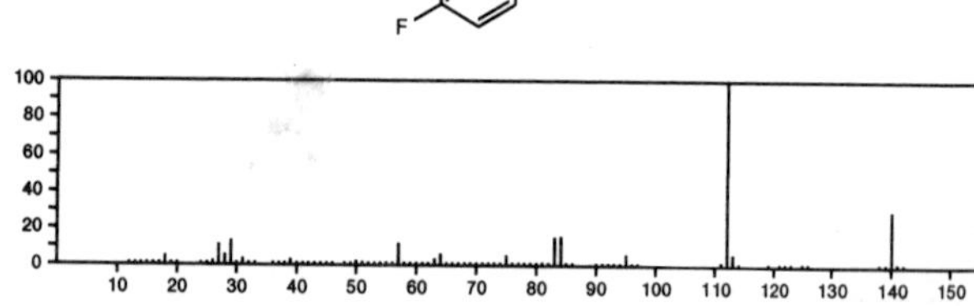

140 C_8H_9FO 7589-27-7
Benzeneethanol, 4-fluoro-

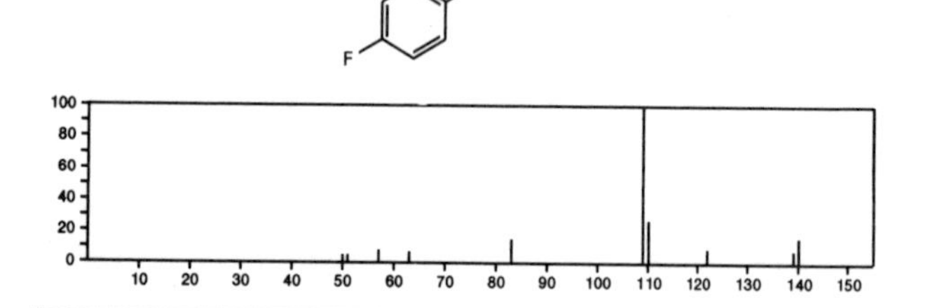

140 C_8H_9FO 52059-53-7
Benzeneethanol, 3-fluoro-

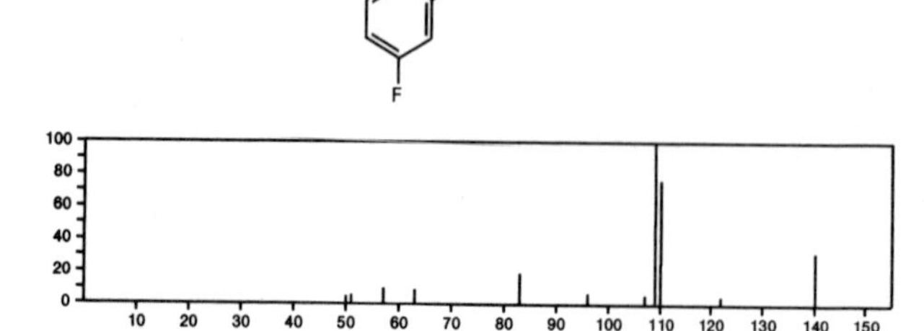

140 $C_8H_{12}O_2$ 698-56-6
Bicyclo[3.1.0]hexan-2-ol, acetate, (1α,2β,5α)-

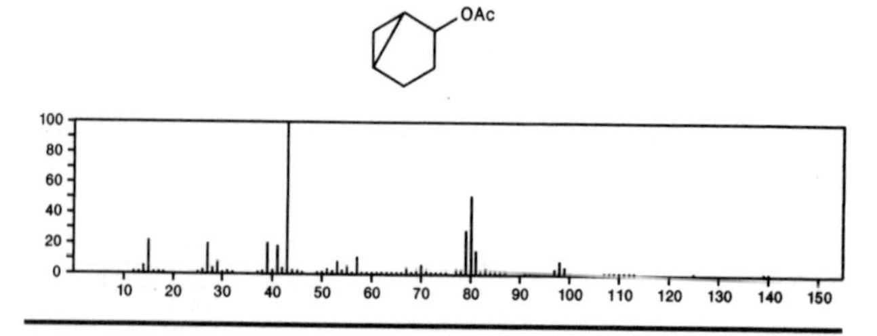

140 $C_8H_{12}O_2$ 874-23-7
Cyclohexanone, 2-acetyl-

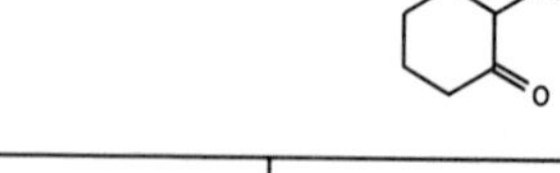

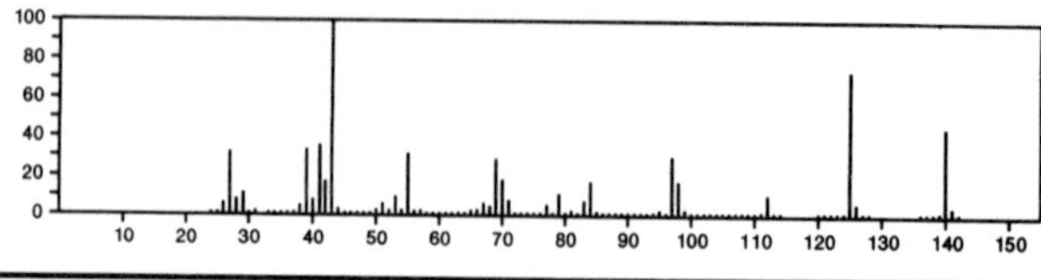

140 $C_8H_{12}O_2$ 1004-58-6
1,4-Dioxaspiro[4.5]dec-6-ene

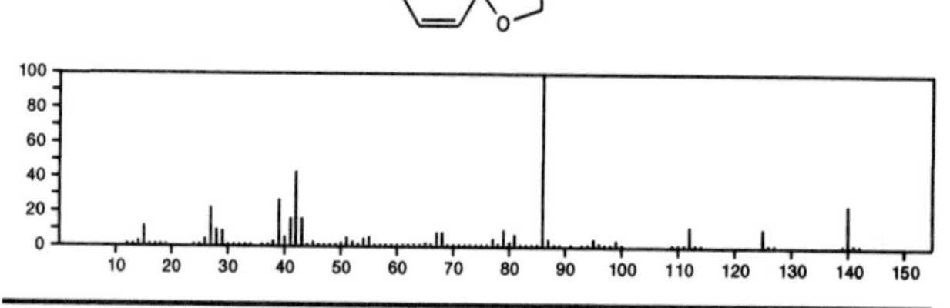

140 $C_8H_{12}O_2$ 1424-22-2
1-Cyclohexen-1-ol, acetate

140 $C_8H_{12}O_2$ 1920-21-4
2*H*-Pyran-2-carboxaldehyde, 3,4-dihydro-2,5-dimethyl-

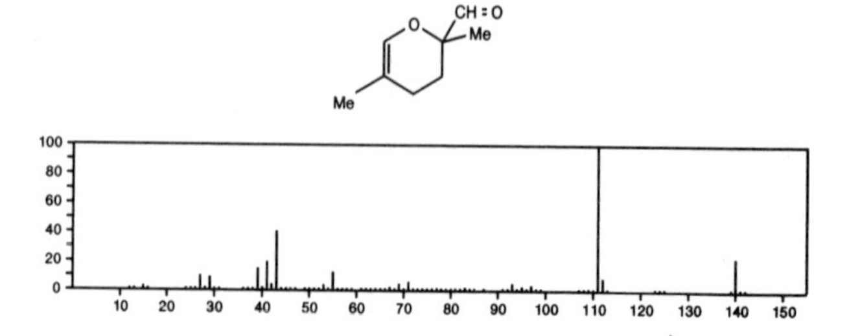

140 $C_8H_{12}O_2$ 3508-78-9
2,4-Pentanedione, 3-(2-propenyl)-

COMe
$H_2C=CHCH_2CHCOMe$

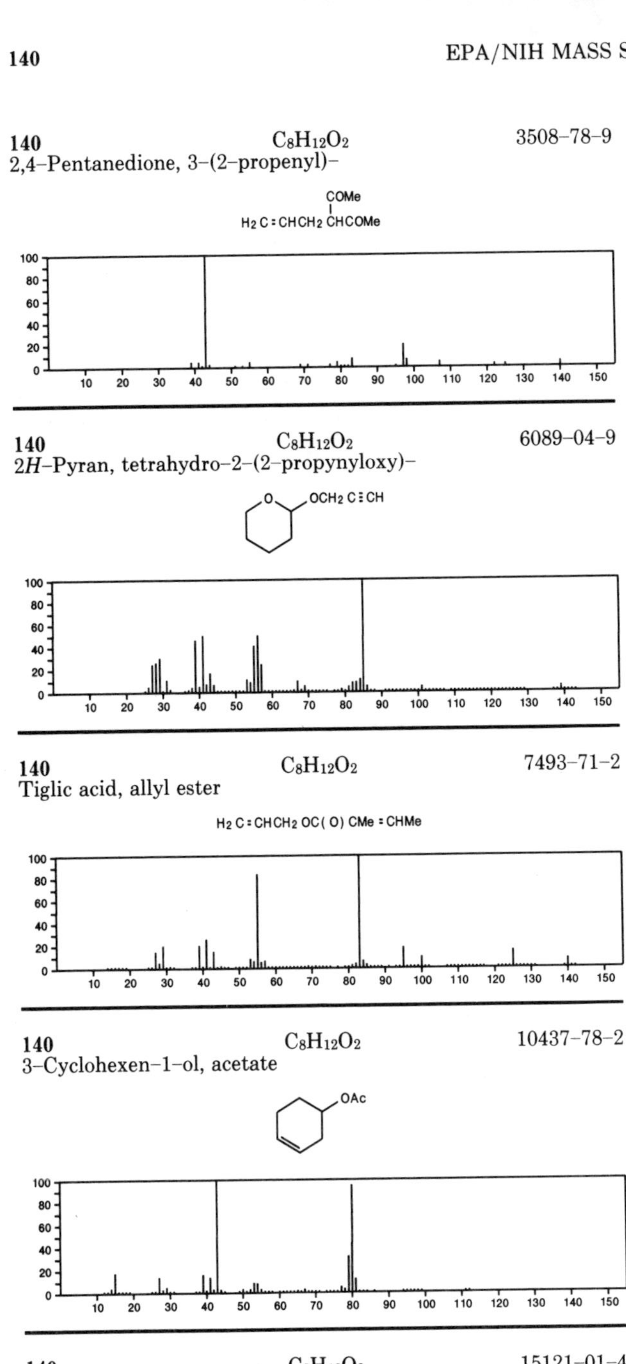

140 $C_8H_{12}O_2$ 6089-04-9
2*H*-Pyran, tetrahydro-2-(2-propynyloxy)-

140 $C_8H_{12}O_2$ 7493-71-2
Tiglic acid, allyl ester

$H_2C=CHCH_2OC(O)CMe=CHMe$

140 $C_8H_{12}O_2$ 10437-78-2
3-Cyclohexen-1-ol, acetate

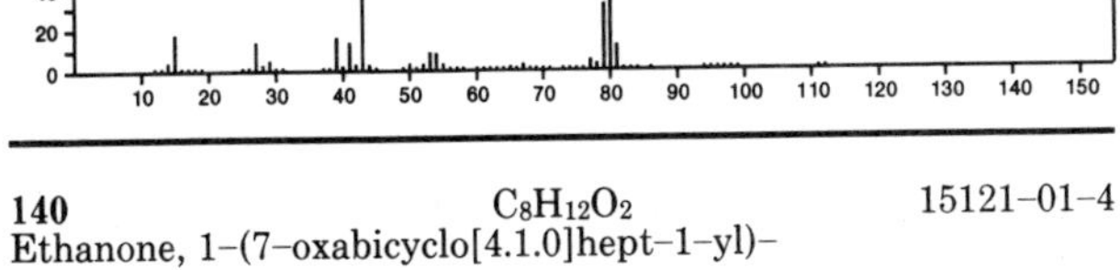

140 $C_8H_{12}O_2$ 15121-01-4
Ethanone, 1-(7-oxabicyclo[4.1.0]hept-1-yl)-

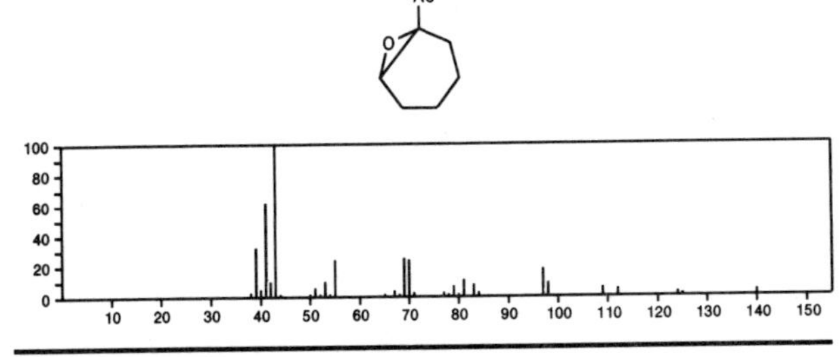

140 $C_8H_{12}O_2$ 16523-06-1
Cyclopentanecarboxylic acid, ethenyl ester

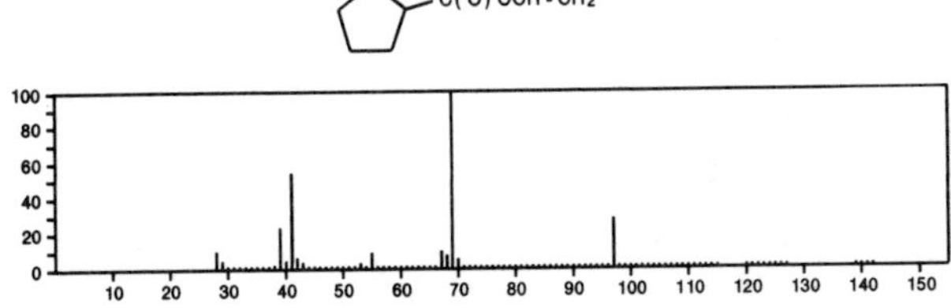

140 $C_8H_{12}O_2$ 21485-51-8
p-Dioxane, 2,5-divinyl-

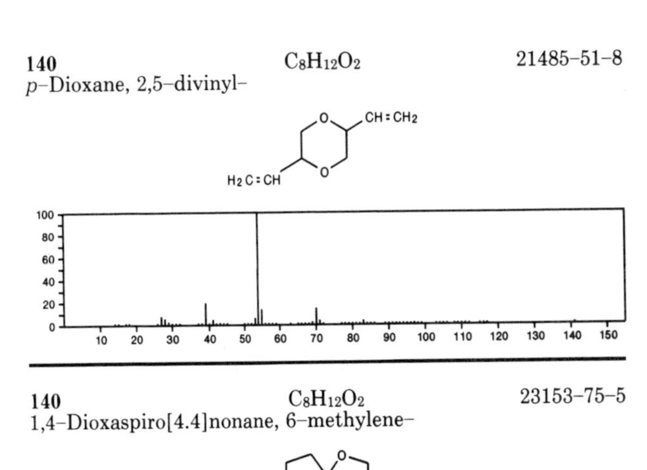

140 $C_8H_{12}O_2$ 23153-75-5
1,4-Dioxaspiro[4.4]nonane, 6-methylene-

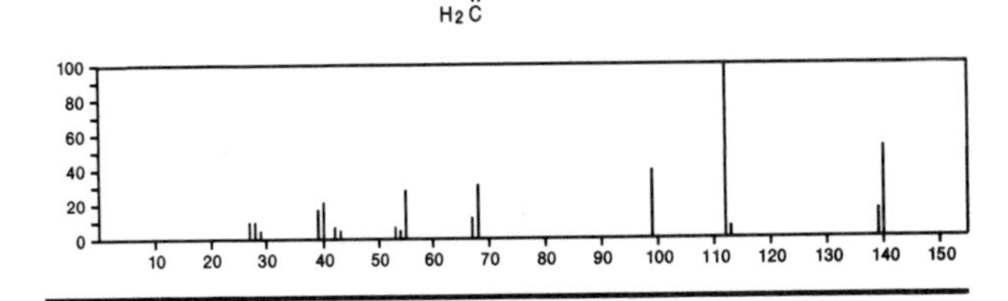

140 $C_8H_{12}O_2$ 25112-87-2
1,3-Cyclopentanedione, 2-ethyl-2-methyl-

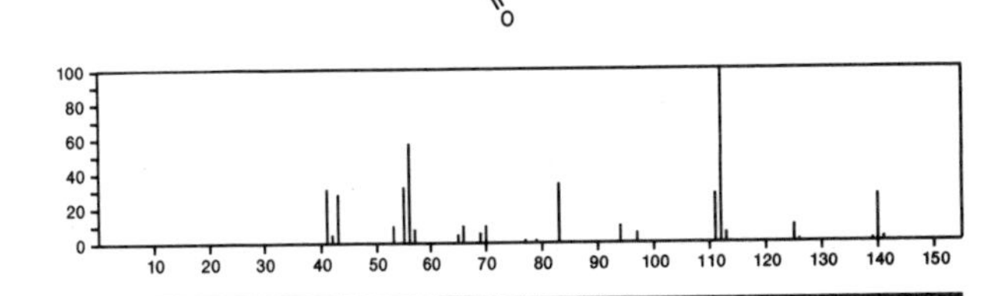

140 $C_8H_{12}O_2$ 34645-17-5
1,3-Pentadien-2-ol, 4-methyl-, acetate

CH_2
$Me_2C=CHCOAc$

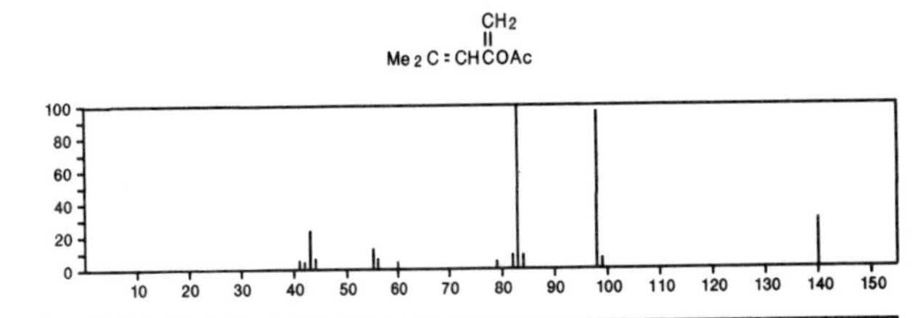

140 $C_8H_{12}O_2$ 38653-28-0
Cyclopentanol, 2-(2-propynyloxy)-, *trans*-

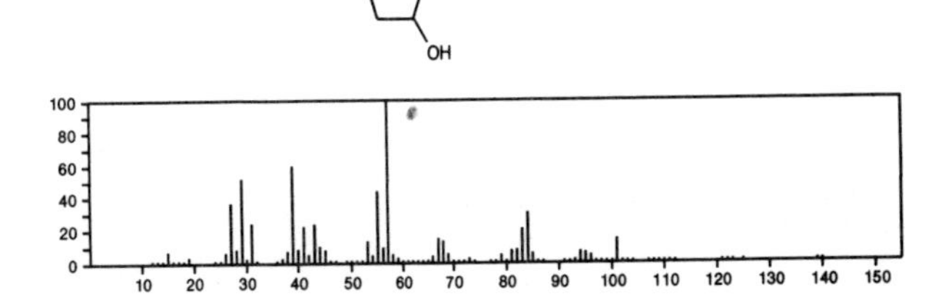

140 $C_8H_{12}O_2$ 38653-39-3
2*H*,6*H*-Cyclopenta[*b*][1,4]dioxepin, 5a,7,8,8a-tetrahydro-, *trans*-

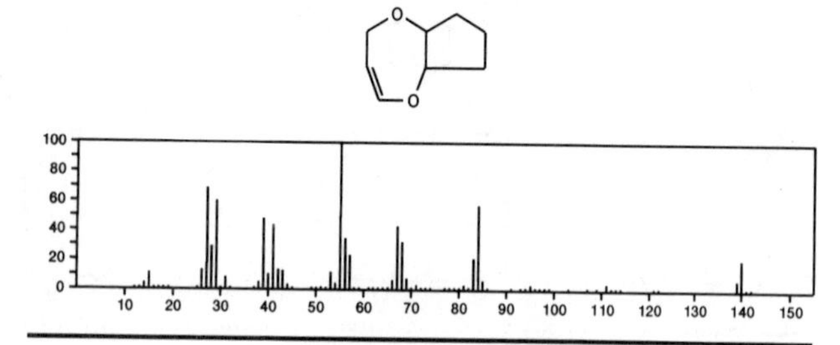

140 $C_8H_{12}O_2$ 38653-47-3
5*H*-Cyclopenta-1,4-dioxin, 4a,6,7,7a-tetrahydro-2-methyl-, *trans*-

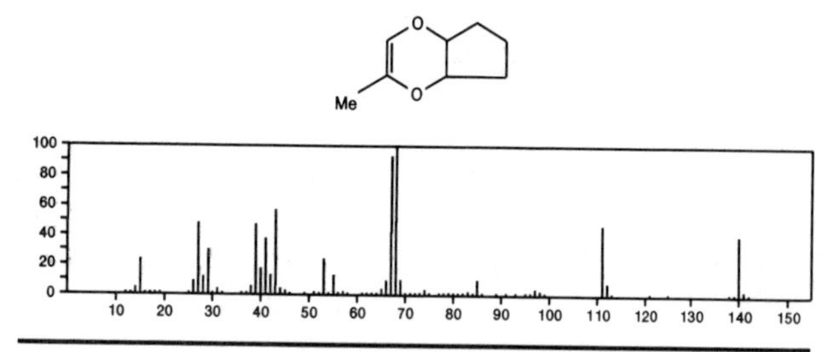

140 $C_8H_{12}O_2$ 55702-69-7
2-Pentanone, 5-(1,2-propadienyloxy)-

MeCO(CH_2) $_3$ OCH = C = CH_2

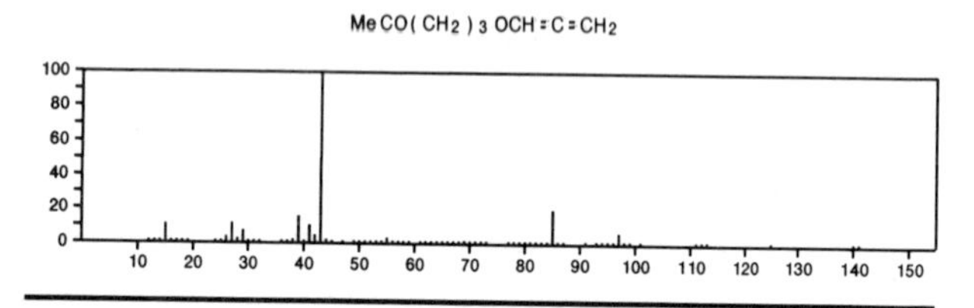

140 $C_8H_{12}O_2$ 55702-70-0
2-Pentanone, 5-(2-propynyloxy)-

MeCO(CH_2) $_3$ OCH_2 C ≡ CH

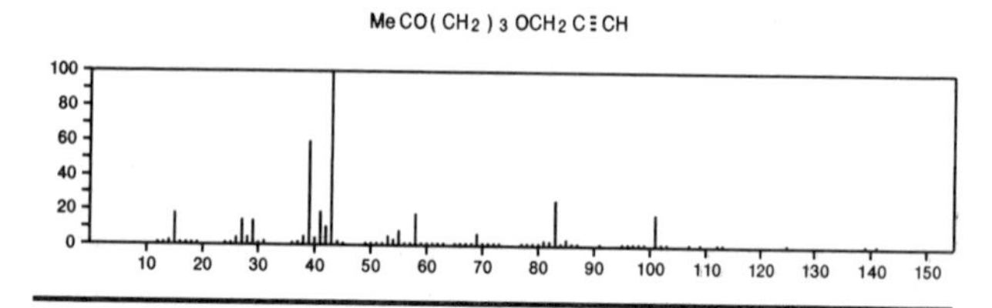

140 $C_8H_{12}O_2$ 55702-71-1
1,4-Benzodioxin, 2,3,4a,5,6,7-hexahydro-

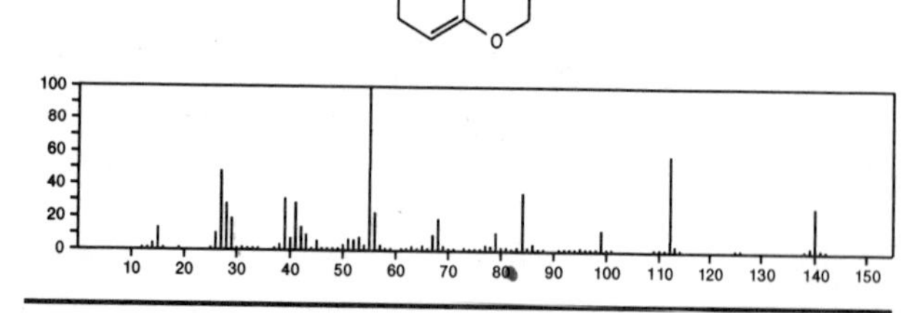

140 $C_8H_{12}O_2$ 56909-02-5
6-Heptynoic acid, methyl ester

HC ≡ C(CH_2) $_4$ C(O) OMe

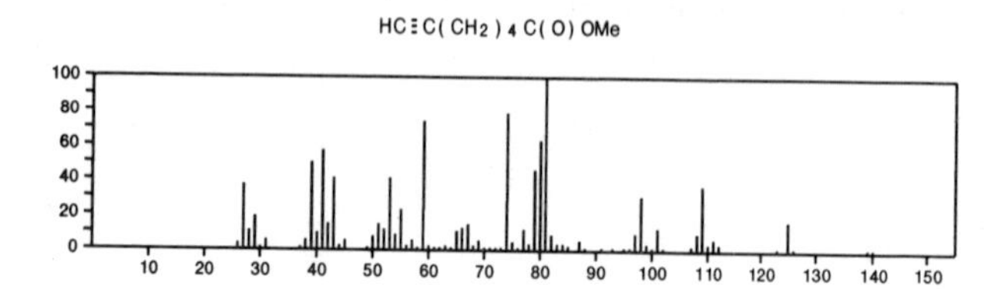

140 $C_8H_{12}S$ 1455-20-5
Thiophene, 2-butyl-

140 $C_8H_{12}S$ 1689-78-7
Thiophene, 2-(1,1-dimethylethyl)-

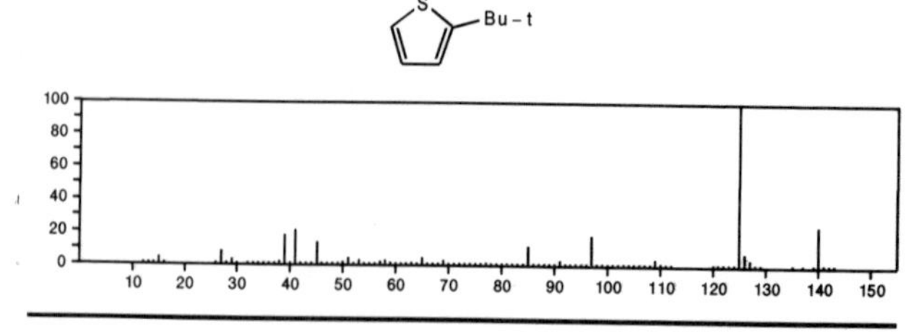

140 $C_8H_{12}S$ 1689-79-8
Thiophene, 3-(1,1-dimethylethyl)-

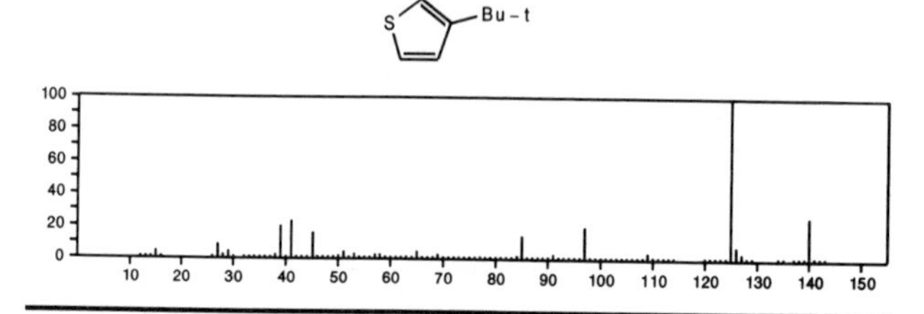

140 $C_8H_{12}S$ 30221-53-5
2-Cyclopentene-1-thione, 3,4,4-trimethyl-

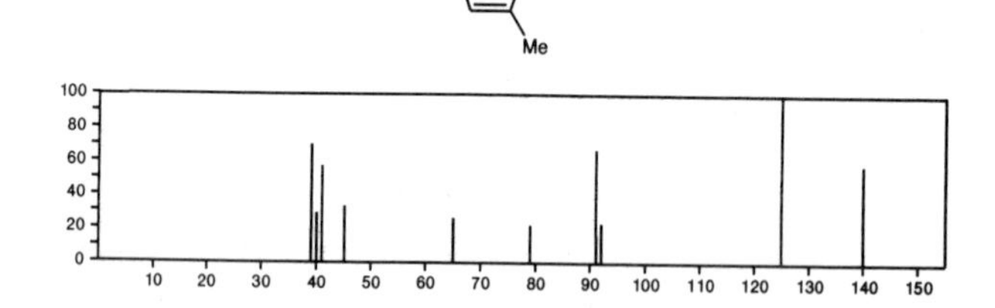

140 $C_8H_{12}S$ 38693-67-3
2-Cyclopentene-1-thione, 2-ethyl-3-methyl-

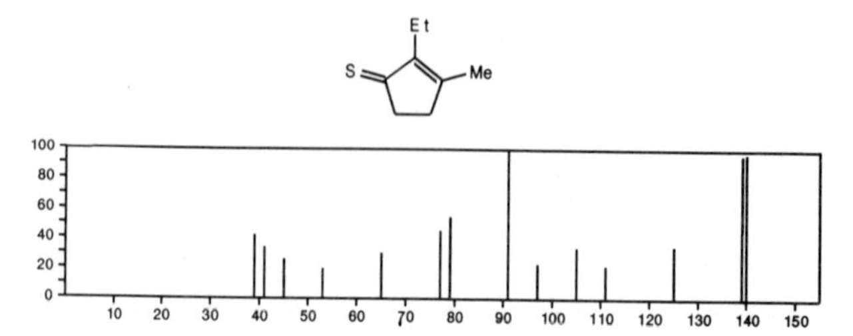

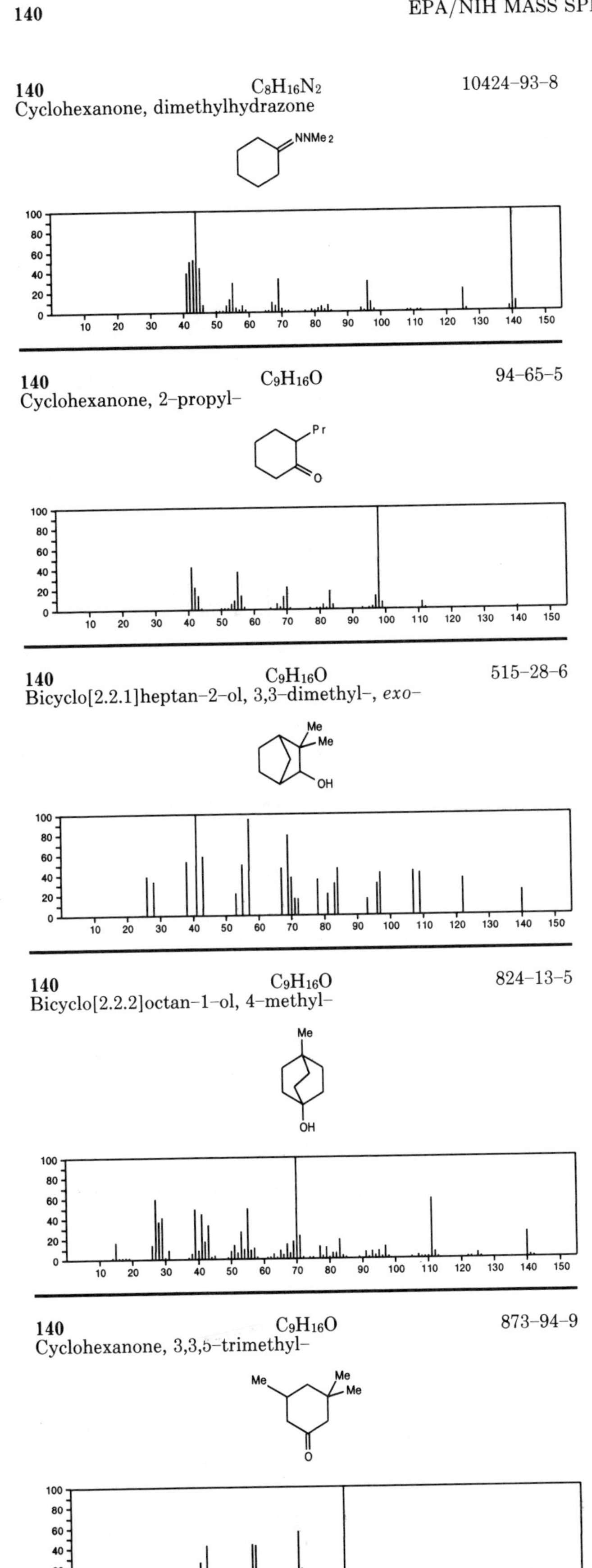

140 $C_8H_{16}N_2$ 10424–93–8
Cyclohexanone, dimethylhydrazone

140 $C_9H_{16}O$ 94–65–5
Cyclohexanone, 2–propyl–

140 $C_9H_{16}O$ 515–28–6
Bicyclo[2.2.1]heptan–2–ol, 3,3–dimethyl–, *exo*–

140 $C_9H_{16}O$ 824–13–5
Bicyclo[2.2.2]octan–1–ol, 4–methyl–

140 $C_9H_{16}O$ 873–94–9
Cyclohexanone, 3,3,5–trimethyl–

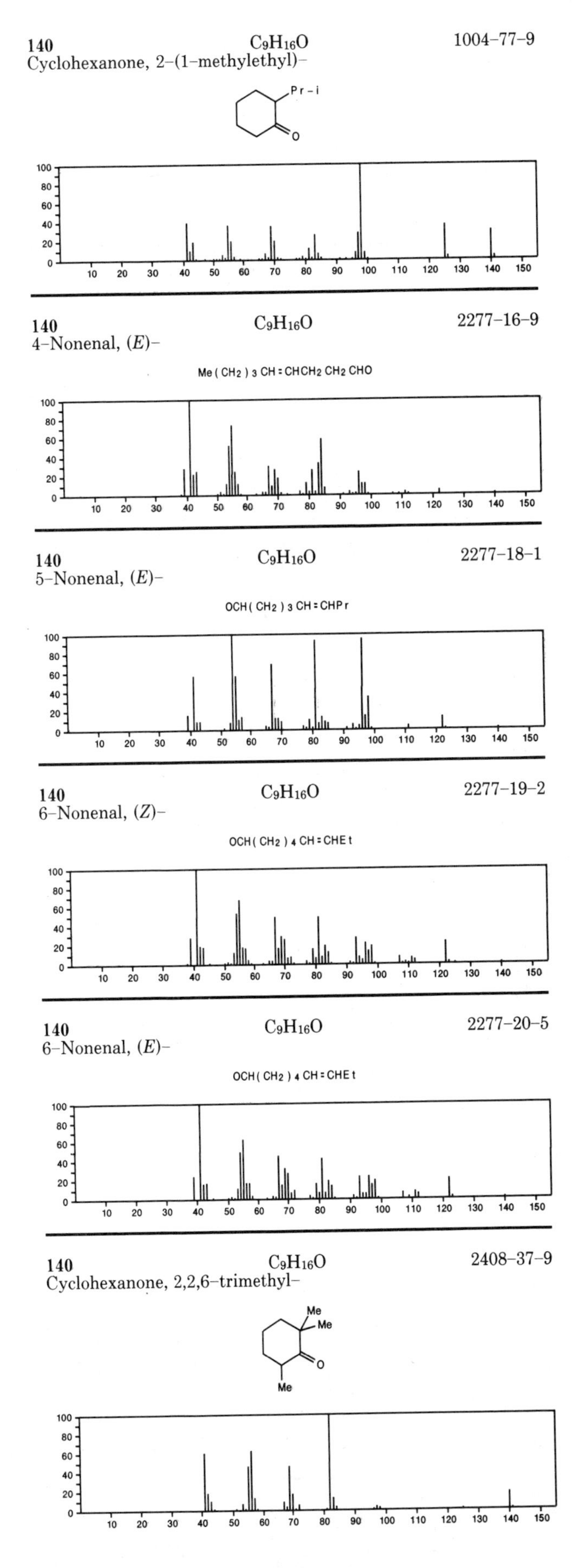

140 $C_9H_{16}O$ 1004–77–9
Cyclohexanone, 2–(1–methylethyl)–

140 $C_9H_{16}O$ 2277–16–9
4–Nonenal, (*E*)–

Me(CH2)3CH=CHCH2CH2CHO

140 $C_9H_{16}O$ 2277–18–1
5–Nonenal, (*E*)–

OCH(CH2)3CH=CHPr

140 $C_9H_{16}O$ 2277–19–2
6–Nonenal, (*Z*)–

OCH(CH2)4CH=CHEt

140 $C_9H_{16}O$ 2277–20–5
6–Nonenal, (*E*)–

OCH(CH2)4CH=CHEt

140 $C_9H_{16}O$ 2408–37–9
Cyclohexanone, 2,2,6–trimethyl–

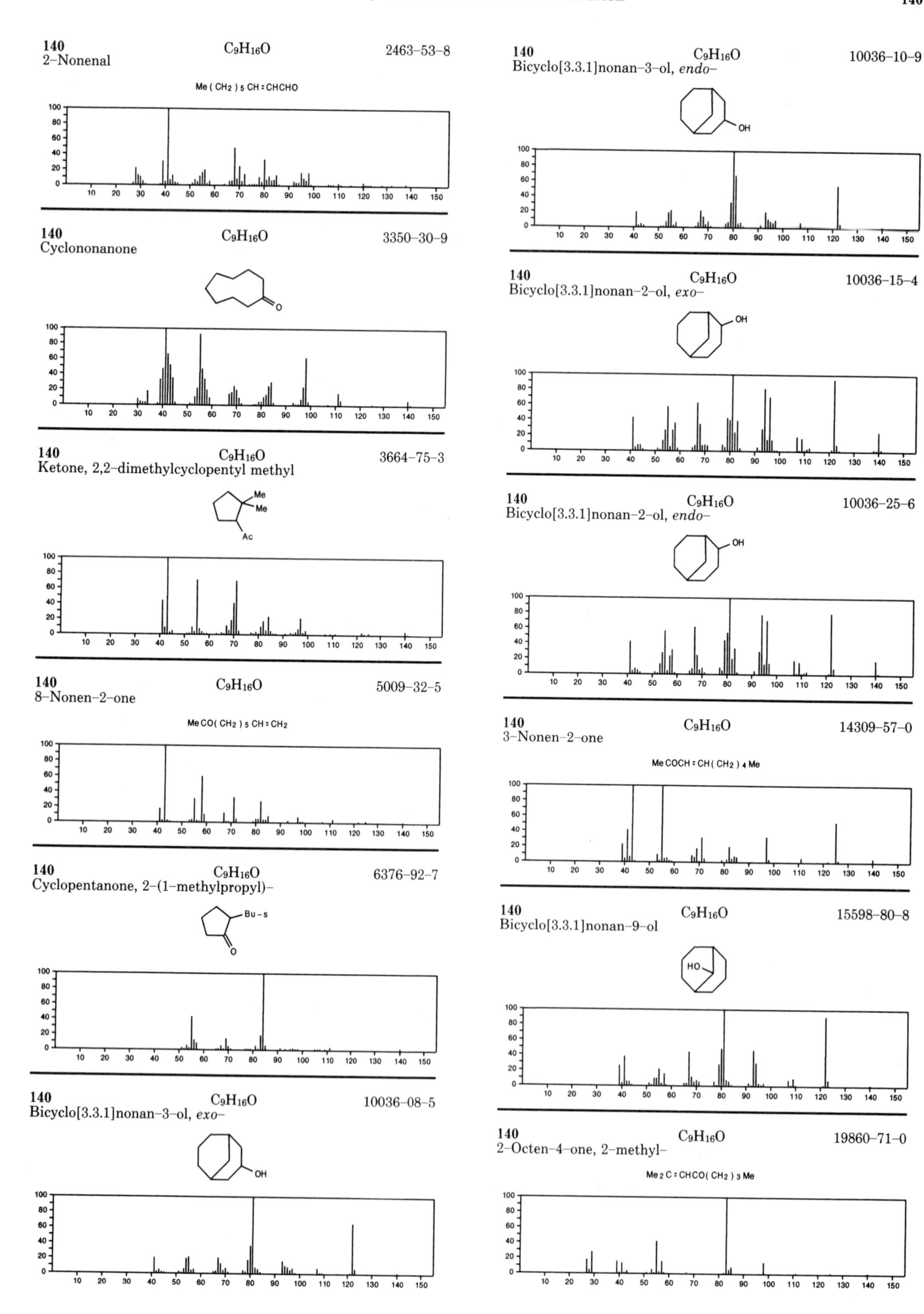
140 2-Nonenal $C_9H_{16}O$ 2463-53-8
Me(CH2)5CH=CHCHO
140 Cyclononanone $C_9H_{16}O$ 3350-30-9
140 Ketone, 2,2-dimethylcyclopentyl methyl $C_9H_{16}O$ 3664-75-3
Me
Me
Ac
140 8-Nonen-2-one $C_9H_{16}O$ 5009-32-5
MeCO(CH2)5CH=CH2
140 Cyclopentanone, 2-(1-methylpropyl)- $C_9H_{16}O$ 6376-92-7
Bu-s
140 Bicyclo[3.3.1]nonan-3-ol, exo- $C_9H_{16}O$ 10036-08-5
OH
140 Bicyclo[3.3.1]nonan-3-ol, endo- $C_9H_{16}O$ 10036-10-9
OH
140 Bicyclo[3.3.1]nonan-2-ol, exo- $C_9H_{16}O$ 10036-15-4
OH
140 Bicyclo[3.3.1]nonan-2-ol, endo- $C_9H_{16}O$ 10036-25-6
OH
140 3-Nonen-2-one $C_9H_{16}O$ 14309-57-0
MeCOCH=CH(CH2)4Me
140 Bicyclo[3.3.1]nonan-9-ol $C_9H_{16}O$ 15598-80-8
HO
140 2-Octen-4-one, 2-methyl- $C_9H_{16}O$ 19860-71-0
Me2C=CHCO(CH2)3Me

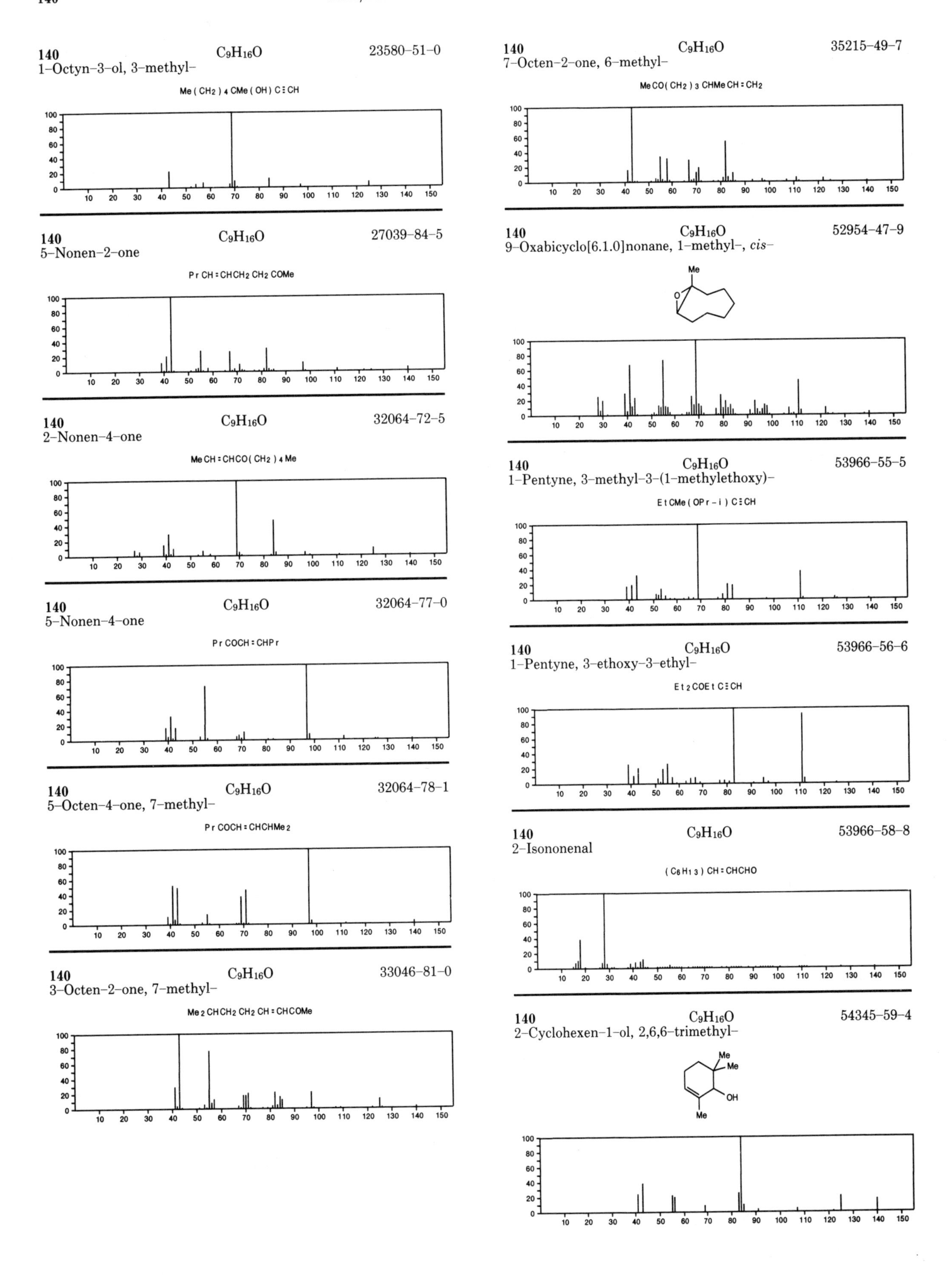

140 C9H16O 23580-51-0
1-Octyn-3-ol, 3-methyl-
Me(CH2)4CMe(OH)C≡CH
140 C9H16O 27039-84-5
5-Nonen-2-one
PrCH=CHCH2CH2COMe
140 C9H16O 32064-72-5
2-Nonen-4-one
MeCH=CHCO(CH2)4Me
140 C9H16O 32064-77-0
5-Nonen-4-one
PrCOCH=CHPr
140 C9H16O 32064-78-1
5-Octen-4-one, 7-methyl-
PrCOCH=CHCHMe2
140 C9H16O 33046-81-0
3-Octen-2-one, 7-methyl-
Me2CHCH2CH2CH=CHCOMe
140 C9H16O 35215-49-7
7-Octen-2-one, 6-methyl-
MeCO(CH2)3CHMeCH=CH2
140 C9H16O 52954-47-9
9-Oxabicyclo[6.1.0]nonane, 1-methyl-, cis-
Me
O
140 C9H16O 53966-55-5
1-Pentyne, 3-methyl-3-(1-methylethoxy)-
EtCMe(OPr-i)C≡CH
140 C9H16O 53966-56-6
1-Pentyne, 3-ethoxy-3-ethyl-
Et2COEtC≡CH
140 C9H16O 53966-58-8
2-Isononenal
(C6H13)CH=CHCHO
140 C9H16O 54345-59-4
2-Cyclohexen-1-ol, 2,6,6-trimethyl-
Me
Me
OH
Me

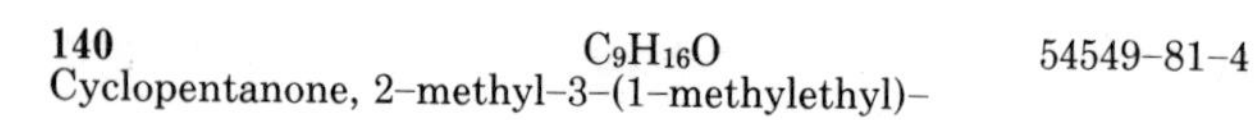

140 $C_9H_{16}O$ 54549-81-4
Cyclopentanone, 2-methyl-3-(1-methylethyl)-

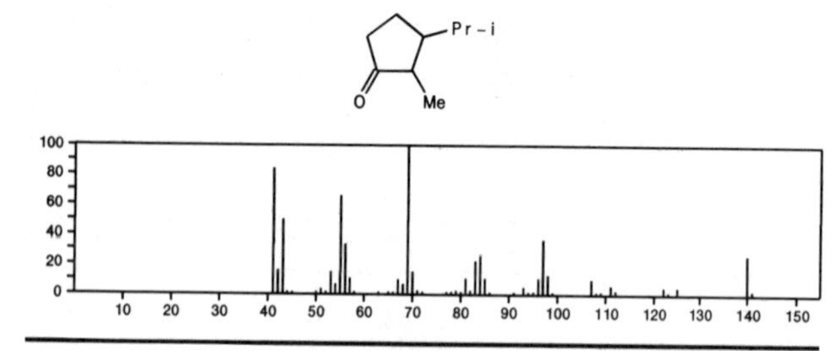

140 $C_9H_{16}O$ 56259-14-4
4-Hepten-3-one, 2,6-dimethyl-

Me2 CHCOCH = CHCHMe2

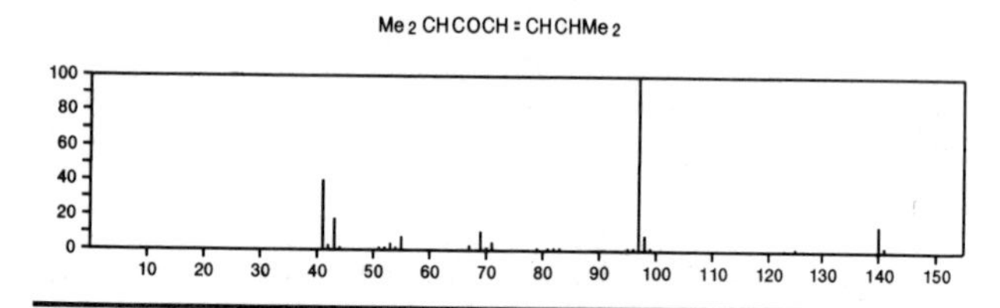

140 $C_9H_{16}O$ 56805-23-3
3,6-Nonadien-1-ol, (*E*,*Z*)-

Et CH = CHCH2 CH = CHCH2 CH2 OH

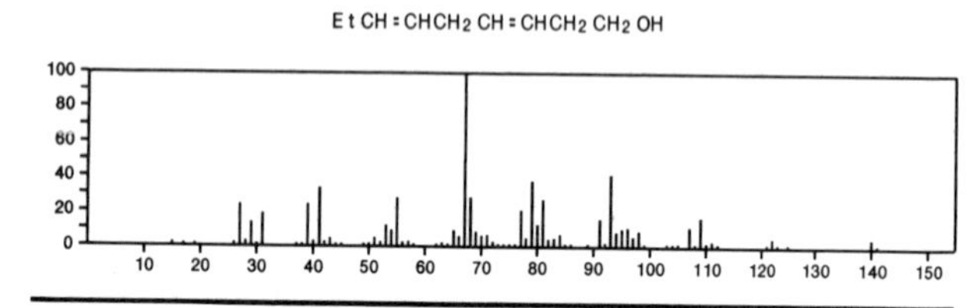

140 $C_{10}H_{20}$ 692-47-7
3-Hexene, 2,2,5,5-tetramethyl-, (*Z*)-

Me3 CCH = CHCMe3

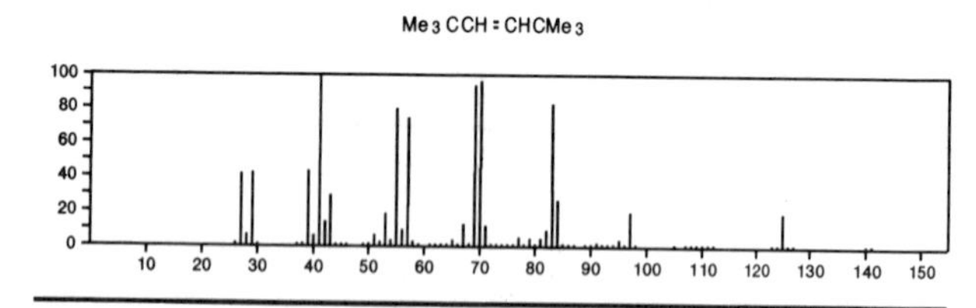

140 $C_{10}H_{20}$ 872-05-9
1-Decene

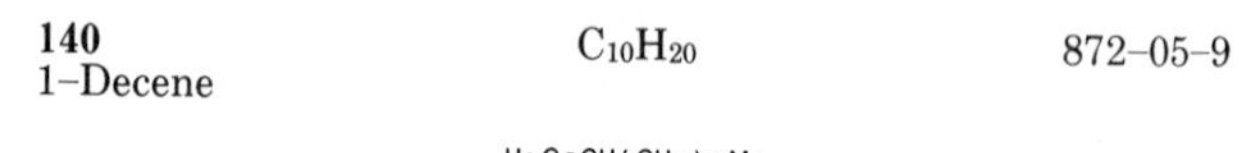

H2 C = CH(CH2) 7 Me

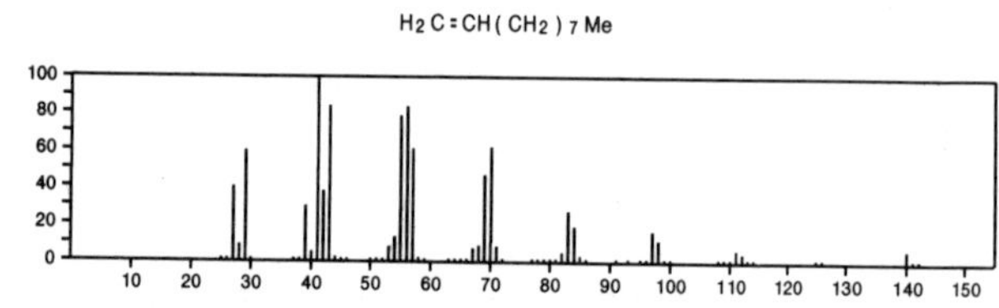

140 $C_{10}H_{20}$ 1331-43-7
Cyclohexane, diethyl-

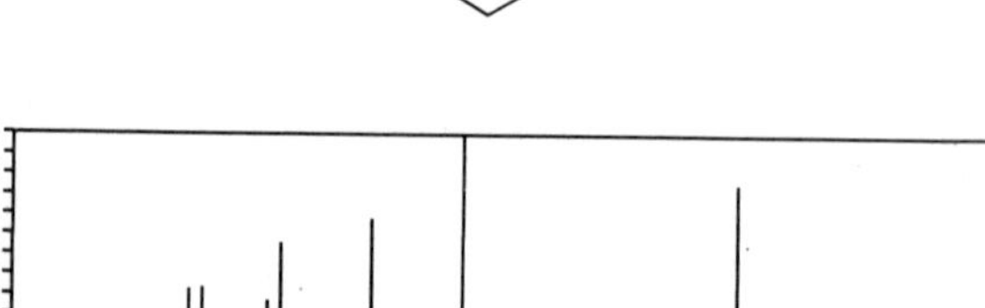

140 $C_{10}H_{20}$ 1678-82-6
Cyclohexane, 1-methyl-4-(1-methylethyl)-, *trans*-

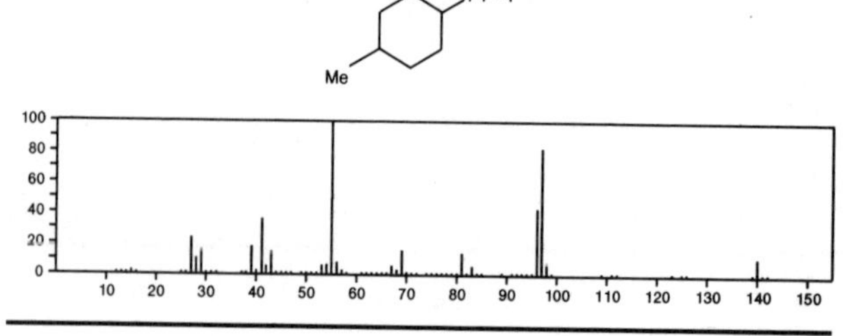

140 $C_{10}H_{20}$ 1678-93-9
Cyclohexane, butyl-

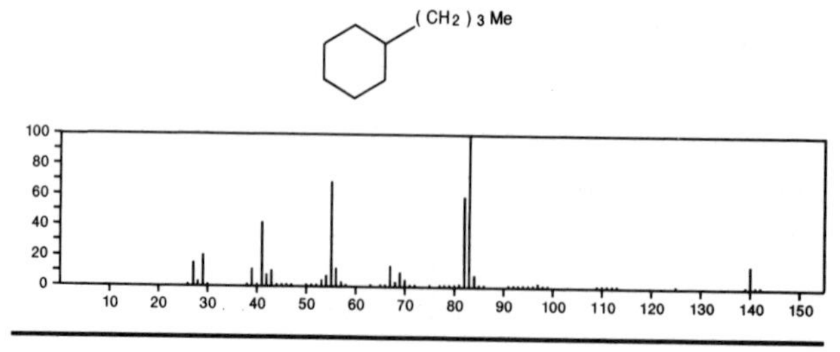

140 $C_{10}H_{20}$ 1678-98-4
Cyclohexane, (2-methylpropyl)-

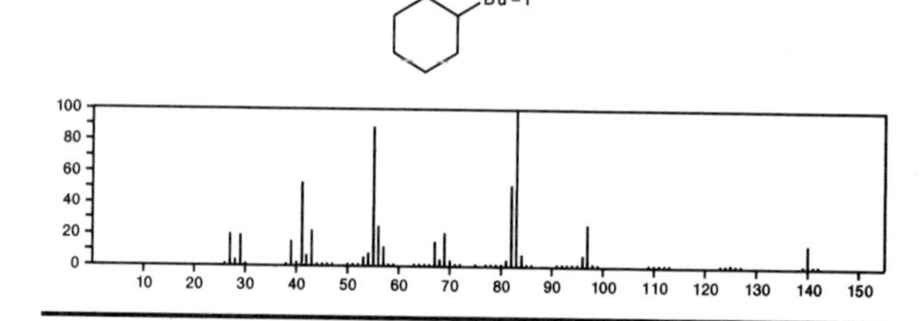

140 $C_{10}H_{20}$ 3178-22-1
Cyclohexane, (1,1-dimethylethyl)-

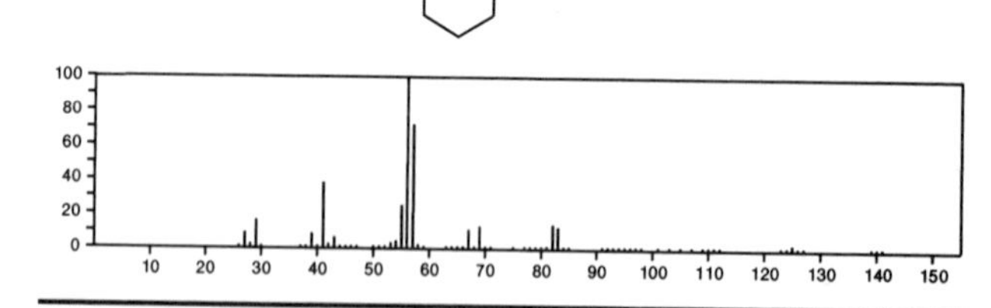

140 $C_{10}H_{20}$ 4485-13-6
3-Heptene, 4-propyl-

Et CH = CPr 2

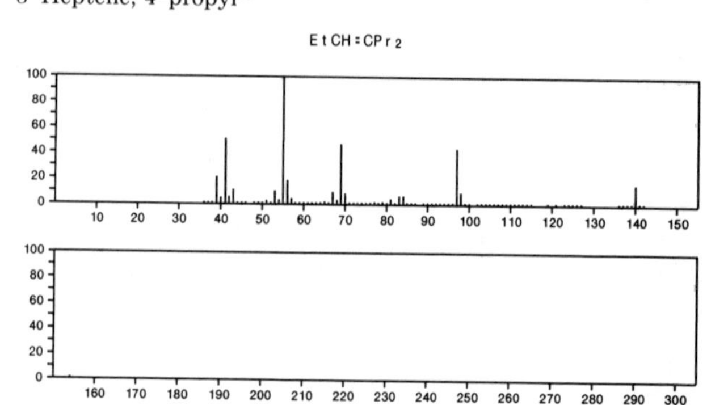

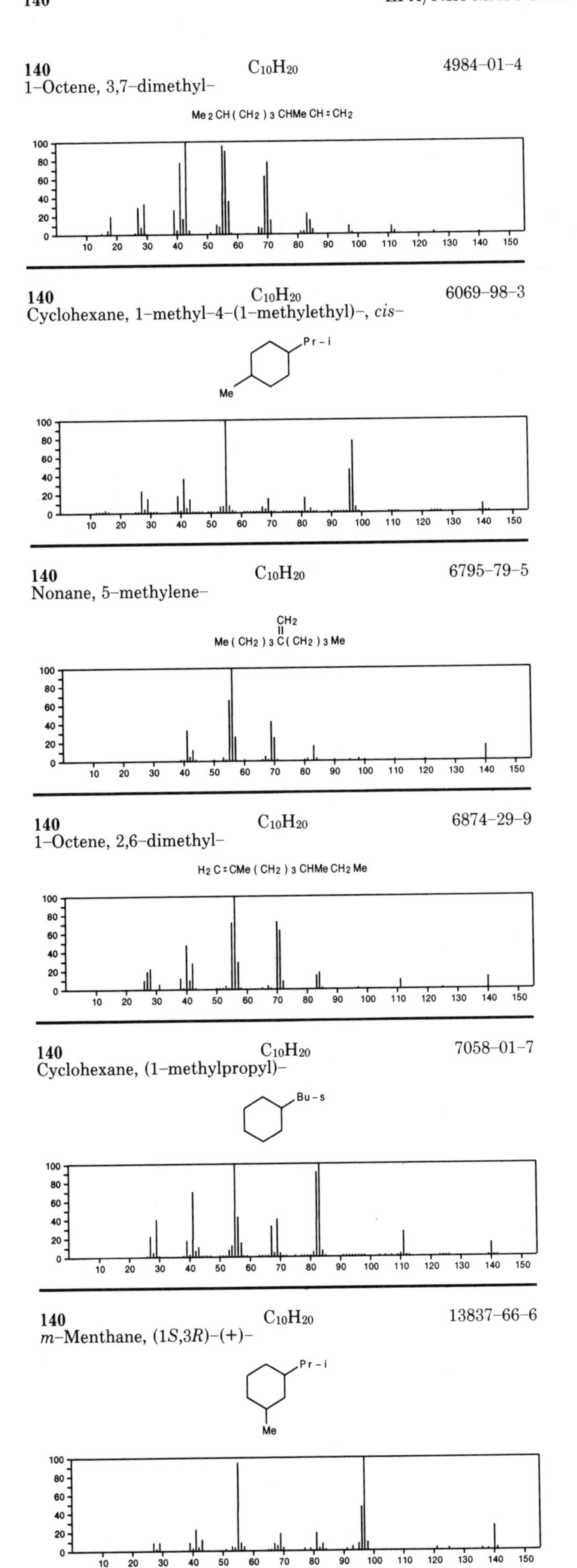

140 $C_{10}H_{20}$ 4984-01-4
1-Octene, 3,7-dimethyl-

140 $C_{10}H_{20}$ 6069-98-3
Cyclohexane, 1-methyl-4-(1-methylethyl)-, *cis*-

140 $C_{10}H_{20}$ 6795-79-5
Nonane, 5-methylene-

140 $C_{10}H_{20}$ 6874-29-9
1-Octene, 2,6-dimethyl-

140 $C_{10}H_{20}$ 7058-01-7
Cyclohexane, (1-methylpropyl)-

140 $C_{10}H_{20}$ 13837-66-6
m-Menthane, (1*S*,3*R*)-(+)-

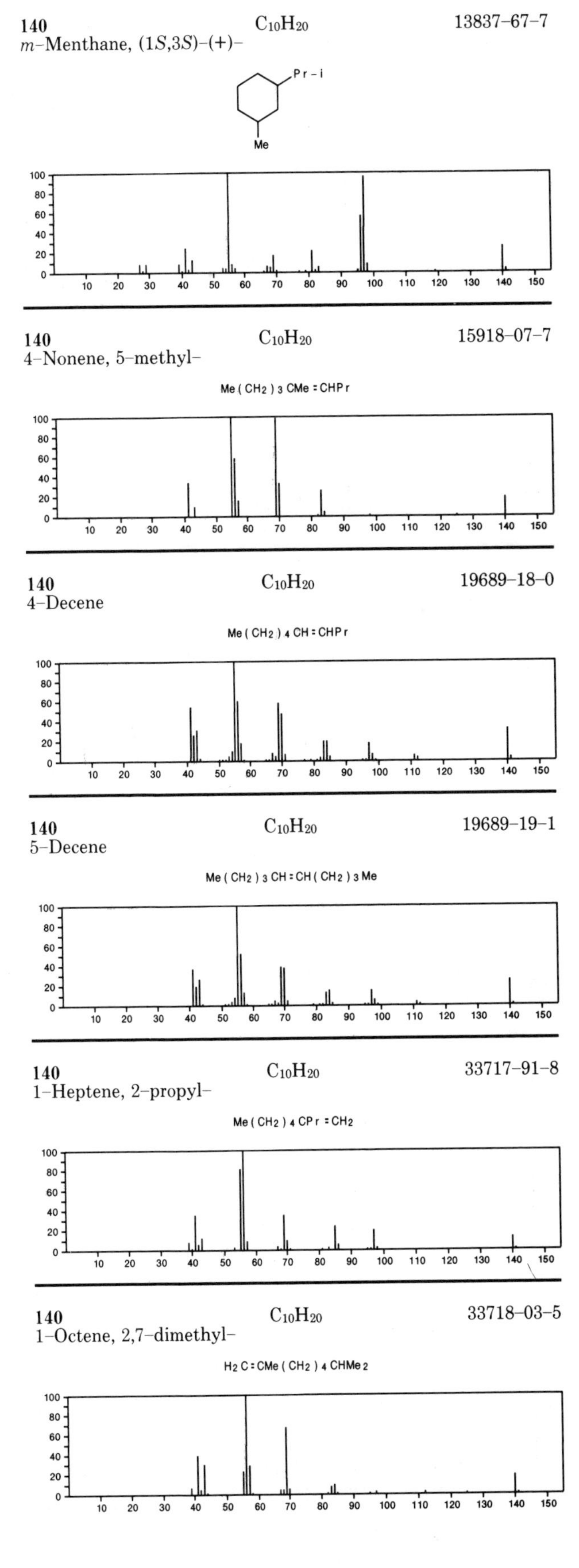

140 $C_{10}H_{20}$ 13837-67-7
m-Menthane, (1*S*,3*S*)-(+)-

140 $C_{10}H_{20}$ 15918-07-7
4-Nonene, 5-methyl-

140 $C_{10}H_{20}$ 19689-18-0
4-Decene

140 $C_{10}H_{20}$ 19689-19-1
5-Decene

140 $C_{10}H_{20}$ 33717-91-8
1-Heptene, 2-propyl-

140 $C_{10}H_{20}$ 33718-03-5
1-Octene, 2,7-dimethyl-

140 $C_{10}H_{20}$ 41977-43-9
Cyclopropane, 1,1,2-trimethyl-3-(2-methylpropyl)-

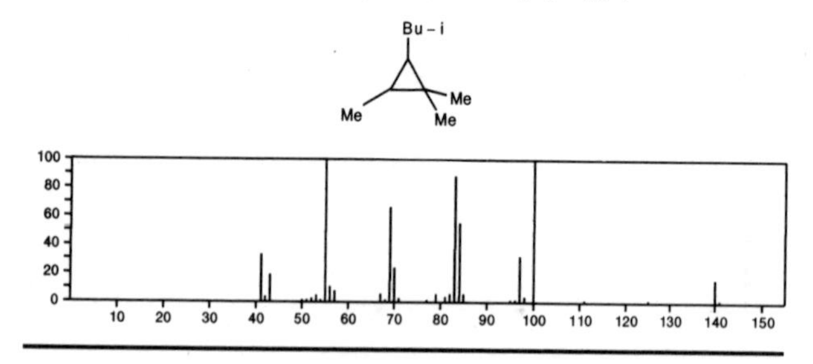

140 $C_{10}H_{20}$ 53966-51-1
3-Octene, 4-ethyl-

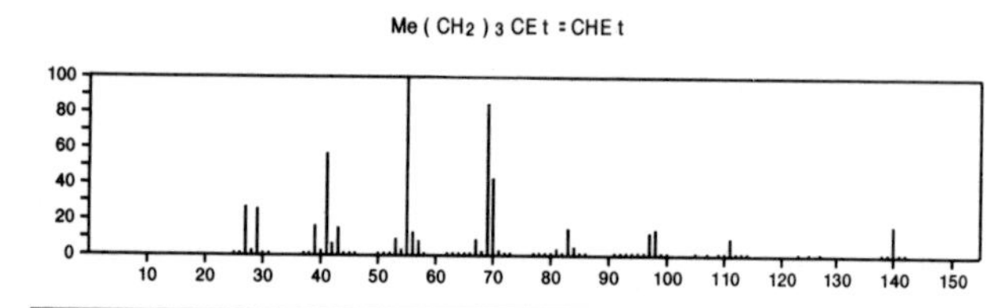

140 $C_{10}H_{20}$ 53966-52-2
2-Octene, 4-ethyl-

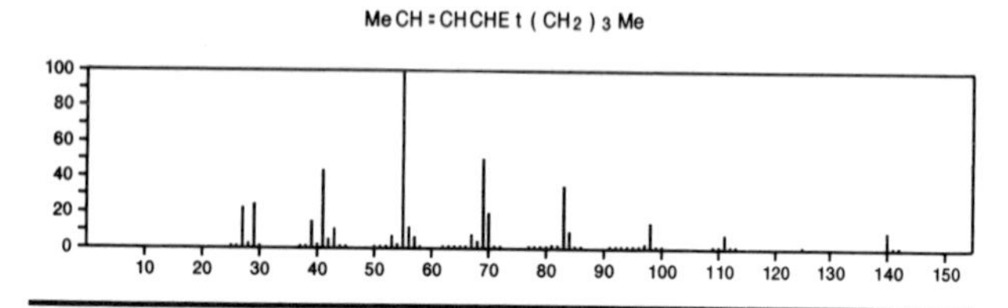

140 $C_{10}H_{20}$ 53966-53-3
3-Nonene, 2-methyl-

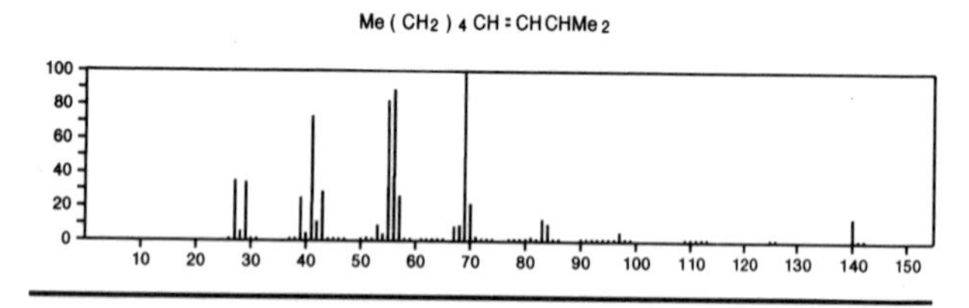

140 $C_{10}H_{20}$ 55724-84-0
4-Nonene, 2-methyl-

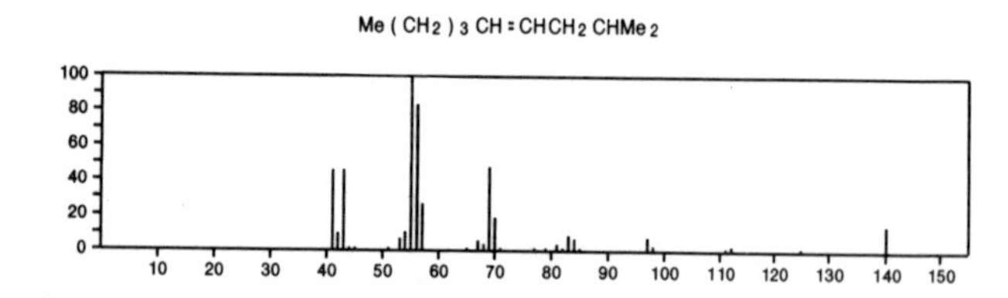

140 $C_{10}H_{20}$ 56728-11-1
1-Octene, 3,4-dimethyl-

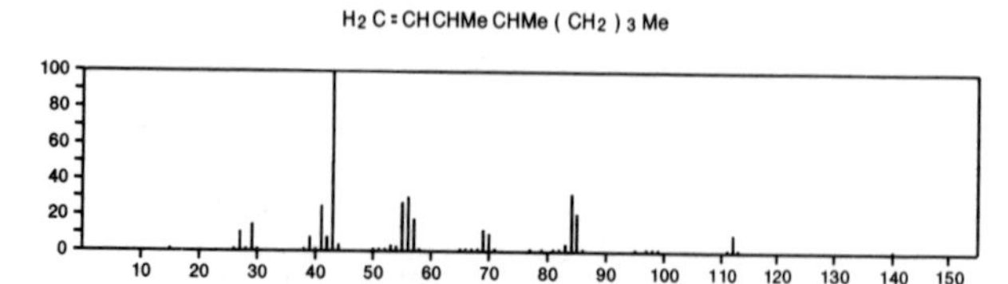

141 $C_4H_3F_4N$ 5522-59-8
Butyronitrile, 2,4,4,4-tetrafluoro-

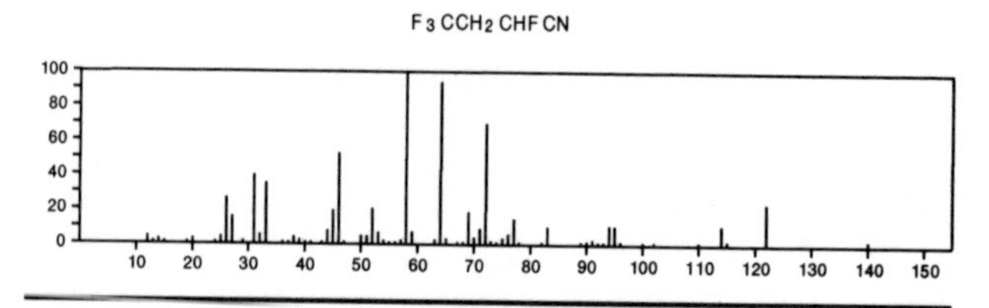

141 $C_4H_6F_3NO$ 1682-66-2
Acetamide, *N*-ethyl-2,2,2-trifluoro-

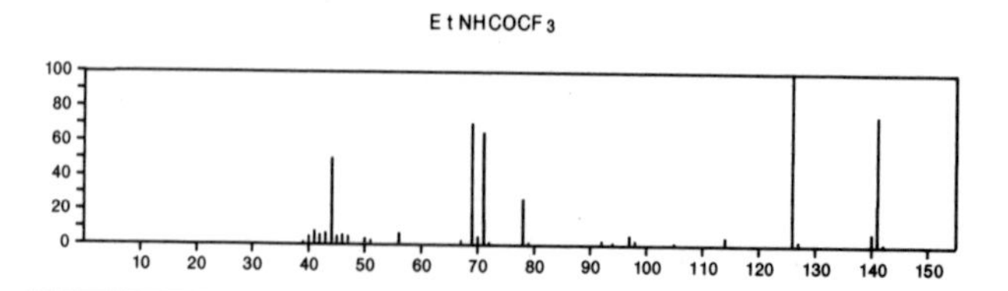

141 $C_5H_7N_3S$ 2183-66-6
4-Pyrimidinamine, 2-(methylthio)-

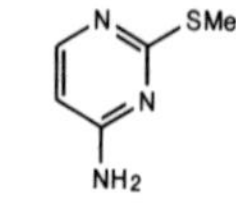

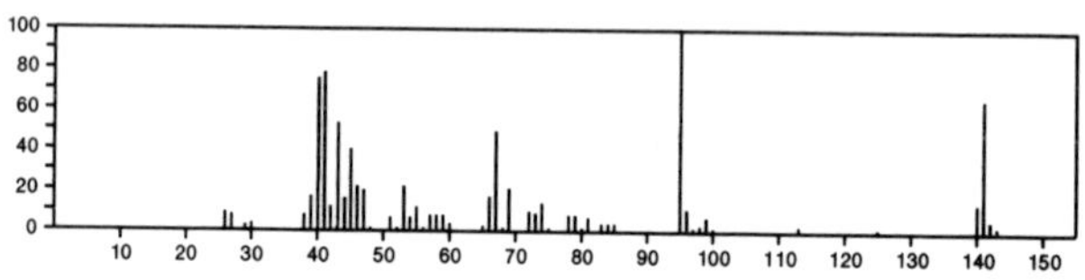

141 $C_6H_4FNO_2$ 350-46-9
Benzene, 1-fluoro-4-nitro-

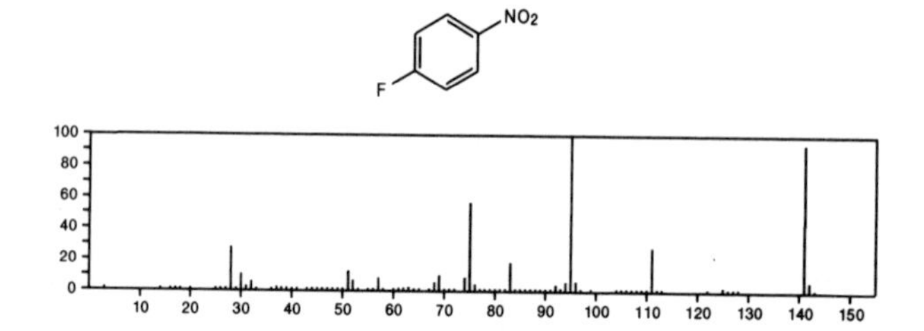

141 $C_6H_4FNO_2$ 402-67-5
Benzene, 1-fluoro-3-nitro-

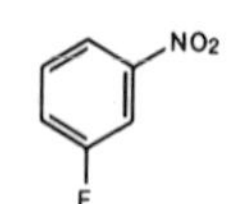

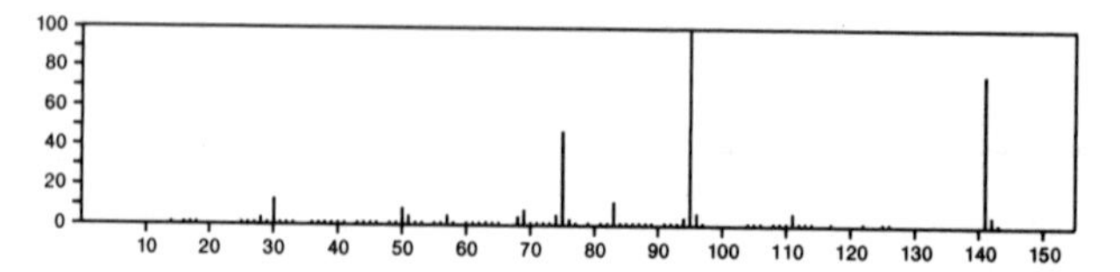

141 C_6H_7NOS 22989-67-9
2(1*H*)-Pyridinethione, 3-hydroxy-6-methyl-

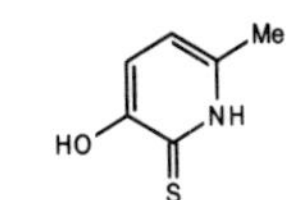

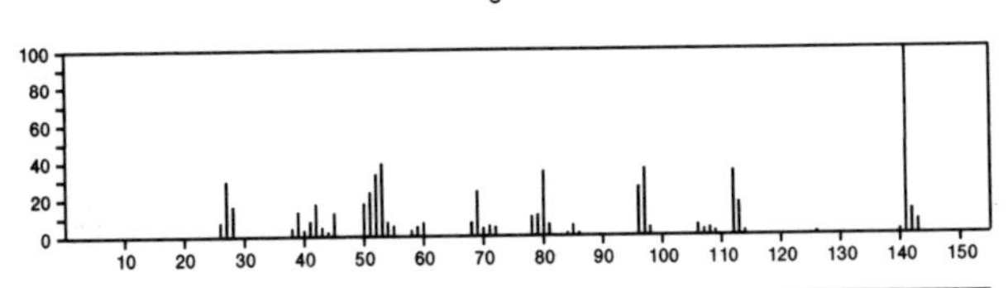

141 C_6H_7NOS 32637-37-9
3-Pyridinol, 2-(methylthio)-

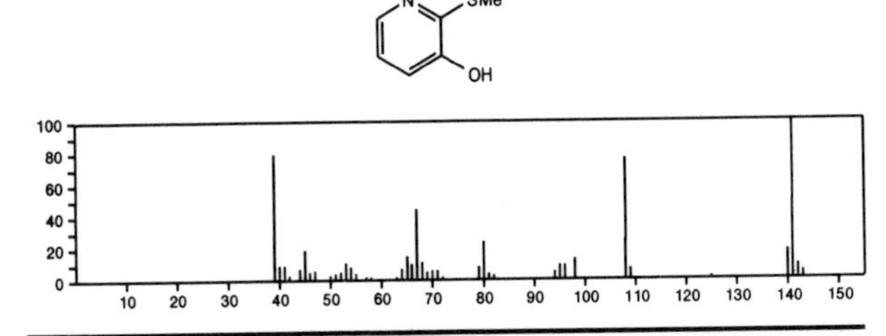

141 $C_6H_7NO_3$ 19788-35-3
3-Isoxazolecarboxylic acid, 5-methyl-, methyl ester

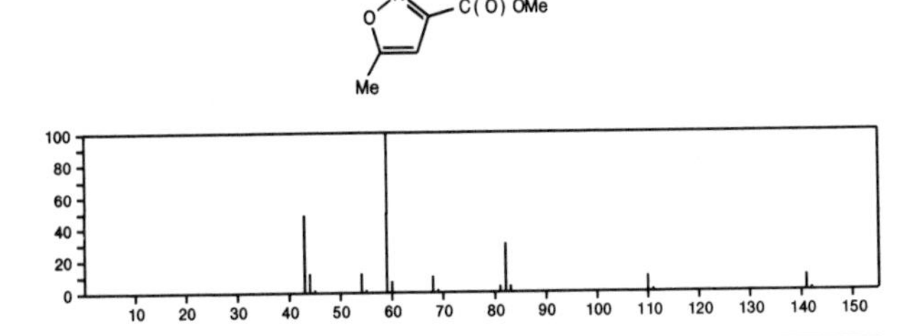

141 $C_6H_{11}N_3O$ 5459-00-7
Hydrazinecarboxamide, 2-cyclopentylidene-

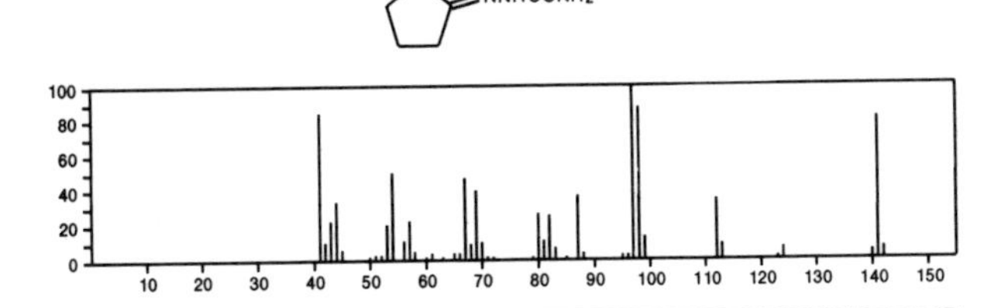

141 C_7H_8ClN 615-65-6
Benzenamine, 2-chloro-4-methyl-

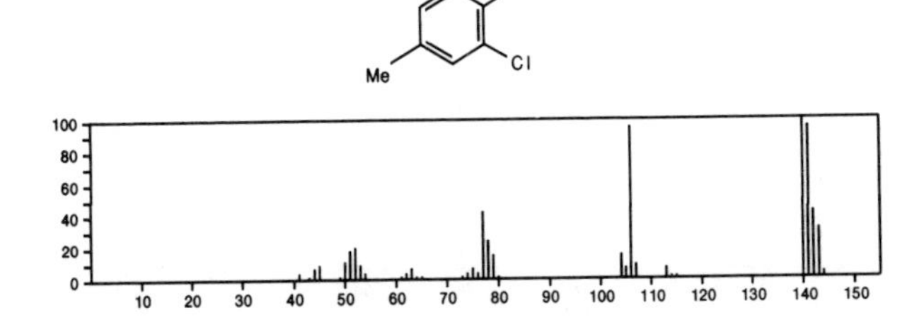

141 $C_7H_{11}NO_2$ 77-67-8
2,5-Pyrrolidinedione, 3-ethyl-3-methyl-

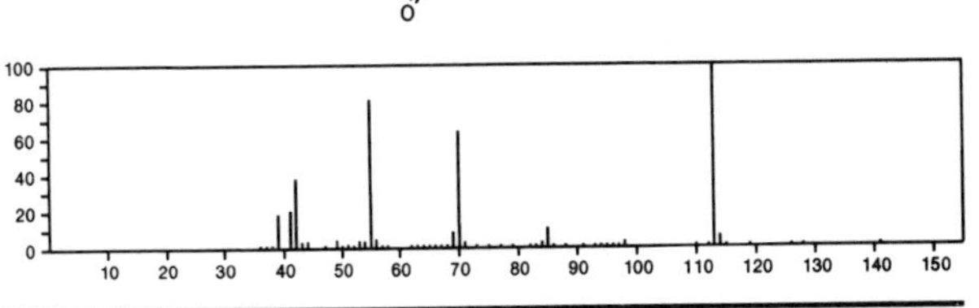

141 $C_7H_{11}NO_2$ 1123-40-6
2,6-Piperidinedione, 4,4-dimethyl-

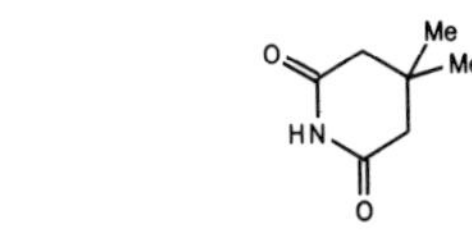

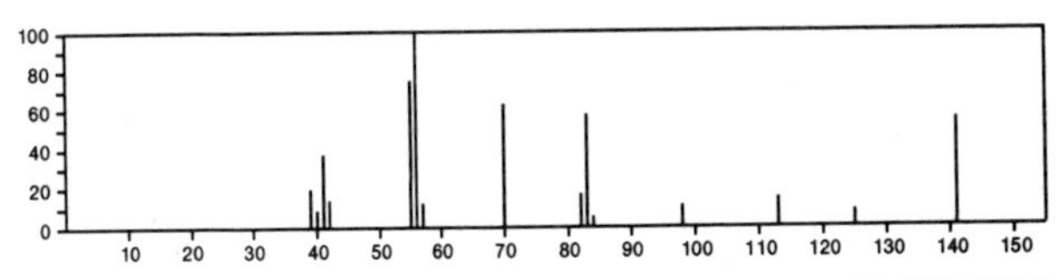

141 $C_7H_{11}NO_2$ 3009-88-9
Pentanoic acid, 5-cyano-, methyl ester

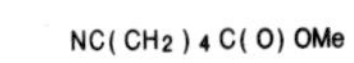

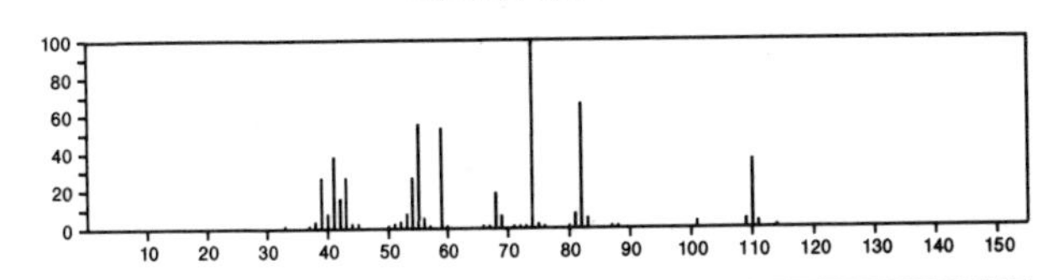

141 $C_7H_{11}NO_2$ 3470-97-1
2,5-Pyrrolidinedione, 1-propyl-

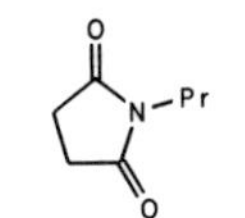

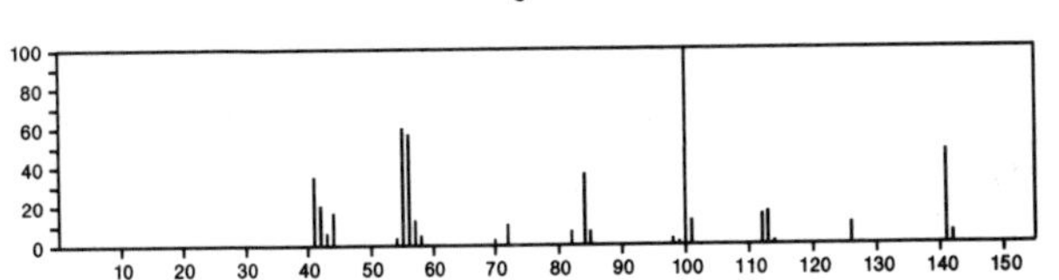

141 $C_7H_{11}NO_2$ 5602-19-7
Hexanoic acid, 6-cyano-

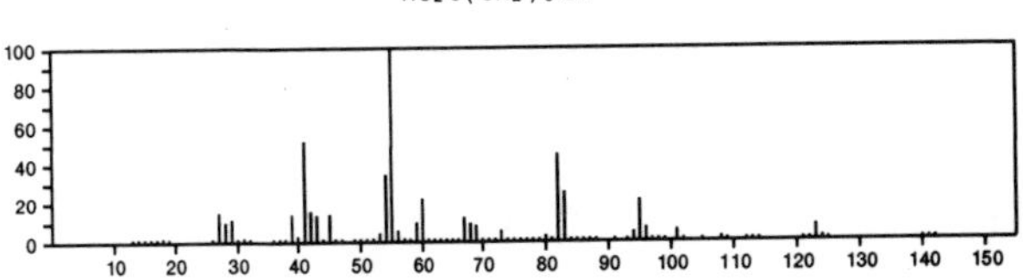

141 $C_7H_{11}NO_2$ 13917–74–3
2*H*–Pyrrol–2–one, 1,5–dihydro–5–methoxy–3,5–dimethyl–

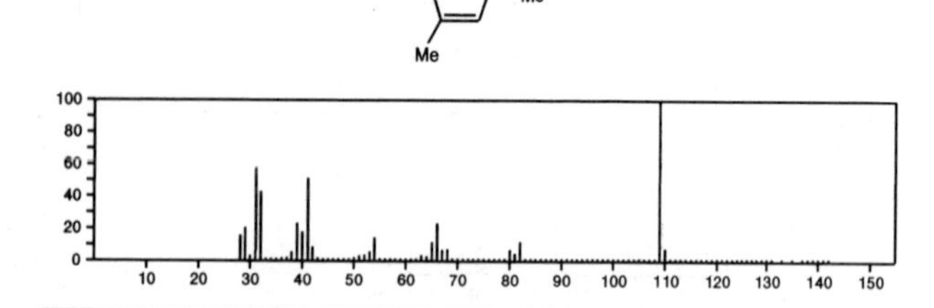

141 $C_7H_{11}NO_2$ 14778–37–1
Diacetamide, *N*–allyl–

H2C=CHCH2N(Ac)2

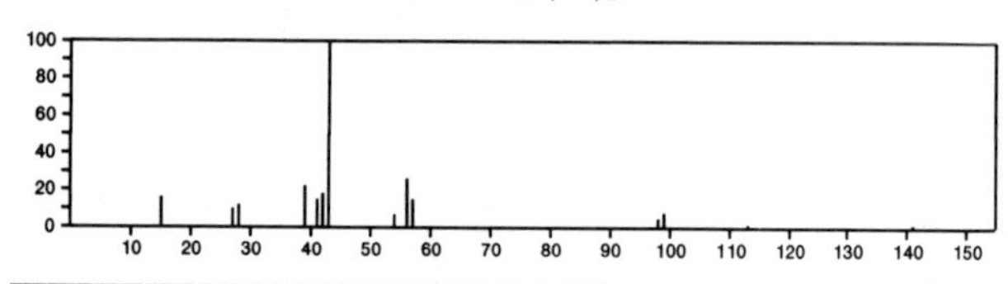

141 $C_7H_{11}NO_2$ 23840–12–2
5–Hexynoic acid, 2–amino–4–methyl–

HO2CCH(NH2)CH2CHMeC≡CH

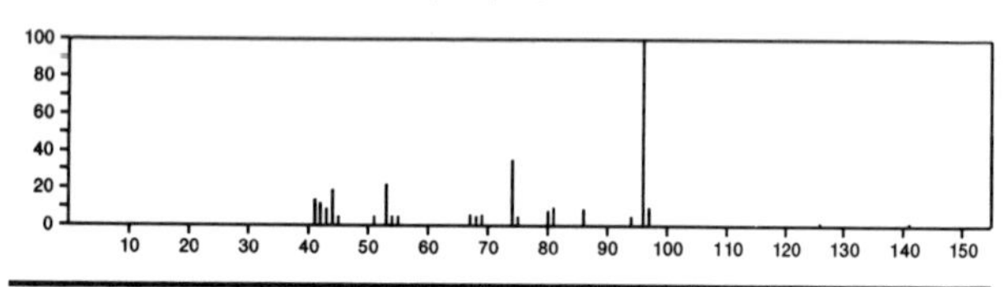

141 $C_7H_{11}NO_2$ 25115–69–9
2,6–Piperidinedione, 3–ethyl–

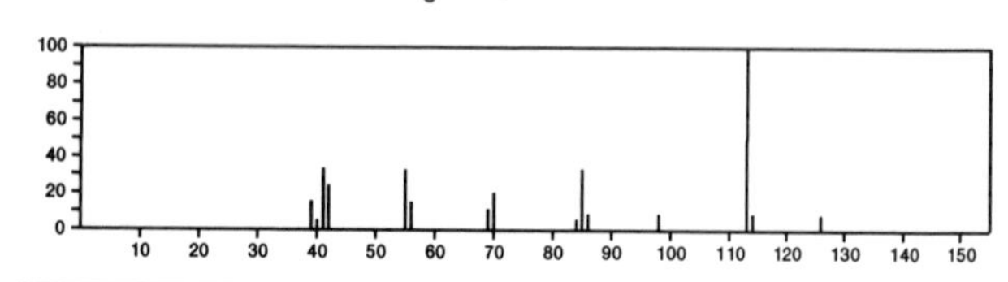

141 $C_7H_{11}NO_2$ 52752–25–7
Butanoic acid, 2–cyano–3–methyl–, methyl ester

Me2CHCH(CN)C(O)OMe

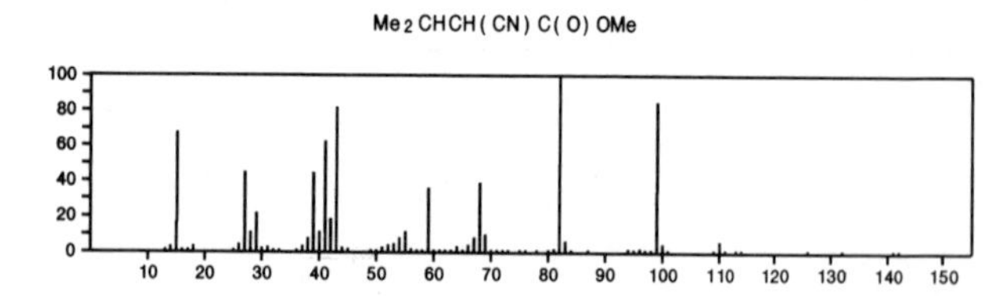

141 $C_7H_{11}NO_2$ 55956–20–2
2–Oxazolidinone, 5–methyl–3–(2–propenyl)–

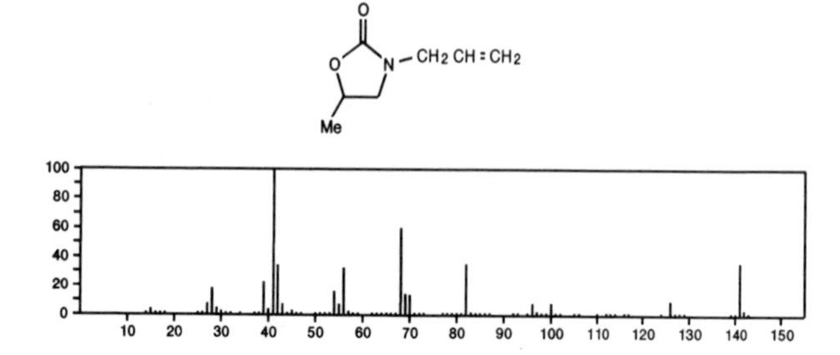

141 $C_7H_{11}NS$ 873–64–3
Thiazole, 2–ethyl–4,5–dimethyl–

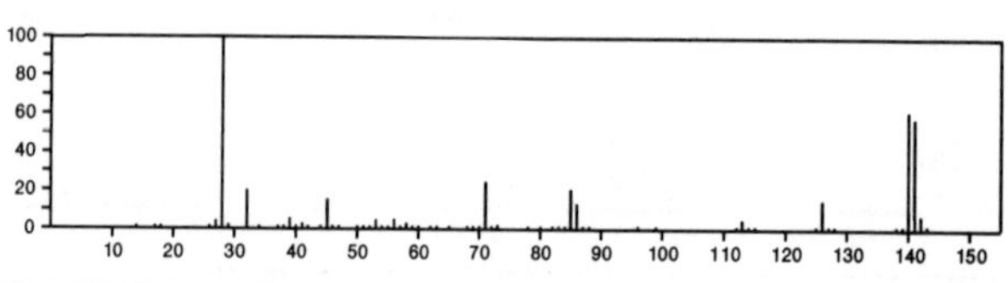

141 $C_7H_{11}NS$ 6081–24–9
Thiazole, 4–(1,1–dimethylethyl)–

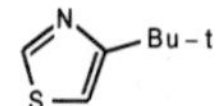

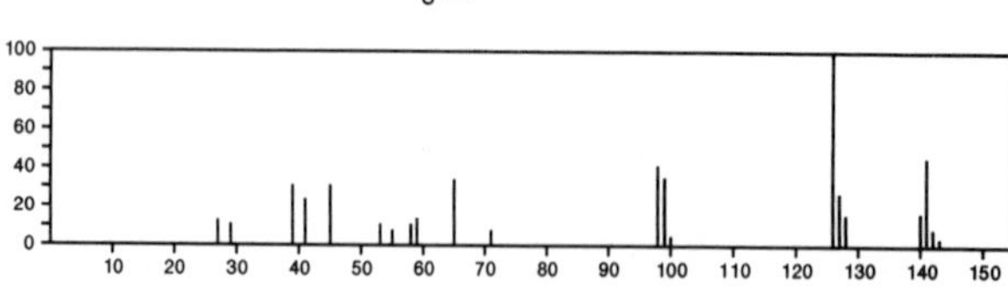

141 $C_7H_{11}NS$ 13623–12–6
Thiazole, 2–(1,1–dimethylethyl)–

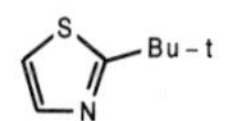

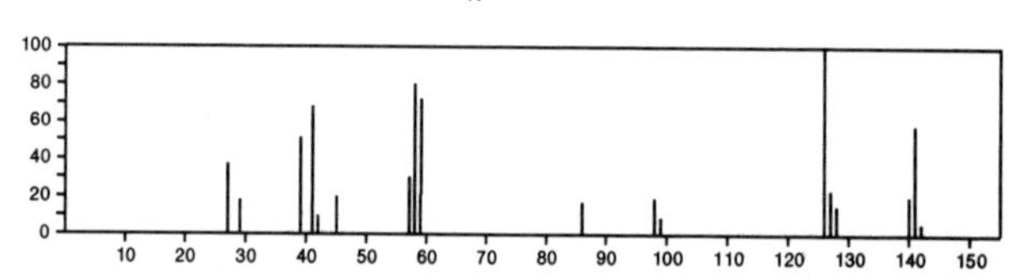

141 $C_7H_{11}NS$ 15679–13–7
Thiazole, 4–methyl–2–(1–methylethyl)–

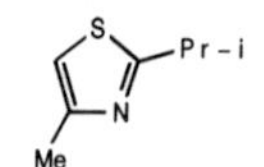

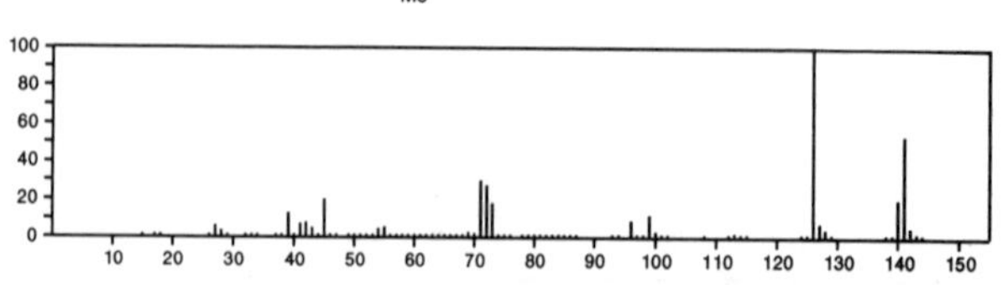

141 $C_7H_{11}NS$ 15729–76–7
Thiazole, 2,5–diethyl–

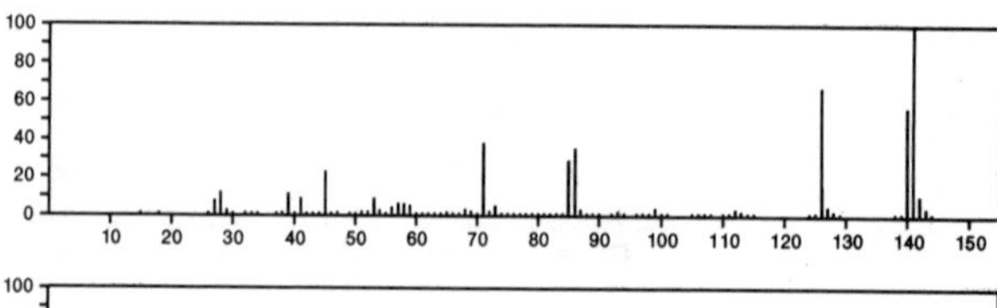

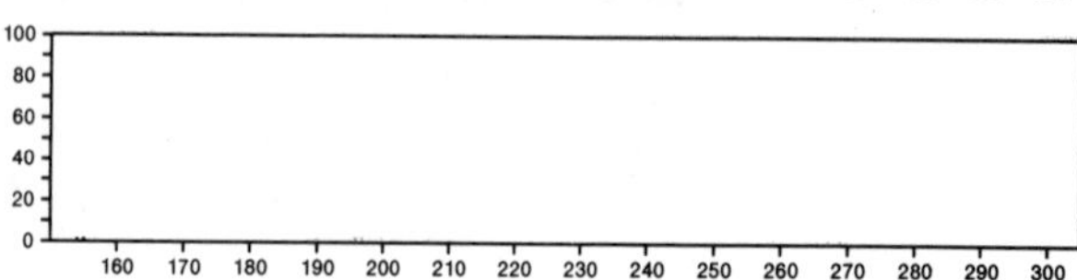

141 $C_7H_{11}NS$ 18640-74-9
Thiazole, 2-(2-methylpropyl)-

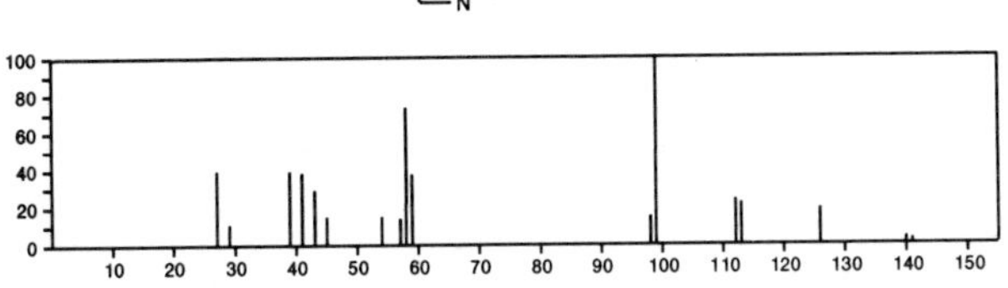

141 $C_7H_{11}NS$ 32272-49-4
Thiazole, 2,4-diethyl-

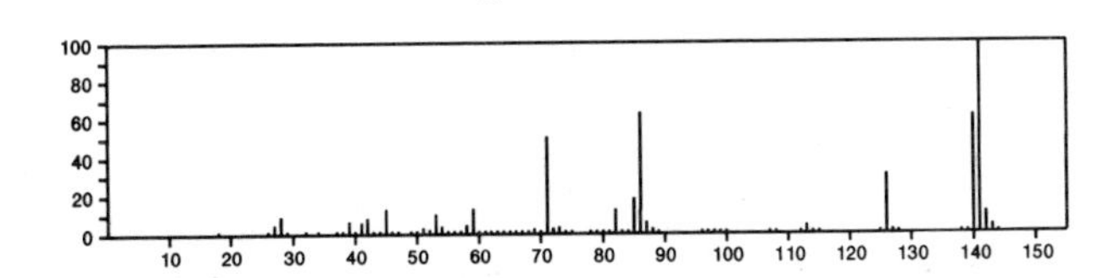

141 $C_7H_{11}NS$ 32272-52-9
Thiazole, 2-methyl-4-(1-methylethyl)-

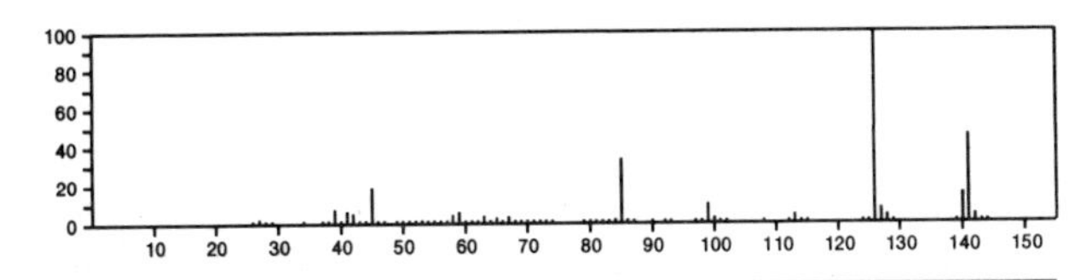

141 $C_7H_{11}NS$ 32272-57-4
Thiazole, 4-ethyl-2,5-dimethyl-

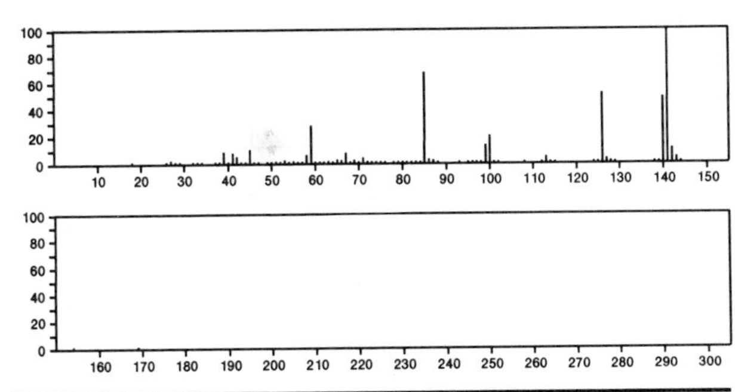

141 $C_7H_{11}NS$ 38205-61-7
Thiazole, 5-ethyl-2,4-dimethyl-

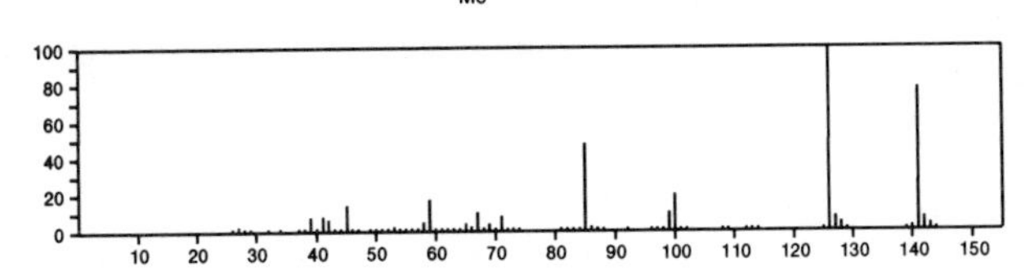

141 $C_7H_{11}NS$ 41981-63-9
Thiazole, 2-methyl-4-propyl-

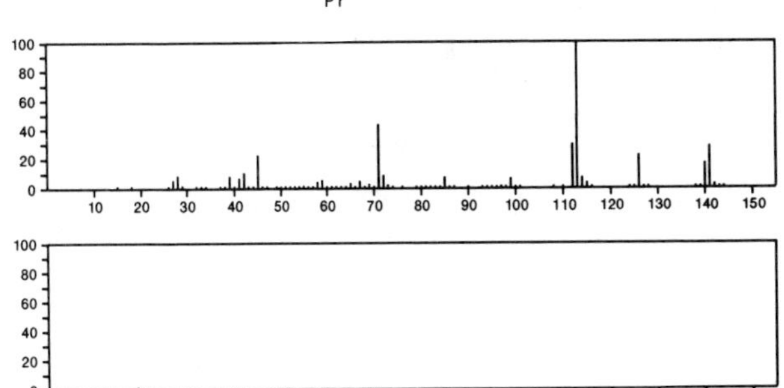

141 $C_7H_{11}NS$ 52414-87-6
Thiazole, 4-methyl-2-propyl-

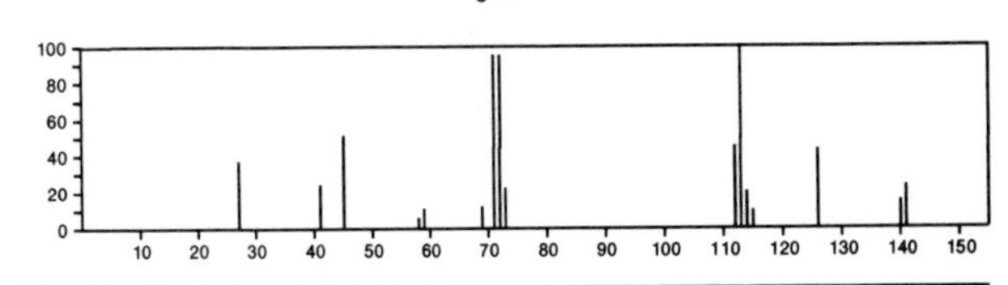

141 $C_7H_{11}NS$ 52414-90-1
Thiazole, 5-butyl-

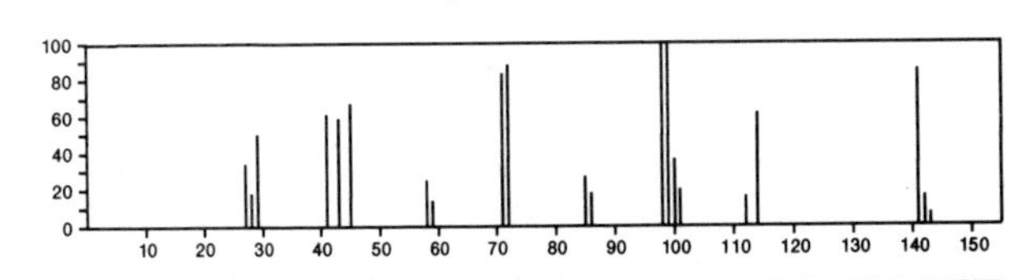

141 $C_7H_{11}NS$ 54031-27-5
2-Propanethione, 1-(2-pyrrolidinylidene)-

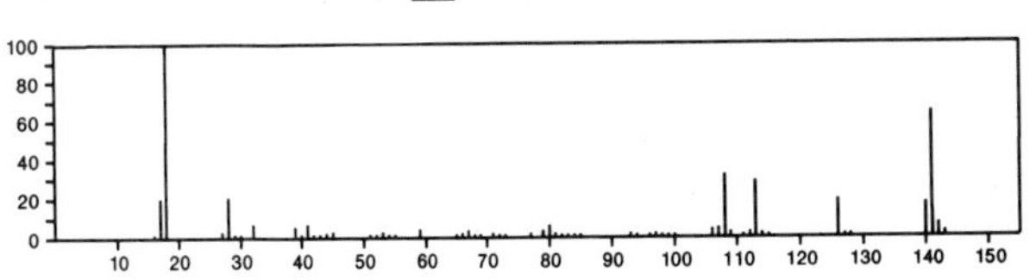

141 $C_7H_{15}N_3$ 27126-22-3
Heptane, 4-azido-

Pr_2CHN_3

141 $C_7H_{15}N_3$ 51677–41–9
Butane, 2–azido–2,3,3–trimethyl–

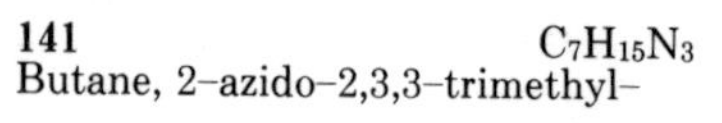

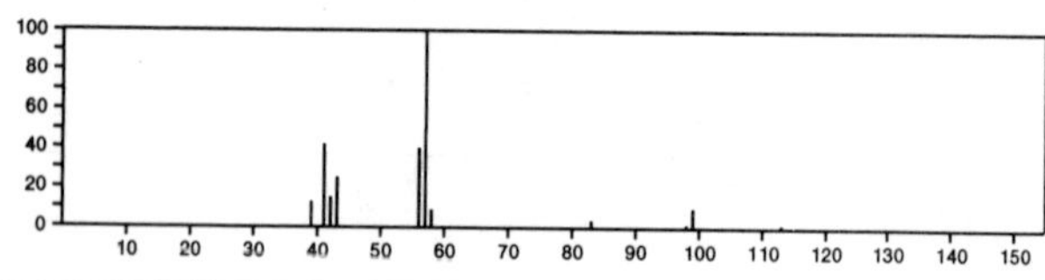

141 $C_7H_{16}BNO$ 24372–00–7
1,3,2–Oxazaborinane, 2–butyl–

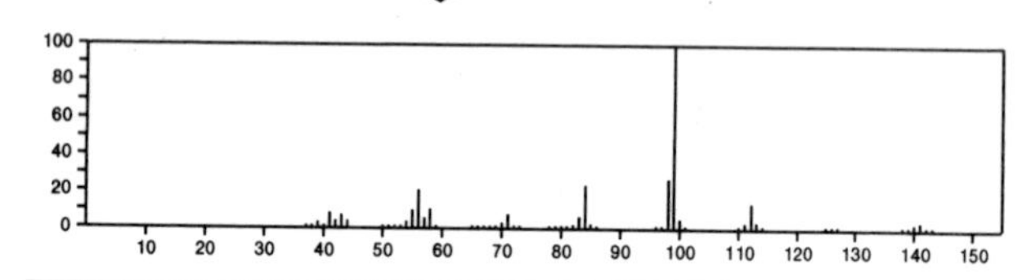

141 $C_7H_{16}BNO$ 31748–11–5
1,3,2–Oxazaborolane, 2–butyl–4–methyl–

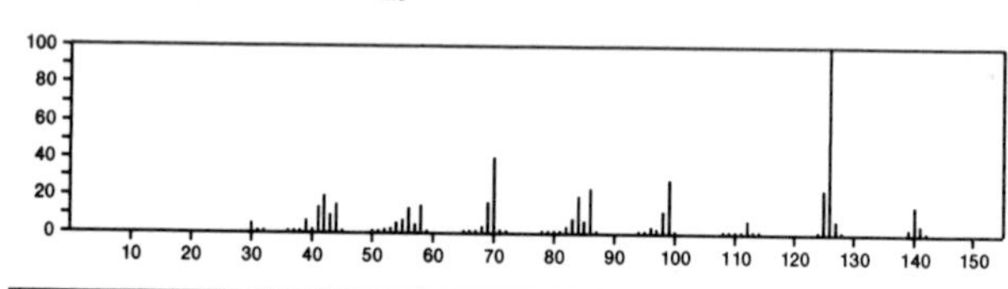

141 $C_8H_{15}NO$ 120–29–6
8–Azabicyclo[3.2.1]octan–3–ol, 8–methyl–, *endo*–

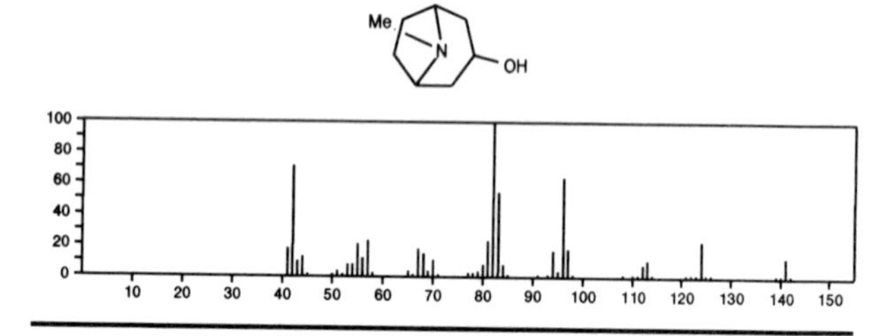

141 $C_8H_{15}NO$ 1074–51–7
Cyclooctanone, oxime

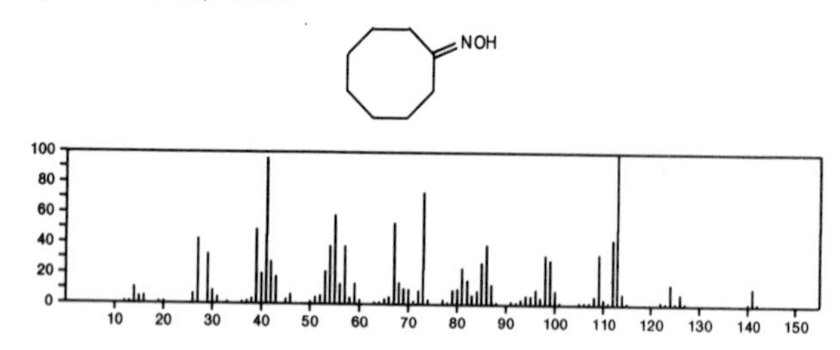

141 $C_8H_{15}NO$ 1124–53–4
Acetamide, *N*–cyclohexyl–

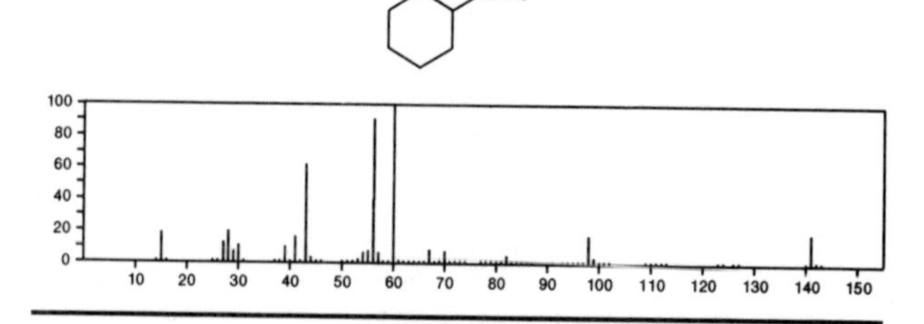

141 $C_8H_{15}NO$ 2403–55–6
Morpholine, 4–(2–methyl–1–propenyl)–

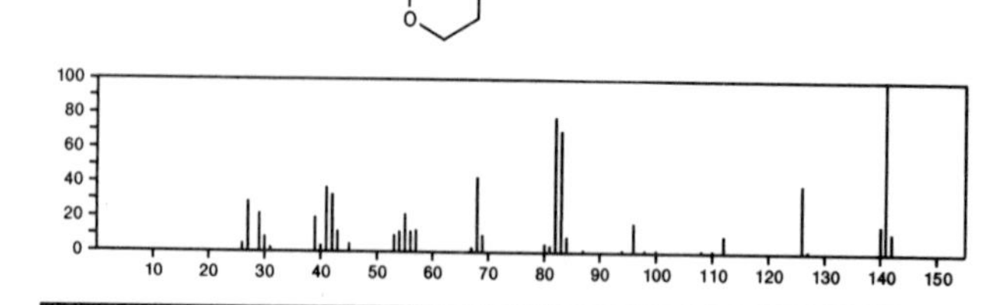

141 $C_8H_{15}NO$ 3376–38–3
Cyclohexanone, *O*–ethyloxime

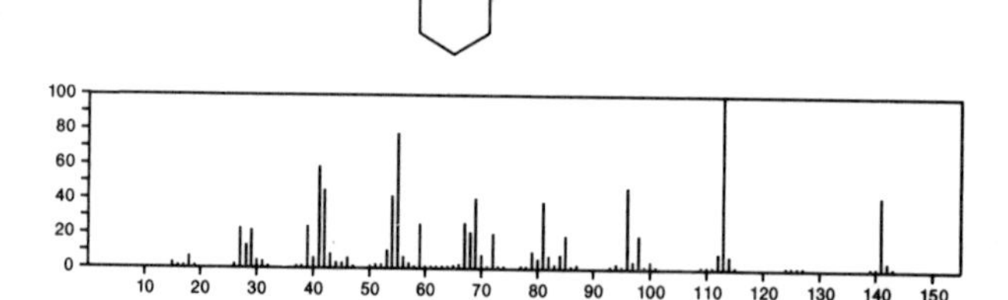

141 $C_8H_{15}NO$ 3470–98–2
2–Pyrrolidinone, 1–butyl–

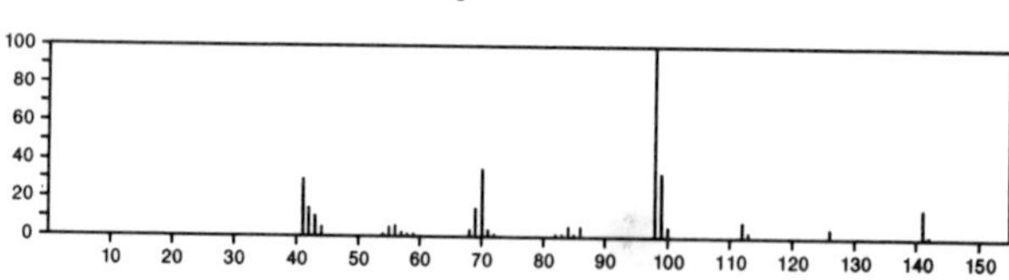

141 $C_8H_{15}NO$ 4396–01–4
2–Propanone, 1–(2–piperidinyl)–

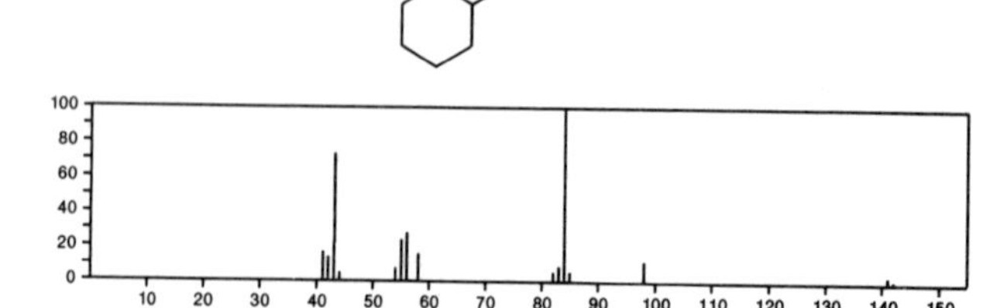

141 $C_8H_{15}NO$ 5809-41-6
1*H*-Azepine, 1-acetylhexahydro-

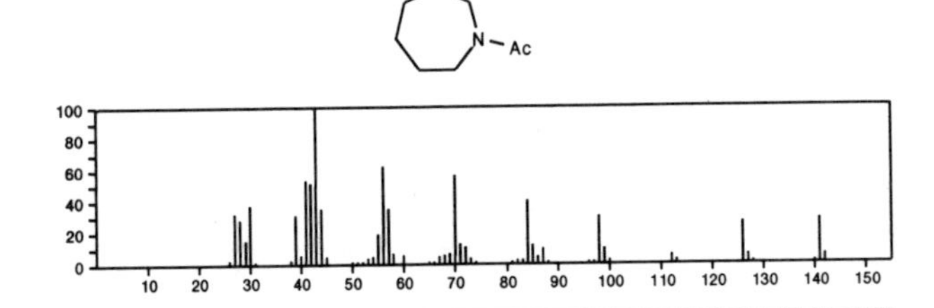

141 $C_8H_{15}NO$ 7432-10-2
8-Azabicyclo[3.2.1]octan-3-ol, 8-methyl-

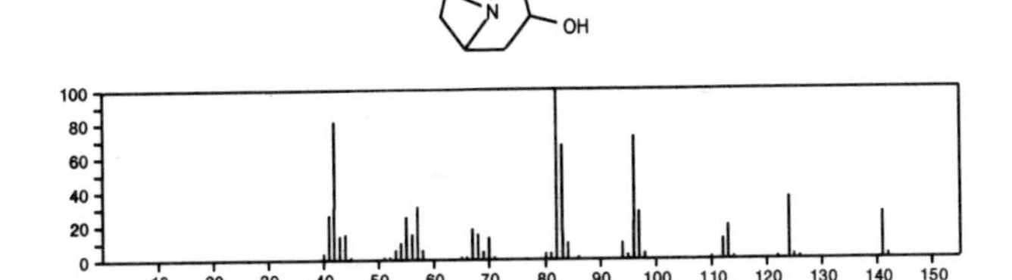

141 $C_8H_{15}NO$ 13493-38-4
3-Azabicyclo[3.2.1]octan-8α-ol, 3-methyl-, *anti*-

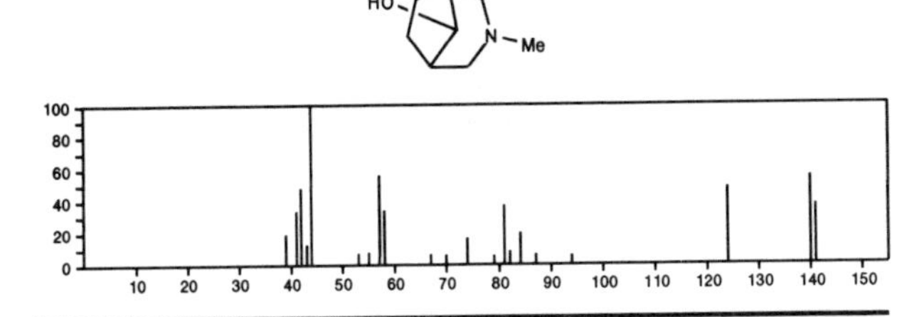

141 $C_8H_{15}NO$ 13493-39-5
3-Azabicyclo[3.2.1]octan-8β-ol, 3-methyl-, *syn*-

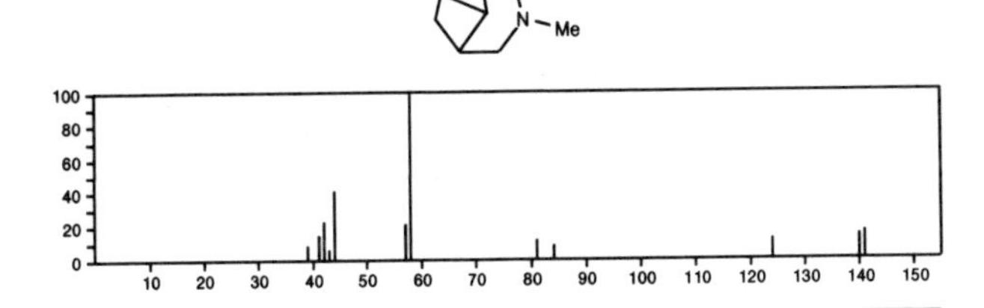

141 $C_8H_{15}NO$ 14387-85-0
Aziridinone, 1-(1,1-dimethylethyl)-3,3-dimethyl-

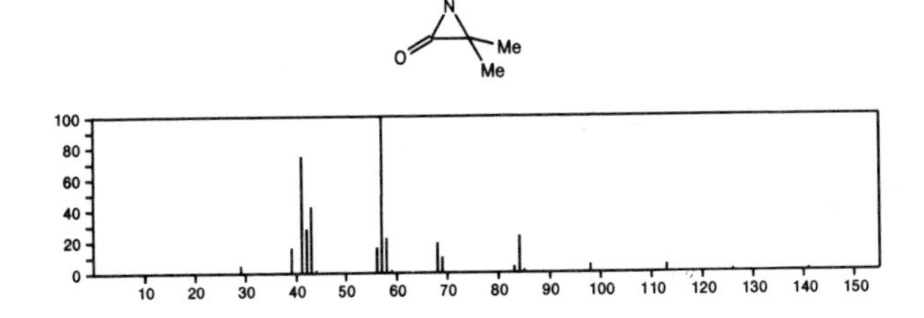

141 $C_8H_{15}NO$ 15431-03-5
Morpholine, 4-(1-butenyl)-

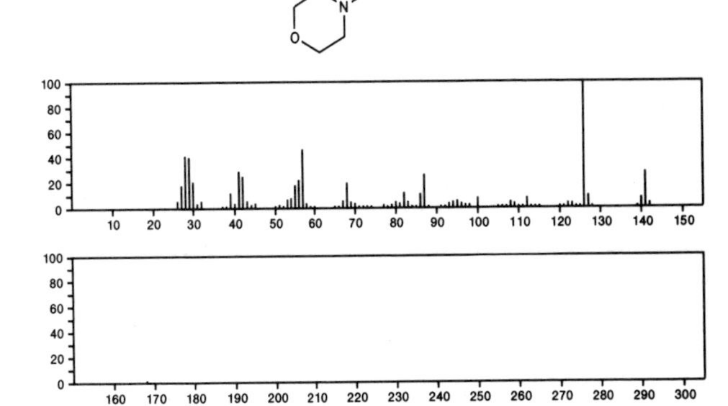

141 $C_8H_{15}NO$ 17719-79-8
Ethanol, 2-(di-2-propenylamino)-

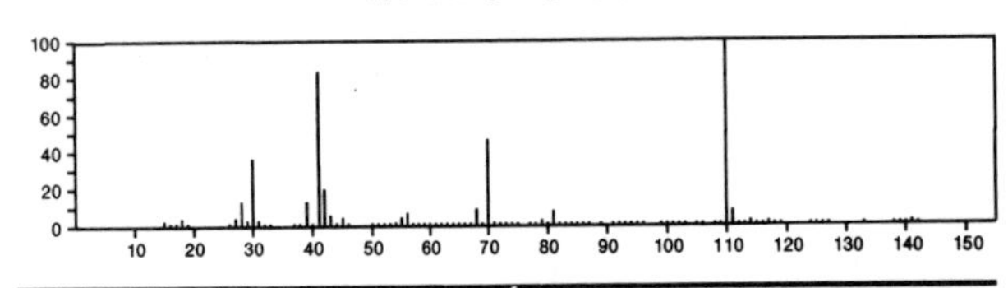

141 $C_8H_{15}NO$ 25943-07-1
2-Oxazoline, 4,5-diethyl-2-methyl-, *cis*-

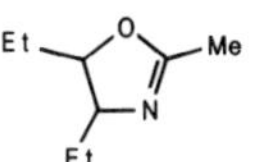

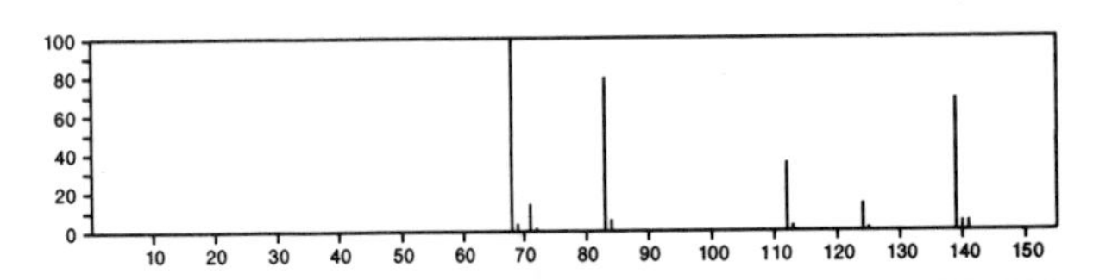

141 $C_8H_{15}NO$ 25943-13-9
2-Oxazoline, 4,5-diethyl-2-methyl-, *trans*-

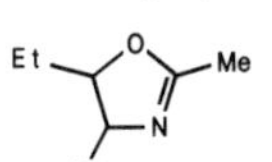

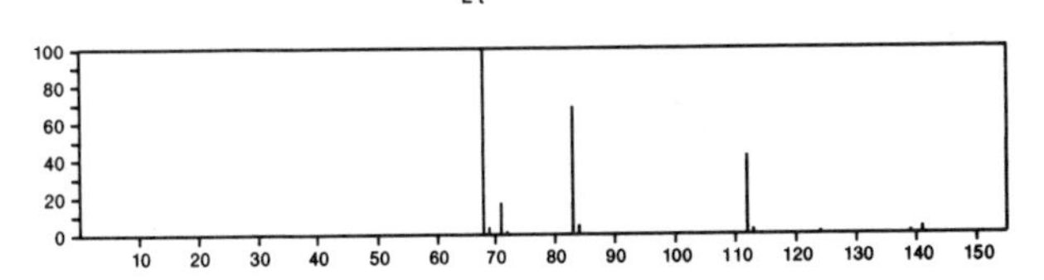

141 $C_8H_{15}NO$ 33527-93-4
Pyrrolidine, 1-(1-oxobutyl)-

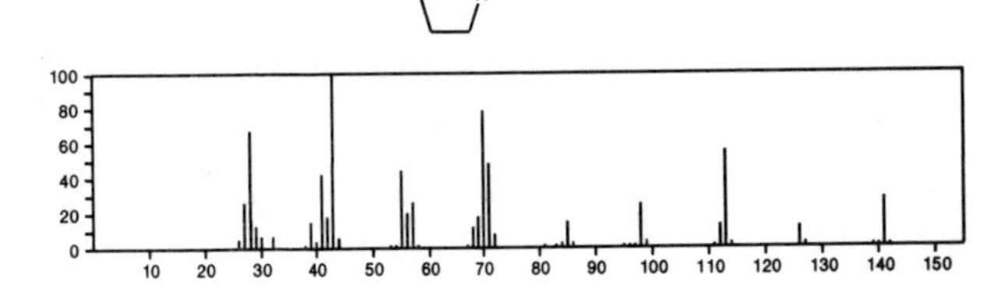

141 $C_8H_{15}NO$ 39209-07-9
Cyclohexanone, 2-methyl-, *O*-methyloxime

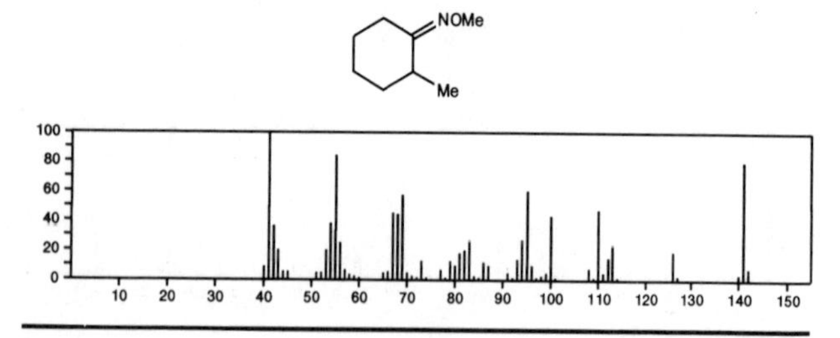

141 $C_8H_{15}NO$ 39477-43-5
Cyclohexanone, 4-methyl-, *O*-methyloxime

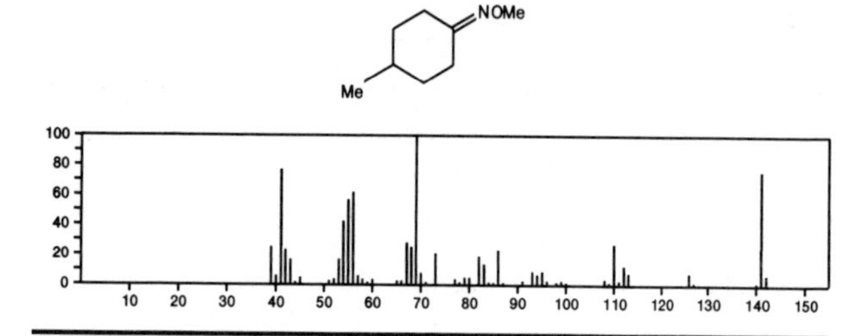

141 $C_8H_{15}NO$ 43152-94-9
3-Piperidinone, 1-ethyl-6-methyl-

141 $C_8H_{15}NO$ 50454-94-9
6-Oxa-1-azabicyclo[3.1.0]hexane, 2,2,4,4-tetramethyl-

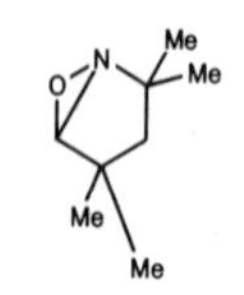

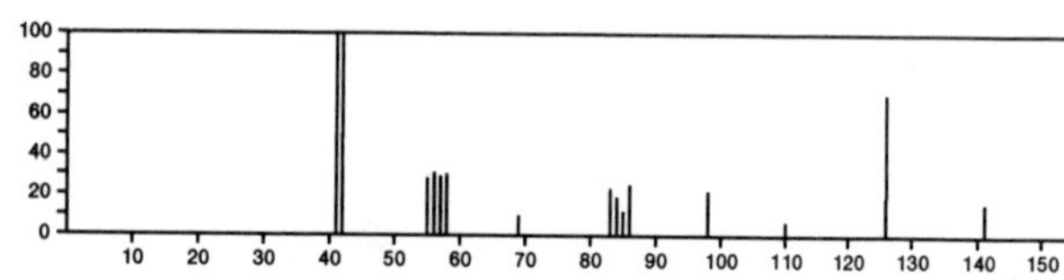

141 $C_8H_{15}NO$ 50455-46-4
1-Azetidinecarboxaldehyde, 2,2,4,4-tetramethyl-

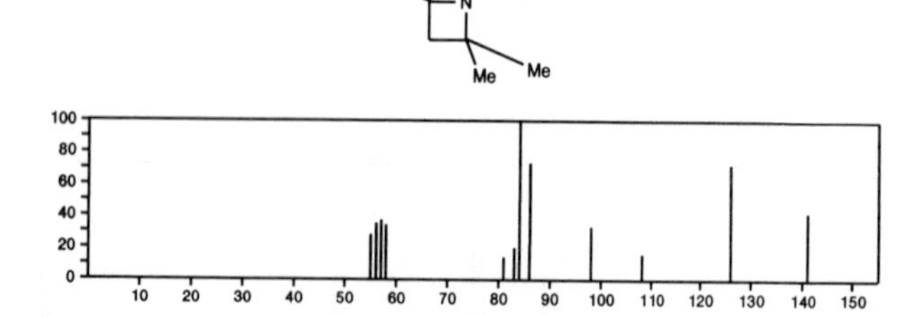

141 $C_8H_{15}NO$ 56336-01-7
3-Hepten-2-one, *O*-methyloxime

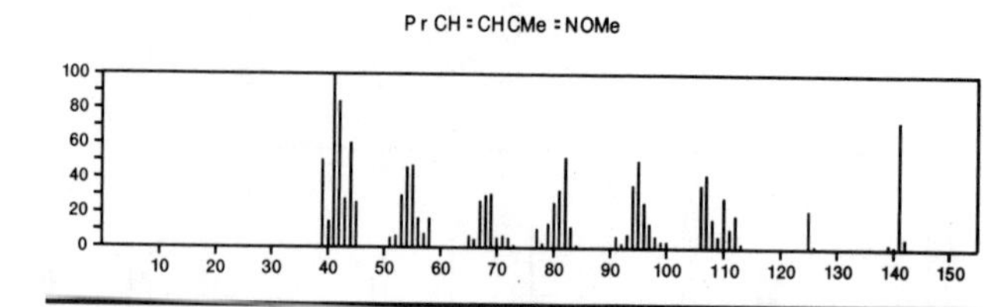

141 $C_8H_{20}BN$ 4023-39-6
Boranamine, *N*,*N*,1,1-tetraethyl-

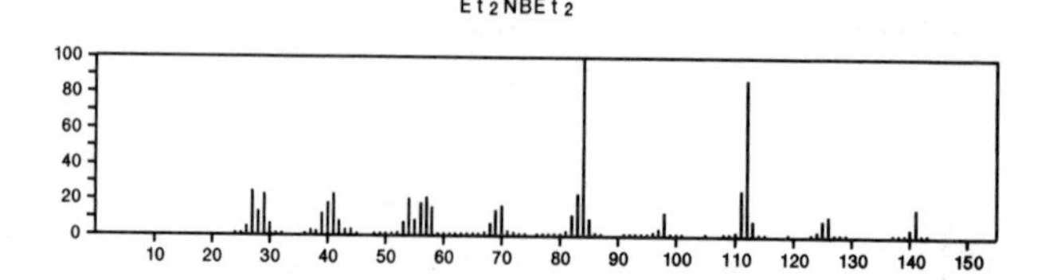

141 $C_9H_{19}N$ 1432-48-0
2-Propanamine, *N*-(2,2-dimethylpropylidene)-2-methyl-

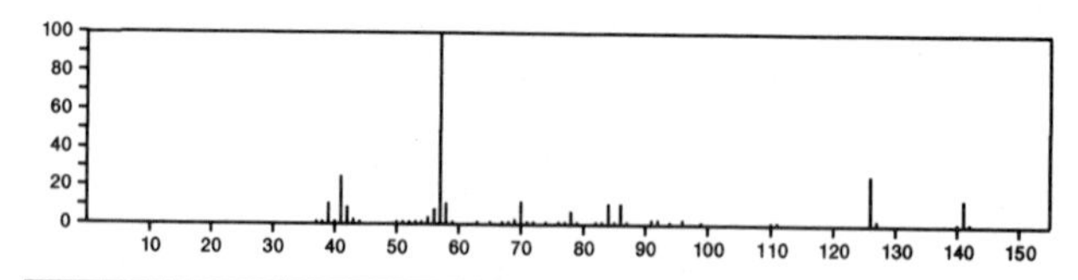

141 $C_9H_{19}N$ 3447-03-8
Pyrrolidine, 2-butyl-1-methyl-

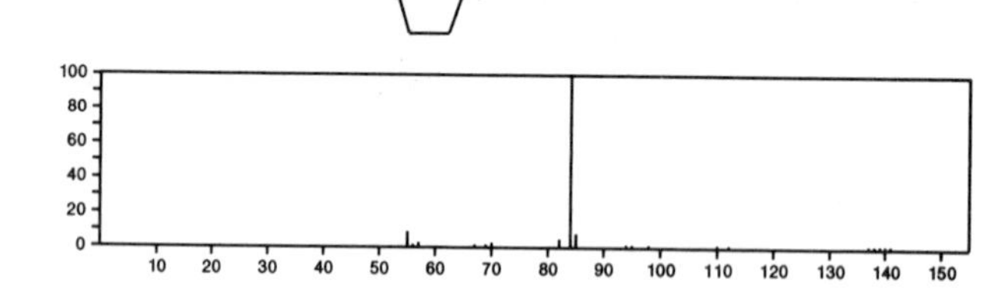

141 $C_9H_{19}N$ 4945-48-6
Piperidine, 1-butyl-

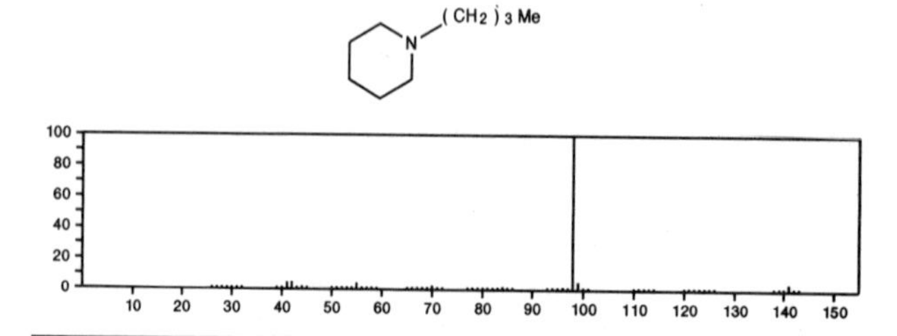

141 $C_9H_{19}N$ 6124-84-1
Aziridine, 1,2-diisopropyl-3-methyl-, *trans*-

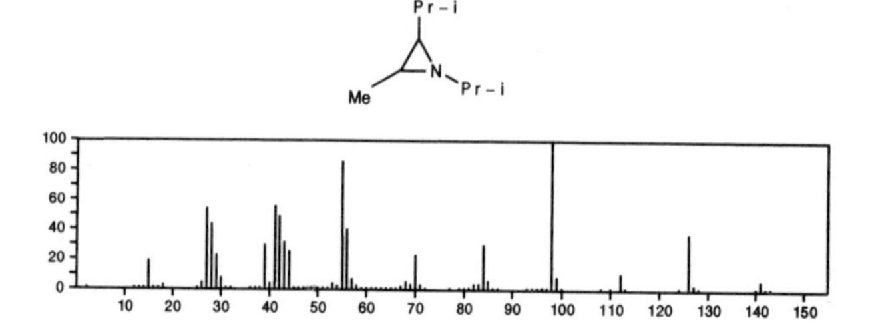

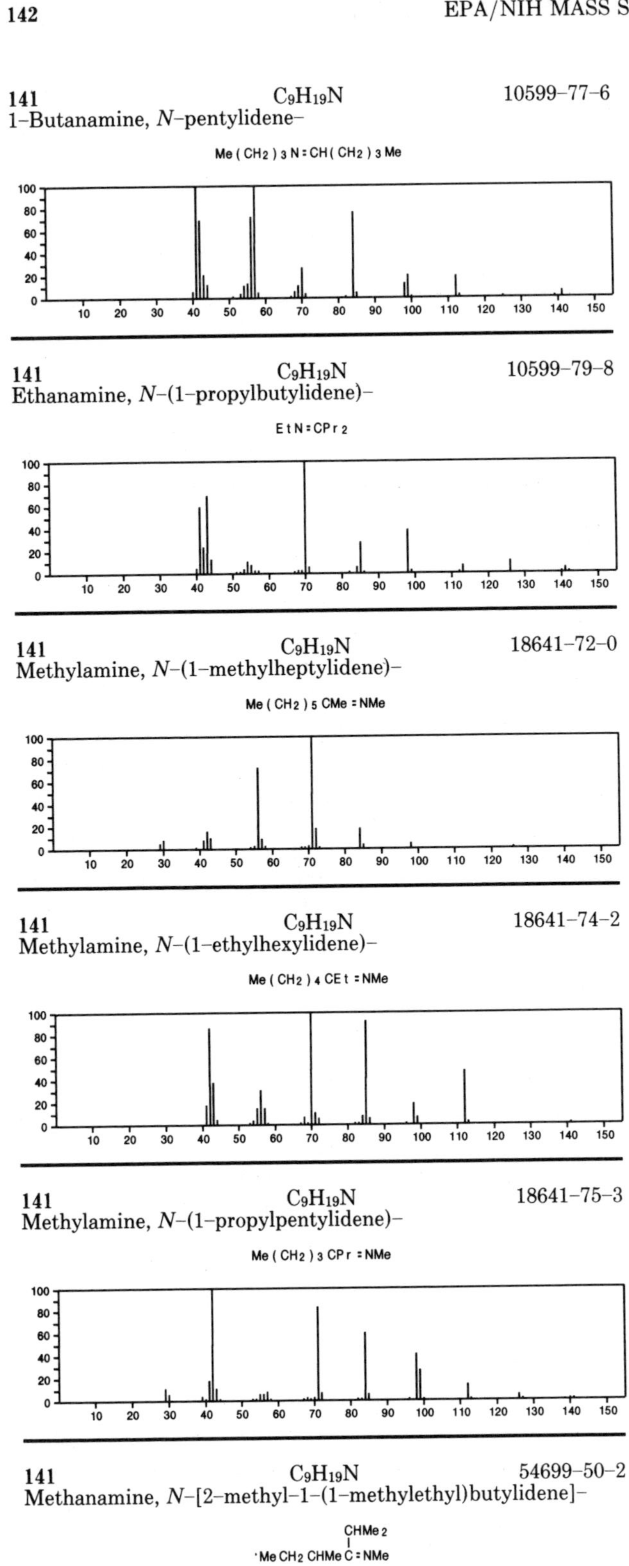

141 $C_9H_{19}N$ 10599-77-6
1-Butanamine, *N*-pentylidene-
Me(CH2)3N=CH(CH2)3Me

141 $C_9H_{19}N$ 10599-79-8
Ethanamine, *N*-(1-propylbutylidene)-
EtN=CPr2

141 $C_9H_{19}N$ 18641-72-0
Methylamine, *N*-(1-methylheptylidene)-
Me(CH2)5CMe=NMe

141 $C_9H_{19}N$ 18641-74-2
Methylamine, *N*-(1-ethylhexylidene)-
Me(CH2)4CEt=NMe

141 $C_9H_{19}N$ 18641-75-3
Methylamine, *N*-(1-propylpentylidene)-
Me(CH2)3CPr=NMe

141 $C_9H_{19}N$ 54699-50-2
Methanamine, *N*-[2-methyl-1-(1-methylethyl)butylidene]-
CHMe2
MeCH2CHMeC=NMe

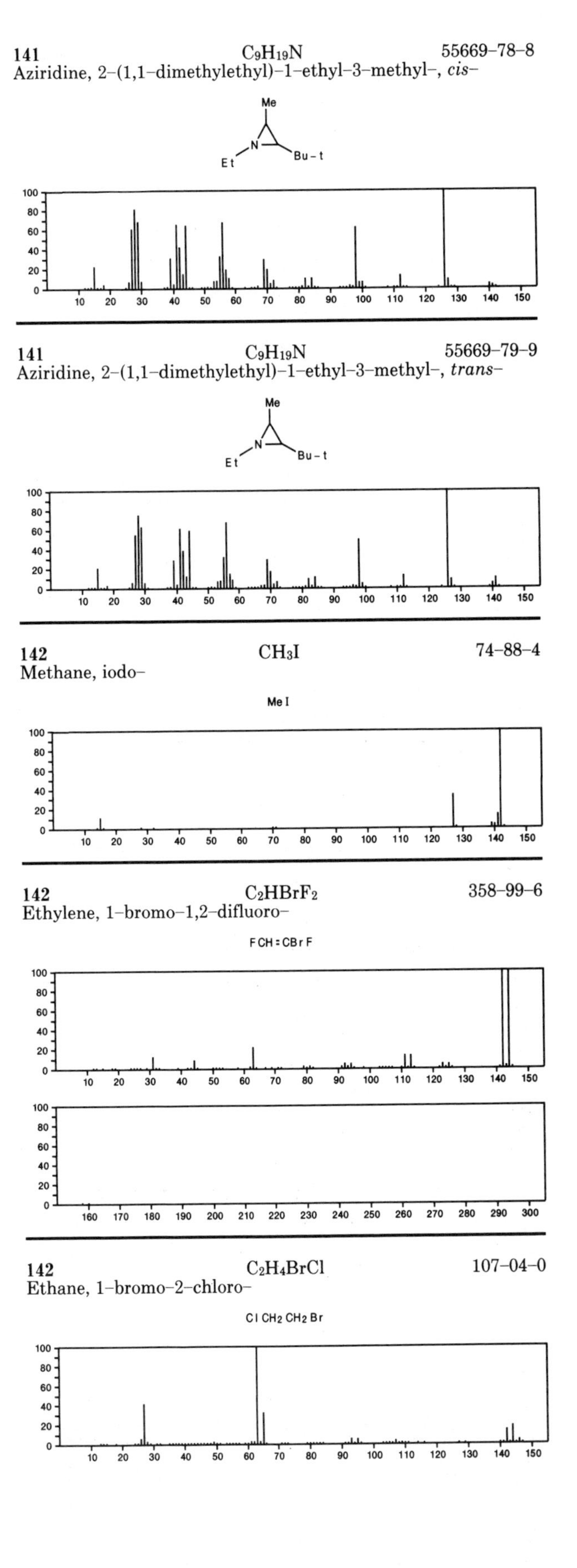

141 $C_9H_{19}N$ 55669-78-8
Aziridine, 2-(1,1-dimethylethyl)-1-ethyl-3-methyl-, *cis*-

141 $C_9H_{19}N$ 55669-79-9
Aziridine, 2-(1,1-dimethylethyl)-1-ethyl-3-methyl-, *trans*-

142 CH_3I 74-88-4
Methane, iodo-
MeI

142 C_2HBrF_2 358-99-6
Ethylene, 1-bromo-1,2-difluoro-
FCH=CBrF

142 C_2H_4BrCl 107-04-0
Ethane, 1-bromo-2-chloro-
ClCH2CH2Br

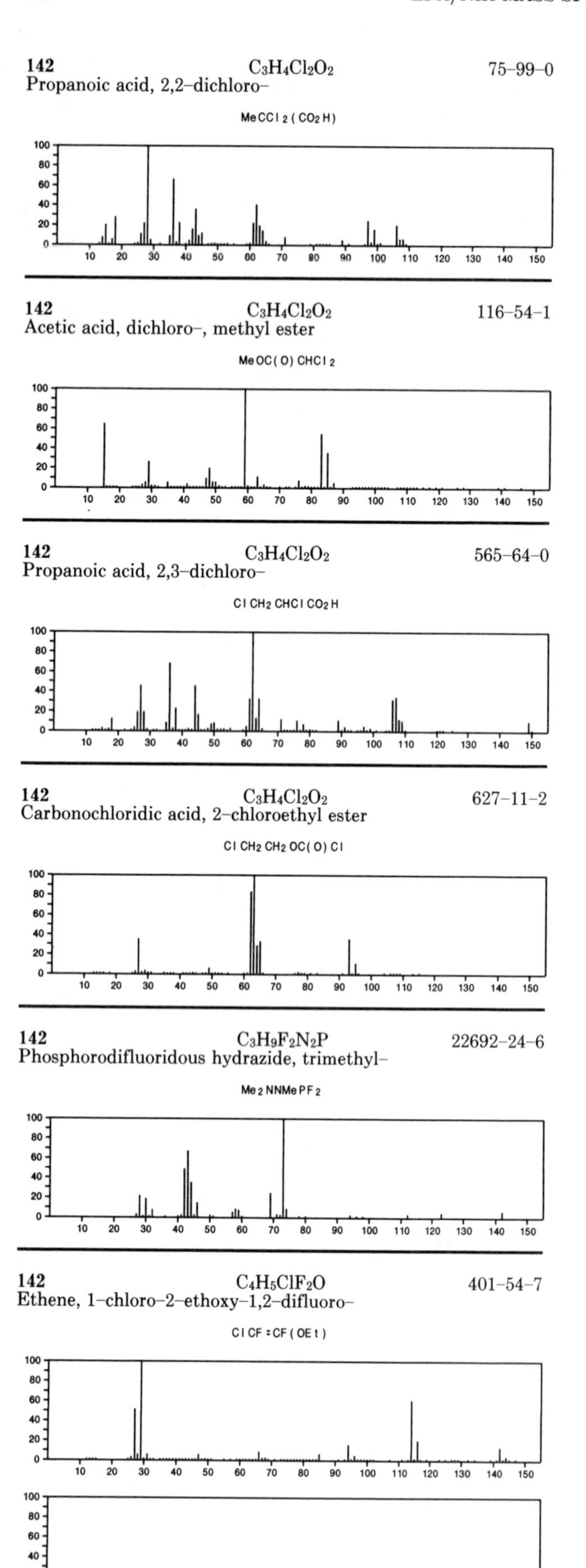
142 $C_3H_4Cl_2O_2$ 75-99-0
Propanoic acid, 2,2-dichloro-
MeCCl2(CO2H)
142 $C_3H_4Cl_2O_2$ 116-54-1
Acetic acid, dichloro-, methyl ester
MeOC(O)CHCl2
142 $C_3H_4Cl_2O_2$ 565-64-0
Propanoic acid, 2,3-dichloro-
ClCH2CHClCO2H
142 $C_3H_4Cl_2O_2$ 627-11-2
Carbonochloridic acid, 2-chloroethyl ester
ClCH2CH2OC(O)Cl
142 $C_3H_9F_2N_2P$ 22692-24-6
Phosphorodifluoridous hydrazide, trimethyl-
Me2NNMePF2
142 $C_4H_5ClF_2O$ 401-54-7
Ethene, 1-chloro-2-ethoxy-1,2-difluoro-
ClCF=CF(OEt)

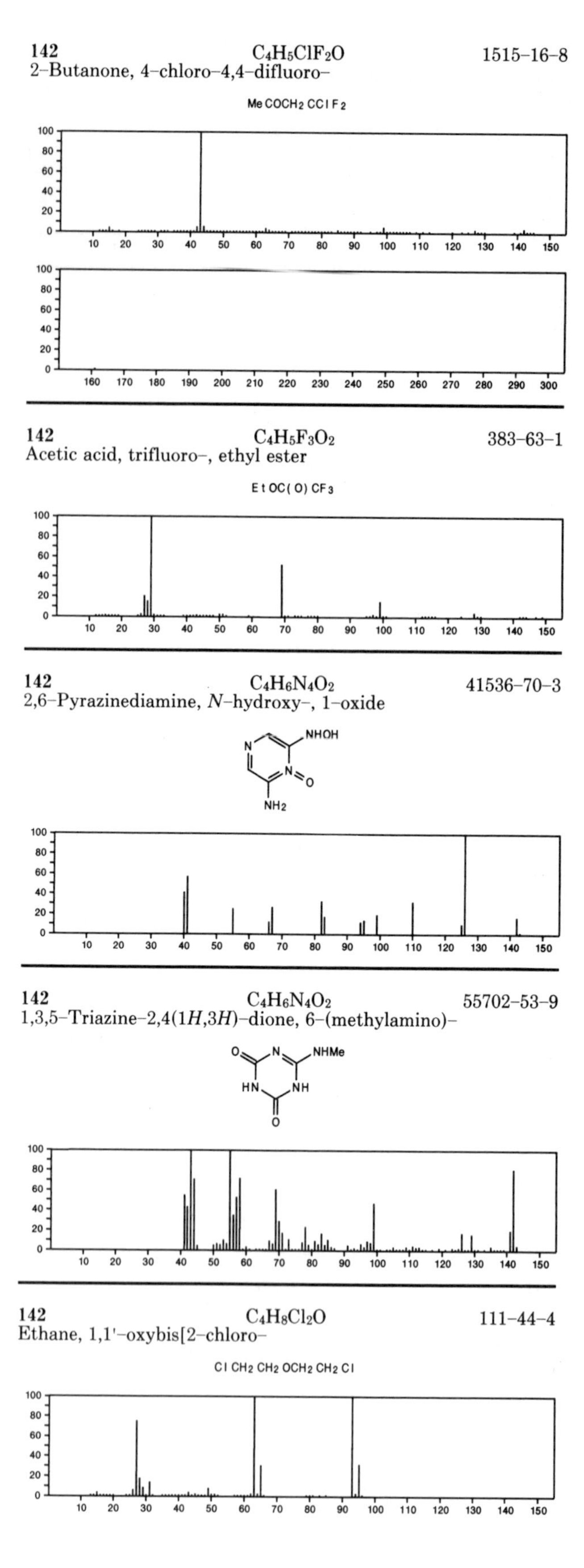
142 $C_4H_5ClF_2O$ 1515-16-8
2-Butanone, 4-chloro-4,4-difluoro-
MeCOCH2CClF2
142 $C_4H_5F_3O_2$ 383-63-1
Acetic acid, trifluoro-, ethyl ester
EtOC(O)CF3
142 $C_4H_6N_4O_2$ 41536-70-3
2,6-Pyrazinediamine, N-hydroxy-, 1-oxide
NHOH
NH2
142 $C_4H_6N_4O_2$ 55702-53-9
1,3,5-Triazine-2,4(1H,3H)-dione, 6-(methylamino)-
NHMe
HN
NH
142 $C_4H_8Cl_2O$ 111-44-4
Ethane, 1,1'-oxybis[2-chloro-
ClCH2CH2OCH2CH2Cl

142 $C_4H_8Cl_2O$ 623-46-1
Ethane, 1,2-dichloro-1-ethoxy-

ClCH2CHCl(OEt)

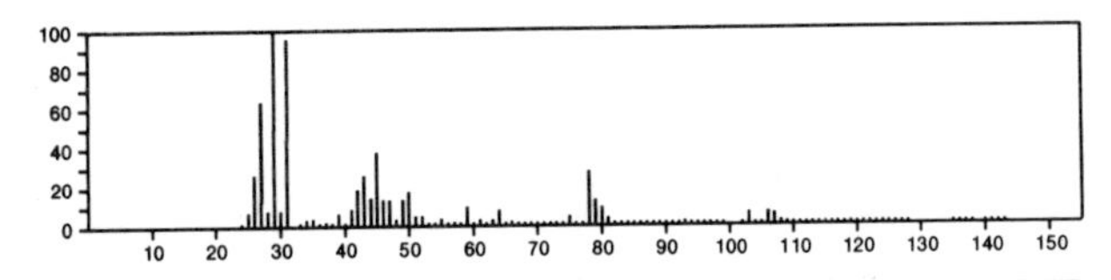

142 $C_5H_2O_5$ 488-86-8
4-Cyclopentene-1,2,3-trione, 4,5-dihydroxy-

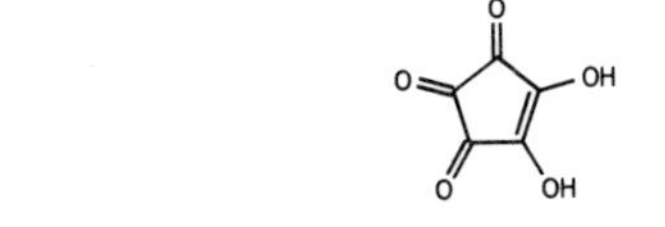

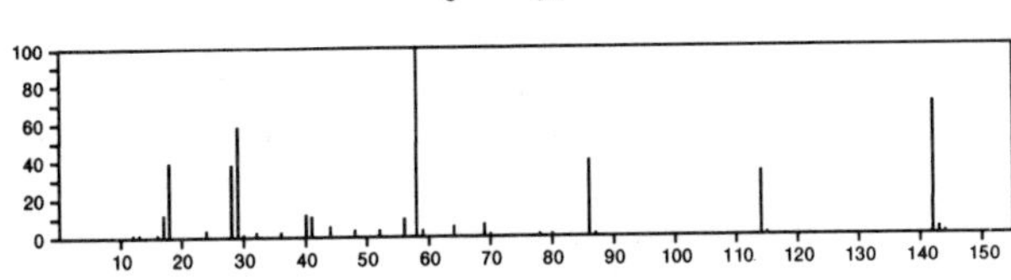

142 $C_5H_6N_2OS$ 636-26-0
4(1*H*)-Pyrimidinone, 2,3-dihydro-5-methyl-2-thioxo-

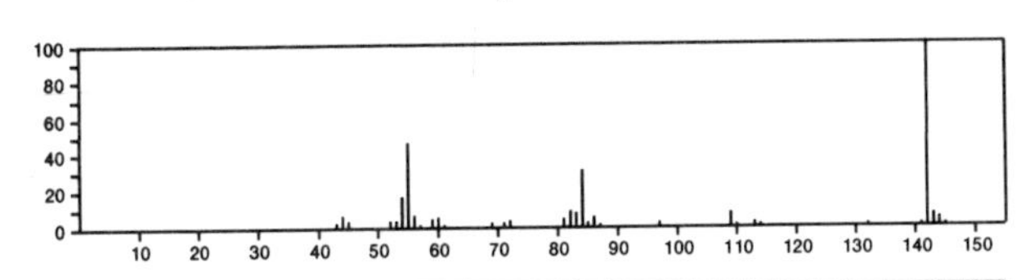

142 $C_5H_6N_2OS$ 5751-20-2
4(1*H*)-Pyrimidinone, 2-(methylthio)-

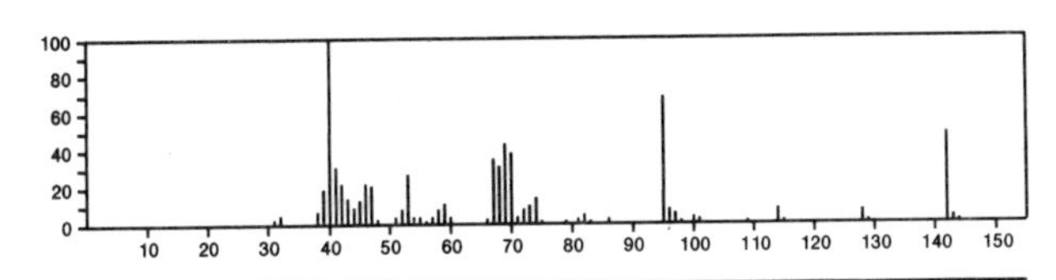

142 $C_5H_6N_2OS$ 24611-14-1
4(1*H*)-Pyrimidinethione, 5-hydroxy-2-methyl-

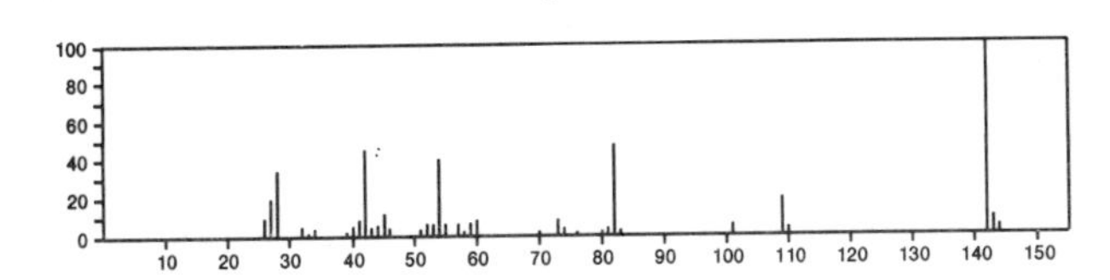

142 $C_5H_6N_2OS$ 35455-86-8
2(1*H*)-Pyrimidinone, 3,4-dihydro-1-methyl-4-thioxo-

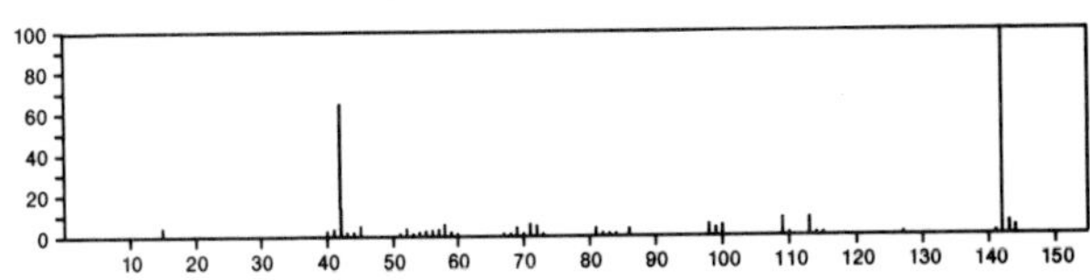

142 $C_5H_6N_2OS$ 40757-61-7
Ethanone, 1-(5-methyl-1,2,3-thiadiazol-4-yl)-

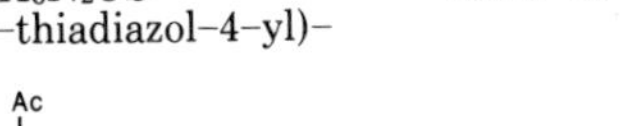

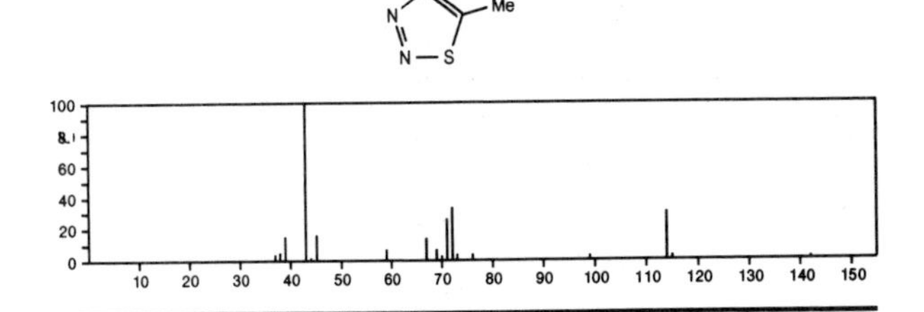

142 $C_5H_6N_2OS$ 42956-80-9
4(1*H*)-Pyrimidinone, 6-mercapto-2-methyl-

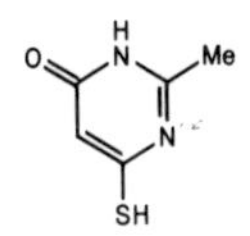

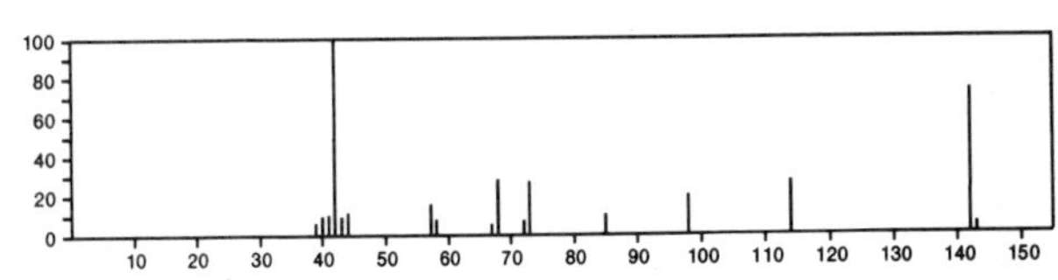

142 $C_5H_6N_2O_3$ 1123-49-5
Isoxazole, 3,5-dimethyl-4-nitro-

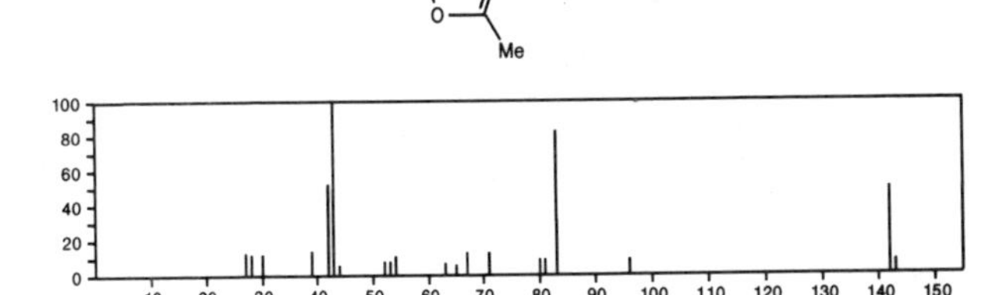

142 $C_6H_3ClO_2$ 695-99-8
2,5-Cyclohexadiene-1,4-dione, 2-chloro-

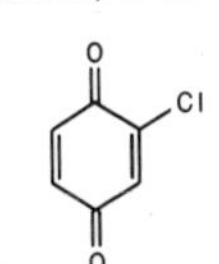

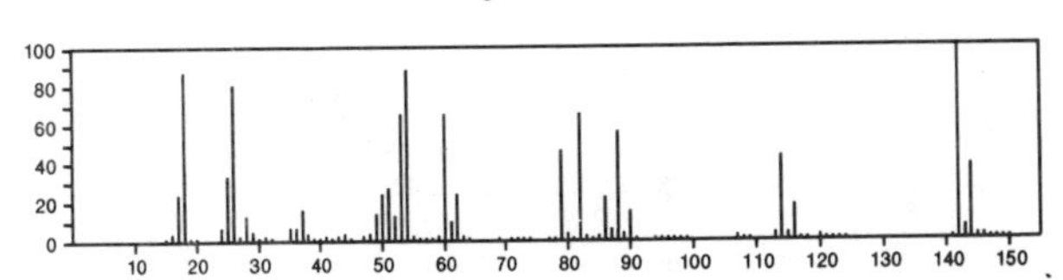

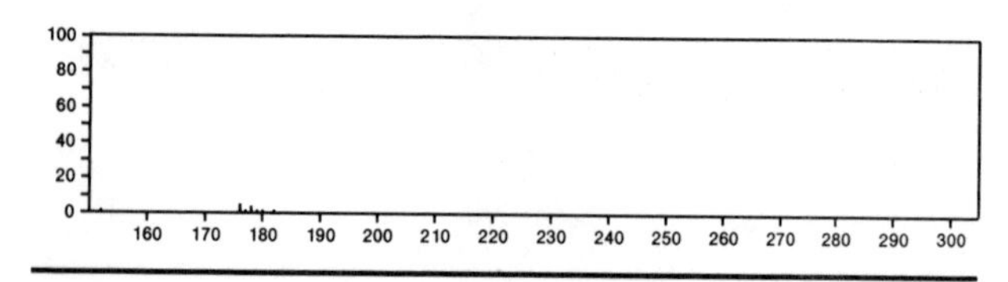

142 $C_6H_6O_4$ 501-30-4
4*H*-Pyran-4-one, 5-hydroxy-2-(hydroxymethyl)-

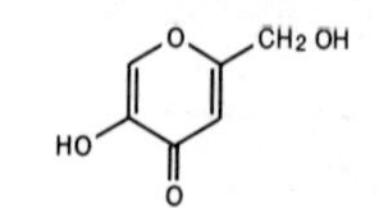

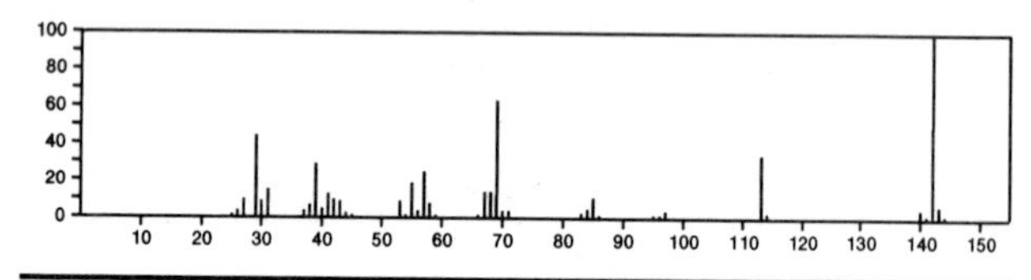

142 $C_6H_6O_4$ 1073-96-7
4*H*-Pyran-4-one, 3,5-dihydroxy-2-methyl-

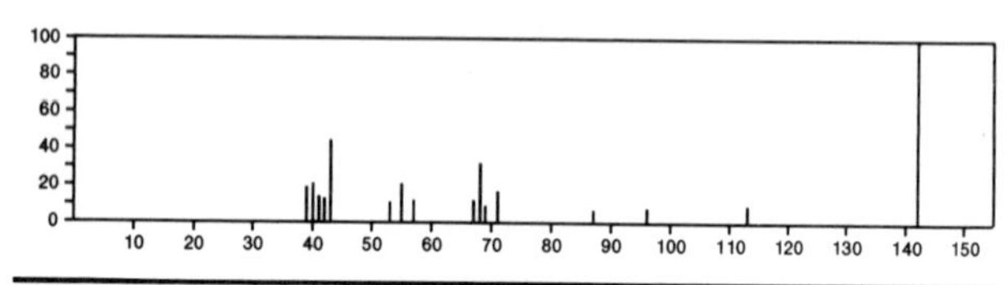

142 $C_6H_6O_4$ 1119-72-8
2,4-Hexadienedioic acid, (*Z*,*Z*)-

$HO_2CCH=CHCH=CHCO_2H$

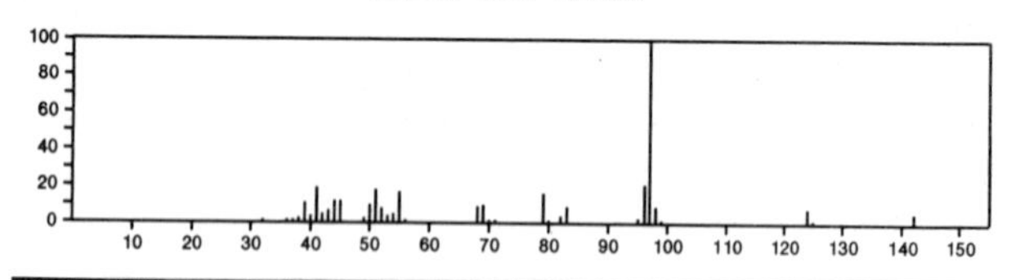

142 $C_6H_6O_4$ 6338-41-6
2-Furancarboxylic acid, 5-(hydroxymethyl)-

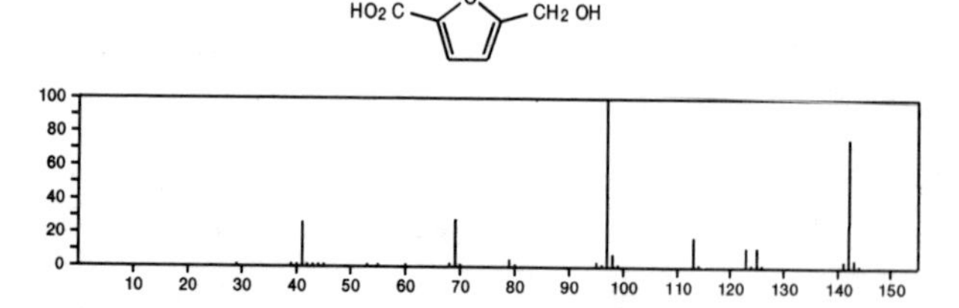

142 $C_6H_6O_4$ 16508-05-7
1-Cyclobutene-1,2-dicarboxylic acid

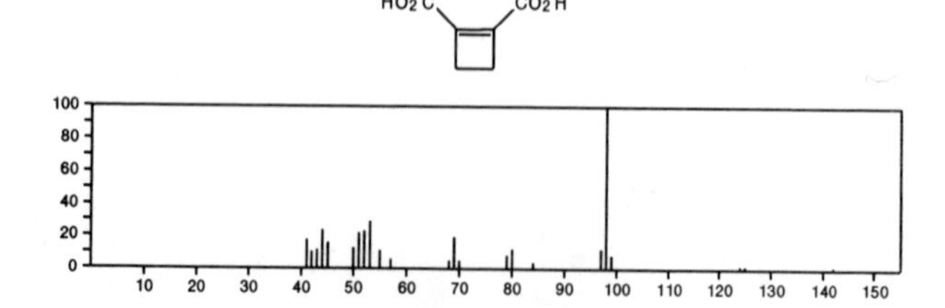

142 $C_6H_6O_4$ 36677-73-3
2-Butyne-1,4-diol, diformate

$O=CHOCH_2C\equiv CCH_2OCH=O$

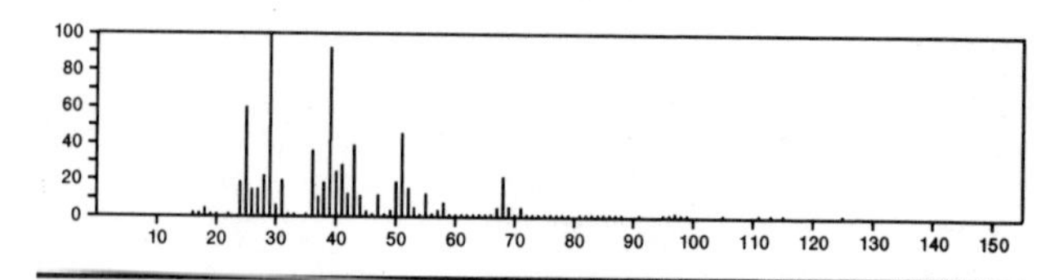

142 $C_6H_7O_2P$ 1779-48-2
Phosphinic acid, phenyl-

PhPH(OH)=O

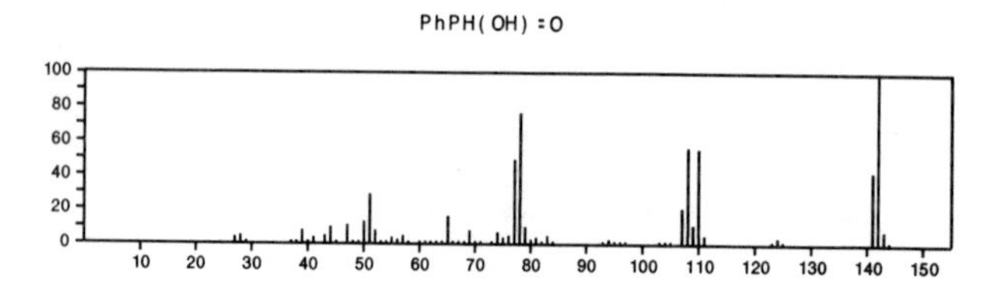

142 $C_6H_{10}N_2O_2$ 6345-19-3
2,4-Imidazolidinedione, 3,5,5-trimethyl-

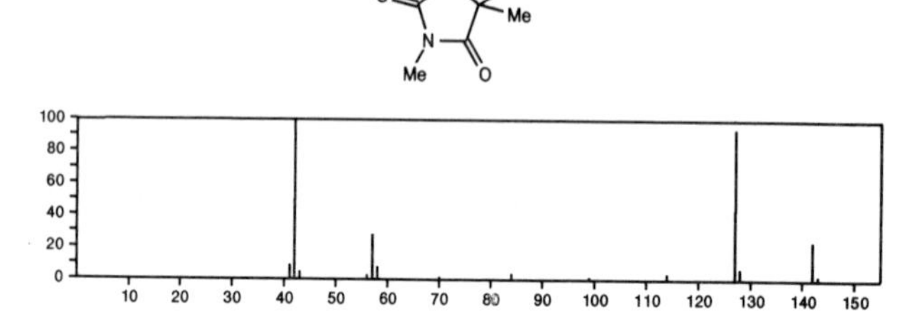

142 $C_6H_{10}N_2O_2$ 6851-81-6
2,4-Imidazolidinedione, 1,5,5-trimethyl-

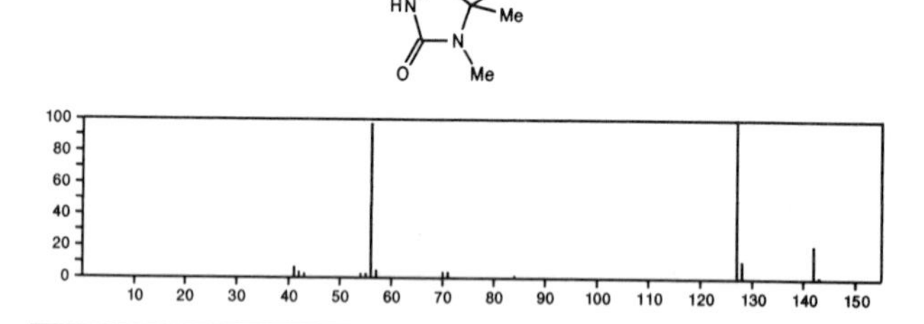

142 $C_6H_{10}N_2O_2$ 6939-25-9
Sydnone, 3-(1,1-dimethylethyl)-

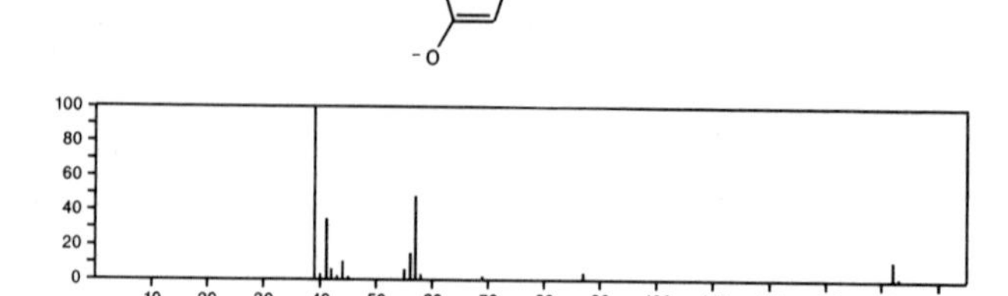

142 C_7H_7ClO 59-50-7
Phenol, 4-chloro-3-methyl-

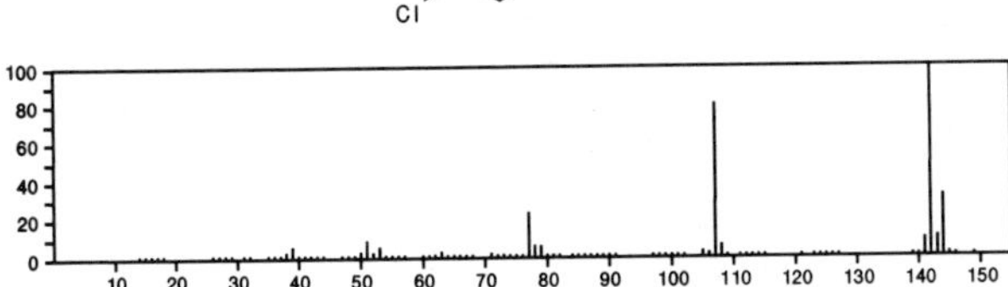

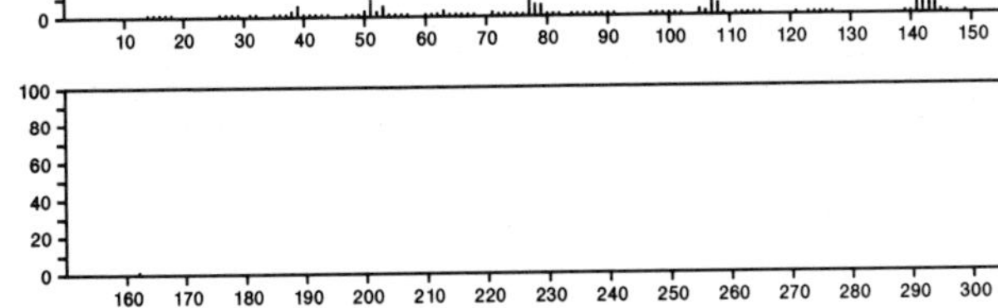

142 C_7H_7ClO 87-64-9
Phenol, 2-chloro-6-methyl-

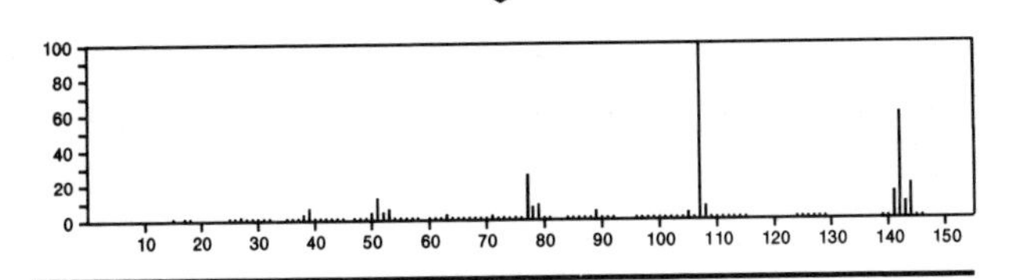

142 C_7H_7ClO 623-12-1
Benzene, 1-chloro-4-methoxy-

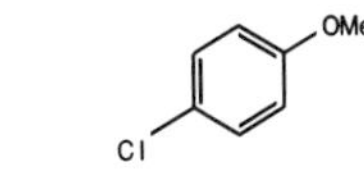

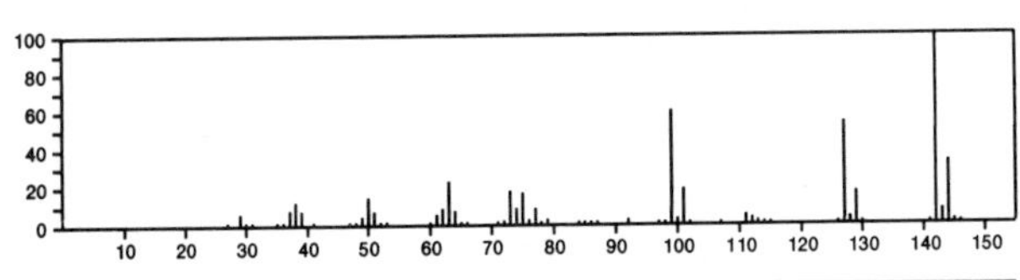

142 C_7H_7ClO 1570-64-5
Phenol, 4-chloro-2-methyl-

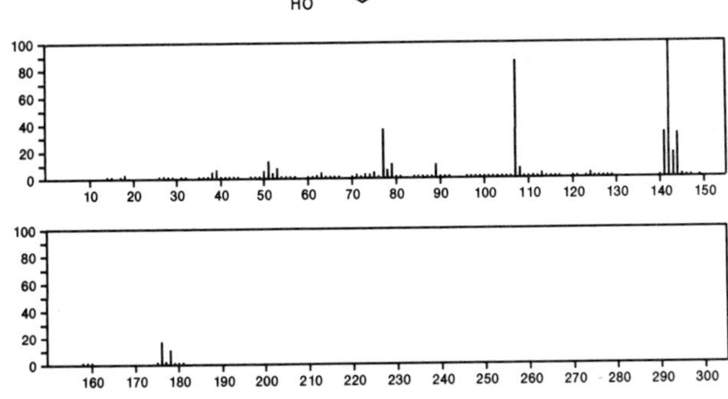

142 C_7H_7ClO 17849-38-6
Benzenemethanol, 2-chloro-

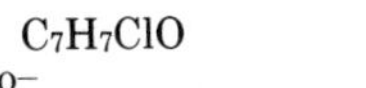

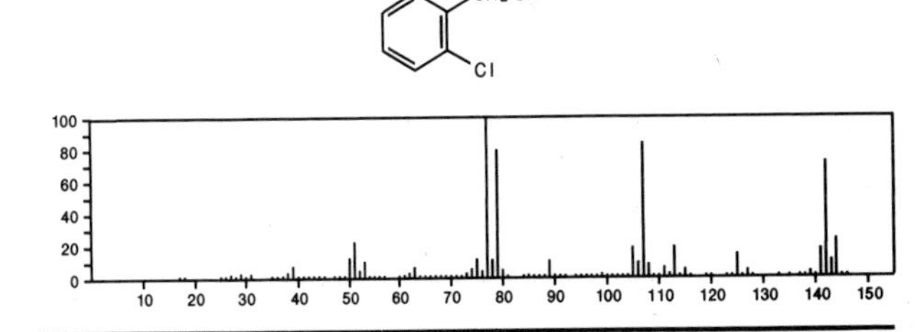

142 $C_7H_{10}O_3$ 19456-20-3
1,3-Benzodioxol-2-one, hexahydro-, *cis*-

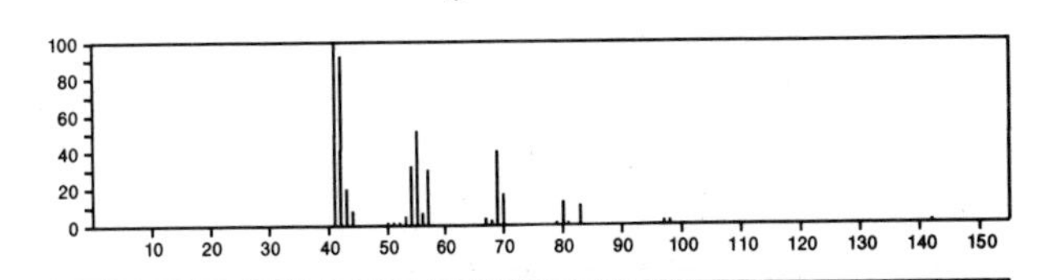

142 $C_7H_{10}O_3$ 20192-66-9
1,3-Benzodioxol-2-one, hexahydro-, *trans*-

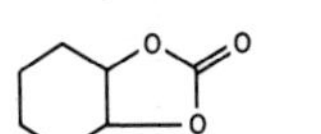

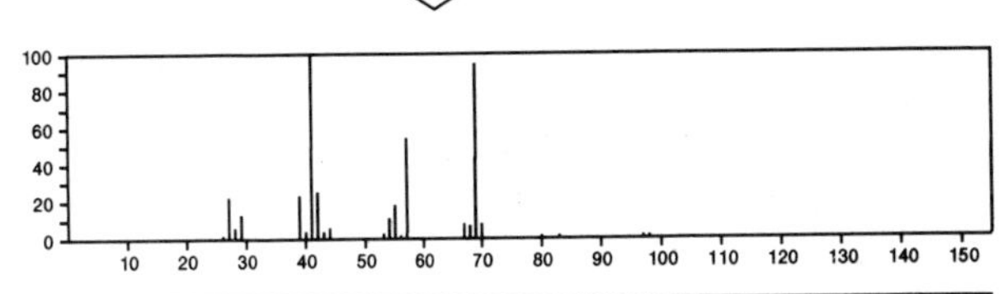

142 $C_7H_{14}N_2O$ 5432-28-0
Cyclohexanamine, *N*-methyl-*N*-nitroso-

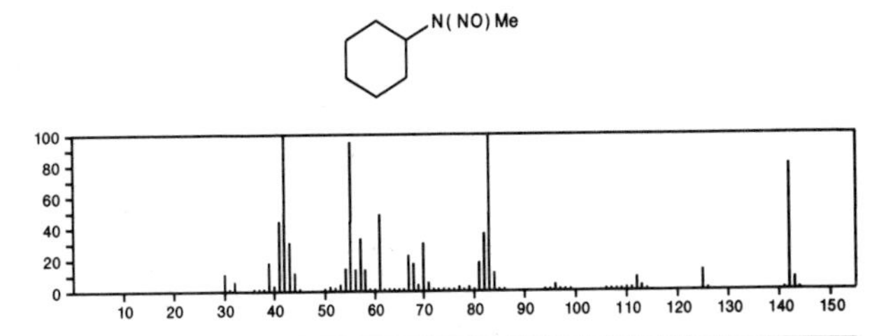

142 $C_7H_{14}N_2O$ 17721-95-8
Piperidine, 2,6-dimethyl-1-nitroso-

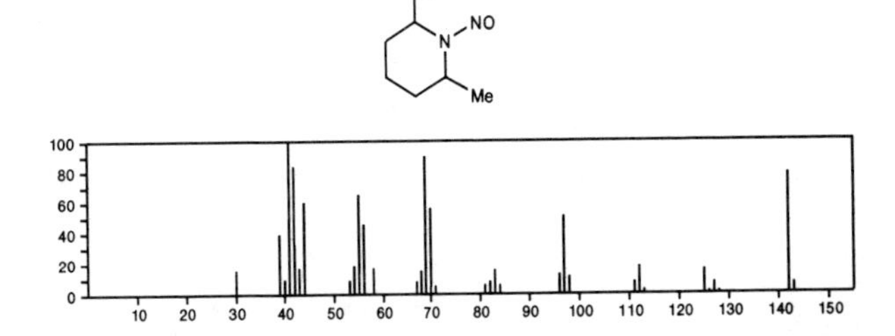

142 $C_7H_{14}N_2O$ 20917-49-1
Azocine, octahydro-1-nitroso-

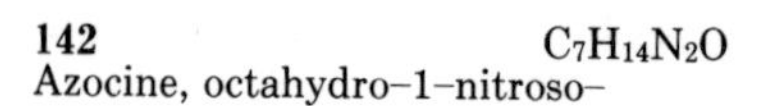

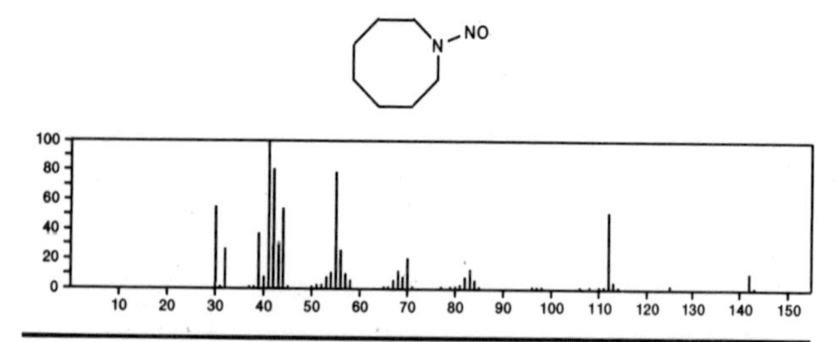

142 $C_7H_{14}N_2O$ 35214-90-5
Diazene, cyclohexylmethyl-, 1-oxide, (*E*)-

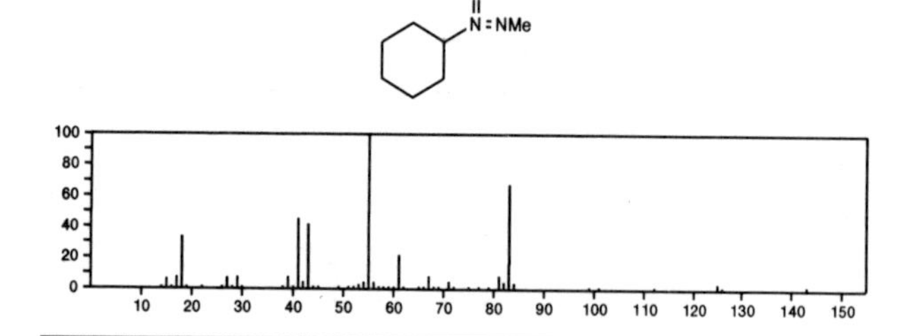

142 $C_7H_{14}N_2O$ 40347-20-4
3-Buten-2-one, 4-(dimethylamino)-4-(n.ethylamino)-

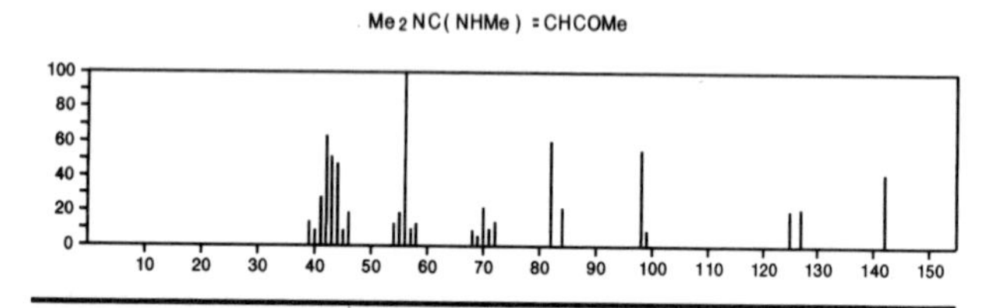

142 $C_7H_{14}N_2O$ 42145-18-6
2-Propenal, 2,3-bis(dimethylamino)-

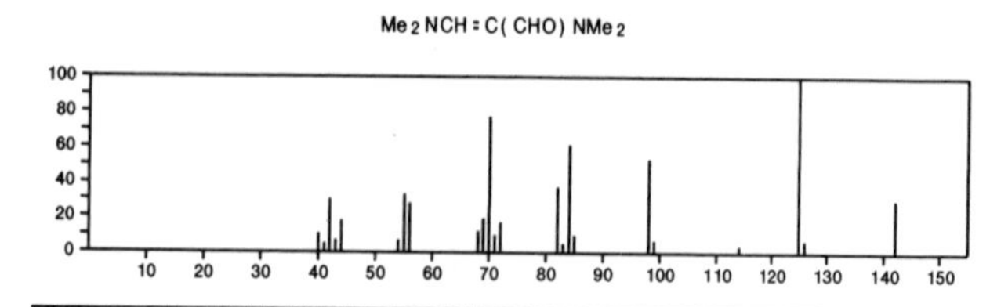

142 $C_7H_{14}N_2O$ 49582-45-8
2-Propenal, 3,3-bis(dimethylamino)-

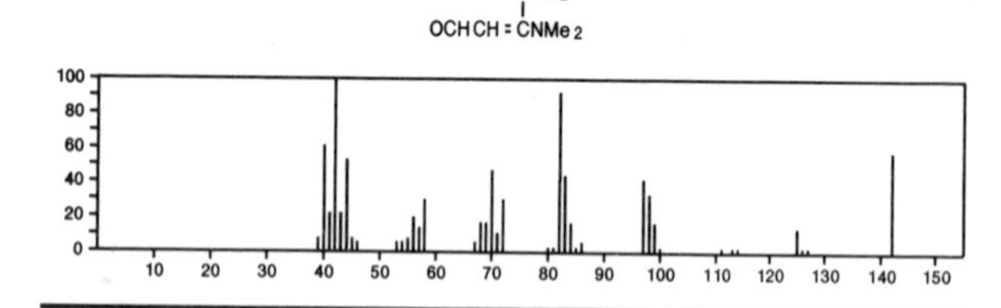

142 $C_7H_{14}N_2O$ 49582-63-0
3-Buten-2-one, 4-(dimethylamino)-3-(methylamino)-

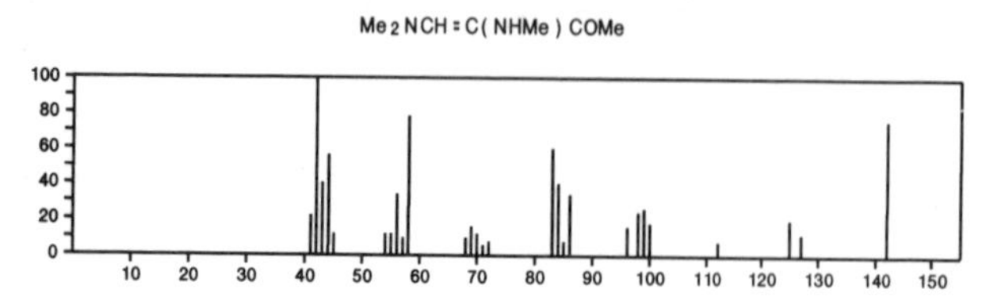

142 $C_7H_{14}N_2O$ 56666-79-6
3-Pyrazolidinone, 1,2,4,5-tetramethyl-

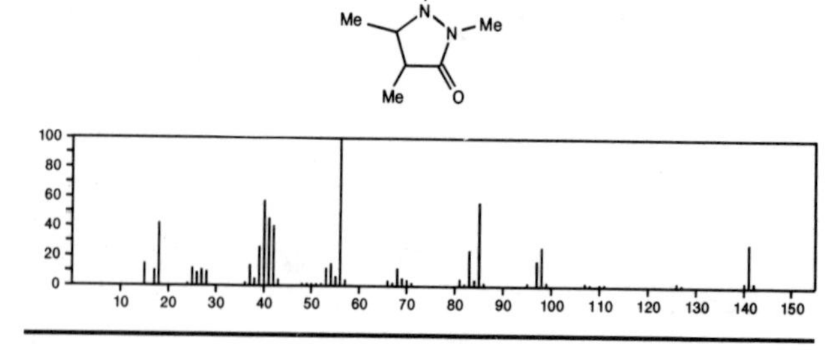

142 $C_7H_{14}N_2O$ 56666-82-1
3-Pyrazolidinone, 1,2,4,4-tetramethyl-

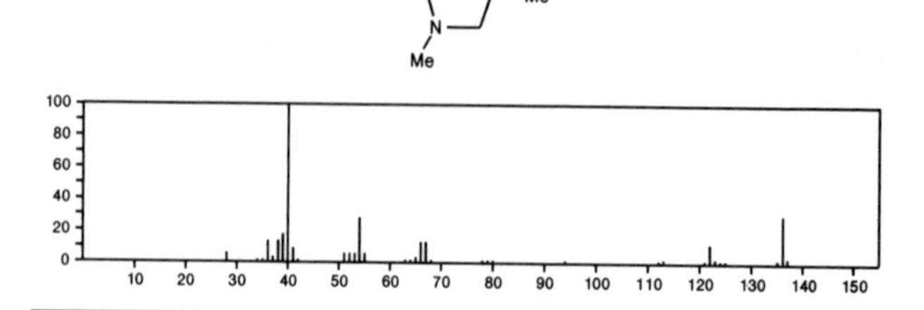

142 $C_8H_{11}Cl$ 23804-47-9
Bicyclo[2.2.2]oct-2-ene, 2-chloro-

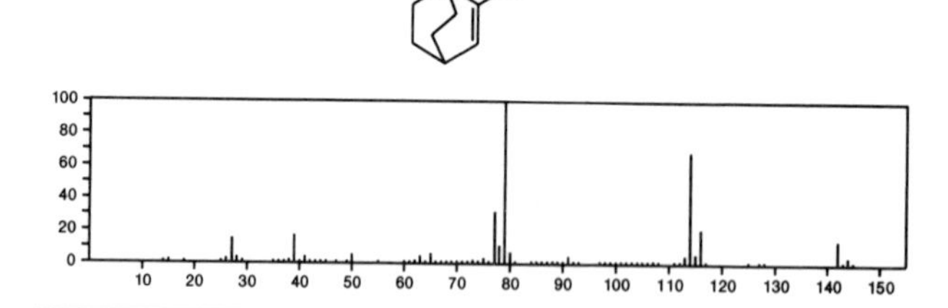

142 $C_8H_{11}Cl$ 35242-17-2
Bicyclo[3.2.1]oct-2-ene, 3-chloro-

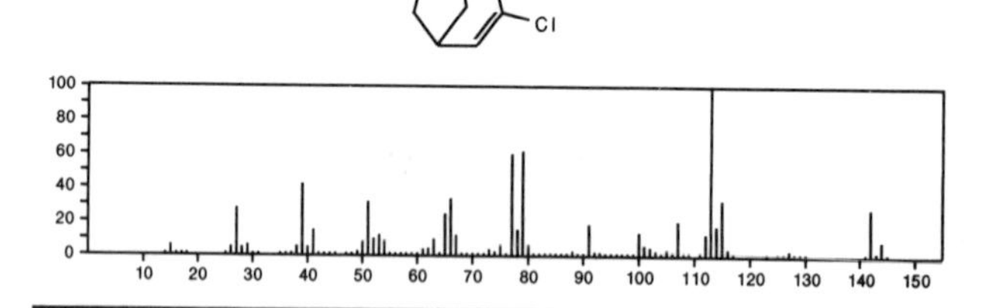

142 $C_8H_{11}Cl$ 49826-39-3
Bicyclo[3.2.1]oct-2-ene, 2-chloro-

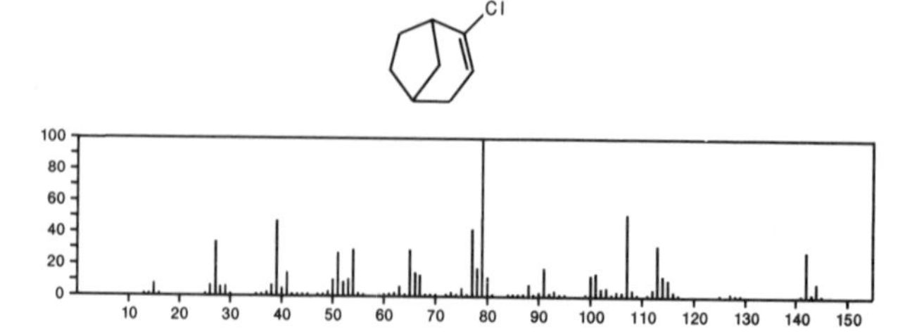

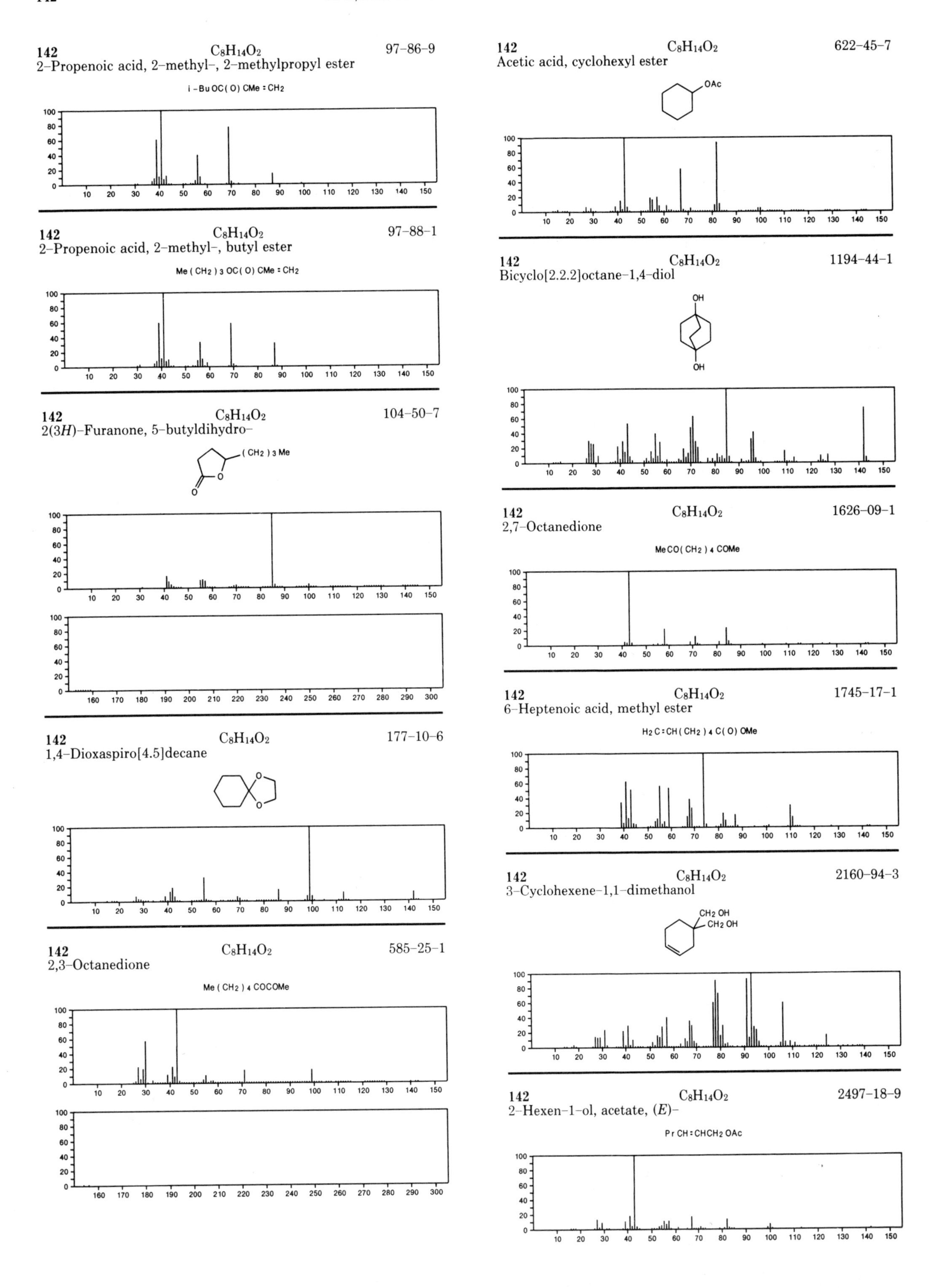
142 $C_8H_{14}O_2$ 97–86–9
2–Propenoic acid, 2–methyl–, 2–methylpropyl ester
i–BuOC(O)CMe=CH2
142 $C_8H_{14}O_2$ 97–88–1
2–Propenoic acid, 2–methyl–, butyl ester
Me(CH2)3OC(O)CMe=CH2
142 $C_8H_{14}O_2$ 104–50–7
2(3H)–Furanone, 5–butyldihydro–
(CH2)3Me
142 $C_8H_{14}O_2$ 177–10–6
1,4–Dioxaspiro[4.5]decane
142 $C_8H_{14}O_2$ 585–25–1
2,3–Octanedione
Me(CH2)4COCOMe
142 $C_8H_{14}O_2$ 622–45–7
Acetic acid, cyclohexyl ester
OAc
142 $C_8H_{14}O_2$ 1194–44–1
Bicyclo[2.2.2]octane–1,4–diol
OH
OH
142 $C_8H_{14}O_2$ 1626–09–1
2,7–Octanedione
MeCO(CH2)4COMe
142 $C_8H_{14}O_2$ 1745–17–1
6–Heptenoic acid, methyl ester
H2C=CH(CH2)4C(O)OMe
142 $C_8H_{14}O_2$ 2160–94–3
3–Cyclohexene–1,1–dimethanol
CH2OH
CH2OH
142 $C_8H_{14}O_2$ 2497–18–9
2–Hexen–1–ol, acetate, (E)–
PrCH=CHCH2OAc

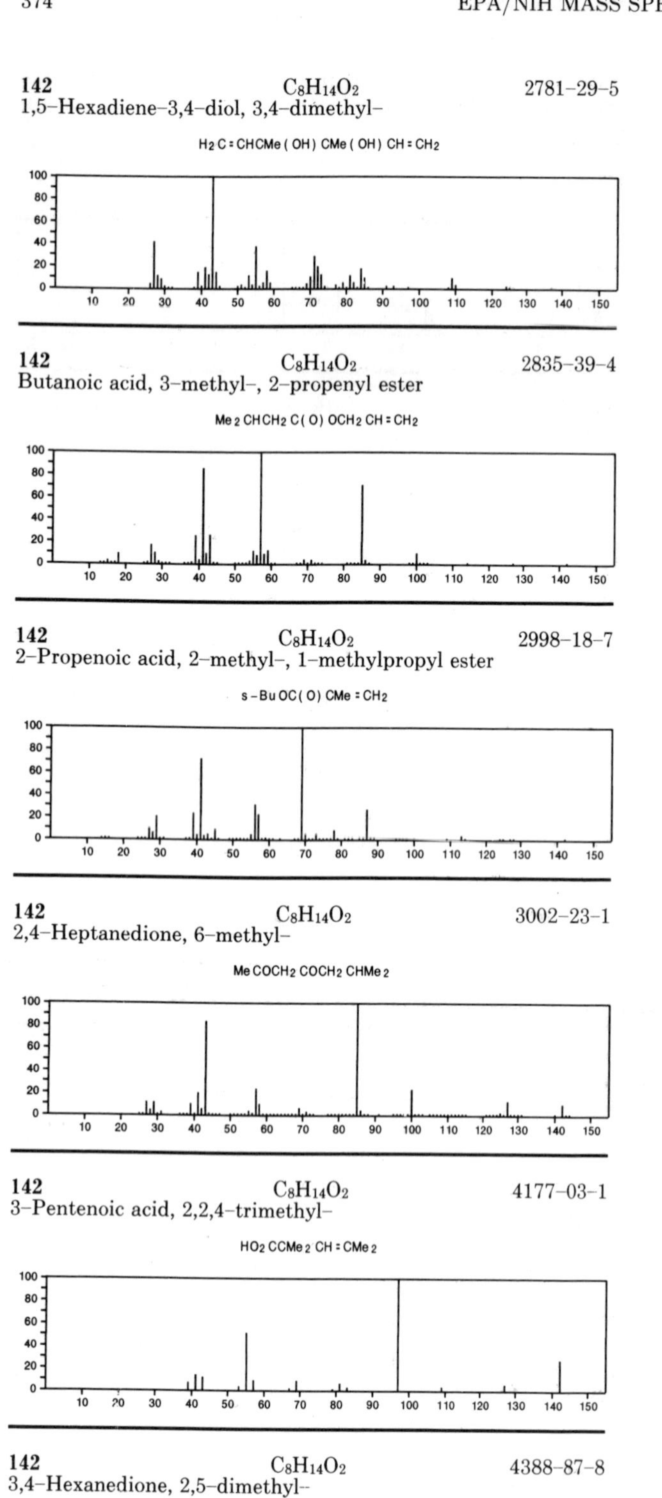
142 C8H14O2 2781-29-5
1,5-Hexadiene-3,4-diol, 3,4-dimethyl-
H2C=CHCMe(OH)CMe(OH)CH=CH2
142 C8H14O2 2835-39-4
Butanoic acid, 3-methyl-, 2-propenyl ester
Me2CHCH2C(O)OCH2CH=CH2
142 C8H14O2 2998-18-7
2-Propenoic acid, 2-methyl-, 1-methylpropyl ester
s-BuOC(O)CMe=CH2
142 C8H14O2 3002-23-1
2,4-Heptanedione, 6-methyl-
MeCOCH2COCH2CHMe2
142 C8H14O2 4177-03-1
3-Pentenoic acid, 2,2,4-trimethyl-
HO2CCMe2CH=CMe2
142 C8H14O2 4388-87-8
3,4-Hexanedione, 2,5-dimethyl-
Me2CHCOCOCHMe2

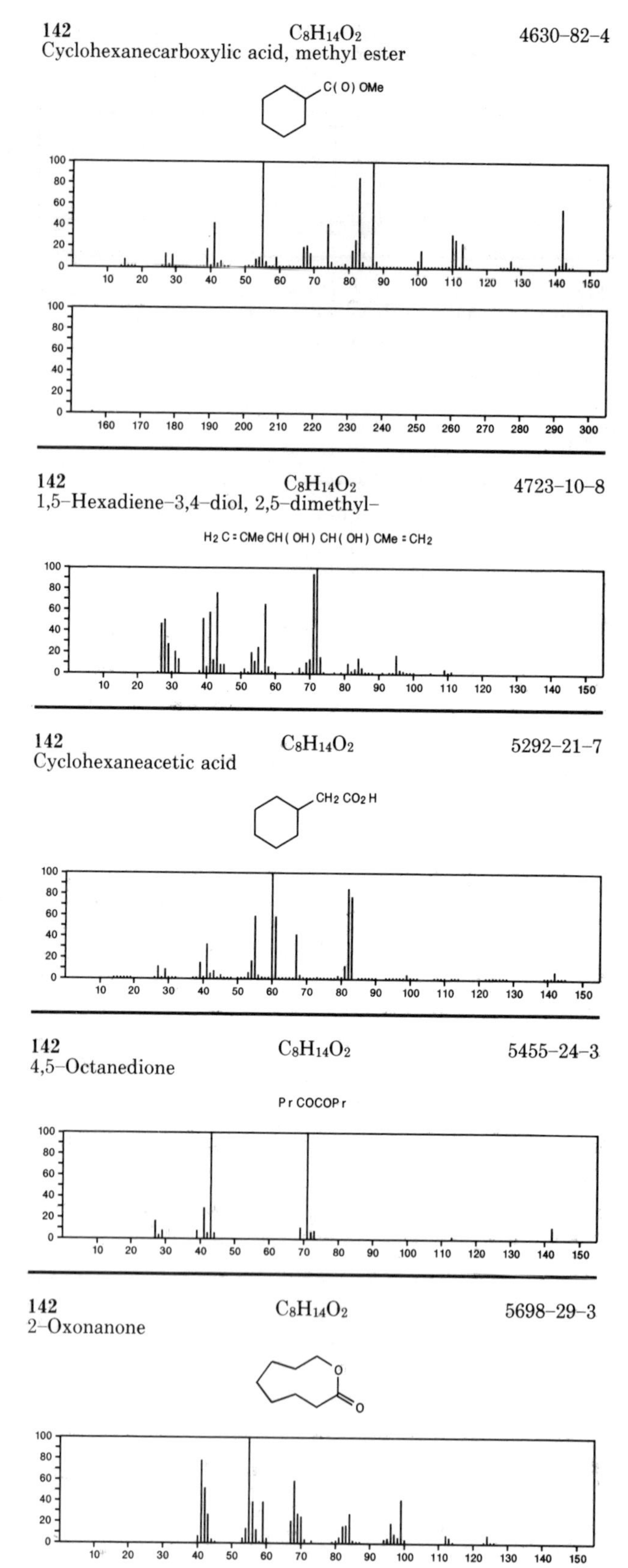
142 C8H14O2 4630-82-4
Cyclohexanecarboxylic acid, methyl ester
C(O)OMe
142 C8H14O2 4723-10-8
1,5-Hexadiene-3,4-diol, 2,5-dimethyl-
H2C=CMeCH(OH)CH(OH)CMe=CH2
142 C8H14O2 5292-21-7
Cyclohexaneacetic acid
CH2CO2H
142 C8H14O2 5455-24-3
4,5-Octanedione
PrCOCOPr
142 C8H14O2 5698-29-3
2-Oxonanone
O
O

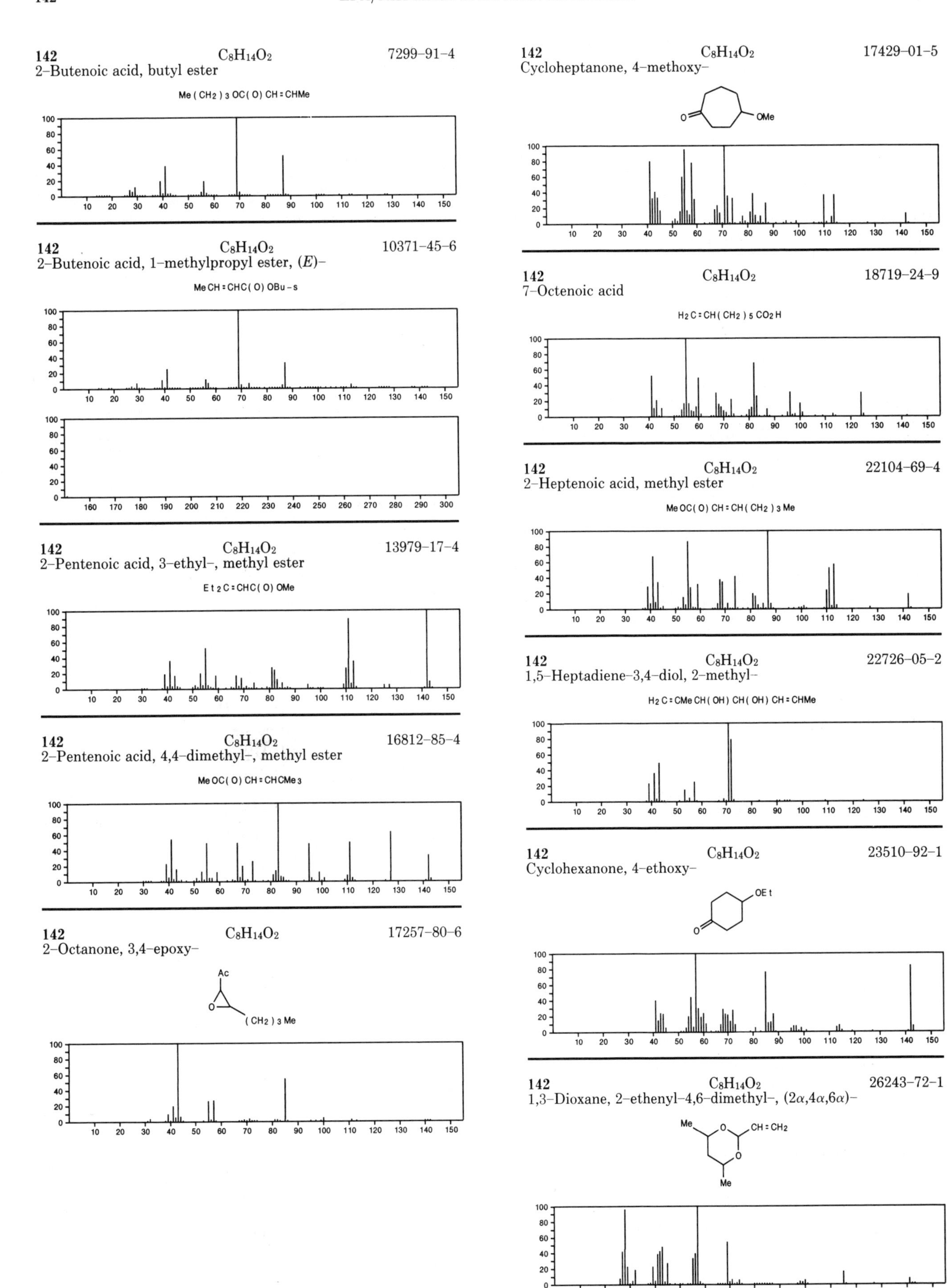
142 $C_8H_{14}O_2$ 7299-91-4
2-Butenoic acid, butyl ester
Me(CH2)3OC(O)CH=CHMe
142 $C_8H_{14}O_2$ 10371-45-6
2-Butenoic acid, 1-methylpropyl ester, (*E*)-
MeCH=CHC(O)OBu-s
142 $C_8H_{14}O_2$ 13979-17-4
2-Pentenoic acid, 3-ethyl-, methyl ester
Et2C=CHC(O)OMe
142 $C_8H_{14}O_2$ 16812-85-4
2-Pentenoic acid, 4,4-dimethyl-, methyl ester
MeOC(O)CH=CHCMe3
142 $C_8H_{14}O_2$ 17257-80-6
2-Octanone, 3,4-epoxy-
Ac
O
(CH2)3Me
142 $C_8H_{14}O_2$ 17429-01-5
Cycloheptanone, 4-methoxy-
O
OMe
142 $C_8H_{14}O_2$ 18719-24-9
7-Octenoic acid
H2C=CH(CH2)5CO2H
142 $C_8H_{14}O_2$ 22104-69-4
2-Heptenoic acid, methyl ester
MeOC(O)CH=CH(CH2)3Me
142 $C_8H_{14}O_2$ 22726-05-2
1,5-Heptadiene-3,4-diol, 2-methyl-
H2C=CMeCH(OH)CH(OH)CH=CHMe
142 $C_8H_{14}O_2$ 23510-92-1
Cyclohexanone, 4-ethoxy-
OEt
O
142 $C_8H_{14}O_2$ 26243-72-1
1,3-Dioxane, 2-ethenyl-4,6-dimethyl-, (2α,4α,6α)-
Me
O
CH=CH2
O
Me

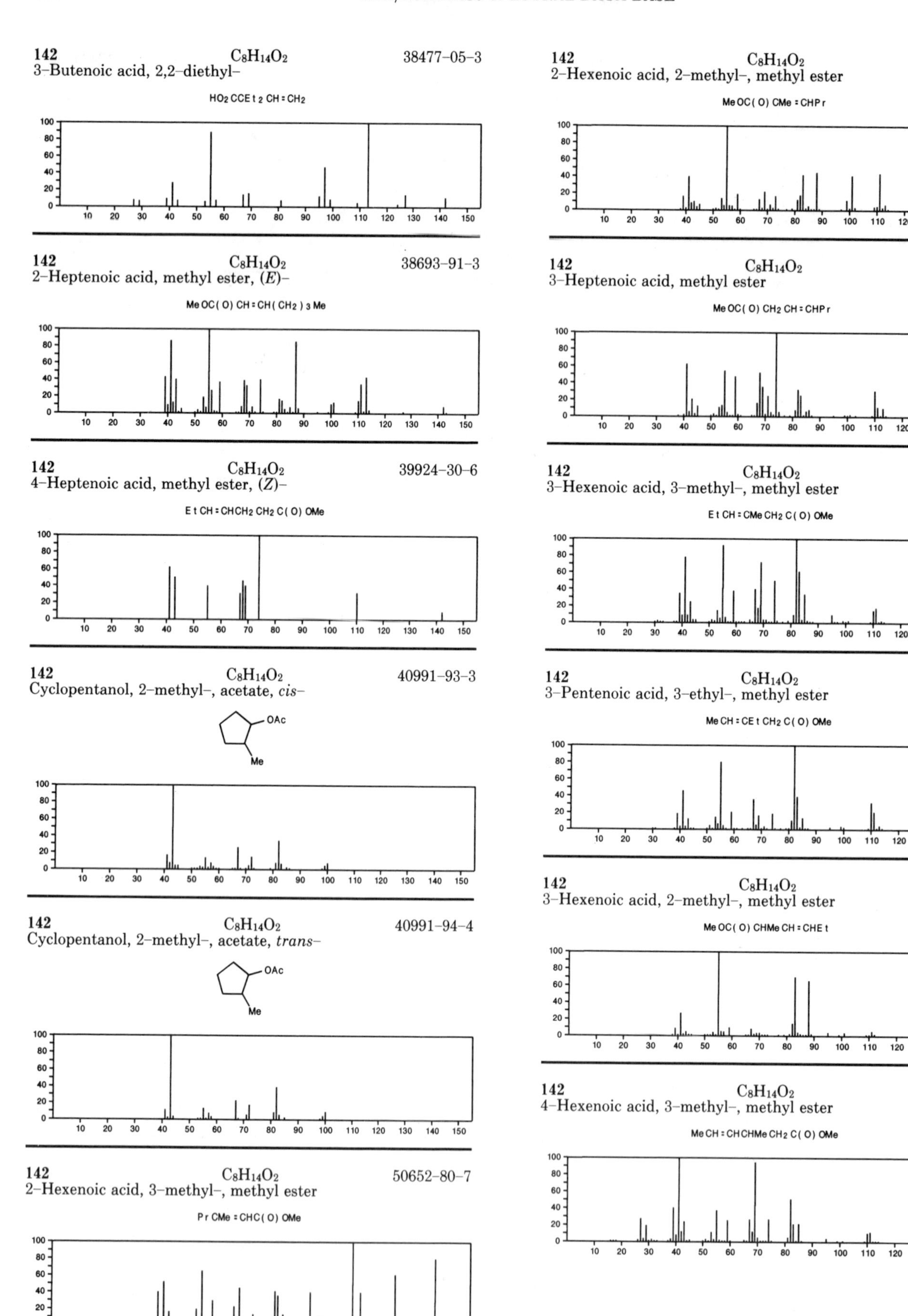

142 $C_8H_{14}O_2$ 38477-05-3
3-Butenoic acid, 2,2-diethyl-
$HO_2CCEt_2CH{=}CH_2$

142 $C_8H_{14}O_2$ 38693-91-3
2-Heptenoic acid, methyl ester, (*E*)-
$MeOC(O)CH{=}CH(CH_2)_3Me$

142 $C_8H_{14}O_2$ 39924-30-6
4-Heptenoic acid, methyl ester, (*Z*)-
$EtCH{=}CHCH_2CH_2C(O)OMe$

142 $C_8H_{14}O_2$ 40991-93-3
Cyclopentanol, 2-methyl-, acetate, *cis-*

142 $C_8H_{14}O_2$ 40991-94-4
Cyclopentanol, 2-methyl-, acetate, *trans-*

142 $C_8H_{14}O_2$ 50652-80-7
2-Hexenoic acid, 3-methyl-, methyl ester
$PrCMe{=}CHC(O)OMe$

142 $C_8H_{14}O_2$ 50652-82-9
2-Hexenoic acid, 2-methyl-, methyl ester
$MeOC(O)CMe{=}CHPr$

142 $C_8H_{14}O_2$ 50652-83-0
3-Heptenoic acid, methyl ester
$MeOC(O)CH_2CH{=}CHPr$

142 $C_8H_{14}O_2$ 50652-84-1
3-Hexenoic acid, 3-methyl-, methyl ester
$EtCH{=}CMeCH_2C(O)OMe$

142 $C_8H_{14}O_2$ 50652-85-2
3-Pentenoic acid, 3-ethyl-, methyl ester
$MeCH{=}CEtCH_2C(O)OMe$

142 $C_8H_{14}O_2$ 50652-86-3
3-Hexenoic acid, 2-methyl-, methyl ester
$MeOC(O)CHMeCH{=}CHEt$

142 $C_8H_{14}O_2$ 54004-27-2
4-Hexenoic acid, 3-methyl-, methyl ester
$MeCH{=}CHCHMeCH_2C(O)OMe$

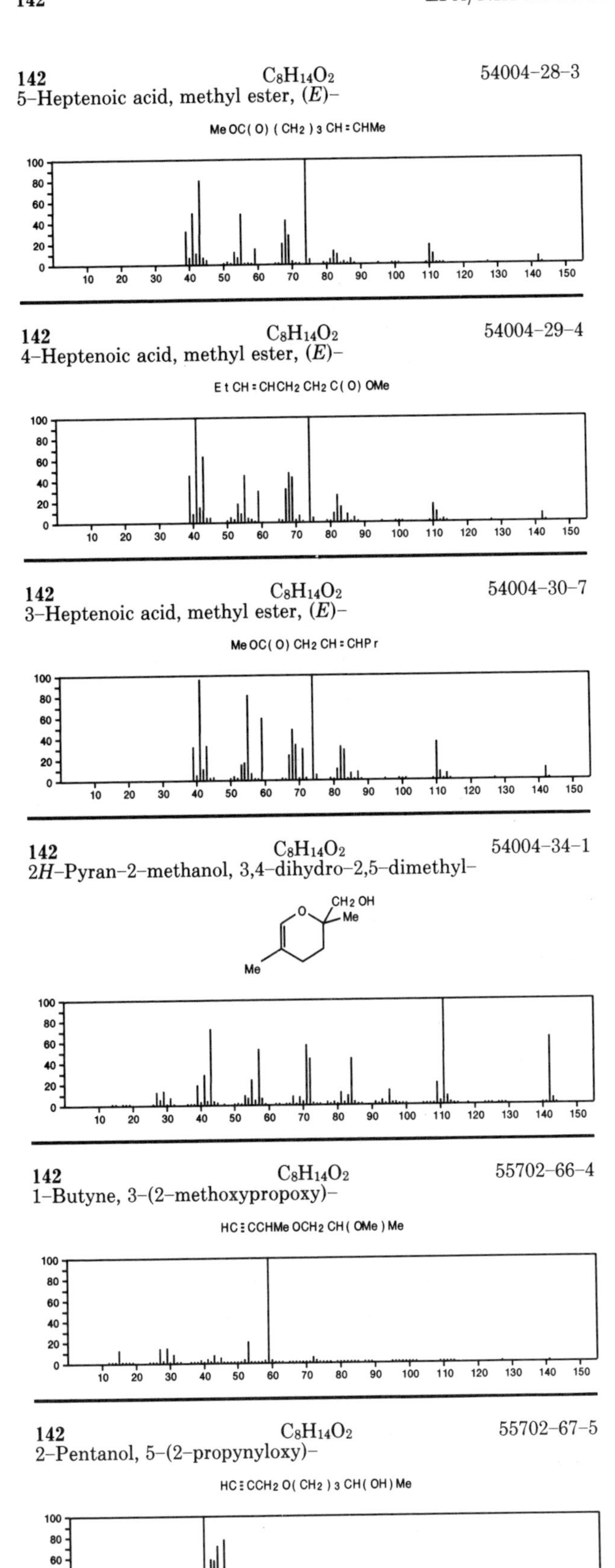

142 $C_8H_{14}O_2$ 54004-28-3
5-Heptenoic acid, methyl ester, (*E*)-
MeOC(O)(CH2)3CH=CHMe
142 $C_8H_{14}O_2$ 54004-29-4
4-Heptenoic acid, methyl ester, (*E*)-
EtCH=CHCH2CH2C(O)OMe
142 $C_8H_{14}O_2$ 54004-30-7
3-Heptenoic acid, methyl ester, (*E*)-
MeOC(O)CH2CH=CHPr
142 $C_8H_{14}O_2$ 54004-34-1
2*H*-Pyran-2-methanol, 3,4-dihydro-2,5-dimethyl-
CH2OH
Me
O
Me
142 $C_8H_{14}O_2$ 55702-66-4
1-Butyne, 3-(2-methoxypropoxy)-
HC≡CCHMeOCH2CH(OMe)Me
142 $C_8H_{14}O_2$ 55702-67-5
2-Pentanol, 5-(2-propynyloxy)-
HC≡CCH2O(CH2)3CH(OH)Me

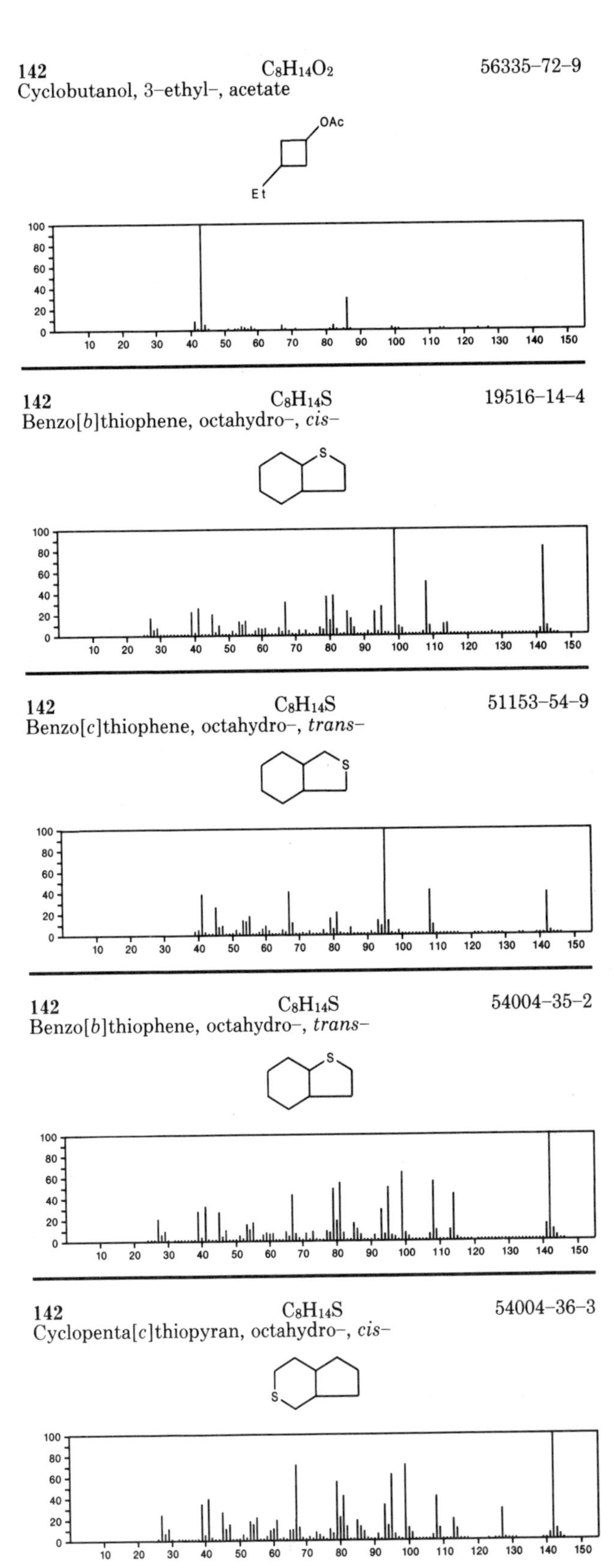

142 $C_8H_{14}O_2$ 56335-72-9
Cyclobutanol, 3-ethyl-, acetate
OAc
Et
142 $C_8H_{14}S$ 19516-14-4
Benzo[*b*]thiophene, octahydro-, *cis*-
S
142 $C_8H_{14}S$ 51153-54-9
Benzo[*c*]thiophene, octahydro-, *trans*-
S
142 $C_8H_{14}S$ 54004-35-2
Benzo[*b*]thiophene, octahydro-, *trans*-
S
142 $C_8H_{14}S$ 54004-36-3
Cyclopenta[*c*]thiopyran, octahydro-, *cis*-
S

142 $C_8H_{14}S$ 54004-37-4
Cyclopenta[*c*]thiopyran, octahydro-, *trans-*

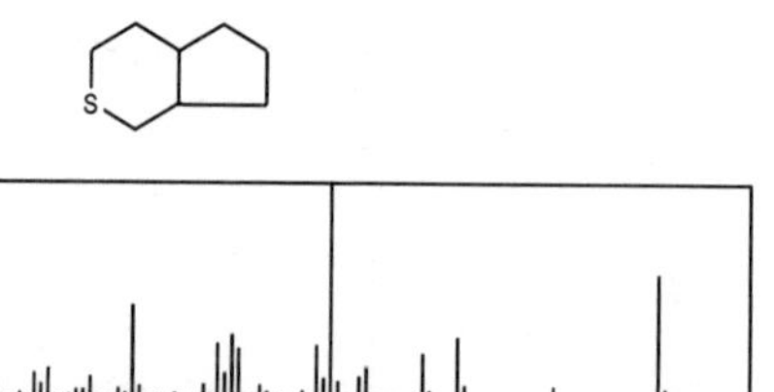

142 $C_8H_{14}S$ 54053-76-8
Benzo[*c*]thiophene, octahydro-, *cis-*

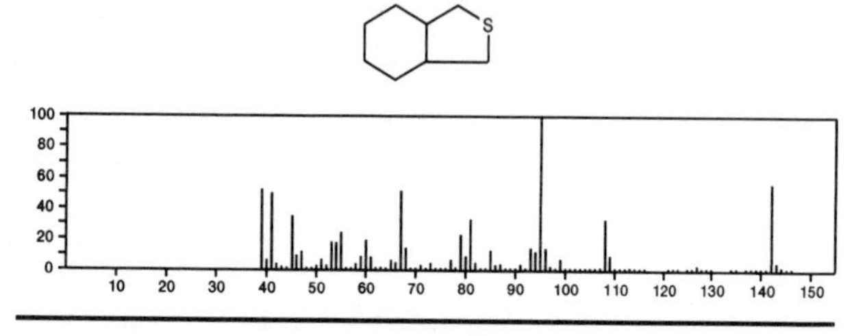

142 $C_8H_{18}N_2$ 927-83-3
Diazene, bis(1,1-dimethylethyl)-

t-BuN=NBu-t

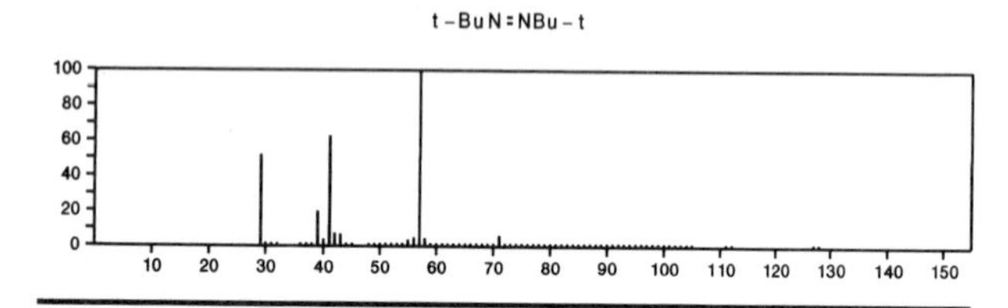

142 $C_8H_{18}Si$ 55956-01-9
Silacyclopentane, 1,1,2,5-tetramethyl-

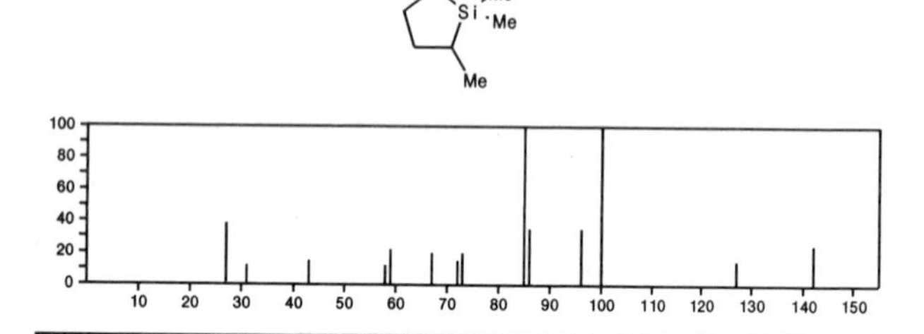

142 $C_9H_6N_2$ 876-31-3
Benzeneacetonitrile, 4-cyano-

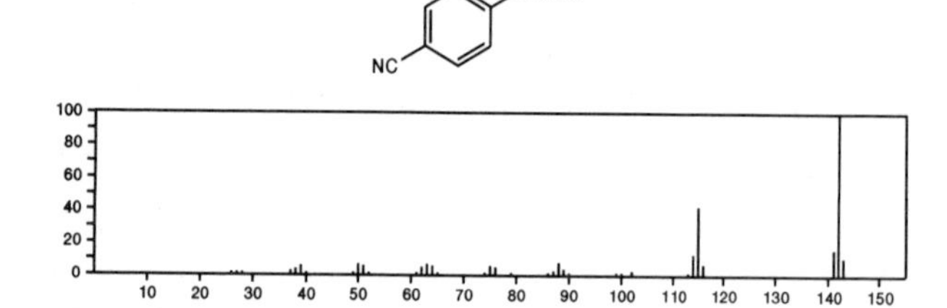

142 $C_9H_6N_2$ 3759-28-2
Benzeneacetonitrile, 2-cyano-

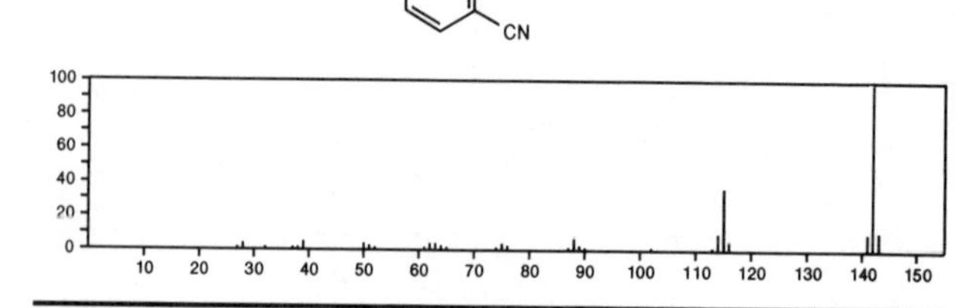

142 $C_9H_{15}F$ 20417-60-1
Bicyclo[2.2.2]octane, 1-fluoro-4-methyl-

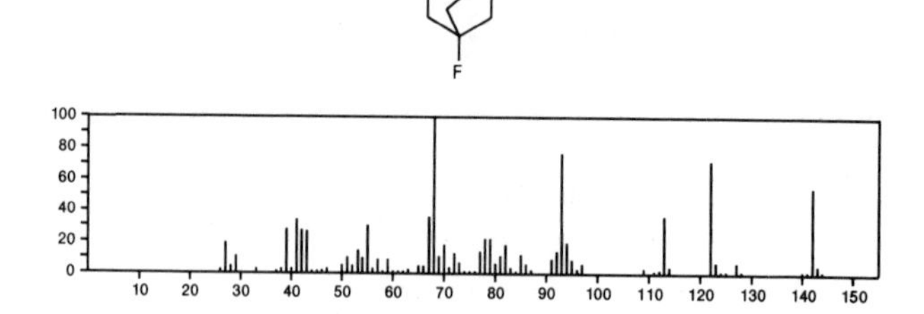

142 $C_9H_{15}F$ 55402-11-4
3-Heptyne, 7-fluoro-2,2-dimethyl-

$Me_3CC\equiv C(CH_2)_3F$

142 $C_9H_{18}O$ 96-07-1
Cyclohexanol, 2-(1-methylethyl)-

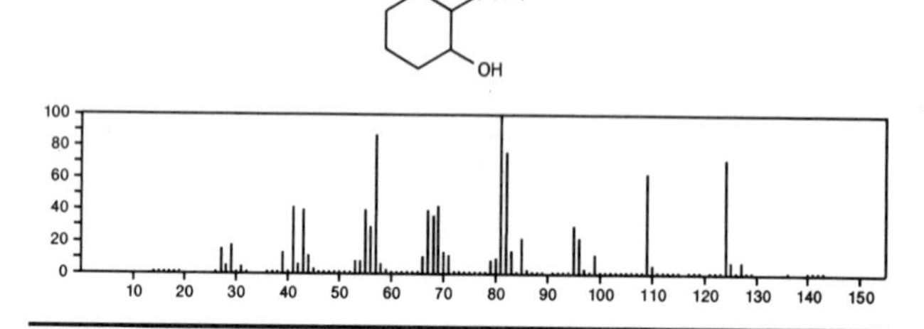

142 $C_9H_{18}O$ 108-83-8
4-Heptanone, 2,6-dimethyl-

$Me_2CHCH_2COCH_2CHMe_2$

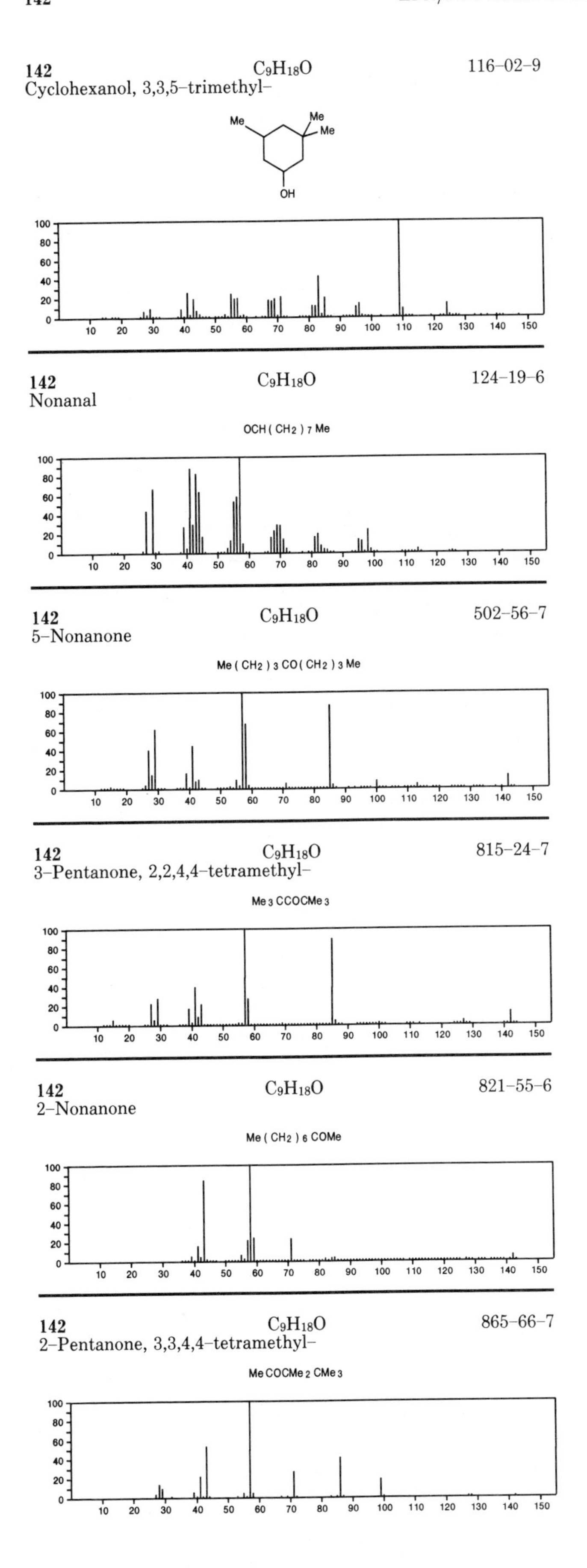
142 $C_9H_{18}O$ 116-02-9
Cyclohexanol, 3,3,5-trimethyl-
142 $C_9H_{18}O$ 124-19-6
Nonanal
OCH(CH2)7Me
142 $C_9H_{18}O$ 502-56-7
5-Nonanone
Me(CH2)3CO(CH2)3Me
142 $C_9H_{18}O$ 815-24-7
3-Pentanone, 2,2,4,4-tetramethyl-
Me3CCOCMe3
142 $C_9H_{18}O$ 821-55-6
2-Nonanone
Me(CH2)6COMe
142 $C_9H_{18}O$ 865-66-7
2-Pentanone, 3,3,4,4-tetramethyl-
MeCOCMe2CMe3

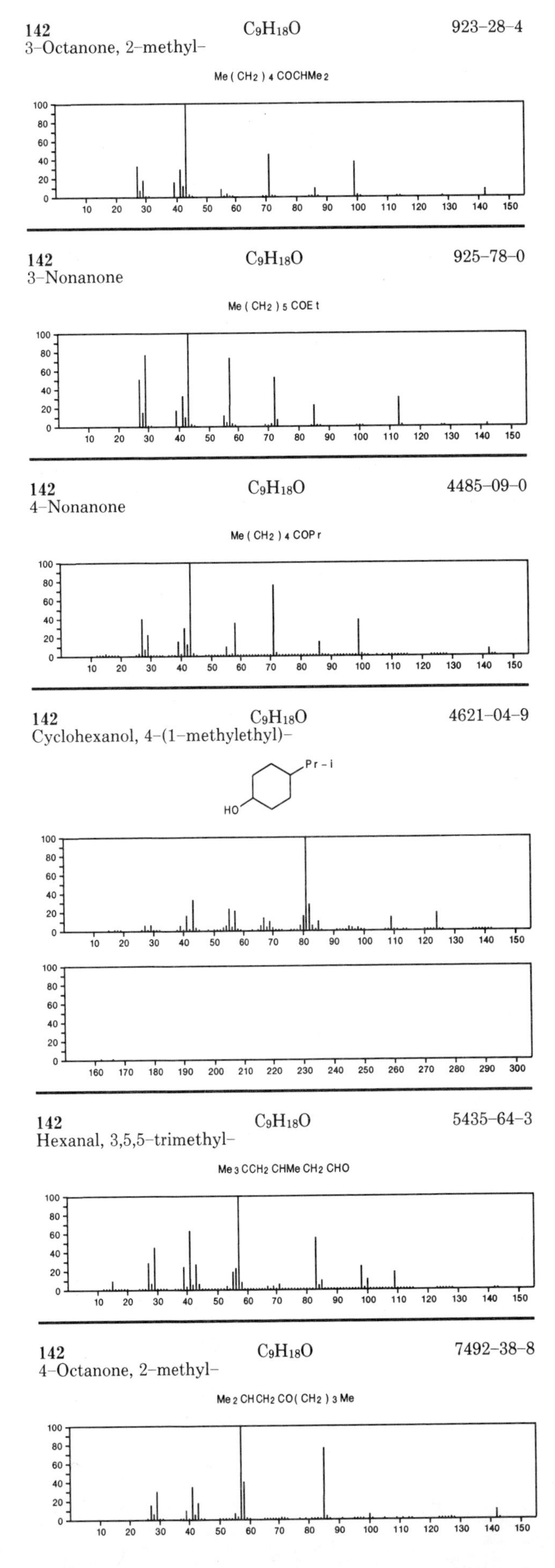
142 $C_9H_{18}O$ 923-28-4
3-Octanone, 2-methyl-
Me(CH2)4COCHMe2
142 $C_9H_{18}O$ 925-78-0
3-Nonanone
Me(CH2)5COEt
142 $C_9H_{18}O$ 4485-09-0
4-Nonanone
Me(CH2)4COPr
142 $C_9H_{18}O$ 4621-04-9
Cyclohexanol, 4-(1-methylethyl)-
142 $C_9H_{18}O$ 5435-64-3
Hexanal, 3,5,5-trimethyl-
Me3CCH2CHMeCH2CHO
142 $C_9H_{18}O$ 7492-38-8
4-Octanone, 2-methyl-
Me2CHCH2CO(CH2)3Me

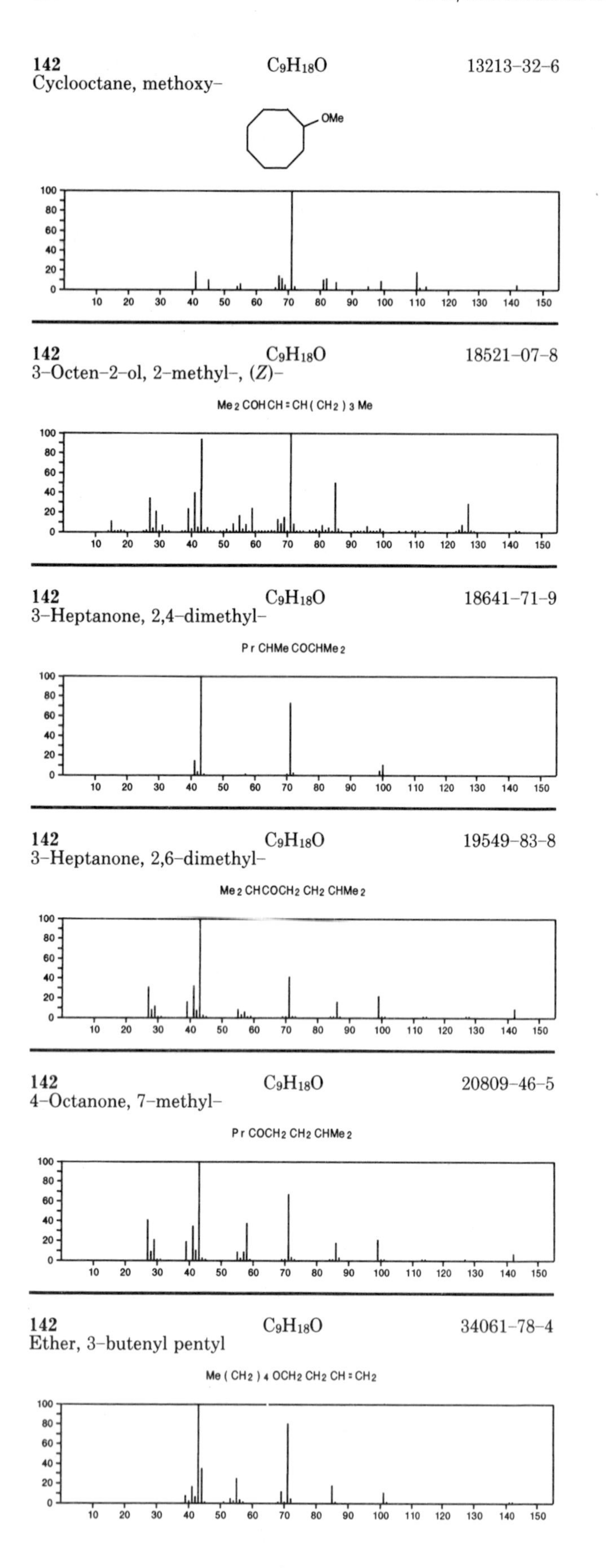

142 $C_9H_{18}O$ 13213-32-6
Cyclooctane, methoxy-

142 $C_9H_{18}O$ 18521-07-8
3-Octen-2-ol, 2-methyl-, (*Z*)-
Me2COHCH=CH(CH2)3Me

142 $C_9H_{18}O$ 18641-71-9
3-Heptanone, 2,4-dimethyl-
PrCHMeCOCHMe2

142 $C_9H_{18}O$ 19549-83-8
3-Heptanone, 2,6-dimethyl-
Me2CHCOCH2CH2CHMe2

142 $C_9H_{18}O$ 20809-46-5
4-Octanone, 7-methyl-
PrCOCH2CH2CHMe2

142 $C_9H_{18}O$ 34061-78-4
Ether, 3-butenyl pentyl
Me(CH2)4OCH2CH2CH=CH2

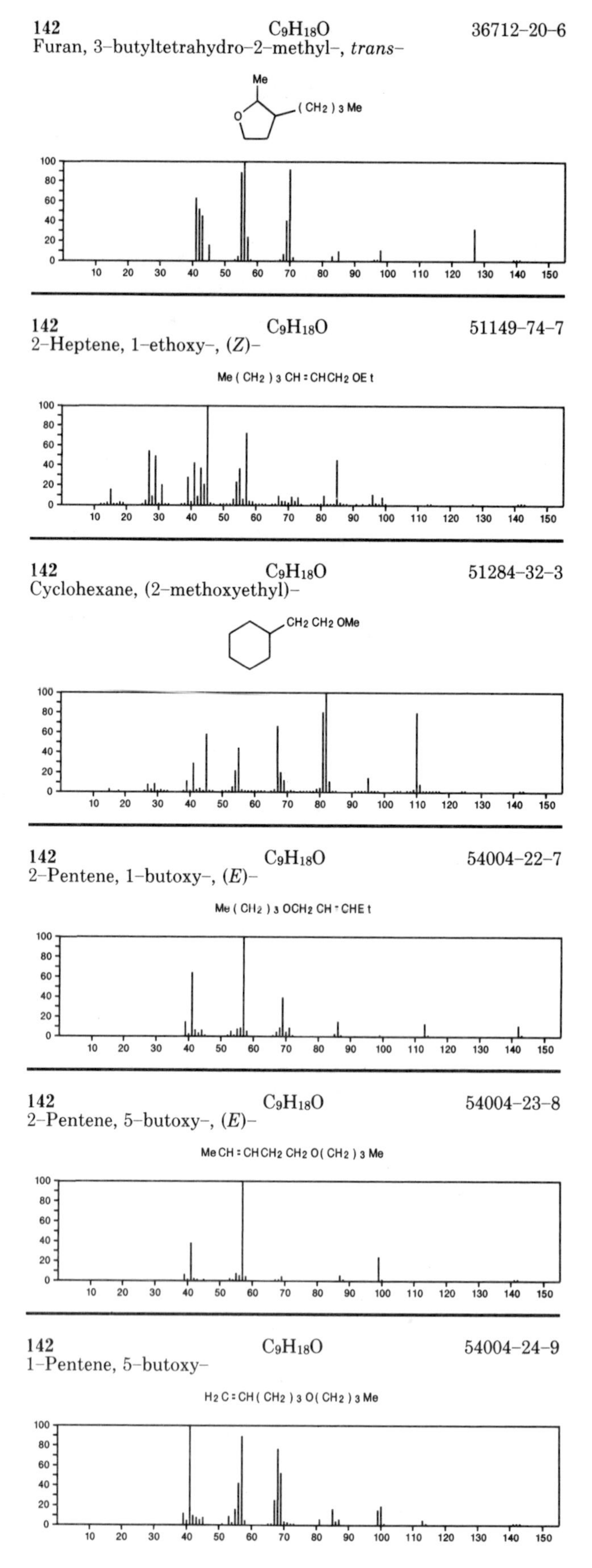

142 $C_9H_{18}O$ 36712-20-6
Furan, 3-butyltetrahydro-2-methyl-, *trans*-

142 $C_9H_{18}O$ 51149-74-7
2-Heptene, 1-ethoxy-, (*Z*)-
Me(CH2)3CH=CHCH2OEt

142 $C_9H_{18}O$ 51284-32-3
Cyclohexane, (2-methoxyethyl)-

142 $C_9H_{18}O$ 54004-22-7
2-Pentene, 1-butoxy-, (*E*)-
Me(CH2)3OCH2CH=CHEt

142 $C_9H_{18}O$ 54004-23-8
2-Pentene, 5-butoxy-, (*E*)-
MeCH=CHCH2CH2O(CH2)3Me

142 $C_9H_{18}O$ 54004-24-9
1-Pentene, 5-butoxy-
H2C=CH(CH2)3O(CH2)3Me

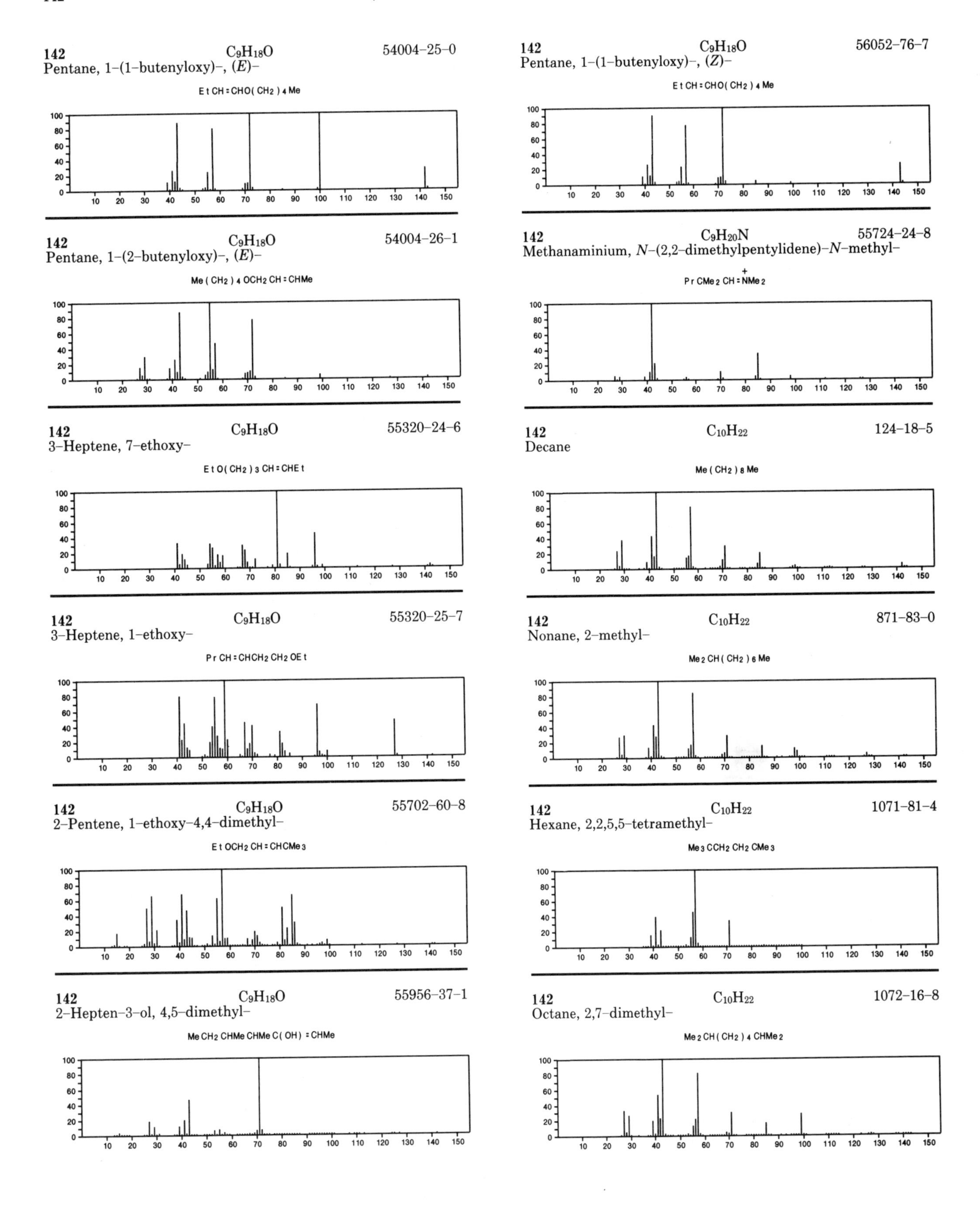
142 $C_9H_{18}O$ 54004-25-0
Pentane, 1-(1-butenyloxy)-, (*E*)-
Et CH = CHO(CH2) 4 Me
142 $C_9H_{18}O$ 56052-76-7
Pentane, 1-(1-butenyloxy)-, (*Z*)-
Et CH = CHO(CH2) 4 Me
142 $C_9H_{18}O$ 54004-26-1
Pentane, 1-(2-butenyloxy)-, (*E*)-
Me (CH2) 4 OCH2 CH = CHMe
142 $C_9H_{20}N$ 55724-24-8
Methanaminium, *N*-(2,2-dimethylpentylidene)-*N*-methyl-
Pr CMe2 CH = N+Me2
142 $C_9H_{18}O$ 55320-24-6
3-Heptene, 7-ethoxy-
Et O(CH2) 3 CH = CHEt
142 $C_{10}H_{22}$ 124-18-5
Decane
Me (CH2) 8 Me
142 $C_9H_{18}O$ 55320-25-7
3-Heptene, 1-ethoxy-
Pr CH = CHCH2 CH2 OEt
142 $C_{10}H_{22}$ 871-83-0
Nonane, 2-methyl-
Me2 CH (CH2) 6 Me
142 $C_9H_{18}O$ 55702-60-8
2-Pentene, 1-ethoxy-4,4-dimethyl-
Et OCH2 CH = CHCMe3
142 $C_{10}H_{22}$ 1071-81-4
Hexane, 2,2,5,5-tetramethyl-
Me3 CCH2 CH2 CMe3
142 $C_9H_{18}O$ 55956-37-1
2-Hepten-3-ol, 4,5-dimethyl-
Me CH2 CHMe CHMe C(OH) = CHMe
142 $C_{10}H_{22}$ 1072-16-8
Octane, 2,7-dimethyl-
Me2 CH (CH2) 4 CHMe2

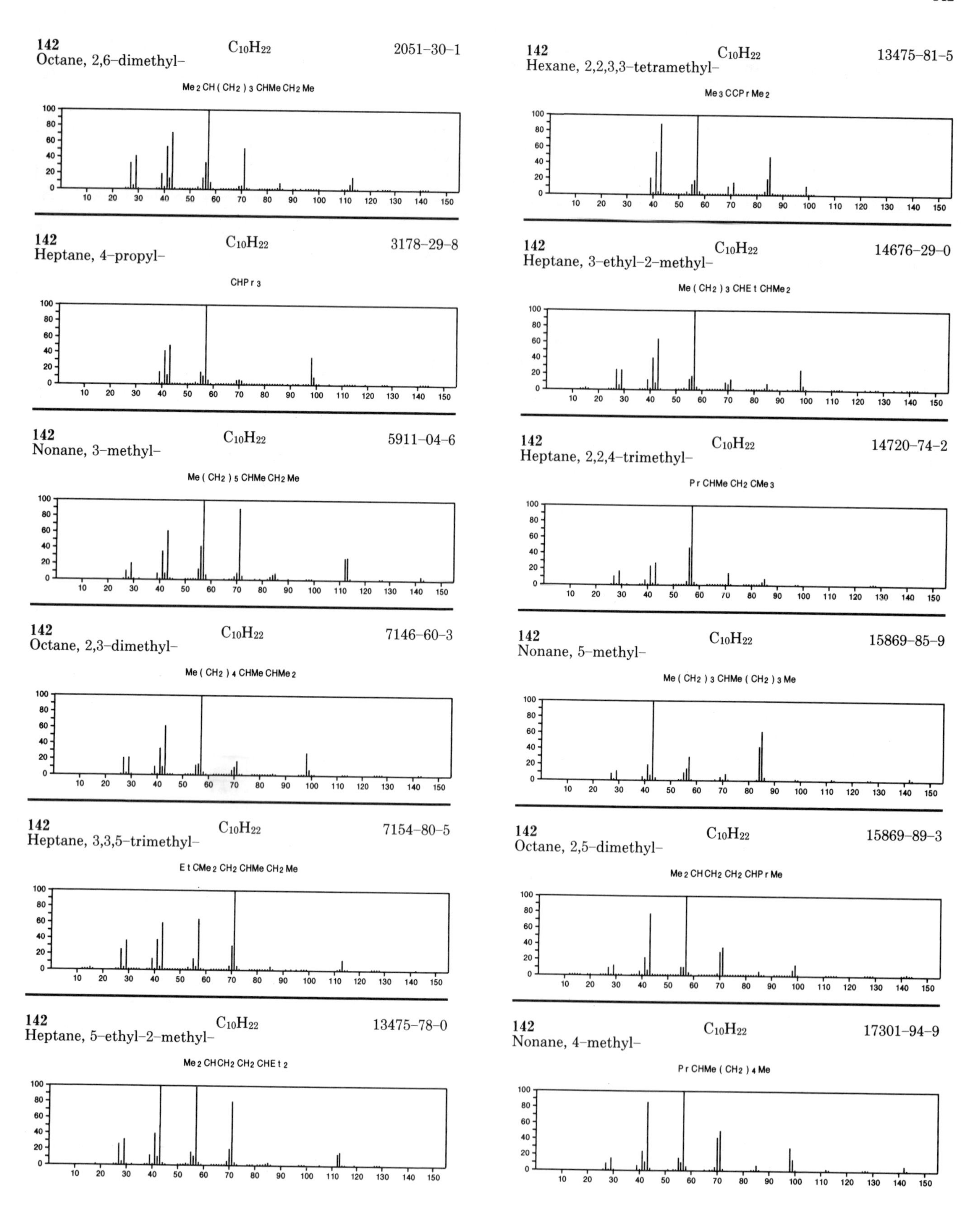
142
Octane, 2,6-dimethyl-
C10H22
2051-30-1
Me2CH(CH2)3CHMeCH2Me
142
Hexane, 2,2,3,3-tetramethyl-
C10H22
13475-81-5
Me3CCPrMe2
142
Heptane, 4-propyl-
C10H22
3178-29-8
CHPr3
142
Heptane, 3-ethyl-2-methyl-
C10H22
14676-29-0
Me(CH2)3CHEtCHMe2
142
Nonane, 3-methyl-
C10H22
5911-04-6
Me(CH2)5CHMeCH2Me
142
Heptane, 2,2,4-trimethyl-
C10H22
14720-74-2
PrCHMeCH2CMe3
142
Octane, 2,3-dimethyl-
C10H22
7146-60-3
Me(CH2)4CHMeCHMe2
142
Nonane, 5-methyl-
C10H22
15869-85-9
Me(CH2)3CHMe(CH2)3Me
142
Heptane, 3,3,5-trimethyl-
C10H22
7154-80-5
EtCMe2CH2CHMeCH2Me
142
Octane, 2,5-dimethyl-
C10H22
15869-89-3
Me2CHCH2CH2CHPrMe
142
Heptane, 5-ethyl-2-methyl-
C10H22
13475-78-0
Me2CHCH2CH2CHEt2
142
Nonane, 4-methyl-
C10H22
17301-94-9
PrCHMe(CH2)4Me

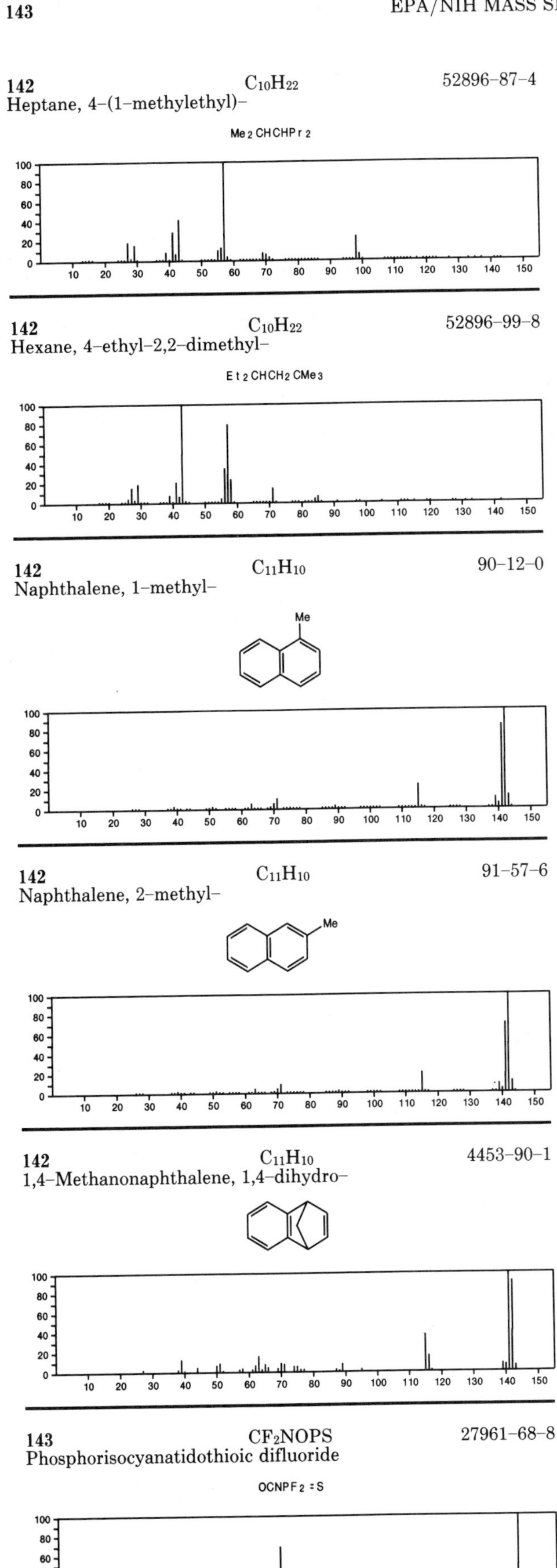

142 $C_{10}H_{22}$ 52896-87-4
Heptane, 4-(1-methylethyl)-

$Me_2CHCHPr_2$

142 $C_{10}H_{22}$ 52896-99-8
Hexane, 4-ethyl-2,2-dimethyl-

$Et_2CHCH_2CMe_3$

142 $C_{11}H_{10}$ 90-12-0
Naphthalene, 1-methyl-

142 $C_{11}H_{10}$ 91-57-6
Naphthalene, 2-methyl-

142 $C_{11}H_{10}$ 4453-90-1
1,4-Methanonaphthalene, 1,4-dihydro-

143 CF_2NOPS 27961-68-8
Phosphorisocyanatidothioic difluoride

$OCNPF_2=S$

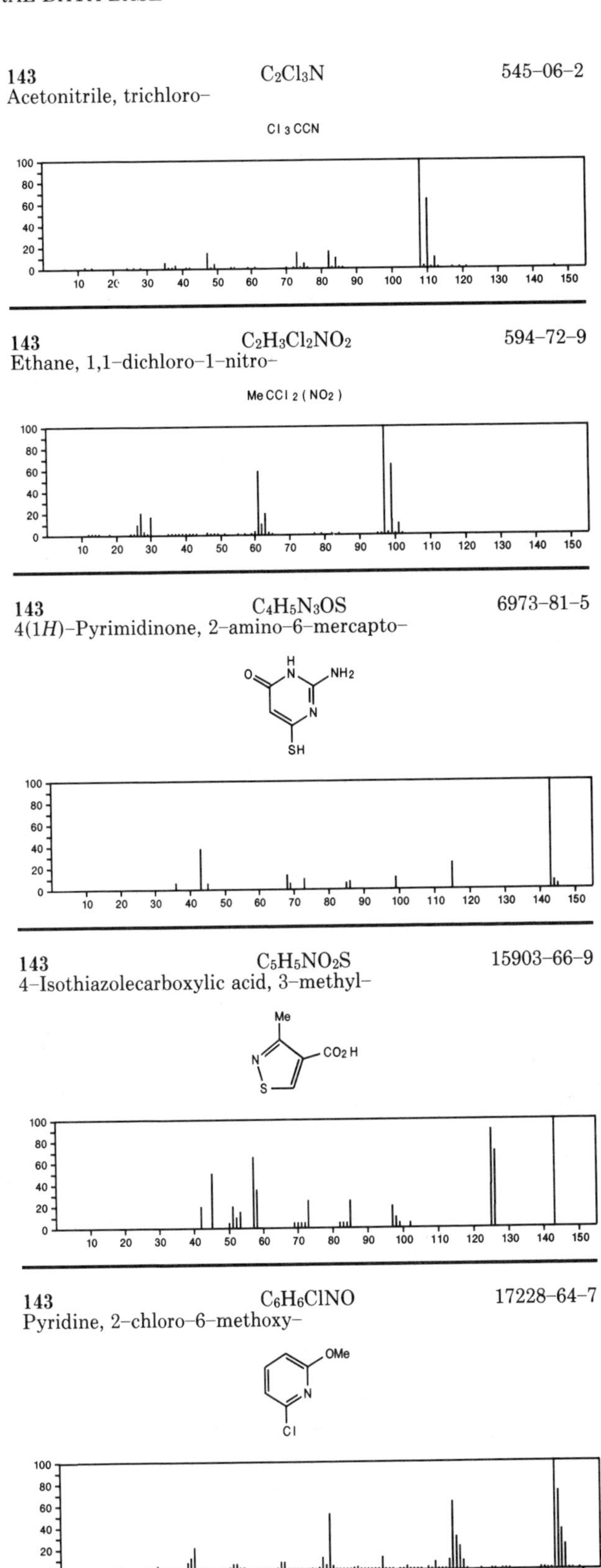

143 C_2Cl_3N 545-06-2
Acetonitrile, trichloro-

Cl_3CCN

143 $C_2H_3Cl_2NO_2$ 594-72-9
Ethane, 1,1-dichloro-1-nitro-

$MeCCl_2(NO_2)$

143 $C_4H_5N_3OS$ 6973-81-5
4(1*H*)-Pyrimidinone, 2-amino-6-mercapto-

143 $C_5H_5NO_2S$ 15903-66-9
4-Isothiazolecarboxylic acid, 3-methyl-

143 C_6H_6ClNO 17228-64-7
Pyridine, 2-chloro-6-methoxy-

143 C_6H_9NOS 137–00–8
5–Thiazoleethanol, 4–methyl–

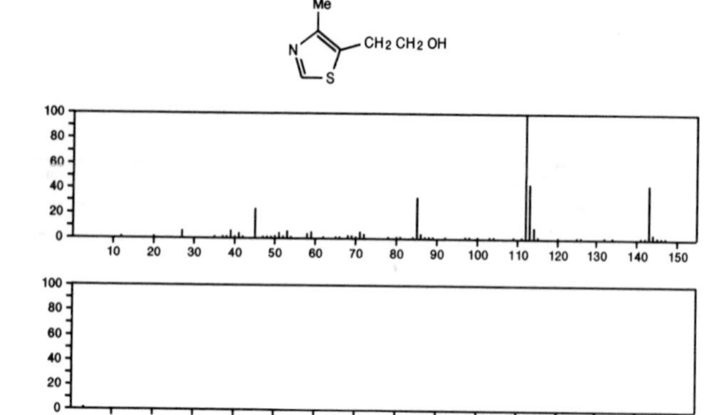

143 $C_6H_9NO_3$ 127–48–0
2,4–Oxazolidinedione, 3,5,5–trimethyl–

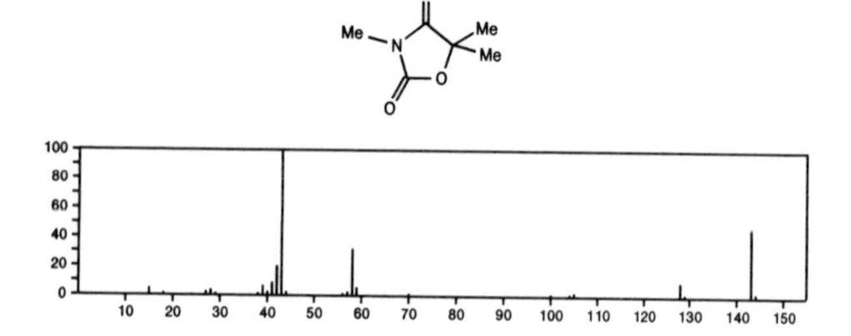

143 $C_6H_9NO_3$ 4931–66–2
L–Proline, 5–oxo–, methyl ester

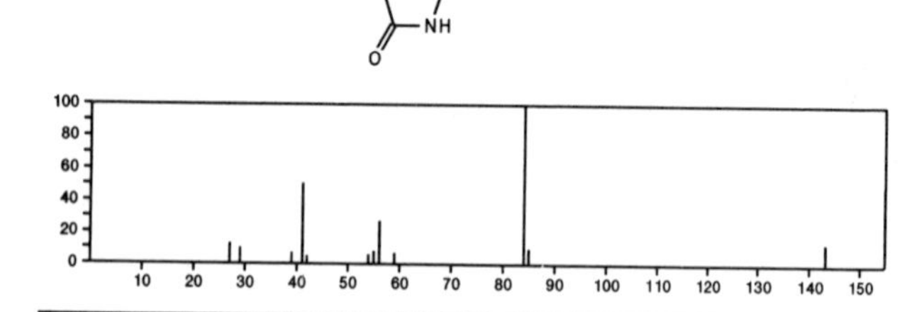

143 $C_6H_9NO_3$ 55649–71–3
2–Butenoic acid, 2–(acetylamino)–

$AcNHC(CO_2H)=CHMe$

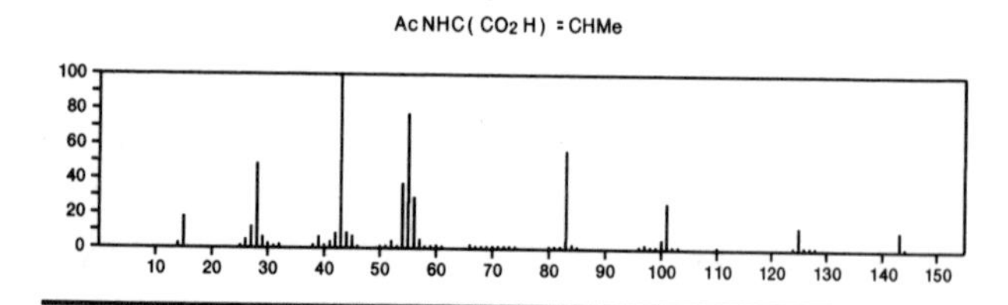

143 $C_6H_{13}N_3O$ 13183–22–7
Valeraldehyde, semicarbazone

$Me(CH_2)_3CH=NNHCONH_2$

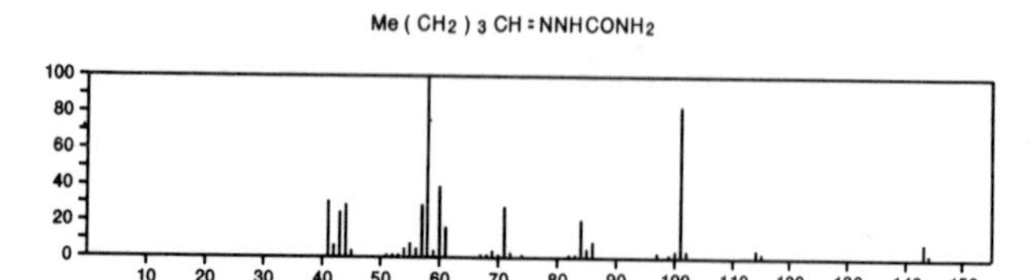

143 $C_6H_{13}N_3O$ 23809–33–8
Pivalaldehyde, semicarbazone

$H_2NCONHN=CHCMe_3$

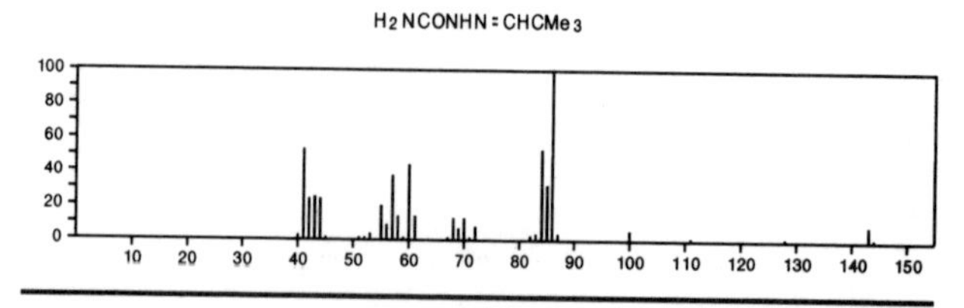

143 $C_7H_9N.ClH$ 540–23–8
Benzenamine, 4–methyl–, hydrochloride

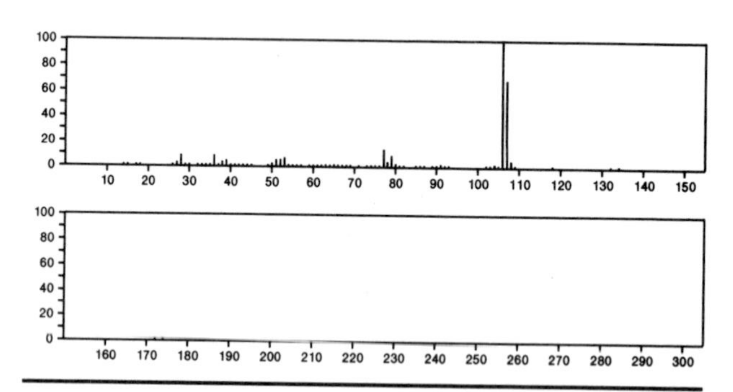

143 $C_7H_9N.ClH$ 636–21–5
Benzenamine, 2–methyl–, hydrochloride

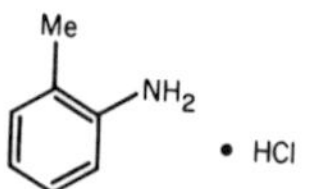

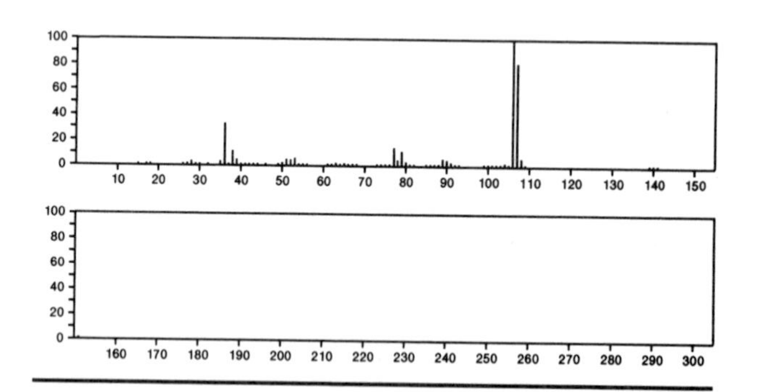

143 $C_7H_9N.ClH$ 638–03–9
Benzenamine, 3–methyl–, hydrochloride

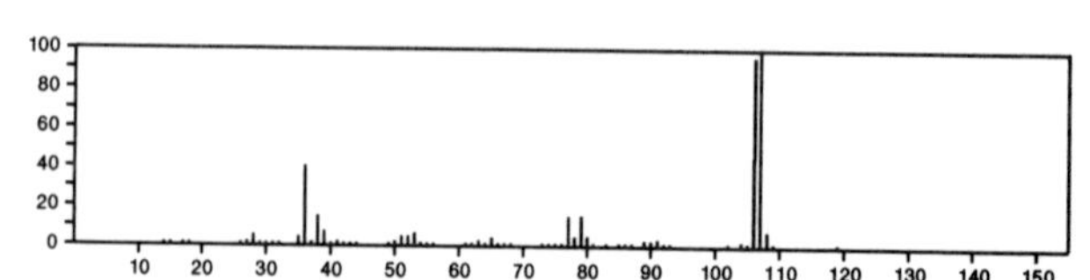

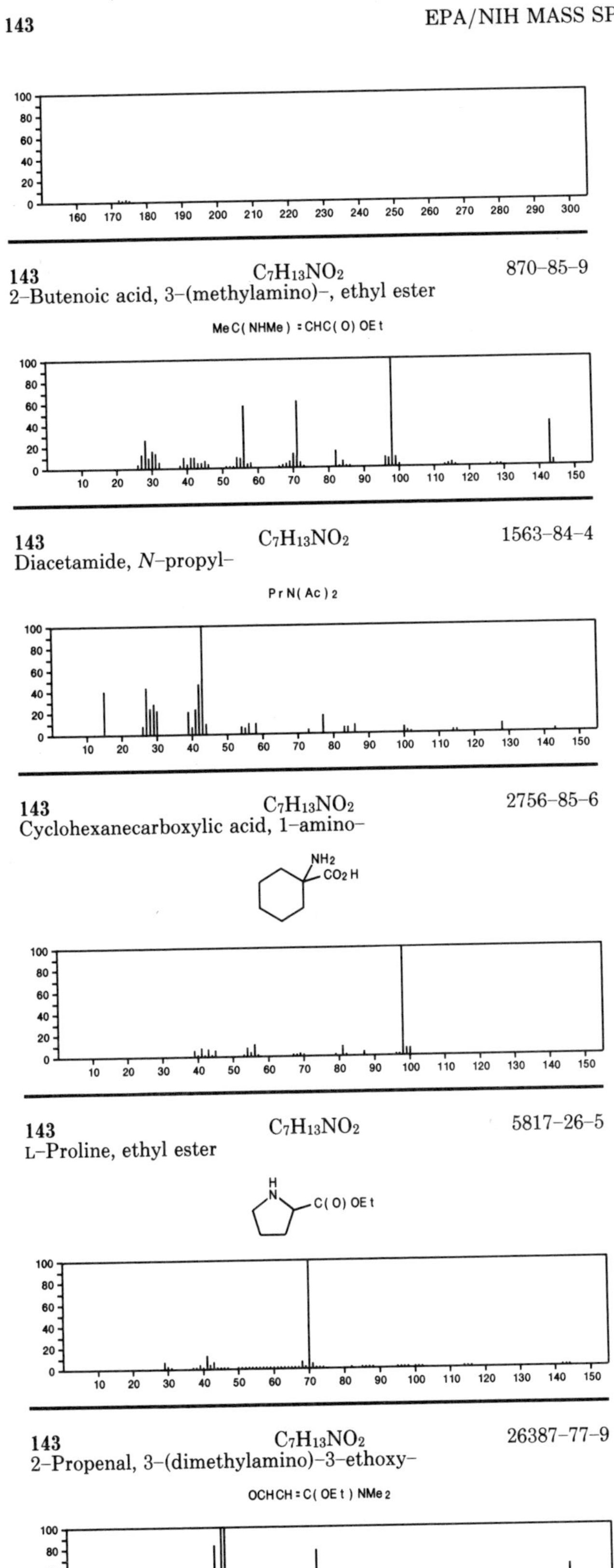

143 $C_7H_{13}NO_2$ 870-85-9
2-Butenoic acid, 3-(methylamino)-, ethyl ester
MeC(NHMe)=CHC(O)OEt
143 $C_7H_{13}NO_2$ 1563-84-4
Diacetamide, N-propyl-
PrN(Ac)2
143 $C_7H_{13}NO_2$ 2756-85-6
Cyclohexanecarboxylic acid, 1-amino-
NH2
CO2H
143 $C_7H_{13}NO_2$ 5817-26-5
L-Proline, ethyl ester
H
N
C(O)OEt
143 $C_7H_{13}NO_2$ 26387-77-9
2-Propenal, 3-(dimethylamino)-3-ethoxy-
OCHCH=C(OEt)NMe2

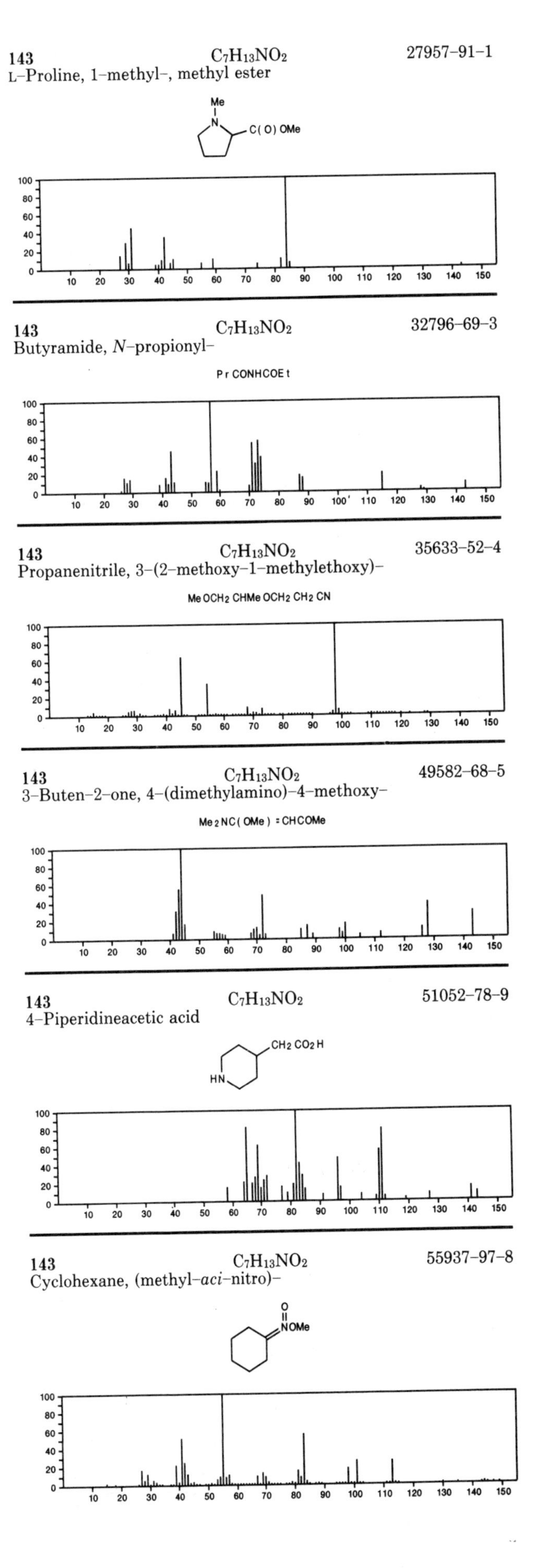

143 $C_7H_{13}NO_2$ 27957-91-1
L-Proline, 1-methyl-, methyl ester
Me
N
C(O)OMe
143 $C_7H_{13}NO_2$ 32796-69-3
Butyramide, N-propionyl-
PrCONHCOEt
143 $C_7H_{13}NO_2$ 35633-52-4
Propanenitrile, 3-(2-methoxy-1-methylethoxy)-
MeOCH2CHMeOCH2CH2CN
143 $C_7H_{13}NO_2$ 49582-68-5
3-Buten-2-one, 4-(dimethylamino)-4-methoxy-
Me2NC(OMe)=CHCOMe
143 $C_7H_{13}NO_2$ 51052-78-9
4-Piperidineacetic acid
CH2CO2H
HN
143 $C_7H_{13}NO_2$ 55937-97-8
Cyclohexane, (methyl-aci-nitro)-
O
NOMe

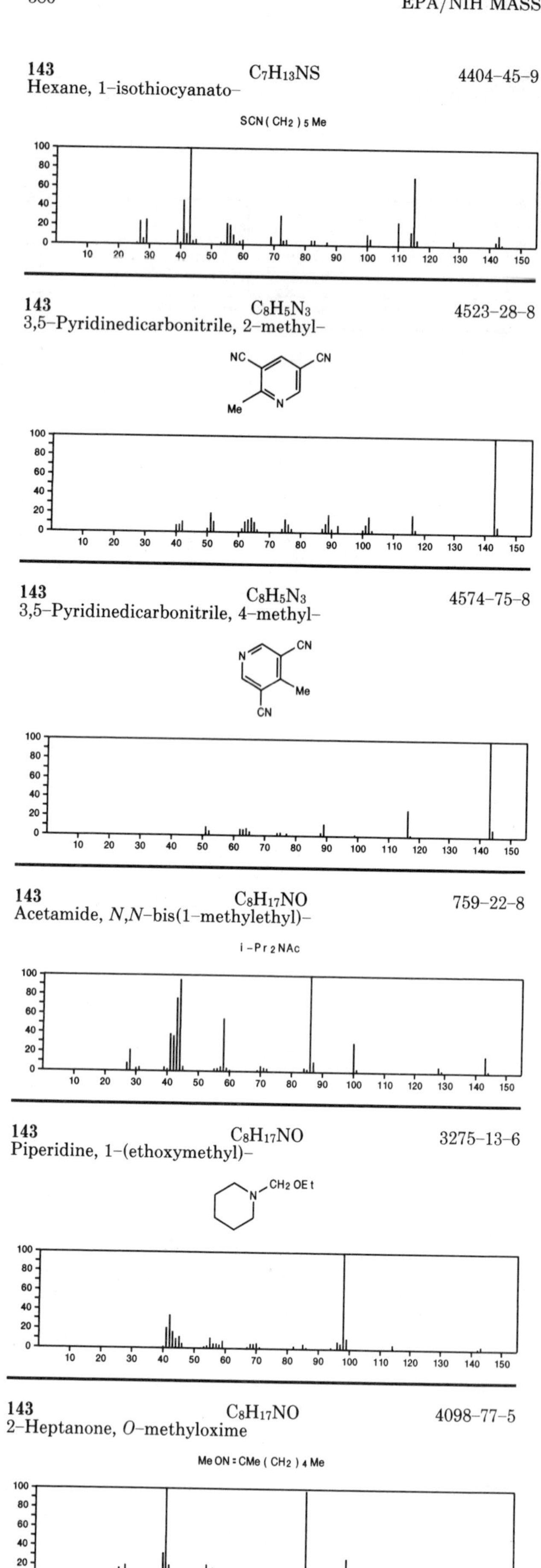

143 $C_7H_{13}NS$ 4404-45-9
Hexane, 1-isothiocyanato-
SCN(CH2)5Me

143 $C_8H_5N_3$ 4523-28-8
3,5-Pyridinedicarbonitrile, 2-methyl-

143 $C_8H_5N_3$ 4574-75-8
3,5-Pyridinedicarbonitrile, 4-methyl-

143 $C_8H_{17}NO$ 759-22-8
Acetamide, *N,N*-bis(1-methylethyl)-
i-Pr2NAc

143 $C_8H_{17}NO$ 3275-13-6
Piperidine, 1-(ethoxymethyl)-

143 $C_8H_{17}NO$ 4098-77-5
2-Heptanone, *O*-methyloxime
MeON=CMe(CH2)4Me

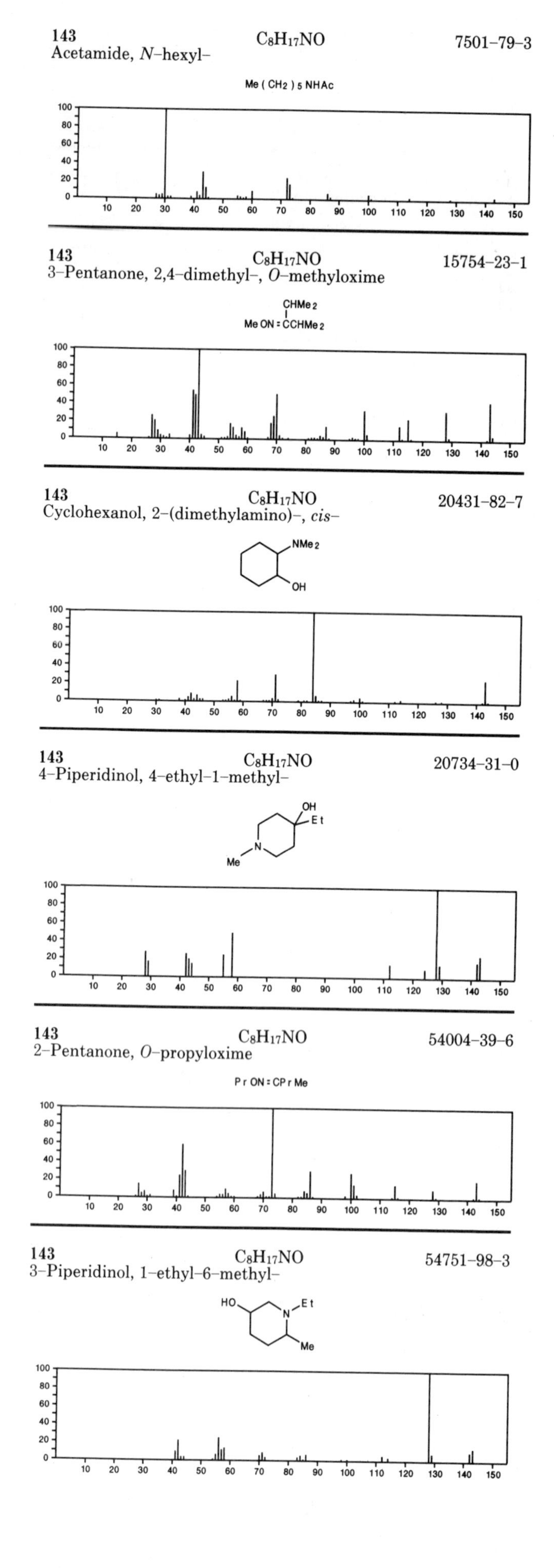

143 $C_8H_{17}NO$ 7501-79-3
Acetamide, *N*-hexyl-
Me(CH2)5NHAc

143 $C_8H_{17}NO$ 15754-23-1
3-Pentanone, 2,4-dimethyl-, *O*-methyloxime
MeON=C(CHMe2)CHMe2

143 $C_8H_{17}NO$ 20431-82-7
Cyclohexanol, 2-(dimethylamino)-, *cis*-

143 $C_8H_{17}NO$ 20734-31-0
4-Piperidinol, 4-ethyl-1-methyl-

143 $C_8H_{17}NO$ 54004-39-6
2-Pentanone, *O*-propyloxime
PrON=CPrMe

143 $C_8H_{17}NO$ 54751-98-3
3-Piperidinol, 1-ethyl-6-methyl-

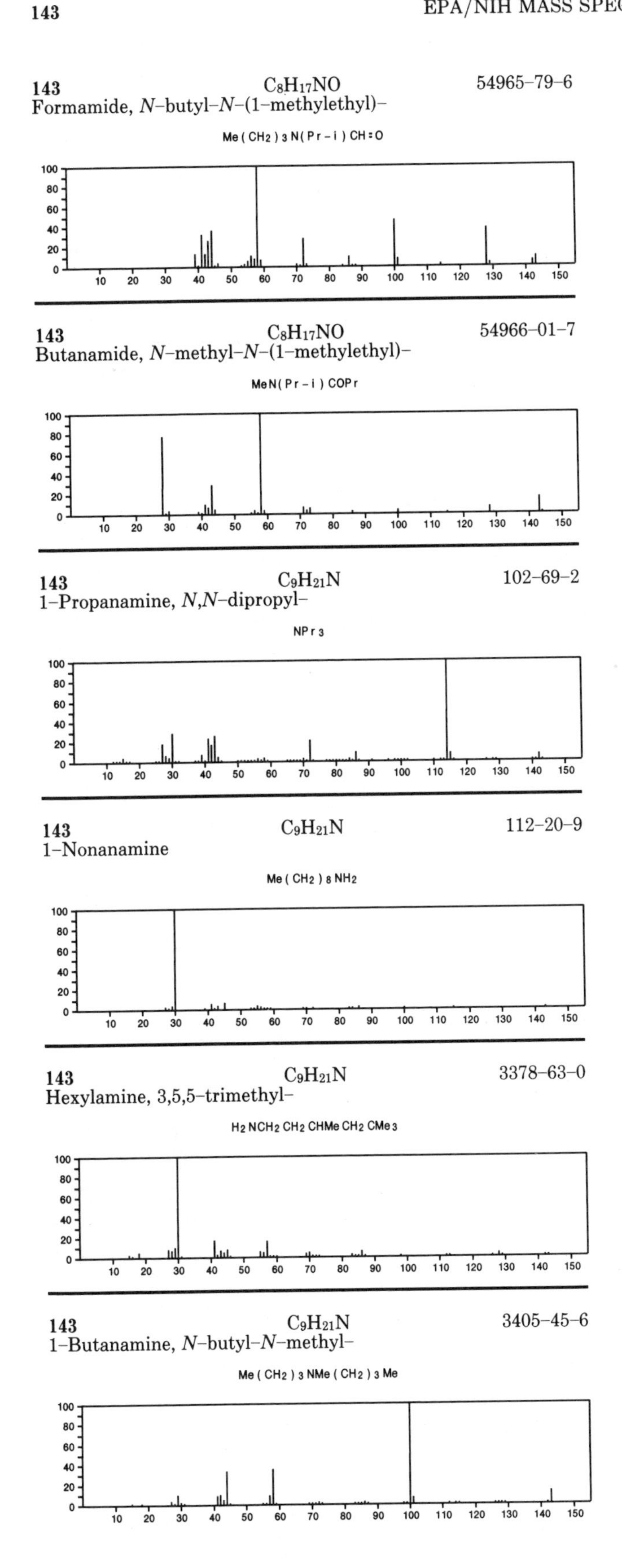

143 $C_8H_{17}NO$ 54965-79-6
Formamide, *N*-butyl-*N*-(1-methylethyl)-
$Me(CH_2)_3N(Pr\text{-}i)CH{=}O$

143 $C_8H_{17}NO$ 54966-01-7
Butanamide, *N*-methyl-*N*-(1-methylethyl)-
$MeN(Pr\text{-}i)COPr$

143 $C_9H_{21}N$ 102-69-2
1-Propanamine, *N*,*N*-dipropyl-
NPr_3

143 $C_9H_{21}N$ 112-20-9
1-Nonanamine
$Me(CH_2)_8NH_2$

143 $C_9H_{21}N$ 3378-63-0
Hexylamine, 3,5,5-trimethyl-
$H_2NCH_2CH_2CHMeCH_2CMe_3$

143 $C_9H_{21}N$ 3405-45-6
1-Butanamine, *N*-butyl-*N*-methyl-
$Me(CH_2)_3NMe(CH_2)_3Me$

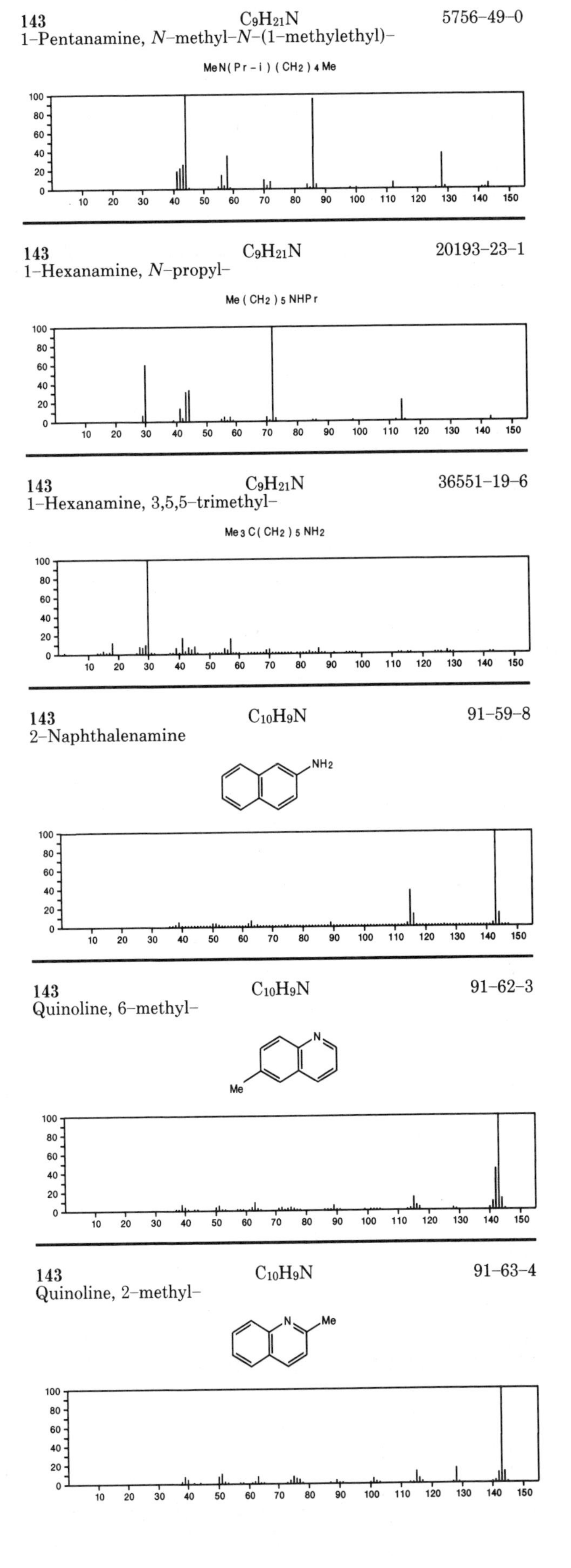

143 $C_9H_{21}N$ 5756-49-0
1-Pentanamine, *N*-methyl-*N*-(1-methylethyl)-
$MeN(Pr\text{-}i)(CH_2)_4Me$

143 $C_9H_{21}N$ 20193-23-1
1-Hexanamine, *N*-propyl-
$Me(CH_2)_5NHPr$

143 $C_9H_{21}N$ 36551-19-6
1-Hexanamine, 3,5,5-trimethyl-
$Me_3C(CH_2)_5NH_2$

143 $C_{10}H_9N$ 91-59-8
2-Naphthalenamine
NH_2

143 $C_{10}H_9N$ 91-62-3
Quinoline, 6-methyl-
N, Me

143 $C_{10}H_9N$ 91-63-4
Quinoline, 2-methyl-
N, Me

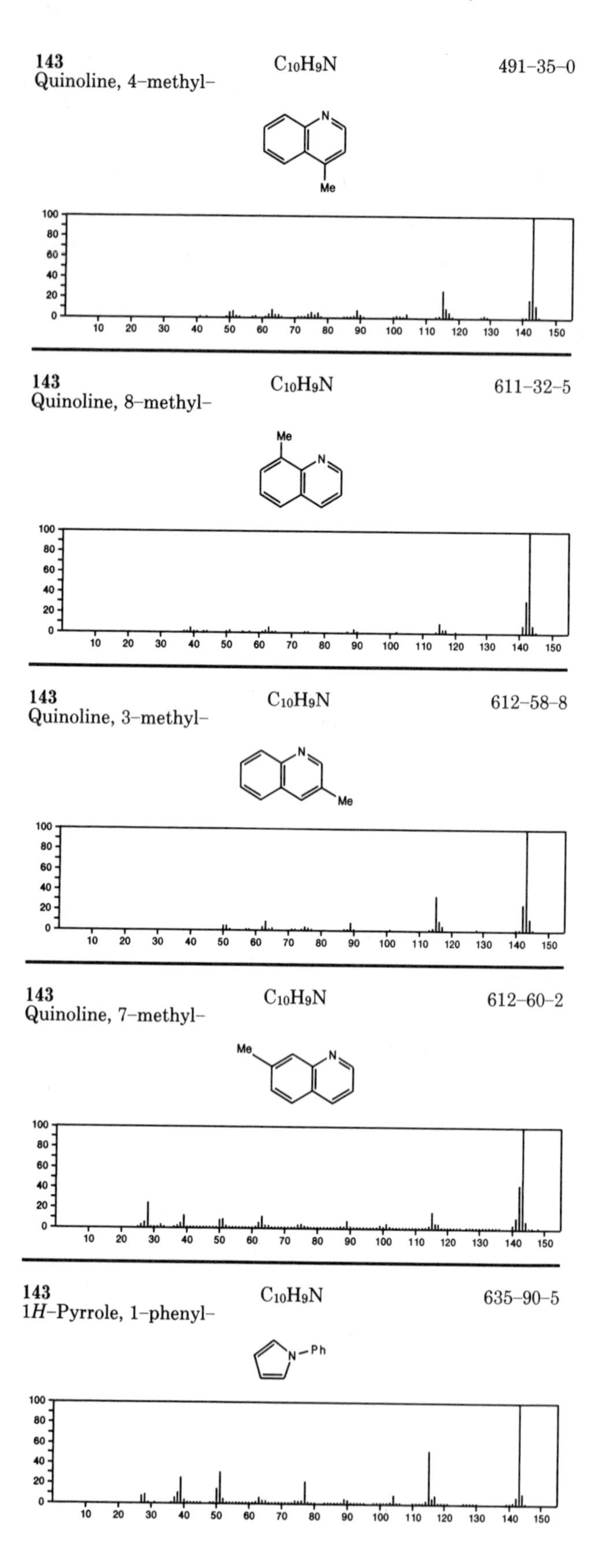
143
Quinoline, 4-methyl-
$C_{10}H_9N$
491-35-0
Me
N
143
Quinoline, 8-methyl-
$C_{10}H_9N$
611-32-5
Me
N
143
Quinoline, 3-methyl-
$C_{10}H_9N$
612-58-8
N
Me
143
Quinoline, 7-methyl-
$C_{10}H_9N$
612-60-2
Me
N
143
1H-Pyrrole, 1-phenyl-
$C_{10}H_9N$
635-90-5
N—Ph
100
80
60
40
20
0
10 20 30 40 50 60 70 80 90 100 110 120 130 140 150

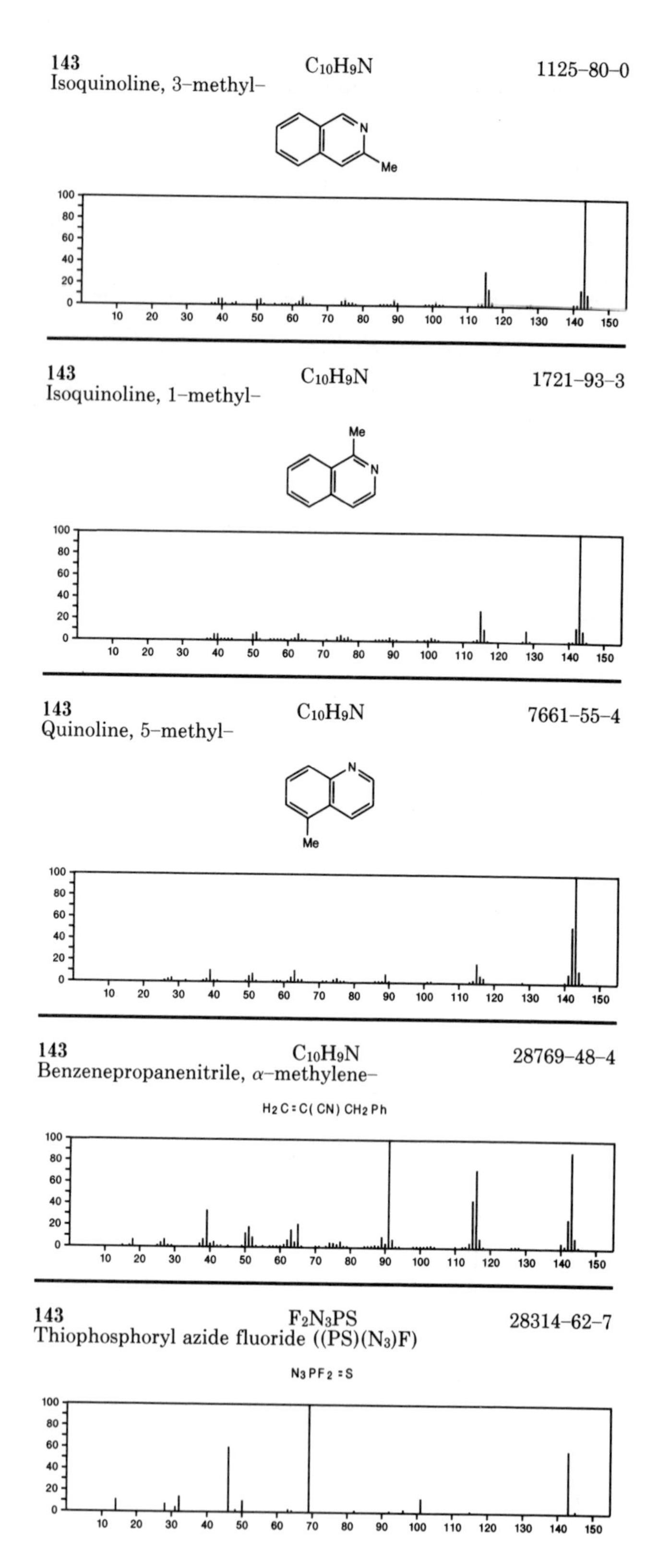
143
Isoquinoline, 3-methyl-
$C_{10}H_9N$
1125-80-0
N
Me
143
Isoquinoline, 1-methyl-
$C_{10}H_9N$
1721-93-3
Me
N
143
Quinoline, 5-methyl-
$C_{10}H_9N$
7661-55-4
N
Me
143
Benzenepropanenitrile, α-methylene-
$C_{10}H_9N$
28769-48-4
$H_2C=C(CN)CH_2Ph$
143
Thiophosphoryl azide fluoride ((PS)(N3)F)
F_2N_3PS
28314-62-7
$N_3PF_2=S$
100
80
60
40
20
0
10 20 30 40 50 60 70 80 90 100 110 120 130 140 150

143 F_5H_2NS 15192-28-6
Sulfur amide fluoride ($S(NH_2)F_5$), (*OC*-6-21)-

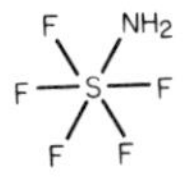

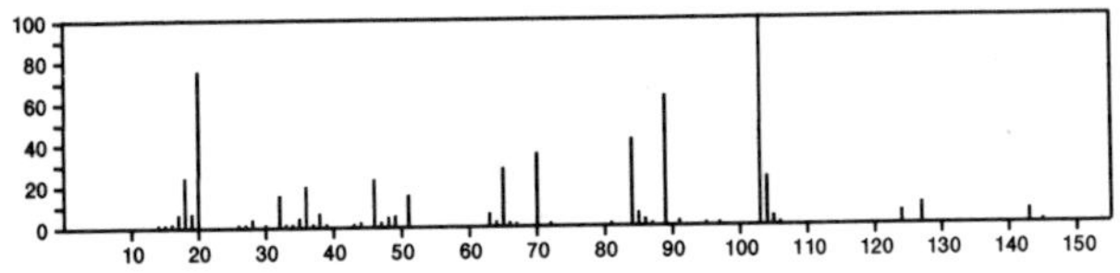

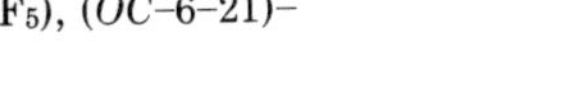

144 $C_2H_6ClO_3P$ 813-77-4
Phosphorochloridic acid, dimethyl ester

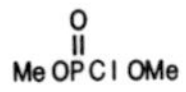

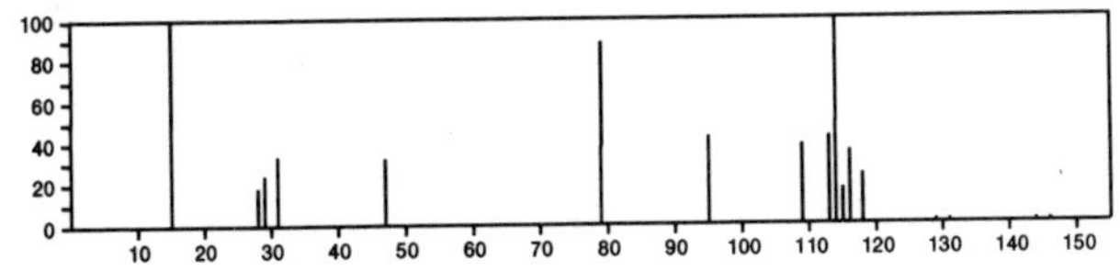

144 $C_3H_3Cl_3$ 2233-00-3
1-Propene, 3,3,3-trichloro-

$Cl_3CCH=CH_2$

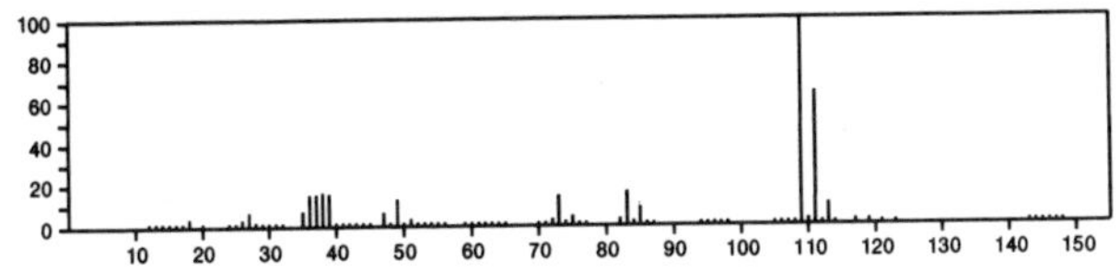

144 $C_3H_3Cl_3$ 2567-14-8
1-Propene, 1,1,3-trichloro-

$ClCH_2CH=CCl_2$

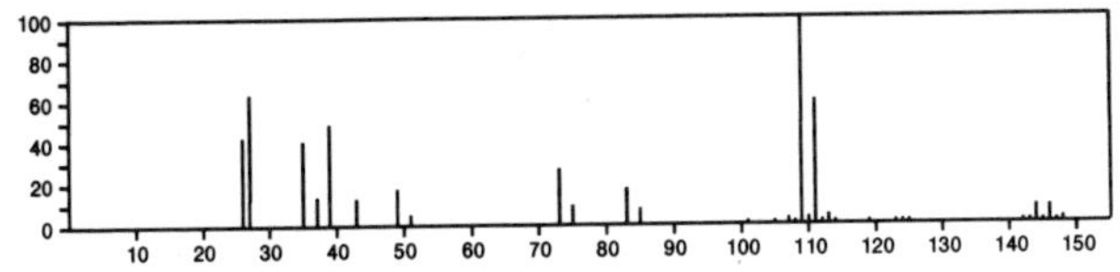

144 $C_3H_3Cl_3$ 21400-25-9
1-Propene, 1,1,2-trichloro-

$Cl_2C=CClMe$

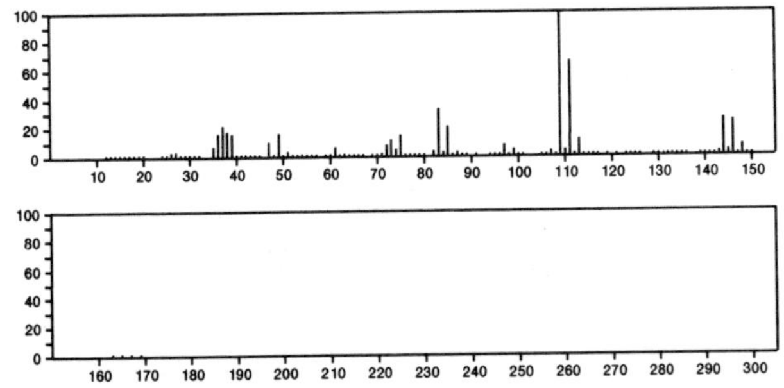

144 $C_4H_4N_2O_2S$ 22097-10-5
1,2,3-Thiadiazole-4-carboxylic acid, 5-methyl-

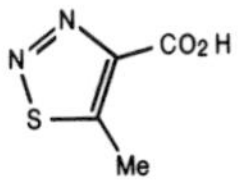

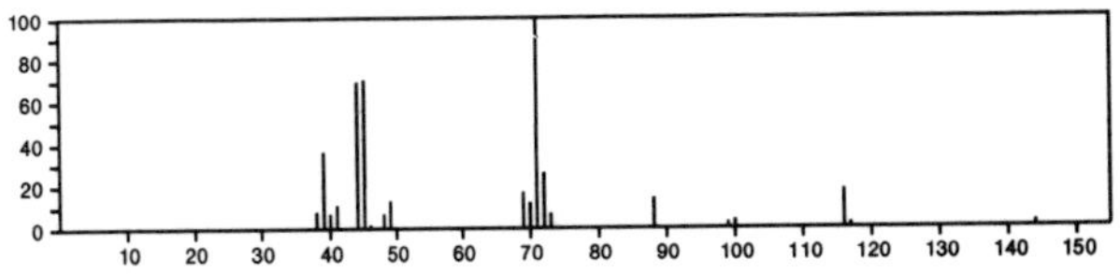

144 $C_4H_4N_2O_4$ 26537-53-1
Sydnone, 3-(carboxymethyl)-

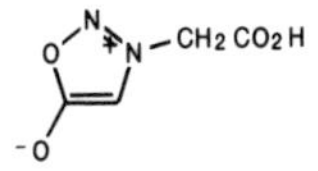

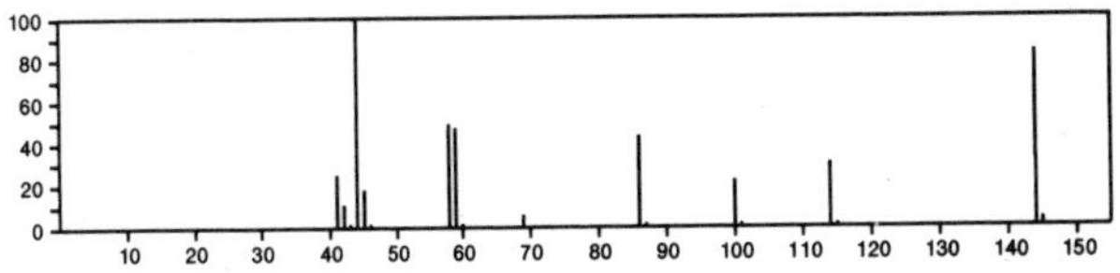

144 $C_4H_5ClN_4$ 14631-08-4
4,5-Pyrimidinediamine, 2-chloro-

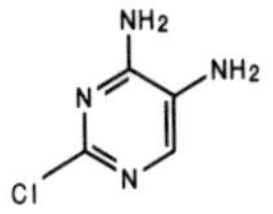

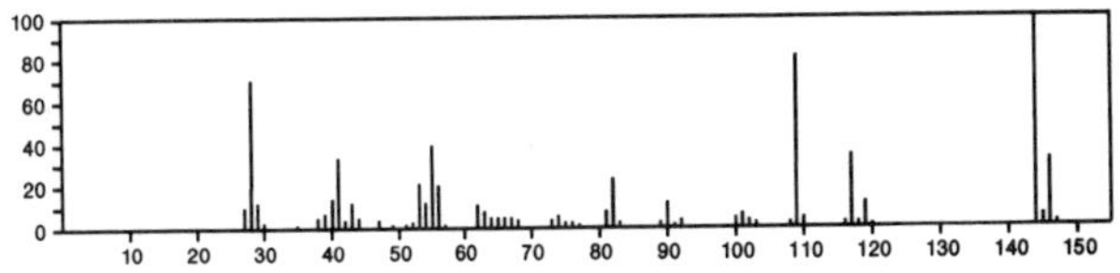

144 $C_4H_8N_4O_2$ 140-79-4
Piperazine, 1,4-dinitroso-

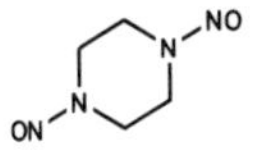

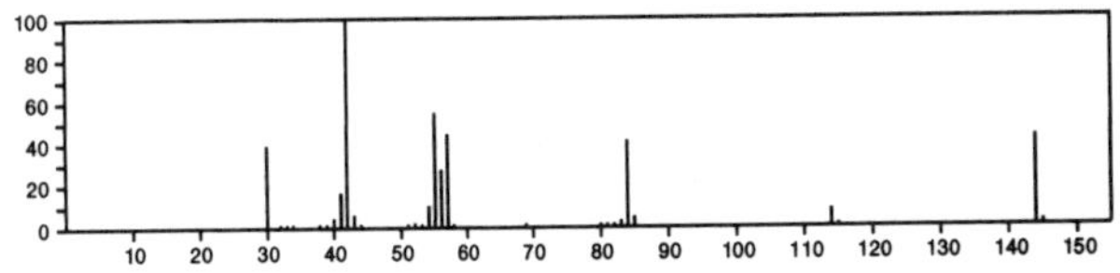

144 $C_4H_8N_4O_2$ 35975-29-2
2,6-Piperazinedione, dioxime

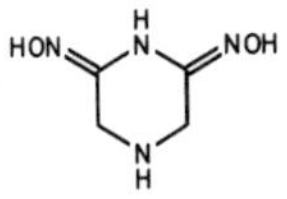

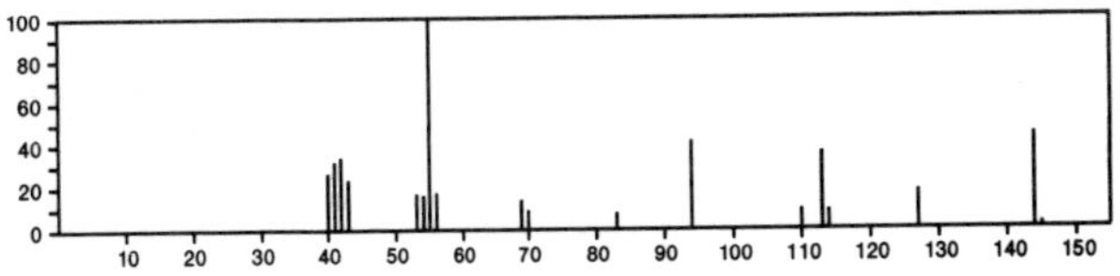

144 $C_4H_8N_4S$ 10455-64-8
3-Thiazolidinecarboxamidine, 2-imino-

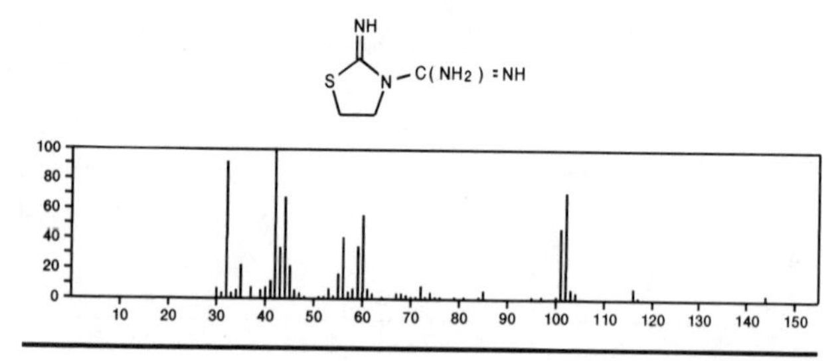

144 $C_5H_4OS_2$ 5694-59-7
Acetaldehyde, 3*H*-1,2-dithiol-3-ylidene-

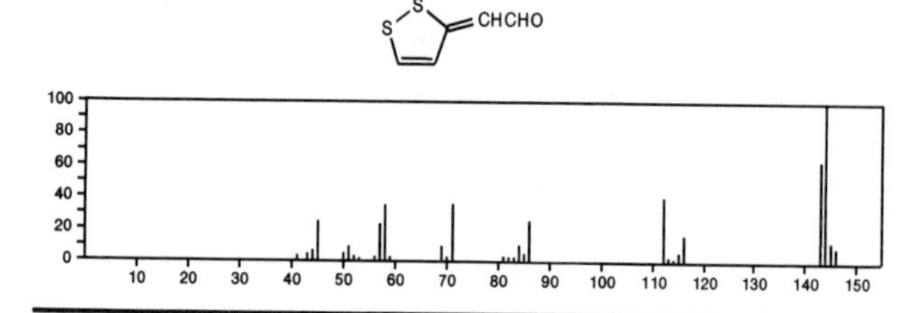

144 $C_5H_4O_3S$ 17396-38-2
4*H*-Thiopyran-4-one, 1,1-dioxide

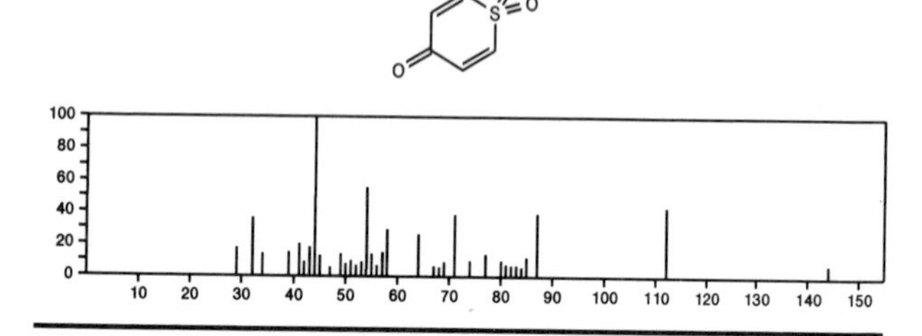

144 $C_5H_5ClN_2O$ 1722-10-7
Pyridazine, 3-chloro-6-methoxy-

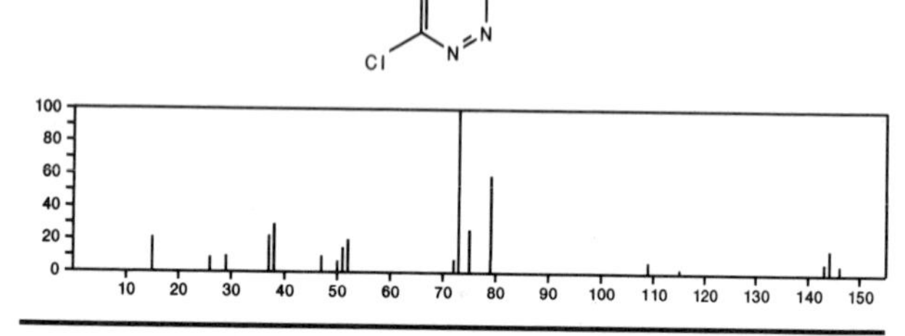

144 $C_5H_8N_2O_3$ 638-20-0
Acetamide, *N*,*N'*-carbonylbis-

AcNHCONHAc

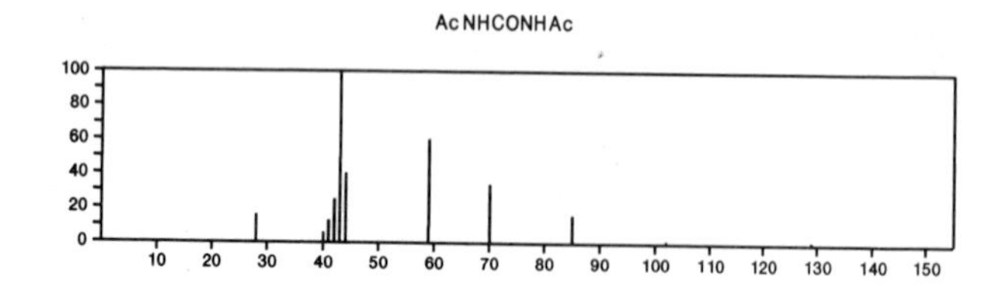

144 $C_5H_8N_2O_3$ 1123-21-3
2,4(1*H*,3*H*)-Pyrimidinedione, dihydro-5-hydroxy-5-methyl-

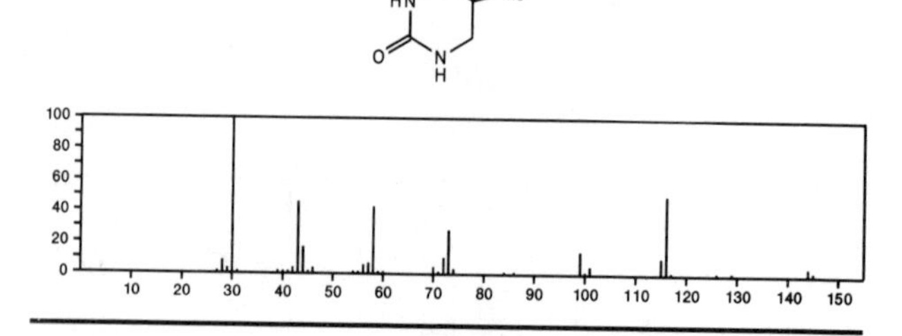

144 $C_5H_8N_2O_3$ 7519-36-0
L-Proline, 1-nitroso-

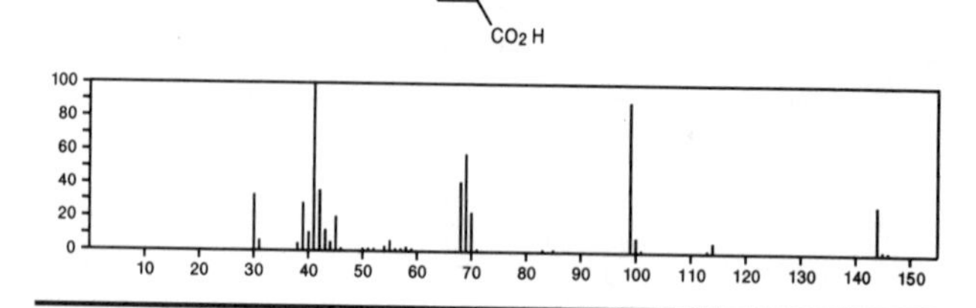

144 $C_6H_8O_4$ 95-96-5
1,4-Dioxane-2,5-dione, 3,6-dimethyl-

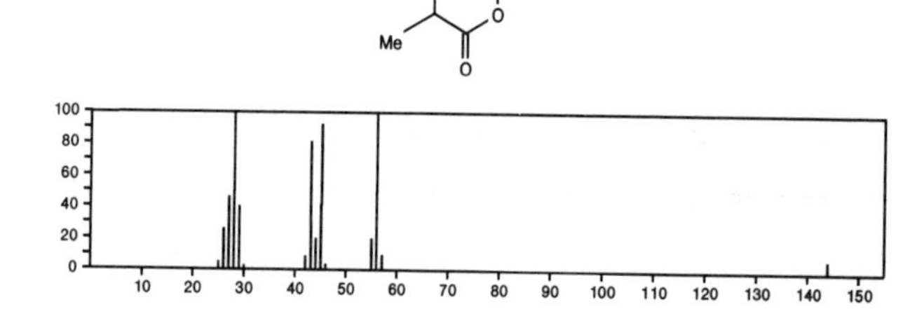

144 $C_6H_8O_4$ 624-48-6
2-Butenedioic acid (*Z*)-, dimethyl ester

MeOC(O)CH=CHC(O)OMe

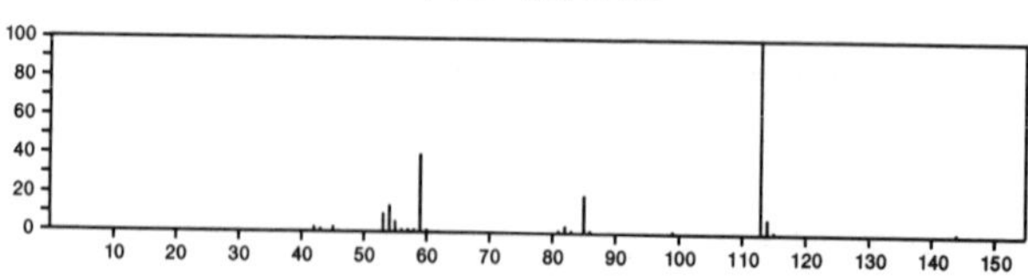

144 $C_6H_8O_4$ 624-49-7
2-Butenedioic acid (*E*)-, dimethyl ester

MeOC(O)CH=CHC(O)OMe

144 $C_6H_8O_4$ 1124–13–6
1,2–Cyclobutanedicarboxylic acid, *trans*–

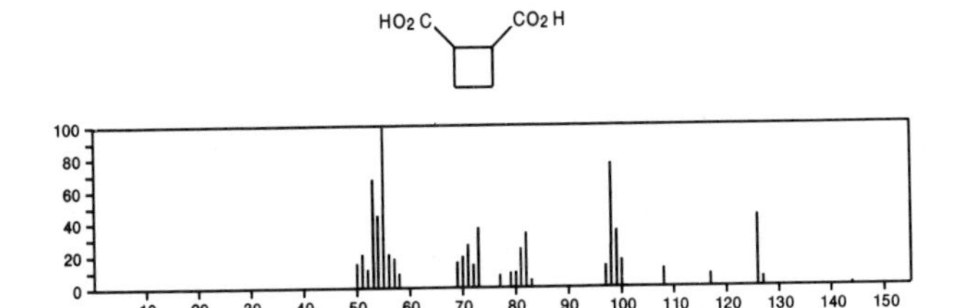

144 $C_6H_8O_4$ 1461–94–5
1,2–Cyclobutanedicarboxylic acid, *cis*–

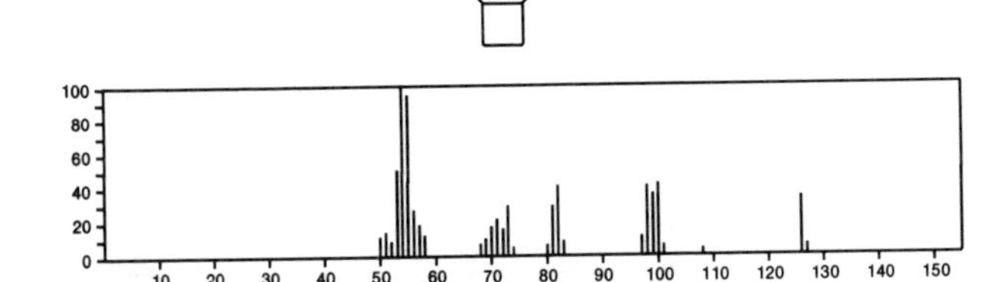

144 $C_6H_8O_4$ 28564–83–2
4H–Pyran–4–one, 2,3–dihydro–3,5–dihydroxy–6–methyl–

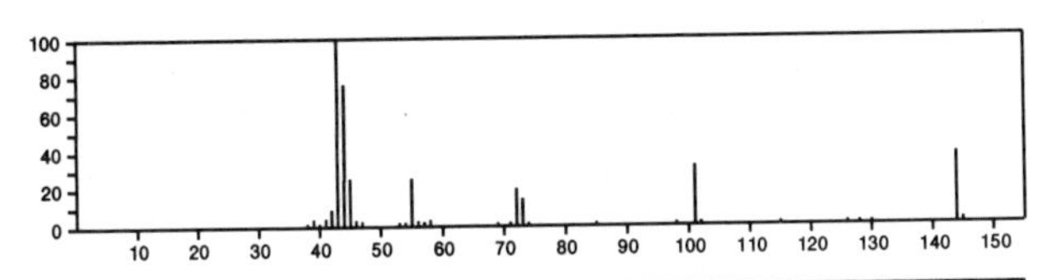

144 $C_6H_8O_4$ 29619–56–5
2–Butene–1,4–diol, diformate

O=CHOCH2 CH=CHCH2 OCH=O

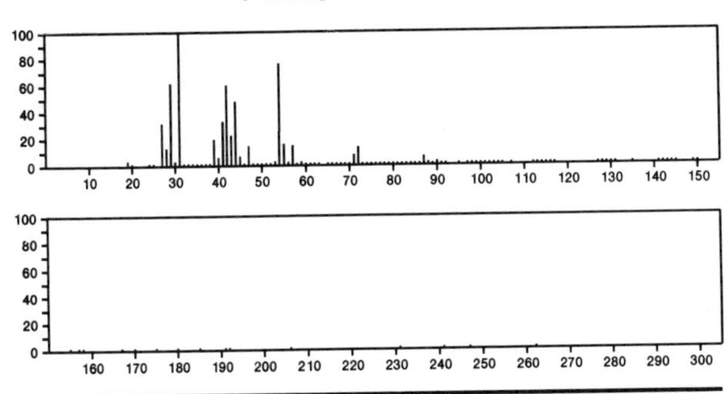

144 $C_6H_8O_4$ 36568–10–2
Pentanoic acid, 3,5–dioxo–, methyl ester

OCHCH2 COCH2 C(O) OMe

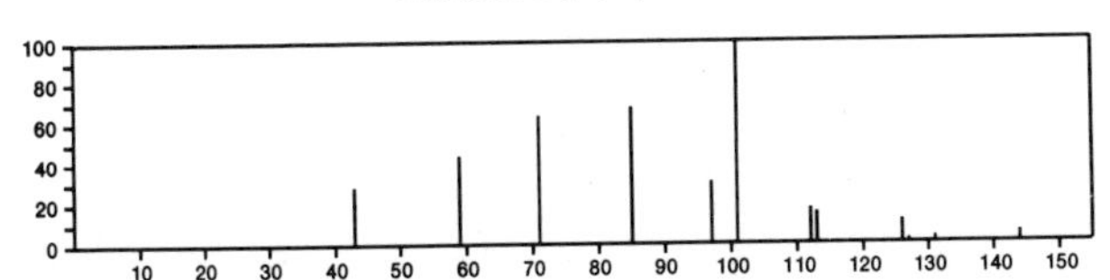

144 $C_6H_9N_2.Cl$ 34061–83–1
Pyridinium, 1–amino–4–methyl–, chloride

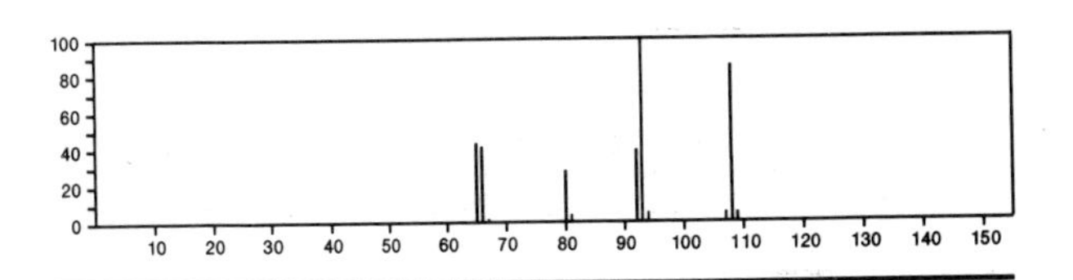

144 $C_6H_9N_2.Cl$ 34061–84–2
Pyridinium, 1–amino–2–methyl–, chloride

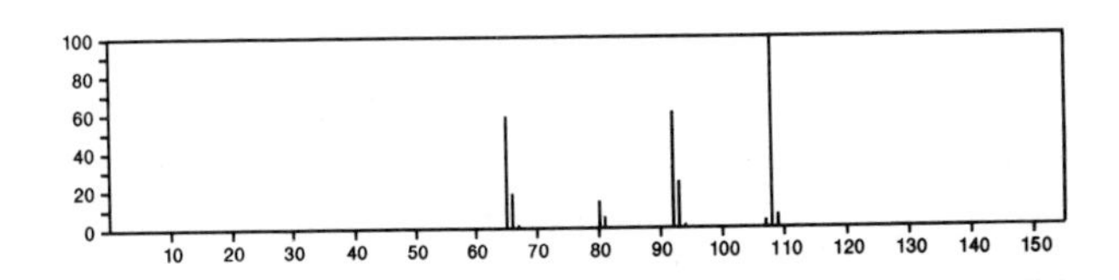

144 $C_6H_{12}N_2O_2$ 55401–87–1
Hydrazinecarboxylic acid, butylidene–, methyl ester

MeOC(O) NHN=CHPr

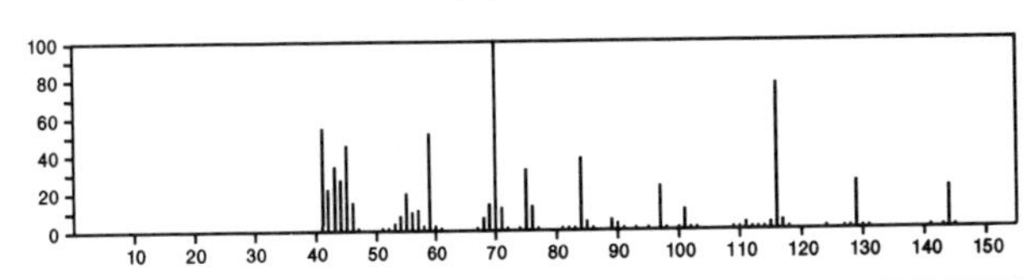

144 $C_6H_{12}N_2O_2$ 55556–87–1
Morpholine, 3,5–dimethyl–4–nitroso–

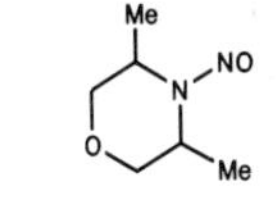

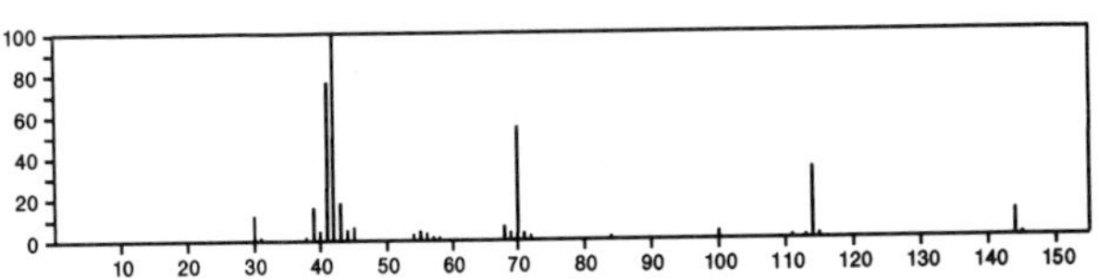

144 $C_6H_{12}O_2Si$ 13688–55–6
2–Propenoic acid, trimethylsilyl ester

Me3 Si OC(O) CH=CH2

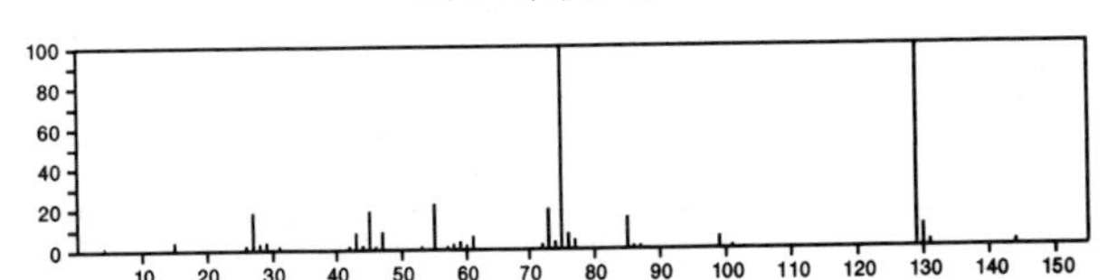

144 $C_6H_{16}N_4$ 35035–69–9
1,2,4,5–Tetrazine, 1,4–diethylhexahydro–

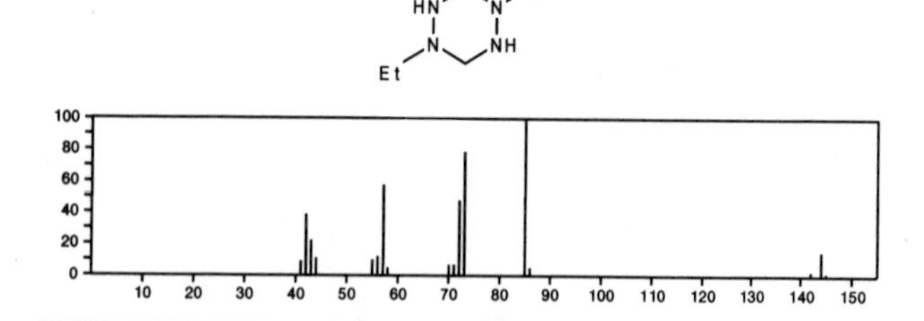

144 $C_6H_{16}Si_2$ 1627–98–1
1,3–Disilacyclobutane, 1,1,3,3–tetramethyl–

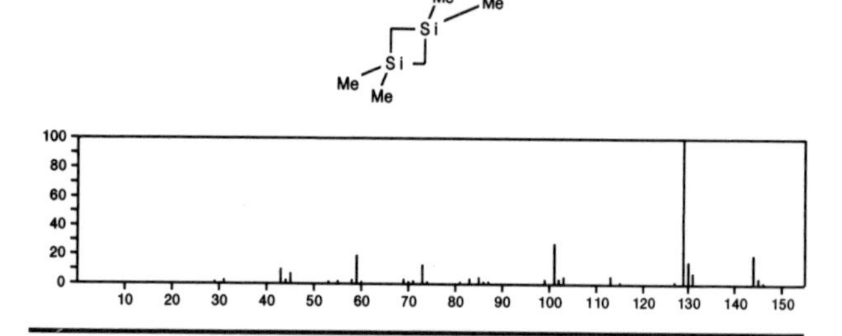

144 C_7H_9ClO 113–18–8
1–Penten–4–yn–3–ol, 1–chloro–3–ethyl–

ClCH=CHCEt(OH)C≡CH

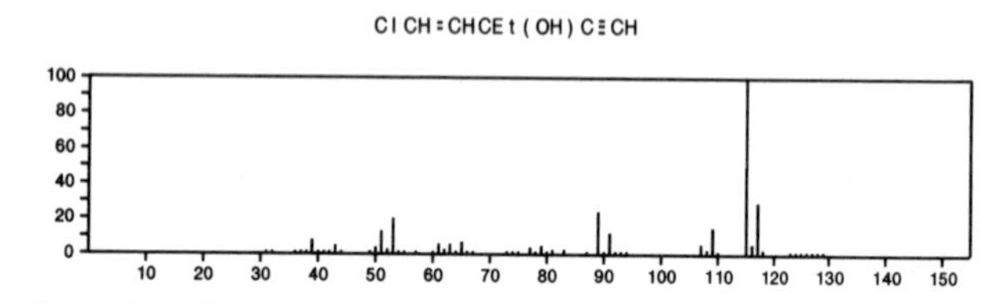

144 $C_7H_{12}OS$ 176–38–5
1–Oxa–4–thiaspiro[4.4]nonane

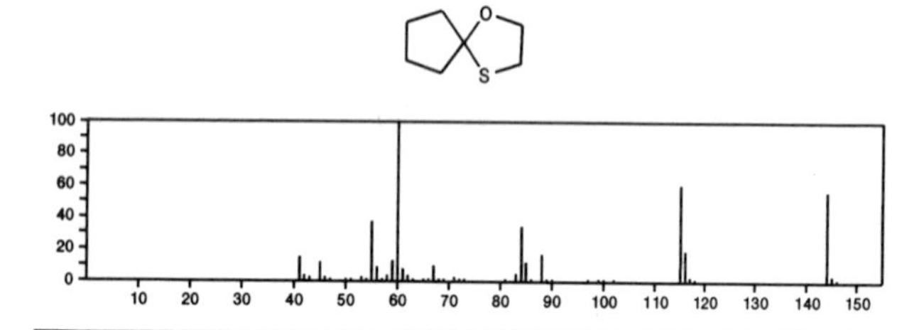

144 $C_7H_{12}OS$ 23510–98–7
Cyclohexanone, 4–(methylthio)–

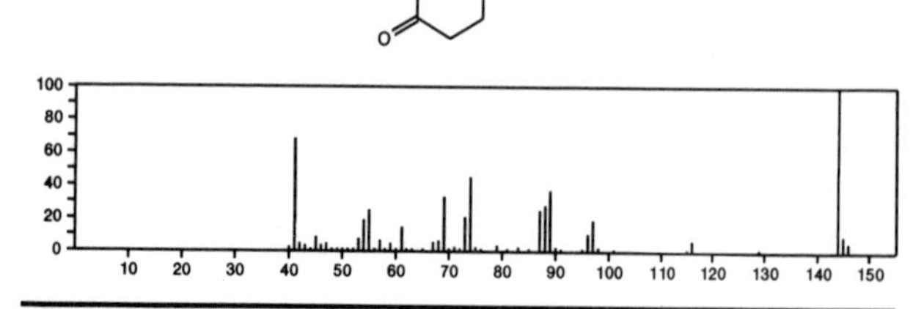

144 $C_7H_{12}O_3$ 539–88–8
Pentanoic acid, 4–oxo–, ethyl ester

EtOC(O)CH2CH2COMe

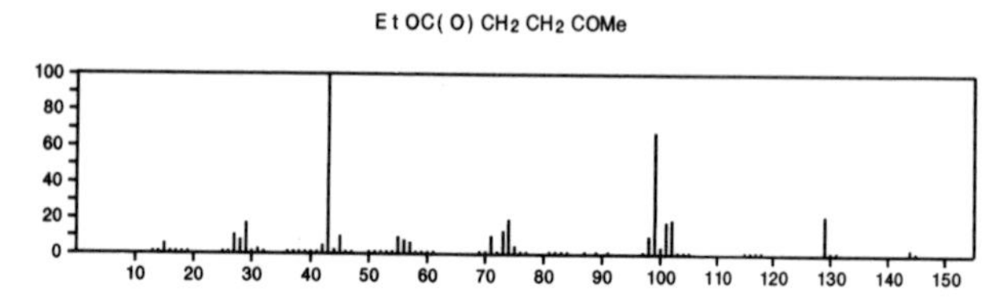

144 $C_7H_{12}O_3$ 609–69–8
Cyclohexanecarboxylic acid, 2–hydroxy–

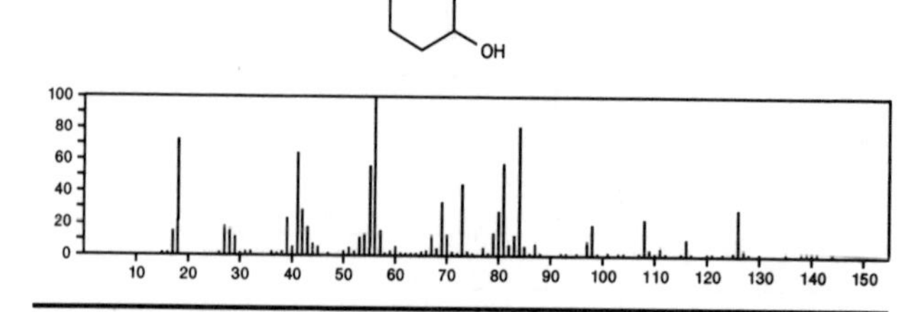

144 $C_7H_{12}O_3$ 637–64–9
2–Furanmethanol, tetrahydro–, acetate

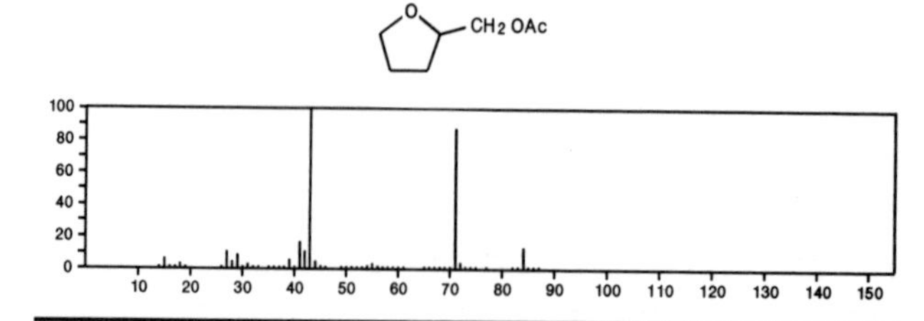

144 $C_7H_{12}O_3$ 3682–42–6
Valeric acid, 3–methyl–2–oxo–, methyl ester

MeCH2CHMeCOC(O)OMe

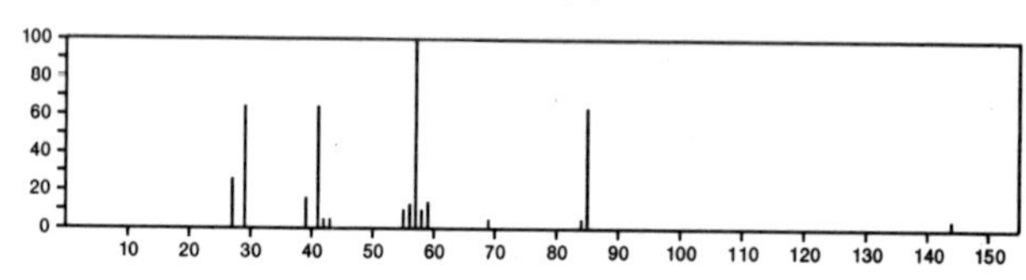

144 $C_7H_{12}O_3$ 3682–43–7
Pentanoic acid, 4–methyl–2–oxo–, methyl ester

MeOC(O)COCH2CHMe2

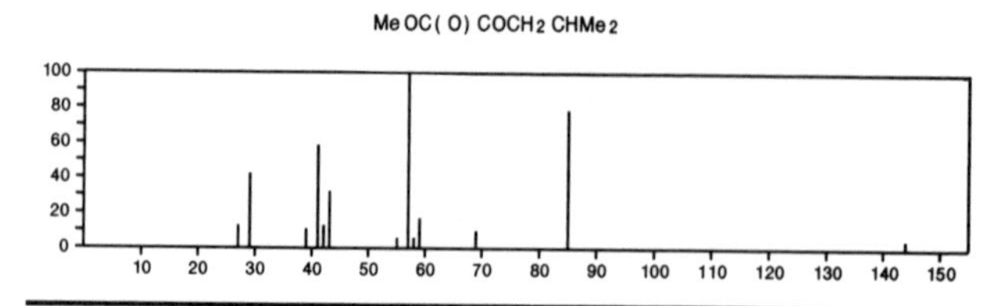

144 $C_7H_{12}O_3$ 5185–97–7
2–Pentanone, 5–(acetyloxy)–

AcO(CH2)3COMe

144 $C_7H_{12}O_3$ 19424–29–4
1,3–Dioxolan–2–one, 4,4,5,5–tetramethyl–

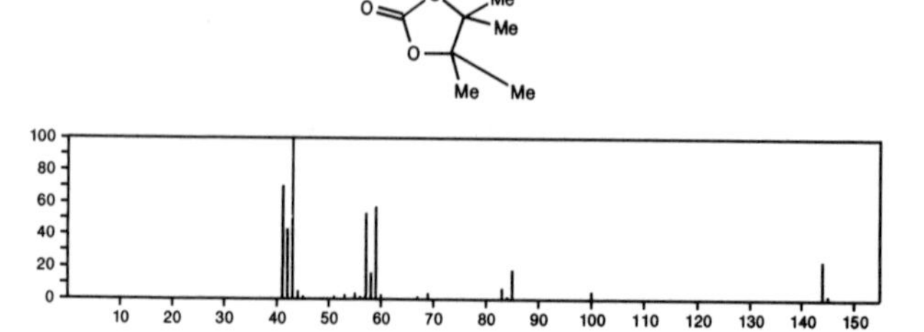

144 $C_7H_{12}O_3$ 24108–29–0
1,3–Dioxolane–2–propanal, 2–methyl–

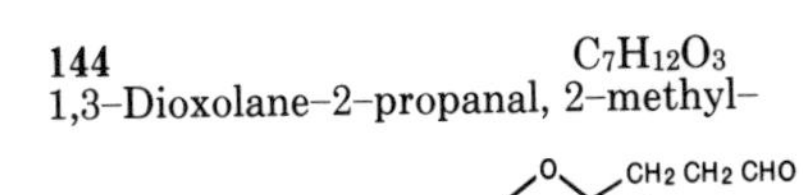
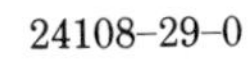
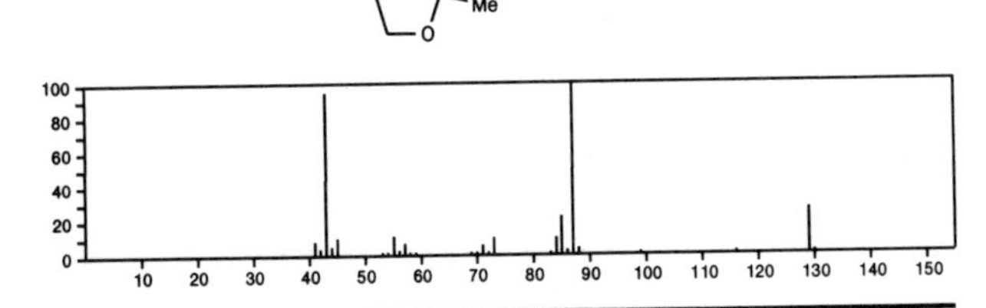

144 $C_7H_{12}O_3$ 51756–08–2
Butanoic acid, 2–ethyl–3–oxo–, methyl ester

MeOC(O)CHEtCOMe

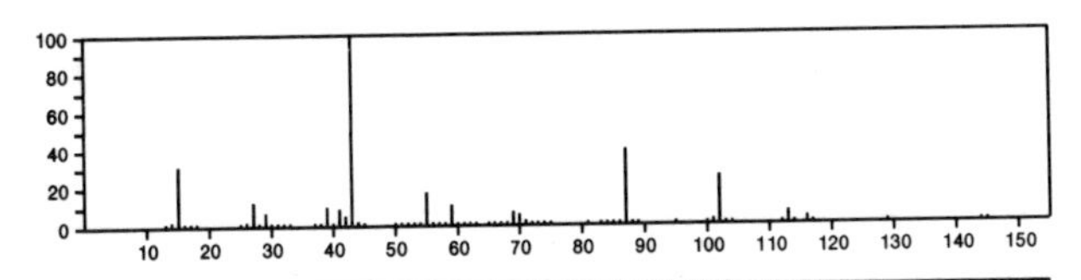

144 $C_7H_{12}O_3$ 56009–32–6
2–Butenoic acid, 2–methoxy–3–methyl–, methyl ester

$Me_2C{:}C(OMe)C(O)OMe$

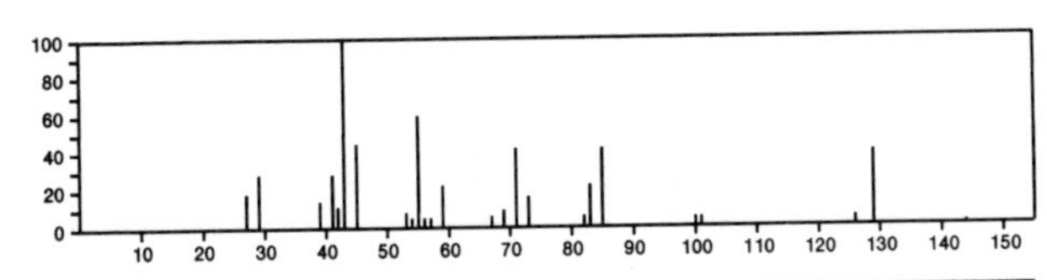

144 $C_7H_{16}N_2O$ 123–00–2
4–Morpholinepropanamine

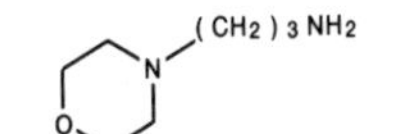
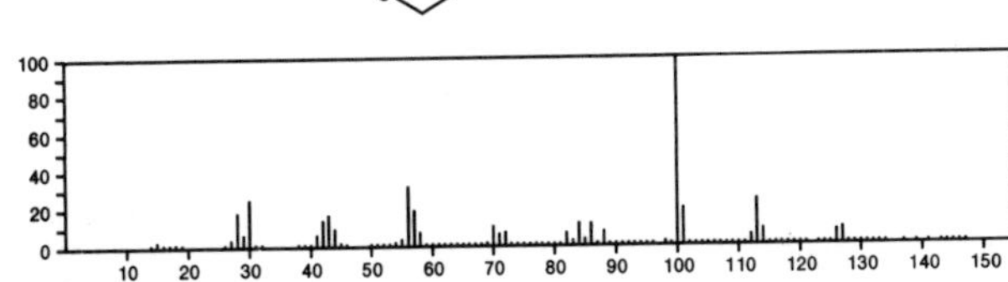

144 $C_7H_{16}N_2O$ 4128–37–4
Urea, *N,N'*–bis(1–methylethyl)–

i–PrNHCONHPr–i

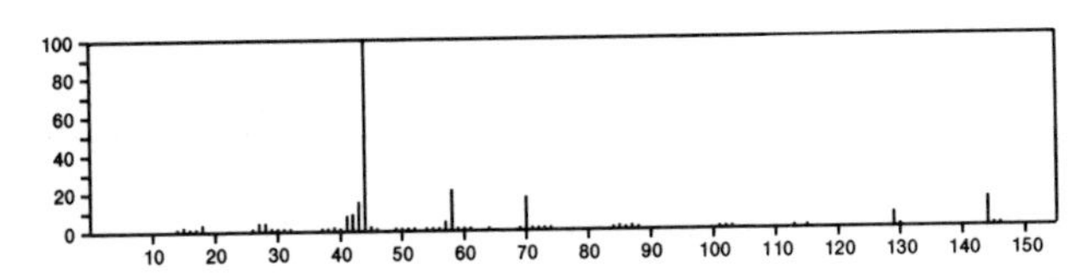

144 $C_7H_{16}N_2O$ 25413–63–2
Pentylamine, *N*–ethyl–*N*–nitroso–

$EtN(NO)(CH_2)_4Me$

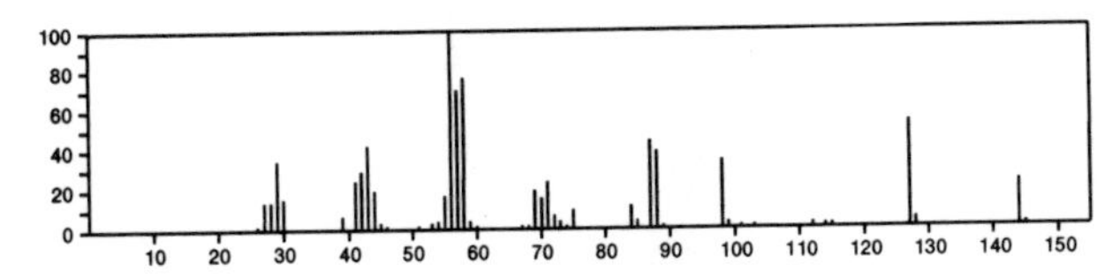

144 $C_7H_{16}N_2O$ 25413–64–3
1–Butanamine, *N*–nitroso–*N*–propyl–

$PrN(NO)(CH_2)_3Me$

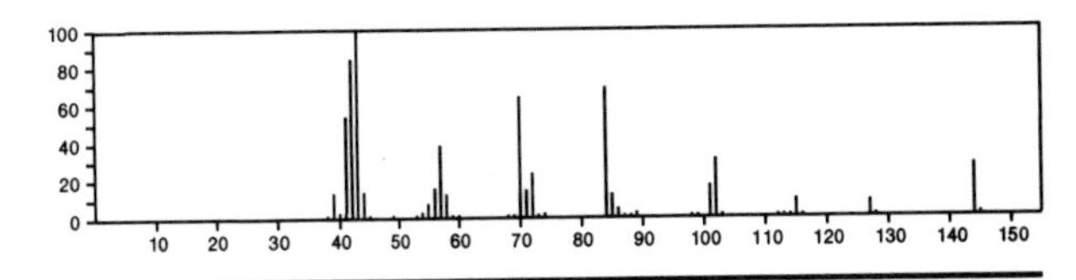

144 $C_8H_{13}Cl$ 823–83–6
Cyclohexane, 1–(chloromethyl)–4–methylene–

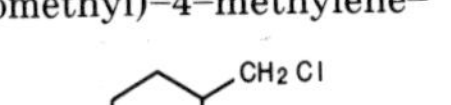
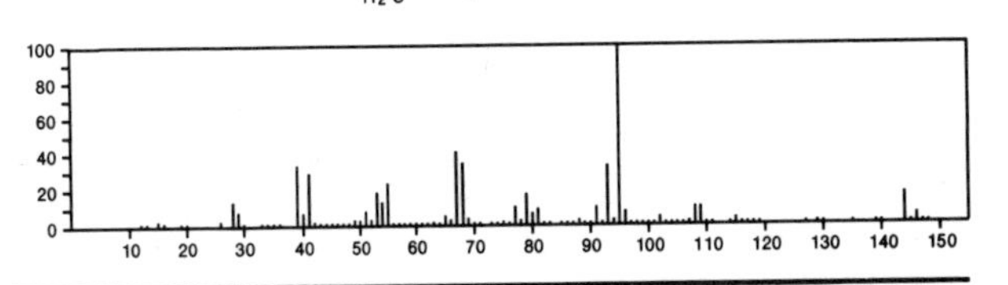

144 $C_8H_{13}Cl$ 2064–03–1
Bicyclo[2.2.2]octane, 1–chloro–

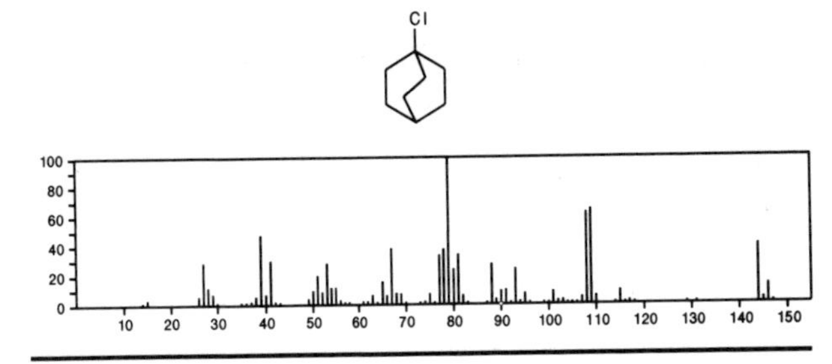

144 $C_8H_{13}Cl$ 15963–69–6
Cyclohexane, (2–chloroethenyl)–

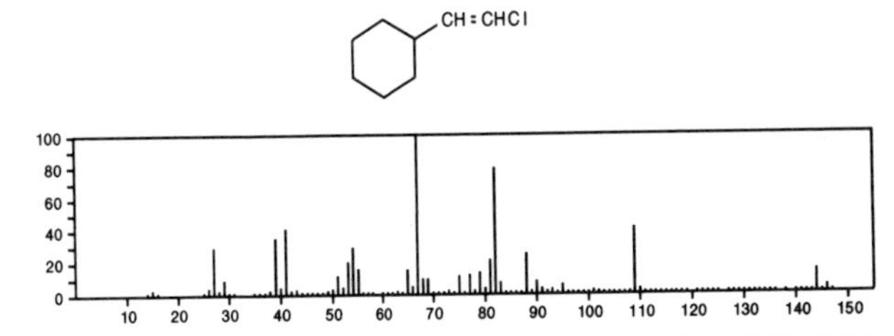

144 $C_8H_{13}Cl$ 19138–54–6
Bicyclo[2.2.1]heptane, 2–chloro–2–methyl–, *exo*–

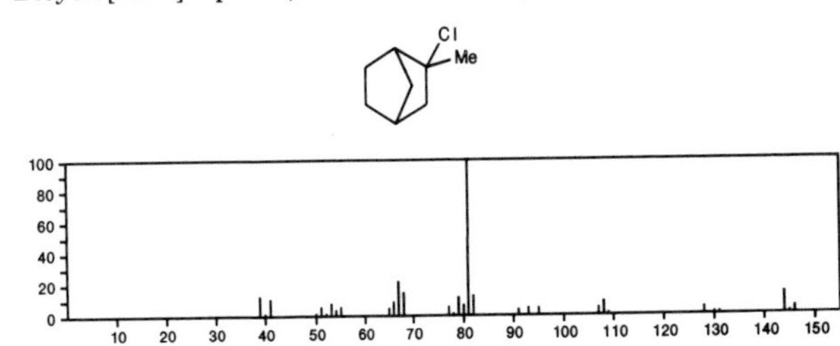

144 $C_8H_{13}Cl$ 22768–96–3
Bicyclo[2.2.1]heptane, 2–chloro–1–methyl–, *exo*–

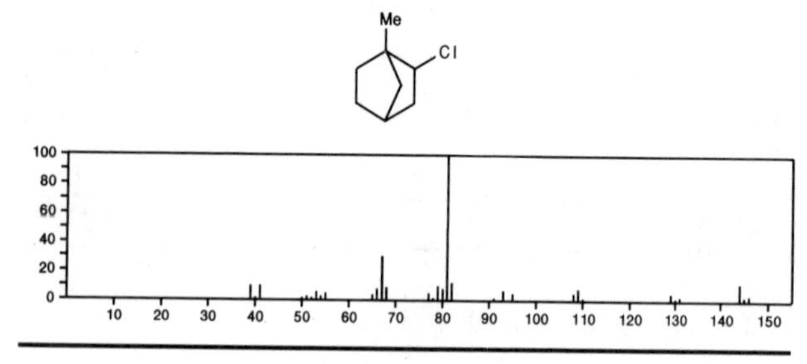

144 $C_8H_{13}Cl$ 33649–79–5
Bicyclo[2.2.2]octane, 2–chloro–

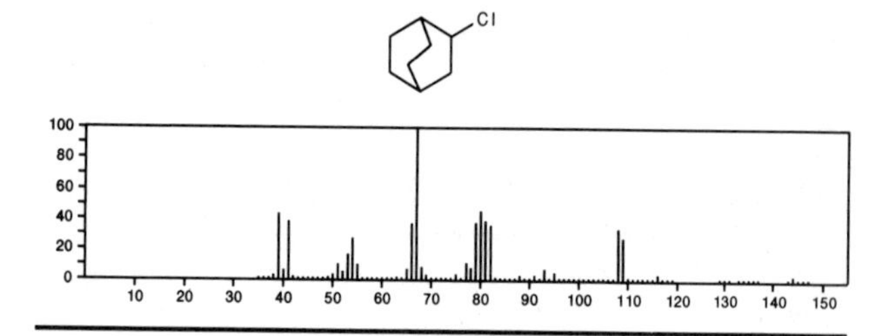

144 $C_8H_{16}O_2$ 97–85–8
Propanoic acid, 2–methyl–, 2–methylpropyl ester

i-BuOC(O)CHMe2

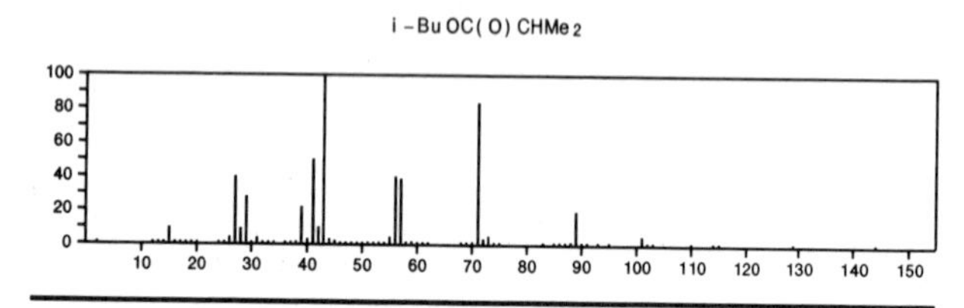

144 $C_8H_{16}O_2$ 97–87–0
Propanoic acid, 2–methyl–, butyl ester

Me2CHC(O)O(CH2)3Me

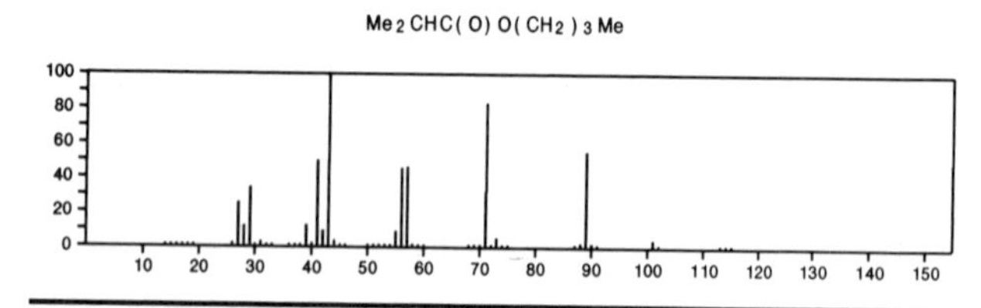

144 $C_8H_{16}O_2$ 105–68–0
1–Butanol, 3–methyl–, propanoate

Me2CHCH2CH2OC(O)Et

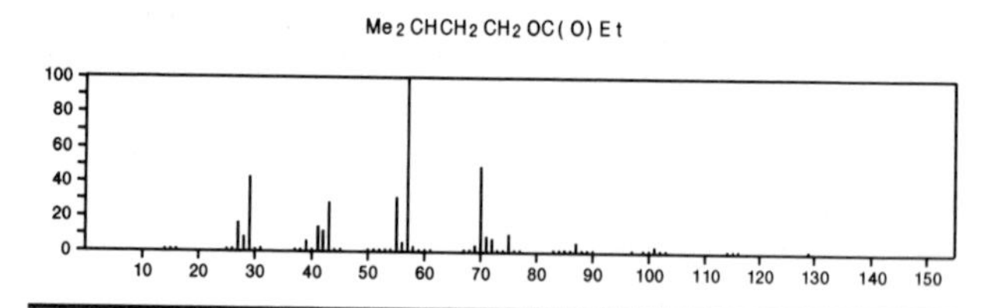

144 $C_8H_{16}O_2$ 106–73–0
Heptanoic acid, methyl ester

Me(CH2)5C(O)OMe

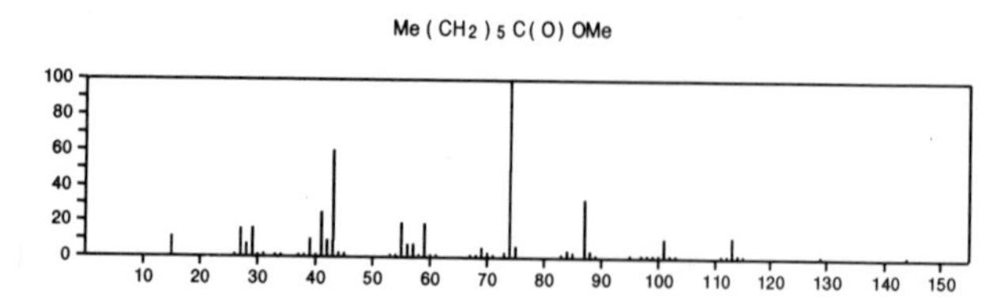

144 $C_8H_{16}O_2$ 109–21–7
Butanoic acid, butyl ester

PrC(O)O(CH2)3Me

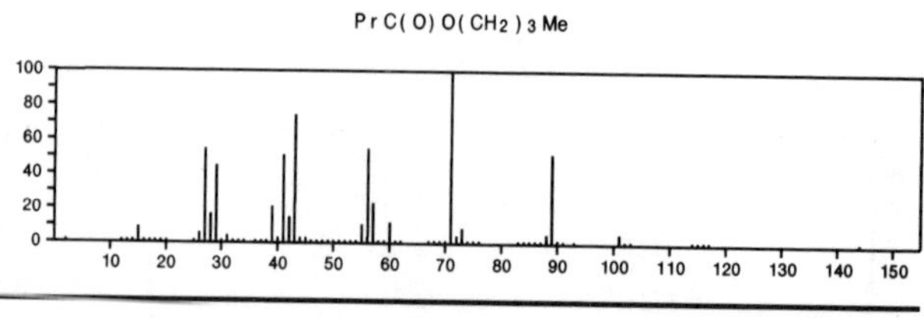

144 $C_8H_{16}O_2$ 124–07–2
Octanoic acid

HO2C(CH2)6Me

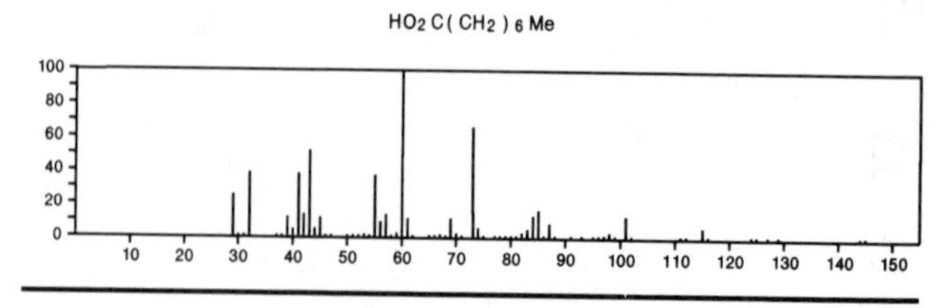

144 $C_8H_{16}O_2$ 142–92–7
Acetic acid, hexyl ester

Me(CH2)5OAc

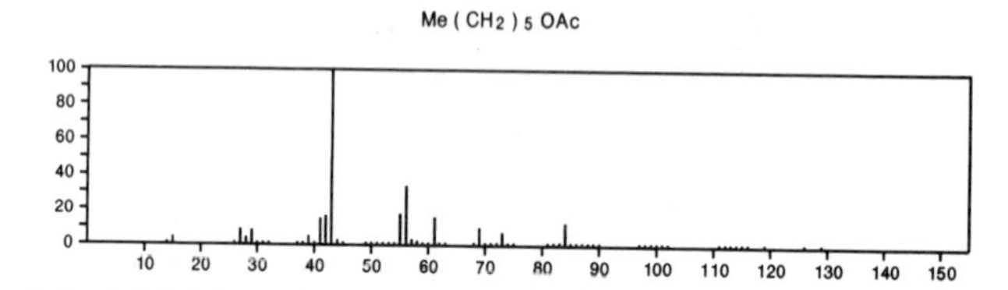

144 $C_8H_{16}O_2$ 149–57–5
Hexanoic acid, 2–ethyl–

HO2CCHEt(CH2)3Me

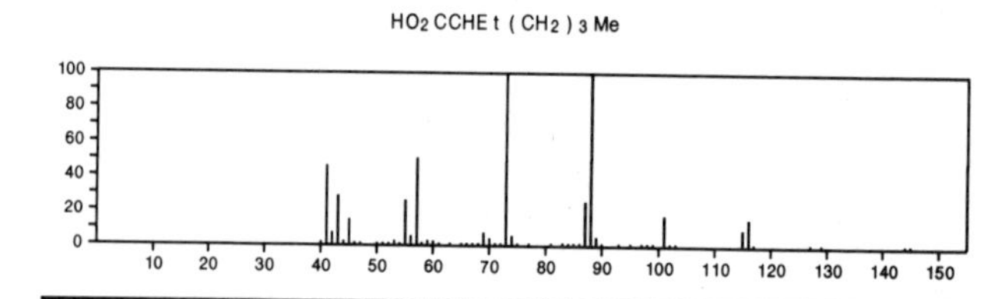

144 $C_8H_{16}O_2$ 624–54–4
Propanoic acid, pentyl ester

EtC(O)O(CH2)4Me

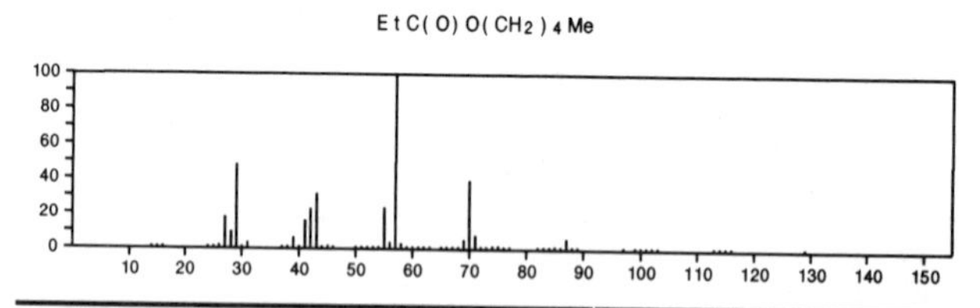

144 $C_8H_{16}O_2$ 819–97–6
Butanoic acid, 1–methylpropyl ester

PrC(O)OBu-s

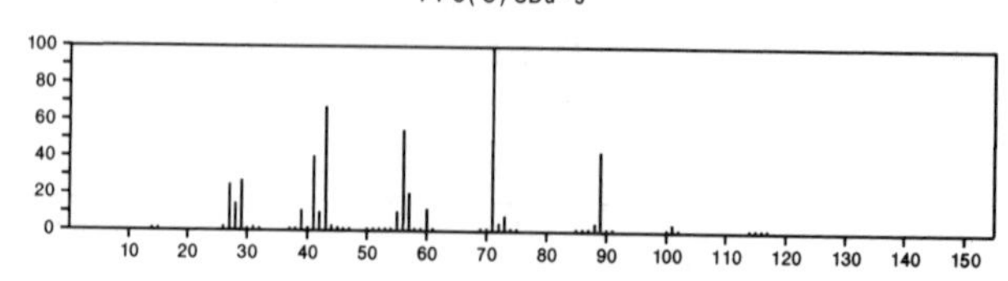

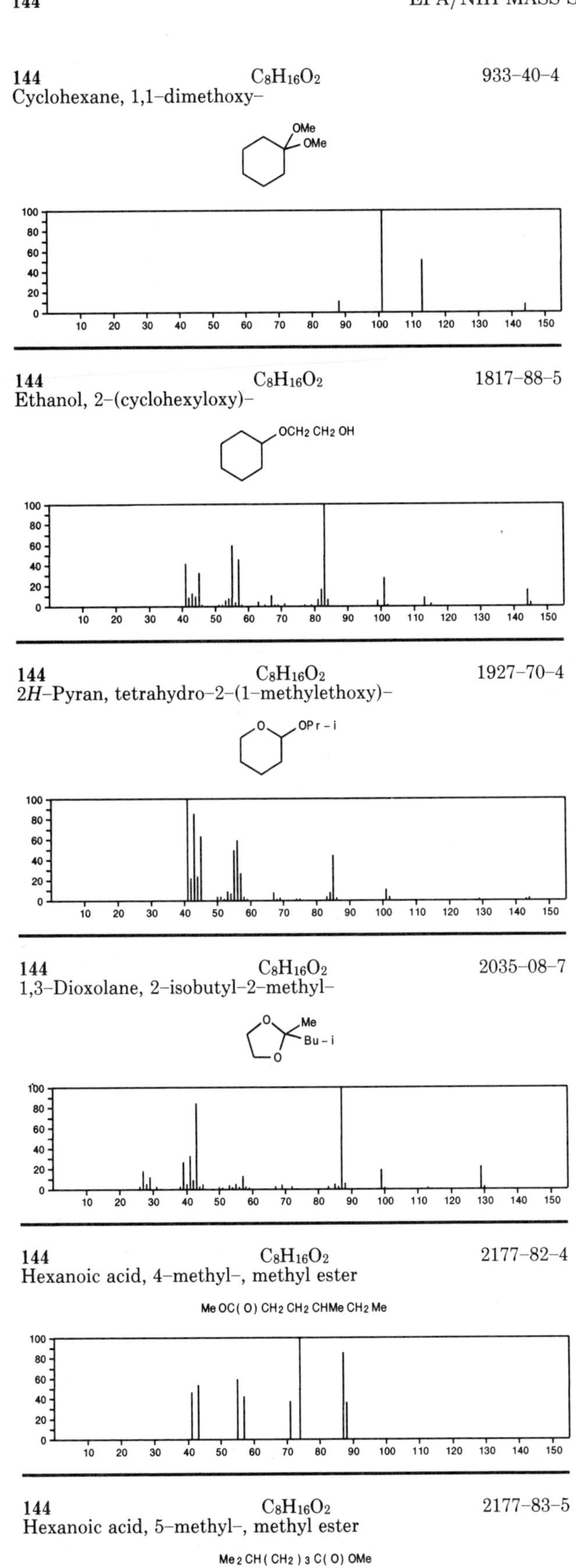

144 $C_8H_{16}O_2$ 933-40-4
Cyclohexane, 1,1-dimethoxy-

144 $C_8H_{16}O_2$ 1817-88-5
Ethanol, 2-(cyclohexyloxy)-

144 $C_8H_{16}O_2$ 1927-70-4
2H-Pyran, tetrahydro-2-(1-methylethoxy)-

144 $C_8H_{16}O_2$ 2035-08-7
1,3-Dioxolane, 2-isobutyl-2-methyl-

144 $C_8H_{16}O_2$ 2177-82-4
Hexanoic acid, 4-methyl-, methyl ester

144 $C_8H_{16}O_2$ 2177-83-5
Hexanoic acid, 5-methyl-, methyl ester

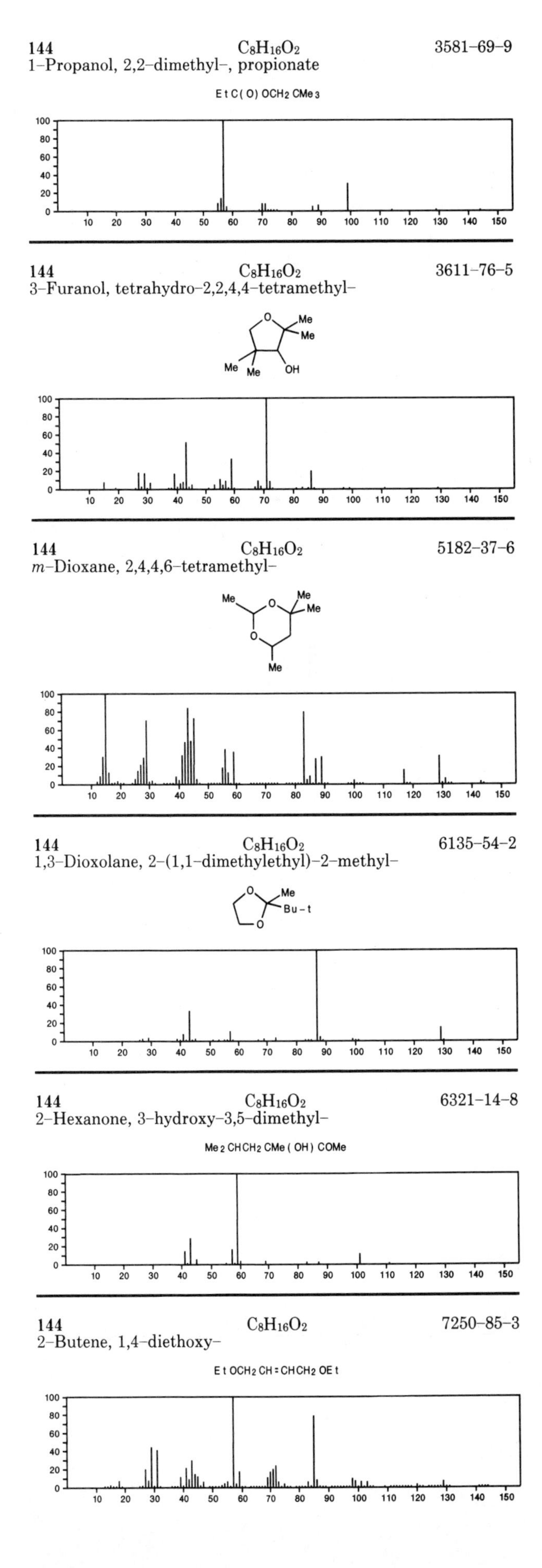

144 $C_8H_{16}O_2$ 3581-69-9
1-Propanol, 2,2-dimethyl-, propionate

144 $C_8H_{16}O_2$ 3611-76-5
3-Furanol, tetrahydro-2,2,4,4-tetramethyl-

144 $C_8H_{16}O_2$ 5182-37-6
m-Dioxane, 2,4,4,6-tetramethyl-

144 $C_8H_{16}O_2$ 6135-54-2
1,3-Dioxolane, 2-(1,1-dimethylethyl)-2-methyl-

144 $C_8H_{16}O_2$ 6321-14-8
2-Hexanone, 3-hydroxy-3,5-dimethyl-

144 $C_8H_{16}O_2$ 7250-85-3
2-Butene, 1,4-diethoxy-

144 $C_8H_{16}O_2$ 13757-91-0
2-Heptanone, 3-hydroxy-3-methyl-

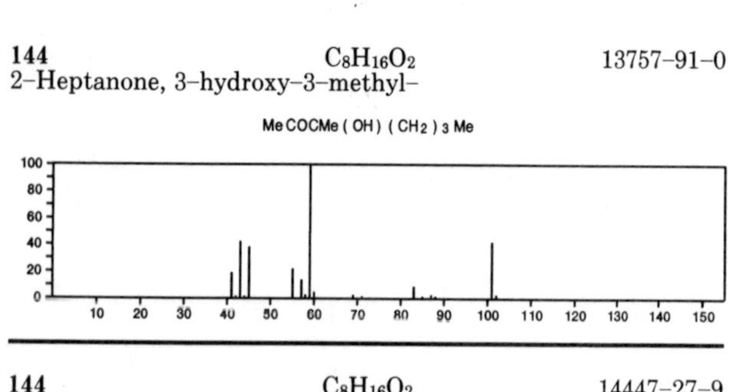

144 $C_8H_{16}O_2$ 14447-27-9
1,3-Dioxolane, 2-butyl-2-methyl-

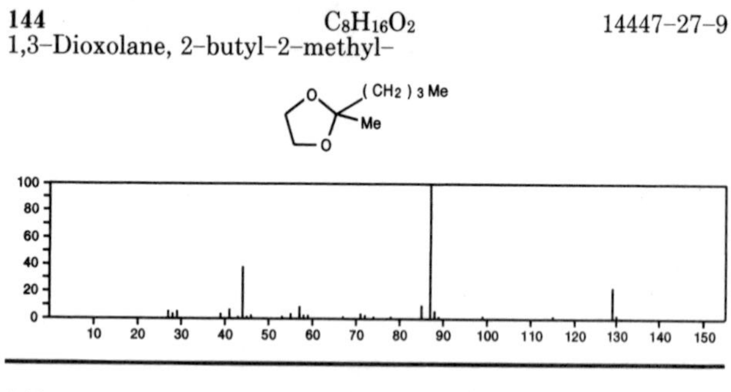

144 $C_8H_{16}O_2$ 14869-38-6
2-Pentanone, 1-ethoxy-4-methyl-

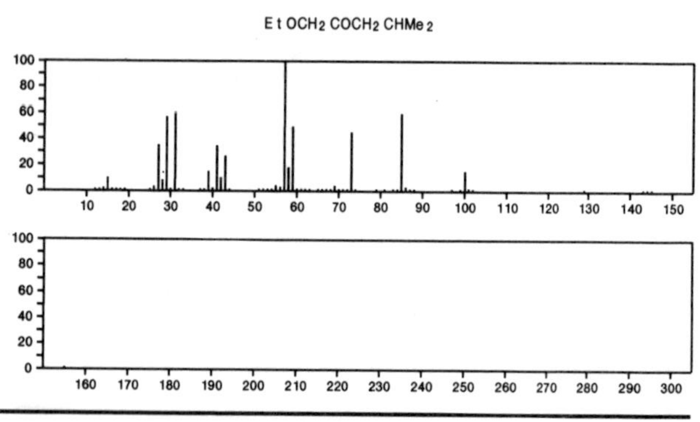

144 $C_8H_{16}O_2$ 17429-05-9
3-Hexanone, 6-methoxy-2-methyl-

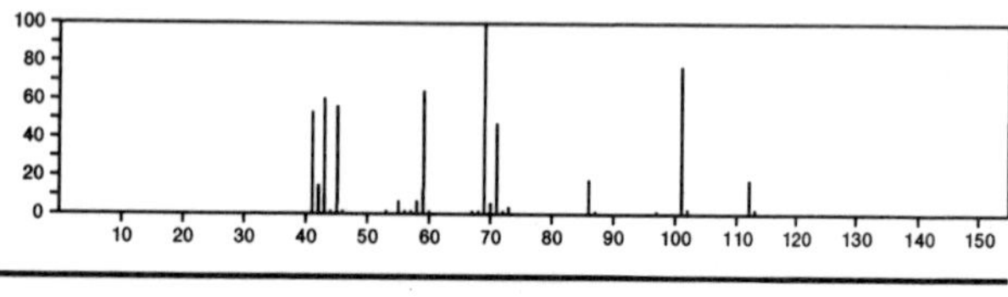

144 $C_8H_{16}O_2$ 20268-00-2
1,3-Dioxane, 2,2,4,6-tetramethyl-, *trans-*

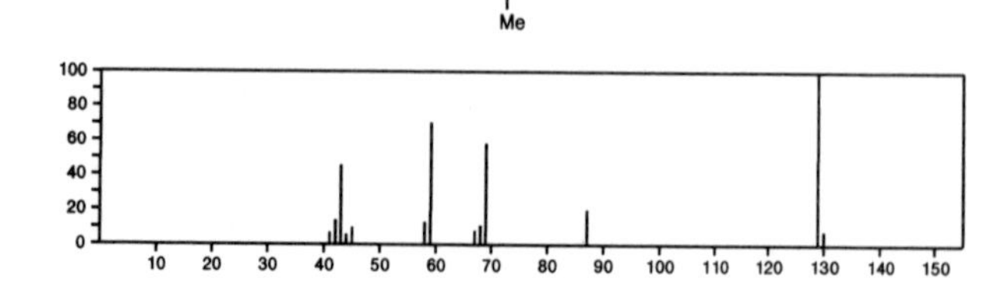

144 $C_8H_{16}O_2$ 28046-68-6
Cyclohexane, 1,4-dimethoxy-, *cis-*

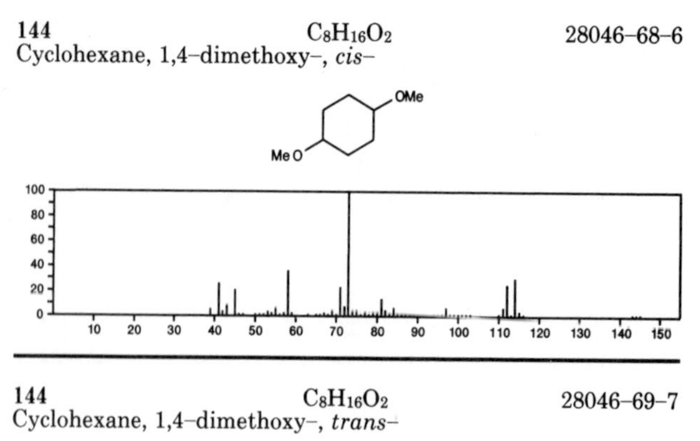

144 $C_8H_{16}O_2$ 28046-69-7
Cyclohexane, 1,4-dimethoxy-, *trans-*

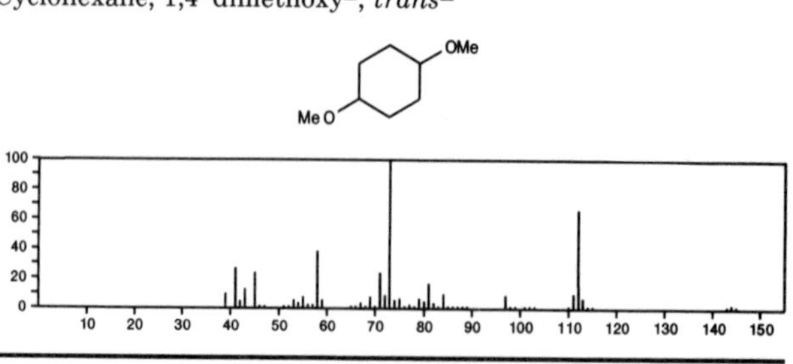

144 $C_8H_{16}O_2$ 29887-60-3
Cyclohexane, 1,2-dimethoxy-, *trans-*

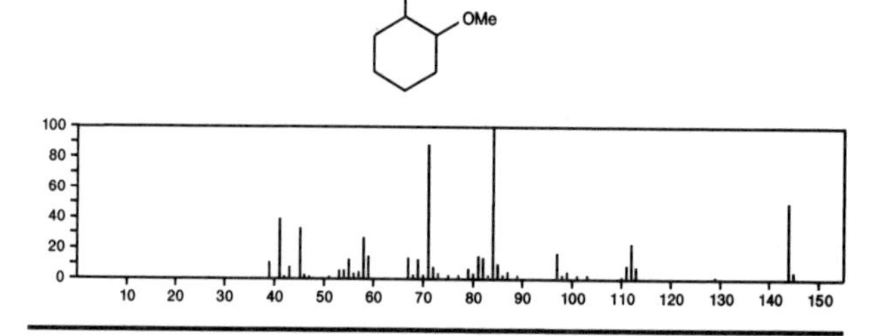

144 $C_8H_{16}O_2$ 29887-61-4
Cyclohexane, 1,3-dimethoxy-, *trans-*

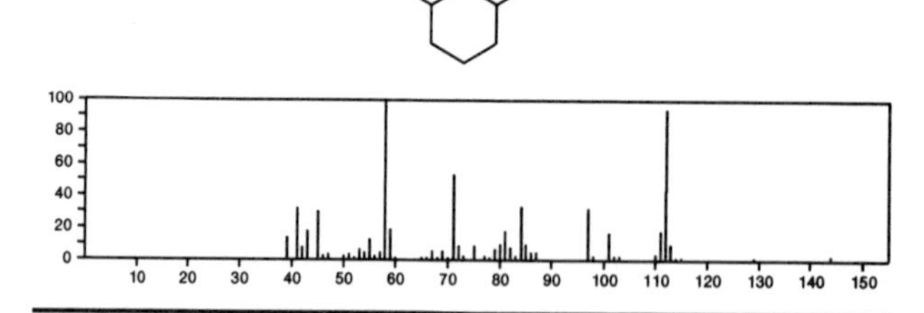

144 $C_8H_{16}O_2$ 30363-80-5
Cyclohexane, 1,2-dimethoxy-, *cis-*

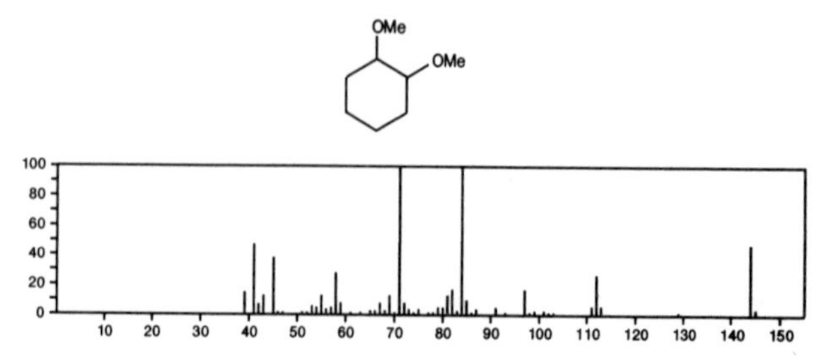

144 $C_8H_{16}O_2$ 30363-81-6
Cyclohexane, 1,3-dimethoxy-, *cis-*

144 $C_8H_{16}O_2$ 54004-45-4
1-Propanol, 2-methyl-2-[(2-methyl-2-propenyl)oxy]-

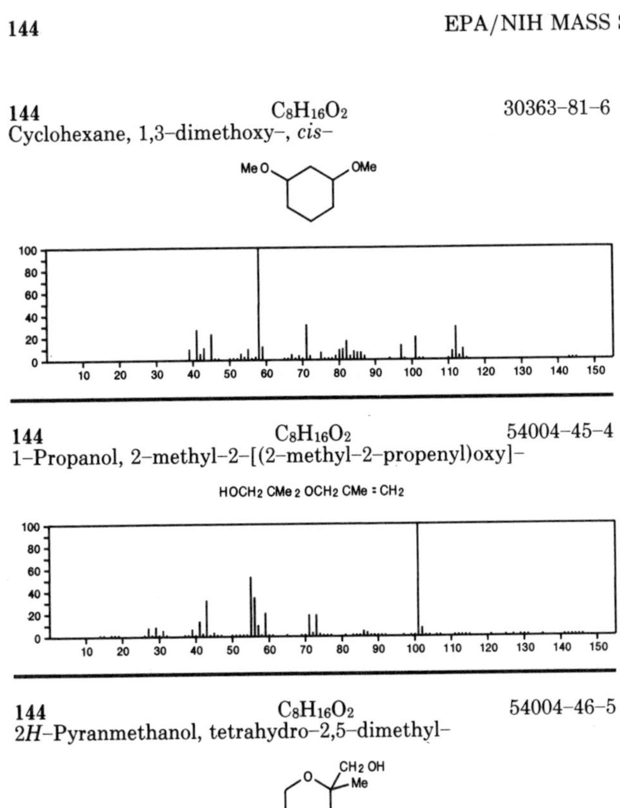

144 $C_8H_{16}O_2$ 54004-46-5
2*H*-Pyranmethanol, tetrahydro-2,5-dimethyl-

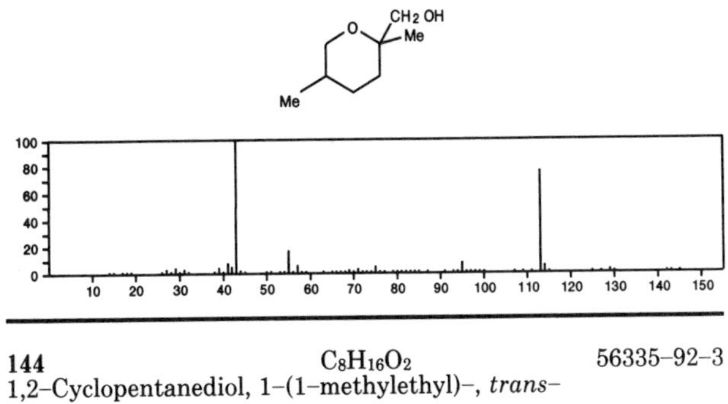

144 $C_8H_{16}O_2$ 56335-92-3
1,2-Cyclopentanediol, 1-(1-methylethyl)-, *trans-*

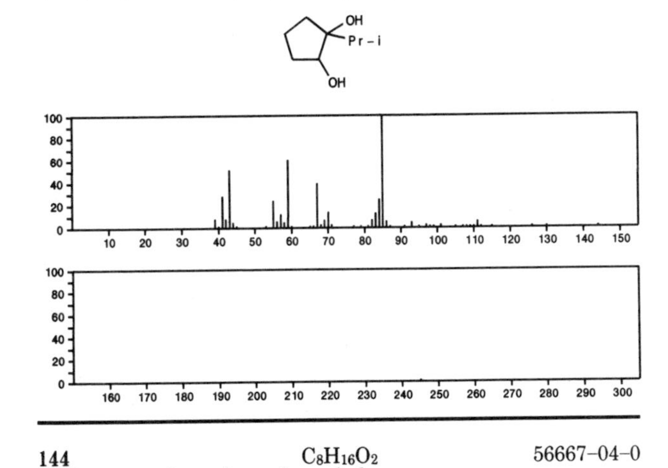

144 $C_8H_{16}O_2$ 56667-04-0
2-Hexanone, 3-methoxy-5-methyl-

Me2 CHCH2 CH(OMe) COMe

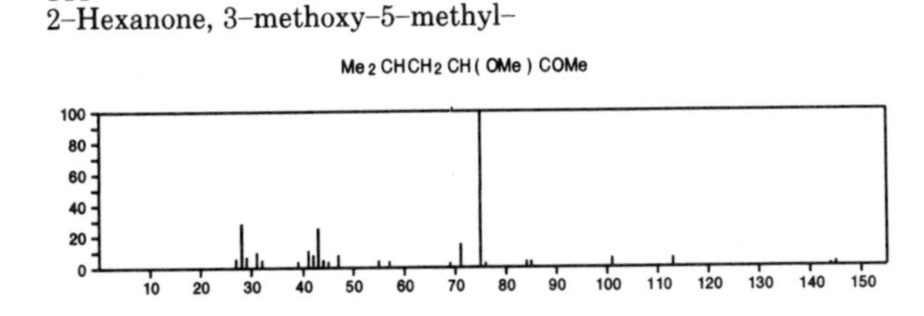

144 $C_8H_{16}O_2$ 56805-30-2
Hexanal, 3-(hydroxymethyl)-4-methyl-

144 $C_8H_{16}O_2$ 56805-31-3
Pentanal, 3-(hydroxymethyl)-4,4-dimethyl-

144 $C_8H_{16}S$ 1613-49-6
Thiophene, 2-butyltetrahydro-

144 $C_8H_{16}S$ 7133-14-4
Sulfide, cyclopentyl propyl

144 $C_8H_{16}S$ 7133-15-5
Sulfide, cyclopentyl isopropyl

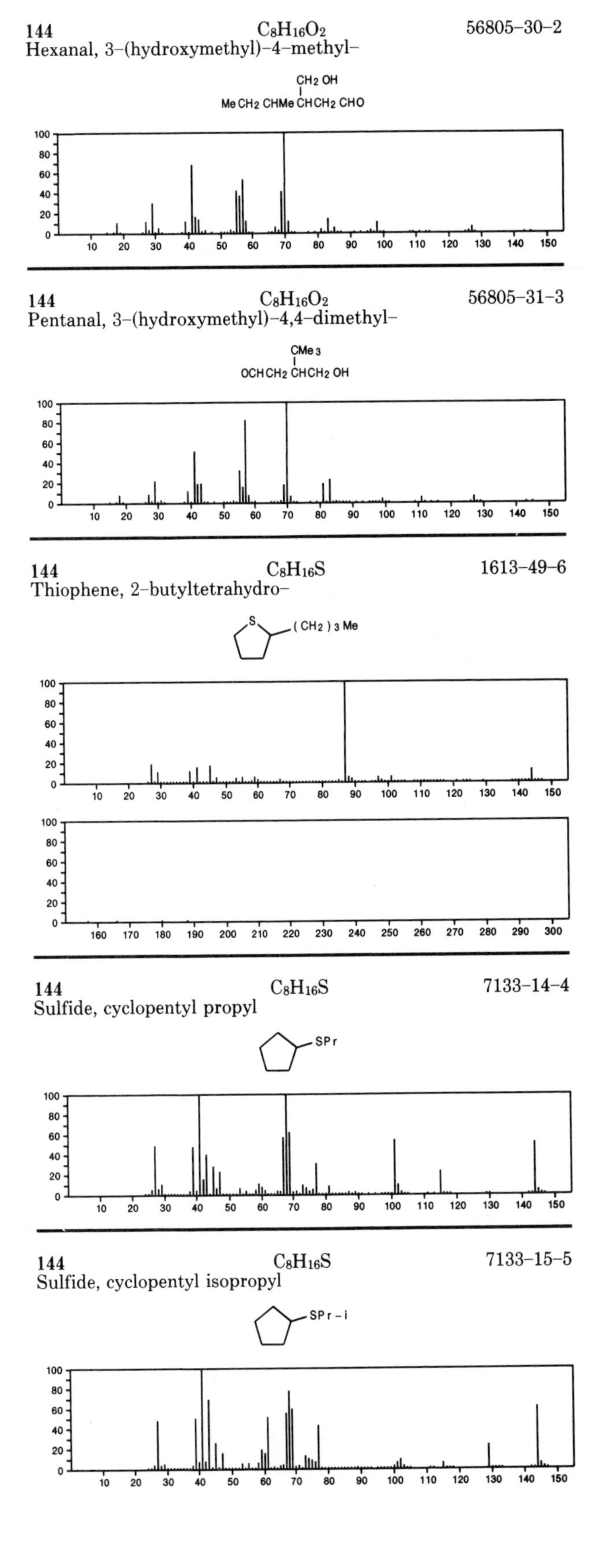

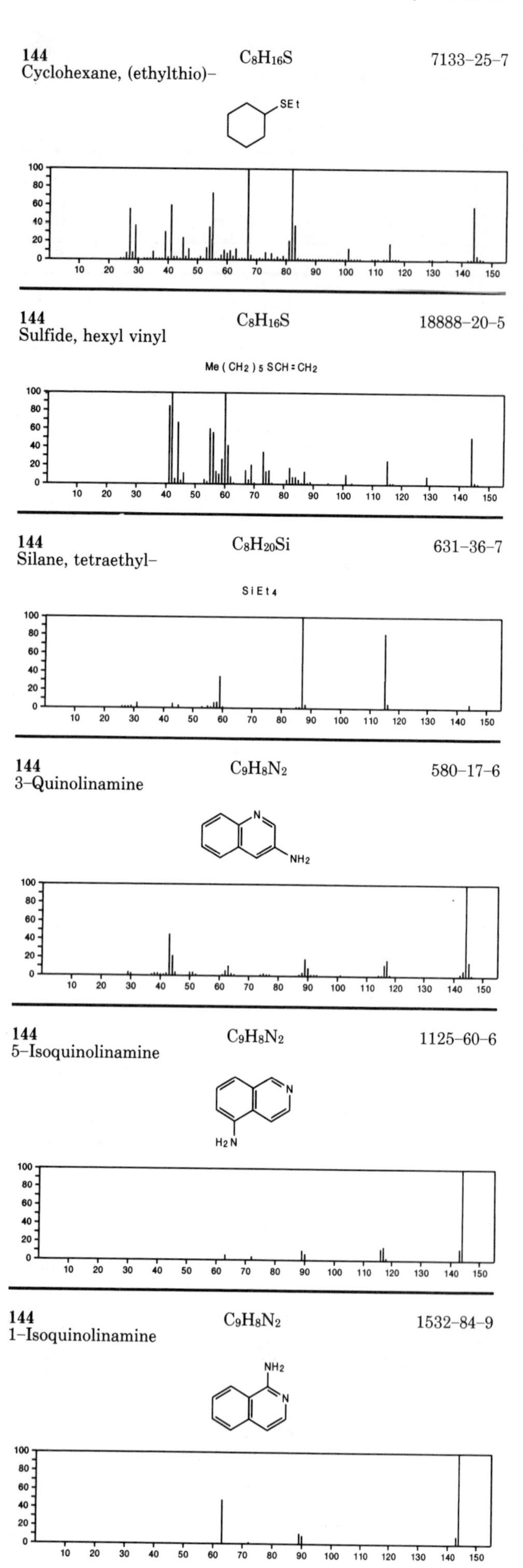

144 Cyclohexane, (ethylthio)- $C_8H_{16}S$ 7133-25-7
SEt
144 Sulfide, hexyl vinyl $C_8H_{16}S$ 18888-20-5
Me (CH2) 5 SCH = CH2
144 Silane, tetraethyl- $C_8H_{20}Si$ 631-36-7
SiEt4
144 3-Quinolinamine $C_9H_8N_2$ 580-17-6
NH2
144 5-Isoquinolinamine $C_9H_8N_2$ 1125-60-6
H2N
144 1-Isoquinolinamine $C_9H_8N_2$ 1532-84-9
NH2

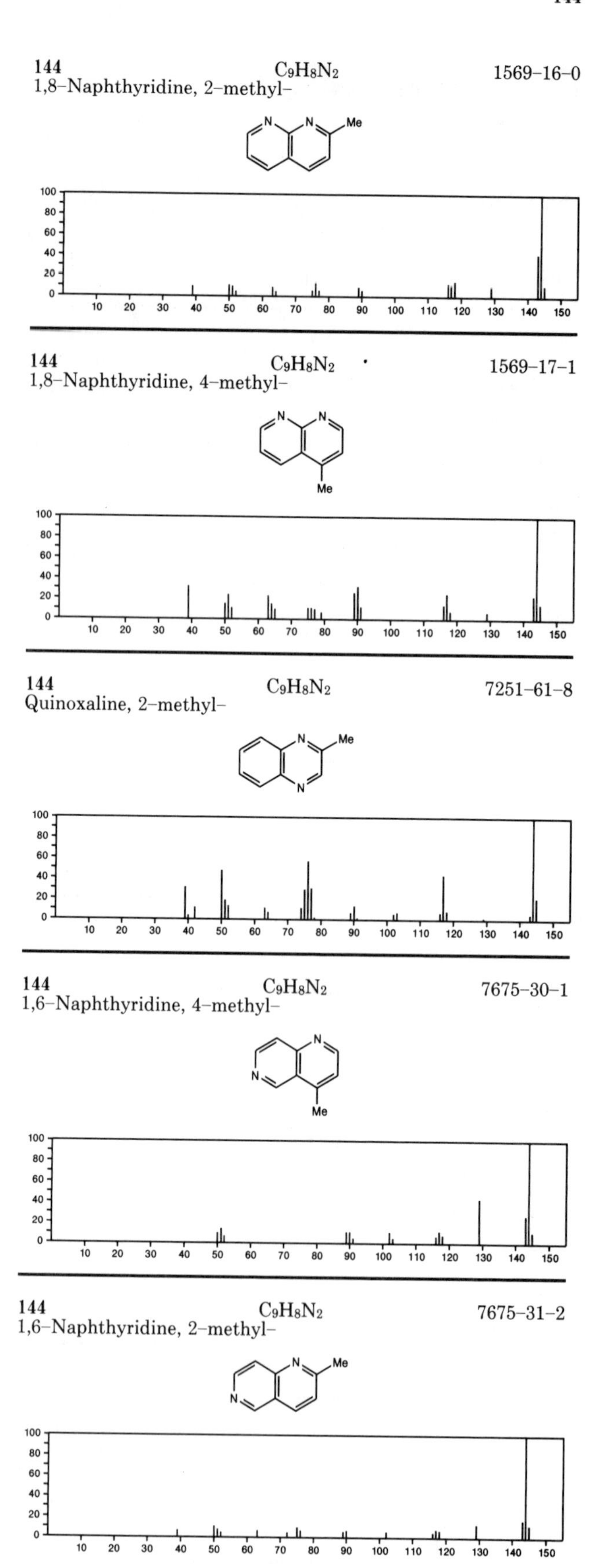

144 1,8-Naphthyridine, 2-methyl- $C_9H_8N_2$ 1569-16-0
Me
144 1,8-Naphthyridine, 4-methyl- $C_9H_8N_2$ 1569-17-1
Me
144 Quinoxaline, 2-methyl- $C_9H_8N_2$ 7251-61-8
Me
144 1,6-Naphthyridine, 4-methyl- $C_9H_8N_2$ 7675-30-1
Me
144 1,6-Naphthyridine, 2-methyl- $C_9H_8N_2$ 7675-31-2
Me

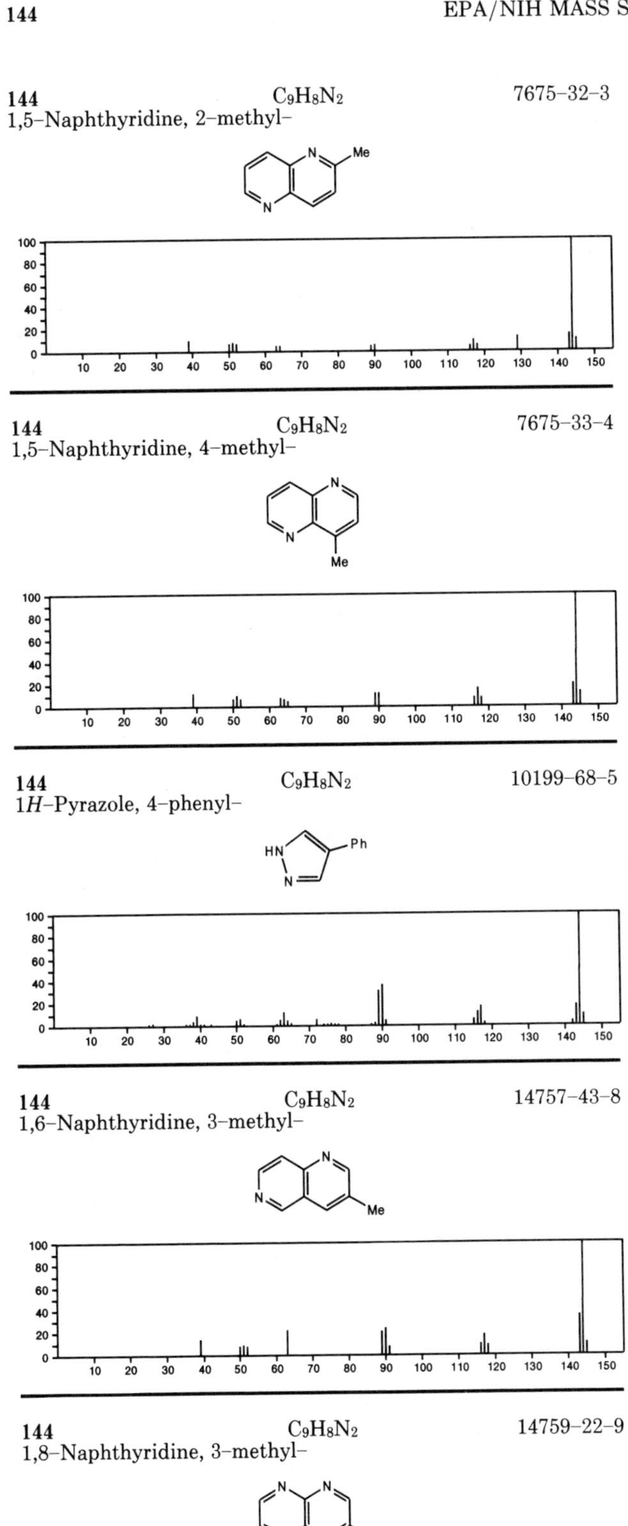

144 $C_9H_8N_2$ 7675–32–3
1,5–Naphthyridine, 2–methyl–

144 $C_9H_8N_2$ 7675–33–4
1,5–Naphthyridine, 4–methyl–

144 $C_9H_8N_2$ 10199–68–5
1*H*–Pyrazole, 4–phenyl–

144 $C_9H_8N_2$ 14757–43–8
1,6–Naphthyridine, 3–methyl–

144 $C_9H_8N_2$ 14759–22–9
1,8–Naphthyridine, 3–methyl–

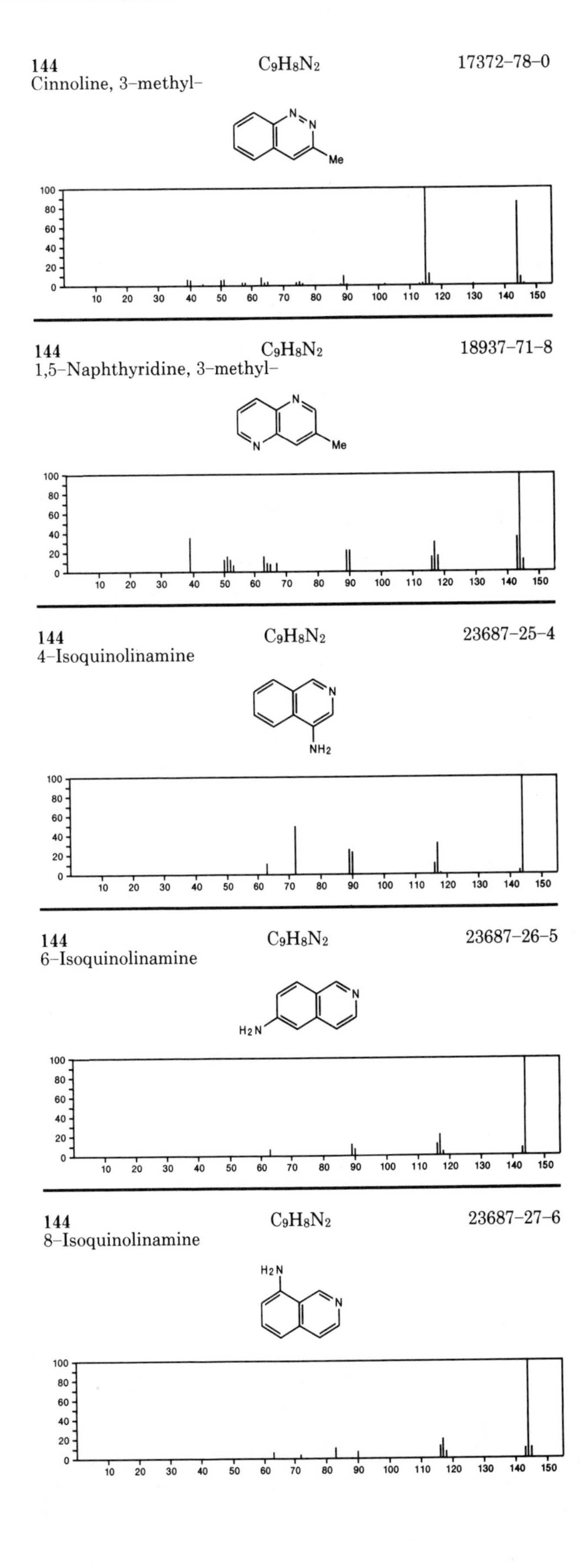

144 $C_9H_8N_2$ 17372–78–0
Cinnoline, 3–methyl–

144 $C_9H_8N_2$ 18937–71–8
1,5–Naphthyridine, 3–methyl–

144 $C_9H_8N_2$ 23687–25–4
4–Isoquinolinamine

144 $C_9H_8N_2$ 23687–26–5
6–Isoquinolinamine

144 $C_9H_8N_2$ 23687–27–6
8–Isoquinolinamine

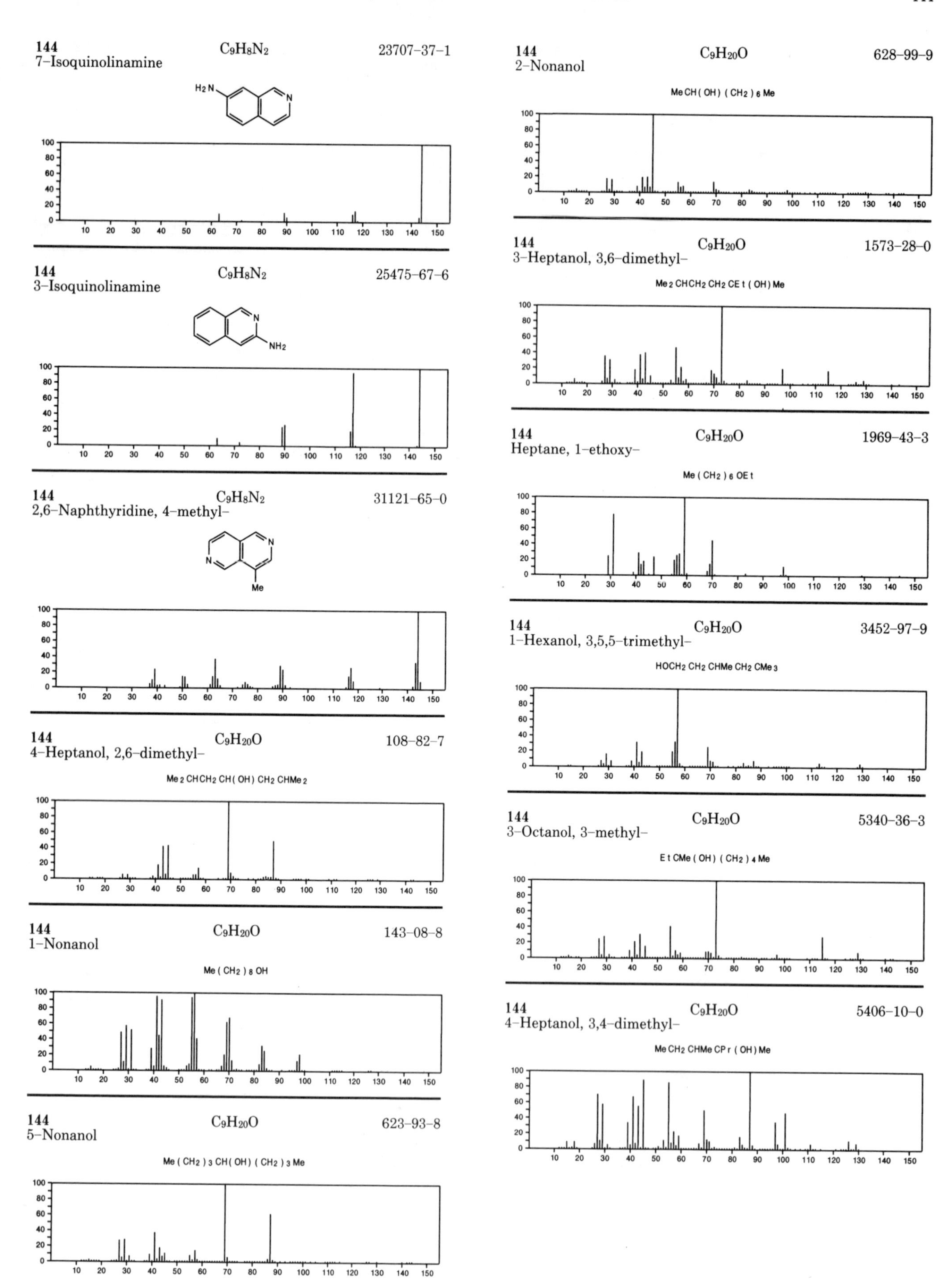
144
7-Isoquinolinamine
C9H8N2
23707-37-1
H2N
N
144
3-Isoquinolinamine
C9H8N2
25475-67-6
N
NH2
144
2,6-Naphthyridine, 4-methyl-
C9H8N2
31121-65-0
N
N
Me
144
4-Heptanol, 2,6-dimethyl-
C9H20O
108-82-7
Me2CHCH2CH(OH)CH2CHMe2
144
1-Nonanol
C9H20O
143-08-8
Me(CH2)8OH
144
5-Nonanol
C9H20O
623-93-8
Me(CH2)3CH(OH)(CH2)3Me
144
2-Nonanol
C9H20O
628-99-9
MeCH(OH)(CH2)6Me
144
3-Heptanol, 3,6-dimethyl-
C9H20O
1573-28-0
Me2CHCH2CH2CEt(OH)Me
144
Heptane, 1-ethoxy-
C9H20O
1969-43-3
Me(CH2)6OEt
144
1-Hexanol, 3,5,5-trimethyl-
C9H20O
3452-97-9
HOCH2CH2CHMeCH2CMe3
144
3-Octanol, 3-methyl-
C9H20O
5340-36-3
EtCMe(OH)(CH2)4Me
144
4-Heptanol, 3,4-dimethyl-
C9H20O
5406-10-0
MeCH2CHMeCPr(OH)Me

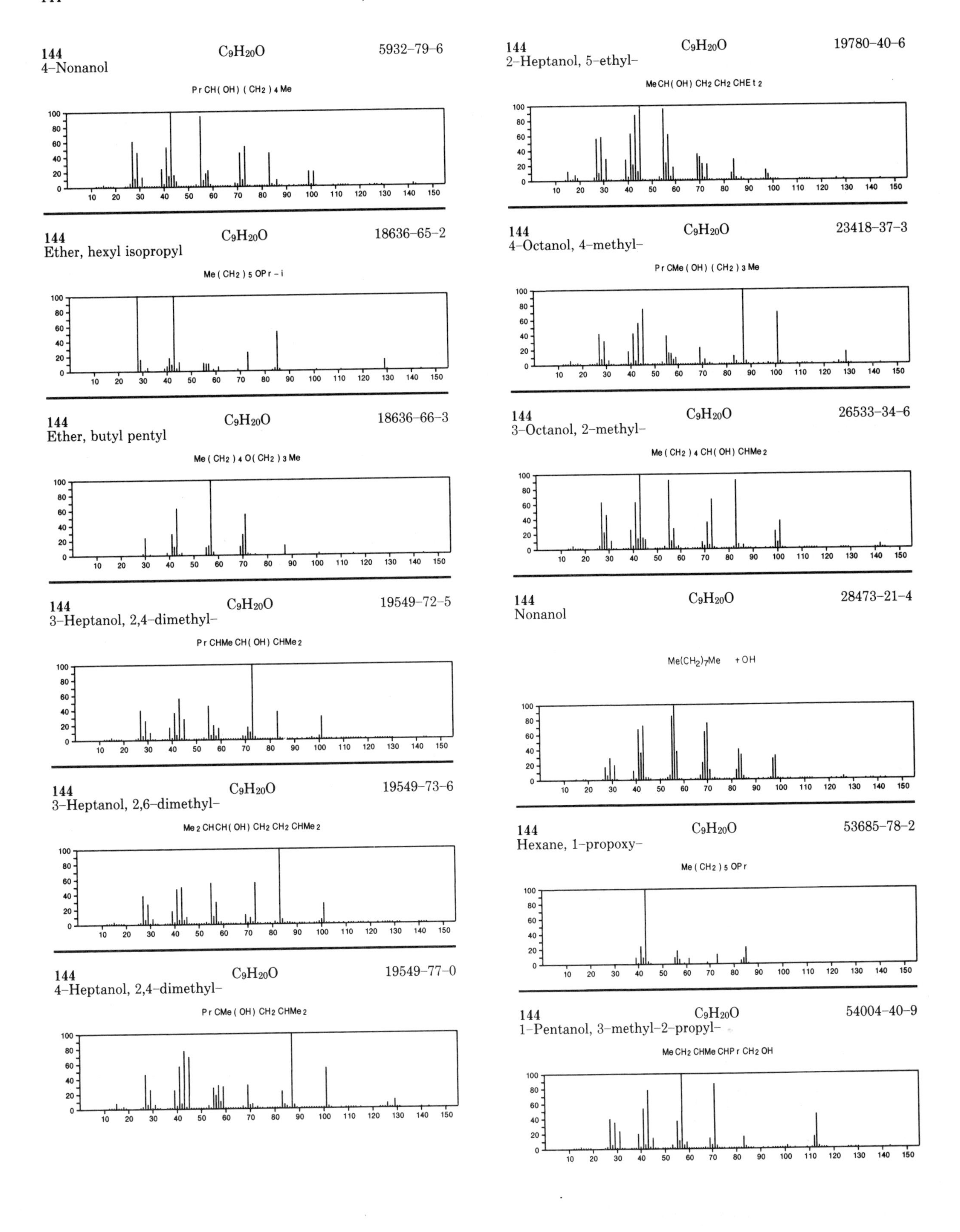
144 C9H20O 5932-79-6
4-Nonanol
PrCH(OH)(CH2)4Me
144 C9H20O 18636-65-2
Ether, hexyl isopropyl
Me(CH2)5OPr-i
144 C9H20O 18636-66-3
Ether, butyl pentyl
Me(CH2)4O(CH2)3Me
144 C9H20O 19549-72-5
3-Heptanol, 2,4-dimethyl-
PrCHMeCH(OH)CHMe2
144 C9H20O 19549-73-6
3-Heptanol, 2,6-dimethyl-
Me2CHCH(OH)CH2CH2CHMe2
144 C9H20O 19549-77-0
4-Heptanol, 2,4-dimethyl-
PrCMe(OH)CH2CHMe2
144 C9H20O 19780-40-6
2-Heptanol, 5-ethyl-
MeCH(OH)CH2CH2CHEt2
144 C9H20O 23418-37-3
4-Octanol, 4-methyl-
PrCMe(OH)(CH2)3Me
144 C9H20O 26533-34-6
3-Octanol, 2-methyl-
Me(CH2)4CH(OH)CHMe2
144 C9H20O 28473-21-4
Nonanol
Me(CH2)7Me +OH
144 C9H20O 53685-78-2
Hexane, 1-propoxy-
Me(CH2)5OPr
144 C9H20O 54004-40-9
1-Pentanol, 3-methyl-2-propyl-
MeCH2CHMeCHPrCH2OH

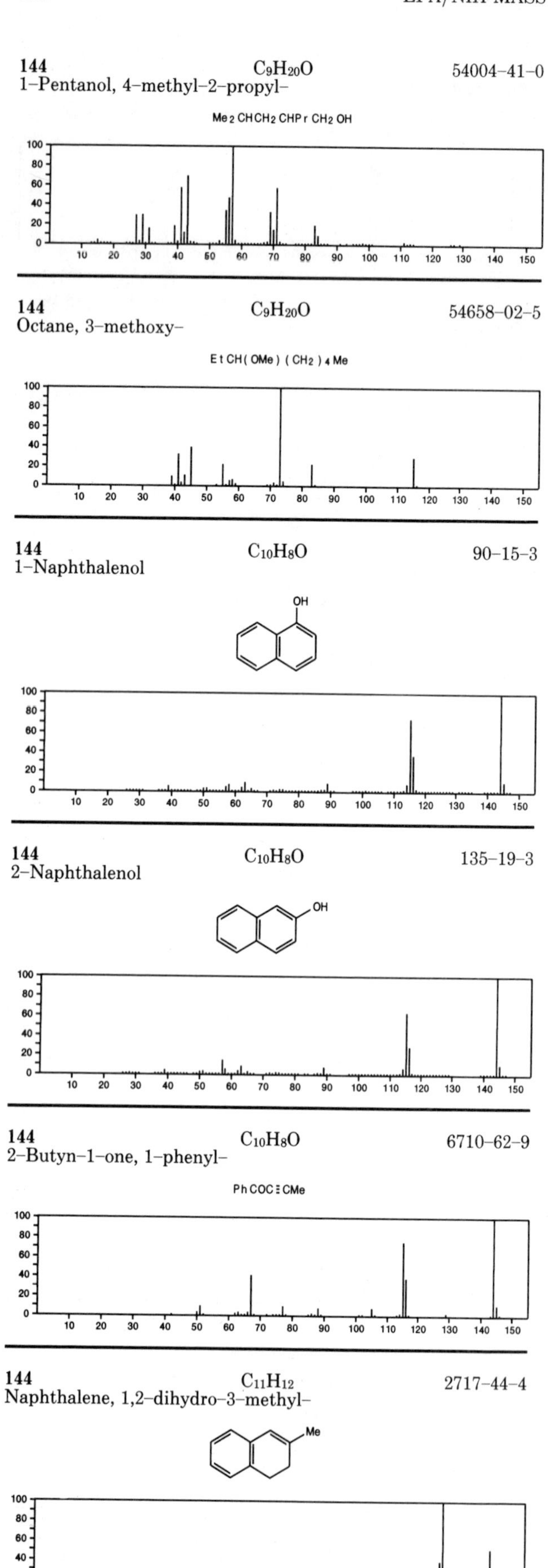

144 $C_9H_{20}O$ 54004-41-0
1-Pentanol, 4-methyl-2-propyl-

Me2 CHCH2 CHPr CH2 OH

144 $C_9H_{20}O$ 54658-02-5
Octane, 3-methoxy-

Et CH(OMe) (CH2) 4 Me

144 $C_{10}H_8O$ 90-15-3
1-Naphthalenol

144 $C_{10}H_8O$ 135-19-3
2-Naphthalenol

144 $C_{10}H_8O$ 6710-62-9
2-Butyn-1-one, 1-phenyl-

PhCOC≡CMe

144 $C_{11}H_{12}$ 2717-44-4
Naphthalene, 1,2-dihydro-3-methyl-

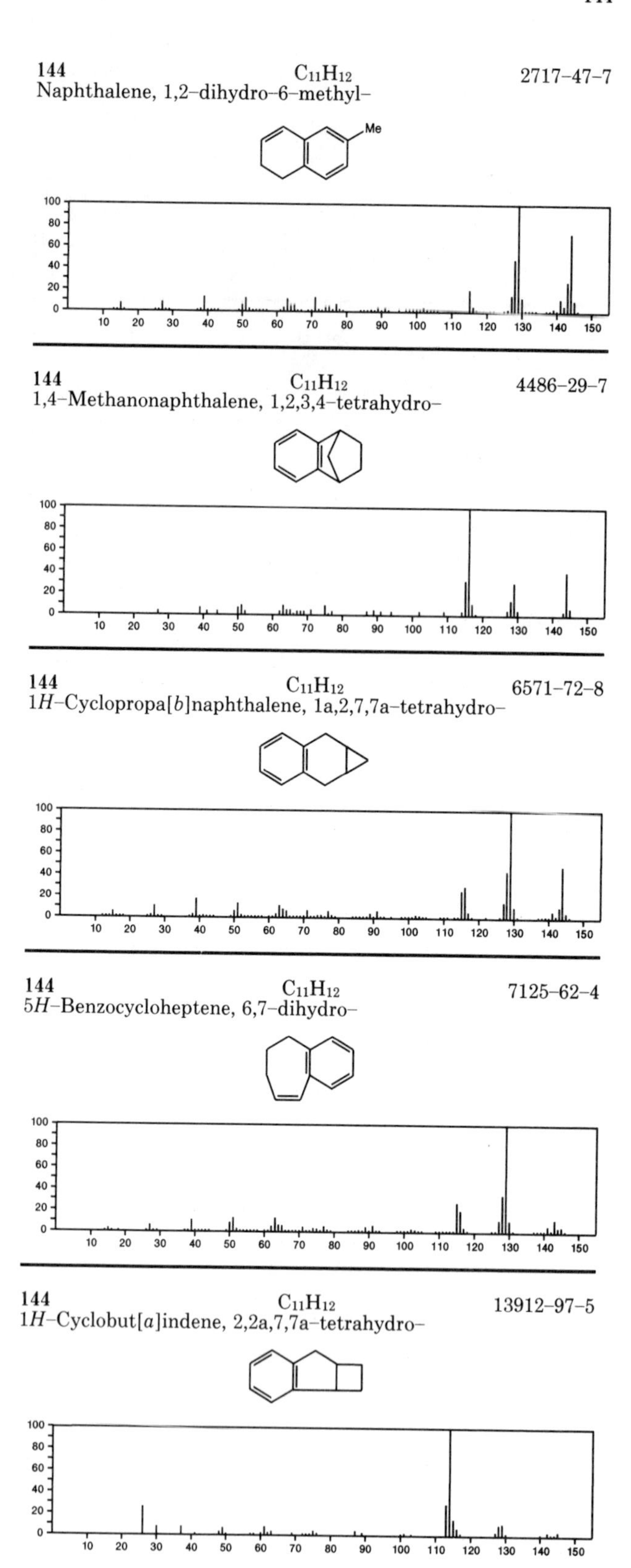

144 $C_{11}H_{12}$ 2717-47-7
Naphthalene, 1,2-dihydro-6-methyl-

144 $C_{11}H_{12}$ 4486-29-7
1,4-Methanonaphthalene, 1,2,3,4-tetrahydro-

144 $C_{11}H_{12}$ 6571-72-8
1*H*-Cyclopropa[*b*]naphthalene, 1a,2,7,7a-tetrahydro-

144 $C_{11}H_{12}$ 7125-62-4
5*H*-Benzocycloheptene, 6,7-dihydro-

144 $C_{11}H_{12}$ 13912-97-5
1*H*-Cyclobut[*a*]indene, 2,2a,7,7a-tetrahydro-

144 $C_{11}H_{12}$ 18636-55-0
1*H*-Indene, 1,1-dimethyl-

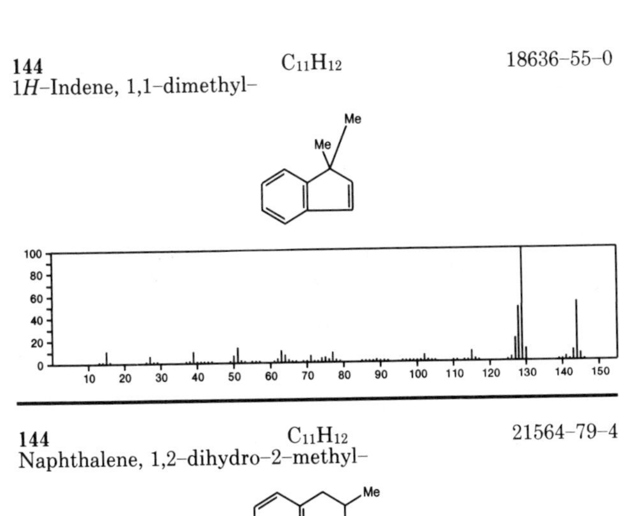

144 $C_{11}H_{12}$ 21564-79-4
Naphthalene, 1,2-dihydro-2-methyl-

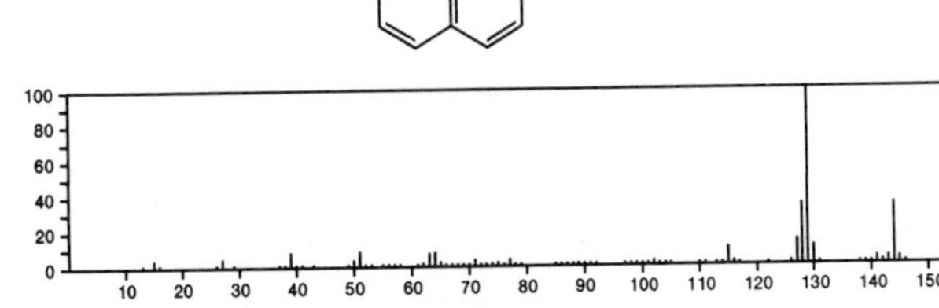

145 $C_2H_6Cl_2NP$ 683-85-2
Phosphoramidous dichloride, dimethyl-

Cl_2PNMe_2

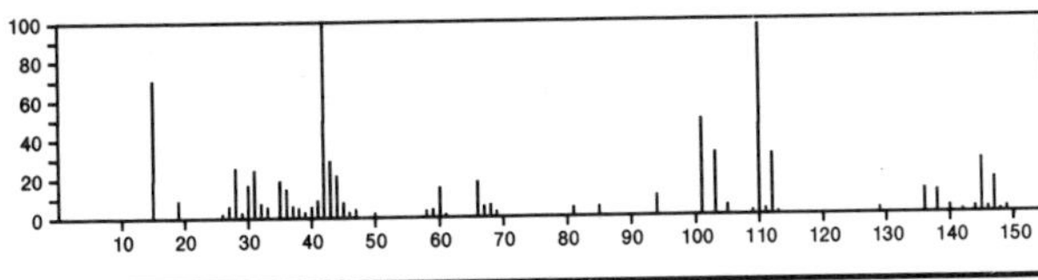

145 $C_3H_3N_3O_2S$ 121-66-4
2-Thiazolamine, 5-nitro-

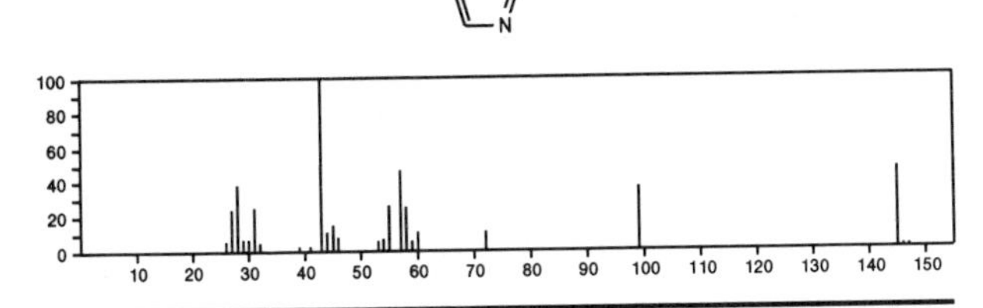

145 $C_3H_4ClN_5$ 3397-62-4
1,3,5-Triazine-2,4-diamine, 6-chloro-

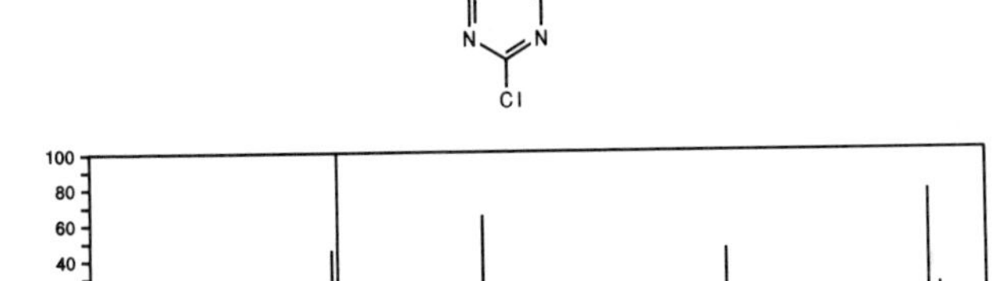

145 $C_5H_{11}N_3O_2$ 50285-71-7
Urea, *N*-ethyl-*N'*,*N'*-dimethyl-*N*-nitroso-

$EtN(NO)CONMe_2$

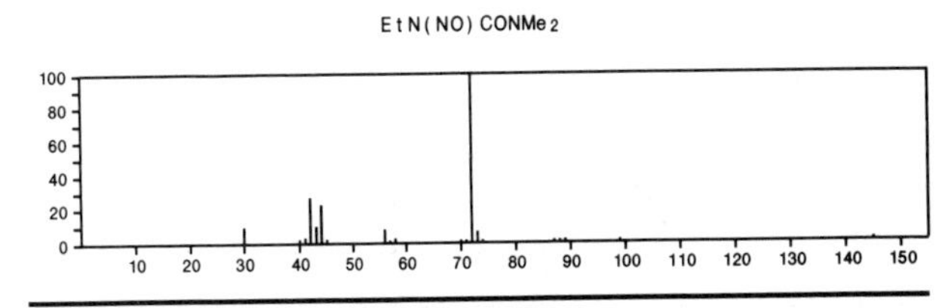

145 $C_5H_{11}N_3S$ 20812-04-8
Hydrazinecarbothioamide, 2-butylidene-

$H_2NCSNHN{=}CHPr$

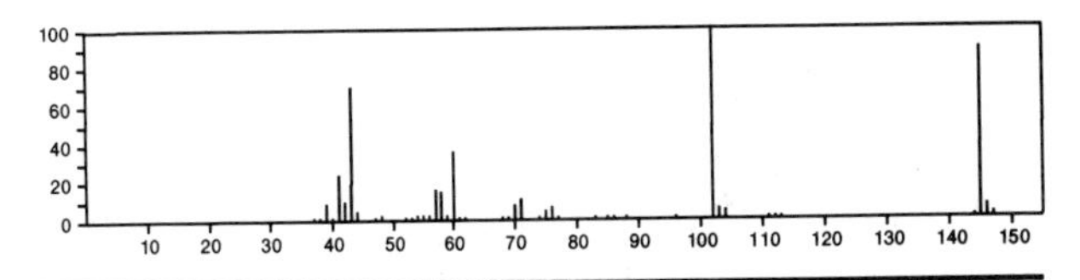

145 C_6H_8ClNO 19788-37-5
Isoxazole, 4-(chloromethyl)-3,5-dimethyl-

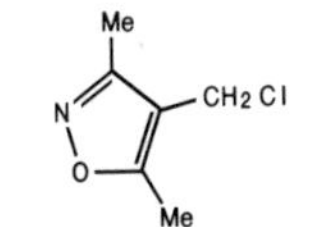

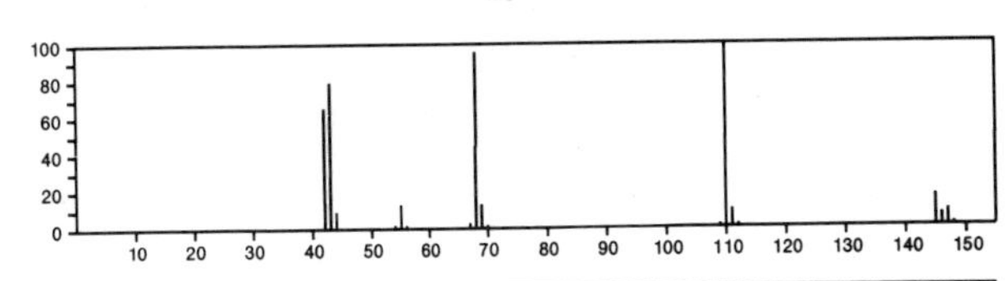

145 $C_6H_{11}NO_3$ 1906-82-7
Glycine, *N*-acetyl-, ethyl ester

$AcNHCH_2C(O)OEt$

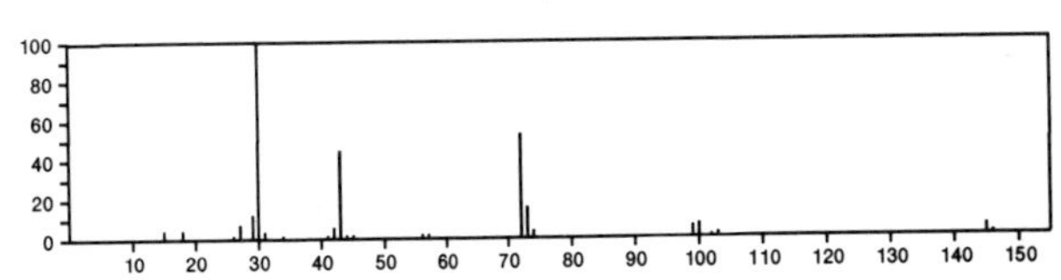

145 $C_6H_{11}NO_3$ 7211-57-6
Butanoic acid, 2-(acetylamino)-

$AcNHCHEtCO_2H$

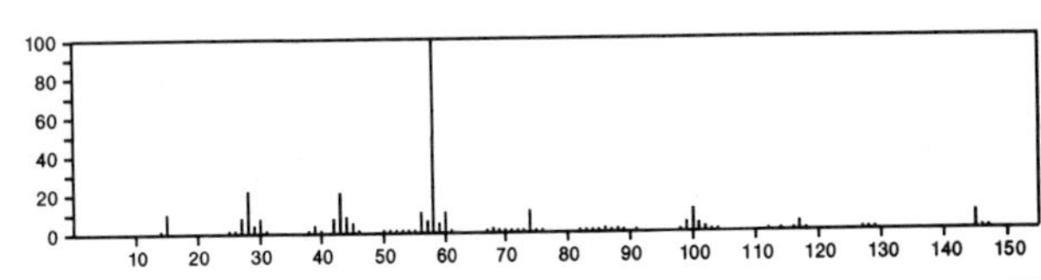

145 $C_6H_{11}NO_3$ 15166-66-2
5-Isoxazolidinecarboxylic acid, 5-methyl-, methyl ester

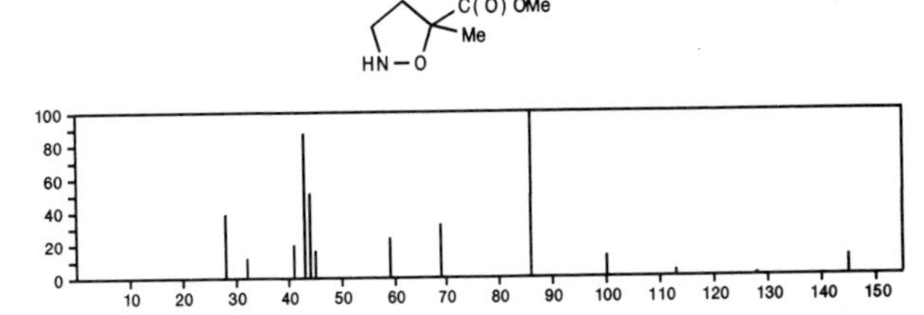

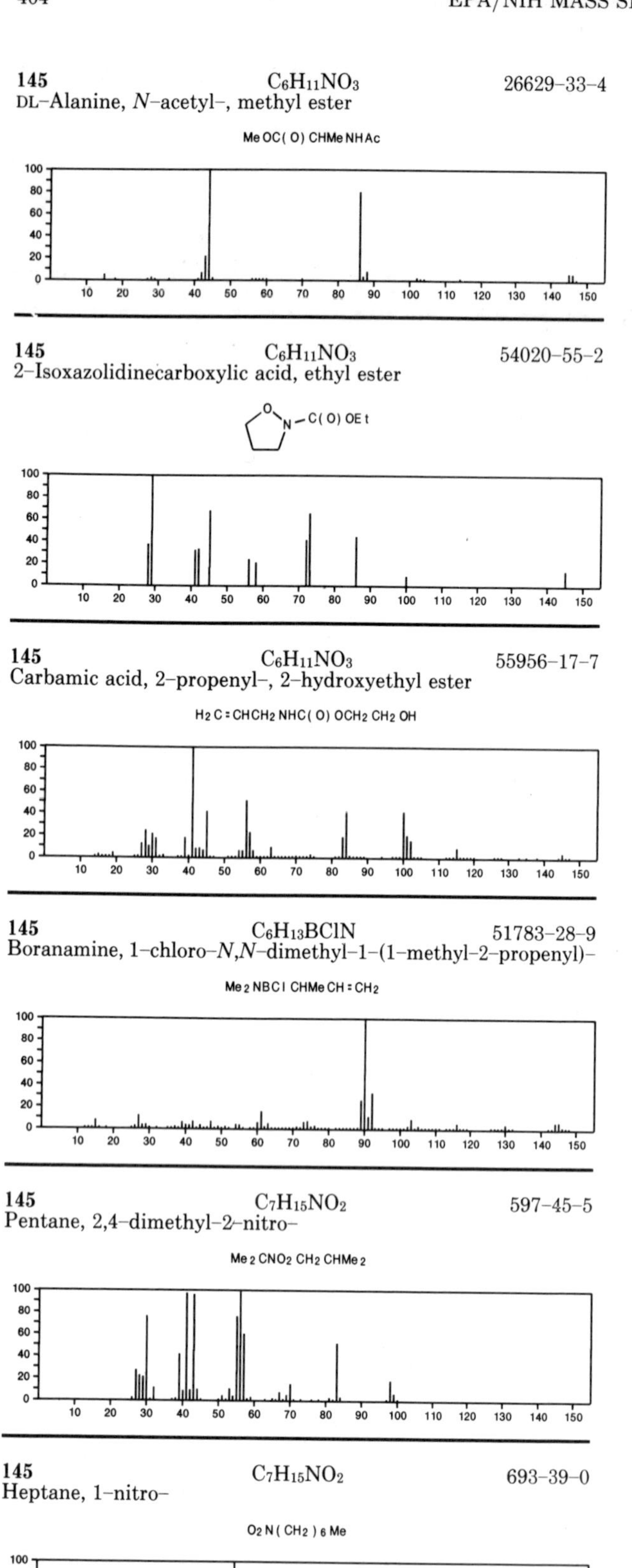

145 $C_6H_{11}NO_3$ 26629-33-4
DL-Alanine, *N*-acetyl-, methyl ester
MeOC(O)CHMeNHAc

145 $C_6H_{11}NO_3$ 54020-55-2
2-Isoxazolidinecarboxylic acid, ethyl ester
N-C(O)OEt

145 $C_6H_{11}NO_3$ 55956-17-7
Carbamic acid, 2-propenyl-, 2-hydroxyethyl ester
H2C=CHCH2NHC(O)OCH2CH2OH

145 $C_6H_{13}BClN$ 51783-28-9
Boranamine, 1-chloro-*N*,*N*-dimethyl-1-(1-methyl-2-propenyl)-
Me2NBClCHMeCH=CH2

145 $C_7H_{15}NO_2$ 597-45-5
Pentane, 2,4-dimethyl-2-nitro-
Me2CNO2CH2CHMe2

145 $C_7H_{15}NO_2$ 693-39-0
Heptane, 1-nitro-
O2N(CH2)6Me

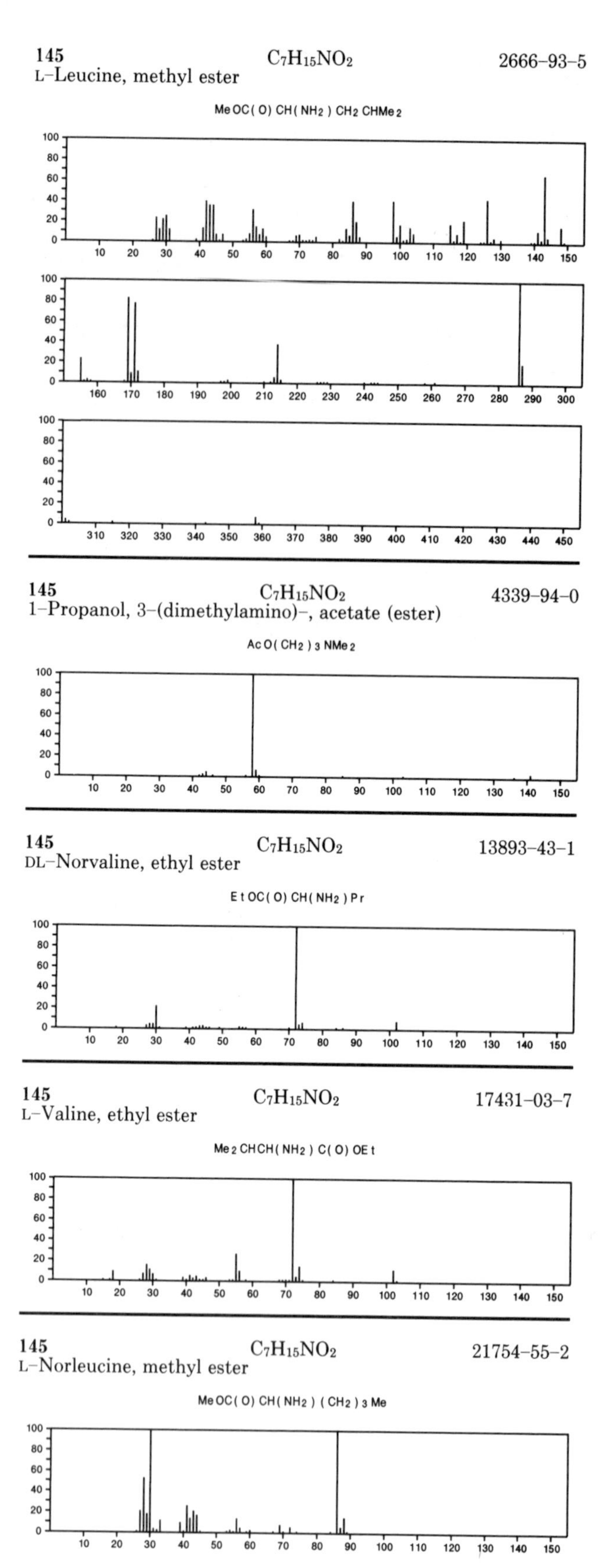

145 $C_7H_{15}NO_2$ 2666-93-5
L-Leucine, methyl ester
MeOC(O)CH(NH2)CH2CHMe2

145 $C_7H_{15}NO_2$ 4339-94-0
1-Propanol, 3-(dimethylamino)-, acetate (ester)
AcO(CH2)3NMe2

145 $C_7H_{15}NO_2$ 13893-43-1
DL-Norvaline, ethyl ester
EtOC(O)CH(NH2)Pr

145 $C_7H_{15}NO_2$ 17431-03-7
L-Valine, ethyl ester
Me2CHCH(NH2)C(O)OEt

145 $C_7H_{15}NO_2$ 21754-55-2
L-Norleucine, methyl ester
MeOC(O)CH(NH2)(CH2)3Me

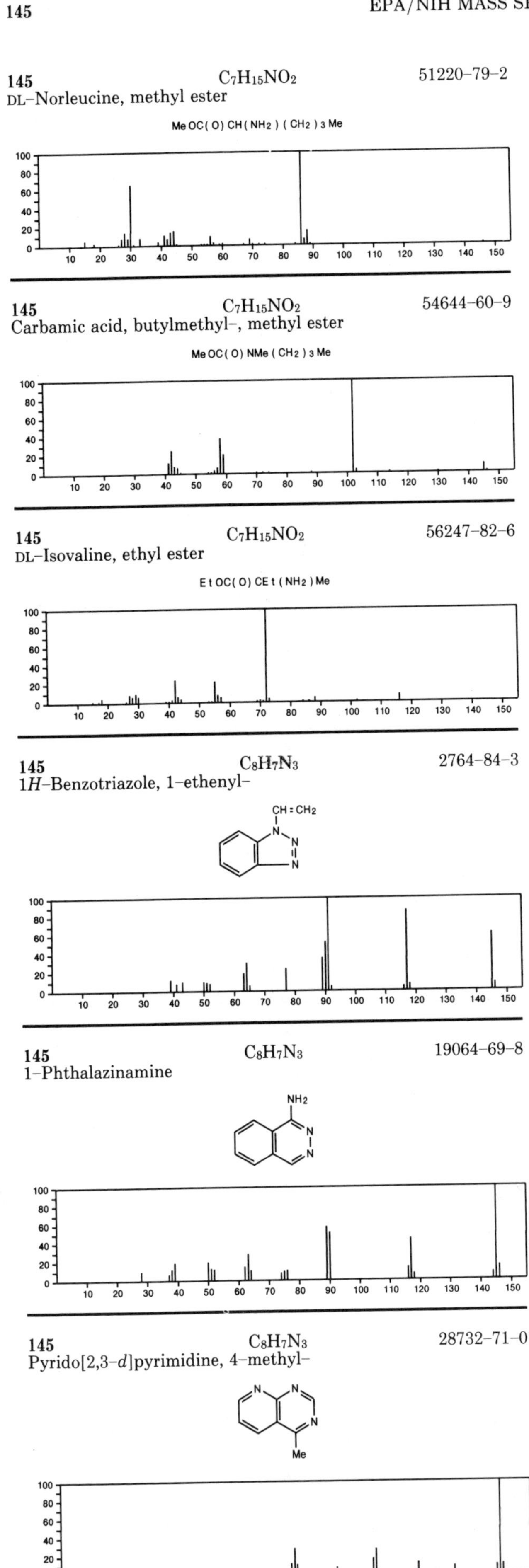

145 $C_7H_{15}NO_2$ 51220-79-2
DL-Norleucine, methyl ester
MeOC(O)CH(NH2)(CH2)3Me

145 $C_7H_{15}NO_2$ 54644-60-9
Carbamic acid, butylmethyl-, methyl ester
MeOC(O)NMe(CH2)3Me

145 $C_7H_{15}NO_2$ 56247-82-6
DL-Isovaline, ethyl ester
EtOC(O)CEt(NH2)Me

145 $C_8H_7N_3$ 2764-84-3
1*H*-Benzotriazole, 1-ethenyl-

145 $C_8H_7N_3$ 19064-69-8
1-Phthalazinamine

145 $C_8H_7N_3$ 28732-71-0
Pyrido[2,3-*d*]pyrimidine, 4-methyl-

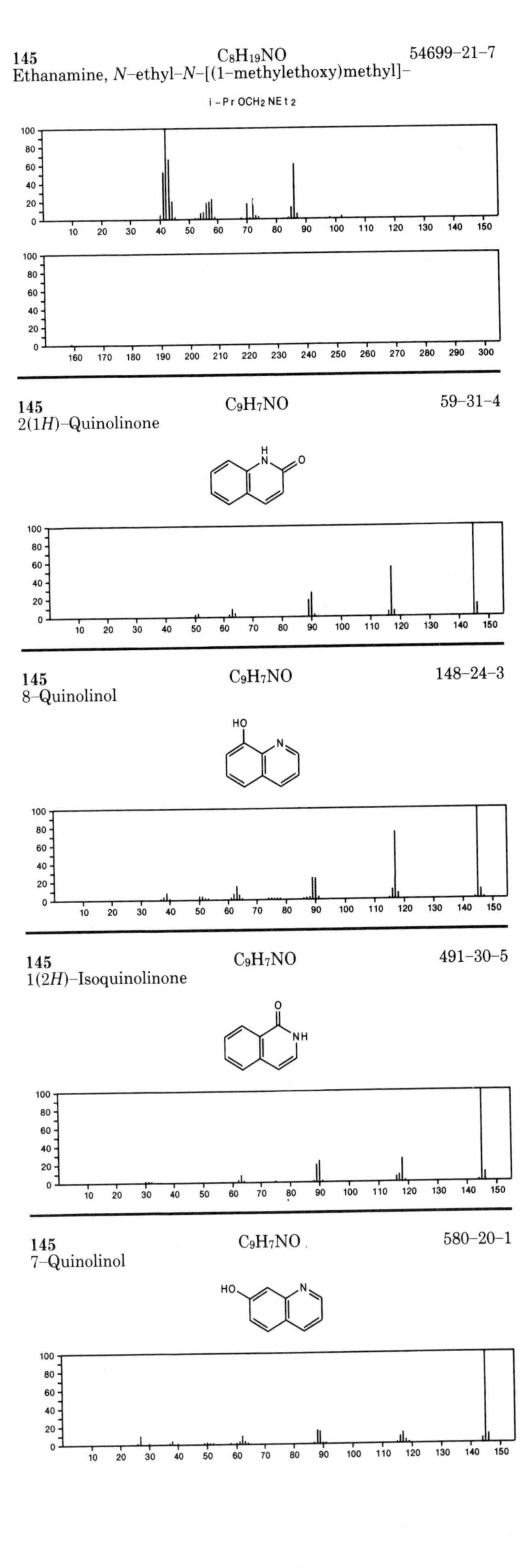

145 $C_8H_{19}NO$ 54699-21-7
Ethanamine, *N*-ethyl-*N*-[(1-methylethoxy)methyl]-
i-PrOCH2NEt2

145 C_9H_7NO 59-31-4
2(1*H*)-Quinolinone

145 C_9H_7NO 148-24-3
8-Quinolinol

145 C_9H_7NO 491-30-5
1(2*H*)-Isoquinolinone

145 C_9H_7NO 580-20-1
7-Quinolinol

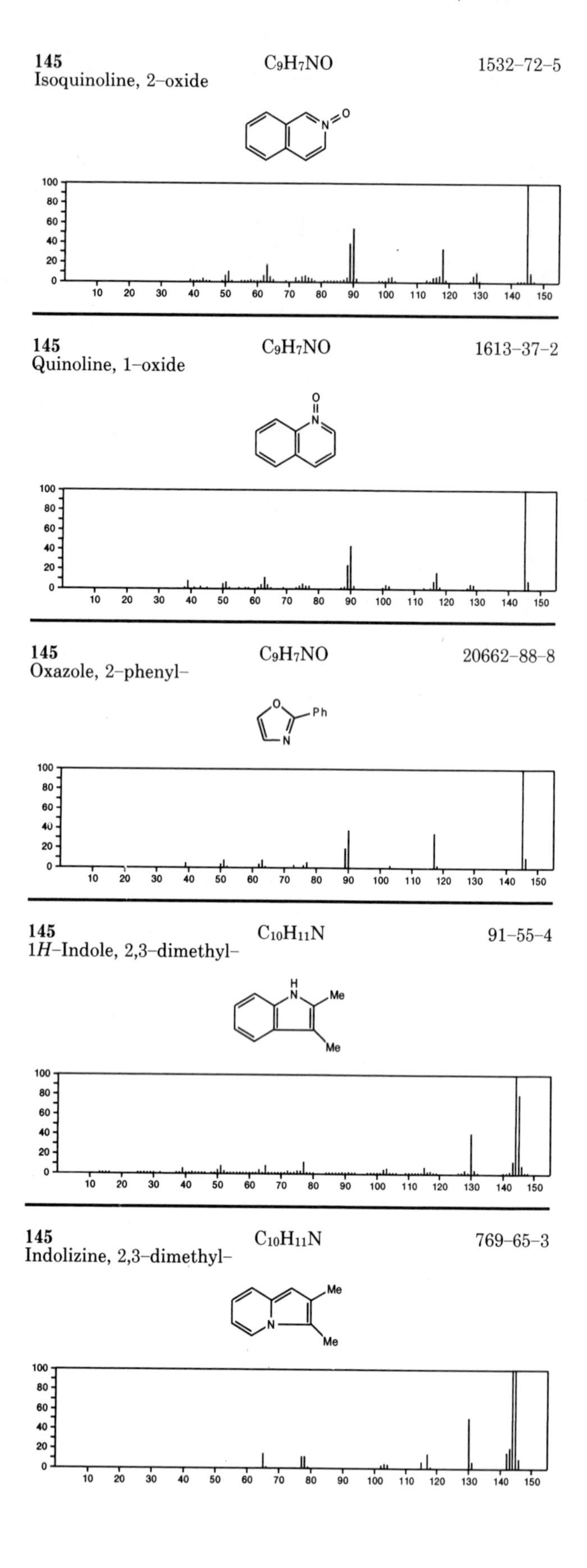

145
Isoquinoline, 2–oxide
C_9H_7NO
1532–72–5

145
Quinoline, 1–oxide
C_9H_7NO
1613–37–2

145
Oxazole, 2–phenyl–
C_9H_7NO
20662–88–8

145
1*H*–Indole, 2,3–dimethyl–
$C_{10}H_{11}N$
91–55–4

145
Indolizine, 2,3–dimethyl–
$C_{10}H_{11}N$
769–65–3

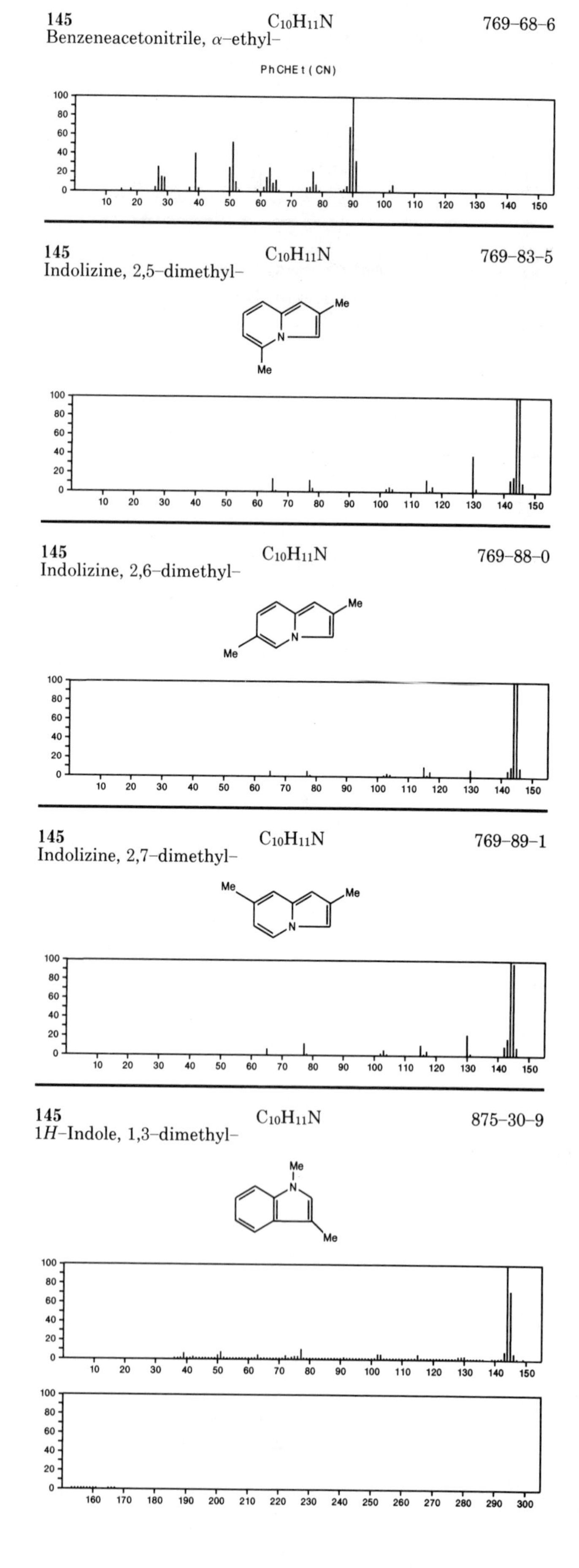

145
Benzeneacetonitrile, α–ethyl–
$C_{10}H_{11}N$
769–68–6

PhCHEt(CN)

145
Indolizine, 2,5–dimethyl–
$C_{10}H_{11}N$
769–83–5

145
Indolizine, 2,6–dimethyl–
$C_{10}H_{11}N$
769–88–0

145
Indolizine, 2,7–dimethyl–
$C_{10}H_{11}N$
769–89–1

145
1*H*–Indole, 1,3–dimethyl–
$C_{10}H_{11}N$
875–30–9

145 $C_{10}H_{11}N$ 1125–77–5
Indolizine, 1,2–dimethyl–

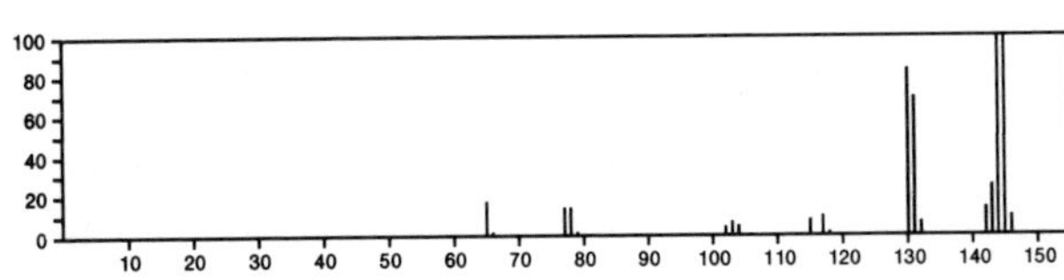

145 $C_{10}H_{11}N$ 1196–79–8
1*H*–Indole, 2,5–dimethyl–

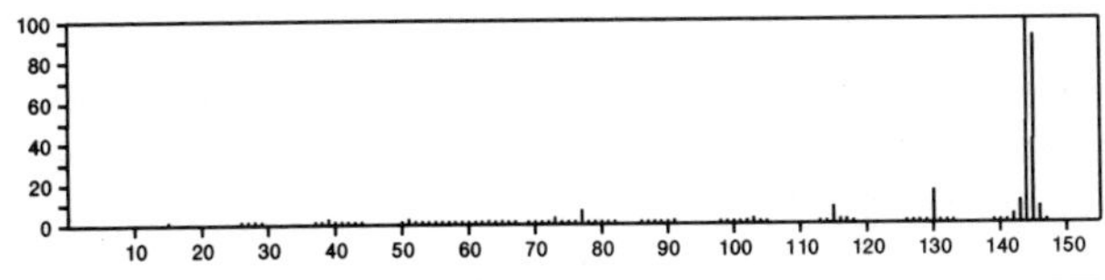

145 $C_{10}H_{11}N$ 1761–13–3
Indolizine, 3,5–dimethyl–

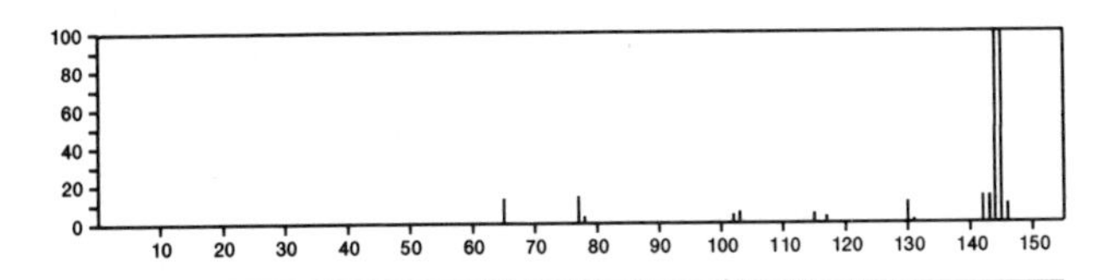

145 $C_{10}H_{11}N$ 2046–18–6
Benzenebutanenitrile

$Ph(CH_2)_3CN$

145 $C_{10}H_{11}N$ 4282–82–0
Benzenamine, *N*–methyl–*N*–2–propynyl–

$HC\equiv CCH_2NMePh$

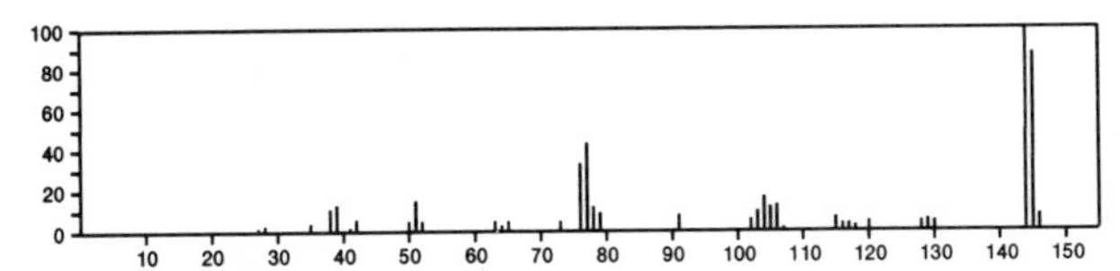

145 $C_{10}H_{11}N$ 5621–16–9
Indole, 1,7–dimethyl–

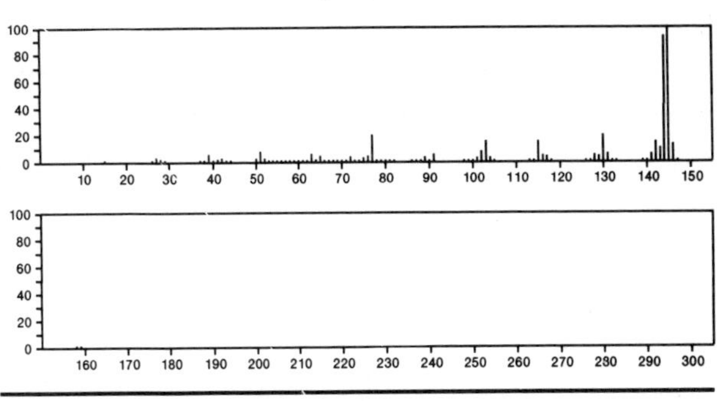

145 $C_{10}H_{11}N$ 5649–36–5
1*H*–Indole, 2,6–dimethyl–

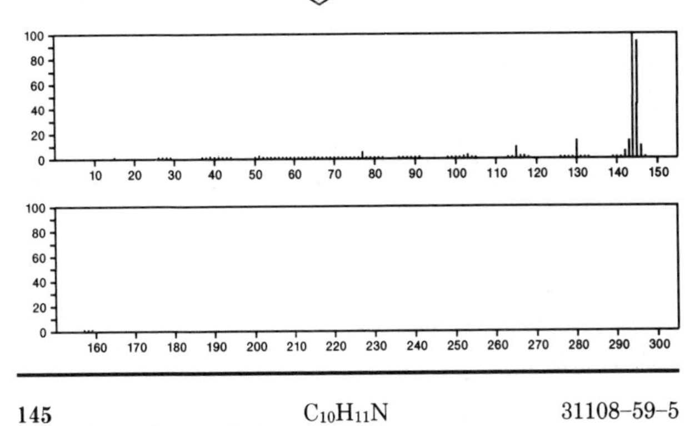

145 $C_{10}H_{11}N$ 31108–59–5
Indolizine, 2,8–dimethyl–

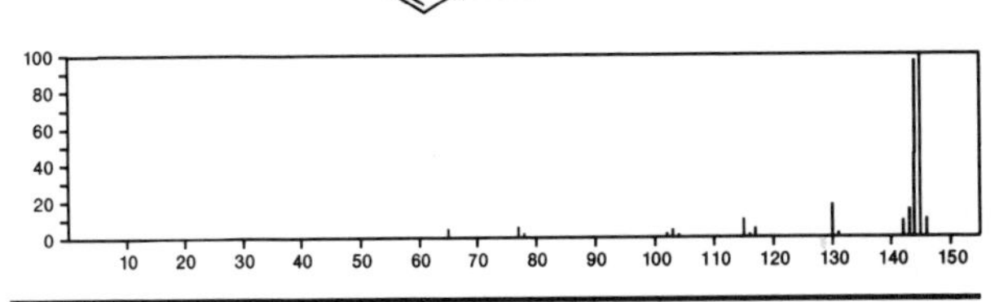

145 $C_{10}H_{11}N$ 54020–53–0
1*H*–Indole, 5,7–dimethyl–

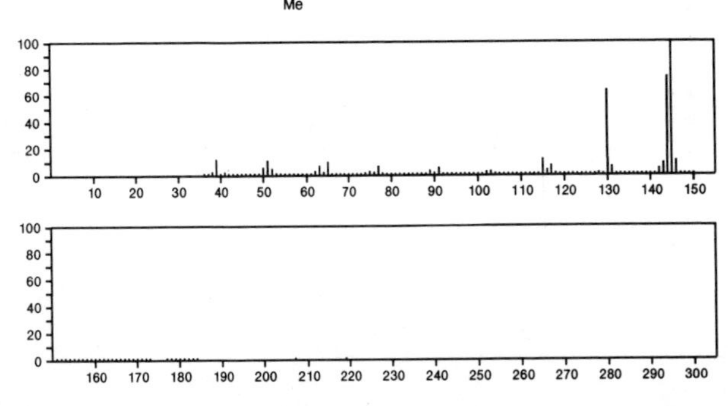

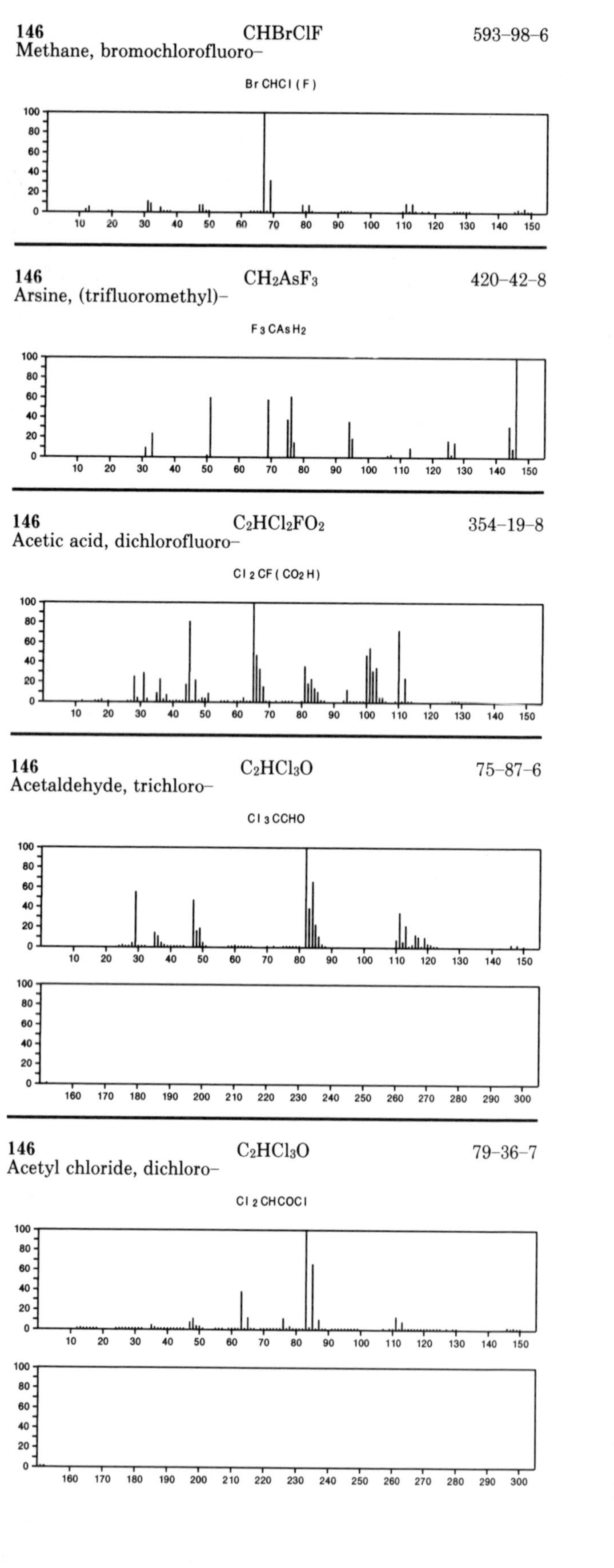

146 $CHBrClF$ 593–98–6
Methane, bromochlorofluoro–

Br CHCl (F)

146 CH_2AsF_3 420–42–8
Arsine, (trifluoromethyl)–

F3 CAsH2

146 $C_2HCl_2FO_2$ 354–19–8
Acetic acid, dichlorofluoro–

Cl 2 CF (CO2 H)

146 C_2HCl_3O 75–87–6
Acetaldehyde, trichloro–

Cl 3 CCHO

146 C_2HCl_3O 79–36–7
Acetyl chloride, dichloro–

Cl 2 CHCOCl

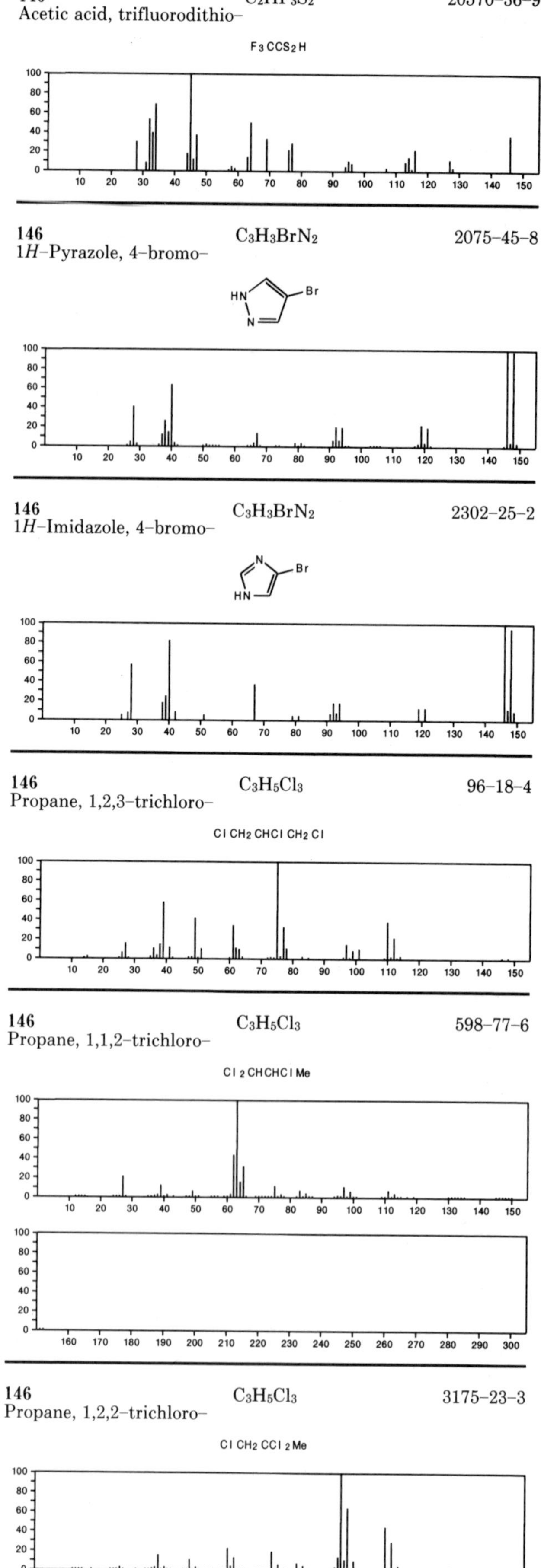

146 $C_2HF_3S_2$ 20570–36–9
Acetic acid, trifluorodithio–

F3 CCS2 H

146 $C_3H_3BrN_2$ 2075–45–8
1*H*–Pyrazole, 4–bromo–

146 $C_3H_3BrN_2$ 2302–25–2
1*H*–Imidazole, 4–bromo–

146 $C_3H_5Cl_3$ 96–18–4
Propane, 1,2,3–trichloro–

Cl CH2 CHCl CH2 Cl

146 $C_3H_5Cl_3$ 598–77–6
Propane, 1,1,2–trichloro–

Cl 2 CHCHCl Me

146 $C_3H_5Cl_3$ 3175–23–3
Propane, 1,2,2–trichloro–

Cl CH2 CCl 2 Me

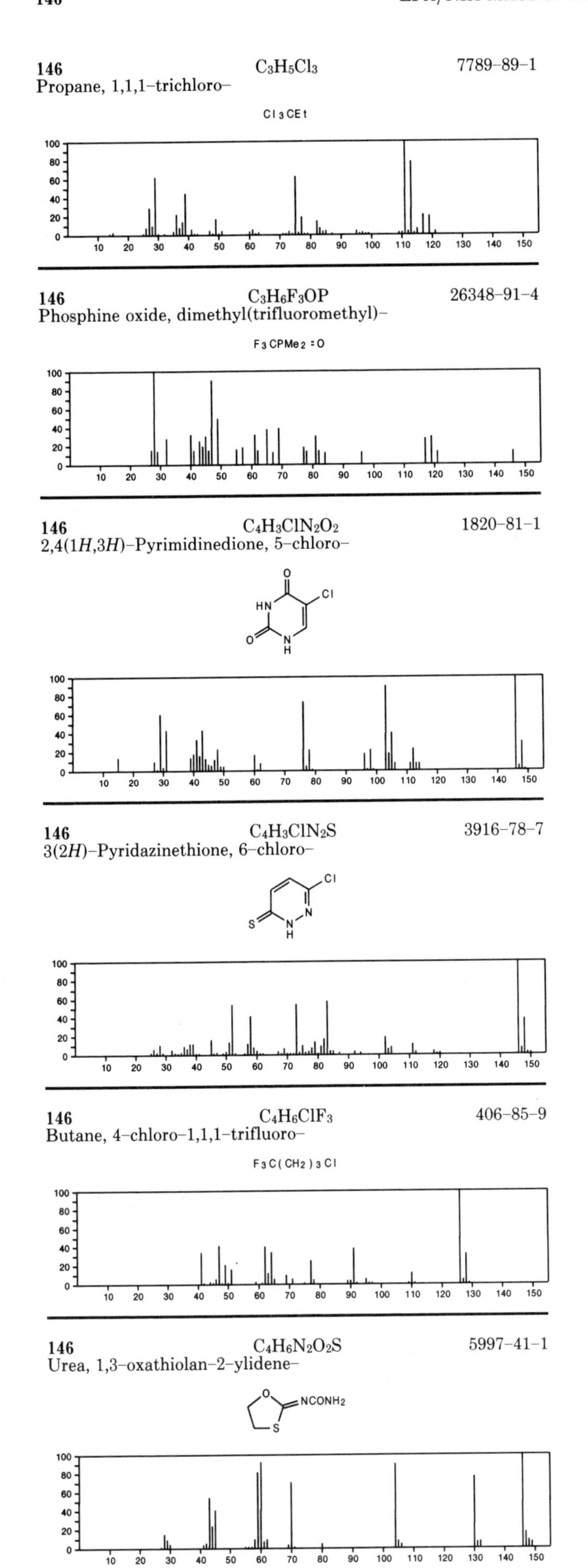
146 $C_3H_5Cl_3$ 7789-89-1
Propane, 1,1,1-trichloro-
Cl3CEt
146 $C_3H_6F_3OP$ 26348-91-4
Phosphine oxide, dimethyl(trifluoromethyl)-
F3CPMe2=O
146 $C_4H_3ClN_2O_2$ 1820-81-1
2,4(1H,3H)-Pyrimidinedione, 5-chloro-
146 $C_4H_3ClN_2S$ 3916-78-7
3(2H)-Pyridazinethione, 6-chloro-
146 $C_4H_6ClF_3$ 406-85-9
Butane, 4-chloro-1,1,1-trifluoro-
F3C(CH2)3Cl
146 $C_4H_6N_2O_2S$ 5997-41-1
Urea, 1,3-oxathiolan-2-ylidene-
NCONH2

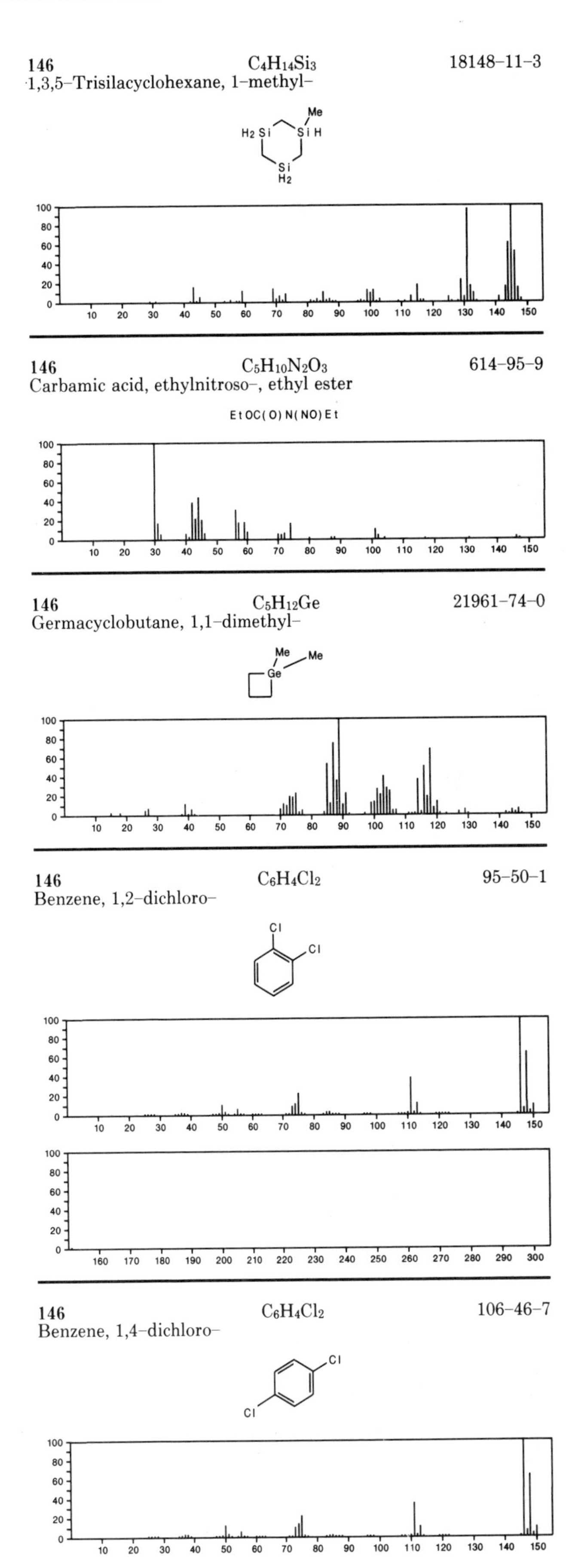
146 $C_4H_{14}Si_3$ 18148-11-3
1,3,5-Trisilacyclohexane, 1-methyl-
146 $C_5H_{10}N_2O_3$ 614-95-9
Carbamic acid, ethylnitroso-, ethyl ester
EtOC(O)N(NO)Et
146 $C_5H_{12}Ge$ 21961-74-0
Germacyclobutane, 1,1-dimethyl-
146 $C_6H_4Cl_2$ 95-50-1
Benzene, 1,2-dichloro-
146 $C_6H_4Cl_2$ 106-46-7
Benzene, 1,4-dichloro-

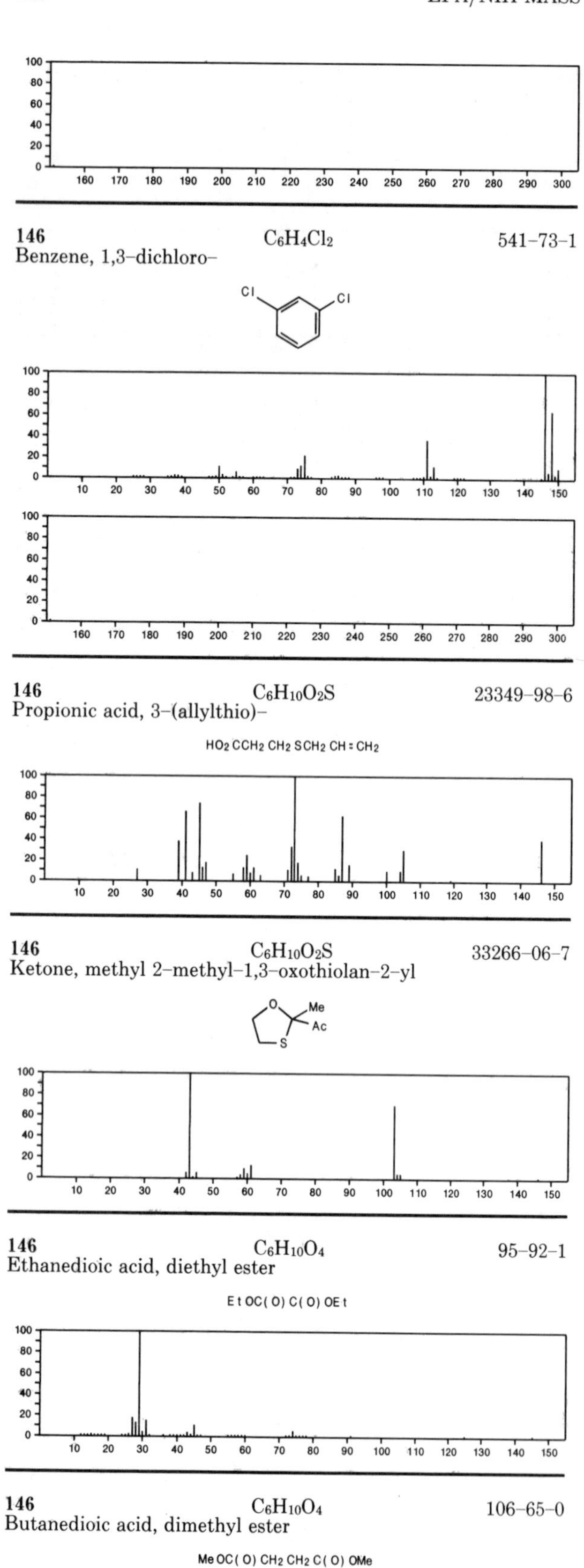

146
Benzene, 1,3-dichloro–
$C_6H_4Cl_2$
541–73–1
146
Propionic acid, 3–(allylthio)–
$C_6H_{10}O_2S$
23349–98–6
HO2CCH2CH2SCH2CH=CH2
146
Ketone, methyl 2–methyl–1,3–oxothiolan–2–yl
$C_6H_{10}O_2S$
33266–06–7
146
Ethanedioic acid, diethyl ester
$C_6H_{10}O_4$
95–92–1
EtOC(O)C(O)OEt
146
Butanedioic acid, dimethyl ester
$C_6H_{10}O_4$
106–65–0
MeOC(O)CH2CH2C(O)OMe

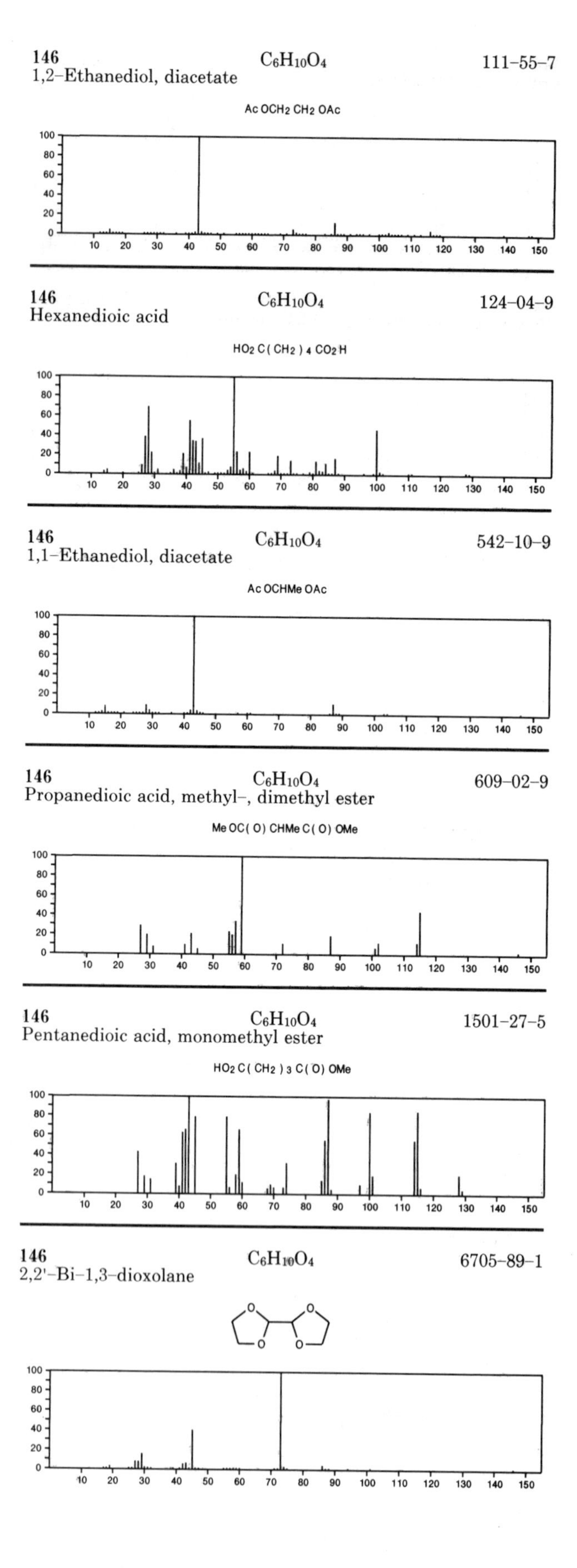

146
1,2–Ethanediol, diacetate
$C_6H_{10}O_4$
111–55–7
AcOCH2CH2OAc
146
Hexanedioic acid
$C_6H_{10}O_4$
124–04–9
HO2C(CH2)4CO2H
146
1,1–Ethanediol, diacetate
$C_6H_{10}O_4$
542–10–9
AcOCHMeOAc
146
Propanedioic acid, methyl–, dimethyl ester
$C_6H_{10}O_4$
609–02–9
MeOC(O)CHMeC(O)OMe
146
Pentanedioic acid, monomethyl ester
$C_6H_{10}O_4$
1501–27–5
HO2C(CH2)3C(O)OMe
146
2,2'–Bi–1,3–dioxolane
$C_6H_{10}O_4$
6705–89–1

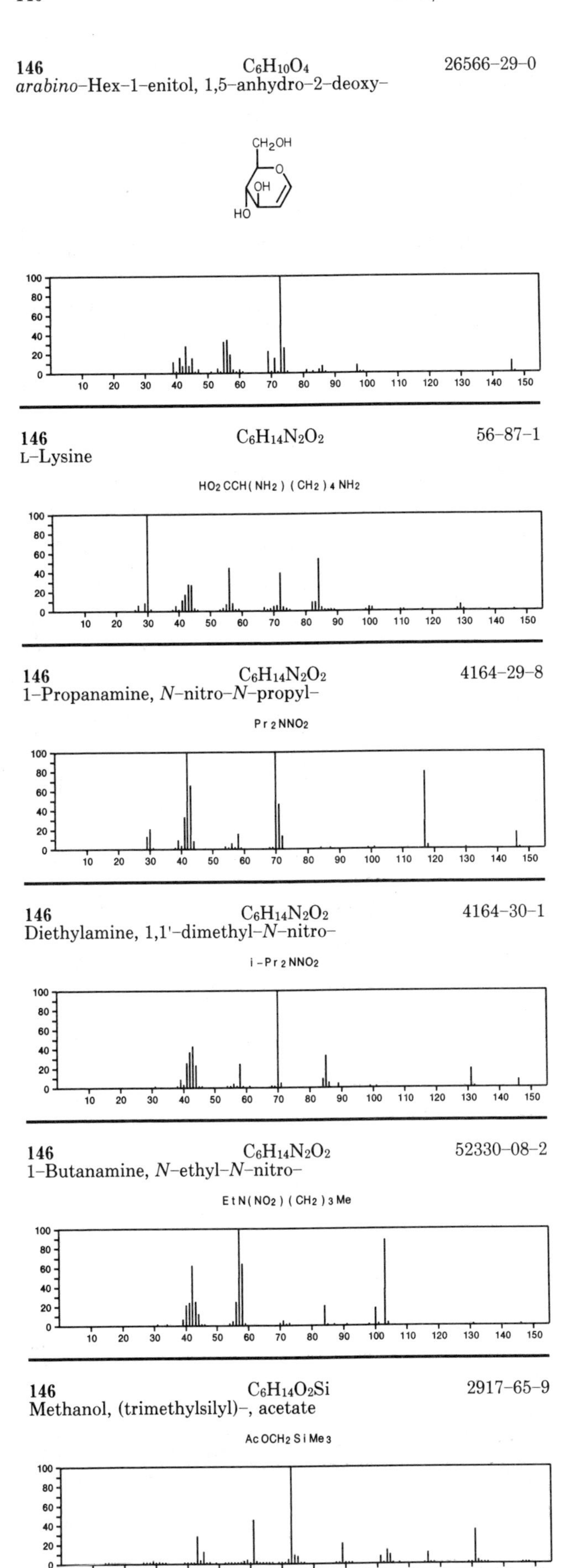

146 $C_6H_{10}O_4$ 26566-29-0
arabino-Hex-1-enitol, 1,5-anhydro-2-deoxy-

146 $C_6H_{14}N_2O_2$ 56-87-1
L-Lysine
$HO_2CCH(NH_2)(CH_2)_4NH_2$

146 $C_6H_{14}N_2O_2$ 4164-29-8
1-Propanamine, *N*-nitro-*N*-propyl-
Pr_2NNO_2

146 $C_6H_{14}N_2O_2$ 4164-30-1
Diethylamine, 1,1'-dimethyl-*N*-nitro-
$i\text{-}Pr_2NNO_2$

146 $C_6H_{14}N_2O_2$ 52330-08-2
1-Butanamine, *N*-ethyl-*N*-nitro-
$EtN(NO_2)(CH_2)_3Me$

146 $C_6H_{14}O_2Si$ 2917-65-9
Methanol, (trimethylsilyl)-, acetate
$AcOCH_2SiMe_3$

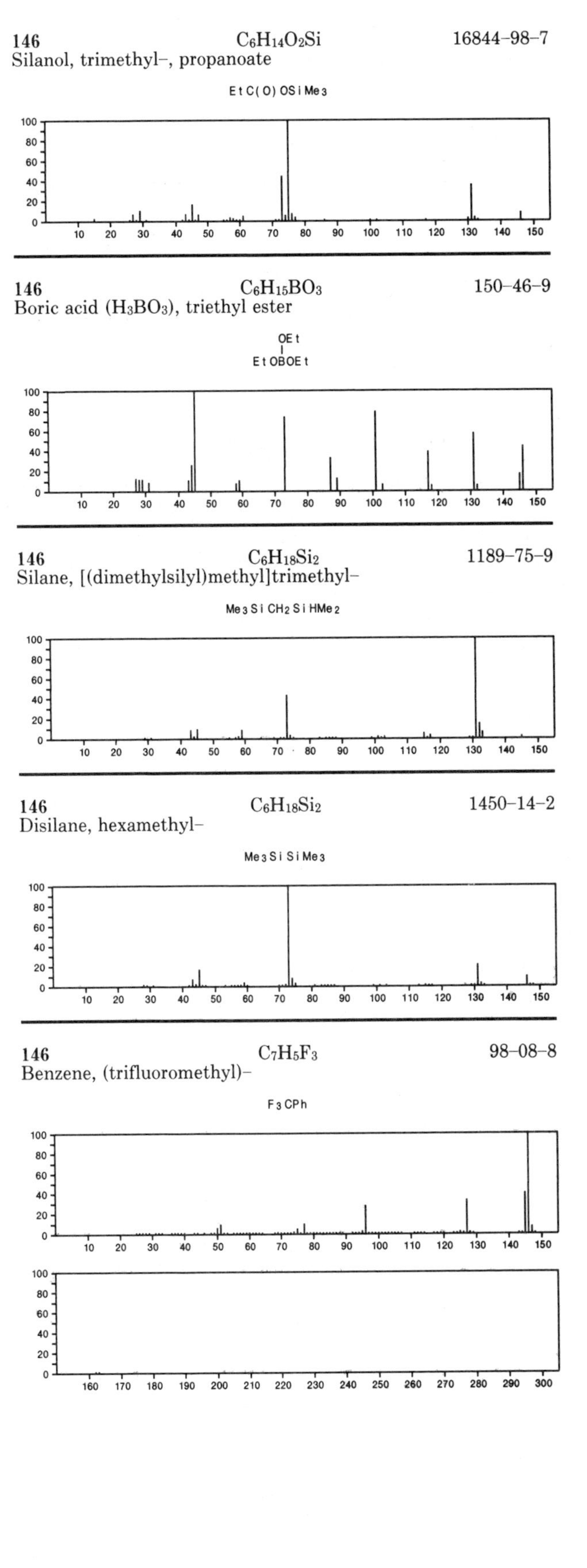

146 $C_6H_{14}O_2Si$ 16844-98-7
Silanol, trimethyl-, propanoate
$EtC(O)OSiMe_3$

146 $C_6H_{15}BO_3$ 150-46-9
Boric acid (H_3BO_3), triethyl ester
$EtOB(OEt)OEt$

146 $C_6H_{18}Si_2$ 1189-75-9
Silane, [(dimethylsilyl)methyl]trimethyl-
$Me_3SiCH_2SiHMe_2$

146 $C_6H_{18}Si_2$ 1450-14-2
Disilane, hexamethyl-
$Me_3SiSiMe_3$

146 $C_7H_5F_3$ 98-08-8
Benzene, (trifluoromethyl)-
F_3CPh

146 $C_7H_5F_3$ 27359-10-0
Benzene, methyl-, trifluoro deriv.

PhMe + 3F

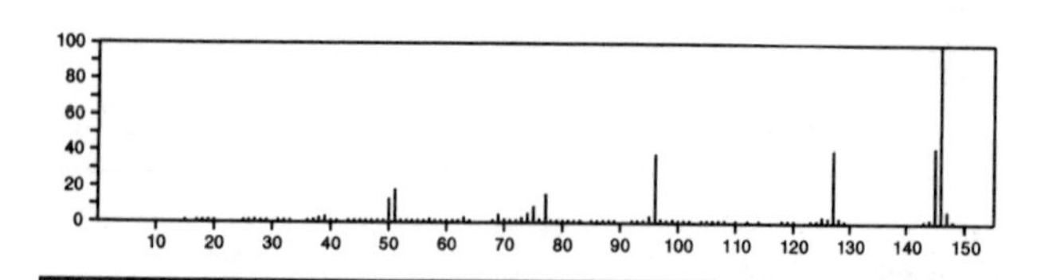

146 $C_7H_6N_4$ 936-40-3
Pteridine, 7-methyl-

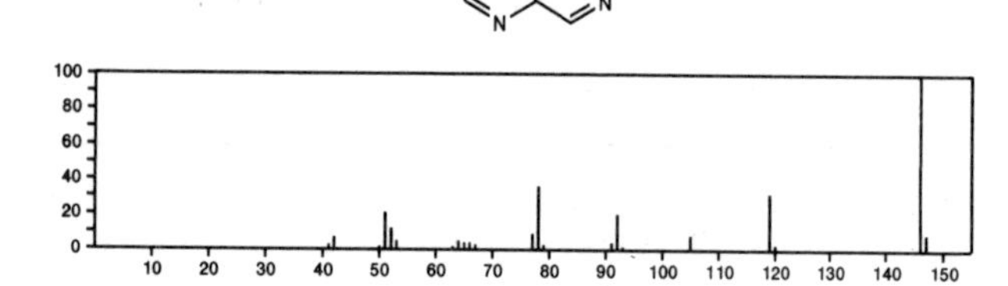

146 $C_7H_6N_4$ 2432-20-4
Pteridine, 2-methyl-

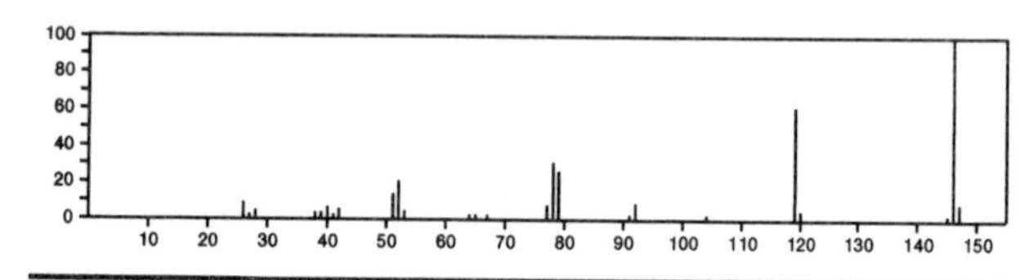

146 $C_7H_6N_4$ 2432-21-5
Pteridine, 4-methyl-

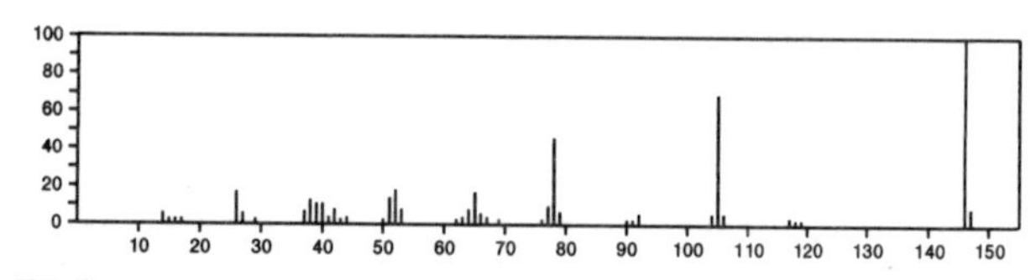

146 $C_7H_{11}ClO$ 56816-12-7
Bicyclo[2.2.1]heptan-2-ol, 3-chloro-

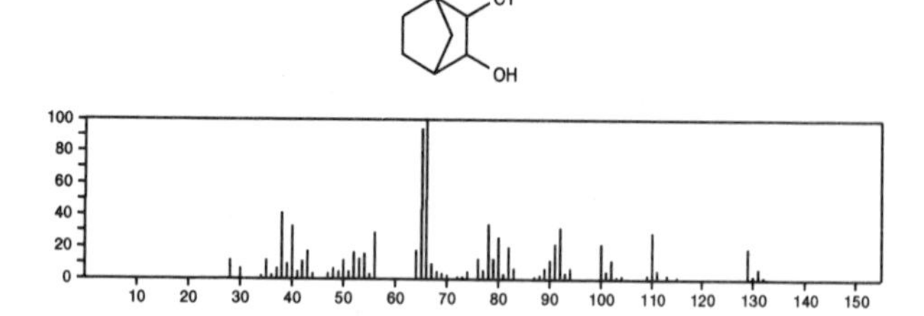

146 $C_7H_{14}OS$ 2432-32-8
Ethanethioic acid, *S*-pentyl ester

Me(CH2)4SAc

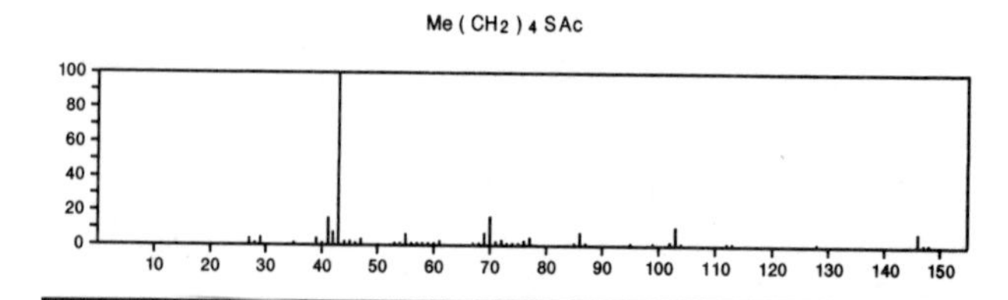

146 $C_7H_{14}OS$ 2432-38-4
Acetic acid, thio-, *S*-isopentyl ester

Me2CHCH2CH2SAc

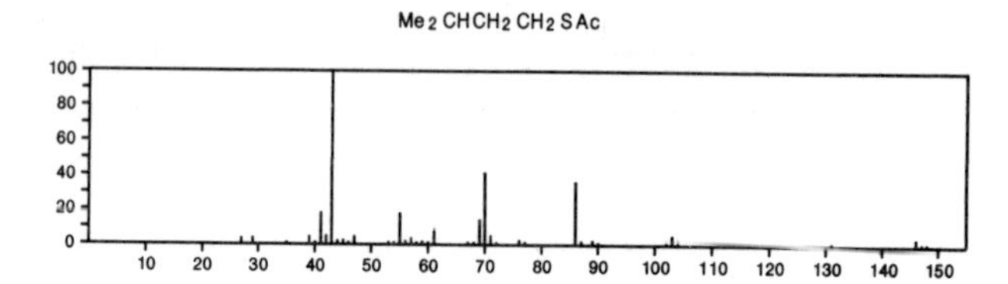

146 $C_7H_{14}OS$ 2432-40-8
Acetic acid, thio-, *S*-1-methylbutyl ester

PrCHMeSAc

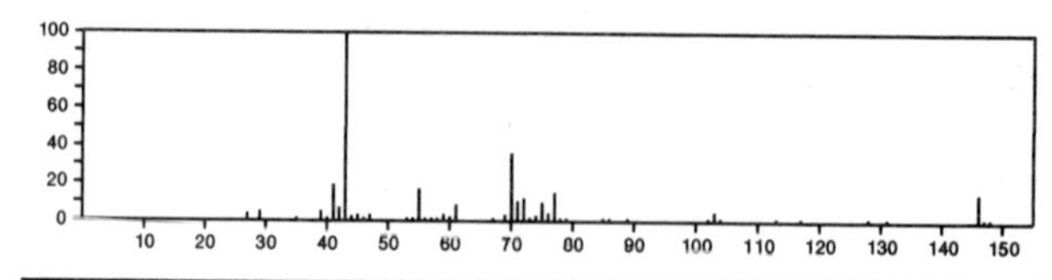

146 $C_7H_{14}OS$ 2432-44-2
Propionic acid, thio-, *S*-*sec*-butyl ester

EtC(O)S(CH2)3Me

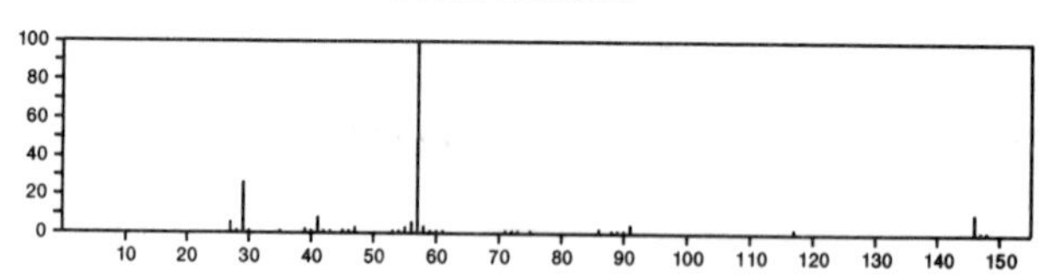

146 $C_7H_{14}OS$ 2432-48-6
Propionic acid, thio-, *S*-isobutyl ester

i-BuSC(O)Et

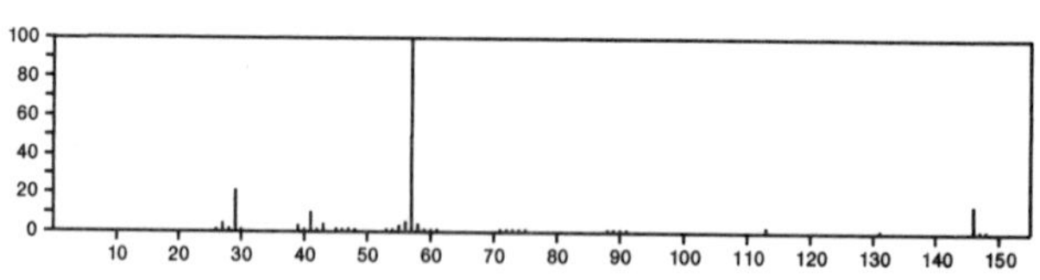

146 $C_7H_{14}OS$ 2432-57-7
Butanethioic acid, 3-methyl-, *S*-ethyl ester

Me2CHCH2C(O)SEt

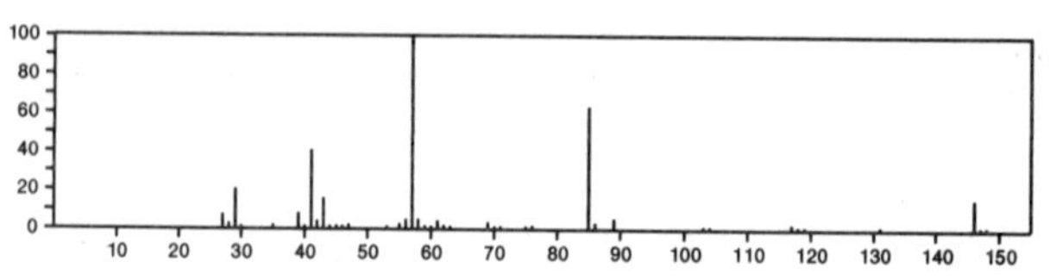

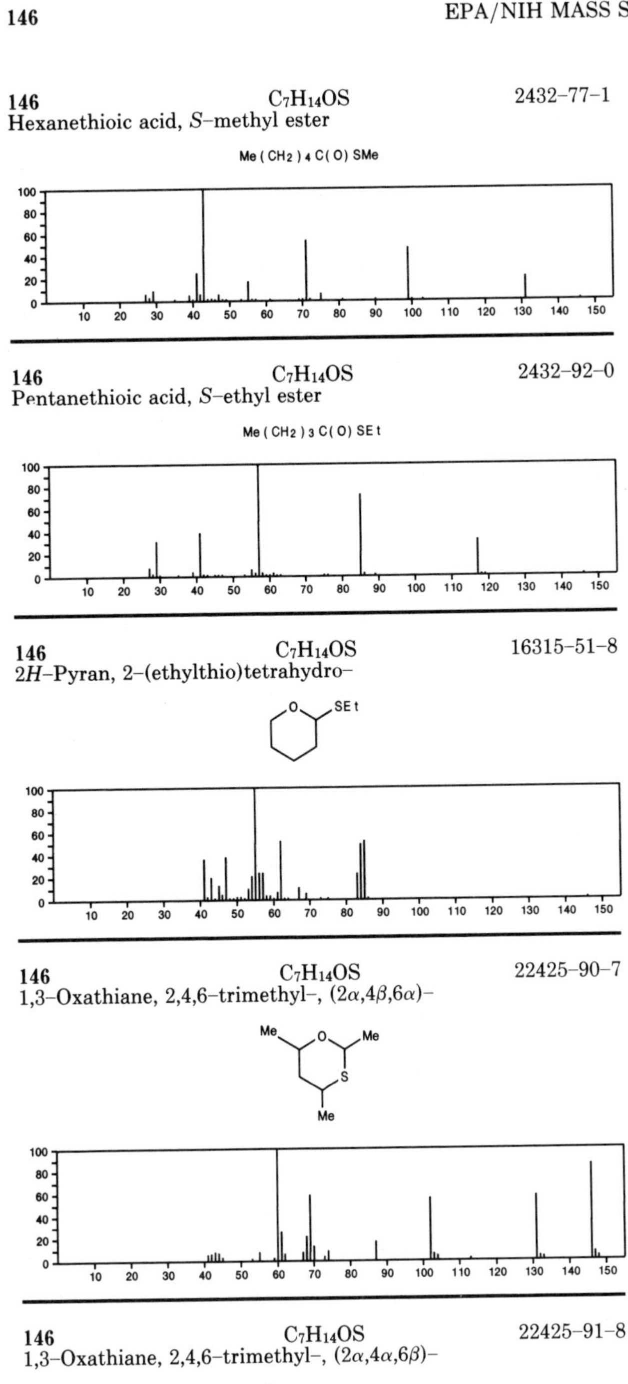
146 $C_7H_{14}OS$ 2432-77-1
Hexanethioic acid, *S*-methyl ester
Me(CH2)4C(O)SMe
146 $C_7H_{14}OS$ 2432-92-0
Pentanethioic acid, *S*-ethyl ester
Me(CH2)3C(O)SEt
146 $C_7H_{14}OS$ 16315-51-8
2*H*-Pyran, 2-(ethylthio)tetrahydro-
SEt
146 $C_7H_{14}OS$ 22425-90-7
1,3-Oxathiane, 2,4,6-trimethyl-, (2α,4β,6α)-
146 $C_7H_{14}OS$ 22425-91-8
1,3-Oxathiane, 2,4,6-trimethyl-, (2α,4α,6β)-

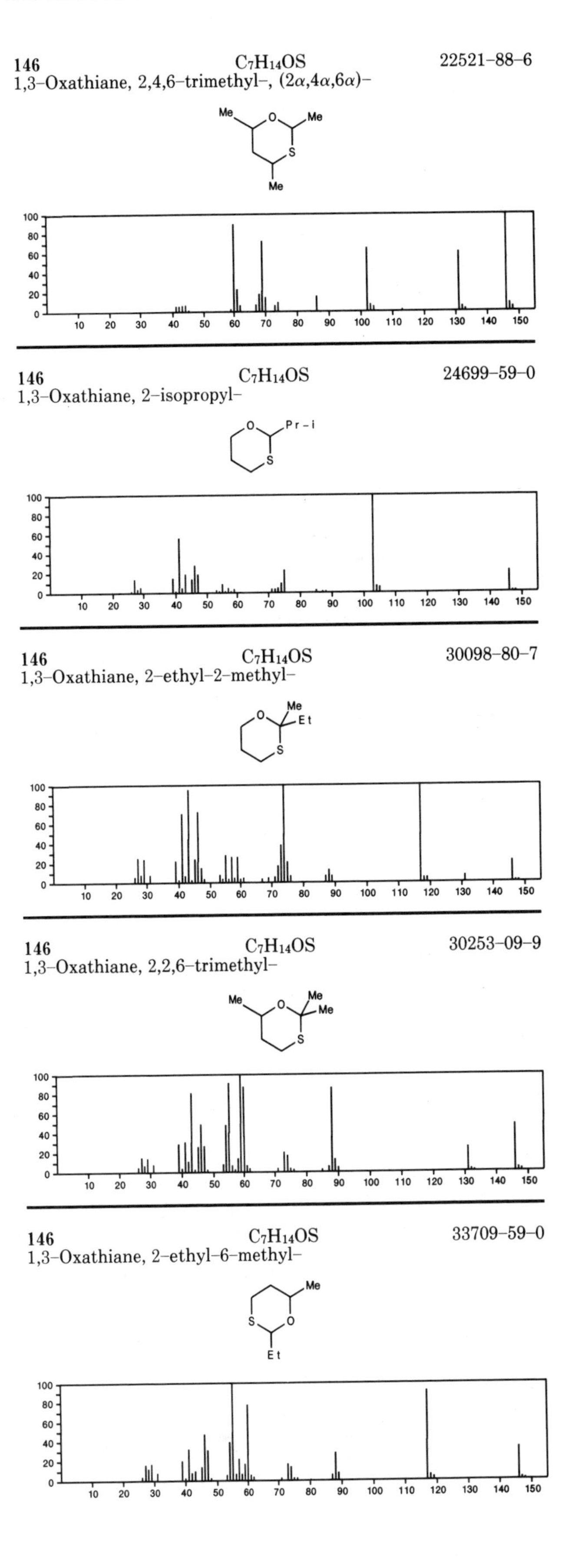
146 $C_7H_{14}OS$ 22521-88-6
1,3-Oxathiane, 2,4,6-trimethyl-, (2α,4α,6α)-
146 $C_7H_{14}OS$ 24699-59-0
1,3-Oxathiane, 2-isopropyl-
Pr-i
146 $C_7H_{14}OS$ 30098-80-7
1,3-Oxathiane, 2-ethyl-2-methyl-
146 $C_7H_{14}OS$ 30253-09-9
1,3-Oxathiane, 2,2,6-trimethyl-
146 $C_7H_{14}OS$ 33709-59-0
1,3-Oxathiane, 2-ethyl-6-methyl-

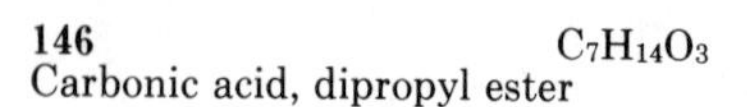

146 $C_7H_{14}O_3$ 623–96–1

Carbonic acid, dipropyl ester

PrOC(O)OPr

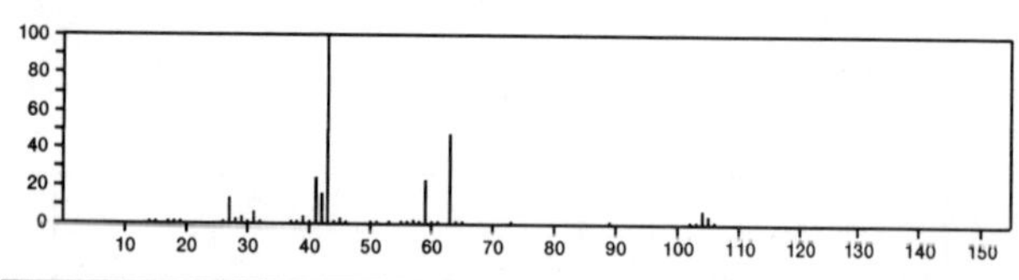

146 $C_7H_{14}O_3$ 763–69–9

Propanoic acid, 3–ethoxy–, ethyl ester

EtOC(O)CH_2CH_2OEt

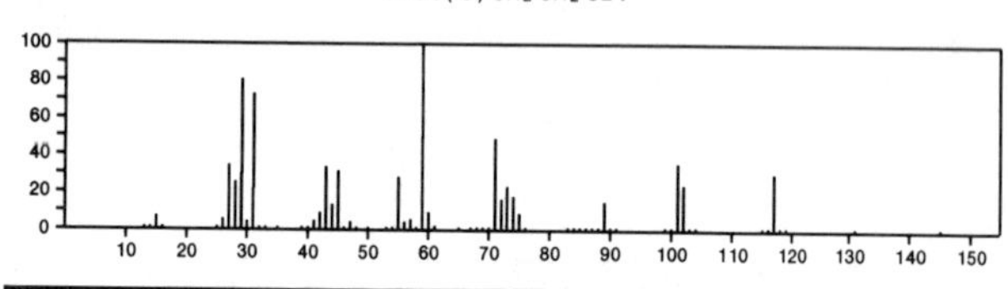

146 $C_7H_{14}O_3$ 2441–06–7

Butanoic acid, 2–hydroxy–3–methyl–, ethyl ester

Me_2CHCH(OH)C(O)OEt

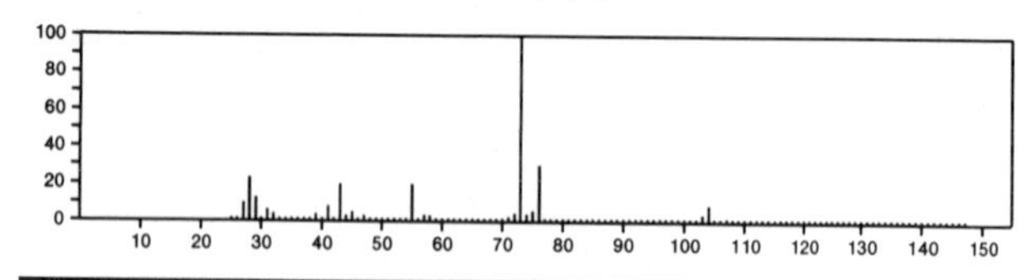

146 $C_7H_{14}O_3$ 5774–26–5

2–Propanone, 1,1–diethoxy–

$(EtO)_2$CHCOMe

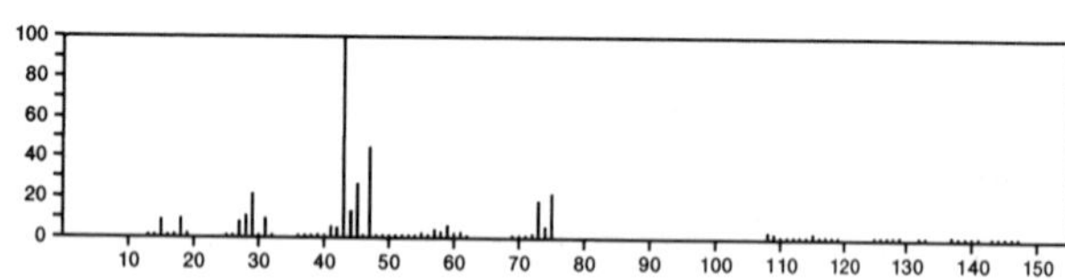

146 $C_7H_{14}O_3$ 6482–34–4

Carbonic acid, bis(1–methylethyl) ester

i–PrOC(O)OPr–i

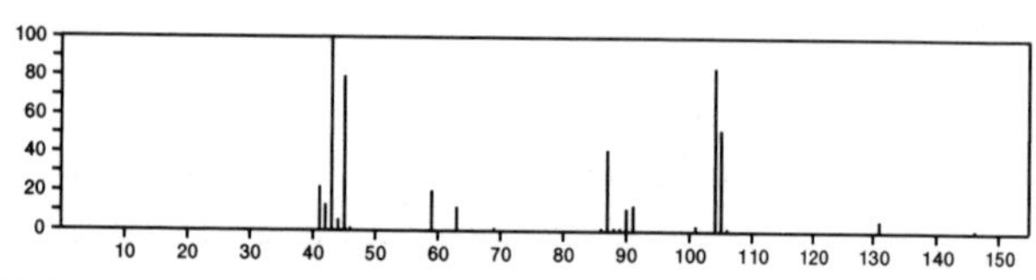

146 $C_7H_{14}O_3$ 6938–26–7

Pentanoic acid, 2–hydroxy–, ethyl ester

EtOC(O)CH(OH)Pr

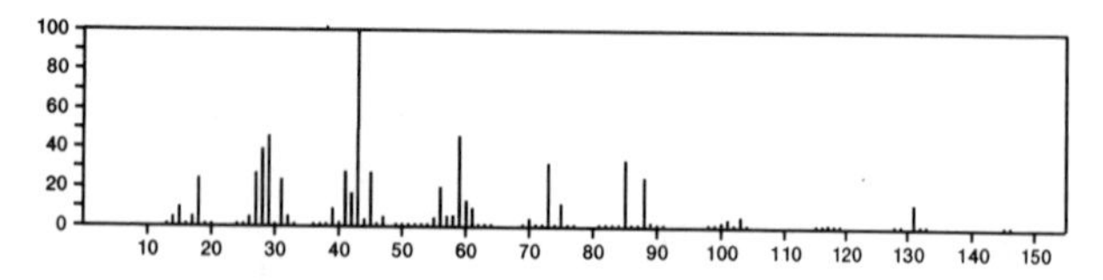

146 $C_7H_{14}O_3$ 21317–50–0

2*H*–Pyran–3–ol, tetrahydro–6–methoxy–2–methyl–, [2*R*–(2α,3β,6β)]–

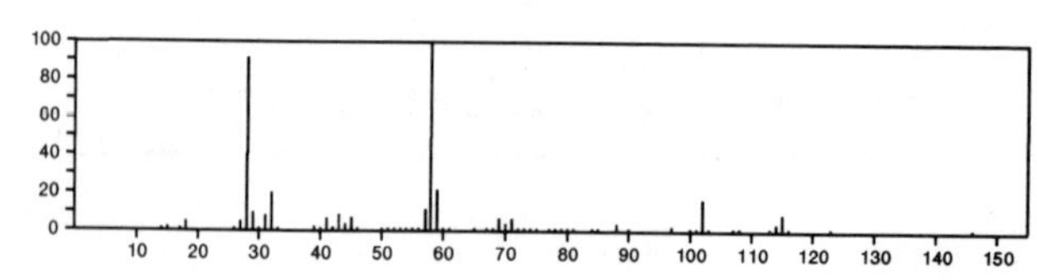

146 $C_7H_{14}O_3$ 29006–04–0

Butanoic acid, 4–ethoxy–, methyl ester

EtO$(CH_2)_3$C(O)OMe

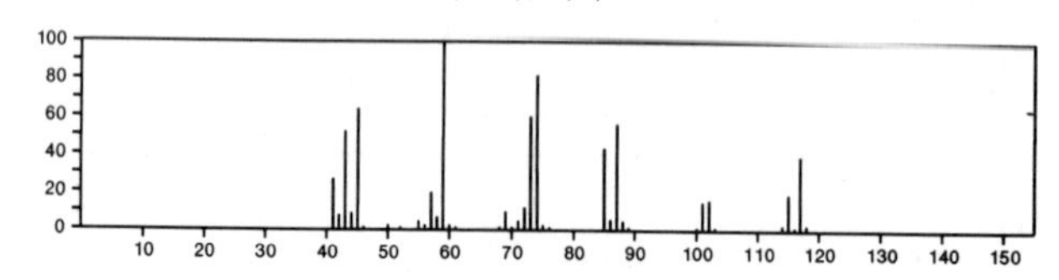

146 $C_7H_{14}O_3$ 29021–98–5

1,3–Dioxolane–2–propanol, 2–methyl–

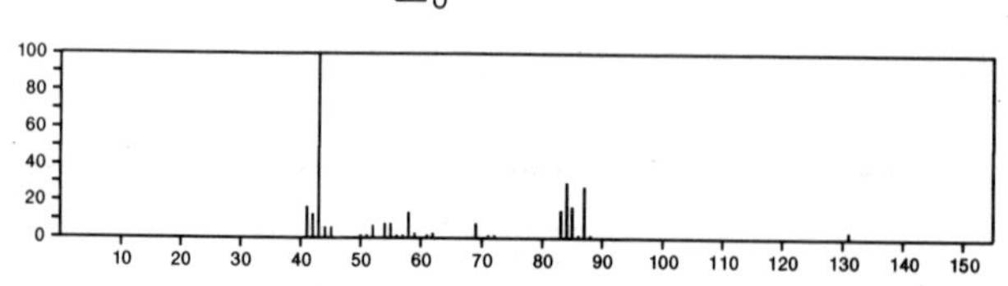

146 $C_7H_{14}O_3$ 40348–72–9

Pentanoic acid, 2–hydroxy–4–methyl–, methyl ester

MeOC(O)CH(OH)CH_2CHMe_2

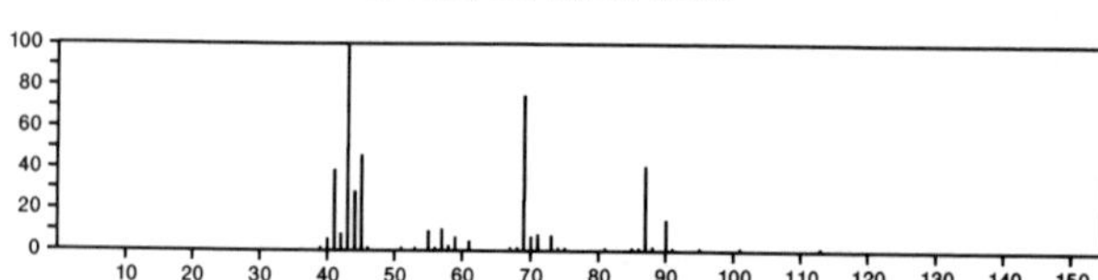

146 $C_7H_{14}O_3$ 41654–19–7

Pentanoic acid, 2–hydroxy–3–methyl–, methyl ester

MeCH_2CHMeCH(OH)C(O)OMe

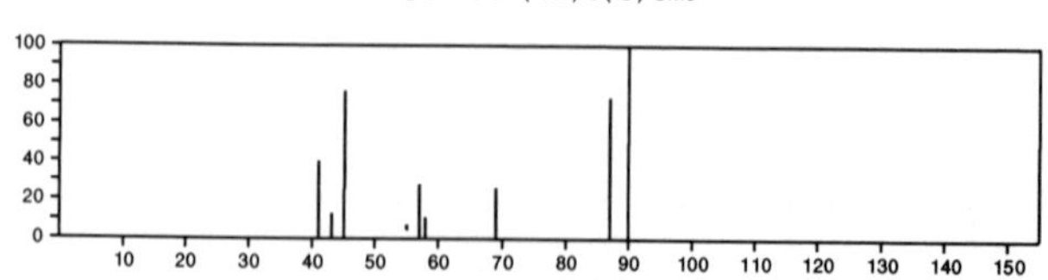

146 $C_7H_{14}O_3$ 42415–64–5

Acetic acid, (1–methylethoxy)–, ethyl ester

i–PrOCH_2C(O)OEt

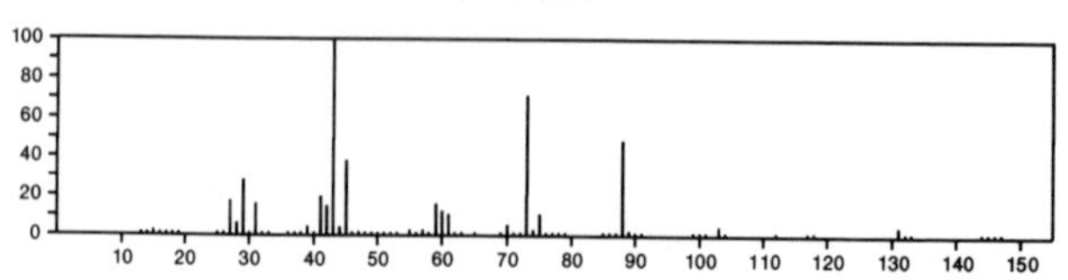

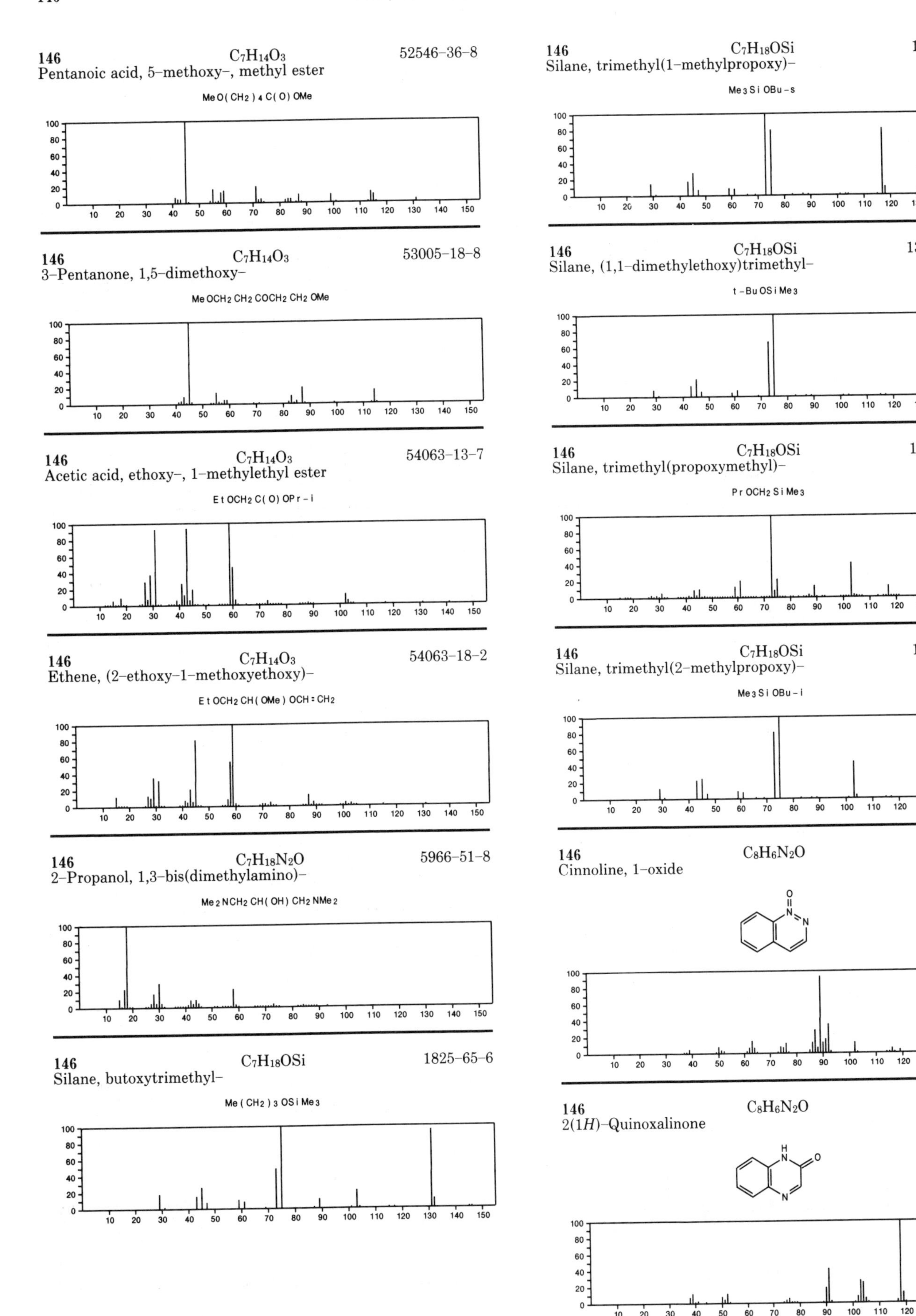

146 $C_7H_{14}O_3$ 52546-36-8
Pentanoic acid, 5-methoxy-, methyl ester
MeO(CH2)4C(O)OMe

146 $C_7H_{14}O_3$ 53005-18-8
3-Pentanone, 1,5-dimethoxy-
MeOCH2CH2COCH2CH2OMe

146 $C_7H_{14}O_3$ 54063-13-7
Acetic acid, ethoxy-, 1-methylethyl ester
EtOCH2C(O)OPr-i

146 $C_7H_{14}O_3$ 54063-18-2
Ethene, (2-ethoxy-1-methoxyethoxy)-
EtOCH2CH(OMe)OCH=CH2

146 $C_7H_{18}N_2O$ 5966-51-8
2-Propanol, 1,3-bis(dimethylamino)-
Me2NCH2CH(OH)CH2NMe2

146 $C_7H_{18}OSi$ 1825-65-6
Silane, butoxytrimethyl-
Me(CH2)3OSiMe3

146 $C_7H_{18}OSi$ 1825-66-7
Silane, trimethyl(1-methylpropoxy)-
Me3SiOBu-s

146 $C_7H_{18}OSi$ 13058-24-7
Silane, (1,1-dimethylethoxy)trimethyl-
t-BuOSiMe3

146 $C_7H_{18}OSi$ 17348-62-8
Silane, trimethyl(propoxymethyl)-
PrOCH2SiMe3

146 $C_7H_{18}OSi$ 18269-50-6
Silane, trimethyl(2-methylpropoxy)-
Me3SiOBu-i

146 $C_8H_6N_2O$ 1125-61-7
Cinnoline, 1-oxide

146 $C_8H_6N_2O$ 1196-57-2
2(1*H*)-Quinoxalinone

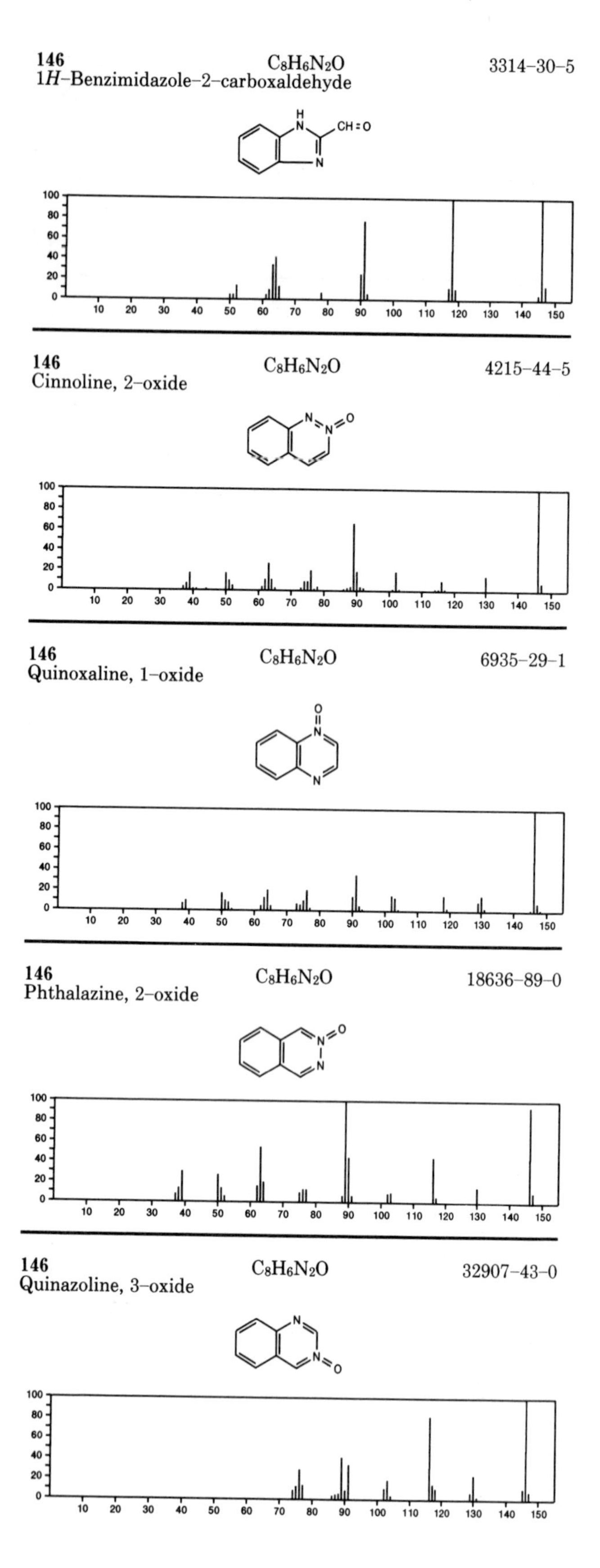

146 $C_8H_6N_2O$ 3314-30-5
1*H*-Benzimidazole-2-carboxaldehyde

146 $C_8H_6N_2O$ 4215-44-5
Cinnoline, 2-oxide

146 $C_8H_6N_2O$ 6935-29-1
Quinoxaline, 1-oxide

146 $C_8H_6N_2O$ 18636-89-0
Phthalazine, 2-oxide

146 $C_8H_6N_2O$ 32907-43-0
Quinazoline, 3-oxide

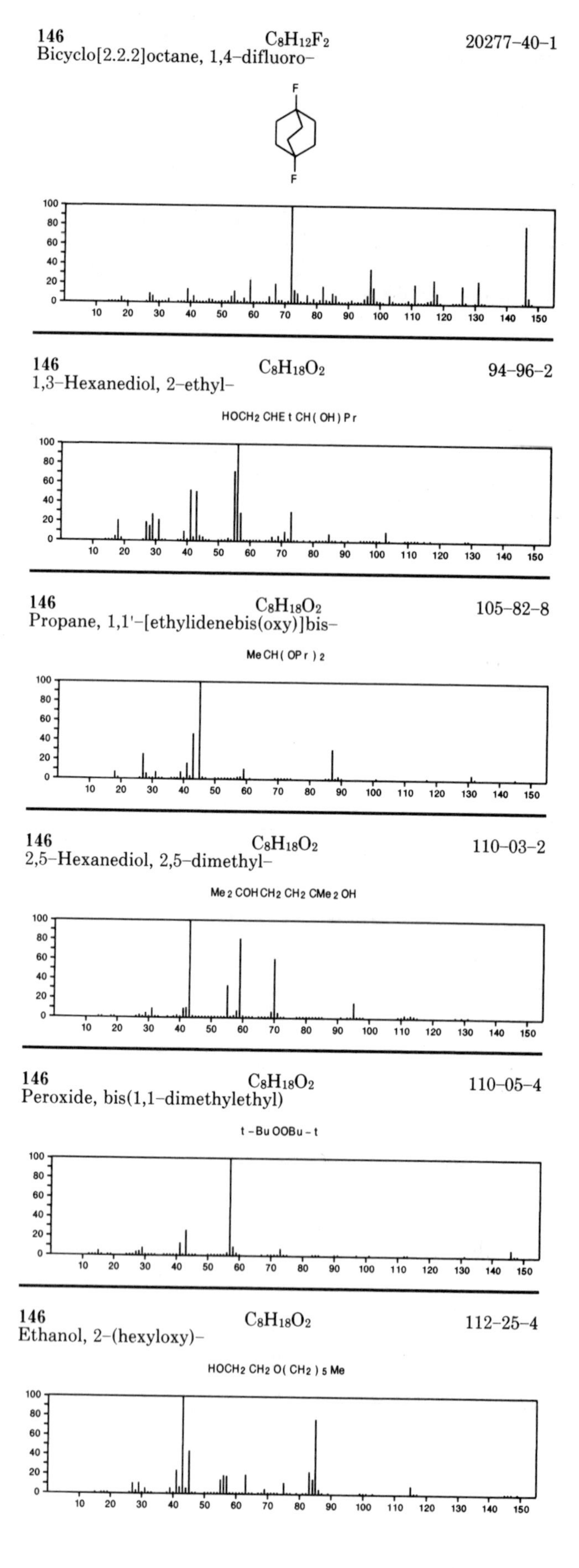

146 $C_8H_{12}F_2$ 20277-40-1
Bicyclo[2.2.2]octane, 1,4-difluoro-

146 $C_8H_{18}O_2$ 94-96-2
1,3-Hexanediol, 2-ethyl-
HOCH2 CHEt CH(OH) Pr

146 $C_8H_{18}O_2$ 105-82-8
Propane, 1,1'-[ethylidenebis(oxy)]bis-
MeCH(OPr)2

146 $C_8H_{18}O_2$ 110-03-2
2,5-Hexanediol, 2,5-dimethyl-
Me2 COHCH2 CH2 CMe2 OH

146 $C_8H_{18}O_2$ 110-05-4
Peroxide, bis(1,1-dimethylethyl)
t-Bu OOBu-t

146 $C_8H_{18}O_2$ 112-25-4
Ethanol, 2-(hexyloxy)-
HOCH2 CH2 O(CH2)5 Me

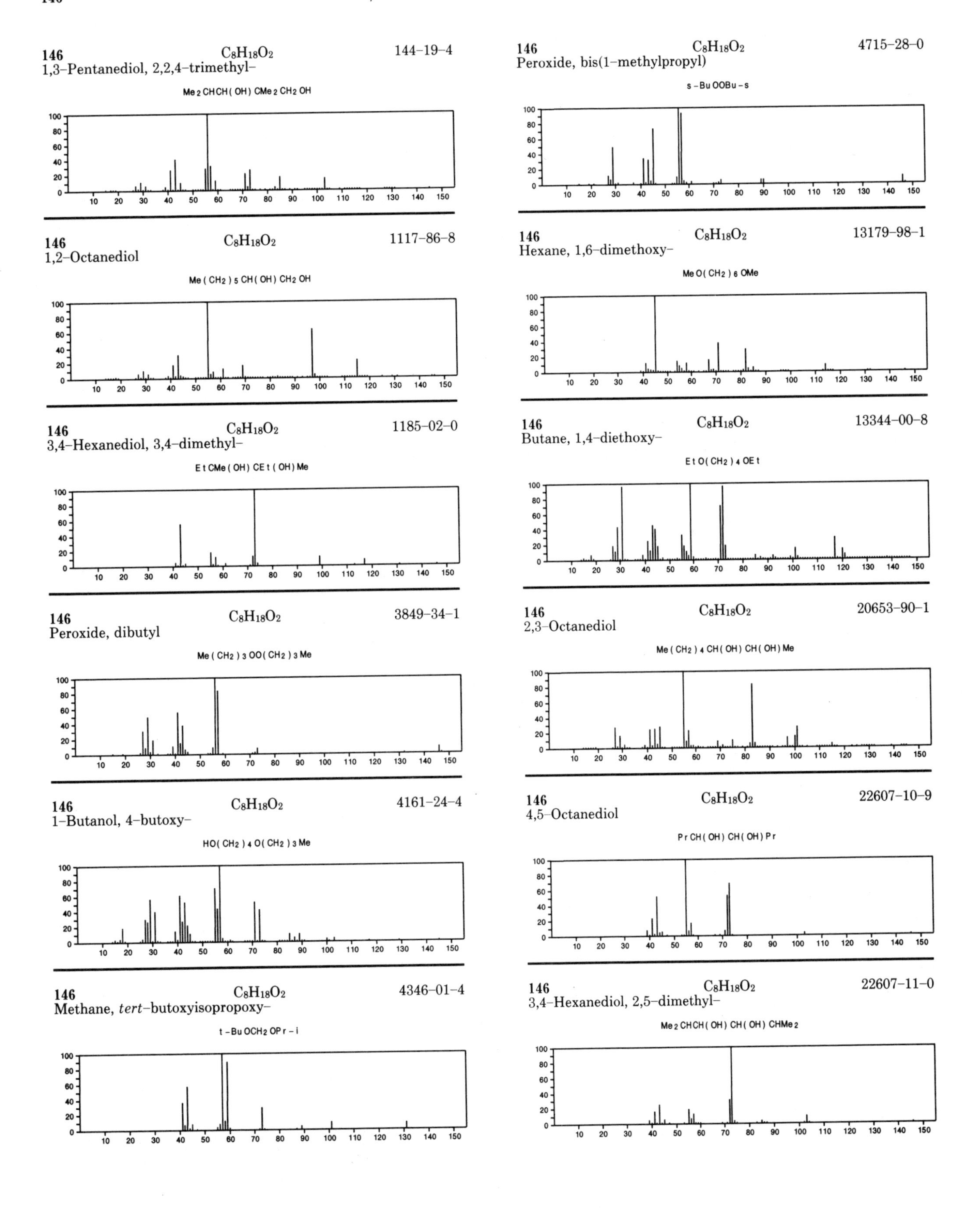
146 C8H18O2 144-19-4
1,3-Pentanediol, 2,2,4-trimethyl-
Me2CHCH(OH)CMe2CH2OH
146 C8H18O2 1117-86-8
1,2-Octanediol
Me(CH2)5CH(OH)CH2OH
146 C8H18O2 1185-02-0
3,4-Hexanediol, 3,4-dimethyl-
EtCMe(OH)CEt(OH)Me
146 C8H18O2 3849-34-1
Peroxide, dibutyl
Me(CH2)3OO(CH2)3Me
146 C8H18O2 4161-24-4
1-Butanol, 4-butoxy-
HO(CH2)4O(CH2)3Me
146 C8H18O2 4346-01-4
Methane, tert-butoxyisopropoxy-
t-BuOCH2OPr-i
146 C8H18O2 4715-28-0
Peroxide, bis(1-methylpropyl)
s-BuOOBu-s
146 C8H18O2 13179-98-1
Hexane, 1,6-dimethoxy-
MeO(CH2)6OMe
146 C8H18O2 13344-00-8
Butane, 1,4-diethoxy-
EtO(CH2)4OEt
146 C8H18O2 20653-90-1
2,3-Octanediol
Me(CH2)4CH(OH)CH(OH)Me
146 C8H18O2 22607-10-9
4,5-Octanediol
PrCH(OH)CH(OH)Pr
146 C8H18O2 22607-11-0
3,4-Hexanediol, 2,5-dimethyl-
Me2CHCH(OH)CH(OH)CHMe2

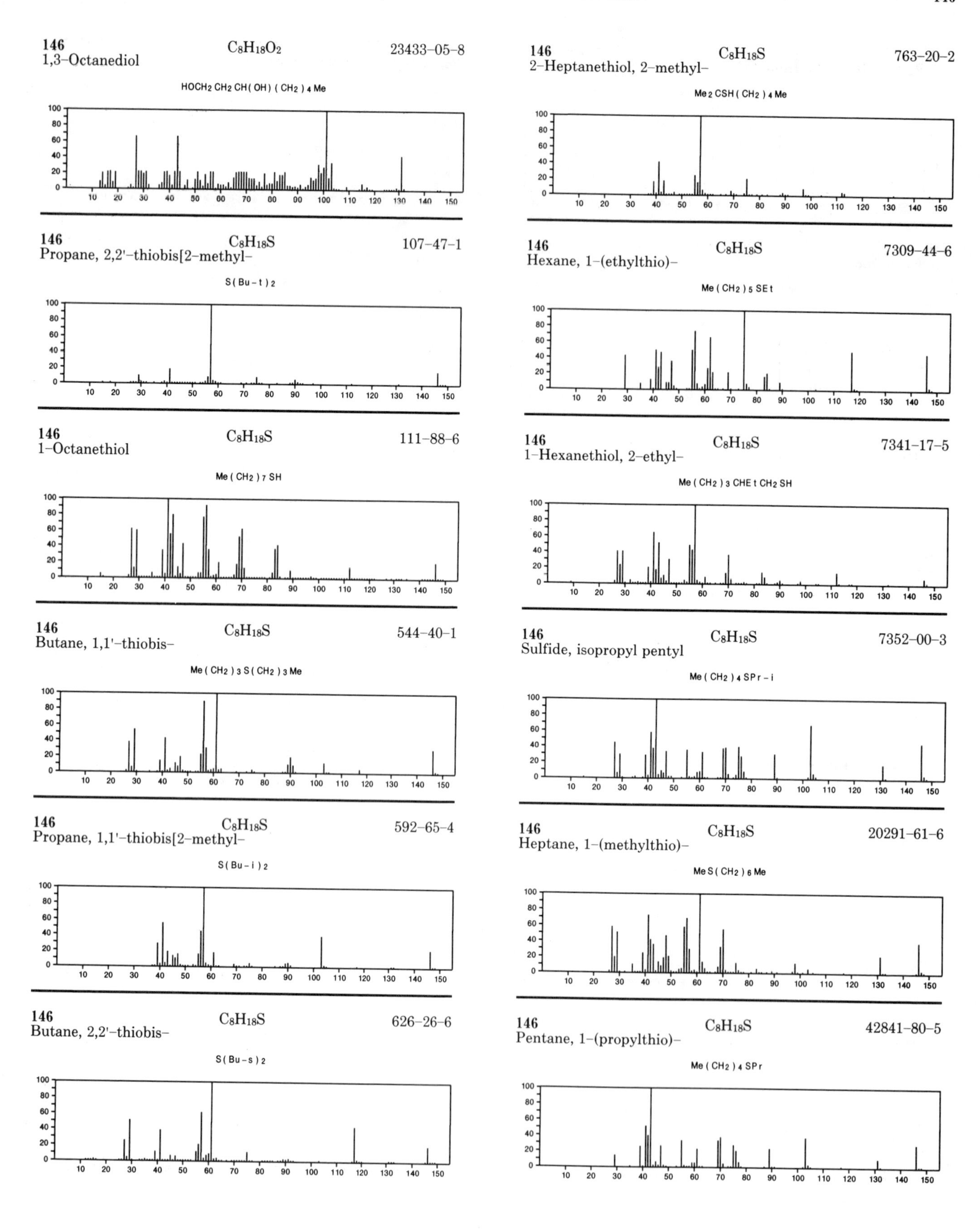
146
1,3-Octanediol
C8H18O2
23433-05-8
HOCH2 CH2 CH(OH) (CH2)4 Me
146
2-Heptanethiol, 2-methyl-
C8H18S
763-20-2
Me2 CSH(CH2)4 Me
146
Propane, 2,2'-thiobis[2-methyl-
C8H18S
107-47-1
S(Bu-t)2
146
Hexane, 1-(ethylthio)-
C8H18S
7309-44-6
Me(CH2)5 SEt
146
1-Octanethiol
C8H18S
111-88-6
Me(CH2)7 SH
146
1-Hexanethiol, 2-ethyl-
C8H18S
7341-17-5
Me(CH2)3 CHEt CH2 SH
146
Butane, 1,1'-thiobis-
C8H18S
544-40-1
Me(CH2)3 S(CH2)3 Me
146
Sulfide, isopropyl pentyl
C8H18S
7352-00-3
Me(CH2)4 SPr-i
146
Propane, 1,1'-thiobis[2-methyl-
C8H18S
592-65-4
S(Bu-i)2
146
Heptane, 1-(methylthio)-
C8H18S
20291-61-6
MeS(CH2)6 Me
146
Butane, 2,2'-thiobis-
C8H18S
626-26-6
S(Bu-s)2
146
Pentane, 1-(propylthio)-
C8H18S
42841-80-5
Me(CH2)4 SPr

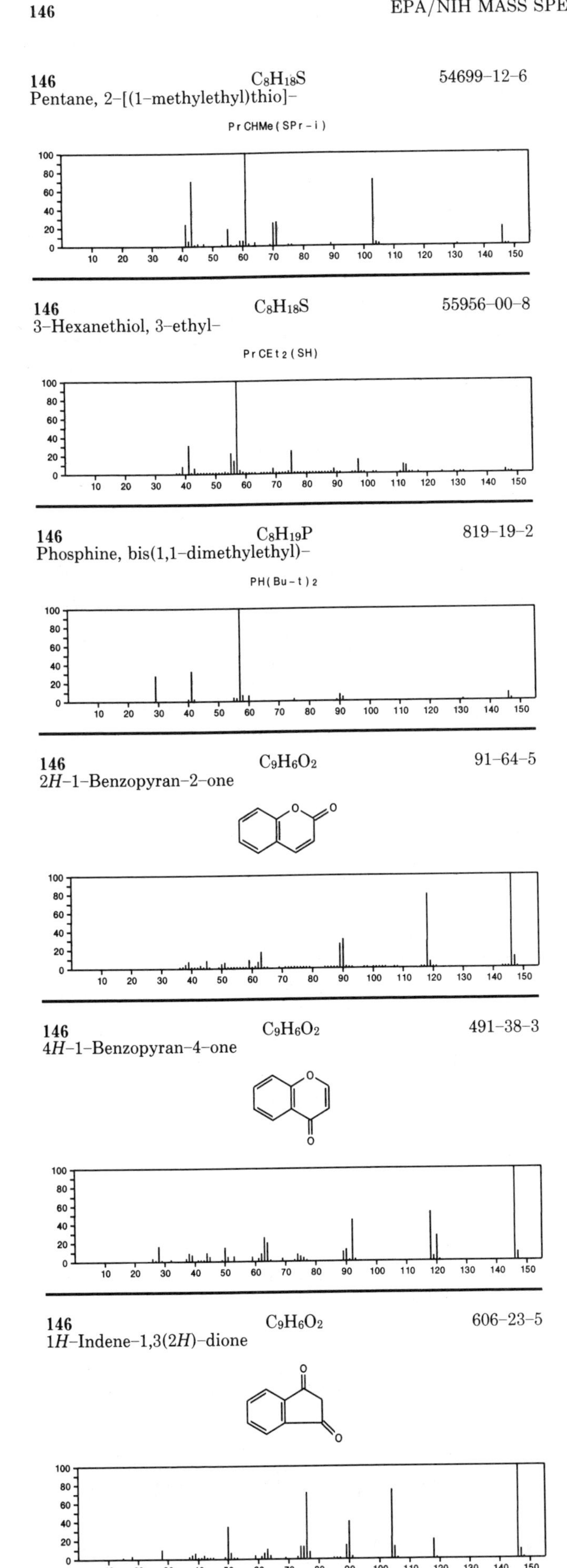

146 $C_8H_{18}S$ 54699–12–6
Pentane, 2–[(1–methylethyl)thio]–

PrCHMe(SPr-i)

146 $C_8H_{18}S$ 55956–00–8
3–Hexanethiol, 3–ethyl–

PrCEt2(SH)

146 $C_8H_{19}P$ 819–19–2
Phosphine, bis(1,1–dimethylethyl)–

PH(Bu-t)2

146 $C_9H_6O_2$ 91–64–5
2*H*–1–Benzopyran–2–one

146 $C_9H_6O_2$ 491–38–3
4*H*–1–Benzopyran–4–one

146 $C_9H_6O_2$ 606–23–5
1*H*–Indene–1,3(2*H*)–dione

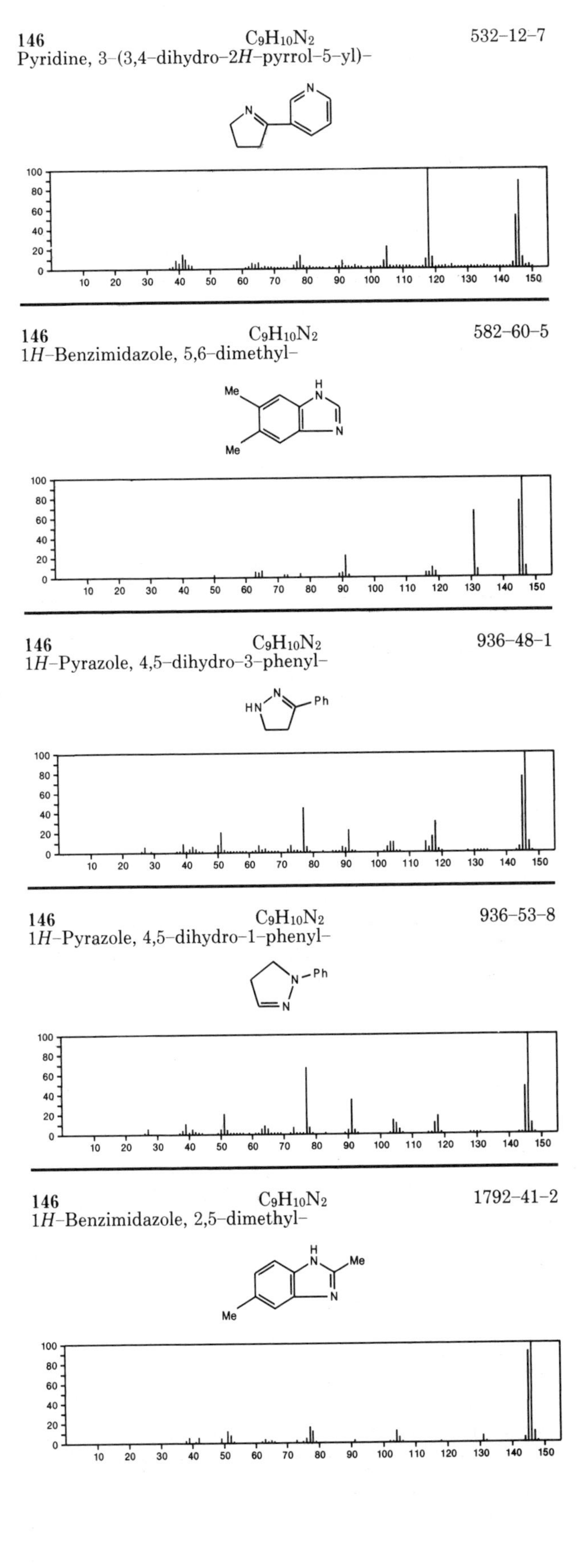

146 $C_9H_{10}N_2$ 532–12–7
Pyridine, 3–(3,4–dihydro–2*H*–pyrrol–5–yl)–

146 $C_9H_{10}N_2$ 582–60–5
1*H*–Benzimidazole, 5,6–dimethyl–

146 $C_9H_{10}N_2$ 936–48–1
1*H*–Pyrazole, 4,5–dihydro–3–phenyl–

146 $C_9H_{10}N_2$ 936–53–8
1*H*–Pyrazole, 4,5–dihydro–1–phenyl–

146 $C_9H_{10}N_2$ 1792–41–2
1*H*–Benzimidazole, 2,5–dimethyl–

146 $C_9H_{10}N_2$ 1848–84–6
1*H*–Benzimidazole, 2–ethyl–

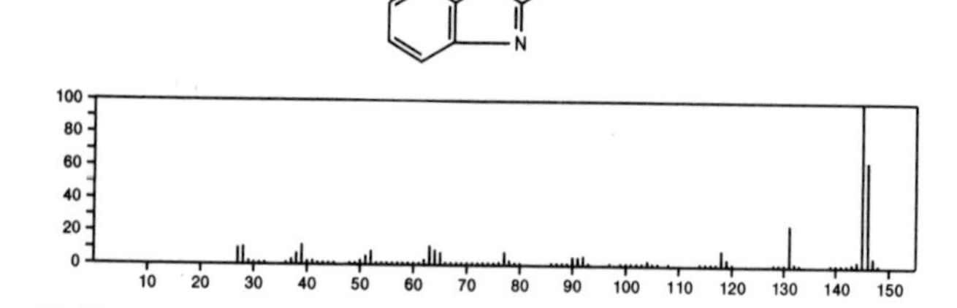

146 $C_9H_{10}N_2$ 23612–49–9
1*H*–Pyrrolo[2,3–*b*]pyridine, 2–ethyl–

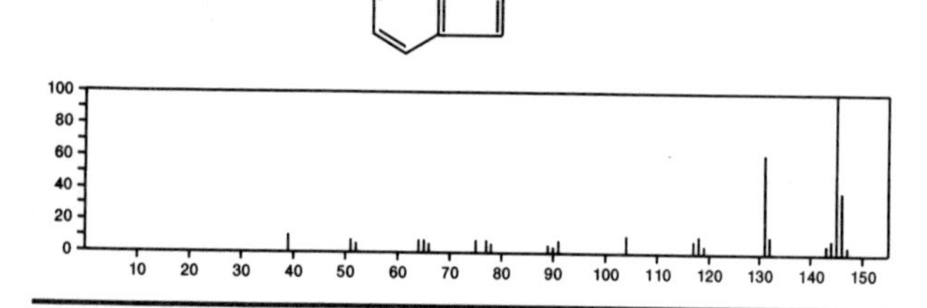

146 $C_9H_{10}N_2$ 23612–70–6
1*H*–Pyrrolo[2,3–*b*]pyridine, 3,4–dimethyl–

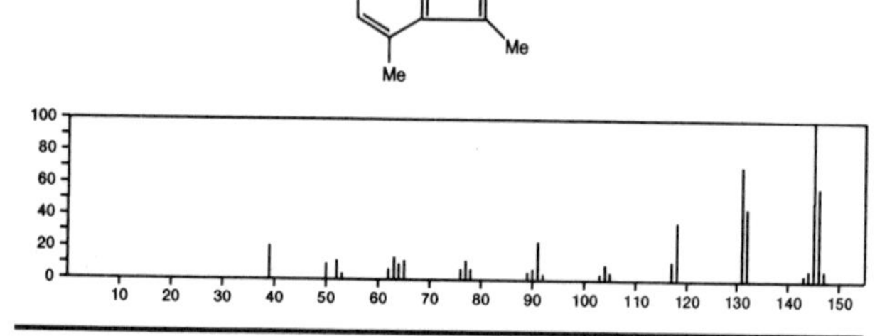

146 $C_9H_{10}N_2$ 28537–56–6
Pyrazolo[1,5–*a*]pyridine, 2,3–dimethyl–

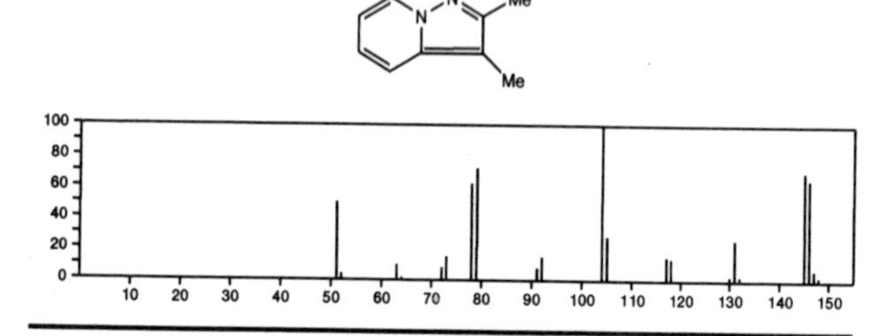

146 $C_9H_{10}N_2$ 54063–11–5
1*H*–Pyrrole, 1,1'–methylenebis–

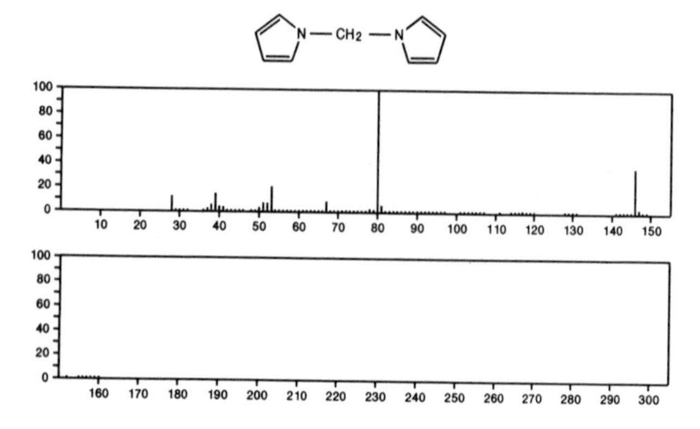

146 $C_{10}H_7F$ 321–38–0
Naphthalene, 1–fluoro–

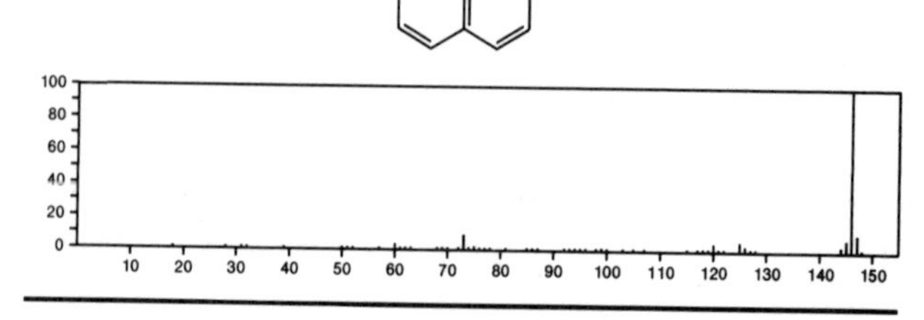

146 $C_{10}H_7F$ 323–09–1
Naphthalene, 2–fluoro–

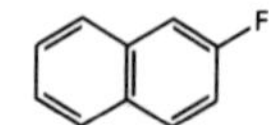

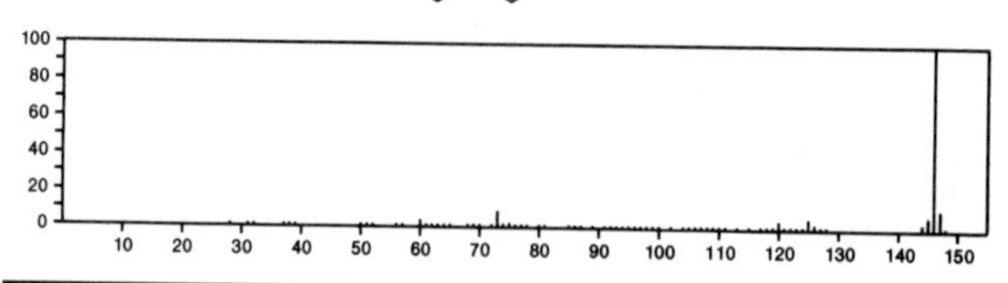

146 $C_{10}H_{10}O$ 122–57–6
3–Buten–2–one, 4–phenyl–

PhCH=CHCOMe

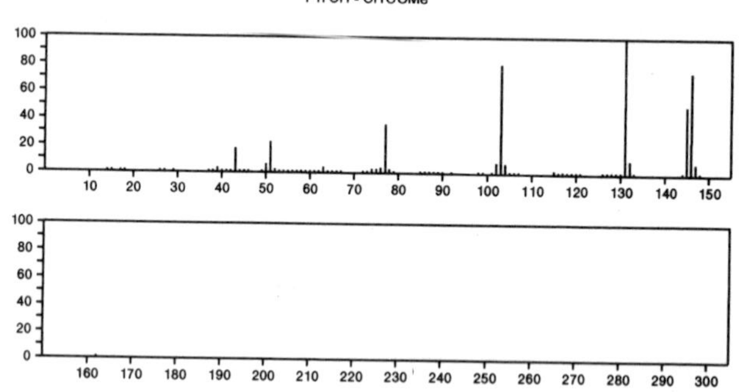

146 $C_{10}H_{10}O$ 529–34–0
1(2*H*)–Naphthalenone, 3,4–dihydro–

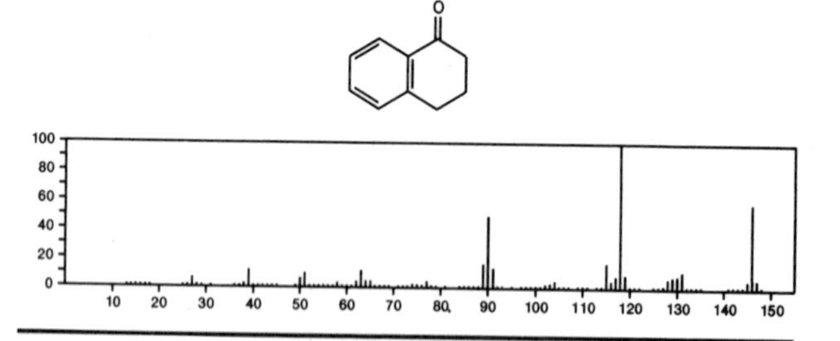

146 $C_{10}H_{10}O$ 530–93–8
2(1*H*)–Naphthalenone, 3,4–dihydro–

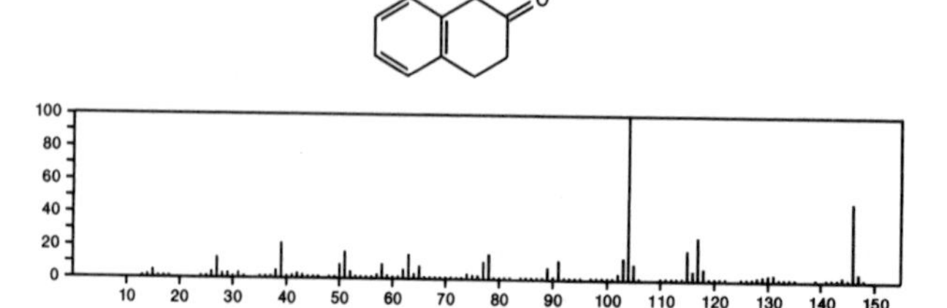

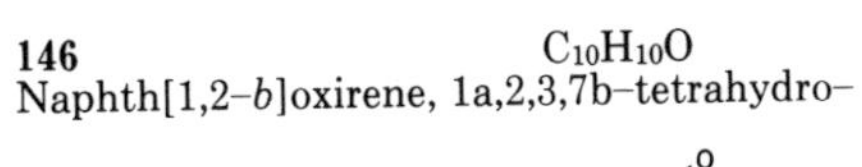

146 $C_{10}H_{10}O$ 2461-34-9
Naphth[1,2-*b*]oxirene, 1a,2,3,7b-tetrahydro-

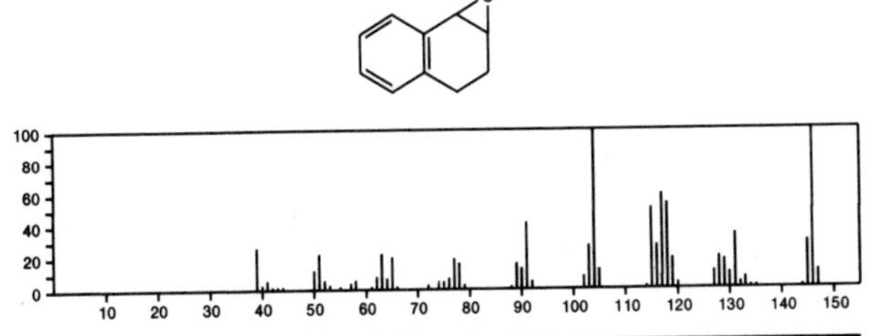

146 $C_{10}H_{10}O$ 3481-02-5
Methanone, cyclopropylphenyl-

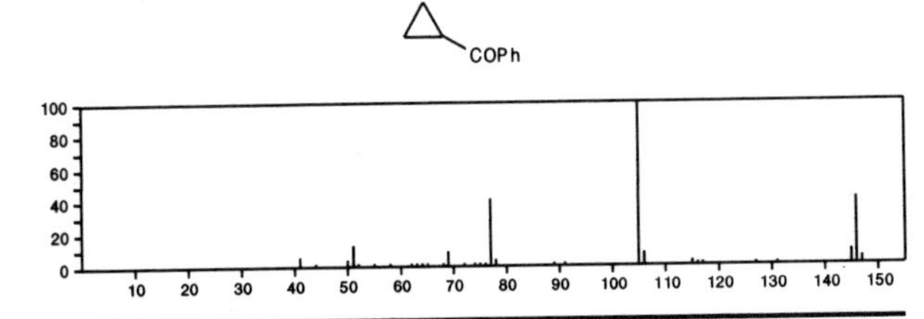

146 $C_{10}H_{10}O$ 6072-57-7
1*H*-Inden-1-one, 2,3-dihydro-3-methyl-

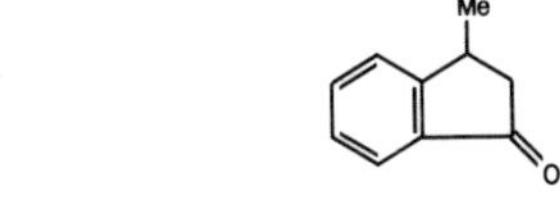

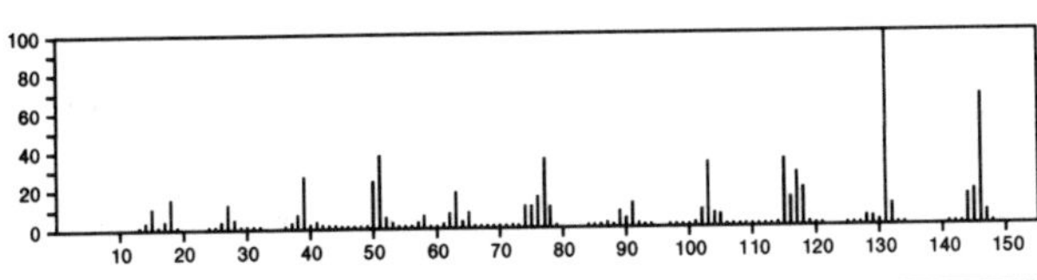

146 $C_{10}H_{10}O$ 16017-24-6
2-Propyn-1-ol, 3-*p*-tolyl-

146 $C_{10}H_{10}O$ 22228-27-9
Cycloprop[*a*]inden-6-ol, 1,1a,6,6a-tetrahydro-

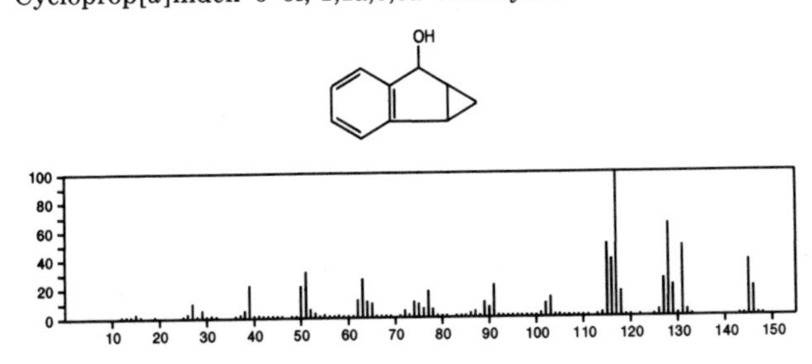

146 $C_{10}H_{10}O$ 28715-26-6
Benzofuran, 4,7-dimethyl-

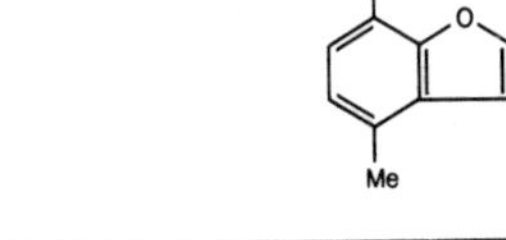

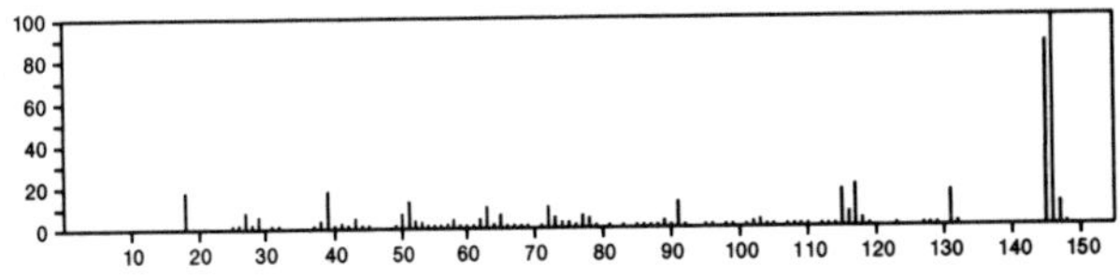

146 $C_{11}H_{14}$ 700-88-9
Benzene, cyclopentyl-

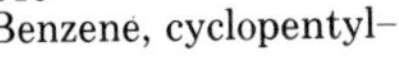

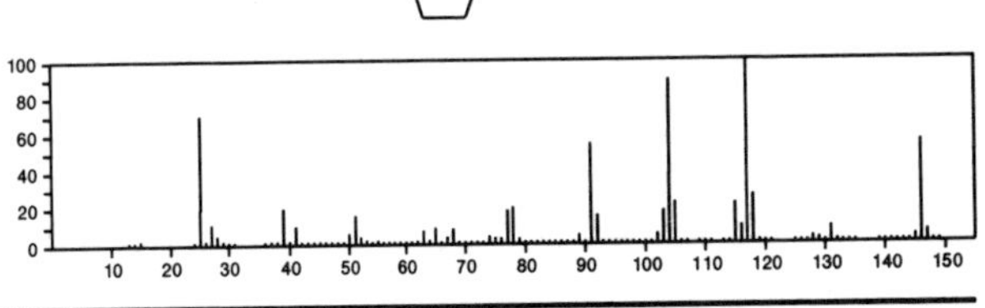

146 $C_{11}H_{14}$ 1075-22-5
Indan, 5,6-dimethyl-

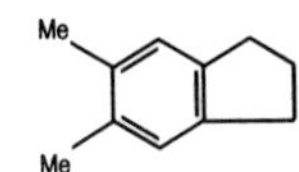

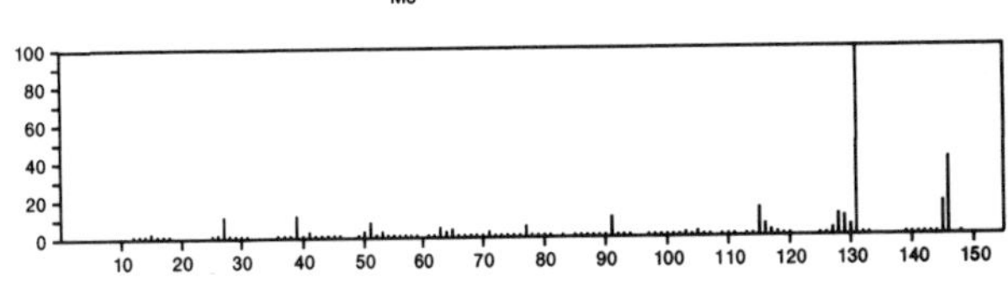

146 $C_{11}H_{14}$ 1559-81-5
Naphthalene, 1,2,3,4-tetrahydro-1-methyl-

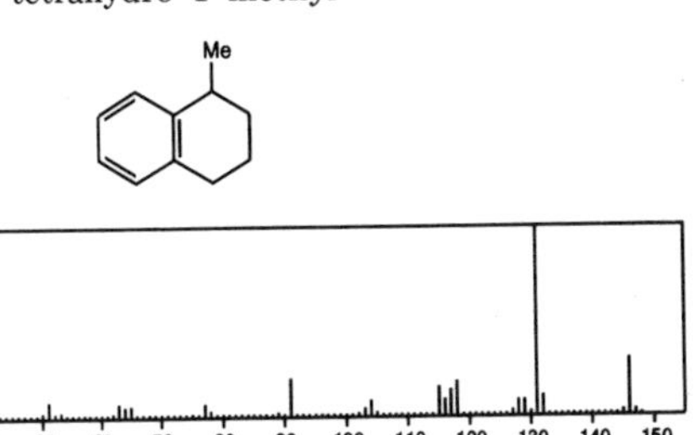

146 $C_{11}H_{14}$ 1647-06-9
Benzene, (2-methyl-3-butenyl)-

PhCH2CHMeCH=CH2

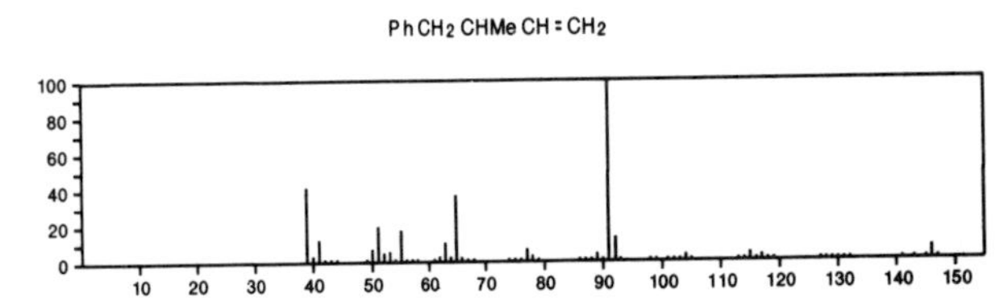

146 $C_{11}H_{14}$ 1680–51–9
Naphthalene, 1,2,3,4–tetrahydro–6–methyl–

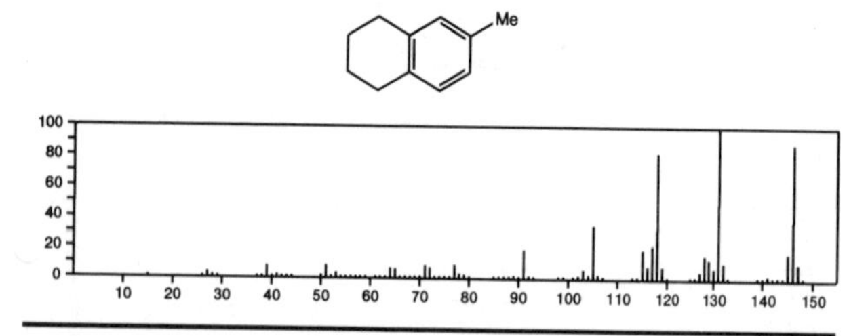

146 $C_{11}H_{14}$ 1685–82–1
1*H*–Indene, 2,3–dihydro–4,6–dimethyl–

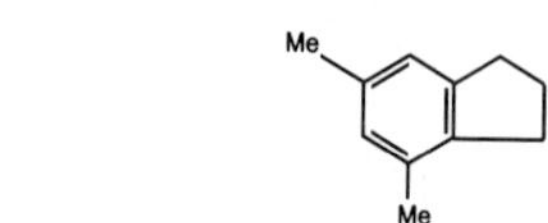

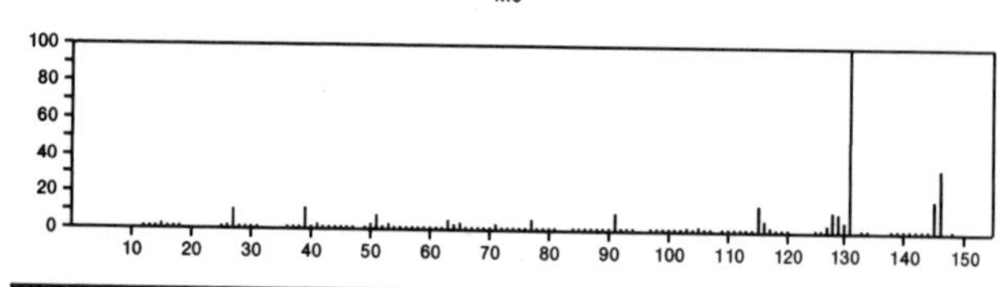

146 $C_{11}H_{14}$ 2809–64–5
Naphthalene, 1,2,3,4–tetrahydro–5–methyl–

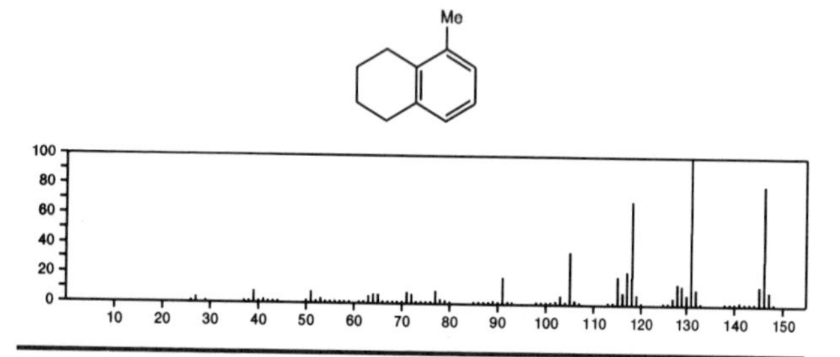

146 $C_{11}H_{14}$ 3877–19–8
Naphthalene, 1,2,3,4–tetrahydro–2–methyl–

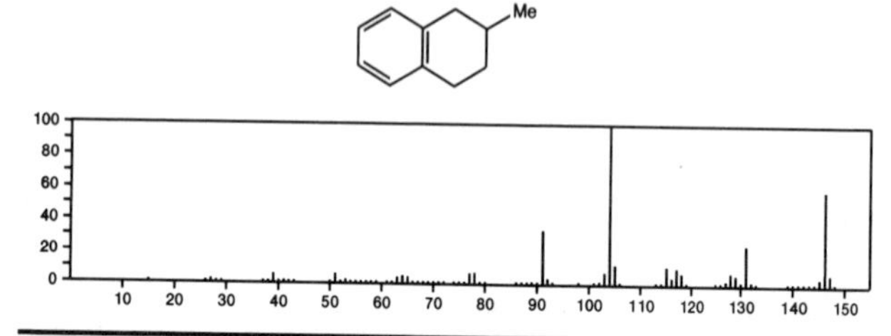

146 $C_{11}H_{14}$ 4175–53–5
1*H*–Indene, 2,3–dihydro–1,3–dimethyl–

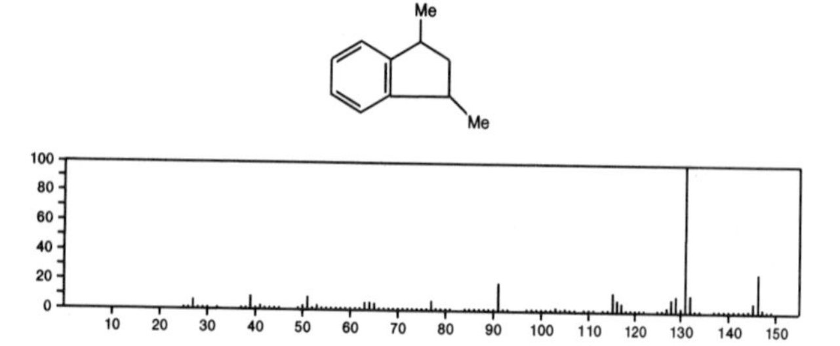

146 $C_{11}H_{14}$ 4489–84–3
Benzene, (3–methyl–2–butenyl)–

$Me_2C=CHCH_2Ph$

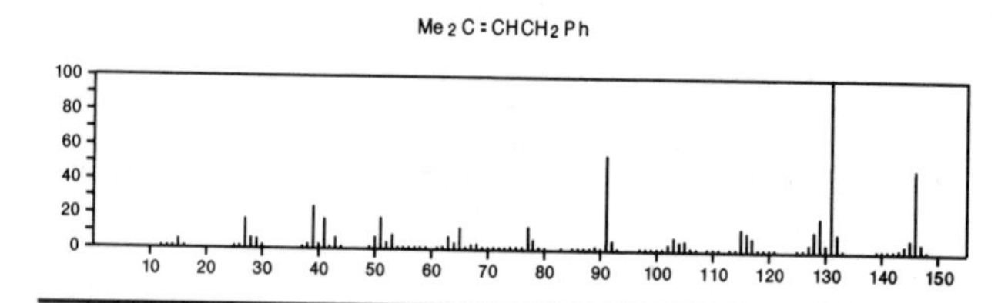

146 $C_{11}H_{14}$ 4830–99–3
1*H*–Indene, 1–ethyl–2,3–dihydro–

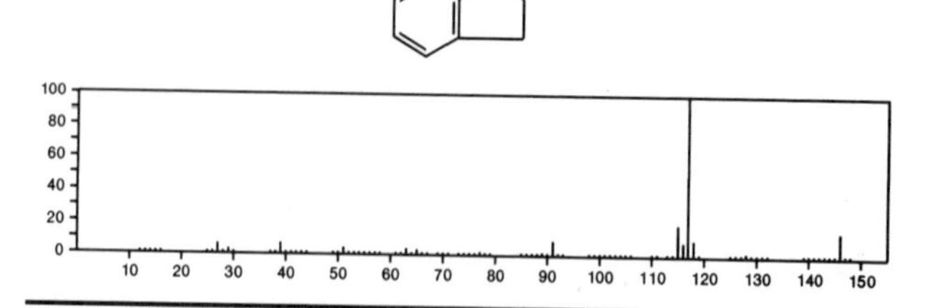

146 $C_{11}H_{14}$ 4912–92–9
1*H*–Indene, 2,3–dihydro–1,1–dimethyl–

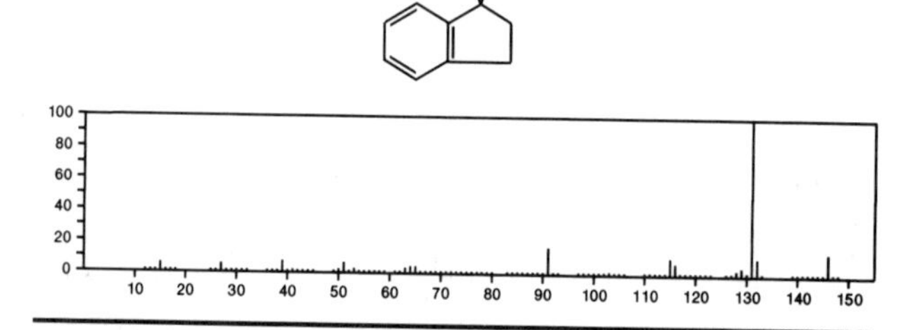

146 $C_{11}H_{14}$ 6682–71–9
1*H*–Indene, 2,3–dihydro–4,7–dimethyl–

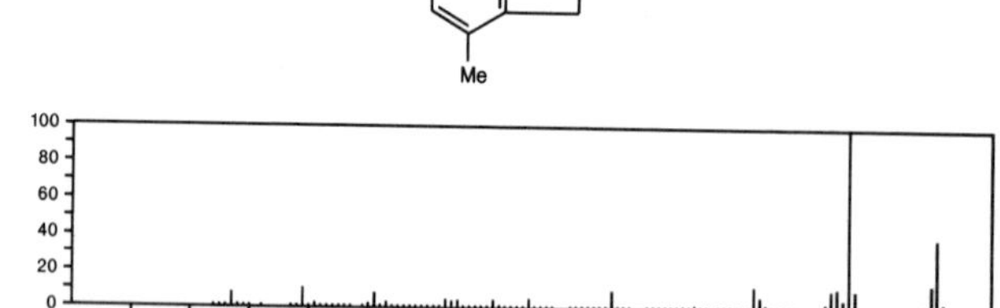

146 $C_{11}H_{14}$ 6683–51–8
Benzene, (3–methyl–3–butenyl)–

$PhCH_2CH_2CMe=CH_2$

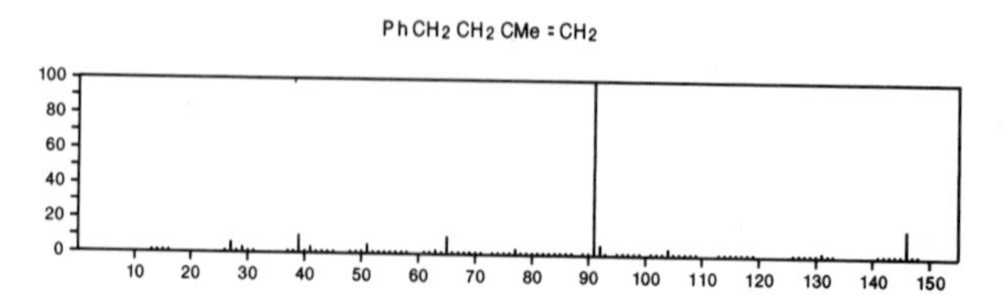

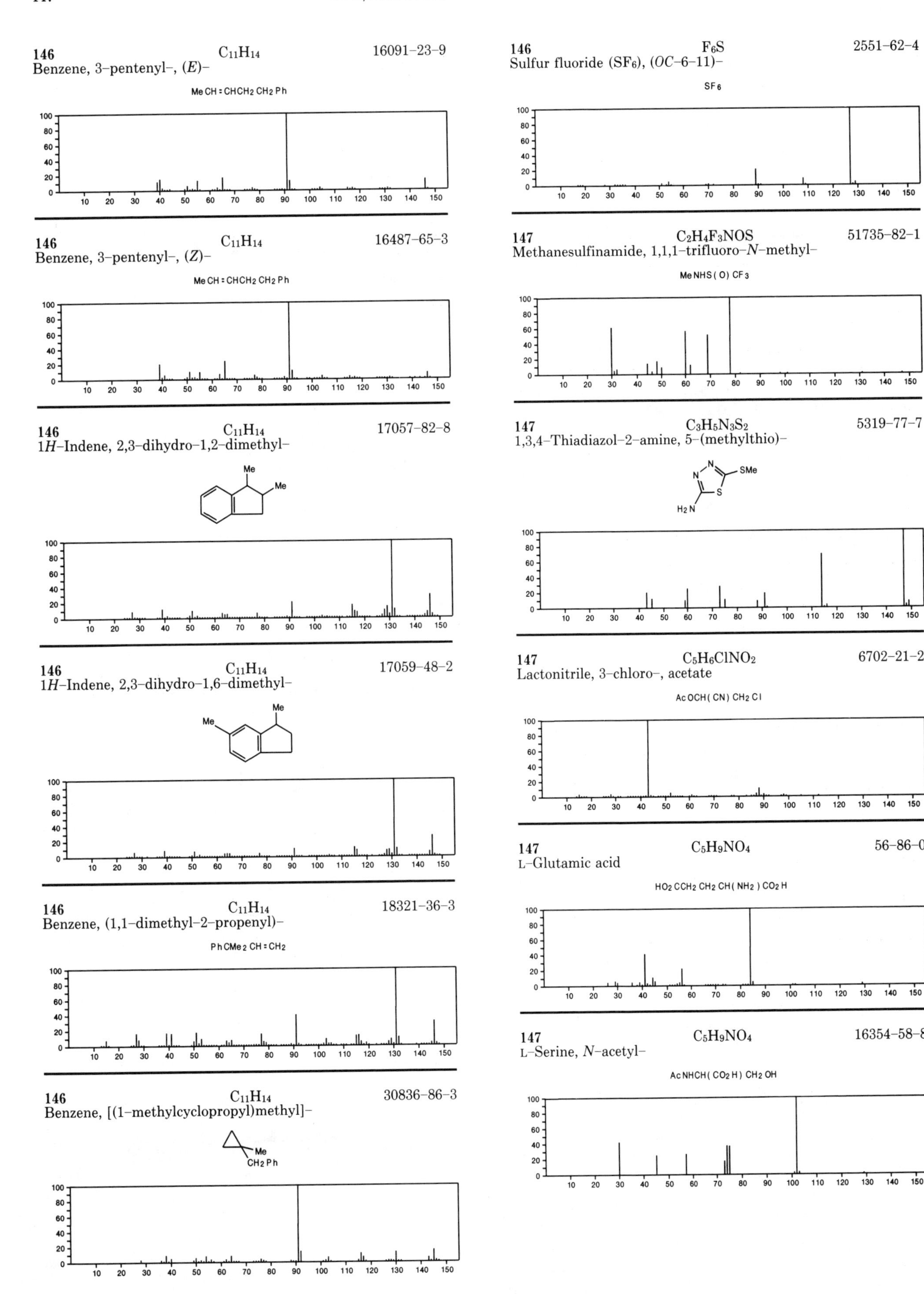
146 C11H14 16091-23-9
Benzene, 3-pentenyl-, (E)-
MeCH=CHCH2CH2Ph
146 C11H14 16487-65-3
Benzene, 3-pentenyl-, (Z)-
MeCH=CHCH2CH2Ph
146 C11H14 17057-82-8
1H-Indene, 2,3-dihydro-1,2-dimethyl-
Me
Me
146 C11H14 17059-48-2
1H-Indene, 2,3-dihydro-1,6-dimethyl-
Me
Me
146 C11H14 18321-36-3
Benzene, (1,1-dimethyl-2-propenyl)-
PhCMe2CH=CH2
146 C11H14 30836-86-3
Benzene, [(1-methylcyclopropyl)methyl]-
Me
CH2Ph
146 F6S 2551-62-4
Sulfur fluoride (SF6), (OC-6-11)-
SF6
147 C2H4F3NOS 51735-82-1
Methanesulfinamide, 1,1,1-trifluoro-N-methyl-
MeNHS(O)CF3
147 C3H5N3S2 5319-77-7
1,3,4-Thiadiazol-2-amine, 5-(methylthio)-
SMe
H2N
147 C5H6ClNO2 6702-21-2
Lactonitrile, 3-chloro-, acetate
AcOCH(CN)CH2Cl
147 C5H9NO4 56-86-0
L-Glutamic acid
HO2CCH2CH2CH(NH2)CO2H
147 C5H9NO4 16354-58-8
L-Serine, N-acetyl-
AcNHCH(CO2H)CH2OH

147 $C_6H_{10}ClNO$ 16580-31-7
Cyclohexane, 1-chloro-2-nitroso-

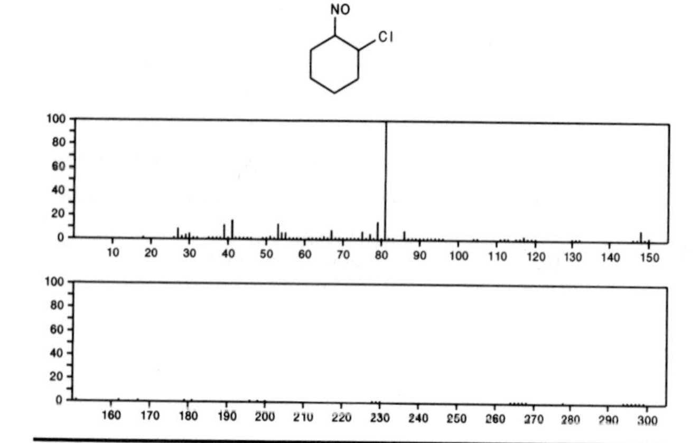

147 $C_6H_{13}NOS$ 32805-46-2
Sulfonium, (acetylamino)diethyl-, hydroxide, inner salt

$Ac\bar{N}—\overset{+}{S}Et_2$

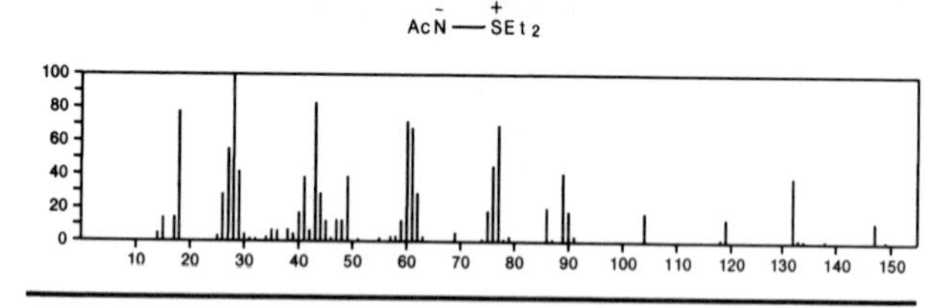

147 $C_6H_{13}NO_3$ 3688-46-8
Butanoic acid, 4-(dimethylamino)-3-hydroxy-

$HO_2CCH_2CH(OH)CH_2NMe_2$

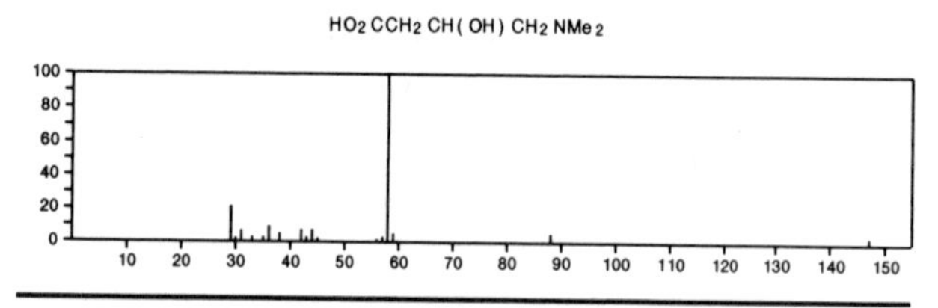

147 $C_6H_{13}NO_3$ 20633-11-8
Nitric acid, hexyl ester

$Me(CH_2)_5ONO_2$

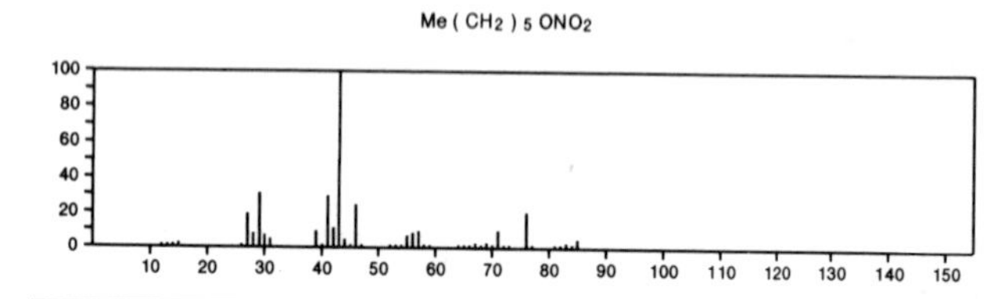

147 $C_6H_{13}NO_3$ 55956-18-8
Acetamide, 2-methoxy-*N*-(2-methoxyethyl)-

$MeOCH_2CONHCH_2CH_2OMe$

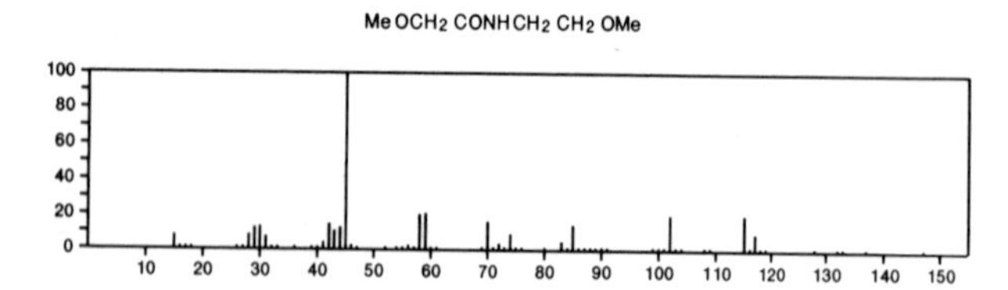

147 $C_7H_5N_3O$ 90-16-4
1,2,3-Benzotriazin-4(1*H*)-one

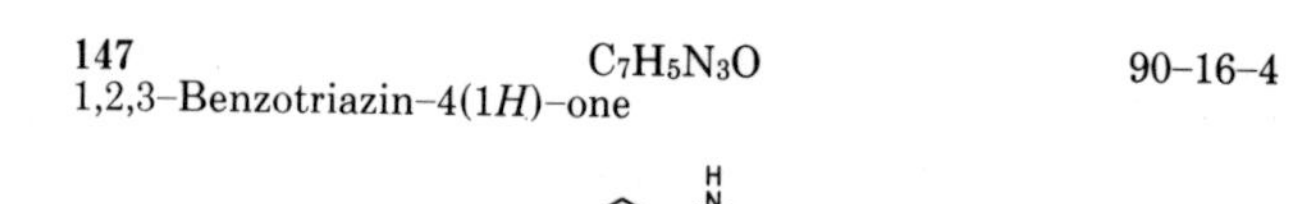

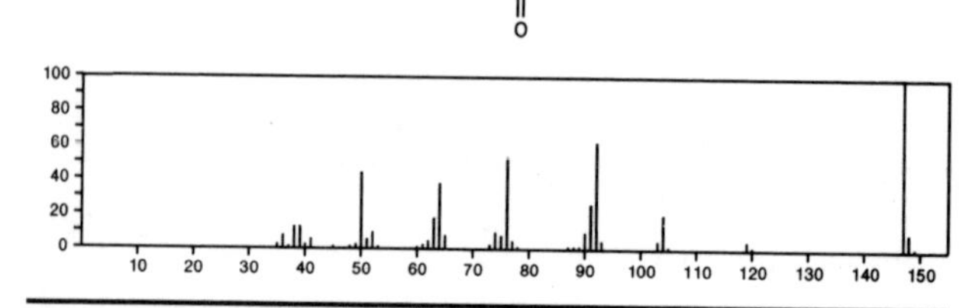

147 $C_7H_5N_3O$ 582-61-6
Benzoyl azide

N_3COPh

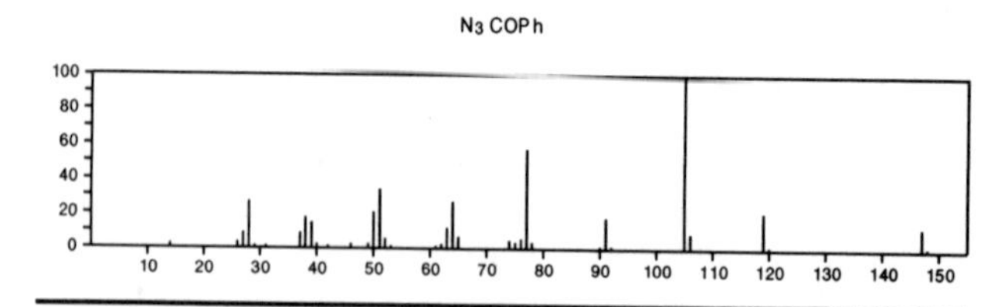

147 $C_7H_5N_3O$ 13389-59-8
Pyridine, 2-(1,2,4-oxadiazol-3-yl)-

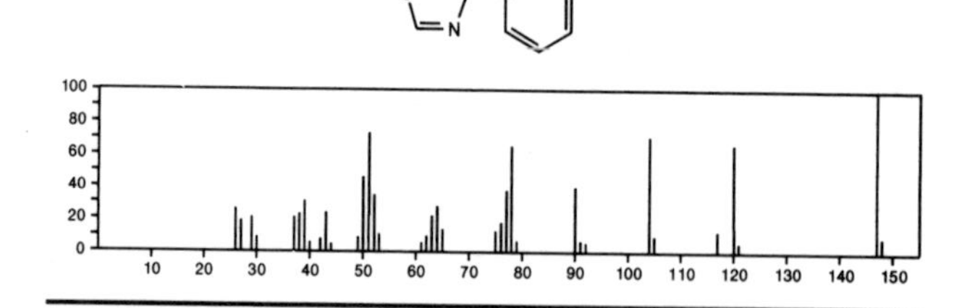

147 $C_7H_5N_3O$ 13428-22-3
Pyridine, 2-(1,3,4-oxadiazol-2-yl)-

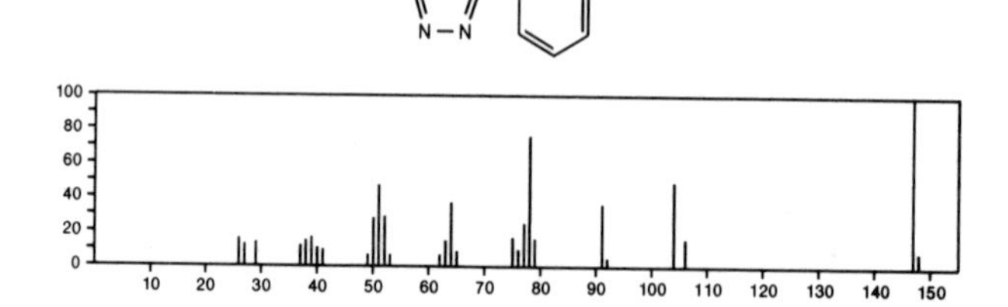

147 $C_7H_5N_3O$ 16952-64-0
Pyrido[4,3-*d*]pyrimidin-4(3*H*)-one

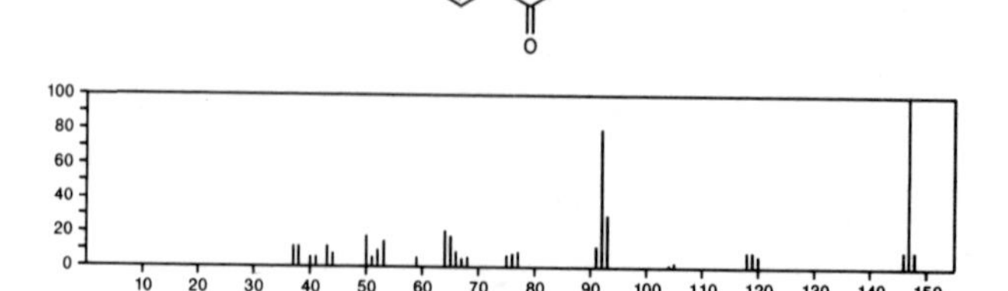

147 $C_7H_5N_3O$ 19178-25-7
Pyrido[3,4-*d*]pyrimidin-4(3*H*)-one

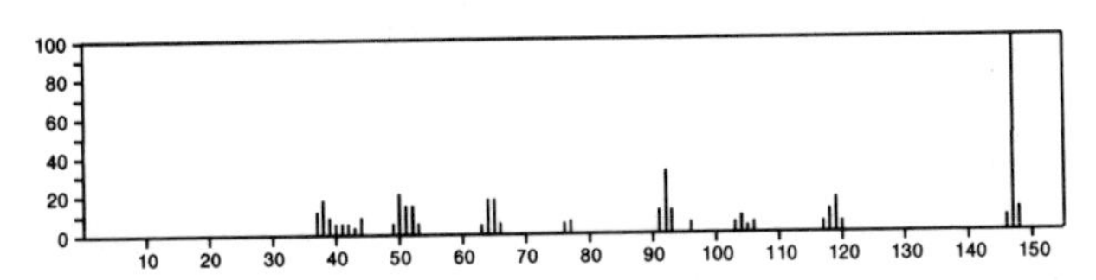

---

147 $C_7H_5N_3O$ 24410-19-3
Pyrido[2,3-*d*]pyrimidin-4(1*H*)-one

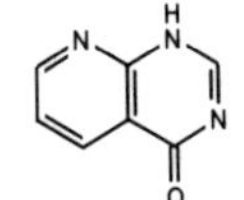

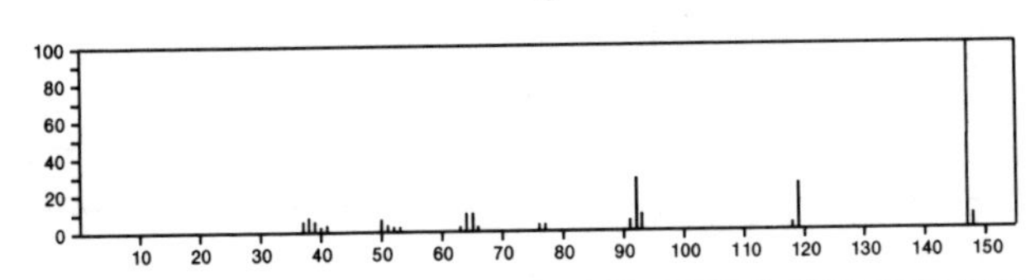

---

147 $C_7H_5N_3O$ 37538-67-3
Pyrido[3,2-*d*]pyrimidin-4-ol

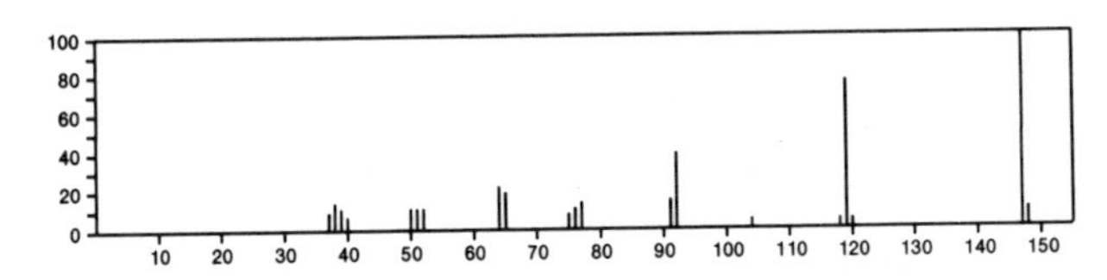

---

147 $C_7H_5N_3O$ 56805-24-4
1*H*-Imidazo[4,5-*b*]pyridine-2-carboxaldehyde

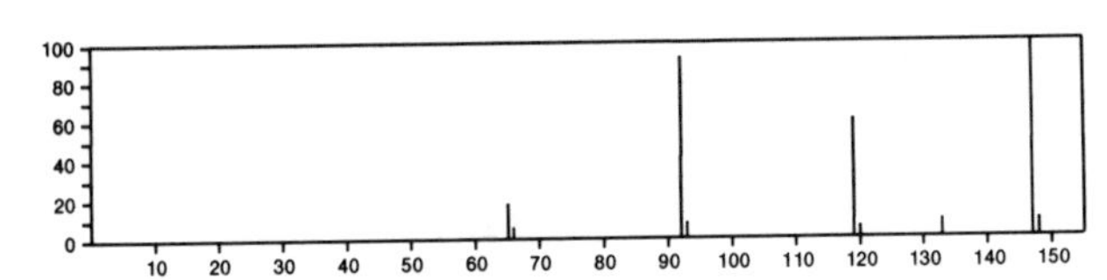

---

147 $C_7H_5N_3O$ 56805-25-5
1*H*-Imidazo[4,5-*c*]pyridine-2-carboxaldehyde

---

147 $C_7H_{14}ClN$ 2167-11-5
Piperidine, 3-chloro-1-ethyl-

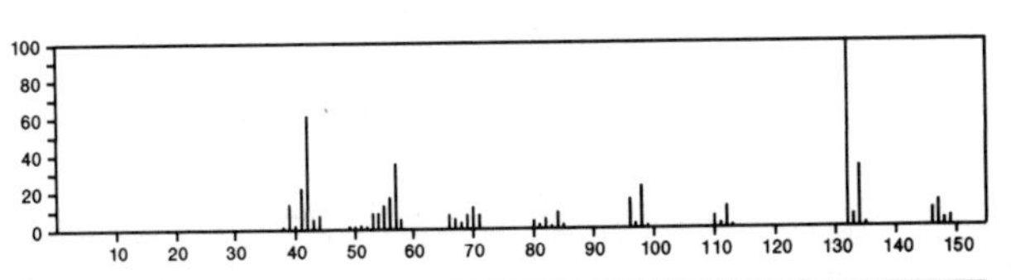

---

147 $C_7H_{14}ClN$ 23240-43-9
Allylamine, 3-chloro-*N*-isopropyl-2-methyl-, (*Z*)-

i-Pr $NHCH_2CMe=CHCl$

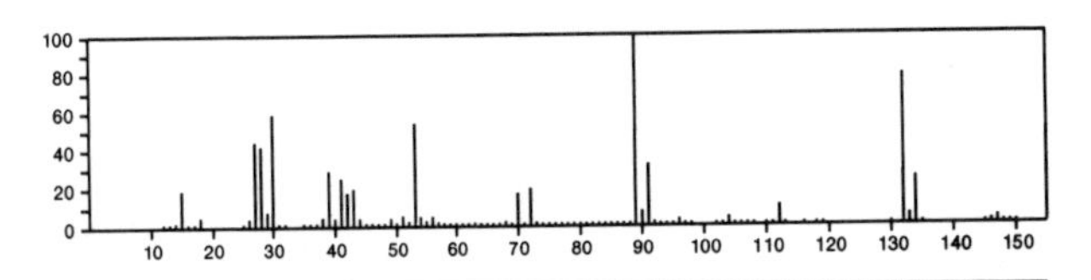

---

147 $C_7H_{14}ClN$ 23240-44-0
Allylamine, 3-chloro-*N*-isopropyl-2-methyl-, (*E*)-

i-Pr $NHCH_2CMe=CHCl$

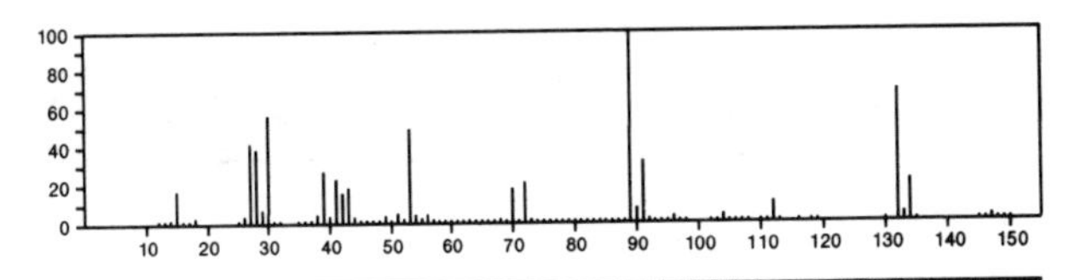

---

147 $C_7H_{17}NO_2$ 55759-85-8
1-Propanamine, 3-(2-methoxy-1-methylethoxy)-

$H_2N(CH_2)_3OCHMeCH_2OMe$

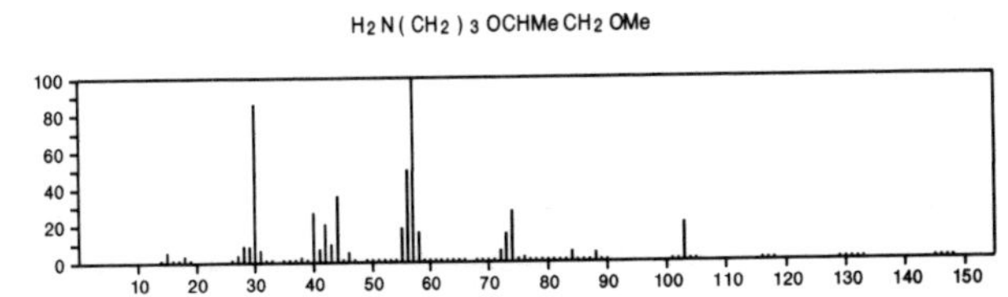

---

147 $C_7H_{17}NS$ 3492-79-3
Ethanamine, *N*-ethyl-*N*-[(ethylthio)methyl]-

$EtSCH_2NEt_2$

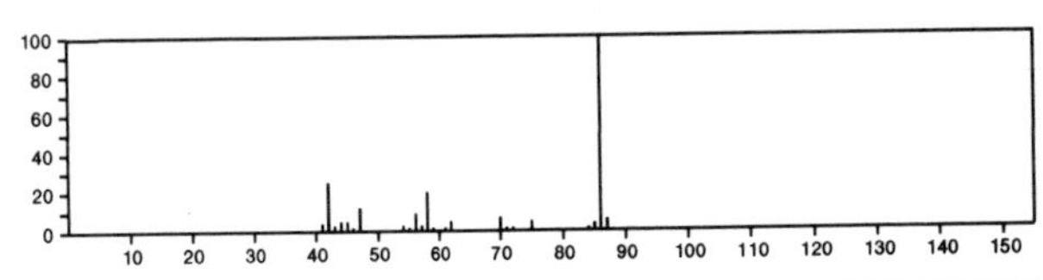

---

147 $C_8H_5NO_2$ 85-41-6
1*H*-Isoindole-1,3(2*H*)-dione

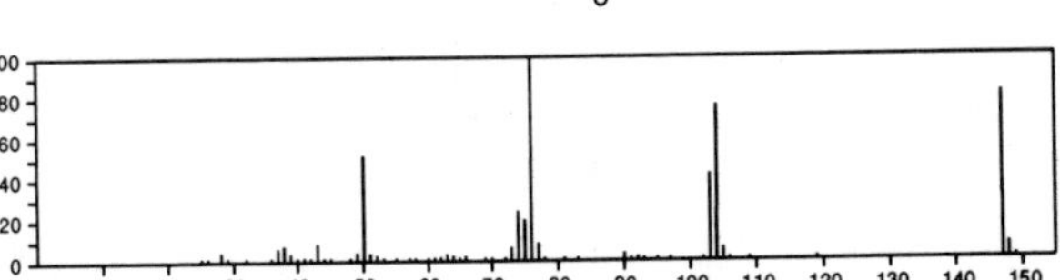

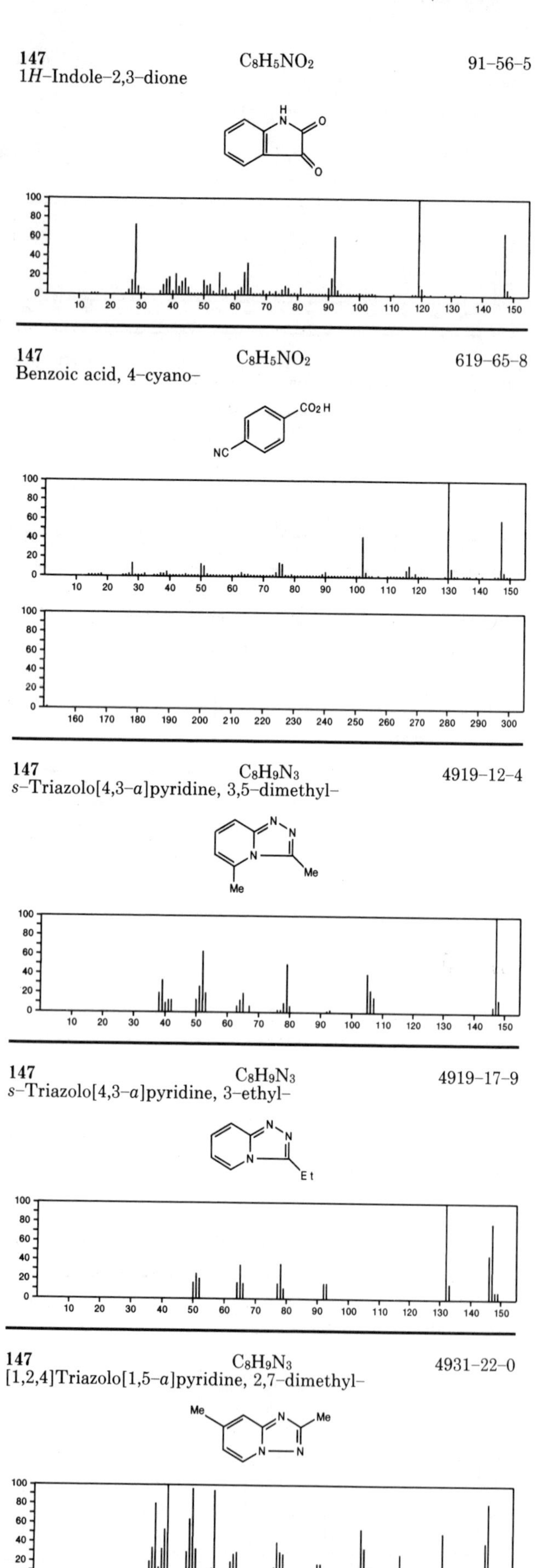

147 $C_8H_5NO_2$ 91-56-5
1*H*-Indole-2,3-dione

147 $C_8H_5NO_2$ 619-65-8
Benzoic acid, 4-cyano-

147 $C_8H_9N_3$ 4919-12-4
s-Triazolo[4,3-*a*]pyridine, 3,5-dimethyl-

147 $C_8H_9N_3$ 4919-17-9
s-Triazolo[4,3-*a*]pyridine, 3-ethyl-

147 $C_8H_9N_3$ 4931-22-0
[1,2,4]Triazolo[1,5-*a*]pyridine, 2,7-dimethyl-

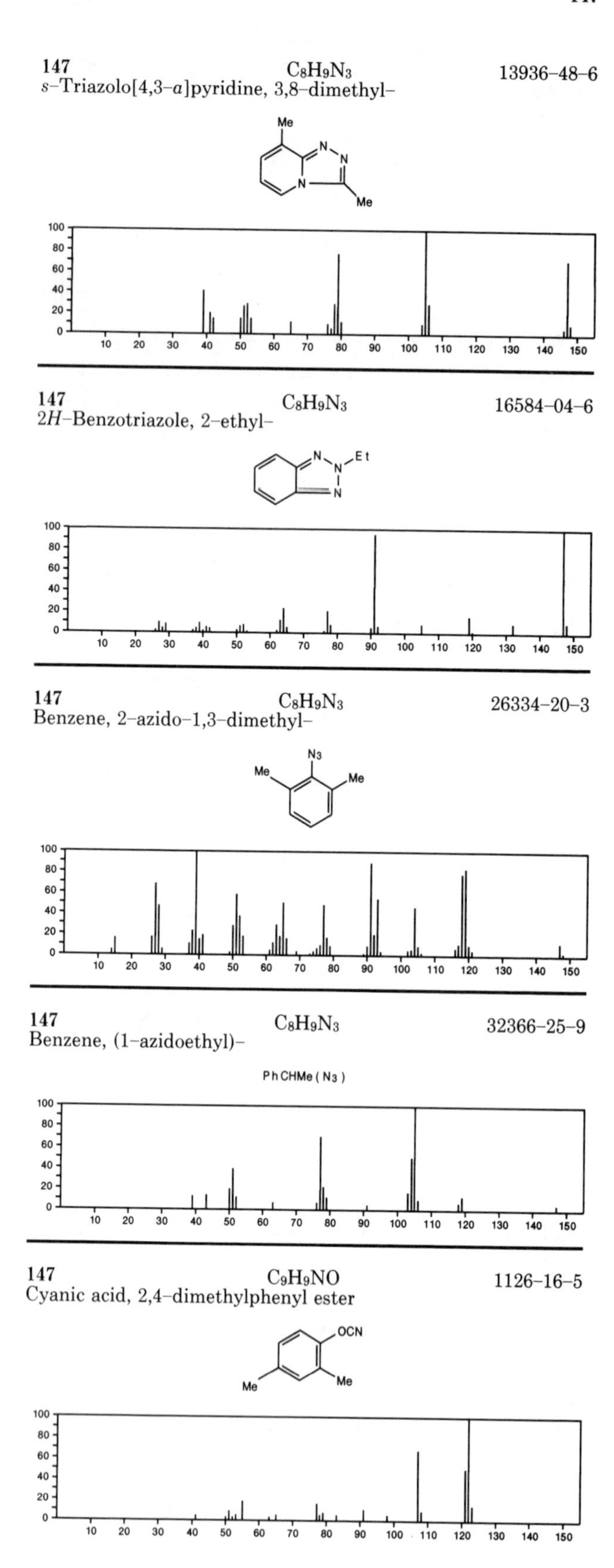

147 $C_8H_9N_3$ 13936-48-6
s-Triazolo[4,3-*a*]pyridine, 3,8-dimethyl-

147 $C_8H_9N_3$ 16584-04-6
2*H*-Benzotriazole, 2-ethyl-

147 $C_8H_9N_3$ 26334-20-3
Benzene, 2-azido-1,3-dimethyl-

147 $C_8H_9N_3$ 32366-25-9
Benzene, (1-azidoethyl)-

PhCHMe(N3)

147 C_9H_9NO 1126-16-5
Cyanic acid, 2,4-dimethylphenyl ester

147 C_9H_9NO 2861–59–8
1*H*–Indole–1–carboxaldehyde, 2,3–dihydro–

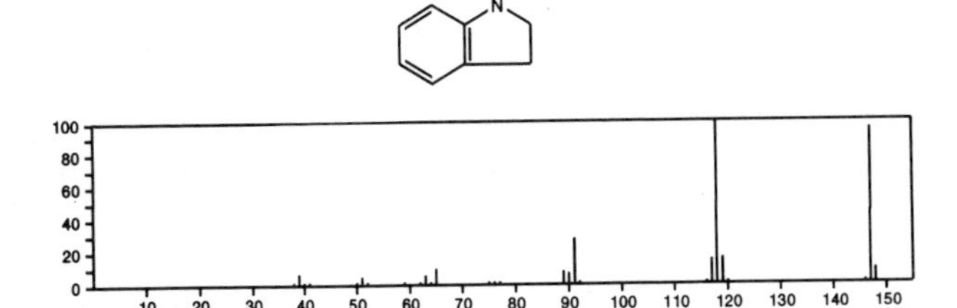

147 C_9H_9NO 17190–29–3
Benzenepropanenitrile, β–hydroxy–

NCCH2 CH(OH) Ph

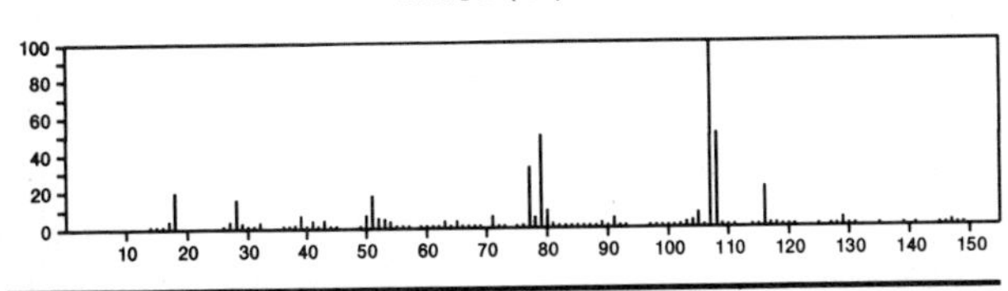

147 C_9H_9NO 22320–21–4
2–Indolizinemethanol

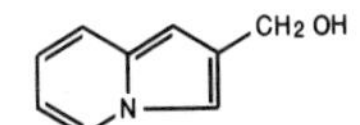

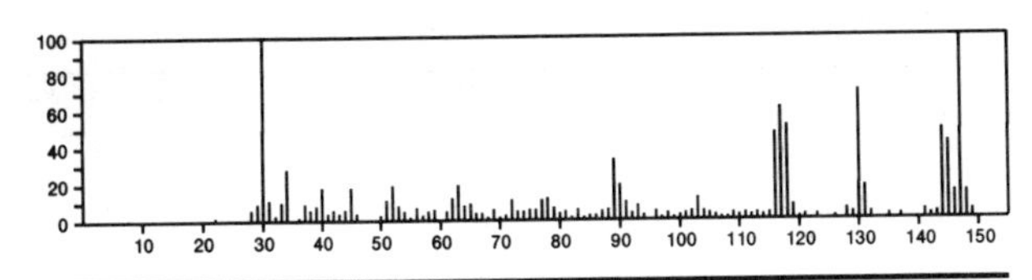

147 $C_9H_{14}BN$ 3519–71–9
Boranamine, *N*,*N*,1–trimethyl–1–phenyl–

Me2NBMePh

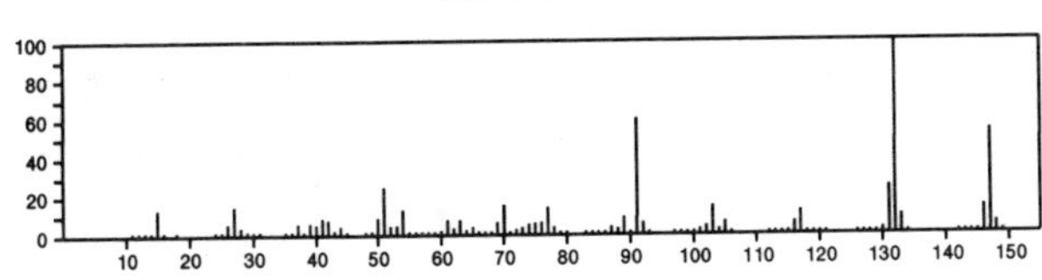

147 $C_{10}H_{13}N$ 936–43–6
Aziridine, 1,2–dimethyl–3–phenyl–, *trans*–

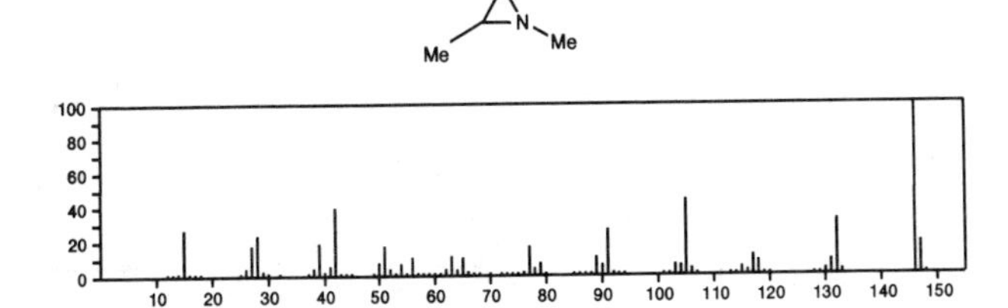

147 $C_{10}H_{13}N$ 2217–41–6
1–Naphthalenamine, 5,6,7,8–tetrahydro–

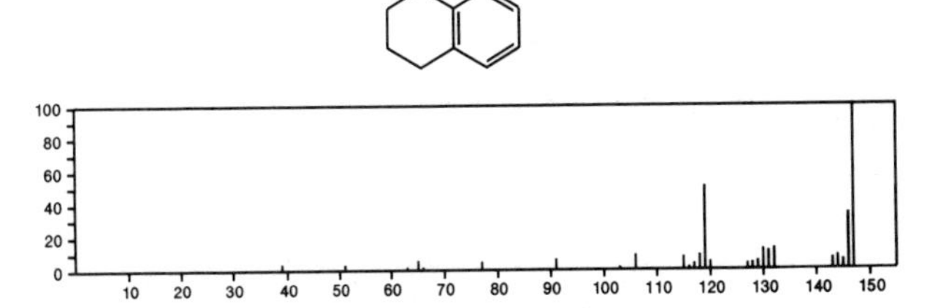

147 $C_{10}H_{13}N$ 2217–43–8
2–Naphthalenamine, 5,6,7,8–tetrahydro–

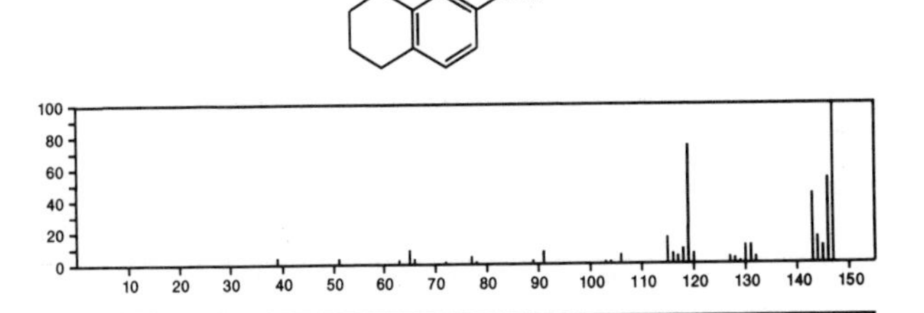

147 $C_{10}H_{13}N$ 3164–46–3
Aziridine, 1–(2–phenylethyl)–

147 $C_{10}H_{13}N$ 4965–09–7
Isoquinoline, 1,2,3,4–tetrahydro–1–methyl–

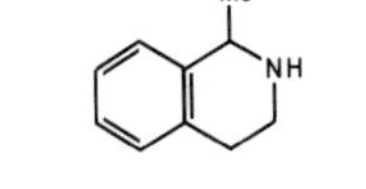

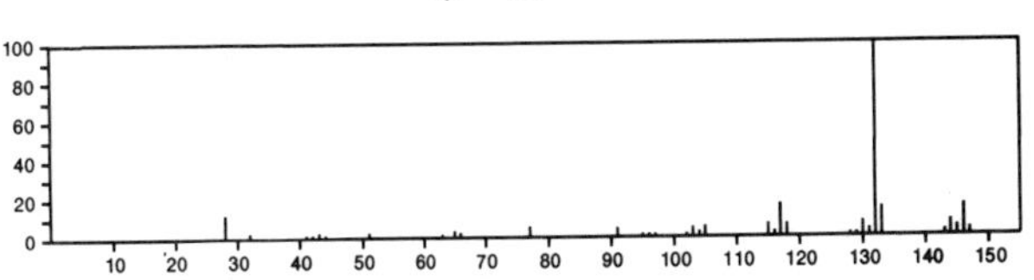

147 $C_{10}H_{13}N$ 6852–55–7
1–Propanamine, *N*–(phenylmethylene)–

PrN=CHPh

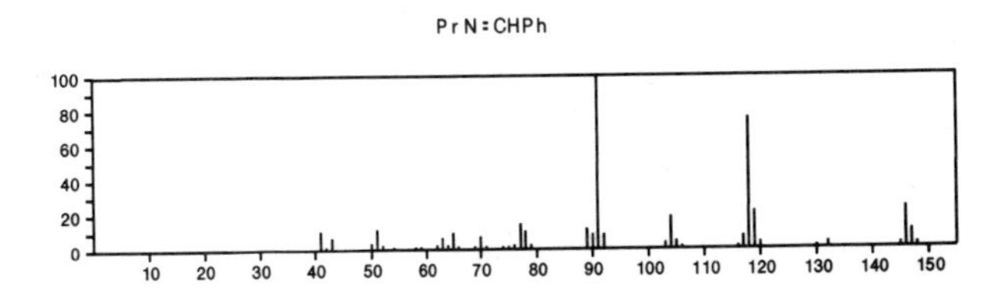

147 $C_{10}H_{13}N$ 6852–56–8
2–Propanamine, *N*–(phenylmethylene)–

i-PrN=CHPh

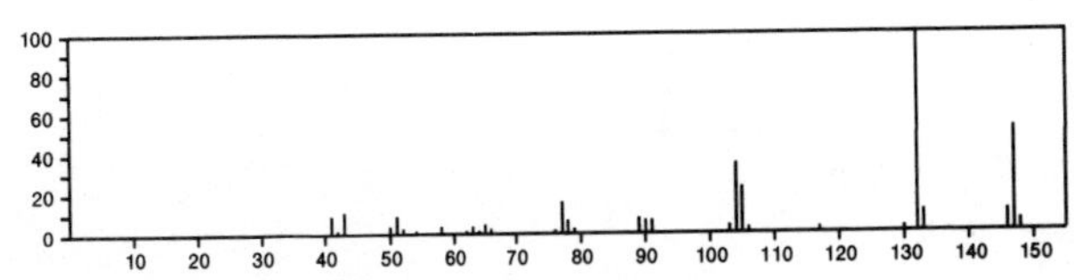

147 $C_{10}H_{13}N$ 26216-93-3
1*H*-Indole, 2,3-dihydro-1,2-dimethyl-

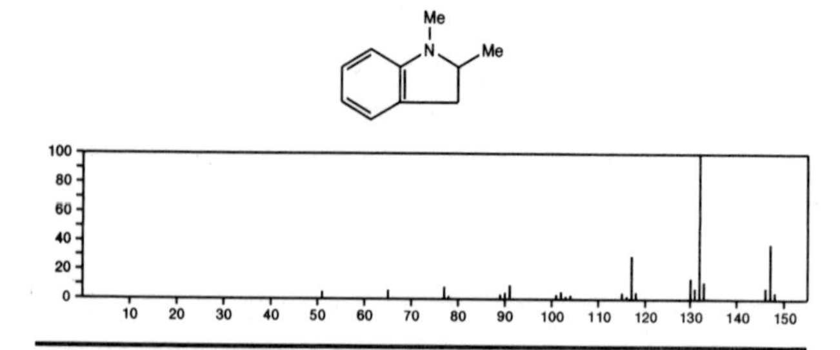

147 $C_{10}H_{13}N$ 29726-60-1
Isoquinoline, 1,2,3,4-tetrahydro-3-methyl-

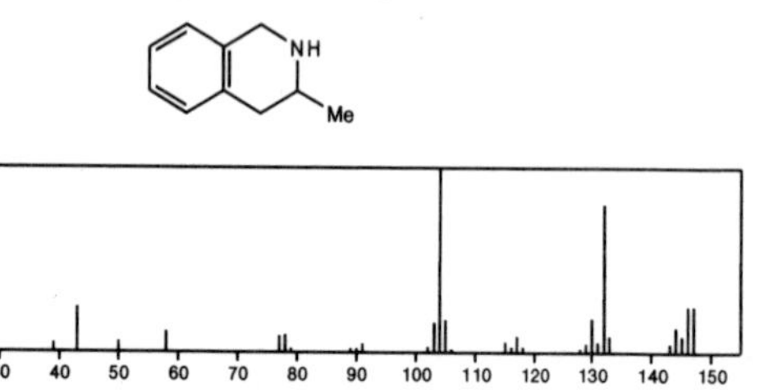

147 $C_{10}H_{13}N$ 55702-57-3
Azetidine, 2-methyl-1-phenyl-

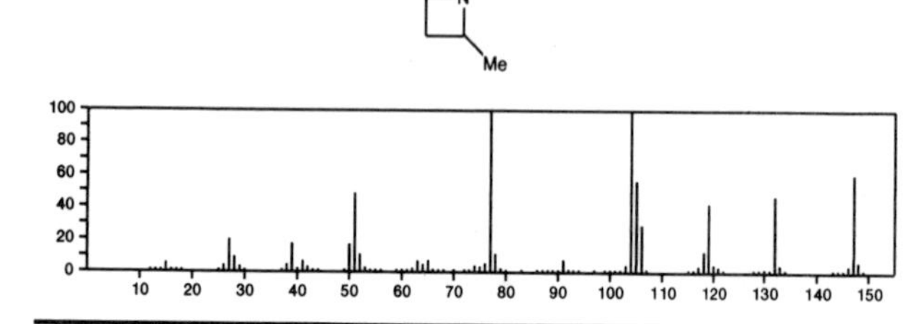

147 $C_{10}H_{13}N$ 55702-58-4
Azetidine, 1-methyl-2-phenyl-

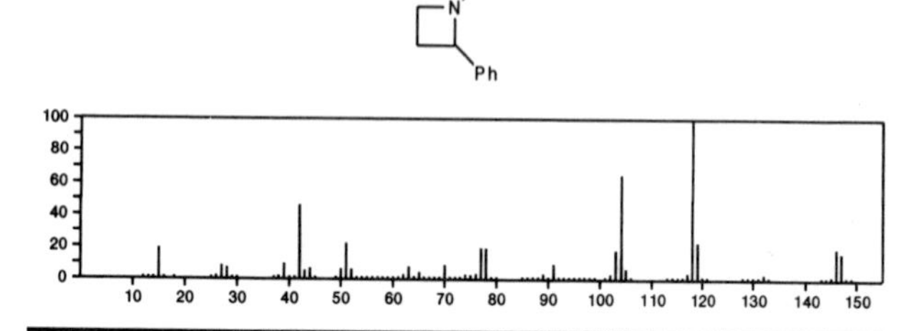

147 $C_{10}H_{13}N$ 56771-48-3
Cyclopropanamine, *N*-methyl-1-phenyl-

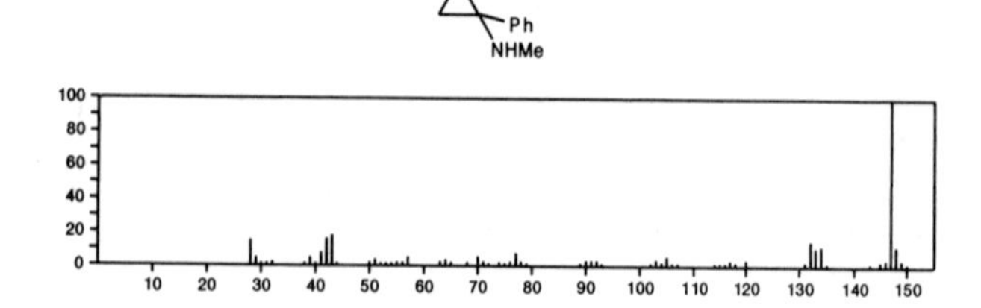

148 $CBrF_3$ 75-63-8
Methane, bromotrifluoro-

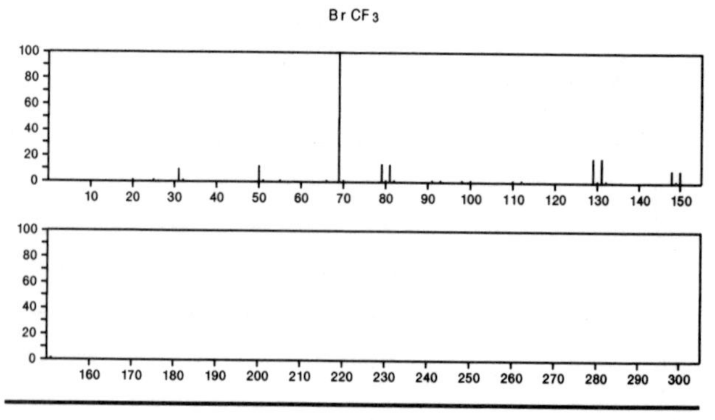

148 C_2Cl_3F 359-29-5
Ethene, trichlorofluoro-

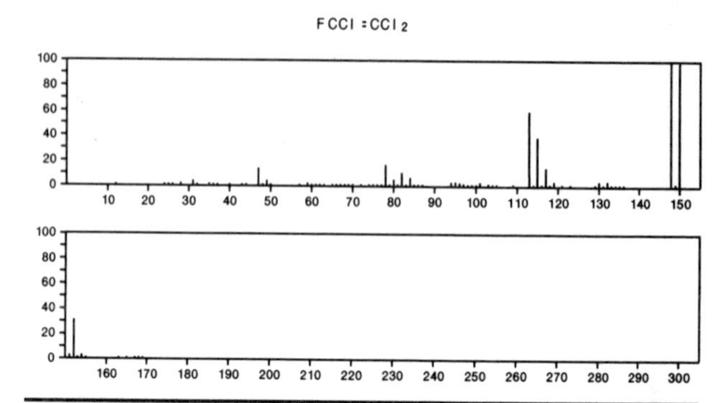

148 $C_2H_3Cl_3O$ 115-20-8
Ethanol, 2,2,2-trichloro-

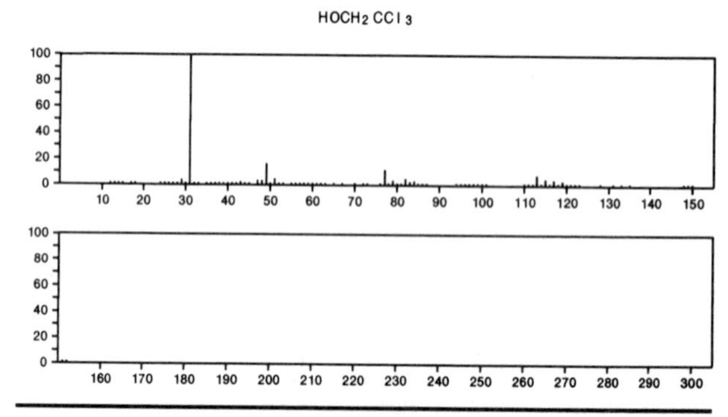

148 $C_2H_3Cl_3O$ 2799-32-8
Ether, chloromethyl dichloromethyl

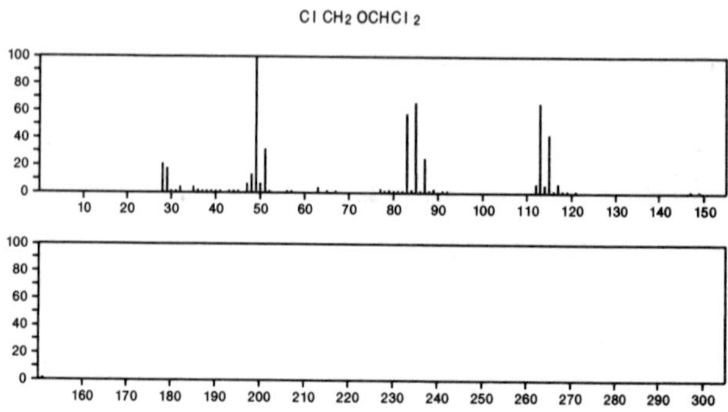

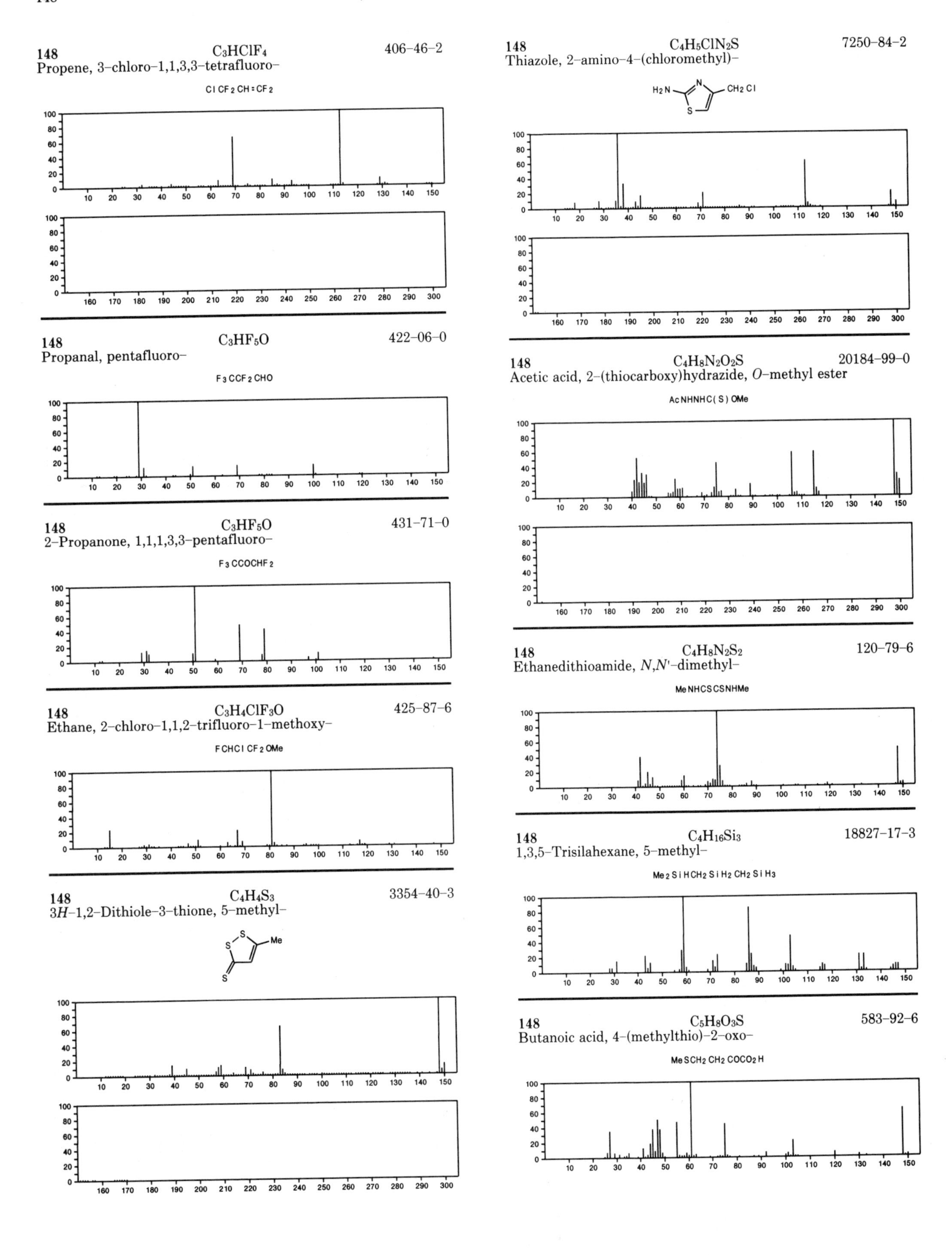
148 C3HClF4 406-46-2
Propene, 3-chloro-1,1,3,3-tetrafluoro-
Cl CF2 CH=CF2
148 C3HF5O 422-06-0
Propanal, pentafluoro-
F3 CCF2 CHO
148 C3HF5O 431-71-0
2-Propanone, 1,1,1,3,3-pentafluoro-
F3 CCOCHF2
148 C3H4ClF3O 425-87-6
Ethane, 2-chloro-1,1,2-trifluoro-1-methoxy-
FCHCl CF2 OMe
148 C4H4S3 3354-40-3
3H-1,2-Dithiole-3-thione, 5-methyl-
148 C4H5ClN2S 7250-84-2
Thiazole, 2-amino-4-(chloromethyl)-
H2N CH2 Cl
148 C4H8N2O2S 20184-99-0
Acetic acid, 2-(thiocarboxy)hydrazide, O-methyl ester
AcNHNHC(S)OMe
148 C4H8N2S2 120-79-6
Ethanedithioamide, N,N'-dimethyl-
MeNHCSCSNHMe
148 C4H16Si3 18827-17-3
1,3,5-Trisilahexane, 5-methyl-
Me2SiHCH2SiH2CH2SiH3
148 C5H8O3S 583-92-6
Butanoic acid, 4-(methylthio)-2-oxo-
MeSCH2CH2COCO2H

148 $C_5H_8O_3S$ 17396–35–9
4*H*–Thiopyran–4–one, tetrahydro–, 1,1–dioxide

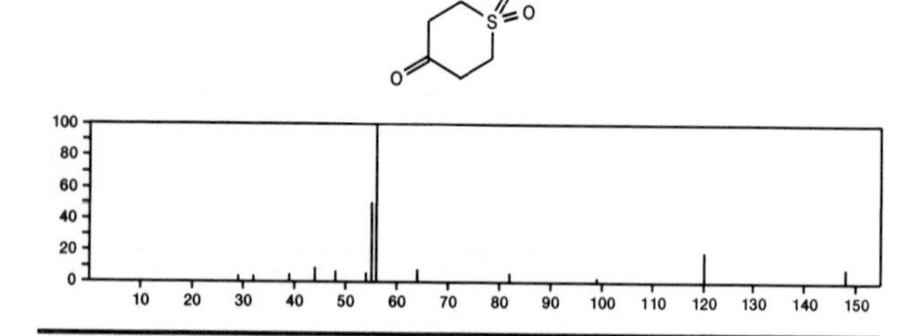

148 C_5H_9Br 137–43–9
Cyclopentane, bromo–

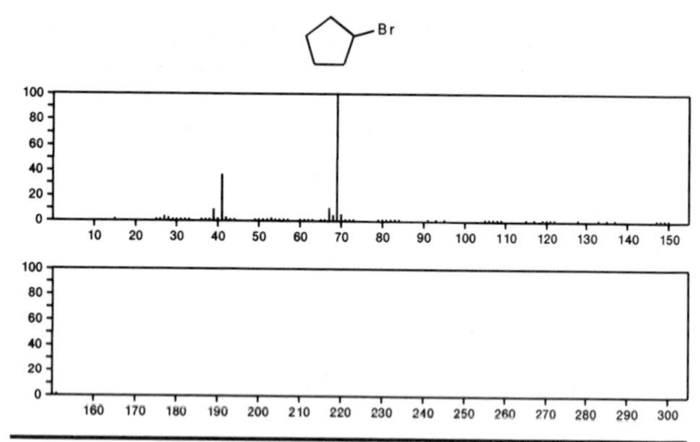

148 $C_5H_9O_3P$ 1449–91–8
2,6,7–Trioxa–1–phosphabicyclo[2.2.2]octane, 4–methyl–

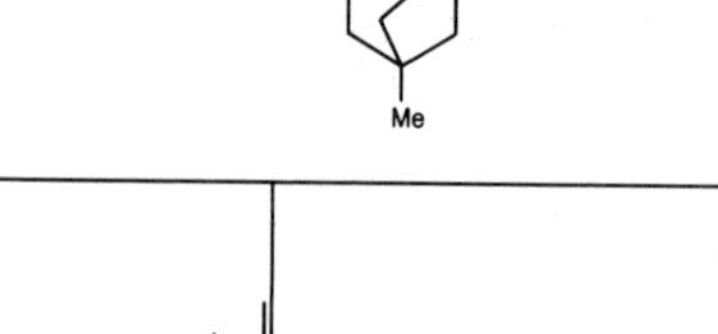

148 $C_5H_{12}N_2O_3$ 15438–70–7
Urea, *N,N'*–bis(2–hydroxyethyl)–

HOCH2 CH2 NHCONHCH2 CH2 OH

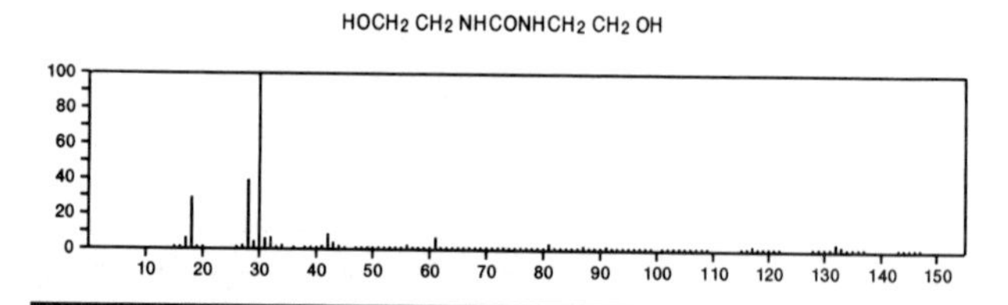

148 $C_6H_4N_4O$ 700–47–0
4(1*H*)–Pteridinone

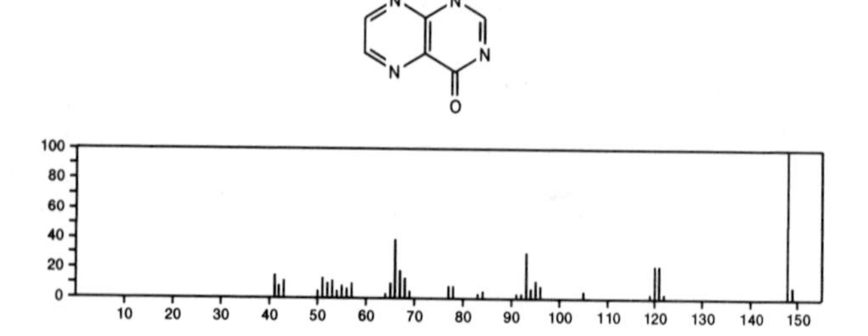

148 $C_6H_4N_4O$ 2432–24–8
2(1*H*)–Pteridinone

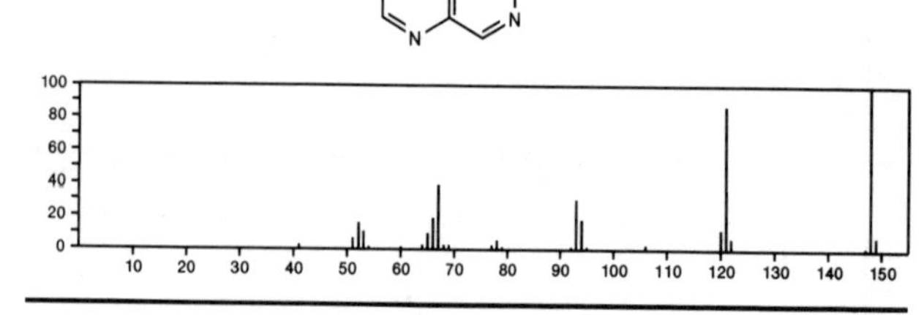

148 $C_6H_4N_4O$ 2432–26–0
6(5*H*)–Pteridinone

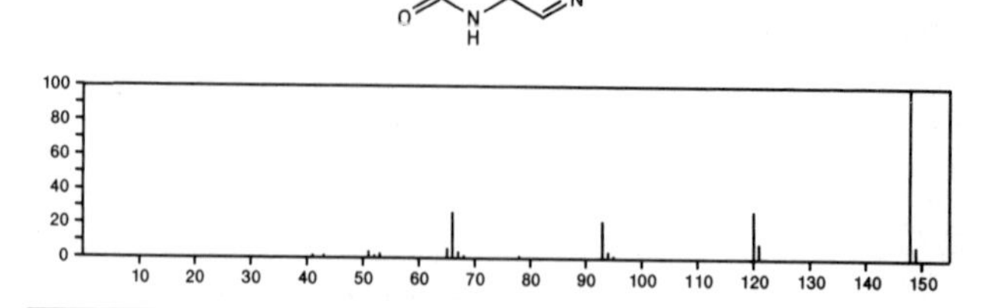

148 $C_6H_4N_4O$ 2432–27–1
7(1*H*)–Pteridinone

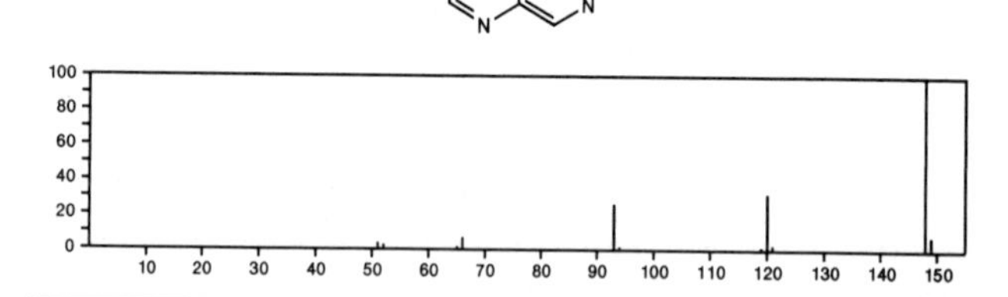

148 $C_6H_4N_4O$ 56805–26–6
1*H*–Purine–8–carboxaldehyde

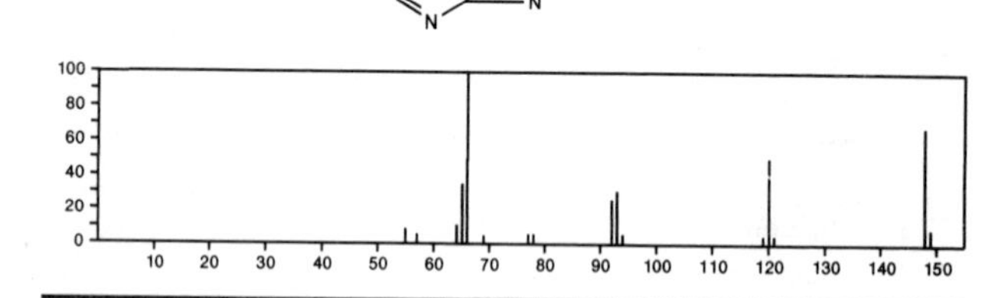

148 $C_6H_{12}O_2S$ 6251–33–8
Thiepane, 1,1–dioxide

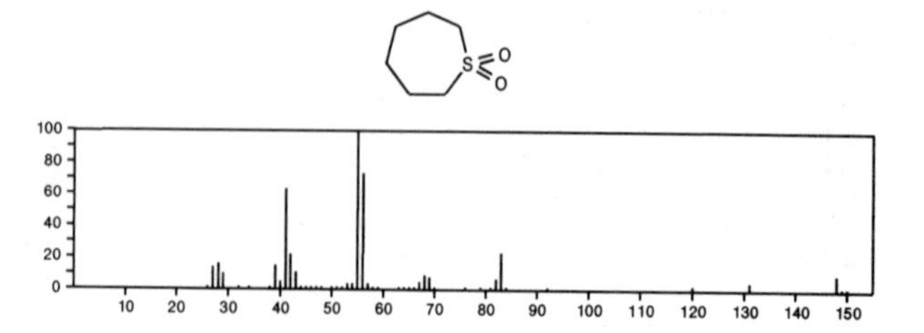

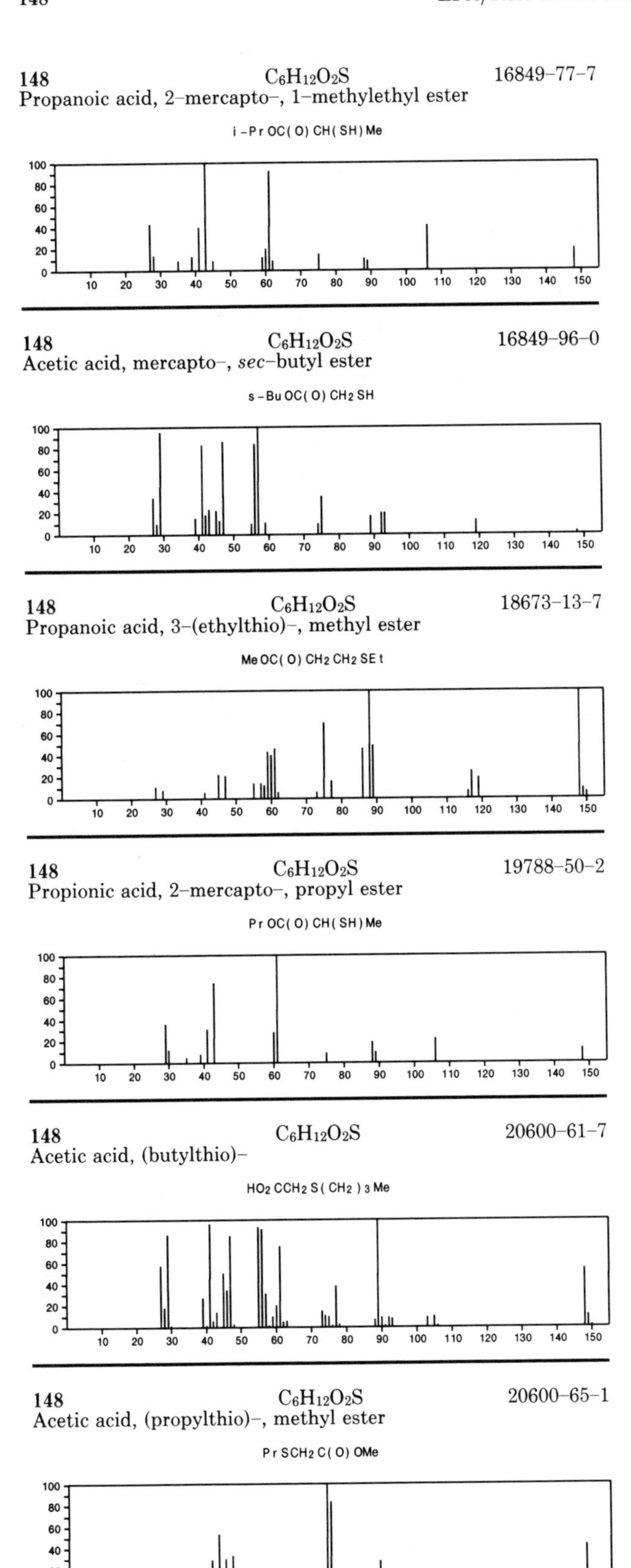

148 $C_6H_{12}O_2S$ 16849-77-7
Propanoic acid, 2-mercapto-, 1-methylethyl ester
i-PrOC(O)CH(SH)Me

148 $C_6H_{12}O_2S$ 16849-96-0
Acetic acid, mercapto-, *sec*-butyl ester
s-BuOC(O)CH2SH

148 $C_6H_{12}O_2S$ 18673-13-7
Propanoic acid, 3-(ethylthio)-, methyl ester
MeOC(O)CH2CH2SEt

148 $C_6H_{12}O_2S$ 19788-50-2
Propionic acid, 2-mercapto-, propyl ester
PrOC(O)CH(SH)Me

148 $C_6H_{12}O_2S$ 20600-61-7
Acetic acid, (butylthio)-
HO2CCH2S(CH2)3Me

148 $C_6H_{12}O_2S$ 20600-65-1
Acetic acid, (propylthio)-, methyl ester
PrSCH2C(O)OMe

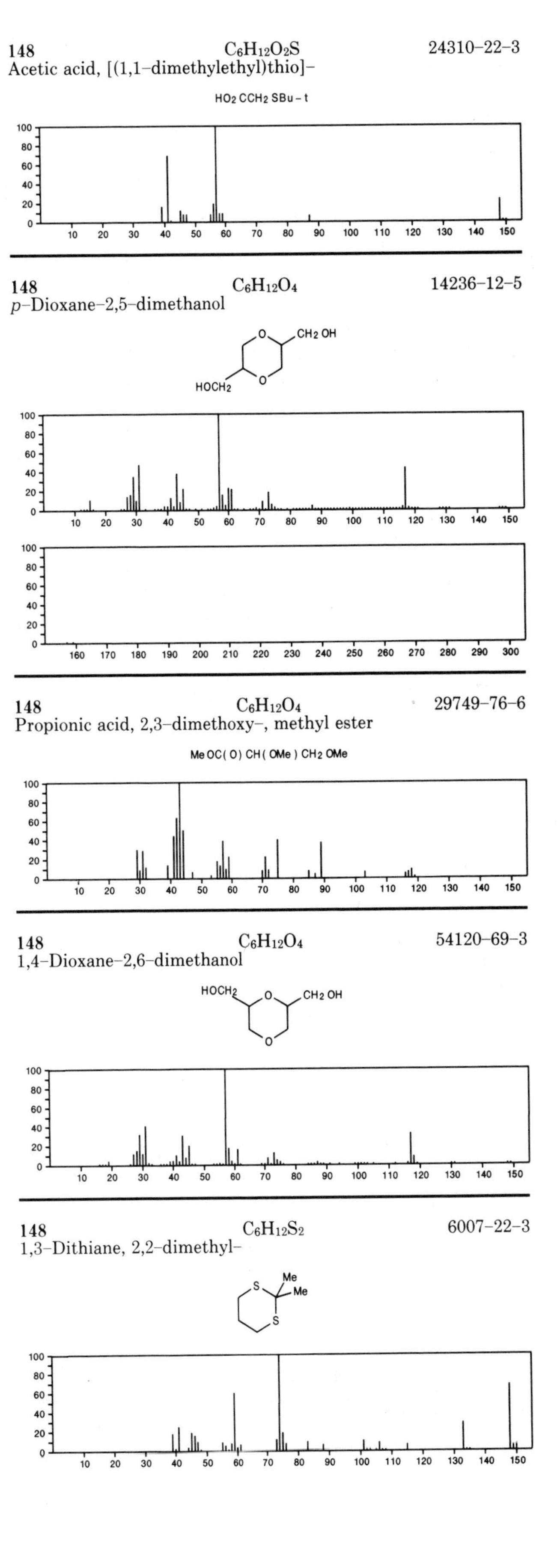

148 $C_6H_{12}O_2S$ 24310-22-3
Acetic acid, [(1,1-dimethylethyl)thio]-
HO2CCH2SBu-t

148 $C_6H_{12}O_4$ 14236-12-5
p-Dioxane-2,5-dimethanol

148 $C_6H_{12}O_4$ 29749-76-6
Propionic acid, 2,3-dimethoxy-, methyl ester
MeOC(O)CH(OMe)CH2OMe

148 $C_6H_{12}O_4$ 54120-69-3
1,4-Dioxane-2,6-dimethanol

148 $C_6H_{12}S_2$ 6007-22-3
1,3-Dithiane, 2,2-dimethyl-

148 $C_6H_{12}S_2$ 13105-10-7
Ethene, 1,2-bis(ethylthio)-

EtSCH=CHSEt

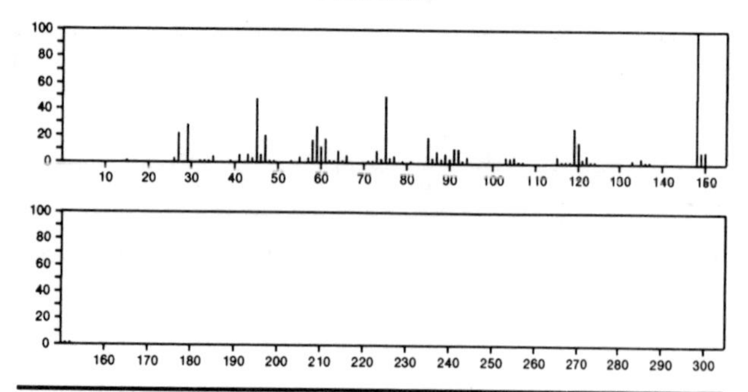

148 $C_6H_{16}N_2O_2$ 4439-20-7
Ethanol, 2,2'-(1,2-ethanediyldiimino)bis-

$HOCH_2CH_2NHCH_2CH_2NHCH_2CH_2OH$

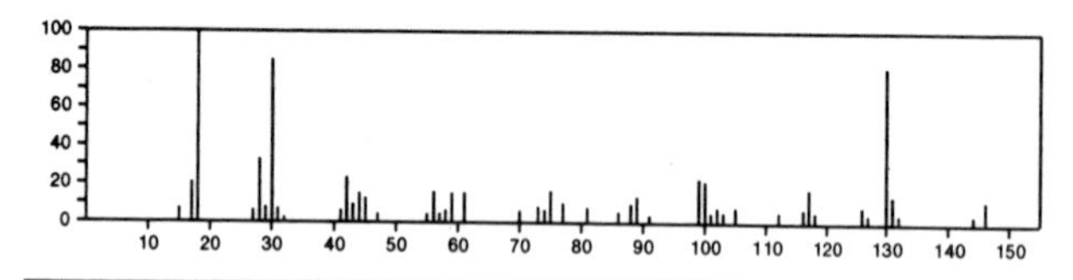

148 $C_6H_{16}O_2Si$ 18173-74-5
Silane, (2-methoxyethoxy)trimethyl-

$MeOCH_2CH_2OSiMe_3$

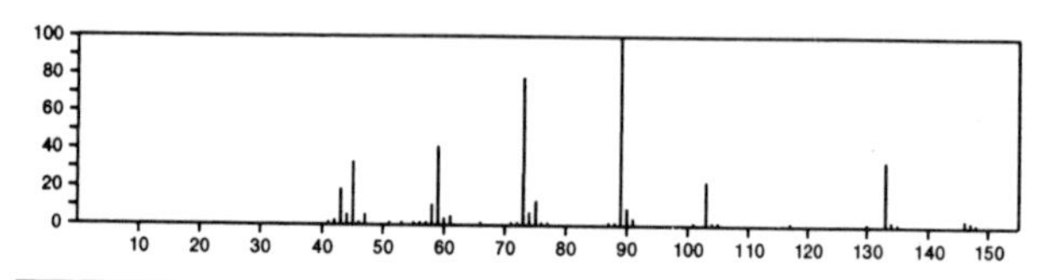

148 $C_7H_8N_4$ 5006-56-4
1,2,4-Triazolo[4,3-*a*]pyridin-3-amine, 7-methyl-

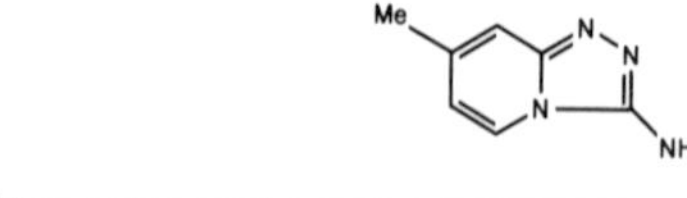

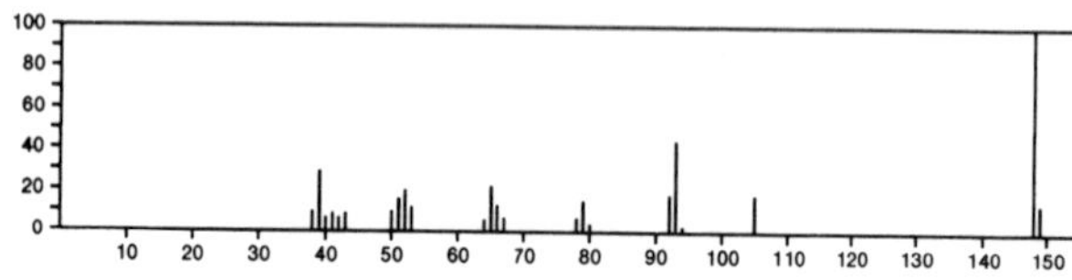

148 $C_7H_8N_4$ 5528-60-9
s-Triazolo[4,3-*a*]pyridine, 3-amino-6-methyl-

148 $C_7H_8N_4$ 5595-15-3
s-Triazolo[4,3-*a*]pyridine, 3-amino-5-methyl-

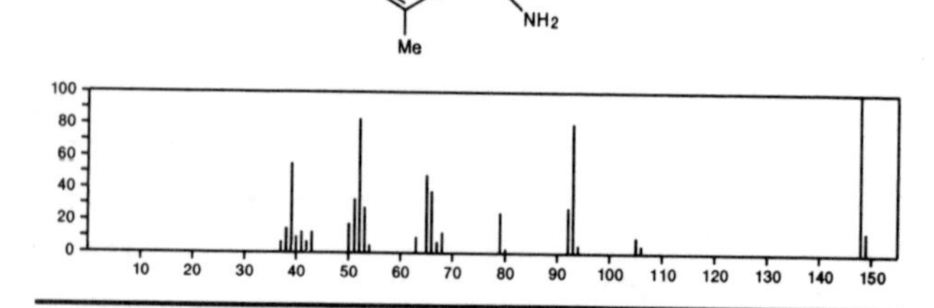

148 $C_7H_8N_4$ 6726-49-4
Pyrazolo[5,1-*c*][1,2,4]triazine, 3,4-dimethyl-

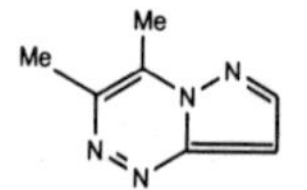

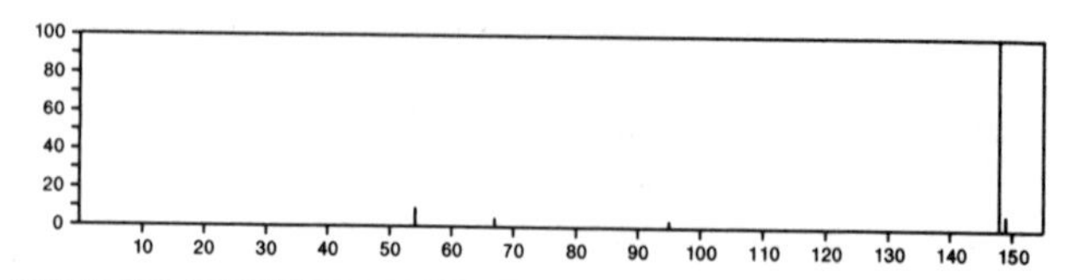

148 $C_7H_8N_4$ 13183-01-2
1*H*-Benzotriazole, 7-amino-1-methyl-

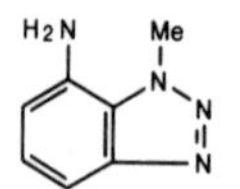

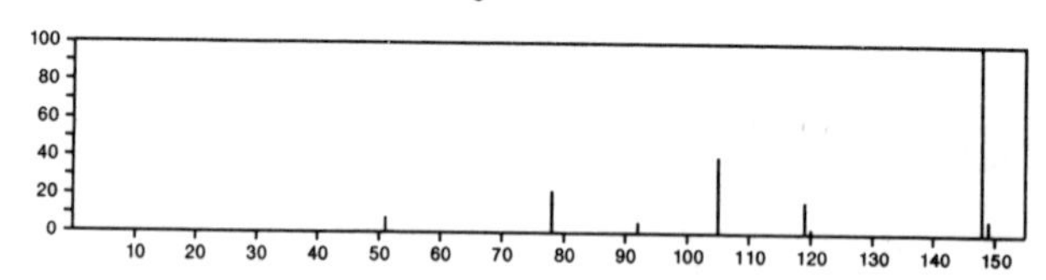

148 $C_7H_8N_4$ 19848-77-2
s-Triazolo[4,3-*a*]pyrazine, 5,8-dimethyl-

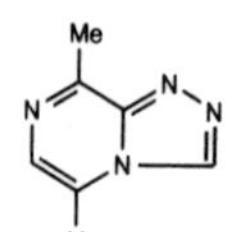

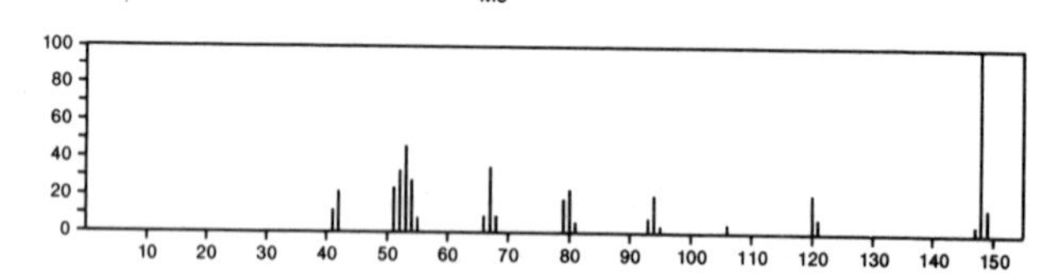

148 $C_7H_8N_4$ 19848-78-3
s-Triazolo[4,3-*a*]pyrazine, 3,8-dimethyl-

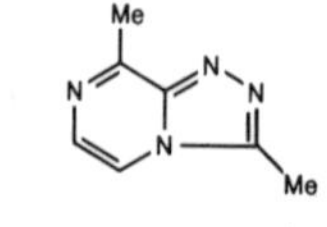

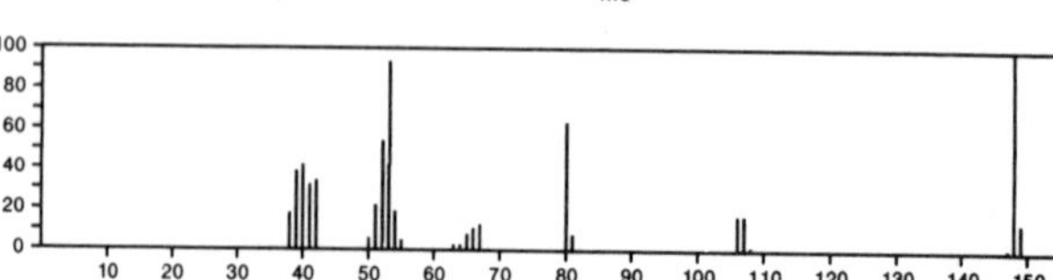

148 $C_7H_8N_4$ 26861-23-4
1*H*-Benzotriazole, 6-amino-1-methyl-

148 $C_7H_8N_4$ 27799-82-2
1*H*-Benzotriazole, 4-amino-1-methyl-

148 $C_7H_8N_4$ 27799-83-3
1*H*-Benzotriazol-5-amine, 1-methyl-

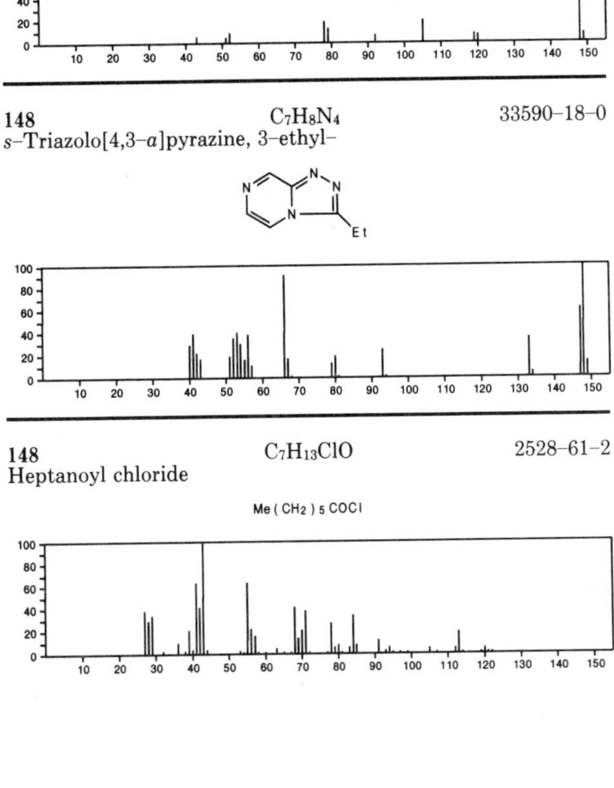

148 $C_7H_8N_4$ 33590-18-0
s-Triazolo[4,3-*a*]pyrazine, 3-ethyl-

148 $C_7H_{13}ClO$ 2528-61-2
Heptanoyl chloride

Me(CH2)5COCl

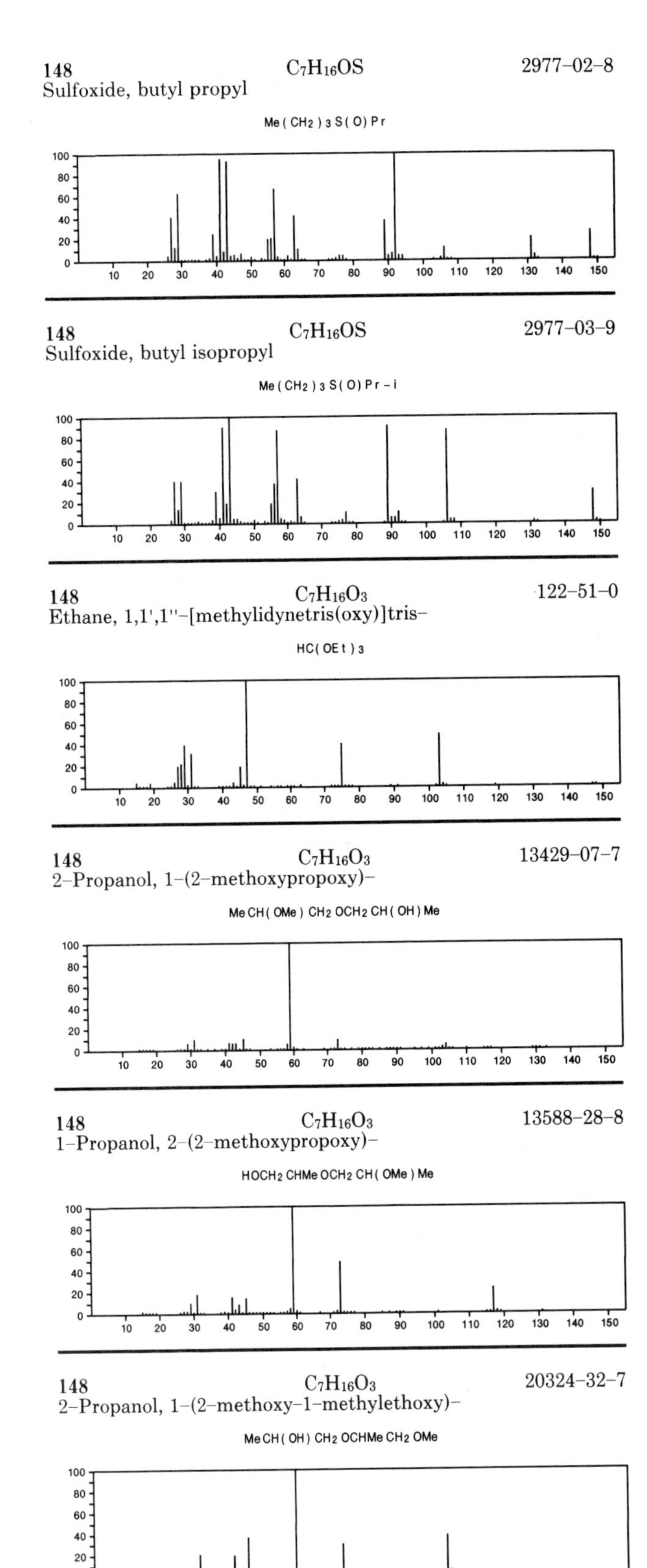

148 $C_7H_{16}OS$ 2977-02-8
Sulfoxide, butyl propyl

Me(CH2)3S(O)Pr

148 $C_7H_{16}OS$ 2977-03-9
Sulfoxide, butyl isopropyl

Me(CH2)3S(O)Pr-i

148 $C_7H_{16}O_3$ 122-51-0
Ethane, 1,1',1''-[methylidynetris(oxy)]tris-

HC(OEt)3

148 $C_7H_{16}O_3$ 13429-07-7
2-Propanol, 1-(2-methoxypropoxy)-

MeCH(OMe)CH2OCH2CH(OH)Me

148 $C_7H_{16}O_3$ 13588-28-8
1-Propanol, 2-(2-methoxypropoxy)-

HOCH2CHMeOCH2CH(OMe)Me

148 $C_7H_{16}O_3$ 20324-32-7
2-Propanol, 1-(2-methoxy-1-methylethoxy)-

MeCH(OH)CH2OCHMeCH2OMe

148 $C_7H_{16}O_3$ 20637-34-7
1-Propanol, 3-methoxy-2-(methoxymethyl)-2-methyl-

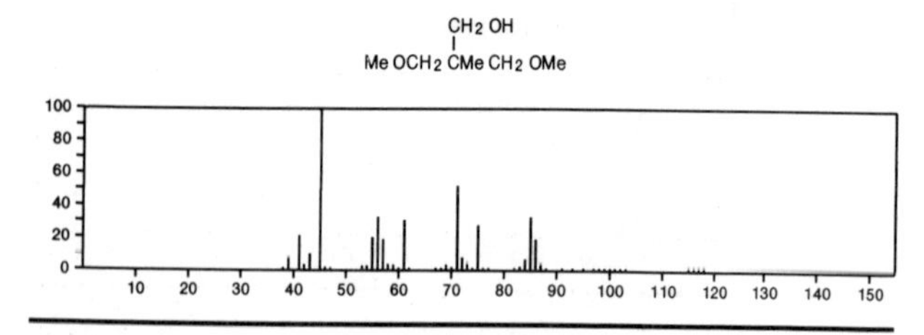

148 $C_7H_{16}O_3$ 20637-48-3
Butane, 1,2,4-trimethoxy-

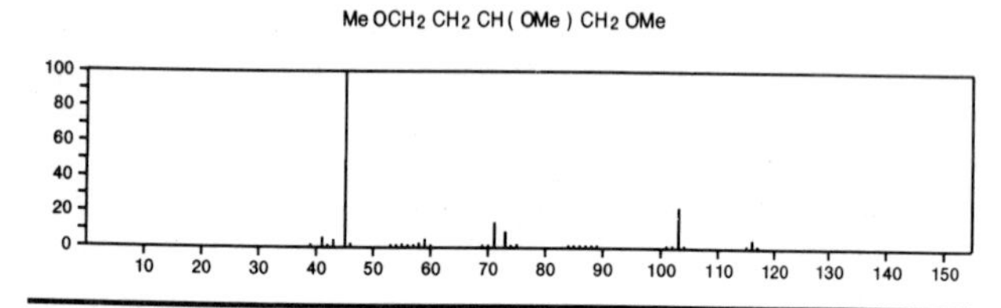

148 $C_7H_{16}O_3$ 55956-21-3
1-Propanol, 2-(2-methoxy-1-methylethoxy)-

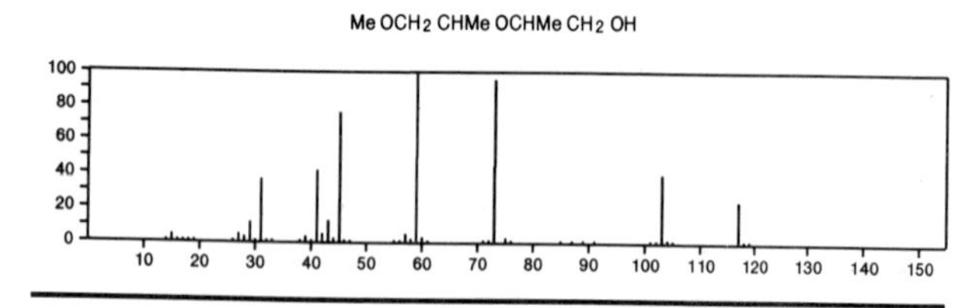

148 $C_8H_4O_3$ 85-44-9
1,3-Isobenzofurandione

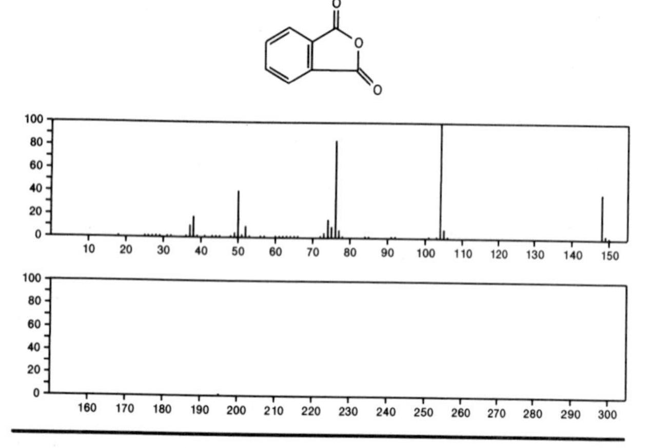

148 $C_8H_8N_2O$ 769-28-8
3-Pyridinecarbonitrile, 1,2-dihydro-4,6-dimethyl-2-oxo-

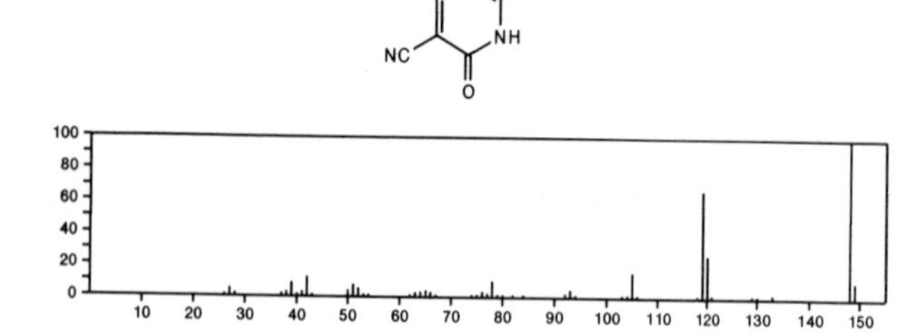

148 $C_8H_8N_2O$ 1006-19-5
3*H*-Indazol-3-one, 1,2-dihydro-1-methyl-

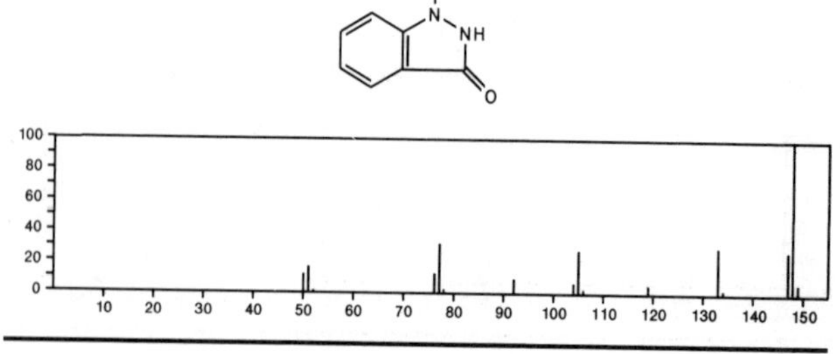

148 $C_8H_8N_2O$ 1848-40-4
3-Indazolinone, 2-methyl-

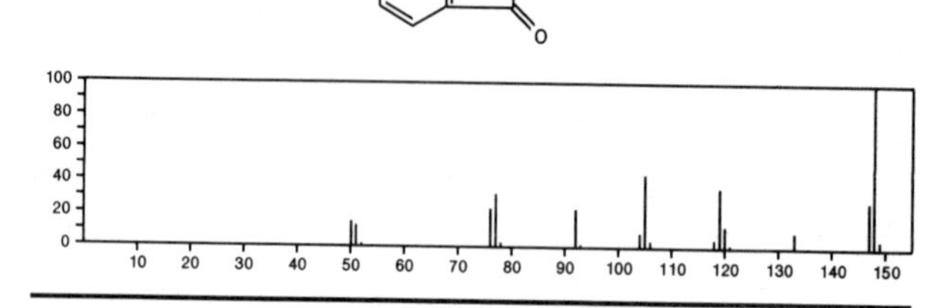

148 $C_8H_8N_2O$ 1848-41-5
1*H*-Indazole, 3-methoxy-

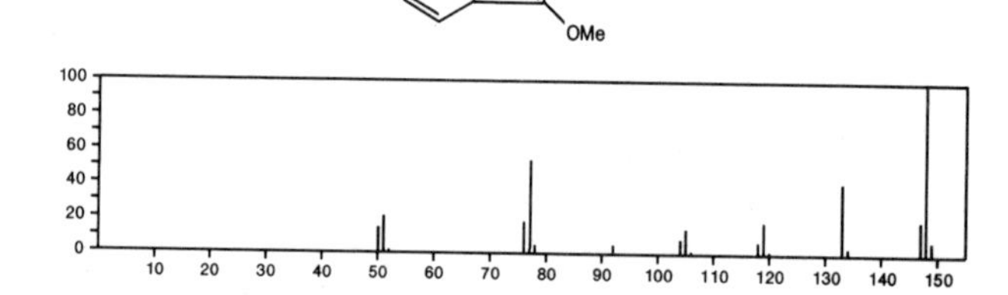

148 $C_8H_8N_2O$ 5400-75-9
2*H*-Benzimidazol-2-one, 1,3-dihydro-5-methyl-

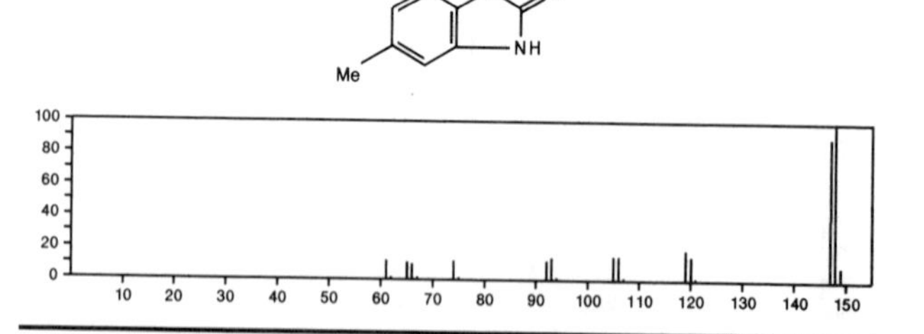

148 $C_8H_8N_2O$ 16007-52-6
Benzimidazole, 2-methyl-, 3-oxide

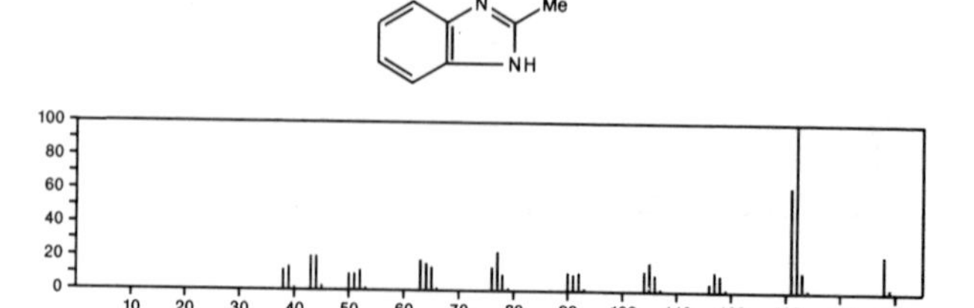

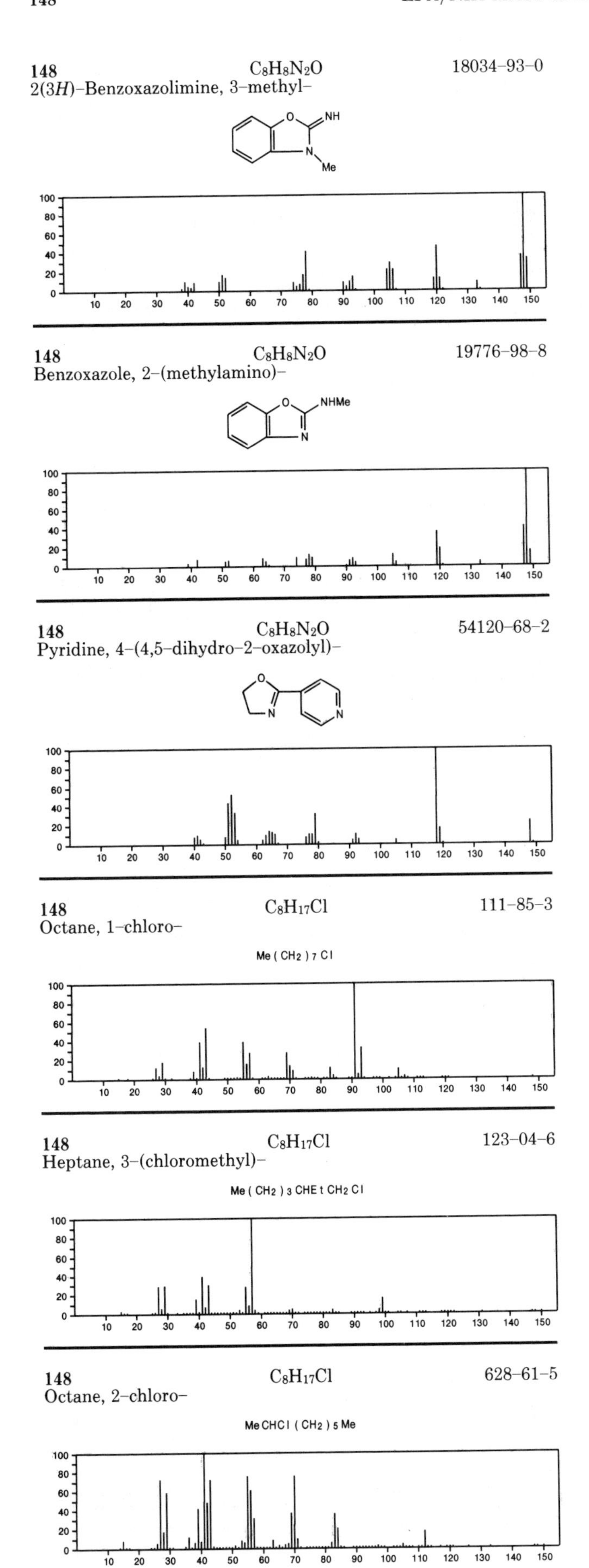
148 $C_8H_8N_2O$ 18034-93-0
2(3*H*)-Benzoxazolimine, 3-methyl-
148 $C_8H_8N_2O$ 19776-98-8
Benzoxazole, 2-(methylamino)-
NHMe
148 $C_8H_8N_2O$ 54120-68-2
Pyridine, 4-(4,5-dihydro-2-oxazolyl)-
148 $C_8H_{17}Cl$ 111-85-3
Octane, 1-chloro-
Me (CH2) 7 Cl
148 $C_8H_{17}Cl$ 123-04-6
Heptane, 3-(chloromethyl)-
Me (CH2) 3 CHEt CH2 Cl
148 $C_8H_{17}Cl$ 628-61-5
Octane, 2-chloro-
Me CHCl (CH2) 5 Me

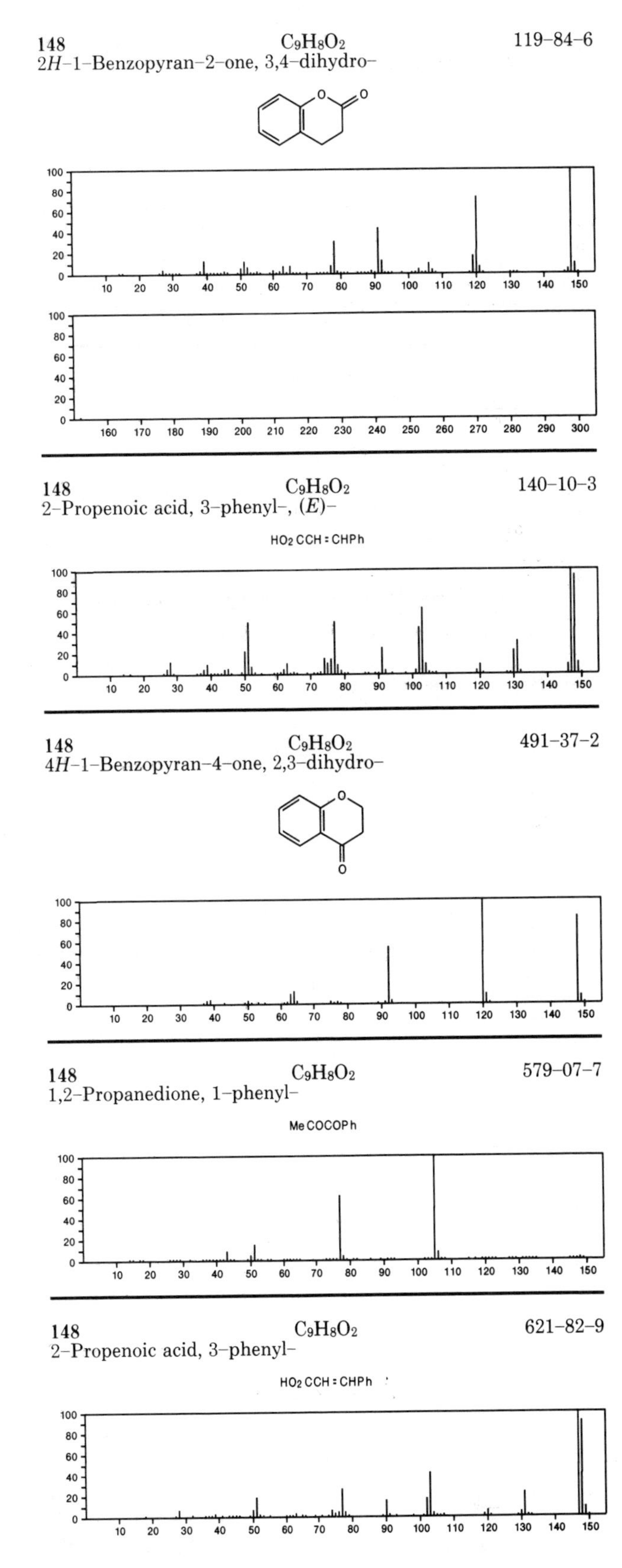
148 $C_9H_8O_2$ 119-84-6
2*H*-1-Benzopyran-2-one, 3,4-dihydro-
148 $C_9H_8O_2$ 140-10-3
2-Propenoic acid, 3-phenyl-, (*E*)-
HO2CCH = CHPh
148 $C_9H_8O_2$ 491-37-2
4*H*-1-Benzopyran-4-one, 2,3-dihydro-
148 $C_9H_8O_2$ 579-07-7
1,2-Propanedione, 1-phenyl-
MeCOCOPh
148 $C_9H_8O_2$ 621-82-9
2-Propenoic acid, 3-phenyl-
HO2CCH = CHPh

148 $C_9H_8O_2$ 669-04-5
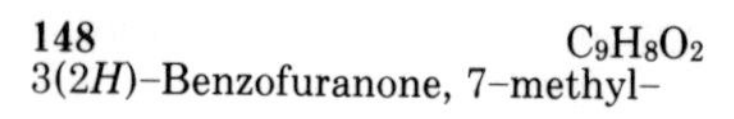
3(2*H*)-Benzofuranone, 7-methyl-

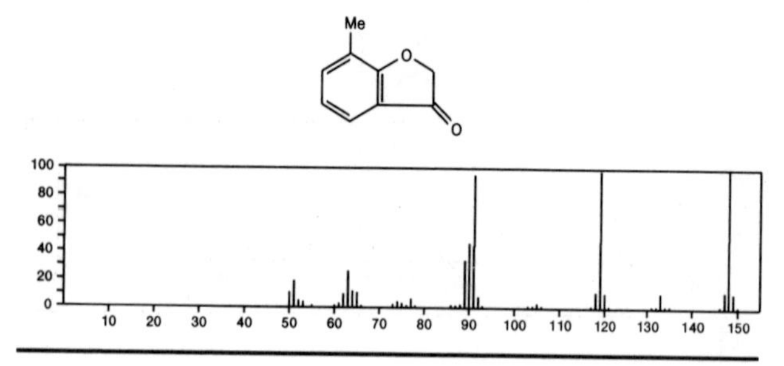

148 $C_9H_8O_2$ 1197-40-6
Furan, 2,2'-methylenebis-

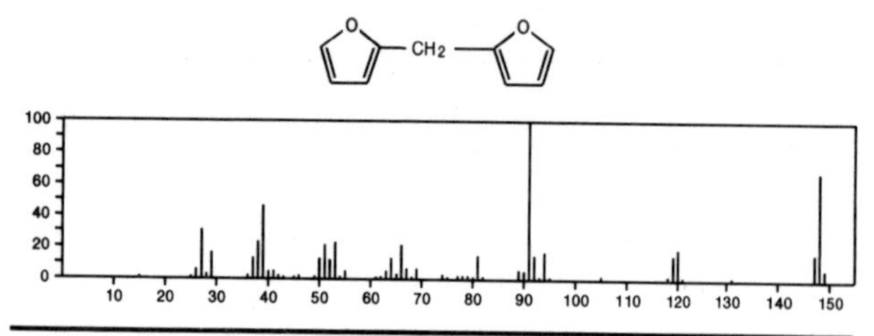

148 $C_9H_8O_2$ 20895-41-4
3(2*H*)-Benzofuranone, 6-methyl-

148 $C_9H_8O_2$ 27587-17-3
1,4-Benzenedicarboxaldehyde, 2-methyl-

148 $C_9H_8O_2$ 32267-71-3
2(3*H*)-Benzofuranone, 3-methyl-

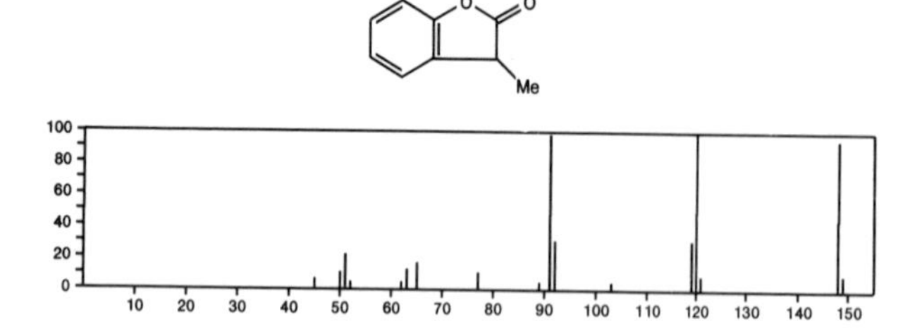

148 $C_9H_8O_2$ 35567-59-0
3(2*H*)-Benzofuranone, 2-methyl-

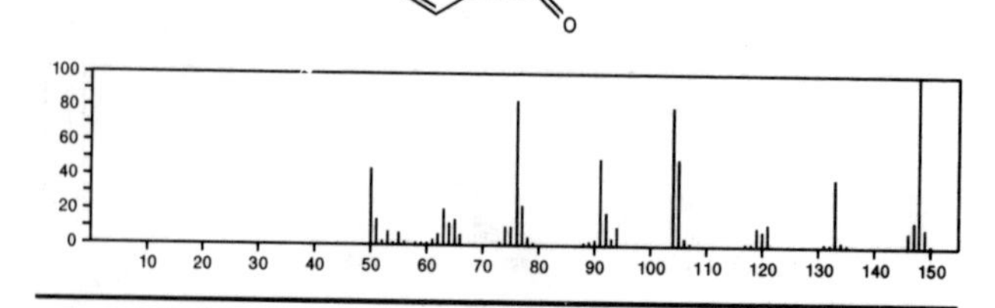

148 $C_9H_8O_2$ 54120-64-8
1(3*H*)-Isobenzofuranone, 5-methyl-

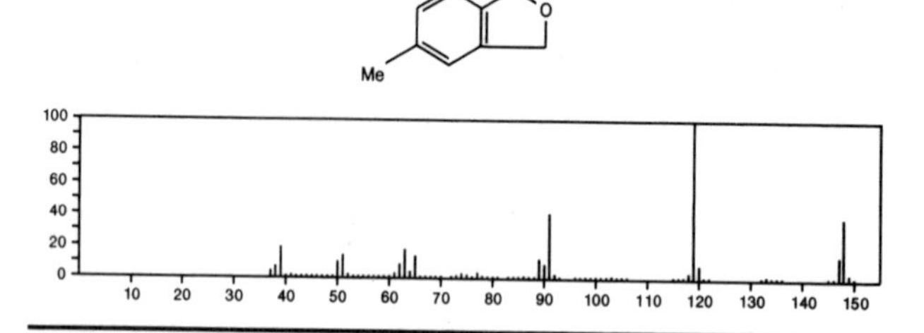

148 $C_9H_8O_2$ 54120-65-9
3(2*H*)-Benzofuranone, 4-methyl-

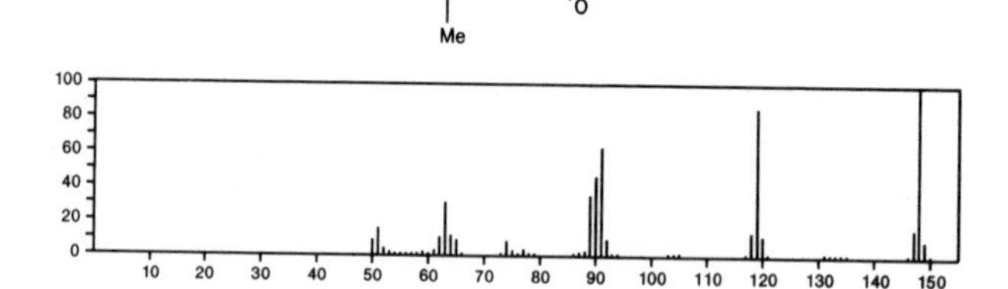

148 $C_9H_8O_2$ 54120-66-0
3(2*H*)-Benzofuranone, 5-methyl-

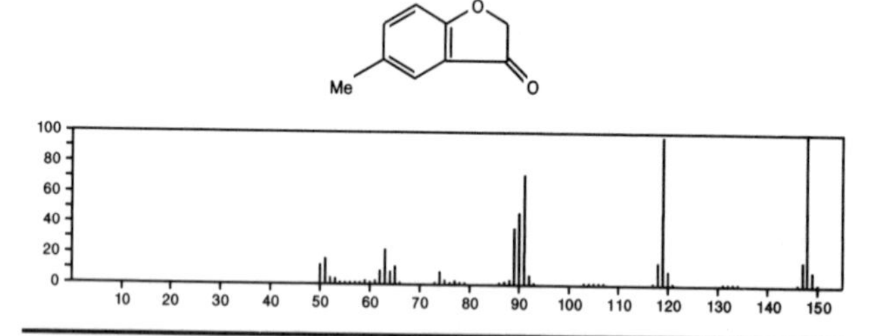

148 C_9H_8S 14315-11-8
Benzo[*b*]thiophene, 4-methyl-

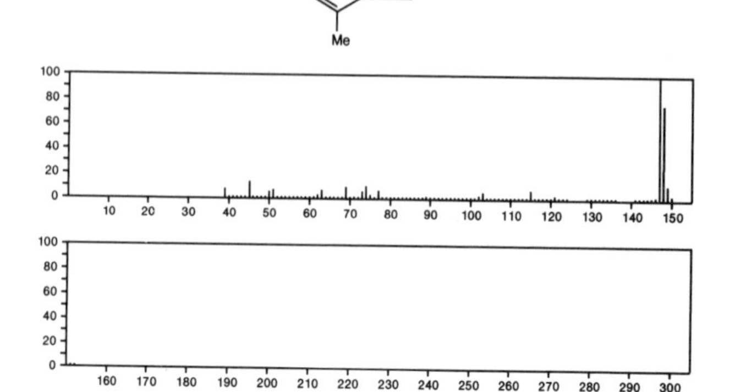

148 C_9H_8S 14315–14–1
Benzo[*b*]thiophene, 5–methyl–

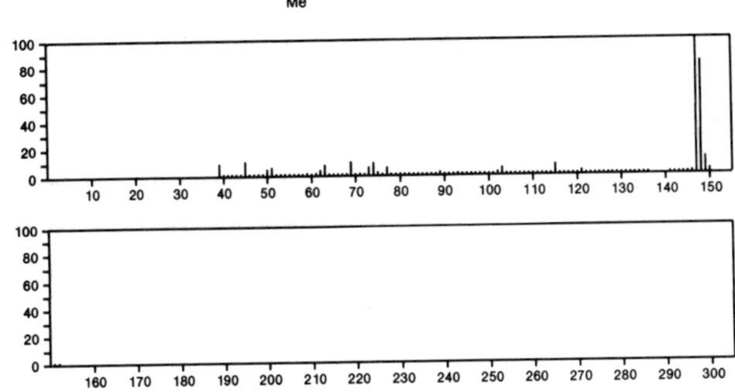

148 C_9H_8S 16587–47–6
Benzo[*b*]thiophene, 6–methyl–

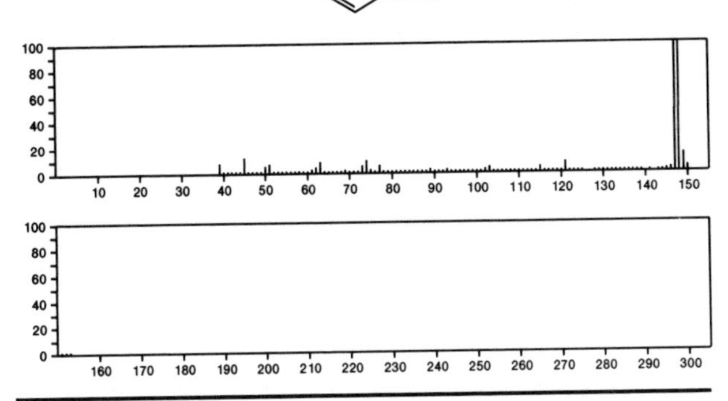

148 $C_9H_{12}N_2$ 494–97–3
Pyridine, 3–(2–pyrrolidinyl)–, (*S*)–

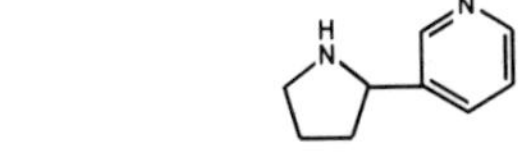

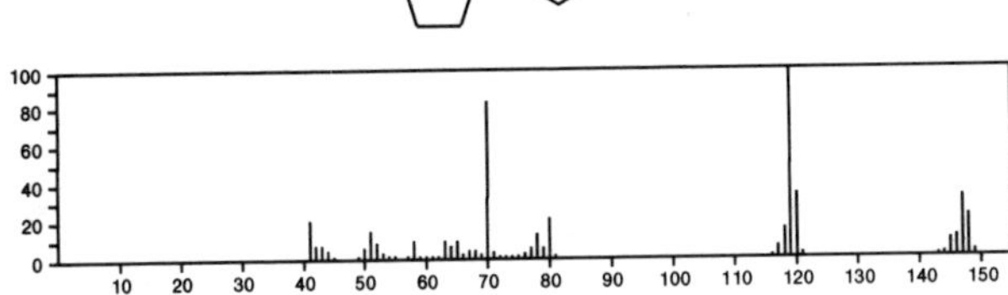

148 $C_9H_{12}N_2$ 1783–25–1
Methanimidamide, *N*,*N*–dimethyl–*N*'–phenyl–

Me2NCH=NPh

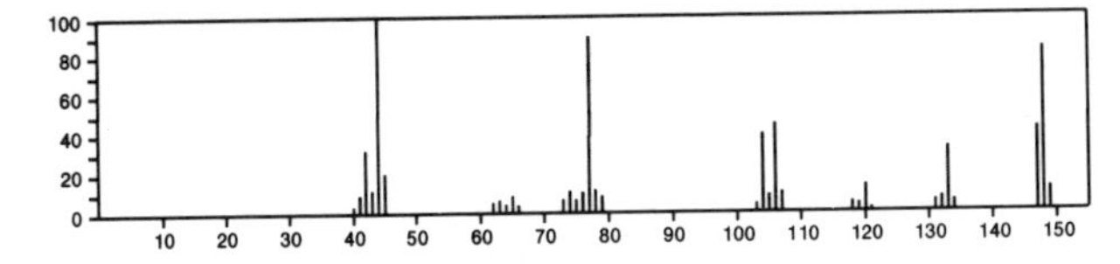

148 $C_9H_{12}N_2$ 16738–88–8
1,1–Cyclopropanedicarbonitrile, 2–methyl–2–propyl–

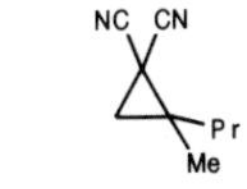

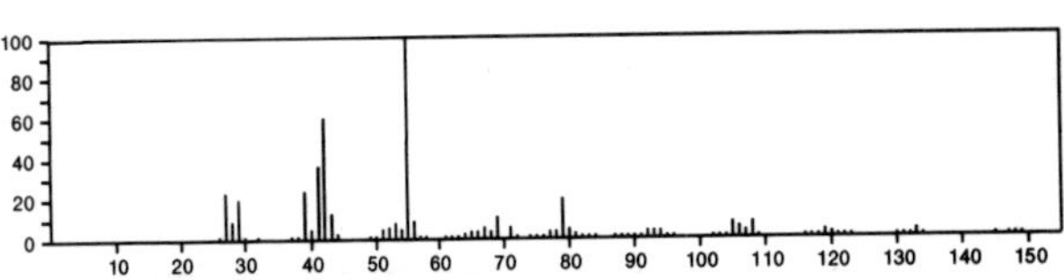

148 $C_{10}H_{12}O$ 89–74–7
Ethanone, 1–(2,4–dimethylphenyl)–

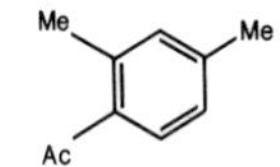

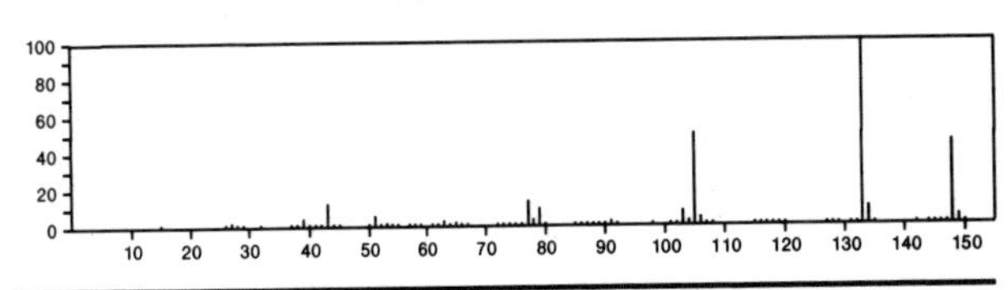

148 $C_{10}H_{12}O$ 104–46–1
Benzene, 1–methoxy–4–(1–propenyl)–

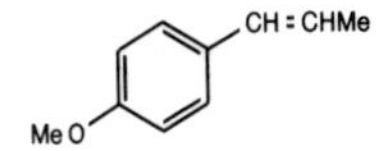

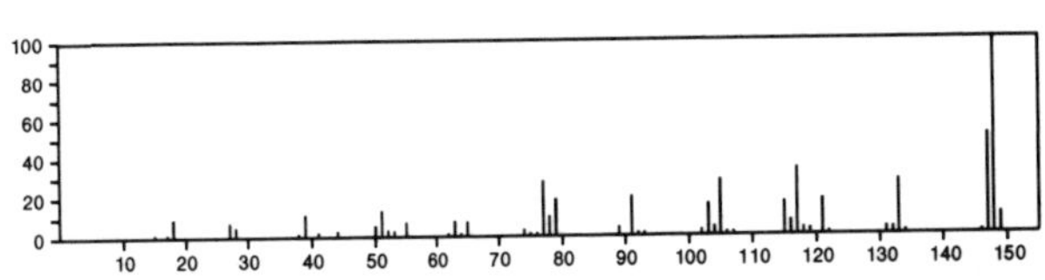

148 $C_{10}H_{12}O$ 122–03–2
Benzaldehyde, 4–(1–methylethyl)–

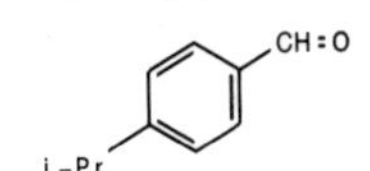

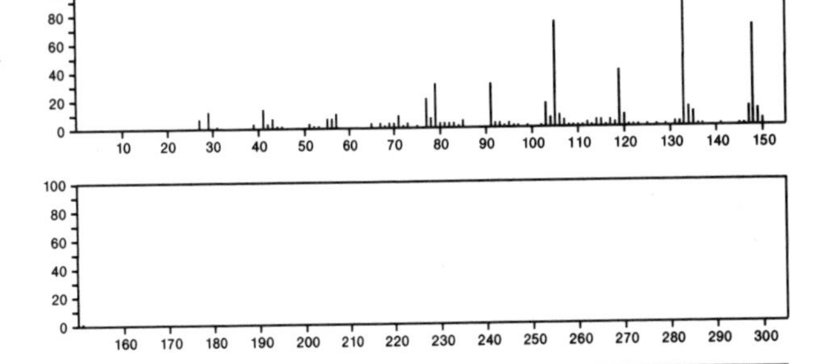

148 $C_{10}H_{12}O$ 140–67–0
Benzene, 1–methoxy–4–(2–propenyl)–

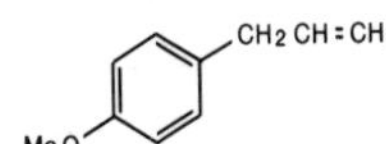

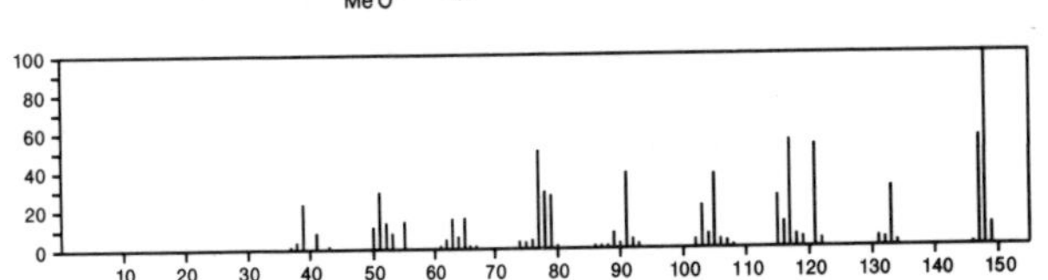

148 $C_{10}H_{12}O$ 487-68-3
Benzaldehyde, 2,4,6-trimethyl-

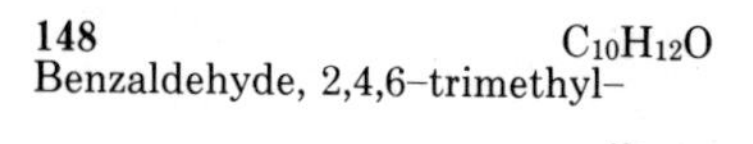

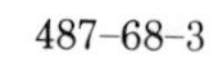

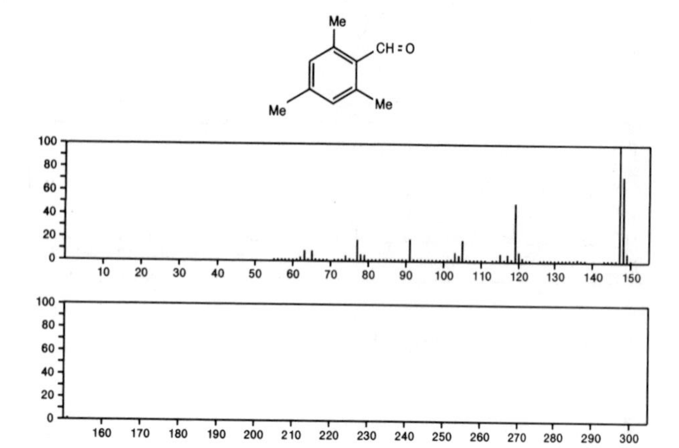

148 $C_{10}H_{12}O$ 495-40-9
1-Butanone, 1-phenyl-

PrCOPh

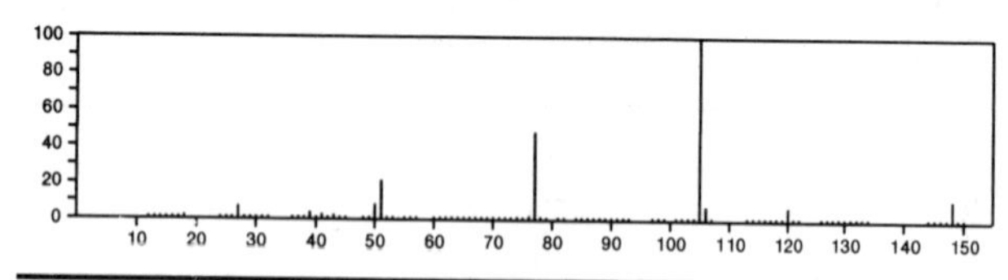

148 $C_{10}H_{12}O$ 529-33-9
1-Naphthalenol, 1,2,3,4-tetrahydro-

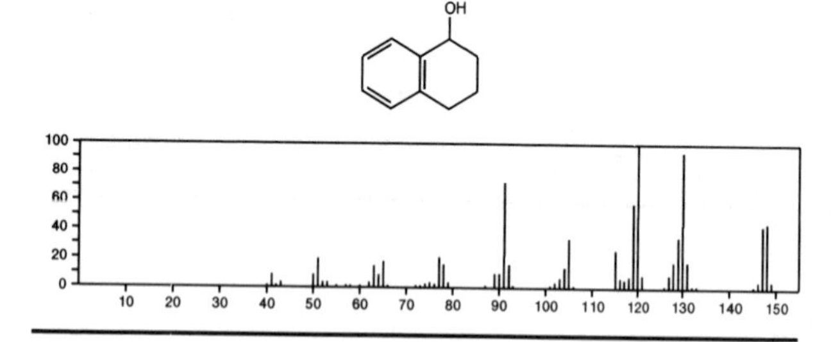

148 $C_{10}H_{12}O$ 529-35-1
1-Naphthalenol, 5,6,7,8-tetrahydro-

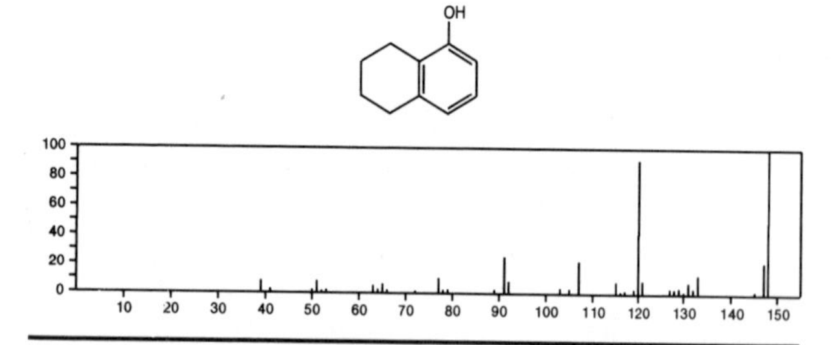

148 $C_{10}H_{12}O$ 611-70-1
1-Propanone, 2-methyl-1-phenyl-

$Me_2CHCOPh$

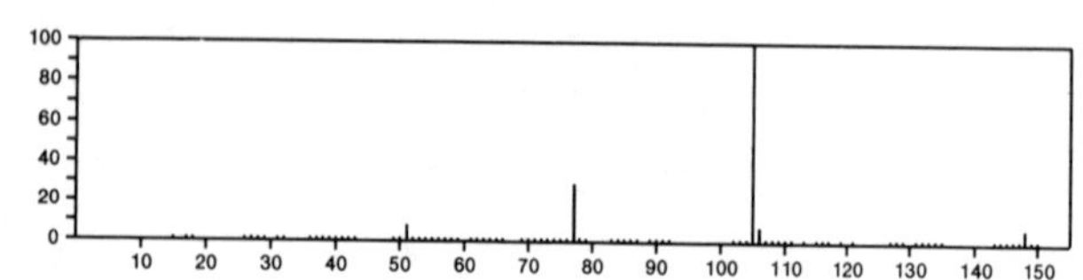

148 $C_{10}H_{12}O$ 937-30-4
Ethanone, 1-(4-ethylphenyl)-

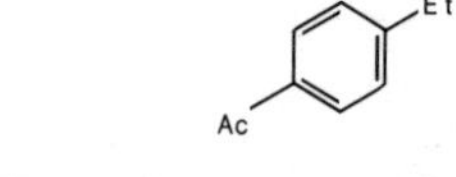

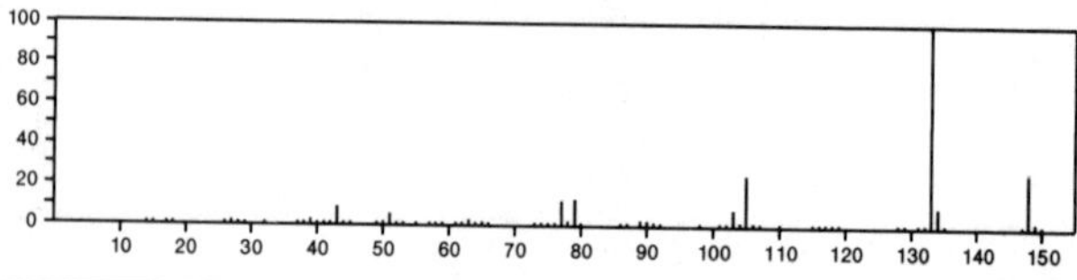

148 $C_{10}H_{12}O$ 1125-78-6
2-Naphthalenol, 5,6,7,8-tetrahydro-

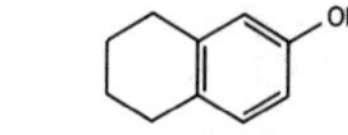

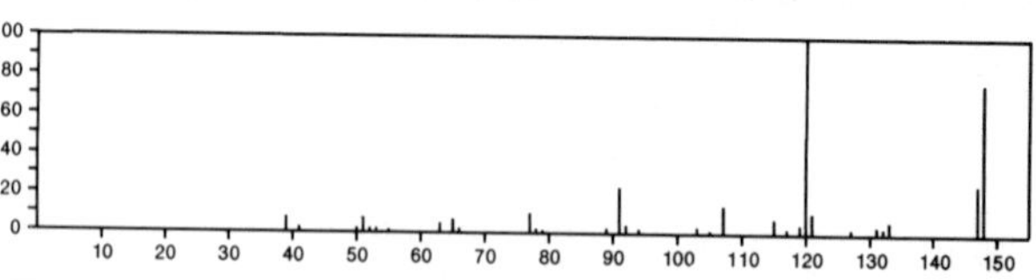

148 $C_{10}H_{12}O$ 5779-72-6
Benzaldehyde, 2,4,5-trimethyl-

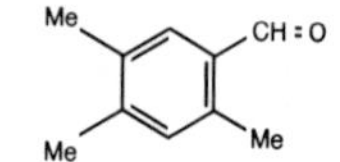

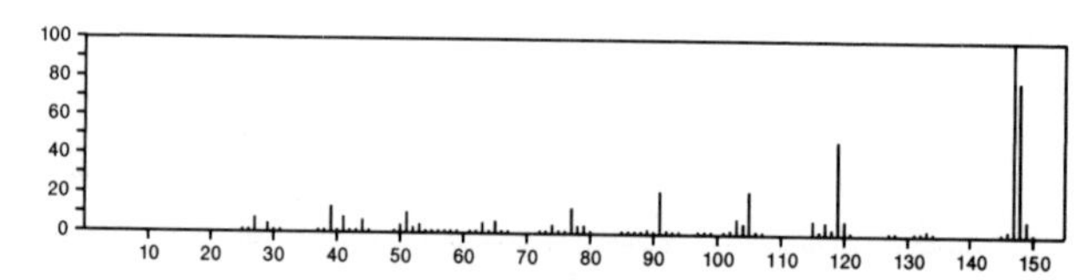

148 $C_{10}H_{12}O$ 5820-22-4
Benzene, [(2-methyl-2-propenyl)oxy]-

$PhOCH_2CMe=CH_2$

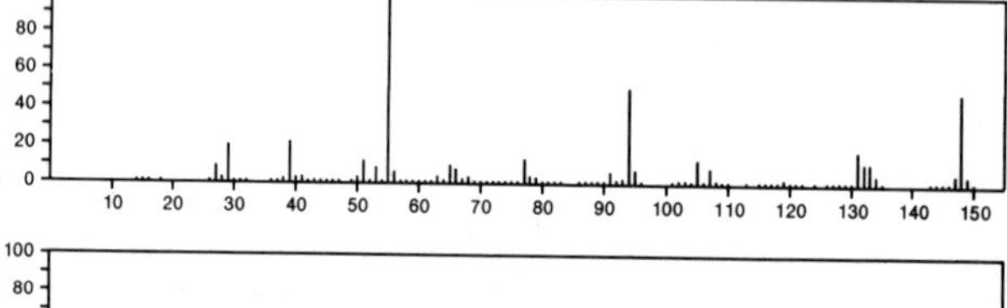

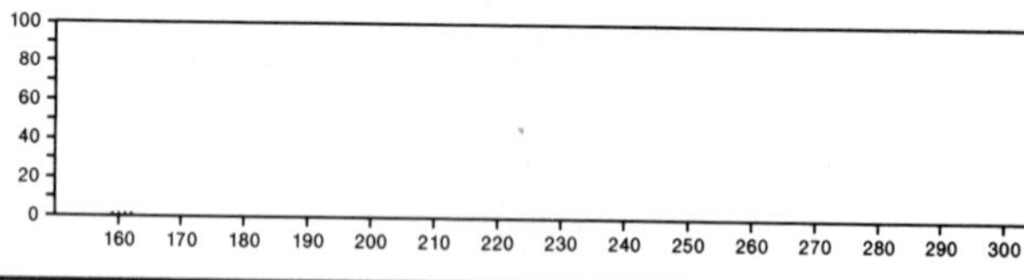

148 $C_{10}H_{12}O$ 6169-78-4
1-Benzoxepin, 2,3,4,5-tetrahydro-

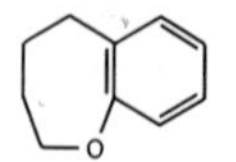

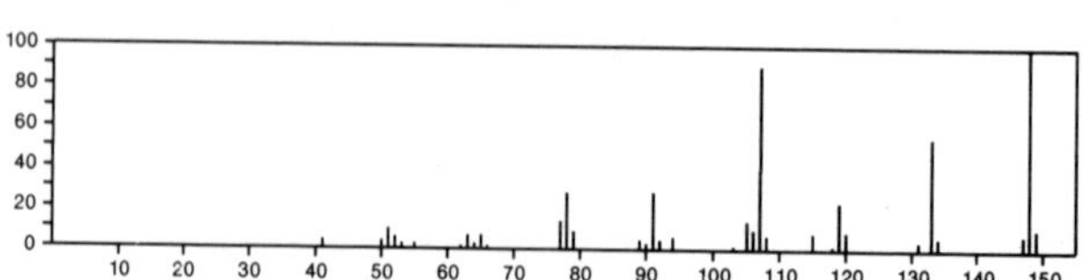

148 $C_{10}H_{12}O$ 13030-26-7
2*H*-1-Benzopyran, 3,4-dihydro-2-methyl-

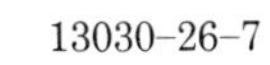

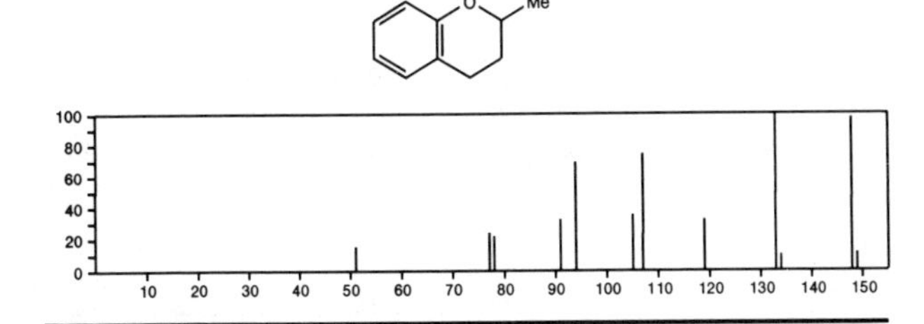

---

148 $C_{10}H_{12}O$ 13351-15-0
1,2,4-Metheno-1*H*-cyclobuta[*cd*]pentalen-3-ol, octahydro-, (1α,1aβ,2α,3α,3aβ,4α,4aβ,5bβ,6*S**)-

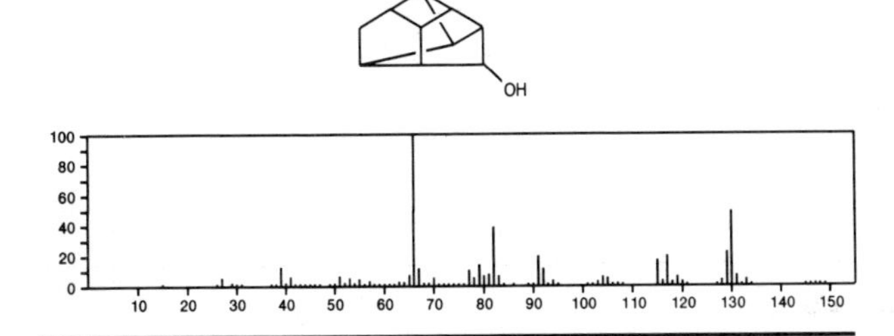

---

148 $C_{10}H_{12}O$ 13656-81-0
2,4,6-Cycloheptatrien-1-one, 4-(1-methylethyl)-

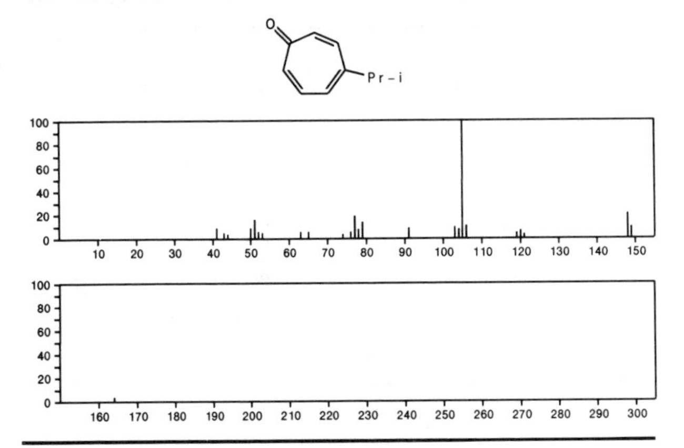

---

148 $C_{10}H_{12}O$ 14593-43-2
Benzene, [(2-propenyloxy)methyl]-

$H_2C=CHCH_2OCH_2Ph$

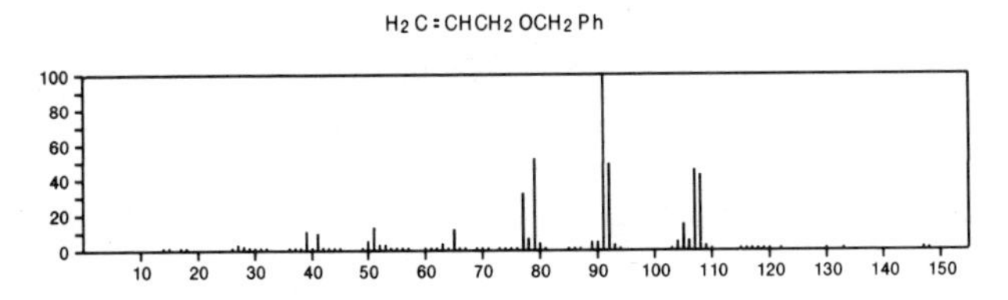

---

148 $C_{10}H_{12}O$ 15776-05-3
1,2,4-Metheno-1*H*-cyclobuta[*cd*]pentalen-3-ol, octahydro-, (1α,1aβ,2α,3β,3aβ,4α,5aβ,5bβ,6*S**)-

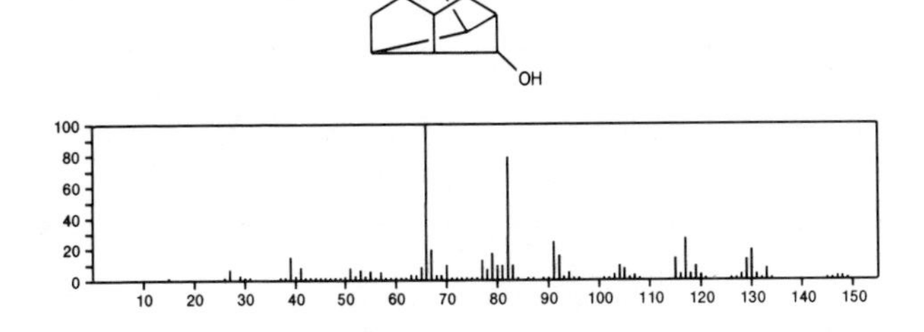

---

148 $C_{10}H_{12}O$ 16277-67-1
Benzene, (3-methoxy-1-propenyl)-

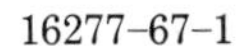

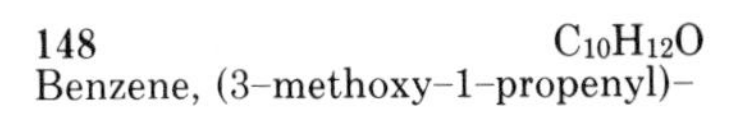

$MeOCH_2CH=CHPh$

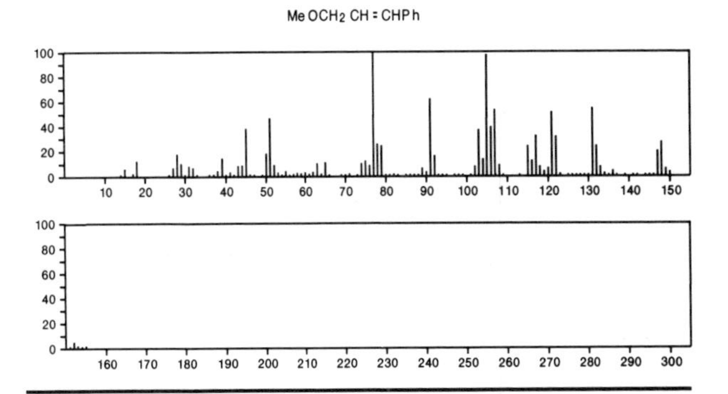

---

148 $C_{10}H_{12}O$ 18328-11-5
Benzenebutanal

$Ph(CH_2)_3CHO$

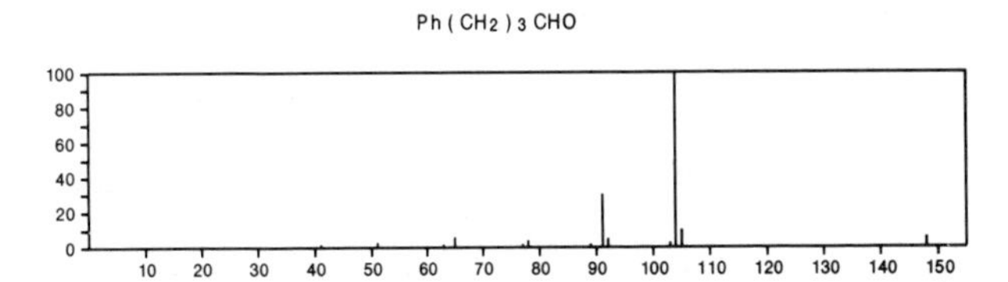

---

148 $C_{10}H_{12}O$ 20944-88-1
Phenol, 2-(2-methyl-2-propenyl)-

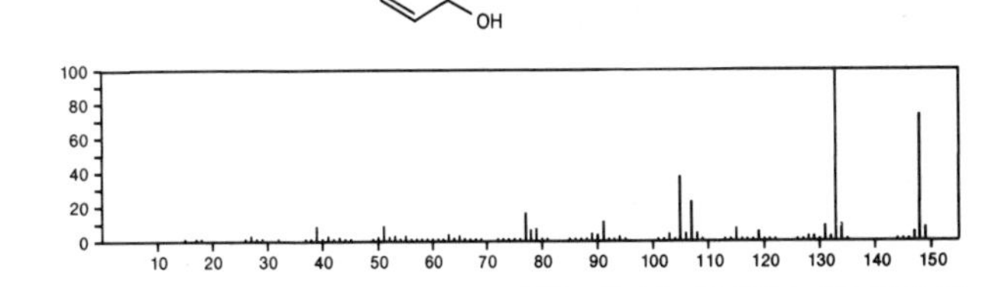

---

148 $C_{10}H_{12}O$ 30844-13-4
1,3,5,7-Cyclooctatetraene, 1-(methoxymethyl)-

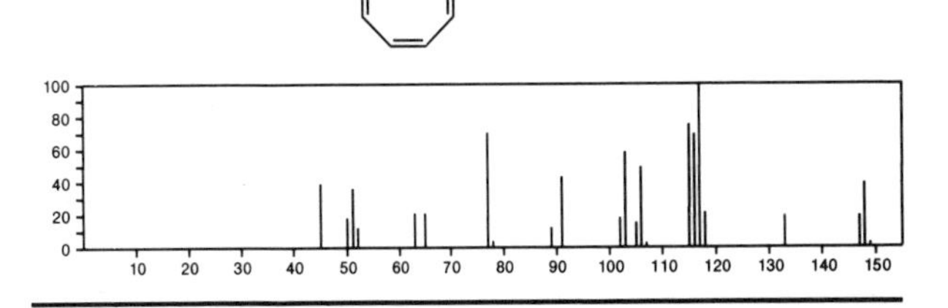

---

148 $C_{10}H_{12}O$ 33641-78-0
Phenol, *p*-(2-methylallyl)-

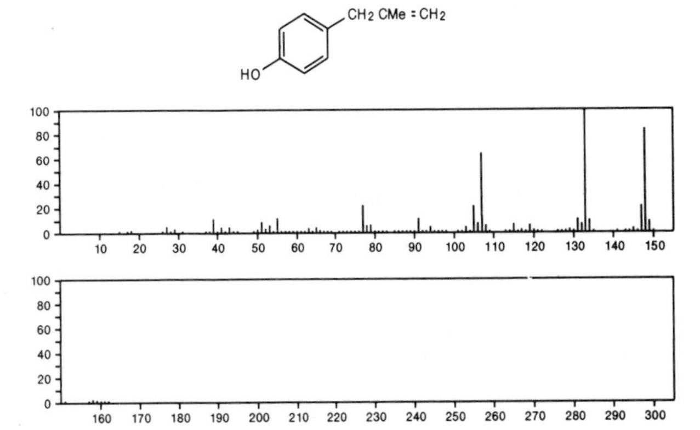

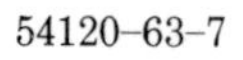

148 $C_{10}H_{12}O$ 54120-63-7
Benzene, ethenyl(methoxymethyl)-

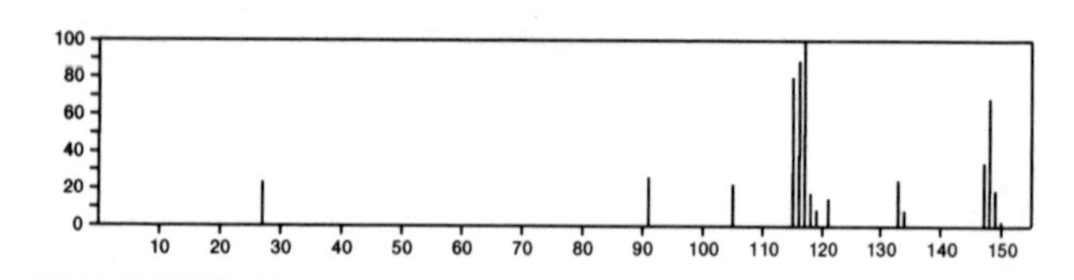

148 $C_{11}H_{16}$ 98-51-1
Benzene, 1-(1,1-dimethylethyl)-4-methyl-

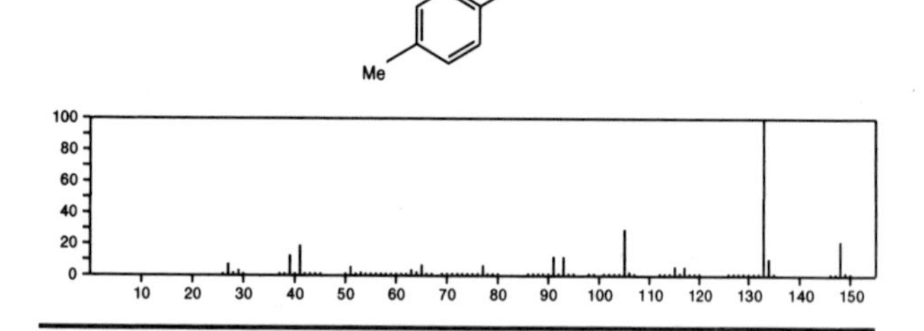

148 $C_{11}H_{16}$ 538-68-1
Benzene, pentyl-

Ph(CH2)4Me

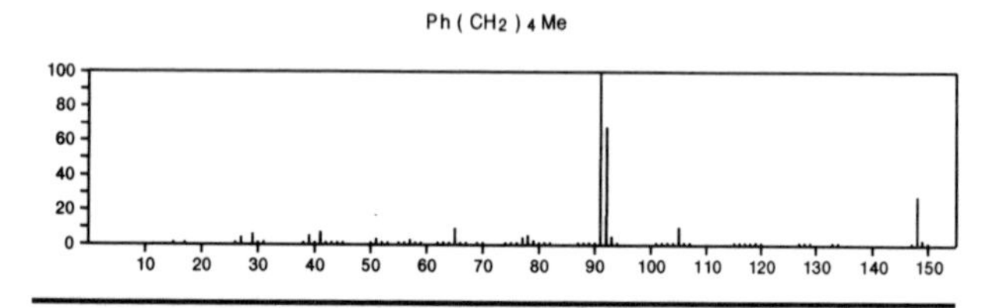

148 $C_{11}H_{16}$ 700-12-9
Benzene, pentamethyl-

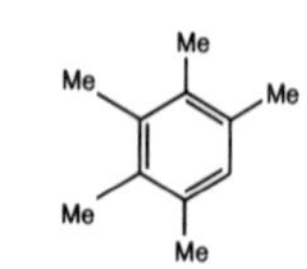

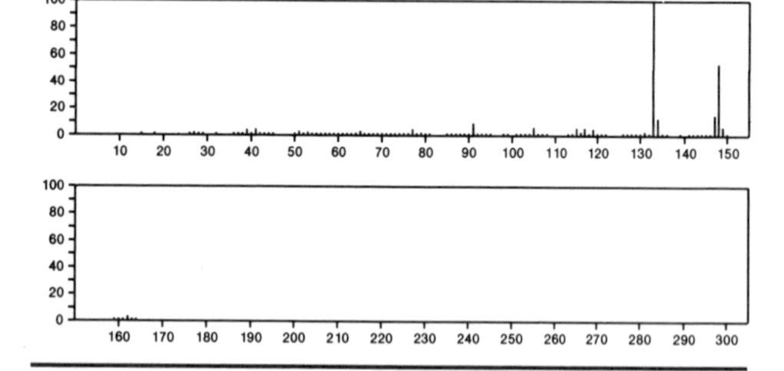

148 $C_{11}H_{16}$ 1007-26-7
Benzene, (2,2-dimethylpropyl)-

Me3CCH2Ph

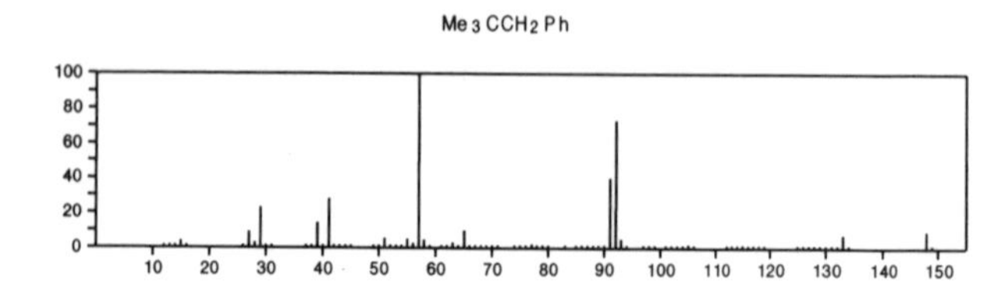

148 $C_{11}H_{16}$ 1075-38-3
Benzene, 1-(1,1-dimethylethyl)-3-methyl-

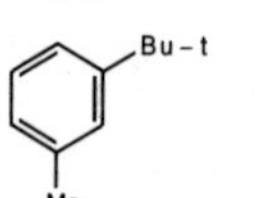

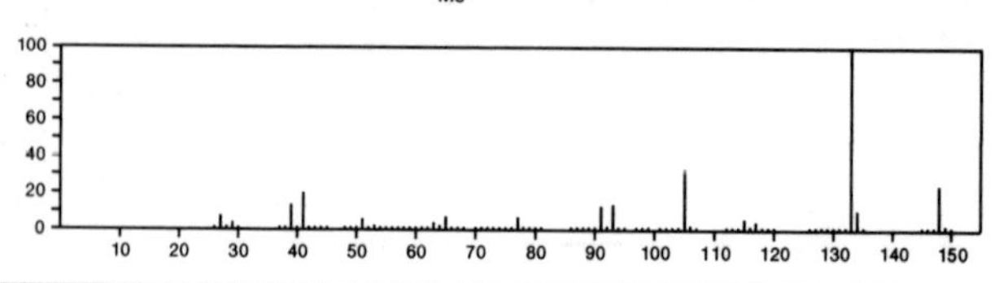

148 $C_{11}H_{16}$ 1196-58-3
Benzene, (1-ethylpropyl)-

Et2CHPh

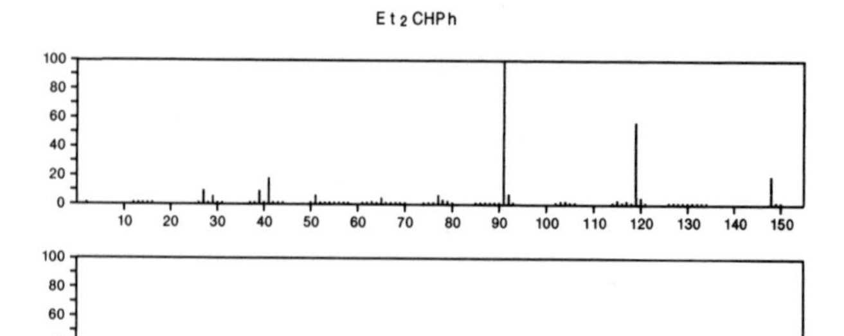

148 $C_{11}H_{16}$ 2049-94-7
Benzene, (3-methylbutyl)-

Me2CHCH2CH2Ph

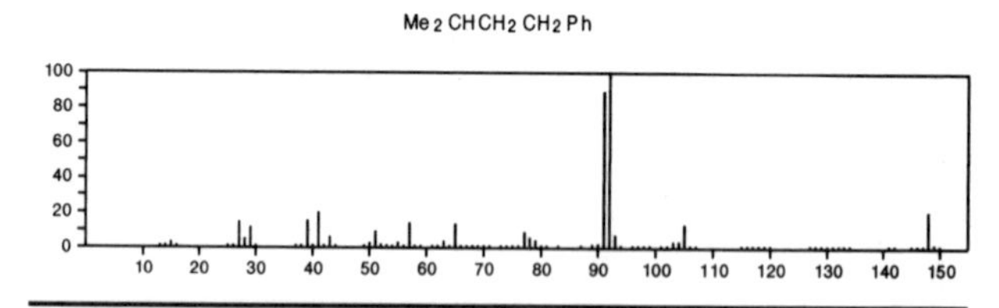

148 $C_{11}H_{16}$ 2049-95-8
Benzene, (1,1-dimethylpropyl)-

PhCEt(Ph)

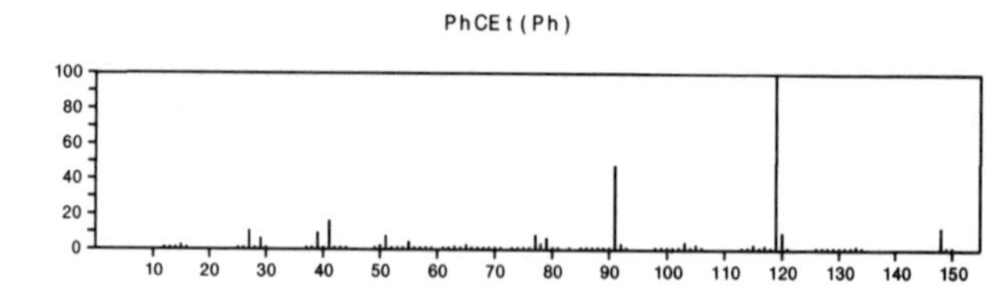

148 $C_{11}H_{16}$ 2719-52-0
Benzene, (1-methylbutyl)-

PhCHPr(Ph)

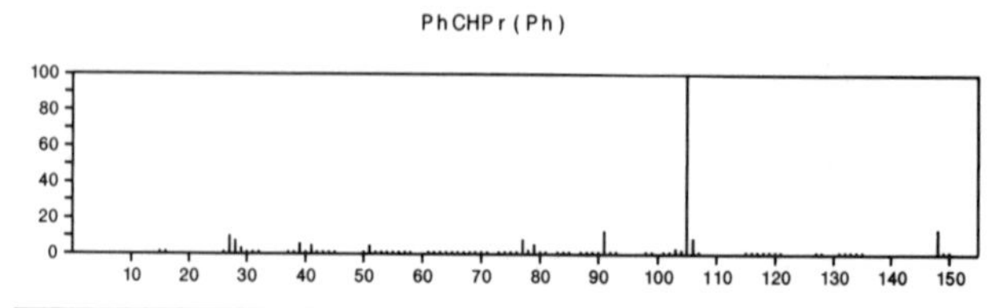

148 $C_{11}H_{16}$ 3968-85-2
Benzene, (2-methylbutyl)-

PhCH2CHMeCH2Me

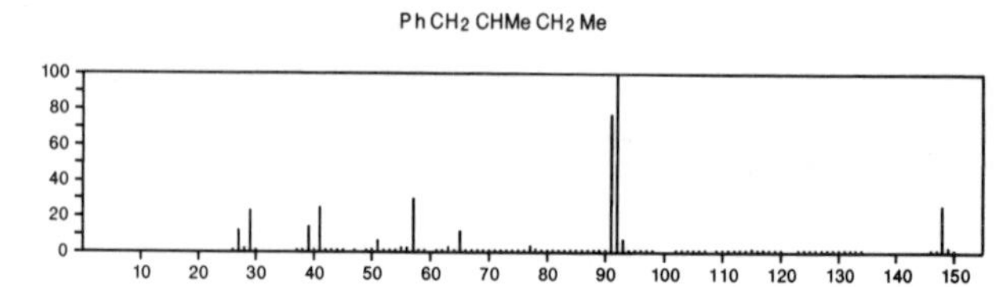

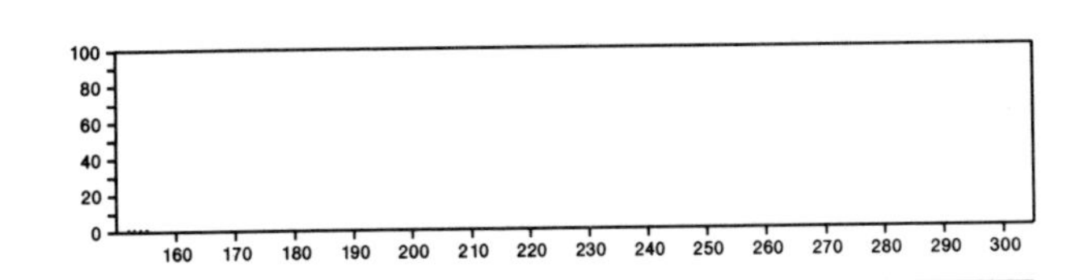

148 $C_{11}H_{16}$ 4132-72-3
Benzene, 1,4-dimethyl-2-(1-methylethyl)-

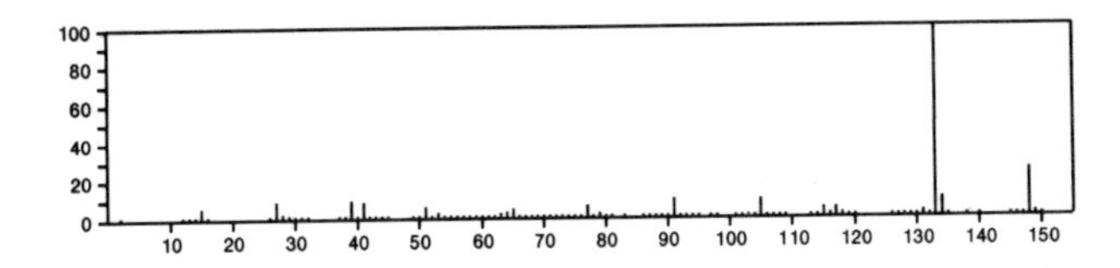

148 $C_{11}H_{16}$ 4218-48-8
Benzene, 1-ethyl-4-(1-methylethyl)-

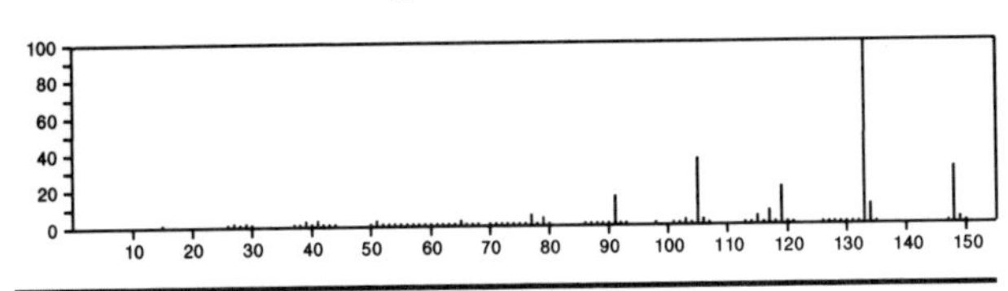

148 $C_{11}H_{16}$ 4481-30-5
Benzene, (1,2-dimethylpropyl)-

$Me_2CHCHMePh$

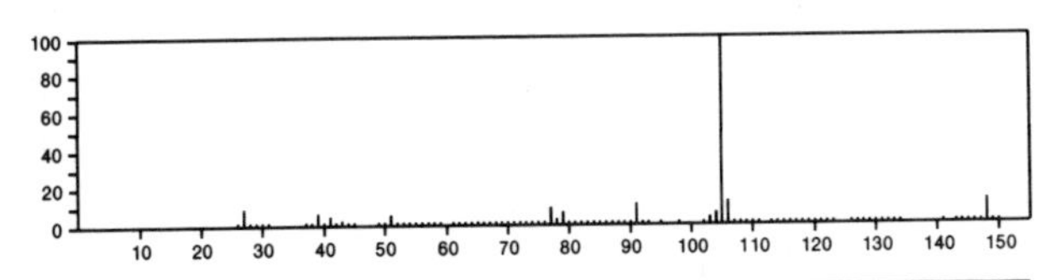

148 $C_{11}H_{16}$ 4706-89-2
Benzene, 2,4-dimethyl-1-(1-methylethyl)-

148 $C_{11}H_{16}$ 4706-90-5
Benzene, 1,3-dimethyl-5-(1-methylethyl)-

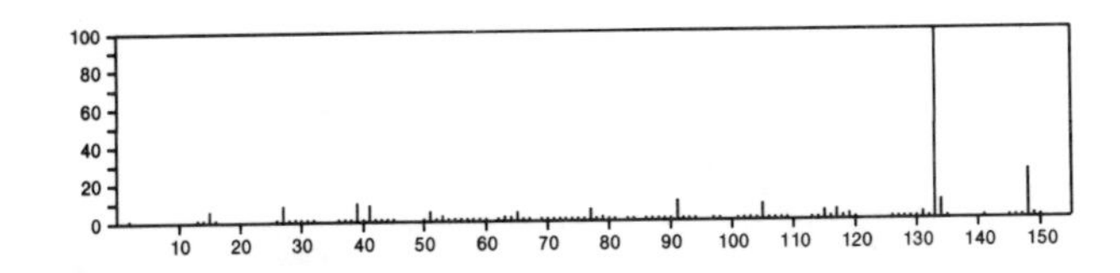

148 $C_{11}H_{16}$ 4920-99-4
Benzene, 1-ethyl-3-(1-methylethyl)-

148 $C_{11}H_{16}$ 5161-04-6
Toluene, *p*-isobutyl-

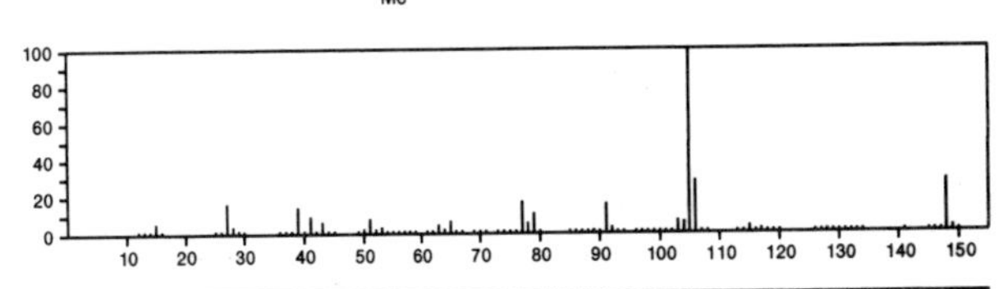

148 $C_{11}H_{16}$ 17851-27-3
Benzene, 1-ethyl-2,4,5-trimethyl-

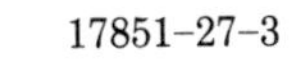

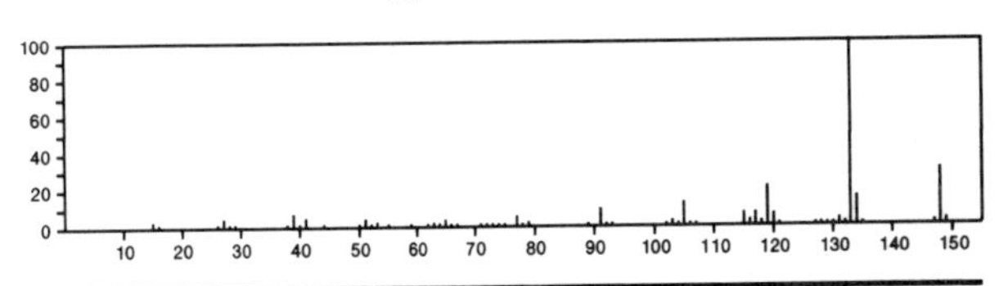

148 $C_{11}H_{16}$ 25550-13-4
Benzene, diethylmethyl-

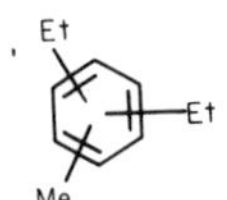

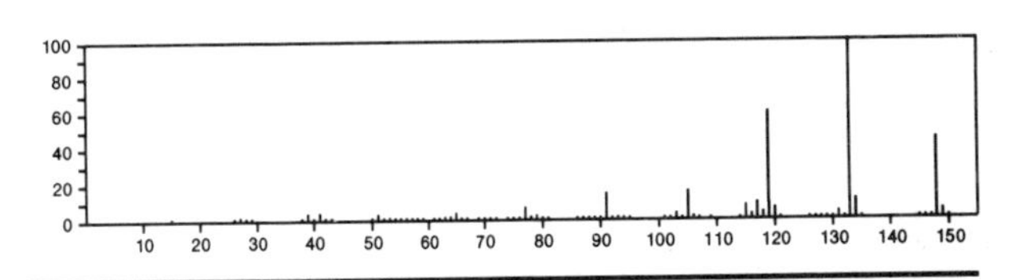

148 $C_{11}H_{16}$ 27138-21-2
Benzene, (1,1-dimethylethyl)methyl-

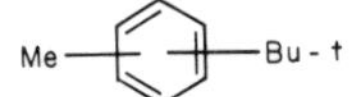

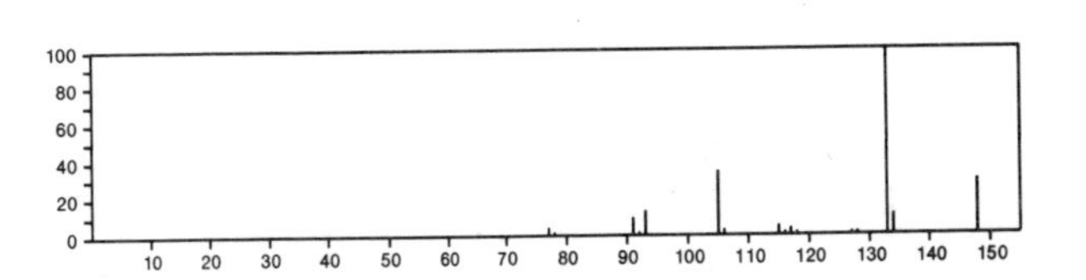

148 $C_{11}H_{16}$ 33156–92–2

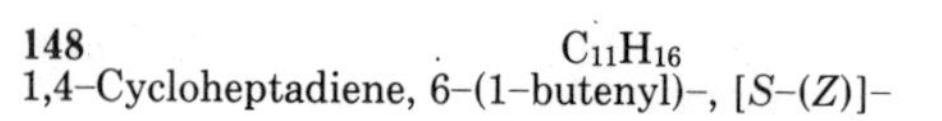
1,4–Cycloheptadiene, 6–(1–butenyl)–, [*S*–(*Z*)]–

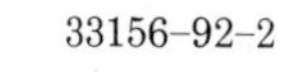

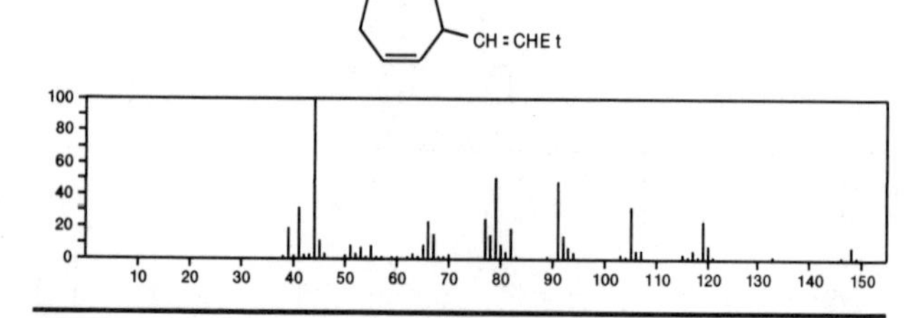

148 $C_{11}H_{16}$ 54120–62–6
Benzene, ethyl–1,2,4–trimethyl–

Me Me Et Me

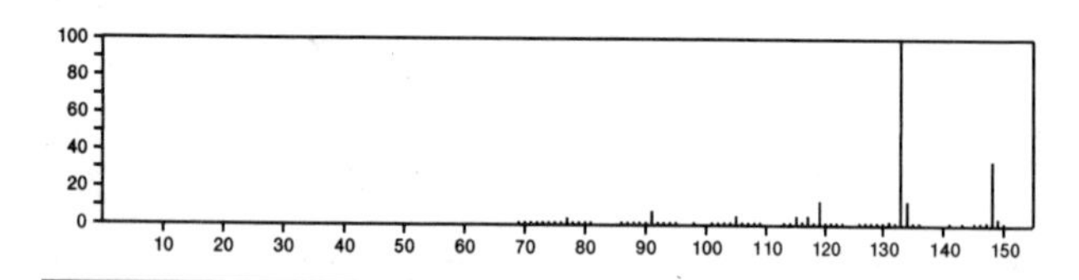

149 CF_5N_3 17224–08–7
Diaziridine, 3–(difluoroamino)trifluoro–

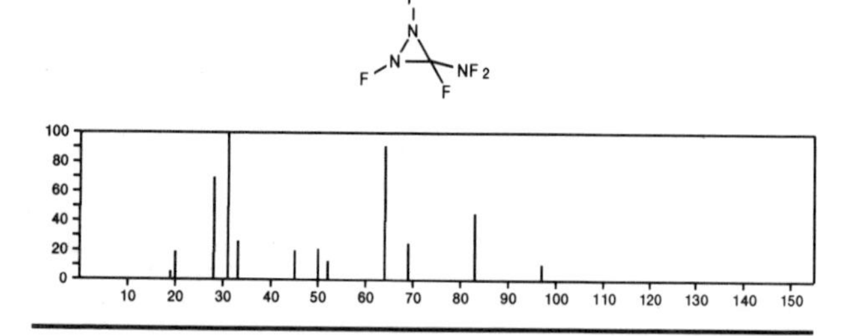

149 $C_4H_7NOS_2$ 16696–88–1
Carbamic acid, acetyldithio–, methyl ester

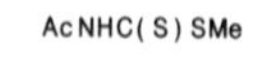

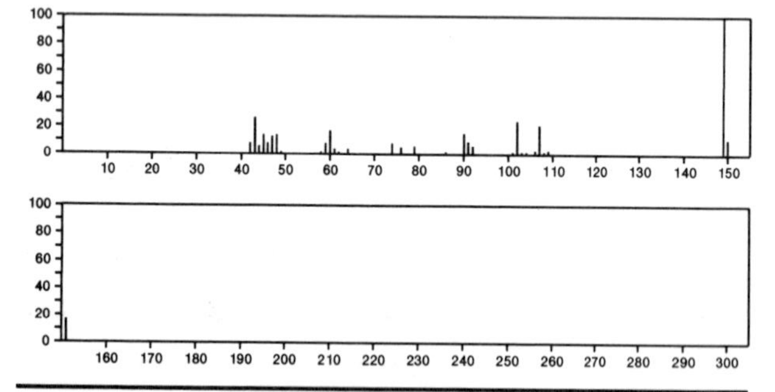

149 C_4H_8BrN 38455–32–2
Azetidine, 1–bromo–2–methyl–

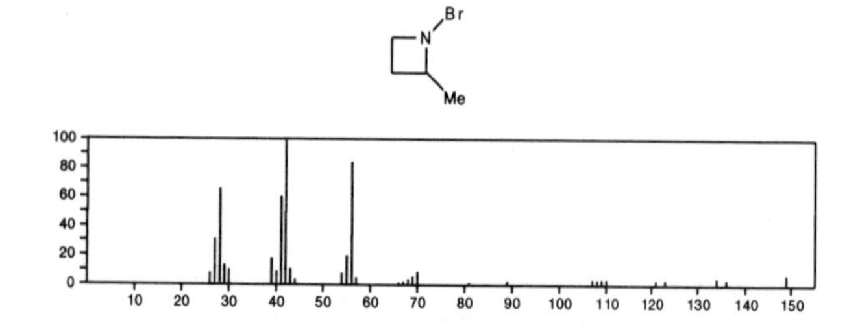

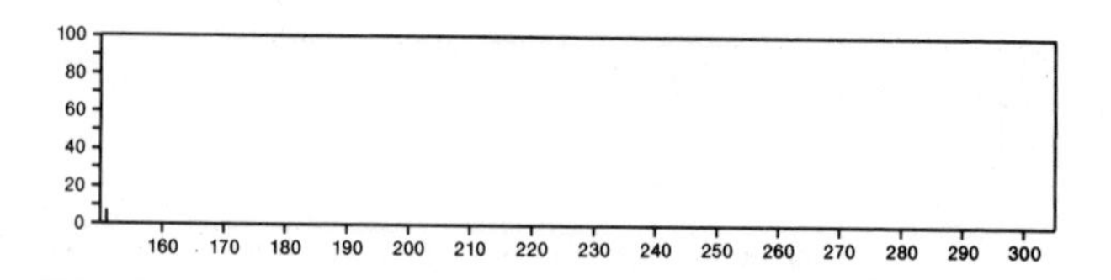

149 $C_5H_{11}NO_2S$ 63–68–3
L–Methionine

HO2CCH(NH2)CH2CH2SMe

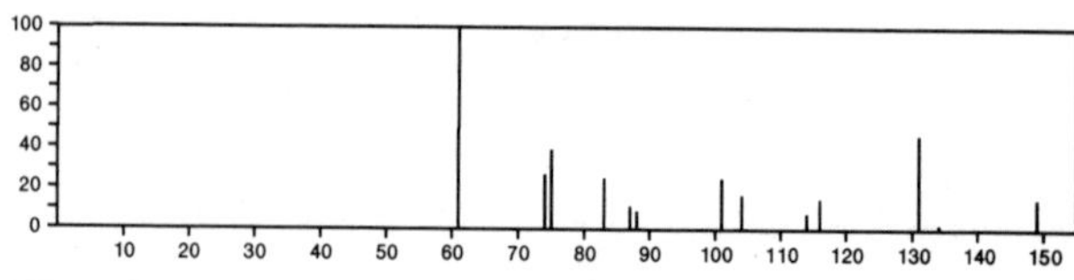

149 $C_5H_{11}NS_2$ 617–38–9
Carbamodithioic acid, dimethyl–, ethyl ester

EtSC(S)NMe2

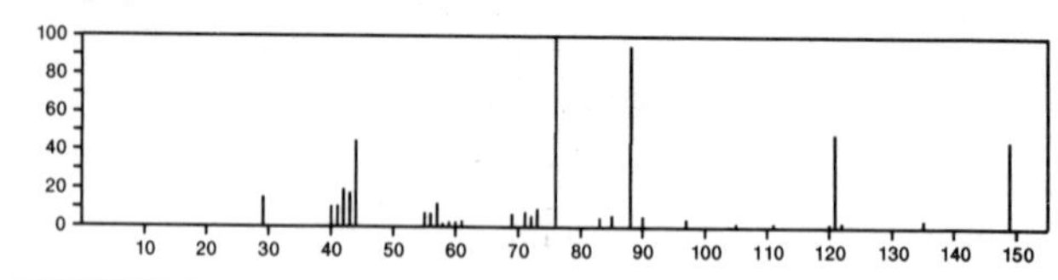

149 $C_6H_7N_5$ 443–72–1
1*H*–Purin–6–amine, *N*–methyl–

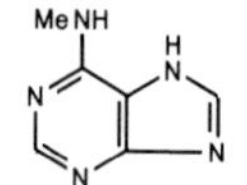

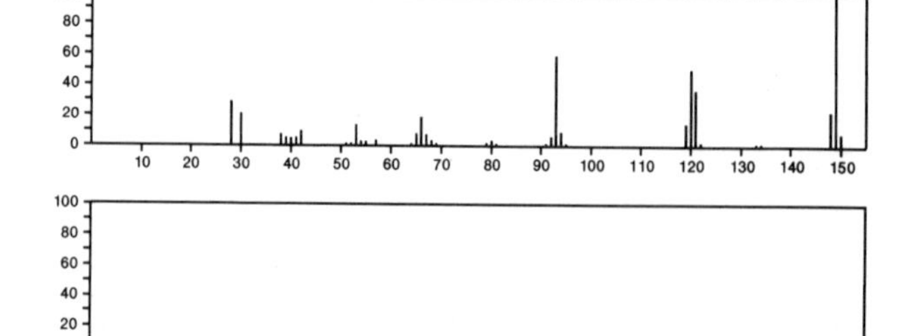

149 $C_6H_7N_5$ 935–69–3
7*H*–Purin–6–amine, 7–methyl–

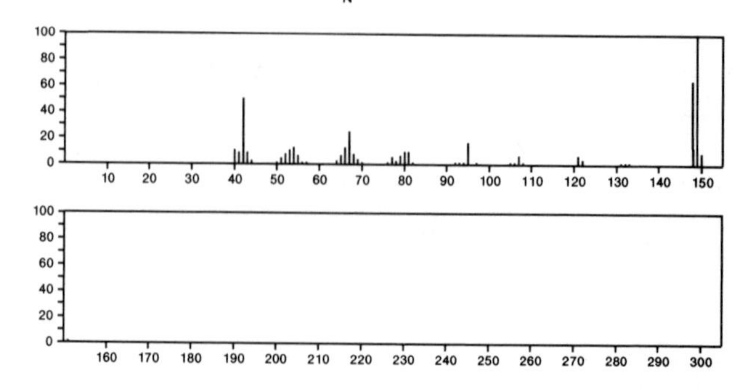

149 $C_7H_7N_3O$ 2101-87-3
Benzene, 1-azido-4-methoxy-

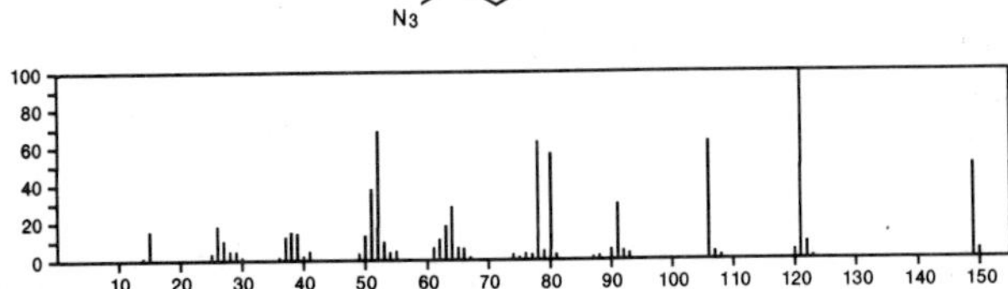

149 $C_7H_7N_3O$ 13980-64-8
1*H*-*s*-Triazolo[1,5-*a*]pyridin-4-ium, 2-hydroxy-1-methyl-, hy=droxide, inner salt

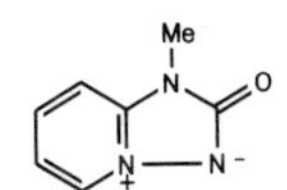

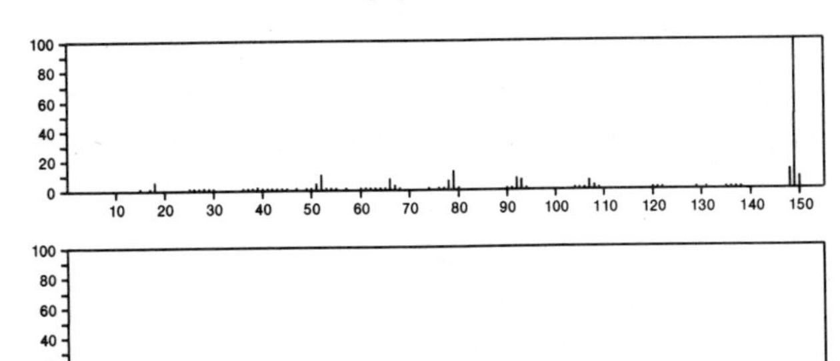

149 $C_7H_7N_3O$ 20442-97-1
Benzene, 1-azido-2-methoxy-

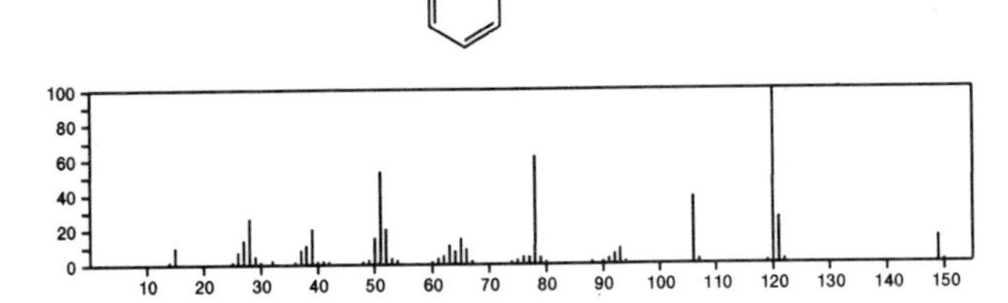

149 $C_7H_7N_3O$ 27799-90-2
1*H*-Benzotriazole, 4-methoxy-

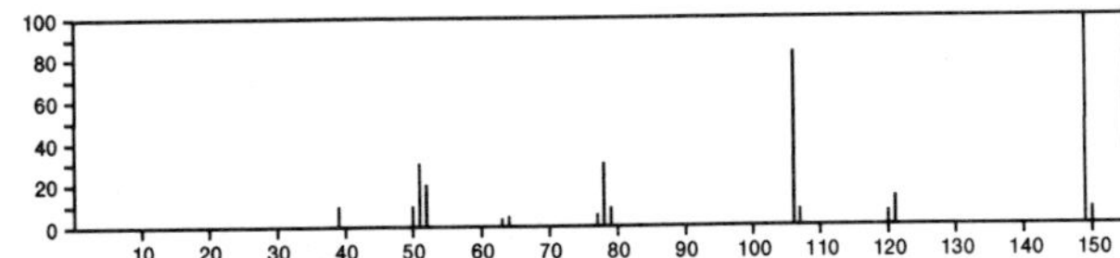

149 $C_7H_7N_3O$ 27799-91-3
1*H*-Benzotriazole, 5-methoxy-

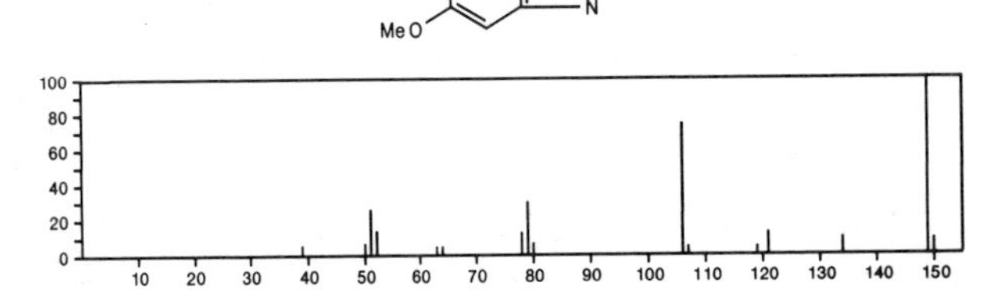

149 $C_8H_7NO_2$ 5466-88-6
2*H*-1,4-Benzoxazin-3(4*H*)-one

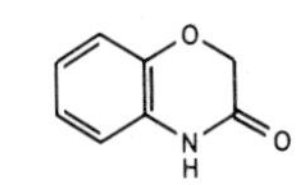

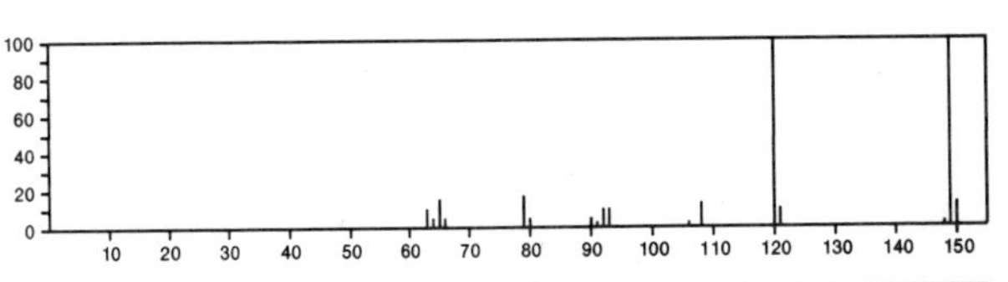

149 $C_8H_7NO_2$ 16877-22-8
1,3-Oxazetidin-2-one, 3-phenyl-

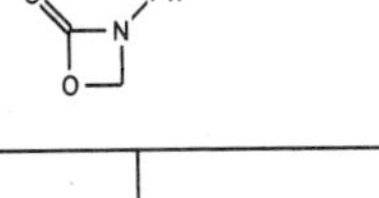

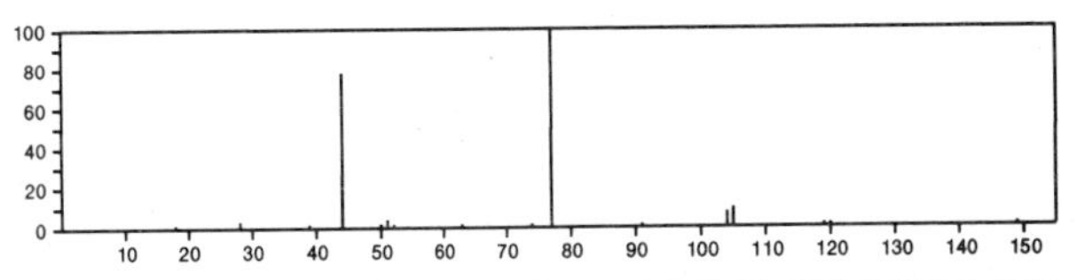

149 C_8H_7NS 120-75-2
Benzothiazole, 2-methyl-

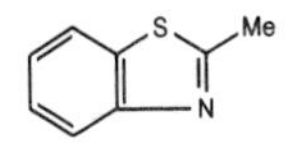

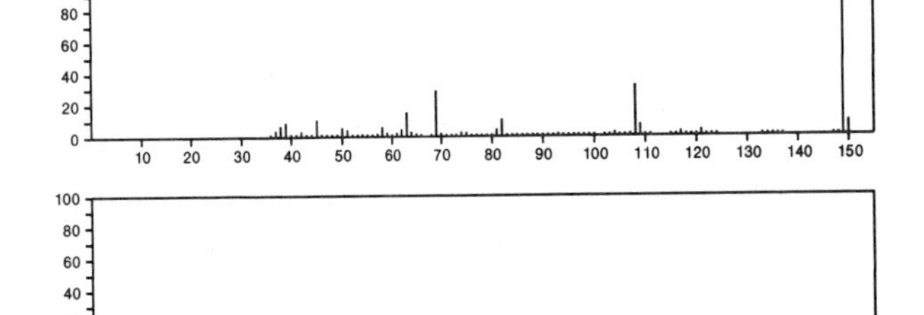

149 C_8H_7NS 6187-89-9
1,2-Benzisothiazole, 3-methyl-

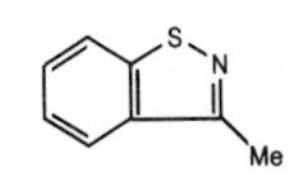

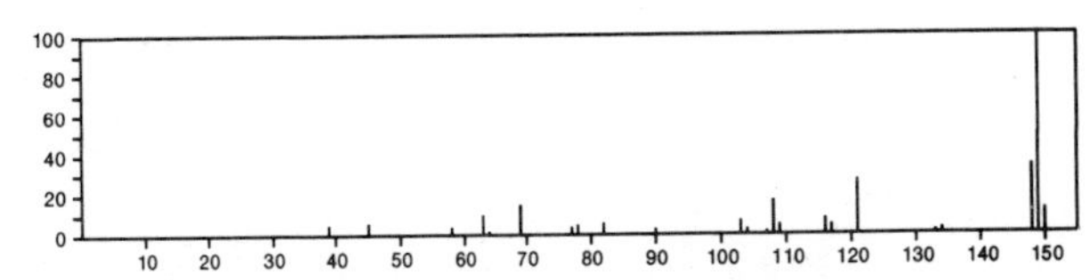

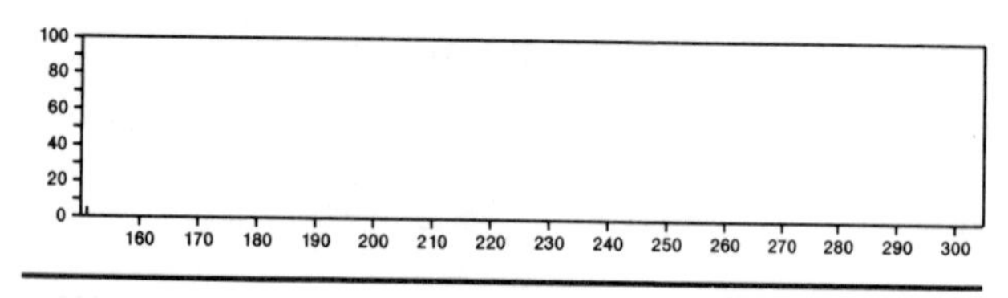

149 $C_8H_{11}N_3$ 42544-37-6
Guanidine, *N*-methyl-*N'*-phenyl-

HN=C(NHMe)NHPh

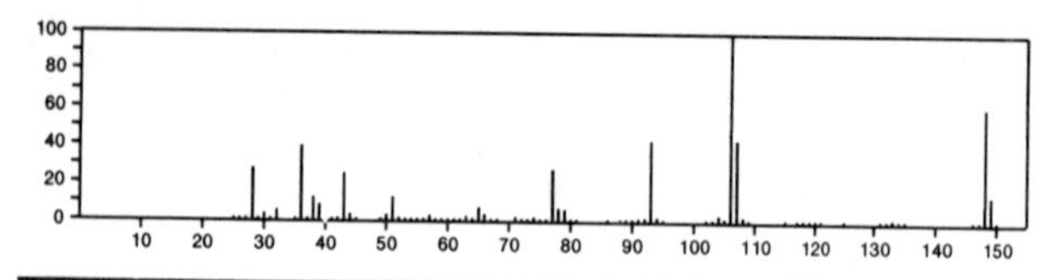

149 $C_9H_{11}NO$ 100-10-7
Benzaldehyde, 4-(dimethylamino)-

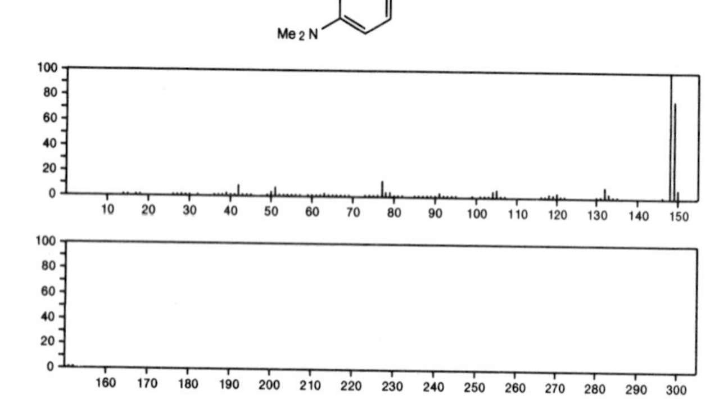

149 $C_9H_{11}NO$ 103-89-9
Acetamide, *N*-(4-methylphenyl)-

149 $C_9H_{11}NO$ 120-66-1
Acetamide, *N*-(2-methylphenyl)-

149 $C_9H_{11}NO$ 537-92-8
Acetamide, *N*-(3-methylphenyl)-

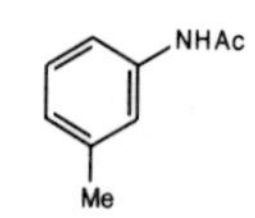

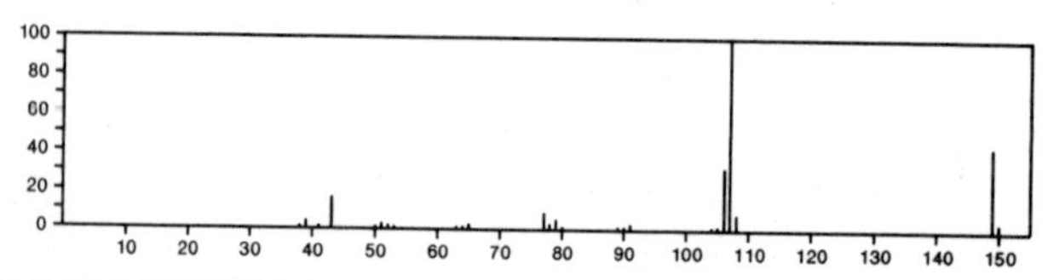

149 $C_9H_{11}NO$ 579-10-2
Acetamide, *N*-methyl-*N*-phenyl-

PhNMe(Ac)

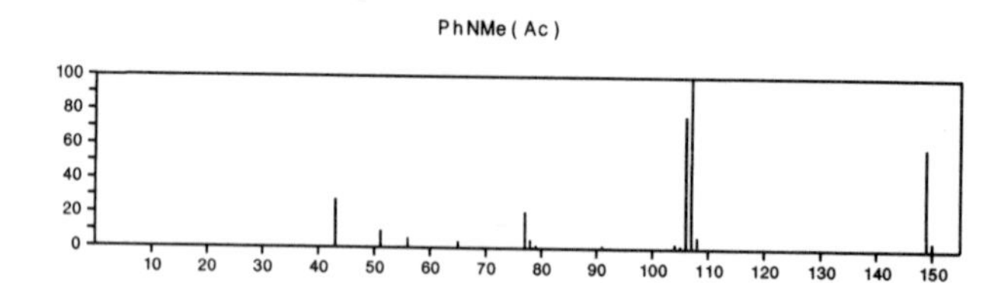

149 $C_9H_{11}NO$ 588-46-5
Acetamide, *N*-(phenylmethyl)-

AcNHCH2Ph

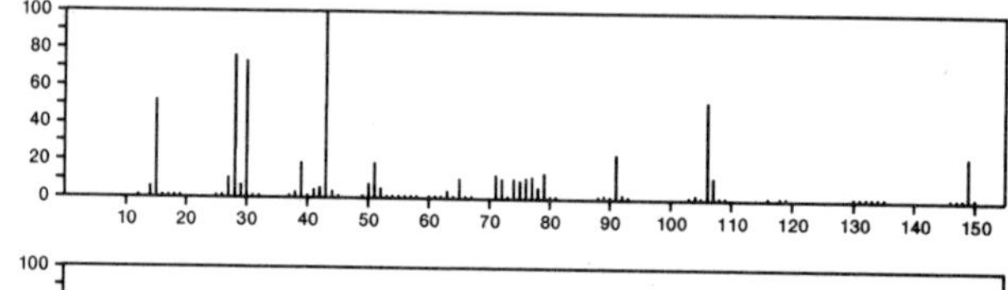

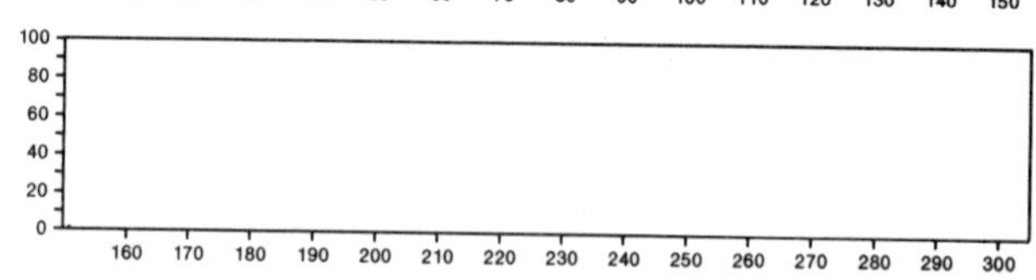

149 $C_9H_{11}NO$ 611-74-5
Benzamide, *N,N*-dimethyl-

Me2NCOPh

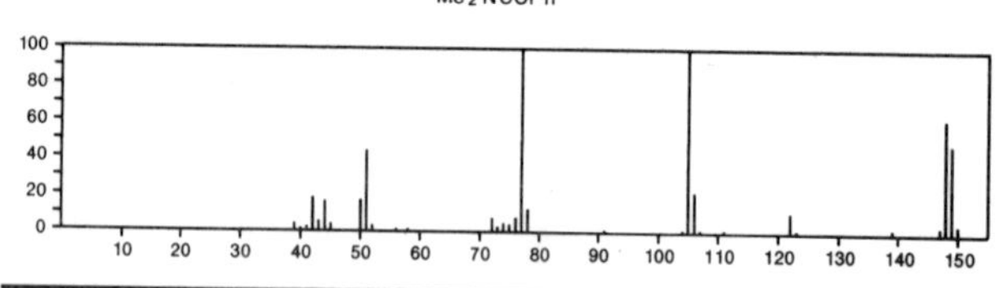

149 $C_9H_{11}NO$ 614-17-5
Benzamide, *N*-ethyl-

EtNHCOPh

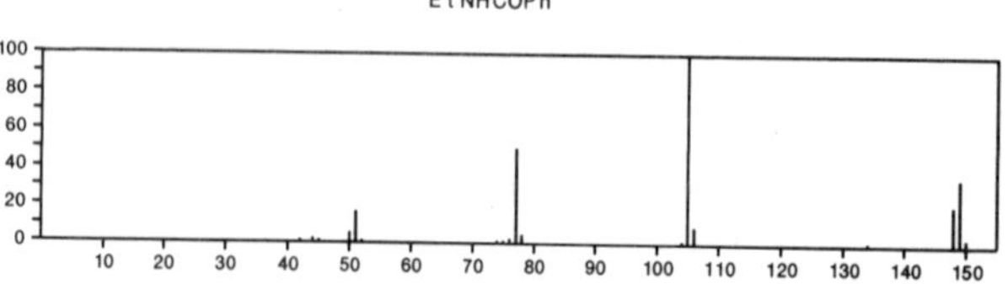

149 $C_9H_{11}NO$ 825-60-5
Benzenecarboximidic acid, ethyl ester

PhC(OEt)=NH

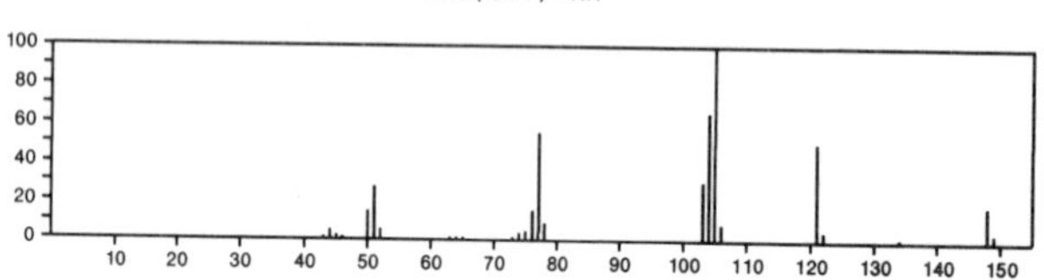

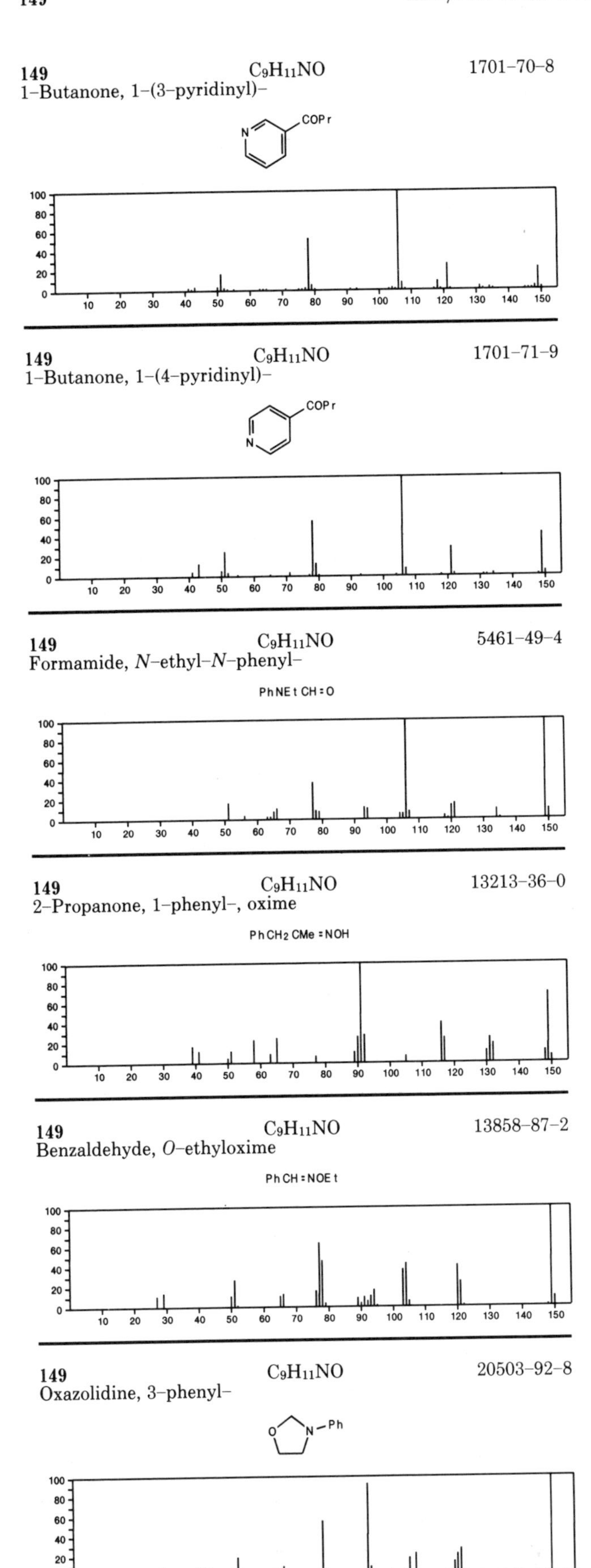

149 $C_9H_{11}NO$ 1701-70-8
1-Butanone, 1-(3-pyridinyl)-

149 $C_9H_{11}NO$ 1701-71-9
1-Butanone, 1-(4-pyridinyl)-

149 $C_9H_{11}NO$ 5461-49-4
Formamide, *N*-ethyl-*N*-phenyl-

149 $C_9H_{11}NO$ 13213-36-0
2-Propanone, 1-phenyl-, oxime

149 $C_9H_{11}NO$ 13858-87-2
Benzaldehyde, *O*-ethyloxime

149 $C_9H_{11}NO$ 20503-92-8
Oxazolidine, 3-phenyl-

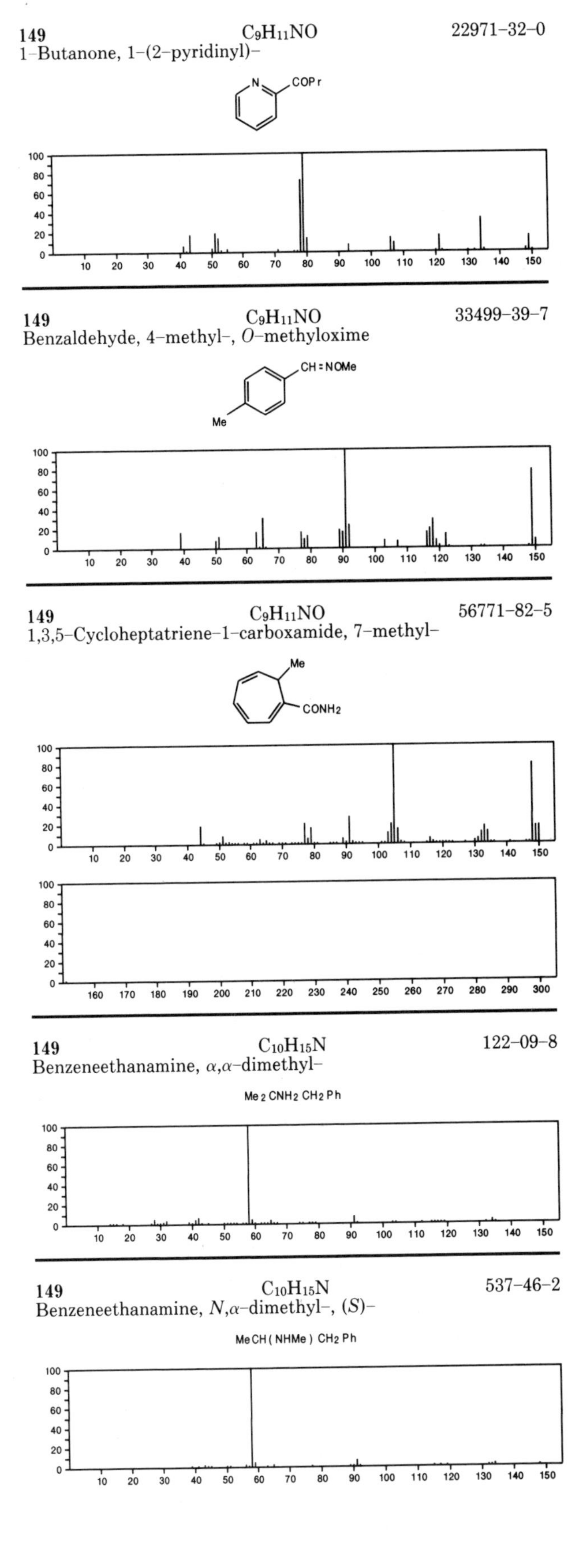

149 $C_9H_{11}NO$ 22971-32-0
1-Butanone, 1-(2-pyridinyl)-

149 $C_9H_{11}NO$ 33499-39-7
Benzaldehyde, 4-methyl-, *O*-methyloxime

149 $C_9H_{11}NO$ 56771-82-5
1,3,5-Cycloheptatriene-1-carboxamide, 7-methyl-

149 $C_{10}H_{15}N$ 122-09-8
Benzeneethanamine, α,α-dimethyl-

149 $C_{10}H_{15}N$ 537-46-2
Benzeneethanamine, *N*,α-dimethyl-, (*S*)-

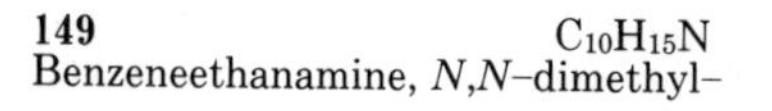

149 $C_{10}H_{15}N$ 1126–71–2
Benzeneethanamine, *N,N*–dimethyl–

Me2NCH2CH2Ph

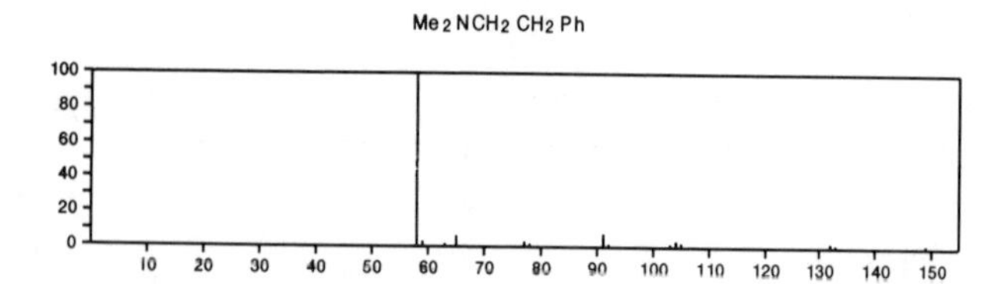

149 $C_{10}H_{15}N$ 4052–88–4
Benzenemethanamine, *N,N*,4–trimethyl–

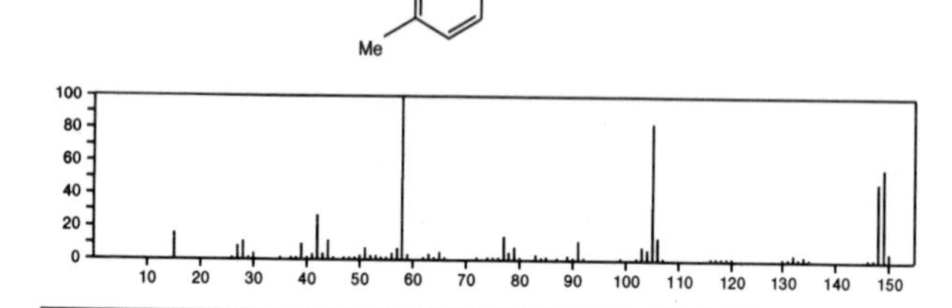

149 $C_{10}H_{15}N$ 7632–10–2
Benzeneethanamine, *N*,α–dimethyl–

MeCH(NHMe)CH2Ph

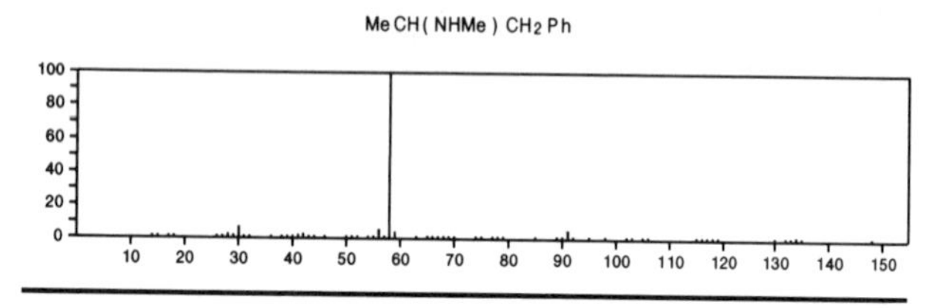

149 $C_{10}H_{15}N$ 31590–84–8
Pyridine, 2–neopentyl–

N CH2CMe3

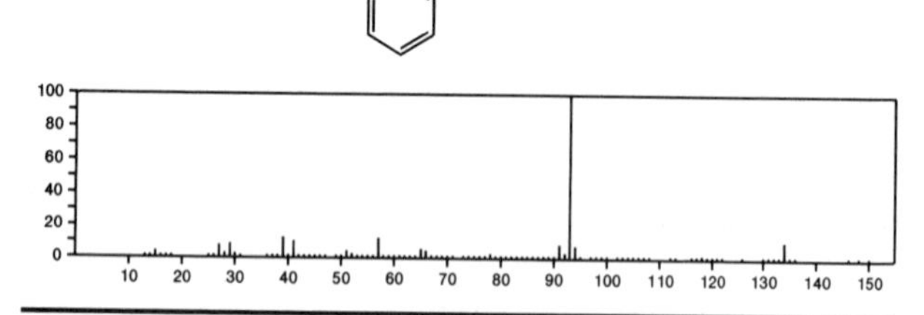

149 $C_{10}H_{15}N$ 33817–09–3
Benzeneethanamine, *N*,α–dimethyl–, (*R*)–

MeCH(NHMe)CH2Ph

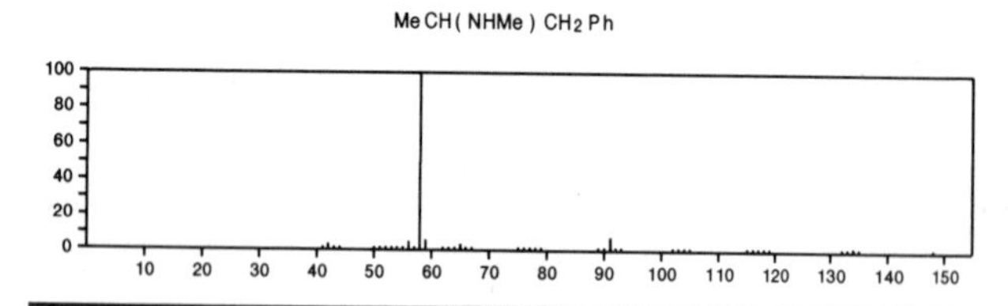

149 $C_{10}H_{15}N$ 55760–14–0
Cyclobutaneacetonitrile, 1–methyl–2–(1–methylethylidene)–

CH2CN Me CMe2

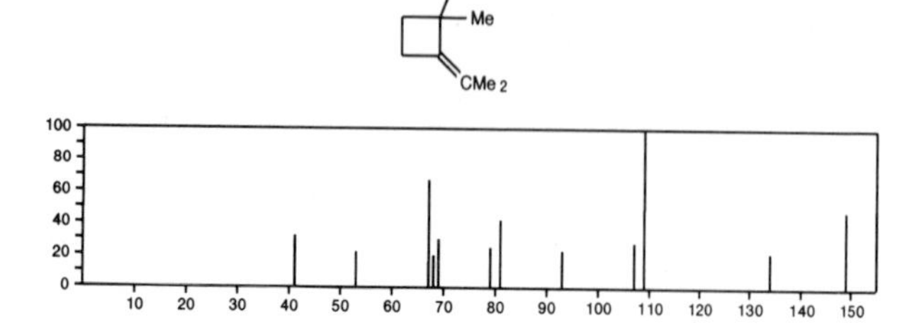

149 $C_{10}H_{15}N$ 55760–15–1
Cyclobutaneacetonitrile, 1–methyl–2–(1–methylethenyl)–

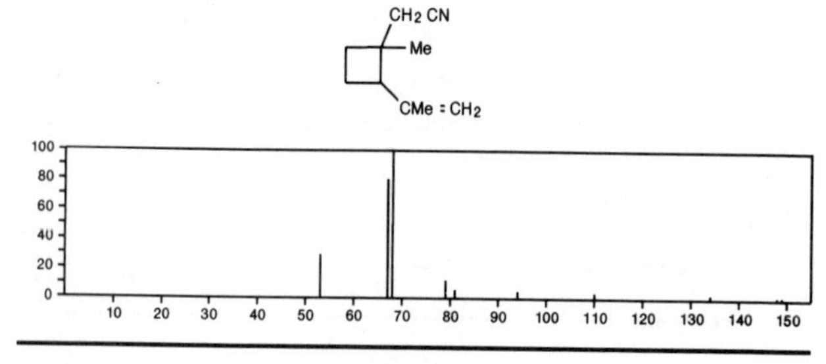

150 $C_2H_2Cl_3F$ 359–28–4
Ethane, 1,1,2–trichloro–2–fluoro–

FCHClCHCl2

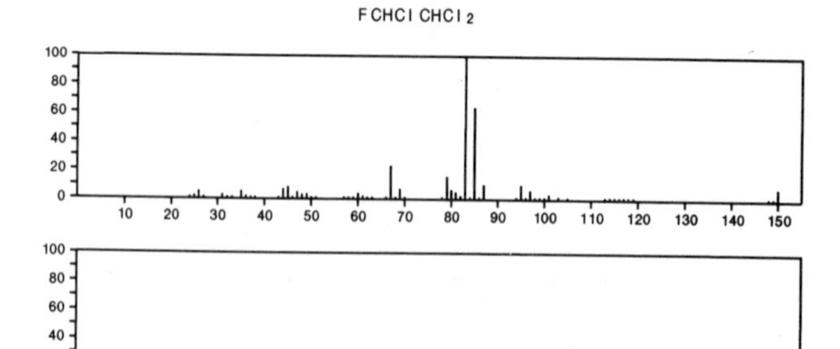

150 $C_2H_8N_2.C_2H_2O_4$ 6629–60–3
Hydrazine, ethyl–, ethanedioate (1:1)

MeCH2NHNH2 • HO2CCO2H

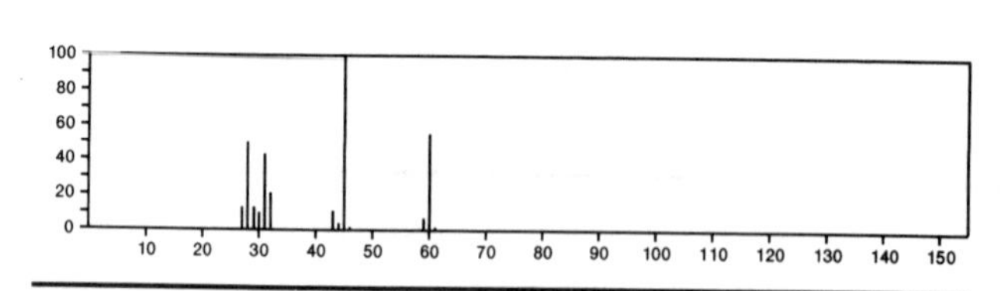

150 C_3F_6 116–15–4
1–Propene, 1,1,2,3,3,3–hexafluoro–

F3CCF=CF2

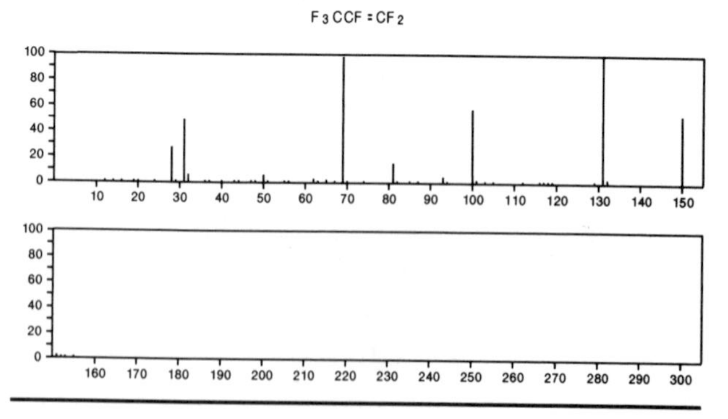

150 $C_4H_6O_2S_2$ 6629–12–5
1,2–Dithiolane–3–carboxylic acid

S S CO2H

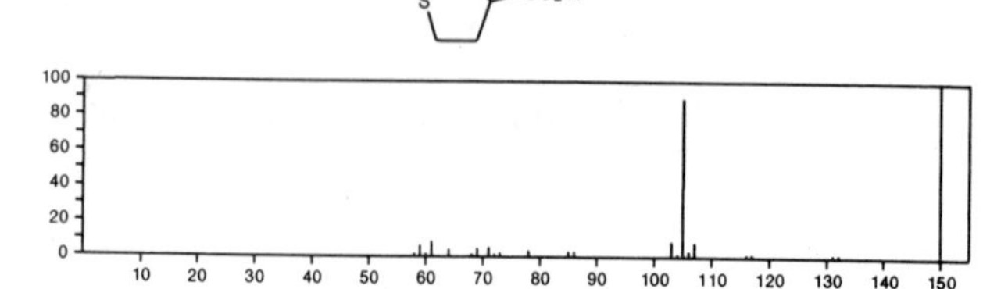

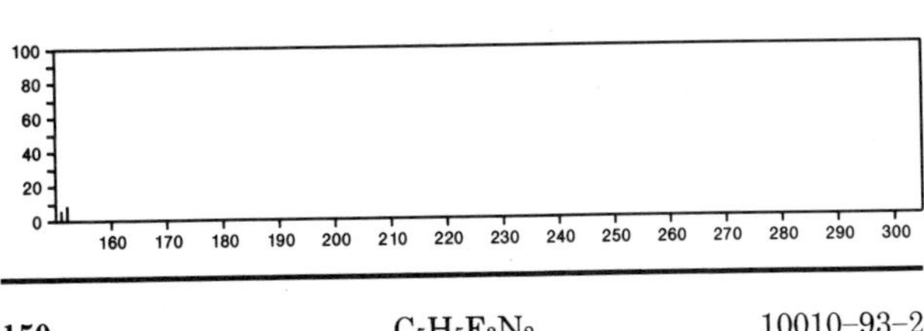

150 $C_5H_5F_3N_2$ 10010-93-2
1*H*-Pyrazole, 3-methyl-5-(trifluoromethyl)-

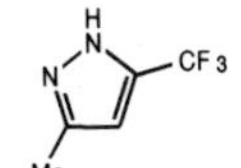

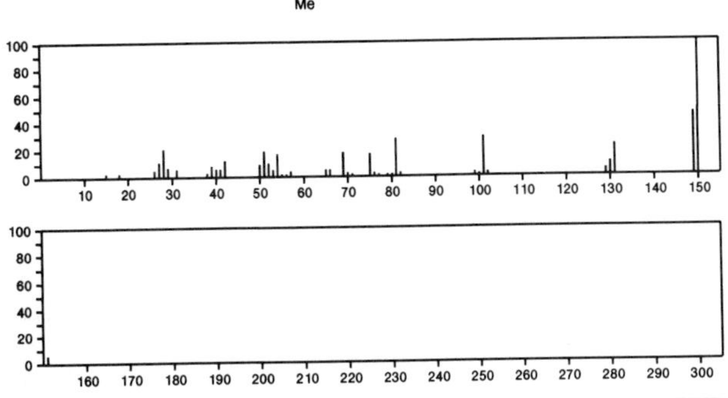

150 $C_5H_6N_6$ 51292-20-7
Imidazo[5,1-*f*][1,2,4]triazine-2,7-diamine

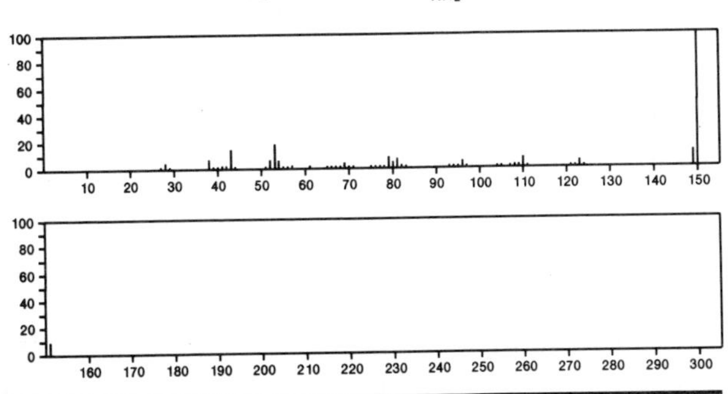

150 $C_5H_{10}OS_2$ 623-79-0
Carbonodithioic acid, *O,S*-diethyl ester

EtOC(S)SEt

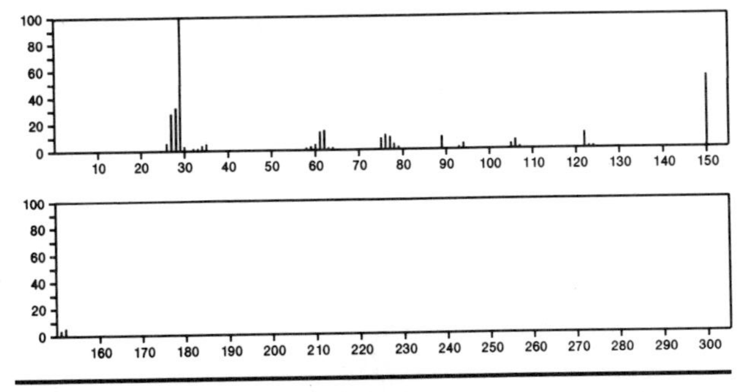

150 $C_5H_{10}OS_2$ 623-80-3
Carbonodithioic acid, *S,S*-diethyl ester

EtSC(O)SEt

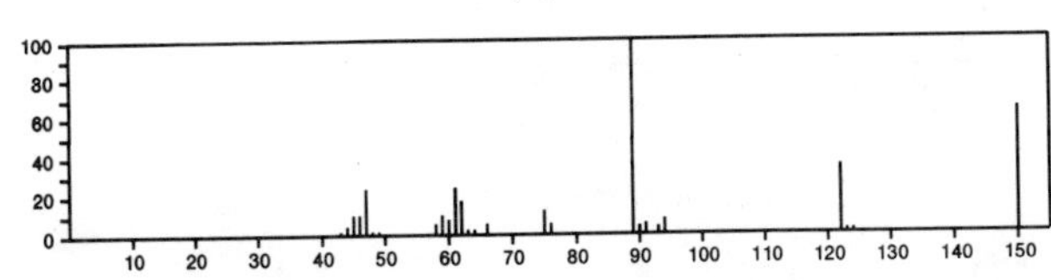

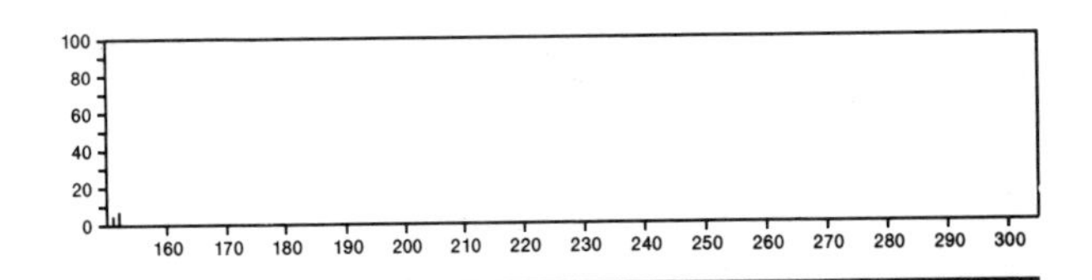

150 $C_5H_{10}O_3S$ 19788-48-8
Acetic acid, mercapto-, 2-methoxyethyl ester

$MeOCH_2CH_2OC(O)CH_2SH$

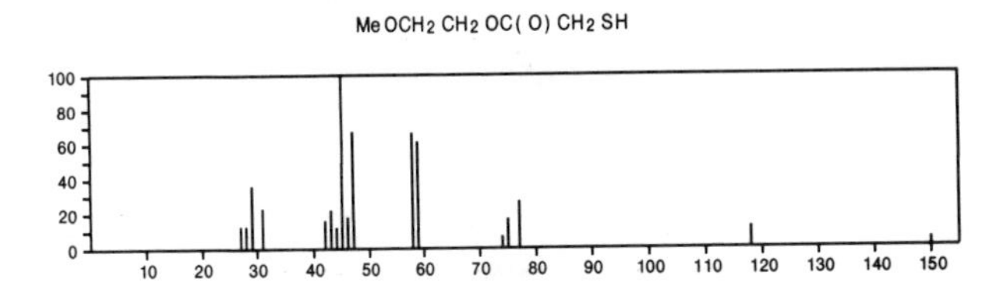

150 $C_5H_{10}O_5$ 50-69-1
D-Ribose

$OCHCH(OH)CH(OH)CH(OH)CH_2OH$

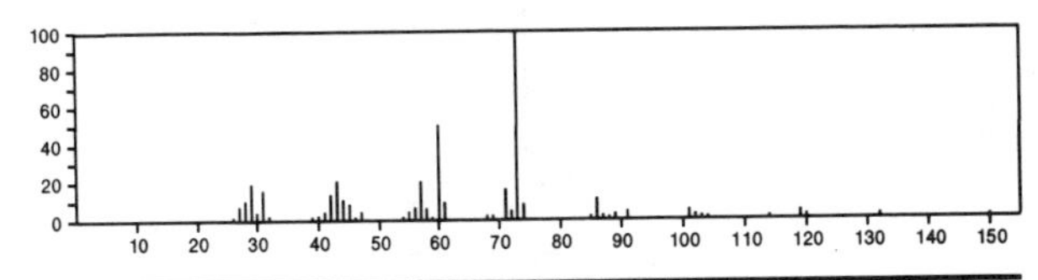

150 $C_5H_{10}O_5$ 56772-25-9
1,2,3,4,5-Cyclopentanepentol

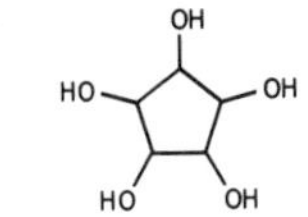

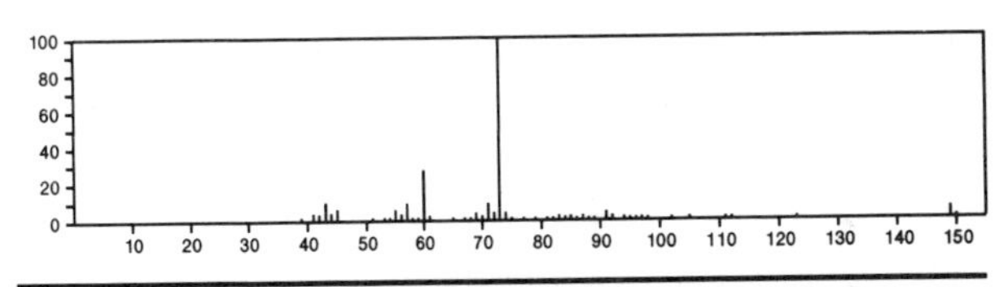

150 $C_5H_{11}Br$ 107-81-3
Pentane, 2-bromo-

PrCHBr(Me)

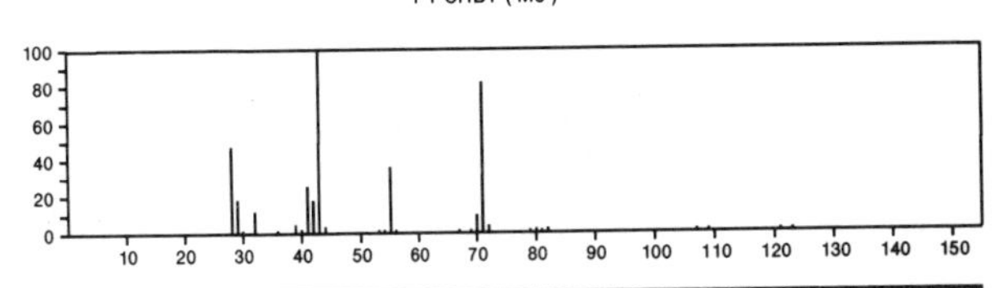

150 $C_5H_{11}Br$ 107-82-4
Butane, 1-bromo-3-methyl-

$BrCH_2CH_2CHMe_2$

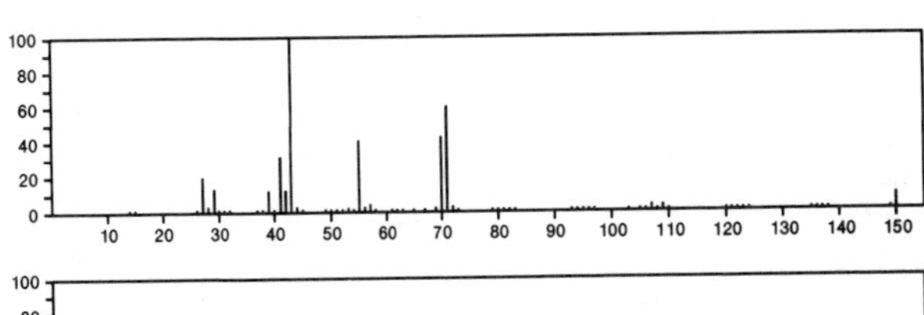

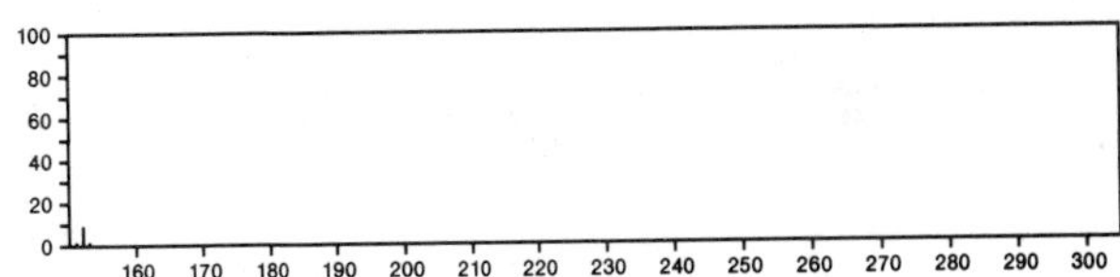

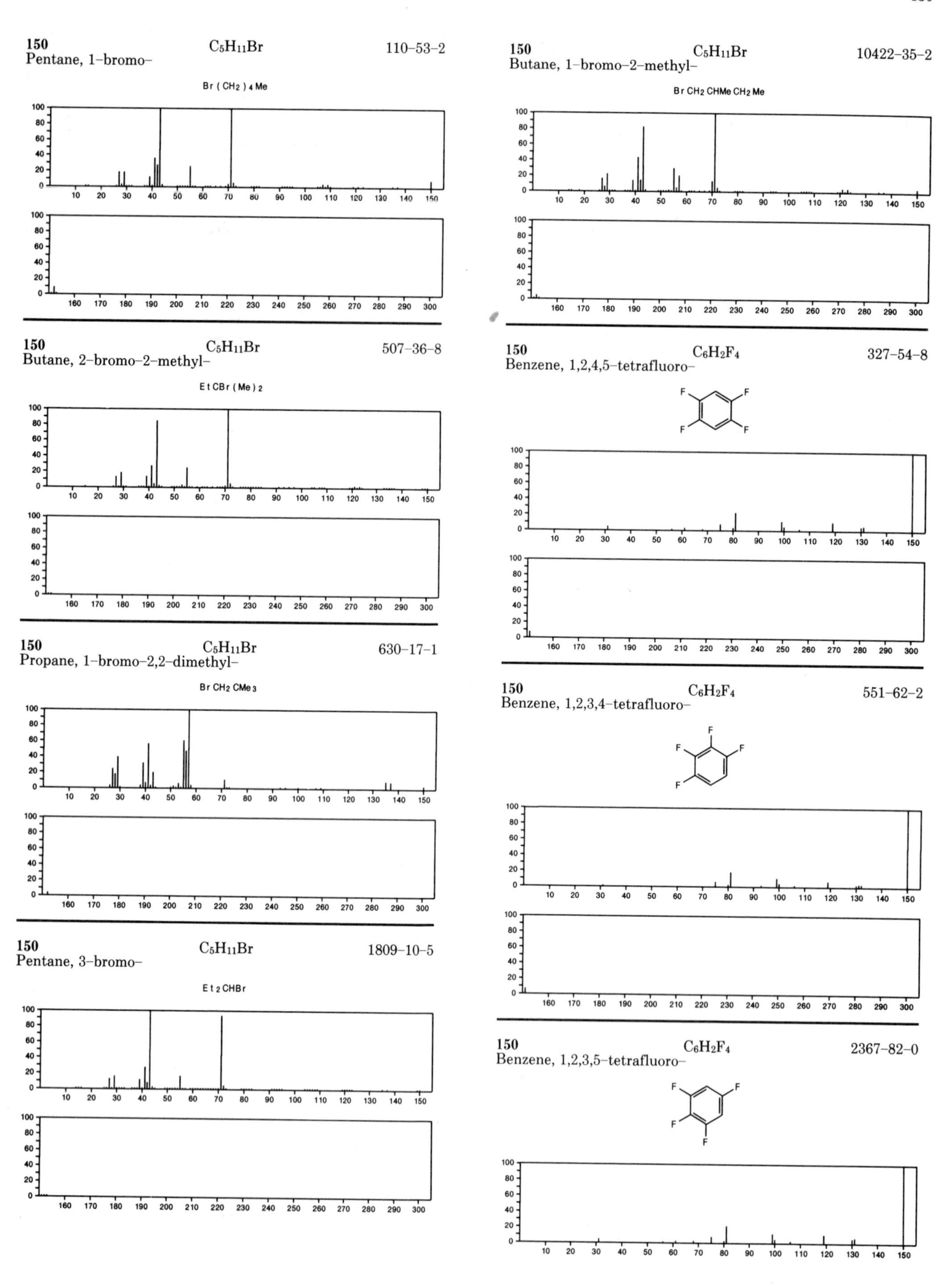
150
Pentane, 1-bromo-
$C_5H_{11}Br$
110-53-2
Br (CH2) 4 Me
150
Butane, 1-bromo-2-methyl-
$C_5H_{11}Br$
10422-35-2
Br CH2 CHMe CH2 Me
150
Butane, 2-bromo-2-methyl-
$C_5H_{11}Br$
507-36-8
Et CBr (Me) 2
150
Benzene, 1,2,4,5-tetrafluoro-
$C_6H_2F_4$
327-54-8
150
Propane, 1-bromo-2,2-dimethyl-
$C_5H_{11}Br$
630-17-1
Br CH2 CMe 3
150
Benzene, 1,2,3,4-tetrafluoro-
$C_6H_2F_4$
551-62-2
150
Pentane, 3-bromo-
$C_5H_{11}Br$
1809-10-5
Et 2 CHBr
150
Benzene, 1,2,3,5-tetrafluoro-
$C_6H_2F_4$
2367-82-0

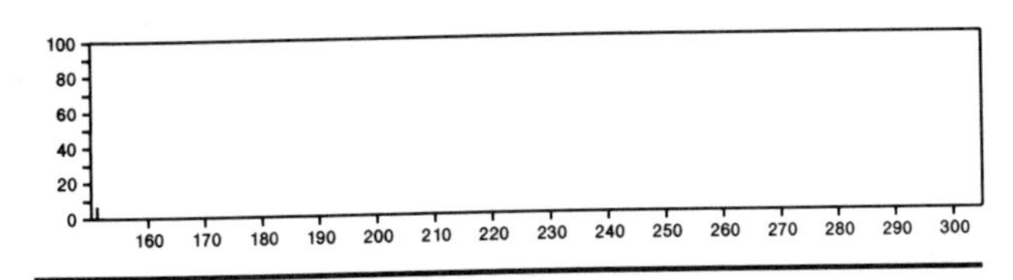

150 $C_6H_6N_4O$ 1006–08–2

6*H*–Purin–6–one, 1,7–dihydro–7–methyl–

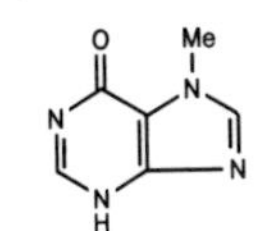

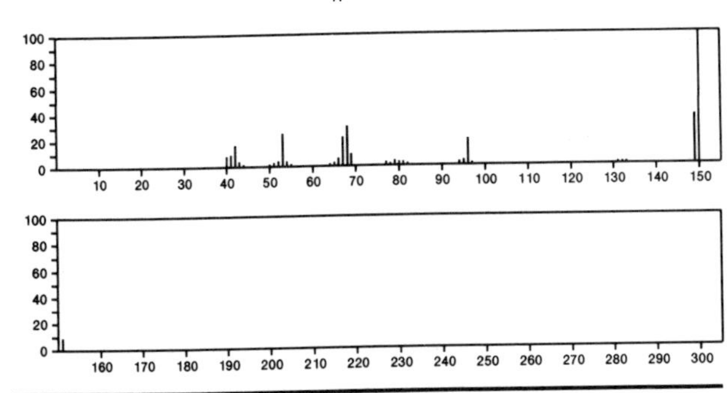

150 $C_6H_6N_4O$ 1006–11–7

6*H*–Purin–6–one, 3,7–dihydro–3–methyl–

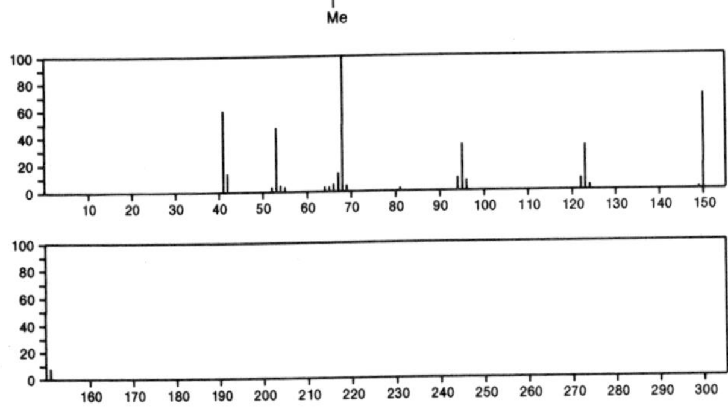

150 $C_6H_6N_4O$ 1074–89–1

1*H*–Purine, 6–methoxy–

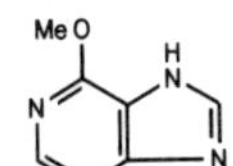

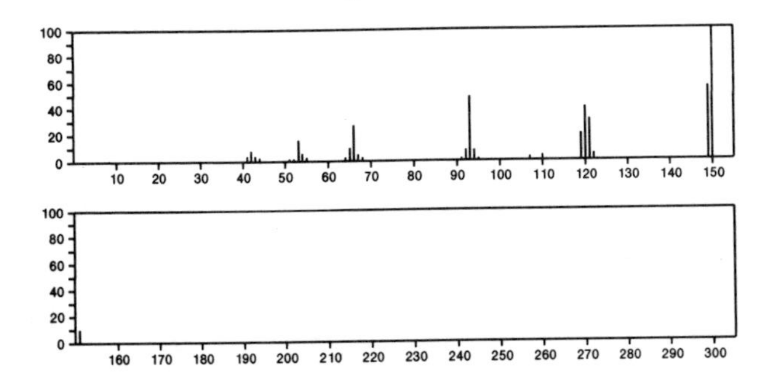

150 $C_6H_6N_4O$ 1125–39–9

6*H*–Purin–6–one, 1,7–dihydro–1–methyl–

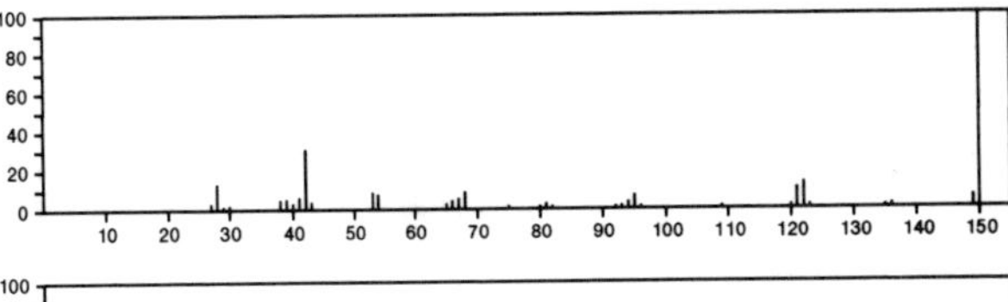

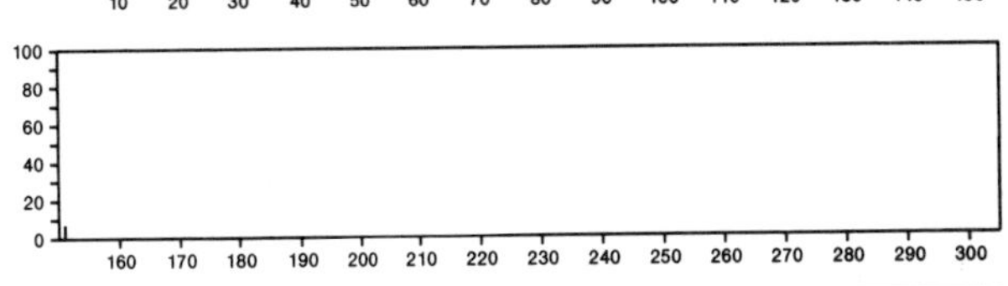

150 $C_6H_{11}ClO_2$ 590–02–3

Acetic acid, chloro–, butyl ester

$Me(CH_2)_3OC(O)CH_2Cl$

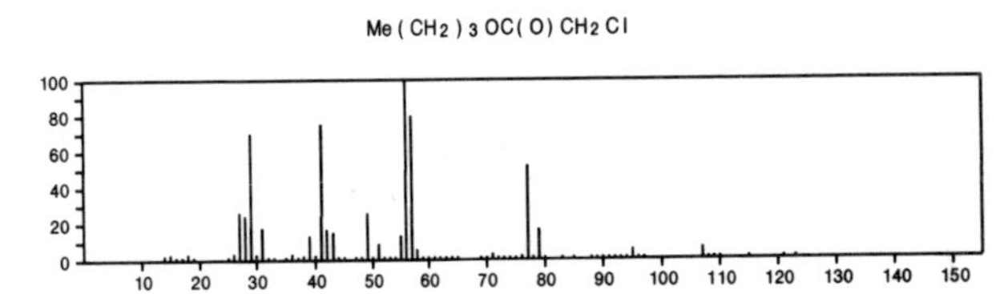

150 $C_6H_{11}ClO_2$ 638–41–5

Carbonochloridic acid, pentyl ester

$Me(CH_2)_4OC(O)Cl$

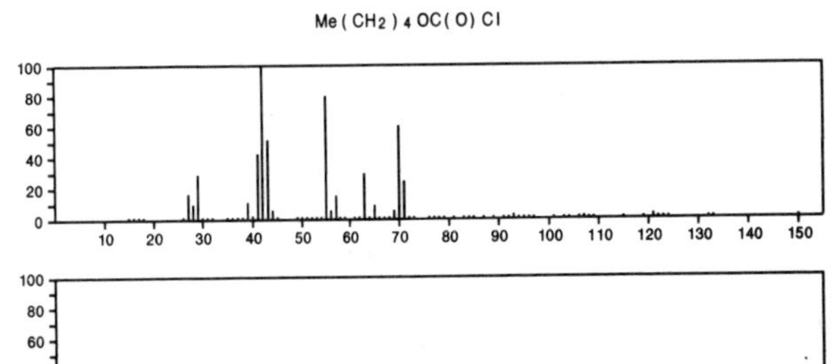

150 $C_6H_{11}ClO_2$ 691–93–0

Propanoic acid, 3–chloro–, 1–methylethyl ester

$ClCH_2CH_2C(O)OPr-i$

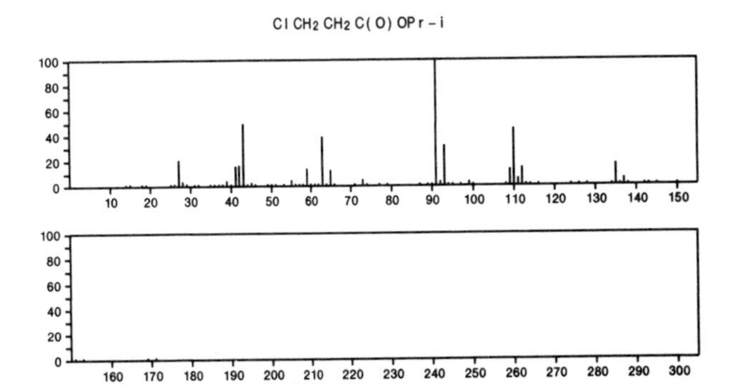

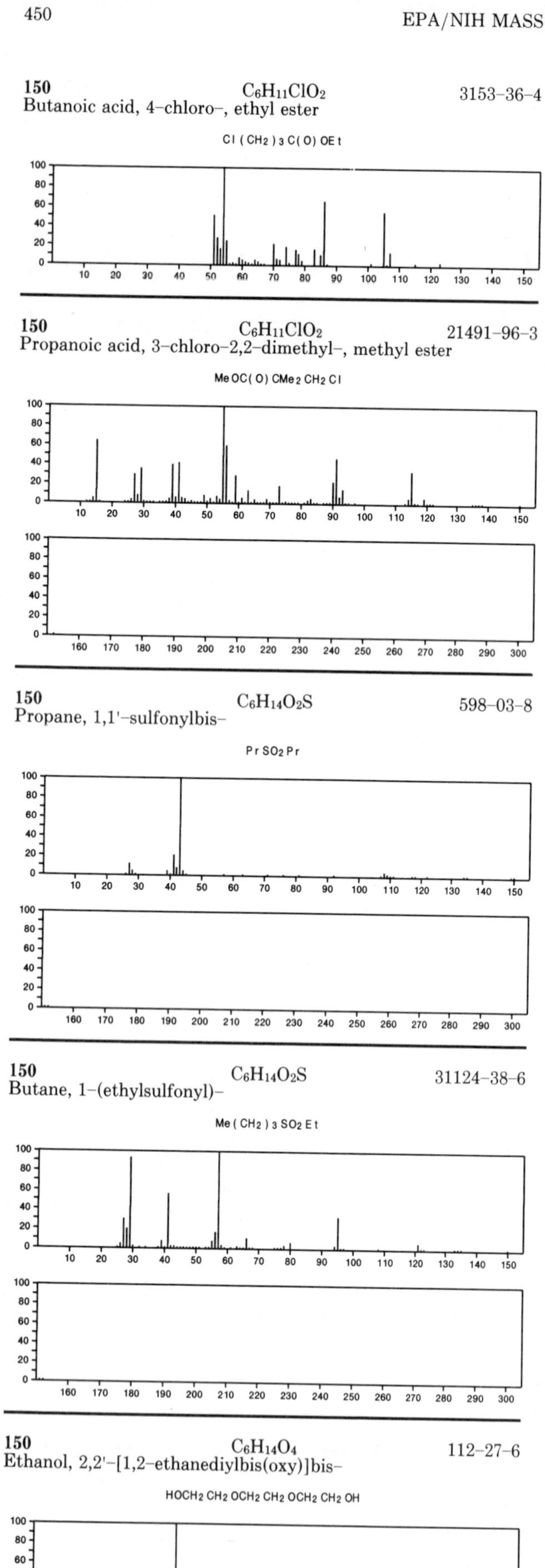

150 $C_6H_{11}ClO_2$ 3153-36-4
Butanoic acid, 4-chloro-, ethyl ester
Cl(CH2)3C(O)OEt

150 $C_6H_{11}ClO_2$ 21491-96-3
Propanoic acid, 3-chloro-2,2-dimethyl-, methyl ester
MeOC(O)CMe2CH2Cl

150 $C_6H_{14}O_2S$ 598-03-8
Propane, 1,1'-sulfonylbis-
PrSO2Pr

150 $C_6H_{14}O_2S$ 31124-38-6
Butane, 1-(ethylsulfonyl)-
Me(CH2)3SO2Et

150 $C_6H_{14}O_4$ 112-27-6
Ethanol, 2,2'-[1,2-ethanediylbis(oxy)]bis-
HOCH2CH2OCH2CH2OCH2CH2OH

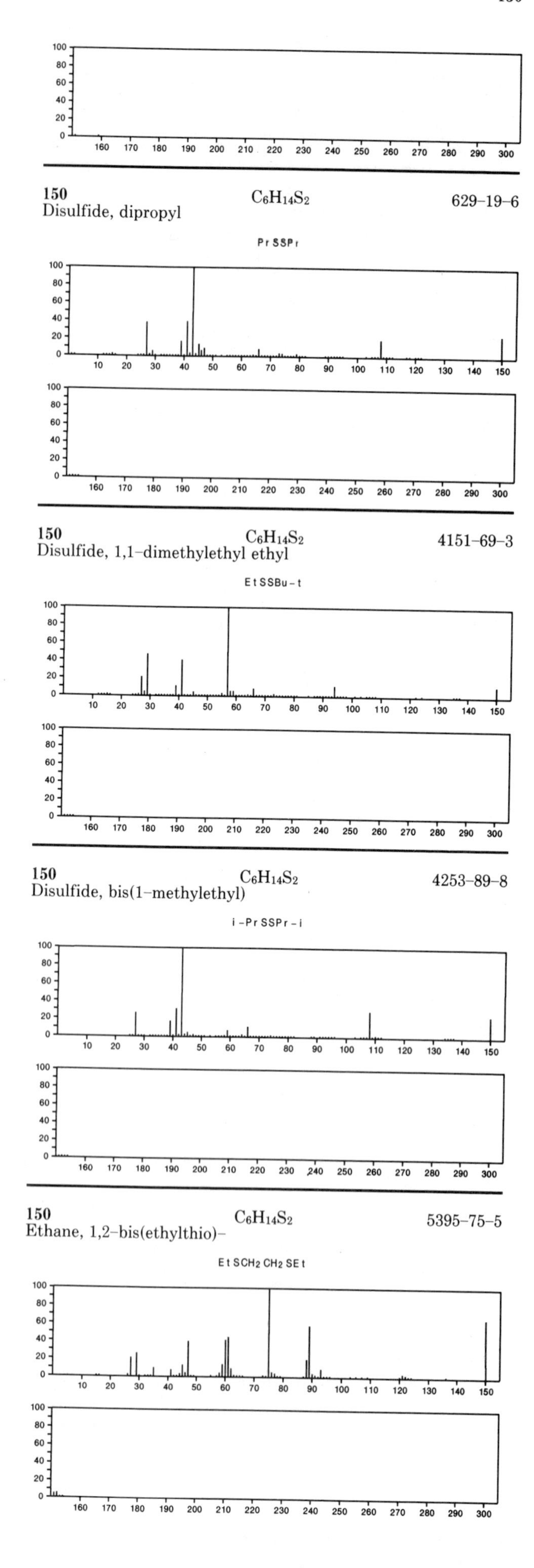

150 $C_6H_{14}S_2$ 629-19-6
Disulfide, dipropyl
PrSSPr

150 $C_6H_{14}S_2$ 4151-69-3
Disulfide, 1,1-dimethylethyl ethyl
EtSSBu-t

150 $C_6H_{14}S_2$ 4253-89-8
Disulfide, bis(1-methylethyl)
i-PrSSPr-i

150 $C_6H_{14}S_2$ 5395-75-5
Ethane, 1,2-bis(ethylthio)-
EtSCH2CH2SEt

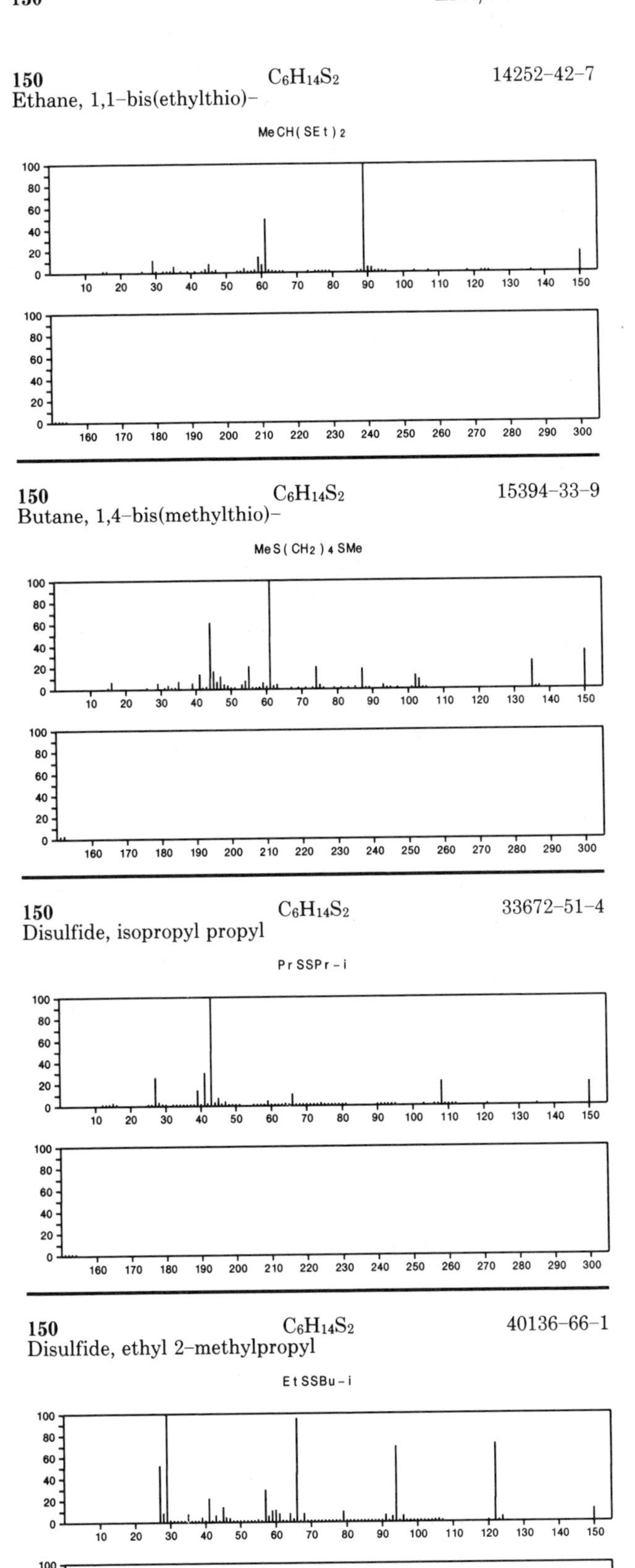

150 $C_6H_{14}S_2$ 14252-42-7
Ethane, 1,1-bis(ethylthio)-
MeCH(SEt)2

150 $C_6H_{14}S_2$ 15394-33-9
Butane, 1,4-bis(methylthio)-
MeS(CH2)4SMe

150 $C_6H_{14}S_2$ 33672-51-4
Disulfide, isopropyl propyl
PrSSPr-i

150 $C_6H_{14}S_2$ 40136-66-1
Disulfide, ethyl 2-methylpropyl
EtSSBu-i

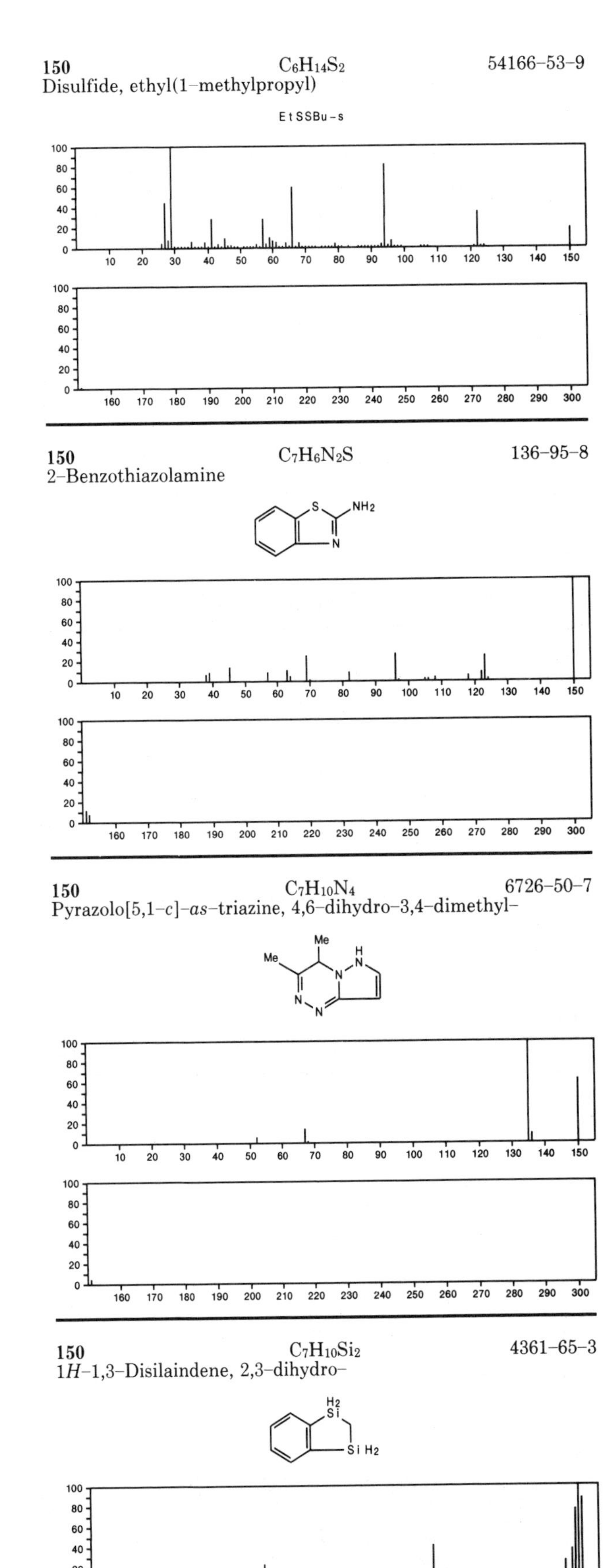

150 $C_6H_{14}S_2$ 54166-53-9
Disulfide, ethyl(1-methylpropyl)
EtSSBu-s

150 $C_7H_6N_2S$ 136-95-8
2-Benzothiazolamine

150 $C_7H_{10}N_4$ 6726-50-7
Pyrazolo[5,1-*c*]-*as*-triazine, 4,6-dihydro-3,4-dimethyl-

150 $C_7H_{10}Si_2$ 4361-65-3
1*H*-1,3-Disilaindene, 2,3-dihydro-

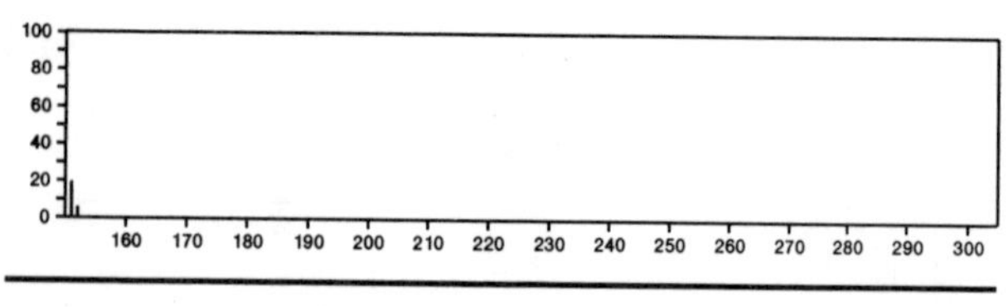

150 $C_8H_6O_3$ 119-67-5
Benzoic acid, 2-formyl-

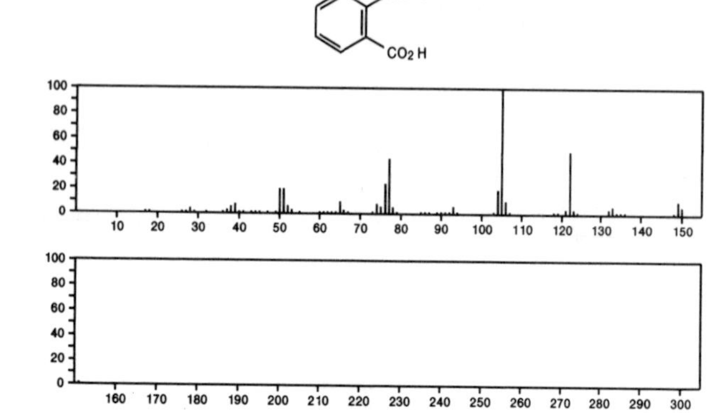

150 $C_8H_6O_3$ 120-57-0
1,3-Benzodioxole-5-carboxaldehyde

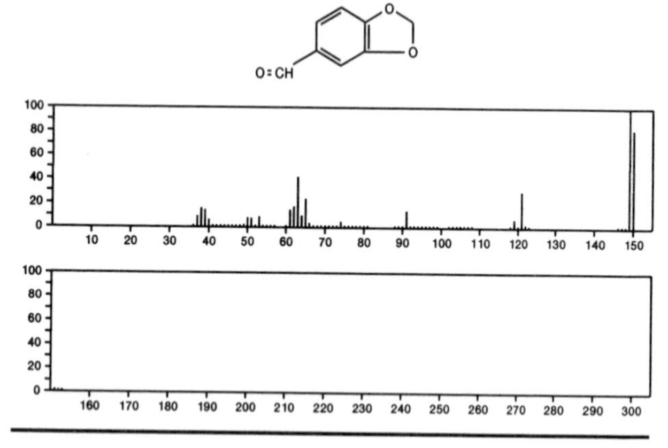

150 $C_8H_6O_3$ 619-66-9
Benzoic acid, 4-formyl-

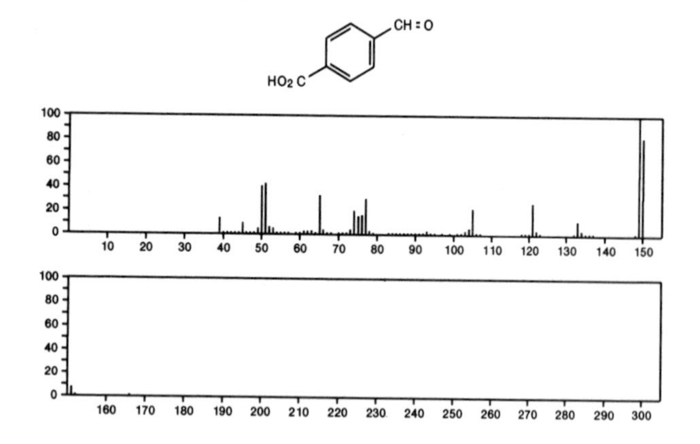

150 $C_8H_{10}N_2O$ 138-89-6
Benzenamine, *N*,*N*-dimethyl-4-nitroso-

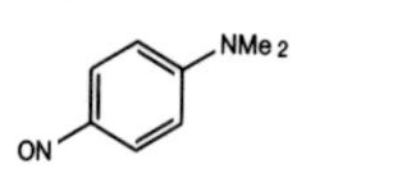

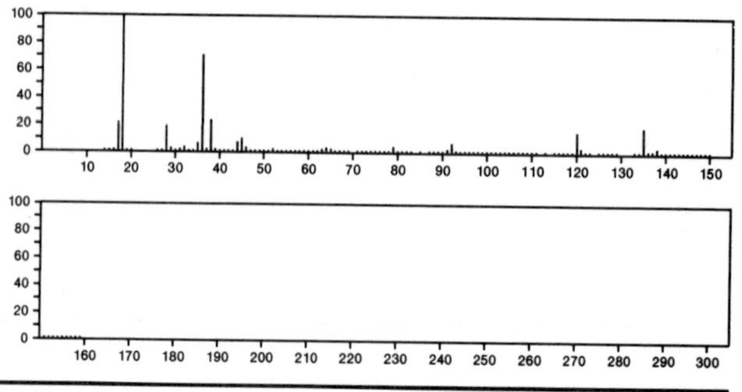

150 $C_8H_{10}N_2O$ 937-40-6
Benzenemethanamine, *N*-methyl-*N*-nitroso-

MeN(NO)CH2Ph

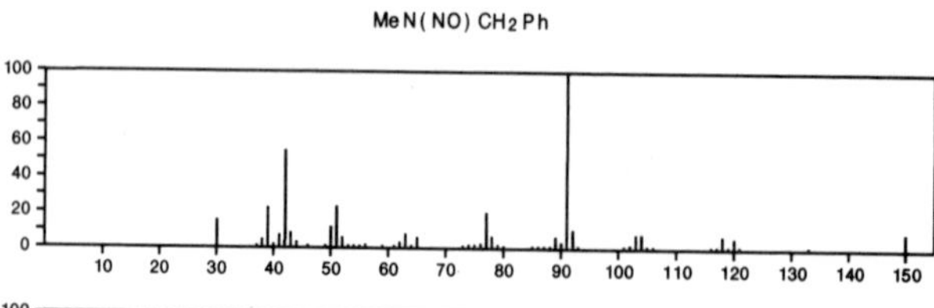

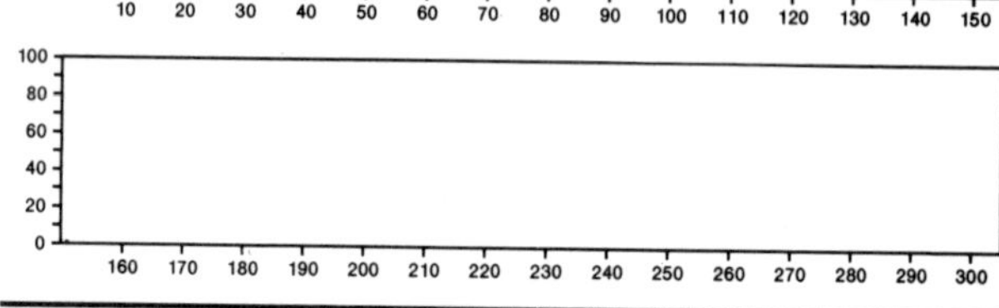

150 $C_8H_{10}N_2O$ 1660-24-8
Benzoic acid, 2-methylhydrazide

PhCONHNHMe

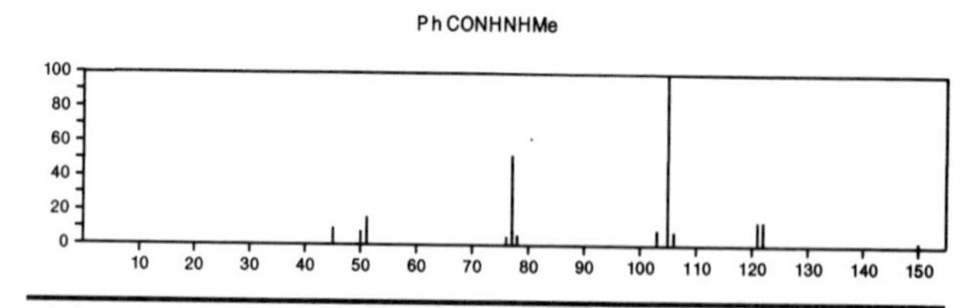

150 $C_8H_{10}N_2O$ 7584-27-2
Pyridinium, 1-(acetylamino)-2-methyl-, hydroxide, inner salt

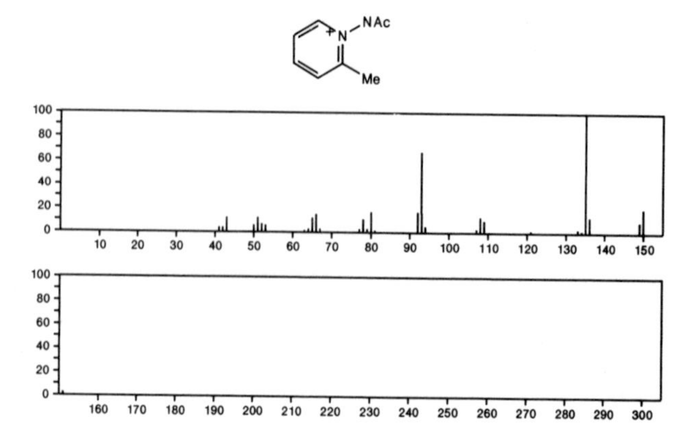

150 $C_8H_{10}N_2O$ 7584–29–4
4–Picolinium, 1–acetamido–, hydroxide, inner salt

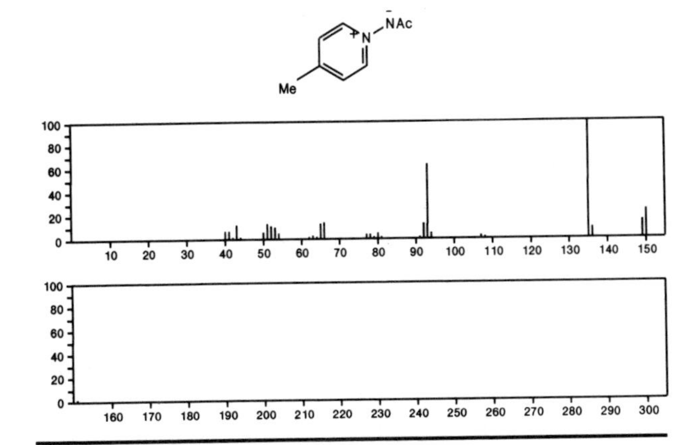

150 $C_8H_{10}N_2O$ 35150–72–2
Diazene, ethylphenyl–, 2–oxide, (*Z*)–

EtN=NPh=O

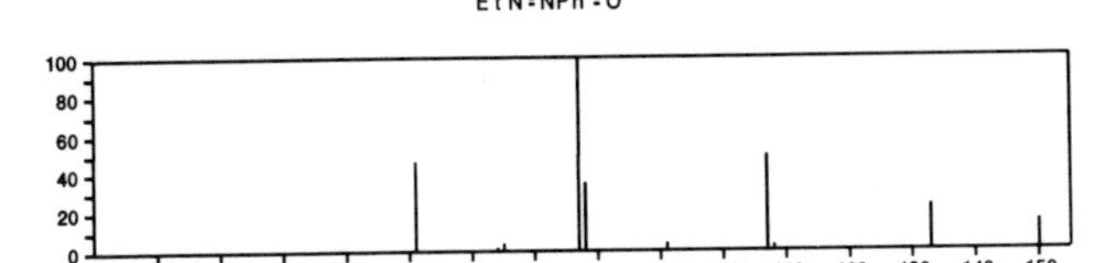

150 $C_8H_{10}N_2O$ 35150–76–6
Diazene, ethylphenyl–, 2–oxide, (*E*)–

EtN=NPh=O

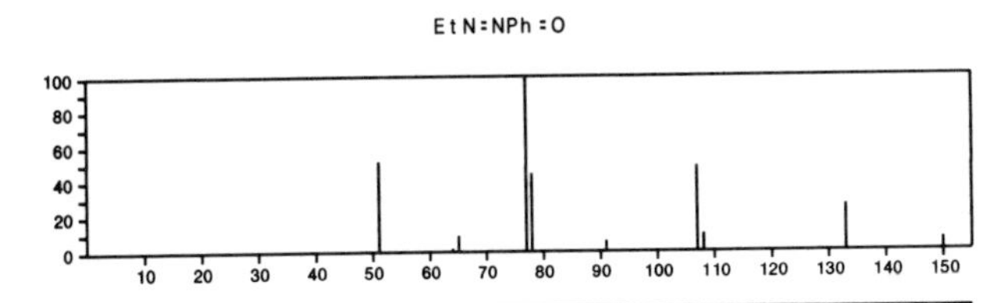

150 $C_8H_{10}N_2O$ 35150–77–7
Diazene, ethylphenyl–, 1–oxide, (*Z*)–

O=NEt=NPh

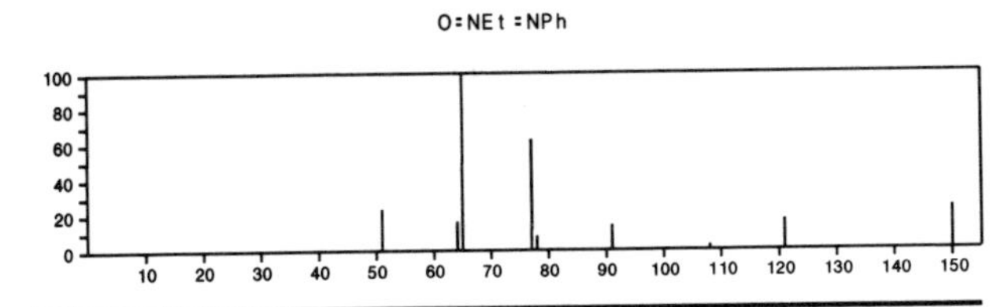

150 $C_8H_{10}N_2O$ 35150–78–8
Diazene, ethylphenyl–, 1–oxide, (*E*)–

O=NEt=NPh

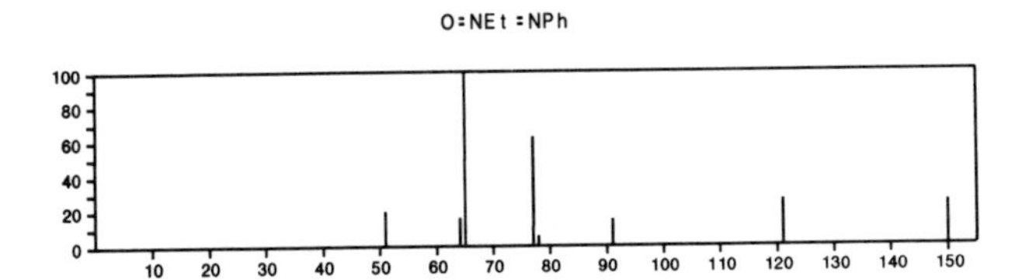

150 $C_9H_{10}O_2$ 89–71–4
Benzoic acid, 2–methyl–, methyl ester

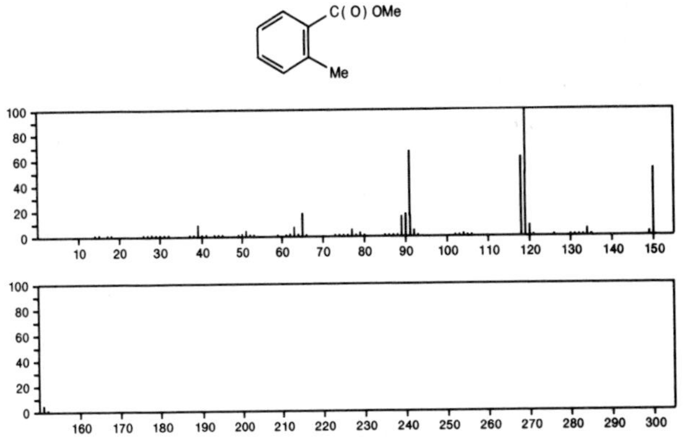

150 $C_9H_{10}O_2$ 93–89–0
Benzoic acid, ethyl ester

EtOC(O)Ph

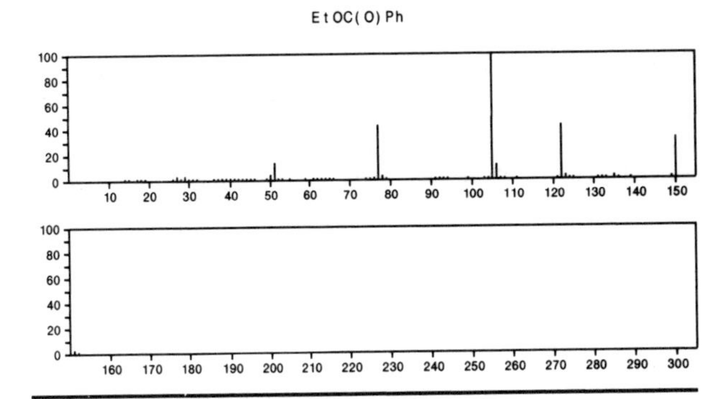

150 $C_9H_{10}O_2$ 99–36–5
Benzoic acid, 3–methyl–, methyl ester

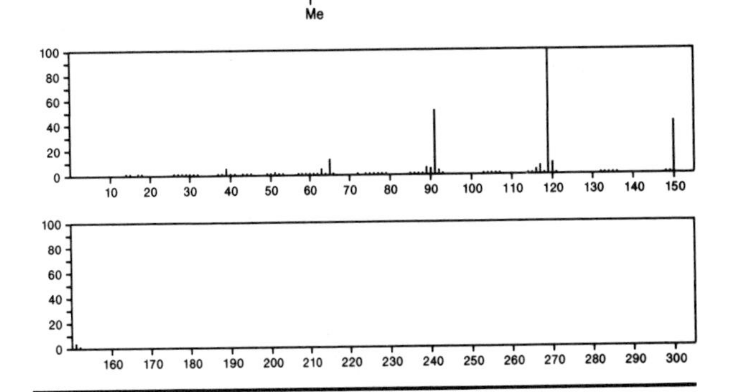

150 $C_9H_{10}O_2$ 99–75–2
Benzoic acid, 4–methyl–, methyl ester

C(O)OMe

Me

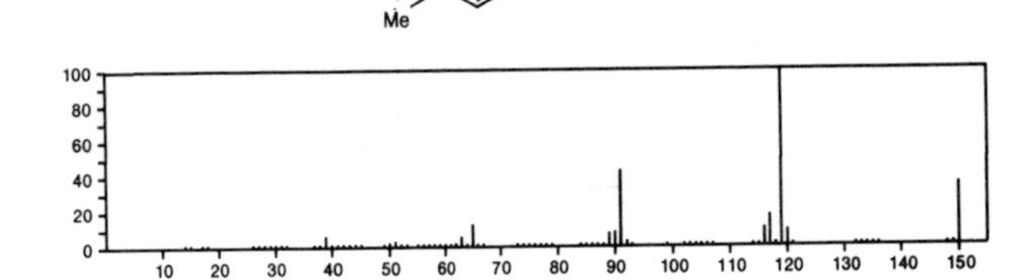

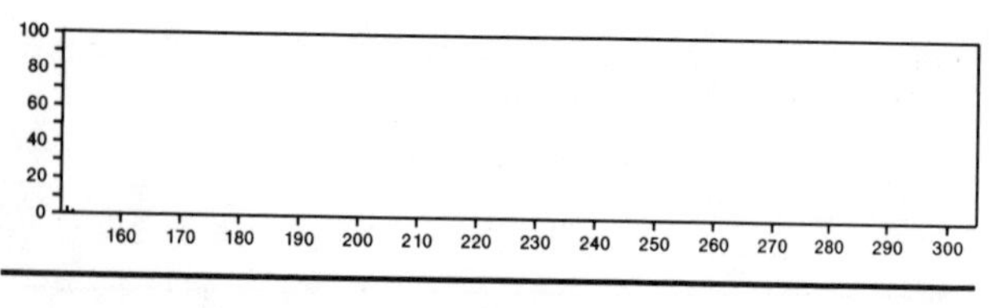

150 $C_9H_{10}O_2$ 100-06-1
Ethanone, 1-(4-methoxyphenyl)-

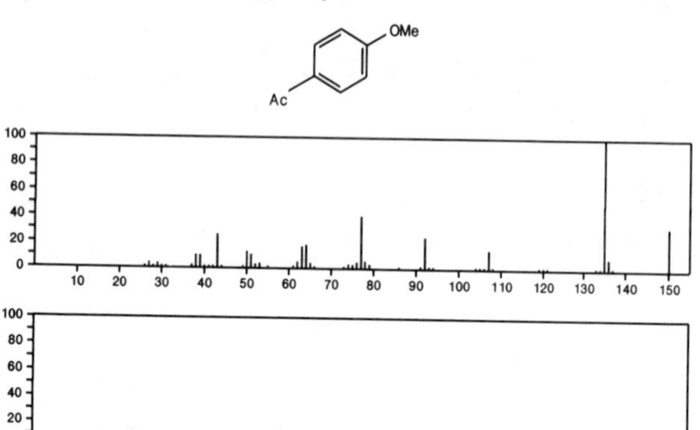

150 $C_9H_{10}O_2$ 101-41-7
Benzeneacetic acid, methyl ester

PhCH2C(O)OMe

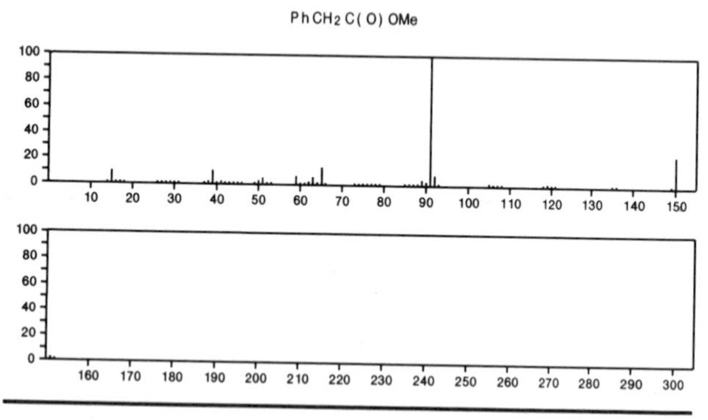

150 $C_9H_{10}O_2$ 104-62-1
Formic acid, 2-phenylethyl ester

O=CHOCH2CH2Ph

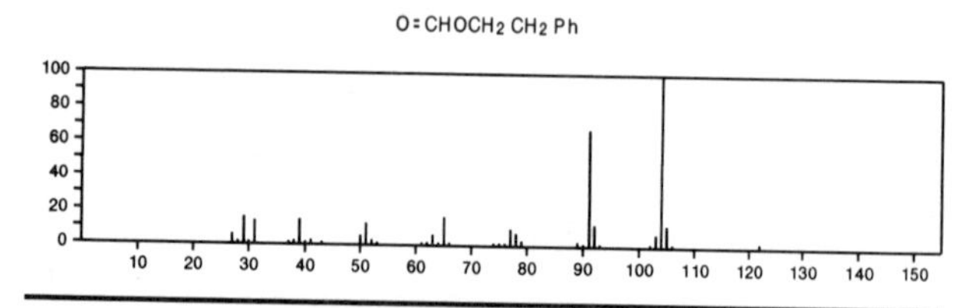

150 $C_9H_{10}O_2$ 122-60-1
Oxirane, (phenoxymethyl)-

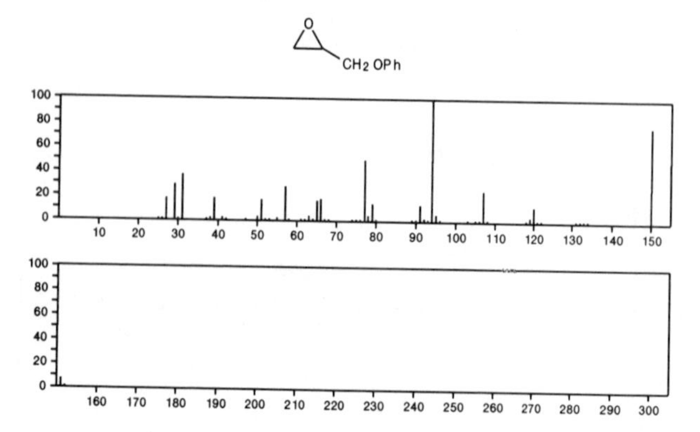

150 $C_9H_{10}O_2$ 140-11-4
Acetic acid, phenylmethyl ester

AcOCH2Ph

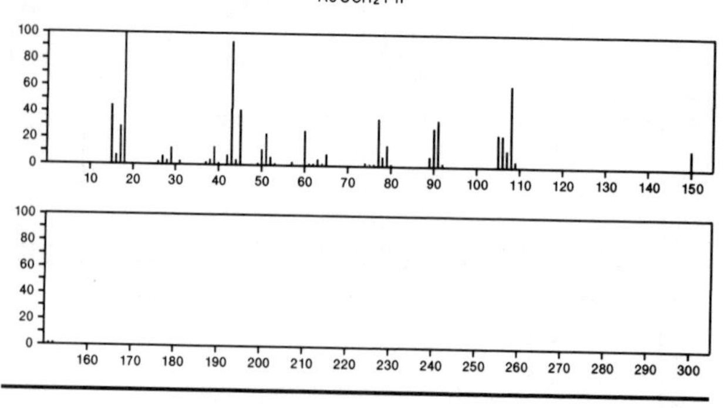

150 $C_9H_{10}O_2$ 499-06-9
Benzoic acid, 3,5-dimethyl-

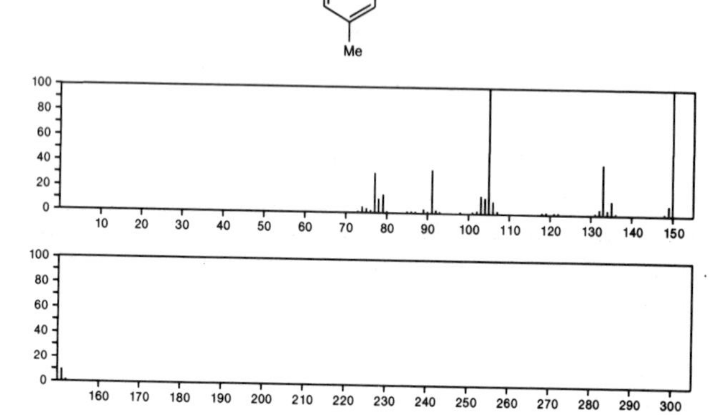

150 $C_9H_{10}O_2$ 501-52-0
Benzenepropanoic acid

HO2CCH2CH2Ph

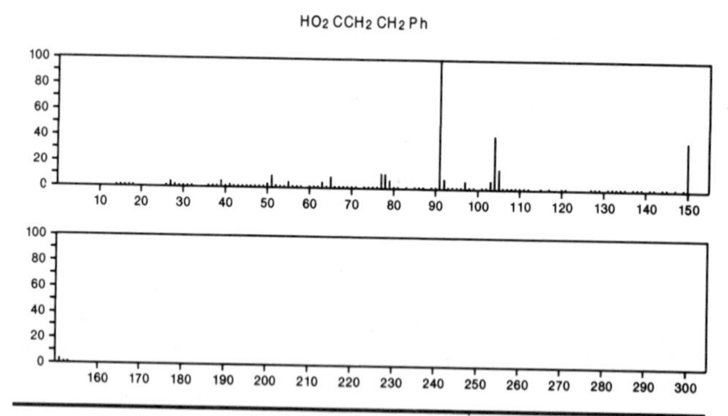

150 $C_9H_{10}O_2$ 533-18-6
Acetic acid, 2-methylphenyl ester

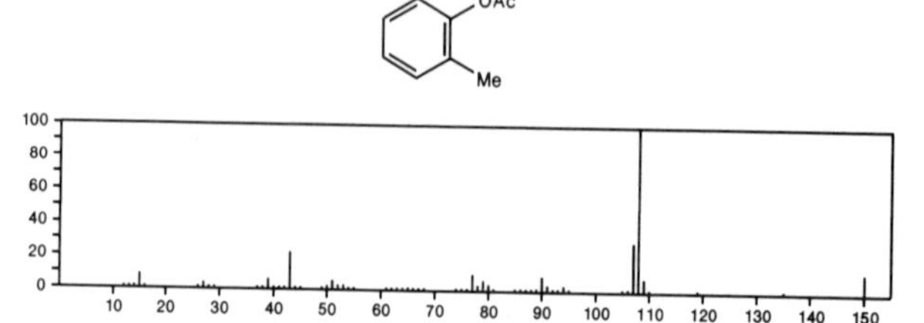

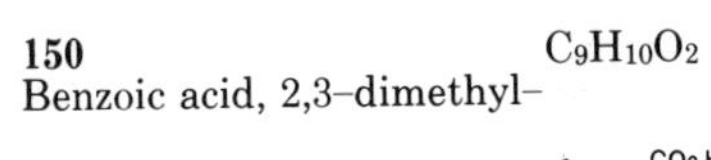

150 $C_9H_{10}O_2$ 603–79–2
Benzoic acid, 2,3–dimethyl–

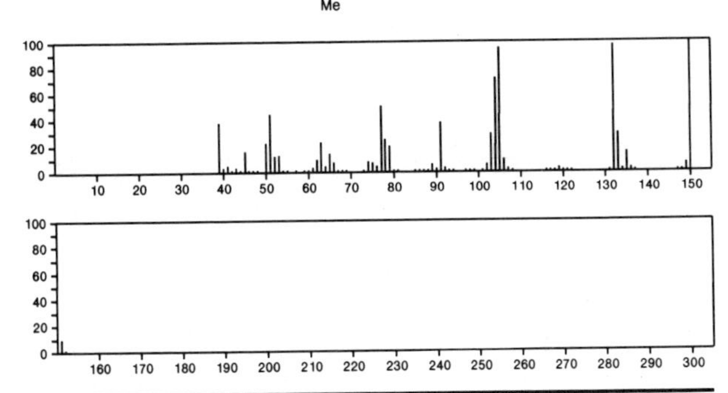

150 $C_9H_{10}O_2$ 610–72–0
Benzoic acid, 2,5–dimethyl–

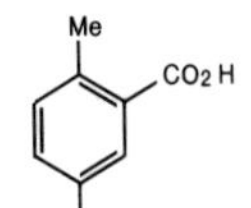

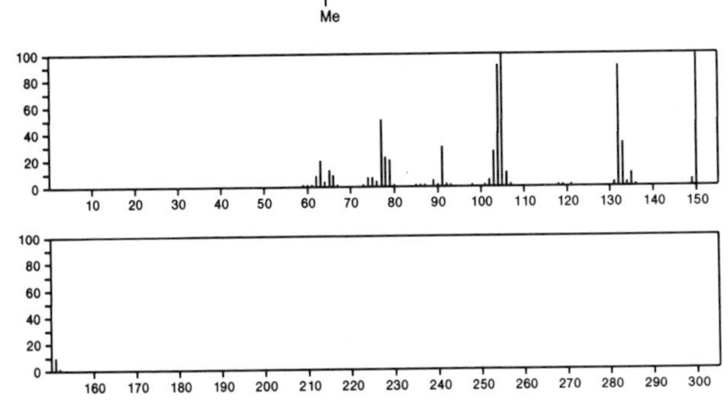

150 $C_9H_{10}O_2$ 611–01–8
Benzoic acid, 2,4–dimethyl–

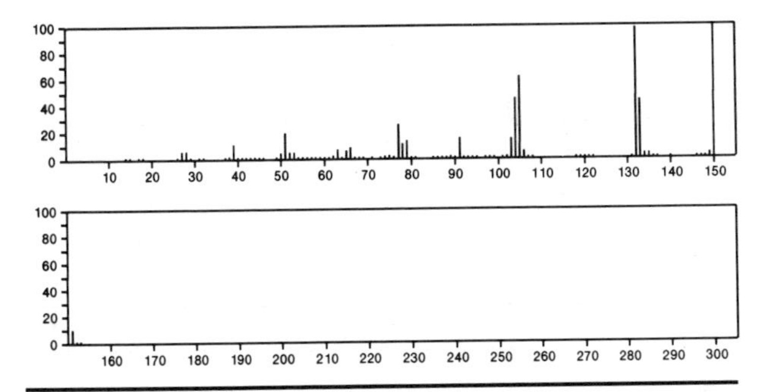

150 $C_9H_{10}O_2$ 619–04–5
Benzoic acid, 3,4–dimethyl–

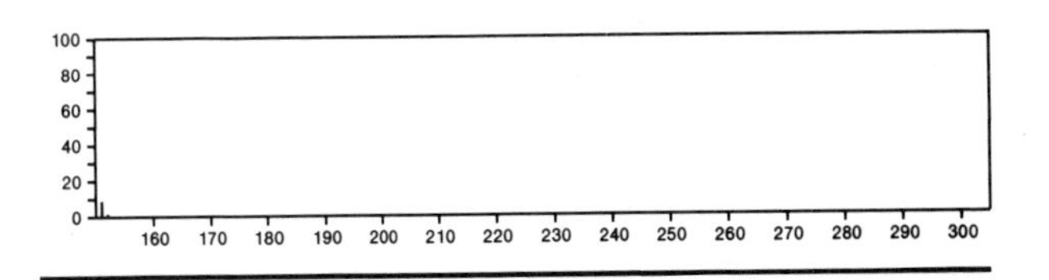

150 $C_9H_{10}O_2$ 621–87–4
2–Propanone, 1–phenoxy–

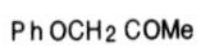

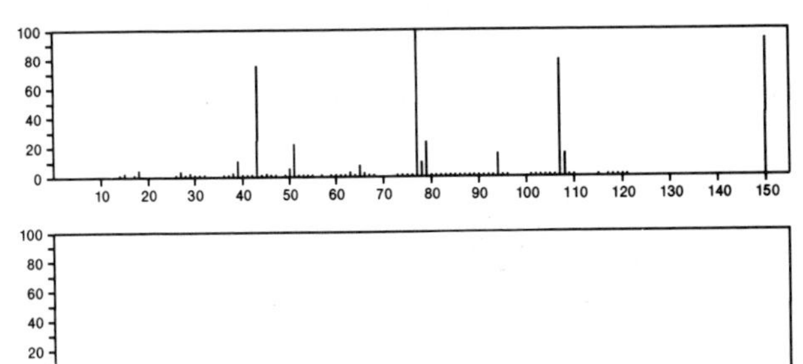

150 $C_9H_{10}O_2$ 632–46–2
Benzoic acid, 2,6–dimethyl–

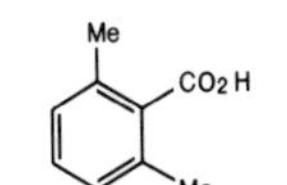

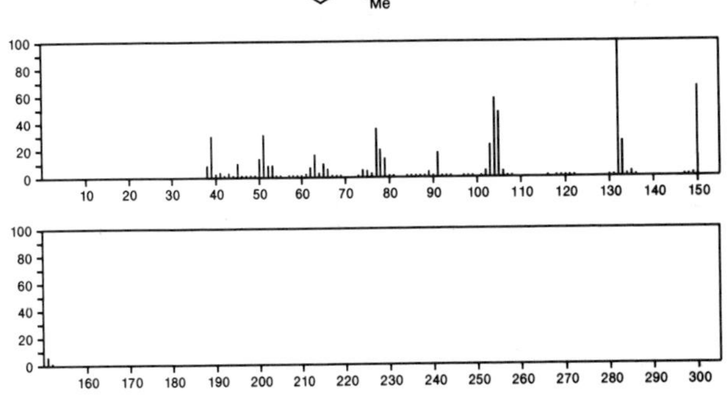

150 $C_9H_{10}O_2$ 637–27–4
Propanoic acid, phenyl ester

EtC(O)OPh

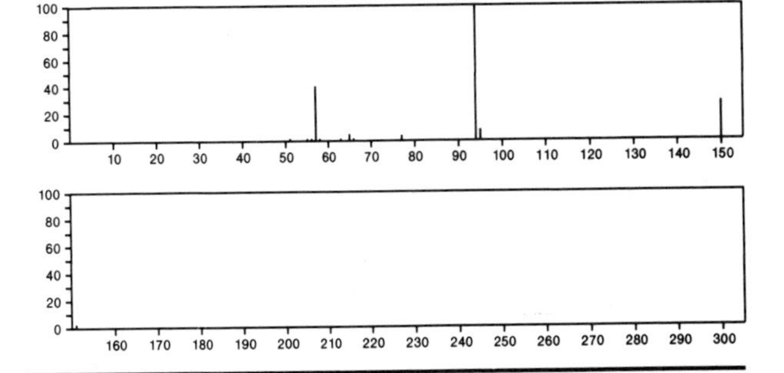

150 $C_9H_{10}O_2$ 936–51–6
1,3–Dioxolane, 2–phenyl–

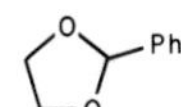

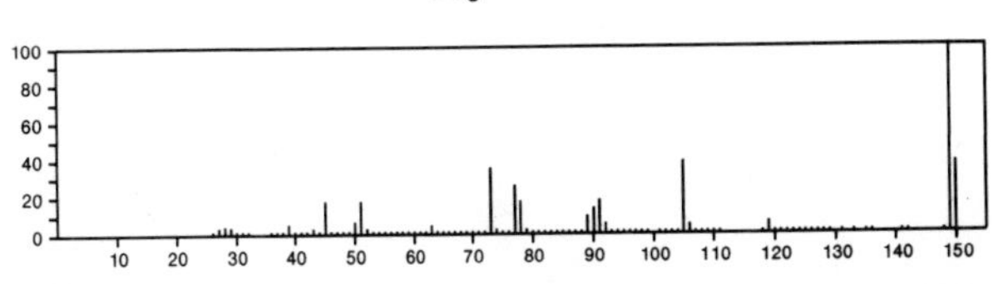

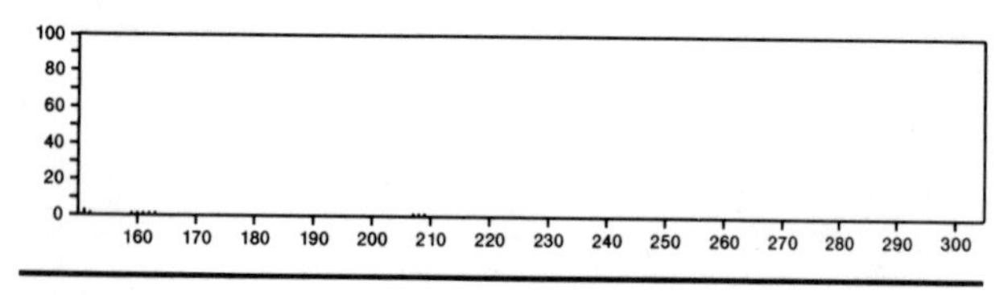

150 $C_9H_{10}O_2$ 3690-05-9
Phenol, 4-(3-hydroxy-1-propenyl)-

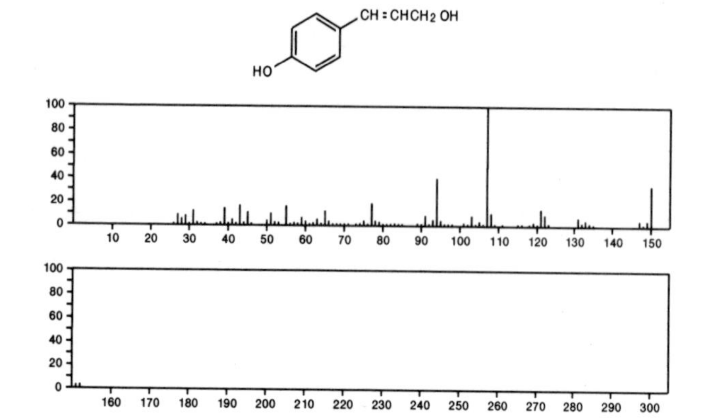

150 $C_9H_{10}O_2$ 4647-42-1
1*H*-Indene-1,2-diol, 2,3-dihydro-, *cis*-

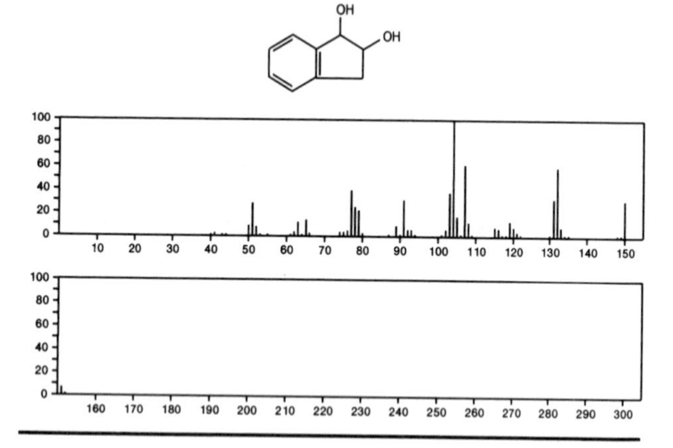

150 $C_9H_{10}O_2$ 4647-43-2
1*H*-Indene-1,2-diol, 2,3-dihydro-, *trans*-

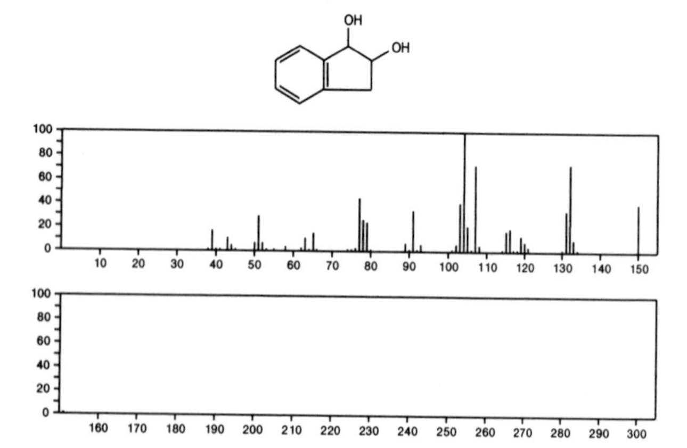

150 $C_9H_{10}O_2$ 7216-18-4
2*H*-1,5-Benzodioxepin, 3,4-dihydro-

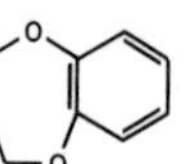

150 $C_9H_{10}O_2$ 28134-31-8
Benzoic acid, ethyl-

Et CO_2H

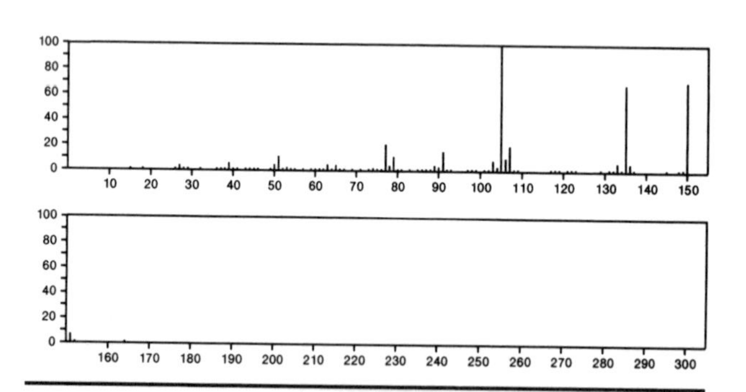

150 $C_9H_{10}O_2$ 31681-28-4
2-Furanacetaldehyde, α-isopropylidene-

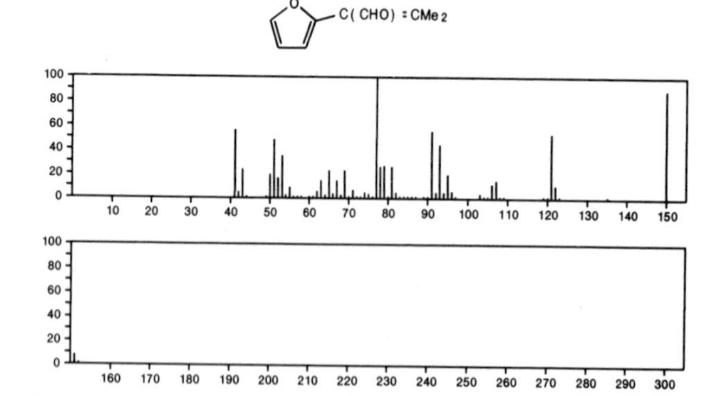

150 $C_9H_{10}O_2$ 31776-28-0
2-Furanacetaldehyde, α-methyl-α-vinyl-

O CMe(CHO)CH=CH2

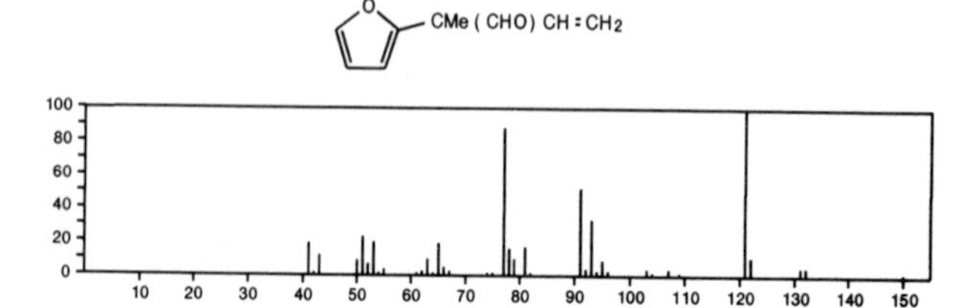

150 $C_9H_{10}O_2$ 56335-77-4

3-Penten-2-one, 3-(2-furanyl)-

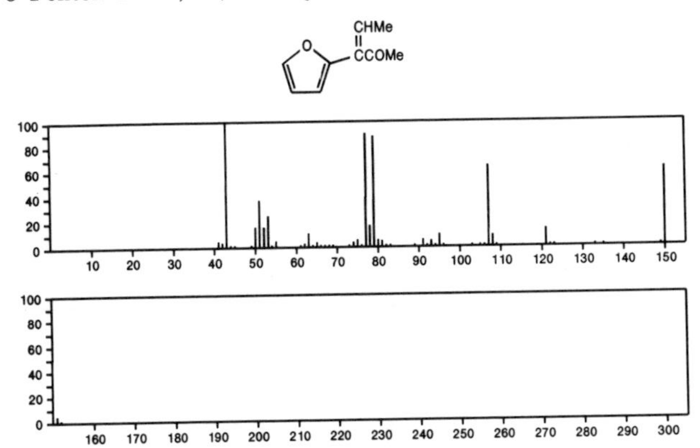

150 $C_9H_{10}O_2$ 56771-81-4

7-Oxabicyclo[4.2.1]nona-2,4-dien-8-one, 9-methyl-

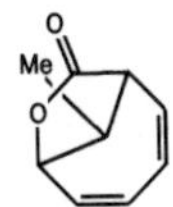

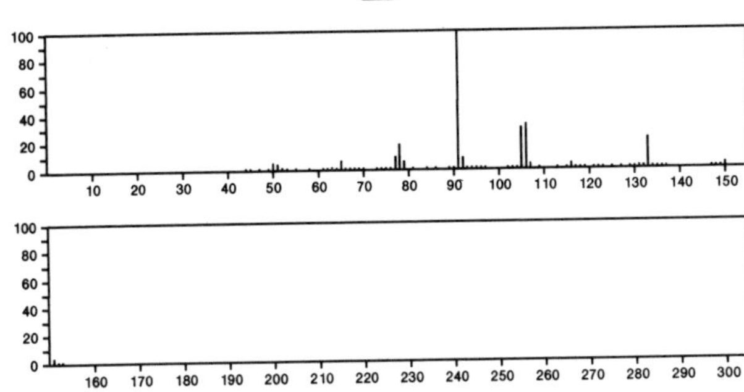

150 $C_9H_{10}S$ 4426-75-9

1*H*-2-Benzothiopyran, 3,4-dihydro-

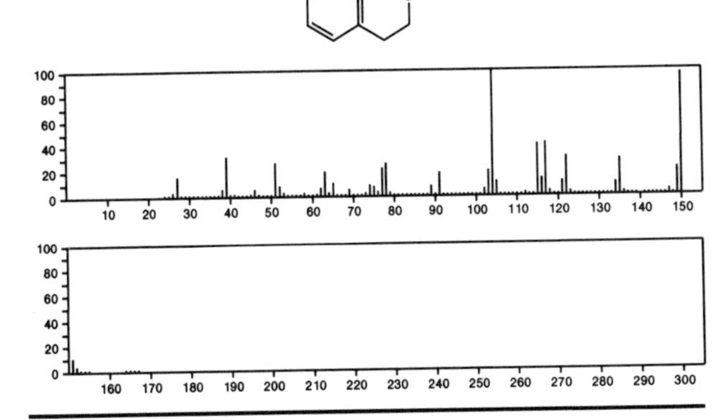

150 $C_9H_{14}N_2$ 15984-10-8

1*H*-Cyclooctapyrazole, 4,5,6,7,8,9-hexahydro-

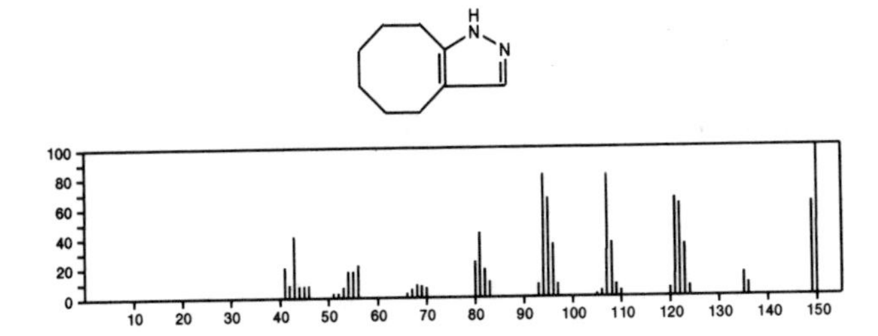

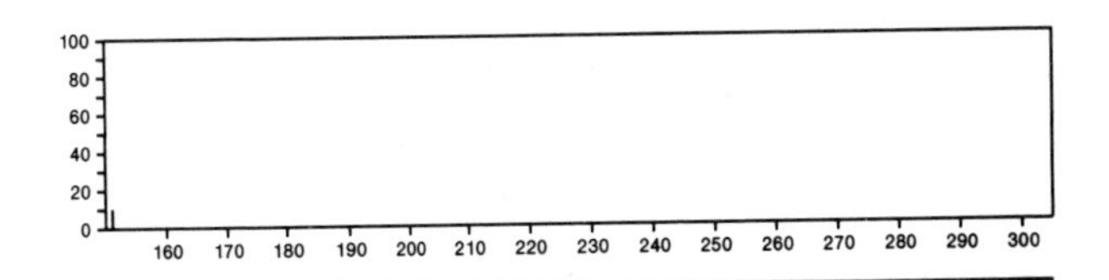

150 $C_9H_{14}N_2$ 18433-97-1

Pyrazine, 2,5-dimethyl-3-propyl-

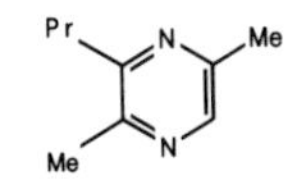

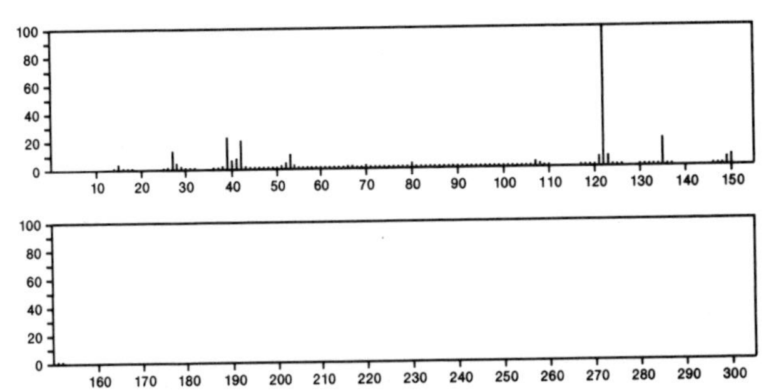

150 $C_9H_{14}N_2$ 54966-12-0

Piperidine, 2-(1*H*-pyrrol-2-yl)-

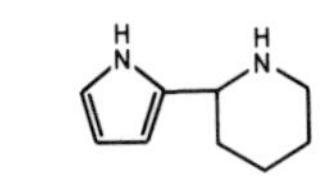

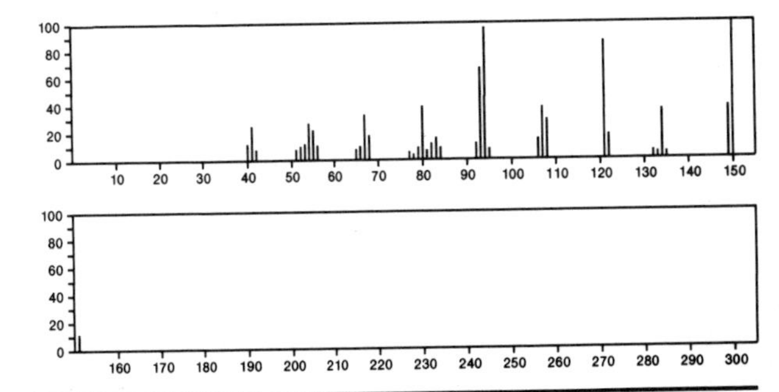

150 $C_9H_{14}Si$ 768-32-1

Silane, trimethylphenyl-

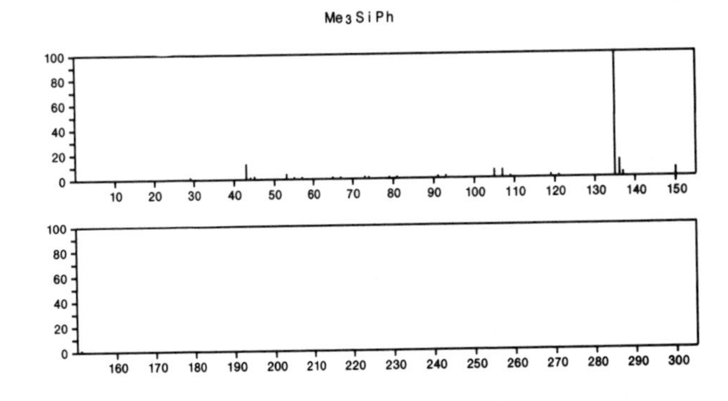

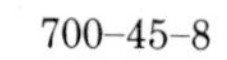

150 $C_{10}H_{11}F$ 700-45-8
Naphthalene, 5-fluoro-1,2,3,4-tetrahydro-

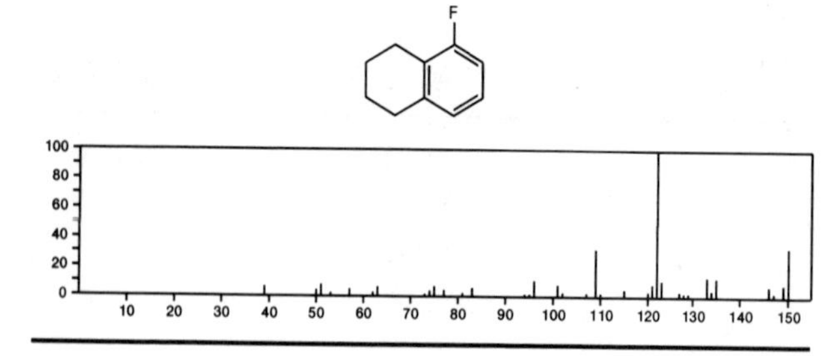

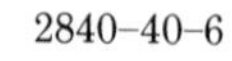

150 $C_{10}H_{11}F$ 2840-40-6
Naphthalene, 6-fluoro-1,2,3,4-tetrahydro-

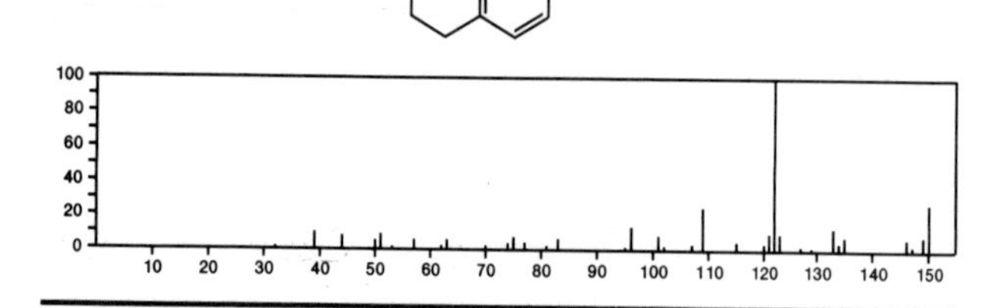

150 $C_{10}H_{14}O$ 88-18-6
Phenol, 2-(1,1-dimethylethyl)-

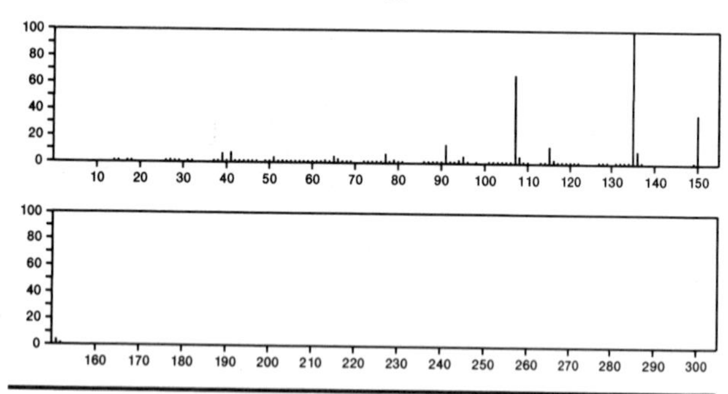

150 $C_{10}H_{14}O$ 89-72-5
Phenol, 2-(1-methylpropyl)-

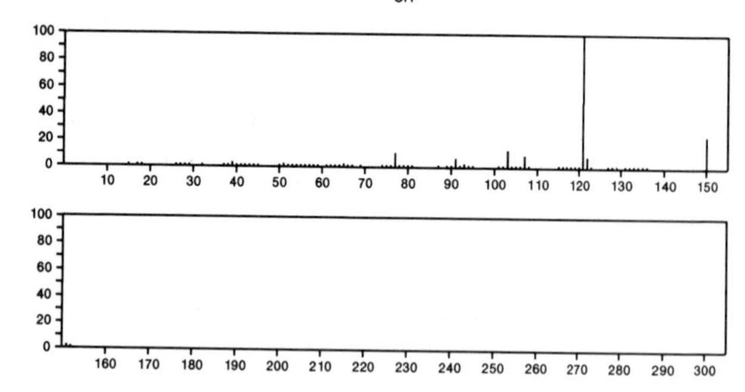

150 $C_{10}H_{14}O$ 89-83-8
Phenol, 5-methyl-2-(1-methylethyl)-

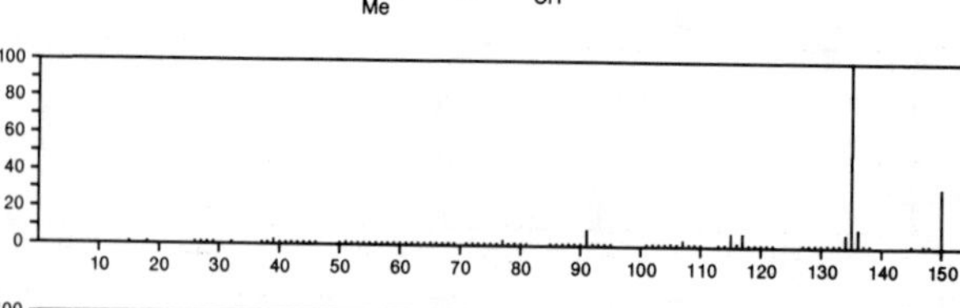

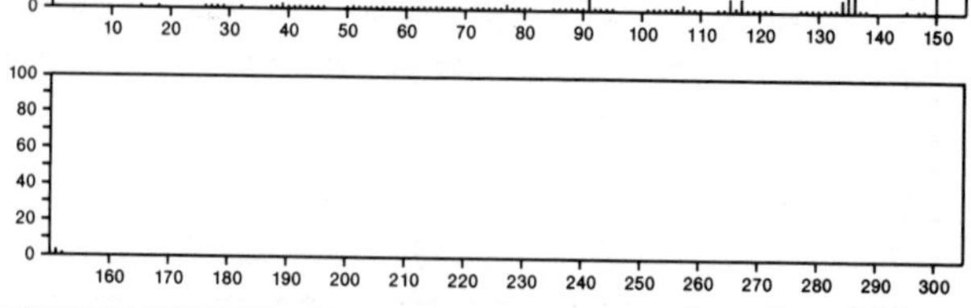

150 $C_{10}H_{14}O$ 98-54-4
Phenol, 4-(1,1-dimethylethyl)-

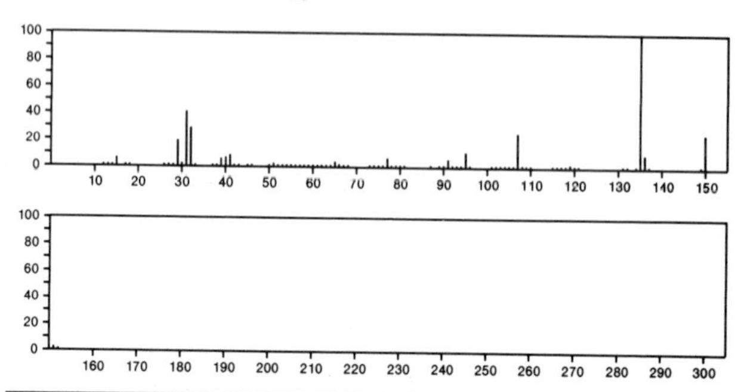

150 $C_{10}H_{14}O$ 99-49-0
2-Cyclohexen-1-one, 2-methyl-5-(1-methylethenyl)-

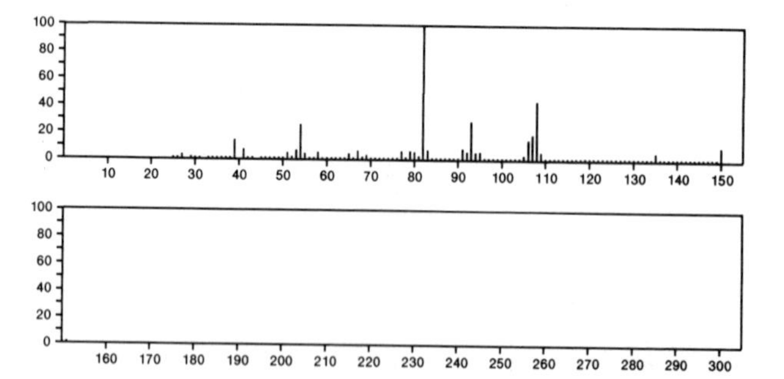

150 $C_{10}H_{14}O$ 99-71-8
Phenol, 4-(1-methylpropyl)-

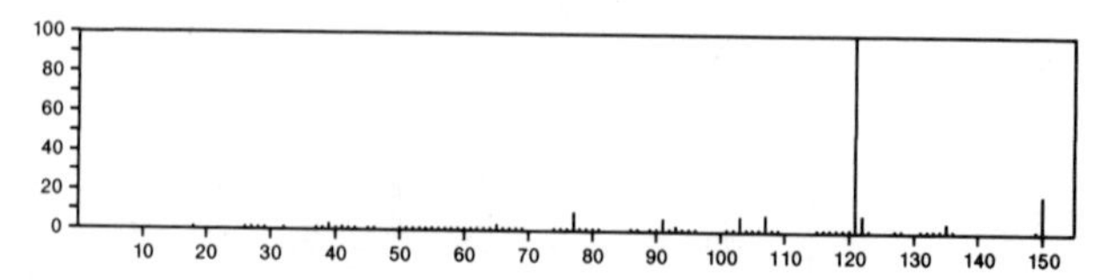

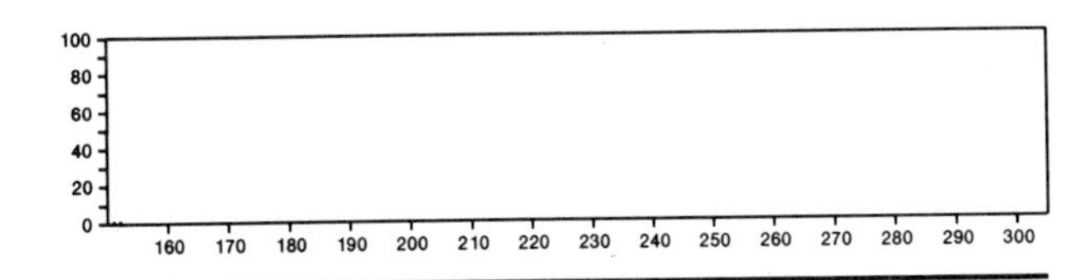

150 $C_{10}H_{14}O$ 100-86-7
Benzeneethanol, α,α-dimethyl-

Me2 COH CH2 Ph

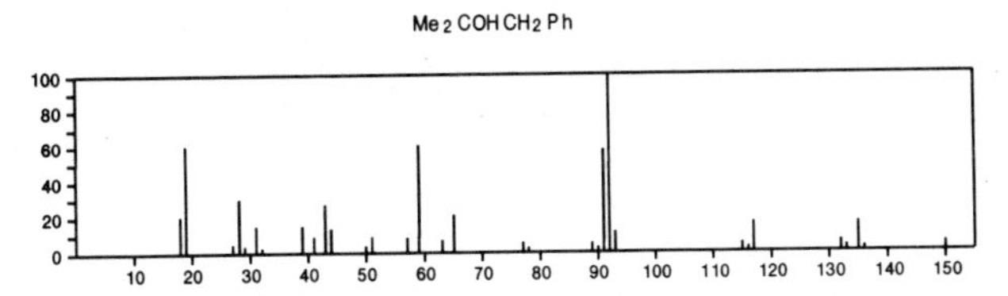

150 $C_{10}H_{14}O$ 104-45-0
Benzene, 1-methoxy-4-propyl-

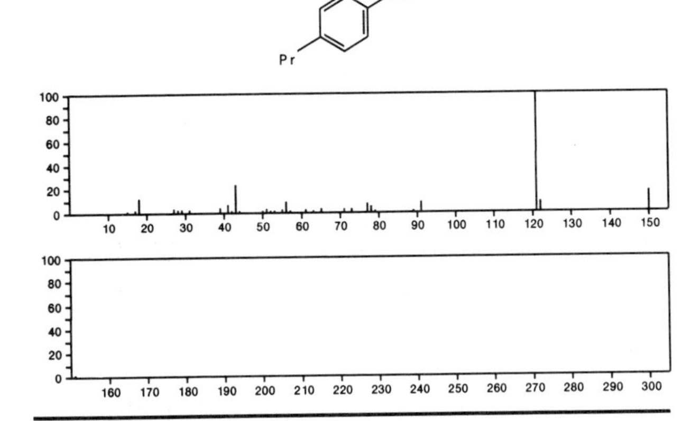

150 $C_{10}H_{14}O$ 491-09-8
2-Cyclohexen-1-one, 3-methyl-6-(1-methylethylidene)-

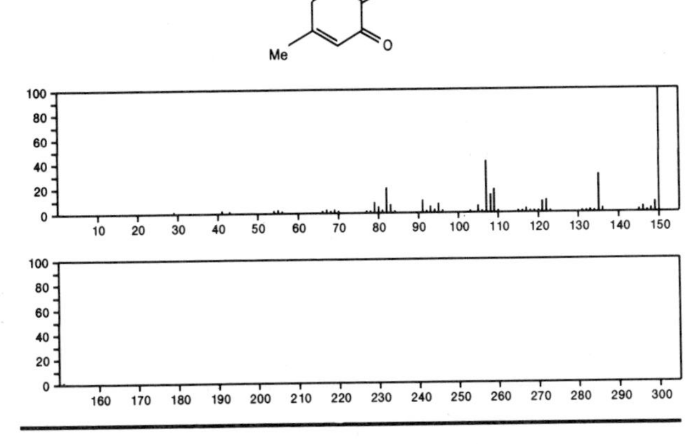

150 $C_{10}H_{14}O$ 494-90-6
Benzofuran, 4,5,6,7-tetrahydro-3,6-dimethyl-

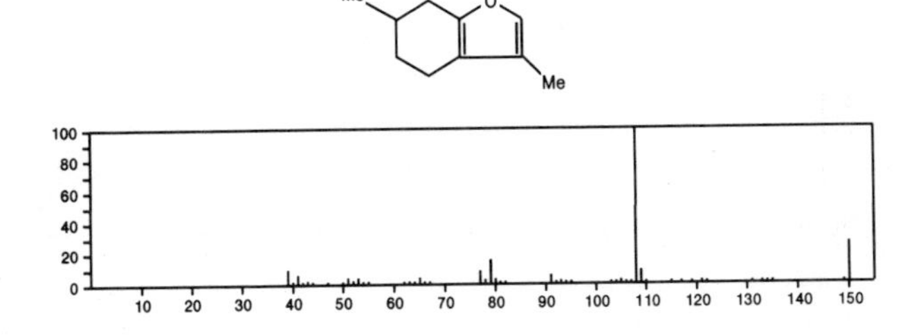

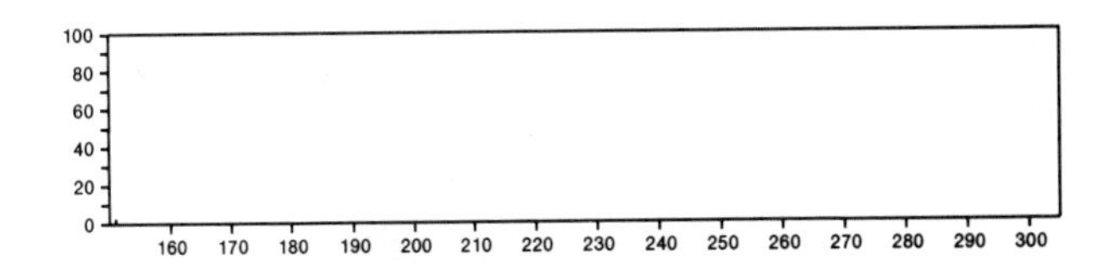

150 $C_{10}H_{14}O$ 499-75-2
Phenol, 2-methyl-5-(1-methylethyl)-

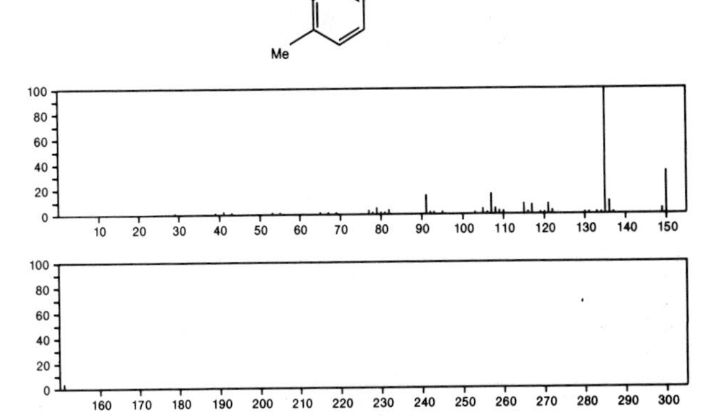

150 $C_{10}H_{14}O$ 539-52-6
Furan, 3-(4-methyl-3-pentenyl)-

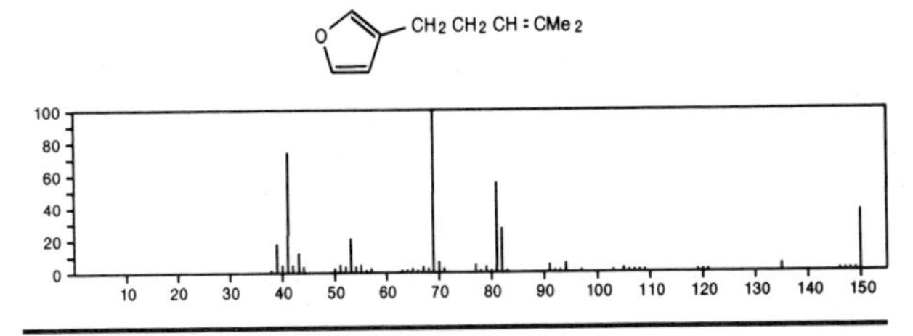

150 $C_{10}H_{14}O$ 585-34-2
Phenol, 3-(1,1-dimethylethyl)-

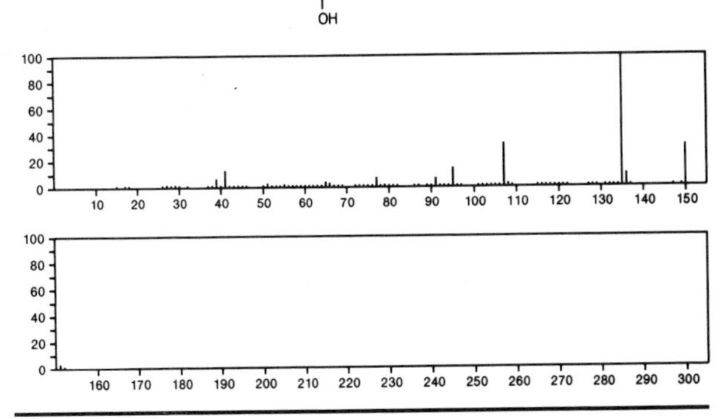

150 $C_{10}H_{14}O$ 700-58-3
Tricyclo[3.3.1.1^{3,7}]decanone

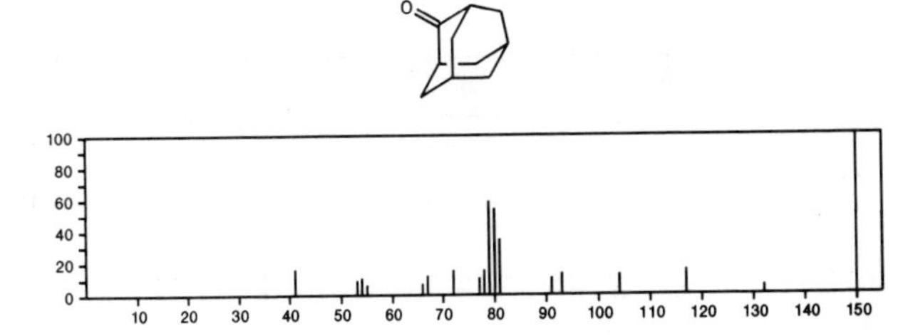

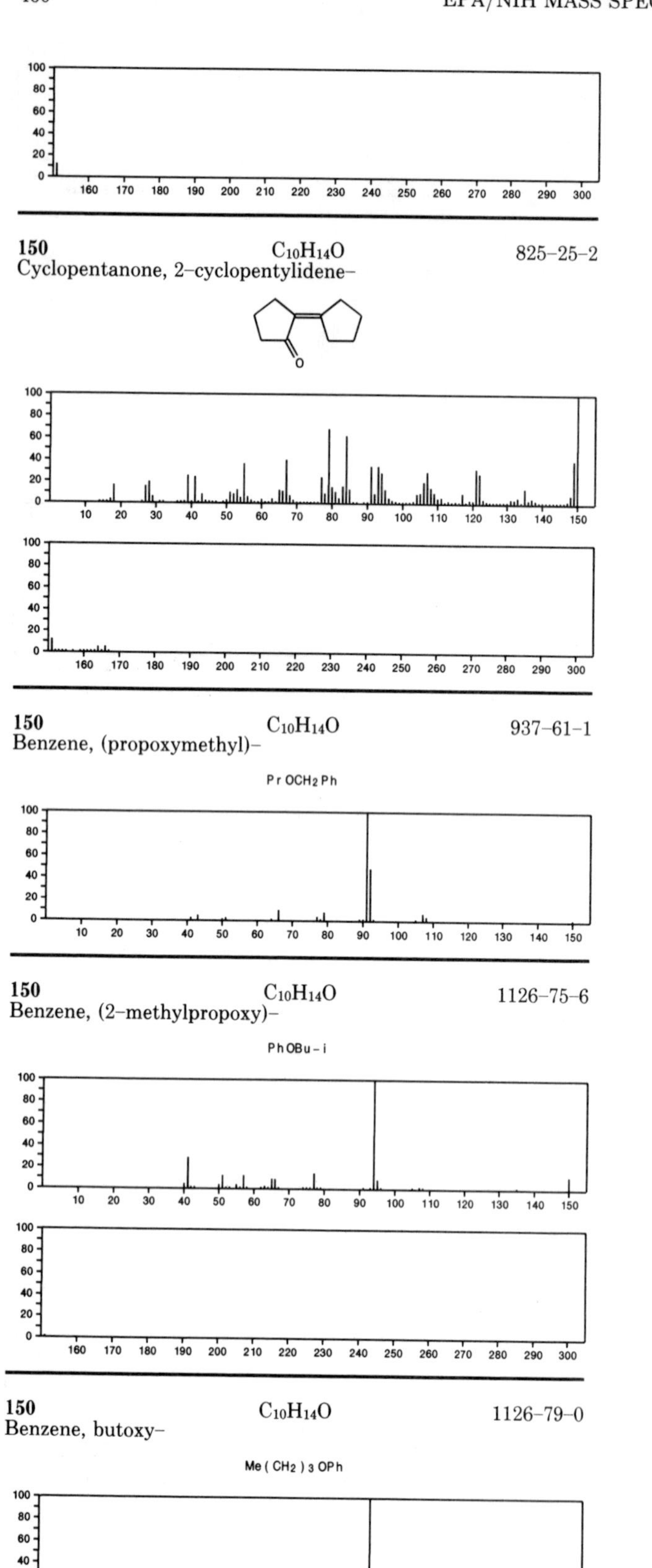

150 $C_{10}H_{14}O$ 825–25–2
Cyclopentanone, 2–cyclopentylidene–

150 $C_{10}H_{14}O$ 937–61–1
Benzene, (propoxymethyl)–

150 $C_{10}H_{14}O$ 1126–75–6
Benzene, (2–methylpropoxy)–

150 $C_{10}H_{14}O$ 1126–79–0
Benzene, butoxy–

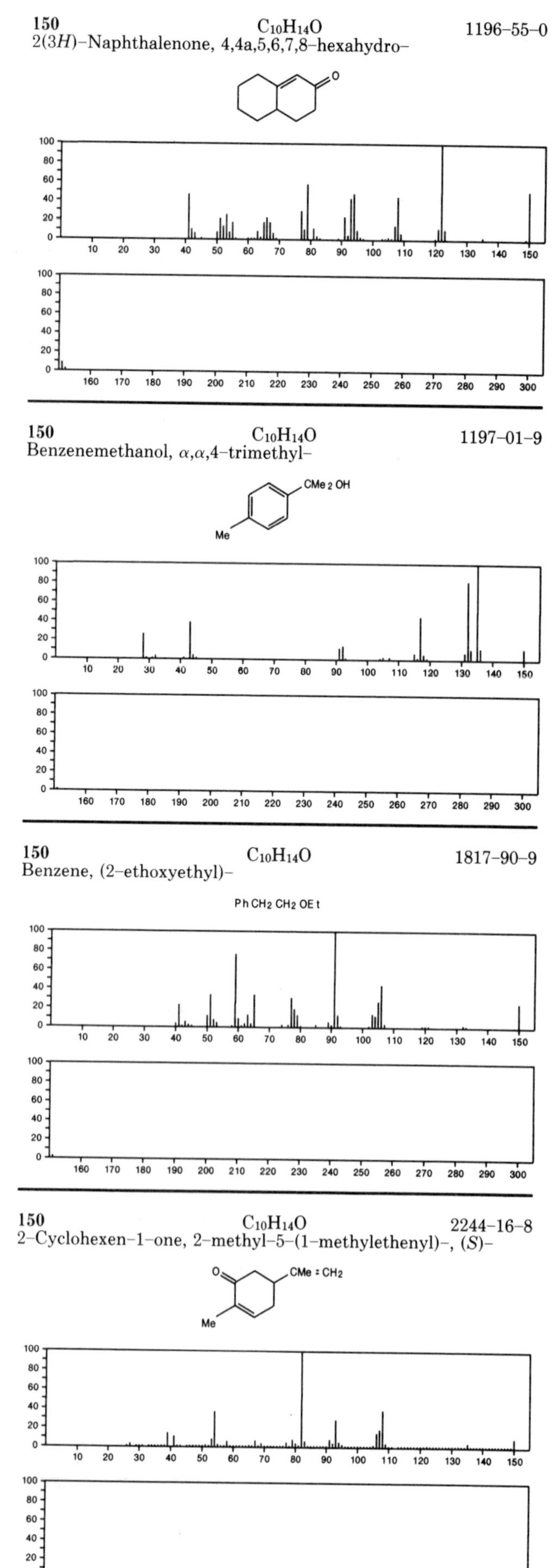

150 $C_{10}H_{14}O$ 1196–55–0
2(3*H*)–Naphthalenone, 4,4a,5,6,7,8–hexahydro–

150 $C_{10}H_{14}O$ 1197–01–9
Benzenemethanol, α,α,4–trimethyl–

150 $C_{10}H_{14}O$ 1817–90–9
Benzene, (2–ethoxyethyl)–

150 $C_{10}H_{14}O$ 2244–16–8
2–Cyclohexen–1–one, 2–methyl–5–(1–methylethenyl)–, (*S*)–

150 $C_{10}H_{14}O$ 2817-96-1

3-Buten-2-one, 4-(3-cyclohexen-1-yl)-

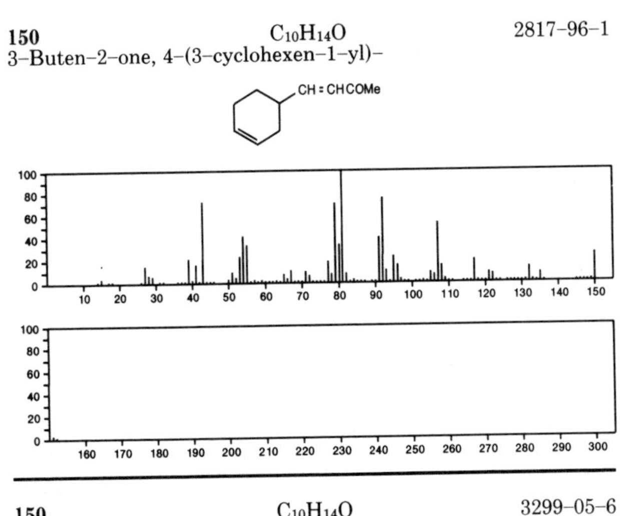

150 $C_{10}H_{14}O$ 3299-05-6

Benzene, (1-ethoxyethyl)-

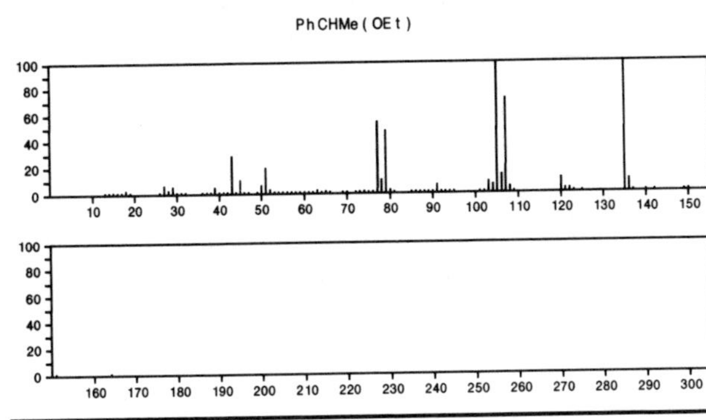

150 $C_{10}H_{14}O$ 4028-66-4

Benzene, 2-methoxy-1,3,5-trimethyl-

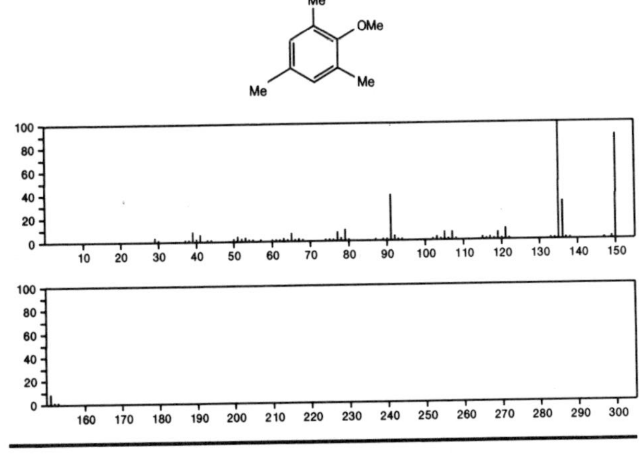

150 $C_{10}H_{14}O$ 4132-48-3

Benzene, 1-methoxy-4-(1-methylethyl)-

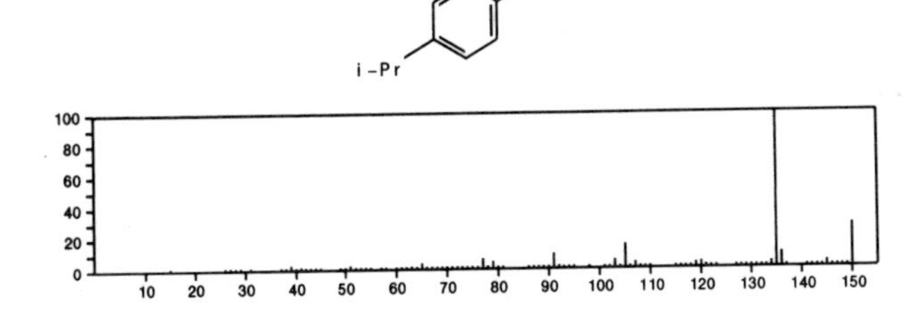

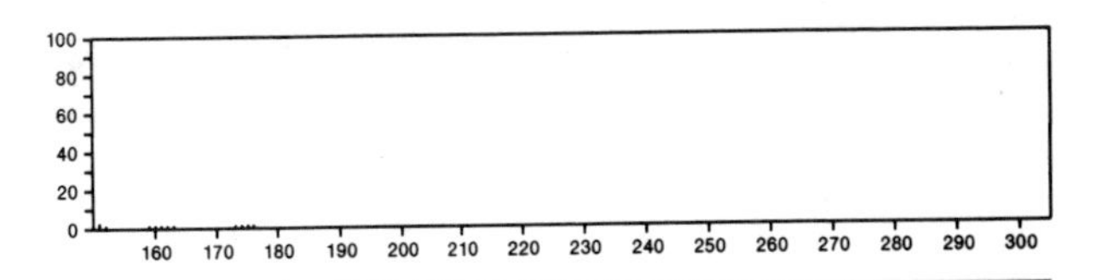

150 $C_{10}H_{14}O$ 4371-50-0

Benzeneethanol, β,4-dimethyl-

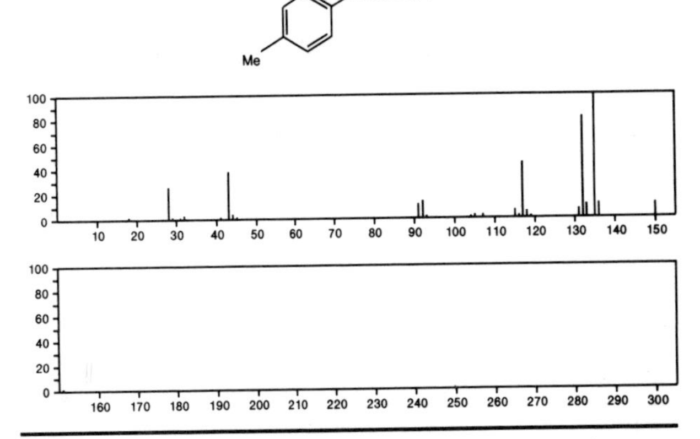

150 $C_{10}H_{14}O$ 4393-05-9

Benzenemethanol, 2,4,5-trimethyl-

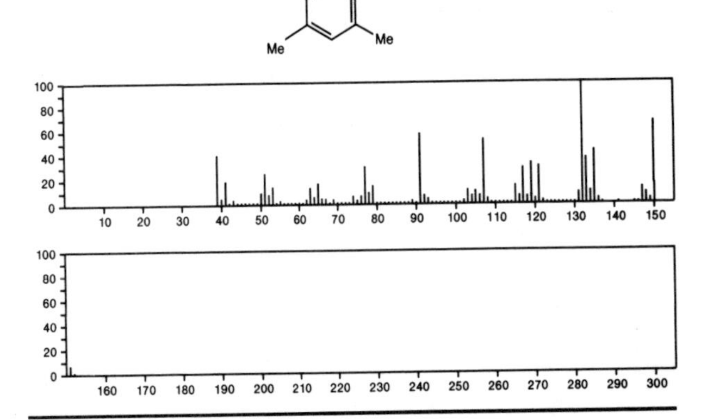

150 $C_{10}H_{14}O$ 6485-40-1

2-Cyclohexen-1-one, 2-methyl-5-(1-methylethenyl)-, (*R*)-

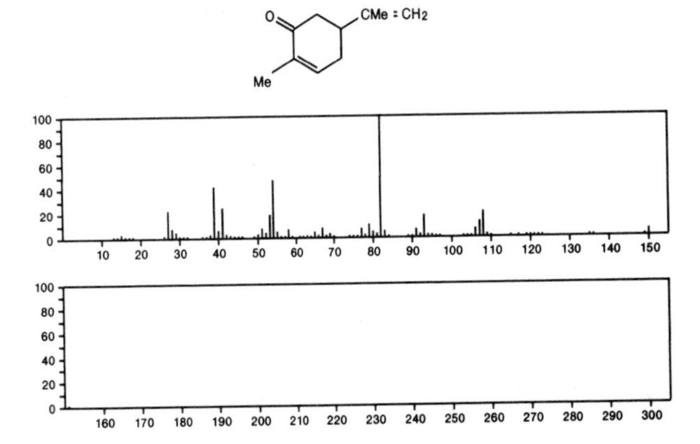

150 $C_{10}H_{14}O$ 6669–13–2
Benzene, (1,1-dimethylethoxy)-

PhOBu-t

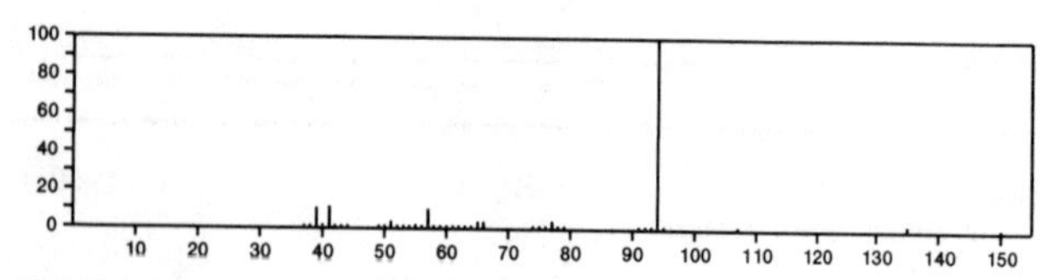

150 $C_{10}H_{14}O$ 17190–71–5
2-Cyclopenten-1-one, 2-(2-butenyl)-3-methyl-, (*Z*)-

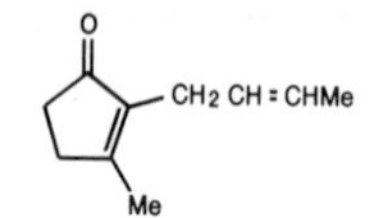

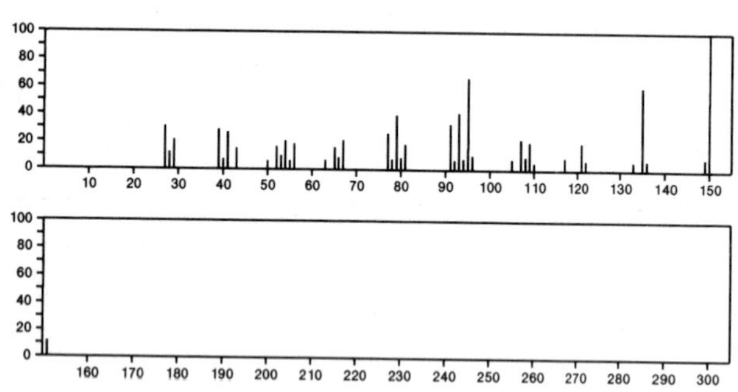

150 $C_{10}H_{14}O$ 17428–89–6
1-Indanone, 3a,4,7,7a-tetrahydro-7a-methyl-, *cis*-

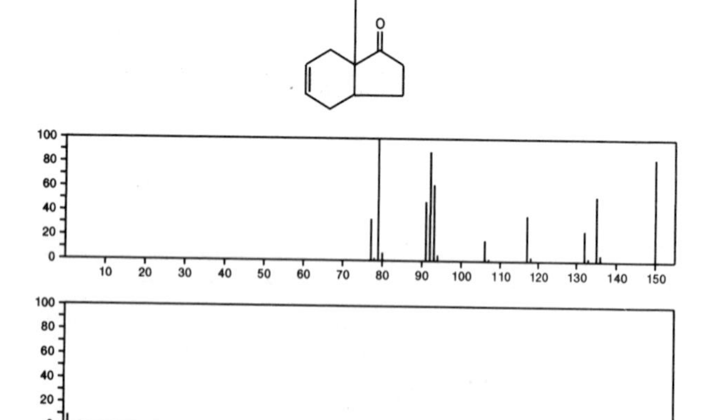

150 $C_{10}H_{14}O$ 17429–25–3
5(4*H*)-Indanone, 3a,7a-dihydro-7a-methyl-, *trans*-

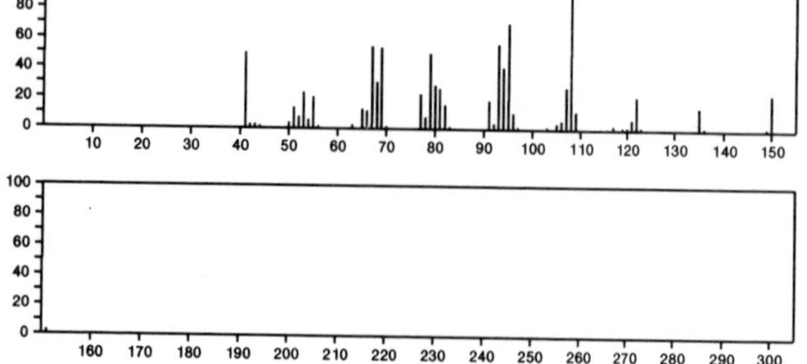

150 $C_{10}H_{14}O$ 17429–35–5
2,5-Cyclohexadien-1-one, 4-ethyl-3,4-dimethyl-

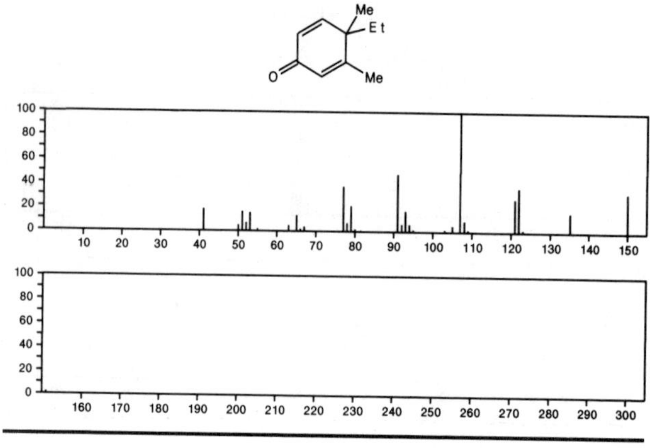

150 $C_{10}H_{14}O$ 18031–40–8
1-Cyclohexene-1-carboxaldehyde, 4-(1-methylethenyl)-, (*S*)-

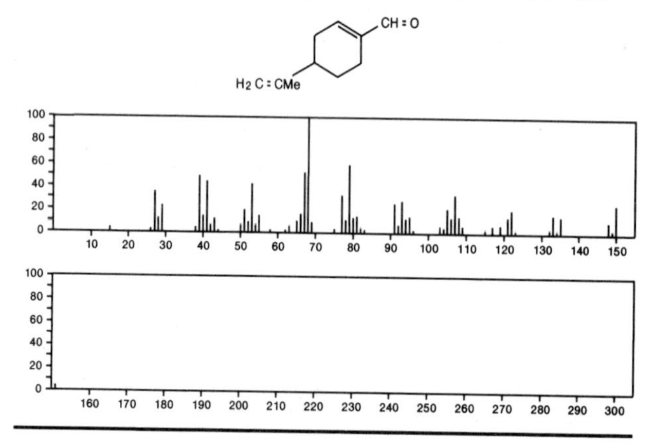

150 $C_{10}H_{14}O$ 18309–32–5
Bicyclo[3.1.1]hept-3-en-2-one, 4,6,6-trimethyl-, (1*R*-*cis*)-

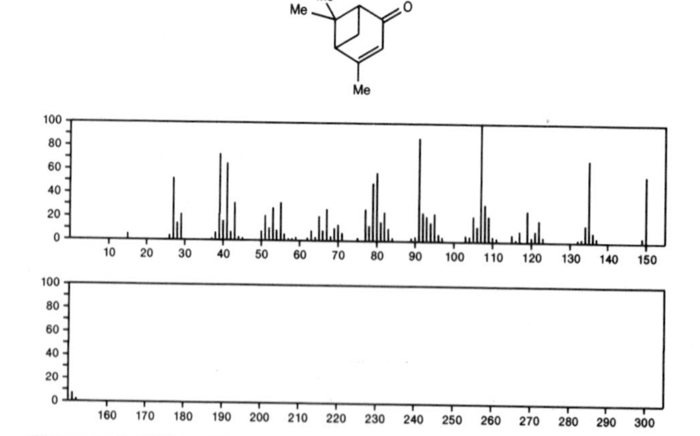

150 $C_{10}H_{14}O$ 21573–36–4
Anisole, 2,3,6-trimethyl-

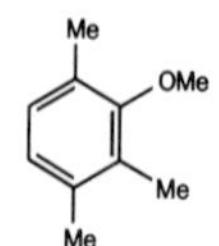

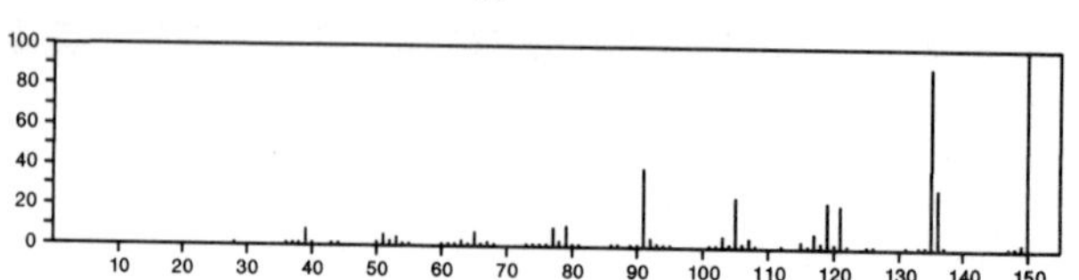

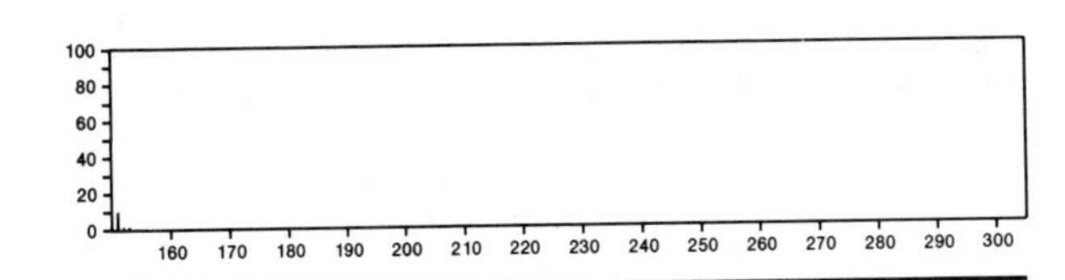

150 $C_{10}H_{14}O$ 23727–16–4
Bicyclo[3.1.1]hept–2–ene–2–carboxaldehyde, 6,6–dimethyl–, (1S)–

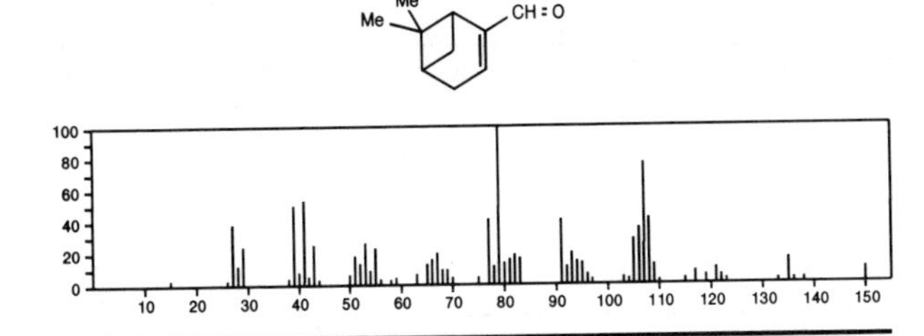

150 $C_{10}H_{14}O$ 24545–81–1
Bicyclo[3.1.0]hex–3–en–2–one, 4–methyl–1–(1–methylethyl)–

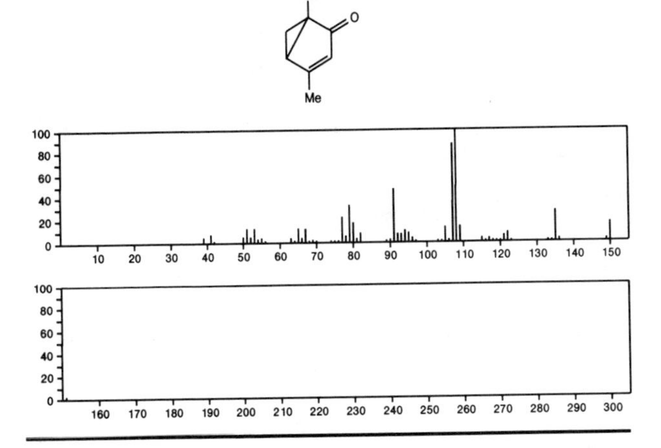

150 $C_{10}H_{14}O$ 26967–65–7
Phenol, diethyl–

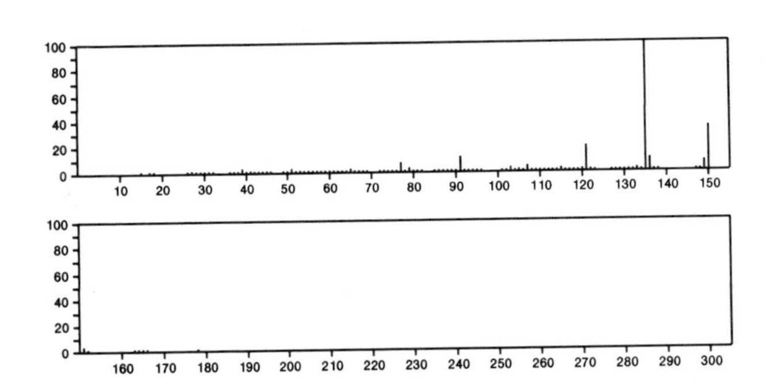

150 $C_{10}H_{14}O$ 27567–85–7
2,5–Methano–1*H*–inden–7(4*H*)–one, hexahydro–

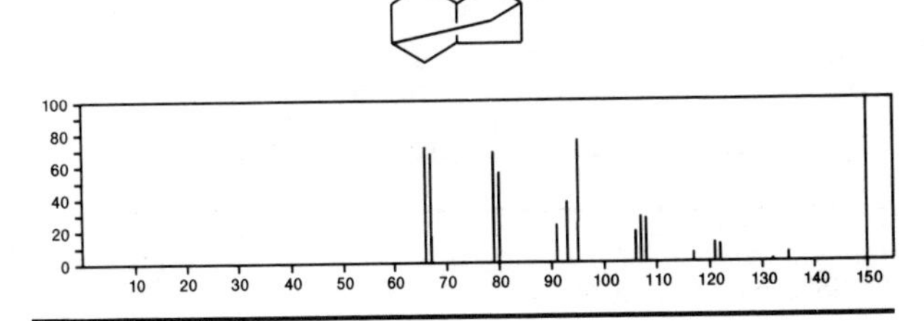

150 $C_{10}H_{14}O$ 27577–96–4
Ethanol, 2–xylyl–

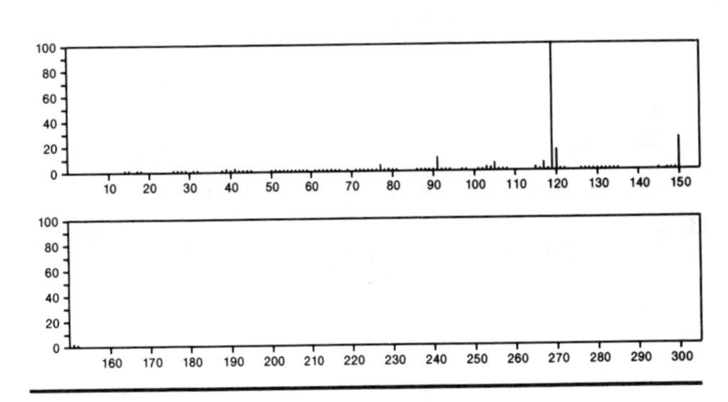

150 $C_{10}H_{14}O$ 41673–72–7
Benzeneethanol, *ar*–ethyl–

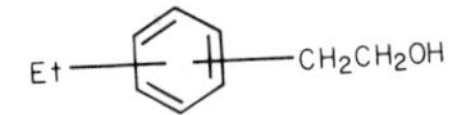

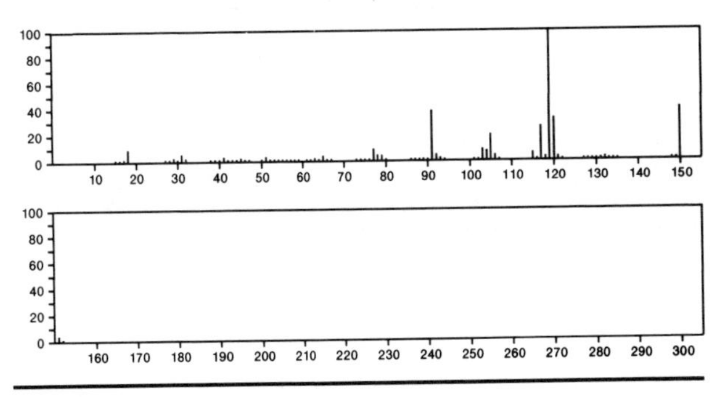

150 $C_{10}H_{14}O$ 41702–60–7
1,7–Octadien–3–one, 2–methyl–6–methylene–

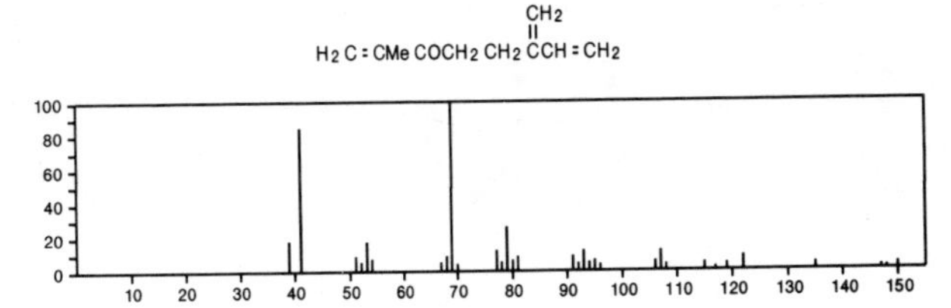

150 $C_{10}H_{14}O$ 52089-32-4
Benzeneethanol, α,β-dimethyl-

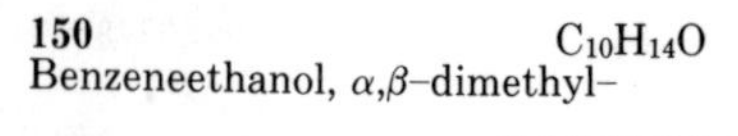
MeCH(OH)CHMePh

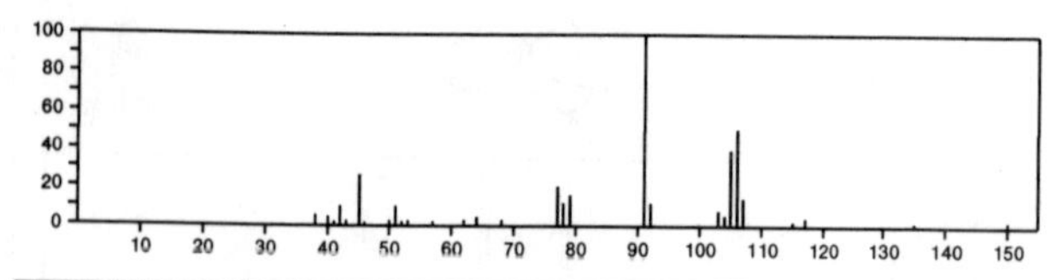

150 $C_{10}H_{14}O$ 54166-48-2
Cyclohexanone, 2-(2-butynyl)-

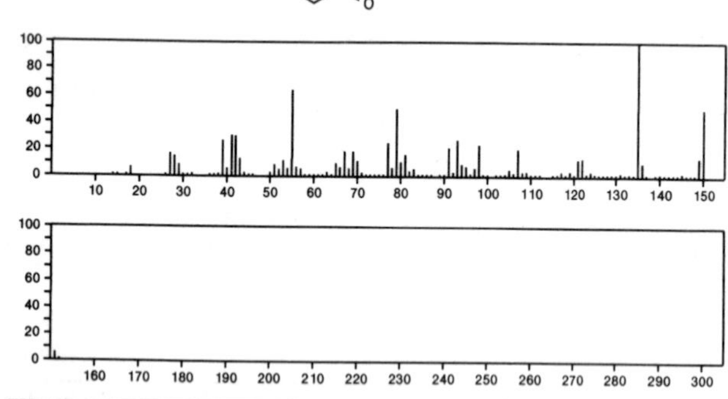

150 $C_{10}H_{14}O$ 54166-49-3
Benzenemethanol, *ar*,*ar*,α-trimethyl-

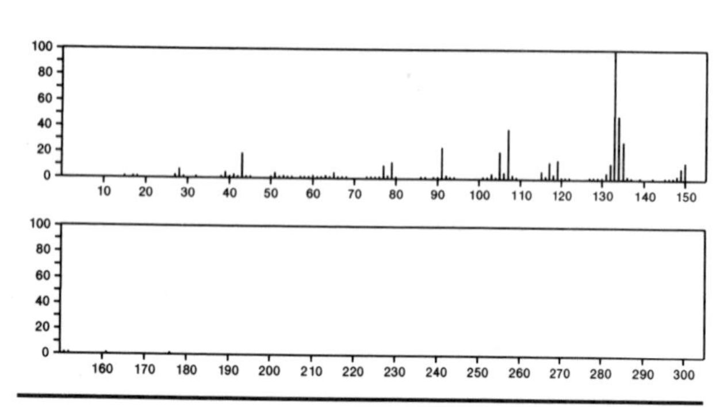

150 $C_{10}H_{14}O$ 54345-60-7
1-Oxaspiro[2.5]oct-5-ene, 8,8-dimethyl-4-methylene-

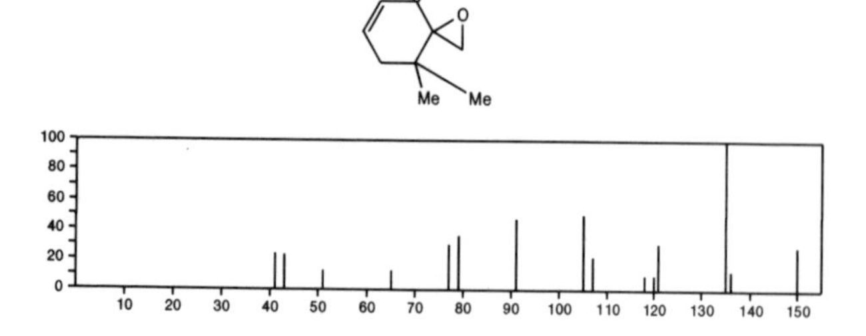

150 $C_{10}H_{14}O$ 54725-16-5
2*H*-Inden-2-one, 1,4,5,6,7,7a-hexahydro-7a-methyl-, (*S*)-

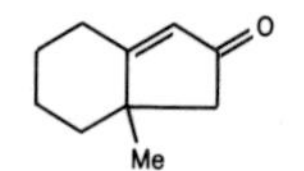

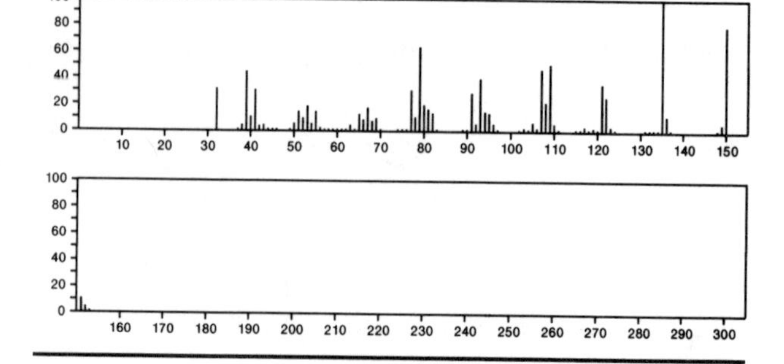

150 $C_{10}H_{14}O$ 55712-51-1
1,5,7-Octatrien-3-one, 2,6-dimethyl-, (*E*)-

H2C:CHCMe:CHCH2COCMe:CH2

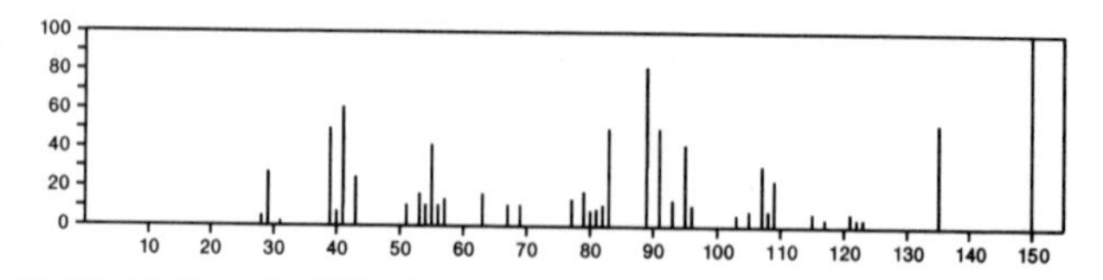

150 $C_{11}H_{18}$ 281-46-9
Tricyclo[4.3.1.1^{3,8}]undecane

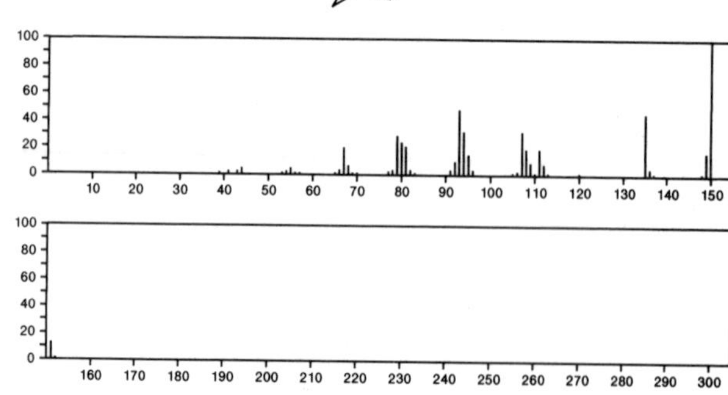

150 $C_{11}H_{18}$ 21789-56-0
Naphthalene, 1,2,3,4,4a,5,8,8a-octahydro-4a-methyl-, *trans*-

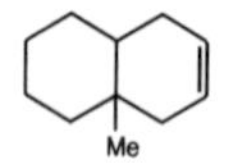

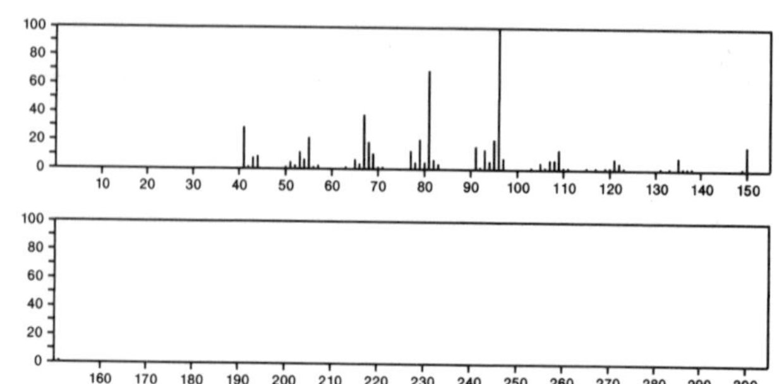

150 $C_{11}H_{18}$ 22822-99-7
Cyclopropane, 1-ethenyl-2-hexenyl-, [1α,2β(*E*)]-(±)-

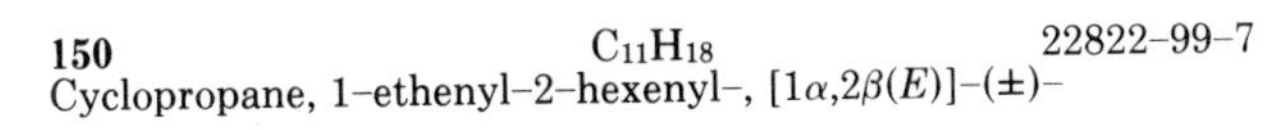

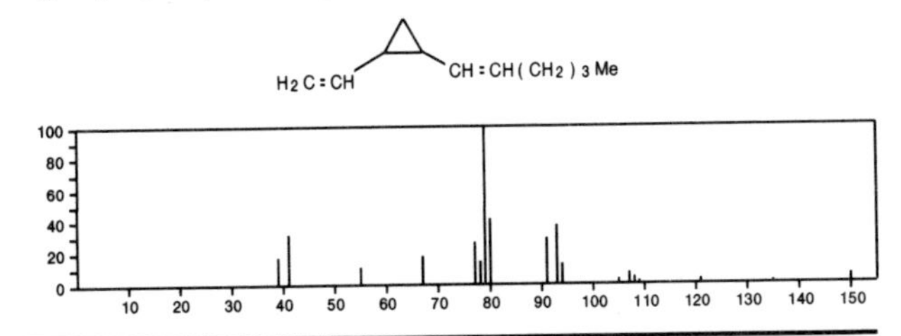

150 $C_{11}H_{18}$ 54166-47-1
Bicyclo[5.1.0]octane, 8-(1-methylethylidene)-

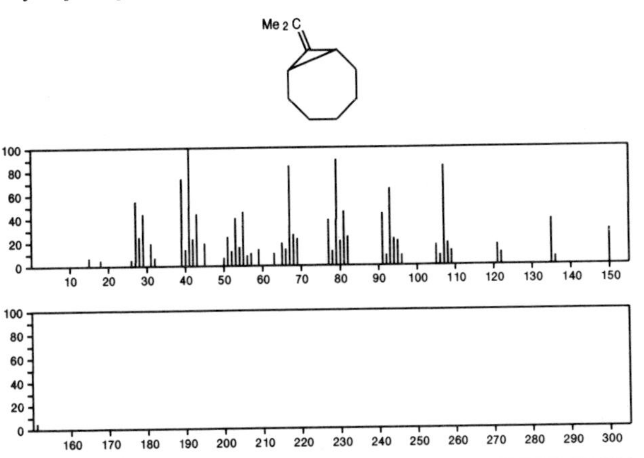

150 $C_{11}H_{18}$ 56030-49-0
Cyclohexene, 3-(3-methyl-1-butenyl)-, (*E*)-

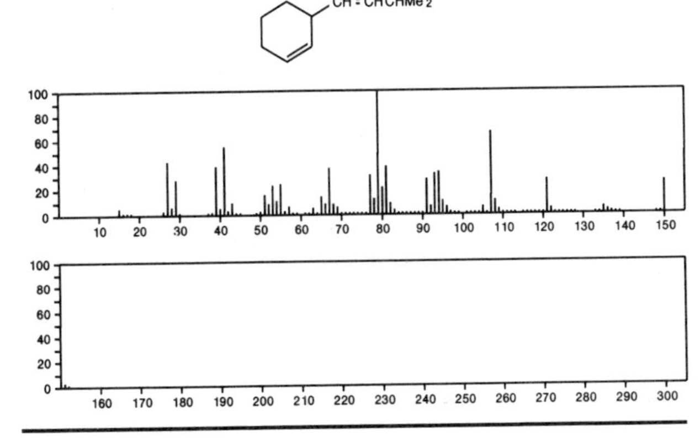

150 $C_{11}H_{18}$ 56324-70-0
1*H*-Indene, 1-ethylideneoctahydro-, *trans*-

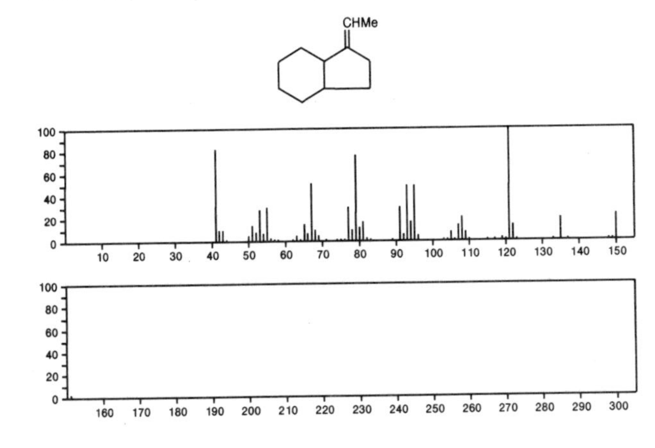

151 CHN_3O_6 517-25-9
Methane, trinitro-

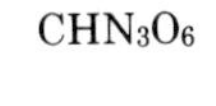

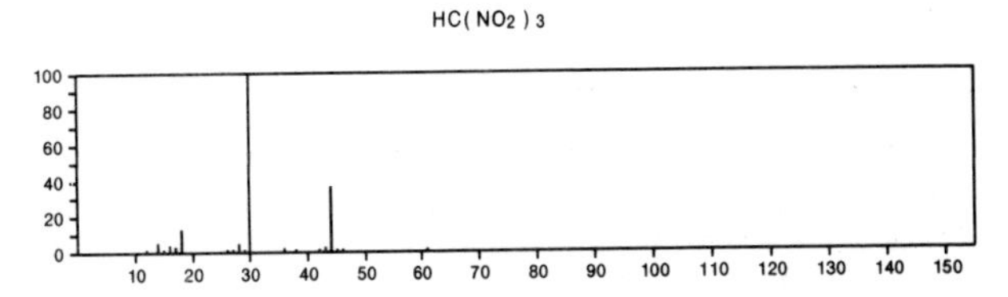

151 $C_4H_9NO_5$ 20633-16-3
Ethanol, 2,2'-oxybis-, mononitrate

HOCH2CH2OCH2CH2ONO2

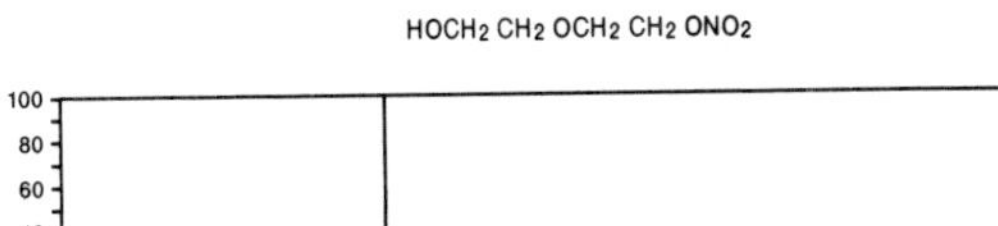

151 C_5HF_4N 2875-18-5
Pyridine, 2,3,5,6-tetrafluoro-

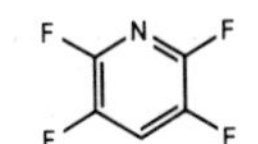

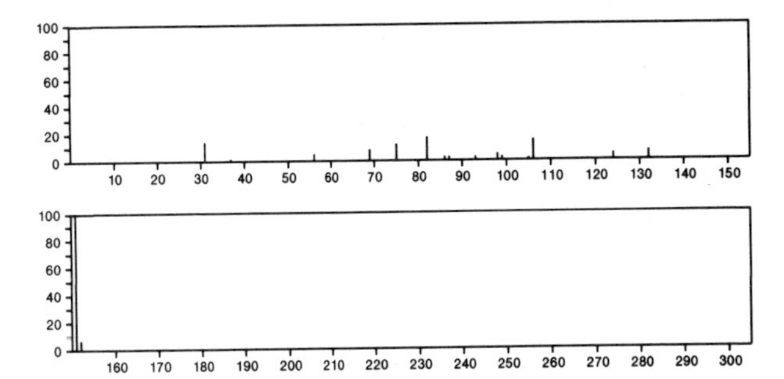

151 $C_5H_5N_5O$ 73-40-5
6*H*-Purin-6-one, 2-amino-1,7-dihydro-

O H N N N H2N N H

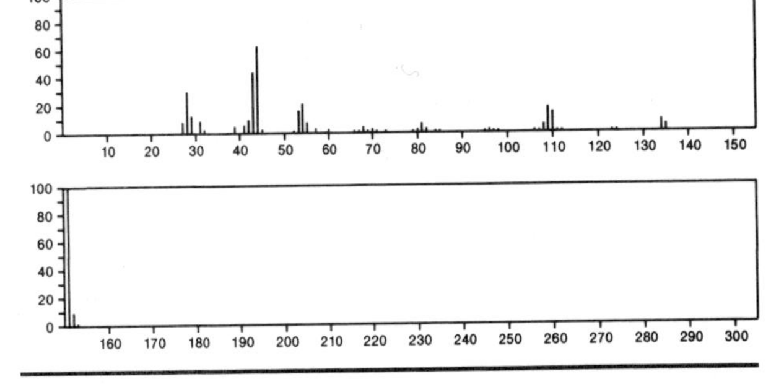

151 $C_6H_5N_3S$ 6952-68-7
1,2,4-Triazolo[4,3-*a*]pyridine-3(2*H*)-thione

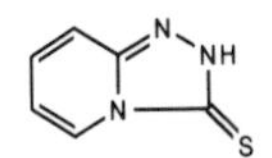

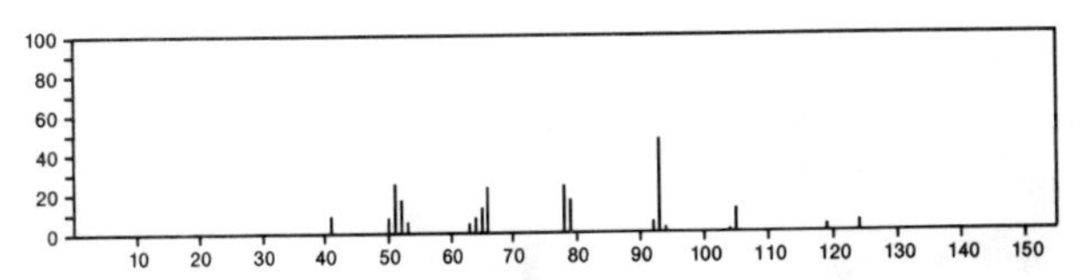

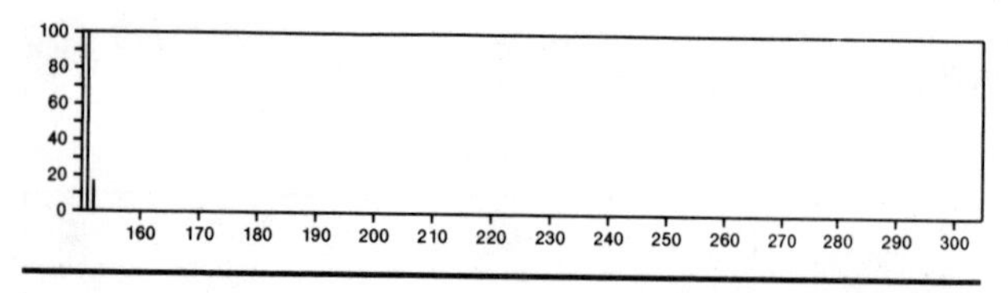

151 $C_6H_5N_3S$ 13316-06-8
Thiazolo[5,4-*d*]pyrimidine, 7-methyl-

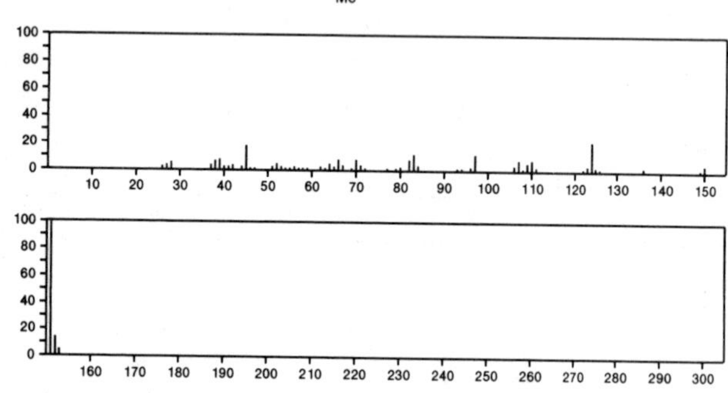

151 $C_6H_5N_3S$ 13554-88-6
Thiazolo[5,4-*d*]pyrimidine, 2-methyl-

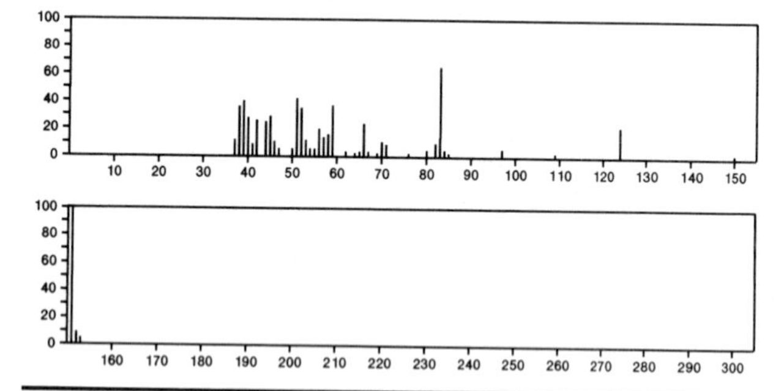

151 $C_6H_5N_3S$ 13554-89-7
Thiazolo[5,4-*d*]pyrimidine, 5-methyl-

151 $C_6H_9N_5$ 35975-34-9
1(2*H*)-Pyrazineacetonitrile, 5-amino-3,6-dihydro-3-imino-

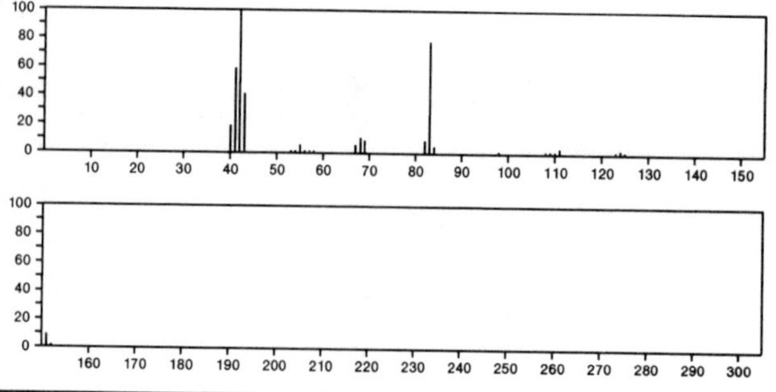

151 C_7H_5NOS 2634-33-5
1,2-Benzisothiazol-3(2*H*)-one

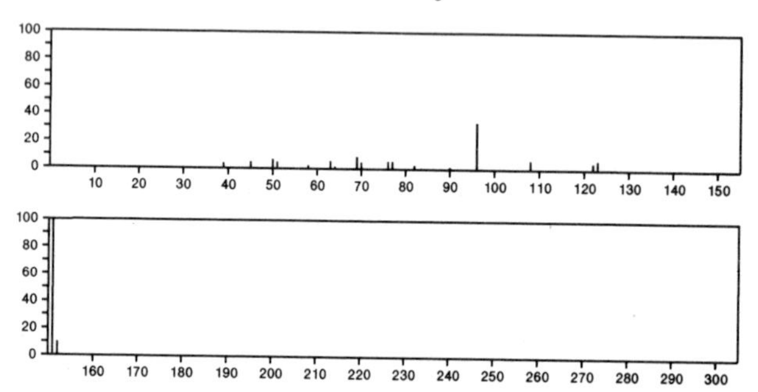

151 $C_7H_5NO_3$ 99-61-6
Benzaldehyde, 3-nitro-

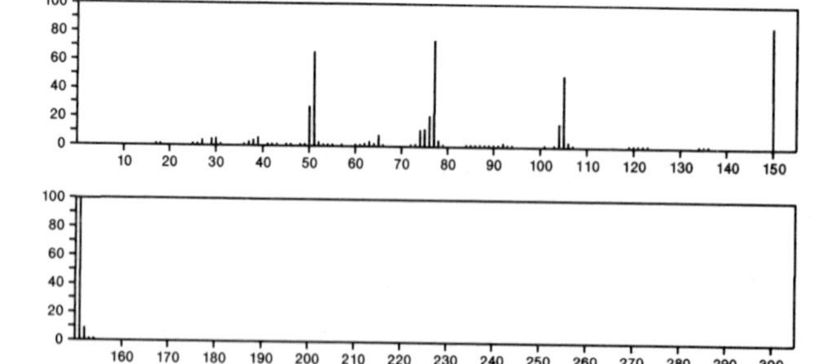

151 $C_7H_5NO_3$ 552-89-6
Benzaldehyde, 2-nitro-

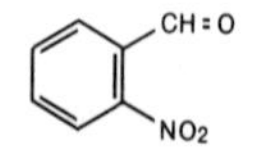

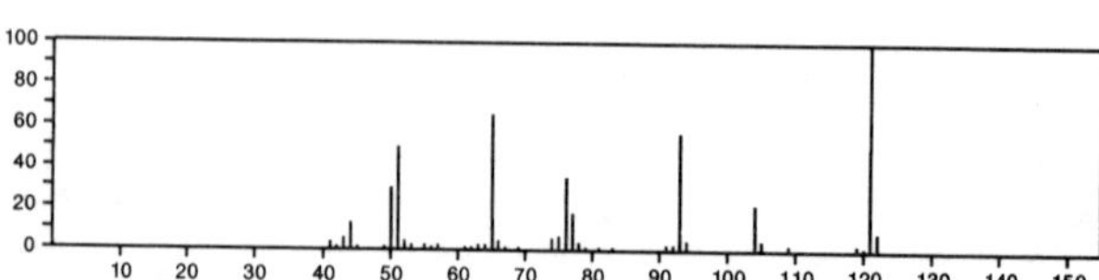

151 $C_7H_5NO_3$ 555–16–8
Benzaldehyde, 4–nitro–

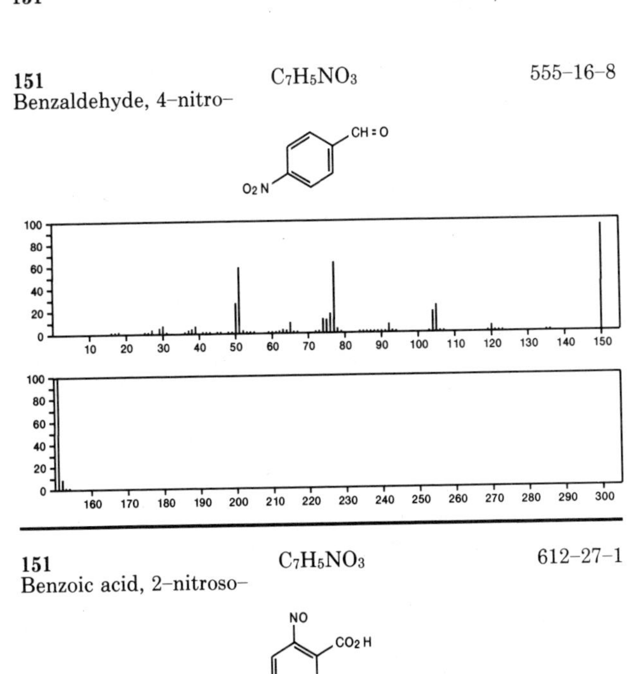

151 $C_7H_5NO_3$ 612–27–1
Benzoic acid, 2–nitroso–

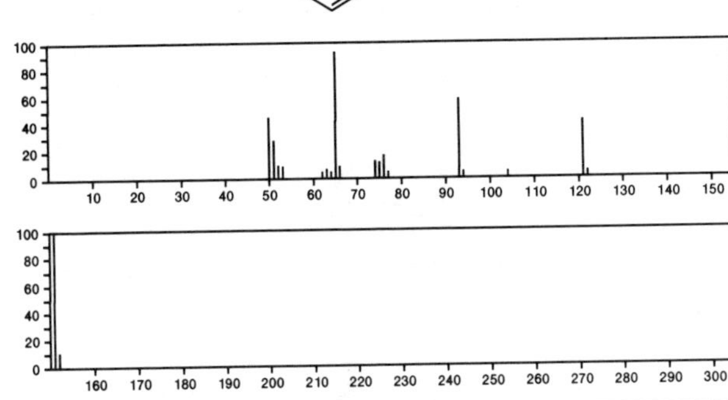

151 $C_7H_9N_3O$ 5351–17–7
Benzoic acid, 4–amino–, hydrazide

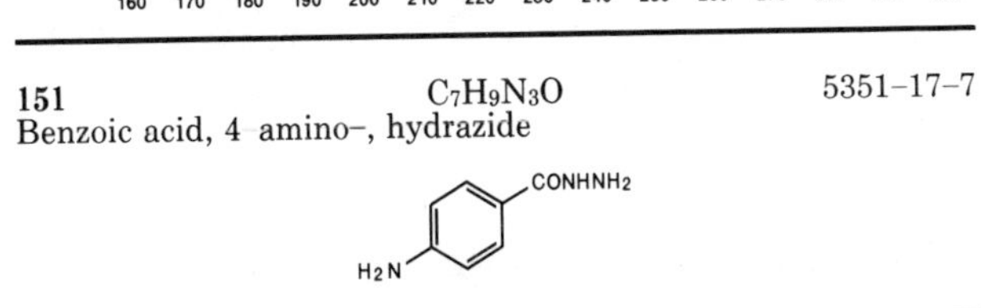

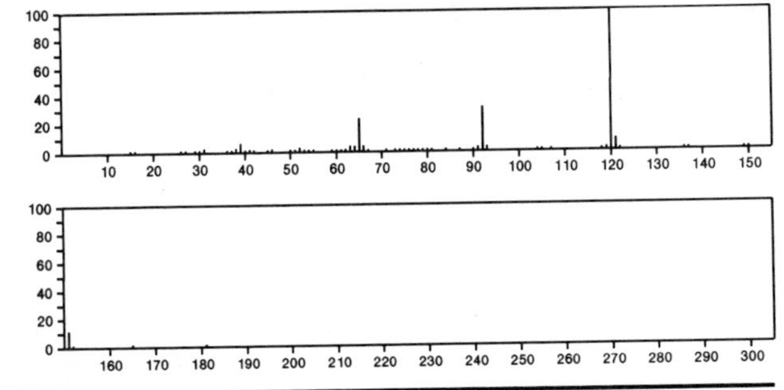

151 $C_7H_9N_3O$ 31479–64–8
3–Picolinium, 1–ureido–, hydroxide, inner salt

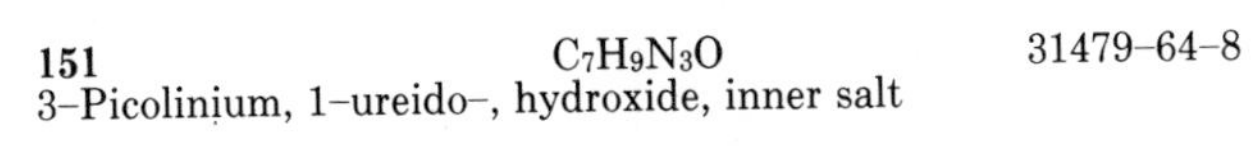

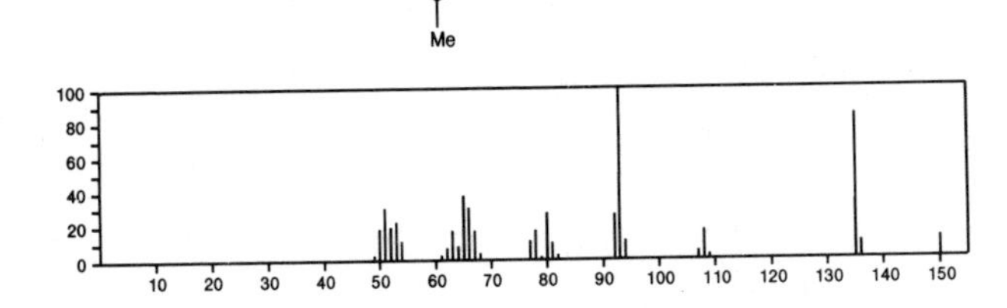

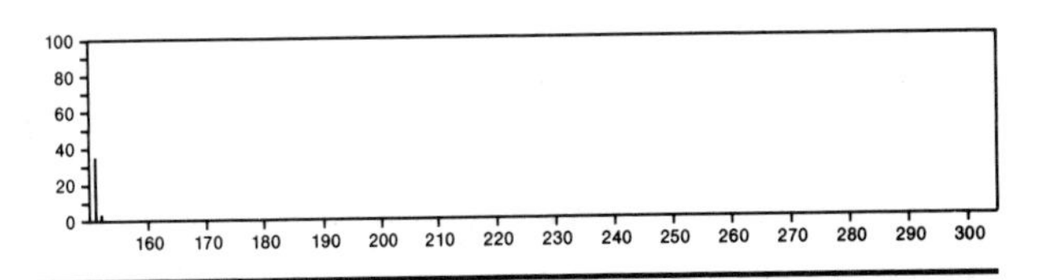

151 $C_8H_9NO_2$ 100–12–9
Benzene, 1–ethyl–4–nitro–

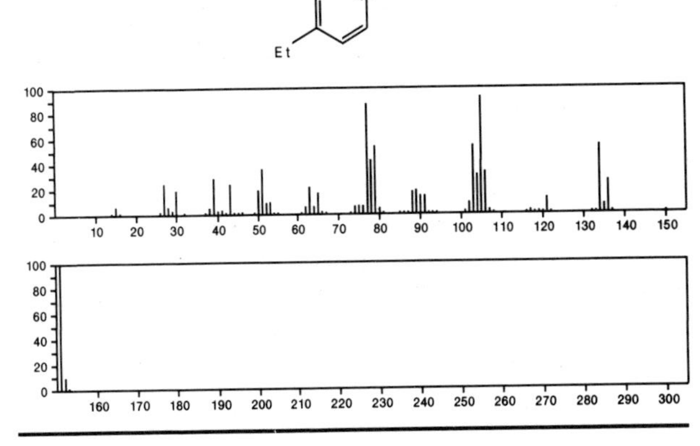

151 $C_8H_9NO_2$ 103–90–2
Acetamide, *N*–(4–hydroxyphenyl)–

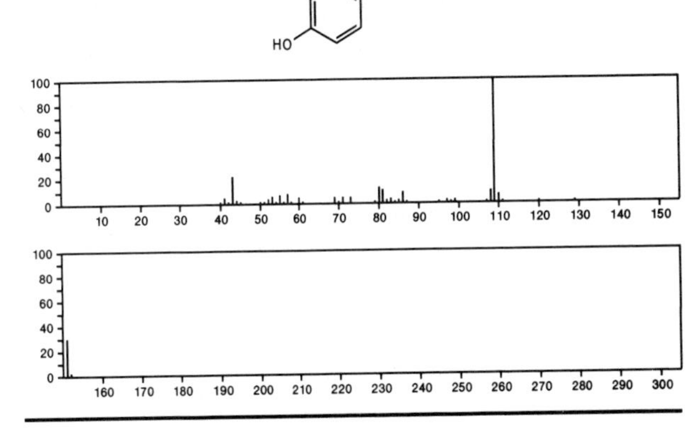

151 $C_8H_9NO_2$ 134–20–3
Benzoic acid, 2–amino–, methyl ester

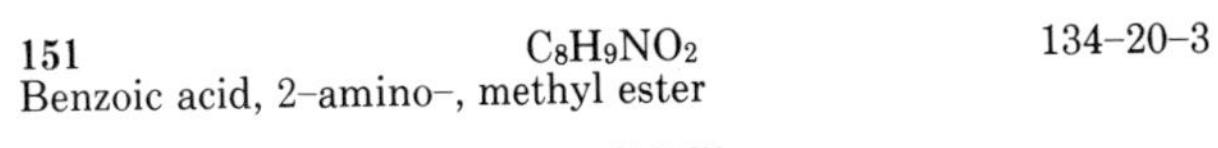

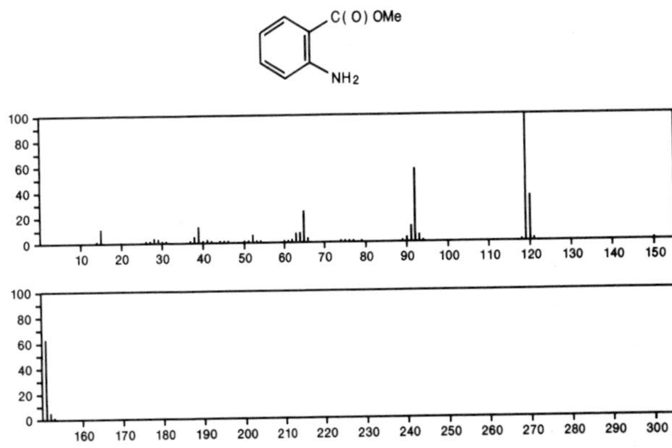

151 $C_8H_9NO_2$ 614-80-2
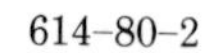
Acetamide, *N*-(2-hydroxyphenyl)-
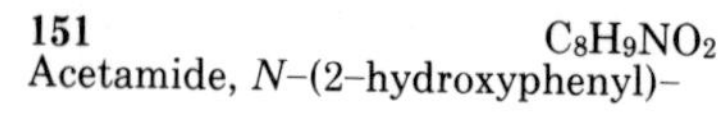

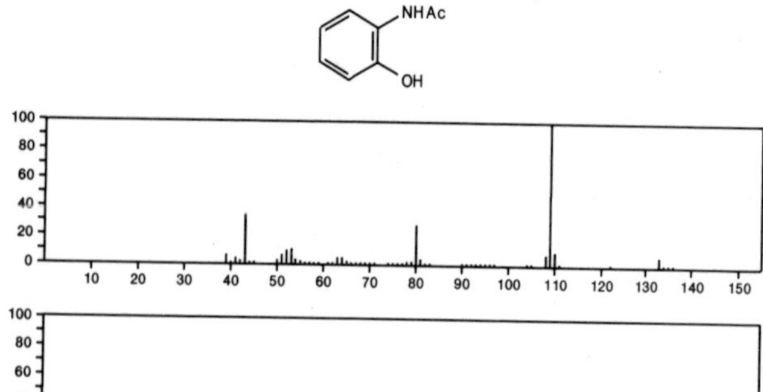

151 $C_8H_9NO_2$ 619-45-4
Benzoic acid, 4-amino-, methyl ester

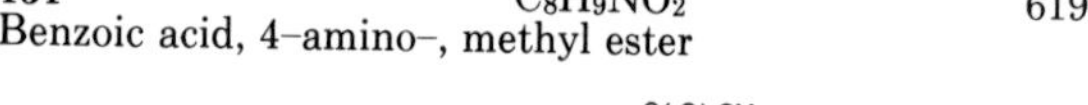

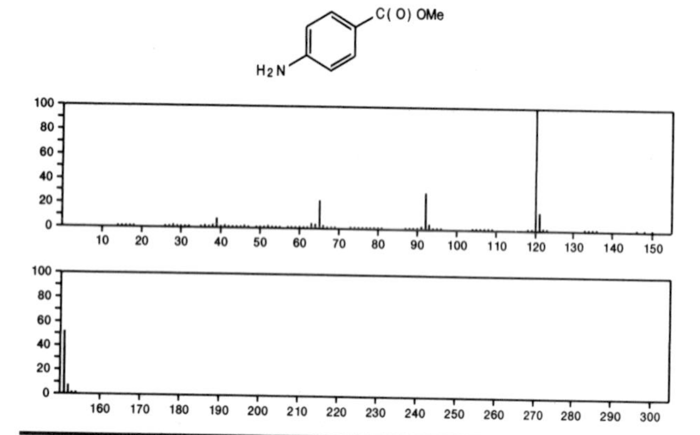

151 $C_8H_9NO_2$ 1006-96-8
3-Pyridinol, 4-methyl-, acetate (ester)

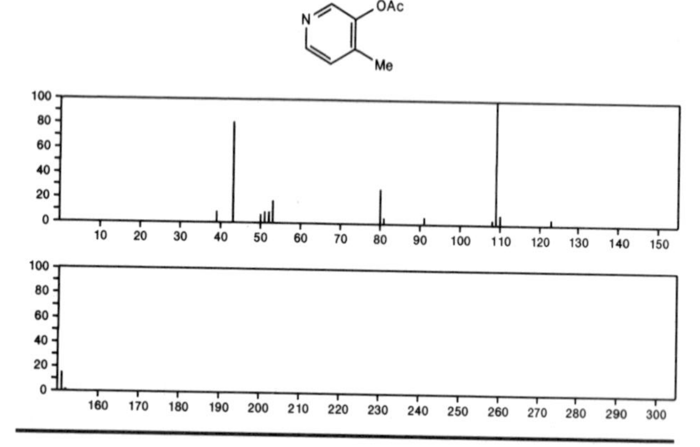

151 $C_8H_9NO_2$ 1007-48-3
4-Pyridinemethanol, acetate (ester)

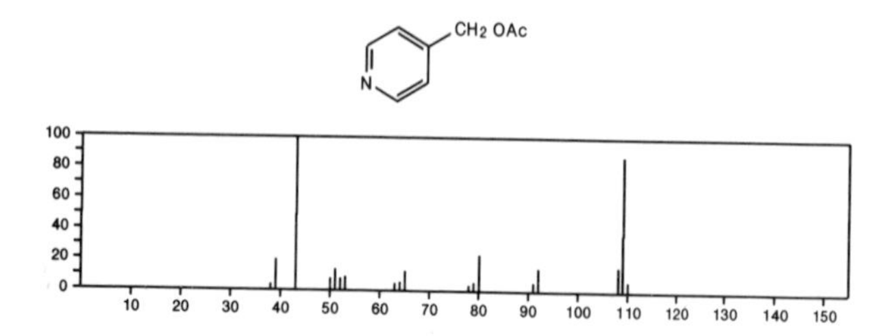

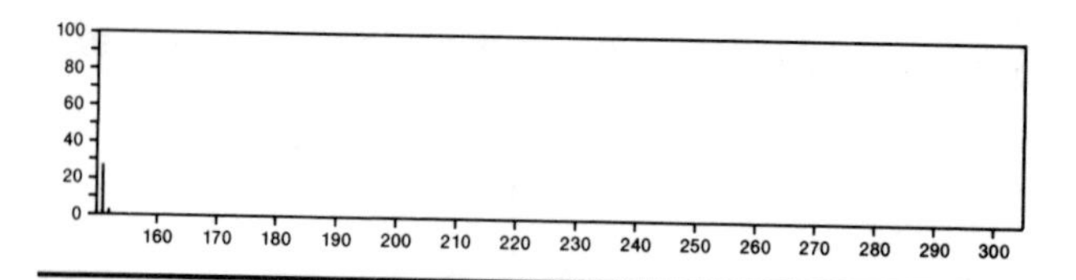

151 $C_8H_9NO_2$ 1007-49-4
2-Pyridinemethanol, acetate (ester)

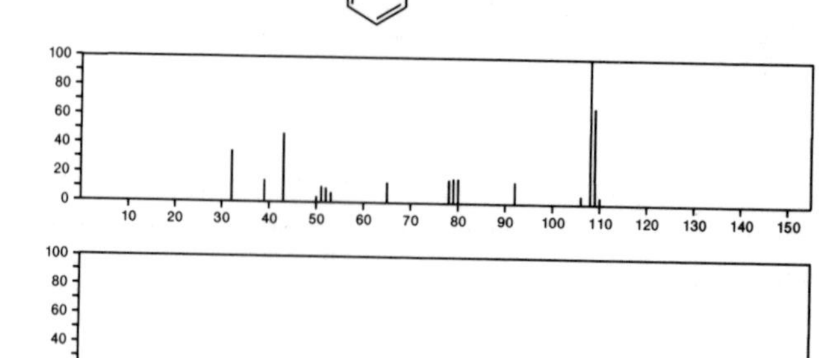

151 $C_8H_9NO_2$ 1469-48-3
1*H*-Isoindole-1,3(2*H*)-dione, 3a,4,7,7a-tetrahydro-, *cis*-

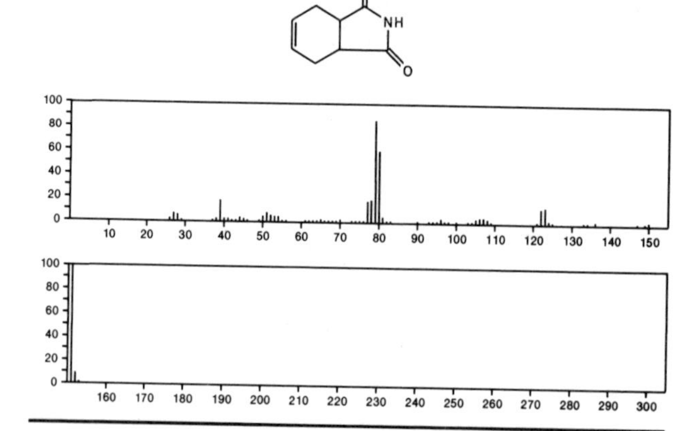

151 $C_8H_9NO_2$ 1943-79-9
Carbamic acid, methyl-, phenyl ester

PhOC(O)NHMe

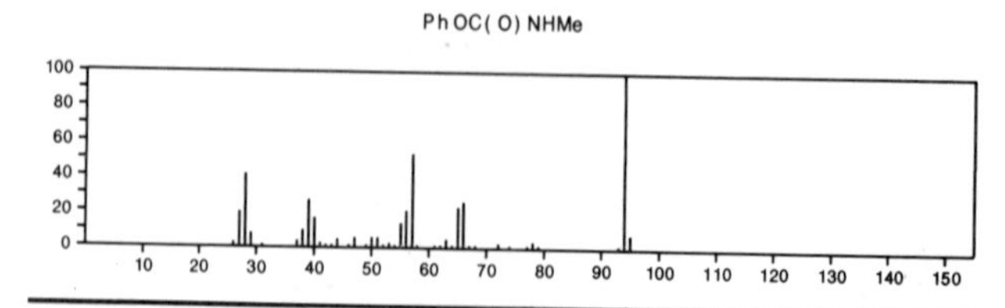

151 $C_8H_9NO_2$ 2603-10-3
Carbamic acid, phenyl-, methyl ester

PhNHC(O)OMe

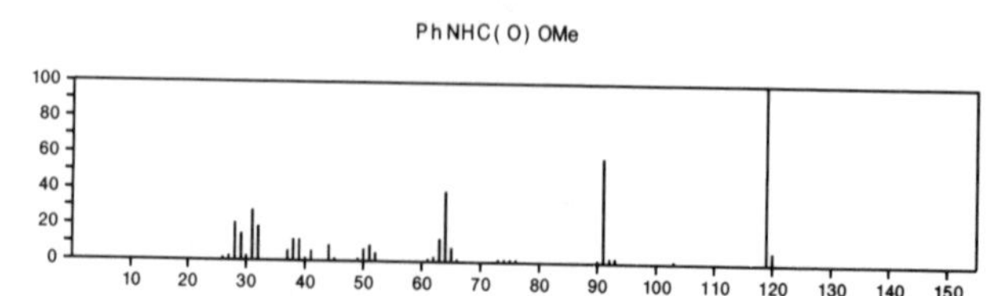

151 $C_8H_9NO_2$ 3235-04-9
Benzaldehyde, 4-methoxy-, oxime

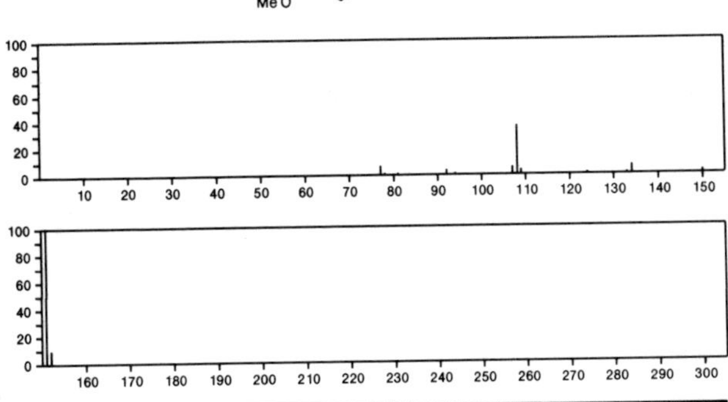

151 $C_8H_9NO_2$ 3717-21-3
Benzaldehyde, 4-methoxy-, oxime, (*E*)-

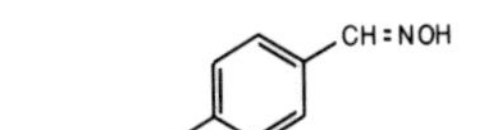

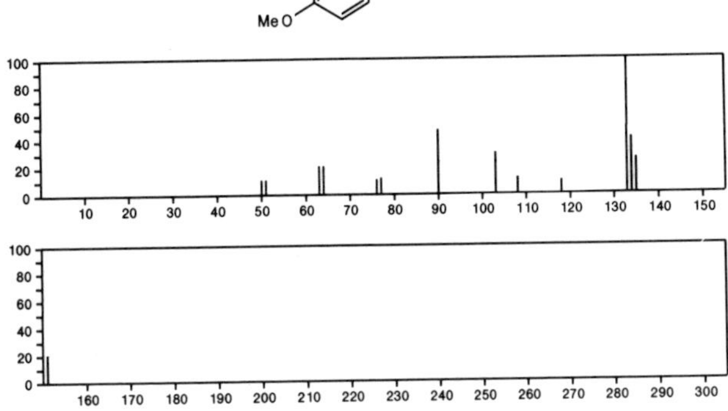

151 $C_8H_9NO_2$ 3717-22-4
Benzaldehyde, 4-methoxy-, oxime, (*Z*)-

CH=NOH
MeO

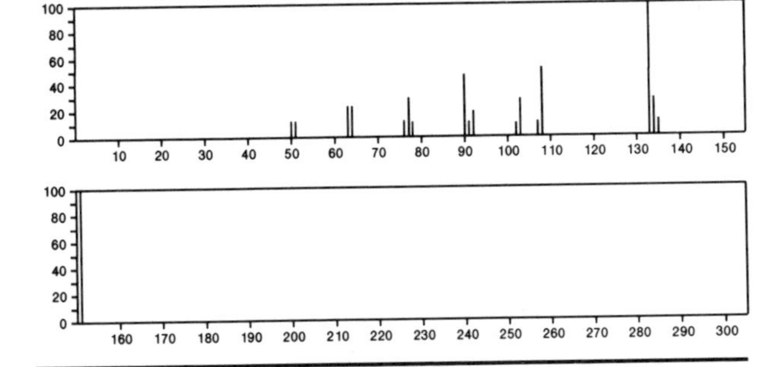

151 $C_8H_9NO_2$ 4746-61-6
Acetamide, 2-hydroxy-*N*-phenyl-

PhNHCOCH2OH

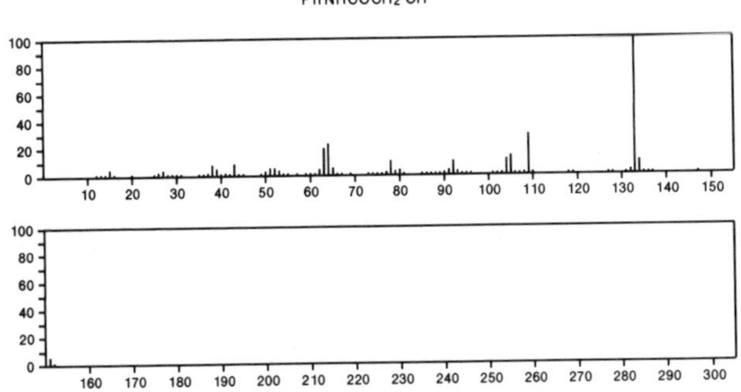

151 $C_8H_9NO_2$ 4842-89-1
3-Pyridinol, 6-methyl-, acetate (ester)

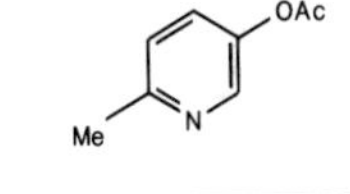

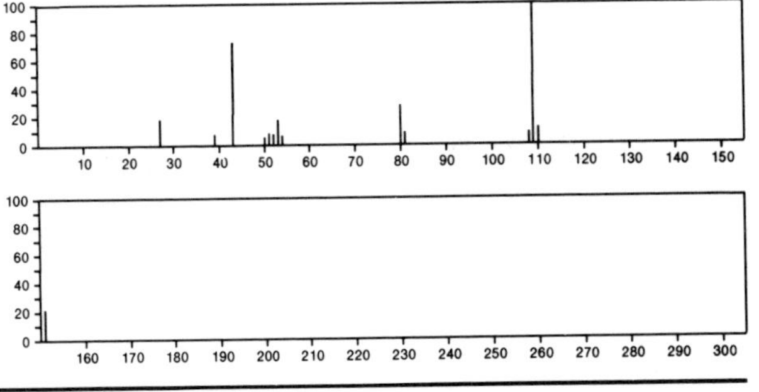

151 $C_8H_9NO_2$ 7214-61-1
Benzene, (1-nitroethyl)-

PhCHMe(NO2)

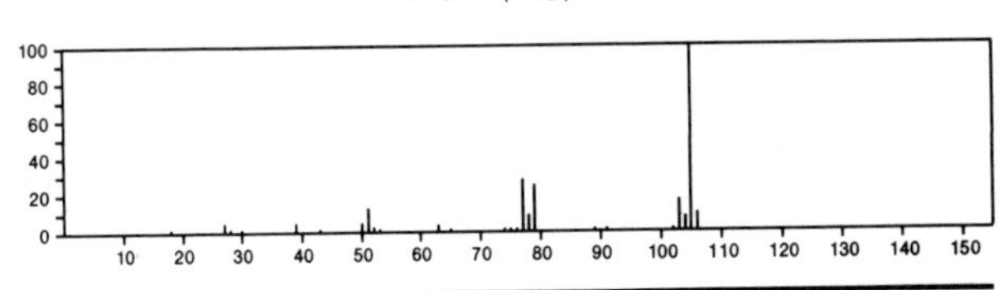

151 $C_8H_9NO_2$ 19628-76-3
Phenol, 3,5-dimethyl-4-nitroso-

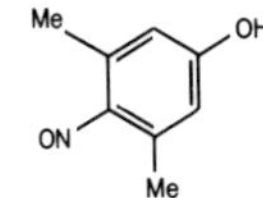

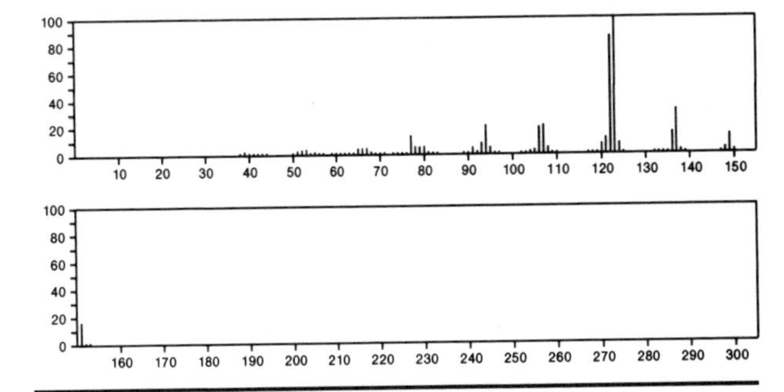

151 $C_8H_9NO_2$ 24935-26-0
Pyridinium, 1-(2-carboxyethyl)-, hydroxide, inner salt

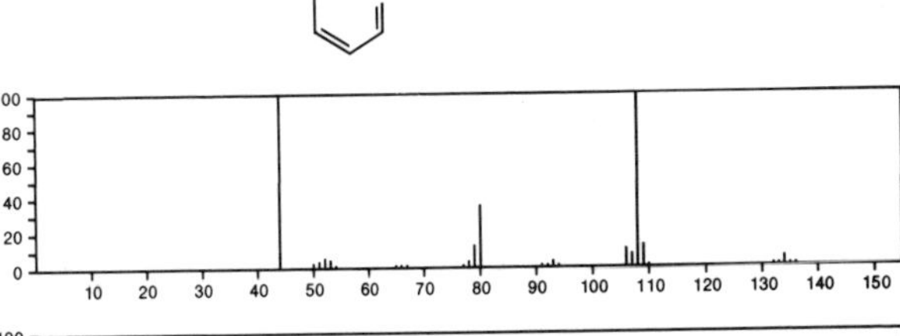

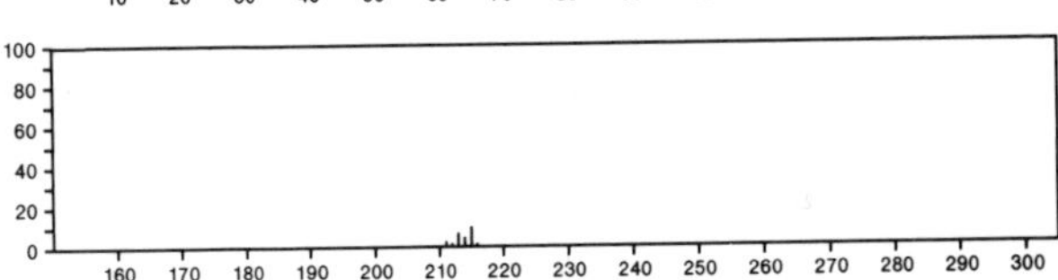

151 $C_8H_9NO_2$ 29559-27-1
Benzene, 1-methyl-4-(nitromethyl)-

151 $C_8H_9NO_2$ 34600-54-9
1*H*-Pyrrole-2-carboxylic acid, 1-(1-propenyl)-, (*E*)-

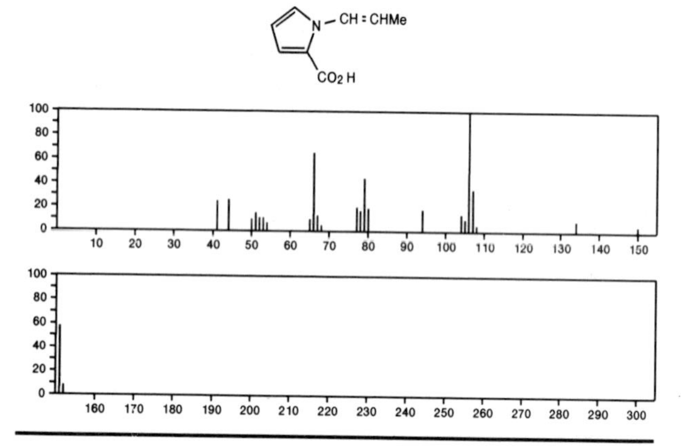

151 $C_8H_9NO_2$ 36880-54-3
Pyridinium, 1-(1-carboxyethyl)-, hydroxide, inner salt

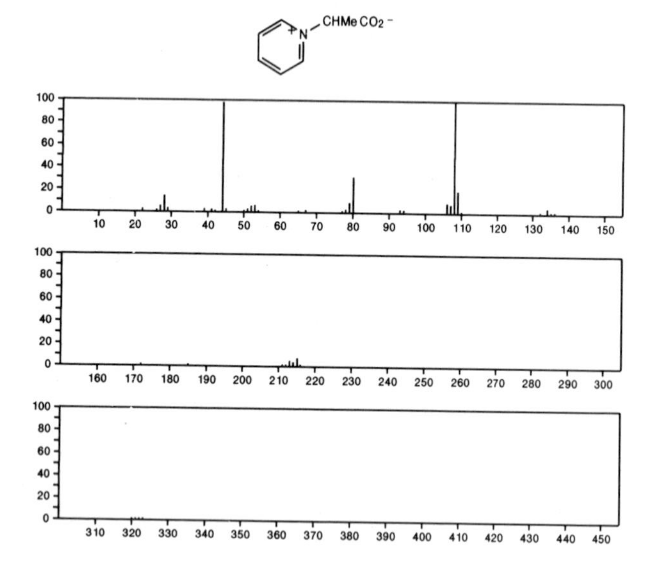

151 $C_8H_9NO_2$ 38489-80-4
Benzaldehyde, 3-methoxy-, oxime

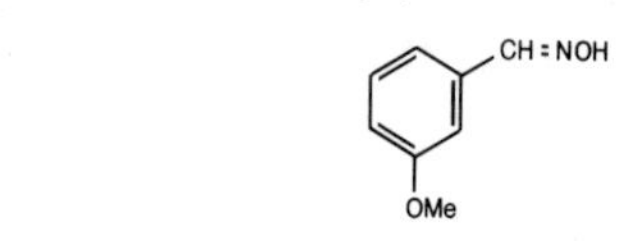

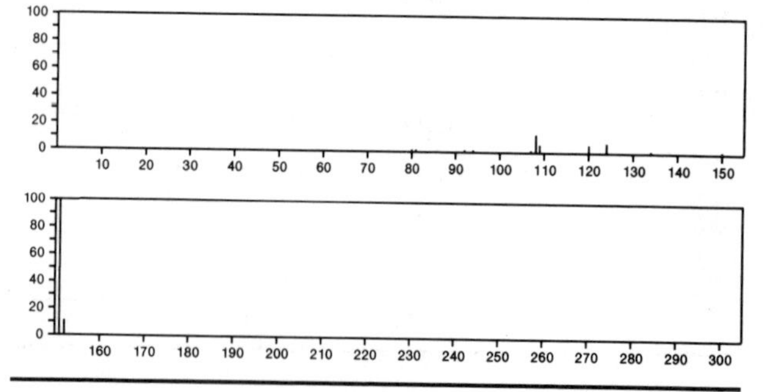

151 C_8H_9NS 645-54-5
Benzeneethanethioamide

H_2NCSCH_2Ph

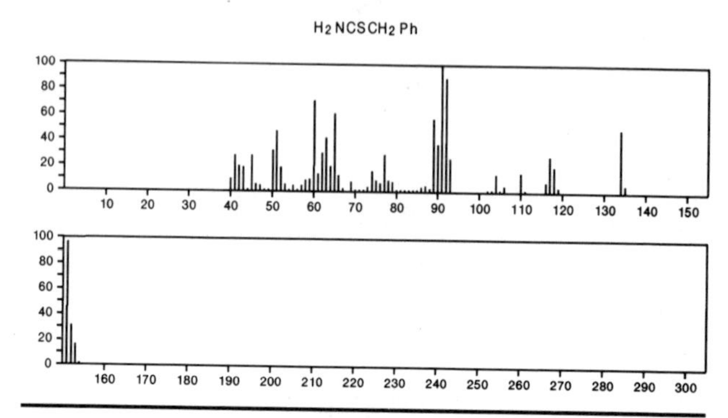

151 $C_9H_{13}NO$ 55-81-2
Benzeneethanamine, 4-methoxy-

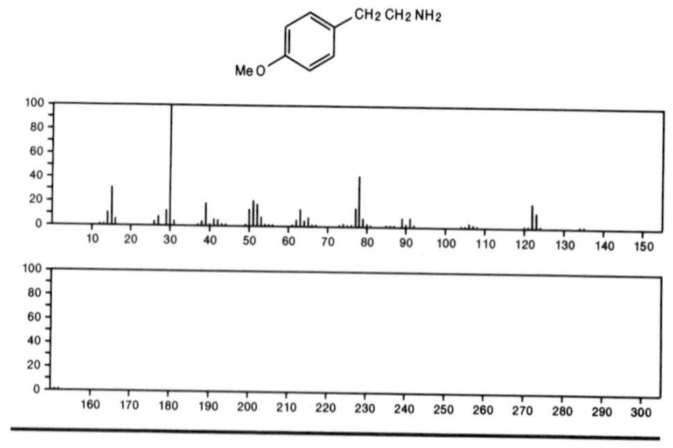

151 $C_9H_{13}NO$ 102-41-0
Ethanol, 2-[(3-methylphenyl)amino]-

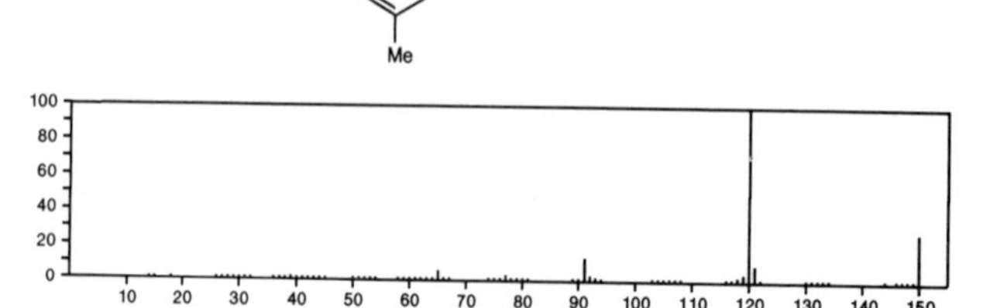

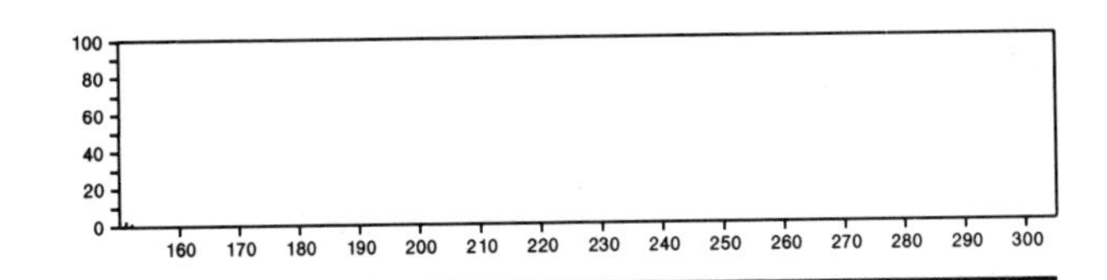

151 $C_9H_{13}NO$ 370–98–9
Phenol, 4–[2–(methylamino)ethyl]–

HO CH2 CH2 NHMe

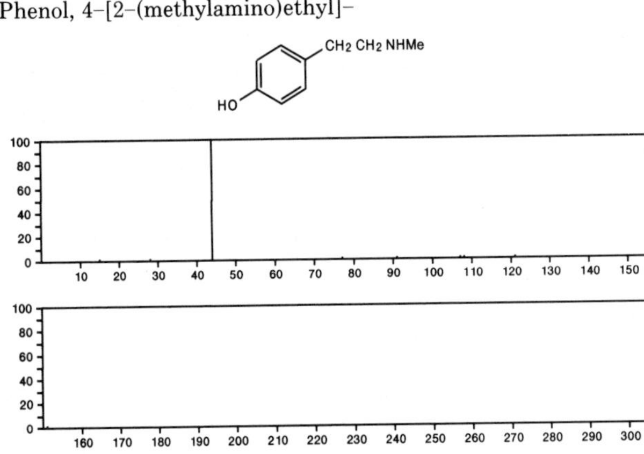

151 $C_9H_{13}NO$ 4594–78–9
Cyclohexanepropanenitrile, 2–oxo–

CH2 CH2 CN O

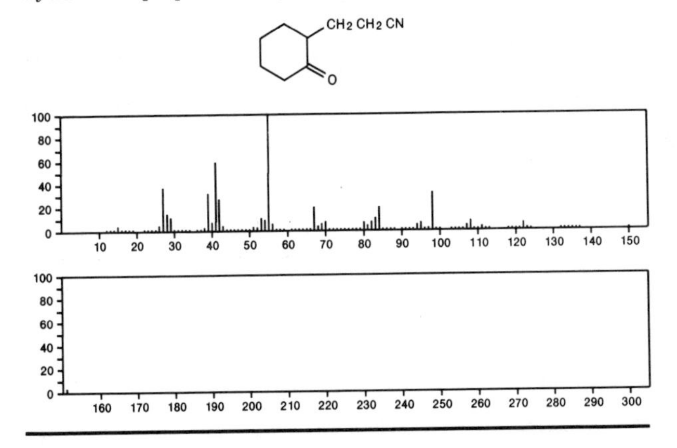

151 $C_9H_{13}NO$ 4707–56–6
Benzenemethanol, 2–(dimethylamino)–

CH2 OH NMe2

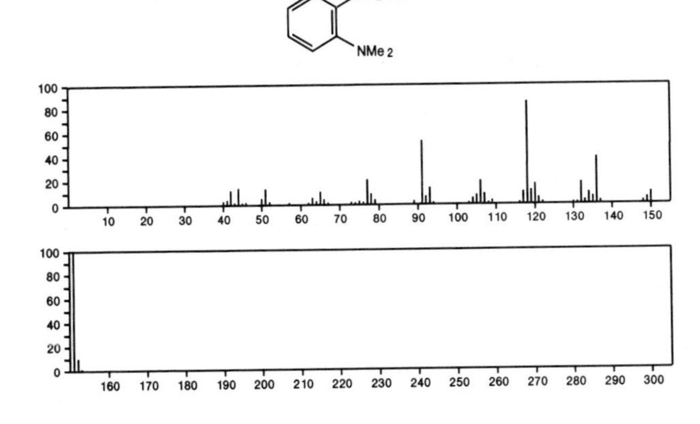

151 $C_9H_{13}NO$ 14098–17–0
Acetamide, *N*–bicyclo[2.2.1]hept–2–en–7–yl–, *anti*–

AcNH

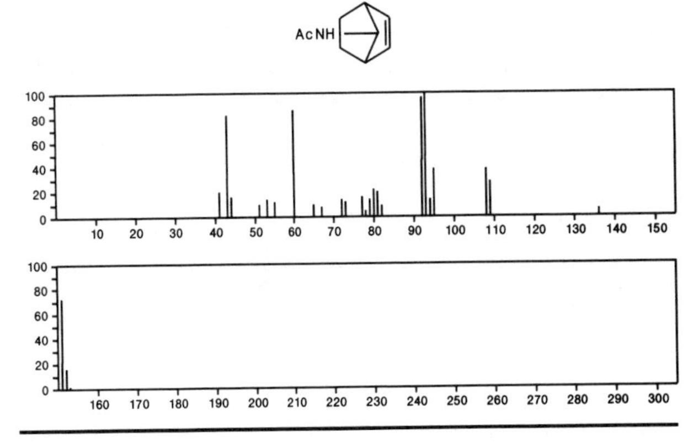

151 $C_9H_{13}NO$ 14174–01–7
Acetamide, *N*–2–norbornen–7–yl–, *syn*–

AcNH

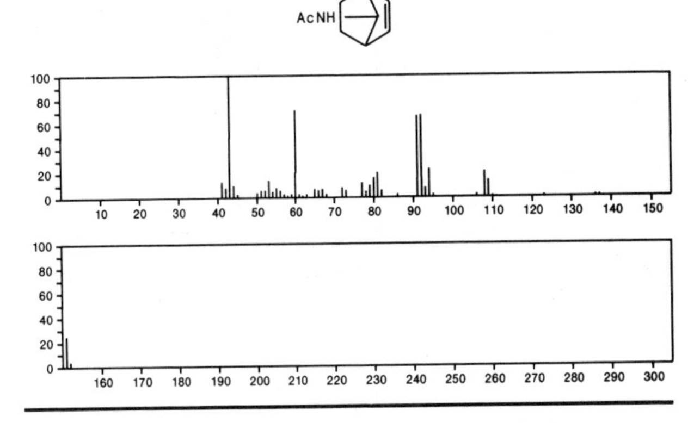

151 $C_9H_{13}NO$ 23501–93–1
Benzenemethanol, 3–(dimethylamino)–

Me2N CH2 OH

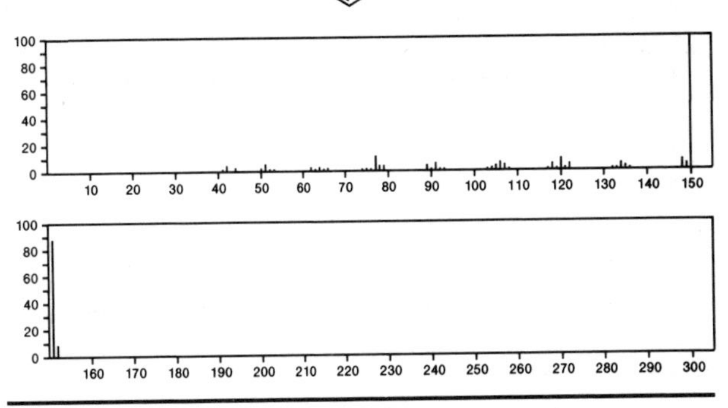

151 $C_9H_{13}NO$ 31061–58–2
Benzenaminium, 2–hydroxy–*N*,*N*,*N*–trimethyl–, hydroxide, inner salt

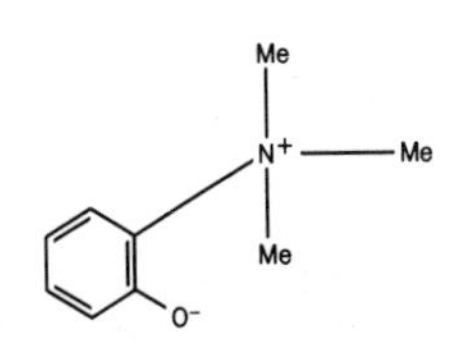

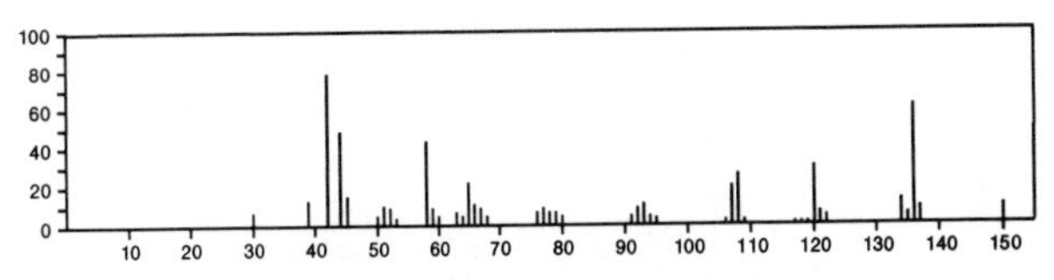

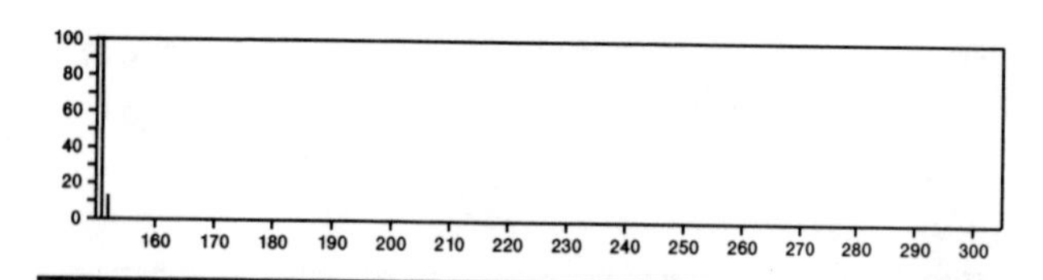

151 $C_9H_{13}NO$ 31061-59-3
Benzenaminium, 3-hydroxy-*N*,*N*,*N*-trimethyl-, hydroxide, inner salt

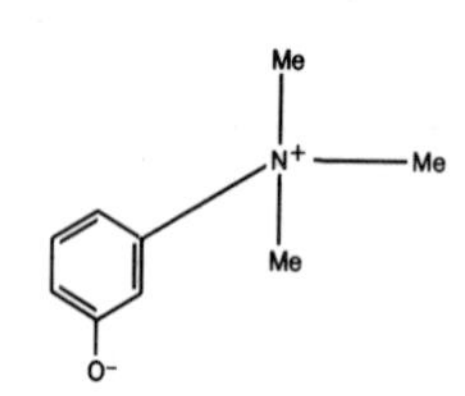

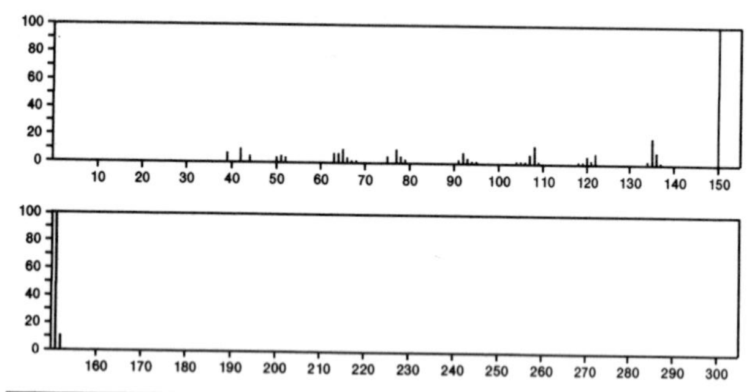

151 $C_9H_{13}NO$ 31061-60-6
Ammonium, (*p*-hydroxyphenyl)trimethyl-, hydroxide, inner salt

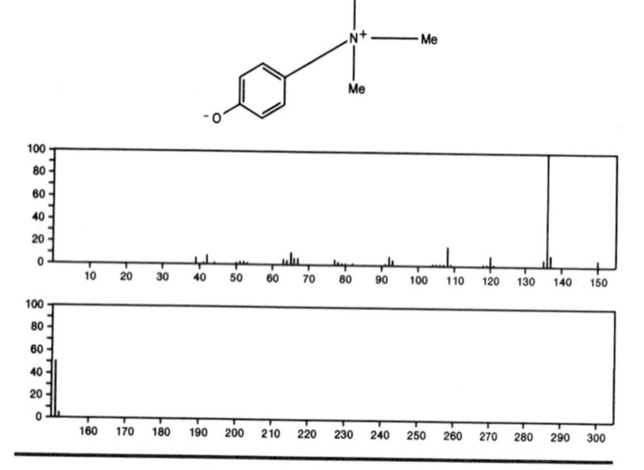

151 $C_9H_{13}NO$ 31396-32-4
Pyridine, 2-butyl-, 1-oxide

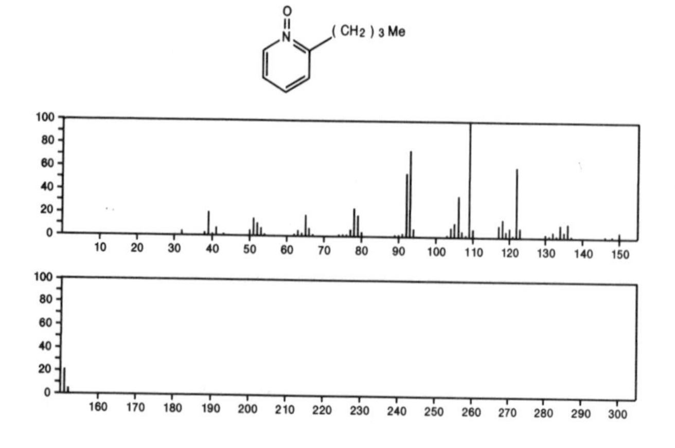

151 $C_9H_{13}NO$ 31396-33-5
Pyridine, 3-butyl-, 1-oxide

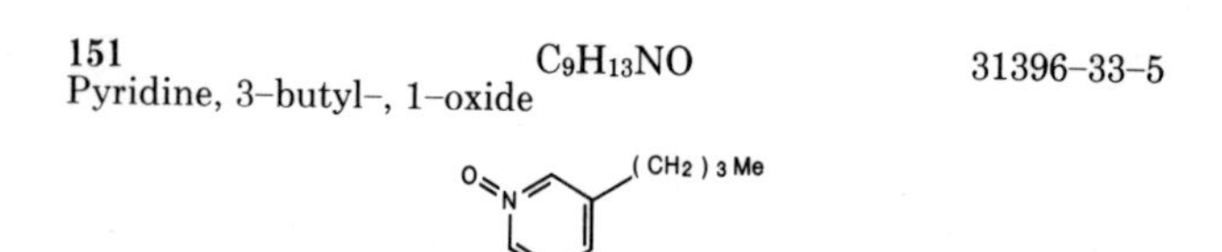

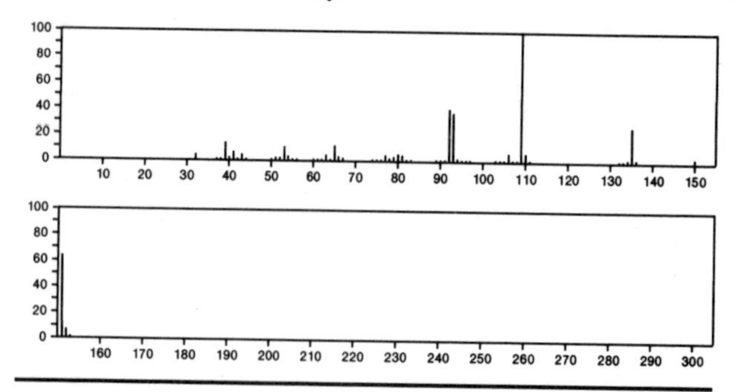

151 $C_9H_{13}NO$ 31396-34-6
Pyridine, 4-butyl-, 1-oxide

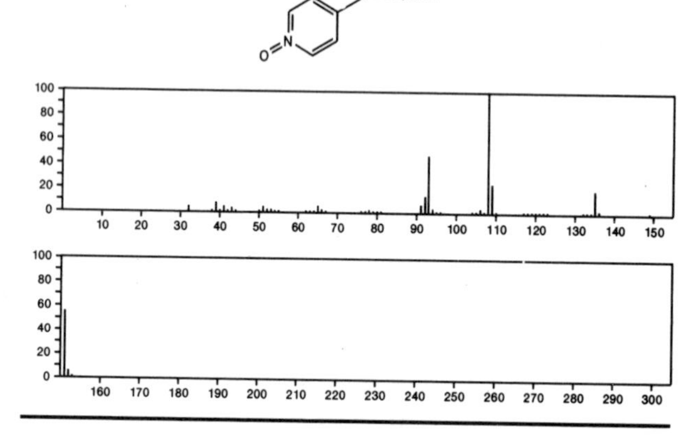

151 $C_{10}H_{17}N$ 1125-99-1
Pyrrolidine, 1-(1-cyclohexen-1-yl)-

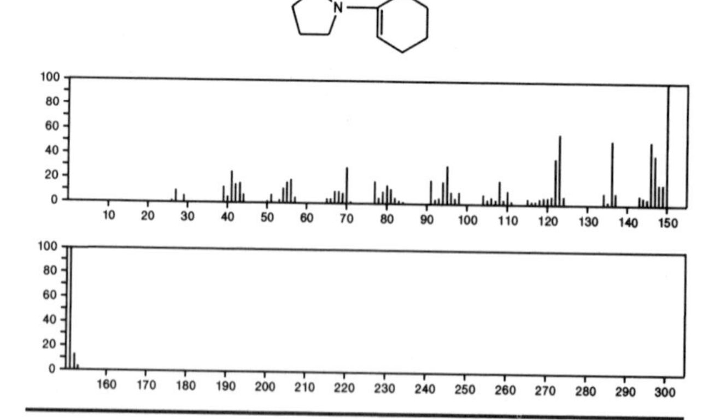

151 $C_{10}H_{17}N$ 1614-92-2
Piperidine, 1-(1-cyclopenten-1-yl)-

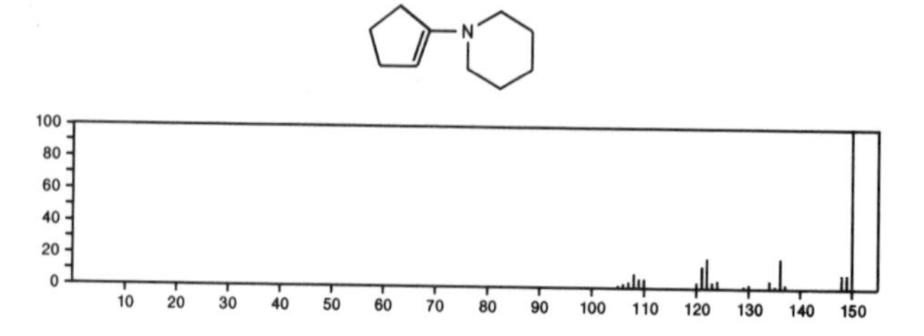

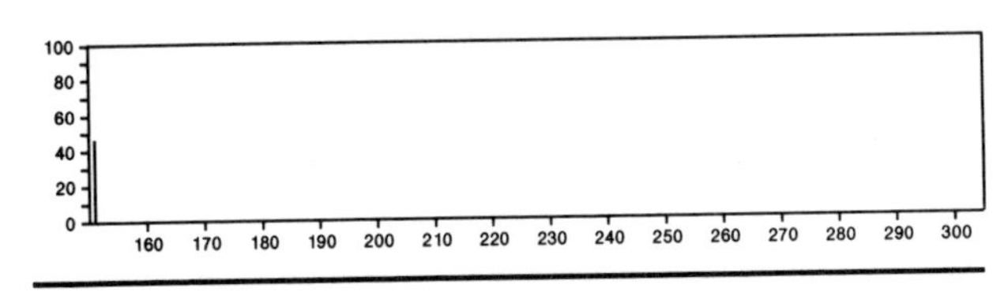

151 $C_{10}H_{17}N$ 5735-21-7
Naphthalen-4a,8a-imine, octahydro-

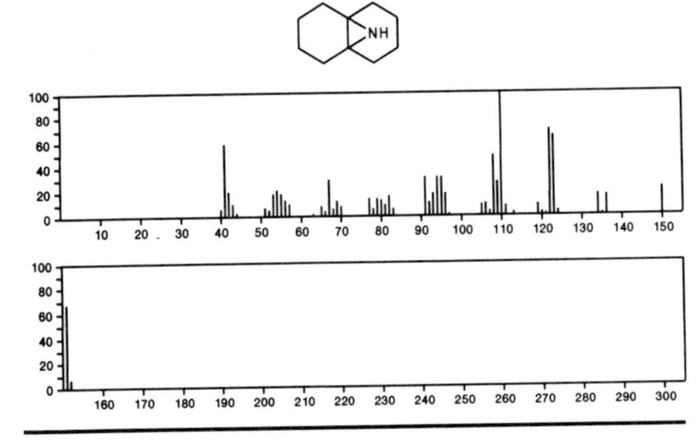

152 CCl_4 56-23-5
Methane, tetrachloro-

CCl_4

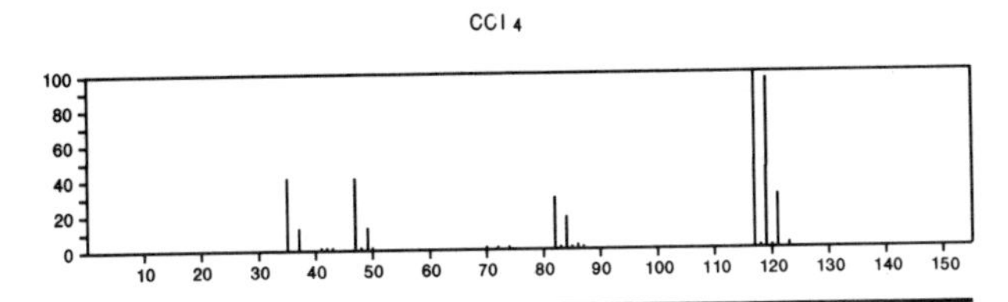

152 $C_2HCl_2F_3$ 306-83-2
Ethane, 2,2-dichloro-1,1,1-trifluoro-

Cl_2CHCF_3

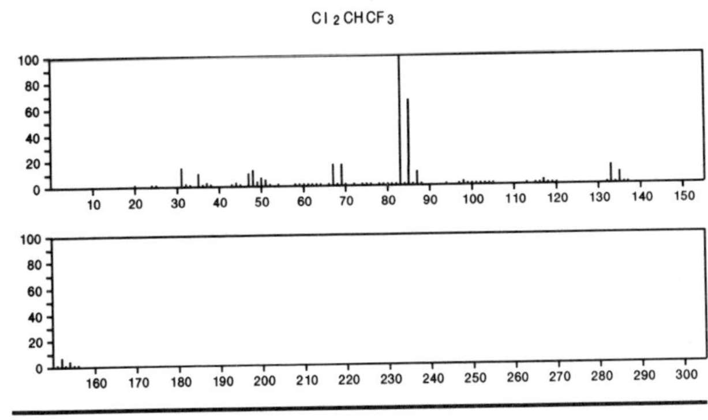

152 $C_2HCl_2F_3$ 354-23-4
Ethane, 1,2-dichloro-1,1,2-trifluoro-

$FCHClCClF_2$

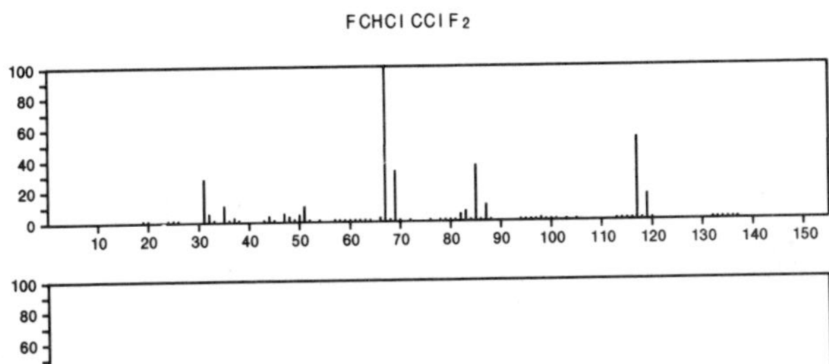

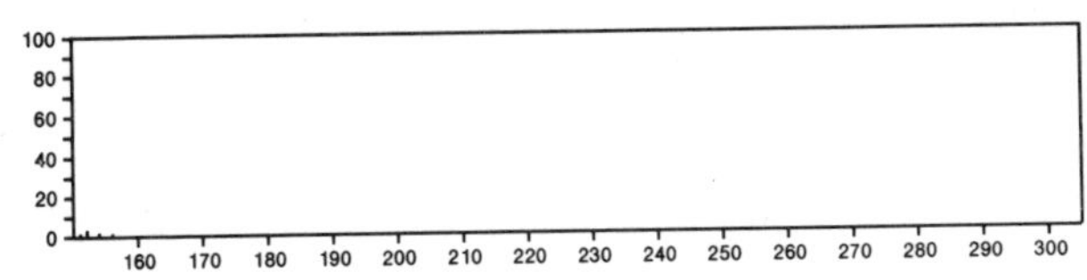

152 $C_2H_4N_2O_6$ 628-96-6
1,2-Ethanediol, dinitrate

$O_2NOCH_2CH_2ONO_2$

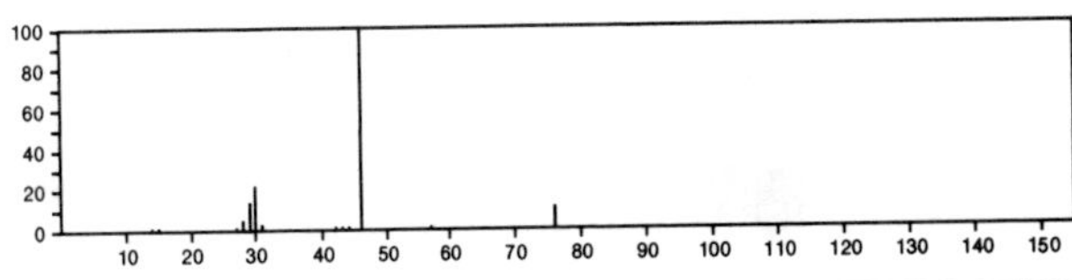

152 $C_3H_3F_3N_4$ 697-94-9
1*H*-Tetrazole, 1-methyl-5-(trifluoromethyl)-

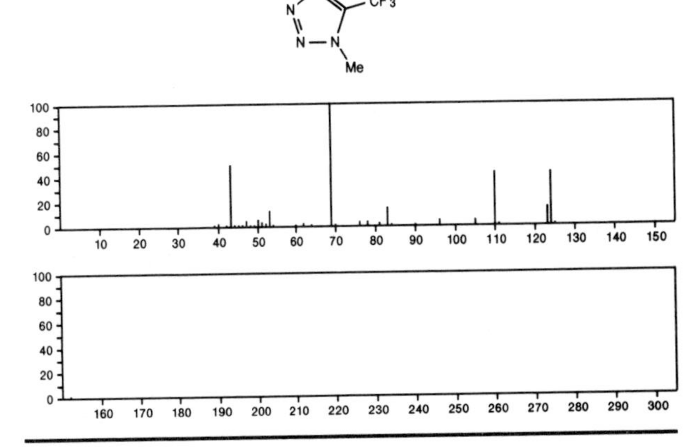

152 $C_3H_3F_3N_4$ 768-27-4
2*H*-Tetrazole, 2-methyl-5-(trifluoromethyl)-

N CF3 N N N Me

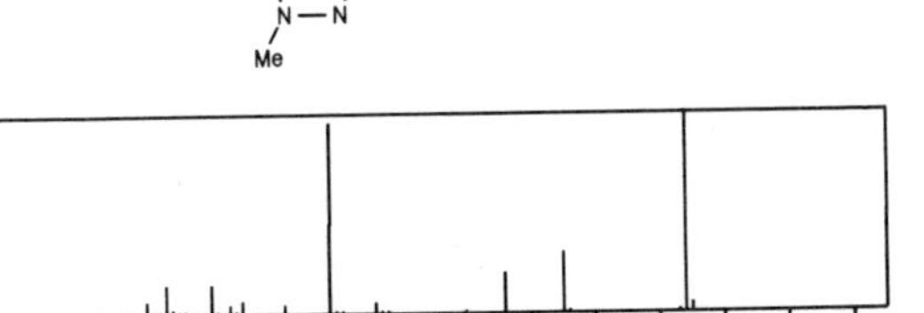

152 $C_3H_5BrO_2$ 96-32-2
Acetic acid, bromo-, methyl ester

$BrCH_2C(O)OMe$

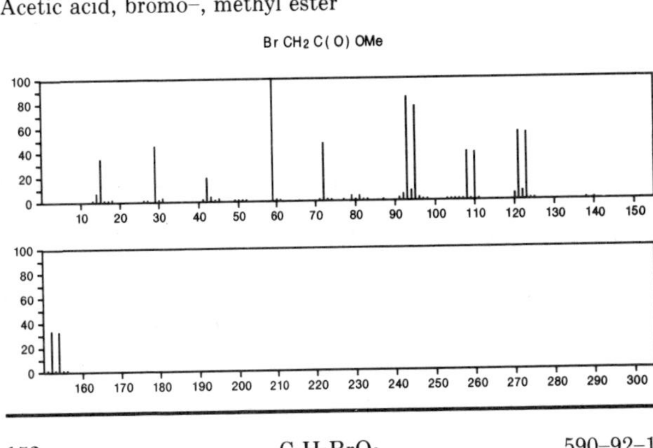

152 $C_3H_5BrO_2$ 590-92-1
Propanoic acid, 3-bromo-

$BrCH_2CH_2CO_2H$

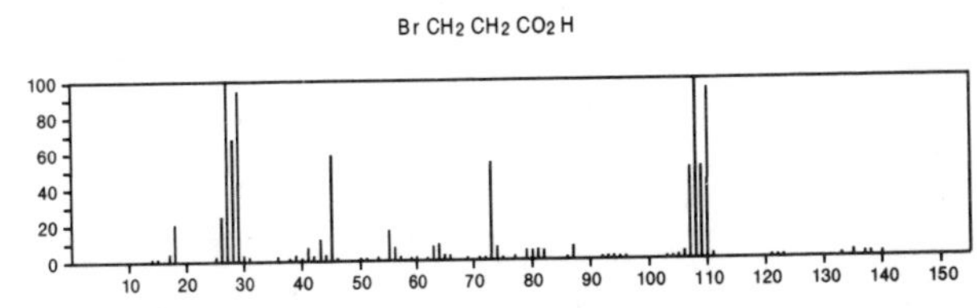

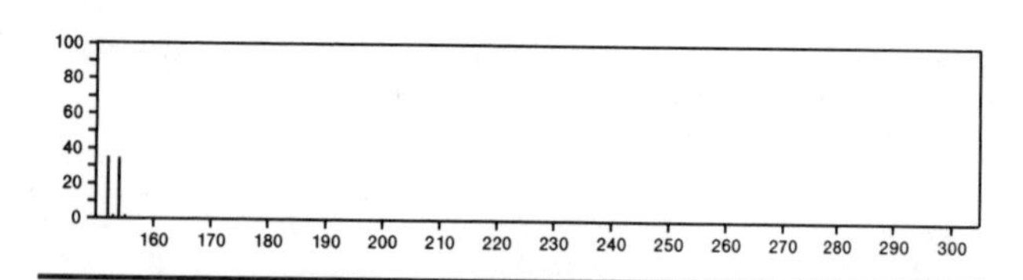

152 $C_3H_9BS_3$ 997-49-9
Thioboric acid (H_3BS_3), trimethyl ester

SMe
MeSBSMe

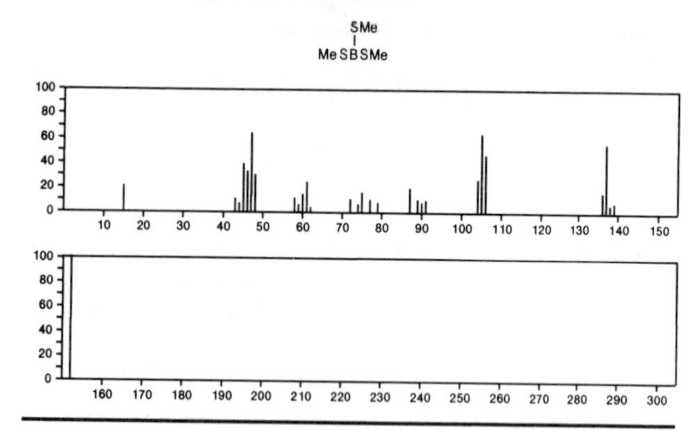

152 C_4H_9BrO 592-55-2
Ethane, 1-bromo-2-ethoxy-

EtOCH2CH2Br

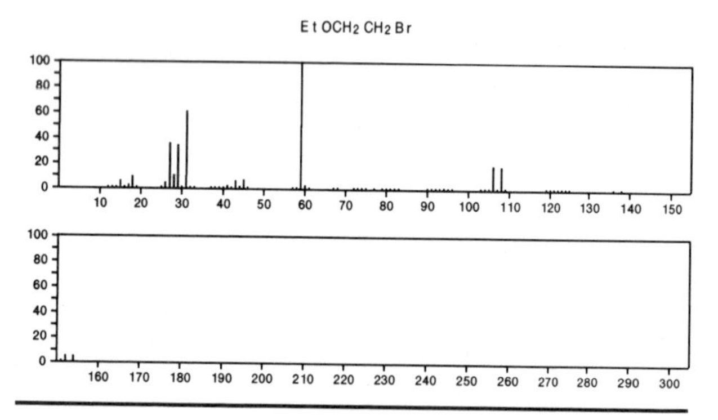

152 $C_4H_9O_4P$ 10429-10-4
Phosphoric acid, ethenyl dimethyl ester

OMe
H2C=CHOPOMe
O

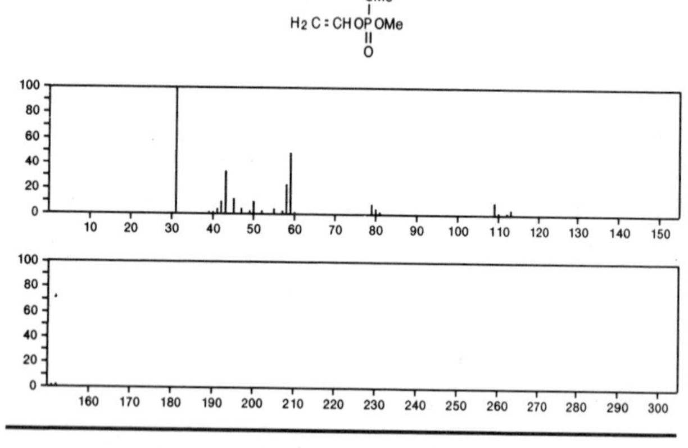

152 $C_4H_{12}O_2SSi$ 54193-90-7
Methanesulfinic acid, trimethylsilyl ester

MeS(O)OSiMe3

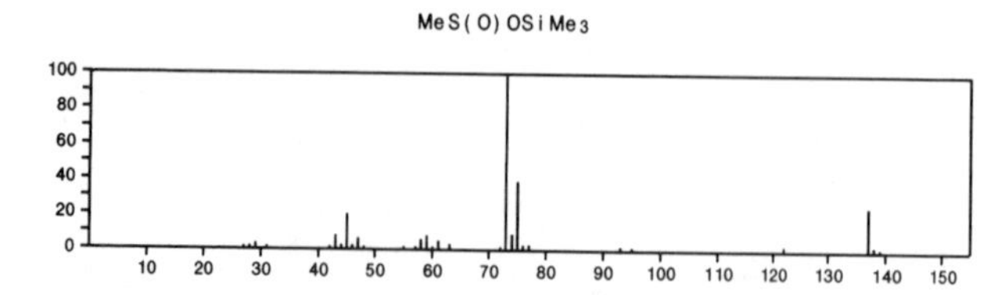

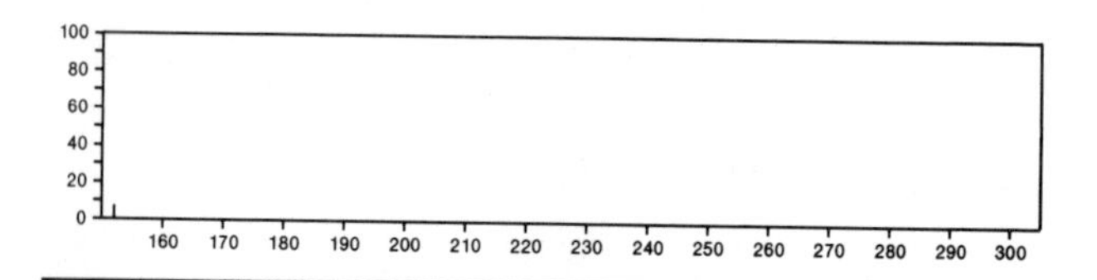

152 $C_4H_{12}O_4Si$ 681-84-5
Silicic acid (H_4SiO_4), tetramethyl ester

OMe
MeOSiOMe
MeO

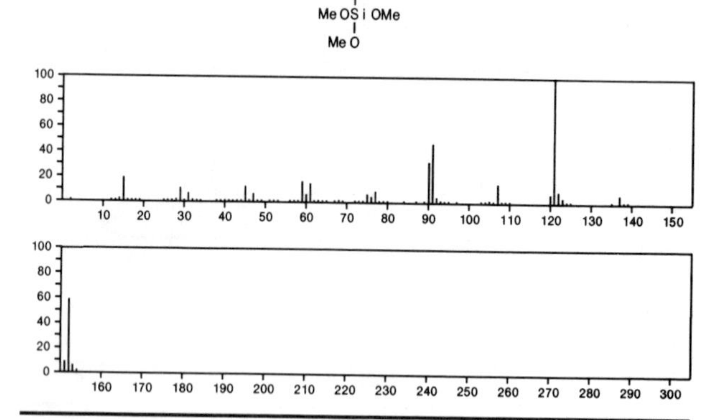

152 $C_4H_{16}B_8$ 31566-09-3
9,10-Dicarbadecaborane(12), 9,10-dimethyl-

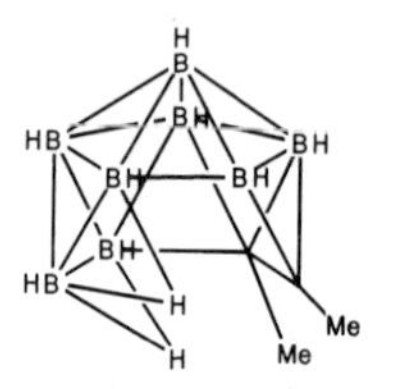

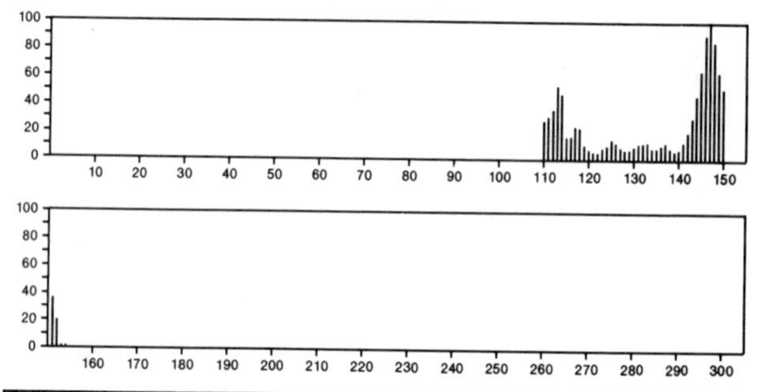

152 $C_5H_4N_4O_2$ 69-89-6
1*H*-Purine-2,6-dione, 3,7-dihydro-

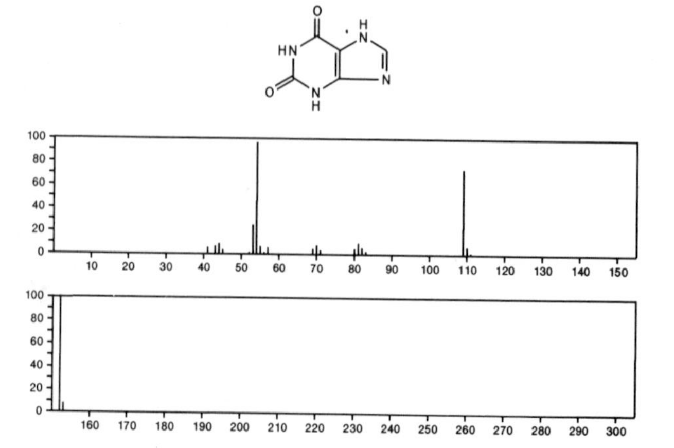

152 $C_5H_4N_4O_2$ 2465-59-0

1*H*-Pyrazolo[3,4-*d*]pyrimidine-4,6(5*H*,7*H*)-dione

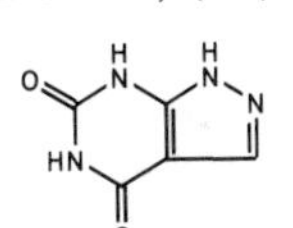

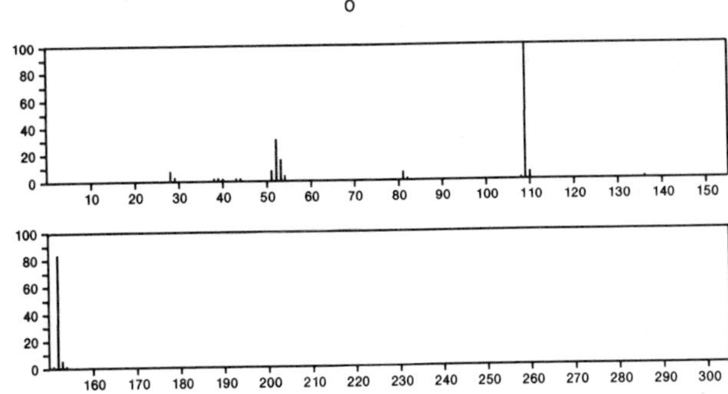

152 $C_5H_4N_4S$ 50-44-2

6*H*-Purine-6-thione, 1,7-dihydro-

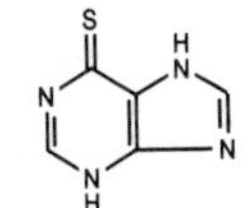

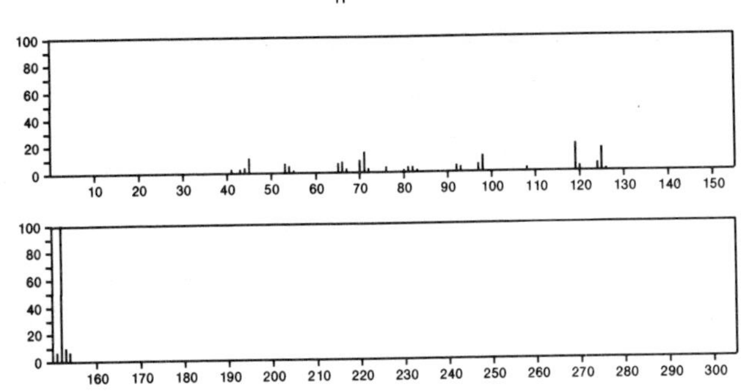

152 $C_6H_8N_4O$ 36361-68-9

1*H*-Purine-6-methanol, 6,7-dihydro-

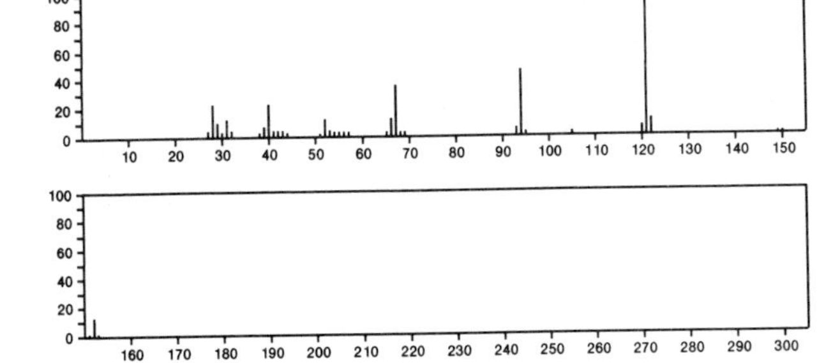

152 $C_6H_{10}Cl_2$ 822-86-6

Cyclohexane, 1,2-dichloro-, *trans*-

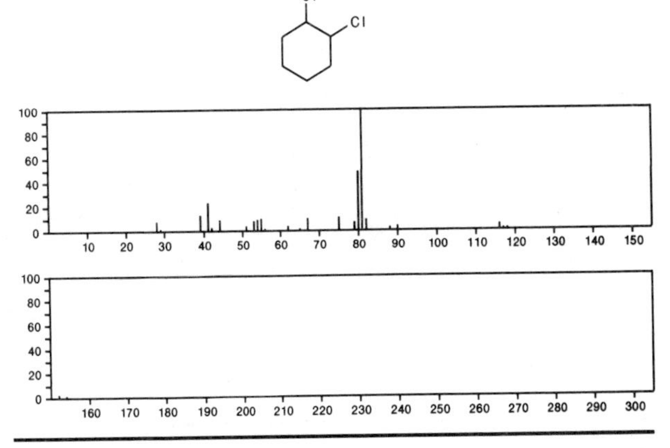

152 $C_6H_{10}Cl_2$ 1121-21-7

Cyclohexane, 1,2-dichloro-

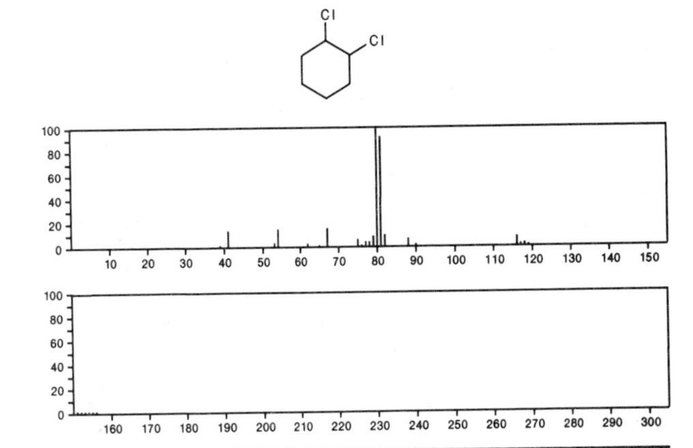

152 $C_6H_{10}Cl_2$ 2108-92-1

Cyclohexane, 1,1-dichloro-

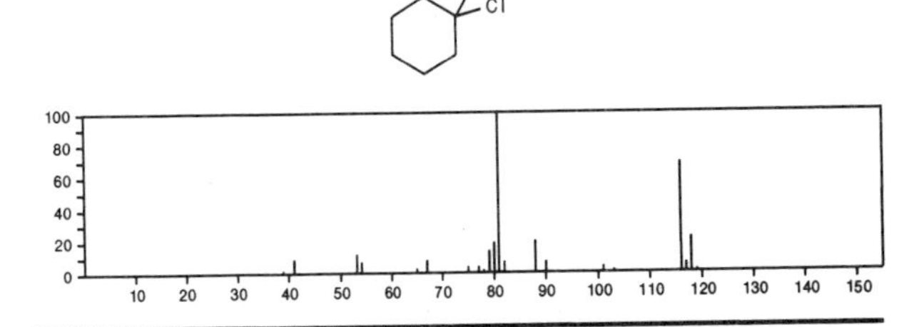

152 $C_6H_{10}Cl_2$ 10498-35-8

Cyclohexane, 1,2-dichloro-, *cis*-

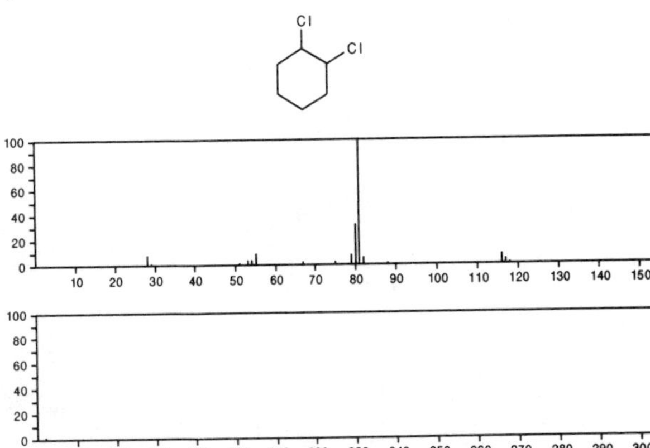

152 $C_6H_{10}Cl_2$ 16749-11-4
Cyclohexane, 1,4-dichloro-, *cis-*

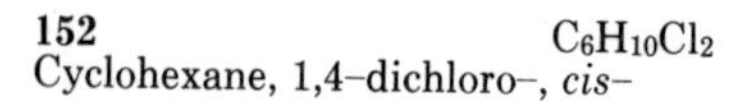

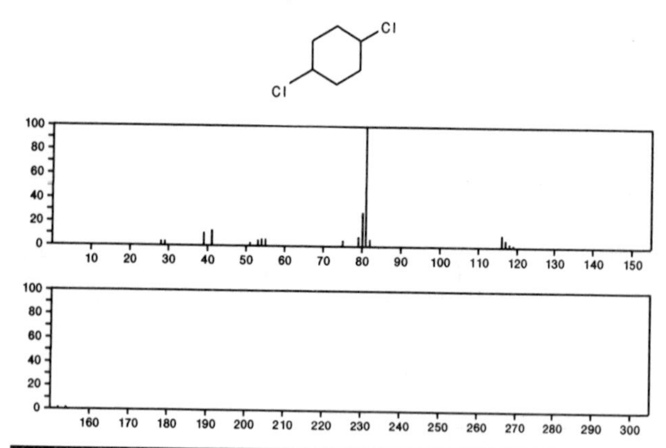

152 $C_6H_{10}Cl_2$ 16890-91-8
Cyclohexane, 1,4-dichloro-, *trans-*

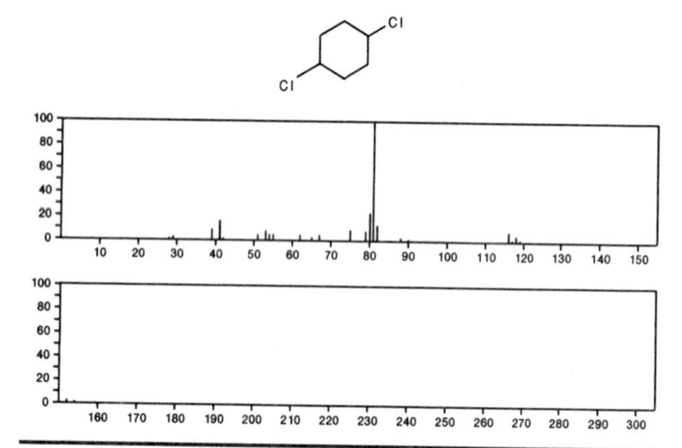

152 $C_6H_{10}Cl_2$ 19398-57-3
Cyclohexane, 1,4-dichloro-

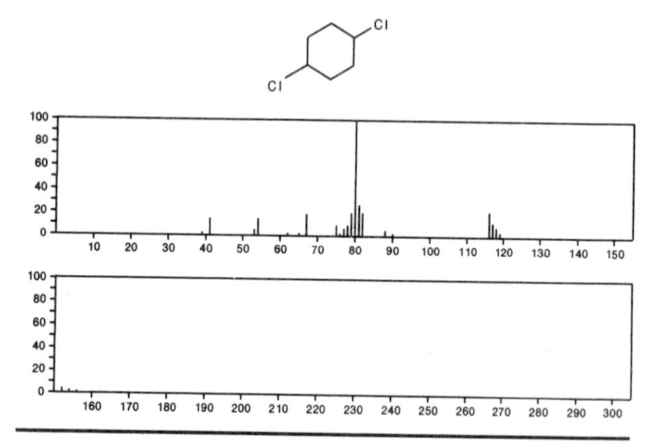

152 $C_6H_{10}Cl_2$ 24955-62-2
Cyclohexane, 1,3-dichloro-, *trans-*

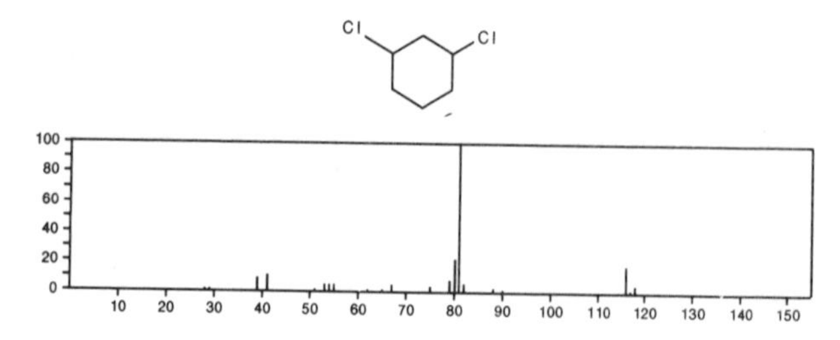

152 $C_6H_{10}Cl_2$ 24955-63-3
Cyclohexane, 1,3-dichloro-, *cis-*

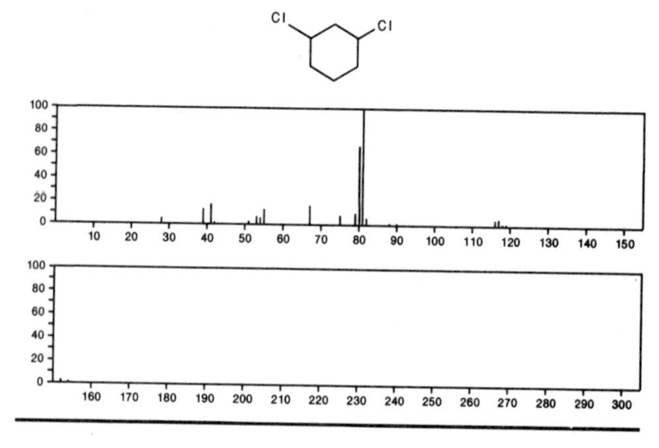

152 $C_6H_{10}Cl_2$ 27990-74-5
1-Pentene, 5-chloro-4-(chloromethyl)-

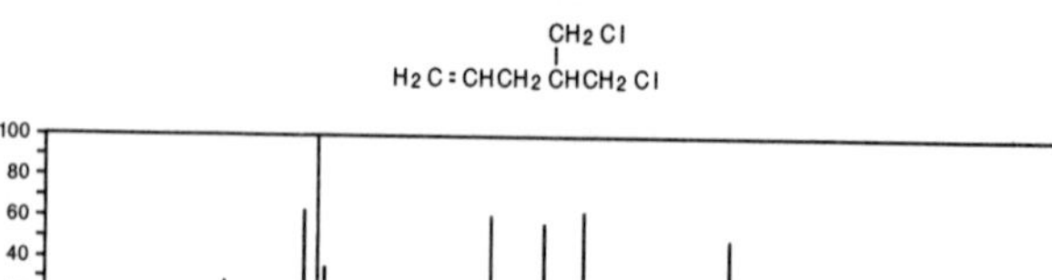

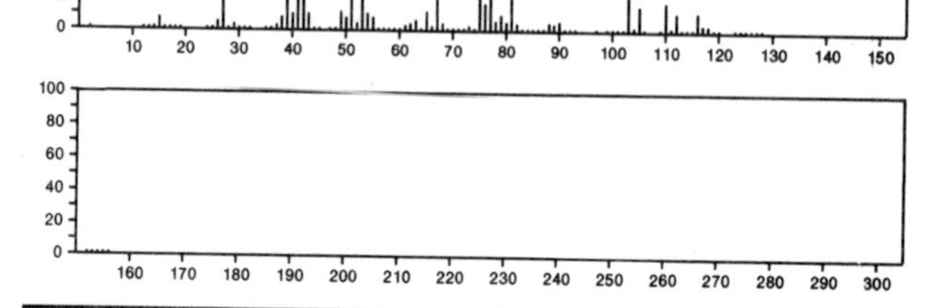

152 $C_6H_{10}Cl_2$ 55887-78-0
Cyclohexane, 1,3-dichloro-

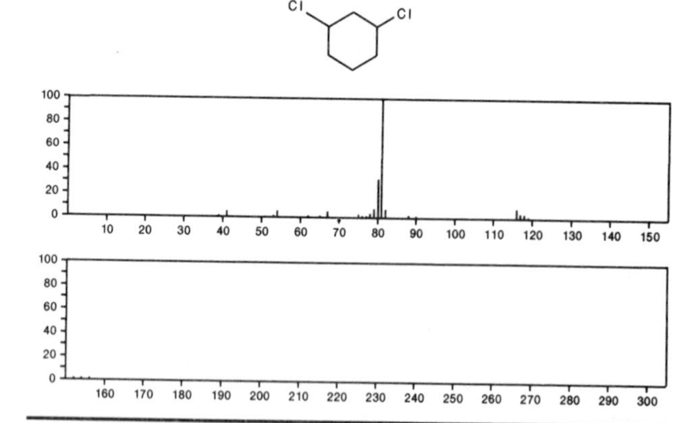

152 $C_6H_{14}ClP$ 40244-90-4
Phosphinous chloride, bis(1-methylethyl)-

ClP(Pr-i)2

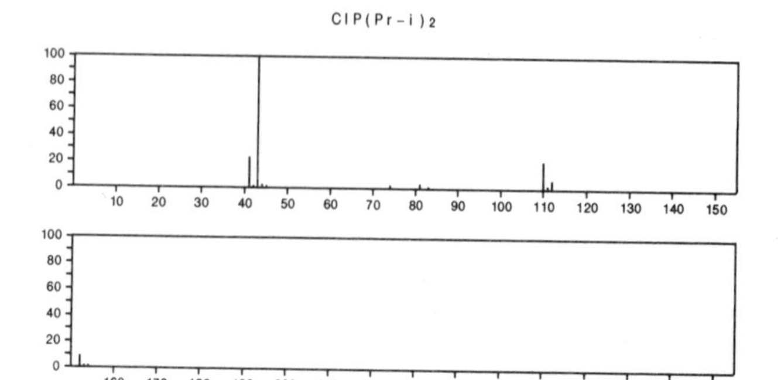

152 $C_7H_8N_2O_2$ 99–52–5
Benzenamine, 2–methyl–4–nitro–

152 $C_7H_8N_2O_2$ 5351–23–5
Benzoic acid, 4–hydroxy–, hydrazide

152 $C_7H_8N_2S$ 103–85–5
Thiourea, phenyl–

152 $C_7H_8N_2S$ 20605–40–7
Benzenecarbothioic acid, hydrazide

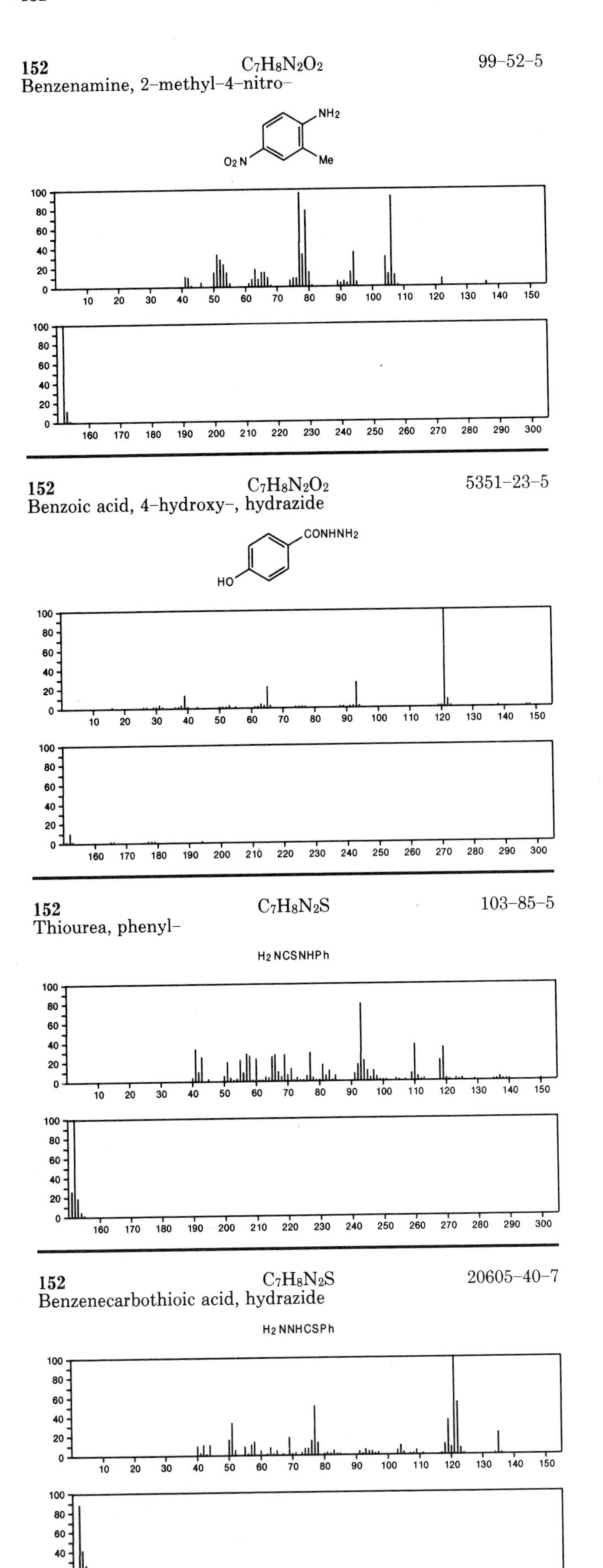

152 C_8H_8OS 5873–86–9
Benzenecarbothioic acid, *O*–methyl ester

152 C_8H_8OS 5925–68–8
Benzenecarbothioic acid, *S*–methyl ester

152 $C_8H_8O_3$ 83–40–9
Benzoic acid, 2–hydroxy–3–methyl–

152 $C_8H_8O_3$ 85–43–8
1,3–Isobenzofurandione, 3a,4,7,7a–tetrahydro–

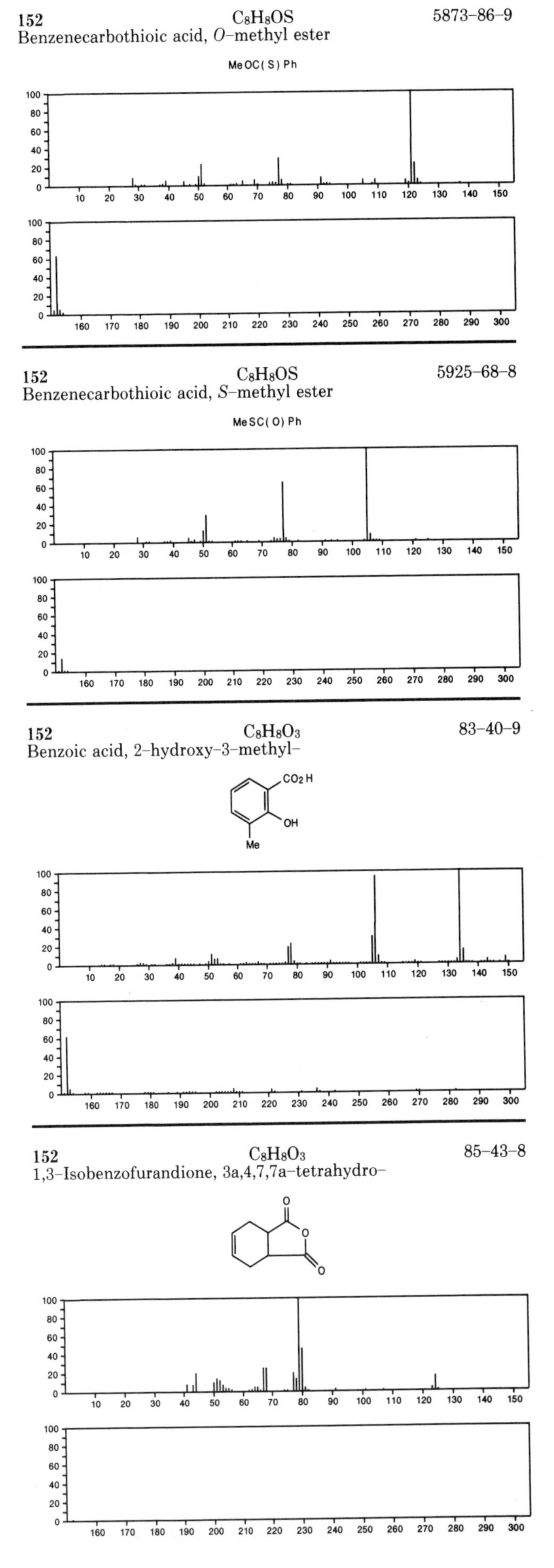

152 $C_8H_8O_3$ 89-84-9
Ethanone, 1-(2,4-dihydroxyphenyl)-

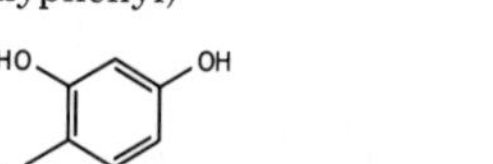

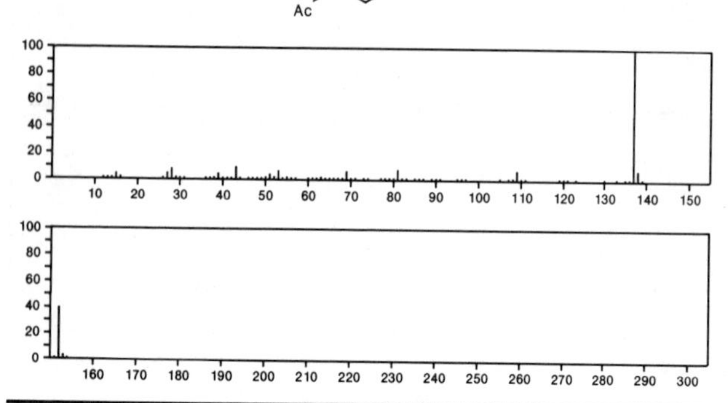

152 $C_8H_8O_3$ 119-36-8
Benzoic acid, 2-hydroxy-, methyl ester

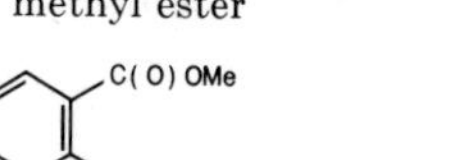

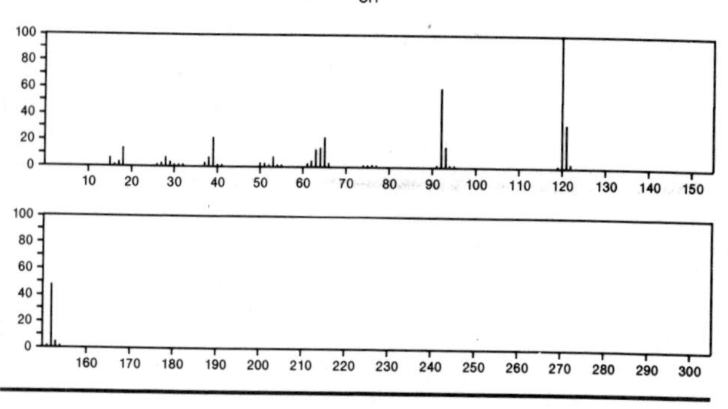

152 $C_8H_8O_3$ 90-64-2
Benzeneacetic acid, α-hydroxy-

PhCH(OH)CO2H

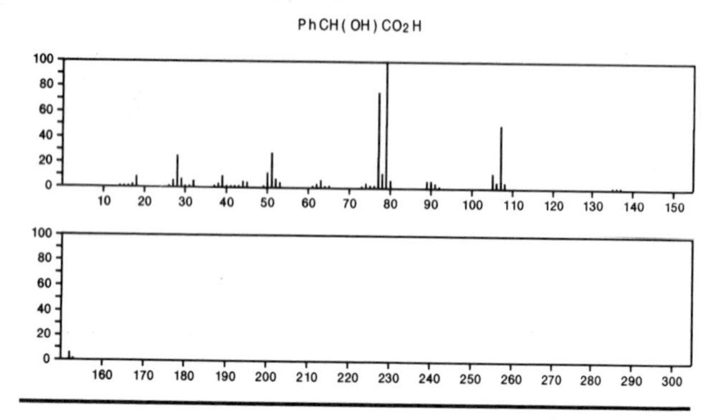

152 $C_8H_8O_3$ 121-33-5
Benzaldehyde, 4-hydroxy-3-methoxy-

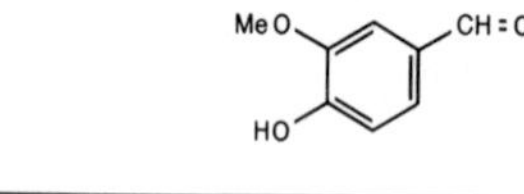

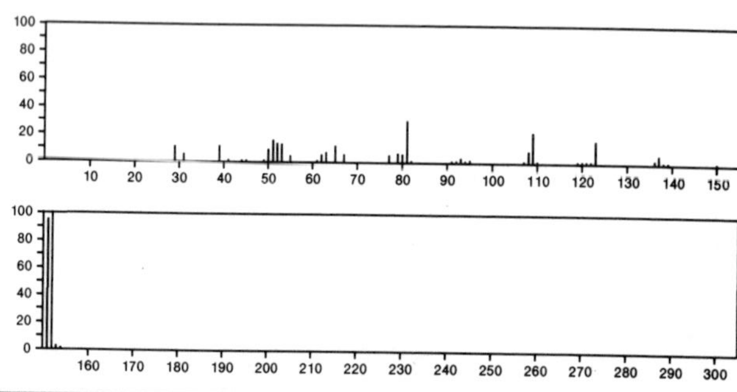

152 $C_8H_8O_3$ 99-76-3
Benzoic acid, 4-hydroxy-, methyl ester

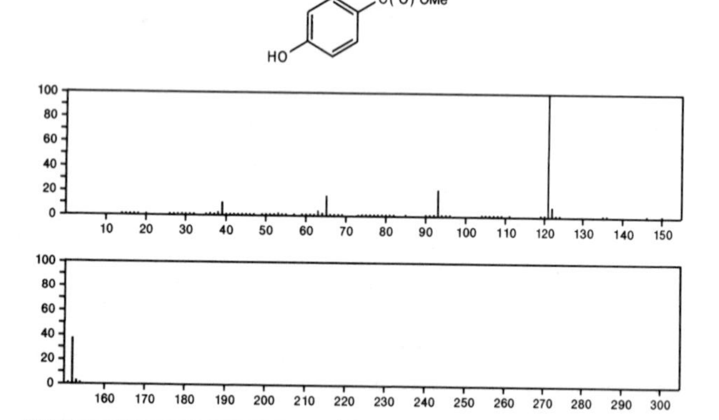

152 $C_8H_8O_3$ 122-59-8
Acetic acid, phenoxy-

HO2CCH2OPh

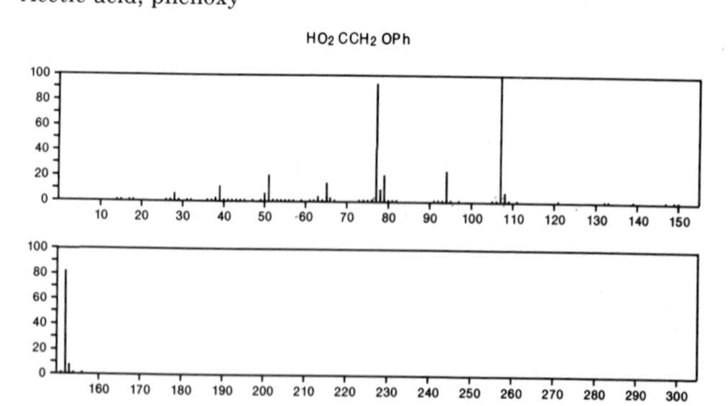

152 $C_8H_8O_3$ 100-09-4
Benzoic acid, 4-methoxy-

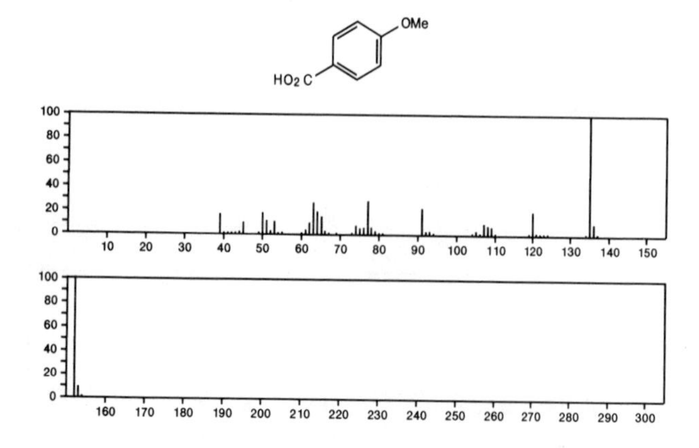

152 $C_8H_8O_3$ 156-38-7
Benzeneacetic acid, 4-hydroxy-

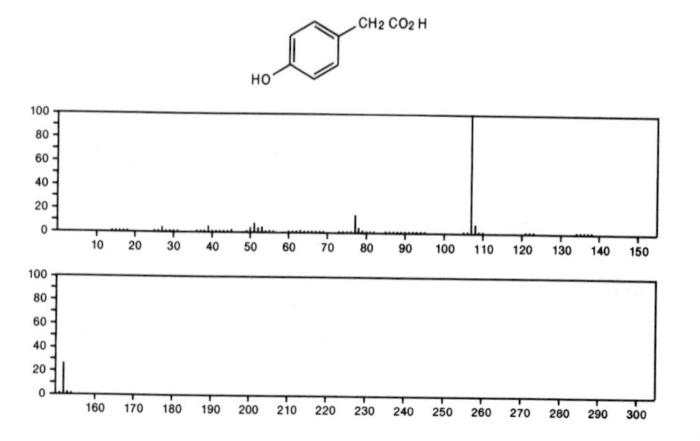

152 $C_8H_8O_3$ 487-69-4
Benzaldehyde, 2,4-dihydroxy-6-methyl-

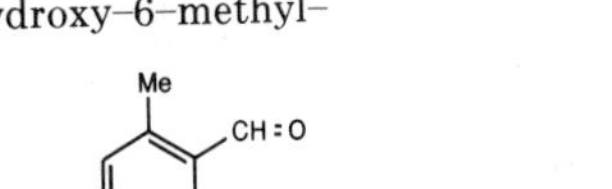

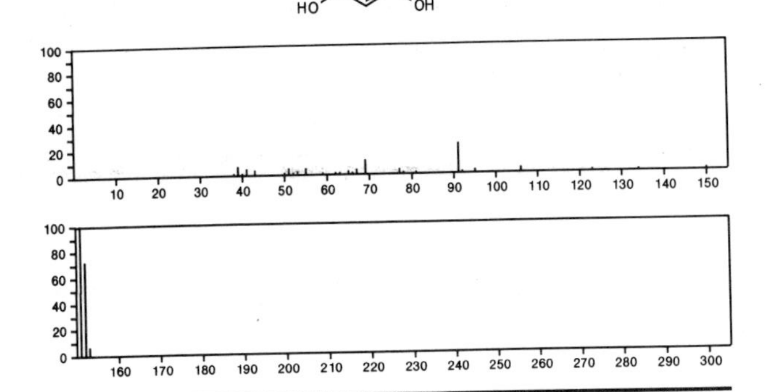

152 $C_8H_8O_3$ 567-61-3
Benzoic acid, 2-hydroxy-6-methyl-

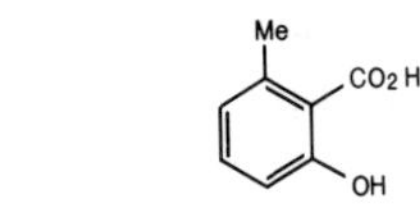

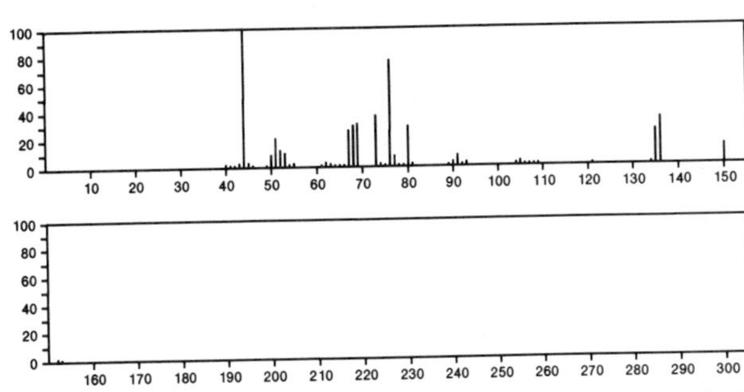

152 $C_8H_8O_3$ 579-75-9
Benzoic acid, 2-methoxy-

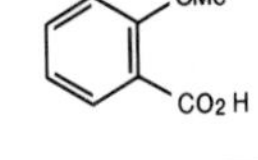

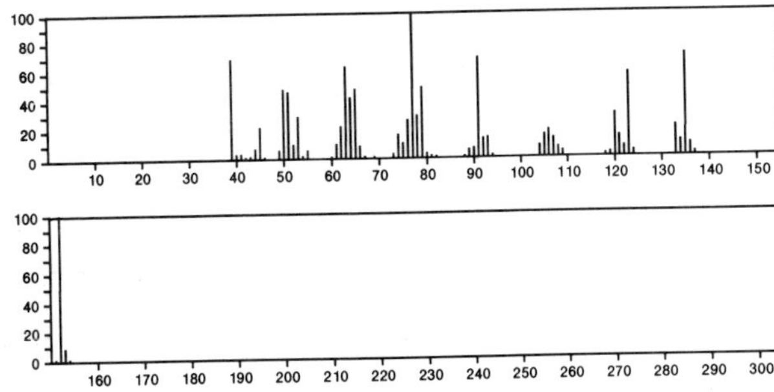

152 $C_8H_8O_3$ 586-38-9
Benzoic acid, 3-methoxy-

HO2C OMe

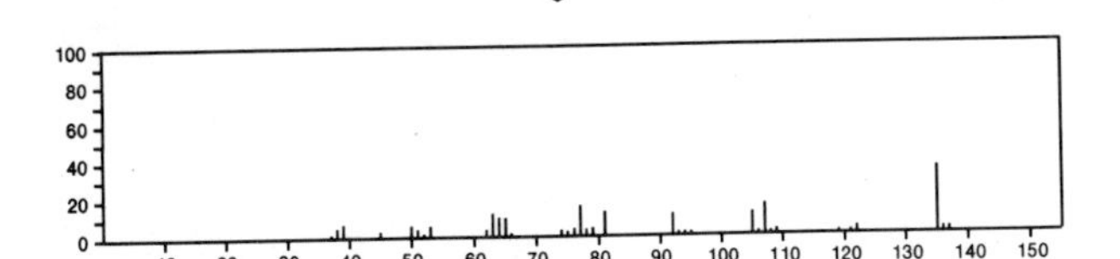

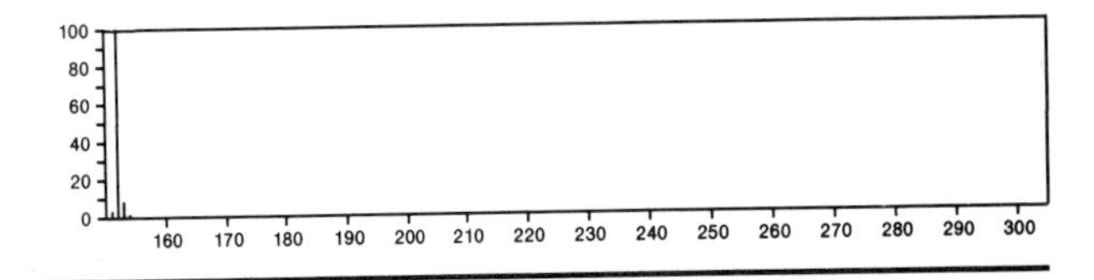

152 $C_8H_8O_3$ 612-20-4
Benzoic acid, 2-(hydroxymethyl)-

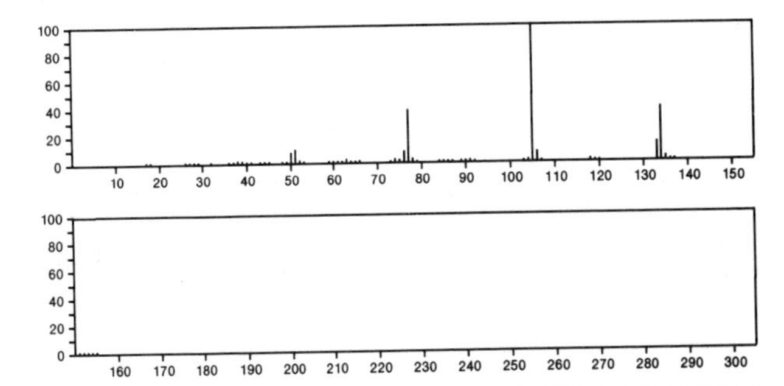

152 $C_8H_8O_3$ 614-13-1
2,5-Cyclohexadiene-1,4-dione, 2-methoxy-5-methyl-

O OMe Me O

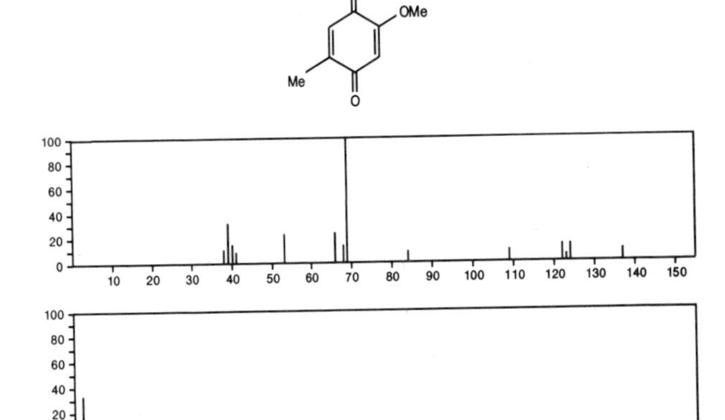

152 $C_8H_8O_3$ 614-75-5
Benzeneacetic acid, 2-hydroxy-

CH2 CO2 H
OH

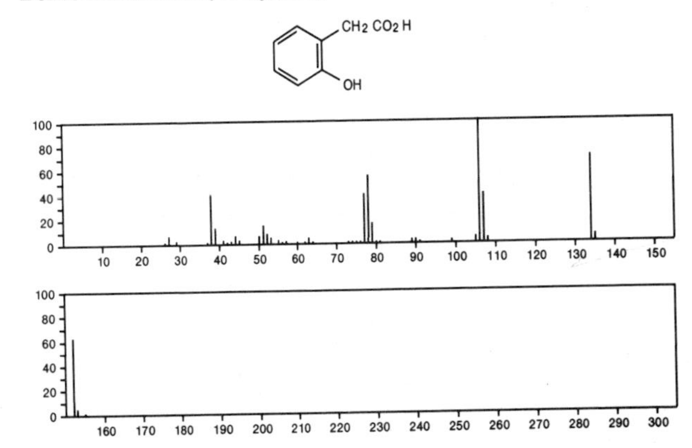

152 $C_8H_8O_3$ 621–37–4
Benzeneacetic acid, 3–hydroxy–

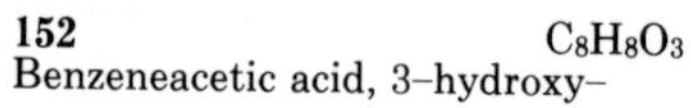

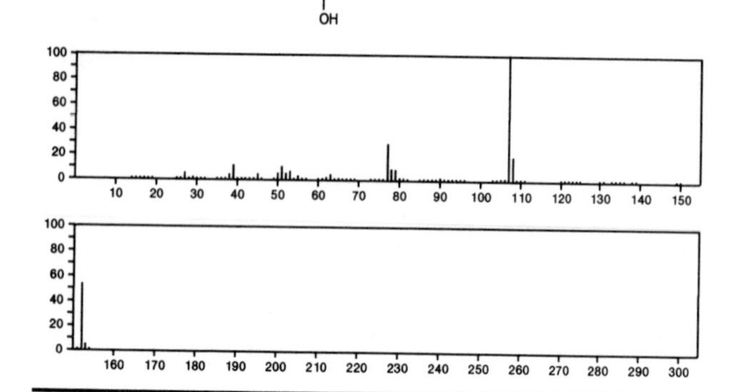

152 $C_8H_8O_3$ 4208–49–5
2–Furancarboxylic acid, 2–propenyl ester

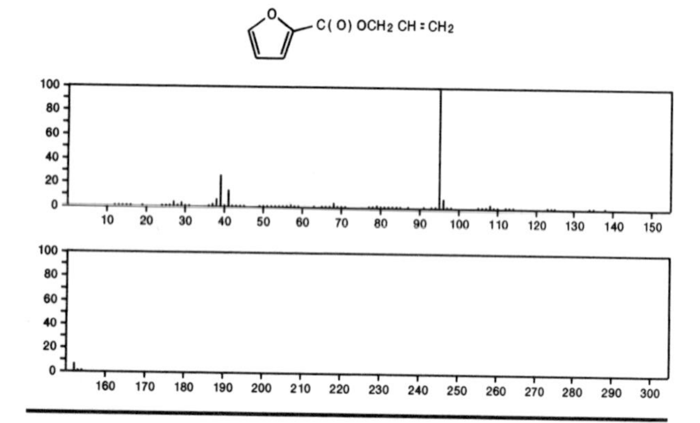

152 $C_8H_8O_3$ 5770–59–2
1,4–Benzodioxan–2–ol

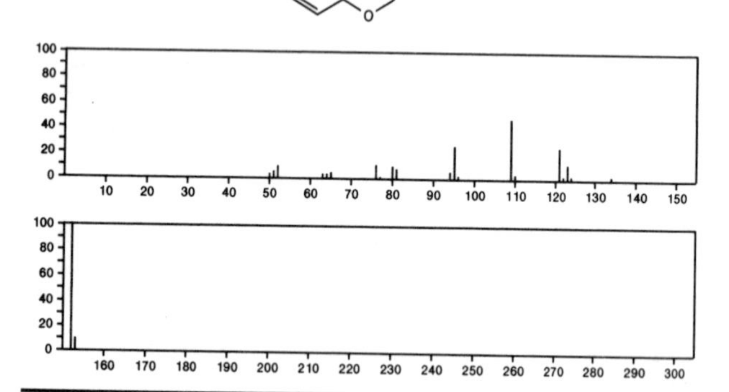

152 $C_8H_8O_3$ 13149–03–6
1,3–Isobenzofurandione, 3a,4,7,7a–tetrahydro–, *trans*–

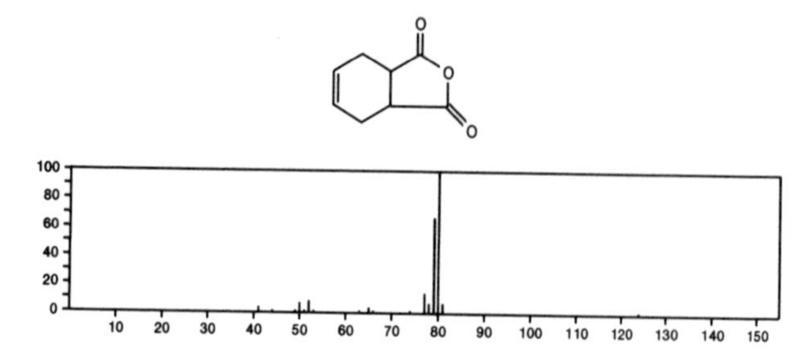

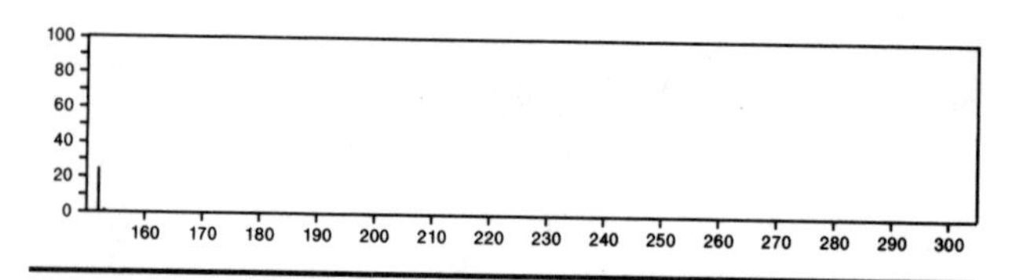

152 $C_8H_8O_3$ 13509–27–8
Carbonic acid, methyl phenyl ester

PhOC(O)OMe

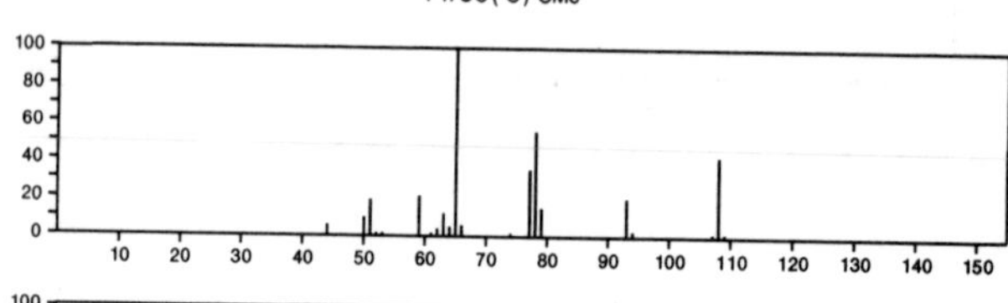

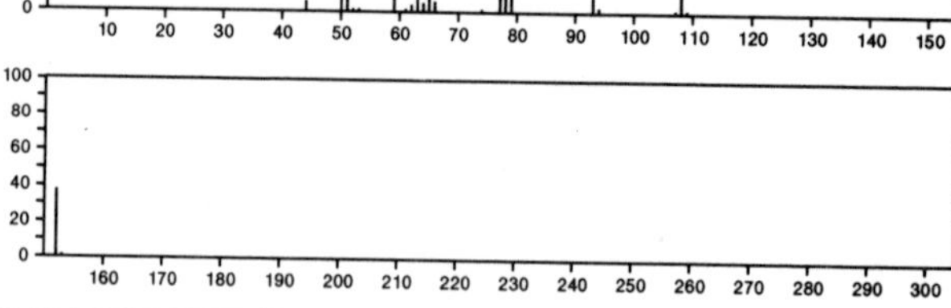

152 $C_8H_8O_3$ 19438–10–9
Benzoic acid, 3–hydroxy–, methyl ester

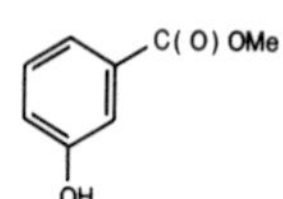

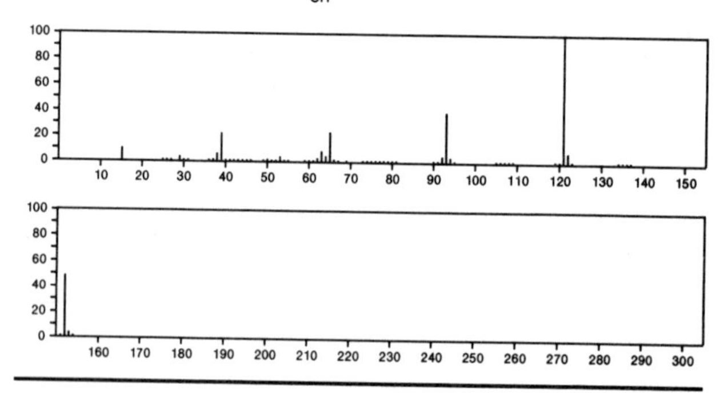

152 $C_8H_8O_3$ 26266–63–7
1,3–Isobenzofurandione, tetrahydro–

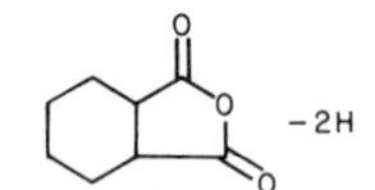

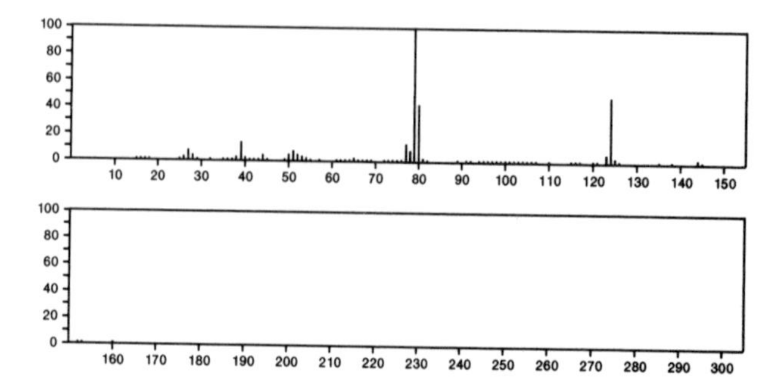

152 $C_8H_{12}N_2O$ 2814-20-2
4(1*H*)-Pyrimidinone, 6-methyl-2-(1-methylethyl)-

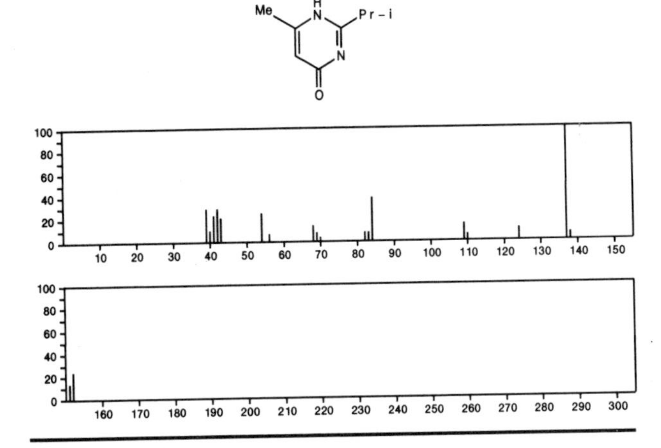

152 $C_8H_{12}N_2O$ 7781-21-7
Pyrimidine, 2-ethoxy-4,6-dimethyl-

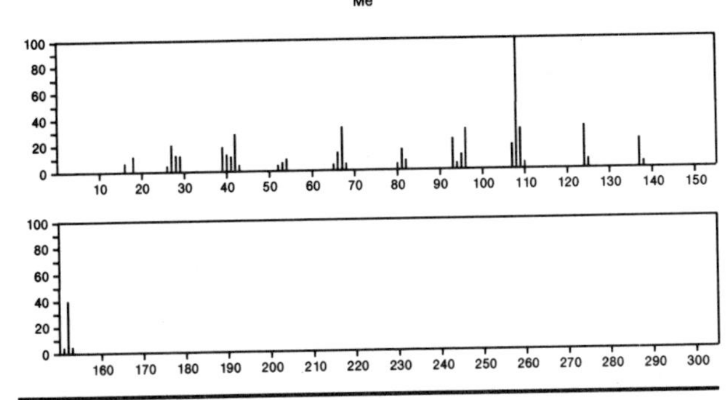

152 $C_8H_{12}N_2O$ 16858-16-5
4(1*H*)-Pyrimidinone, 6-methyl-2-propyl-

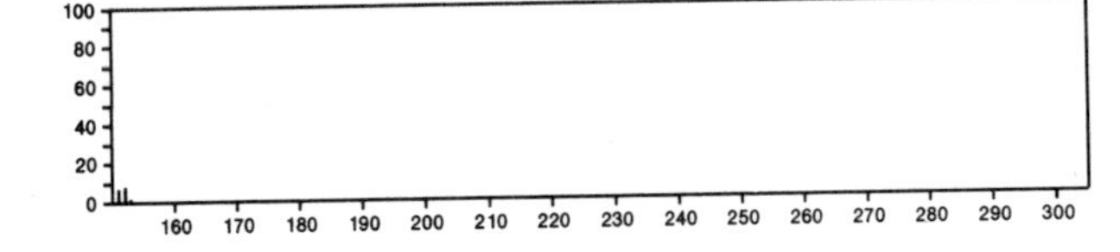

152 $C_8H_{12}N_2O$ 32363-52-3
4(3*H*)-Pyrimidinone, 3-ethyl-2,6-dimethyl-

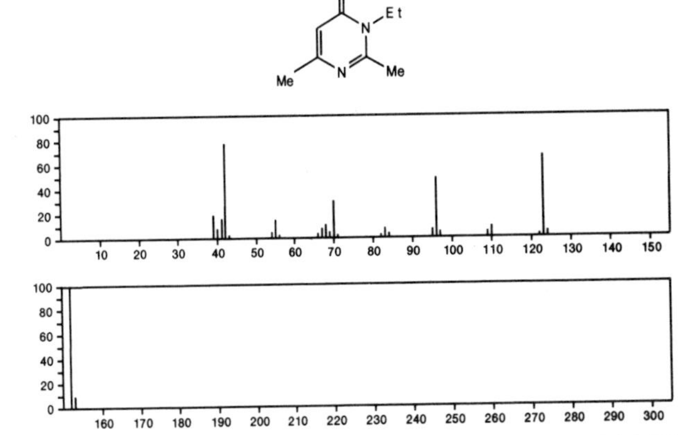

152 $C_8H_{12}N_2O$ 32363-54-5
4(3*H*)-Pyrimidinone, 2-ethyl-3,6-dimethyl-

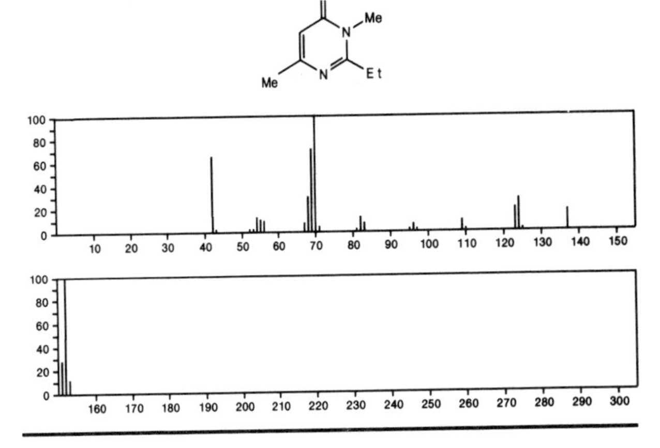

152 $C_8H_{12}N_2O$ 49540-59-2
Ethanol, 2-(1-phenylhydrazino)-

$HOCH_2CH_2N(NH_2)Ph$

152 C_9H_9Cl 2687-12-9
Benzene, (3-chloro-1-propenyl)-

$PhCH=CHCH_2Cl$

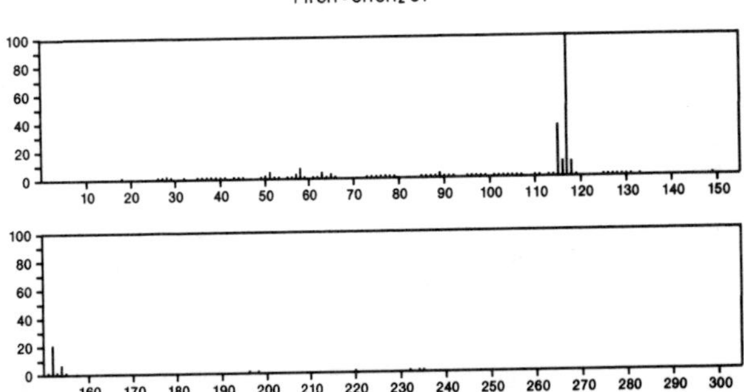

152 C_9H_9Cl 6268-37-7
Benzene, (3-chloroallyl)-

PhCH2 CH=CHCl

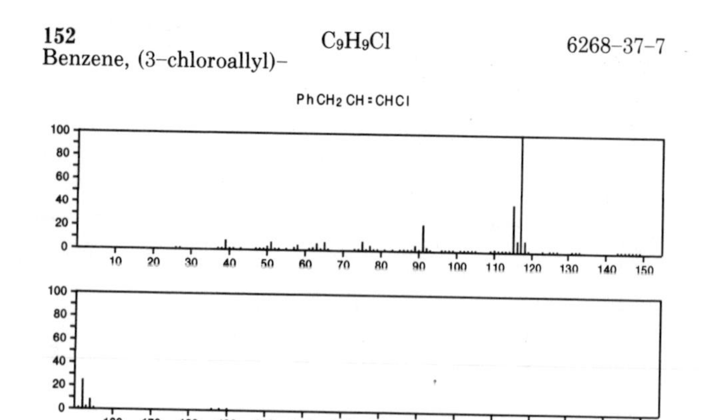

152 C_9H_9Cl 30030-25-2
Benzene, (chloromethyl)ethenyl-

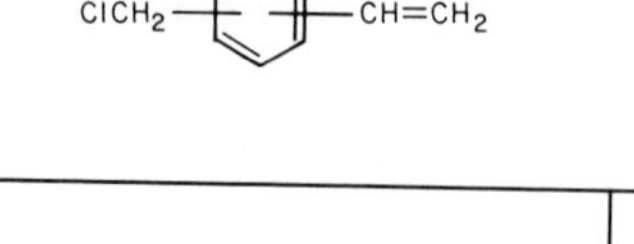

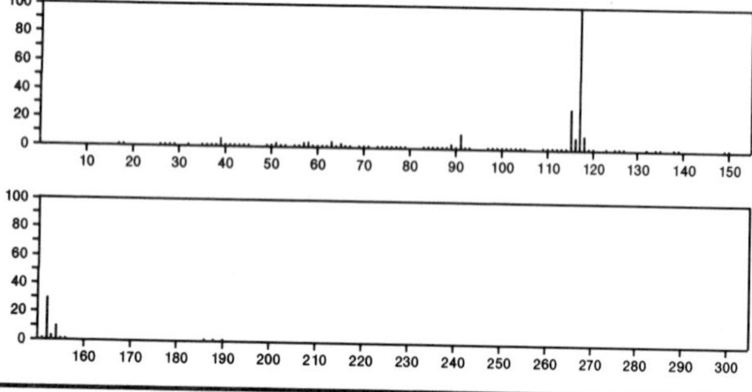

152 C_9H_9Cl 35275-62-8
1*H*-Indene, 1-chloro-2,3-dihydro-

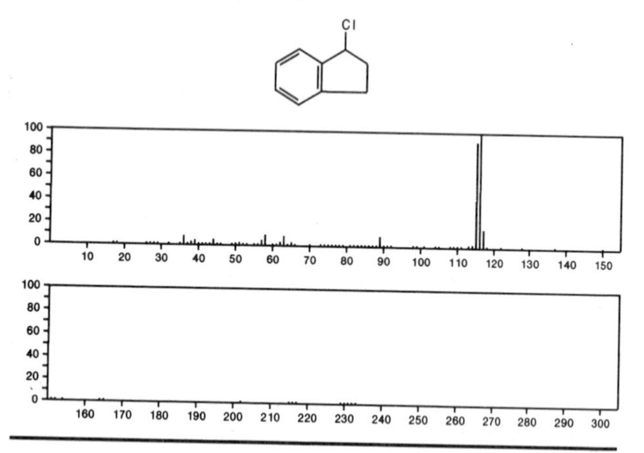

152 C_9H_9FO 456-03-1
1-Propanone, 1-(4-fluorophenyl)-

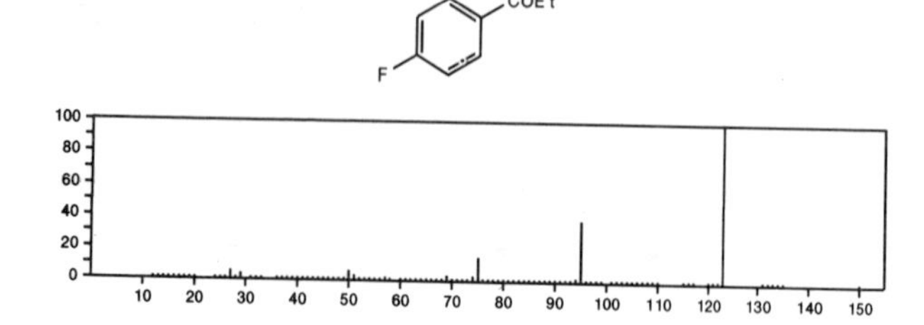

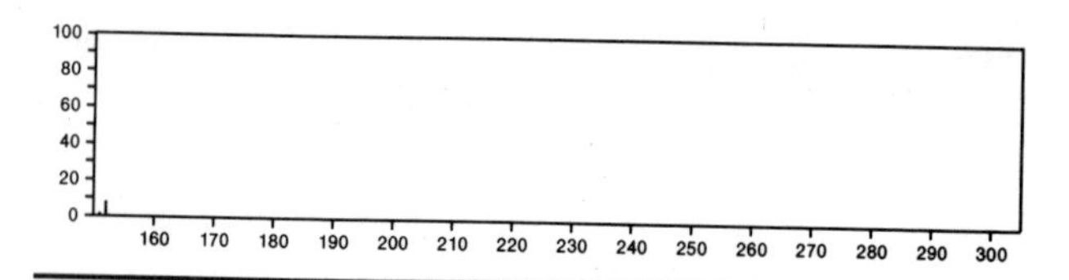

152 $C_9H_{12}O_2$ 80-15-9
Hydroperoxide, 1-methyl-1-phenylethyl

HOOCMe2 Ph

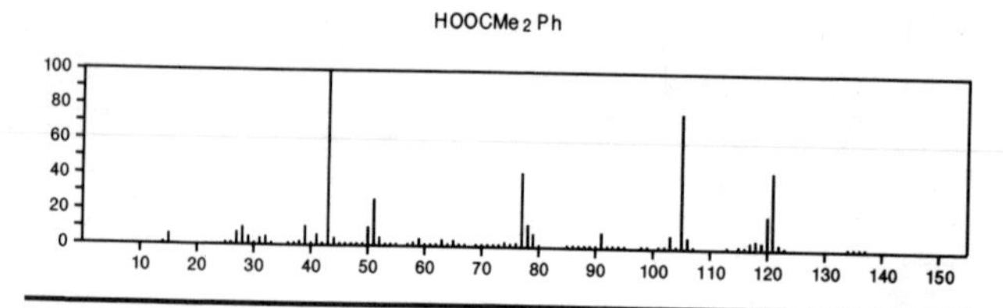

152 $C_9H_{12}O_2$ 551-45-1
2-Cyclopenten-1-one, 4-hydroxy-3-methyl-2-(2-propenyl)-

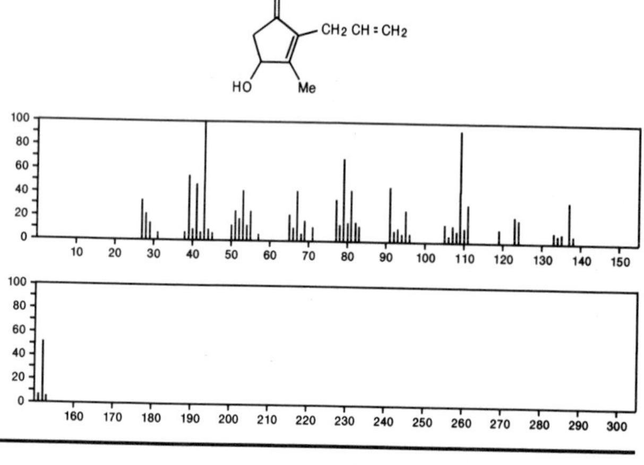

152 $C_9H_{12}O_2$ 622-08-2
Ethanol, 2-(phenylmethoxy)-

HOCH2 CH2 OCH2 Ph

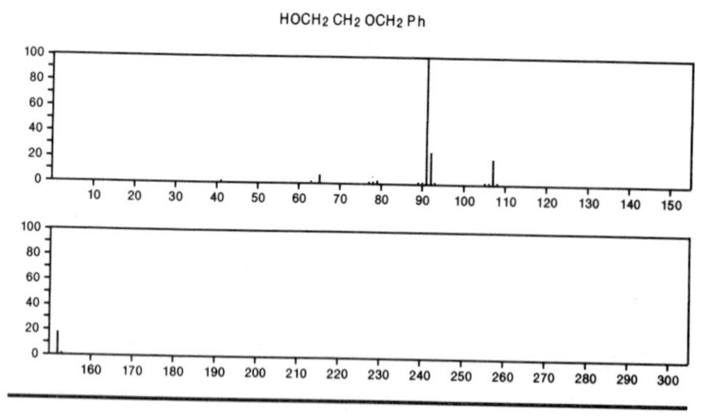

152 $C_9H_{12}O_2$ 1481-92-1
Benzenepropanol, 2-hydroxy-

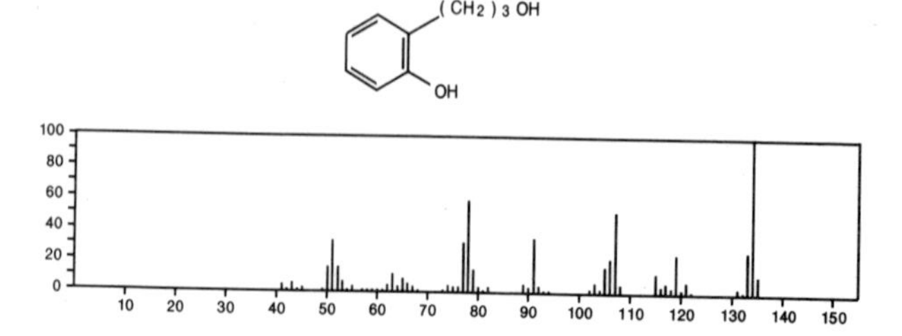

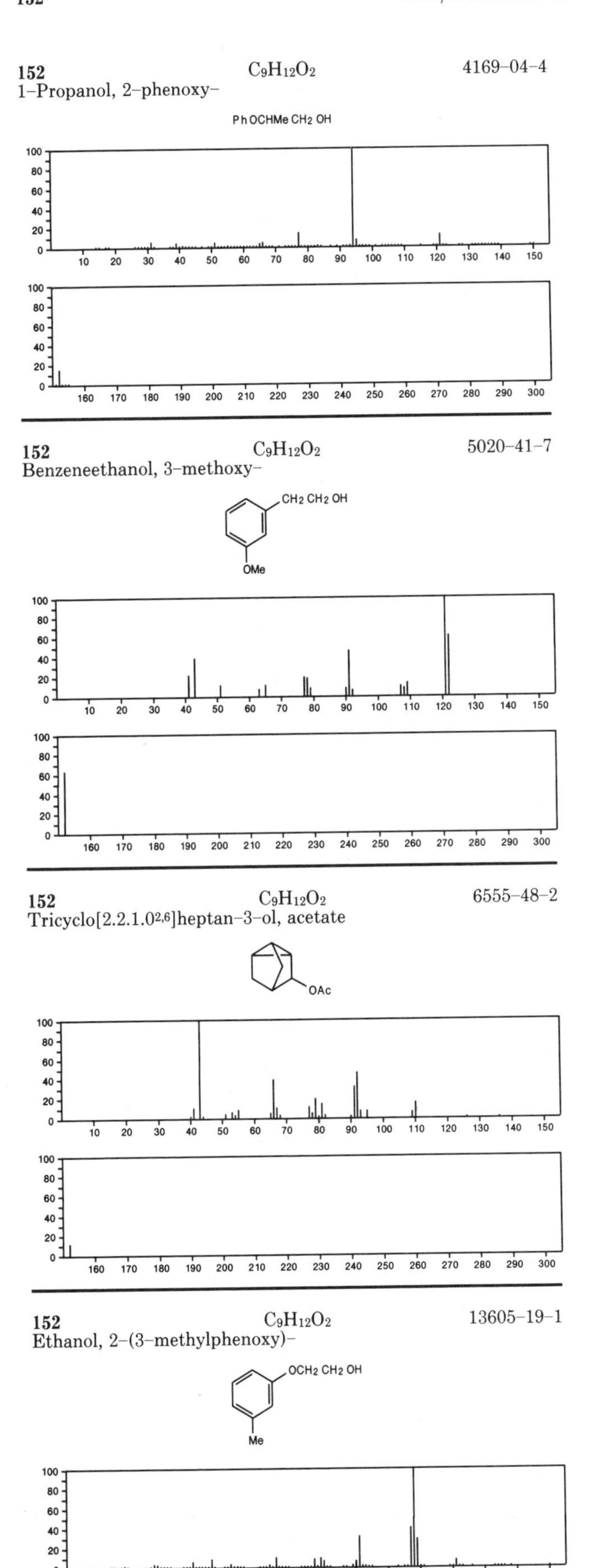

152 $C_9H_{12}O_2$ 4169-04-4
1-Propanol, 2-phenoxy-
PhOCHMe CH2 OH
152 $C_9H_{12}O_2$ 5020-41-7
Benzeneethanol, 3-methoxy-
CH2 CH2 OH
OMe
152 $C_9H_{12}O_2$ 6555-48-2
Tricyclo[2.2.1.0^{2,6}]heptan-3-ol, acetate
OAc
152 $C_9H_{12}O_2$ 13605-19-1
Ethanol, 2-(3-methylphenoxy)-
OCH2 CH2 OH
Me

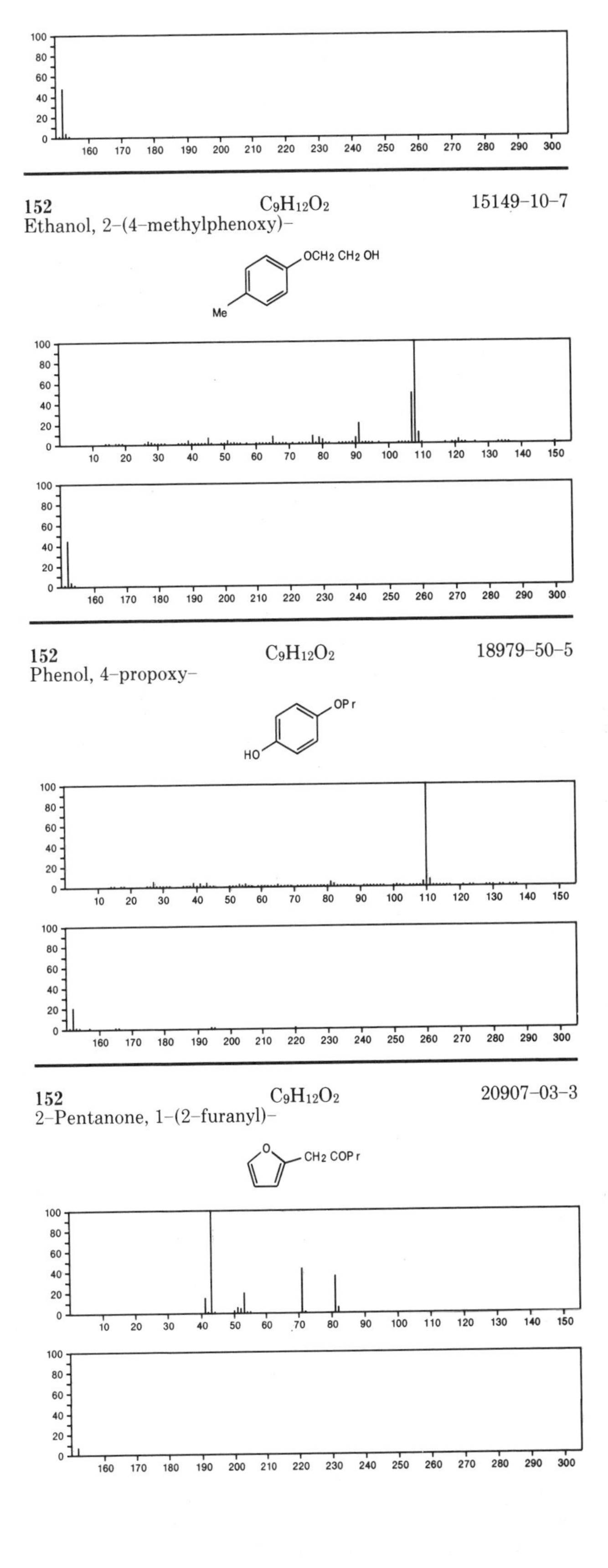

152 $C_9H_{12}O_2$ 15149-10-7
Ethanol, 2-(4-methylphenoxy)-
OCH2 CH2 OH
Me
152 $C_9H_{12}O_2$ 18979-50-5
Phenol, 4-propoxy-
OPr
HO
152 $C_9H_{12}O_2$ 20907-03-3
2-Pentanone, 1-(2-furanyl)-
CH2 COPr

152 $C_9H_{12}O_2$ 20907-04-4
2-Butanone, 1-(2-furanyl)-3-methyl-

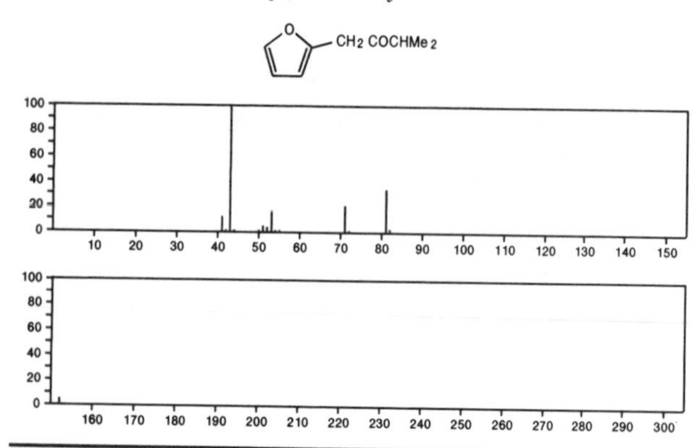

152 $C_9H_{12}O_2$ 21998-86-7
Benzene, 1-methoxy-2-(methoxymethyl)-

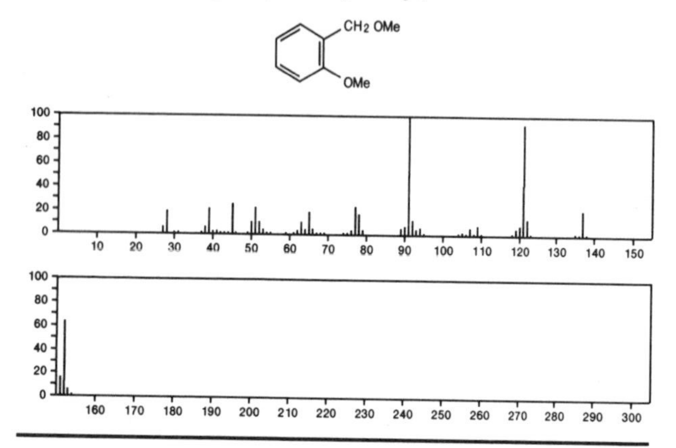

152 $C_9H_{12}O_2$ 22805-42-1
Benzenemethanol, α-ethyl-4-hydroxy-

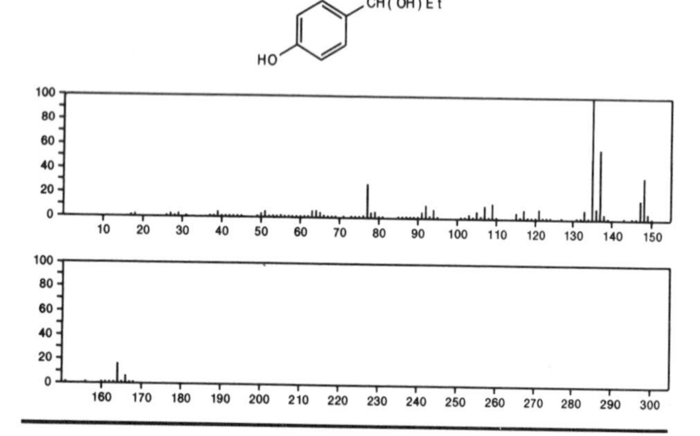

152 $C_9H_{12}O_2$ 31600-55-2
Benzene, [(methoxymethoxy)methyl]-

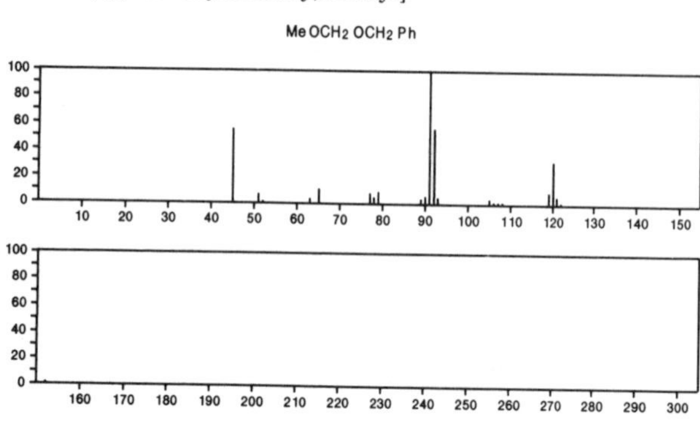

152 $C_9H_{12}O_2$ 31681-26-2
2-Furanacetaldehyde, α-propyl-

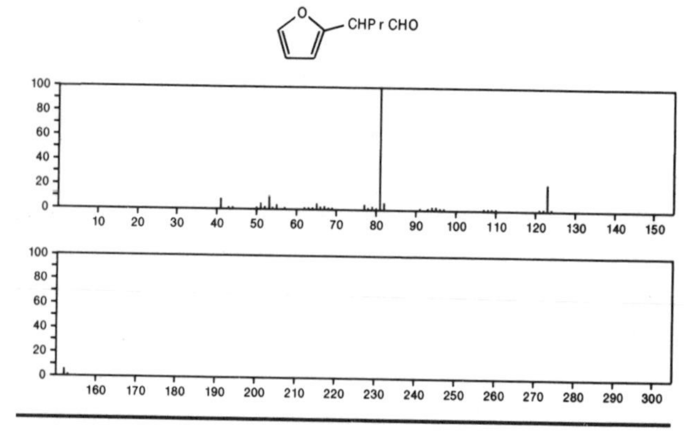

152 $C_9H_{12}O_2$ 31681-30-8
2-Furanacetaldehyde, α-isopropyl-

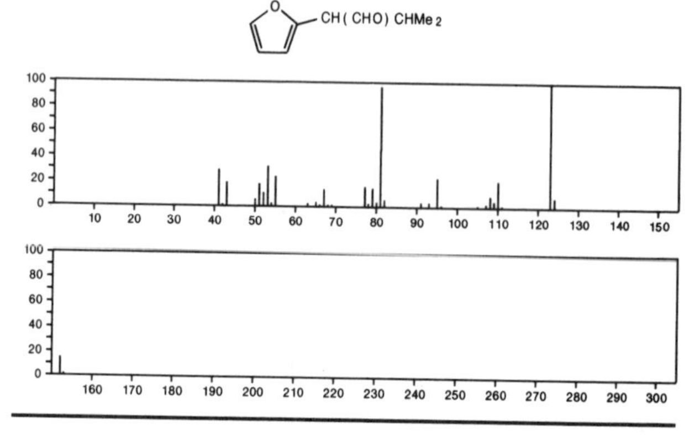

152 $C_9H_{12}O_2$ 34883-01-7
Phenol, 5-methoxy-2,3-dimethyl-

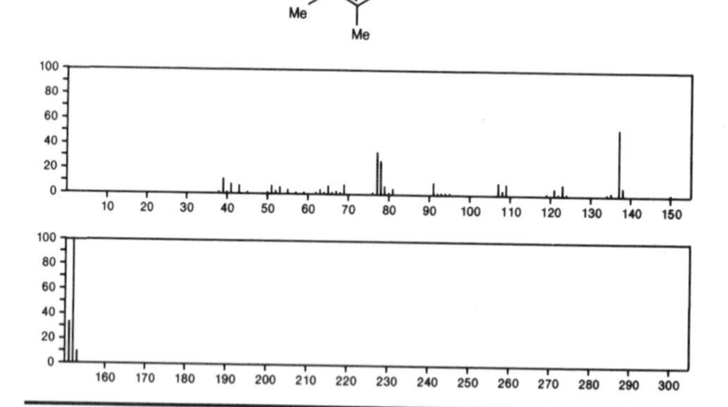

152 $C_9H_{12}O_2$ 41532-81-4
Benzene, (2-methoxyethoxy)-

Me OCH2 CH2 OPh

152 $C_9H_{12}O_2$ 53387-38-5
2(3*H*)-Benzofuranone, hexahydro-3-methylene-

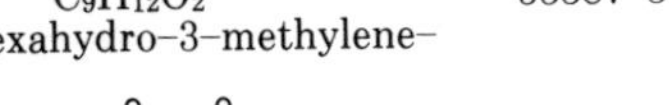

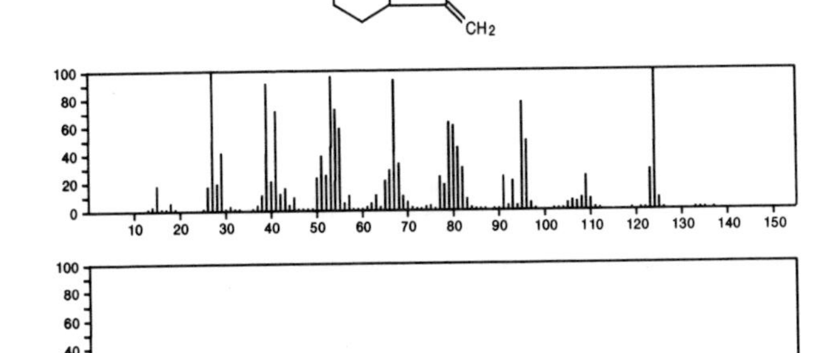

152 $C_9H_{12}O_2$ 56051-94-6
2-Cyclohexen-1-one, 4-(2-oxopropyl)-

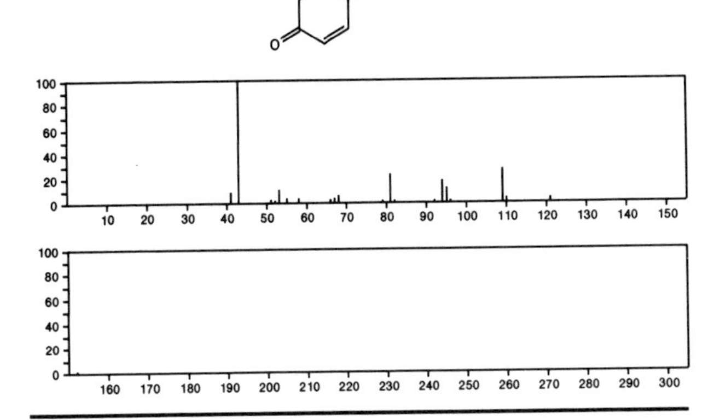

152 $C_9H_{12}O_2$ 56701-05-4
Tricyclo[3.2.0.0^{2,7}]heptane-3-carboxylic acid, methyl ester, (1α,2α,3α,5β,7α)-

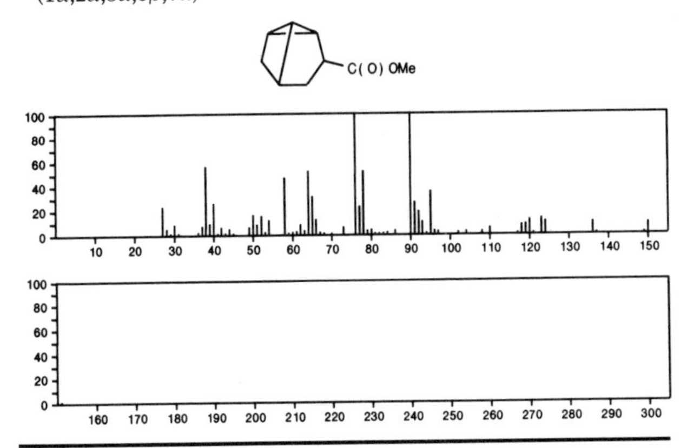

152 $C_9H_{12}O_2$ 56760-80-6
Tricyclo[3.2.0.0^{2,7}]heptane-3-carboxylic acid, methyl ester, (1α,2α,3β,5β,7α)-

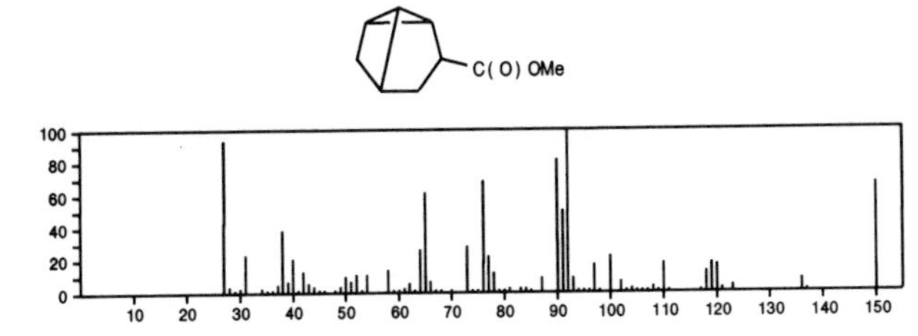

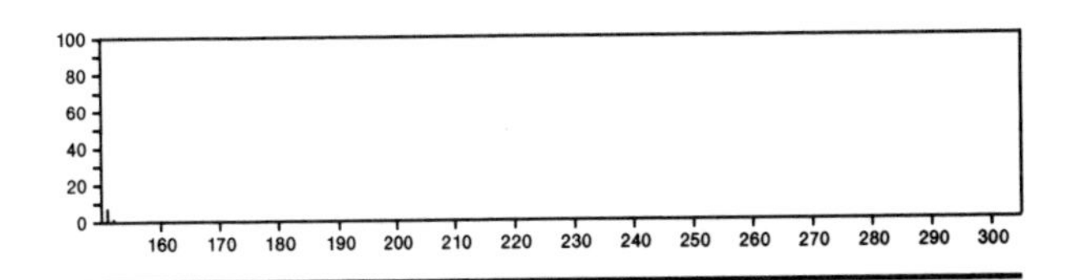

152 $C_9H_{12}S$ 622-63-9
Benzene, 1-(ethylthio)-4-methyl-

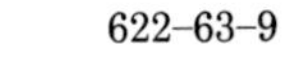

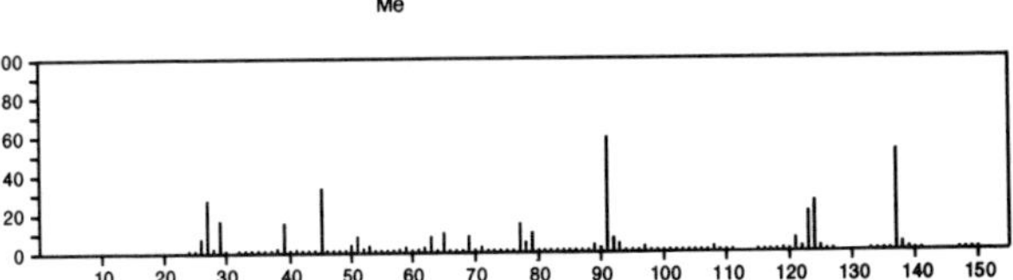

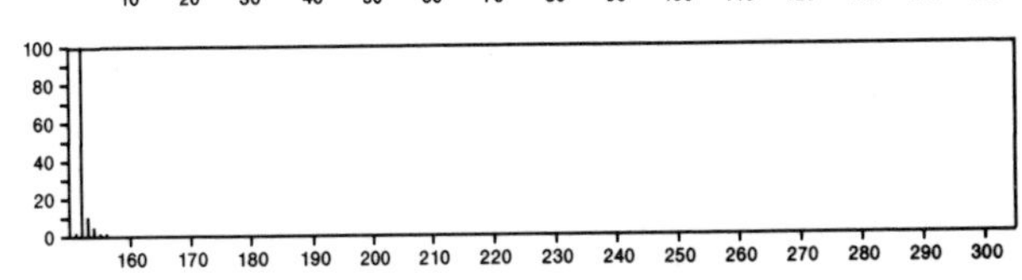

152 $C_9H_{12}S$ 874-79-3
Benzene, (propylthio)-

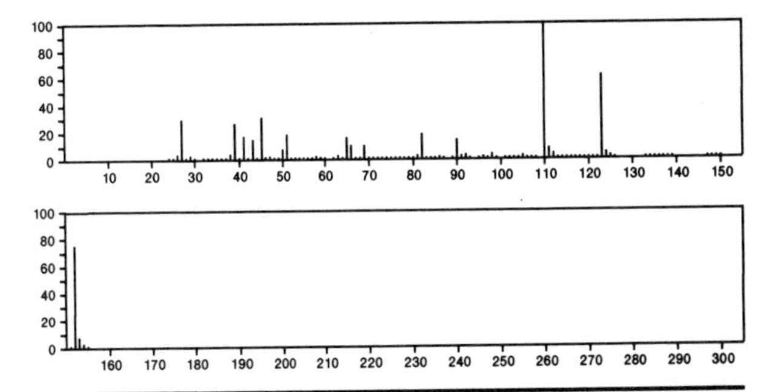

152 $C_9H_{12}S$ 3695-36-1
Benzene, 1-(ethylthio)-2-methyl-

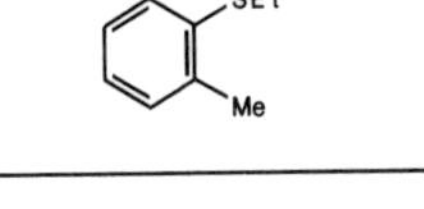

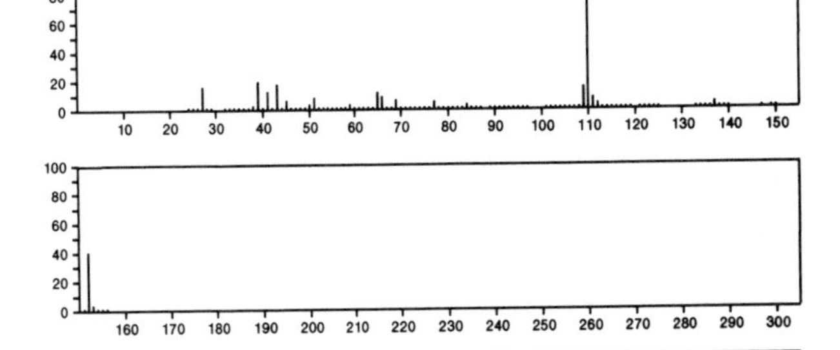

152 $C_9H_{12}S$ 6263-62-3
Benzene, [(ethylthio)methyl]-

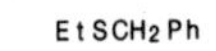

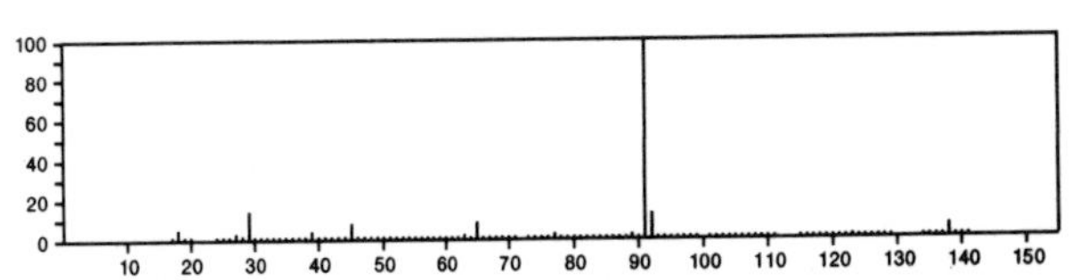

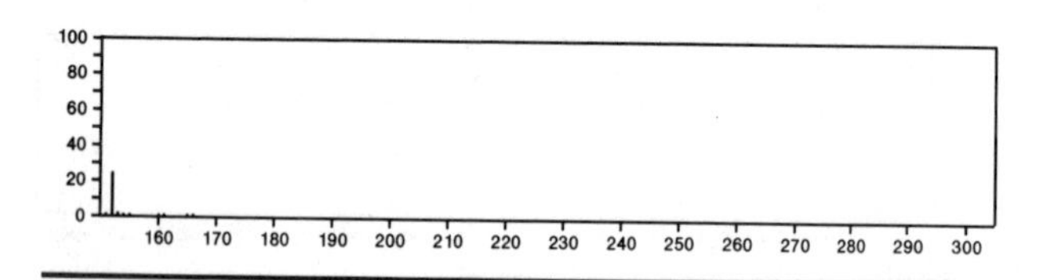

152 $C_9H_{12}S$ 34786-24-8
Benzene, 1-(ethylthio)-3-methyl-

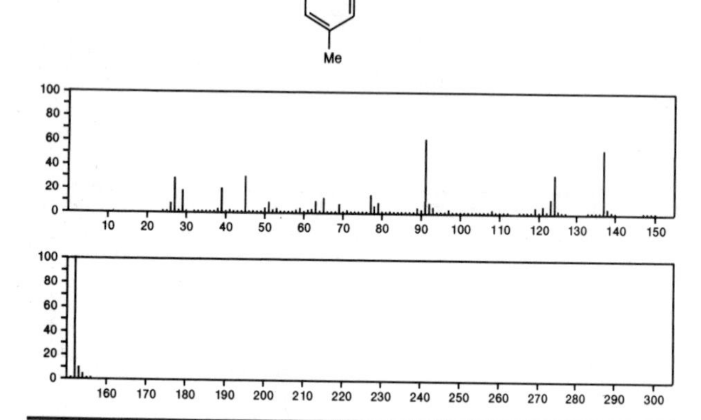

152 $C_{10}H_{16}O$ 76-22-2
Bicyclo[2.2.1]heptan-2-one, 1,7,7-trimethyl-

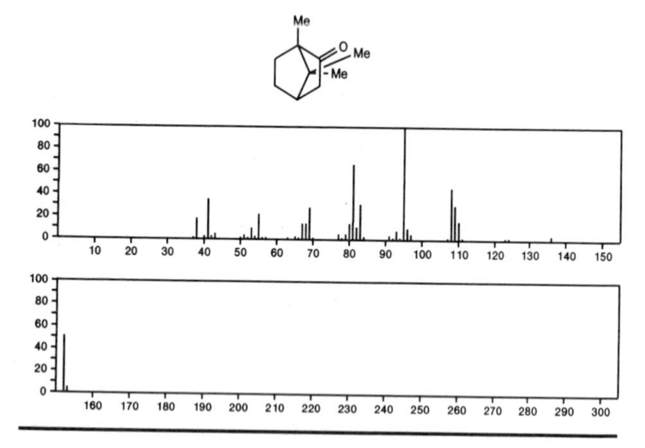

152 $C_{10}H_{16}O$ 89-81-6
2-Cyclohexen-1-one, 3-methyl-6-(1-methylethyl)-

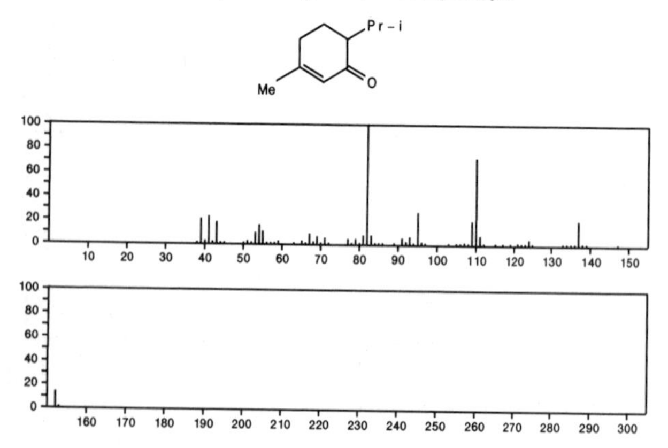

152 $C_{10}H_{16}O$ 89-82-7
Cyclohexanone, 5-methyl-2-(1-methylethylidene)-, (*R*)-

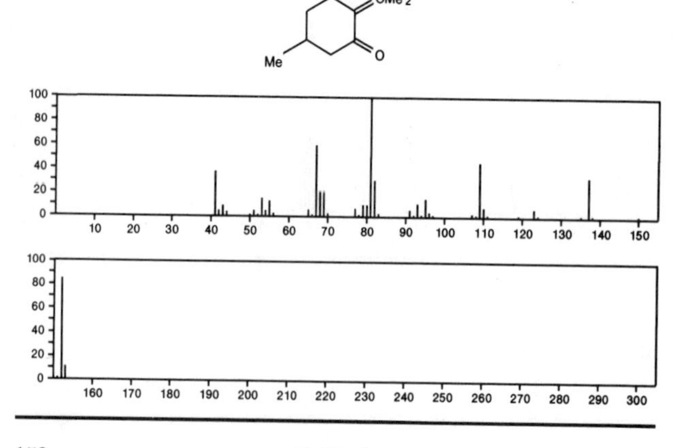

152 $C_{10}H_{16}O$ 99-48-9
2-Cyclohexen-1-ol, 2-methyl-5-(1-methylethenyl)-

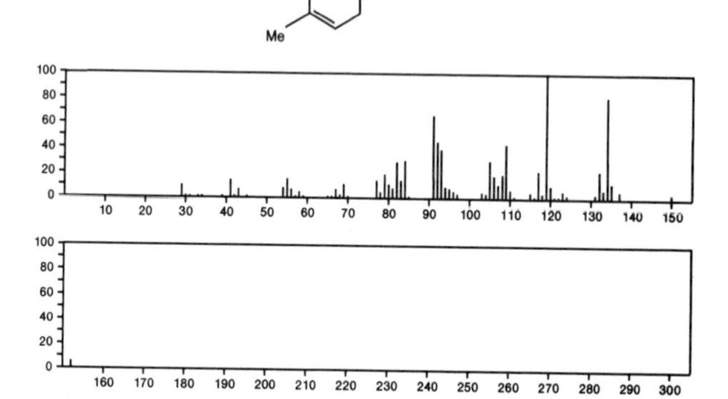

152 $C_{10}H_{16}O$ 106-26-3
2,6-Octadienal, 3,7-dimethyl-, (*Z*)-

OCHCH = CMe CH2 CH2 CH = CMe2

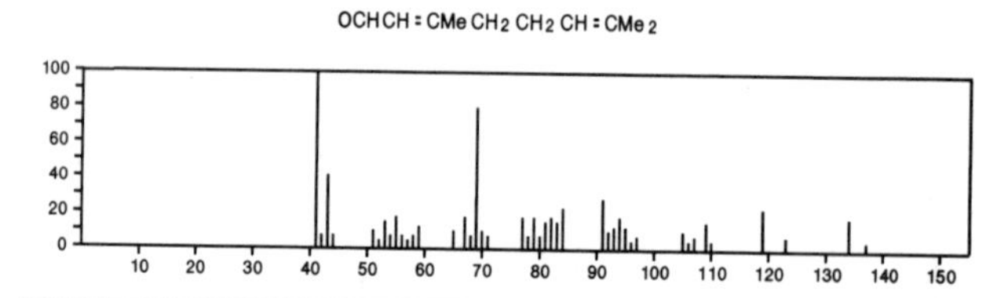

152 $C_{10}H_{16}O$ 141-27-5
2,6-Octadienal, 3,7-dimethyl-, (*E*)-

OCHCH = CMe CH2 CH2 CH = CMe2

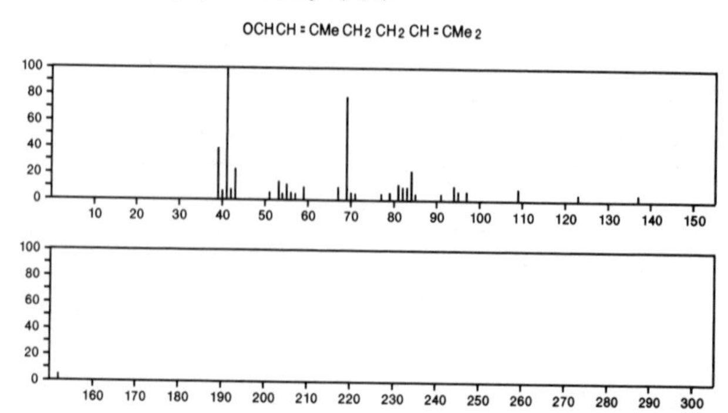

152 $C_{10}H_{16}O$ 432-25-7
1-Cyclohexene-1-carboxaldehyde, 2,6,6-trimethyl-

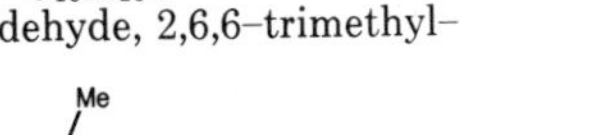

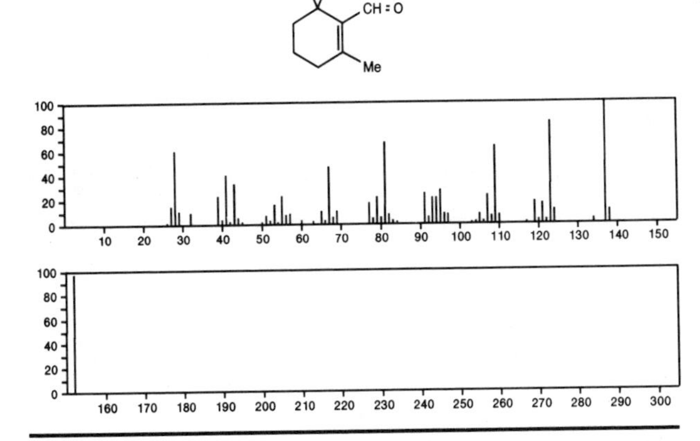

152 $C_{10}H_{16}O$ 471-15-8
Bicyclo[3.1.0]hexan-3-one, 4-methyl-1-(1-methylethyl)-, [1*S*-(1α,4β,5α)]-

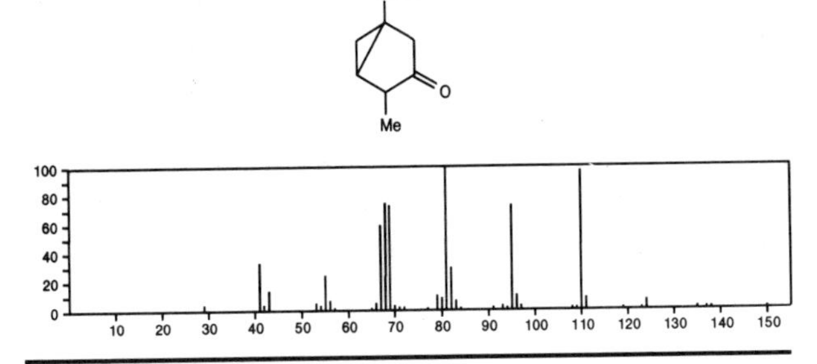

152 $C_{10}H_{16}O$ 471-16-9
Bicyclo[3.1.0]hexan-3-ol, 4-methylene-1-(1-methylethyl)-, [1*S*-(1α,3β,5α)]-

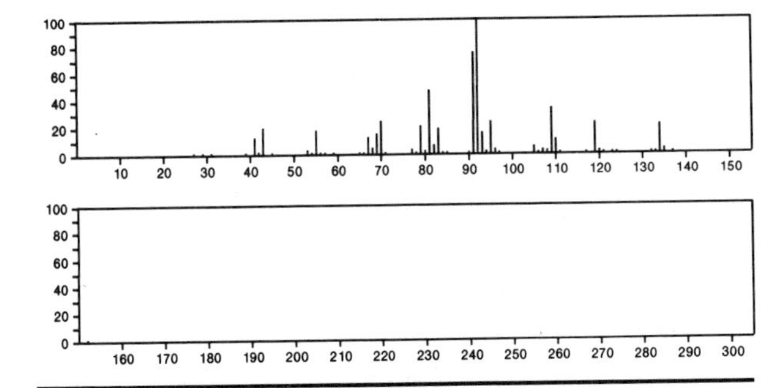

152 $C_{10}H_{16}O$ 497-62-1
Bicyclo[4.1.0]heptan-2-one, 3,7,7-trimethyl-

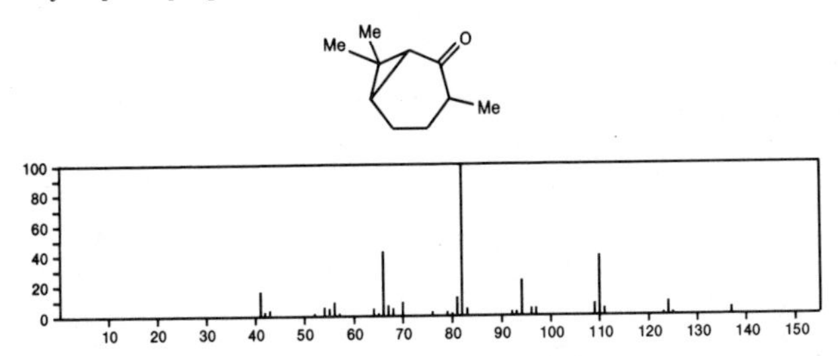

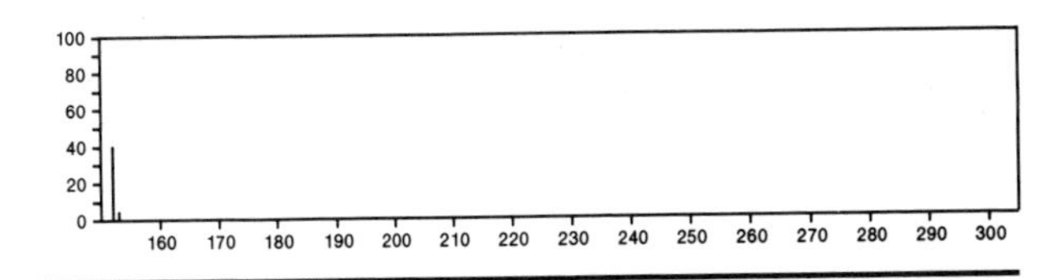

152 $C_{10}H_{16}O$ 515-00-4
Bicyclo[3.1.1]hept-2-ene-2-methanol, 6,6-dimethyl-

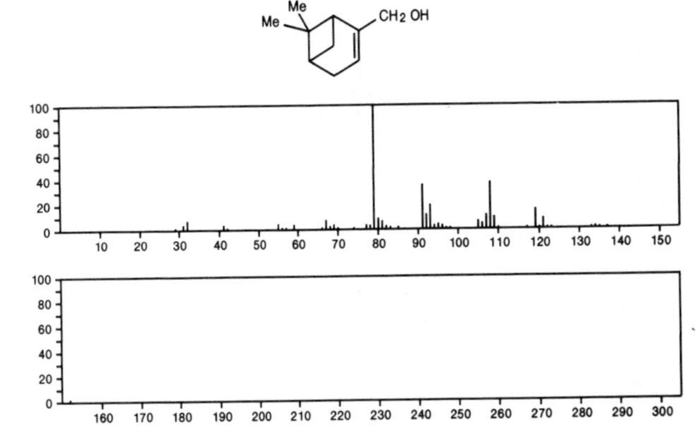

152 $C_{10}H_{16}O$ 536-59-4
1-Cyclohexene-1-methanol, 4-(1-methylethenyl)-

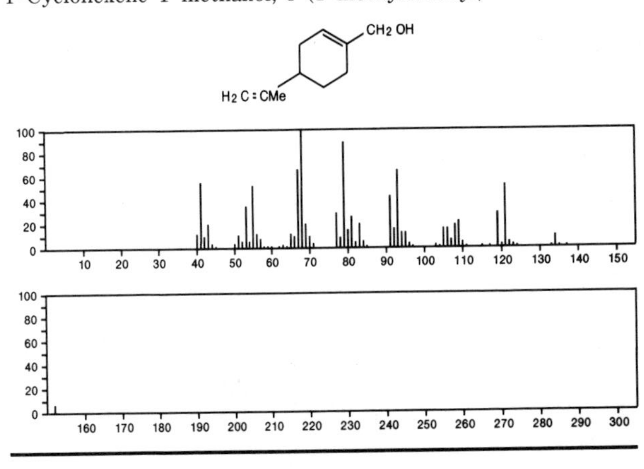

152 $C_{10}H_{16}O$ 546-49-6
1,5-Heptadien-4-one, 3,3,6-trimethyl-

$Me_2C:CHCOCMe_2CH:CH_2$

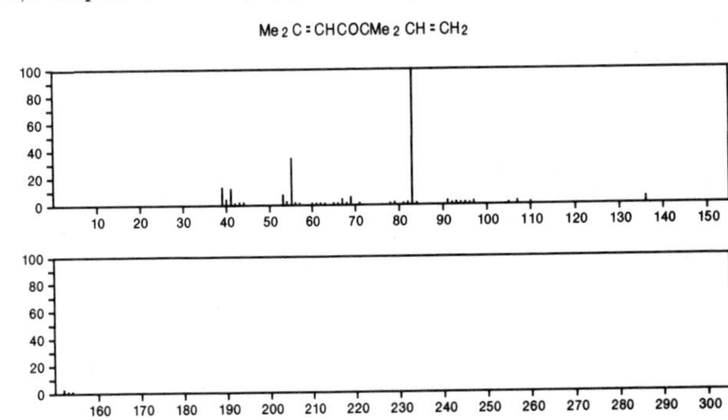

152 $C_{10}H_{16}O$ 546-80-5
Bicyclo[3.1.0]hexan-3-one, 4-methyl-1-(1-methylethyl)-, [1*S*-(1α,4α,5α)]-

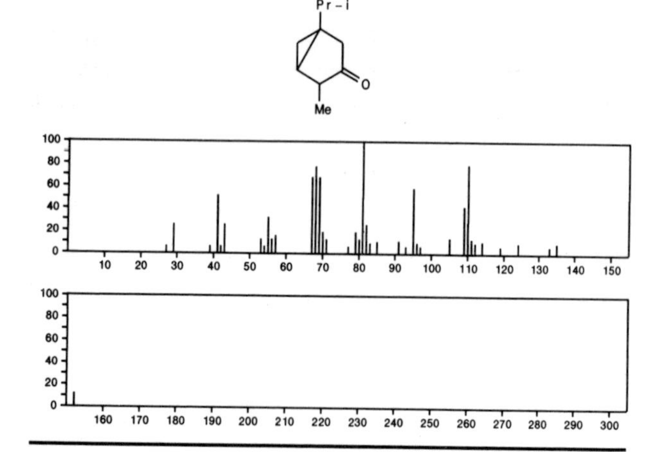

152 $C_{10}H_{16}O$ 1195-79-5
Bicyclo[2.2.1]heptan-2-one, 1,3,3-trimethyl-

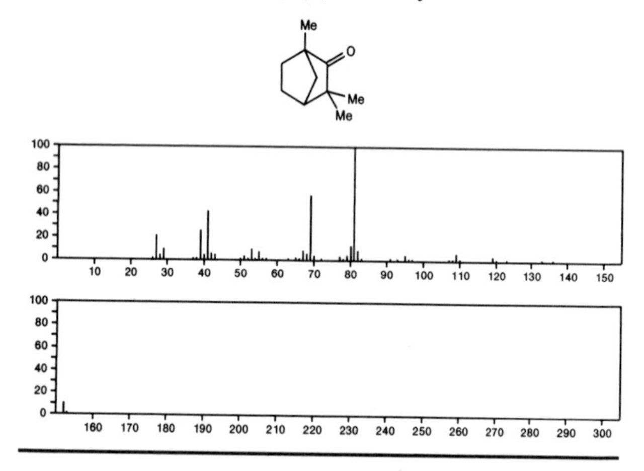

152 $C_{10}H_{16}O$ 1197-06-4
2-Cyclohexen-1-ol, 2-methyl-5-(1-methylethenyl)-, *cis*-

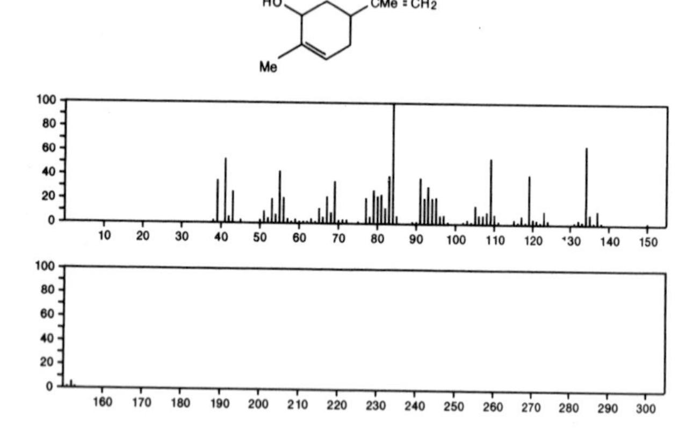

152 $C_{10}H_{16}O$ 1197-07-5
2-Cyclohexen-1-ol, 2-methyl-5-(1-methylethenyl)-, *trans*-

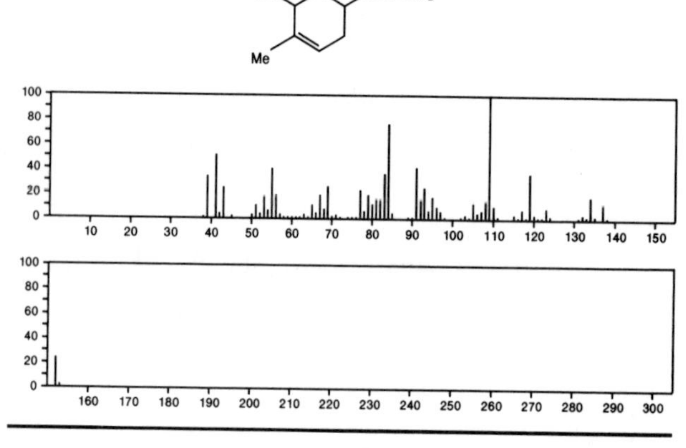

152 $C_{10}H_{16}O$ 4696-15-5
Bicyclo[5.2.1]decan-10-one

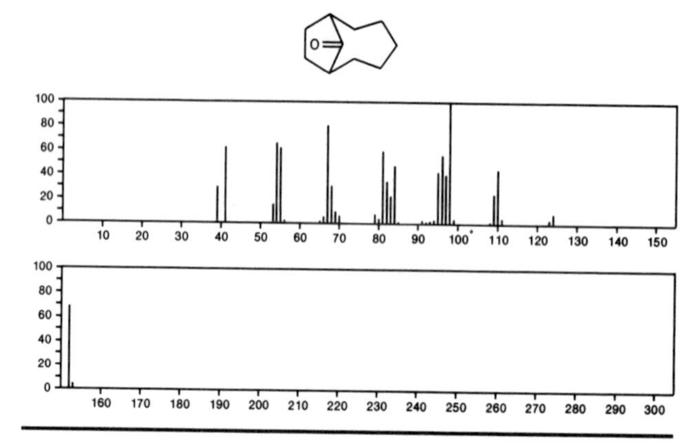

152 $C_{10}H_{16}O$ 4884-24-6
[1,1'-Bicyclopentyl]-2-one

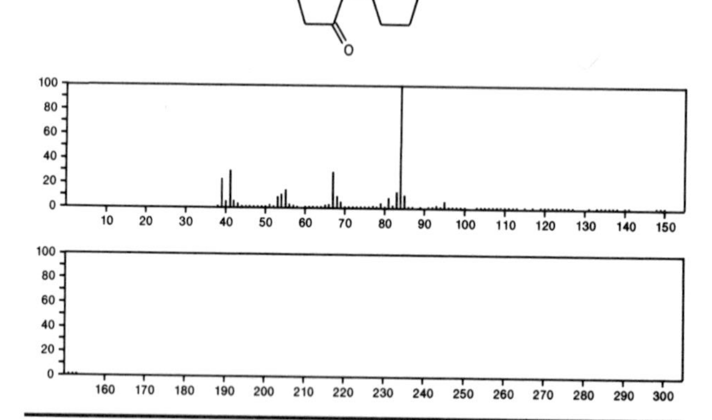

152 $C_{10}H_{16}O$ 5113-66-6
2-Cyclohexen-1-one, 5-methyl-2-(1-methylethyl)-

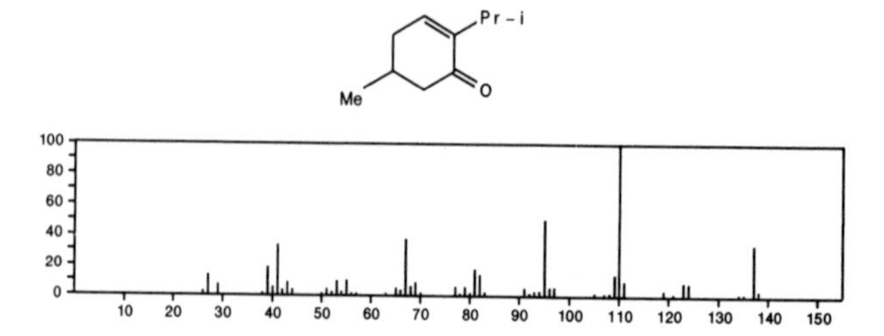

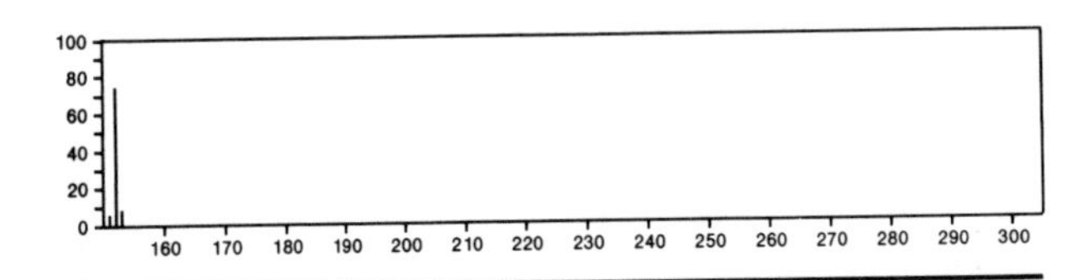

152 $C_{10}H_{16}O$ 5392-40-5

2,6-Octadienal, 3,7-dimethyl-

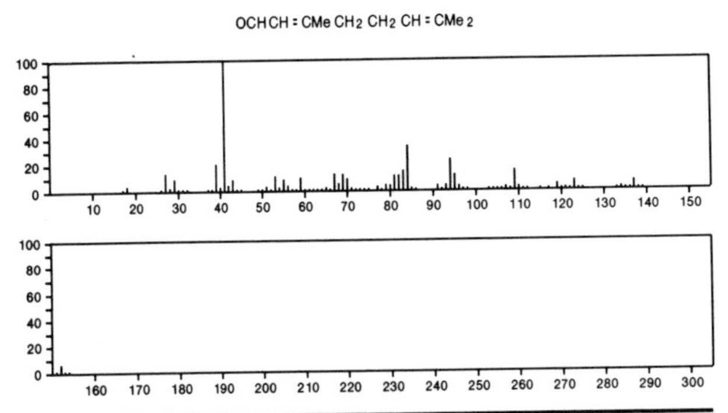

152 $C_{10}H_{16}O$ 6508-22-1

2-Oxatricyclo[3.3.1.1^{3,7}]decane, 1-methyl-

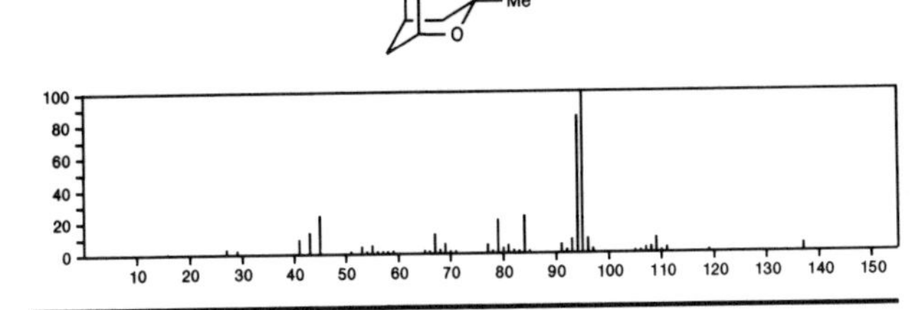

152 $C_{10}H_{16}O$ 7764-50-3

Cyclohexanone, 2-methyl-5-(1-methylethenyl)-

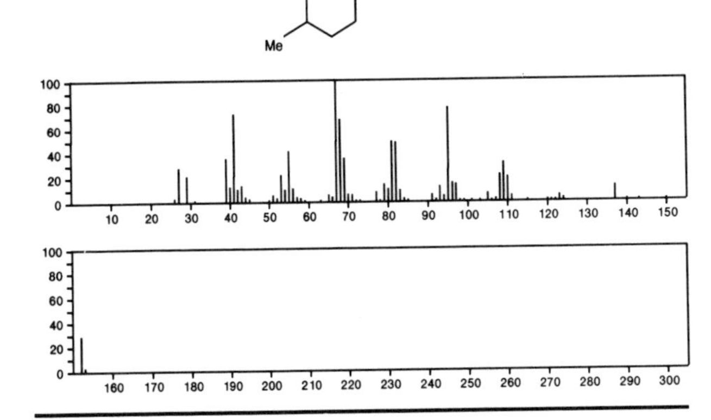

152 $C_{10}H_{16}O$ 10292-98-5

Bicyclo[2.2.1]heptan-2-one, 4,7,7-trimethyl-, (1*S*)-

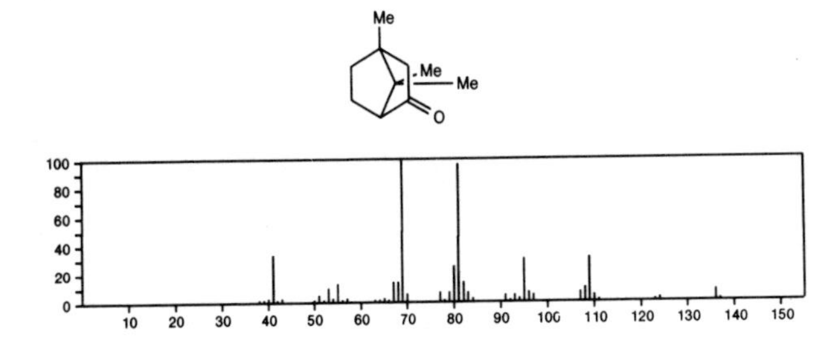

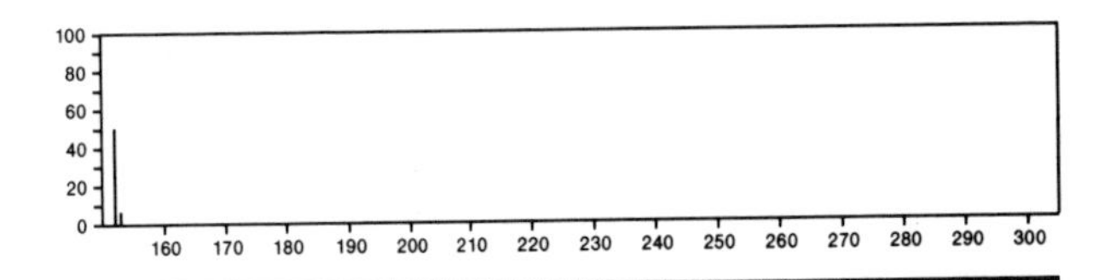

152 $C_{10}H_{16}O$ 13025-91-7

1*H*-Inden-1-one, octahydro-7a-methyl-, *cis*-

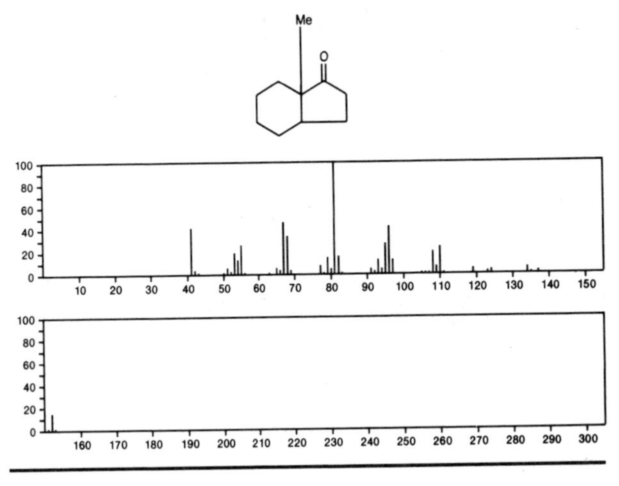

152 $C_{10}H_{16}O$ 13351-29-6

2*H*-Inden-2-one, octahydro-3a-methyl-, *cis*-

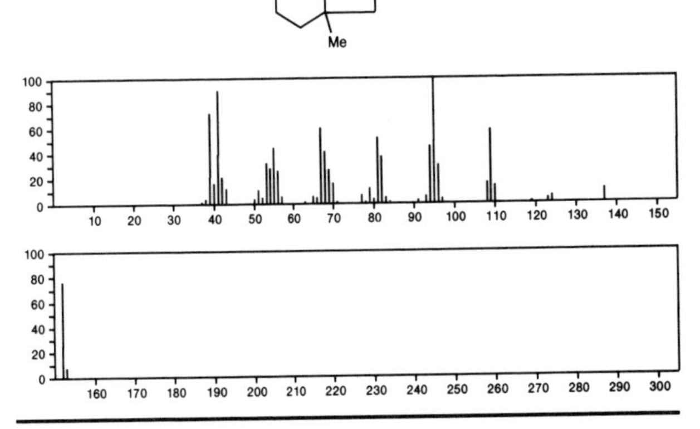

152 $C_{10}H_{16}O$ 13854-85-8

Bicyclo[2.2.1]heptan-2-one, 4,7,7-trimethyl-, (1*R*)-

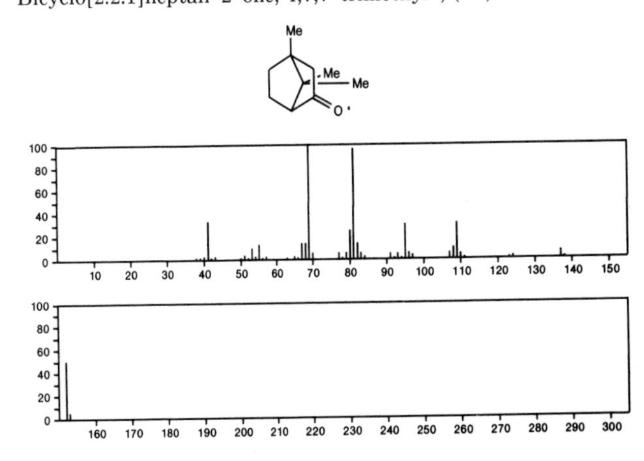

152 $C_{10}H_{16}O$ 15358-88-0
Bicyclo[3.1.1]heptan-3-one, 2,6,6-trimethyl-, (1α,2β,5α)-

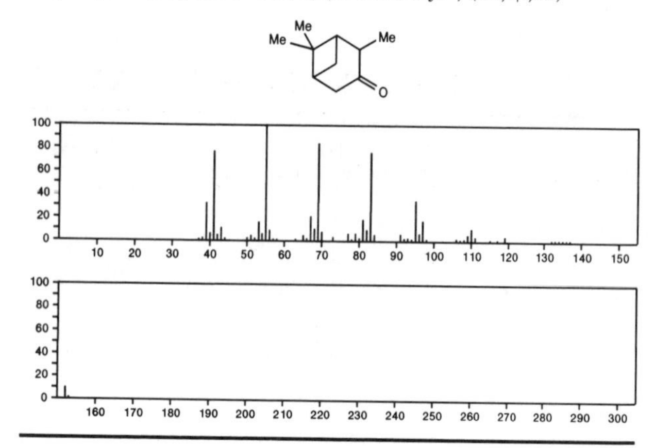

152 $C_{10}H_{16}O$ 15932-80-6
Cyclohexanone, 5-methyl-2-(1-methylethylidene)-

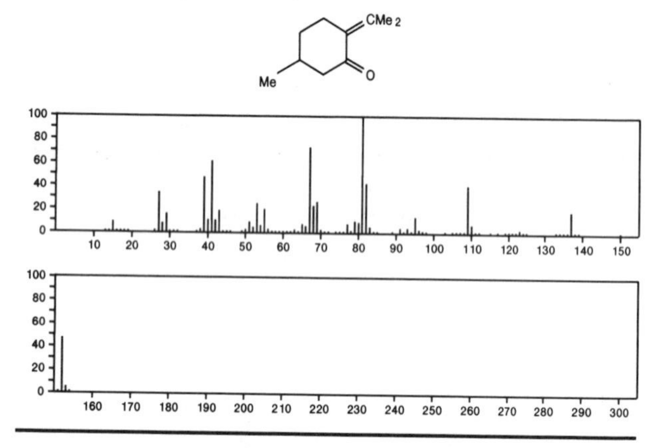

152 $C_{10}H_{16}O$ 16021-08-2
2(1*H*)-Naphthalenone, octahydro-, *trans*-

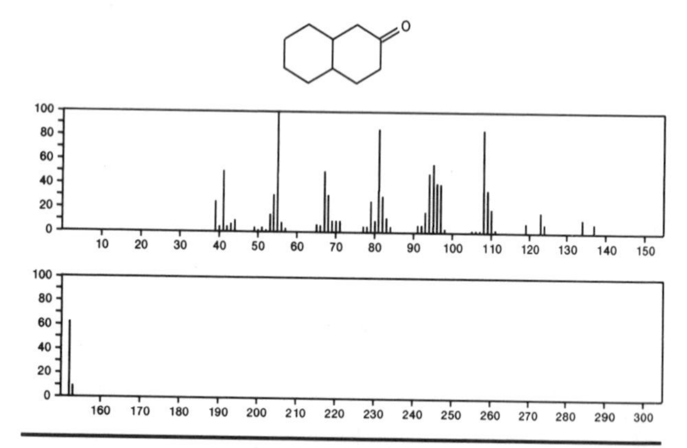

152 $C_{10}H_{16}O$ 17428-83-0
1*H*-Inden-1-one, octahydro-7a-methyl-, *trans*-

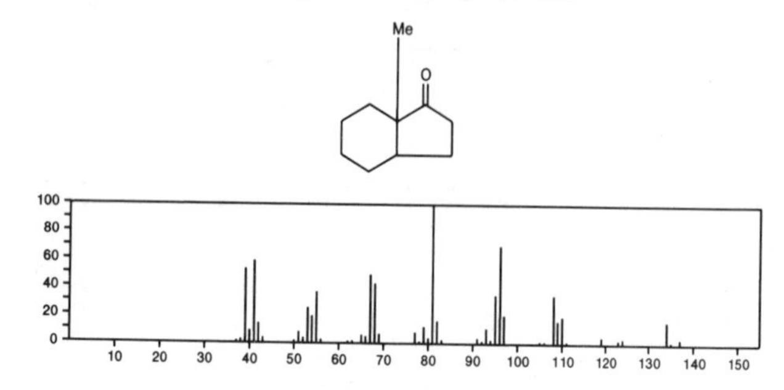

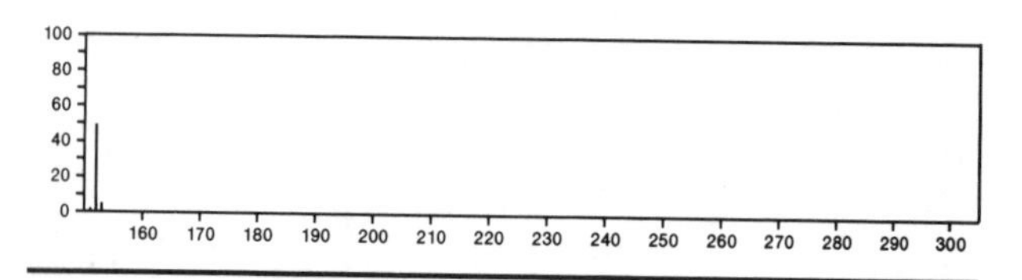

152 $C_{10}H_{16}O$ 17429-33-3
2-Cyclohexen-1-one, 4-ethyl-4,5-dimethyl-

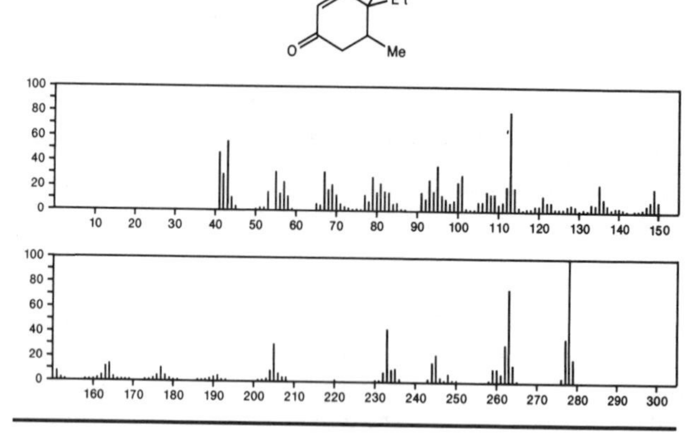

152 $C_{10}H_{16}O$ 17622-46-7
2-Cyclohexen-1-one, 4-ethyl-3,4-dimethyl-

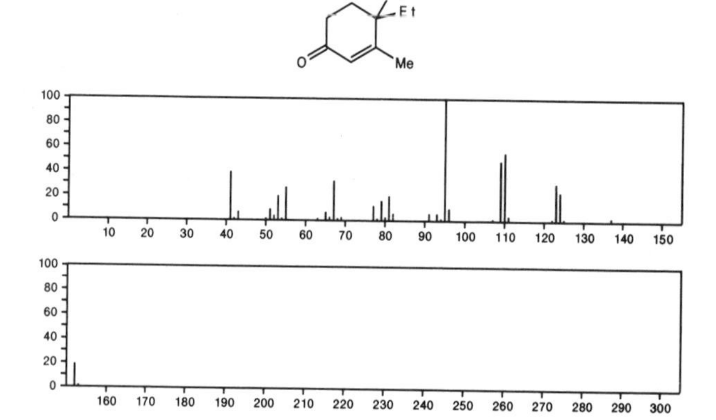

152 $C_{10}H_{16}O$ 20379-99-1
2-Indanone, hexahydro-3a-methyl-, *trans*-

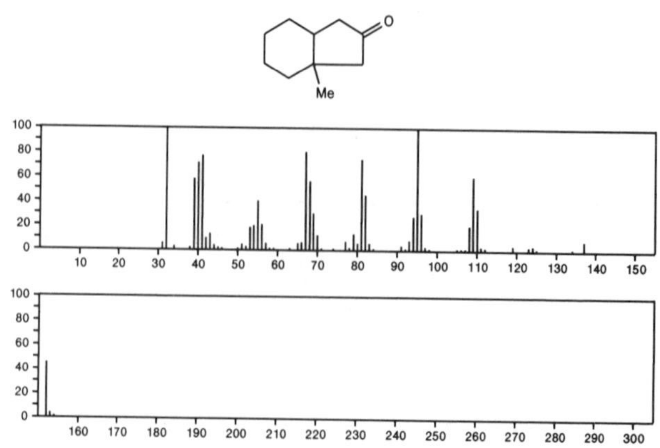

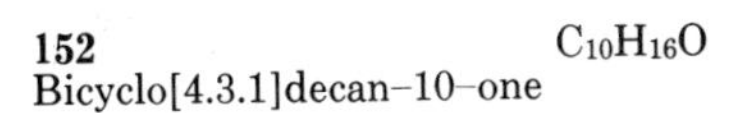

152 $C_{10}H_{16}O$ 20440-21-5
Bicyclo[4.3.1]decan-10-one

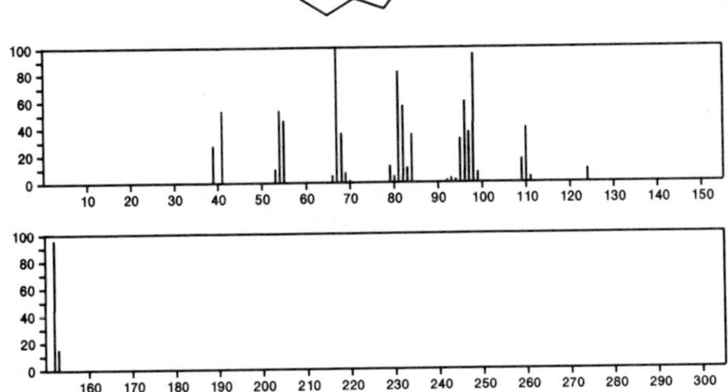

152 $C_{10}H_{16}O$ 21399-51-9
Naphth[2,3-*b*]oxirene, decahydro-

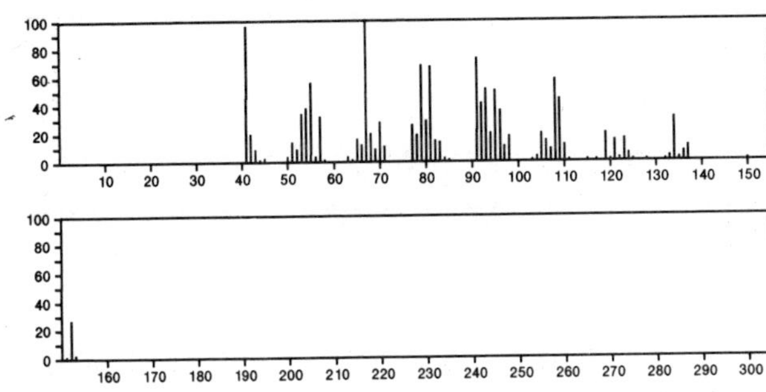

152 $C_{10}H_{16}O$ 26532-24-1
Acetaldehyde, (3,3-dimethylcyclohexylidene)-, (*Z*)-

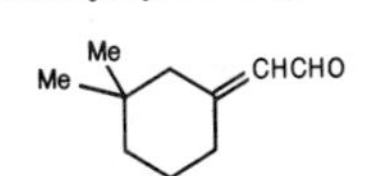

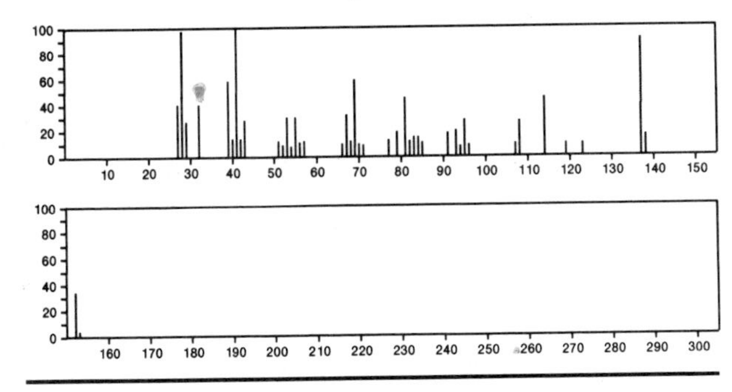

152 $C_{10}H_{16}O$ 26532-25-2
Acetaldehyde, (3,3-dimethylcyclohexylidene)-, (*E*)-

Me
Me
CHCHO

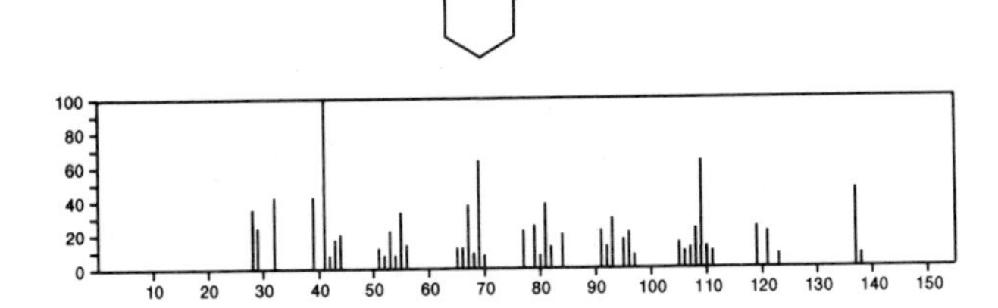

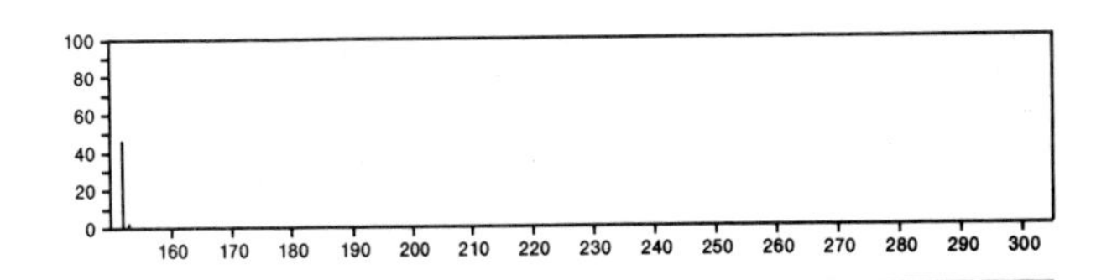

152 $C_{10}H_{16}O$ 28840-87-1
2,5-Methano-1*H*-inden-7-ol, octahydro-, (2α,3aβ,5α,7α,7aβ)-

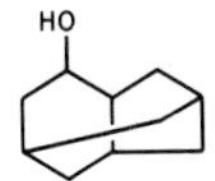

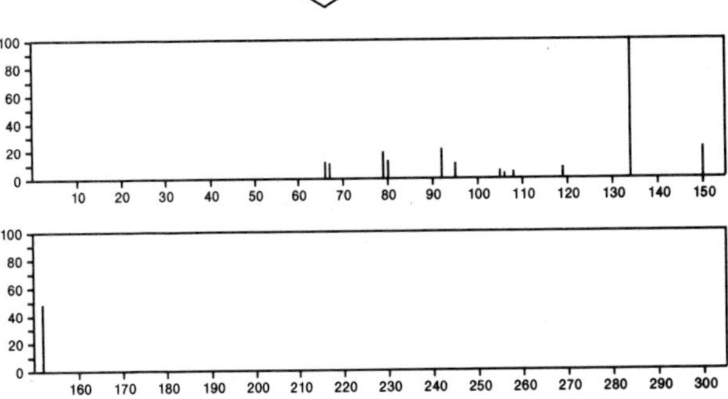

152 $C_{10}H_{16}O$ 28840-88-2
2,5-Methano-1*H*-inden-7-ol, octahydro-, (2α,3aβ,5α,7β,7aβ)-

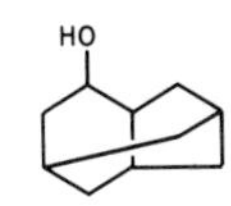

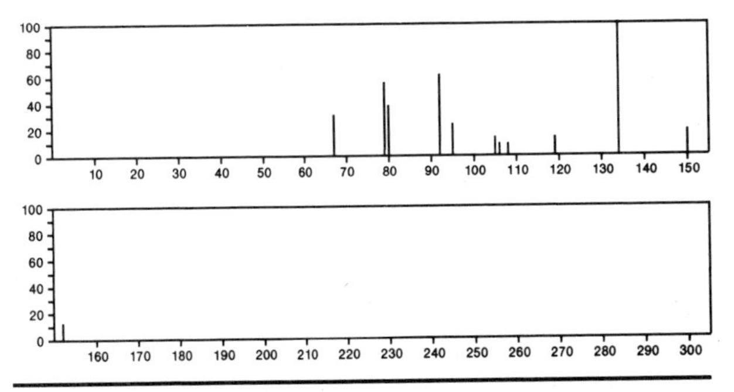

152 $C_{10}H_{16}O$ 29030-74-8
5-Octyn-4-one, 2,7-dimethyl-

$Me_2CHC{\equiv}CCOCH_2CHMe_2$

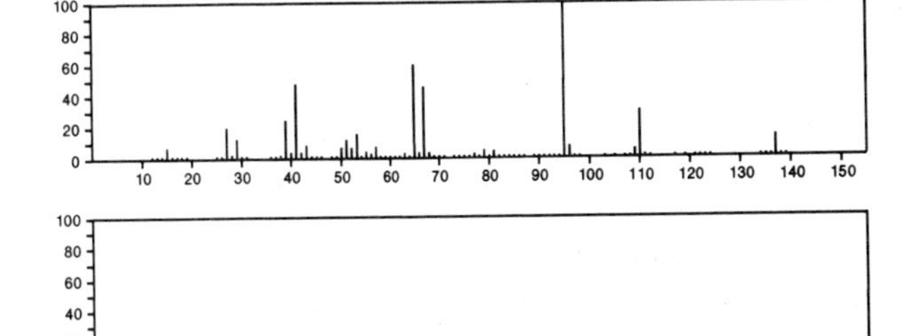

152 $C_{10}H_{16}O$ 29414-56-0
1,5,7-Octatrien-3-ol, 2,6-dimethyl-

$H_2C{:}CHCMe{:}CHCH_2CH(OH)CMe{:}CH_2$

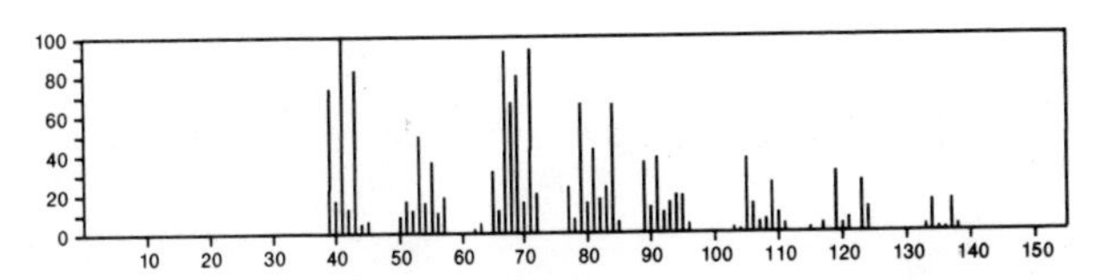

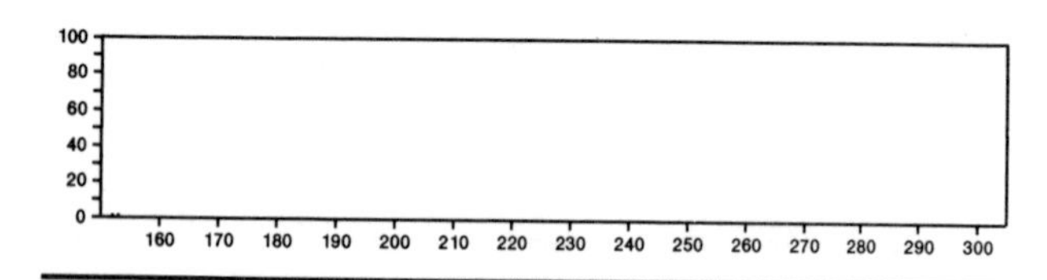

152 $C_{10}H_{16}O$ 29548-13-8
p-Mentha-1(7),8(10)-dien-9-ol

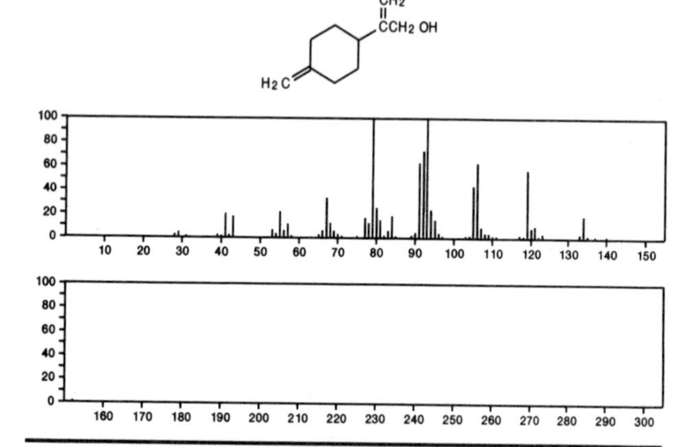

152 $C_{10}H_{16}O$ 29550-55-8
Tricyclo[2.2.1.0^{2,6}]heptane-3-methanol, 2,3-dimethyl-

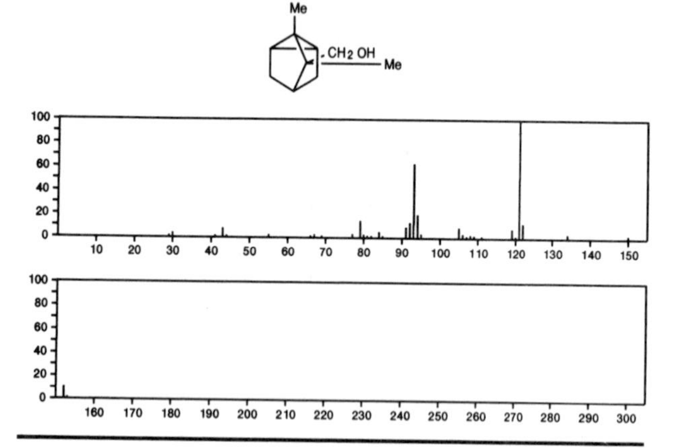

152 $C_{10}H_{16}O$ 29606-79-9
Cyclohexanone, 5-methyl-2-(1-methylethenyl)-, *trans*-

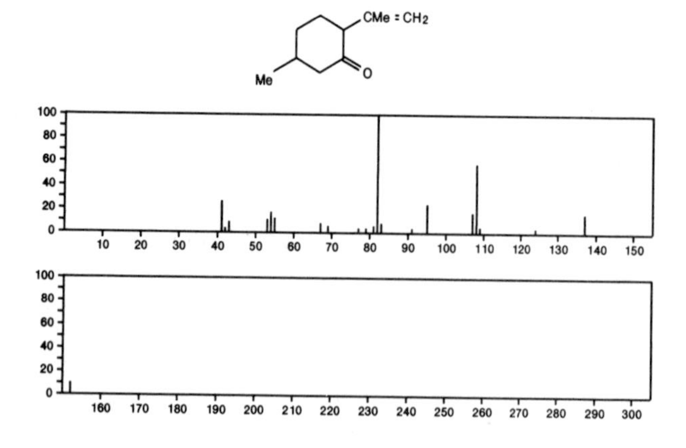

152 $C_{10}H_{16}O$ 29750-24-1
Bicyclo[4.1.0]heptan-2-one, 3,5,5-trimethyl-

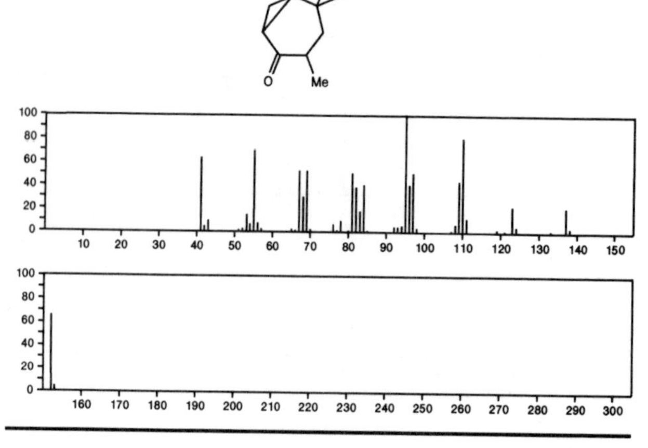

152 $C_{10}H_{16}O$ 35408-14-1
3,7-Nonadien-2-one, 8-methyl-, (*E*)-

MeCOCH = CHCH2 CH2 CH = CMe2

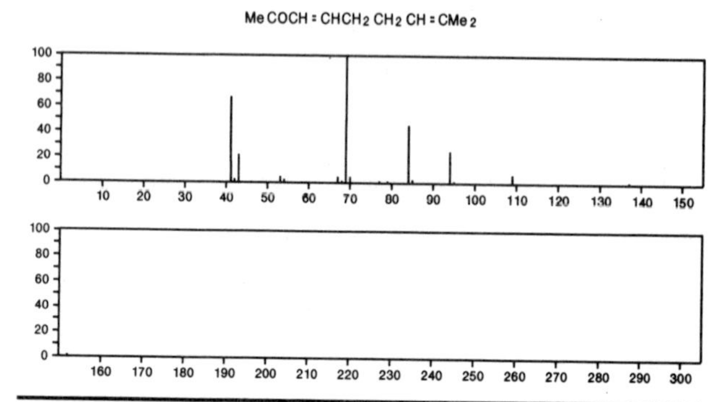

152 $C_{10}H_{16}O$ 40702-26-9
3-Cyclohexene-1-carboxaldehyde, 1,3,4-trimethyl-

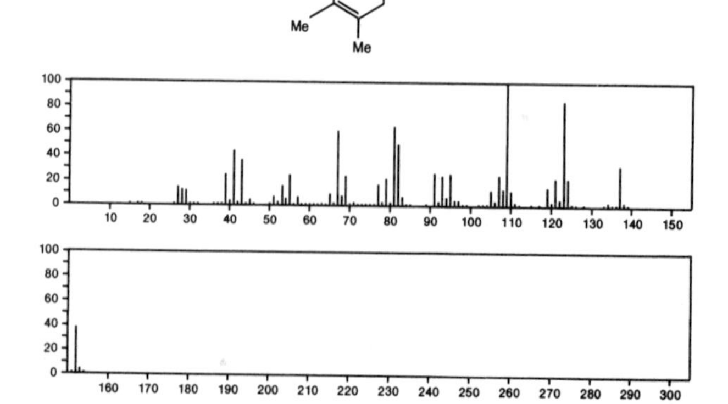

152 $C_{10}H_{16}O$ 43219-68-7
Ethanone, 1-(1,4-dimethyl-3-cyclohexen-1-yl)-

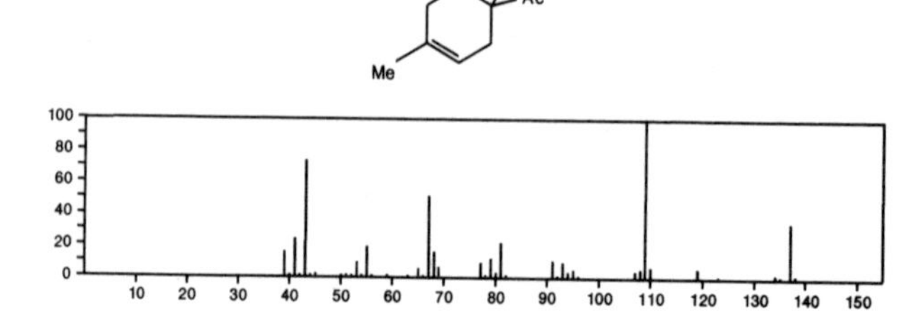

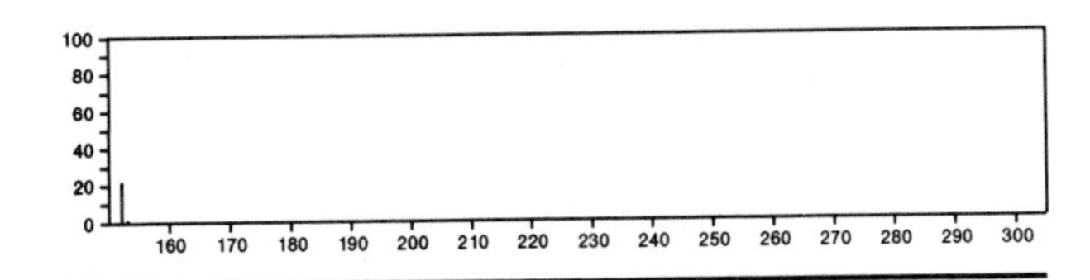

152 $C_{10}H_{16}O$ 49833-93-4

Cyclohexane, 1-methoxy-1-(1,2-propadienyl)-

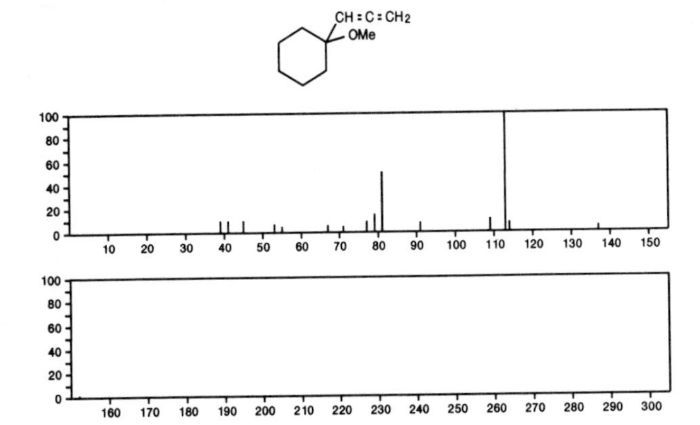

152 $C_{10}H_{16}O$ 51733-68-7

Ethanone, 1-(1,3-dimethyl-3-cyclohexen-1-yl)-

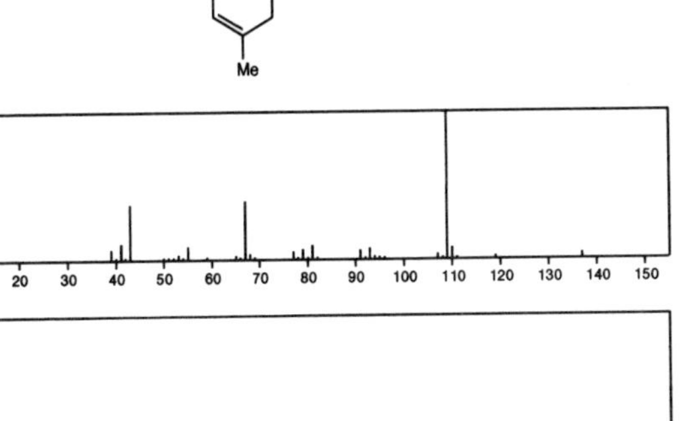

152 $C_{10}H_{16}O$ 54345-56-1

1-Oxaspiro[2.5]octane, 4,4-dimethyl-8-methylene-

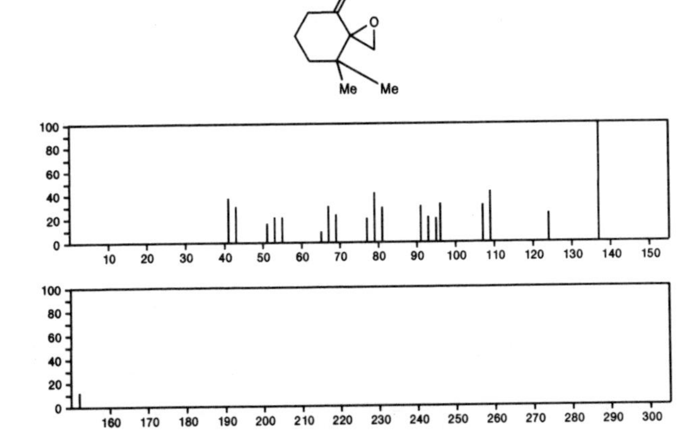

152 $C_{10}H_{16}O$ 55759-84-7

Bicyclo[3.2.0]heptan-2-one, 1,4,4-trimethyl-

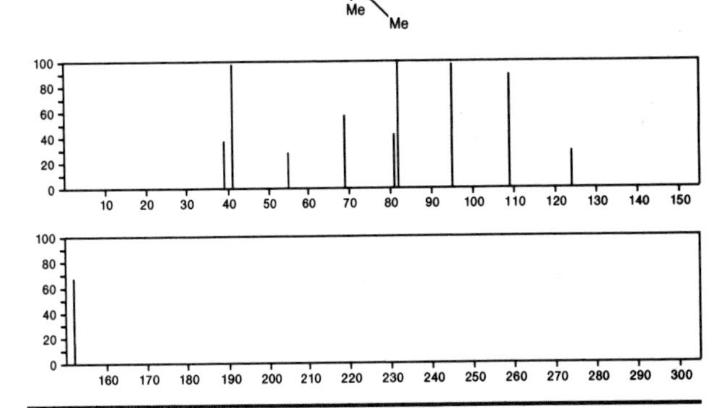

152 $C_{11}H_{20}$ 180-43-8

Spiro[5.5]undecane

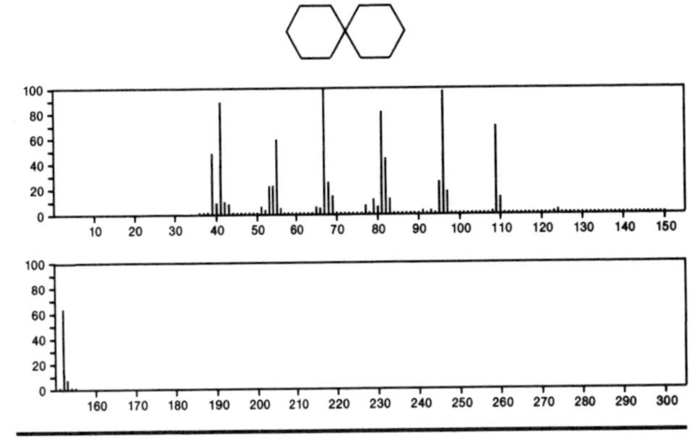

152 $C_{11}H_{20}$ 2243-98-3

1-Undecyne

$HC\equiv C(CH_2)_8Me$

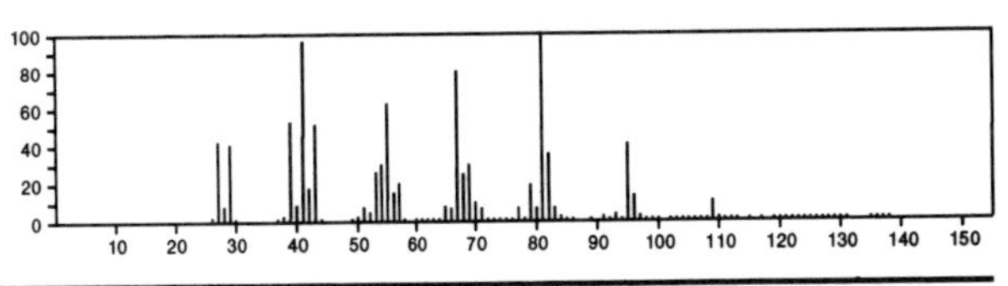

152 $C_{11}H_{20}$ 2958-76-1

Naphthalene, decahydro-2-methyl-

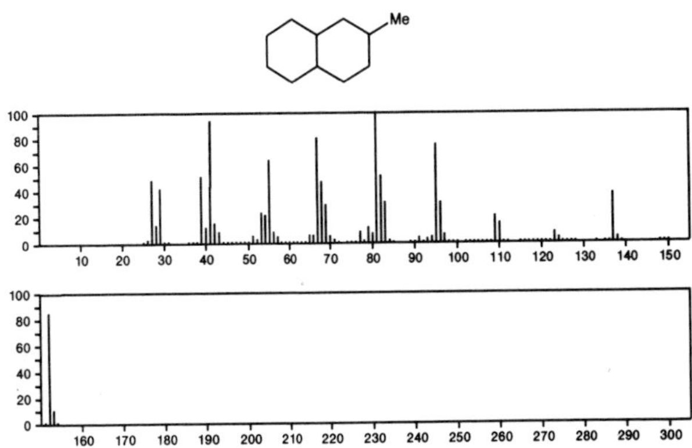

152 Cyclohexene, 1-isopentyl- $C_{11}H_{20}$ 3983-04-8

152 Cyclohexene, 1-pentyl- $C_{11}H_{20}$ 15232-85-6

152 Bicyclo[4.1.0]heptane, 7-butyl- $C_{11}H_{20}$ 18645-10-8

152 3-Octyne, 2,2,7-trimethyl- $C_{11}H_{20}$ 55402-13-6

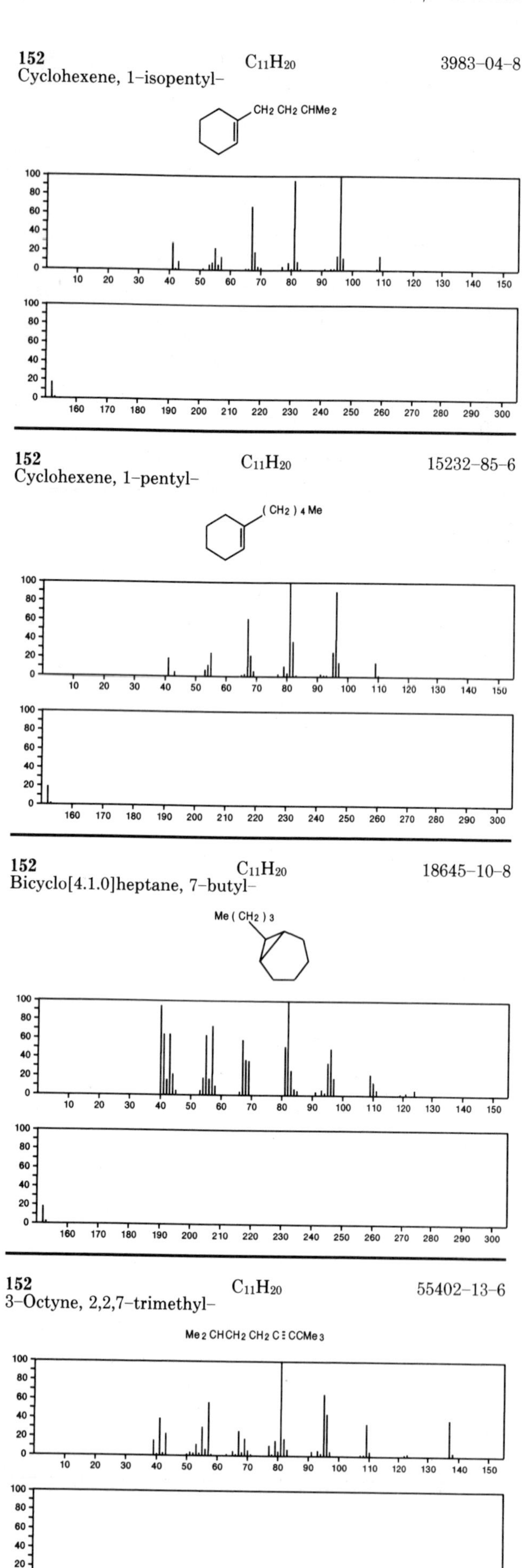

152 1,4-Undecadiene, (*E*)- $C_{11}H_{20}$ 55976-13-1

152 1,4-Undecadiene, (*Z*)- $C_{11}H_{20}$ 55976-14-2

152 Cyclopropane, 1-(1-methylethyl)-2-(2-methyl-1-methylenepropyl)- $C_{11}H_{20}$ 56259-17-7

152 Cyclopropane, 1,1-dimethyl-2-(1-methylethyl)-3-(1-methyl=ethylidene)-, (±)- $C_{11}H_{20}$ 56701-50-9

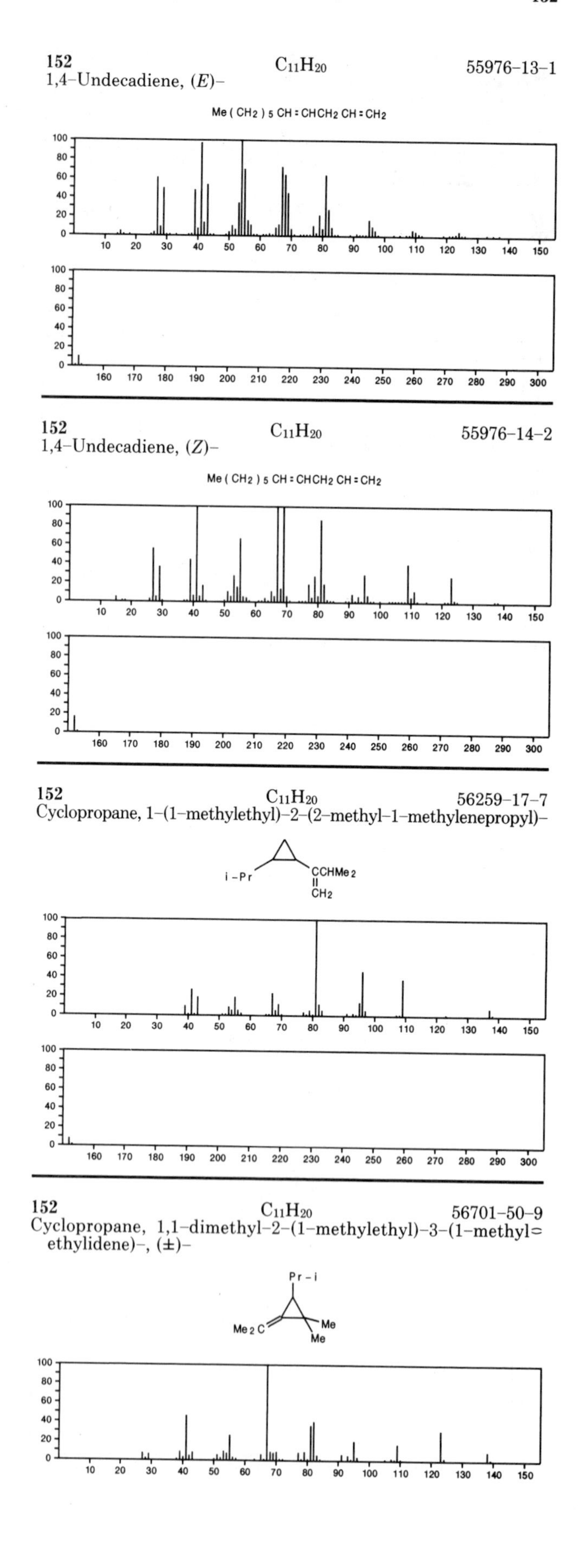

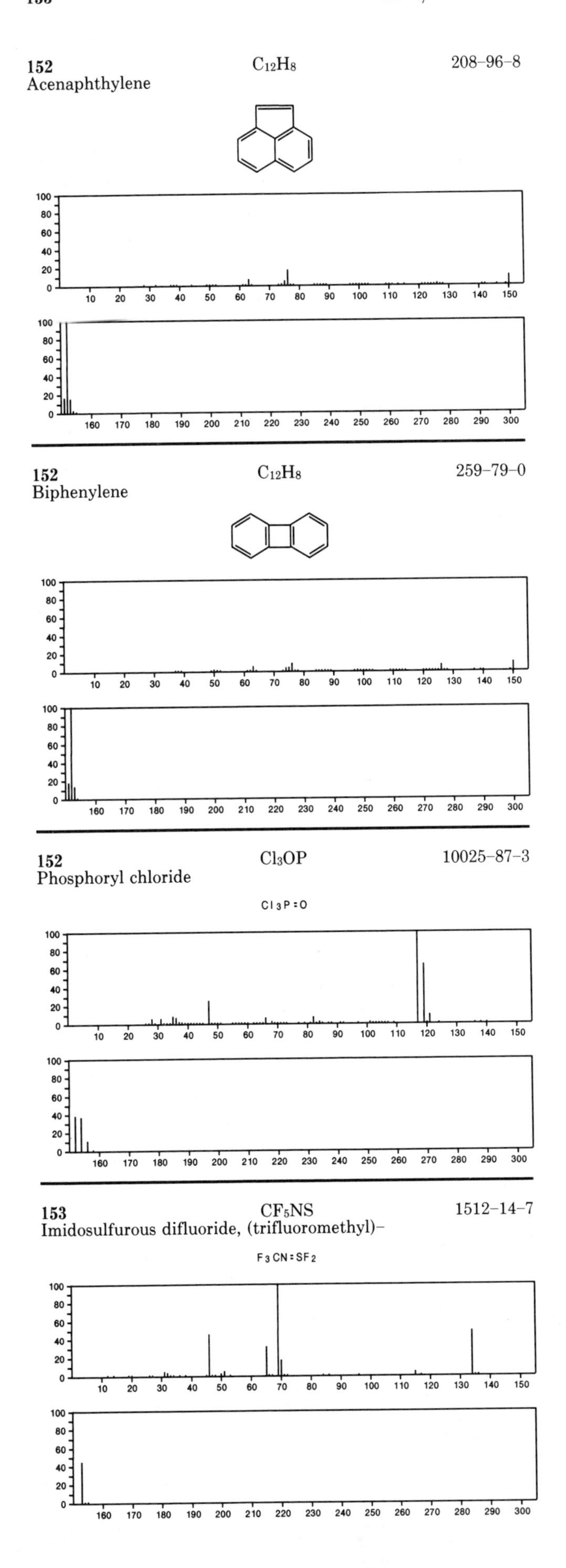

152 $C_{12}H_8$ 208–96–8
Acenaphthylene

152 $C_{12}H_8$ 259–79–0
Biphenylene

152 Cl_3OP 10025–87–3
Phosphoryl chloride

$Cl_3P{=}O$

153 CF_5NS 1512–14–7
Imidosulfurous difluoride, (trifluoromethyl)–

$F_3CN{=}SF_2$

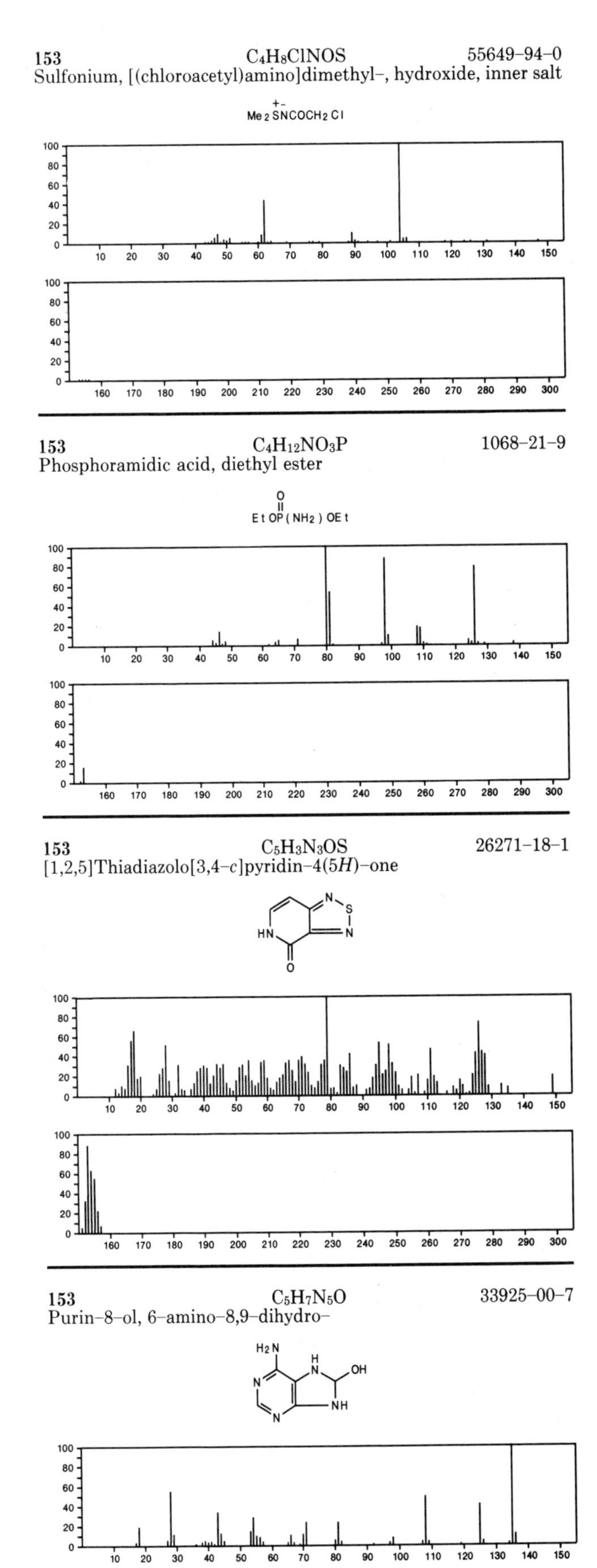

153 C_4H_8ClNOS 55649–94–0
Sulfonium, [(chloroacetyl)amino]dimethyl–, hydroxide, inner salt

$Me_2\overset{+-}{S}NCOCH_2Cl$

153 $C_4H_{12}NO_3P$ 1068–21–9
Phosphoramidic acid, diethyl ester

EtOP(O)(NH2)OEt

153 $C_5H_3N_3OS$ 26271–18–1
[1,2,5]Thiadiazolo[3,4–*c*]pyridin–4(5*H*)–one

153 $C_5H_7N_5O$ 33925–00–7
Purin–8–ol, 6–amino–8,9–dihydro–

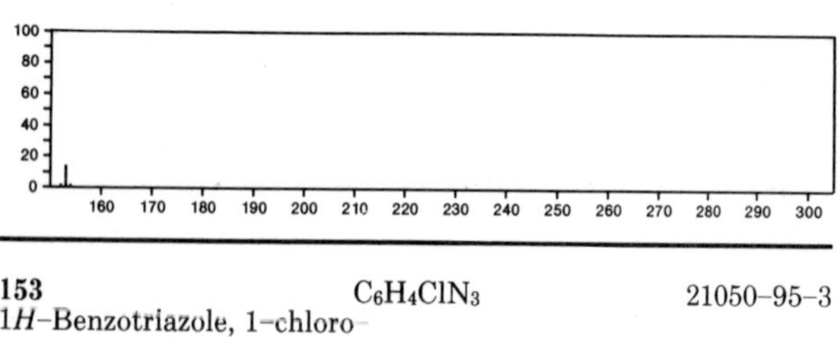

153 $C_6H_4ClN_3$ 21050-95-3
1*H*-Benzotriazole, 1-chloro-

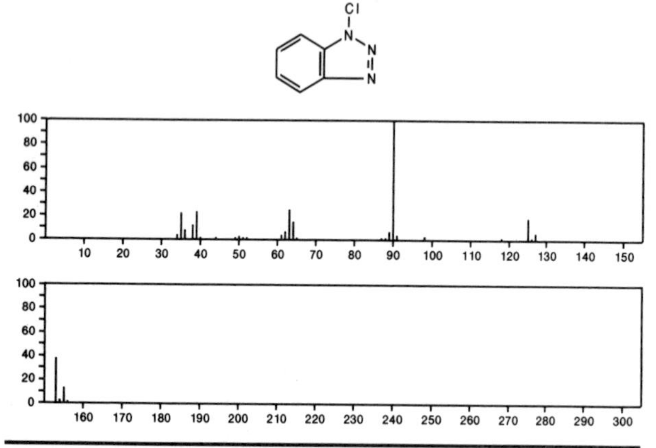

153 $C_6H_7N_3O_2$ 100-16-3
Hydrazine, (4-nitrophenyl)-

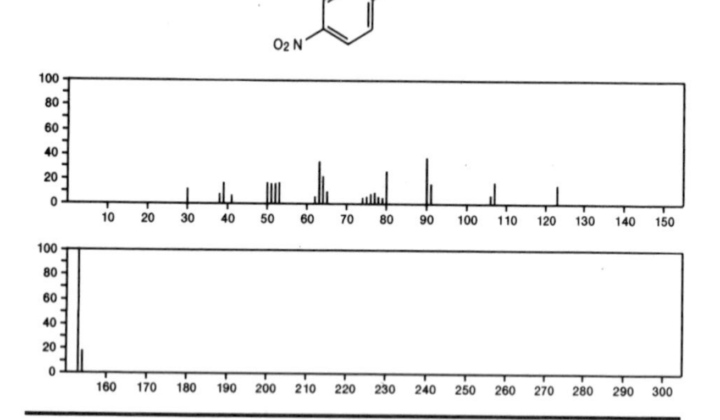

153 $C_6H_7N_3O_2$ 3034-19-3
Hydrazine, (2-nitrophenyl)-

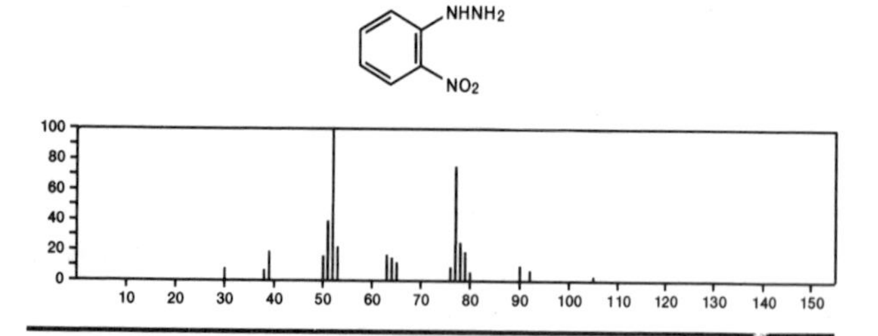

153 $C_6H_7N_3O_2$ 3694-52-8
1,2-Benzenediamine, 3-nitro-

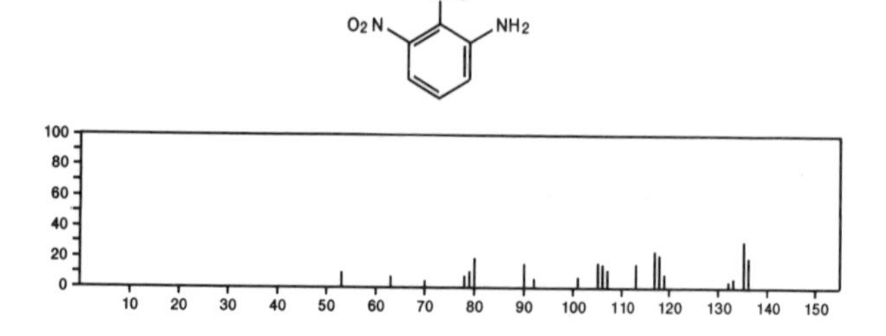

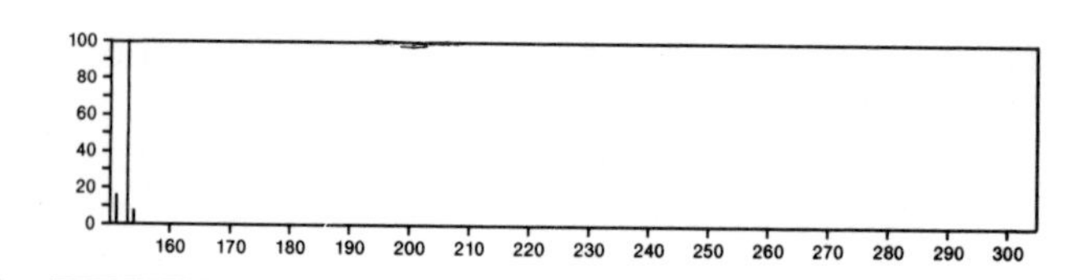

153 $C_6H_7N_3O_2$ 18344-53-1
2-Pyridinamine, 3-methyl-*N*-nitro-

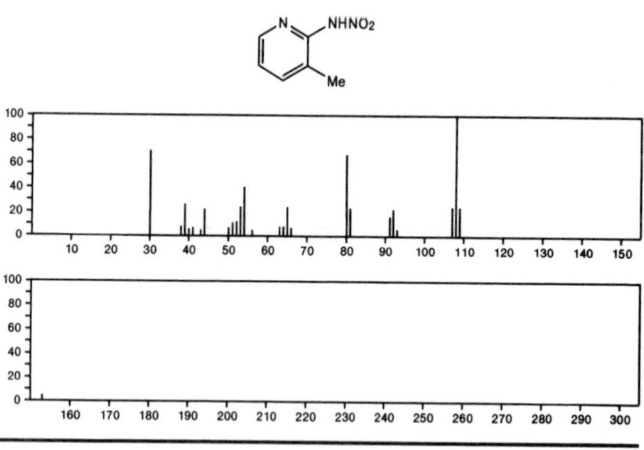

153 $C_6H_7N_3O_2$ 31396-29-9
2(1*H*)-Pyridinimine, 1-methyl-*N*-nitro-

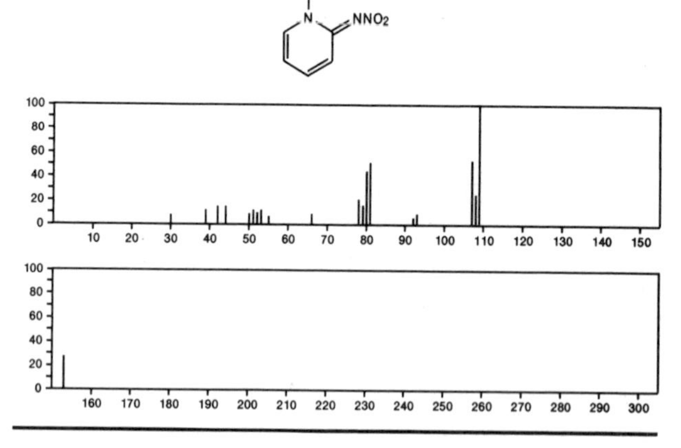

153 $C_6H_7N_3O_2$ 55760-13-9
3*H*-Pyrazolo[4,3-*b*]pyridine-3,3-diol, 1,2-dihydro-

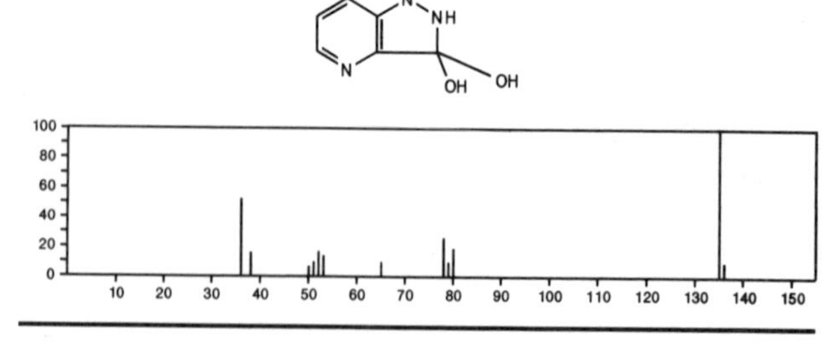

153 $C_6H_{15}NO.ClH$ 14426-20-1
Ethanol, 2-(diethylamino)-, hydrochloride

$HOCH_2CH_2NEt_2 \cdot HCl$

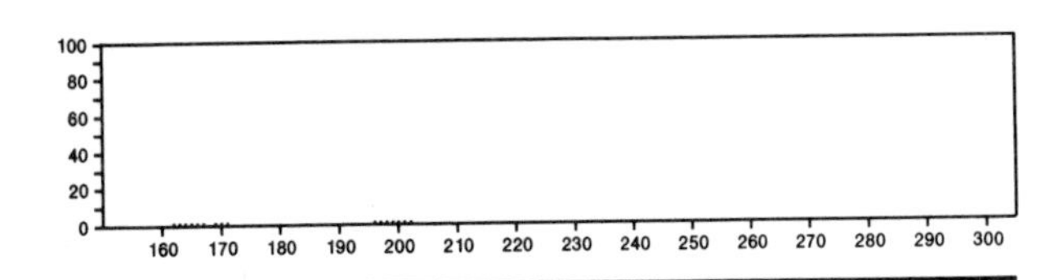

153 C_7H_4ClNO 104-12-1
Benzene, 1-chloro-4-isocyanato-

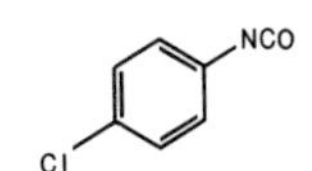

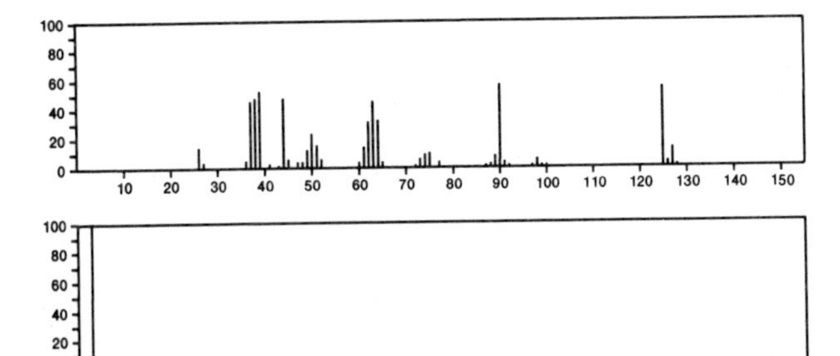

153 C_7H_4ClNO 51134-03-3
Benzene, chloroisocyanato-

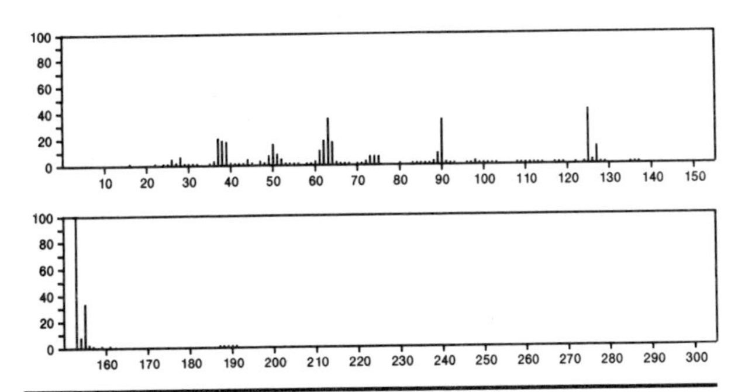

153 C_7H_7NOS 15795-43-4
Benzenamine, 3-methyl-*N*-sulfinyl-

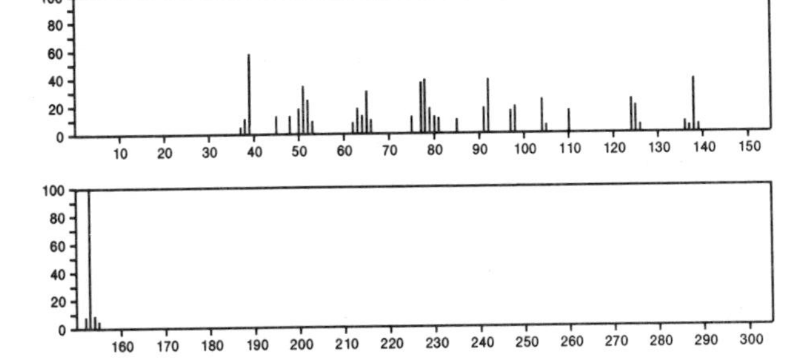

153 C_7H_7NOS 23003-45-4
Thiazolo[3,2-*a*]pyridinium, 2,3-dihydro-8-hydroxy-, hydroxide, inner salt

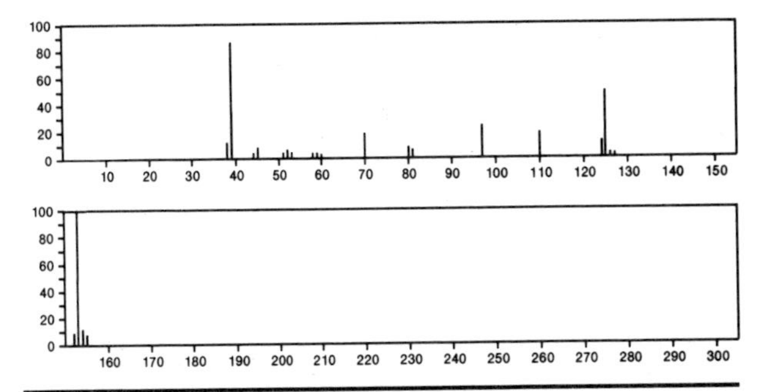

153 $C_7H_7NO_3$ 91-23-6
Benzene, 1-methoxy-2-nitro-

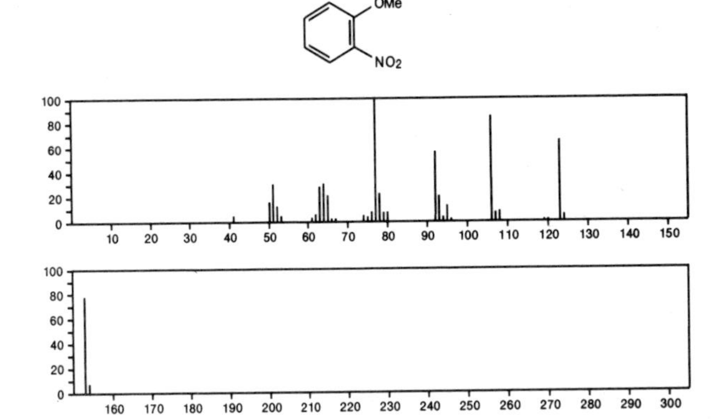

153 $C_7H_7NO_3$ 100-17-4
Benzene, 1-methoxy-4-nitro-

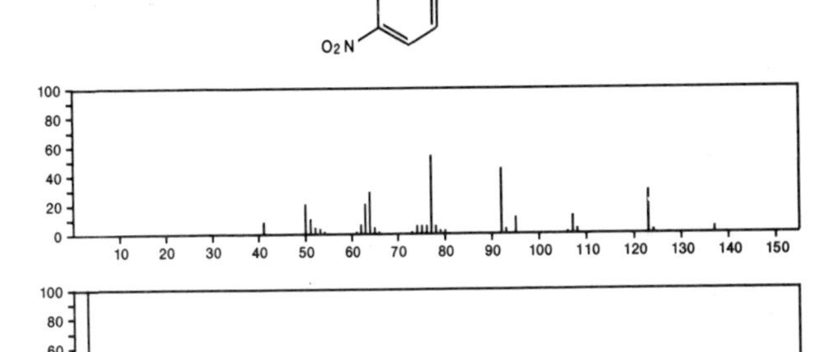

153 $C_7H_7NO_3$ 119-33-5
Phenol, 4-methyl-2-nitro-

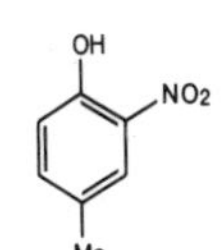

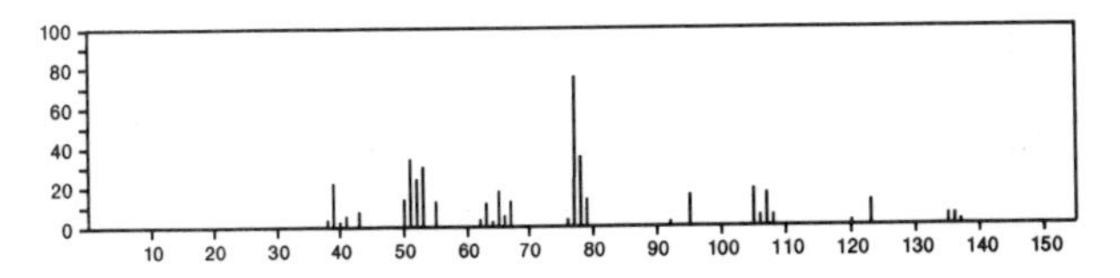

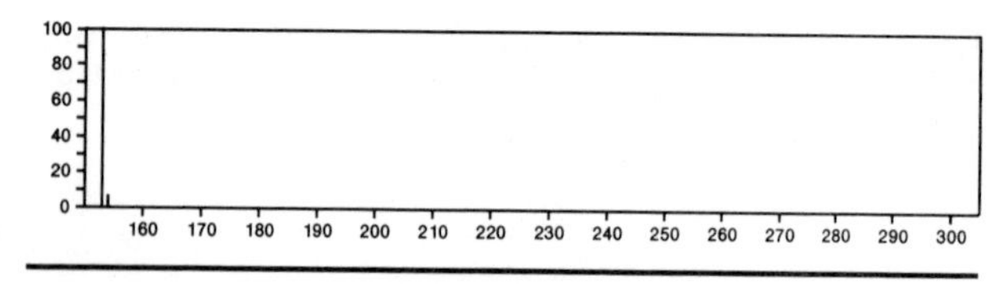

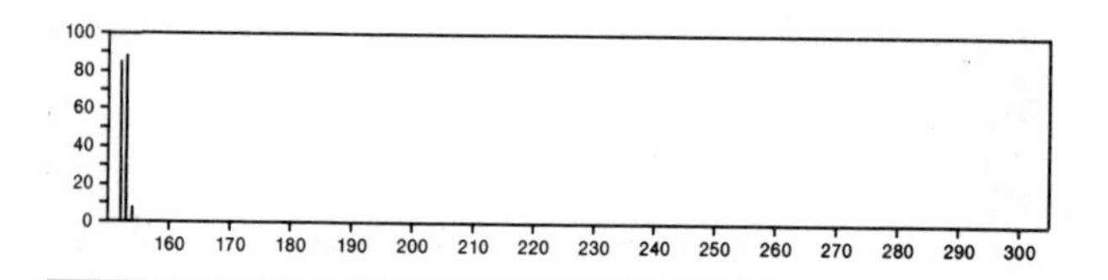

153 $C_7H_7NO_3$ 548-93-6
Benzoic acid, 2-amino-3-hydroxy-

CO_2H, NH_2, OH

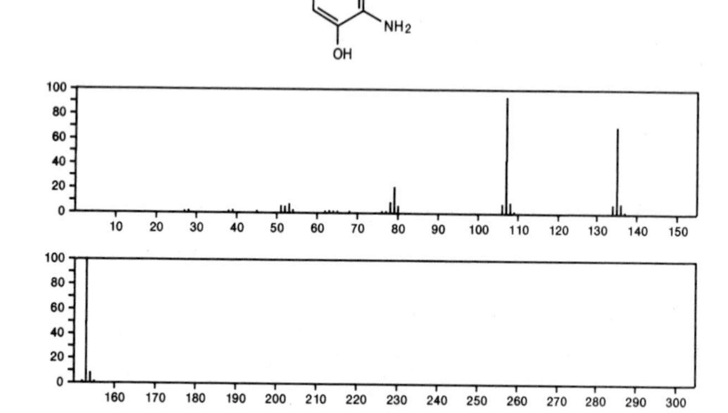

153 $C_7H_7NO_3$ 555-03-3
Benzene, 1-methoxy-3-nitro-

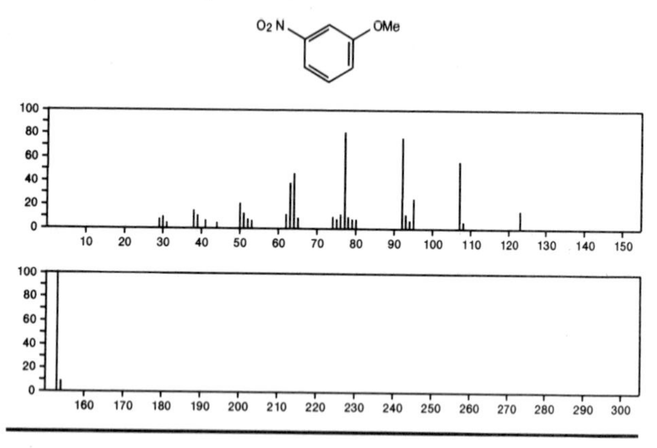

153 $C_7H_7NO_3$ 612-25-9
Benzenemethanol, 2-nitro-

CH_2OH, NO_2

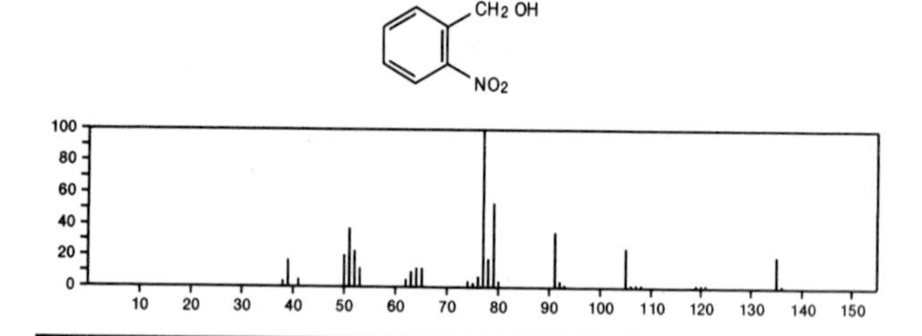

153 $C_7H_7NO_3$ 26893-73-2
2-Pyridinecarboxylic acid, 6-methoxy-

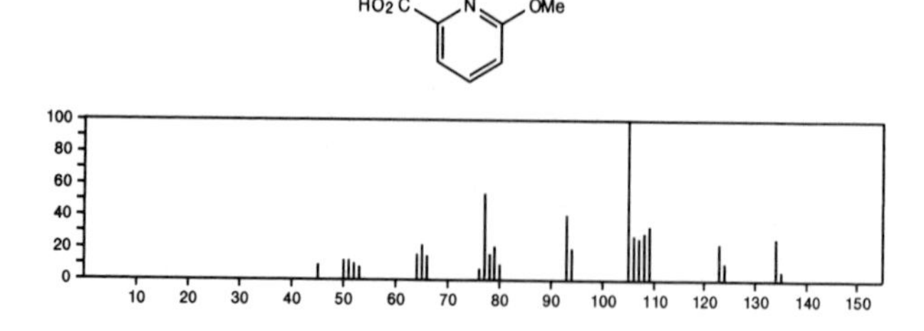

153 $C_7H_7NO_3$ 29082-92-6
2-Pyridinecarboxylic acid, 5-methoxy-

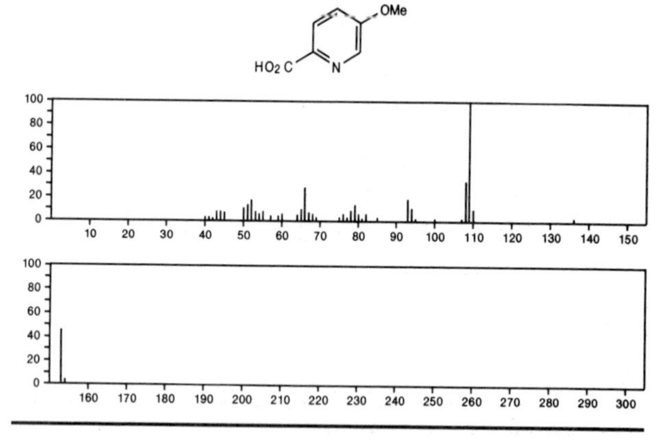

153 $C_7H_7NO_3$ 56666-50-3
Tricyclo[2.2.1.0^{1,4}]heptan-2-one, 6-nitro-

O_2N, O

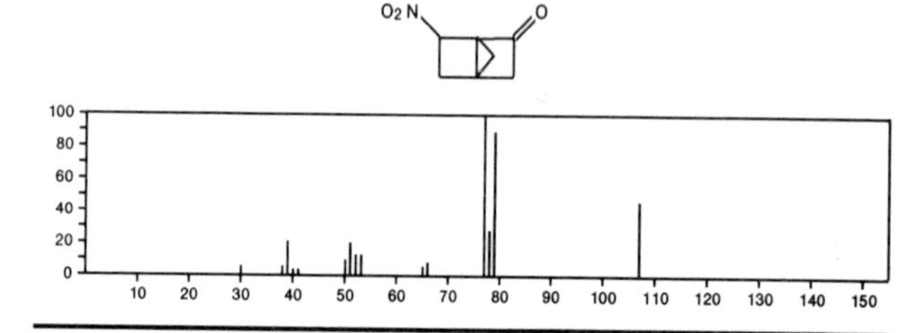

153 $C_7H_{11}N_3O$ 7749-48-6
2-Pyrimidinamine, 4-ethoxy-6-methyl-

Me, OEt, N, N, NH_2

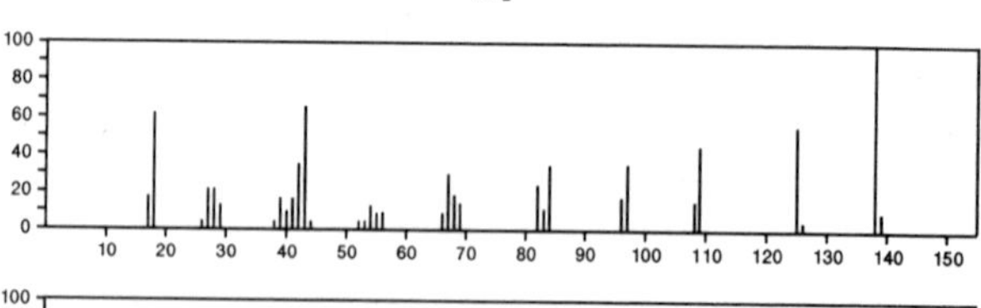

153 $C_8H_{11}NO_2$ 51-61-6
1,2-Benzenediol, 4-(2-aminoethyl)-

HO, HO, $CH_2CH_2NH_2$

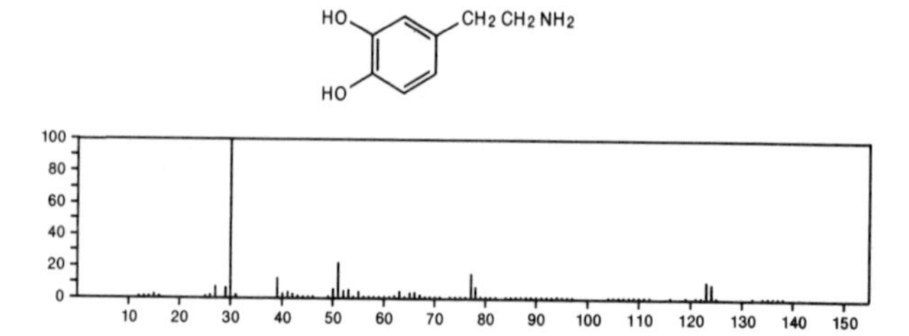

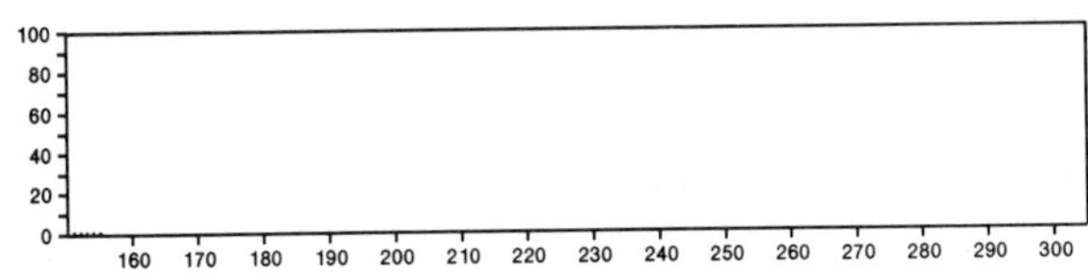

153 $C_8H_{11}NO_2$ 102-56-7
Benzenamine, 2,5-dimethoxy-

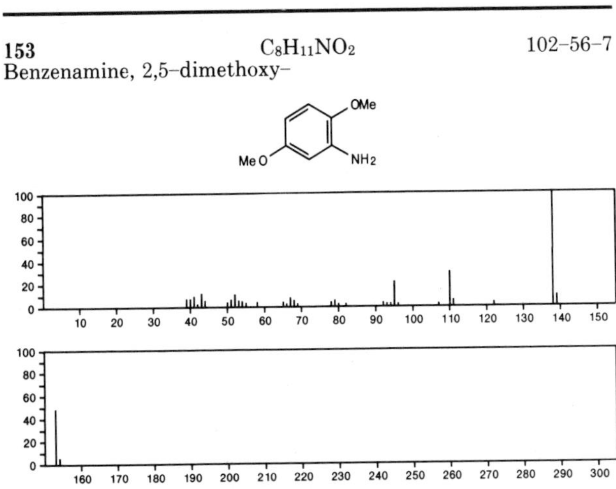

153 $C_8H_{11}NO_2$ 1196-92-5
Phenol, 4-(aminomethyl)-2-methoxy-

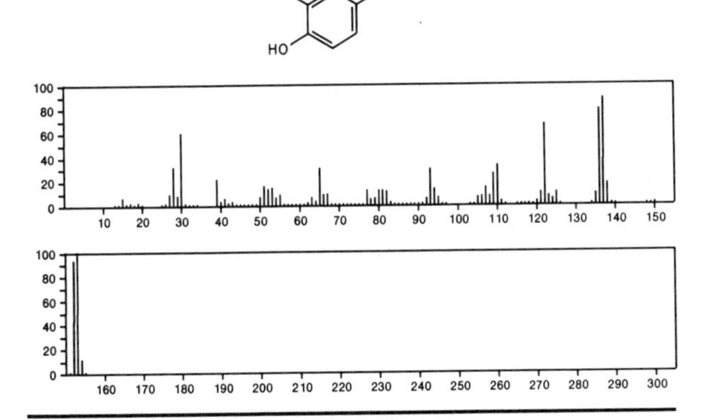

153 $C_8H_{11}NO_2$ 1444-94-6
1*H*-Isoindole-1,3(2*H*)-dione, hexahydro-

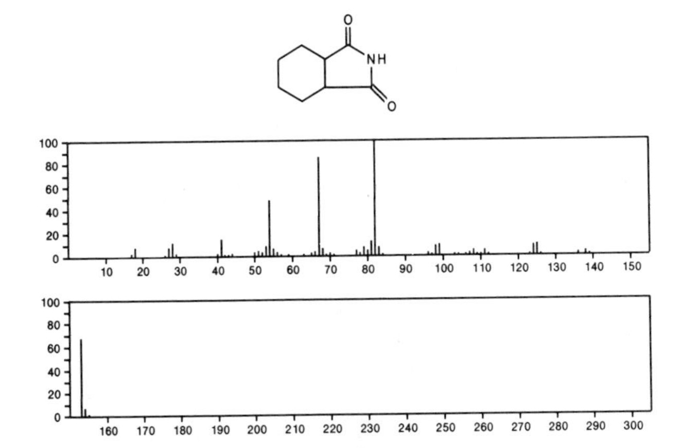

153 $C_8H_{11}NO_2$ 1915-83-9
Benzenemethanol, α-(aminomethyl)-4-hydroxy-, (±)-

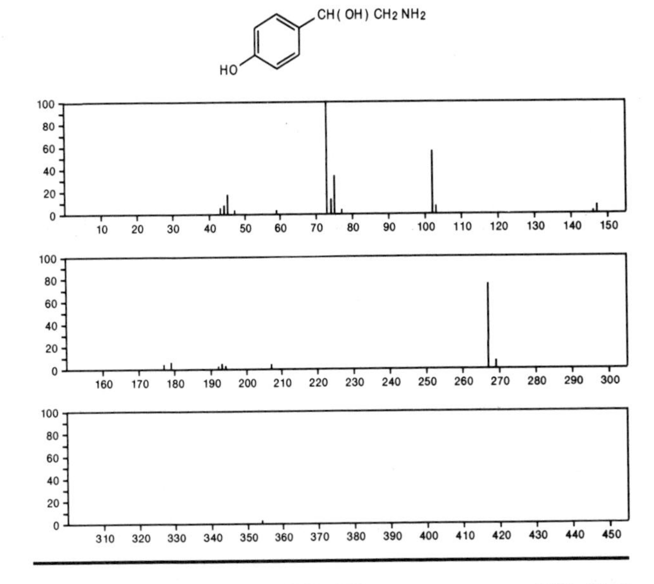

153 $C_8H_{11}NO_2$ 2973-09-3
1*H*-Pyrrole-2,5-dione, 1-butyl-

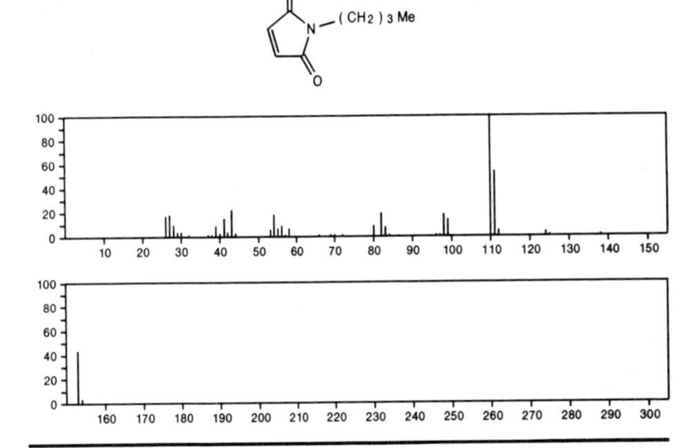

153 $C_8H_{11}NO_2$ 25471-69-6
Furo[2,3,4-*gh*]pyrrolizin-2(2a*H*)-one, hexahydro-, (2aα,7aα,7bα)-

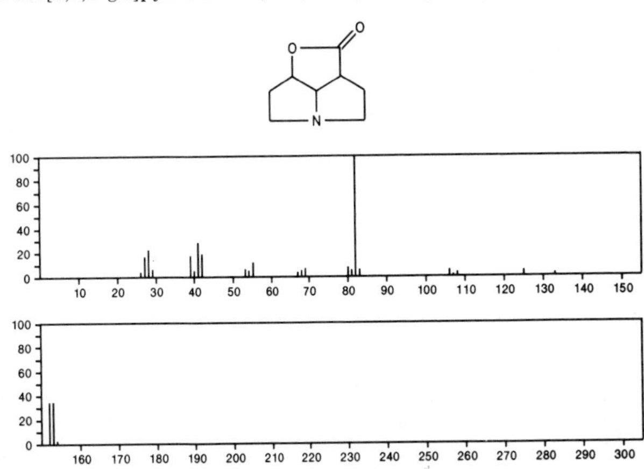

153 $C_8H_{11}NO_2$ 27396-39-0
2(5*H*)-Furanone, 5-(butylimino)-

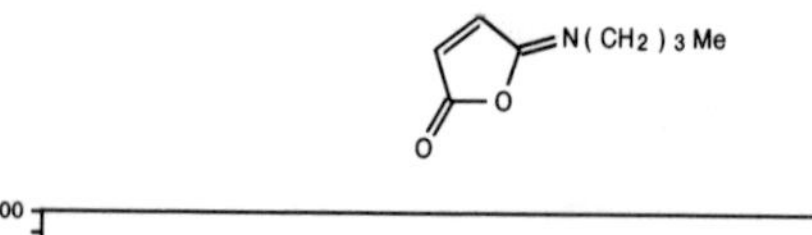

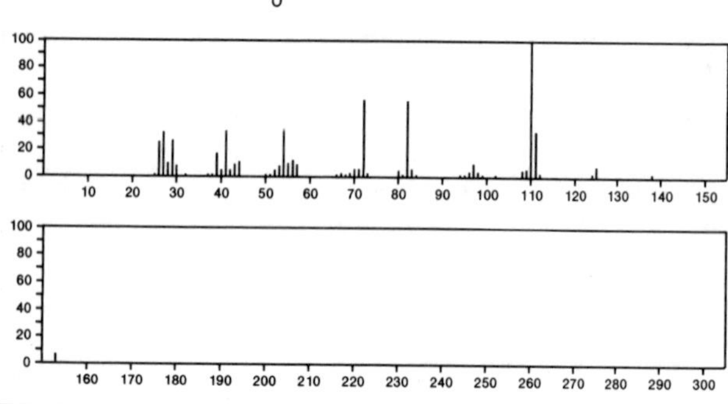

153 $C_8H_{11}NO_2$ 54244-77-8
2-Butenoic acid, 2-(cyanomethyl)-, ethyl ester

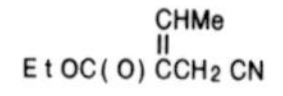

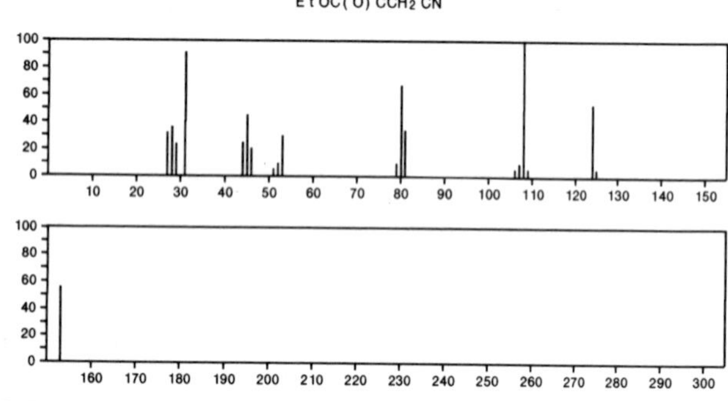

153 $C_8H_{11}NO_2$ 56701-03-2
Carbamic acid, 1,3,5-hexatrienyl-, methyl ester

$H_2C{=}CHCH{=}CHCH{=}CHNHC(O)OMe$

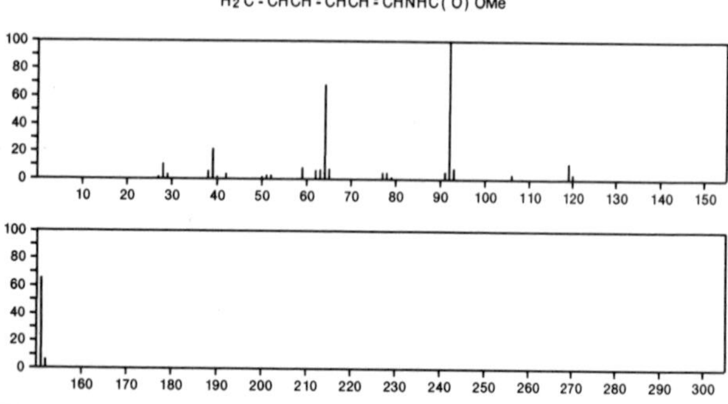

153 $C_8H_{11}NS$ 19006-70-3
2(1*H*)-Pyridinethione, 1,4,6-trimethyl-

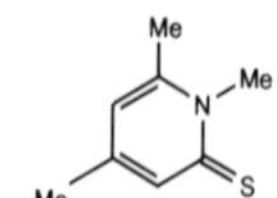

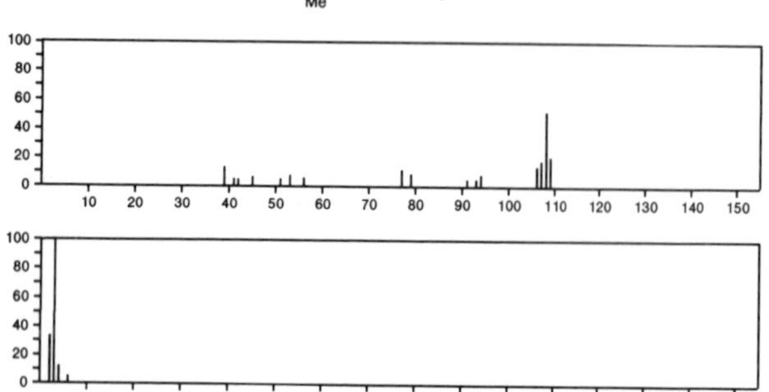

153 $C_8H_{11}NS$ 19006-72-5
2(1*H*)-Pyridinethione, 1-ethyl-4-methyl-

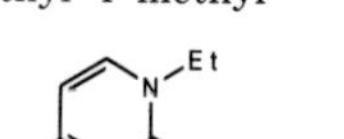

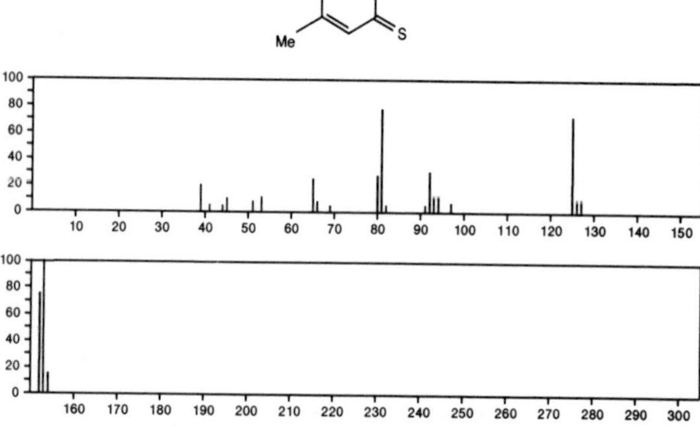

153 $C_8H_{11}NS$ 19006-73-6
2(1*H*)-Pyridinethione, 1-ethyl-6-methyl-

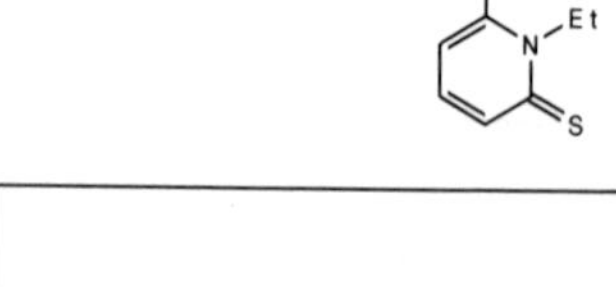

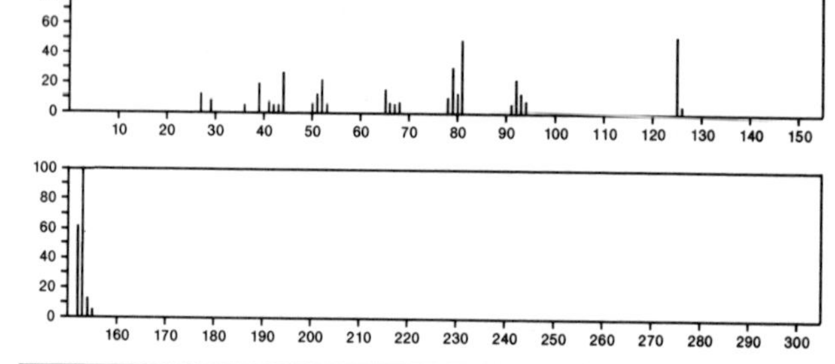

153 $C_8H_{11}NS$ 19006-74-7
2(1*H*)-Pyridinethione, 1-propyl-

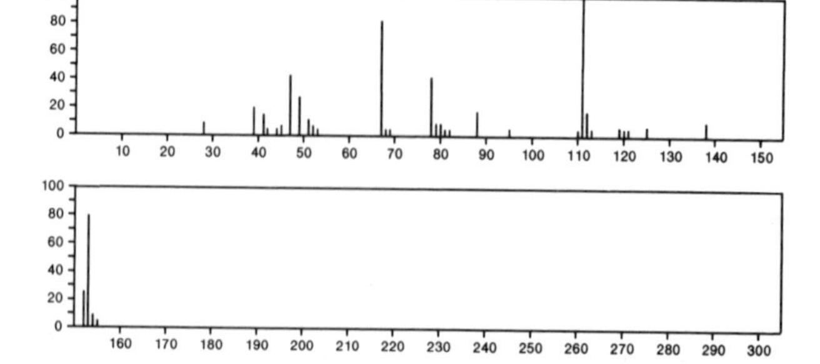

153 $C_8H_{11}NS$ 19006-78-1
4-Picoline, 2-(ethylthio)-

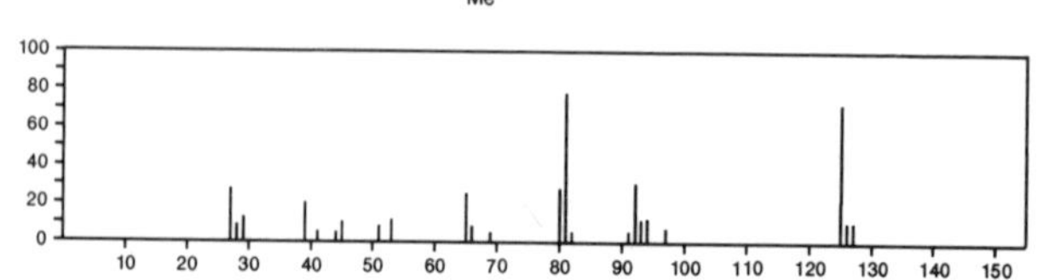

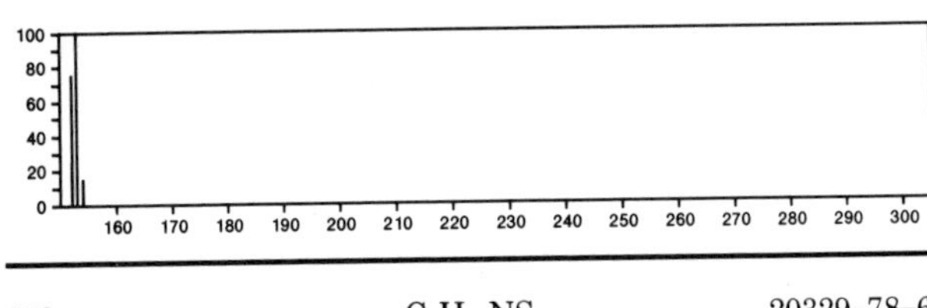

153 $C_8H_{11}NS$ 20329-78-6

2-Picoline, 6-(ethylthio)-

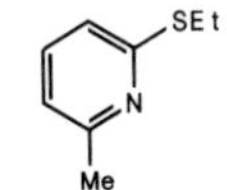

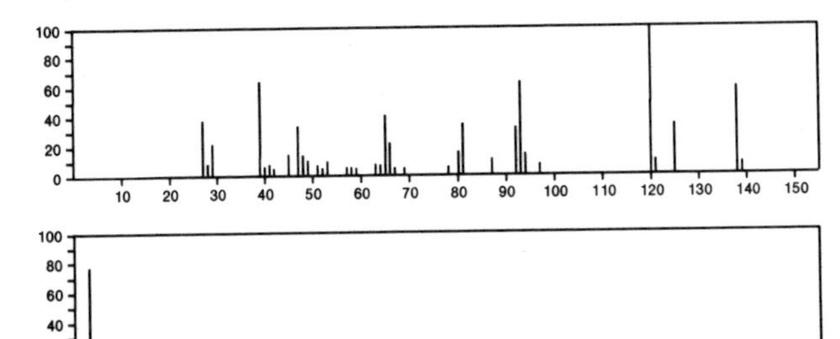

153 $C_8H_{15}N_3$ 45967-45-1

1*H*-Imidazole-4-ethanamine, *N*,*N*,2-trimethyl-

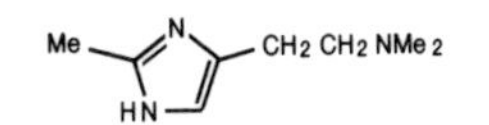

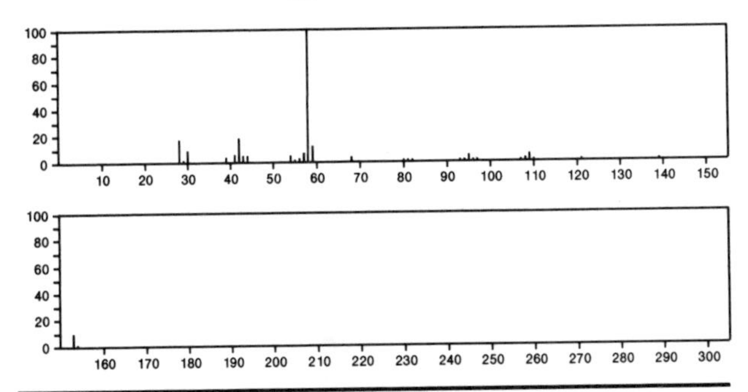

153 $C_9H_{15}NO$ 552-70-5

9-Azabicyclo[3.3.1]nonan-3-one, 9-methyl-

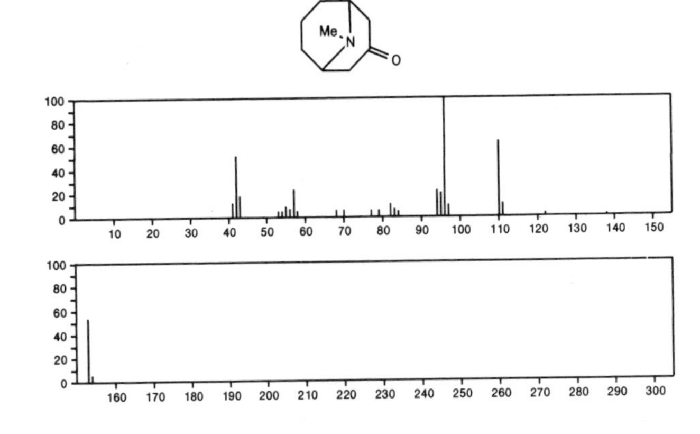

153 $C_9H_{15}NO$ 769-04-0

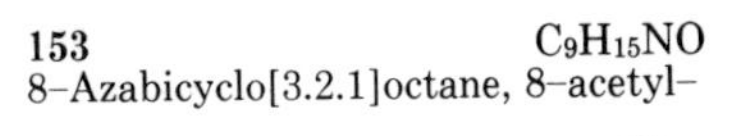

8-Azabicyclo[3.2.1]octane, 8-acetyl-

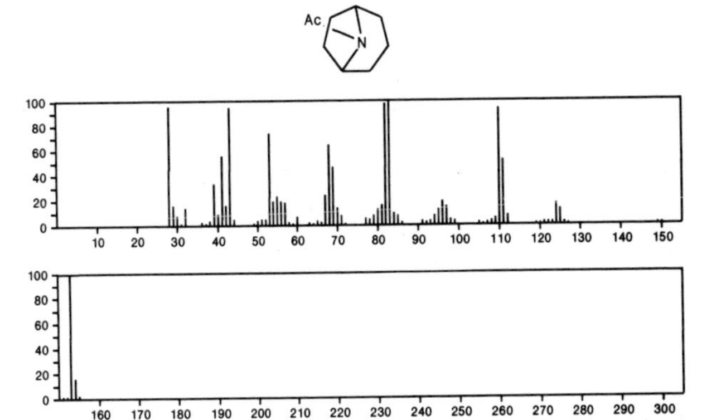

153 $C_9H_{15}NO$ 936-52-7

Morpholine, 4-(1-cyclopenten-1-yl)-

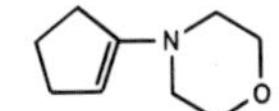

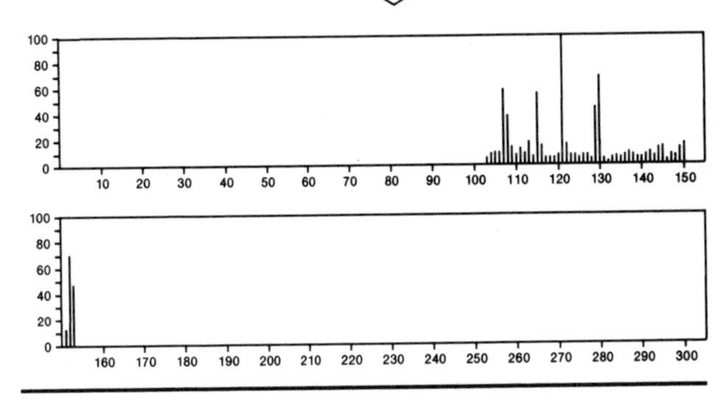

153 $C_9H_{15}NO$ 1809-57-0

3-Buten-2-one, 4-(1-piperidinyl)-

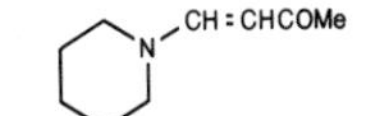

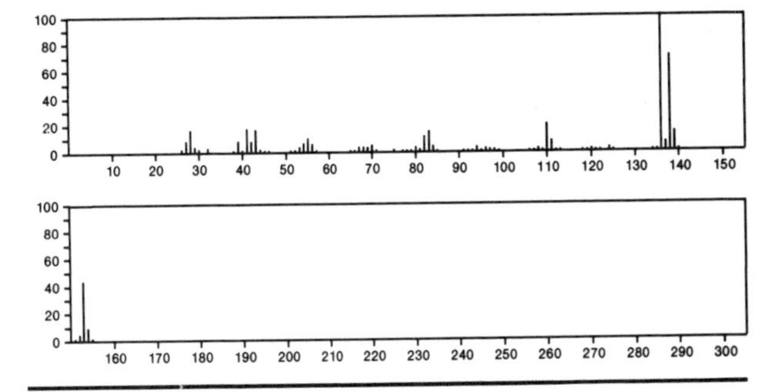

153 $C_9H_{15}NO$ 4146-35-4

3-Azabicyclo[3.3.1]nonan-9-one, 3-methyl-

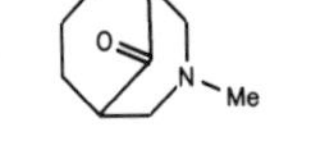

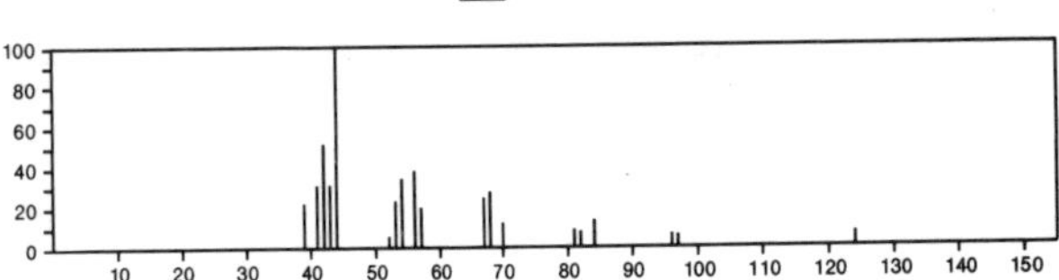

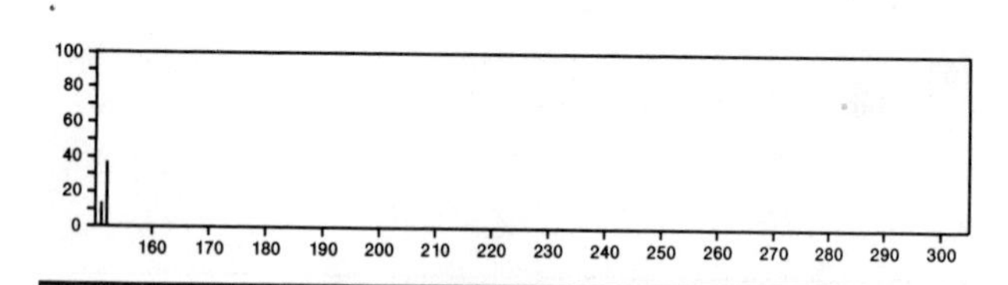

153 $C_9H_{15}NO$ 32363-15-8
Morpholine, 4-(3-methyl-1,3-butadienyl)-

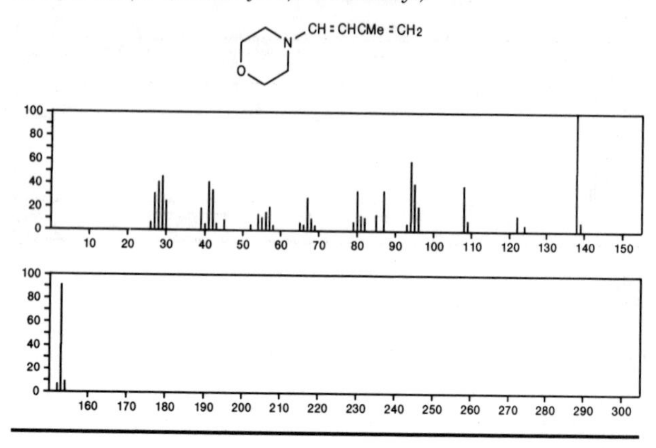

153 $C_9H_{15}NO$ 49656-54-4
9-Azabicyclo[3.3.1]non-6-en-2-ol, 9-methyl-, *endo*-

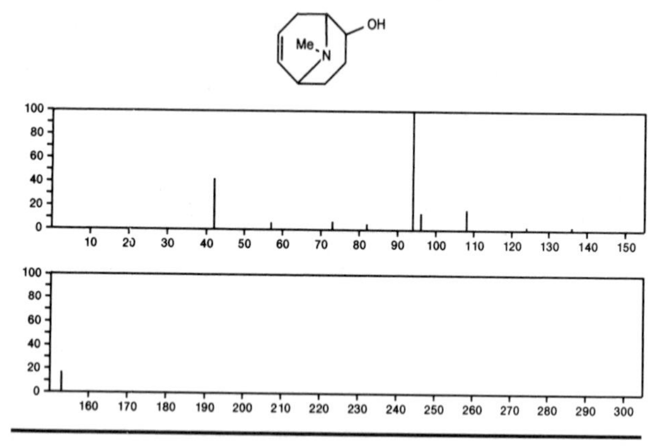

153 $C_9H_{15}NO$ 50483-91-5
2-Azetidinone, 1,3,3-trimethyl-4-(1-methylethylidene)-

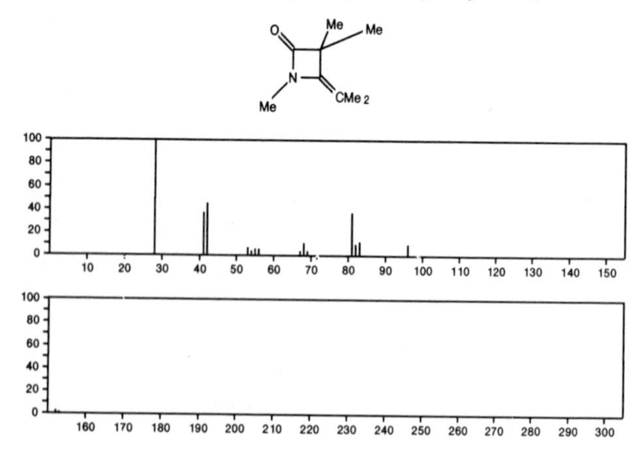

153 $C_9H_{15}NO$ 55320-41-7
Heptanenitrile, 4-acetyl-

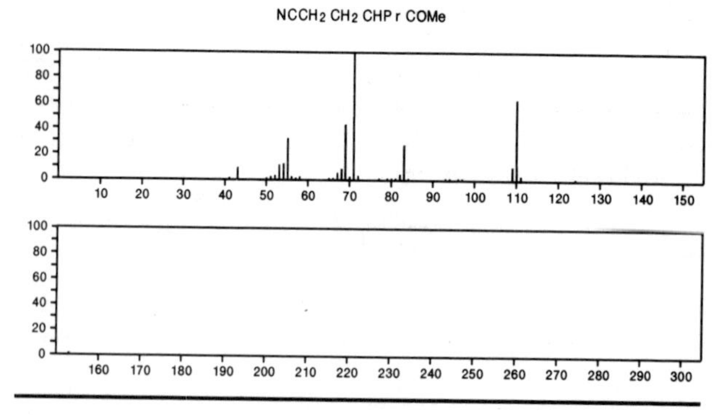

153 $C_9H_{15}NO$ 56336-06-2
2-Cyclohexen-1-one, 3,5-dimethyl-, *O*-methyloxime

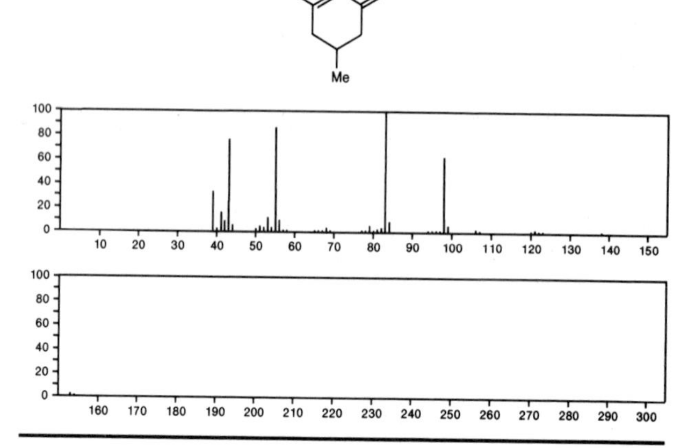

153 $C_{10}H_{19}N$ 464-42-6
Bicyclo[2.2.1]heptan-2-amine, 1,7,7-trimethyl-, *endo*-

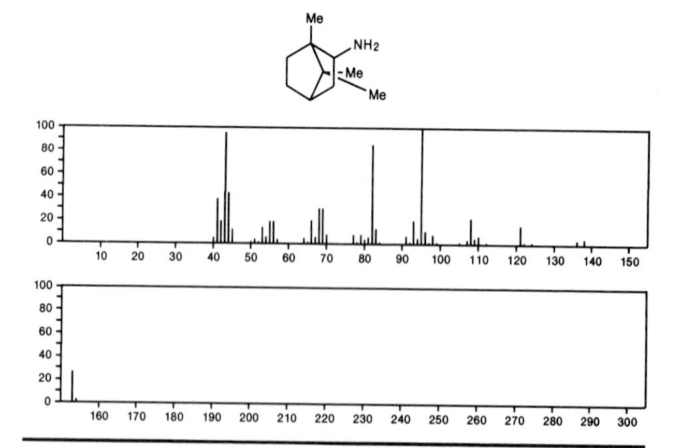

153 $C_{10}H_{19}N$ 465-49-6
3-Azabicyclo[3.2.1]octane, 1,8,8-trimethyl-, (1*R*)-

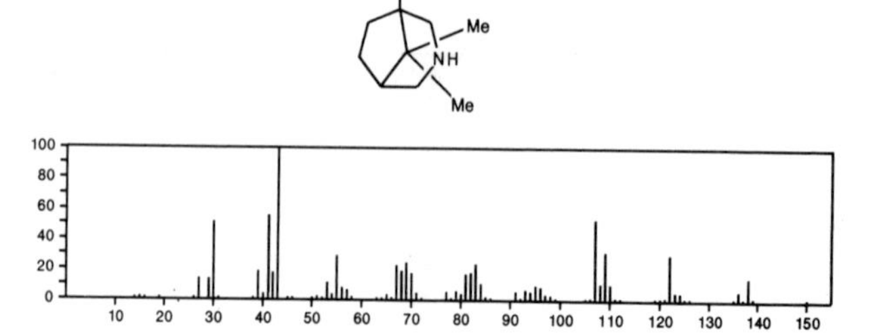

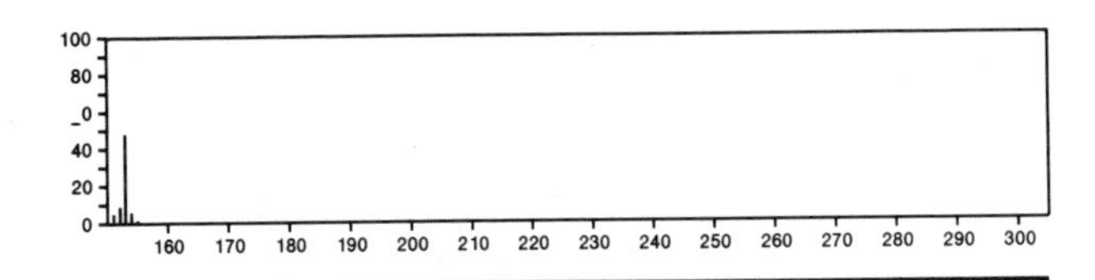

153 $C_{10}H_{19}N$ 875-63-8
Quinoline, decahydro-1-methyl-, *trans-*

Me N

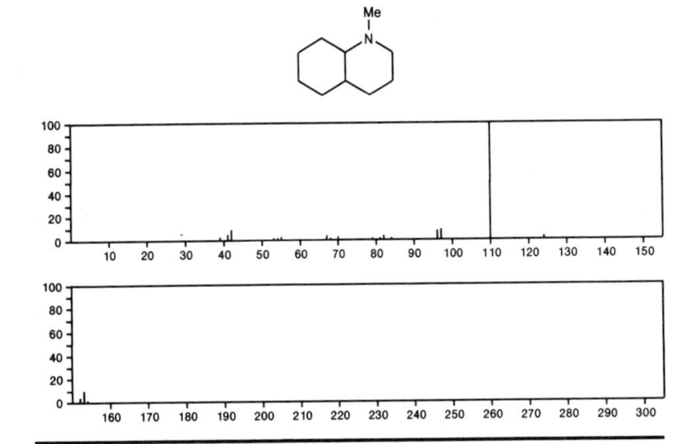

153 $C_{10}H_{19}N$ 16726-25-3
Quinoline, decahydro-1-methyl-, *cis-*

Me N

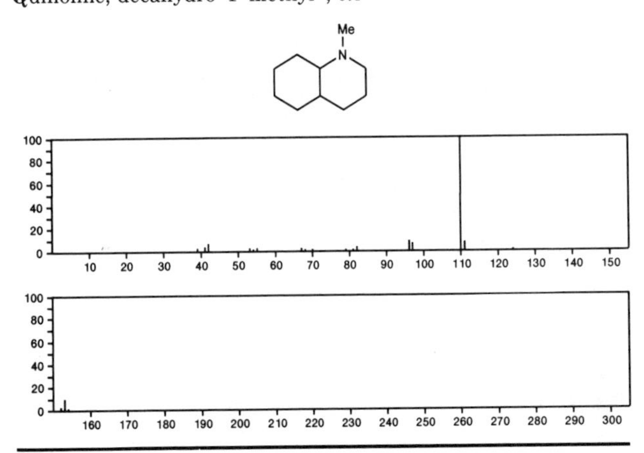

153 $C_{10}H_{19}N$ 35155-43-2
Piperidine, 1-(2-methyl-1-butenyl)-

N CH = CEt Me

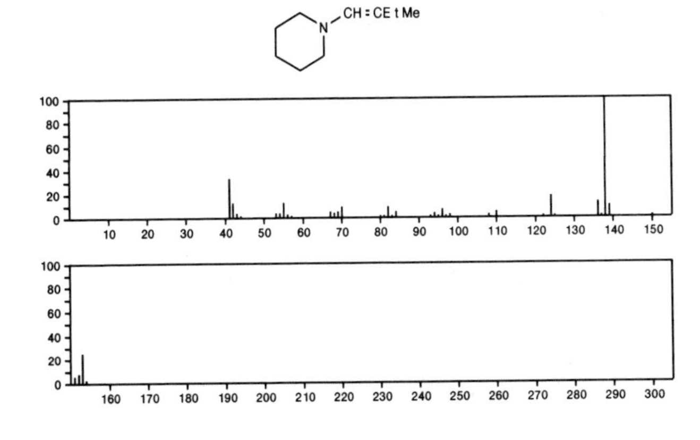

153 $C_{10}H_{19}N$ 49845-25-2
Piperidine, 1-(1-pentenyl)-

N CH = CHPr

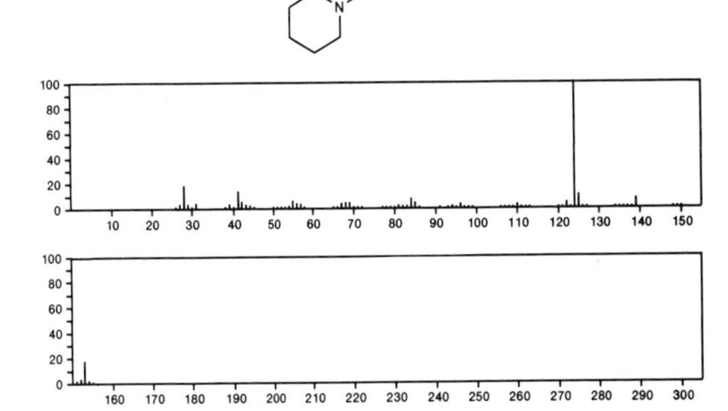

153 $C_{10}H_{19}N$ 55669-80-2
Aziridine, 2-(1,1-dimethylethyl)-3-methyl-1-(2-propenyl)-, *trans-*

Bu - t
Me N $CH_2CH=CH_2$

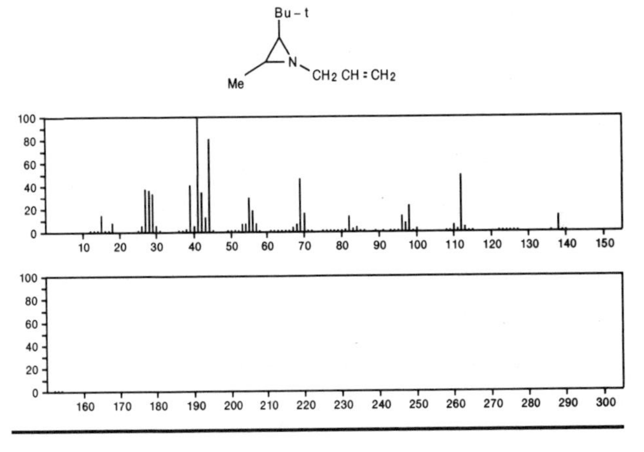

153 $C_{11}H_7N$ 86-53-3
1-Naphthalenecarbonitrile

CN

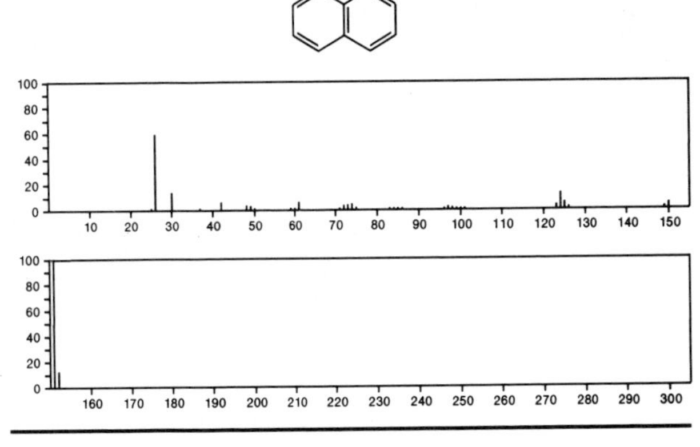

154 C_2ClF_5 76-15-3
Ethane, chloropentafluoro-

F_3CCClF_2

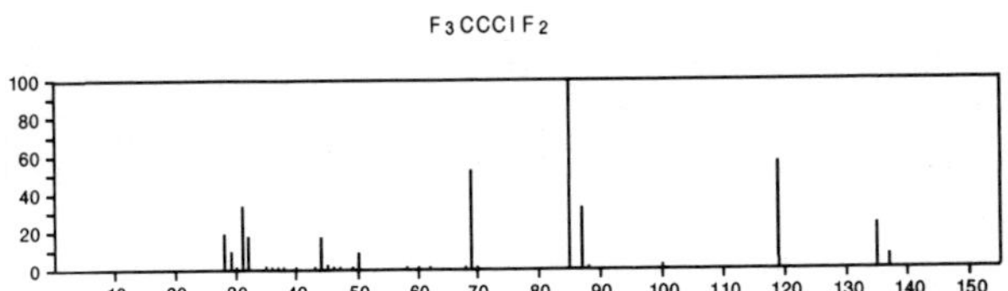

154 $C_4H_4Cl_2O_2$ 1588–60–9
2(3*H*)-Furanone, 3,3-dichlorodihydro-

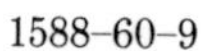

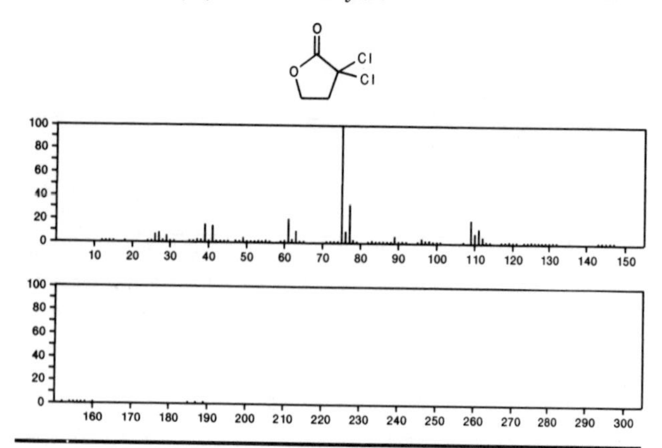

154 $C_4H_{10}S_3$ 5418–86–0
Methane, tris(methylthio)-

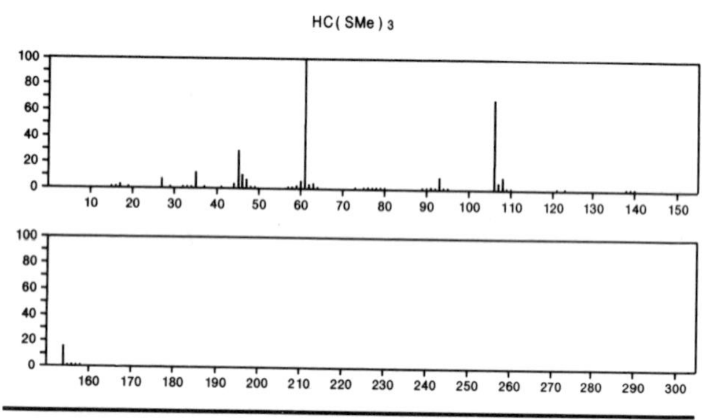

154 $C_4H_{12}ClN_2P$ 3348–44–5
Phosphorodiamidous chloride, tetramethyl-

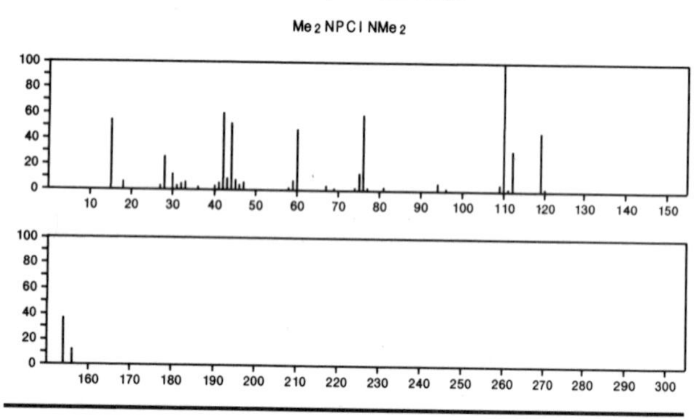

154 $C_5H_5F_3O_2$ 367–57–7
2,4-Pentanedione, 1,1,1-trifluoro-

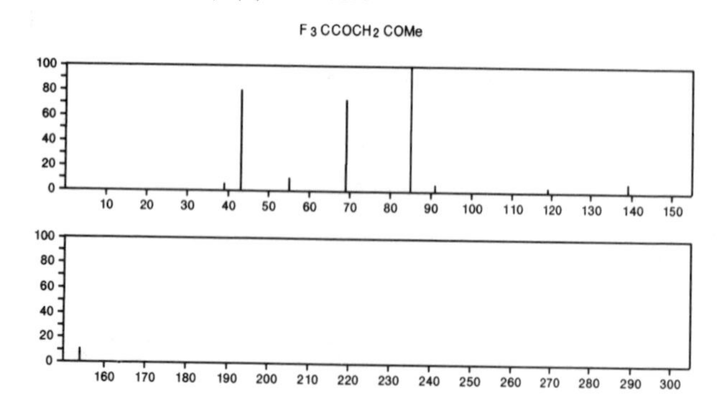

154 $C_5H_5F_3O_2$ 400–39–5
Acetic acid, trifluoro-, 1-methylethenyl ester

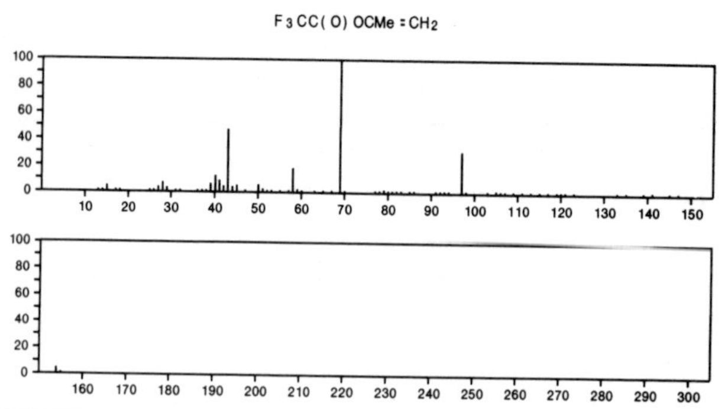

154 $C_5H_9F_3Si$ 51676–19–8
Silane, trifluoro(2-methyl-2-butenyl)-

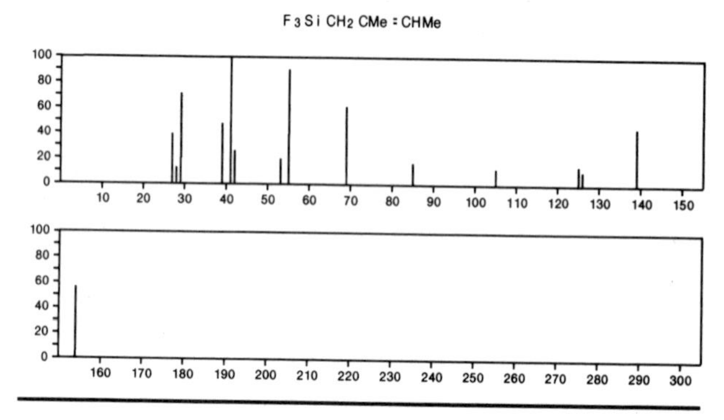

154 $C_6H_6F_4$ 1763–21–9
1,5-Hexadiene, 3,3,4,4-tetrafluoro-

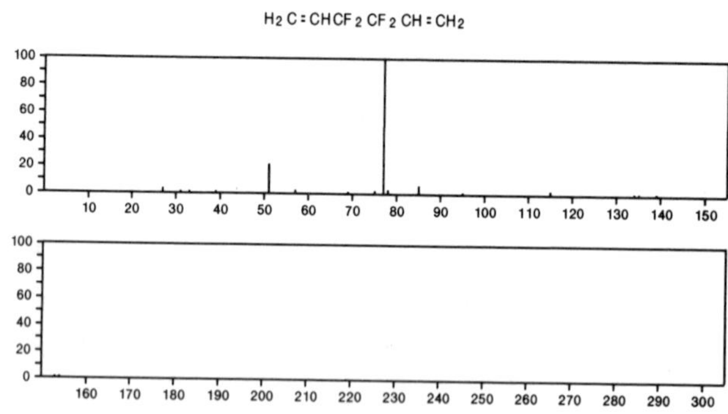

154 $C_6H_6N_2OS$ 17420–03–0
Hydrazine, phenylsulfinyl-

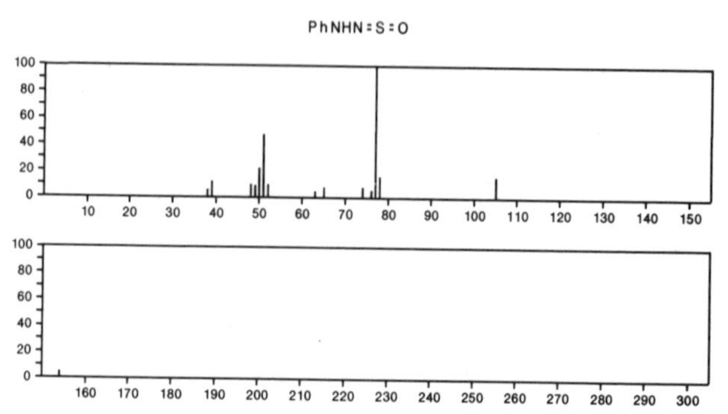

154 $C_6H_6N_2OS$ 56196-65-7
Acetonitrile, (3-methyl-4-oxo-2-thiazolidinylidene)-

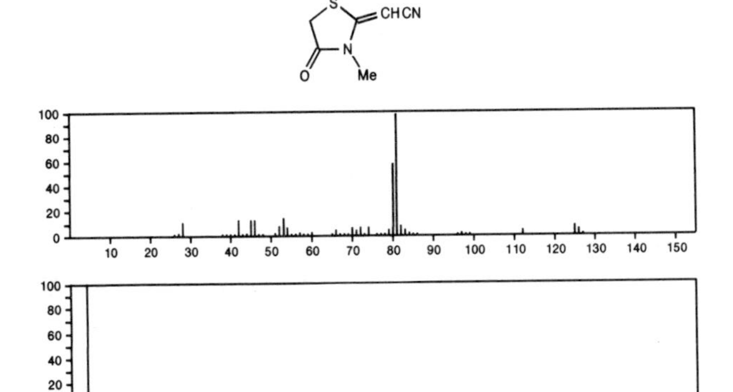

154 $C_6H_9F_3O$ 360-34-9
2-Hexanone, 1,1,1-trifluoro-

$Me(CH_2)_3COCF_3$

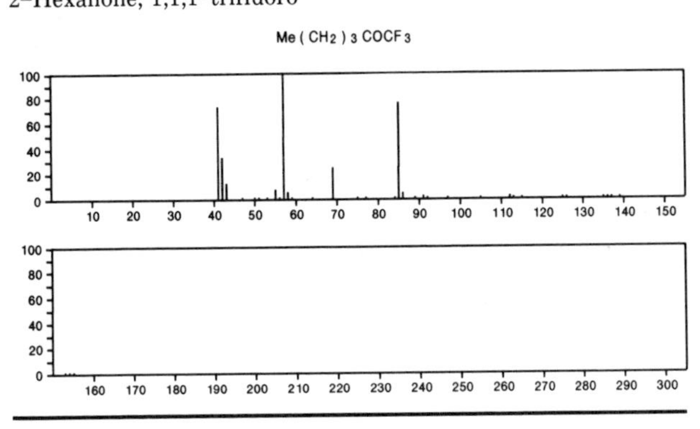

154 $C_6H_{10}N_4O$ 37034-43-8
1,3,5-Triazin-2-amine, *N*-ethyl-4-methoxy-

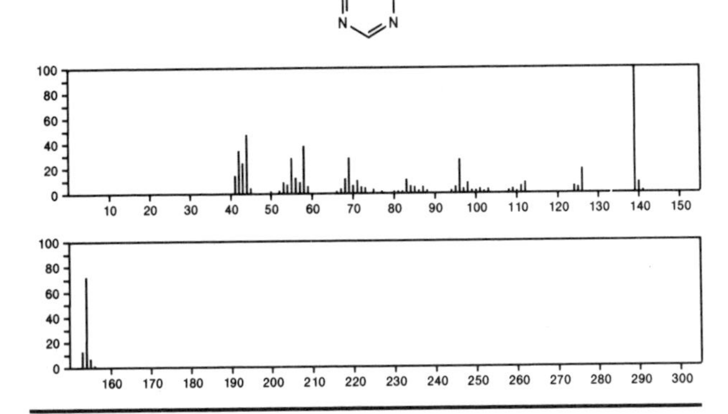

154 $C_6H_{15}BN_4$ 20534-04-7
Δ^2-Tetrazaboroline, 1,4,5-triethyl-

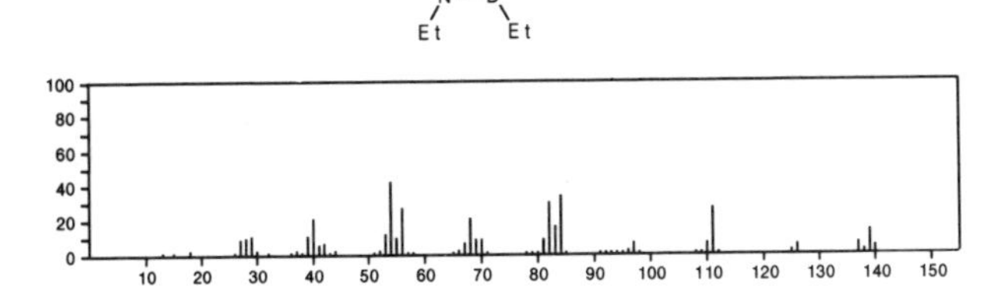

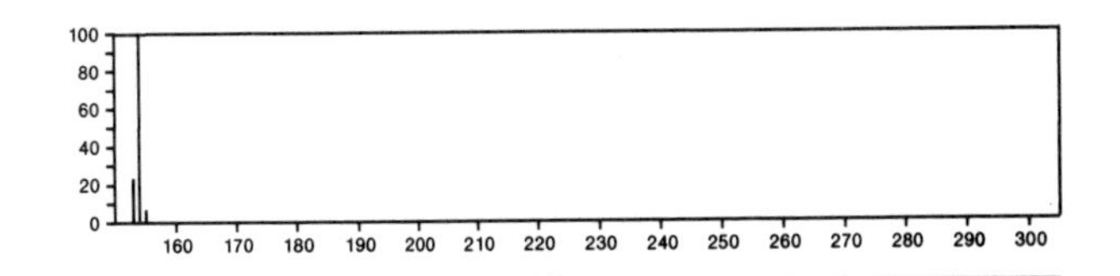

154 $C_7H_6O_2S$ 147-93-3
Benzoic acid, 2-mercapto-

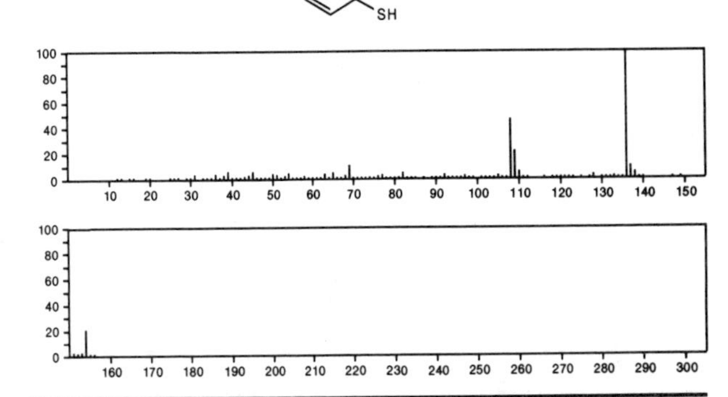

154 $C_7H_6O_2S$ 7283-41-2
Benzenecarbothioic acid, 2-hydroxy-

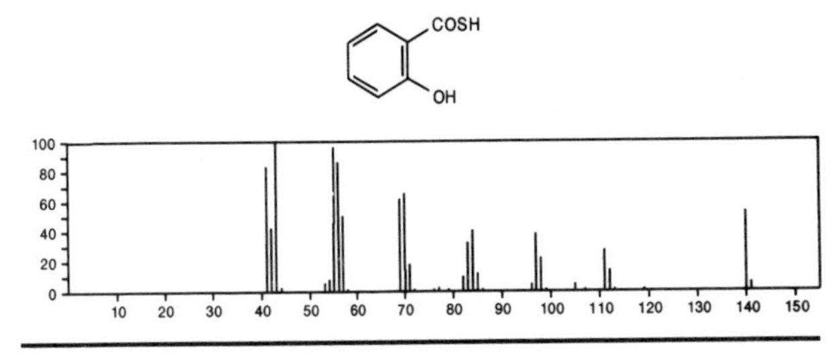

154 $C_7H_6O_4$ 89-86-1
Benzoic acid, 2,4-dihydroxy-

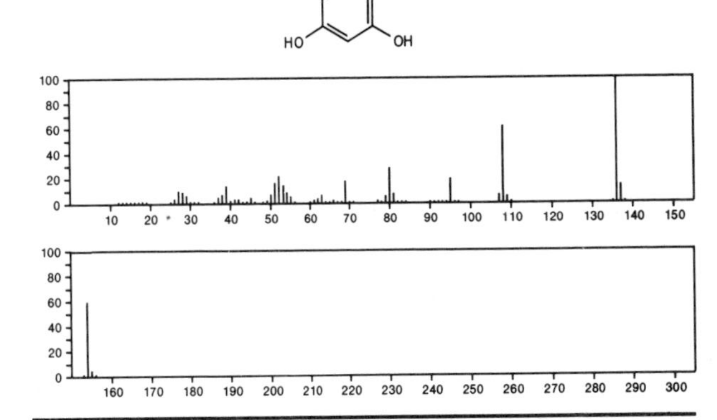

154 $C_7H_6O_4$ 99-10-5
Benzoic acid, 3,5-dihydroxy-

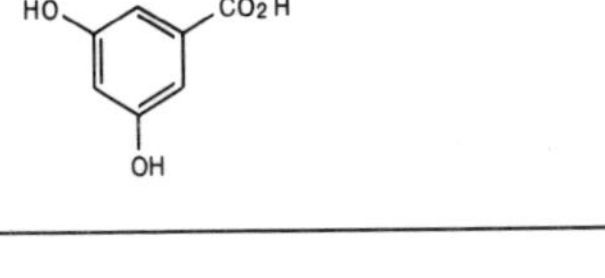

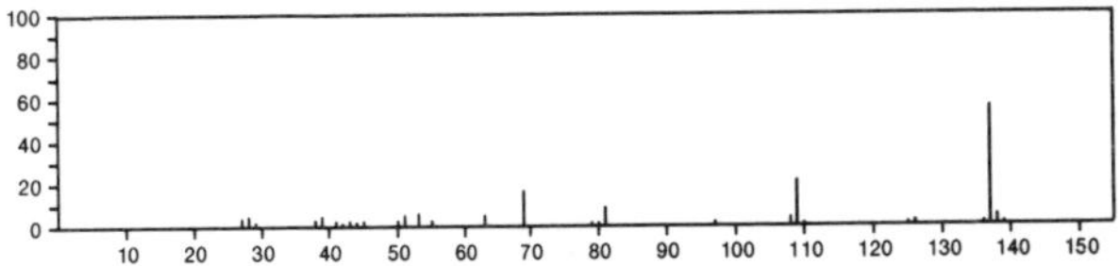

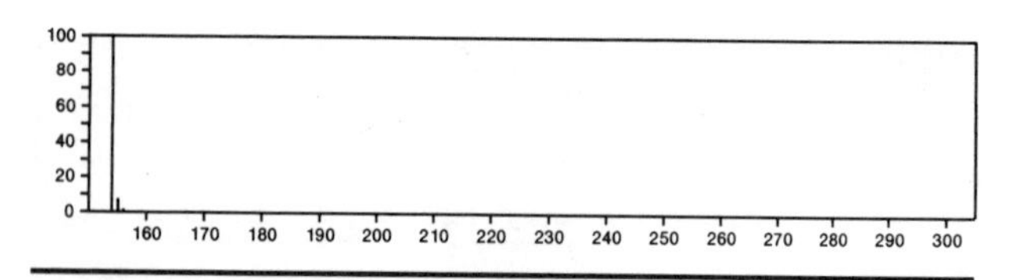

154 $C_7H_6O_4$ 99-50-3
Benzoic acid, 3,4-dihydroxy-

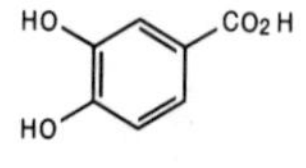

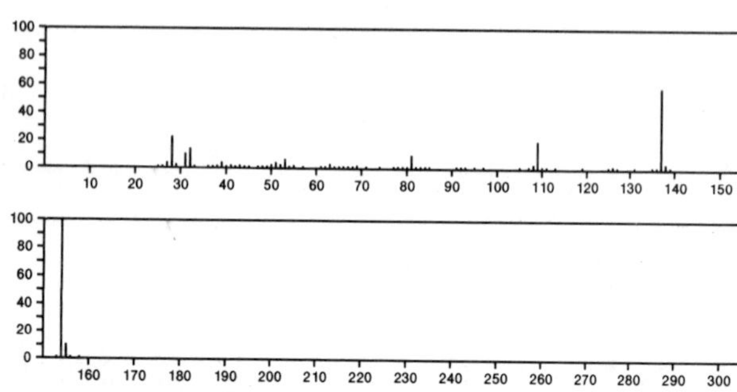

154 $C_7H_6O_4$ 303-07-1
Benzoic acid, 2,6-dihydroxy-

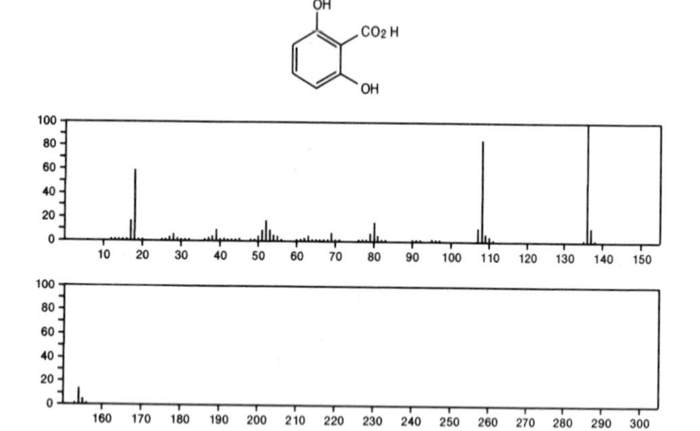

154 $C_7H_6O_4$ 303-38-8
Benzoic acid, 2,3-dihydroxy-

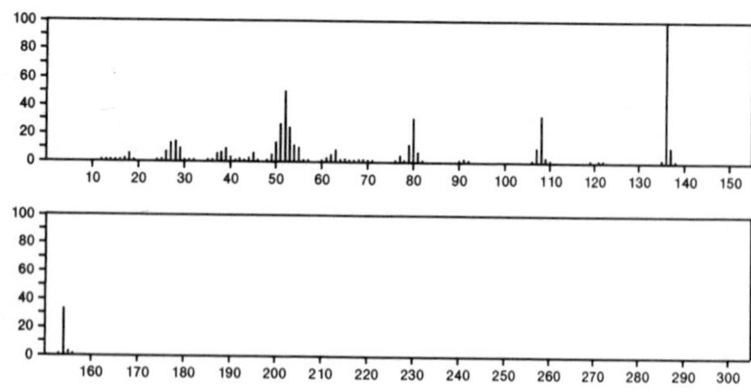

154 $C_7H_6O_4$ 490-79-9
Benzoic acid, 2,5-dihydroxy-

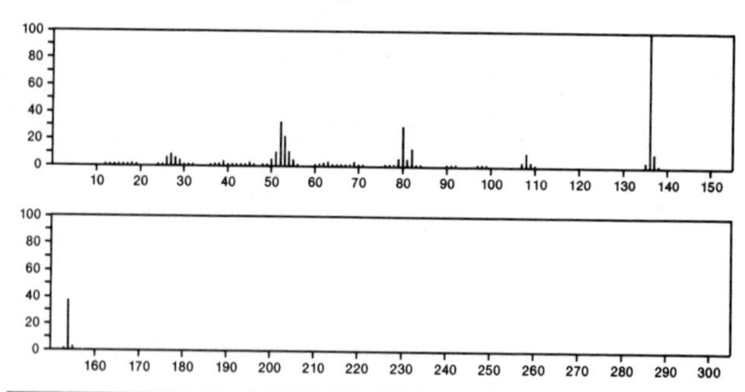

154 $C_7H_6O_4$ 32180-43-1
7-Oxabicyclo[4.1.0]hept-3-ene-2,5-dione, 3-(hydroxymethyl)-, (±)-

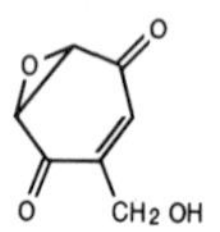

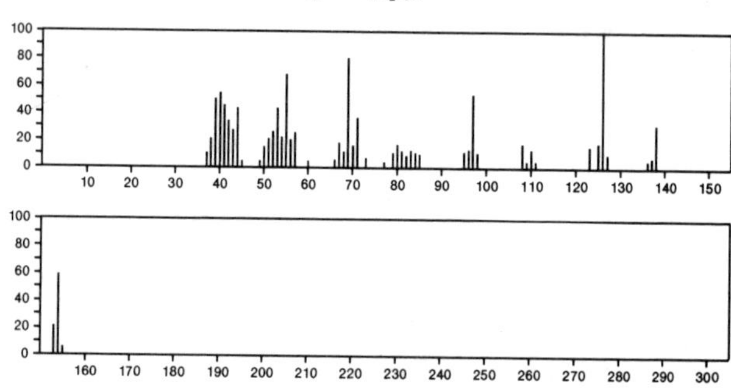

154 $C_7H_6S_2$ 21505-25-9
2,4,6-Cycloheptatriene-1-thione, 2-mercapto-

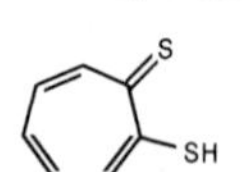

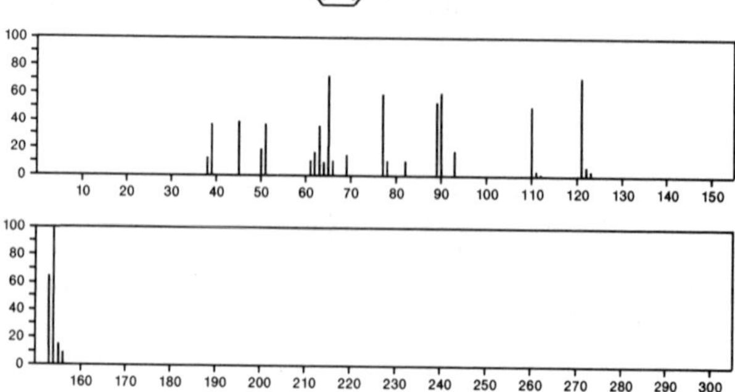

154 $C_7H_{10}N_2O_2$ 1006–24–2
2,4(1*H*,3*H*)–Pyrimidinedione, 3–ethyl–6–methyl–

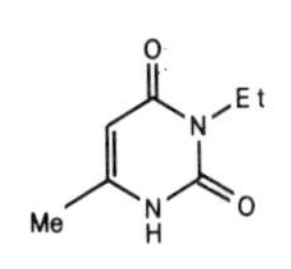
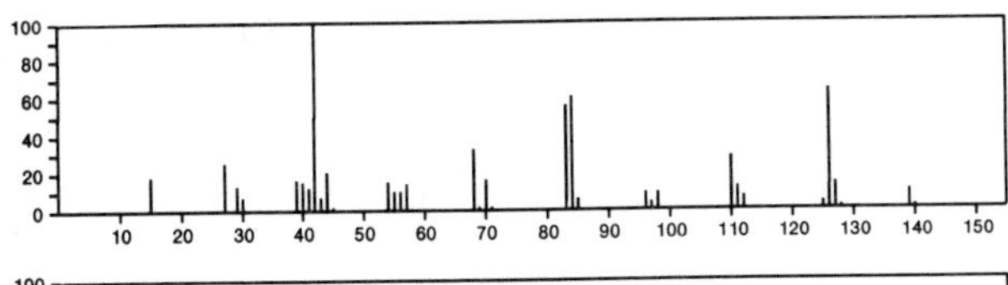
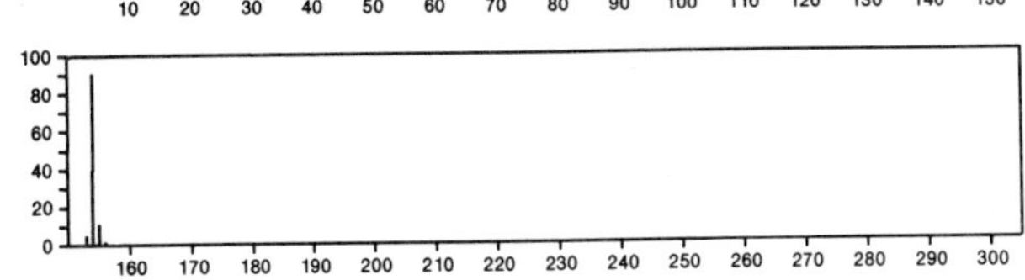

154 $C_7H_{10}N_2O_2$ 13509–52–9
2,4(1*H*,3*H*)–Pyrimidinedione, 1,3,6–trimethyl–

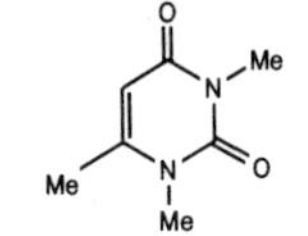
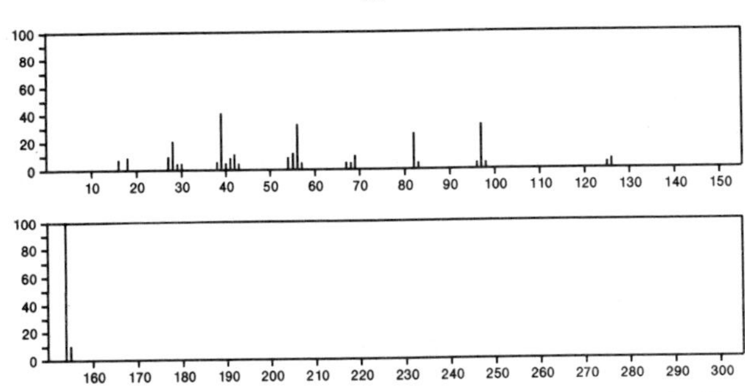

154 $C_7H_{10}N_2O_2$ 24611–11–8
4(1*H*)–Pyrimidinone, 5–ethoxy–2–methyl–

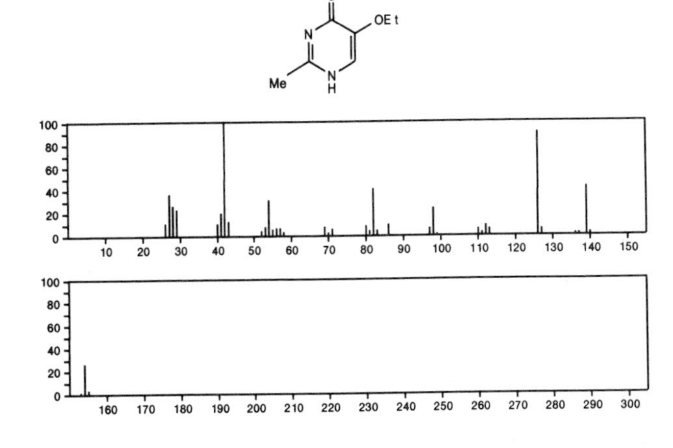

154 $C_7H_{10}N_2O_2$ 38249–34–2
4(1*H*)–Pyrimidinone, 6–ethoxy–2–methyl–

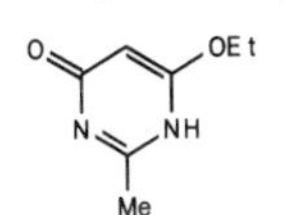
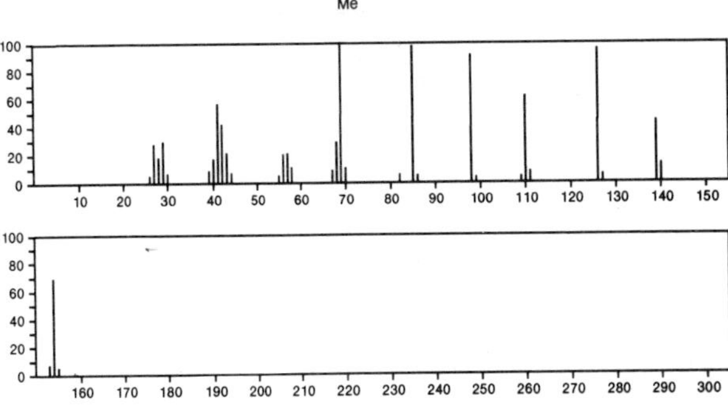

154 C_8H_7ClO 99–02–5
Ethanone, 1–(3–chlorophenyl)–

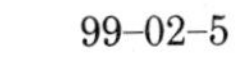
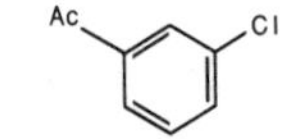
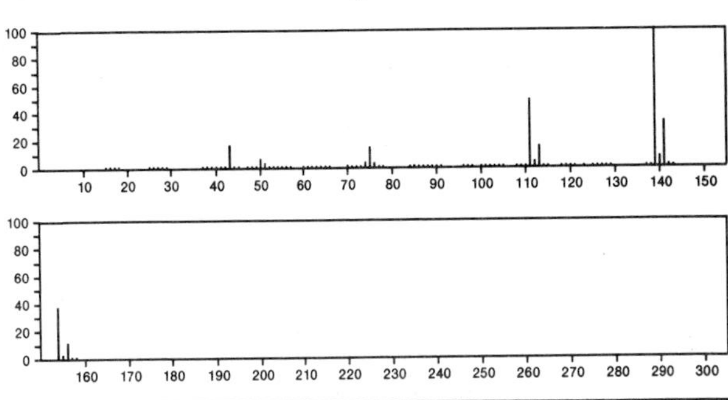

154 C_8H_7ClO 99–91–2
Ethanone, 1–(4–chlorophenyl)–

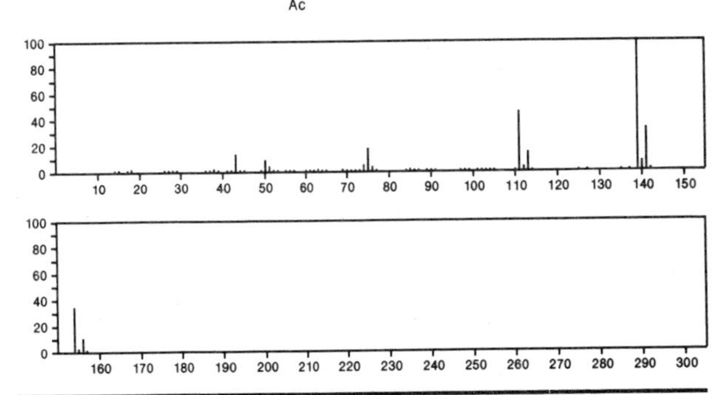

154 C_8H_7ClO 532–27–4
Ethanone, 2–chloro–1–phenyl–

Cl CH2 COPh

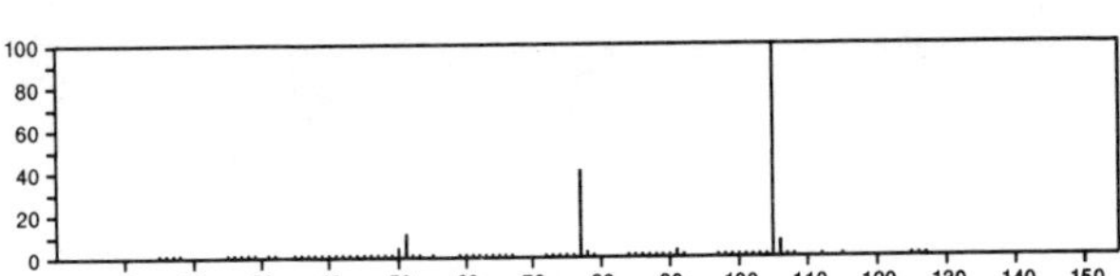

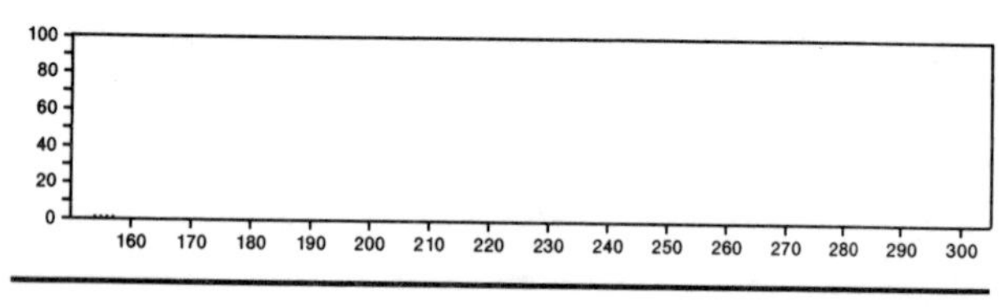

154 C_8H_7ClO 933-88-0
Benzoyl chloride, 2-methyl-

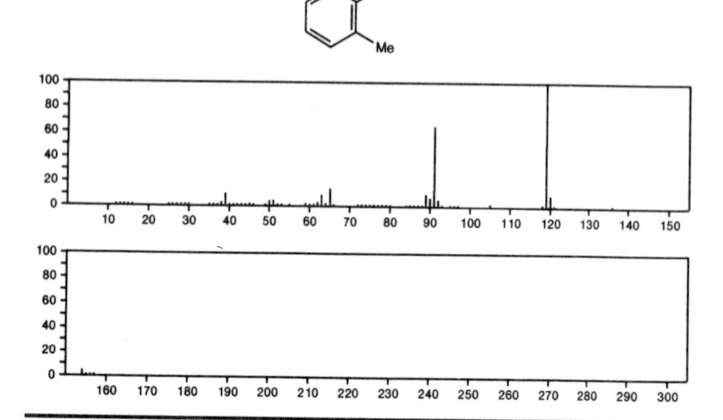

154 C_8H_7ClO 2142-68-9
Ethanone, 1-(2-chlorophenyl)-

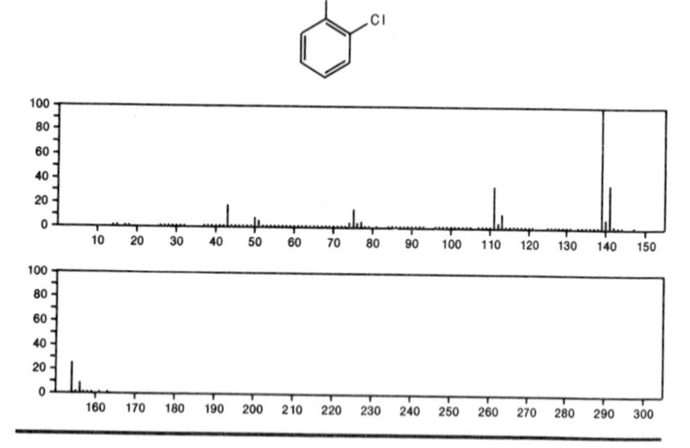

154 $C_8H_{10}OS$ 3120-74-9
Phenol, 3-methyl-4-(methylthio)-

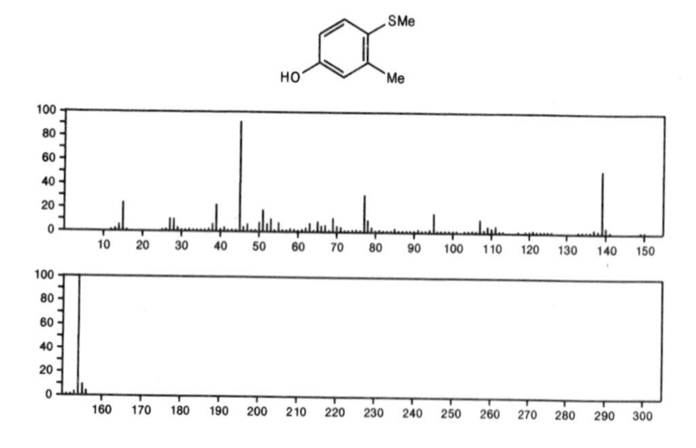

154 $C_8H_{10}OS$ 5333-83-5
1-Butanone, 1-(2-thienyl)-

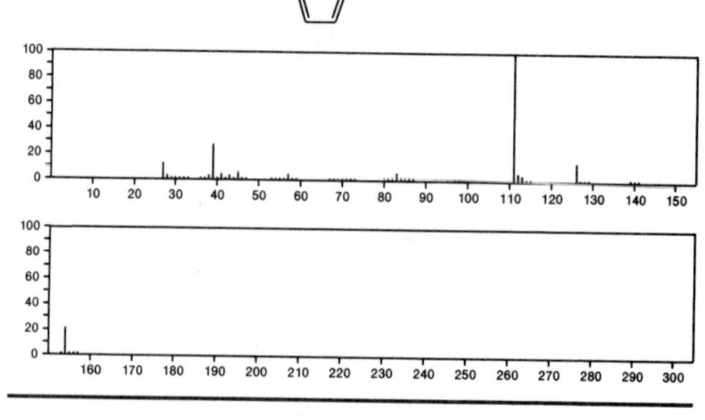

154 $C_8H_{10}OS$ 6338-63-2
Ethanethiol, 2-phenoxy-

PhOCH2 CH2 SH

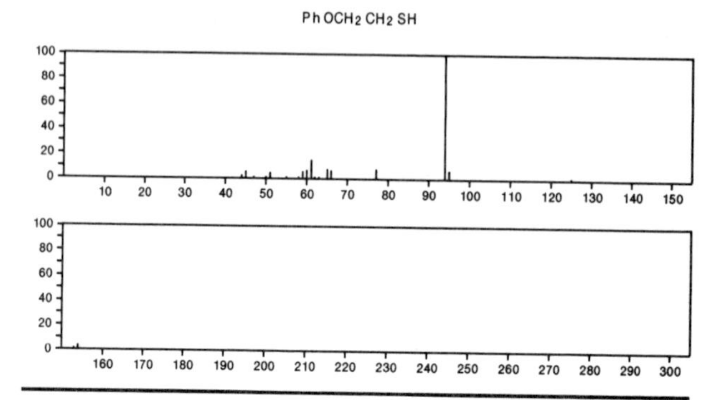

154 $C_8H_{10}OS$ 29549-60-8
Phenol, 2-(ethylthio)-

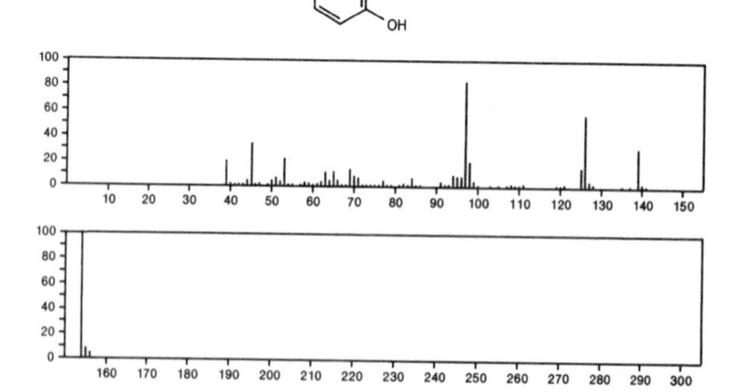

154 $C_8H_{10}O_3$ 85-42-7
1,3-Isobenzofurandione, hexahydro-

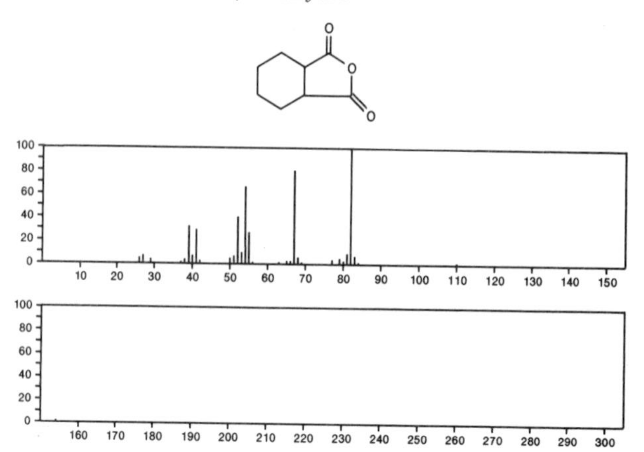

154 $C_8H_{10}O_3$ 500-99-2
Phenol, 3,5-dimethoxy-

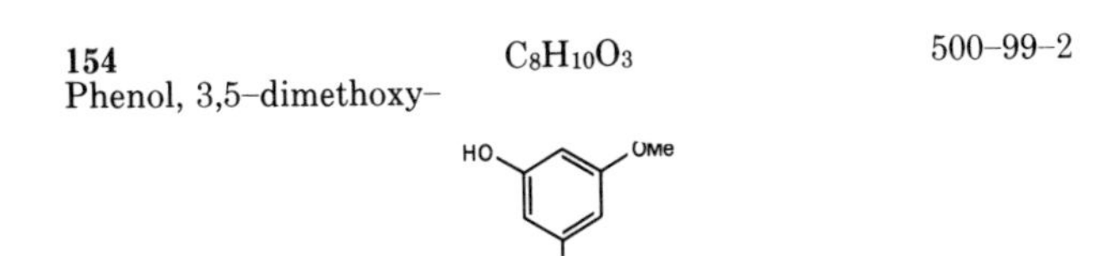

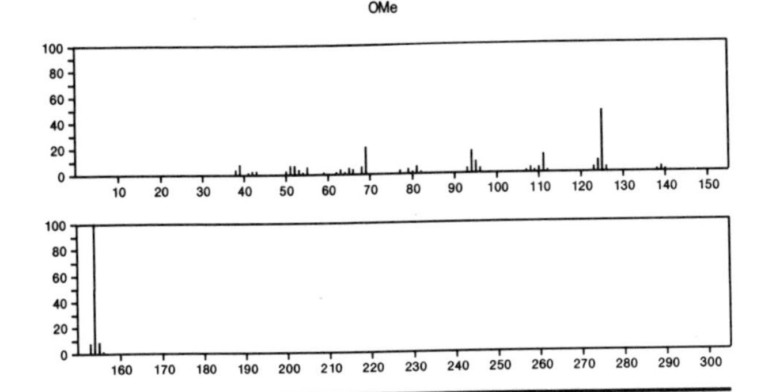

154 $C_8H_{10}O_3$ 615-10-1
2-Furancarboxylic acid, propyl ester

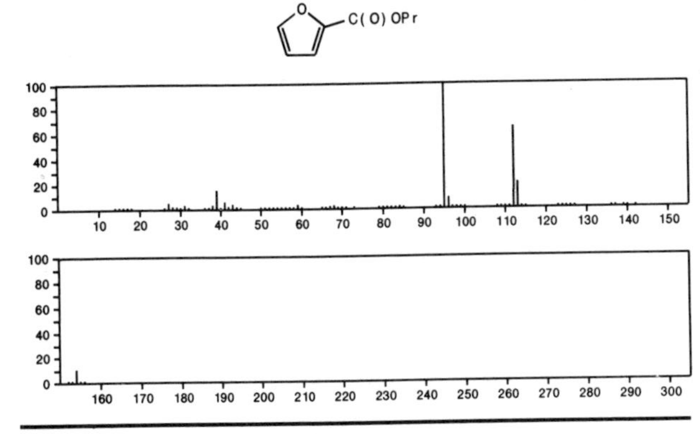

154 $C_8H_{10}O_3$ 623-19-8
2-Furanmethanol, propanoate

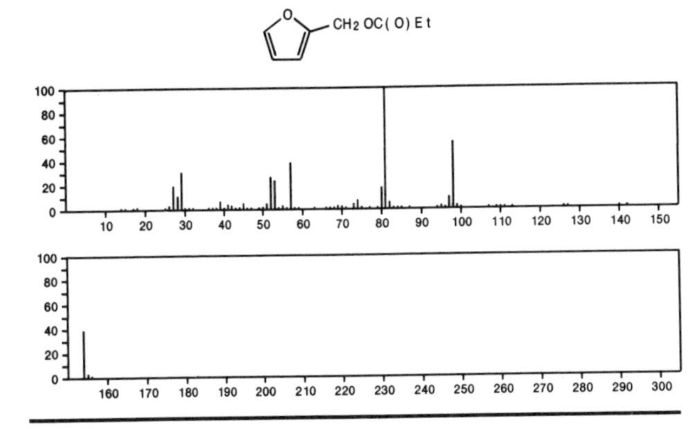

154 $C_8H_{10}O_3$ 2033-89-8
Phenol, 3,4-dimethoxy-

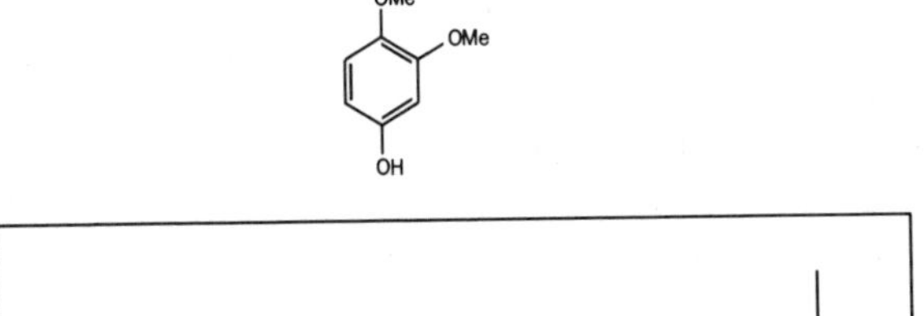

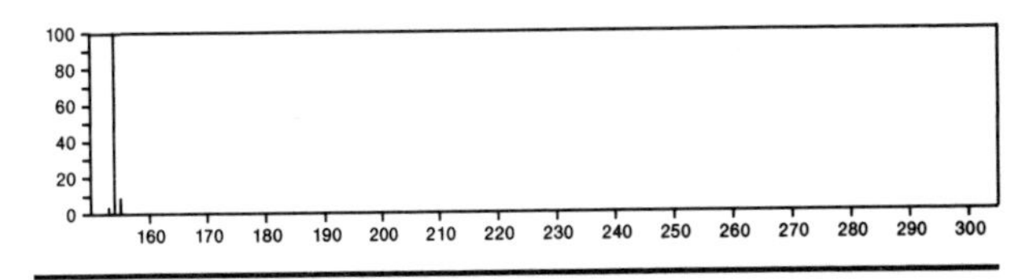

154 $C_8H_{10}O_3$ 4056-69-3
1,3-Cyclopentanedione, 2-acetyl-4-methyl-

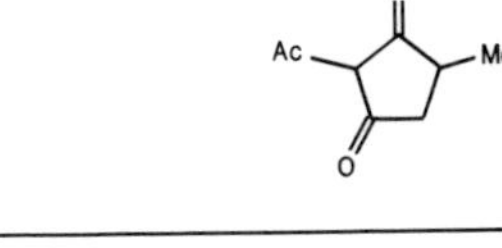

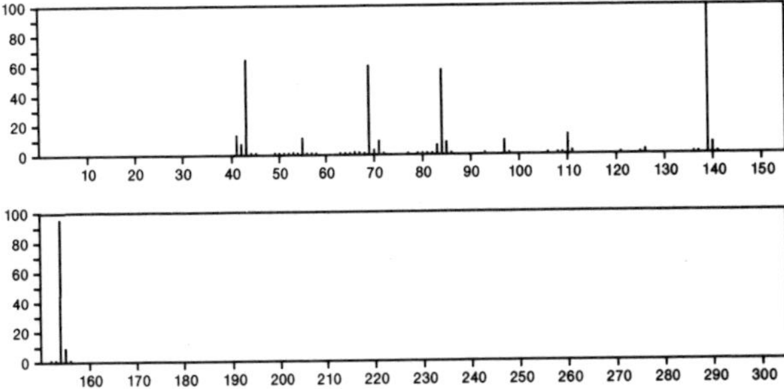

154 $C_8H_{10}O_3$ 10597-60-1
1,2-Benzenediol, 4-(2-hydroxyethyl)-

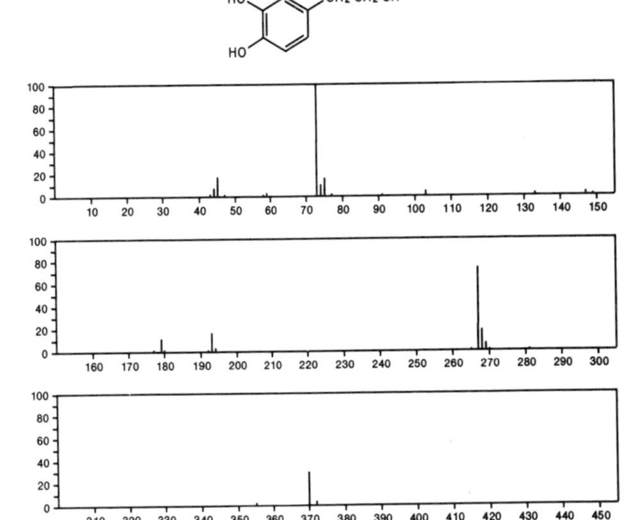

154 $C_8H_{10}O_3$ 19261-13-3
3-Butene-1,2-diol, 1-(2-furanyl)-

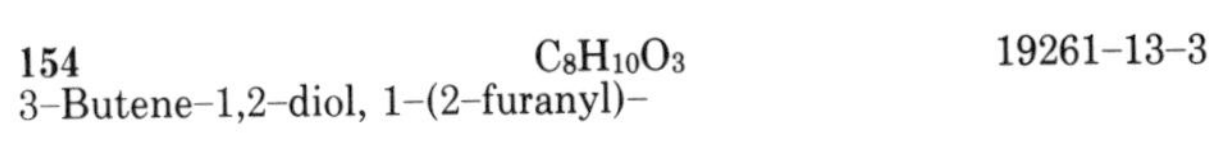

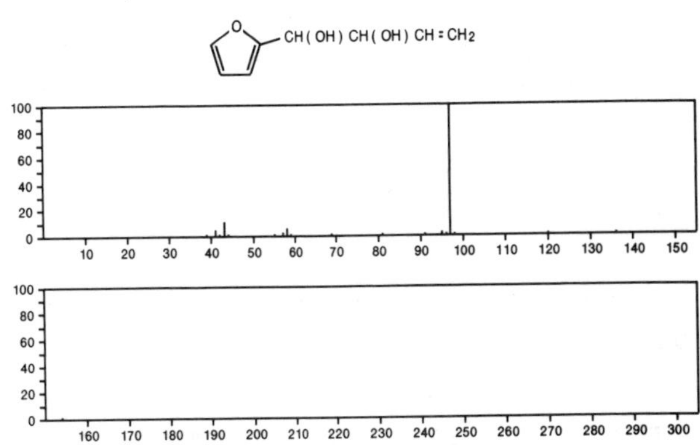

154 $C_8H_{10}O_3$ 50267–11–3
3,8,11–Trioxatetracyclo[4.4.1.$0^{2,4}$.$0^{7,9}$]undecane, (1α,2α,4α,6α,7β,9β)–

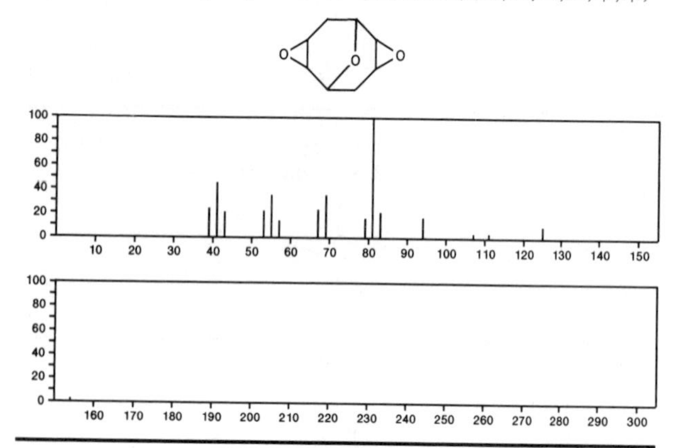

154 $C_8H_{10}O_3$ 50267–13–5
3,8,11–Trioxatetracyclo[4.4.1.$0^{2,4}$.$0^{7,9}$]undecane, (1α,2β,4β,6α,7β,9β)–

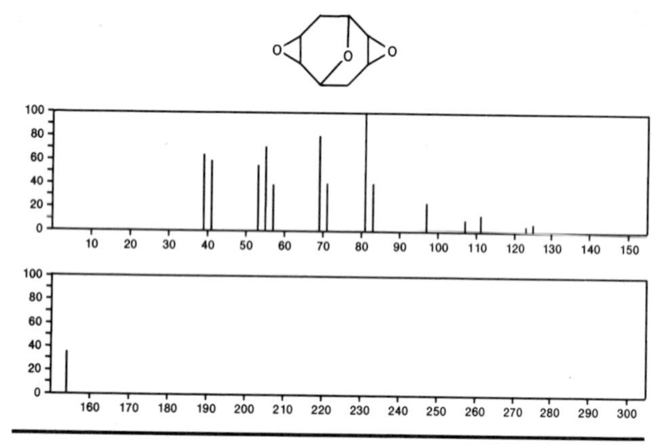

154 $C_8H_{10}O_3$ 50607–35–7
2*H*–Pyran–2–one, 3–ethyl–4–hydroxy–6–methyl–

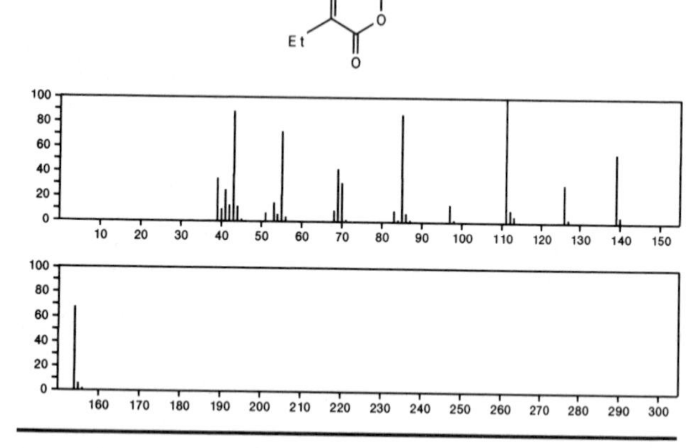

154 $C_8H_{10}O_3$ 56666–97–8
1,4–Dioxane, 2–(2–furanyl)–

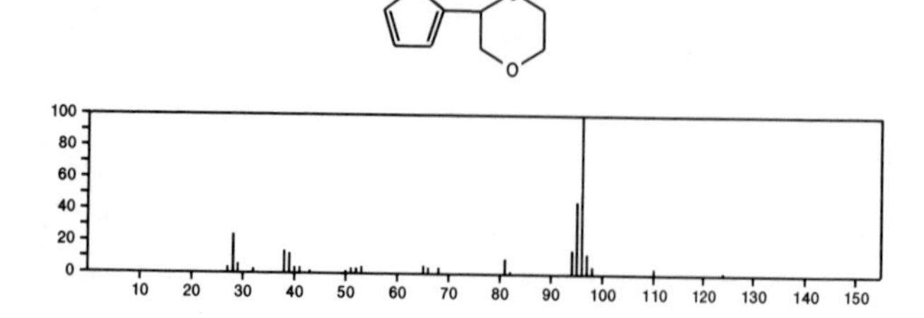

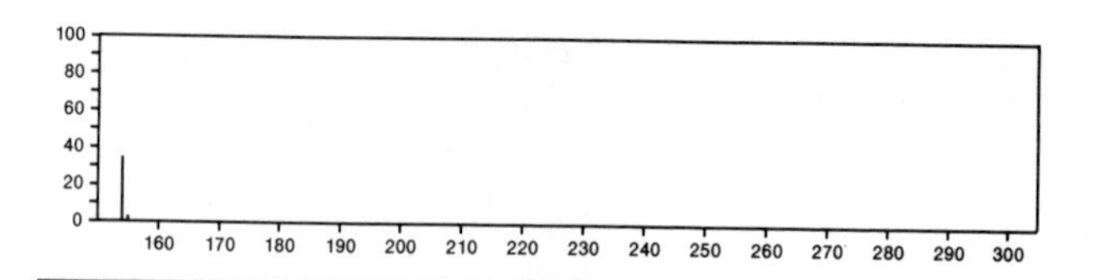

154 $C_8H_{14}N_2O$ 1522–09–4
3–Azabicyclo[3.2.2]nonane, 3–nitroso–

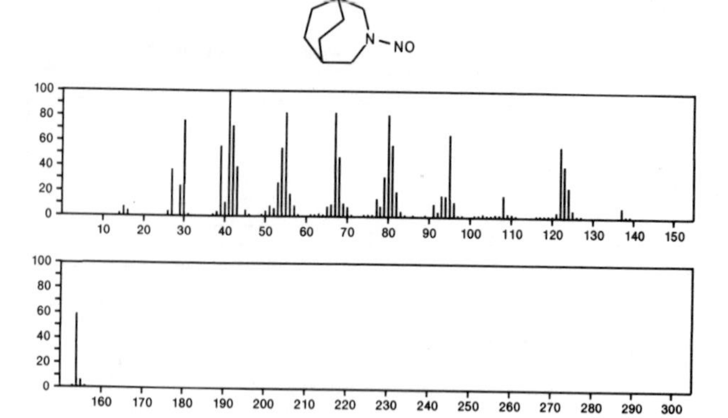

154 $C_8H_{14}N_2O$ 49582–43–6
3–Buten–2–one, 4–(1–aziridinyl)–4–(dimethylamino)–

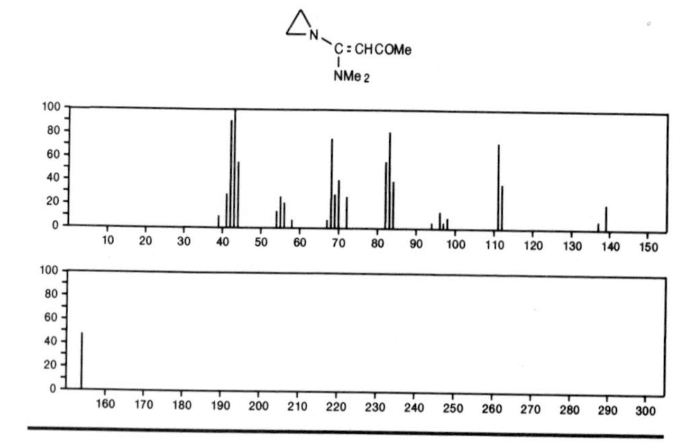

154 $C_9H_{11}Cl$ 104–52–9
Benzene, (3–chloropropyl)–

Ph(CH_2)$_3$Cl

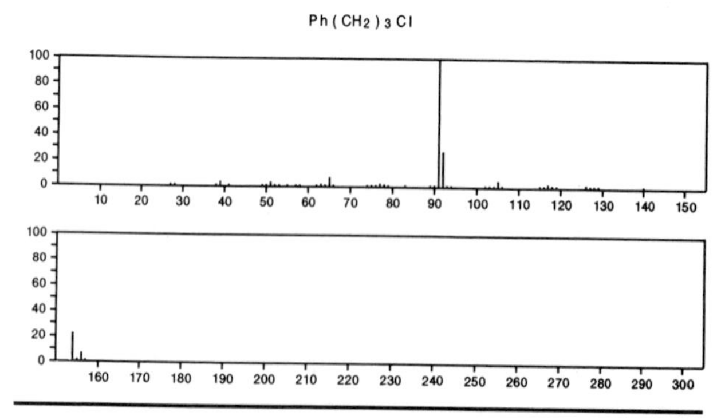

154 $C_9H_{11}Cl$ 824–47–5
Benzene, (2–chloro–1–methylethyl)–

ClCH_2CHMePh

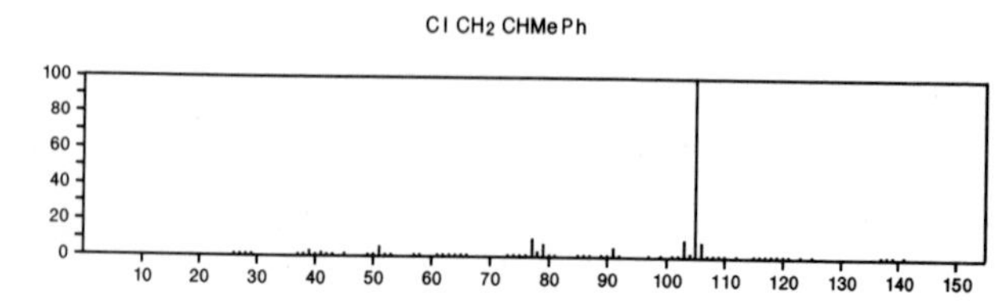

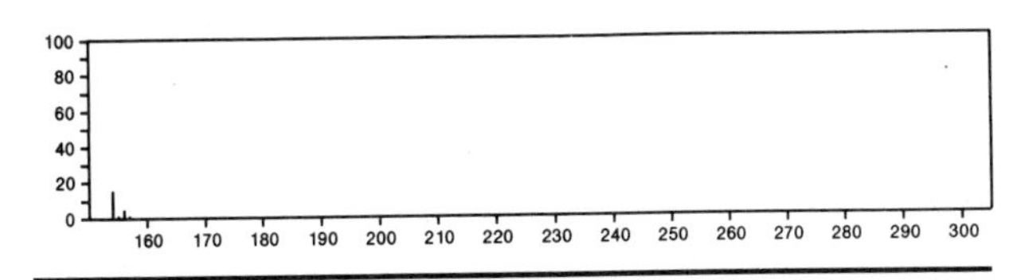

154 $C_9H_{11}Cl$ 1667–04–5
Benzene, 2–chloro–1,3,5–trimethyl–

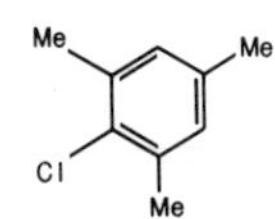

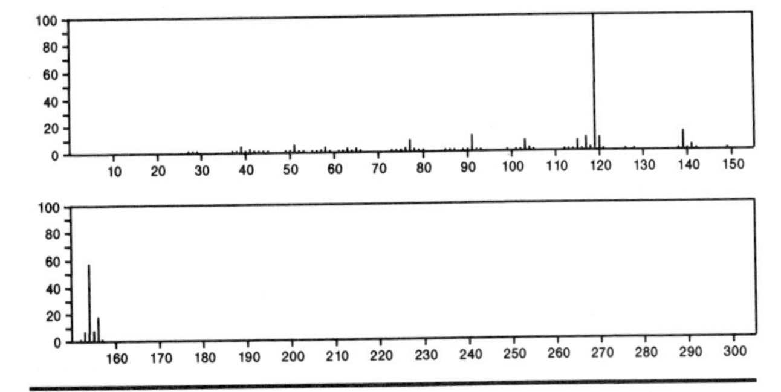

154 $C_9H_{11}Cl$ 1730–86–5
Benzene, 1–chloro–2–propyl–

Pr
Cl

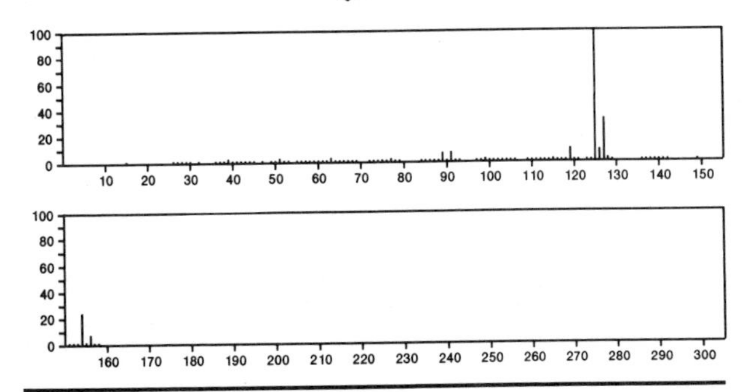

154 $C_9H_{11}Cl$ 2077–13–6
Benzene, 1–chloro–2–(1–methylethyl)–

Pr-i
Cl

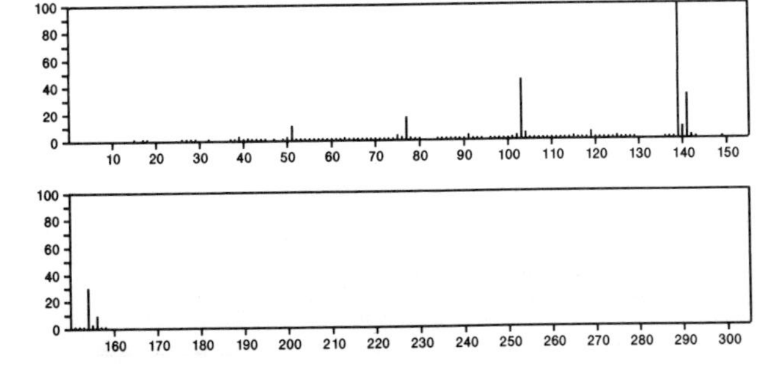

154 $C_9H_{11}Cl$ 2621–46–7
Benzene, 1–chloro–4–(1–methylethyl)–

Pr-i
Cl

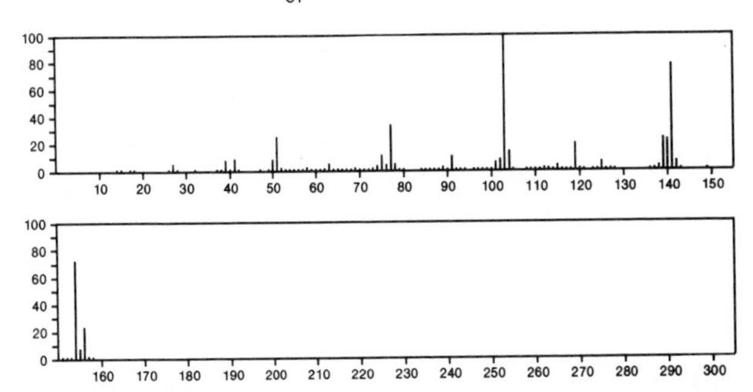

154 $C_9H_{11}Cl$ 10304–81–1
Benzene, (2–chloropropyl)–

MeCHCl CH_2Ph

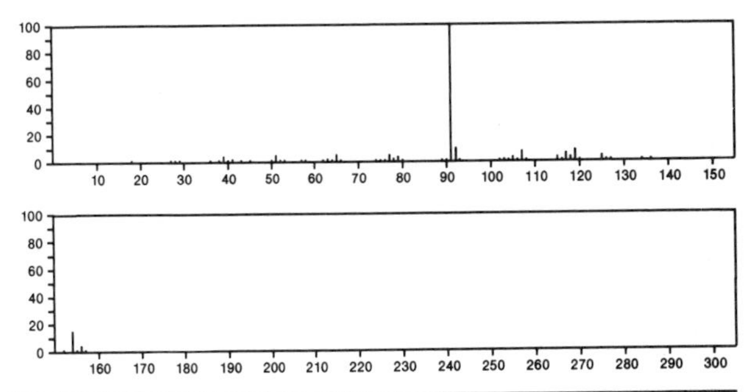

154 $C_9H_{11}Cl$ 26968–58–1
Benzene, (chloromethyl)ethyl–

Et — CH_2Cl

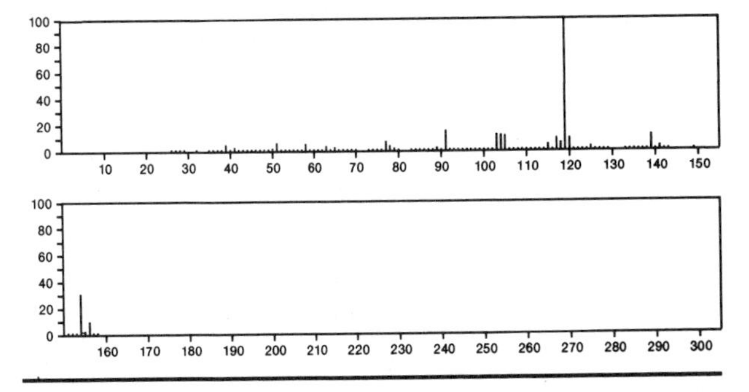

154 $C_9H_{14}O_2$ 699–55–8
Bicyclo[2.2.2]octane–1–carboxylic acid

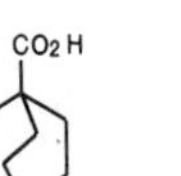

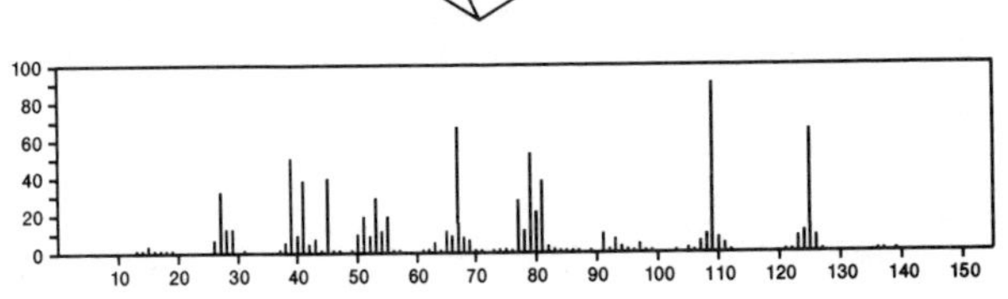

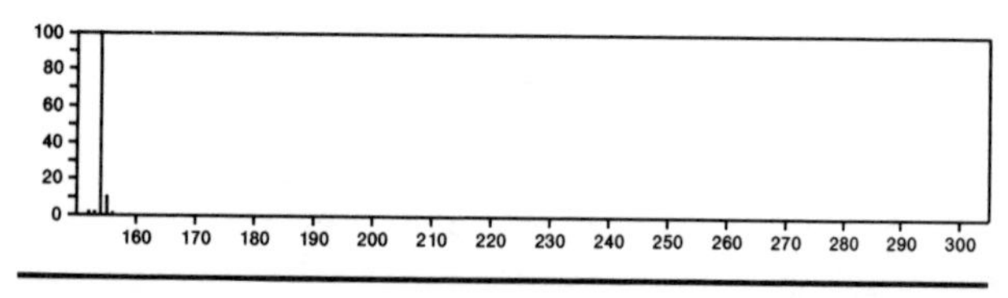

154 $C_9H_{14}O_2$ 825-31-0
1,3-Cyclopentanedione, 2-butyl-

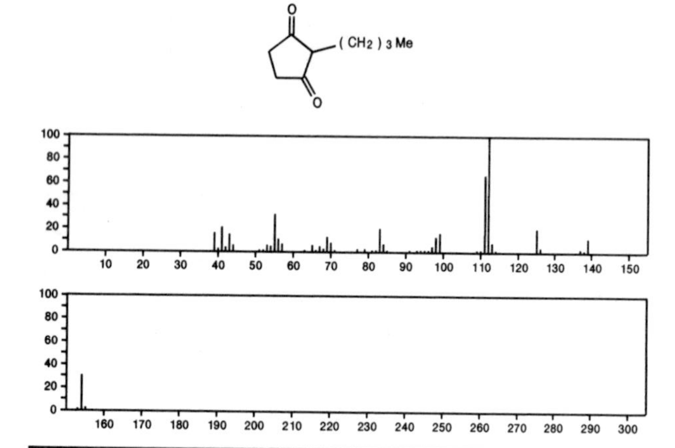

154 $C_9H_{14}O_2$ 1728-24-1
1,4-Dioxaspiro[4.6]undec-6-ene

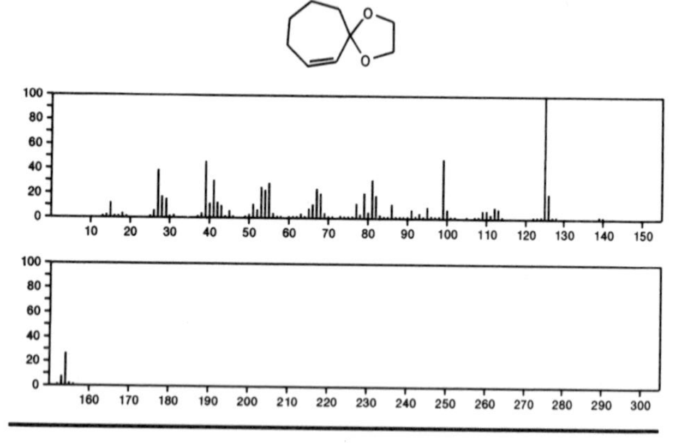

154 $C_9H_{14}O_2$ 1846-70-4
2-Nonynoic acid

Me (CH2) 5 C≡CCO2 H

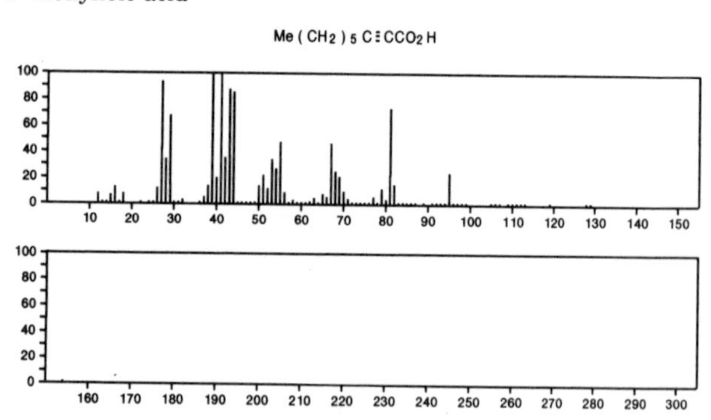

154 $C_9H_{14}O_2$ 3066-71-5
2-Propenoic acid, cyclohexyl ester

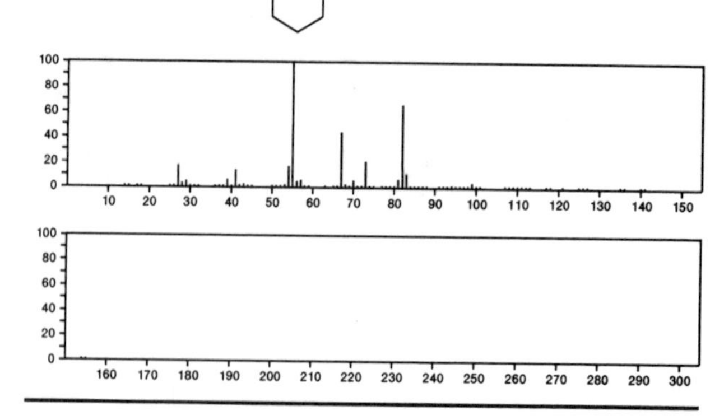

154 $C_9H_{14}O_2$ 4362-67-8
1,3-Dioxolane, 4,5-diethenyl-2,2-dimethyl-

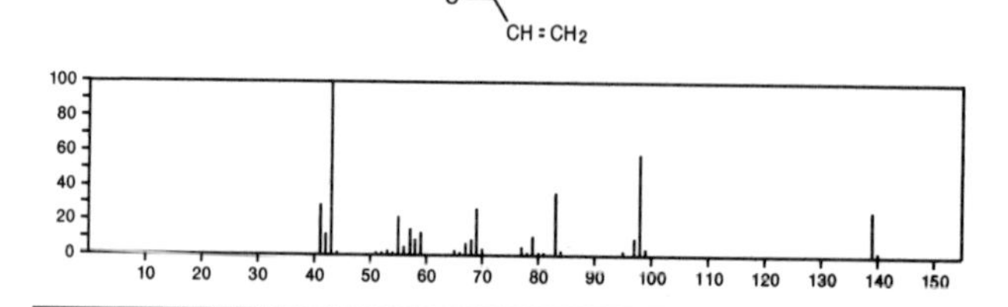

154 $C_9H_{14}O_2$ 4840-76-0
Cyclohexanecarboxylic acid, ethenyl ester

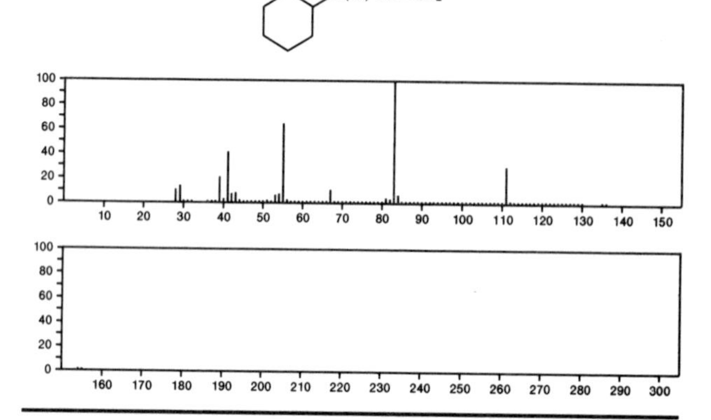

154 $C_9H_{14}O_2$ 4893-16-7
Bicyclo[2.2.2]octanone, 4-methoxy-

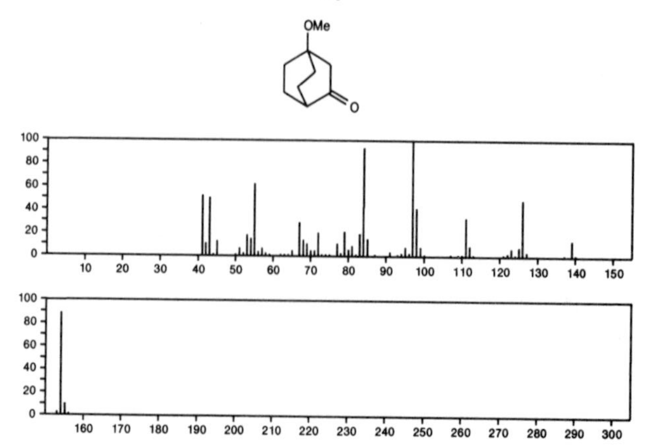

154 $C_9H_{14}O_2$ 6493-79-4

3-Cyclohexene-1-carboxylic acid, 4-methyl-, methyl ester

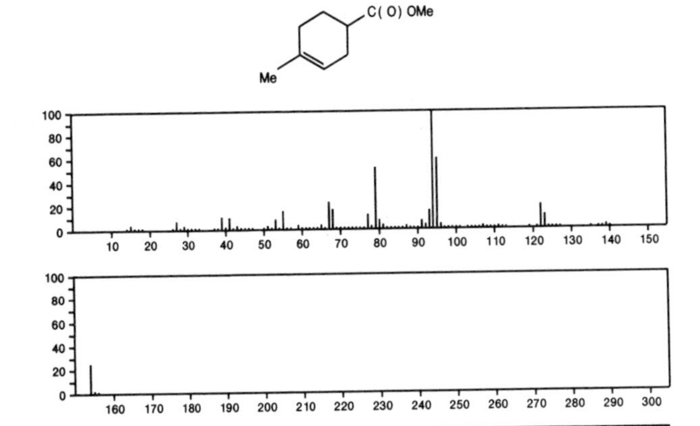

154 $C_9H_{14}O_2$ 7140-60-5

1,4-Dioxaspiro[4.6]undec-7-ene

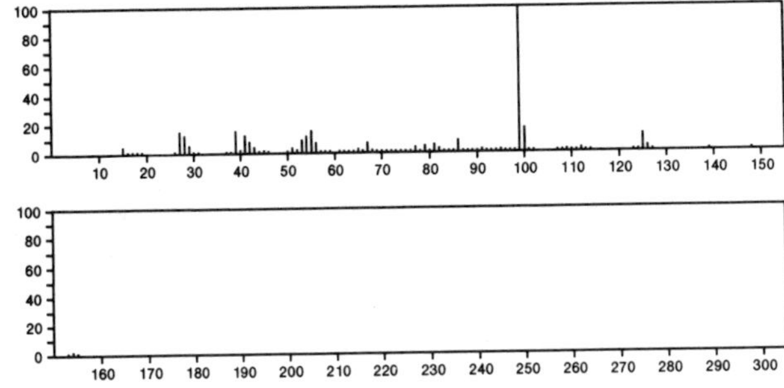

154 $C_9H_{14}O_2$ 7196-96-5

1,4-Benzodioxin, 4a,5,6,7,8,8a-hexahydro-2-methyl-, *trans-*

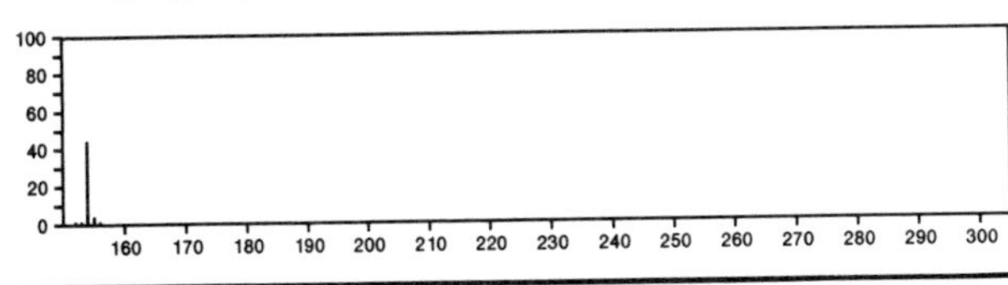

154 $C_9H_{14}O_2$ 7229-32-5

Cyclohexanol, 2-(2-propynyloxy)-, *trans-*

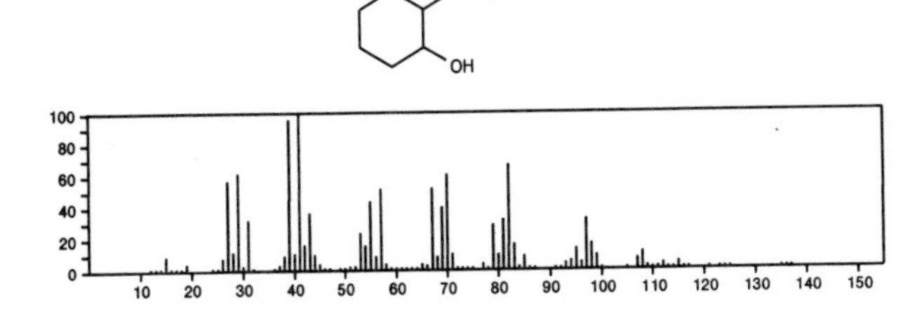

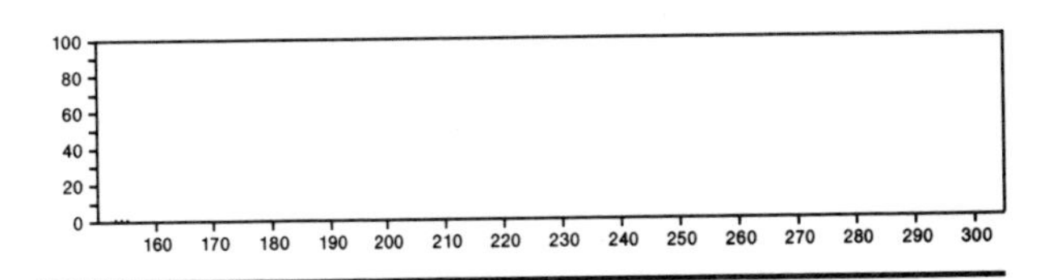

154 $C_9H_{14}O_2$ 10276-21-8

7-Oxabicyclo[4.1.0]heptan-2-one, 4,4,6-trimethyl-

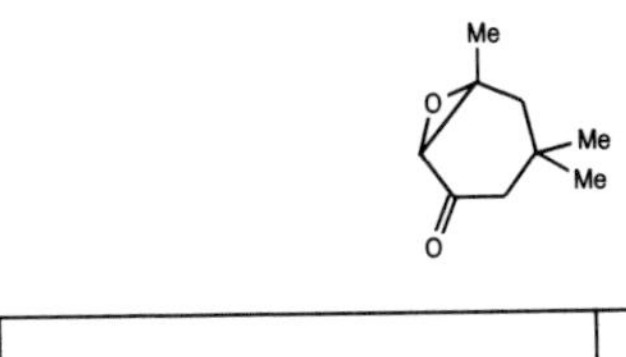

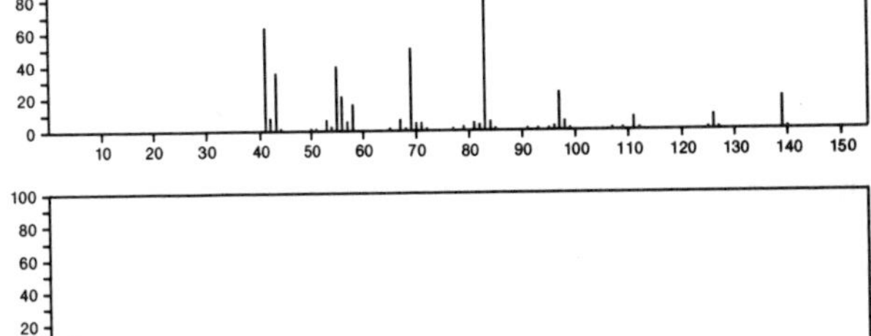

154 $C_9H_{14}O_2$ 18458-50-9

7-Octynoic acid, methyl ester

HC≡C(CH2) 5 C(O) OMe

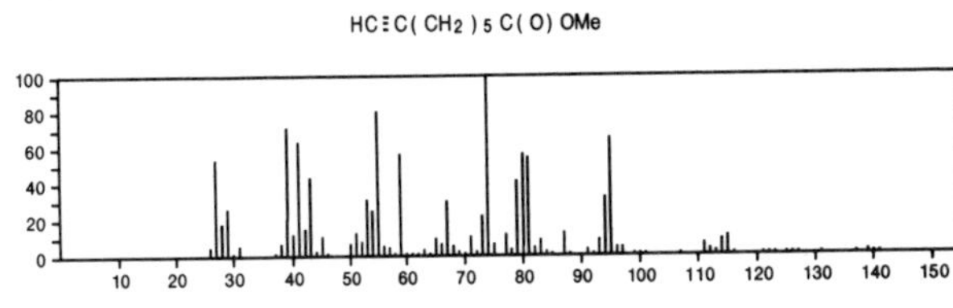

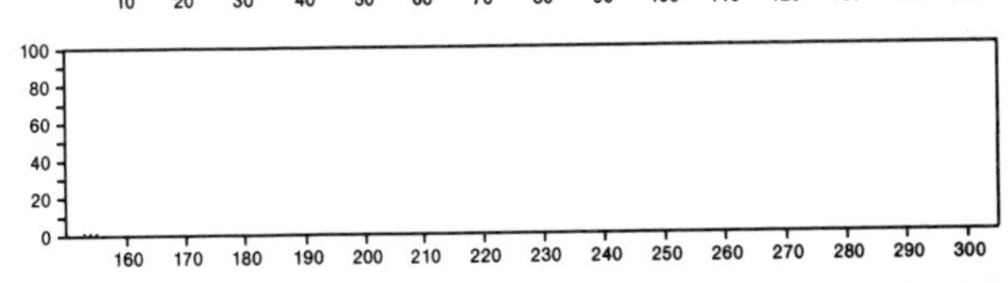

154 $C_9H_{14}O_2$ 20547-99-3

1,4-Cyclohexanedione, 2,2,6-trimethyl-

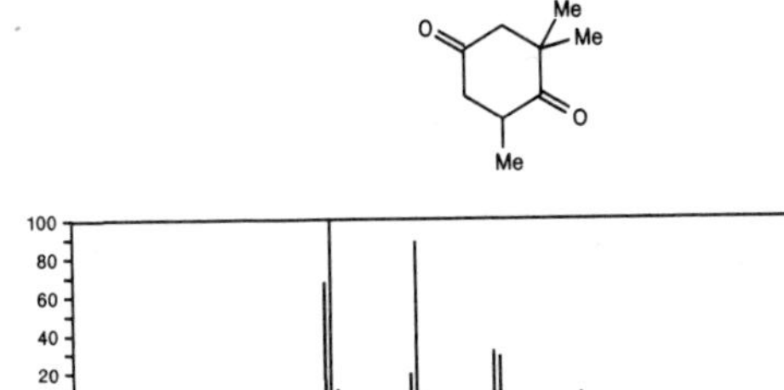

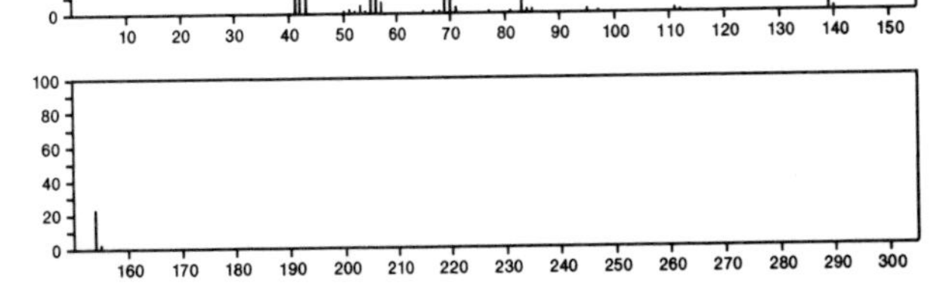

154 $C_9H_{14}O_2$ 23153-81-3
1,4-Dioxaspiro[4.5]decane, 6-methylene-

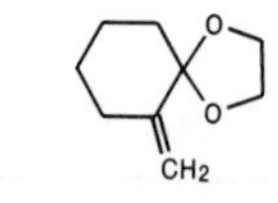

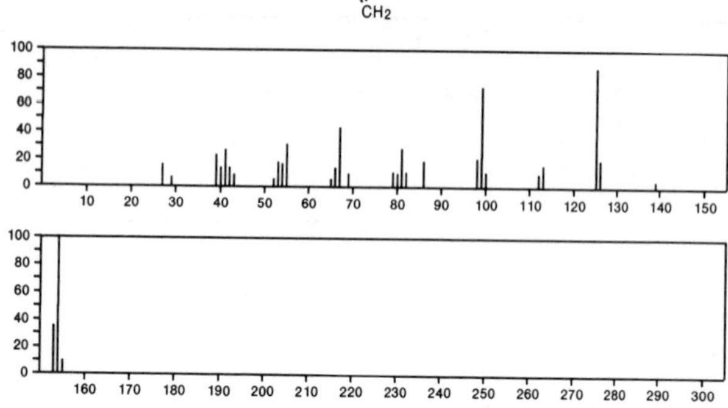

154 $C_9H_{14}O_2$ 30964-01-3
8-Nonynoic acid

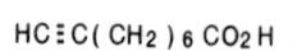

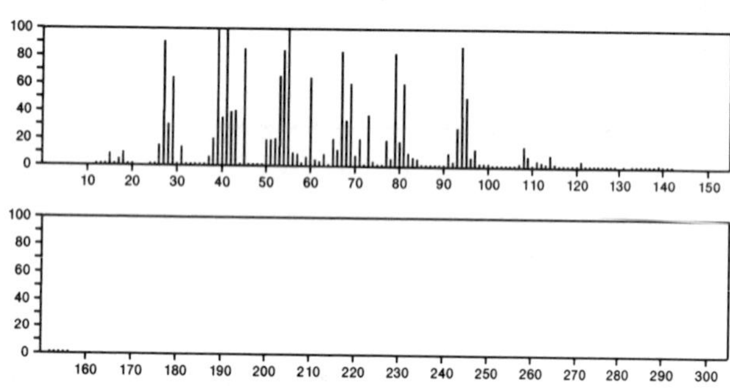

154 $C_9H_{14}O_2$ 34640-76-1
Bicyclo[2.2.1]heptan-2-ol, acetate

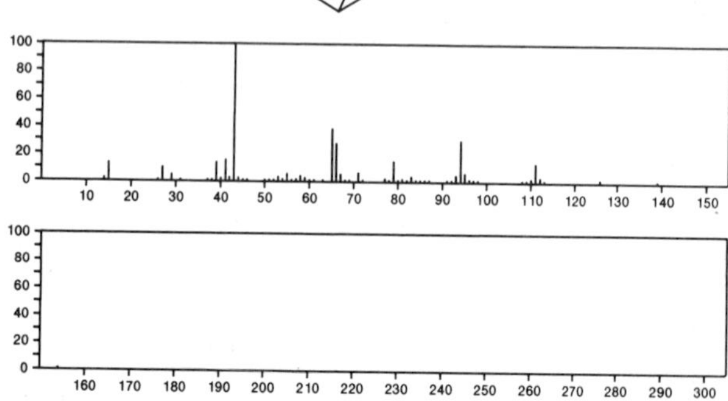

154 $C_9H_{14}O_2$ 38653-34-8
1,4-Benzodioxin, octahydro-2-methylene-, *trans*-

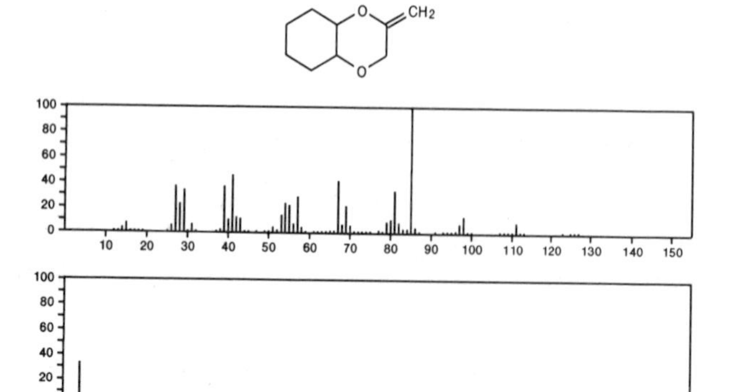

154 $C_9H_{14}O_2$ 40365-61-5
2H-Pyran, 2-(3-butynyloxy)tetrahydro-

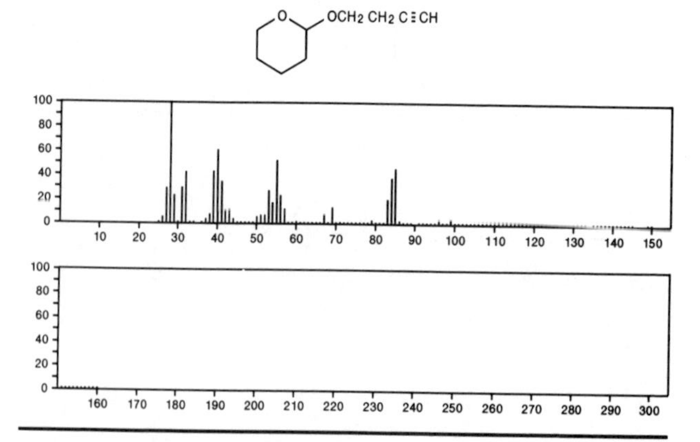

154 $C_9H_{14}O_2$ 41977-59-7
Cyclopropanecarboxylic acid, 3-ethenyl-2,2-dimethyl-, methyl ester, *trans*-

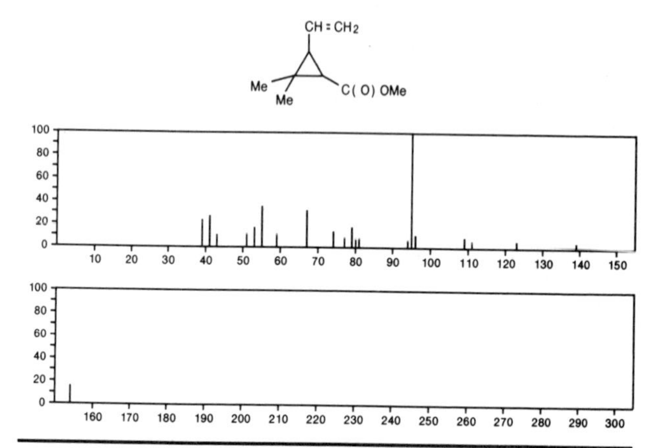

154 $C_9H_{14}O_2$ 45955-66-6
Cyclopentaneacetic acid, ethenyl ester

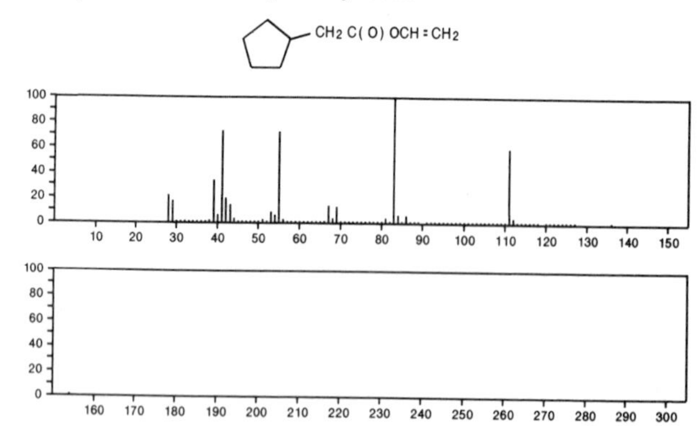

154 $C_9H_{14}O_2$ 50786-09-9
1-Oxaspiro[2.5]octan-4-one, 2,2-dimethyl-

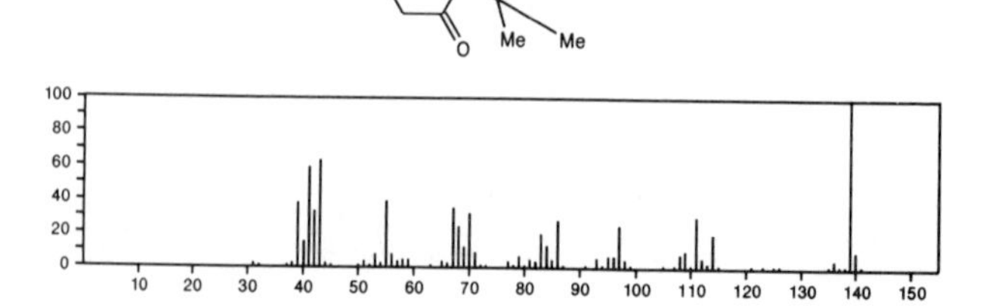

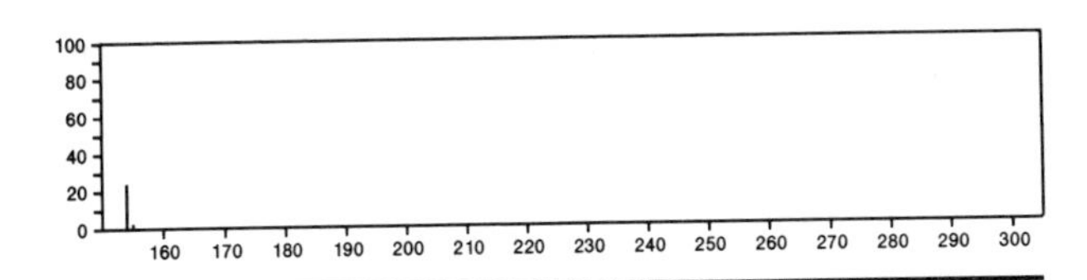

154 $C_9H_{14}O_2$ 54244-72-3
1,3-Cyclopentanedione, 4-butyl-

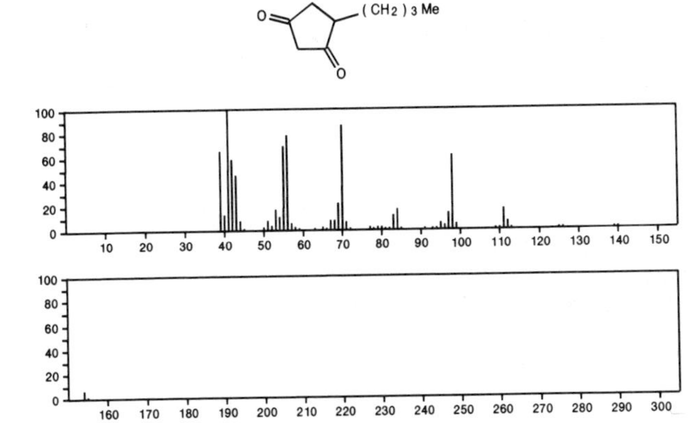

154 $C_9H_{14}O_2$ 55702-63-1
1,3-Benzodioxole, 2-ethenylhexahydro-

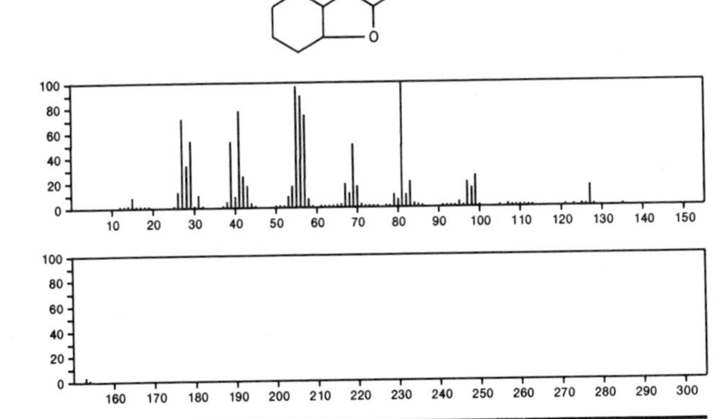

154 $C_9H_{14}O_2$ 55956-38-2
5*H*-Cyclohepta-1,4-dioxin, 2,3,6,7,8,9-hexahydro-

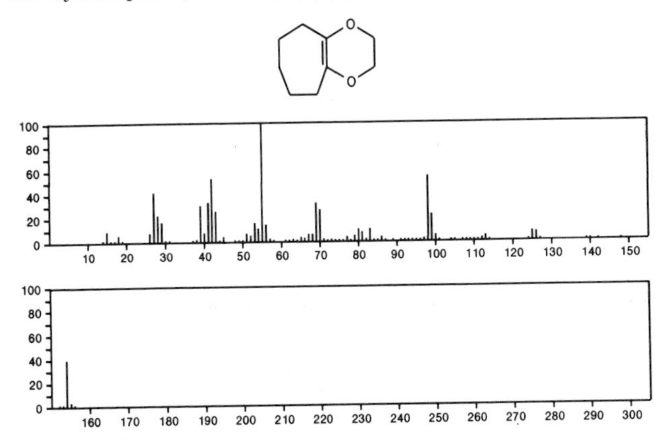

154 $C_9H_{14}O_2$ 55956-39-3
5*H*-Cyclohepta-1,4-dioxin, 2,3,4a,6,7,9a-hexahydro-, *cis*-

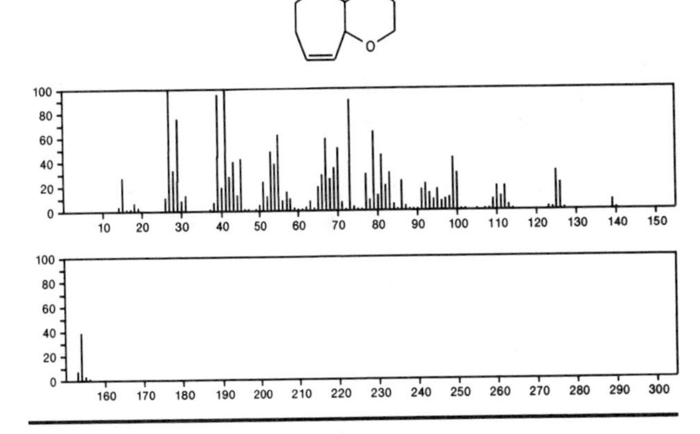

154 $C_9H_{14}O_2$ 55956-40-6
5-Hexen-2-one, 3-ethylidene-1-methoxy-

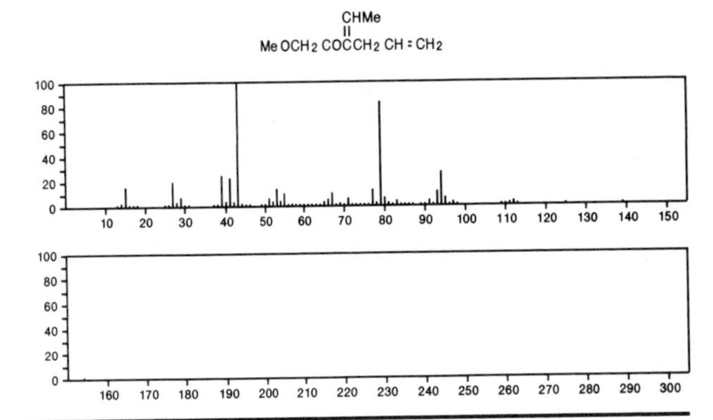

154 $C_9H_{14}O_2$ 55975-33-2
5*H*-Cyclohepta-1,4-dioxin, 2,3,4a,6,7,9a-hexahydro-, *trans*-

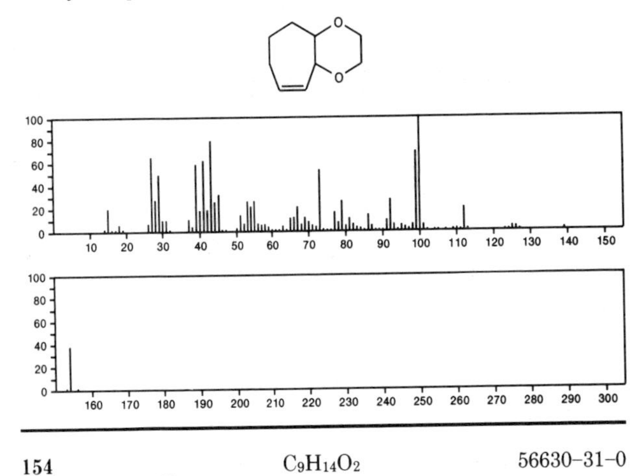

154 $C_9H_{14}O_2$ 56630-31-0
6-Nonynoic acid

HO2C(CH2)4C≡CEt

154 $C_9H_{14}O_2$ 56630–32–1
7–Nonynoic acid

154 $C_9H_{14}O_2$ 56630–33–2
3–Nonynoic acid

154 $C_9H_{14}O_2$ 56630–34–3
5–Nonynoic acid

154 $C_9H_{14}S$ 281–25–4
2–Thiatricyclo[3.3.1.1^{3,7}]decane

154 $C_9H_{14}S$ 30221–54–6
2–Cyclopentene–1–thione, 2,3,4,4–tetramethyl–

154 $C_9H_{14}S$ 30221–55–7
2–Cyclohexene–1–thione, 3,5,5–trimethyl–

154 $C_9H_{14}S$ 33312–98–0
Bicyclo[2.2.1]heptane–2–thione, 3,3–dimethyl–

154 $C_9H_{14}S$ 38693–71–9
2–Cyclohexene–1–thione, 3,6,6–trimethyl–

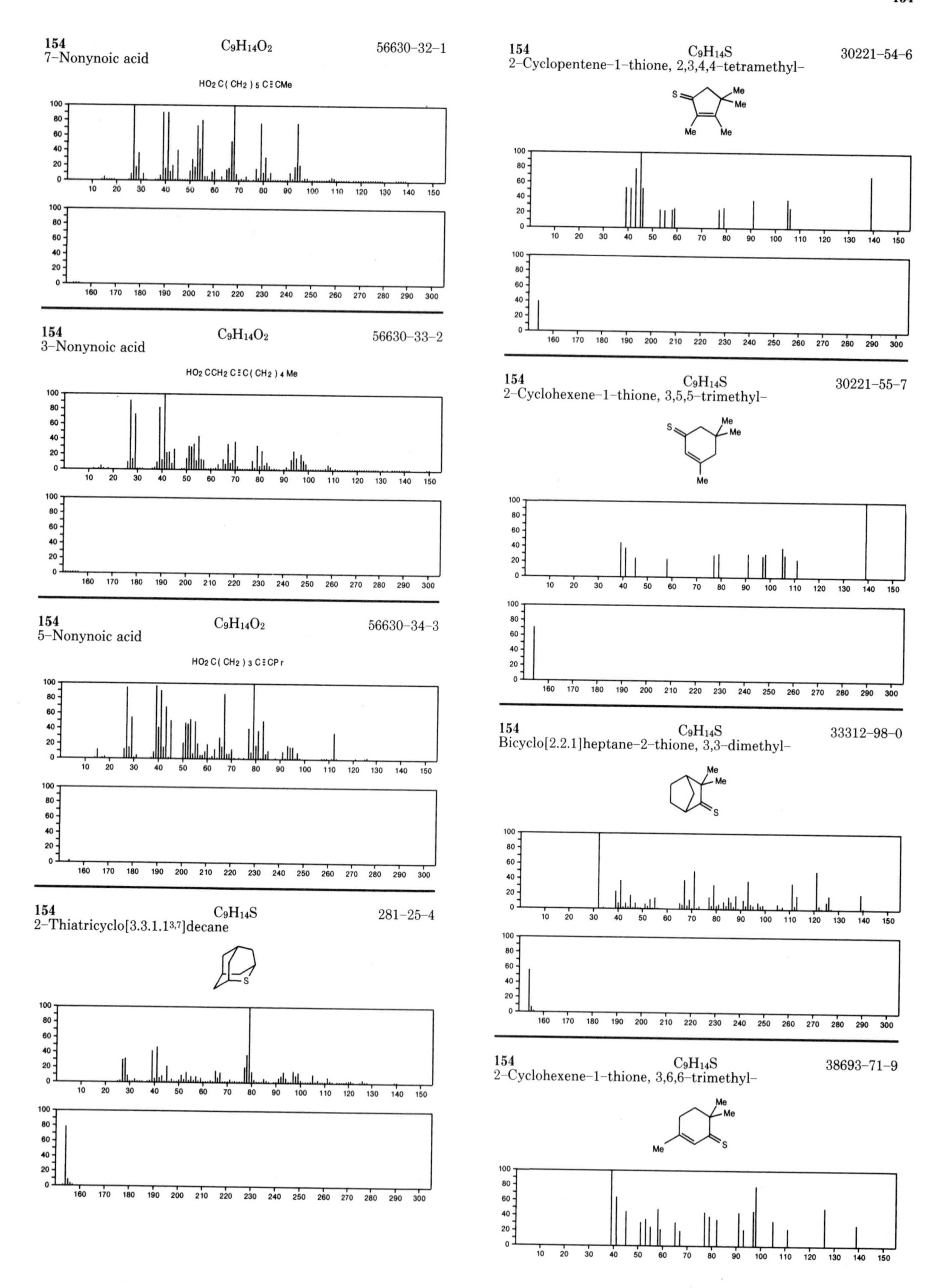

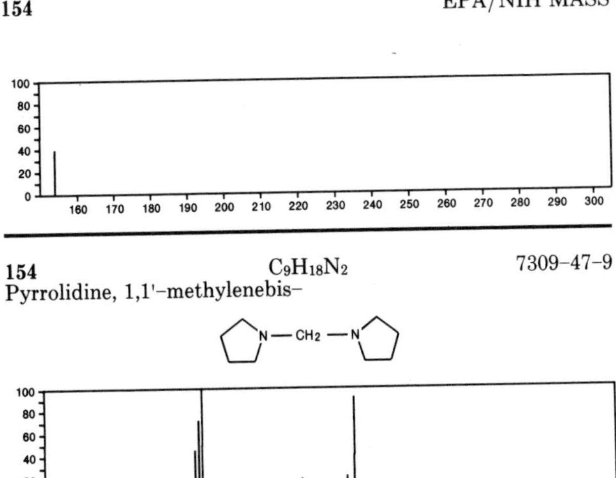

154 $C_9H_{18}N_2$ 7309-47-9
Pyrrolidine, 1,1'-methylenebis-

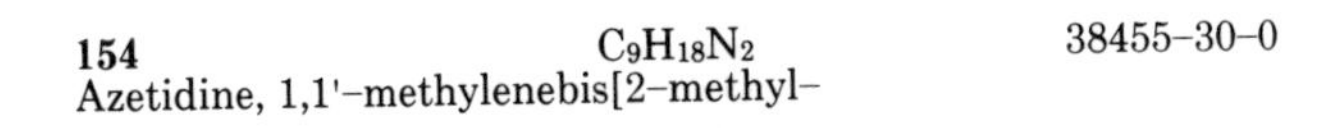

154 $C_9H_{18}N_2$ 38455-30-0
Azetidine, 1,1'-methylenebis[2-methyl-

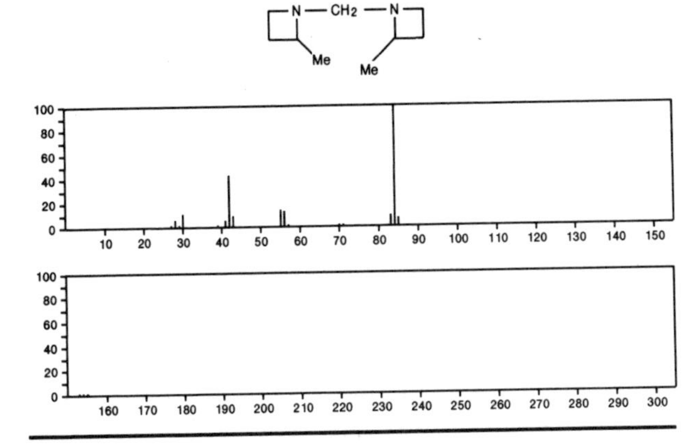

154 $C_9H_{18}Si$ 3844-94-8
Silane, 1-hexynyltrimethyl-

Me3SiC≡C(CH2)3Me

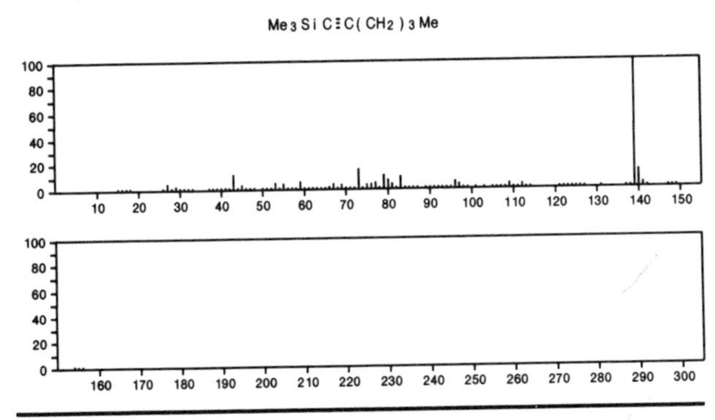

154 $C_9H_{18}Si$ 17874-17-8
Silane, 1-cyclohexen-1-yltrimethyl-

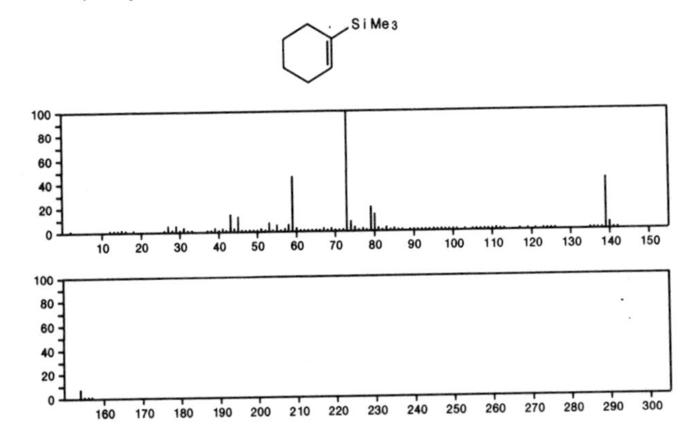

154 $C_9H_{18}Si$ 40934-71-2
Silane, 2-cyclohexen-1-yltrimethyl-

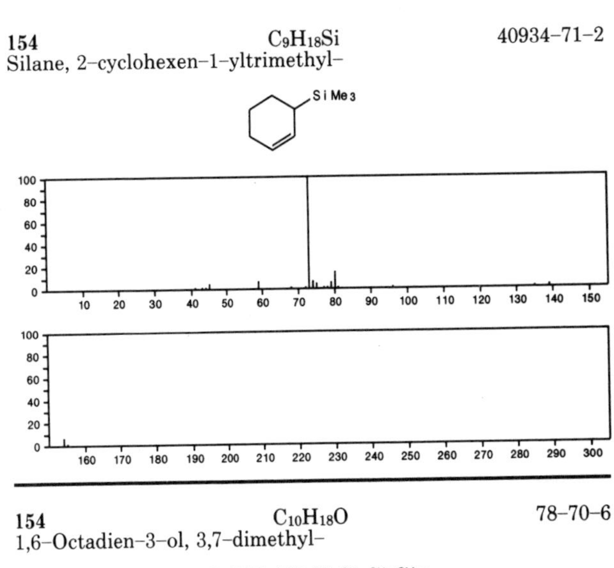

154 $C_{10}H_{18}O$ 78-70-6
1,6-Octadien-3-ol, 3,7-dimethyl-

H2C=CHCMe(OH)CH2CH2CH=CMe2

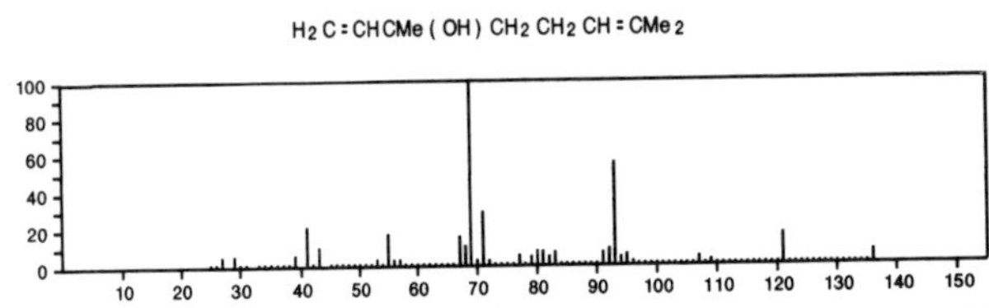

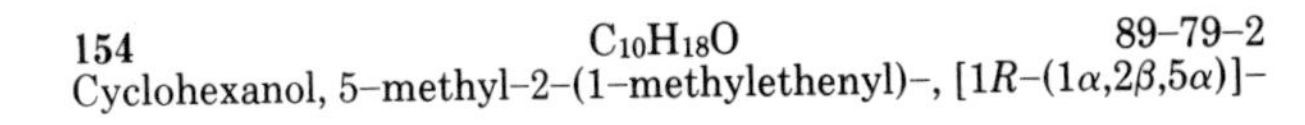

154 $C_{10}H_{18}O$ 89-79-2
Cyclohexanol, 5-methyl-2-(1-methylethenyl)-, [1*R*-(1α,2β,5α)]-

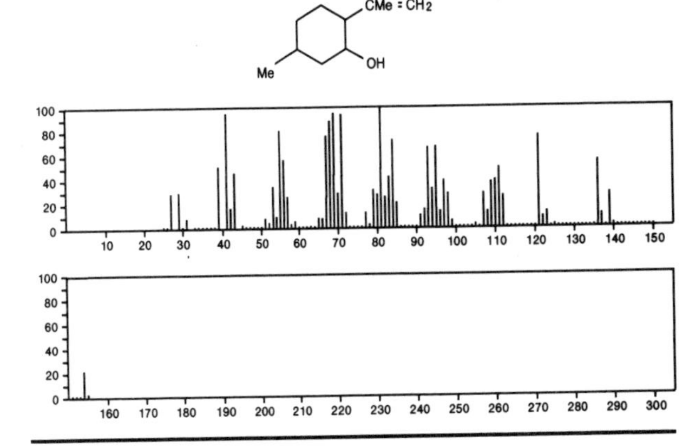

154 $C_{10}H_{18}O$ 89-80-5
Cyclohexanone, 5-methyl-2-(1-methylethyl)-, *trans*-

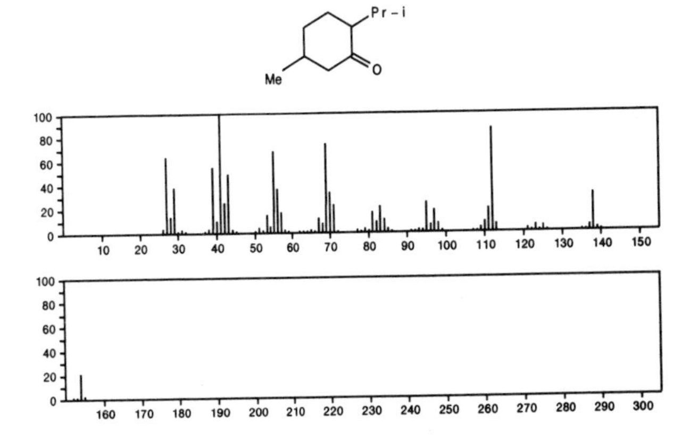

154 $C_{10}H_{18}O$ 98-53-3
Cyclohexanone, 4-(1,1-dimethylethyl)-

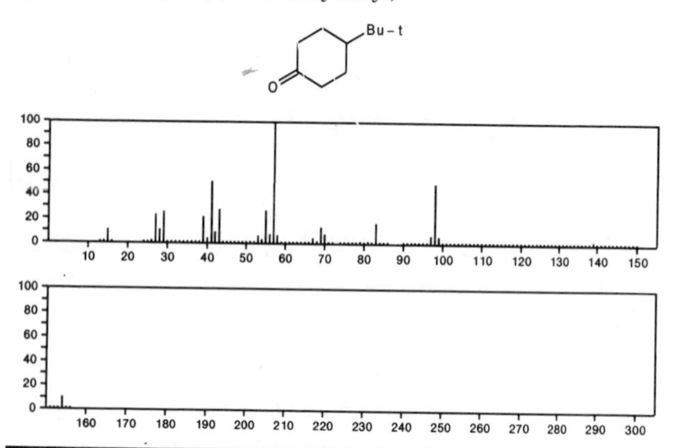

154 $C_{10}H_{18}O$ 98-55-5
3-Cyclohexene-1-methanol, α,α,4-trimethyl-

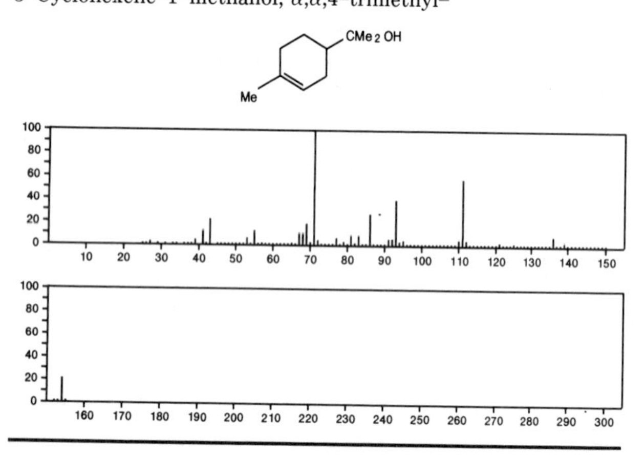

154 $C_{10}H_{18}O$ 106-23-0
6-Octenal, 3,7-dimethyl-

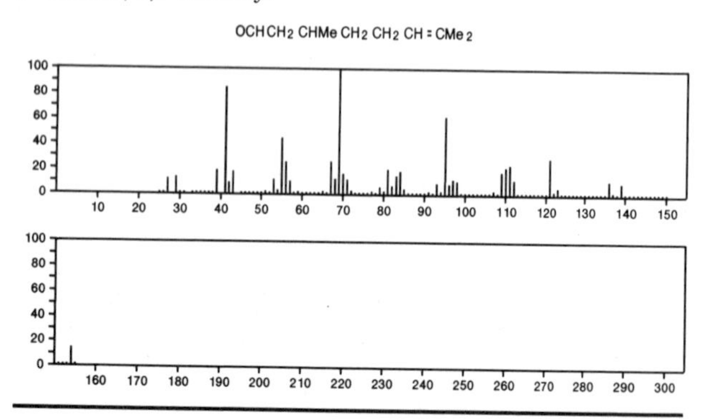

154 $C_{10}H_{18}O$ 106-24-1
2,6-Octadien-1-ol, 3,7-dimethyl-, (*E*)-

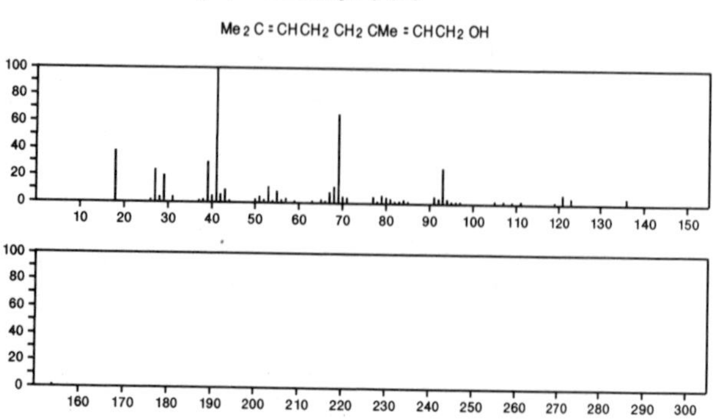

154 $C_{10}H_{18}O$ 106-25-2
2,6-Octadien-1-ol, 3,7-dimethyl-, (*Z*)-

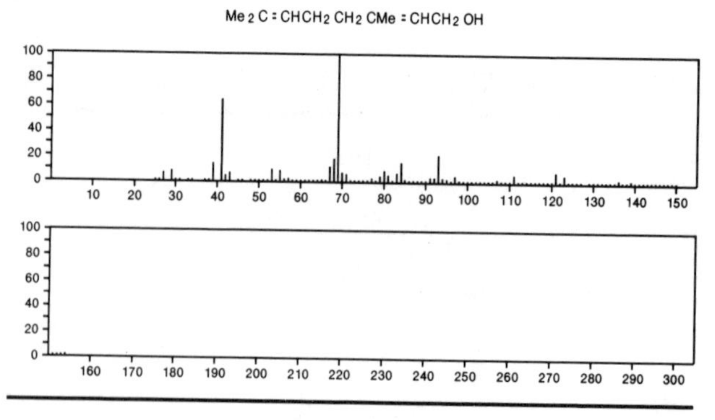

154 $C_{10}H_{18}O$ 124-76-5
Bicyclo[2.2.1]heptan-2-ol, 1,7,7-trimethyl-, *exo*-

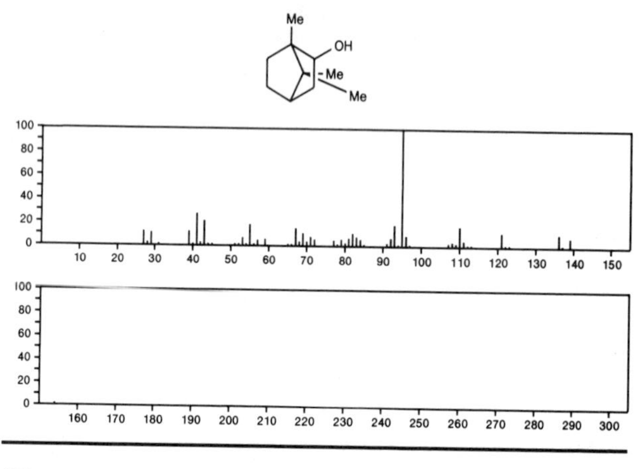

154 $C_{10}H_{18}O$ 138-87-4
Cyclohexanol, 1-methyl-4-(1-methylethenyl)-

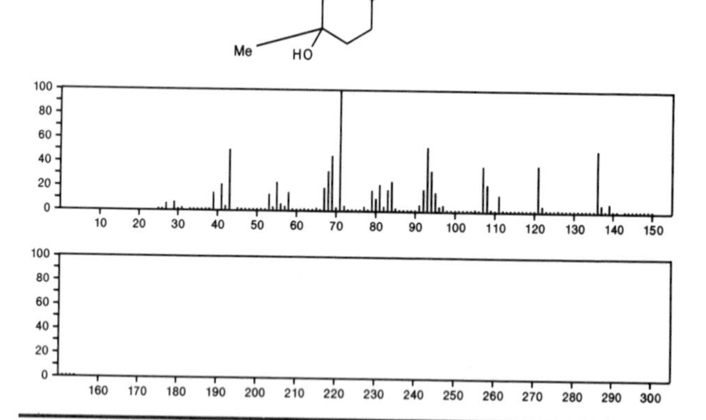

154 $C_{10}H_{18}O$ 464-45-9
Bicyclo[2.2.1]heptan-2-ol, 1,7,7-trimethyl-, (1*S*-*endo*)-

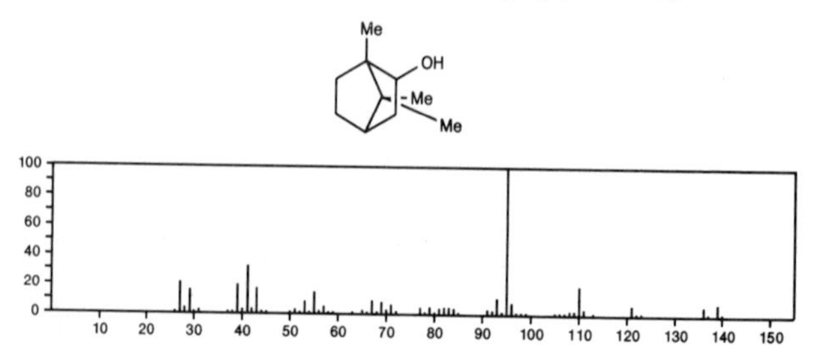

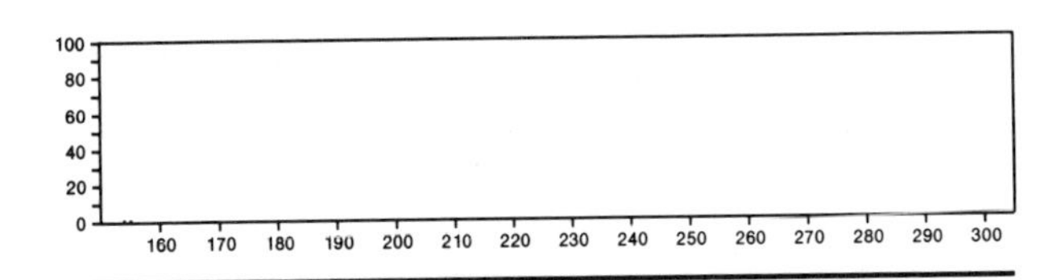

154 $C_{10}H_{18}O$ 470-67-7

7-Oxabicyclo[2.2.1]heptane, 1-methyl-4-(1-methylethyl)-

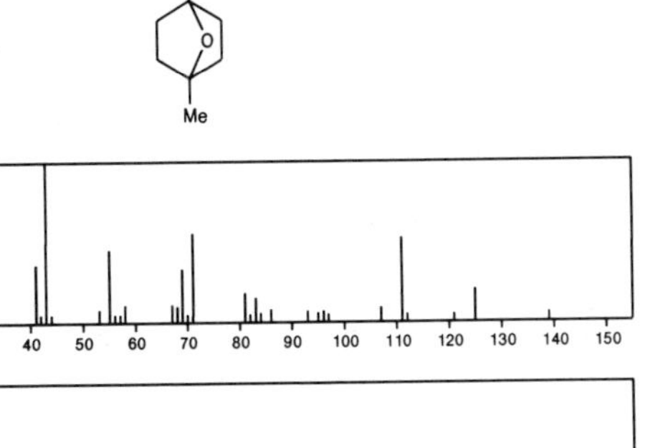

154 $C_{10}H_{18}O$ 470-82-6

2-Oxabicyclo[2.2.2]octane, 1,3,3-trimethyl-

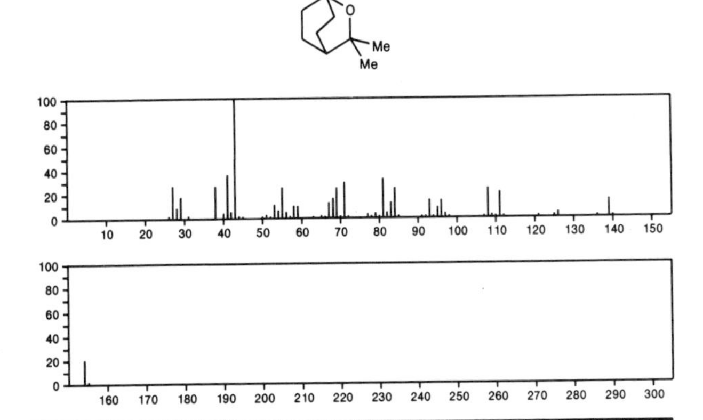

154 $C_{10}H_{18}O$ 491-07-6

Cyclohexanone, 5-methyl-2-(1-methylethyl)-, *cis*-

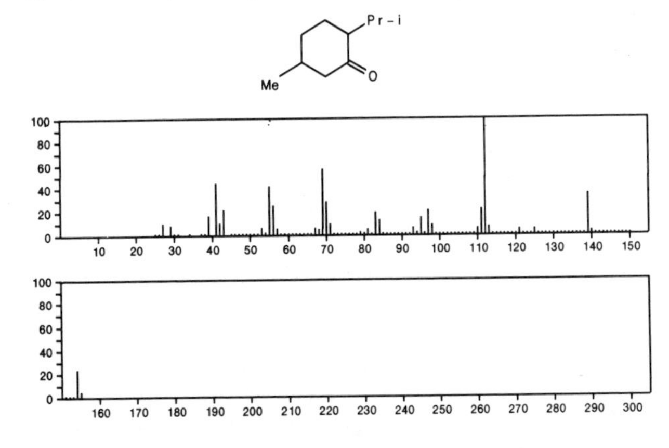

154 $C_{10}H_{18}O$ 498-16-8

4-Hexen-1-ol, 5-methyl-2-(1-methylethenyl)-, (*R*)-

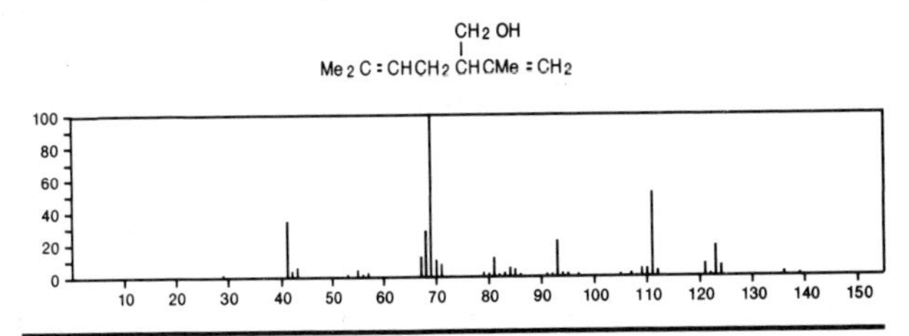

154 $C_{10}H_{18}O$ 507-70-0

Bicyclo[2.2.1]heptan-2-ol, 1,7,7-trimethyl-, *endo*-

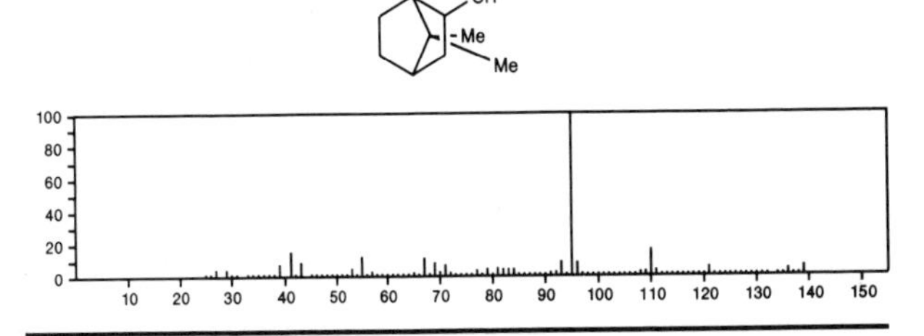

154 $C_{10}H_{18}O$ 513-23-5

Bicyclo[3.1.0]hexan-3-ol, 4-methyl-1-(1-methylethyl)-

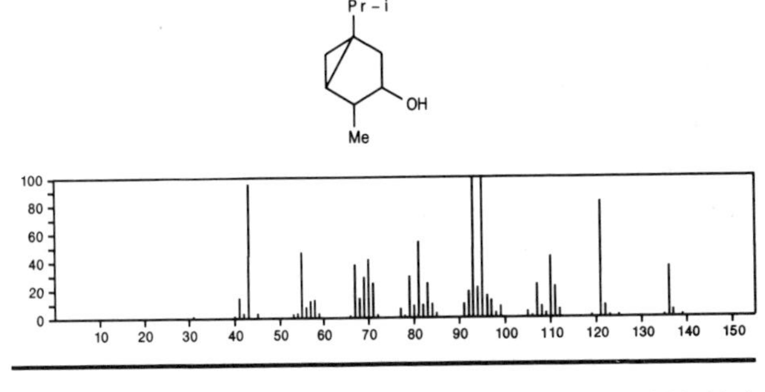

154 $C_{10}H_{18}O$ 536-30-1

2-Cyclohexen-1-ol, 2-methyl-5-(1-methylethyl)-, (1*S*-*cis*)-

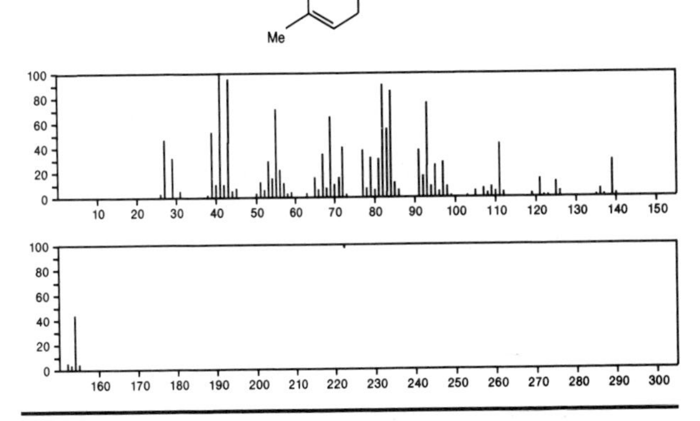

154 $C_{10}H_{18}O$ 543-39-5

7-Octen-2-ol, 2-methyl-6-methylene-

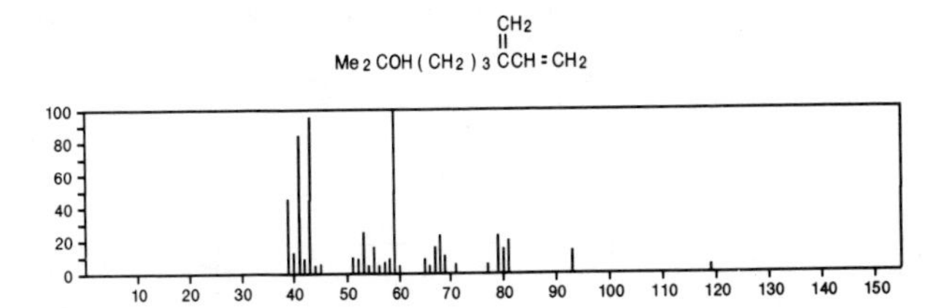

154 $C_{10}H_{18}O$ 546–79–2
Bicyclo[3.1.0]hexan–2–ol, 2–methyl–5–(1–methylethyl)–

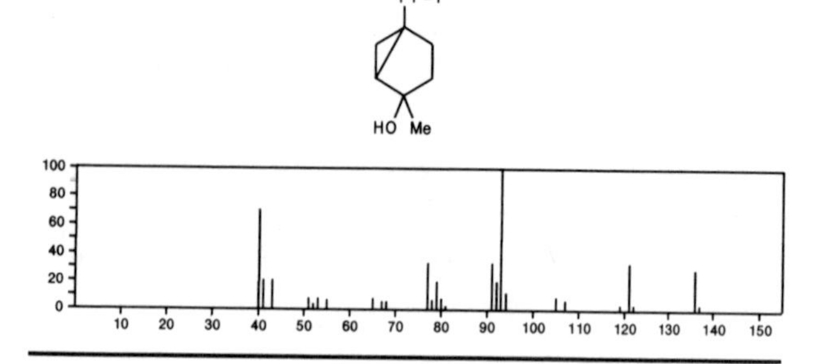

154 $C_{10}H_{18}O$ 562–74–3
3–Cyclohexen–1–ol, 4–methyl–1–(1–methylethyl)–

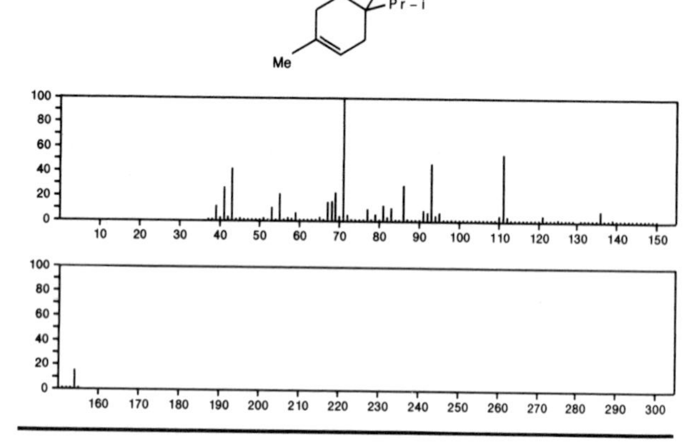

154 $C_{10}H_{18}O$ 624–15–7
2,6–Octadien–1–ol, 3,7–dimethyl–

$Me_2C:CHCH_2CH_2CMe:CHCH_2OH$

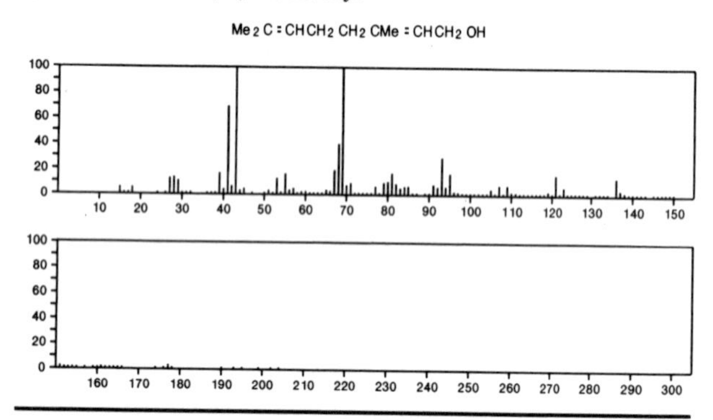

154 $C_{10}H_{18}O$ 1196–31–2
Cyclohexanone, 5–methyl–2–(1–methylethyl)–, (2*R*–*cis*)–

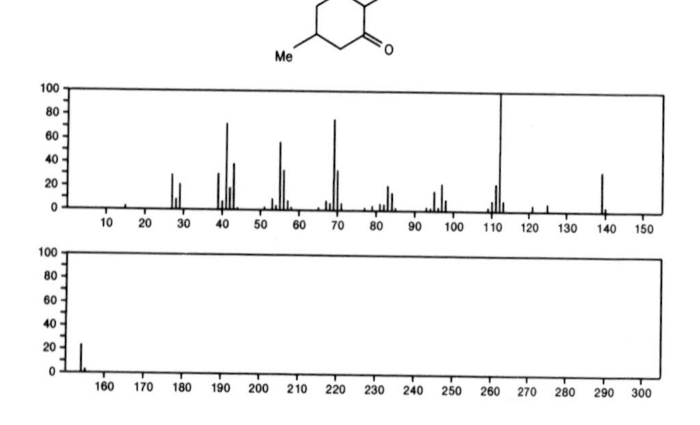

154 $C_{10}H_{18}O$ 1502–06–3
Cyclodecanone

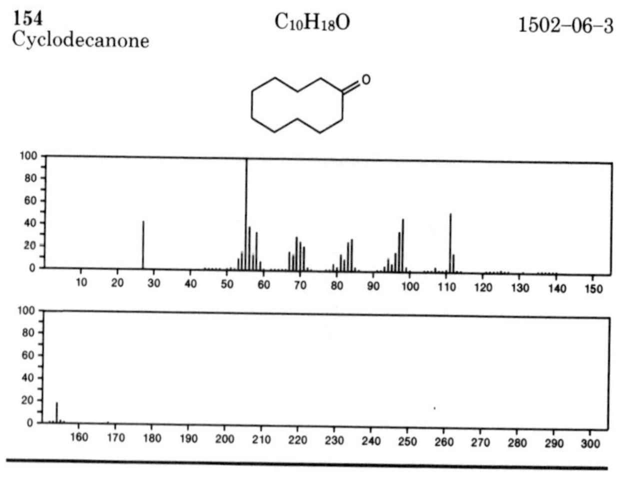

154 $C_{10}H_{18}O$ 1632–73–1
Bicyclo[2.2.1]heptan–2–ol, 1,3,3–trimethyl–

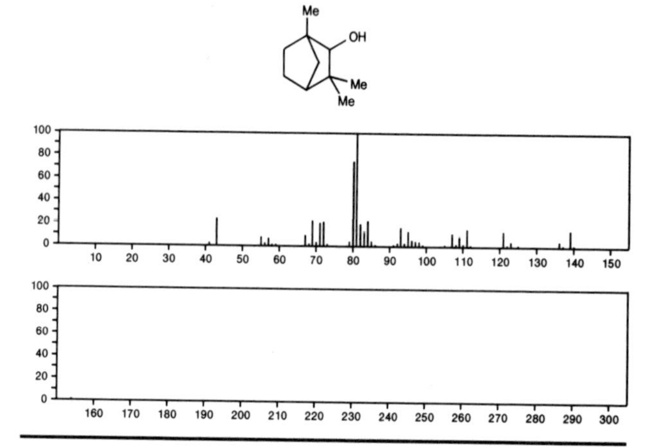

154 $C_{10}H_{18}O$ 2316–85–0
2–Butanone, 4–cyclohexyl–

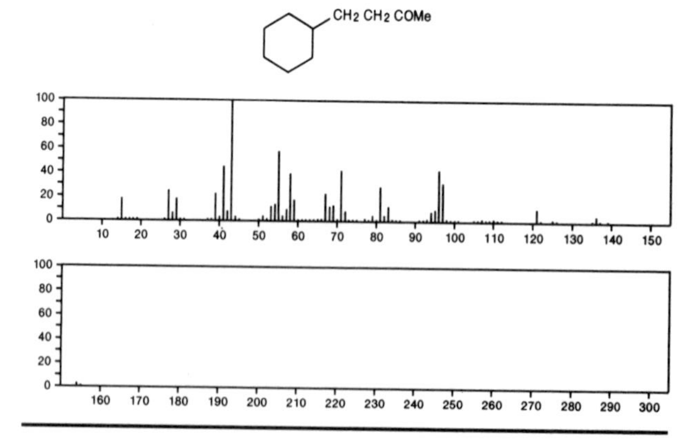

154 $C_{10}H_{18}O$ 2903–23–3
2–Nonen–4–one, 2–methyl–

$Me(CH_2)_4COCH:CMe_2$

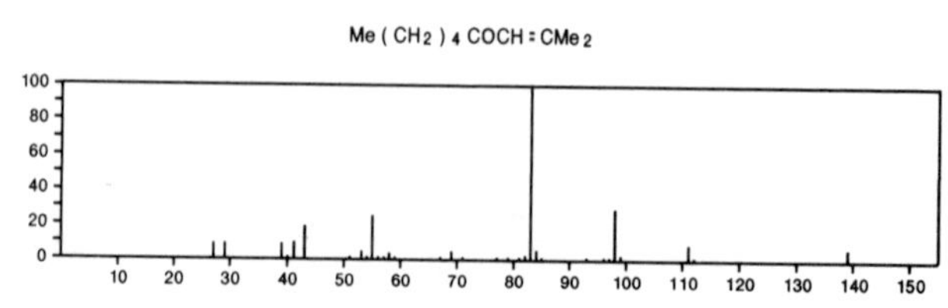

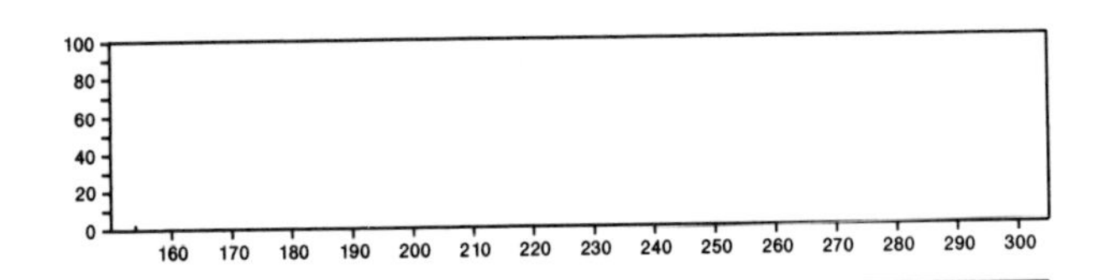

154 $C_{10}H_{18}O$ 4028-58-4
Cyclopentanol, 1,2-dimethyl-3-(1-methylethenyl)-, [1*R*-(1α,2β,3β)]-

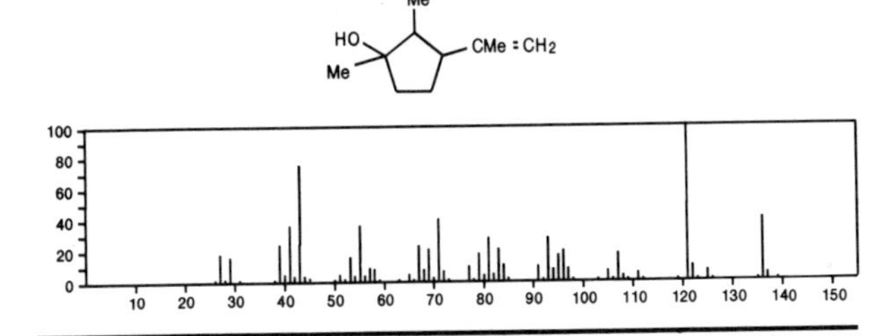

154 $C_{10}H_{18}O$ 4028-59-5
Cyclopentanol, 1,2-dimethyl-3-(1-methylethenyl)-, [1*R*-(1α,2α,3β)]-

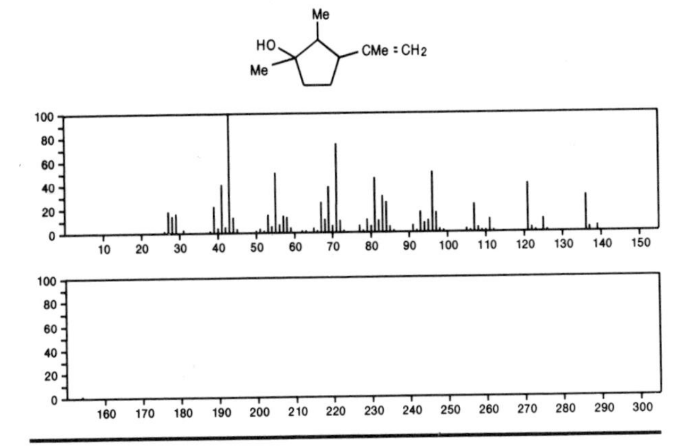

154 $C_{10}H_{18}O$ 4028-60-8
Cyclopentanol, 1,2-dimethyl-3-(1-methylethenyl)-, [1*R*-(1α,2α,3α)]-

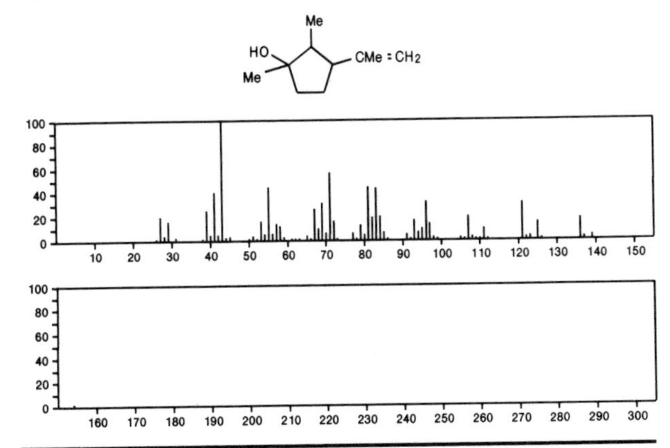

154 $C_{10}H_{18}O$ 4099-07-4
Cyclopentanol, 1,2-dimethyl-3-(1-methylethenyl)-, [1*R*-(1α,2β,3α)]-

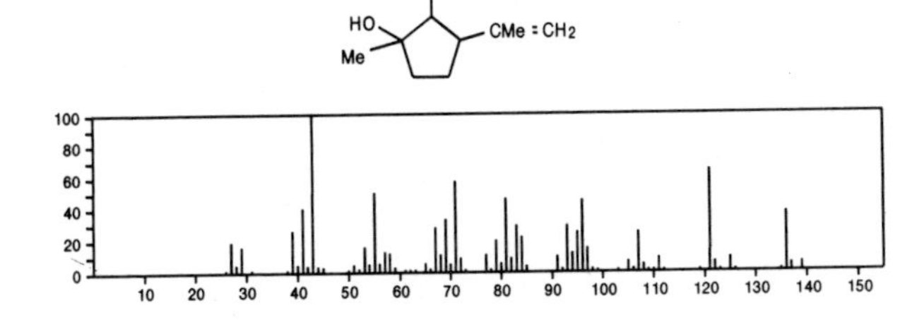

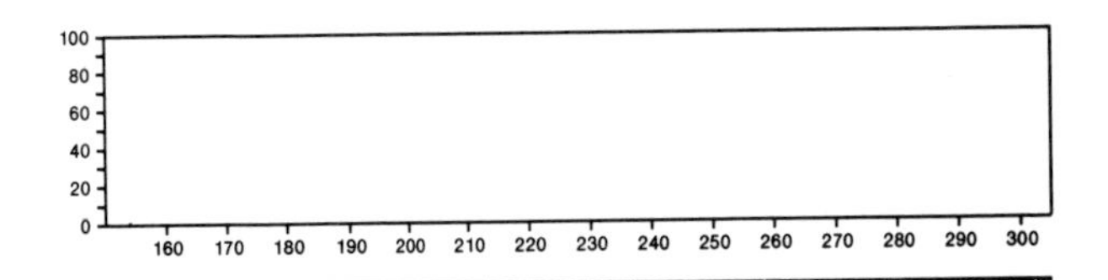

154 $C_{10}H_{18}O$ 4884-25-7
[1,1'-Bicyclopentyl]-2-ol

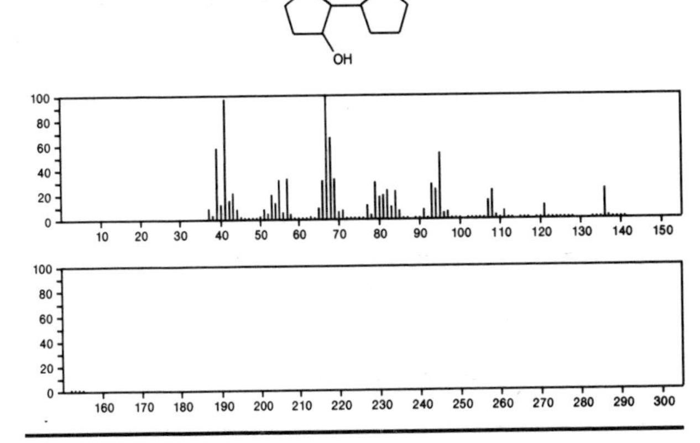

154 $C_{10}H_{18}O$ 4884-28-0
[1,1'-Bicyclopentyl]-1-ol

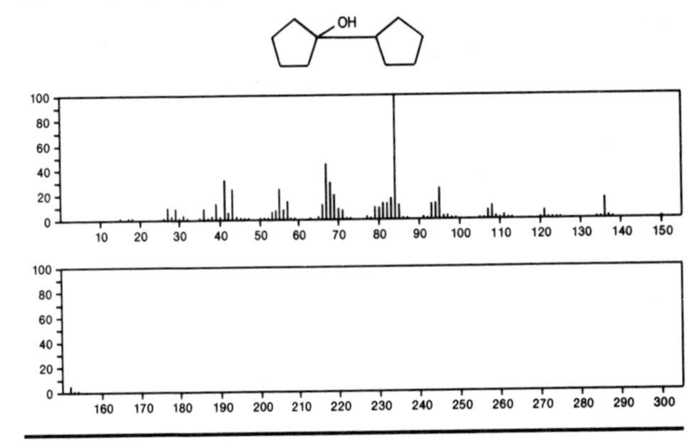

154 $C_{10}H_{18}O$ 5502-99-8
Cyclohexaneethanol, 4-methyl-β-methylene-

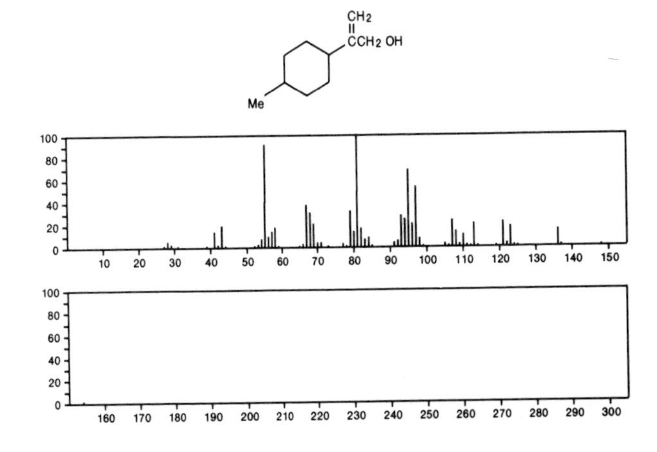

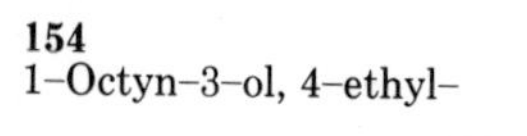

154 $C_{10}H_{18}O$ 5877–42–9
1–Octyn–3–ol, 4–ethyl–

HC≡CCH(OH)CHEt(CH2)3Me

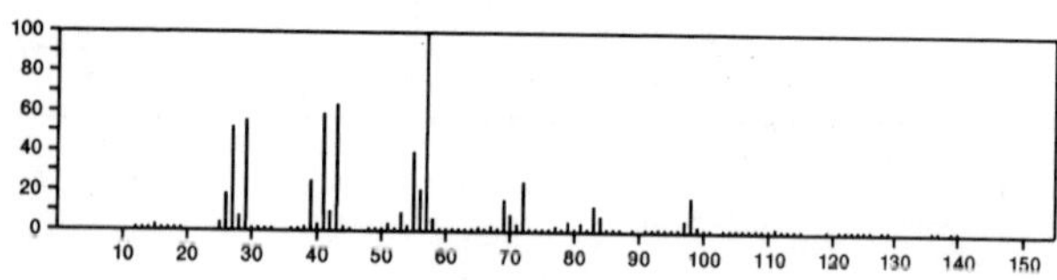

154 $C_{10}H_{18}O$ 6555–95–9
Bicyclo[2.2.2]octane, 1–methoxy–4–methyl–

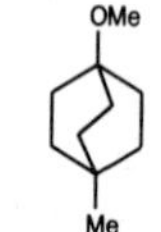

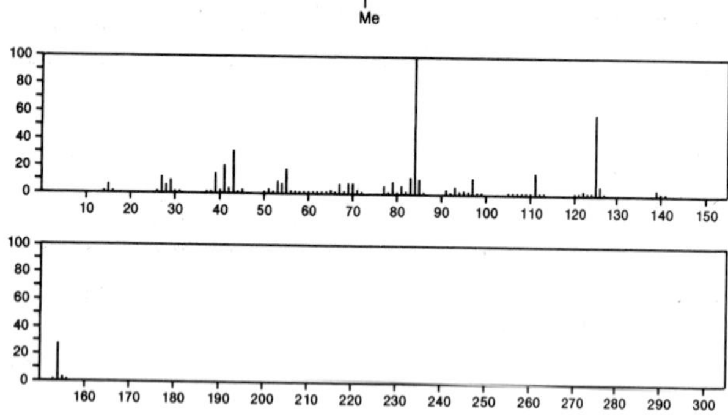

154 $C_{10}H_{18}O$ 7036–98–8
5–Nonen–4–one, 6–methyl–

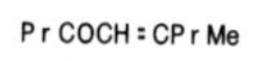

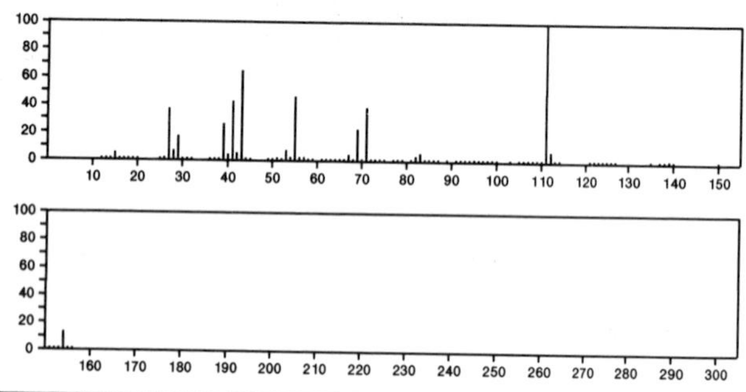

154 $C_{10}H_{18}O$ 10458–14–7
Cyclohexanone, 5–methyl–2–(1–methylethyl)–

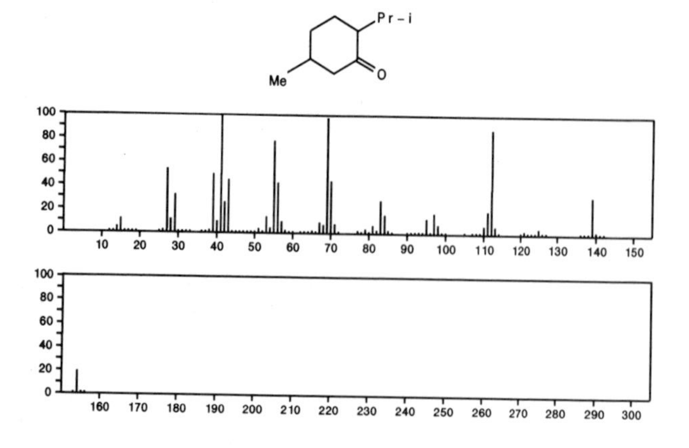

154 $C_{10}H_{18}O$ 10482–56–1
3–Cyclohexene–1–methanol, α,α,4–trimethyl–, (*S*)–

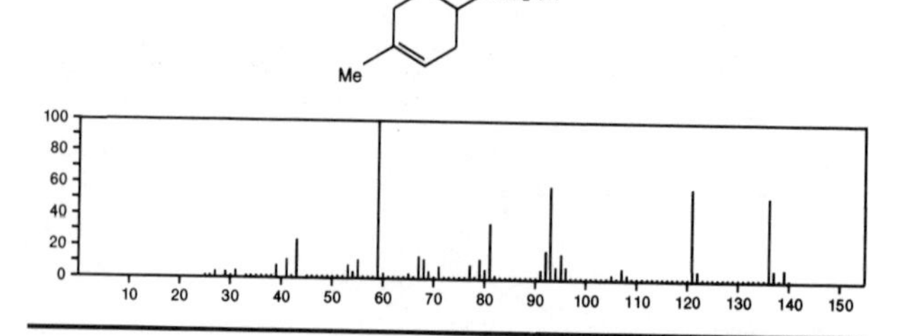

154 $C_{10}H_{18}O$ 10519–33–2
3–Decen–2–one

Me COCH = CH(CH2)5 Me

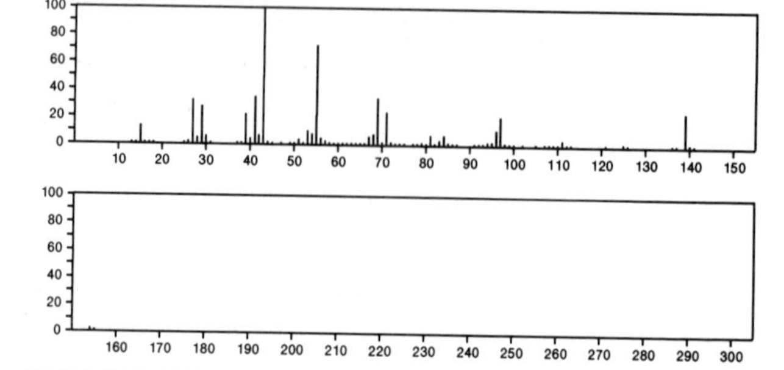

154 $C_{10}H_{18}O$ 13835–30–8
3–Cyclohexene–1–ethanol, β,4–dimethyl–, [*R*–(*R**,*R**)]–

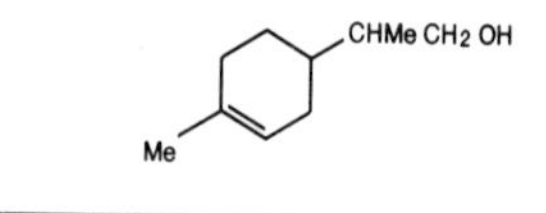

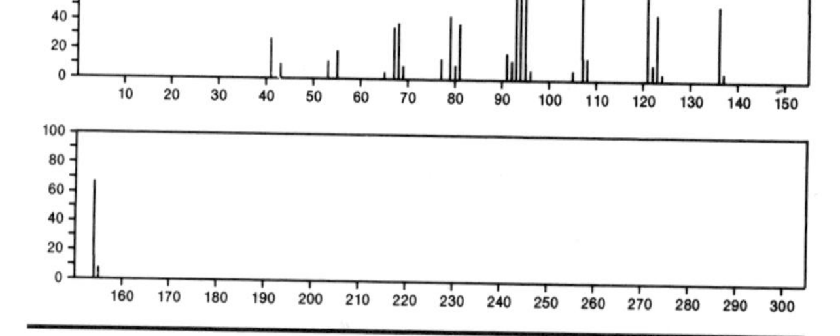

154 $C_{10}H_{18}O$ 15404–56–5
2–Oxabicyclo[2.2.1]heptane, 1,3,3,7–tetramethyl–, (1*R*,4*S*,7*R*)–(+)–

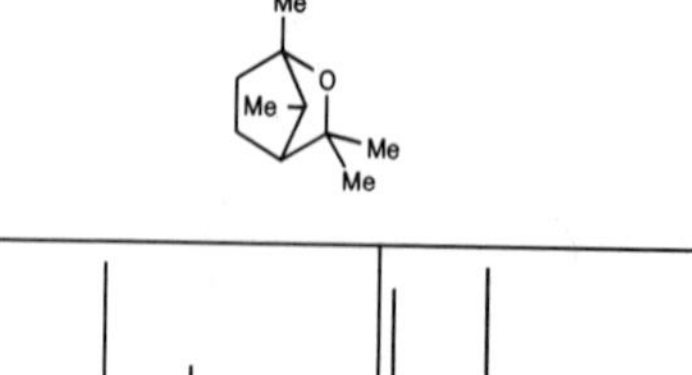

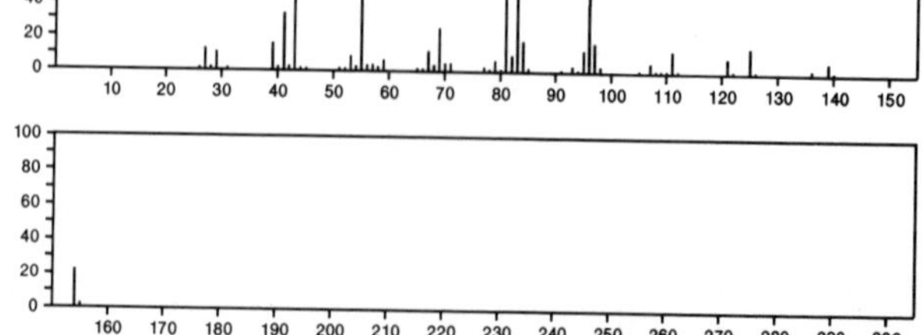

154 $C_{10}H_{18}O$ 15404-57-6
2-Oxabicyclo[2.2.1]heptane, 1,3,3,7-tetramethyl-, (1*R*,4*S*,7*S*)-(+)-

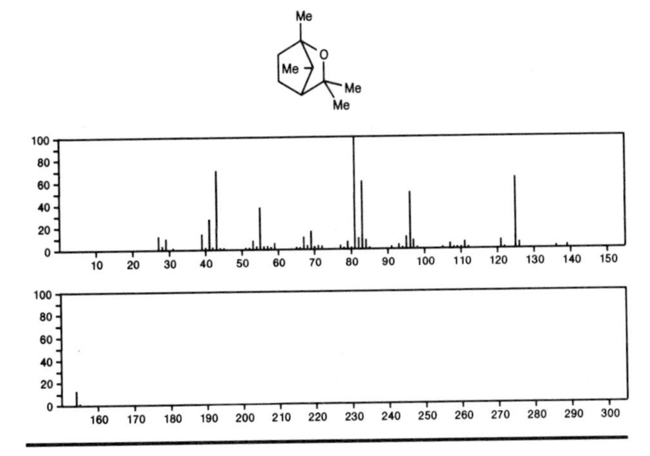

154 $C_{10}H_{18}O$ 16409-43-1
2*H*-Pyran, tetrahydro-4-methyl-2-(2-methyl-1-propenyl)-

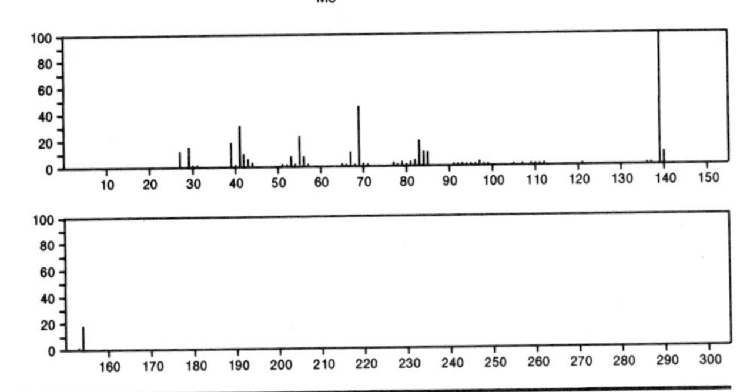

154 $C_{10}H_{18}O$ 16519-67-8
Cyclohexanone, 2,2-diethyl-

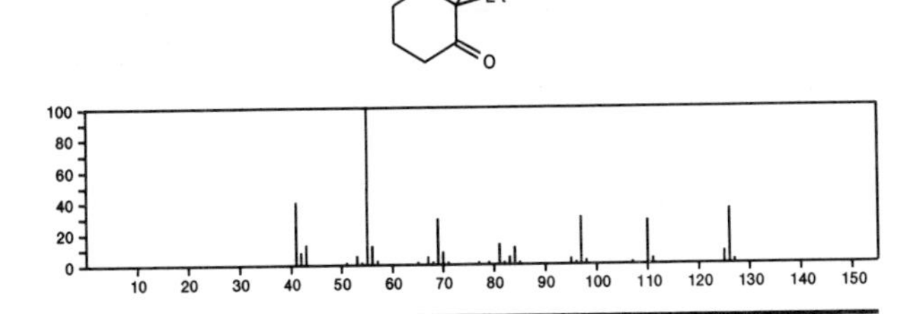

154 $C_{10}H_{18}O$ 16519-68-9
Cyclohexanone, 2,6-diethyl-

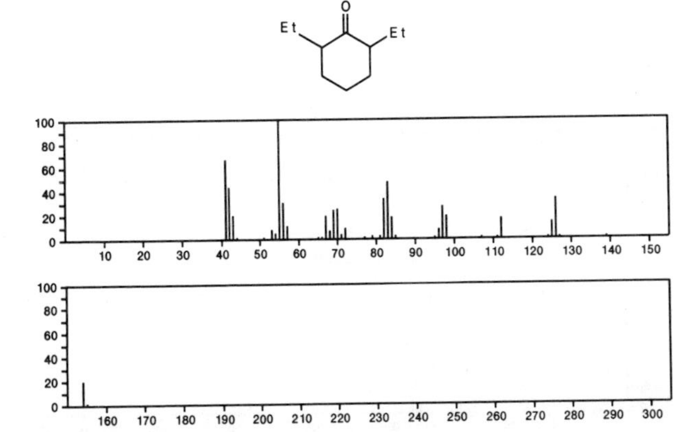

154 $C_{10}H_{18}O$ 17429-42-4
Cyclohexanone, 4-ethyl-3,4-dimethyl-

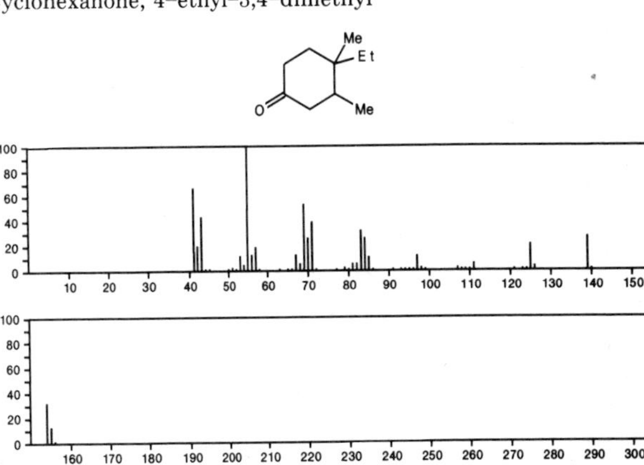

154 $C_{10}H_{18}O$ 17983-22-1
Ketone, methyl 2,2,3-trimethylcyclopentyl

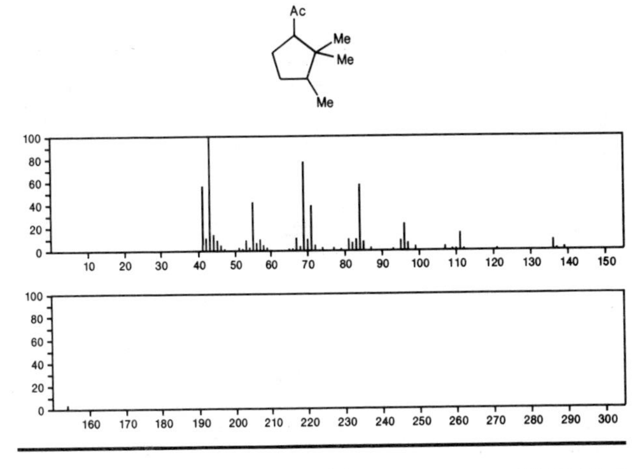

154 $C_{10}H_{18}O$ 17983-26-5
Ketone, 2,2-dimethylcyclohexyl methyl

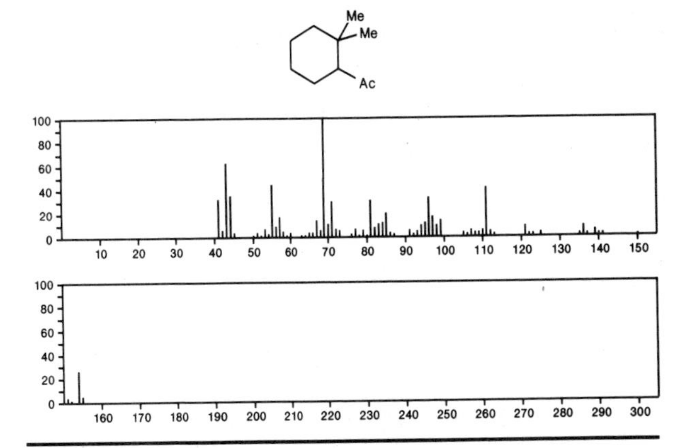

154 $C_{10}H_{18}O$ 18479-68-0
3-Cyclohexene-1-ethanol, β,4-dimethyl-

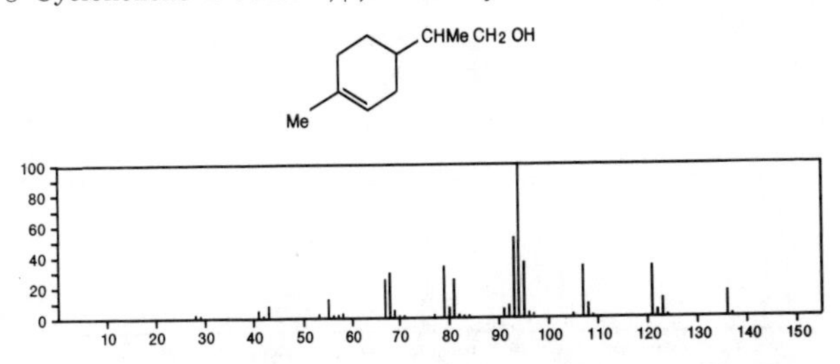

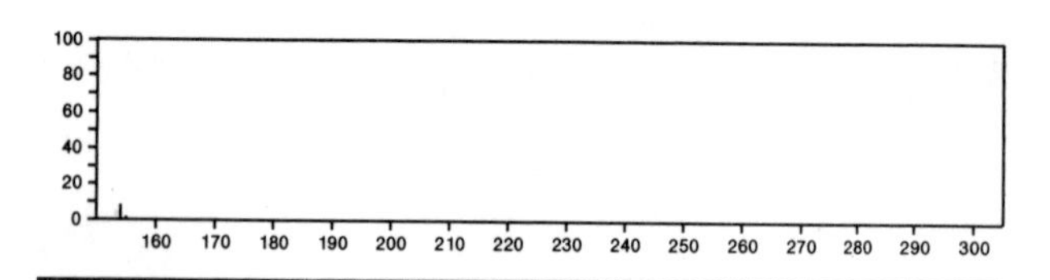

154 $C_{10}H_{18}O$ 18675-33-7
Cyclohexanol, 2-methyl-5-(1-methylethenyl)-, (1α,2α,5β)-

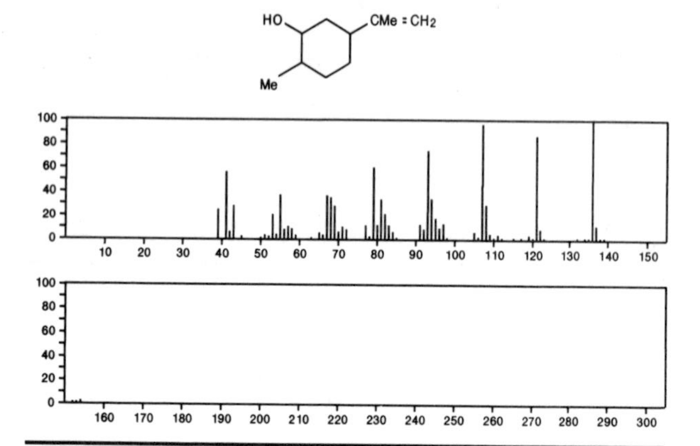

154 $C_{10}H_{18}O$ 22336-76-1
Bicyclo[2.2.1]heptan-2-ol, 4,7,7-trimethyl-

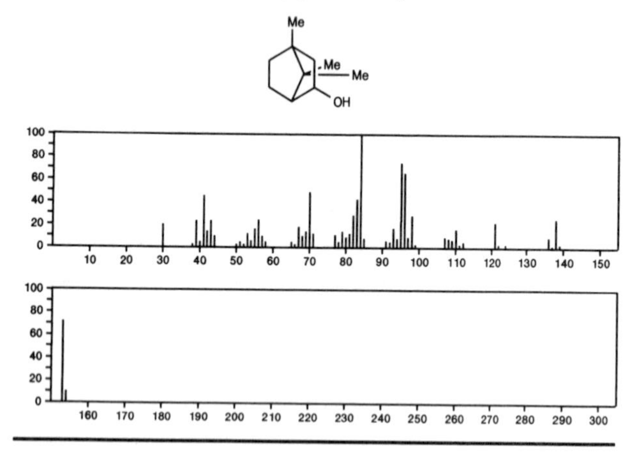

154 $C_{10}H_{18}O$ 24524-56-9
Ether, *tert*-butyl isopropylidenecyclopropyl

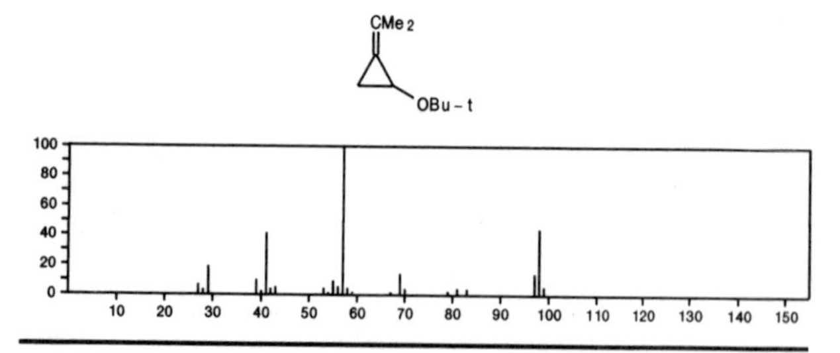

154 $C_{10}H_{18}O$ 26532-22-9
Cyclobutaneethanol, 1-methyl-2-(1-methylethenyl)-, (1*R*-*cis*)-

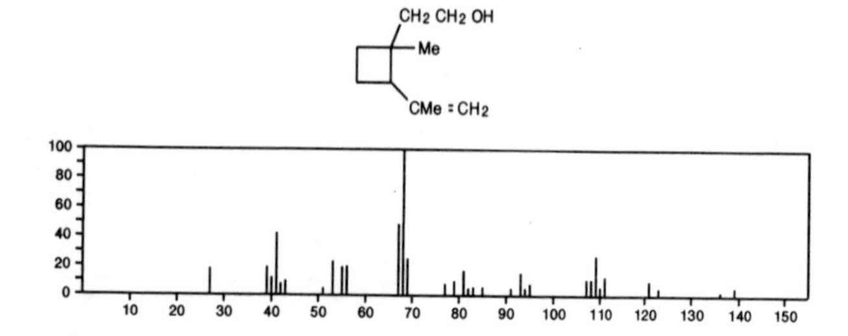

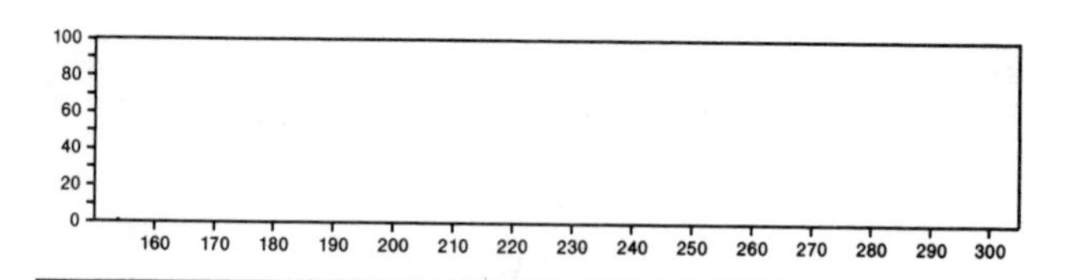

154 $C_{10}H_{18}O$ 30346-21-5
Cyclobutaneethanol, 1-methyl-2-(1-methylethenyl)-, *trans*-

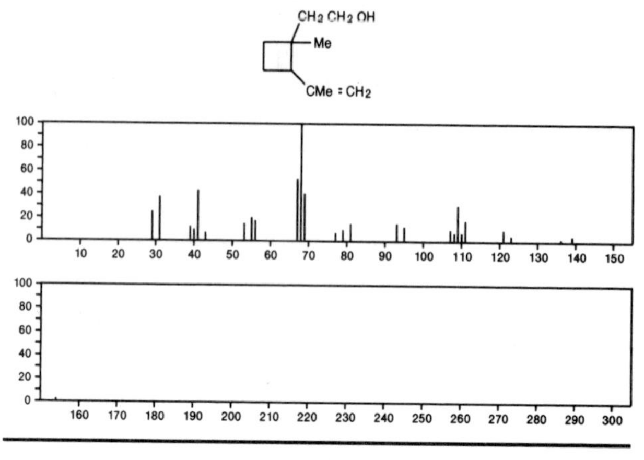

154 $C_{10}H_{18}O$ 31053-82-4
Cyclohexene, 1-(1,1-dimethylethoxy)-

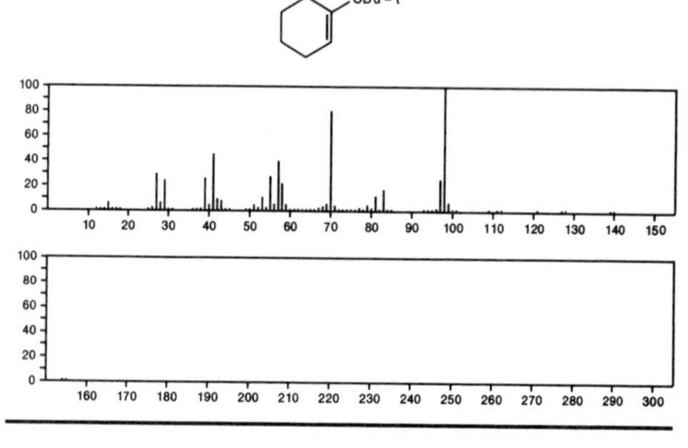

154 $C_{10}H_{18}O$ 32064-69-0
3-Hepten-2-one, 3-propyl-

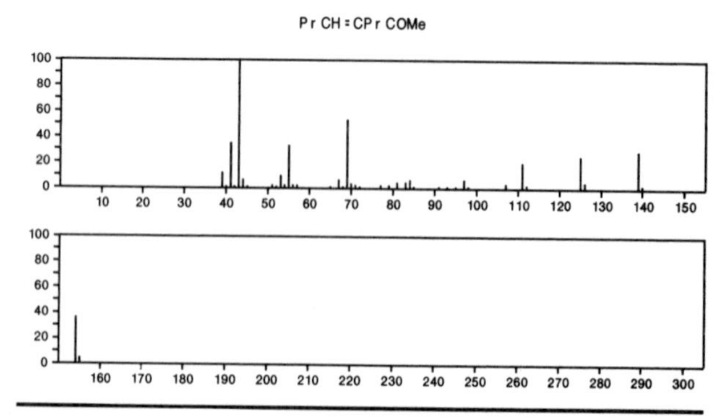

154 $C_{10}H_{18}O$ 32064-70-3
2-Heptanone, 3-propylidene-

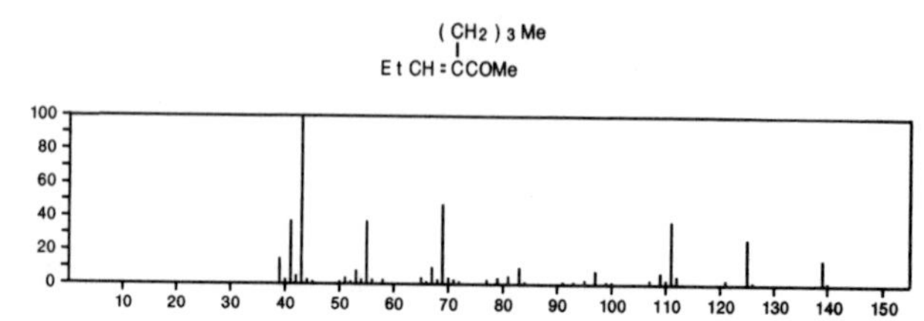

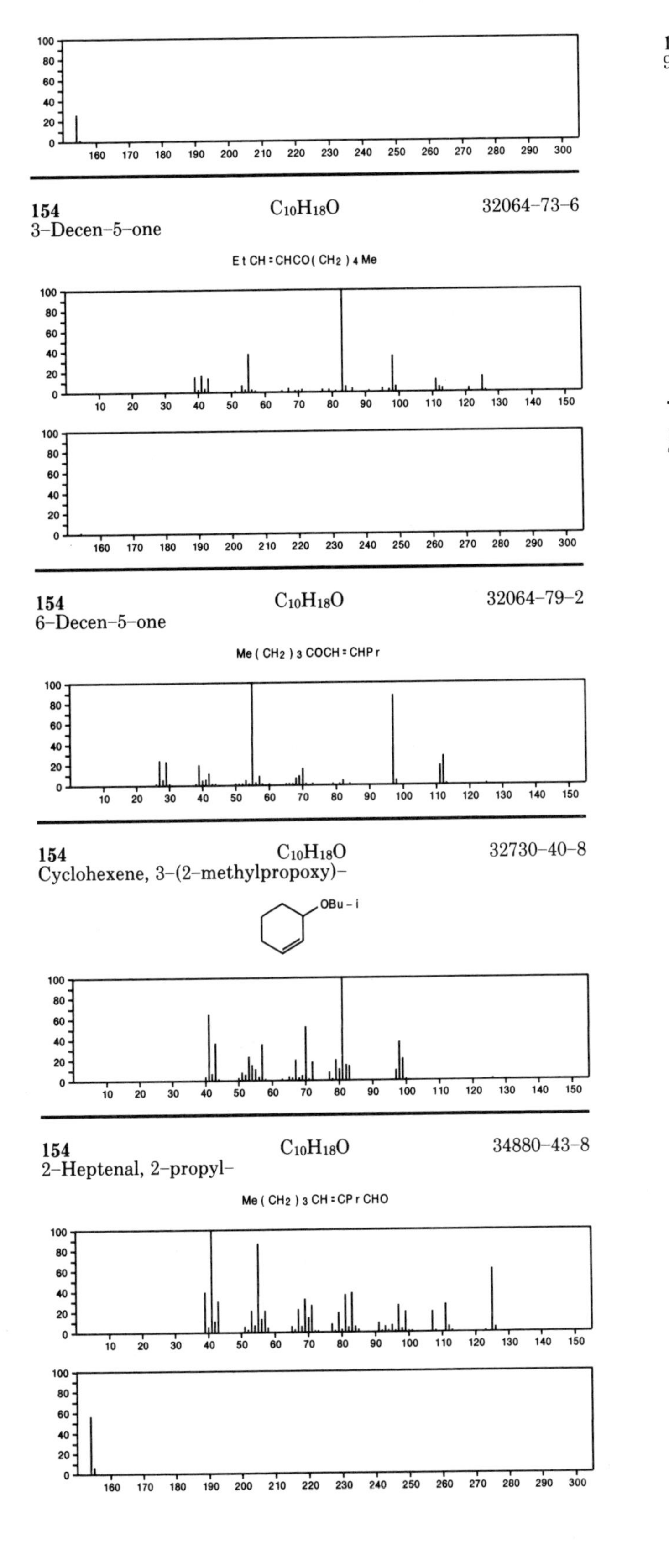
154
$C_{10}H_{18}O$
32064-73-6
3-Decen-5-one
Et CH=CHCO(CH2)4 Me
154
$C_{10}H_{18}O$
32064-79-2
6-Decen-5-one
Me(CH2)3 COCH=CHPr
154
$C_{10}H_{18}O$
32730-40-8
Cyclohexene, 3-(2-methylpropoxy)-
OBu-i
154
$C_{10}H_{18}O$
34880-43-8
2-Heptenal, 2-propyl-
Me(CH2)3 CH=CPr CHO

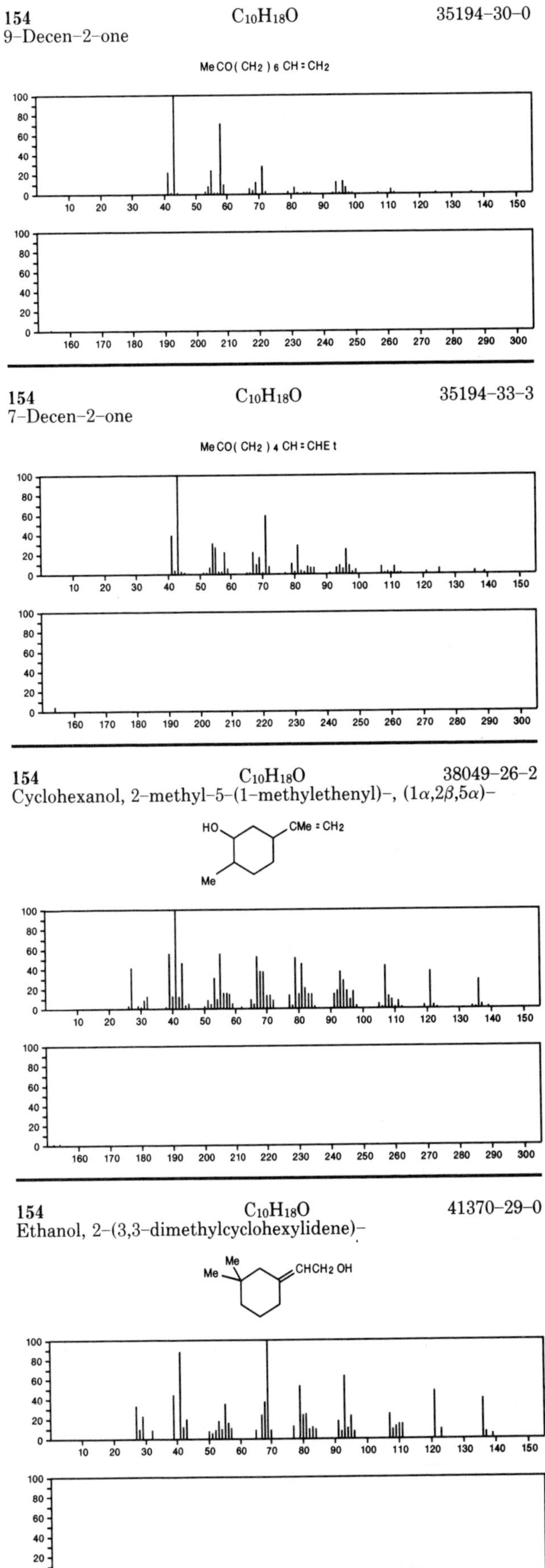
154
$C_{10}H_{18}O$
35194-30-0
9-Decen-2-one
MeCO(CH2)6 CH=CH2
154
$C_{10}H_{18}O$
35194-33-3
7-Decen-2-one
MeCO(CH2)4 CH=CHEt
154
$C_{10}H_{18}O$
38049-26-2
Cyclohexanol, 2-methyl-5-(1-methylethenyl)-, (1α,2β,5α)-
HO
CMe=CH2
Me
154
$C_{10}H_{18}O$
41370-29-0
Ethanol, 2-(3,3-dimethylcyclohexylidene)-
Me
Me
CHCH2 OH

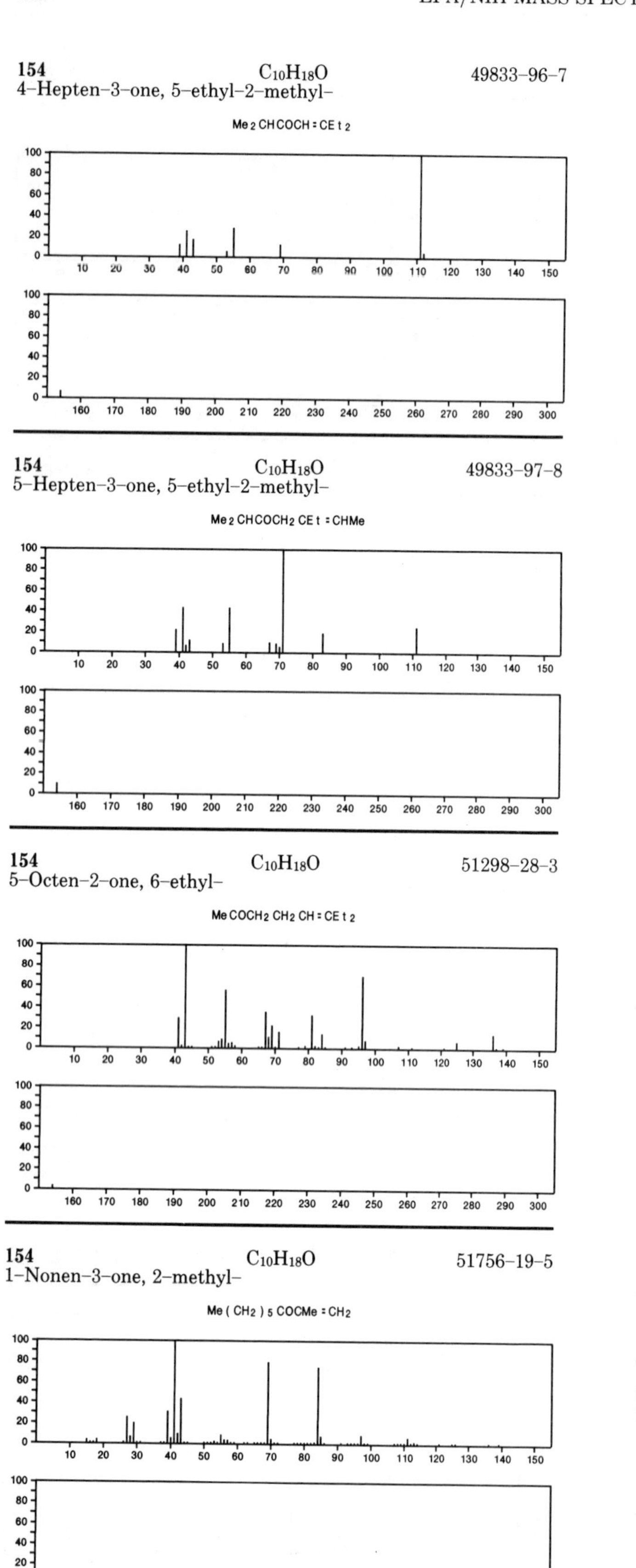

154 $C_{10}H_{18}O$ 49833-96-7
4-Hepten-3-one, 5-ethyl-2-methyl-

Me2CHCOCH=CEt2

154 $C_{10}H_{18}O$ 49833-97-8
5-Hepten-3-one, 5-ethyl-2-methyl-

Me2CHCOCH2CEt=CHMe

154 $C_{10}H_{18}O$ 51298-28-3
5-Octen-2-one, 6-ethyl-

MeCOCH2CH2CH=CEt2

154 $C_{10}H_{18}O$ 51756-19-5
1-Nonen-3-one, 2-methyl-

Me(CH2)5COCMe=CH2

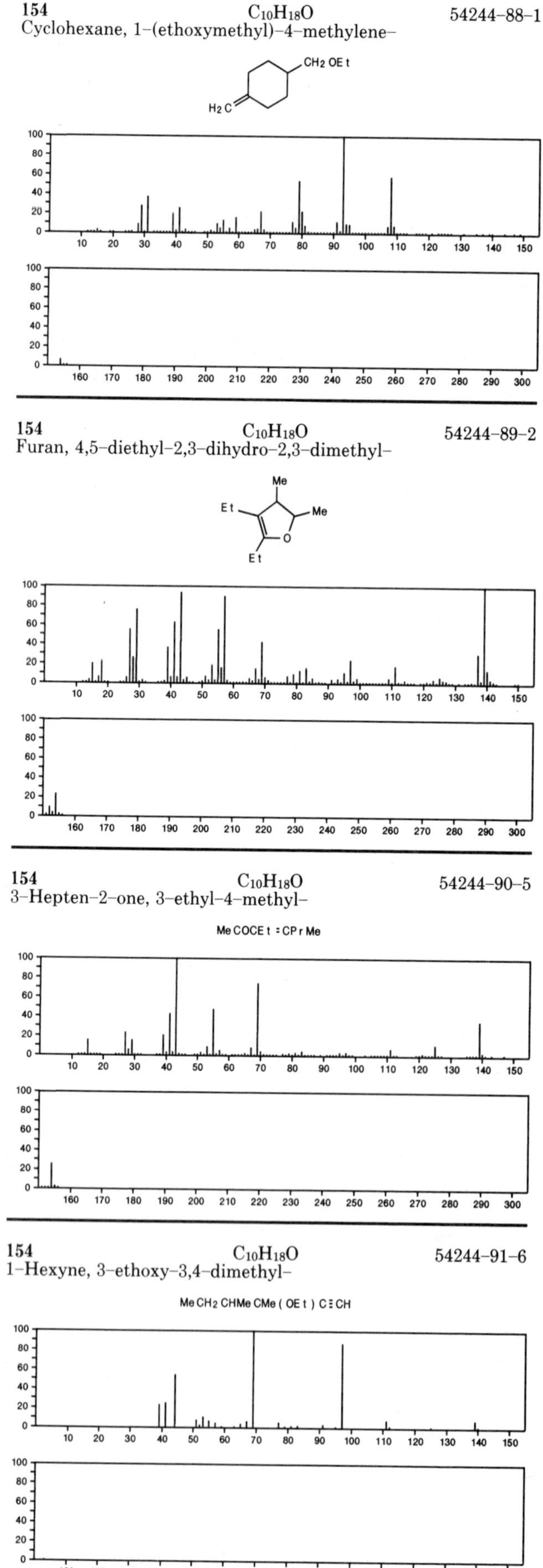

154 $C_{10}H_{18}O$ 54244-88-1
Cyclohexane, 1-(ethoxymethyl)-4-methylene-

154 $C_{10}H_{18}O$ 54244-89-2
Furan, 4,5-diethyl-2,3-dihydro-2,3-dimethyl-

154 $C_{10}H_{18}O$ 54244-90-5
3-Hepten-2-one, 3-ethyl-4-methyl-

MeCOCEt=CPrMe

154 $C_{10}H_{18}O$ 54244-91-6
1-Hexyne, 3-ethoxy-3,4-dimethyl-

MeCH2CHMeCMe(OEt)C≡CH

154 $C_{10}H_{18}O$ 54244-92-7
1-Heptyne, 3-methoxy-3,4-dimethyl-

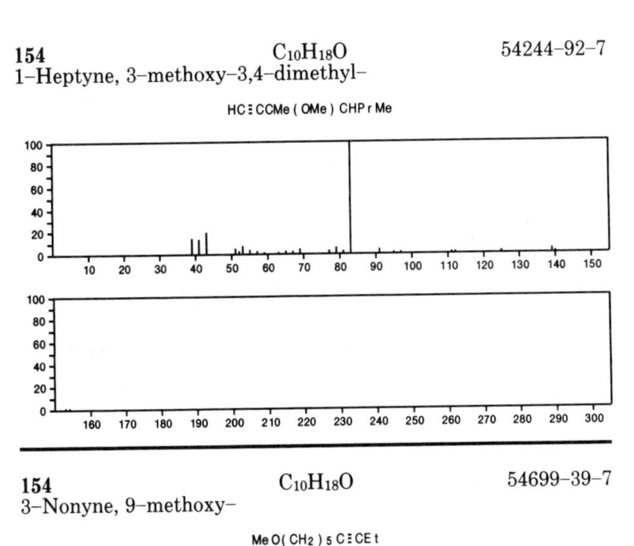

154 $C_{10}H_{18}O$ 54699-39-7
3-Nonyne, 9-methoxy-

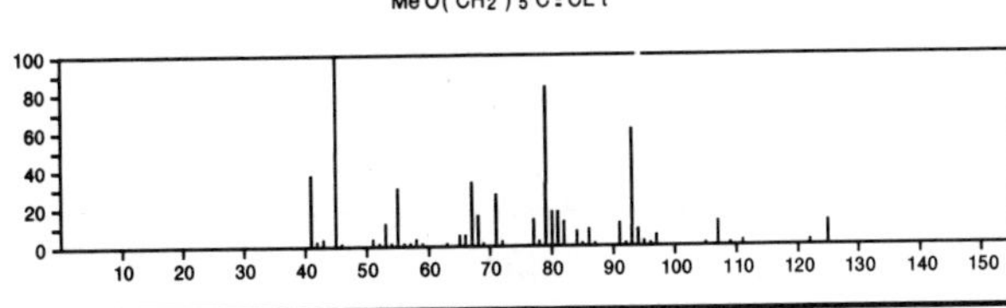

154 $C_{10}H_{18}O$ 55162-55-5
2-Cyclohexen-1-ol, 4-ethyl-1,4-dimethyl-

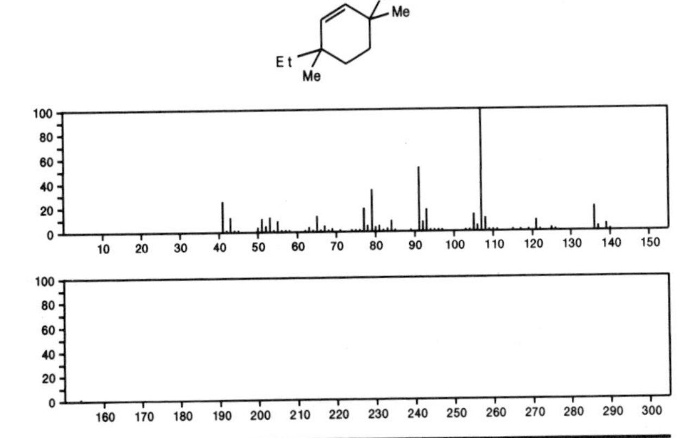

154 $C_{10}H_{18}O$ 56259-15-5
1-Propanone, 2-methyl-1-[2-(1-methylethyl)cyclopropyl]-

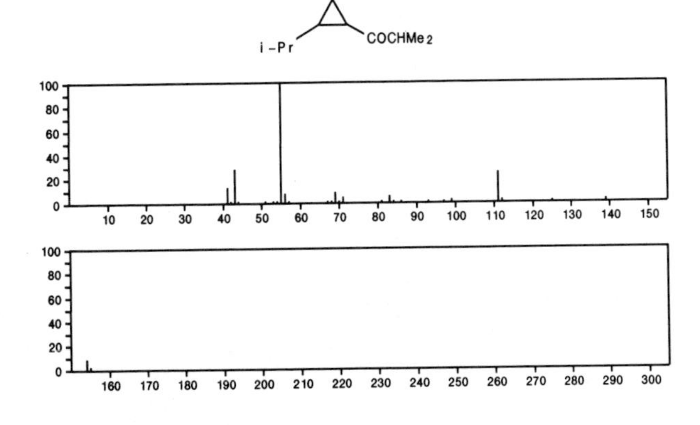

154 $C_{10}H_{18}O$ 56666-68-3
1-Cyclopentene-1-methanol, α,α,4,5-tetramethyl-, *cis*-

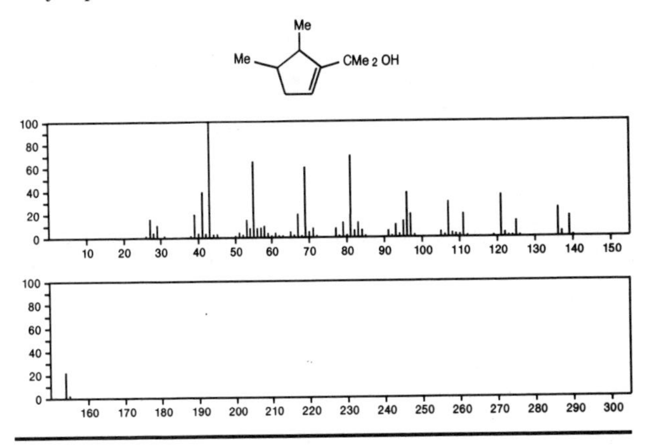

154 $C_{10}H_{18}O$ 56666-69-4
1-Cyclopentene-1-methanol, α,α,4,5-tetramethyl-, *trans*-

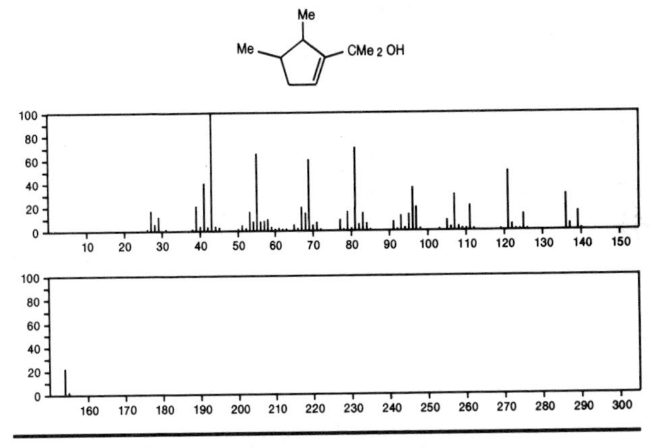

154 $C_{10}H_{18}O$ 56701-51-0
Cyclopropane, (1,1-dimethylethoxy)(1-methylethylidene)-

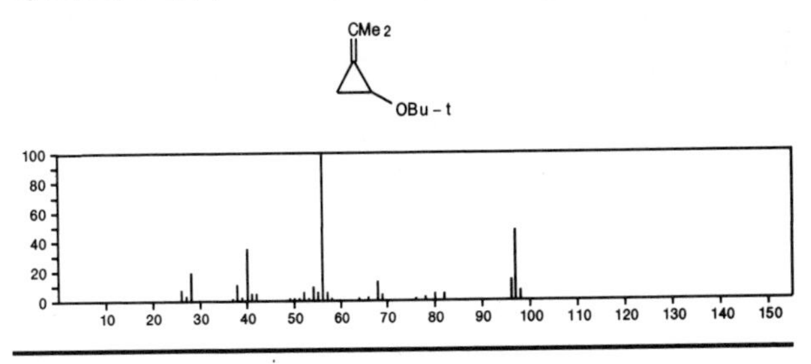

154 $C_{11}H_{6}O$ 277-96-3
2*H*-Cyclopropa[3,4]naphth[1,2-*b*]oxirene

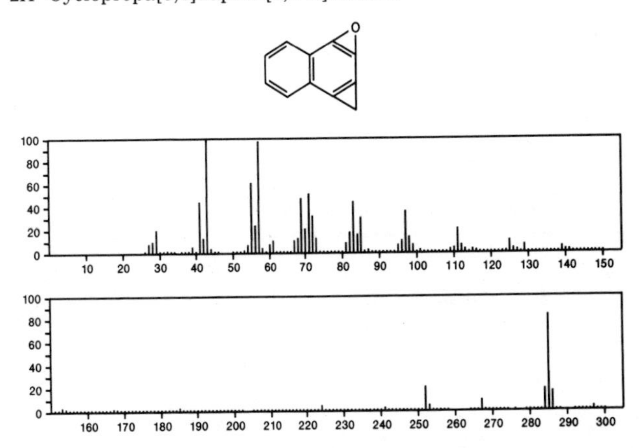

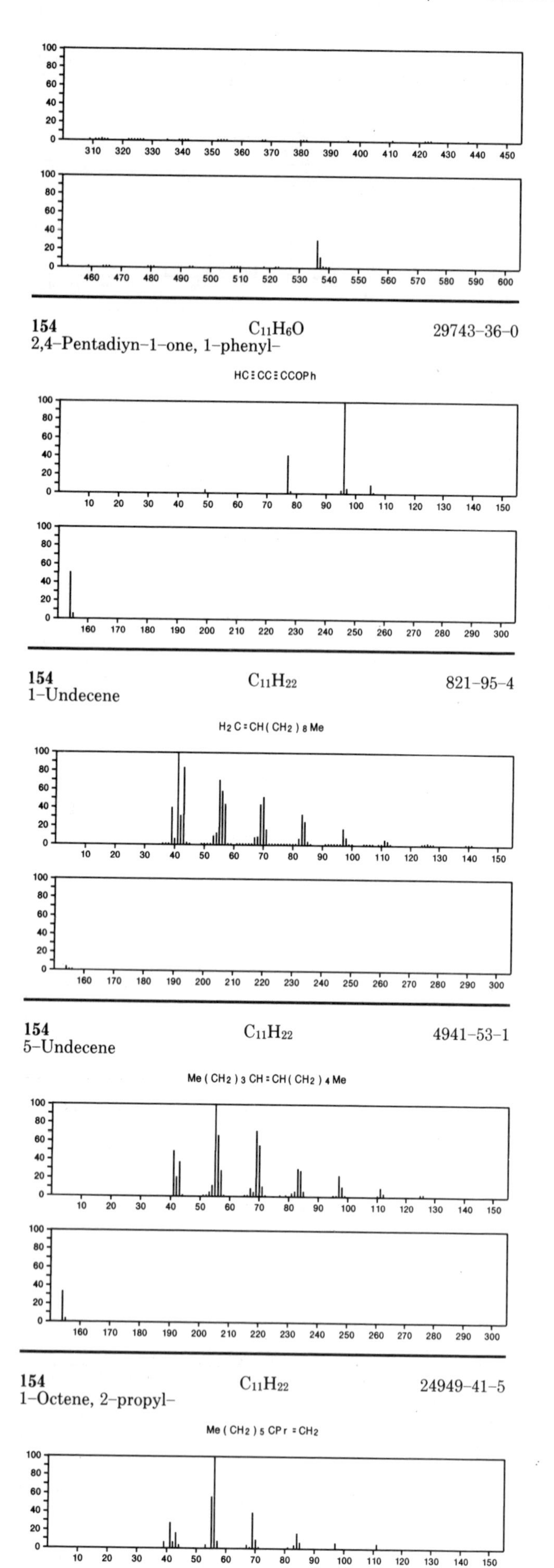

154 $C_{11}H_6O$ 29743-36-0
2,4-Pentadiyn-1-one, 1-phenyl-

HC≡CC≡CCOPh

154 $C_{11}H_{22}$ 821-95-4
1-Undecene

$H_2C=CH(CH_2)_8Me$

154 $C_{11}H_{22}$ 4941-53-1
5-Undecene

$Me(CH_2)_3CH=CH(CH_2)_4Me$

154 $C_{11}H_{22}$ 24949-41-5
1-Octene, 2-propyl-

$Me(CH_2)_5CPr=CH_2$

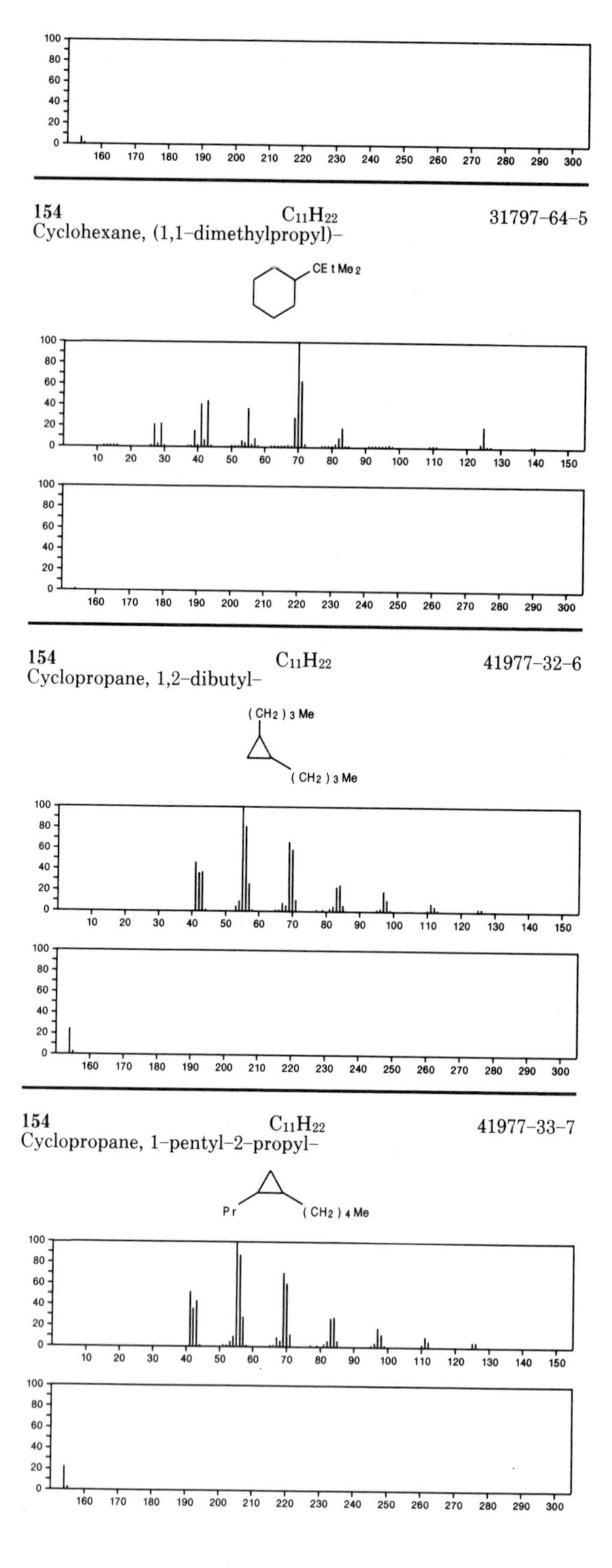

154 $C_{11}H_{22}$ 31797-64-5
Cyclohexane, (1,1-dimethylpropyl)-

CEtMe2

154 $C_{11}H_{22}$ 41977-32-6
Cyclopropane, 1,2-dibutyl-

$(CH_2)_3Me$
$(CH_2)_3Me$

154 $C_{11}H_{22}$ 41977-33-7
Cyclopropane, 1-pentyl-2-propyl-

Pr
$(CH_2)_4Me$

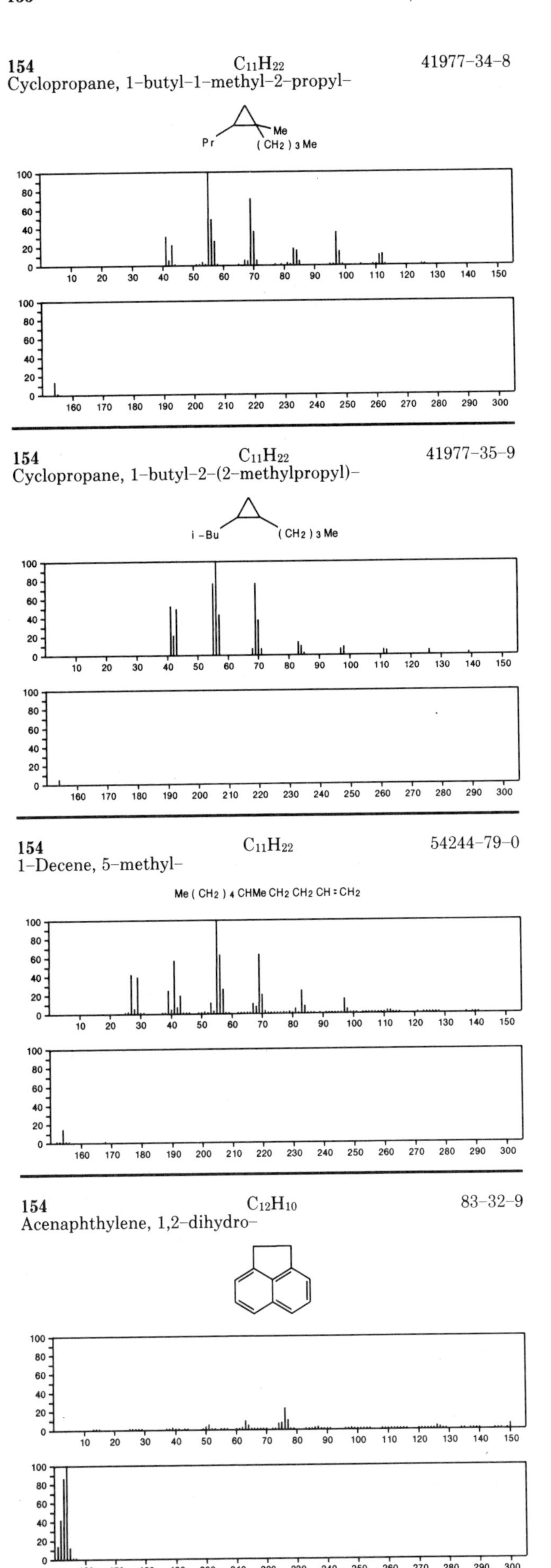
154 $C_{11}H_{22}$ 41977-34-8
Cyclopropane, 1-butyl-1-methyl-2-propyl-
Me
Pr (CH2)3Me
154 $C_{11}H_{22}$ 41977-35-9
Cyclopropane, 1-butyl-2-(2-methylpropyl)-
i-Bu (CH2)3Me
154 $C_{11}H_{22}$ 54244-79-0
1-Decene, 5-methyl-
Me(CH2)4CHMeCH2CH2CH=CH2
154 $C_{12}H_{10}$ 83-32-9
Acenaphthylene, 1,2-dihydro-

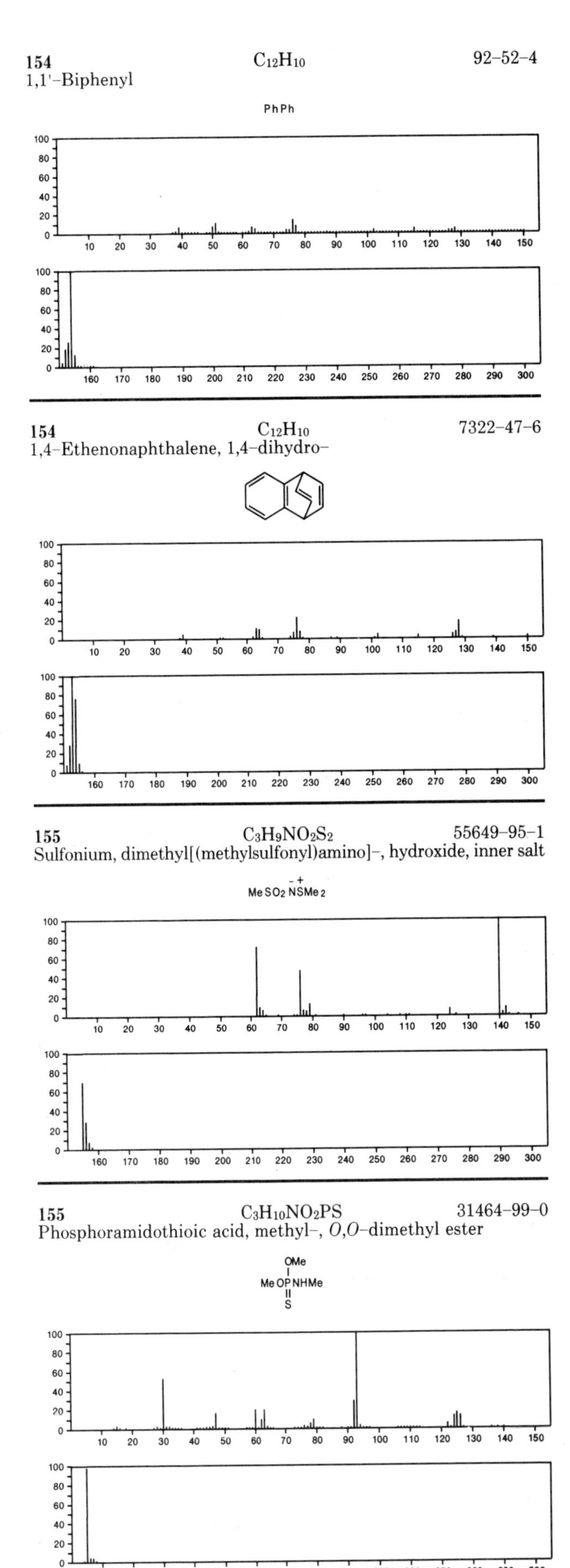
154 $C_{12}H_{10}$ 92-52-4
1,1'-Biphenyl
PhPh
154 $C_{12}H_{10}$ 7322-47-6
1,4-Ethenonaphthalene, 1,4-dihydro-
155 $C_3H_9NO_2S_2$ 55649-95-1
Sulfonium, dimethyl[(methylsulfonyl)amino]-, hydroxide, inner salt
MeSO2NSMe2
155 $C_3H_{10}NO_2PS$ 31464-99-0
Phosphoramidothioic acid, methyl-, O,O-dimethyl ester
OMe
MeOPNHMe
S

155 $C_5H_8F_3NO$ 10056-69-6
Acetamide, 2,2,2-trifluoro-*N*-propyl-

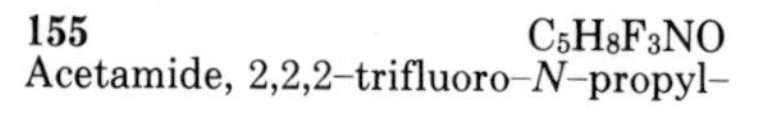

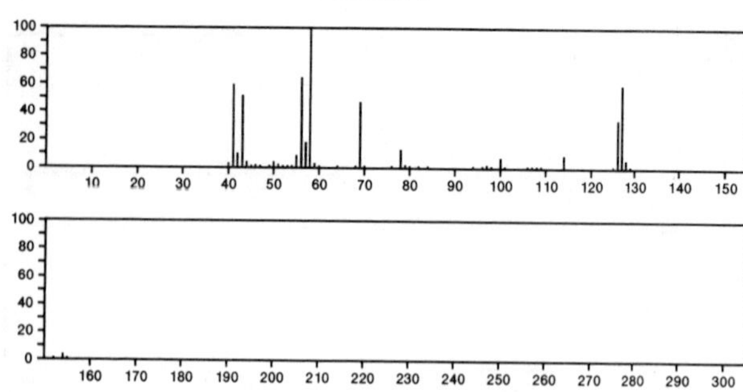

155 $C_5H_9N_5O$ 7313-54-4
1,3,5-Triazin-2(1*H*)-one, 4-amino-6-(ethylamino)-

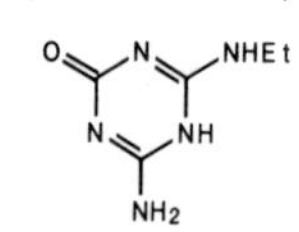

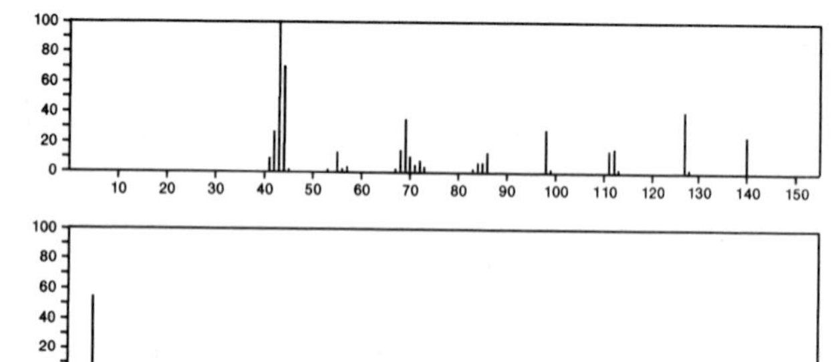

155 $C_5H_9N_5O$ 55702-52-8
1,3,5-Triazin-2(1*H*)-one, 4,6-bis(methylamino)-

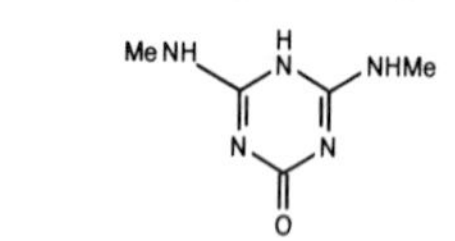

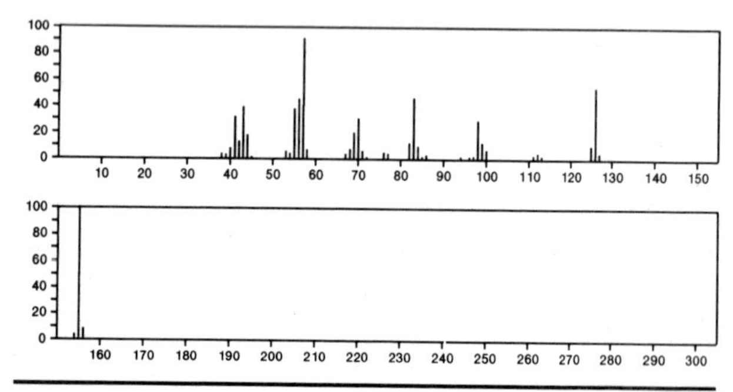

155 $C_6H_9N_3O_2$ 71-00-1
L-Histidine

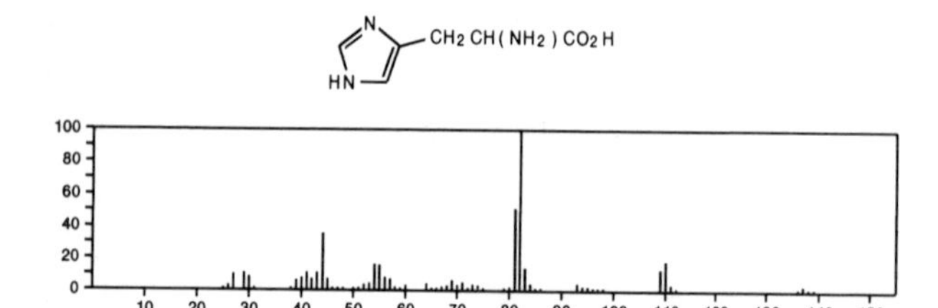

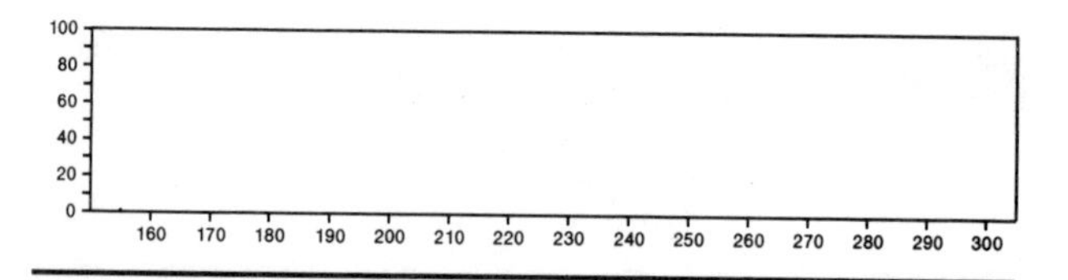

155 $C_6H_9N_3O_2$ 42956-82-1
4(1*H*)-Pyrimidinone, 2-amino-6-ethoxy-

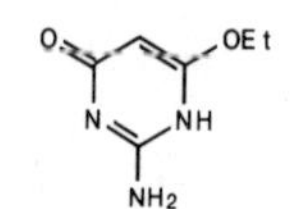

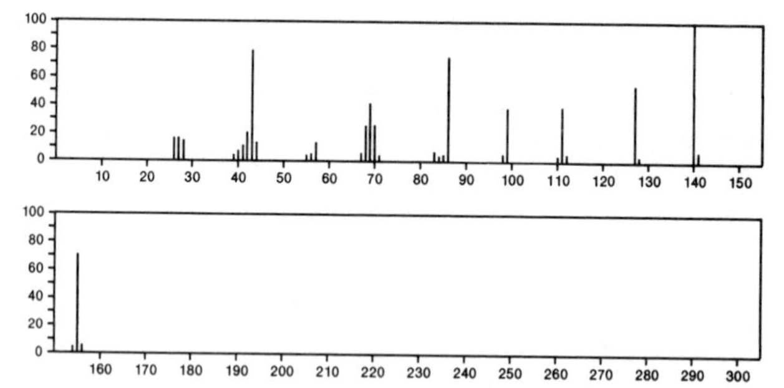

155 $C_6H_9N_3S$ 54308-63-3
4-Pyrimidinamine, 2-(ethylthio)-

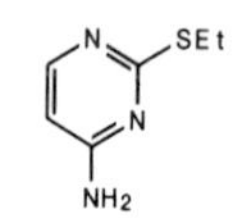

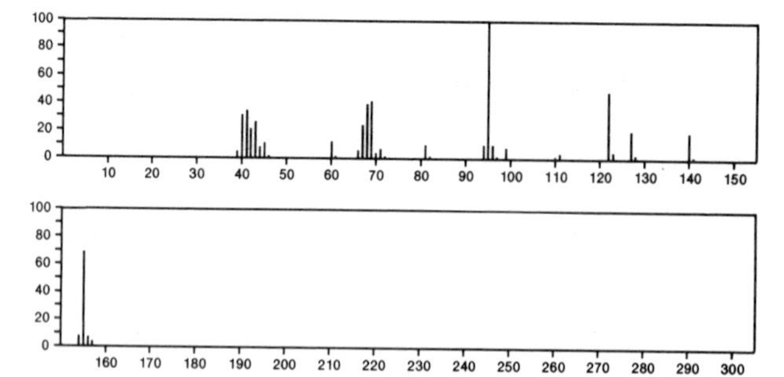

155 $C_6H_9N_3S$ 54308-64-4
4-Pyrimidinamine, 5-methyl-2-(methylthio)-

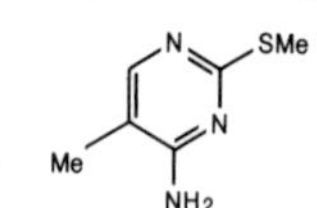

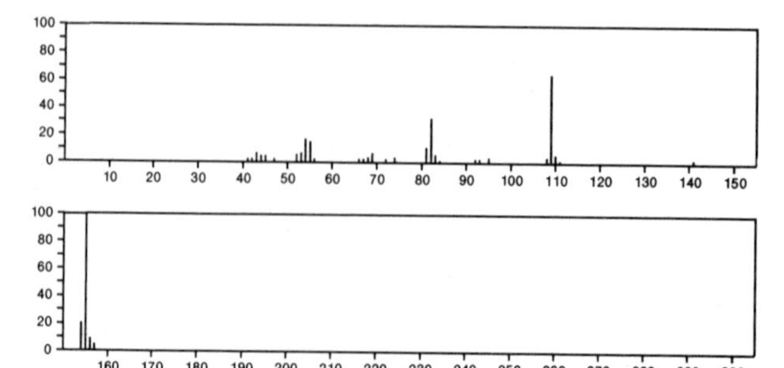

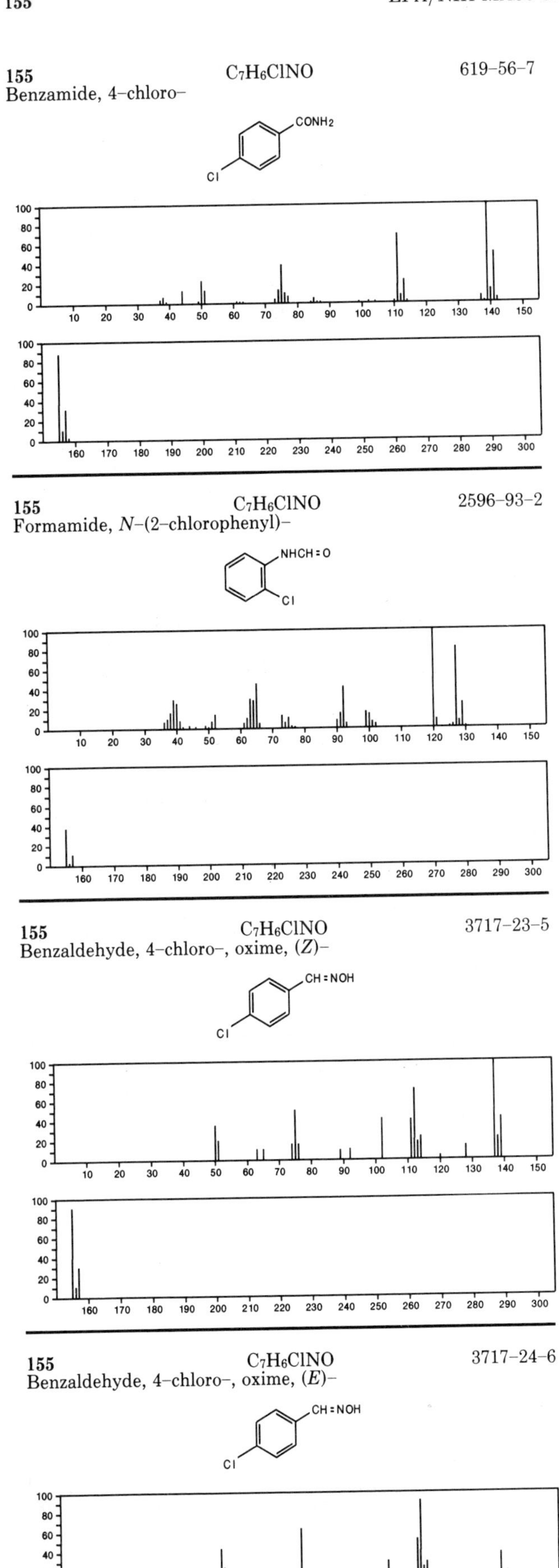

155 C_7H_6ClNO 619-56-7
Benzamide, 4-chloro-

155 C_7H_6ClNO 2596-93-2
Formamide, *N*-(2-chlorophenyl)-

155 C_7H_6ClNO 3717-23-5
Benzaldehyde, 4-chloro-, oxime, (*Z*)-

155 C_7H_6ClNO 3717-24-6
Benzaldehyde, 4-chloro-, oxime, (*E*)-

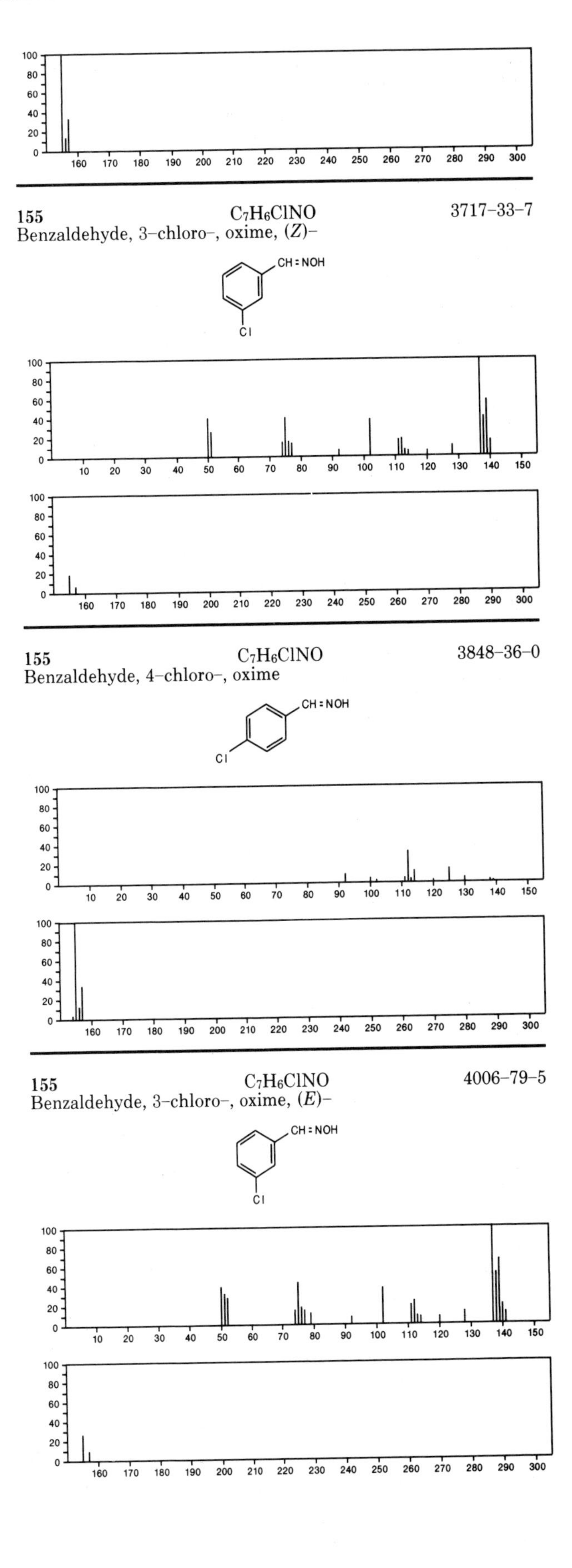

155 C_7H_6ClNO 3717-33-7
Benzaldehyde, 3-chloro-, oxime, (*Z*)-

155 C_7H_6ClNO 3848-36-0
Benzaldehyde, 4-chloro-, oxime

155 C_7H_6ClNO 4006-79-5
Benzaldehyde, 3-chloro-, oxime, (*E*)-

155 C_7H_6ClNO 34158–71–9
Benzaldehyde, 3–chloro–, oxime

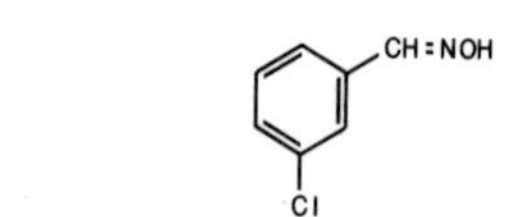

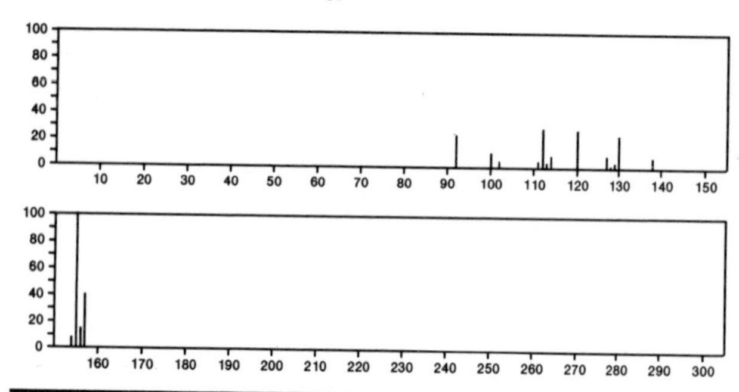

155 C_7H_9NOS 4381–25–3
Sulfoximine, *S*–methyl–*S*–phenyl–

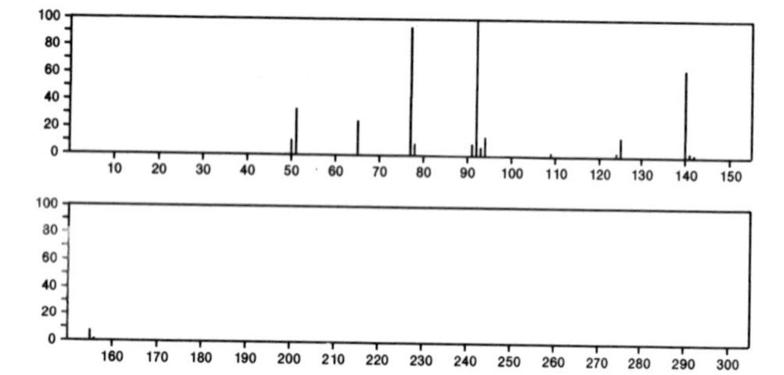

155 C_7H_9NOS 23003–25–0
3–Pyridinol, 6–methyl–2–(methylthio)–

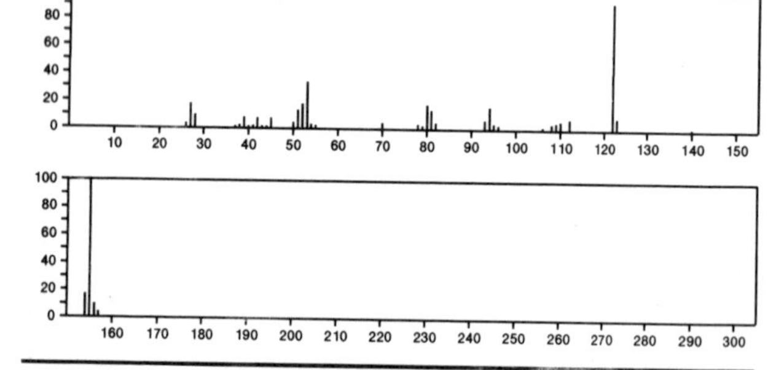

155 $C_7H_{13}N_3O$ 1589–61–3
Hydrazinecarboxamide, 2–cyclohexylidene–

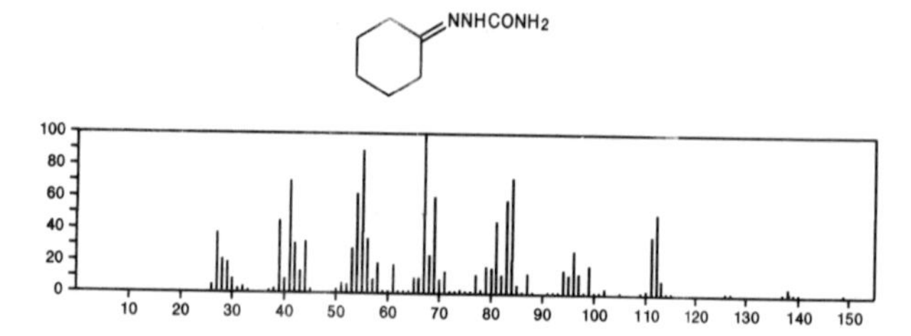

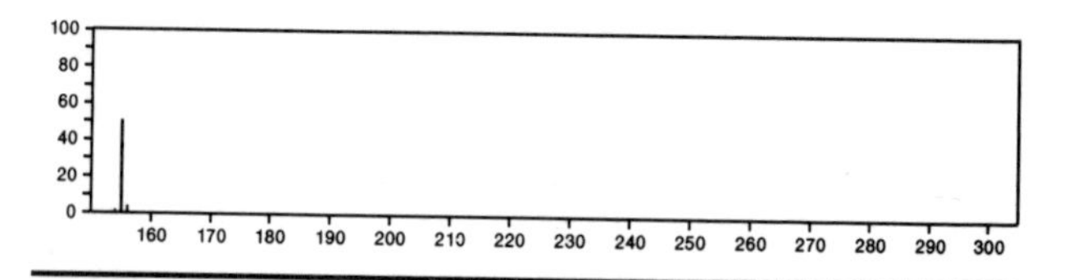

155 $C_8H_{10}ClN$ 13519–74–9
Benzenamine, 2–chloro–*N*–ethyl–

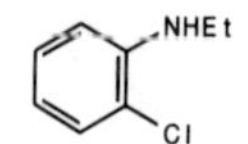

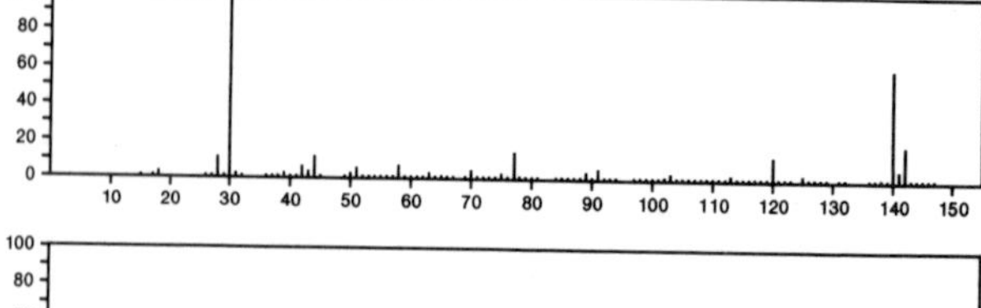

155 $C_8H_{13}NO_2$ 63–75–2
3–Pyridinecarboxylic acid, 1,2,5,6–tetrahydro–1–methyl–, methyl ester

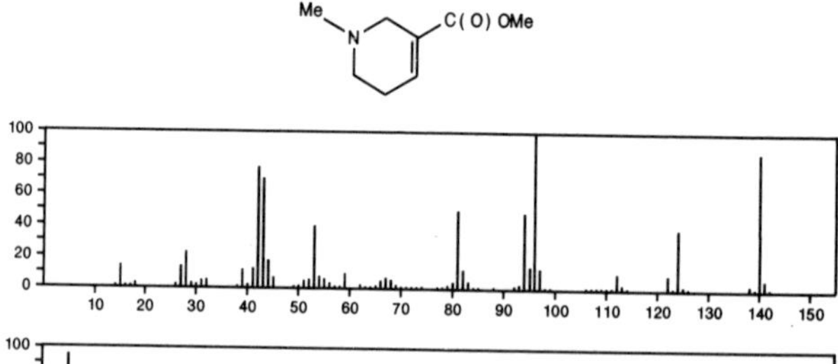

155 $C_8H_{13}NO_2$ 520–63–8
1*H*–Pyrrolizine–7–methanol, 2,3,5,7a–tetrahydro–1–hydroxy–, (1*S*–*cis*)–

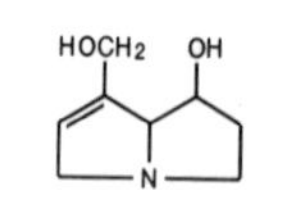

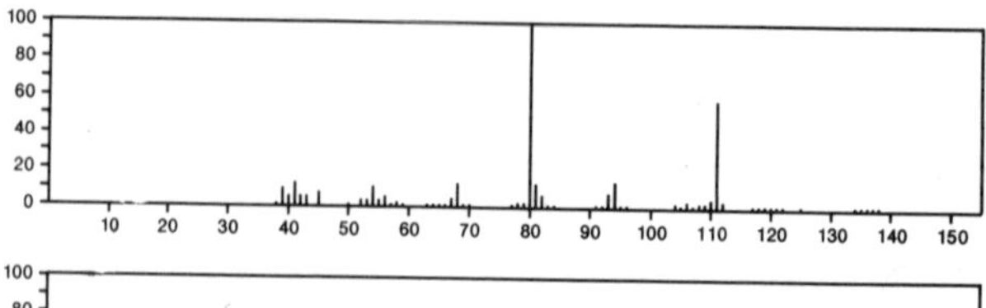

155 $C_8H_{13}NO_2$ 3213-49-8
Butanoic acid, 2-cyano-3-methyl-, ethyl ester

Me2CHCH(CN)C(O)OEt

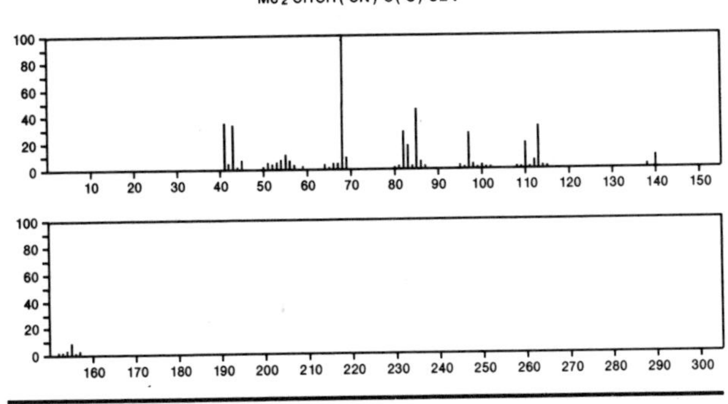

155 $C_8H_{13}NO_2$ 7309-46-8
Hexanoic acid, 2-cyano-, methyl ester

MeOC(O)CH(CN)(CH2)3Me

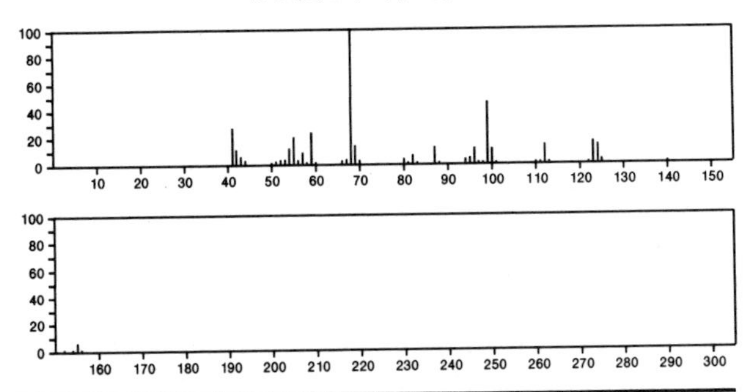

155 $C_8H_{13}NO_2$ 13861-99-9
Succinimide, 2-ethyl-*N*,2-dimethyl-

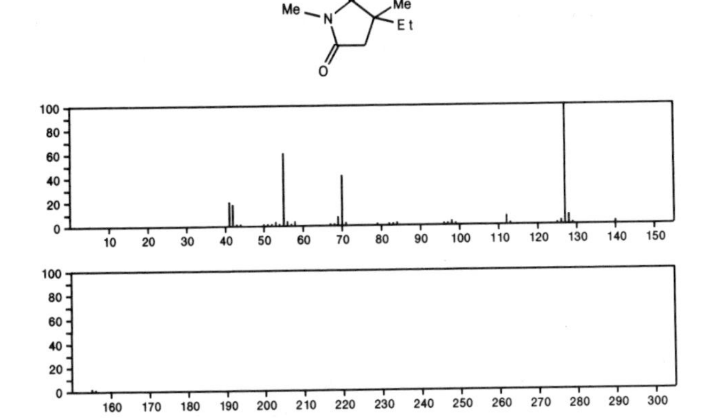

155 $C_8H_{13}NO_2$ 19822-83-4
Cyclopropanepropionic acid, α-amino-β-methyl-2-methylene-

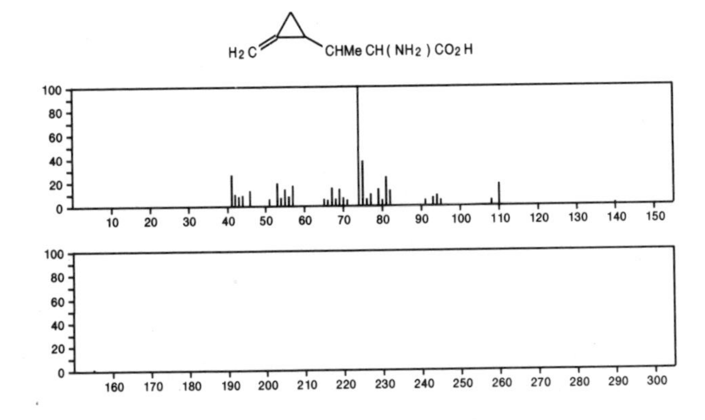

155 $C_8H_{13}NO_2$ 25115-67-7
Glutarimide, *N*,3,3-trimethyl-

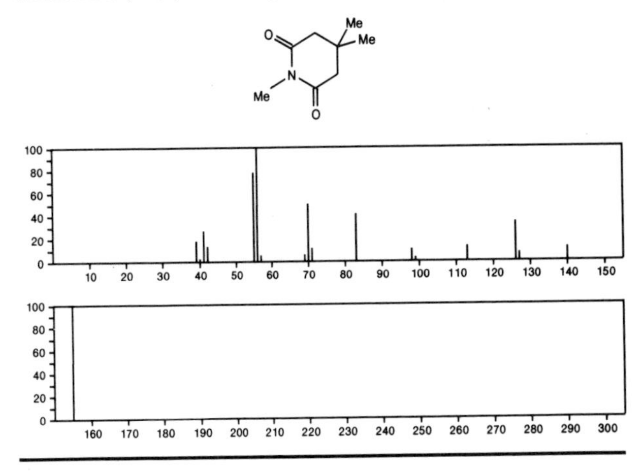

155 $C_8H_{13}NO_2$ 34291-62-8
1-Azabicyclo[2.2.2]octan-3-one, 6-(hydroxymethyl)-

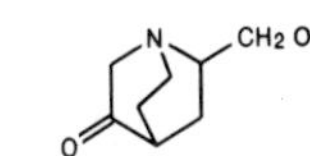

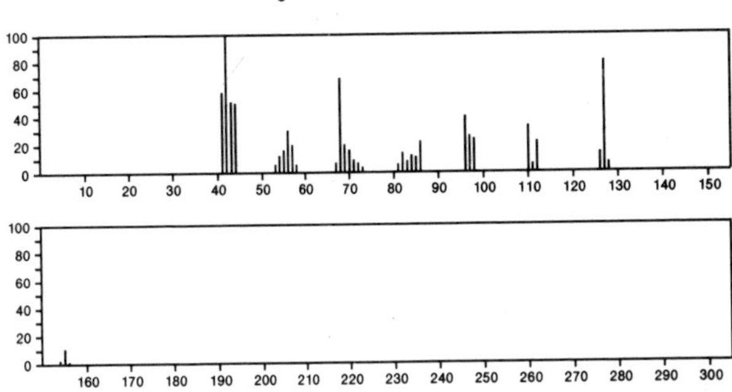

155 $C_8H_{13}NO_2$ 54751-97-2
3-Piperidinone, 1-acetyl-6-methyl-

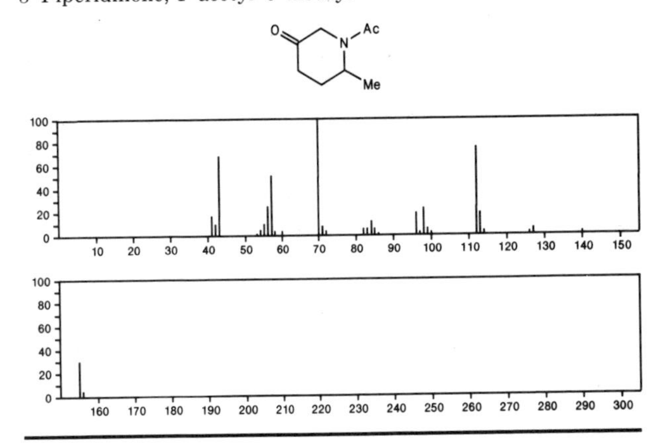

155 $C_8H_{13}NS$ 15679-14-8
Thiazole, 2-(1,1-dimethylethyl)-4-methyl-

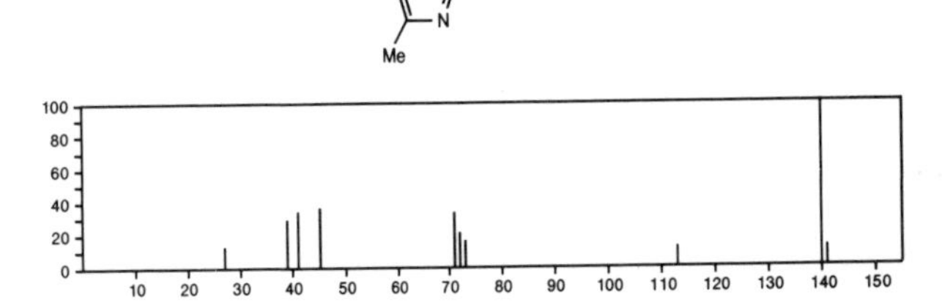

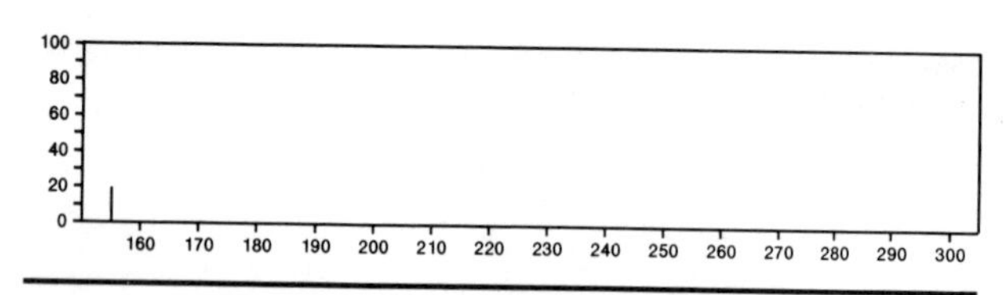

155 $C_8H_{13}NS$ 27149-25-3
Thiazole, 2-ethyl-5-propyl-

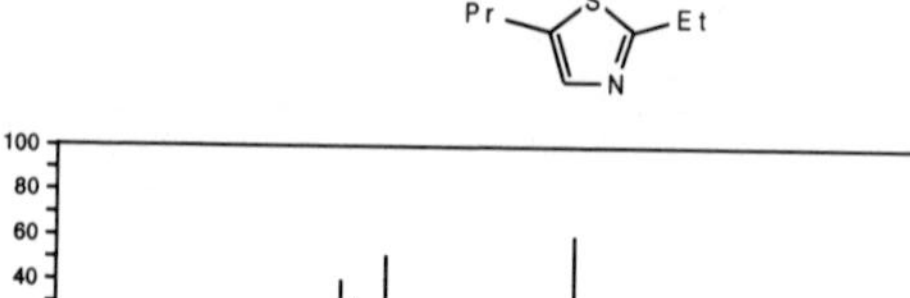

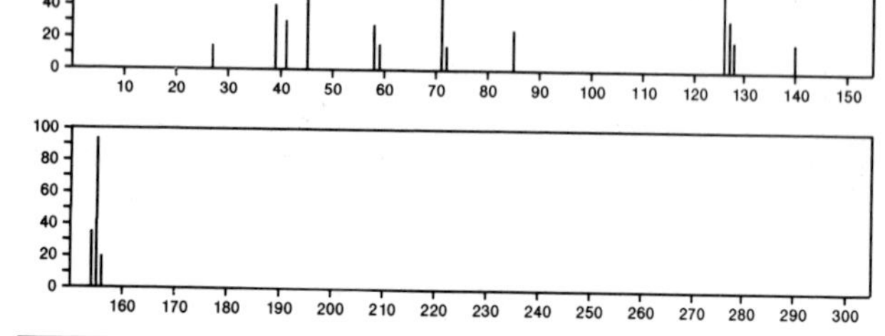

155 $C_8H_{13}NS$ 41981-68-4
Thiazole, 4-ethyl-2-propyl-

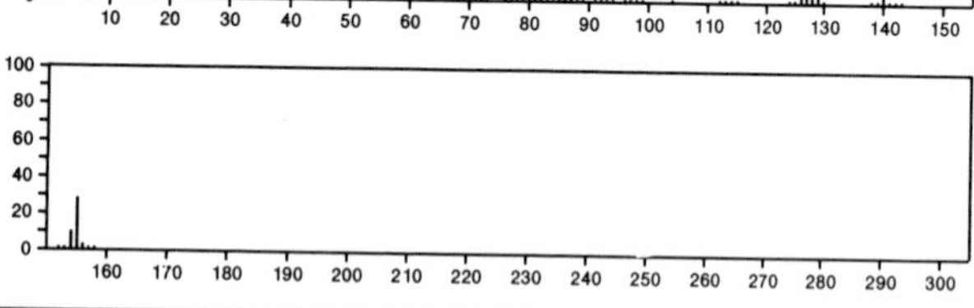

155 $C_8H_{13}NS$ 41981-69-5
Thiazole, 4-butyl-2-methyl-

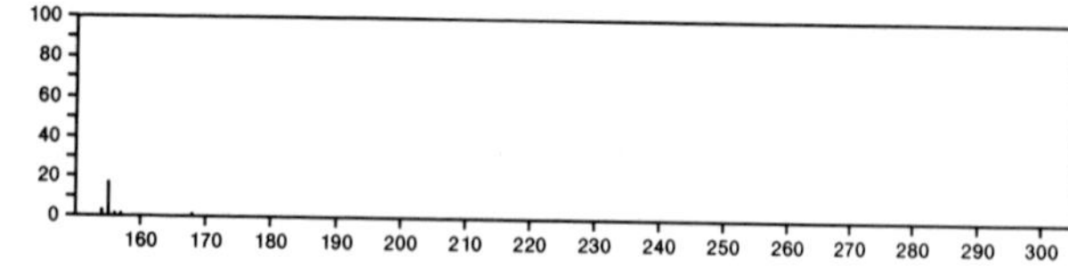

155 $C_8H_{13}NS$ 41981-70-8
Thiazole, 2,4-dimethyl-5-propyl-

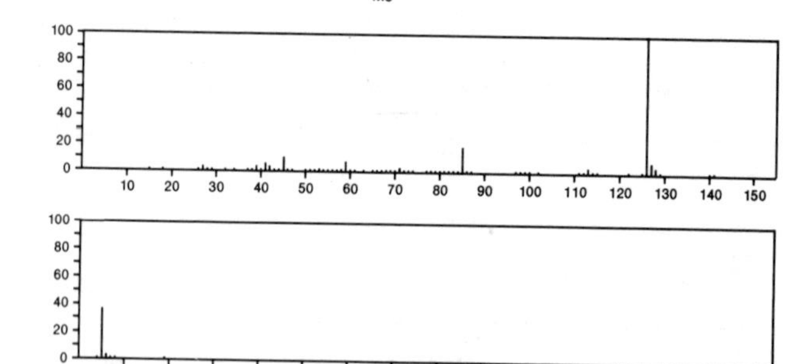

155 $C_8H_{13}NS$ 41981-71-9
Thiazole, 2,5-diethyl-4-methyl-

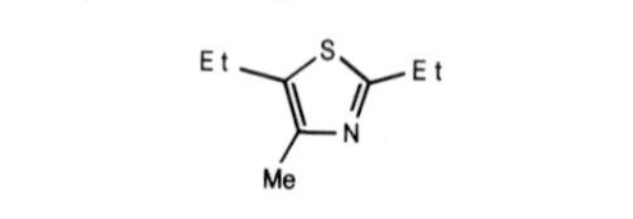

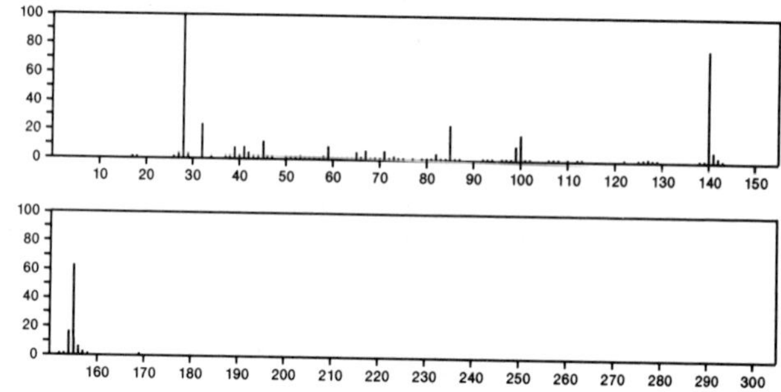

155 $C_8H_{13}NS$ 41981-72-0
Thiazole, 4,5-dimethyl-2-propyl-

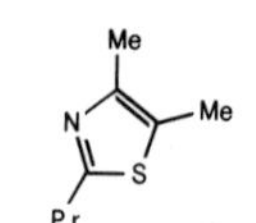

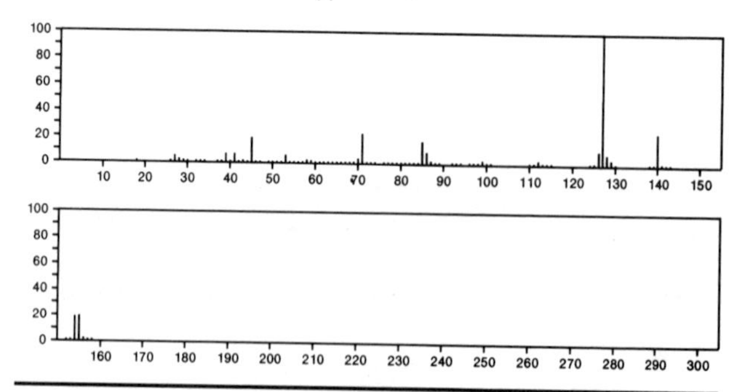

155 $C_8H_{13}NS$ 52414-85-4
Thiazole, 2-butyl-5-methyl-

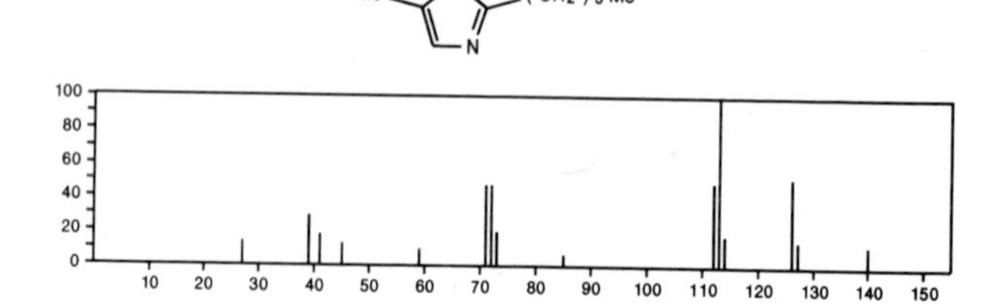

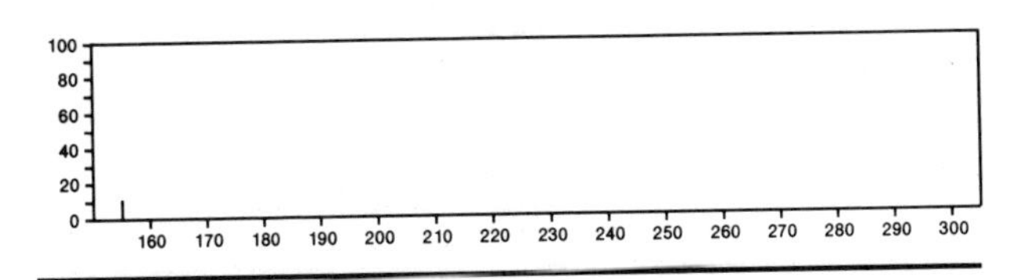

155 $C_8H_{13}NS$ 52414-89-8
Thiazole, 2,4-diethyl-5-methyl-

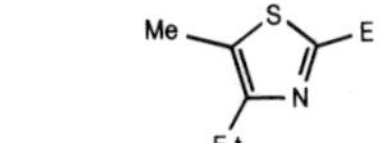

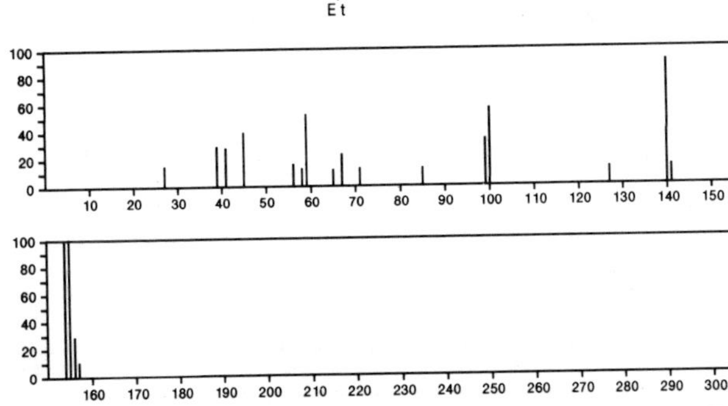

155 $C_8H_{13}NS$ 54031-34-4
2H-Pyrrole, 3,4-dihydro-5-[2-(methylthio)-1-propenyl]-

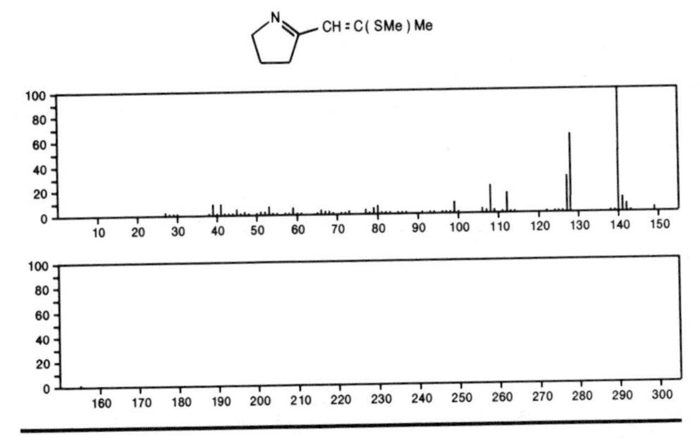

155 $C_9H_{17}NO$ 553-77-5
2-Propanone, 1-(1-methyl-2-piperidinyl)-

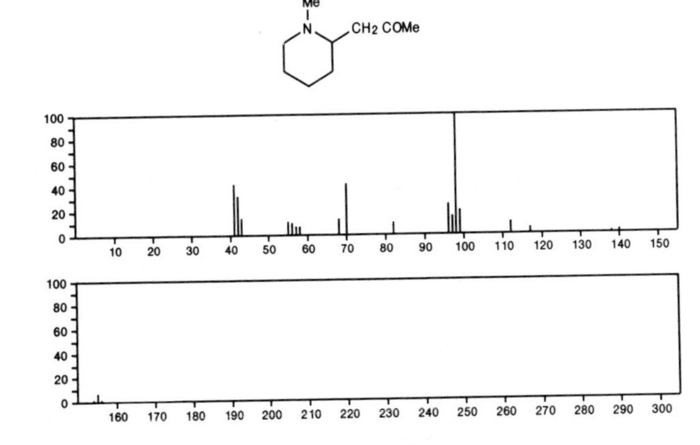

155 $C_9H_{17}NO$ 2972-02-3
Cyclononanone, oxime

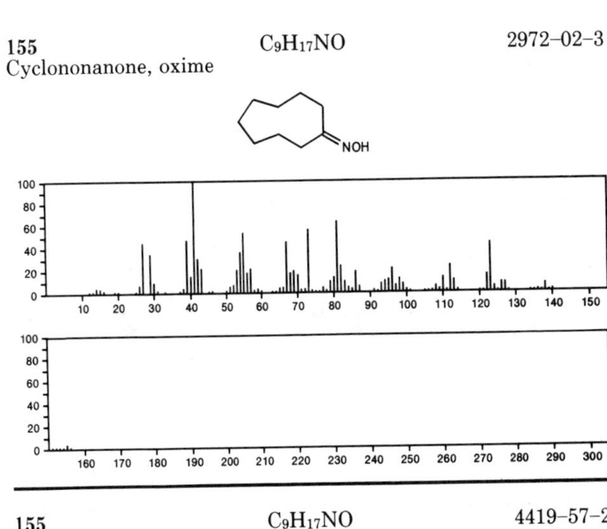

155 $C_9H_{17}NO$ 4419-57-2
Pyrrolidine, 1-valeryl-

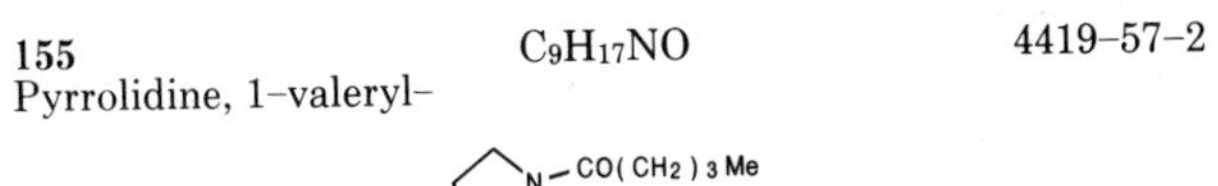

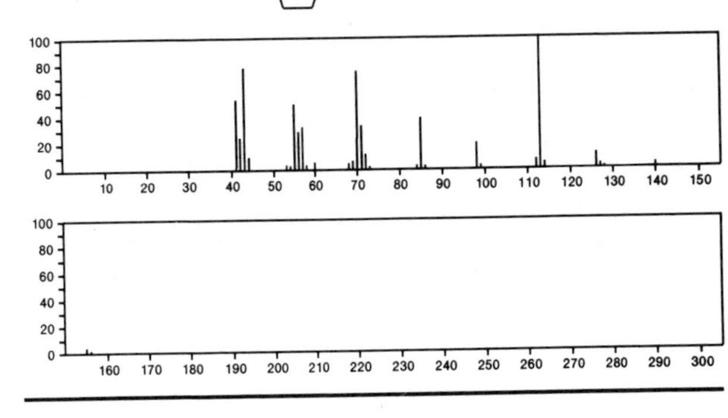

155 $C_9H_{17}NO$ 5327-02-6
Propanenitrile, 3-(hexyloxy)-

$NCCH_2CH_2O(CH_2)_5Me$

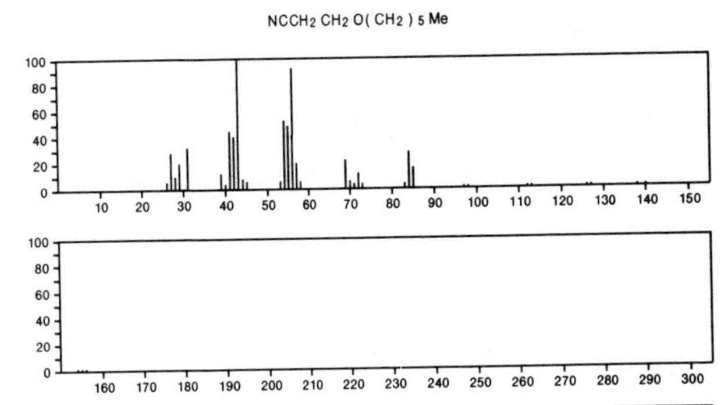

155 $C_9H_{17}NO$ 10447-19-5
2H-Quinolizin-1-ol, octahydro-, *cis*-

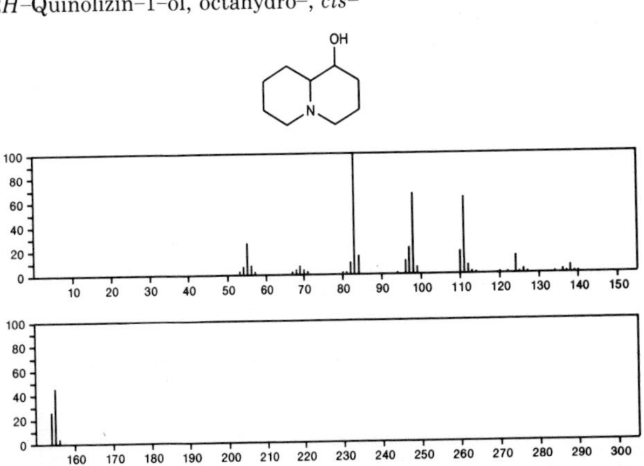

155 $C_9H_{17}NO$ 10447-20-8
2*H*-Quinolizin-1-ol, octahydro-, *trans*-

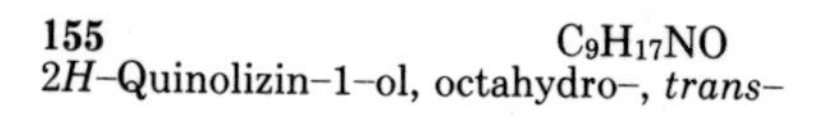

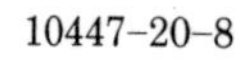

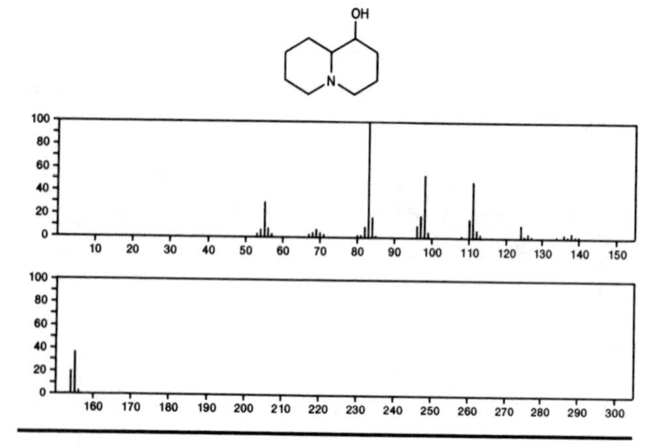

155 $C_9H_{17}NO$ 13493-40-8
3-Azabicyclo[3.3.1]nonan-9-ol, 3-methyl-, *syn*-

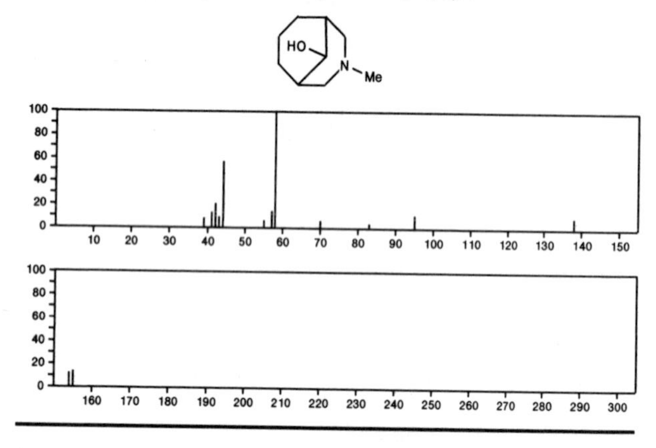

155 $C_9H_{17}NO$ 13962-79-3
3-Azabicyclo[3.3.1]nonan-9-ol, 3-methyl-, *anti*-

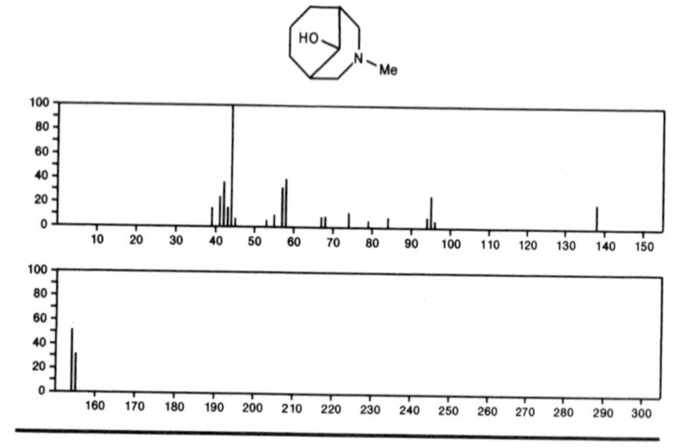

155 $C_9H_{17}NO$ 15409-60-6
Cyclohexanone, 2-[(dimethylamino)methyl]-

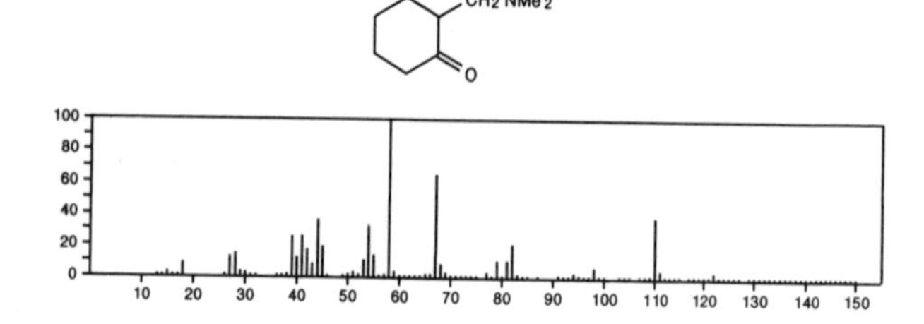

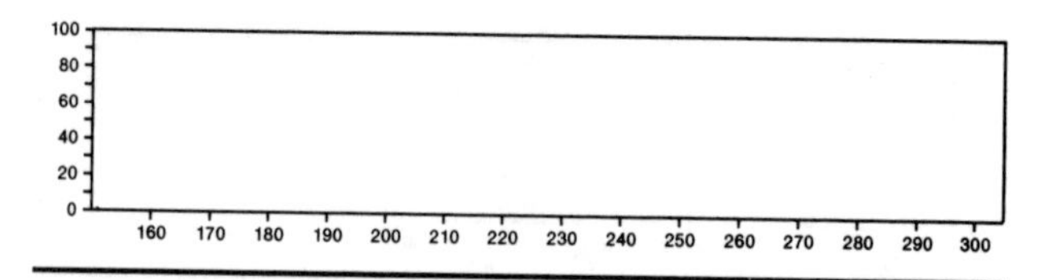

155 $C_9H_{17}NO$ 15769-36-5
2*H*-Quinolizin-3-ol, octahydro-, *trans*-

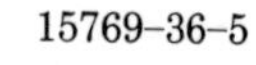

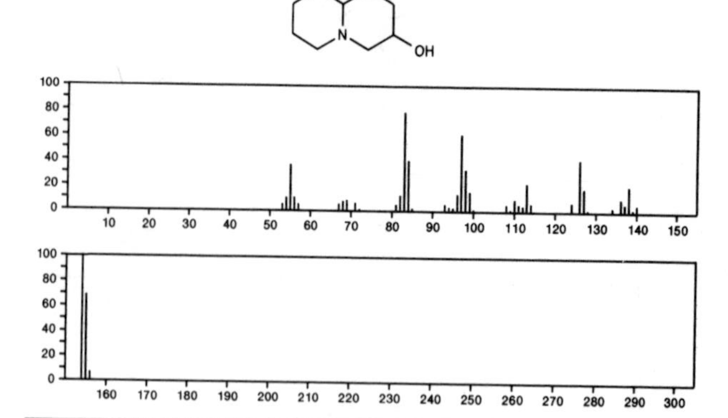

155 $C_9H_{17}NO$ 22525-60-6
2*H*-Quinolizin-1-ol, octahydro-

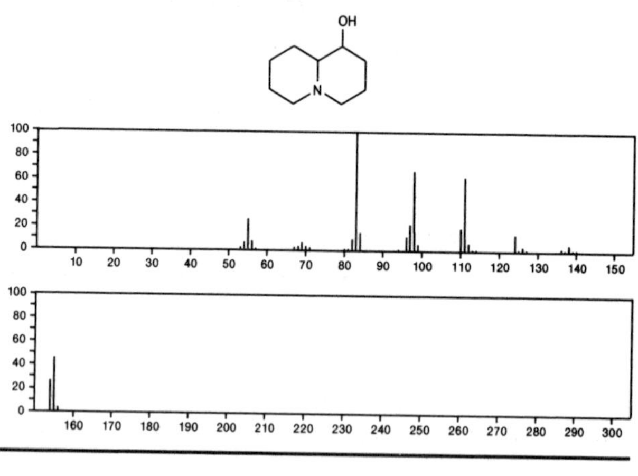

155 $C_9H_{17}NO$ 24985-57-7
2-Octen-4-one, 2-(methylamino)-

MeC(NHMe)=CHCO(CH2)3Me

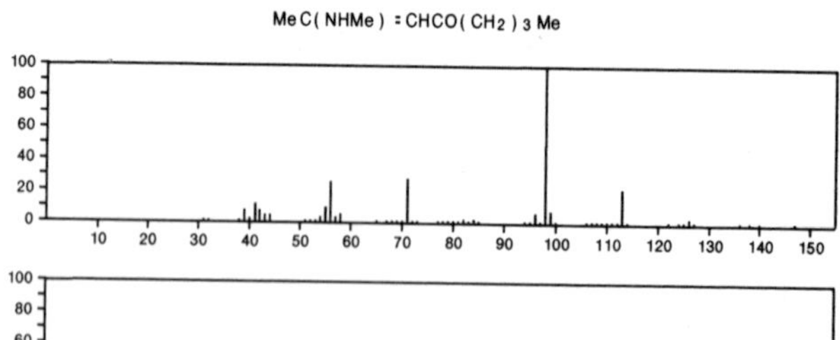

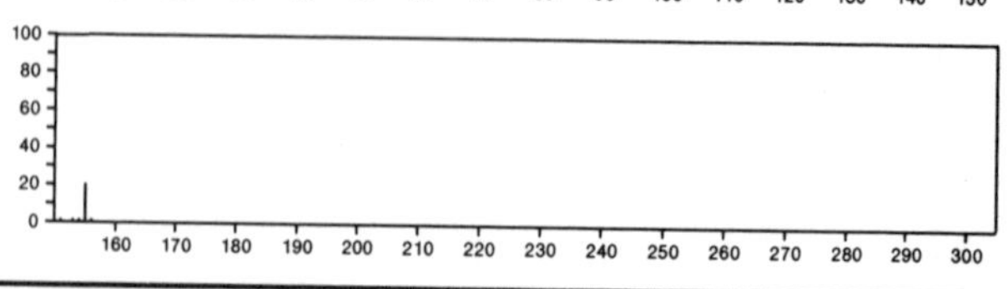

155 $C_9H_{17}NO$ 24985-58-8
3-Octen-2-one, 4-(methylamino)-

MeCOCH=C(NHMe)(CH2)3Me

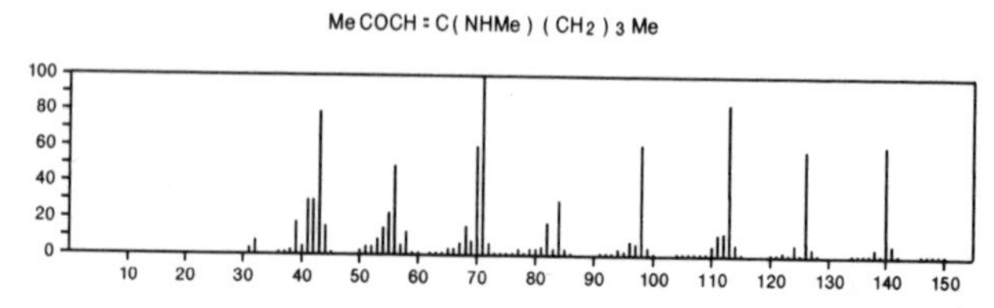

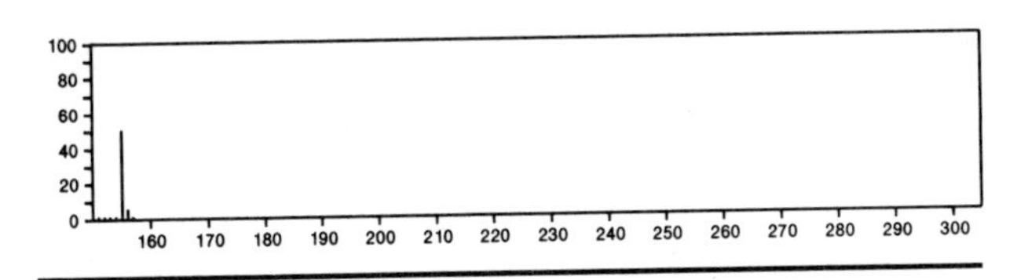

155 $C_9H_{17}NO$ 31172-60-8

2*H*-Quinolizin-2-ol, octahydro-, *trans*-

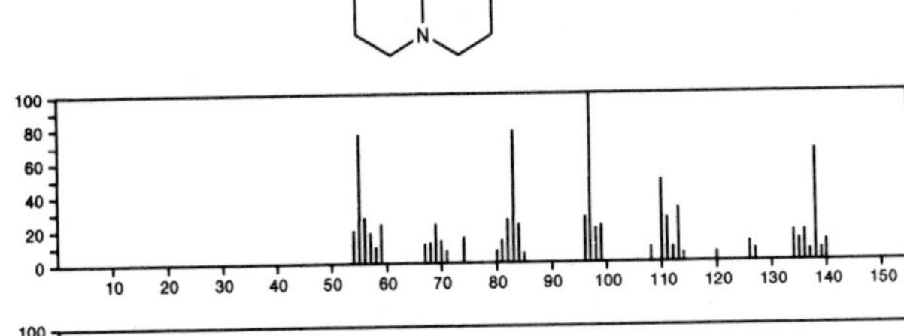

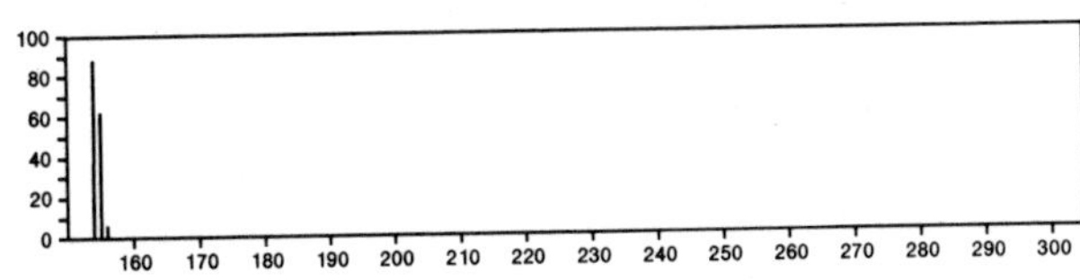

155 $C_9H_{17}NO$ 35790-99-9

6-Azabicyclo[3.2.1]octane, 5-methoxy-6-methyl-

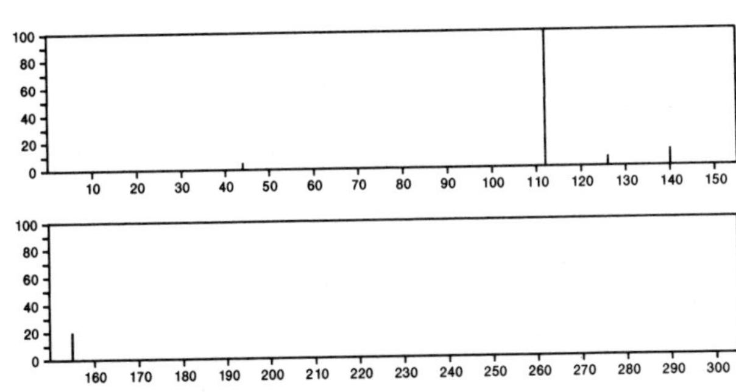

155 $C_9H_{17}NO$ 37835-57-7

4-Pyridinemethanol, 1-ethyl-1,2,3,6-tetrahydro-α-methyl-

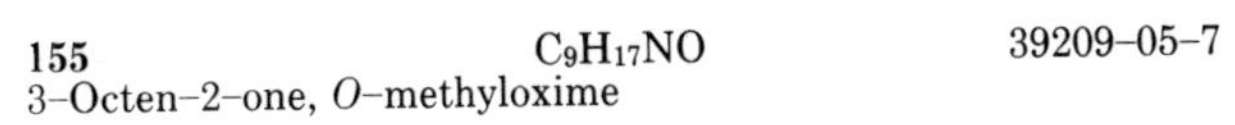

155 $C_9H_{17}NO$ 39209-05-7

3-Octen-2-one, *O*-methyloxime

Me(CH2)3CH=CHCMe=NOMe

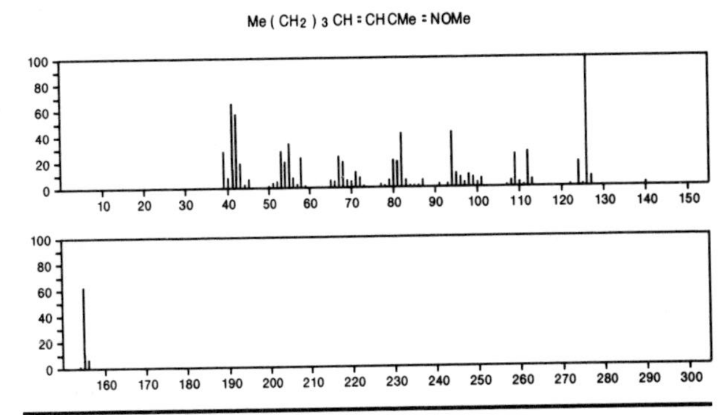

155 $C_9H_{17}NO$ 54244-76-7

Acetamide, *N*-cyclopentyl-*N*-ethyl-

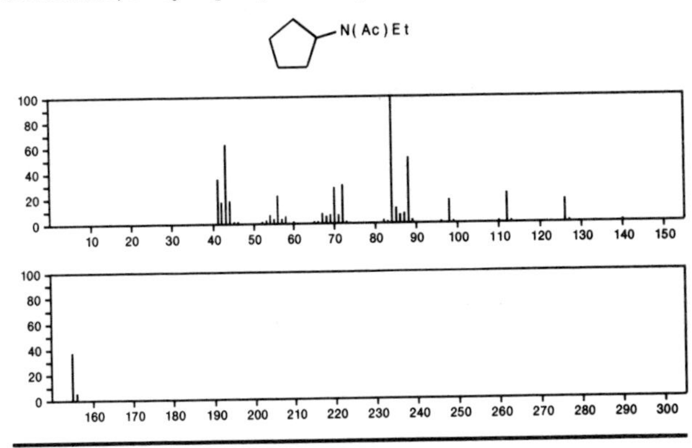

155 $C_9H_{17}NO$ 54308-61-1

2*H*-Quinolizin-3-ol, octahydro-

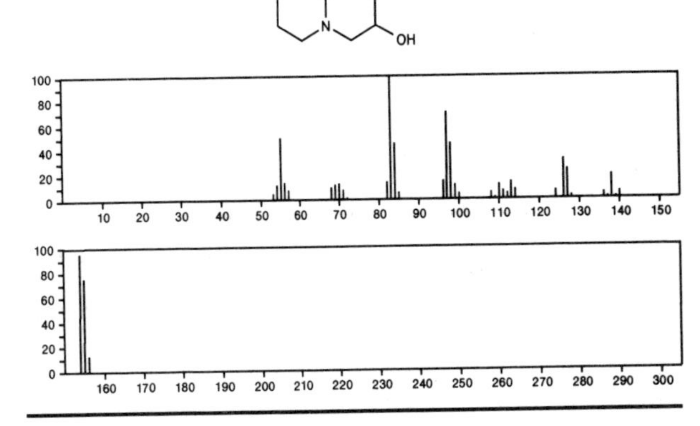

155 $C_9H_{17}NO$ 54308-62-2

2*H*-Quinolizin-2-ol, octahydro-

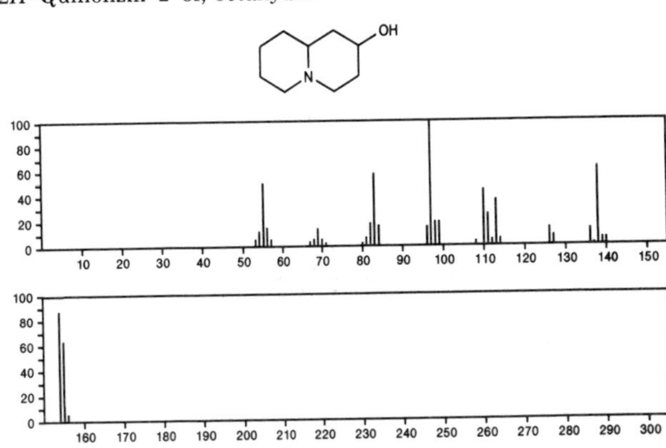

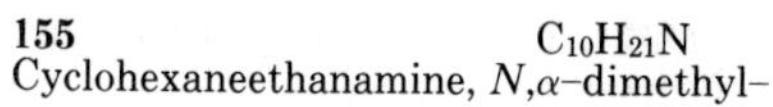

155 $C_{10}H_{21}N$ 101-40-6
Cyclohexaneethanamine, *N*,α-dimethyl-

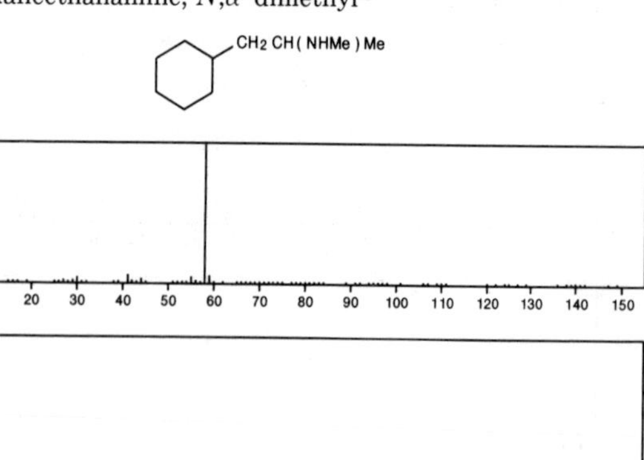

155 $C_{10}H_{21}N$ 10324-58-0
Piperidine, 1-pentyl-

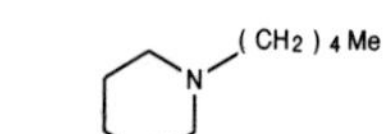

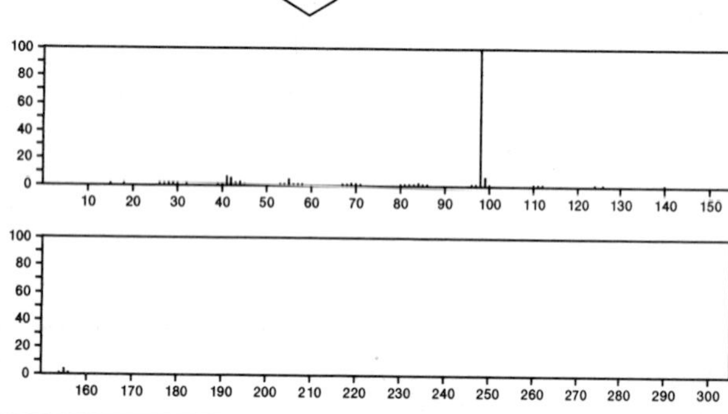

155 $C_{10}H_{21}N$ 10599-81-2
Methylamine, *N*-(1-butylpentylidene)-

(CH2) 3 Me
Me (CH2) 3 C = NMe

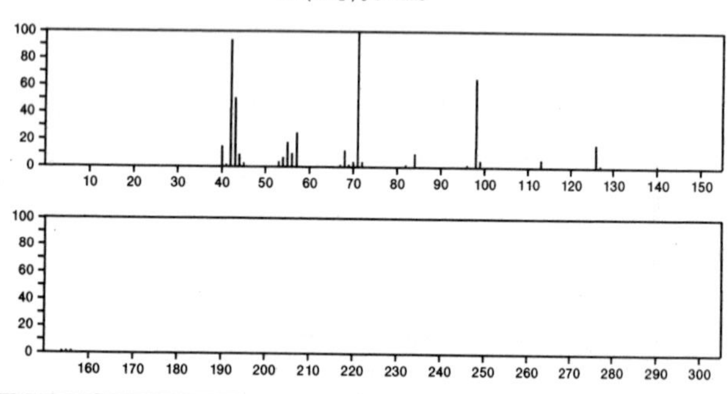

155 $C_{10}H_{21}N$ 18641-77-5
Methylamine, *N*-(1-isopropylhexylidene)-

CHMe 2
Me (CH2) 4 C = NMe

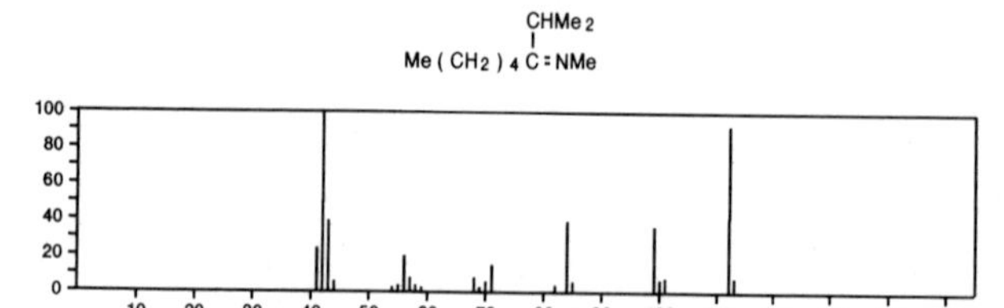

155 $C_{10}H_{21}N$ 23399-21-5
Cyclohexanamine, 5-methyl-2-(1-methylethyl)-, (1α,2α,5β)-

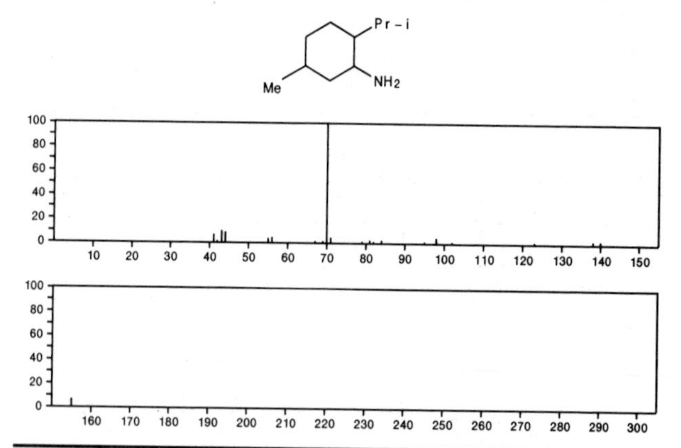

155 $C_{10}H_{21}N$ 54518-97-7
1-Butanamine, 2-methyl-*N*-(2-methylbutylidene)-

Me CH2 CHMe CH2 N = CHCHMe CH2 Me

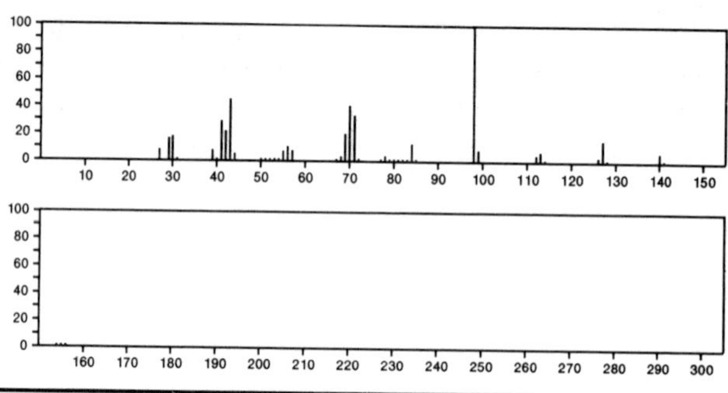

155 $C_{10}H_{21}N$ 55956-31-5
3-Octen-2-amine, *N*,*N*-dimethyl-, (*E*)-

Me (CH2) 3 CH = CHCHMe NMe 2

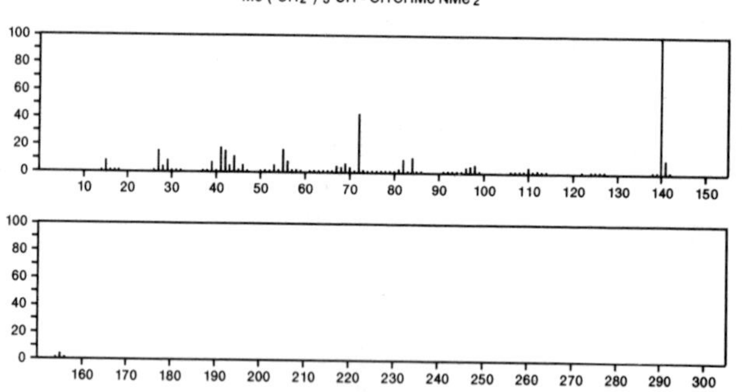

156 C_2H_5I 75-03-6
Ethane, iodo-

EtI

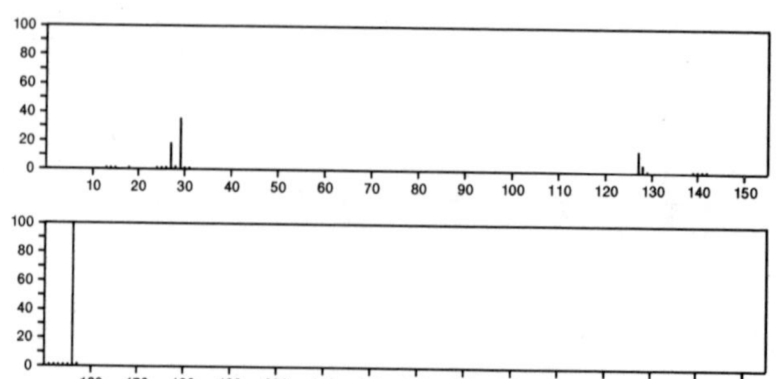

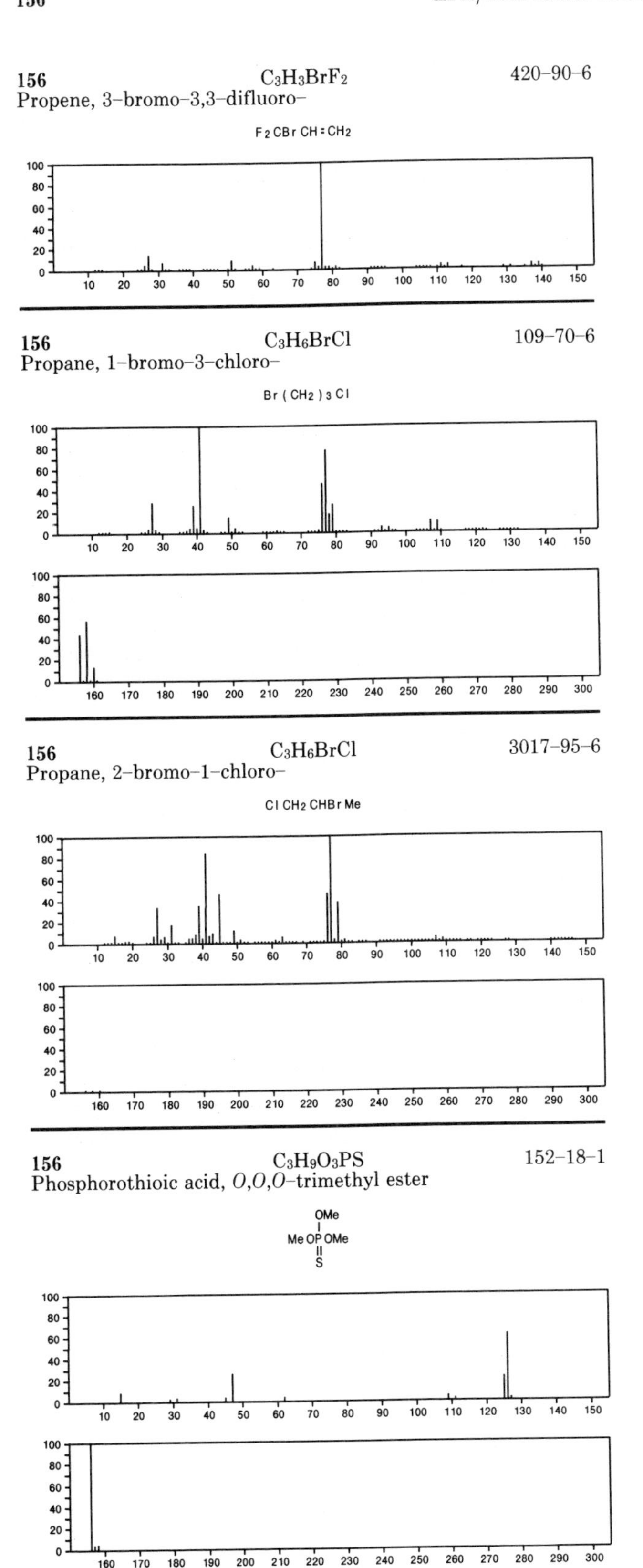
156 C3H3BrF2 420-90-6
Propene, 3-bromo-3,3-difluoro-
F2CBr CH=CH2
156 C3H6BrCl 109-70-6
Propane, 1-bromo-3-chloro-
Br (CH2)3 Cl
156 C3H6BrCl 3017-95-6
Propane, 2-bromo-1-chloro-
Cl CH2 CHBr Me
156 C3H9O3PS 152-18-1
Phosphorothioic acid, O,O,O-trimethyl ester
OMe
Me OP OMe
S

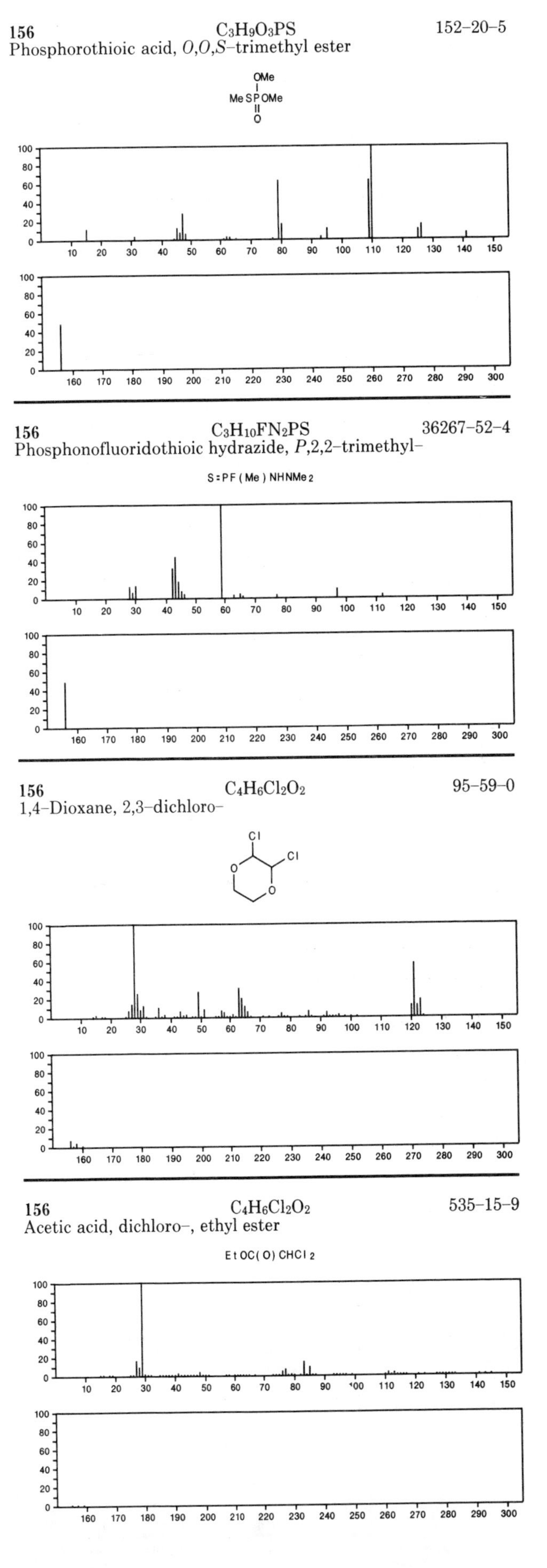
156 C3H9O3PS 152-20-5
Phosphorothioic acid, O,O,S-trimethyl ester
OMe
Me SP OMe
O
156 C3H10FN2PS 36267-52-4
Phosphonofluoridothioic hydrazide, P,2,2-trimethyl-
S=PF(Me) NHNMe2
156 C4H6Cl2O2 95-59-0
1,4-Dioxane, 2,3-dichloro-
Cl
O
156 C4H6Cl2O2 535-15-9
Acetic acid, dichloro-, ethyl ester
Et OC(O) CHCl2

156 $C_4H_6Cl_2O_2$ 600–32–8
Butyric acid, 2,3–dichloro–

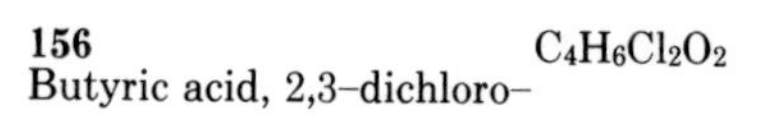

$MeCHCl\,CHCl\,CO_2H$

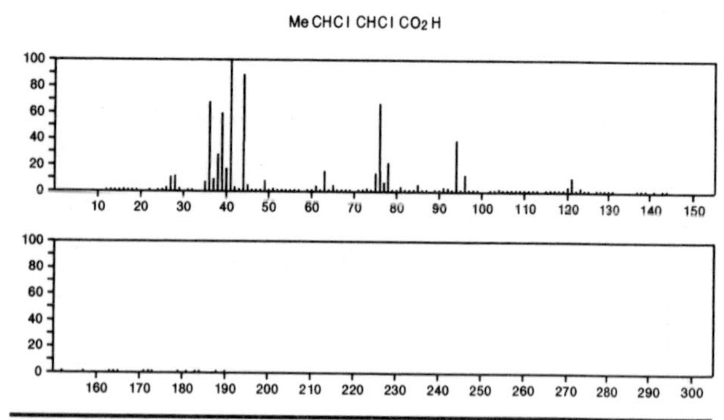

156 $C_4H_6Cl_2O_2$ 817–80–1
Carbonochloridic acid, 2–chloro–1–methylethyl ester

$ClC(O)OCHMeCH_2Cl$

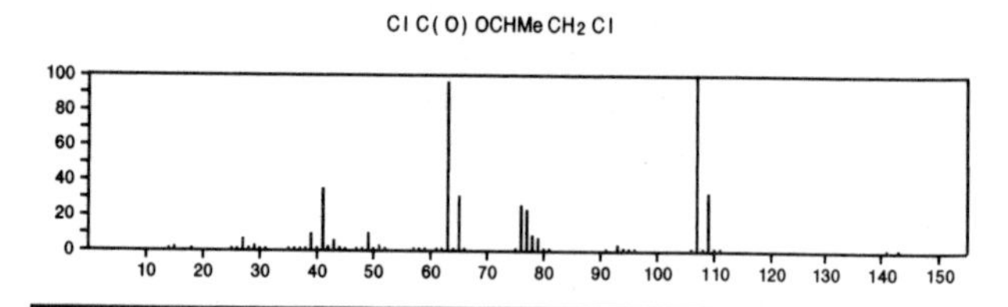

156 $C_4H_6Cl_2O_2$ 2612–35–3
1,3–Dioxolane, 2–(dichloromethyl)–

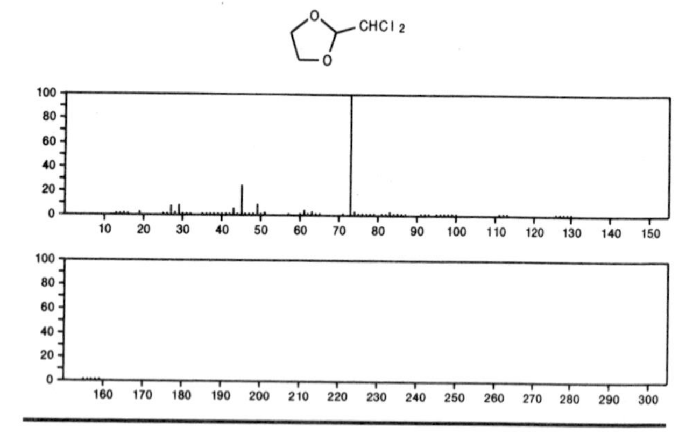

156 $C_4H_6Cl_2O_2$ 3674–09–7
Propanoic acid, 2,3–dichloro–, methyl ester

$MeOC(O)CHCl\,CH_2Cl$

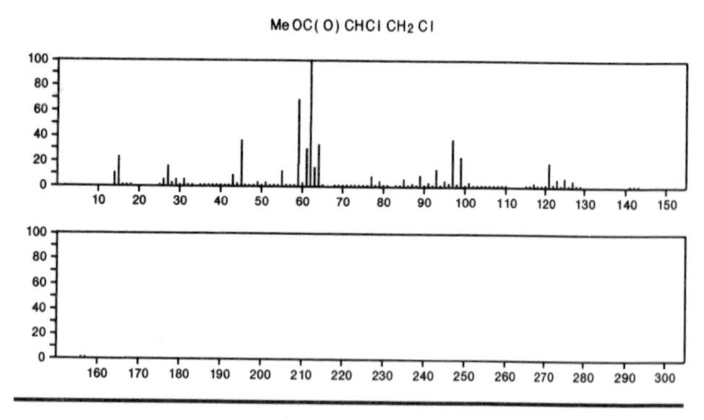

156 $C_4H_6Cl_2O_2$ 13023–00–2
Butyric acid, 2,2–dichloro–

$EtCCl_2(CO_2H)$

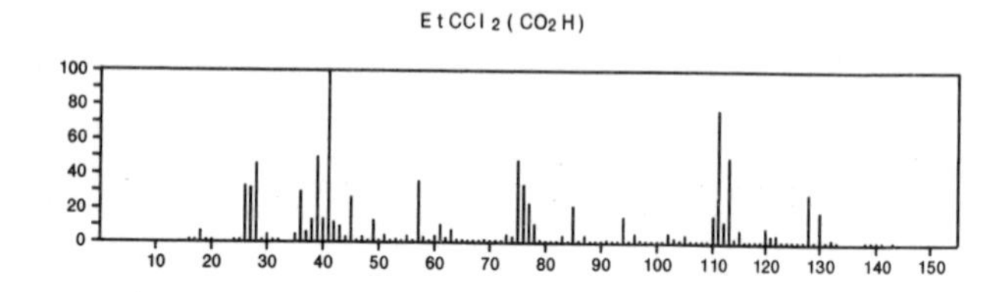

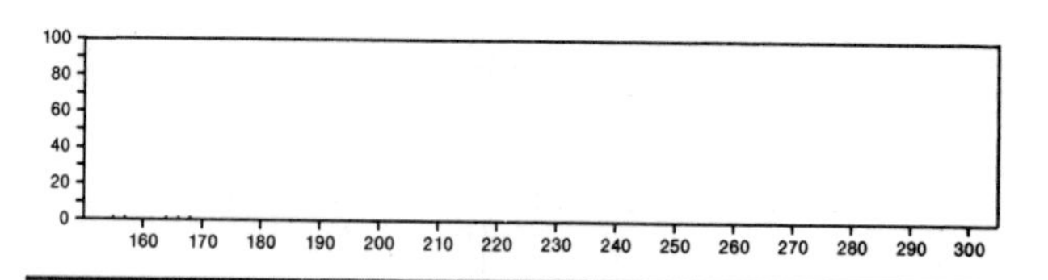

156 $C_4H_6Cl_2O_2$ 17640–02–7
Propanoic acid, 2,2–dichloro–, methyl ester

$MeOC(O)CCl_2Me$

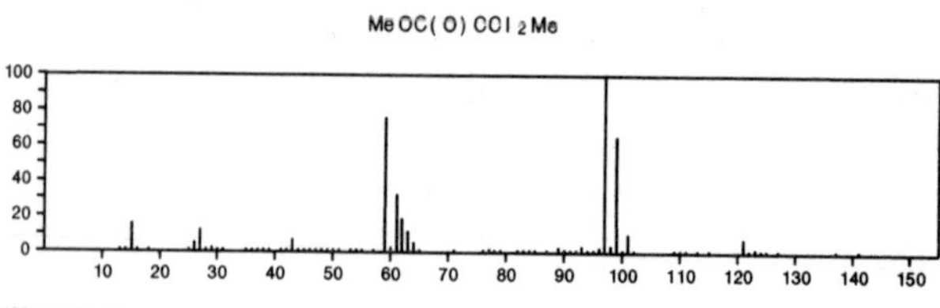

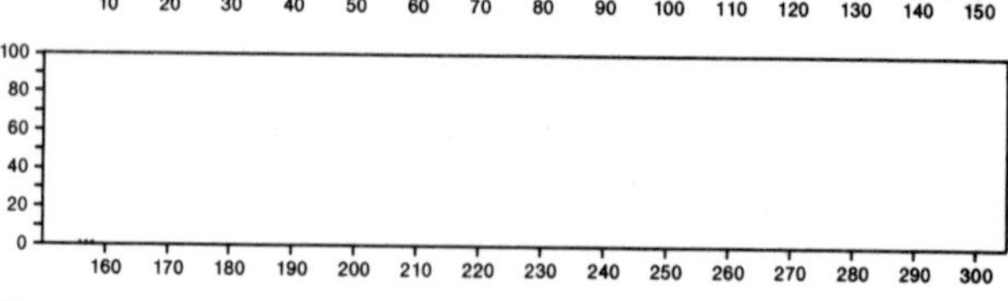

156 $C_5H_4N_2O_4$ 65–86–1
4–Pyrimidinecarboxylic acid, 1,2,3,6–tetrahydro–2,6–dioxo–

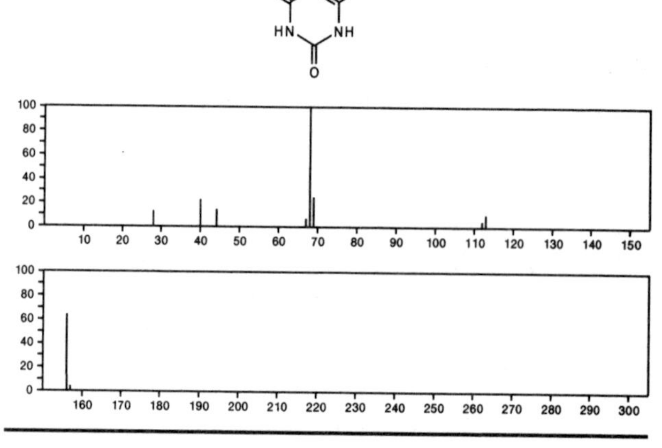

156 $C_6H_4O_5$ 3238–40–2
2,5–Furandicarboxylic acid

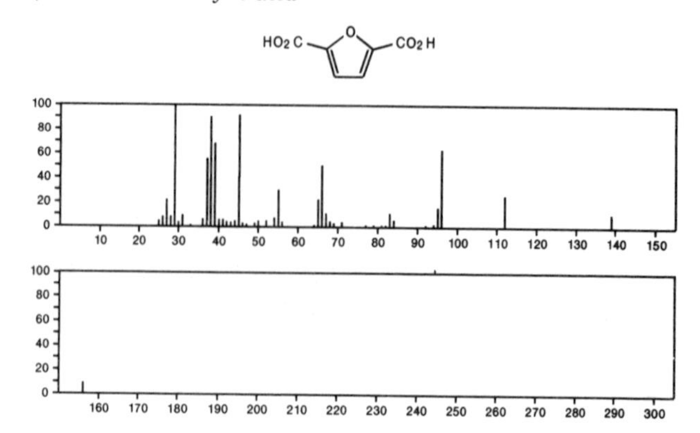

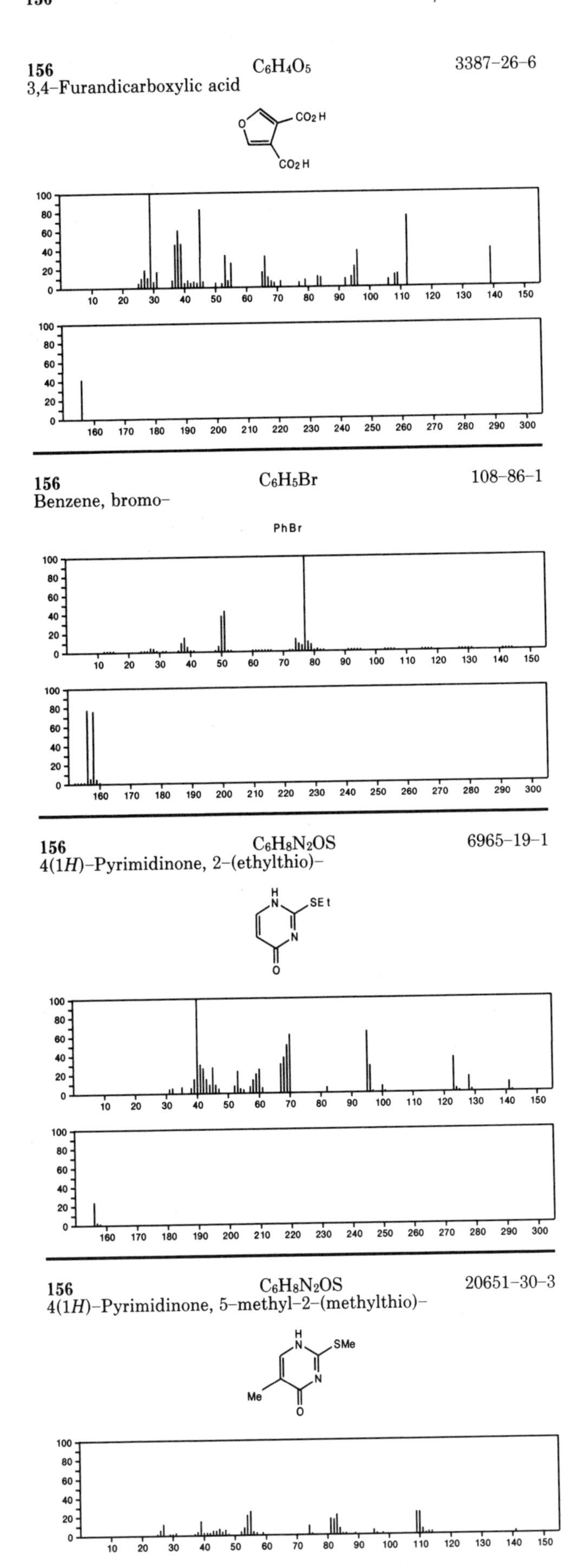

156 $C_6H_4O_5$ 3387-26-6
3,4-Furandicarboxylic acid

156 C_6H_5Br 108-86-1
Benzene, bromo-

156 $C_6H_8N_2OS$ 6965-19-1
4(1*H*)-Pyrimidinone, 2-(ethylthio)-

156 $C_6H_8N_2OS$ 20651-30-3
4(1*H*)-Pyrimidinone, 5-methyl-2-(methylthio)-

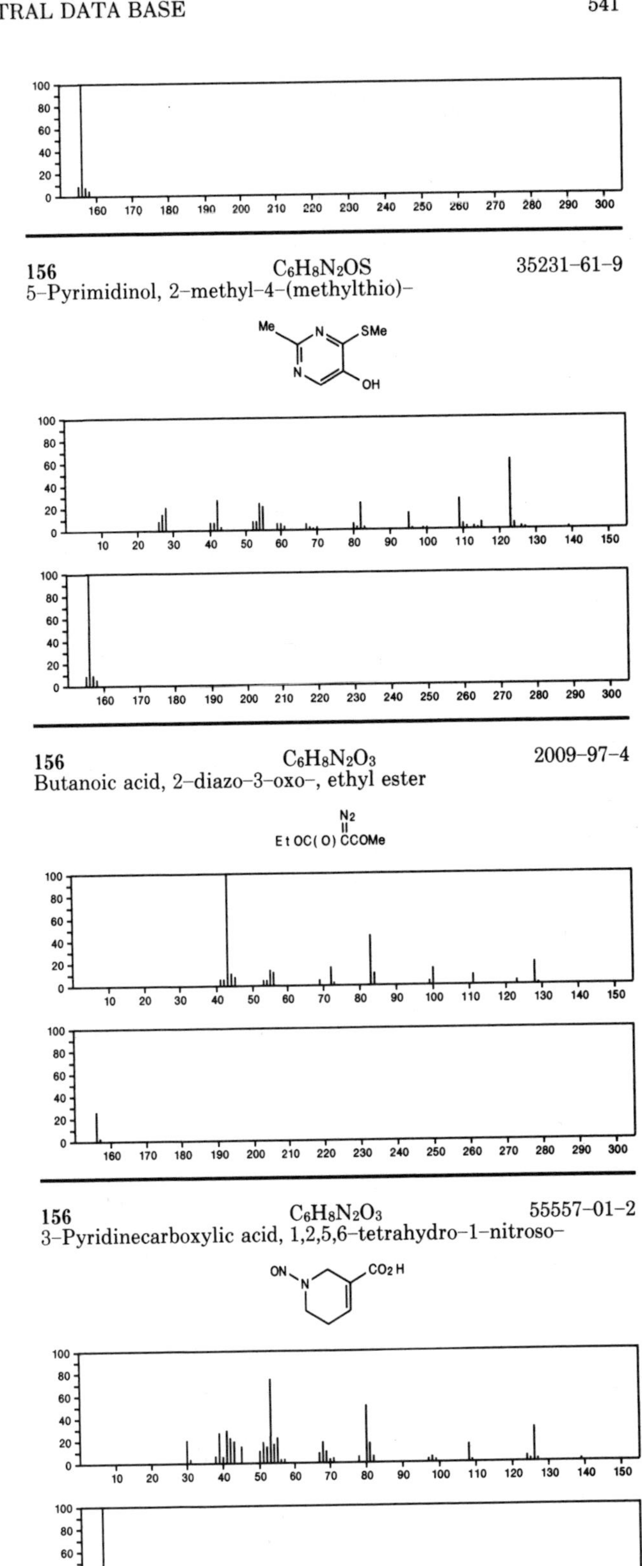

156 $C_6H_8N_2OS$ 35231-61-9
5-Pyrimidinol, 2-methyl-4-(methylthio)-

156 $C_6H_8N_2O_3$ 2009-97-4
Butanoic acid, 2-diazo-3-oxo-, ethyl ester

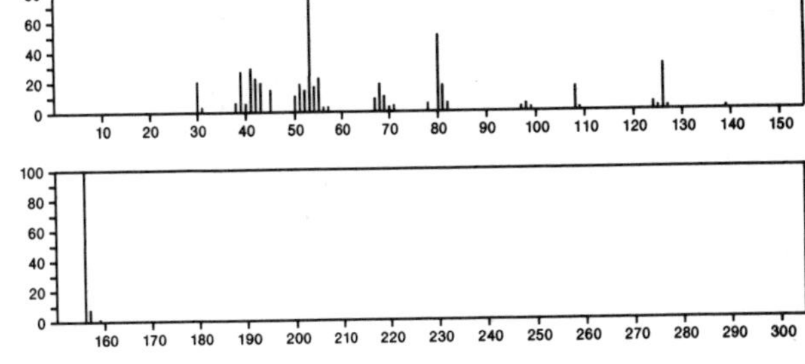

156 $C_6H_8N_2O_3$ 55557-01-2
3-Pyridinecarboxylic acid, 1,2,5,6-tetrahydro-1-nitroso-

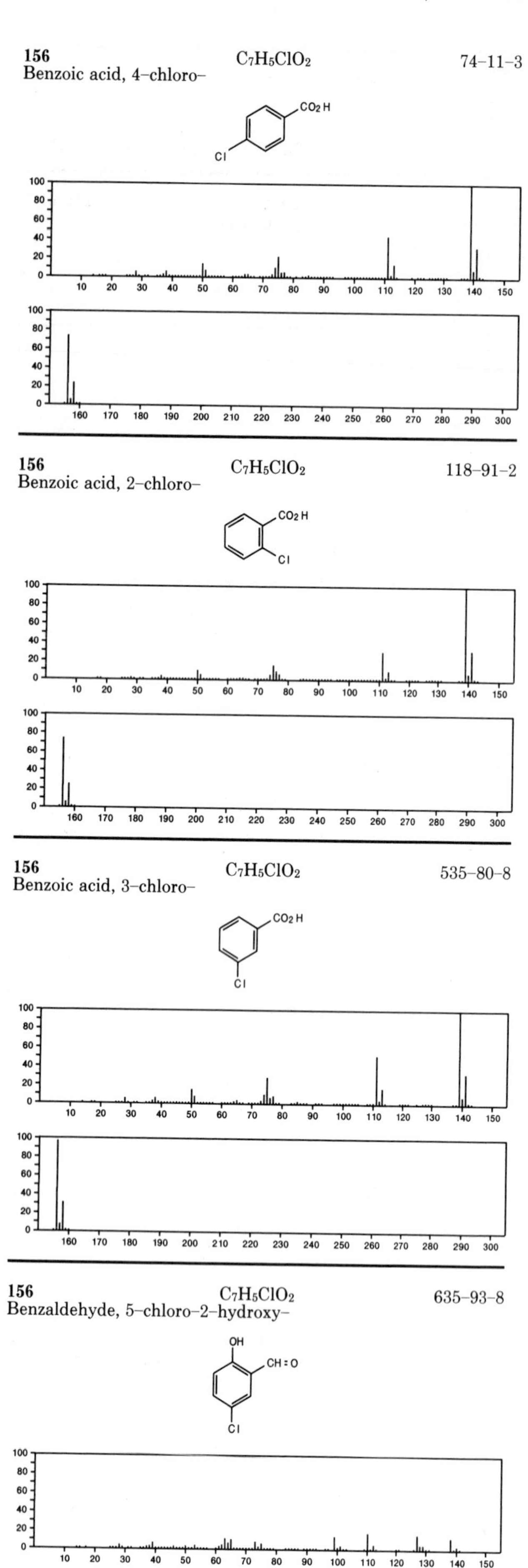

156
$C_7H_5ClO_2$
74-11-3
Benzoic acid, 4-chloro-
CO2H
Cl
156
$C_7H_5ClO_2$
118-91-2
Benzoic acid, 2-chloro-
CO2H
Cl
156
$C_7H_5ClO_2$
535-80-8
Benzoic acid, 3-chloro-
CO2H
Cl
156
$C_7H_5ClO_2$
635-93-8
Benzaldehyde, 5-chloro-2-hydroxy-
OH
CH:O
Cl

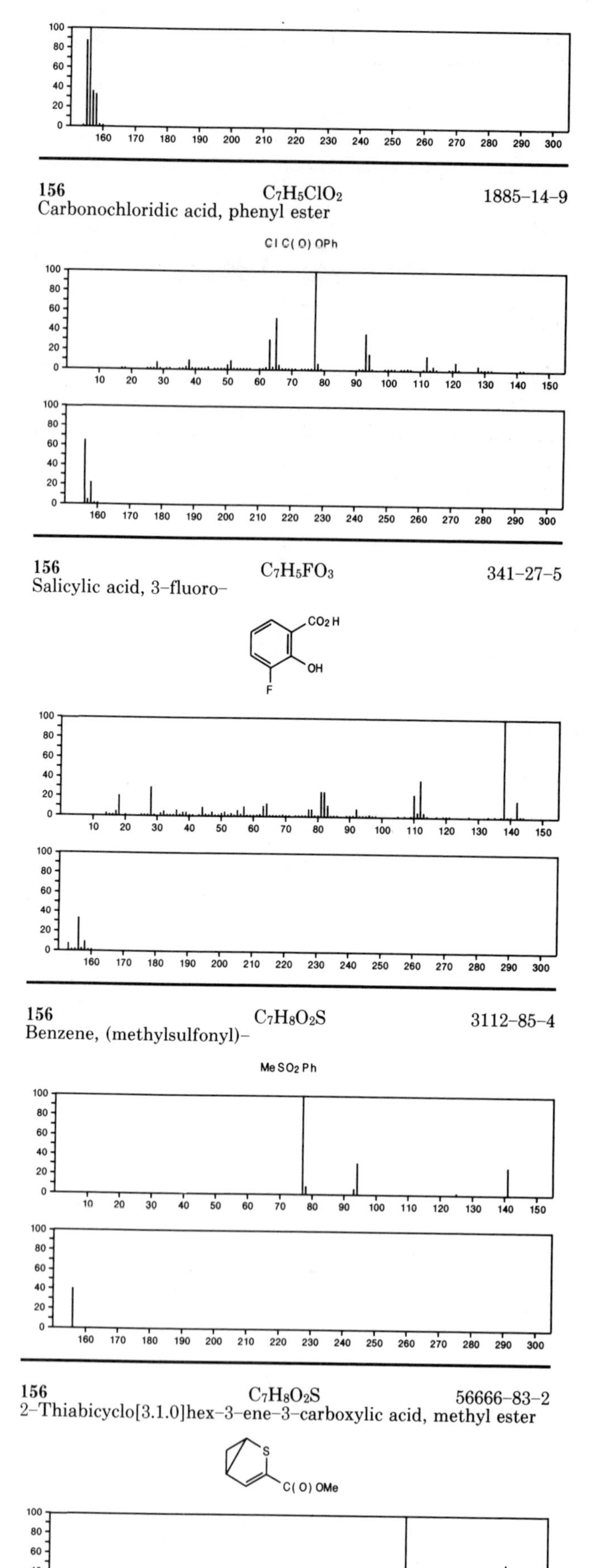

156
$C_7H_5ClO_2$
1885-14-9
Carbonochloridic acid, phenyl ester
Cl C(O) OPh
156
$C_7H_5FO_3$
341-27-5
Salicylic acid, 3-fluoro-
CO2H
OH
F
156
$C_7H_8O_2S$
3112-85-4
Benzene, (methylsulfonyl)-
Me SO2 Ph
156
$C_7H_8O_2S$
56666-83-2
2-Thiabicyclo[3.1.0]hex-3-ene-3-carboxylic acid, methyl ester
S
C(O) OMe

156 $C_7H_8O_4$ 3128–15–2
1–Cyclopentene–1,2–dicarboxylic acid

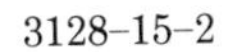

CO_2H CO_2H

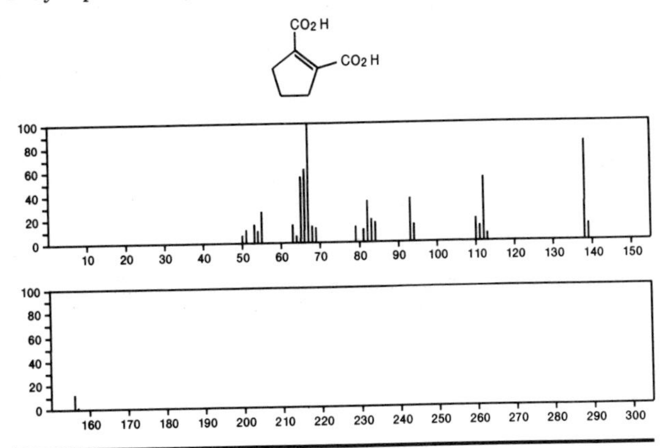

156 $C_7H_8O_4$ 24554–00–5
1,5–Cyclohexadiene–1–carboxylic acid, 3,4–dihydroxy–

HO CO_2H HO

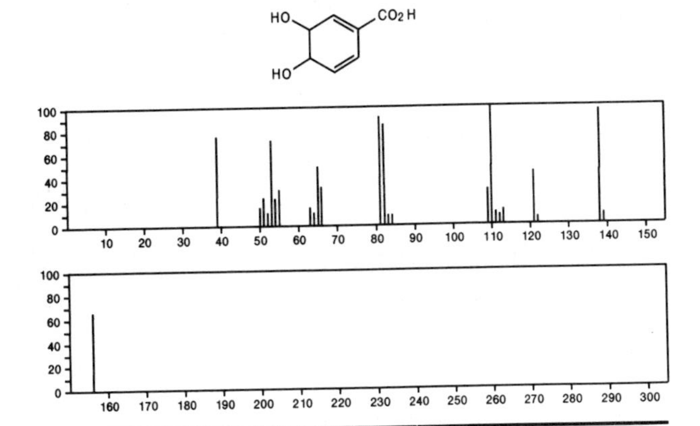

156 $C_7H_8O_4$ 36802–01–4
2–Furancarboxylic acid, 5–(hydroxymethyl)–, methyl ester

C(O)OMe $HOCH_2$

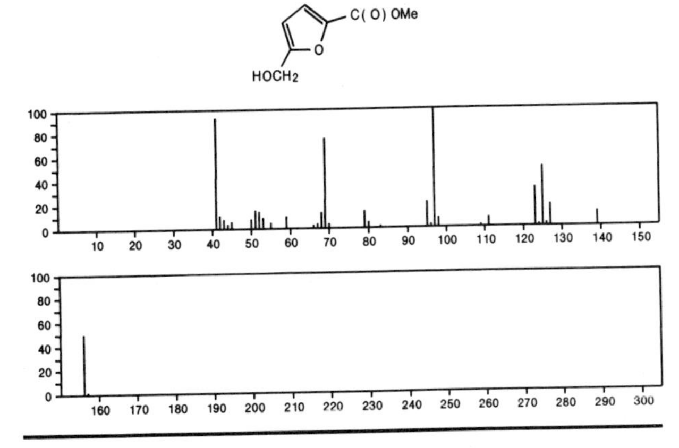

156 $C_7H_8O_4$ 56771–77–8
4–Hexenoic acid, 3–methyl–2,6–dioxo–

$OCHCH{=}CHCHMe\,COCO_2H$

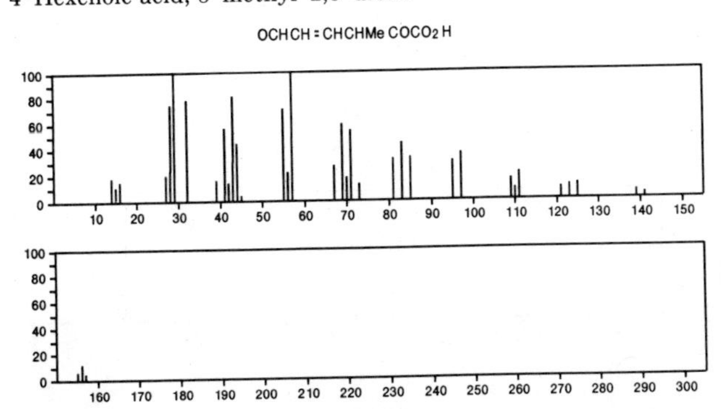

156 $C_7H_{12}N_2O_2$ 5455–34–5
2,4–Imidazolidinedione, 5,5–diethyl–

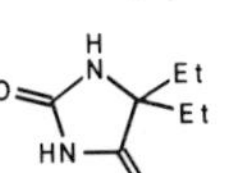

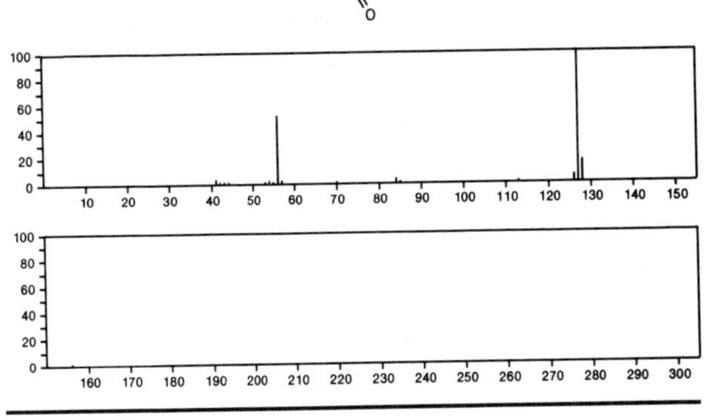

156 $C_7H_{12}N_2O_2$ 14702–41–1
Carbazic acid, 3–cyclopentylidene–, methyl ester

NNHC(O)OMe

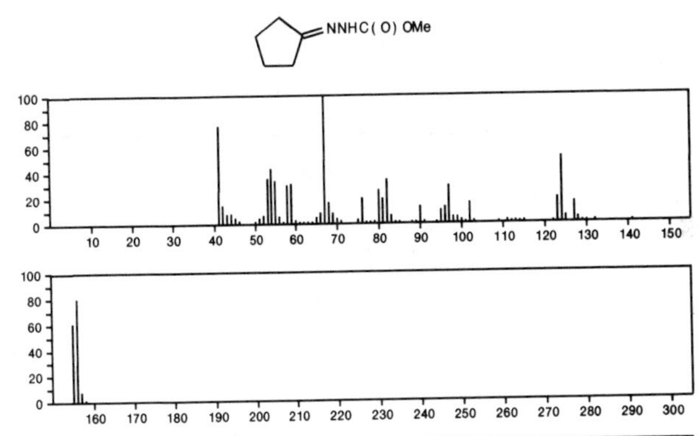

156 $C_7H_{12}N_2O_2$ 26537–48–4
Sydnone, 3–neopentyl–

CH_2CMe_3

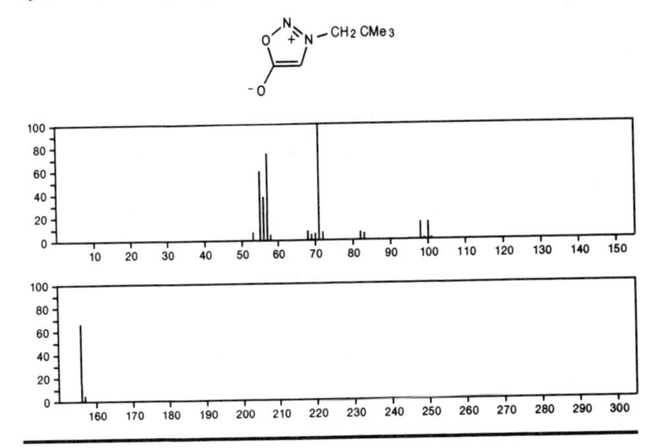

156 $C_7H_{12}N_2O_2$ 26537–51–9
Sydnone, 3–isopentyl–

$CH_2CH_2CHMe_2$

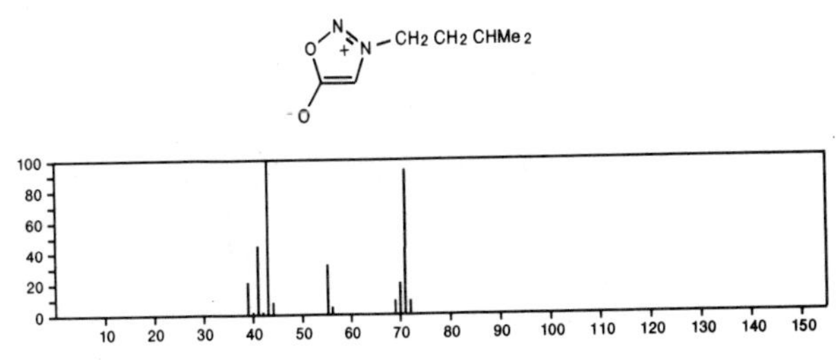

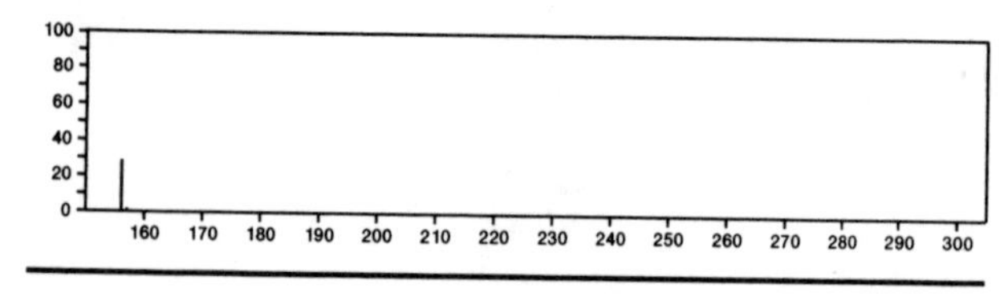

156 $C_7H_{12}N_2O_2$ 33599-31-4
Hydantoin, 3-butyl-

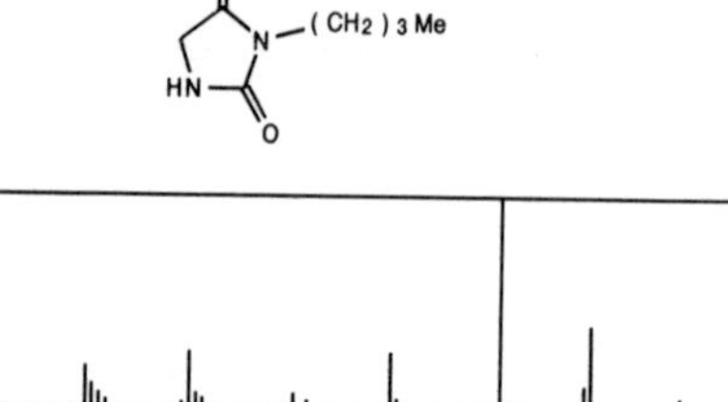

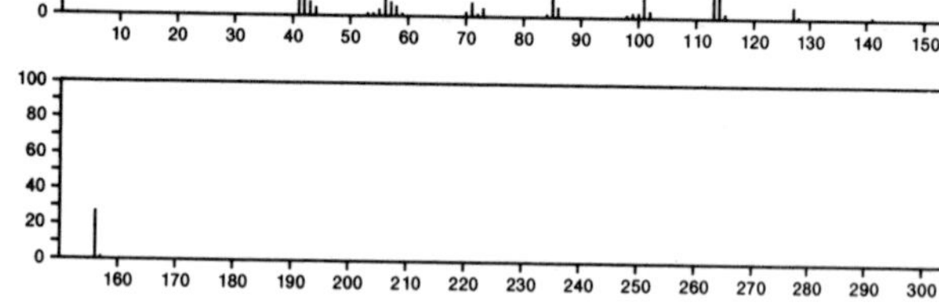

156 $C_7H_{12}N_2O_2$ 33599-32-5
Hydantoin, 1-butyl-

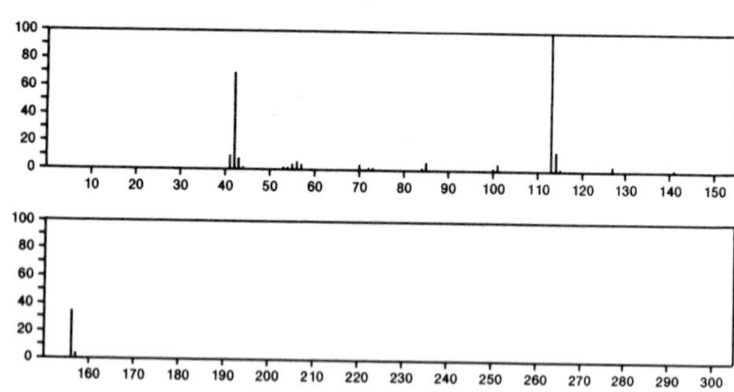

156 $C_7H_{13}BO_3$ 24372-01-8
Lactic acid, monoanhydride with 1-butaneboronic acid, cyclic ester

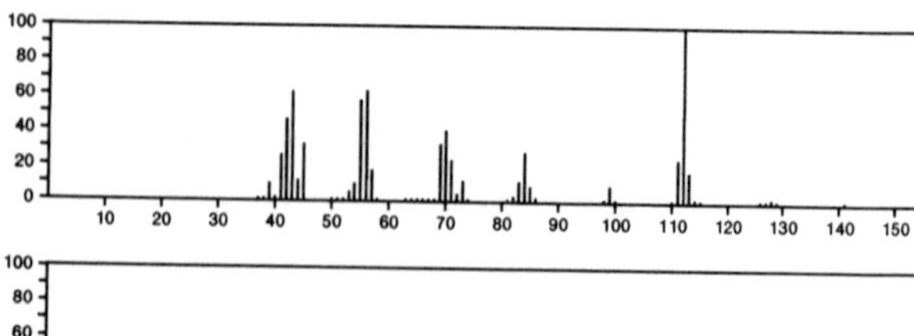

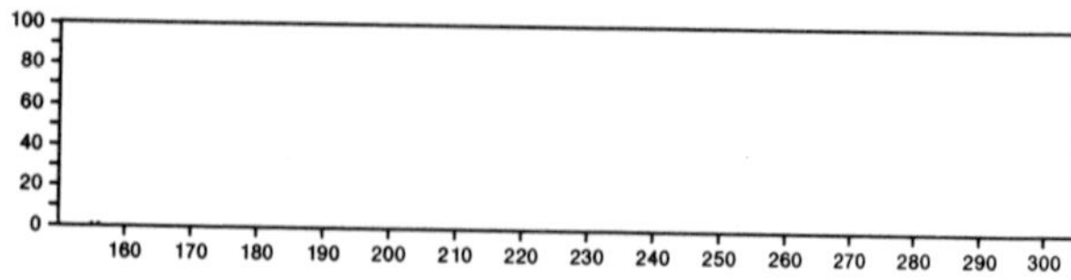

156 $C_7H_{13}BO_3$ 33823-94-8
Hydracrylic acid, monoanhydride with 1-butaneboronic acid, cyclic ester

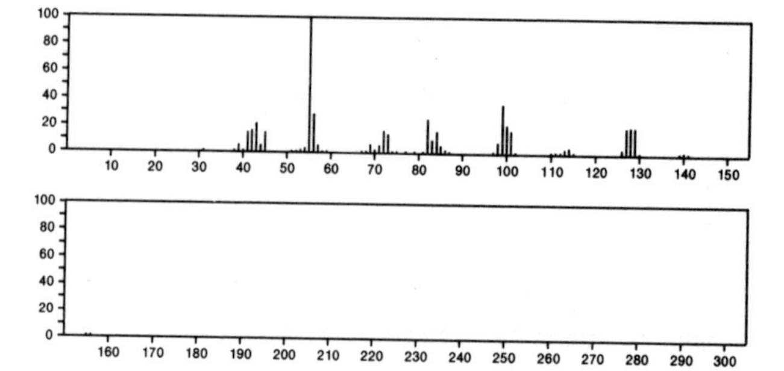

156 $C_7H_{16}Si_2$ 5927-28-6
1,3-Disilacyclopent-4-ene, 1,1,3,3-tetramethyl-

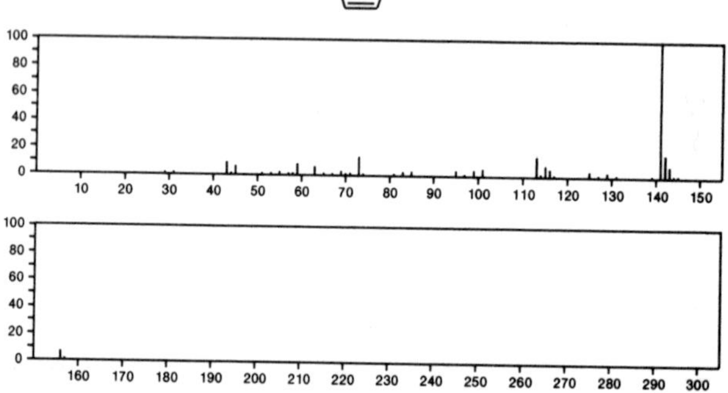

156 $C_8H_4N_4$ 29482-47-1
Pyrido[2,3-*d*]pyrimidine-4-carbonitrile

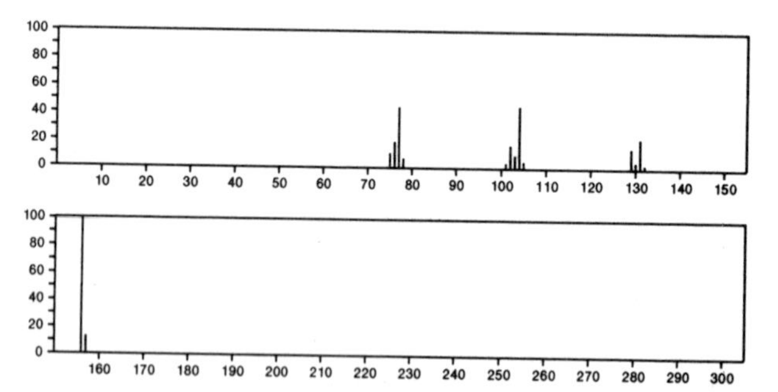

156 C_8H_9ClO 614-72-2
Benzene, 1-chloro-2-ethoxy-

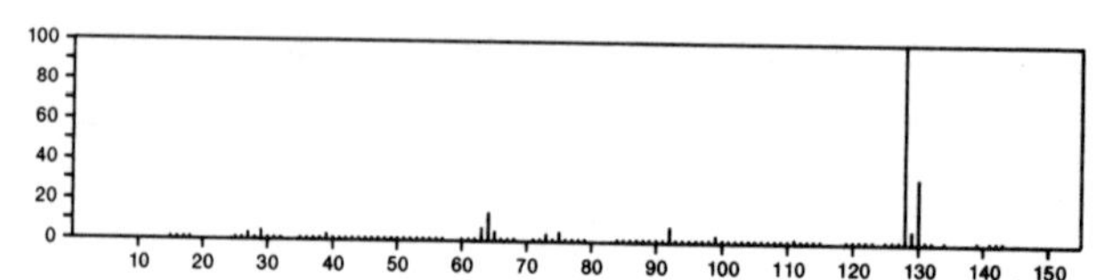

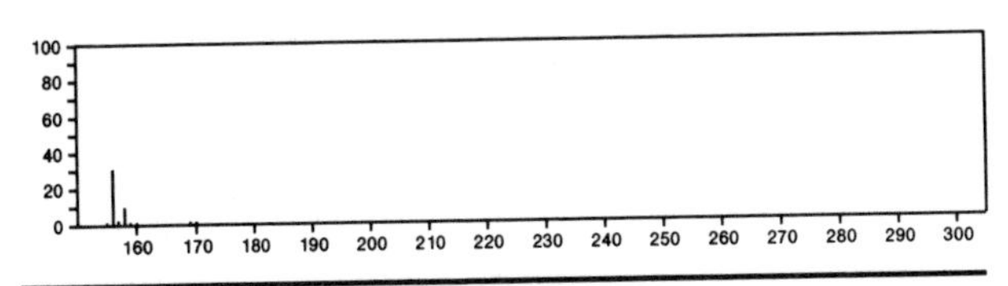

156 C_8H_9ClO 622–61–7

Benzene, 1–chloro–4–ethoxy–

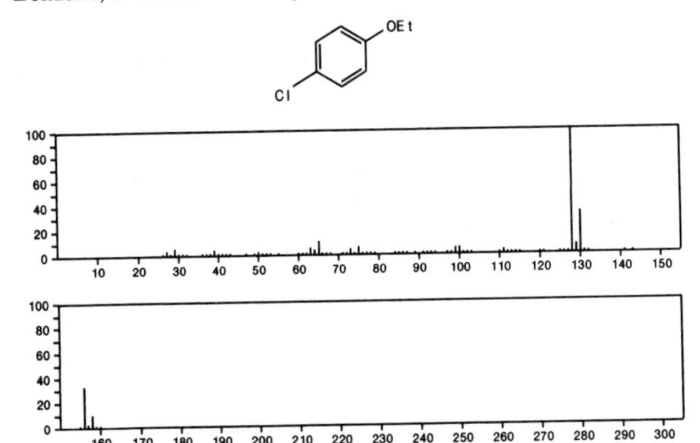

156 C_8H_9ClO 622–86–6

Benzene, (2–chloroethoxy)–

Ph OCH2 CH2 Cl

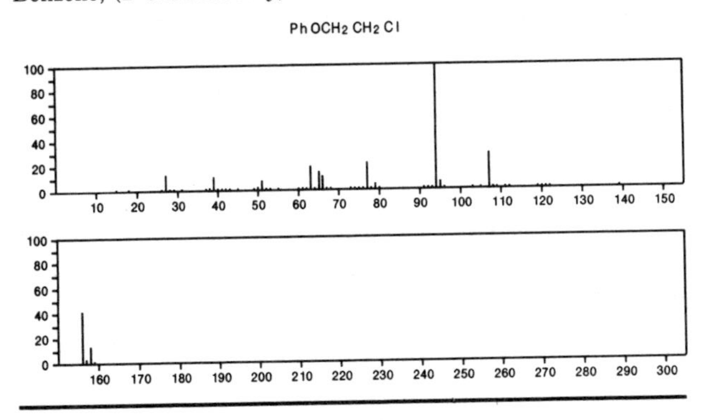

156 C_8H_9ClO 1674–30–2

Benzenemethanol, α–(chloromethyl)–

Cl CH2 CH(OH) Ph

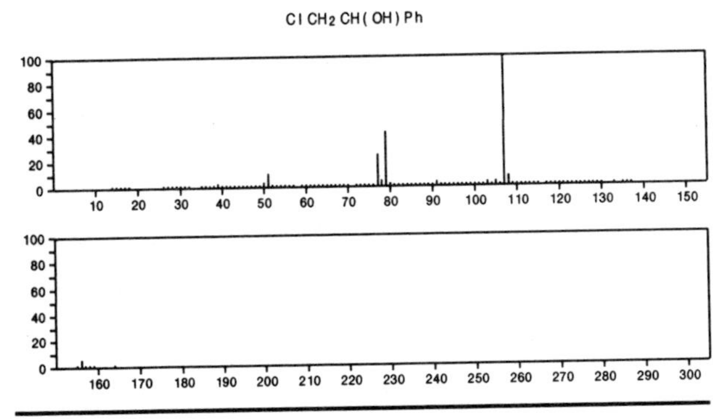

156 C_8H_9ClO 3391–10–4

Benzenemethanol, 4–chloro–α–methyl–

CH(OH) Me

Cl

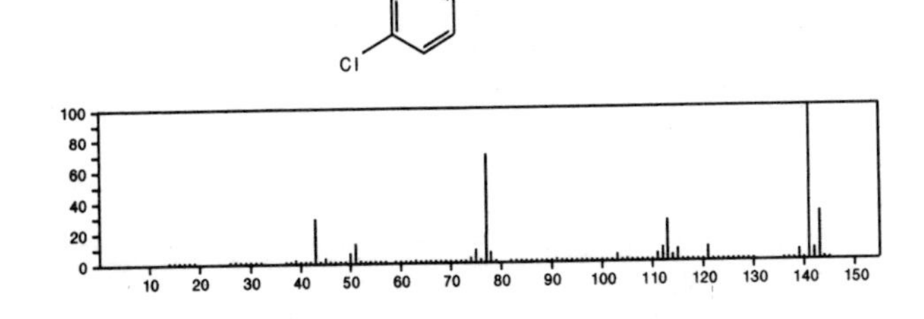

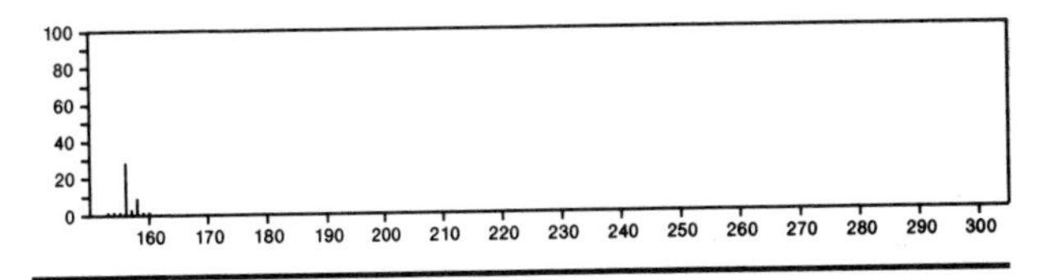

156 C_8H_9ClO 5182–44–5

Benzeneethanol, 3–chloro–

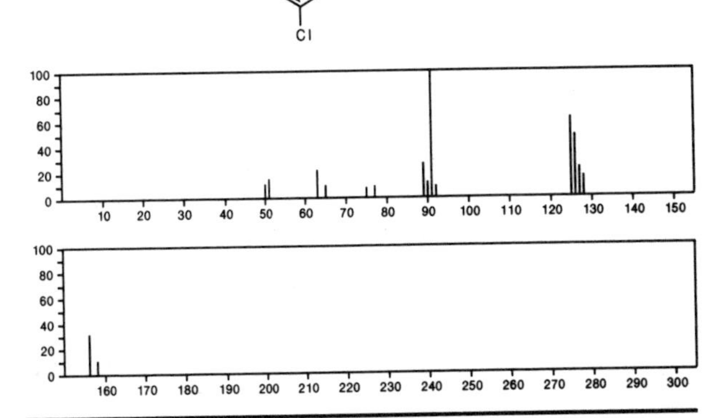

156 C_8H_9ClO 13524–04–4

Benzenemethanol, 2–chloro–α–methyl–

CH(OH) Me

Cl

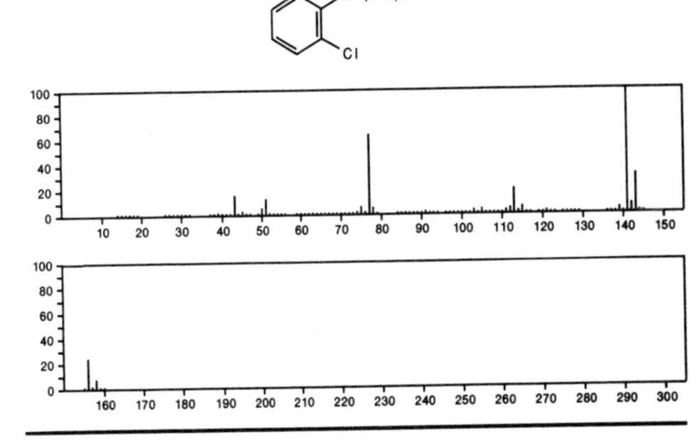

156 $C_8H_{12}O_3$ 611–10–9

Cyclopentanecarboxylic acid, 2–oxo–, ethyl ester

C(O) OEt

O

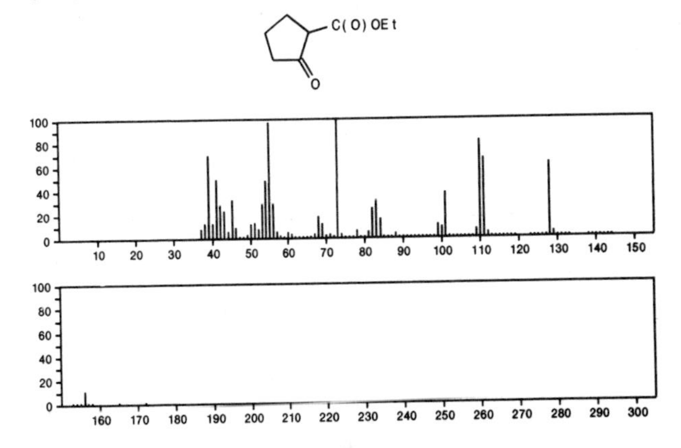

156 $C_8H_{12}O_3$ 4208-60-0
1,2-Butanediol, 1-(2-furyl)-

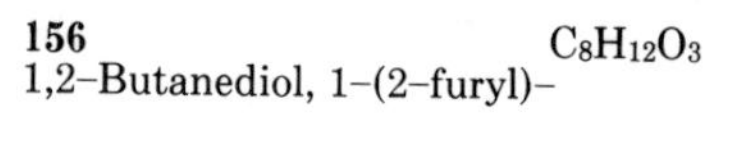

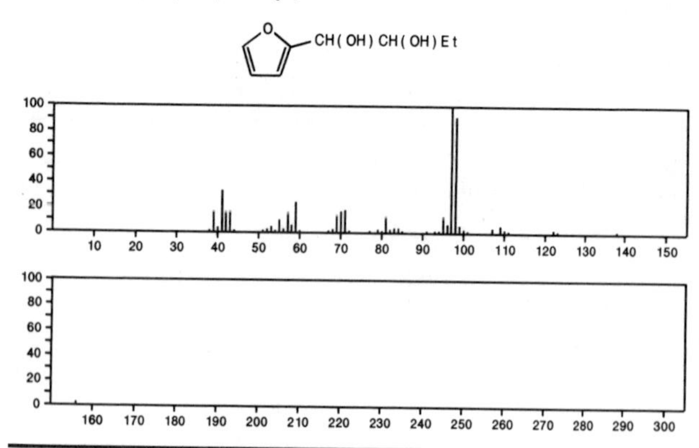

156 $C_8H_{12}O_3$ 4746-97-8
1,4-Dioxaspiro[4.5]decan-8-one

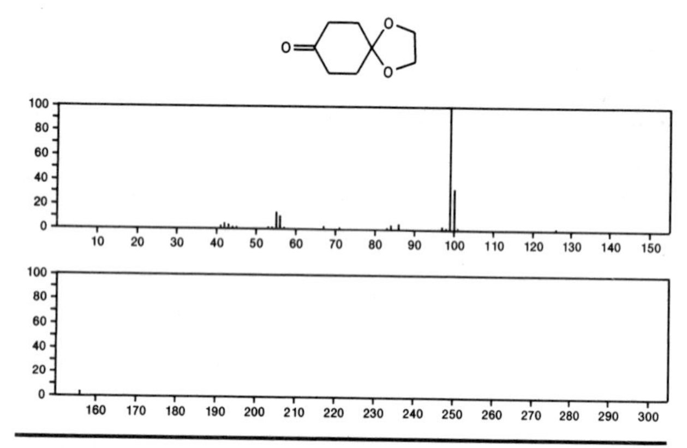

156 $C_8H_{12}O_3$ 14161-46-7
7-Oxabicyclo[4.1.0]heptan-1-ol, acetate

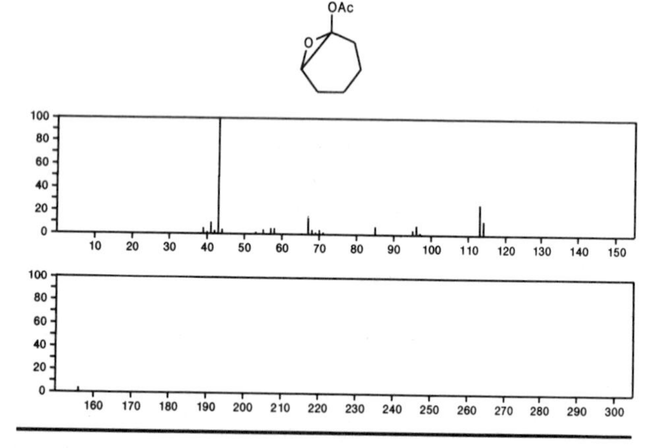

156 $C_8H_{12}O_3$ 14744-18-4
3,4(2*H*,5*H*)-Furandione, 2,2,5,5-tetramethyl-

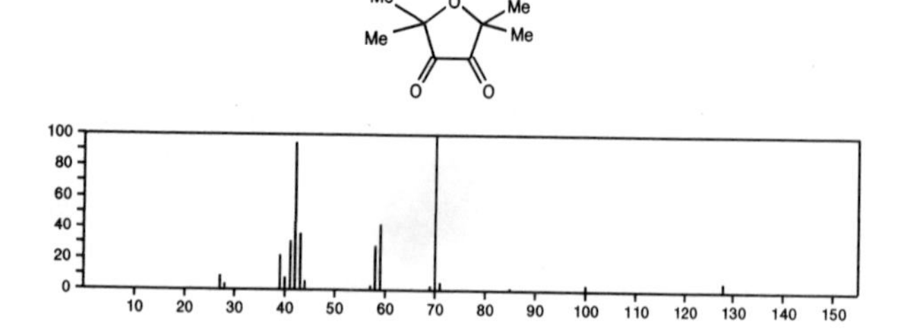

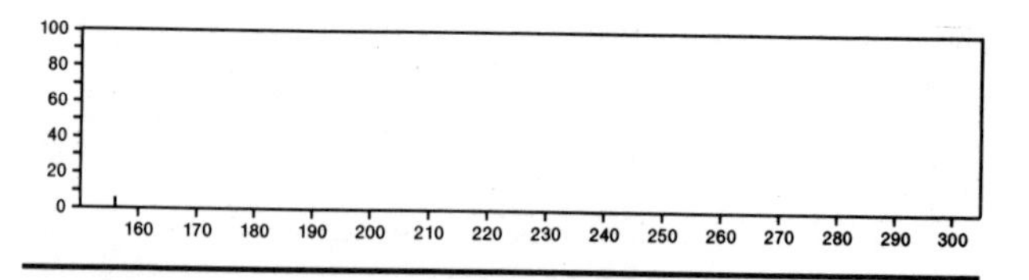

156 $C_8H_{12}O_3$ 55702-68-6
Propanoic acid, 3-(2-propynyloxy)-, ethyl ester

EtOC(O)CH2CH2OCH2C≡CH

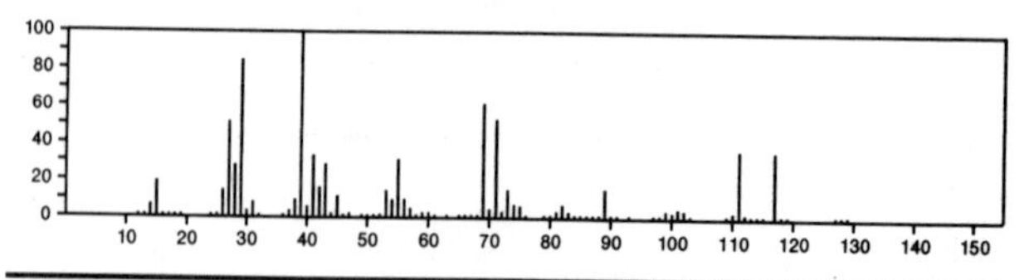

156 $C_8H_{14}NO_2$ 2154-34-9
1-Pyrrolidinyloxy, 2,2,5,5-tetramethyl-3-oxo-

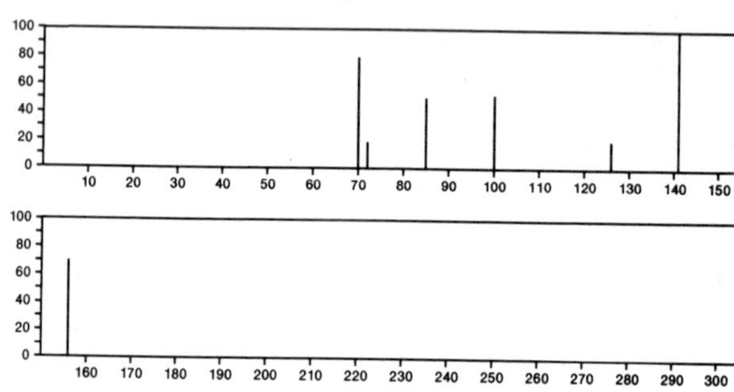

156 $C_8H_{16}N_2O$ 20917-50-4
1*H*-Azonine, octahydro-1-nitroso-

156 $C_8H_{16}N_2O$ 42801-00-3
2-Propenal, 3-(dimethylamino)-3-[(1-methylethyl)amino]-

OCHCH=C(NHPr-i)NMe2

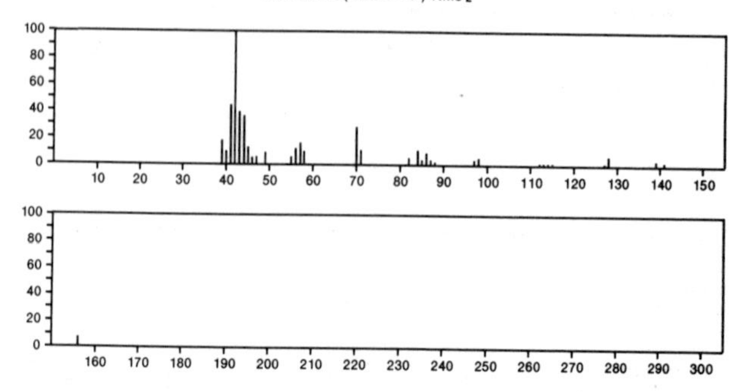

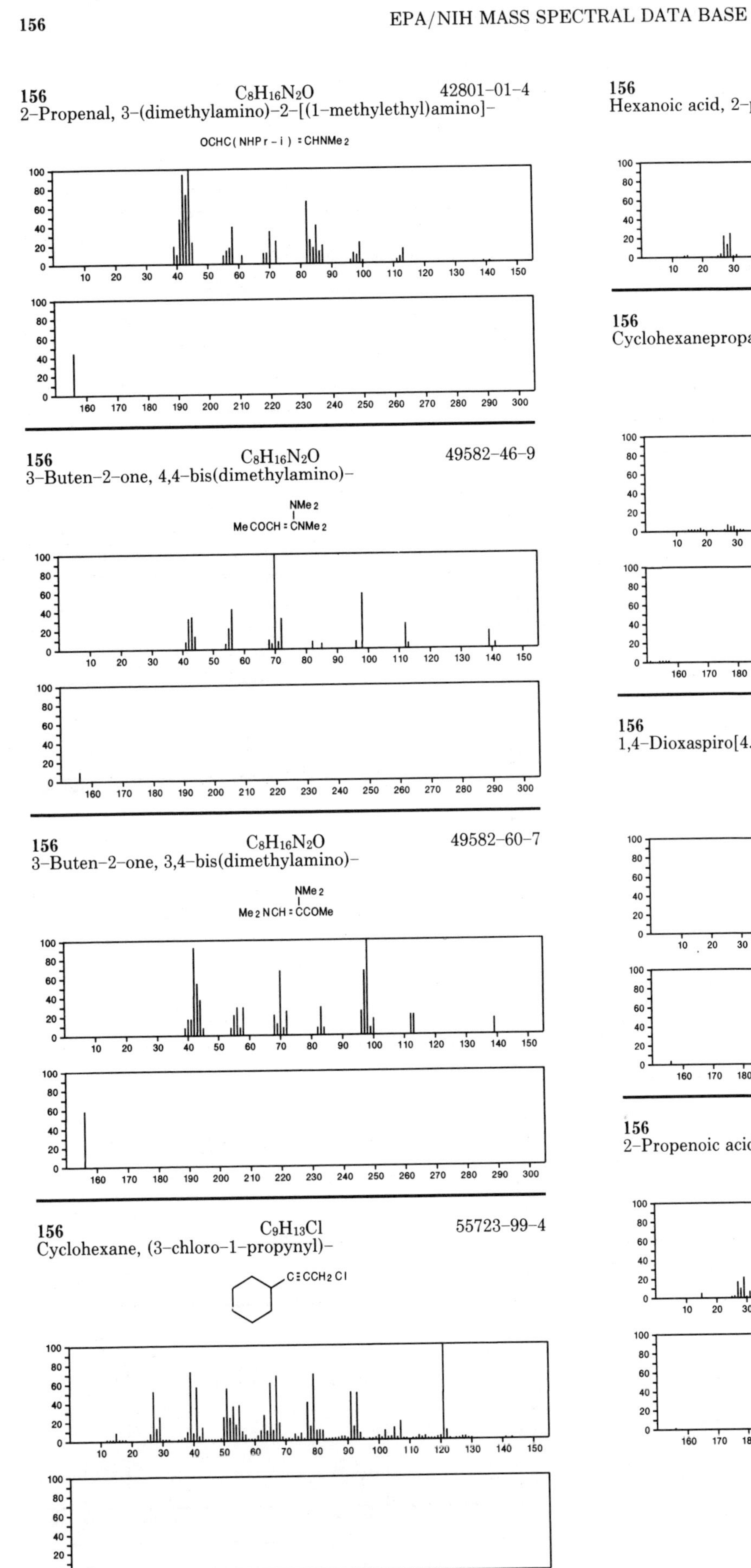

156 $C_8H_{16}N_2O$ 42801-01-4
2-Propenal, 3-(dimethylamino)-2-[(1-methylethyl)amino]-

156 $C_8H_{16}N_2O$ 49582-46-9
3-Buten-2-one, 4,4-bis(dimethylamino)-

156 $C_8H_{16}N_2O$ 49582-60-7
3-Buten-2-one, 3,4-bis(dimethylamino)-

156 $C_9H_{13}Cl$ 55723-99-4
Cyclohexane, (3-chloro-1-propynyl)-

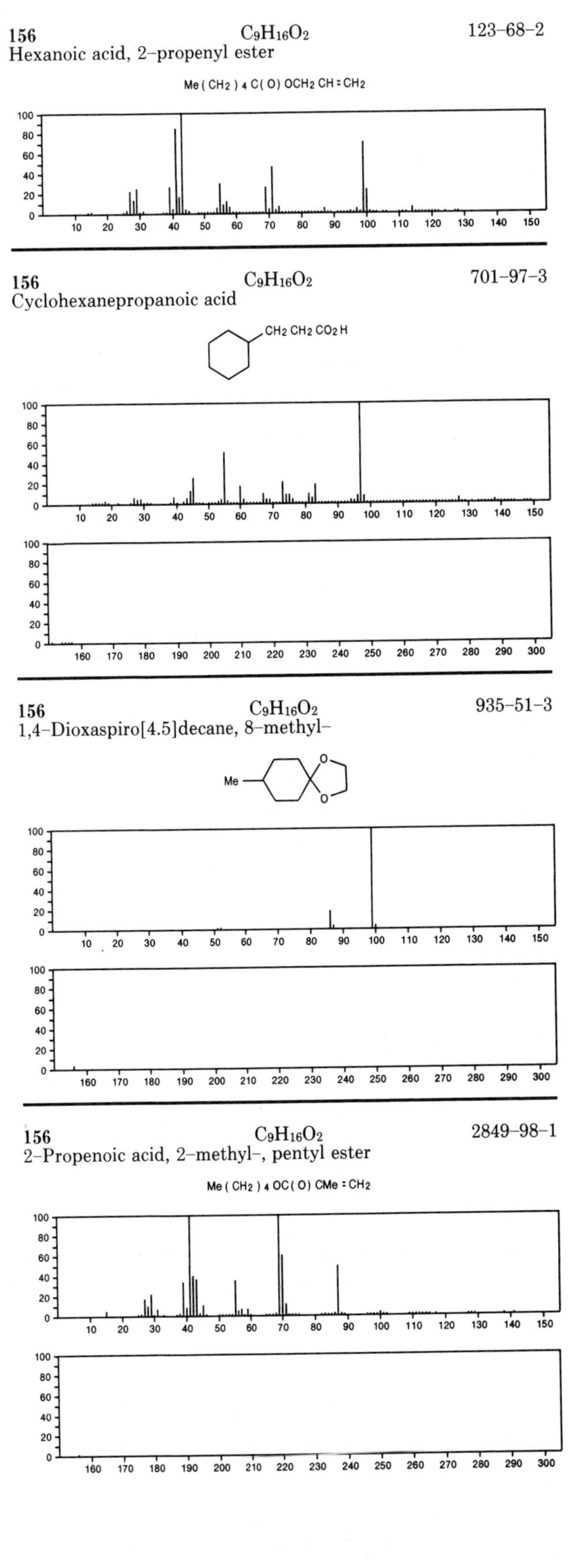

156 $C_9H_{16}O_2$ 123-68-2
Hexanoic acid, 2-propenyl ester

156 $C_9H_{16}O_2$ 701-97-3
Cyclohexanepropanoic acid

156 $C_9H_{16}O_2$ 935-51-3
1,4-Dioxaspiro[4.5]decane, 8-methyl-

156 $C_9H_{16}O_2$ 2849-98-1
2-Propenoic acid, 2-methyl-, pentyl ester

156 $C_9H_{16}O_2$ 3431-87-6
3-Penten-2-one, 4-butoxy-

Me(CH2)3OCMe=CHCOMe

156 $C_9H_{16}O_2$ 7560-65-8
Cyclohexene, 4-(dimethoxymethyl)-

156 $C_9H_{16}O_2$ 13482-27-4
Cyclohexanone, 2-ethyl-4-methoxy-

156 $C_9H_{16}O_2$ 17257-83-9
4-Octanone, 2,3-epoxy-2-methyl-

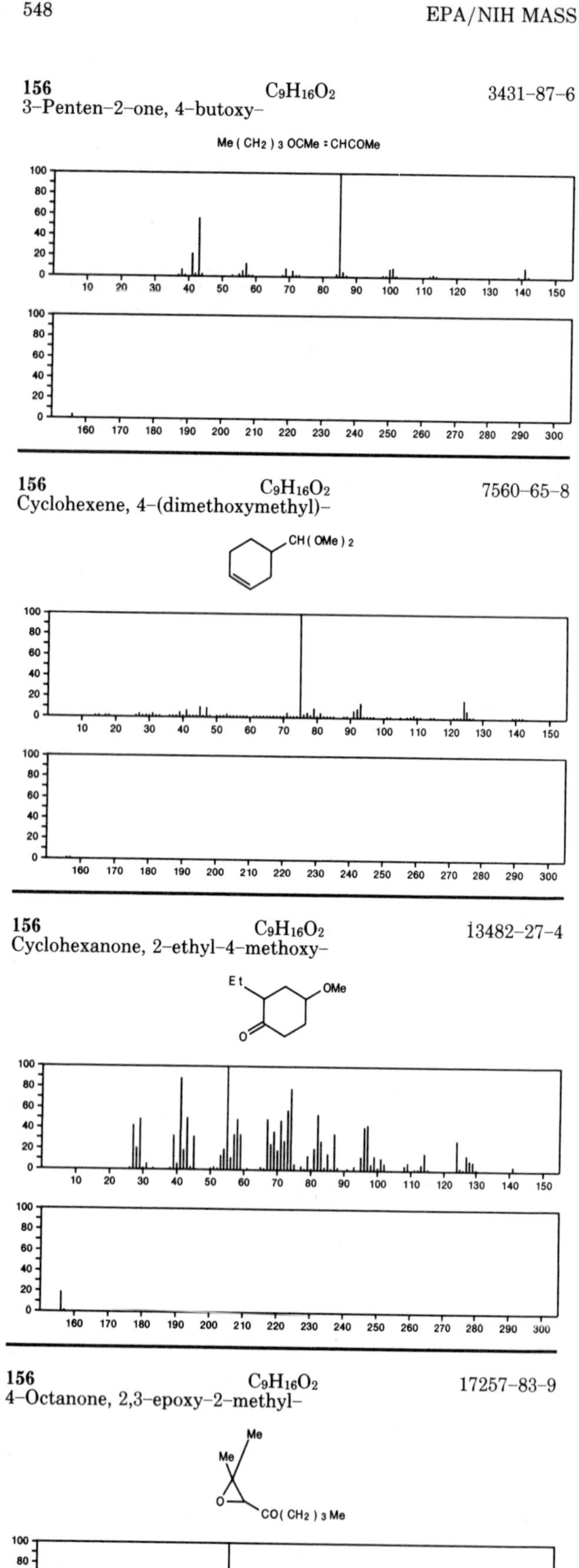

156 $C_9H_{16}O_2$ 21063-71-8
4-Octenoic acid, methyl ester, (*Z*)-

MeOC(O)CH2CH2CH=CHPr

156 $C_9H_{16}O_2$ 22607-16-5
1,5-Heptadiene-3,4-diol, 2,5-dimethyl-

H2C=CMeCH(OH)CH(OH)CMe=CHMe

156 $C_9H_{16}O_2$ 24985-48-6
2-Octen-4-one, 2-methoxy-

MeC(OMe)=CHCO(CH2)3Me

156 $C_9H_{16}O_2$ 24985-52-2
3-Octen-2-one, 4-methoxy-

MeCOCH=C(OMe)(CH2)3Me

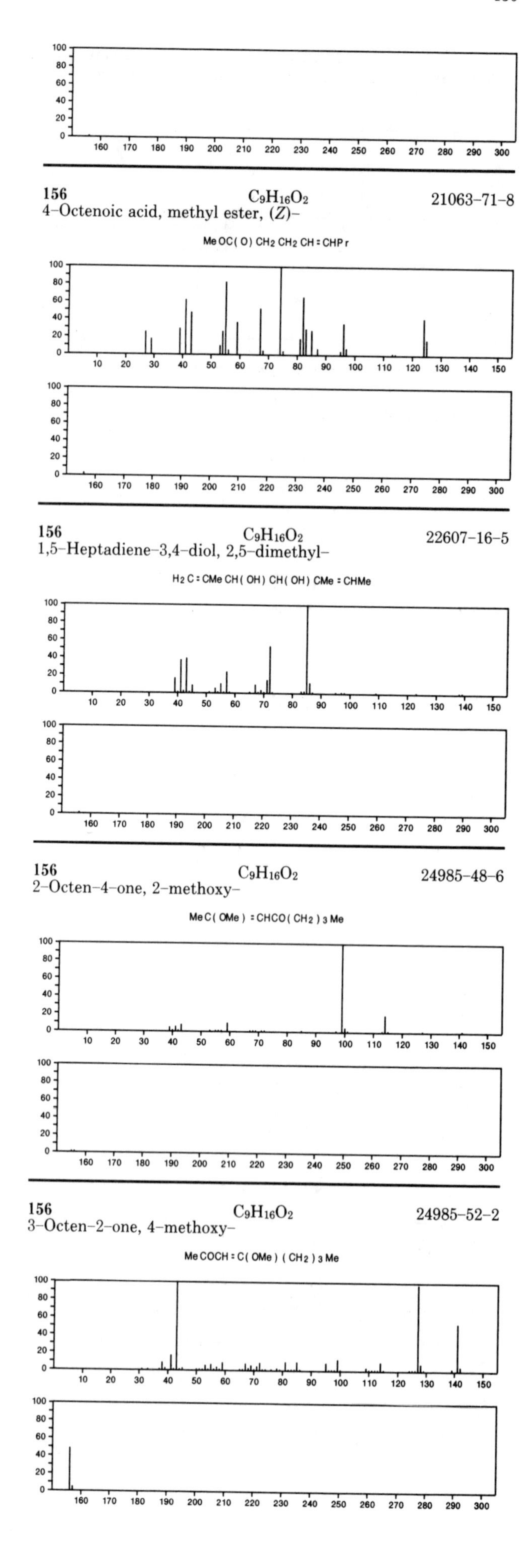

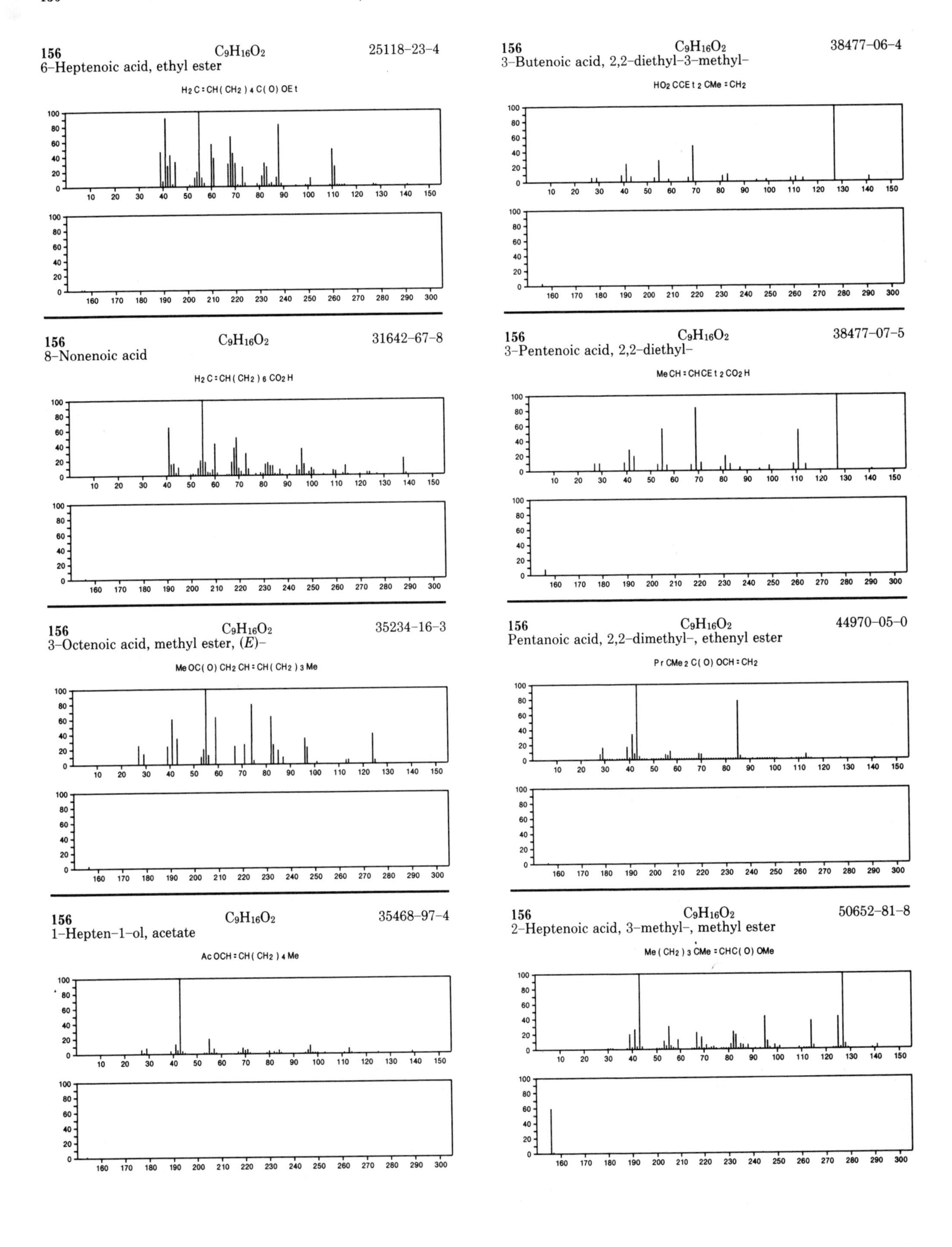
156 $C_9H_{16}O_2$ 25118-23-4
6-Heptenoic acid, ethyl ester
H2C=CH(CH2)4C(O)OEt
156 $C_9H_{16}O_2$ 38477-06-4
3-Butenoic acid, 2,2-diethyl-3-methyl-
HO2CCEt2CMe=CH2
156 $C_9H_{16}O_2$ 31642-67-8
8-Nonenoic acid
H2C=CH(CH2)6CO2H
156 $C_9H_{16}O_2$ 38477-07-5
3-Pentenoic acid, 2,2-diethyl-
MeCH=CHCEt2CO2H
156 $C_9H_{16}O_2$ 35234-16-3
3-Octenoic acid, methyl ester, (E)-
MeOC(O)CH2CH=CH(CH2)3Me
156 $C_9H_{16}O_2$ 44970-05-0
Pentanoic acid, 2,2-dimethyl-, ethenyl ester
PrCMe2C(O)OCH=CH2
156 $C_9H_{16}O_2$ 35468-97-4
1-Hepten-1-ol, acetate
AcOCH=CH(CH2)4Me
156 $C_9H_{16}O_2$ 50652-81-8
2-Heptenoic acid, 3-methyl-, methyl ester
Me(CH2)3CMe=CHC(O)OMe

156 $C_9H_{16}O_2$ 51181-40-9
Cyclohexanecarboxylic acid, 4-methyl-, methyl ester

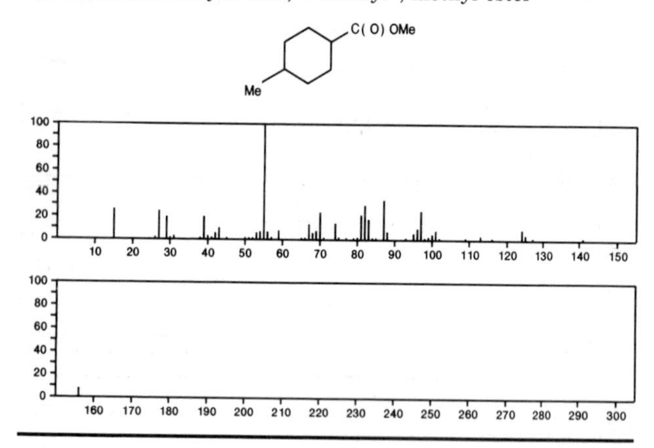

156 $C_9H_{16}O_2$ 54340-69-1
5-Heptenoic acid, ethyl ester, (*E*)-

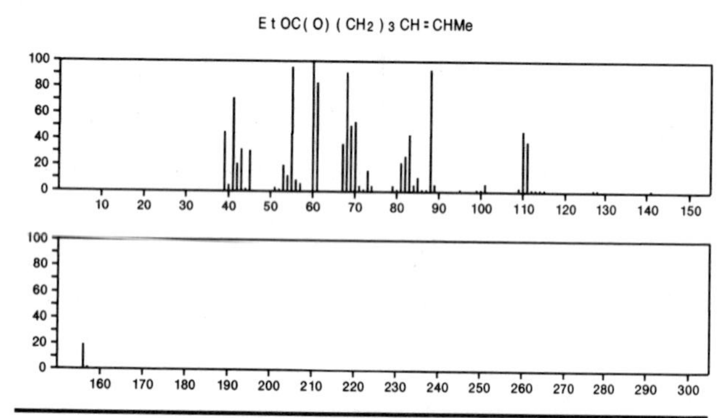

156 $C_9H_{16}O_2$ 54340-70-4
4-Heptenoic acid, ethyl ester, (*E*)-

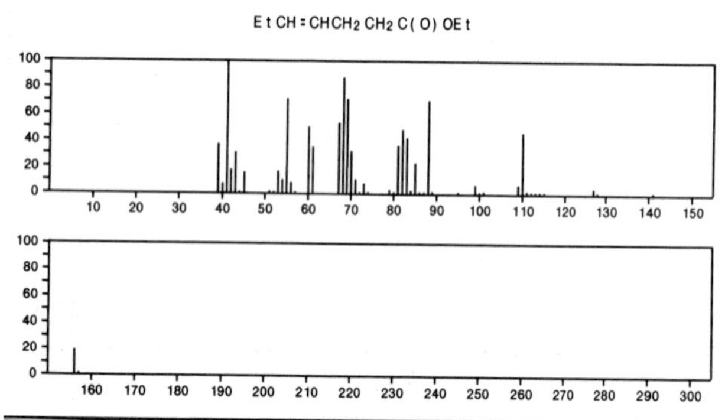

156 $C_9H_{16}O_2$ 54340-71-5
3-Heptenoic acid, ethyl ester, (*E*)-

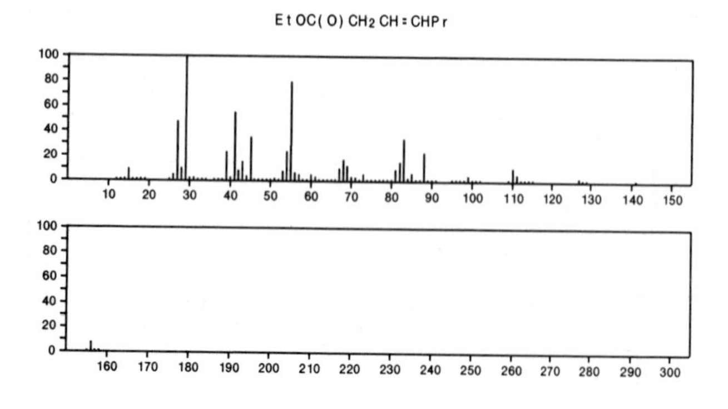

156 $C_9H_{16}O_2$ 54340-72-6
2-Heptenoic acid, ethyl ester, (*E*)-

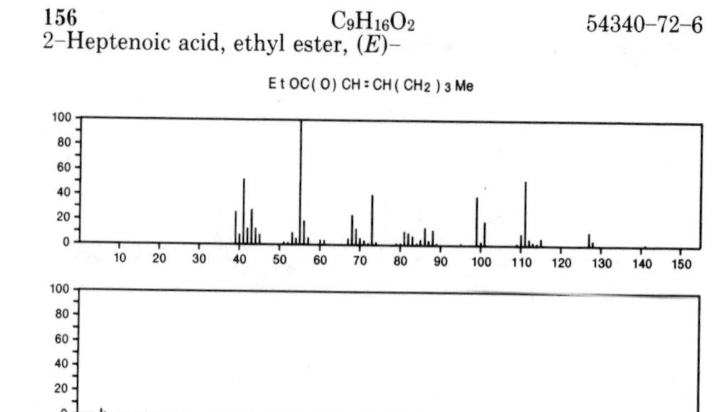

156 $C_9H_{16}O_2$ 54714-33-9
Cyclohexanol, 2-methyl-, acetate, *cis*-

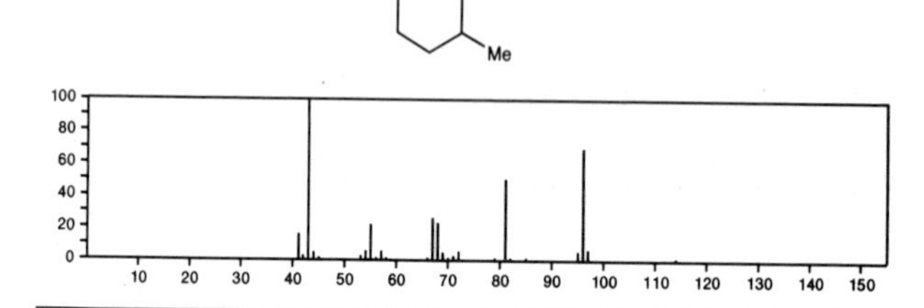

156 $C_9H_{16}O_2$ 54714-34-0
Cyclohexanol, 2-methyl-, acetate, *trans*-

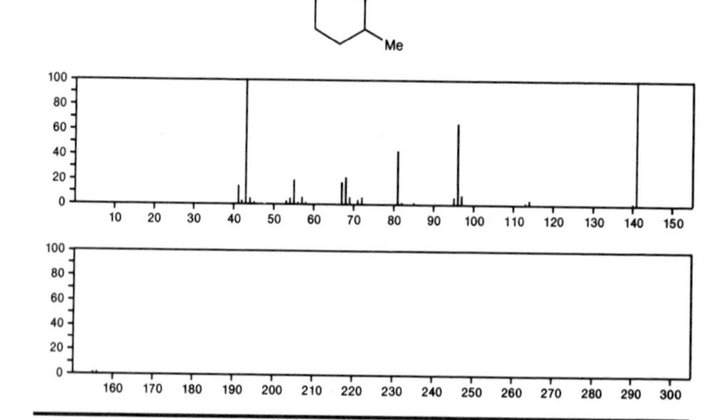

156 $C_9H_{16}O_2$ 55013-32-6
2(3*H*)-Furanone, 5-butyldihydro-4-methyl-, *cis*-

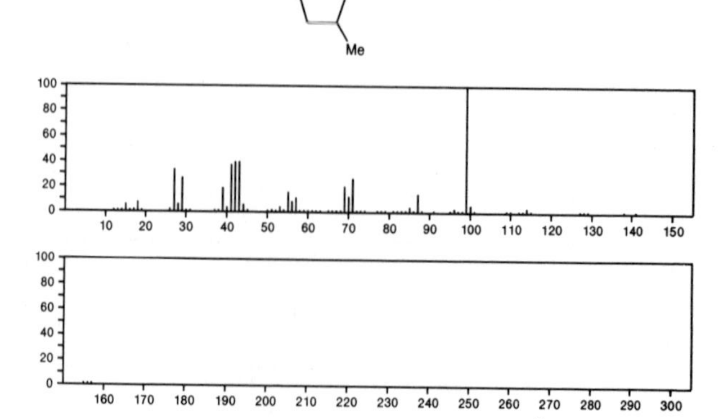

156 $C_9H_{16}O_2$ 56335-74-1
2,6-Octadiene-4,5-diol, 4-methyl-

MeCH:CHCH(OH)CMe(OH)CH:CHMe

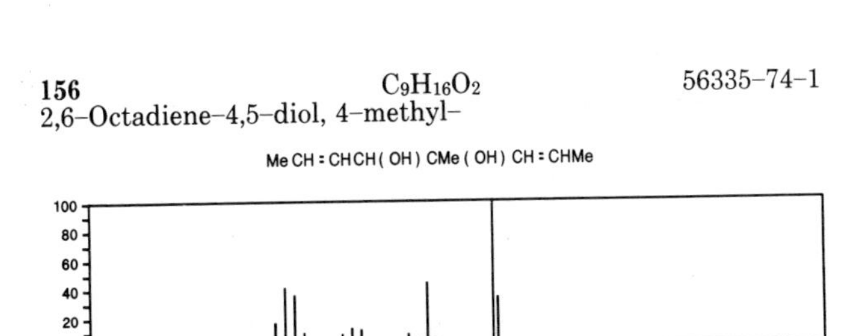

156 $C_9H_{16}O_2$ 56943-71-6
Ethanone, 1-(hexahydro-2*H*-oxocin-3-yl)-

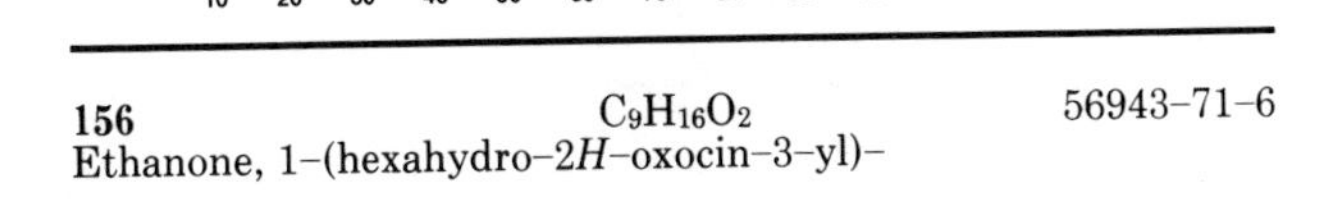

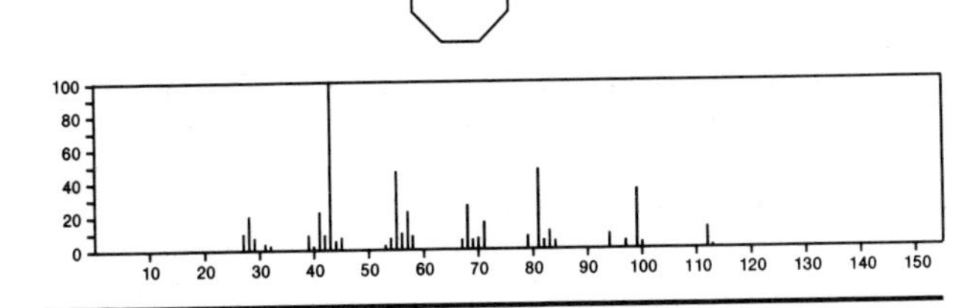

156 $C_9H_{16}S$ 39825-77-9
Bicyclo[2.2.2]octane-1-thiol, 4-methyl-

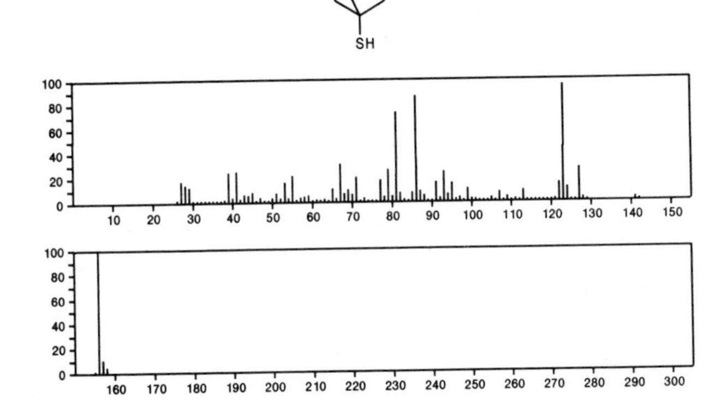

156 $C_9H_{16}S$ 54340-73-7
2*H*-1-Benzothiopyran, octahydro-, *trans*-

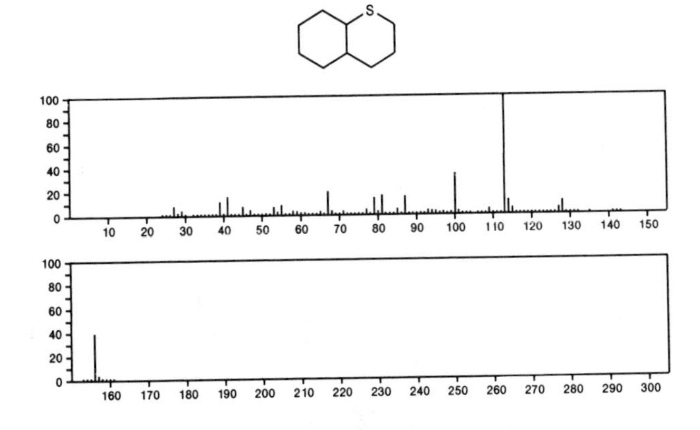

156 $C_9H_{16}S$ 54340-74-8
1*H*-2-Benzothiopyran, octahydro-, *cis*-

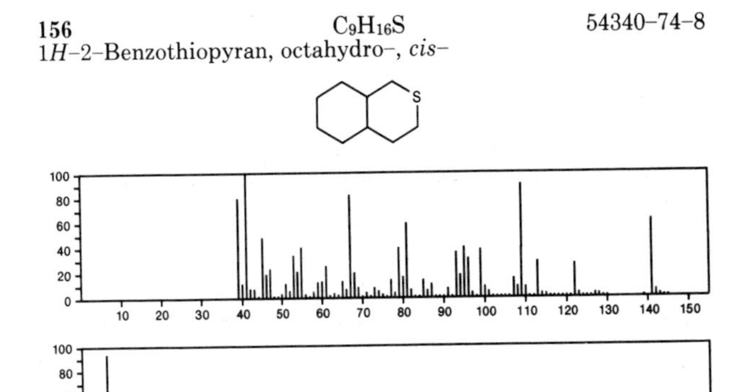

156 $C_9H_{18}NO$ 2564-83-2
1-Piperidinyloxy, 2,2,6,6-tetramethyl-

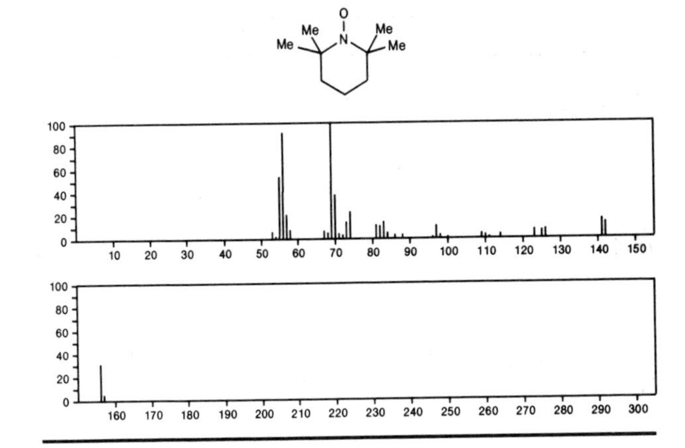

156 $C_9H_{20}N_2$ 14090-58-5
4-Heptanone, dimethylhydrazone

Pr₂C:NNMe₂

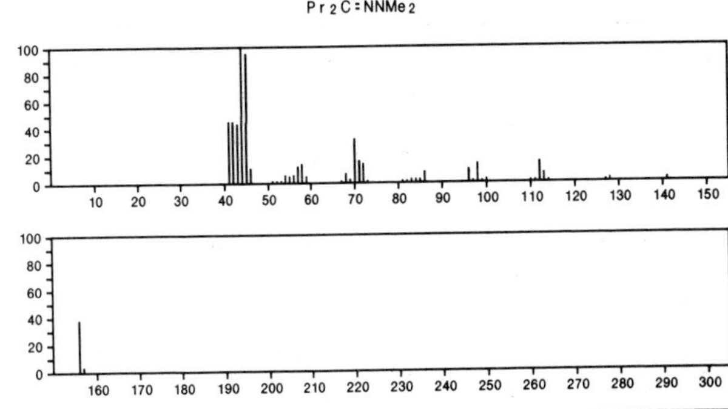

156 $C_9H_{20}N_2$ 54365-81-0
Piperazine, 2,5-dimethyl-3-propyl-

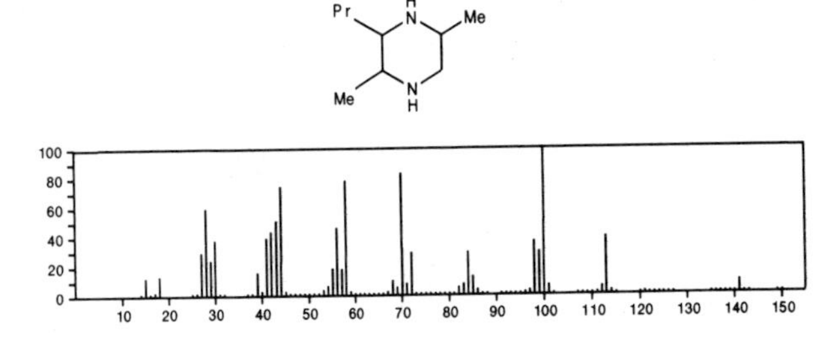

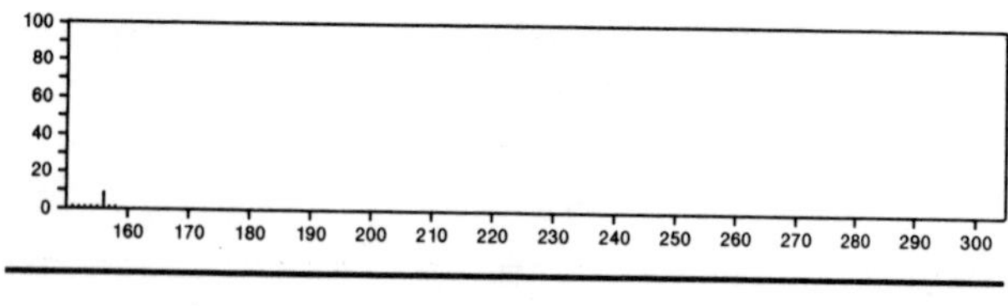

156 $C_9H_{20}Si$ 52835-06-0
Silane, 1-hexenyltrimethyl-, (*Z*)-

$Me_3SiCH=CH(CH_2)_3Me$

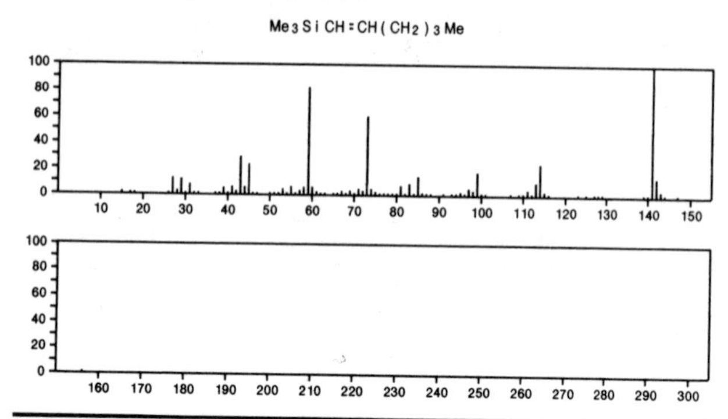

156 $C_{10}H_8N_2$ 366-18-7
2,2'-Bipyridine

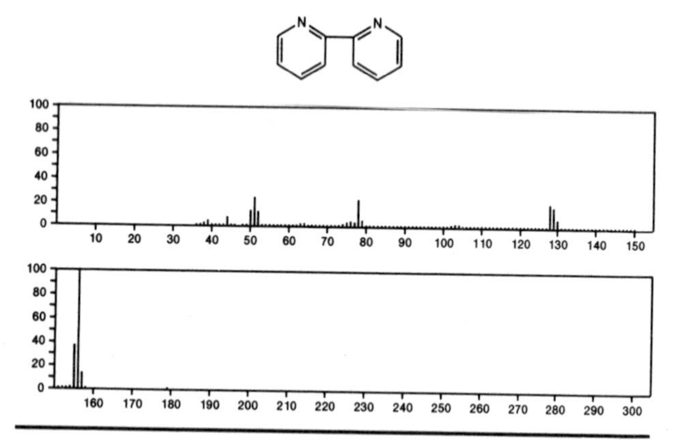

156 $C_{10}H_8N_2$ 22320-36-1
6-Indolizinecarbonitrile, 2-methyl-

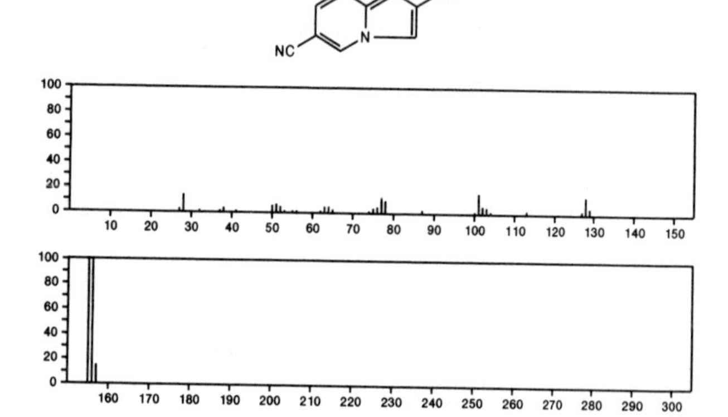

156 $C_{10}H_{20}O$ 89-78-1
Cyclohexanol, 5-methyl-2-(1-methylethyl)-, (1α,2β,5α)-

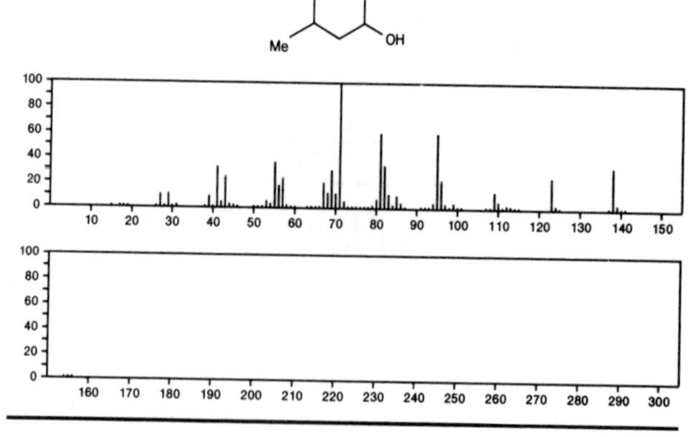

156 $C_{10}H_{20}O$ 98-52-2
Cyclohexanol, 4-(1,1-dimethylethyl)-

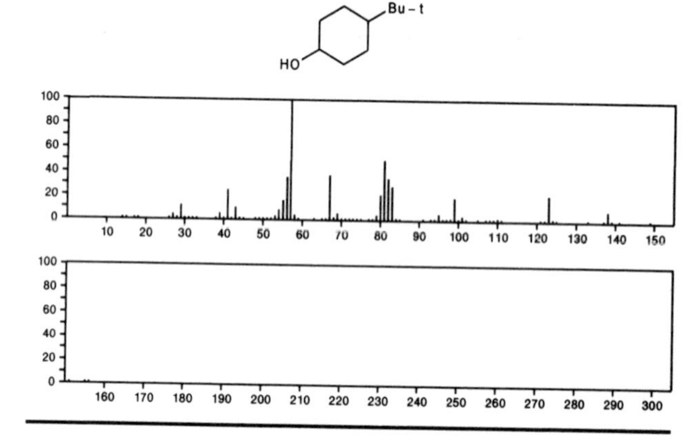

156 $C_{10}H_{20}O$ 103-44-6
Heptane, 3-[(ethenyloxy)methyl]-

$Me(CH_2)_3CHEtCH_2OCH=CH_2$

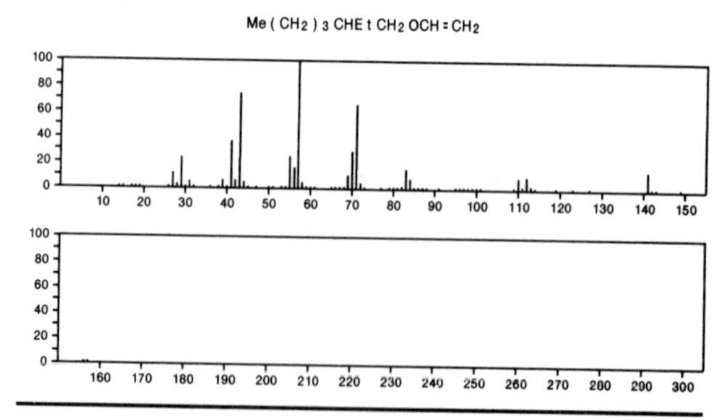

156 $C_{10}H_{20}O$ 106-22-9
6-Octen-1-ol, 3,7-dimethyl-

$Me_2C=CHCH_2CH_2CHMeCH_2CH_2OH$

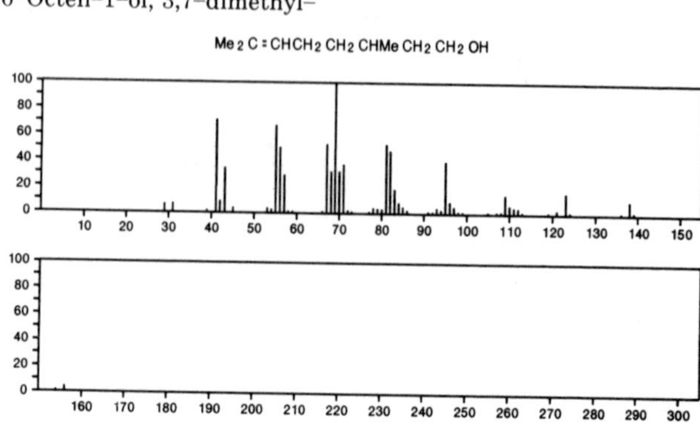

156 $C_{10}H_{20}O$ 112–31–2
Decanal

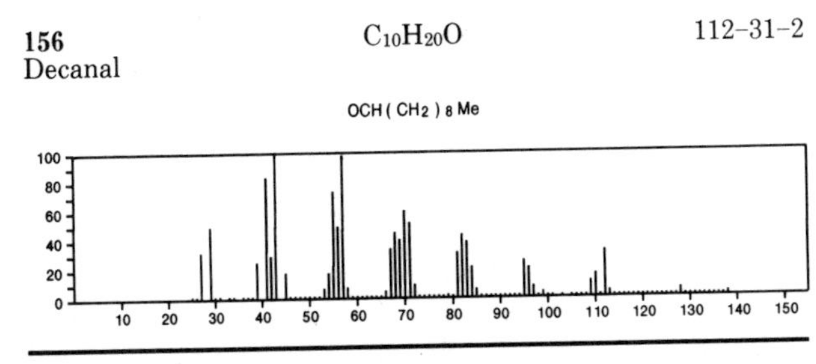

156 $C_{10}H_{20}O$ 490–99–3
Cyclohexanol, 5–methyl–2–(1–methylethyl)–, (1α,2β,5β)–

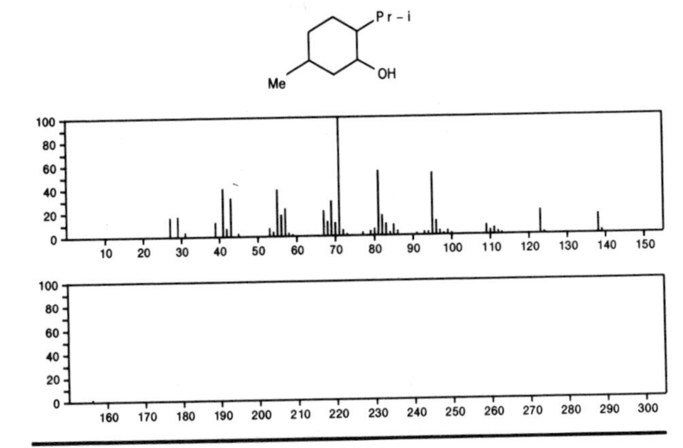

156 $C_{10}H_{20}O$ 491–01–0
Cyclohexanol, 5–methyl–2–(1–methylethyl)–, (1α,2α,5β)–

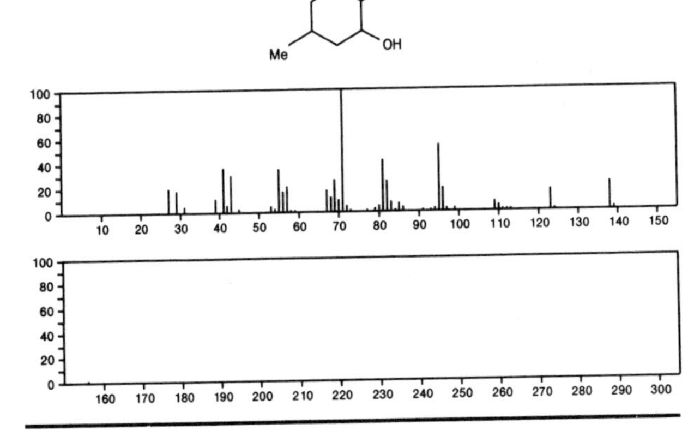

156 $C_{10}H_{20}O$ 491–02–1
Cyclohexanol, 5–methyl–2–(1–methylethyl)–, (1α,2α,5α)–

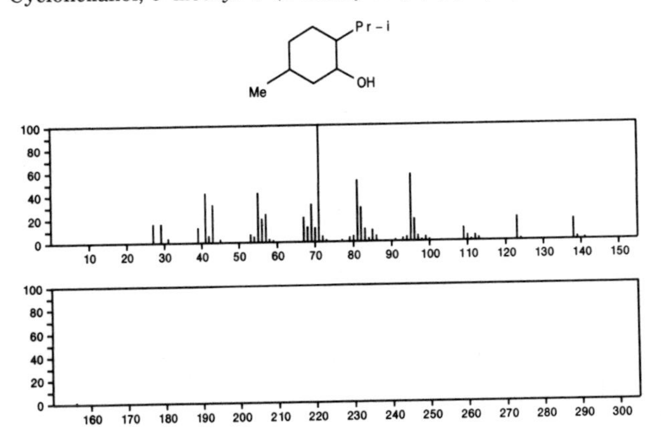

156 $C_{10}H_{20}O$ 499–69–4
Cyclohexanol, 2–methyl–5–(1–methylethyl)–, (1α,2β,5α)–

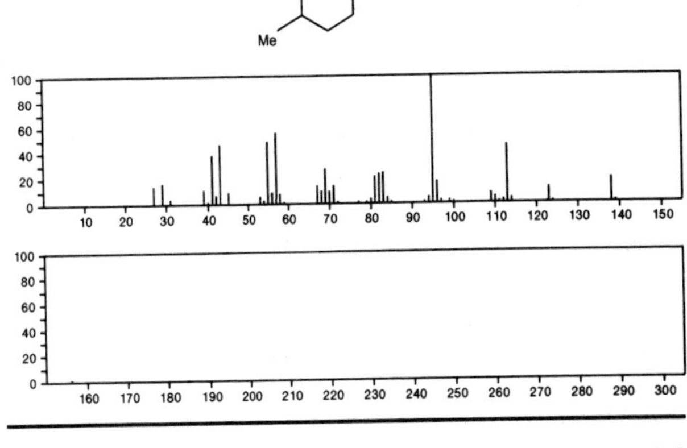

156 $C_{10}H_{20}O$ 624–16–8
4–Decanone

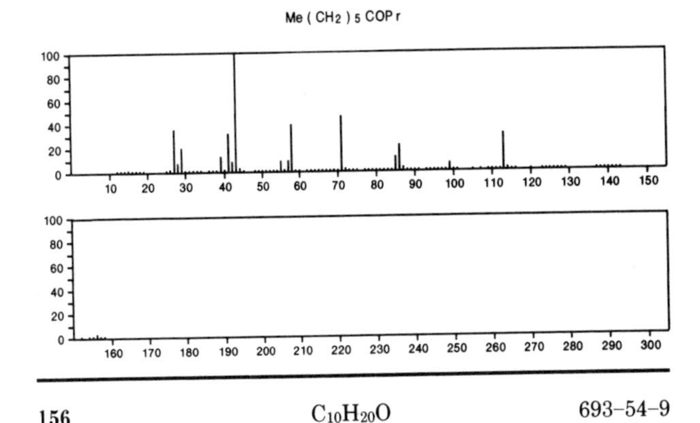

156 $C_{10}H_{20}O$ 693–54–9
2–Decanone

Me (CH2) 7 COMe

156 $C_{10}H_{20}O$ 820–29–1
5–Decanone

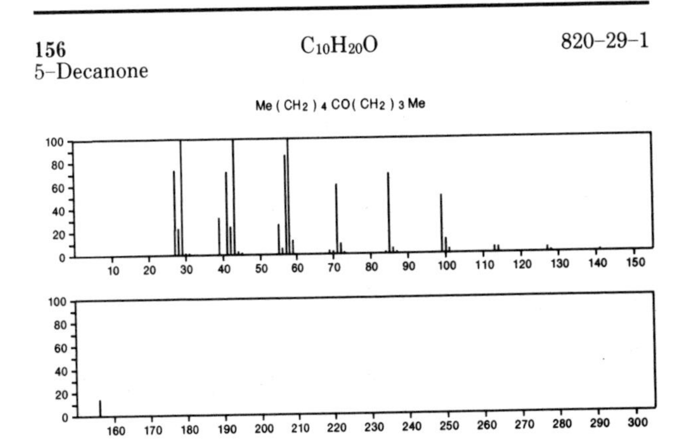

156
3-Decanone
$C_{10}H_{20}O$
928-80-3

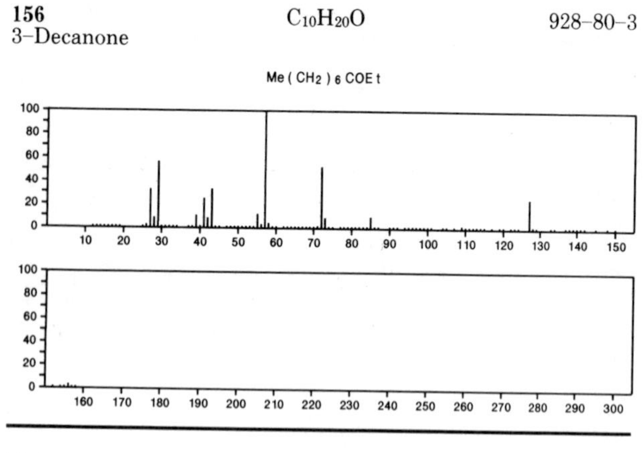

156
Cyclohexanol, 4-(1,1-dimethylethyl)-, *cis-*
$C_{10}H_{20}O$
937-05-3

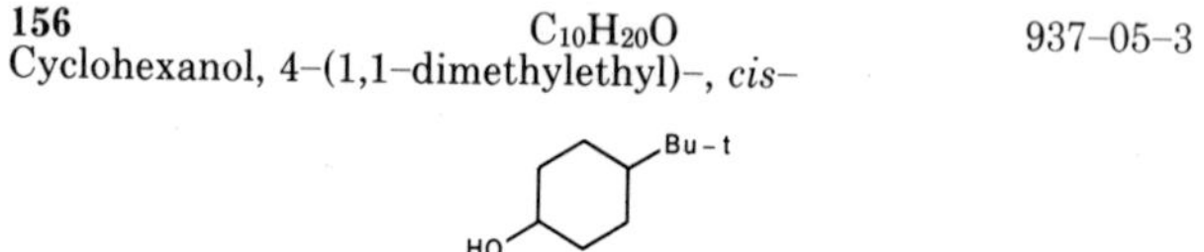

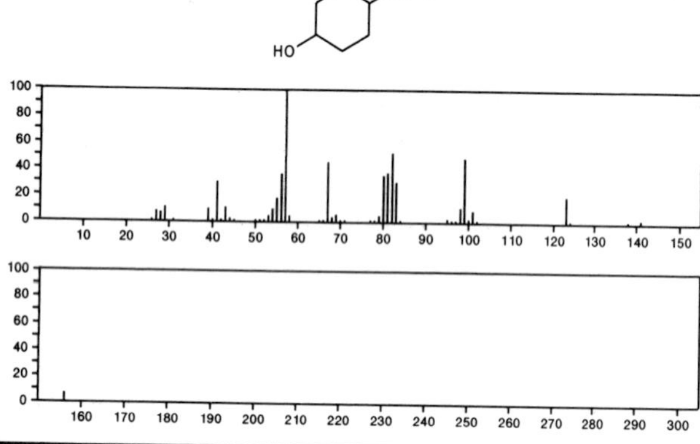

156
Cyclohexanol, 2-methyl-5-(1-methylethyl)-, (1α,2α,5β)-
$C_{10}H_{20}O$
1126-40-5

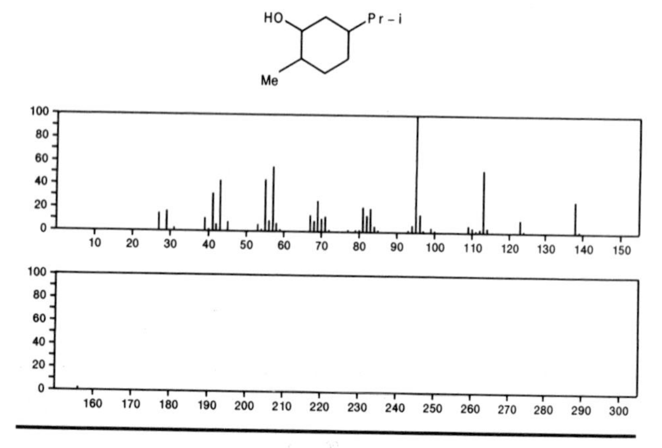

156
Cyclohexanol, 5-methyl-2-(1-methylethyl)-
$C_{10}H_{20}O$
1490-04-6

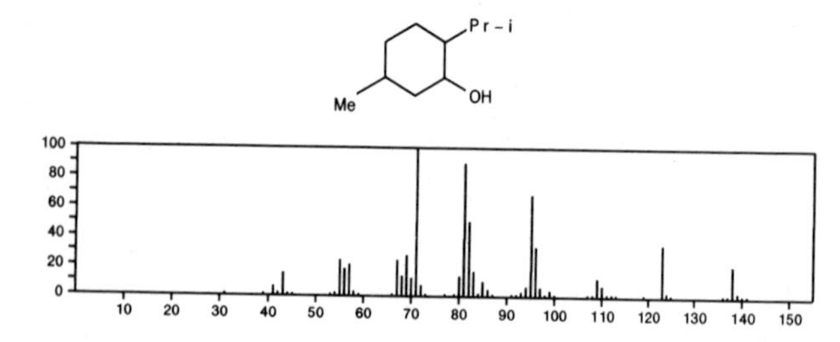

156
Cyclodecanol
$C_{10}H_{20}O$
1502-05-2

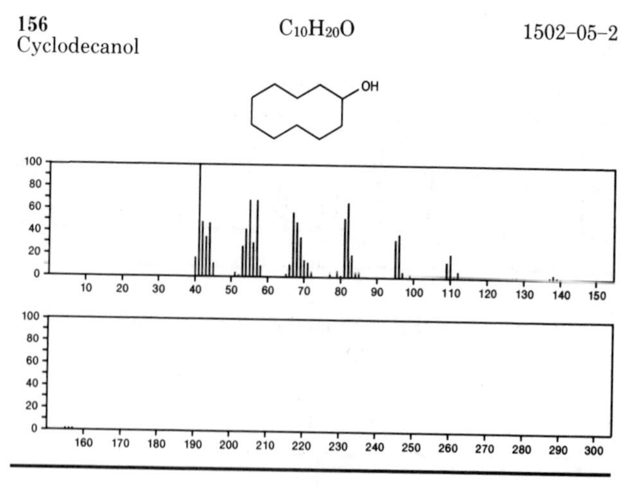

156
Cyclohexanol, 2-methyl-5-(1-methylethyl)-, (1α,2β,5β)-
$C_{10}H_{20}O$
3127-80-8

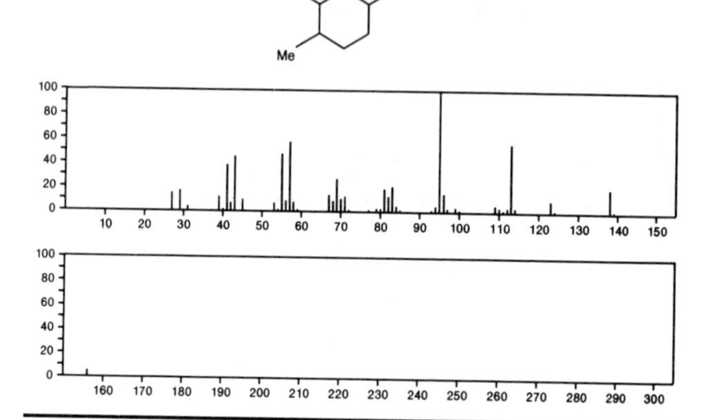

156
p-Menthan-4-ol, *cis-*
$C_{10}H_{20}O$
3239-02-9

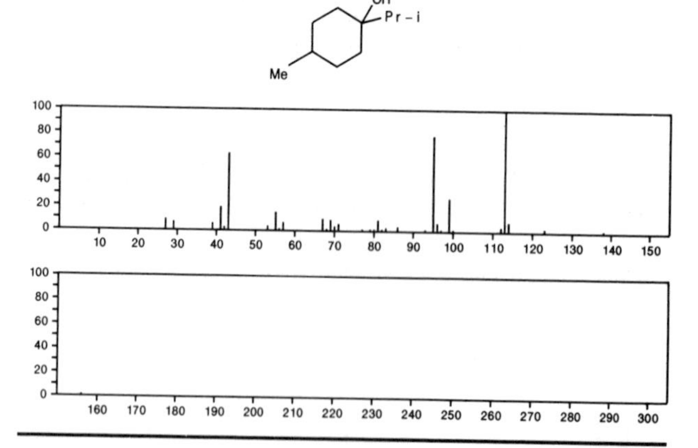

156
Cyclohexanol, 1-methyl-4-(1-methylethyl)-, *trans-*
$C_{10}H_{20}O$
3901-93-7

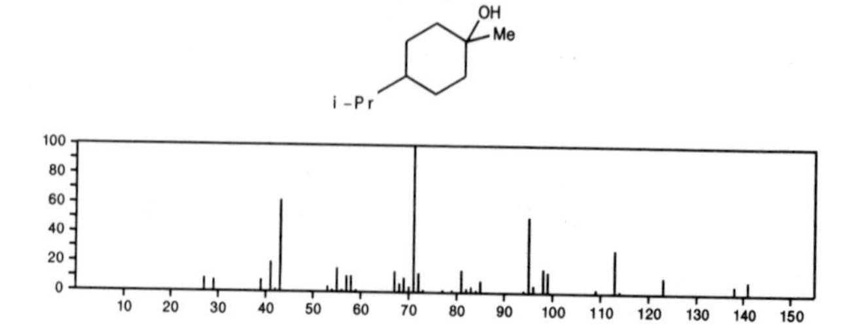

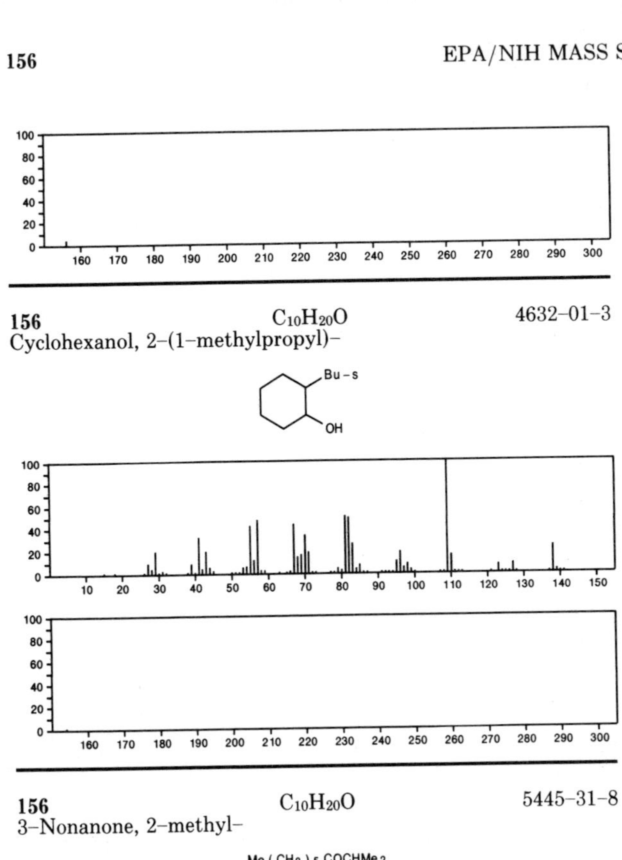

156 $C_{10}H_{20}O$ 4632-01-3
Cyclohexanol, 2-(1-methylpropyl)-

156 $C_{10}H_{20}O$ 5445-31-8
3-Nonanone, 2-methyl-

Me(CH2)5COCHMe2

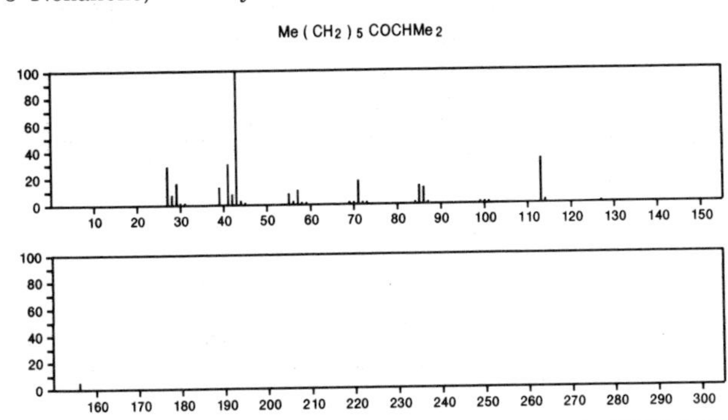

156 $C_{10}H_{20}O$ 6137-29-7
4-Nonanone, 8-methyl-

Me2CH(CH2)3COPr

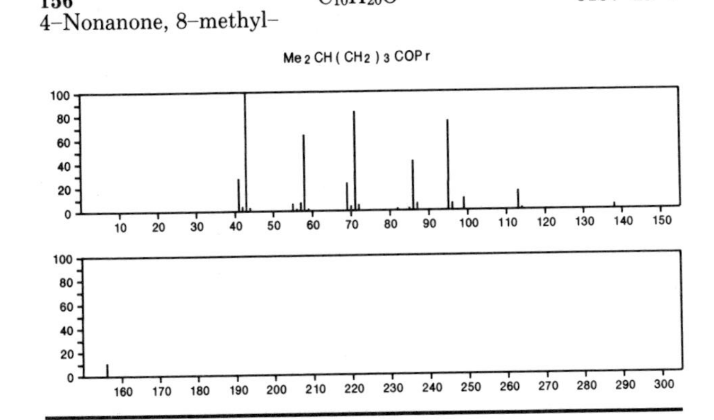

156 $C_{10}H_{20}O$ 6292-20-2
Cyclohexanol, 4-*sec*-butyl-

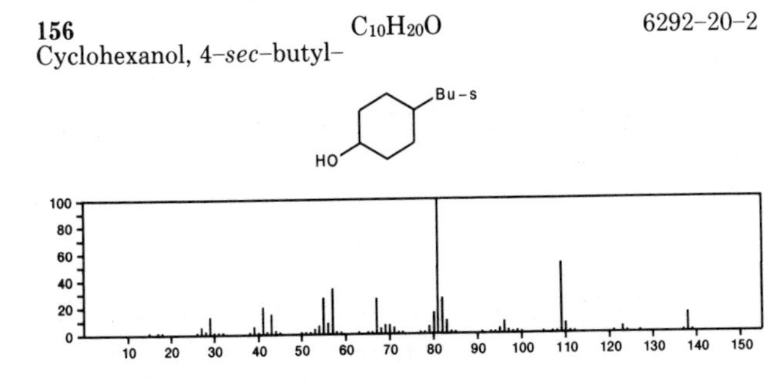

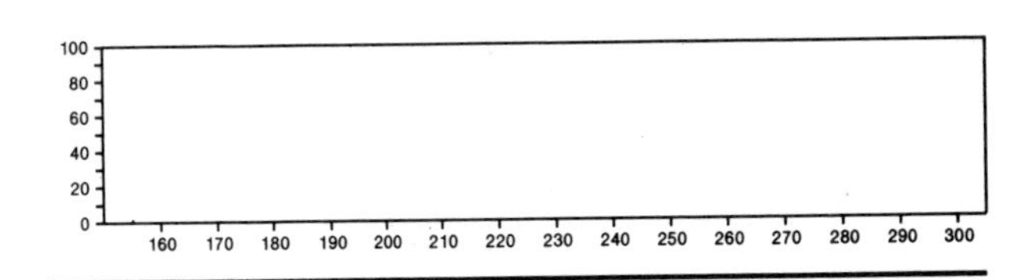

156 $C_{10}H_{20}O$ 10528-67-3
Cyclohexanepropanol, α-methyl-

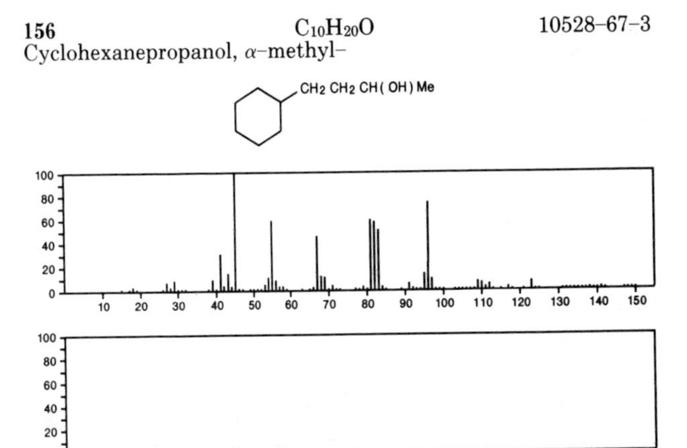

156 $C_{10}H_{20}O$ 13491-79-7
Cyclohexanol, 2-(1,1-dimethylethyl)-

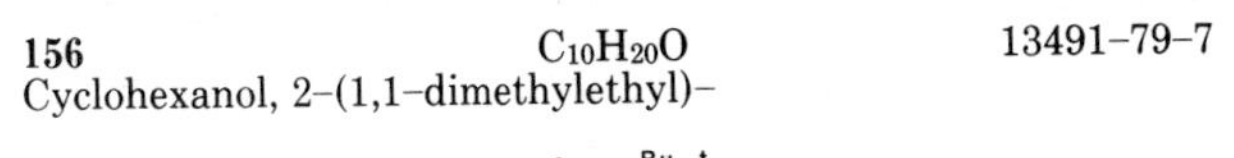

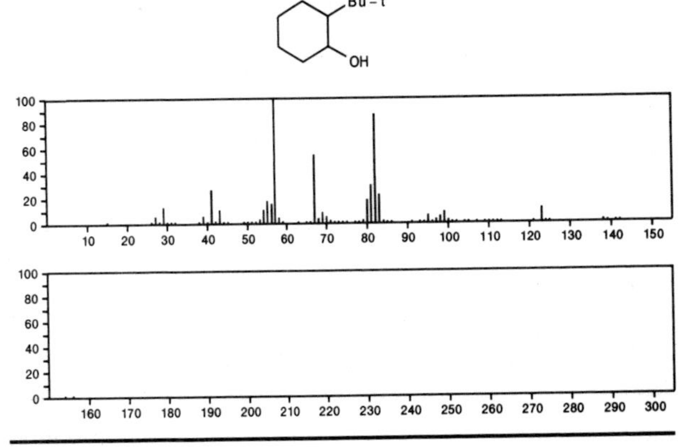

156 $C_{10}H_{20}O$ 16519-24-7
Heptane, 1-(2-propenyloxy)-

H2C=CHCH2O(CH2)6Me

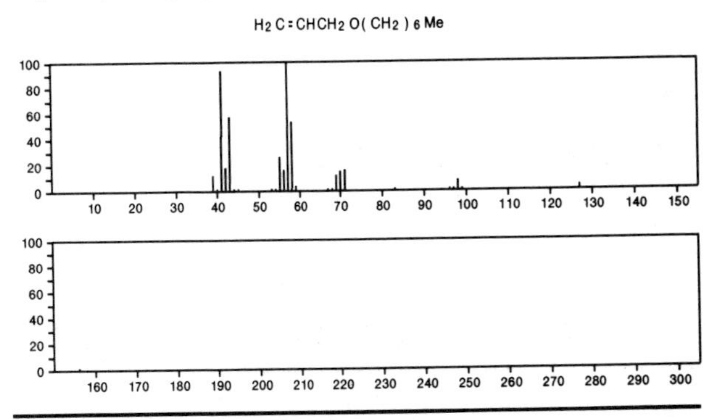

156 $C_{10}H_{20}O$ 26489-01-0
6-Octen-1-ol, 3,7-dimethyl-, (±)-

Me2C=CHCH2CH2CHMeCH2CH2OH

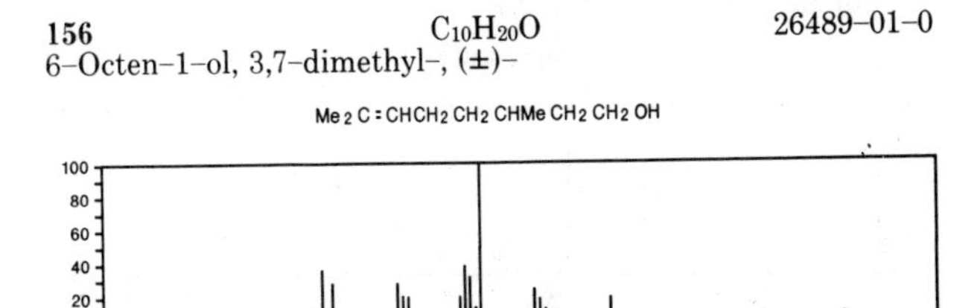

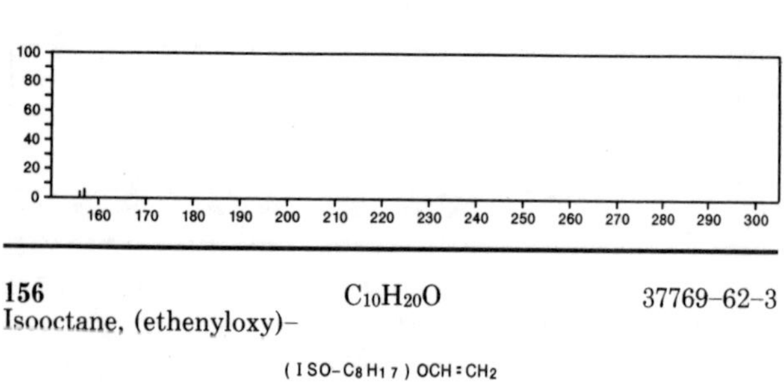

156 $C_{10}H_{20}O$ 37769–62–3
Isooctane, (ethenyloxy)–

(ISO–C_8H_{17}) OCH=CH_2

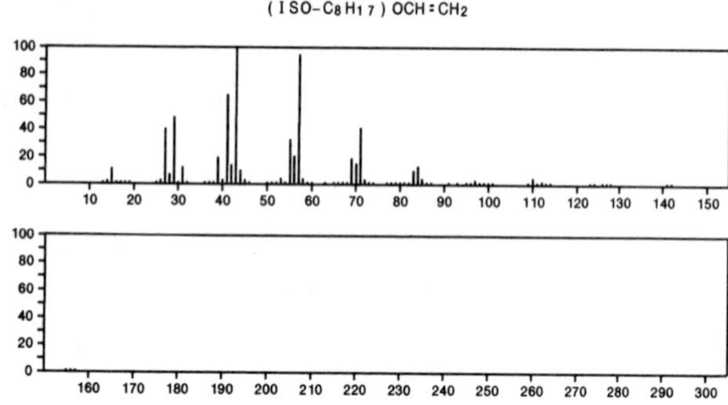

156 $C_{10}H_{20}O$ 42846–32–2
Cyclohexanol, 2–methyl–5–(1–methylethyl)–, (1α,2α,5α)–

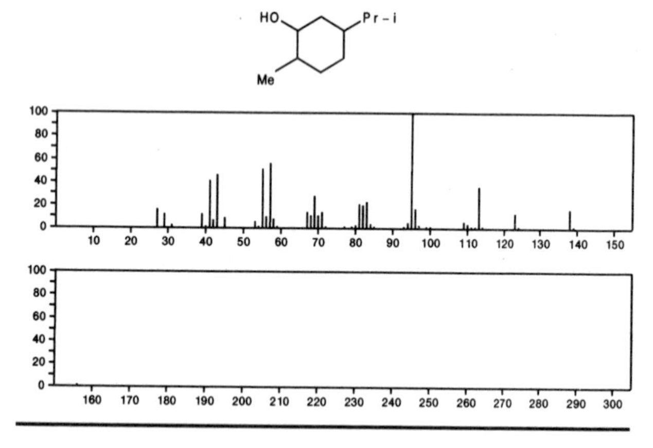

156 $C_{10}H_{20}O$ 54340–67–9
2–Hexene, 1–butoxy–, (*E*)–

Pr CH=CHCH$_2$ O(CH$_2$)$_3$ Me

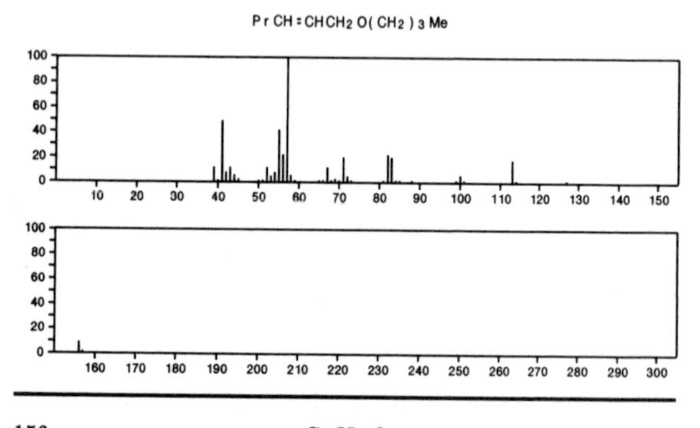

156 $C_{10}H_{20}O$ 54340–68–0
2–Pentene, 1–(pentyloxy)–, (*E*)–

Me(CH$_2$)$_4$ OCH$_2$ CH=CHEt

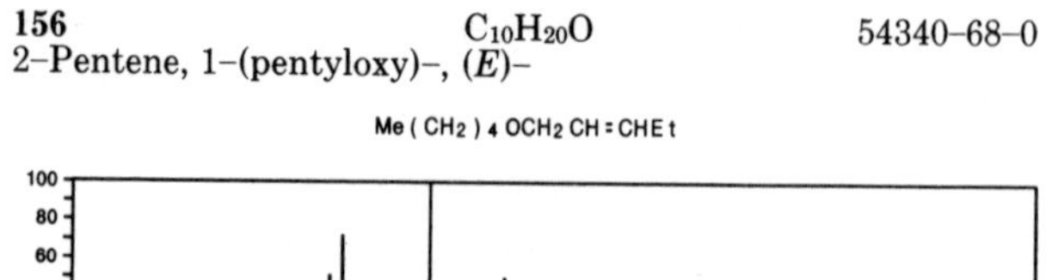

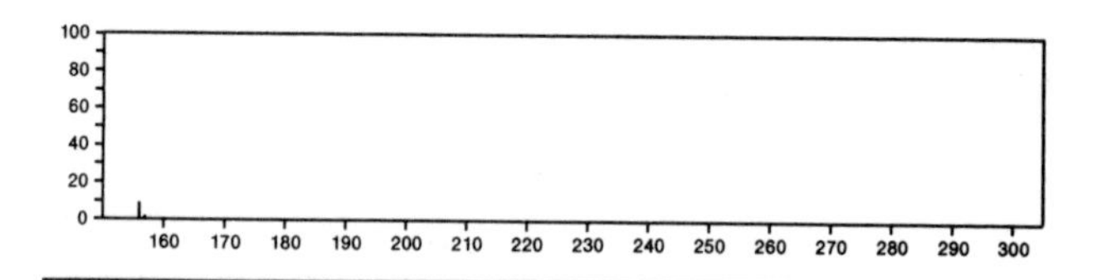

156 $C_{10}H_{20}O$ 56052–85–8
2–Pentene, 5–(pentyloxy)–, (*E*)–

MeCH=CHCH$_2$ CH$_2$ O(CH$_2$)$_4$ Me

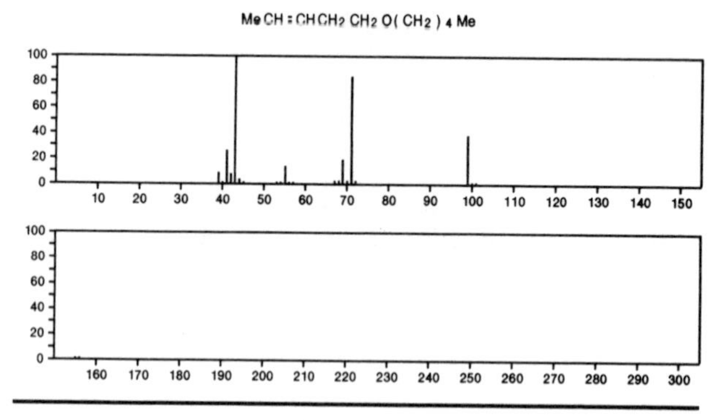

156 $C_{10}H_{20}O$ 56052–88–1
1–Pentene, 5–(pentyloxy)–

H$_2$C=CH(CH$_2$)$_3$ O(CH$_2$)$_4$ Me

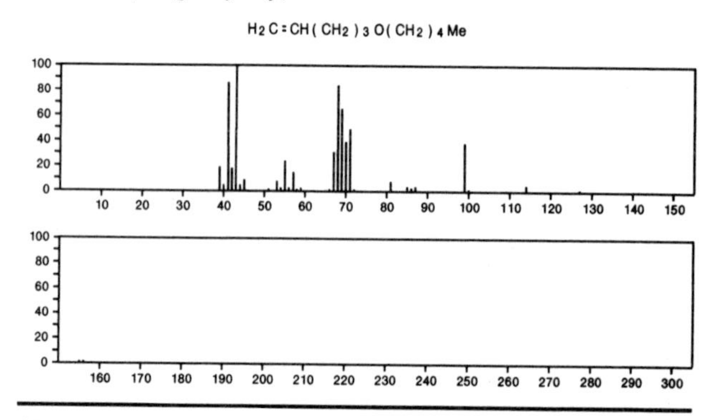

156 $C_{10}H_{20}O$ 56259–16–6
Cyclopropanemethanol, α,2–bis(1–methylethyl)–

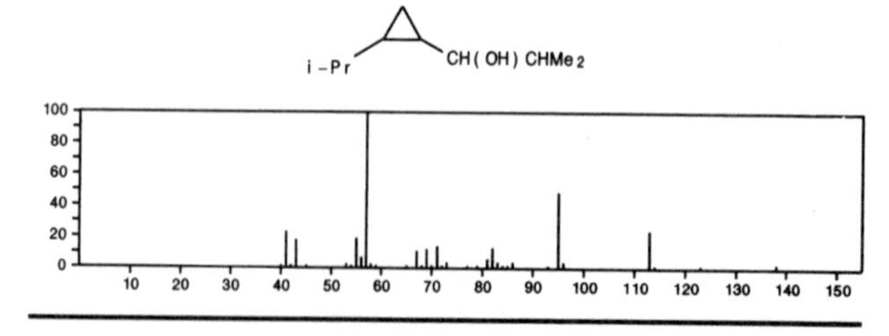

156 $C_{11}H_8O$ 36628–80–5
Bicyclo[4.4.1]undeca–1,3,5,7,9–pentaen–11–one

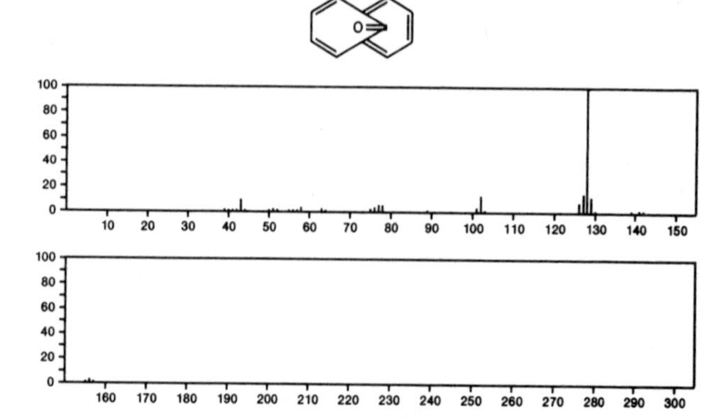

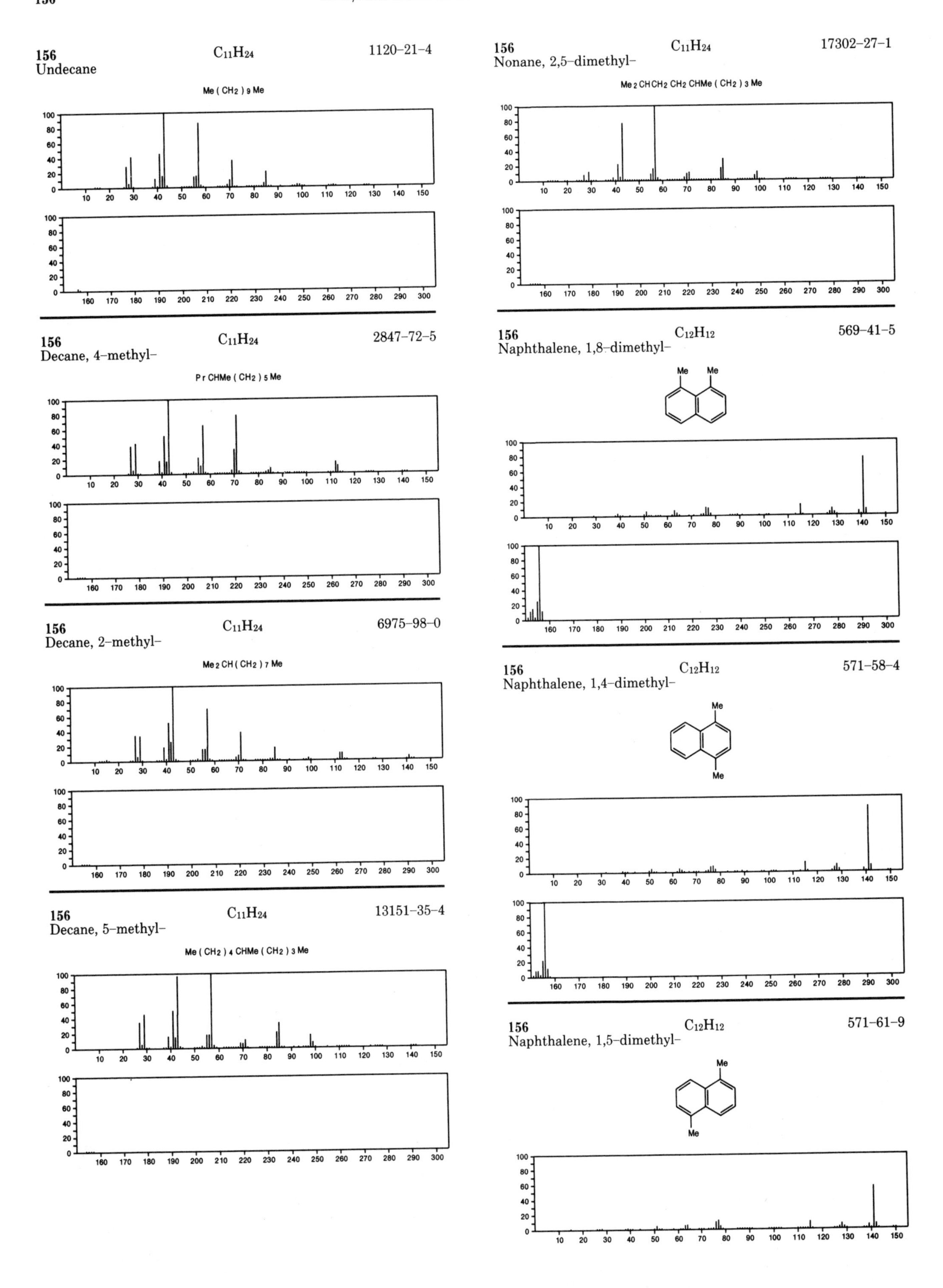

156
Undecane
C11H24
1120-21-4
Me (CH2) 9 Me
156
Decane, 4-methyl-
C11H24
2847-72-5
Pr CHMe (CH2) 5 Me
156
Decane, 2-methyl-
C11H24
6975-98-0
Me2 CH (CH2) 7 Me
156
Decane, 5-methyl-
C11H24
13151-35-4
Me (CH2) 4 CHMe (CH2) 3 Me
156
Nonane, 2,5-dimethyl-
C11H24
17302-27-1
Me2 CHCH2 CH2 CHMe (CH2) 3 Me
156
Naphthalene, 1,8-dimethyl-
C12H12
569-41-5
Me
Me
156
Naphthalene, 1,4-dimethyl-
C12H12
571-58-4
Me
Me
156
Naphthalene, 1,5-dimethyl-
C12H12
571-61-9
Me
Me

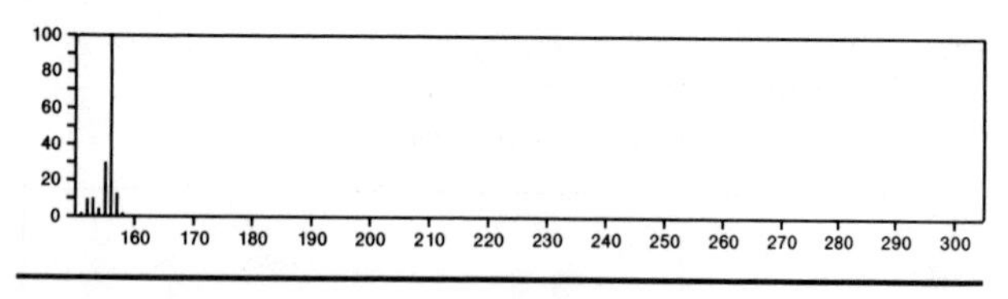

156 $C_{12}H_{12}$ 573–98–8
Naphthalene, 1,2–dimethyl–

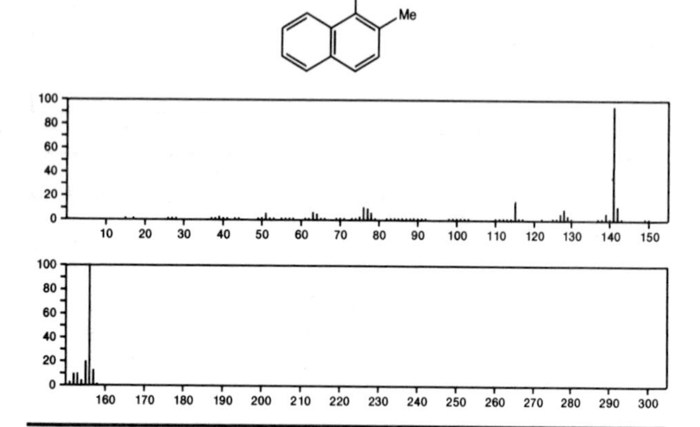

156 $C_{12}H_{12}$ 575–37–1
Naphthalene, 1,7–dimethyl–

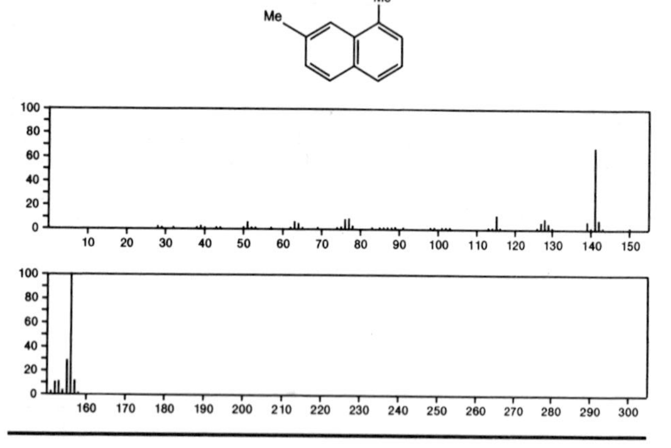

156 $C_{12}H_{12}$ 575–41–7
Naphthalene, 1,3–dimethyl–

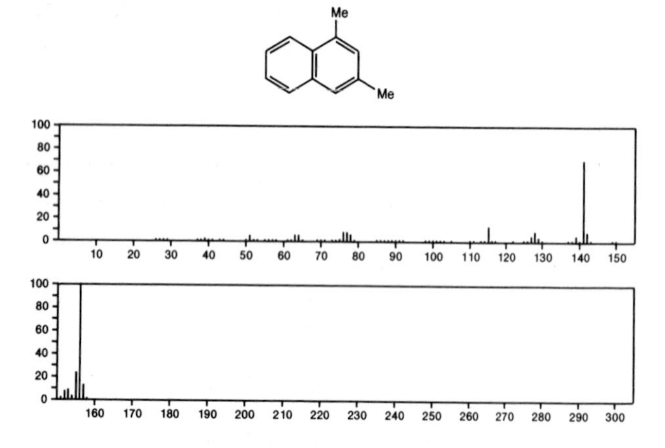

156 $C_{12}H_{12}$ 575–43–9
Naphthalene, 1,6–dimethyl–

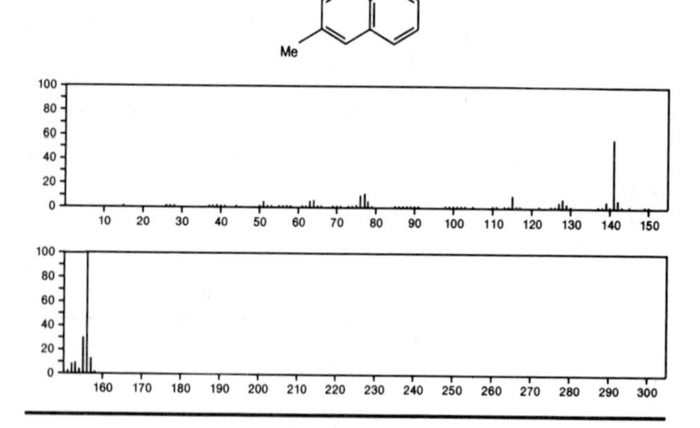

156 $C_{12}H_{12}$ 581–40–8
Naphthalene, 2,3–dimethyl–

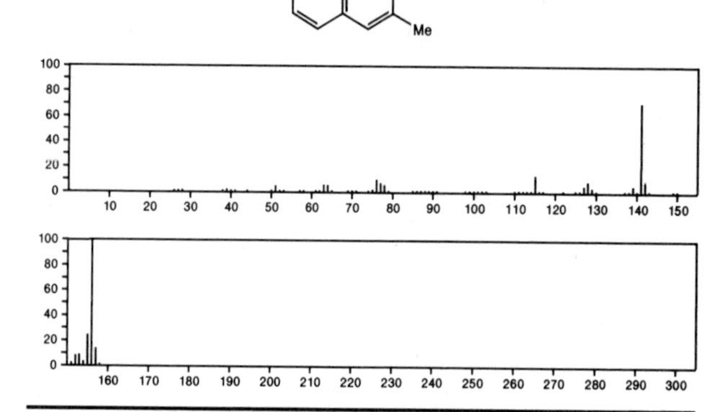

156 $C_{12}H_{12}$ 581–42–0
Naphthalene, 2,6–dimethyl–

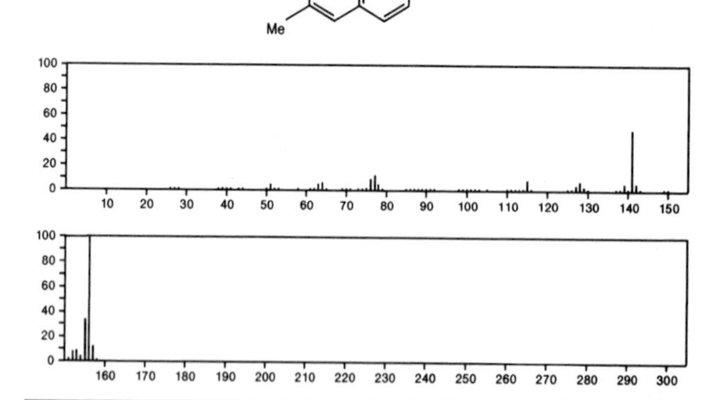

156 $C_{12}H_{12}$ 582–16–1
Naphthalene, 2,7–dimethyl–

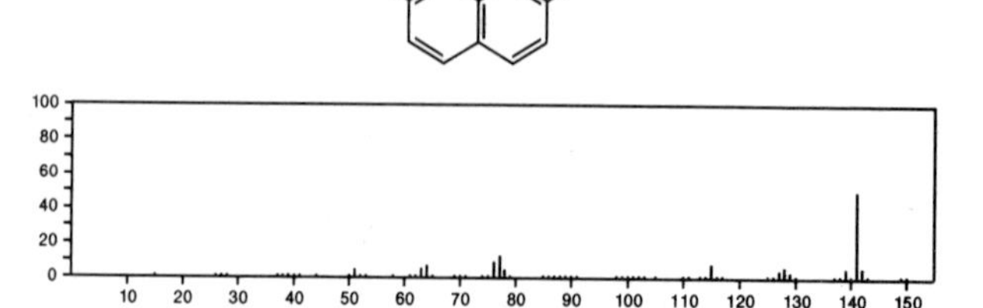

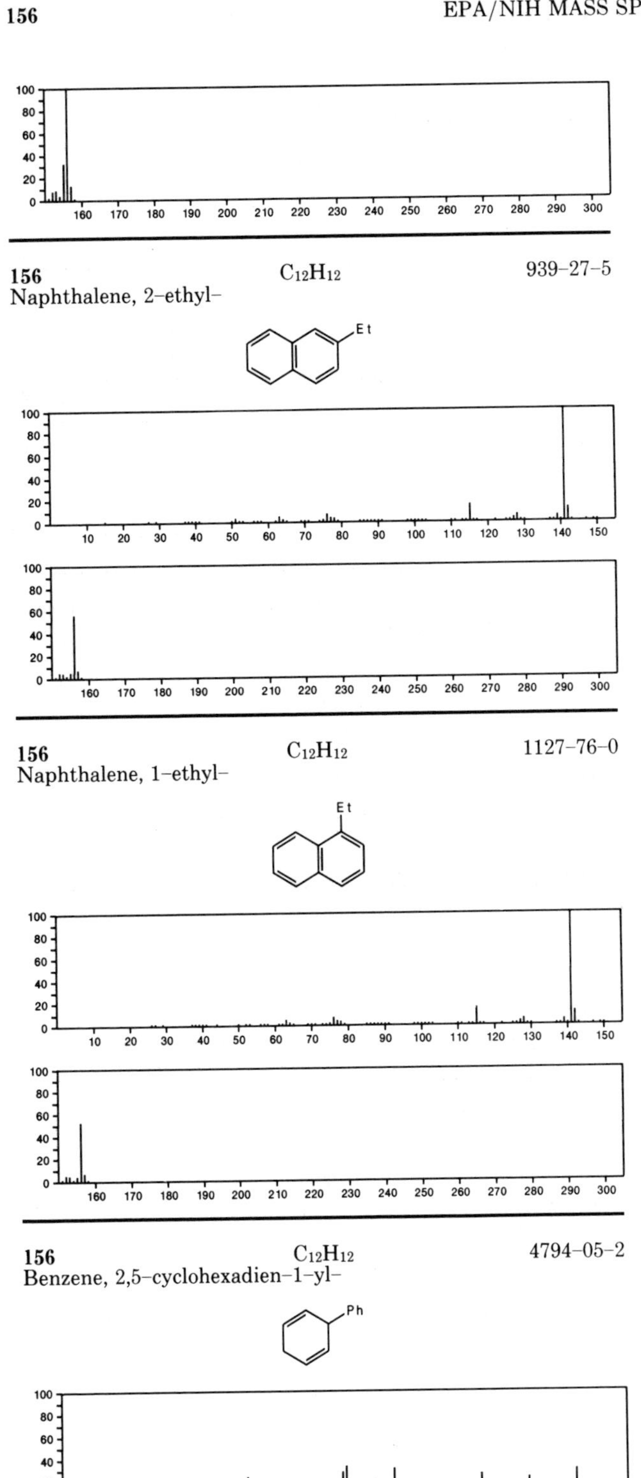

156 $C_{12}H_{12}$ 939-27-5
Naphthalene, 2-ethyl-

156 $C_{12}H_{12}$ 1127-76-0
Naphthalene, 1-ethyl-

156 $C_{12}H_{12}$ 4794-05-2
Benzene, 2,5-cyclohexadien-1-yl-

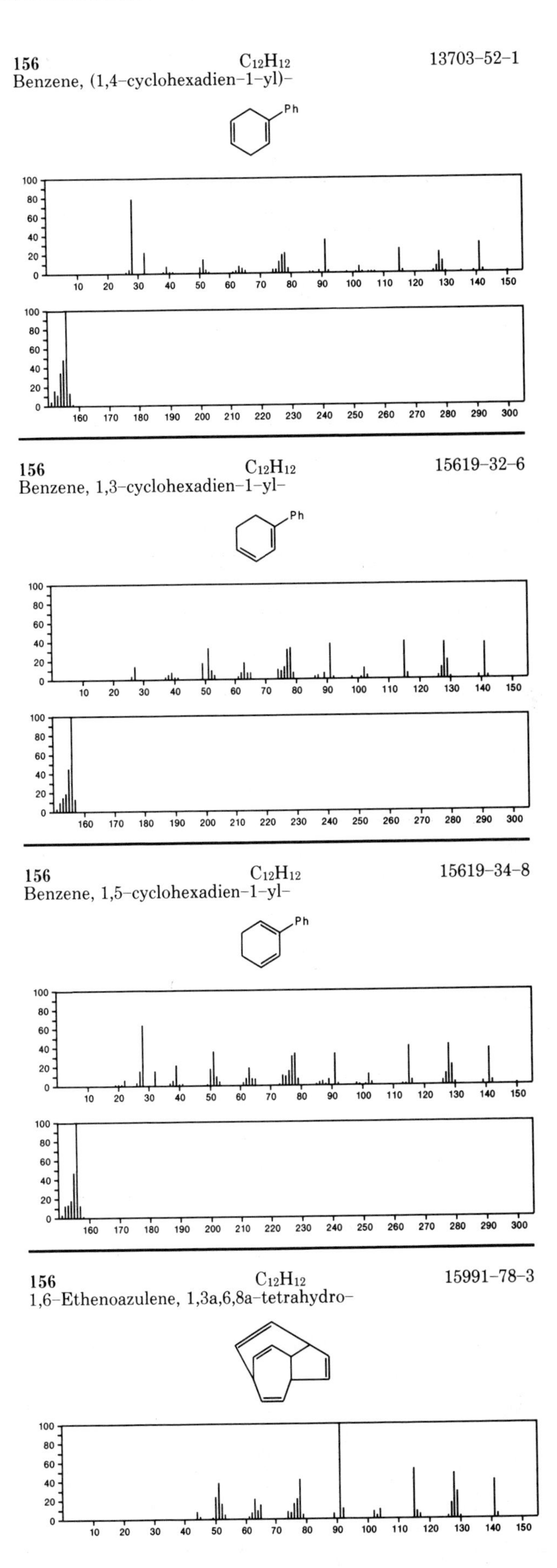

156 $C_{12}H_{12}$ 13703-52-1
Benzene, (1,4-cyclohexadien-1-yl)-

156 $C_{12}H_{12}$ 15619-32-6
Benzene, 1,3-cyclohexadien-1-yl-

156 $C_{12}H_{12}$ 15619-34-8
Benzene, 1,5-cyclohexadien-1-yl-

156 $C_{12}H_{12}$ 15991-78-3
1,6-Ethenoazulene, 1,3a,6,8a-tetrahydro-

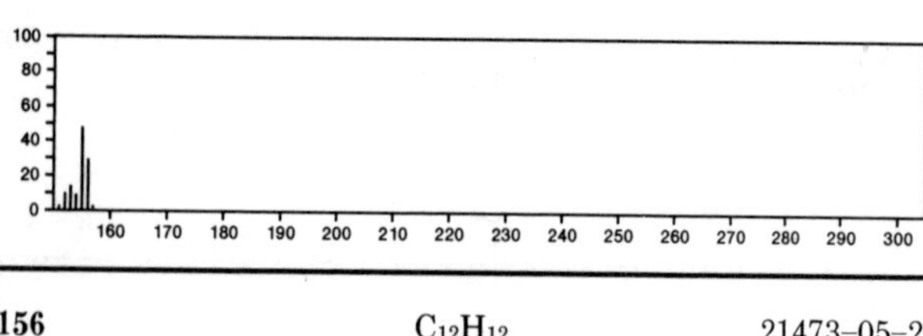

156 $C_{12}H_{12}$ 21473-05-2
Benzene, 2,4-cyclohexadien-1-yl-

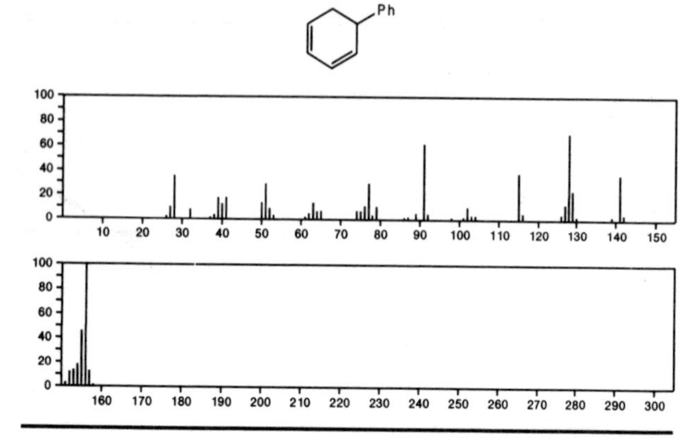

157 $C_3H_2Cl_3N$ 813-74-1
Propanenitrile, 2,2,3-trichloro-

$ClCH_2CCl_2CN$

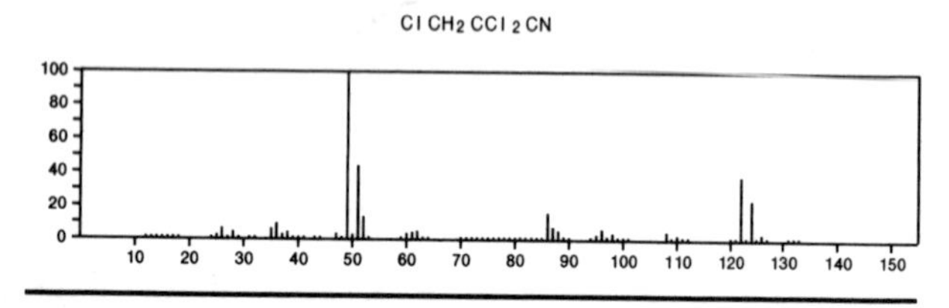

157 C_5H_4BrN 109-04-6
Pyridine, 2-bromo-

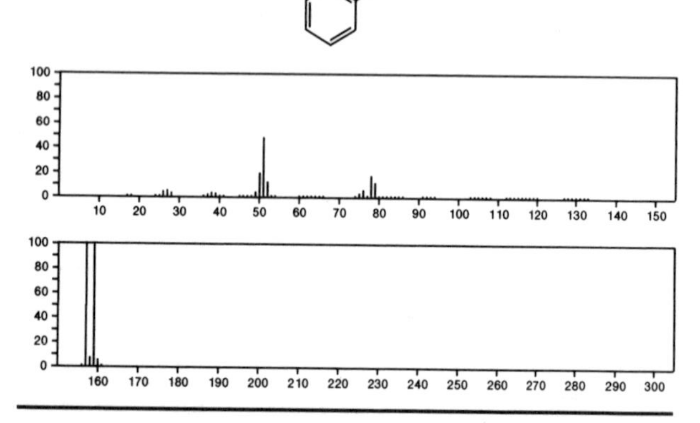

157 C_5H_4BrN 626-55-1
Pyridine, 3-bromo-

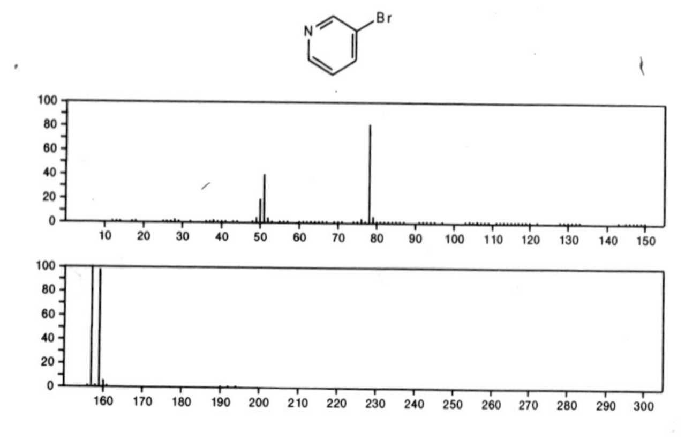

157 $C_6H_4ClNO_2$ 121-73-3
Benzene, 1-chloro-3-nitro-

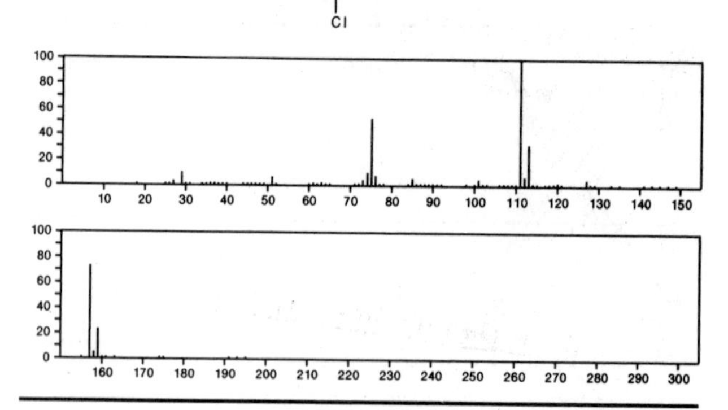

157 $C_6H_4ClNO_2$ 4684-94-0
2-Pyridinecarboxylic acid, 6-chloro-

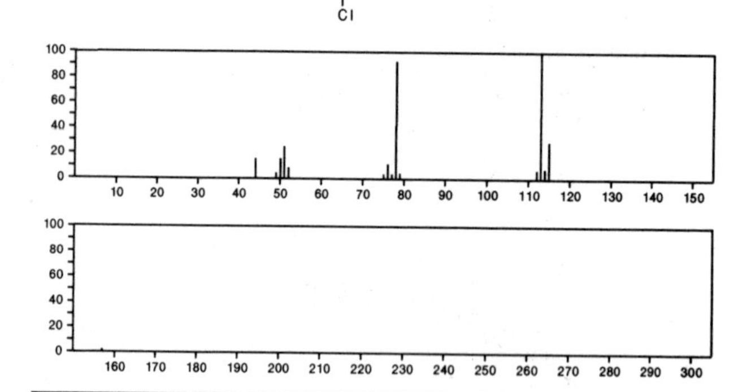

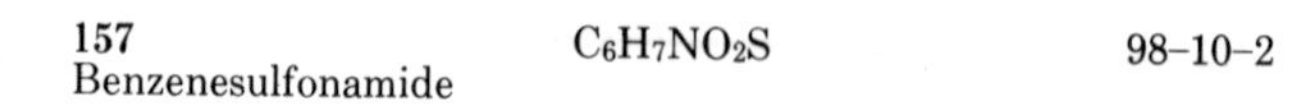

157 $C_6H_7NO_2S$ 98-10-2
Benzenesulfonamide

H_2NSO_2Ph

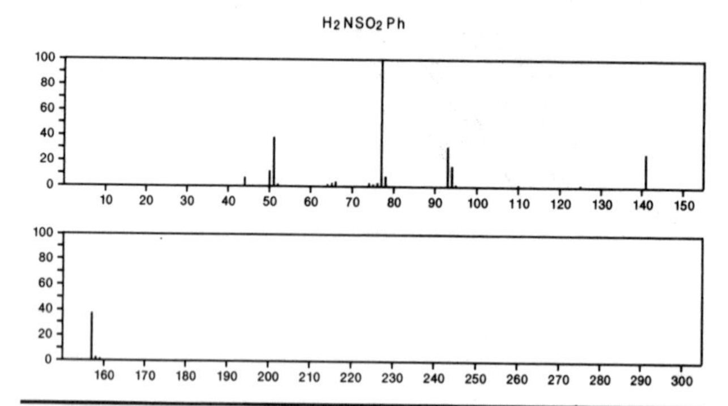

157 $C_6H_7NO_2S$ 5255-33-4
5-Thiazoleacetic acid, 4-methyl-

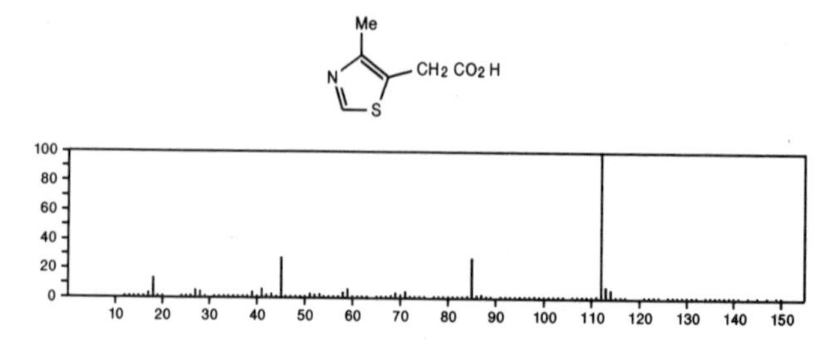

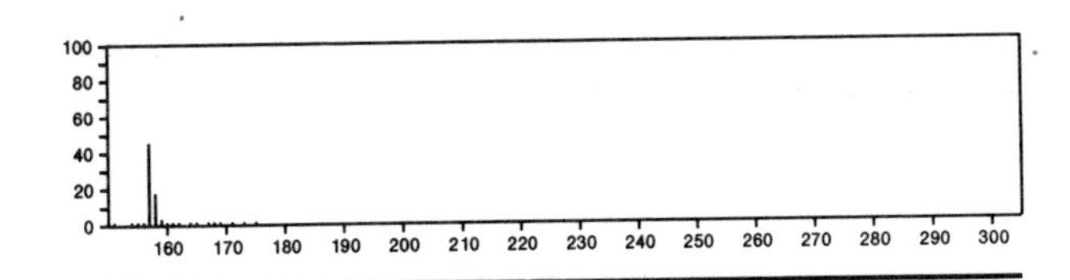

157 $C_6H_7NO_2S$ 23244-32-8
3-Isothiazolecarboxylic acid, ethyl ester

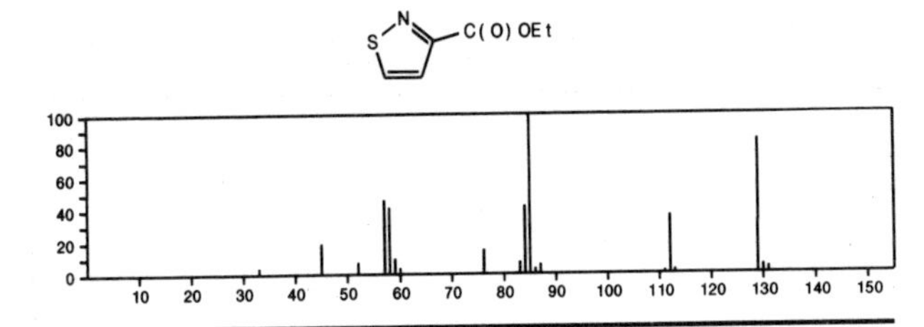

157 $C_6H_{11}N_3S$ 7283-39-8
Hydrazinecarbothioamide, 2-cyclopentylidene-

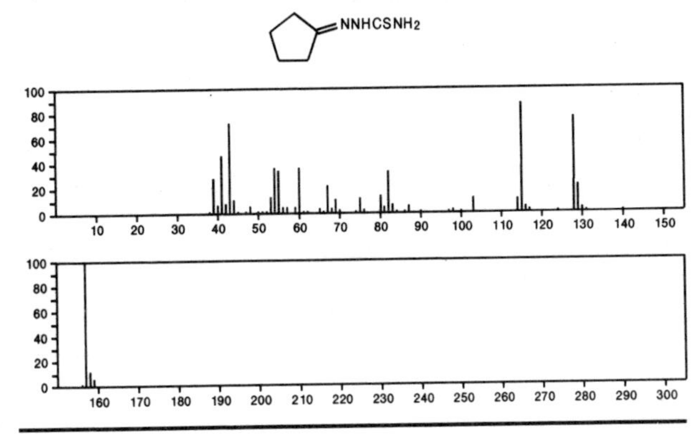

157 $C_7H_{11}NO_3$ 68-95-1
L-Proline, 1-acetyl-

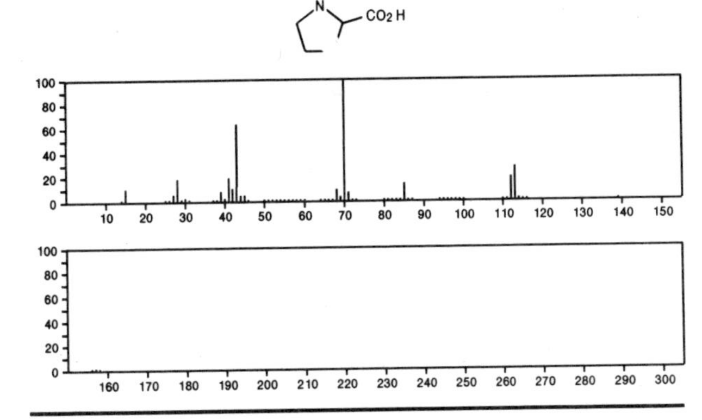

157 $C_7H_{11}NO_3$ 115-67-3
2,4-Oxazolidinedione, 5-ethyl-3,5-dimethyl-

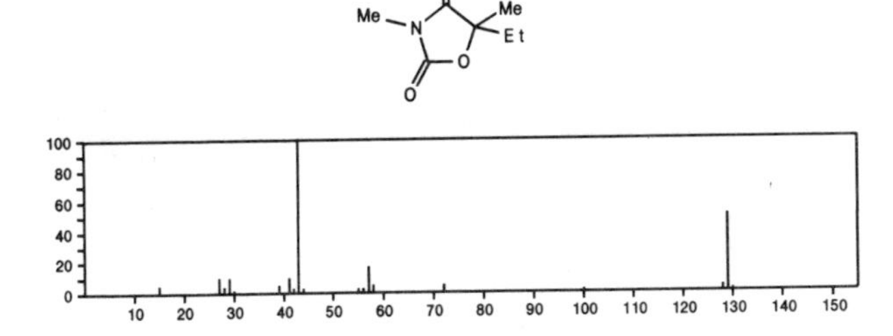

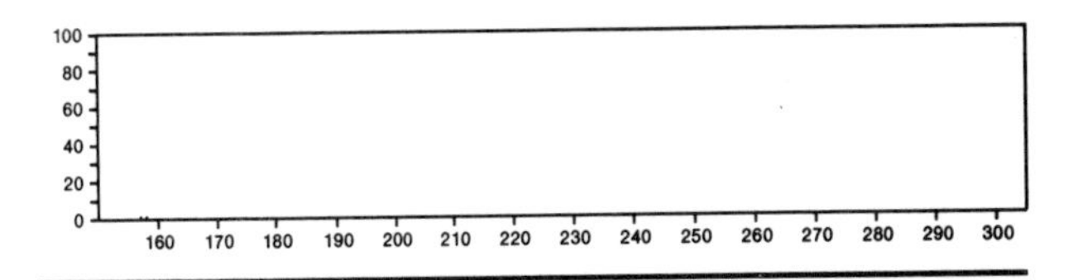

157 $C_7H_{11}NO_3$ 23840-13-3
5-Hexynoic acid, 2-amino-4-(hydroxymethyl)-

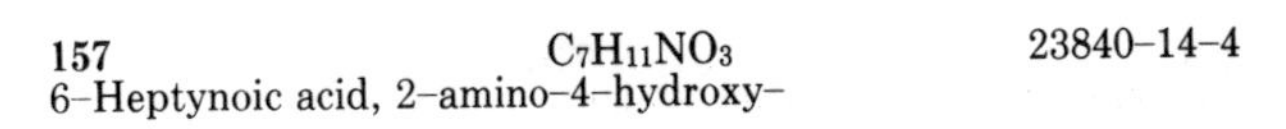

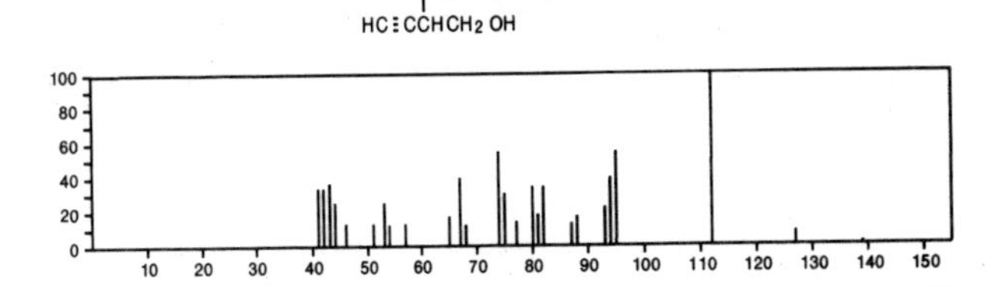

157 $C_7H_{11}NO_3$ 23840-14-4
6-Heptynoic acid, 2-amino-4-hydroxy-

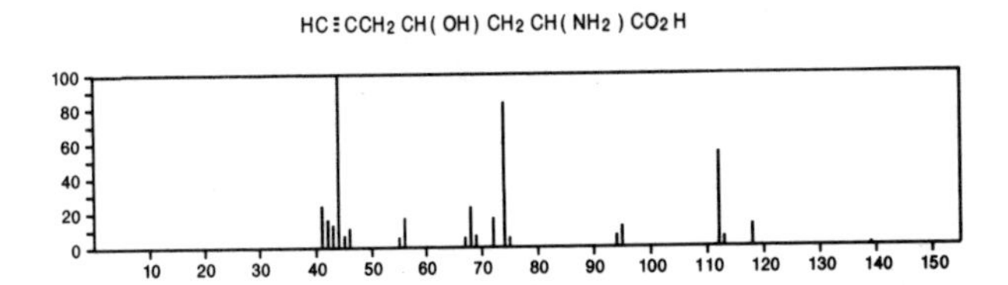

157 $C_7H_{11}NO_3$ 42435-88-1
L-Proline, 1-methyl-5-oxo-, methyl ester

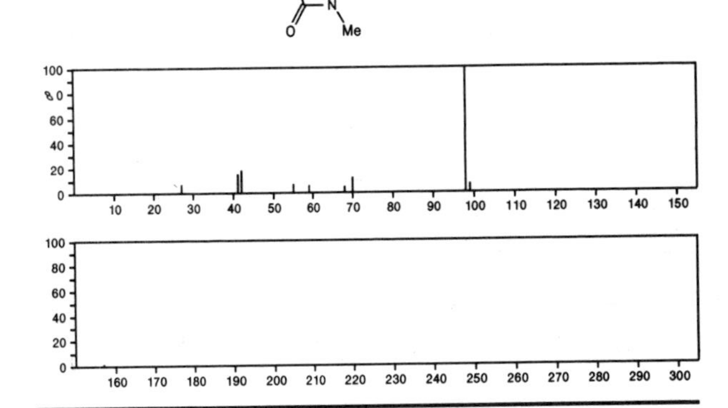

157 $C_7H_{11}NO_3$ 52812-86-9
4-Hexenoic acid, 5-amino-3-oxo-, methyl ester

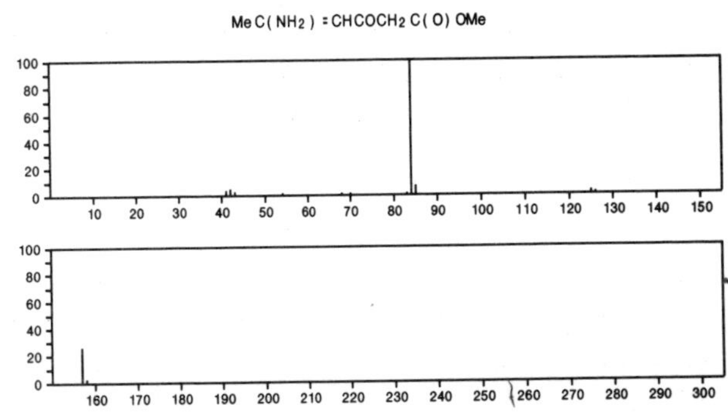

157 $C_7H_{11}NO_3$ 56145-24-5
2-Pyrrolidinecarboxylic acid, 2-methyl-5-oxo-, methyl ester

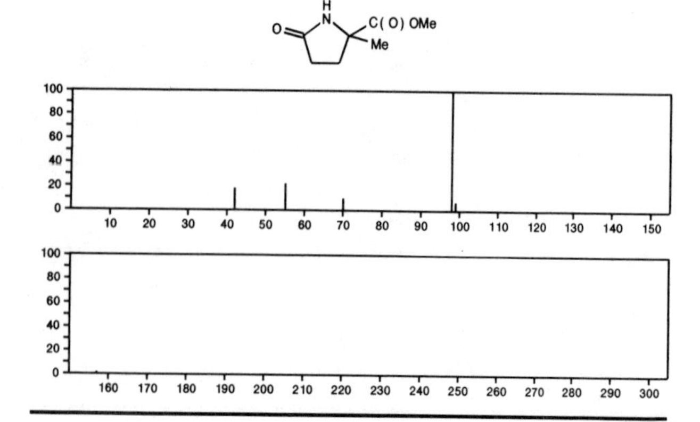

157 $C_8H_{11}N.ClH$ 156-28-5
Benzeneethanamine, hydrochloride

$H_2NCH_2CH_2Ph \cdot HCl$

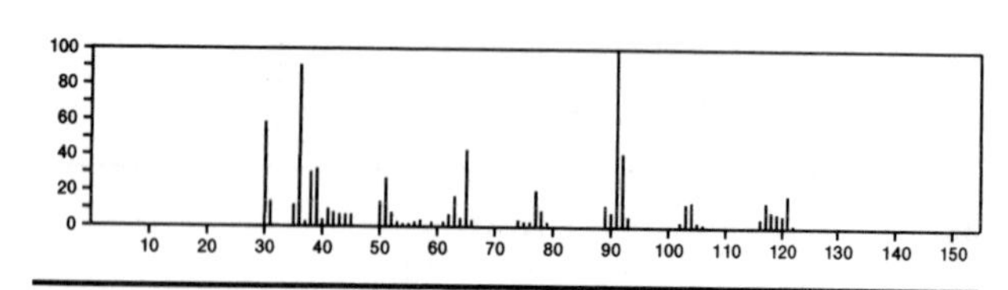

157 $C_8H_{15}NO_2$ 520-62-7
1*H*-Pyrrolizine-1-methanol, hexahydro-7-hydroxy-, [1*S*-(1α,7α,7aβ)]-

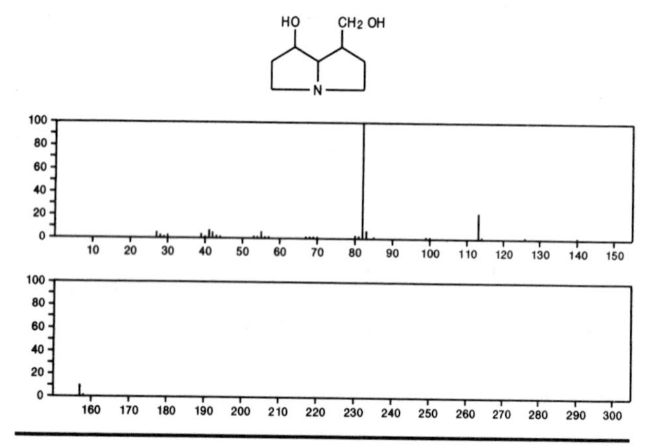

157 $C_8H_{15}NO_2$ 1787-52-6
Diacetamide, *N*-isobutyl-

i-BuN(Ac)2

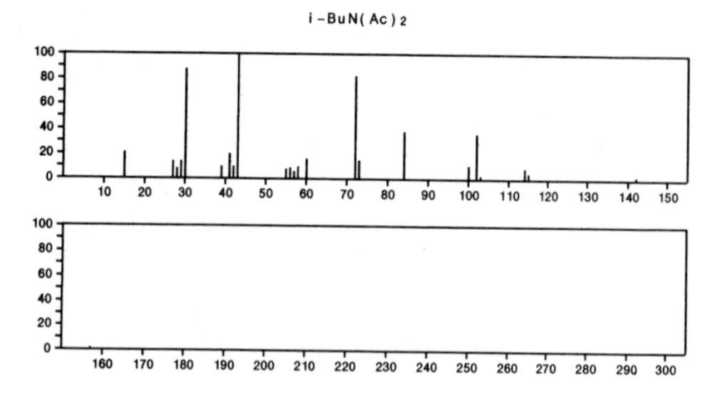

157 $C_8H_{15}NO_2$ 6949-77-5
Cycloheptanecarboxylic acid, 1-amino-

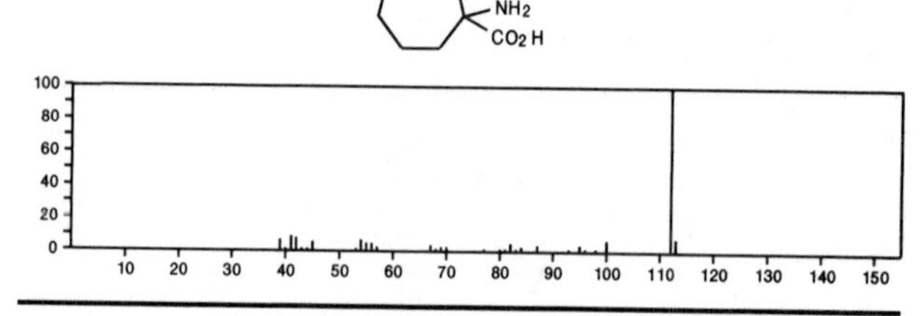

157 $C_8H_{15}NO_2$ 14205-43-7
2-Butenoic acid, 3-amino-, 1,1-dimethylethyl ester

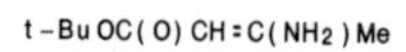

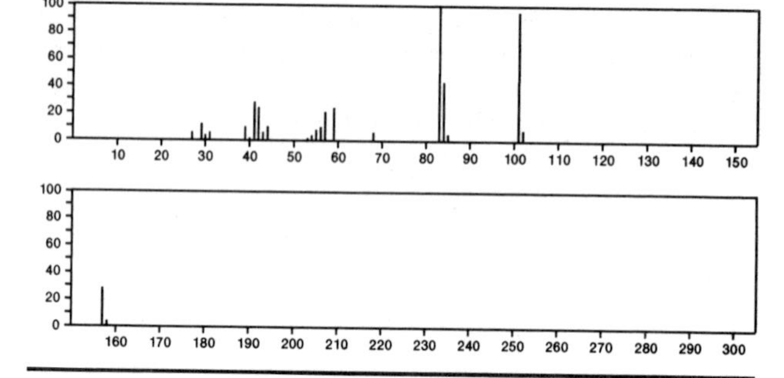

157 $C_8H_{15}NO_2$ 19264-30-3
Diacetamide, *N*-*sec*-butyl-

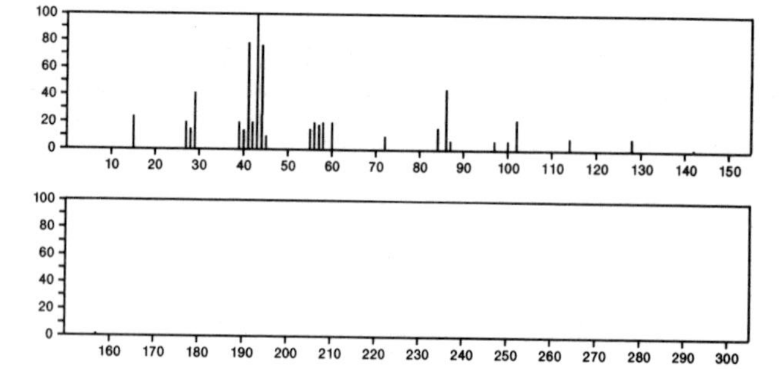

157 $C_8H_{15}NO_2$ 23435-07-
2*H*-Azonin-2-one, octahydro-6-hydroxy-

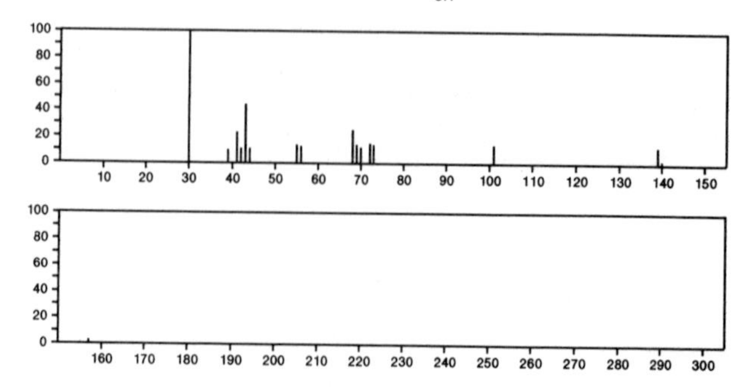

157 $C_8H_{15}NO_2$ 32663-70-0
8-Azabicyclo[3.2.1]octan-3-ol, 8-methyl-, 8-oxide, (*endo,anti*)-

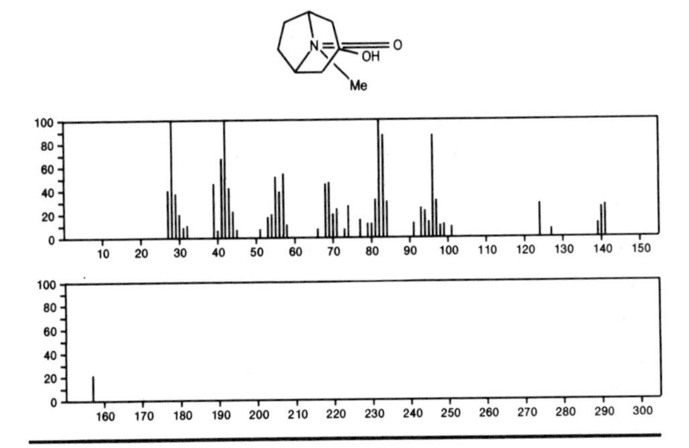

157 $C_8H_{15}NO_2$ 32663-71-1
8-Azabicyclo[3.2.1]octan-3-ol, 8-methyl-, 8-oxide, (*endo,syn*)-

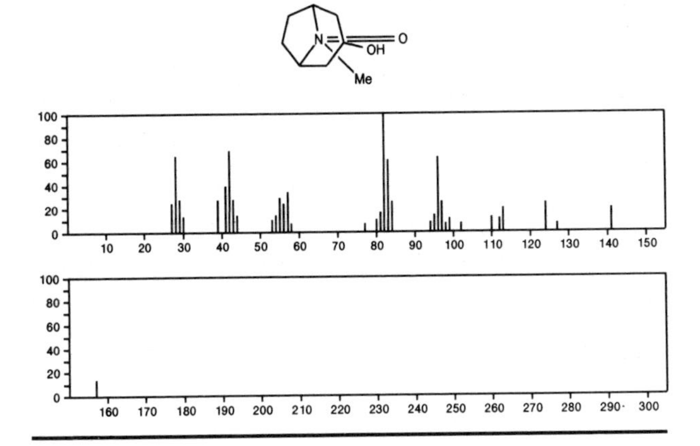

157 $C_8H_{15}NO_2$ 49582-71-0
3-Buten-2-one, 4-(dimethylamino)-4-ethoxy-

Me2NC(OEt)=CHCOMe

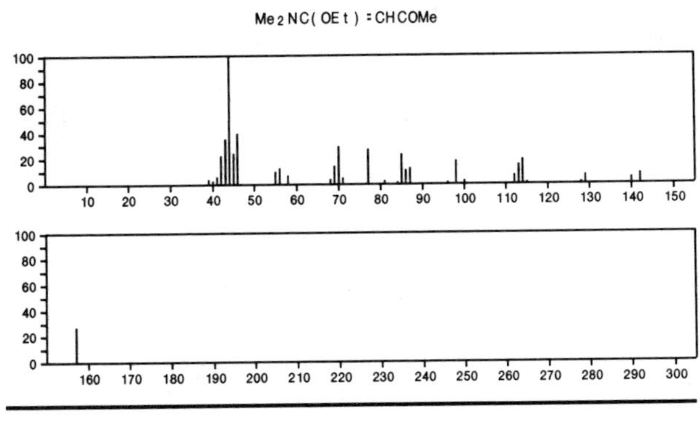

157 $C_8H_{15}NO_2$ 54751-96-1
3-Piperidinol, 1-acetyl-6-methyl-

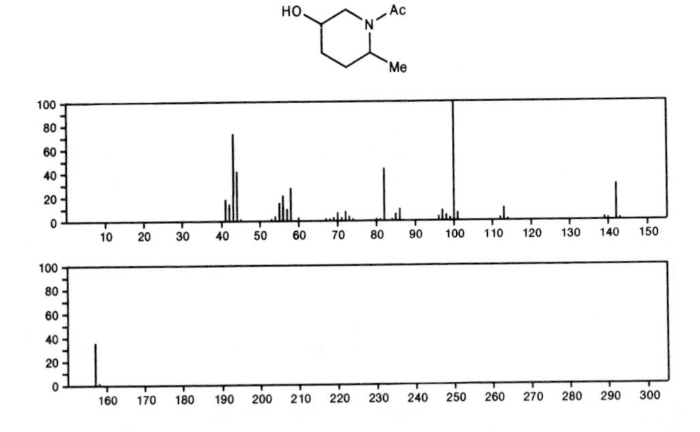

157 $C_8H_{15}NS$ 35418-38-3
2-Pyrrolidinethione, 3,3,5,5-tetramethyl-

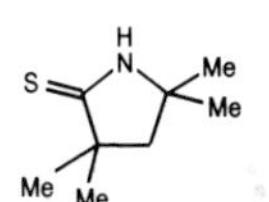

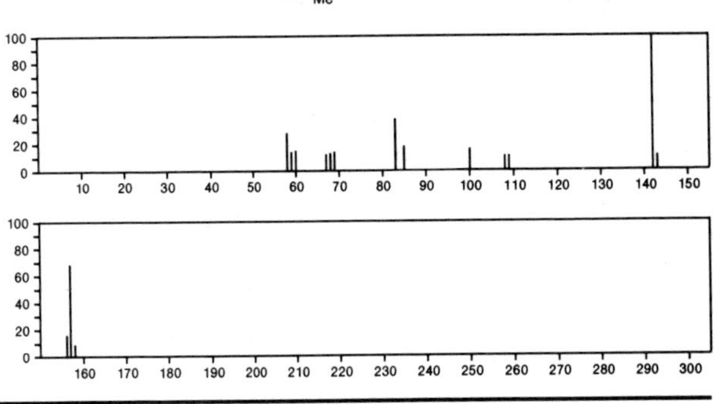

157 $C_8H_{17}N_2O$ 34272-83-8
1-Pyrrolidinyloxy, 3-amino-2,2,5,5-tetramethyl-

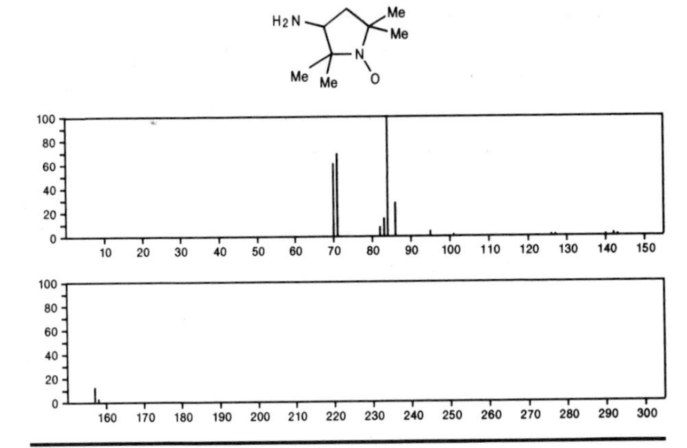

157 $C_9H_7N_3$ 1656-95-7
3,5-Pyridinedicarbonitrile, 2,6-dimethyl-

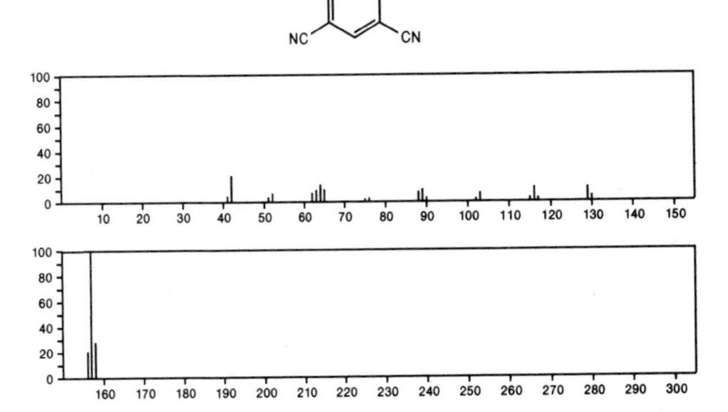

157 $C_9H_{19}NO$ 686-96-4
Propanamide, *N*-(1,1-dimethylethyl)-2,2-dimethyl-

t-BuNHCOCMe3

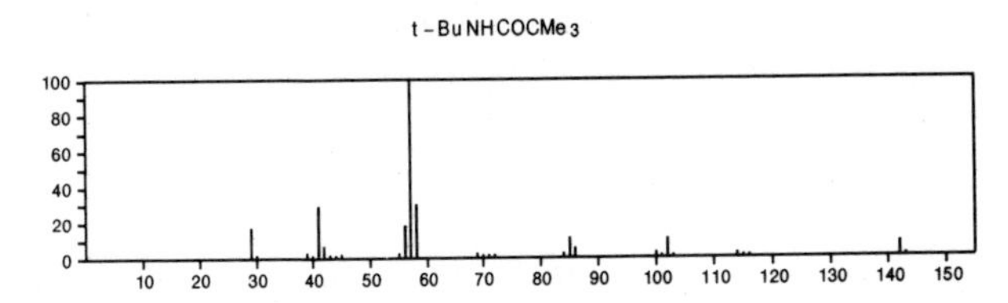

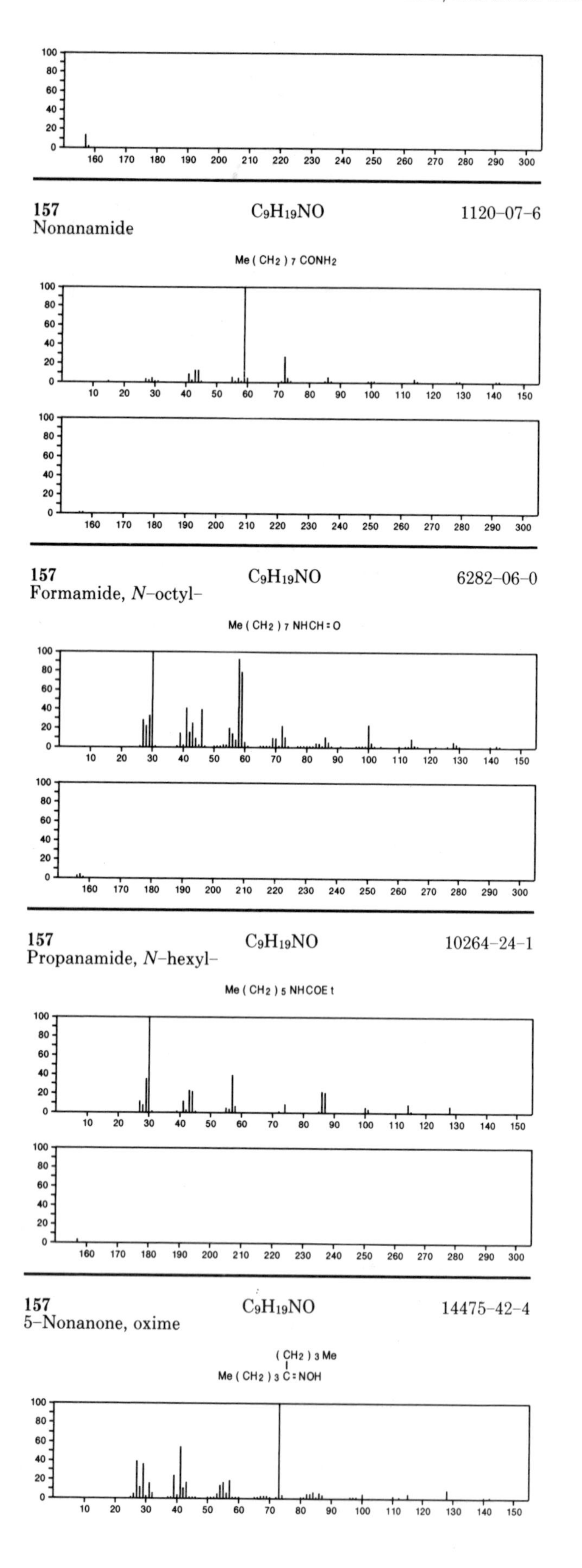

157 C9H19NO 1120-07-6
Nonanamide
Me (CH2) 7 CONH2
157 C9H19NO 6282-06-0
Formamide, N-octyl-
Me (CH2) 7 NHCH = O
157 C9H19NO 10264-24-1
Propanamide, N-hexyl-
Me (CH2) 5 NHCOEt
157 C9H19NO 14475-42-4
5-Nonanone, oxime
(CH2) 3 Me
Me (CH2) 3 C = NOH

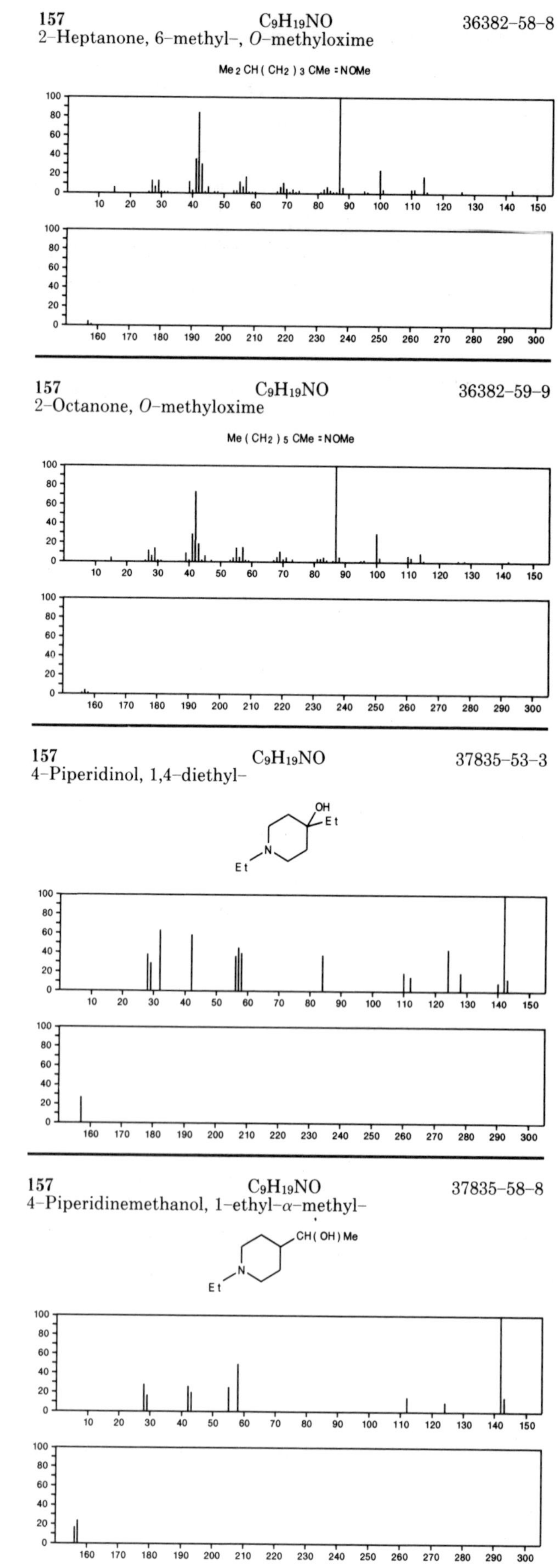

157 C9H19NO 36382-58-8
2-Heptanone, 6-methyl-, O-methyloxime
Me2 CH (CH2) 3 CMe = NOMe
157 C9H19NO 36382-59-9
2-Octanone, O-methyloxime
Me (CH2) 5 CMe = NOMe
157 C9H19NO 37835-53-3
4-Piperidinol, 1,4-diethyl-
OH
Et
N
Et
157 C9H19NO 37835-58-8
4-Piperidinemethanol, 1-ethyl-α-methyl-
CH(OH) Me
N
Et

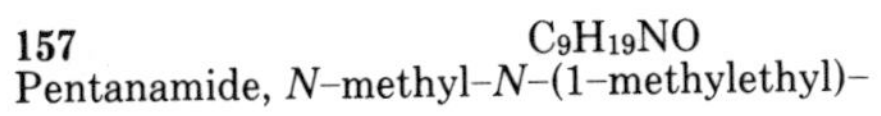

157 $C_9H_{19}NO$ 54965-74-1
Pentanamide, *N*-methyl-*N*-(1-methylethyl)-

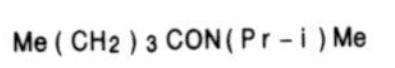

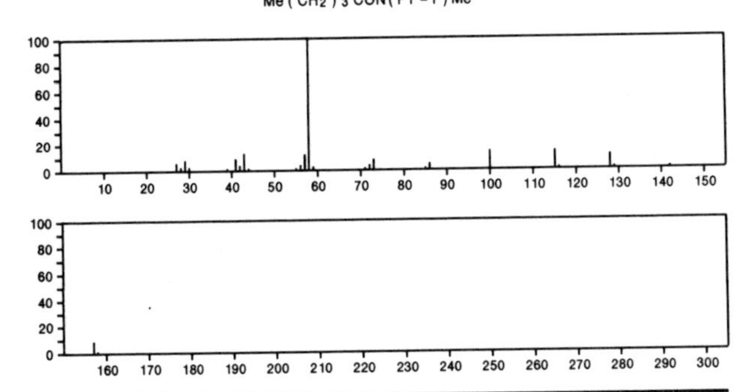

157 $C_9H_{19}NO$ 55669-83-5
1-Aziridinepropanol, 2-methyl-3-(1-methylethyl)-, *trans*-

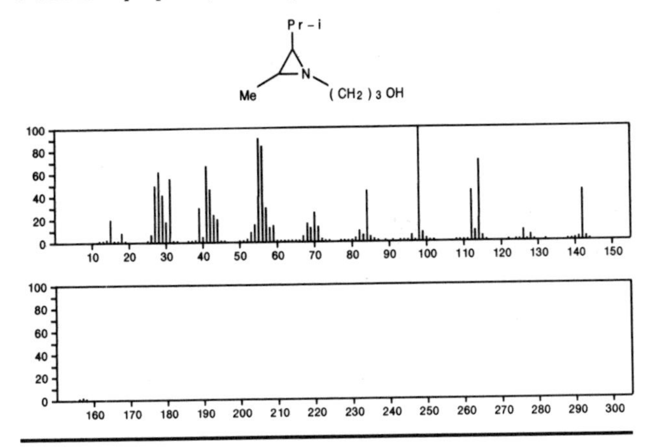

157 $C_9H_{19}NO$ 55955-98-1
Oxazolidine, 2,2,5-trimethyl-3-propyl-

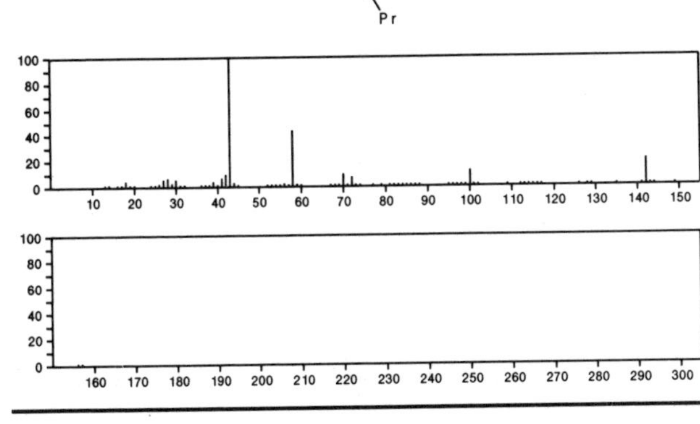

157 $C_{10}H_{23}N$ 544-00-3
1-Butanamine, 3-methyl-*N*-(3-methylbutyl)-

Me2 CHCH2 CH2 NHCH2 CH2 CHMe2

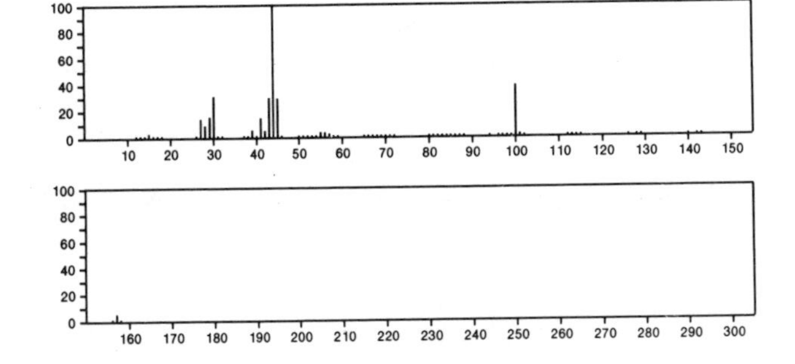

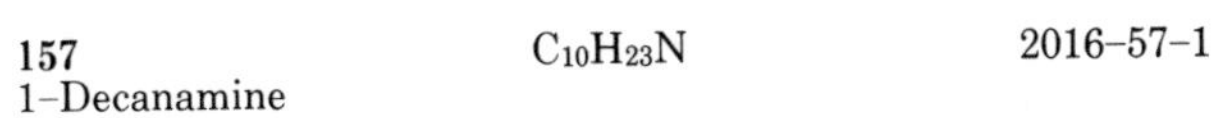

157 $C_{10}H_{23}N$ 2016-57-1
1-Decanamine

Me (CH2) 9 NH2

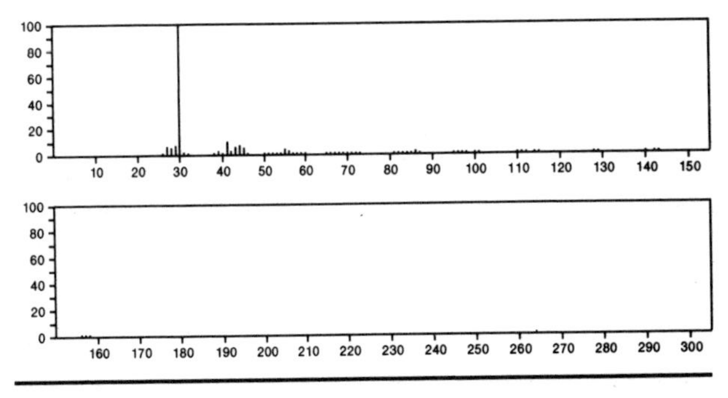

157 $C_{10}H_{23}N$ 2050-92-2
1-Pentanamine, *N*-pentyl-

Me (CH2) 4 NH (CH2) 4 Me

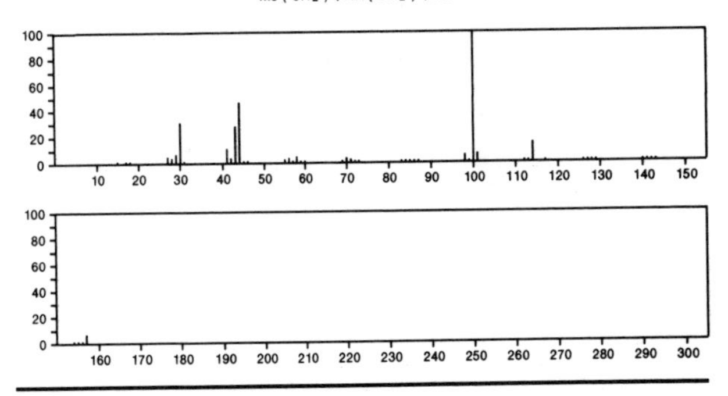

157 $C_{10}H_{23}N$ 24539-82-0
3-Octanamine, *N*,*N*-dimethyl-

Me (CH2) 4 CHE t NMe 2

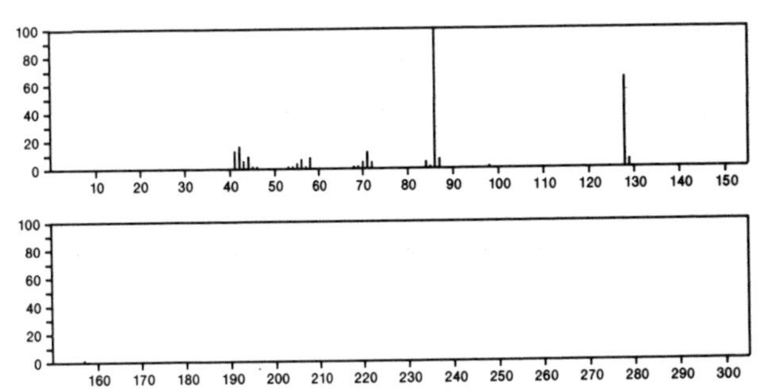

157 $C_{10}H_{23}N$ 24552-00-9
Hexylamine, *N*-methyl-*N*-propyl-

Pr NMe (CH2) 5 Me

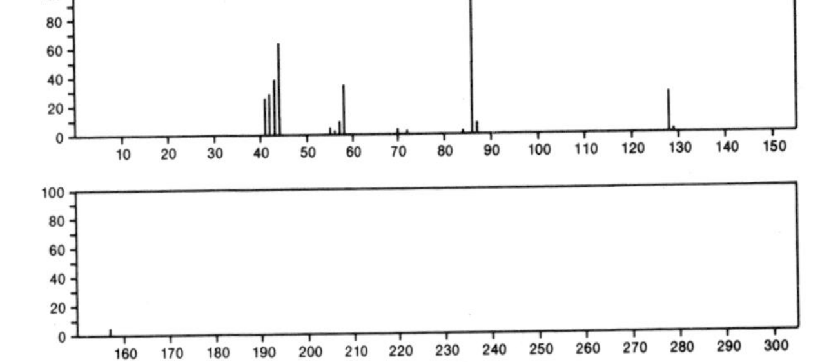

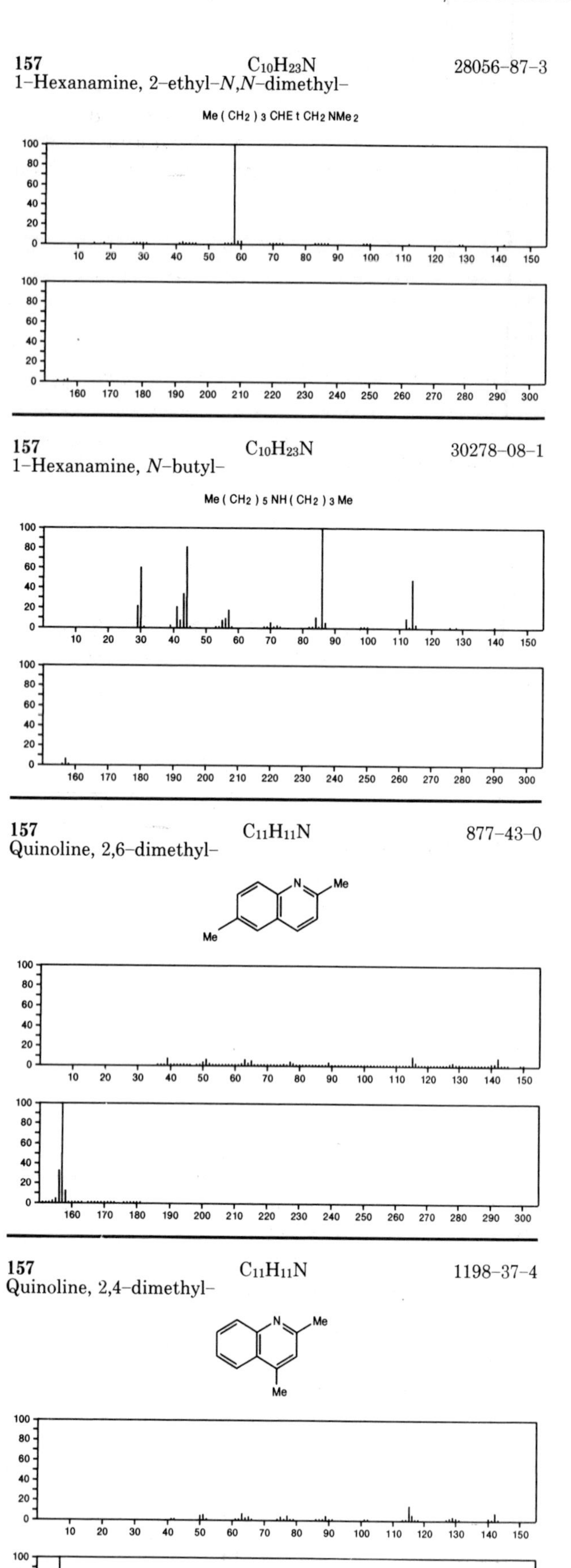
157
C10H23N
28056-87-3
1-Hexanamine, 2-ethyl-N,N-dimethyl-
Me (CH2) 3 CHE t CH2 NMe 2
157
C10H23N
30278-08-1
1-Hexanamine, N-butyl-
Me (CH2) 5 NH (CH2) 3 Me
157
C11H11N
877-43-0
Quinoline, 2,6-dimethyl-
157
C11H11N
1198-37-4
Quinoline, 2,4-dimethyl-

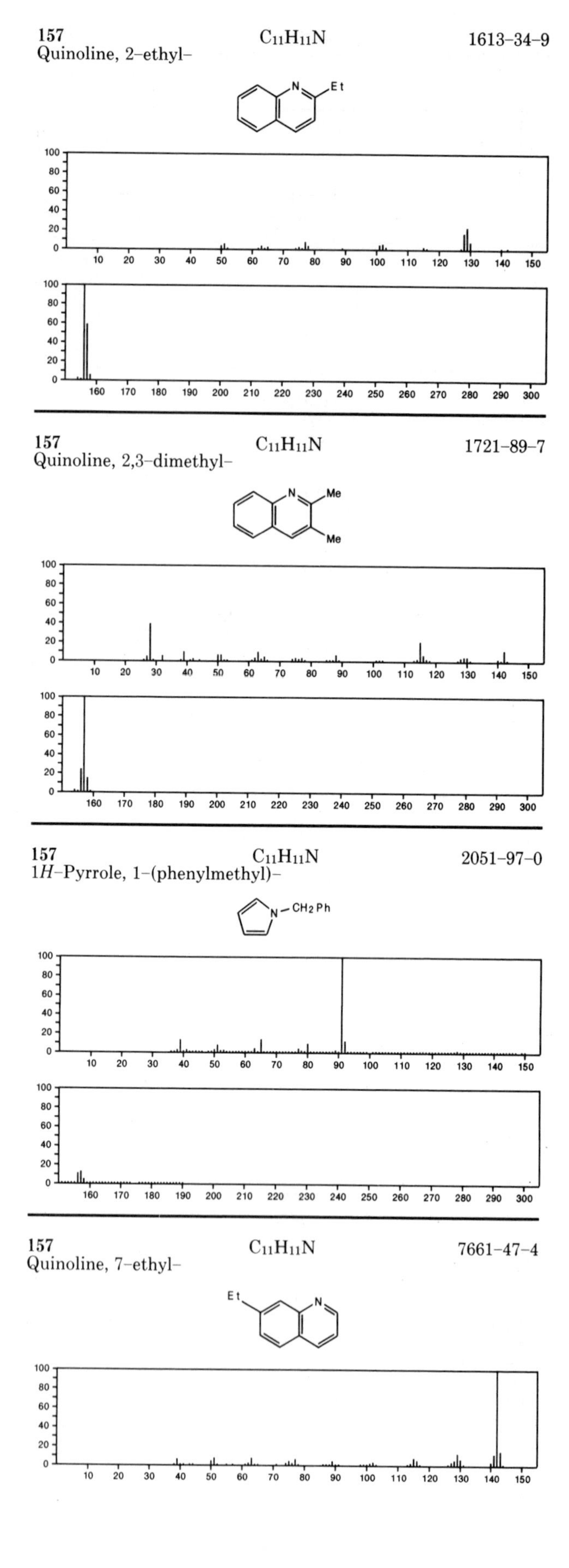
157
C11H11N
1613-34-9
Quinoline, 2-ethyl-
157
C11H11N
1721-89-7
Quinoline, 2,3-dimethyl-
157
C11H11N
2051-97-0
1H-Pyrrole, 1-(phenylmethyl)-
157
C11H11N
7661-47-4
Quinoline, 7-ethyl-

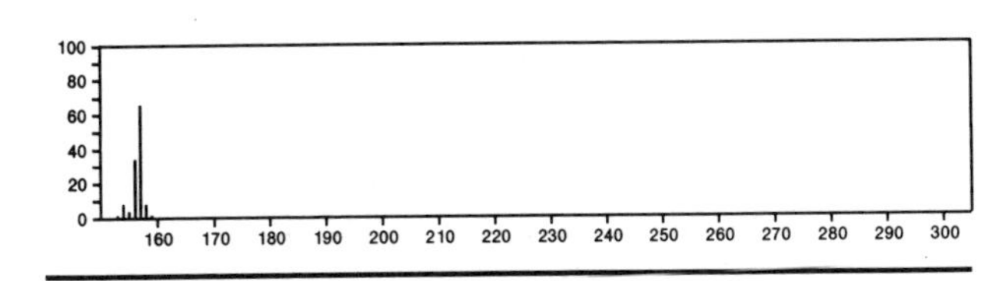

157 $C_{11}H_{11}N$ 7661-60-1
Isoquinoline, 1-ethyl-

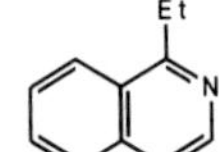

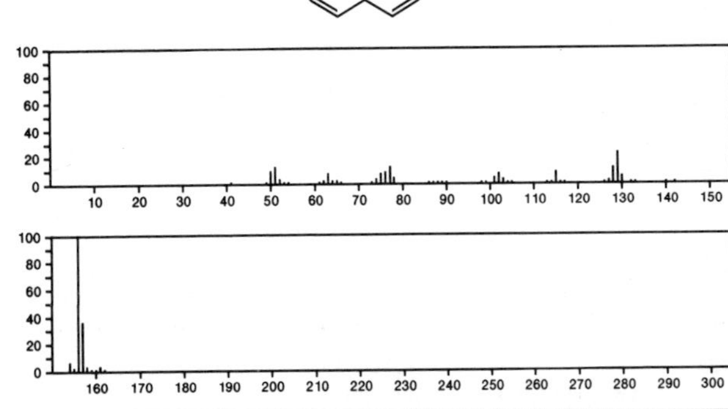

157 $C_{11}H_{11}N$ 17104-67-5
2-Naphthonitrile, 5,6,7,8-tetrahydro-

CN

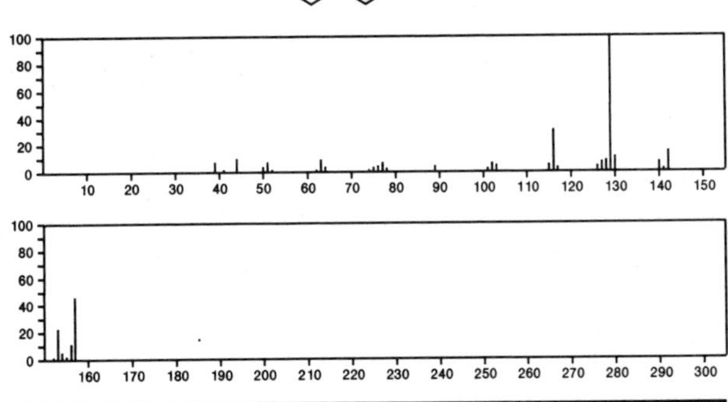

157 $C_{11}H_{11}N$ 29809-13-0
1-Naphthonitrile, 5,6,7,8-tetrahydro-

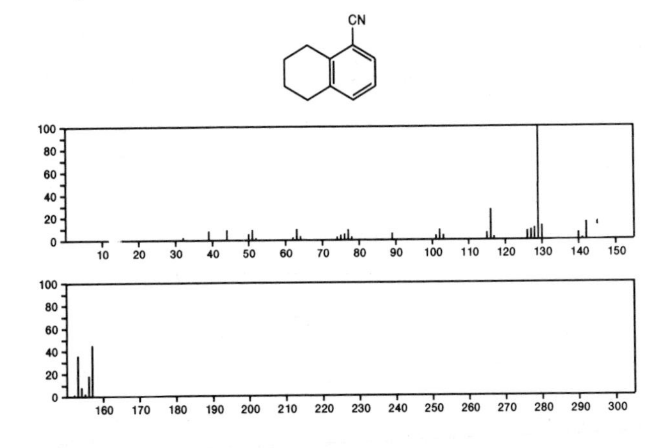

158 CF_6S 374-10-7
Sulfur, trifluoro(trifluoromethyl)-

F_3SCF_3

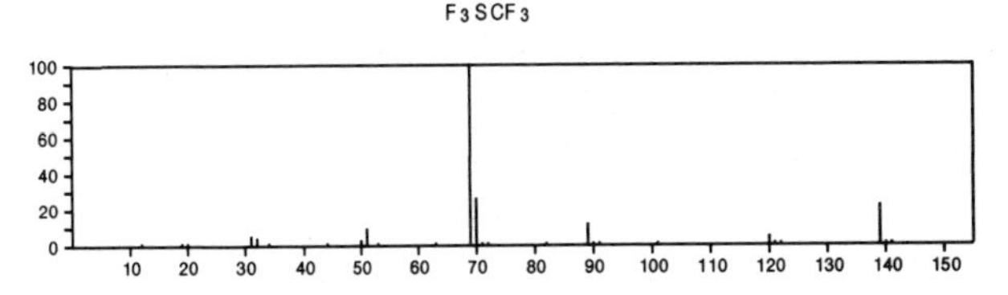

158 $C_4H_5Cl_3$ 4749-27-3
1-Propene, 3,3,3-trichloro-2-methyl-

$Cl_3CCMe=CH_2$

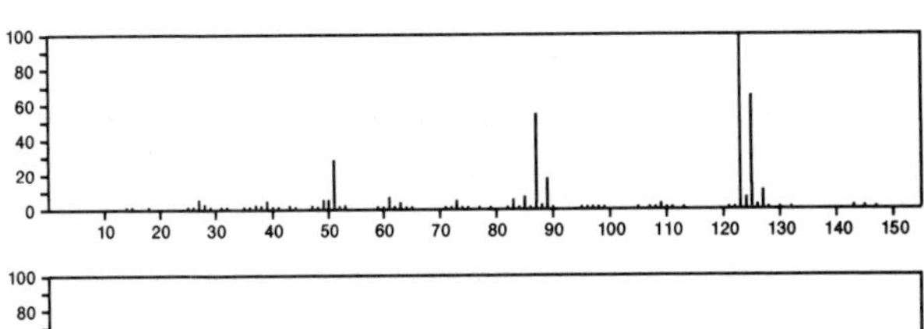

158 $C_4H_5Cl_3$ 31702-33-7
1-Propene, 1,1,3-trichloro-2-methyl-

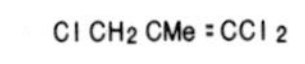

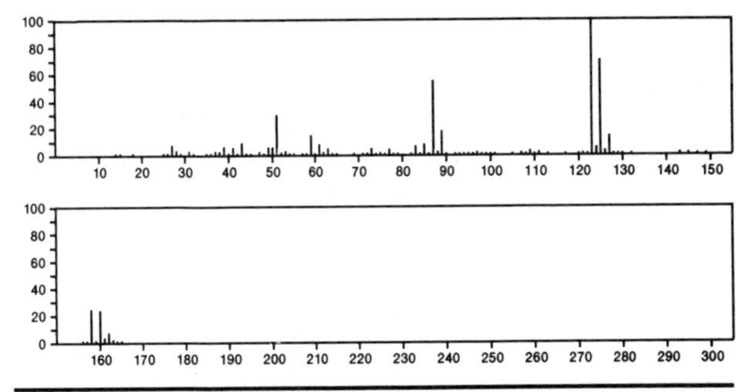

158 $C_5H_6N_2O_2S$ 584-26-9
4-Imidazolidinone, 1-acetyl-2-thioxo-

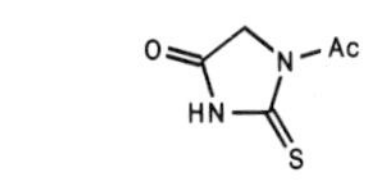

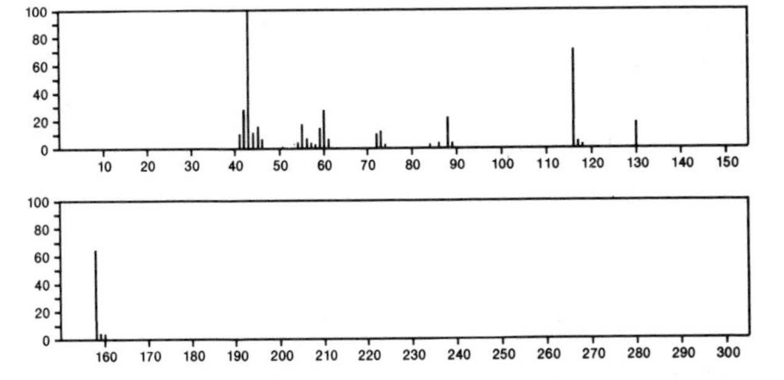

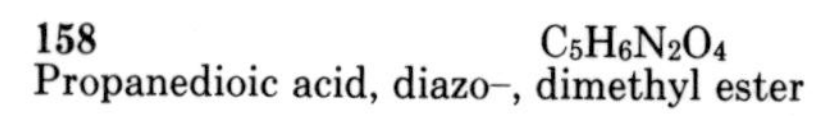

158 $C_5H_6N_2O_4$ 6773–29–1

Propanedioic acid, diazo–, dimethyl ester

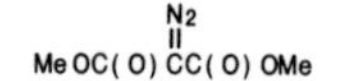

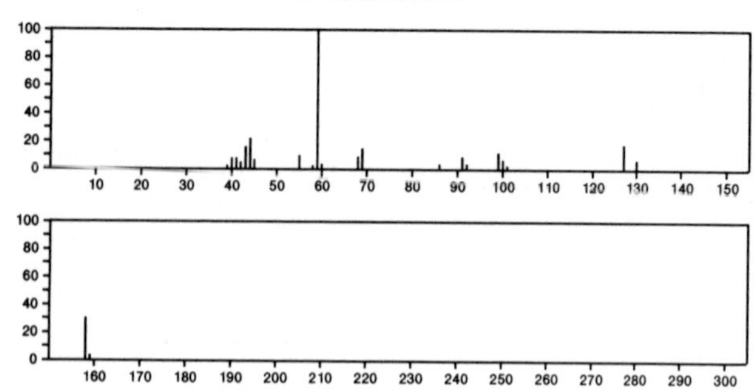

158 $C_5H_6N_2O_4$ 26574–32–3

Sydnone, 3–(2–carboxyethyl)–

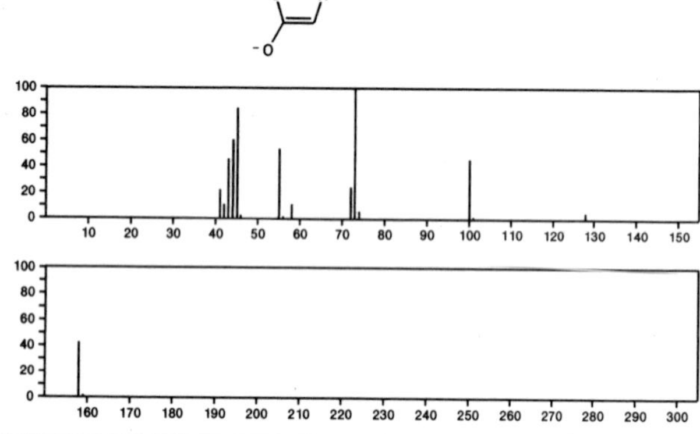

158 $C_5H_{10}N_4O_2$ 55556–94–0

Piperazine, 2–methyl–1,4–dinitroso–

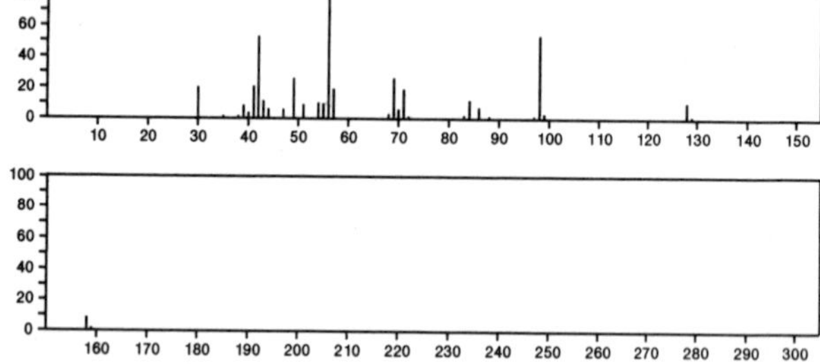

158 $C_5H_{10}N_4O_2$ 55557–00–1

1*H*–1,4–Diazepine, hexahydro–1,4–dinitroso–

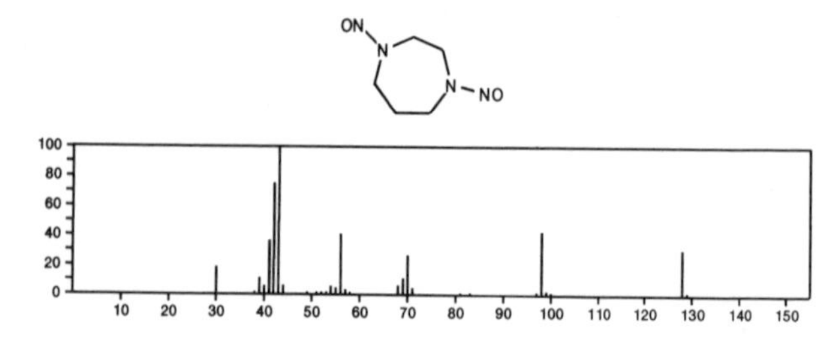

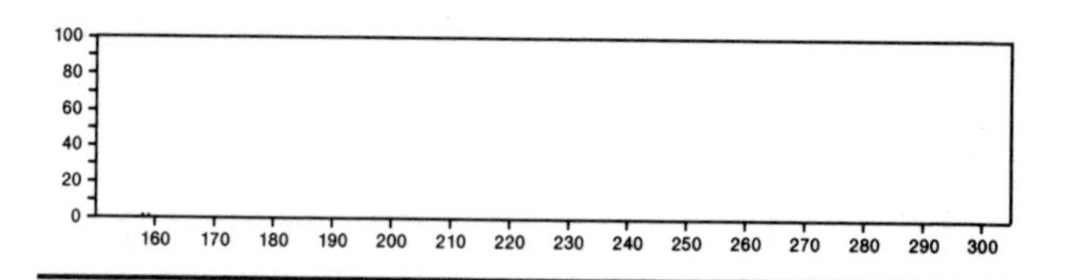

158 $C_6H_5BCl_2$ 873–51–8

Borane, dichlorophenyl–

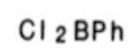

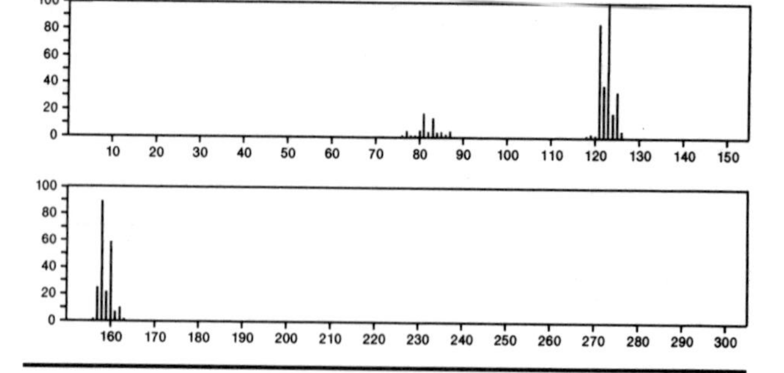

158 $C_6H_6OS_2$ 35972–85–1

2–Furancarbodithioic acid, methyl ester

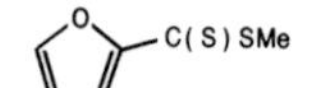

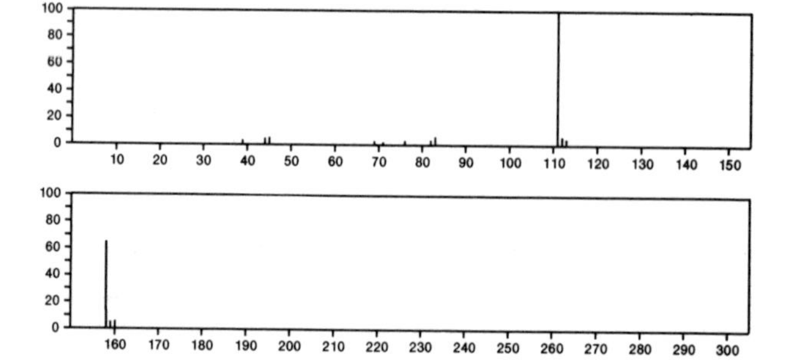

158 $C_6H_6O_3S$ 98–11–3

Benzenesulfonic acid

PhSO3H

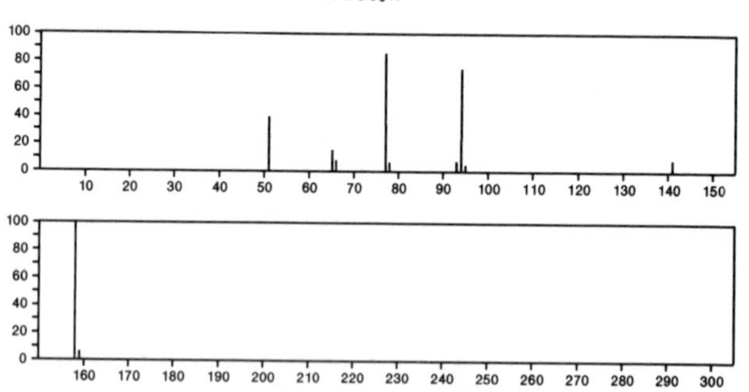

158 $C_6H_7ClN_2O$ 20551–34–2

4(3*H*)–Pyrimidinone, 5–chloro–2,6–dimethyl–

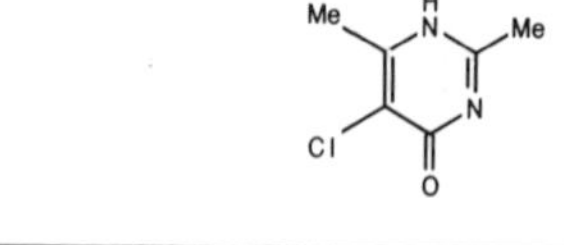

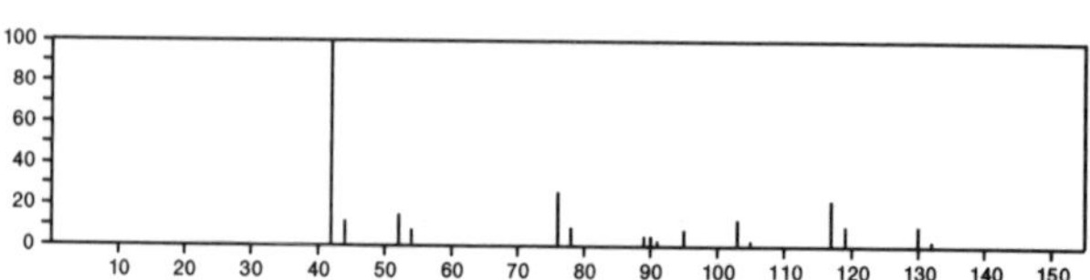

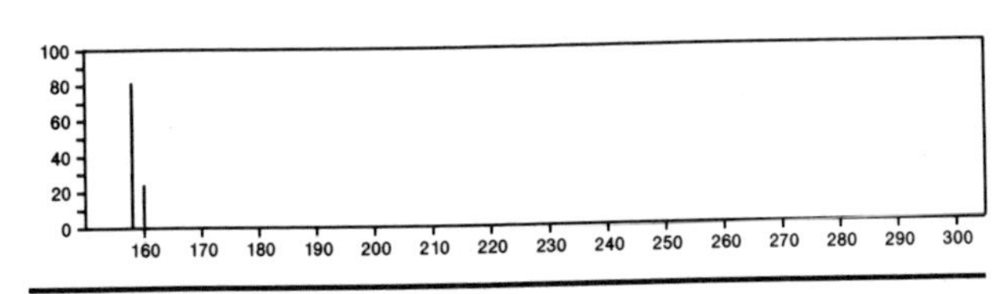

158 $C_6H_7O_3P$ 1571-33-1
Phosphonic acid, phenyl-

$PhPO_3H_2$

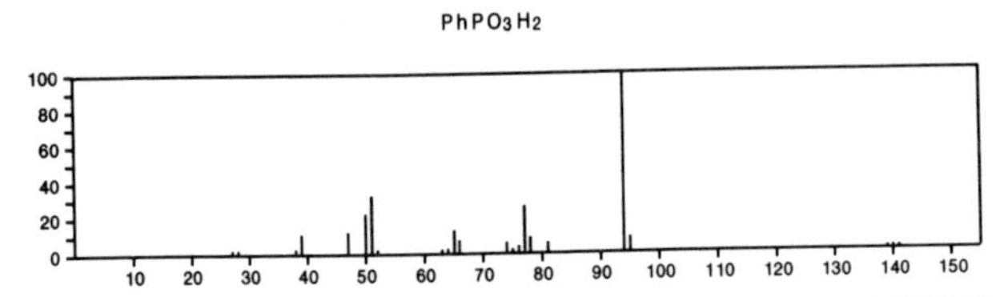

158 $C_6H_{10}F_4$ 648-36-2
Hexane, 3,3,4,4-tetrafluoro-

$EtCF_2CF_2Et$

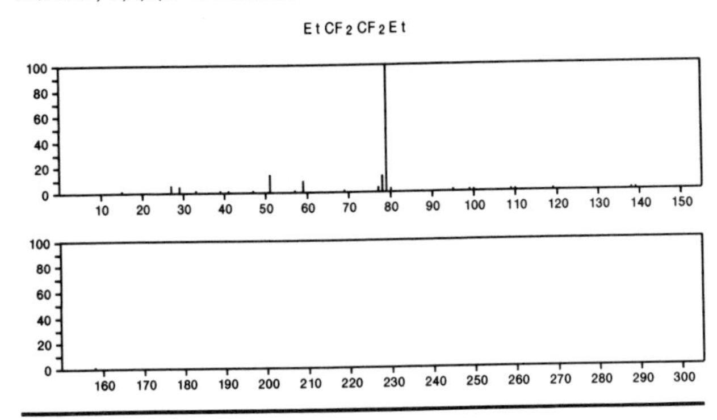

158 $C_6H_{10}N_2OS$ 56805-20-0
4-Imidazolidinone, 5-(1-methylethyl)-2-thioxo-

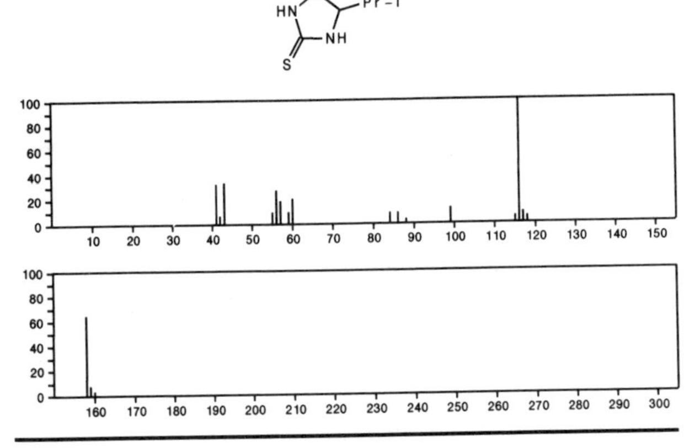

158 $C_6H_{10}N_2O_3$ 4515-18-8
2-Piperidinecarboxylic acid, 1-nitroso-

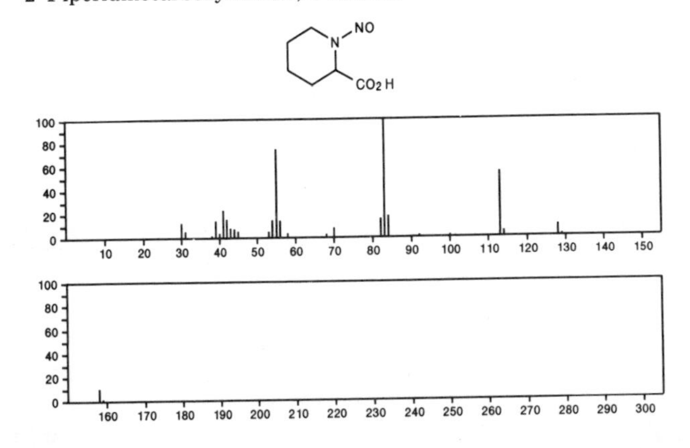

158 $C_6H_{10}N_2O_3$ 6238-69-3
Isonipecotic acid, 1-nitroso-

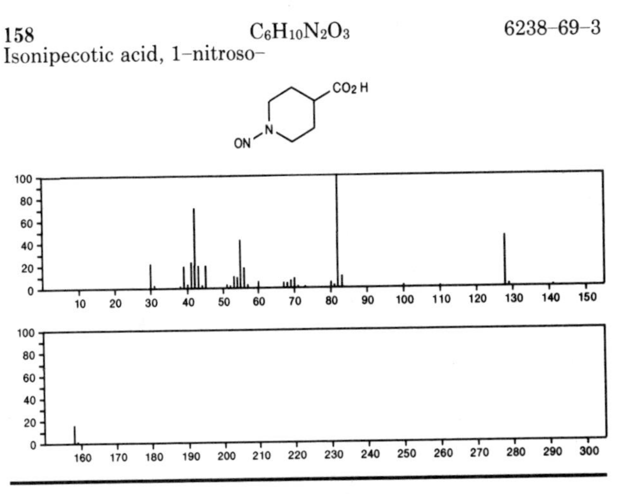

158 $C_6H_{10}N_2O_3$ 16228-00-5
1-Imidazolidinemethanol, 4,4-dimethyl-2,5-dioxo-

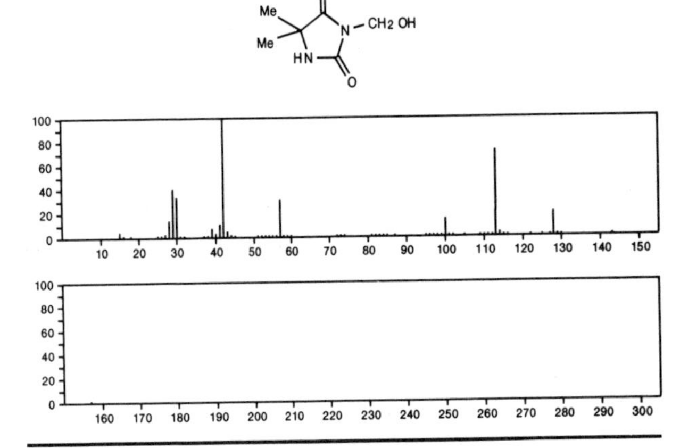

158 $C_6H_{10}N_2O_3$ 30310-83-9
2-Piperidinecarboxylic acid, 1-nitroso-, (*S*)-

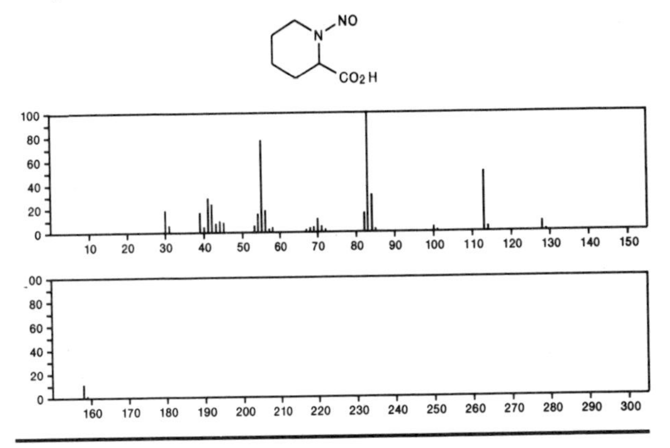

158 $C_7H_{10}N_2.ClH$ 637-60-5
Hydrazine, (4-methylphenyl)-, monohydrochloride

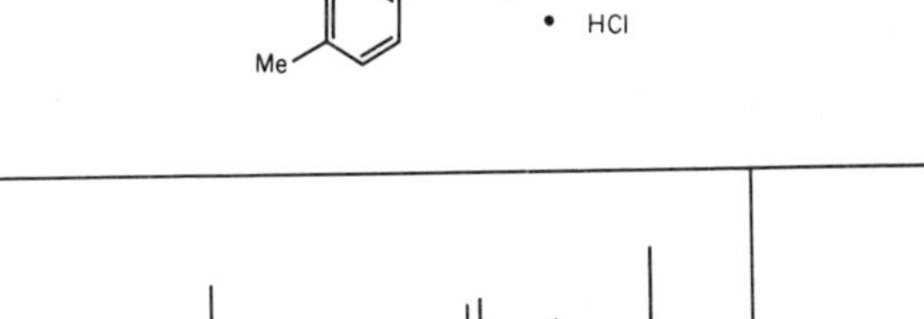

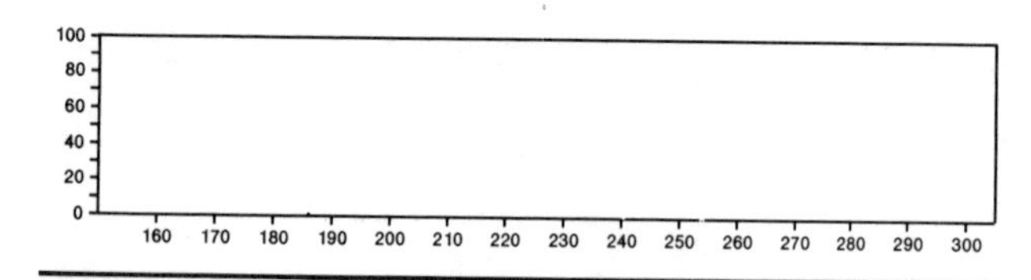

158 $C_7H_{10}N_2.ClH$ 1073–62–7
Hydrazine, (phenylmethyl)–, monohydrochloride

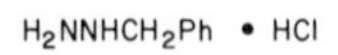

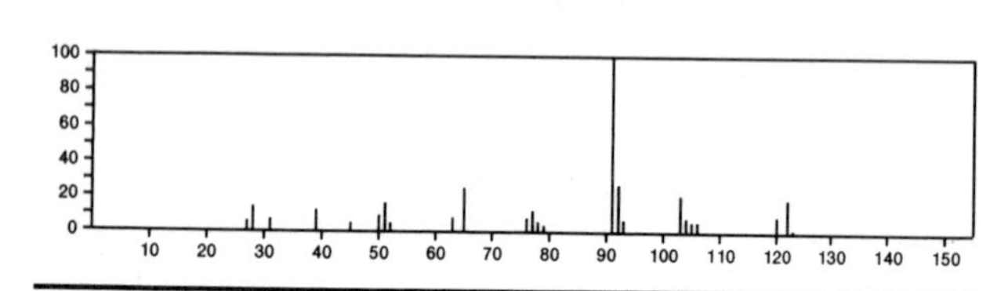

158 $C_7H_{10}O_4$ 869–29–4
2–Propene–1,1–diol, diacetate

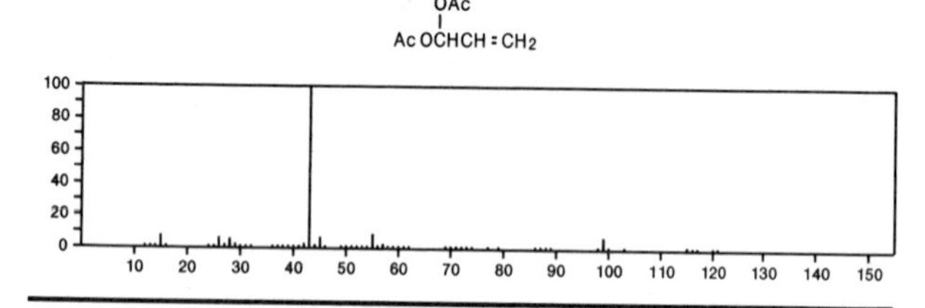

158 $C_7H_{10}O_4$ 29736–80–9
Hexanoic acid, 3,5–dioxo–, methyl ester

MeOC(O)CH2COCH2COMe

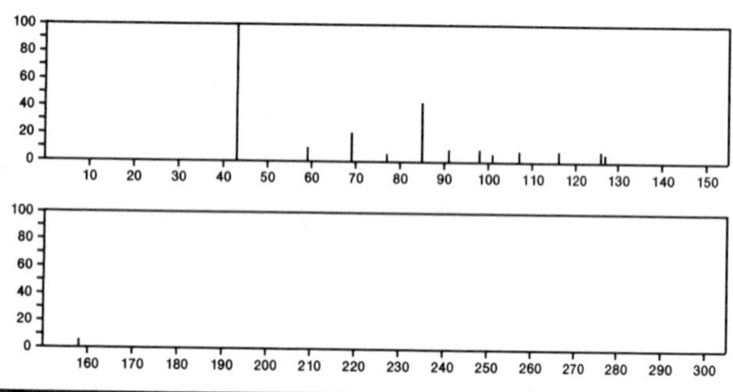

158 $C_7H_{10}S_2$ 50878–65–4
Thiophene, 2–[(1–methylethyl)thio]–

SPr–i

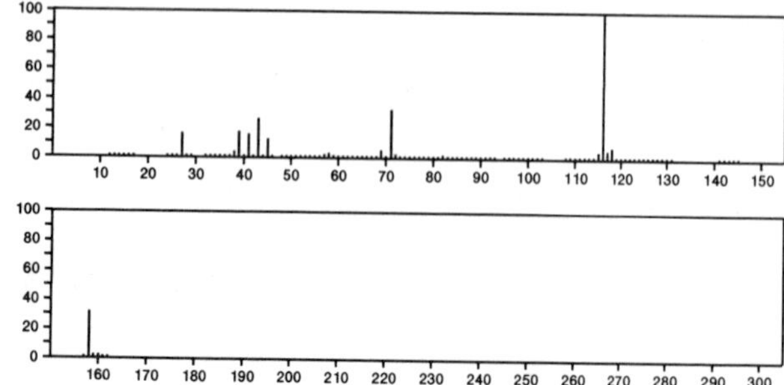

158 $C_7H_{11}N_2.Cl$ 34061–85–3
Pyridinium, 1–amino–2,6–dimethyl–, chloride

Me NH2 Me • Cl⁻

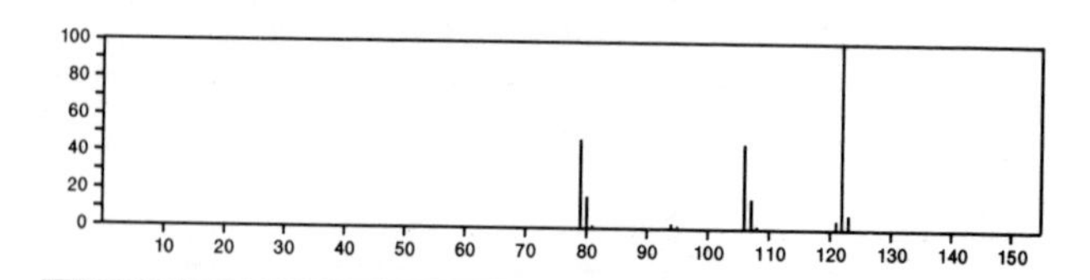

158 $C_7H_{14}N_2O_2$ 7400–28–4
Hydrazinecarboxylic acid, butylidene–, ethyl ester

EtOC(O)NHN=CHPr

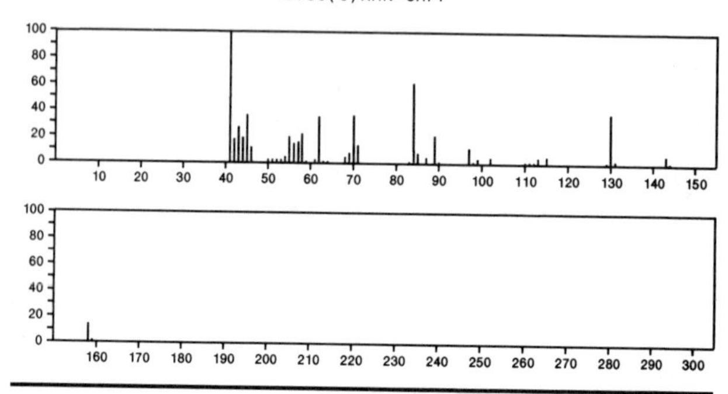

158 $C_7H_{14}N_2O_2$ 14702–36–4
Carbazic acid, 3–pentylidene–, methyl ester

MeOC(O)NHN=CH(CH2)3Me

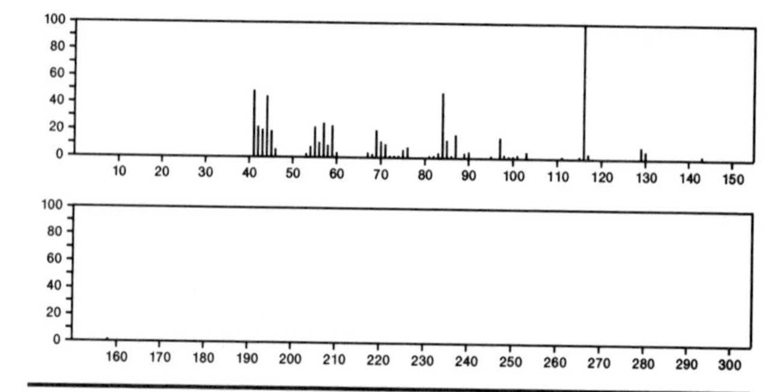

158 $C_7H_{14}N_2O_2$ 49582–53–8
2–Propenoic acid, 3–(dimethylamino)–3–(methylamino)–, methyl ester

MeOC(O)CH=C(NHMe)NMe2

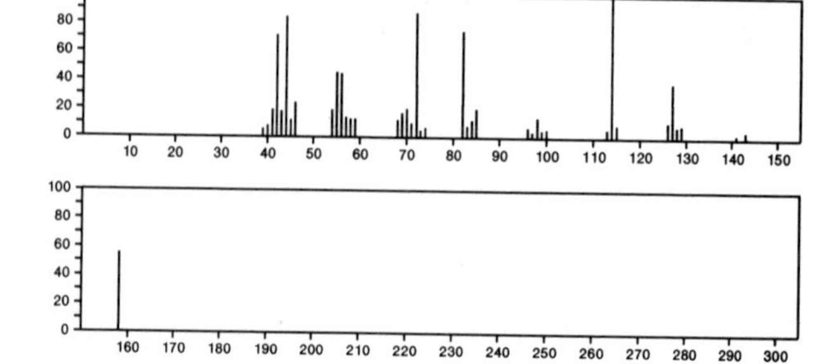

158 $C_7H_{14}N_2S$ 17709-98-7
Diethylamine, *N*-isothiocyanato-1,1'-dimethyl-

i-Pr₂NNCS

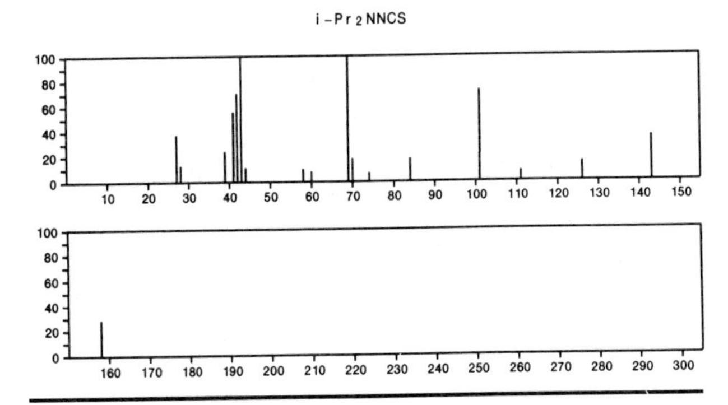

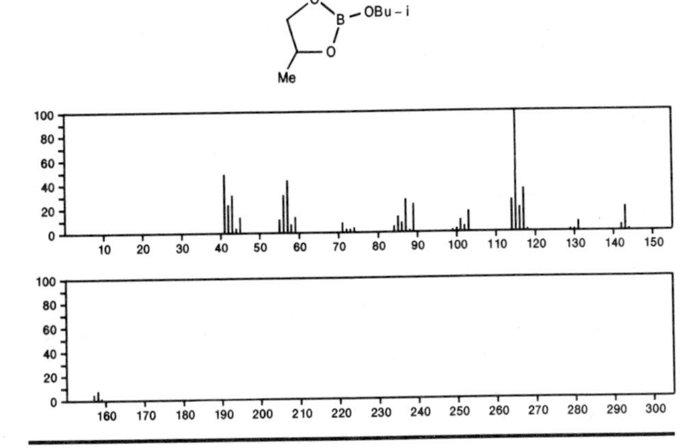

158 $C_7H_{15}BO_3$ 52910-21-1
1,3,2-Dioxaborolane, 4-methyl-2-(2-methylpropoxy)-

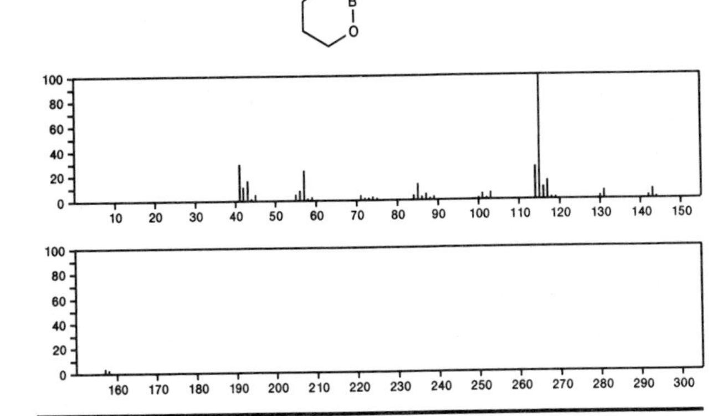

158 $C_7H_{15}BO_3$ 55162-67-9
1,3,2-Dioxaborinane, 2-(2-methylpropoxy)-

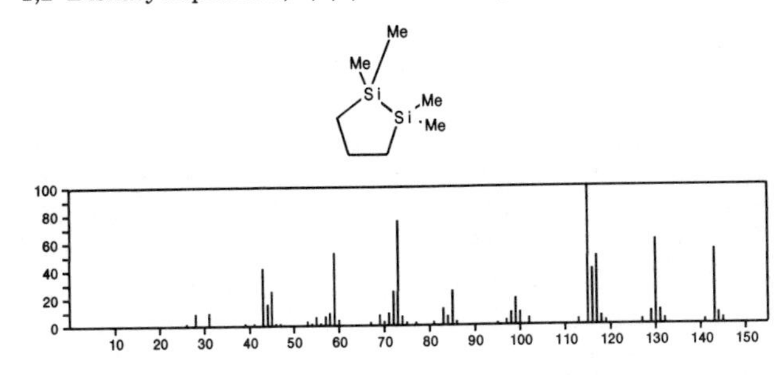

158 $C_7H_{18}Si_2$ 15003-82-4
1,2-Disilacyclopentane, 1,1,2,2-tetramethyl-

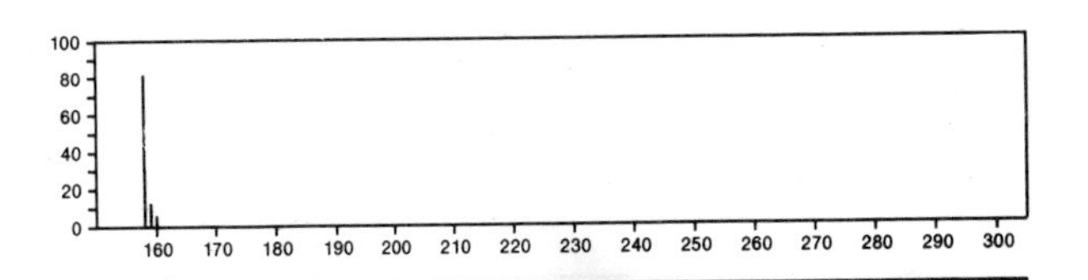

158 $C_8H_{11}ClO$ 7697-11-2
Bicyclo[2.2.2]octanone, 4-chloro-

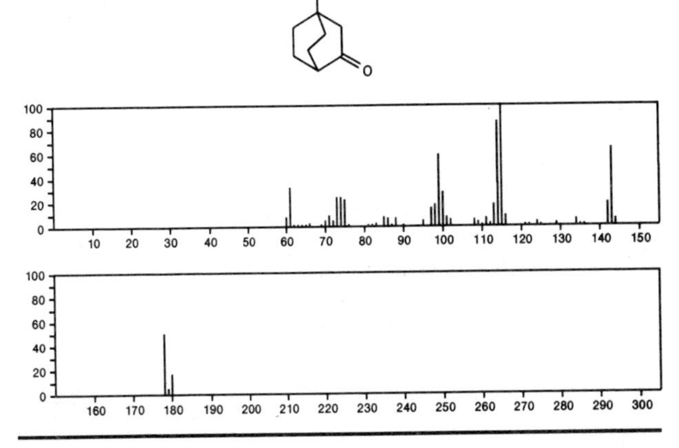

158 $C_8H_{11}ClO$ 23804-48-0
Bicyclo[2.2.2]octanone, 3-chloro-

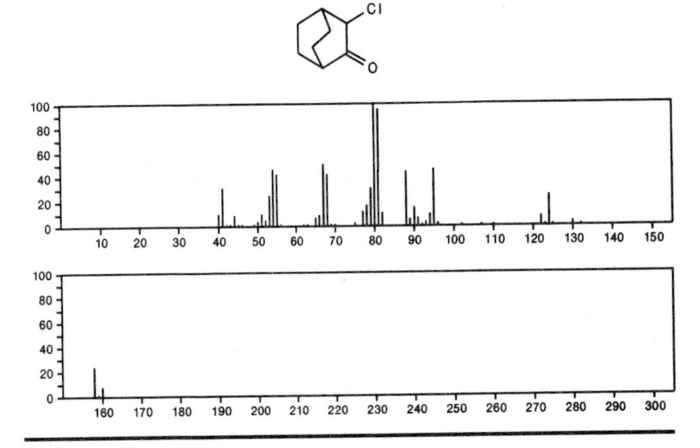

158 $C_8H_{11}ClO$ 56324-75-5
Bicyclo[2.2.2]octan-2-one, 5-chloro-, *exo*-

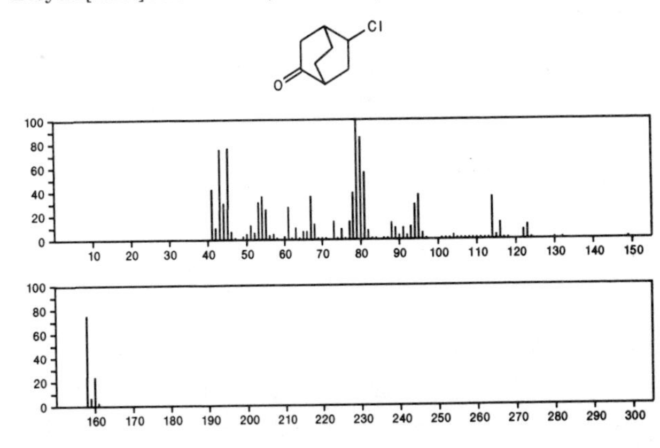

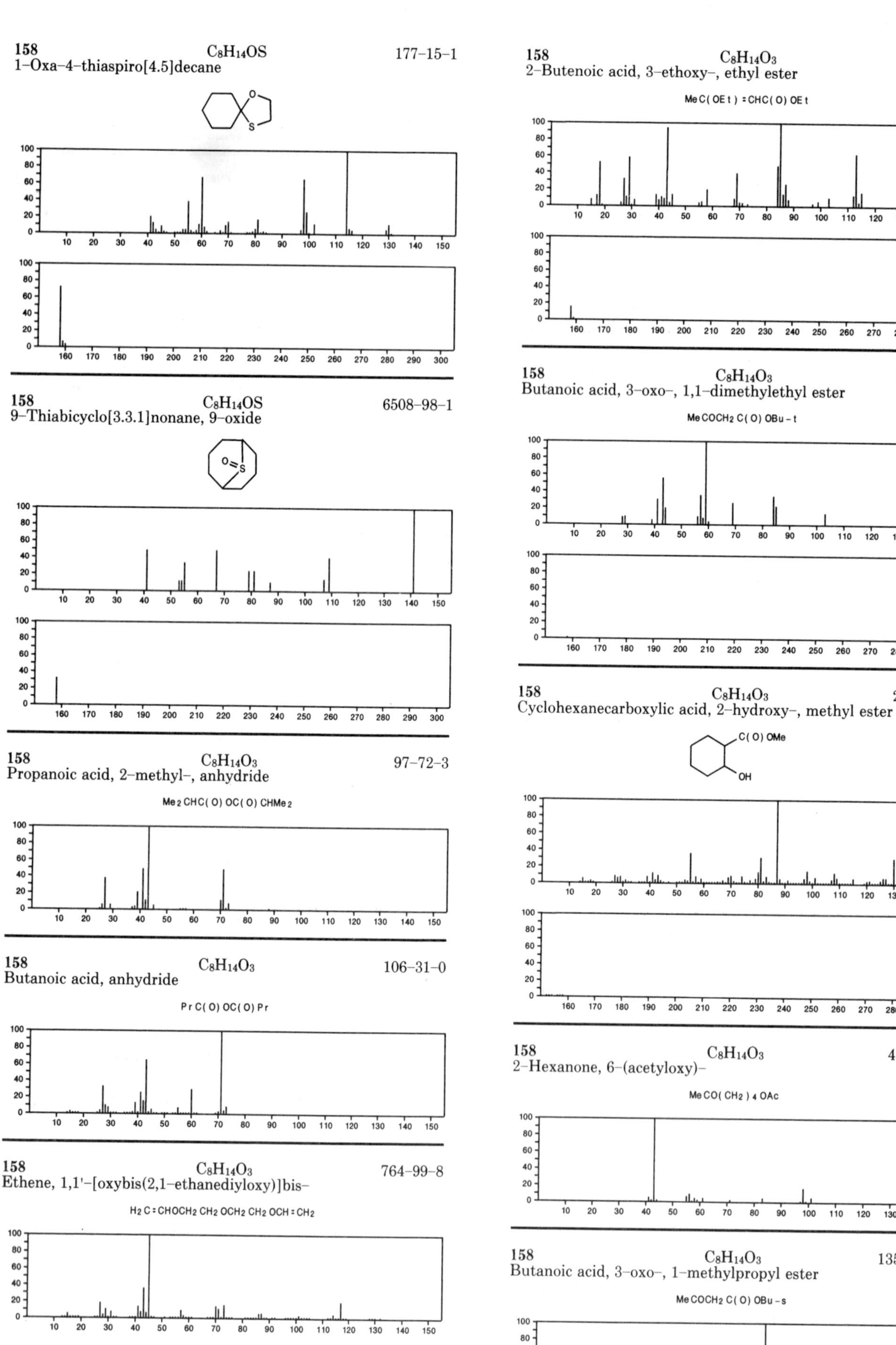

158 $C_8H_{14}O_3$ 998-91-4
2-Butenoic acid, 3-ethoxy-, ethyl ester

MeC(OEt)=CHC(O)OEt

158 $C_8H_{14}O_3$ 1694-31-1
Butanoic acid, 3-oxo-, 1,1-dimethylethyl ester

MeCOCH2 C(O) OBu-t

158 $C_8H_{14}O_3$ 2236-11-5
Cyclohexanecarboxylic acid, 2-hydroxy-, methyl ester

C(O) OMe
OH

158 $C_8H_{14}O_3$ 4305-26-4
2-Hexanone, 6-(acetyloxy)-

MeCO(CH2)4 OAc

158 $C_8H_{14}O_3$ 13562-76-0
Butanoic acid, 3-oxo-, 1-methylpropyl ester

MeCOCH2 C(O) OBu-s

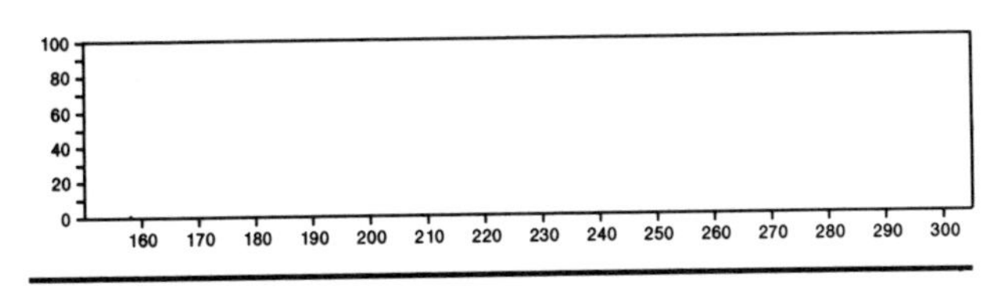

158 $C_8H_{14}O_3$ 13984-57-1

Hexanoic acid, 5-oxo-, ethyl ester

MeCO(CH2)3C(O)OEt

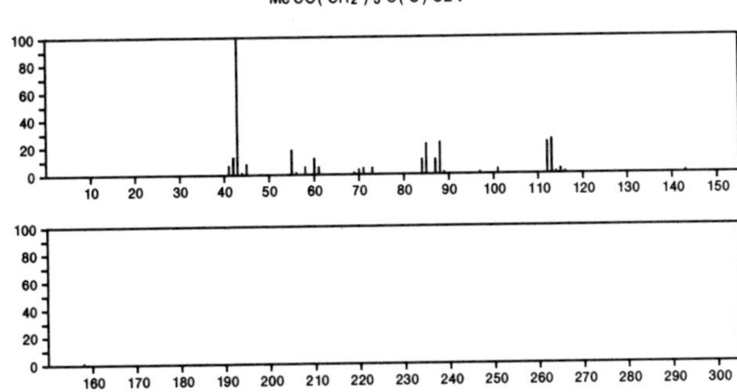

158 $C_8H_{14}O_3$ 22428-87-1

1,4-Dioxaspiro[4.5]decan-8-ol

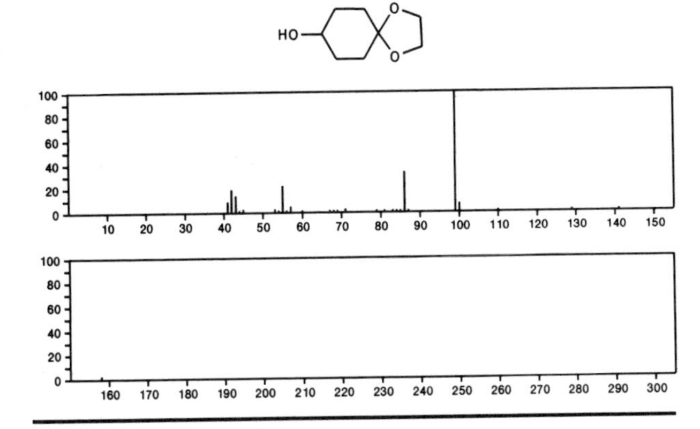

158 $C_8H_{14}O_3$ 29887-64-7

Cyclohexanone, 3,5-dimethoxy-, *trans-*

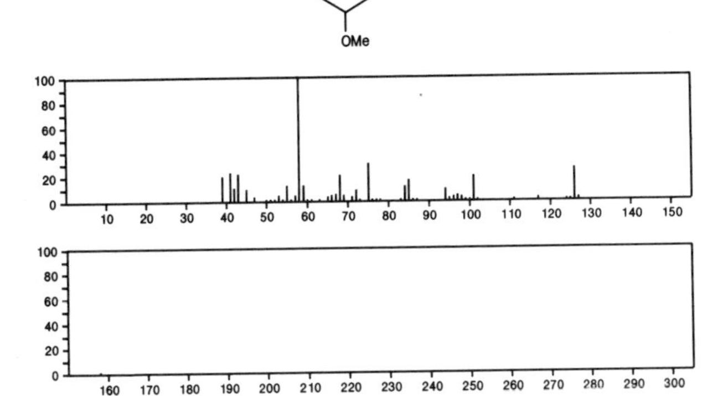

158 $C_8H_{14}O_3$ 30363-85-0

Cyclohexanone, 3,5-dimethoxy-, *cis-*

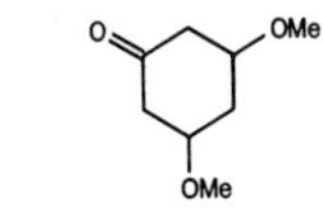

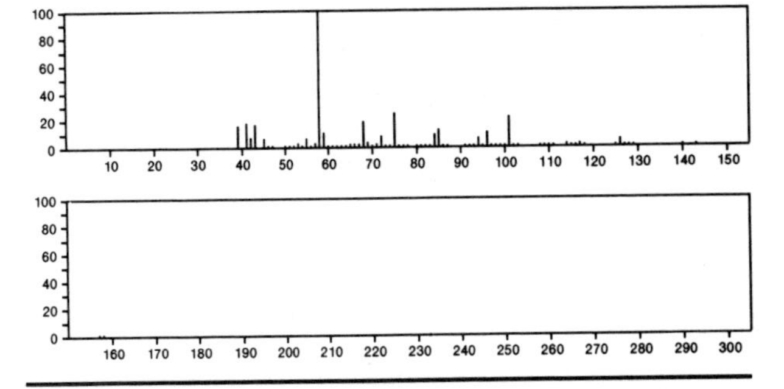

158 $C_8H_{14}O_3$ 33528-35-7

2-Butanone, 4-(2-methyl-1,3-dioxolan-2-yl)-

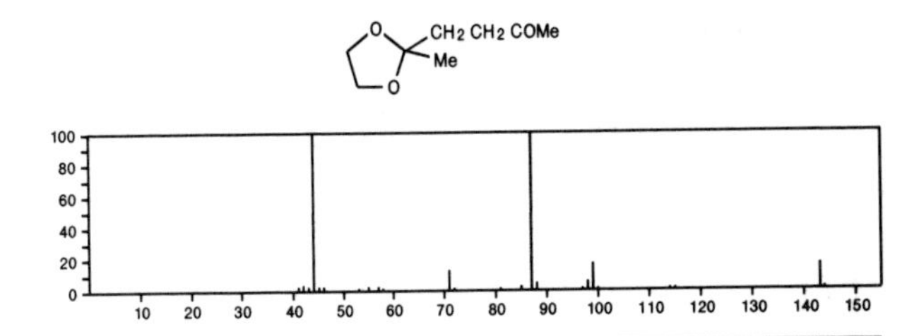

158 $C_8H_{14}O_3$ 34553-37-2

Hexanoic acid, 5-methyl-4-oxo-, methyl ester

Me2CHCOCH2CH2C(O)OMe

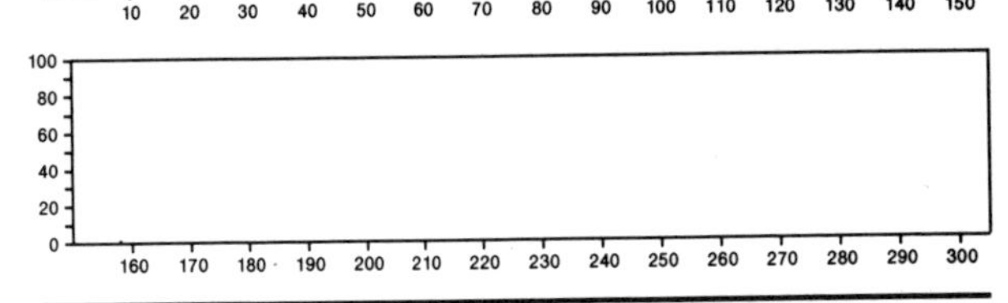

158 $C_8H_{14}O_3$ 51756-10-6

Butanoic acid, 2-acetyl-3-methyl-, methyl ester

CHMe2
MeOC(O)CHCOMe

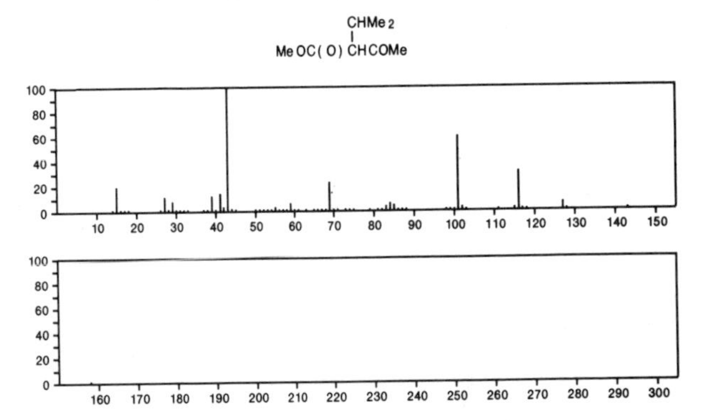

158 $C_8H_{14}O_3$ 56009-35-9

2-Pentenoic acid, 2-methoxy-3-methyl-, methyl ester

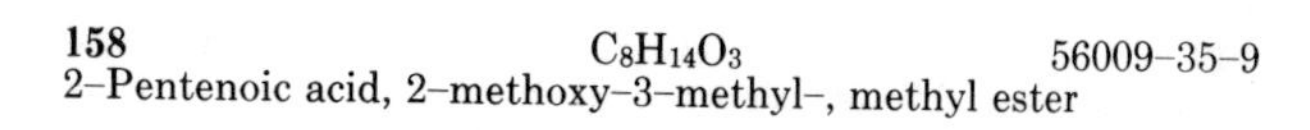

MeOC(O)C(OMe)=CEtMe

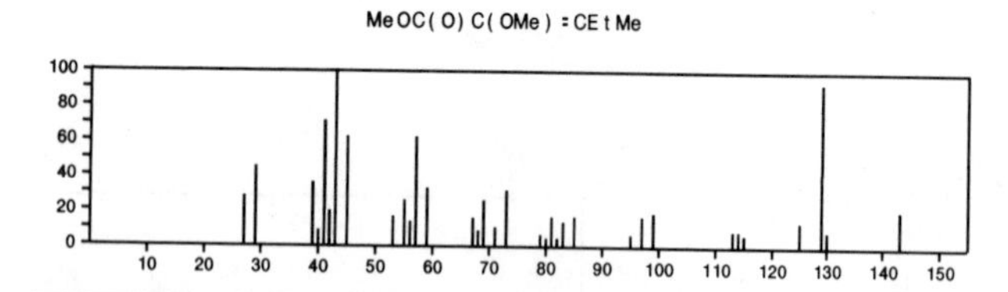

158 $C_8H_{14}O_3$ 56009-36-0

2-Pentenoic acid, 2-methoxy-4-methyl-, methyl ester

MeOC(O)C(OMe)=CHCHMe2

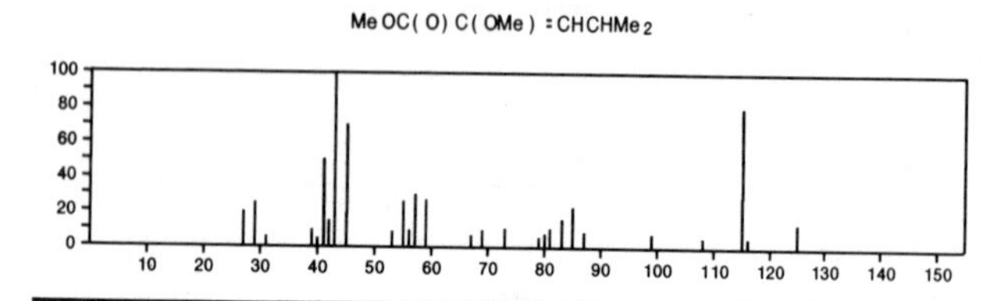

158 $C_8H_{16}NO_2$ 2154-37-2

1-Pyrrolidinyloxy, 3-hydroxy-2,2,5,5-tetramethyl-

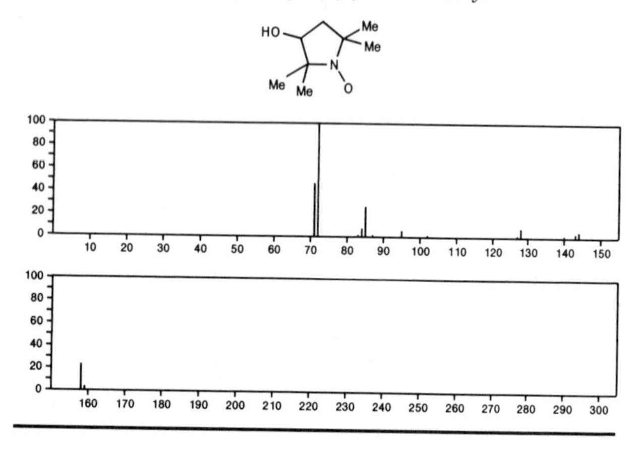

158 $C_8H_{18}N_2O$ 924-16-3

1-Butanamine, *N*-butyl-*N*-nitroso-

Me(CH2)3N(NO)(CH2)3Me

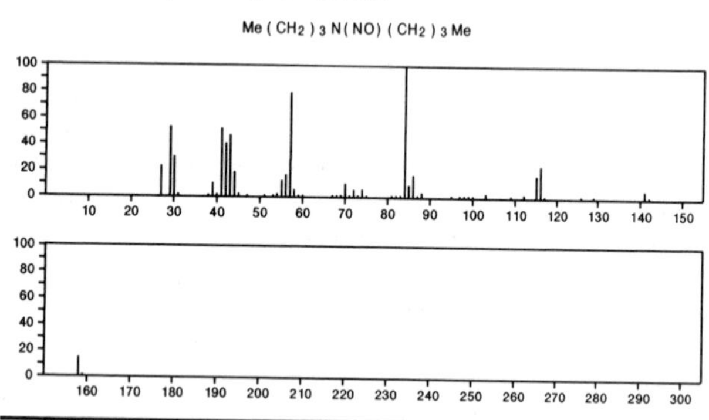

158 $C_8H_{18}N_2O$ 997-95-5

1-Propanamine, 2-methyl-*N*-(2-methylpropyl)-*N*-nitroso-

i-Bu2NNO

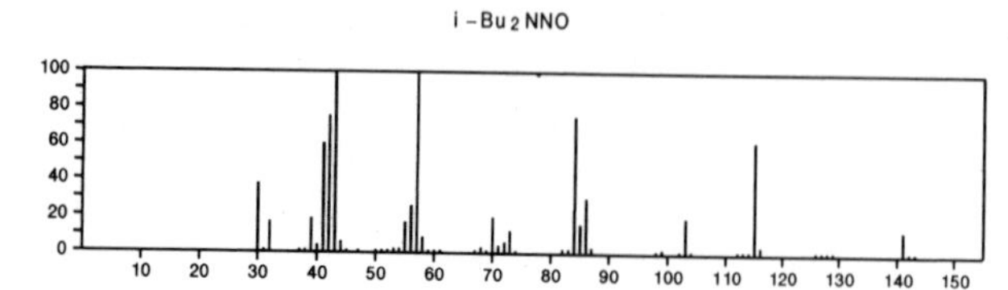

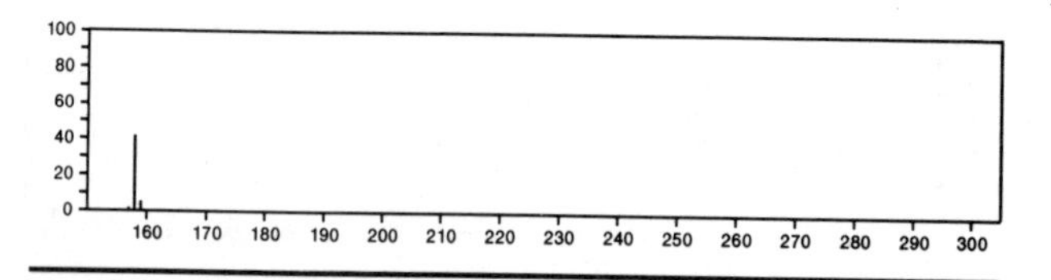

158 $C_8H_{18}N_2O$ 28023-77-0

1-Pentanamine, *N*-(1-methylethyl)-*N*-nitroso-

i-PrN(NO)(CH2)4Me

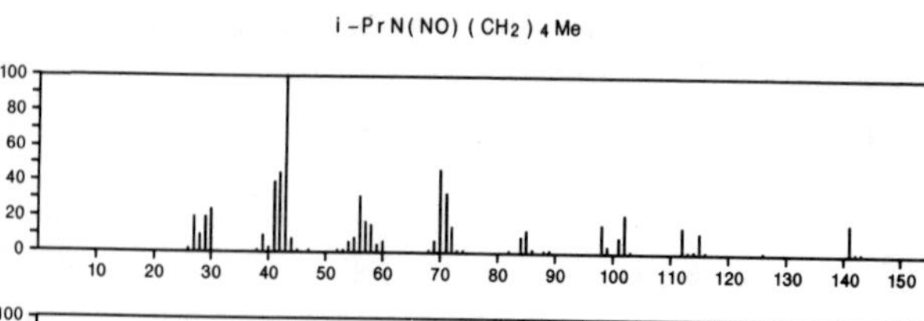

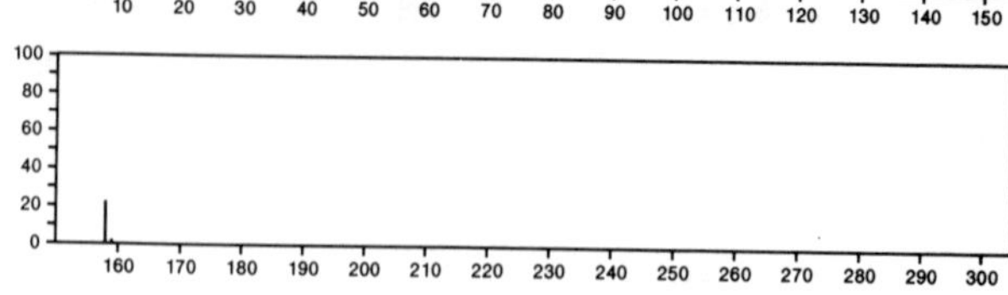

158 $C_8H_{18}N_2O$ 28023-78-1

1-Pentanamine, *N*-nitroso-*N*-propyl-

PrN(NO)(CH2)4Me

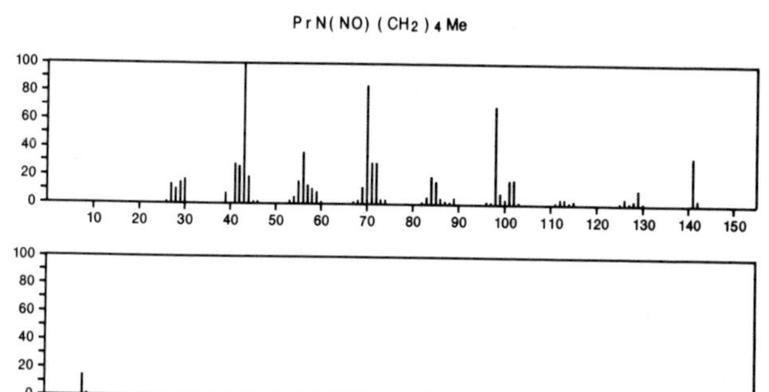

158 $C_8H_{18}N_2O$ 28023-81-6

1-Butanamine, *N*-(1-methylpropyl)-*N*-nitroso-

s-BuN(NO)(CH2)3Me

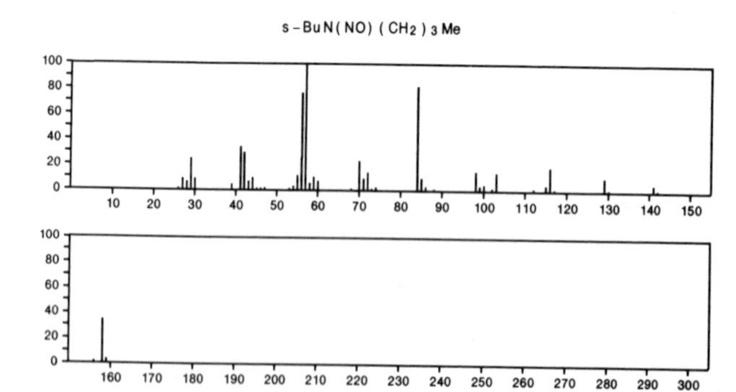

158 $C_9H_6N_2O$ 1775-23-1

1*H*-Inden-1-one, 2-diazo-2,3-dihydro-

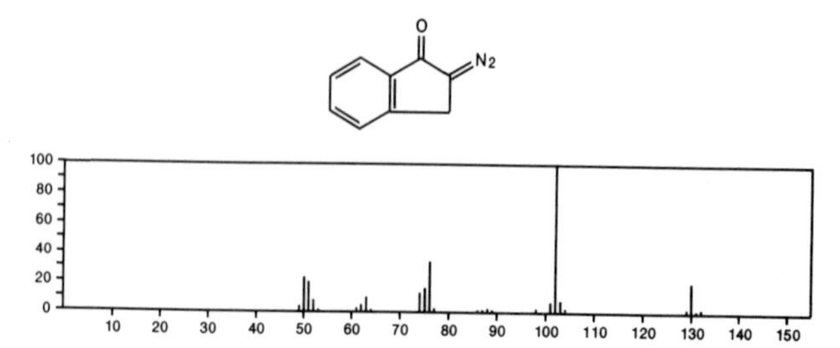

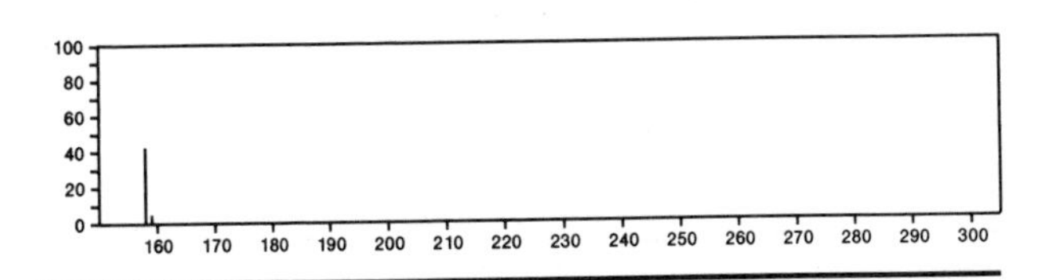

158 $C_9H_{15}Cl$ 7697-06-5
Bicyclo[2.2.2]octane, 1-chloro-4-methyl-

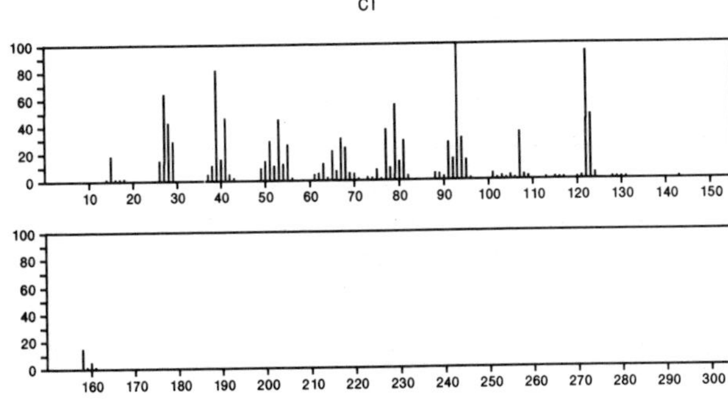

158 $C_9H_{15}Cl$ 22768-97-4
Bicyclo[2.2.1]heptane, 3-chloro-2,2-dimethyl-, *exo*-

Me Me Cl

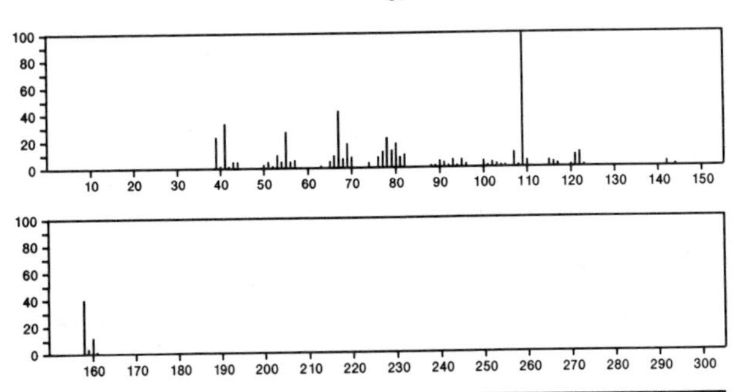

158 $C_9H_{15}Cl$ 22768-98-5
Bicyclo[2.2.1]heptane, 2-chloro-7,7-dimethyl-, *exo*-

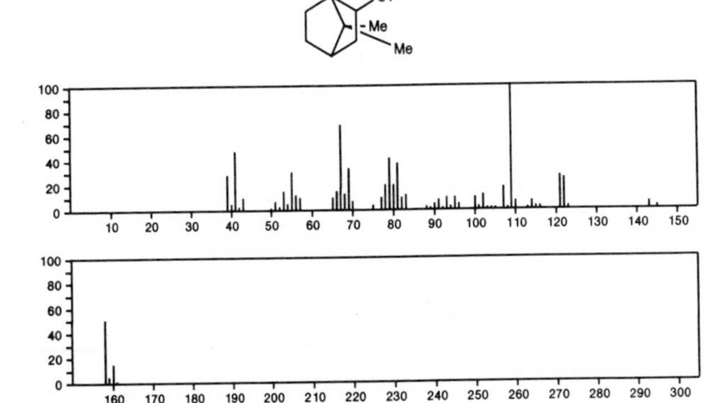

158 $C_9H_{15}Cl$ 55402-10-3
3-Heptyne, 7-chloro-2,2-dimethyl-

$Me_3CC{\equiv}C(CH_2)_3Cl$

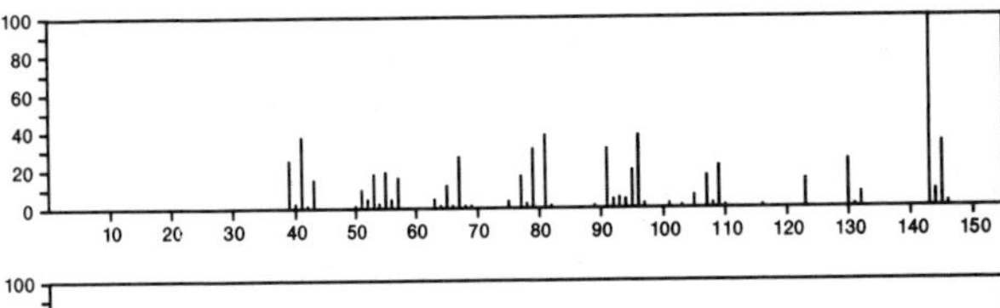

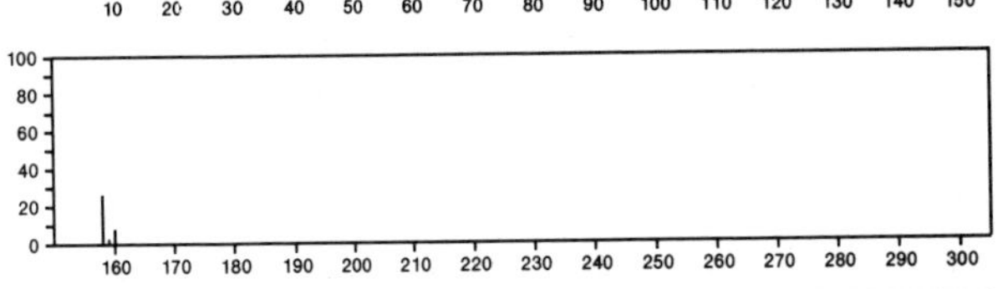

158 $C_9H_{18}O_2$ 106-30-9
Heptanoic acid, ethyl ester

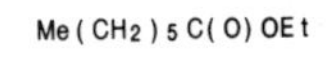

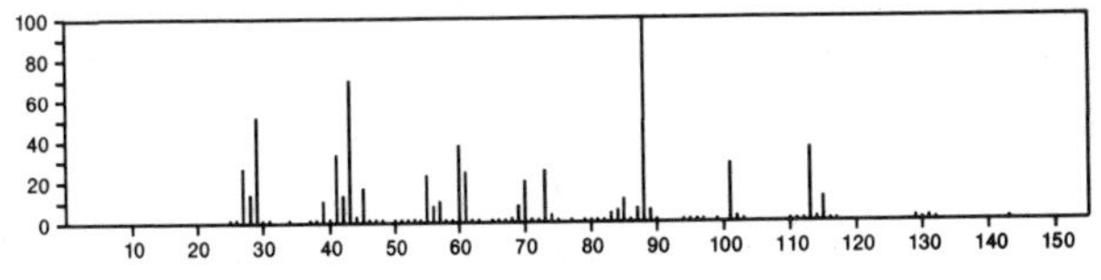

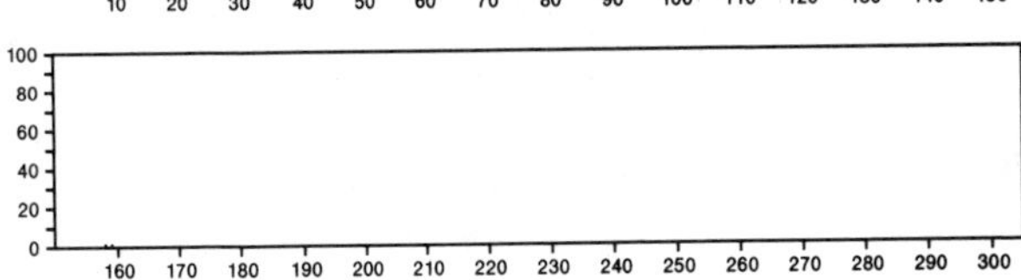

158 $C_9H_{18}O_2$ 109-19-3
Butanoic acid, 3-methyl-, butyl ester

$Me_2CHCH_2C(O)O(CH_2)_3Me$

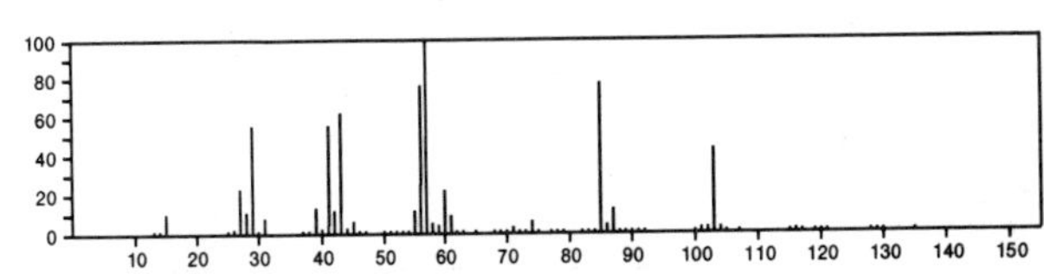

158 $C_9H_{18}O_2$ 111-11-5
Octanoic acid, methyl ester

$Me(CH_2)_6C(O)OMe$

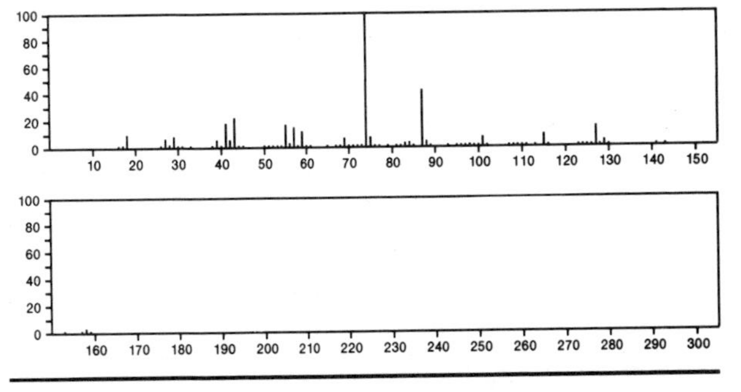

158 $C_9H_{18}O_2$ 112-05-0
Nonanoic acid

$HO_2C(CH_2)_7Me$

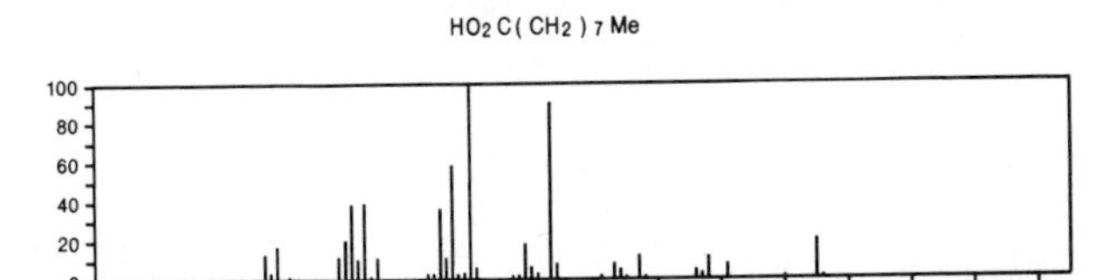

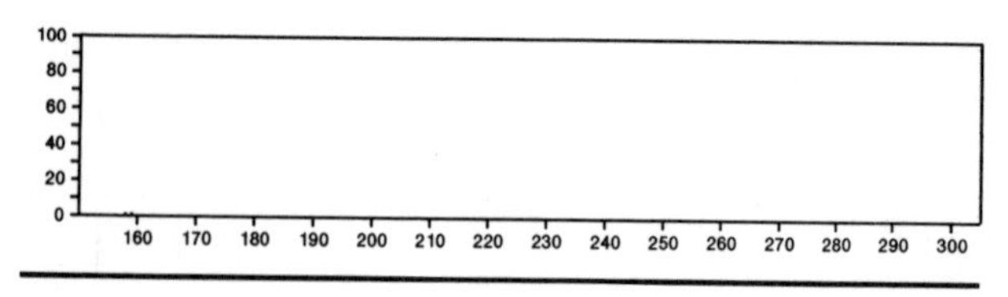

158 $C_9H_{18}O_2$ 112–06–1
Acetic acid, heptyl ester

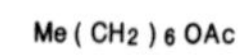

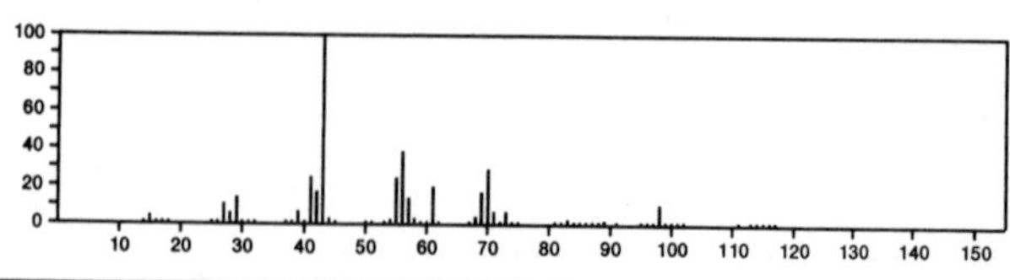

158 $C_9H_{18}O_2$ 540–18–1
Butanoic acid, pentyl ester

Pr C(O) O(CH2) 4 Me

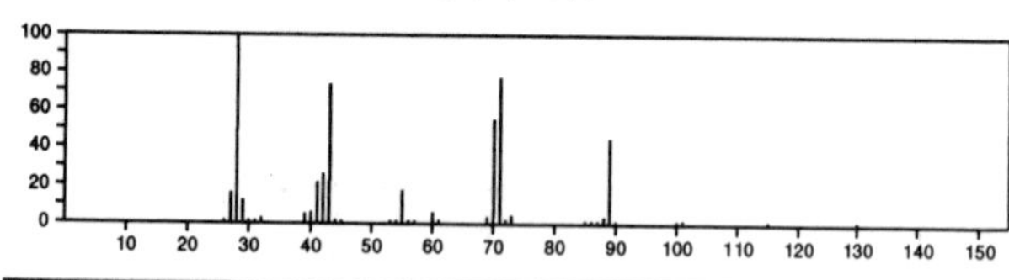

158 $C_9H_{18}O_2$ 589–59–3
Butanoic acid, 3–methyl–, 2–methylpropyl ester

i –Bu OC(O) CH2 CHMe 2

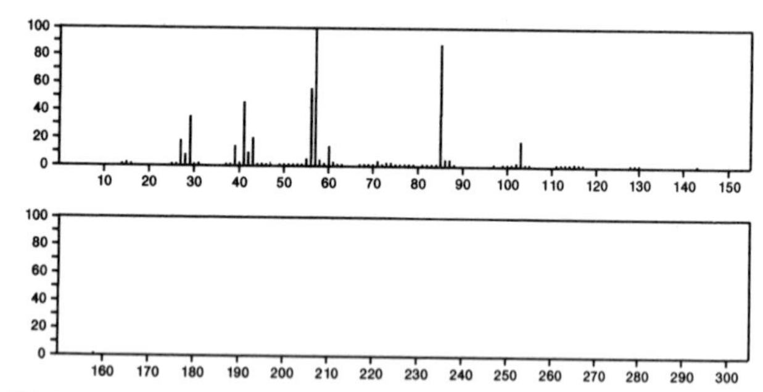

158 $C_9H_{18}O_2$ 816–19–3
Hexanoic acid, 2–ethyl–, methyl ester

Me OC(O) CHE t (CH2) 3 Me

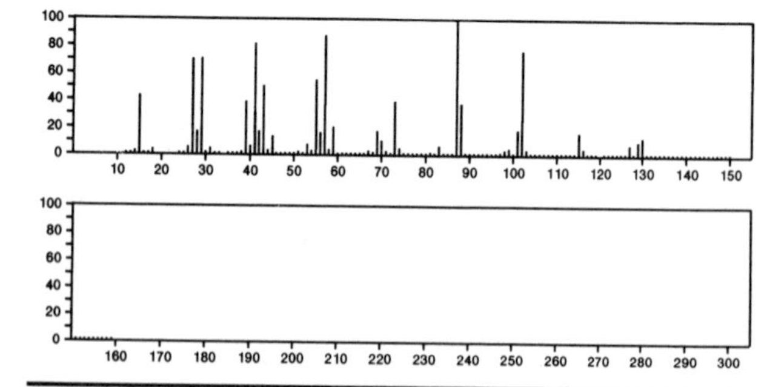

158 $C_9H_{18}O_2$ 869–08–9
Butanoic acid, 2–methyl–, 1–methylpropyl ester

Me CH2 CHMe C(O) OBu – s

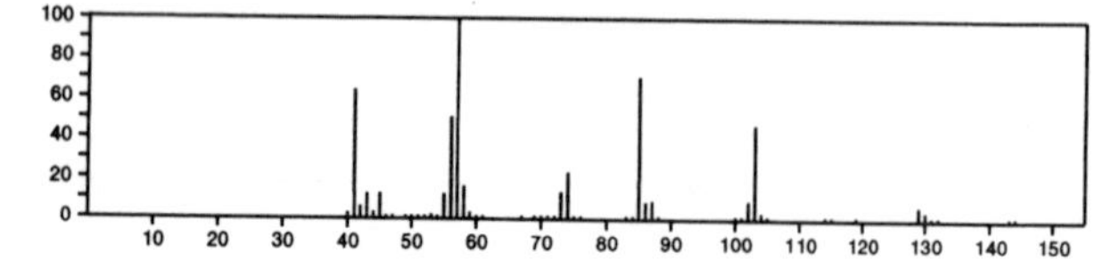

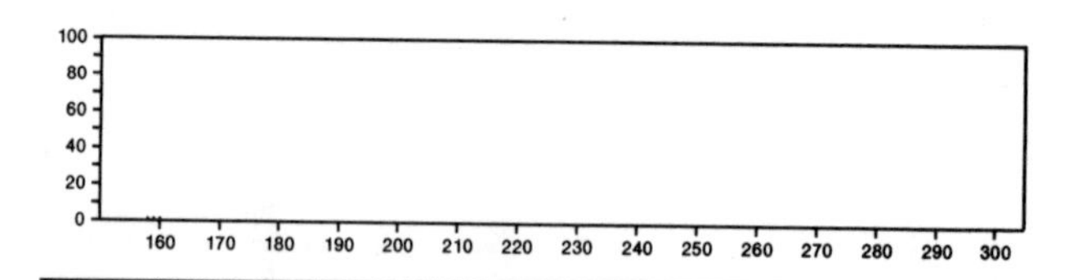

158 $C_9H_{18}O_2$ 935–45–5
1,3–Dioxolane, 2–ethyl–2–isobutyl–

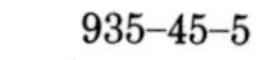

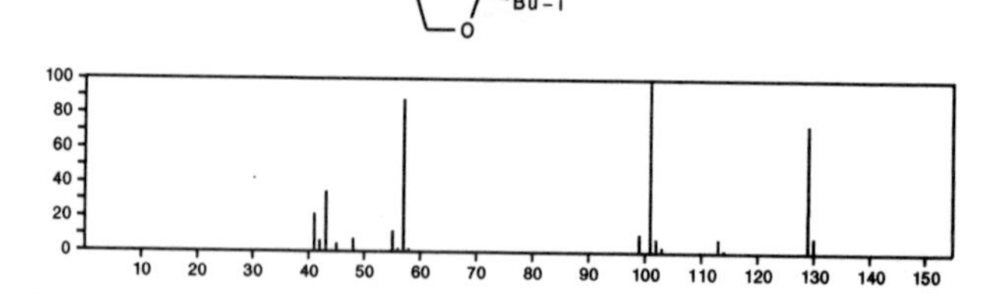

158 $C_9H_{18}O_2$ 935–49–9
1,3–Dioxolane, 2–butyl–2–ethyl–

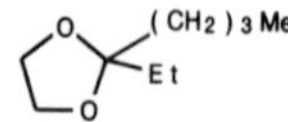

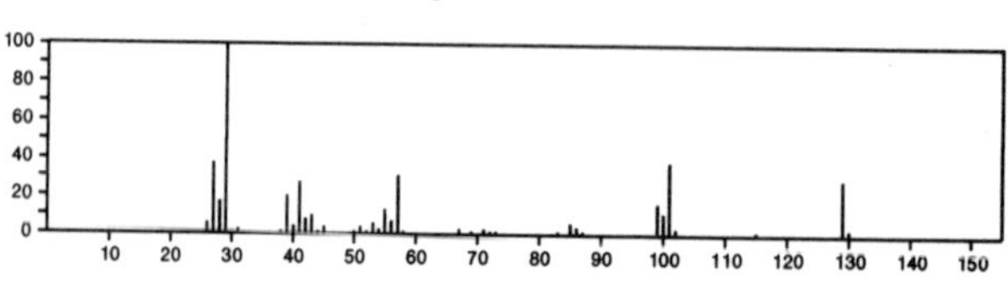

158 $C_9H_{18}O_2$ 1927–68–0
2H–Pyran, 2–butoxytetrahydro–

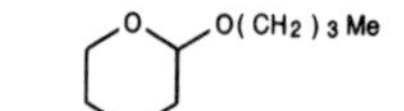

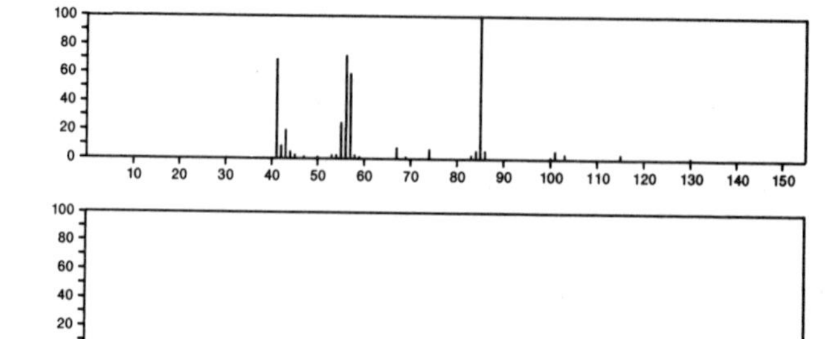

158 $C_9H_{18}O_2$ 1927–69–1
2H–Pyran, 2–(1,1–dimethylethoxy)tetrahydro–

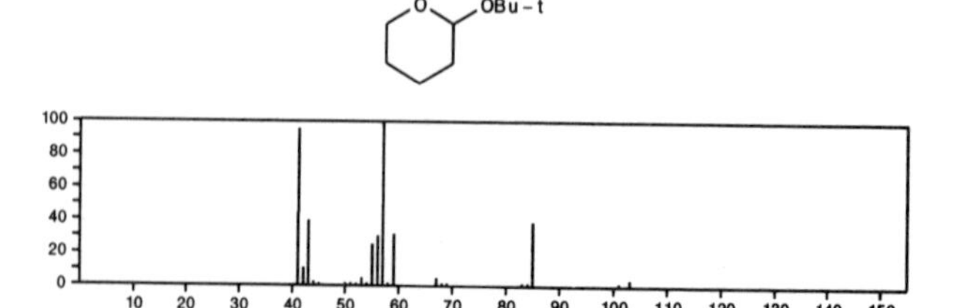

158 $C_9H_{18}O_2$ 2445–67–2
Butanoic acid, 2–methyl–, 2–methylpropyl ester

Me CH2 CHMe C(O) OBu – i

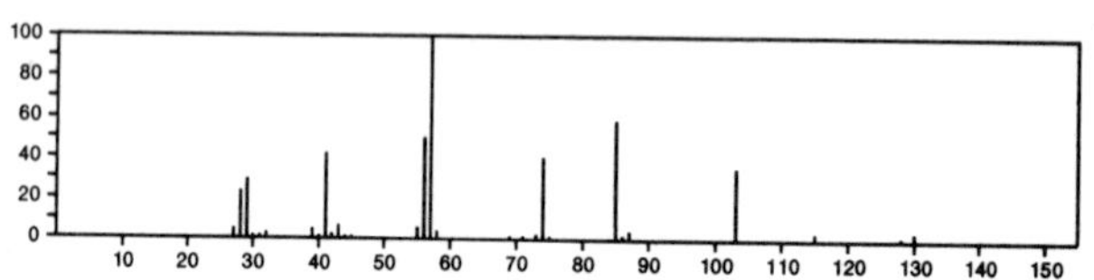

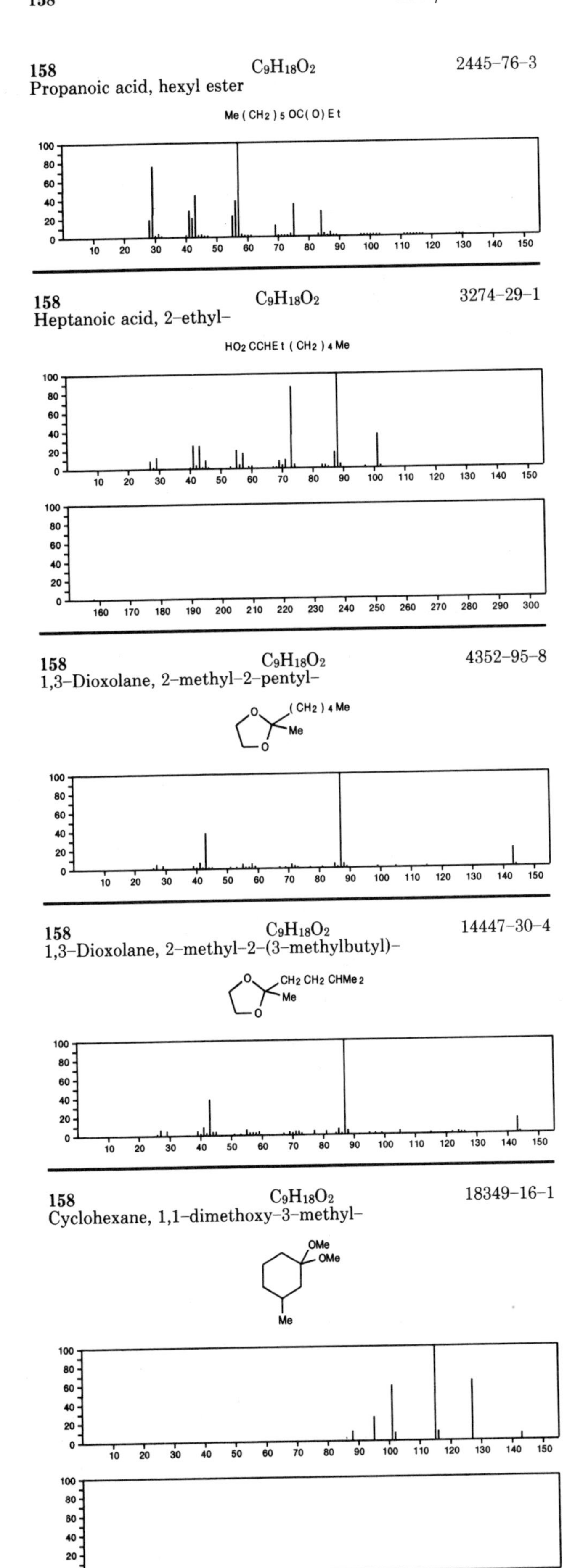
158 $C_9H_{18}O_2$ 2445-76-3
Propanoic acid, hexyl ester
Me(CH2)5OC(O)Et
158 $C_9H_{18}O_2$ 3274-29-1
Heptanoic acid, 2-ethyl-
HO2CCHEt(CH2)4Me
158 $C_9H_{18}O_2$ 4352-95-8
1,3-Dioxolane, 2-methyl-2-pentyl-
(CH2)4Me
158 $C_9H_{18}O_2$ 14447-30-4
1,3-Dioxolane, 2-methyl-2-(3-methylbutyl)-
CH2CH2CHMe2
158 $C_9H_{18}O_2$ 18349-16-1
Cyclohexane, 1,1-dimethoxy-3-methyl-
OMe
OMe
Me

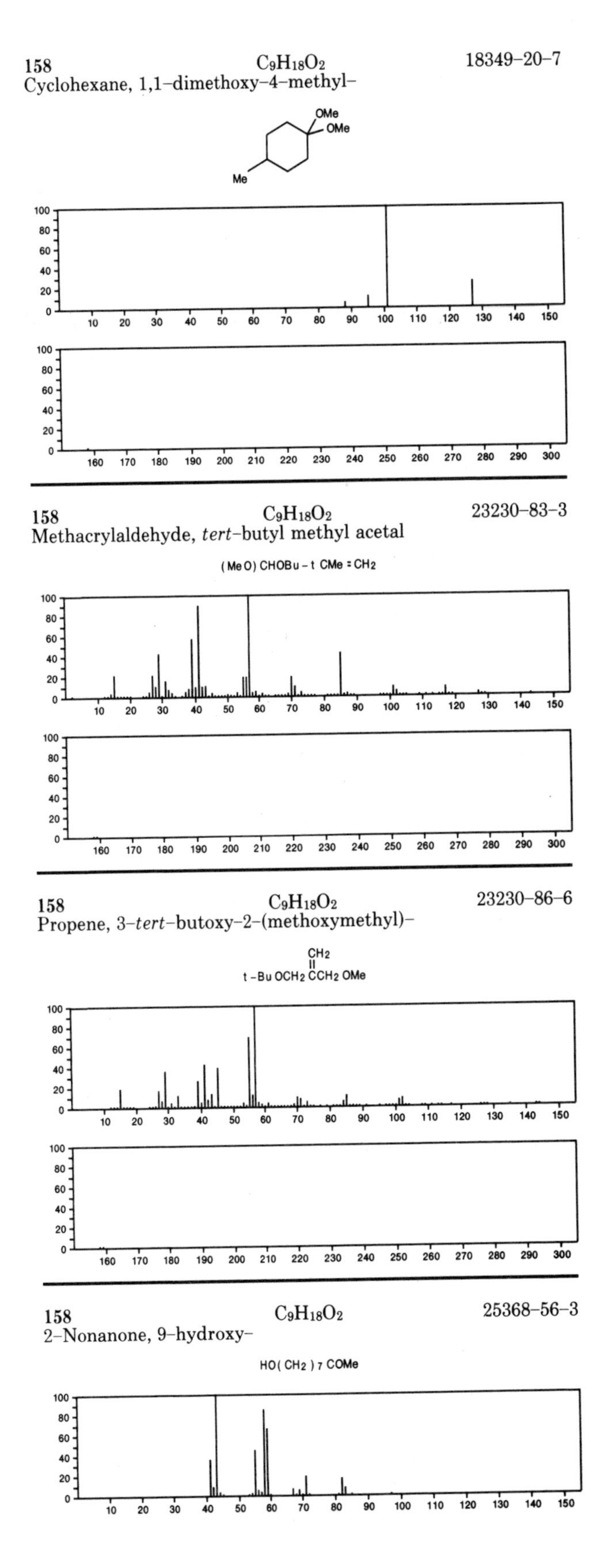
158 $C_9H_{18}O_2$ 18349-20-7
Cyclohexane, 1,1-dimethoxy-4-methyl-
OMe
OMe
Me
158 $C_9H_{18}O_2$ 23230-83-3
Methacrylaldehyde, *tert*-butyl methyl acetal
(MeO)CHOBu-t CMe=CH2
158 $C_9H_{18}O_2$ 23230-86-6
Propene, 3-*tert*-butoxy-2-(methoxymethyl)-
CH2
t-BuOCH2CCH2OMe
158 $C_9H_{18}O_2$ 25368-56-3
2-Nonanone, 9-hydroxy-
HO(CH2)7COMe

158 $C_9H_{18}O_2$ 29887–62–5
Cyclohexane, 1,3–dimethoxy–5–methyl–, (1α,3β,5α)–

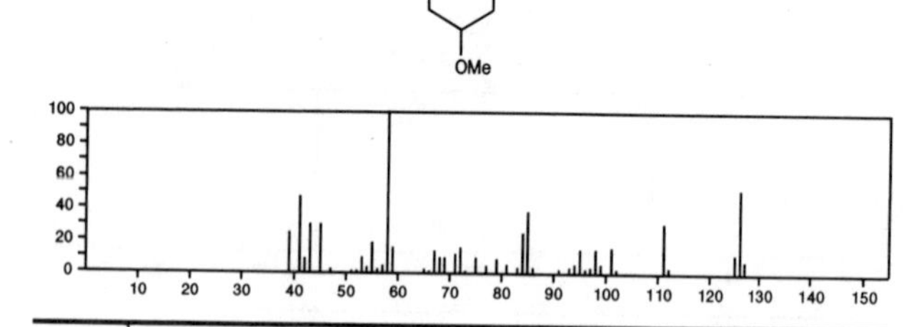

158 $C_9H_{18}O_2$ 29887–66–9
Cyclohexane, 1,4–dimethoxy–2–methyl–, stereoisomer

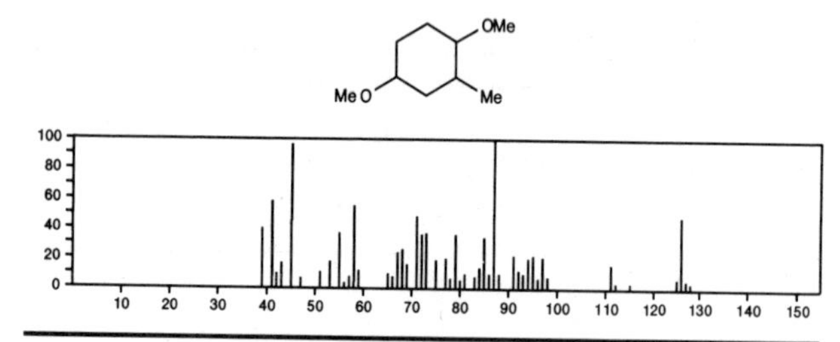

158 $C_9H_{18}O_2$ 29887–78–3
Cycloheptane, 1,2–dimethoxy–, *trans*–

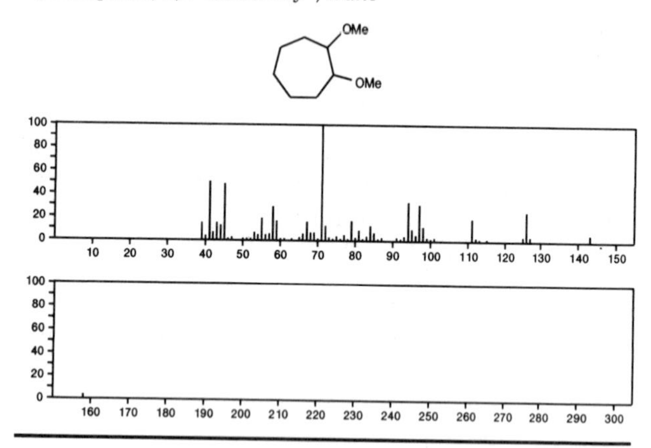

158 $C_9H_{18}O_2$ 29887–79–4
Cycloheptane, 1,3–dimethoxy–, *trans*–

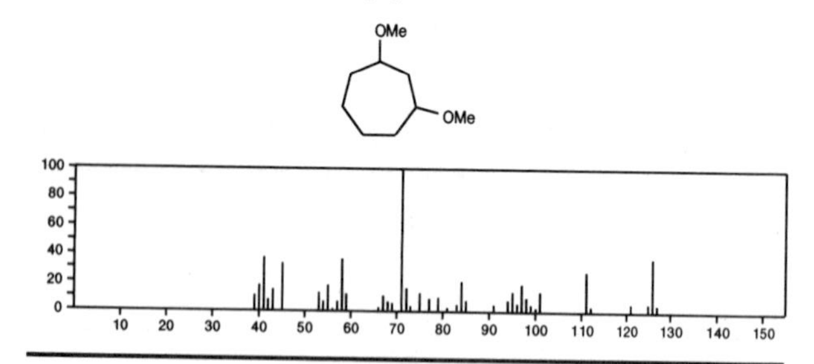

158 $C_9H_{18}O_2$ 29887–80–7
Cycloheptane, 1,4–dimethoxy–, *trans*–

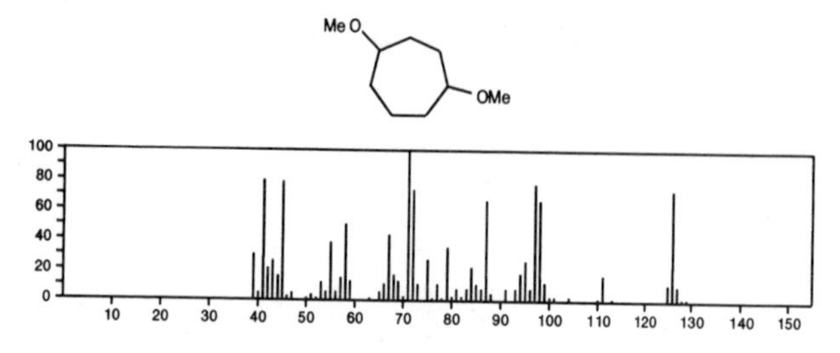

158 $C_9H_{18}O_2$ 30363–82–7
Cyclohexane, 1,3–dimethoxy–5–methyl–, stereoisomer

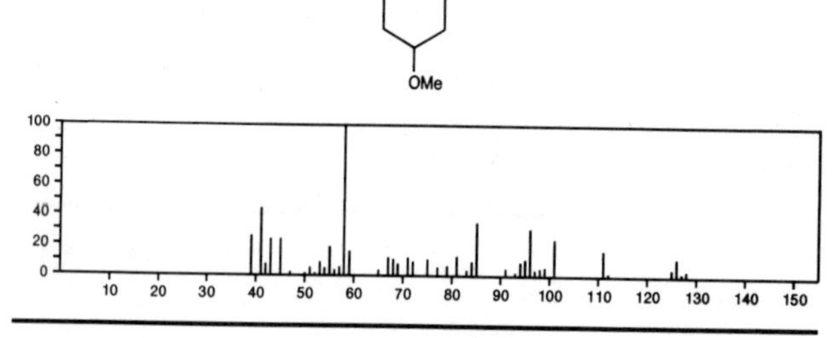

158 $C_9H_{18}O_2$ 30363–88–3
Cyclohexane, 1,4–dimethoxy–2–methyl–, stereoisomer

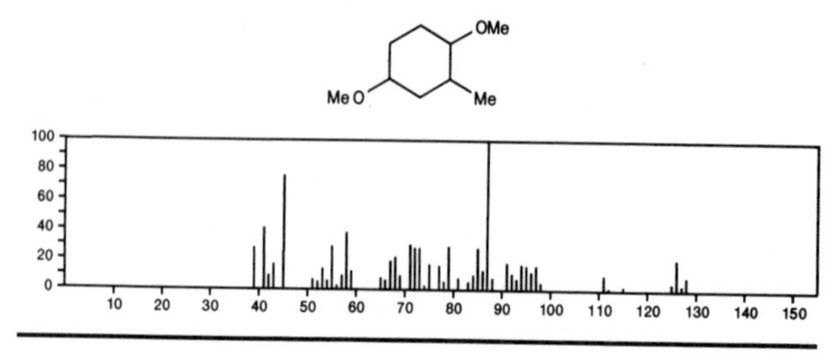

158 $C_9H_{18}O_2$ 30363–90–7
Cycloheptane, 1,3–dimethoxy–, *cis*–

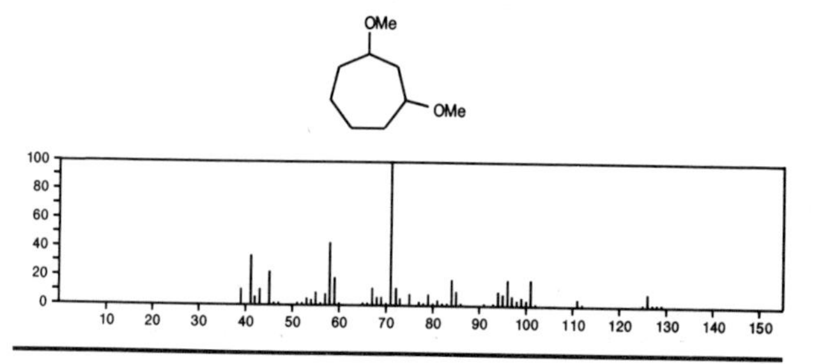

158 $C_9H_{18}O_2$ 30363–91–8
Cycloheptane, 1,4–dimethoxy–, *cis*–

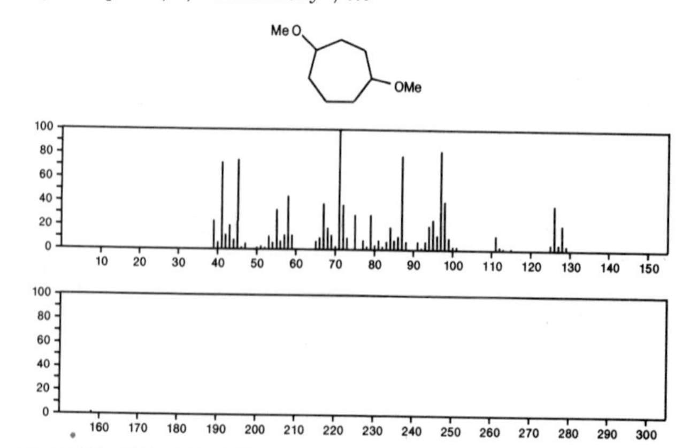

158 $C_9H_{18}O_2$ 38574–09–3
Cyclohexane, 1,1–dimethoxy–2–methyl–

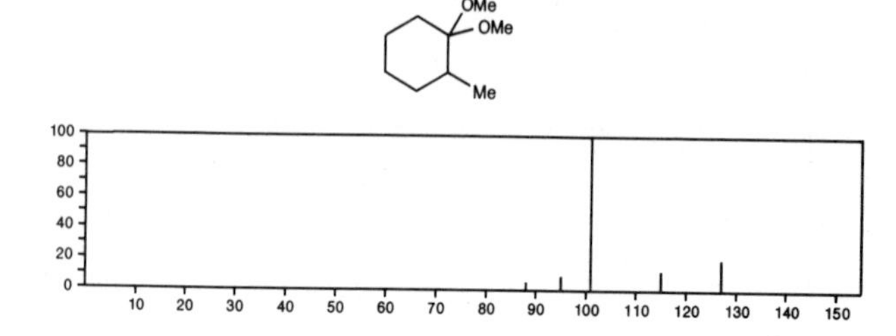

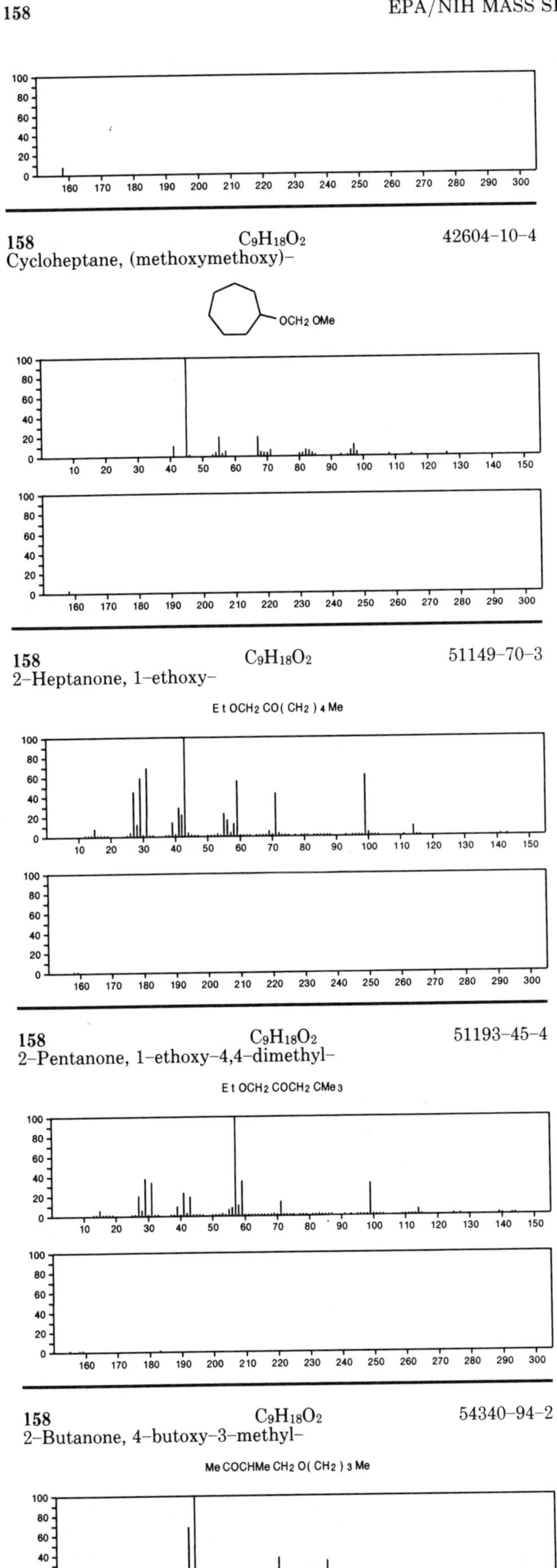
158 $C_9H_{18}O_2$ 42604-10-4
Cycloheptane, (methoxymethoxy)-
OCH2 OMe
158 $C_9H_{18}O_2$ 51149-70-3
2-Heptanone, 1-ethoxy-
Et OCH2 CO(CH2)4 Me
158 $C_9H_{18}O_2$ 51193-45-4
2-Pentanone, 1-ethoxy-4,4-dimethyl-
Et OCH2 COCH2 CMe3
158 $C_9H_{18}O_2$ 54340-94-2
2-Butanone, 4-butoxy-3-methyl-
Me COCHMe CH2 O(CH2)3 Me

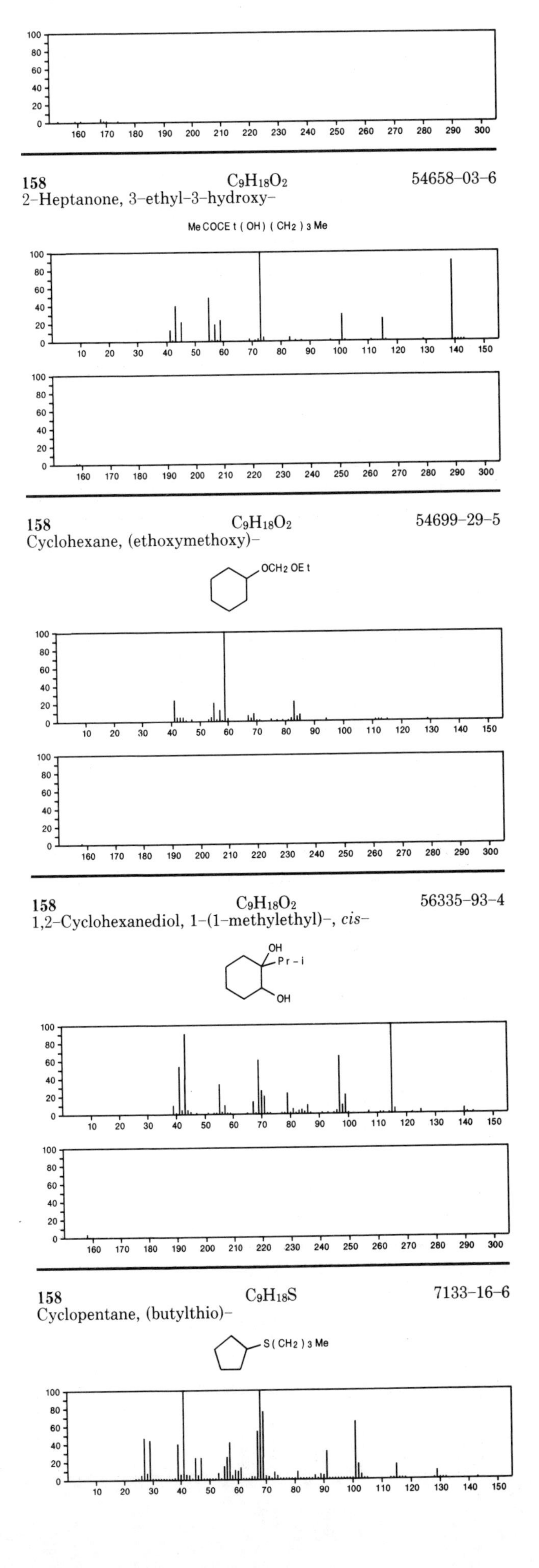
158 $C_9H_{18}O_2$ 54658-03-6
2-Heptanone, 3-ethyl-3-hydroxy-
Me COCEt (OH) (CH2)3 Me
158 $C_9H_{18}O_2$ 54699-29-5
Cyclohexane, (ethoxymethoxy)-
OCH2 OEt
158 $C_9H_{18}O_2$ 56335-93-4
1,2-Cyclohexanediol, 1-(1-methylethyl)-, cis-
OH
Pr-i
OH
158 $C_9H_{18}S$ 7133-16-6
Cyclopentane, (butylthio)-
S(CH2)3 Me

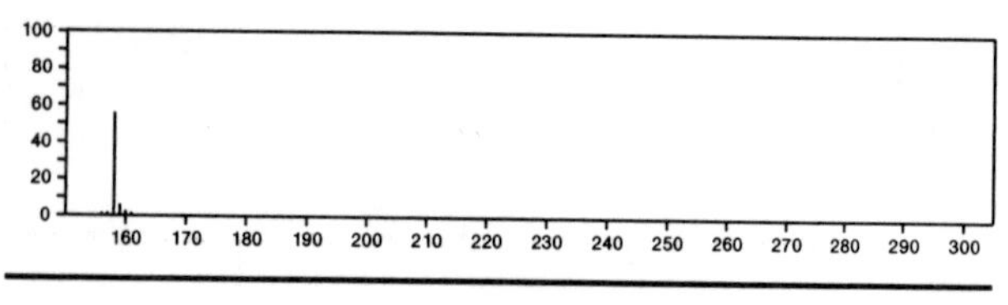

158 $C_9H_{18}S$ 7133–17–7
Sulfide, cyclopentyl isobutyl

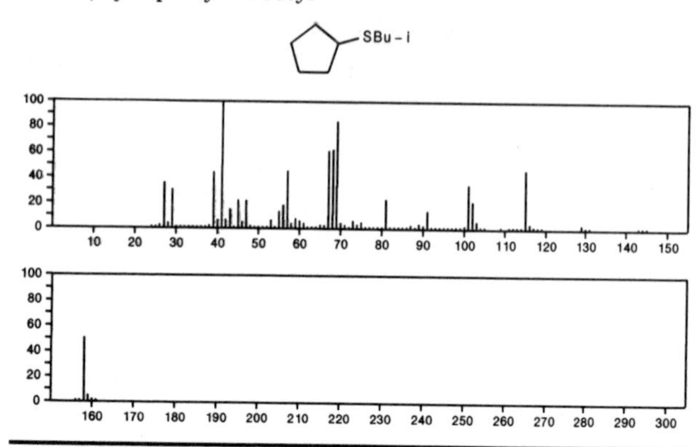

158 $C_9H_{18}S$ 7133–18–8
Sulfide, *sec*–butyl cyclopentyl

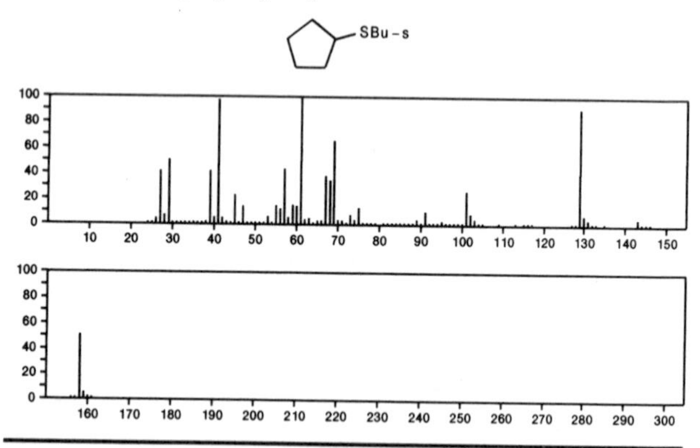

158 $C_9H_{18}S$ 7133–19–9
Sulfide, *tert*–butyl cyclopentyl

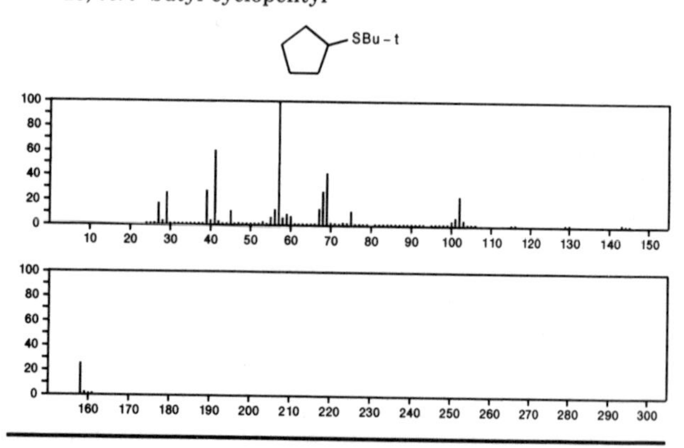

158 $C_9H_{18}S$ 7133–38–2
Sulfide, cyclohexyl propyl

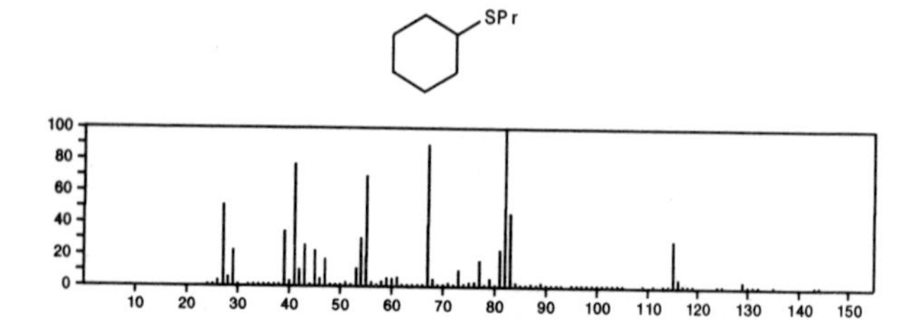

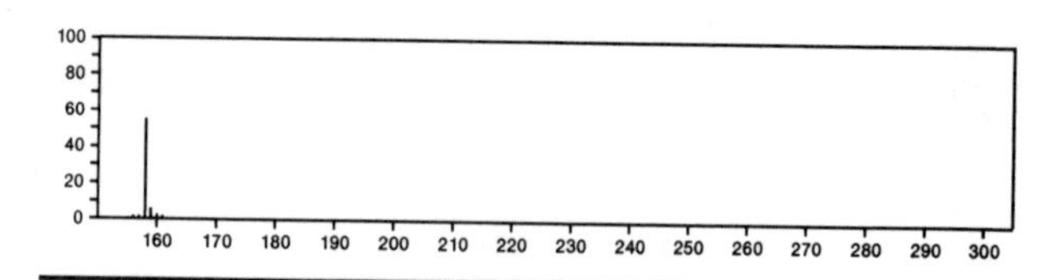

158 $C_9H_{18}S$ 7133–39–3
Cyclohexane, [(1–methylethyl)thio]–

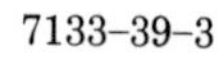

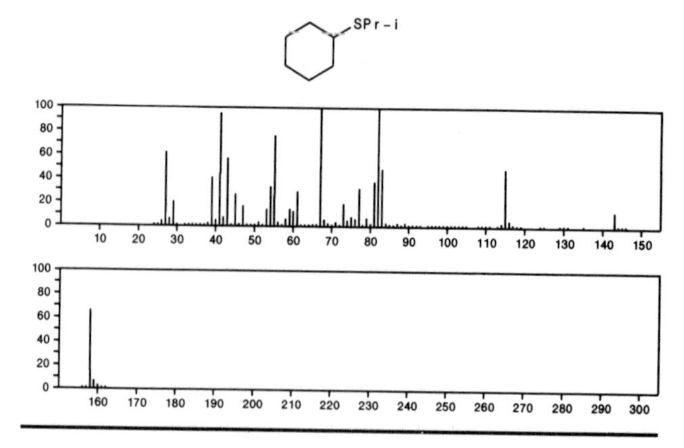

158 $C_9H_{18}S$ 21961–05–7
Heptane, 1–(ethenylthio)–

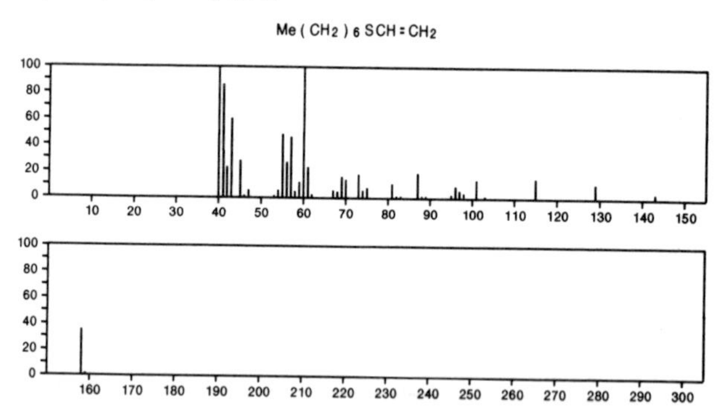

158 $C_9H_{18}S$ 55320–20–2
3–Heptene, 7–(ethylthio)–

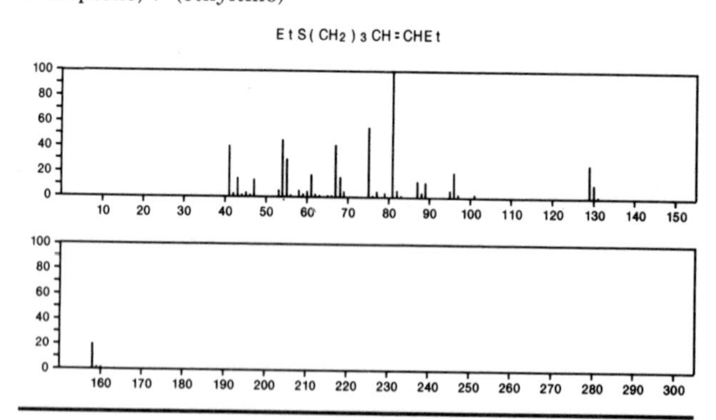

158 $C_9H_{18}S$ 55320–22–4
Thiophene, 3–butyltetrahydro–2–methyl–

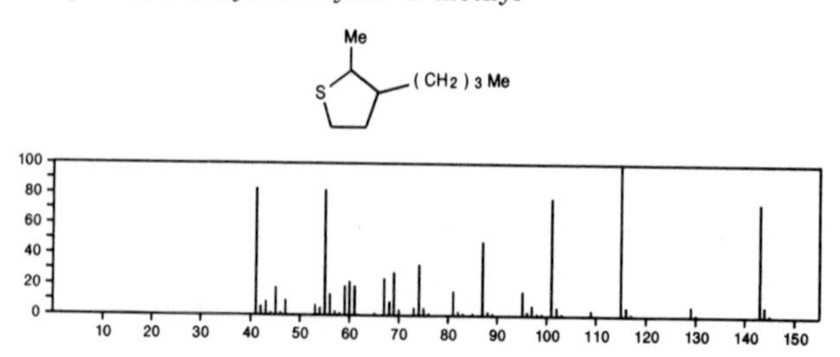

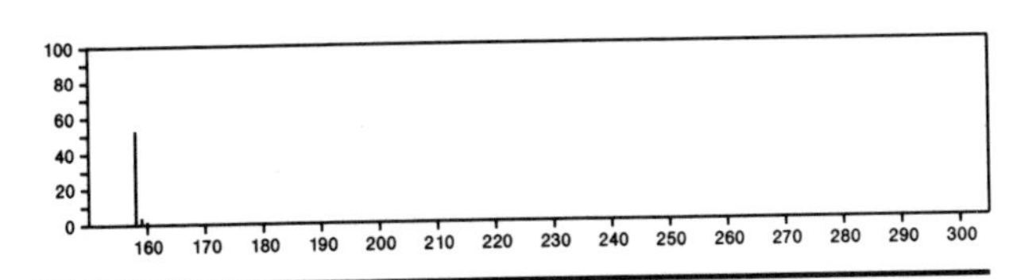

158 $C_9H_{18}S$ 55320-23-5

2*H*-Thiopyran, tetrahydro-2-methyl-3-propyl-

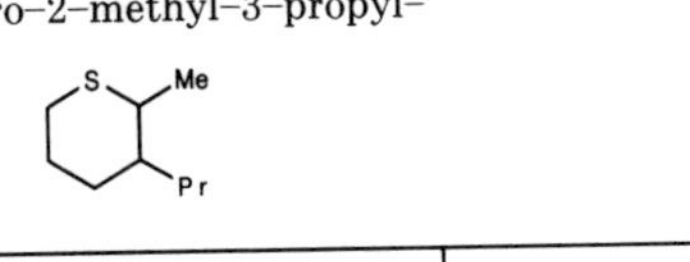

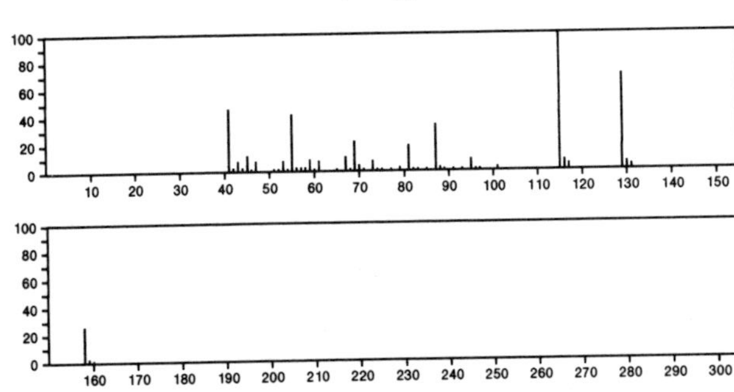

158 $C_9H_{22}N_2$ 102-53-4

Methanediamine, *N*,*N*,*N'*,*N'*-tetraethyl-

$Et_2NCH_2NEt_2$

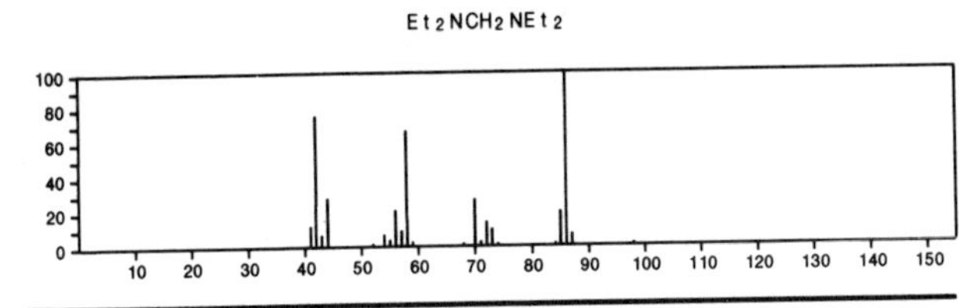

158 $C_{10}H_6O_2$ 130-15-4

1,4-Naphthalenedione

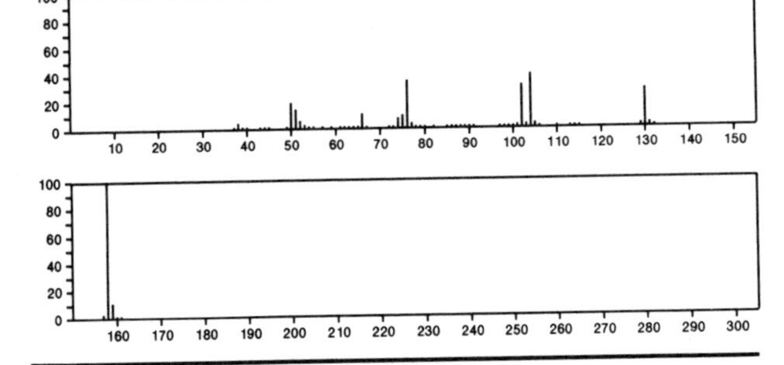

158 $C_{10}H_6O_2$ 524-42-5

1,2-Naphthalenedione

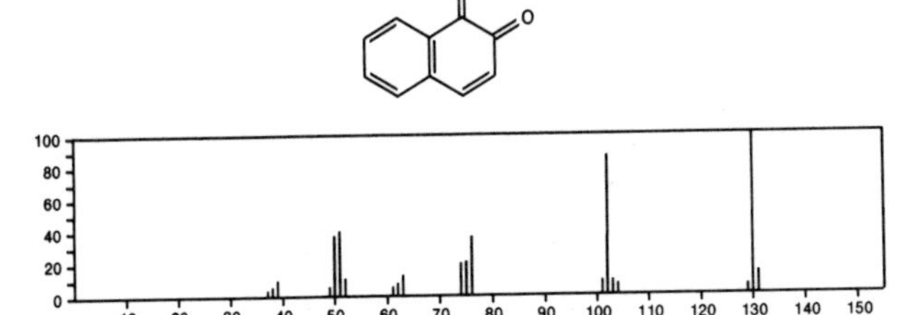

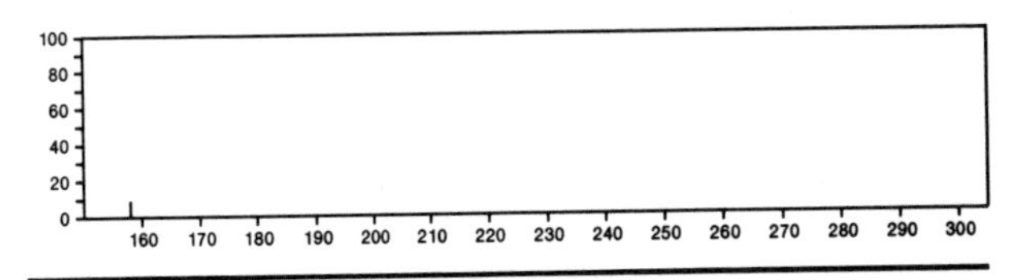

158 $C_{10}H_{10}N_2$ 670-91-7

1*H*-Imidazole, 4-(4-methylphenyl)-

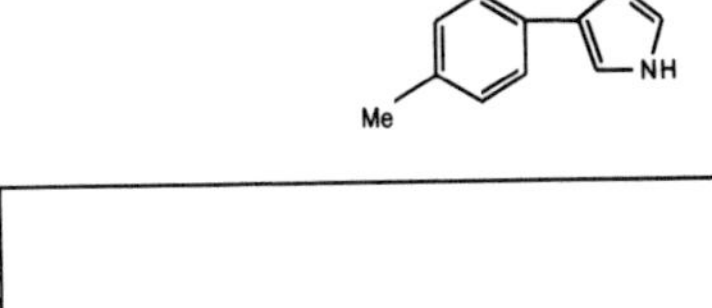

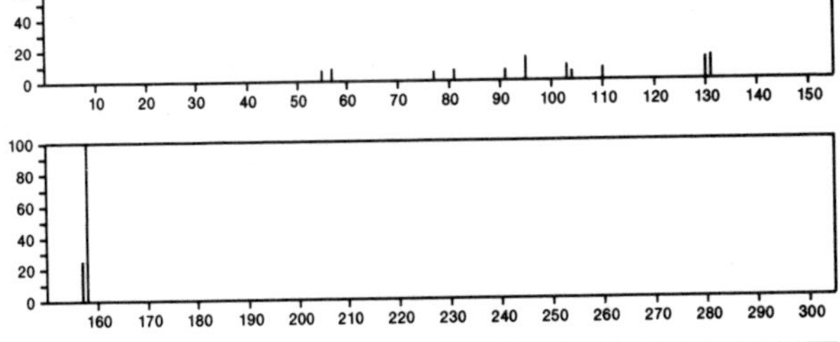

158 $C_{10}H_{10}N_2$ 1199-13-9

1,8-Naphthyridine, 3,6-dimethyl-

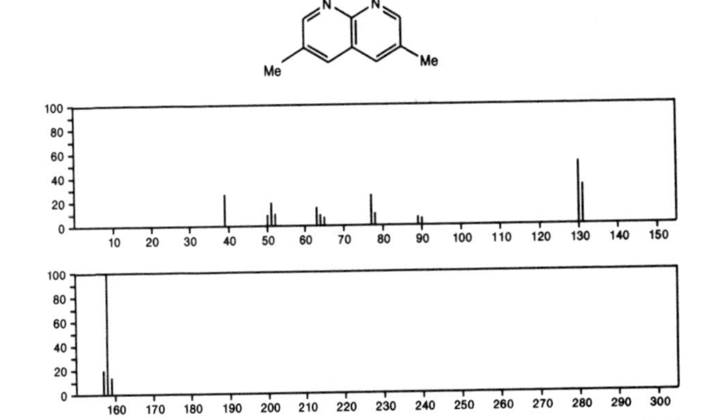

158 $C_{10}H_{10}N_2$ 2379-55-7

Quinoxaline, 2,3-dimethyl-

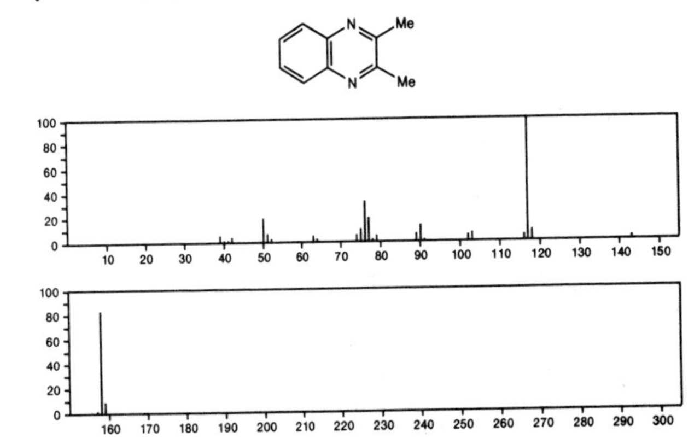

158 $C_{10}H_{10}N_2$ 3347-62-4
1*H*-Pyrazole, 3-methyl-5-phenyl-

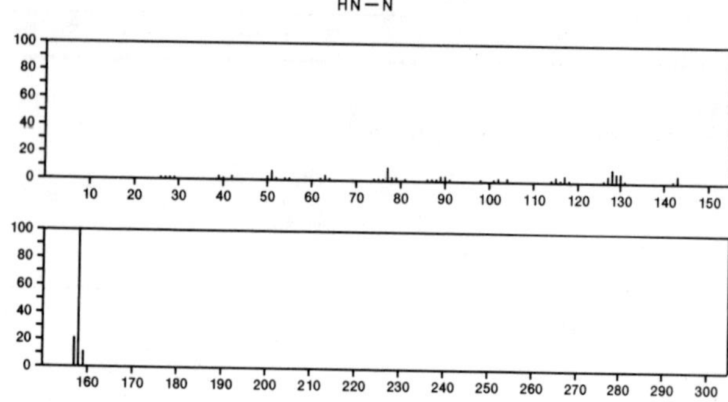

158 $C_{10}H_{10}N_2$ 3929-83-7
Cinnoline, 3,4-dimethyl-

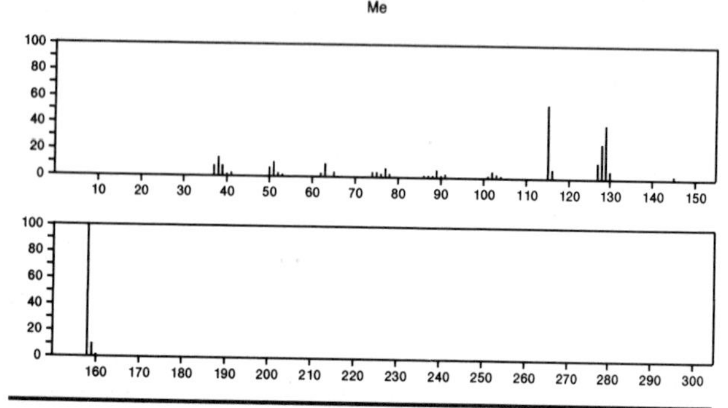

158 $C_{10}H_{10}N_2$ 7544-64-1
1,8-Naphthyridine, 2,4-dimethyl-

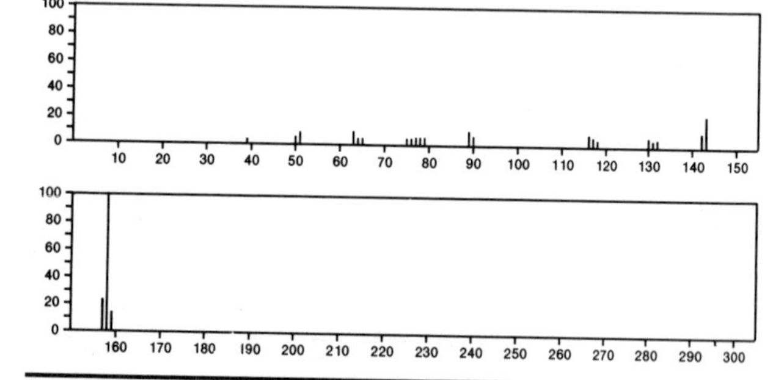

158 $C_{10}H_{10}N_2$ 14757-45-0
1,8-Naphthyridine, 2,6-dimethyl-

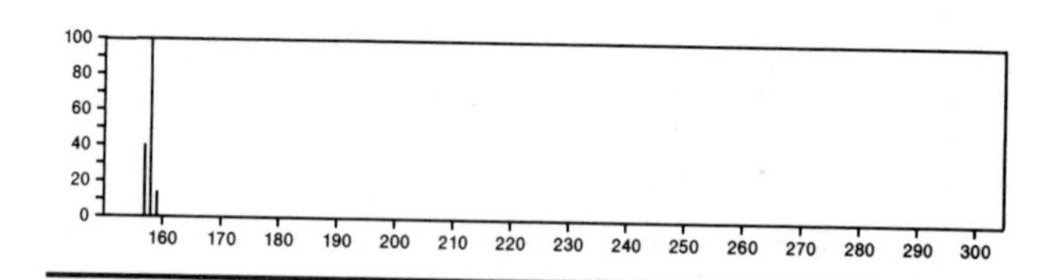

158 $C_{10}H_{10}N_2$ 14757-46-1
1,8-Naphthyridine, 3,5-dimethyl-

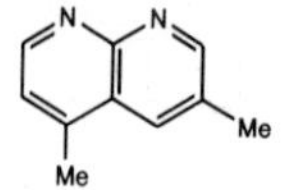

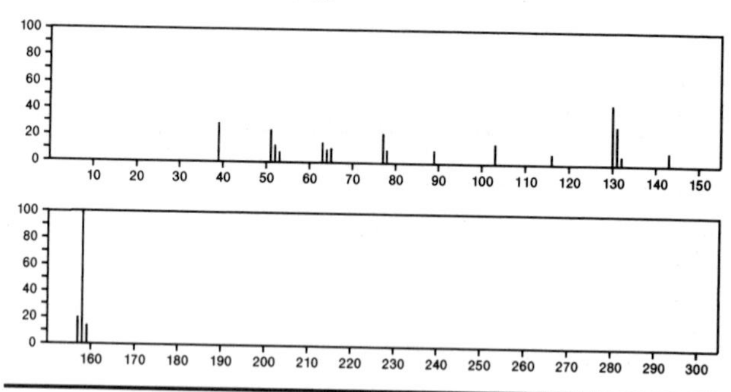

158 $C_{10}H_{10}N_2$ 14759-23-0
1,8-Naphthyridine, 2,5-dimethyl-

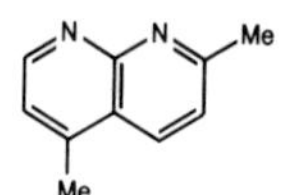

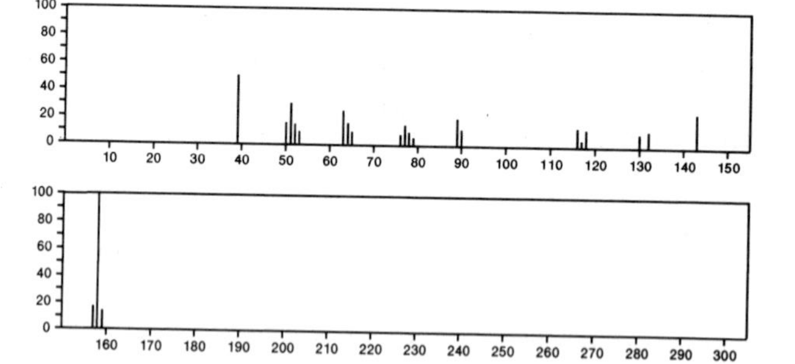

158 $C_{10}H_{10}N_2$ 14903-77-6
1,8-Naphthyridine, 4,5-dimethyl-

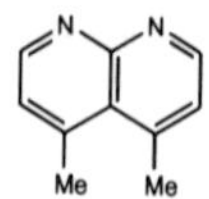

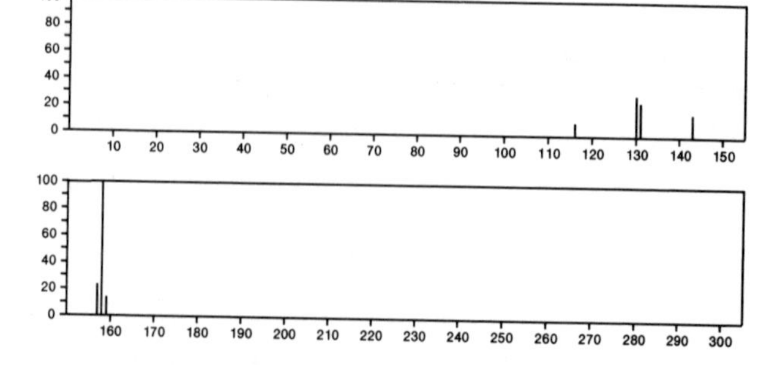

158 $C_{10}H_{10}N_2$ 14903–78–7
1,8–Naphthyridine, 2,7–dimethyl–

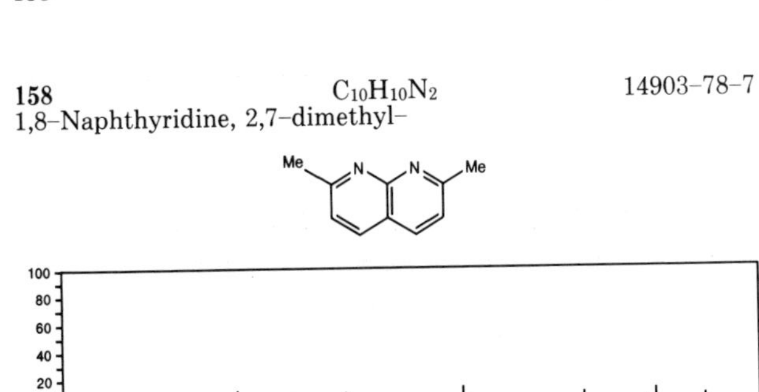

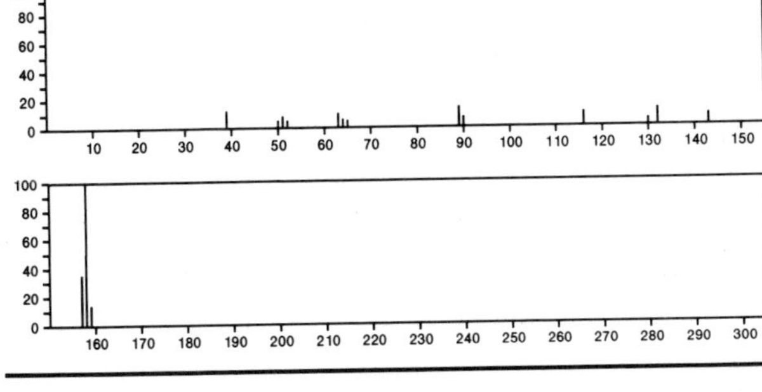

158 $C_{10}H_{10}N_2$ 19018–22–5
Benzimidazole, 1–allyl–

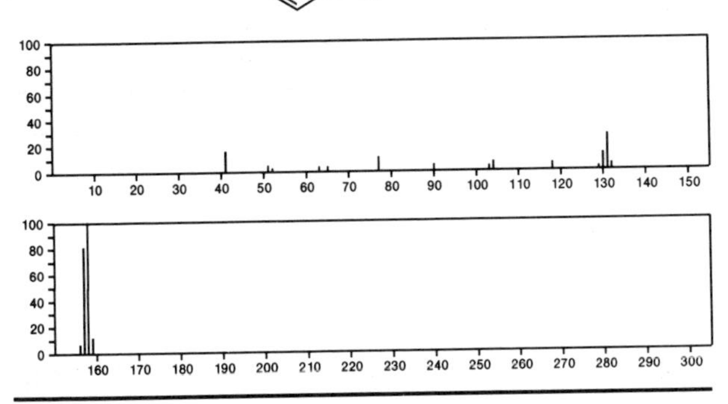

158 $C_{10}H_{10}N_2$ 29100–32–1
1*H*–2,3–Benzodiazepine, 1–methyl–

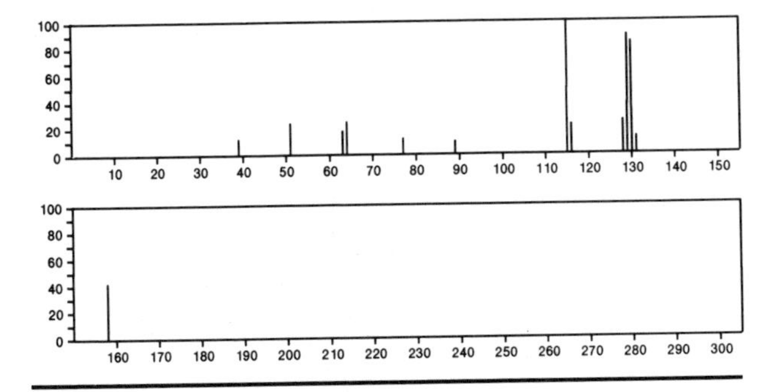

158 $C_{10}H_{10}N_2$ 37122–50–2
1*H*–Imidazole, 2–(4–methylphenyl)–

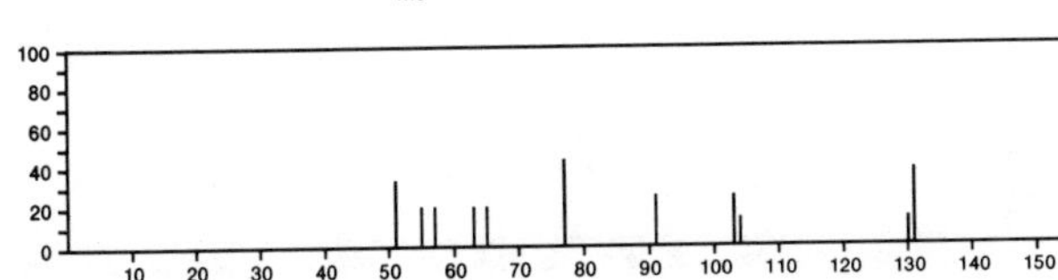

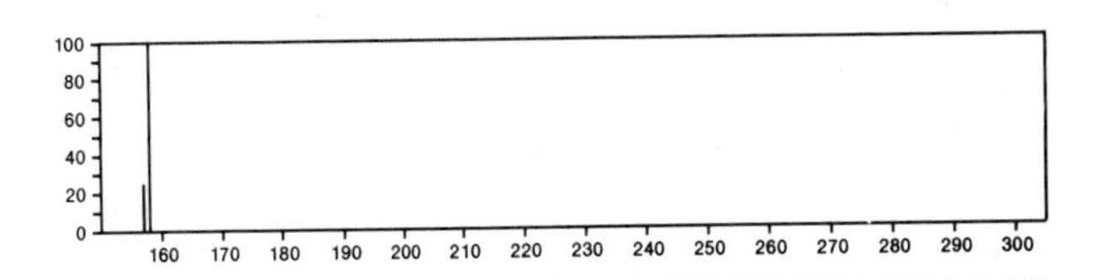

158 $C_{10}H_{22}O$ 78–69–3
3–Octanol, 3,7–dimethyl–

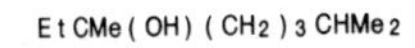

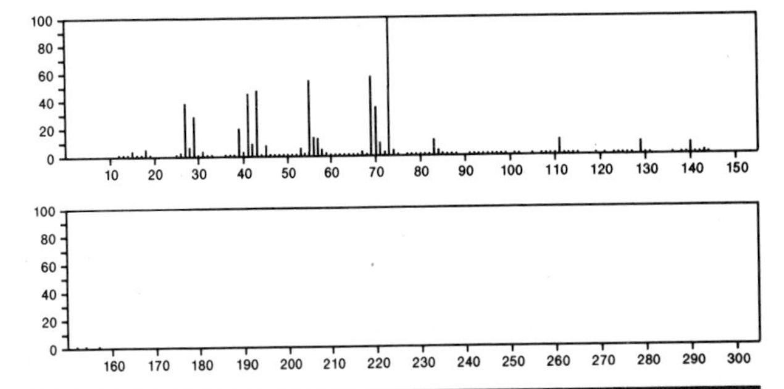

158 $C_{10}H_{22}O$ 106–21–8
1–Octanol, 3,7–dimethyl–

HOCH2 CH2 CHMe (CH2) 3 CHMe 2

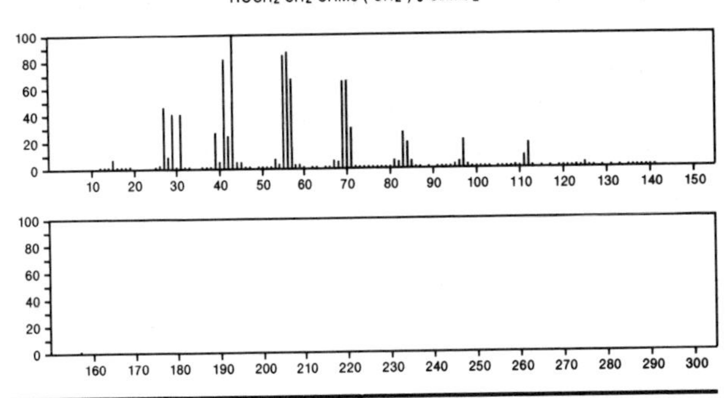

158 $C_{10}H_{22}O$ 112–30–1
1–Decanol

Me (CH2) 9 OH

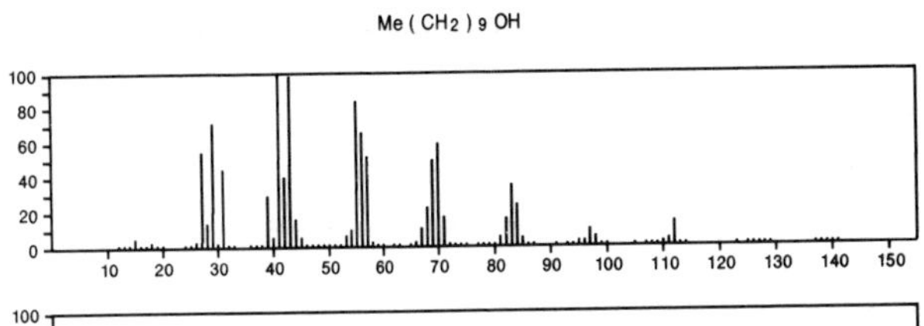

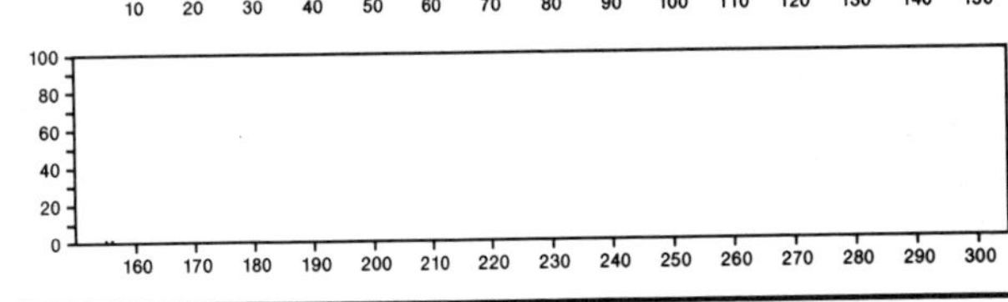

158 $C_{10}H_{22}O$ 151–19–9
3–Octanol, 3,6–dimethyl–

Me CH2 CHMe CH2 CH2 CEt (OH) Me

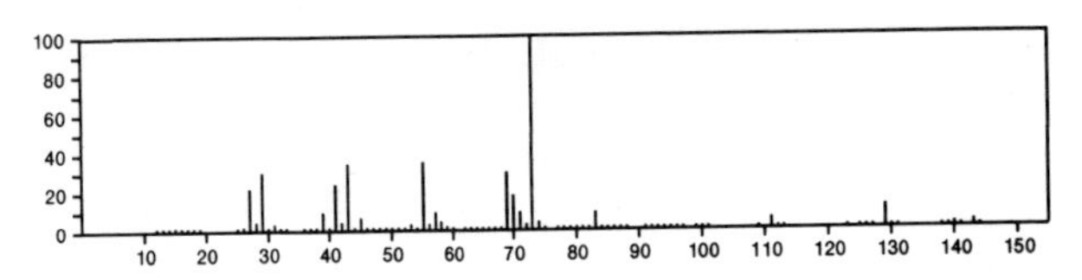

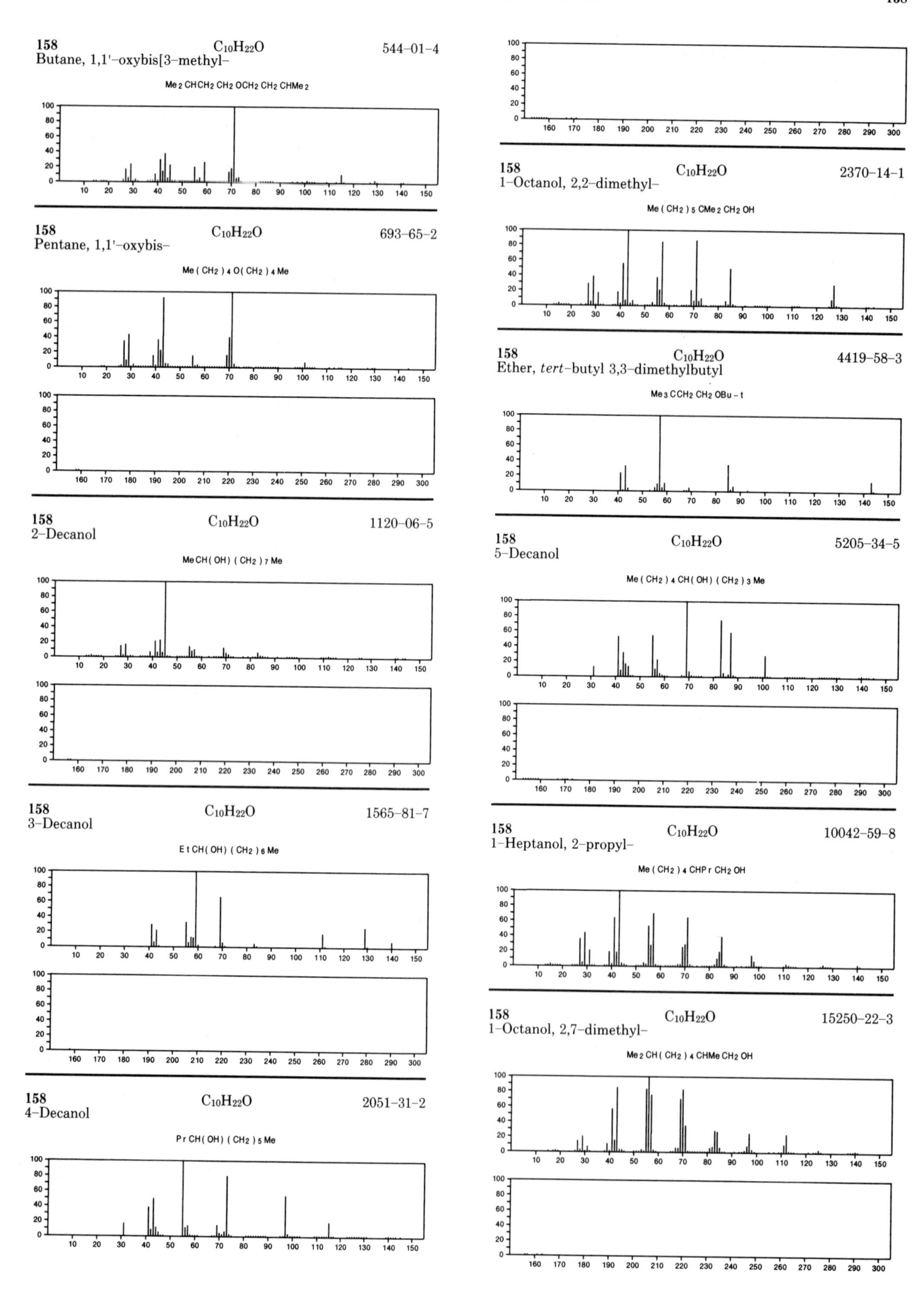

158 C10H22O 544-01-4
Butane, 1,1'-oxybis[3-methyl-
Me 2 CHCH2 CH2 OCH2 CH2 CHMe 2
158 C10H22O 693-65-2
Pentane, 1,1'-oxybis-
Me (CH2) 4 O(CH2) 4 Me
158 C10H22O 1120-06-5
2-Decanol
Me CH(OH) (CH2) 7 Me
158 C10H22O 1565-81-7
3-Decanol
E t CH(OH) (CH2) 6 Me
158 C10H22O 2051-31-2
4-Decanol
P r CH(OH) (CH2) 5 Me
158 C10H22O 2370-14-1
1-Octanol, 2,2-dimethyl-
Me (CH2) 5 CMe 2 CH2 OH
158 C10H22O 4419-58-3
Ether, tert-butyl 3,3-dimethylbutyl
Me 3 C CH2 CH2 OBu - t
158 C10H22O 5205-34-5
5-Decanol
Me (CH2) 4 CH(OH) (CH2) 3 Me
158 C10H22O 10042-59-8
1-Heptanol, 2-propyl-
Me (CH2) 4 CHP r CH2 OH
158 C10H22O 15250-22-3
1-Octanol, 2,7-dimethyl-
Me 2 CH(CH2) 4 CHMe CH2 OH

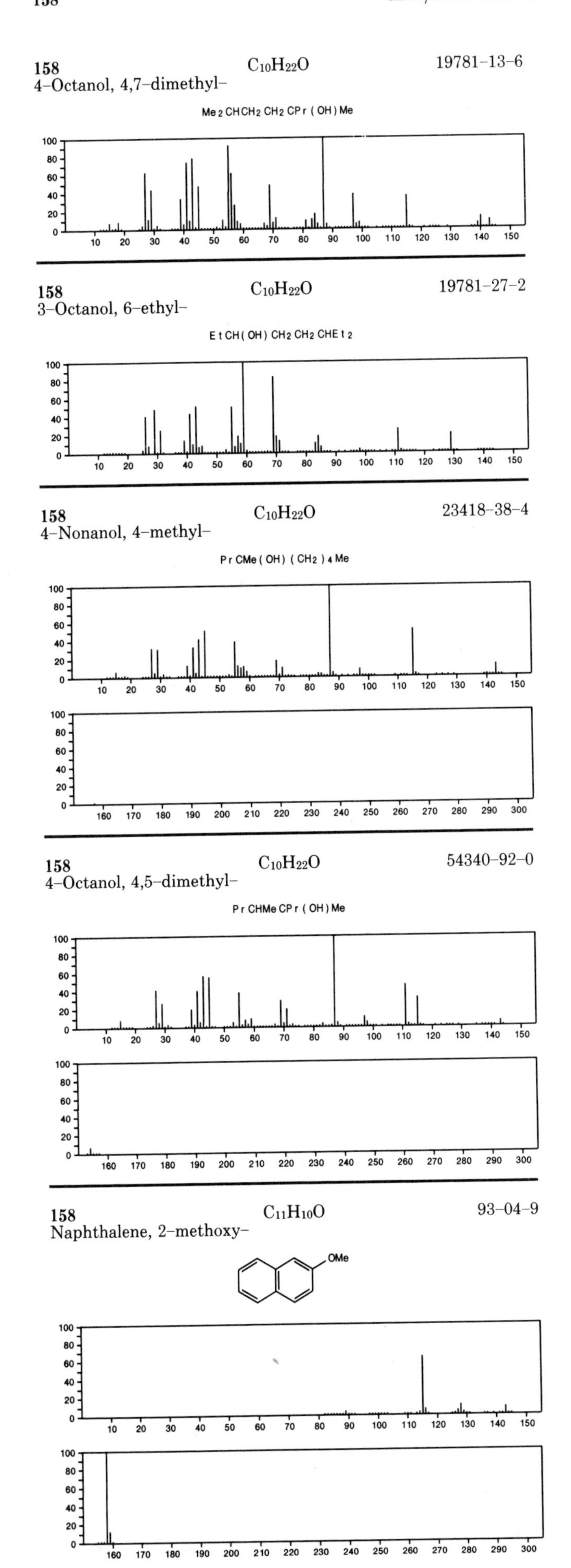
158 $C_{10}H_{22}O$ 19781–13–6
4–Octanol, 4,7–dimethyl–
Me2CHCH2CH2CPr(OH)Me
158 $C_{10}H_{22}O$ 19781–27–2
3–Octanol, 6–ethyl–
EtCH(OH)CH2CH2CHEt2
158 $C_{10}H_{22}O$ 23418–38–4
4–Nonanol, 4–methyl–
PrCMe(OH)(CH2)4Me
158 $C_{10}H_{22}O$ 54340–92–0
4–Octanol, 4,5–dimethyl–
PrCHMeCPr(OH)Me
158 $C_{11}H_{10}O$ 93–04–9
Naphthalene, 2–methoxy–
OMe

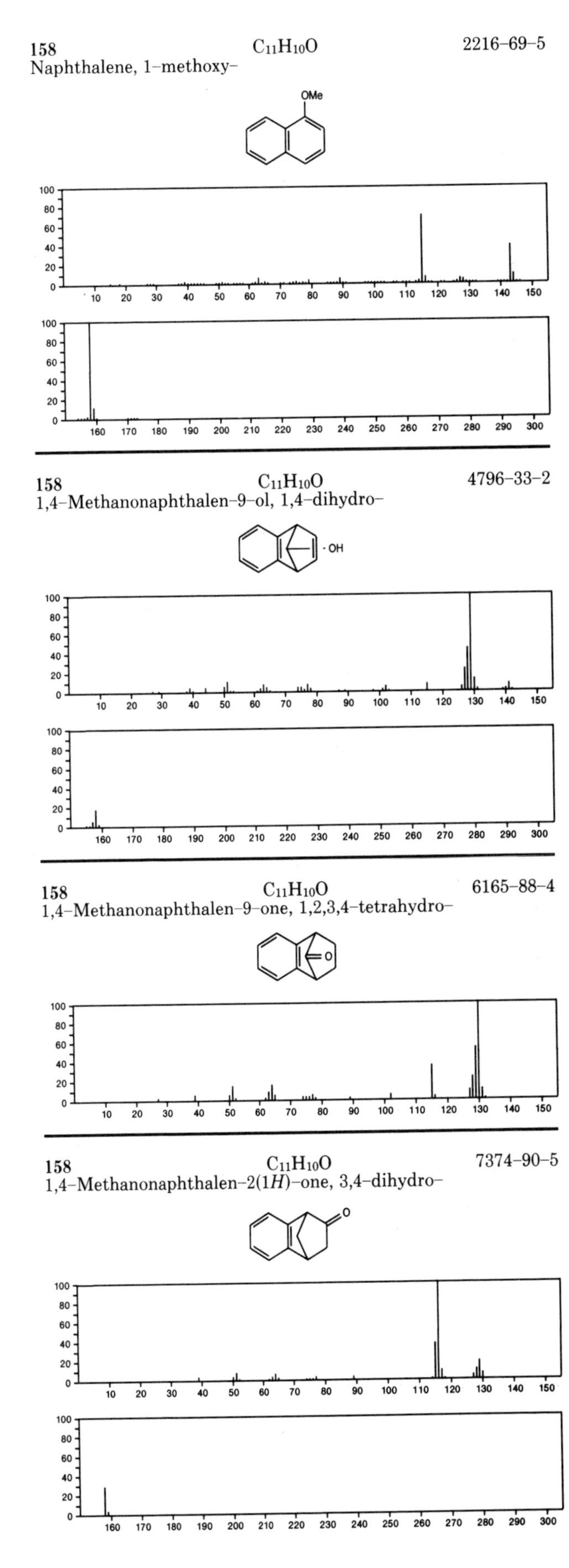
158 $C_{11}H_{10}O$ 2216–69–5
Naphthalene, 1–methoxy–
OMe
158 $C_{11}H_{10}O$ 4796–33–2
1,4–Methanonaphthalen–9–ol, 1,4–dihydro–
OH
158 $C_{11}H_{10}O$ 6165–88–4
1,4–Methanonaphthalen–9–one, 1,2,3,4–tetrahydro–
O
158 $C_{11}H_{10}O$ 7374–90–5
1,4–Methanonaphthalen–2(1*H*)–one, 3,4–dihydro–
O

158 $C_{11}H_{10}O$ 7469–77–4
1–Naphthalenol, 2–methyl–

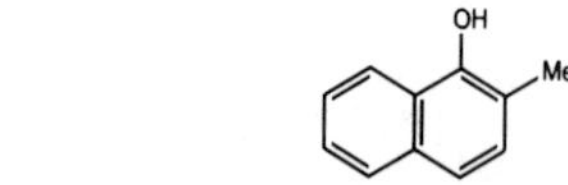

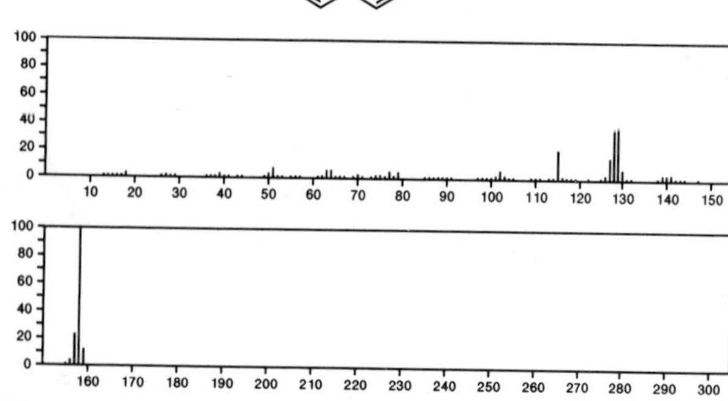

158 $C_{11}H_{10}O$ 10240–08–1
1–Naphthalenol, 4–methyl–

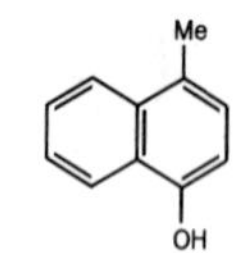

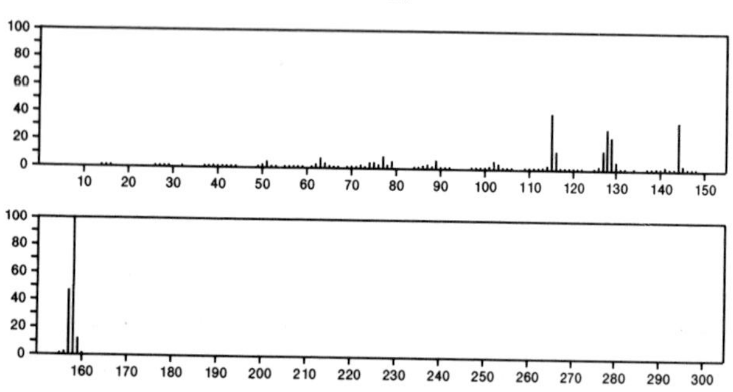

158 $C_{11}H_{10}O$ 13137–34–3
2,7–Methanonaphth[2,3–*b*]oxirene, 1a,2,7,7a–tetrahydro–, (1aα,2β,7β,7aα)–

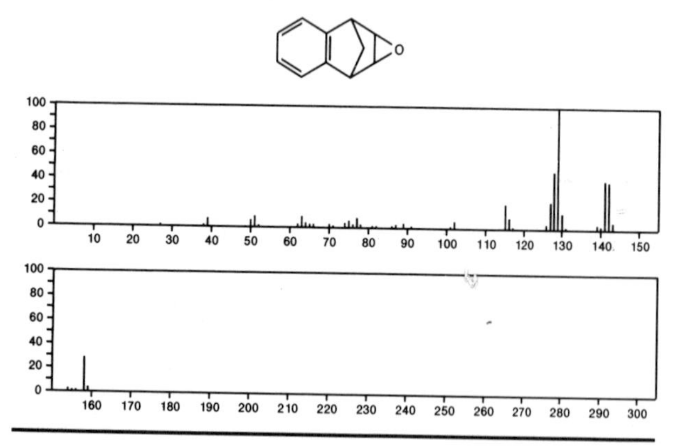

158 $C_{11}H_{10}O$ 13615–40–2
1–Naphthalenol, 3–methyl–

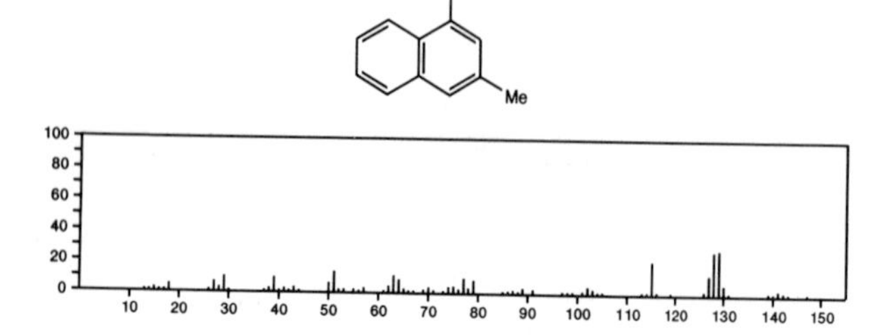

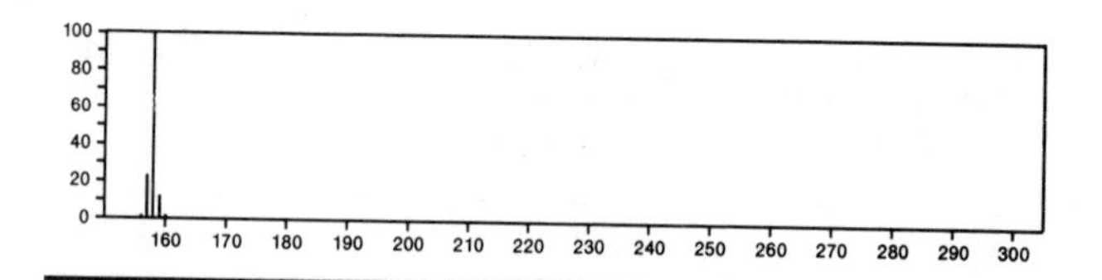

158 $C_{12}H_{14}$ 480–72–8
Acenaphthylene, 1,2,2a,3,4,5–hexahydro–

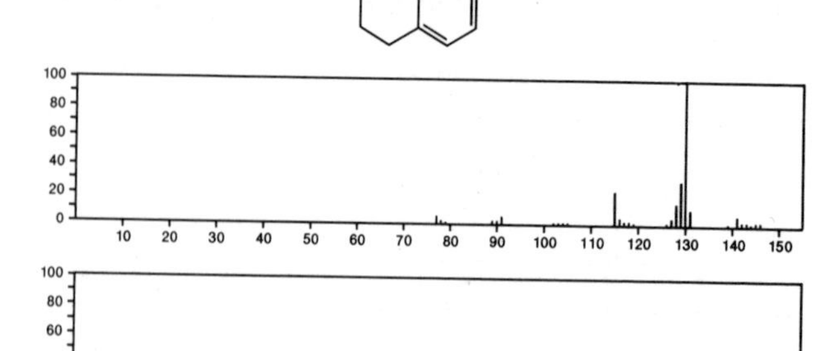

158 $C_{12}H_{14}$ 771–98–2
Benzene, 1–cyclohexen–1–yl–

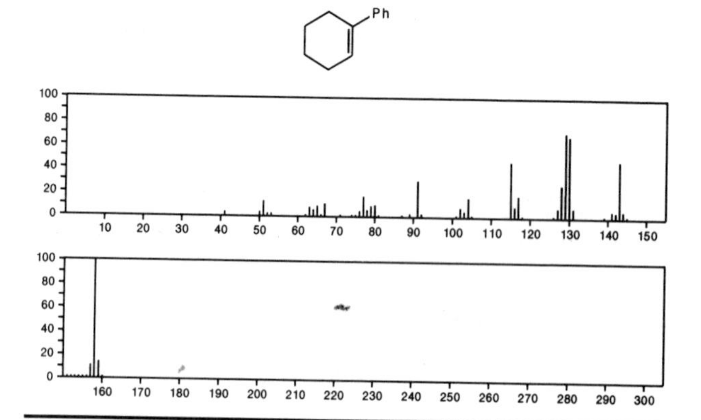

158 $C_{12}H_{14}$ 1129–65–3
Benzene, 1–hexynyl–

PhC≡C(CH2)3Me

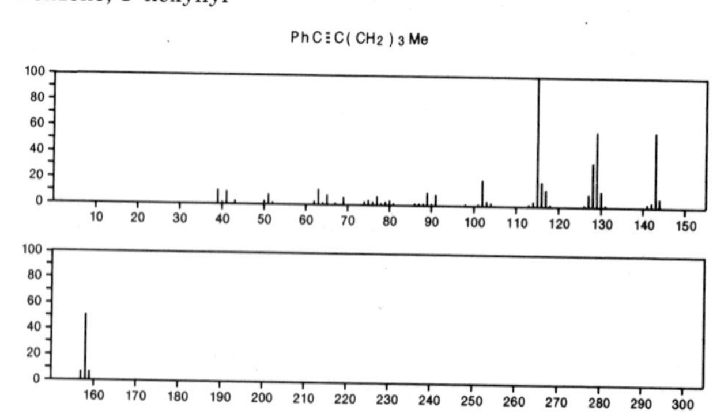

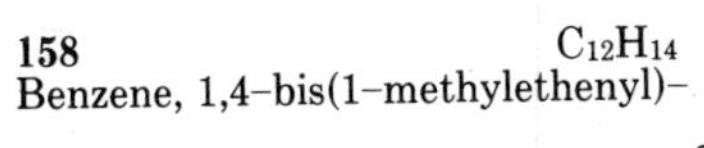

158 $C_{12}H_{14}$ 1605–18–1
Benzene, 1,4-bis(1-methylethenyl)–

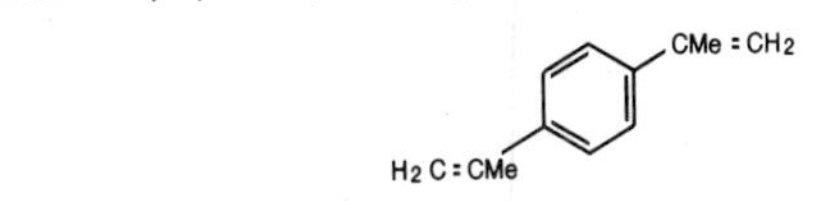

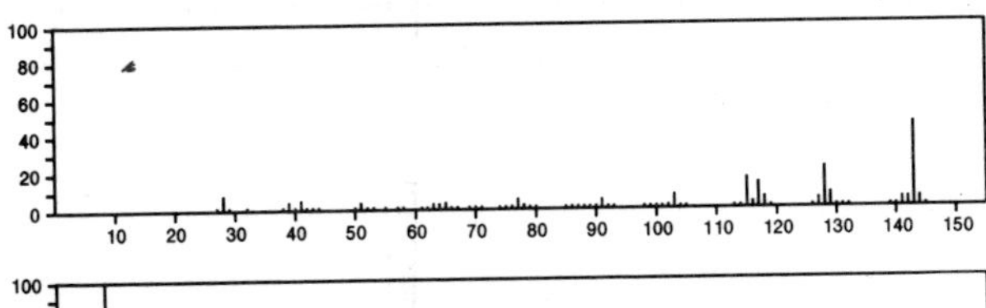

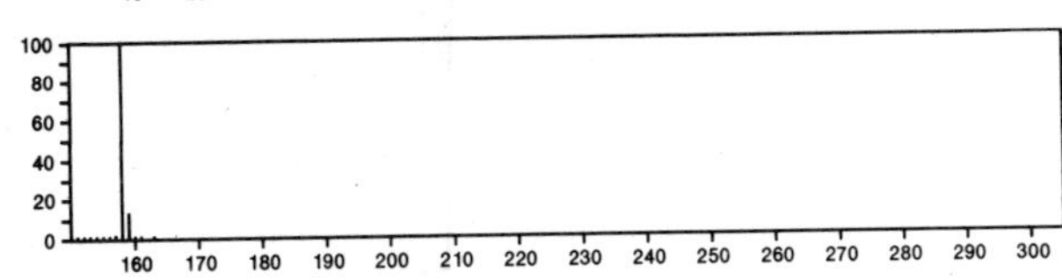

158 $C_{12}H_{14}$ 3748–13–8
Benzene, 1,3-bis(1-methylethenyl)–

H2 C = CMe CMe = CH2

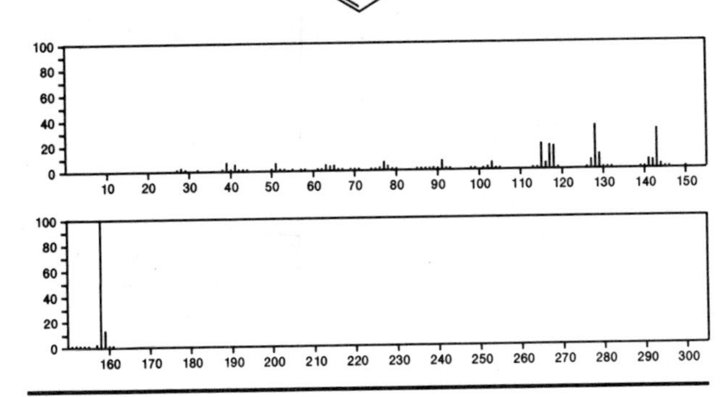

158 $C_{12}H_{14}$ 4175–52–4
1,4-Ethanonaphthalene, 1,2,3,4-tetrahydro–

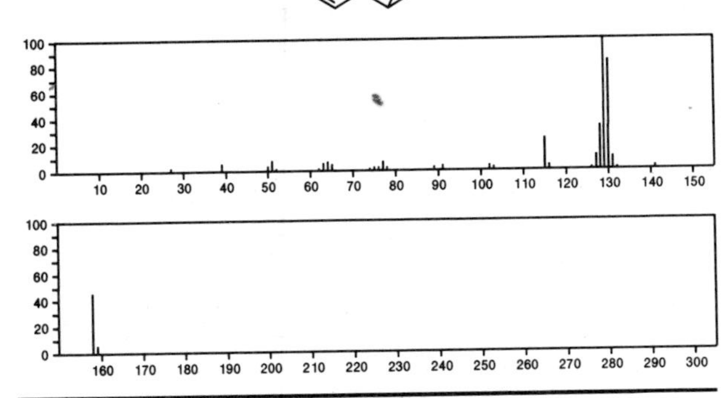

158 $C_{12}H_{14}$ 4994–16–5
Benzene, 3-cyclohexen-1-yl–

Ph

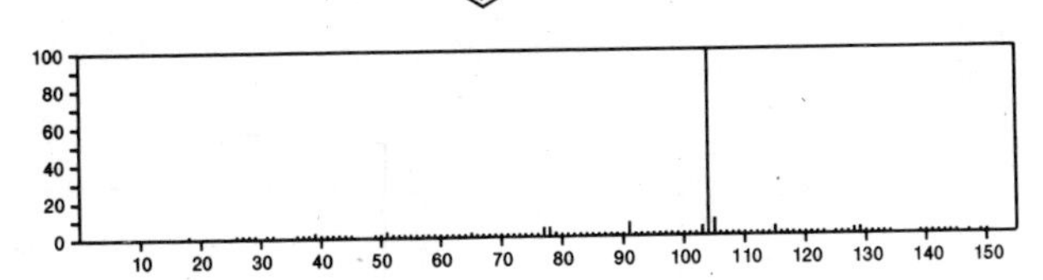

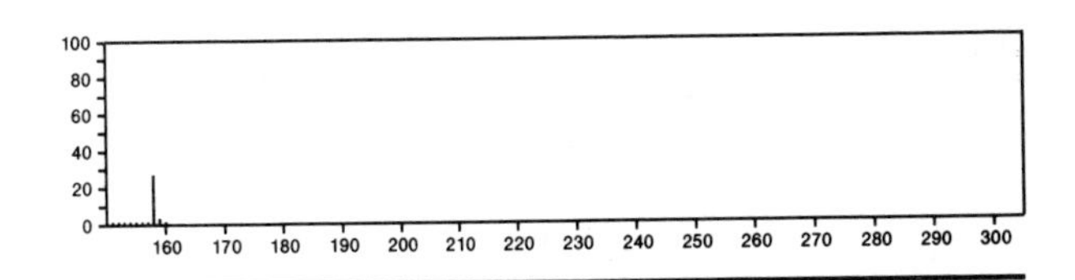

158 $C_{12}H_{14}$ 15232–96–9
Benzene, 2-cyclohexen-1-yl–

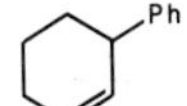

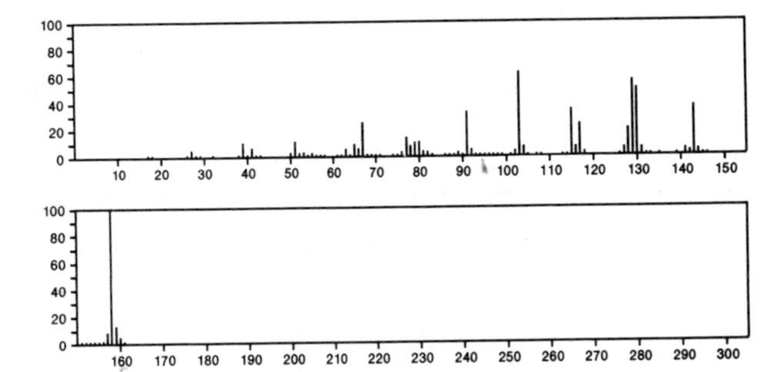

158 $C_{12}H_{14}$ 20295–17–4
4a,8a-Ethenonaphthalene, 1,4,5,8-tetrahydro–

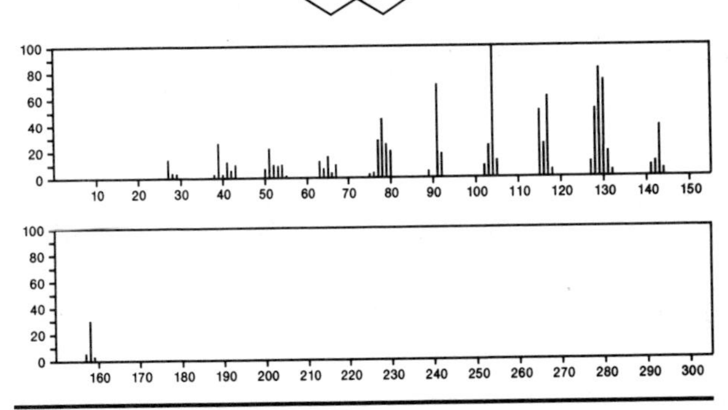

158 $C_{12}H_{14}$ 24329–97–3
Indan, 2-allyl–

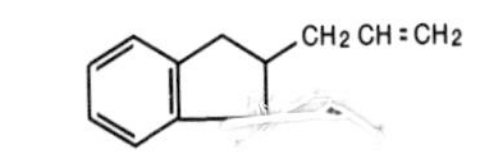

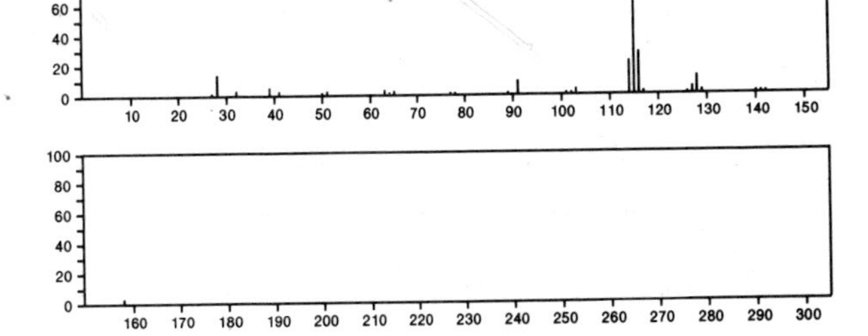

158 $C_{12}H_{14}$ 24524–55–8
Benzene, [(1–methylethylidene)cyclopropyl]–, (*R*)–

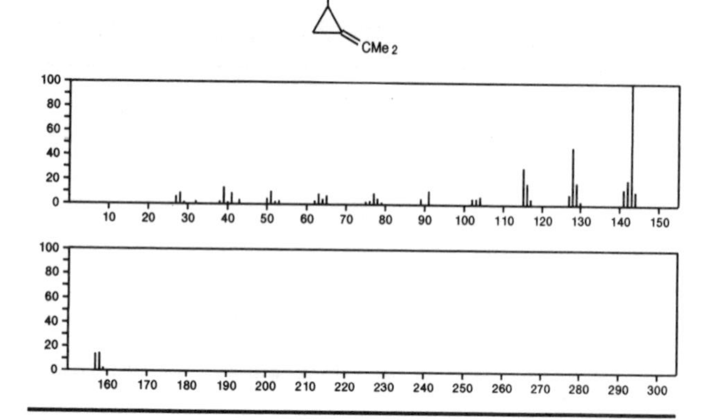

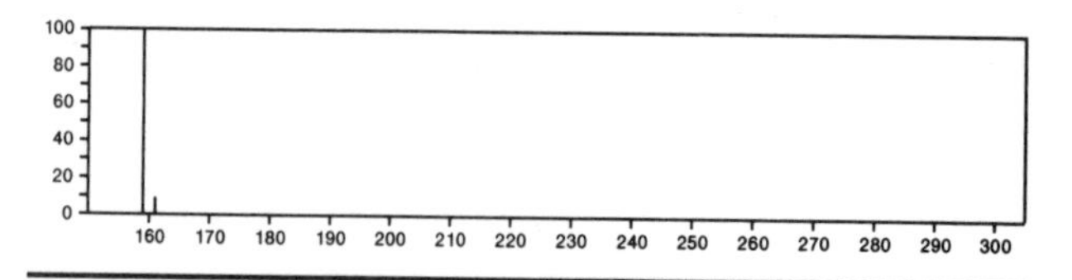

158 $C_{12}H_{14}$ 29555–07–5
Benzene, 1,3,5–trimethyl–2–(1,2–propadienyl)–

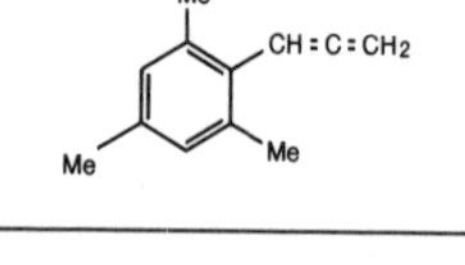

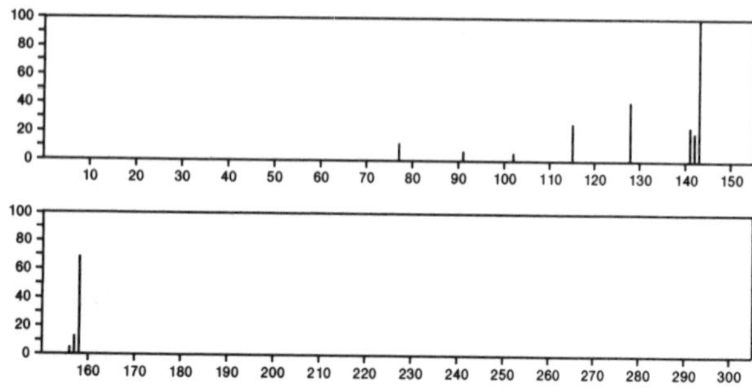

158 $C_{12}H_{14}$ 56701–47–4
Benzene, [(1–methylethylidene)cyclopropyl]–

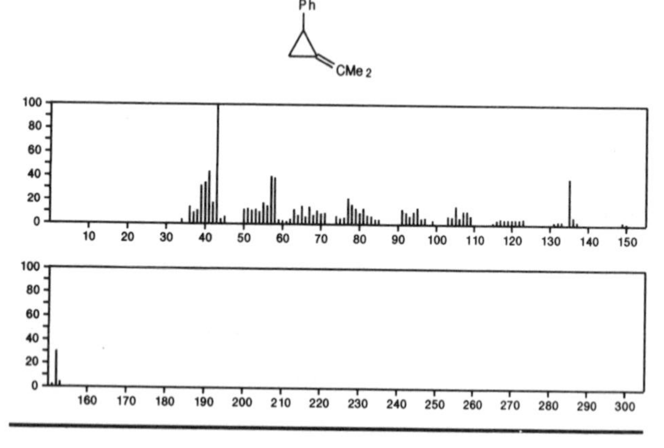

159 CF_2NPS_2 14526–12–6
Phosphor(isothiocyanatido)thioic difluoride

SCNPF2 = S

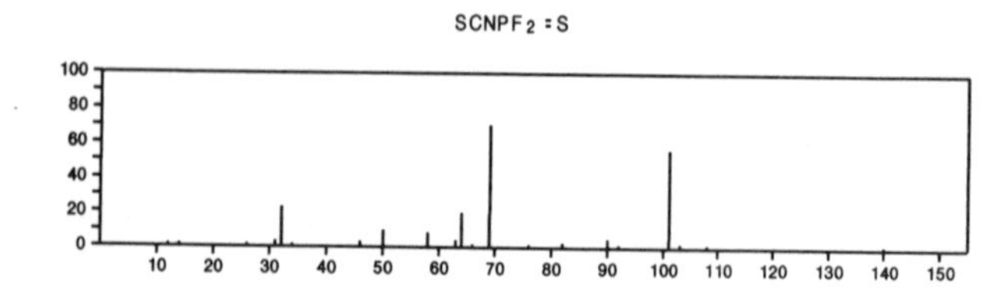

159 $C_2F_3NO_2S$ 51735–89–8
Methanesulfinyl isocyanate, trifluoro–

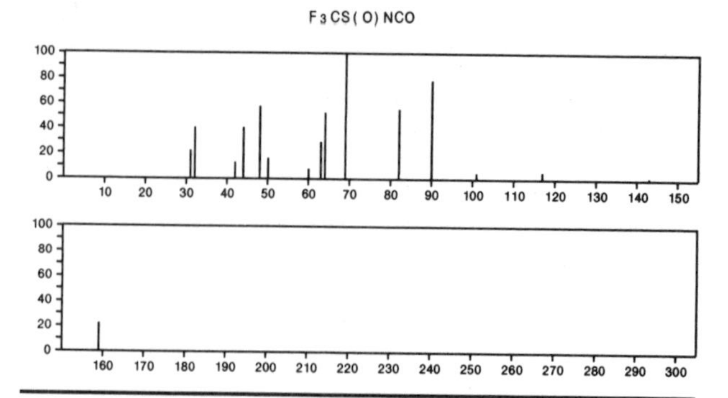

159 $C_2H_7ClNOPS$ 681–04–9
Phosphoramidochloridothioic acid, methyl–, *O*–methyl ester

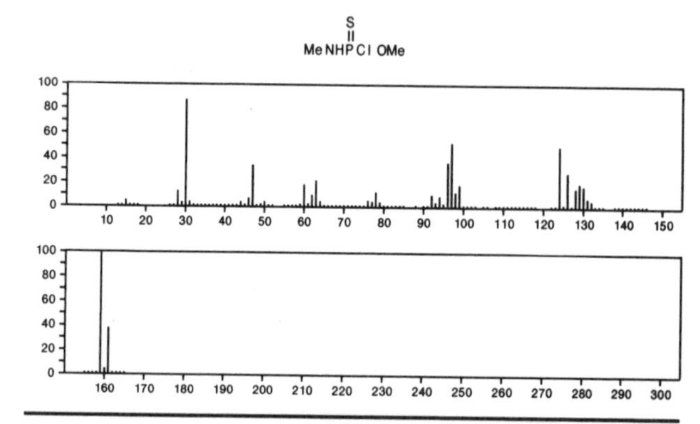

159 $C_5H_6ClN_3O$ 5734–64–5
2–Pyrimidinamine, 4–chloro–6–methoxy–

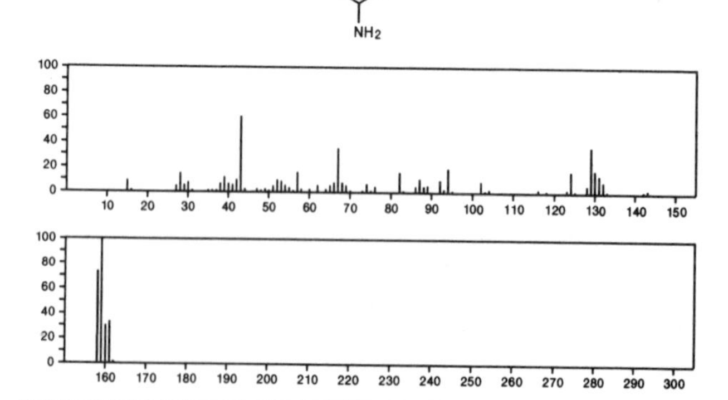

159 $C_6H_9NS_2$ 51598–96–0
1–Butene, 4–isothiocyanato–1–(methylthio)–

SCNCH2 CH2 CH=CHSMe

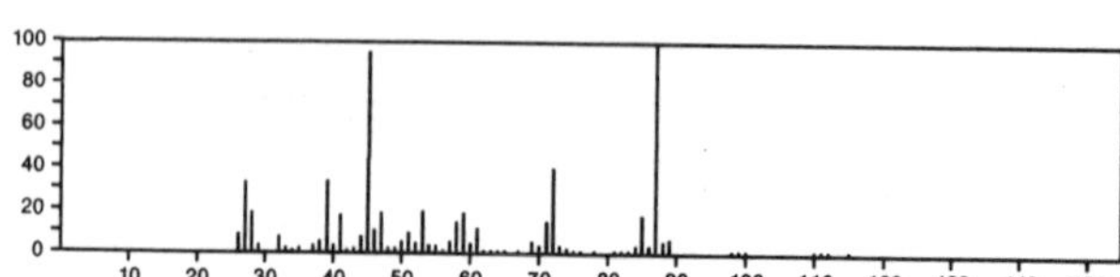

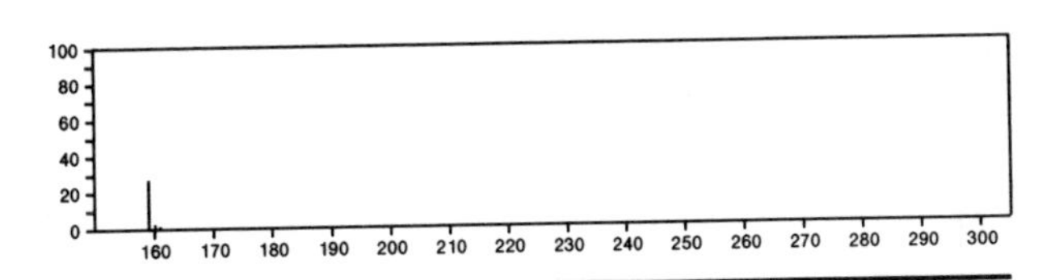

159 $C_6H_{13}N_3O_2$ 50285-72-8
Urea, *N*,*N*-diethyl-*N'*-methyl-*N'*-nitroso-

$Et_2NCON(NO)Me$

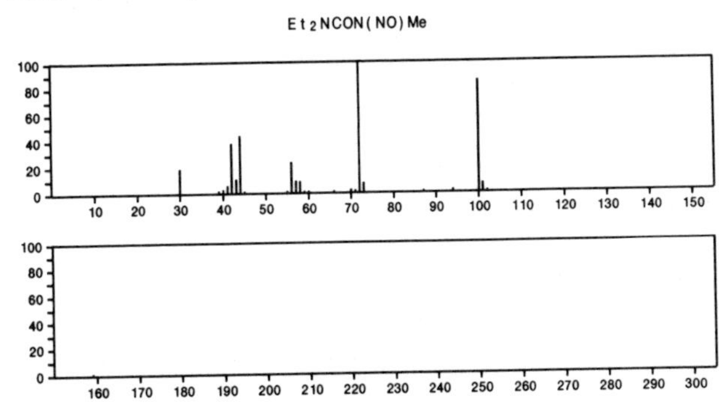

159 $C_6H_{13}N_3S$ 1752-39-2
Hydrazinecarbothioamide, 2-(1-methylbutylidene)-

$PrCMe{=}NNHCSNH_2$

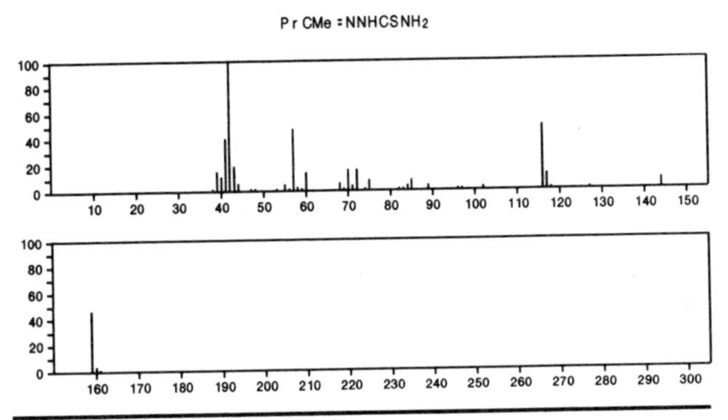

159 $C_6H_{13}N_3S$ 22397-19-9
Hydrazinecarbothioamide, 2-pentylidene-

$Me(CH_2)_3CH{=}NNHCSNH_2$

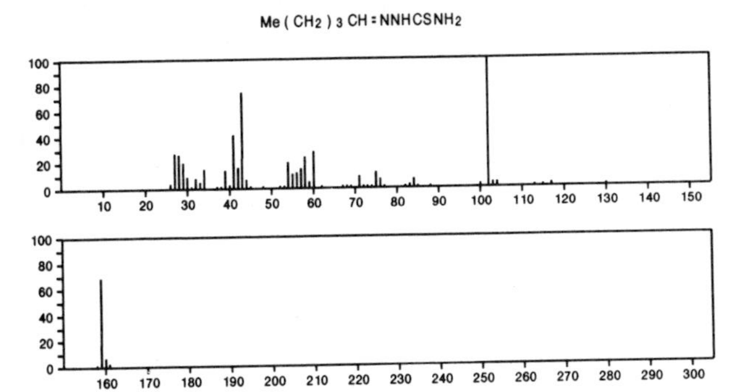

159 $C_7H_{13}NO_3$ 96-81-1
L-Valine, *N*-acetyl-

$AcNHCH(CO_2H)CHMe_2$

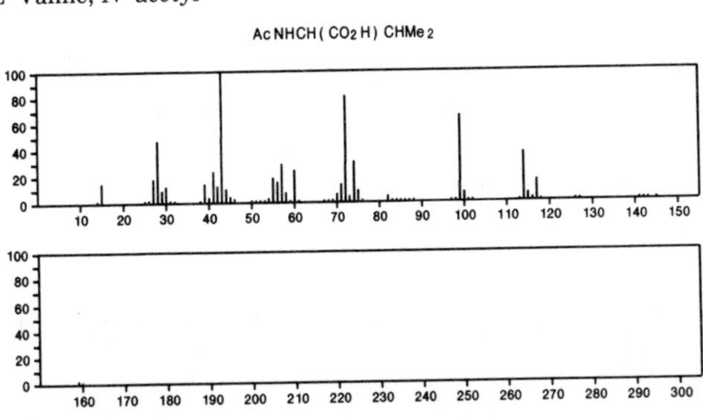

159 $C_7H_{13}NO_3$ 3375-84-6
2-Oxazolidinone, 3-(2-hydroxypropyl)-5-methyl-

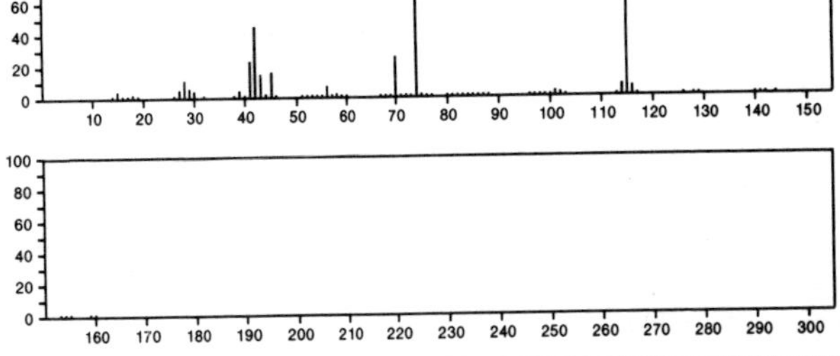

159 $C_7H_{13}NO_3$ 49582-69-6
2-Propenoic acid, 3-(dimethylamino)-3-methoxy-, methyl ester

$MeOC(O)CH{=}C(OMe)NMe_2$

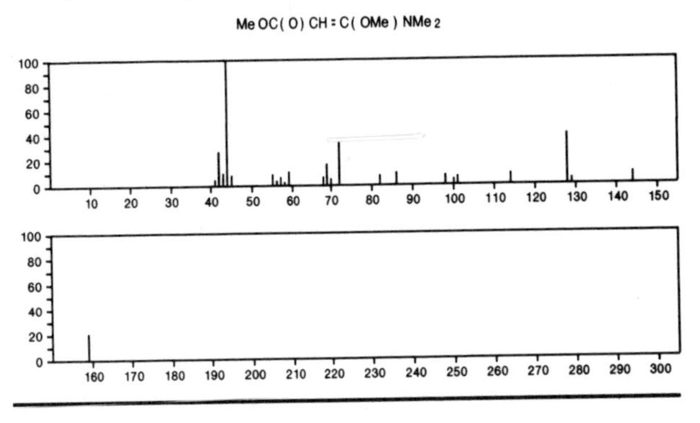

159 $C_7H_{13}NO_3$ 54340-82-8
4-Hexenoic acid, 2-amino-6-hydroxy-4-methyl-

$HOCH_2CH{=}CMeCH_2CH(NH_2)CO_2H$

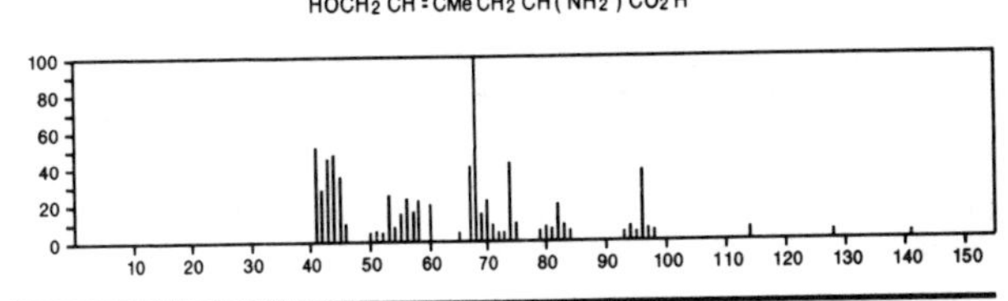

159 $C_8H_5N_3O$ 3265-29-0
2*H*-Indol-2-one, 3-diazo-1,3-dihydro-

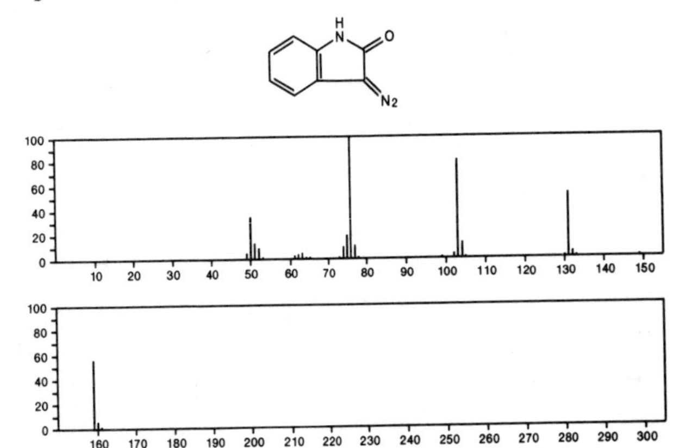

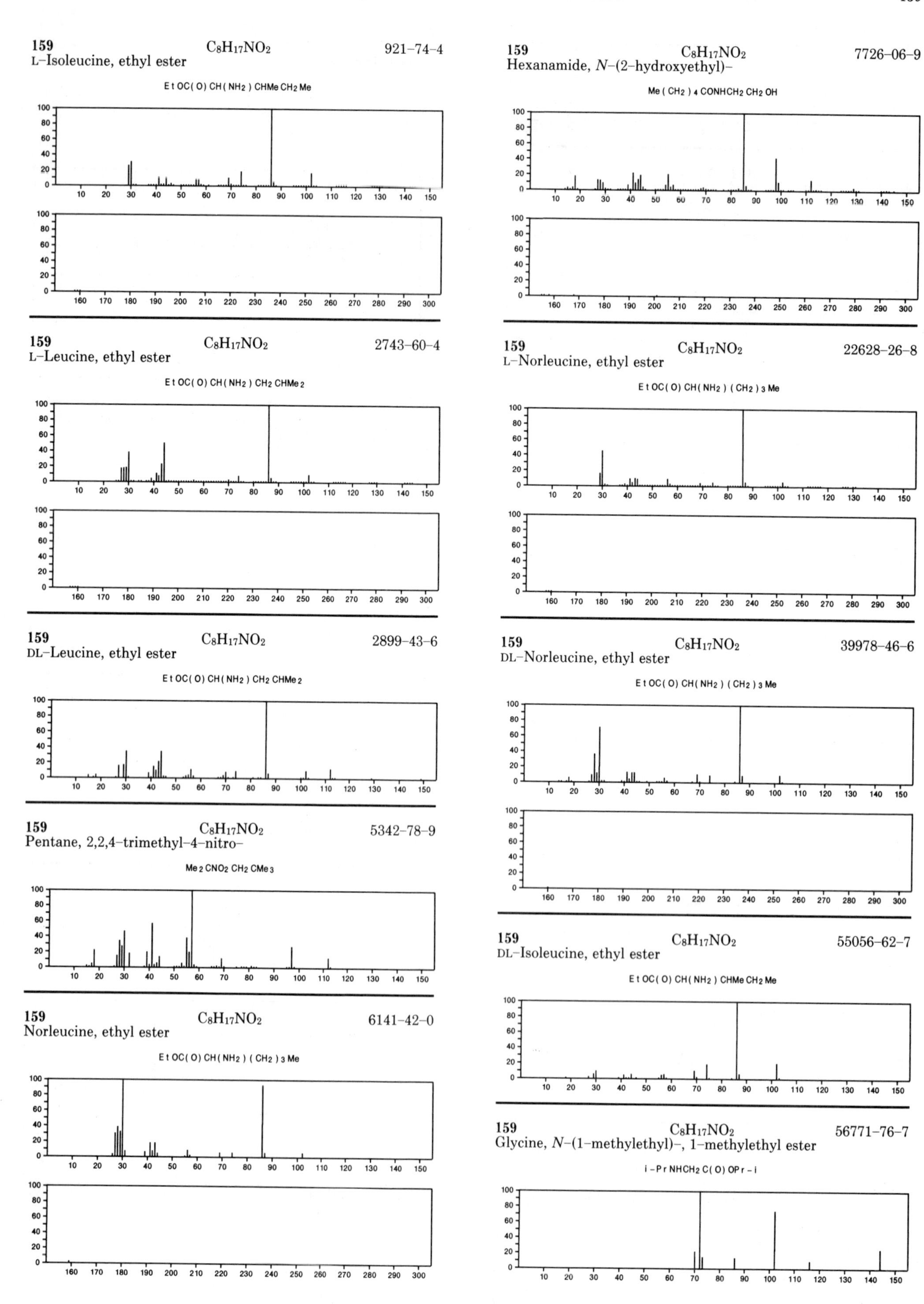
159 $C_8H_{17}NO_2$ 921-74-4
L-Isoleucine, ethyl ester
EtOC(O)CH(NH2)CHMeCH2Me
159 $C_8H_{17}NO_2$ 7726-06-9
Hexanamide, N-(2-hydroxyethyl)-
Me(CH2)4CONHCH2CH2OH
159 $C_8H_{17}NO_2$ 2743-60-4
L-Leucine, ethyl ester
EtOC(O)CH(NH2)CH2CHMe2
159 $C_8H_{17}NO_2$ 22628-26-8
L-Norleucine, ethyl ester
EtOC(O)CH(NH2)(CH2)3Me
159 $C_8H_{17}NO_2$ 2899-43-6
DL-Leucine, ethyl ester
EtOC(O)CH(NH2)CH2CHMe2
159 $C_8H_{17}NO_2$ 39978-46-6
DL-Norleucine, ethyl ester
EtOC(O)CH(NH2)(CH2)3Me
159 $C_8H_{17}NO_2$ 5342-78-9
Pentane, 2,2,4-trimethyl-4-nitro-
Me2CNO2CH2CMe3
159 $C_8H_{17}NO_2$ 55056-62-7
DL-Isoleucine, ethyl ester
EtOC(O)CH(NH2)CHMeCH2Me
159 $C_8H_{17}NO_2$ 6141-42-0
Norleucine, ethyl ester
EtOC(O)CH(NH2)(CH2)3Me
159 $C_8H_{17}NO_2$ 56771-76-7
Glycine, N-(1-methylethyl)-, 1-methylethyl ester
i-PrNHCH2C(O)OPr-i

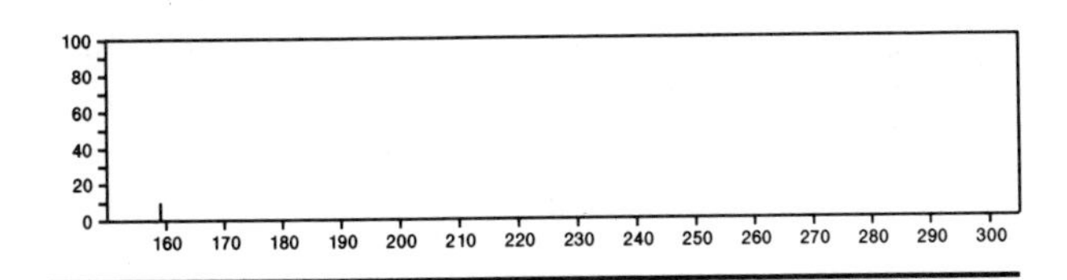

159 $C_9H_9N_3$ 6085–94–5
1*H*–1,2,4–Triazole, 1–(phenylmethyl)–

N N–CH2 Ph N

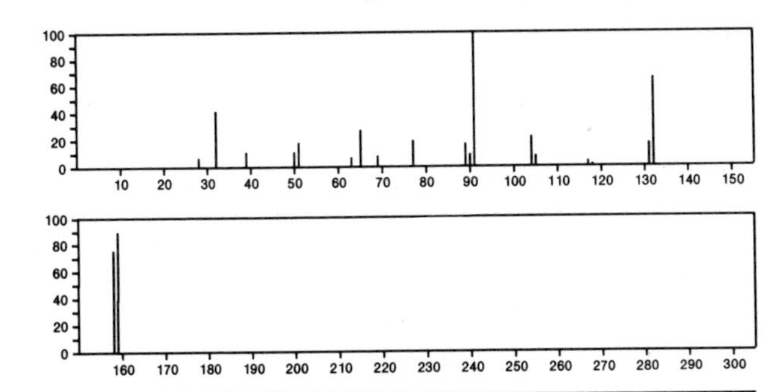

159 $C_9H_9N_3$ 28732–68–5
Pyrido[2,3–*d*]pyrimidine, 4–ethyl–

N N N Et

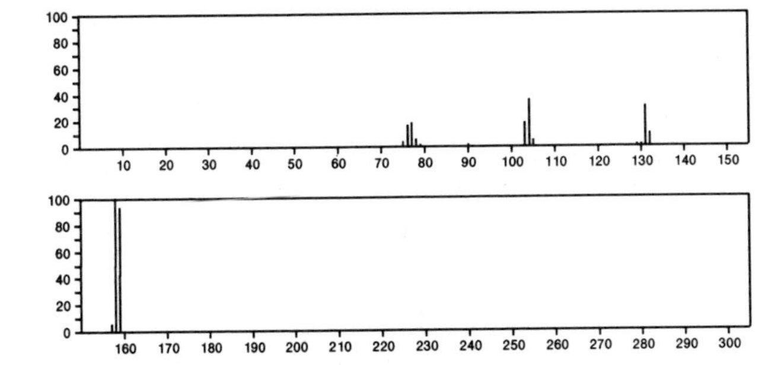

159 $C_{10}H_9NO$ 606–43–9
2(1*H*)–Quinolinone, 1–methyl–

Me N O

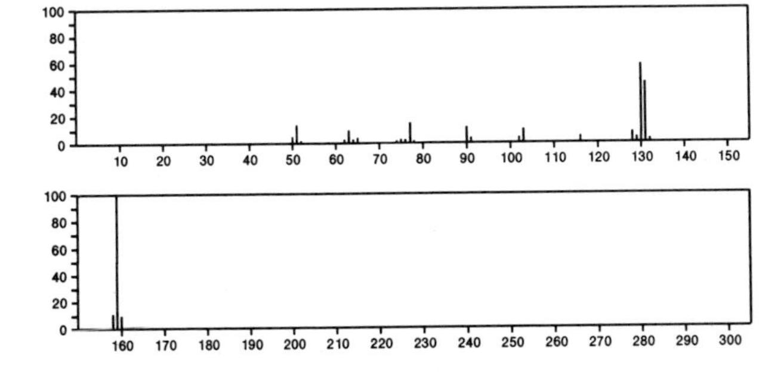

159 $C_{10}H_9NO$ 607–66–9
2(1*H*)–Quinolinone, 4–methyl–

H N O Me

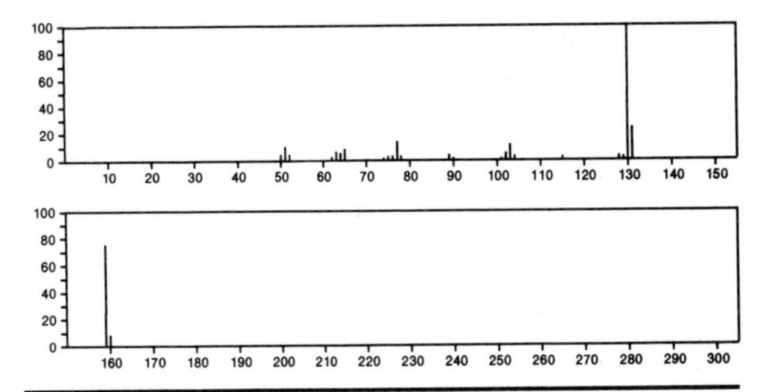

159 $C_{10}H_9NO$ 607–67–0
4–Quinolinol, 2–methyl–

N Me OH

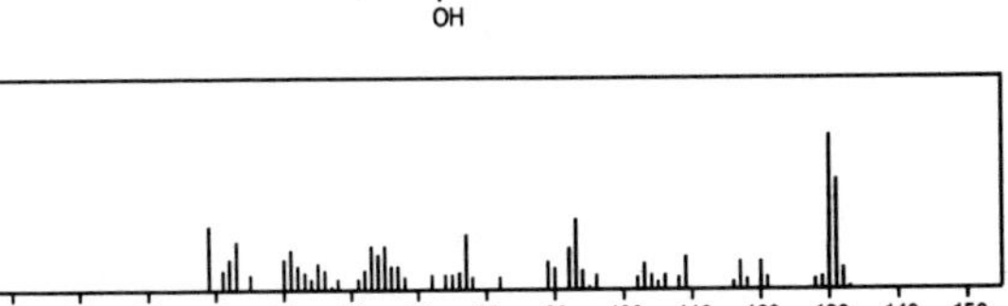

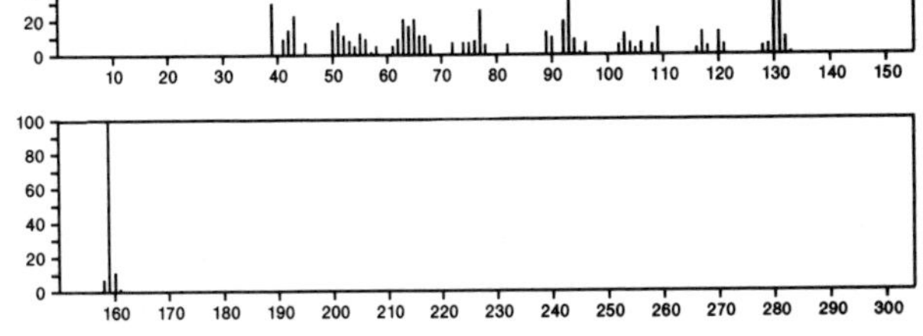

159 $C_{10}H_9NO$ 703–80–0
Ethanone, 1–(1*H*–indol–3–yl)–

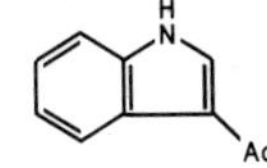

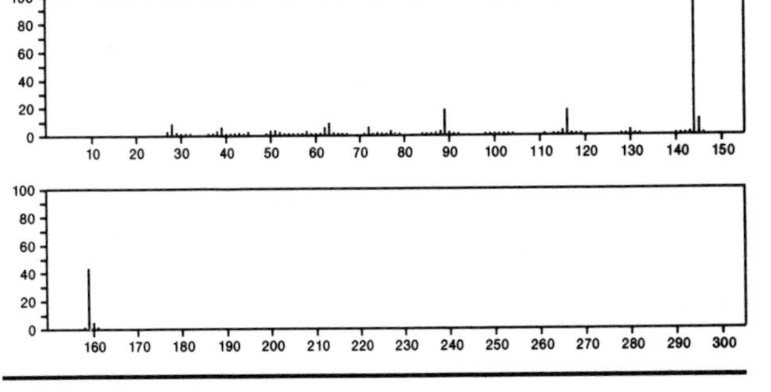

159 $C_{10}H_9NO$ 826–81–3
8–Quinolinol, 2–methyl–

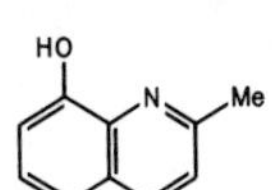

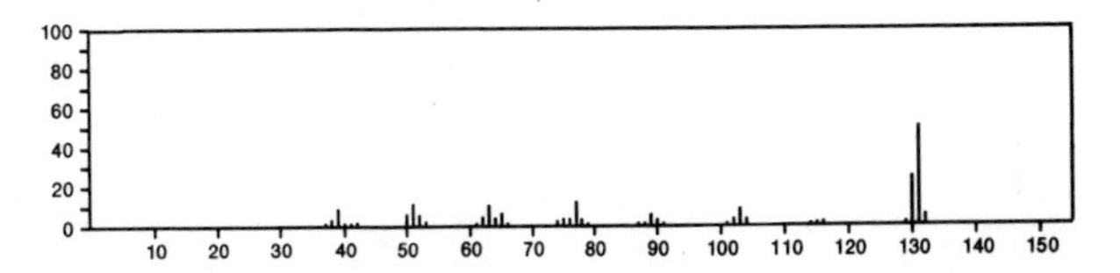

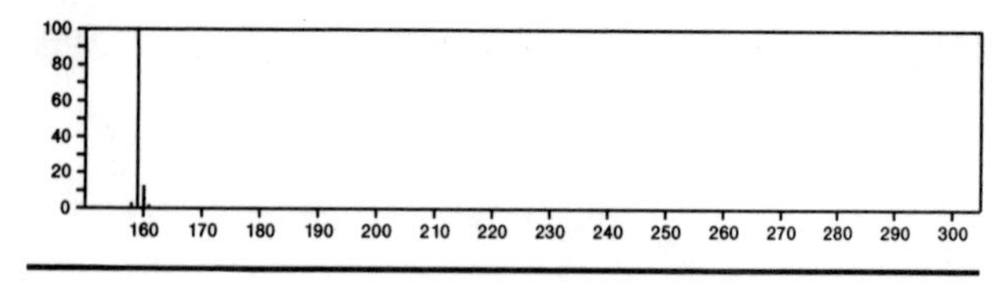

159 $C_{10}H_9NO$ 877-39-4
Oxazole, 4-methyl-2-phenyl-

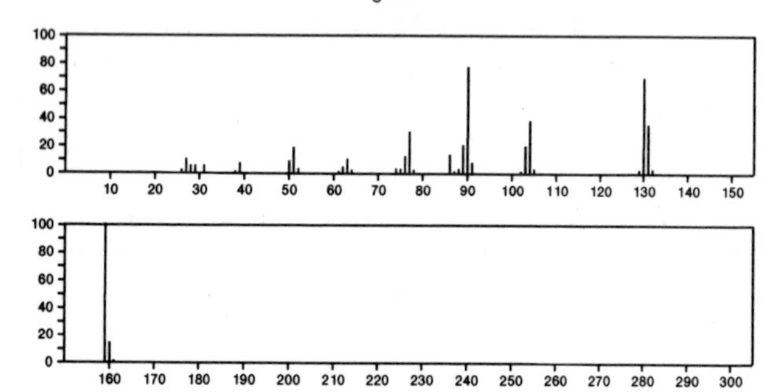

159 $C_{10}H_9NO$ 1076-28-4
Quinoline, 2-methyl-, 1-oxide

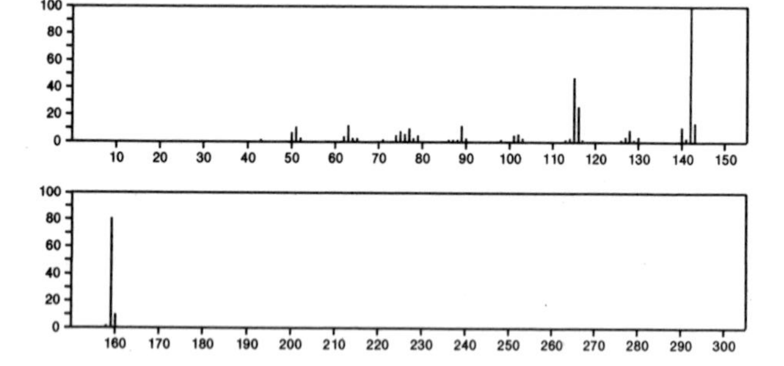

159 $C_{10}H_9NO$ 1780-17-2
2-Quinolinemethanol

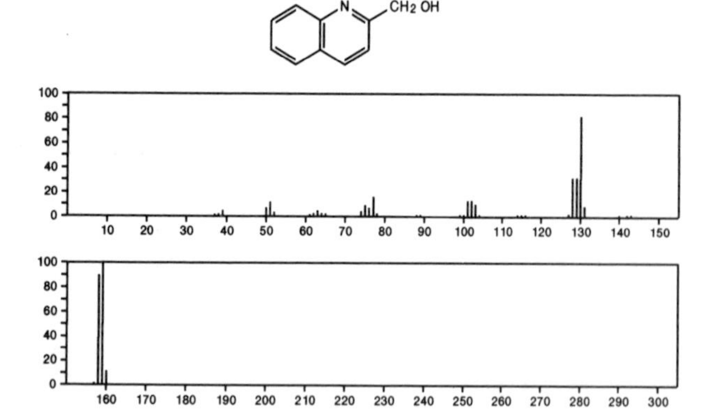

159 $C_{10}H_9NO$ 2721-59-7
2(1*H*)-Quinolinone, 3-methyl-

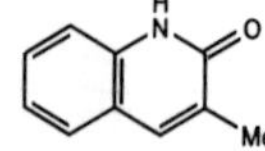

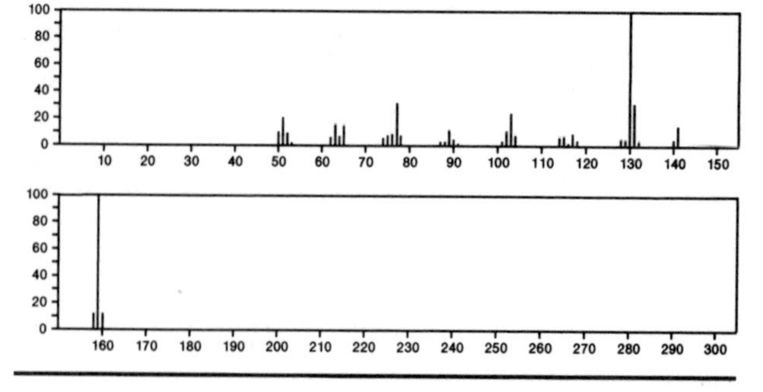

159 $C_{10}H_9NO$ 3222-65-9
Isoquinoline, 1-methyl-, 2-oxide

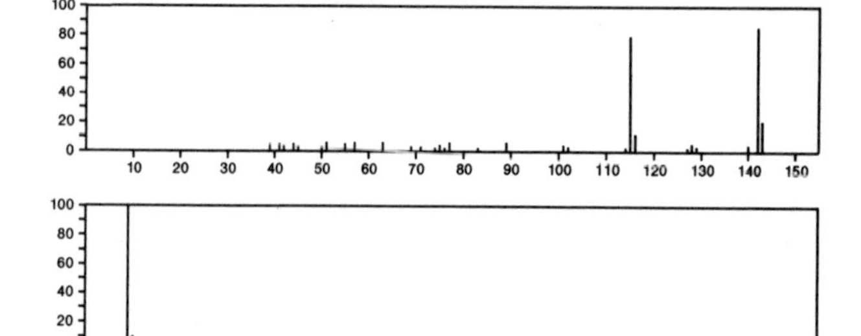

159 $C_{10}H_9NO$ 3846-73-9
8-Quinolinol, 4-methyl-

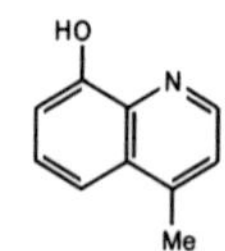

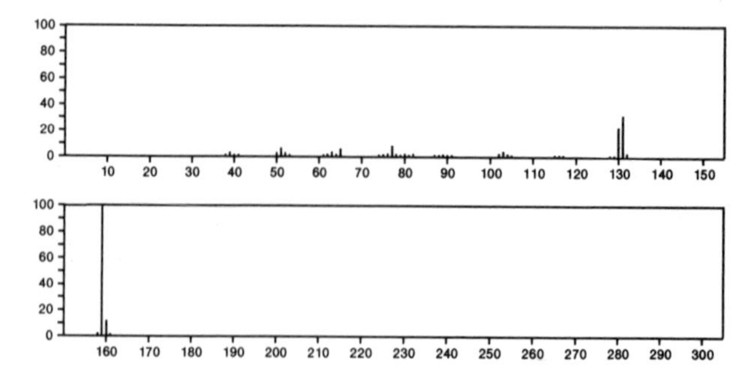

159 $C_{10}H_9NO$ 4053–40–1
Quinoline, 4-methyl-, 1-oxide

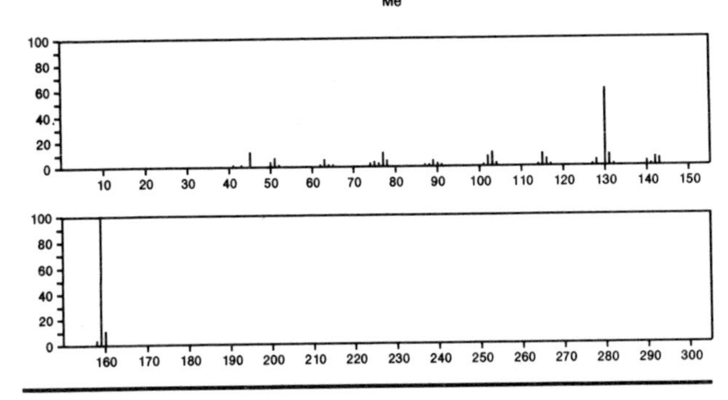

159 $C_{10}H_9NO$ 4053–42–3
Quinoline, 6-methyl-, 1-oxide

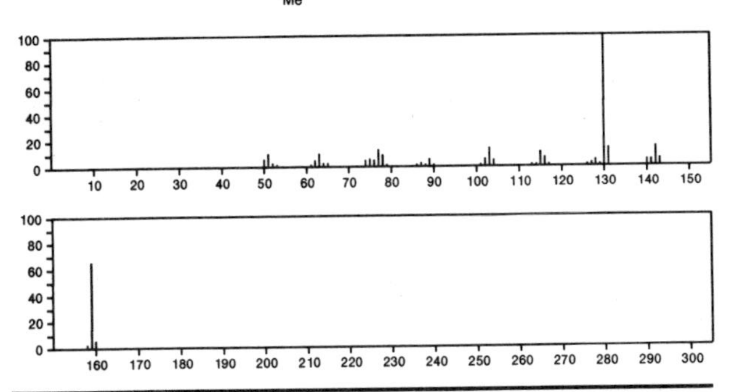

159 $C_{10}H_9NO$ 5221–67–0
Oxazole, 5-methyl-2-phenyl-

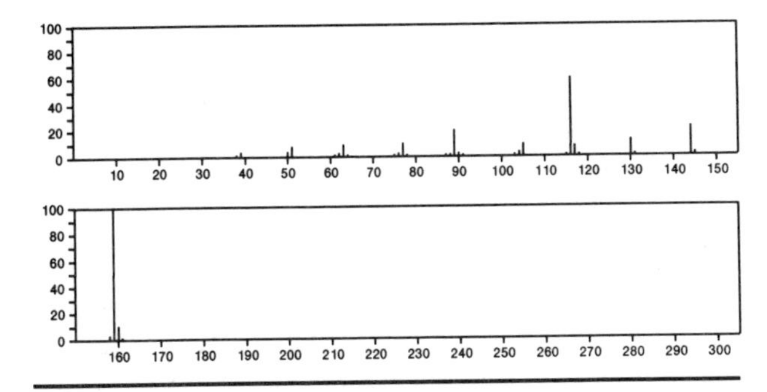

159 $C_{10}H_9NO$ 5541–68–4
8-Quinolinol, 7-methyl-

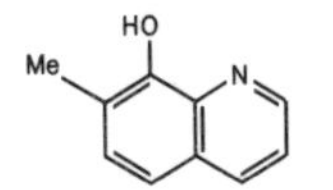

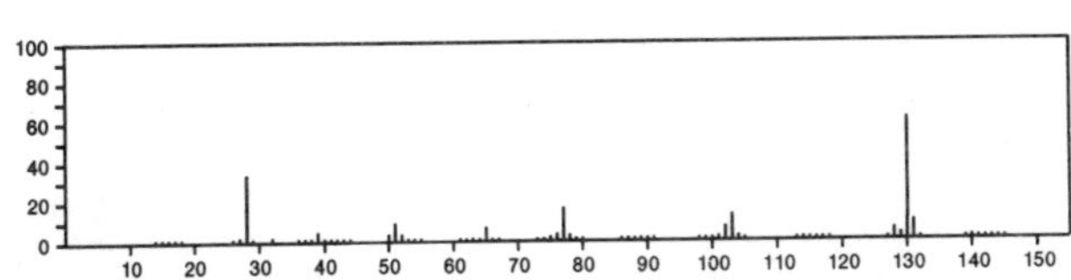

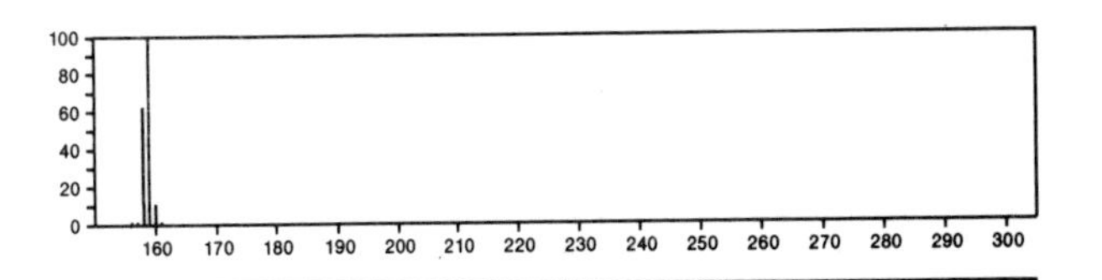

159 $C_{10}H_9NO$ 6931–16–4
Quinoline, 2-methoxy-

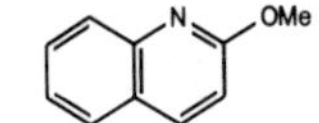

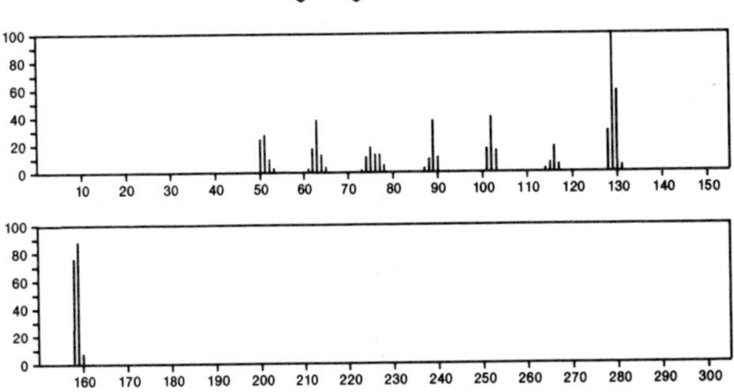

159 $C_{10}H_9NO$ 14548–00–6
Isoquinoline, 3-methyl-, 2-oxide

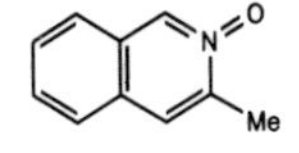

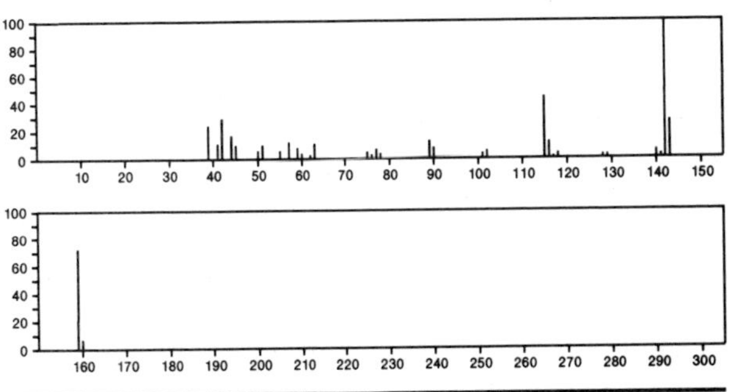

159 $C_{10}H_9NO$ 16032–35–2
8-Quinolinemethanol

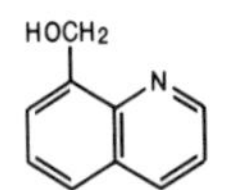

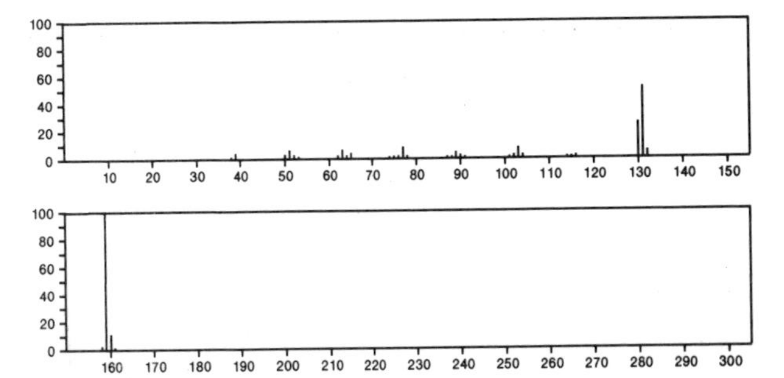

159 $C_{10}H_9NO$ 25314-91-4
Ethanone, 1-(3-indolizinyl)-

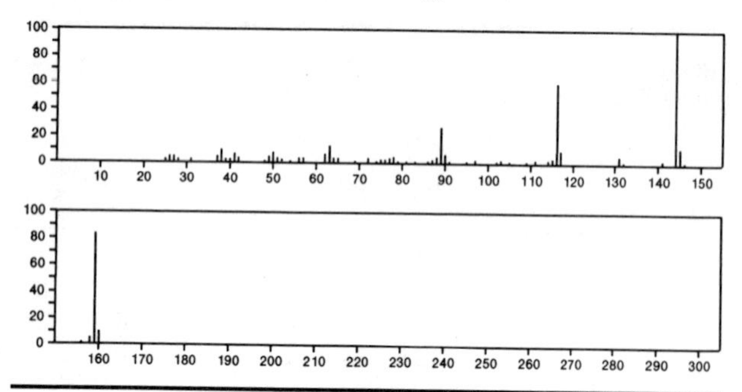

159 $C_{10}H_9NO$ 55887-59-7
Benzene, (1-isocyanato-2-propenyl)-

PhCH(NCO) CH=CH2

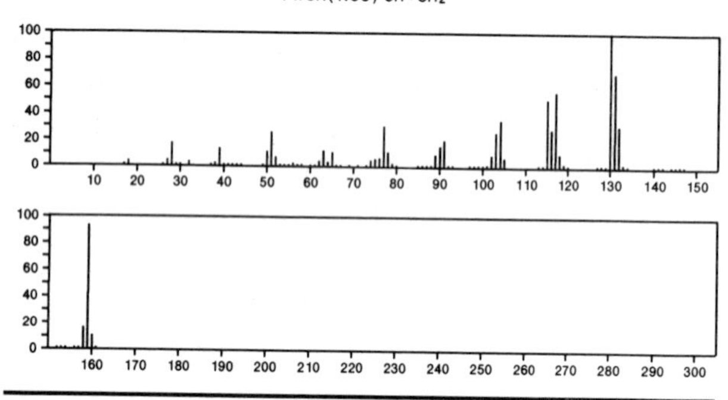

159 $C_{11}H_{13}N$ 555-57-7
Benzenemethanamine, *N*-methyl-*N*-2-propynyl-

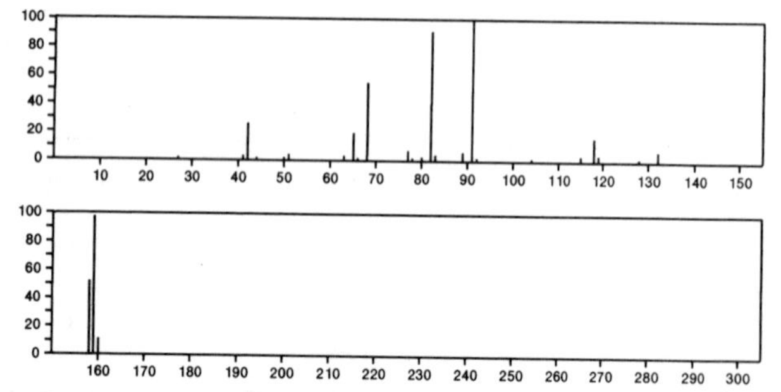

159 $C_{11}H_{13}N$ 1640-39-7
3*H*-Indole, 2,3,3-trimethyl-

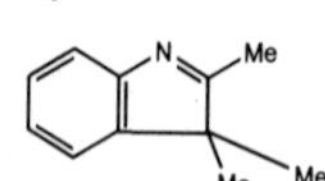

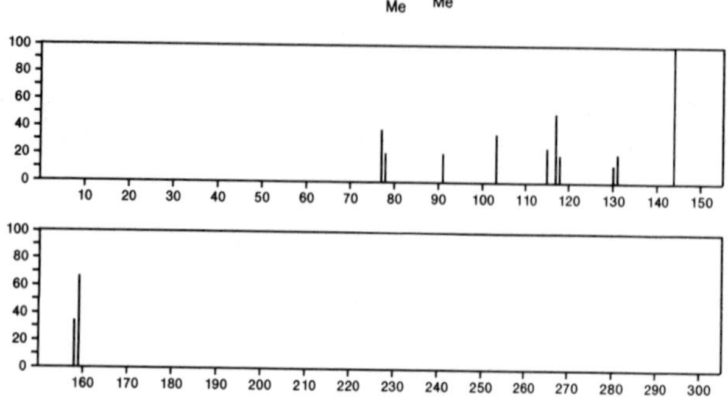

159 $C_{11}H_{13}N$ 1971-46-6
1*H*-Indole, 1,2,3-trimethyl-

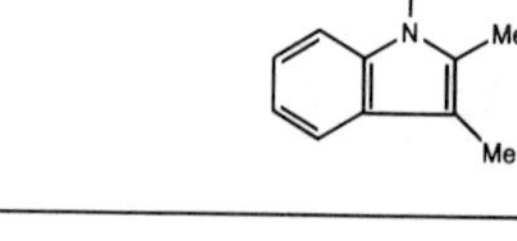

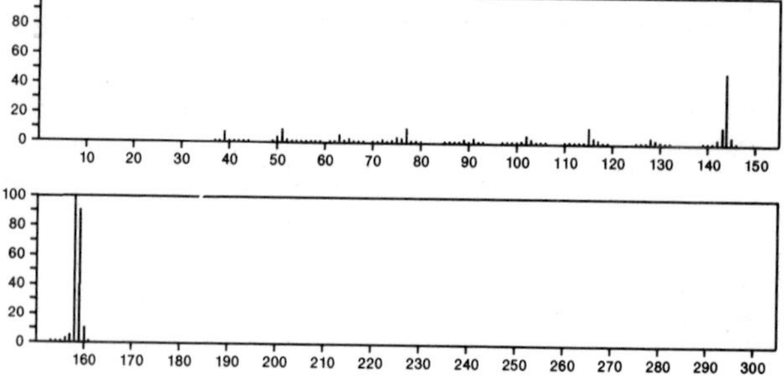

159 $C_{11}H_{13}N$ 4363-25-1
2*H*-1,4-Ethanoquinoline, 3,4-dihydro-

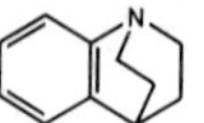

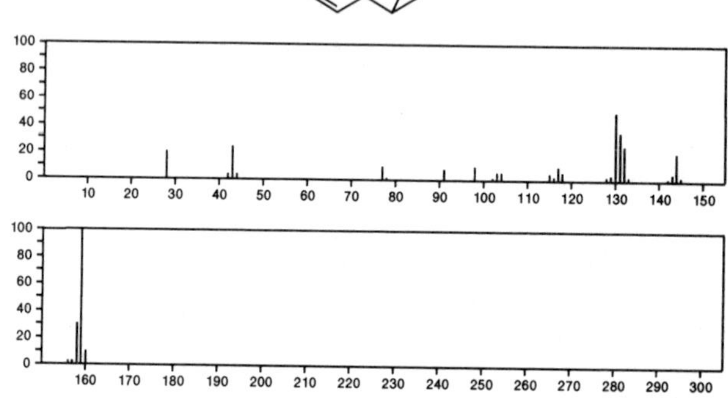

159 $C_{11}H_{13}N$ 54340-99-7
1*H*-Indole, 5,6,7-trimethyl-

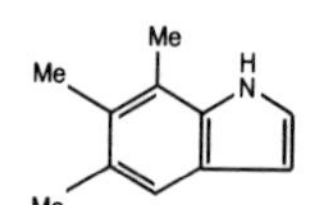

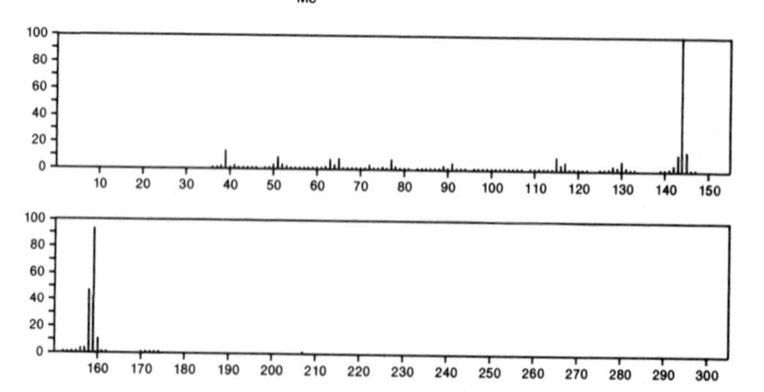

160 $C_2H_6ClO_2PS$ 2524-03-0
Phosphorochloridothioic acid, *O,O*-dimethyl ester

S
‖
Me OPCl OMe

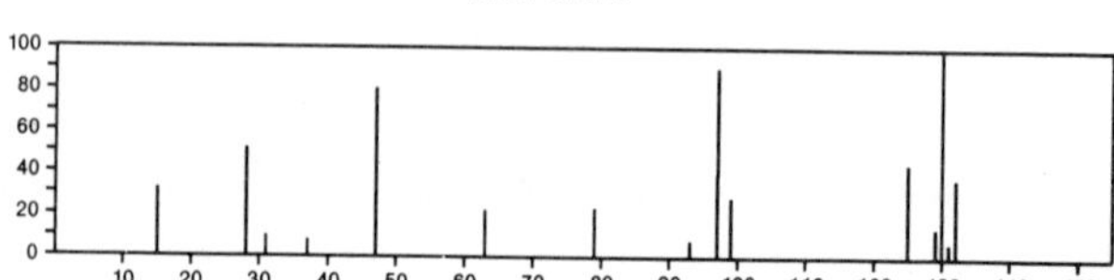

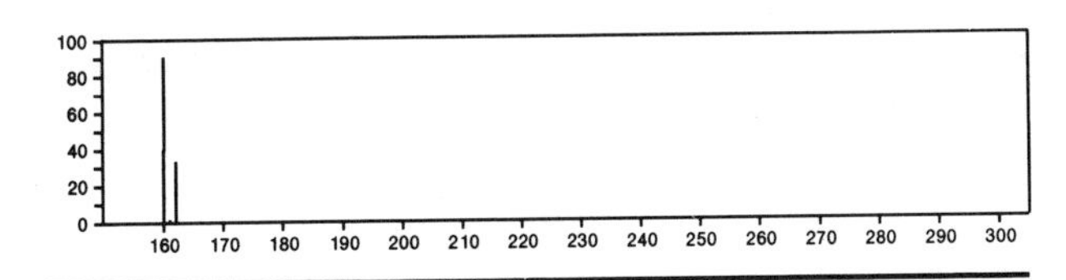

160 $C_2H_7F_2N_2PS$ 36267-50-2
Phosphorodifluoridothioic hydrazide, 2,2-dimethyl-

S=PF2NHNMe2

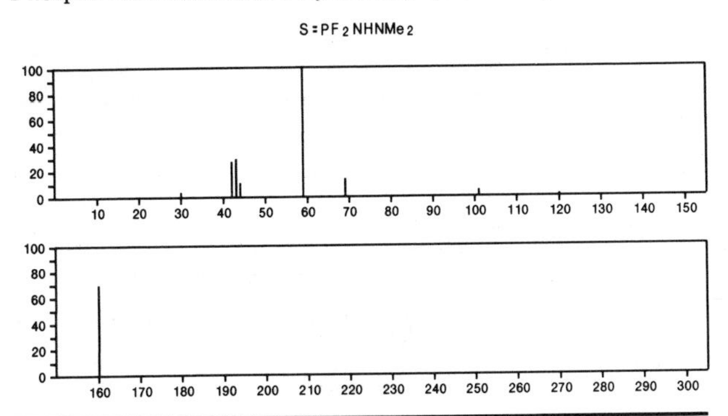

160 $C_3H_3Cl_3O$ 918-00-3
2-Propanone, 1,1,1-trichloro-

Cl3CCOMe

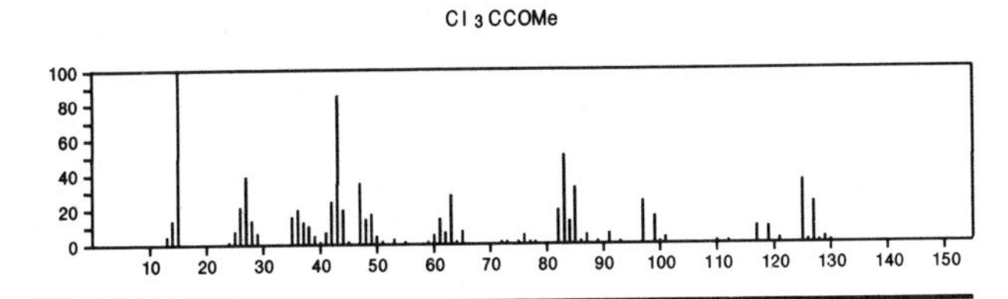

160 $C_3H_3Cl_3O$ 26073-26-7
Propanoyl chloride, 2,2-dichloro-

MeCCl2COCl

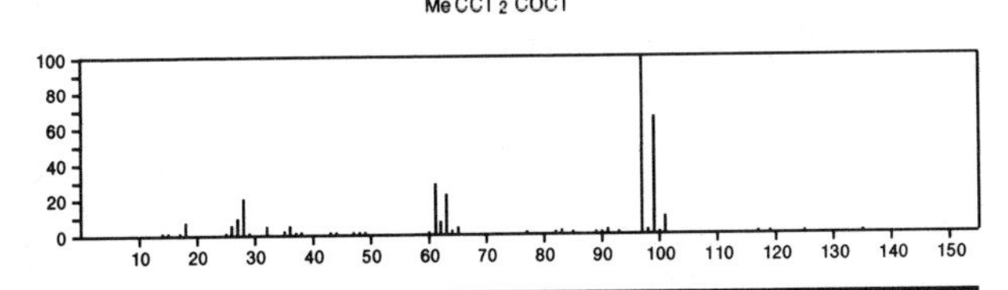

160 $C_5H_4S_3$ 252-09-5
[1,2]Dithiolo[1,5-*b*][1,2]dithiole-7-S^{IV}

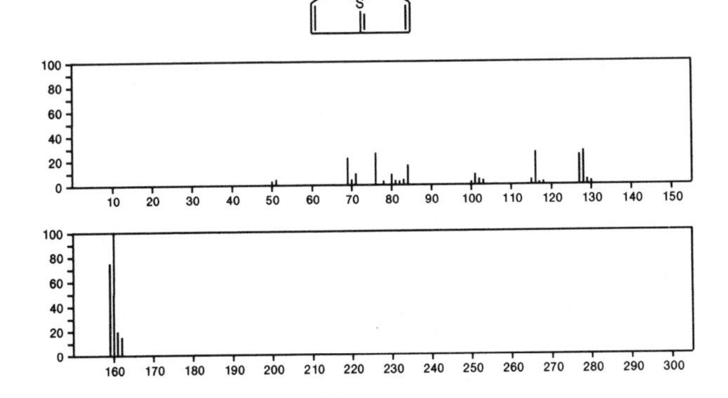

160 $C_5H_5ClN_2S$ 7145-61-1
Pyridazine, 3-chloro-6-(methylthio)-

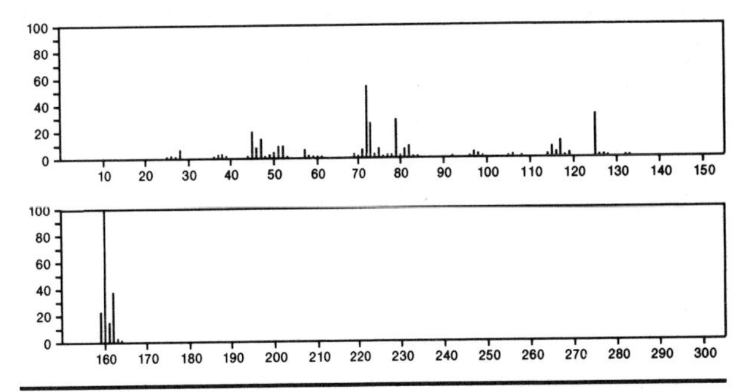

160 $C_5H_8N_2O_4$ 2943-56-8
2,4(1*H*,3*H*)-Pyrimidinedione, dihydro-5,6-dihydroxy-5-methyl-

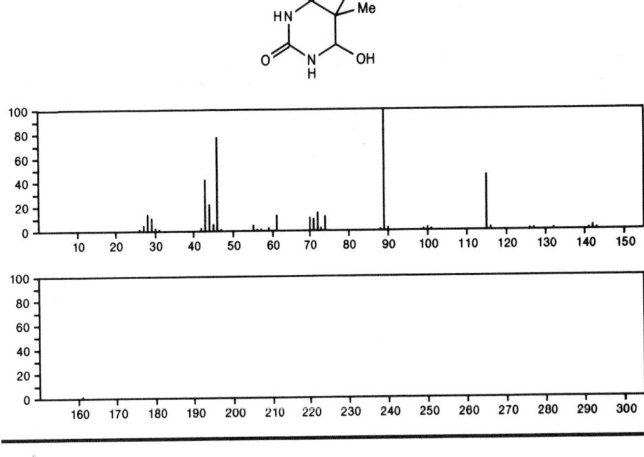

160 $C_5H_8N_2O_4$ 30310-80-6
L-Proline, 4-hydroxy-1-nitroso-, *trans*-

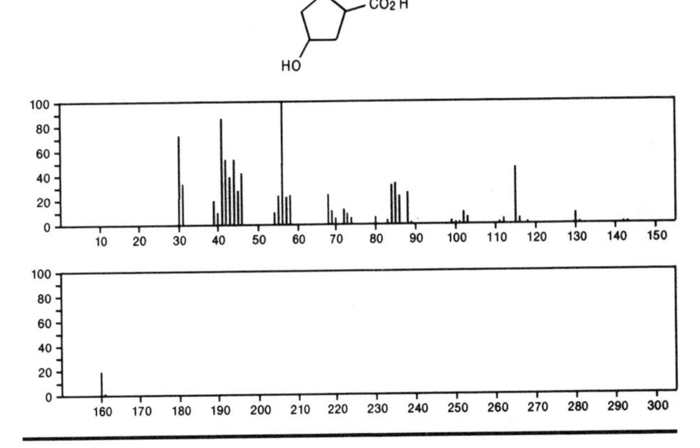

160 $C_5H_8N_2S_2$ 56701-32-7
1,2,3-Thiadiazolium, 4-ethyl-5-mercapto-3-methyl-, hydroxide, inner salt

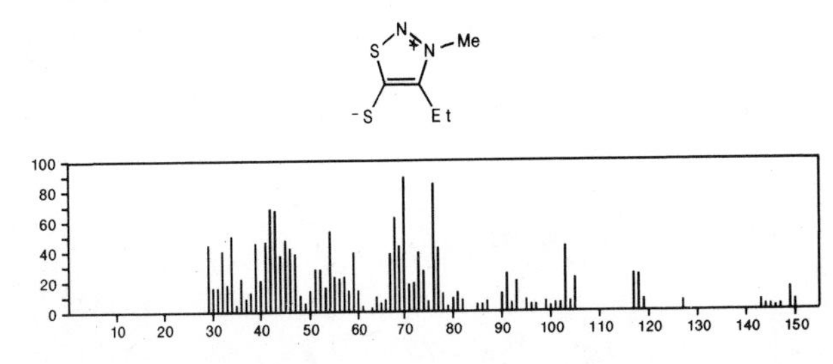

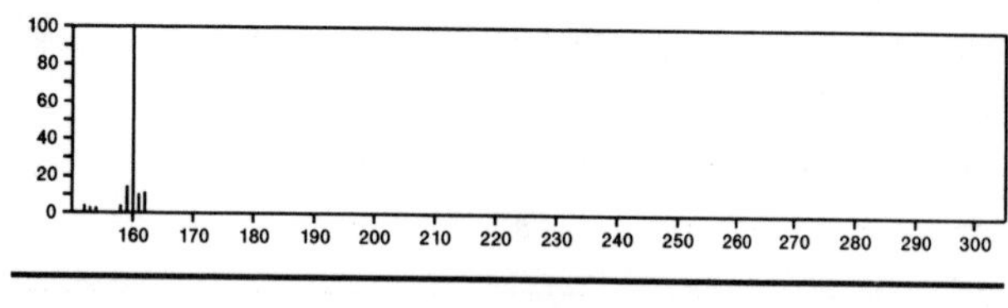

160 $C_5H_{16}Si_3$ 54424-15-6
1,3,5-Trisilacyclohexane, 1,1-dimethyl-

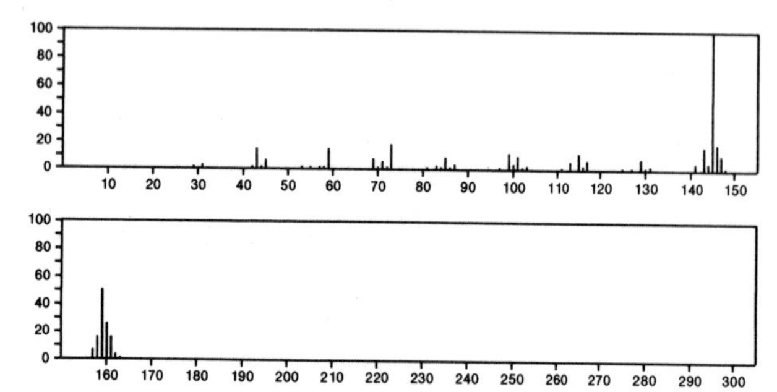

160 $C_6H_5ClO_3$ 7559-81-1
4*H*-Pyran-4-one, 2-(chloromethyl)-5-hydroxy-

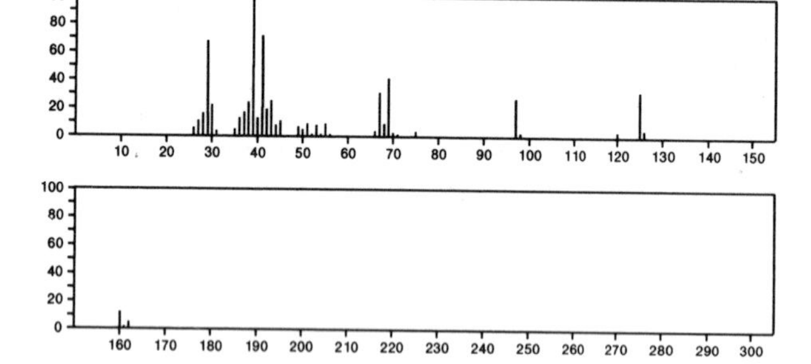

160 $C_6H_5FO_4$ 10318-01-1
2,4-Hexadienedioic acid, 2-fluoro-

$HO_2CCF=CHCH=CHCO_2H$

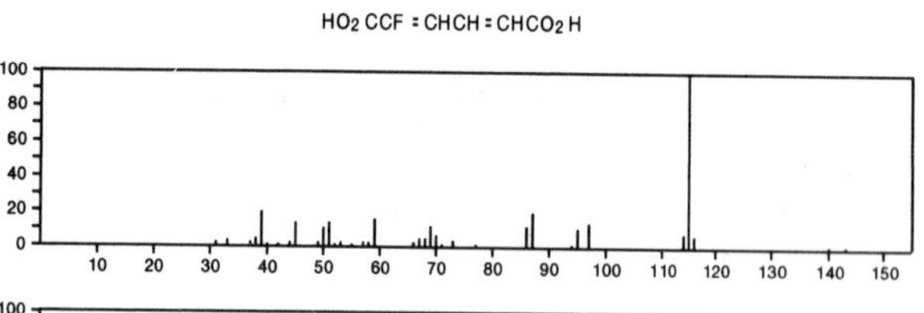

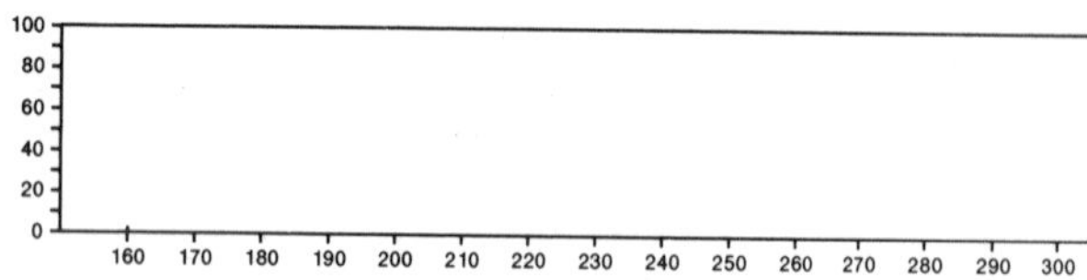

160 C_6H_9Br 1521-51-3
Cyclohexene, 3-bromo-

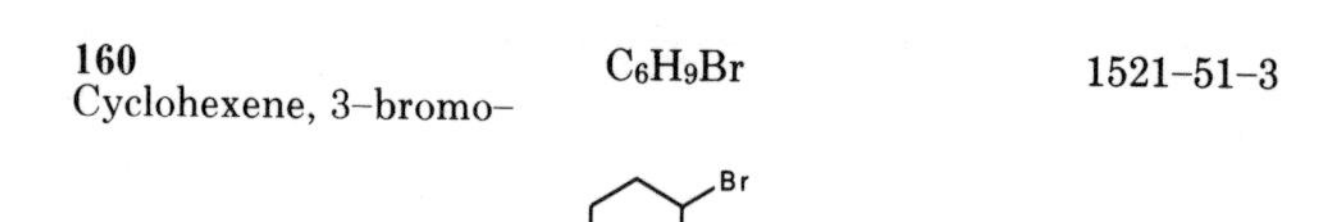

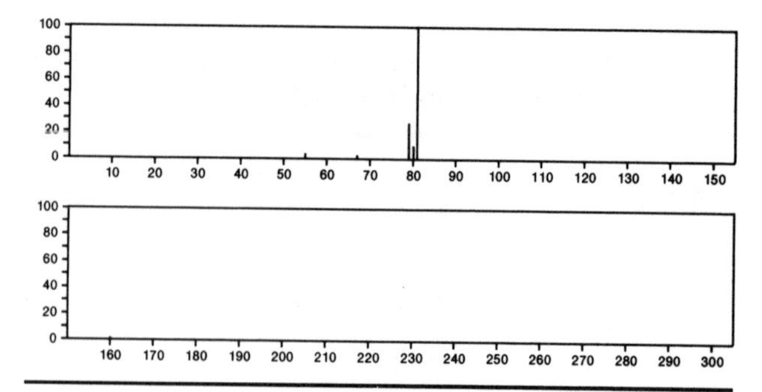

160 C_6H_9Br 17645-61-3
Cyclopentene, 3-(bromomethyl)-

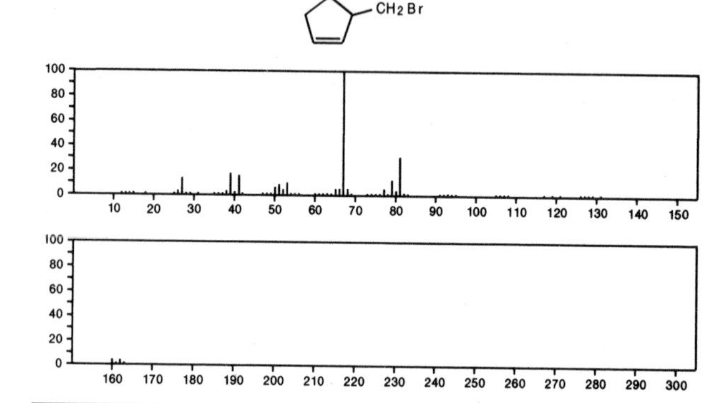

160 C_6H_9Br 55402-12-5
2-Hexyne, 6-bromo-

$Br(CH_2)_3C\equiv CMe$

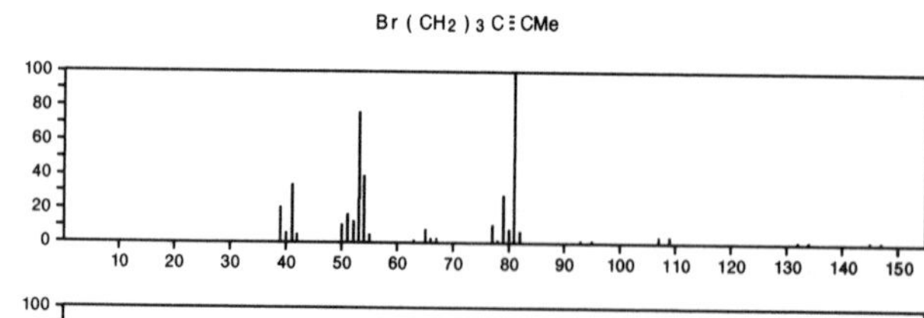

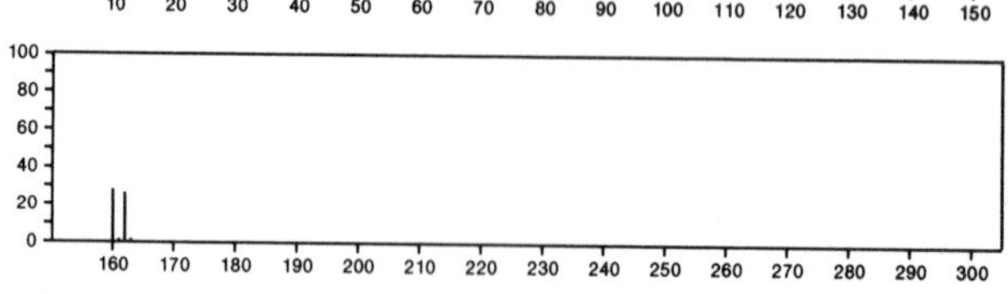

160 $C_6H_{12}N_2O_3$ 1948-31-8
L-Alanine, *N*-L-alanyl-

$MeCH(NH_2)CONHCHMeCO_2H$

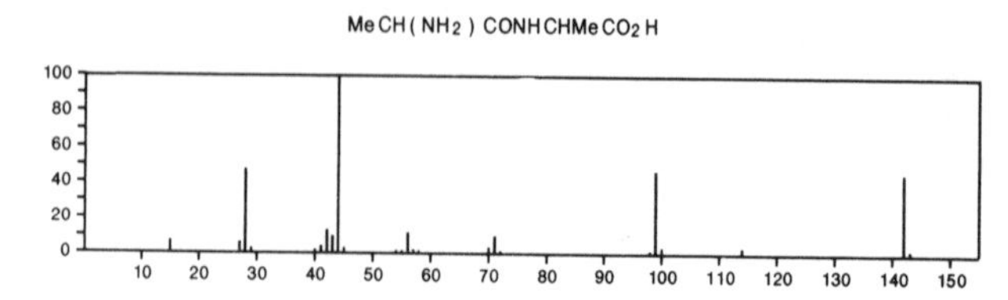

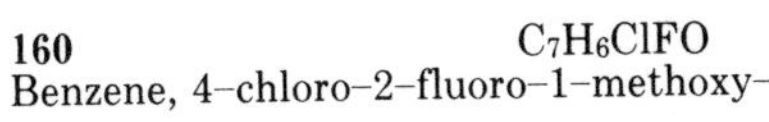

160 C_7H_6ClFO 452–09–5
Benzene, 4–chloro–2–fluoro–1–methoxy–

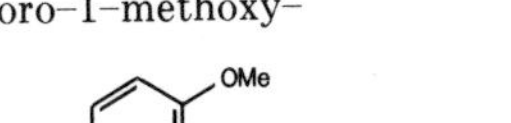

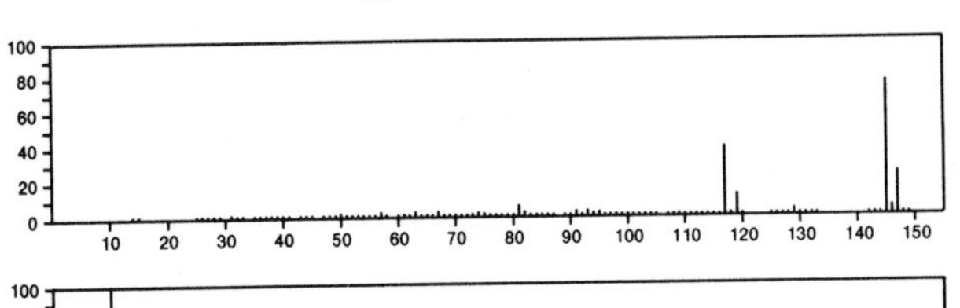

160 $C_7H_6Cl_2$ 95–73–8
Benzene, 2,4–dichloro–1–methyl–

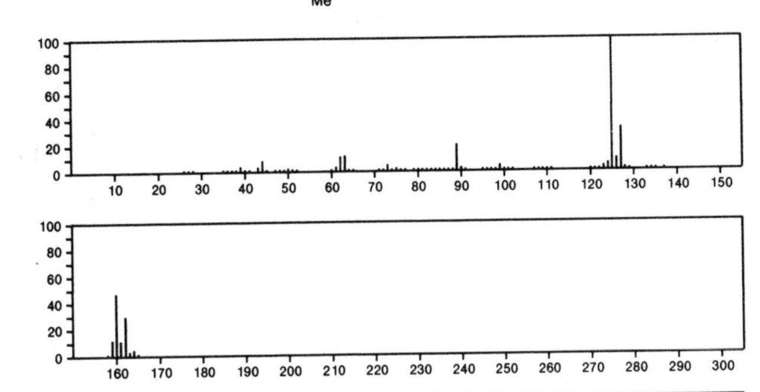

160 $C_7H_6Cl_2$ 95–75–0
Benzene, 1,2–dichloro–4–methyl–

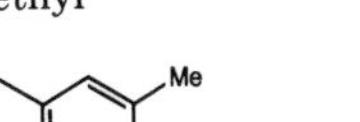

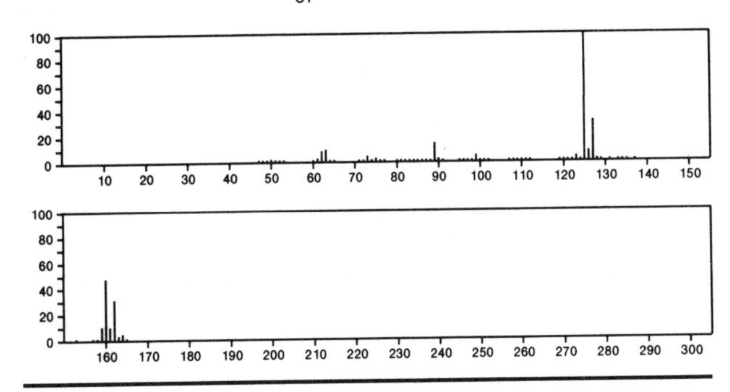

160 $C_7H_6Cl_2$ 98–87–3
Benzene, (dichloromethyl)–

Cl_2CHPh

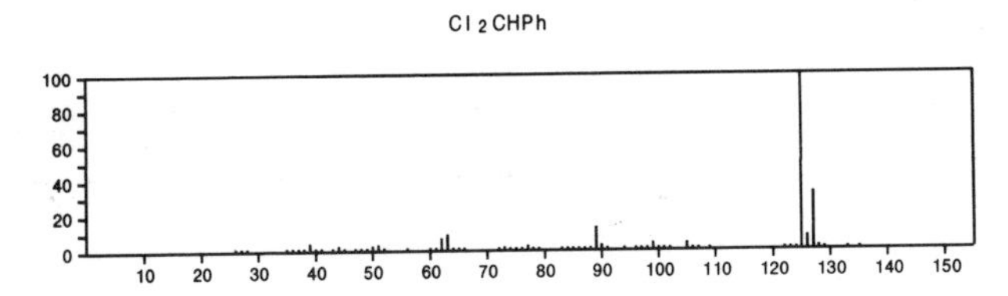

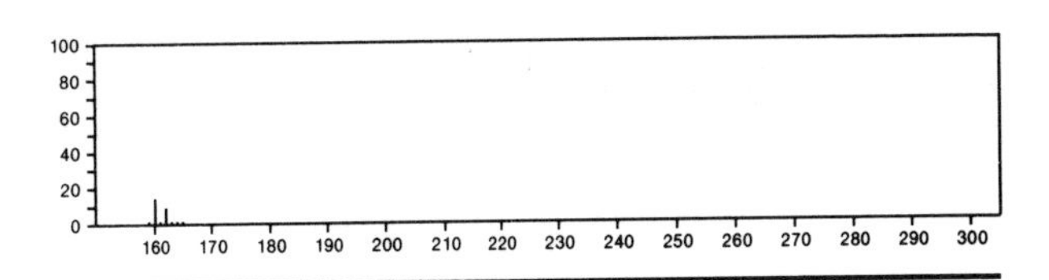

160 $C_7H_6Cl_2$ 118–69–4
Benzene, 1,3–dichloro–2–methyl–

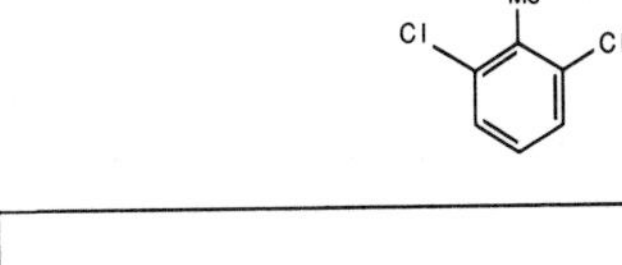

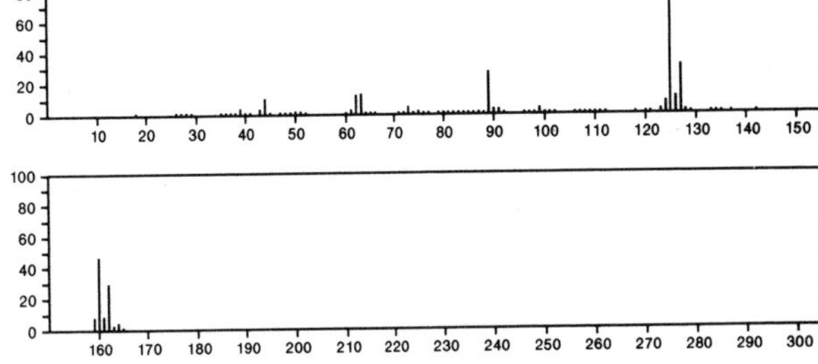

160 $C_7H_6Cl_2$ 611–19–8
Benzene, 1–chloro–2–(chloromethyl)–

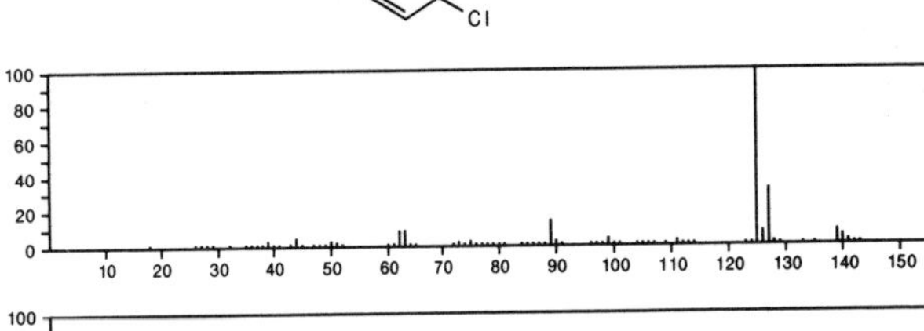

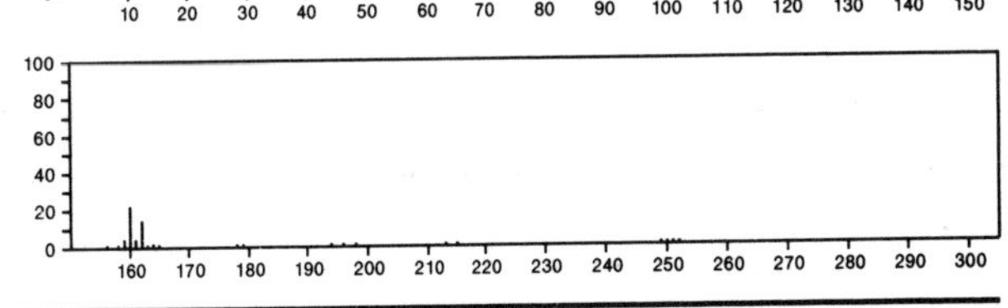

160 $C_7H_6Cl_2$ 620–20–2
Benzene, 1–chloro–3–(chloromethyl)–

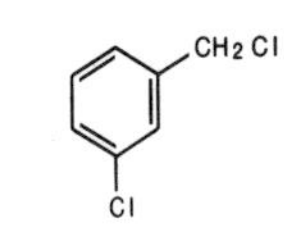

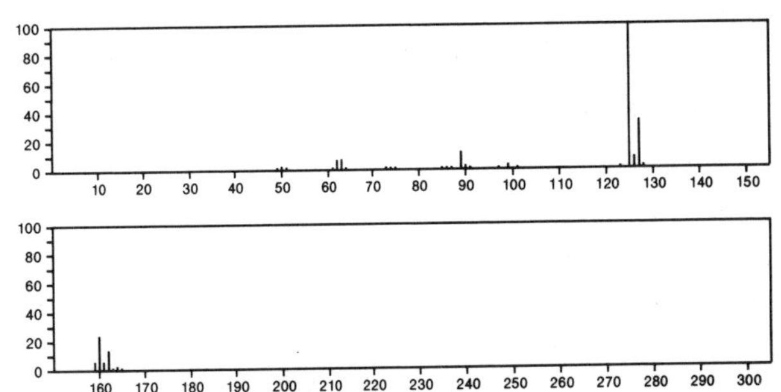

160 $C_7H_6Cl_2$ 19398-61-9
Benzene, 1,4-dichloro-2-methyl-

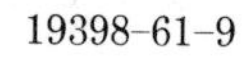

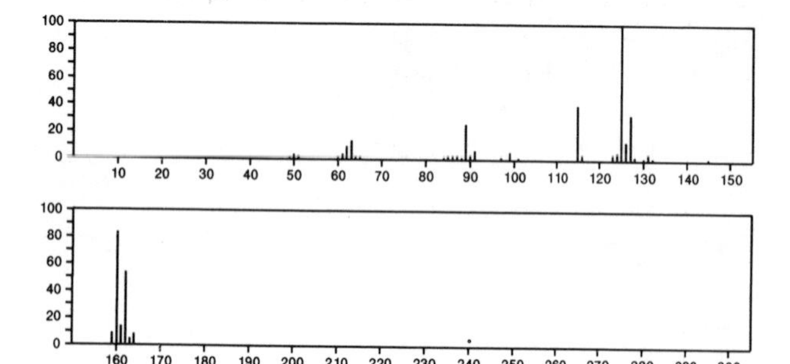

160 $C_7H_6Cl_2$ 29797-40-8
Benzene, dichloromethyl-

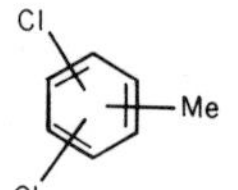

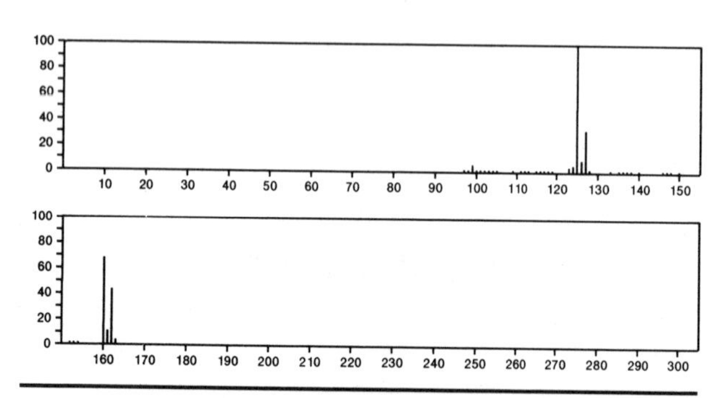

160 $C_7H_9BN_4$ 6982-52-1
Δ2-Tetrazaboroline, 1-methyl-4-phenyl-

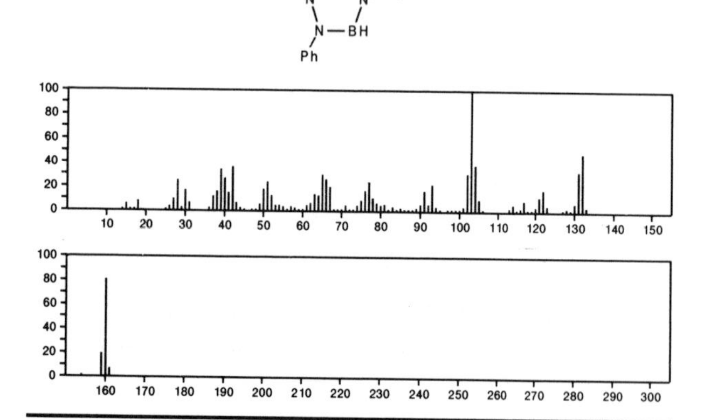

160 $C_7H_{12}O_2S$ 23246-23-3
Propionic acid, 3-(allylthio)-, methyl ester

MeOC(O)CH2CH2SCH2CH=CH2

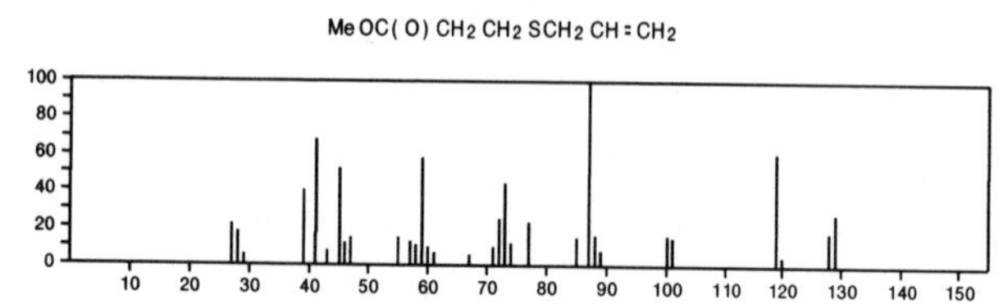

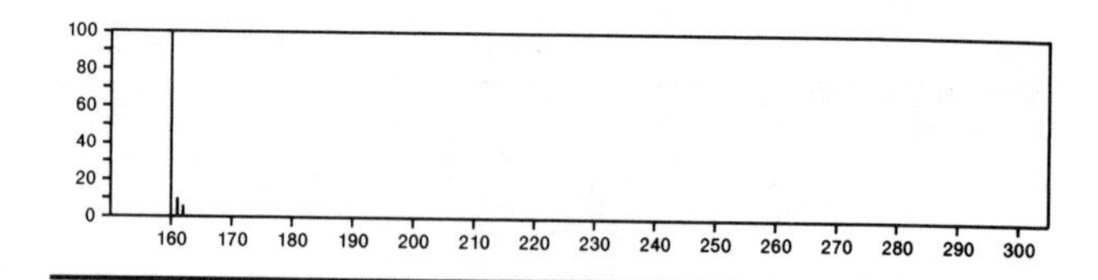

160 $C_7H_{12}O_2S$ 35562-79-9
1,4-Oxathiepan-2-one, 3,3-dimethyl-

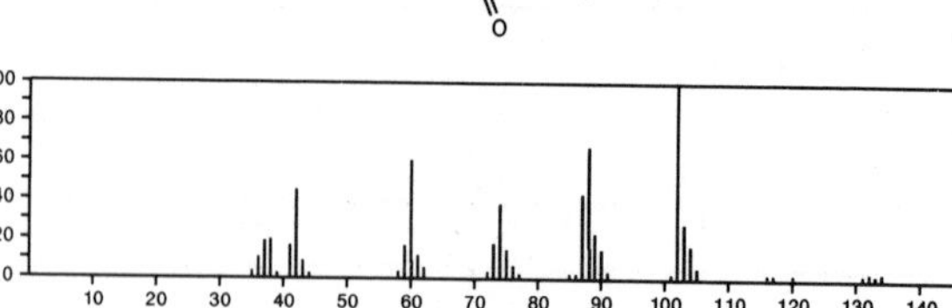

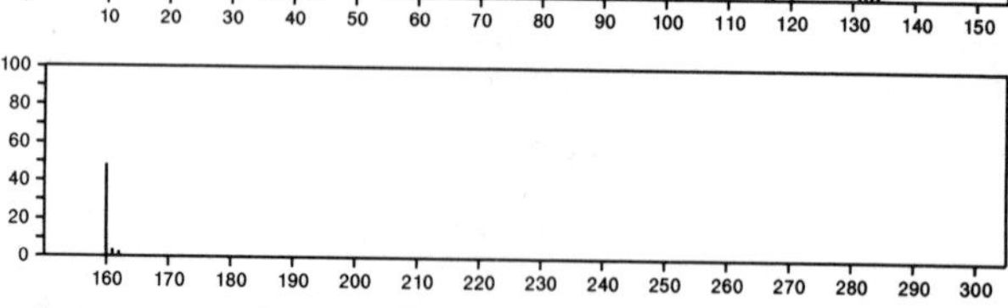

160 $C_7H_{12}O_4$ 105-53-3
Propanedioic acid, diethyl ester

EtOC(O)CH2C(O)OEt

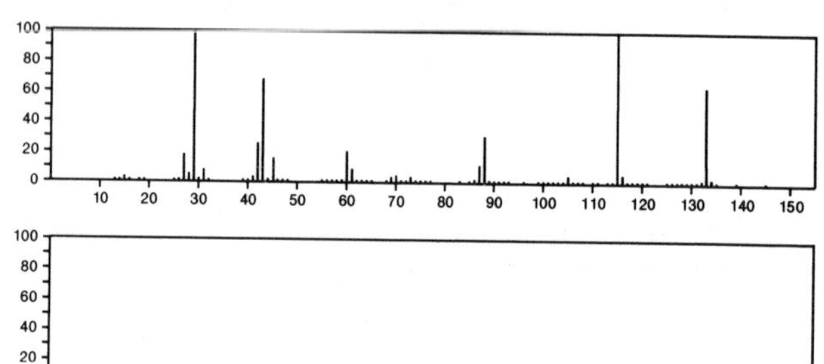

160 $C_7H_{12}O_4$ 111-16-0
Heptanedioic acid

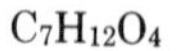

HO2C(CH2)5CO2H

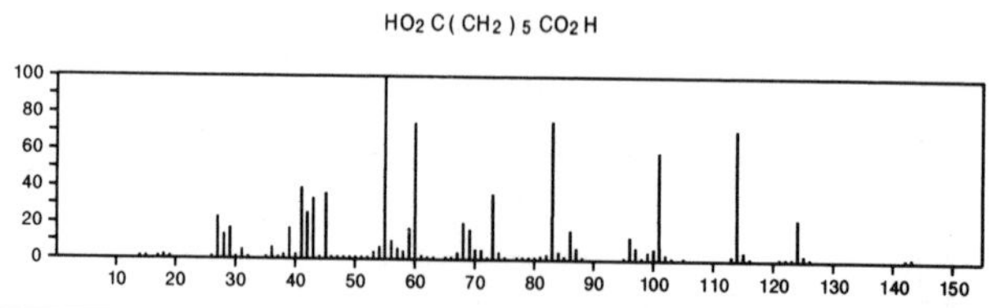

160 $C_7H_{12}O_4$ 628-66-0
1,3-Propanediol, diacetate

AcO(CH2)3OAc

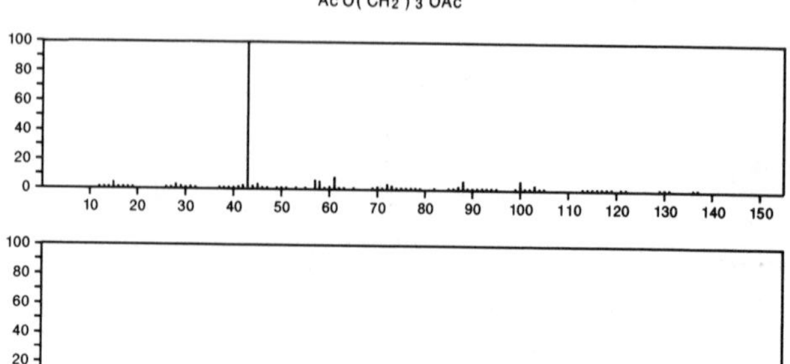

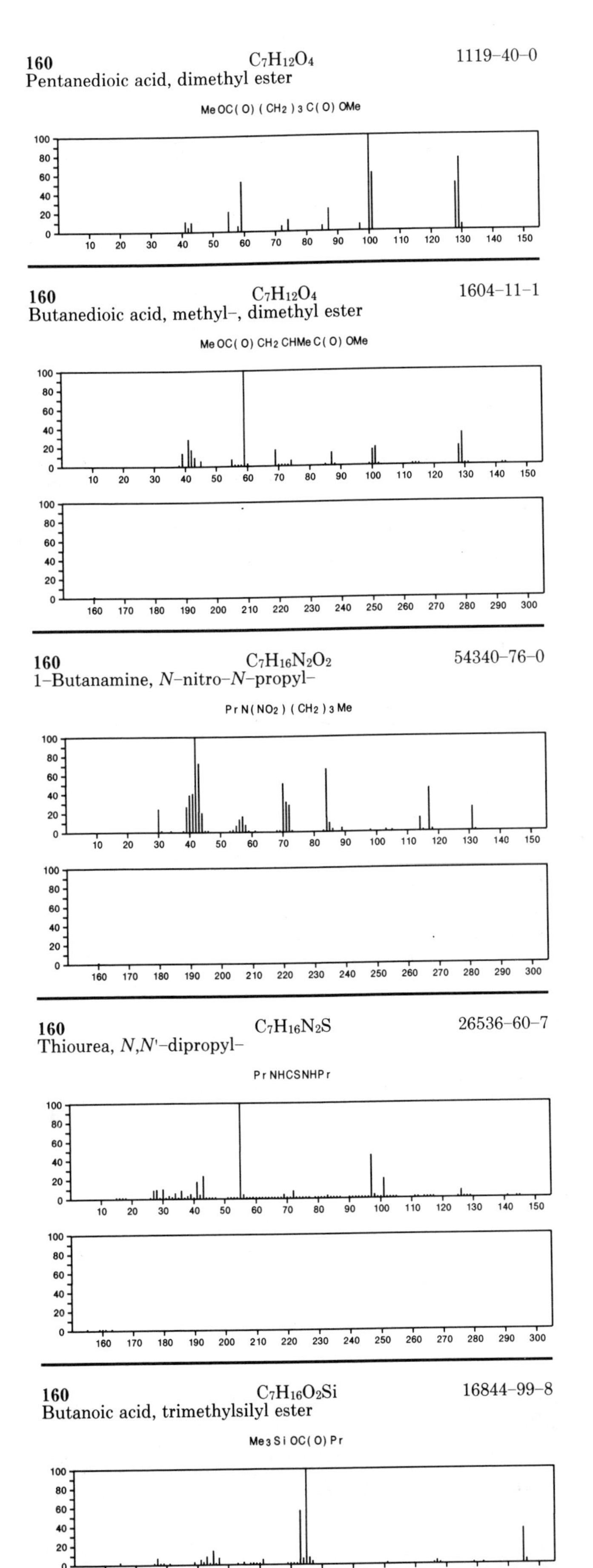

160 $C_7H_{12}O_4$ 1119-40-0
Pentanedioic acid, dimethyl ester
MeOC(O)(CH2)3C(O)OMe

160 $C_7H_{12}O_4$ 1604-11-1
Butanedioic acid, methyl-, dimethyl ester
MeOC(O)CH2CHMeC(O)OMe

160 $C_7H_{16}N_2O_2$ 54340-76-0
1-Butanamine, *N*-nitro-*N*-propyl-
PrN(NO2)(CH2)3Me

160 $C_7H_{16}N_2S$ 26536-60-7
Thiourea, *N*,*N*'-dipropyl-
PrNHCSNHPr

160 $C_7H_{16}O_2Si$ 16844-99-8
Butanoic acid, trimethylsilyl ester
Me3SiOC(O)Pr

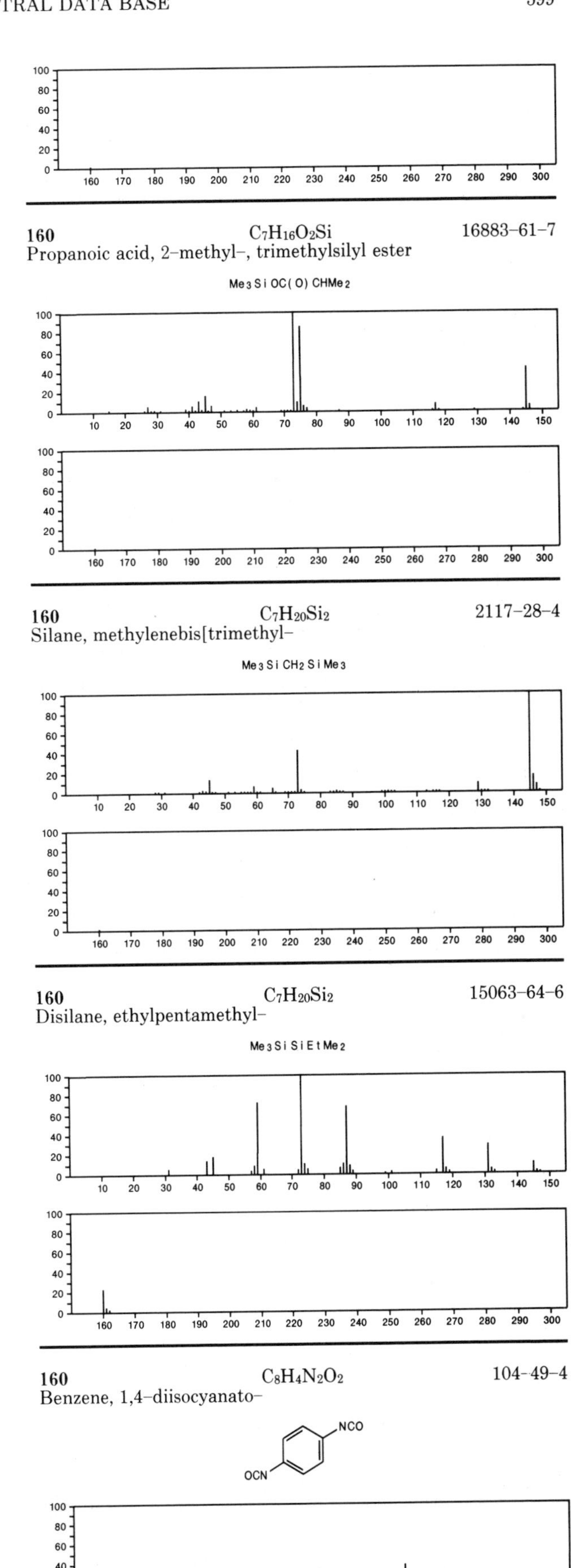

160 $C_7H_{16}O_2Si$ 16883-61-7
Propanoic acid, 2-methyl-, trimethylsilyl ester
Me3SiOC(O)CHMe2

160 $C_7H_{20}Si_2$ 2117-28-4
Silane, methylenebis[trimethyl-
Me3SiCH2SiMe3

160 $C_7H_{20}Si_2$ 15063-64-6
Disilane, ethylpentamethyl-
Me3SiSiEtMe2

160 $C_8H_4N_2O_2$ 104-49-4
Benzene, 1,4-diisocyanato-

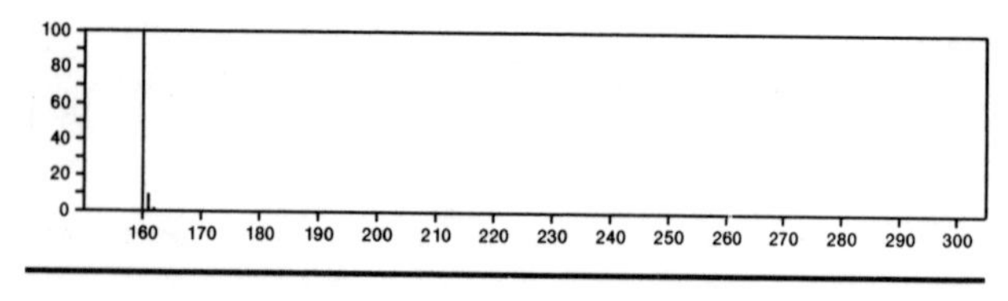

160 $C_8H_7F_3$ 401-79-6
m-Xylene, α,α,α-trifluoro-

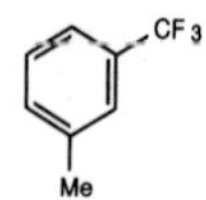

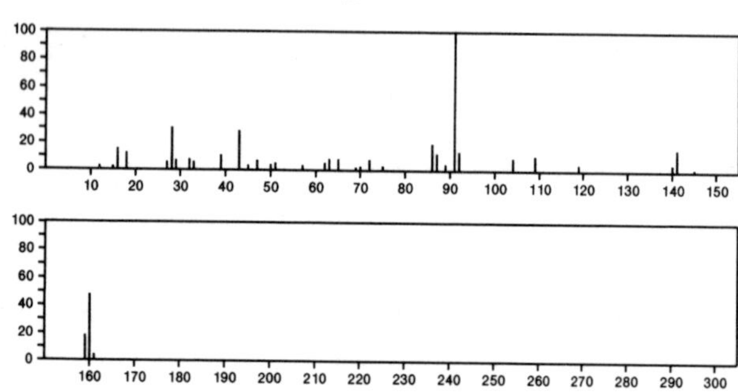

160 $C_8H_7F_3$ 6140-17-6
Benzene, 1-methyl-4-(trifluoromethyl)-

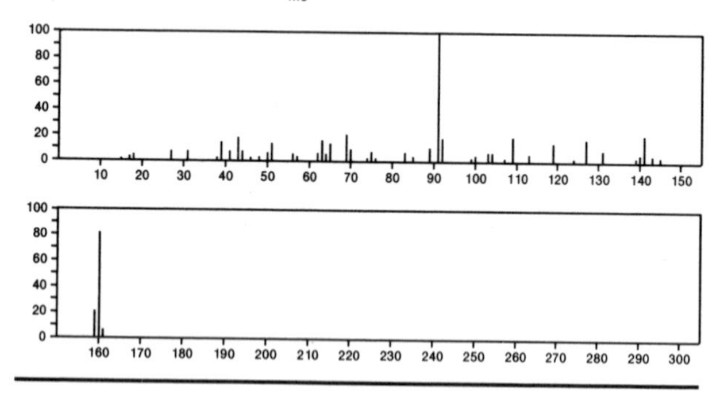

160 $C_8H_8N_4$ 704-61-0
Pteridine, 6,7-dimethyl-

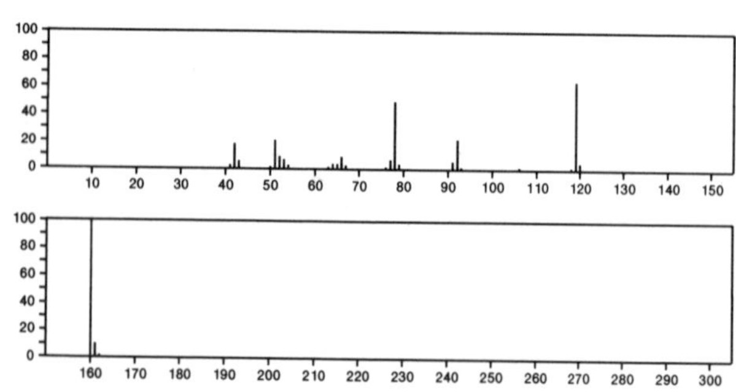

160 $C_8H_8N_4$ 17987-70-1
1,4-Phthalazinediamine

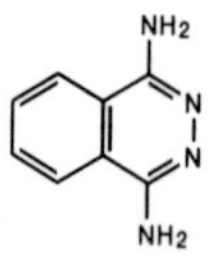

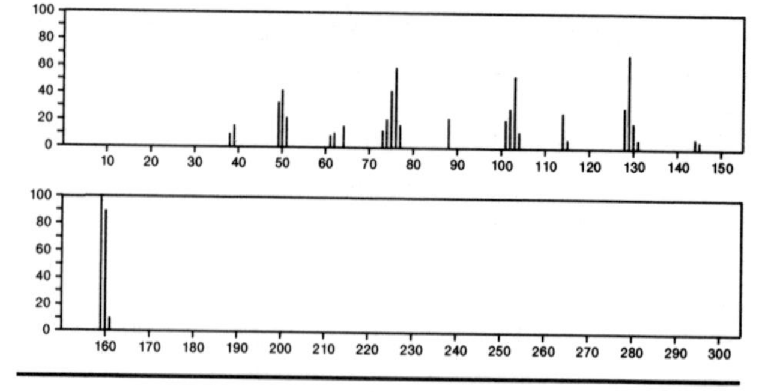

160 $C_8H_{16}OS$ 1927-57-7
2*H*-Pyran, tetrahydro-2-[(1-methylethyl)thio]-

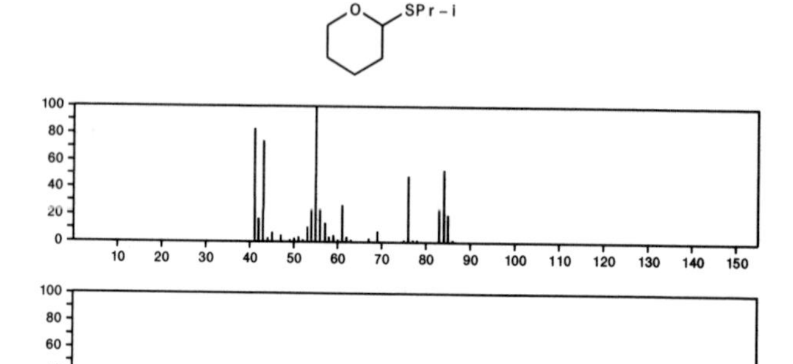

160 $C_8H_{16}OS$ 2307-12-2
Acetic acid, thio-, *S*-hexyl ester

Me(CH2)5SAc

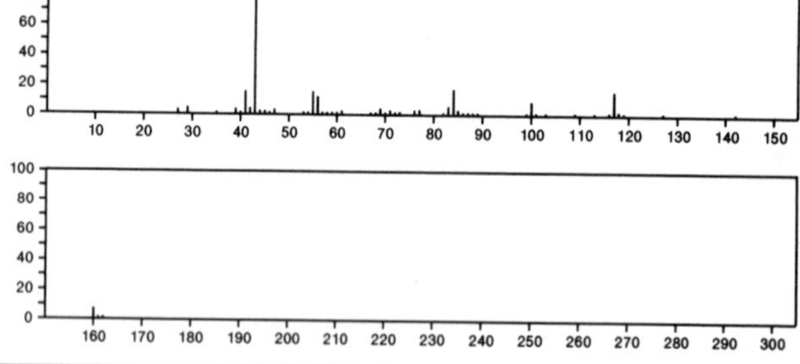

160 $C_8H_{16}OS$ 2432-49-7
Propionic acid, thio-, *S*-isopentyl ester

Me2CHCH2CH2SC(O)Et

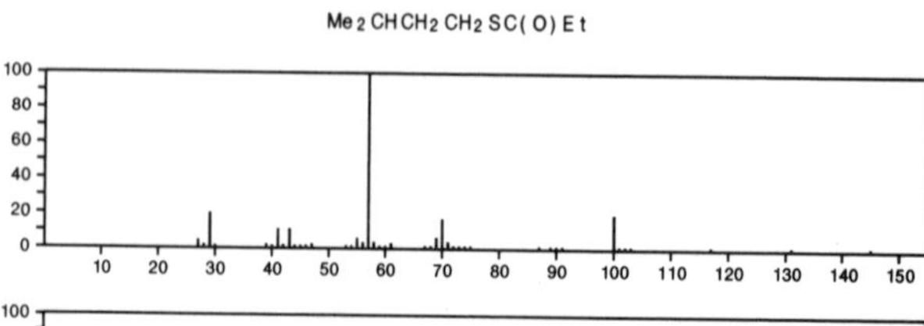

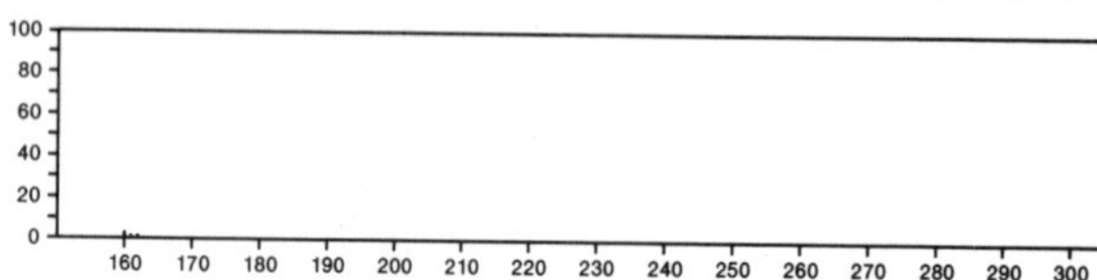

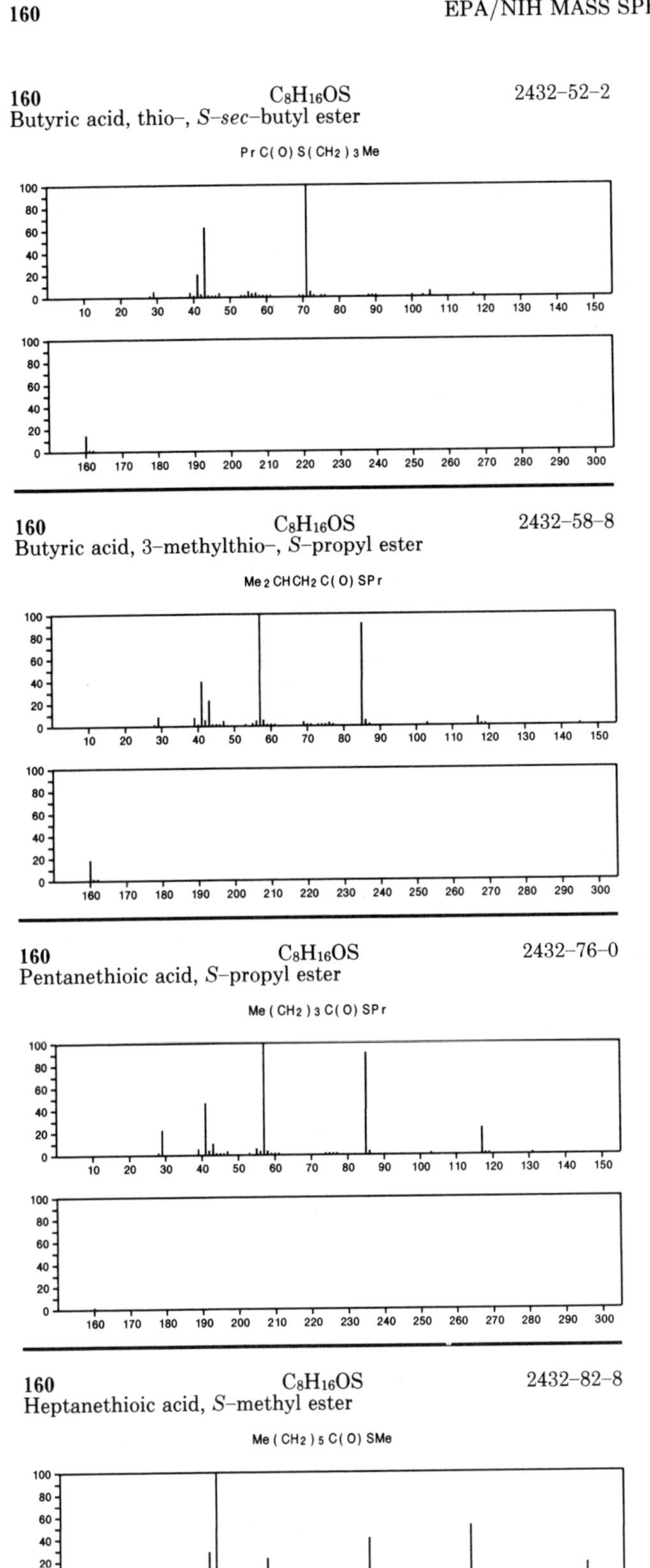

160 $C_8H_{16}OS$ 2432-52-2
Butyric acid, thio-, *S-sec*-butyl ester
PrC(O)S(CH2)3Me

160 $C_8H_{16}OS$ 2432-58-8
Butyric acid, 3-methylthio-, *S*-propyl ester
Me2CHCH2C(O)SPr

160 $C_8H_{16}OS$ 2432-76-0
Pentanethioic acid, *S*-propyl ester
Me(CH2)3C(O)SPr

160 $C_8H_{16}OS$ 2432-82-8
Heptanethioic acid, *S*-methyl ester
Me(CH2)5C(O)SMe

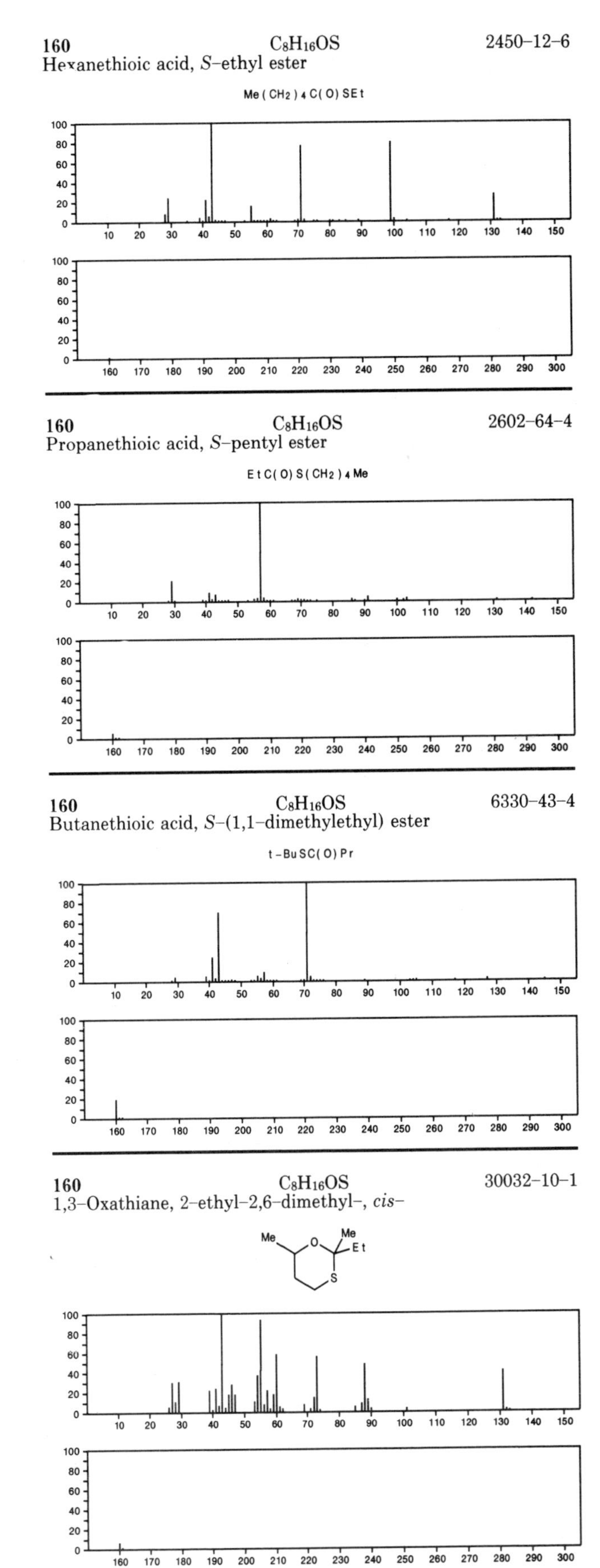

160 $C_8H_{16}OS$ 2450-12-6
Hexanethioic acid, *S*-ethyl ester
Me(CH2)4C(O)SEt

160 $C_8H_{16}OS$ 2602-64-4
Propanethioic acid, *S*-pentyl ester
EtC(O)S(CH2)4Me

160 $C_8H_{16}OS$ 6330-43-4
Butanethioic acid, *S*-(1,1-dimethylethyl) ester
t-BuSC(O)Pr

160 $C_8H_{16}OS$ 30032-10-1
1,3-Oxathiane, 2-ethyl-2,6-dimethyl-, *cis*-

160 $C_8H_{16}OS$ 30032-11-2
1,3-Oxathiane, 2-ethyl-2,6-dimethyl-, *trans*-

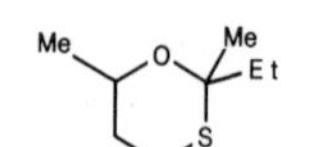

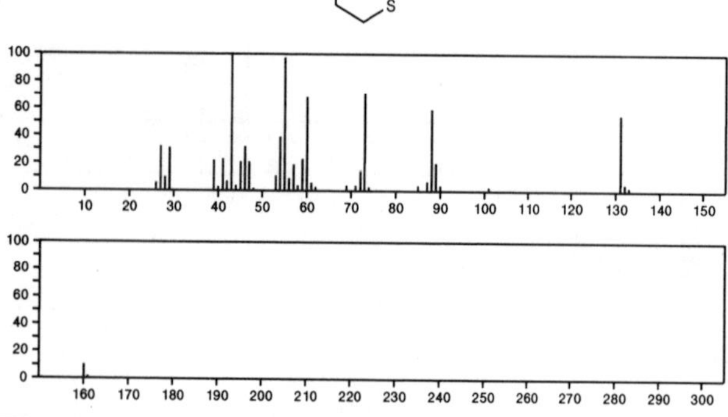

160 $C_8H_{16}OS$ 30098-81-8
1,3-Oxathiane, 2-methyl-2-(1-methylethyl)-

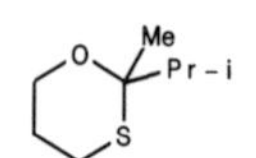

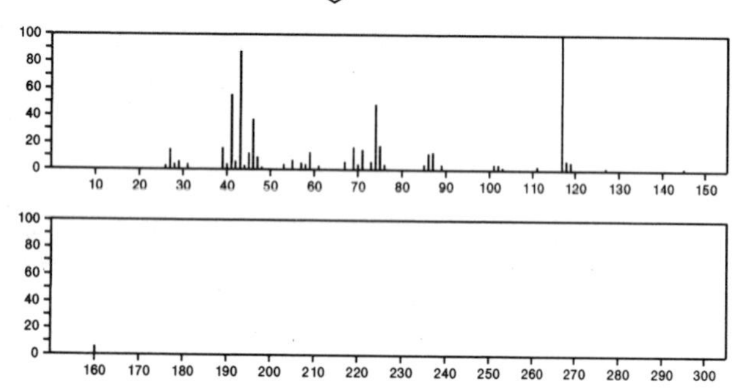

160 $C_8H_{16}OS$ 33709-60-3
1,3-Oxathiane, 2-isopropyl-6-methyl-

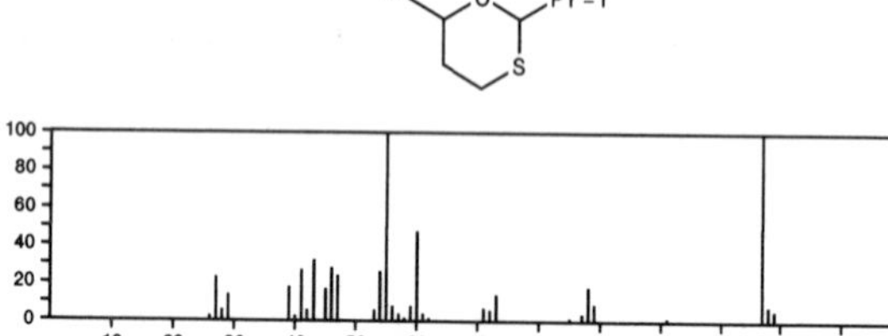

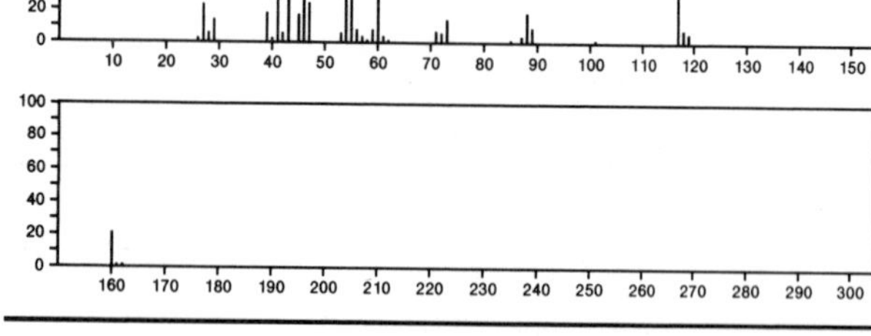

160 $C_8H_{16}OS$ 34560-79-7
1,3-Oxathiane, 2,2,4,6-tetramethyl-, *cis*-

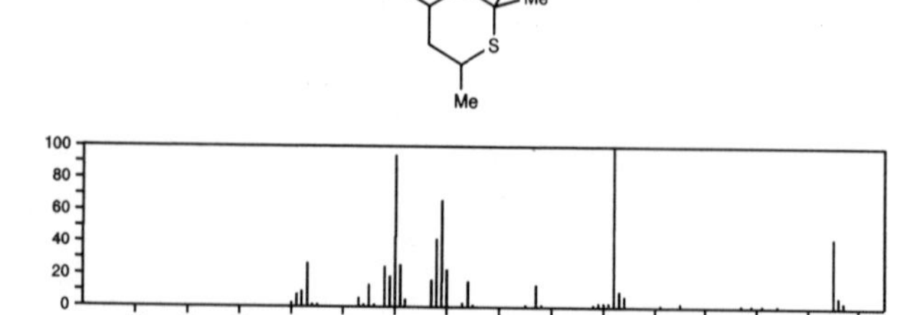

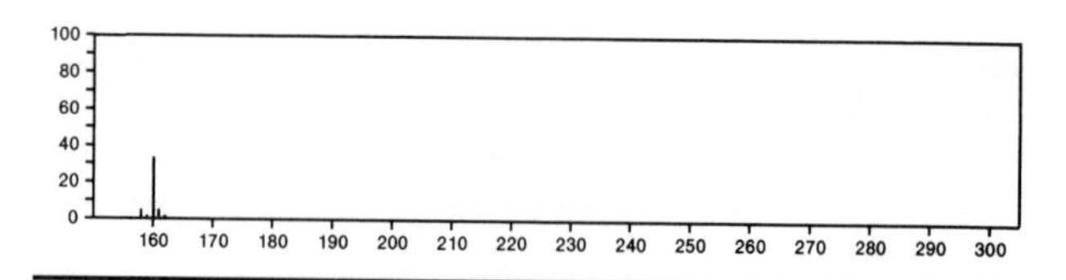

160 $C_8H_{16}OS$ 55590-84-6
Ethanethioic acid, *S*-(1-ethylbutyl) ester

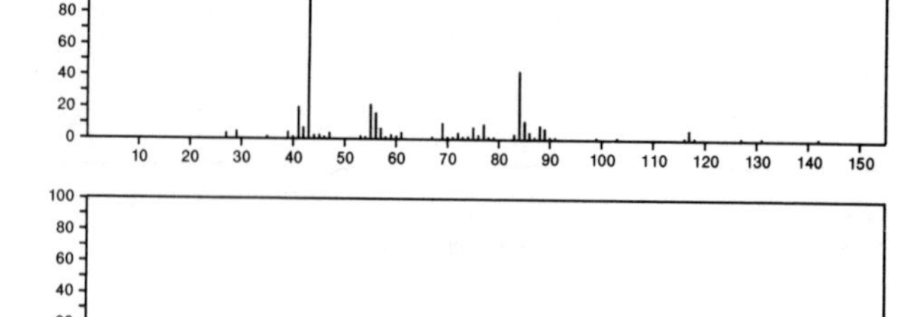

160 $C_8H_{16}OS$ 56052-25-6
2-Propanone, 1-(pentylthio)-

MeCOCH2S(CH2)4Me

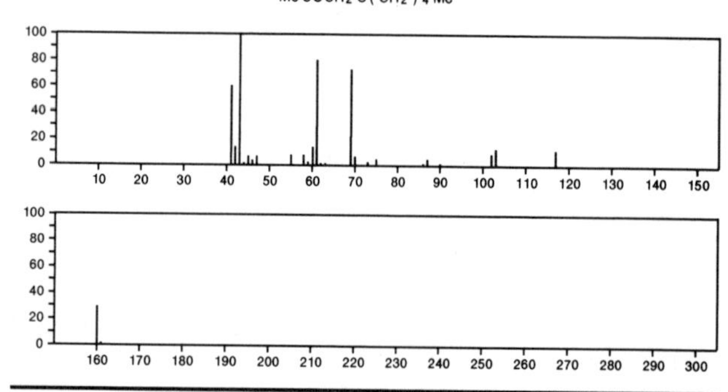

160 $C_8H_{16}O_3$ 764-89-6
Octanoic acid, 8-hydroxy-

HO2C(CH2)7OH

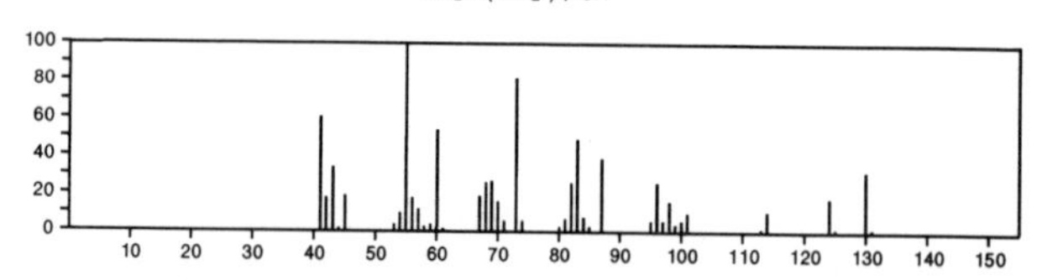

160 $C_8H_{16}O_3$ 3320-90-9
Furan, 2,5-diethoxytetrahydro-

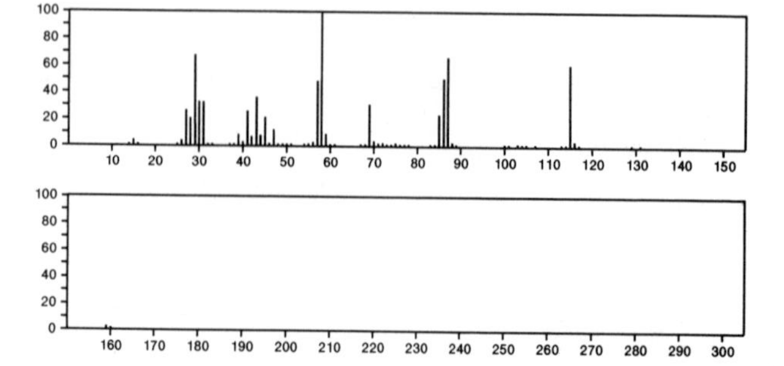

160 $C_8H_{16}O_3$ 5745–75–5

1,3–Dioxolane–2–butanol, 2–methyl–

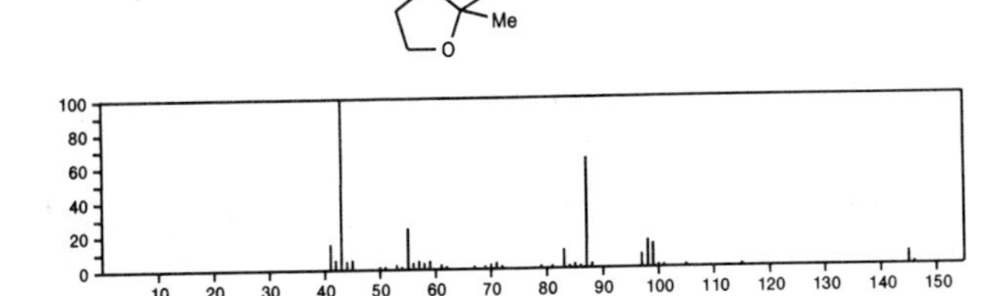

160 $C_8H_{16}O_3$ 17639–74–6

Acetic acid, (1–methylethoxy)–, 1–methylethyl ester

i-Pr $OCH_2C(O)OPr$-i

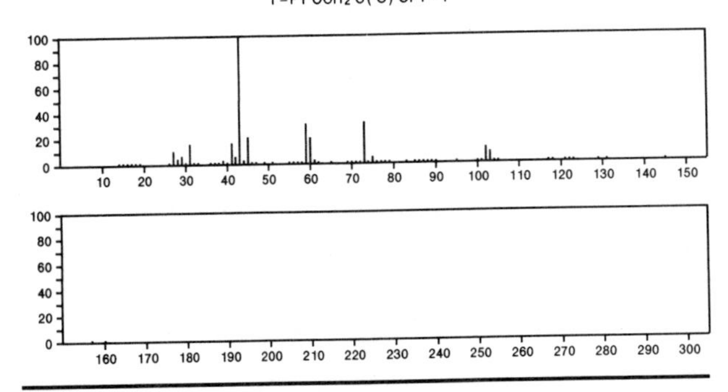

160 $C_8H_{16}O_3$ 29006–05–1

Butyric acid, 4–isopropoxy–, methyl ester

i-Pr $O(CH_2)_3C(O)OMe$

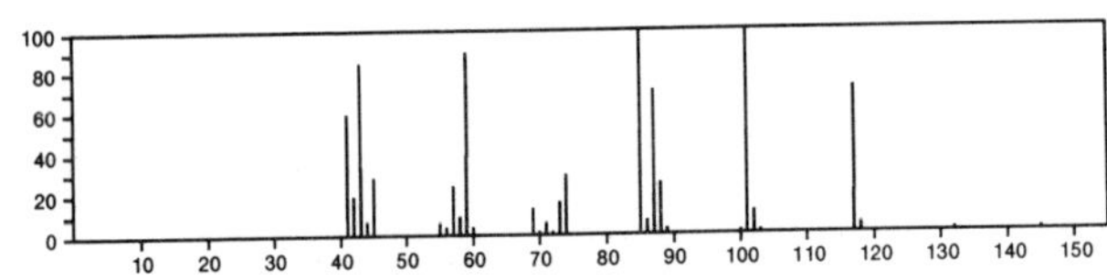

160 $C_8H_{16}O_3$ 29887–58–9

Cyclopentane, 1,2,3–trimethoxy–, stereoisomer

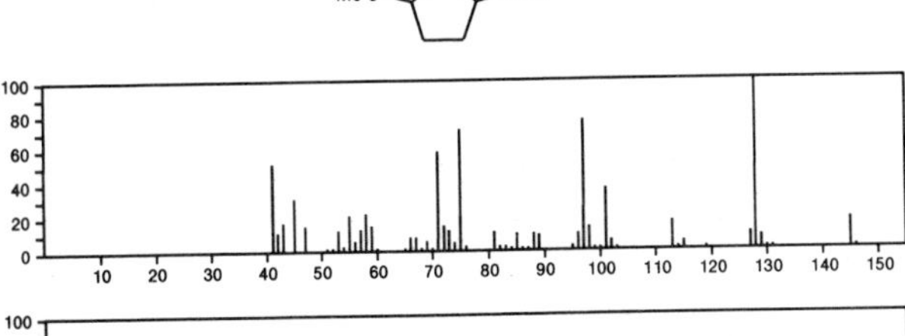

160 $C_8H_{16}O_3$ 29887–63–6

Cyclohexanol, 3,5–dimethoxy–, *cis*–1,3,*trans*–1,5–

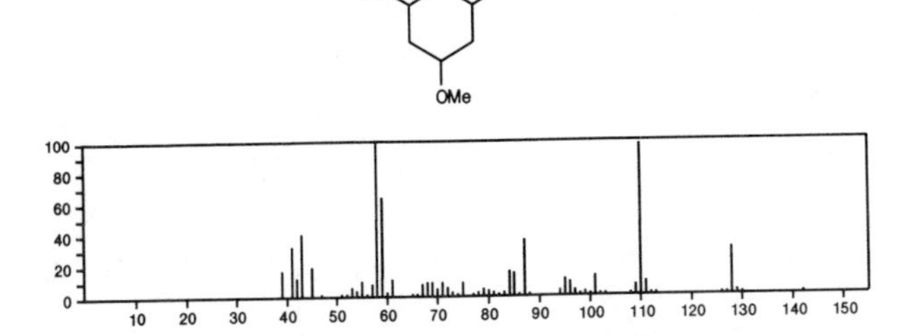

160 $C_8H_{16}O_3$ 30363–64–5

Cyclohexanol, 3,5–dimethoxy–, stereoisomer

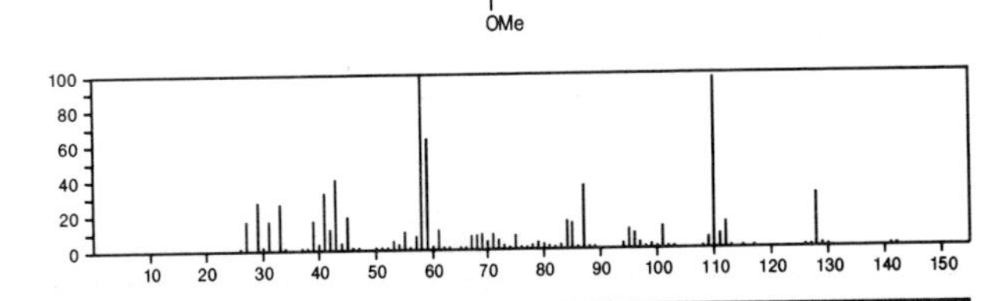

160 $C_8H_{16}O_3$ 30363–83–8

Cyclohexanol, 3,5–dimethoxy–, stereoisomer

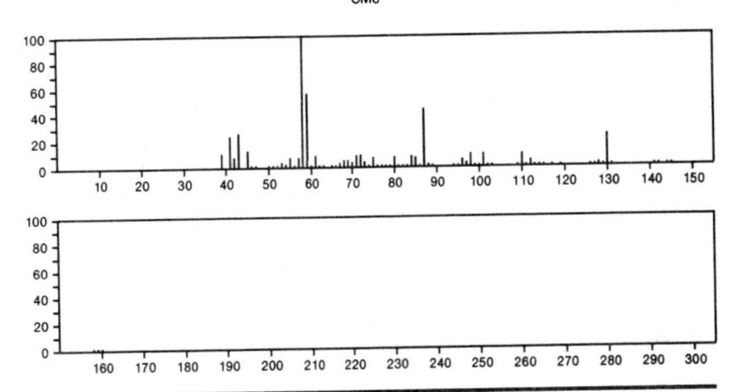

160 $C_8H_{16}O_3$ 30517–18–1

Cyclohexanol, 3,5–dimethoxy–, stereoisomer

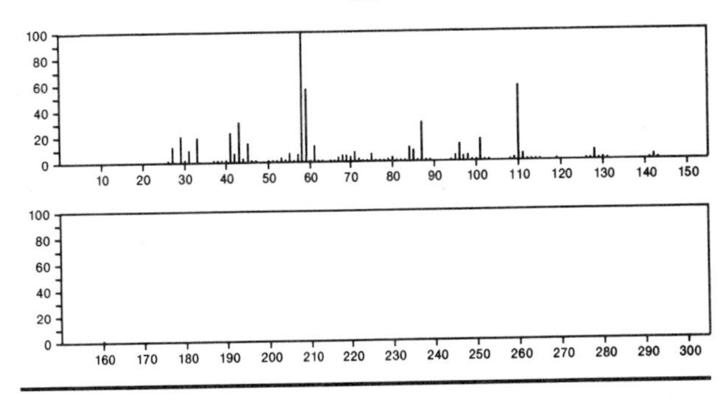

160 $C_8H_{16}O_3$ 36651–29–3

1,3–Dioxolane–2–propanol, 2,4–dimethyl–

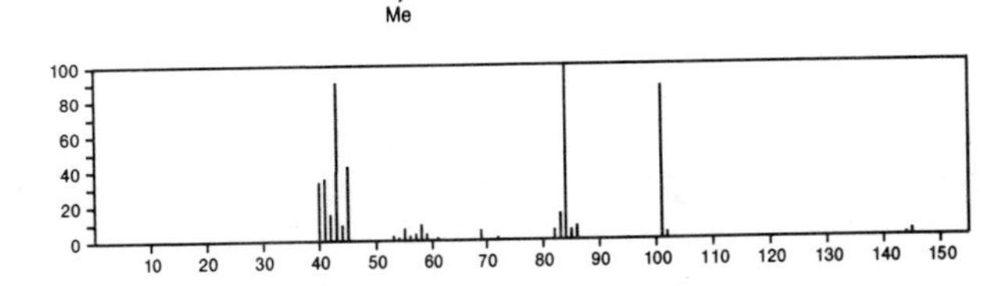

160 $C_8H_{16}O_3$ 36651-31-7
1,3-Dioxane-2-propanol, 2-methyl-

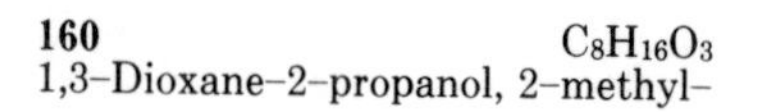

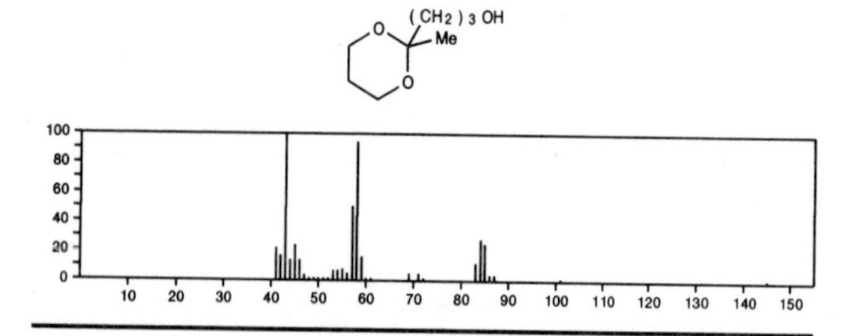

160 $C_8H_{16}O_3$ 54632-67-6
1,3-Dioxolane-2-propanol, α,2-dimethyl-

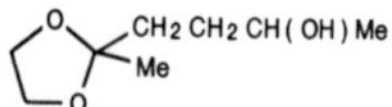

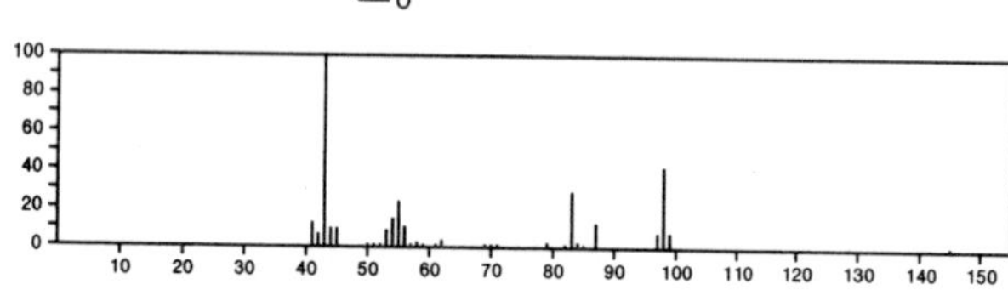

160 $C_8H_{16}O_3$ 54751-80-3
1,3-Dioxolane, 2-(3-methoxypropyl)-2-methyl-

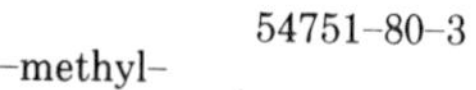

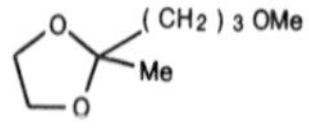

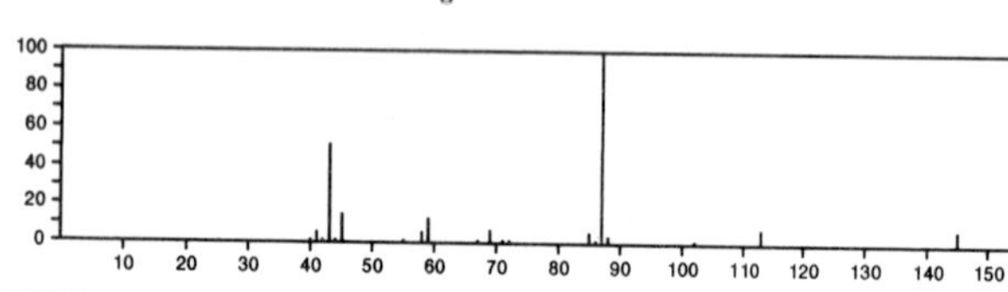

160 $C_8H_{16}O_3$ 55724-73-7
Butanoic acid, 4-butoxy-

$HO_2C(CH_2)_3O(CH_2)_3Me$

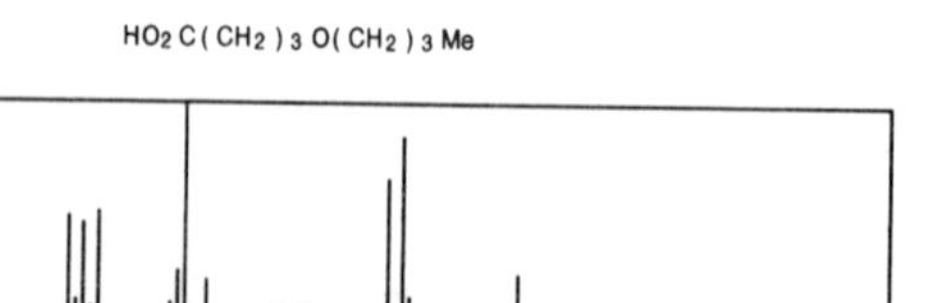

160 $C_8H_{17}OP$ 50837-80-4
Phosphine, acetylbis(1-methylethyl)-

i-Pr_2PAc

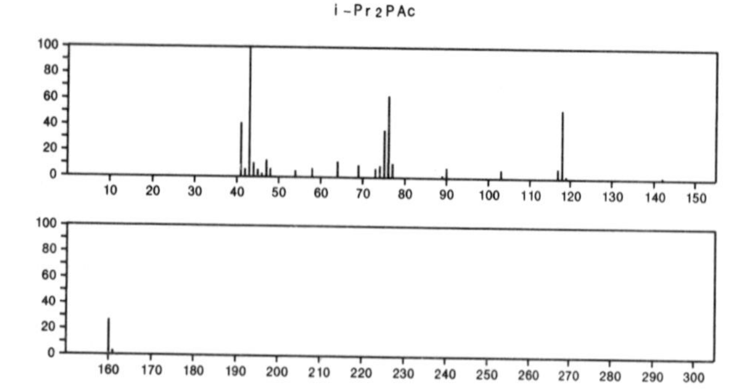

160 $C_8H_{20}N_2O$ 3033-62-3
Ethanamine, 2,2'-oxybis[*N*,*N*-dimethyl-

$Me_2NCH_2CH_2OCH_2CH_2NMe_2$

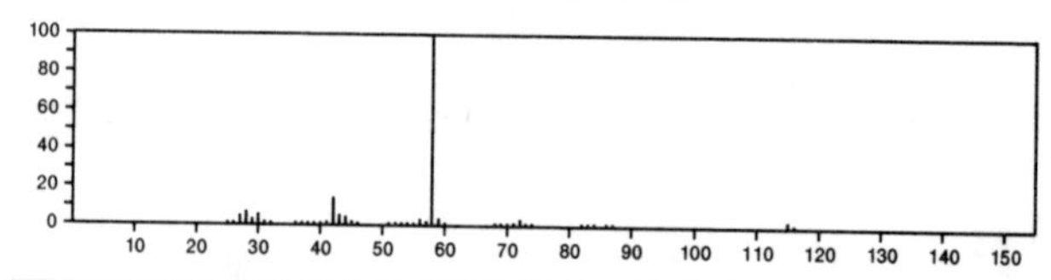

160 $C_8H_{20}OSi$ 1825-67-8
Silane, trimethyl(1-methylbutoxy)-

$PrCHMeOSiMe_3$

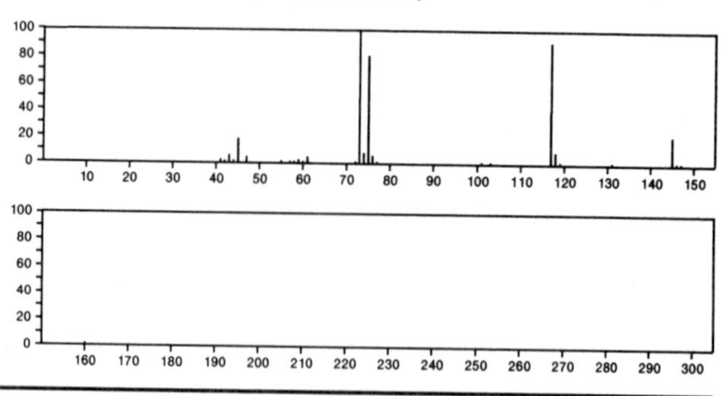

160 $C_8H_{20}OSi$ 6689-16-3
Silane, (1,1-dimethylpropoxy)trimethyl-

$EtCMe_2OSiMe_3$

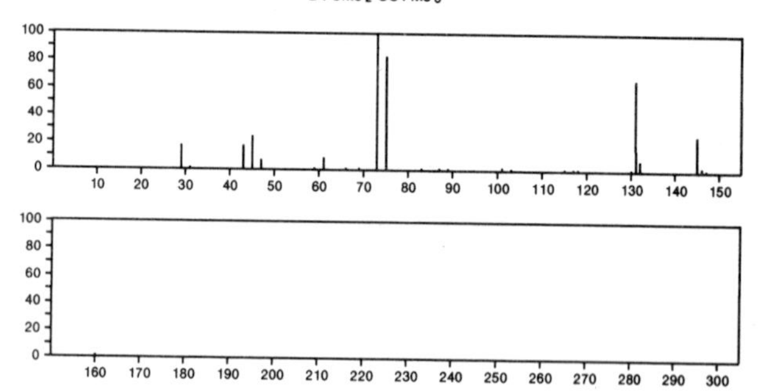

160 $C_8H_{20}OSi$ 14629-45-9
Silane, trimethyl(pentyloxy)-

$Me(CH_2)_4OSiMe_3$

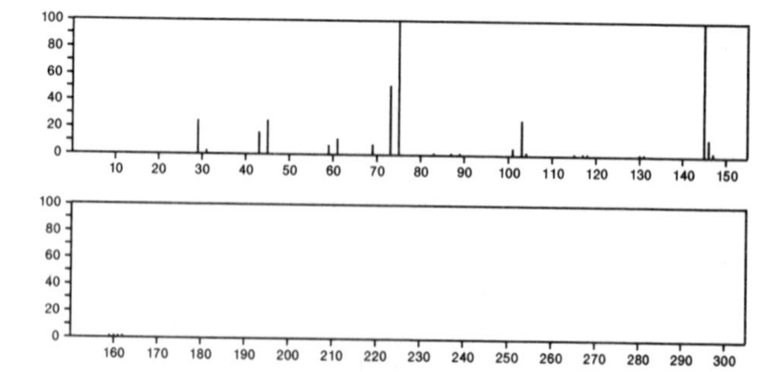

160 $C_8H_{20}OSi$ 17348-64-0
Silane, (*sec*-butoxymethyl)trimethyl-

Me_3SiCH_2OBu-s

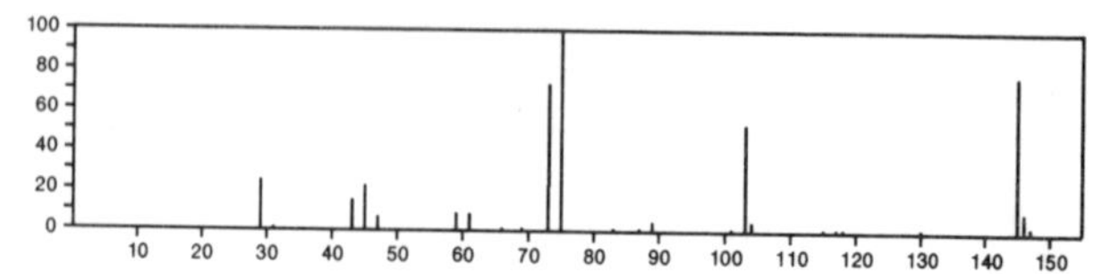

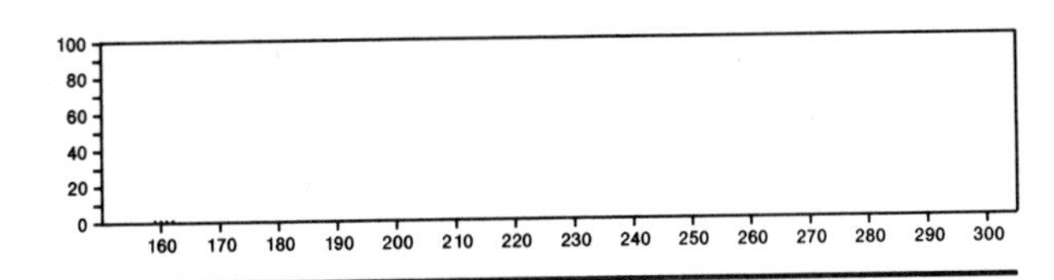

160 $C_8H_{20}OSi$ 18246-52-1
Silane, (butoxymethyl)trimethyl-

$Me_3SiCH_2O(CH_2)_3Me$

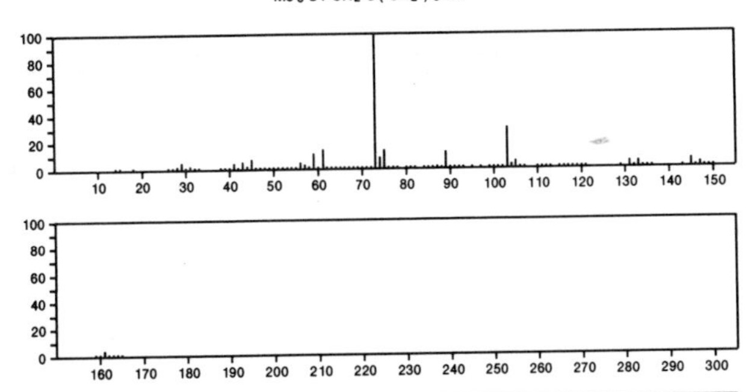

160 $C_8H_{20}OSi$ 18246-56-5
Silane, trimethyl(3-methylbutoxy)-

$Me_3SiOCH_2CH_2CHMe_2$

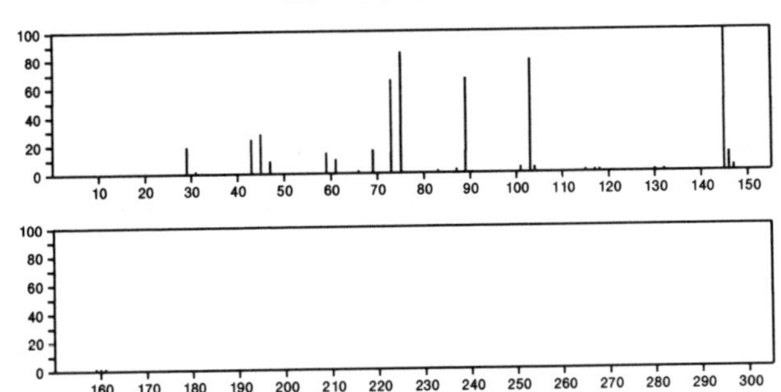

160 $C_8H_{20}OSi$ 18246-63-4
Silane, (2,2-dimethylpropoxy)trimethyl-

$Me_3SiOCH_2CMe_3$

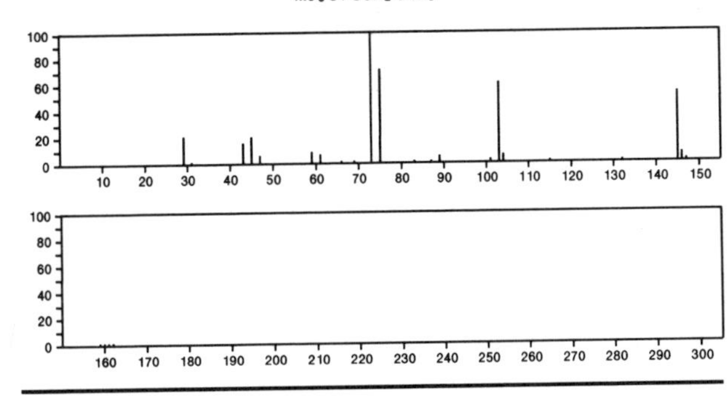

160 $C_8H_{20}OSi$ 18246-76-9
Silane, (1,2-dimethylpropoxy)trimethyl-

$Me_3SiOCHMeCHMe_2$

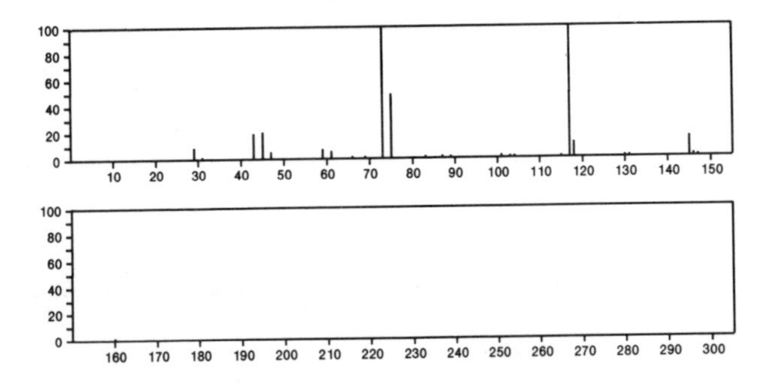

160 $C_9H_4O_3$ 938-24-9
1*H*-Indene-1,2,3-trione

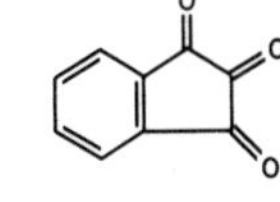

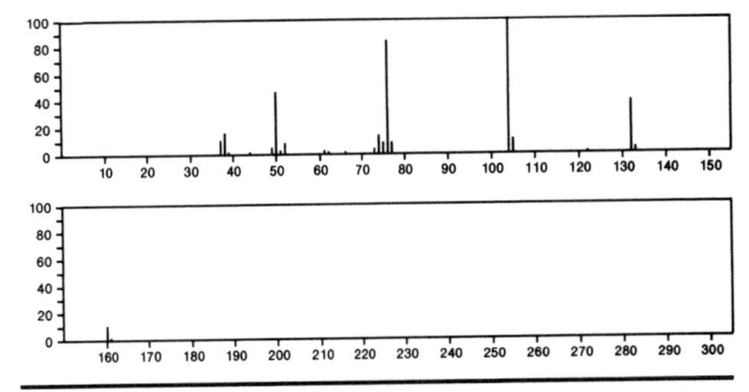

160 $C_9H_8N_2O$ 1693-94-3
4*H*-Pyrido[1,2-*a*]pyrimidin-4-one, 2-methyl-

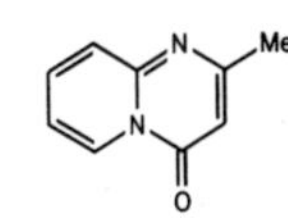

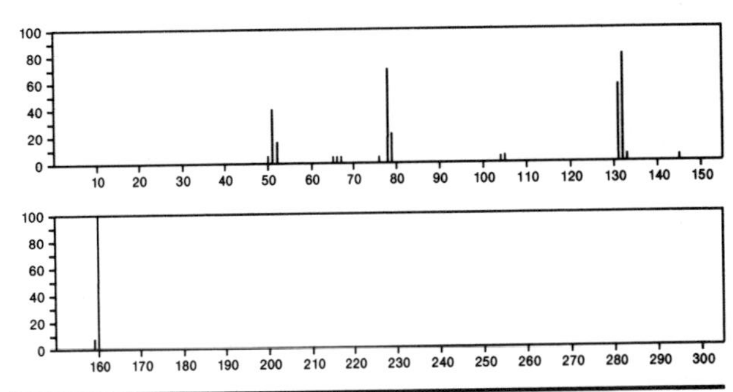

160 $C_9H_8N_2O$ 4369-55-5
5-Isoxazolamine, 3-phenyl-

Ph NH2 N–O

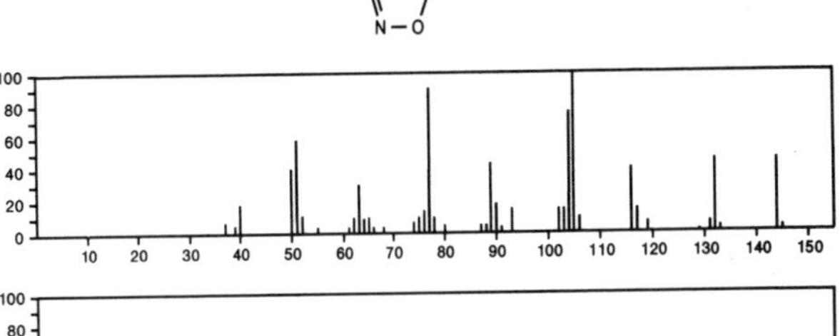

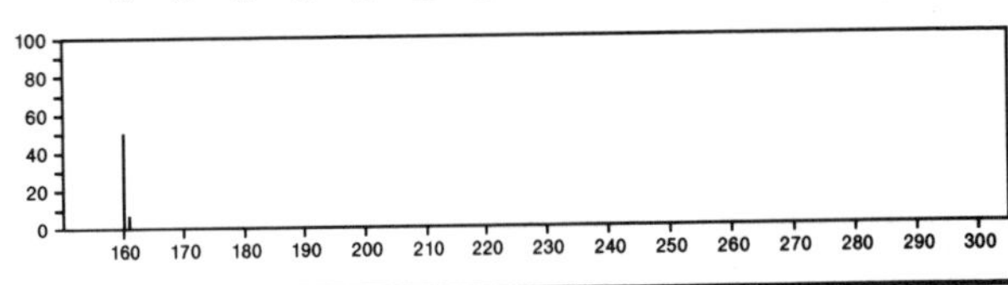

160 $C_9H_8N_2O$ 4860-93-9
3*H*-Pyrazol-3-one, 2,4-dihydro-5-phenyl-

O Ph HN–N

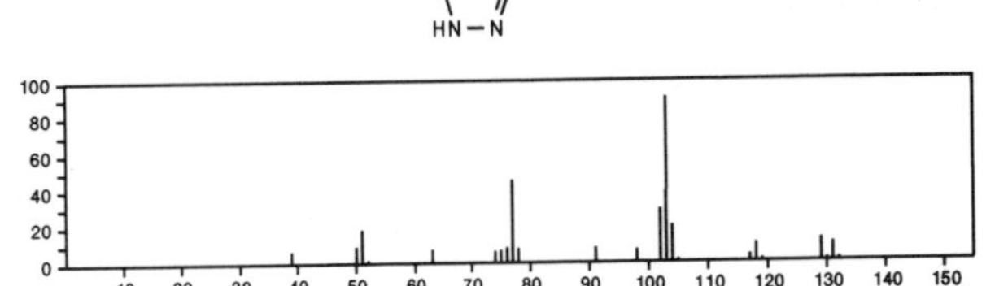

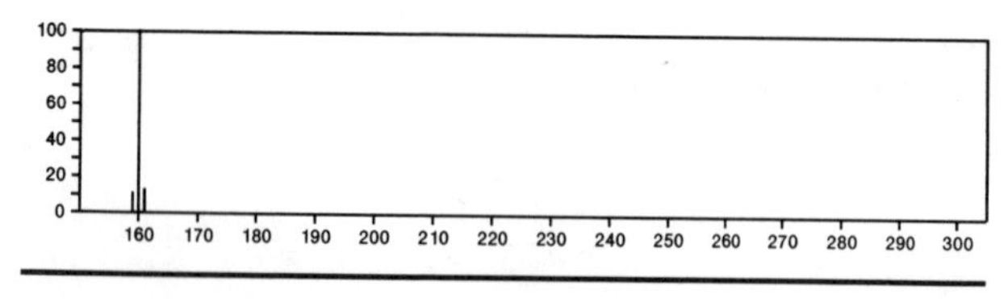

160 $C_9H_8N_2O$ 5004-48-8
1(2*H*)-Phthalazinone, 4-methyl-

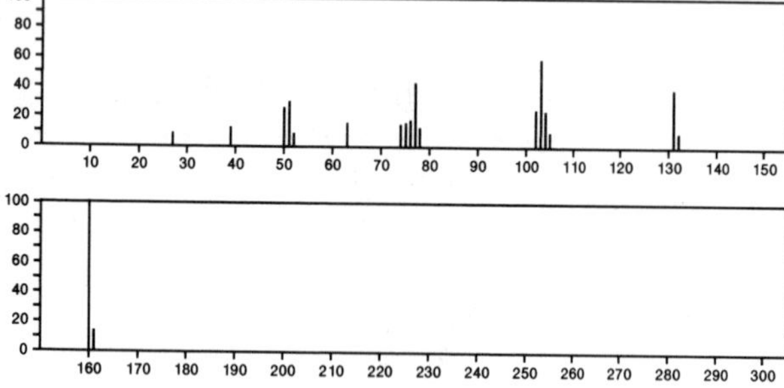

160 $C_9H_8N_2O$ 5580-85-8
Cinnoline, 4-methyl-, 2-oxide

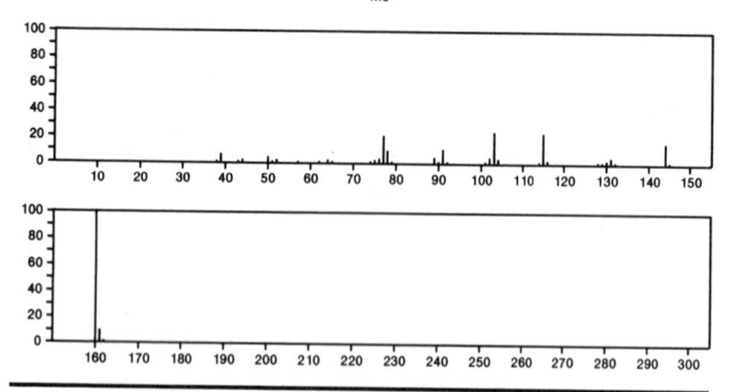

160 $C_9H_8N_2O$ 5580-86-9
Cinnoline, 4-methyl-, 1-oxide

160 $C_9H_8N_2O$ 10501-56-1
Quinazoline, 4-methyl-, 3-oxide

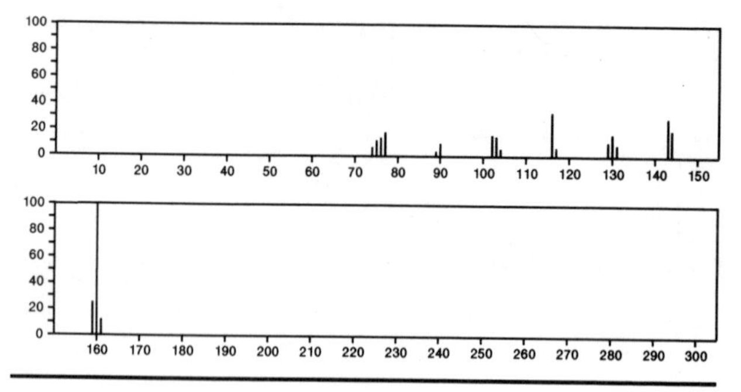

160 $C_9H_8N_2O$ 18916-44-4
Quinoxaline, 2-methyl-, 1-oxide

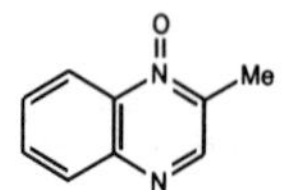

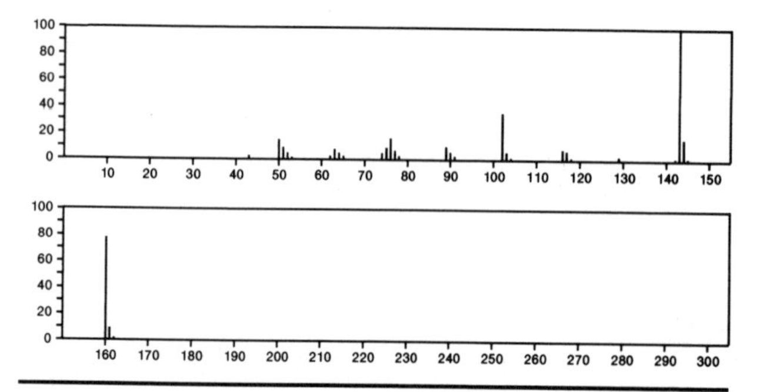

160 $C_9H_8N_2O$ 18916-45-5
Quinoxaline, 2-methyl-, 4-oxide

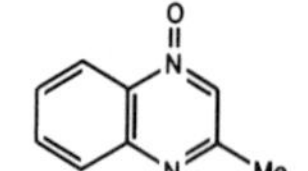

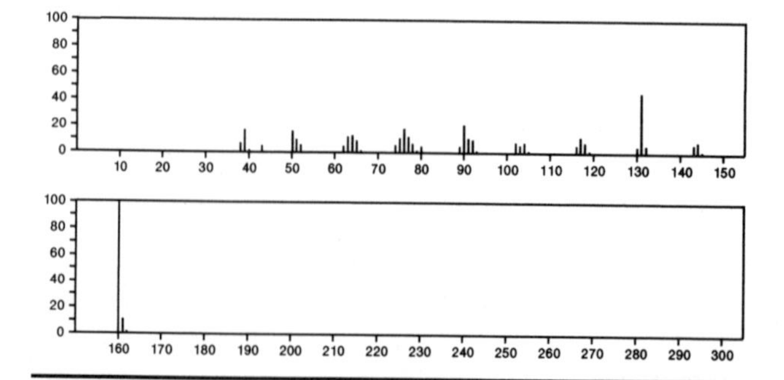

160 $C_9H_8N_2O$ 22320-27-0
3-Indolizinecarboxamide

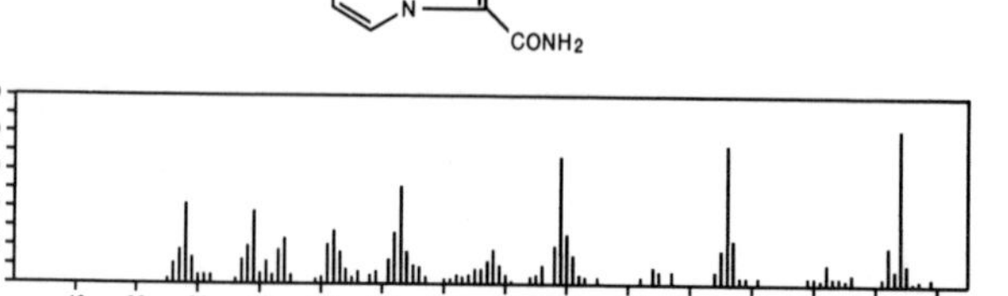

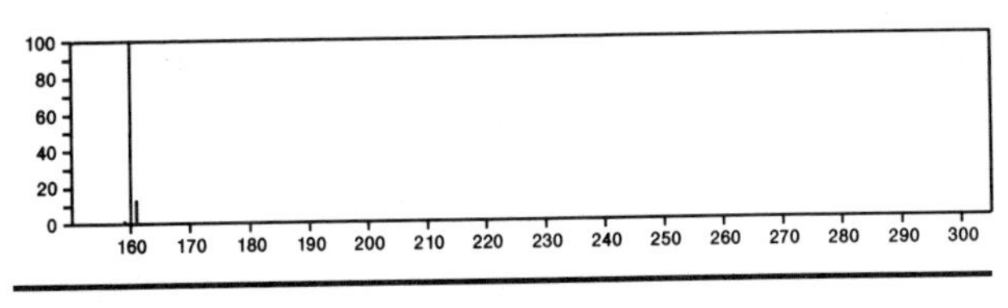

160 $C_9H_8N_2O$ 23350-02-9
4-Isoxazolamine, 3-phenyl-

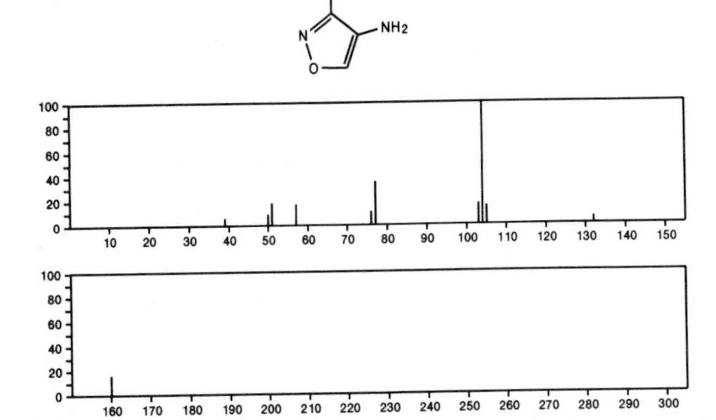

160 $C_9H_8N_2O$ 28883-94-5
2*H*-Azirine-2-carboxamide, 3-phenyl-

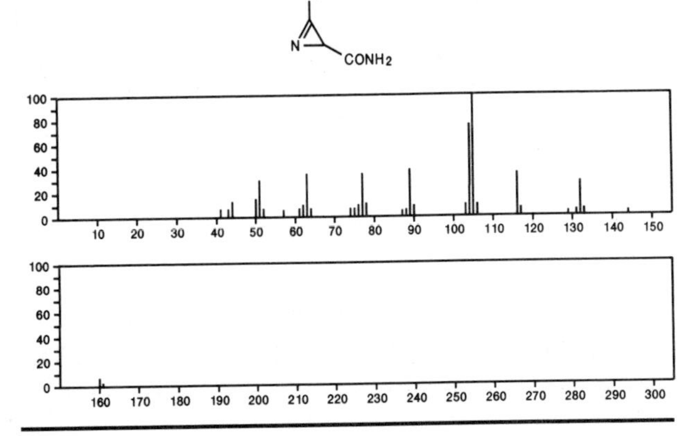

160 $C_9H_8N_2O$ 33499-34-2
Benzonitrile, 4-[(methoxyimino)methyl]-

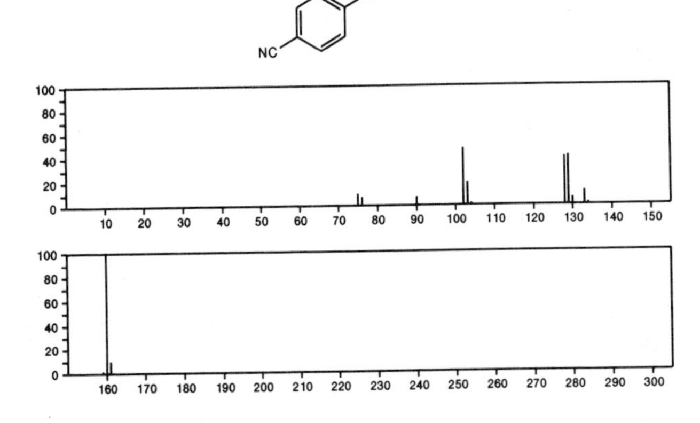

160 $C_9H_8N_2O$ 35549-22-5
2*H*-Pyrido[1,2-*a*]pyrimidin-2-one, 4-methyl-

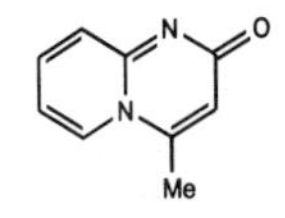

160 $C_9H_8N_2O$ 37920-72-2
Quinazoline, 4-methyl-, 1-oxide

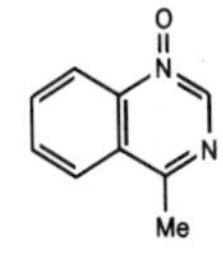

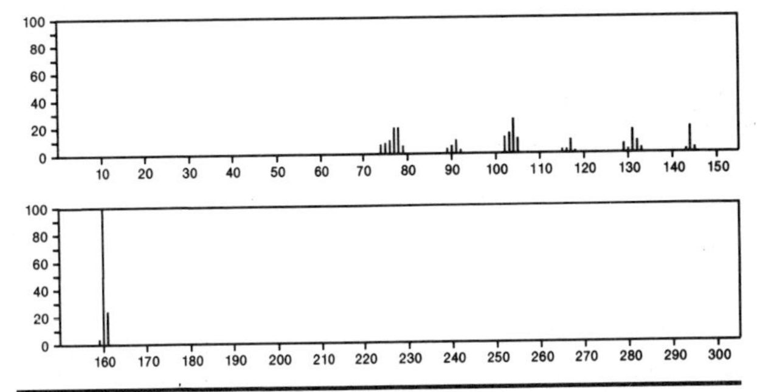

160 $C_9H_{20}O_2$ 115-84-4
1,3-Propanediol, 2-butyl-2-ethyl-

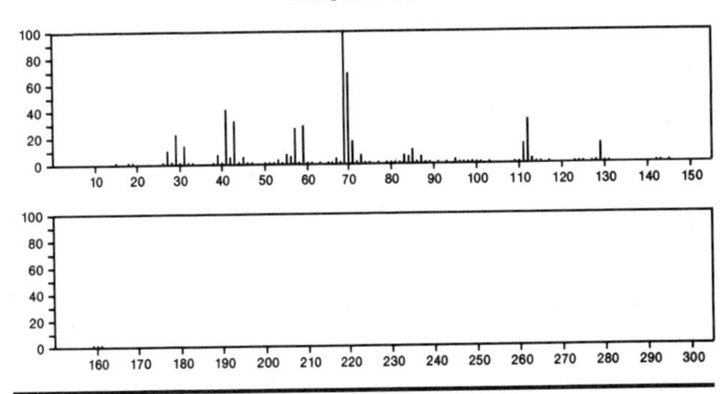

160 $C_9H_{20}O_2$ 2568-90-3
Butane, 1,1'-[methylenebis(oxy)]bis-

Me (CH2) 3 OCH2 O(CH2) 3 Me

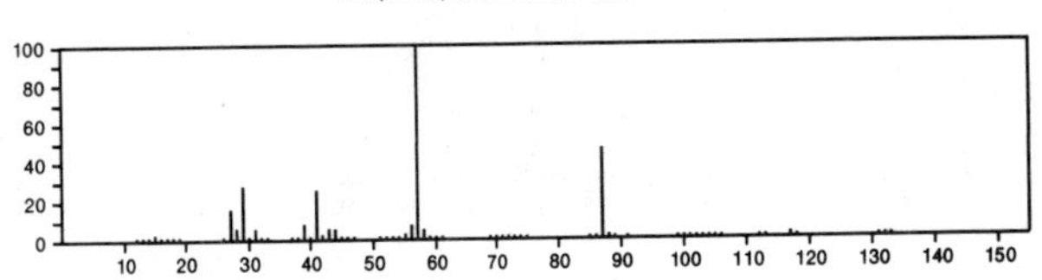

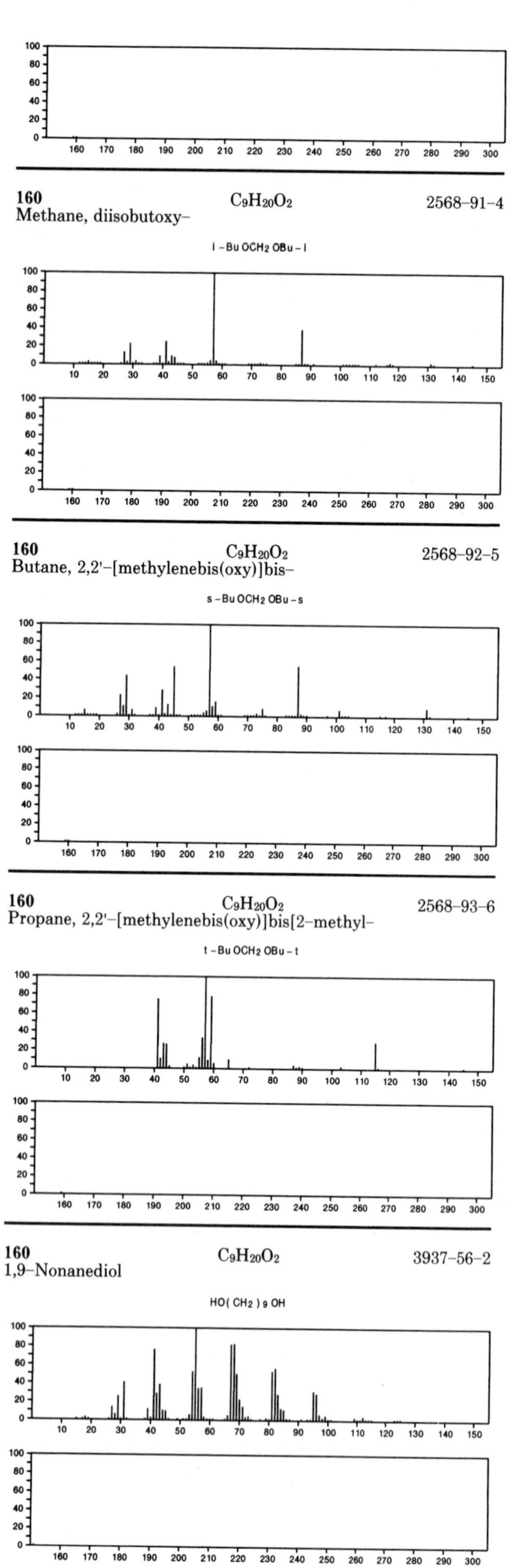
160
Methane, diisobutoxy-
C9H20O2
2568-91-4
i-BuOCH2OBu-i
160
Butane, 2,2'-[methylenebis(oxy)]bis-
C9H20O2
2568-92-5
s-BuOCH2OBu-s
160
Propane, 2,2'-[methylenebis(oxy)]bis[2-methyl-
C9H20O2
2568-93-6
t-BuOCH2OBu-t
160
1,9-Nonanediol
C9H20O2
3937-56-2
HO(CH2)9OH

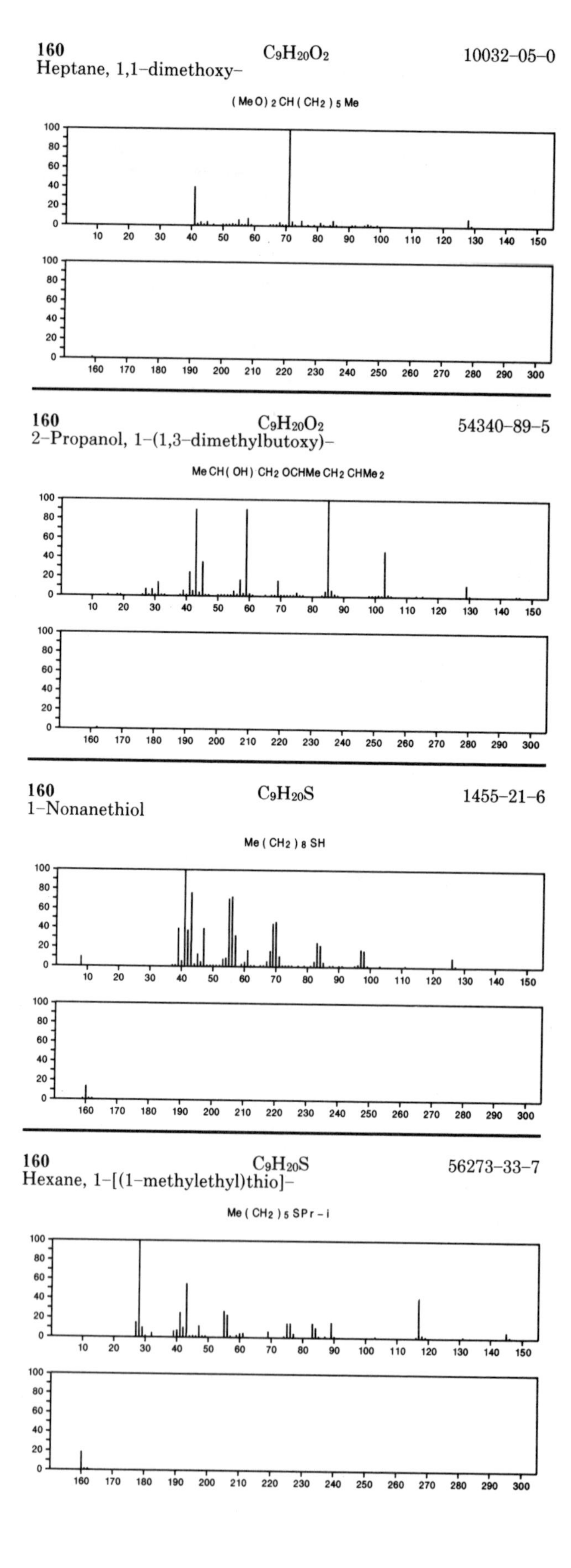
160
Heptane, 1,1-dimethoxy-
C9H20O2
10032-05-0
(MeO)2CH(CH2)5Me
160
2-Propanol, 1-(1,3-dimethylbutoxy)-
C9H20O2
54340-89-5
MeCH(OH)CH2OCHMeCH2CHMe2
160
1-Nonanethiol
C9H20S
1455-21-6
Me(CH2)8SH
160
Hexane, 1-[(1-methylethyl)thio]-
C9H20S
56273-33-7
Me(CH2)5SPr-i

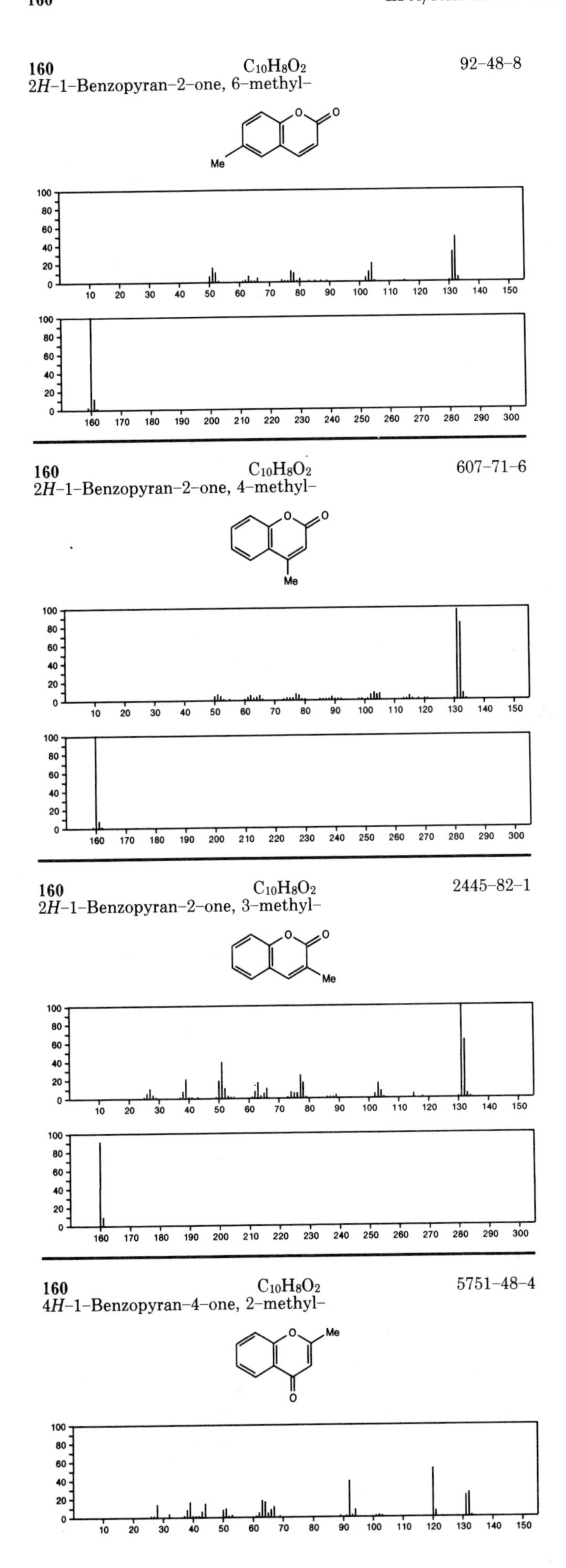
160
$C_{10}H_8O_2$
92–48–8
2*H*–1–Benzopyran–2–one, 6–methyl–
160
$C_{10}H_8O_2$
607–71–6
2*H*–1–Benzopyran–2–one, 4–methyl–
160
$C_{10}H_8O_2$
2445–82–1
2*H*–1–Benzopyran–2–one, 3–methyl–
160
$C_{10}H_8O_2$
5751–48–4
4*H*–1–Benzopyran–4–one, 2–methyl–

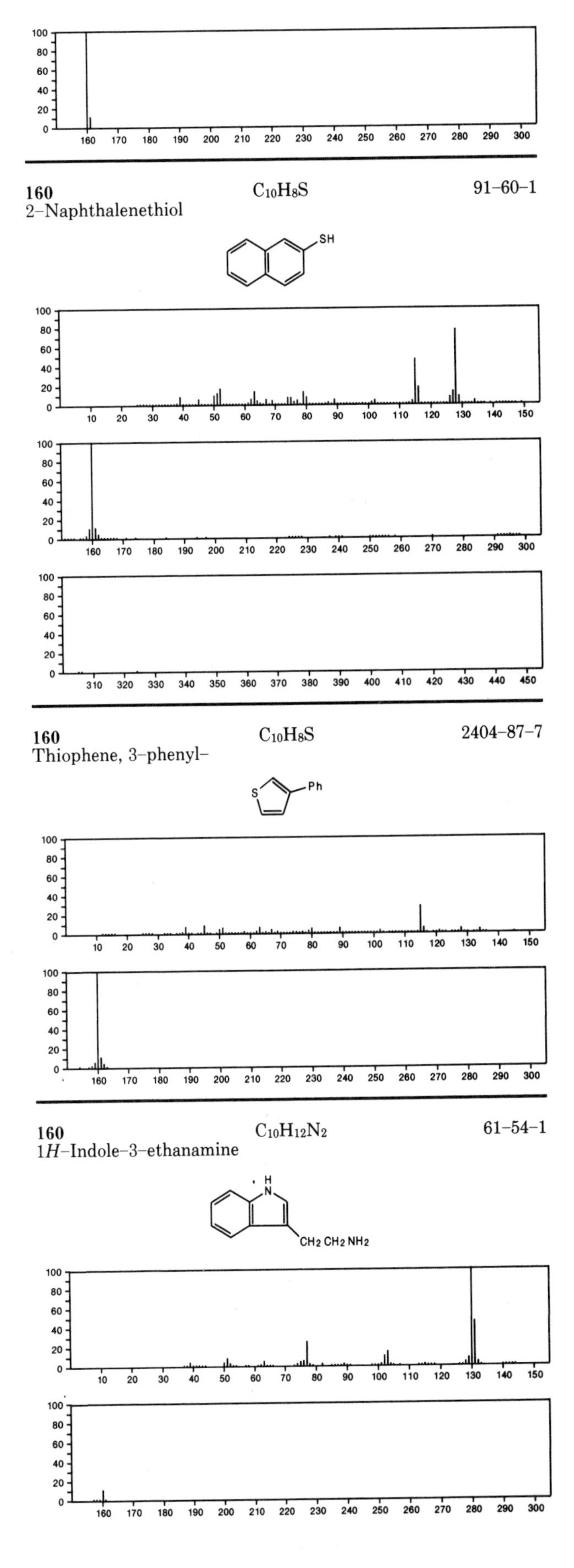
160
$C_{10}H_8S$
91–60–1
2–Naphthalenethiol
160
$C_{10}H_8S$
2404–87–7
Thiophene, 3–phenyl–
160
$C_{10}H_{12}N_2$
61–54–1
1*H*–Indole–3–ethanamine

160 $C_{10}H_{12}N_2$ 939-06-0
1*H*-Imidazole, 4,5-dihydro-4-methyl-2-phenyl-

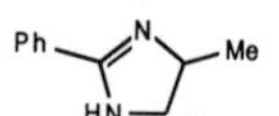

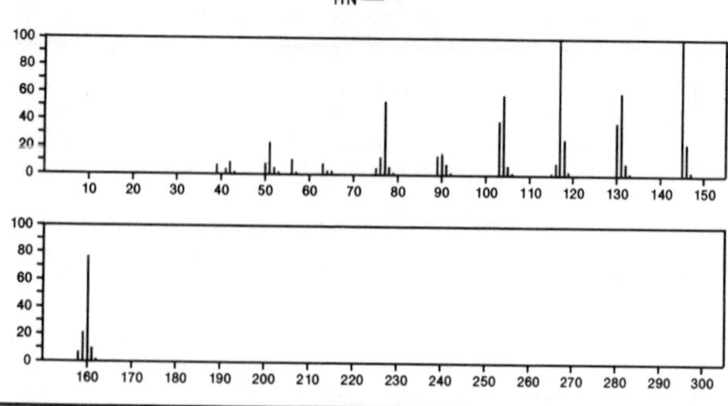

160 $C_{10}H_{12}N_2$ 5465-29-2
1*H*-Benzimidazole, 2-propyl-

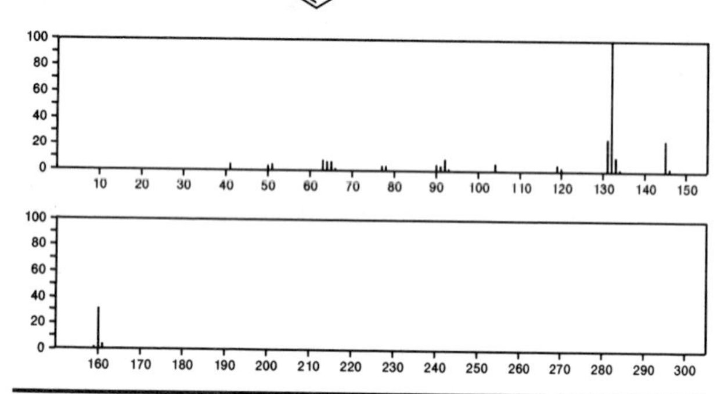

160 $C_{10}H_{12}N_2$ 5851-43-4
1*H*-Benzimidazole, 2-(1-methylethyl)-

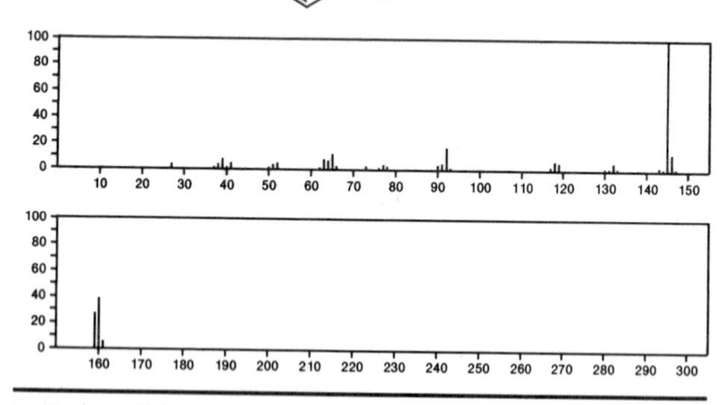

160 $C_{10}H_{12}N_2$ 17408-34-3
Pyrazolo[1,5-*a*]pyridine, 2,3,7-trimethyl-

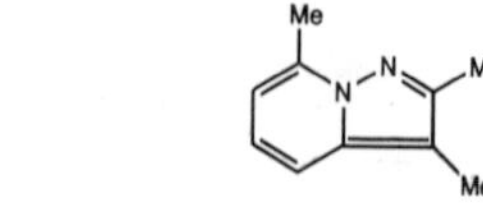

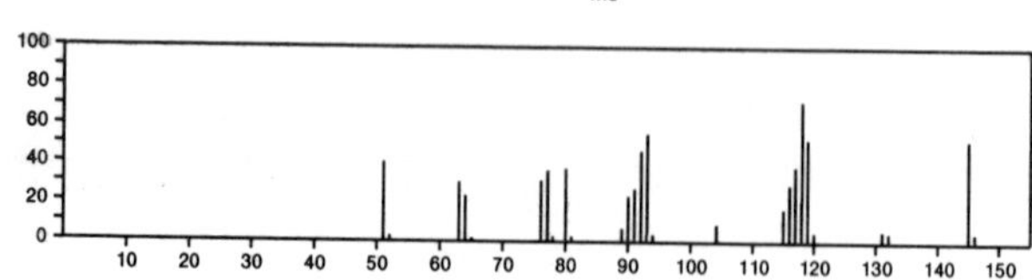

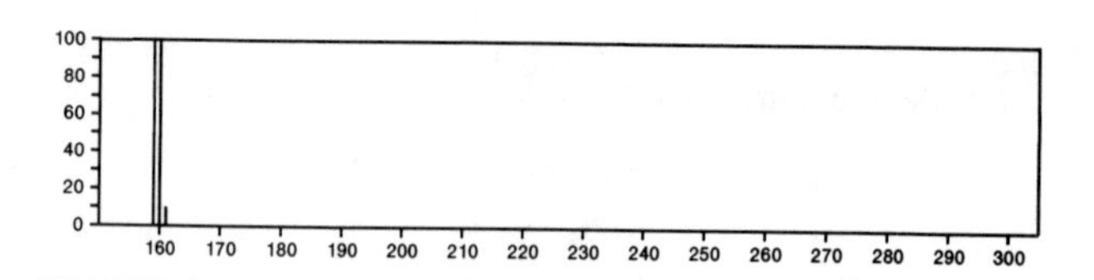

160 $C_{10}H_{12}N_2$ 18233-71-1
1*H*-2,4-Benzodiazepine, 2,5-dihydro-3-methyl-

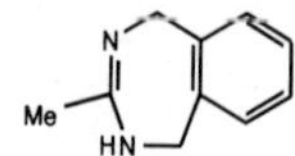

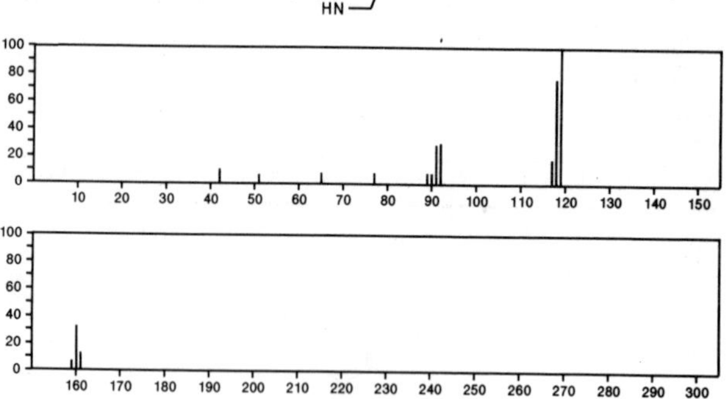

160 $C_{10}H_{12}N_2$ 27257-18-7
1*H*-Pyrrolo[2,3-*b*]pyridine, 2-(1-methylethyl)-

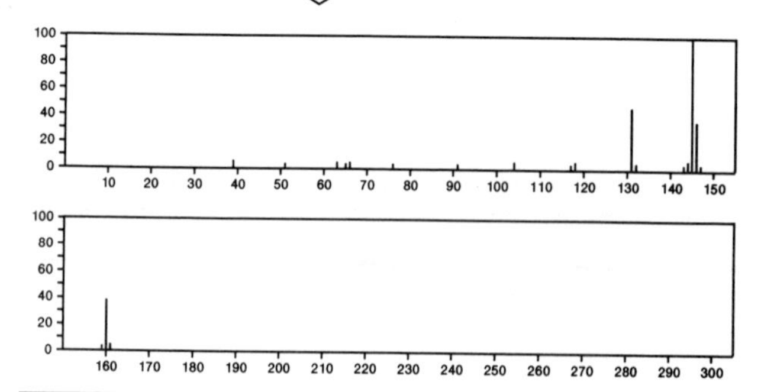

160 $C_{10}H_{12}N_2$ 46035-38-5
3*H*-1,3-Benzodiazepine, 4,5-dihydro-2-methyl-

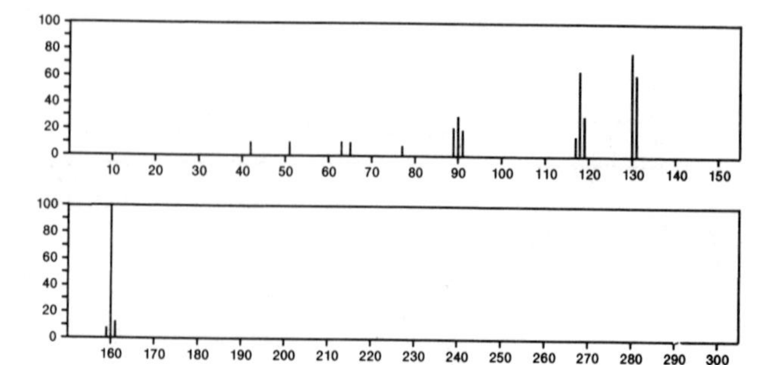

160 $C_{11}H_{12}O$ 1198–20–5
1,4–Methanonaphthalen–9–ol, 1,2,3,4–tetrahydro–, stereoisomer

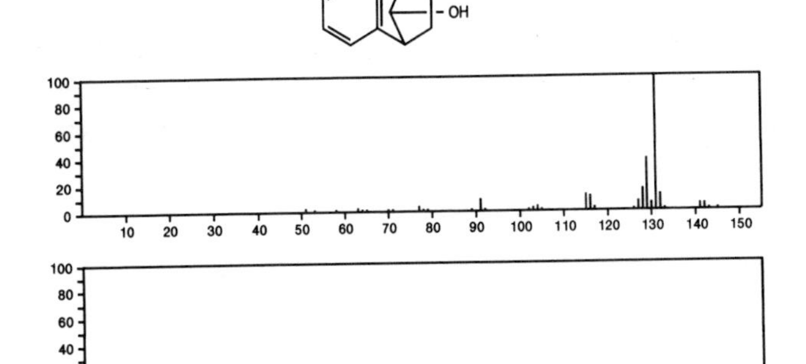

160 $C_{11}H_{12}O$ 1590–08–5
1(2*H*)–Naphthalenone, 3,4–dihydro–2–methyl–

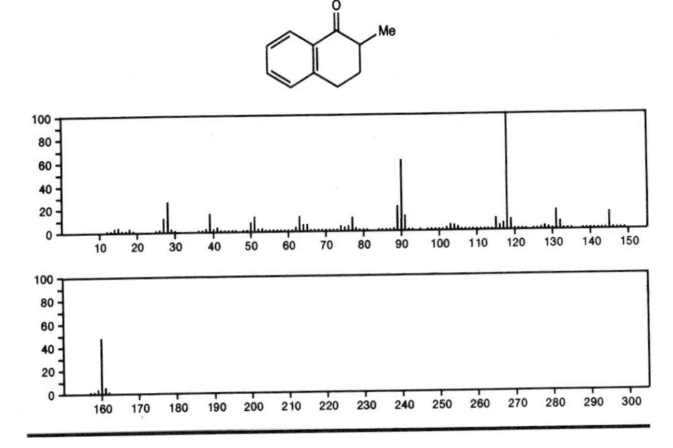

160 $C_{11}H_{12}O$ 1901–26–4
3–Buten–2–one, 3–methyl–4–phenyl–

PhCH=CMeCOMe

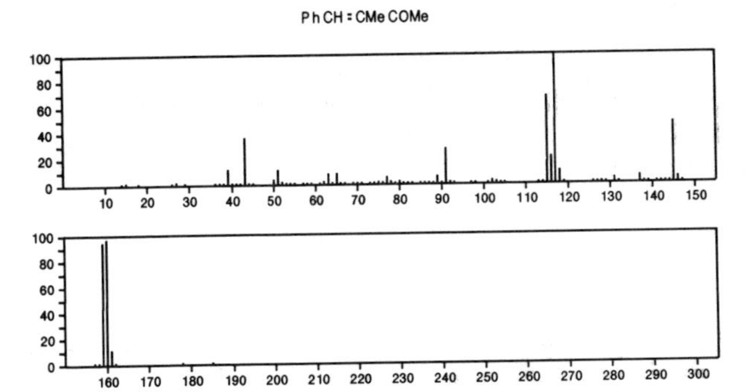

160 $C_{11}H_{12}O$ 5037–60–5
1*H*–Inden–1–one, 2,3–dihydro–4,7–dimethyl–

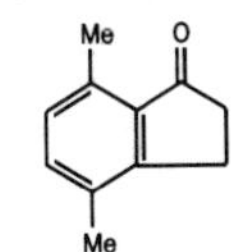

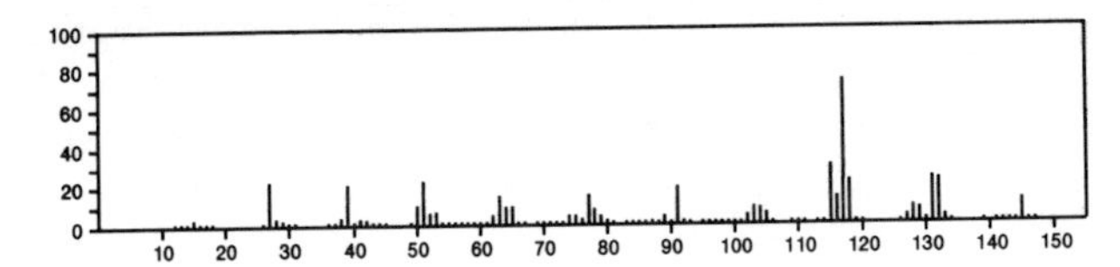

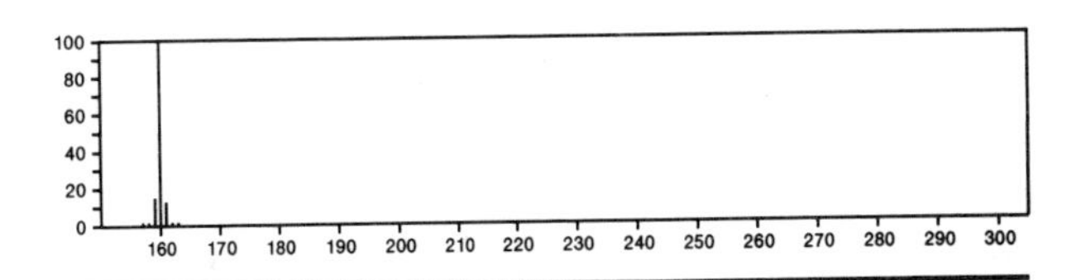

160 $C_{11}H_{12}O$ 5359–04–6
Ethanone, 1–[4–(1–methylethenyl)phenyl]–

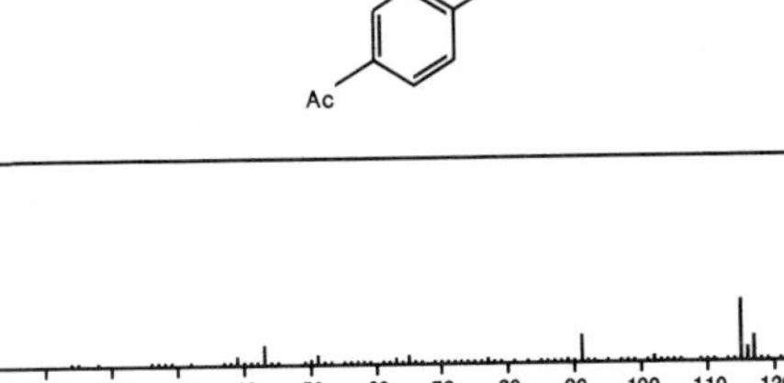

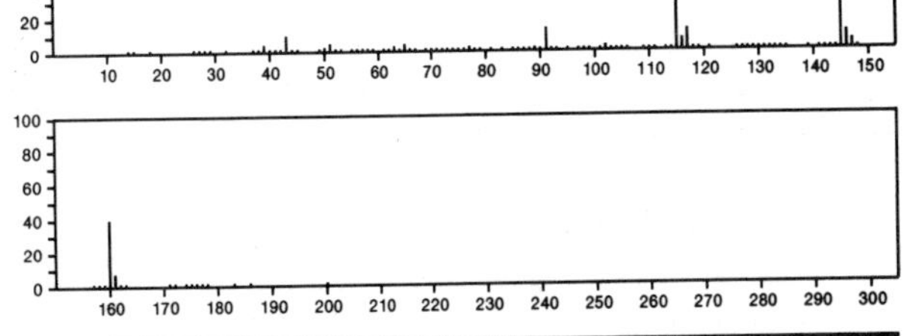

160 $C_{11}H_{12}O$ 6682–69–5
1*H*–Inden–1–one, 2,3–dihydro–5,7–dimethyl–

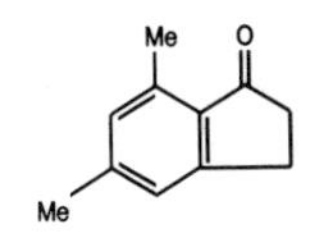

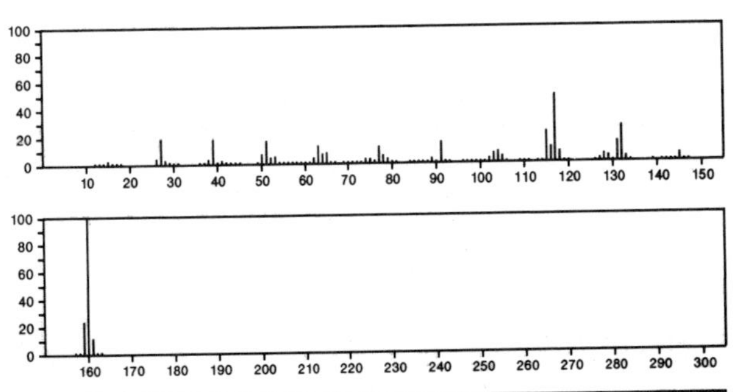

160 $C_{11}H_{12}O$ 6712–35–2
Benzenemethanol, α–methyl–α–2–propynyl–

HC≡CCH2CMe(OH)Ph

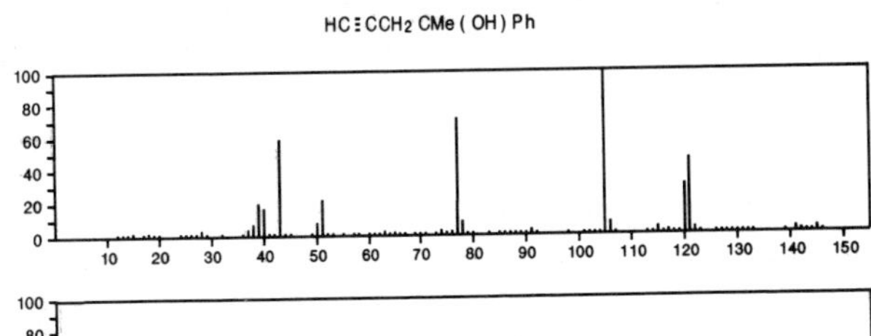

160 $C_{11}H_{12}O$ 6939–35–1
1(2*H*)–Naphthalenone, 3,4–dihydro–5–methyl–

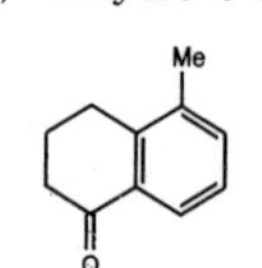

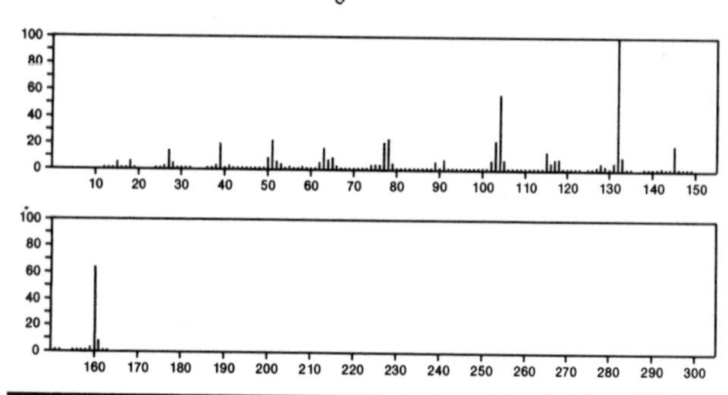

160 $C_{11}H_{12}O$ 10521–97–8
3–Penten–2–one, 5–phenyl–

Me COCH = CHCH2 Ph

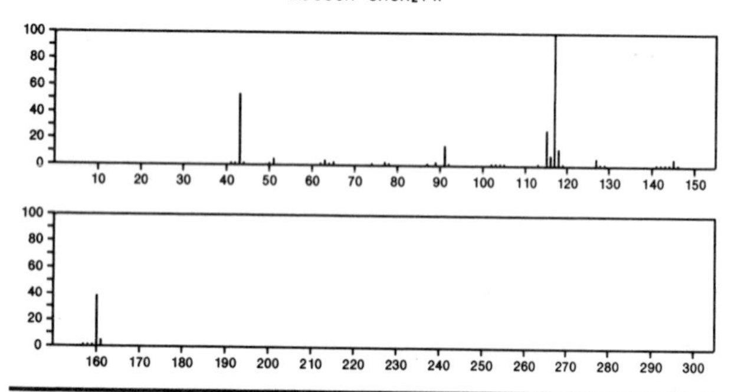

160 $C_{11}H_{12}O$ 13153–75–8
1,4–Methanonaphthalen–2–ol, 1,2,3,4–tetrahydro–, *endo*–

OH

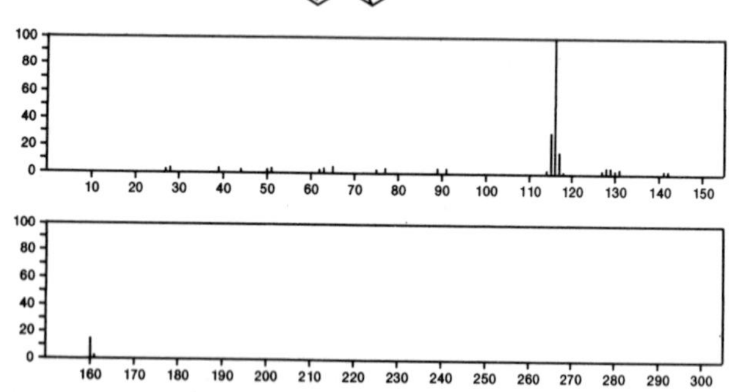

160 $C_{11}H_{12}O$ 13999–10–5
1,4–Methanonaphthalen–9–ol, 1,2,3,4–tetrahydro–, *syn*–

– OH

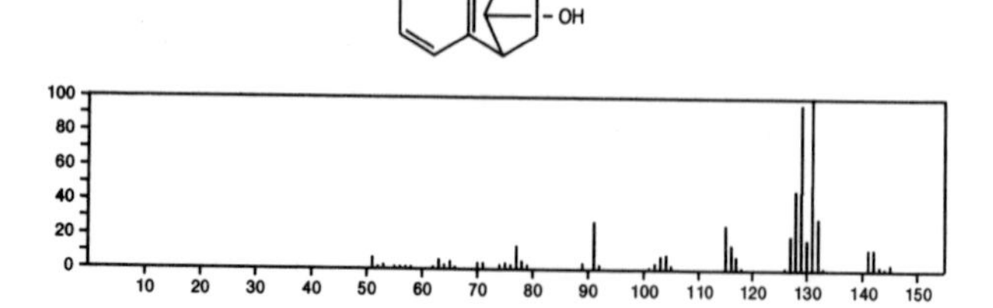

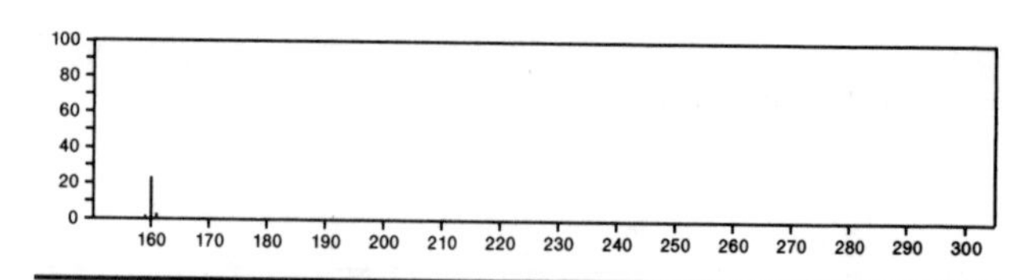

160 $C_{11}H_{12}O$ 14113–94–1
Ethanone, 1–cyclopropyl–2–phenyl–

COCH2 Ph

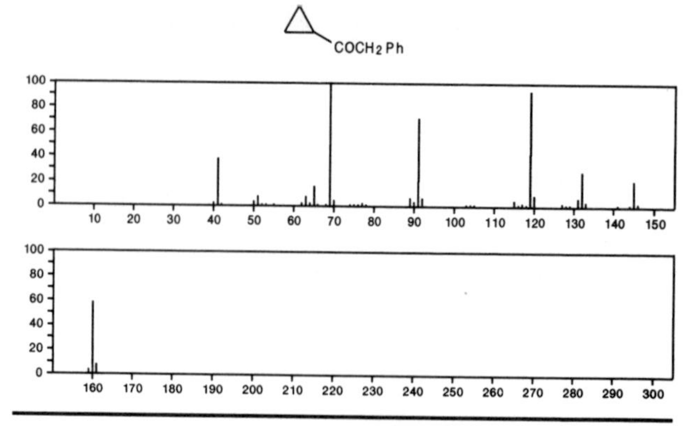

160 $C_{11}H_{12}O$ 14944–23–1
1(2*H*)–Naphthalenone, 3,4–dihydro–3–methyl–

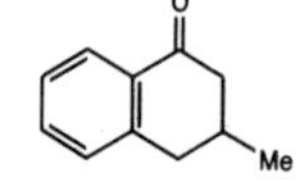

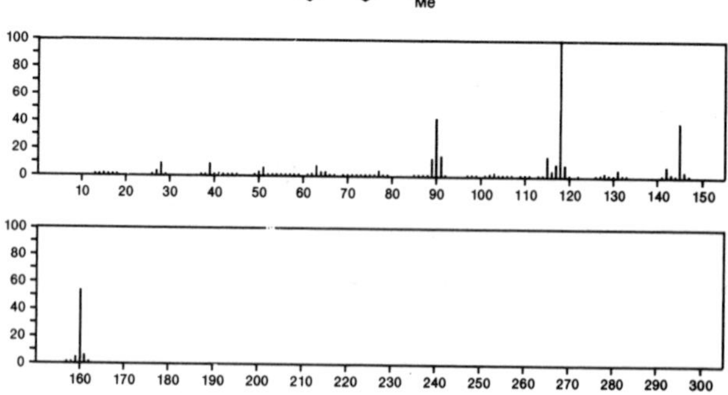

160 $C_{11}H_{12}O$ 16440–97–4
1–Indanone, 5,6–dimethyl–

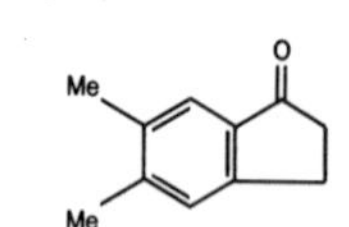

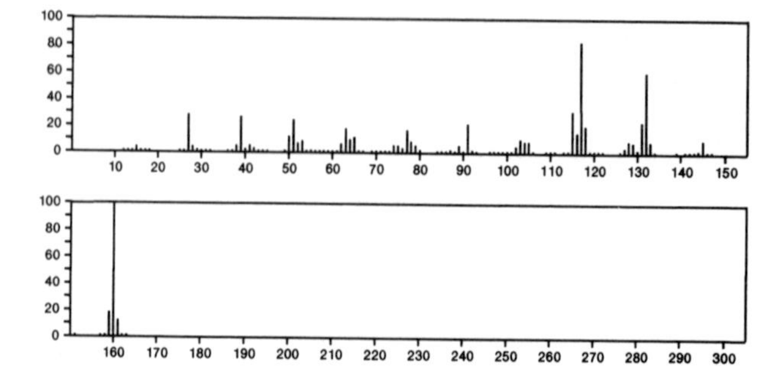

160 $C_{11}H_{12}O$ 19832-98-5
1(2*H*)-Naphthalenone, 3,4-dihydro-4-methyl-

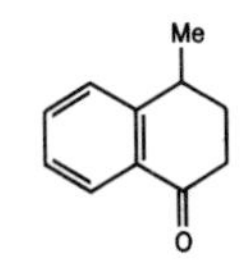

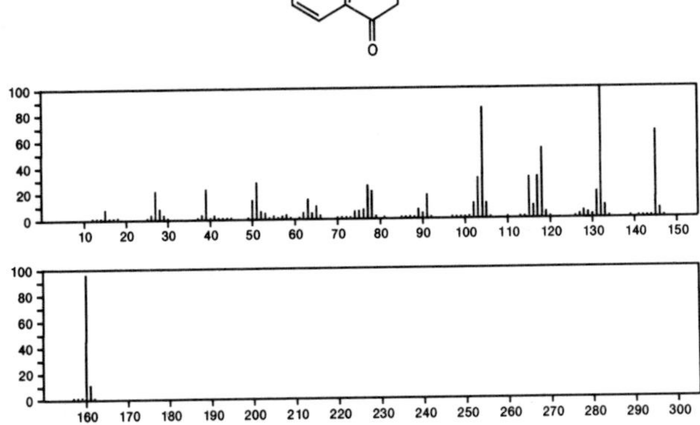

160 $C_{11}H_{12}O$ 26465-81-6
1*H*-Inden-1-one, 2,3-dihydro-3,3-dimethyl-

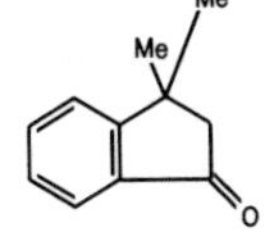

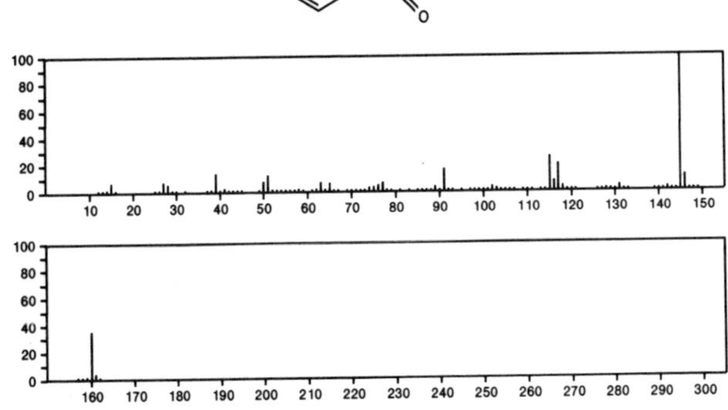

160 $C_{11}H_{12}O$ 33046-84-3
2-Pentenal, 5-phenyl-

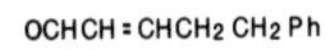

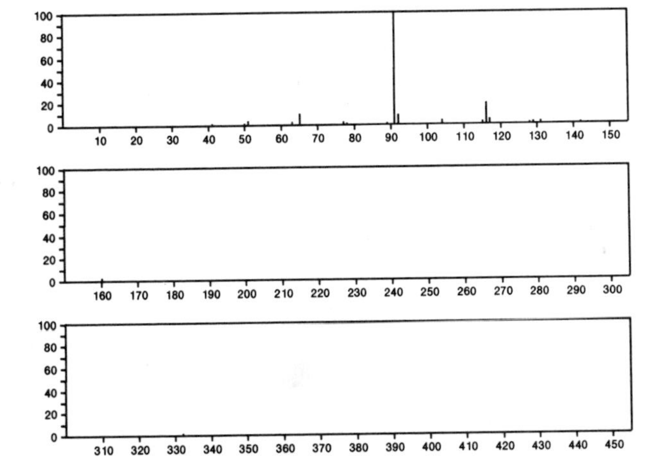

160 $C_{11}H_{12}O$ 51015-28-2
1(2*H*)-Naphthalenone, 3,4-dihydro-8-methyl-

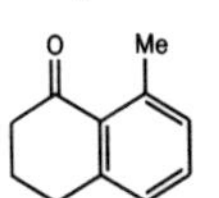

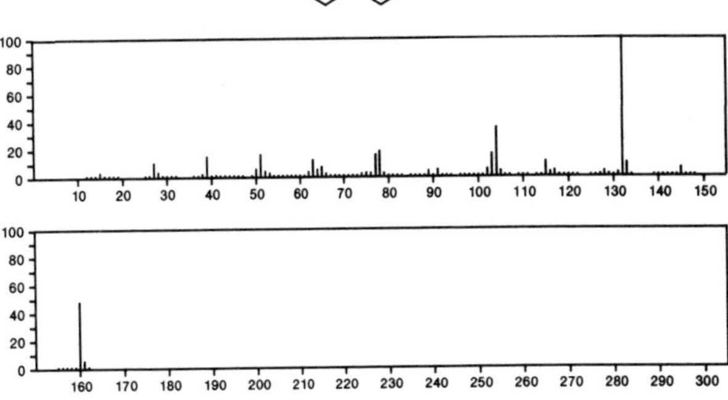

160 $C_{11}H_{12}O$ 51738-14-8
2,4-Cyclohexadien-1-one, 2,6-dimethyl-6-(2-propynyl)-

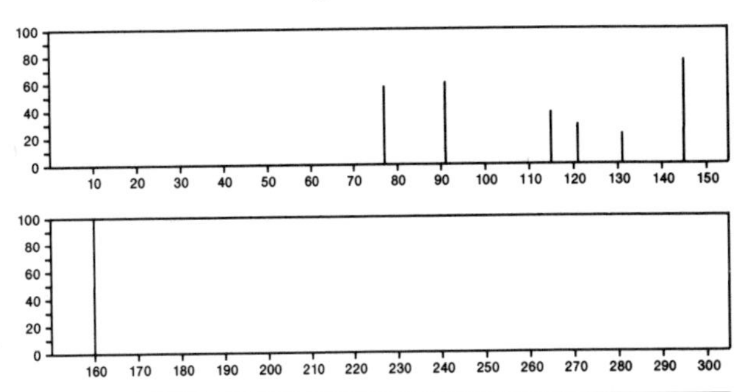

160 $C_{11}H_{12}O$ 51738-15-9
2,4-Cyclohexadien-1-one, 4,6-dimethyl-6-(2-propynyl)-

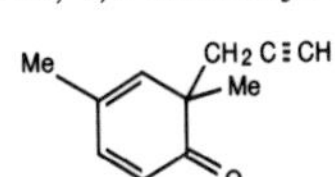

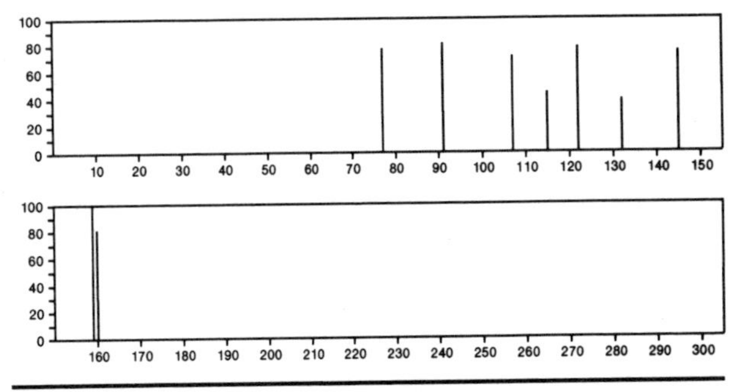

160 $C_{11}H_{12}O$ 55255-94-2
1,4-Methanonaphthalen-9-ol, 1,2,3,4-tetrahydro-

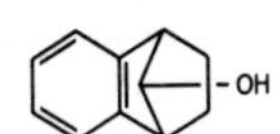

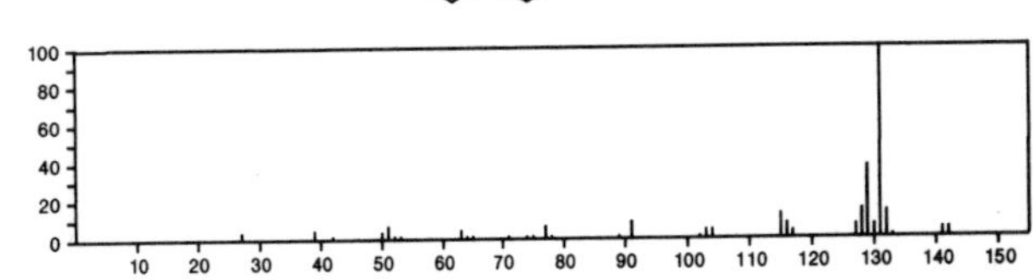

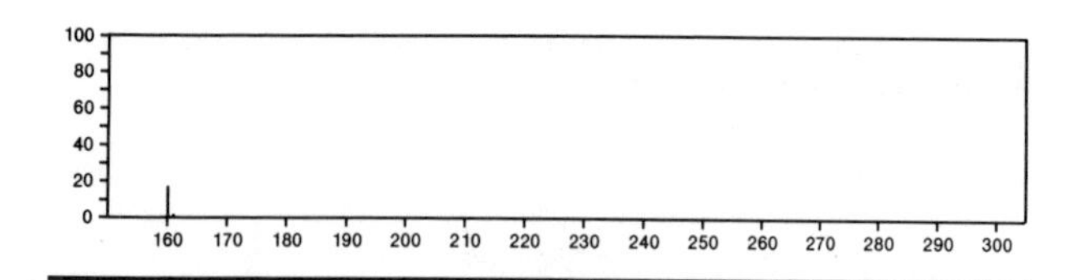

160 $C_{11}H_{12}O$ 55956-30-4
3-Buten-2-one, 3-methyl-1-phenyl-

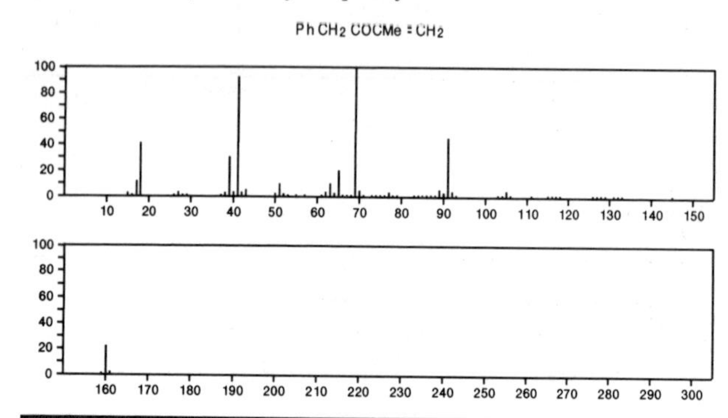

160 $C_{11}H_{12}O$ 56667-10-8
2-Cyclopenten-1-ol, 1-phenyl-

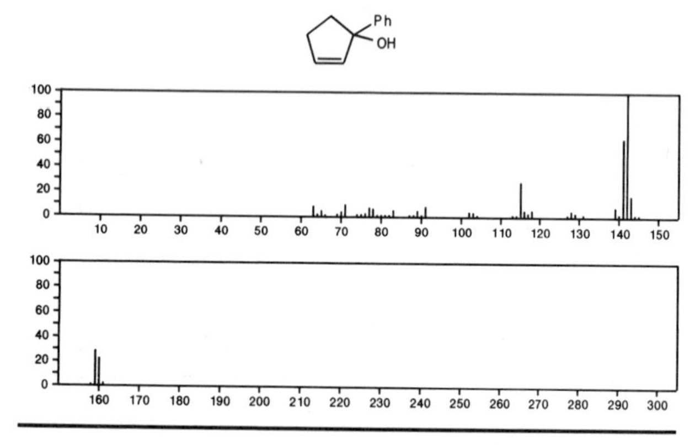

160 $C_{12}H_{16}$ 827-52-1
Benzene, cyclohexyl-

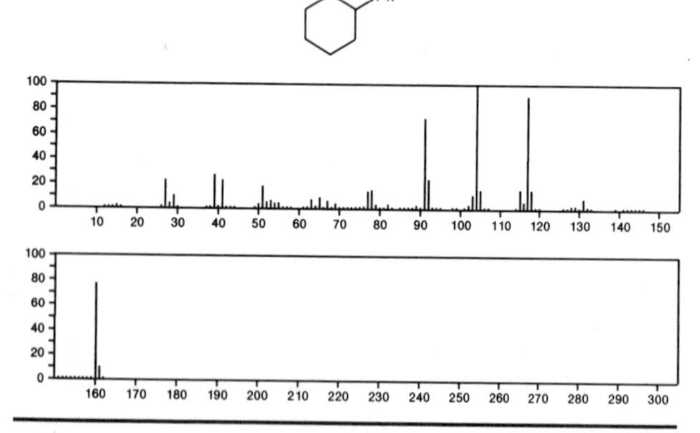

160 $C_{12}H_{16}$ 1076-61-5
Naphthalene, 1,2,3,4-tetrahydro-6,7-dimethyl-

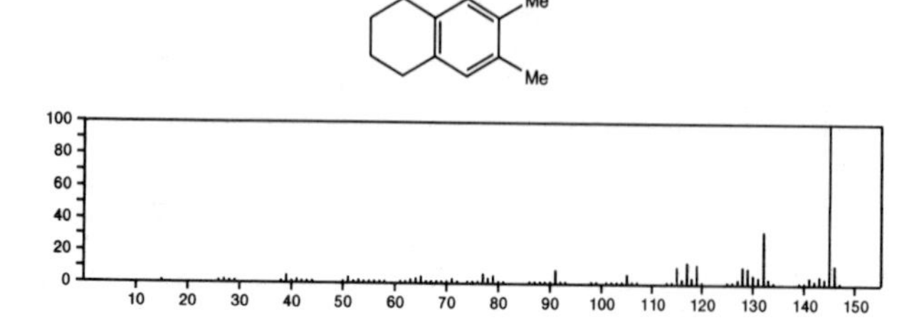

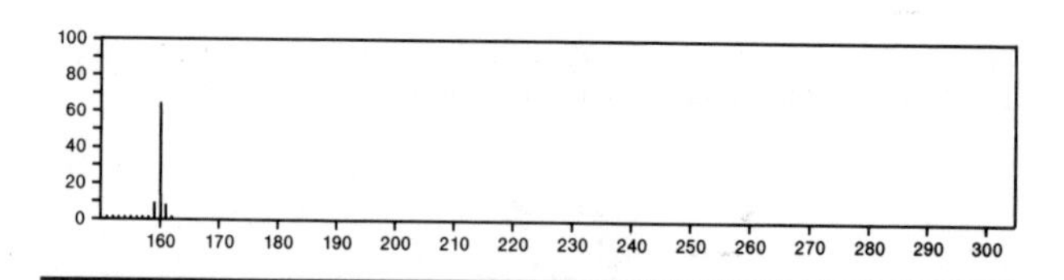

160 $C_{12}H_{16}$ 1129-29-9
Benzene, 1-(1-methylethenyl)-3-(1-methylethyl)-

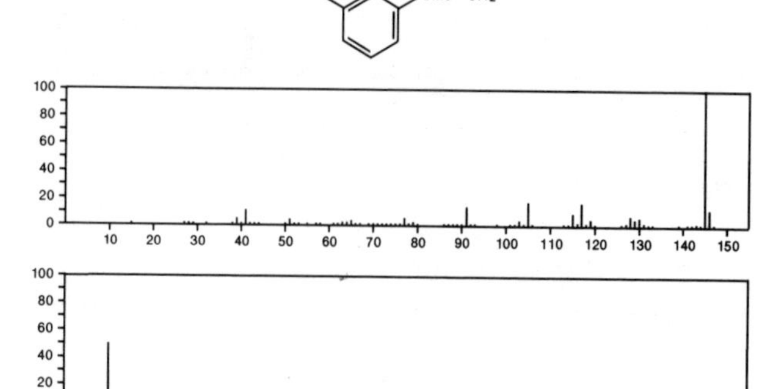

160 $C_{12}H_{16}$ 1746-23-2
Benzene, 1-(1,1-dimethylethyl)-4-ethenyl-

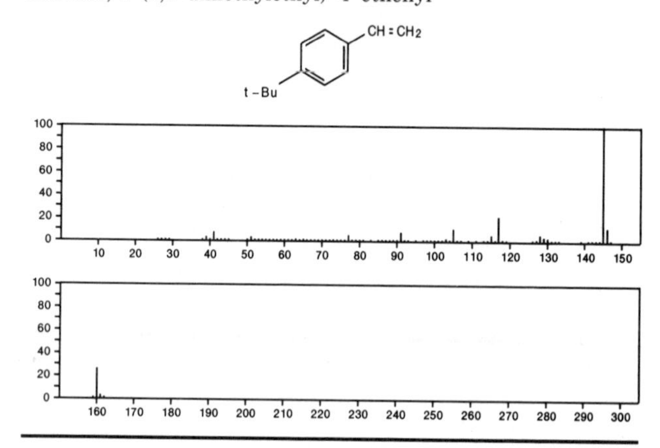

160 $C_{12}H_{16}$ 1985-59-7
Naphthalene, 1,2,3,4-tetrahydro-1,1-dimethyl-

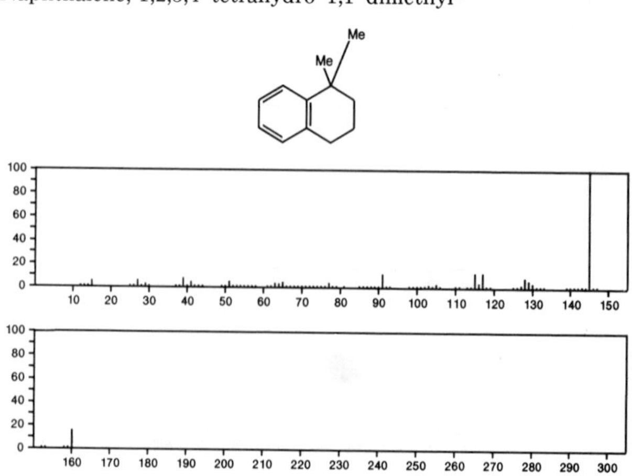

160 $C_{12}H_{16}$ 2388-14-9
Benzene, 1-(1-methylethenyl)-4-(1-methylethyl)-

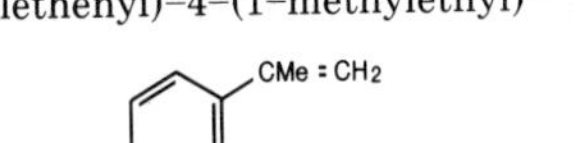

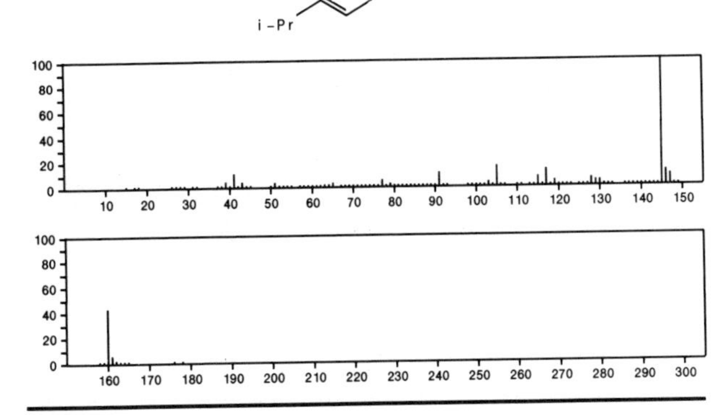

160 $C_{12}H_{16}$ 2613-76-5
1*H*-Indene, 2,3-dihydro-1,1,3-trimethyl-

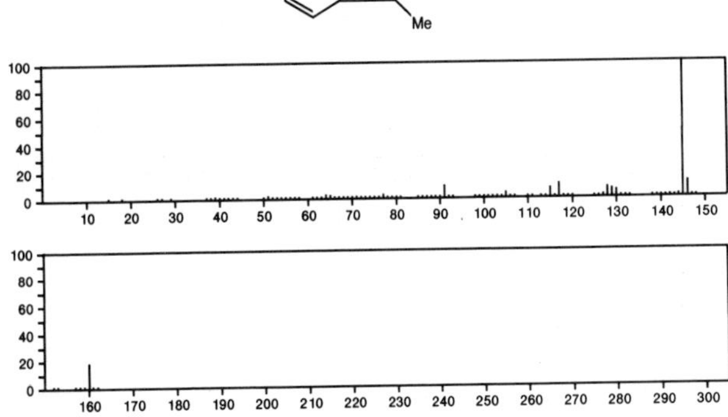

160 $C_{12}H_{16}$ 4175-54-6
Naphthalene, 1,2,3,4-tetrahydro-1,4-dimethyl-

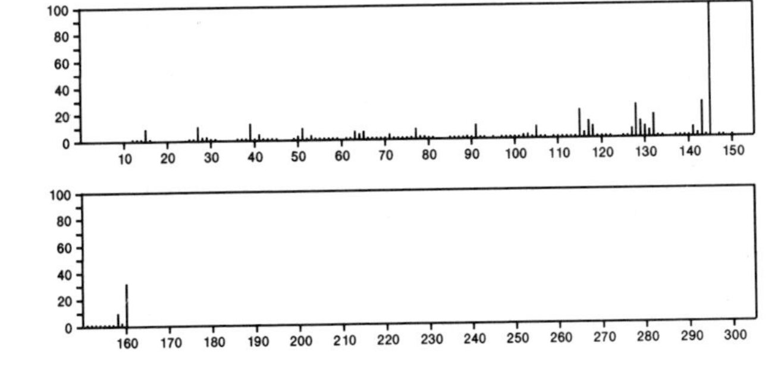

160 $C_{12}H_{16}$ 6682-06-0
Indan, 4,5,7-trimethyl-

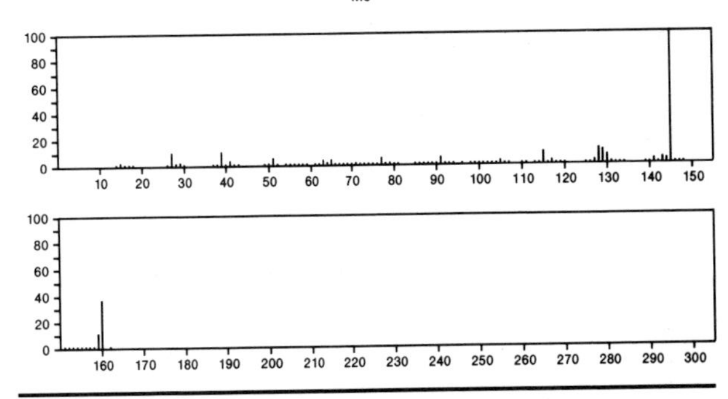

160 $C_{12}H_{16}$ 7524-63-2
Naphthalene, 1,2,3,4-tetrahydro-2,6-dimethyl-

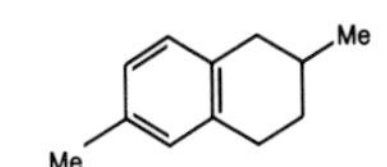

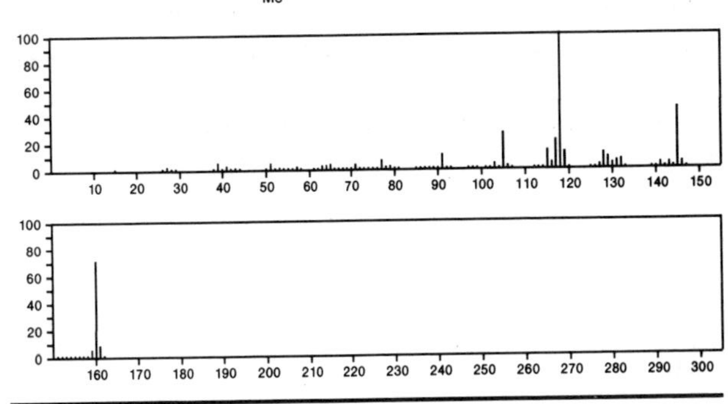

160 $C_{12}H_{16}$ 13065-07-1
Naphthalene, 1,2,3,4-tetrahydro-2,7-dimethyl-

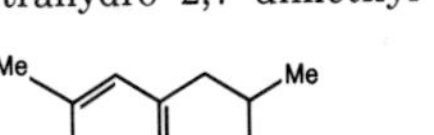

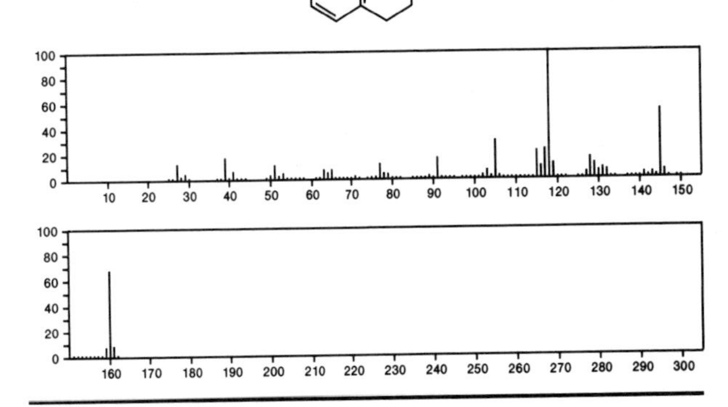

160 $C_{12}H_{16}$ 14276-95-0
1*H*-Indene, 2,3-dihydro-1,1,6-trimethyl-

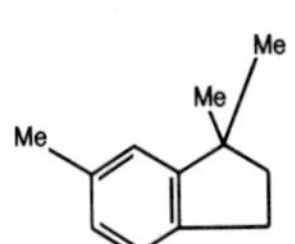

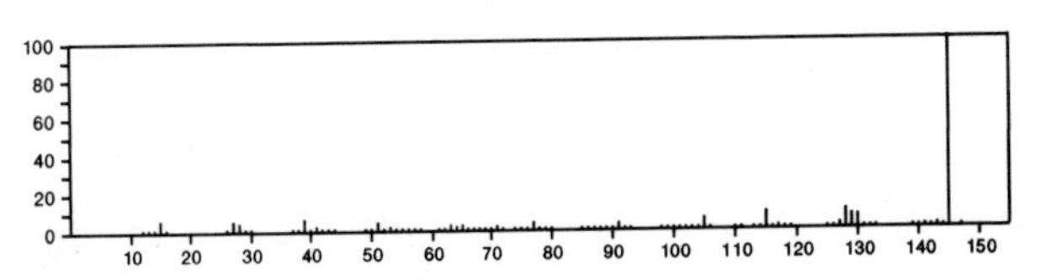

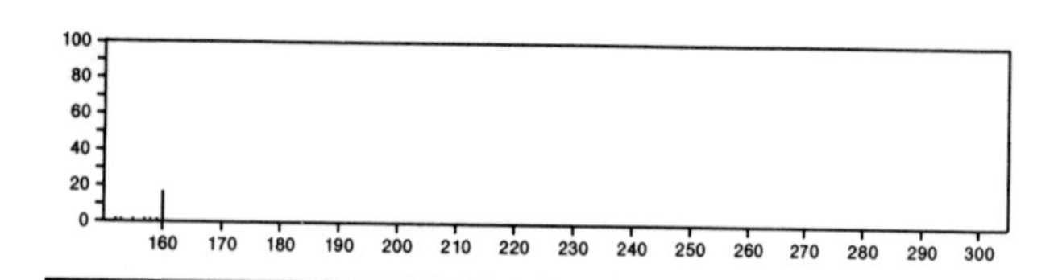

160 $C_{12}H_{16}$ 16204-72-1
1*H*-Indene, 2,3-dihydro-1,1,4-trimethyl-

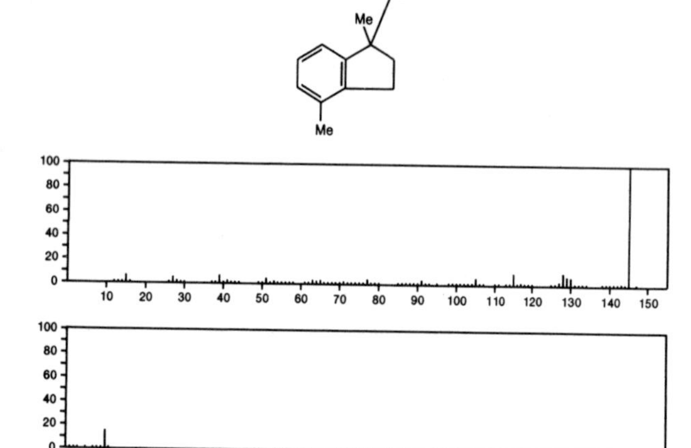

160 $C_{12}H_{16}$ 20027-77-4
Naphthalene, 1,2,3,4-tetrahydro-5,6-dimethyl-

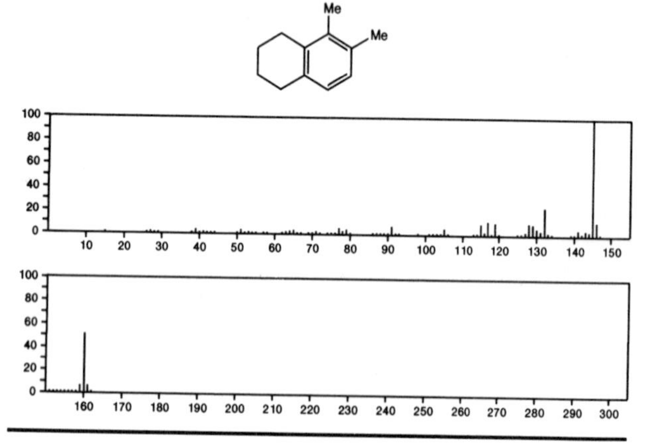

160 $C_{12}H_{16}$ 21564-91-0
Naphthalene, 1,2,3,4-tetrahydro-1,5-dimethyl-

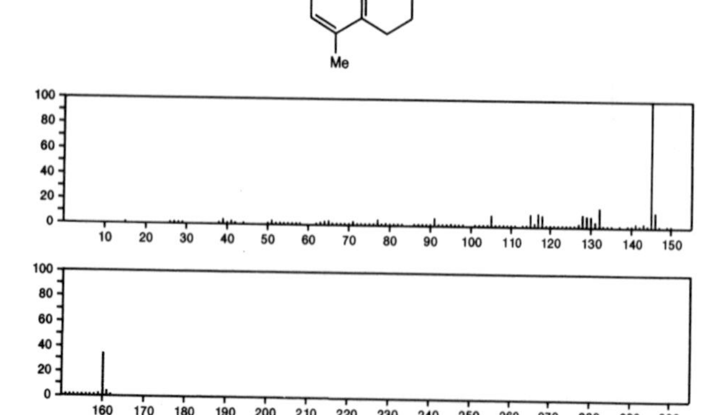

160 $C_{12}H_{16}$ 21564-92-1
Naphthalene, 1,2,3,4-tetrahydro-2,3-dimethyl-

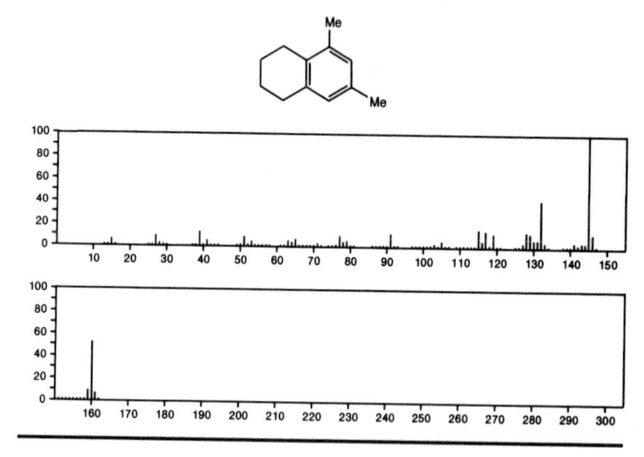

160 $C_{12}H_{16}$ 21693-54-9
Naphthalene, 1,2,3,4-tetrahydro-5,7-dimethyl-

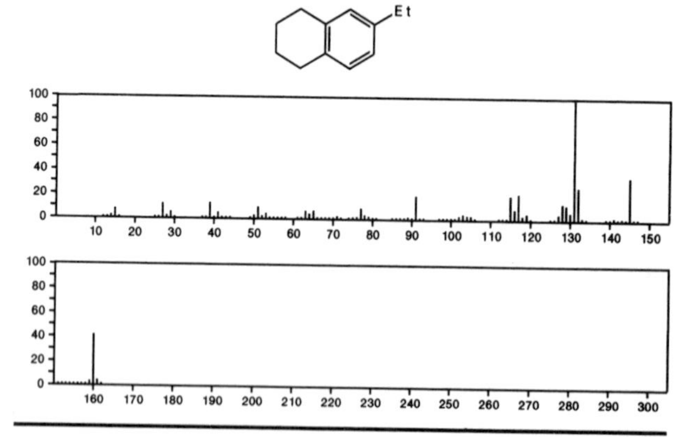

160 $C_{12}H_{16}$ 22531-20-0
Naphthalene, 6-ethyl-1,2,3,4-tetrahydro-

160 $C_{12}H_{16}$ 25419-33-4
Naphthalene, 1,2,3,4-tetrahydro-1,8-dimethyl-

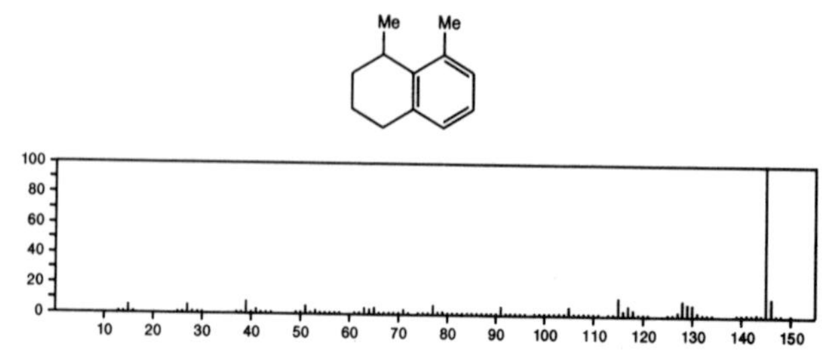

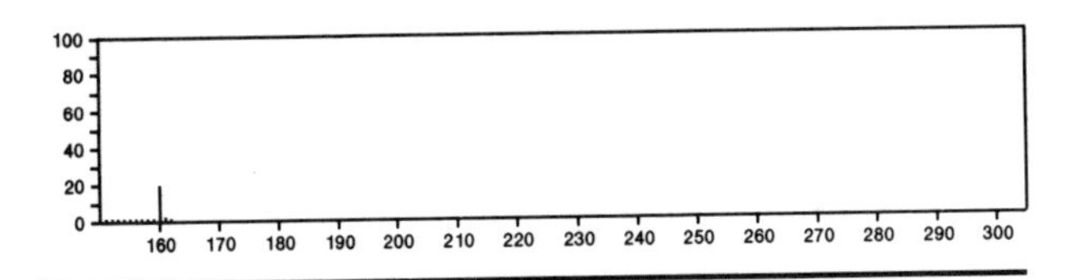

160 $C_{12}H_{16}$ 27193-71-1
Benzene, (1-methylethenyl)(1-methylethyl)-

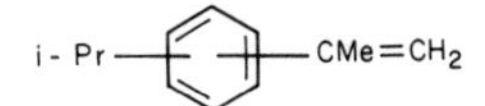

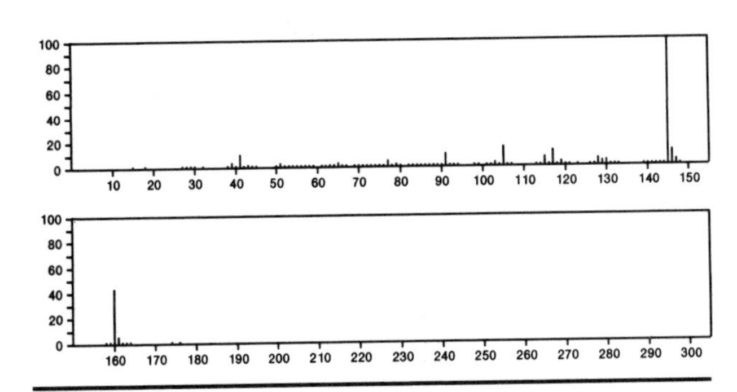

160 $C_{12}H_{16}$ 28229-15-4
Biphenylene, 1,2,3,6,7,8,8a,8b-octahydro-, *trans*-

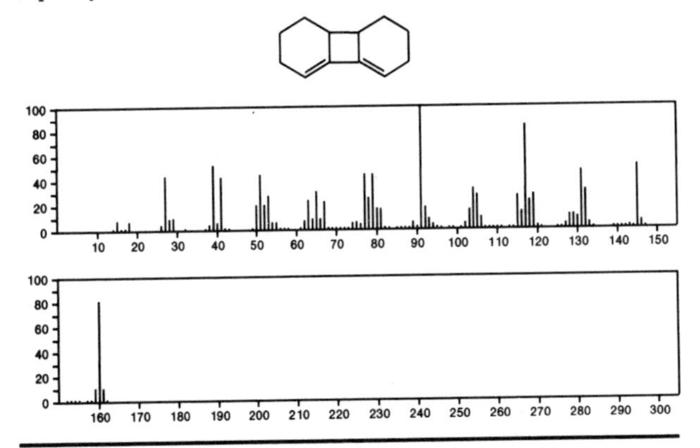

160 $C_{12}H_{16}$ 32367-54-7
Naphthalene, 2-ethyl-1,2,3,4-tetrahydro-

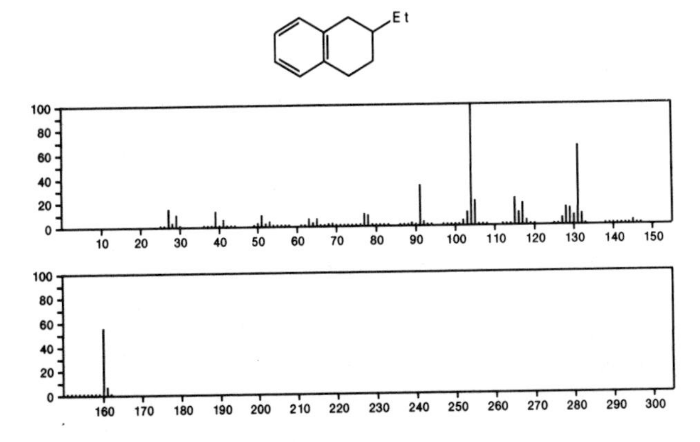

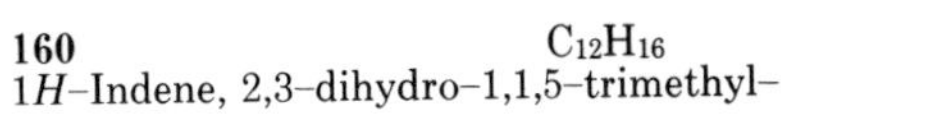

160 $C_{12}H_{16}$ 40650-41-7
1*H*-Indene, 2,3-dihydro-1,1,5-trimethyl-

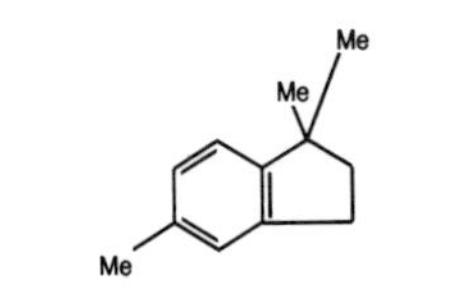

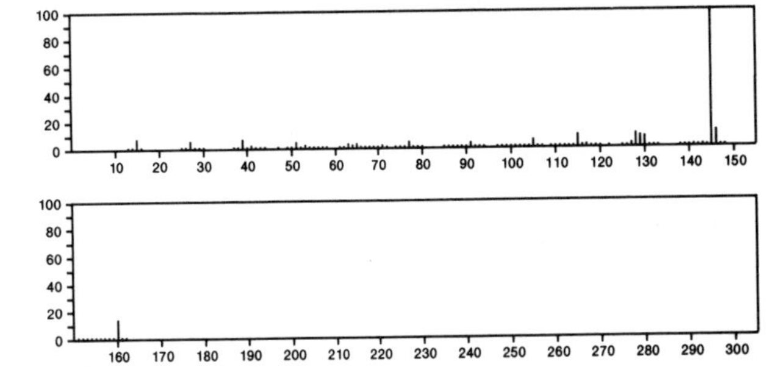

160 $C_{12}H_{16}$ 42524-30-1
Benzene, (3-methyl-4-pentenyl)-

$H_2C:CHCHMeCH_2CH_2Ph$

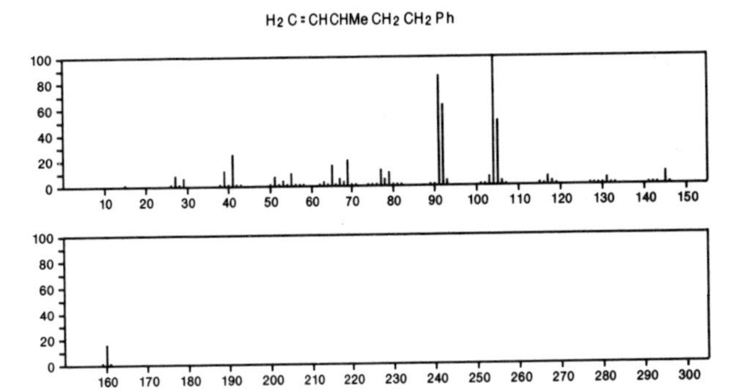

160 $C_{12}H_{16}$ 42775-75-7
Naphthalene, 5-ethyl-1,2,3,4-tetrahydro-

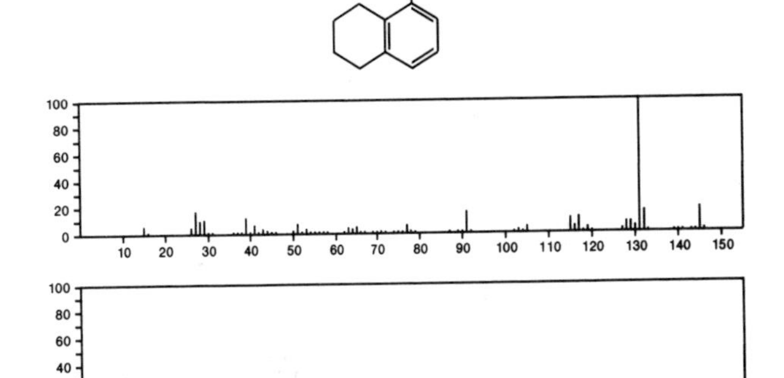

160 $C_{12}H_{16}$ 50871-04-0
Benzene, (1,2-dimethyl-3-butenyl)-

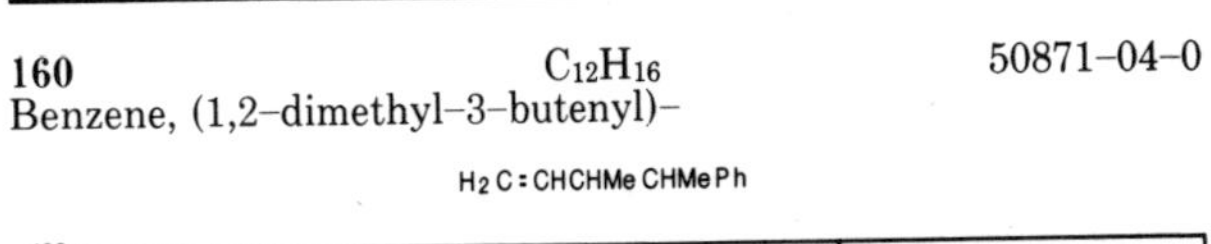

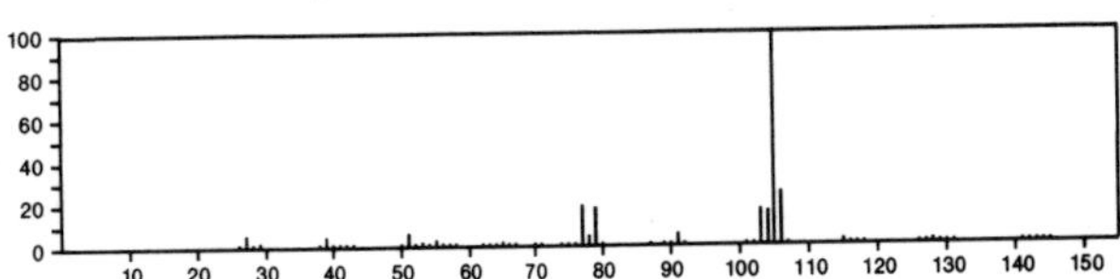

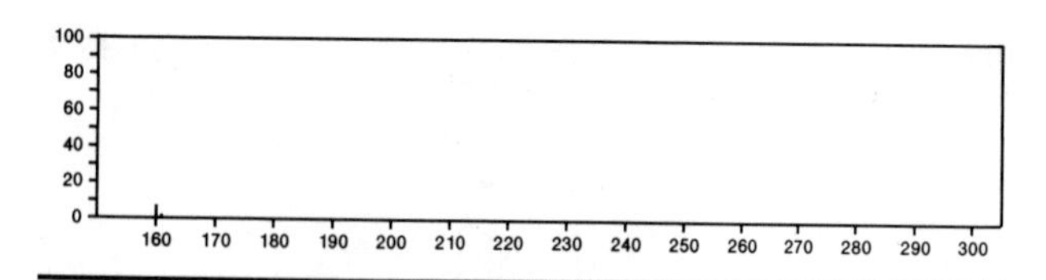

160 $C_{12}H_{16}$ 54340-83-9
Benzene, 1,2,4-trimethyl(1-methylethenyl)-

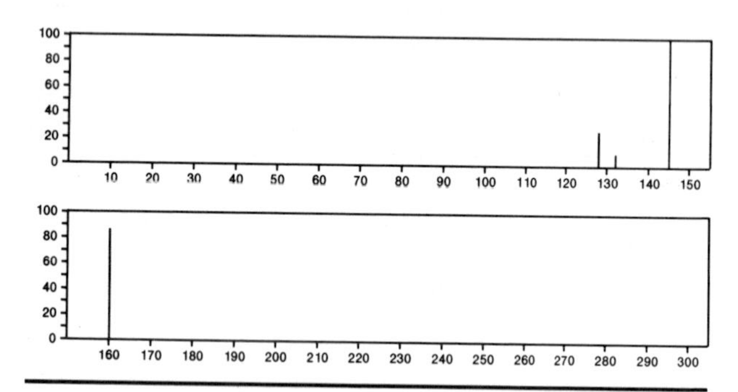

160 $C_{12}H_{16}$ 54340-84-0
Benzene, 1,2,4-trimethyl-5-(1-methylethenyl)-

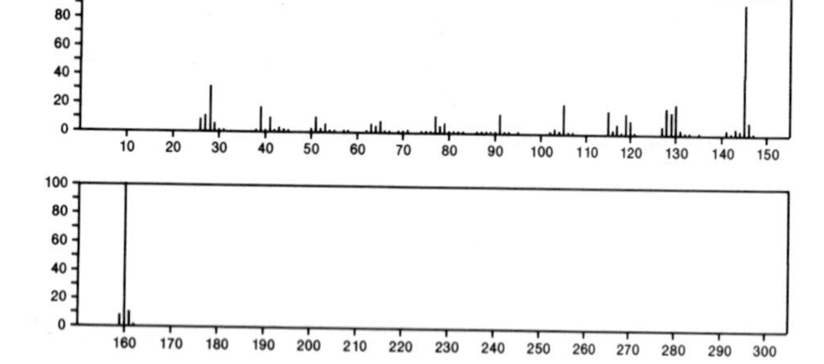

160 $C_{12}H_{16}$ 54340-85-1
Benzene, 1-(2-butenyl)-2,3-dimethyl-

160 $C_{12}H_{16}$ 54340-86-2
Benzene, 4-(2-butenyl)-1,2-dimethyl-, (*E*)-

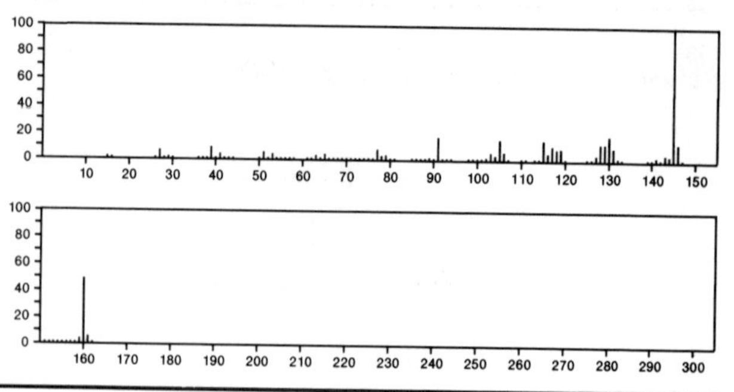

160 $C_{12}H_{16}$ 54340-87-3
1*H*-Indene, 2,3-dihydro-1,4,7-trimethyl-

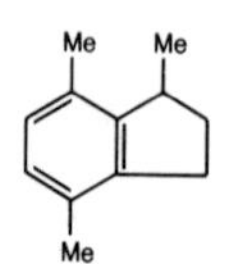

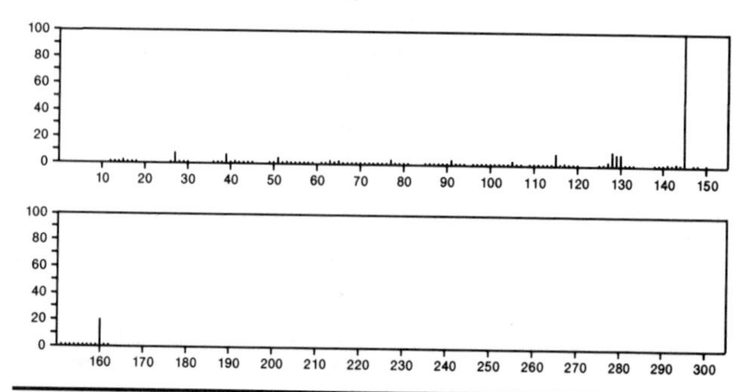

160 $C_{12}H_{16}$ 54340-88-4
1*H*-Indene, 2,3-dihydro-1,5,7-trimethyl-

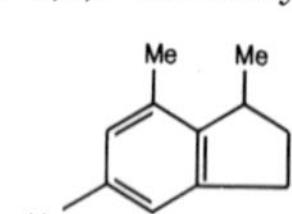

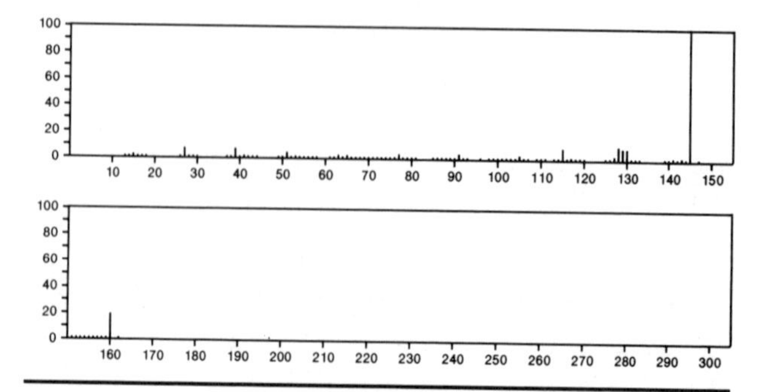

160 $C_{12}H_{16}$ 56282-43-0
Benzene, (1-cyclopropyl-1-methylethyl)-

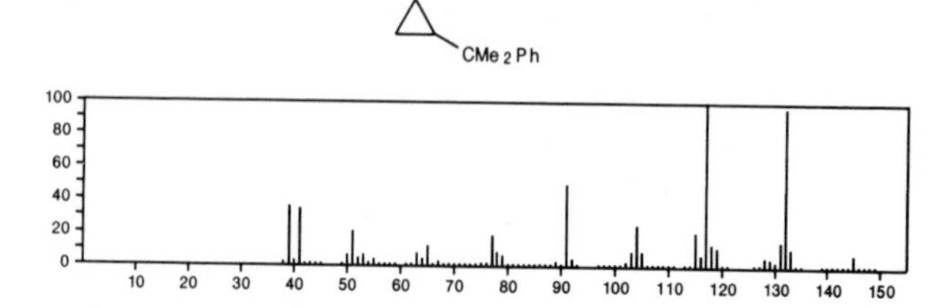

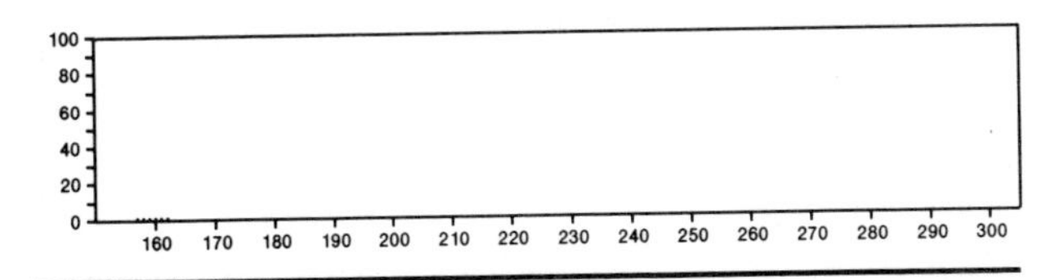

160 $C_{12}H_{16}$ 56298-75-0
1*H*-Indene, 1-ethyl-2,3-dihydro-1-methyl-

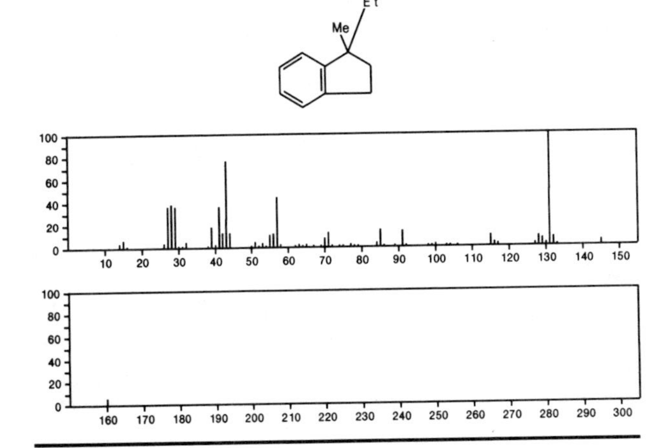

160 $C_{12}H_{16}$ 56818-01-0
Benzene, 1-methyl-4-(3-methyl-3-butenyl)-

161 $C_4H_7N_3S_2$ 25660-70-2
1,3,4-Thiadiazol-2-amine, 5-(ethylthio)-

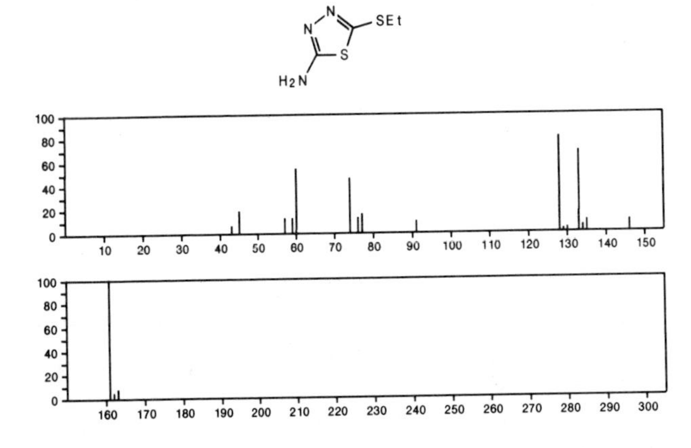

161 $C_4H_{11}N_5O_2$ 20004-00-6
Ethanimidamide, 2,2'-iminobis[*N*-hydroxy-

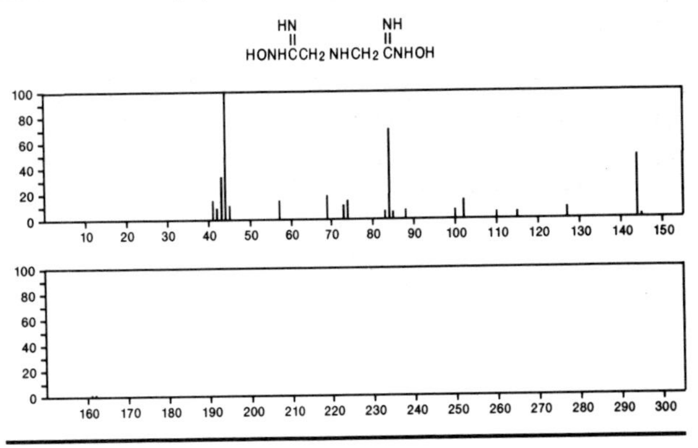

161 $C_6H_5Cl_2N$ 95-76-1
Benzenamine, 3,4-dichloro-

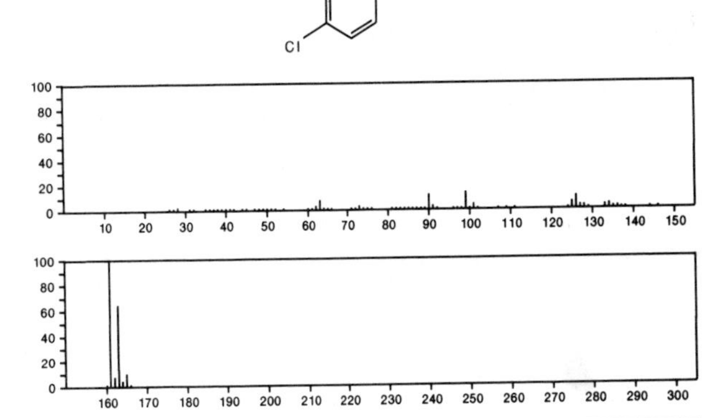

161 $C_6H_5Cl_2N$ 95-82-9
Benzenamine, 2,5-dichloro-

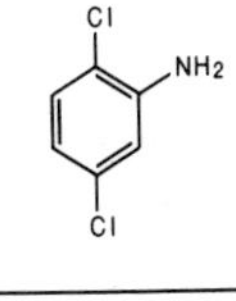

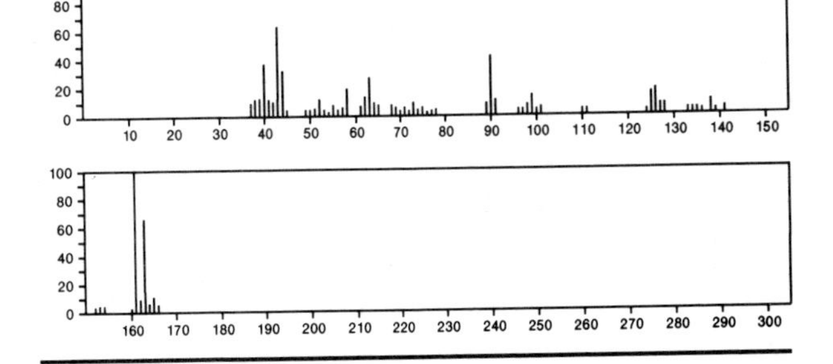

161 $C_6H_5Cl_2N$ 554-00-7
Benzenamine, 2,4-dichloro-

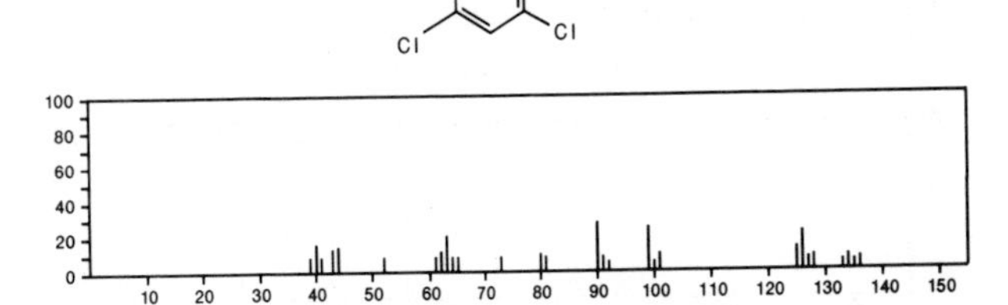

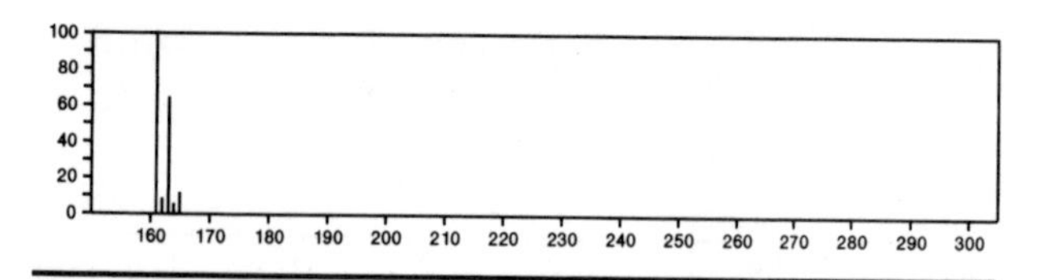

161 $C_6H_5Cl_2N$ 608-27-5
Benzenamine, 2,3-dichloro-

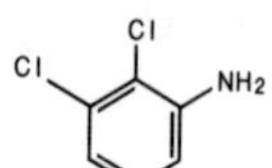

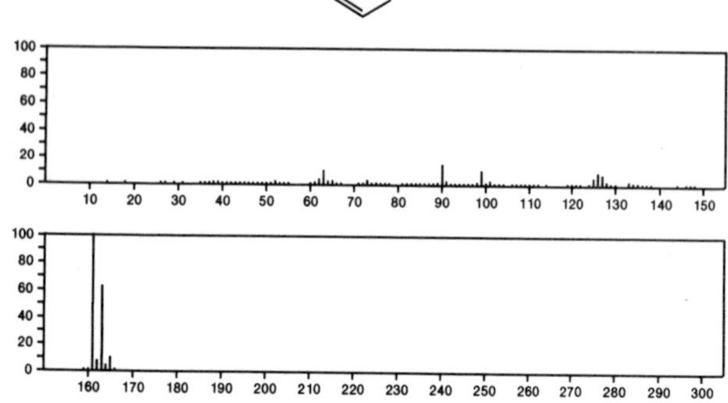

161 $C_6H_{11}NO_2S$ 16703-48-3
Acetic acid, (dimethylamino)thioxo-, ethyl ester

$Me_2NCSC(O)OEt$

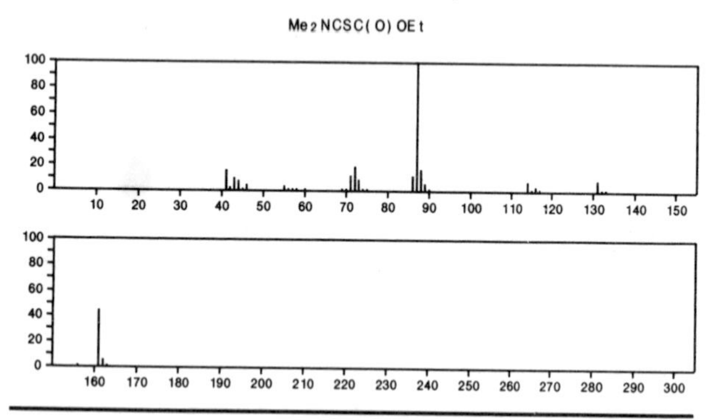

161 $C_6H_{11}NO_4$ 55299-56-4
DL-Serine, *N*-acetyl-, methyl ester

CH_2OH
$MeOC(O)CHNHAc$

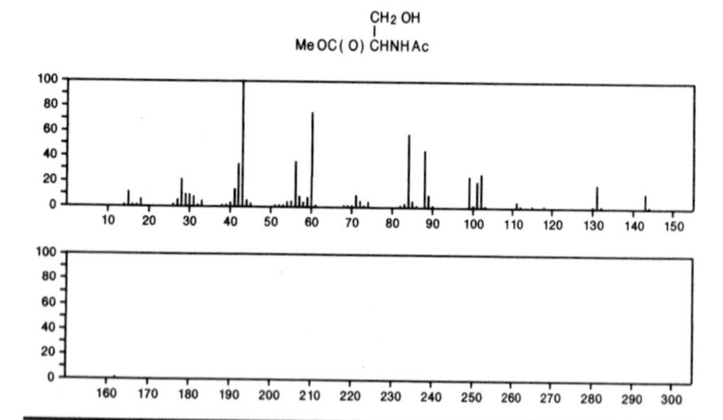

161 $C_6H_{15}NO_2Si$ 25688-72-6
Glycine, *N*-(trimethylsilyl)-, methyl ester

$MeOC(O)CH_2NHSiMe_3$

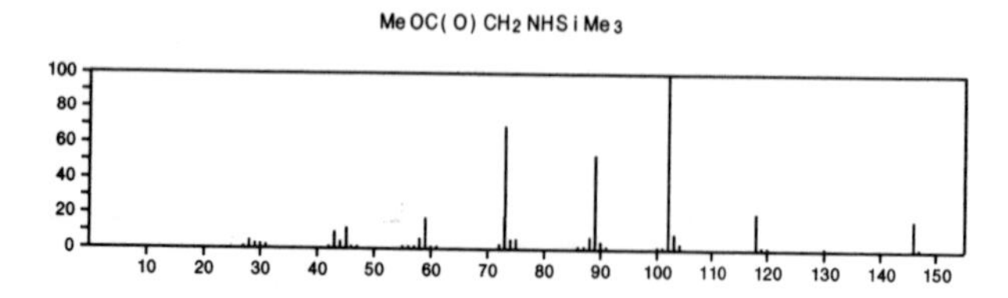

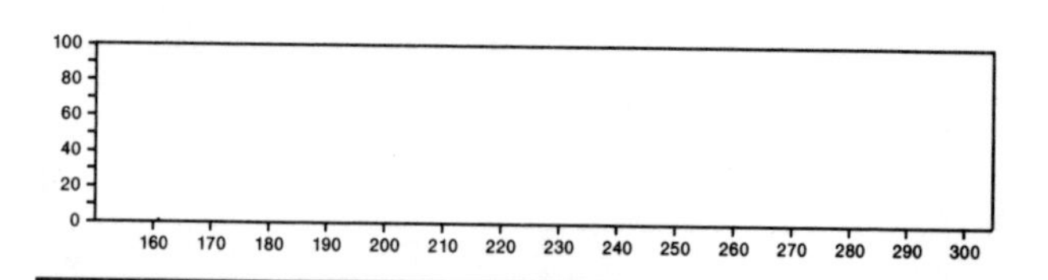

161 $C_6H_{19}NSi_2$ 999-97-3
Silanamine, 1,1,1-trimethyl-*N*-(trimethylsilyl)-

$Me_3SiNHSiMe_3$

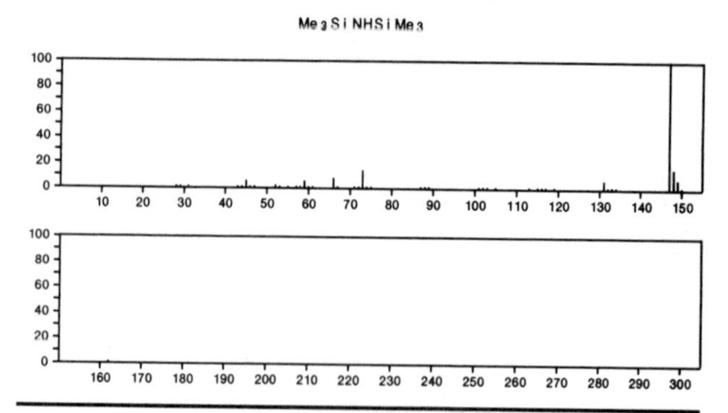

161 $C_6H_{19}NSi_2$ 32713-31-8
Silanamine, *N*-(dimethylsilyl)-1,1,1-trimethyl-, monomethyl deriv.

$NH(SiH_3)_2$ +6Me

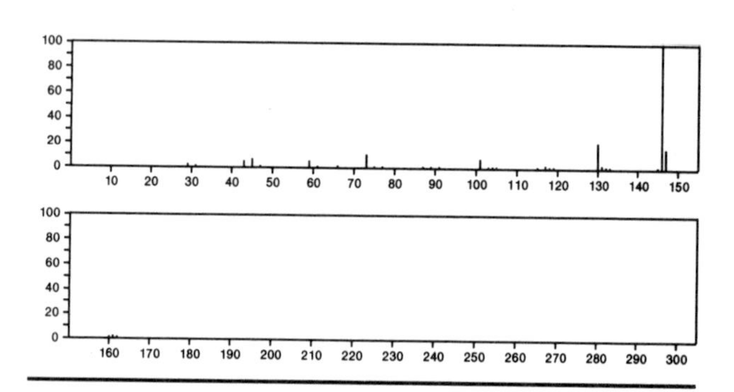

161 $C_7H_6F_3N$ 98-16-8
Benzenamine, 3-(trifluoromethyl)-

F_3C NH_2

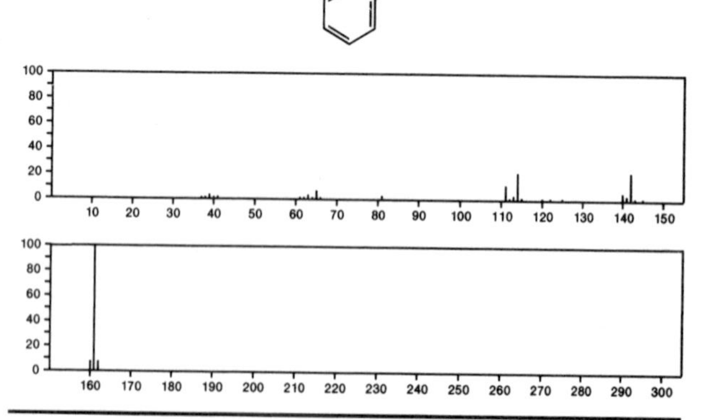

161 $C_7H_{15}NO_3$ 20633-12-9
Nitric acid, heptyl ester

$Me(CH_2)_6ONO_2$

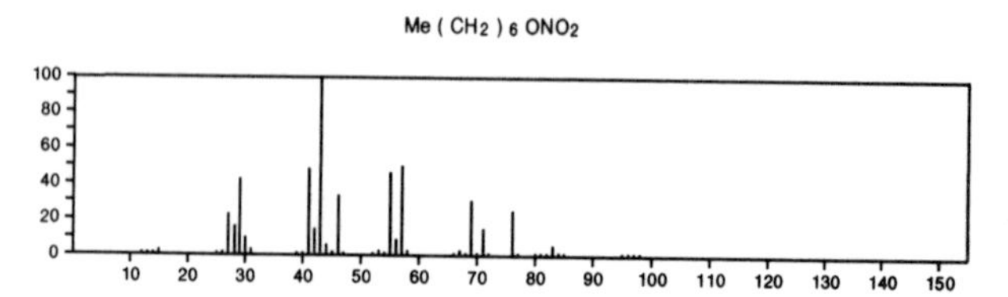

161 $C_7H_{19}NOSi$ 16654-64-1
Ethanamine, *N,N*-dimethyl-2-[(trimethylsilyl)oxy]-

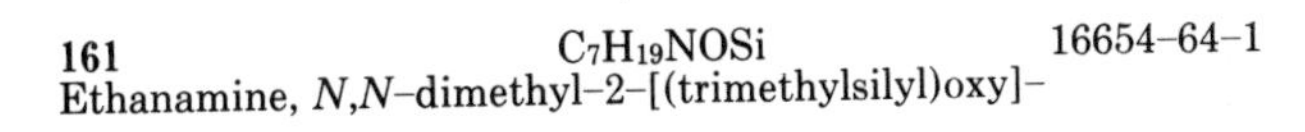

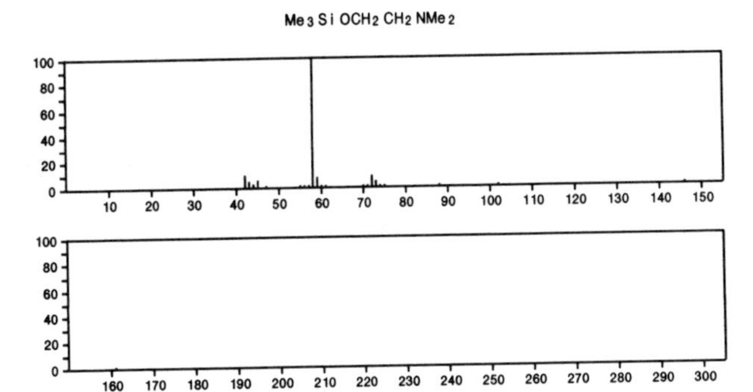

161 $C_8H_7N_3O$ 3303-26-2
Pyrido[3,2-*d*]pyrimidin-4(3*H*)-one, 2-methyl-

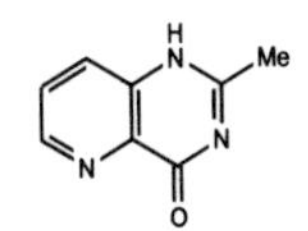

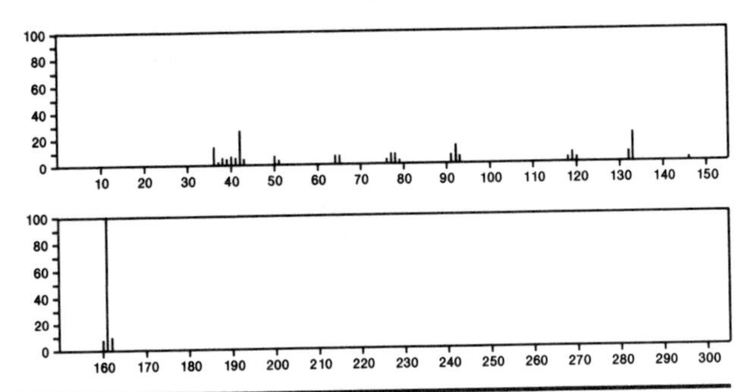

161 $C_8H_7N_3O$ 3663-37-4
1,2,4-Oxadiazol-5-amine, 3-phenyl-

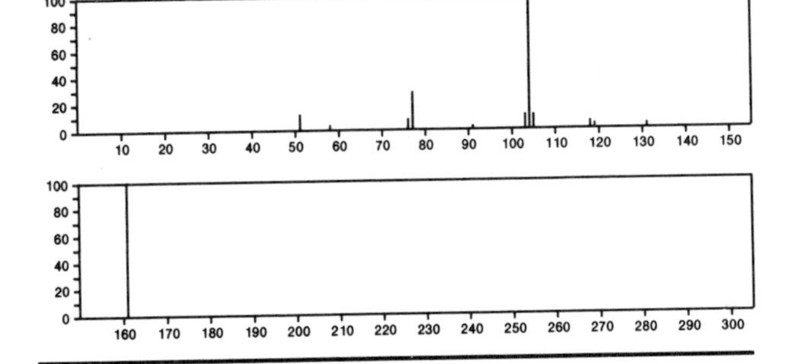

161 $C_8H_7N_3O$ 10350-68-2
Pyridine, 2-(5-methyl-1,2,4-oxadiazol-3-yl)-

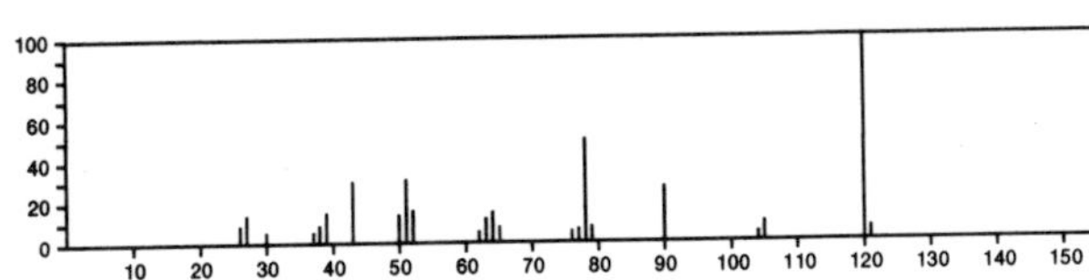

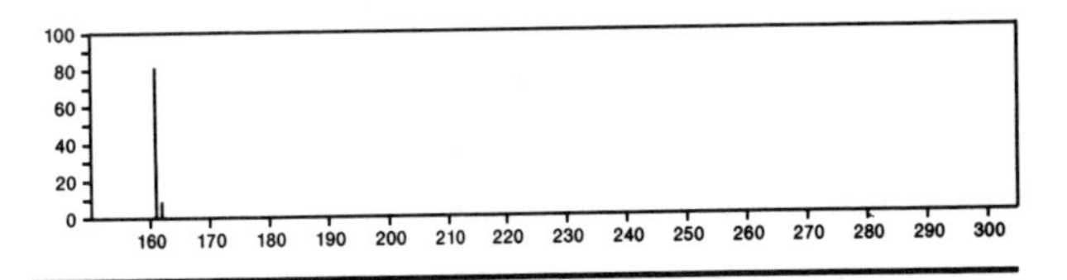

161 $C_8H_7N_3O$ 18773-93-8
1*H*-Benzotriazole, 1-acetyl-

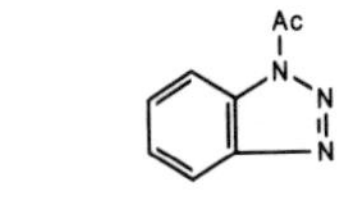

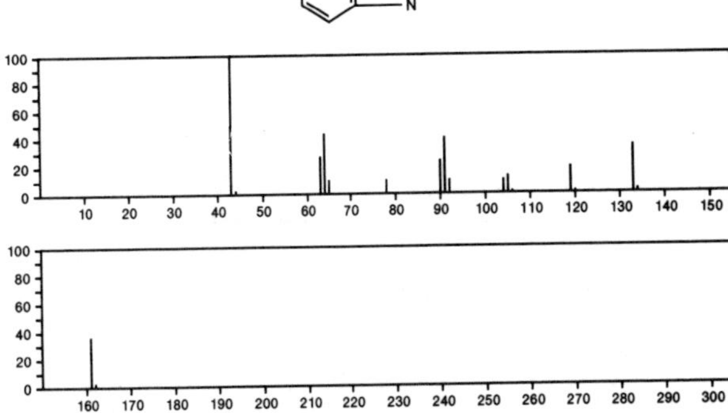

161 $C_8H_7N_3O$ 21434-16-2
3*H*-1,2,4-Triazol-3-one, 2,4-dihydro-2-phenyl-

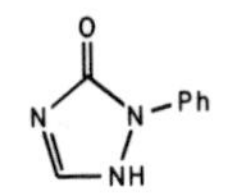

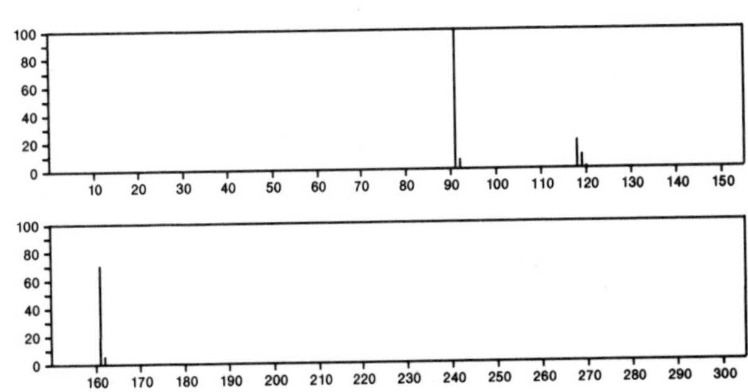

161 $C_8H_7N_3O$ 22389-82-8
Pyrido[3,4-*d*]pyrimidin-4(3*H*)-one, 3-methyl-

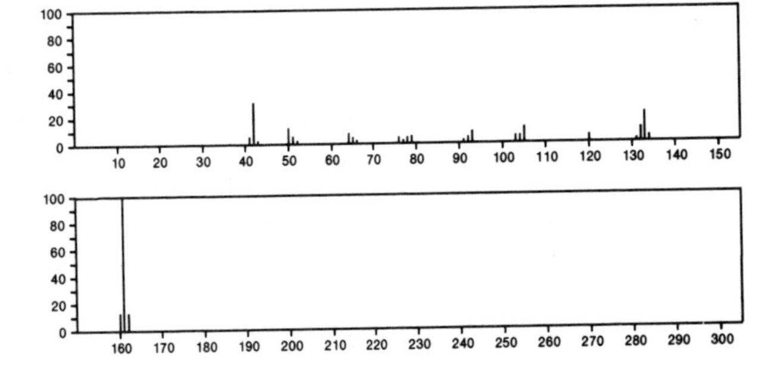

161 $C_8H_7N_3O$ 28732-78-7
Pyrido[2,3-*d*]pyrimidine, 4-methoxy-

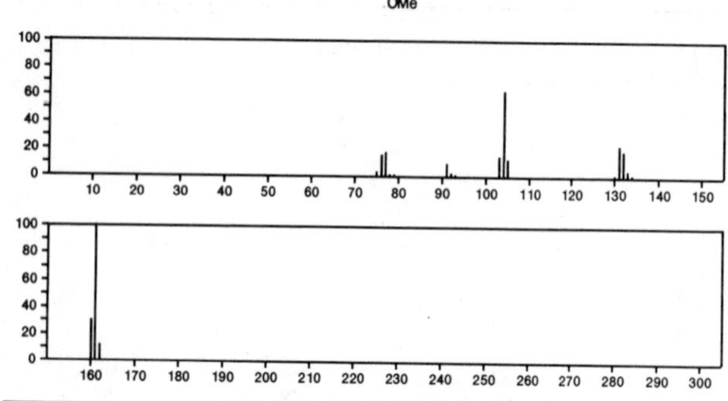

161 $C_9H_7NO_2$ 58-57-1
2(1*H*)-Quinolinone, 1-hydroxy-

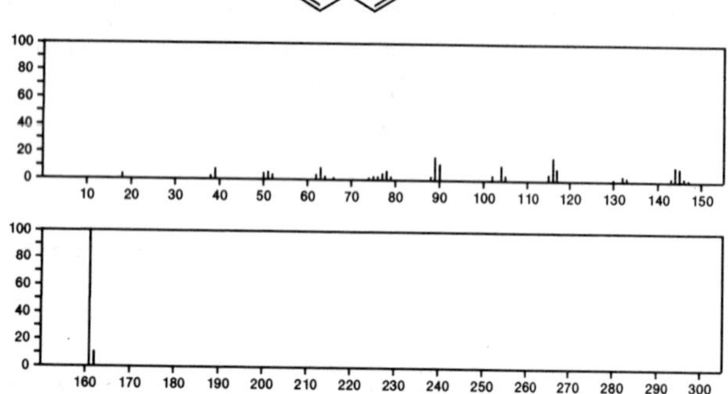

161 $C_9H_7NO_2$ 608-05-9
1*H*-Indole-2,3-dione, 5-methyl-

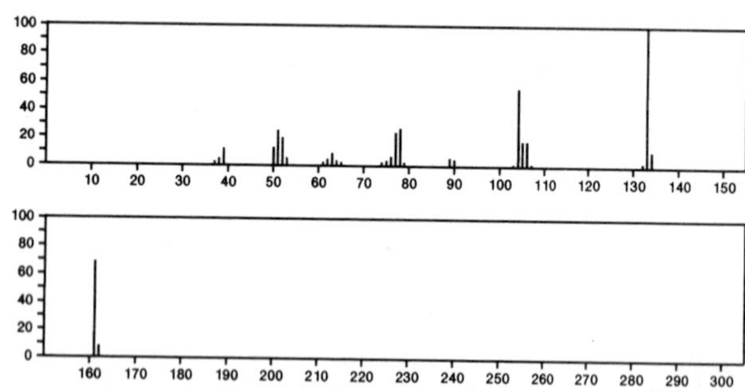

161 $C_9H_7NO_2$ 771-50-6
1*H*-Indole-3-carboxylic acid

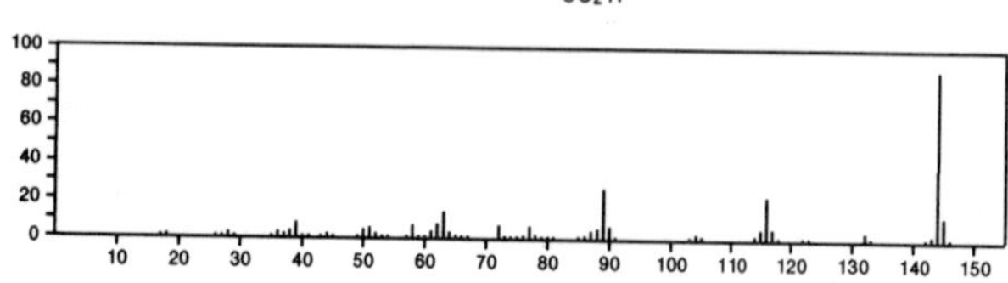

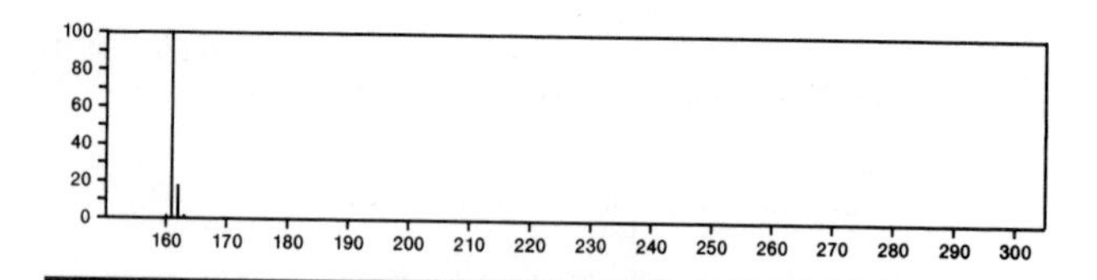

161 $C_9H_7NO_2$ 1076-59-1
5(4*H*)-Isoxazolone, 3-phenyl-

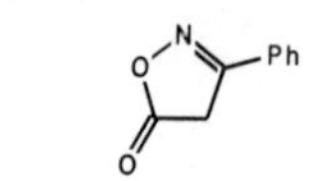

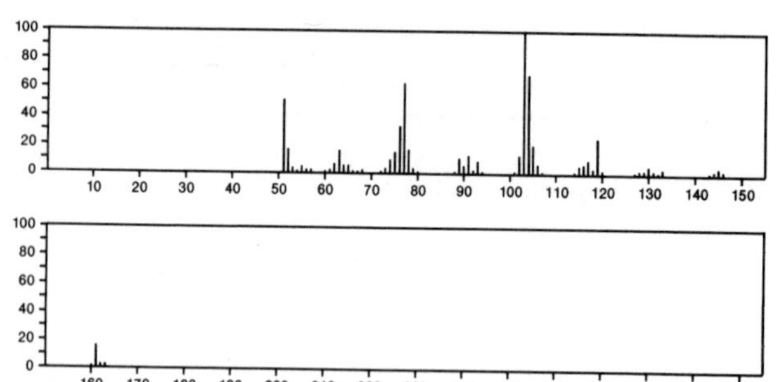

161 $C_9H_7NO_2$ 1127-59-9
1*H*-Indole-2,3-dione, 7-methyl-

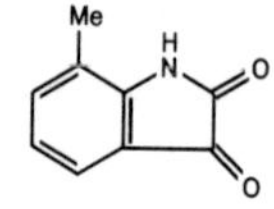

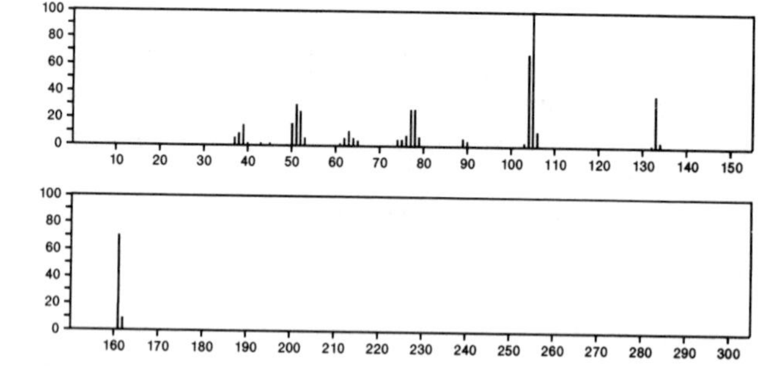

161 $C_9H_7NO_2$ 1128-44-5
1*H*-Indole-2,3-dione, 4-methyl-

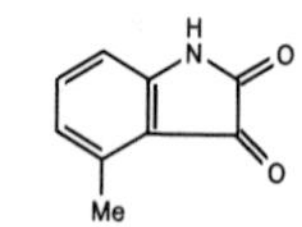

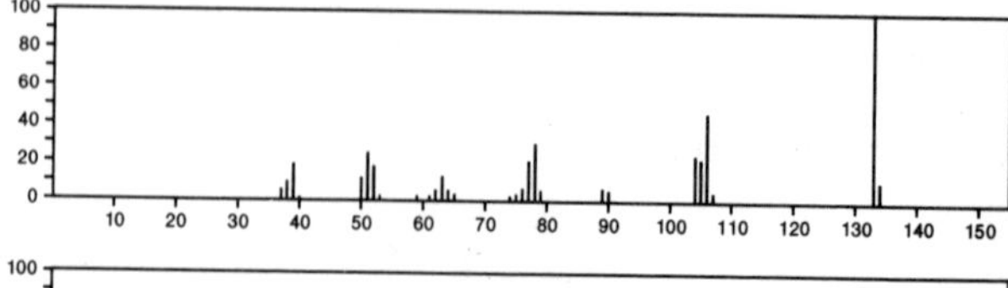

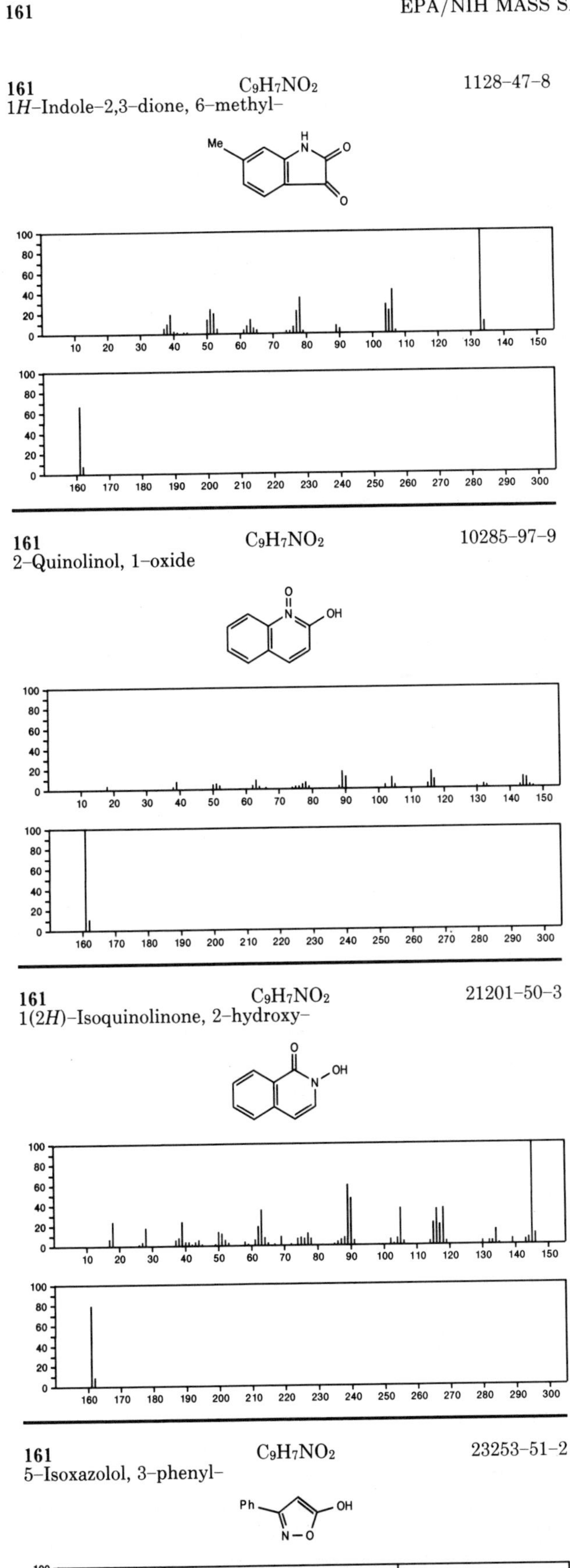

161 $C_9H_7NO_2$ 1128-47-8
1*H*-Indole-2,3-dione, 6-methyl-

161 $C_9H_7NO_2$ 10285-97-9
2-Quinolinol, 1-oxide

161 $C_9H_7NO_2$ 21201-50-3
1(2*H*)-Isoquinolinone, 2-hydroxy-

161 $C_9H_7NO_2$ 23253-51-2
5-Isoxazolol, 3-phenyl-

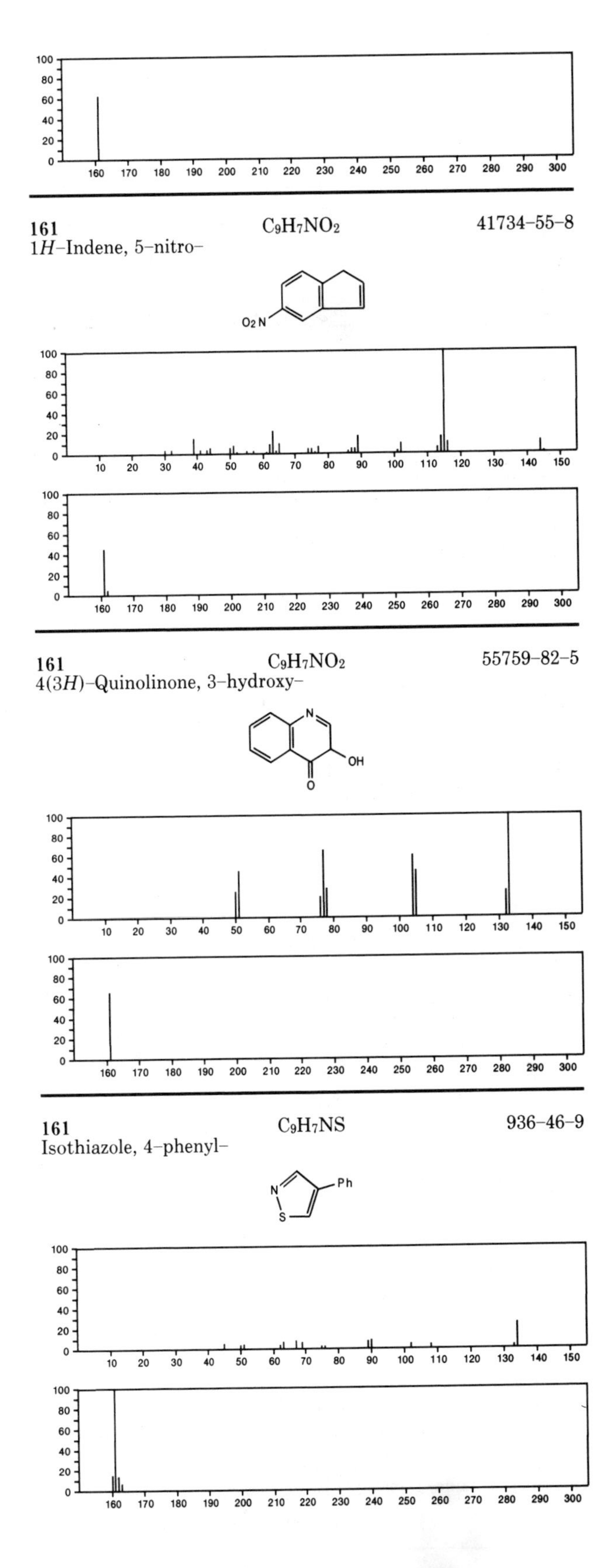

161 $C_9H_7NO_2$ 41734-55-8
1*H*-Indene, 5-nitro-

161 $C_9H_7NO_2$ 55759-82-5
4(3*H*)-Quinolinone, 3-hydroxy-

161 C_9H_7NS 936-46-9
Isothiazole, 4-phenyl-

161 $C_9H_{11}N_3$ 4919-15-7
s-Triazolo[4,3-*a*]pyridine, 3,5,7-trimethyl-

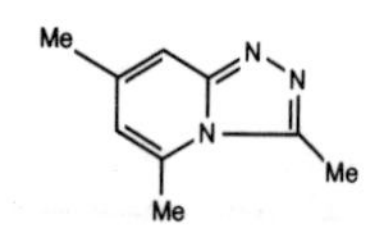

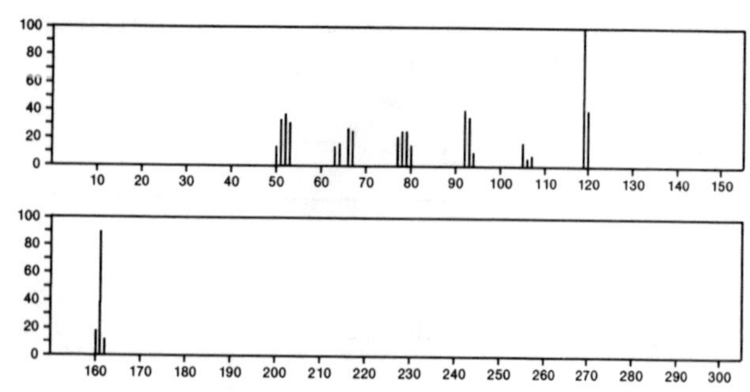

161 $C_9H_{11}N_3$ 4919-18-0
s-Triazolo[4,3-*a*]pyridine, 3-ethyl-5-methyl-

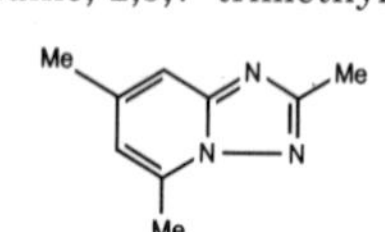

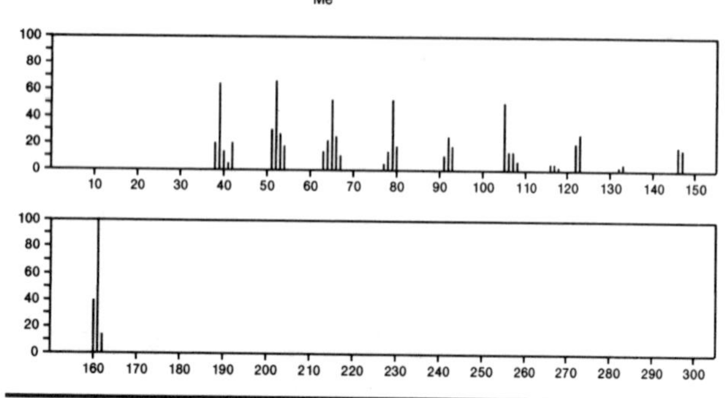

161 $C_9H_{11}N_3$ 4931-30-0
s-Triazolo[1,5-*a*]pyridine, 2,5,7-trimethyl-

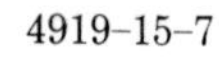

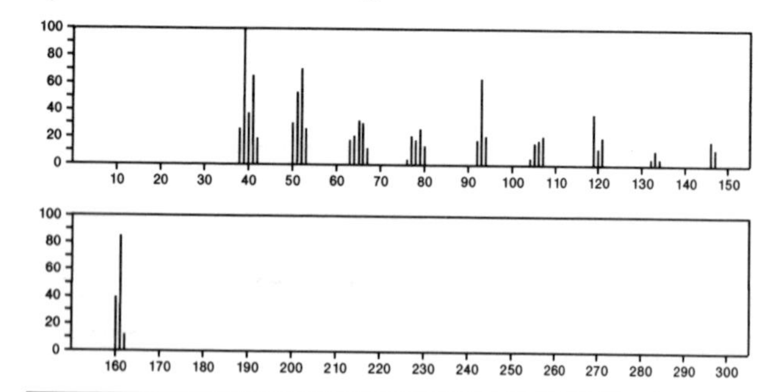

161 $C_9H_{11}N_3$ 32366-26-0
Benzene, (1-azido-1-methylethyl)-

PhCMe2(N3)

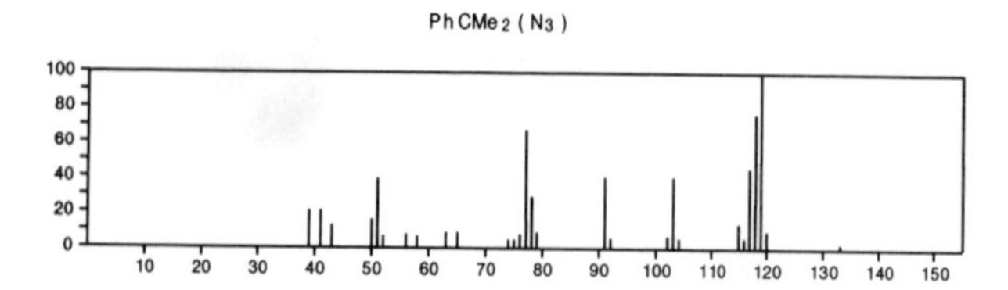

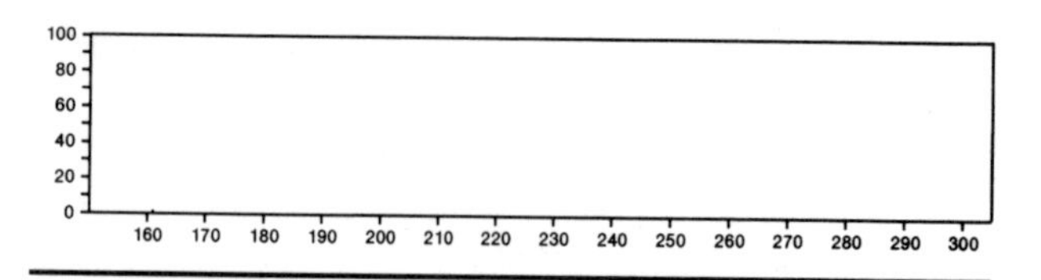

161 $C_9H_{11}N_3$ 50872-97-4
Cyclopenta[4,5]pyrrolo[1,2-*b*][1,2,4]triazole, 4a,5,6,7,7a,8-hexa=hydro-8-methylene-

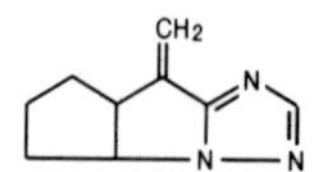

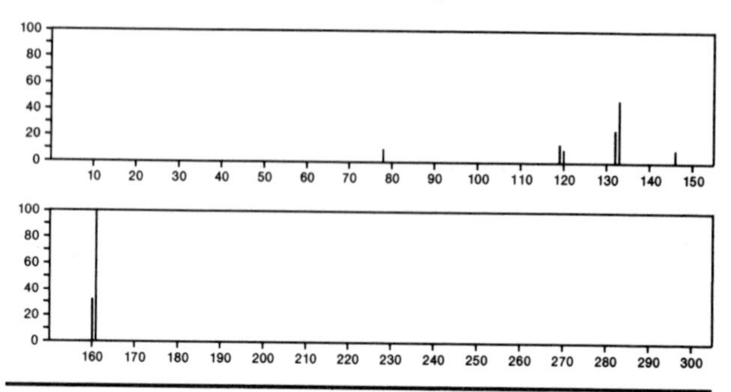

161 $C_{10}H_{11}NO$ 1128-85-4
2-Buten-1-one, 3-amino-1-phenyl-

MeC(NH2)=CHCOPh

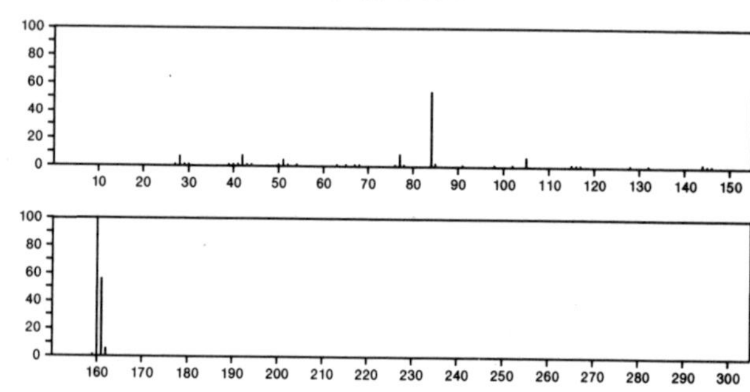

161 $C_{10}H_{11}NO$ 4641-57-0
2-Pyrrolidinone, 1-phenyl-

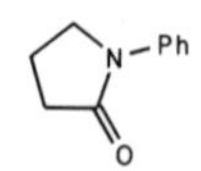

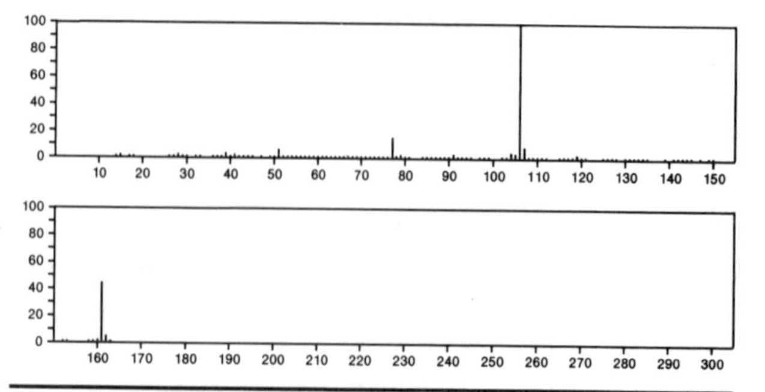

161 $C_{11}H_{15}N$ 1077-18-5
1-Butanamine, *N*-(phenylmethylene)-

PhCH=N(CH2)3Me

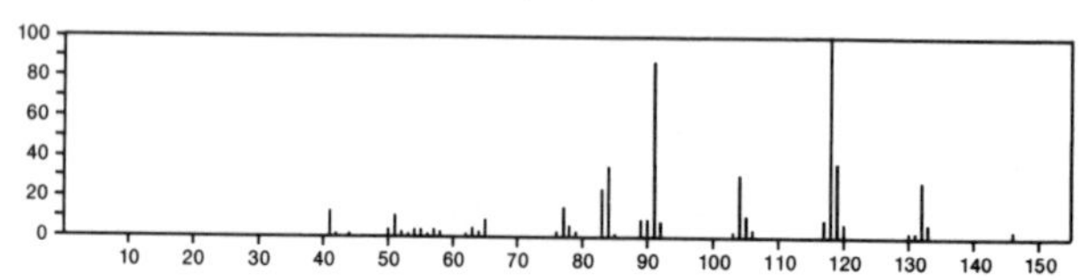

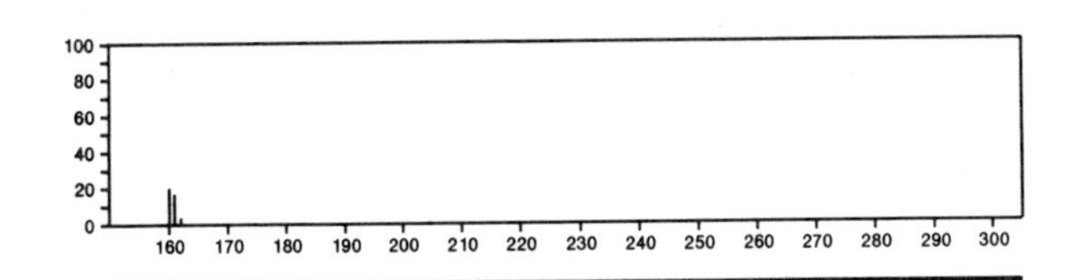

161 $C_{11}H_{15}N$ 6852-58-0
2-Propanamine, 2-methyl-*N*-(phenylmethylene)-

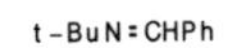

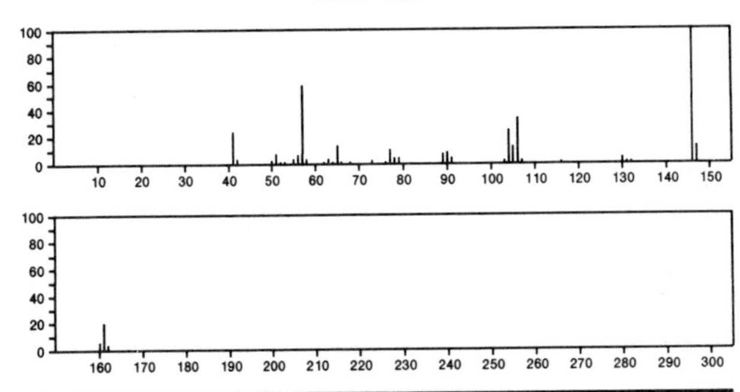

161 $C_{11}H_{15}N$ 7730-40-7
Azetidine, 1-benzyl-2-methyl-

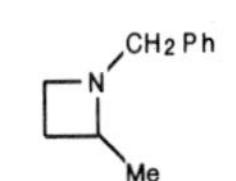

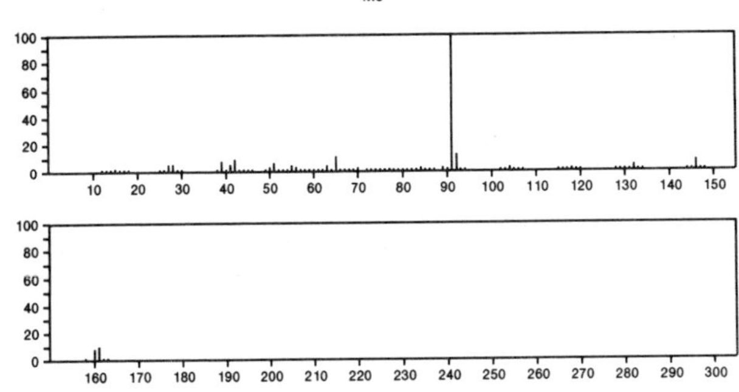

161 $C_{11}H_{15}N$ 10433-34-8
Benzeneethanamine, *N*-(1-methylethylidene)-

PhCH2CH2N=CMe2

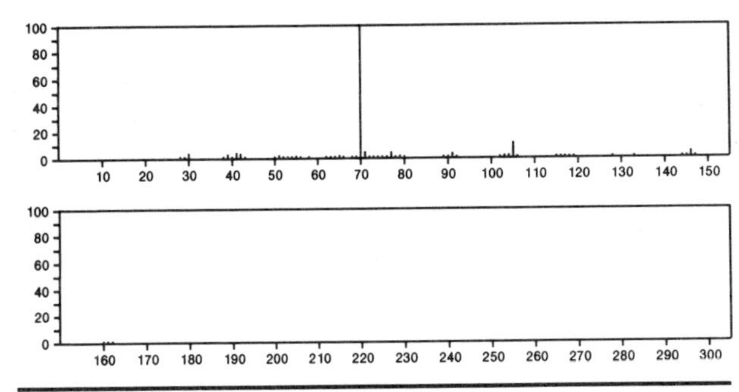

161 $C_{11}H_{15}N$ 14429-09-5
Isoquinoline, 1,2,3,4-tetrahydro-1,2-dimethyl-

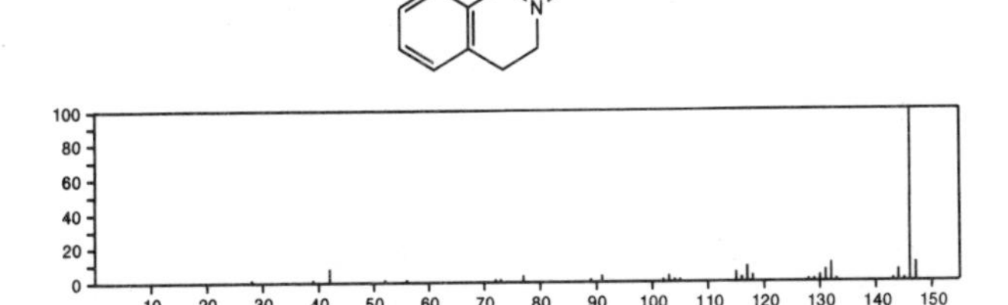

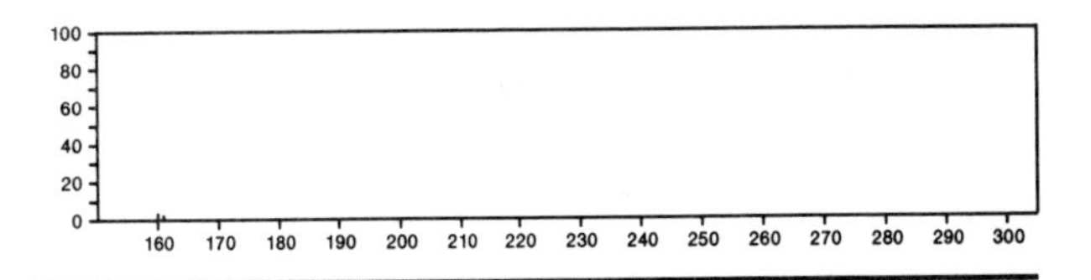

161 $C_{11}H_{15}N$ 18781-58-3
1*H*-Indole, 2,3-dihydro-2,3,3-trimethyl-

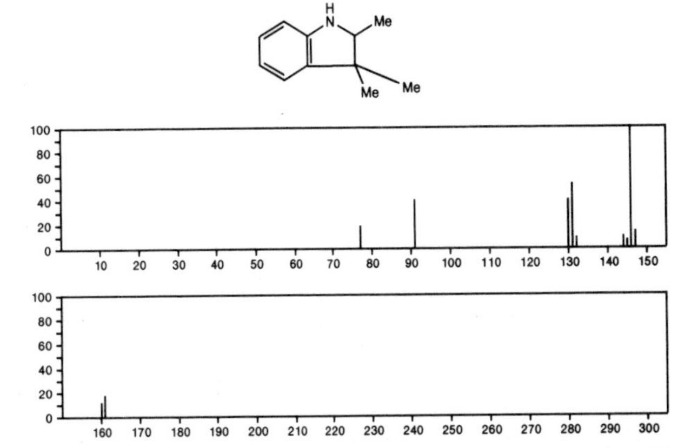

161 $C_{11}H_{15}N$ 23074-42-2
Tricyclo[3.3.1.1^{3,7}]decane-1-carbonitrile

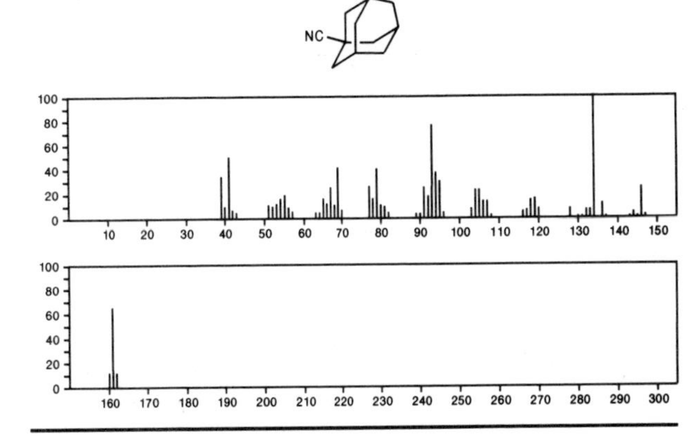

161 $C_{11}H_{15}N$ 24432-52-8
Aziridine, 2,3-dimethyl-1-(phenylmethyl)-, *trans*-

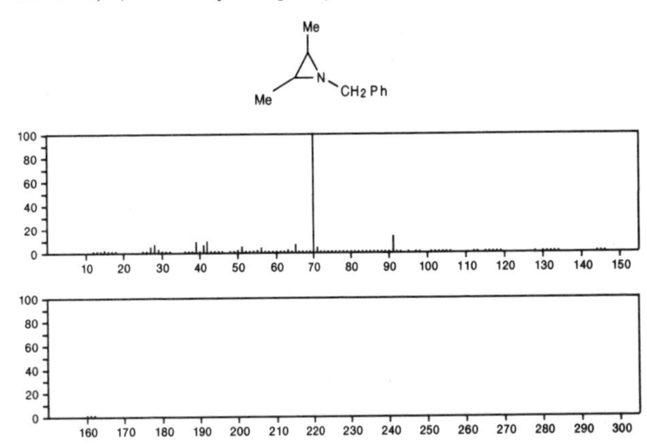

161 $C_{11}H_{15}N$ 29666-60-2
Methanamine, *N*-(1-methyl-3-phenylpropylidene)-

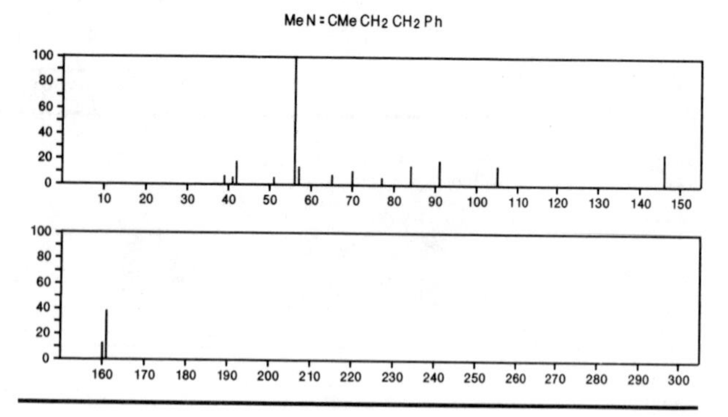

161 $C_{11}H_{15}N$ 33611-54-0
Benzenemethanimine, α-(1,1-dimethylethyl)-

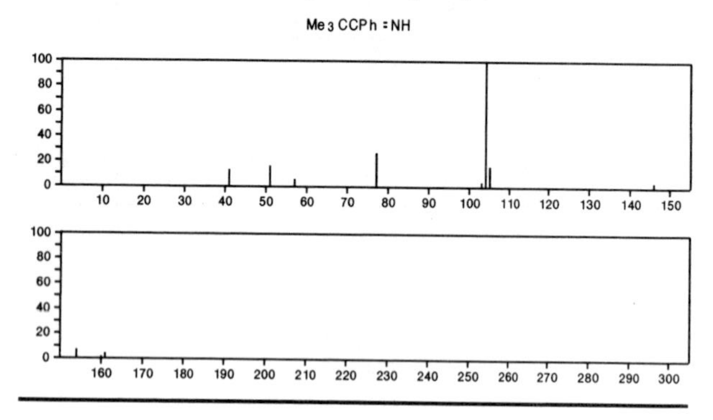

161 $C_{11}H_{15}N$ 40051-50-1
2-Butanamine, *N*-(phenylmethylene)-

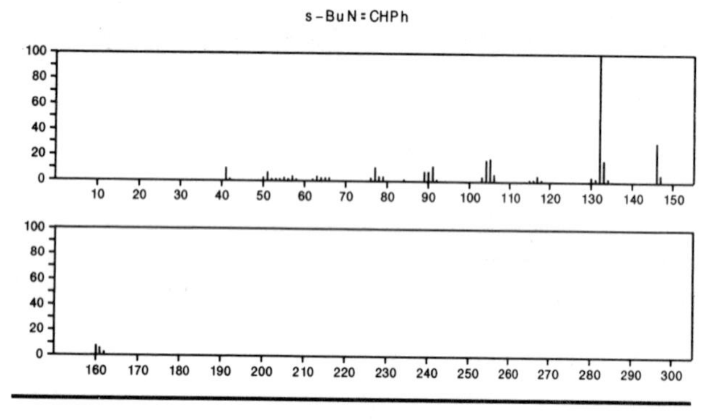

161 $C_{11}H_{15}N$ 54365-72-9
Isoquinoline, 1,2,3,4-tetrahydro-2,3-dimethyl-

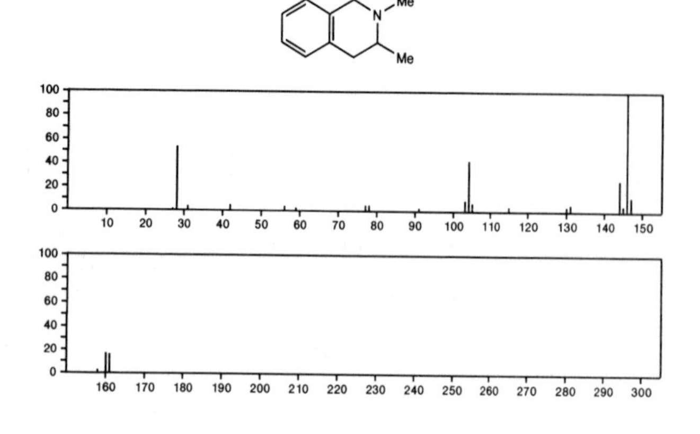

161 $C_{11}H_{15}N$ 55702-31-3
Azetidine, 3-methyl-1-(phenylmethyl)-

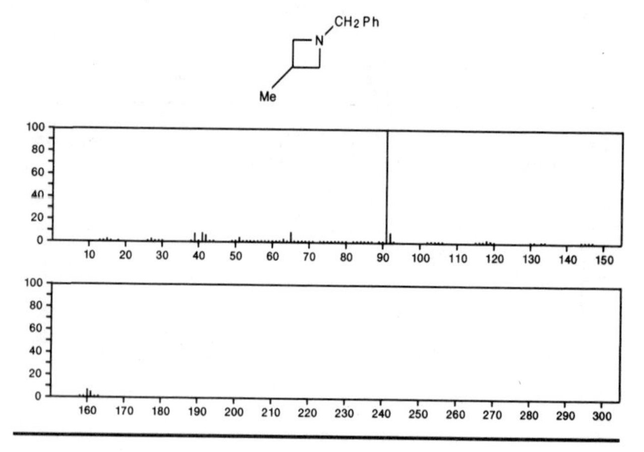

161 $C_{11}H_{15}N$ 56701-44-1
Cyclopropanamine, *N*-methyl-1-(4-methylphenyl)-

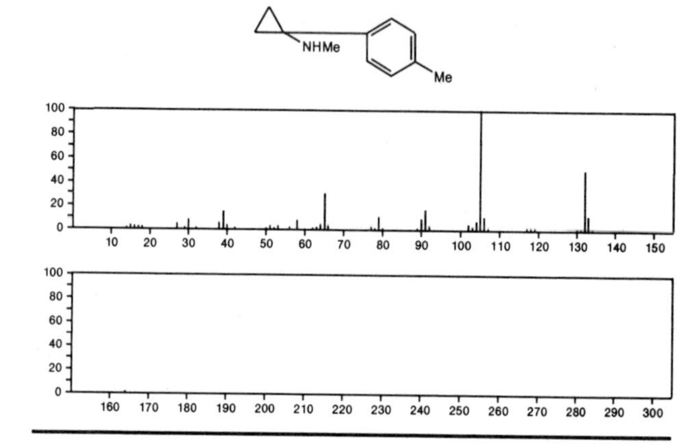

162 $CHBrCl_2$ 75-27-4
Methane, bromodichloro-

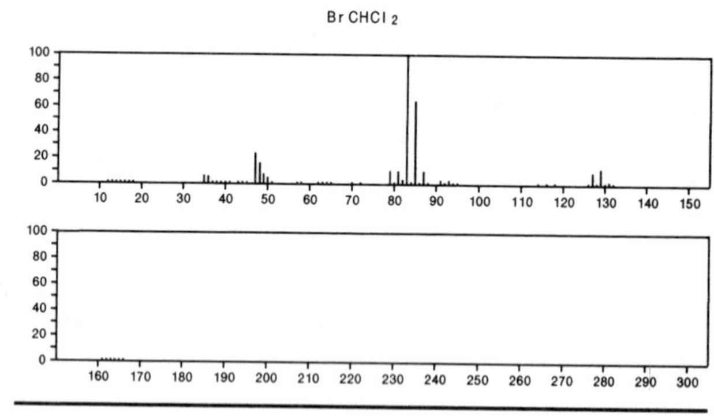

162 $C_2HCl_3O_2$ 76-03-9
Acetic acid, trichloro-

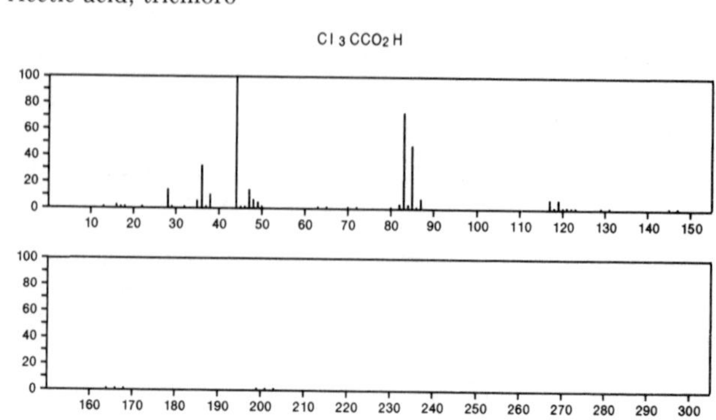

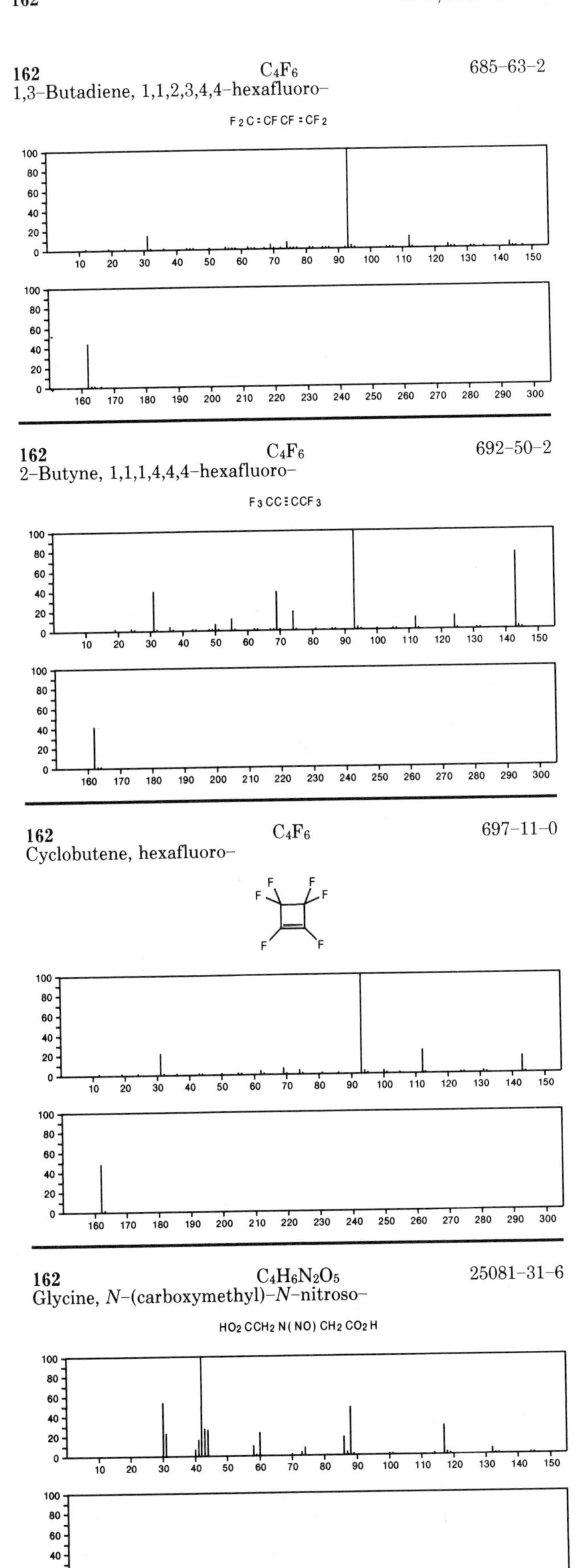

162 C_4F_6 685-63-2
1,3-Butadiene, 1,1,2,3,4,4-hexafluoro-

162 C_4F_6 692-50-2
2-Butyne, 1,1,1,4,4,4-hexafluoro-

162 C_4F_6 697-11-0
Cyclobutene, hexafluoro-

162 $C_4H_6N_2O_5$ 25081-31-6
Glycine, *N*-(carboxymethyl)-*N*-nitroso-

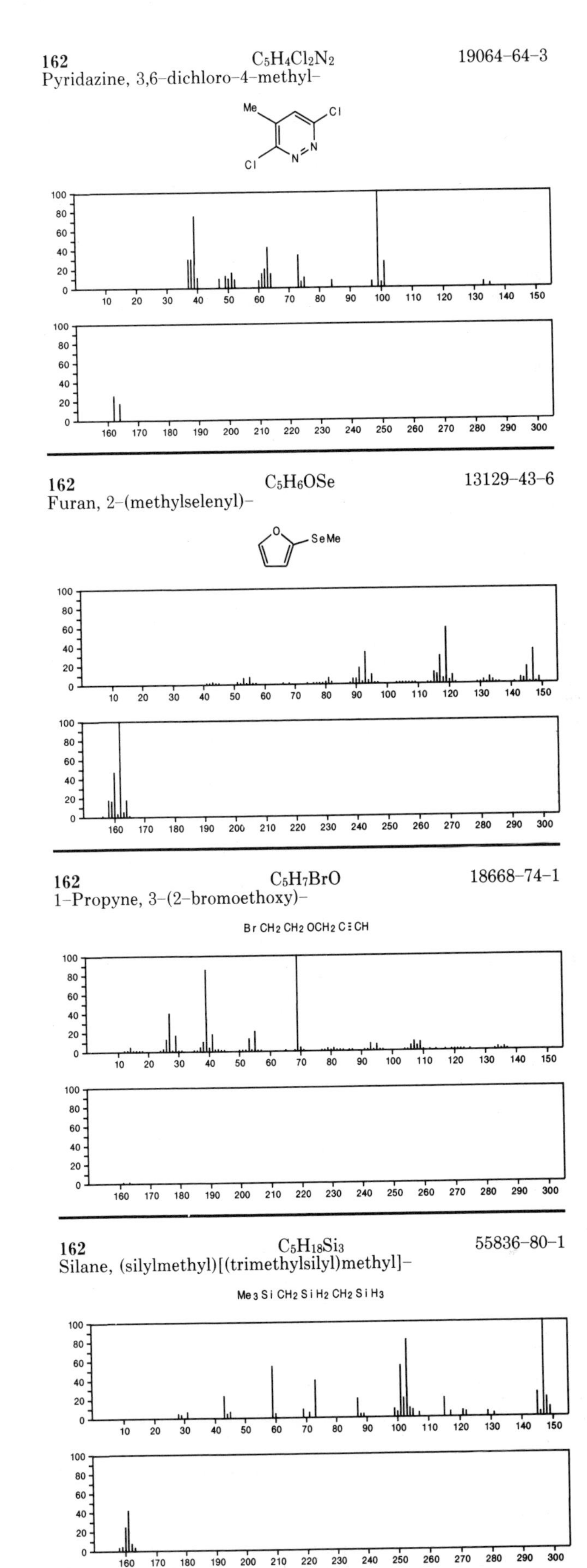

162 $C_5H_4Cl_2N_2$ 19064-64-3
Pyridazine, 3,6-dichloro-4-methyl-

162 C_5H_6OSe 13129-43-6
Furan, 2-(methylselenyl)-

162 C_5H_7BrO 18668-74-1
1-Propyne, 3-(2-bromoethoxy)-

162 $C_5H_{18}Si_3$ 55836-80-1
Silane, (silylmethyl)[(trimethylsilyl)methyl]-

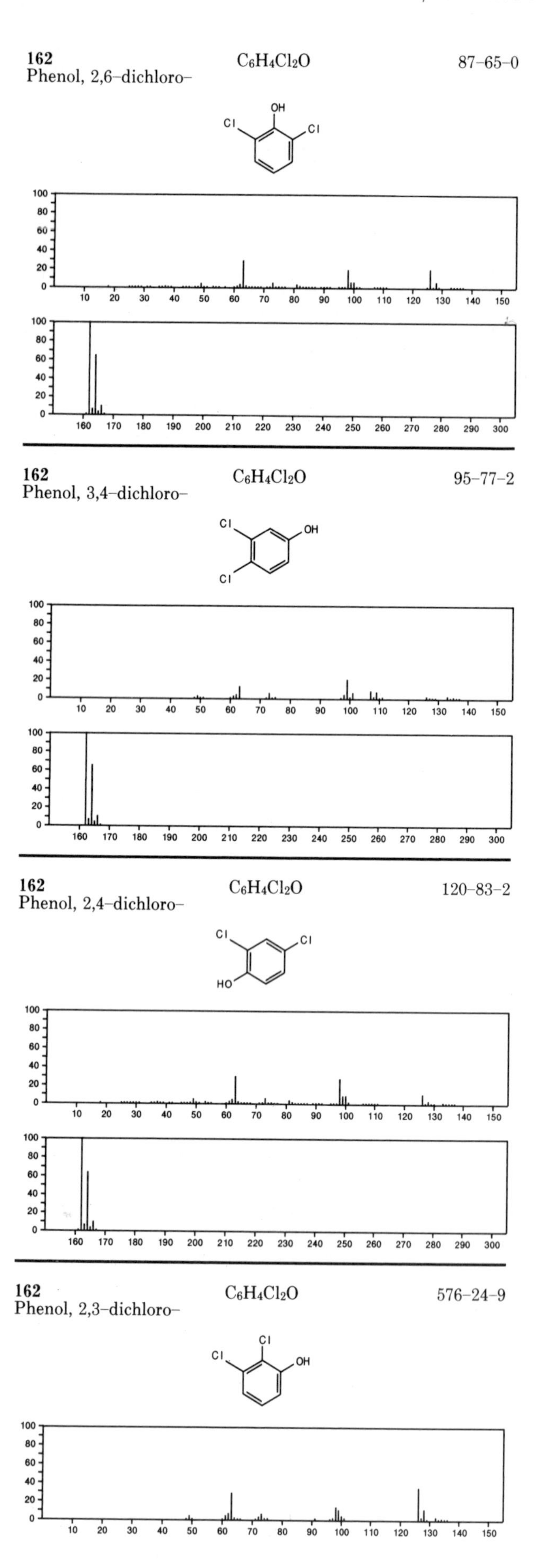

162
Phenol, 2,6-dichloro-
$C_6H_4Cl_2O$
87-65-0
162
Phenol, 3,4-dichloro-
$C_6H_4Cl_2O$
95-77-2
162
Phenol, 2,4-dichloro-
$C_6H_4Cl_2O$
120-83-2
162
Phenol, 2,3-dichloro-
$C_6H_4Cl_2O$
576-24-9

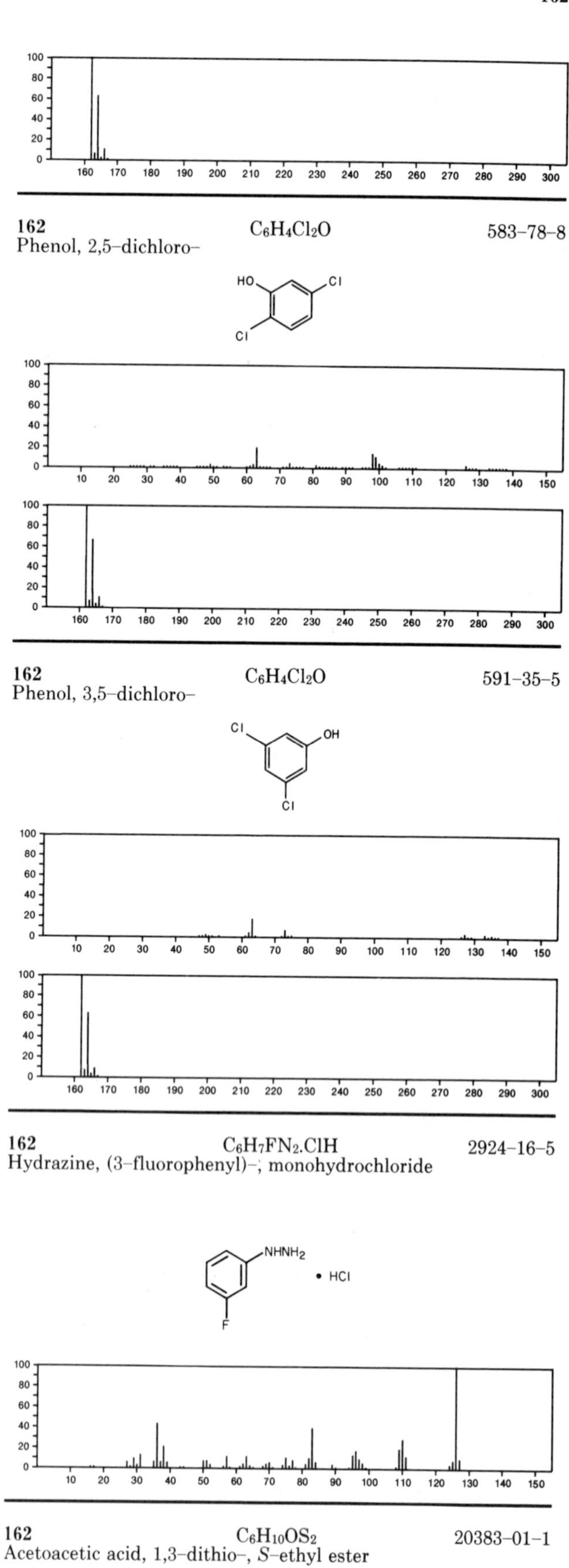

162
Phenol, 2,5-dichloro-
$C_6H_4Cl_2O$
583-78-8
162
Phenol, 3,5-dichloro-
$C_6H_4Cl_2O$
591-35-5
162
Hydrazine, (3-fluorophenyl)-, monohydrochloride
$C_6H_7FN_2.ClH$
2924-16-5
162
Acetoacetic acid, 1,3-dithio-, *S*-ethyl ester
$C_6H_{10}OS_2$
20383-01-1
EtSC(O)CH2CSMe

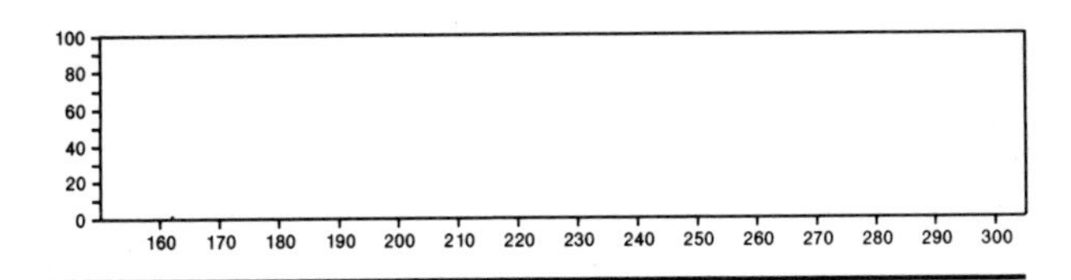

162 $C_6H_{10}OS_2$ 33266-07-8
Ethanone, 1-(2-methyl-1,3-dithiolan-2-yl)-

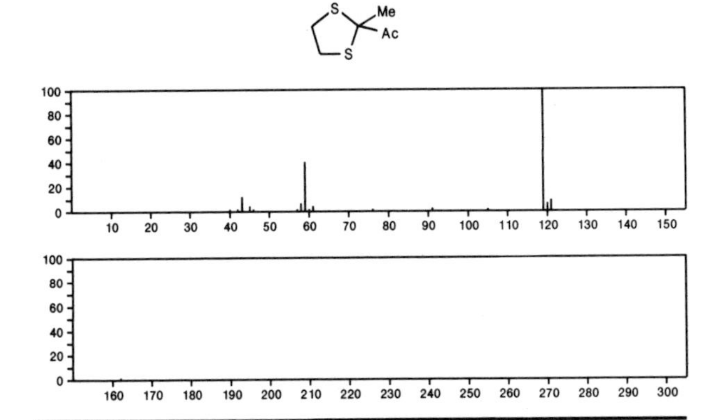

162 $C_6H_{10}O_3S$ 19456-18-9
1,2-Cyclohexanediol, cyclic sulfite, *cis*-

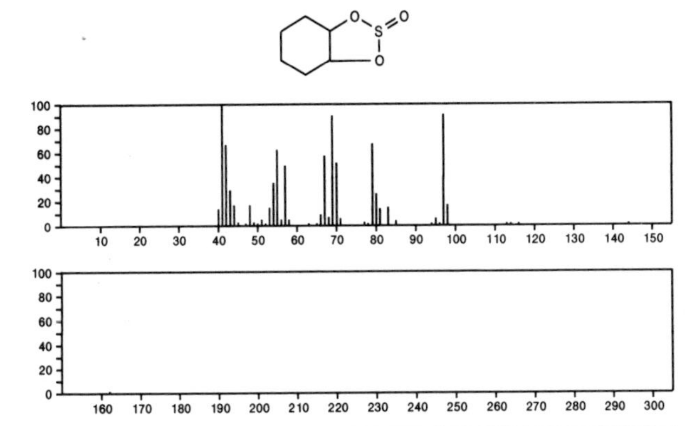

162 $C_6H_{10}O_3S$ 19456-19-0
1,2-Cyclohexanediol, cyclic sulfite, *trans*-

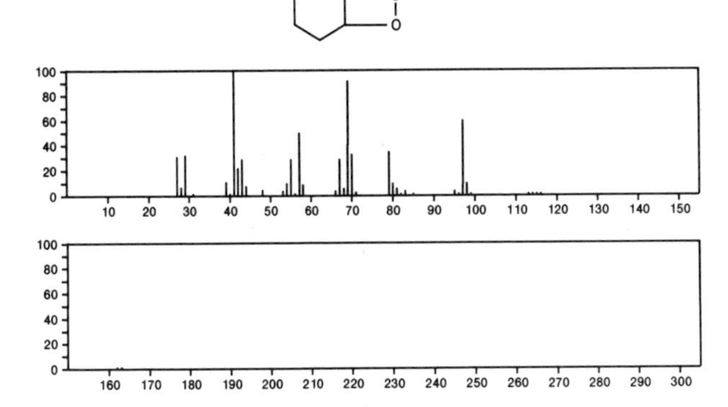

162 $C_6H_{10}O_3S$ 55780-98-8
D-Fructose, 1,3,6-trideoxy-3,6-epithio-

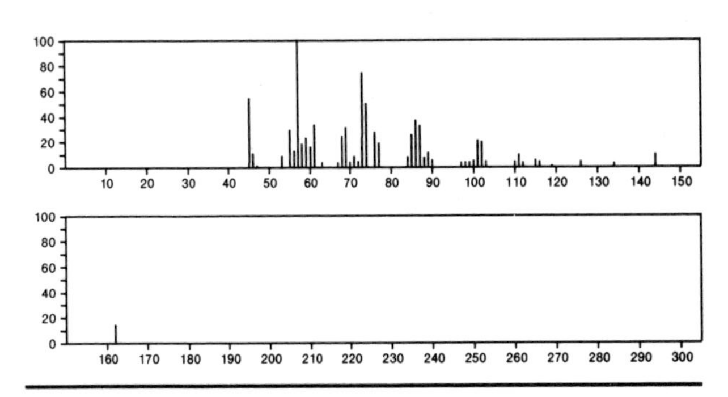

162 $C_6H_{10}O_5$ 1587-15-1
Butanedioic acid, hydroxy-, dimethyl ester

MeOC(O)CH(OH)CH2C(O)OMe

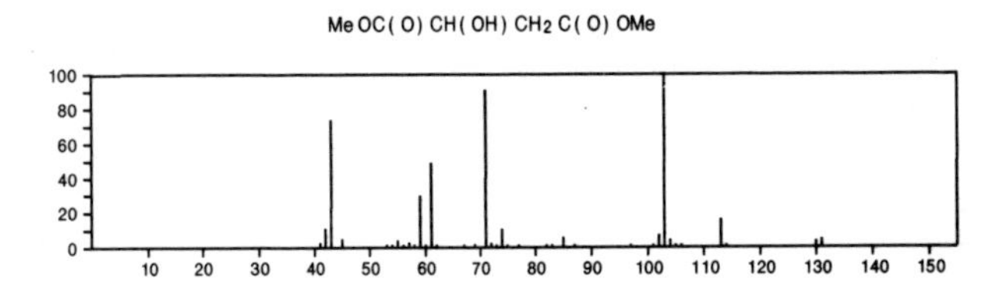

162 $C_6H_{10}O_5$ 19500-95-9
Acetic acid, methoxy-, anhydride

MeOCH2C(O)OC(O)CH2OMe

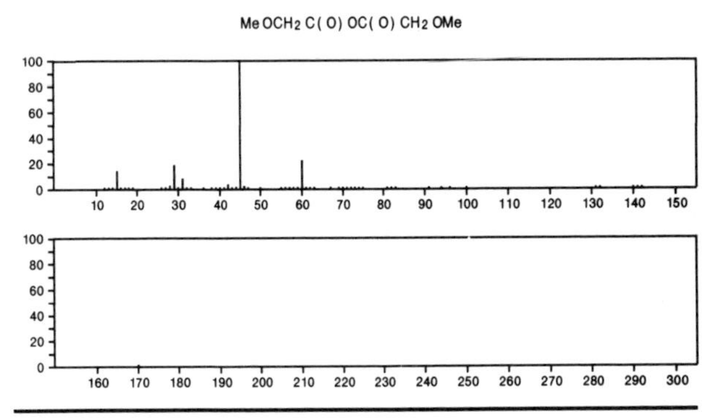

162 $C_6H_{11}Br$ 108-85-0
Cyclohexane, bromo-

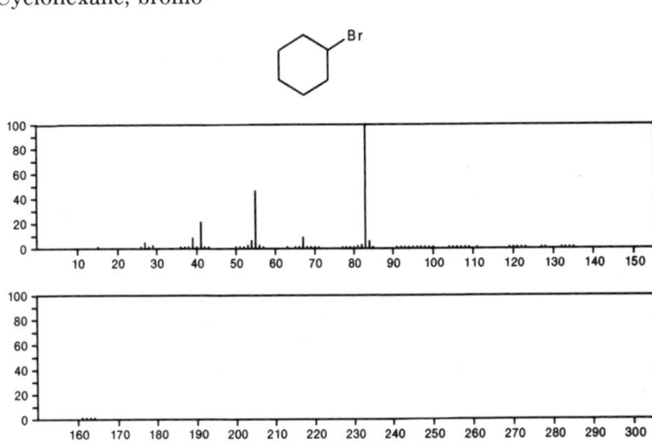

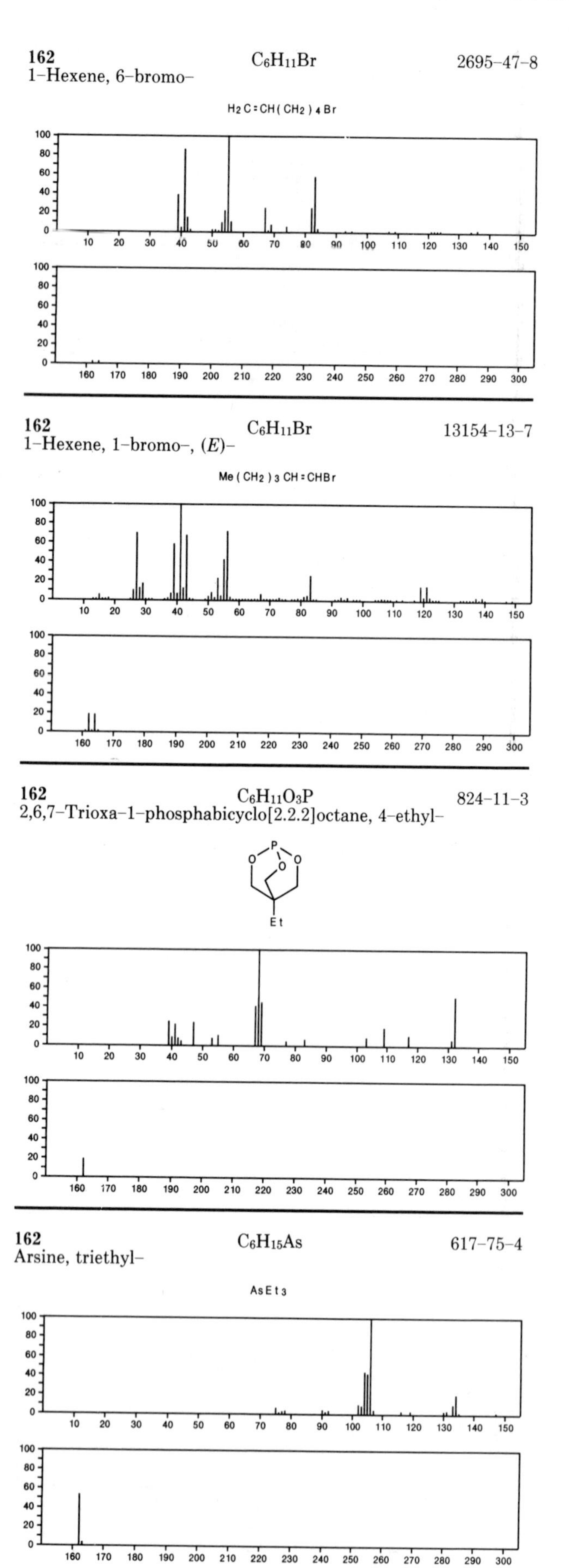

162
1–Hexene, 6–bromo–
$C_6H_{11}Br$
2695–47–8
H2C=CH(CH2)4Br
162
1–Hexene, 1–bromo–, (*E*)–
$C_6H_{11}Br$
13154–13–7
Me(CH2)3CH=CHBr
162
2,6,7–Trioxa–1–phosphabicyclo[2.2.2]octane, 4–ethyl–
$C_6H_{11}O_3P$
824–11–3
162
Arsine, triethyl–
$C_6H_{15}As$
617–75–4
AsEt3

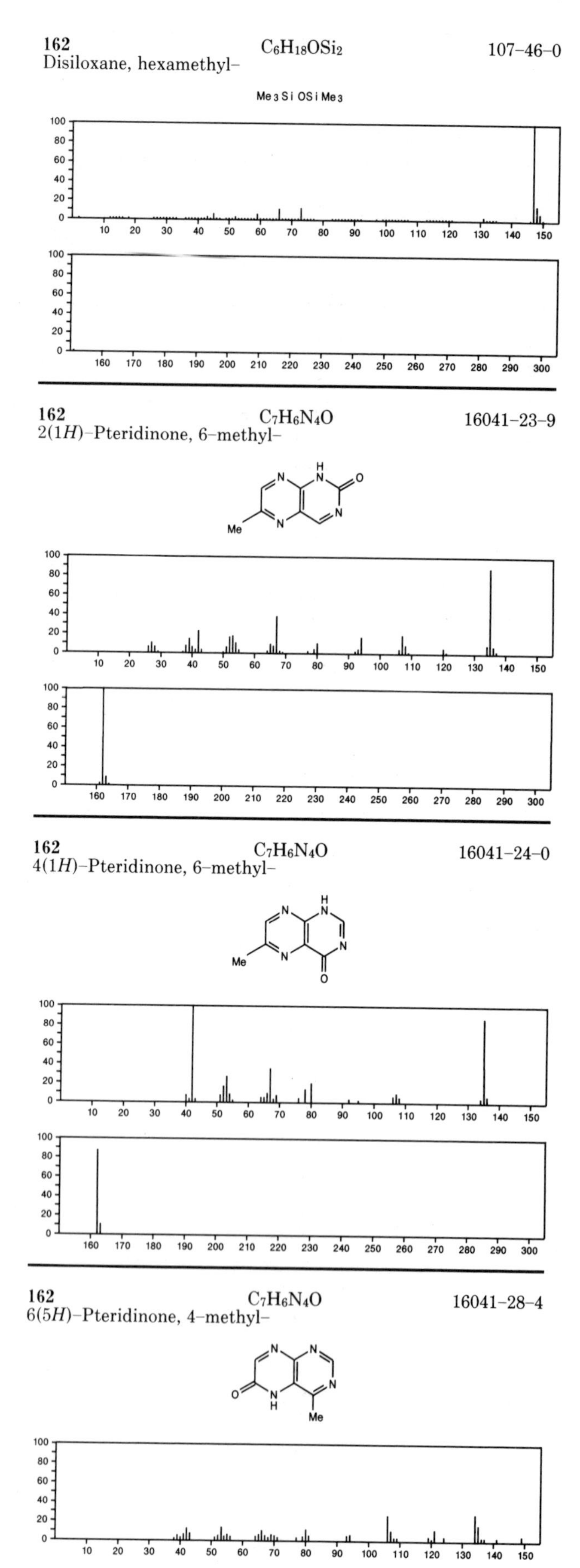

162
Disiloxane, hexamethyl–
$C_6H_{18}OSi_2$
107–46–0
Me3SiOSiMe3
162
2(1*H*)–Pteridinone, 6–methyl–
$C_7H_6N_4O$
16041–23–9
162
4(1*H*)–Pteridinone, 6–methyl–
$C_7H_6N_4O$
16041–24–0
162
6(5*H*)–Pteridinone, 4–methyl–
$C_7H_6N_4O$
16041–28–4

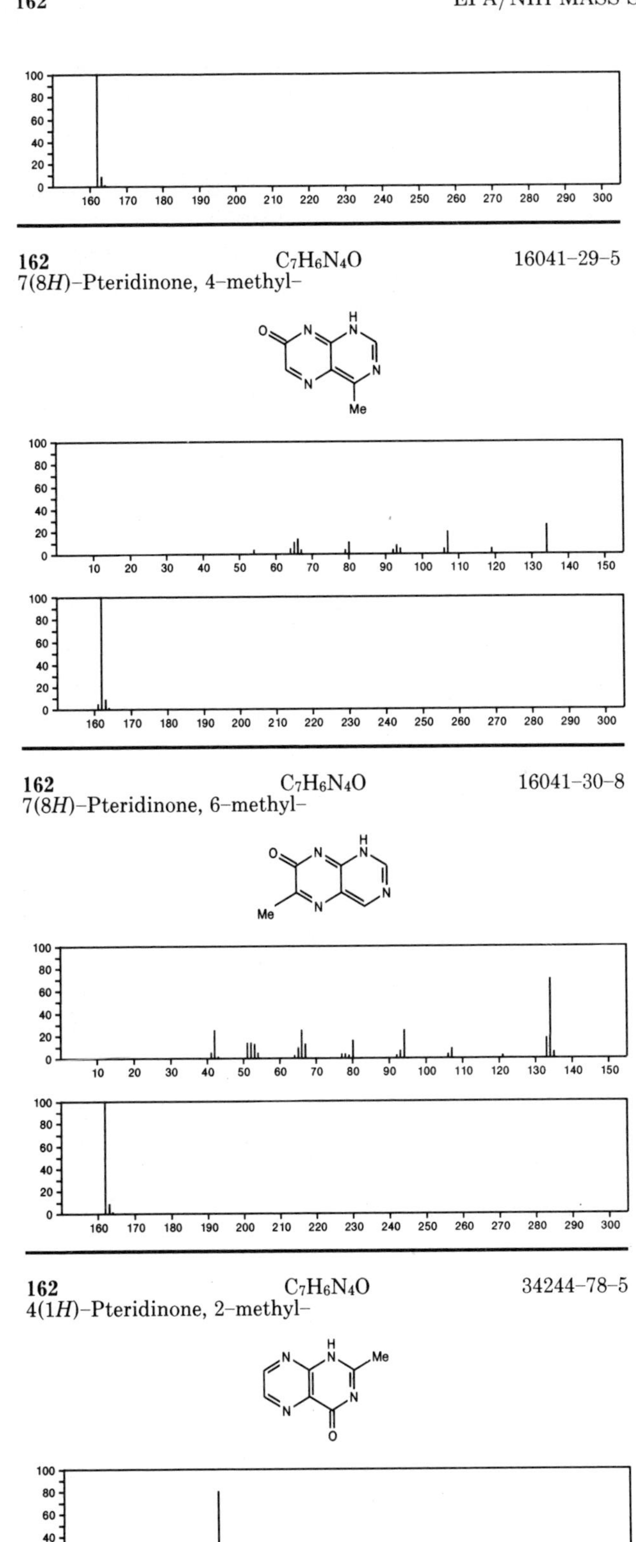

162 $C_7H_6N_4O$ 16041-29-5
7(8*H*)-Pteridinone, 4-methyl-

162 $C_7H_6N_4O$ 16041-30-8
7(8*H*)-Pteridinone, 6-methyl-

162 $C_7H_6N_4O$ 34244-78-5
4(1*H*)-Pteridinone, 2-methyl-

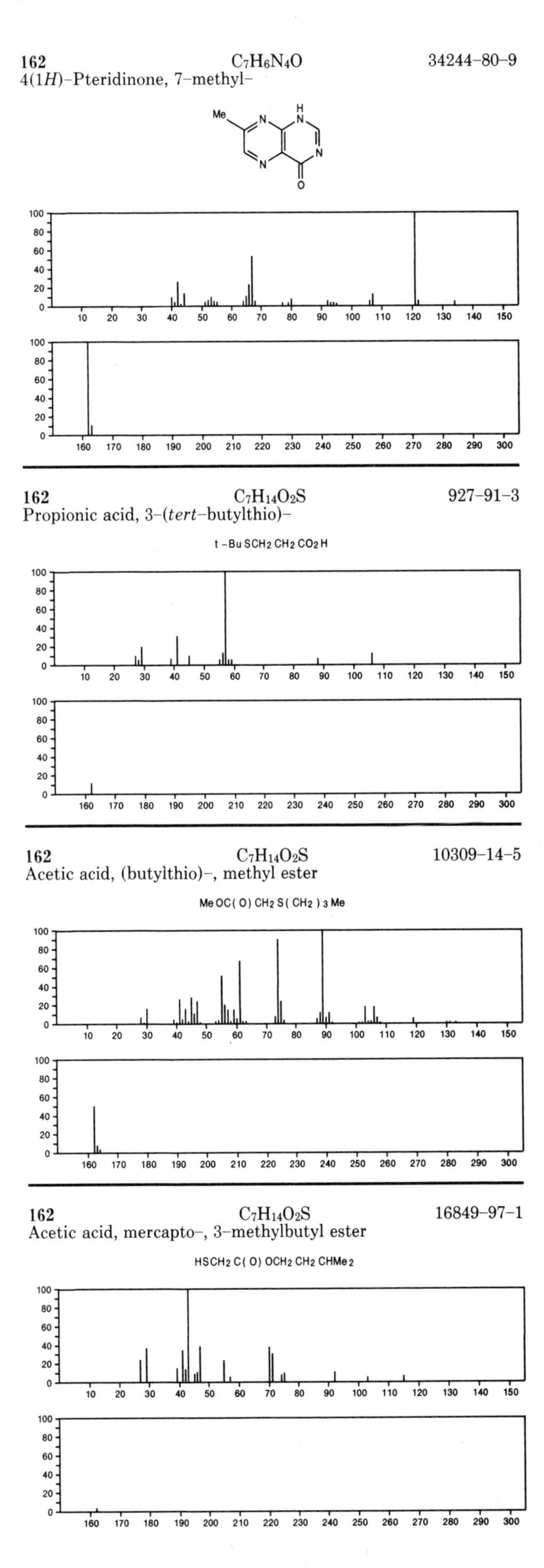

162 $C_7H_6N_4O$ 34244-80-9
4(1*H*)-Pteridinone, 7-methyl-

162 $C_7H_{14}O_2S$ 927-91-3
Propionic acid, 3-(*tert*-butylthio)-

t-Bu SCH2 CH2 CO2 H

162 $C_7H_{14}O_2S$ 10309-14-5
Acetic acid, (butylthio)-, methyl ester

Me OC(O) CH2 S(CH2)3 Me

162 $C_7H_{14}O_2S$ 16849-97-1
Acetic acid, mercapto-, 3-methylbutyl ester

HSCH2 C(O) OCH2 CH2 CHMe2

162 $C_7H_{14}O_2S$ 20600-66-2
Acetic acid, (isobutylthio)-, methyl ester

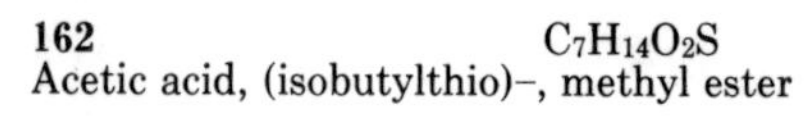

i-BuSCH2C(O)OMe

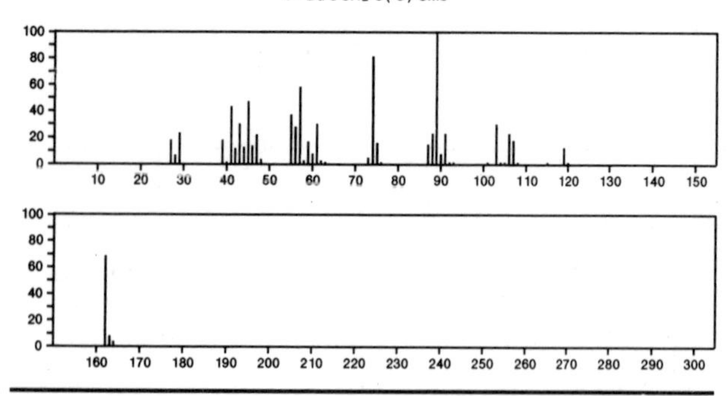

162 $C_7H_{14}O_2S$ 22002-73-9
Propionic acid, 3-(butylthio)-

Me(CH2)3SCH2CH2CO2H

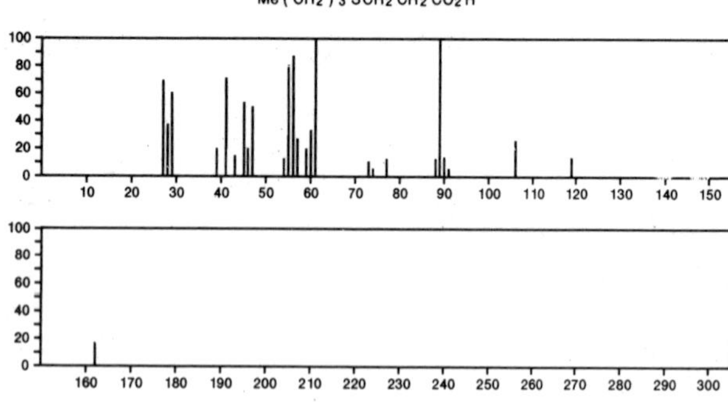

162 $C_7H_{14}O_2S$ 36651-30-6
1,3-Oxathiolane-2-propanol, 2-methyl-

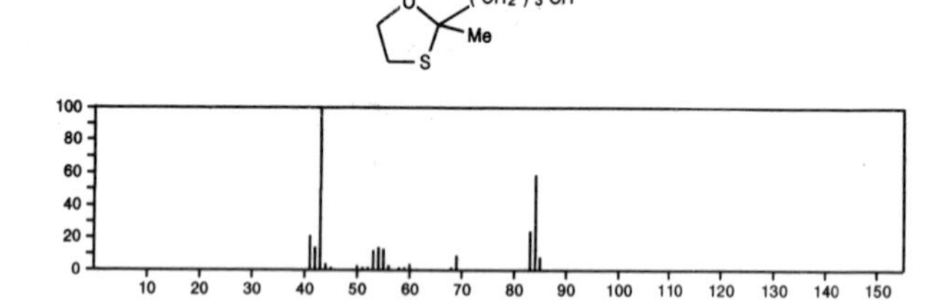

162 $C_7H_{14}O_4$ 13089-76-4
D-*ribo*-Hexose, 2,6-dideoxy-3-*O*-methyl-

MeCH(OH)CH(OH)CH(OMe)CH2CHO

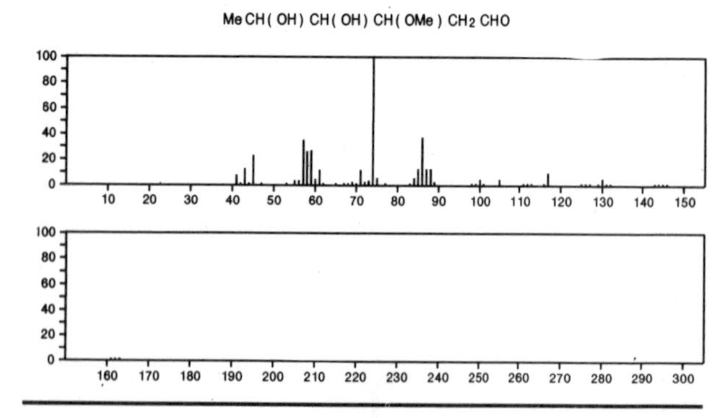

162 $C_7H_{18}O_2Si$ 16654-45-8
Silane, (2-ethoxyethoxy)trimethyl-

EtOCH2CH2OSiMe3

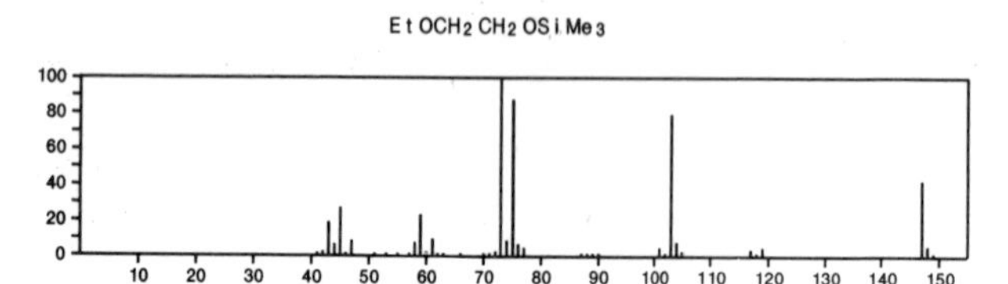

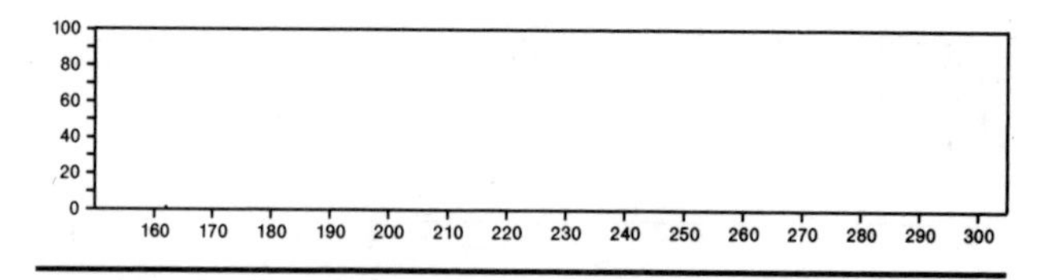

162 $C_7H_{18}SSi$ 3553-78-4
Silane, (butylthio)trimethyl-

Me(CH2)3SSiMe3

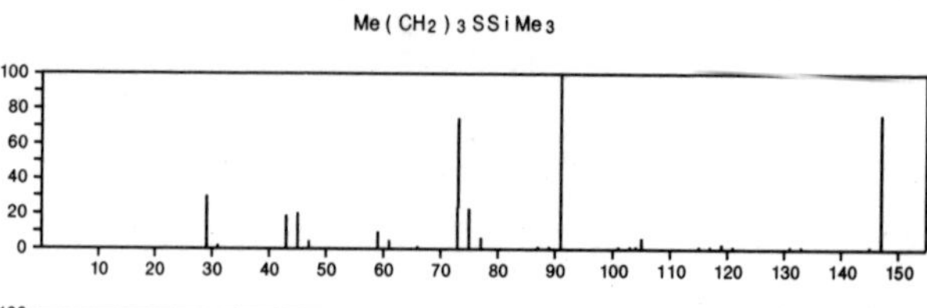

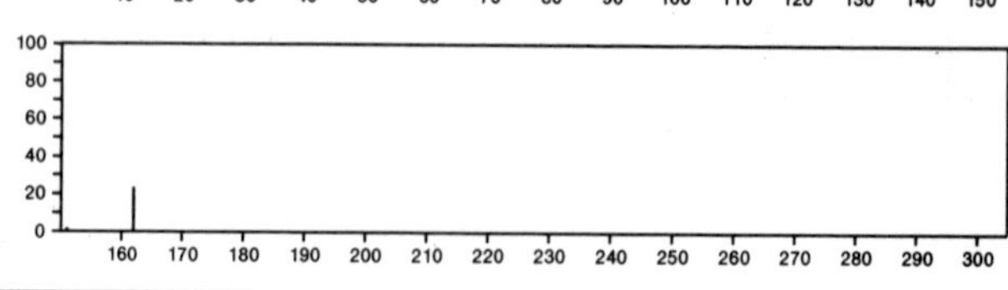

162 $C_8H_6N_2O_2$ 120-06-9
Sydnone, 3-phenyl-

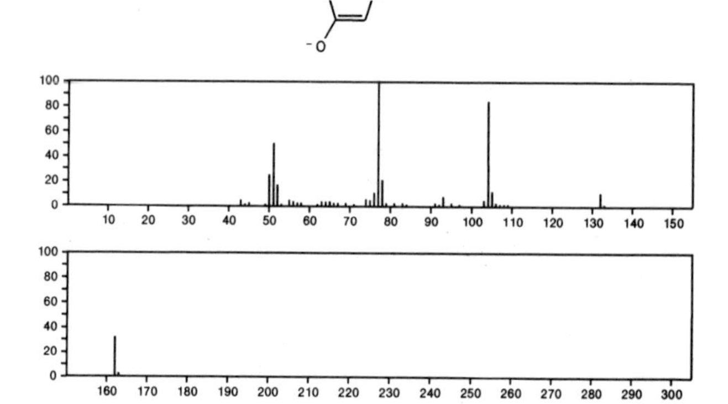

162 $C_8H_6N_2O_2$ 555-21-5
Benzeneacetonitrile, 4-nitro-

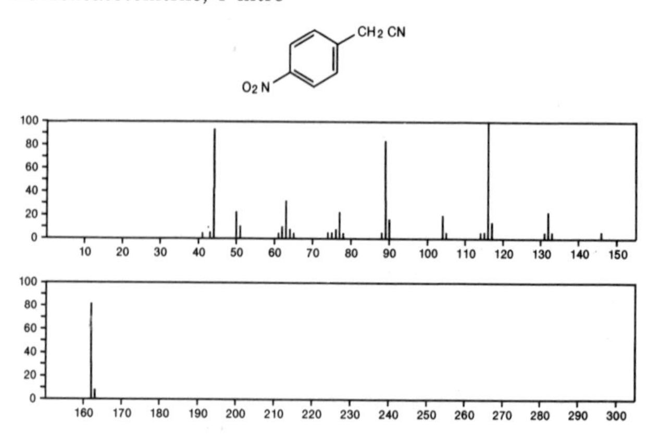

162 $C_8H_6N_2O_2$ 610-66-2
Benzeneacetonitrile, 2-nitro-

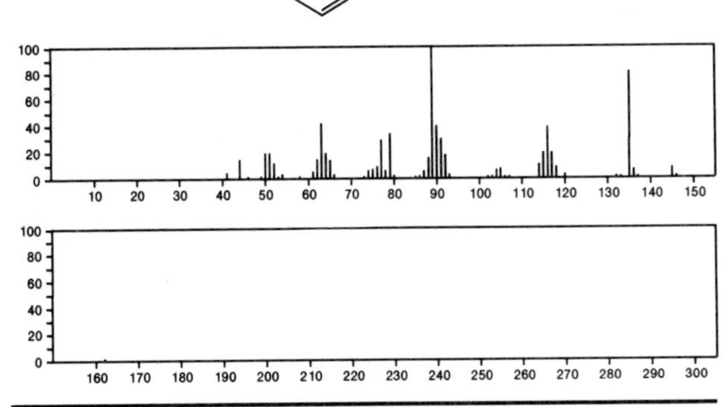

162 $C_8H_6N_2O_2$ 15804-19-0
2,3-Quinoxalinedione, 1,4-dihydro-

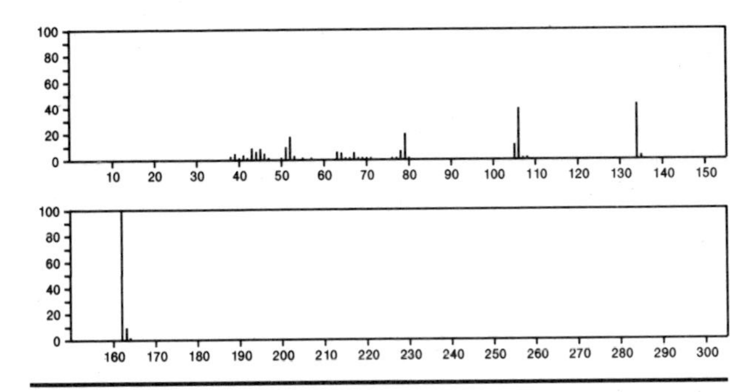

162 $C_8H_6N_2O_2$ 21084-84-4
1,2,4-Oxadiazol-3(2*H*)-one, 5-phenyl-

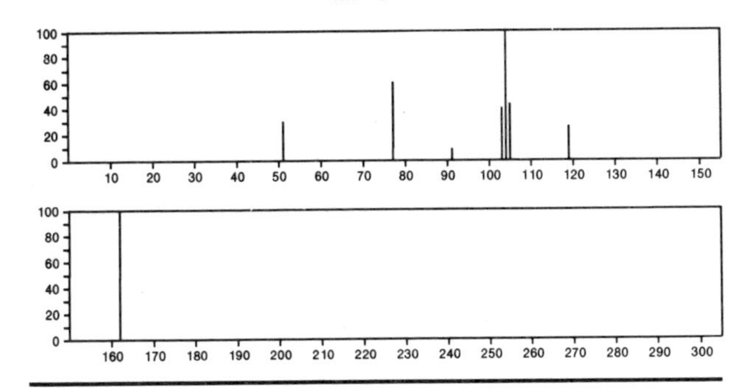

162 $C_8H_6N_2S$ 25445-77-6
1,2,3-Thiadiazole, 4-phenyl-

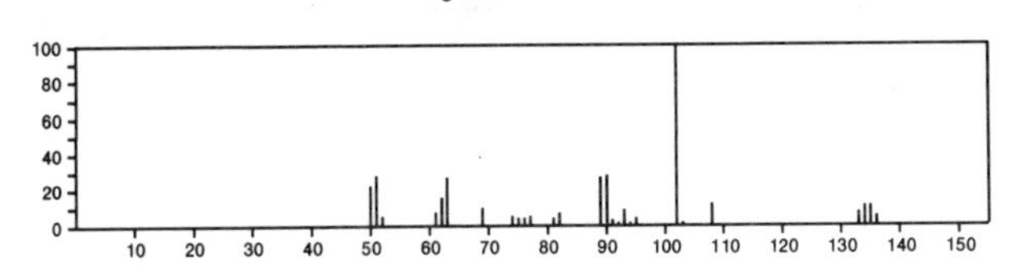

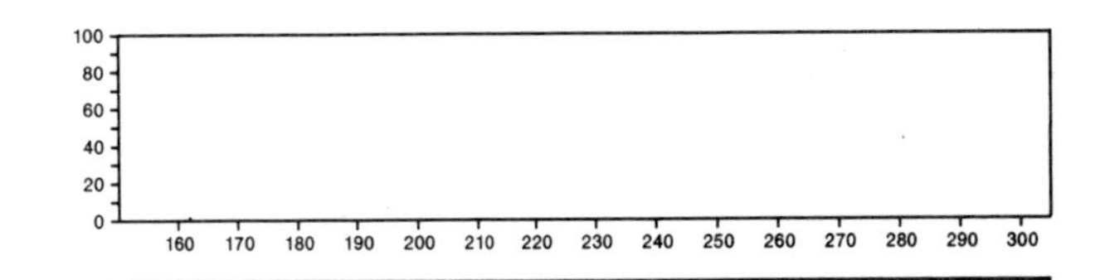

162 $C_8H_7BO_3$ 51901-62-3
4*H*-1,3,2-Benzodioxaborin-4-one, 2-methyl-

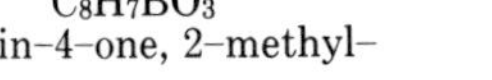

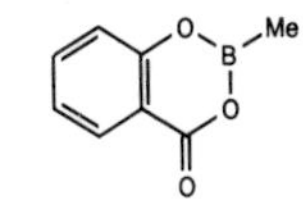

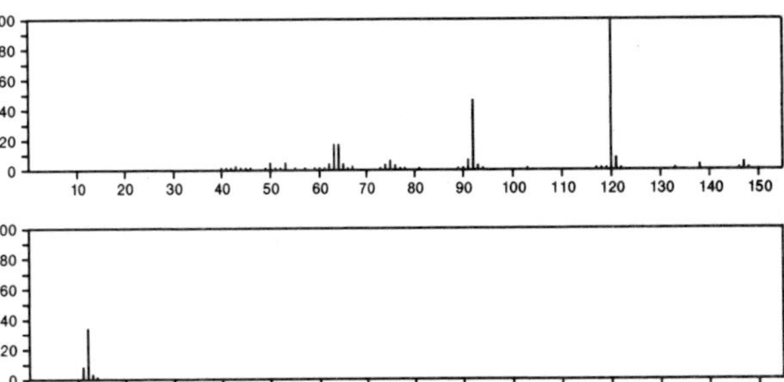

162 $C_8H_{10}N_4$ 19848-79-4
s-Triazolo[4,3-*a*]pyrazine, 3,5,8-trimethyl-

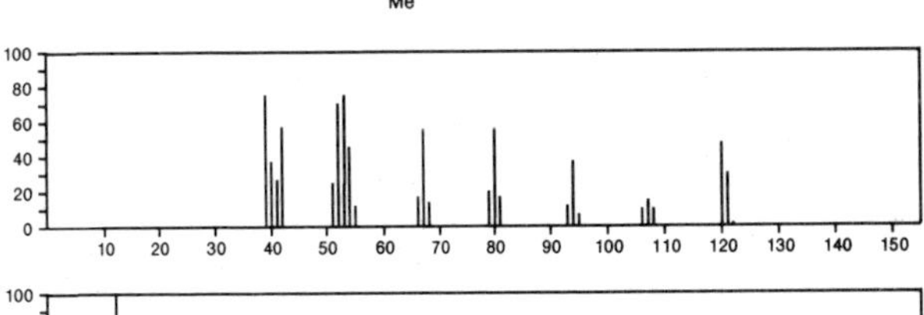

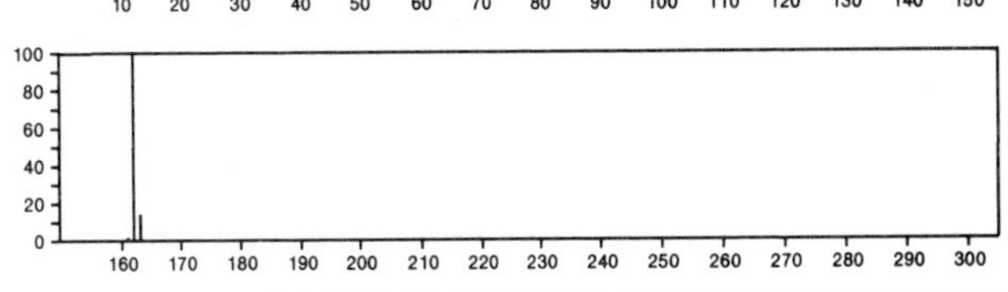

162 $C_8H_{10}N_4$ 19848-80-7
s-Triazolo[4,3-*a*]pyrazine, 3-ethyl-8-methyl-

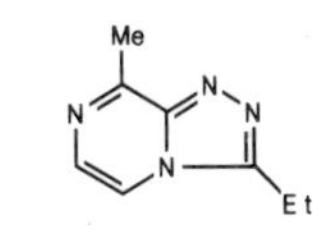

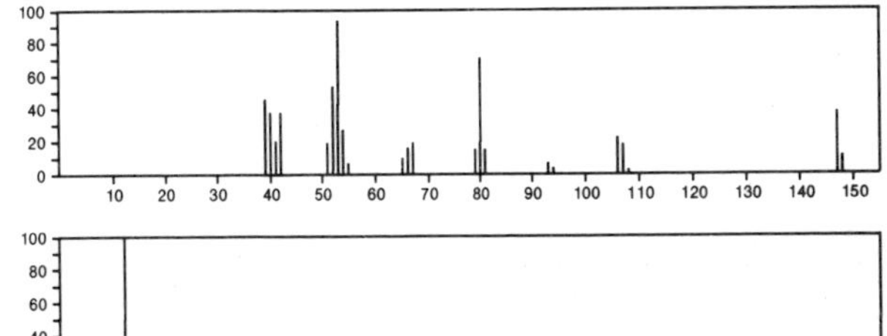

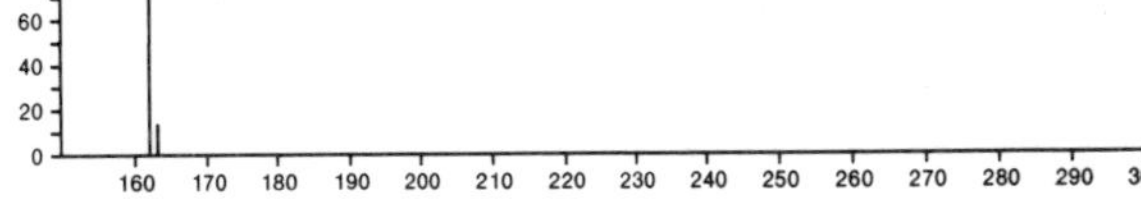

162 $C_8H_{10}N_4$ 54410-76-3
[1,2,4]Triazolo[1,5-*a*]pyrazine, 2,5,8-trimethyl-

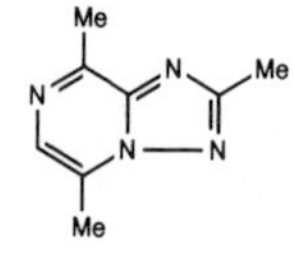

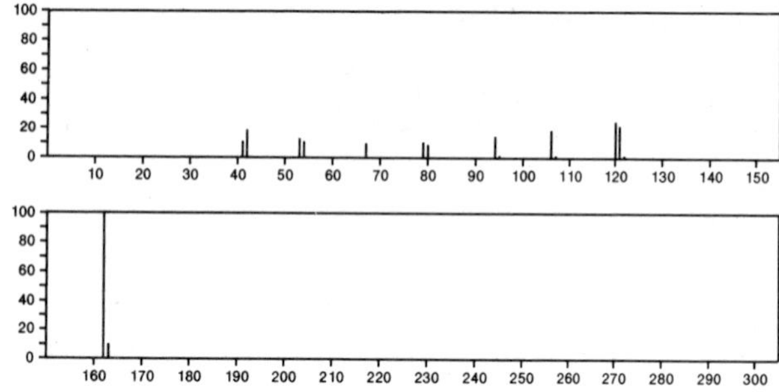

162 $C_8H_{15}ClO$ 111-64-8
Octanoyl chloride

Me(CH2)6COCl

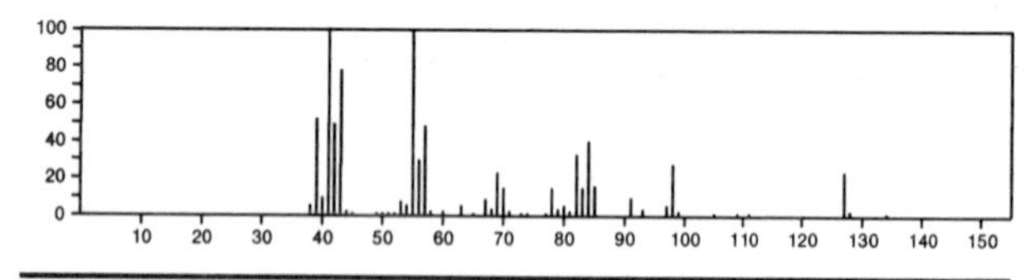

162 $C_8H_{18}OS$ 2168-93-6
Butane, 1,1'-sulfinylbis-

Me(CH2)3S(O)(CH2)3Me

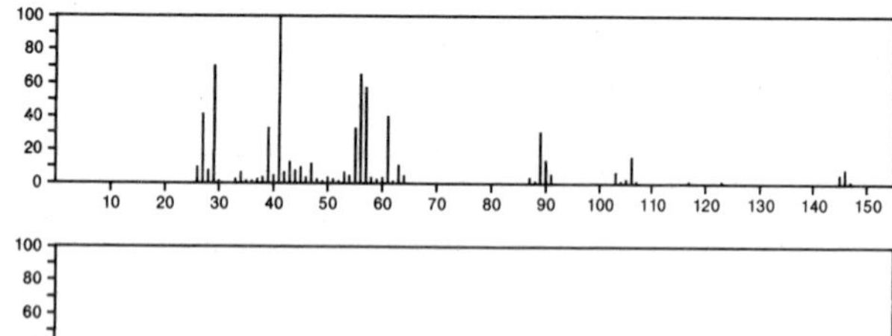

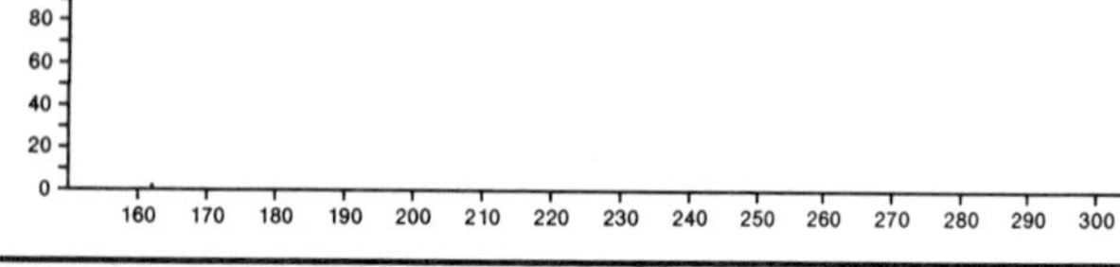

162 $C_8H_{18}OS$ 3085-40-3
Propane, 1,1'-sulfinylbis[2-methyl-

i-BuS(O)Bu-i

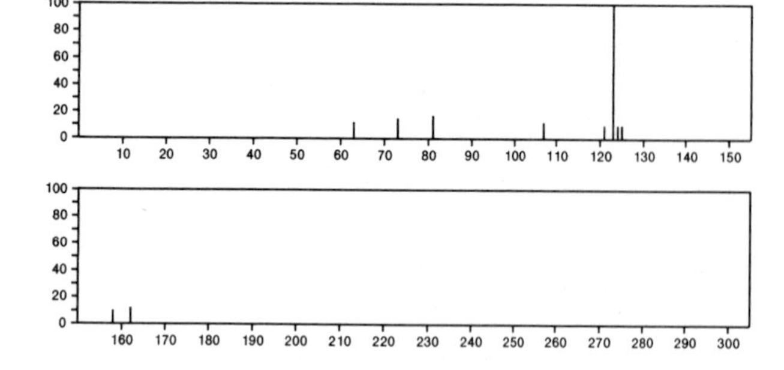

162 $C_8H_{18}O_3$ 78-39-7
Ethane, 1,1,1-triethoxy-

MeC(OEt)3

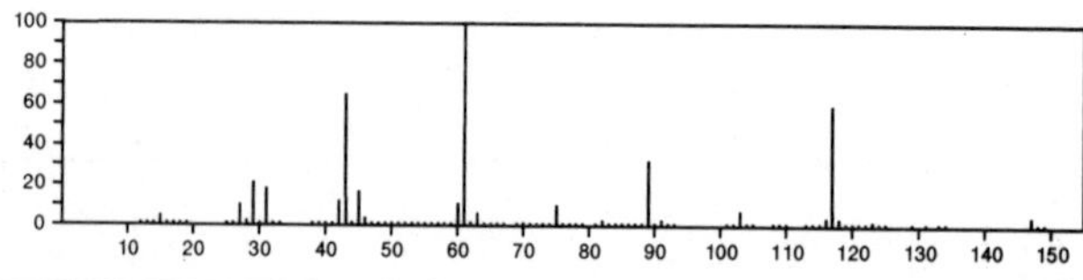

162 $C_8H_{18}O_3$ 112-34-5
Ethanol, 2-(2-butoxyethoxy)-

Me(CH2)3OCH2CH2OCH2CH2OH

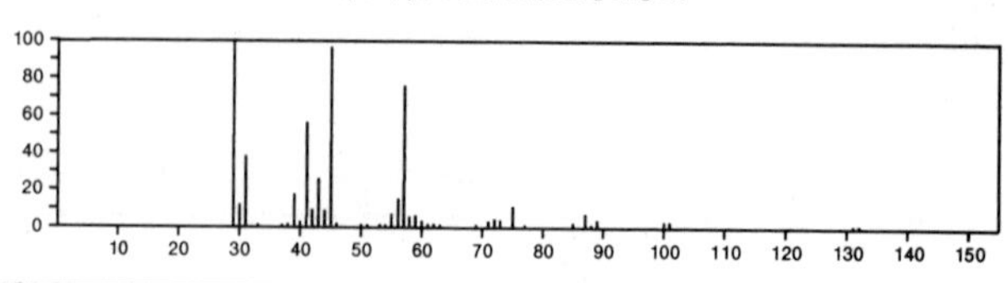

162 $C_8H_{18}O_3$ 112-36-7
Ethane, 1,1'-oxybis[2-ethoxy-

EtOCH2CH2OCH2CH2OEt

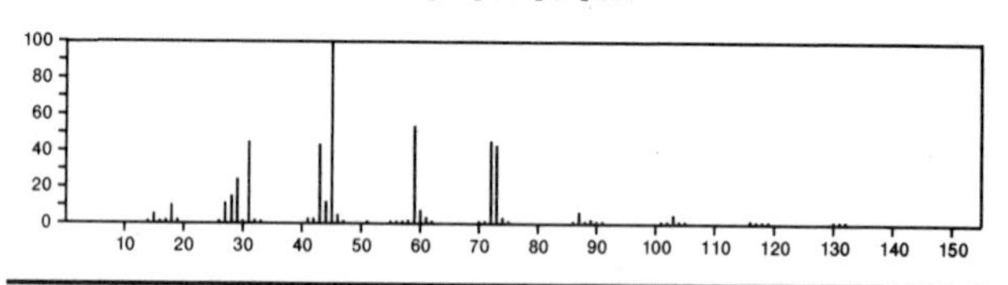

162 $C_8H_{18}O_3$ 10143-32-5
2-Propanol, 1-(2-ethoxypropoxy)-

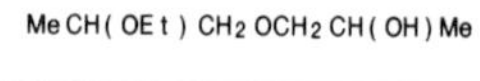

MeCH(OEt)CH2OCH2CH(OH)Me

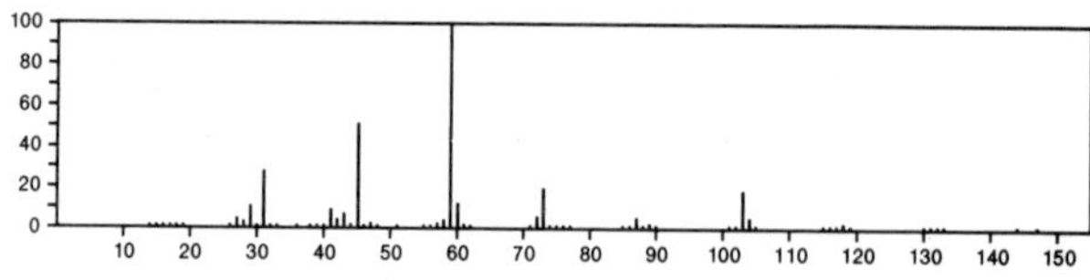

162 $C_8H_{18}O_3$ 13820-09-2
Pentane, 1,1,1-trimethoxy-

(MeO)3C(CH2)3Me

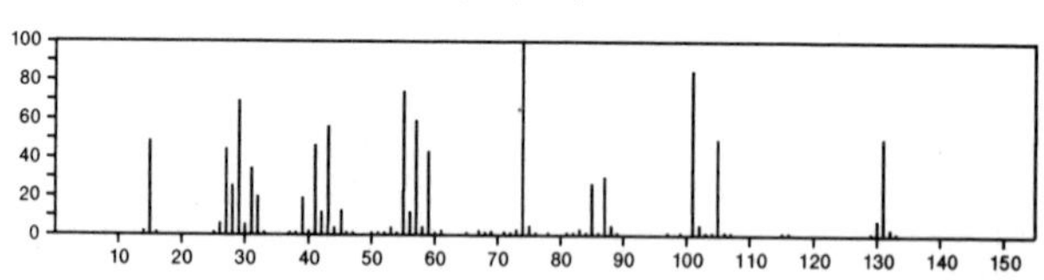

162 $C_8H_{18}O_3$ 15476-20-7
Propane, 1,3-dimethoxy-2-(methoxymethyl)-2-methyl-

MeOCH2C(CH2OMe)MeCH2OMe

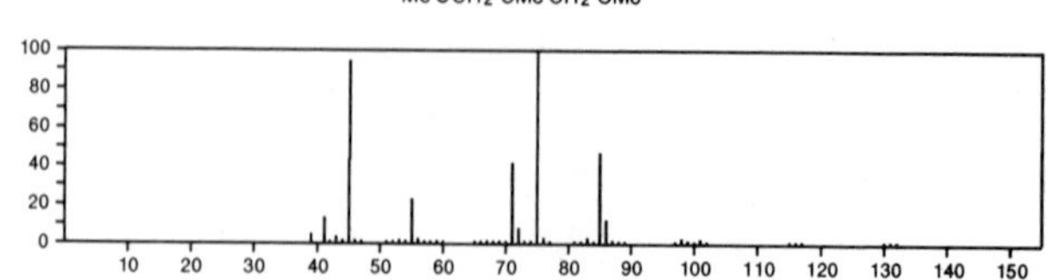

162 $C_8H_{18}O_3$ 20637-28-9
Pentane, 1,2,3-trimethoxy-

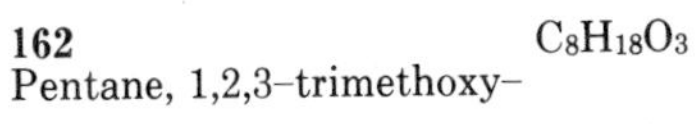

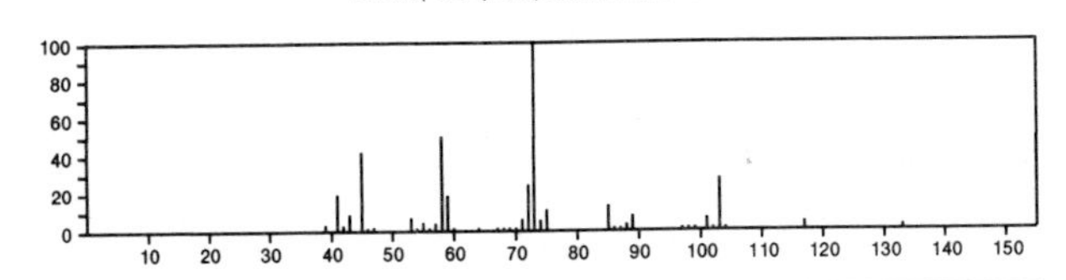

162 $C_8H_{18}O_3$ 54446-78-5
Ethanol, 1-(2-butoxyethoxy)-

Me (CH2) 3 OCH2 CH2 OCH(OH) Me

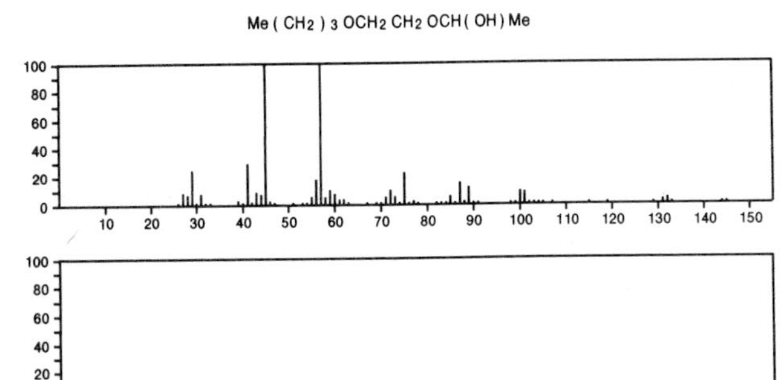

162 $C_8H_{19}OP$ 684-19-5
Phosphine oxide, bis(1,1-dimethylethyl)-

(t-Bu) 2 PH = O

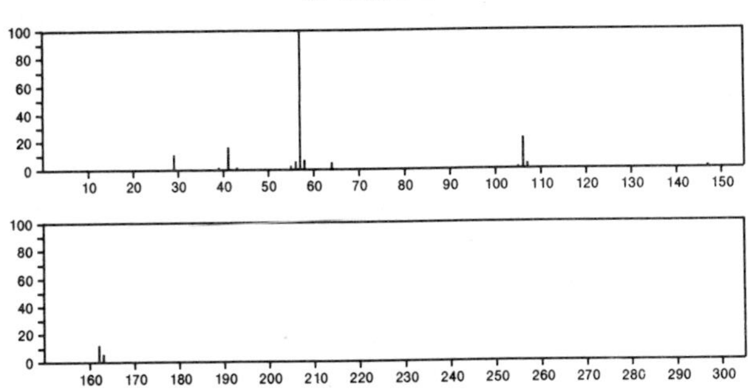

162 $C_9H_6O_3$ 1076-38-6
2*H*-1-Benzopyran-2-one, 4-hydroxy-

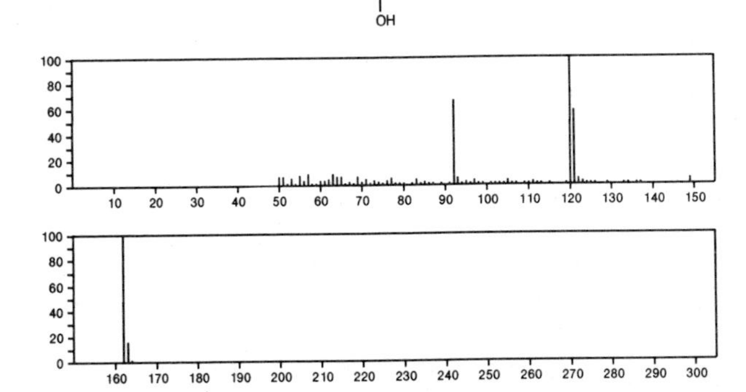

162 $C_9H_{10}N_2O$ 1848-69-7
2-Imidazolidinone, 1-phenyl-

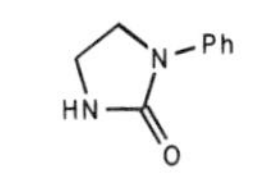

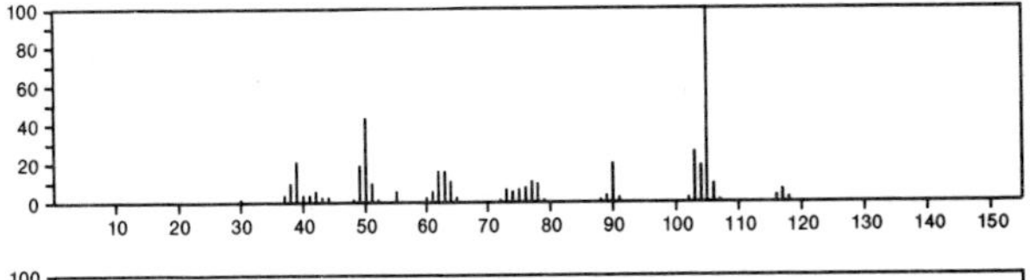

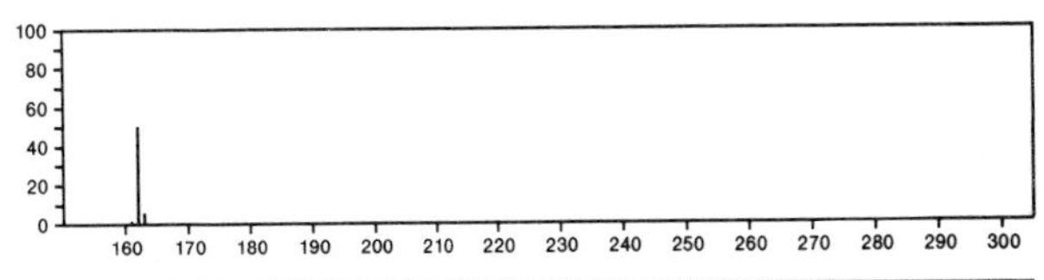

162 $C_9H_{10}N_2O$ 4857-01-6
1*H*-Benzimidazole-2-ethanol

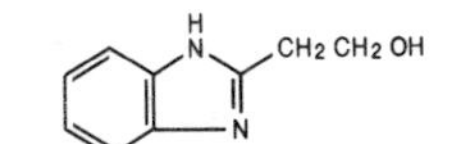

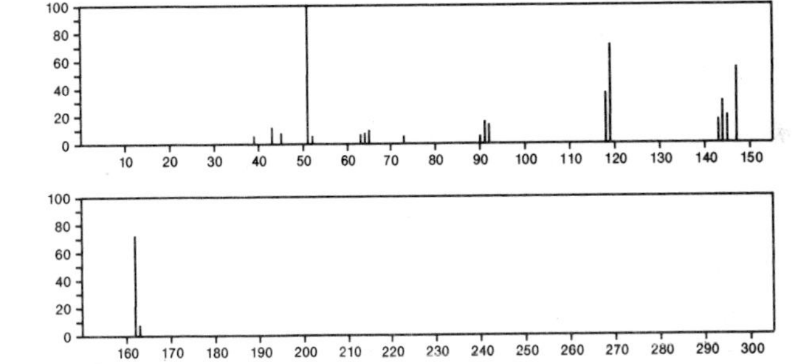

162 $C_9H_{10}N_2O$ 13858-89-4
2-Benzoxazolamine, *N*,*N*-dimethyl-

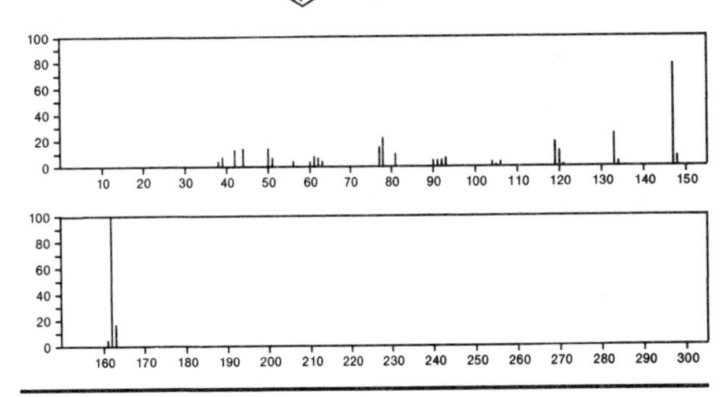

162 $C_9H_{10}N_2O$ 16007-53-7
Benzimidazole, 2-ethyl-, 3-oxide

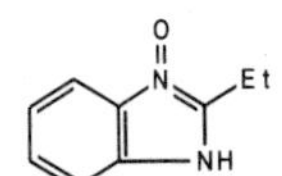

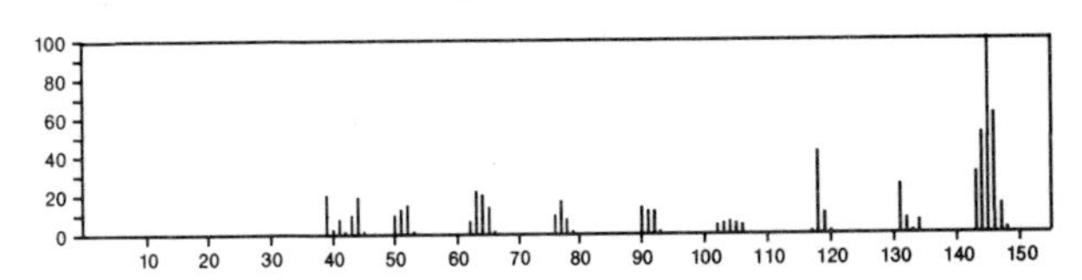

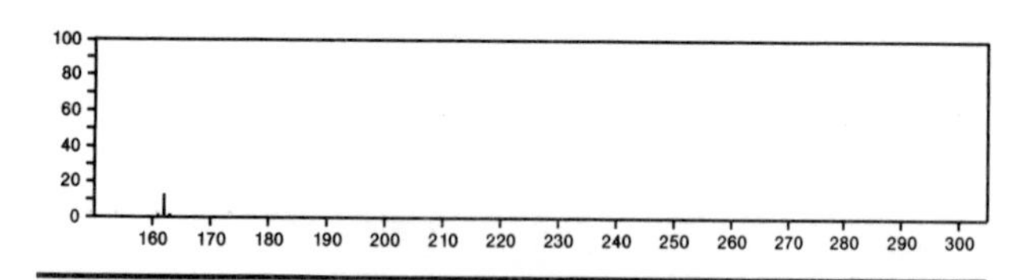

162 $C_9H_{10}N_2O$ 21326-91-0
Benzoxazole, 2-(ethylamino)-

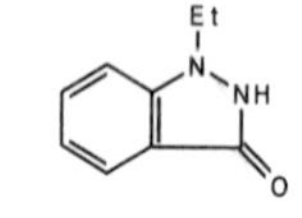

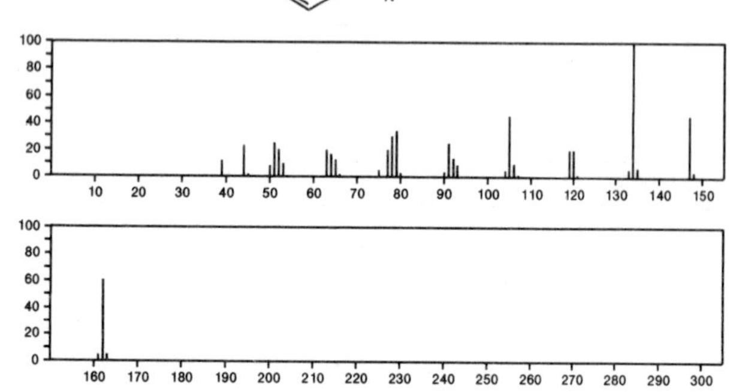

162 $C_9H_{10}N_2O$ 54385-62-5
3H-Indazol-3-one, 1-ethyl-1,2-dihydro-

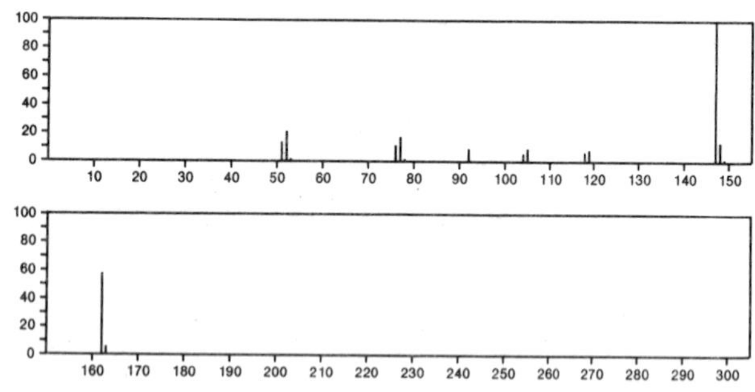

162 $C_9H_{19}Cl$ 2473-01-0
Nonane, 1-chloro-

Me (CH2) 8 Cl

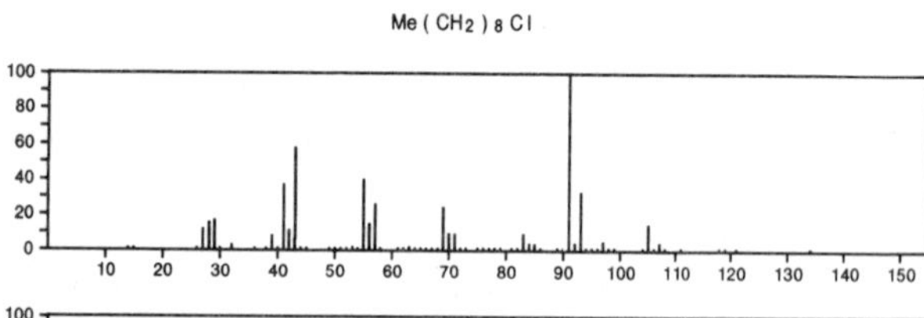

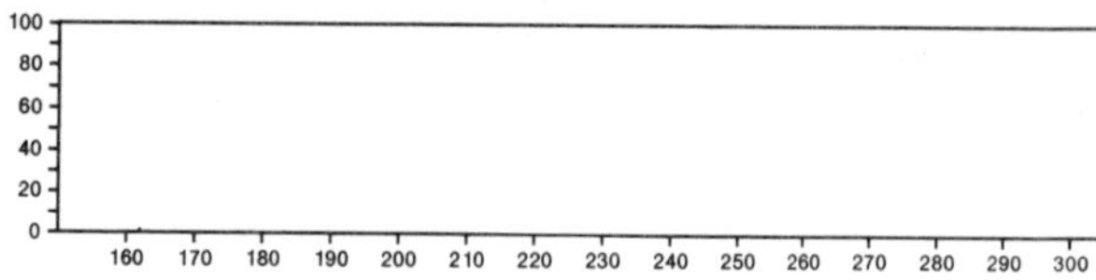

162 $C_{10}H_7Cl$ 90-13-1
Naphthalene, 1-chloro-

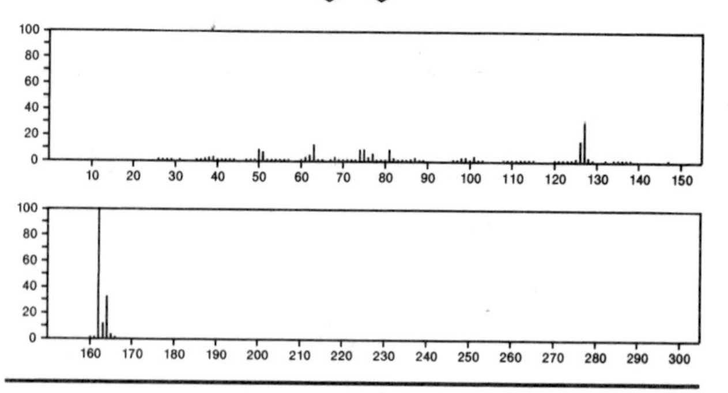

162 $C_{10}H_7Cl$ 91-58-7
Naphthalene, 2-chloro-

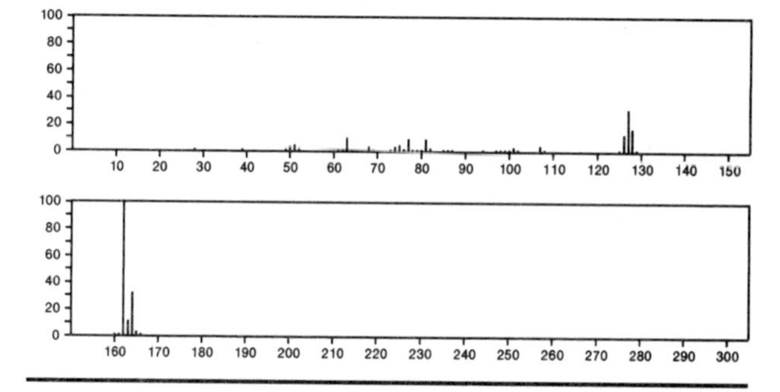

162 $C_{10}H_7Cl$ 25586-43-0
Naphthalene, chloro-

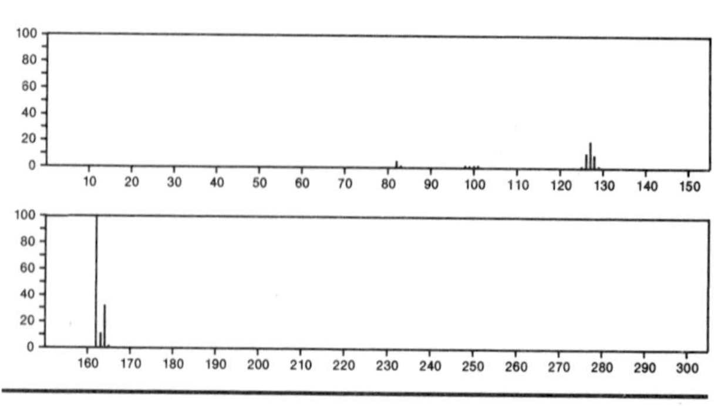

162 $C_{10}H_{10}O_2$ 93-91-4
1,3-Butanedione, 1-phenyl-

PhCOCH2COMe

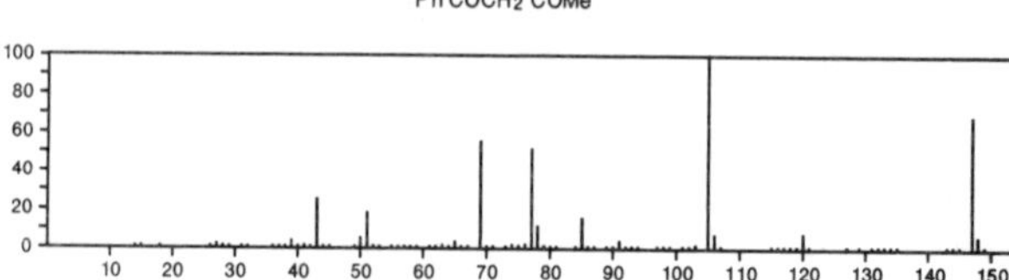

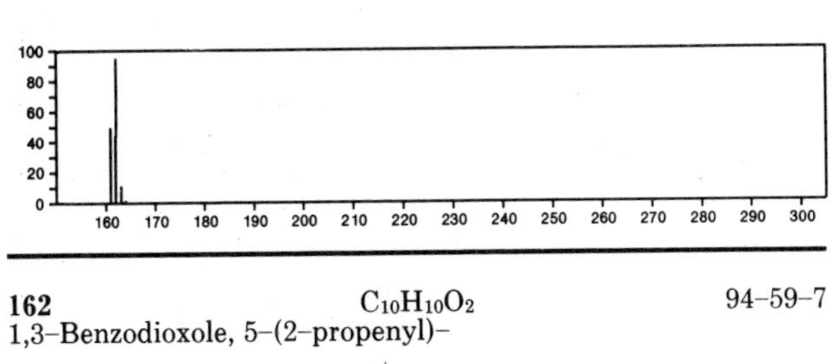

162 $C_{10}H_{10}O_2$ 94-59-7
1,3-Benzodioxole, 5-(2-propenyl)-

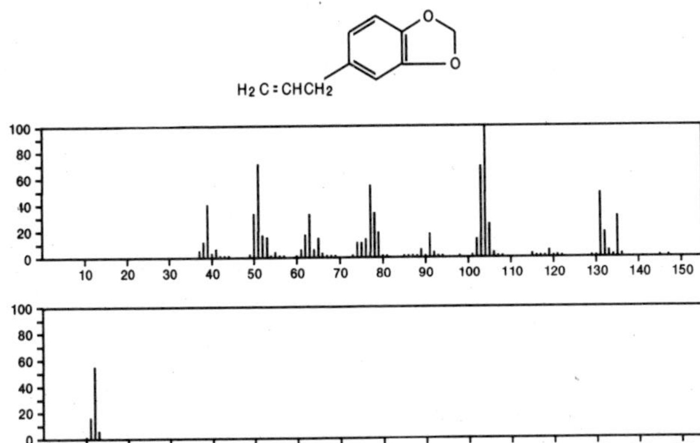

162 $C_{10}H_{10}O_2$ 103-26-4
2-Propenoic acid, 3-phenyl-, methyl ester

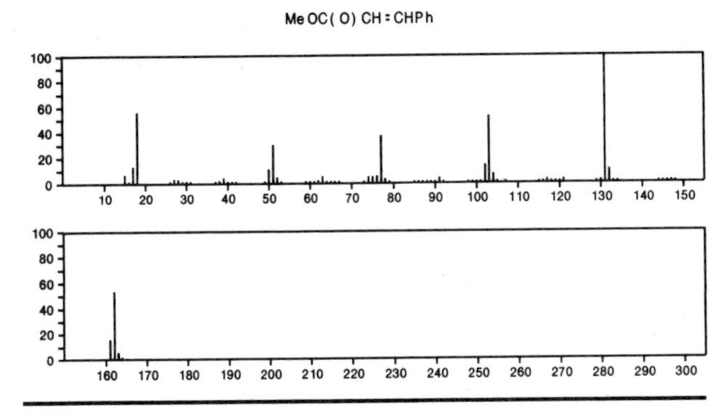

162 $C_{10}H_{10}O_2$ 120-58-1
1,3-Benzodioxole, 5-(1-propenyl)-

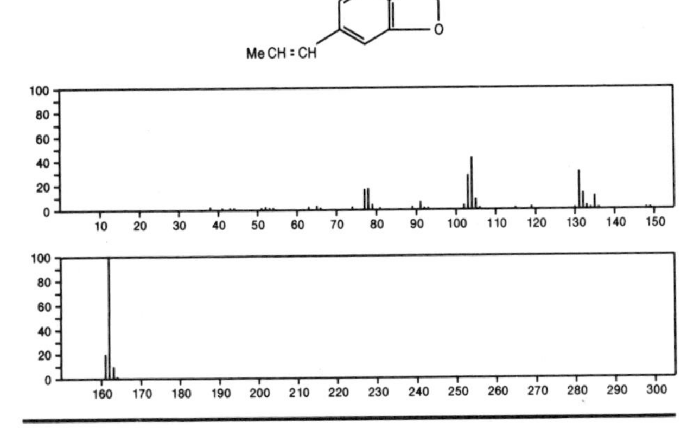

162 $C_{10}H_{10}O_2$ 583-04-0
Benzoic acid, 2-propenyl ester

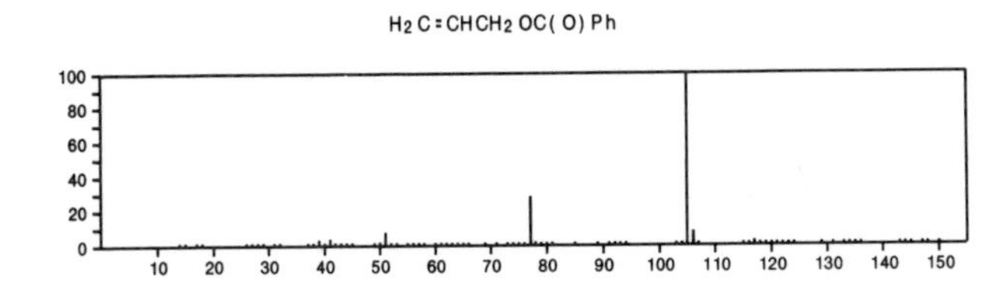

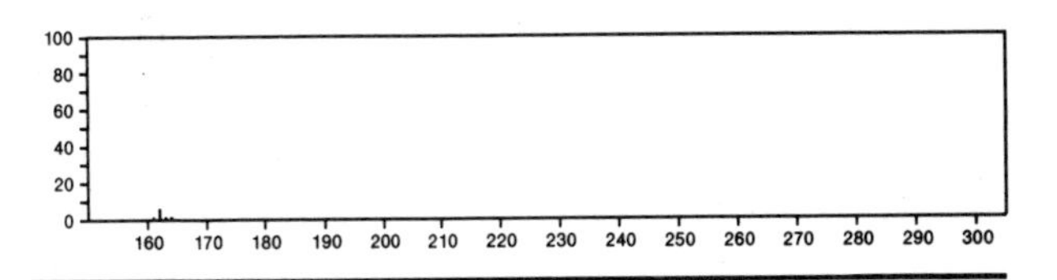

162 $C_{10}H_{10}O_2$ 1008-76-0
2(3*H*)-Furanone, dihydro-5-phenyl-

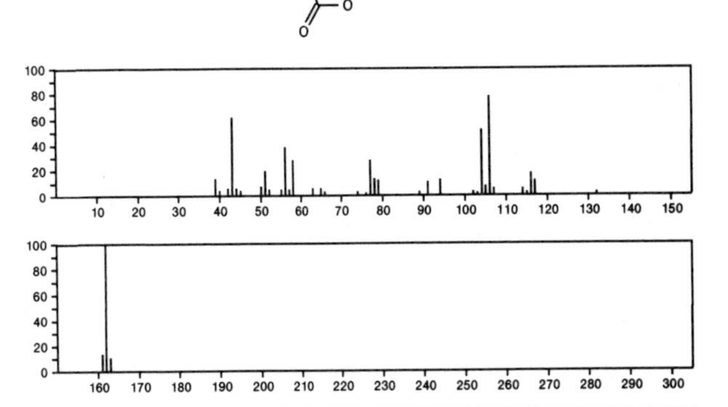

162 $C_{10}H_{10}O_2$ 1009-61-6
Ethanone, 1,1'-(1,4-phenylene)bis-

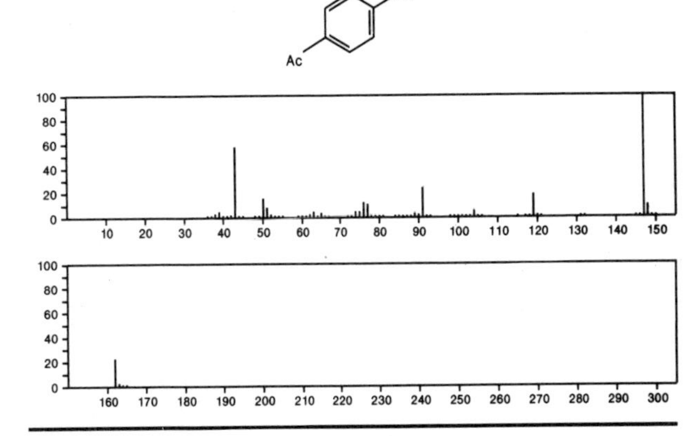

162 $C_{10}H_{10}O_2$ 1076-96-6
Benzoic acid, 4-ethenyl-, methyl ester

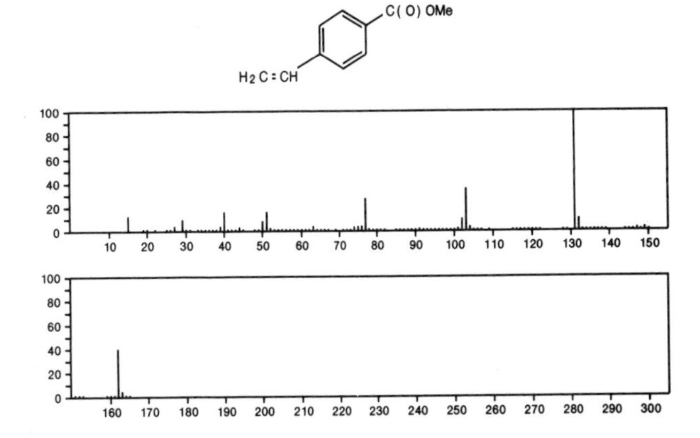

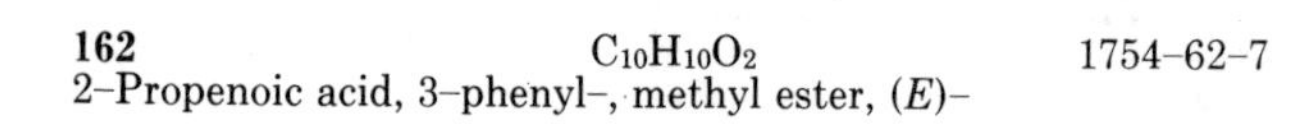

162 $C_{10}H_{10}O_2$ 1754–62–7

2–Propenoic acid, 3–phenyl–, methyl ester, (*E*)–

MeOC(O)CH=CHPh

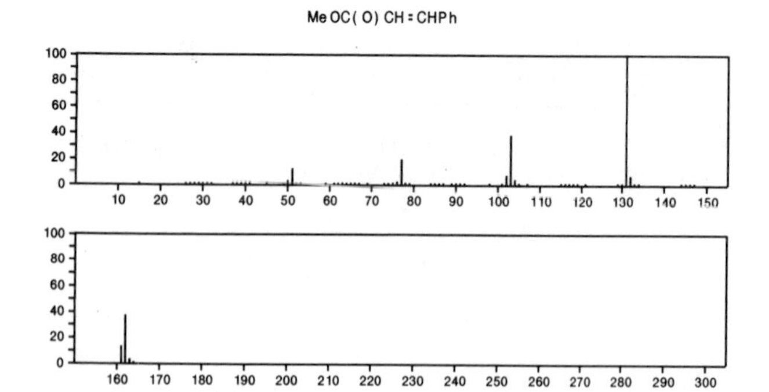

162 $C_{10}H_{10}O_2$ 7044–92–0

Terephthalaldehyde, 2,5–dimethyl–

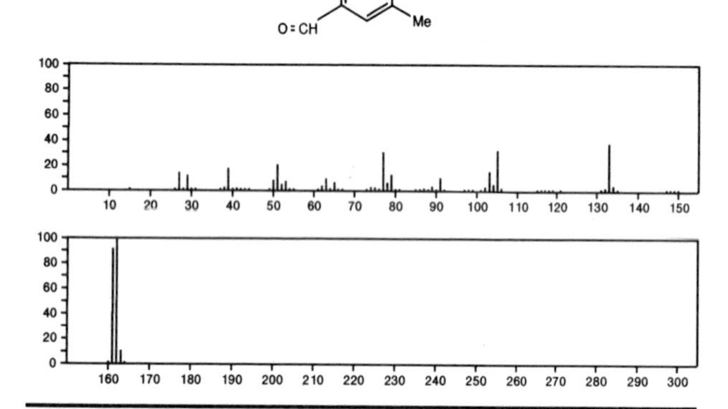

162 $C_{10}H_{10}O_2$ 7234–04–0

1,2–Naphthalenediol, 1,2–dihydro–

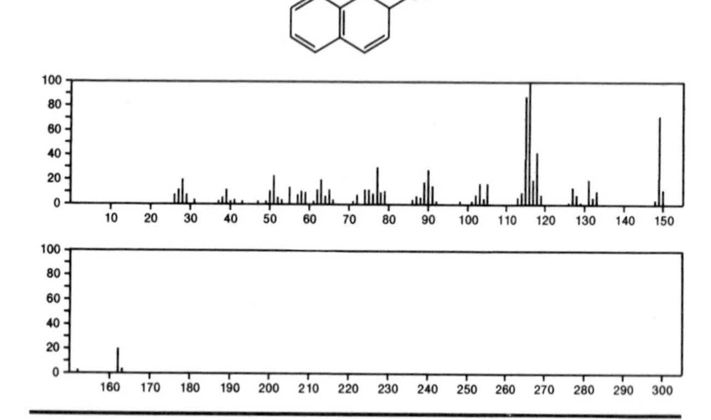

162 $C_{10}H_{10}O_2$ 13524–76–0

2(3*H*)–Benzofuranone, 3,3–dimethyl–

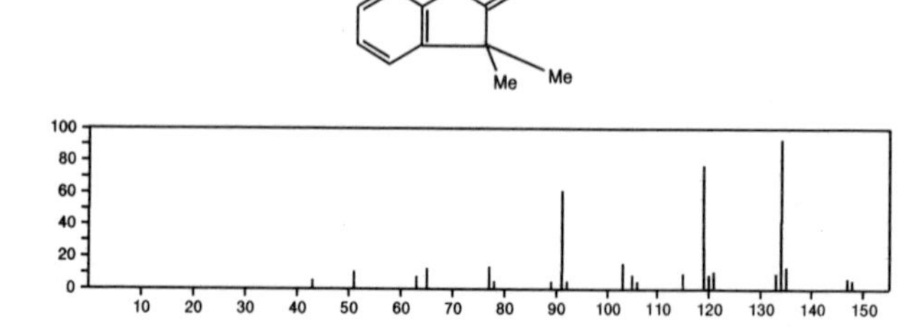

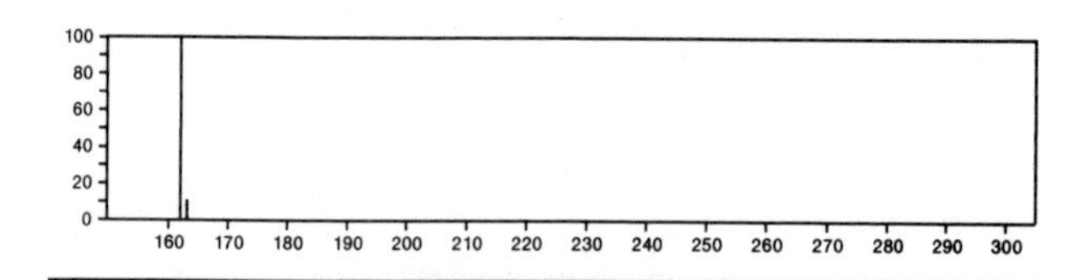

162 $C_{10}H_{10}O_2$ 19560–64–6

2*H*–1,5–Benzodioxepin, 3,4–dihydro–3–methylene–

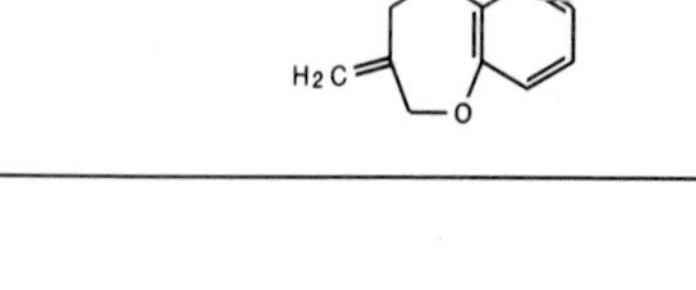

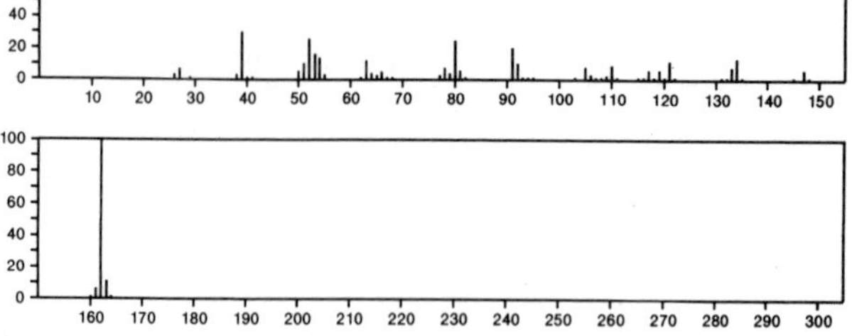

162 $C_{10}H_{10}O_2$ 20895–42–5

3(2*H*)–Benzofuranone, 4,5–dimethyl–

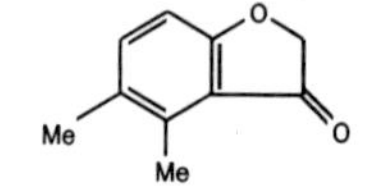

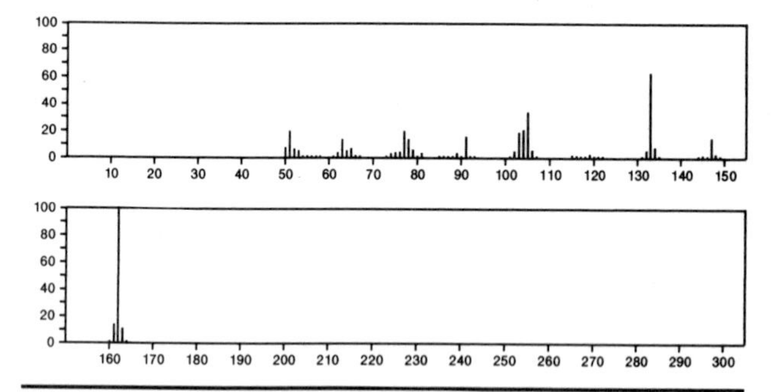

162 $C_{10}H_{10}O_2$ 20895–43–6

3(2*H*)–Benzofuranone, 5,6–dimethyl–

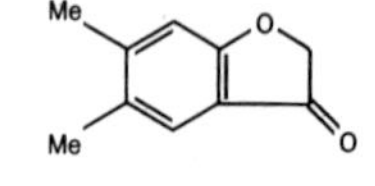

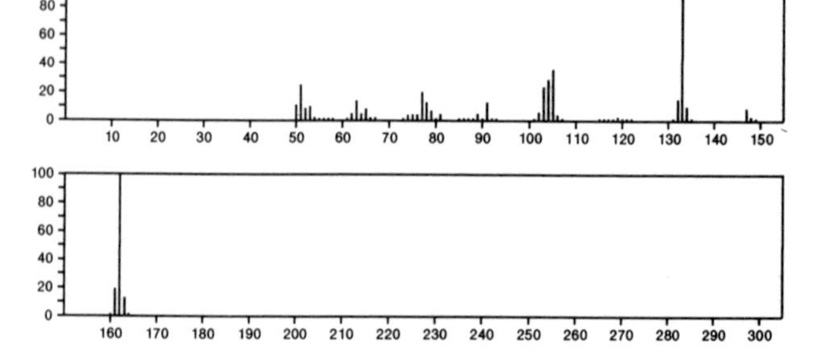

162 $C_{10}H_{10}O_2$ 20895-44-7
3(2*H*)-Benzofuranone, 4,6-dimethyl-

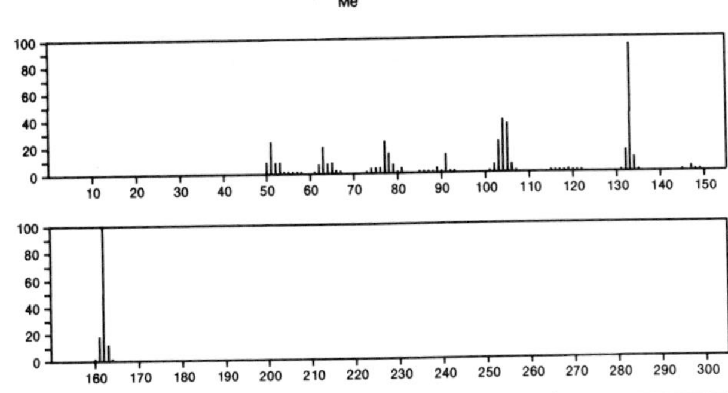

162 $C_{10}H_{10}O_2$ 20895-45-8
3(2*H*)-Benzofuranone, 4,7-dimethyl-

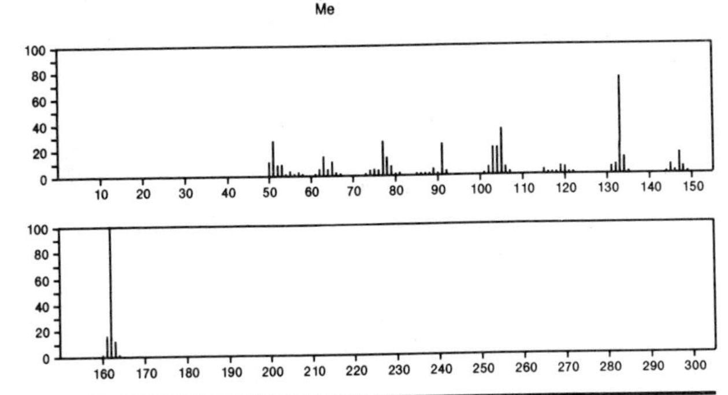

162 $C_{10}H_{10}O_2$ 20895-46-9
3(2*H*)-Benzofuranone, 5,7-dimethyl-

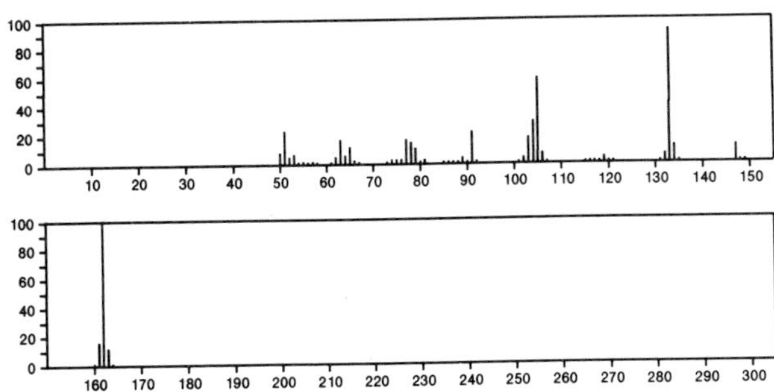

162 $C_{10}H_{10}O_2$ 20895-47-0
3(2*H*)-Benzofuranone, 6,7-dimethyl-

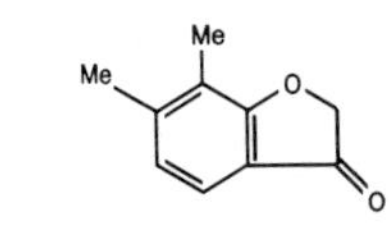

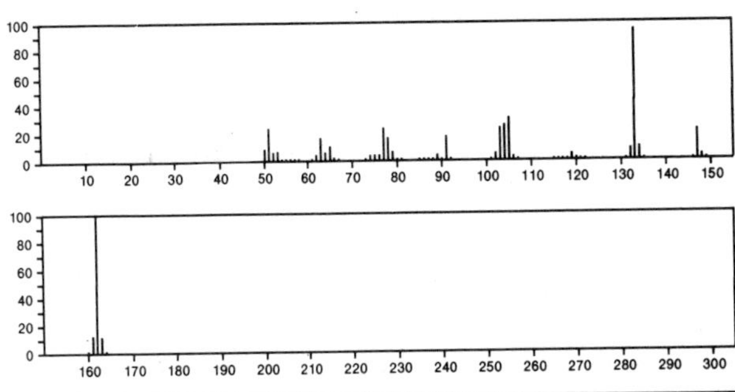

162 $C_{10}H_{10}O_2$ 21149-13-3
2-Cyclopenten-1-one, 2-(4-methyl-2-furyl)-

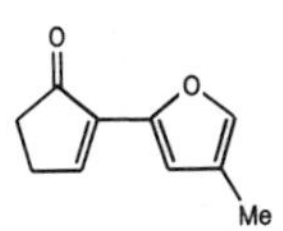

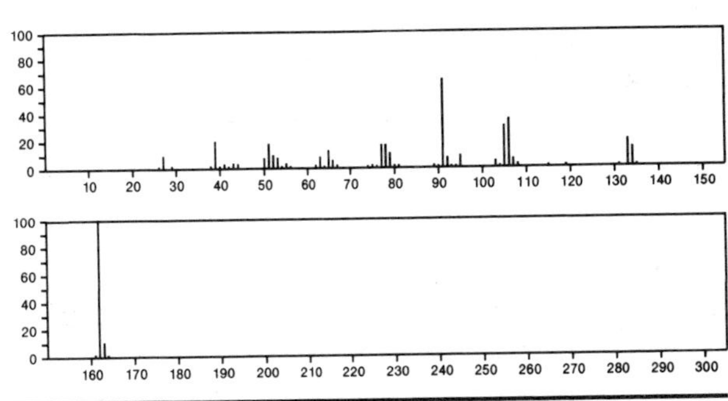

162 $C_{10}H_{10}O_2$ 21149-14-4
Indene-1,7(4*H*)-dione, 3a,7a-dihydro-5-methyl-

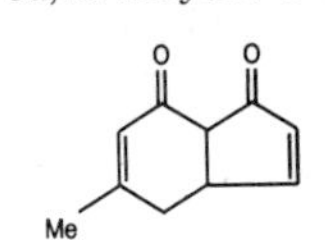

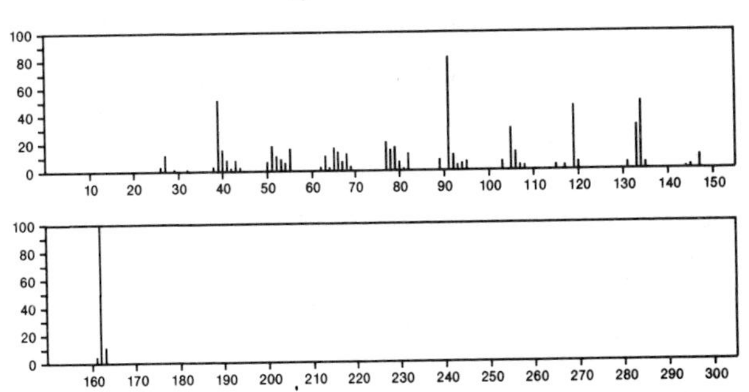

162 $C_{10}H_{10}O_2$ 24019–66–7
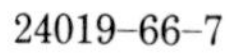
Benzaldehyde, 2–hydroxy–3–(2–propenyl)–

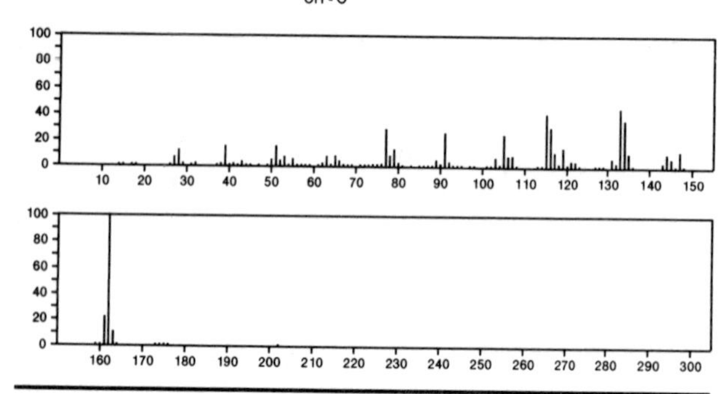

162 $C_{10}H_{10}O_2$ 28752–82–1
Benzaldehyde, 2–(2–propenyloxy)–

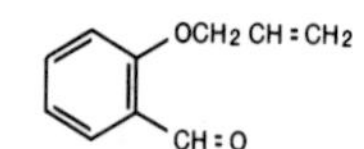

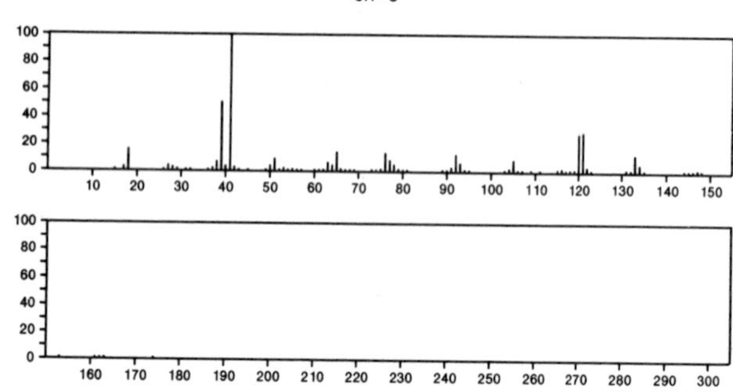

162 $C_{10}H_{10}O_2$ 29800–57–5
Tetracyclo[3.3.0.0^{2,4}.0^{3,6}]oct–7–ene–4–carboxylic acid, methyl ester

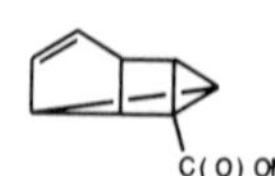

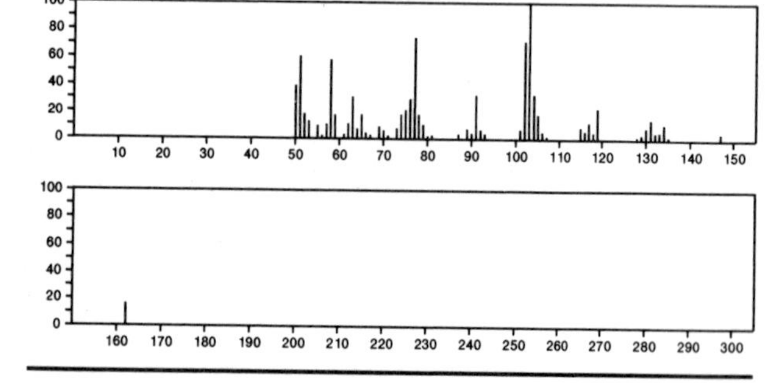

162 $C_{10}H_{10}O_2$ 37570–14–2
Benzeneacetic acid, *ar*–ethenyl–

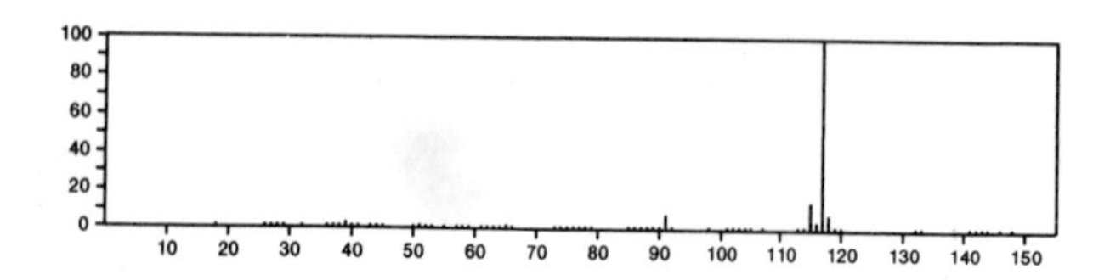

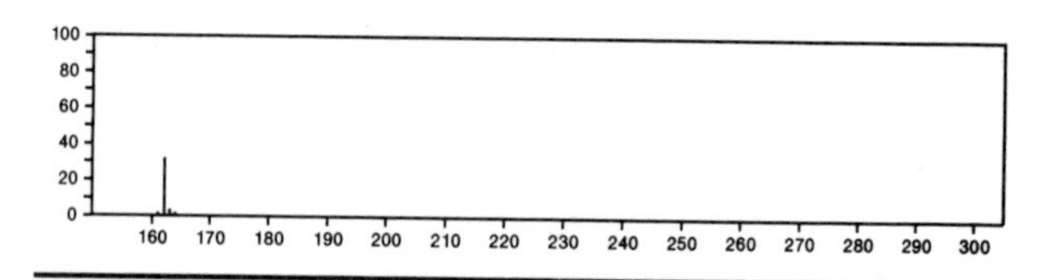

162 $C_{10}H_{10}O_2$ 40663–68–1
Benzaldehyde, 4–(2–propenyloxy)–

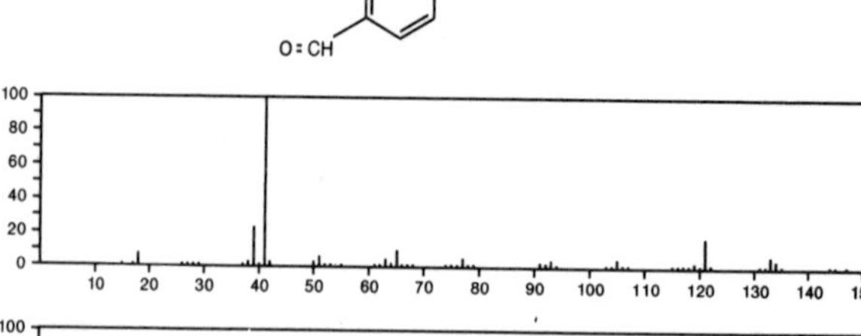

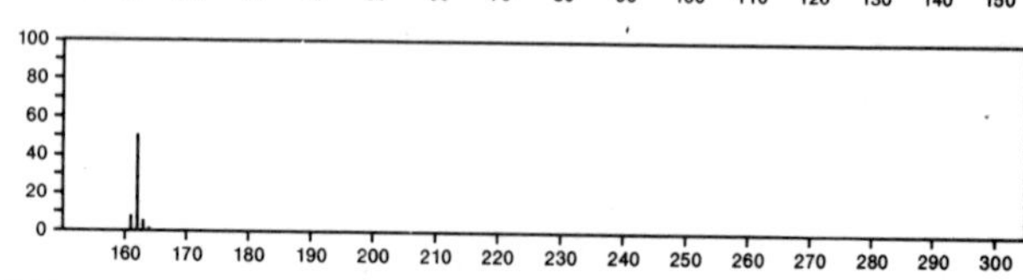

162 $C_{10}H_{10}O_2$ 54365–75–2
5–Benzofurancarboxaldehyde, 2,3–dihydro–2–methyl–

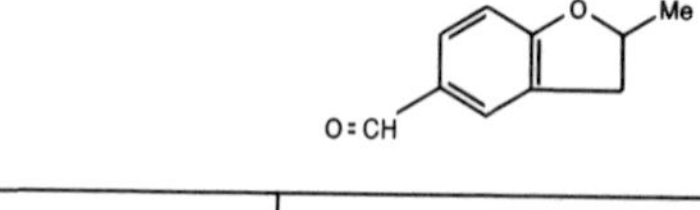

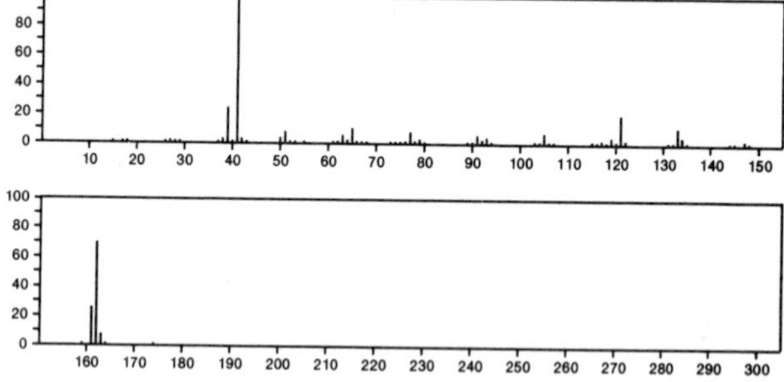

162 $C_{10}H_{10}O_2$ 54365–76–3
3(2*H*)–Benzofuranone, 2,4–dimethyl–

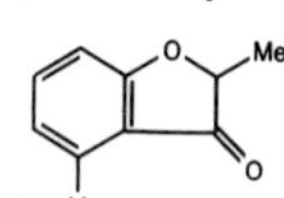

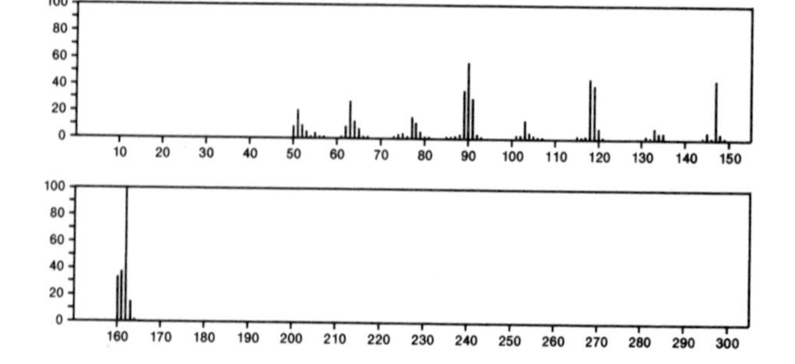

162 $C_{10}H_{10}O_2$ 54365-77-4
3(2*H*)-Benzofuranone, 2,5-dimethyl-

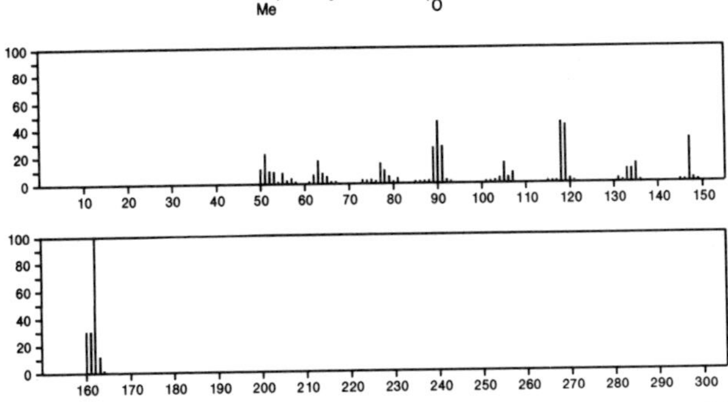

162 $C_{10}H_{10}O_2$ 54365-78-5
3(2*H*)-Benzofuranone, 2,6-dimethyl-

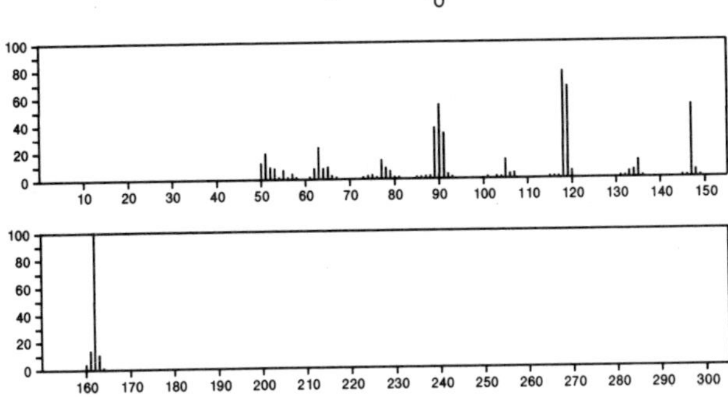

162 $C_{10}H_{10}O_2$ 54365-79-6
3(2*H*)-Benzofuranone, 2,7-dimethyl-

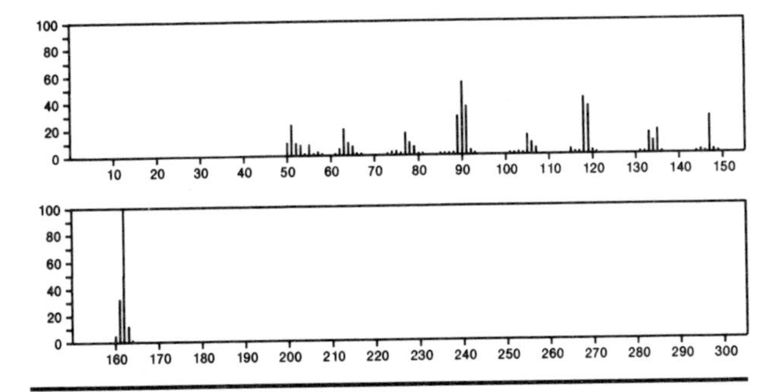

162 $C_{10}H_{10}S$ 16587-40-9
Benzo[*b*]thiophene, 2,7-dimethyl-

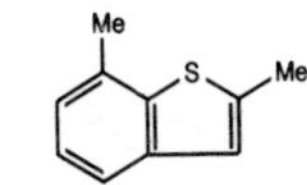

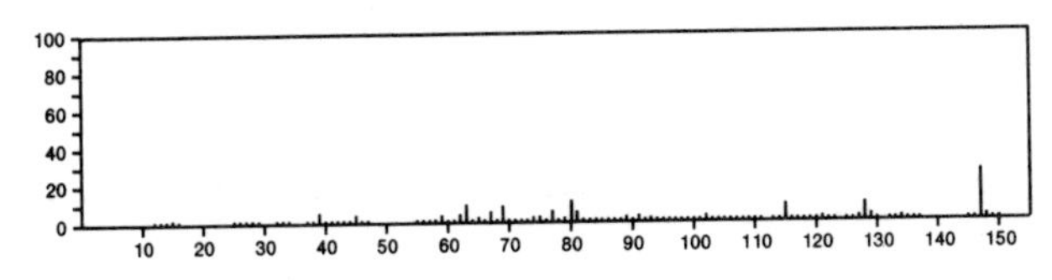

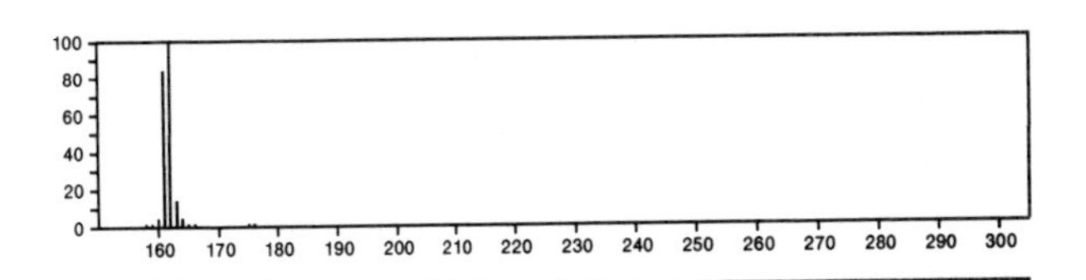

162 $C_{10}H_{10}S$ 16587-42-1
Benzo[*b*]thiophene, 7-ethyl-

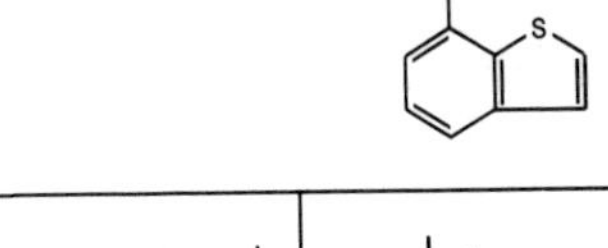

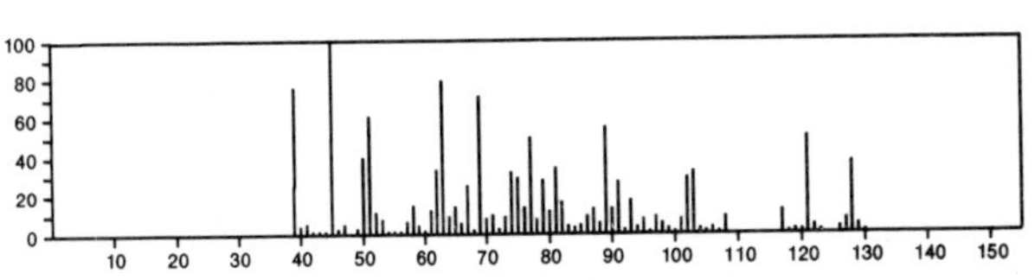

162 $C_{10}H_{10}S$ 16587-48-7
Benzo[*b*]thiophene, 2,5-dimethyl-

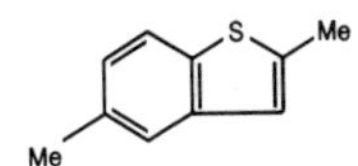

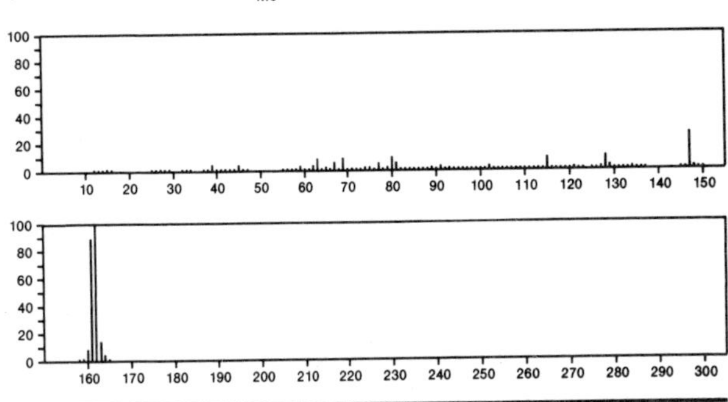

162 $C_{10}H_{10}S$ 16587-50-1
Benzo[*b*]thiophene, 3,6-dimethyl-

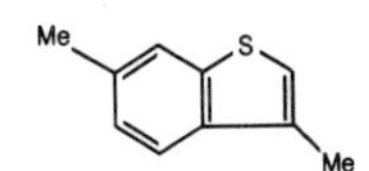

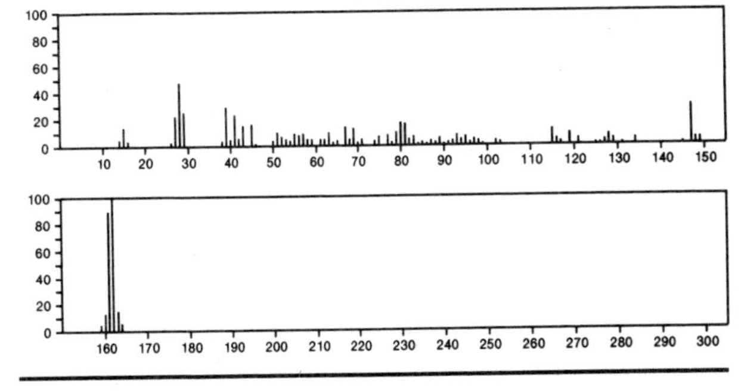

162 $C_{10}H_{10}S$ 54385-63-6
Benzo[*b*]thiophene, ethyl-

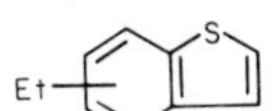

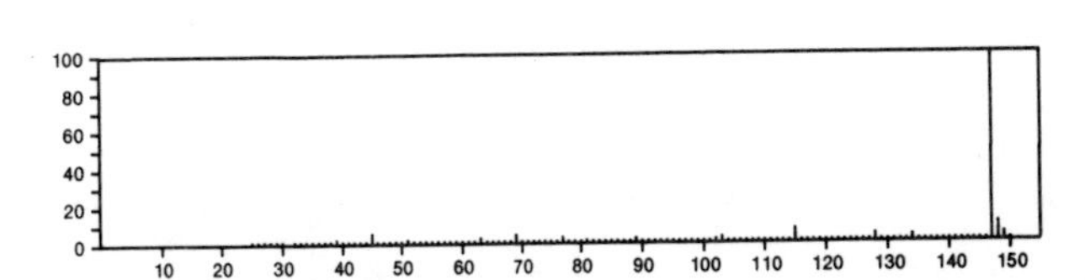

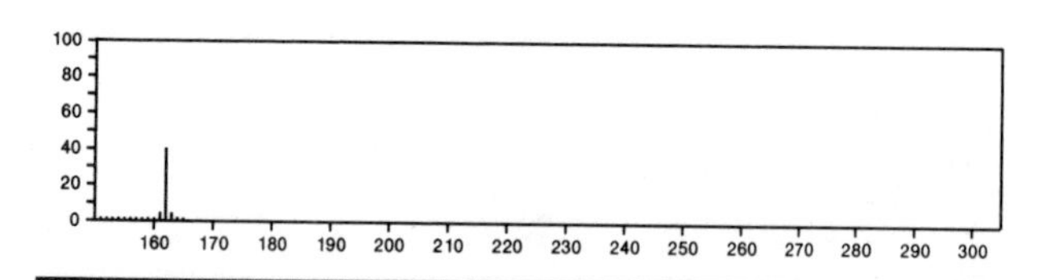

162 $C_{10}H_{14}N_2$ 54-11-5
Pyridine, 3-(1-methyl-2-pyrrolidinyl)-, (*S*)-

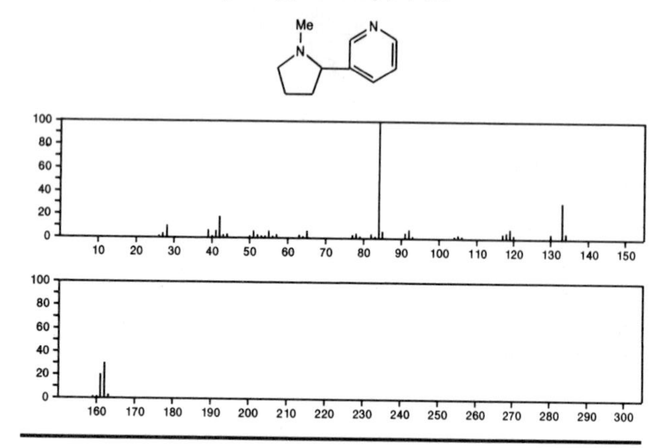

162 $C_{10}H_{14}N_2$ 494-52-0
Pyridine, 3-(2-piperidinyl)-, (*S*)-

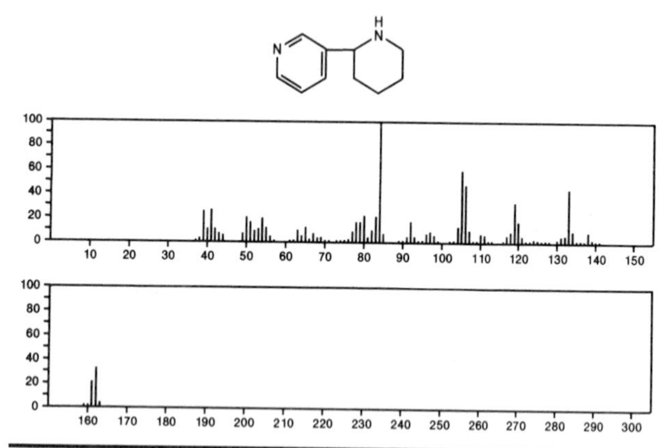

162 $C_{10}H_{14}N_2$ 2305-75-1
Methanimidamide, *N*,*N*-dimethyl-*N'*-(3-methylphenyl)-

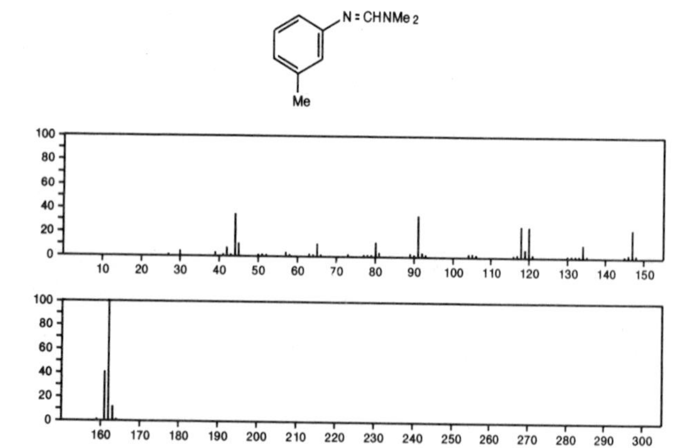

162 $C_{10}H_{14}N_2$ 7549-96-4
Methanimidamide, *N*,*N*-dimethyl-*N'*-(4-methylphenyl)-

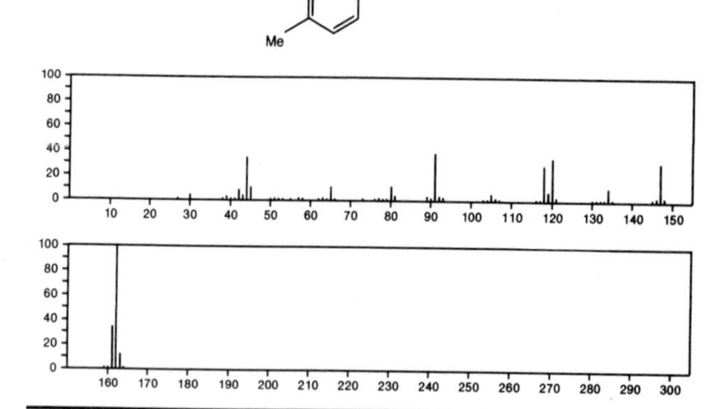

162 $C_{10}H_{14}N_2$ 16738-89-9
1,1-Cyclopropanedicarbonitrile, 2-butyl-2-methyl-

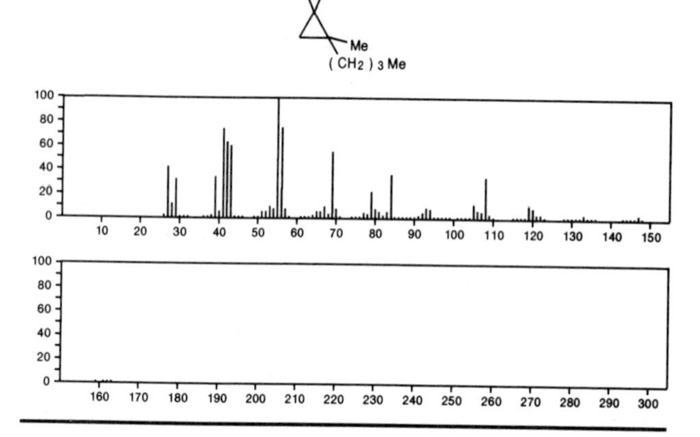

162 $C_{11}H_{14}O$ 645-13-6
Ethanone, 1-[4-(1-methylethyl)phenyl]-

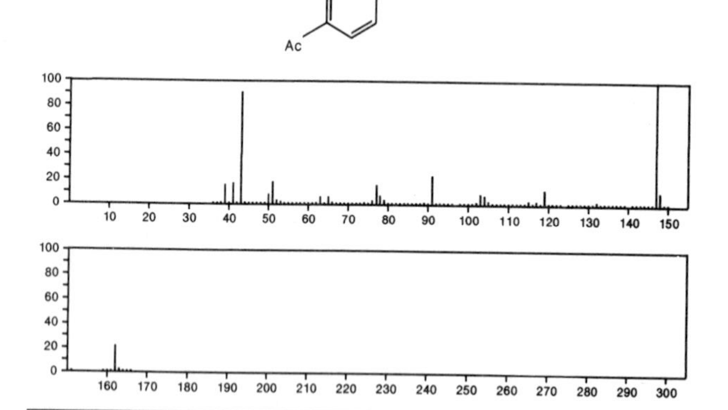

162 $C_{11}H_{14}O$ 1008-19-1
Naphthalene, 1,2,3,4-tetrahydro-5-methoxy-

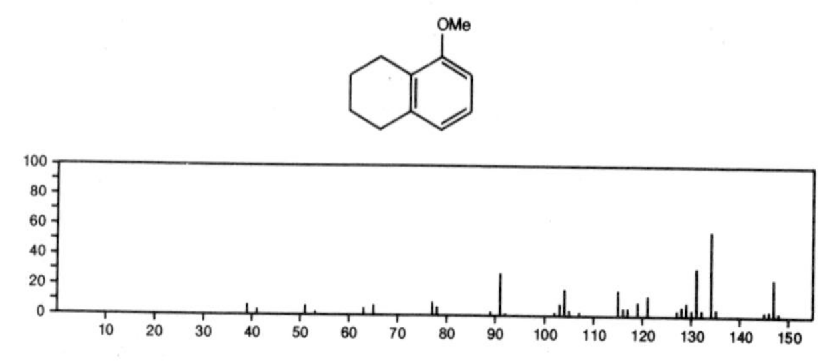

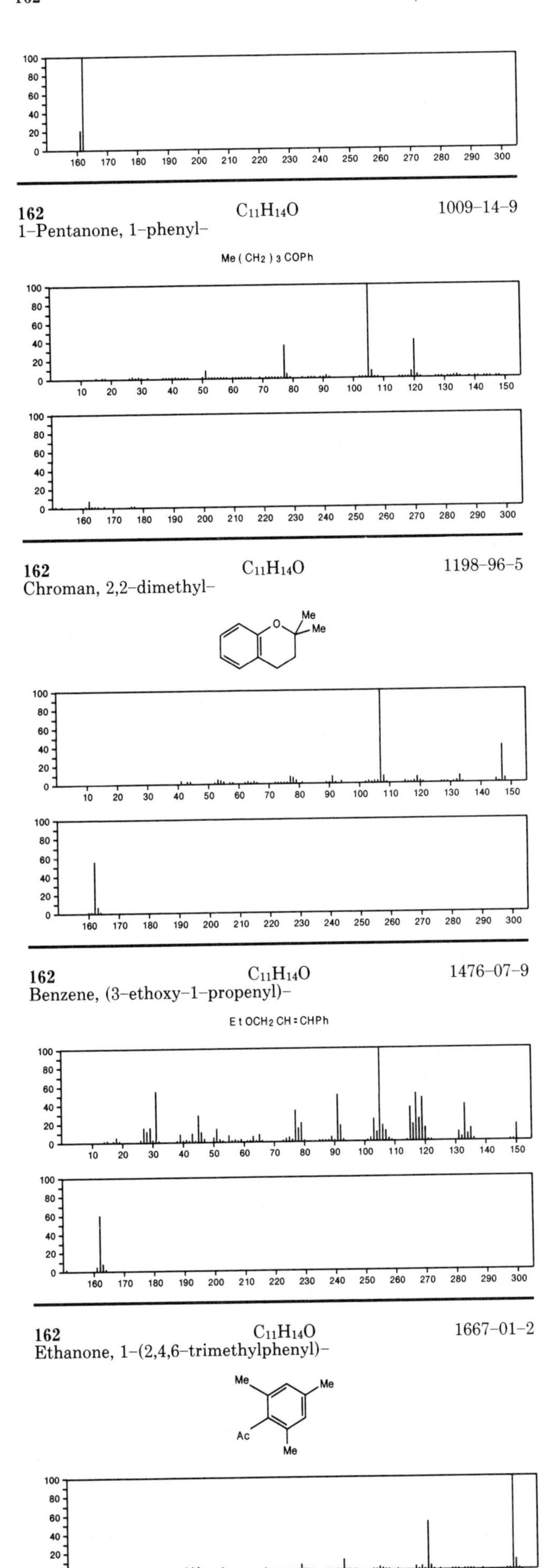

162 $C_{11}H_{14}O$ 1009–14–9
1–Pentanone, 1–phenyl–
Me (CH2) 3 COPh
162 $C_{11}H_{14}O$ 1198–96–5
Chroman, 2,2–dimethyl–
162 $C_{11}H_{14}O$ 1476–07–9
Benzene, (3–ethoxy–1–propenyl)–
E t OCH2 CH = CHPh
162 $C_{11}H_{14}O$ 1667–01–2
Ethanone, 1–(2,4,6–trimethylphenyl)–

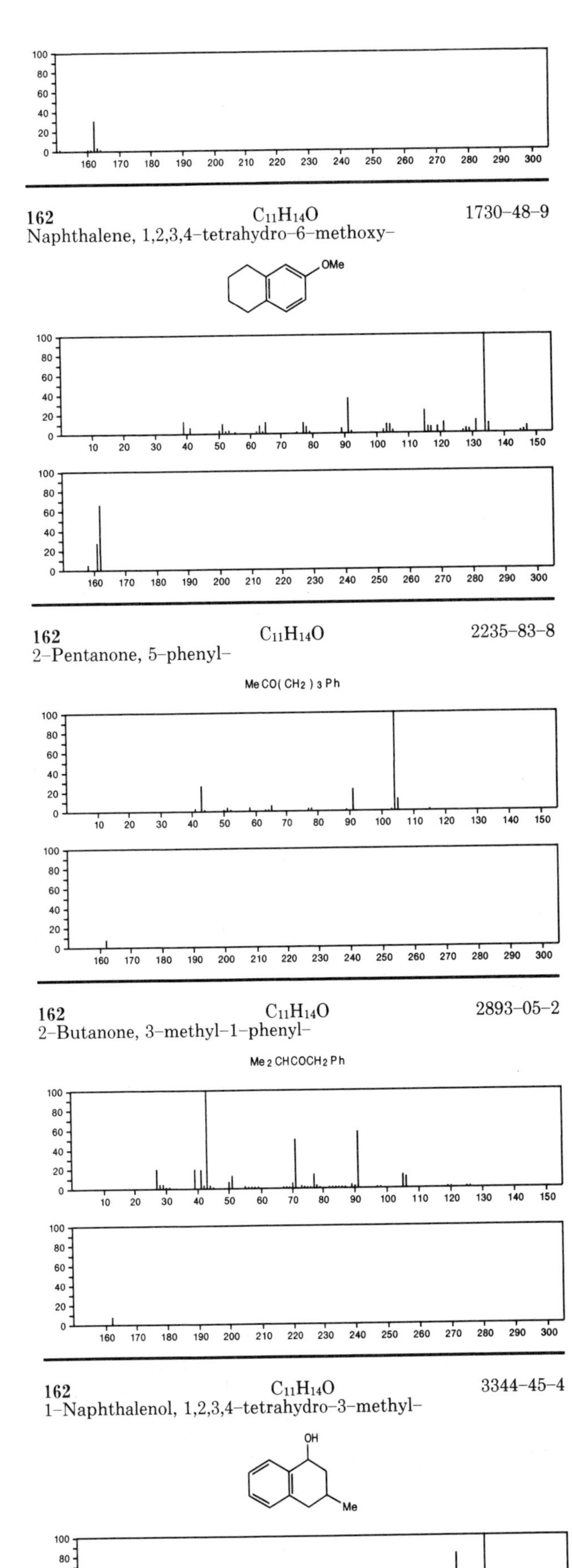

162 $C_{11}H_{14}O$ 1730–48–9
Naphthalene, 1,2,3,4–tetrahydro–6–methoxy–
162 $C_{11}H_{14}O$ 2235–83–8
2–Pentanone, 5–phenyl–
Me CO(CH2) 3 Ph
162 $C_{11}H_{14}O$ 2893–05–2
2–Butanone, 3–methyl–1–phenyl–
Me 2 CHCOCH2 Ph
162 $C_{11}H_{14}O$ 3344–45–4
1–Naphthalenol, 1,2,3,4–tetrahydro–3–methyl–

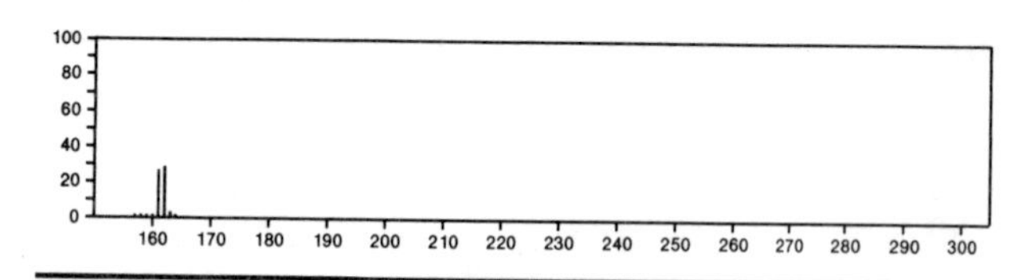

162 $C_{11}H_{14}O$ 6518-50-9
Tricyclo[4.4.0.0^{2,7}]dec-4-en-3-one, 6-methyl-

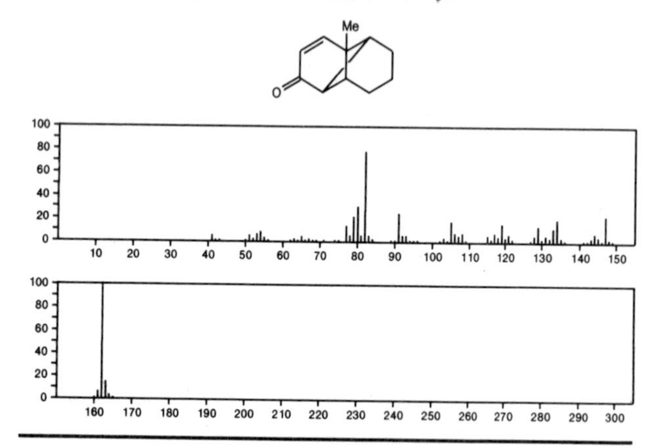

162 $C_{11}H_{14}O$ 6683-92-7
2-Pentanone, 1-phenyl-

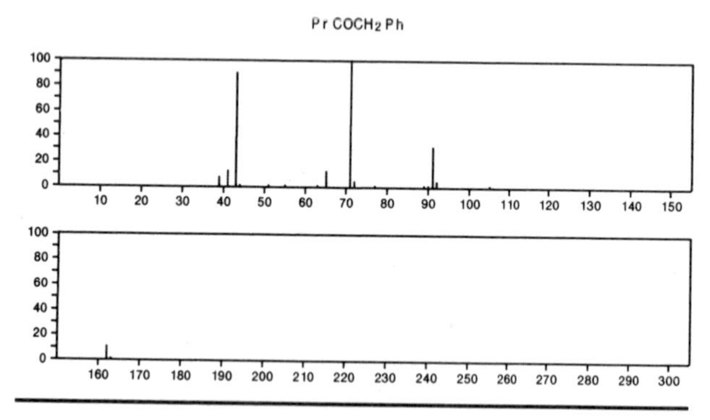

162 $C_{11}H_{14}O$ 14309-15-0
Benzene, [(3-methyl-2-butenyl)oxy]-

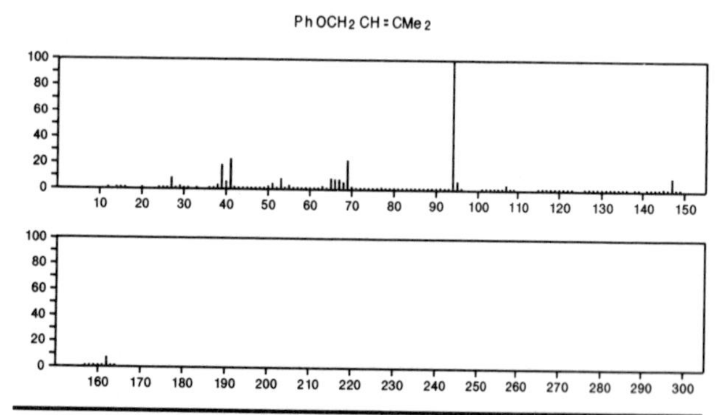

162 $C_{11}H_{14}O$ 17429-21-9
2(1*H*)-Naphthalenone, 4a,5,8,8a-tetrahydro-4a-methyl-, *trans*-

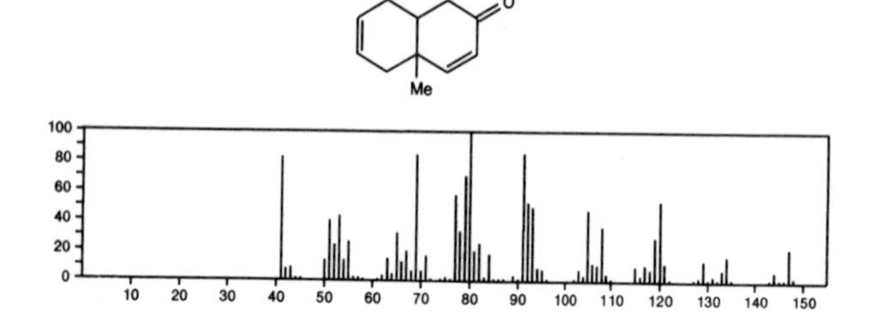

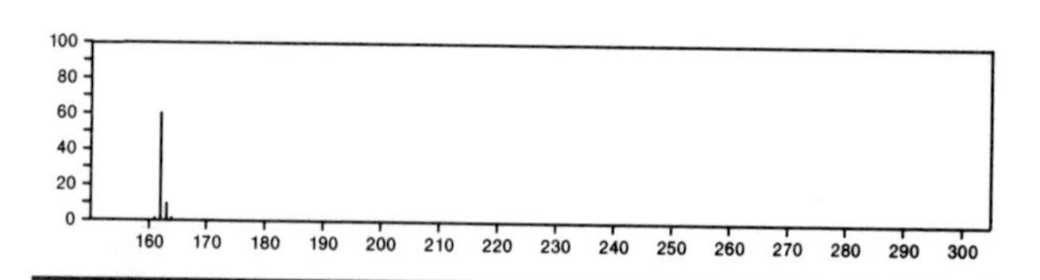

162 $C_{11}H_{14}O$ 22610-79-3
2-Cyclopenten-1-one, 3-methyl-2-(2,4-pentadienyl)-, (*Z*)-

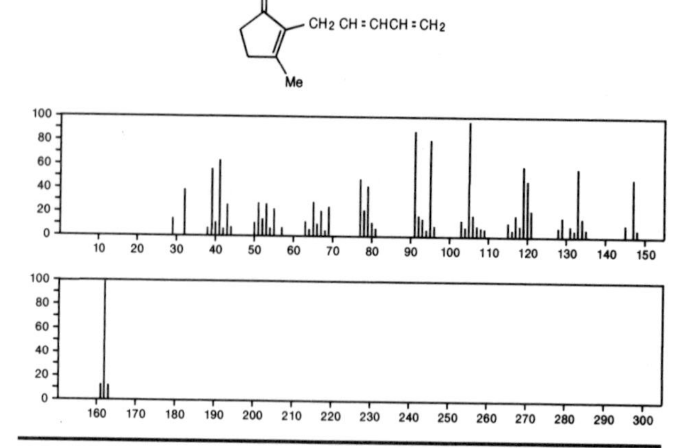

162 $C_{11}H_{14}O$ 22610-80-6
2-Cyclopenten-1-one, 3-methyl-2-(1,3-pentadienyl)-, (*E*,*Z*)-

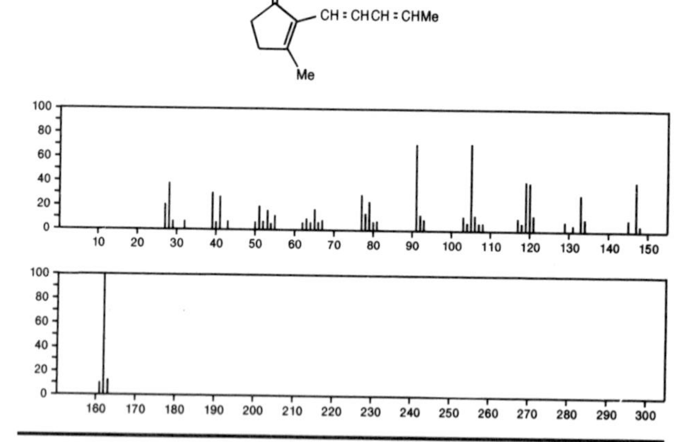

162 $C_{11}H_{14}O$ 32281-70-2
1-Naphthol, 1,2,3,4-tetrahydro-2-methyl-

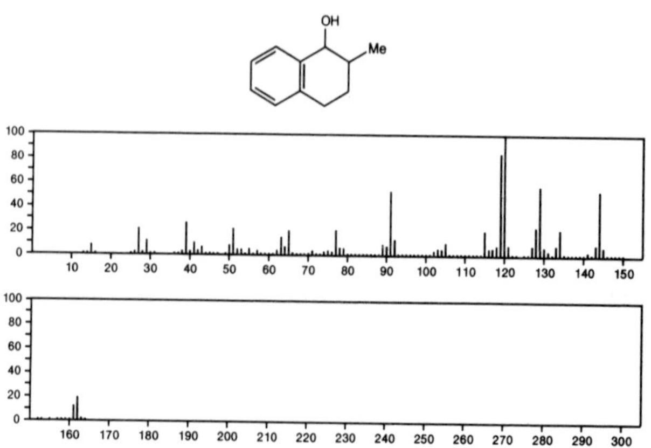

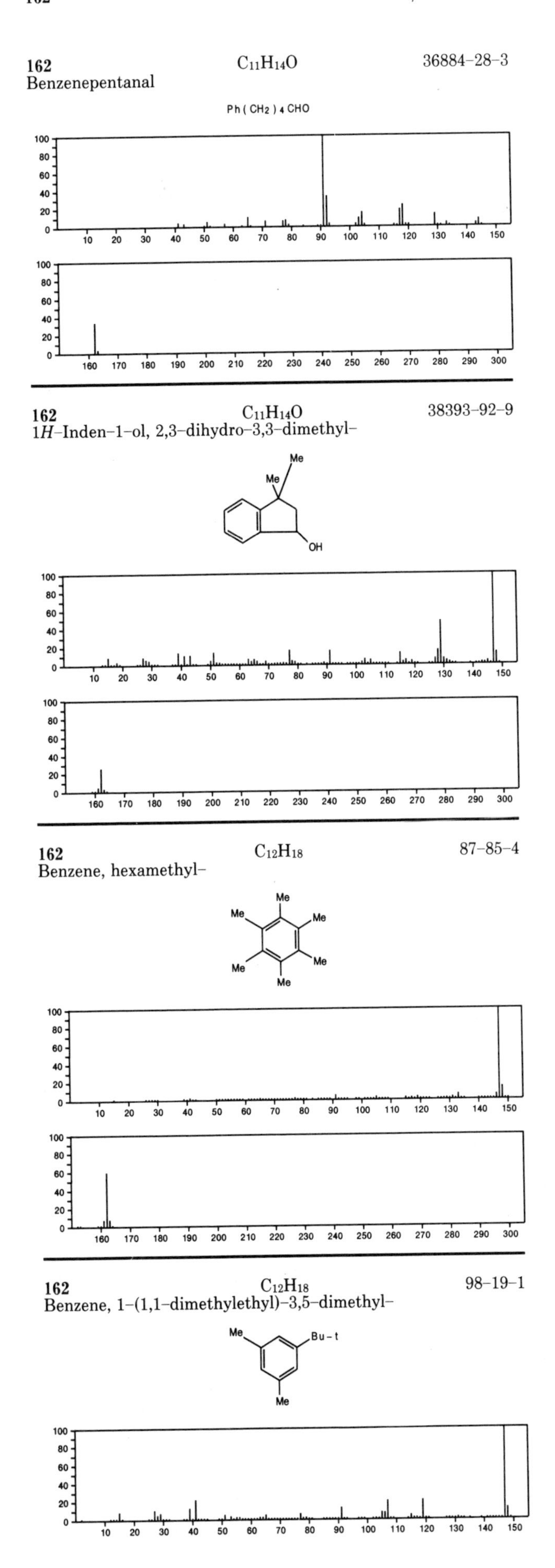

162 $C_{11}H_{14}O$ 36884–28–3
Benzenepentanal

Ph(CH2)4CHO

162 $C_{11}H_{14}O$ 38393–92–9
1*H*–Inden–1–ol, 2,3–dihydro–3,3–dimethyl–

162 $C_{12}H_{18}$ 87–85–4
Benzene, hexamethyl–

162 $C_{12}H_{18}$ 98–19–1
Benzene, 1–(1,1–dimethylethyl)–3,5–dimethyl–

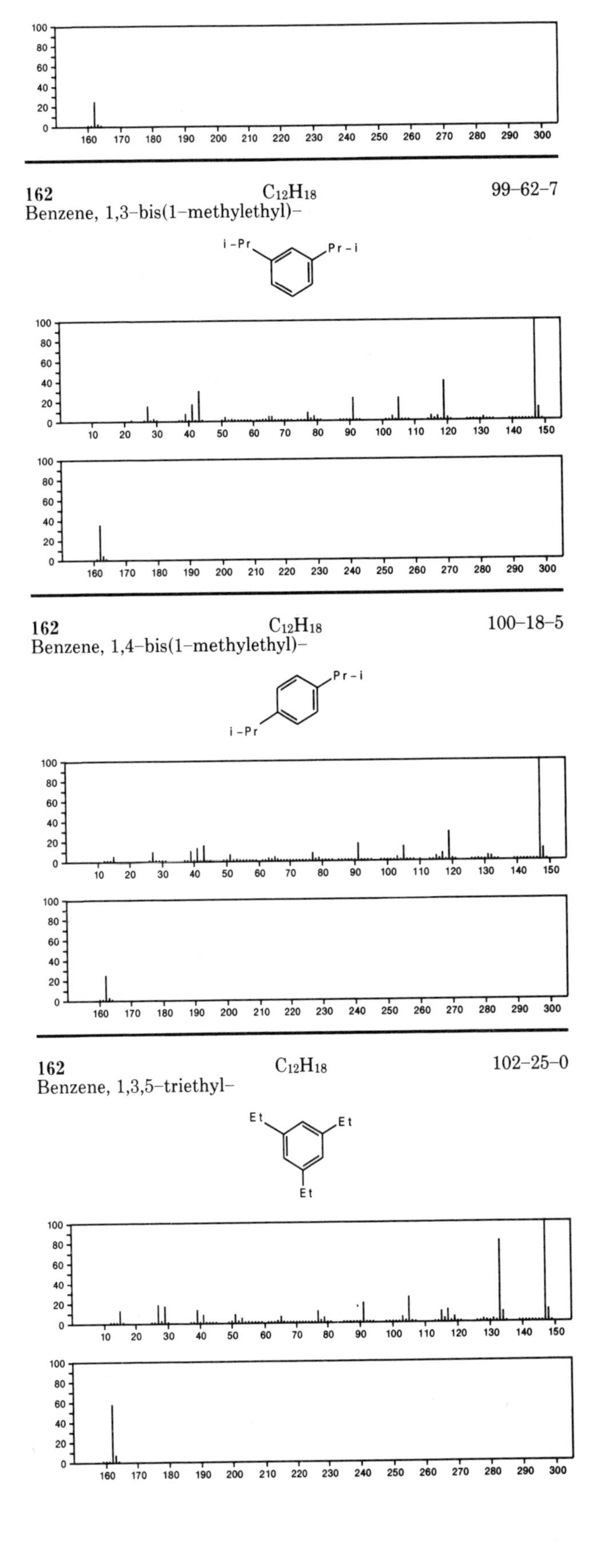

162 $C_{12}H_{18}$ 99–62–7
Benzene, 1,3–bis(1–methylethyl)–

162 $C_{12}H_{18}$ 100–18–5
Benzene, 1,4–bis(1–methylethyl)–

162 $C_{12}H_{18}$ 102–25–0
Benzene, 1,3,5–triethyl–

162 $C_{12}H_{18}$ 577-55-9
Benzene, 1,2-bis(1-methylethyl)-

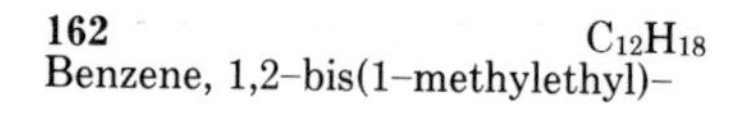

Pr-i
Pr-i

162 $C_{12}H_{18}$ 676-22-2
1,5,9-Cyclododecatriene, (*E,E,E*)-

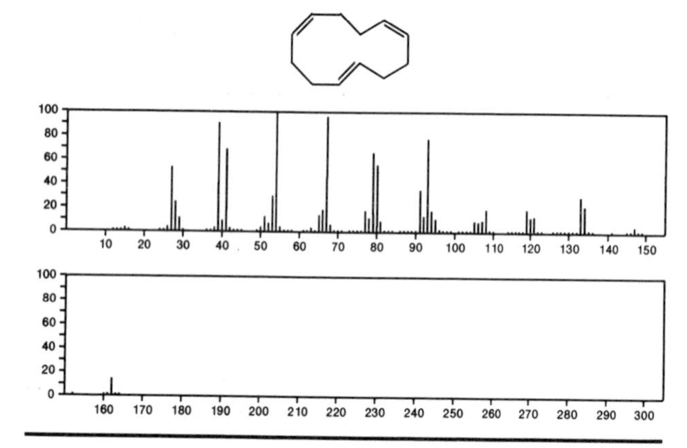

162 $C_{12}H_{18}$ 877-44-1
Benzene, 1,2,4-triethyl-

Et
Et
Et

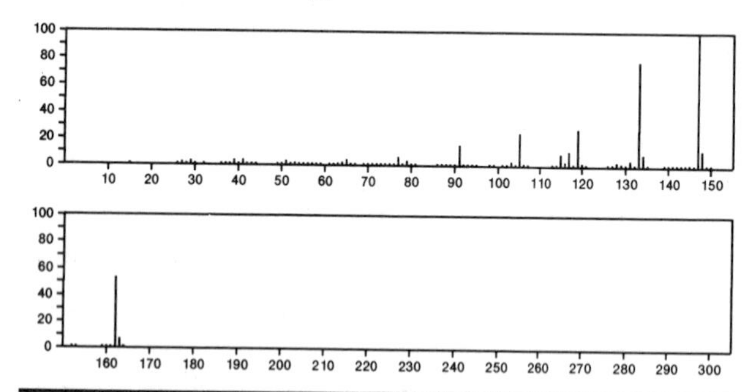

162 $C_{12}H_{18}$ 1077-16-3
Benzene, hexyl-

Ph(CH2)5Me

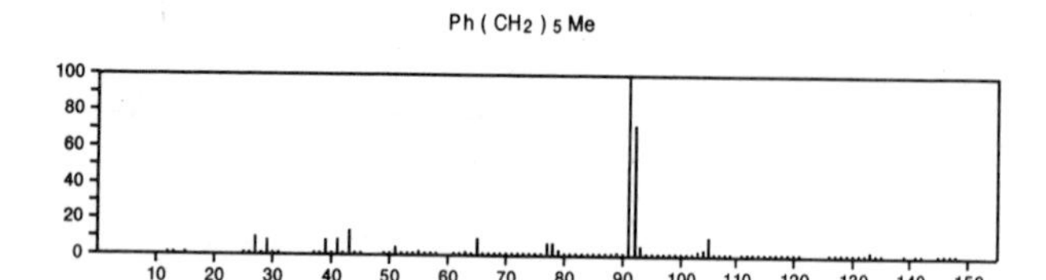

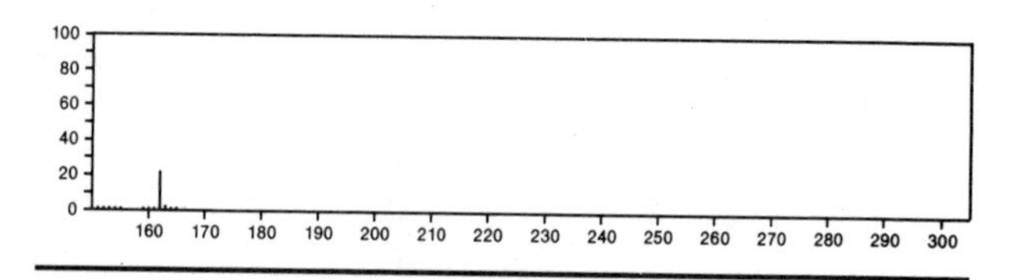

162 $C_{12}H_{18}$ 1483-60-9
Benzene, 2,4-dimethyl-1-(1-methylpropyl)-

Bu-s
Me
Me

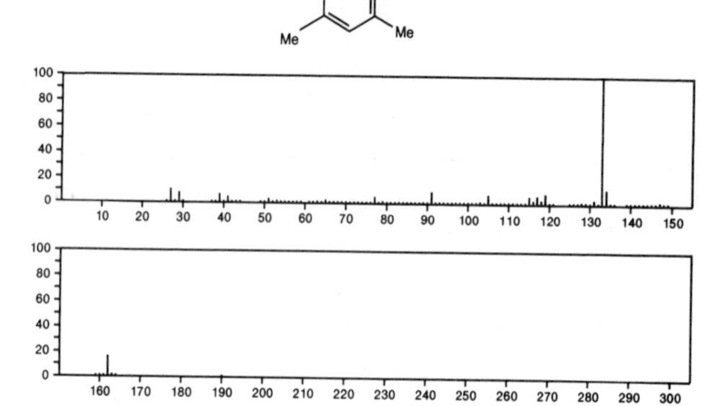

162 $C_{12}H_{18}$ 1541-20-4
Bi-2-cyclohexen-1-yl

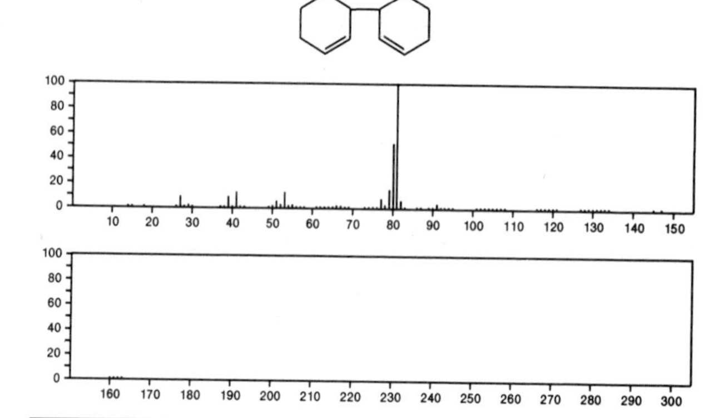

162 $C_{12}H_{18}$ 1985-57-5
Benzene, (1,1-dimethylbutyl)-

PhCPr(Ph)

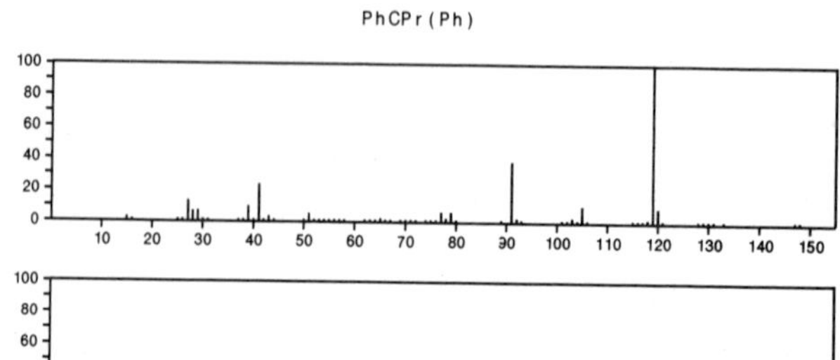

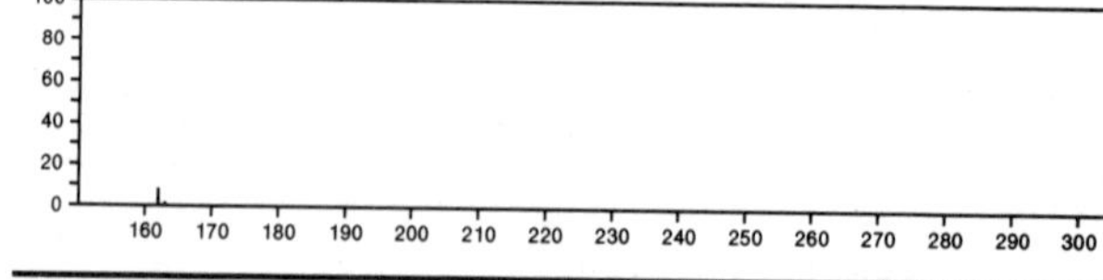

162 $C_{12}H_{18}$ 1985-97-3
Benzene, (1-ethyl-1-methylpropyl)-

PhCEt2(Ph)

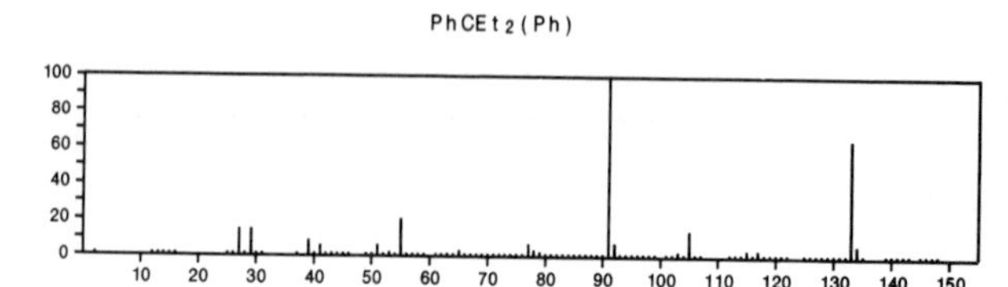

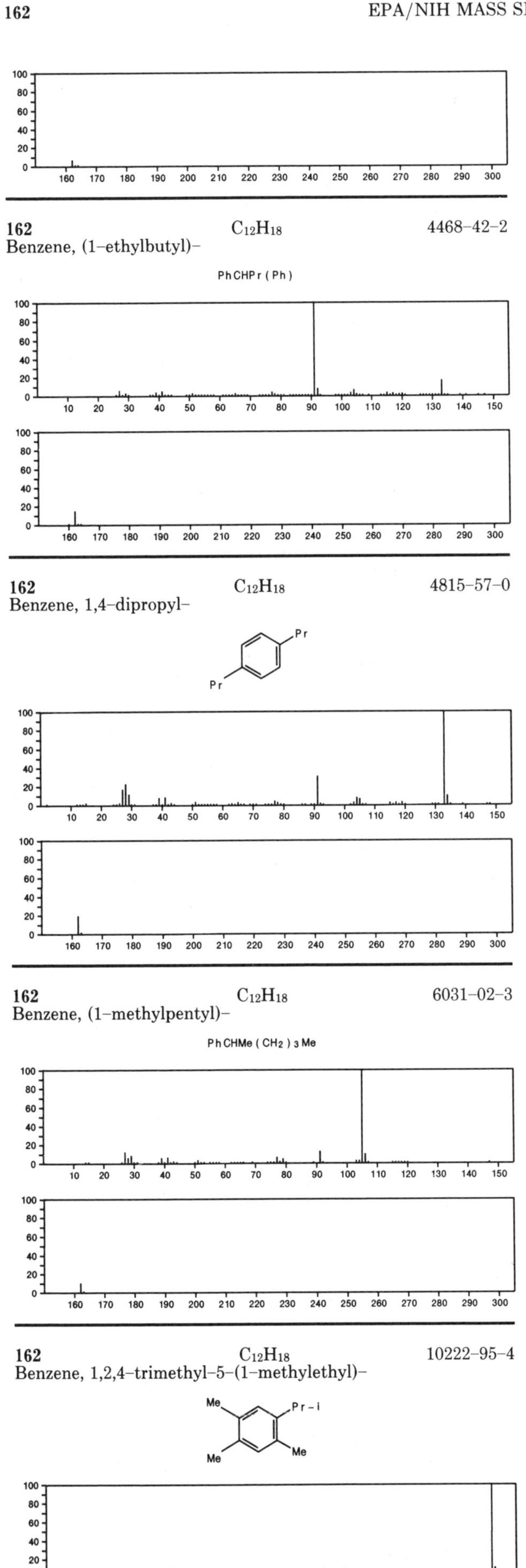
162 C12H18 4468-42-2
Benzene, (1-ethylbutyl)-
PhCHPr(Ph)
162 C12H18 4815-57-0
Benzene, 1,4-dipropyl-
Pr
Pr
162 C12H18 6031-02-3
Benzene, (1-methylpentyl)-
PhCHMe(CH2)3Me
162 C12H18 10222-95-4
Benzene, 1,2,4-trimethyl-5-(1-methylethyl)-
Me
Pr-i
Me
Me

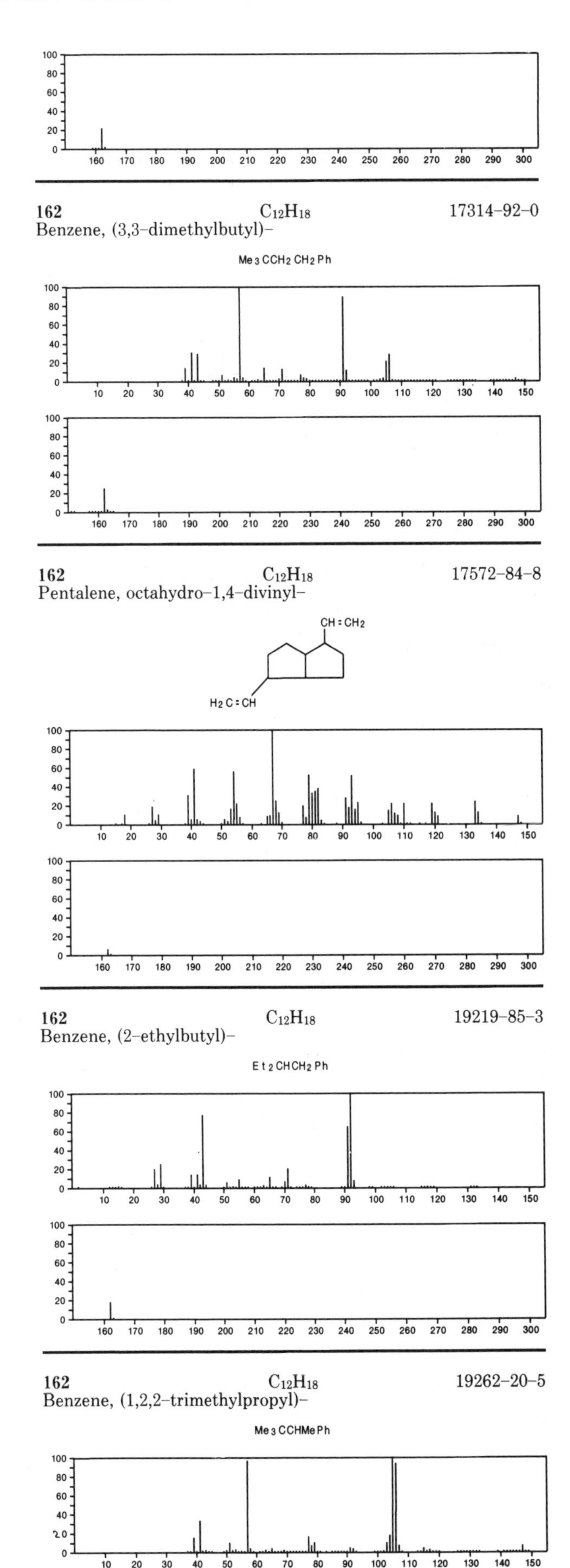
162 C12H18 17314-92-0
Benzene, (3,3-dimethylbutyl)-
Me3CCH2CH2Ph
162 C12H18 17572-84-8
Pentalene, octahydro-1,4-divinyl-
CH=CH2
H2C=CH
162 C12H18 19219-85-3
Benzene, (2-ethylbutyl)-
Et2CHCH2Ph
162 C12H18 19262-20-5
Benzene, (1,2,2-trimethylpropyl)-
Me3CCHMePh

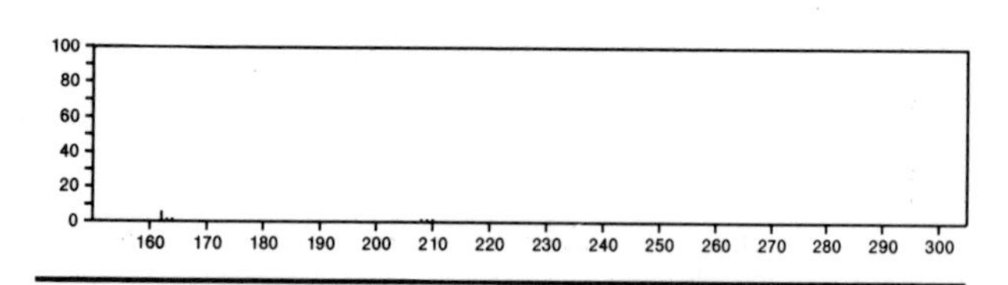

162 $C_{12}H_{18}$ 22975-58-2
Toluene, *p*-(1-ethylpropyl)-

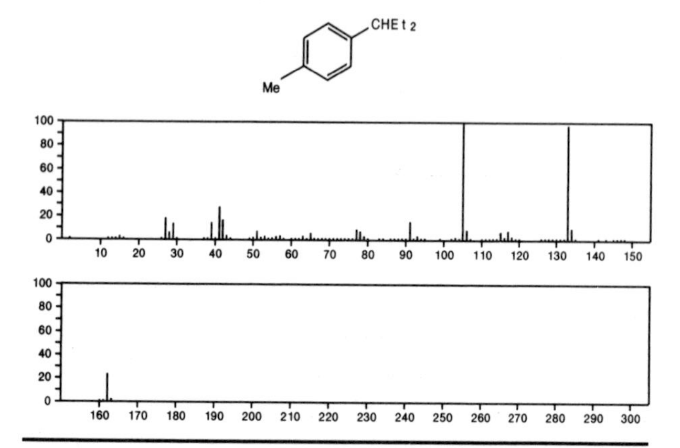

162 $C_{12}H_{18}$ 25321-09-9
Benzene, bis(1-methylethyl)-

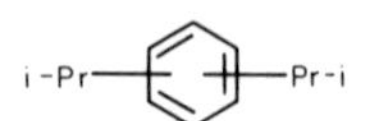

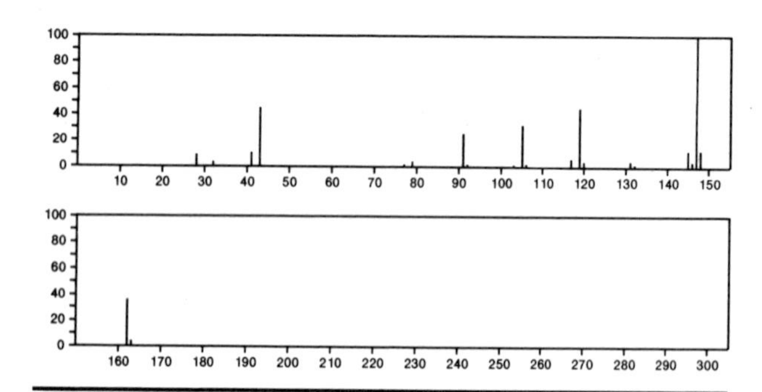

162 $C_{12}H_{18}$ 26356-11-6
Benzene, (1,1,2-trimethylpropyl)-

Me2CHCMe2Ph

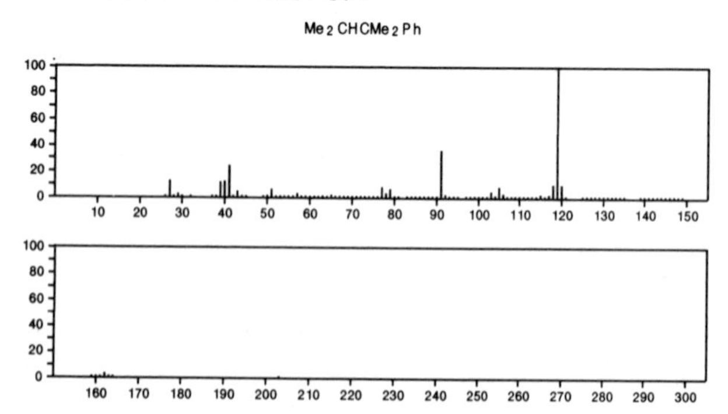

162 $C_{12}H_{18}$ 27070-59-3
Cyclododecatriene

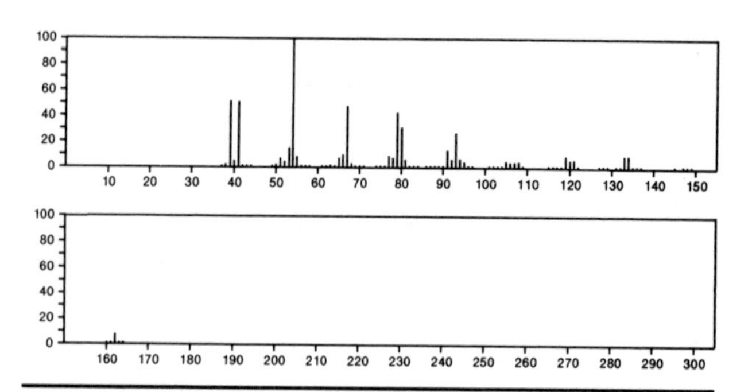

162 $C_{12}H_{18}$ 28080-86-6
Benzene, (2,2-dimethylbutyl)-

EtCMe2CH2Ph

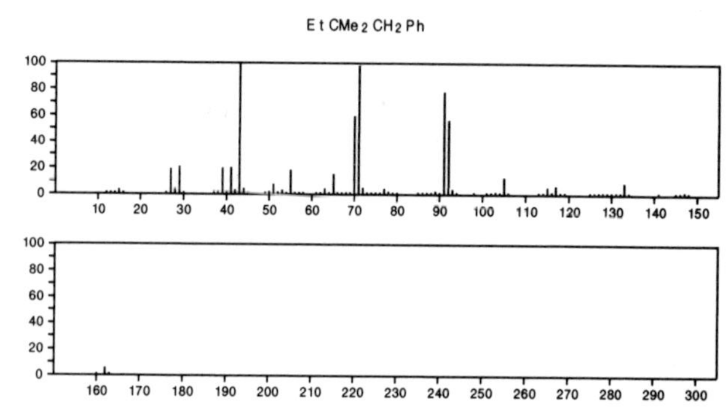

162 $C_{12}H_{18}$ 28654-79-7
Benzene, *sec*-butylethyl-

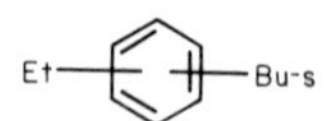

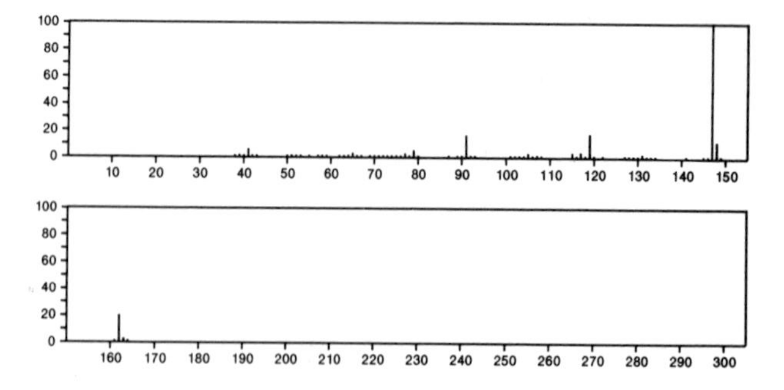

162 $C_{12}H_{18}$ 31365-98-7
Benzene, 3-ethyl-1,2,4,5-tetramethyl-

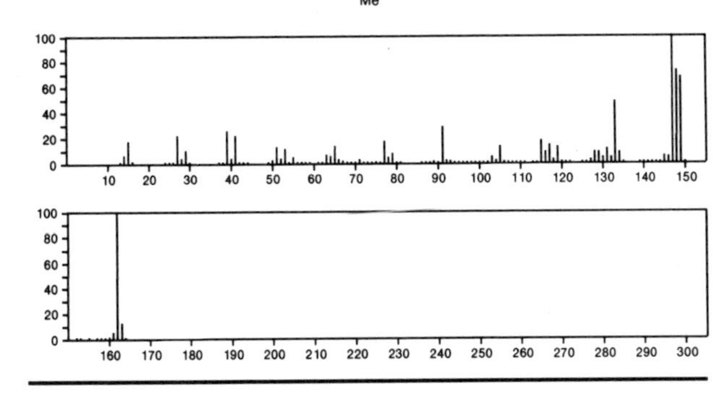

162 $C_{12}H_{18}$ 33991-29-6
Benzene, trimethyl(1-methylethyl)-

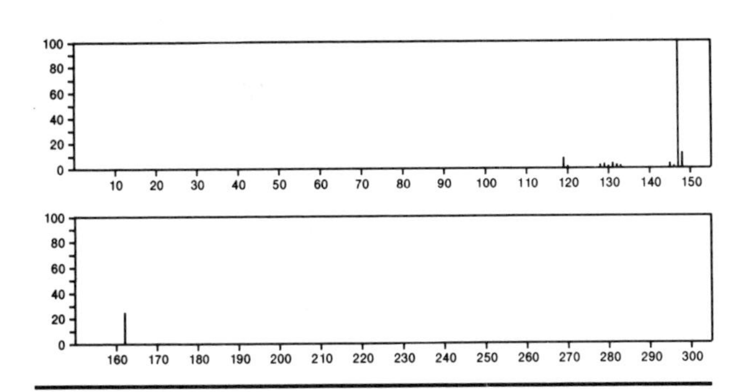

162 $C_{12}H_{18}$ 54410-69-4
Benzene, (3-methylpentyl)-

Me CH2 CHMe CH2 CH2 Ph

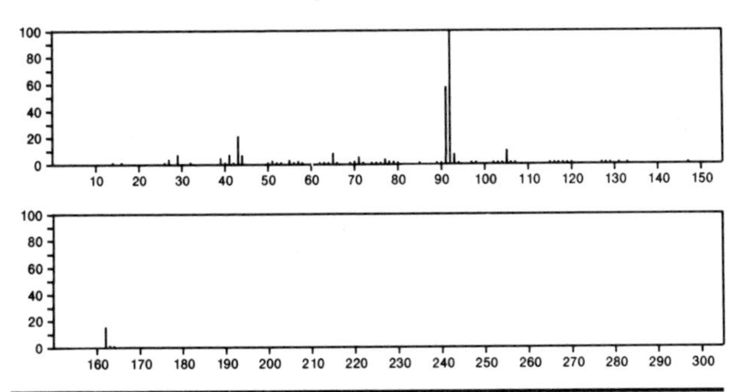

162 $C_{12}H_{18}$ 54410-74-1
Benzene, 1-methyl-2-(1-ethylpropyl)-

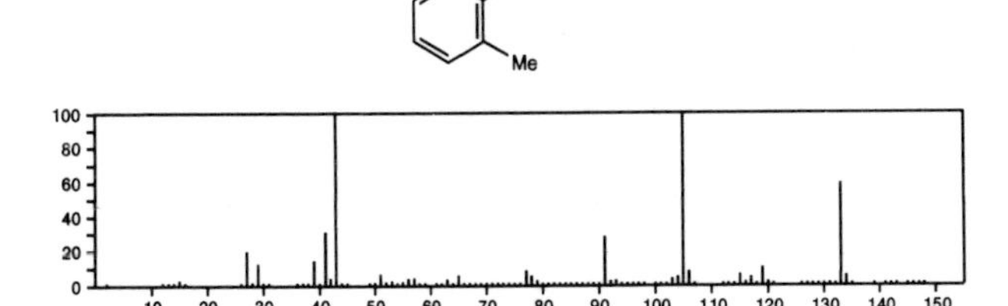

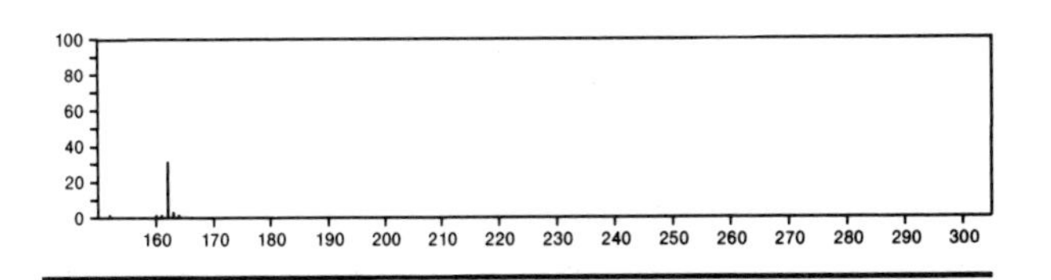

162 $C_{12}H_{18}$ 54410-75-2
Benzene, 1,2-diethyl-3,4-dimethyl-

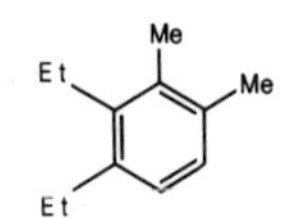

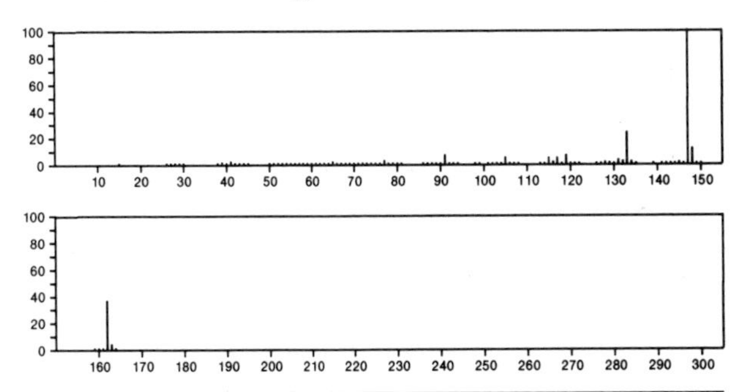

162 $C_{12}H_{18}$ 55669-88-0
Benzene, 1,4-dimethyl-2-(2-methylpropyl)-

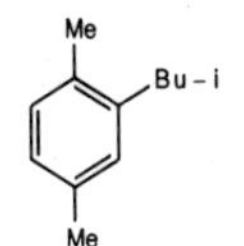

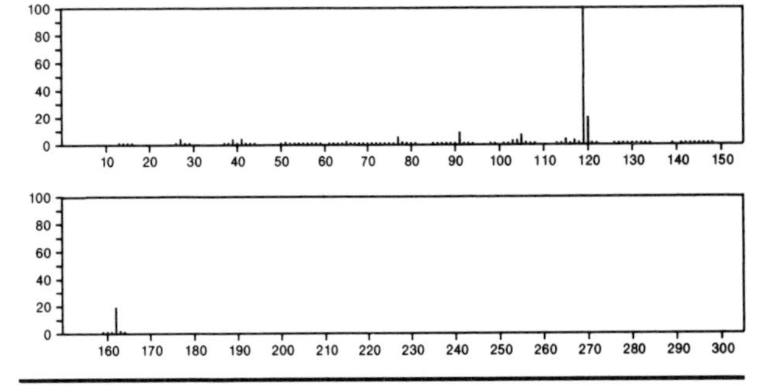

162 $C_{12}H_{18}$ 55682-73-0
3,5-Decadiyne, 2,2-dimethyl-

Me3CC⋮CC⋮C(CH2)3Me

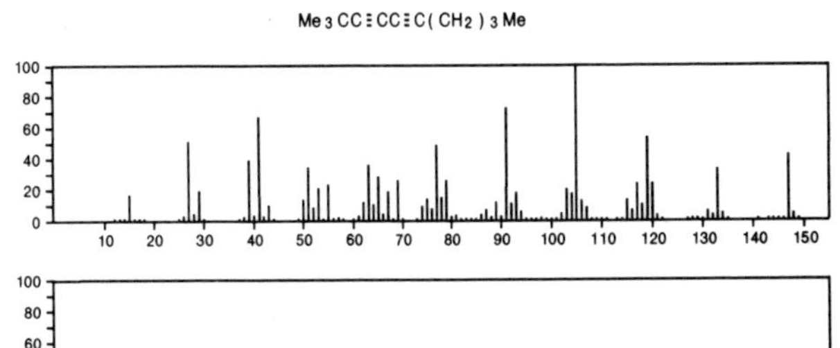

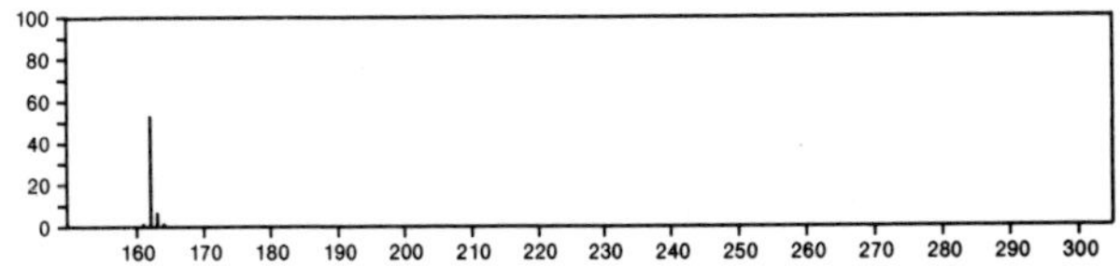

162 $C_{12}H_{18}$ 56248-17-0
Cyclohexene, 1,5,5-trimethyl-6-(2-propenylidene)-

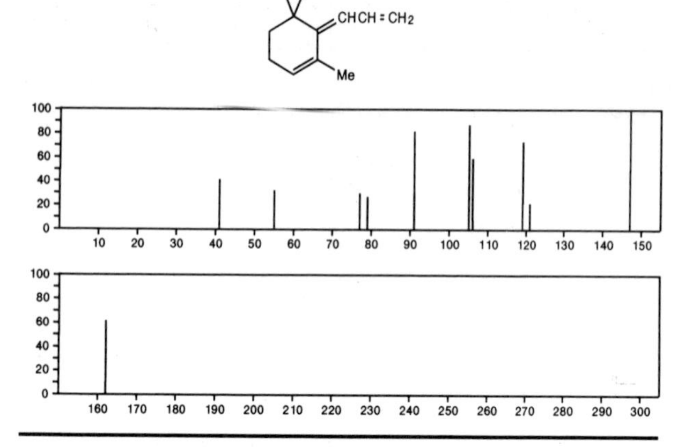

163 CCl_3NO_2 76-06-2
Methane, trichloronitro-

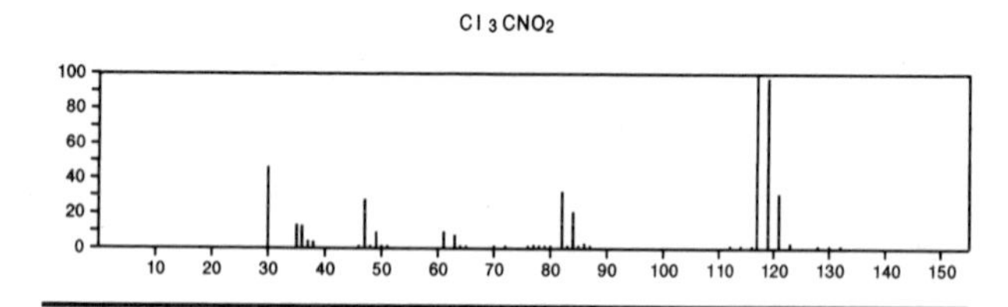

163 C_3H_2BrNS 3034-55-7
Thiazole, 5-bromo-

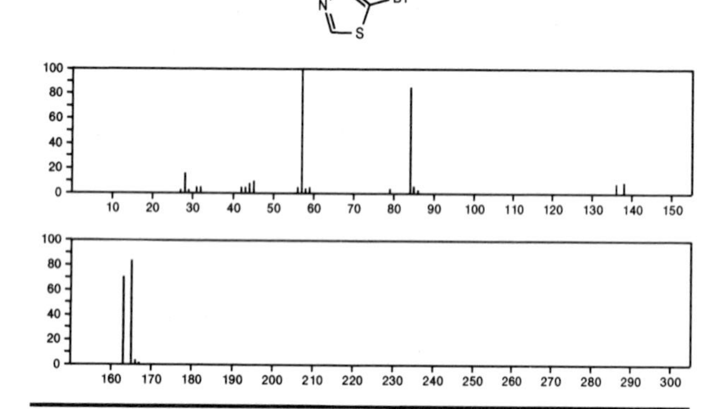

163 $C_4H_3Cl_2N_3$ 56-05-3
2-Pyrimidinamine, 4,6-dichloro-

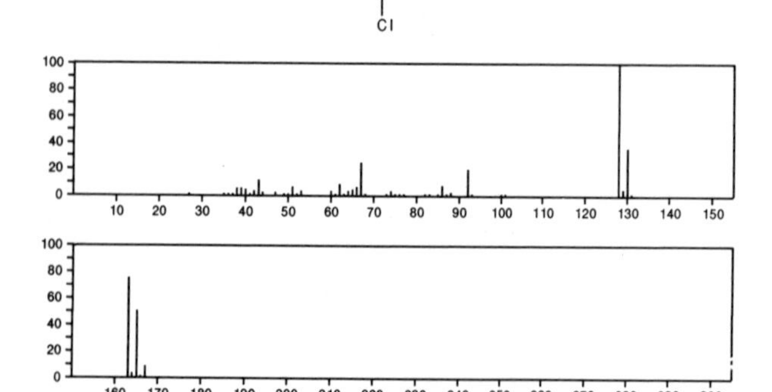

163 $C_5H_4F_3N_3$ 16075-42-6
Pyrimidine, 2-amino-4-(trifluoromethyl)-

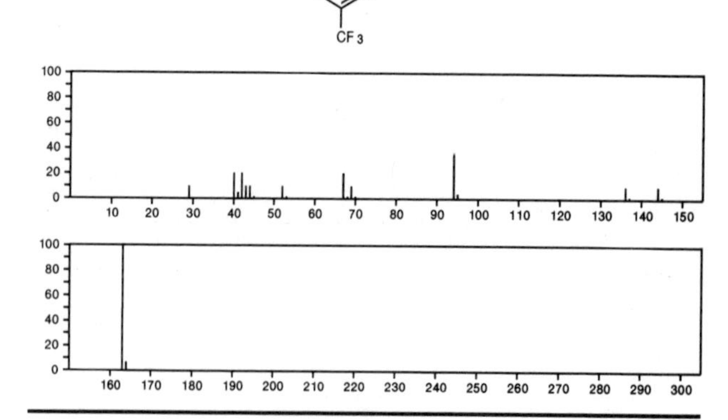

163 $C_5H_9NO_3S$ 55956-23-5
Alanine, *N*-acetyl-2-mercapto-, (±)-

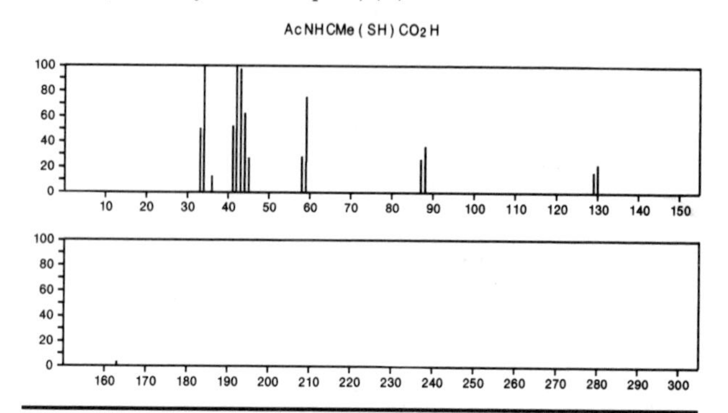

163 $C_6H_5N_5O$ 2236-60-4
4(1*H*)-Pteridinone, 2-amino-

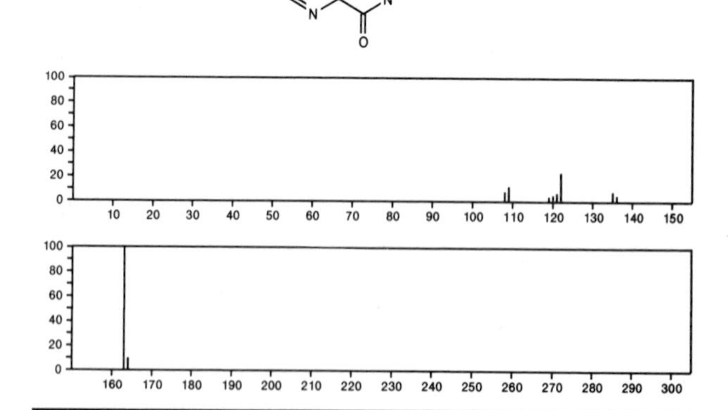

163 $C_6H_{13}NS_2$ 686-07-7
Carbamodithioic acid, diethyl-, methyl ester

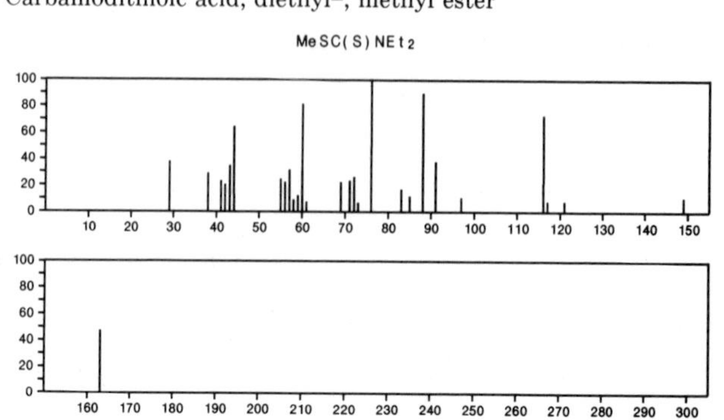

163 $C_6H_{18}N_3P$ 1608-26-0

Phosphorous triamide, hexamethyl-

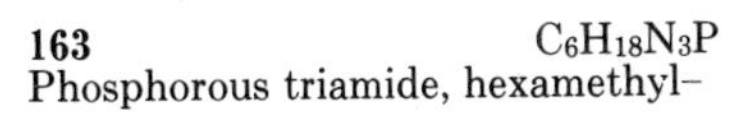

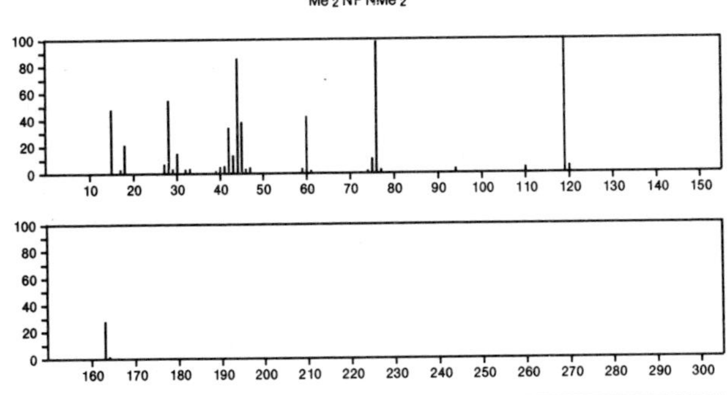

163 $C_7H_5N_3O_2$ 16952-65-1

Pyrido[4,3-*d*]pyrimidine-2,4(1*H*,3*H*)-dione

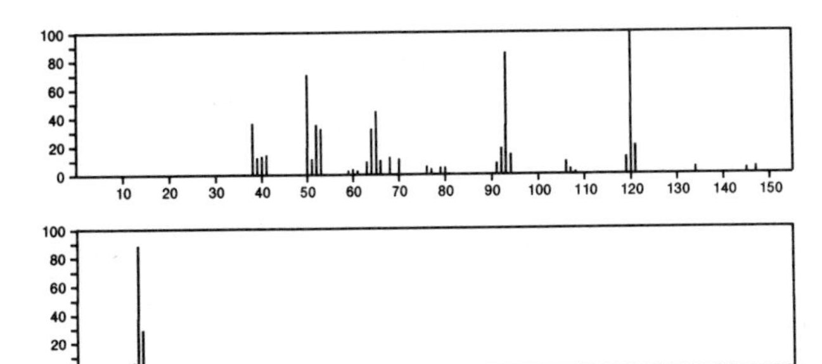

163 $C_7H_5N_3O_2$ 21038-66-4

Pyrido[2,3-*d*]pyrimidine-2,4(1*H*,3*H*)-dione

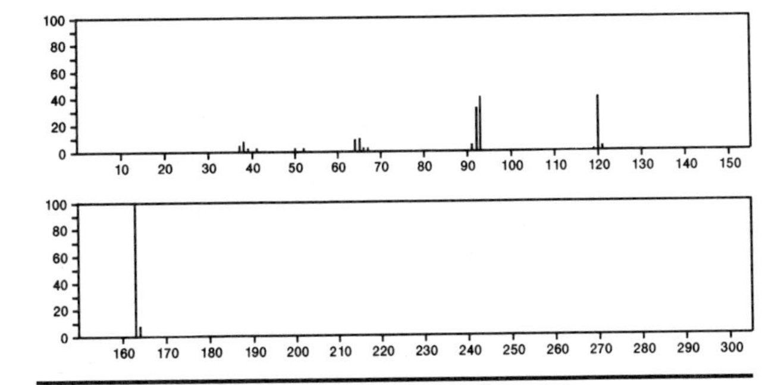

163 $C_7H_5N_3O_2$ 21038-67-5

Pyrido[3,4-*d*]pyrimidine-2,4(1*H*,3*H*)-dione

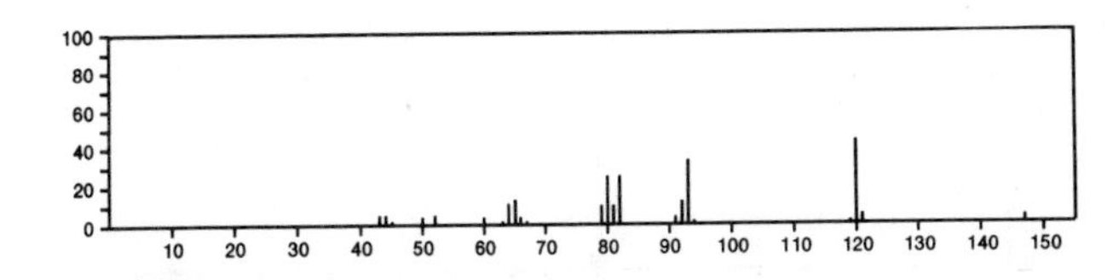

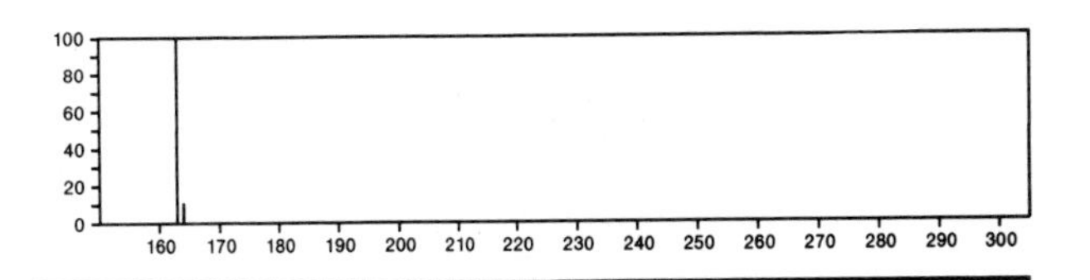

163 $C_7H_5N_3O_2$ 37538-68-4

Pyrido[3,2-*d*]pyrimidine-2,4(1*H*,3*H*)-dione

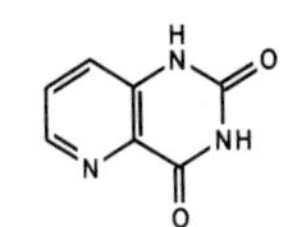

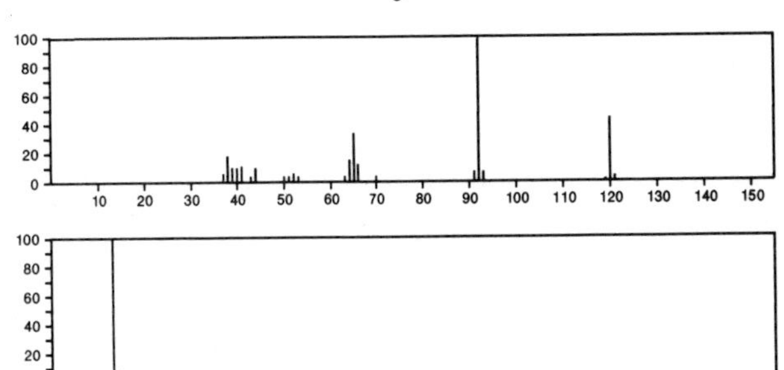

163 $C_7H_5N_3S$ 15370-74-8

Pyrido[2,3-*d*]pyridazine-8(7*H*)-thione

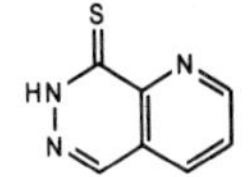

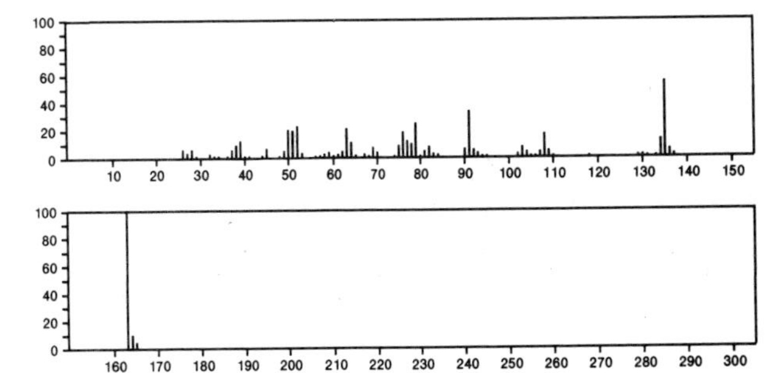

163 $C_7H_5N_3S$ 15370-85-1

Pyrido[2,3-*d*]pyridazine-5(6*H*)-thione

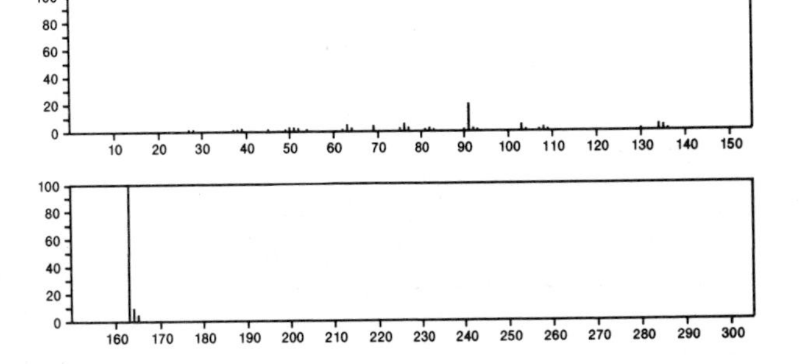

163 $C_7H_9N_5$ 938–55–6
1*H*–Purin–6–amine, *N*,*N*–dimethyl–

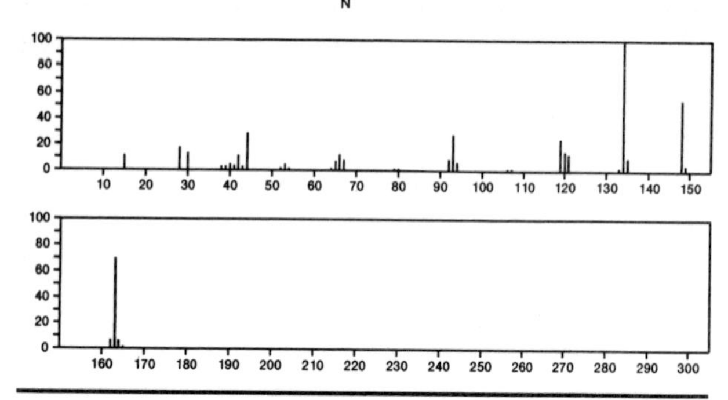

163 $C_7H_9N_5$ 19855–02–8
s–Triazolo[4,3–*a*]pyrazine, 3–amino–5,8–dimethyl–

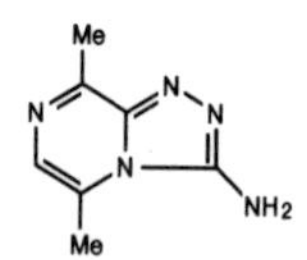

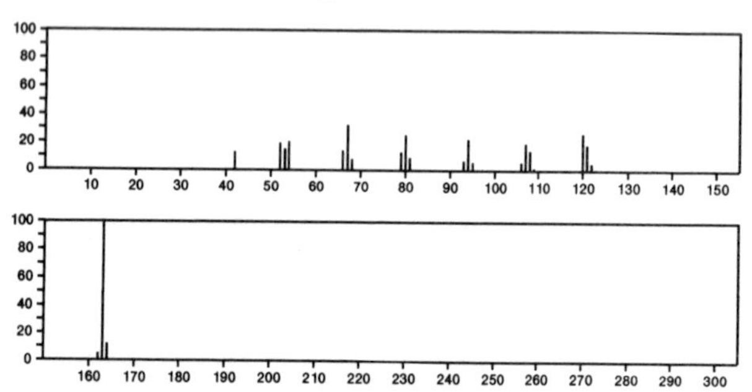

163 $C_8H_9N_3O$ 1574–10–3
Hydrazinecarboxamide, 2–(phenylmethylene)–

$H_2NCONHN{=}CHPh$

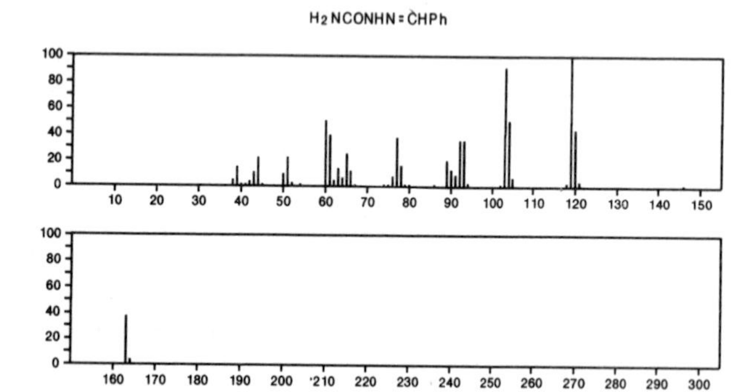

163 C_9H_6ClN 612–62–4
Quinoline, 2–chloro–

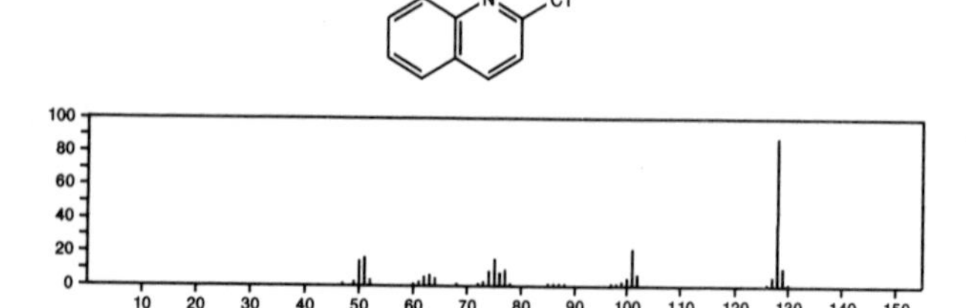

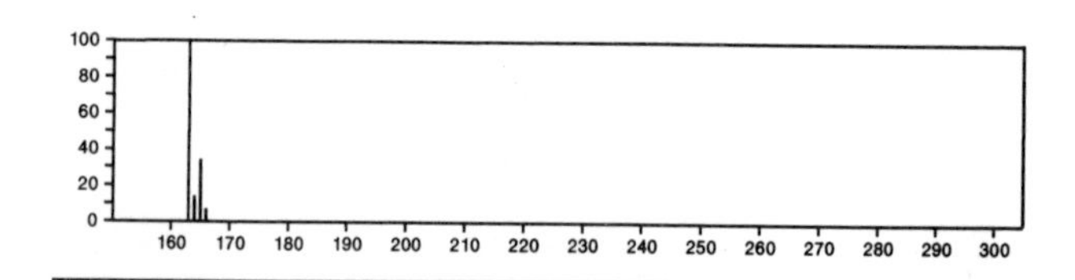

163 $C_9H_9NO_2$ 705–60–2
Benzene, (2–nitro–1–propenyl)–

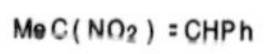

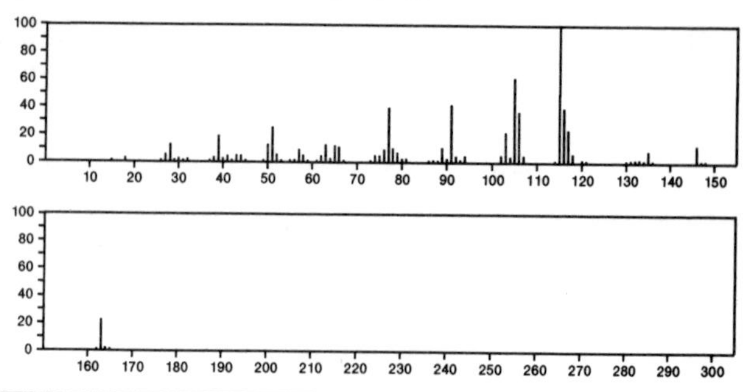

163 $C_9H_9NO_2$ 1575–95–7
Benzamide, *N*–acetyl–

AcNHCOPh

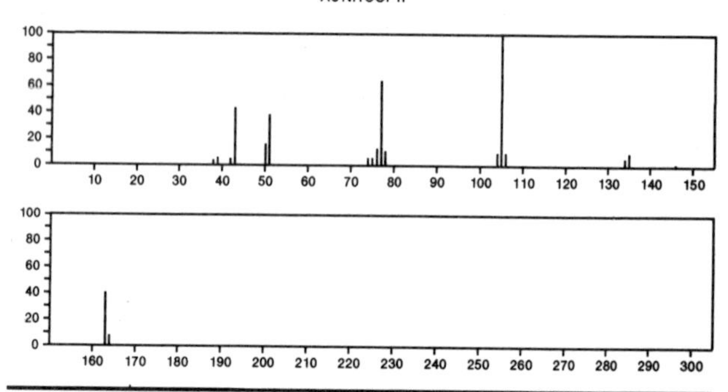

163 $C_9H_9NO_2$ 3594–37–4
1,3–Butanedione, 1–(3–pyridinyl)–

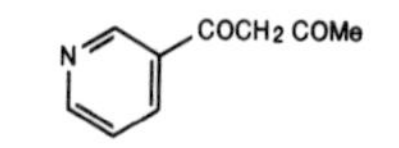

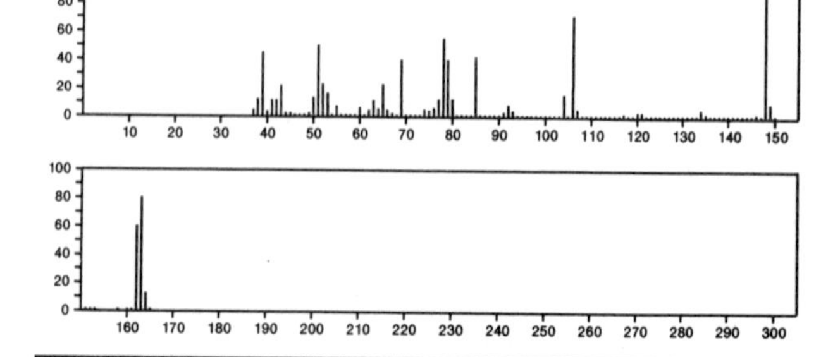

163 $C_9H_9NO_2$ 6265–30–1
4,7–Methano–1*H*–isoindole–1,3(2*H*)–dione, 3a,4,7,7a–tetrahydro–, (3aα,4α,7α,7aα)–

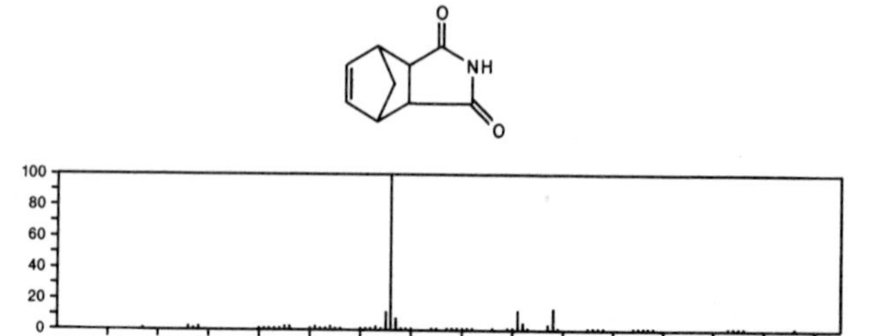

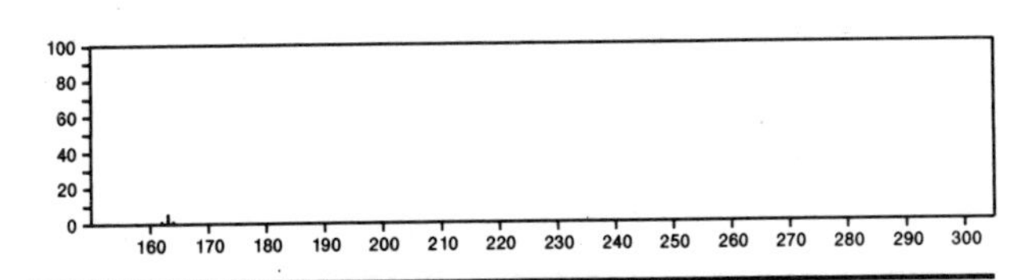

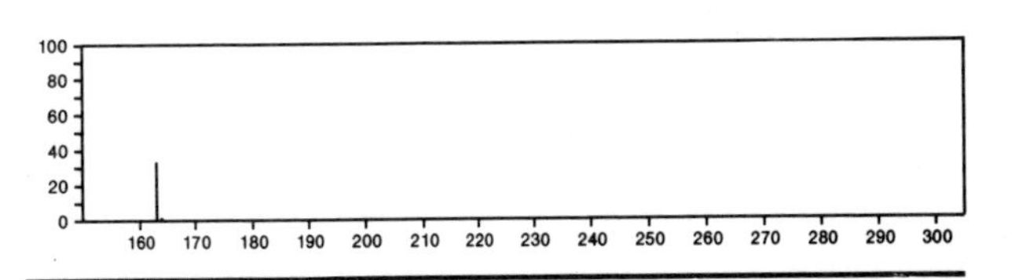

163 $C_9H_9NO_2$ 13303-68-9
1*H*-Indole-1-carboxaldehyde, 2,3-dihydro-2-hydroxy-

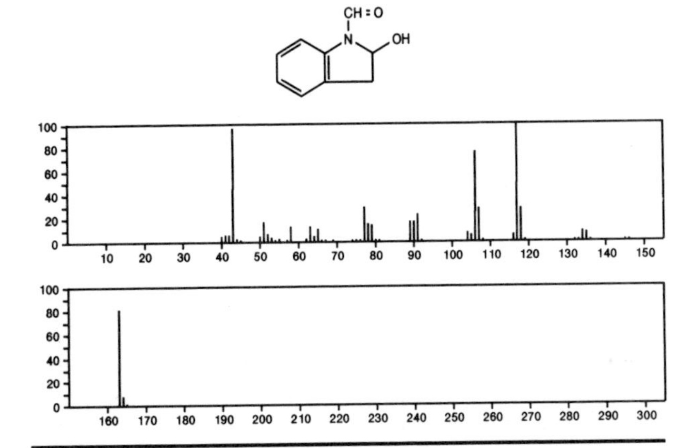

163 $C_9H_9NO_2$ 52479-54-6
Formamide, *N*-(2-formylphenyl)-*N*-methyl-

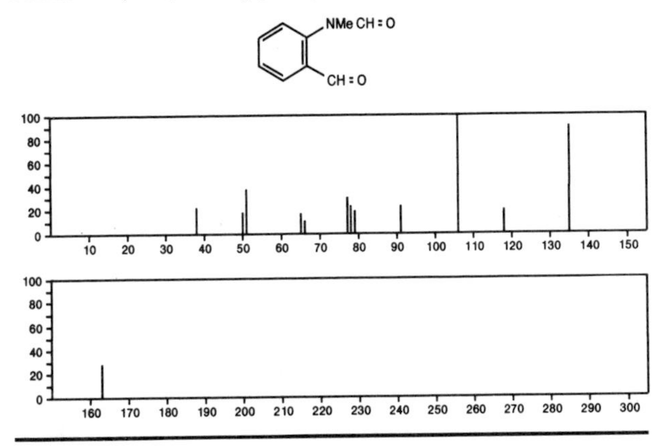

163 $C_{10}H_{13}NO$ 90-26-6
Benzeneacetamide, α-ethyl-

PhCHEt CONH2

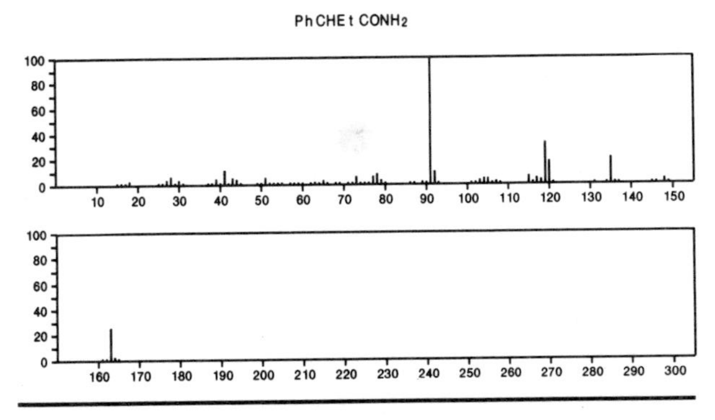

163 $C_{10}H_{13}NO$ 529-65-7
Acetamide, *N*-ethyl-*N*-phenyl-

PhNEt (Ac)

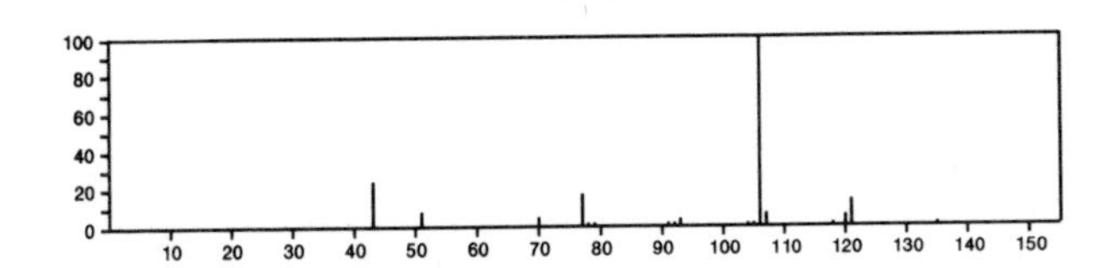

163 $C_{10}H_{13}NO$ 612-03-3
Acetamide, *N*-methyl-*N*-(4-methylphenyl)-

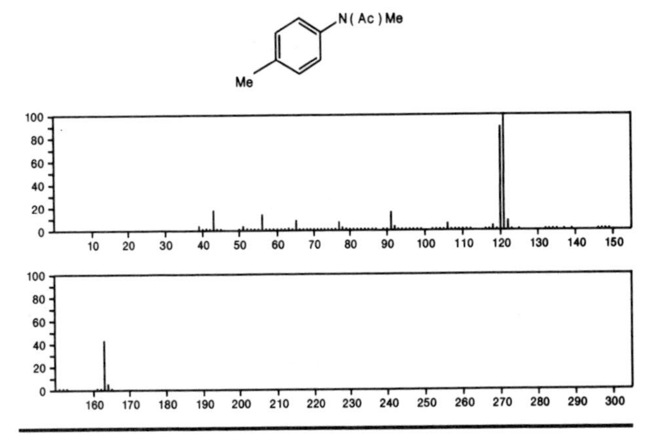

163 $C_{10}H_{13}NO$ 877-95-2
Acetamide, *N*-(2-phenylethyl)-

PhCH2 CH2 NHAc

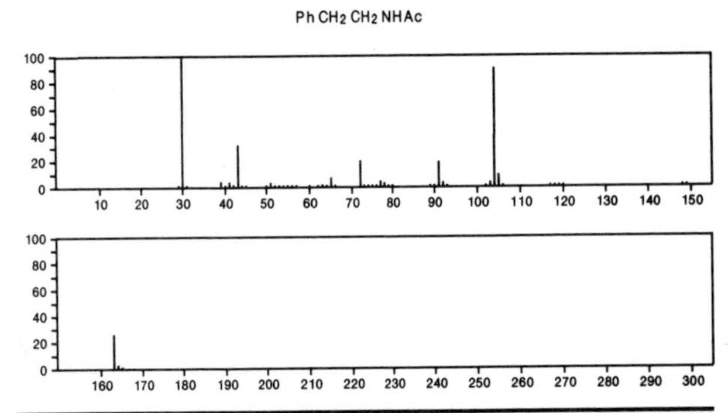

163 $C_{10}H_{13}NO$ 2198-53-0
Acetamide, *N*-(2,6-dimethylphenyl)-

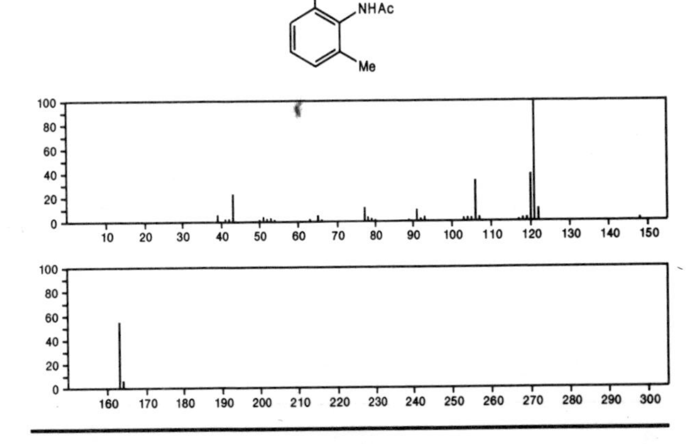

163 $C_{10}H_{13}NO$ 3319-03-7
Ethanone, 2-(dimethylamino)-1-phenyl-

Me2NCH2 COPh

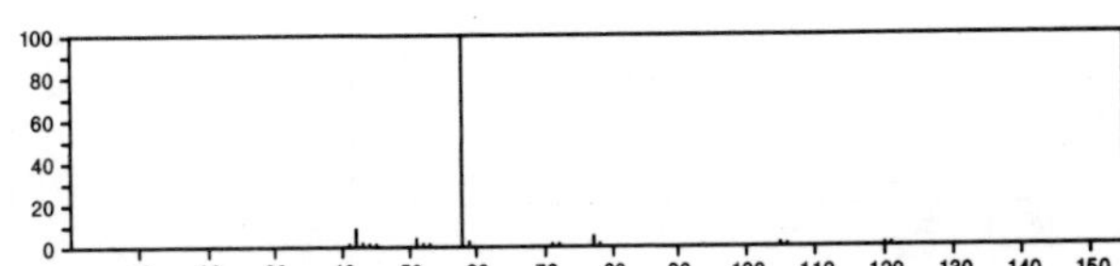

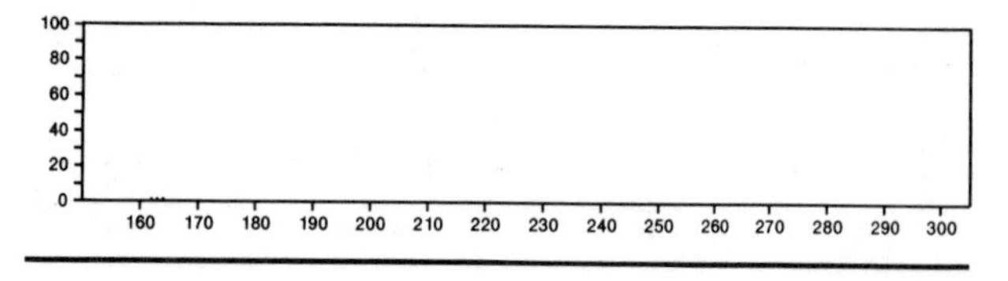

163 $C_{10}H_{13}NO$ 6284-14-6
Acetamide, *N*-(1-phenylethyl)-

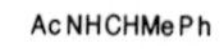

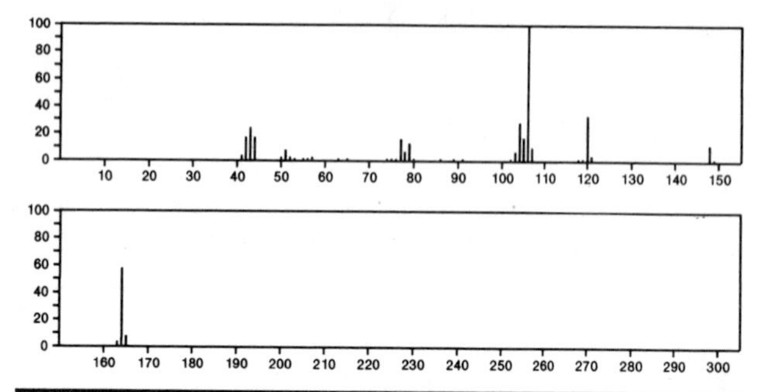

163 $C_{10}H_{13}NO$ 7137-97-5
1-Pentanone, 1-(2-pyridinyl)-

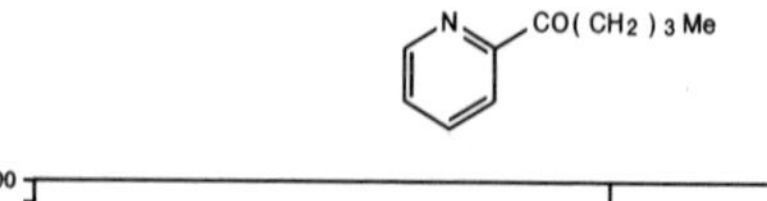

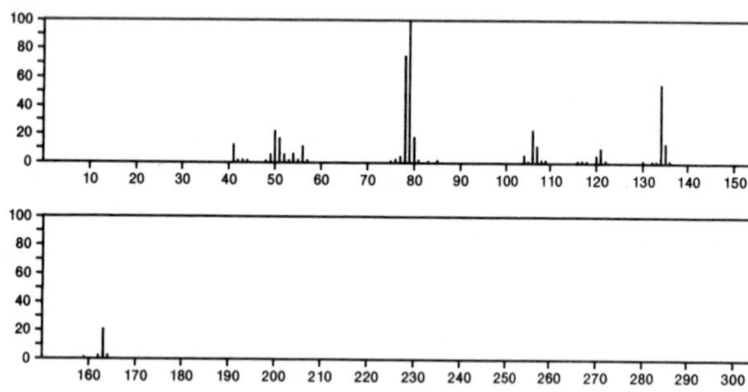

163 $C_{10}H_{13}NO$ 10336-55-7
Ethanone, 1-[2-(dimethylamino)phenyl]-

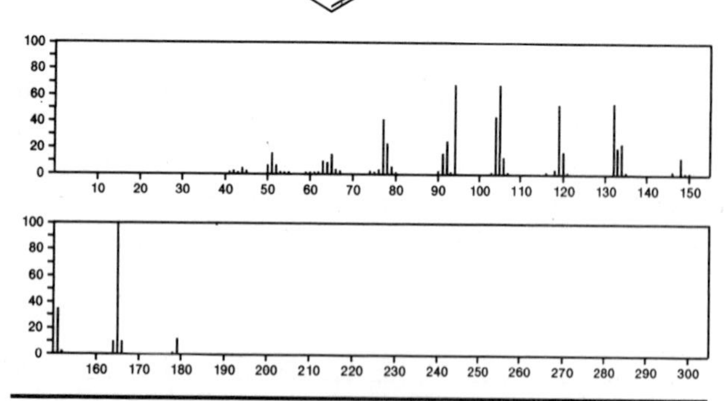

163 $C_{10}H_{13}NO$ 10546-70-0
Benzamide, *N*-propyl-

PrNHCOPh

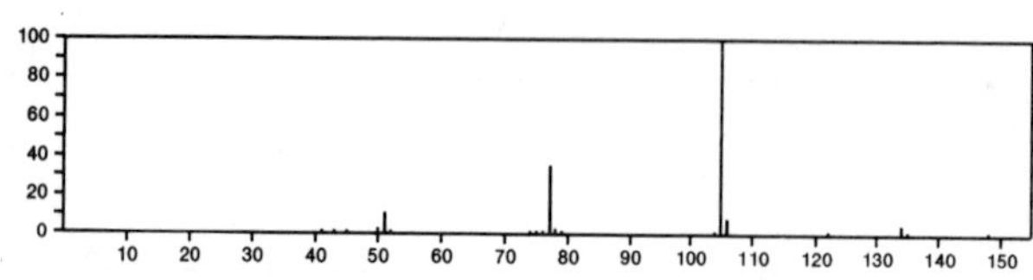

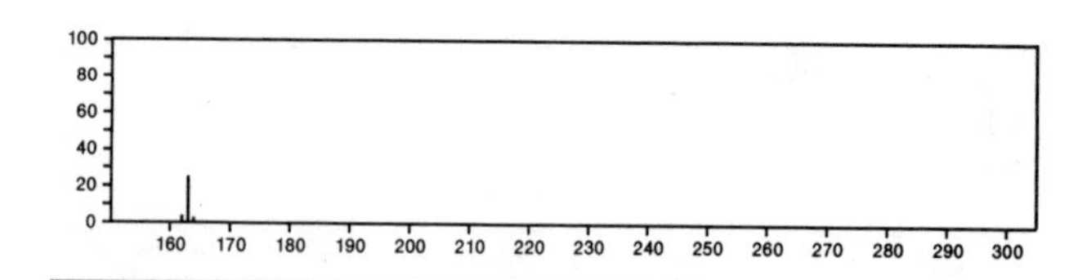

163 $C_{10}H_{13}NO$ 18925-69-4
Benzeneacetamide, *N*,*N*-dimethyl-

Me2NCOCH2Ph

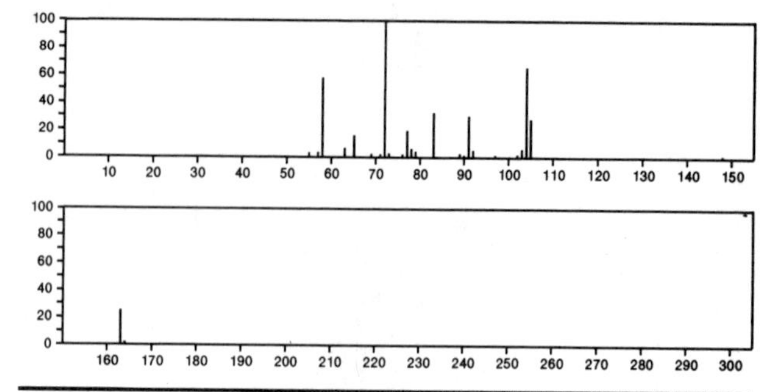

163 $C_{10}H_{13}NO$ 20205-53-2
Azocine, 2-methoxy-3,8-dimethyl-

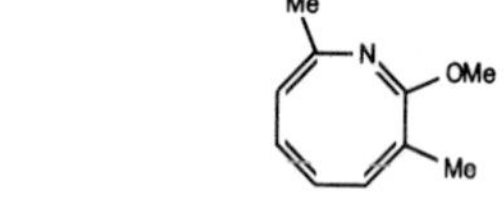

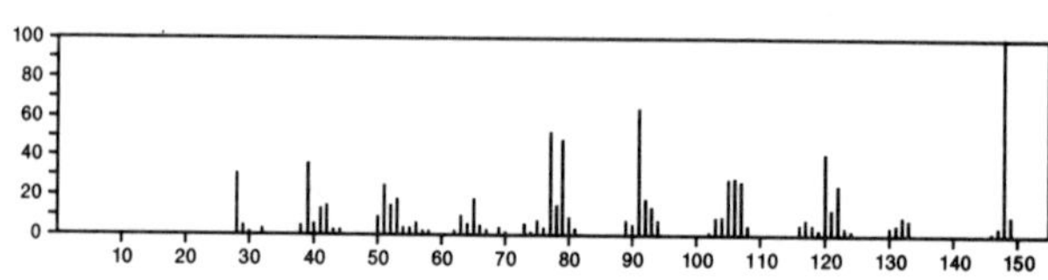

163 $C_{10}H_{13}NO$ 32187-26-1
2-Butanone, 1-amino-1-phenyl-

PhCH(NH2)COEt

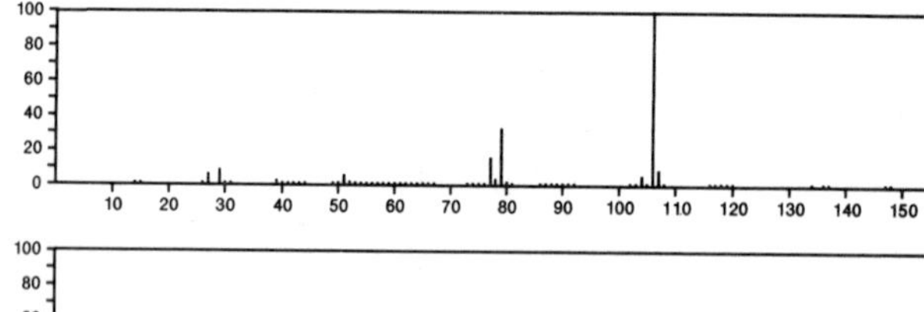

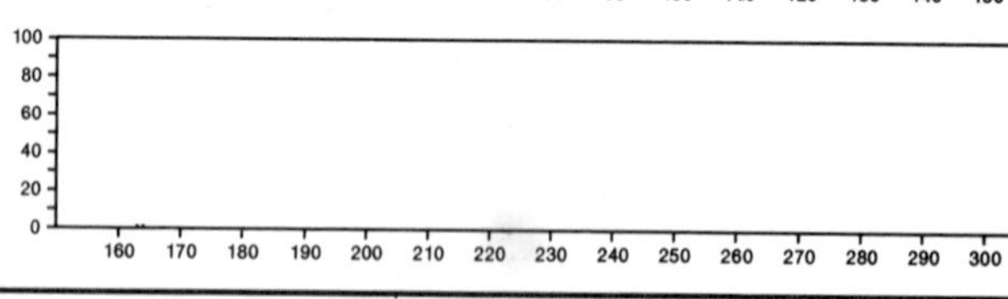

163 $C_{10}H_{13}NO$ 33499-41-1
Benzaldehyde, *O*-isopropyloxime

PhCH=NOPr-i

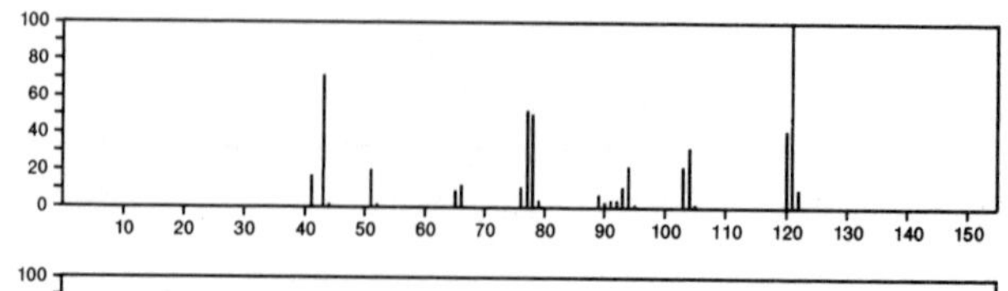

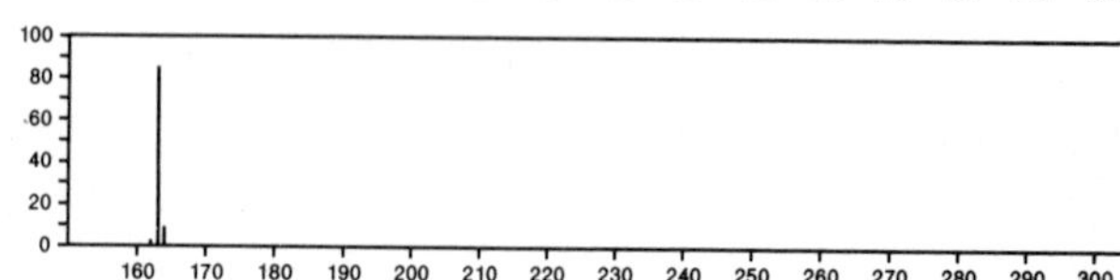

163 $C_{10}H_{13}NO$ 33581-40-7
Benzaldehyde, *O*-propyloxime

163 $C_{10}H_{13}NO$ 40513-35-7
2-Butanone, 3-amino-4-phenyl-

163 $C_{10}H_{13}NO$ 54410-77-4
Formamide, *N*-methyl-*N*-[(4-methylphenyl)methyl]-

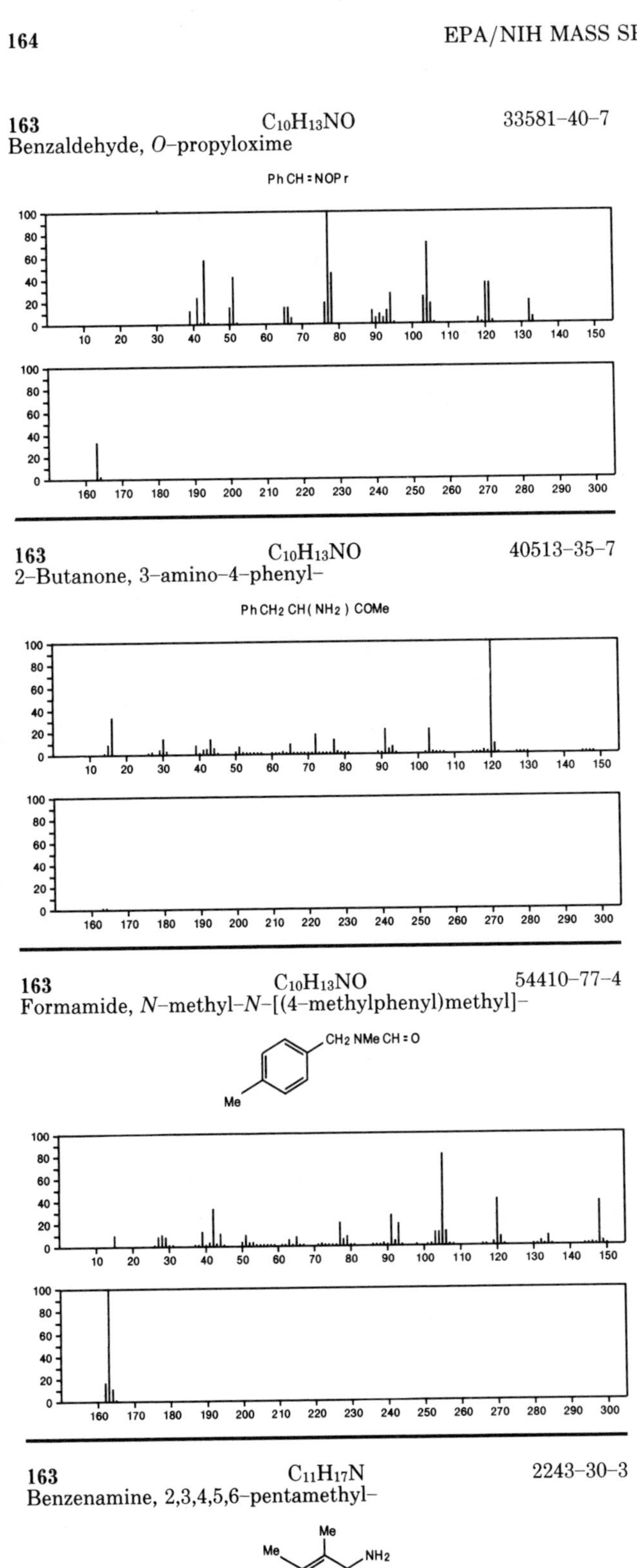

163 $C_{11}H_{17}N$ 2243-30-3
Benzenamine, 2,3,4,5,6-pentamethyl-

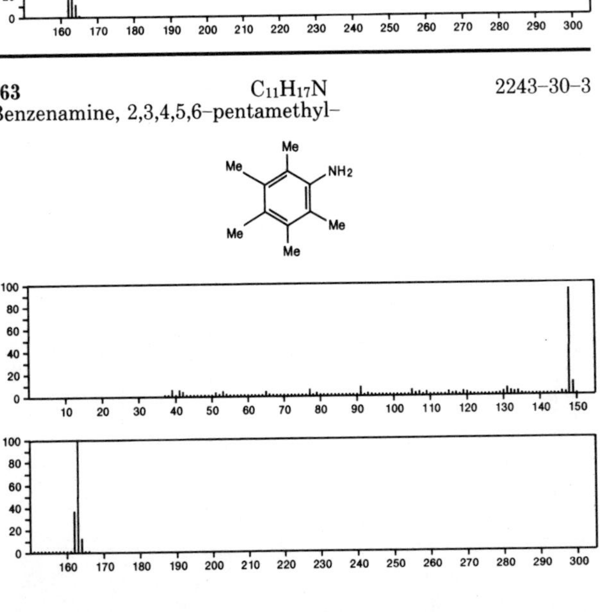

163 $C_{11}H_{17}N$ 7210-81-3
Benzenamine, *N*-(2,2-dimethylpropyl)-

163 $C_{11}H_{17}N$ 18205-60-2
Pyrrolidine, 1-(3-cyclohexen-1-ylidenemethyl)-

163 $C_{11}H_{17}N$ 26462-76-0
p-Menth-1-ene-9-carbonitrile, (4*R*,8*S*)-(+)-

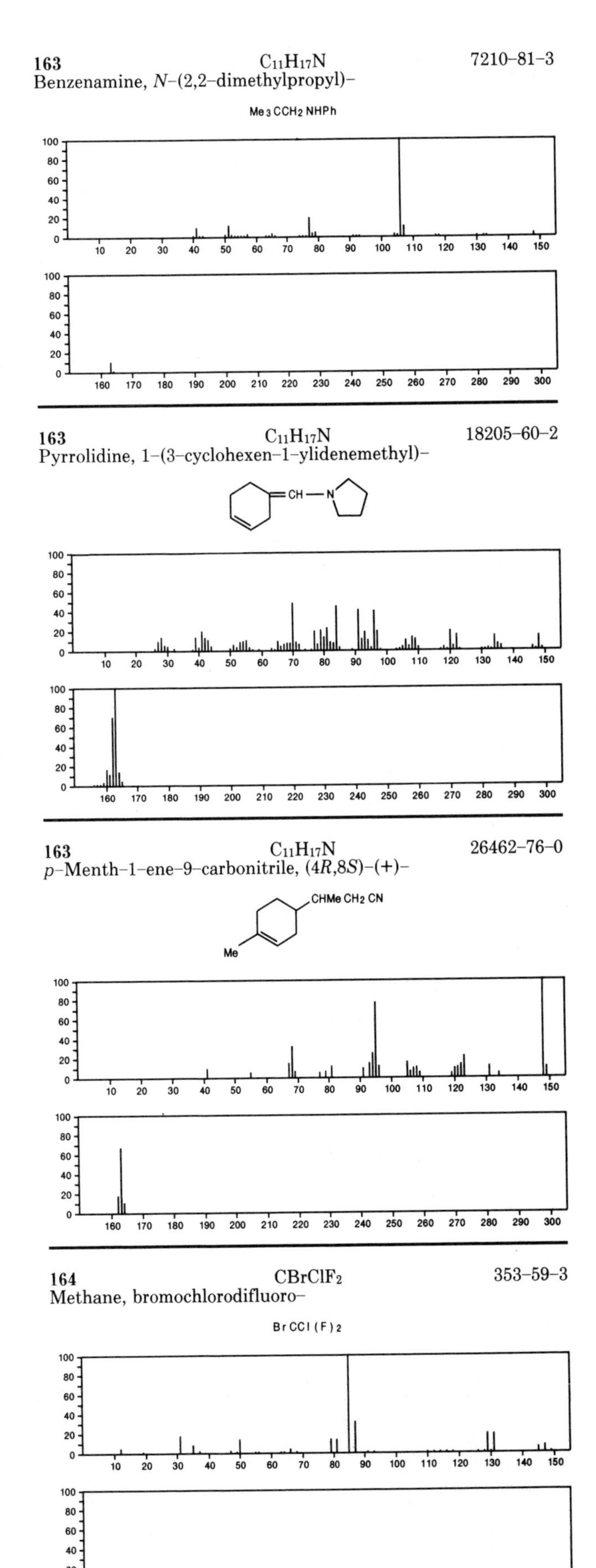

164 $CBrClF_2$ 353-59-3
Methane, bromochlorodifluoro-

BrCCl(F)2

164 CH_3Cl_2OPS 2523–94–6
Phosphorodichloridothioic acid, *O*–methyl ester

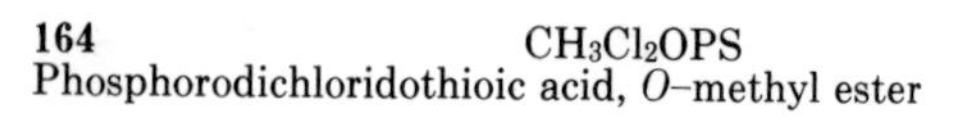

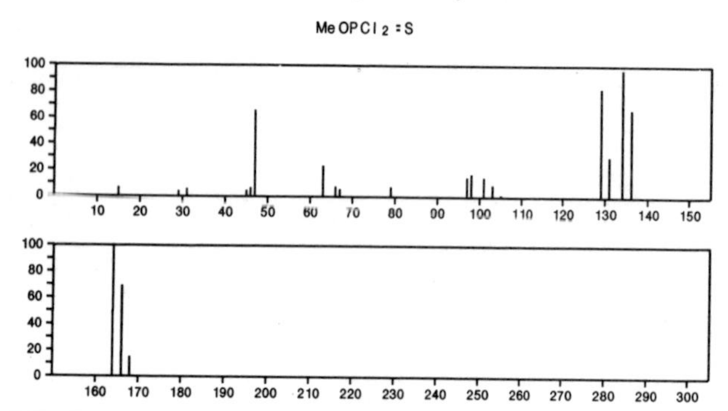

164 C_2Cl_4 127–18–4
Ethene, tetrachloro–

$Cl_2C=CCl_2$

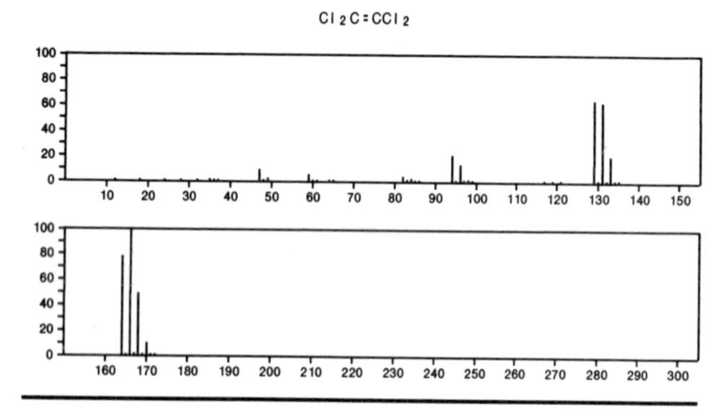

164 $C_2H_3Cl_3O_2$ 302–17–0
1,1–Ethanediol, 2,2,2–trichloro–

$(HO)_2CHCCl_3$

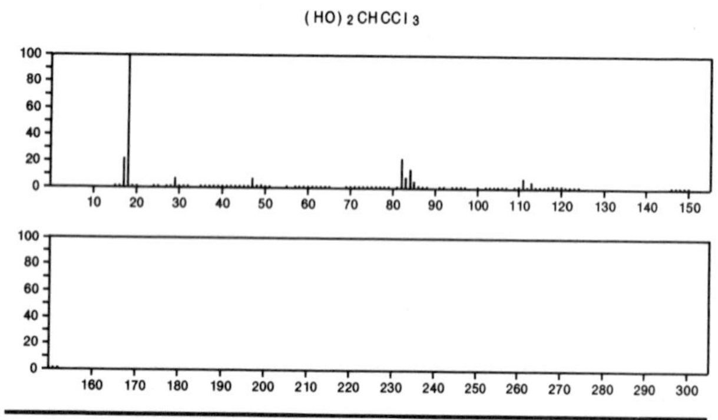

164 $C_3HCl_2F_3$ 431–27–6
Propene, 1,2–dichloro–3,3,3–trifluoro–

$ClCH=CClCF_3$

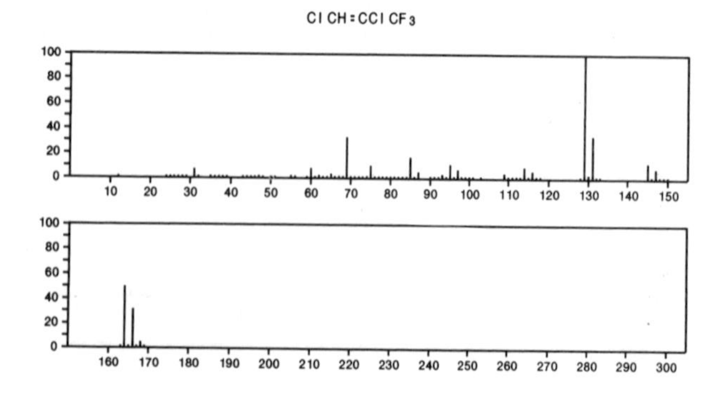

164 $C_3H_2Cl_2N_4$ 933–20–0
1,3,5–Triazin–2–amine, 4,6–dichloro–

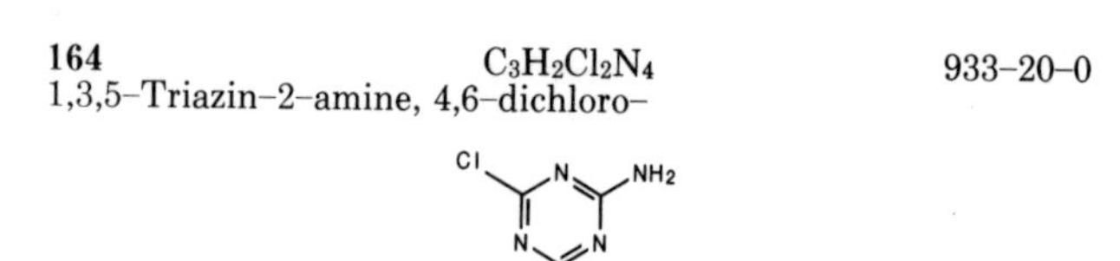

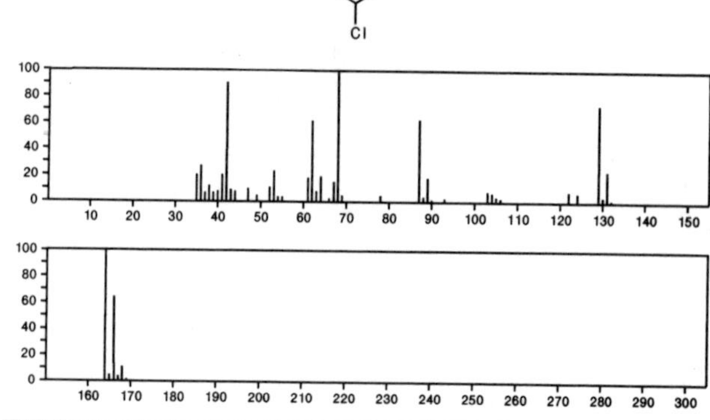

164 $C_3H_4Cl_2F_2O$ 76–38–0
Ethane, 2,2–dichloro–1,1–difluoro–1–methoxy–

Cl_2CHCF_2OMe

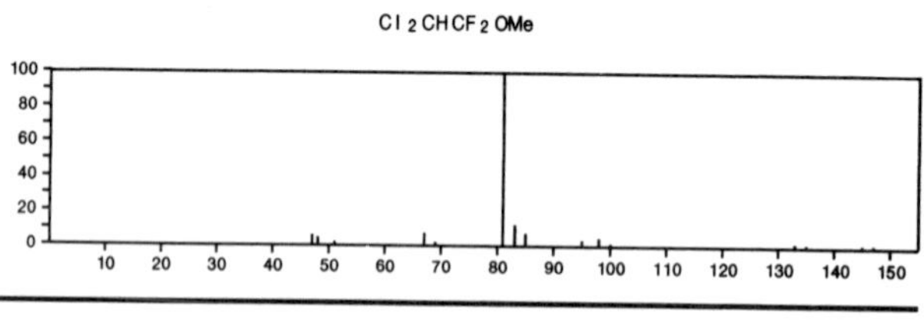

164 $C_3H_{10}N_2.C_2H_2O_4$ 6340–91–6
Hydrazine, propyl–, ethanedioate (1:1)

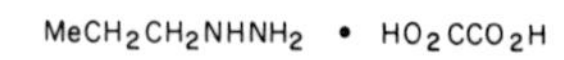

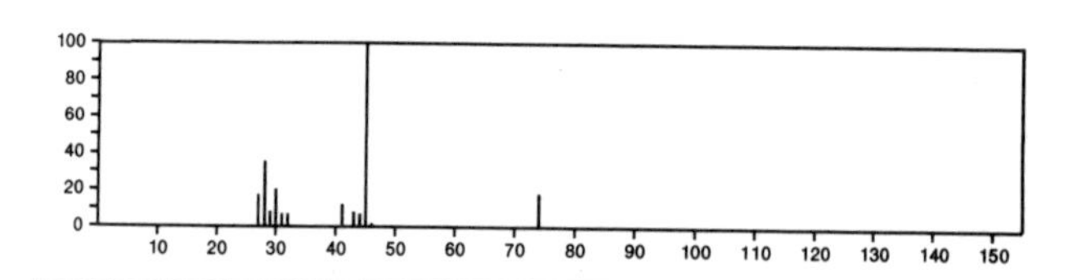

164 $C_3H_{10}N_2.C_2H_2O_4$ 6629–61–4
Hydrazine, (1–methylethyl)–, ethanedioate (1:1)

$Me_2CHNHNH_2 \bullet HO_2CCO_2H$

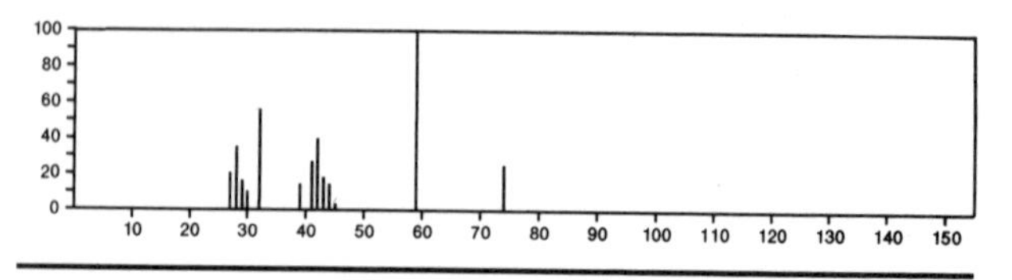

164 $C_4H_9AsO_2$ 24635–97–0
1,3,2–Dioxarsenane, 2–methyl–

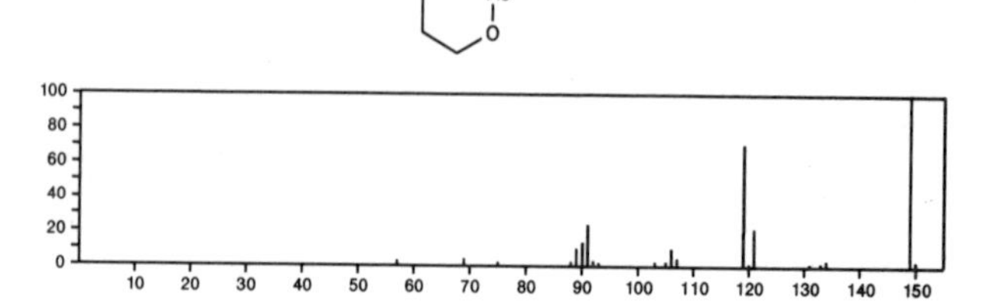

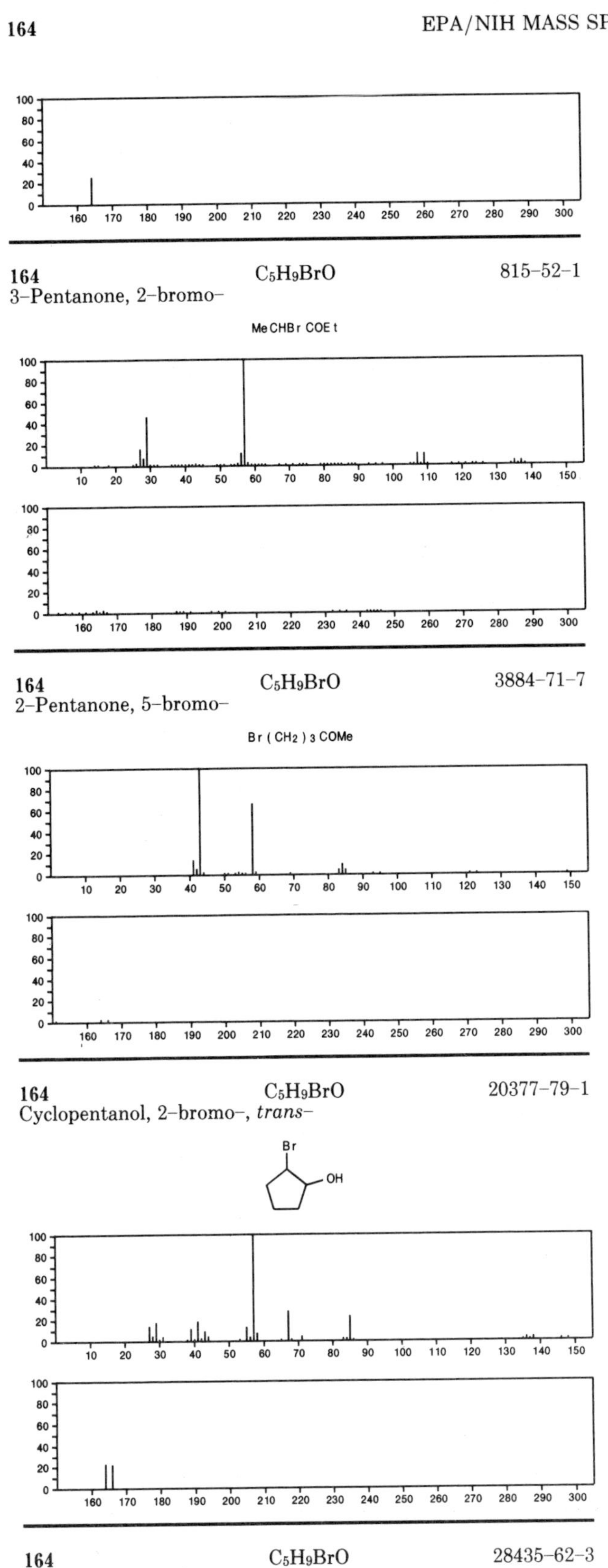

164 C_5H_9BrO 815–52–1
3–Pentanone, 2–bromo–
MeCHBrCOEt
164 C_5H_9BrO 3884–71–7
2–Pentanone, 5–bromo–
Br(CH2)3COMe
164 C_5H_9BrO 20377–79–1
Cyclopentanol, 2–bromo–, trans–
164 C_5H_9BrO 28435–62–3
Cyclopentanol, 2–bromo–, cis–

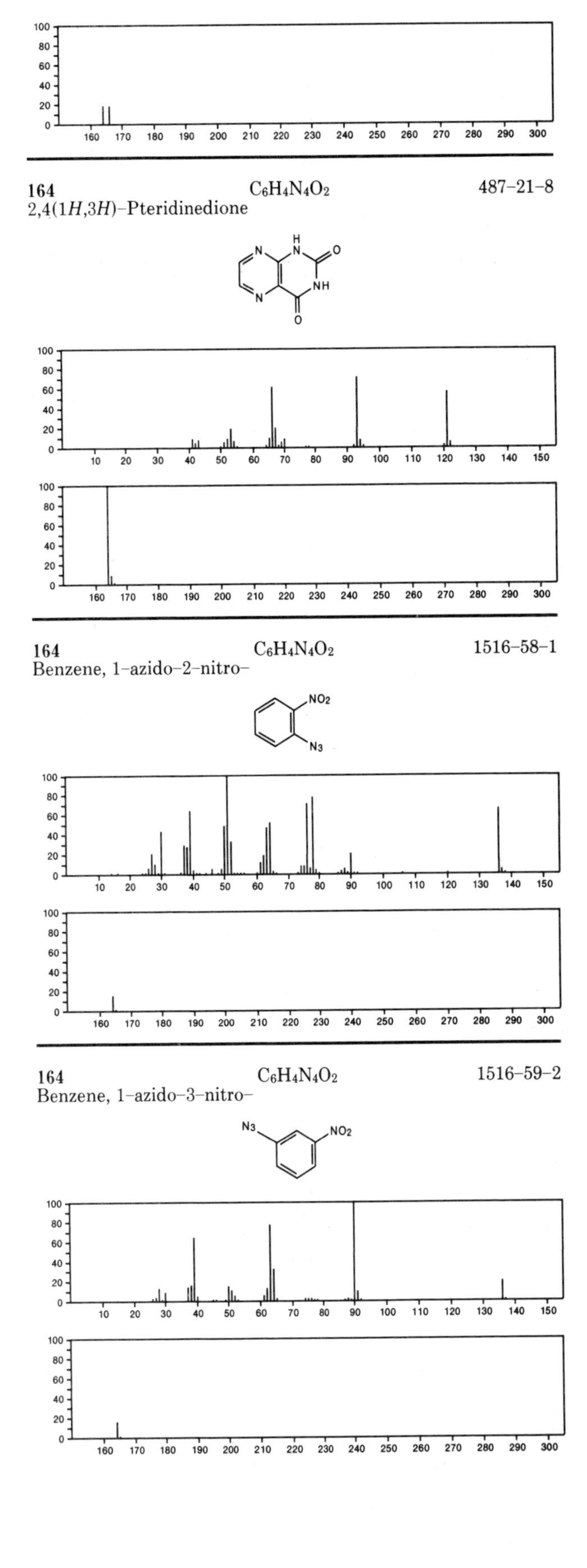

164 $C_6H_4N_4O_2$ 487–21–8
2,4(1H,3H)–Pteridinedione
164 $C_6H_4N_4O_2$ 1516–58–1
Benzene, 1–azido–2–nitro–
164 $C_6H_4N_4O_2$ 1516–59–2
Benzene, 1–azido–3–nitro–

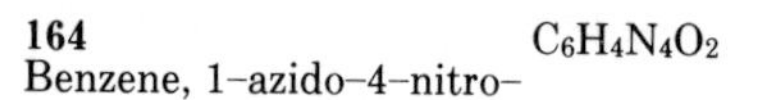

164 $C_6H_4N_4O_2$ 1516-60-5
Benzene, 1-azido-4-nitro-

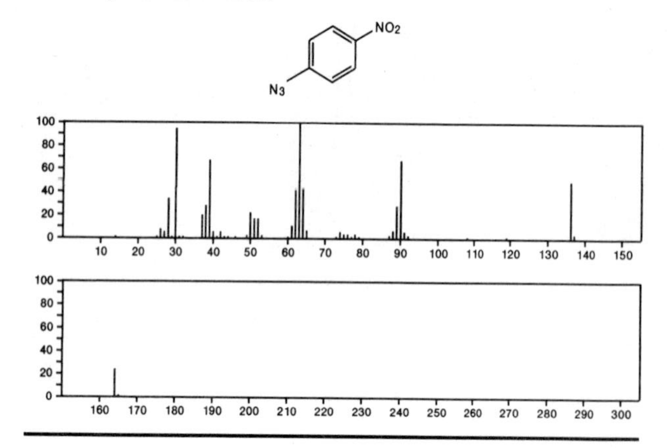

164 $C_6H_4N_4O_2$ 18106-57-5
4(3*H*)-Pteridinone, 3-hydroxy-

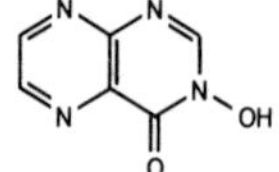

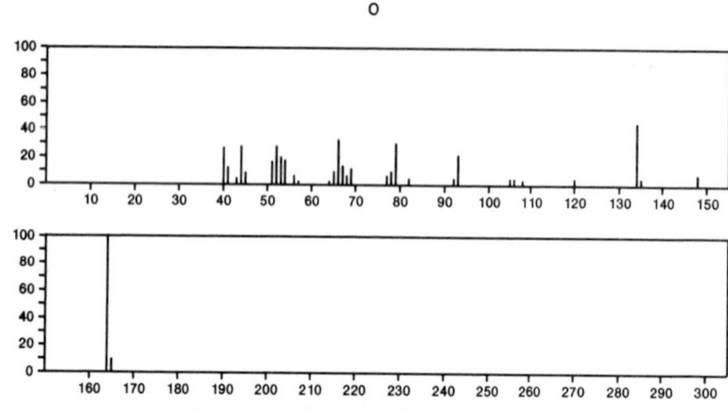

164 $C_6H_9ClO_3$ 609-15-4
Butanoic acid, 2-chloro-3-oxo-, ethyl ester

Et OC(O) CHCl COMe

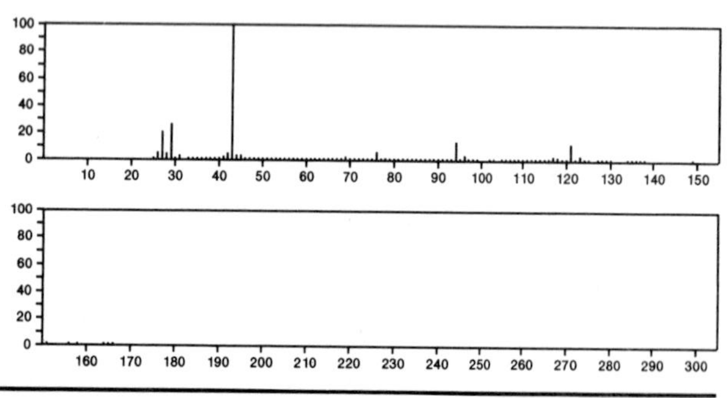

164 $C_6H_{12}OS_2$ 38379-93-0
Carbonodithioic acid, *S*-ethyl *O*-(1-methylethyl) ester

i-Pr OC(S) SEt

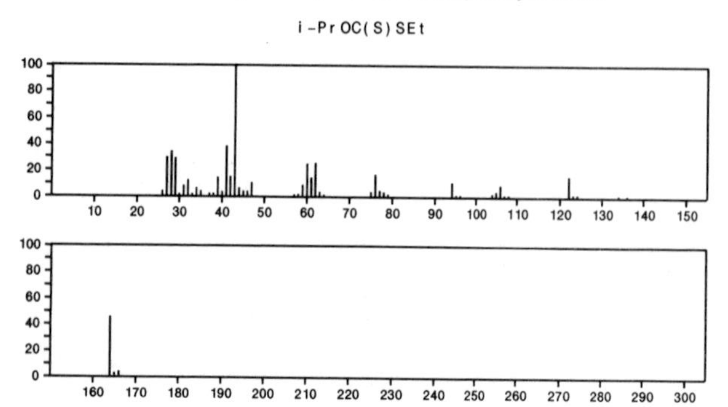

164 $C_6H_{12}O_3S$ 19424-26-1
1,3,2-Dioxathiolane, 4,4,5,5-tetramethyl-, 2-oxide

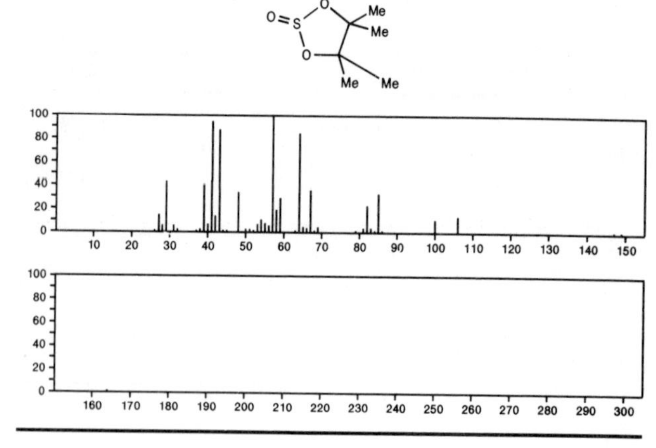

164 $C_6H_{12}O_5$ 1824-96-0
α-D-Xylofuranoside, methyl

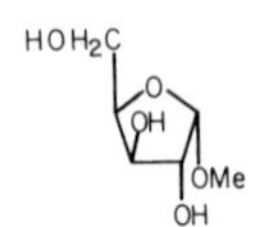

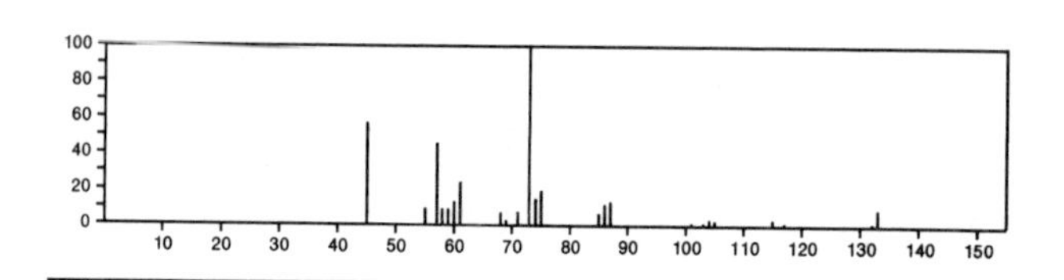

164 $C_6H_{12}O_5$ 1825-00-9
β-L-Arabinopyranoside, methyl

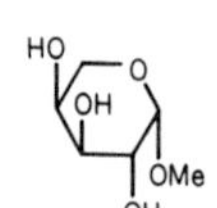

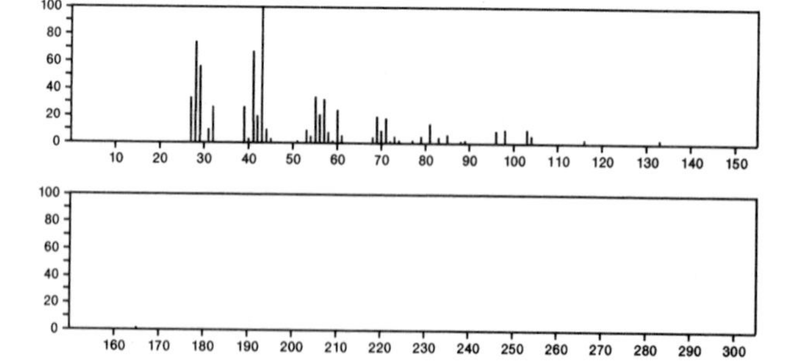

164 $C_6H_{13}Br$ 111-25-1
Hexane, 1-bromo-

Br $(CH_2)_5$ Me

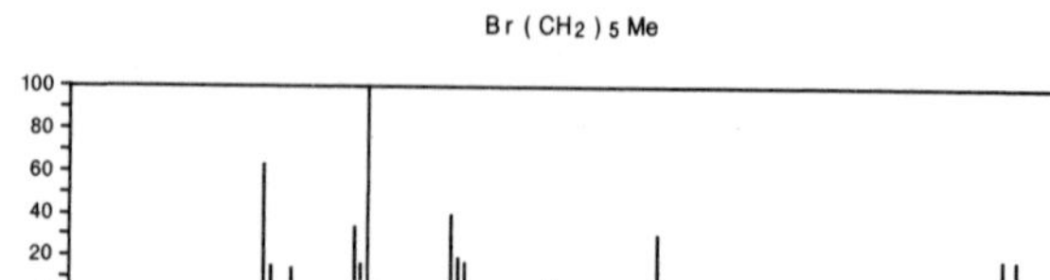

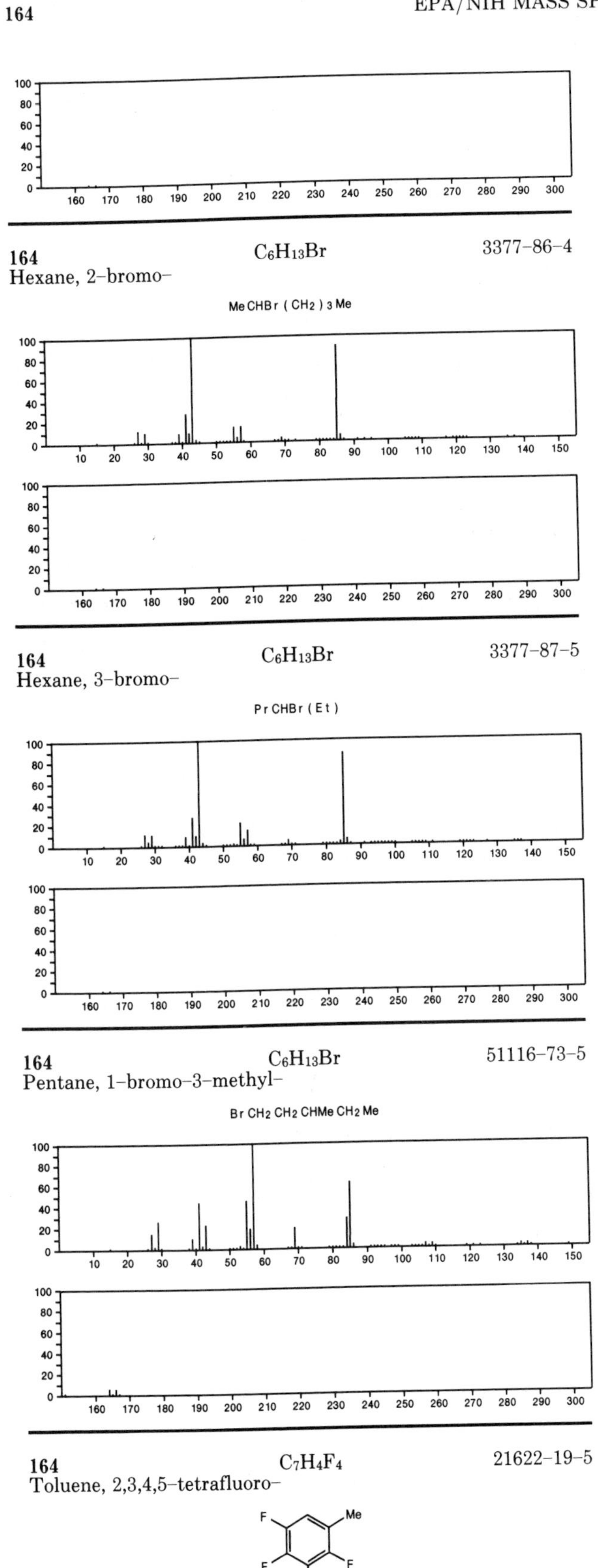

164 $C_6H_{13}Br$ 3377-86-4
Hexane, 2-bromo-

MeCHBr(CH2)3Me

164 $C_6H_{13}Br$ 3377-87-5
Hexane, 3-bromo-

PrCHBr(Et)

164 $C_6H_{13}Br$ 51116-73-5
Pentane, 1-bromo-3-methyl-

BrCH2CH2CHMeCH2Me

164 $C_7H_4F_4$ 21622-19-5
Toluene, 2,3,4,5-tetrafluoro-

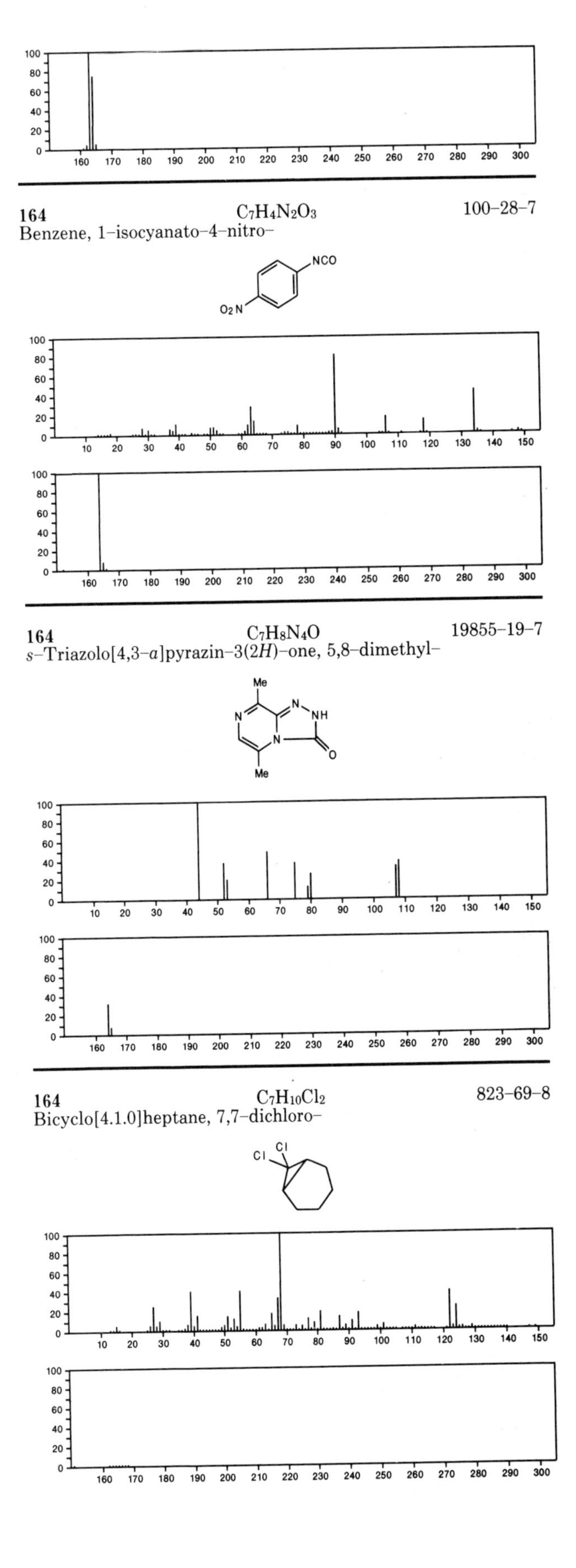

164 $C_7H_4N_2O_3$ 100-28-7
Benzene, 1-isocyanato-4-nitro-

164 $C_7H_8N_4O$ 19855-19-7
s-Triazolo[4,3-*a*]pyrazin-3(2*H*)-one, 5,8-dimethyl-

164 $C_7H_{10}Cl_2$ 823-69-8
Bicyclo[4.1.0]heptane, 7,7-dichloro-

164 $C_7H_{13}ClO_2$ 5978-08-5
1,3-Dioxolane, 2-(3-chloropropyl)-2-methyl-

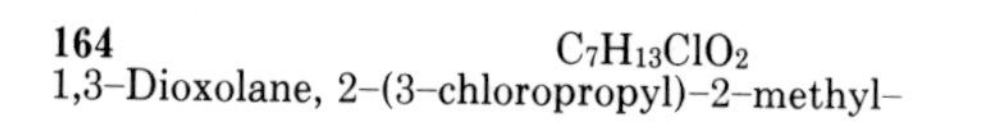

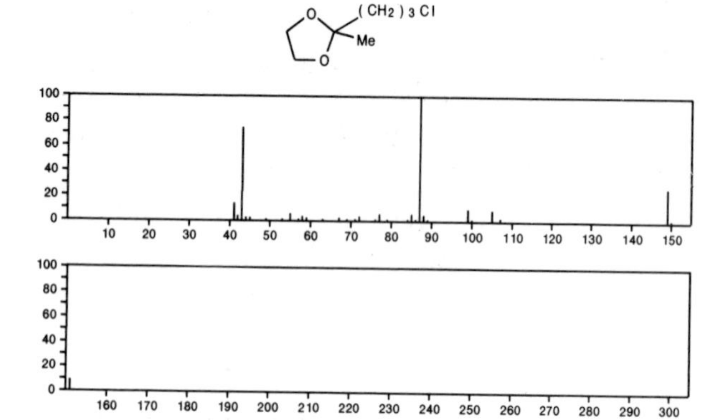

164 $C_7H_{16}O_2S$ 31124-39-7
Sulfone, butyl propyl

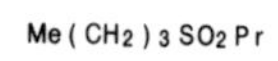

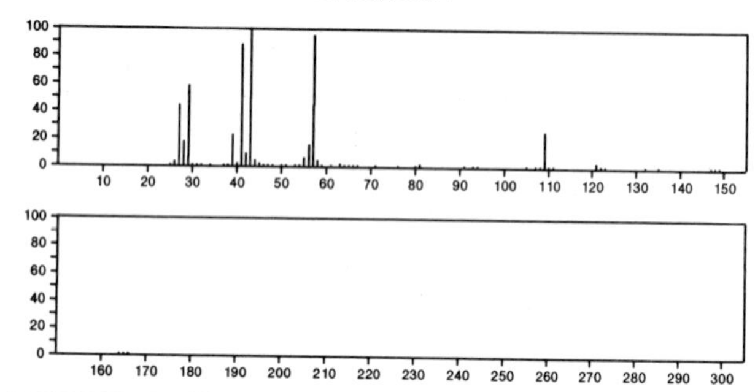

164 $C_7H_{16}O_2S$ 31124-40-0
Sulfone, butyl isopropyl

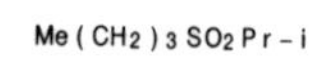

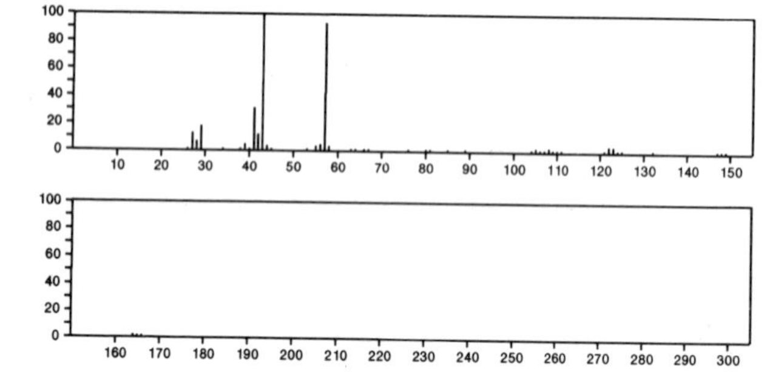

164 $C_7H_{16}O_4$ 112-35-6
Ethanol, 2-[2-(2-methoxyethoxy)ethoxy]-

HOCH2 CH2 OCH2 CH2 OCH2 CH2 OMe

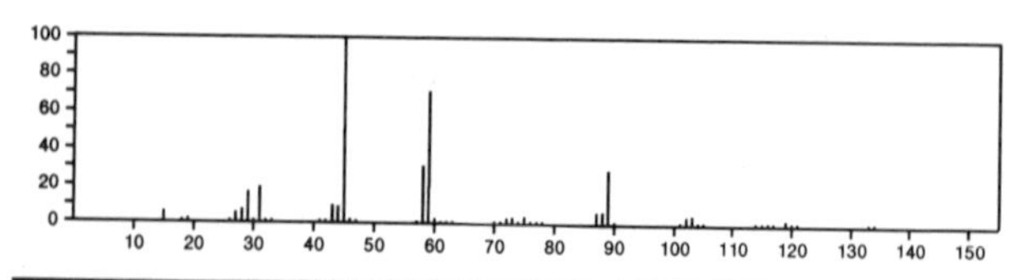

164 $C_7H_{16}S_2$ 14252-45-0
Propane, 2,2-bis(ethylthio)-

Et S2 C(Me) 2

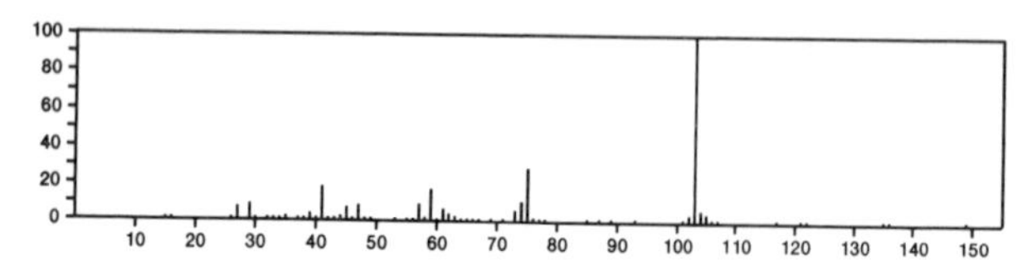

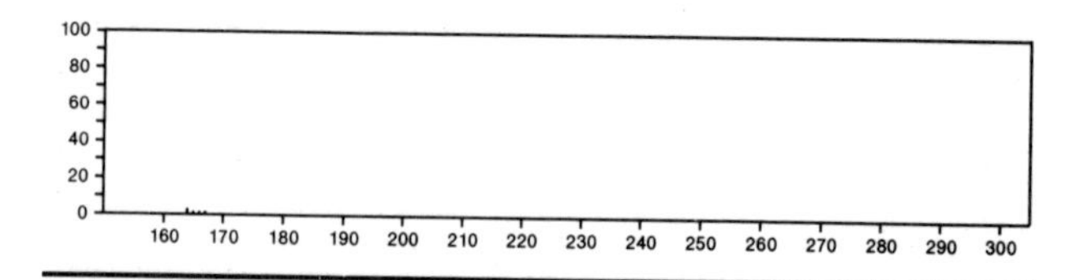

164 $C_7H_{16}S_2$ 33672-52-5
Propane, 1,3-bis(ethylthio)-

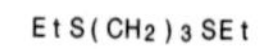

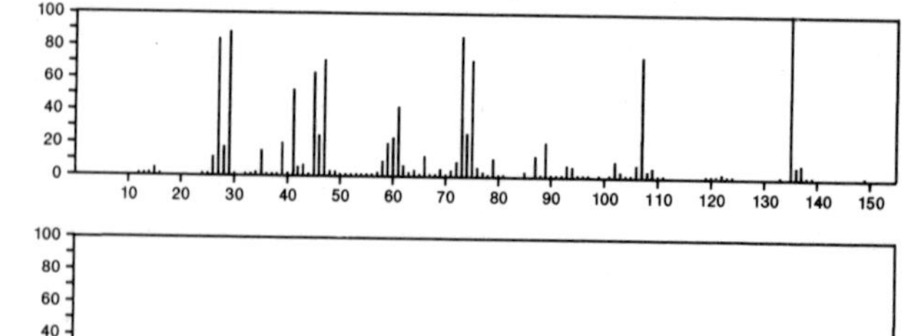

164 $C_7H_{16}S_2$ 54410-62-7
Propane, 1,2-bis(ethylthio)-

Et SCH2 CH(SEt) Me

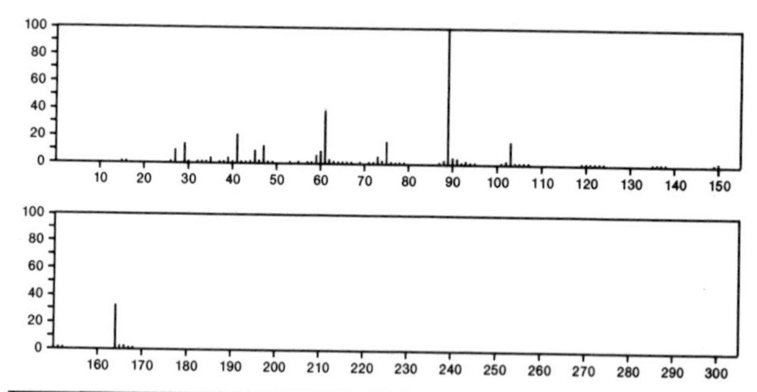

164 $C_7H_{16}S_2$ 54410-63-8
Pentane, 1,5-bis(methylthio)-

Me S (CH2) 5 SMe

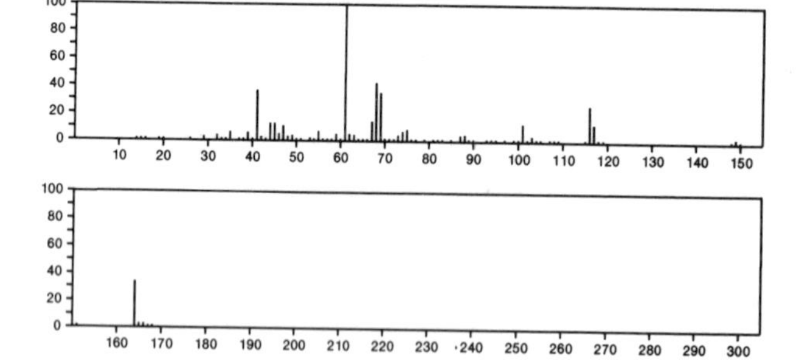

164 $C_8H_8N_2O_2$ 524-40-3
3-Pyridinecarbonitrile, 1,2-dihydro-4-methoxy-1-methyl-2-oxo-

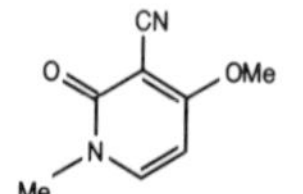

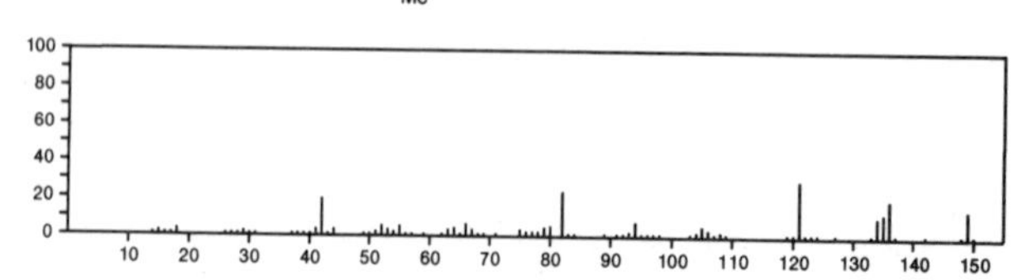

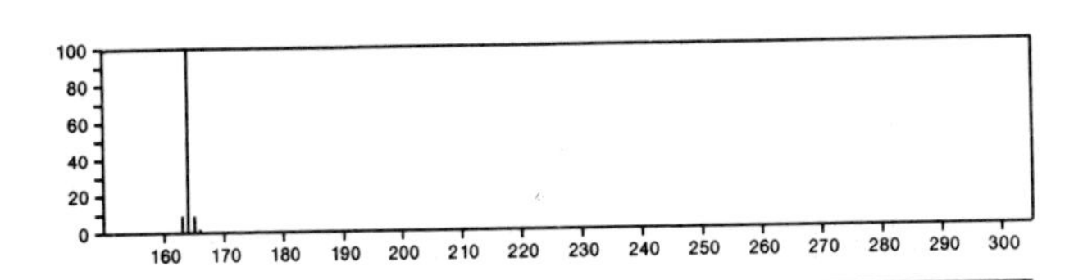

164 $C_8H_8N_2O_2$ 2080–75–3
2*H*–Benzimidazol–2–one, 1,3–dihydro–5–methoxy–

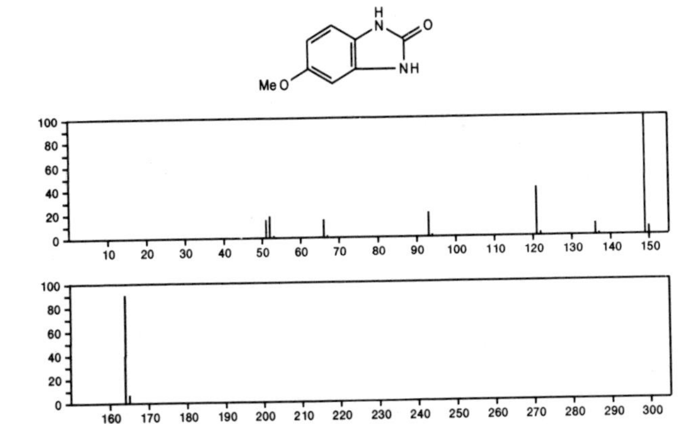

164 $C_8H_8N_2O_2$ 16007–56–0
Benzimidazole, 2–methoxy–, 3–oxide

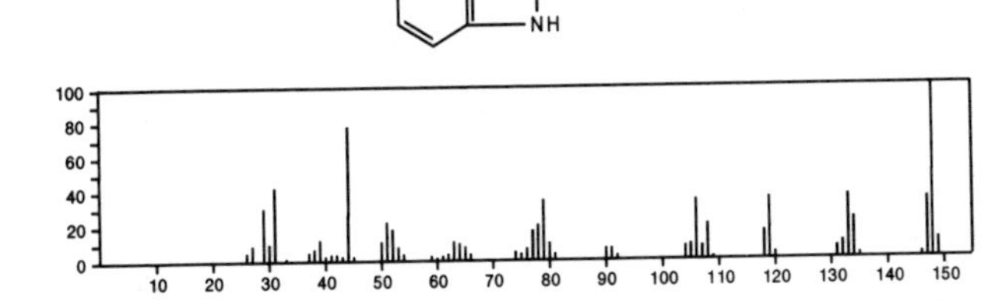

164 $C_8H_8N_2S$ 14779–16–9
2(3*H*)–Benzothiazolimine, 3–methyl–

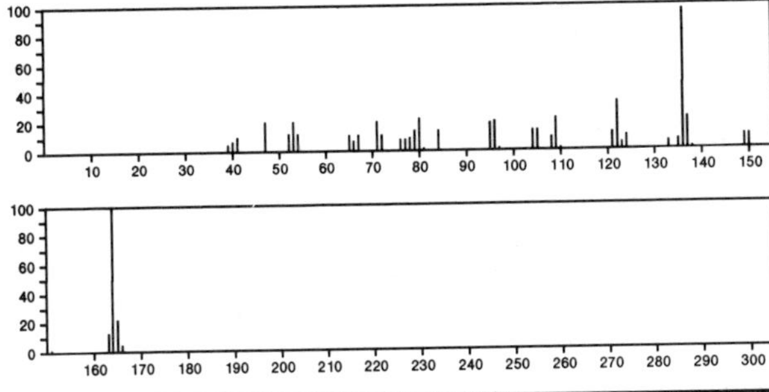

164 $C_8H_8N_2S$ 16954–69–1
2–Benzothiazolamine, *N*–methyl–

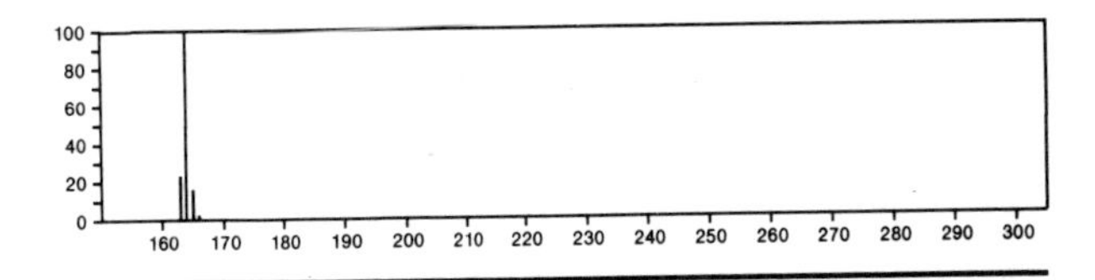

164 $C_8H_{12}N_4$ 51659–18–8
1–Propen–2–amine, *N*,*N*–dimethyl–1–(1,2,4–triazin–5–yl)–

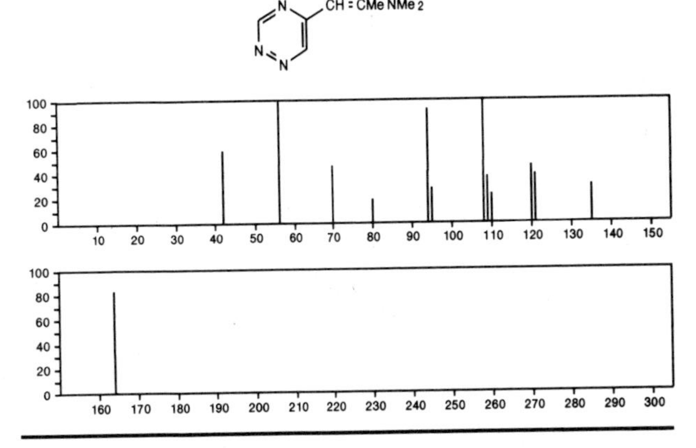

164 $C_8H_{12}Si_2$ 18292–03–0
1,3–Disilaindan, 1–methyl–

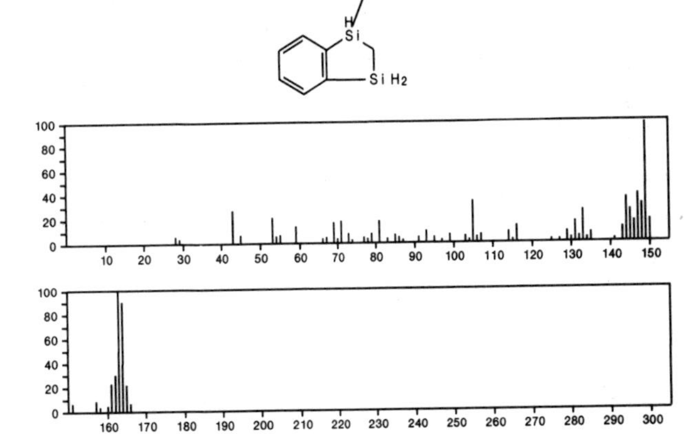

164 $C_9H_8O_3$ 156–06–9
Benzenepropanoic acid, α–oxo–

HO_2CCOCH_2Ph

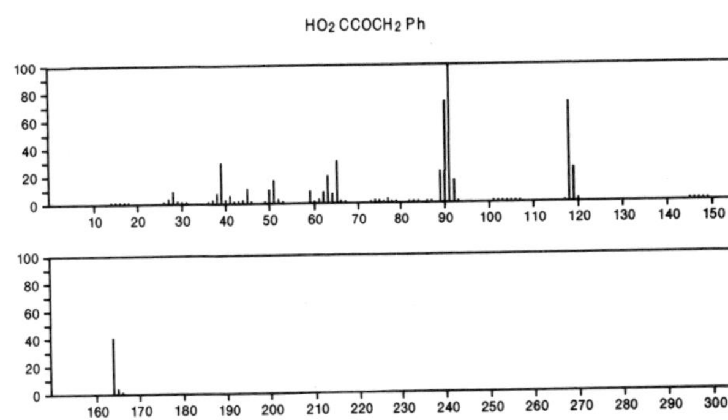

164 $C_9H_8O_3$ 583–17–5
2–Propenoic acid, 3–(2–hydroxyphenyl)–

CH=CHCO2H
OH

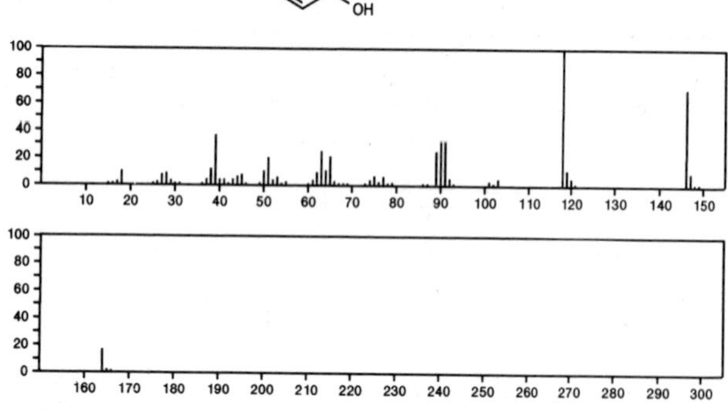

164 $C_9H_8O_3$ 4122–56–9
Benzoic acid, 2–formyl–, methyl ester

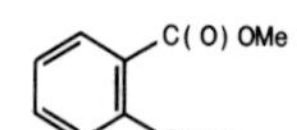

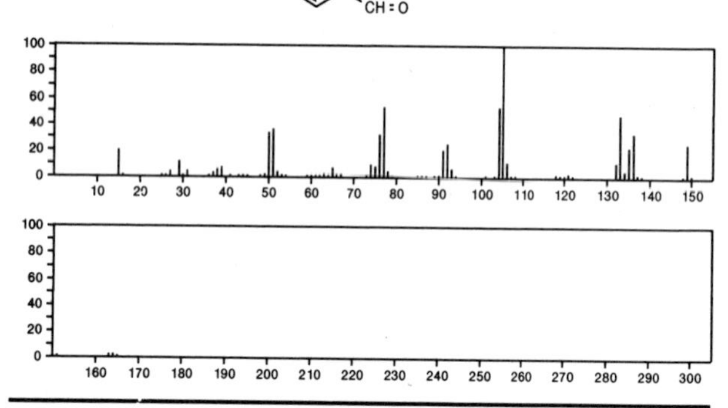

164 $C_9H_8O_3$ 4427–92–3
1,3–Dioxolan–2–one, 4–phenyl–

O O Ph
O

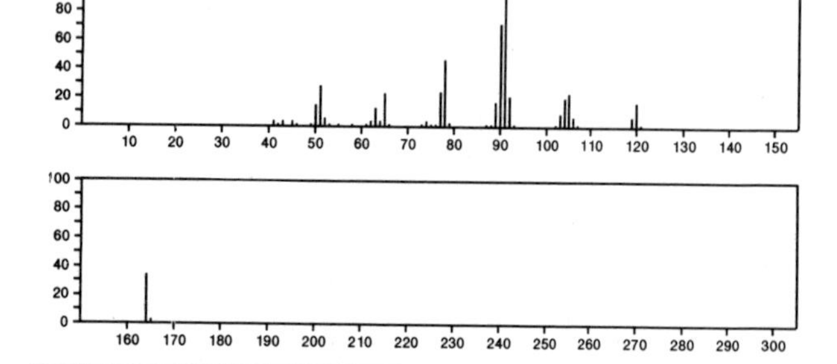

164 $C_9H_8O_3$ 7400–08–0
2–Propenoic acid, 3–(4–hydroxyphenyl)–

CH=CHCO2H
HO

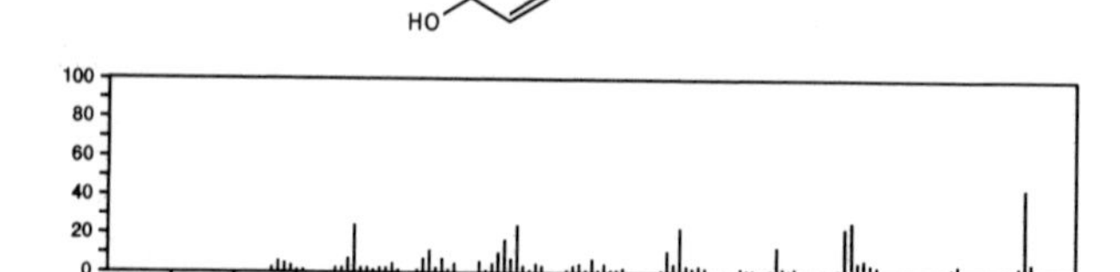

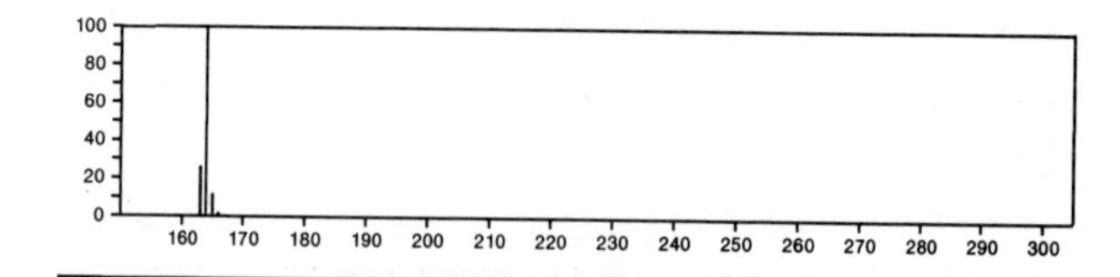

164 $C_9H_{12}N_2O$ 101–42–8
Urea, *N,N*–dimethyl–*N'*–phenyl–

Me2NCONHPh

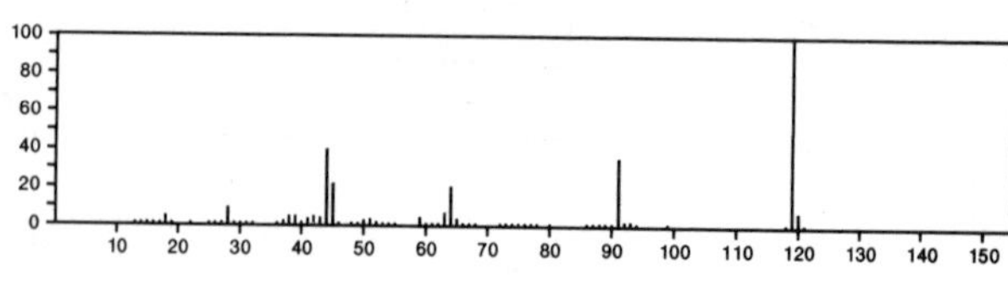

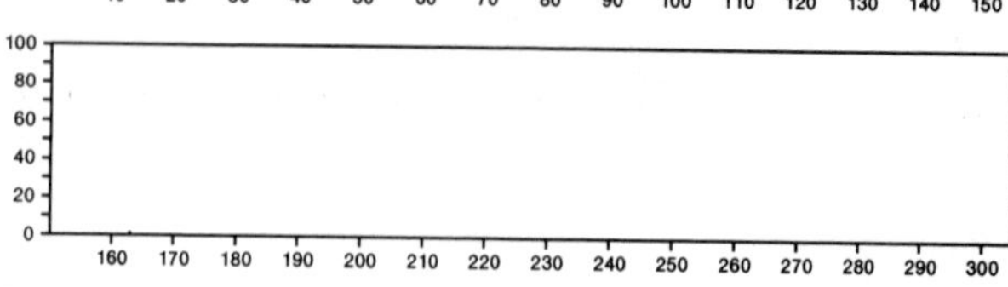

164 $C_9H_{12}N_2O$ 2350–51–8
Formamidine, *N'*–(*p*–hydroxyphenyl)–*N,N*–dimethyl–

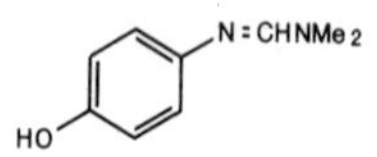

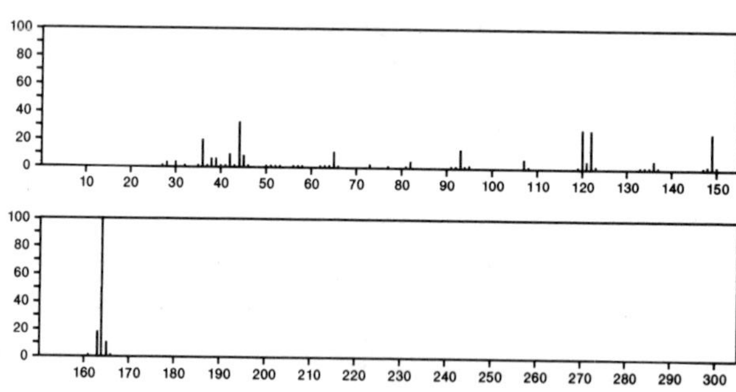

164 $C_9H_{12}N_2O$ 25635–97–6
Methanimidamide, *N'*–(3–hydroxyphenyl)–*N,N*–dimethyl–

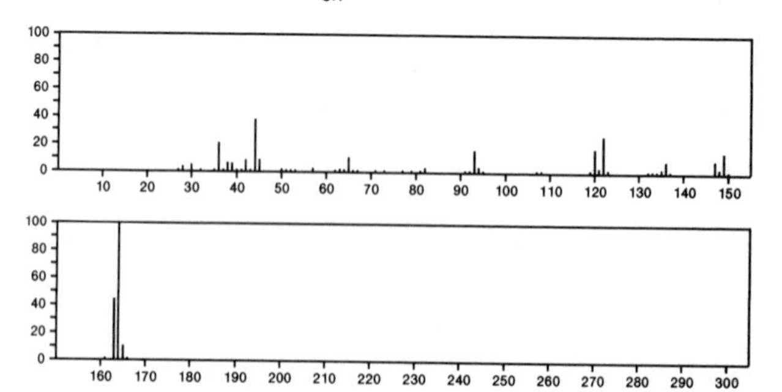

164 $C_9H_{12}N_2O$ 31020-35-6
Pyridinium, 1-(acetylamino)-2,6-dimethyl-, hydroxide, inner salt

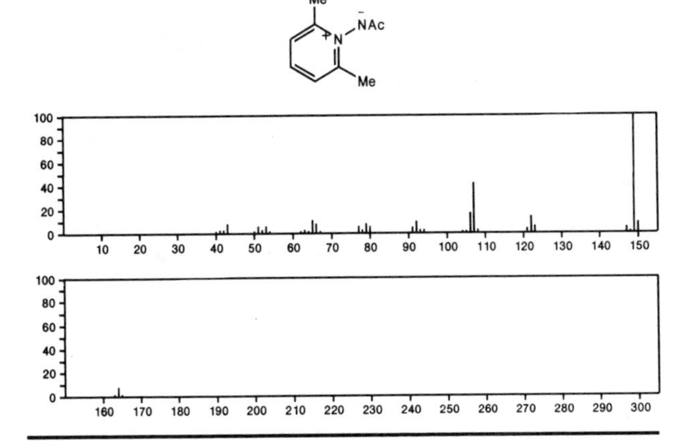

164 $C_{10}H_{12}O_2$ 90-27-7
Benzeneacetic acid, α-ethyl-

PhCHEt(CO2H)

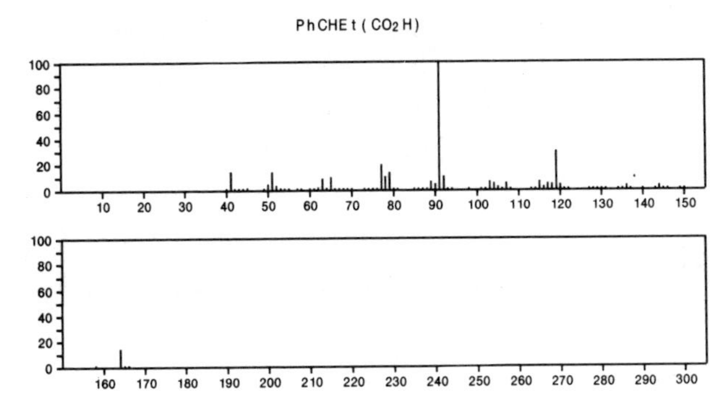

164 $C_{10}H_{12}O_2$ 93-92-5
Benzenemethanol, α-methyl-, acetate

AcOCHMePh

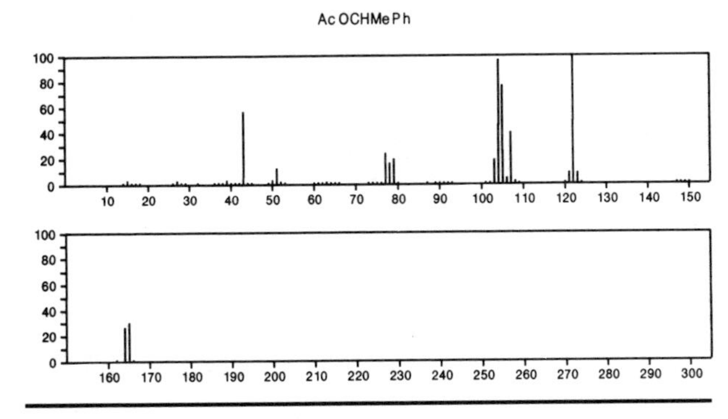

164 $C_{10}H_{12}O_2$ 97-53-0
Phenol, 2-methoxy-4-(2-propenyl)-

MeO CH2CH=CH2 HO

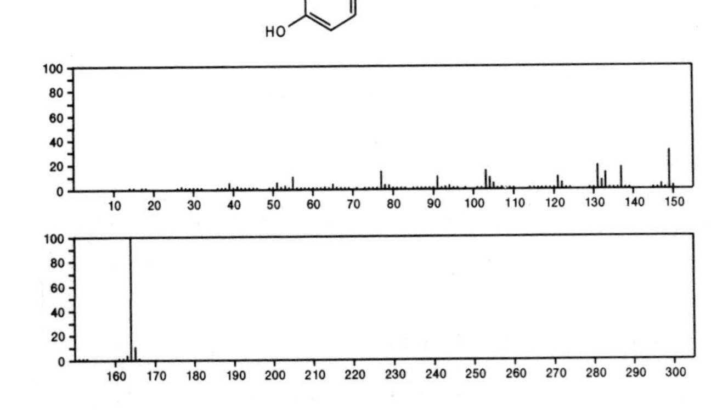

164 $C_{10}H_{12}O_2$ 97-54-1
Phenol, 2-methoxy-4-(1-propenyl)-

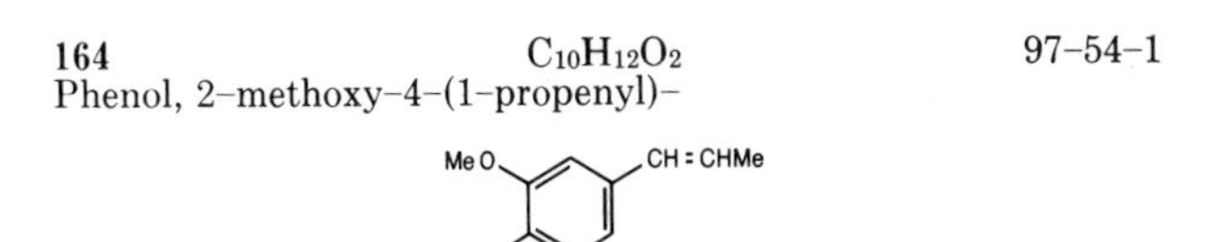

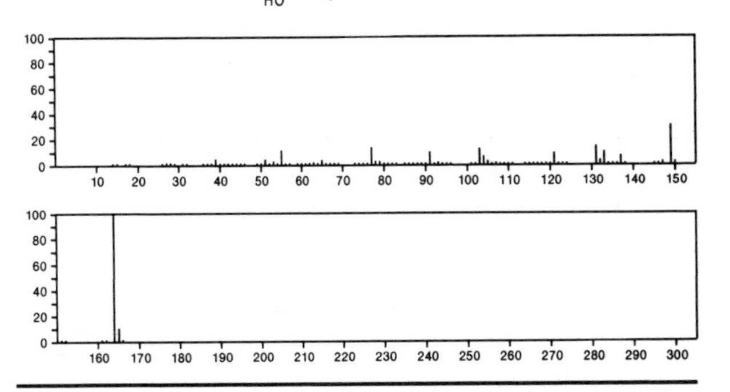

164 $C_{10}H_{12}O_2$ 101-49-5
1,3-Dioxolane, 2-(phenylmethyl)-

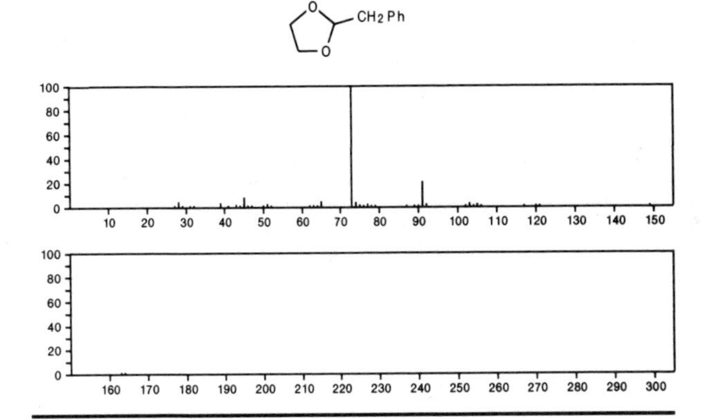

164 $C_{10}H_{12}O_2$ 101-97-3
Benzeneacetic acid, ethyl ester

PhCH2C(O)OEt

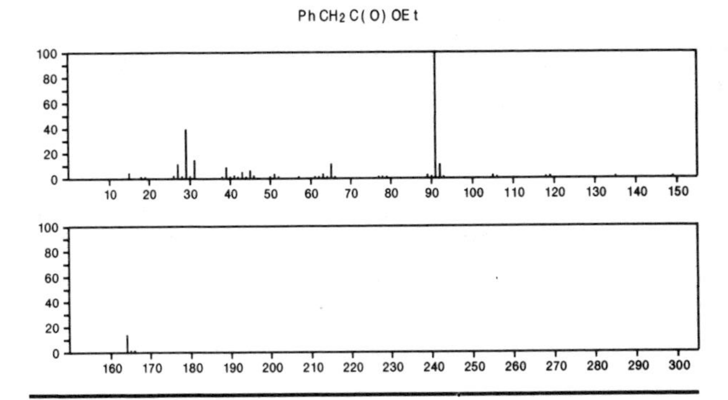

164 $C_{10}H_{12}O_2$ 103-25-3
Benzenepropanoic acid, methyl ester

MeOC(O)CH2CH2Ph

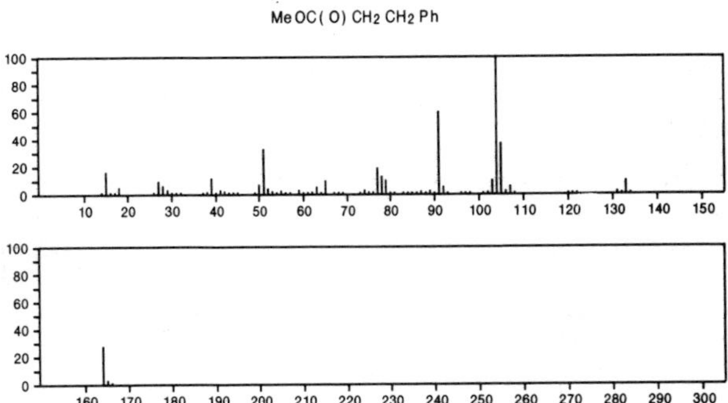

164 $C_{10}H_{12}O_2$ 103-45-7
Acetic acid, 2-phenylethyl ester

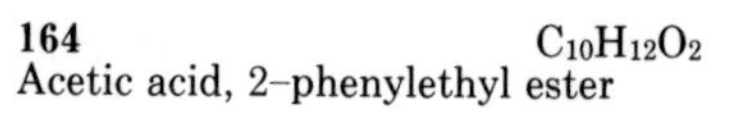

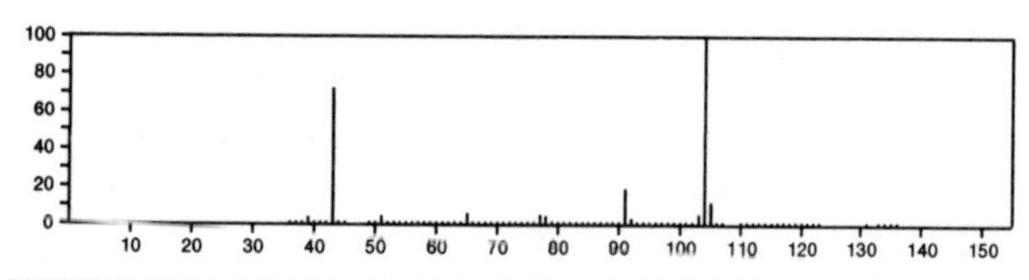

164 $C_{10}H_{12}O_2$ 121-97-1
1-Propanone, 1-(4-methoxyphenyl)-

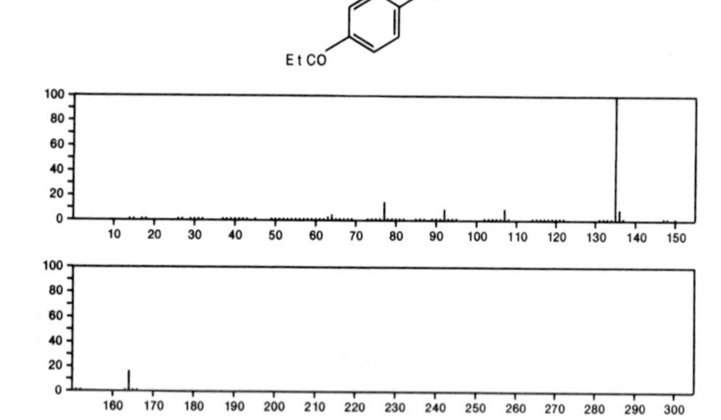

164 $C_{10}H_{12}O_2$ 122-84-9
2-Propanone, 1-(4-methoxyphenyl)-

CH2 COMe
MeO

100 80 60 40 20 0
10 20 30 40 50 60 70 80 90 100 110 120 130 140 150

100 80 60 40 20 0
160 170 180 190 200 210 220 230 240 250 260 270 280 290 300

164 $C_{10}H_{12}O_2$ 480-63-7
Benzoic acid, 2,4,6-trimethyl-

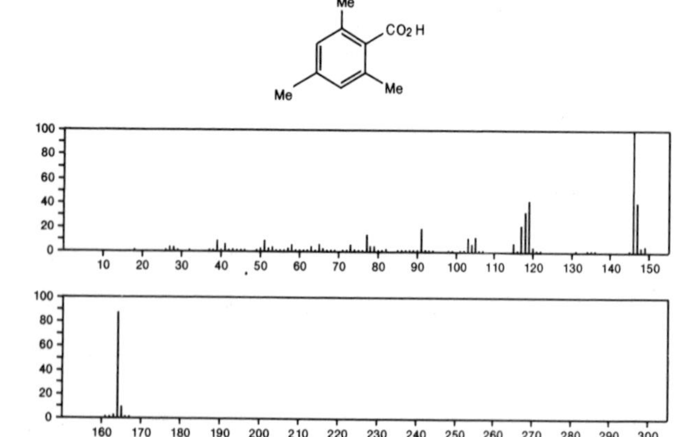

164 $C_{10}H_{12}O_2$ 527-17-3
2,5-Cyclohexadiene-1,4-dione, 2,3,5,6-tetramethyl-

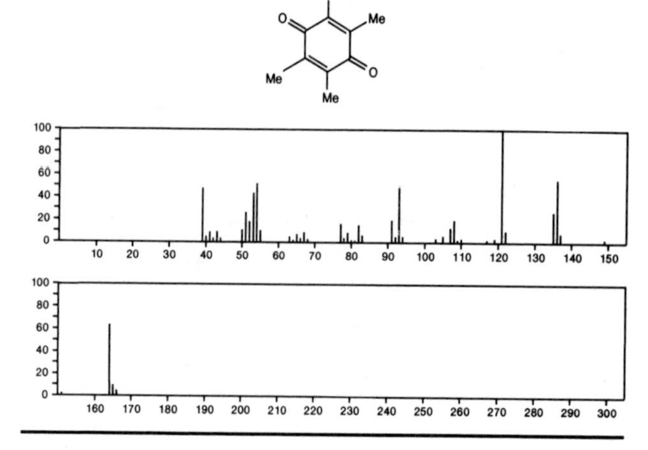

164 $C_{10}H_{12}O_2$ 528-90-5
Benzoic acid, 2,4,5-trimethyl-

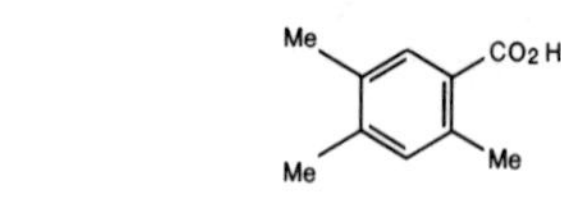

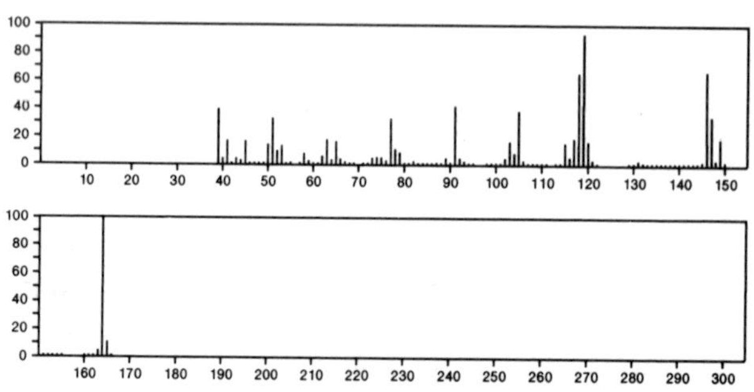

164 $C_{10}H_{12}O_2$ 536-66-3
Benzoic acid, 4-(1-methylethyl)-

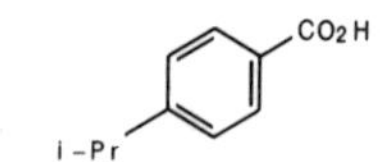

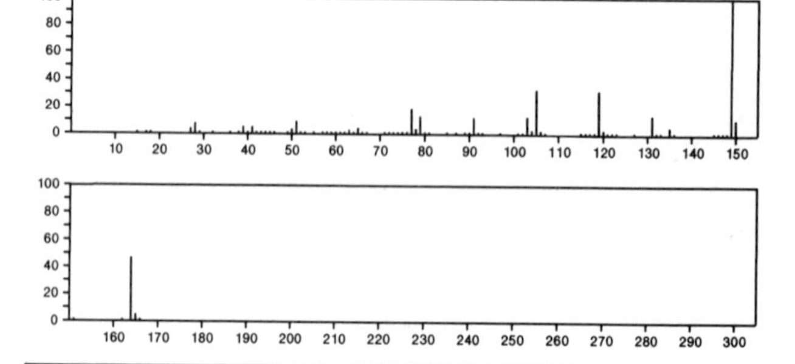

164 $C_{10}H_{12}O_2$ 772-01-0
1,3-Dioxane, 2-phenyl-

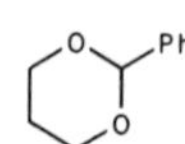

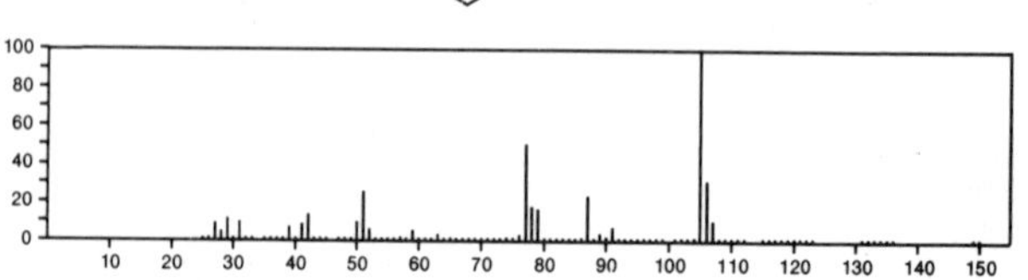

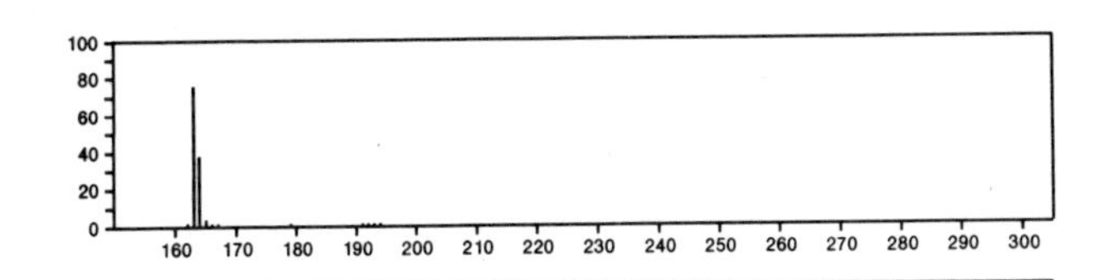

164 $C_{10}H_{12}O_2$ 939-48-0
Benzoic acid, 1-methylethyl ester

i-Pr OC(O) Ph

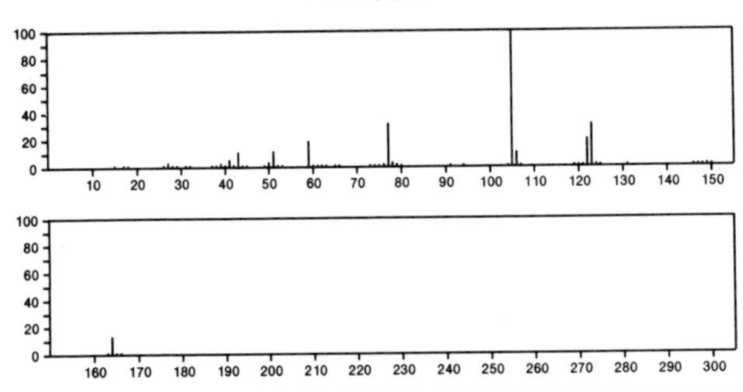

164 $C_{10}H_{12}O_2$ 1076-55-7
Phenol, 2-methoxy-6-(1-propenyl)-

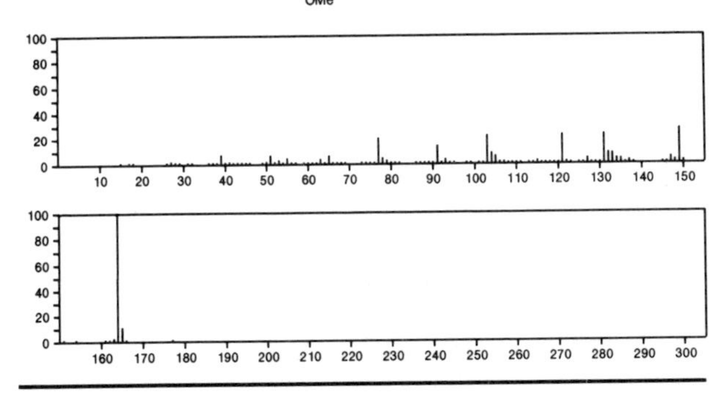

164 $C_{10}H_{12}O_2$ 1821-12-1
Benzenebutanoic acid

Ph(CH_2)$_3$$CO_2$H

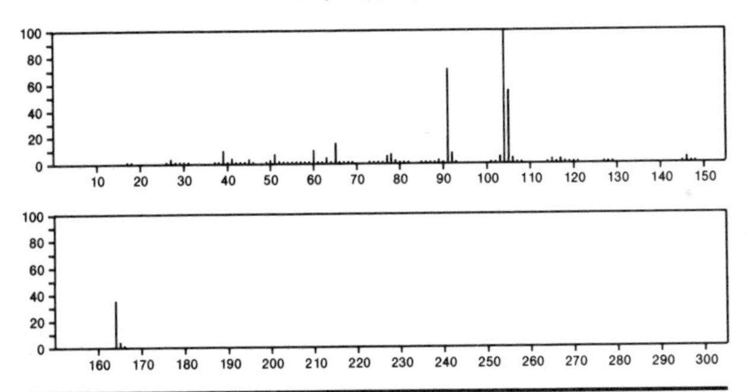

164 $C_{10}H_{12}O_2$ 2403-51-2
1,3-Dioxolane, 2-(4-methylphenyl)-

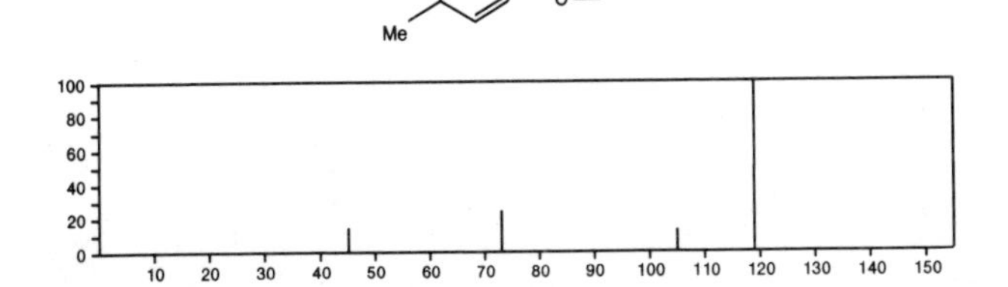

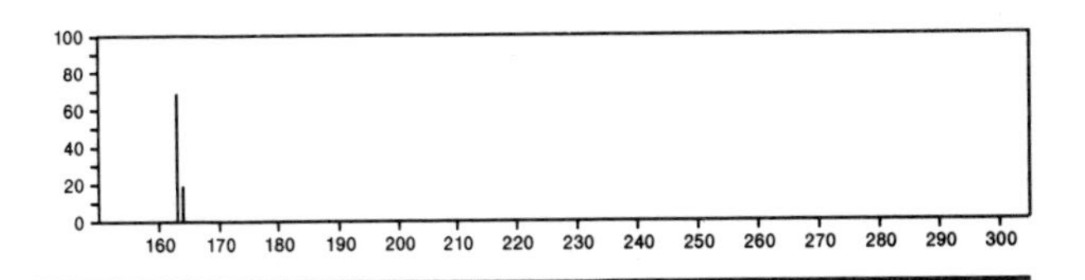

164 $C_{10}H_{12}O_2$ 2568-25-4
1,3-Dioxolane, 4-methyl-2-phenyl-

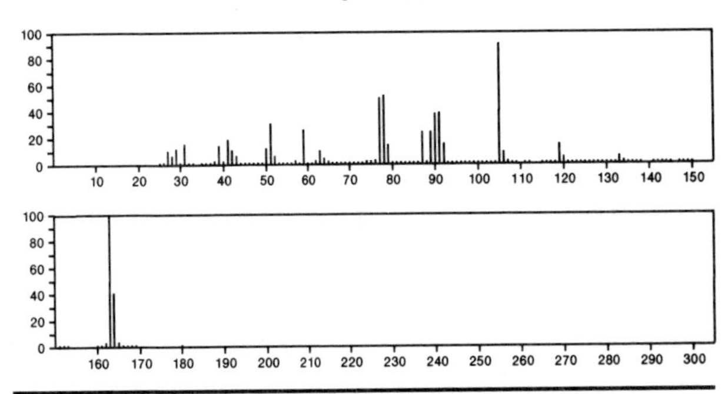

164 $C_{10}H_{12}O_2$ 3056-60-8
Phenol, 3-ethyl-, acetate

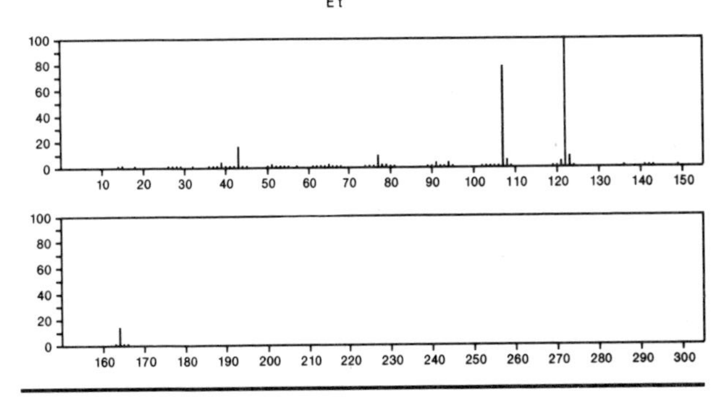

164 $C_{10}H_{12}O_2$ 3602-55-9
2,5-Cyclohexadiene-1,4-dione, 2-(1,1-dimethylethyl)-

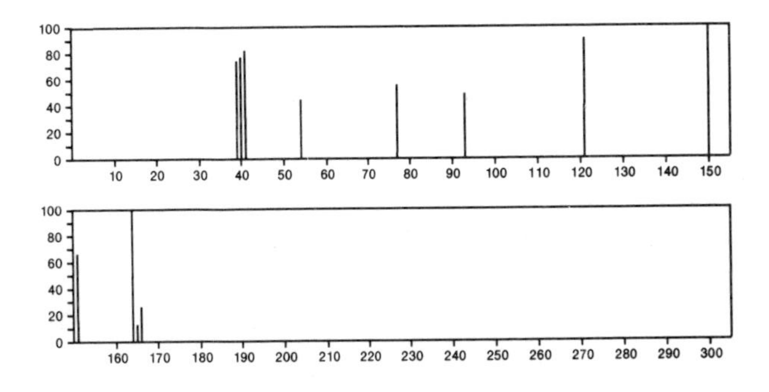

164 $C_{10}H_{12}O_2$ 3674-77-9
1,3-Dioxolane, 2-methyl-2-phenyl-

Ph
Me

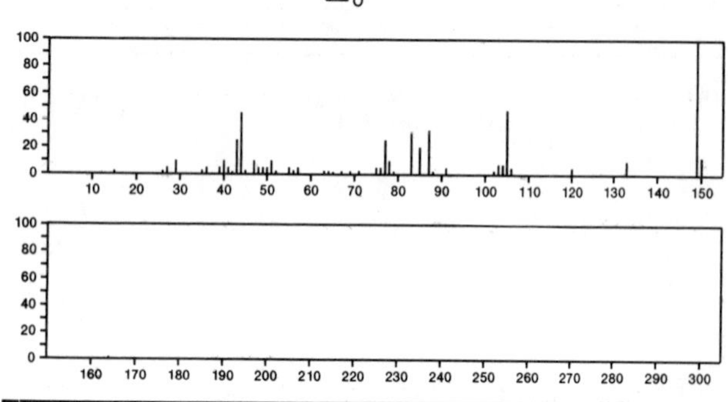

164 $C_{10}H_{12}O_2$ 4346-18-3
Butanoic acid, phenyl ester

PrC(O)OPh

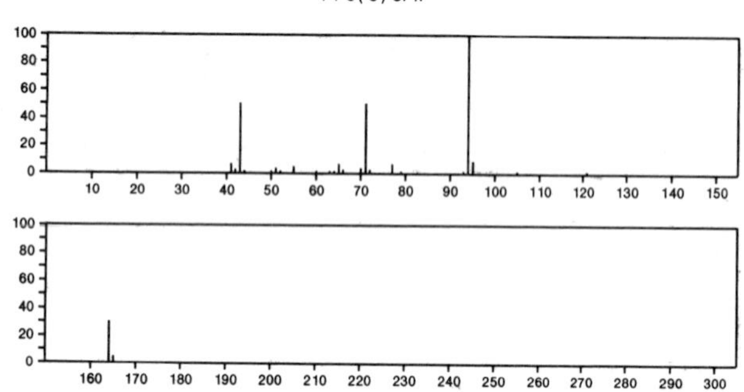

164 $C_{10}H_{12}O_2$ 5471-51-2
2-Butanone, 4-(4-hydroxyphenyl)-

CH2 CH2 COMe
HO

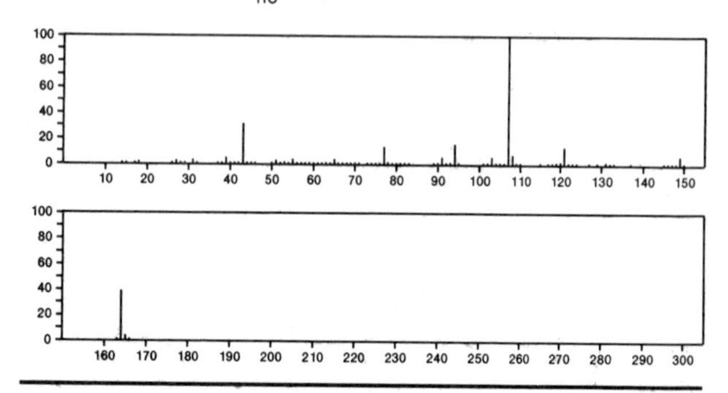

164 $C_{10}H_{12}O_2$ 5932-68-3
Phenol, 2-methoxy-4-(1-propenyl)-, (*E*)-

MeO CH=CHMe
HO

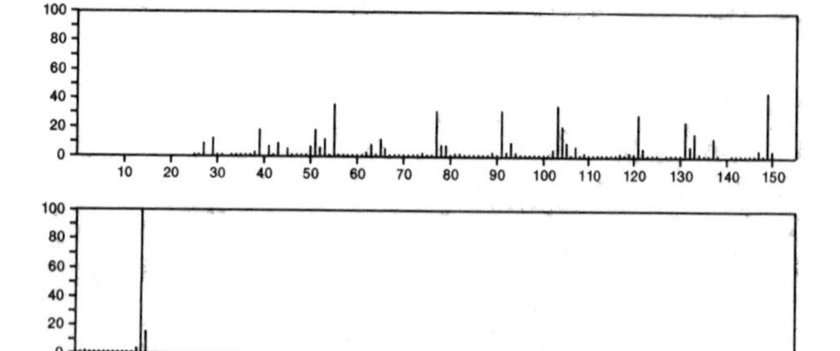

164 $C_{10}H_{12}O_2$ 7124-91-6
1,6-Benzodioxocin, 2,3,4,5-tetrahydro-

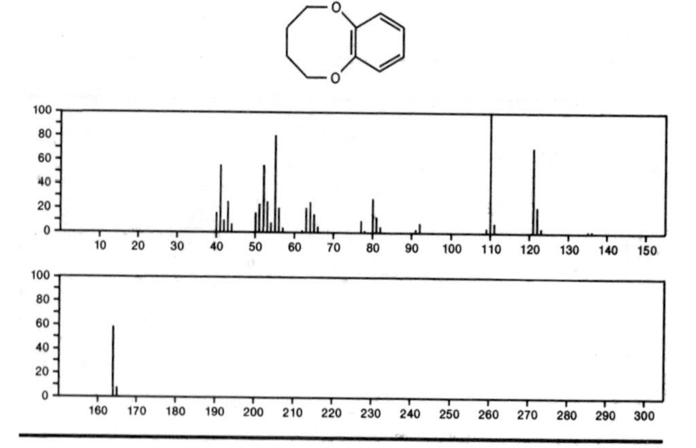

164 $C_{10}H_{12}O_2$ 13730-55-7
Benzoic acid, 2,5-dimethyl-, methyl ester

Me
C(O)OMe
Me

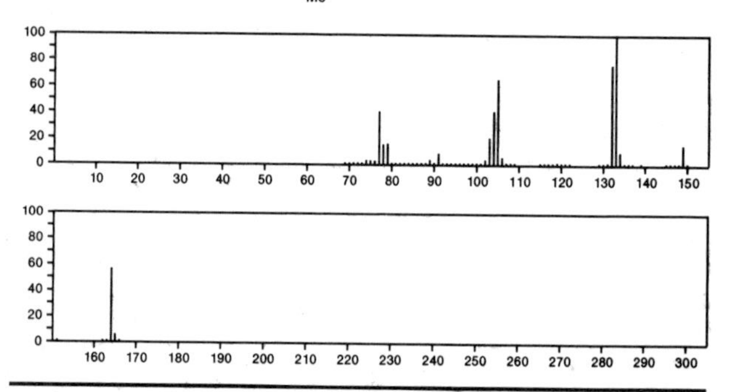

164 $C_{10}H_{12}O_2$ 14211-53-1
1,2-Naphthalenediol, 1,2,3,4-tetrahydro-, *trans*-

OH
OH

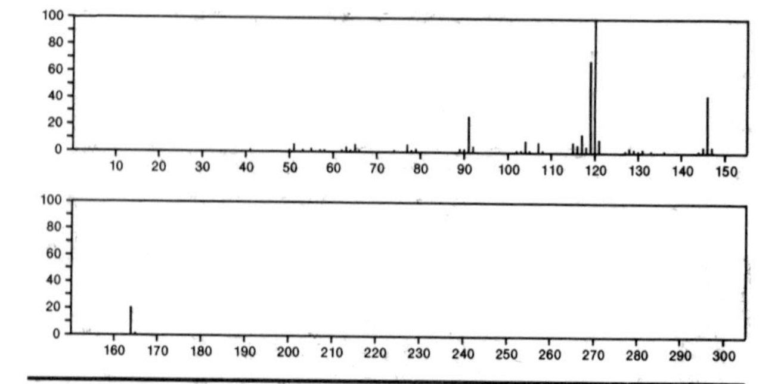

164 $C_{10}H_{12}O_2$ 17373-93-2
Benzenemethanol, 2-methyl-, acetate

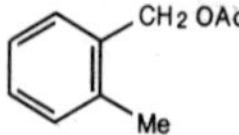

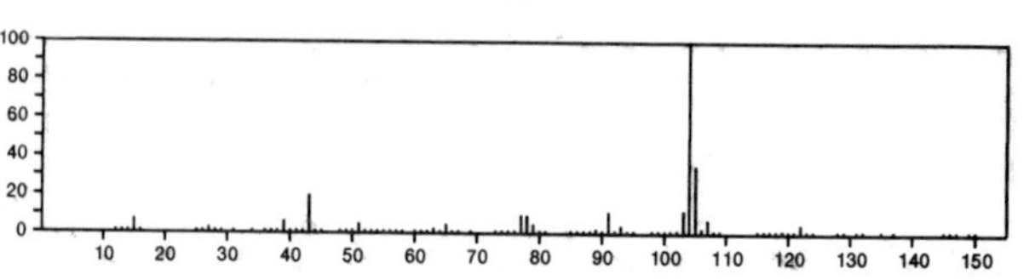

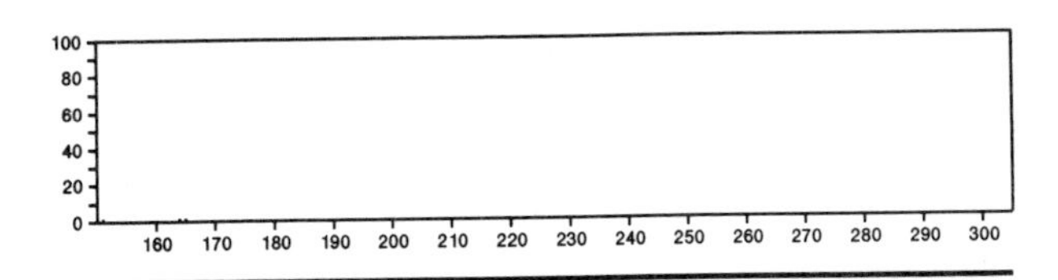

164 $C_{10}H_{12}O_2$ 19784-98-6
Phenol, 2-methoxy-5-(1-propenyl)-, (*E*)-

HO CH=CHMe
MeO

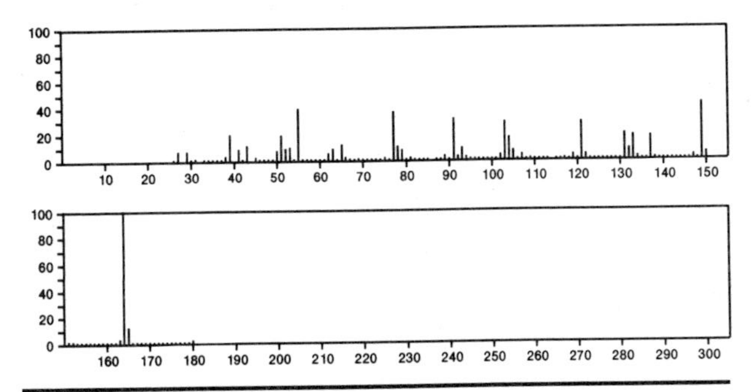

164 $C_{10}H_{12}O_2$ 21016-53-5
1,2-Naphthalenediol, 1,2,3,4-tetrahydro-, *cis*-

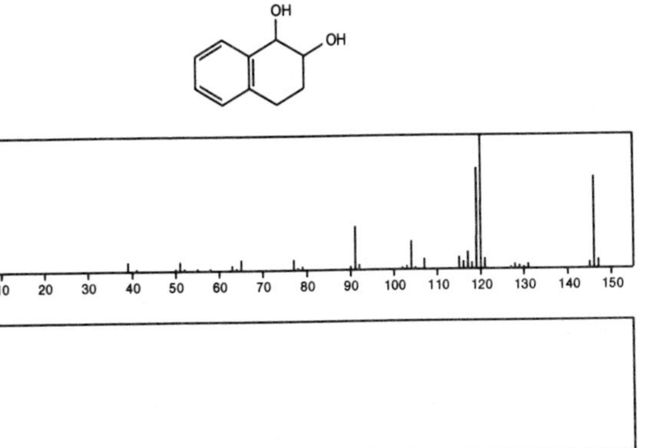

164 $C_{10}H_{12}O_2$ 23617-71-2
Benzoic acid, 2,4-dimethyl-, methyl ester

C(O)OMe
Me Me

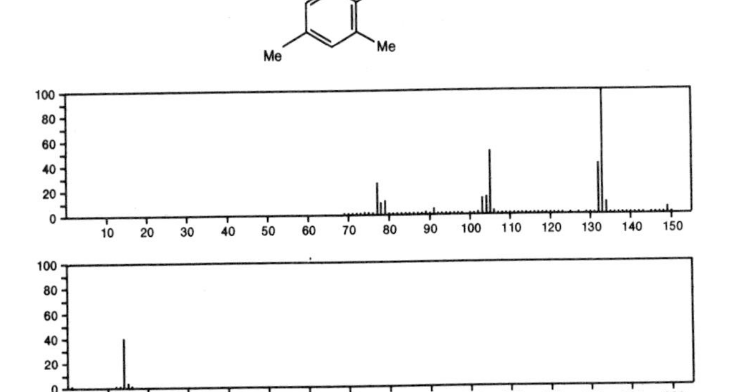

164 $C_{10}H_{12}O_2$ 25081-39-4
Benzoic acid, 3,5-dimethyl-, methyl ester

Me C(O)OMe
Me

164 $C_{10}H_{12}O_2$ 30458-34-5
Benzene, 1,2-(butylidenedioxy)-

O Pr
O

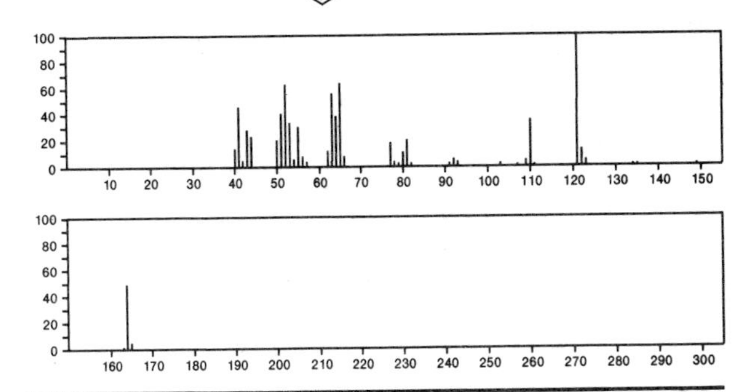

164 $C_{10}H_{12}O_2$ 41137-14-8
1,2-Naphthalenediol, 1,2,3,4-tetrahydro-

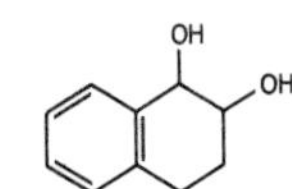

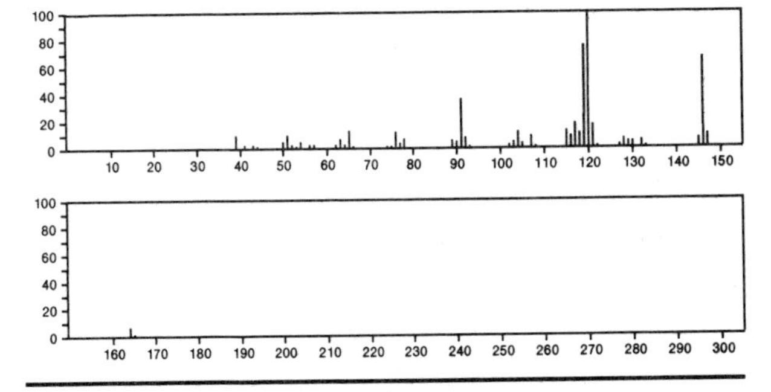

164 $C_{10}H_{12}O_2$ 56588-29-5
1*H*-Indene-1,2-diol, 2,3-dihydro-1-methyl-, *cis*-

HO Me
OH

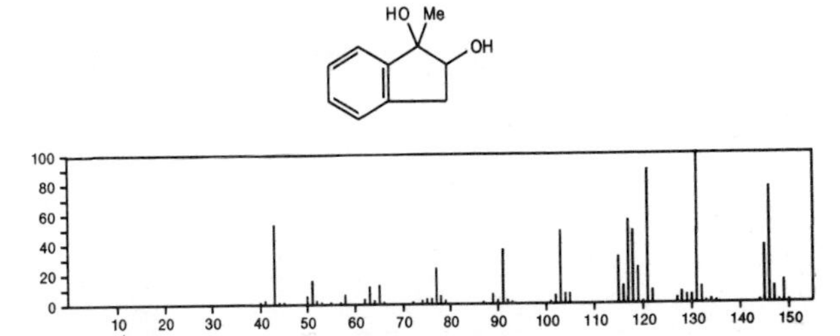

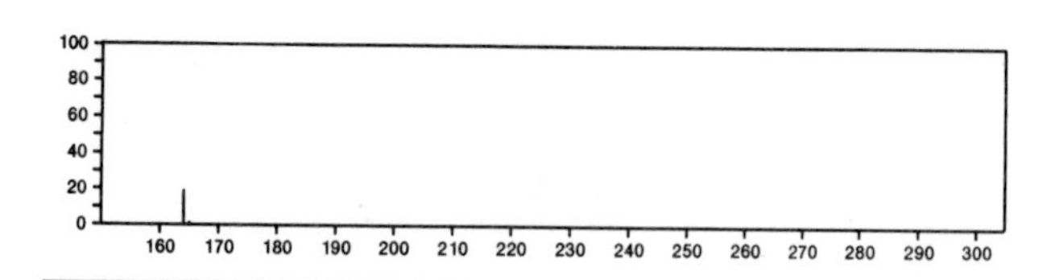

164 $C_{10}H_{12}O_2$ 56588-40-0
1*H*-Indene-1,2-diol, 2,3-dihydro-2-methyl-, *cis*-

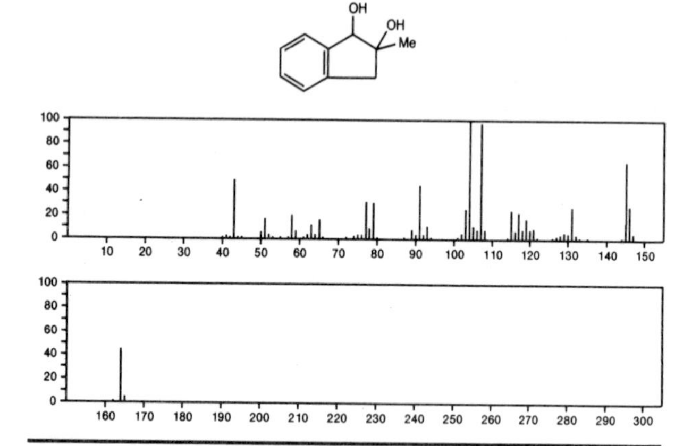

164 $C_{10}H_{16}N_2$ 93-05-0
1,4-Benzenediamine, *N*,*N*-diethyl-

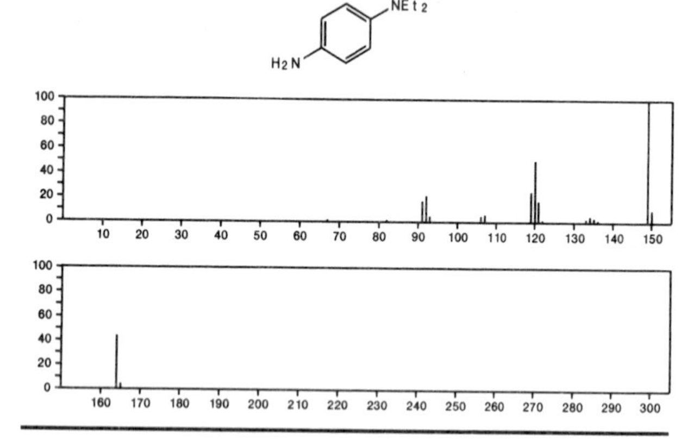

164 $C_{10}H_{16}N_2$ 3010-30-8
1,4-Benzenediamine, *N*,*N*'-diethyl-

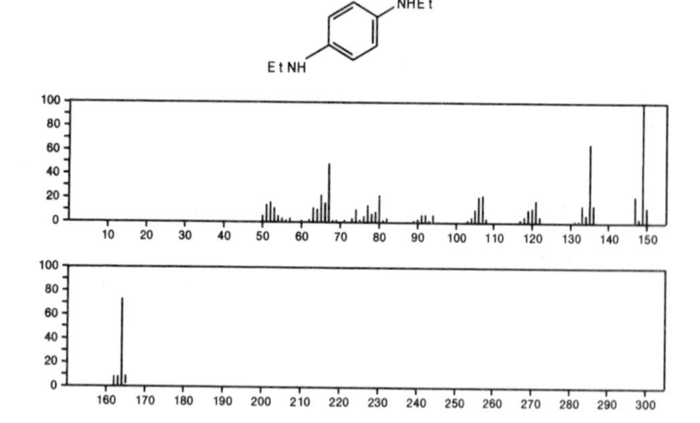

164 $C_{10}H_{16}N_2$ 5857-99-8
m-Phenylenediamine, *N*,*N*'-diethyl-

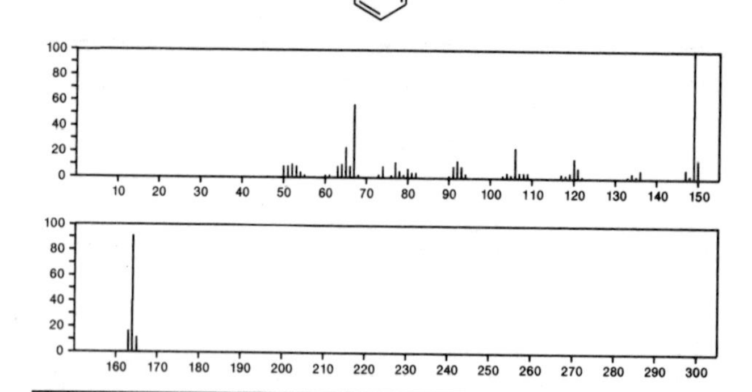

164 $C_{10}H_{16}N_2$ 15834-78-3
Pyrazine, 5-butyl-2,3-dimethyl-

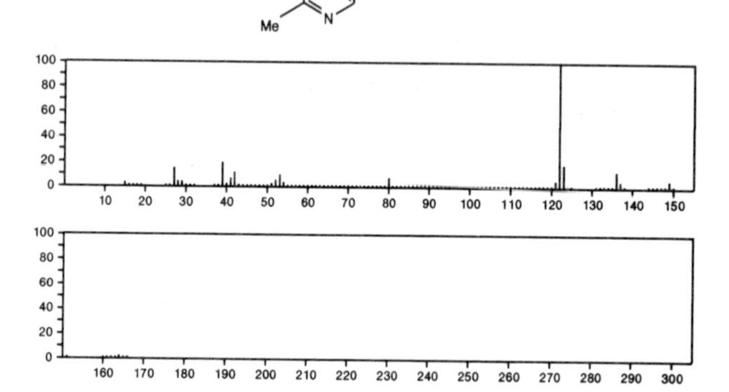

164 $C_{10}H_{16}N_2$ 24340-87-2
o-Phenylenediamine, *N*,*N*'-diethyl-

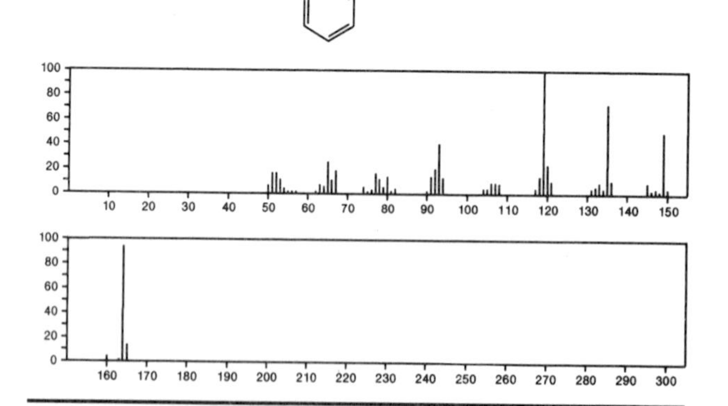

164 $C_{10}H_{16}N_2$ 24340-88-3
p-Phenylenediamine, *N*'-ethyl-*N*,*N*-dimethyl-

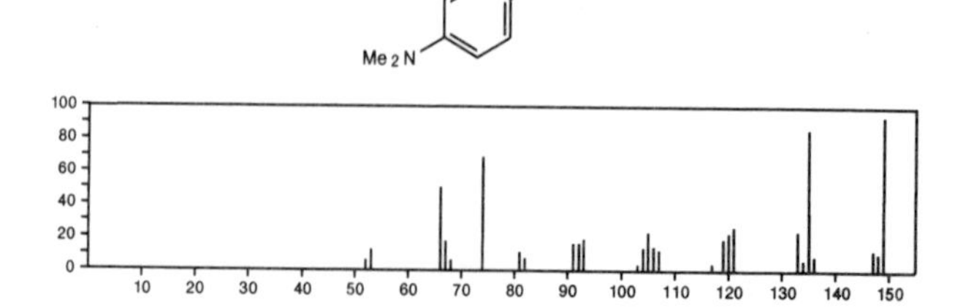

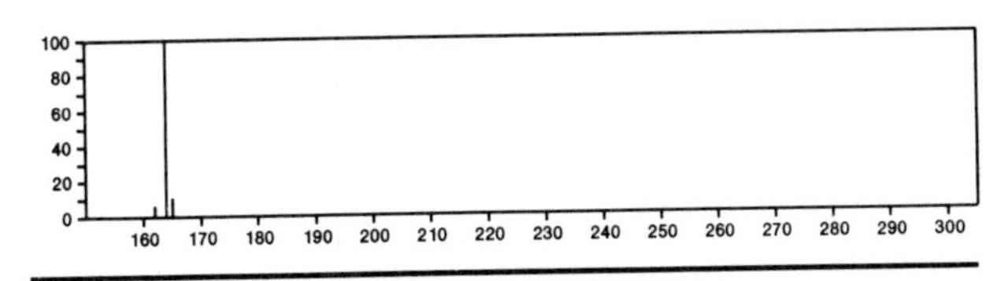

164 $C_{10}H_{16}N_2$ 32263-00-6
Pyrazine, 5-*sec*-butyl-2,3-dimethyl-

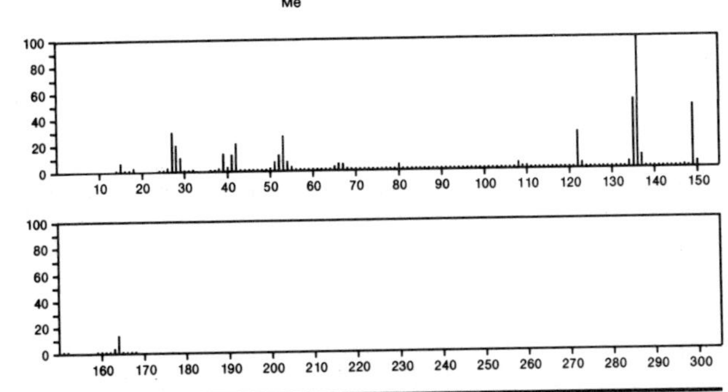

164 $C_{10}H_{16}N_2$ 32736-94-0
Pyrazine, 3-isobutyl-2,5-dimethyl-

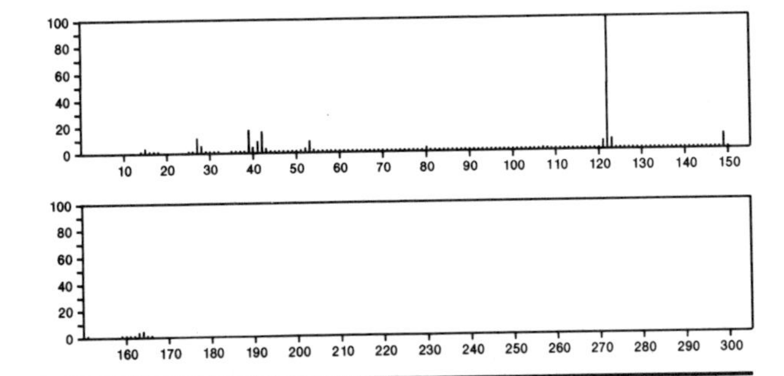

164 $C_{10}H_{16}N_2$ 40790-29-2
Pyrazine, 3-butyl-2,5-dimethyl-

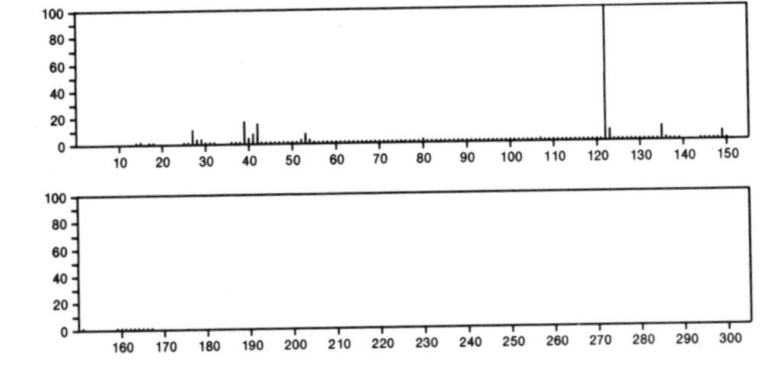

164 $C_{10}H_{16}N_2$ 50888-63-6
Pyrazine, 2-butyl-3,5-dimethyl-

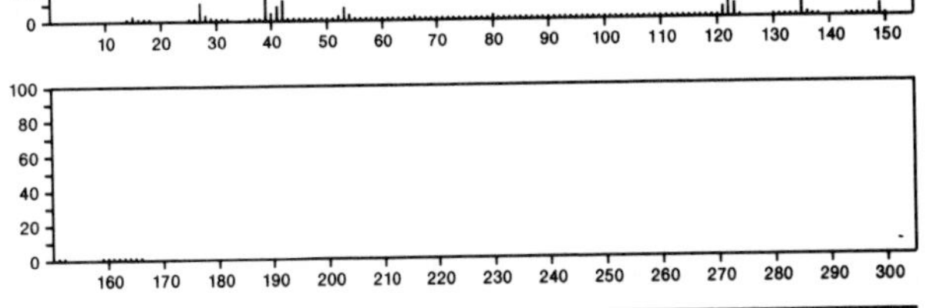

164 $C_{10}H_{16}N_2$ 54410-83-2
Pyrazine, 2,3-dimethyl-5-(2-methylpropyl)-

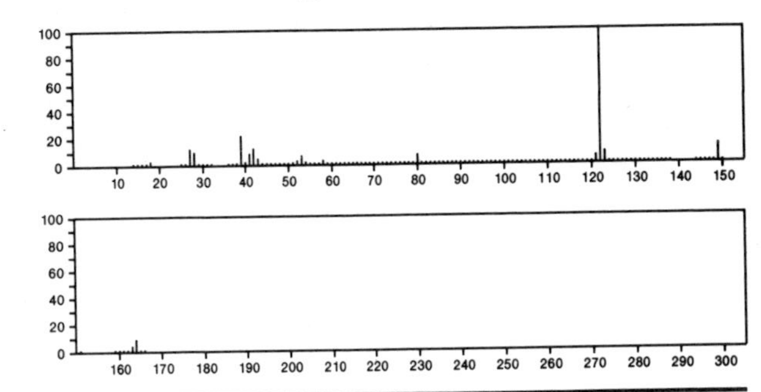

164 $C_{11}H_{16}O$ 80-46-6
Phenol, 4-(1,1-dimethylpropyl)-

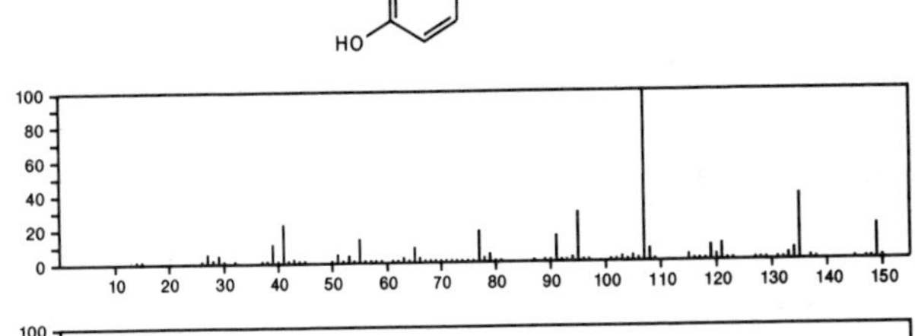

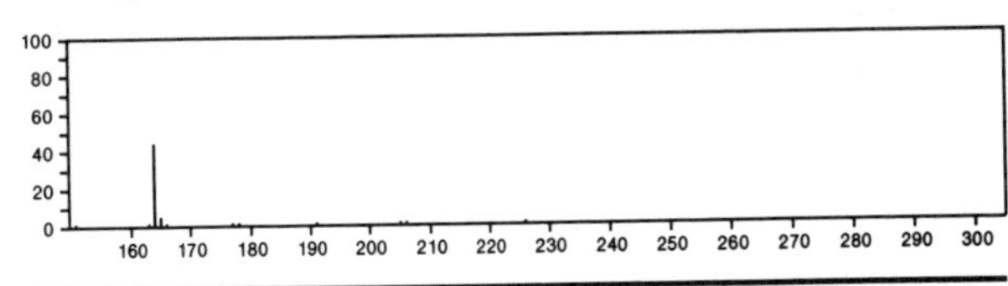

164 $C_{11}H_{16}O$ 98-27-1
Phenol, 4-(1,1-dimethylethyl)-2-methyl-

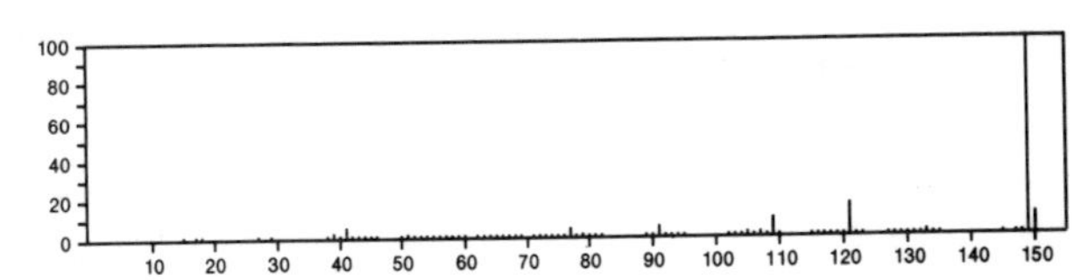

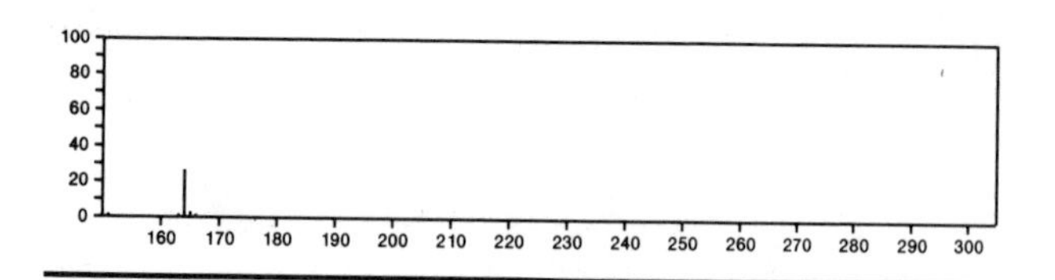

164 $C_{11}H_{16}O$ 136-81-2
Phenol, 2-pentyl-

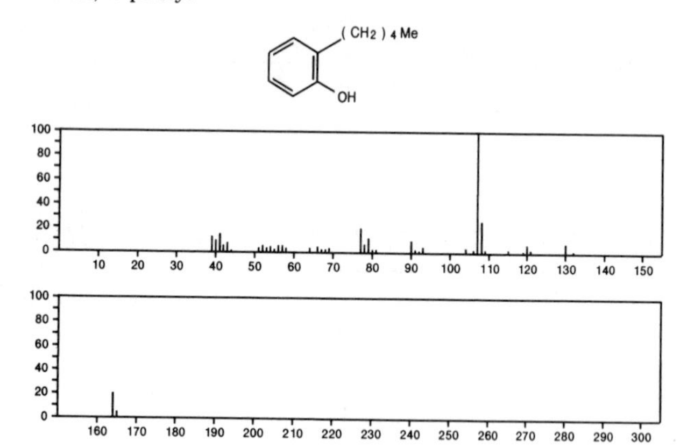

164 $C_{11}H_{16}O$ 488-10-8
2-Cyclopenten-1-one, 3-methyl-2-(2-pentenyl)-, (*Z*)-

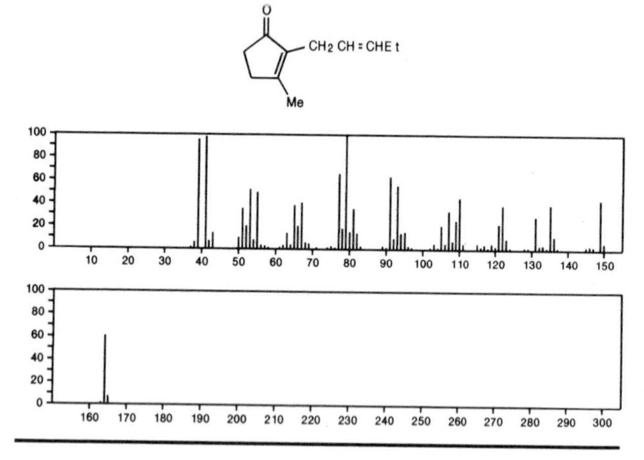

164 $C_{11}H_{16}O$ 588-67-0
Benzene, (butoxymethyl)-

PhCH2O(CH2)3Me

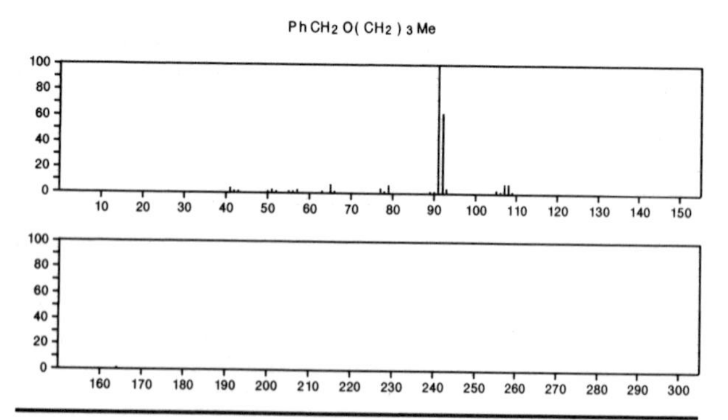

164 $C_{11}H_{16}O$ 2189-88-0
Benzene, (2,2-dimethylpropoxy)-

Me3CCH2OPh

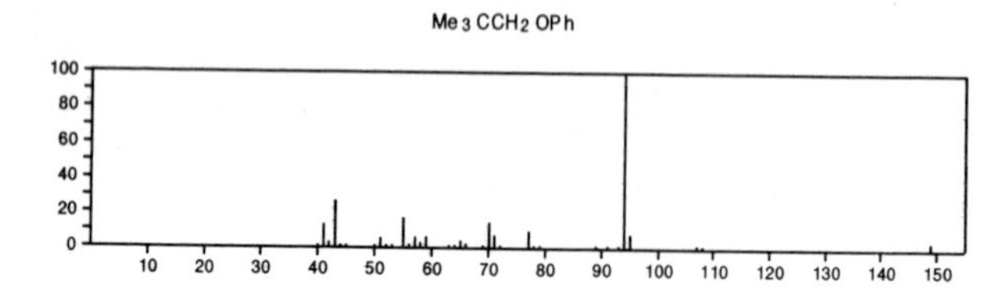

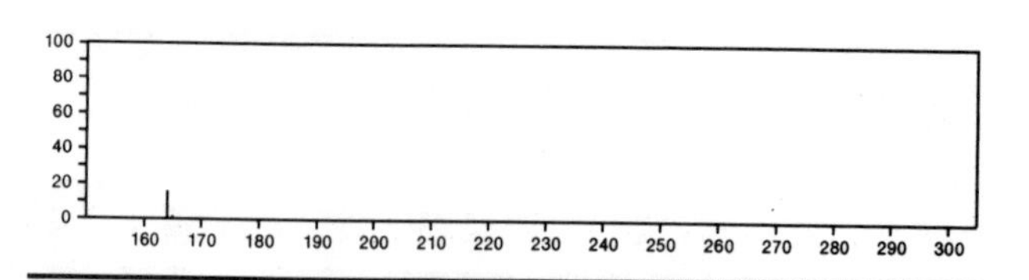

164 $C_{11}H_{16}O$ 2219-82-1
Phenol, 2-(1,1-dimethylethyl)-6-methyl-

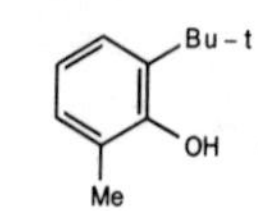

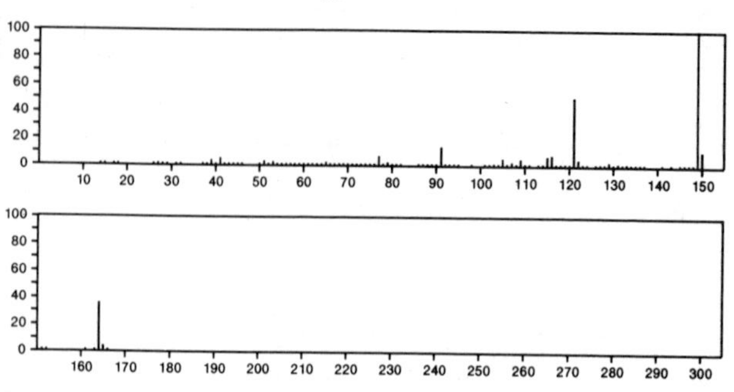

164 $C_{11}H_{16}O$ 2409-55-4
Phenol, 2-(1,1-dimethylethyl)-4-methyl-

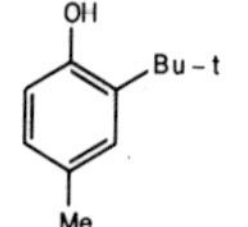

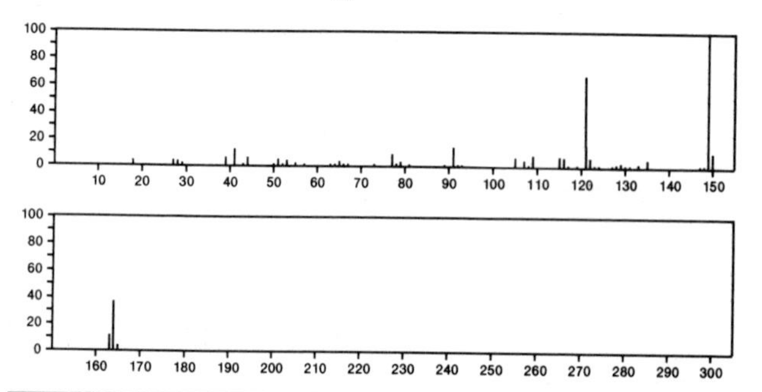

164 $C_{11}H_{16}O$ 3280-08-8
Benzeneethanol, α,α,β-trimethyl-

Me2COHCHMePh

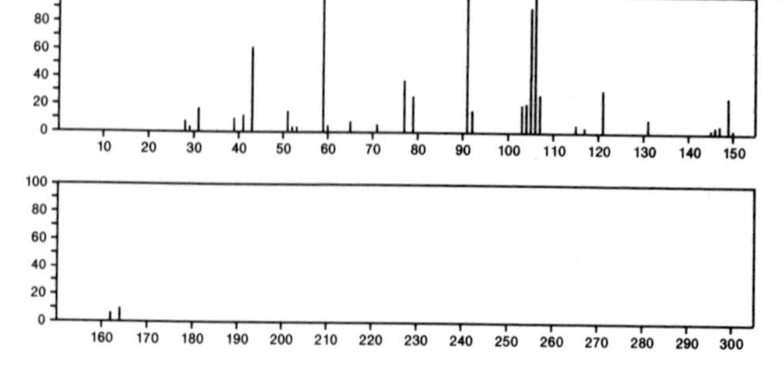

164 $C_{11}H_{16}O$ 3459-80-1
Benzene, [(1,1-dimethylethoxy)methyl]-

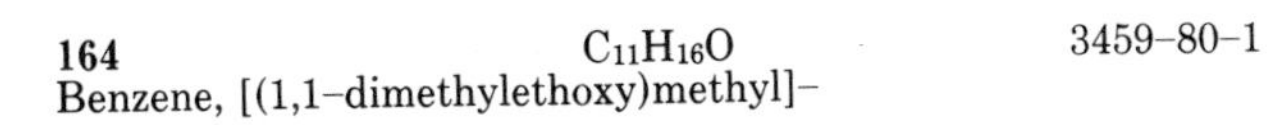

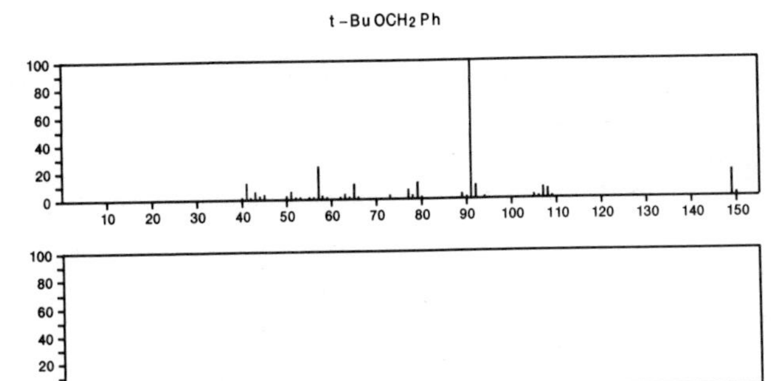

164 $C_{11}H_{16}O$ 4087-39-2
2(3*H*)-Naphthalenone, 4,4a,5,6,7,8-hexahydro-4a-methyl-, (*S*)-

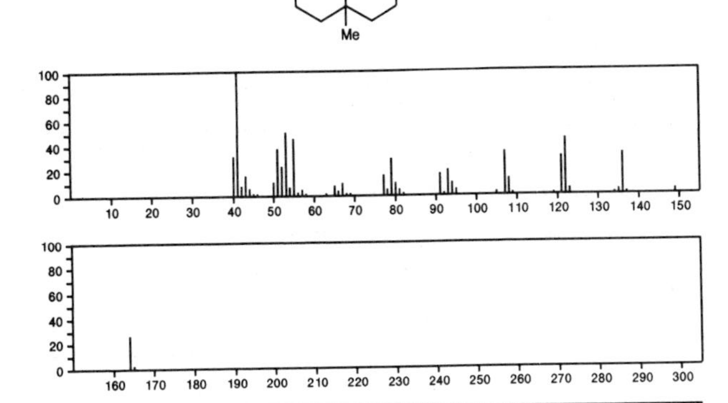

164 $C_{11}H_{16}O$ 5396-38-3
Benzene, 1-(1,1-dimethylethyl)-4-methoxy-

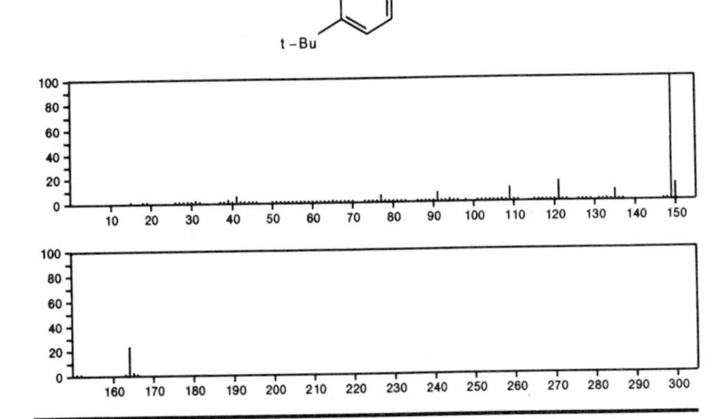

164 $C_{11}H_{16}O$ 13037-79-1
m-Cresol, 2-*tert*-butyl-

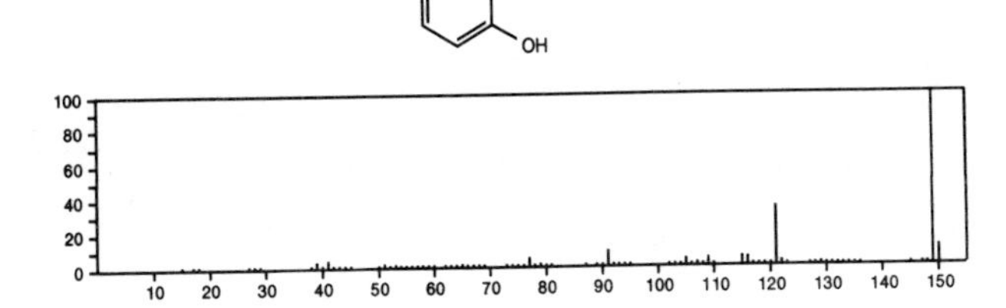

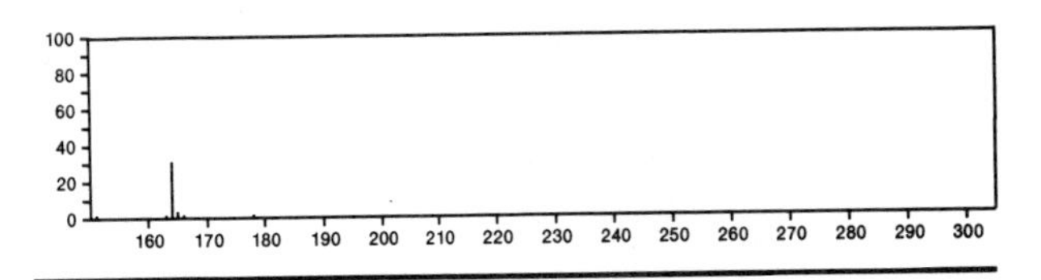

164 $C_{11}H_{16}O$ 14938-35-3
Phenol, 4-pentyl-

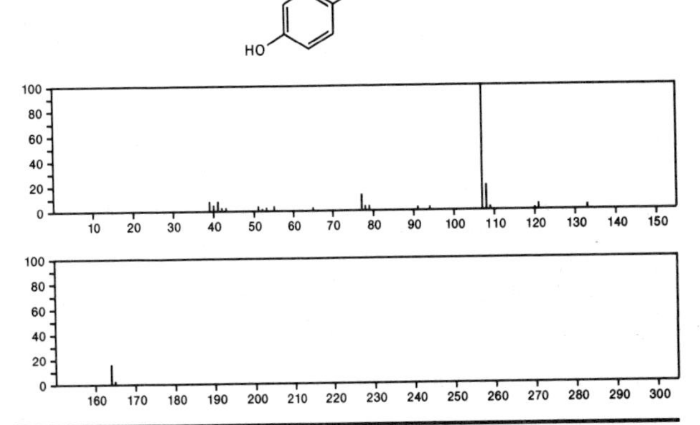

164 $C_{11}H_{16}O$ 18366-35-3
Inden-5(4*H*)-one, 2,6,7,7a-tetrahydro-4,4-dimethyl-

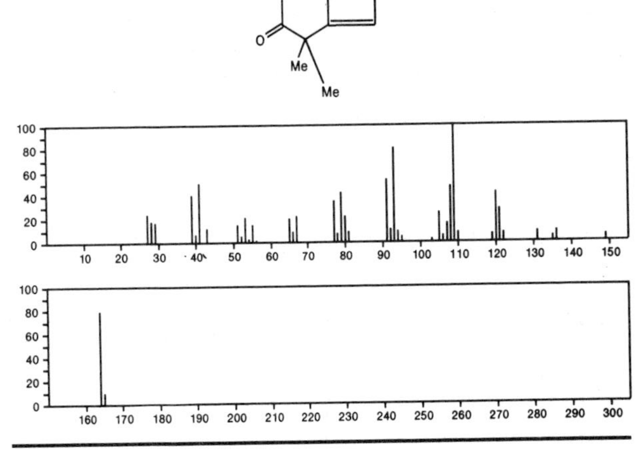

164 $C_{11}H_{16}O$ 20056-66-0
Phenol, 3-pentyl-

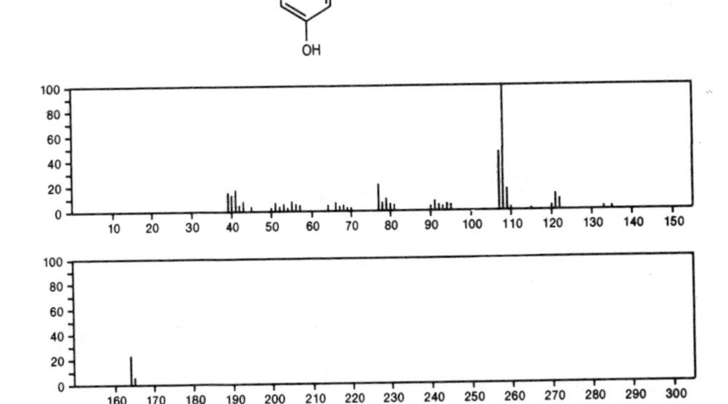

164 $C_{11}H_{16}O$ 20834–59–7
Benzeneethanol, α,α,4–trimethyl–

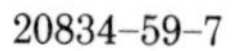

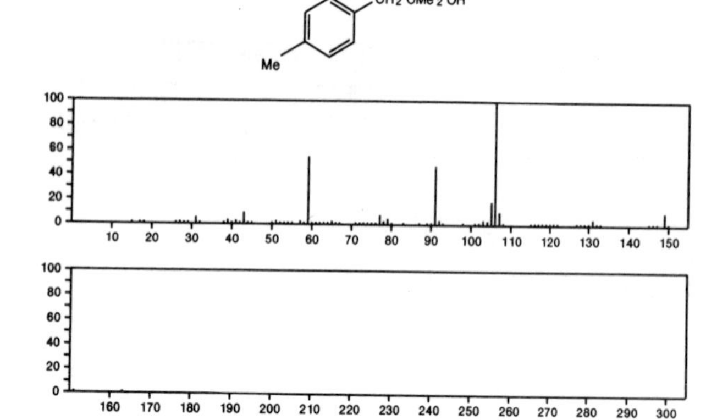

164 $C_{11}H_{16}O$ 21841–29–2
1(2*H*)–Naphthalenone, 3,4,4a,5,8,8a–hexahydro–8a–methyl–, *trans*–

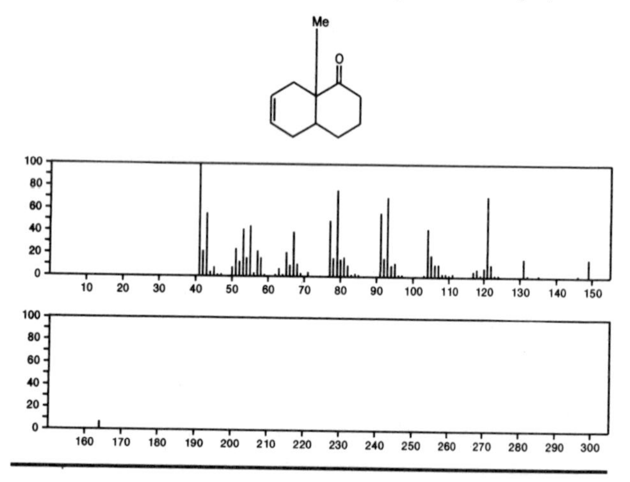

164 $C_{11}H_{16}O$ 22844–34–4
2(1*H*)–Naphthalenone, 4a,5,6,7,8,8a–hexahydro–4a–methyl–, *trans*–

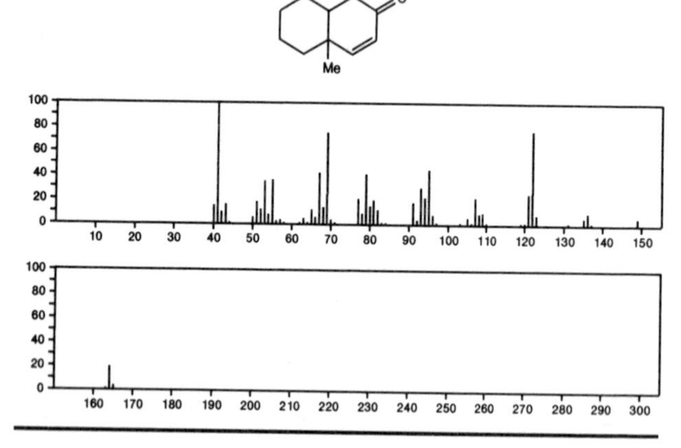

164 $C_{11}H_{16}O$ 55103–73–6
2(1*H*)–Azulenone, 4,5,6,7,8,8a–hexahydro–8a–methyl–, (*S*)–

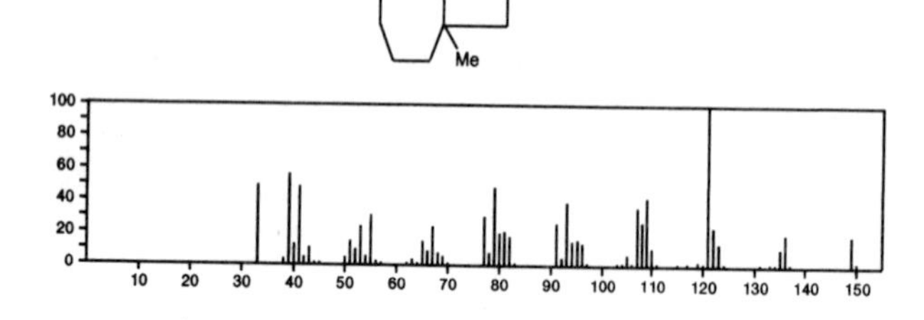

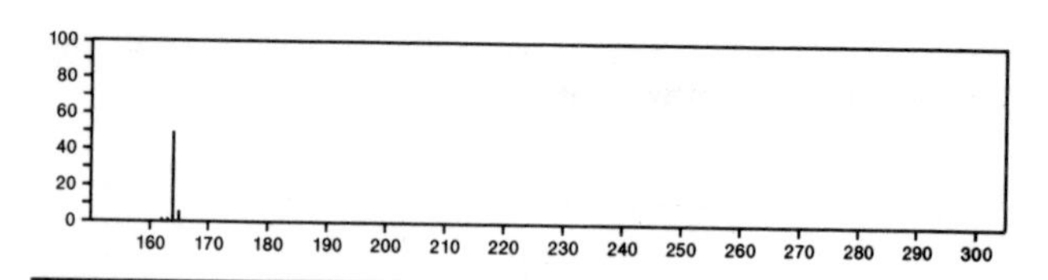

164 $C_{11}H_{16}O$ 55283–48–2
2(1*H*)–Naphthalenone, 3,4,4a,5,8,8a–hexahydro–4a–methyl–, *trans*–

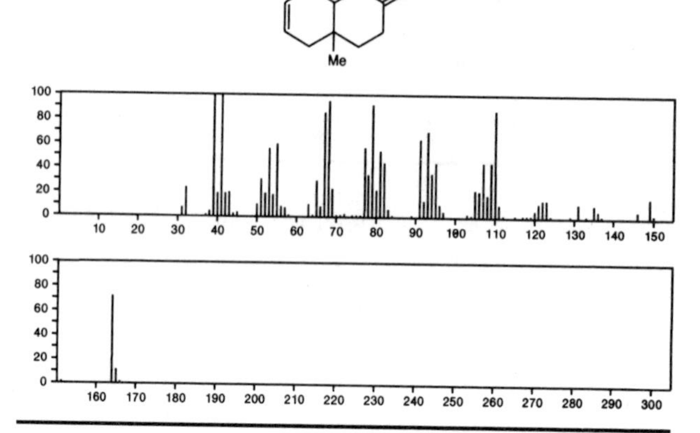

164 $C_{11}H_{16}O$ 56630–96–7
Bicyclo[3.2.1]octan–2–one, 1–(1–propenyl)–

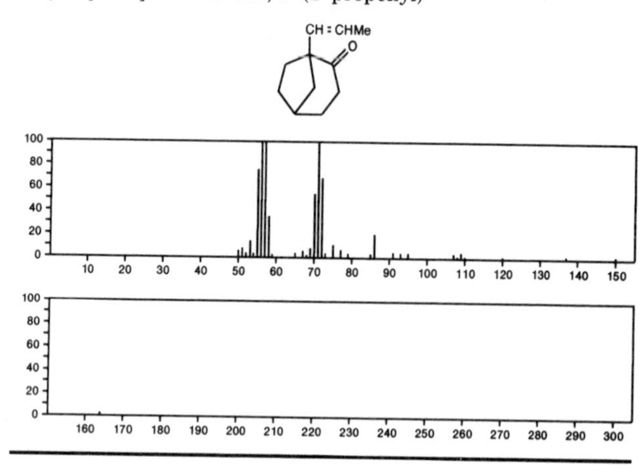

164 $C_{12}H_{20}$ 702–79–4
Tricyclo[3.3.1.1^{3,7}]decane, 1,3–dimethyl–

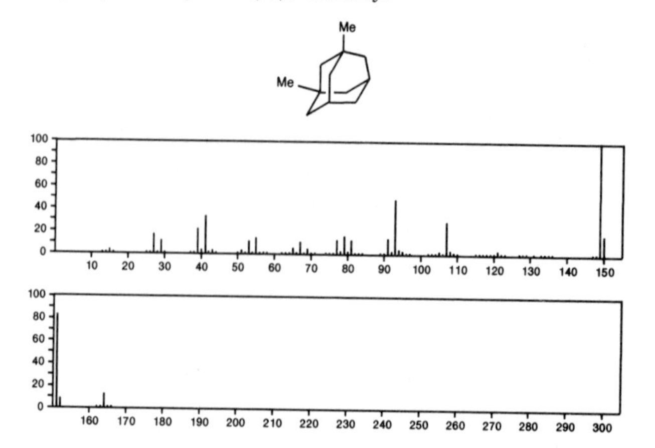

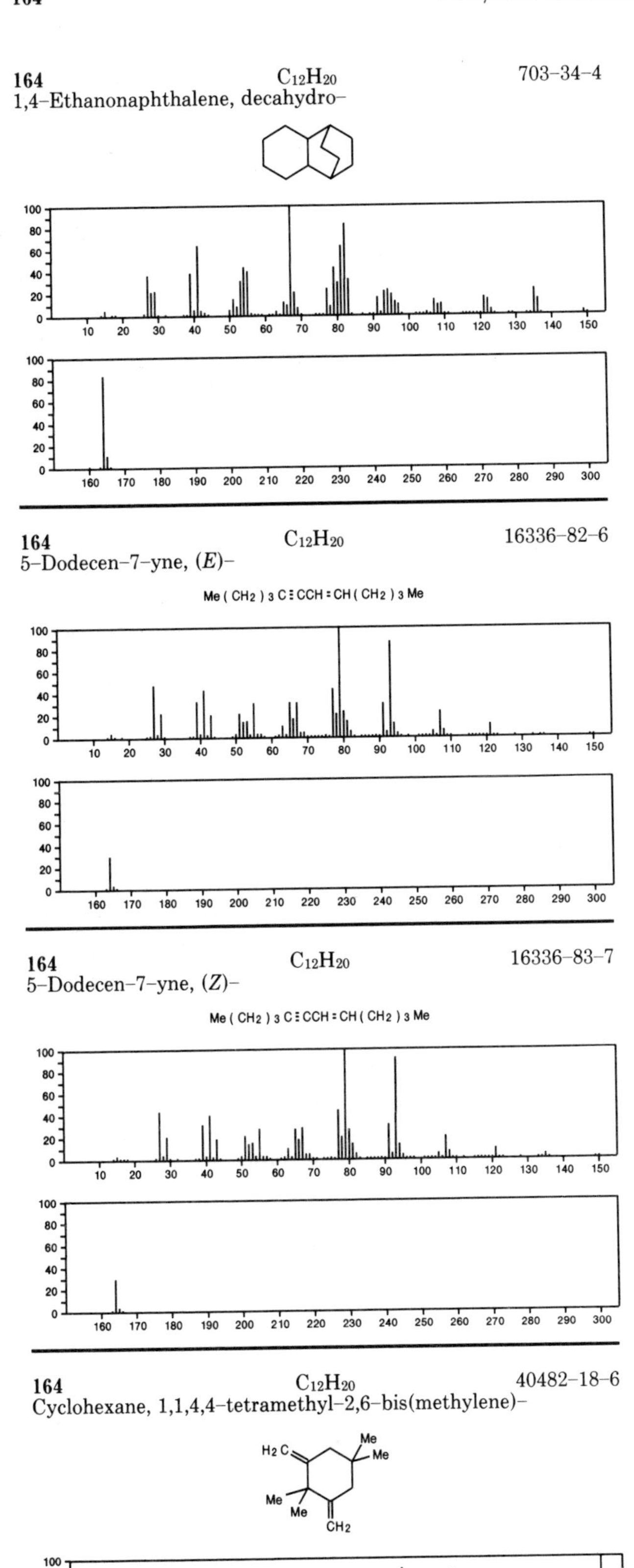

164 $C_{12}H_{20}$ 703-34-4
1,4-Ethanonaphthalene, decahydro-

164 $C_{12}H_{20}$ 16336-82-6
5-Dodecen-7-yne, (*E*)-

Me(CH2)3C≡CCH=CH(CH2)3Me

164 $C_{12}H_{20}$ 16336-83-7
5-Dodecen-7-yne, (*Z*)-

Me(CH2)3C≡CCH=CH(CH2)3Me

164 $C_{12}H_{20}$ 40482-18-6
Cyclohexane, 1,1,4,4-tetramethyl-2,6-bis(methylene)-

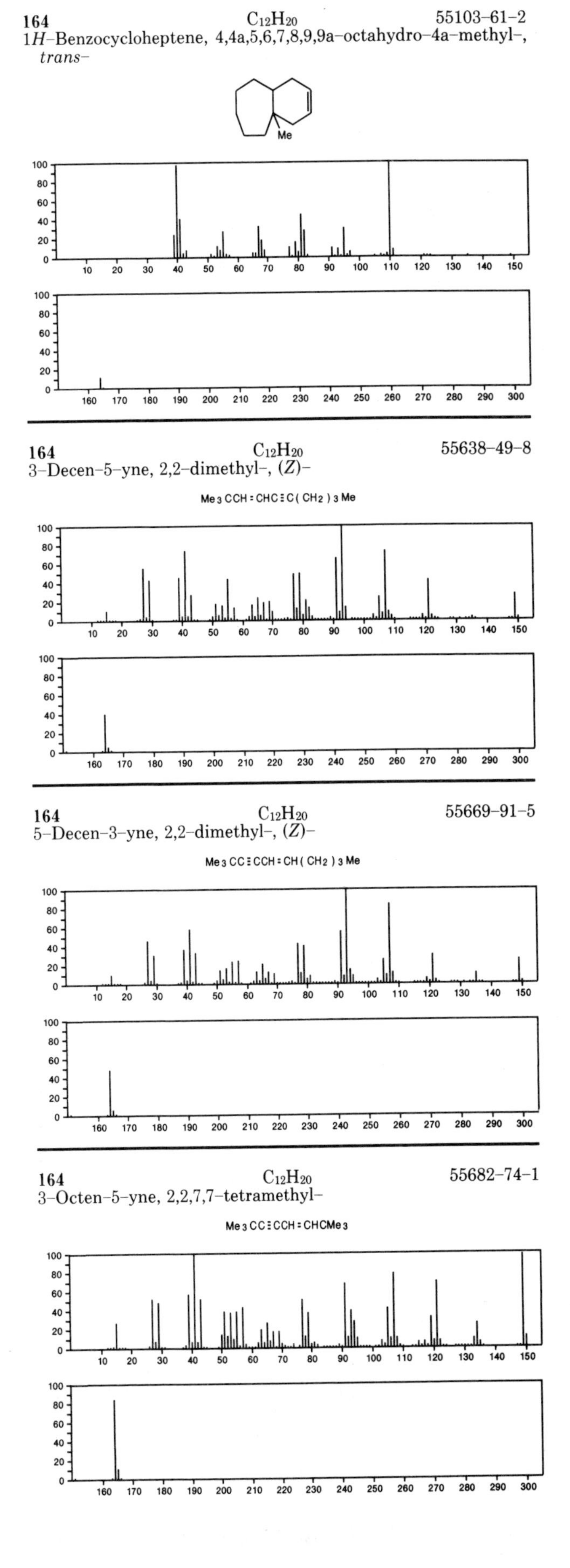

164 $C_{12}H_{20}$ 55103-61-2
1*H*-Benzocycloheptene, 4,4a,5,6,7,8,9,9a-octahydro-4a-methyl-, *trans*-

164 $C_{12}H_{20}$ 55638-49-8
3-Decen-5-yne, 2,2-dimethyl-, (*Z*)-

Me3CCH=CHC≡C(CH2)3Me

164 $C_{12}H_{20}$ 55669-91-5
5-Decen-3-yne, 2,2-dimethyl-, (*Z*)-

Me3CC≡CCH=CH(CH2)3Me

164 $C_{12}H_{20}$ 55682-74-1
3-Octen-5-yne, 2,2,7,7-tetramethyl-

Me3CC≡CCH=CHCMe3

164 $C_{12}H_{20}$ 55976–09–5
Naphthalene, 1,2,3,4,4a,5,6,8a–octahydro–4a,8–dimethyl–

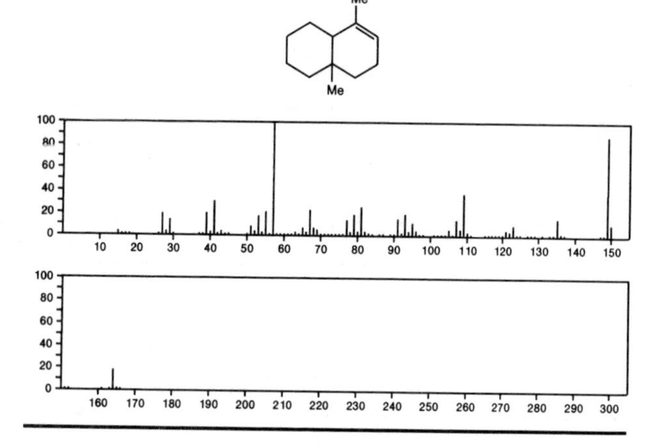

164 $C_{12}H_{20}$ 55976–10–8
5–Decene, 4–ethynyl–, (*E*)–

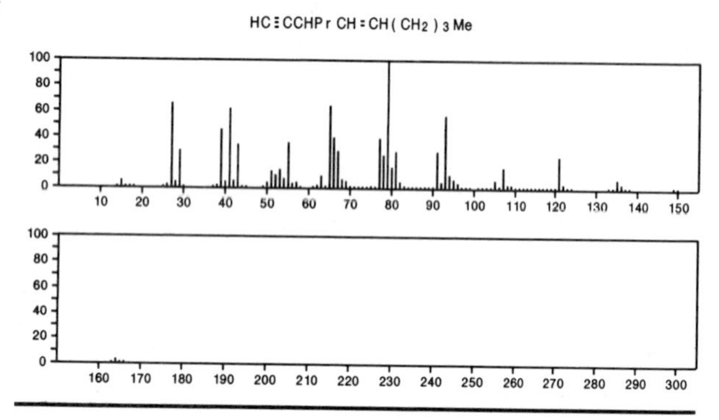

164 $C_{12}H_{20}$ 55976–11–9
Cyclohexene, 3–(1–hexenyl)–, (*E*)–

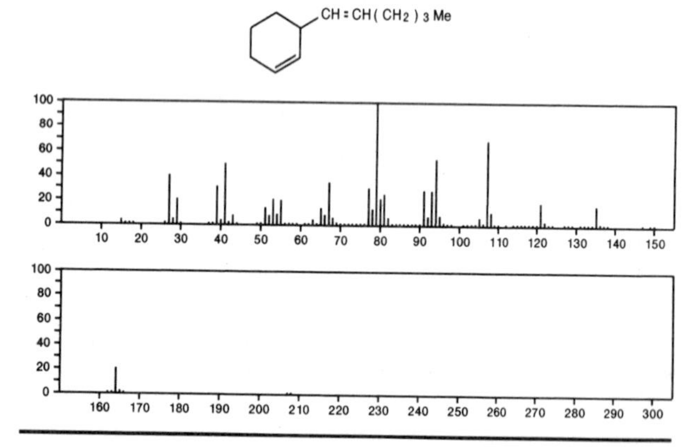

164 $C_{12}H_{20}$ 55976–12–0
3–Octen–5–yne, 2,2,7,7–tetramethyl–, (*E*)–

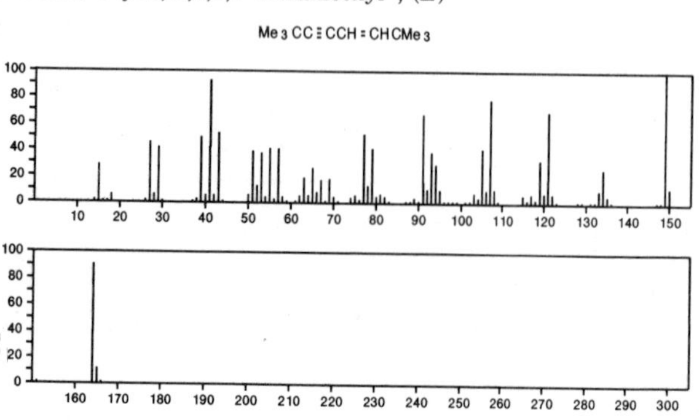

164 $C_{12}H_{20}$ 56324–68–6
1*H*–Indene, 1–ethylideneoctahydro–7a–methyl–, (1*E*,3aα,7aβ)–

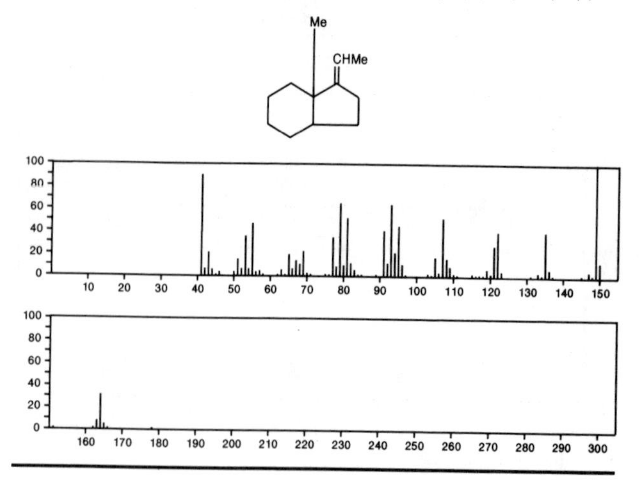

164 $C_{12}H_{20}$ 56324–69–7
1*H*–Indene, 1–ethylideneoctahydro–7a–methyl–, (1*Z*,3aα,7aβ)–

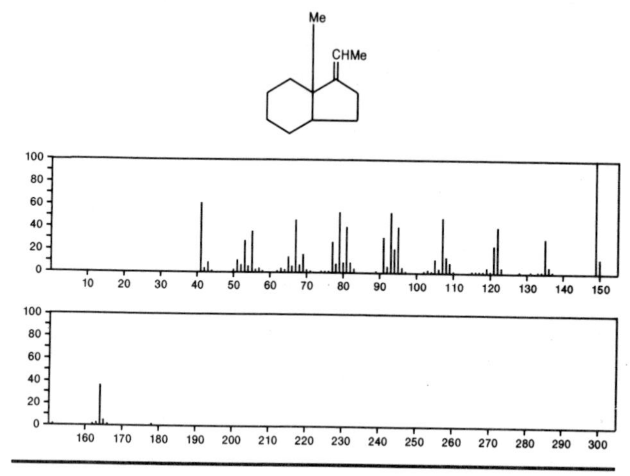

164 $C_{12}H_{20}$ 56362–87–9
1*H*–Indene, 1–ethylideneoctahydro–7a–methyl–, *cis*–

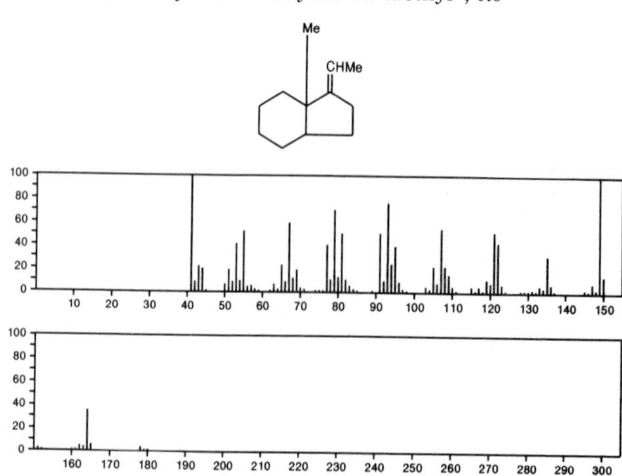

164 $C_{12}H_{20}$ 56666-90-1

Bicyclo[6.1.0]nonane, 9-(1-methylethylidene)-

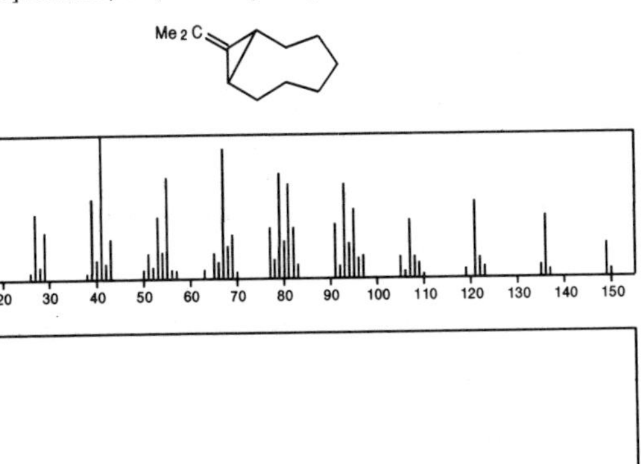

165 $C_2H_3N_3O_6$ 595-86-8

Ethane, 1,1,1-trinitro-

MeC(NO2)3

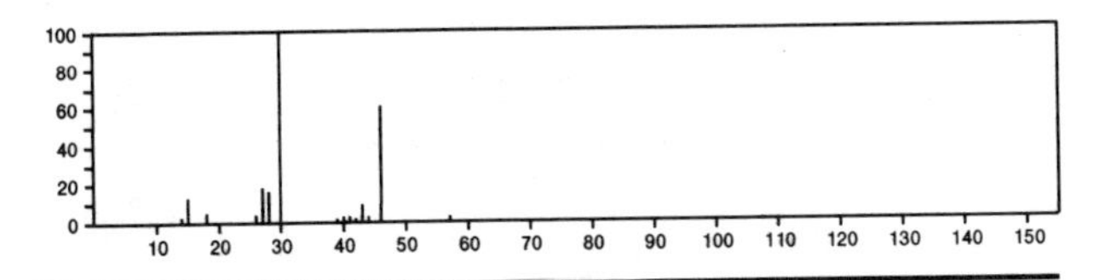

165 $C_5H_{12}NO_3P$ 29727-88-6

1,3,2-Dioxaphospholan-2-amine, 5,5-dimethyl-, 2-oxide

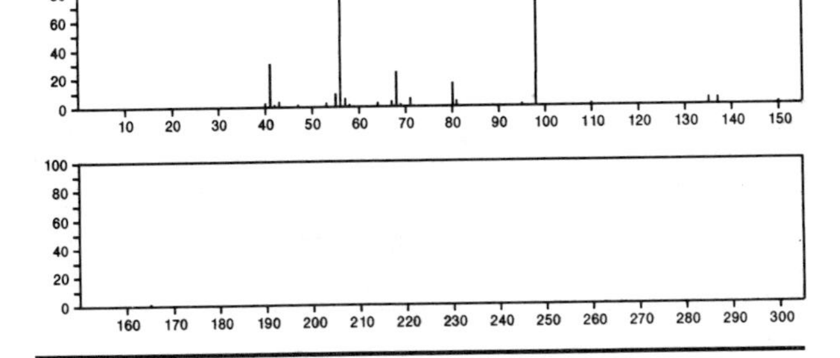

165 $C_6H_3F_4N$ 5580-80-3

Benzenamine, 2,3,4,5-tetrafluoro-

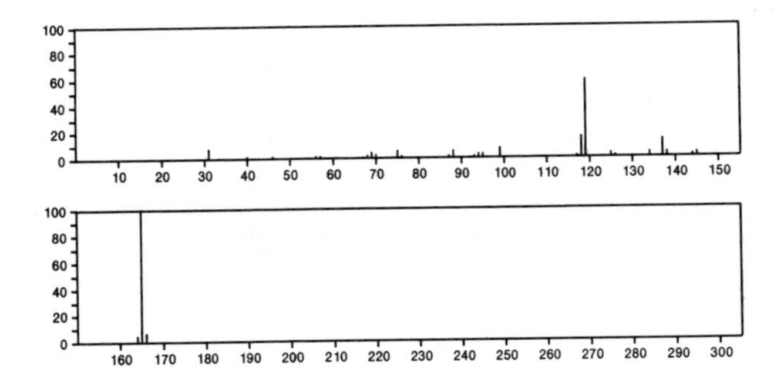

165 $C_6H_7N_5O$ 578-76-7

6*H*-Purin-6-one, 2-amino-1,7-dihydro-7-methyl-

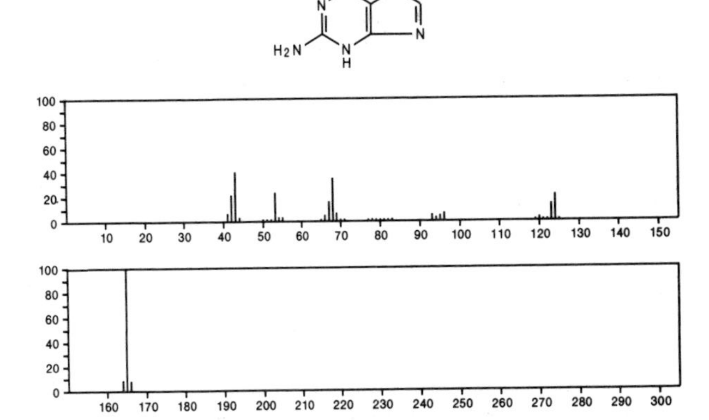

165 $C_6H_7N_5O$ 938-85-2

6*H*-Purin-6-one, 2-amino-1,7-dihydro-1-methyl-

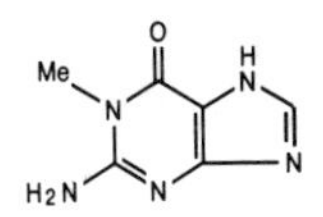

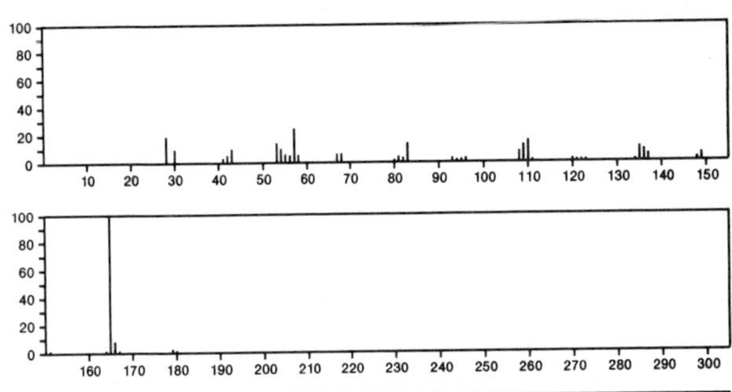

165 $C_6H_7N_5O$ 10030-78-1

6*H*-Purin-6-one, 1,7-dihydro-2-(methylamino)-

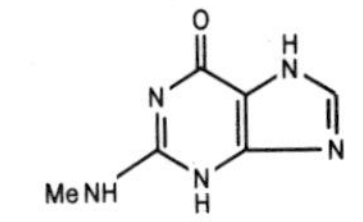

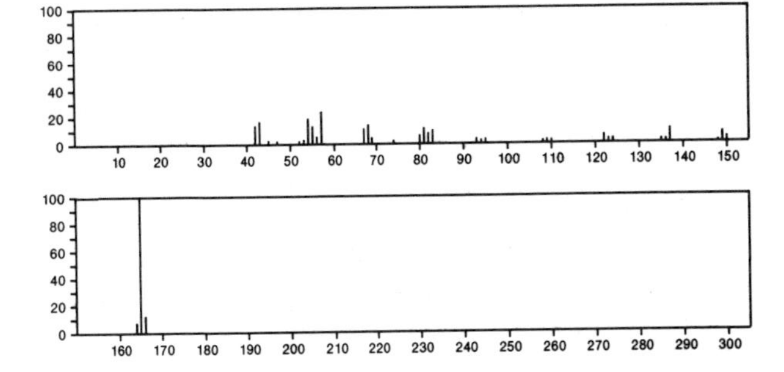

165 $C_6H_7N_5O$ 20535-83-5
1*H*-Purin-2-amine, 6-methoxy-

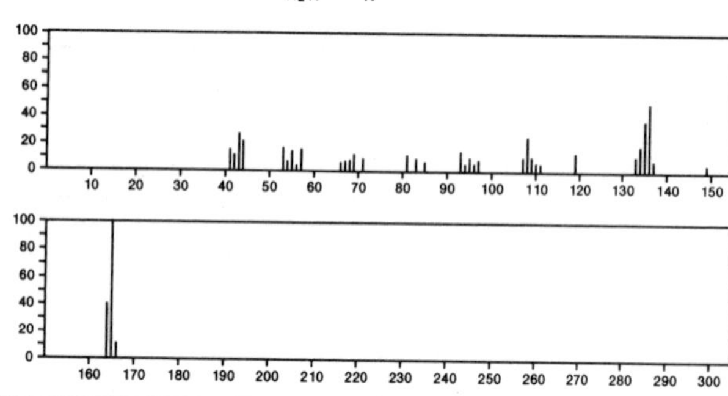

165 $C_7H_4ClN_3$ 28732-79-8
Pyrido[2,3-*d*]pyrimidine, 4-chloro-

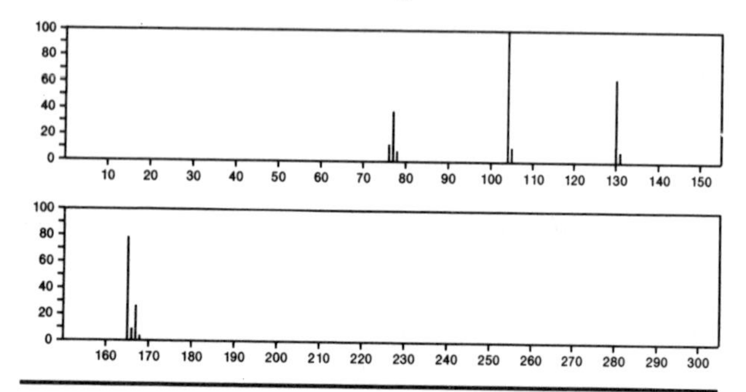

165 $C_7H_7N_3O_2$ 6268-32-2
Urea, *N*-nitroso-*N*-phenyl-

PhN(NO)CONH2

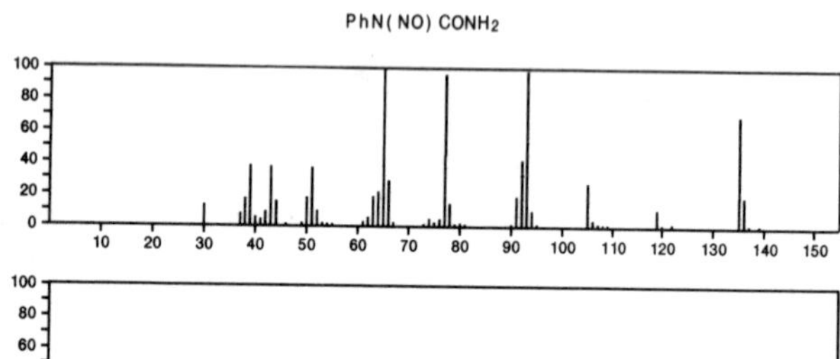

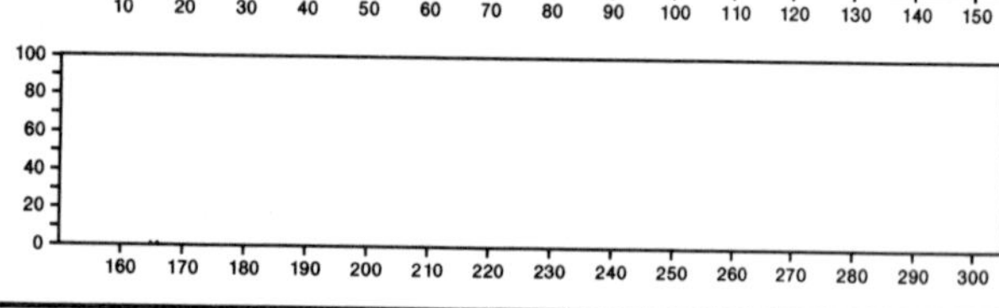

165 $C_7H_7N_3S$ 4926-22-1
s-Triazolo[4,3-*a*]pyridine-3-thiol, 5-methyl-

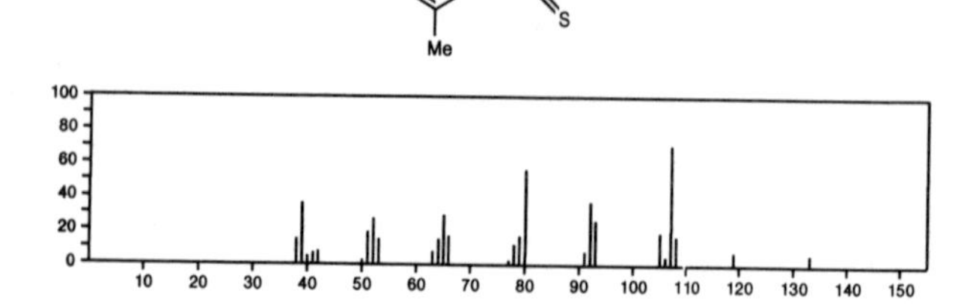

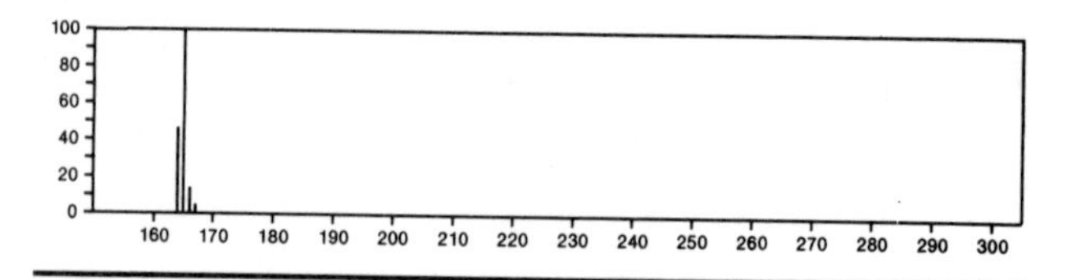

165 $C_7H_7N_3S$ 4926-23-2
s-Triazolo[4,3-*a*]pyridine-3-thiol, 7-methyl-

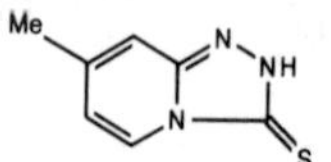

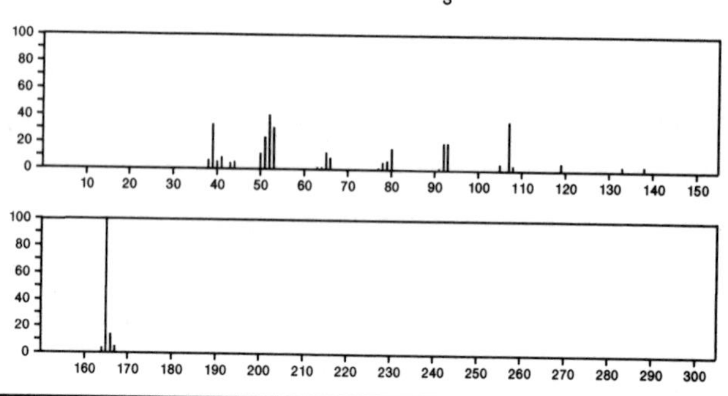

165 C_8H_7NOS 5325-20-2
2*H*-1,4-Benzothiazin-3(4*H*)-one

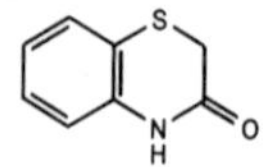

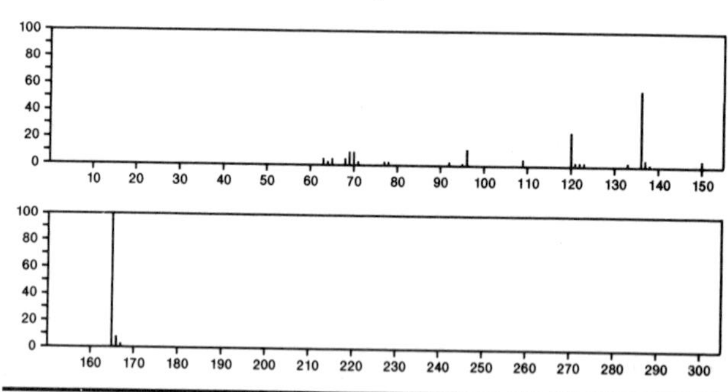

165 C_8H_7NOS 30276-99-4
Thiazolo[3,2-*a*]pyridinium, 8-hydroxy-3-methyl-, hydroxide, inner salt

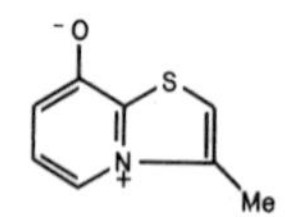

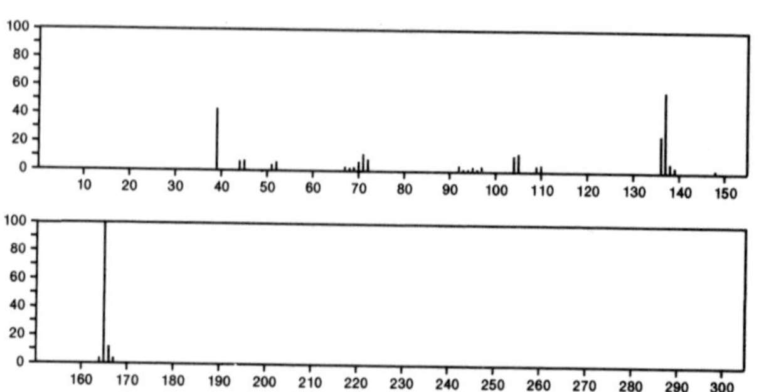

165 C_8H_7NOS 30277-17-9
Thiazolo[3,2-*a*]pyridinium, 8-hydroxy-5-methyl-, hydroxide, inner salt

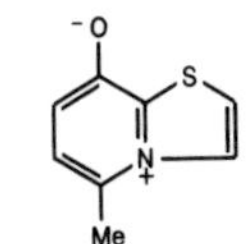

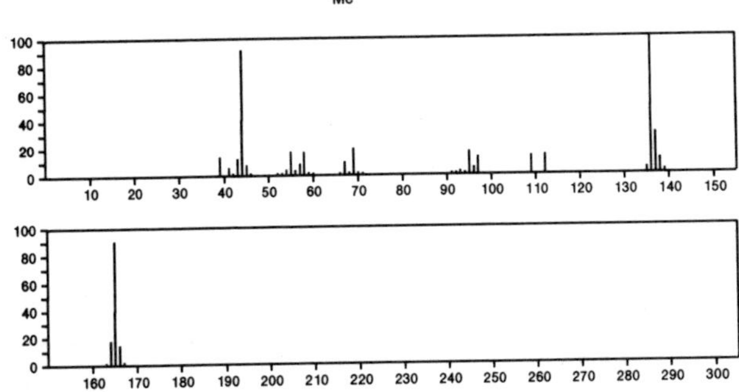

165 C_8H_7NOS 40991-38-6
1,2-Benzisothiazole, 3-methoxy-

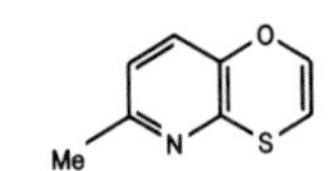

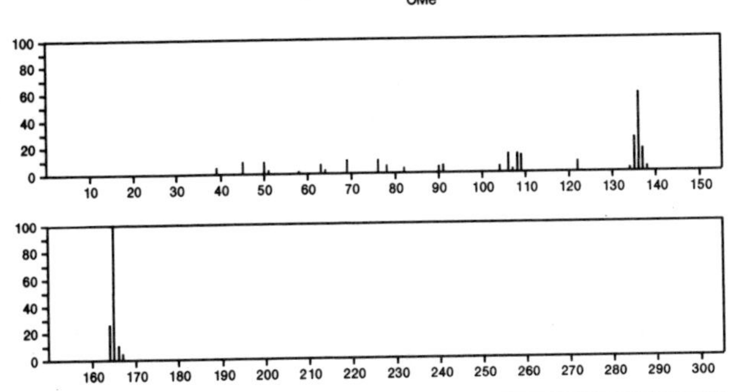

165 C_8H_7NOS 56114-39-7
1,4-Oxathiino[3,2-*b*]pyridine, 6-methyl-

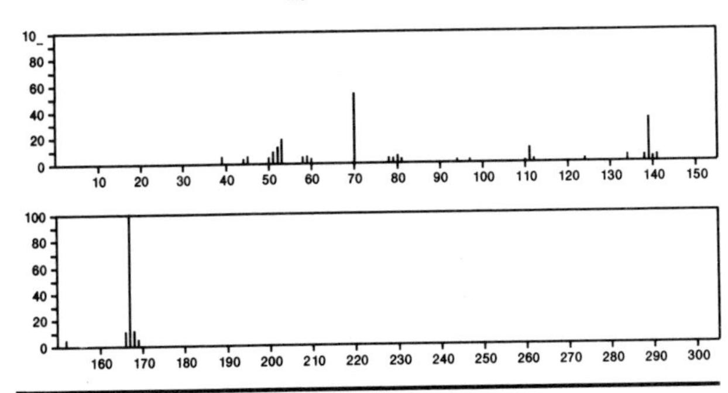

165 $C_8H_7NO_3$ 100-19-6
Ethanone, 1-(4-nitrophenyl)-

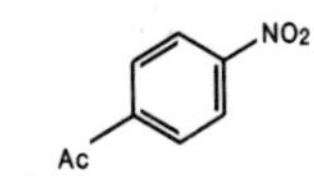

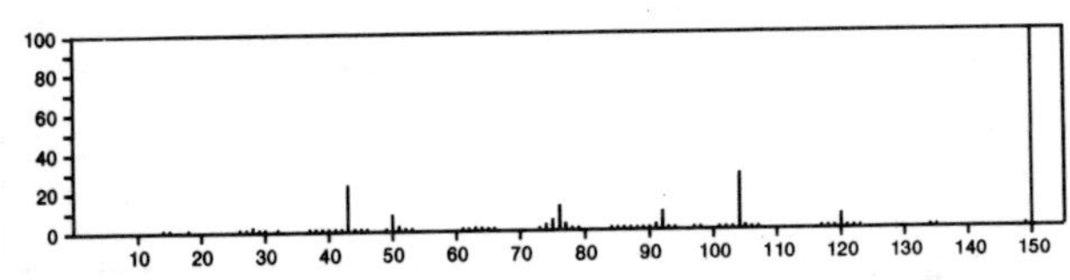

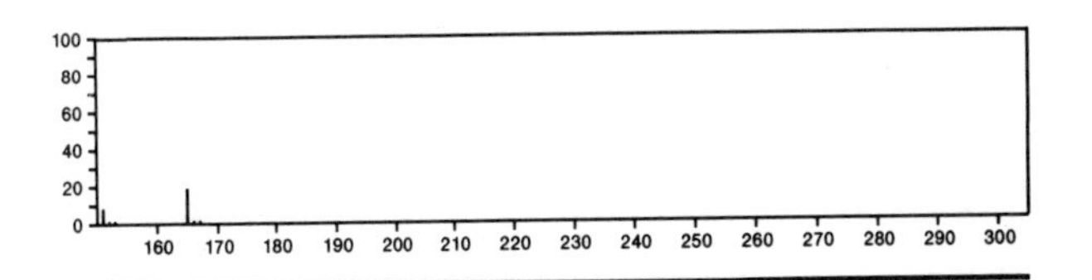

165 $C_8H_7NO_3$ 771-26-6
2*H*-1,4-Benzoxazin-3(4*H*)-one, 4-hydroxy-

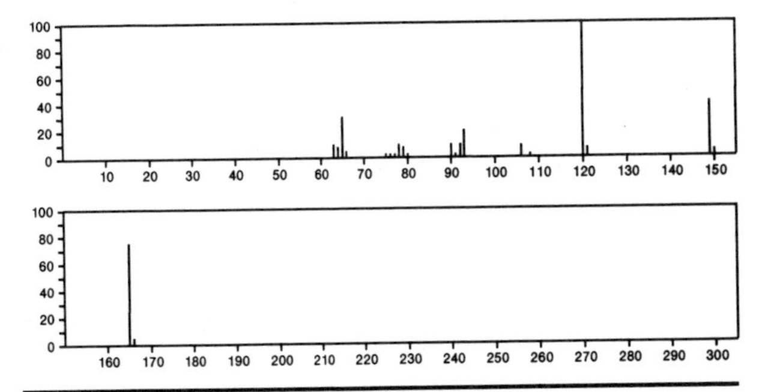

165 $C_8H_7NO_3$ 6383-59-1
Phthalaldehydic acid, oxime

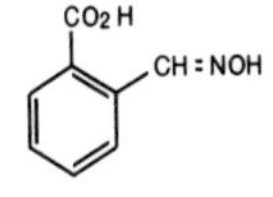

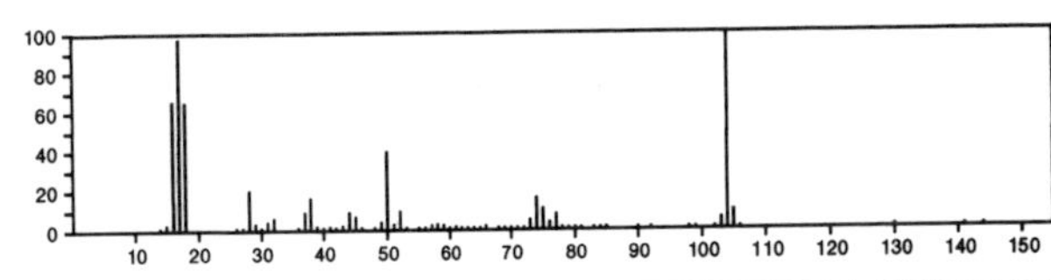

165 $C_8H_7NO_3$ 20357-22-6
Benzaldehyde, 4-methyl-2-nitro-

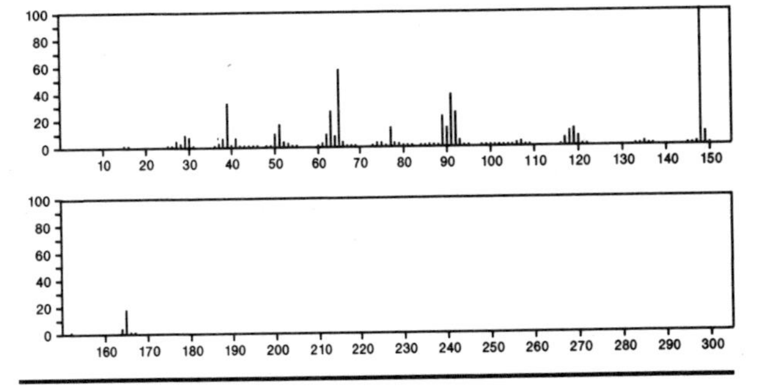

165 $C_8H_{11}N_3O$ 31382-90-8
Pyridinium, 3,5-dimethyl-1-ureido-, hydroxide, inner salt

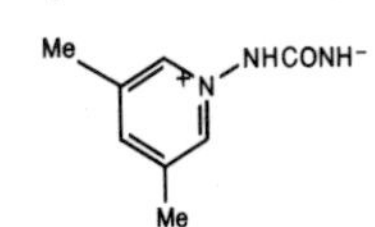

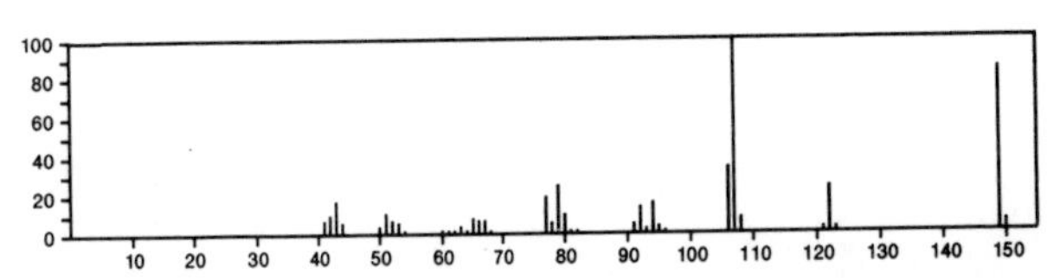

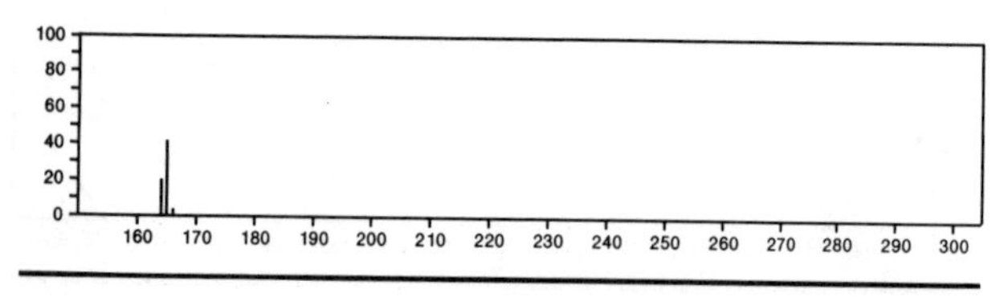

165 $C_9H_{11}NO_2$ 51-66-1
Acetamide, *N*-(4-methoxyphenyl)-

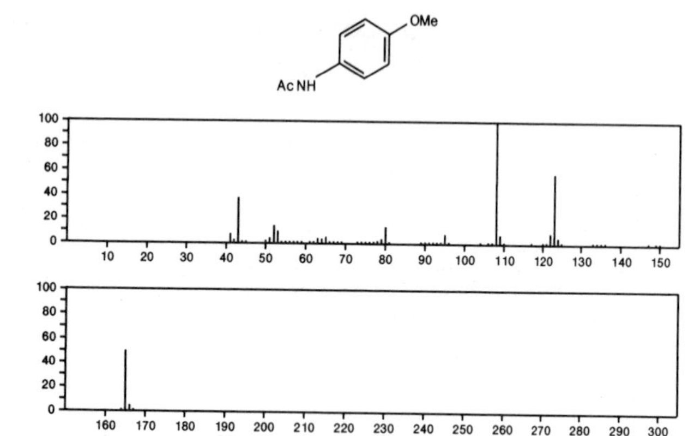

165 $C_9H_{11}NO_2$ 63-91-2
L-Phenylalanine

$HO_2CCH(NH_2)CH_2Ph$

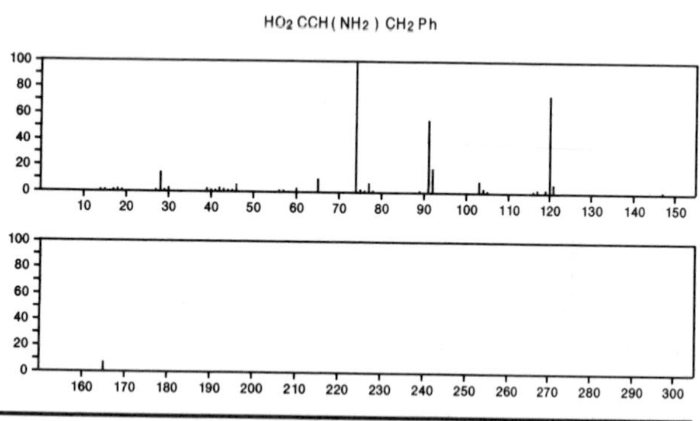

165 $C_9H_{11}NO_2$ 87-25-2
Benzoic acid, 2-amino-, ethyl ester

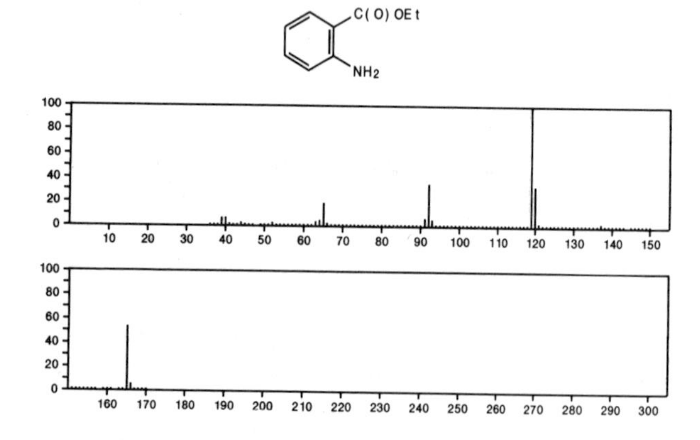

165 $C_9H_{11}NO_2$ 93-26-5
Acetamide, *N*-(2-methoxyphenyl)-

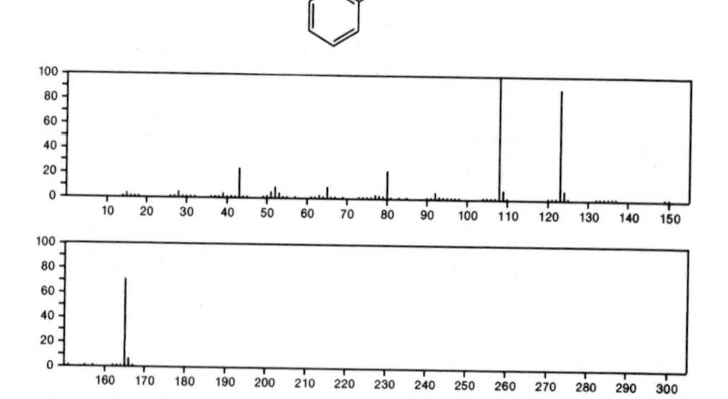

165 $C_9H_{11}NO_2$ 94-09-7
Benzoic acid, 4-amino-, ethyl ester

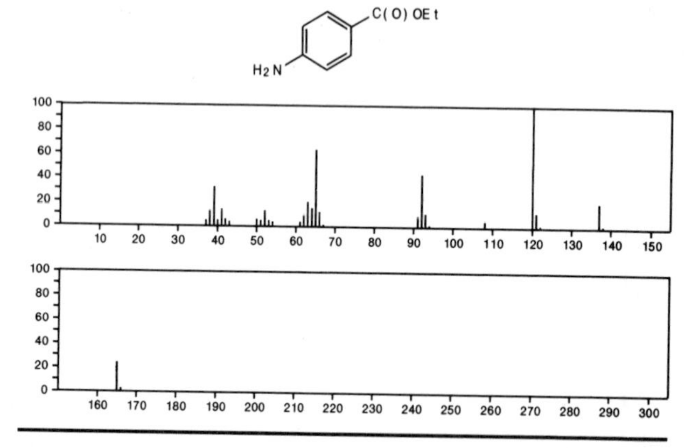

165 $C_9H_{11}NO_2$ 99-64-9
Benzoic acid, 3-(dimethylamino)-

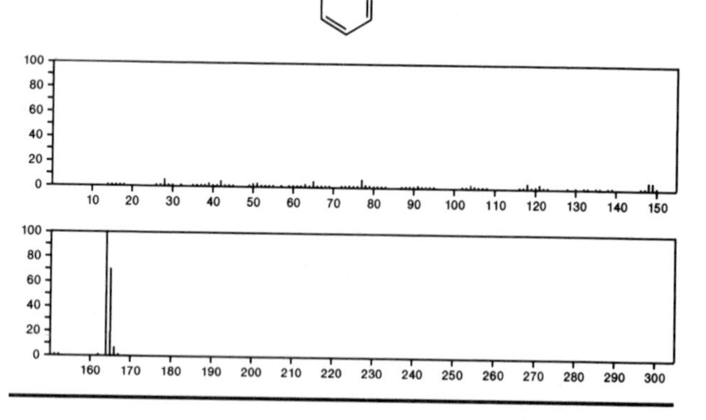

165 $C_9H_{11}NO_2$ 101-99-5
Carbamic acid, phenyl-, ethyl ester

PhNHC(O)OEt

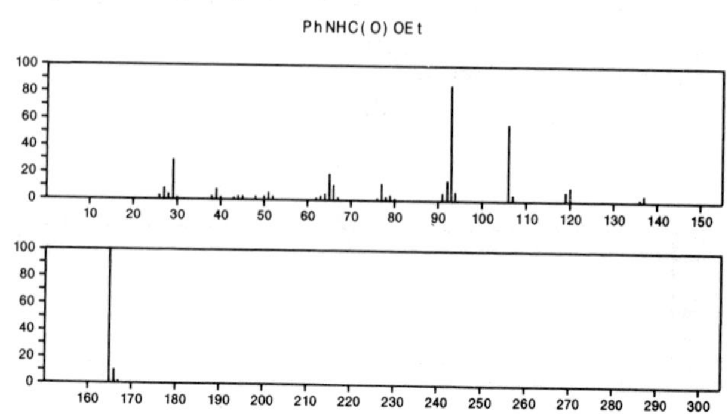

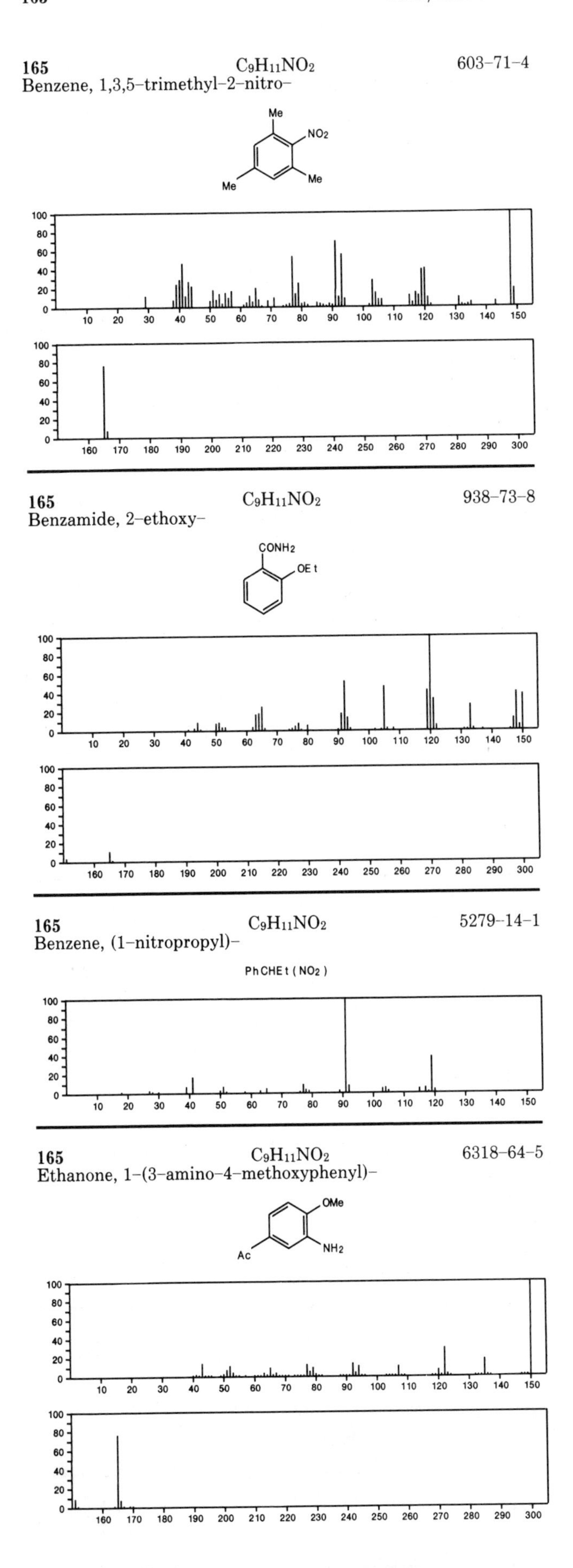

165 $C_9H_{11}NO_2$ 603–71–4
Benzene, 1,3,5–trimethyl–2–nitro–

165 $C_9H_{11}NO_2$ 938–73–8
Benzamide, 2–ethoxy–

165 $C_9H_{11}NO_2$ 5279–14–1
Benzene, (1–nitropropyl)–

PhCHEt(NO2)

165 $C_9H_{11}NO_2$ 6318–64–5
Ethanone, 1–(3–amino–4–methoxyphenyl)–

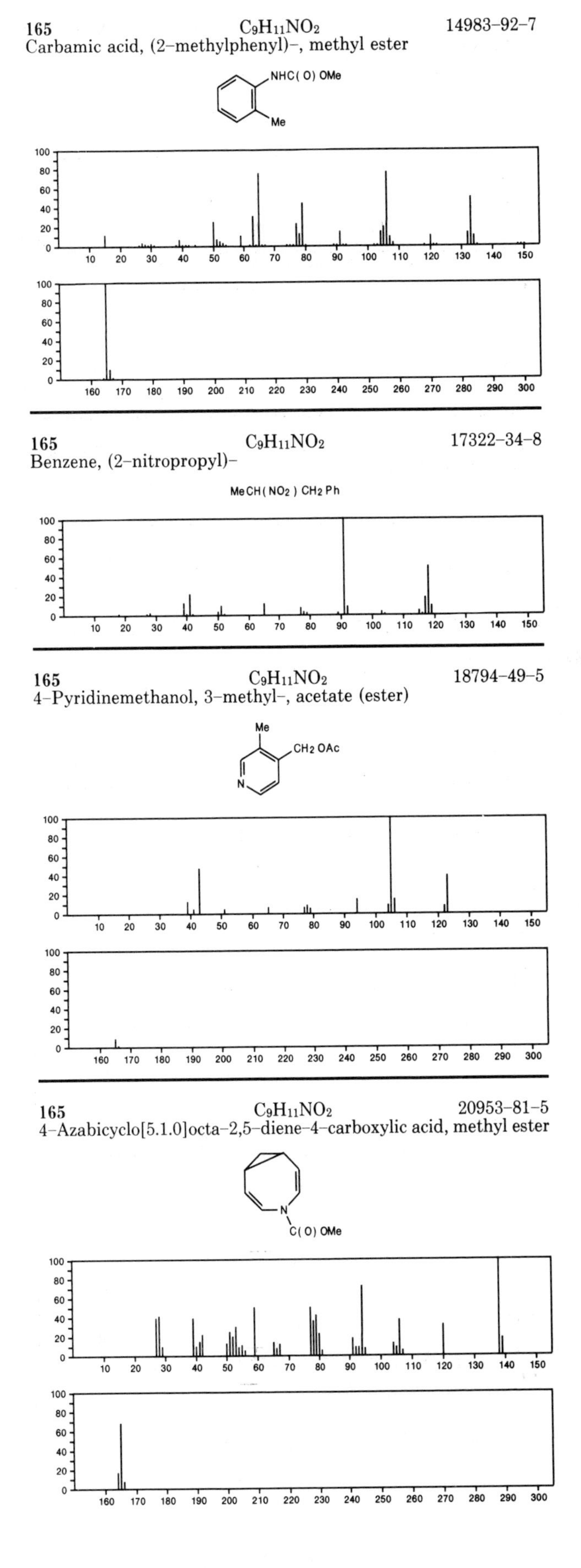

165 $C_9H_{11}NO_2$ 14983–92–7
Carbamic acid, (2–methylphenyl)–, methyl ester

165 $C_9H_{11}NO_2$ 17322–34–8
Benzene, (2–nitropropyl)–

MeCH(NO2)CH2Ph

165 $C_9H_{11}NO_2$ 18794–49–5
4–Pyridinemethanol, 3–methyl–, acetate (ester)

165 $C_9H_{11}NO_2$ 20953–81–5
4–Azabicyclo[5.1.0]octa–2,5–diene–4–carboxylic acid, methyl ester

165 $C_9H_{11}NO_2$ 22139–35–1
5–Azatricyclo[4.2.0.0^{2,4}]oct–7–ene–5–carboxylic acid, methyl ester

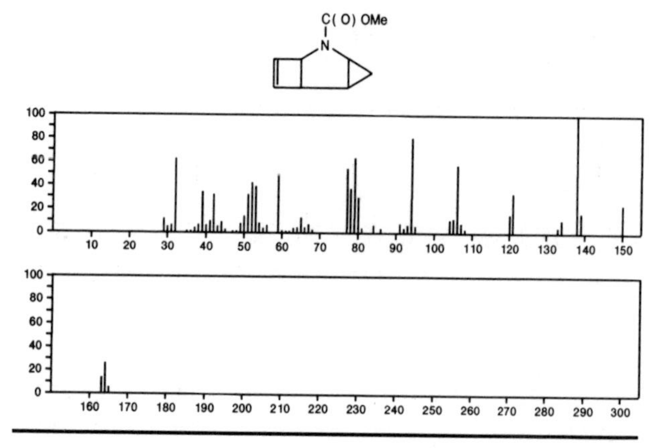

165 $C_9H_{11}NO_2$ 22818–69–5
Benzene, (3–nitropropyl)–

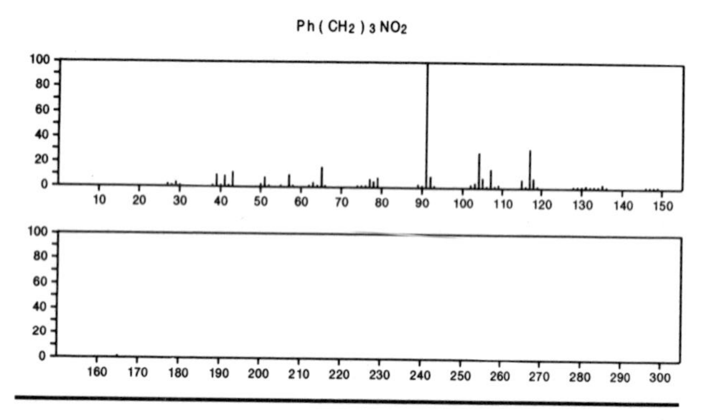

165 $C_9H_{11}NO_2$ 26682–99–5
Benzeneacetic acid, α–amino–, methyl ester

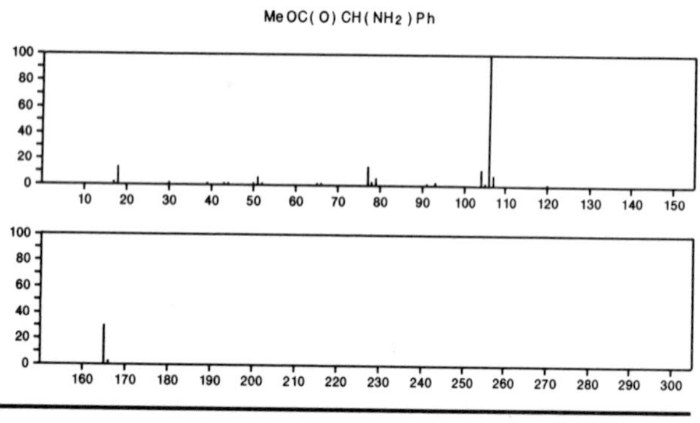

165 $C_9H_{11}NO_2$ 28537–55–5
Benzenepropanol, nitrite

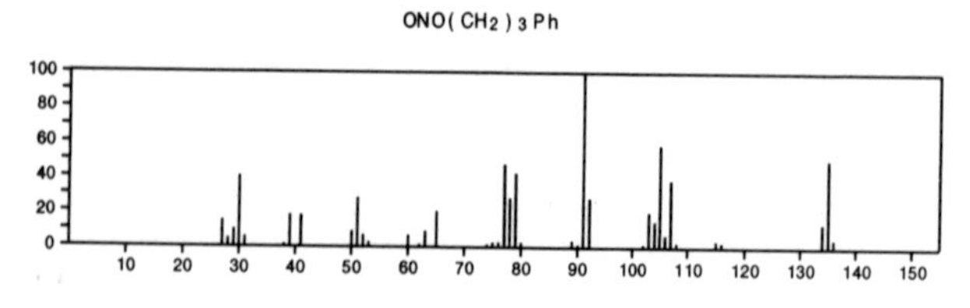

165 $C_9H_{11}NO_2$ 28839–49–8
1*H*–Isoindole–1,3(2*H*)–dione, 4,5,6,7–tetrahydro–2–methyl–

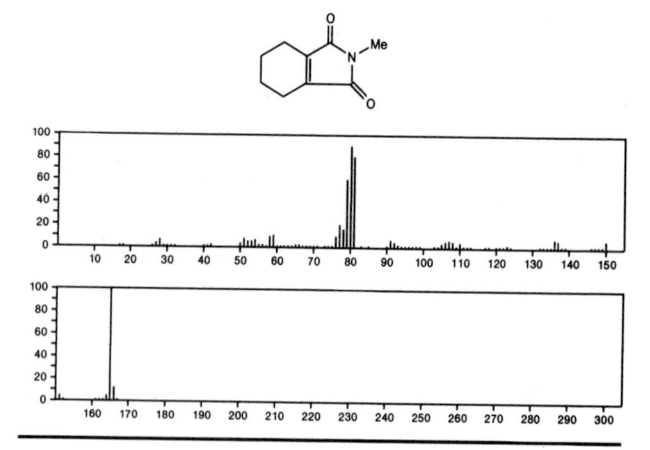

165 $C_9H_{11}NO_2$ 33499–40–0
Benzaldehyde, 4–methoxy–, *O*–methyloxime

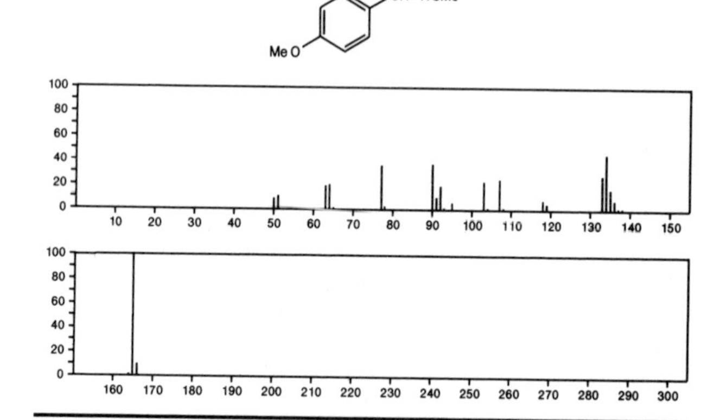

165 $C_9H_{11}NO_2$ 39076–18–1
Carbamic acid, (3–methylphenyl)–, methyl ester

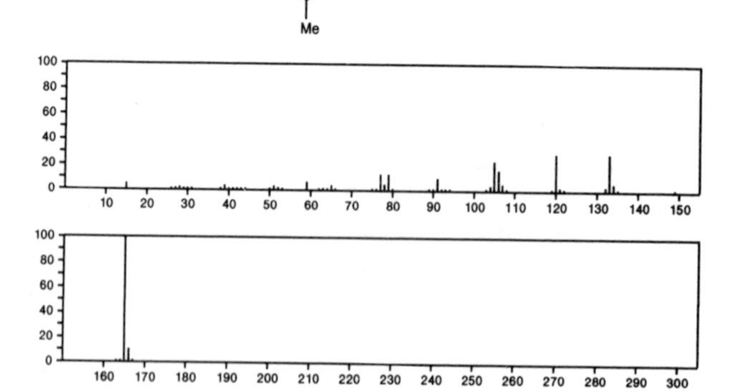

165 $C_9H_{11}NO_2$ 39998–21–5
Acetic acid, (1–methyl–2(1*H*)–pyridinylidene)–, methyl ester

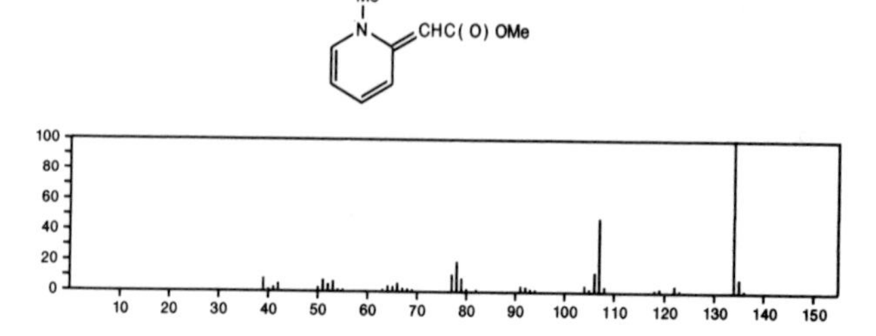

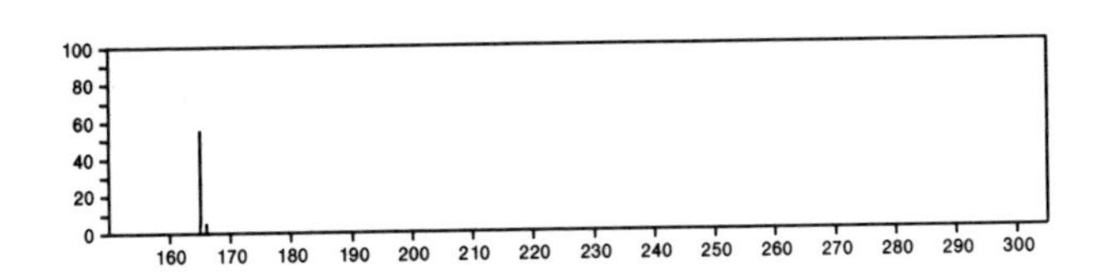

165 $C_9H_{11}NO_2$ 39998–22–6
Acetic acid, (1–methyl–4(1*H*)–pyridinylidene)–, methyl ester

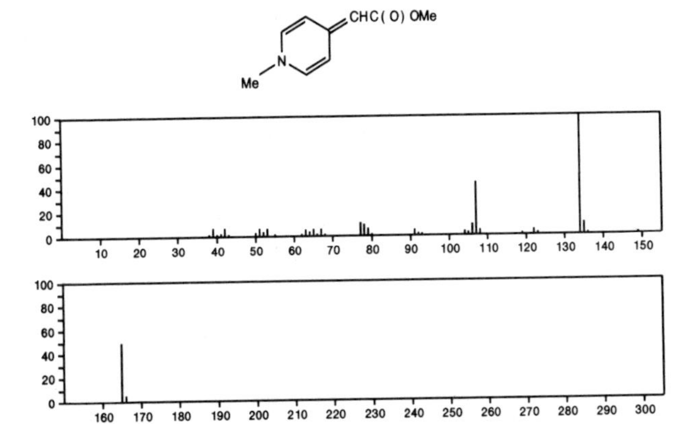

165 $C_9H_{11}NO_2$ 55836–69–6
Benzamide, 3–ethoxy–

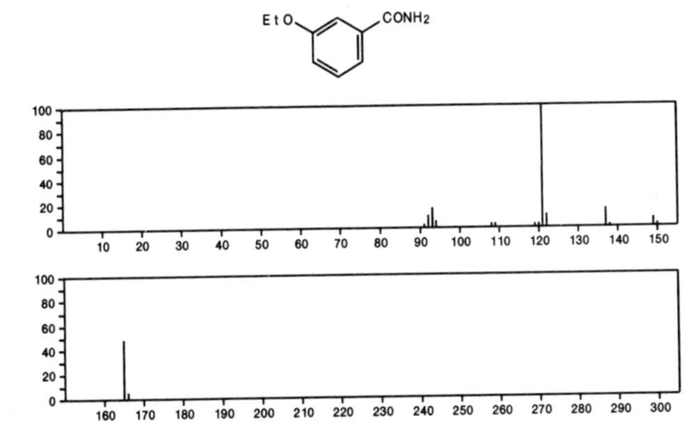

165 $C_9H_{11}NO_2$ 55836–71–0
Benzamide, 4–ethoxy–

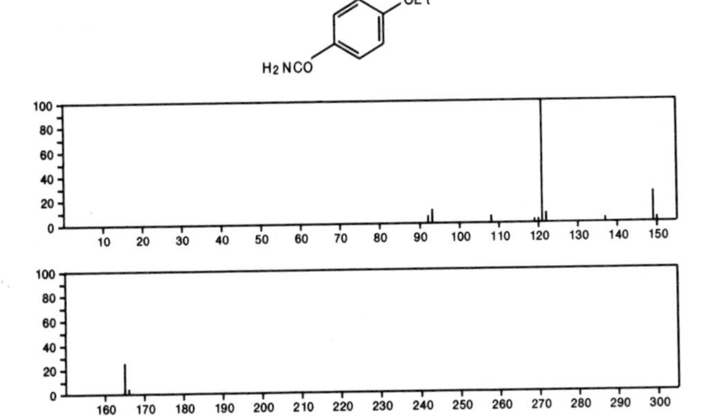

165 $C_9H_{11}NO_2$ 56666–98–9
2–Azabicyclo[3.2.0]hepta–3,6–diene–1–carboxylic acid, 3–methyl–, methyl ester

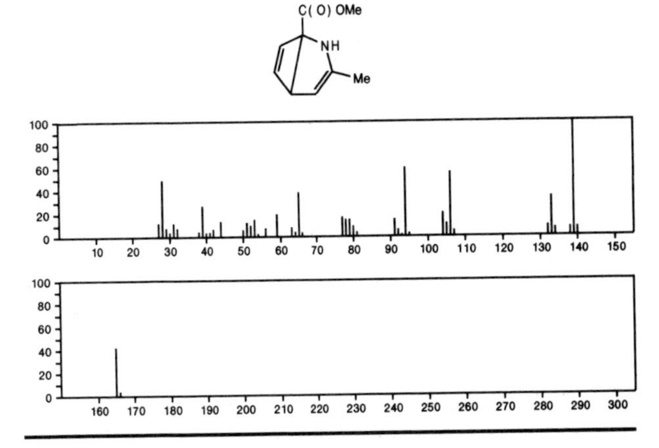

165 $C_9H_{11}NO_2$ 56666–99–0
2–Azabicyclo[3.2.0]hepta–3,6–diene–2–carboxylic acid, 5–methyl–, methyl ester

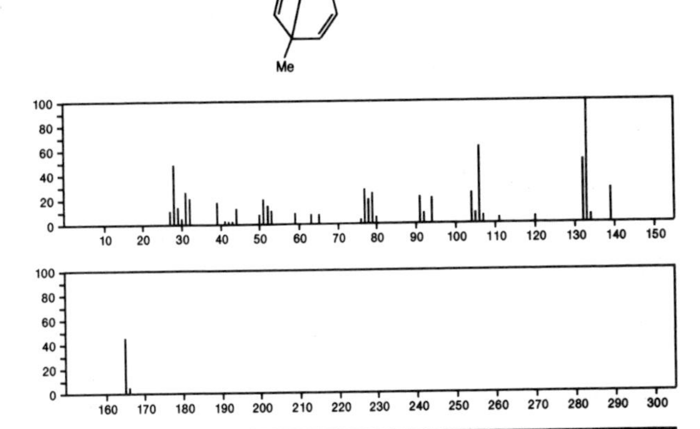

165 $C_9H_{11}NO_2$ 56667–06–2
2–Azabicyclo[3.2.0]hepta–3,6–diene–2–carboxylic acid, 4–methyl–, methyl ester

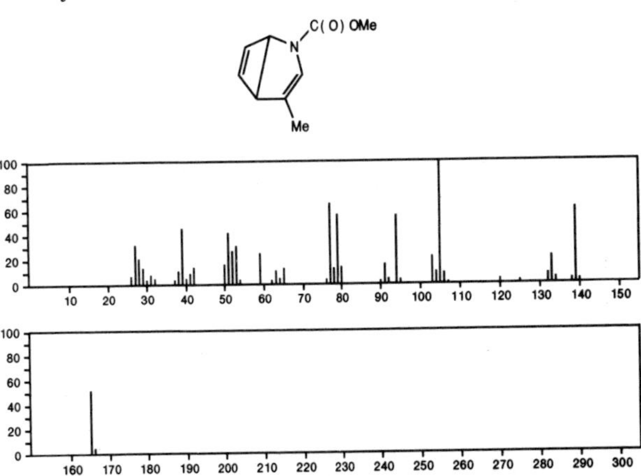

165 $C_9H_{11}NO_2$ 56667–07–3
2–Azabicyclo[3.2.0]hepta–3,6–diene–2–carboxylic acid, 1–methyl–, methyl ester

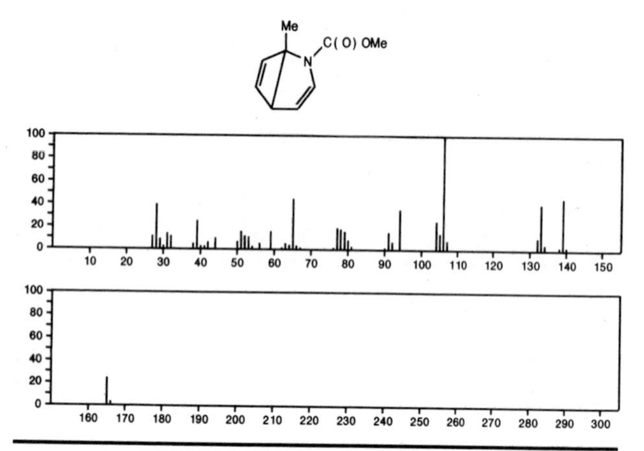

165 $C_9H_{11}NO_2$ 56667–08–4
2–Azabicyclo[3.2.0]hepta–3,6–diene–2–carboxylic acid, 6–methyl–, methyl ester

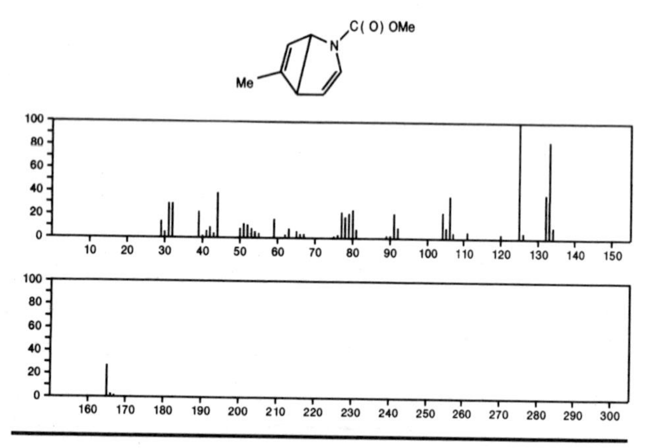

165 $C_9H_{15}NSi$ 17881–80–0
Pyridine, 2–[(trimethylsilyl)methyl]–

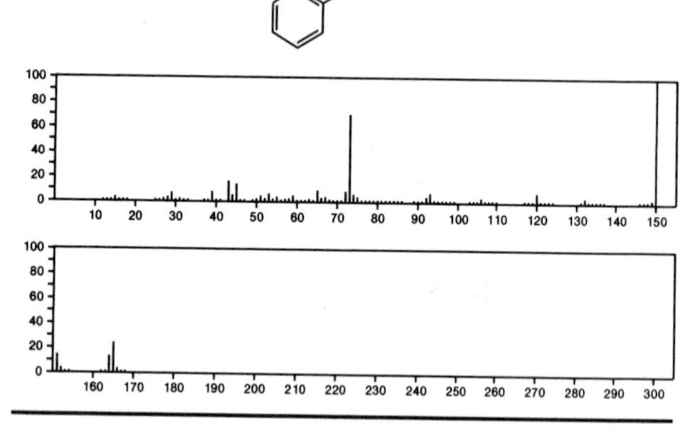

165 $C_{10}H_{15}NO$ 90–82–4
Benzenemethanol, α–[1–(methylamino)ethyl]–, [*S*–(*R**,*R**)]–

MeCH(NHMe)CH(OH)Ph

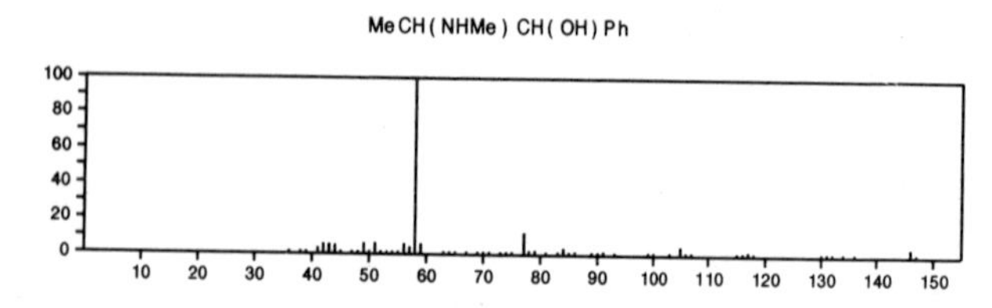

165 $C_{10}H_{15}NO$ 299–42–3
Benzenemethanol, α–[1–(methylamino)ethyl]–, [*R*–(*R**,*S**)]–

MeCH(NHMe)CH(OH)Ph

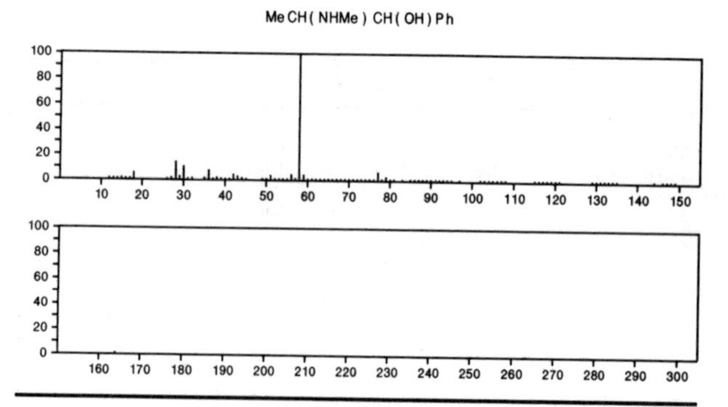

165 $C_{10}H_{15}NO$ 5300–22–1
Benzenemethanol, α–[(ethylamino)methyl]–

EtNHCH2CH(OH)Ph

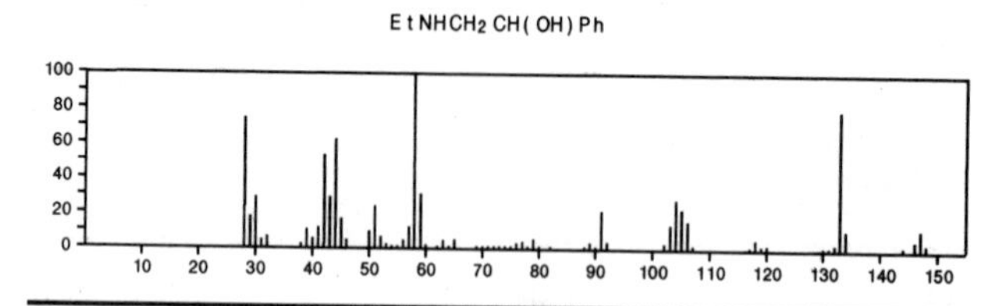

165 $C_{10}H_{15}NO$ 6120–10–1
Phenol, 4–(dimethylamino)–3,5–dimethyl–

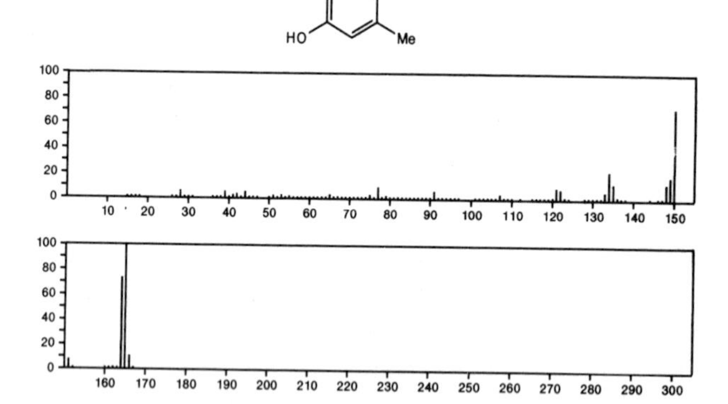

165 $C_{10}H_{15}NO$ 32180–92–0
Phenol, *p*–[3–(methylamino)propyl]–

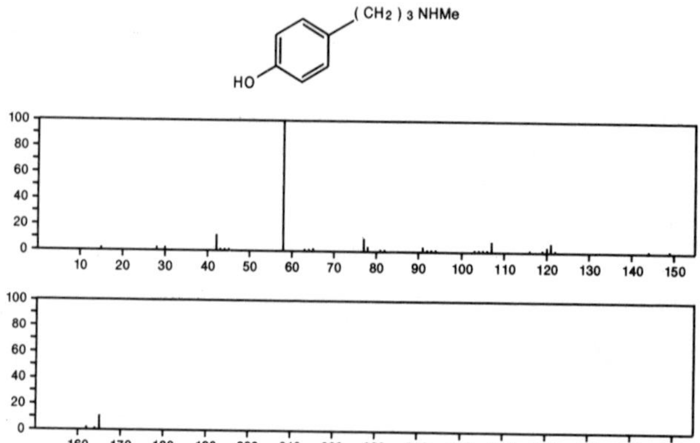

165 $C_{10}H_{15}NO$ 34874-88-9

Benzenamine, 3-methoxy-2,4,6-trimethyl-

165 $C_{10}H_{15}NO$ 55956-29-1

Benzenamine, 4-ethoxyethyl-

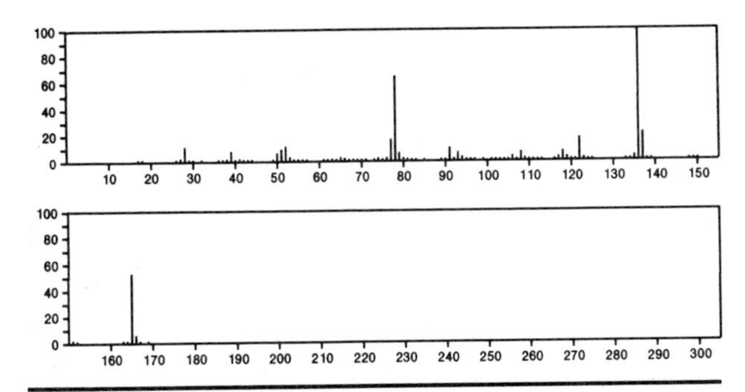

165 $C_{11}H_{19}N$ 2981-10-4

Piperidine, 1-(1-cyclohexen-1-yl)-

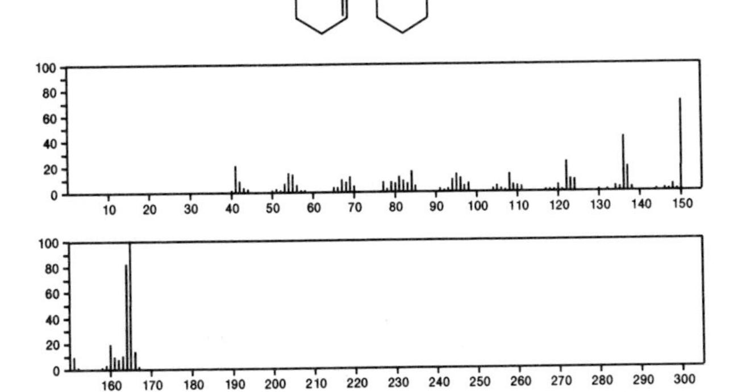

165 $C_{11}H_{19}N$ 3048-63-3

Tricyclo[4.3.1.1^{3,8}]undecan-3-amine

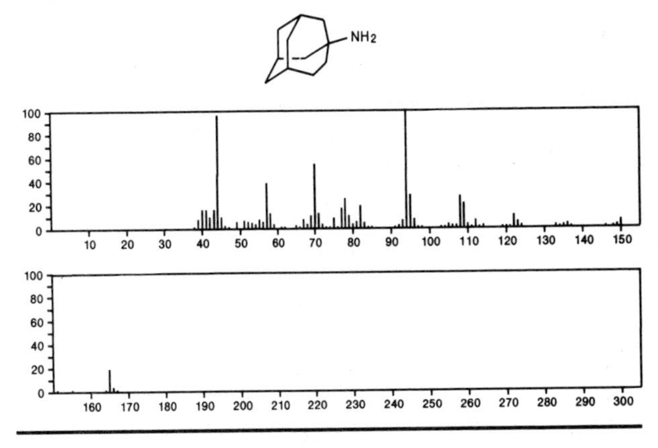

165 $C_{11}H_{19}N$ 5049-51-4

Pyrrolidine, 1-(6-methyl-1-cyclohexen-1-yl)-

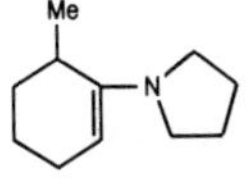

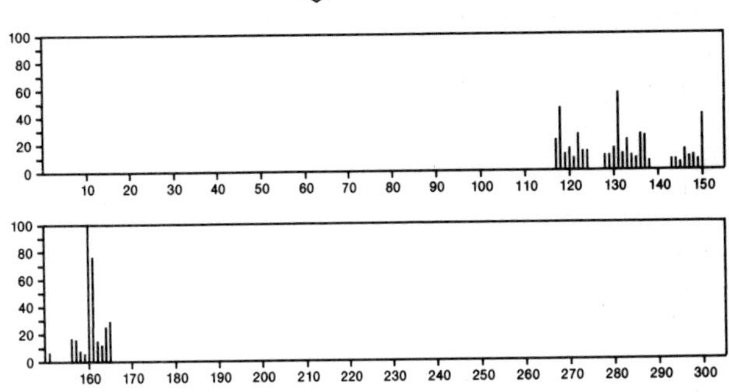

165 $C_{11}H_{19}N$ 14092-11-6

Pyrrolidine, 1-(1-cyclohepten-1-yl)-

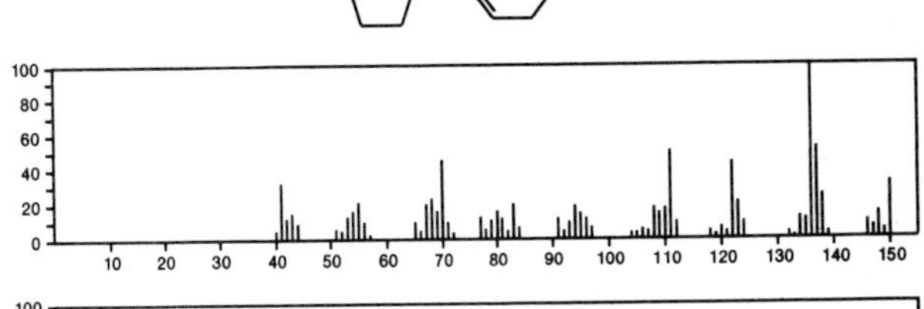

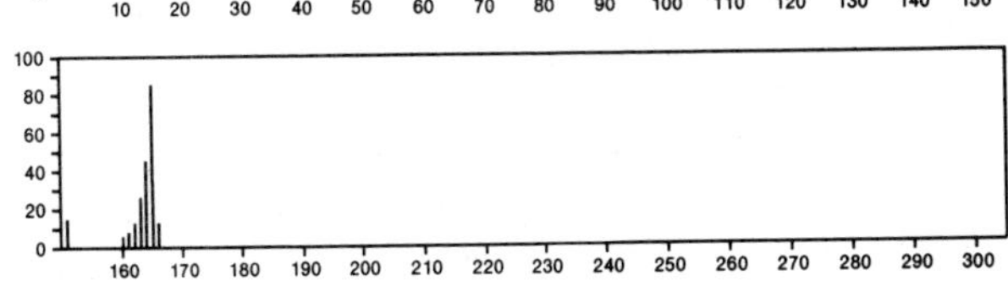

165 $C_{11}H_{19}N$ 31083-61-1

Tricyclo[4.3.1.1^{3,8}]undecan-1-amine

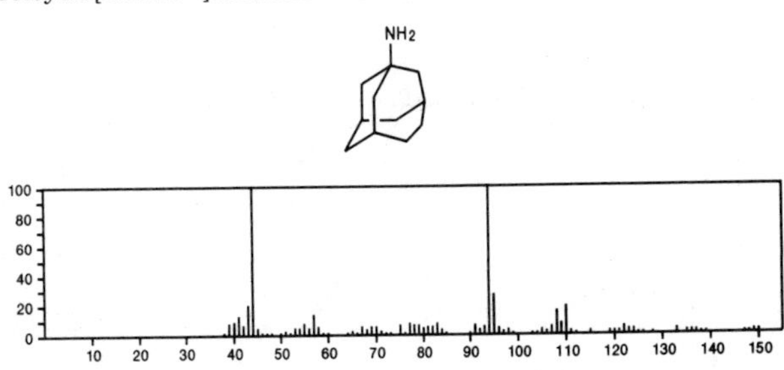

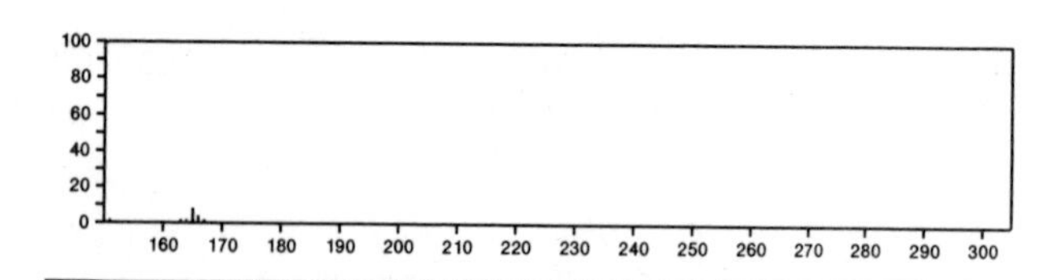

165 $C_{11}H_{19}N$ 56053-03-3
2-Naphthalenamine, 1,2,4a,5,6,7,8,8a-octahydro-4a-methyl-

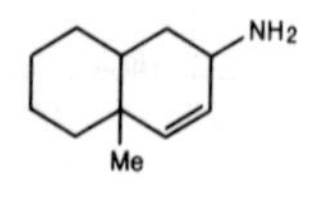

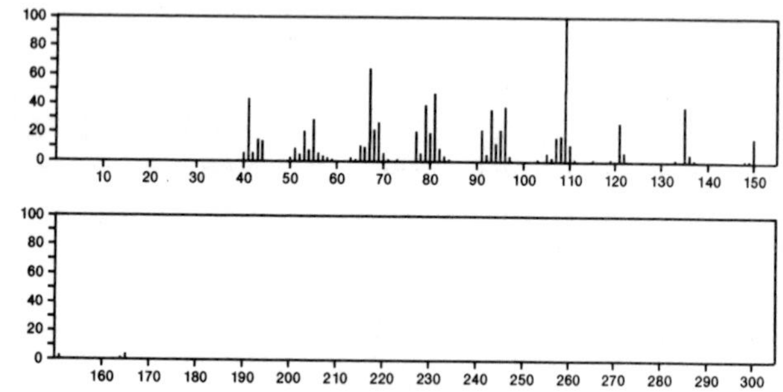

166 $C_2H_2Cl_4$ 79-34-5
Ethane, 1,1,2,2-tetrachloro-

$Cl_2CHCHCl_2$

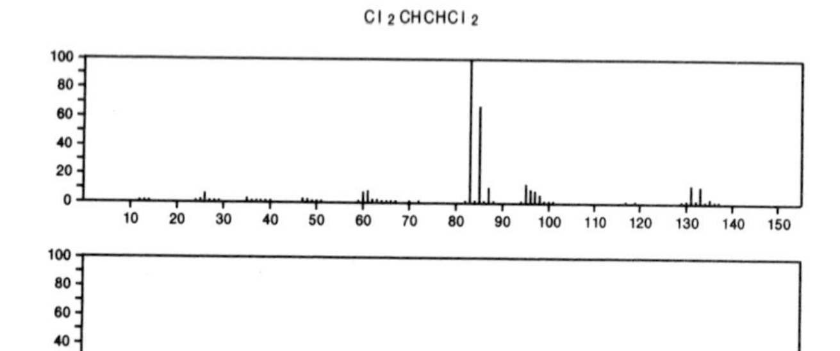

166 $C_2H_2Cl_4$ 630-20-6
Ethane, 1,1,1,2-tetrachloro-

$ClCH_2CCl_3$

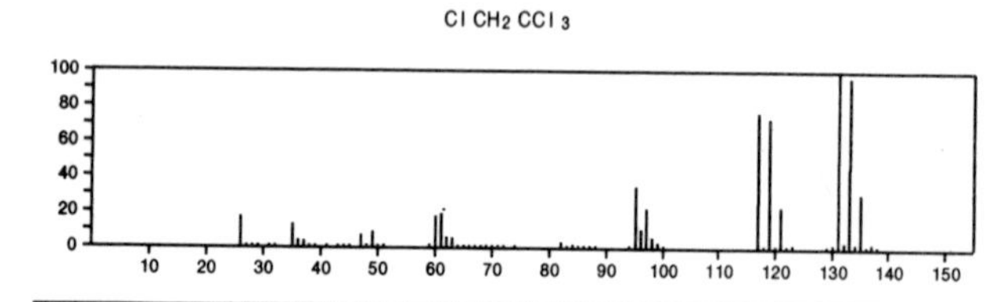

166 $C_2H_4F_4P_2$ 50966-32-0
Phosphonous difluoride, 1,2-ethanediylbis-

$F_2PCH_2CH_2PF_2$

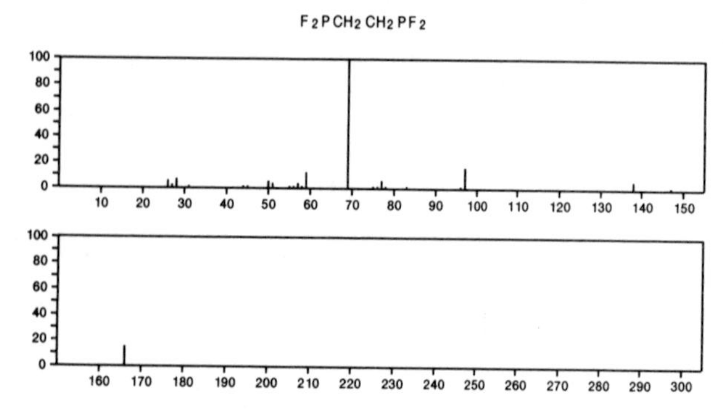

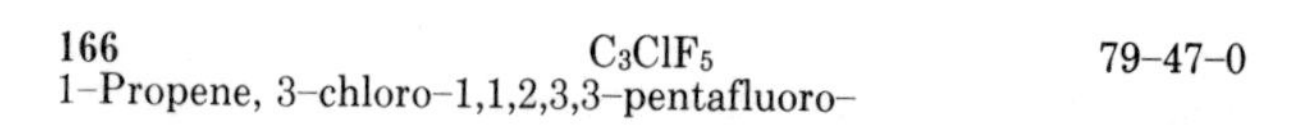

166 C_3ClF_5 79-47-0
1-Propene, 3-chloro-1,1,2,3,3-pentafluoro-

$F_2CClCF{:}CF_2$

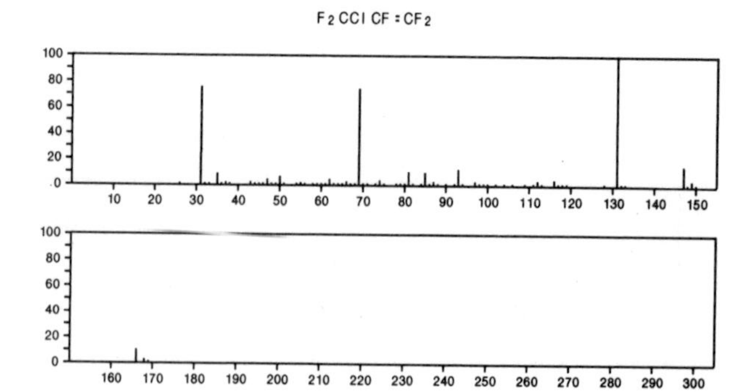

166 C_3ClF_5 2804-50-4
1-Propene, 2-chloro-1,1,3,3,3-pentafluoro-

$F_3CCCl{:}CF_2$

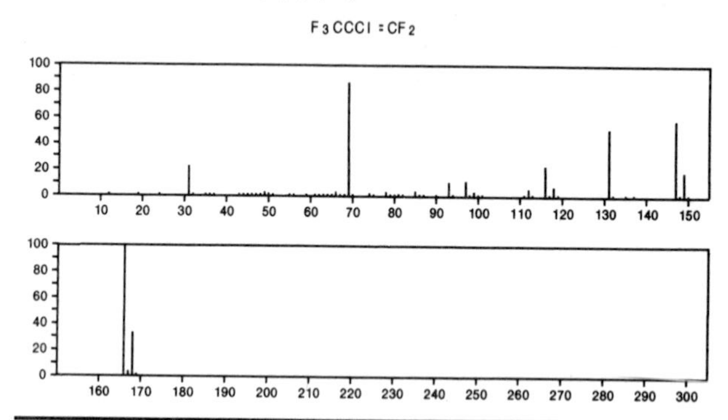

166 C_3F_6O 684-16-2
2-Propanone, 1,1,1,3,3,3-hexafluoro-

F_3CCOCF_3

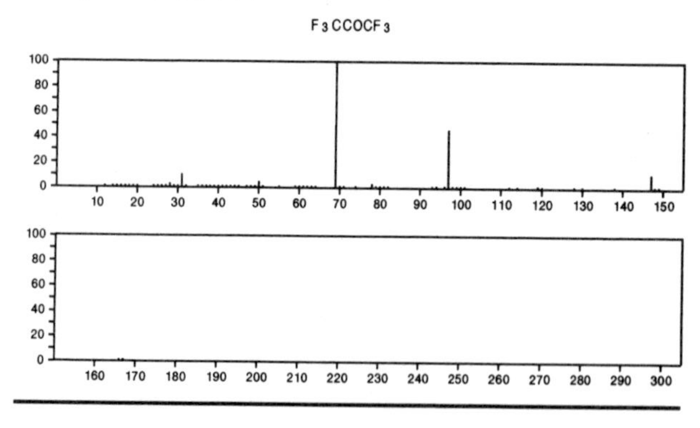

166 $C_3H_3Cl_2F_3$ 338-75-0
Propane, 2,3-dichloro-1,1,1-trifluoro-

$ClCH_2CHClCF_3$

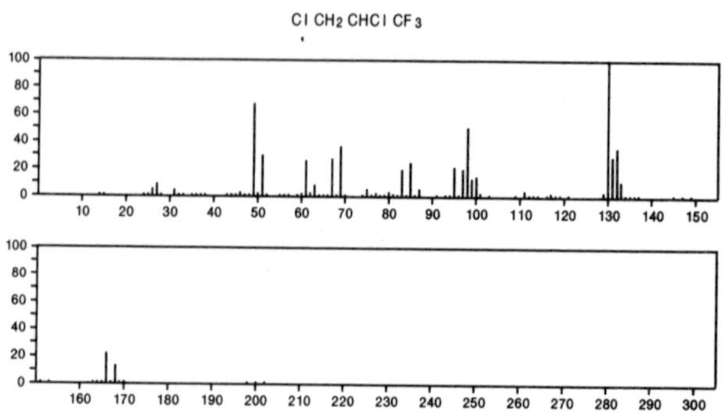

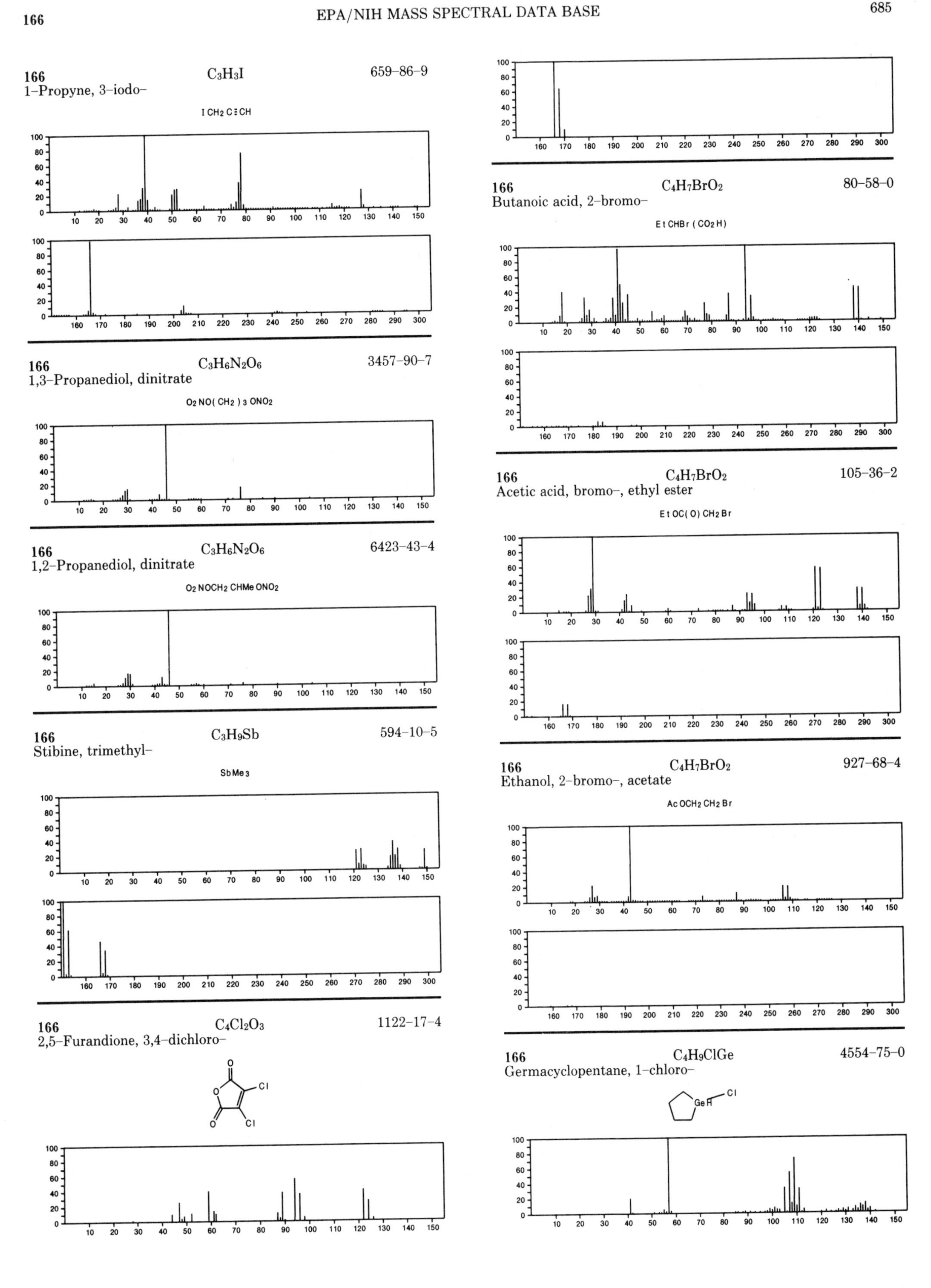

166 C_3H_3I 659–86–9
1–Propyne, 3–iodo–
I CH2 C≡CH

166 $C_3H_6N_2O_6$ 3457–90–7
1,3–Propanediol, dinitrate
O2 NO(CH2) 3 ONO2

166 $C_3H_6N_2O_6$ 6423–43–4
1,2–Propanediol, dinitrate
O2 NOCH2 CHMe ONO2

166 C_3H_9Sb 594–10–5
Stibine, trimethyl–
SbMe3

166 $C_4Cl_2O_3$ 1122–17–4
2,5–Furandione, 3,4–dichloro–

166 $C_4H_7BrO_2$ 80–58–0
Butanoic acid, 2–bromo–
Et CHBr (CO2 H)

166 $C_4H_7BrO_2$ 105–36–2
Acetic acid, bromo–, ethyl ester
Et OC(O) CH2 Br

166 $C_4H_7BrO_2$ 927–68–4
Ethanol, 2–bromo–, acetate
Ac OCH2 CH2 Br

166 C_4H_9ClGe 4554–75–0
Germacyclopentane, 1–chloro–

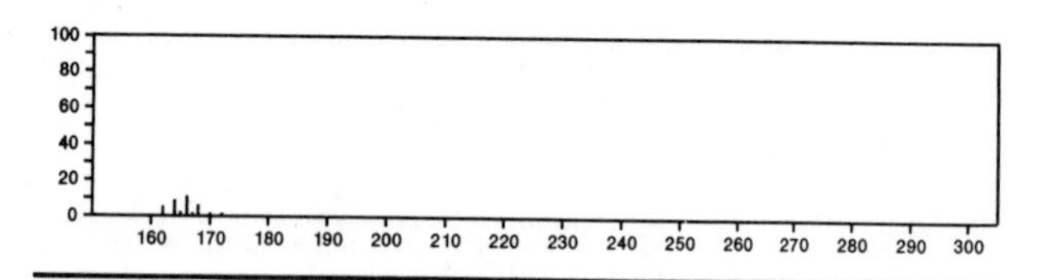

166 $C_5H_4Cl_2O_2$ 14203-21-5
1,3-Cyclopentanedione, 2,2-dichloro-

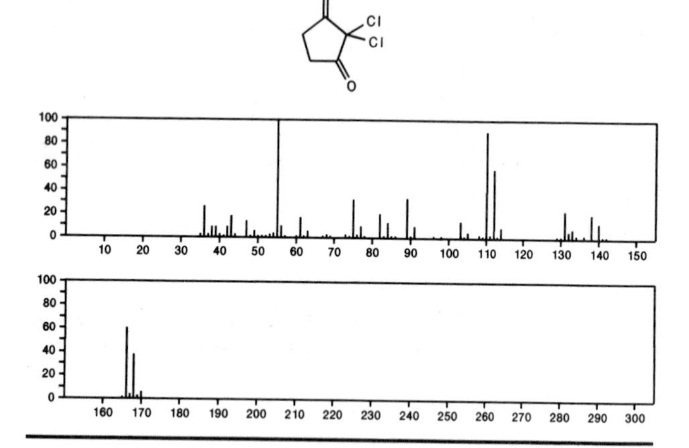

166 C_5H_8BrF 51422-72-1
Cyclopentane, 1-bromo-2-fluoro-, *cis*-

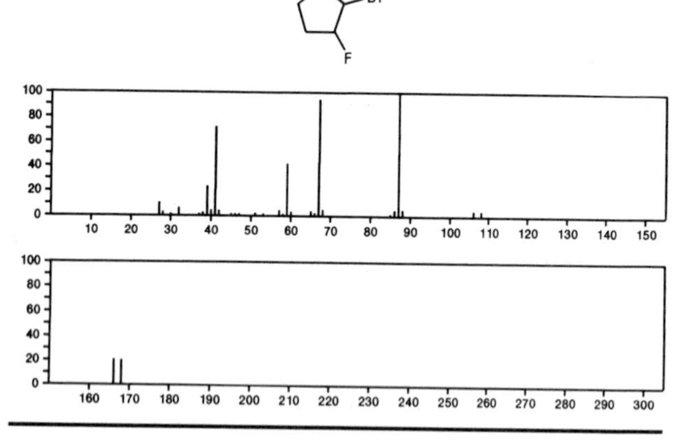

166 C_5H_8BrF 51422-73-2
Cyclopentane, 1-bromo-2-fluoro-, *trans*-

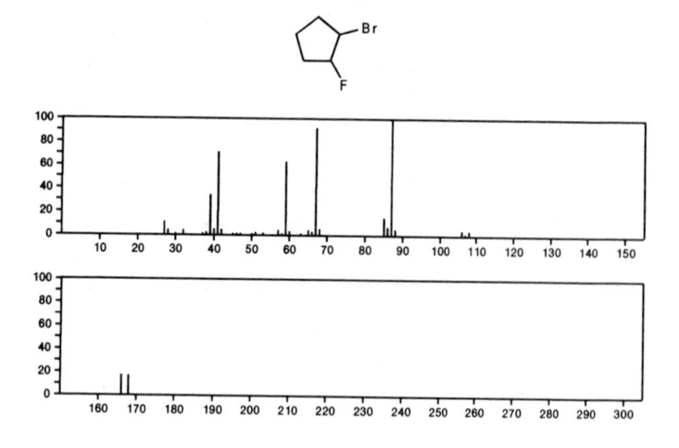

166 $C_5H_{11}BrO$ 4457-67-4
Butane, 1-bromo-4-methoxy-

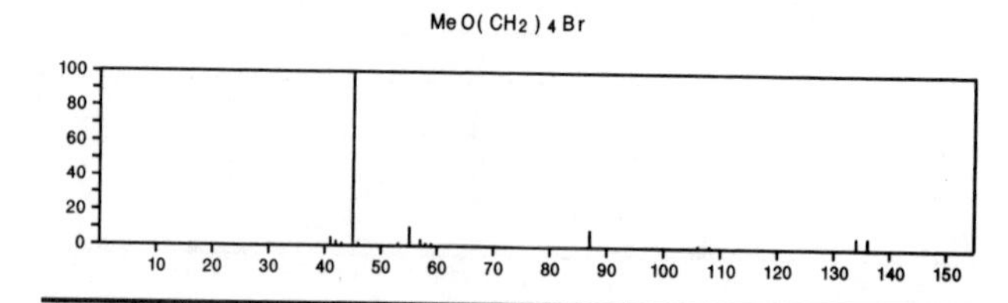

166 $C_5H_{11}O_4P$ 4185-82-4
Phosphoric acid, dimethyl 1-methylethenyl ester

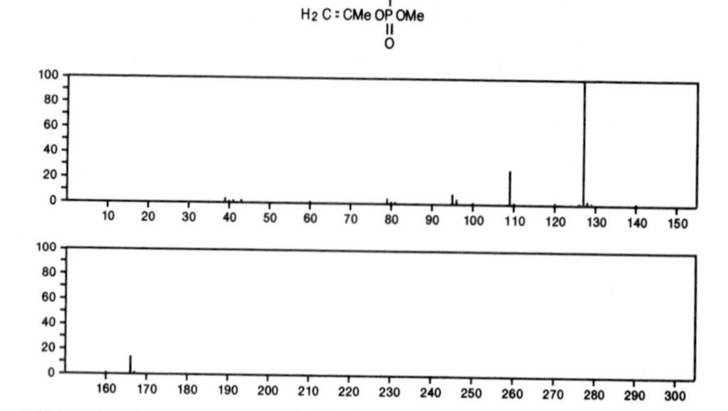

166 $C_5H_{11}O_4P$ 55712-50-0
Phosphoric acid, dimethyl 1-propenyl ester

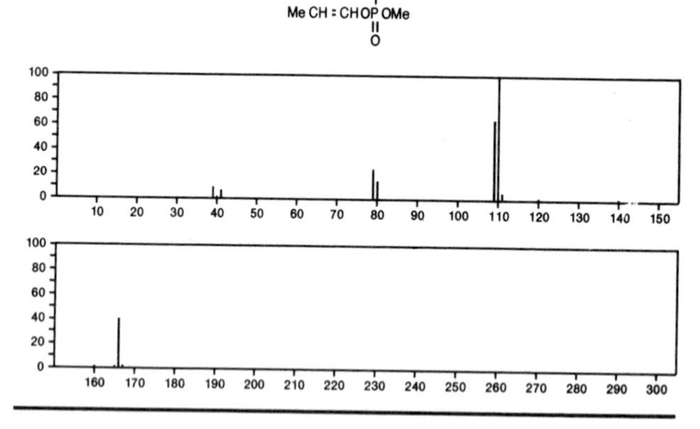

166 $C_6H_6N_4O_2$ 1076-22-8
1*H*-Purine-2,6-dione, 3,7-dihydro-3-methyl-

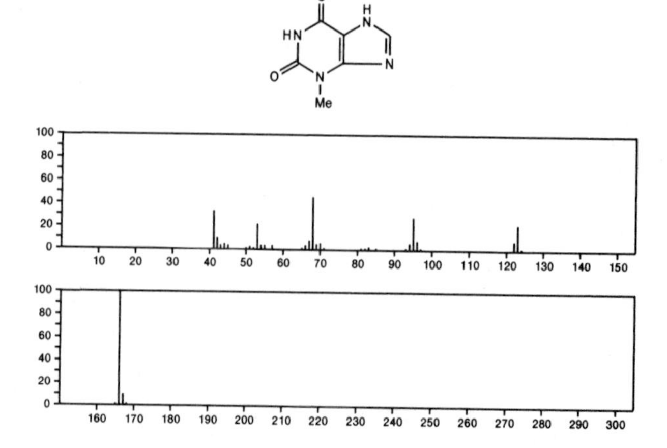

166 $C_6H_6N_4O_2$ 6136-37-4

1*H*-Purine-2,6-dione, 3,7-dihydro-1-methyl-

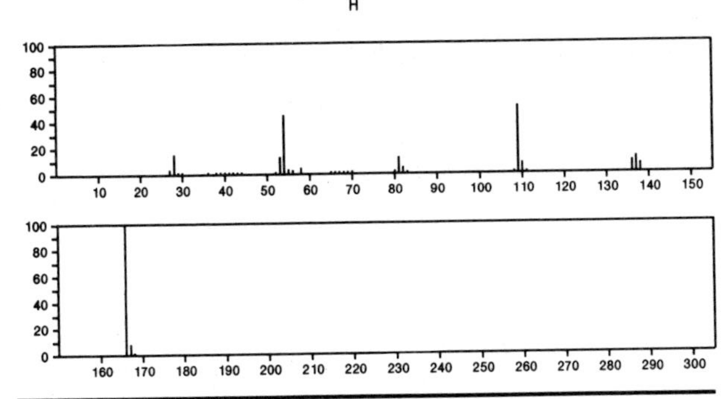

166 $C_6H_6N_4O_2$ 28109-92-4

1*H*-Purine-2,6-dione, 3,7-dihydromethyl-

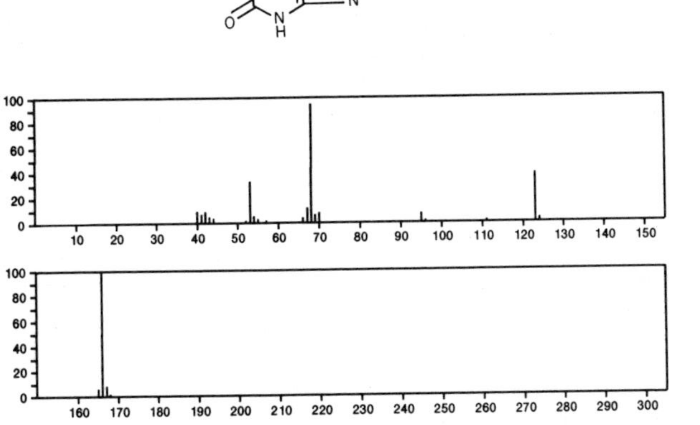

166 $C_6H_6N_4S$ 50-66-8

1*H*-Purine, 6-(methylthio)-

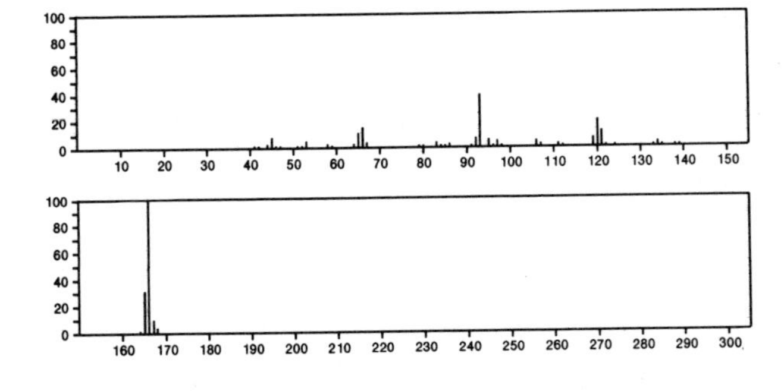

166 $C_6H_6N_4S$ 1006-12-8

6*H*-Purine-6-thione, 3,7-dihydro-3-methyl-

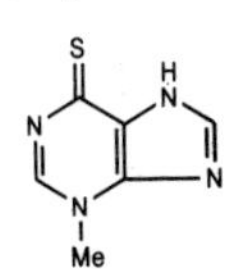

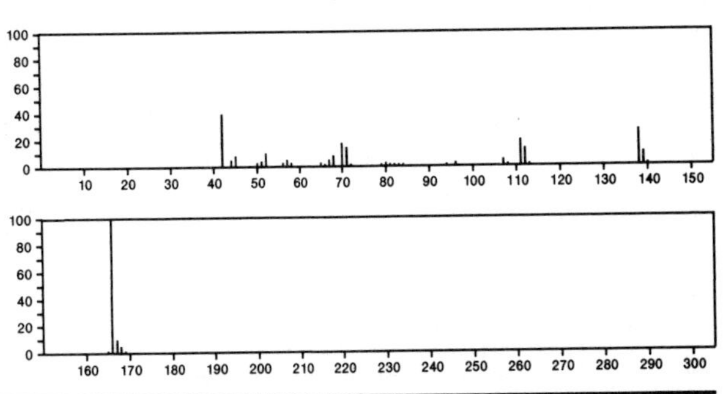

166 $C_6H_6N_4S$ 1006-20-8

6*H*-Purine-6-thione, 1,9-dihydro-9-methyl-

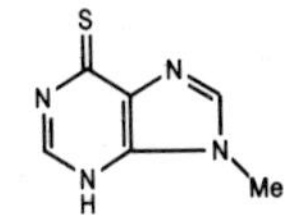

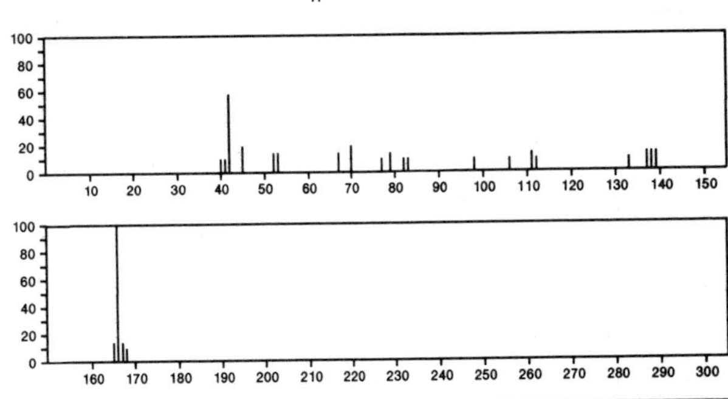

166 $C_6H_6N_4S$ 1006-22-0

6*H*-Purine-6-thione, 1,7-dihydro-1-methyl-

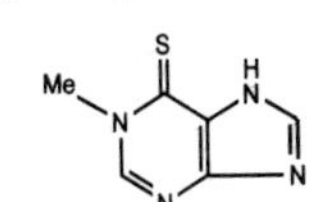

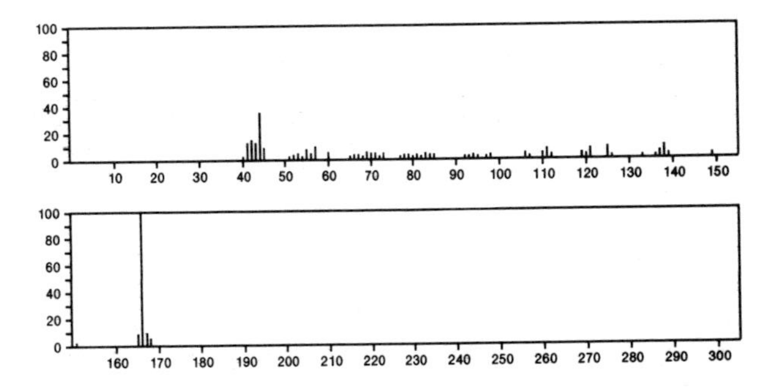

166 $C_6H_6N_4S$ 1126-23-4
6H-Purine-6-thione, 1,7-dihydro-8-methyl-

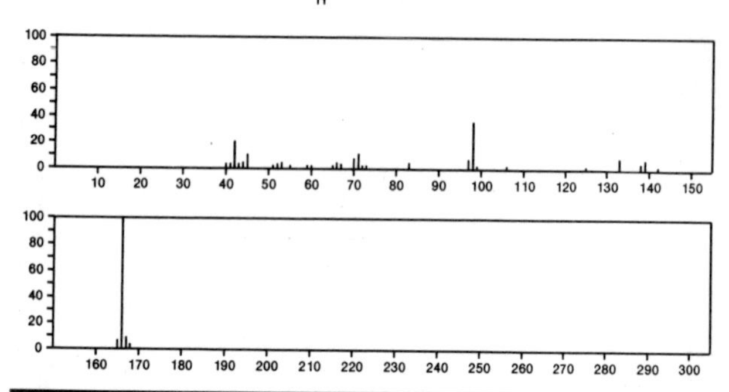

166 $C_6H_6N_4S$ 3324-79-6
6H-Purine-6-thione, 1,7-dihydro-7-methyl-

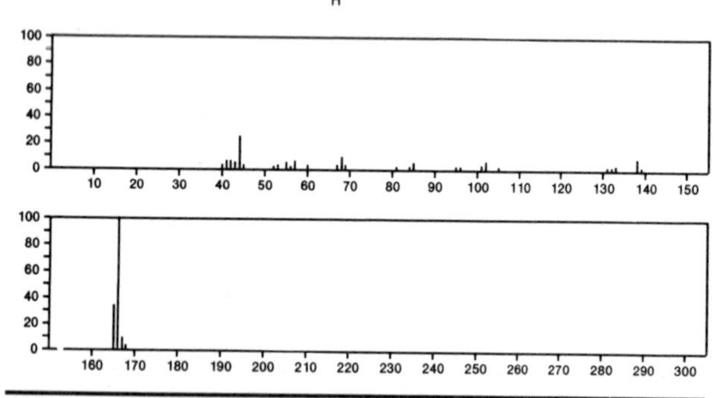

166 $C_6H_6N_4S$ 33426-53-8
1H-Purine, 8-(methylthio)-

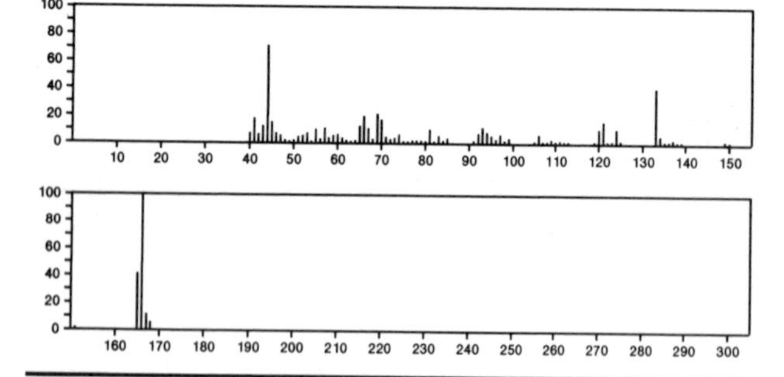

166 $C_6H_6N_4S$ 33512-51-5
1H-Purine, 2-(methylthio)-

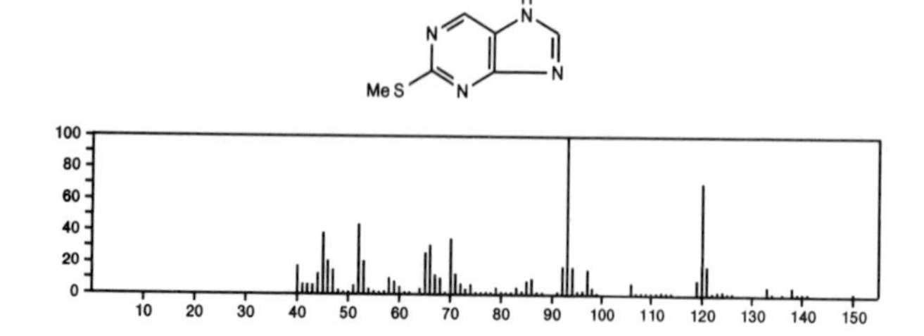

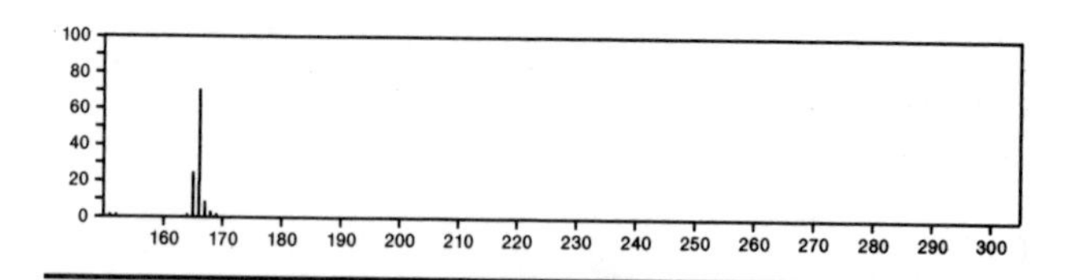

166 $C_6H_6N_4S$ 38917-31-6
6H-Purine-6-thione, 1,7-dihydro-2-methyl-

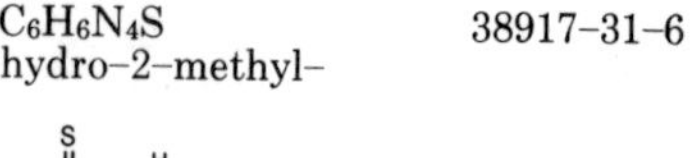

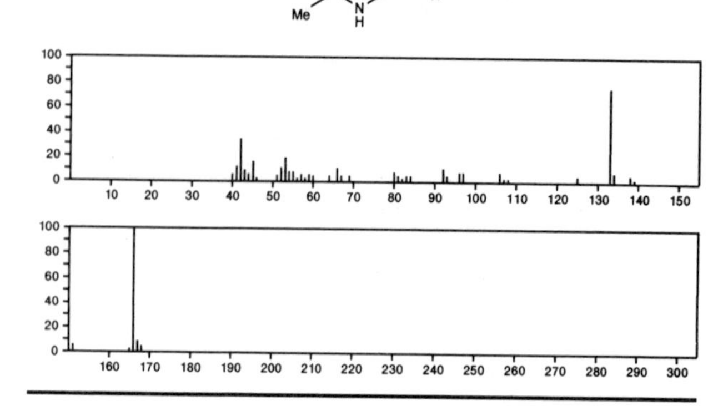

166 $C_6H_8Cl_2O$ 4162-62-3
Ether, bis(2-chloroallyl)

$H_2C=CCl\,CH_2\,OCH_2\,CCl=CH_2$

166 $C_6H_{14}OS_2$ 54411-14-2
Ethane, 1,1'-[oxybis(methylenethio)]bis-

$EtSCH_2OCH_2SEt$

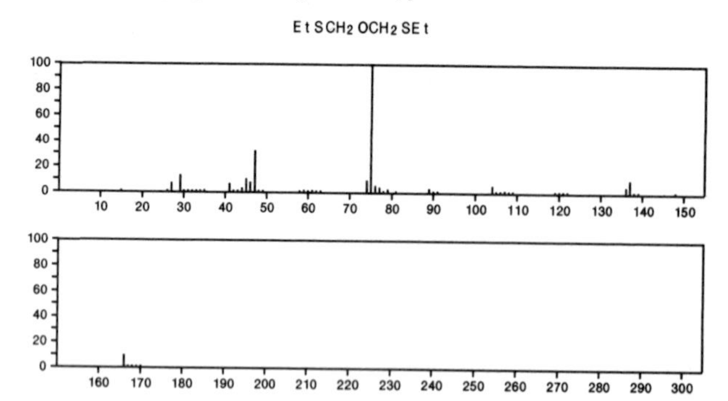

166 $C_6H_{14}O_3S$ 623-98-3
Sulfurous acid, dipropyl ester

PrOS(O)OPr

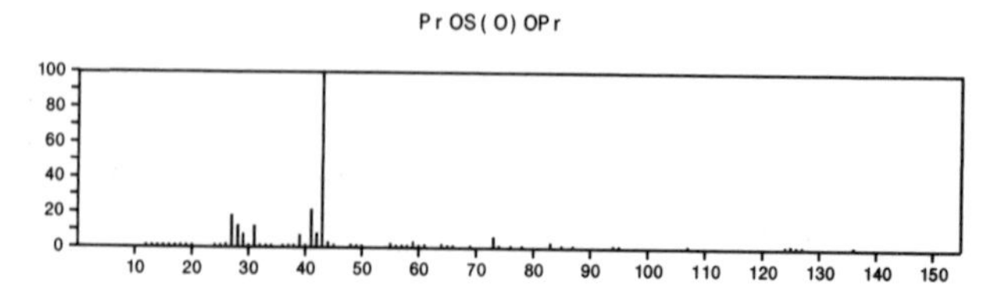

166 $C_6H_{14}O_3S$ 4773-13-1
Sulfurous acid, bis(1-methylethyl) ester

i-PrOS(O)OPr-i

166 $C_6H_{14}O_3S$ 56282-36-1
Ethanol, 2-[2-(2-mercaptoethoxy)ethoxy]-

$HSCH_2CH_2OCH_2CH_2OCH_2CH_2OH$

166 $C_7H_6N_2O_3$ 610-15-1
Benzamide, 2-nitro-

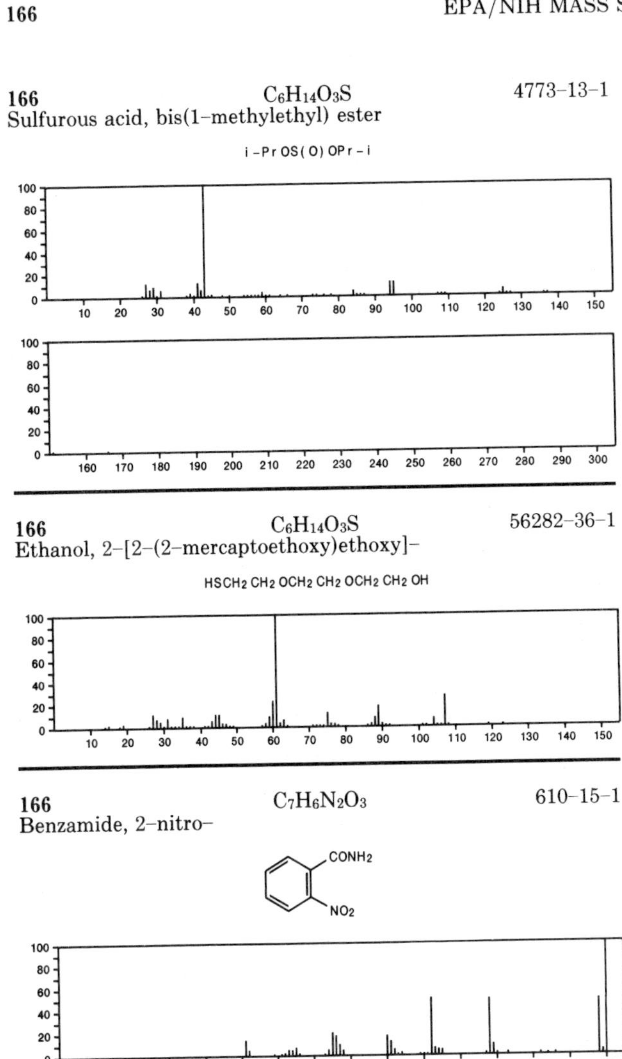

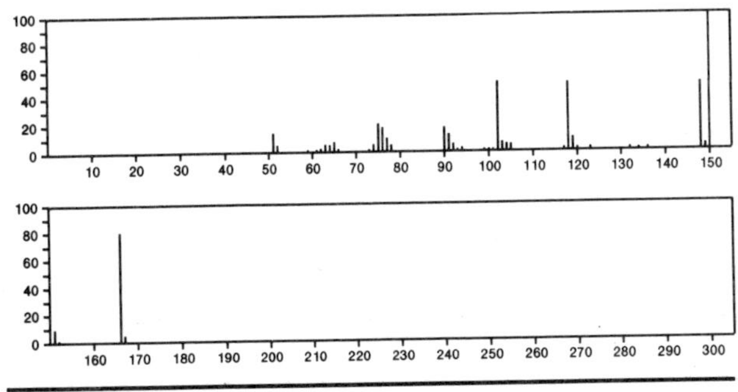

166 $C_7H_6N_2O_3$ 1129-37-9
Benzaldehyde, 4-nitro-, oxime

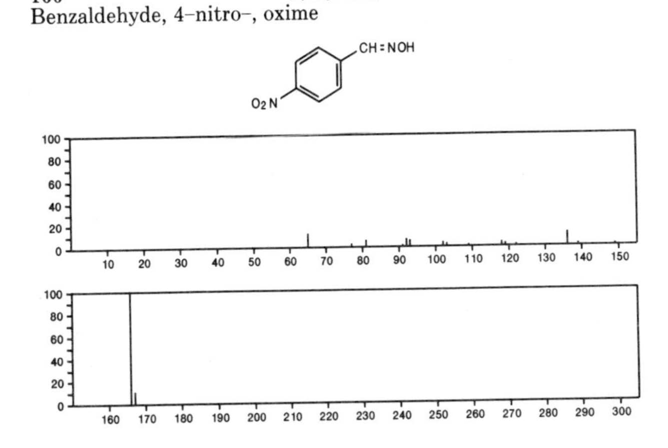

166 $C_7H_6N_2O_3$ 3431-62-7
Benzaldehyde, 3-nitro-, oxime

166 $C_7H_6N_2O_3$ 3717-19-9
Benzaldehyde, 4-nitro-, oxime, (*E*)-

166 $C_7H_6N_2O_3$ 3717-20-2
Benzaldehyde, 4-nitro-, oxime, (*Z*)-

166 $C_7H_6N_2O_3$ 3717-29-1
Benzaldehyde, 3-nitro-, oxime, (*E*)-

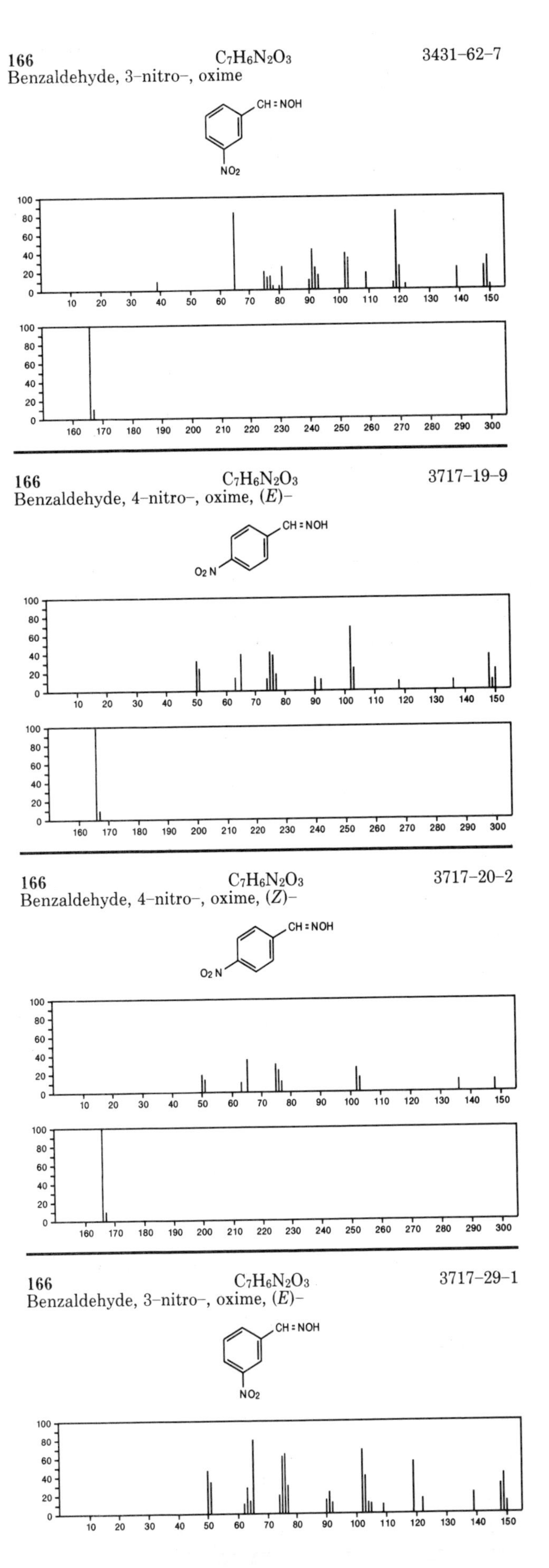

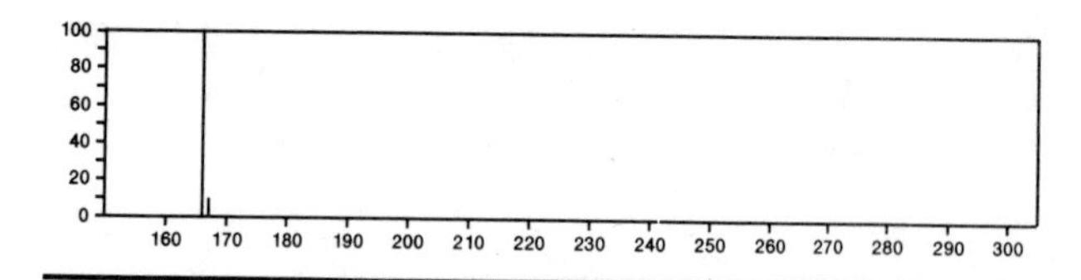

166 $C_7H_6N_2O_3$ 3717-30-4
Benzaldehyde, 3-nitro-, oxime, (*Z*)-

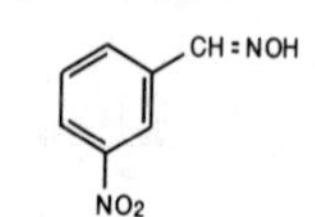

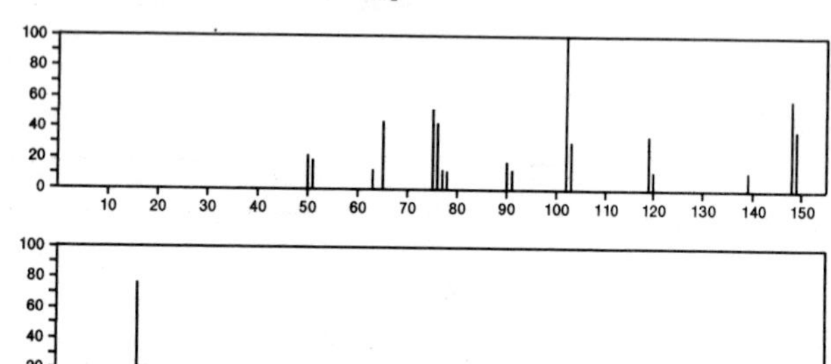

166 $C_7H_{10}N_4O$ 36361-69-0
1*H*-Purine-6-methanol, 6,7-dihydro-α-methyl-

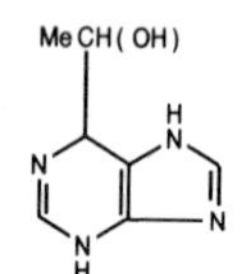

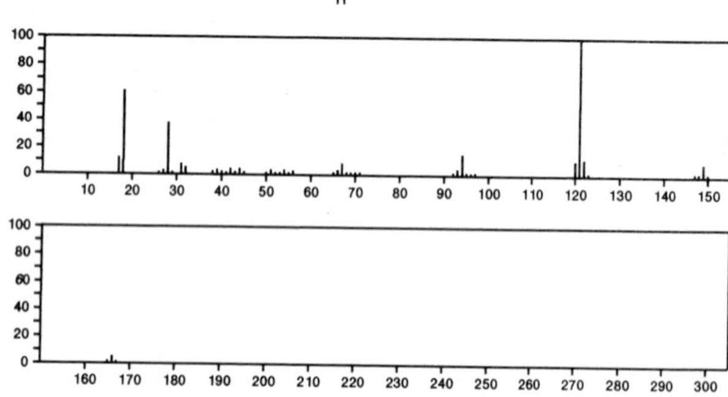

166 $C_7H_{15}ClO_2$ 20637-37-0
Propane, 2-(chloromethyl)-1,3-dimethoxy-2-methyl-

CH2Cl
MeOCH2CMeCH2OMe

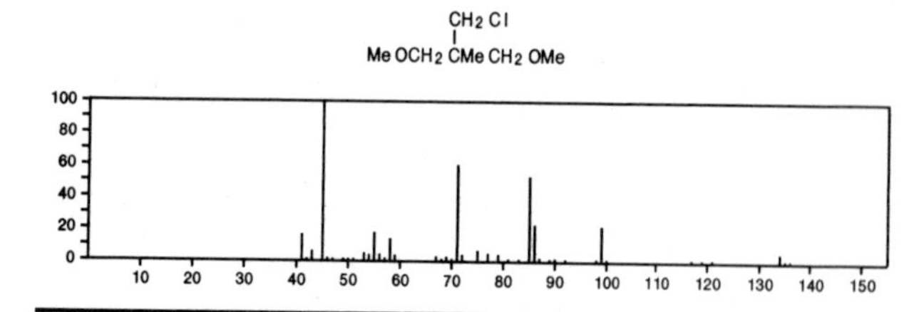

166 $C_8H_6O_4$ 88-99-3
1,2-Benzenedicarboxylic acid

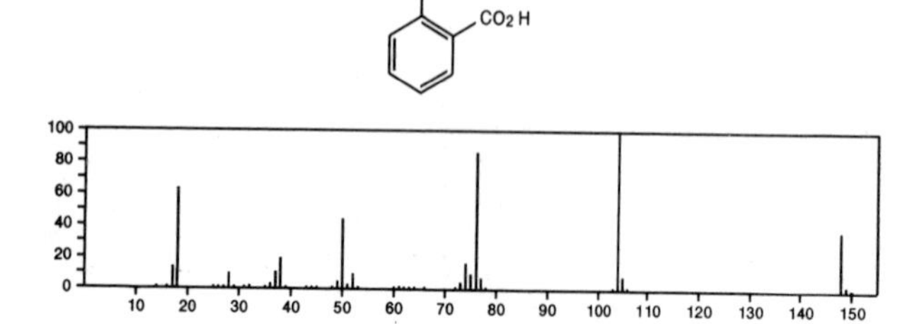

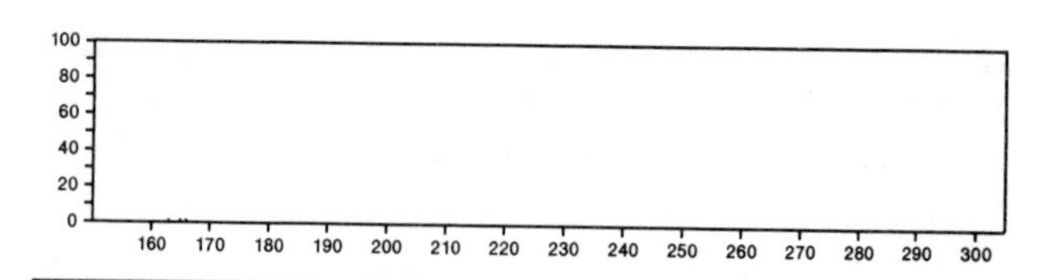

166 $C_8H_6O_4$ 100-21-0
1,4-Benzenedicarboxylic acid

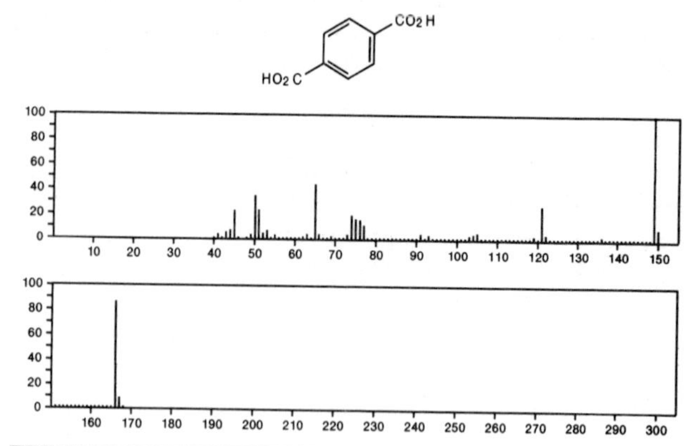

166 $C_8H_6O_4$ 121-91-5
1,3-Benzenedicarboxylic acid

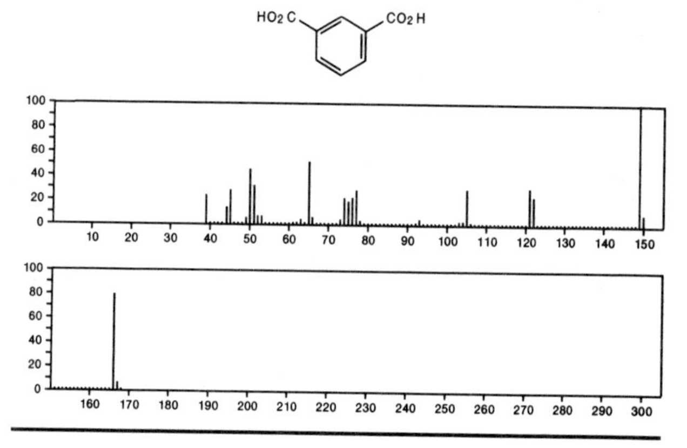

166 $C_8H_6S_2$ 492-97-7
2,2'-Bithiophene

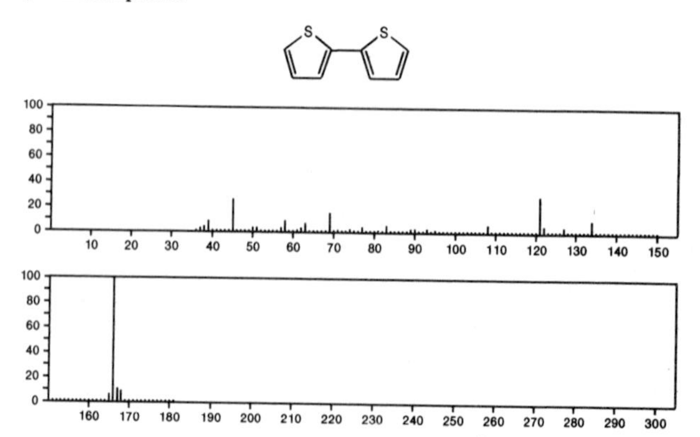

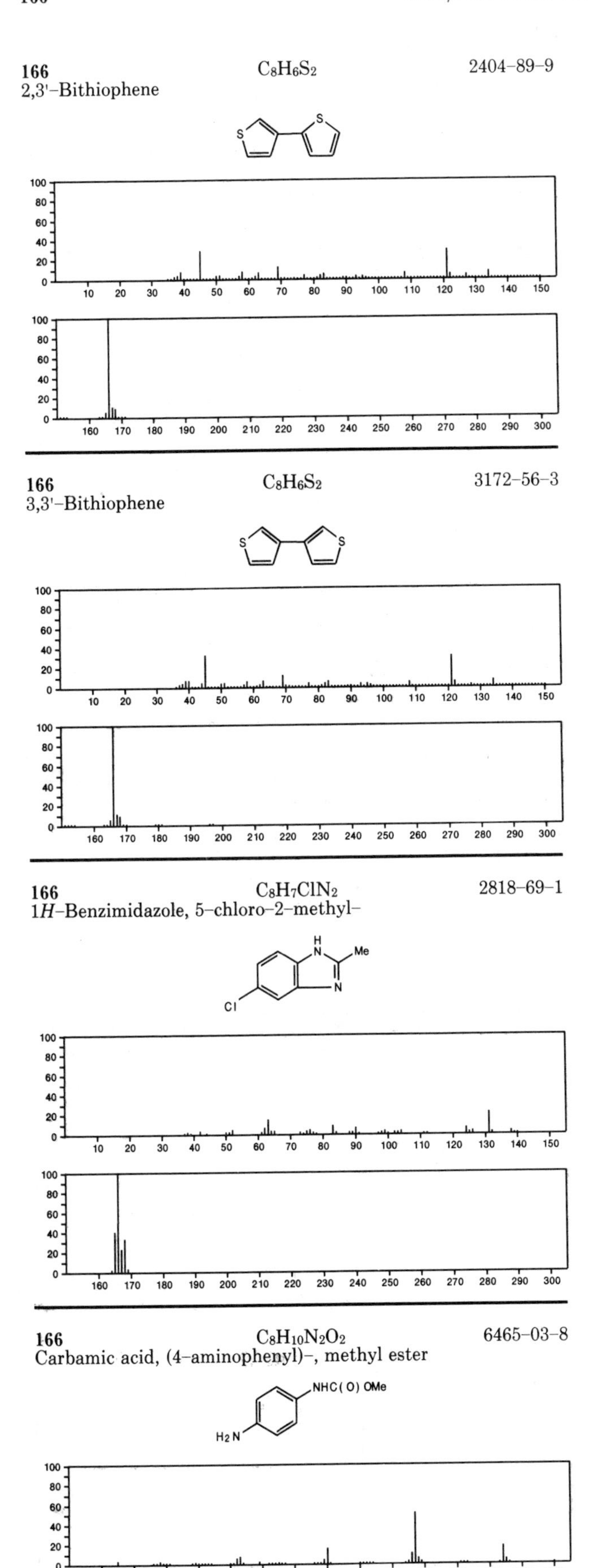

166 $C_8H_6S_2$ 2404-89-9
2,3'-Bithiophene
166 $C_8H_6S_2$ 3172-56-3
3,3'-Bithiophene
166 $C_8H_7ClN_2$ 2818-69-1
1*H*-Benzimidazole, 5-chloro-2-methyl-
166 $C_8H_{10}N_2O_2$ 6465-03-8
Carbamic acid, (4-aminophenyl)-, methyl ester
NHC(O)OMe
H2N

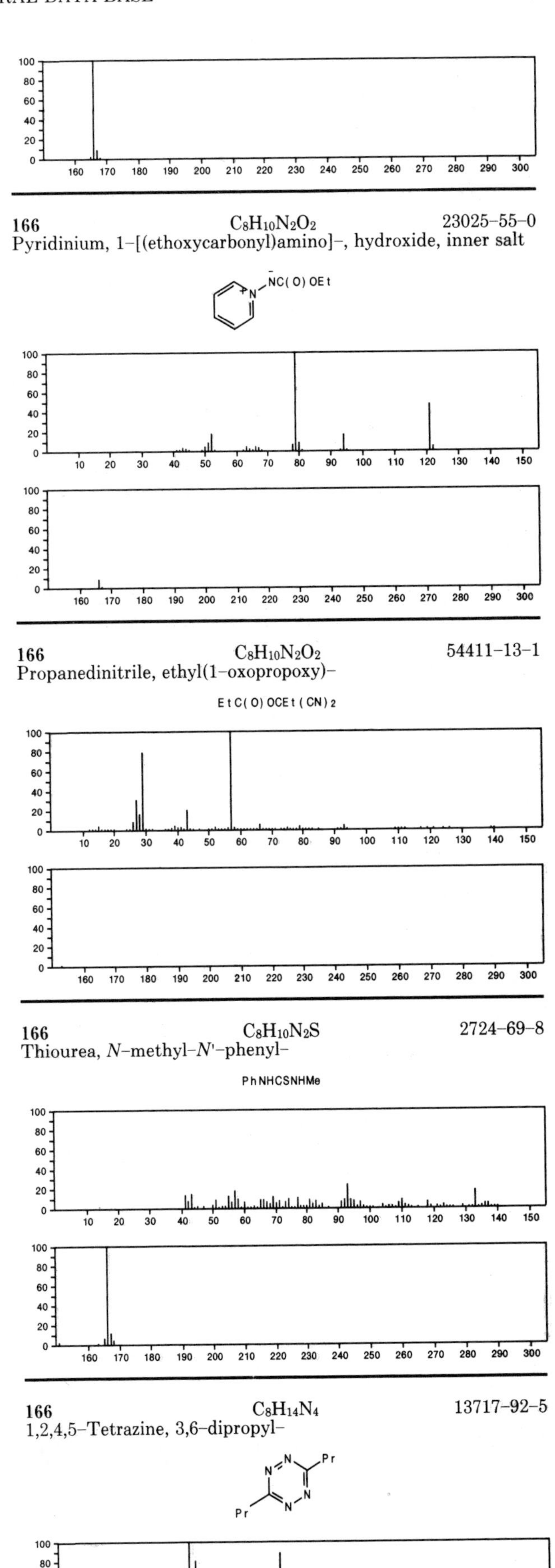

166 $C_8H_{10}N_2O_2$ 23025-55-0
Pyridinium, 1-[(ethoxycarbonyl)amino]-, hydroxide, inner salt
NC(O)OEt
166 $C_8H_{10}N_2O_2$ 54411-13-1
Propanedinitrile, ethyl(1-oxopropoxy)-
EtC(O)OCEt(CN)2
166 $C_8H_{10}N_2S$ 2724-69-8
Thiourea, *N*-methyl-*N'*-phenyl-
PhNHCSNHMe
166 $C_8H_{14}N_4$ 13717-92-5
1,2,4,5-Tetrazine, 3,6-dipropyl-
Pr

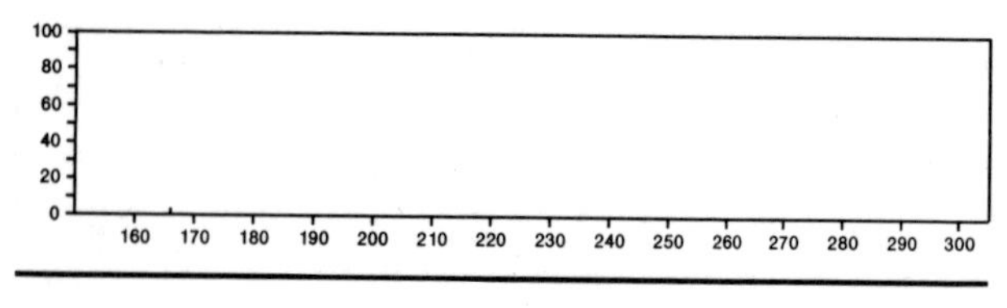

166 $C_8H_{14}N_4$ 13717-93-6
1,2,4,5-Tetrazine, 3,6-bis(1-methylethyl)-

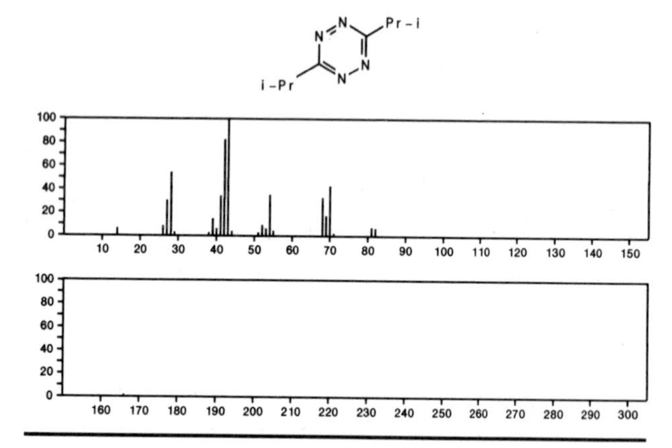

166 C_9H_7ClO 42180-82-5
Benzofuran, 5-chloro-2-methyl-

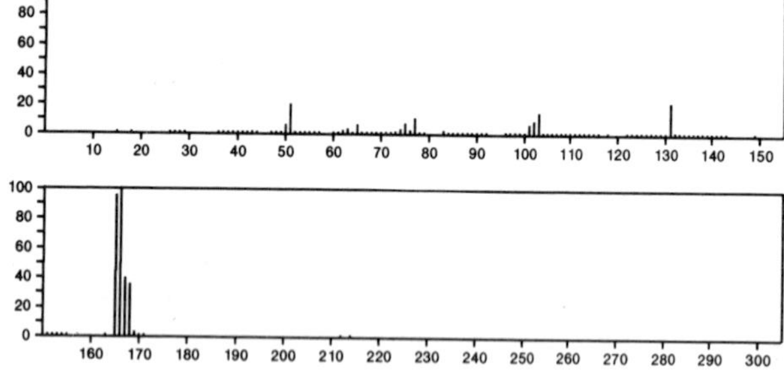

166 $C_9H_7FO_2$ 451-69-4
2-Propenoic acid, 3-(2-fluorophenyl)-

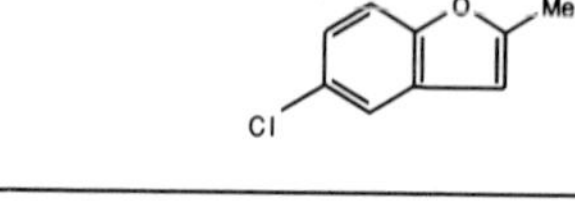

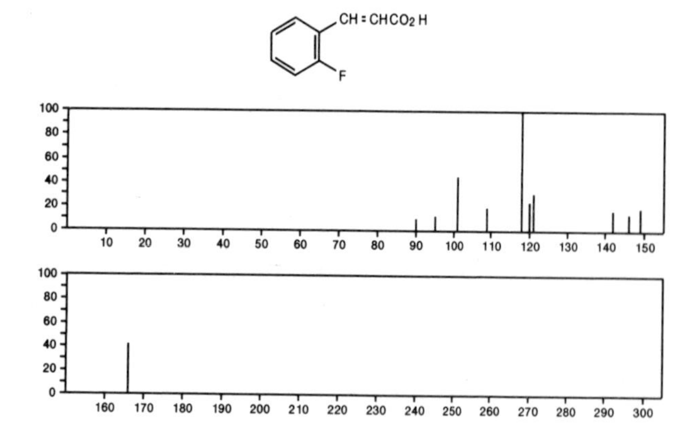

166 $C_9H_{10}OS$ 936-61-8
Benzenecarbothioic acid, *O*-ethyl ester

EtOC(S)Ph

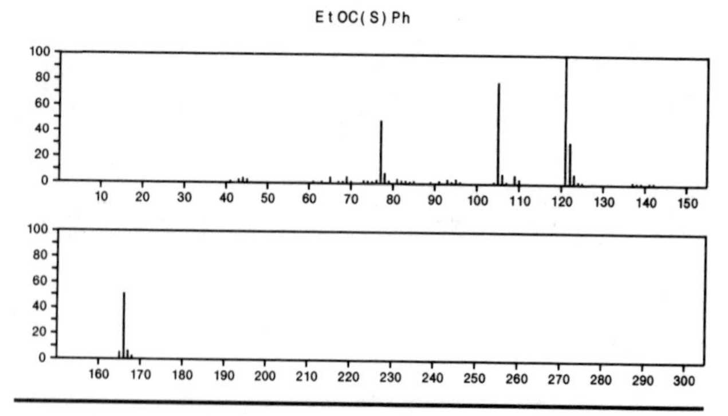

166 $C_9H_{10}OS$ 1484-17-9
Benzenecarbothioic acid, *S*-ethyl ester

EtSC(O)Ph

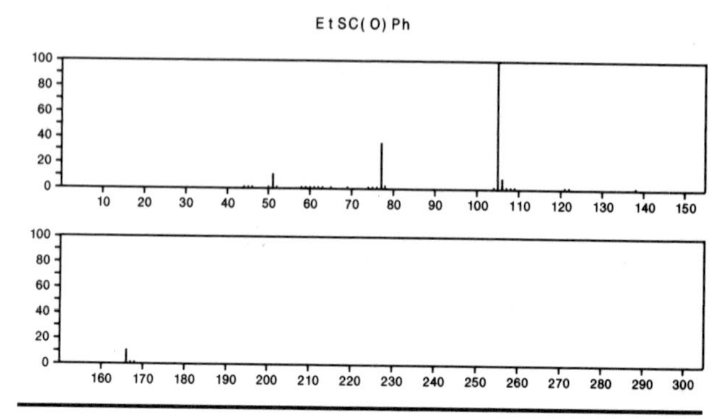

166 $C_9H_{10}OS$ 5925-74-6
Benzeneethanethioic acid, *S*-methyl ester

PhCH2C(O)SMe

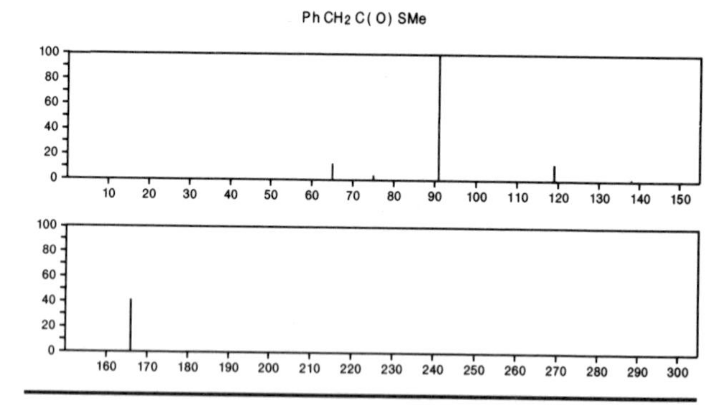

166 $C_9H_{10}OS$ 5925-77-9
Benzenecarbothioic acid, 4-methyl-, *S*-methyl ester

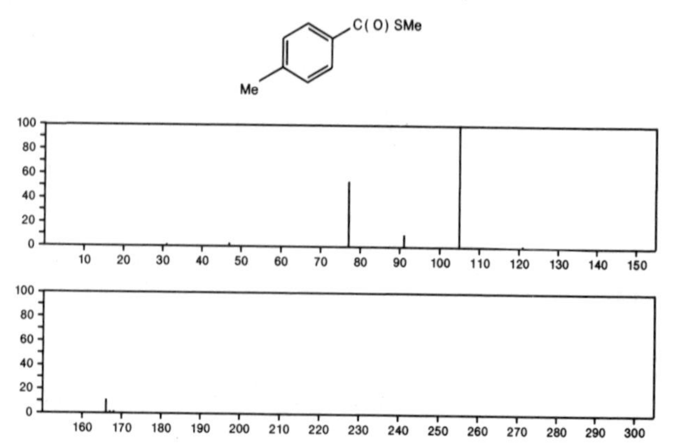

166 $C_9H_{10}OS$ 5977-80-0
Benzenecarbothioic acid, 4-methyl-, *O*-methyl ester

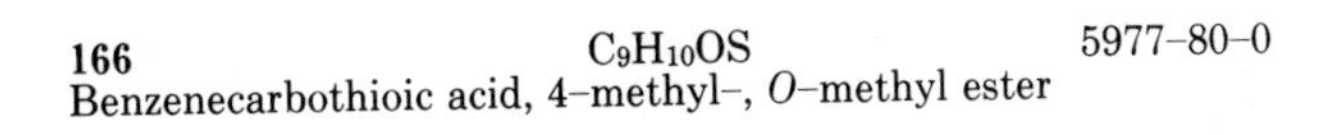

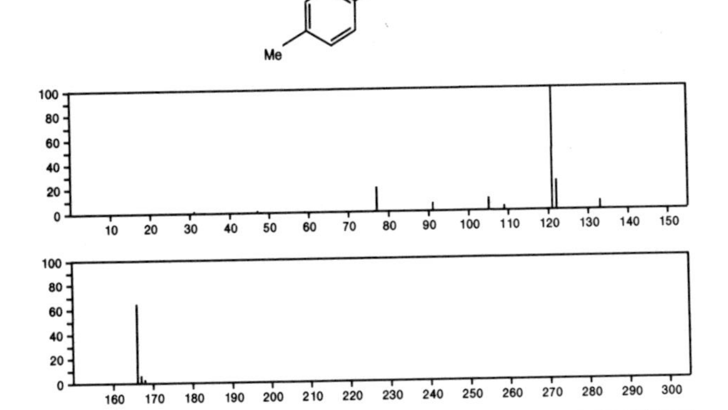

166 $C_9H_{10}OS$ 7715-00-6
Benzene, [2-(methylsulfinyl)ethenyl]-

PhCH=CHS(O)Me

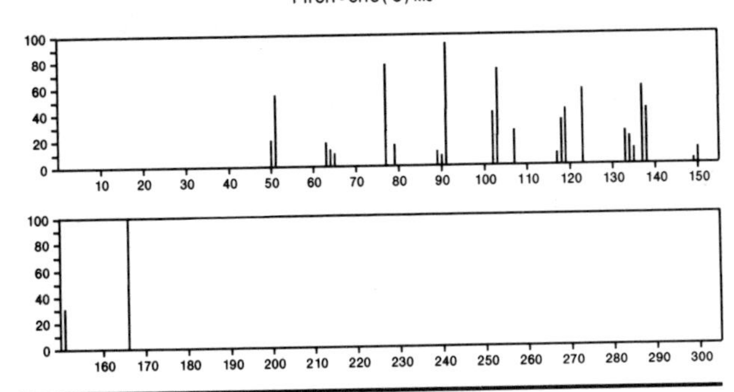

166 $C_9H_{10}OS$ 26027-98-5
Benzenecarbothioic acid, 3-methyl-, *O*-methyl ester

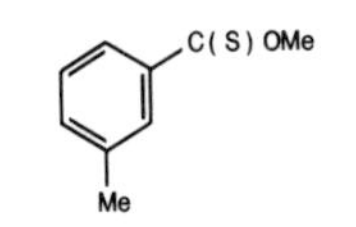

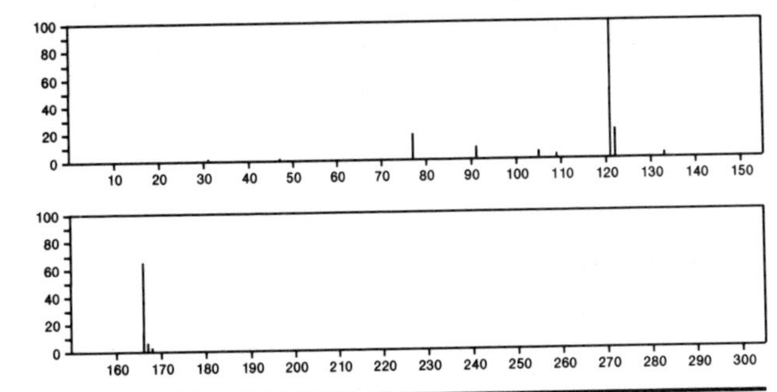

166 $C_9H_{10}OS$ 28145-55-3
m-Toluic acid, thio-, *S*-methyl ester

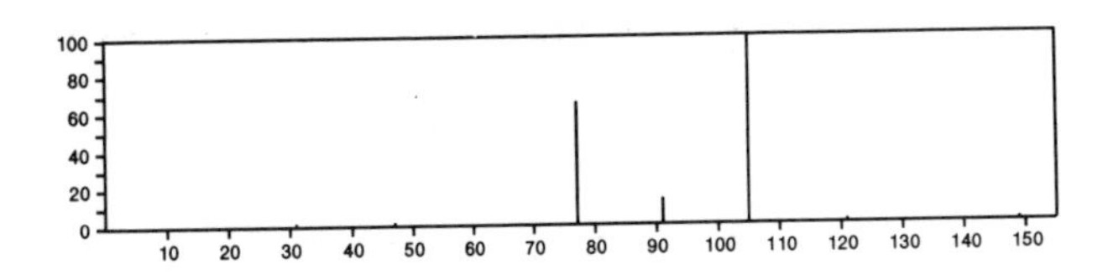

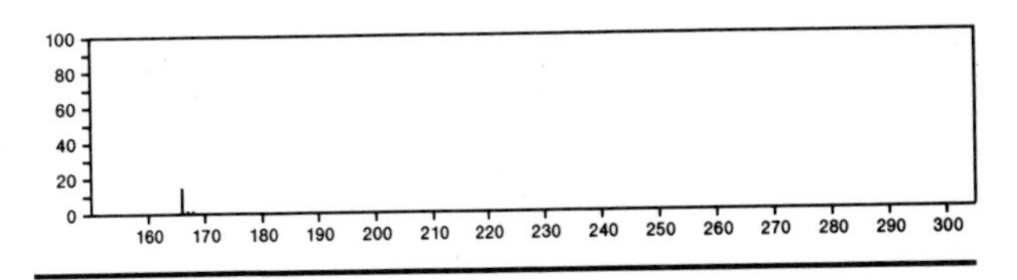

166 $C_9H_{10}OS$ 56817-89-1
Benzenecarbothioic acid, ethyl ester

PhCOSH + EtOH

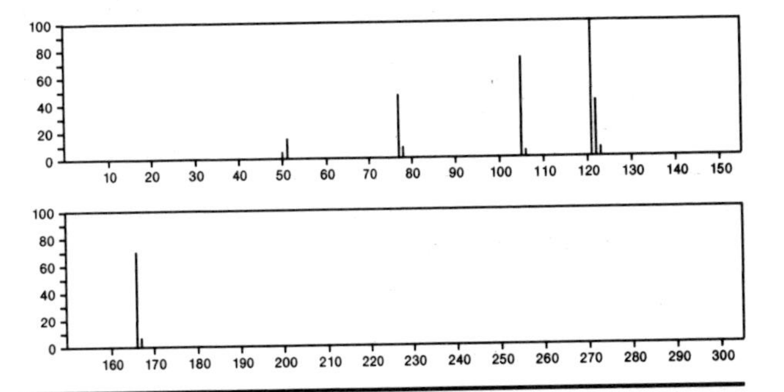

166 $C_9H_{10}O_3$ 93-02-7
Benzaldehyde, 2,5-dimethoxy-

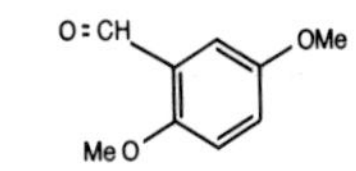

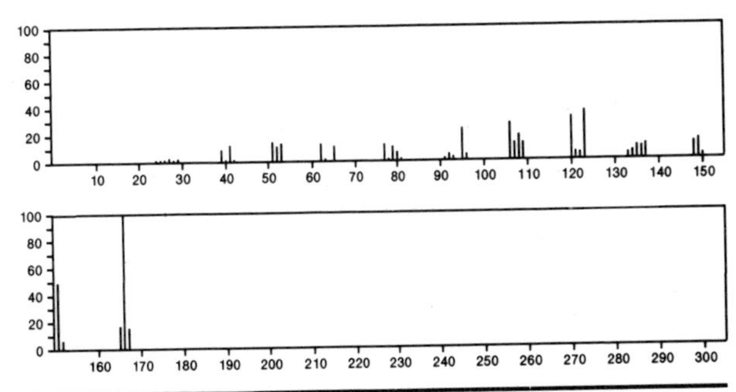

166 $C_9H_{10}O_3$ 93-25-4
Benzeneacetic acid, 2-methoxy-

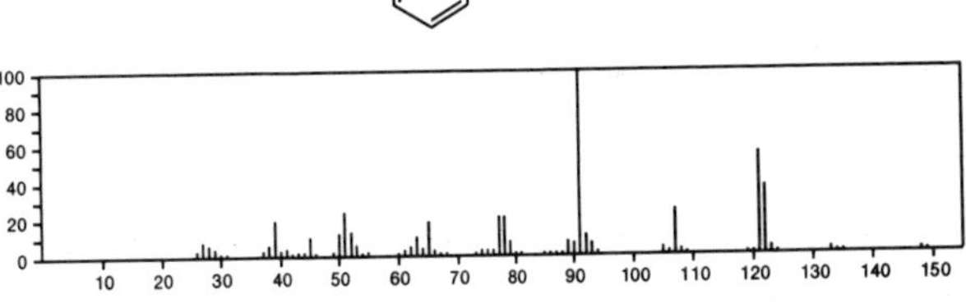

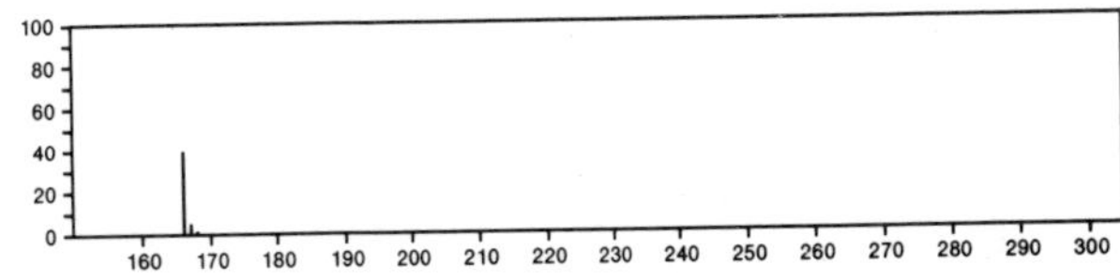

166 $C_9H_{10}O_3$ 104-01-8
Benzeneacetic acid, 4-methoxy-

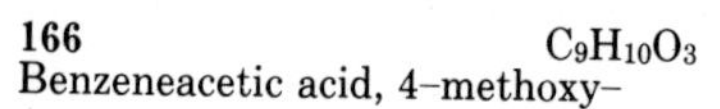

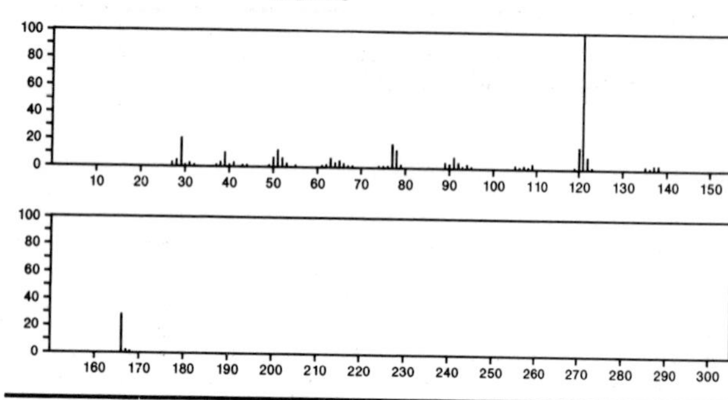

166 $C_9H_{10}O_3$ 120-14-9
Benzaldehyde, 3,4-dimethoxy-

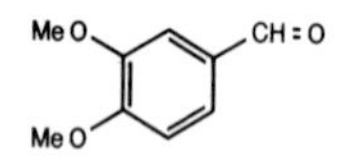

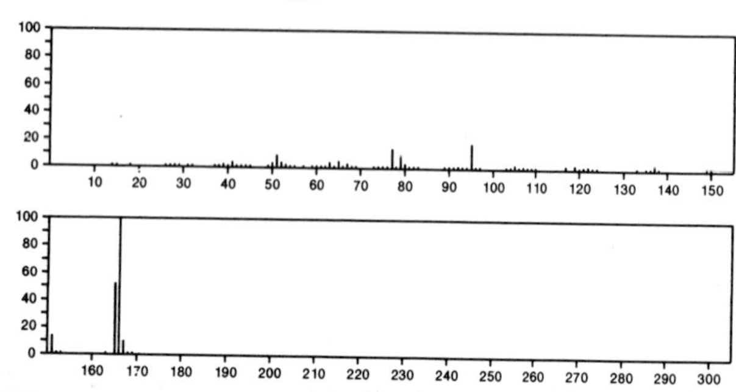

166 $C_9H_{10}O_3$ 120-47-8
Benzoic acid, 4-hydroxy-, ethyl ester

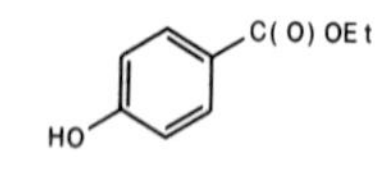

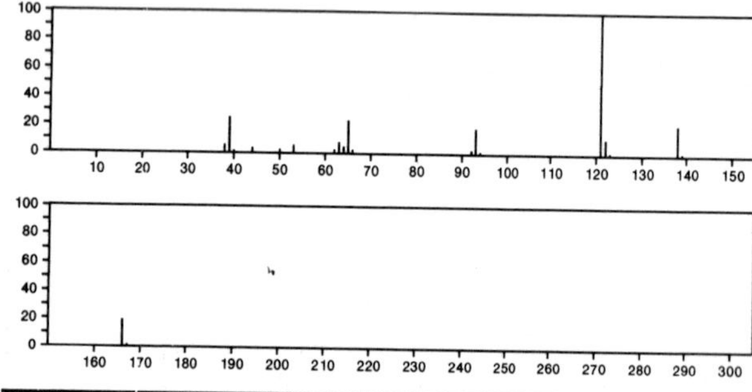

166 $C_9H_{10}O_3$ 121-98-2
Benzoic acid, 4-methoxy-, methyl ester

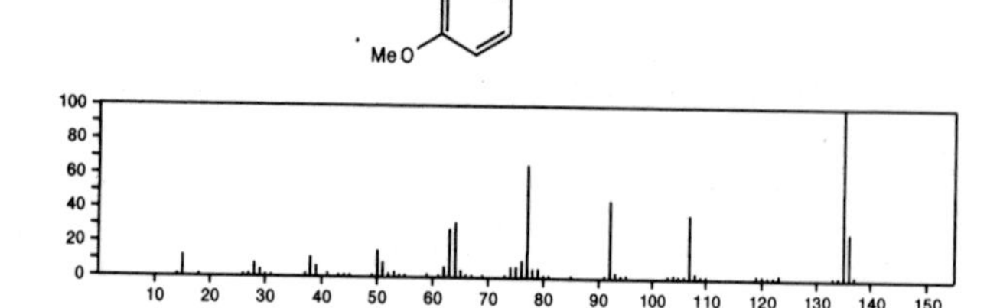

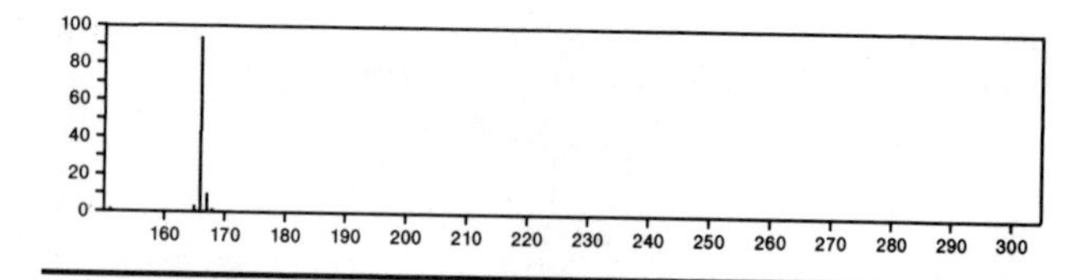

166 $C_9H_{10}O_3$ 156-05-8
Benzenepropanoic acid, α-hydroxy-

$HO_2CCH(OH)CH_2Ph$

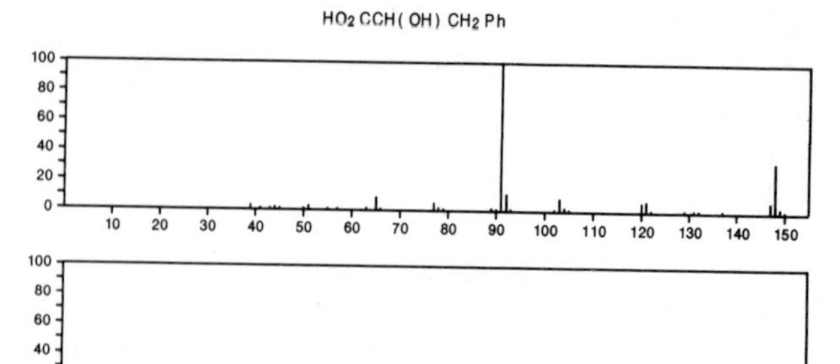

166 $C_9H_{10}O_3$ 498-02-2
Ethanone, 1-(4-hydroxy-3-methoxyphenyl)-

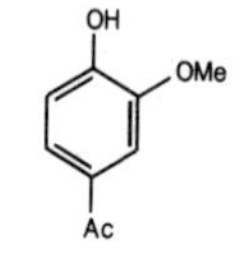

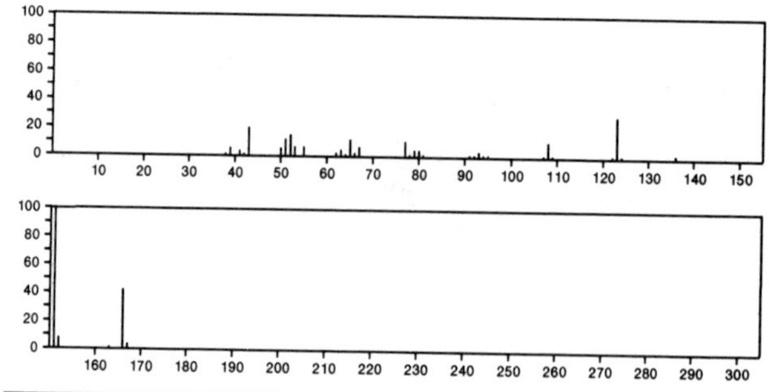

166 $C_9H_{10}O_3$ 501-97-3
Benzenepropanoic acid, 4-hydroxy-

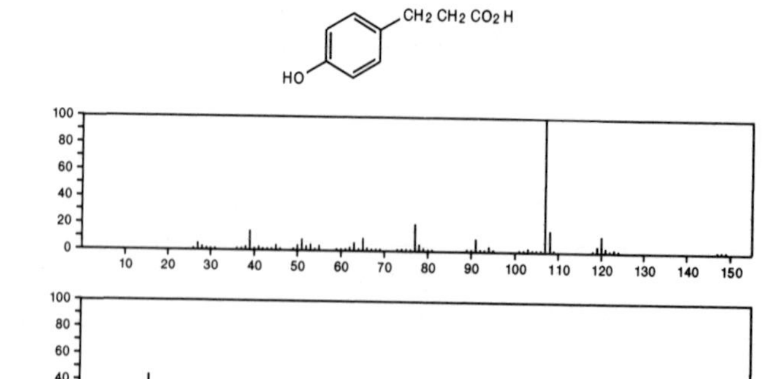

166 $C_9H_{10}O_3$ 515-30-0
Benzeneacetic acid, α-hydroxy-α-methyl-

$CPhMeMeCO_2H$

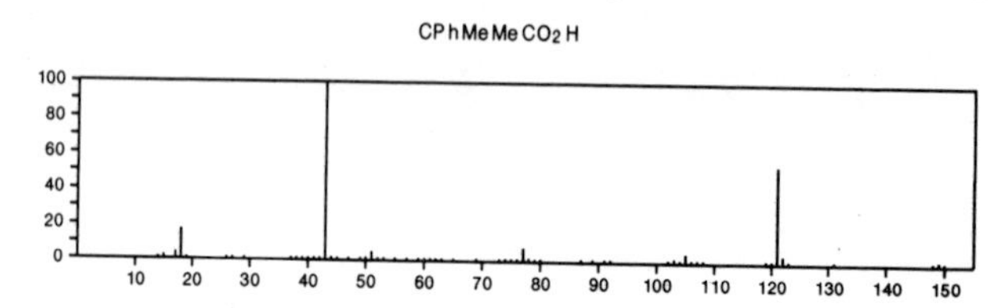

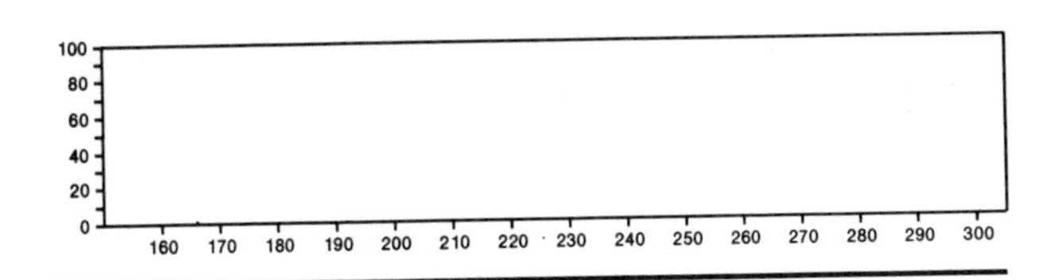

166 $C_9H_{10}O_3$ 552–41–0

Ethanone, 1–(2–hydroxy–4–methoxyphenyl)–

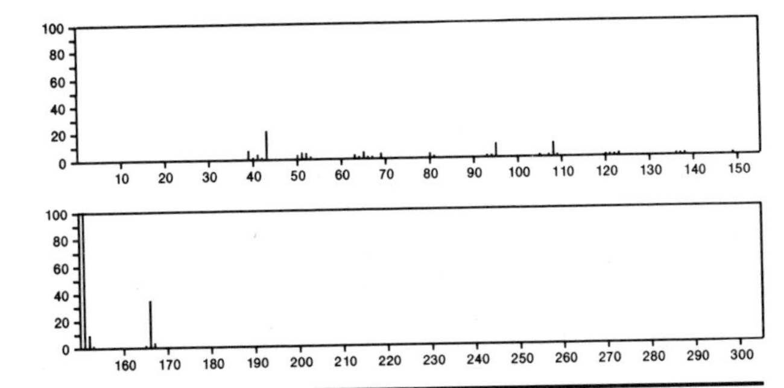

166 $C_9H_{10}O_3$ 606–45–1

Benzoic acid, 2–methoxy–, methyl ester

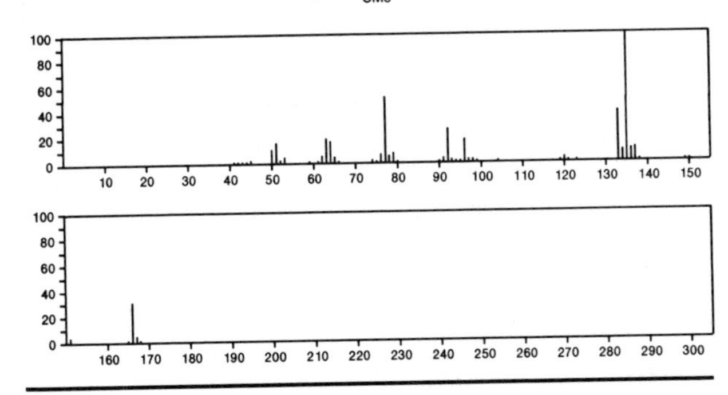

166 $C_9H_{10}O_3$ 613–70–7

Phenol, 2–methoxy–, acetate

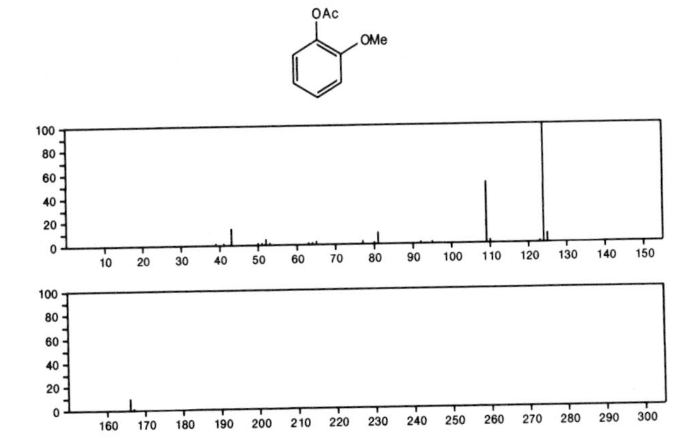

166 $C_9H_{10}O_3$ 623–20–1

2–Propenoic acid, 3–(2–furanyl)–, ethyl ester

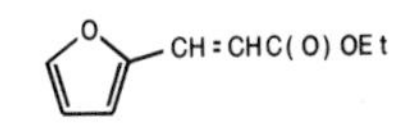

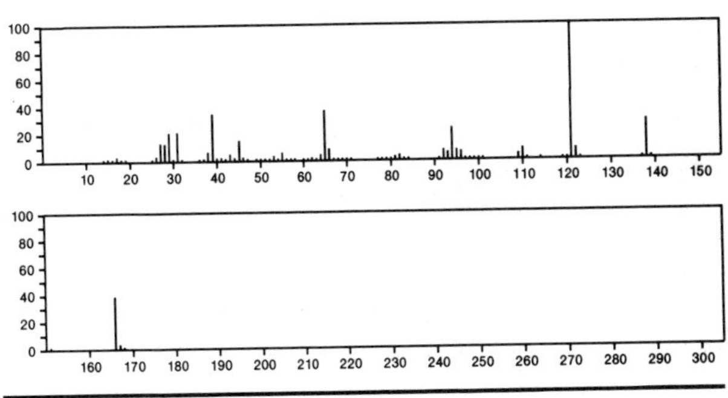

166 $C_9H_{10}O_3$ 703–23–1

Ethanone, 1–(2–hydroxy–6–methoxyphenyl)–

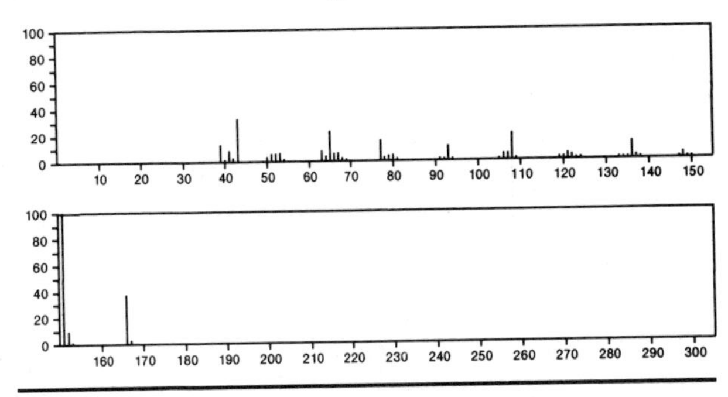

166 $C_9H_{10}O_3$ 705–15–7

Ethanone, 1–(2–hydroxy–5–methoxyphenyl)–

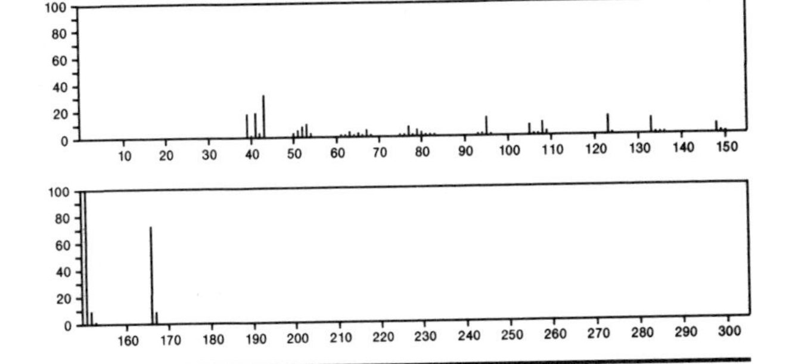

166 $C_9H_{10}O_3$ 771–90–4

Benzeneacetic acid, α–hydroxy–, methyl ester

MeOC(O)CH(OH)Ph

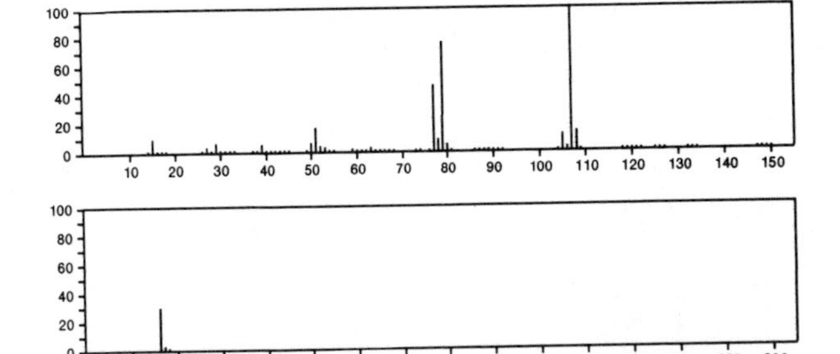

166 $C_9H_{10}O_3$ 940–31–8
Propanoic acid, 2–phenoxy–

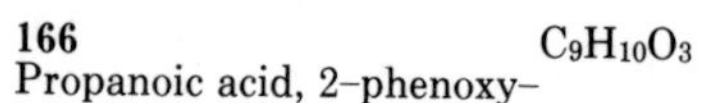

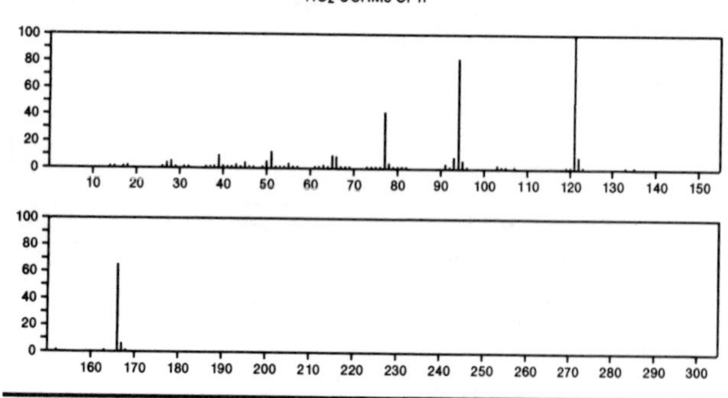

166 $C_9H_{10}O_3$ 1200–06–2
Phenol, 4–methoxy–, acetate

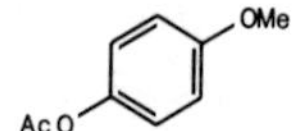

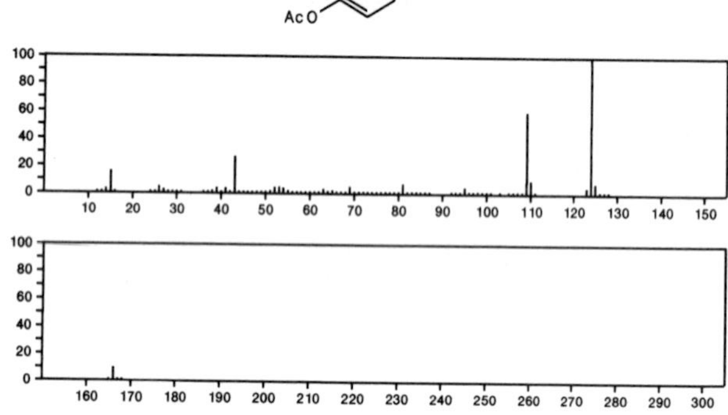

166 $C_9H_{10}O_3$ 1798–09–0
Benzeneacetic acid, 3–methoxy–

HO2 CCH2 OMe

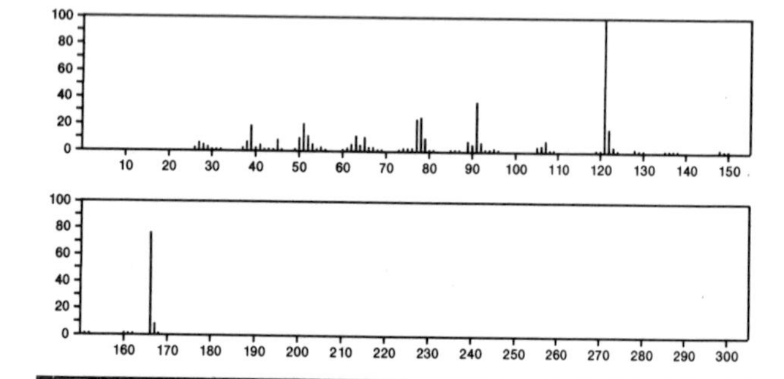

166 $C_9H_{10}O_3$ 1848–01–7
Carbonic acid, methyl 4–methylphenyl ester

OC(O) OMe
Me

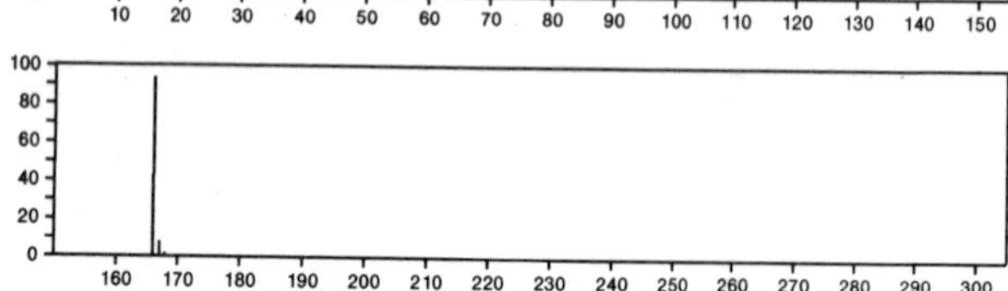

166 $C_9H_{10}O_3$ 1848–02–8
Carbonic acid, methyl *m*–tolyl ester

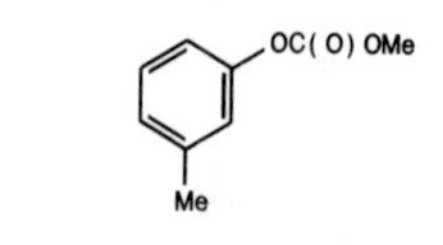

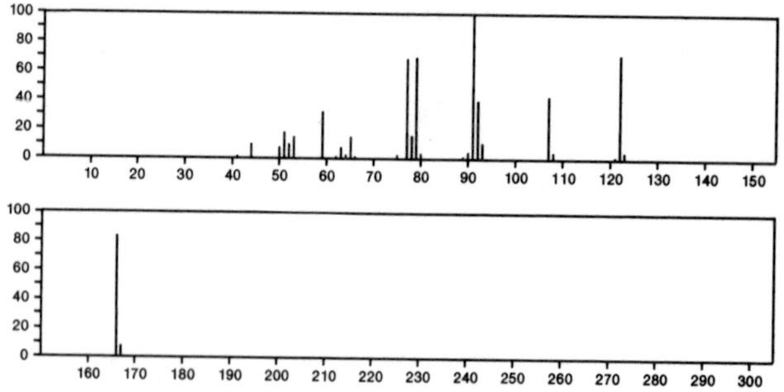

166 $C_9H_{10}O_3$ 2065–23–8
Acetic acid, phenoxy–, methyl ester

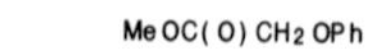

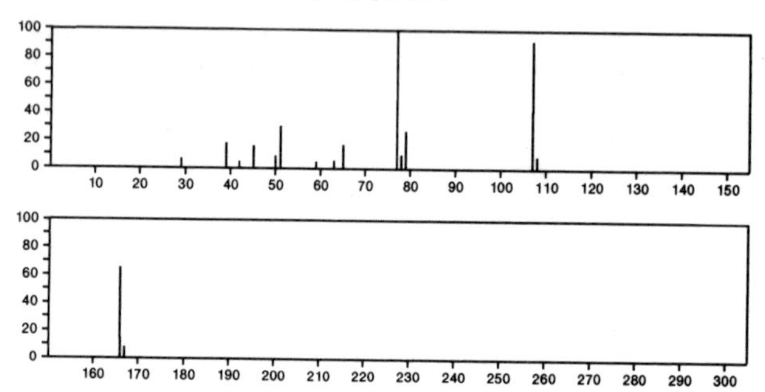

166 $C_9H_{10}O_3$ 2990–31–0
Benzaldehyde, 4,6–dihydroxy–2,3–dimethyl–

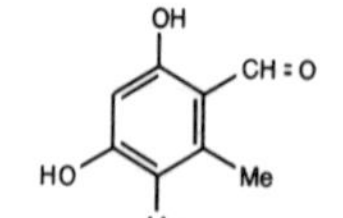

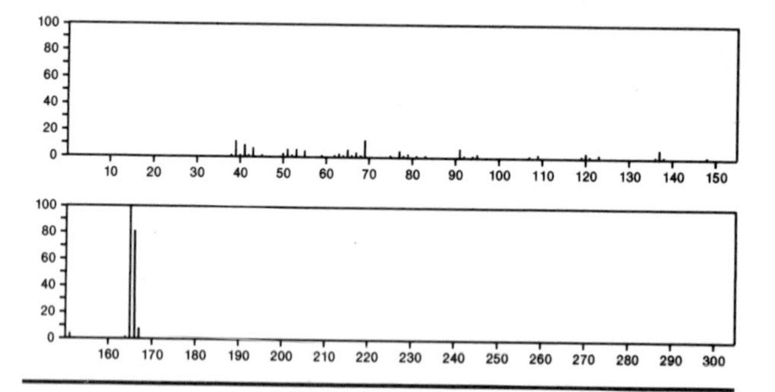

166 $C_9H_{10}O_3$ 3878–46–4
Carbonic acid, ethyl phenyl ester

PhOC(O) OEt

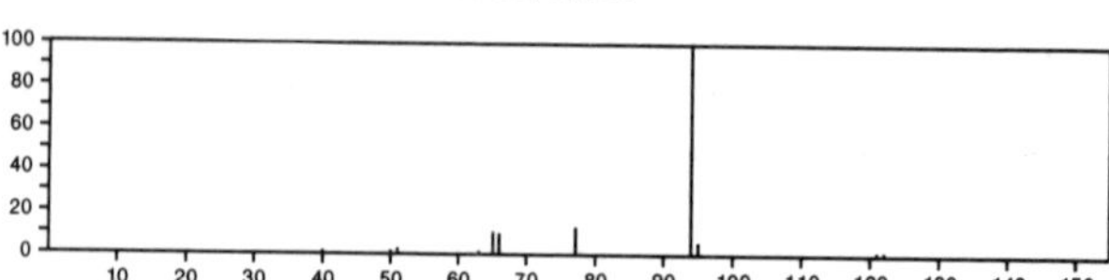

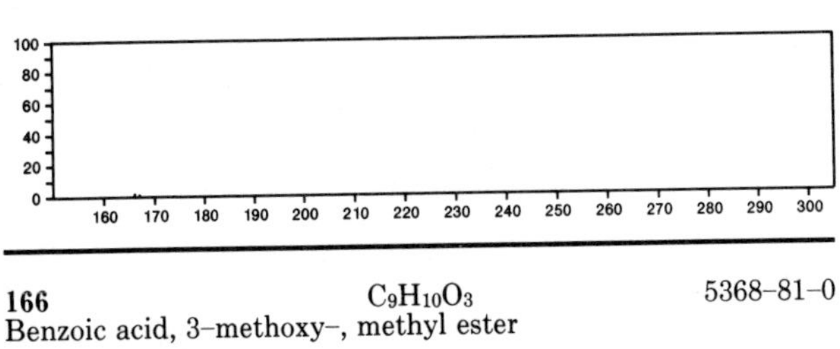

166 $C_9H_{10}O_3$ 5368–81–0
Benzoic acid, 3–methoxy–, methyl ester

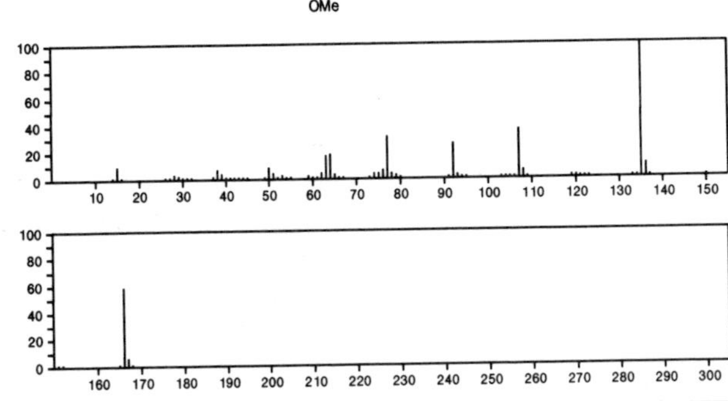

166 $C_9H_{10}O_3$ 5451–83–2
Phenol, 3–methoxy–, acetate

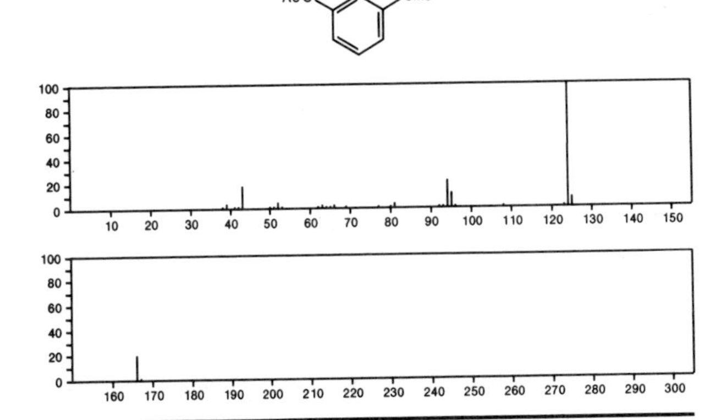

166 $C_9H_{10}O_3$ 7521–38–2
4H–Pyran–4–one, 3–acetyl–2,6–dimethyl–

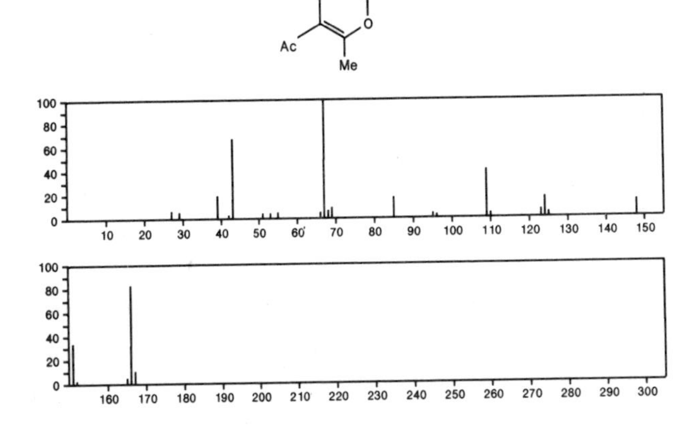

166 $C_9H_{10}O_3$ 14199–15–6
Benzeneacetic acid, 4–hydroxy–, methyl ester

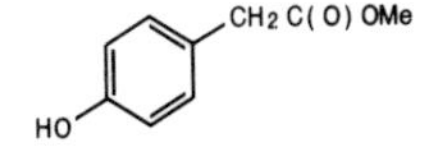

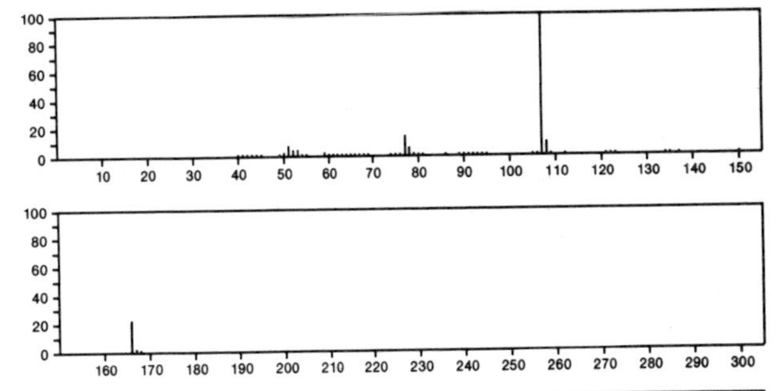

166 $C_9H_{10}O_3$ 22446–37–3
Benzeneacetic acid, 2–hydroxy–, methyl ester

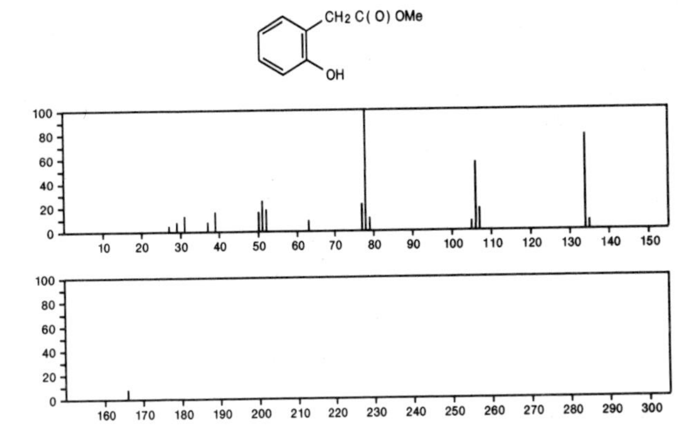

166 $C_9H_{10}O_3$ 23287–26–5
Benzoic acid, 2–hydroxy–3–methyl–, methyl ester

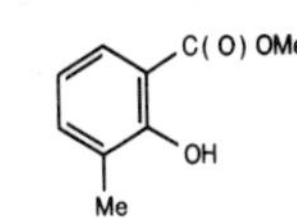

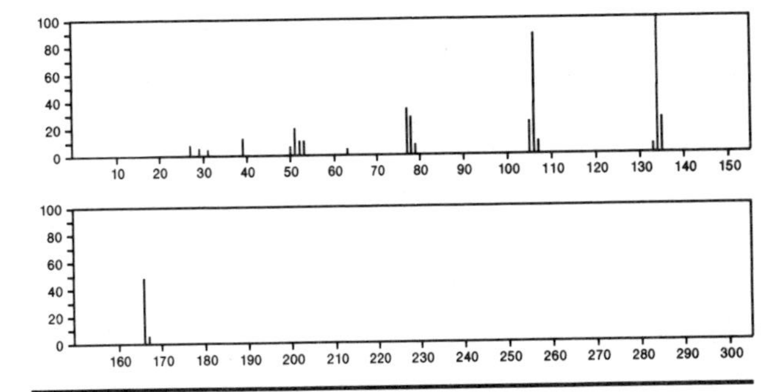

166 $C_9H_{10}O_3$ 33528–09–5
Benzoic acid, 2–hydroxy–6–methyl–, methyl ester

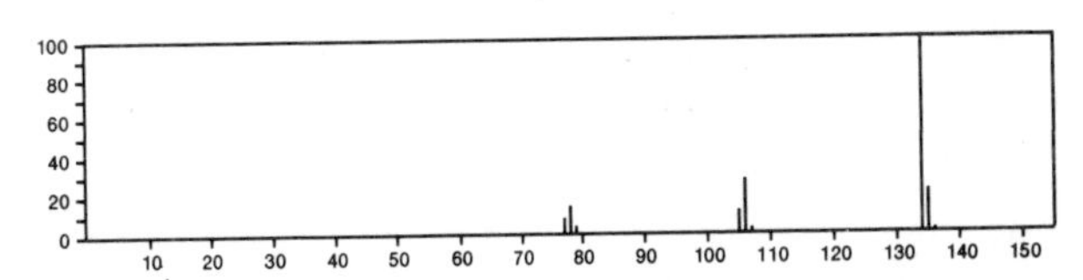

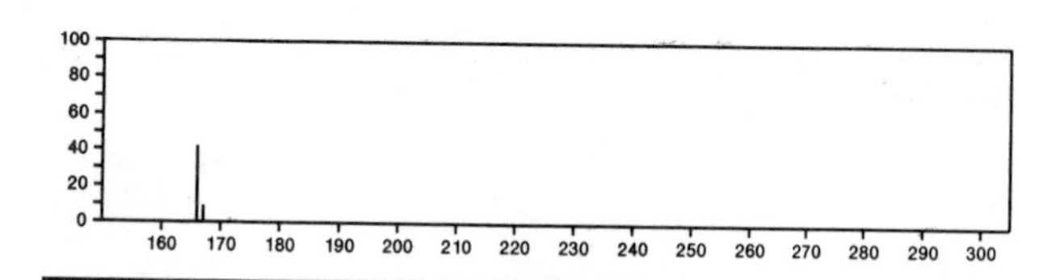

166 $C_9H_{10}O_3$ 34883-08-4
Benzaldehyde, 2-hydroxy-4-methoxy-6-methyl-

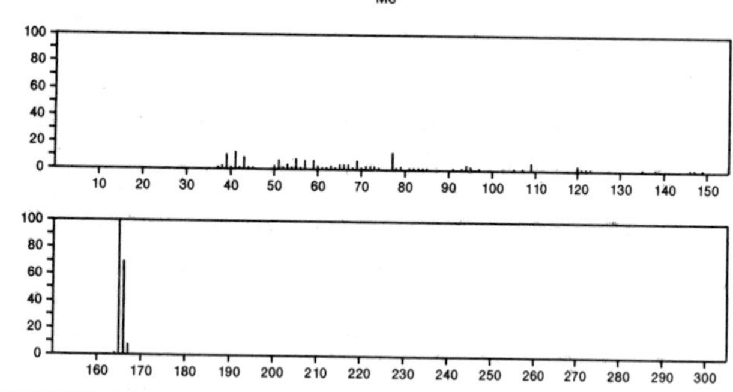

166 $C_9H_{10}O_3$ 34883-14-2
Benzaldehyde, 2,4-dihydroxy-3,6-dimethyl-

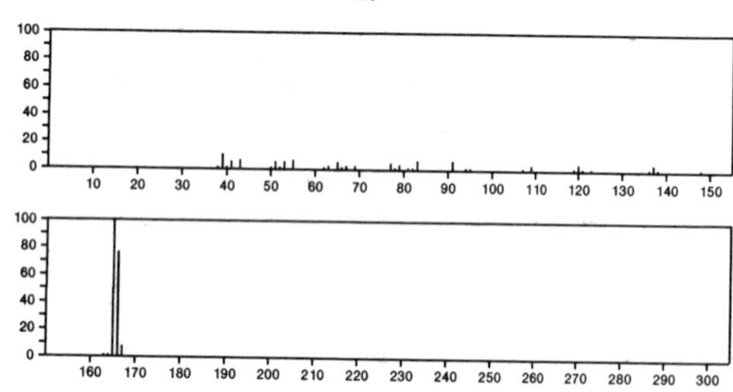

166 $C_9H_{10}O_3$ 42058-59-3
Benzeneacetic acid, 3-hydroxy-, methyl ester

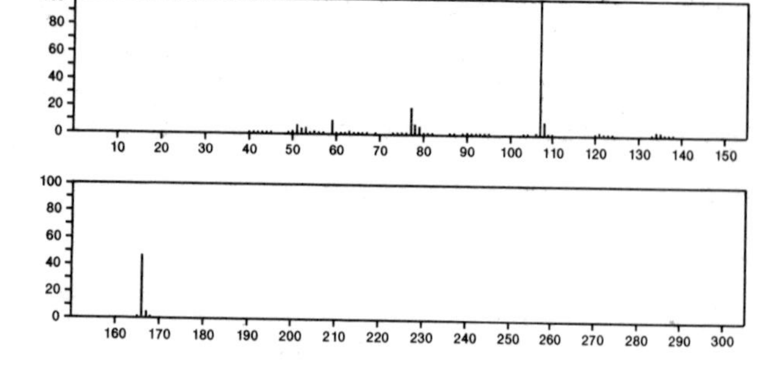

166 $C_9H_{10}O_3$ 54966-48-2
2-Butynoic acid, 4-cyclopropyl-4-oxo-, ethyl ester

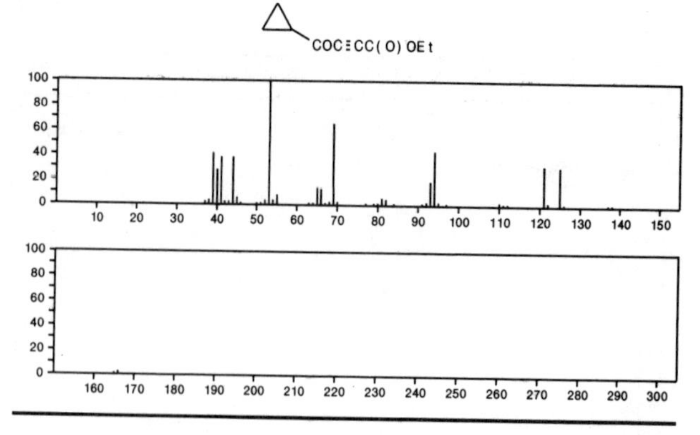

166 $C_9H_{14}OSi$ 1529-17-5
Silane, trimethylphenoxy-

Me3SiOPh

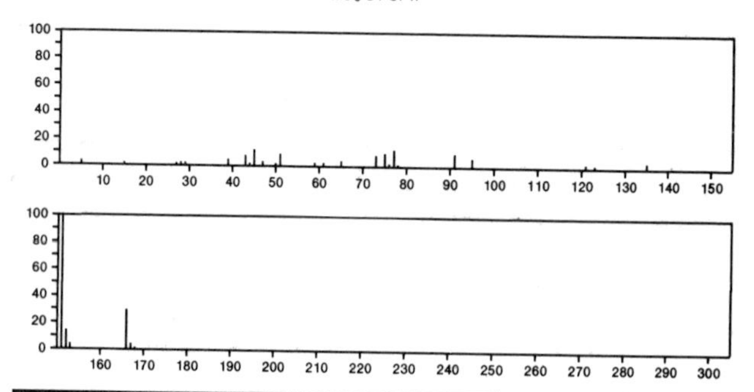

166 $C_{10}H_{11}Cl$ 16608-68-7
2-Butene, 3-chloro-1-phenyl-, (*Z*)-

MeCCl =CHCH2Ph

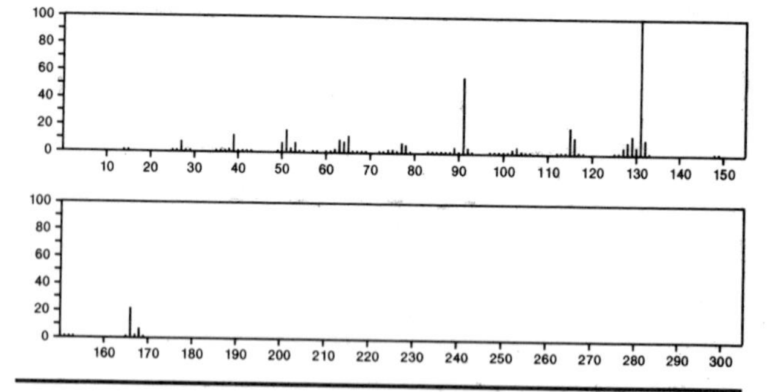

166 $C_{10}H_{11}Cl$ 54411-12-0
Benzene, (2-chloro-2-butenyl)-

PhCH2CCl =CHMe

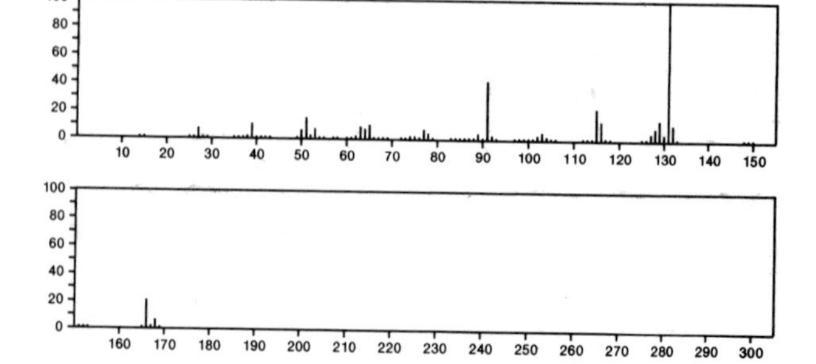

166 $C_{10}H_{11}FO$ 582–83–2
1–Butanone, 1–(4–fluorophenyl)–

166 $C_{10}H_{14}O_2$ 98–29–3
1,2–Benzenediol, 4–(1,1–dimethylethyl)–

166 $C_{10}H_{14}O_2$ 101–48–4
Benzene, (2,2–dimethoxyethyl)–

(MeO)2CHCH2Ph

166 $C_{10}H_{14}O_2$ 122–94–1
Phenol, 4–butoxy–

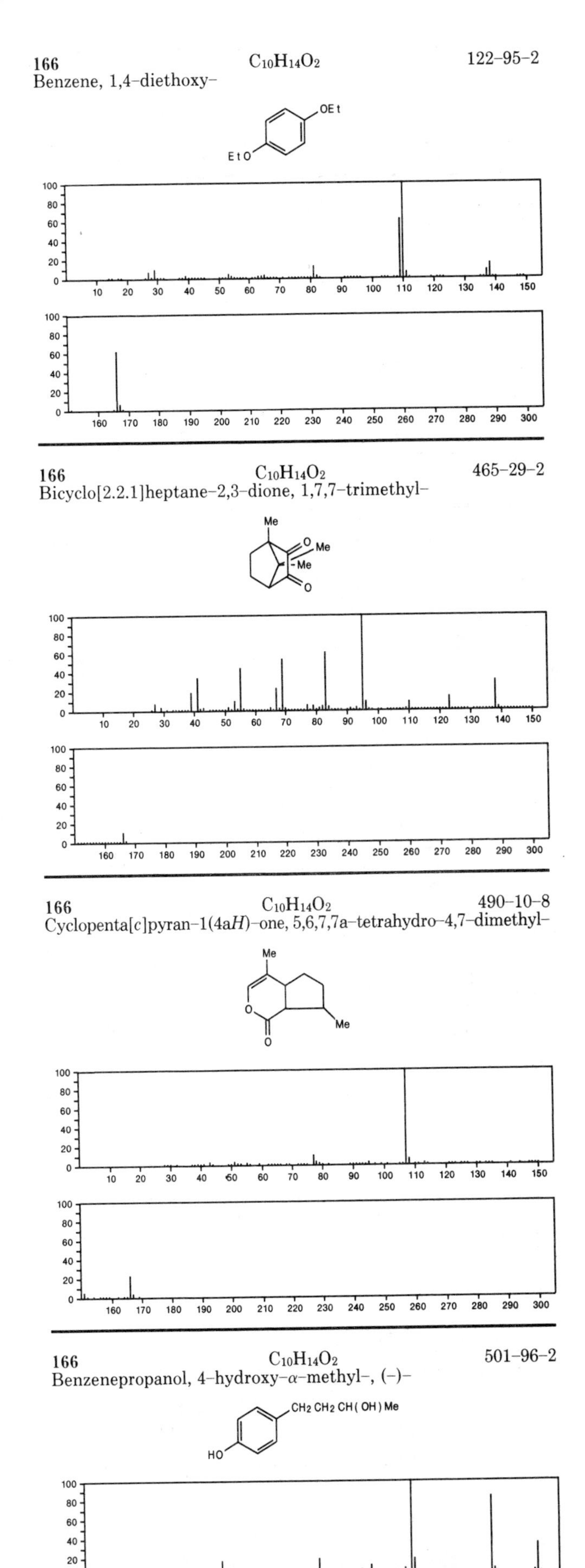

166 $C_{10}H_{14}O_2$ 122–95–2
Benzene, 1,4–diethoxy–

166 $C_{10}H_{14}O_2$ 465–29–2
Bicyclo[2.2.1]heptane–2,3–dione, 1,7,7–trimethyl–

166 $C_{10}H_{14}O_2$ 490–10–8
Cyclopenta[*c*]pyran–1(4a*H*)–one, 5,6,7,7a–tetrahydro–4,7–dimethyl–

166 $C_{10}H_{14}O_2$ 501–96–2
Benzenepropanol, 4–hydroxy–α–methyl–, (–)–

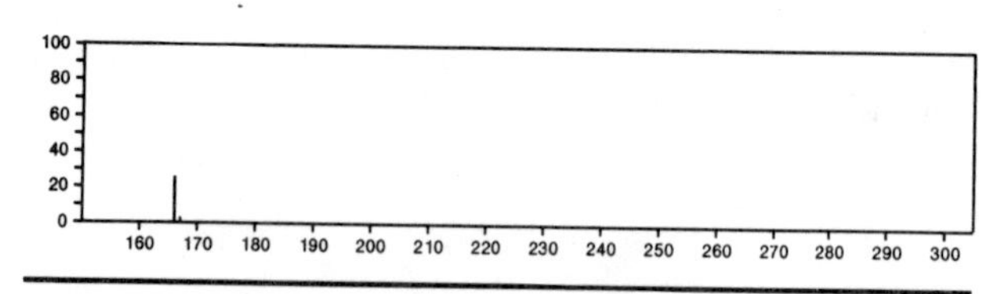

166 $C_{10}H_{14}O_2$ 828–12–6
Methane, isopropoxyphenoxy–

PhOCH2 OPr – i

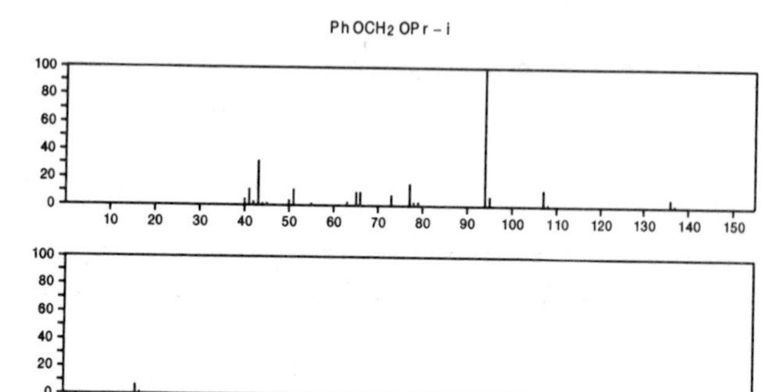

166 $C_{10}H_{14}O_2$ 2785–87–7
Phenol, 2–methoxy–4–propyl–

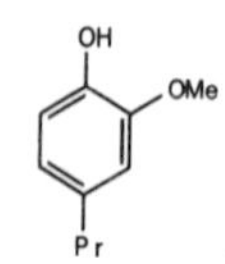

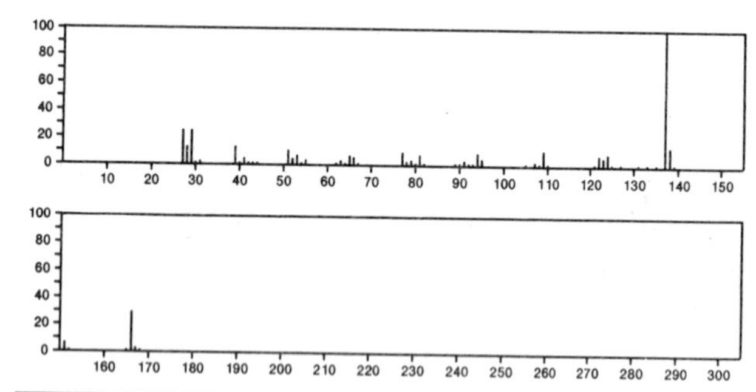

166 $C_{10}H_{14}O_2$ 4013–37–0
Benzene, (1,2–dimethoxyethyl)–

MeOCH2 CH(OMe) Ph

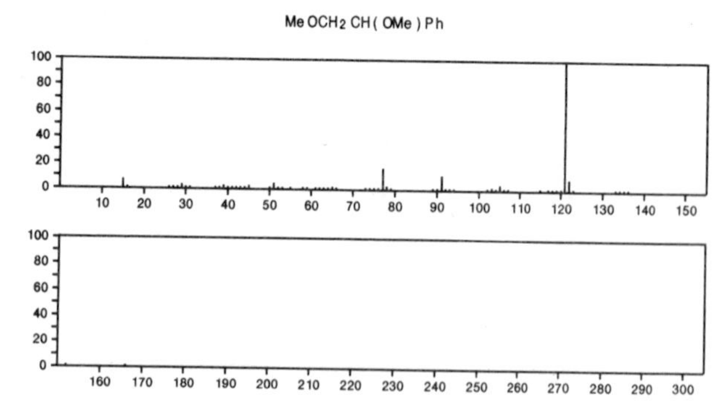

166 $C_{10}H_{14}O_2$ 4230–32–4
Bicyclo[2.2.1]heptane–2,5–dione, 1,7,7–trimethyl–

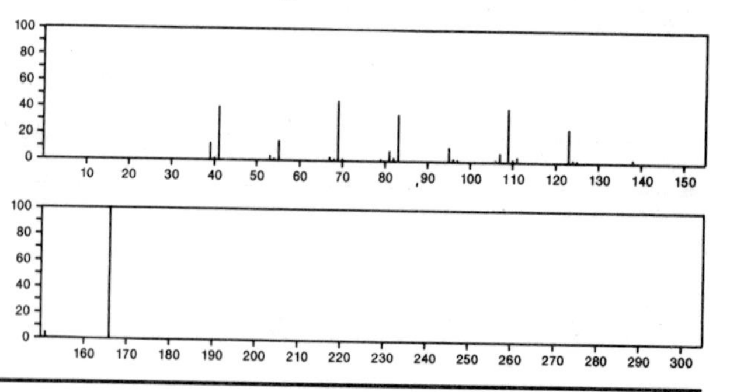

166 $C_{10}H_{14}O_2$ 4457–16–3
Methane, phenoxypropoxy–

PhOCH2 OPr

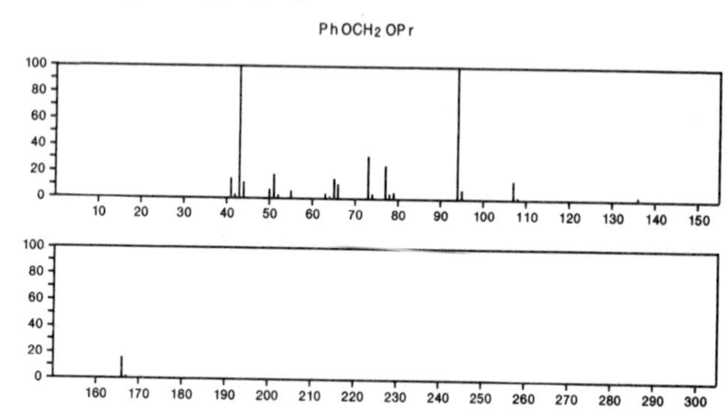

166 $C_{10}H_{14}O_2$ 4799–66–0
Ethanol, 2–(1–phenylethoxy)–

PhCHMe OCH2 CH2 OH

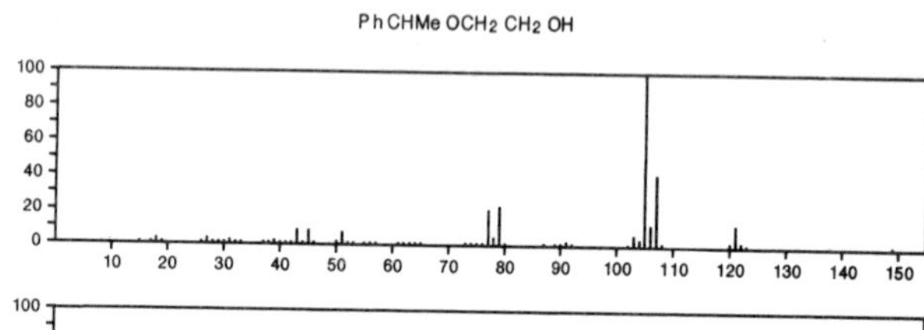

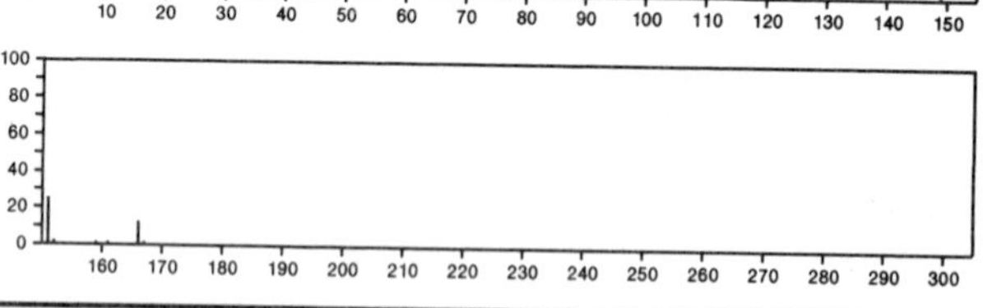

166 $C_{10}H_{14}O_2$ 4799–68–2
1–Propanol, 3–(phenylmethoxy)–

PhCH2 O(CH2)3 OH

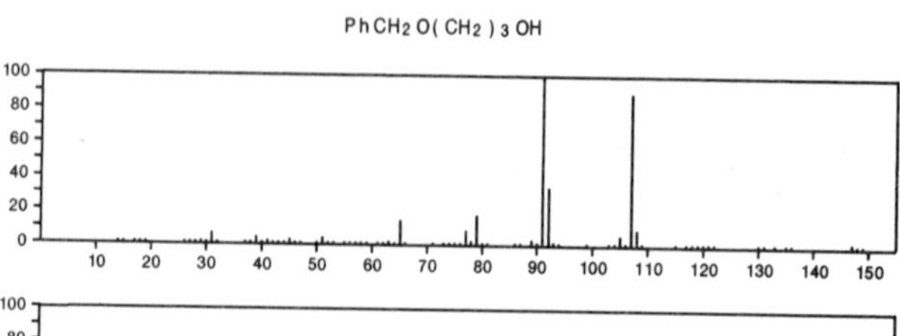

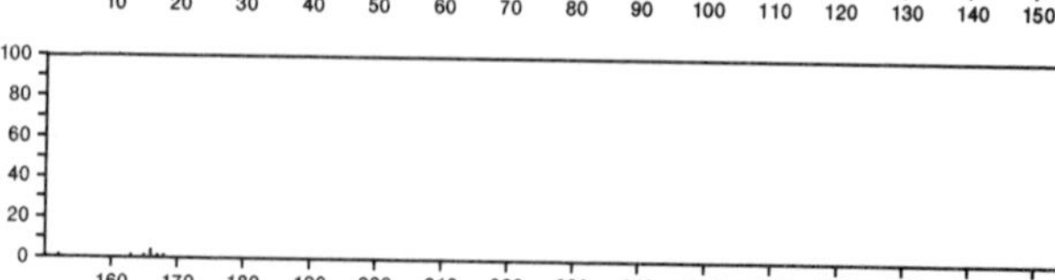

166 $C_{10}H_{14}O_2$ 5349-60-0
Benzenemethanol, α-ethyl-4-methoxy-

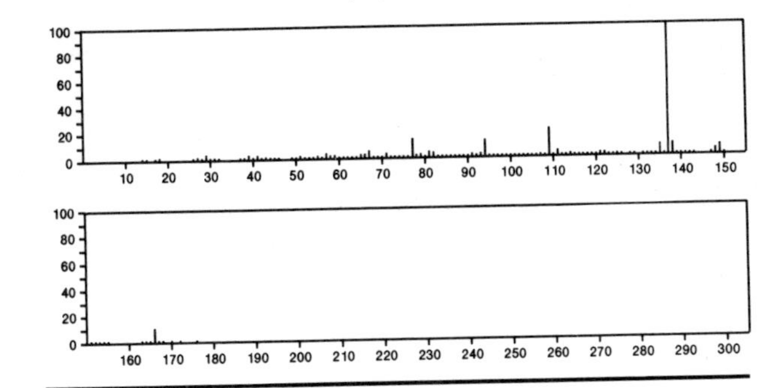

166 $C_{10}H_{14}O_2$ 6684-66-8
Spiro[4.5]decane-6,10-dione

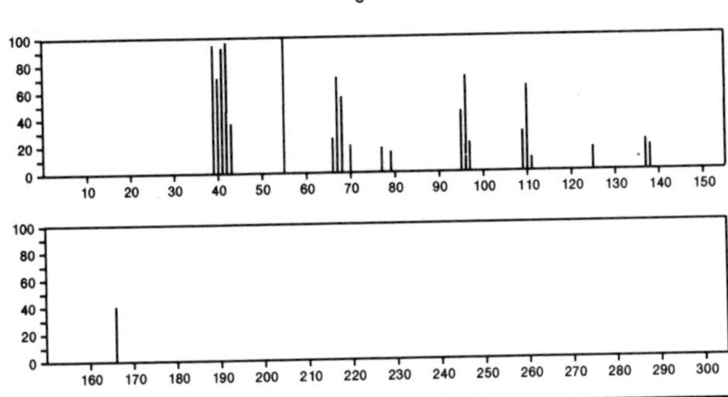

166 $C_{10}H_{14}O_2$ 10493-37-5
Benzenepropanol, 2-methoxy-

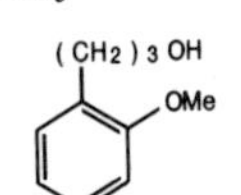

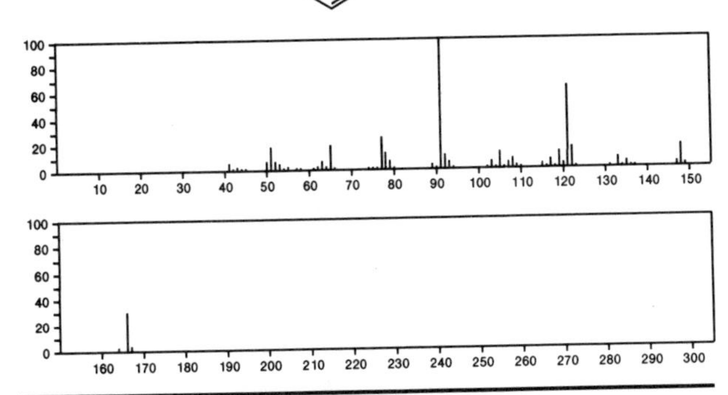

166 $C_{10}H_{14}O_2$ 13807-91-5
2-Propanol, 1-(phenylmethoxy)-

PhCH2OCH2CH(OH)Me

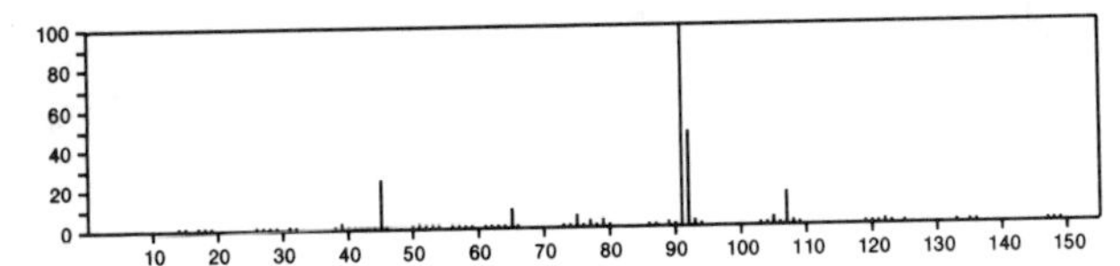

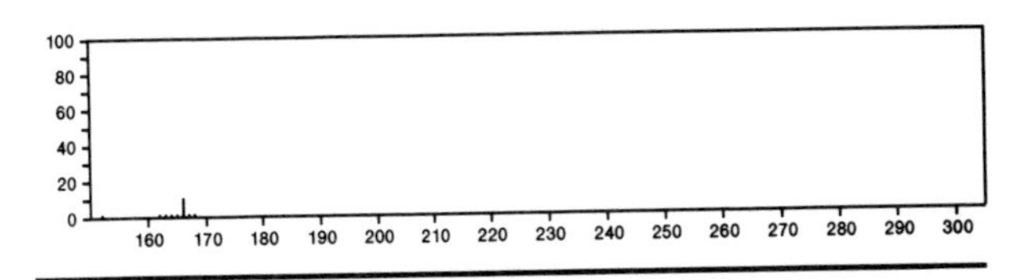

166 $C_{10}H_{14}O_2$ 17190-74-8
2-Cyclopenten-1-one, 2-(2-butenyl)-4-hydroxy-3-methyl-, (*Z*)-

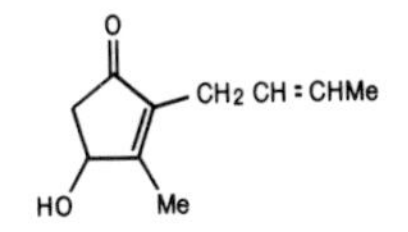

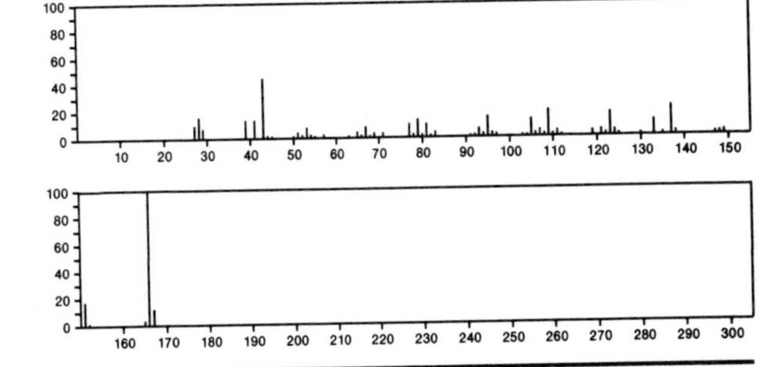

166 $C_{10}H_{14}O_2$ 19594-02-6
Ethane, 1-ethoxy-2-phenoxy-

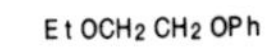

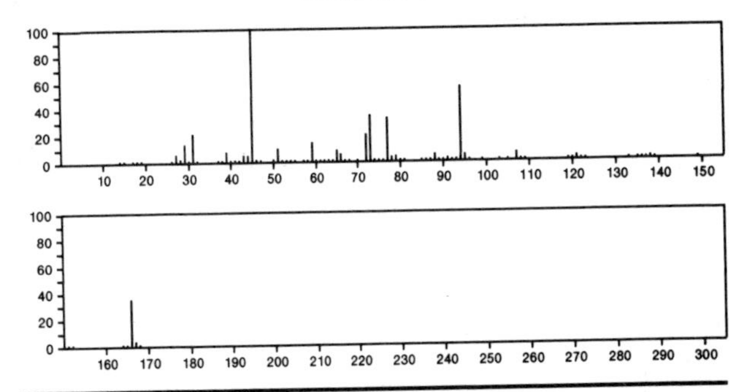

166 $C_{10}H_{14}O_2$ 22607-13-2
1,2-Butanediol, 1-phenyl-

EtCH(OH)CH(OH)Ph

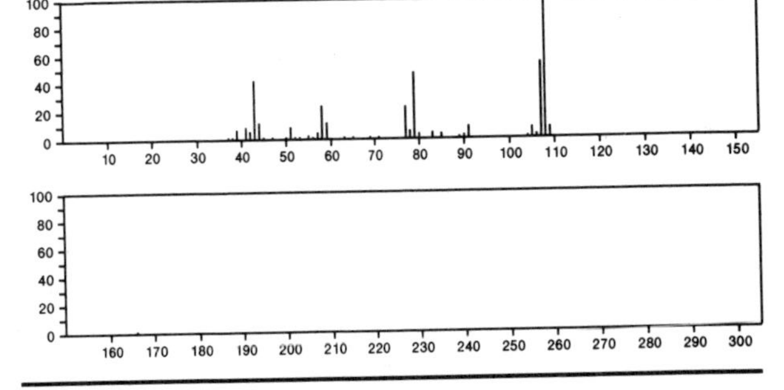

166 $C_{10}H_{14}O_2$ 24765-53-5
1,3-Propanediol, 2-methyl-2-phenyl-

HOCH2CPhMeCH2OH

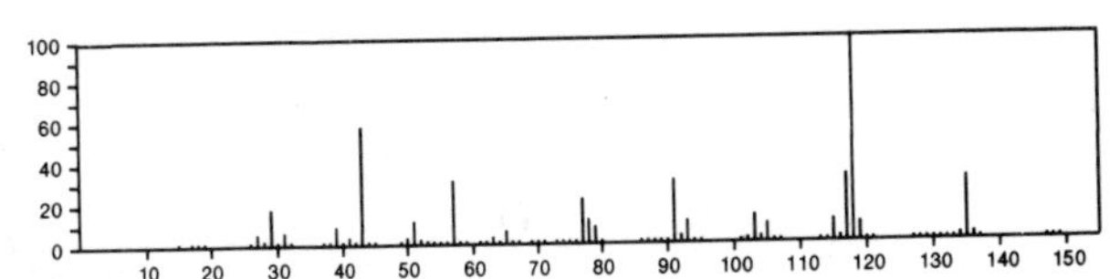

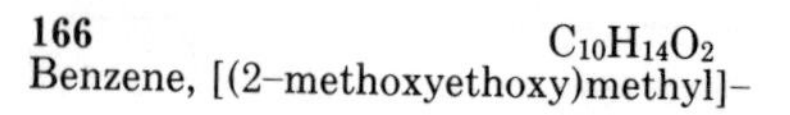

166 $C_{10}H_{14}O_2$ 31600-56-3
Benzene, [(2-methoxyethoxy)methyl]-

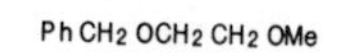

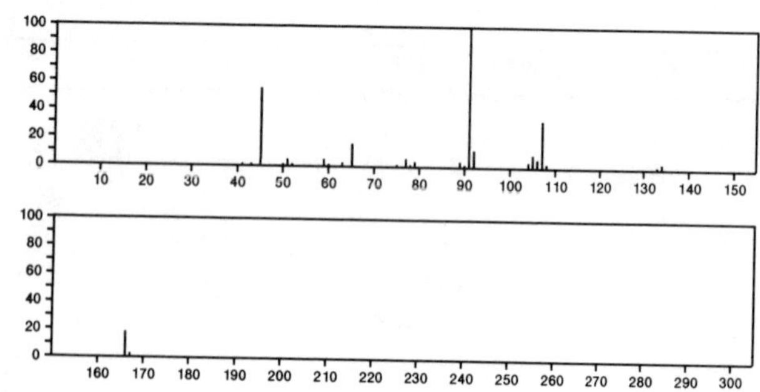

166 $C_{10}H_{14}O_2$ 34883-02-8
Phenol, 5-methoxy-2,3,4-trimethyl-

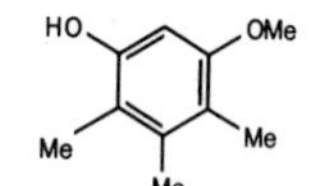

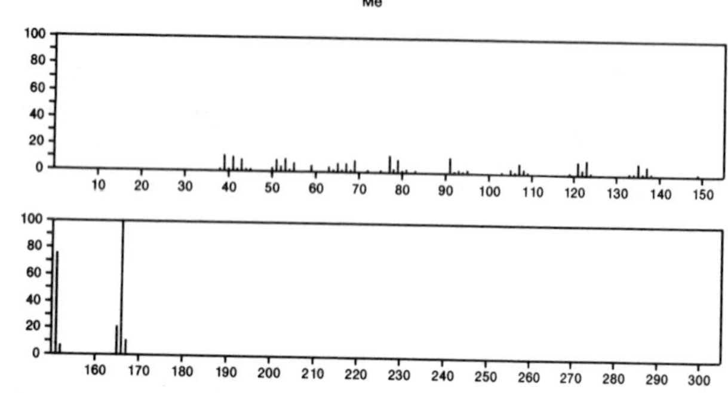

166 $C_{10}H_{14}O_2$ 34883-03-9
Phenol, 3-methoxy-2,5,6-trimethyl-

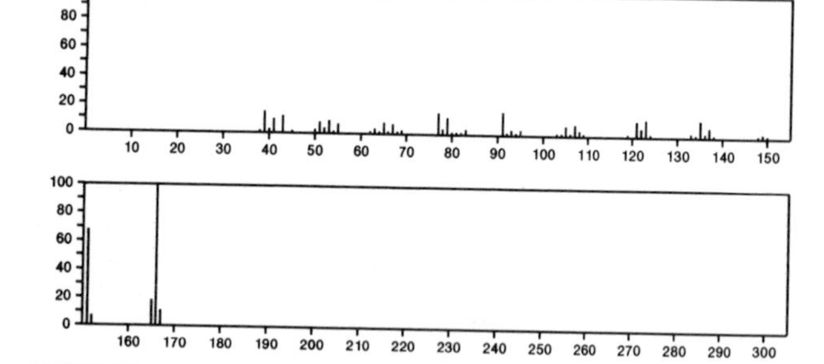

166 $C_{10}H_{14}O_2$ 34883-04-0
Phenol, 3-methoxy-2,4,5-trimethyl-

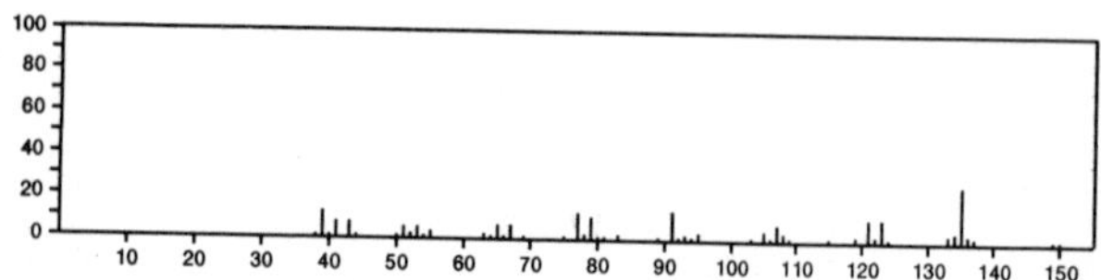

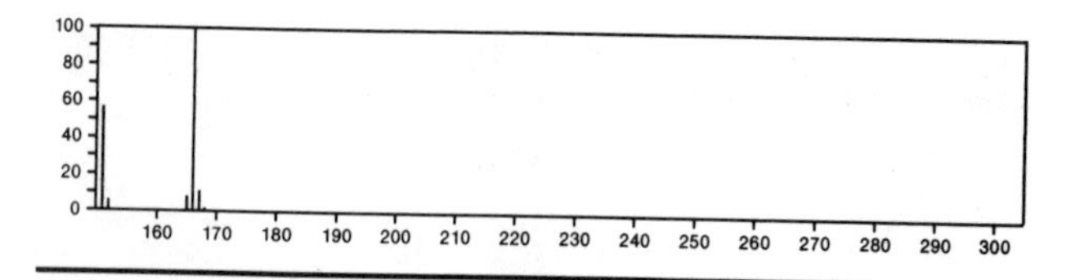

166 $C_{10}H_{14}O_2$ 34883-05-1
Phenol, 3-methoxy-2,4,6-trimethyl-

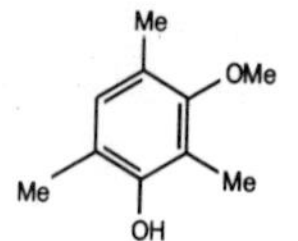

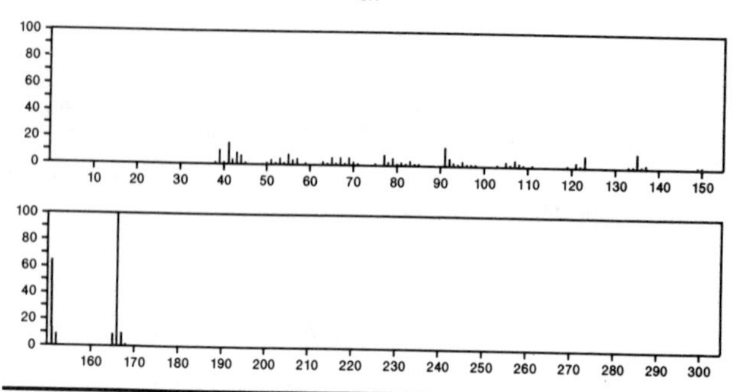

166 $C_{10}H_{14}O_2$ 36803-48-2
Spiro[4.5]decane-1,6-dione

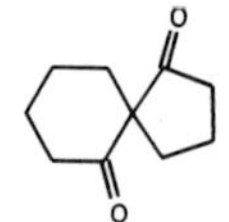

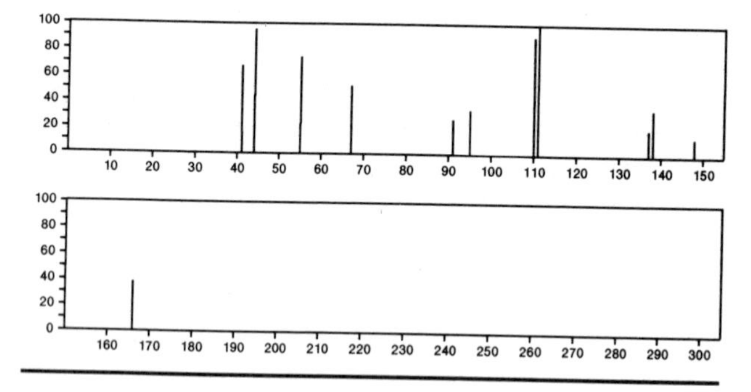

166 $C_{10}H_{14}O_2$ 51020-64-5
Tricyclo[3.3.1.1^{3,7}]decanone, 4-hydroxy-, (1α,3β,4β,5α,7β)-

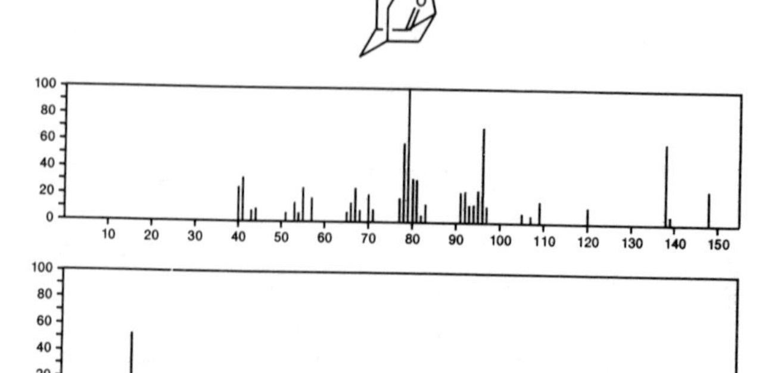

166 $C_{10}H_{14}O_2$ 51020-65-6
Tricyclo[3.3.1.1^{3,7}]decanone, 4-hydroxy-, (1α,3β,4α,5α,7β)-

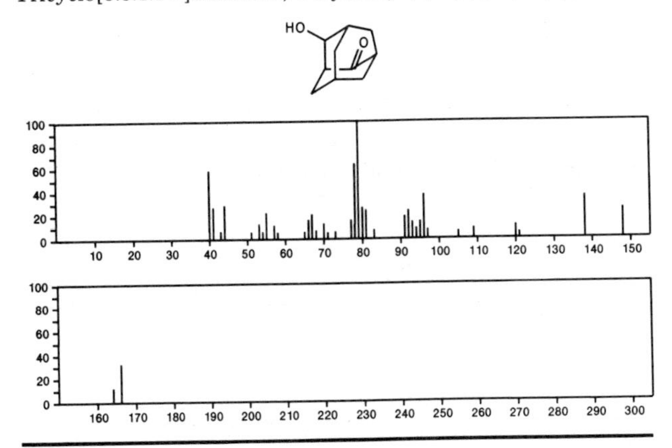

166 $C_{10}H_{14}O_2$ 54411-10-8
Ethanol, 2-(4-ethylphenoxy)-

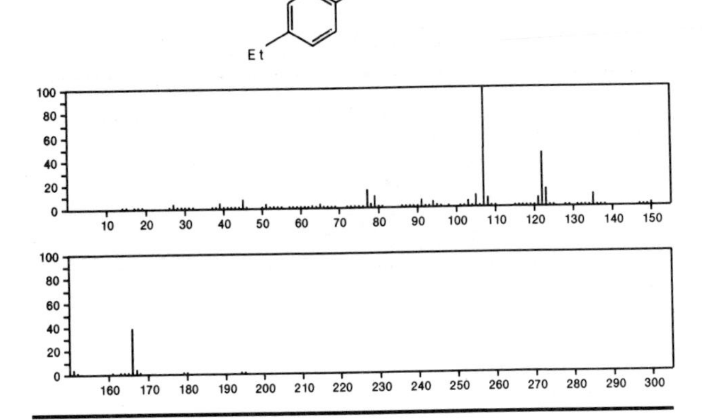

166 $C_{10}H_{14}O_2$ 54411-11-9
Ethanol, 2-[(3-methylphenyl)methoxy]-

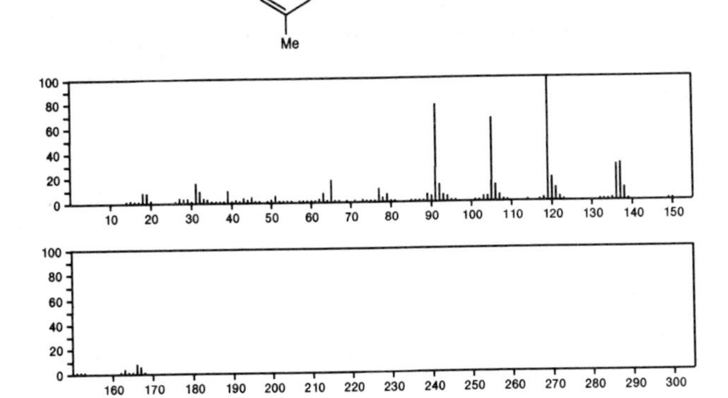

166 $C_{10}H_{14}O_2$ 54411-20-0
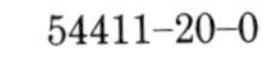
Ethanol, 2-(dimethylphenoxy)-

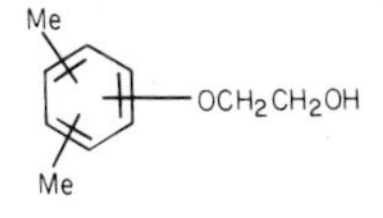

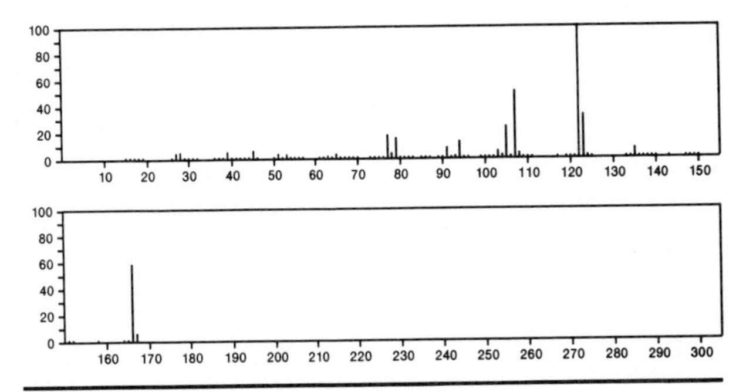

166 $C_{10}H_{14}O_2$ 55956-35-9
Spiro[bicyclo[3.2.1]oct-3-ene-2,2'-[1,3]dioxolane]

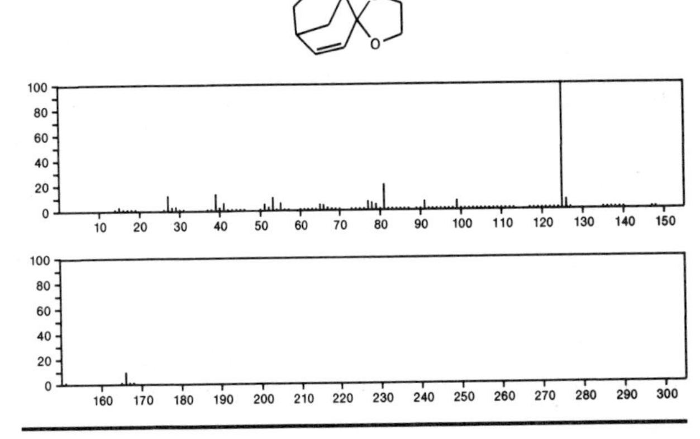

166 $C_{10}H_{14}O_2$ 55956-36-0
5,8-Methano-5*H*-cyclohepta-1,4-dioxin, 2,3,6,7,8,9-hexahydro-

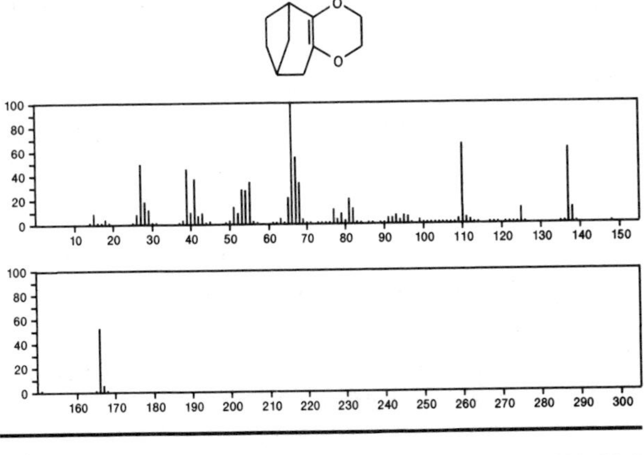

166 $C_{10}H_{14}S$ 1126-80-3
Benzene, (butylthio)-

Me (CH2) 3 SPh

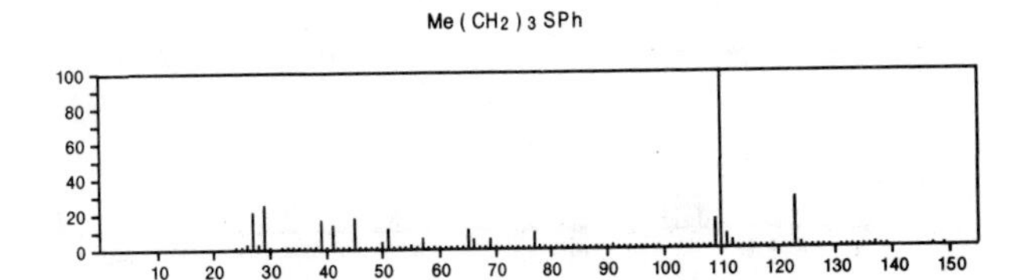

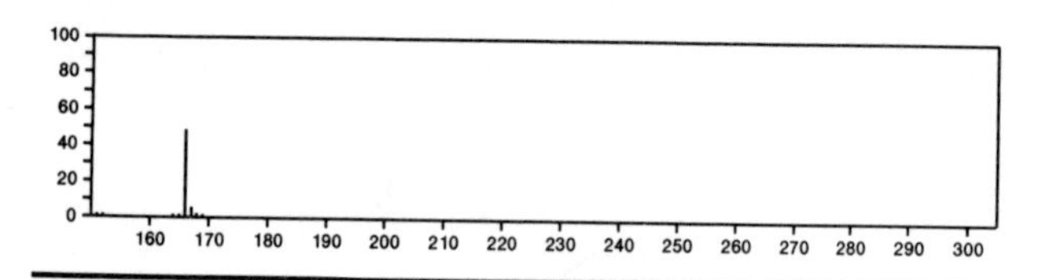

166 $C_{10}H_{14}S$ 3019-19-0
Benzene, [(1,1-dimethylethyl)thio]-

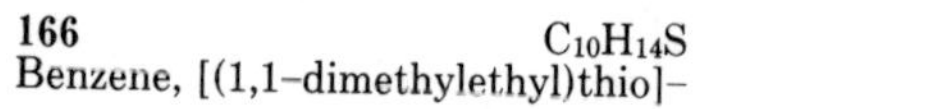

PhSBu-t

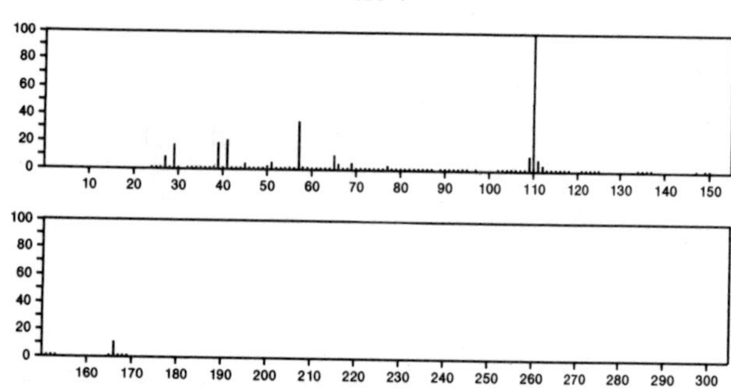

166 $C_{10}H_{14}S$ 13307-61-4
Benzene, [(2-methylpropyl)thio]-

PhSBu-i

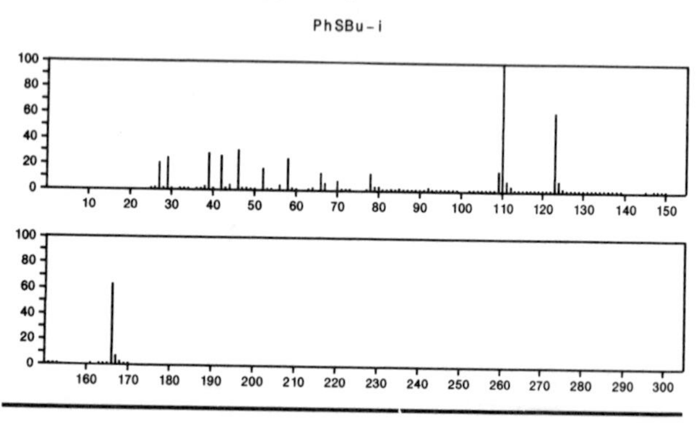

166 $C_{10}H_{14}S$ 14905-79-4
Benzene, [(1-methylpropyl)thio]-

PhSBu-s

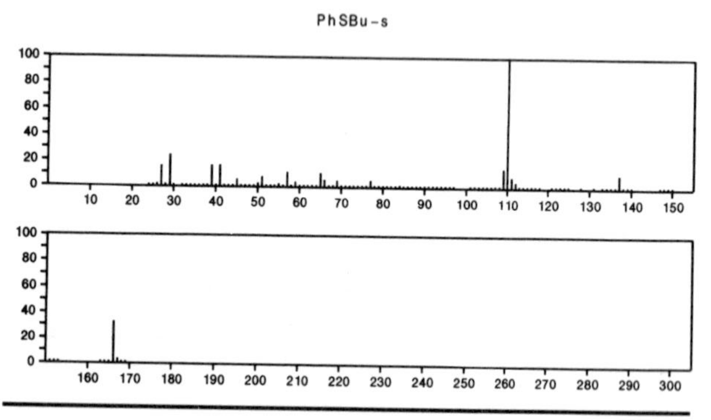

166 $C_{10}H_{14}S$ 14905-80-7
Benzene, 1-methyl-3-[(1-methylethyl)thio]-

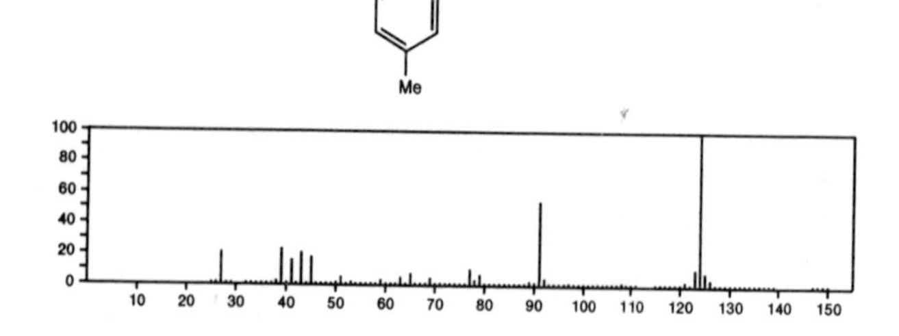

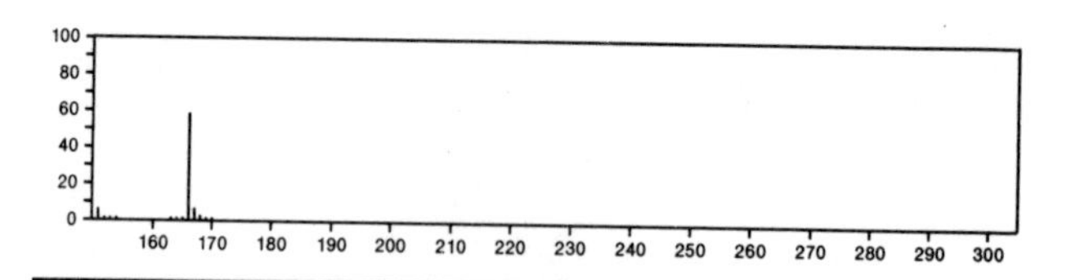

166 $C_{10}H_{14}S$ 14905-81-8
Benzene, 1-methyl-4-[(1-methylethyl)thio]-

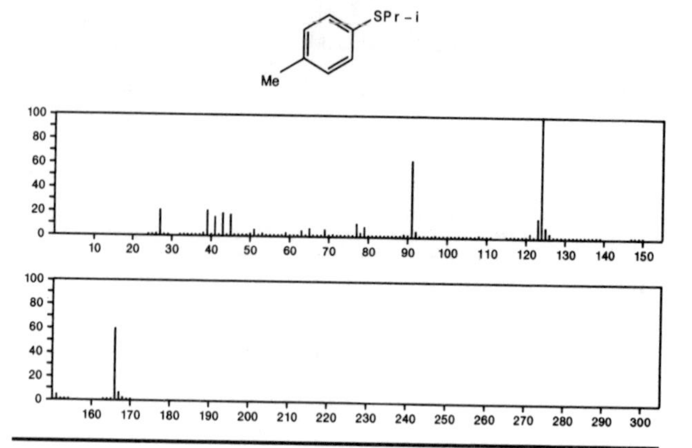

166 $C_{10}H_{14}S$ 15560-97-1
Benzene, 1-methyl-2-(propylthio)-

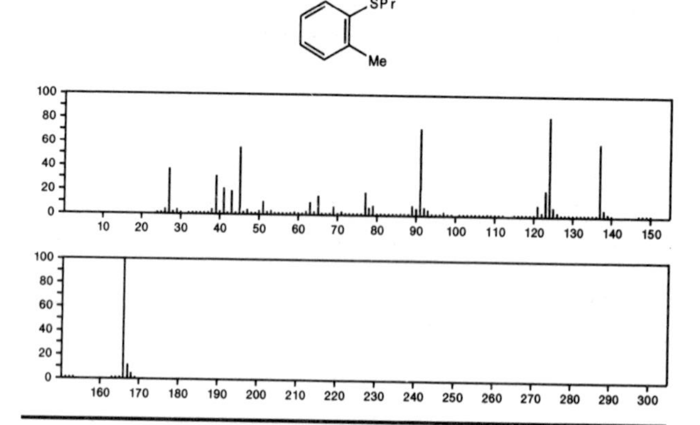

166 $C_{10}H_{14}S$ 15560-98-2
Benzene, 1-methyl-2-[(1-methylethyl)thio]-

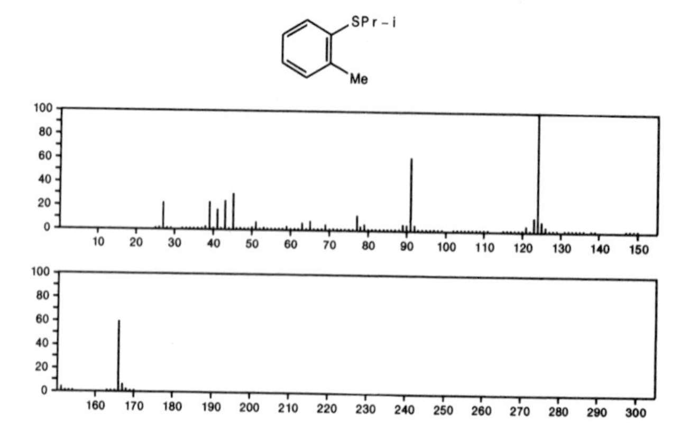

166 $C_{10}H_{14}S$ 22914–08–5
Sulfide, ethyl phenethyl

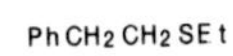

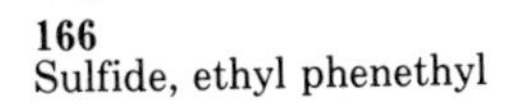

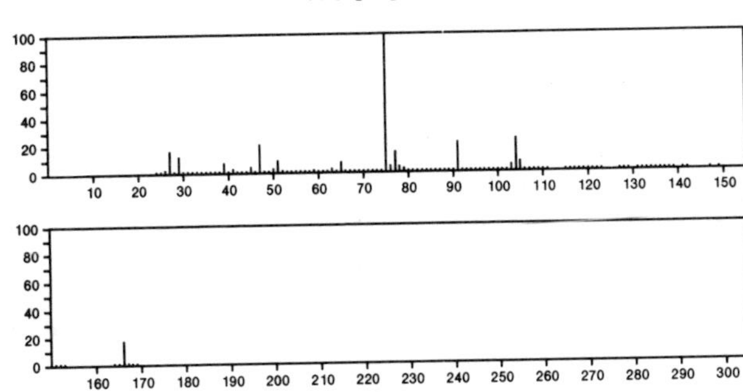

166 $C_{10}H_{14}S$ 24599–52–8
Benzene, 1–methyl–4–(propylthio)–

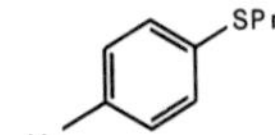

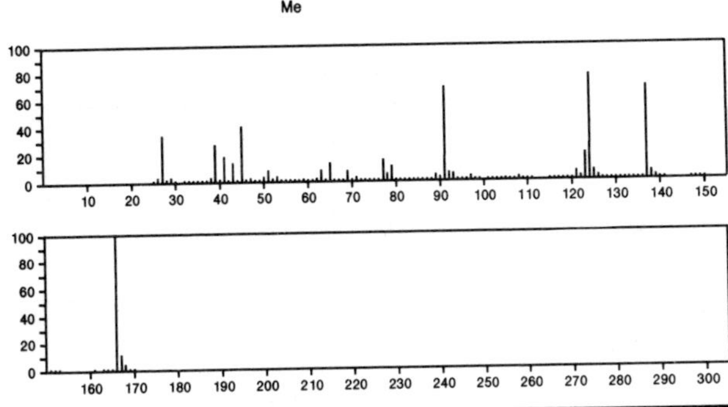

166 $C_{10}H_{14}S$ 24767–95–1
Sulfide, propyl *m*–tolyl

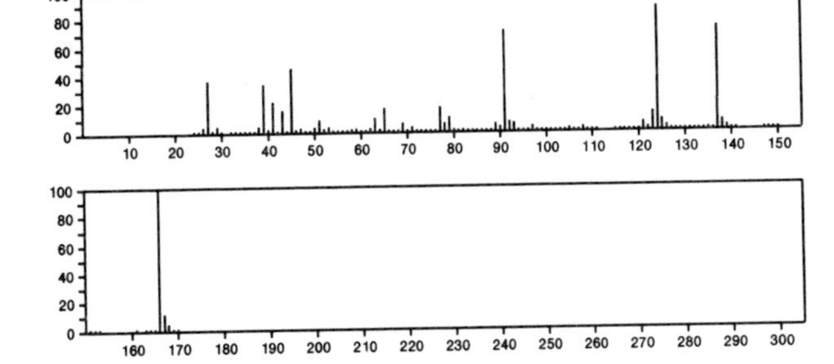

166 $C_{10}H_{18}N_2$ 54966–11–9
1*H*,5*H*–Pyrrolo[1',2':3,4]imidazo[1,5–*a*]pyridine, octahydro–

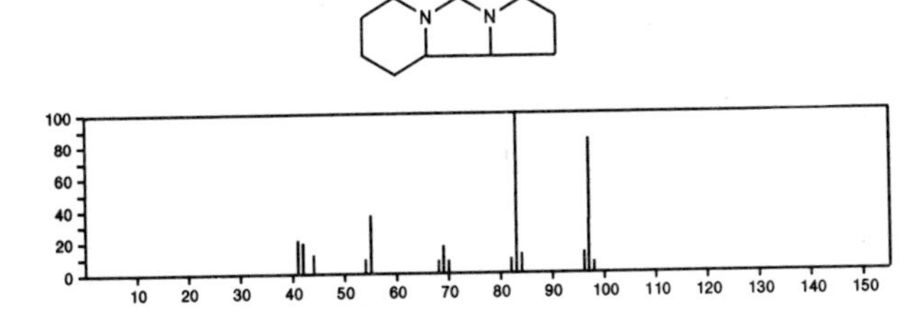

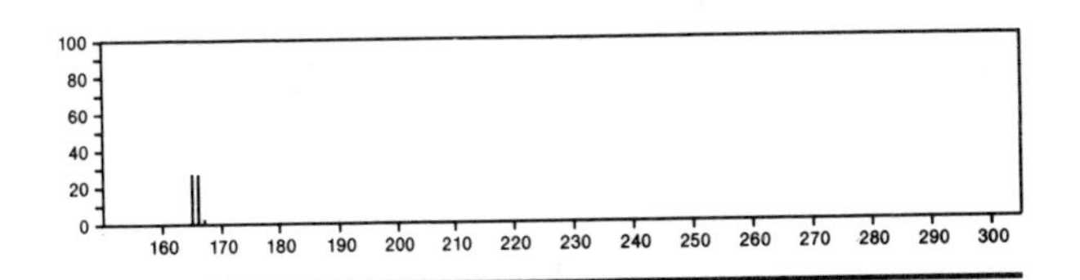

166 $C_{11}H_{18}O$ 128–50–7
Bicyclo[3.1.1]hept–2–ene–2–ethanol, 6,6–dimethyl–

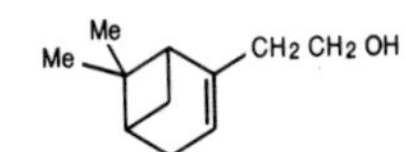

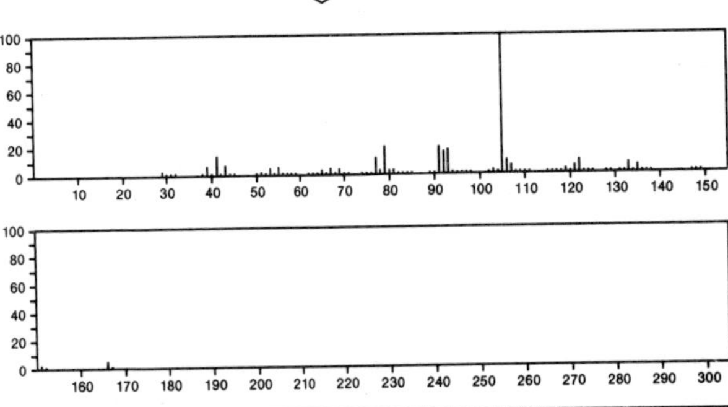

166 $C_{11}H_{18}O$ 770–62–7
1(2*H*)–Naphthalenone, octahydro–8a–methyl–, *cis*–

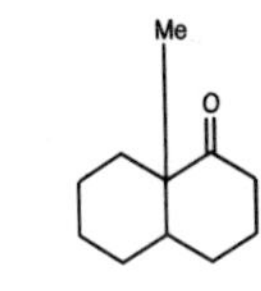

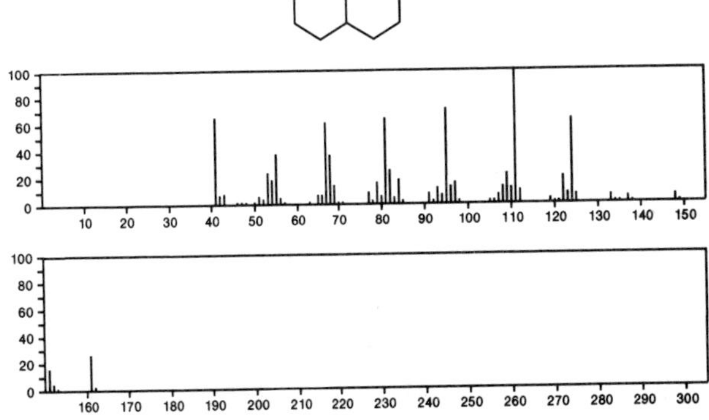

166 $C_{11}H_{18}O$ 937–99–5
1(2*H*)–Naphthalenone, octahydro–4a–methyl–, *trans*–

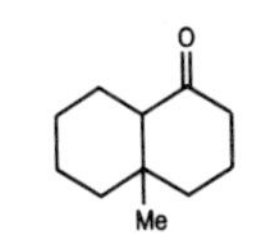

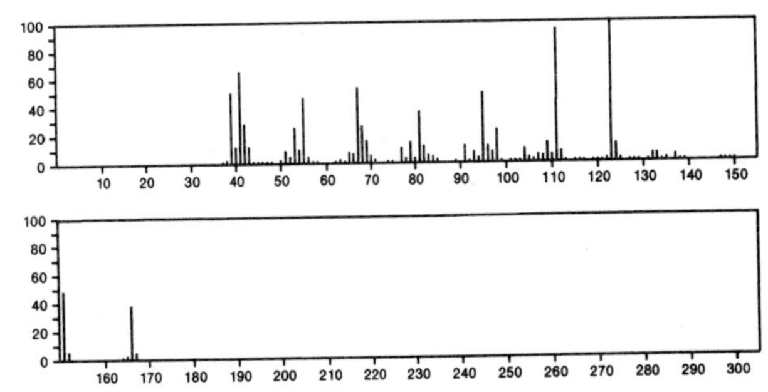

166 $C_{11}H_{18}O$ 938-06-7
2(1*H*)-Naphthalenone, octahydro-4a-methyl-, *cis*-

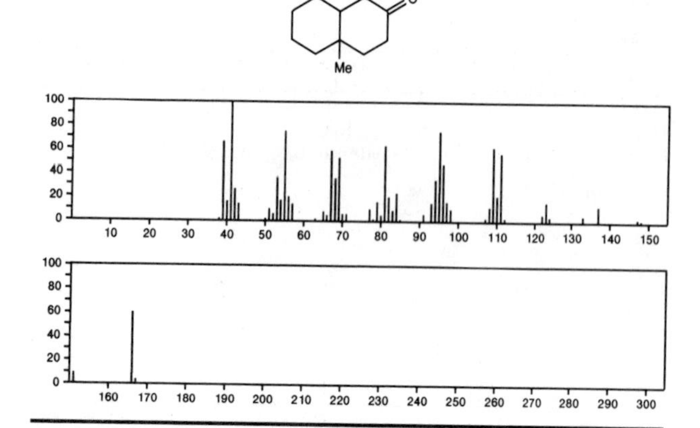

166 $C_{11}H_{18}O$ 938-07-8
2(1*H*)-Naphthalenone, octahydro-4a-methyl-, *trans*-

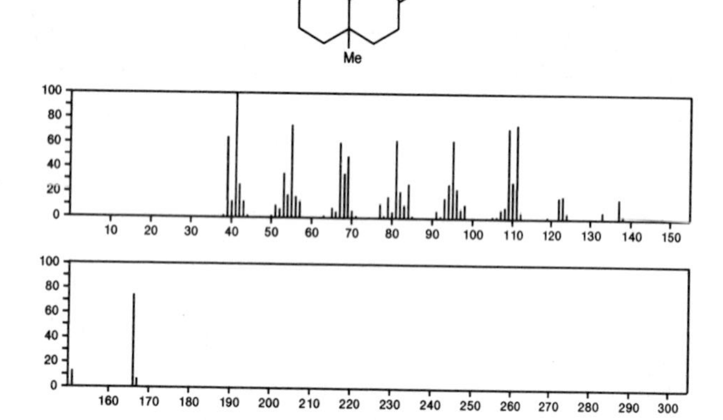

166 $C_{11}H_{18}O$ 1197-95-1
2(1*H*)-Naphthalenone, octahydro-8a-methyl-, *trans*-

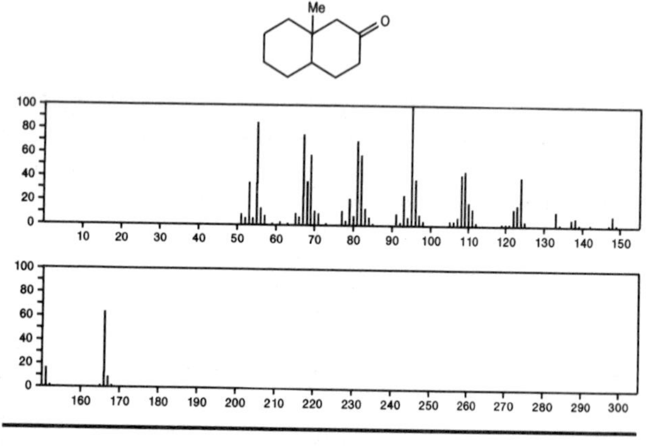

166 $C_{11}H_{18}O$ 1781-81-3
Spiro[5.5]undecan-2-one

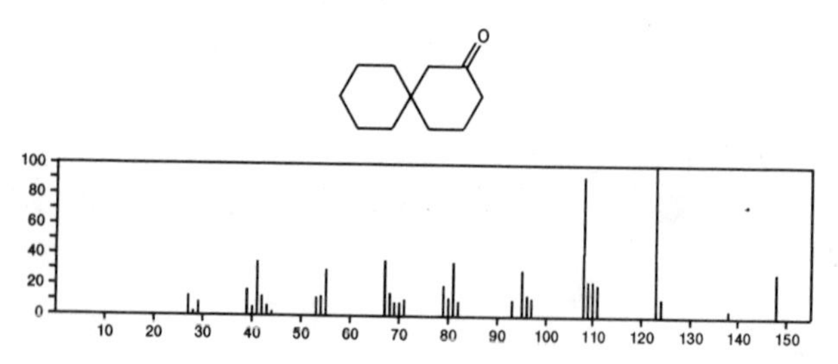

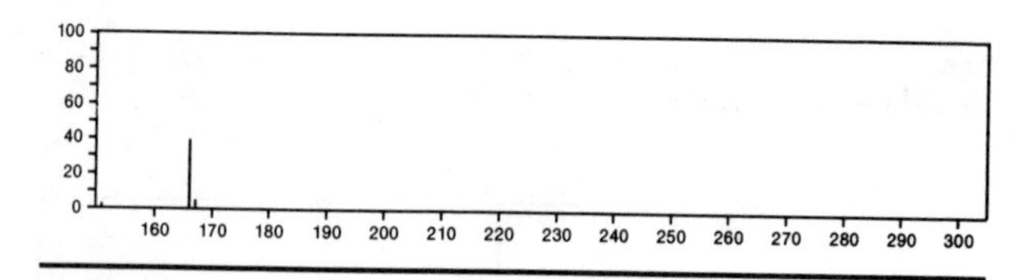

166 $C_{11}H_{18}O$ 1781-83-5
Spiro[5.5]undecan-1-one

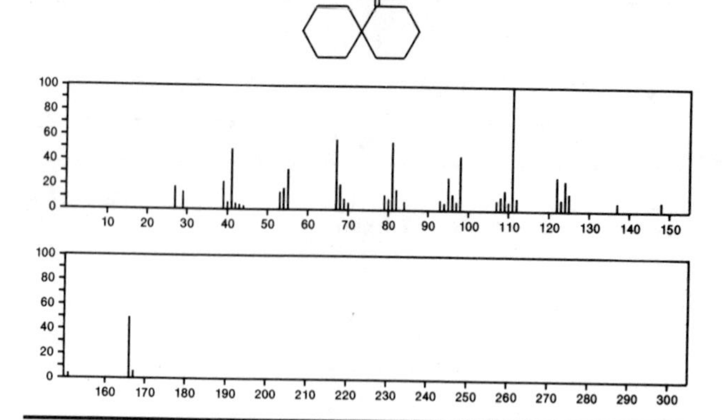

166 $C_{11}H_{18}O$ 1890-25-1
Spiro[5.5]undecan-3-one

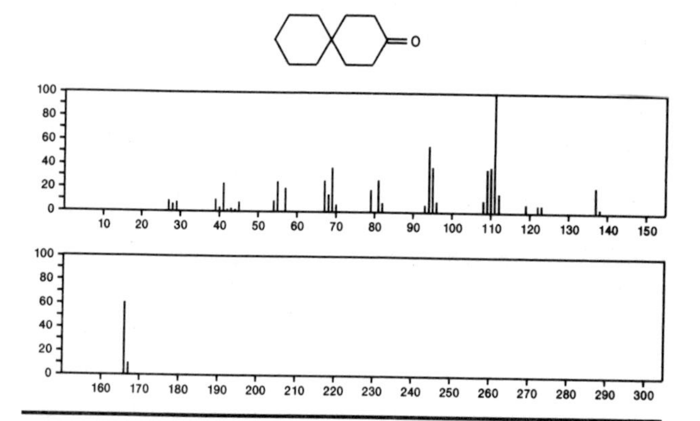

166 $C_{11}H_{18}O$ 2530-17-8
2(1*H*)-Naphthalenone, octahydro-8a-methyl-, *cis*-

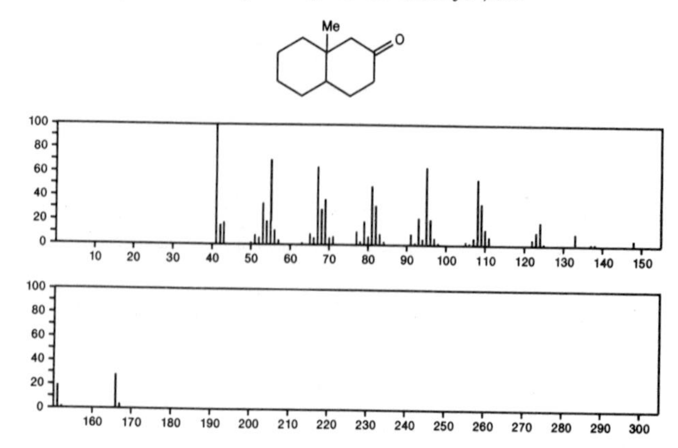

166 $C_{11}H_{18}O$ 5811–48–3
Bicyclo[2.2.1]heptan–2–one, 1,3,7,7–tetramethyl–

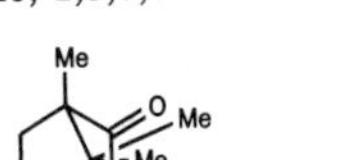
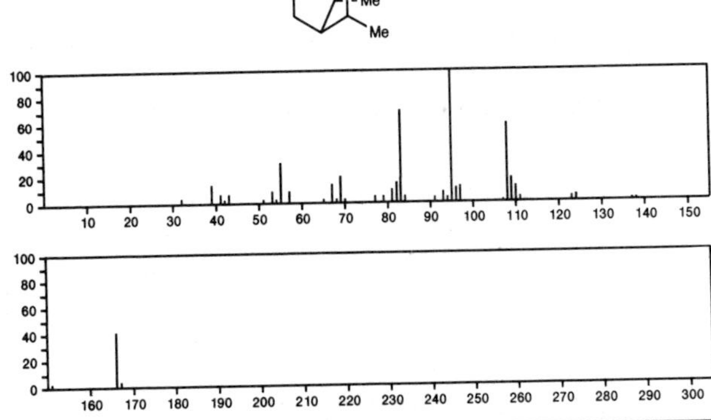

166 $C_{11}H_{18}O$ 6711–26–8
Cyclohexanone, 2,5–dimethyl–2–(1–methylethenyl)–

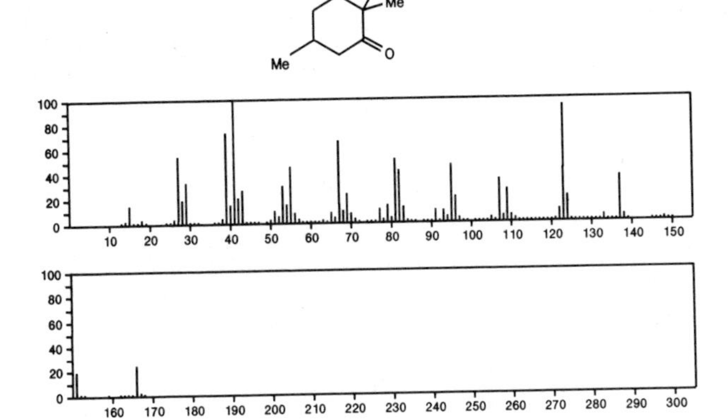

166 $C_{11}H_{18}O$ 13348–11–3
Bicyclo[5.3.1]undecan–11–one

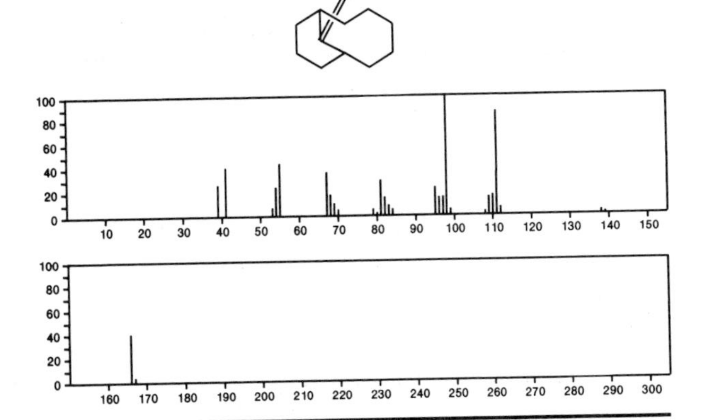

166 $C_{11}H_{18}O$ 14504–80–4
Tricyclo[4.3.1.1^3,8]undecan–3–ol

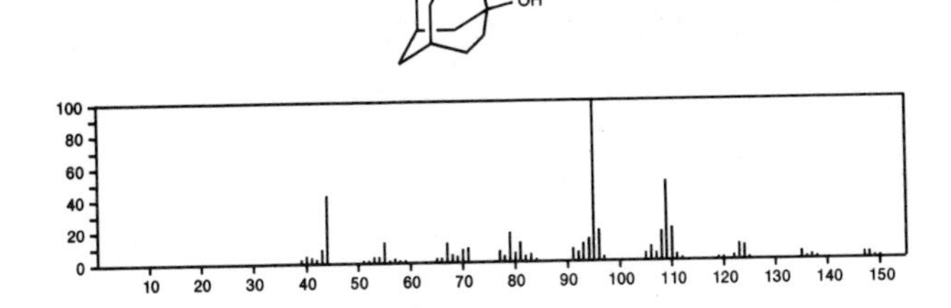
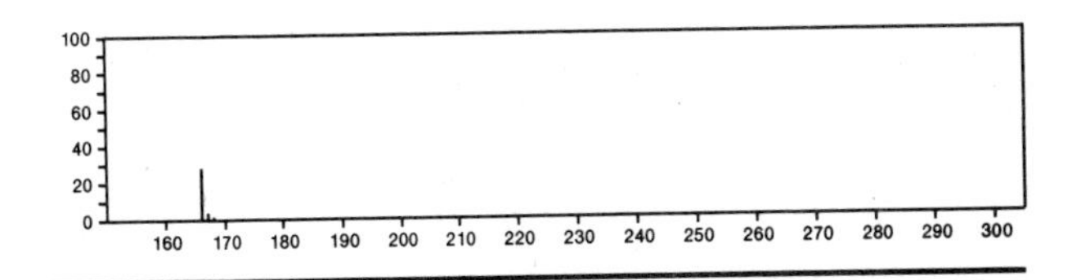

166 $C_{11}H_{18}O$ 21102–88–5
2(1*H*)–Naphthalenone, octahydro–1–methyl–, (1α,4aβ,8aα)–

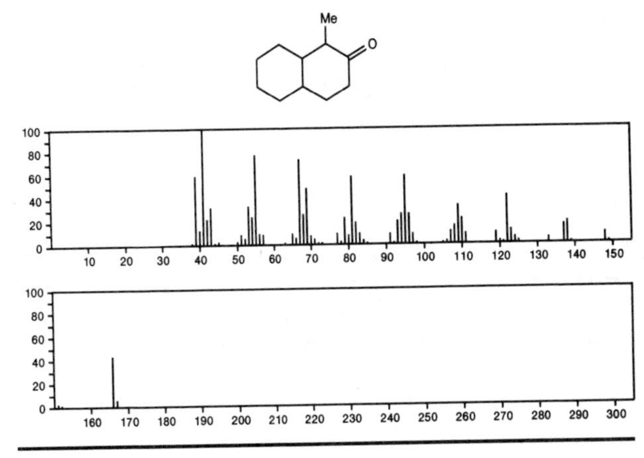

166 $C_{11}H_{18}O$ 28017–79–0
2–Cyclohexen–1–one, 5,5–dimethyl–3–(1–methylethyl)–

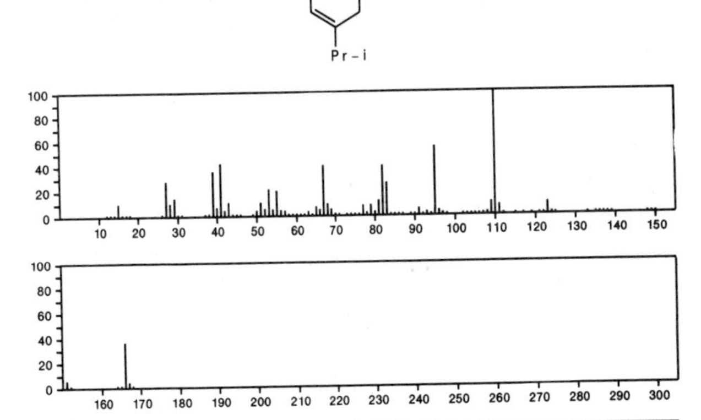

166 $C_{11}H_{18}O$ 31061–64–0
Tricyclo[4.3.1.1^3,8]undecan–1–ol

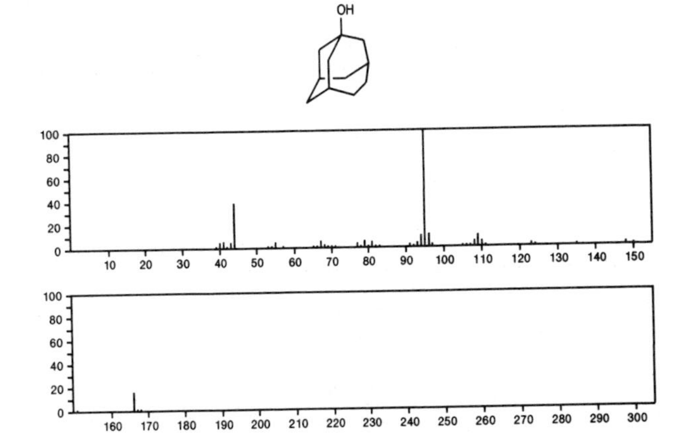

166 $C_{11}H_{18}O$ 35505-97-6
Cyclohexanone, 2,2,5,5-tetramethyl-3-methylene-

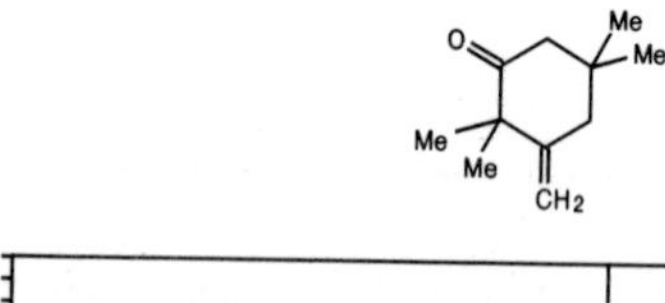

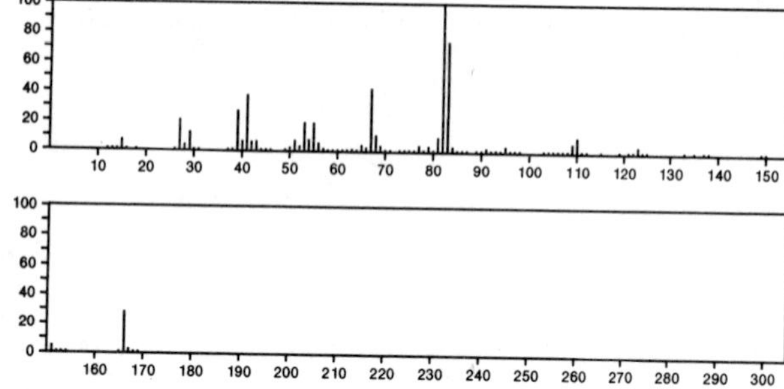

166 $C_{11}H_{18}O$ 51149-78-1
Cyclohexene, 1-(3-ethoxy-1-propenyl)-, (*Z*)-

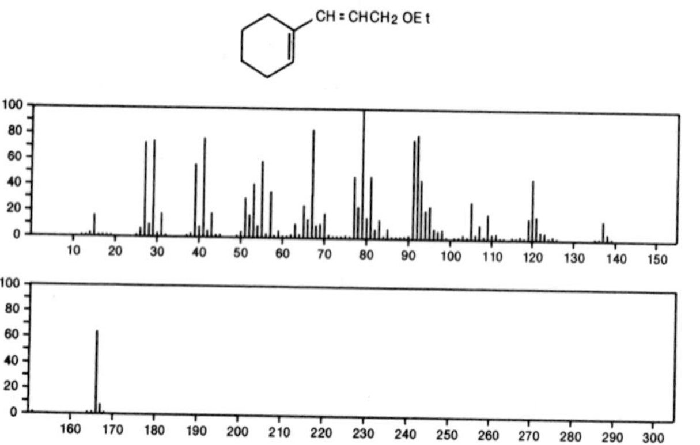

166 $C_{11}H_{18}O$ 54345-62-9
1-Oxaspiro[2.5]octane, 2,4,4-trimethyl-8-methylene-

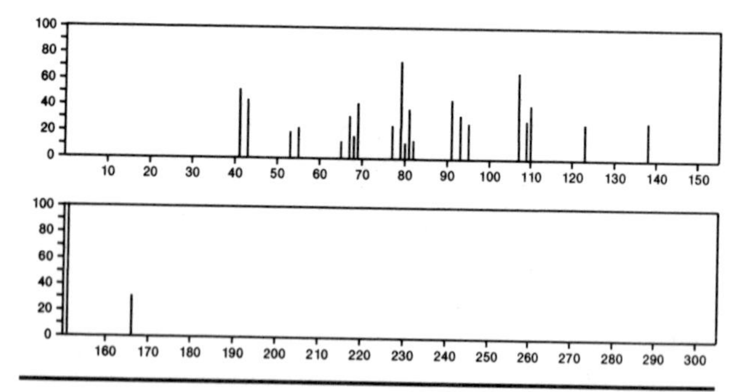

166 $C_{11}H_{18}O$ 54410-58-1
2-Cyclohexen-1-one, 3,6-dimethyl-6-(1-methylethyl)-

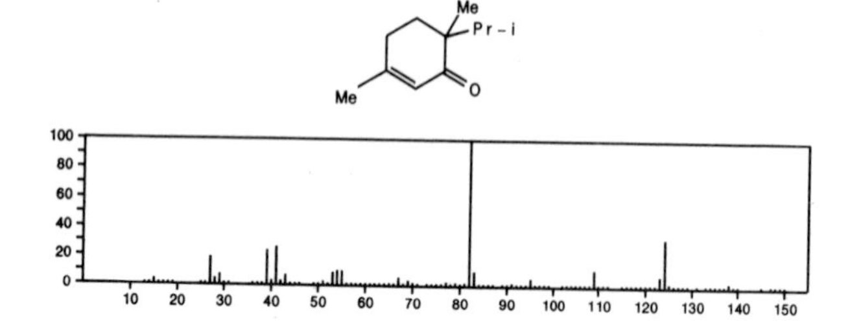

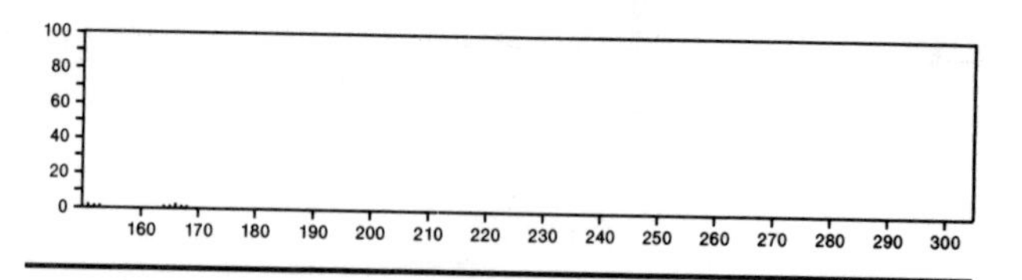

166 $C_{11}H_{18}O$ 54764-62-4
1-Butanone, 1-bicyclo[4.1.0]hept-7-yl-

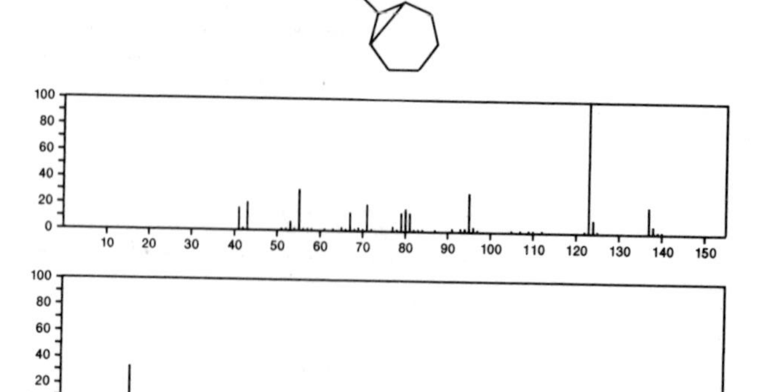

166 $C_{11}H_{18}O$ 55332-01-9
2(1*H*)-Naphthalenone, octahydro-3-methyl-, (3α,4aβ,8aα)-

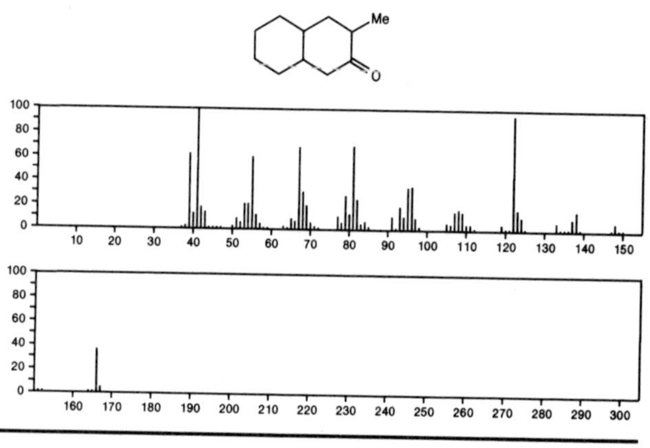

166 $C_{11}H_{18}O$ 55702-54-0
4-Penten-2-one, 3-cyclohexyl-

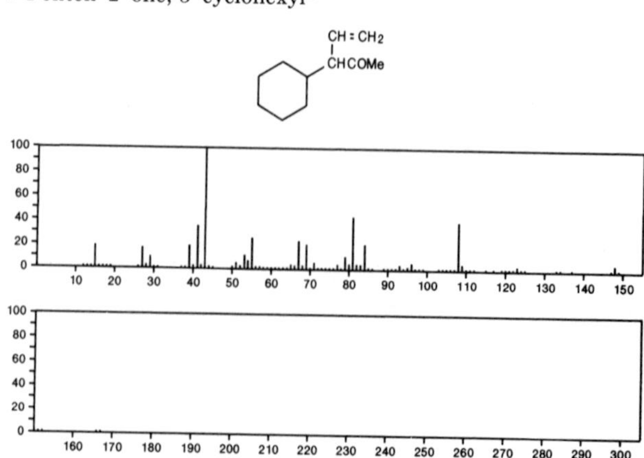

166 $C_{11}H_{18}O$ 55723-95-0
4*H*-Cyclopentacycloocten-4-one, decahydro-

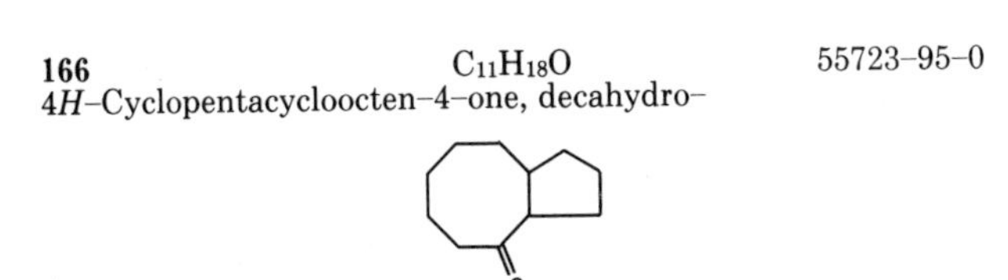

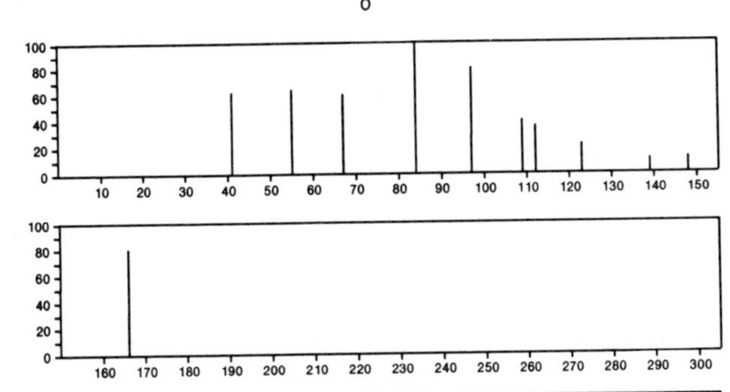

166 $C_{11}H_{18}O$ 56362-32-4
Ethanone, 1-(octahydro-1*H*-inden-1-yl)-, (1α,3aα,7aβ)-

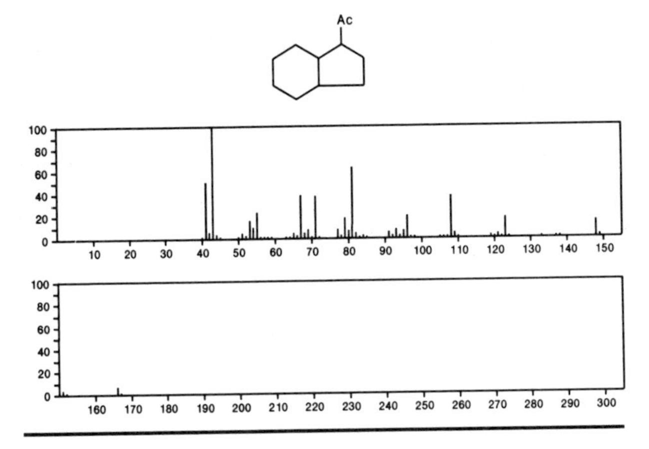

166 $C_{11}H_{18}O$ 56362-33-5
Ethanone, 1-(octahydro-1*H*-inden-1-yl)-, (1α,3aβ,7aα)-

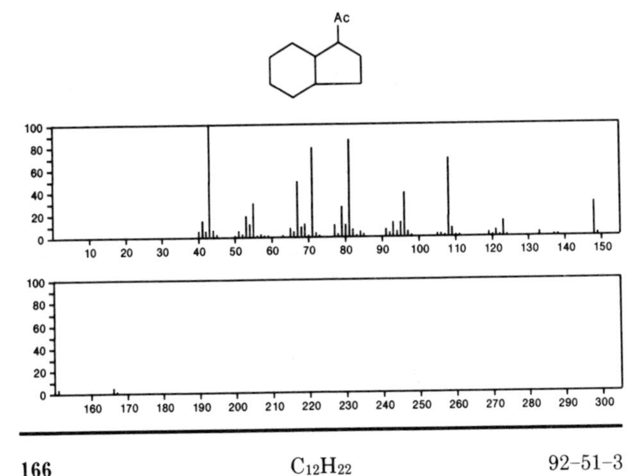

166 $C_{12}H_{22}$ 92-51-3
1,1'-Bicyclohexyl

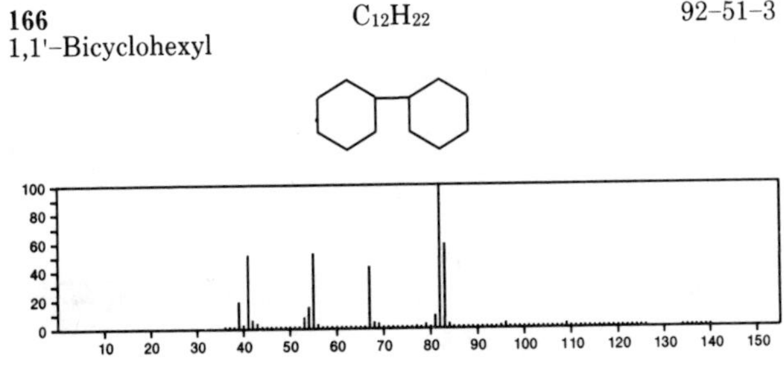

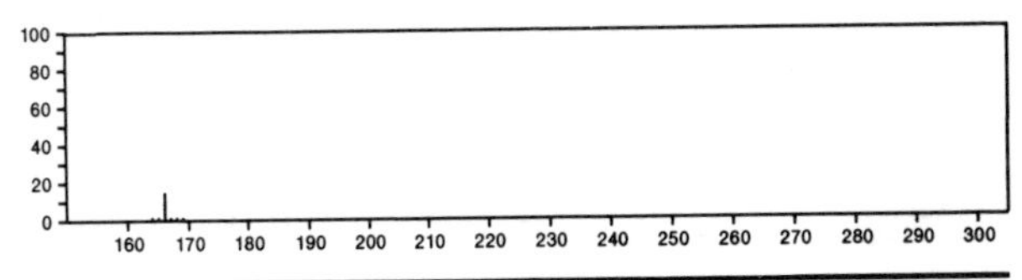

166 $C_{12}H_{22}$ 181-15-7
Spiro[5.6]dodecane

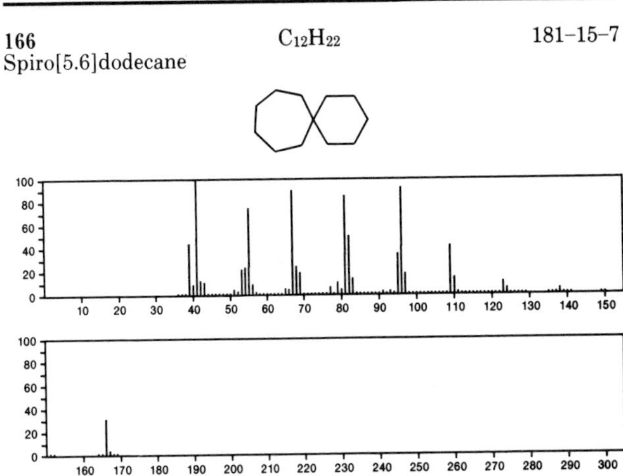

166 $C_{12}H_{22}$ 765-03-7
1-Dodecyne

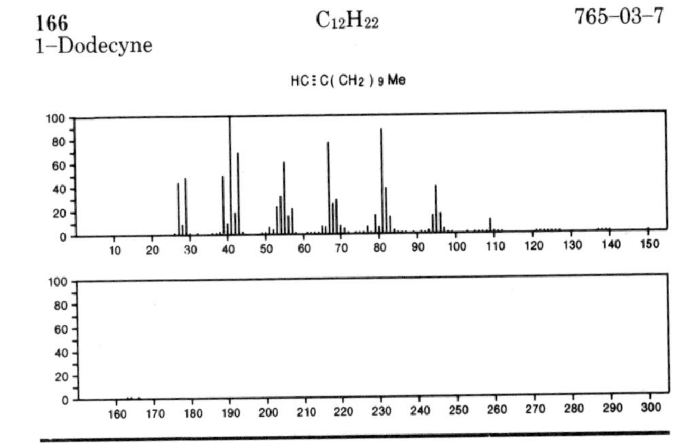

166 $C_{12}H_{22}$ 1008-80-6
Naphthalene, decahydro-2,3-dimethyl-

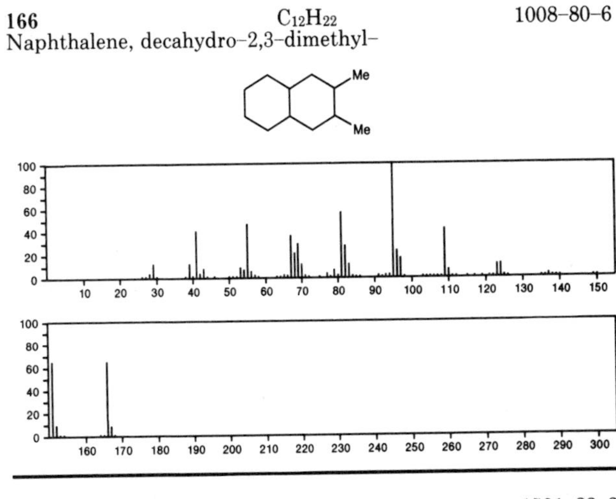

166 $C_{12}H_{22}$ 1501-82-2
Cyclododecene

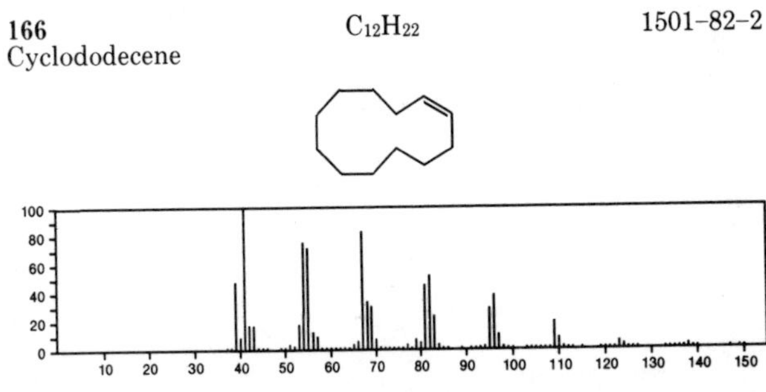

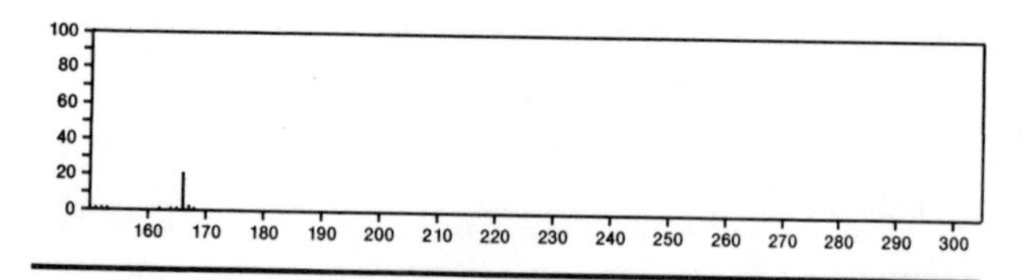

166 $C_{12}H_{22}$ 1618-22-0
Naphthalene, decahydro-2,6-dimethyl-

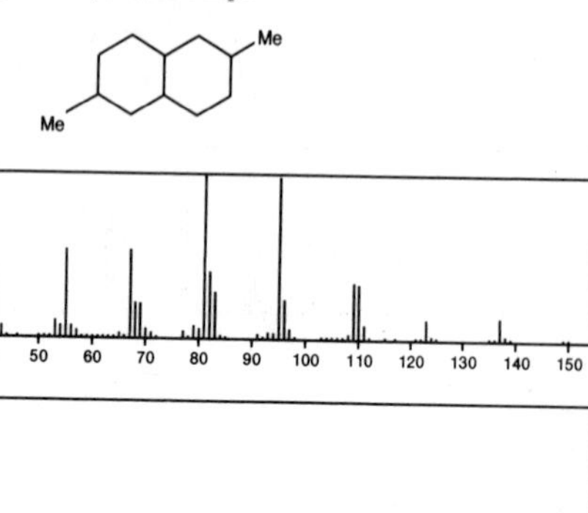

166 $C_{12}H_{22}$ 1750-51-2
Naphthalene, decahydro-1,6-dimethyl-

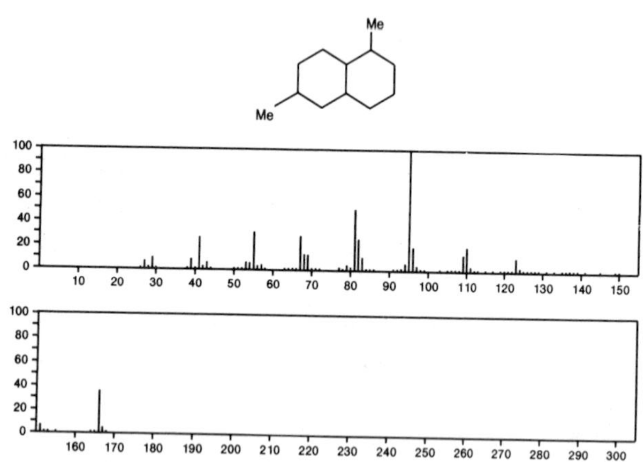

166 $C_{12}H_{22}$ 3604-14-6
Naphthalene, decahydro-1,2-dimethyl-

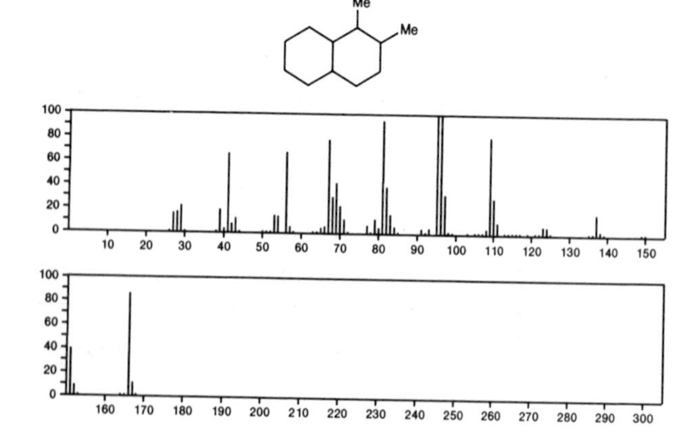

166 $C_{12}H_{22}$ 4413-21-2
Cyclopentane, 1,1'-ethylidenebis-

166 $C_{12}H_{22}$ 4431-89-4
Cyclohexane, (cyclopentylmethyl)-

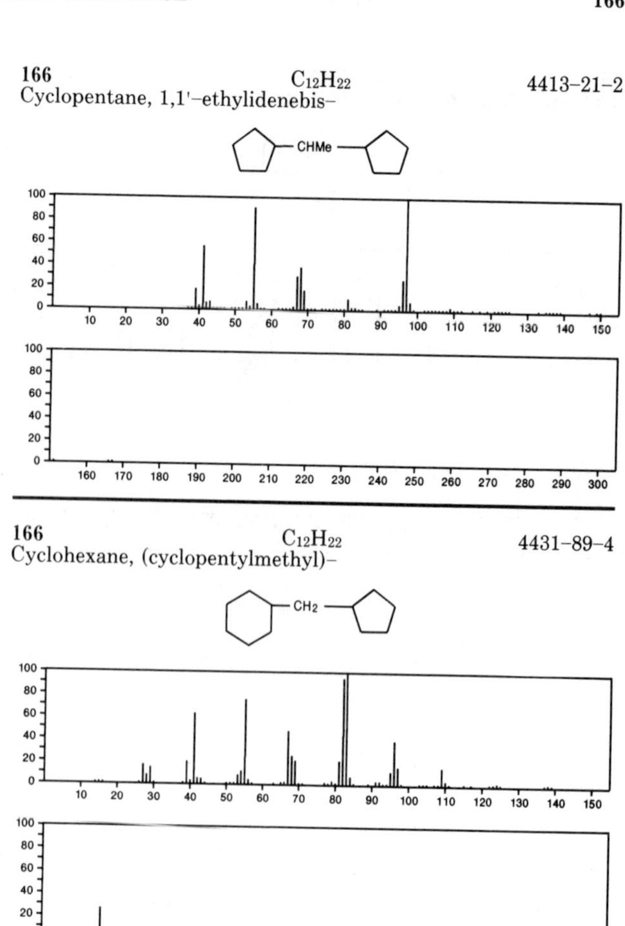

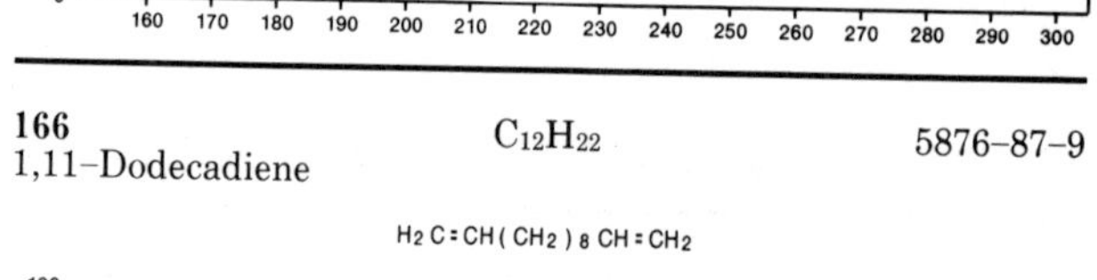

166 $C_{12}H_{22}$ 5876-87-9
1,11-Dodecadiene

$H_2C=CH(CH_2)_8CH=CH_2$

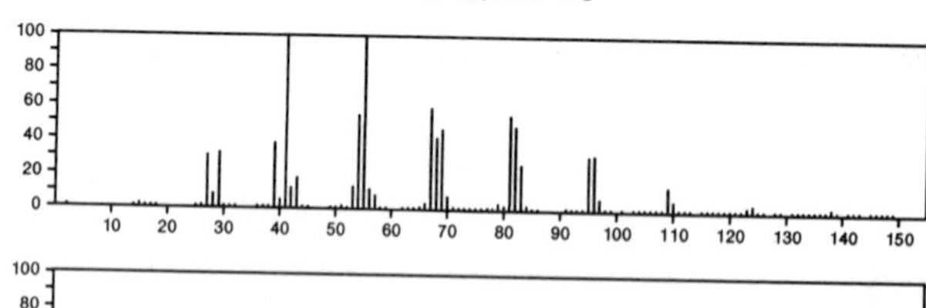

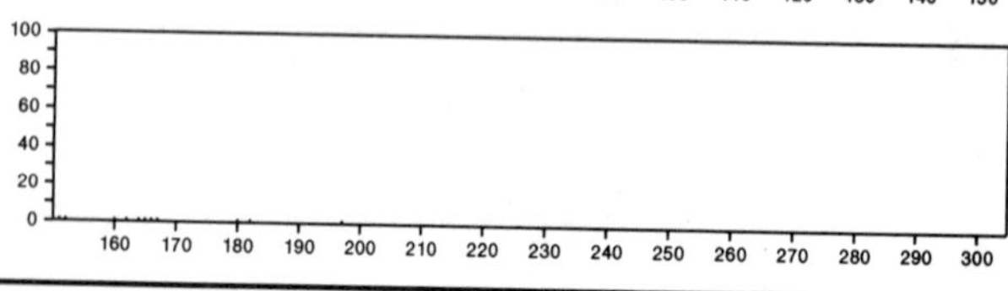

166 $C_{12}H_{22}$ 6108-62-9
5,7-Dodecadiene, (*Z,Z*)-

$Me(CH_2)_3CH=CHCH=CH(CH_2)_3Me$

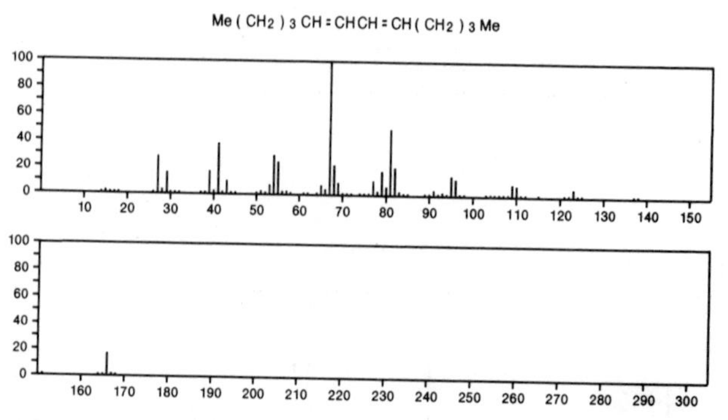

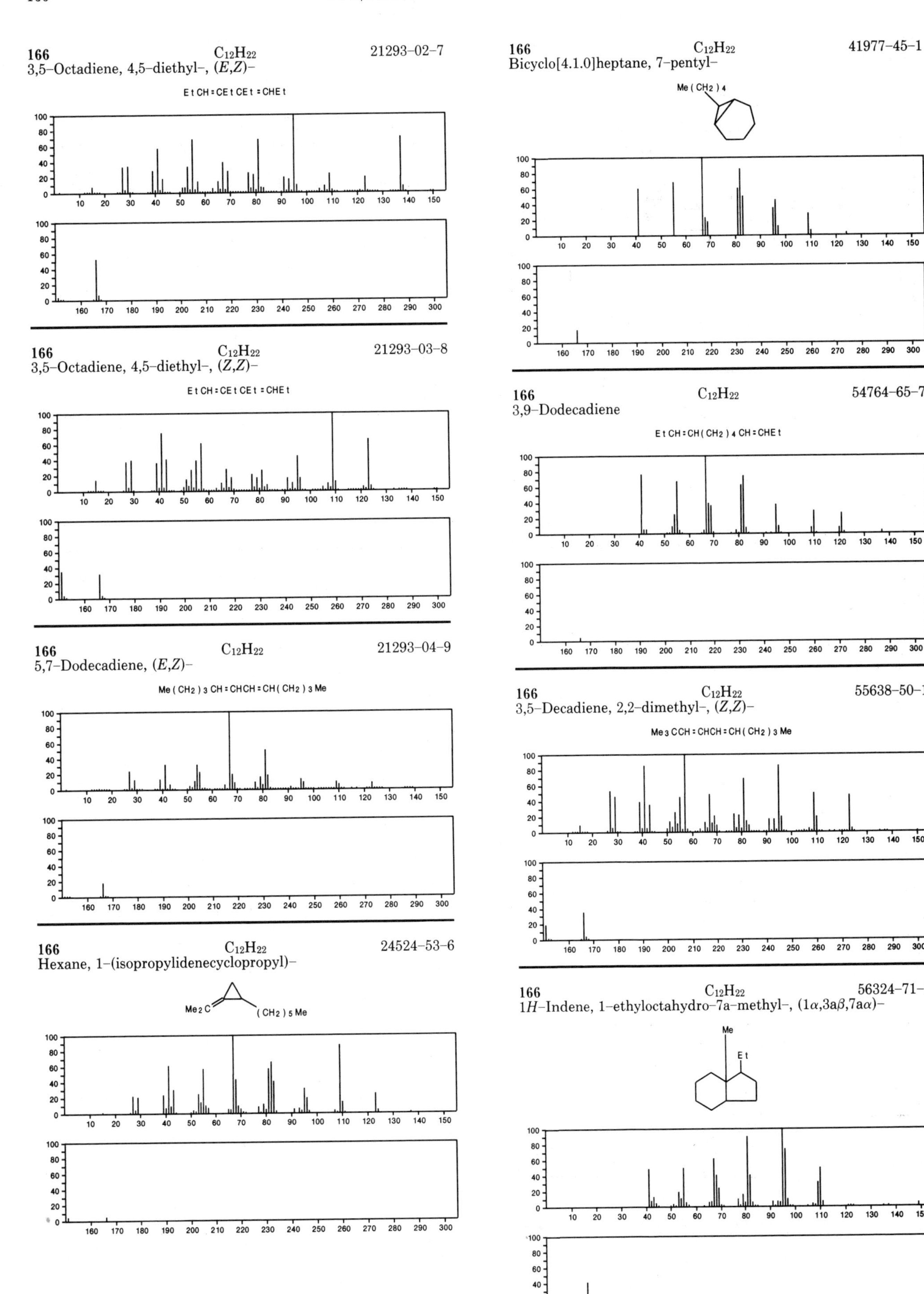
166 $C_{12}H_{22}$ 21293-02-7
3,5-Octadiene, 4,5-diethyl-, (*E*,*Z*)-
EtCH=CEtCEt=CHEt
166 $C_{12}H_{22}$ 21293-03-8
3,5-Octadiene, 4,5-diethyl-, (*Z*,*Z*)-
EtCH=CEtCEt=CHEt
166 $C_{12}H_{22}$ 21293-04-9
5,7-Dodecadiene, (*E*,*Z*)-
Me(CH2)3CH=CHCH=CH(CH2)3Me
166 $C_{12}H_{22}$ 24524-53-6
Hexane, 1-(isopropylidenecyclopropyl)-
Me2C
(CH2)5Me
166 $C_{12}H_{22}$ 41977-45-1
Bicyclo[4.1.0]heptane, 7-pentyl-
Me(CH2)4
166 $C_{12}H_{22}$ 54764-65-7
3,9-Dodecadiene
EtCH=CH(CH2)4CH=CHEt
166 $C_{12}H_{22}$ 55638-50-1
3,5-Decadiene, 2,2-dimethyl-, (*Z*,*Z*)-
Me3CCH=CHCH=CH(CH2)3Me
166 $C_{12}H_{22}$ 56324-71-1
1*H*-Indene, 1-ethyloctahydro-7a-methyl-, (1α,3aβ,7aα)-
Me
Et

166 $C_{12}H_{22}$ 56701-46-3
Cyclopropane, hexyl(1-methylethylidene)-

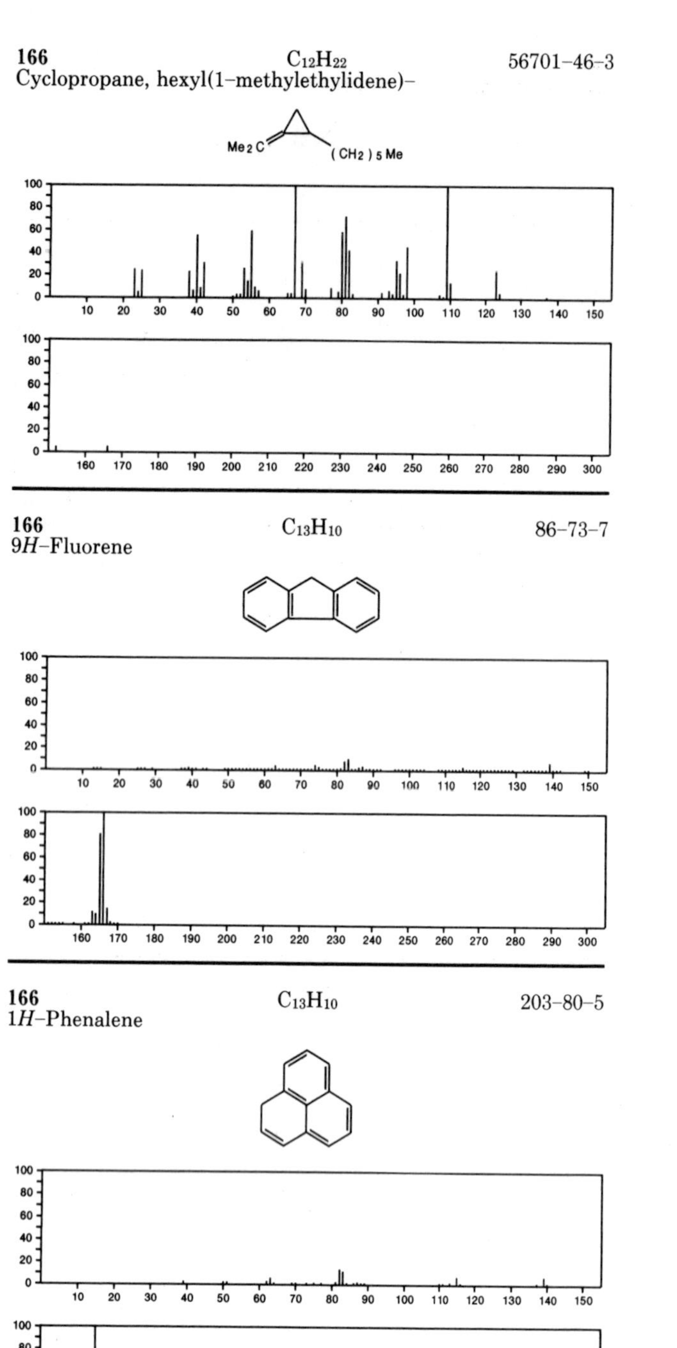

166 $C_{13}H_{10}$ 86-73-7
9*H*-Fluorene

166 $C_{13}H_{10}$ 203-80-5
1*H*-Phenalene

167 $C_4H_{10}NO_2PS$ 7114-53-6
1,3,2-Dioxaphospholan-2-amine, *N*,*N*-dimethyl-, 2-sulfide

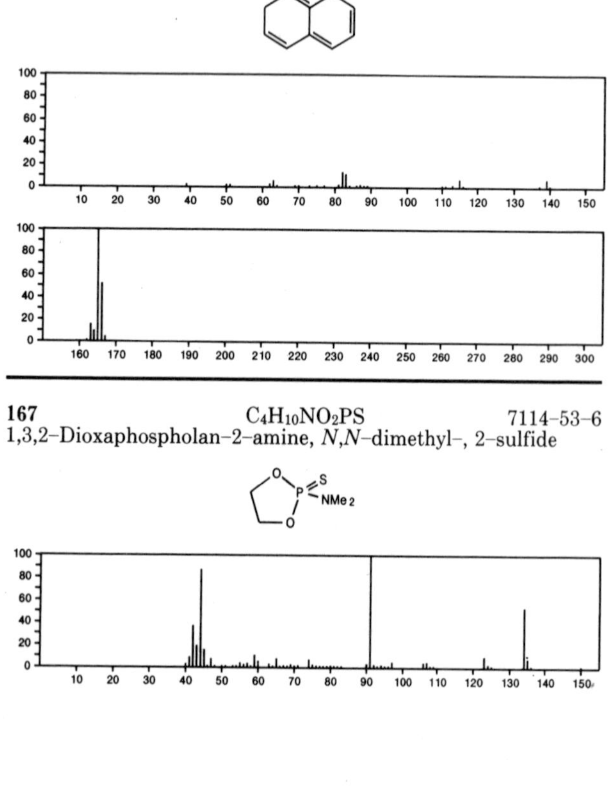

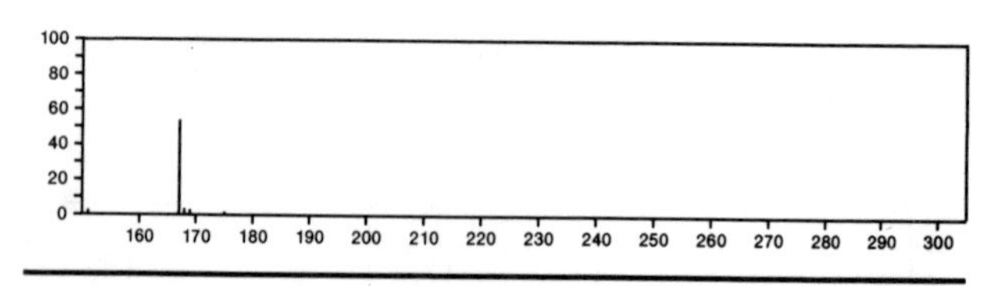

167 $C_5H_{14}N_3O.Cl$ 123-46-6
Ethanaminium, 2-hydrazino-*N*,*N*,*N*-trimethyl-2-oxo-, chloride

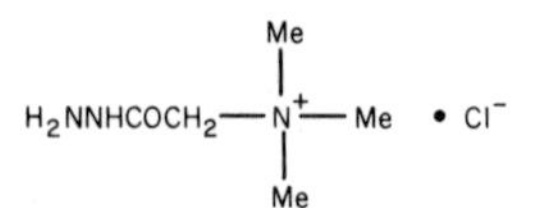

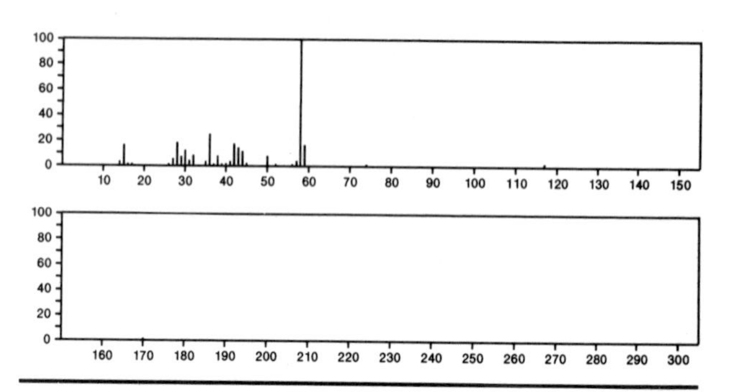

167 $C_7H_5NO_4$ 62-23-7
Benzoic acid, 4-nitro-

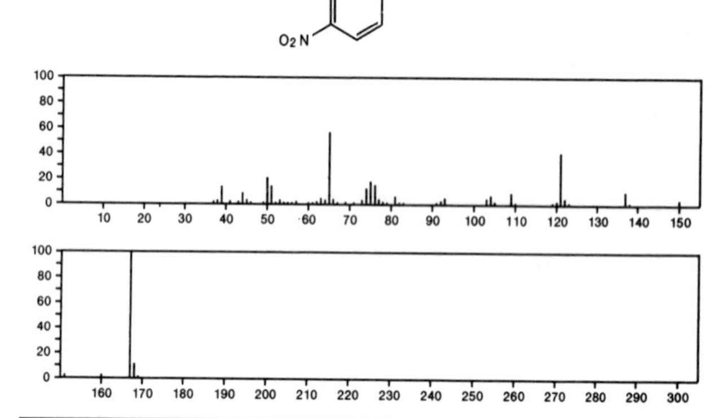

167 $C_7H_5NO_4$ 89-00-9
2,3-Pyridinedicarboxylic acid

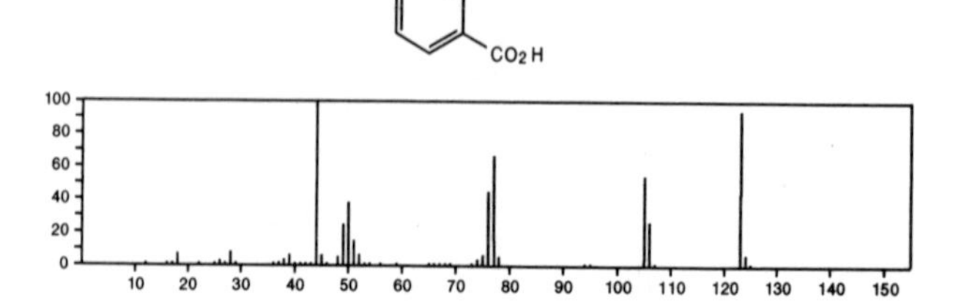

167 $C_7H_5NO_4$ 97–51–8
Benzaldehyde, 2–hydroxy–5–nitro–

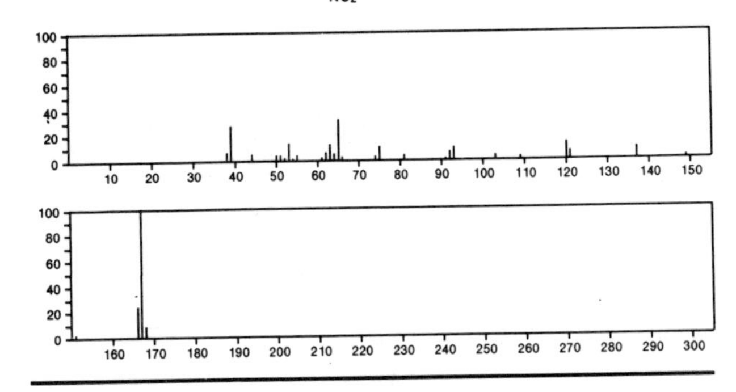

167 $C_7H_5NO_4$ 100–26–5
2,5–Pyridinedicarboxylic acid

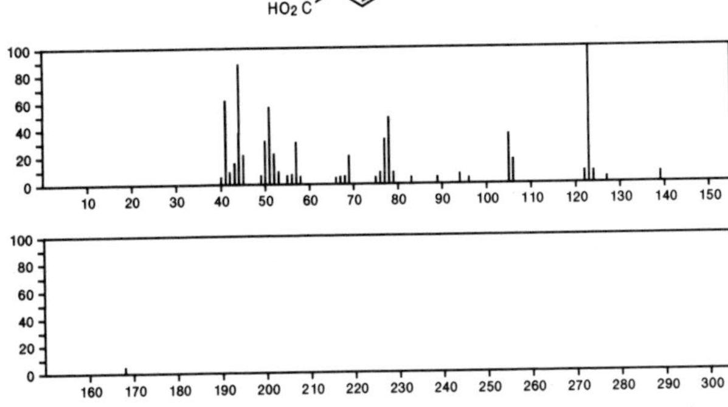

167 $C_7H_5NO_4$ 121–92–6
Benzoic acid, 3–nitro–

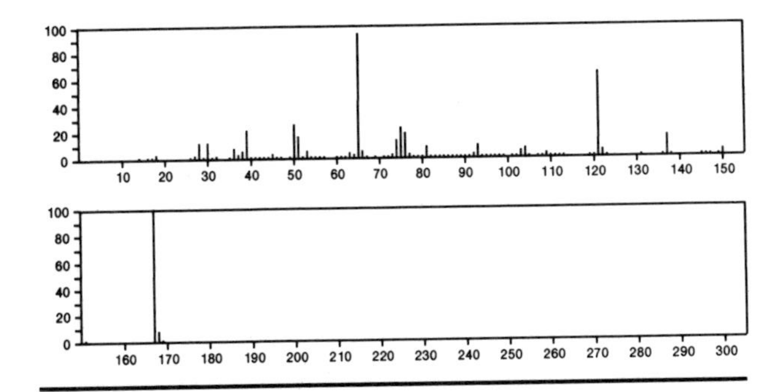

167 $C_7H_5NO_4$ 552–16–9
Benzoic acid, 2–nitro–

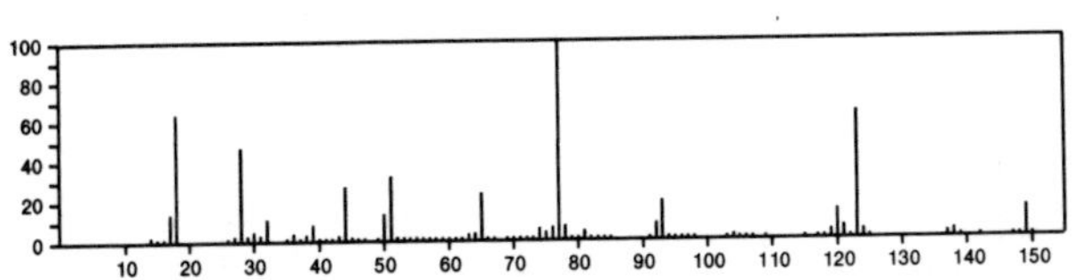

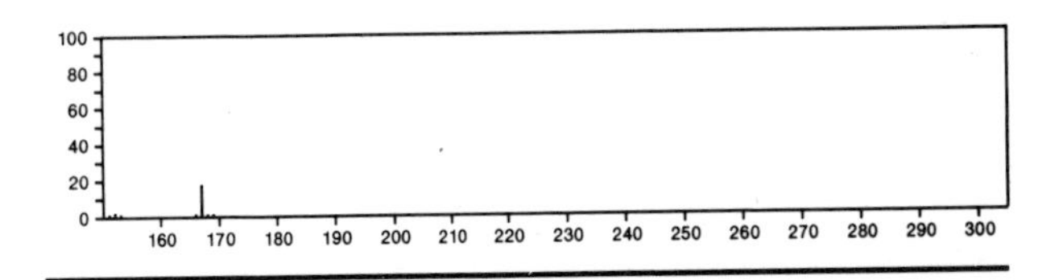

167 $C_7H_5NS_2$ 149–30–4
2(3*H*)–Benzothiazolethione

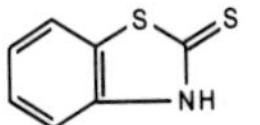

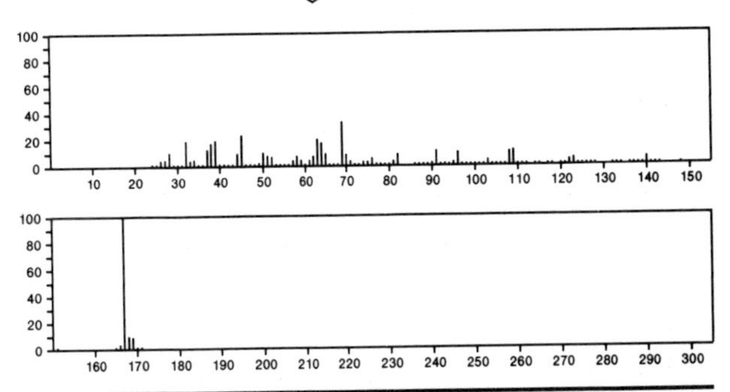

167 $C_7H_{13}N_5$ 4150–59–8
1,3,5–Triazine–2,4–diamine, *N,N'*–diethyl–

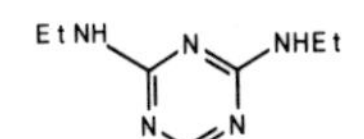

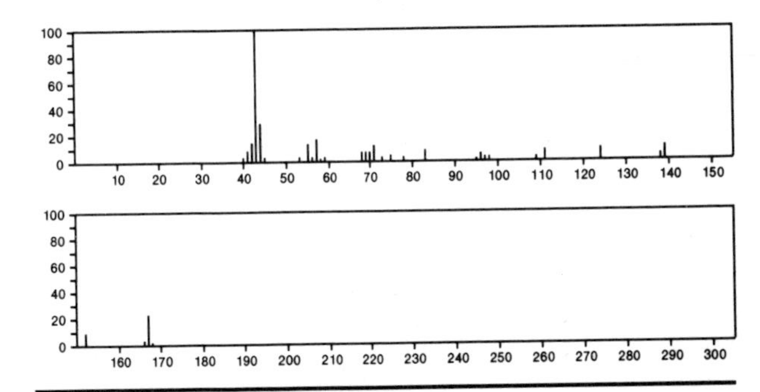

167 C_8H_9NOS 13509–38–1
Carbamothioic acid, phenyl–, *S*–methyl ester

PhNHC(O)SMe

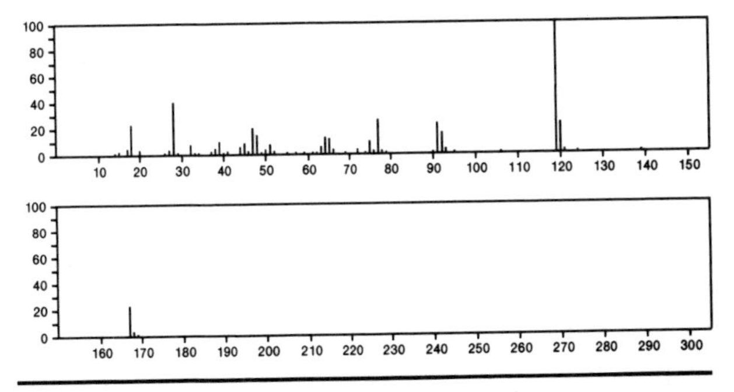

167 C_8H_9NOS 13509–39–2
Carbamothioic acid, methyl–, *S*–phenyl ester

PhSC(O)NHMe

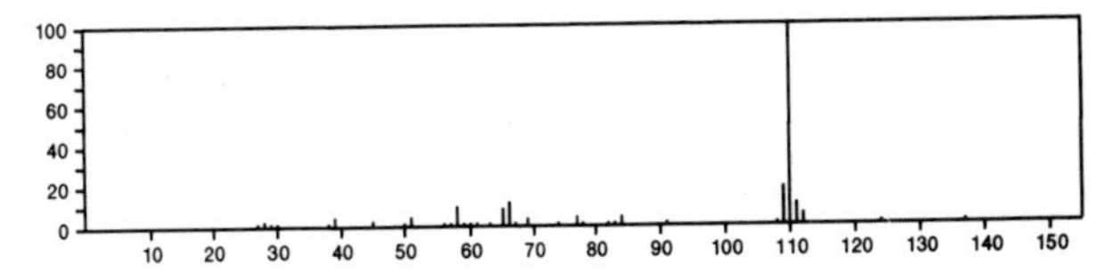

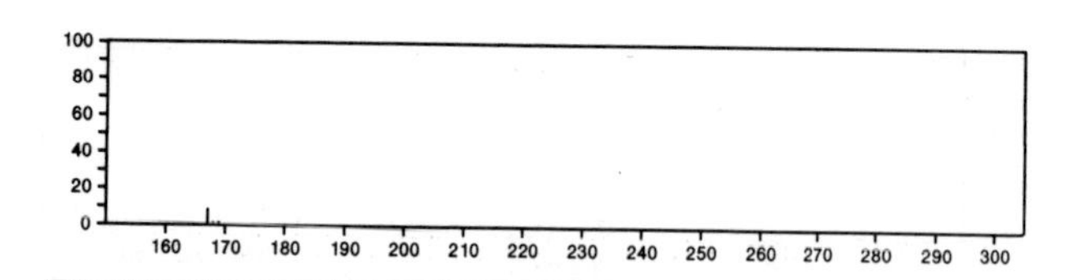

167 C_8H_9NOS 17420-02-9
Benzenamine, 2,6-dimethyl-*N*-sulfinyl-

167 C_8H_9NOS 23003-43-2
Thiazolo[3,2-*a*]pyridinium, 2,3-dihydro-8-hydroxy-5-methyl-, hydroxide, inner salt

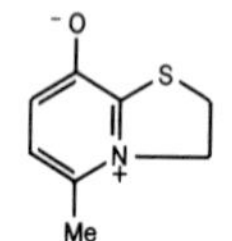

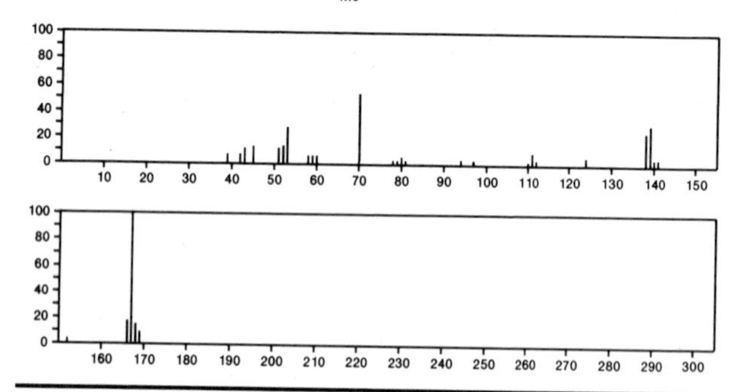

167 $C_8H_9NO_3$ 66-72-8
4-Pyridinecarboxaldehyde, 3-hydroxy-5-(hydroxymethyl)-2-methyl-

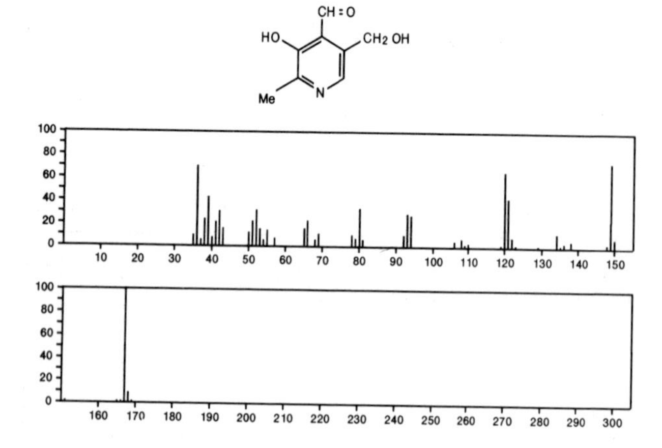

167 $C_8H_9NO_3$ 5501-39-3
2(1*H*)-Pyridinone, 3-acetyl-4-hydroxy-6-methyl-

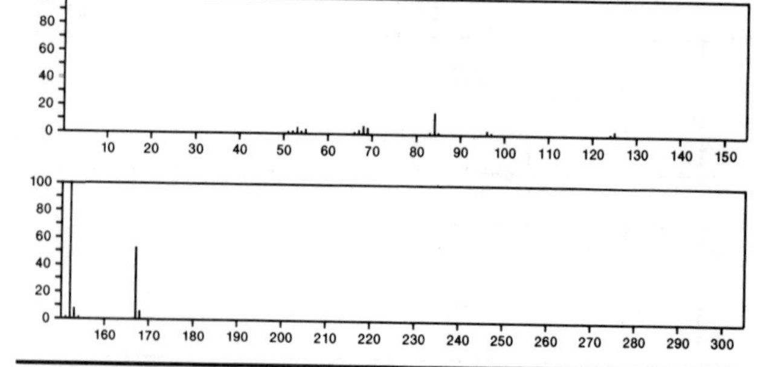

167 $C_8H_9NO_3$ 7292-76-4
Glycine, 2-(*m*-hydroxyphenyl)-, DL-

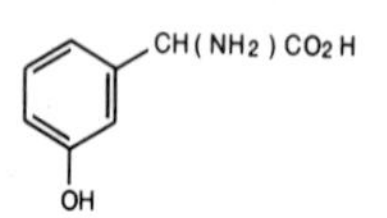

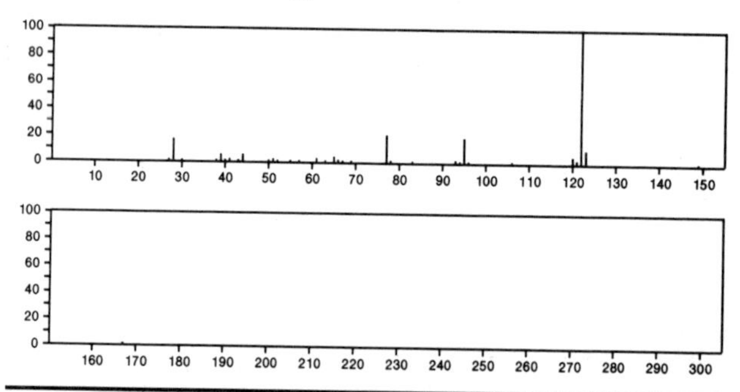

167 $C_8H_9NO_3$ 13959-08-5
2(1*H*)-Pyridinone, 4-(acetyloxy)-6-methyl-

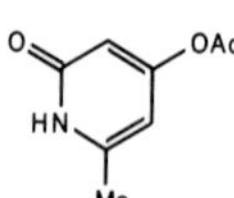

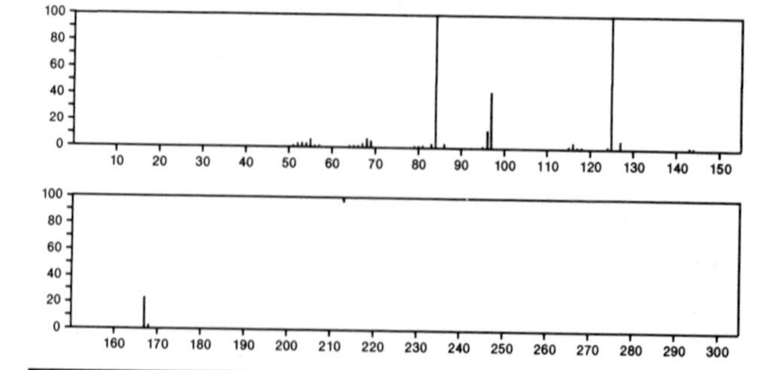

167 $C_8H_9NO_3$ 52022-77-2
Benzeneethanol, 3-nitro-

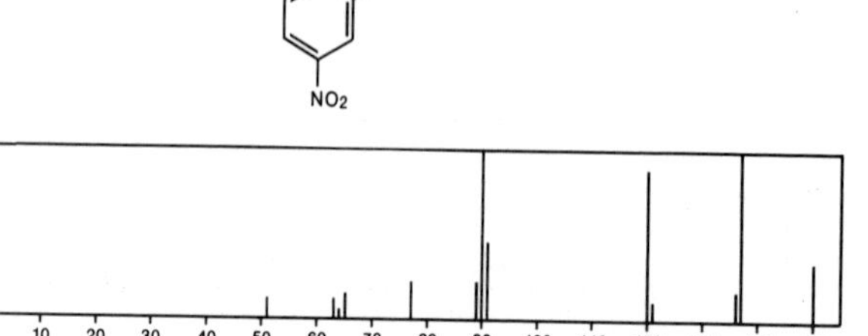

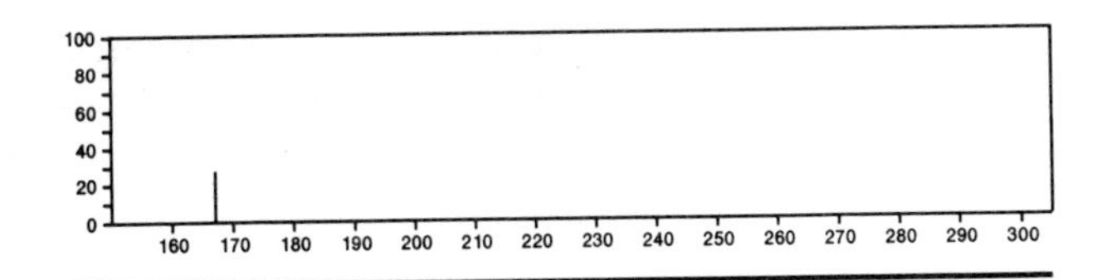

167 $C_8H_{11}BClN$ 1196-44-7
Boranamine, 1-chloro-*N*,*N*-dimethyl-1-phenyl-

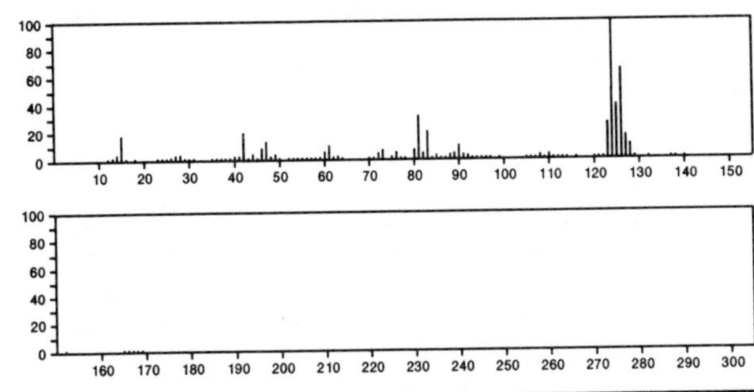

167 $C_8H_{13}NOSi$ 56196-99-7
4(1*H*)-Pyridinone, 1-(trimethylsilyl)-

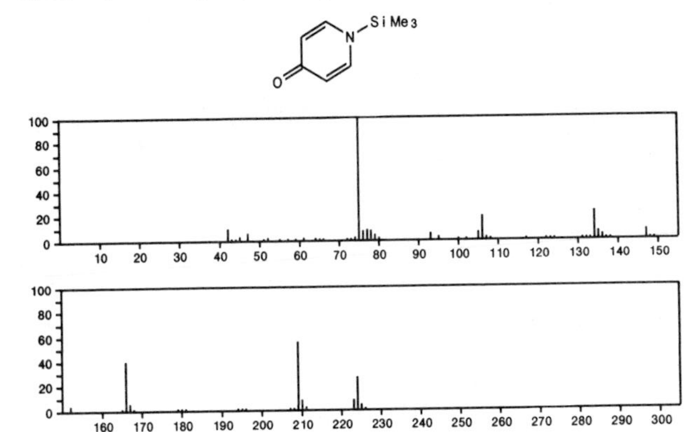

167 $C_9H_{10}ClN$ 30839-64-6
Azetidine, 1-chloro-2-phenyl-

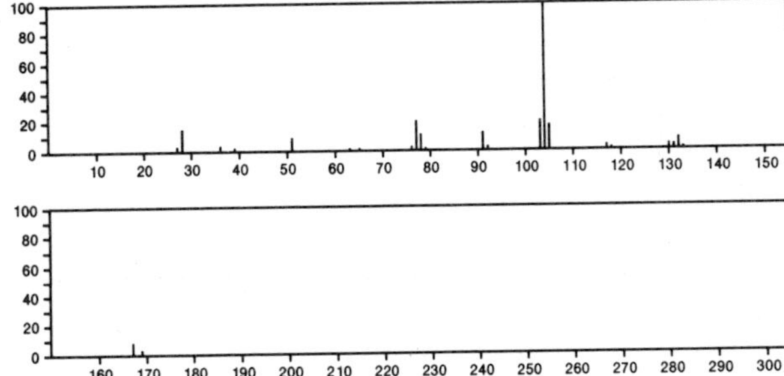

167 $C_9H_{13}NO_2$ 77-04-3
2,4(1*H*,3*H*)-Pyridinedione, 3,3-diethyl-

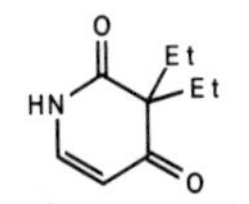

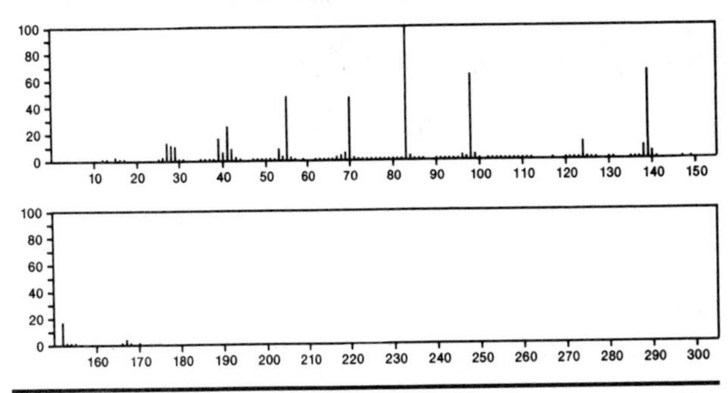

167 $C_9H_{13}NO_2$ 94-07-5
Benzenemethanol, 4-hydroxy-α-[(methylamino)methyl]-

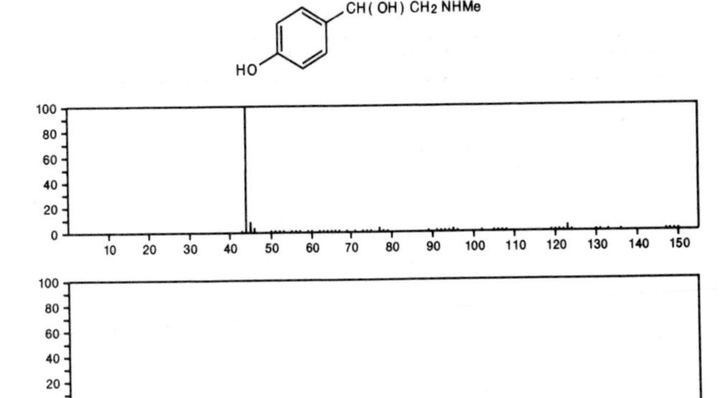

167 $C_9H_{13}NO_2$ 126-52-3
Cyclohexanol, 1-ethynyl-, carbamate

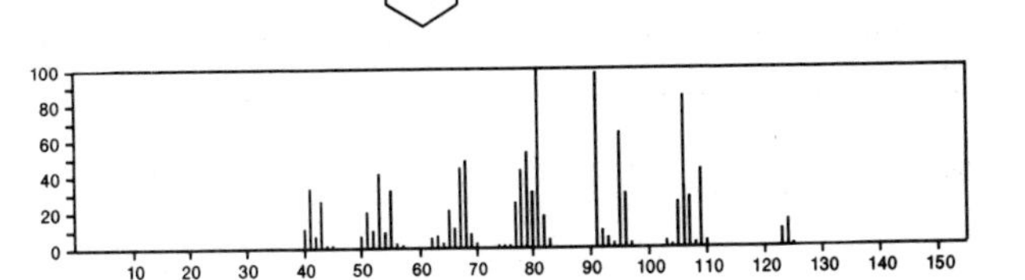

167 $C_9H_{13}NO_2$ 554-52-9
Phenol, 4-(2-aminoethyl)-2-methoxy-

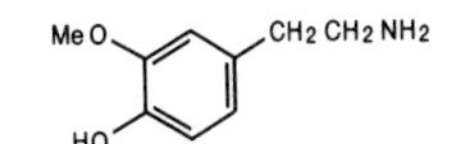

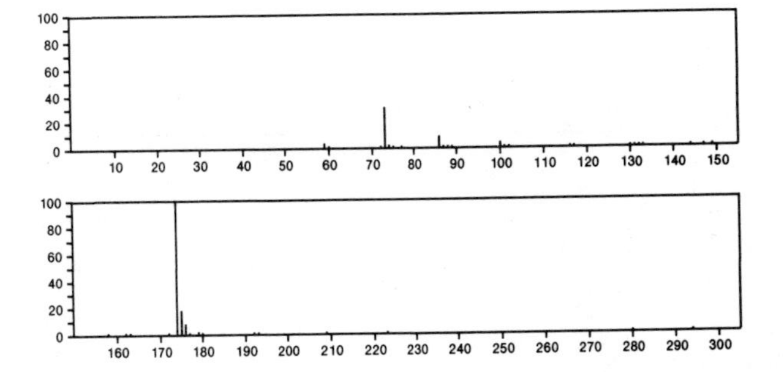

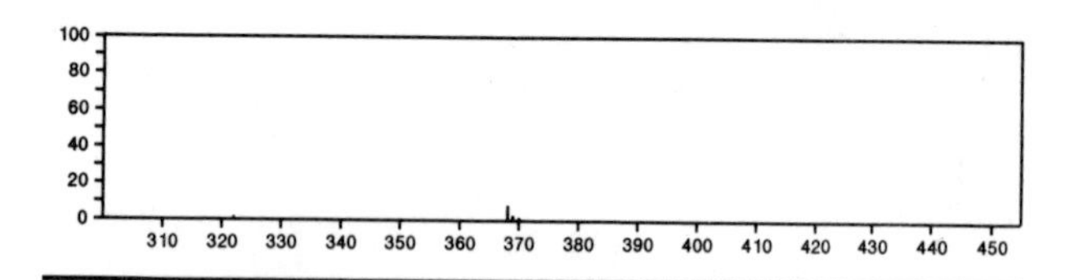

167 $C_9H_{13}NO_2$ 3213-30-7
Phenol, 5-(2-aminoethyl)-2-methoxy-

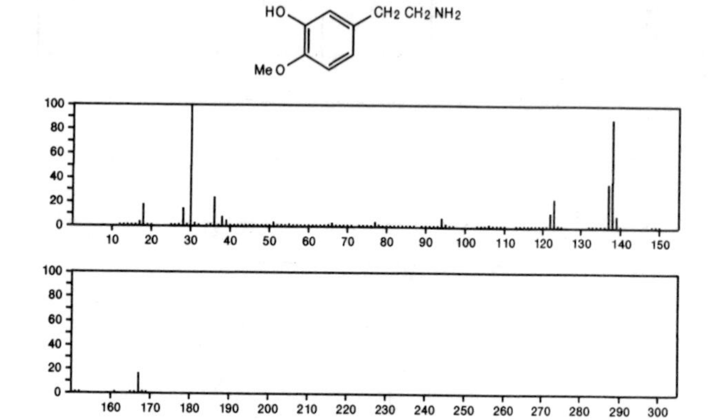

167 $C_9H_{13}NO_2$ 19788-38-6
2-Butanone, 4-(3,5-dimethyl-4-isoxazolyl)-

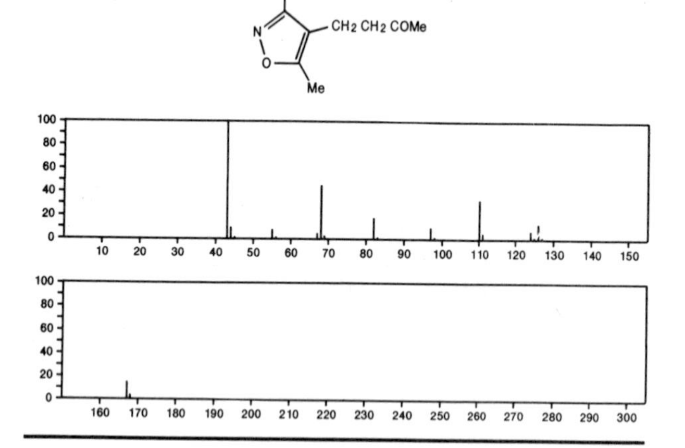

167 $C_9H_{13}NO_2$ 29249-00-1
Phenol, 4-(2-aminoethyl)methoxy-

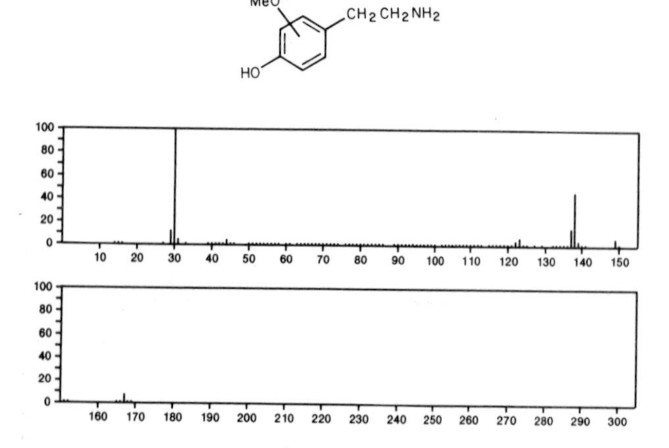

167 $C_9H_{13}NO_2$ 31539-88-5
1-Azabicyclo[2.2.2]oct-2-ene-3-carboxylic acid, methyl ester

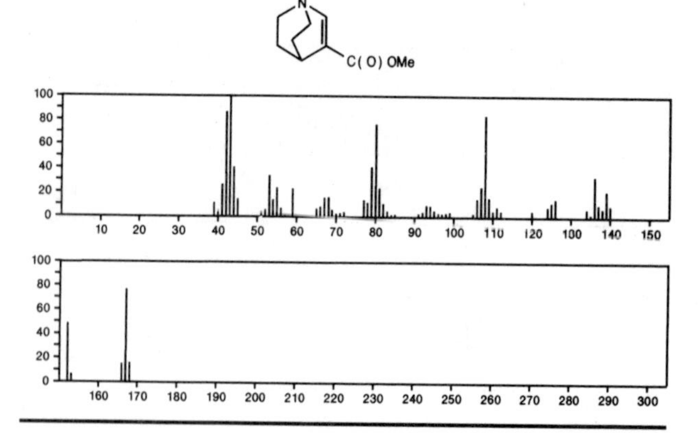

167 $C_9H_{13}NO_2$ 39998-23-7
3-Pyridineacetic acid, 1,4-dihydro-1-methyl-, methyl ester

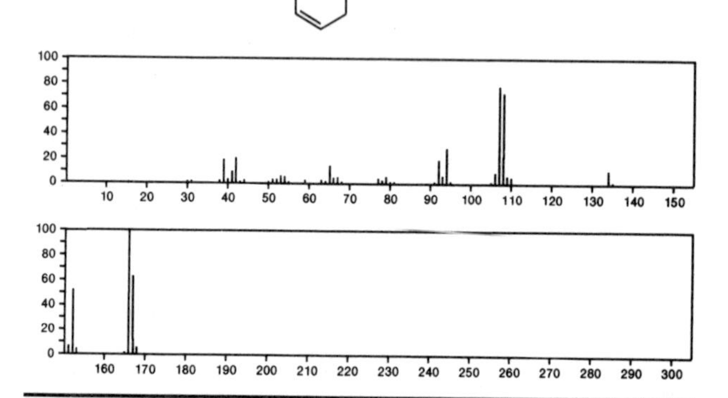

167 $C_9H_{13}NO_2$ 50267-24-8
2-Oxa-6-azatricyclo[3.3.1.1^{3,7}]decane-6-carboxaldehyde

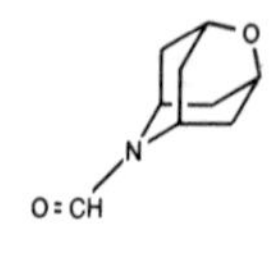

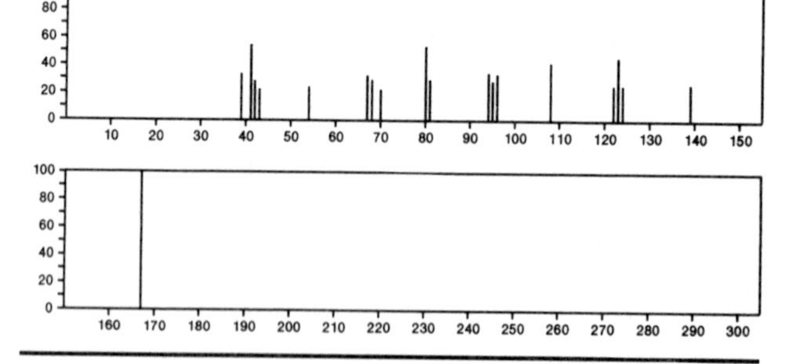

167 $C_9H_{13}NO_2$ 54410-97-8
Phenol, 2-(2-aminoethyl)-5-methoxy-

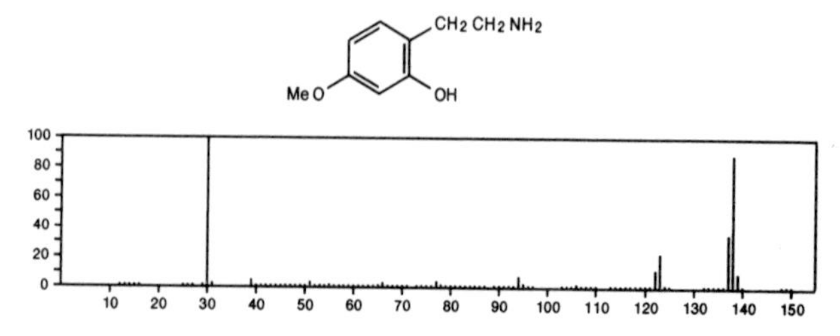

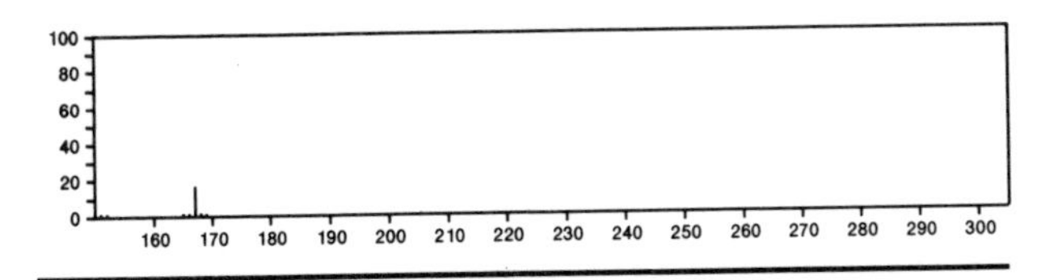

167 $C_9H_{13}NS$ 18794-26-8
Pyridine, 4-(*tert*-butylthio)-

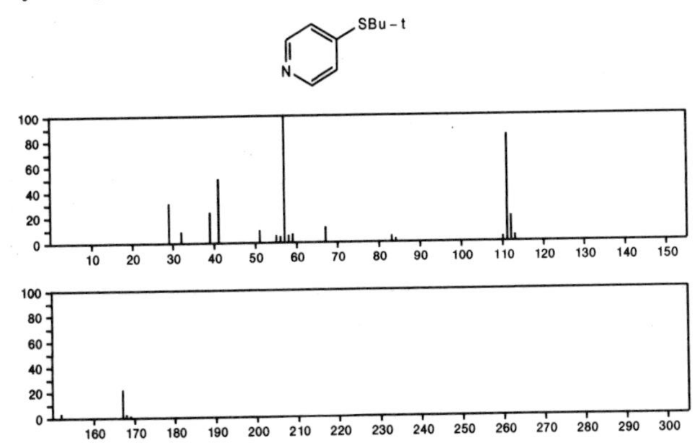

167 $C_{10}H_{17}NO$ 670-80-4
Morpholine, 4-(1-cyclohexen-1-yl)-

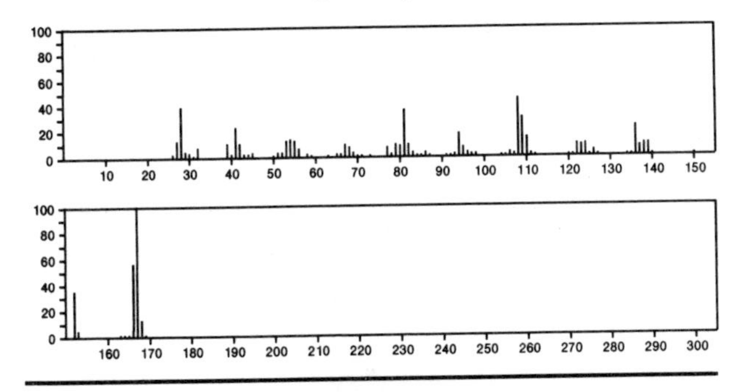

167 $C_{10}H_{17}NO$ 4146-36-5
8-Azabicyclo[4.3.1]decan-10-one, 8-methyl-

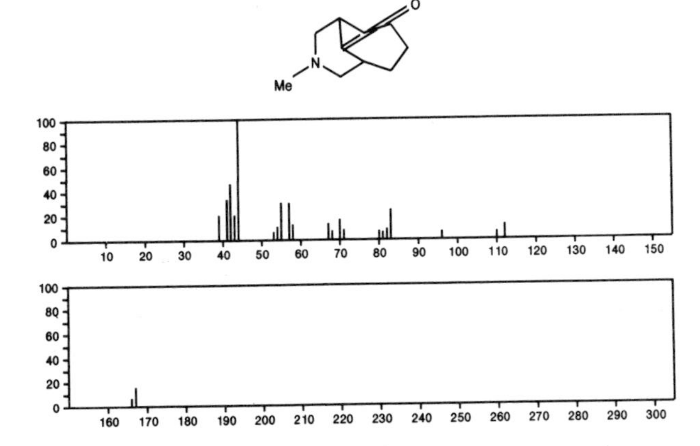

167 $C_{10}H_{17}NO$ 4514-87-8
Bicyclo[2.2.1]heptan-2-one, 4,7,7-trimethyl-, oxime

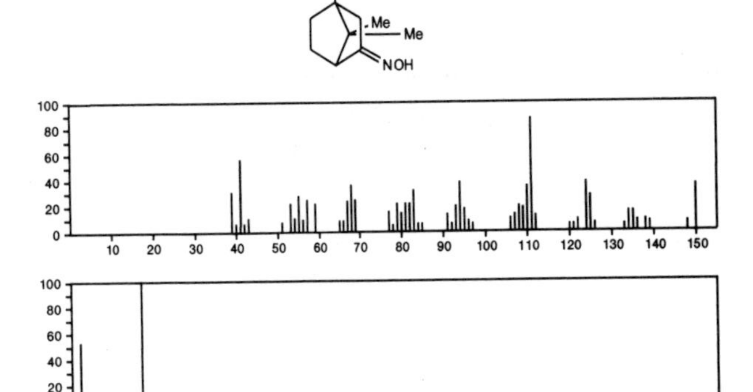

167 $C_{10}H_{17}NO$ 13559-66-5
Bicyclo[2.2.1]heptan-2-one, 1,7,7-trimethyl-, oxime

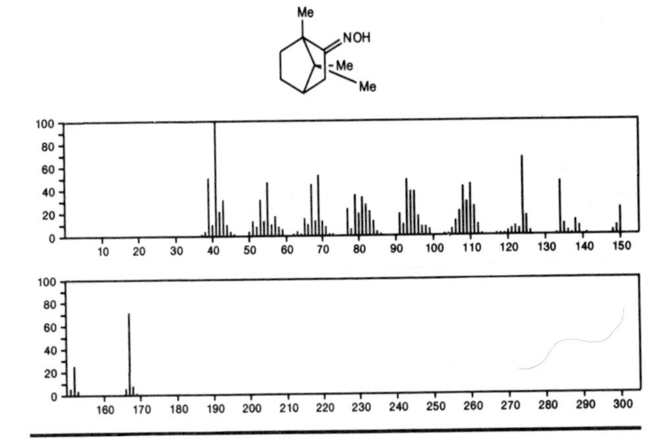

167 $C_{10}H_{17}NO$ 55760-17-3
Bicyclo[3.2.0]heptan-3-one, 1,4,4-trimethyl-, oxime

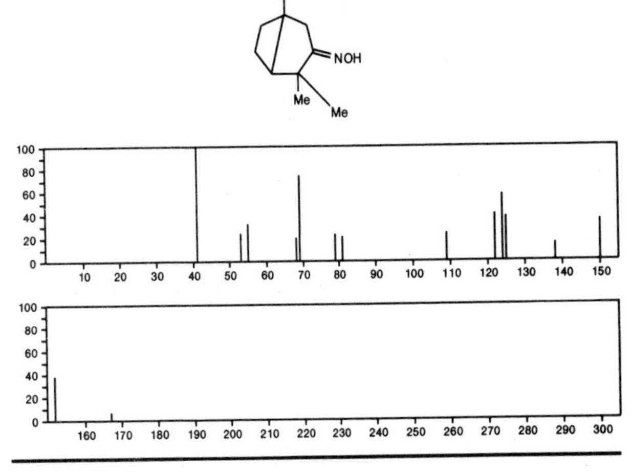

167 $C_{10}H_{17}NO$ 56701-31-6
Bicyclo[3.3.1]nonan-3-one, 7-(aminomethyl)-

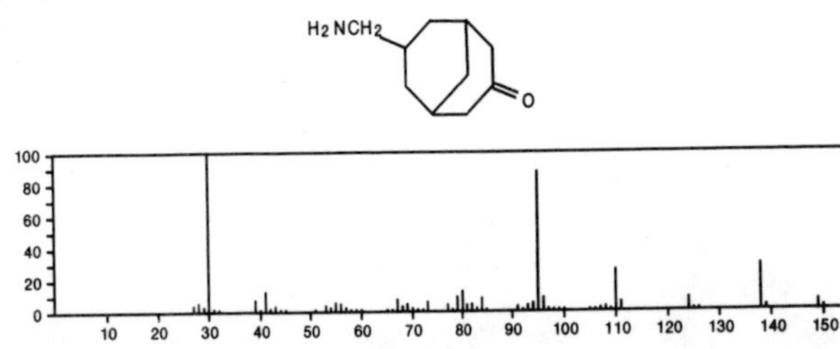

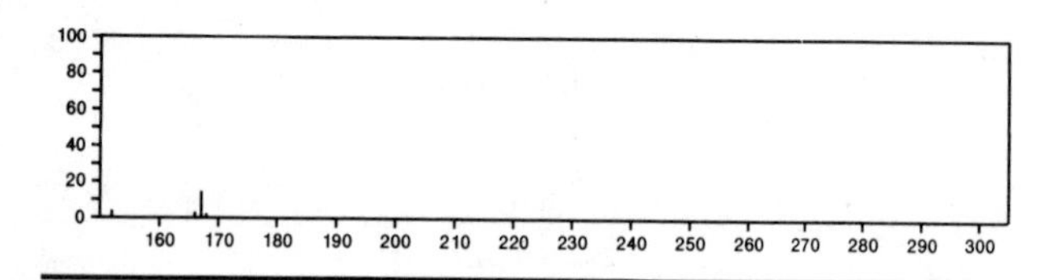

167 $C_{10}H_{17}NO$ 56771-94-9
8-Azabicyclo[3.2.1]octane, 8-(1-oxopropyl)-

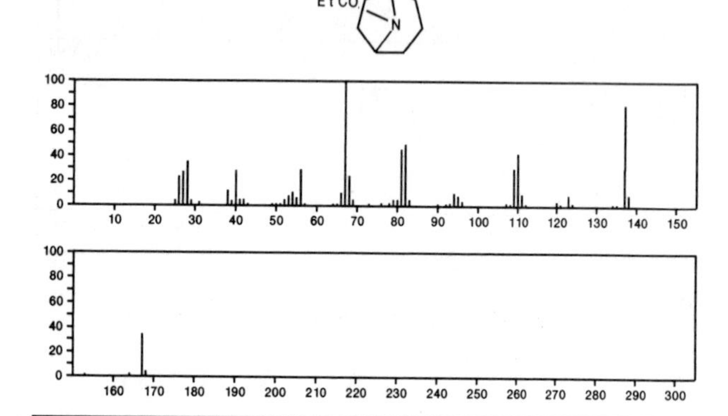

167 $C_{11}H_{21}N$ 2244-07-7
Undecanenitrile

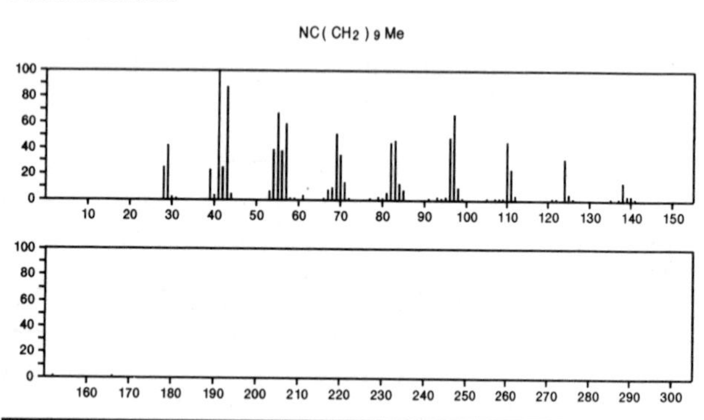

167 $C_{11}H_{21}N$ 22285-82-1
Bicyclo[2.2.1]heptan-2-amine, *N*,1,7,7-tetramethyl-

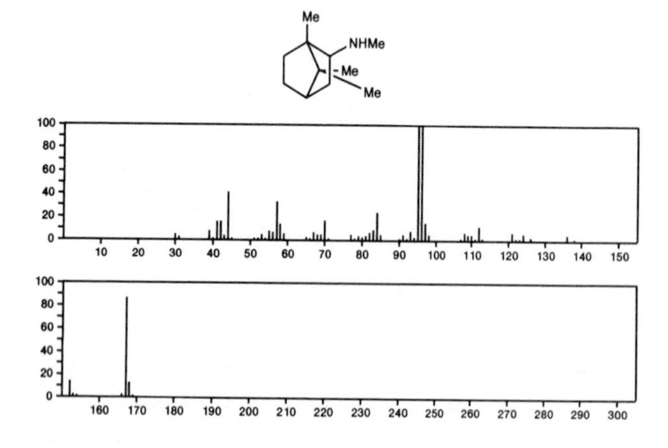

167 $C_{11}H_{21}N$ 32064-85-0
Quinoline, decahydro-1,7-dimethyl-

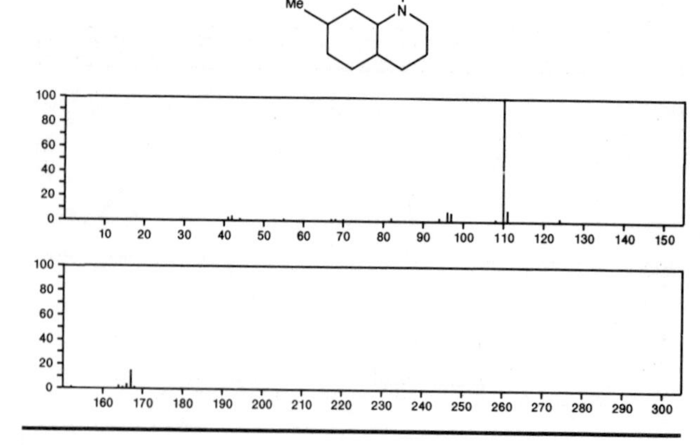

167 $C_{11}H_{21}N$ 35973-44-5
Bicyclo[2.2.1]heptan-2-amine, *N*,4,7,7-tetramethyl-

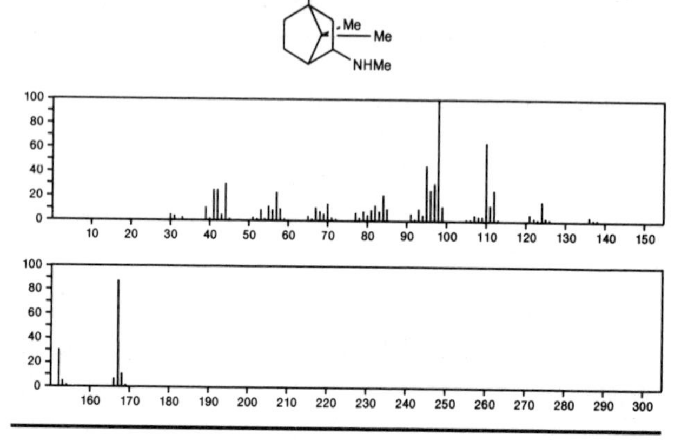

167 $C_{12}H_9N$ 86-74-8
9*H*-Carbazole

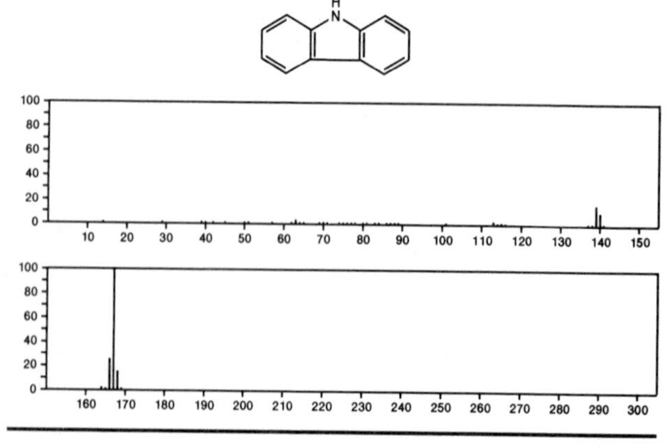

168 $C_2HCl_3F_2$ 354-21-2
Ethane, 1,2,2-trichloro-1,1-difluoro-

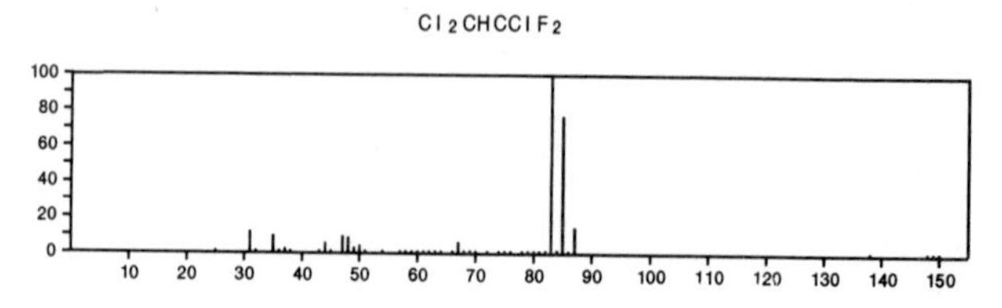

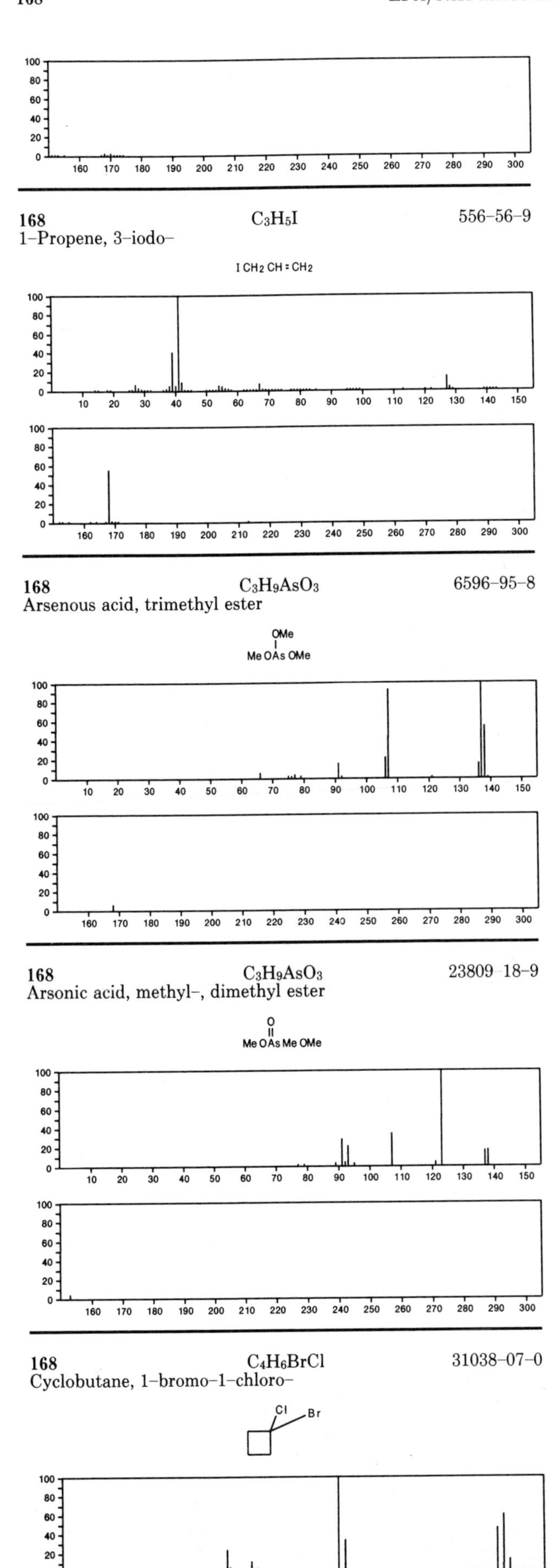

168 C_3H_5I 556-56-9
1-Propene, 3-iodo-

168 $C_3H_9AsO_3$ 6596-95-8
Arsenous acid, trimethyl ester

168 $C_3H_9AsO_3$ 23809-18-9
Arsonic acid, methyl-, dimethyl ester

168 C_4H_6BrCl 31038-07-0
Cyclobutane, 1-bromo-1-chloro-

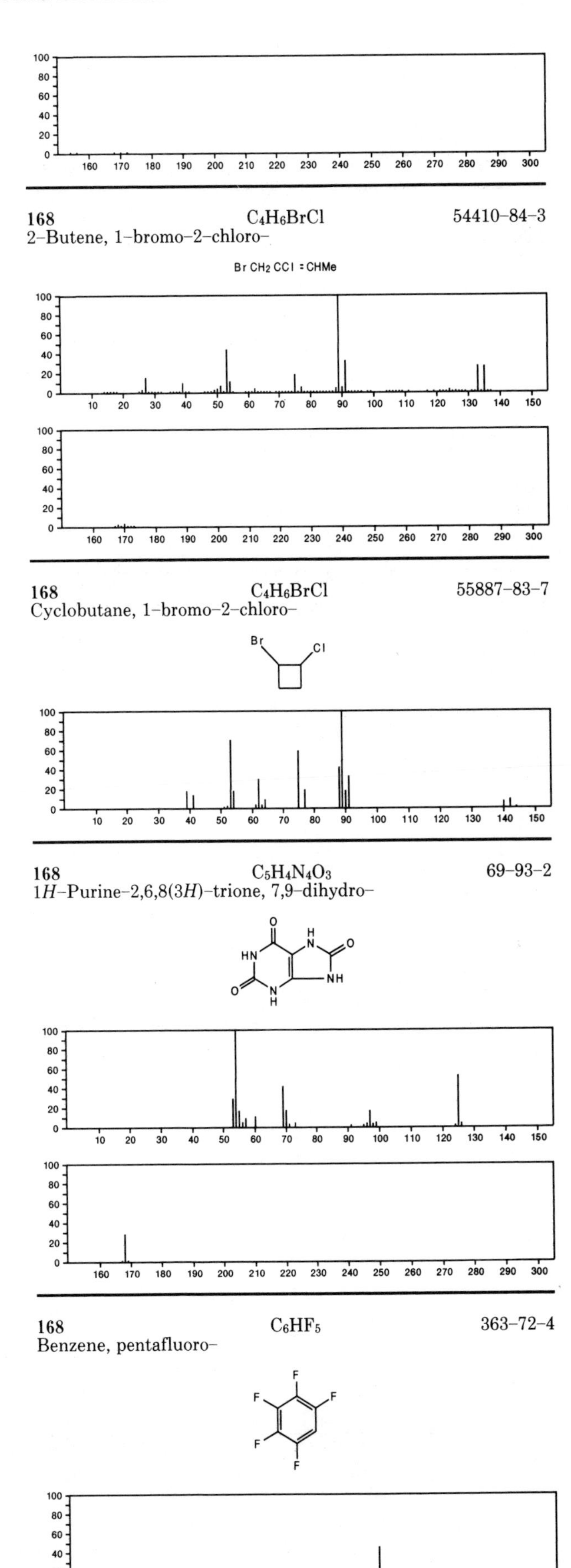

168 C_4H_6BrCl 54410-84-3
2-Butene, 1-bromo-2-chloro-

168 C_4H_6BrCl 55887-83-7
Cyclobutane, 1-bromo-2-chloro-

168 $C_5H_4N_4O_3$ 69-93-2
1*H*-Purine-2,6,8(3*H*)-trione, 7,9-dihydro-

168 C_6HF_5 363-72-4
Benzene, pentafluoro-

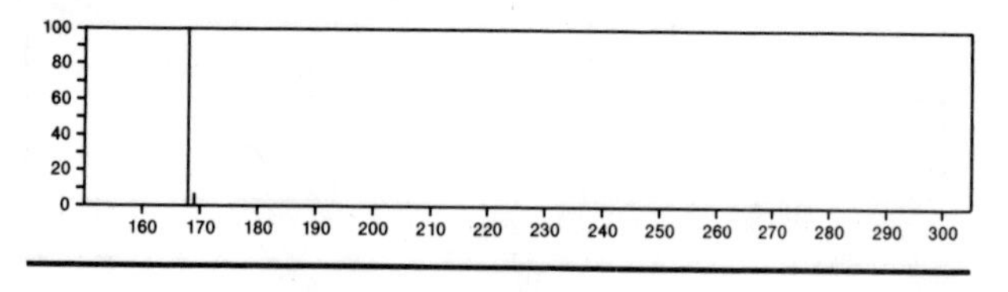

168 $C_6H_4N_2O_4$ 99-65-0
Benzene, 1,3-dinitro-

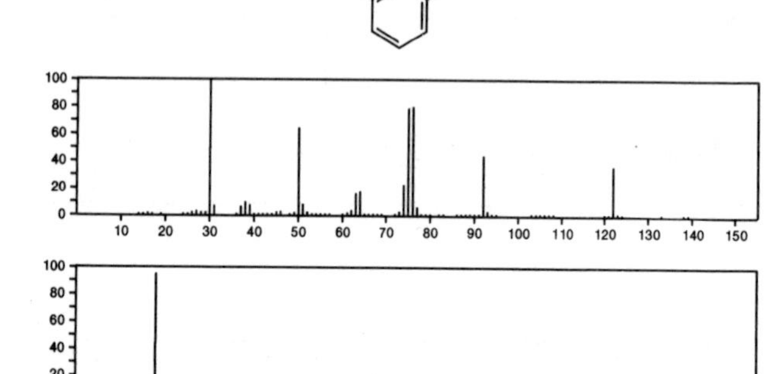

168 $C_6H_4N_2O_4$ 100-25-4
Benzene, 1,4-dinitro-

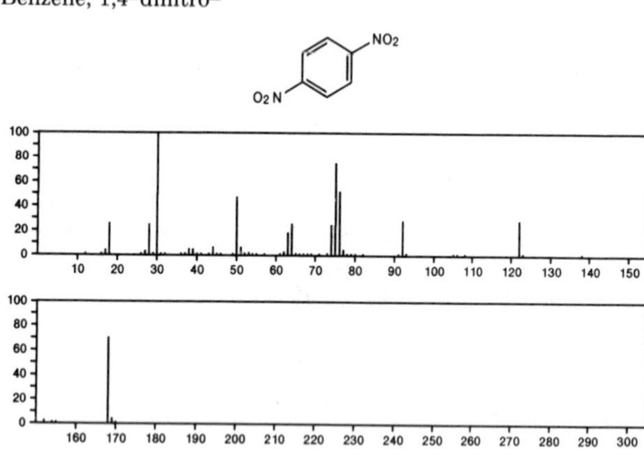

168 $C_6H_4N_2O_4$ 26893-68-5
2-Pyridinecarboxylic acid, 6-nitro-

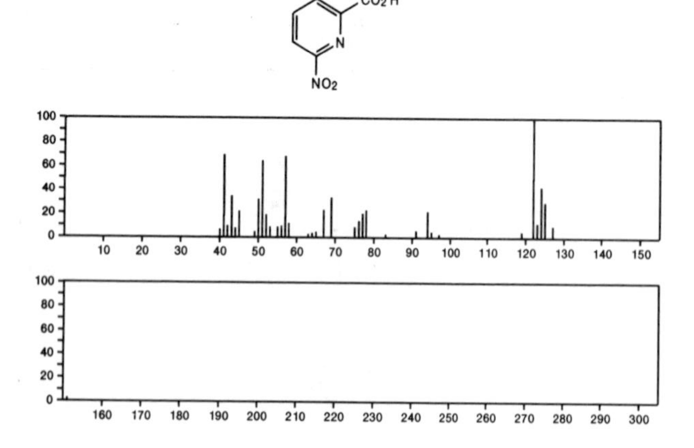

168 $C_6H_4N_2O_4$ 30651-24-2
2-Pyridinecarboxylic acid, 5-nitro-

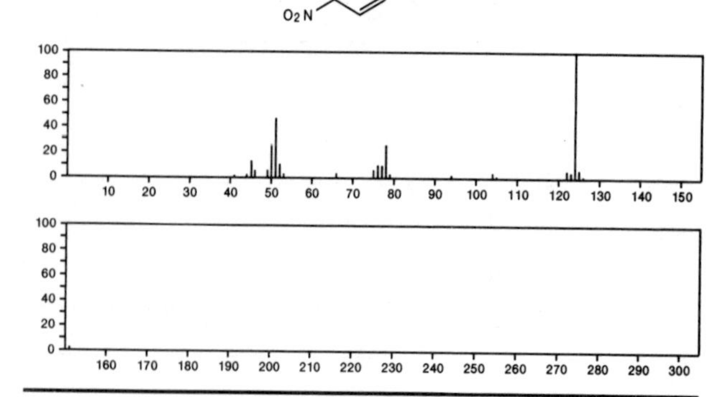

168 $C_6H_7F_3O_2$ 400-54-4
2,4-Hexanedione, 1,1,1-trifluoro-

F_3CCOCH_2COEt

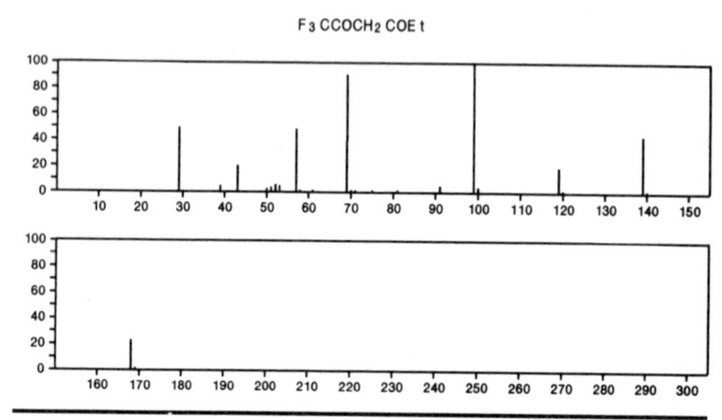

168 $C_6H_8N_4O_2$ 35975-28-1
Formamide, *N,N'*-2,6-piperazinediylidenebis-

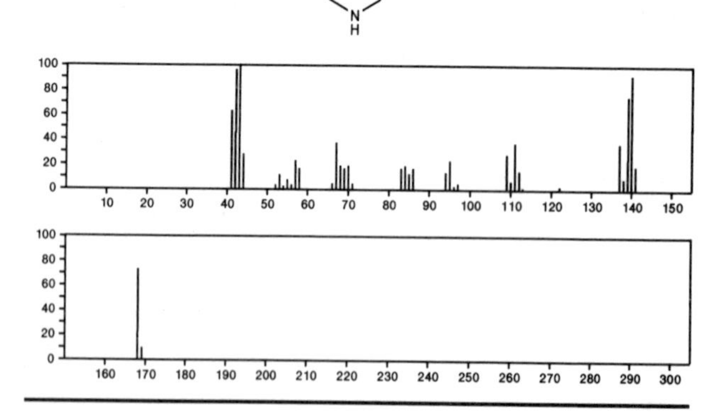

168 $C_6H_{12}N_6$ 877-77-0
s-Tetrazine, 3,6-bis(dimethylamino)-

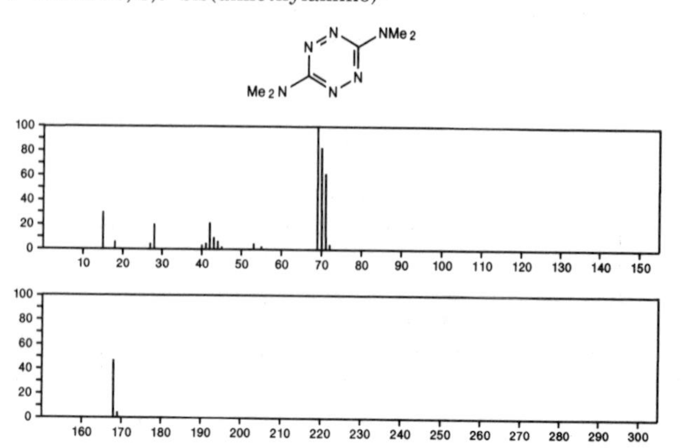

168 $C_6H_{13}ClO_3$ 5197-62-6
Ethanol, 2-[2-(2-chloroethoxy)ethoxy]-

HOCH2 CH2 OCH2 CH2 OCH2 CH2 Cl

168 $C_6H_{15}B_3O_3$ 3043-60-5
Boroxin, triethyl-

168 $C_7H_5ClN_2O$ 61-80-3
2-Benzoxazolamine, 5-chloro-

168 $C_7H_8N_2OS$ 24614-07-1
Thiazolo[3,2-*c*]pyrimidin-4-ium, 2,3-dihydro-8-hydroxy-5-methyl-, hydroxide, inner salt

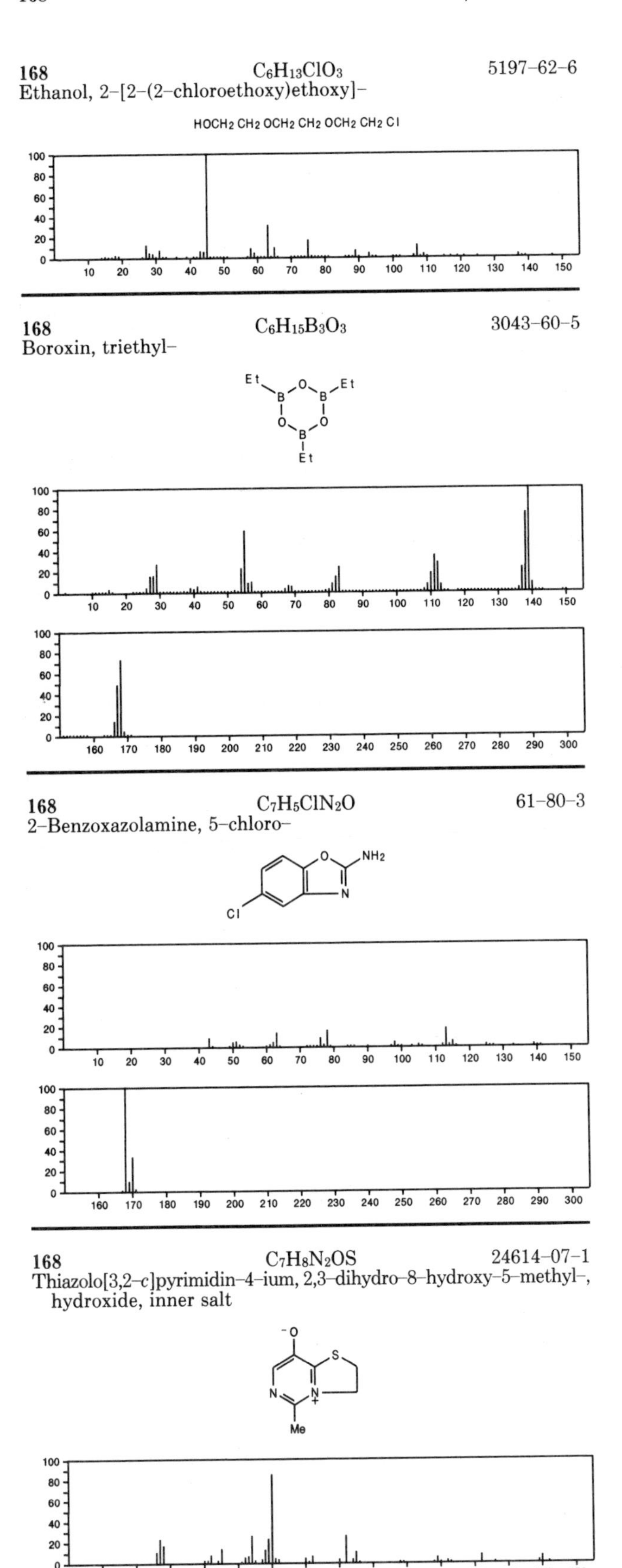

168 $C_7H_{14}Cl_2$ 821-25-0
Heptane, 1,1-dichloro-

Cl 2 CH(CH2) 5 Me

168 $C_7H_{14}Cl_2$ 16703-32-5
Hexane, 1,1-dichloro-3-methyl-

Pr CHMe CH2 CHCl 2

168 C_8H_5ClS 7342-86-1
Benzo[*b*]thiophene, 3-chloro-

168 $C_8H_8O_2S$ 103-04-8
Acetic acid, (phenylthio)-

HO2 CCH2 SPh

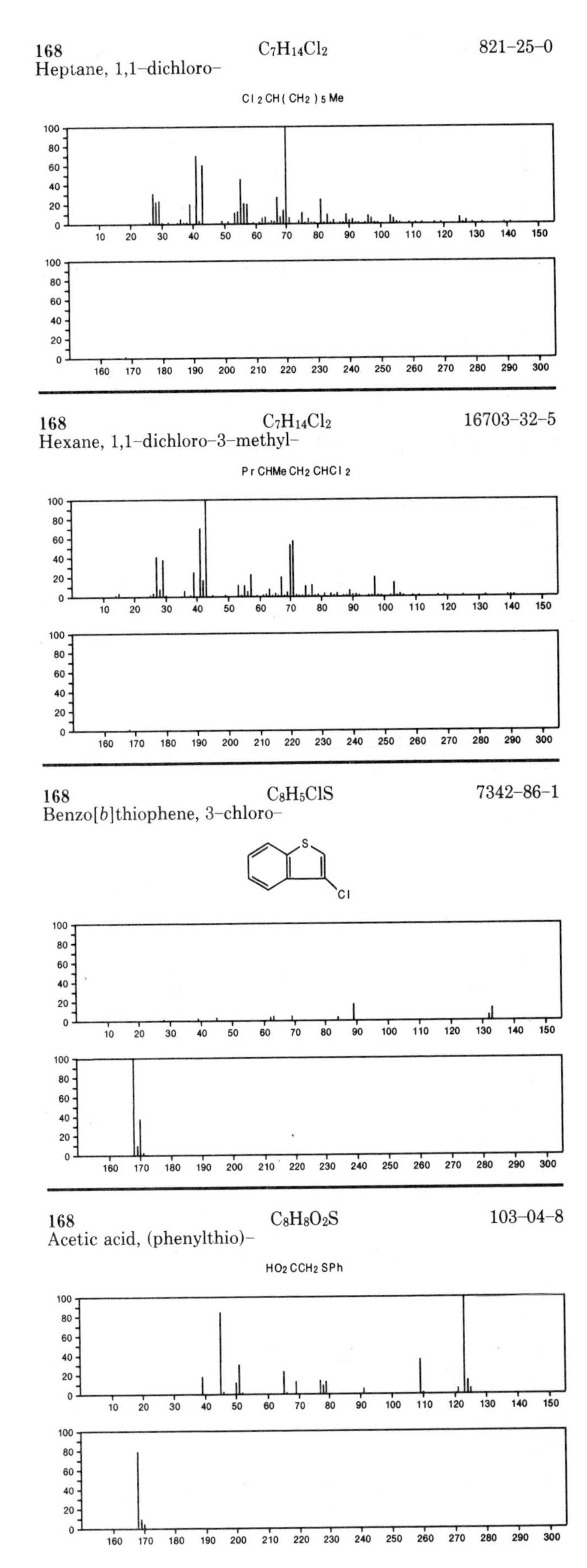

168 $C_8H_8O_2S$ 1007-37-0
Carbonic acid, thio-, *O*-methyl *O*-phenyl ester

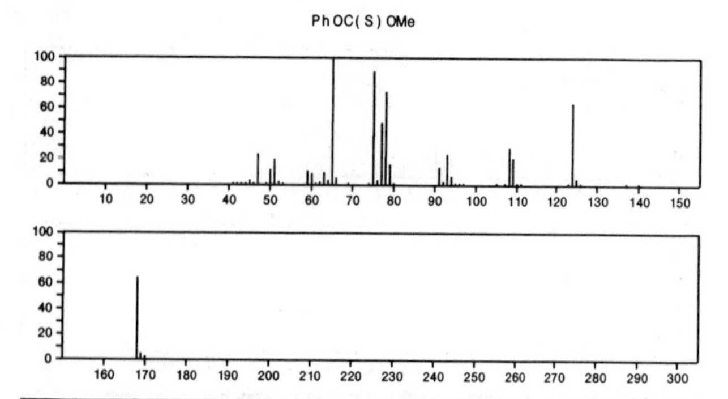

168 $C_8H_8O_2S$ 3186-52-5
Carbonothioic acid, *O*-methyl *S*-phenyl ester

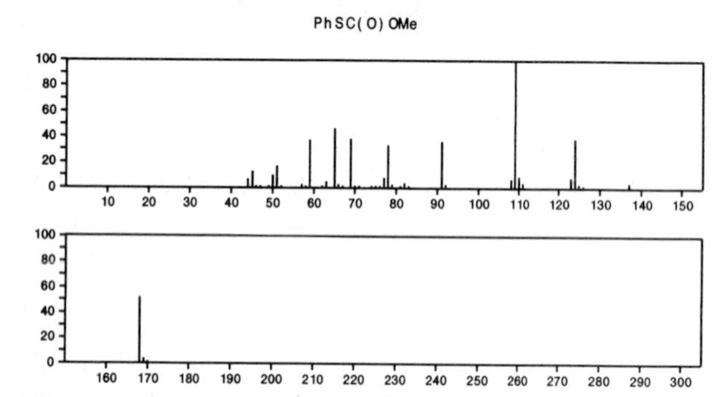

168 $C_8H_8O_2S$ 5535-48-8
Benzene, (ethenylsulfonyl)-

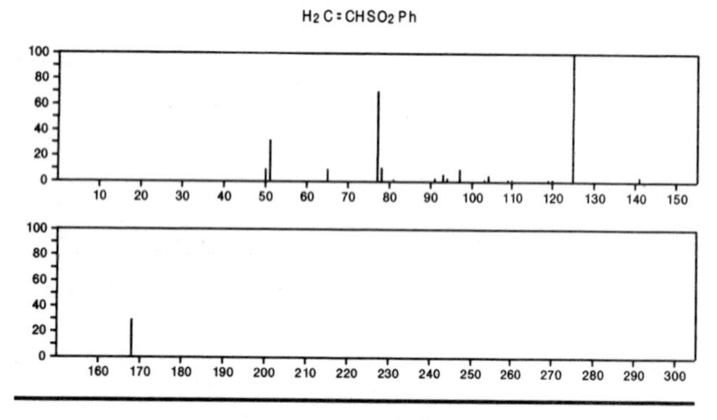

168 $C_8H_8O_2S$ 13509-28-9
Carbonic acid, thio-, *S*-methyl *O*-phenyl ester

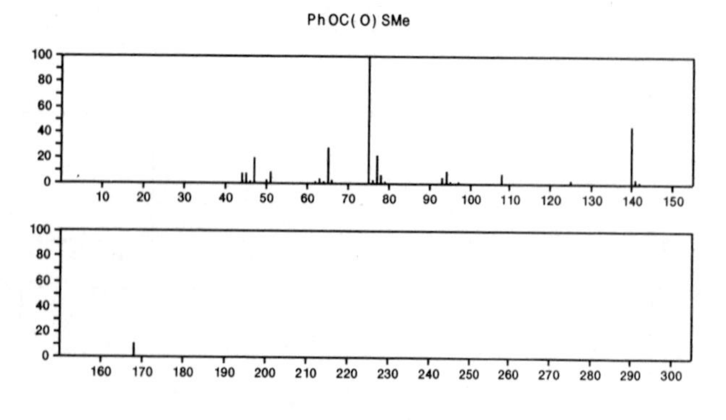

168 $C_8H_8O_4$ 102-32-9
Benzeneacetic acid, 3,4-dihydroxy-

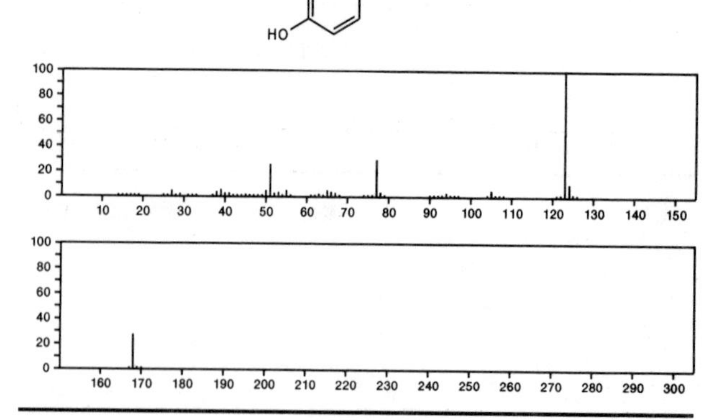

168 $C_8H_8O_4$ 121-34-6
Benzoic acid, 4-hydroxy-3-methoxy-

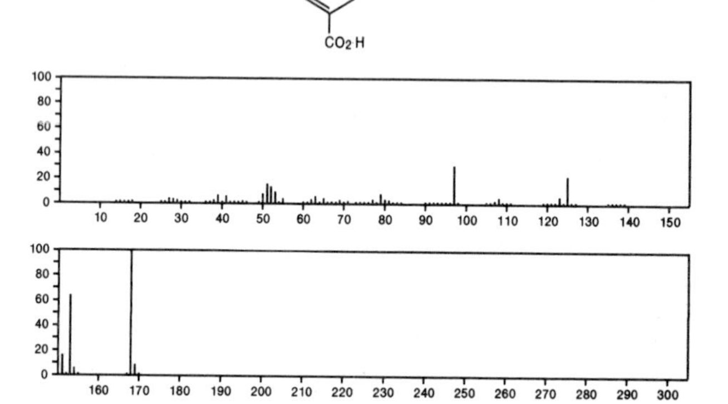

168 $C_8H_8O_4$ 451-13-8
Benzeneacetic acid, 2,5-dihydroxy-

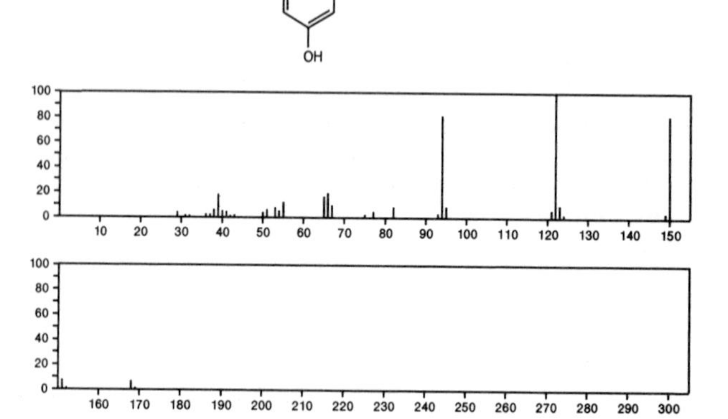

168 $C_8H_8O_4$ 480–64–8
Benzoic acid, 2,4–dihydroxy–6–methyl–

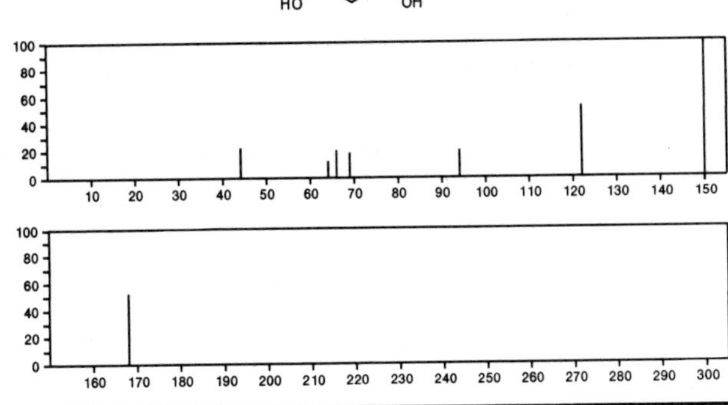

168 $C_8H_8O_4$ 520–45–6
2*H*–Pyran–2,4(3*H*)–dione, 3–acetyl–6–methyl–

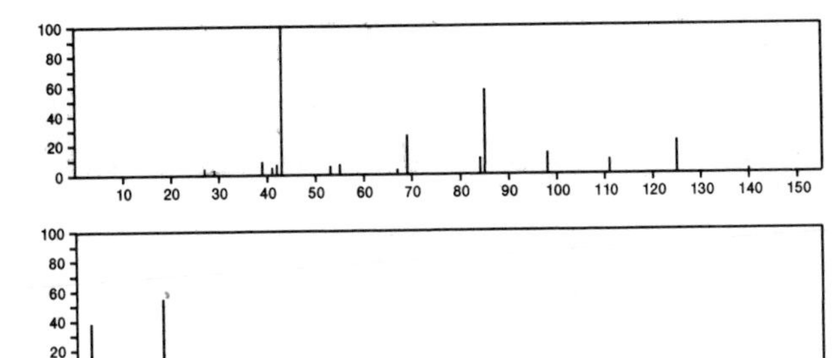

168 $C_8H_8O_4$ 669–40–9
2*H*–Pyran–5–carboxylic acid, 6–methyl–2–oxo–, methyl ester

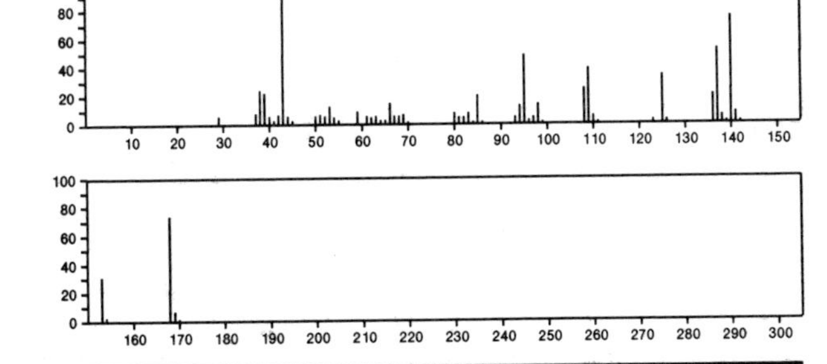

168 $C_8H_8O_4$ 1198–84–1
Benzeneacetic acid, α,4–dihydroxy–

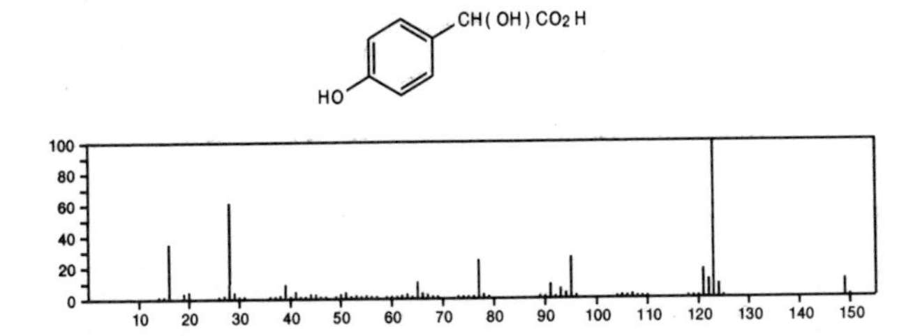

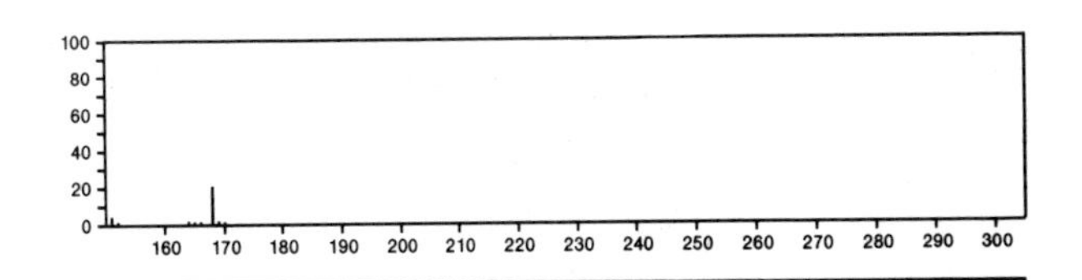

168 $C_8H_8O_4$ 2150–43–8
Benzoic acid, 3,4–dihydroxy–, methyl ester

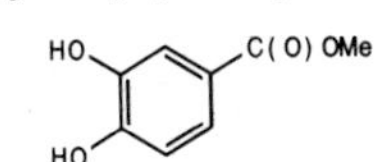

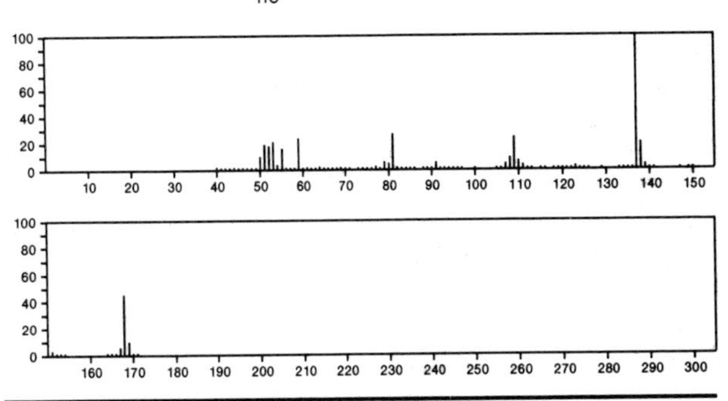

168 $C_8H_8O_4$ 2150–45–0
Benzoic acid, 2,6–dihydroxy–, methyl ester

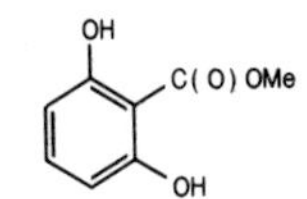

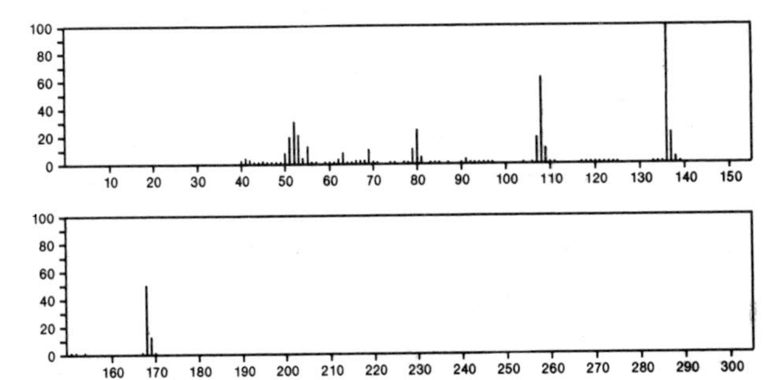

168 $C_8H_8O_4$ 2150–46–1
Benzoic acid, 2,5–dihydroxy–, methyl ester

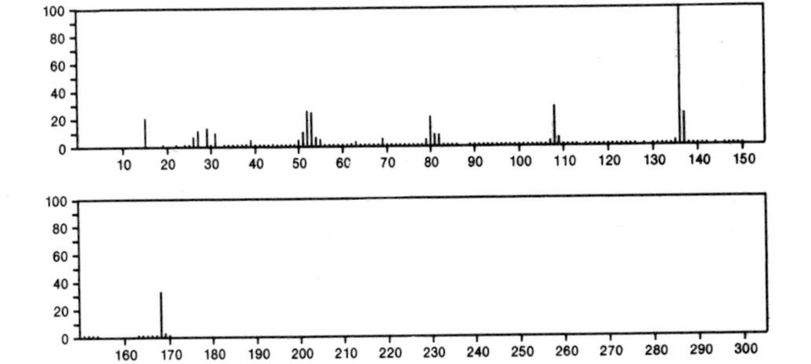

168 $C_8H_8O_4$ 2150-47-2
Benzoic acid, 2,4-dihydroxy-, methyl ester

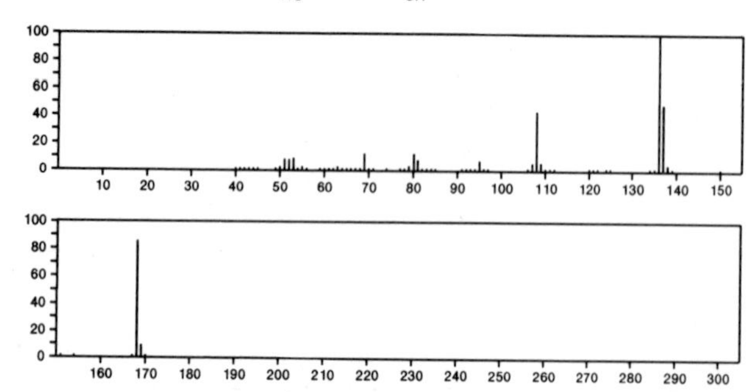

168 $C_8H_8O_4$ 2411-83-8
Benzoic acid, 2,3-dihydroxy-, methyl ester

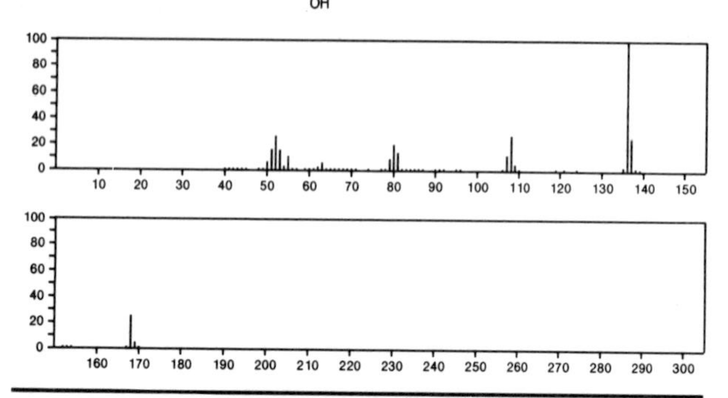

168 $C_8H_8O_4$ 2654-72-0
2,5-Cyclohexadiene-1,4-dione, 2,5-dihydroxy-3,6-dimethyl-

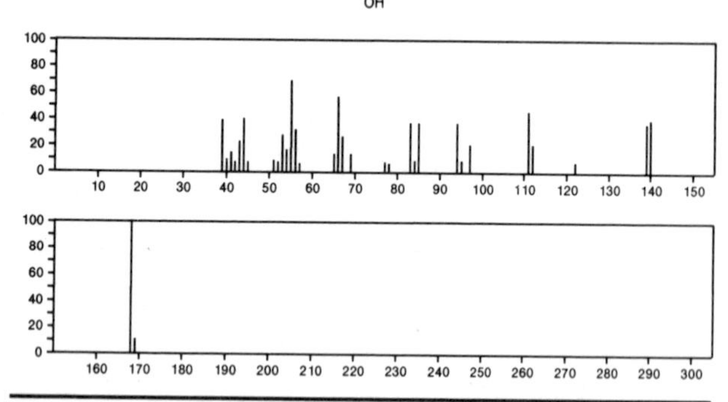

168 $C_8H_8O_4$ 17119-15-2
Benzeneacetic acid, α,3-dihydroxy-

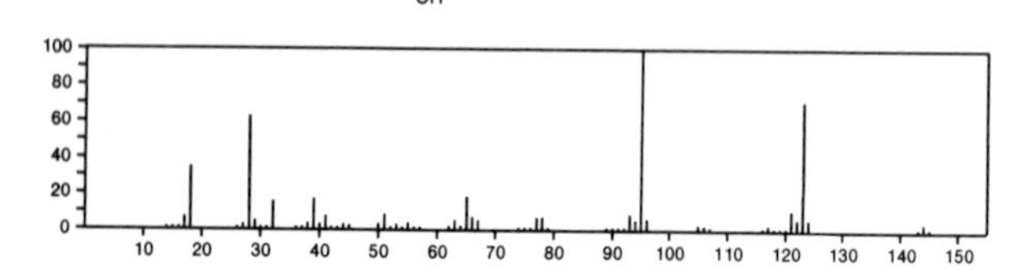

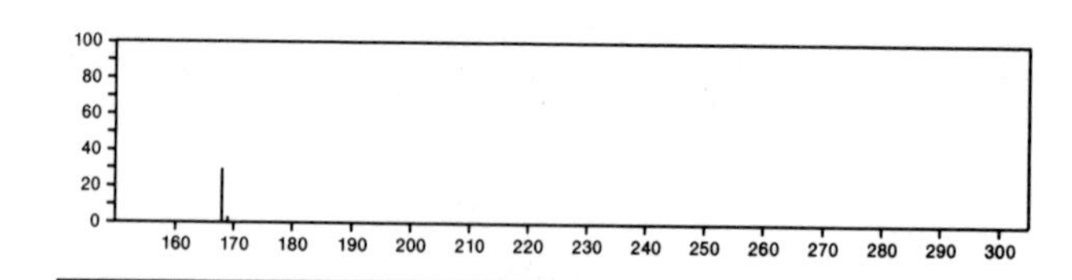

168 $C_8H_8S_2$ 2168-78-7
Benzenecarbodithioic acid, methyl ester

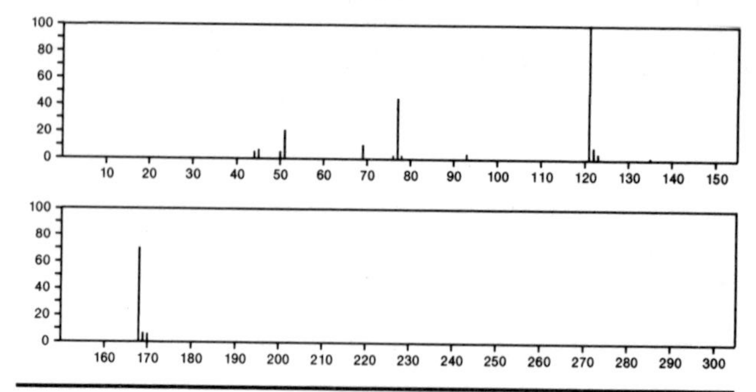

168 $C_8H_{12}N_2O_2$ 7454-99-1
Uracil, 6-methyl-3-propyl-

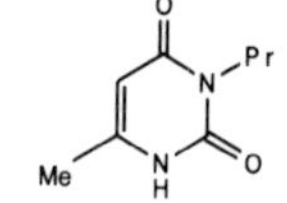

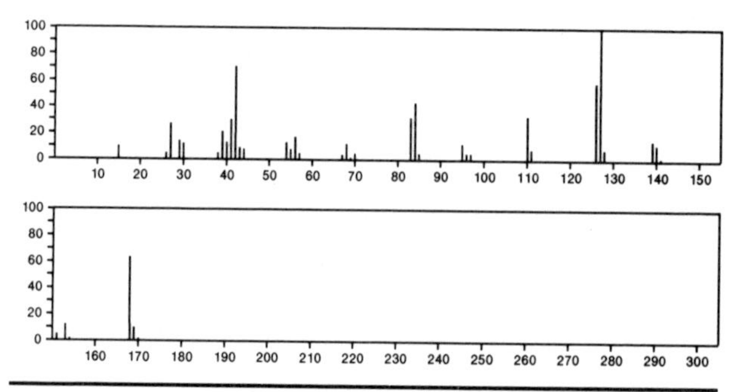

168 $C_8H_{12}N_2O_2$ 20600-69-5
Sydnone, 3-cyclohexyl-

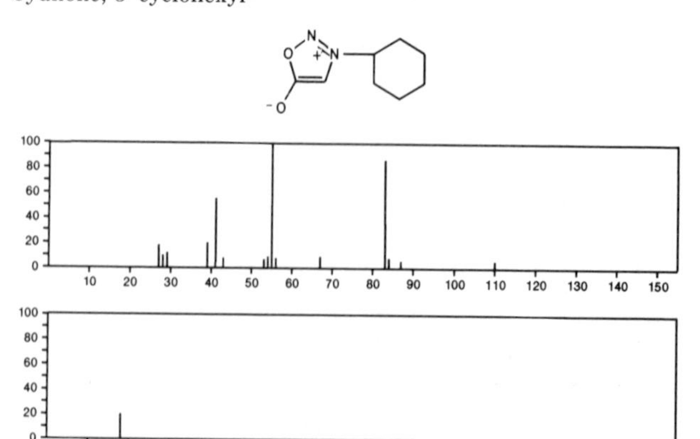

168 $C_8H_{12}N_2O_2$ 24614-11-7
4(3*H*)-Pyrimidinone, 5-ethoxy-2,3-dimethyl-

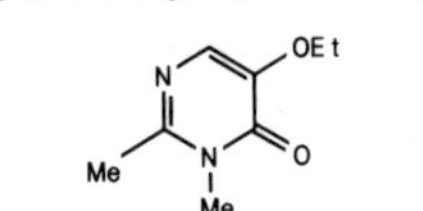

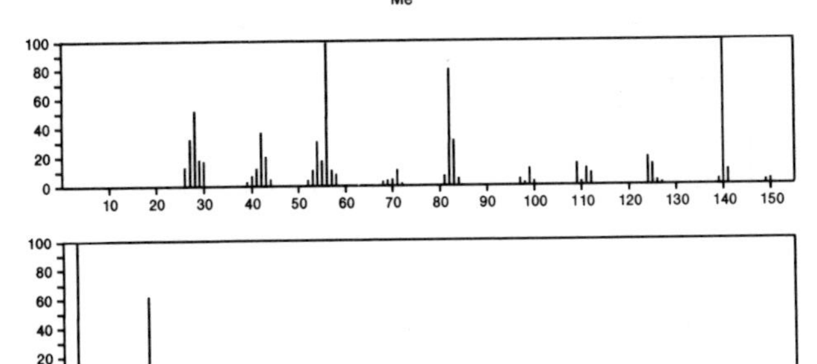

168 $C_8H_{12}N_2O_2$ 24614-12-8
Pyrimidine, 5-ethoxy-4-methoxy-2-methyl-

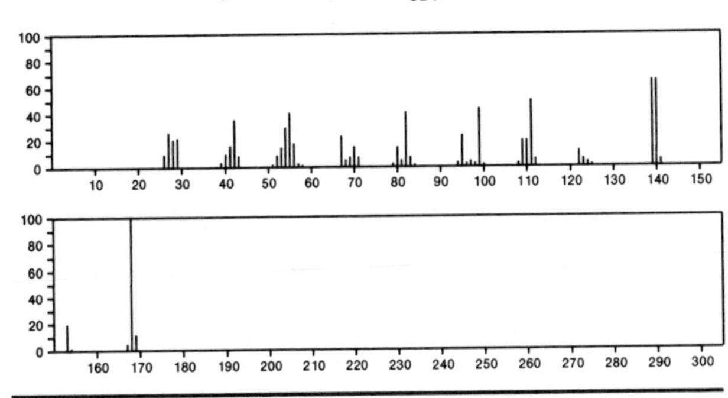

168 $C_8H_{22}B_2N_2$ 19162-23-3
1,2-Diborane(4)diamine, 1,2-diethyl-*N,N,N',N'*-tetramethyl-

$Me_2NBEtBEtNMe_2$

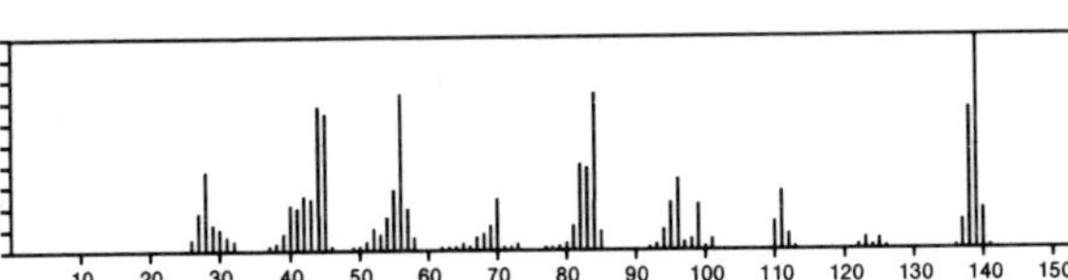

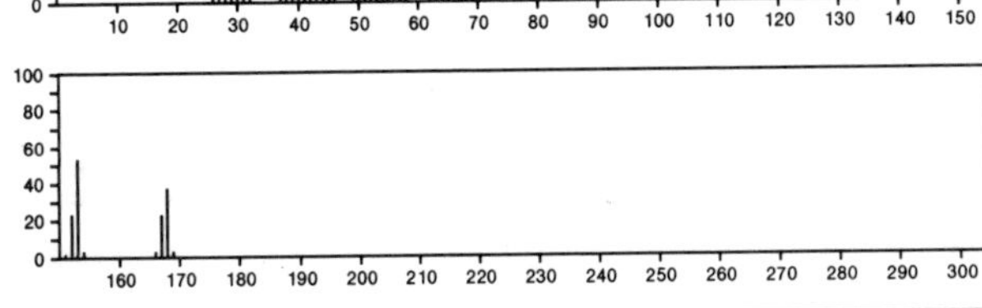

168 C_9H_9ClO 6285-05-8
1-Propanone, 1-(4-chlorophenyl)-

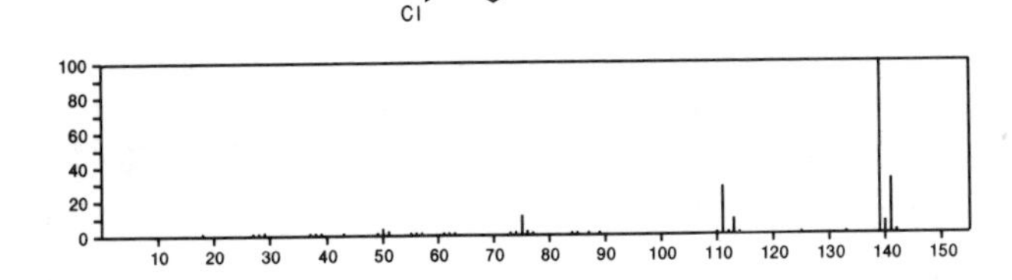

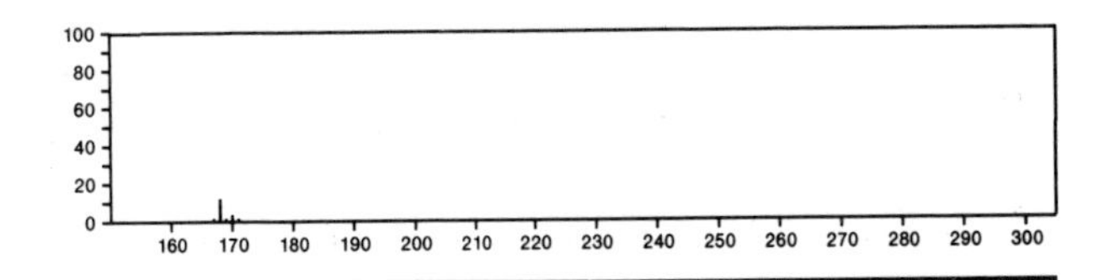

168 C_9H_9ClO 37074-39-8
Ethanone, 1-(4-chloro-3-methylphenyl)-

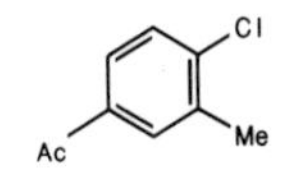

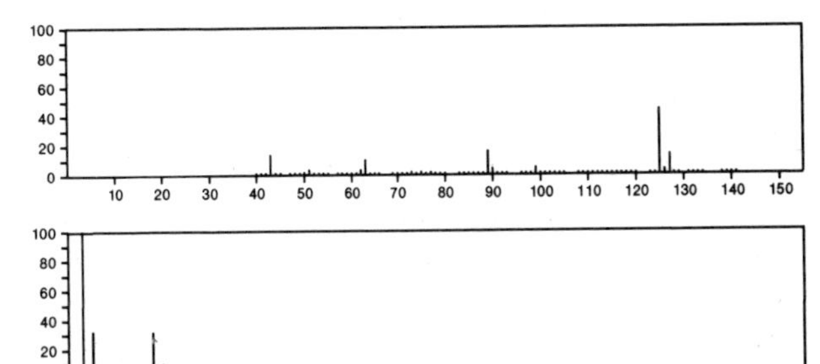

168 C_9H_9ClO 53299-53-9
Benzene, [(2-chloro-2-propenyl)oxy]-

$PhOCH_2CCl:CH_2$

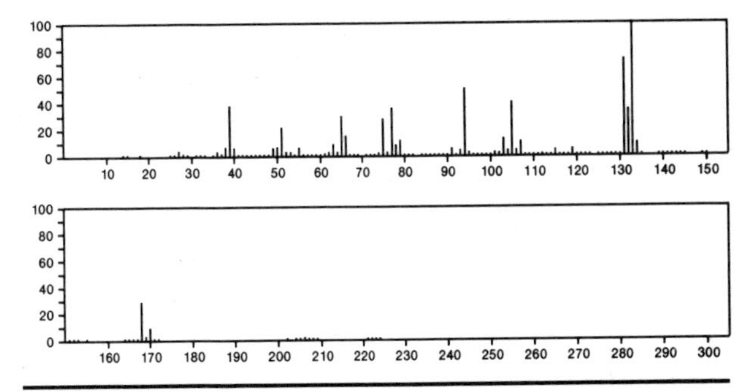

168 C_9H_9ClO 54410-95-6
Benzene, [(3-chloro-2-propenyl)oxy]-

$ClCH:CHCH_2OPh$

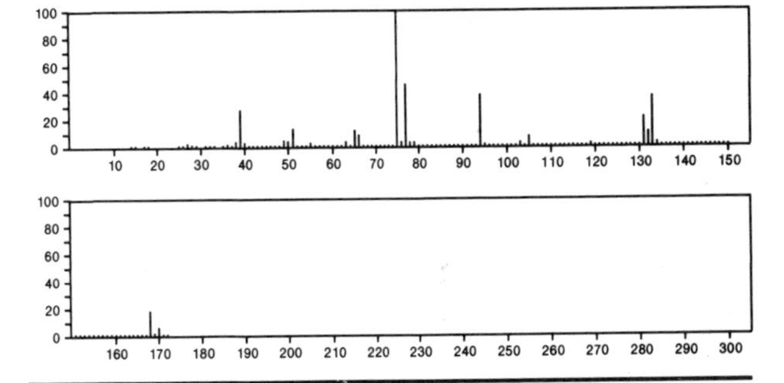

168 C_9H_9ClO 54410-96-7
Benzofuran, 5-chloro-2,3-dihydro-2-methyl-

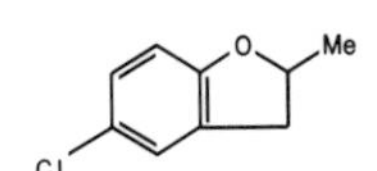

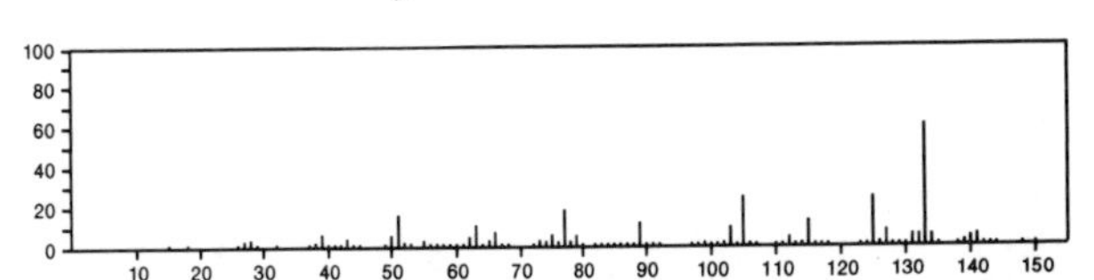

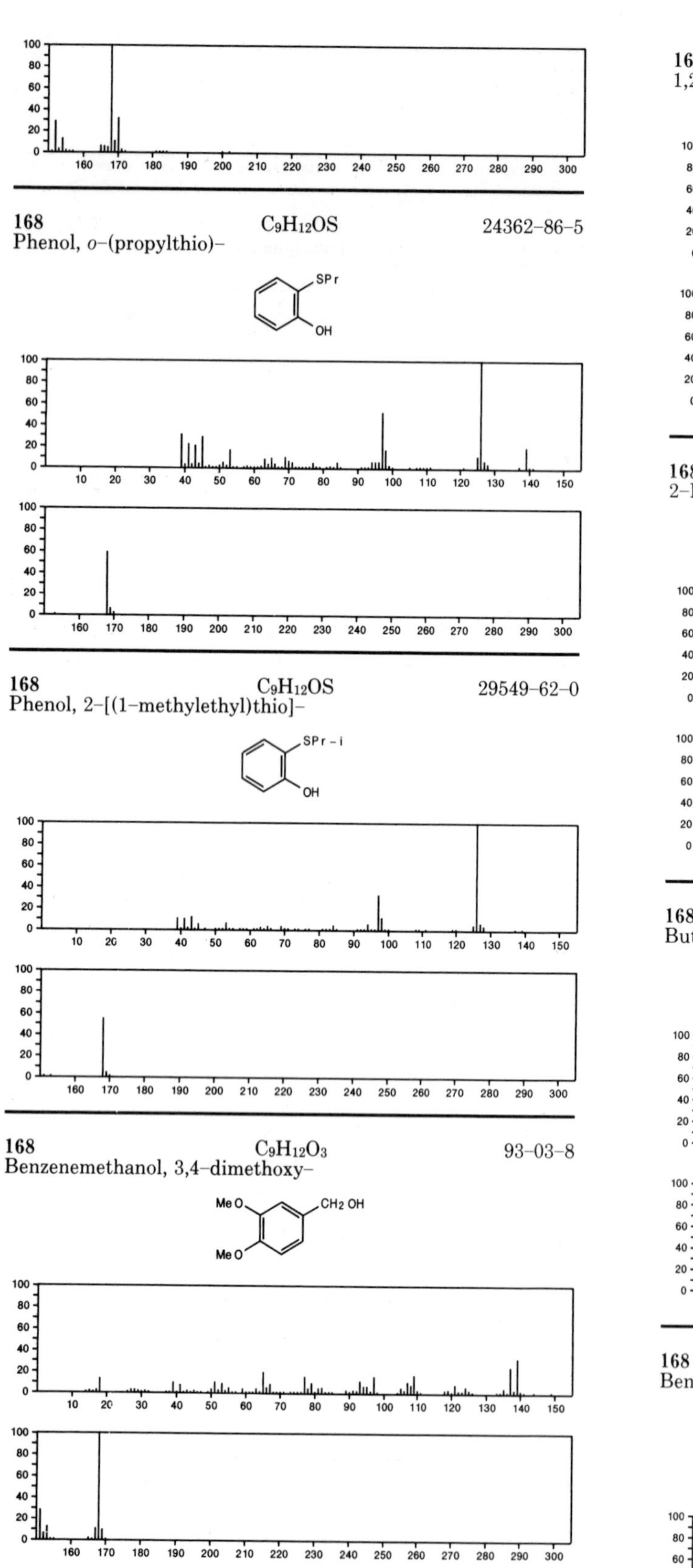

168 $C_9H_{12}OS$ 24362-86-5
Phenol, *o*-(propylthio)-

168 $C_9H_{12}OS$ 29549-62-0
Phenol, 2-[(1-methylethyl)thio]-

168 $C_9H_{12}O_3$ 93-03-8
Benzenemethanol, 3,4-dimethoxy-

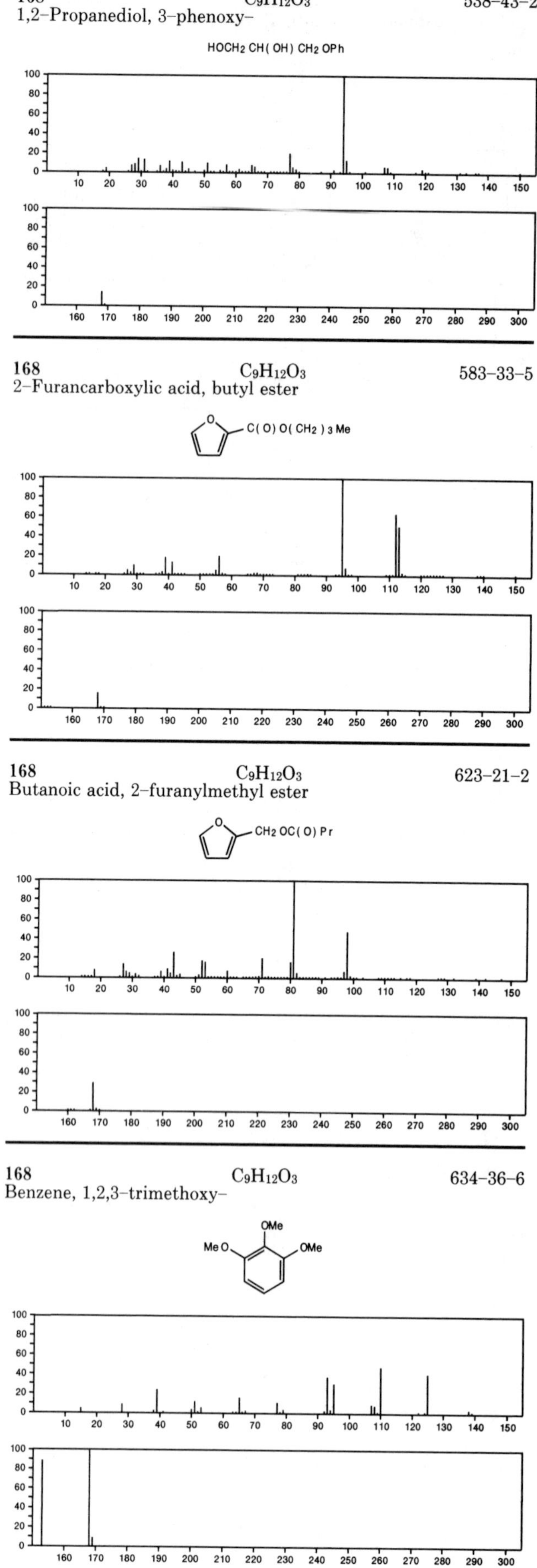

168 $C_9H_{12}O_3$ 538-43-2
1,2-Propanediol, 3-phenoxy-

HOCH2 CH(OH) CH2 OPh

168 $C_9H_{12}O_3$ 583-33-5
2-Furancarboxylic acid, butyl ester

168 $C_9H_{12}O_3$ 623-21-2
Butanoic acid, 2-furanylmethyl ester

168 $C_9H_{12}O_3$ 634-36-6
Benzene, 1,2,3-trimethoxy-

168 $C_9H_{12}O_3$ 2380-78-1
Benzeneethanol, 4-hydroxy-3-methoxy-

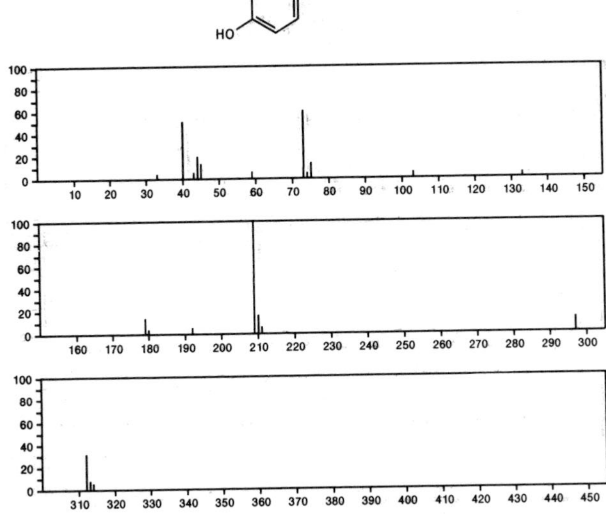

168 $C_9H_{12}O_3$ 14293-26-6
Phenethyl alcohol, 2,5-dihydroxy-α-methyl-, (-)-

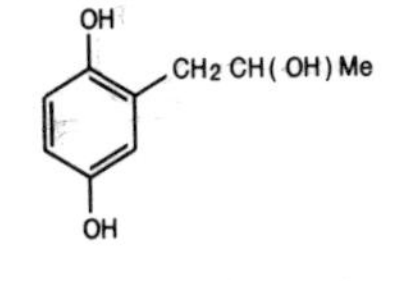

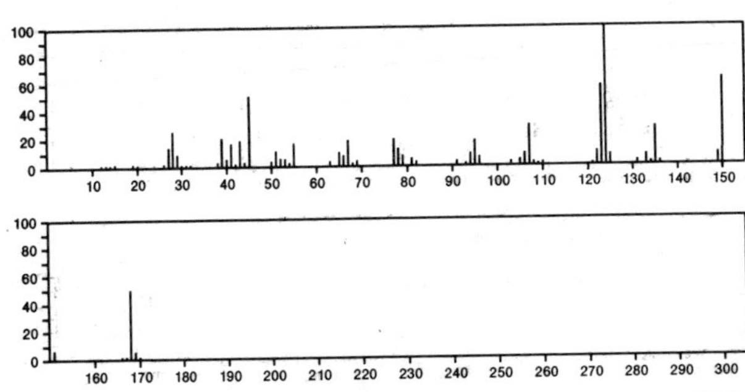

168 $C_9H_{12}O_3$ 18927-20-3
3-Butene-1,2-diol, 1-(2-furanyl)-2-methyl-

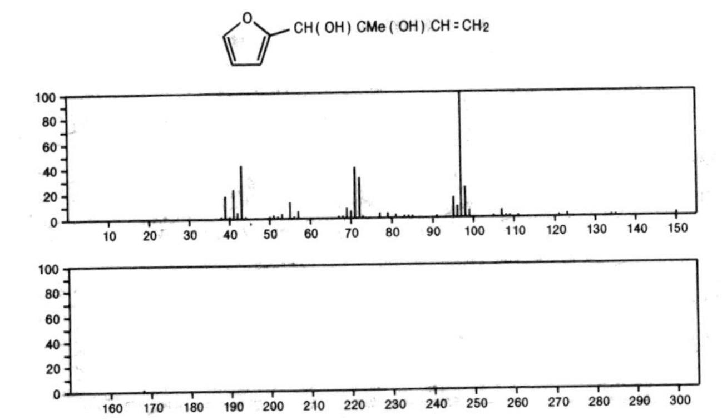

168 $C_9H_{12}O_3$ 21141-71-9
3-Butene-1,2-diol, 1-(2-furanyl)-3-methyl-

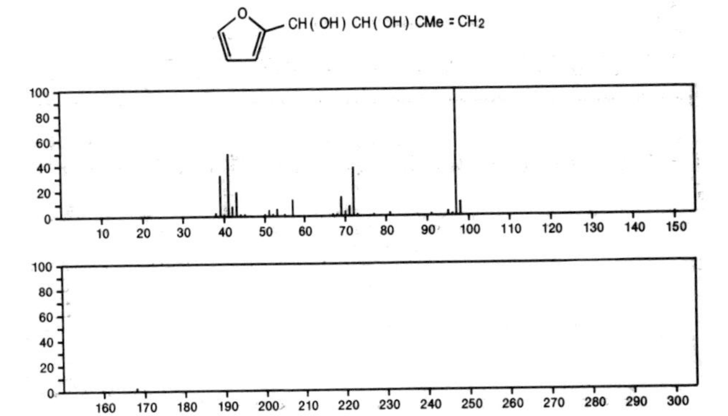

168 $C_9H_{12}O_3$ 46005-09-8
1,2,4-Cyclopentanetrione, 3-butyl-

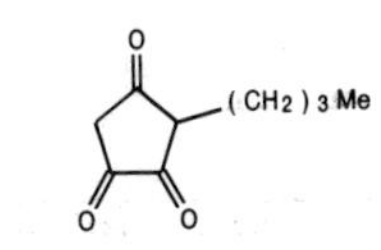

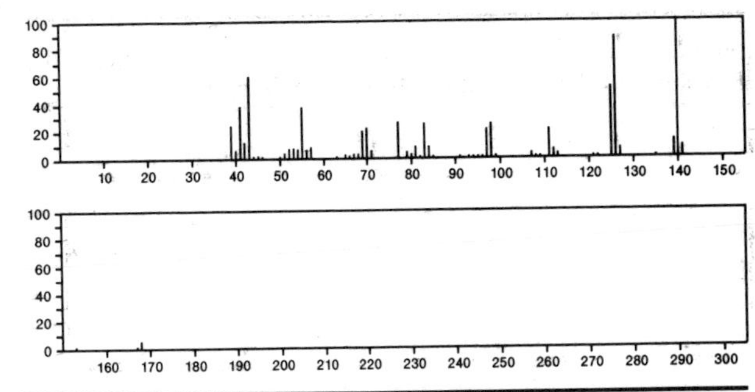

168 $C_9H_{12}O_3$ 55956-42-8
Cyclohexanecarboxylic acid, 3-methylene-2-oxo-, methyl ester

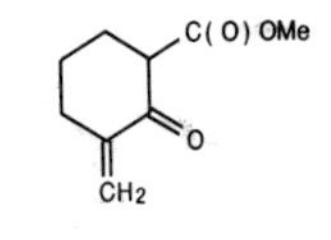

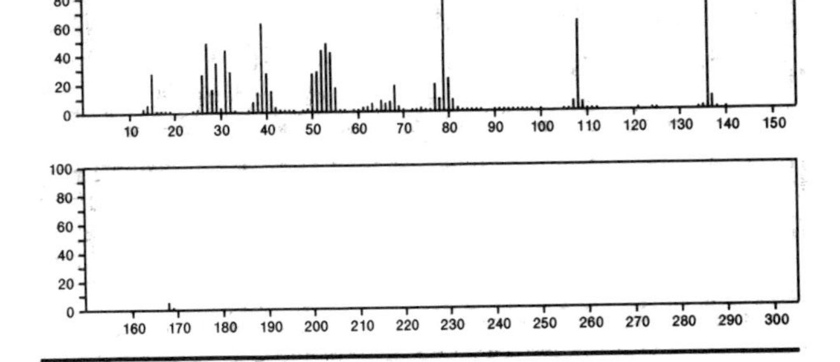

168 $C_9H_{16}N_2O$ 16620-84-1
Pyrido[1,2-*d*][1,4]diazepin-2(3*H*)-one, octahydro-

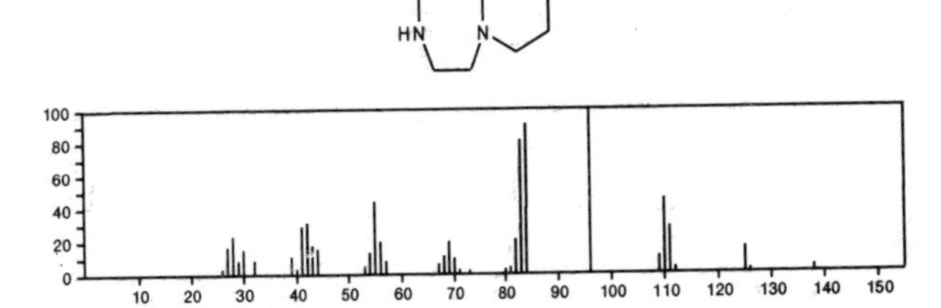

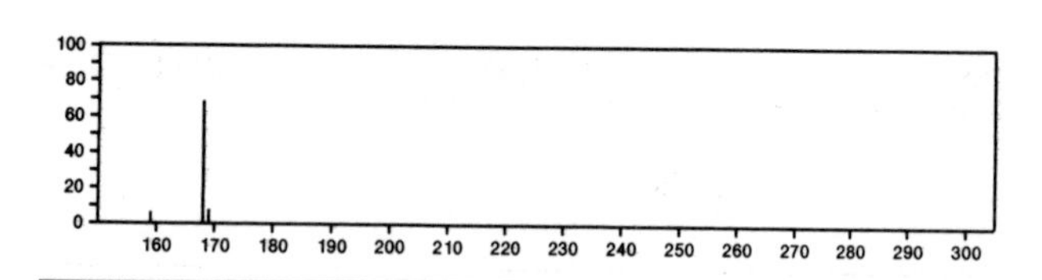

168 $C_9H_{16}N_2O$ 54411-09-5

1*H*-Pyrazole-1-carboxaldehyde, 4-ethyl-4,5-dihydro-5-propyl-

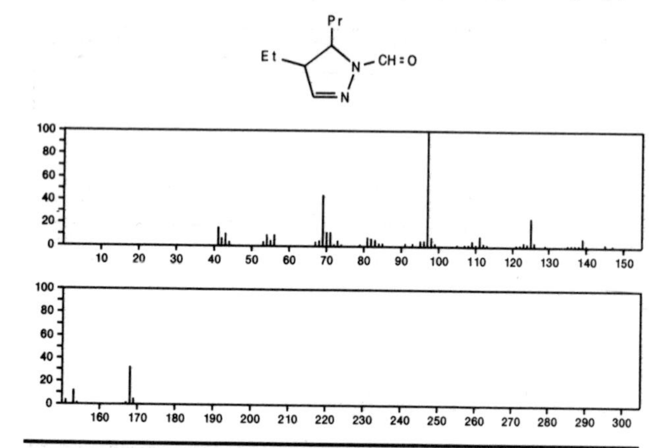

168 $C_{10}H_{13}Cl$ 4395-79-3

Benzene, 2-chloro-1-methyl-4-(1-methylethyl)-

Cl
Pr-i
Me

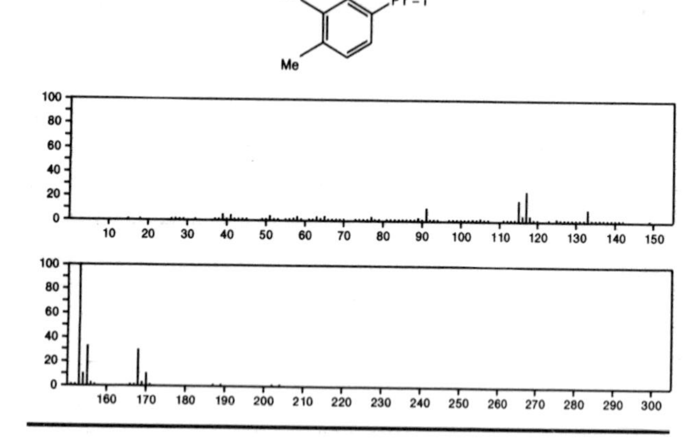

168 $C_{10}H_{13}Cl$ 26937-23-5

Benzene, *sec*-butylchloro-

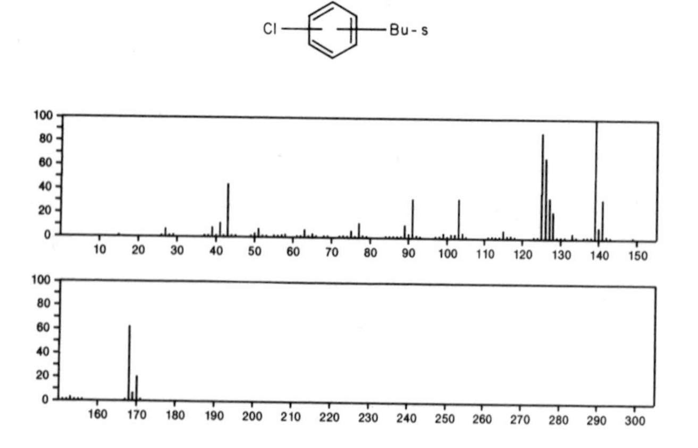

168 $C_{10}H_{13}Cl$

27378-66-1

Benzene, chloro(1,1-dimethylethyl)-

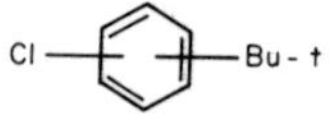

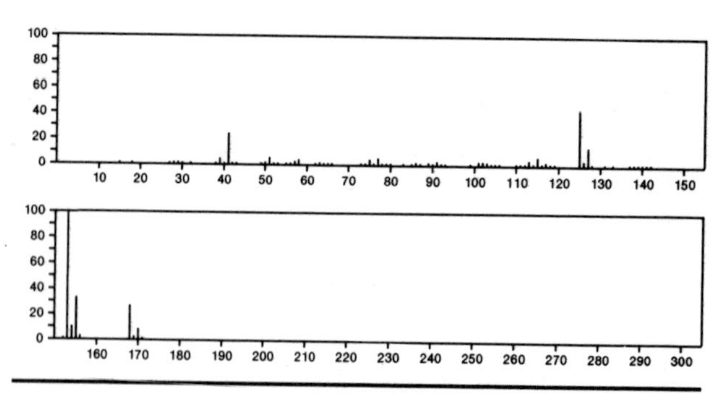

168 $C_{10}H_{13}Cl$ 27755-22-2

Cymene, 7-chloro-

ClCH2
Pr-i

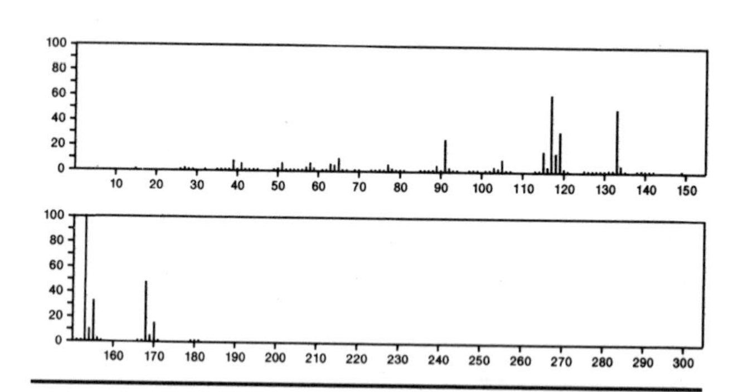

168 $C_{10}H_{13}Cl$ 54411-19-7

Benzene, chloro-1-methyl-4-(1-methylethyl)-

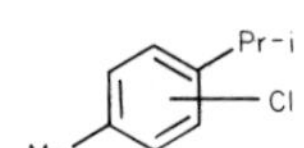

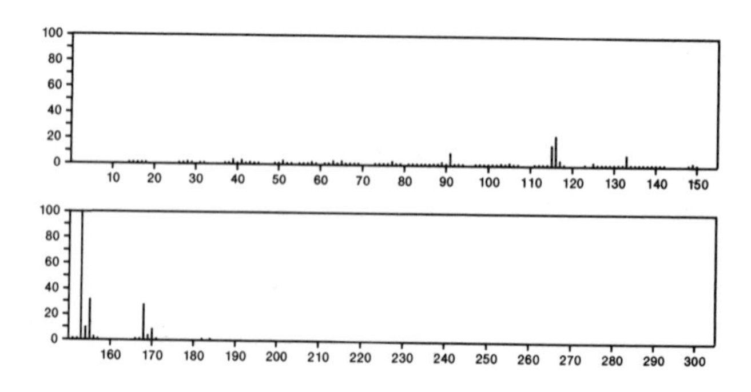

168 $C_{10}H_{13}Cl$ 54411-21-1
Benzene, (1-chloroethyl)dimethyl-

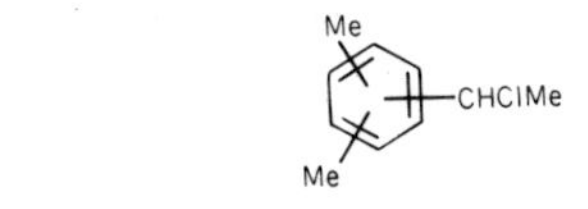

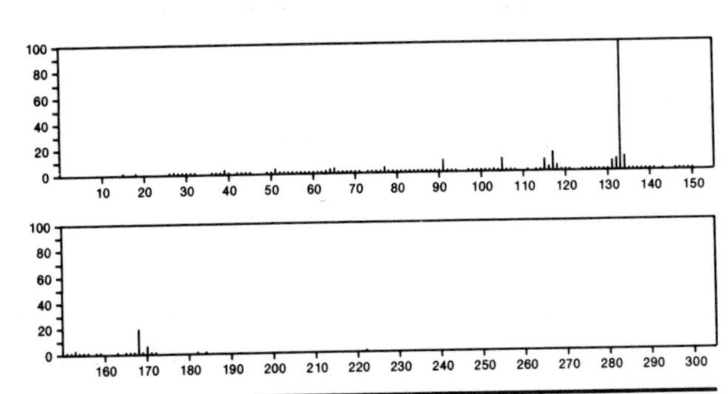

168 $C_{10}H_{13}Cl$ 54868-29-0
Benzene, diethyl-, monochloro deriv.

Et
Et
Cl

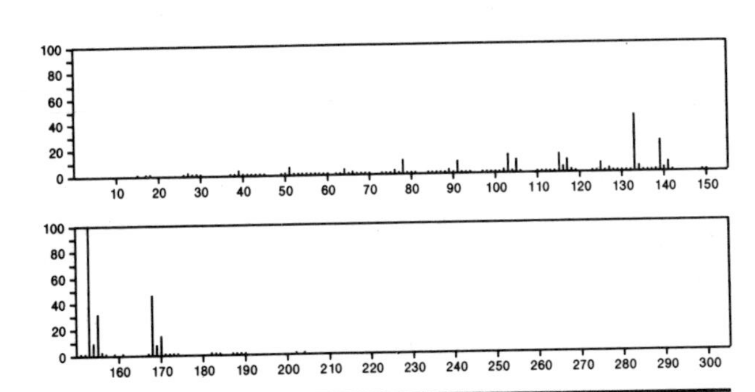

168 $C_{10}H_{13}FO$ 56781-83-0
Tricyclo[3.3.1.1^{3,7}]decan-2-one, 4-fluoro-, (1α,3β,4α,5α,7β)-

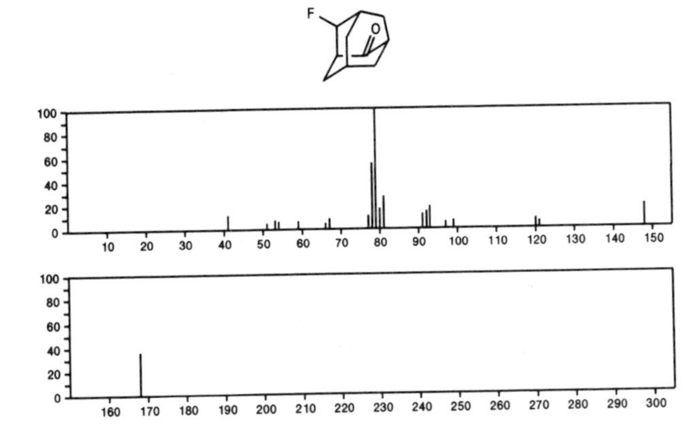

168 $C_{10}H_{13}FO$ 56781-84-1
Tricyclo[3.3.1.1^{3,7}]decan-2-one, 4-fluoro-, (1α,3β,4β,5α,7β)-

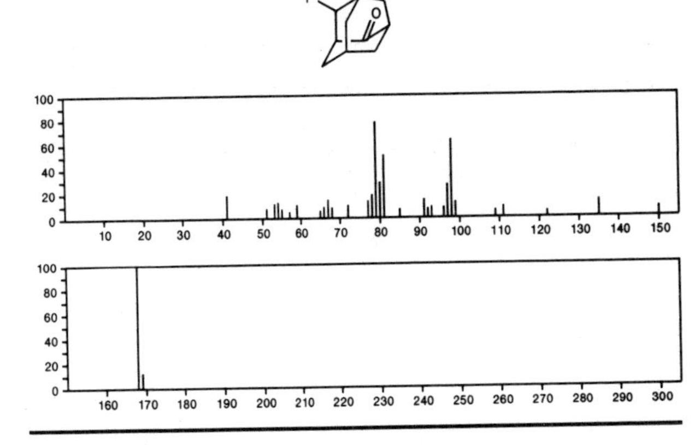

168 $C_{10}H_{16}O_2$ 111-80-8
2-Nonynoic acid, methyl ester

Me(CH$_2$)$_5$C≡CC(O)OMe

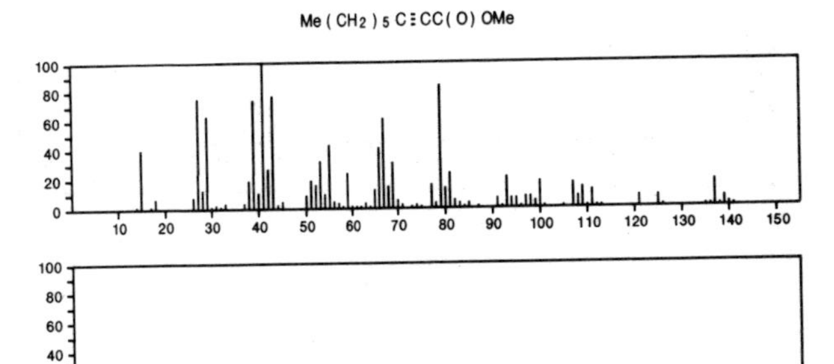

168 $C_{10}H_{16}O_2$ 512-85-6
2,3-Dioxabicyclo[2.2.2]oct-5-ene, 1-methyl-4-(1-methylethyl)-

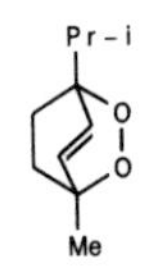

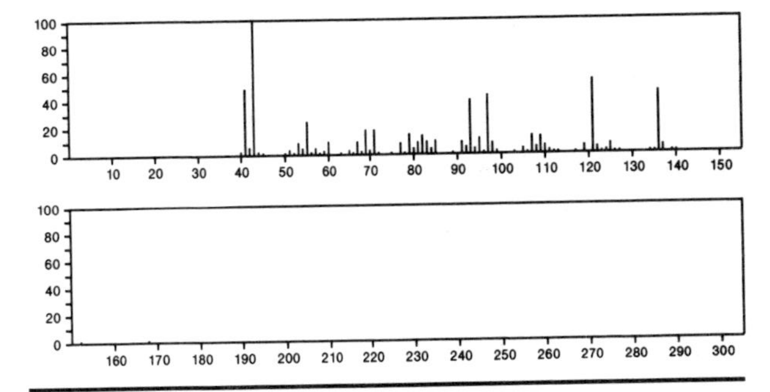

168 $C_{10}H_{16}O_2$ 702-67-0
Bicyclo[2.2.2]octane-1-carboxylic acid, 4-methyl-

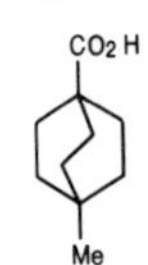

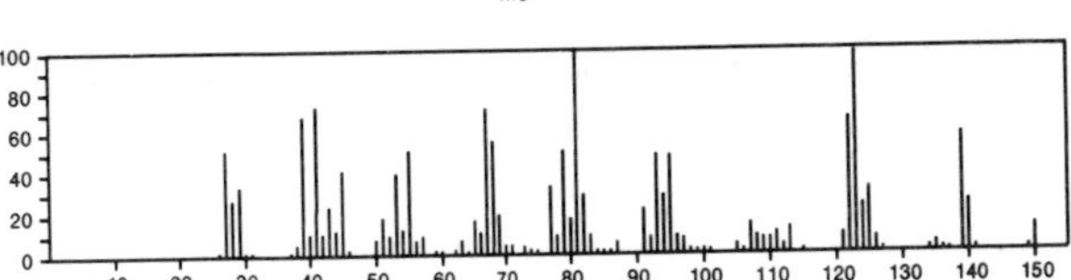

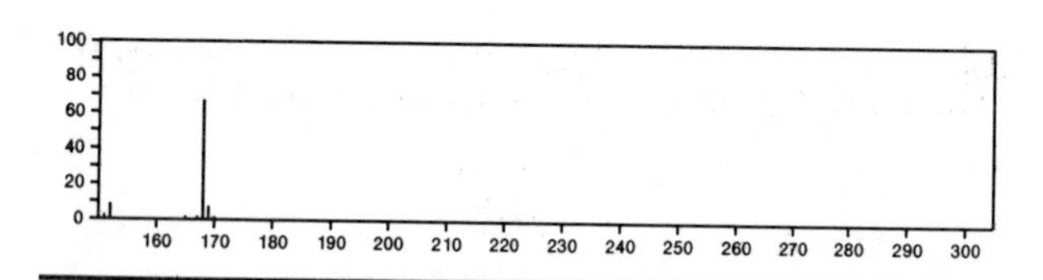

168 $C_{10}H_{16}O_2$ 705-16-8
Cyclopropanecarboxylic acid, 2,2-dimethyl-3-(2-methyl-1-propenyl)-, *trans*-(±)-

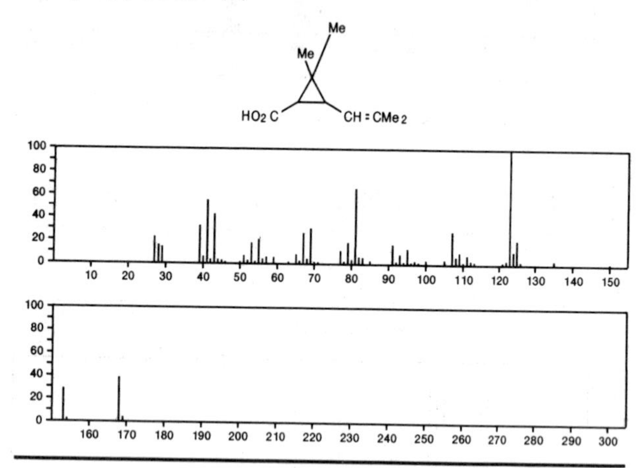

168 $C_{10}H_{16}O_2$ 827-03-2
1,3-Cyclopentanedione, 2-isopentyl-

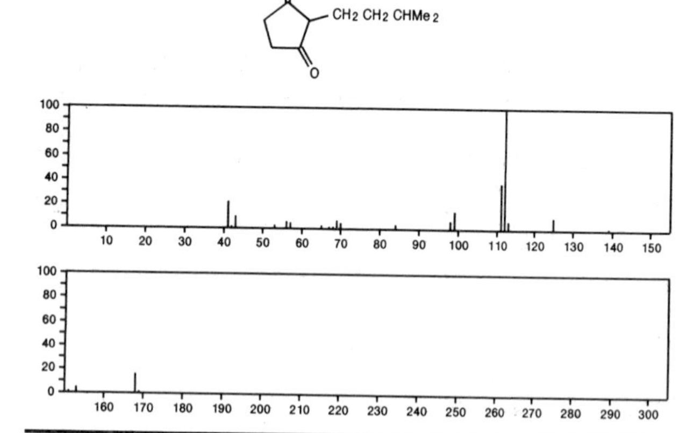

168 $C_{10}H_{16}O_2$ 939-86-6
1,3-Cyclopentanedione, 4-isopentyl-

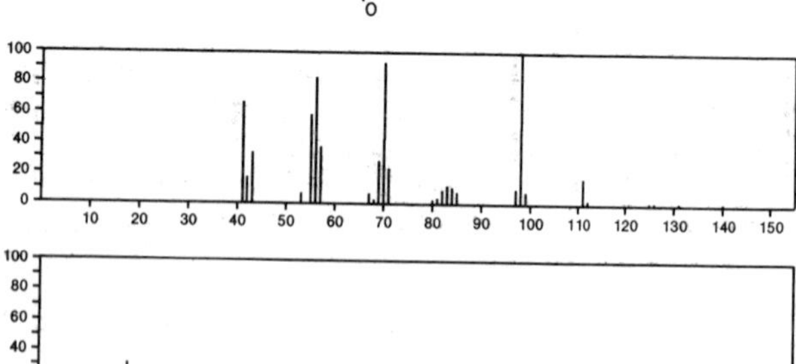

168 $C_{10}H_{16}O_2$ 3907-11-7
Bicyclo[2.2.2]octanone, 4-methoxy-1-methyl-

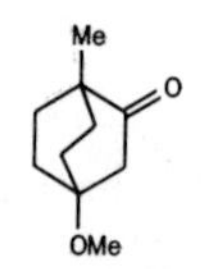

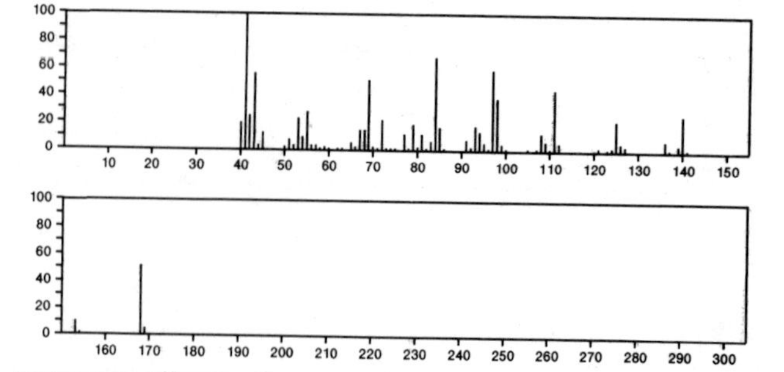

168 $C_{10}H_{16}O_2$ 7003-48-7
8-Nonynoic acid, methyl ester

$HC\equiv C(CH_2)_6C(O)OMe$

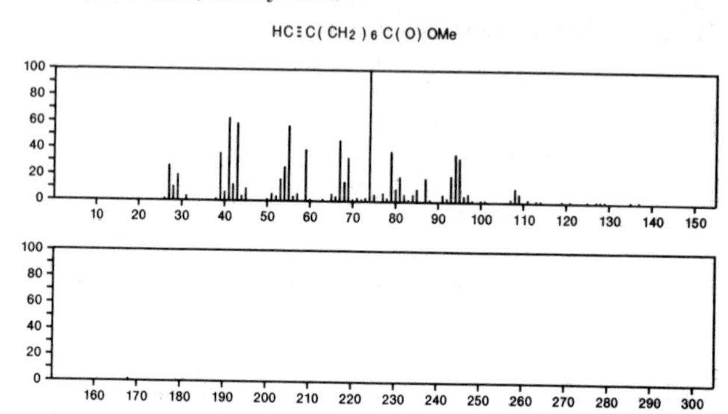

168 $C_{10}H_{16}O_2$ 10453-89-1
Cyclopropanecarboxylic acid, 2,2-dimethyl-3-(2-methyl-1-propenyl)-

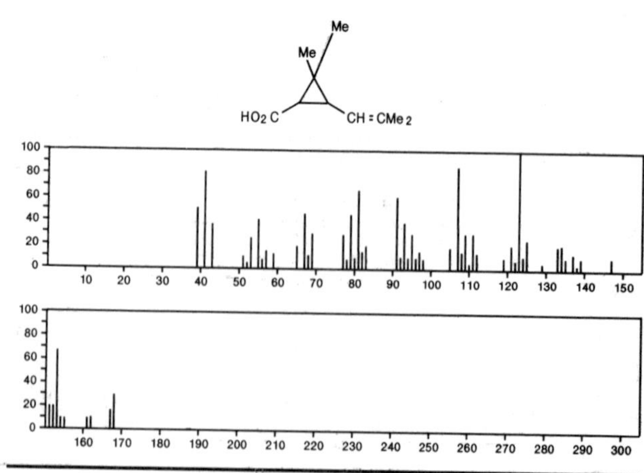

168 $C_{10}H_{16}O_2$ 13080-28-9
1-Oxaspiro[2.5]octan-4-one, 2,2,6-trimethyl-, *trans*-

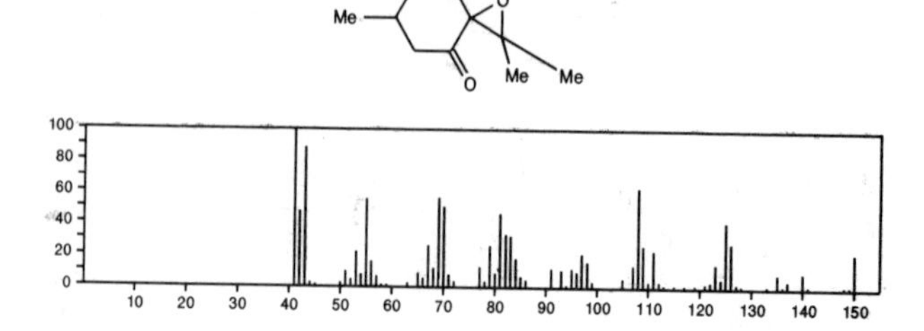

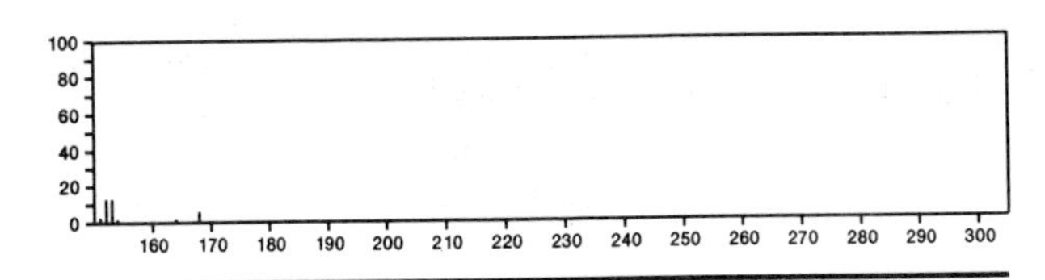

168 $C_{10}H_{16}O_2$ 13080-29-0
1-Oxaspiro[2.5]octan-4-one, 2,2,6-trimethyl-, *cis*-

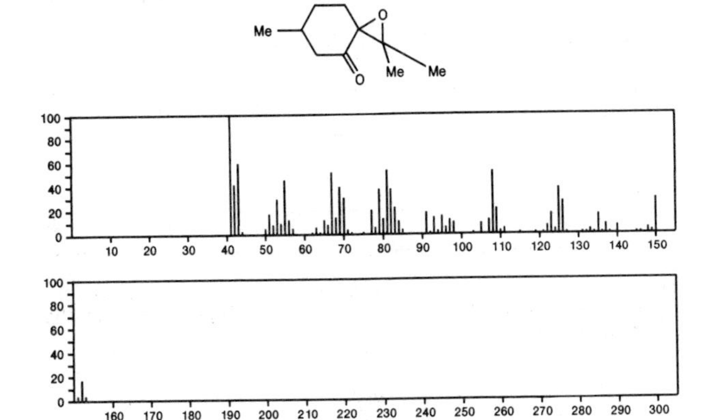

168 $C_{10}H_{16}O_2$ 16491-62-6
2-Butenoic acid, cyclohexyl ester

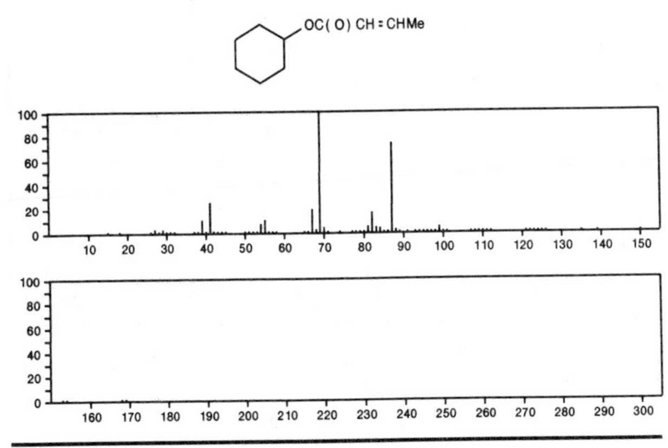

168 $C_{10}H_{16}O_2$ 16642-55-0
1-Cyclohexene-1-acetic acid, α,α-dimethyl-

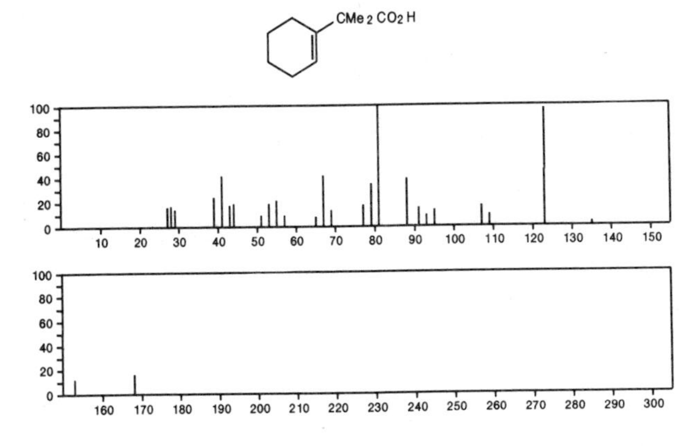

168 $C_{10}H_{16}O_2$ 17672-96-7
Cyclopenta[*c*]pyran-1(3*H*)-one, hexahydro-4,7-dimethyl-, (4α,4aα,7α,7aα)-

168 $C_{10}H_{16}O_2$ 19931-20-5
Bicyclo[1.1.0]butane-1-carboxylic acid, 2,2,4,4-tetramethyl-, methyl ester

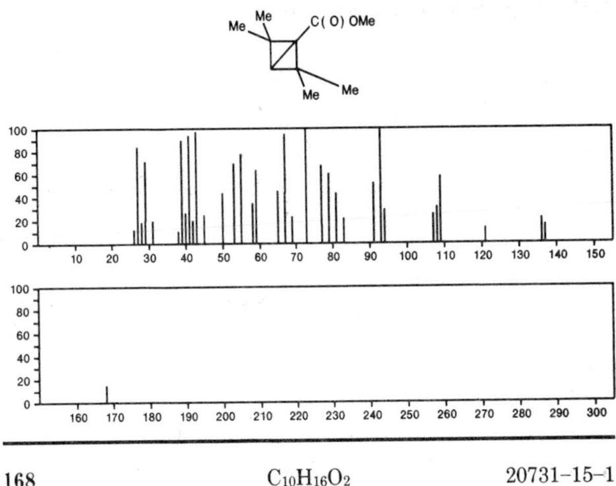

168 $C_{10}H_{16}O_2$ 20731-15-1
4-Nonynoic acid, methyl ester

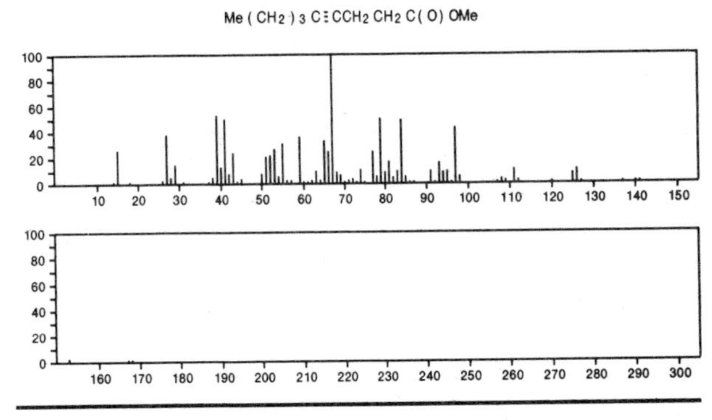

168 $C_{10}H_{16}O_2$ 20731-17-3
6-Nonynoic acid, methyl ester

Et C≡C(CH2) 4 C(O) OMe

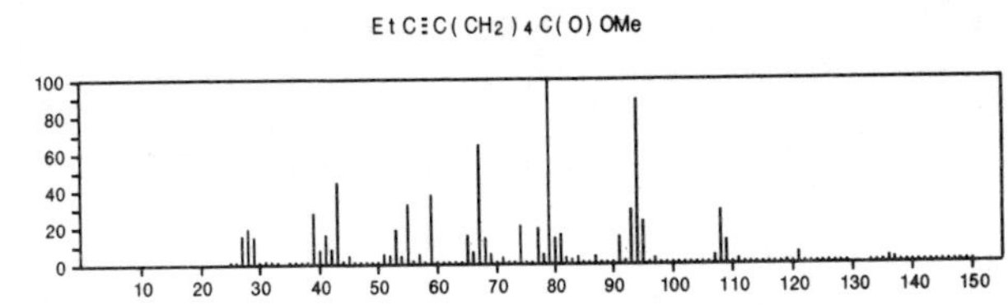

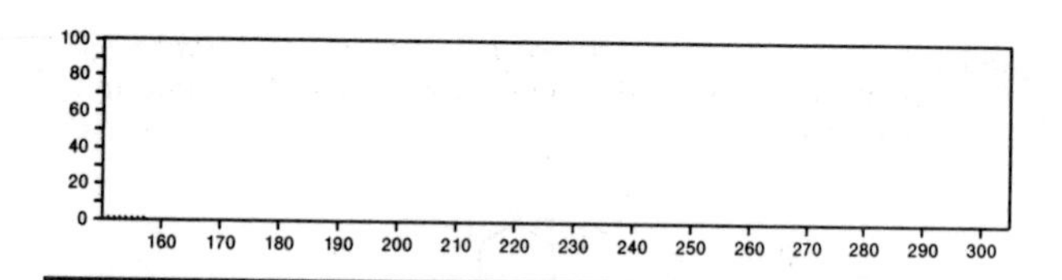

168 $C_{10}H_{16}O_2$ 20731-18-4
7-Nonynoic acid, methyl ester

MeOC(O)(CH2)5C≡CMe

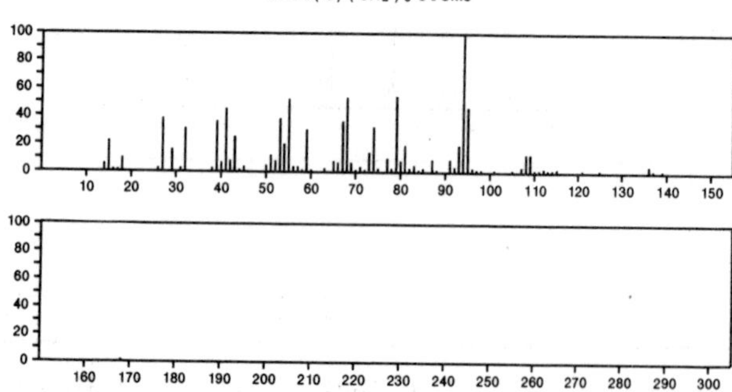

168 $C_{10}H_{16}O_2$ 21766-50-7
1(2*H*)-Naphthalenone, octahydro-4-hydroxy-, *trans*-

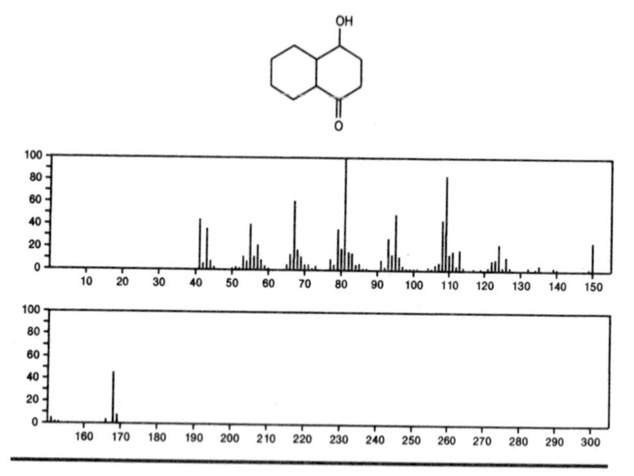

168 $C_{10}H_{16}O_2$ 21950-33-4
Cyclopenta[*c*]pyran-1(3*H*)-one, hexahydro-4,7-dimethyl-

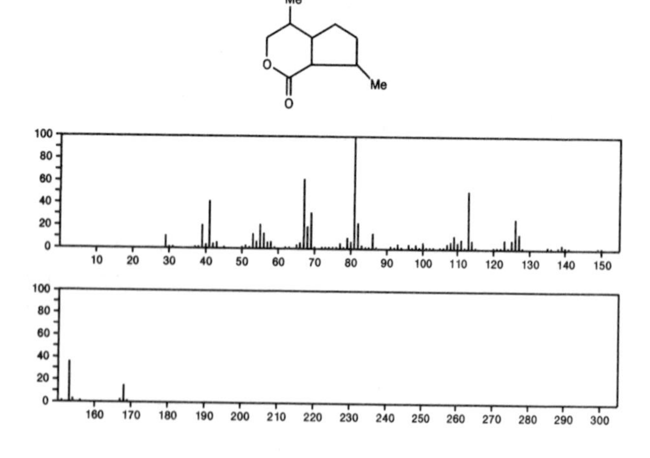

168 $C_{10}H_{16}O_2$ 22975-39-9
2-Cyclopenten-1-one, 2-butyl-3-methoxy-

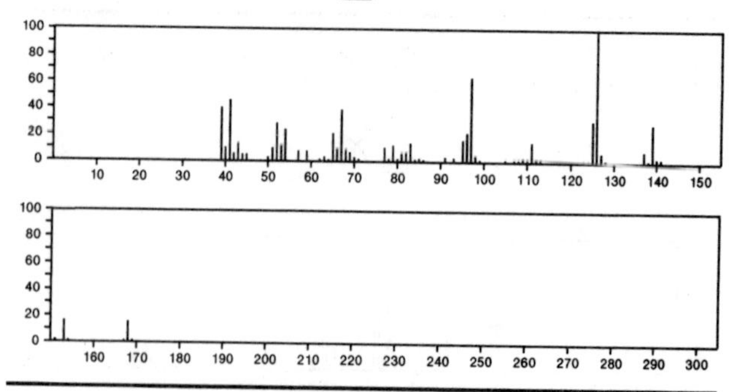

168 $C_{10}H_{16}O_2$ 33383-56-1
Cyclopropanecarboxylic acid, 2,2-dimethyl-3-(1-propenyl)-, methyl ester, [1α,3β(*Z*)]-

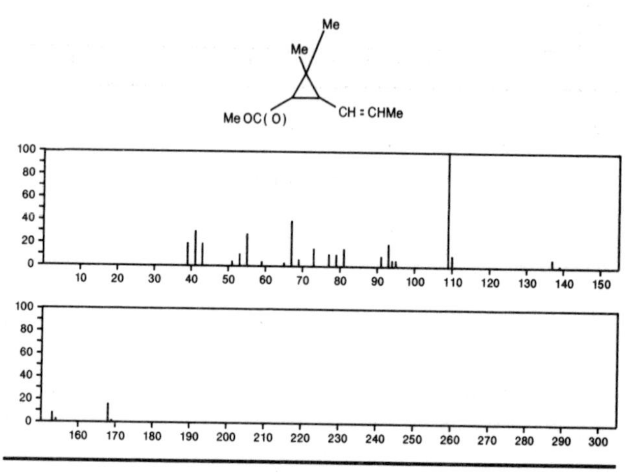

168 $C_{10}H_{16}O_2$ 41654-18-6
2,6-Nonadienoic acid, methyl ester, (*E*,*Z*)-

MeOC(O)CH=CHCH2CH2CH=CHEt

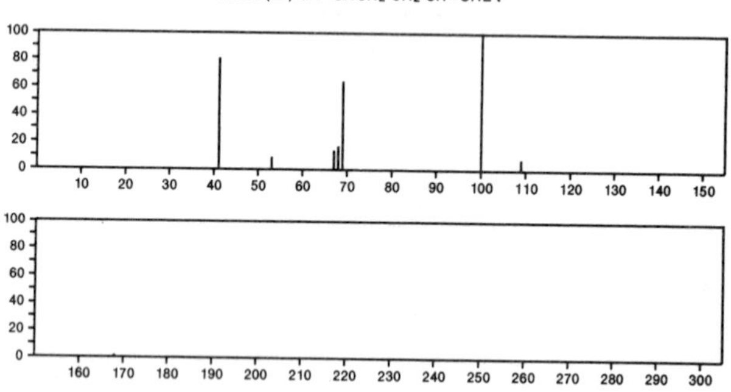

168 $C_{10}H_{16}O_2$ 42569–58–4
4,8–Dioxatricyclo[5.1.0.0^{3,5}]octane, 1–methyl–5–(1–methylethyl)–, (1α,3α,5α,7α)–

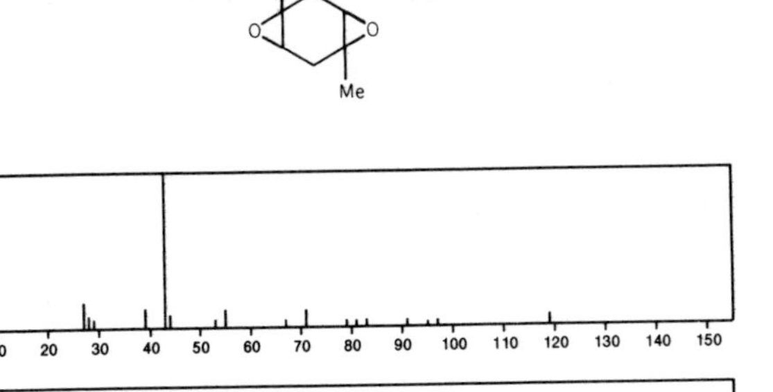

168 $C_{10}H_{16}O_2$ 42569–59–5
4,8–Dioxatricyclo[5.1.0.0^{3,5}]octane, 1–methyl–5–(1–methylethyl)–, (1α,3β,5β,7α)–

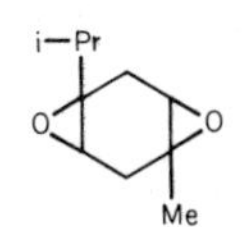

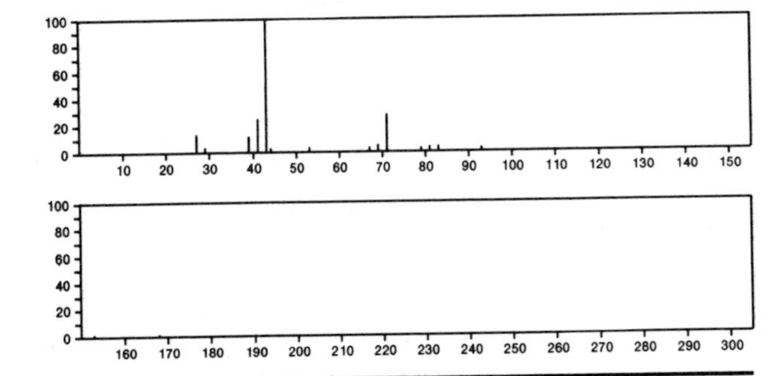

168 $C_{10}H_{16}O_2$ 53690–89–4
2–Cyclopenten–1–one, 5–butyl–3–methoxy–

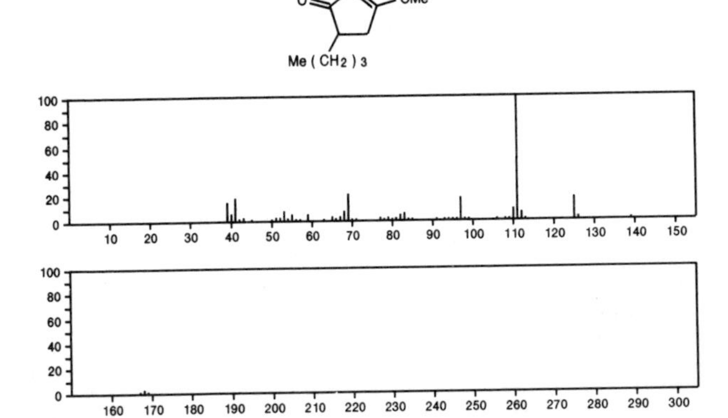

168 $C_{10}H_{16}O_2$ 53690–92–9
2–Cyclopenten–1–one, 4–butyl–3–methoxy–

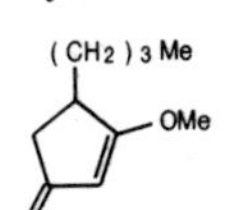

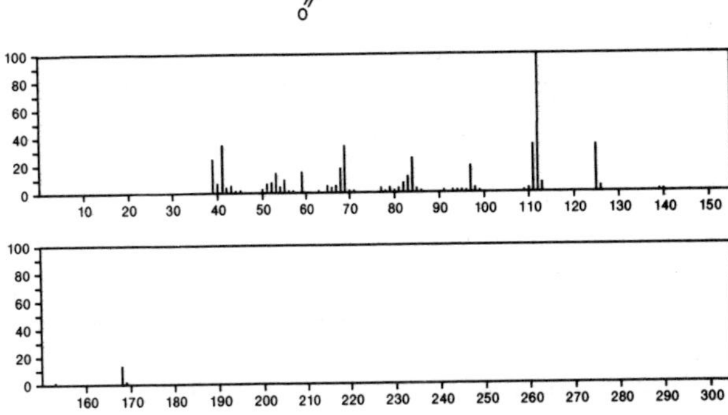

168 $C_{10}H_{16}O_2$ 54346–05–3
4(1*H*)–Isobenzofuranone, hexahydro–3a,7a–dimethyl–, *cis*–(±)–

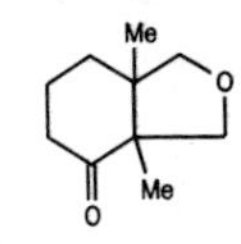

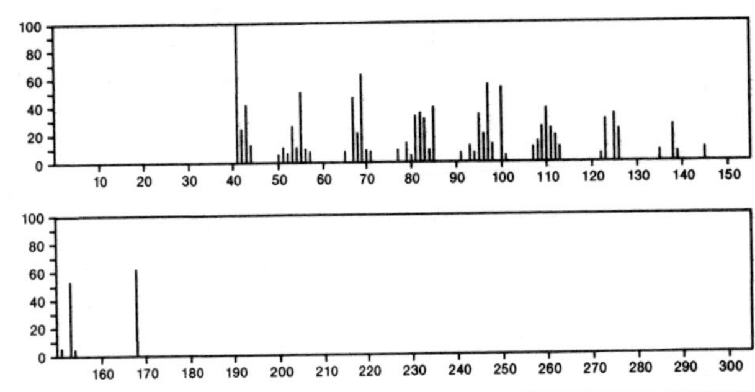

168 $C_{10}H_{16}O_2$ 55282–91–2
2,7–Octadienoic acid, ethyl ester, (*E*)–

H2C=CH(CH2)3CH=CHC(O)OEt

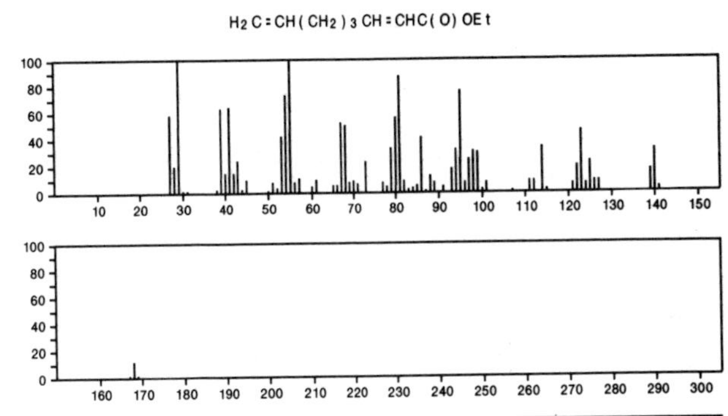

168 $C_{10}H_{16}O_2$ 55283–12–0
2,6–Octadienoic acid, 3–methyl–, methyl ester, (*E,Z*)–

MeCH=CHCH2CH2CMe=CHC(O)OMe

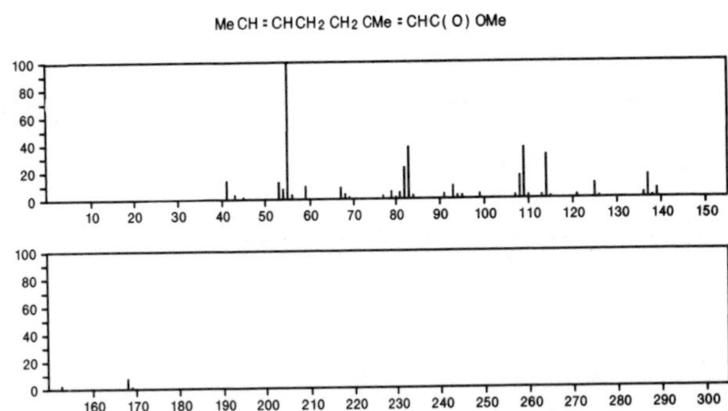

168 $C_{10}H_{16}O_2$ 55283-13-1
2,6-Octadienoic acid, 3-methyl-, methyl ester, (*E,E*)-

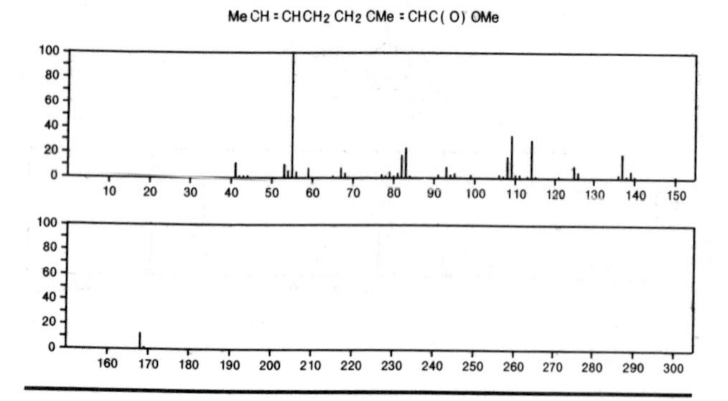

168 $C_{10}H_{16}O_2$ 55760-16-2
Cyclobutaneacetic acid, 1-methyl-2-(1-methylethenyl)-

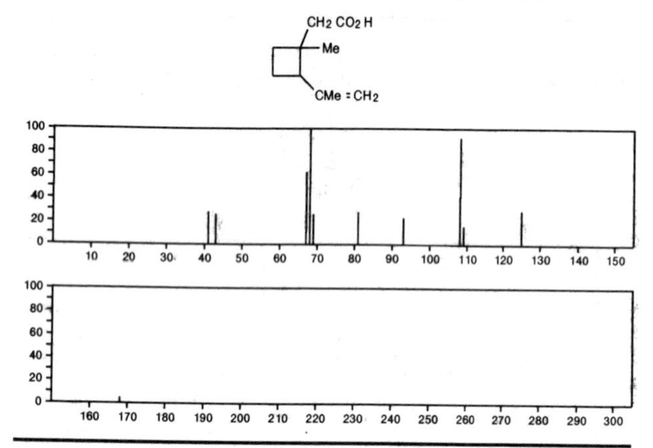

168 $C_{10}H_{16}O_2$ 55955-53-8
2-Cyclohexen-1-one, 4-hydroxy-3-methyl-6-(1-methylethyl)-, *trans-*

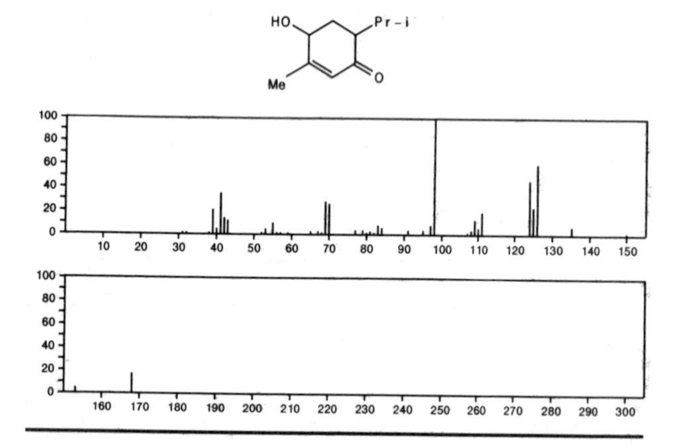

168 $C_{10}H_{16}O_2$ 55955-54-9
2-Cyclohexen-1-one, 3-(hydroxymethyl)-6-(1-methylethyl)-

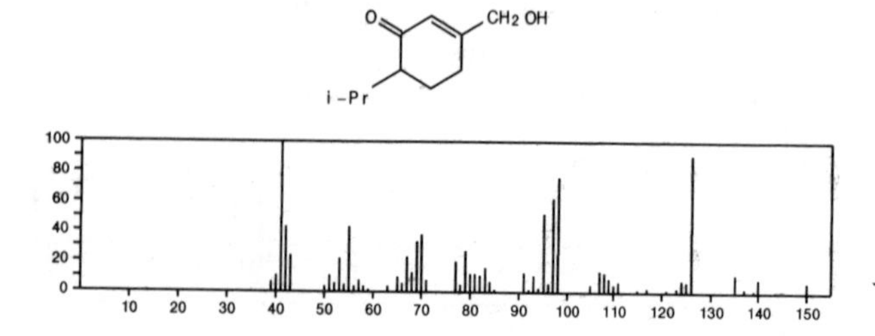

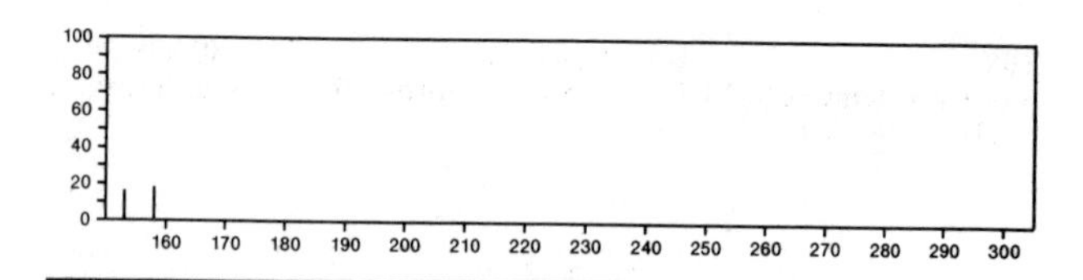

168 $C_{10}H_{16}O_2$ 57157-04-7
1,3-Cyclopentanedione, 2-ethyl-4-propyl-

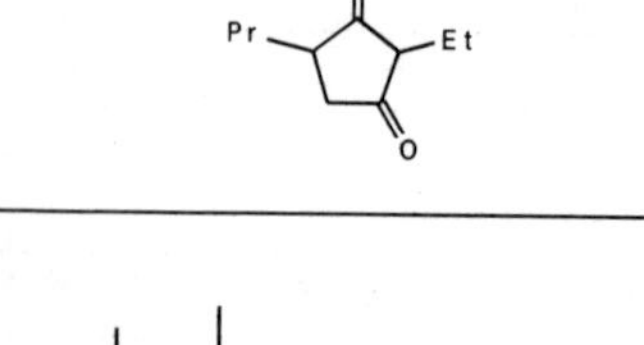

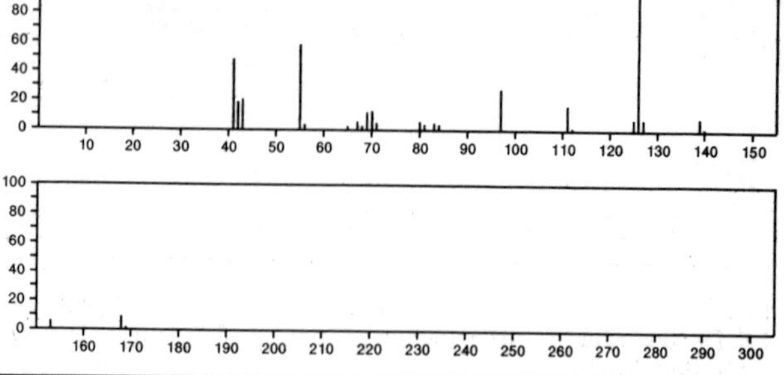

168 $C_{10}H_{16}S$ 875-06-9
Bicyclo[2.2.1]heptane-2-thione, 1,3,3-trimethyl-

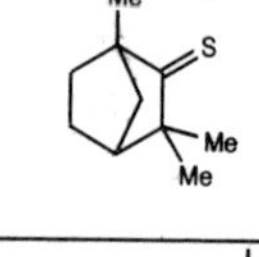

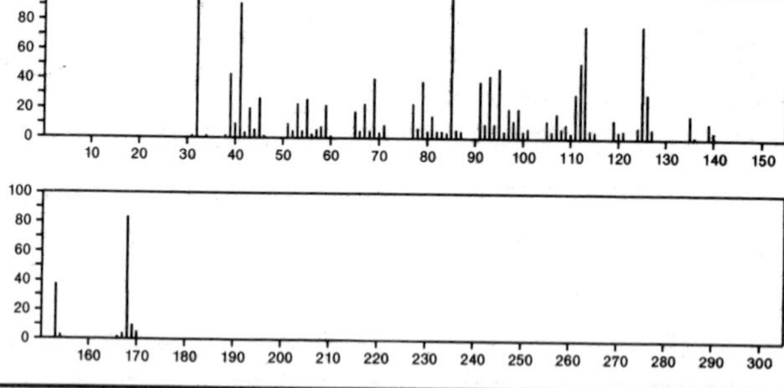

168 $C_{10}H_{16}S$ 7519-74-6
Bicyclo[2.2.1]heptane-2-thione, 1,7,7-trimethyl-

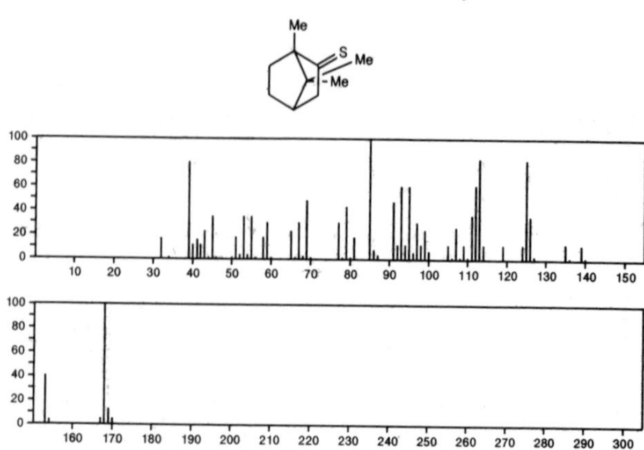

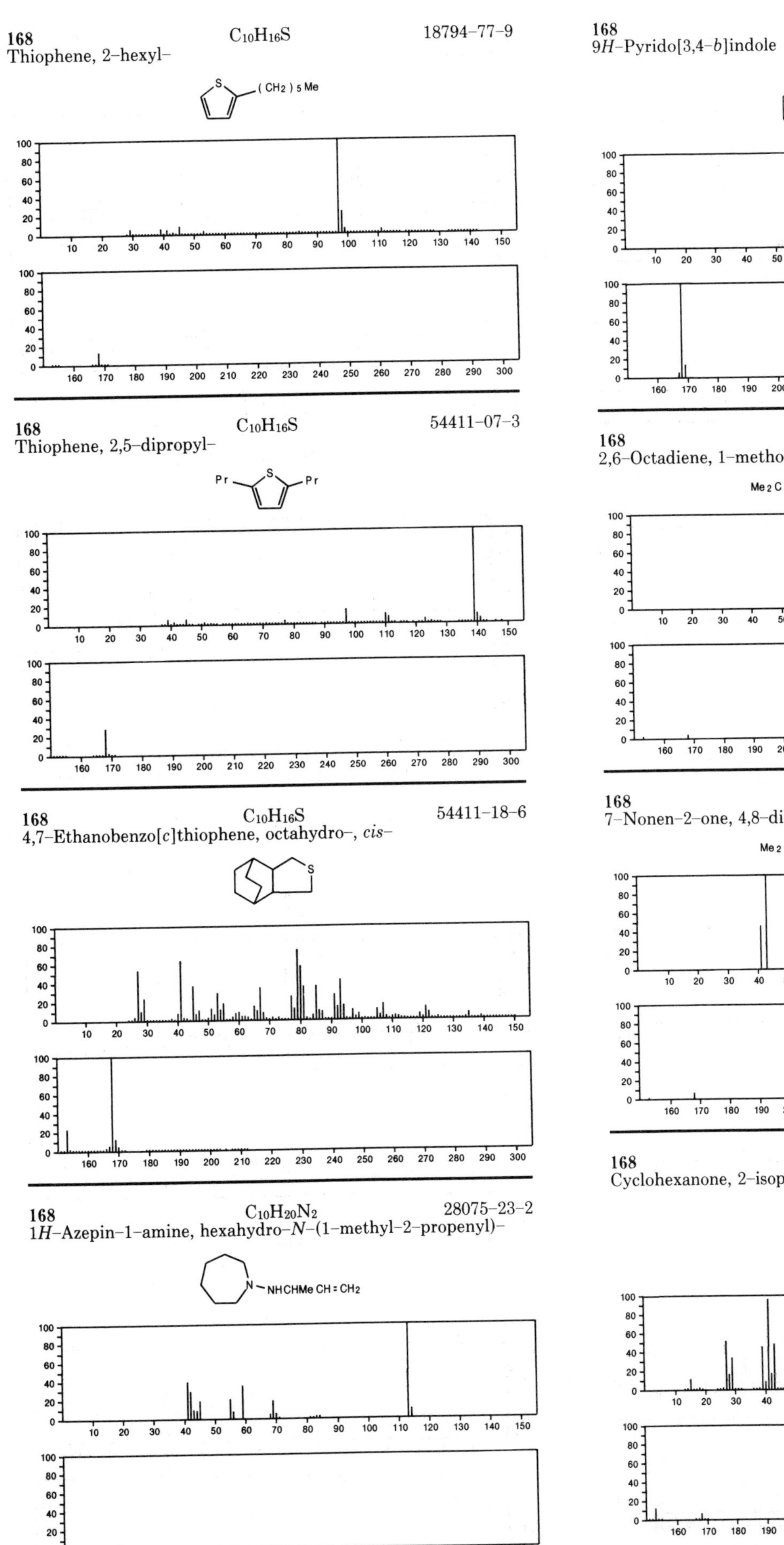

168 $C_{10}H_{16}S$ 18794-77-9
Thiophene, 2-hexyl-

168 $C_{10}H_{16}S$ 54411-07-3
Thiophene, 2,5-dipropyl-

168 $C_{10}H_{16}S$ 54411-18-6
4,7-Ethanobenzo[c]thiophene, octahydro-, *cis*-

168 $C_{10}H_{20}N_2$ 28075-23-2
1*H*-Azepin-1-amine, hexahydro-*N*-(1-methyl-2-propenyl)-

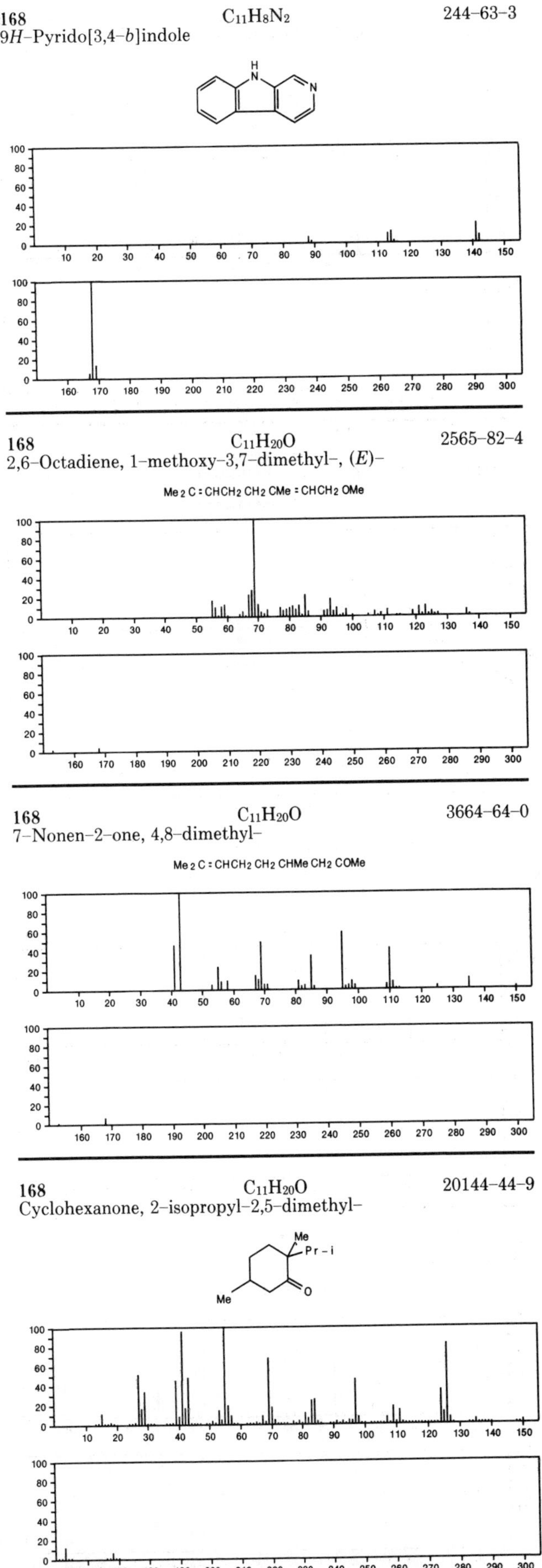

168 $C_{11}H_8N_2$ 244-63-3
9*H*-Pyrido[3,4-*b*]indole

168 $C_{11}H_{20}O$ 2565-82-4
2,6-Octadiene, 1-methoxy-3,7-dimethyl-, (*E*)-

168 $C_{11}H_{20}O$ 3664-64-0
7-Nonen-2-one, 4,8-dimethyl-

168 $C_{11}H_{20}O$ 20144-44-9
Cyclohexanone, 2-isopropyl-2,5-dimethyl-

168 $C_{11}H_{20}O$ 21720-89-8
Naphthalene, decahydro-1-methoxy-

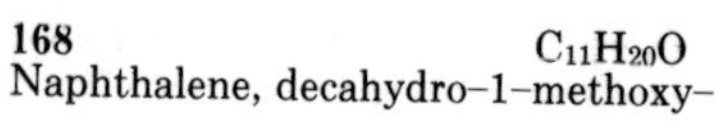
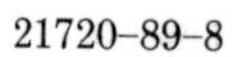
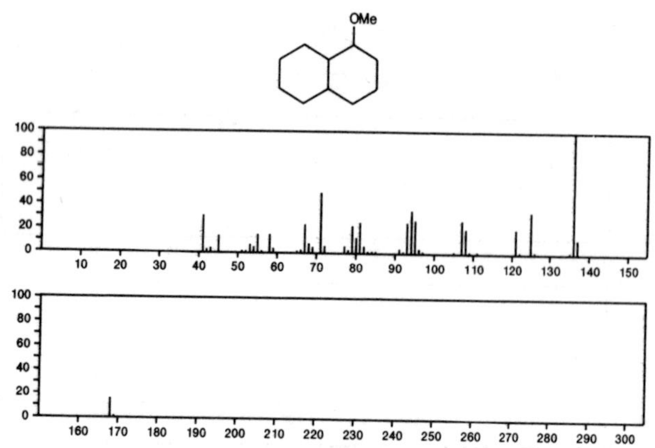

168 $C_{11}H_{20}O$ 22319-29-5
4-Hepten-3-one, 5-ethyl-2,4-dimethyl-

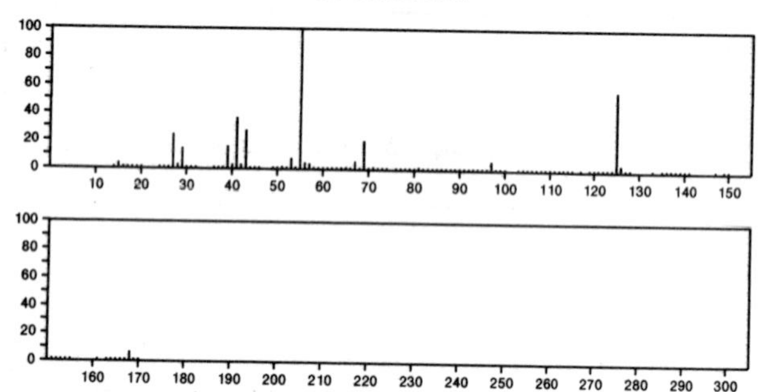

168 $C_{11}H_{20}O$ 31569-55-8
Cyclohexanemethanol, α-(2-methylallyl)-

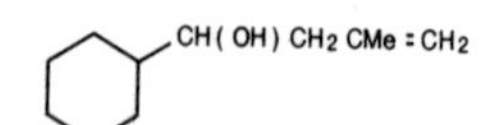

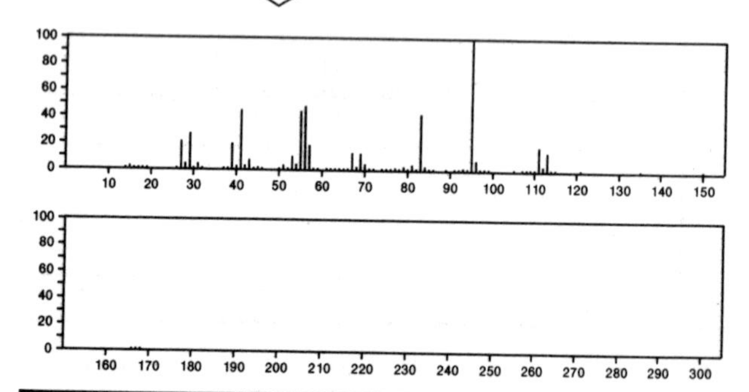

168 $C_{11}H_{20}O$ 32064-74-7
4-Undecen-6-one

Pr CH = CHCO(CH2)4 Me

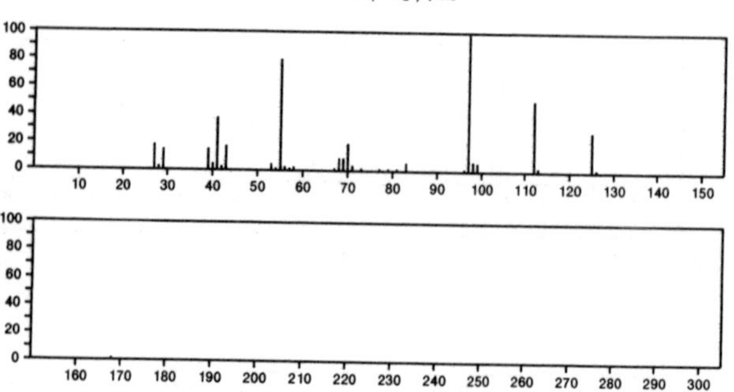

168 $C_{11}H_{20}O$ 32064-75-8
3-Decen-5-one, 2-methyl-

Me2 CHCH = CHCO(CH2)4 Me

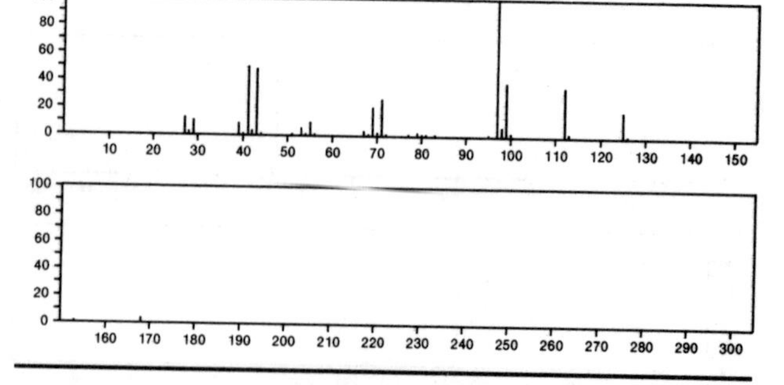

168 $C_{11}H_{20}O$ 40648-24-6
Cyclohexene, 1-(1,1-dimethylethoxy)-3-methyl-

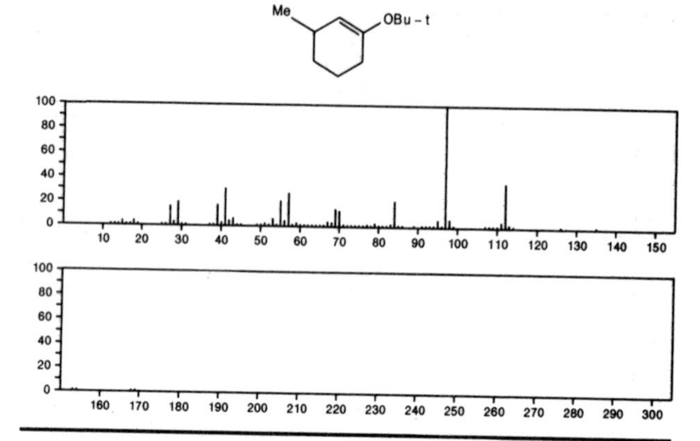

168 $C_{11}H_{20}O$ 40648-25-7
Cyclohexene, 1-(1,1-dimethylethoxy)-6-methyl-

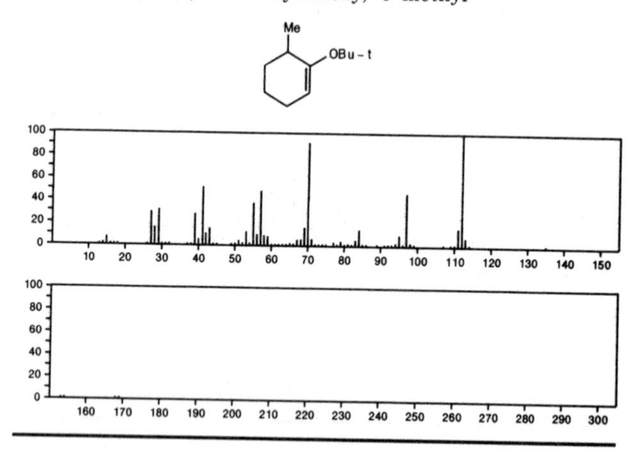

168 $C_{11}H_{20}O$ 40648-26-8
Cyclohexene, 1-(1,1-dimethylethoxy)-2-methyl-

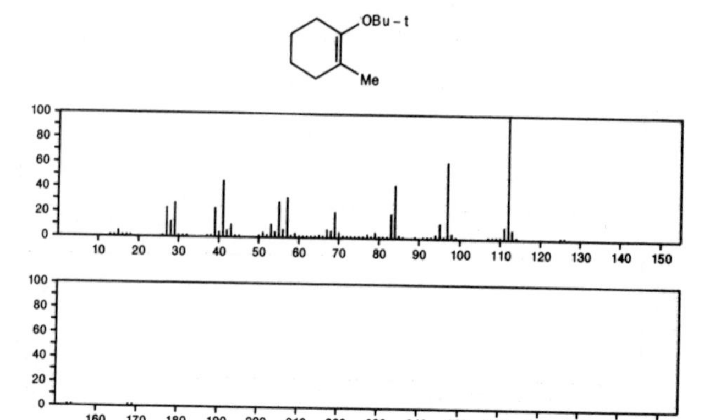

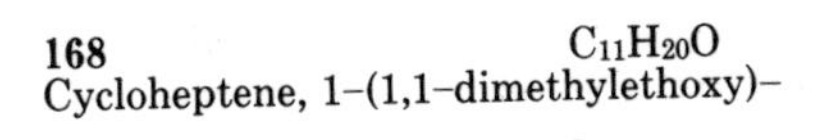

168 $C_{11}H_{20}O$ 49565–07–3
Cycloheptene, 1–(1,1–dimethylethoxy)–

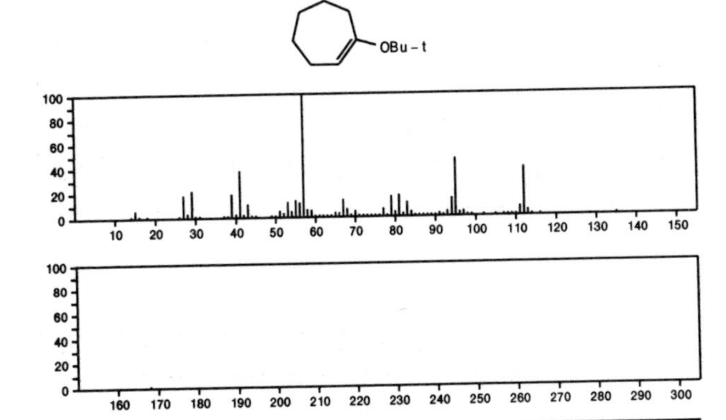

168 $C_{11}H_{20}O$ 54345–61–8
1–Cyclohexene–1–methanol, α,2,6,6–tetramethyl–

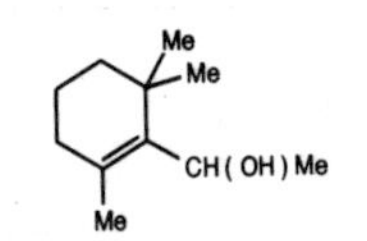

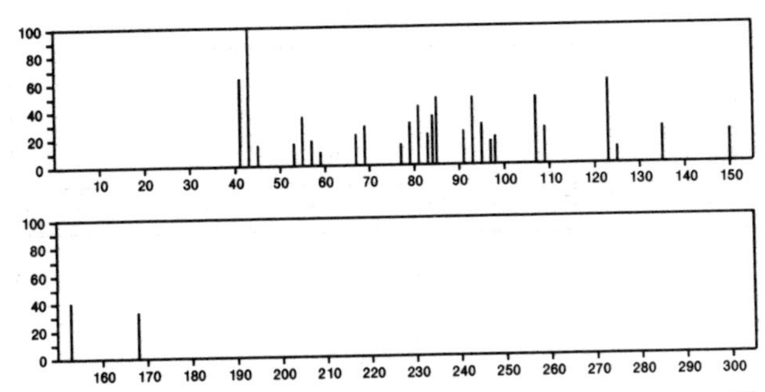

168 $C_{11}H_{20}O$ 54345–64–1
Cyclohexanol, 1–ethyl–2,2–dimethyl–6–methylene–

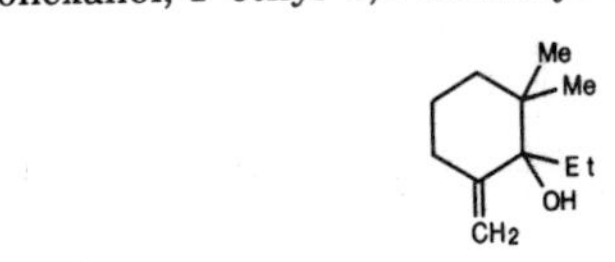

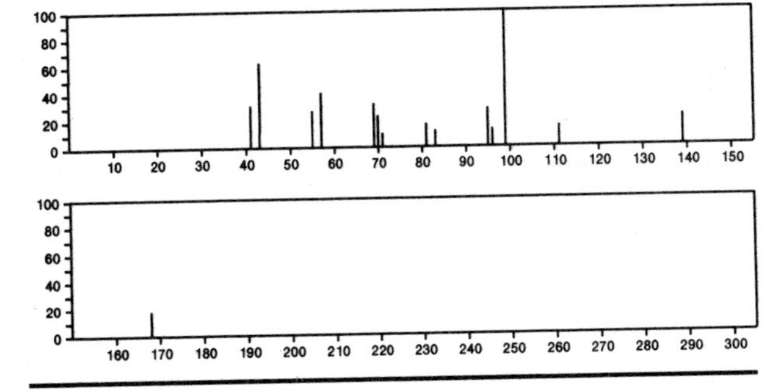

168 $C_{11}H_{20}O$ 54411–03–9
3–Decen–2–one, 3–methyl–

MeCOCMe=CH(CH2)5Me

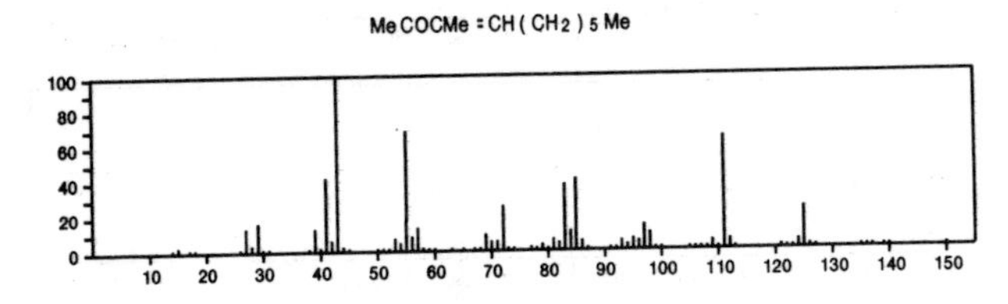

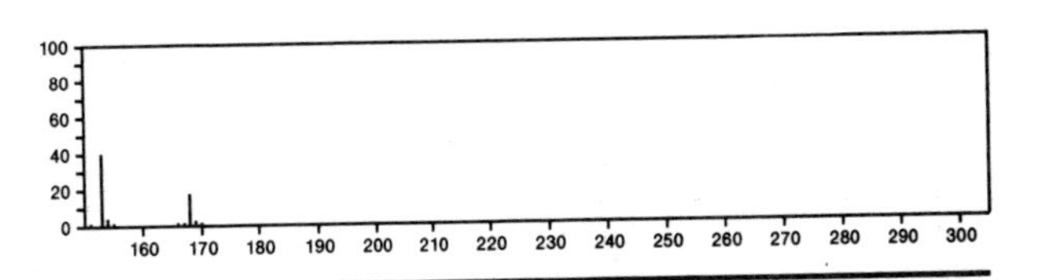

168 $C_{11}H_{20}O$ 54411–04–0
1–Heptyne, 3–ethoxy–3,4–dimethyl–

HC≡CCMe(OEt)CHPrMe

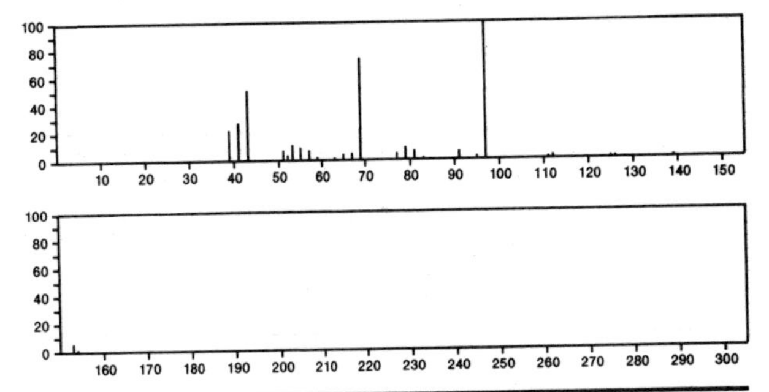

168 $C_{11}H_{20}O$ 54965–85–4
Cyclohexene, 3–(2,2–dimethylpropoxy)–

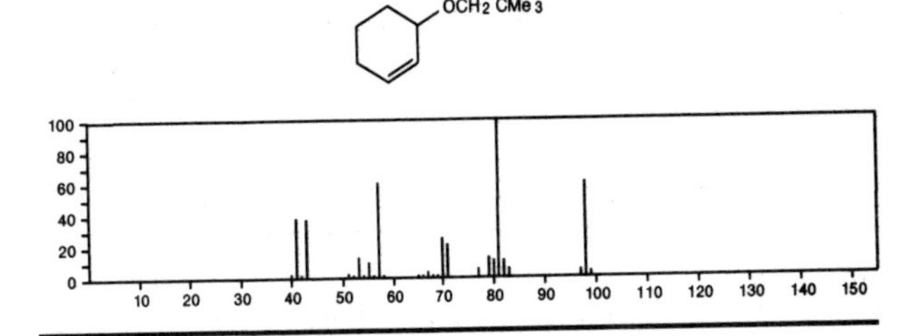

168 $C_{11}H_{20}O$ 55283–56–2
Cyclohexanone, 2–ethyl–2–propyl–

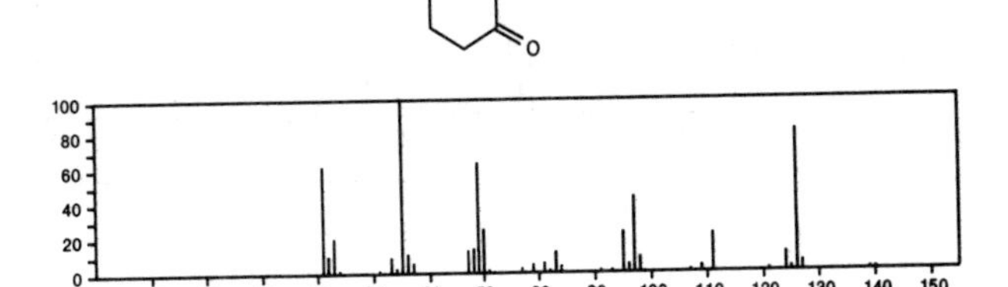

168 $C_{11}H_{20}O$ 55702–30–2
Cyclohexane, (3–ethoxy–1–propenyl)–, (*Z*)–

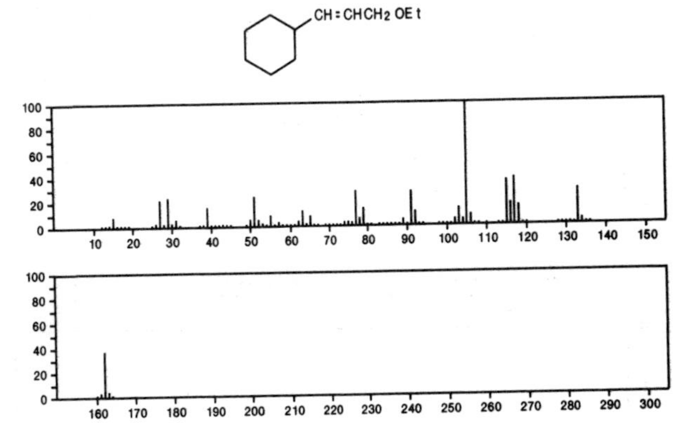

168 $C_{11}H_{20}O$ 56312-55-1
5-Undecen-4-one

168 $C_{11}H_{20}O$ 56312-56-2
3-Nonen-2-one, 3-ethyl-

168 $C_{11}H_{20}O$ 56667-09-5
Cyclopropane, 1-(2-methylbutoxy)-2-(1-methylethylidene)-

168 $C_{12}H_8O$ 132-64-9
Dibenzofuran

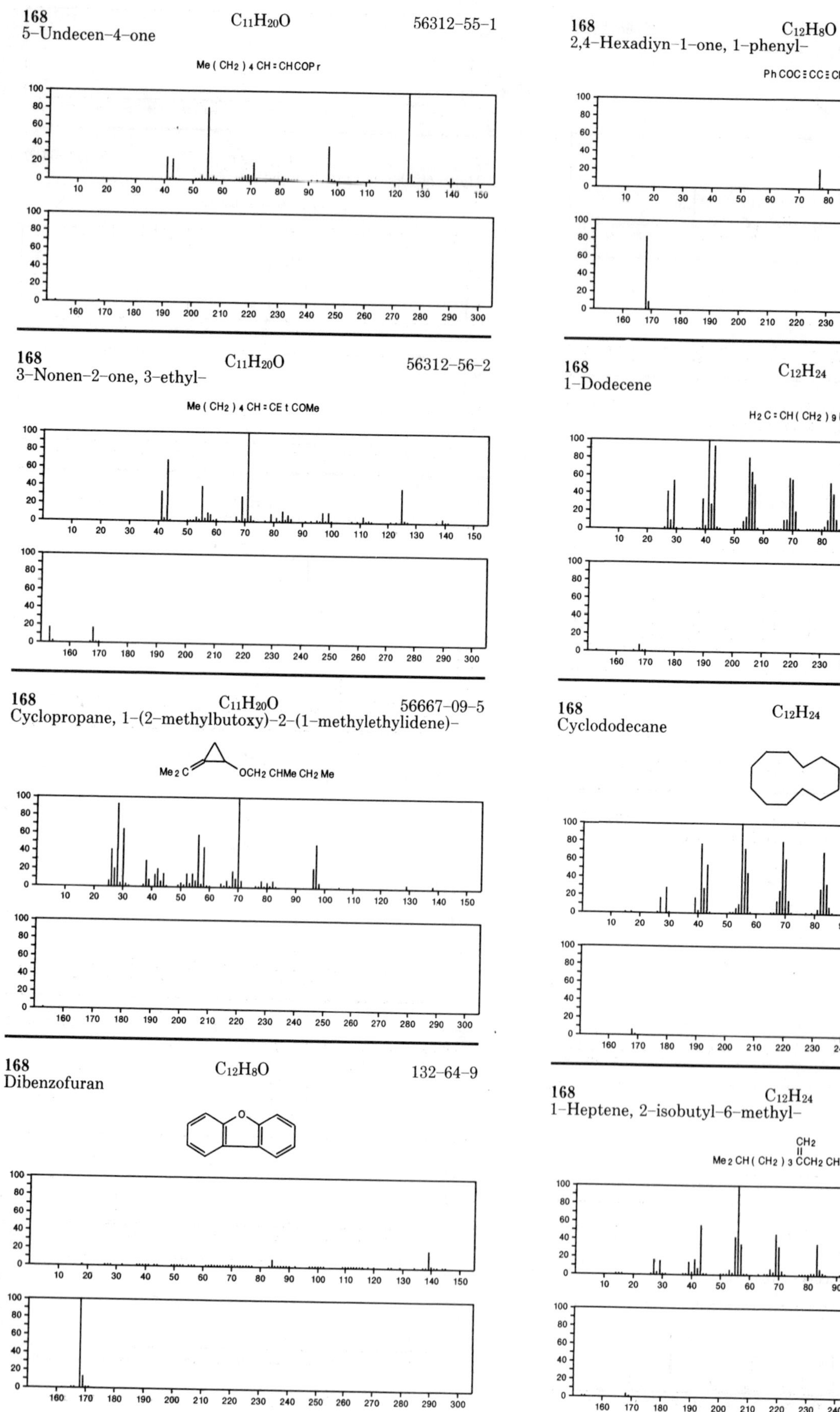

168 $C_{12}H_8O$ 495-74-9
2,4-Hexadiyn-1-one, 1-phenyl-

168 $C_{12}H_{24}$ 112-41-4
1-Dodecene

168 $C_{12}H_{24}$ 294-62-2
Cyclododecane

168 $C_{12}H_{24}$ 7323-15-1
1-Heptene, 2-isobutyl-6-methyl-

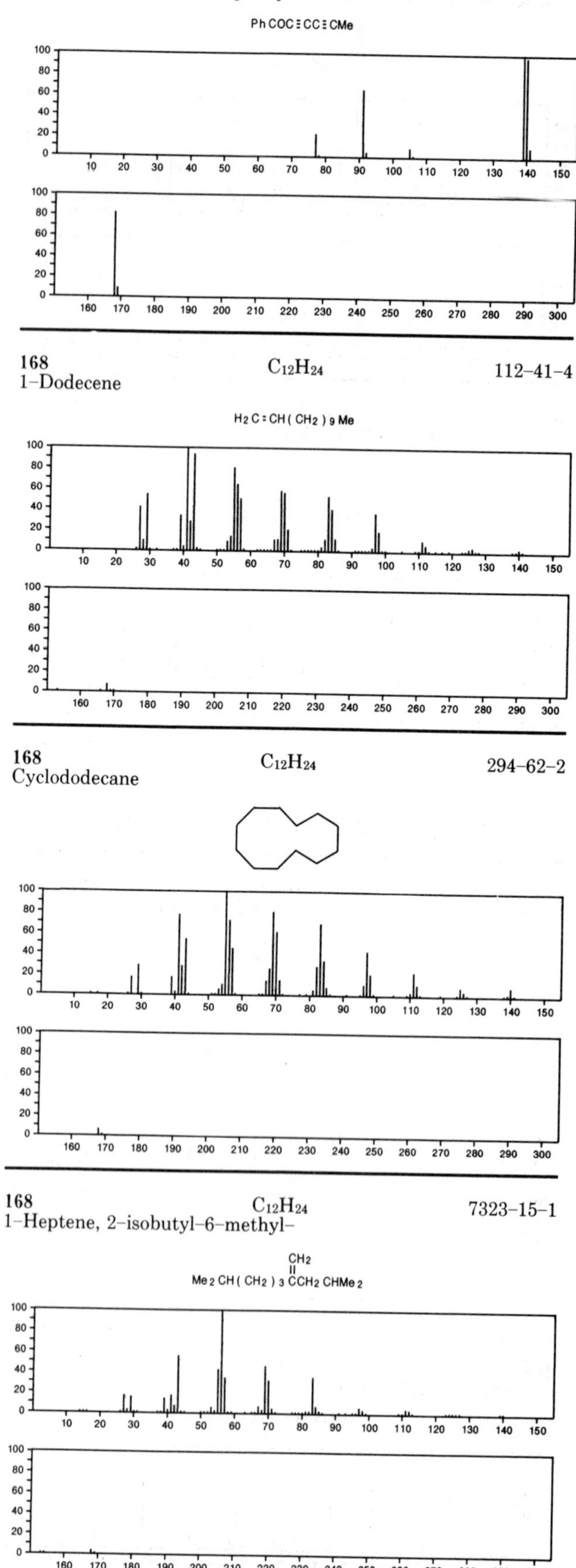

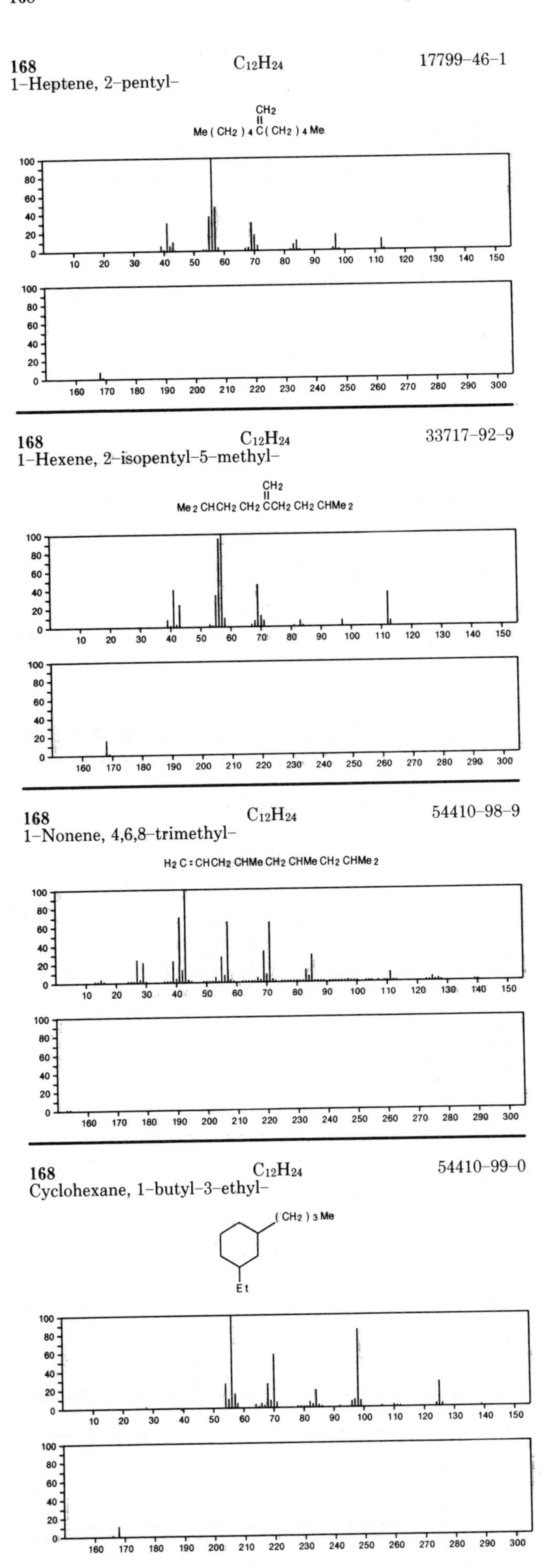
168 $C_{12}H_{24}$ 17799–46–1
1–Heptene, 2–pentyl–
Me(CH2)4C(CH2)4Me
168 $C_{12}H_{24}$ 33717–92–9
1–Hexene, 2–isopentyl–5–methyl–
Me2CHCH2CH2CCH2CH2CHMe2
168 $C_{12}H_{24}$ 54410–98–9
1–Nonene, 4,6,8–trimethyl–
H2C=CHCH2CHMeCH2CHMeCH2CHMe2
168 $C_{12}H_{24}$ 54410–99–0
Cyclohexane, 1–butyl–3–ethyl–
(CH2)3Me
Et

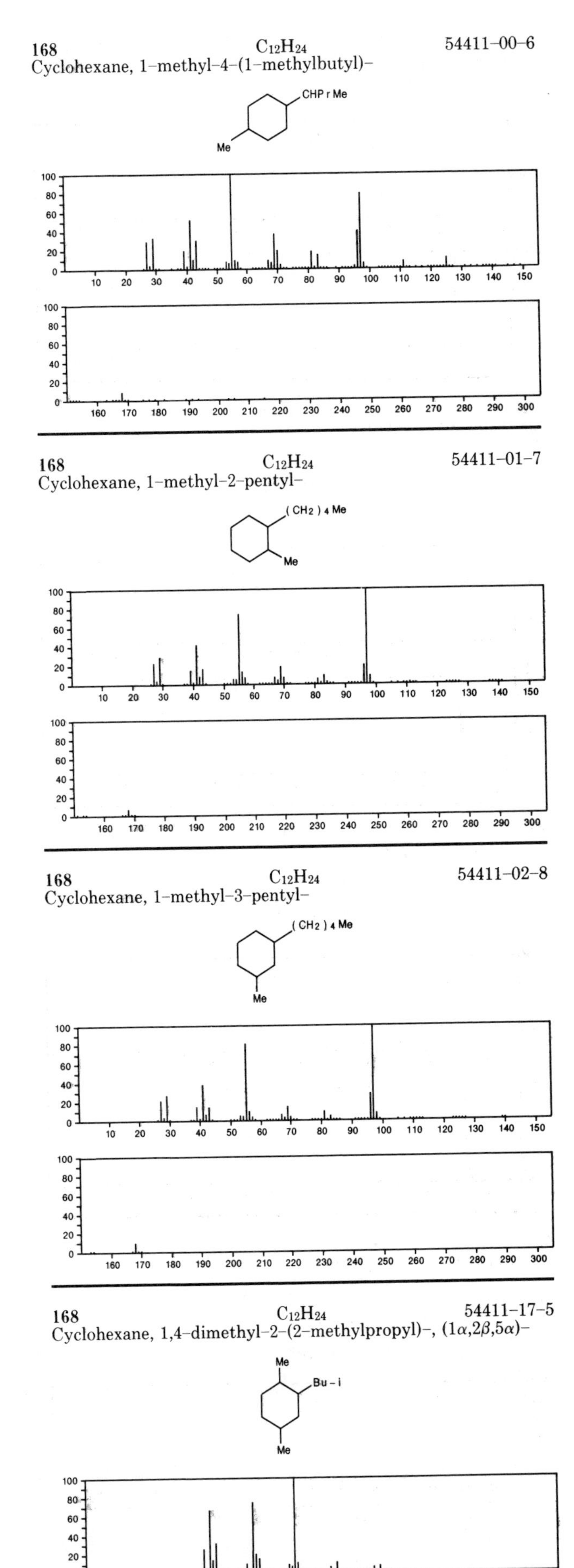
168 $C_{12}H_{24}$ 54411–00–6
Cyclohexane, 1–methyl–4–(1–methylbutyl)–
CHPrMe
Me
168 $C_{12}H_{24}$ 54411–01–7
Cyclohexane, 1–methyl–2–pentyl–
(CH2)4Me
Me
168 $C_{12}H_{24}$ 54411–02–8
Cyclohexane, 1–methyl–3–pentyl–
(CH2)4Me
Me
168 $C_{12}H_{24}$ 54411–17–5
Cyclohexane, 1,4–dimethyl–2–(2–methylpropyl)–, (1α,2β,5α)–
Me
Bu-i
Me

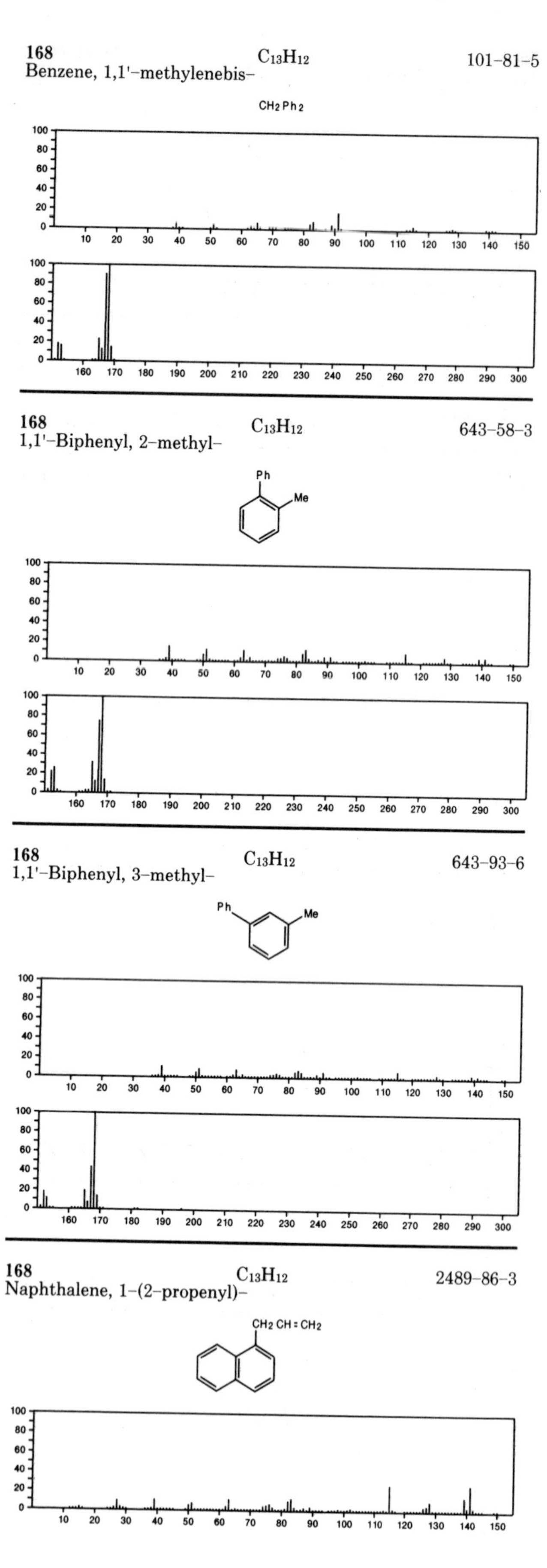

168 Benzene, 1,1'-methylenebis- $C_{13}H_{12}$ 101-81-5

CH_2Ph_2

168 1,1'-Biphenyl, 2-methyl- $C_{13}H_{12}$ 643-58-3

168 1,1'-Biphenyl, 3-methyl- $C_{13}H_{12}$ 643-93-6

168 Naphthalene, 1-(2-propenyl)- $C_{13}H_{12}$ 2489-86-3

$CH_2CH=CH_2$

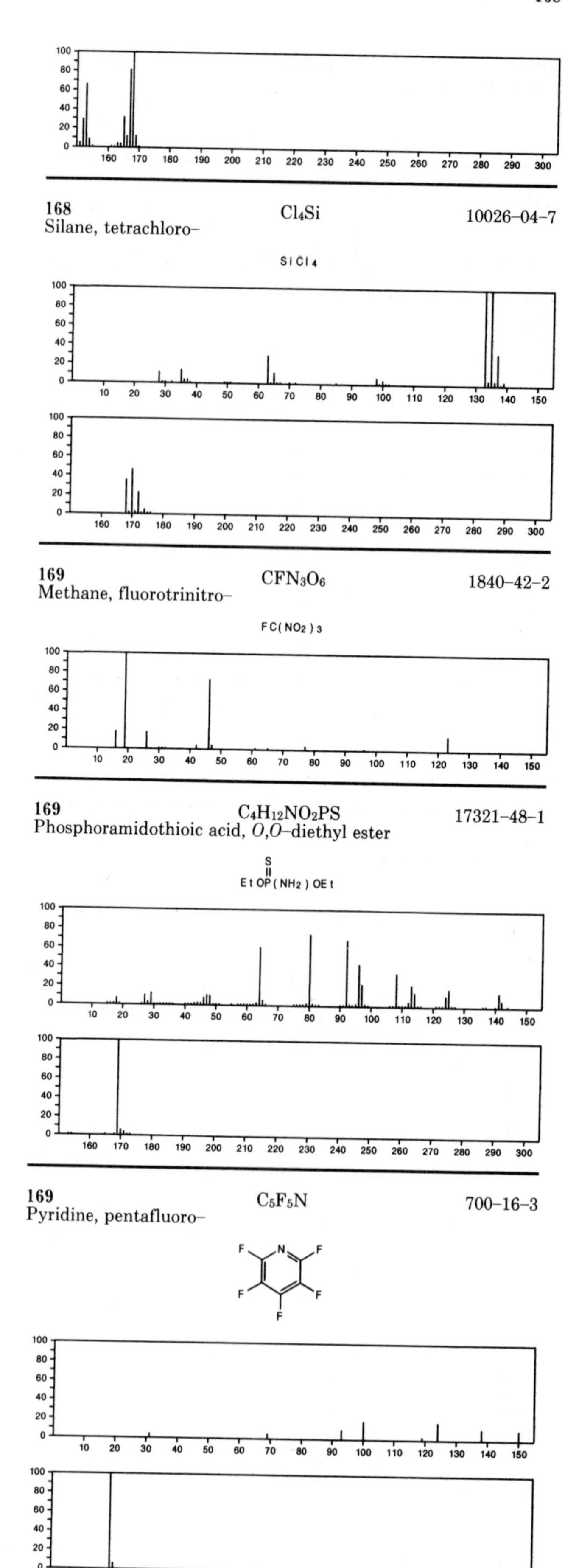

168 Silane, tetrachloro- Cl_4Si 10026-04-7

$SiCl_4$

169 Methane, fluorotrinitro- CFN_3O_6 1840-42-2

$FC(NO_2)_3$

169 Phosphoramidothioic acid, *O,O*-diethyl ester $C_4H_{12}NO_2PS$ 17321-48-1

$EtOP(S)(NH_2)OEt$

169 Pyridine, pentafluoro- C_5F_5N 700-16-3

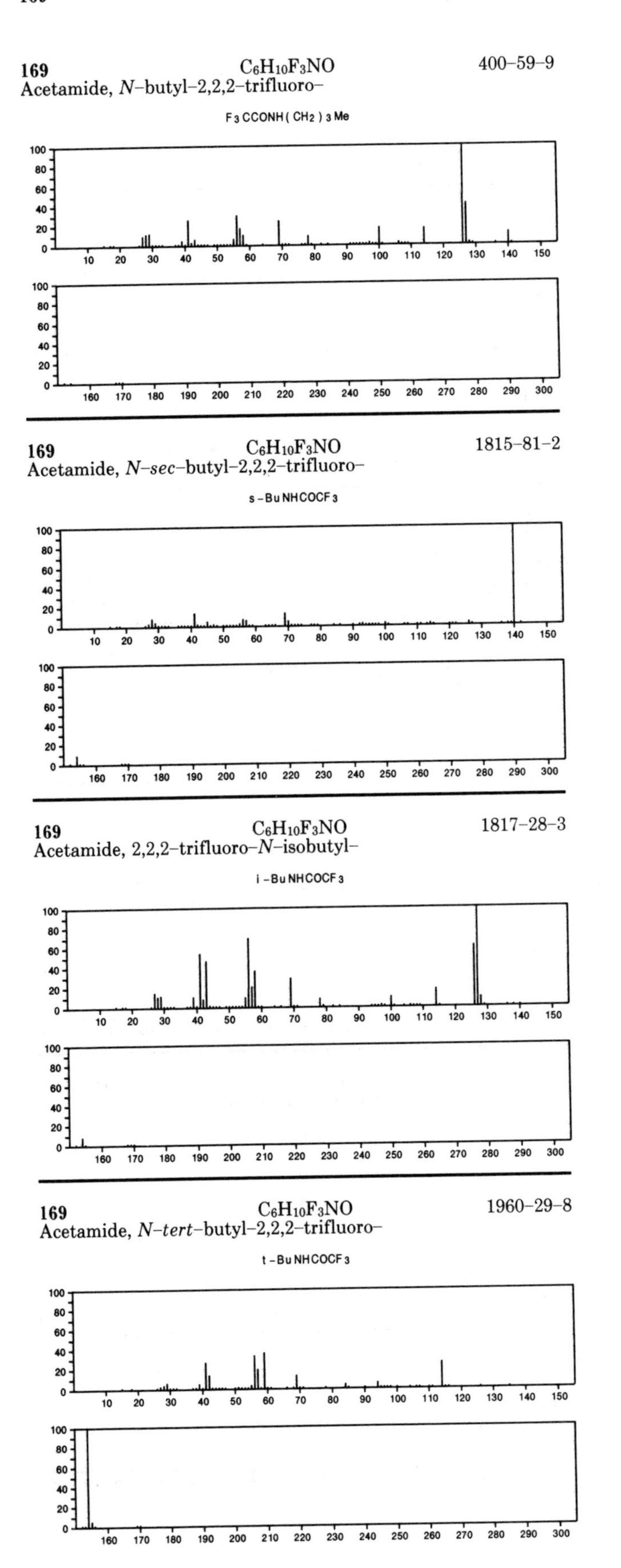

169 $C_6H_{10}F_3NO$ 400-59-9
Acetamide, *N*-butyl-2,2,2-trifluoro-

$F_3CCONH(CH_2)_3Me$

169 $C_6H_{10}F_3NO$ 1815-81-2
Acetamide, *N*-*sec*-butyl-2,2,2-trifluoro-

s-BuNHCOCF$_3$

169 $C_6H_{10}F_3NO$ 1817-28-3
Acetamide, 2,2,2-trifluoro-*N*-isobutyl-

i-BuNHCOCF$_3$

169 $C_6H_{10}F_3NO$ 1960-29-8
Acetamide, *N*-*tert*-butyl-2,2,2-trifluoro-

t-BuNHCOCF$_3$

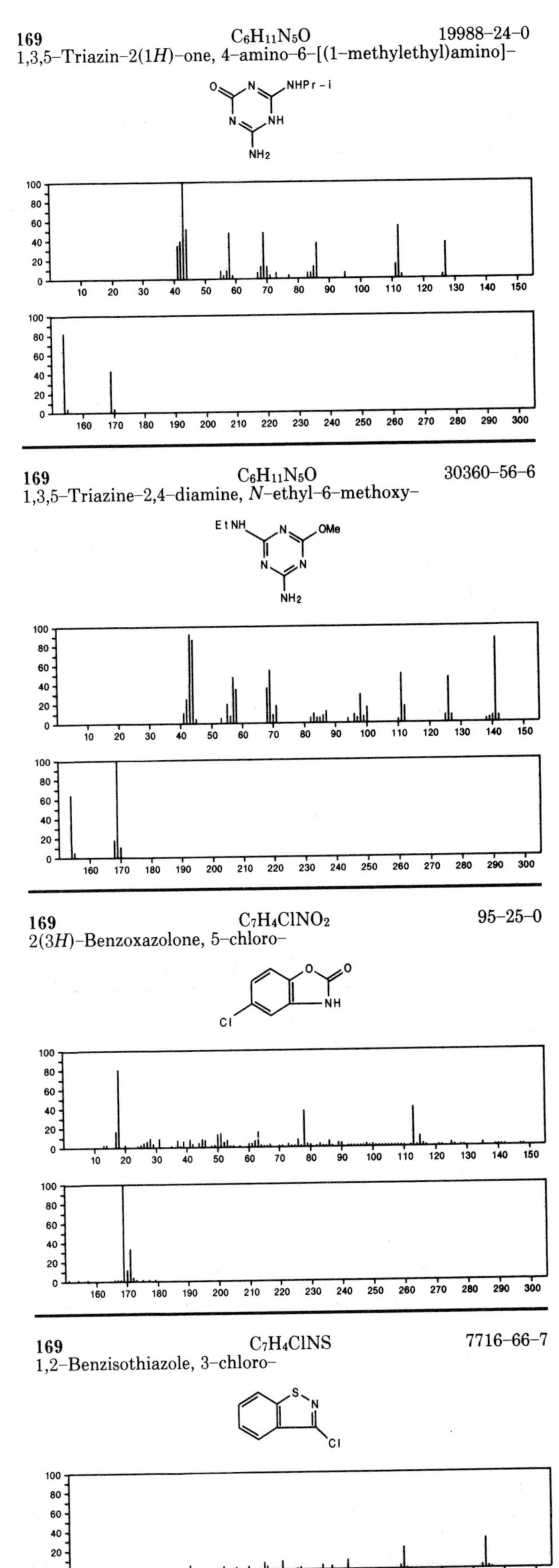

169 $C_6H_{11}N_5O$ 19988-24-0
1,3,5-Triazin-2(1*H*)-one, 4-amino-6-[(1-methylethyl)amino]-

169 $C_6H_{11}N_5O$ 30360-56-6
1,3,5-Triazine-2,4-diamine, *N*-ethyl-6-methoxy-

169 $C_7H_4ClNO_2$ 95-25-0
2(3*H*)-Benzoxazolone, 5-chloro-

169 C_7H_4ClNS 7716-66-7
1,2-Benzisothiazole, 3-chloro-

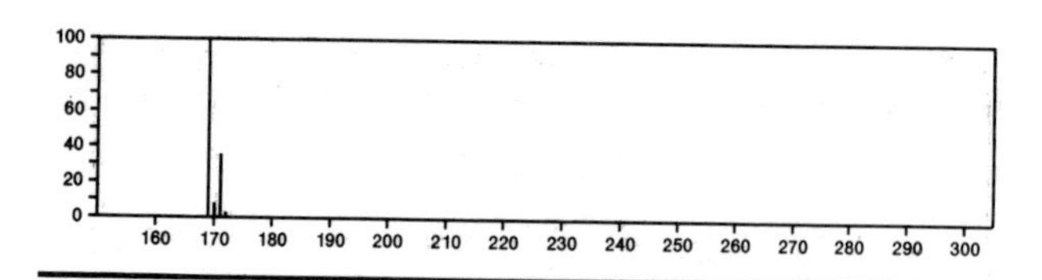

169 C_7H_4ClNS 28783-23-5
Thieno[3,2-*c*]pyridine, 2-chloro-

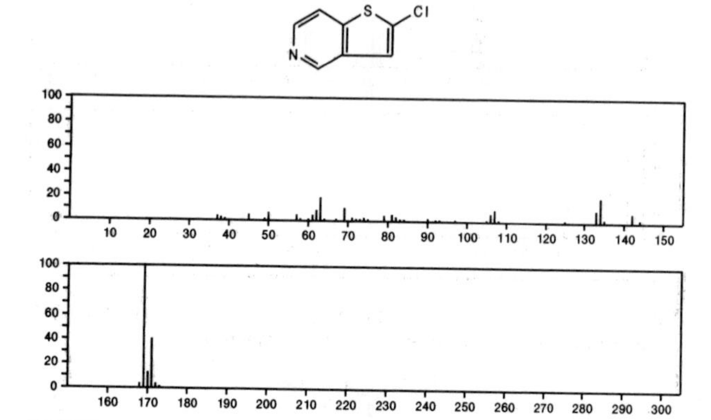

169 C_7H_4ClNS 53399-36-3
Thieno[2,3-*b*]pyridine, 3-chloro-

169 $C_7H_7NO_4$ 7450-68-2
1*H*-Pyrrole-2,5-dione, 1-[(acetyloxy)methyl]-

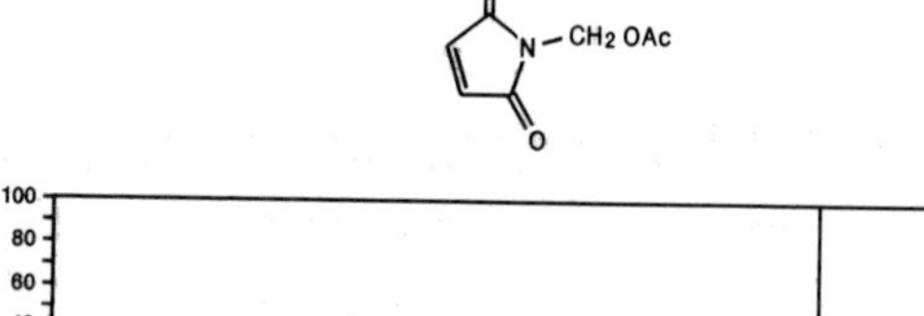

169 $C_7H_{11}N_3O_2$ 332-80-9
L-Histidine, 1-methyl-

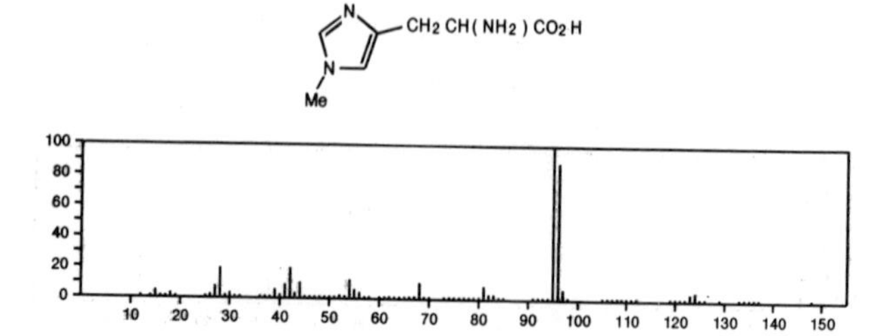

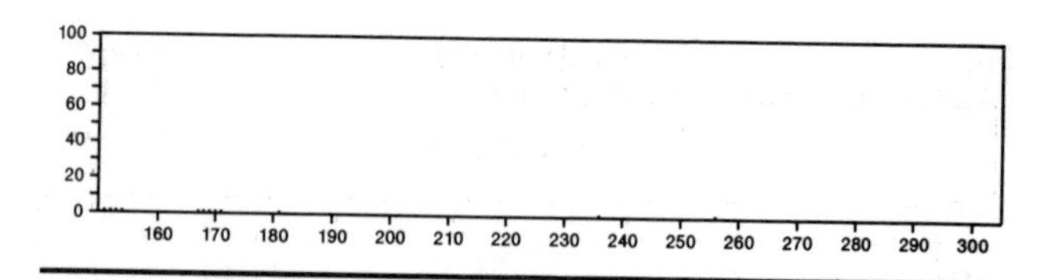

169 $C_7H_{11}N_3O_2$ 368-16-1
L-Histidine, 3-methyl-

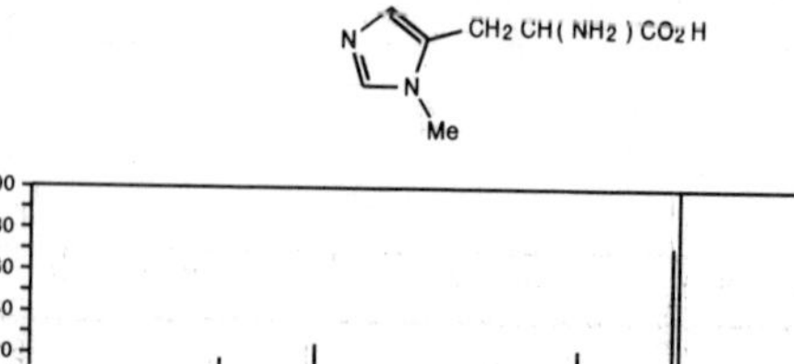

169 $C_7H_{11}N_3O_2$ 29924-76-3
1,2,4-Triazabicyclo[2.2.2]octan-3-one, 2-acetyl-

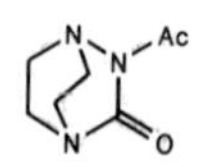

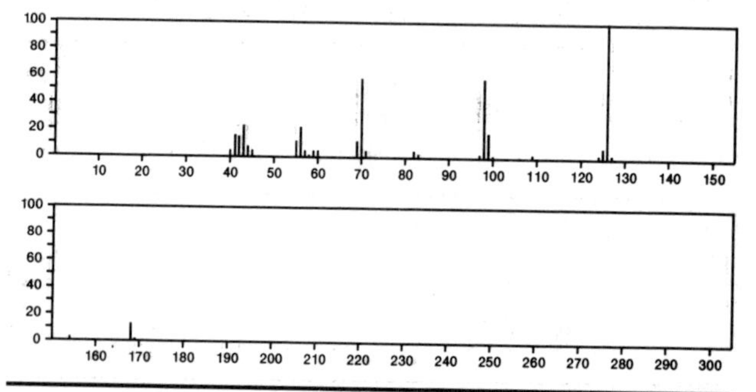

169 $C_7H_{11}N_3S$ 54410-87-6
4-Pyrimidinamine, 2-(ethylthio)-5-methyl-

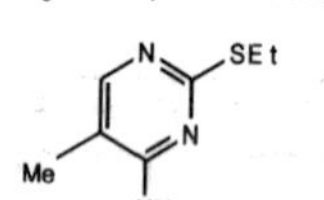

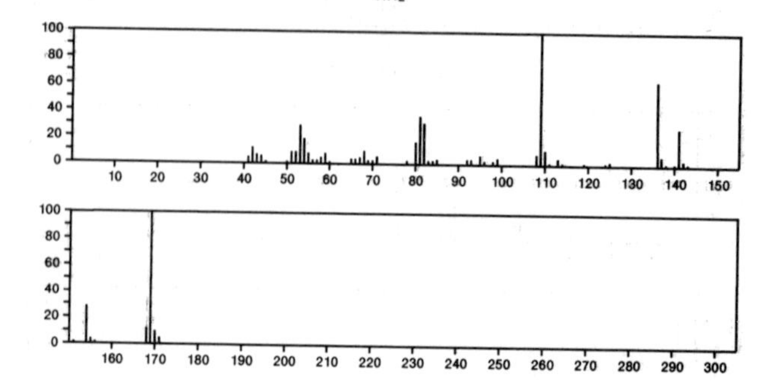

169 $C_7H_{11}N_3S$ 54410-88-7
4-Pyrimidinamine, 2-(propylthio)-

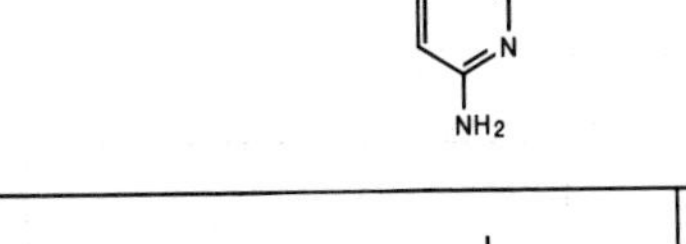

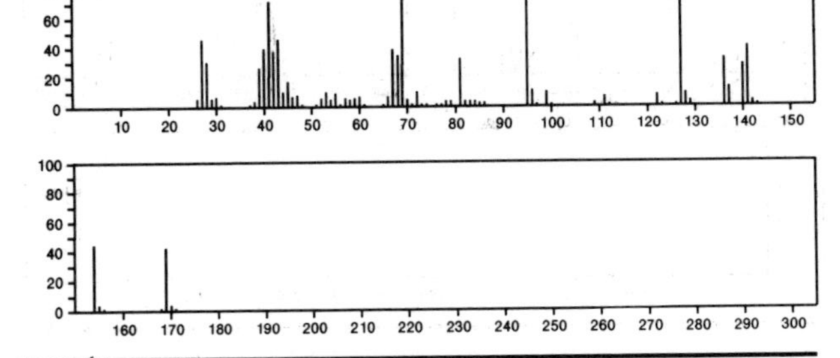

169 $C_8H_5F_2NO$ 1644-16-2
Benzene, (difluoroisocyanatomethyl)-

PhCF2(NCO)

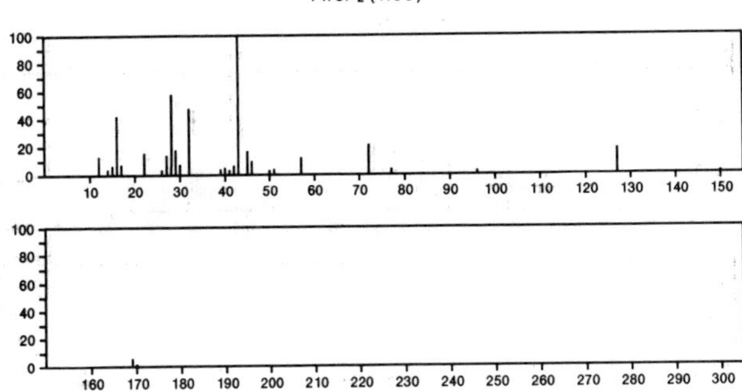

169 C_8H_8ClNO 539-03-7
Acetamide, *N*-(4-chlorophenyl)-

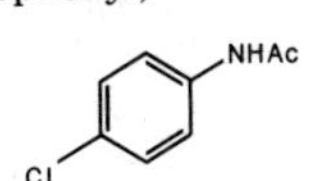

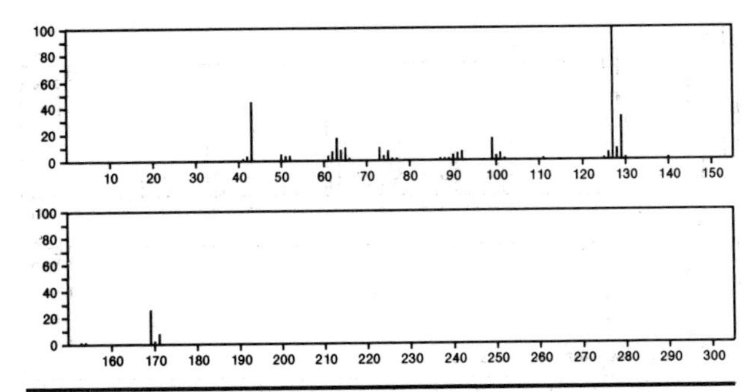

169 C_8H_8ClNO 33499-36-4
Benzaldehyde, *m*-chloro-, *O*-methyloxime

CH=NOMe
Cl

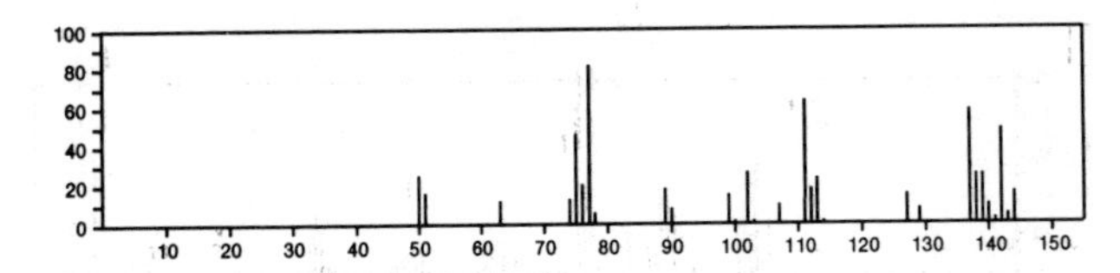

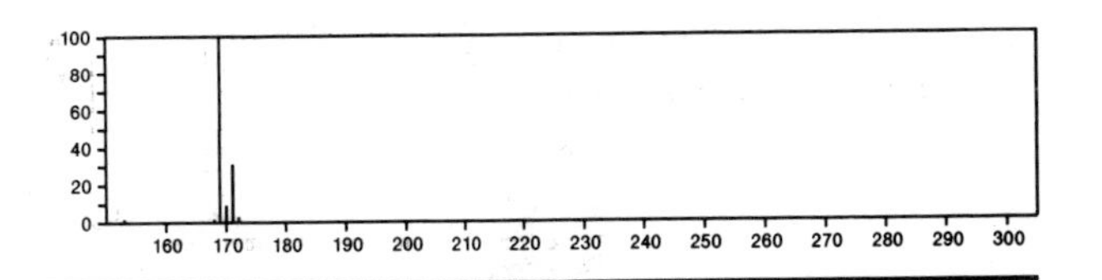

169 C_8H_8ClNO 33499-37-5
Benzaldehyde, 4-chloro-, *O*-methyloxime

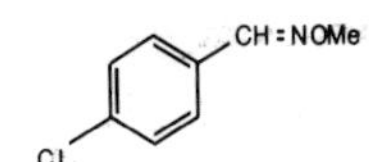

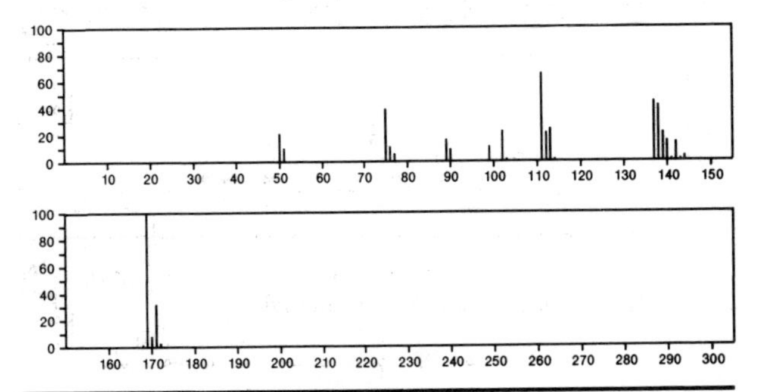

169 C_8H_8ClNO 33513-35-8
Benzaldehyde, *o*-chloro-, *O*-methyloxime

CH=NOMe
Cl

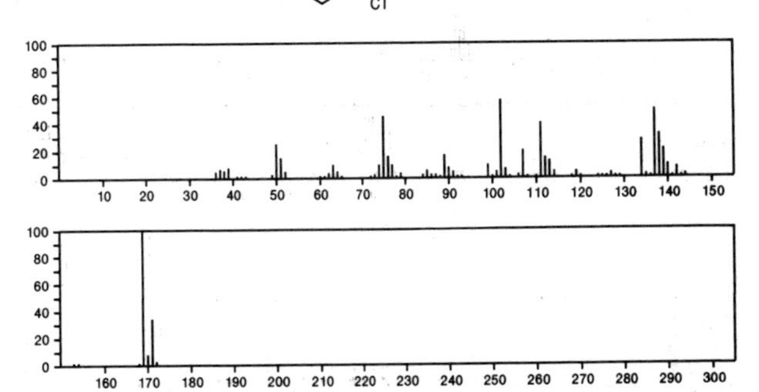

169 $C_8H_{11}NOS$ 23003-26-1
3-Pyridinol, 2-(ethylthio)-6-methyl-

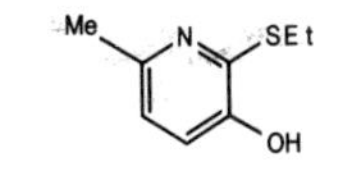

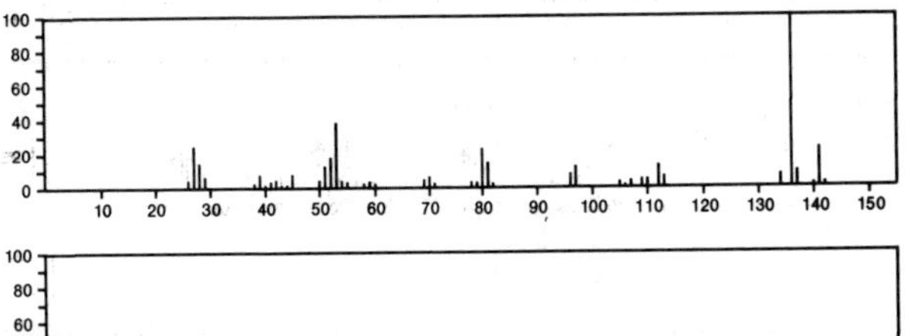

169 $C_8H_{11}NOS$ 24207-15-6
2(1*H*)-Pyridinethione, 1-ethyl-3-hydroxy-6-methyl-

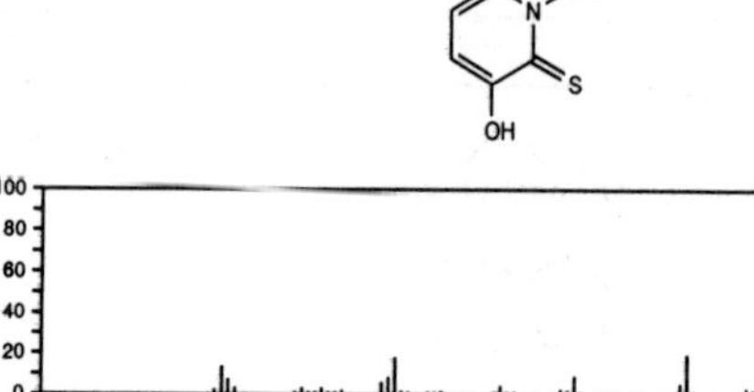

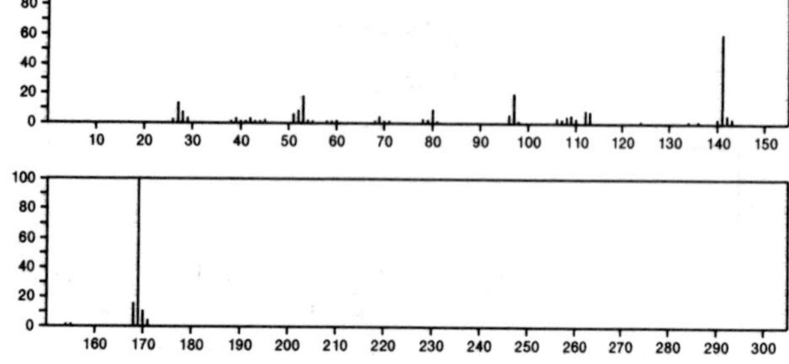

169 $C_8H_{11}NOS$ 40585-12-4
2(1*H*)-Pyridinethione, 3-ethoxy-6-methyl-

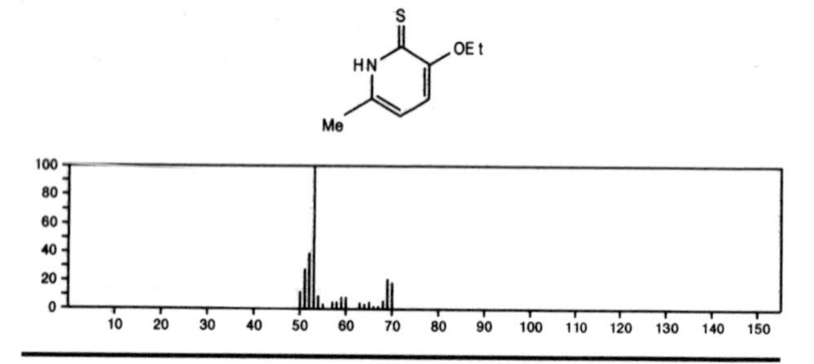

169 $C_8H_{11}NOS$ 55956-24-6
Ethanol, 2-[(methyl-2-thienylmethylene)amino]-

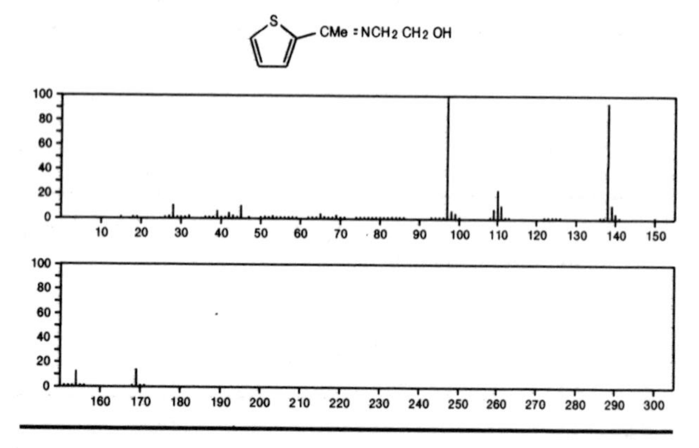

169 $C_8H_{11}NO_3$ 51-41-2
1,2-Benzenediol, 4-(2-amino-1-hydroxyethyl)-, (*R*)-

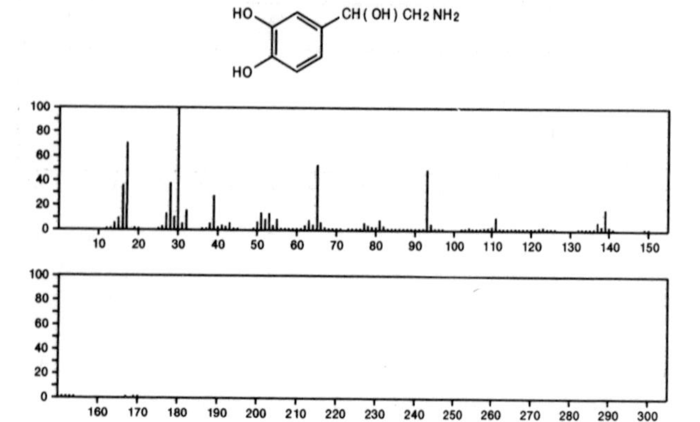

169 $C_8H_{11}NO_3$ 149-95-1
1,2-Benzenediol, 4-(2-amino-1-hydroxyethyl)-, (*S*)-

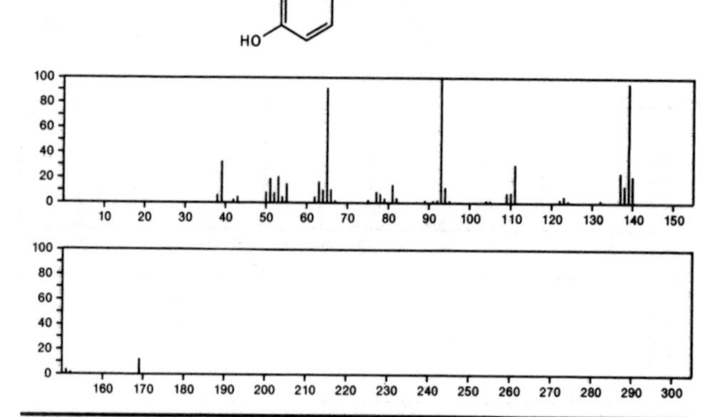

169 $C_8H_{15}N_3O$ 4549-20-6
Hydrazinecarboxamide, 2-(2-methylcyclohexylidene)-

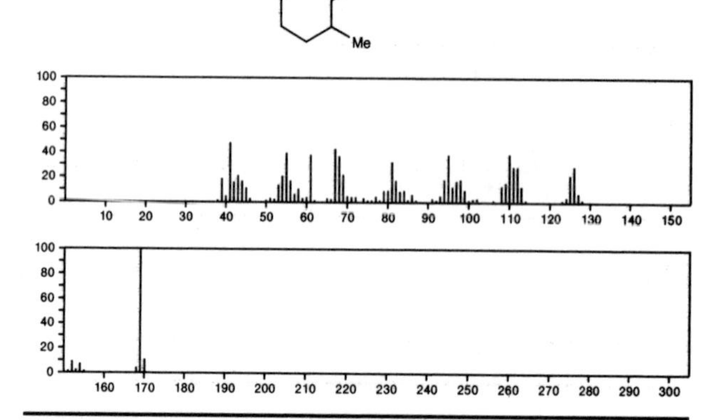

169 $C_8H_{15}N_3O$ 5439-97-4
Cyclohexanone, 4-methyl-, semicarbazone

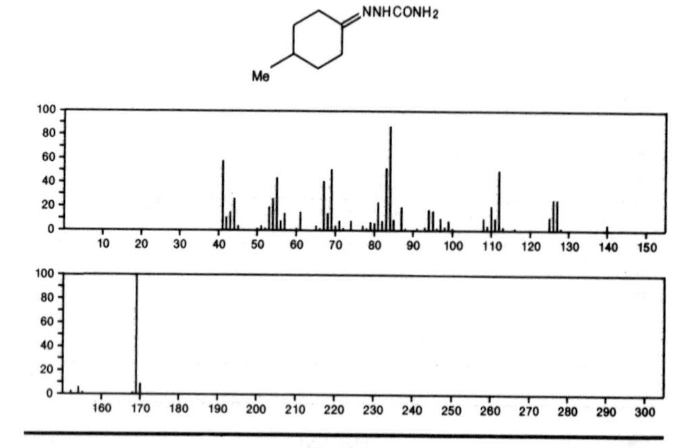

169 $C_8H_{15}N_3O$ 54410-86-5
Hydrazinecarboxamide, 2-(3-methylcyclohexylidene)-

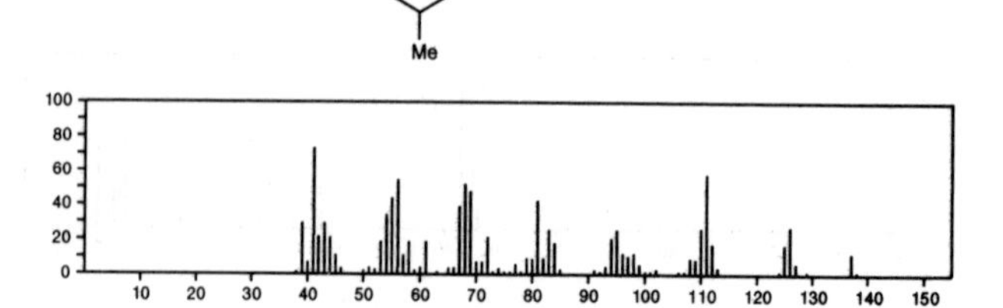

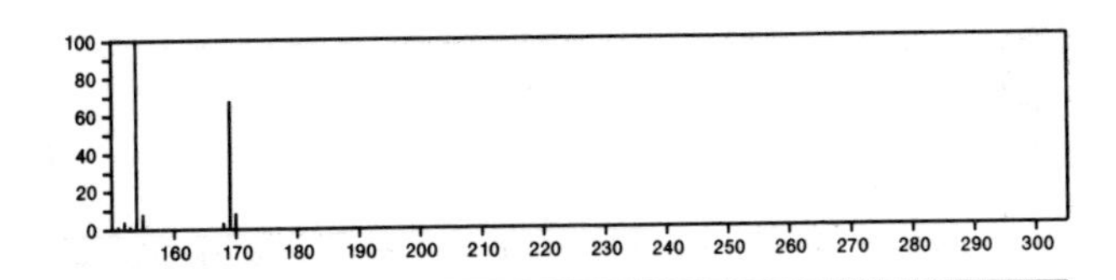

169 $C_9H_{15}NO_2$ 4531-60-6
3-Penten-2-one, 4-(4-morpholinyl)-

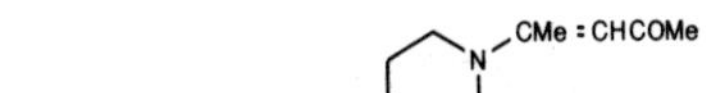

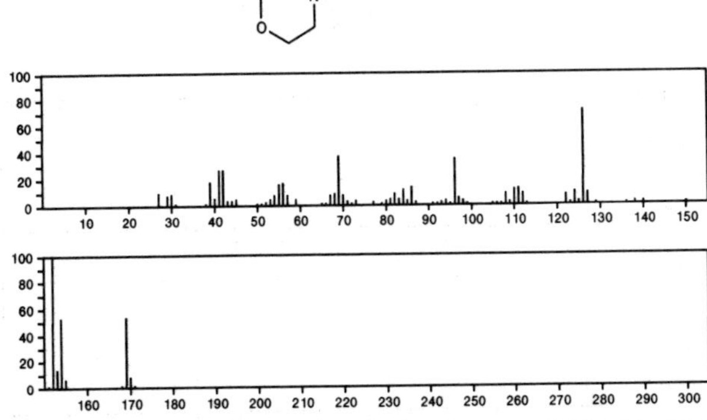

169 $C_9H_{15}NO_2$ 4839-12-7
Tropinone, 6β-methoxy-, (+)-

Me
MeO N O

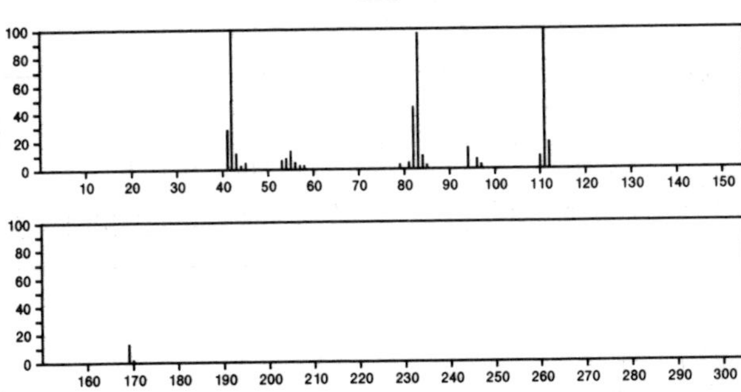

169 $C_9H_{15}NO_2$ 7309-45-7
Pentanoic acid, 2-cyano-3-methyl-, ethyl ester

EtOC(O)CH(CN)CHMeCH$_2$Me

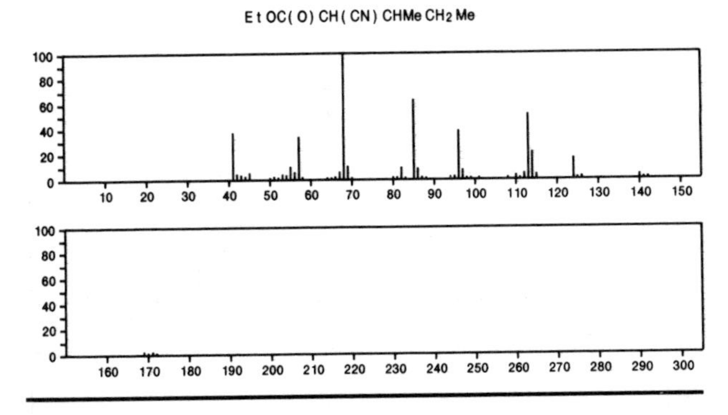

169 $C_9H_{15}NO_2$ 7352-02-5
Valeric acid, 2-cyano-4-methyl-, ethyl ester

EtOC(O)CH(CN)CH$_2$CHMe$_2$

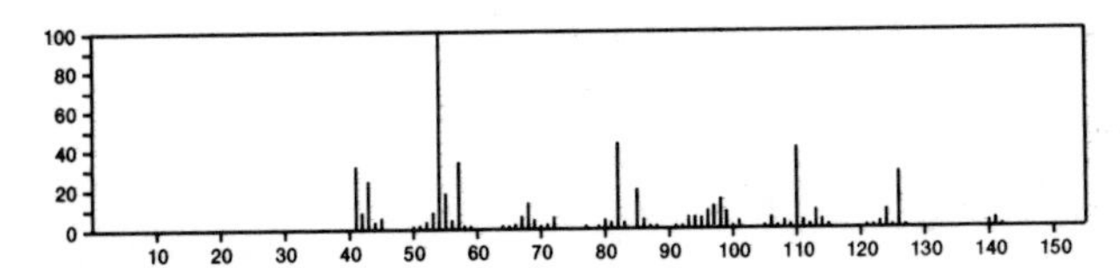

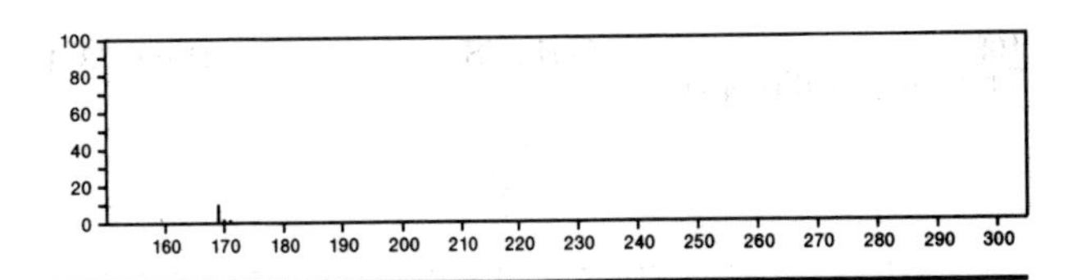

169 $C_9H_{15}NO_2$ 7391-39-1
Hexanoic acid, 2-cyano-, ethyl ester

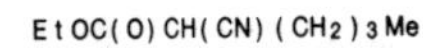

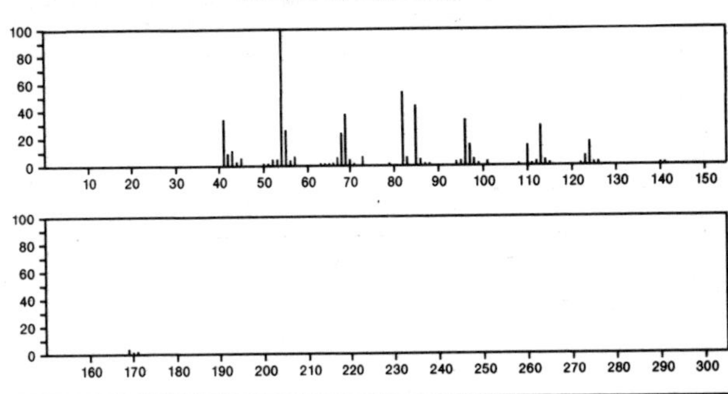

169 $C_9H_{15}NO_2$ 54410-85-4
2,5-Pyrrolidinedione, 1,3-diethyl-3-methyl-

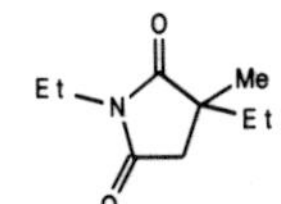

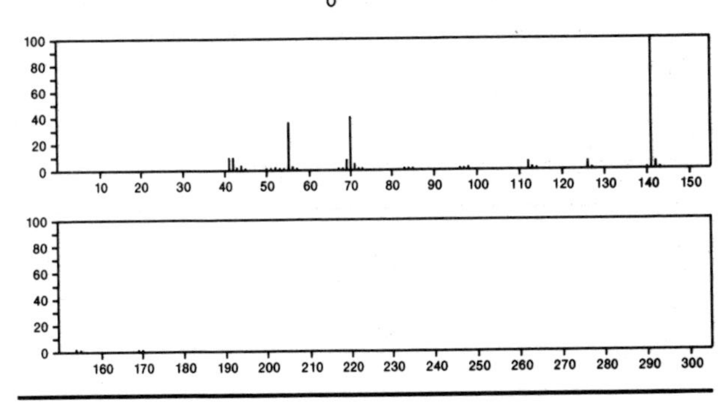

169 $C_9H_{15}NS$ 4276-67-9
Thiazole, 5-ethyl-2-methyl-4-propyl-

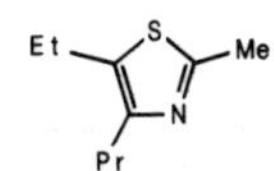

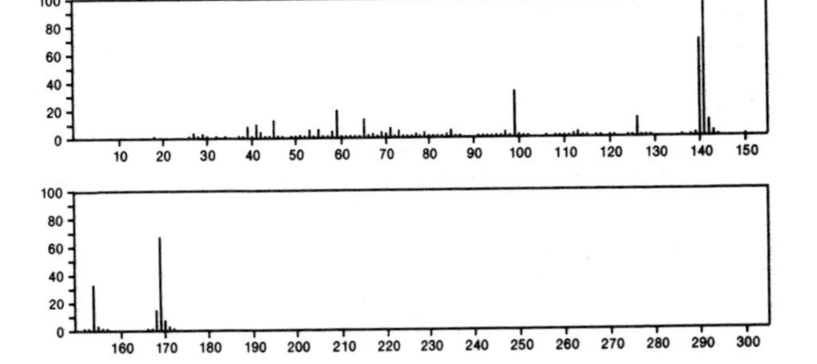

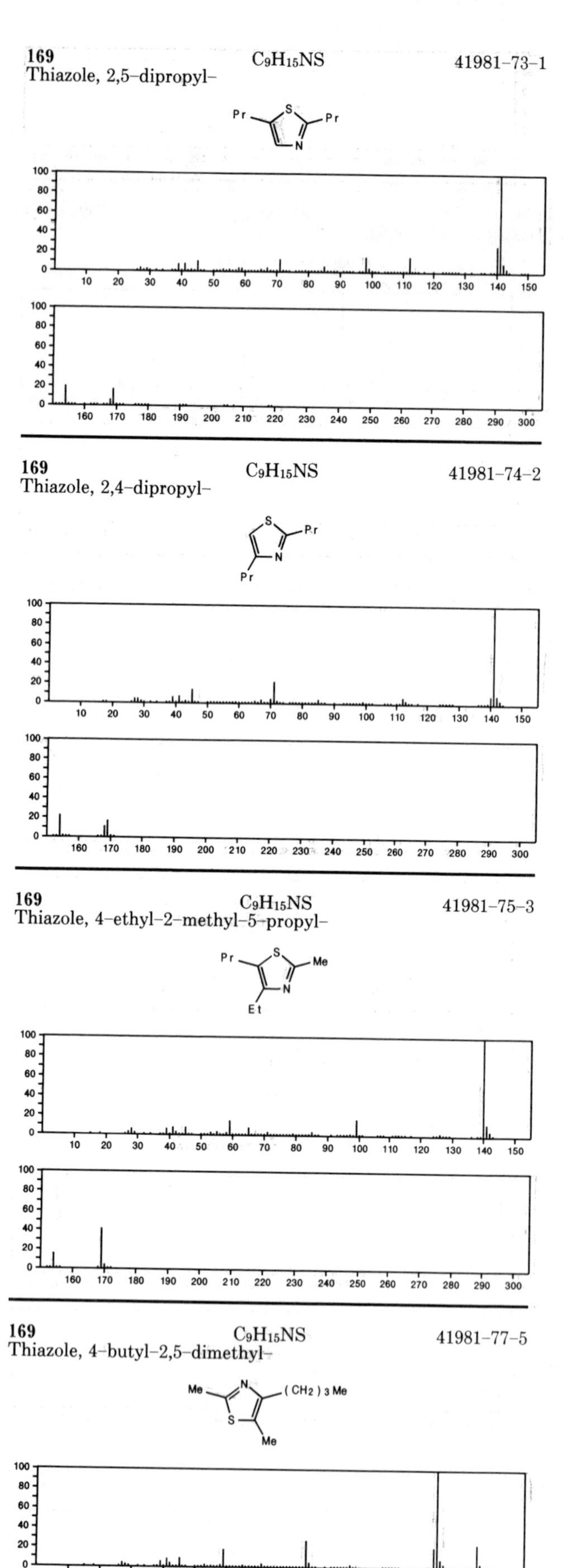

169 Thiazole, 2,5-dipropyl- $C_9H_{15}NS$ 41981-73-1

169 Thiazole, 2,4-dipropyl- $C_9H_{15}NS$ 41981-74-2

169 Thiazole, 4-ethyl-2-methyl-5-propyl- $C_9H_{15}NS$ 41981-75-3

169 Thiazole, 4-butyl-2,5-dimethyl- $C_9H_{15}NS$ 41981-77-5

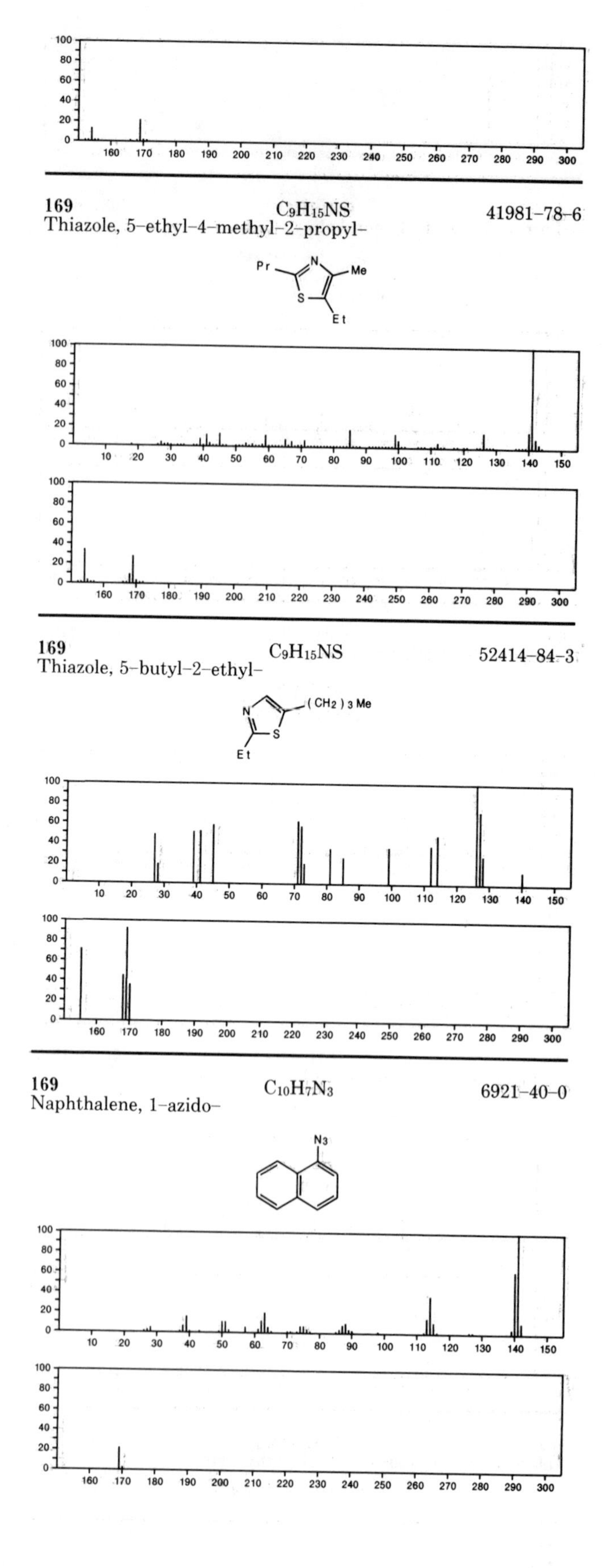

169 Thiazole, 5-ethyl-4-methyl-2-propyl- $C_9H_{15}NS$ 41981-78-6

169 Thiazole, 5-butyl-2-ethyl- $C_9H_{15}NS$ 52414-84-3

169 Naphthalene, 1-azido- $C_{10}H_7N_3$ 6921-40-0

169 $C_{10}H_7N_3$ 17966-00-6
1*H*-Dipyrido[2,3-*b*:3',2'-*d*]pyrrole

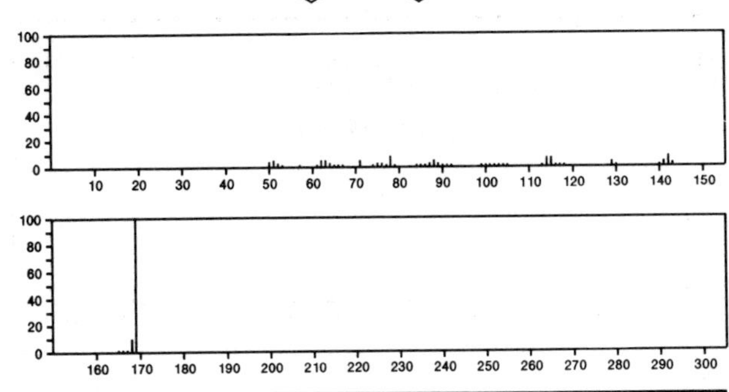

169 $C_{10}H_{19}NO$ 1128-34-3
Acetamide, *N*-cyclohexyl-*N*-ethyl-

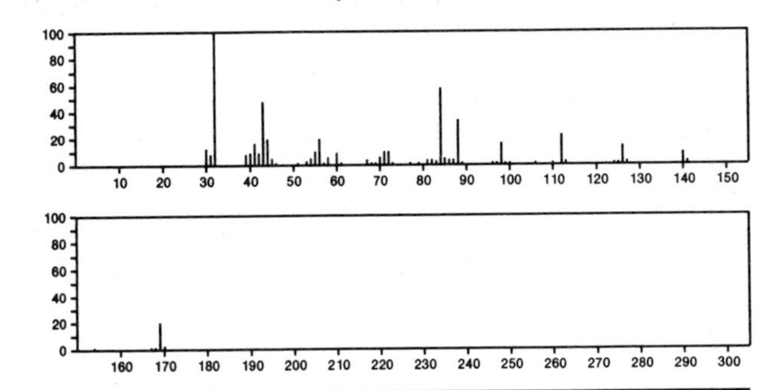

169 $C_{10}H_{19}NO$ 1199-87-7
Butanamide, *N*-cyclohexyl-

169 $C_{10}H_{19}NO$ 1925-44-6
Bicyclo[2.2.1]heptan-2-ol, 3-amino-1,7,7-trimethyl-, (*endo,endo*)-

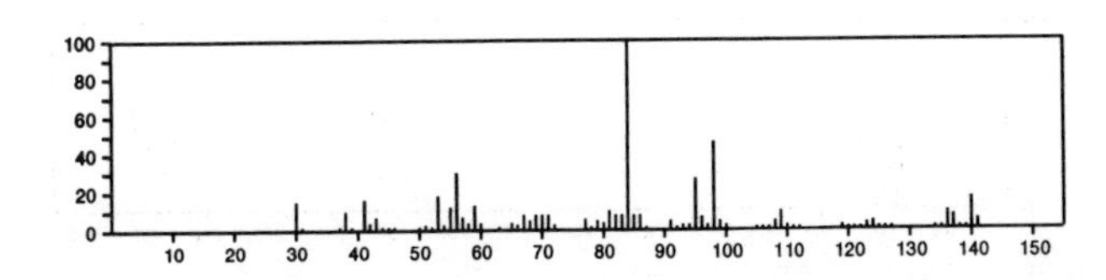

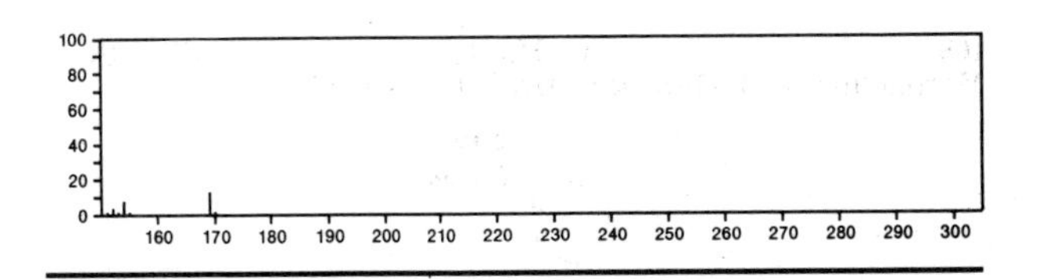

169 $C_{10}H_{19}NO$ 2972-01-2
Cyclodecanone, oxime

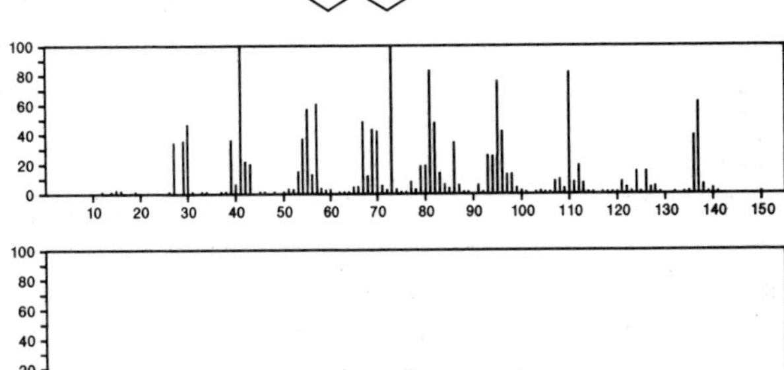

169 $C_{10}H_{19}NO$ 14387-89-4
Aziridinone, 1,3-bis(1,1-dimethylethyl)-

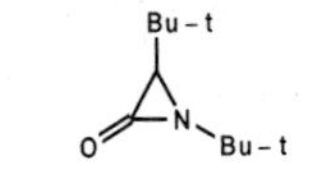

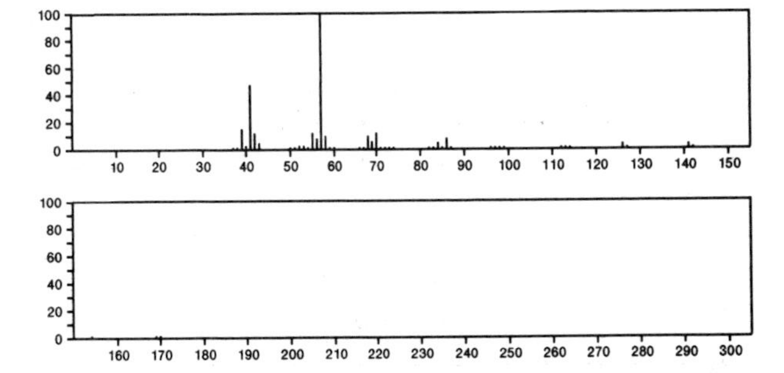

169 $C_{10}H_{19}NO$ 32344-86-8
Bicyclo[2.2.1]heptan-2-ol, 3-amino-4,7,7-trimethyl-, (*endo,endo*)-

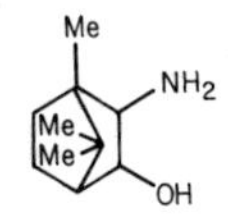

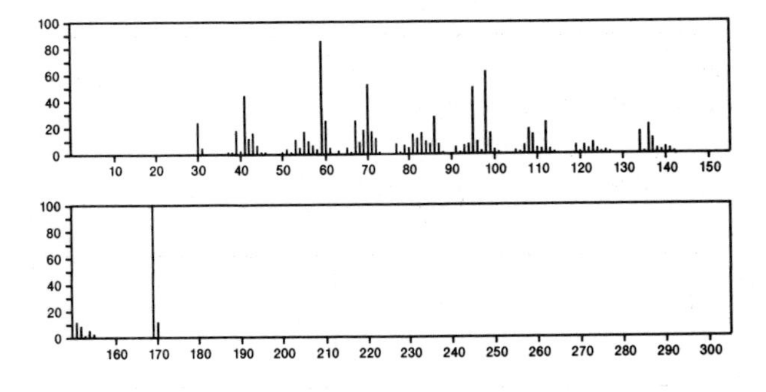

169 $C_{10}H_{19}NO$ 55103-87-2
Morpholine, 4-[1-(1-methylethyl)-1-propenyl]-

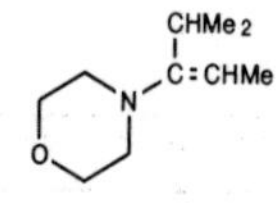

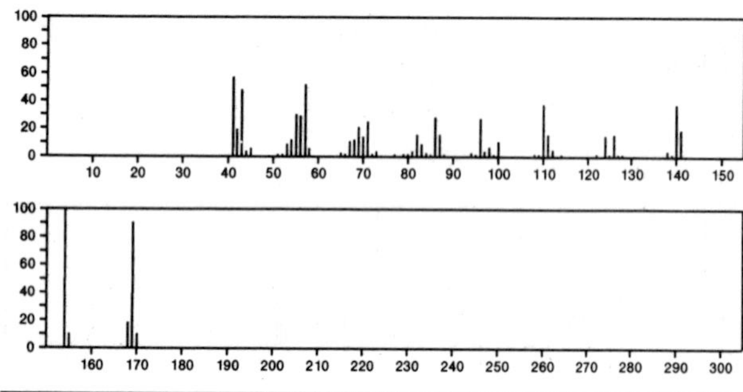

169 $C_{10}H_{19}NO$ 56335-96-7
3-Nonen-2-one, *O*-methyloxime

Me(CH2)4CH=CHCMe=NOMe

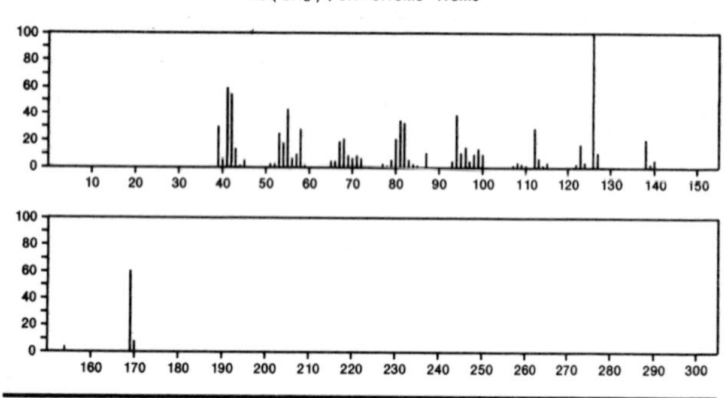

169 $C_{11}H_{7}NO$ 86-84-0
Naphthalene, 1-isocyanato-

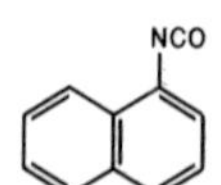

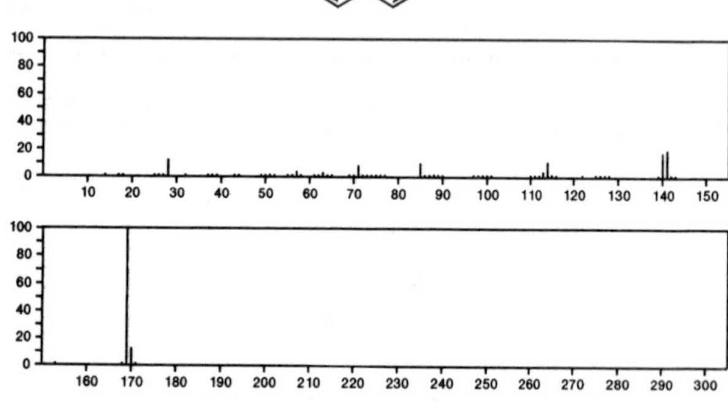

169 $C_{11}H_{23}N$ 3447-05-0
Pyrrolidine, 2-hexyl-1-methyl-

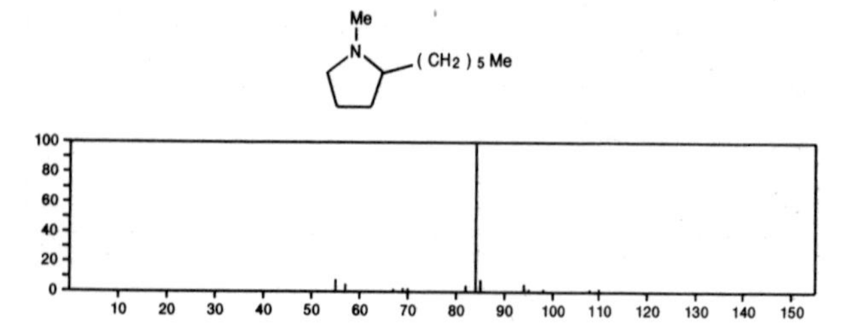

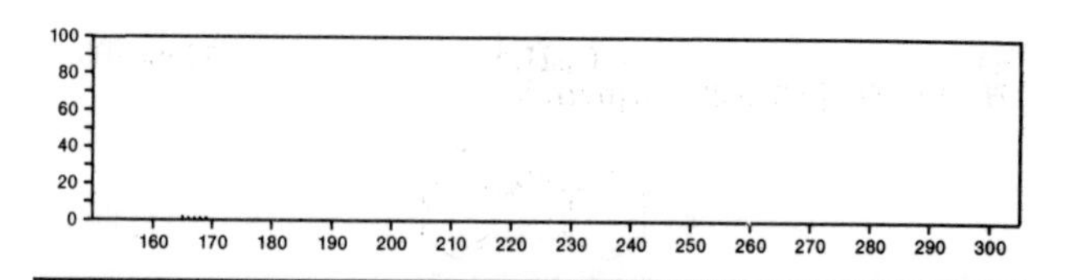

169 $C_{11}H_{23}N$ 10599-80-1
Butylamine, *N*-(1-propylbutylidene)-

Me(CH2)3N=CPr2

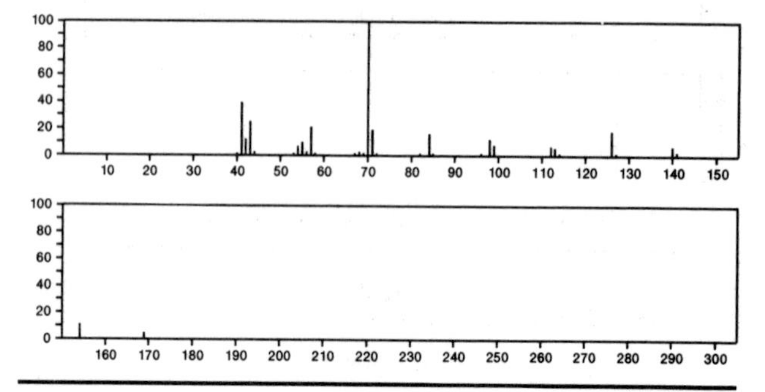

169 $C_{11}H_{23}N$ 10599-82-3
Ethylamine, *N*-(1-butylpentylidene)-

(CH2)3Me
Me(CH2)3C=NEt

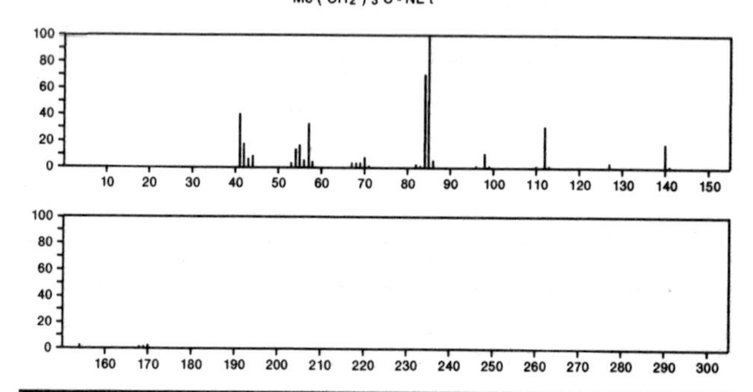

169 $C_{11}H_{23}N$ 18641-76-4
Methylamine, *N*-(1-butylhexylidene)-

(CH2)4Me
Me(CH2)3C=NMe

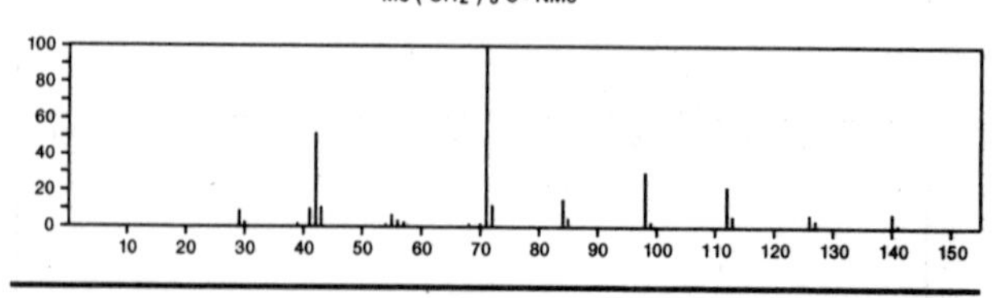

169 $C_{12}H_{11}N$ 90-41-5
[1,1'-Biphenyl]-2-amine

Ph
NH2

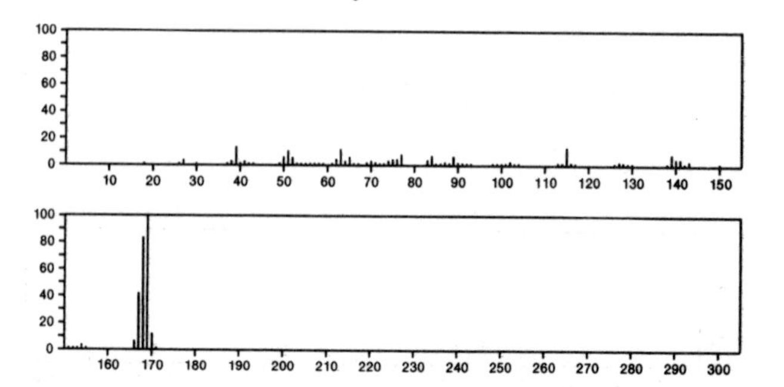

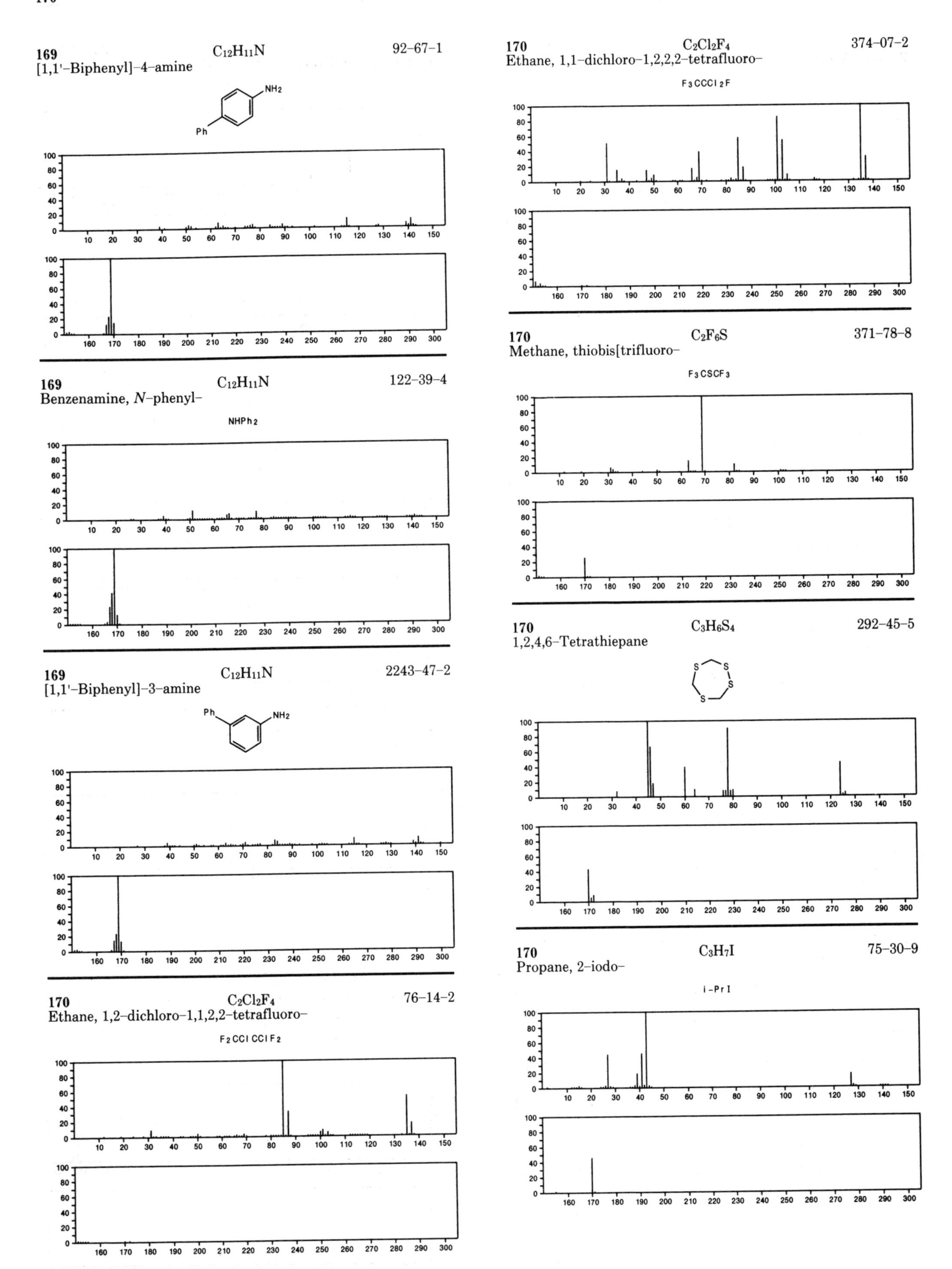
169
C12H11N
92-67-1
[1,1'-Biphenyl]-4-amine
NH2
Ph
169
C12H11N
122-39-4
Benzenamine, N-phenyl-
NHPh2
169
C12H11N
2243-47-2
[1,1'-Biphenyl]-3-amine
Ph
NH2
170
C2Cl2F4
76-14-2
Ethane, 1,2-dichloro-1,1,2,2-tetrafluoro-
F2CCl CClF2
170
C2Cl2F4
374-07-2
Ethane, 1,1-dichloro-1,2,2,2-tetrafluoro-
F3CCCl2F
170
C2F6S
371-78-8
Methane, thiobis[trifluoro-
F3CSCF3
170
C3H6S4
292-45-5
1,2,4,6-Tetrathiepane
S
S
S
S
170
C3H7I
75-30-9
Propane, 2-iodo-
i-PrI

170 C_3H_7I 107–08–4
Propane, 1–iodo–

PrI

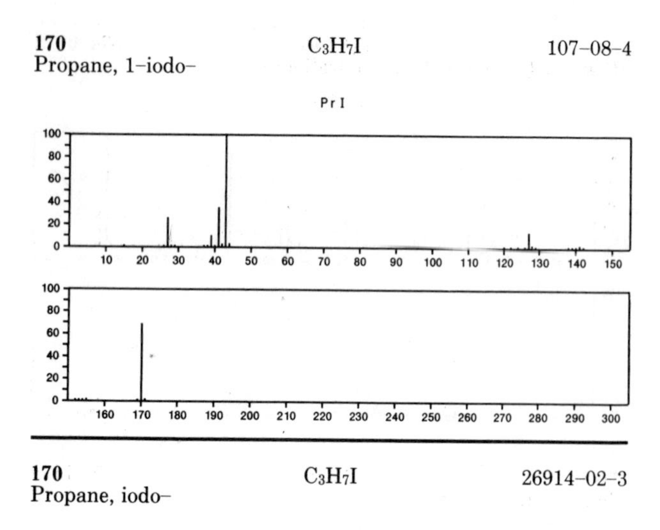

170 C_3H_7I 26914–02–3
Propane, iodo–

EtMe +I

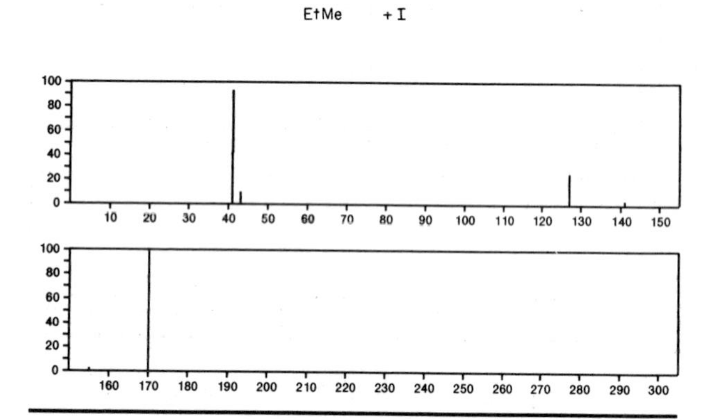

170 $C_4H_4Cl_2O_3$ 56247–52–0
2,3,5–Trioxabicyclo[2.1.0]pentane, 1,4–bis(chloromethyl)–

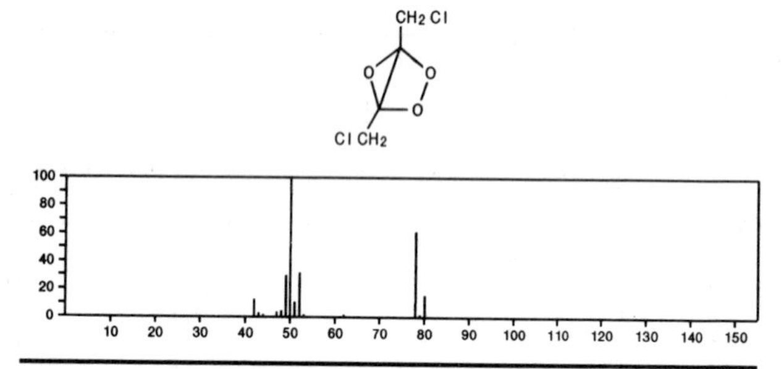

170 $C_4H_{12}FN_2O_2P$ 22692–27–9
Phosphorodiamidous fluoride, *N,N'*–dimethoxy–*N,N'*–dimethyl–

MeONMePFNMeOMe

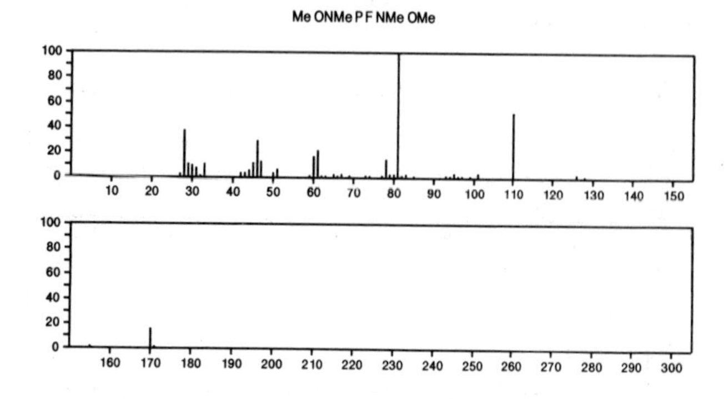

170 $C_4H_{12}FN_2PS$ 36267–53–5
Phosphonofluoridothioic hydrazide, *P*–ethyl–2,2–dimethyl–

S=PF(Et)NHNMe2

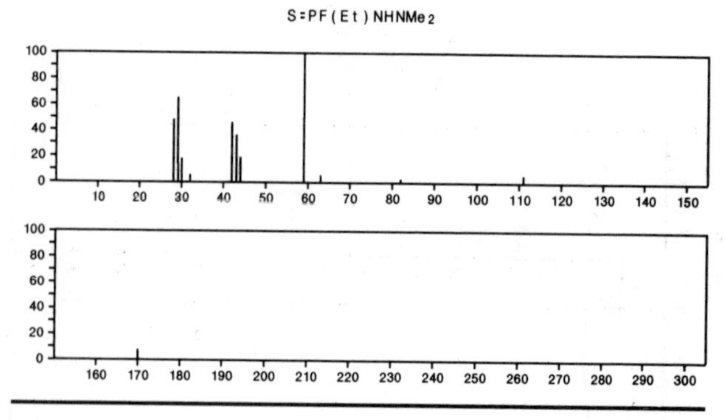

170 C_4NiO_4 13463–39–3
Nickel carbonyl (Ni(CO)4), (*T*–4)–

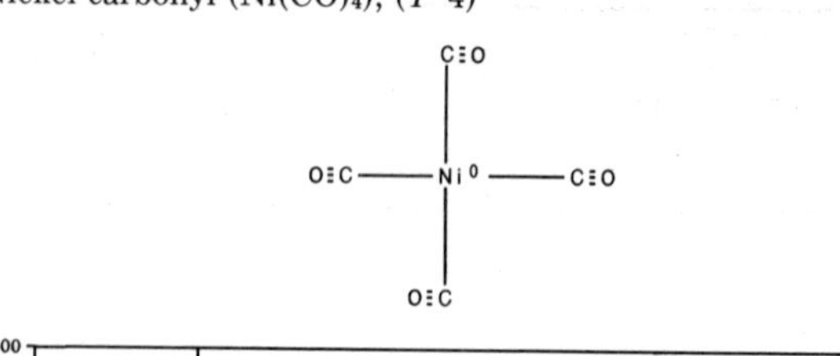

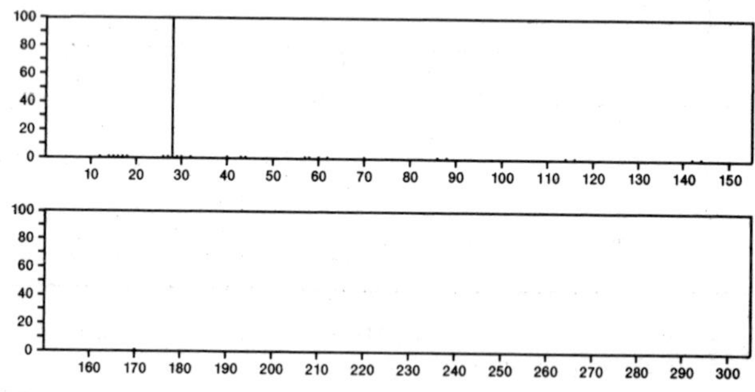

170 $C_5H_8Cl_2O_2$ 589–96–8
1–Propanol, 2,3–dichloro–, acetate

AcOCH2CHClCH2Cl

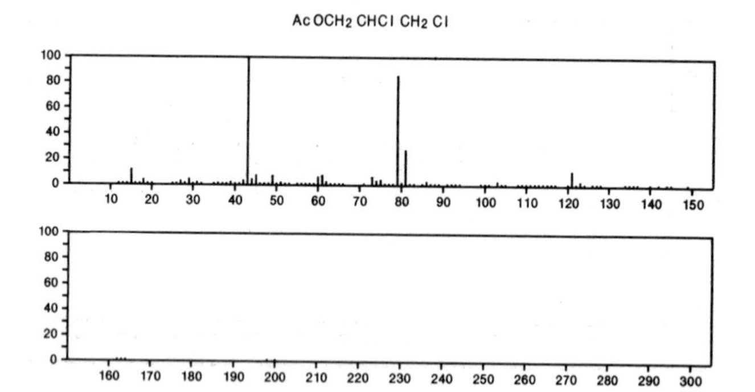

170 $C_5H_8Cl_2O_2$ 18545–44–3
Butyric acid, 2,2–dichloro–, methyl ester

MeOC(O)CCl2Et

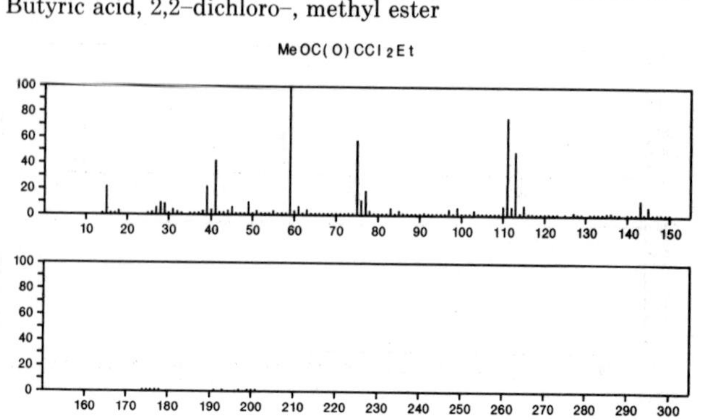

170 $C_5H_8Cl_2O_2$ 54460-97-8
Butanoic acid, 2,3-dichloro-, methyl ester

MeCHCl CHCl C(O) OMe

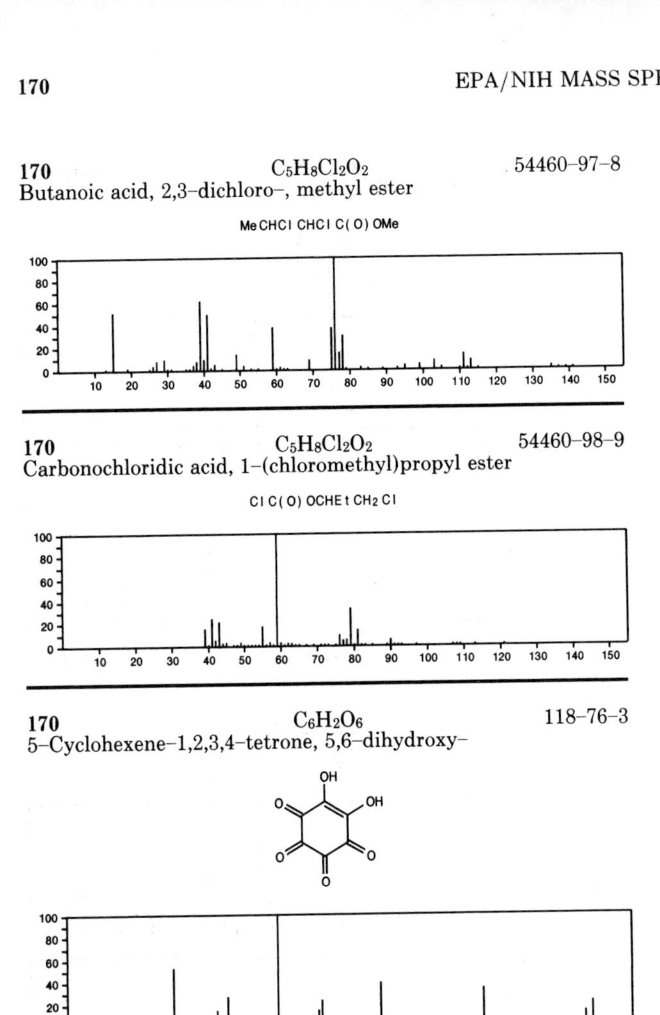

170 $C_5H_8Cl_2O_2$ 54460-98-9
Carbonochloridic acid, 1-(chloromethyl)propyl ester

Cl C(O) OCHEt CH2 Cl

170 $C_6H_2O_6$ 118-76-3
5-Cyclohexene-1,2,3,4-tetrone, 5,6-dihydroxy-

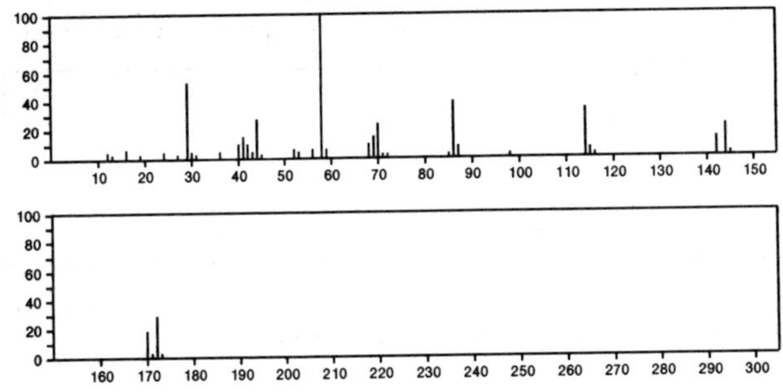

170 $C_6H_6N_2O_2S$ 1615-06-1
2,1,3-Benzothiadiazole, 1,3-dihydro-, 2,2-dioxide

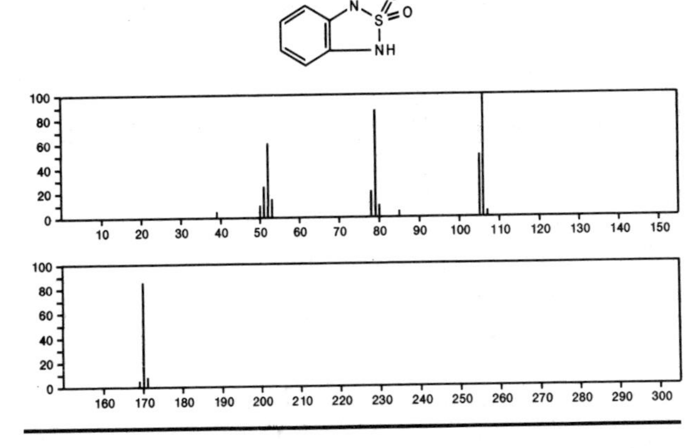

170 $C_6H_9F_3O_2$ 400-52-2
Acetic acid, trifluoro-, 1,1-dimethylethyl ester

t-Bu OC(O) CF3

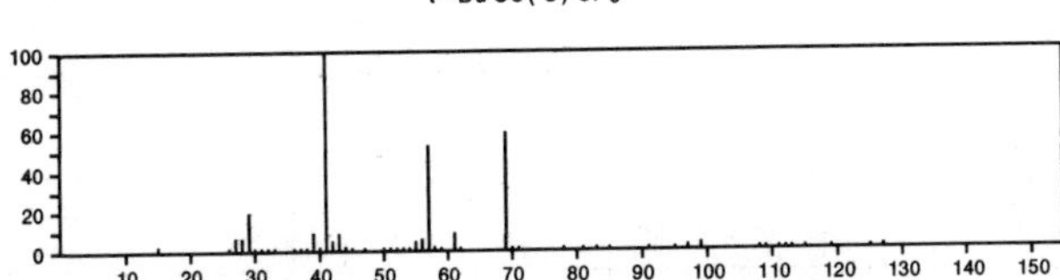

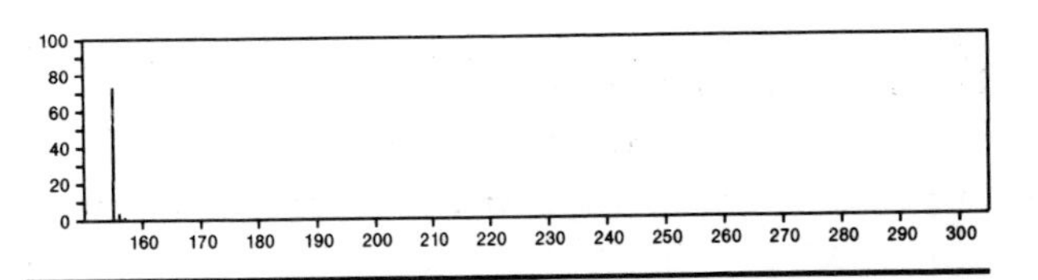

170 $C_6H_{10}N_4O_2$ 35975-36-1
1(2*H*)-Pyrazineacetamide, 5-amino-3,6-dihydro-3-oxo-

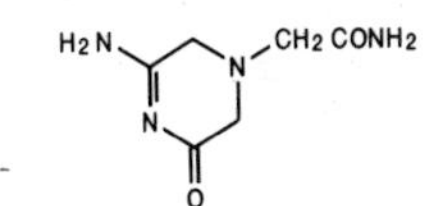

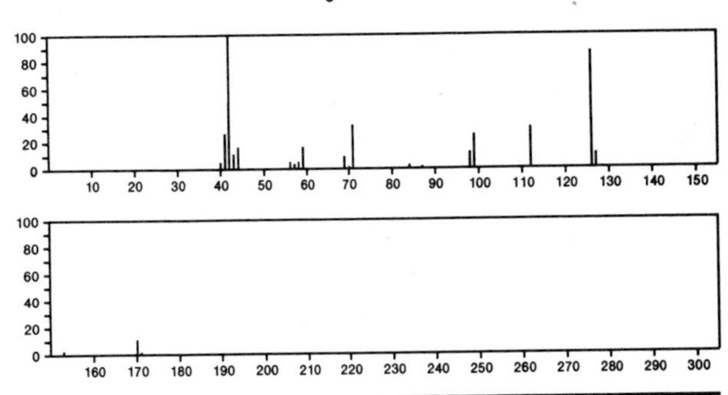

170 $C_6H_{12}Cl_2O$ 108-60-1
Propane, 2,2'-oxybis[1-chloro-

Cl CH2 CHMe OCHMe CH2 Cl

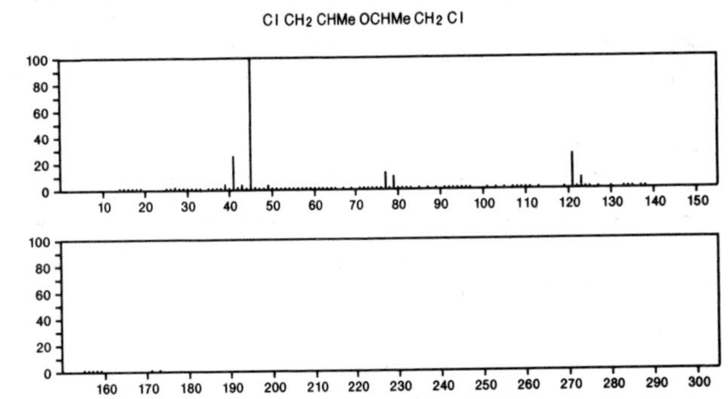

170 $C_6H_{12}Cl_2O$ 629-36-7
Propane, 1,1'-oxybis[3-chloro-

Cl (CH2)3 O(CH2)3 Cl

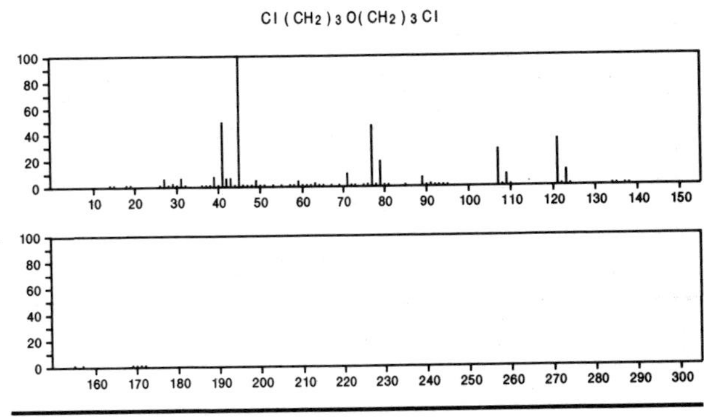

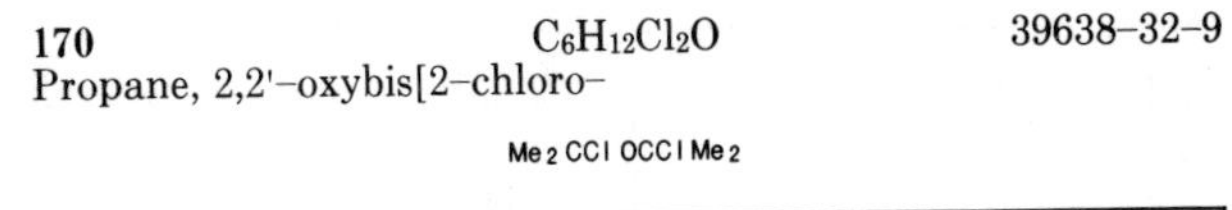

170 $C_6H_{12}Cl_2O$ 39638-32-9
Propane, 2,2'-oxybis[2-chloro-

Me2 CCl OCCl Me2

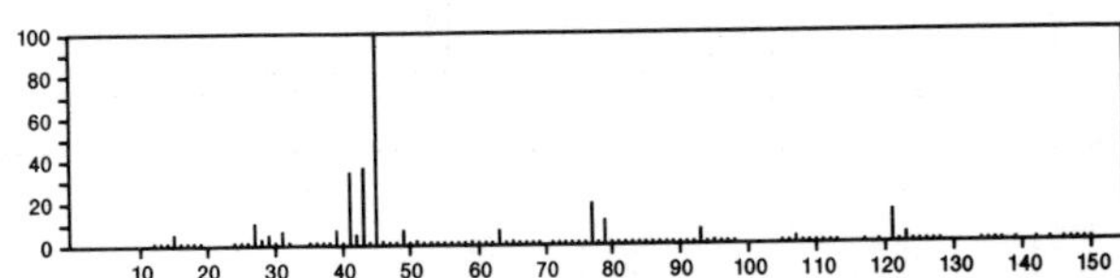

170 $C_6H_{12}Cl_2O$ 54460-96-7
Propane, 1,1'-oxybis[2-chloro-

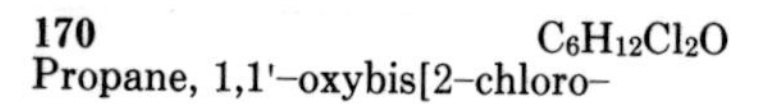

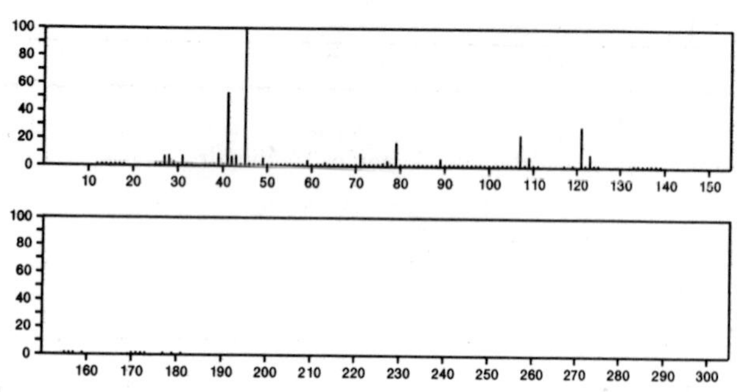

170 $C_7H_6O_5$ 54576-44-2
3,4-Furandicarboxylic acid, 2-methyl-

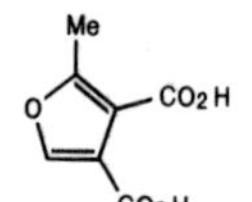

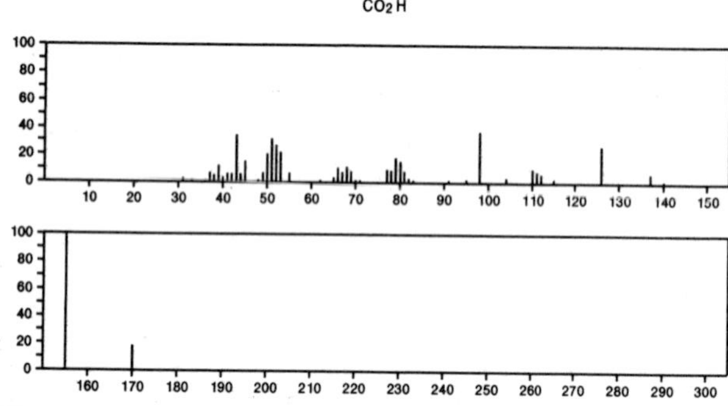

170 C_7H_7Br 95-46-5
Benzene, 1-bromo-2-methyl-

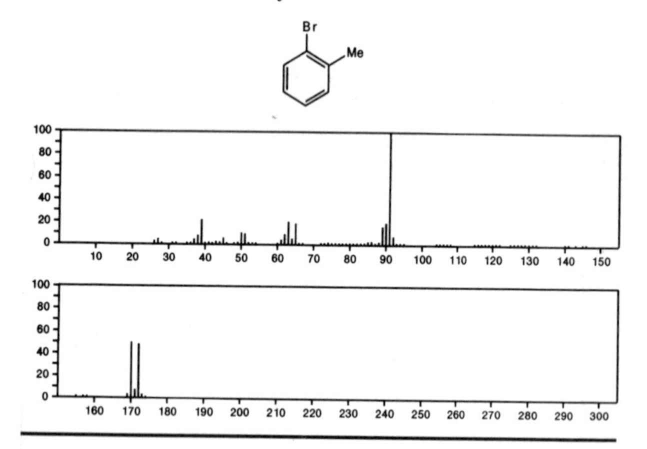

170 C_7H_7Br 100-39-0
Benzene, (bromomethyl)-

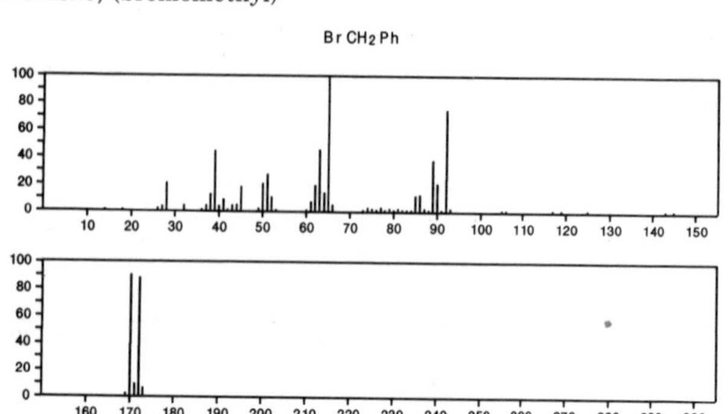

170 C_7H_7Br 106-38-7
Benzene, 1-bromo-4-methyl-

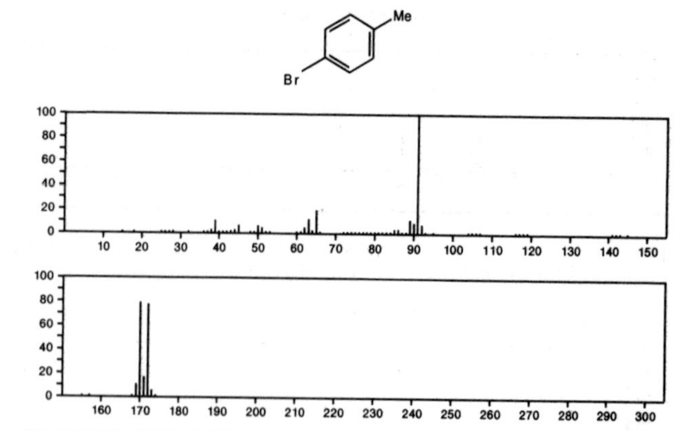

170 C_7H_7Br 591-17-3
Benzene, 1-bromo-3-methyl-

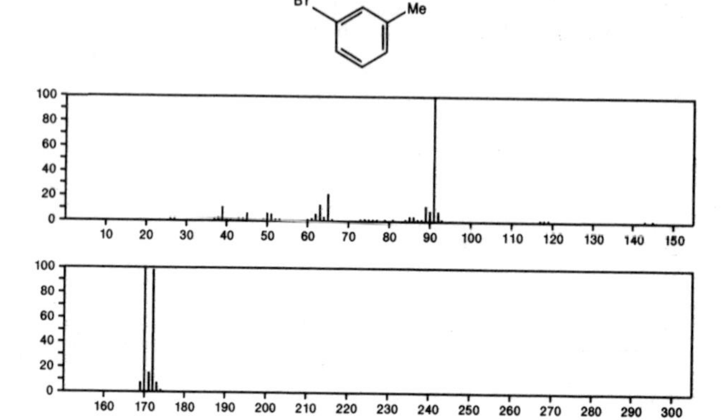

170 $C_7H_{10}N_2OS$ 13480-95-0
4(1*H*)-Pyrimidinone, 2-(ethylthio)-5-methyl-

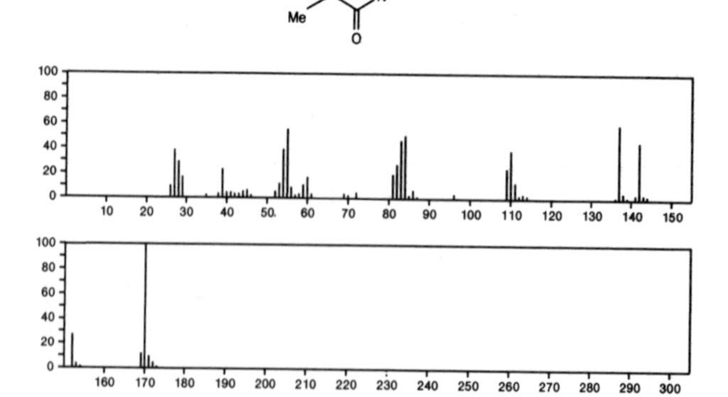

170 $C_7H_{10}N_2OS$ 24611-13-0
4(1*H*)-Pyrimidinethione, 5-ethoxy-2-methyl-

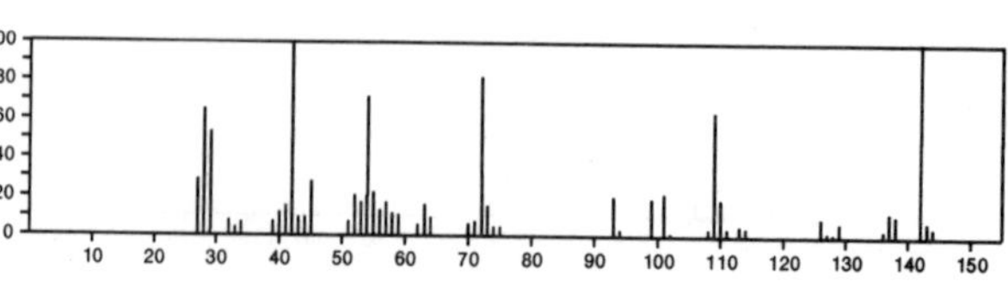

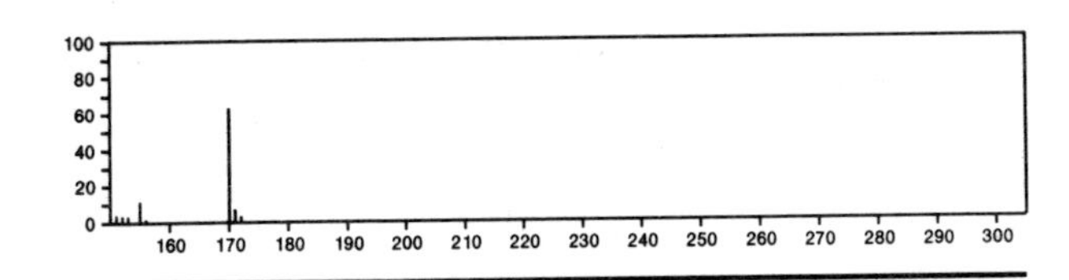

170 $C_7H_{10}N_2OS$ 54460-95-6

4(1*H*)-Pyrimidinone, 2-(propylthio)-

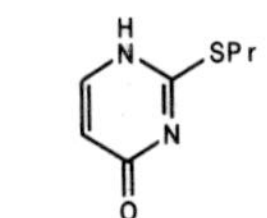

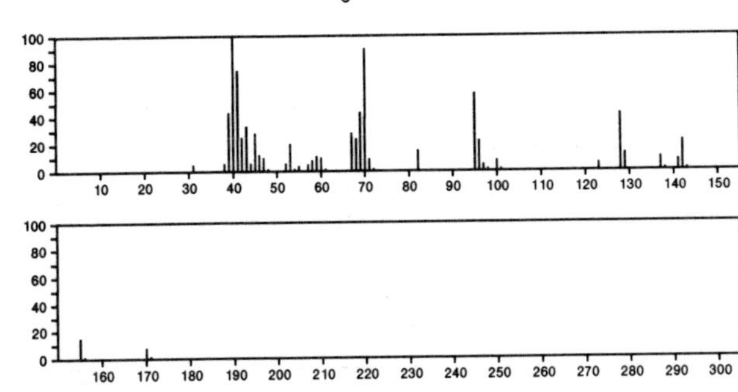

170 $C_7H_{10}N_2O_3$ 55557-02-3

3-Pyridinecarboxylic acid, 1,2,5,6-tetrahydro-1-nitroso-, methyl ester

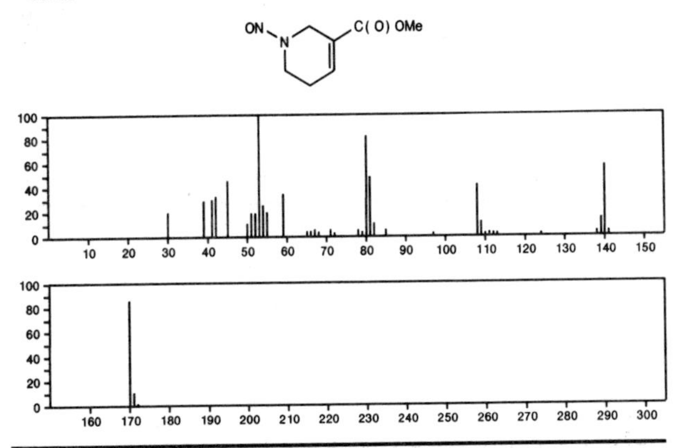

170 $C_8H_7ClO_2$ 610-96-8

Benzoic acid, 2-chloro-, methyl ester

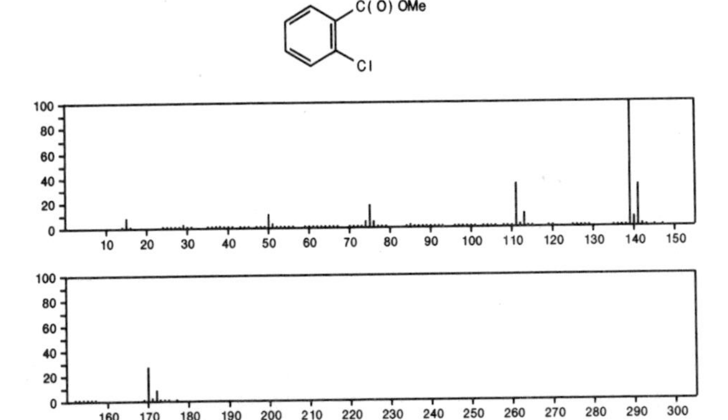

170 $C_8H_7ClO_2$ 701-99-5

Acetyl chloride, phenoxy-

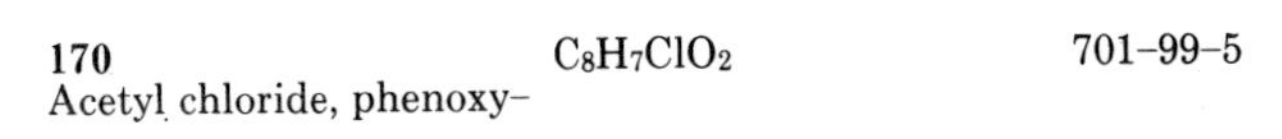

$PhOCH_2COCl$

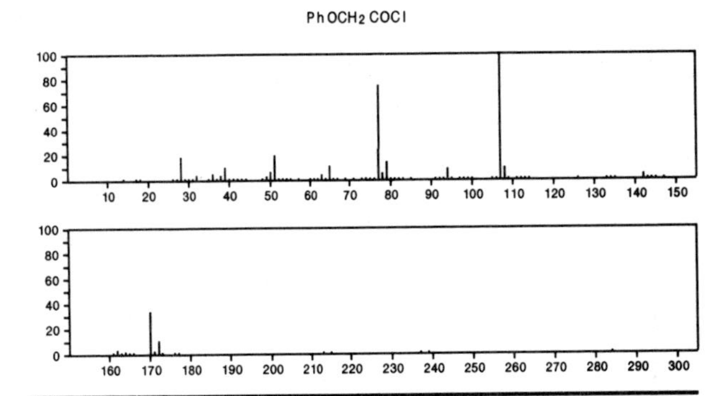

170 $C_8H_7ClO_2$ 876-27-7

Acetic acid, 4-chlorophenyl ester

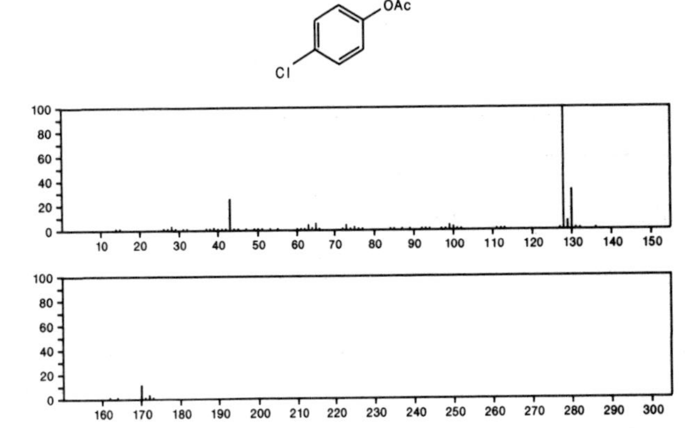

170 $C_8H_7ClO_2$ 1126-46-1

Benzoic acid, 4-chloro-, methyl ester

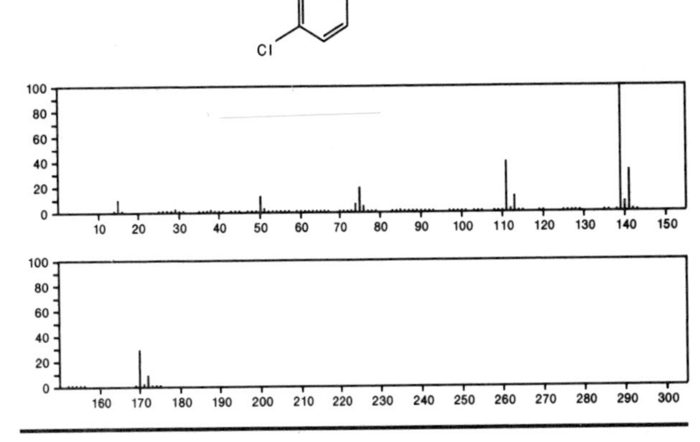

170 $C_8H_7ClO_2$ 1450-74-4

Ethanone, 1-(5-chloro-2-hydroxyphenyl)-

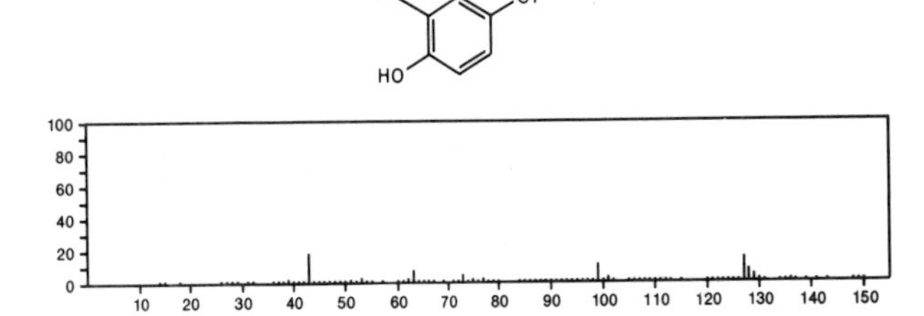

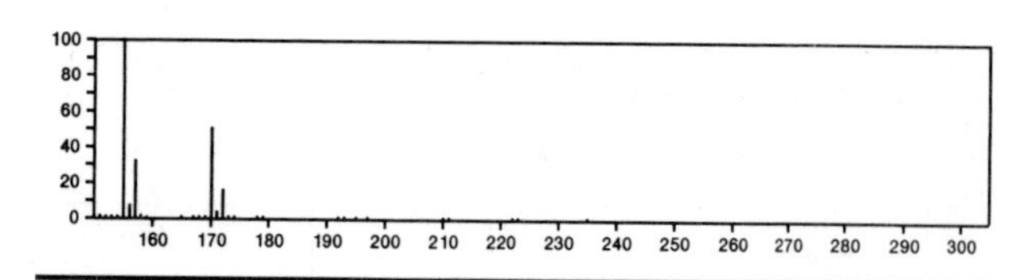

170 $C_8H_7ClO_2$ 1878-66-6
Benzeneacetic acid, 4-chloro-

CH_2CO_2H, Cl

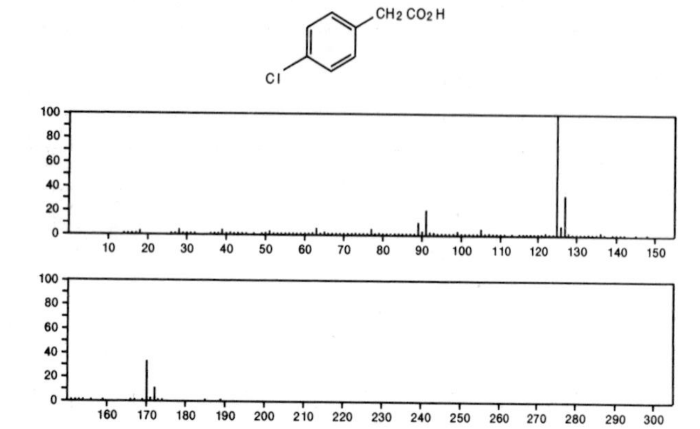

170 $C_8H_7ClO_2$ 2444-36-2
Benzeneacetic acid, 2-chloro-

CH_2CO_2H, Cl

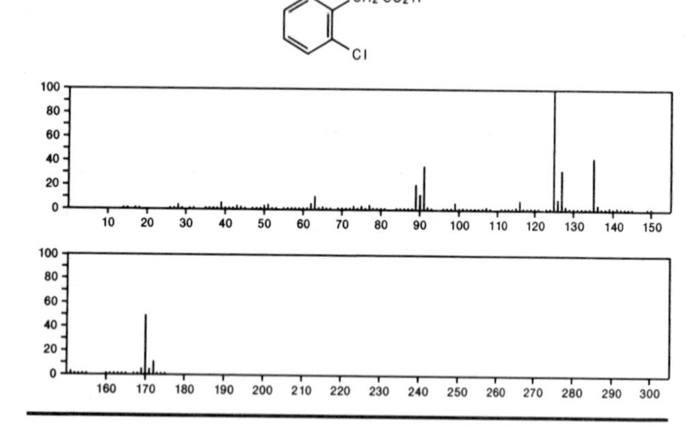

170 $C_8H_7ClO_2$ 2905-65-9
Benzoic acid, 3-chloro-, methyl ester

C(O)OMe, Cl

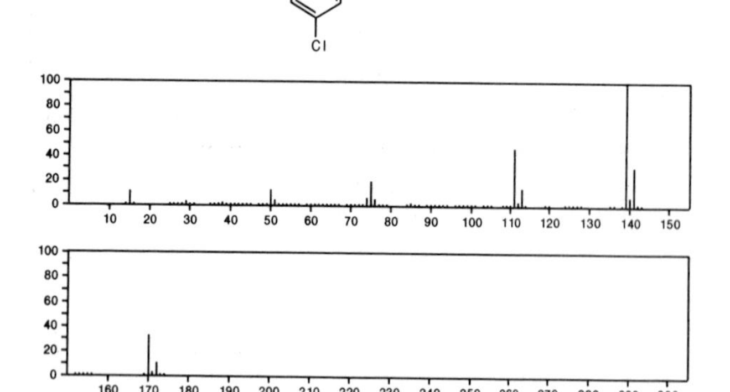

170 $C_8H_7ClO_2$ 23731-06-8
Benzaldehyde, 5-(chloromethyl)-2-hydroxy-

O=CH, CH_2Cl, HO

170 $C_8H_7FO_3$ 1847-98-9
Carbonic acid, *p*-fluorophenyl methyl ester

OC(O)OMe, F

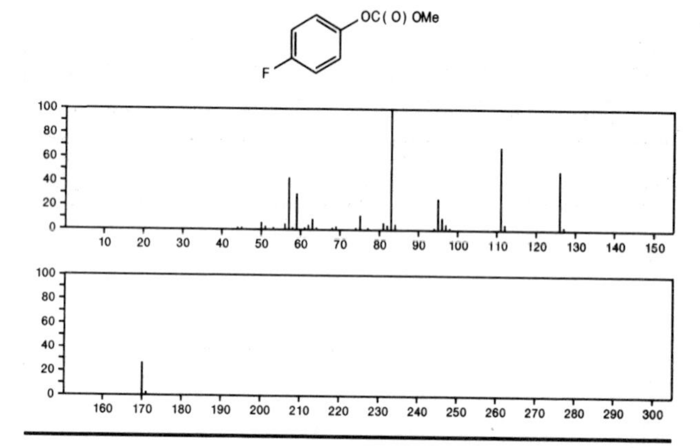

170 $C_8H_7FO_3$ 1847-99-0
Carbonic acid, *m*-fluorophenyl methyl ester

OC(O)OMe, F

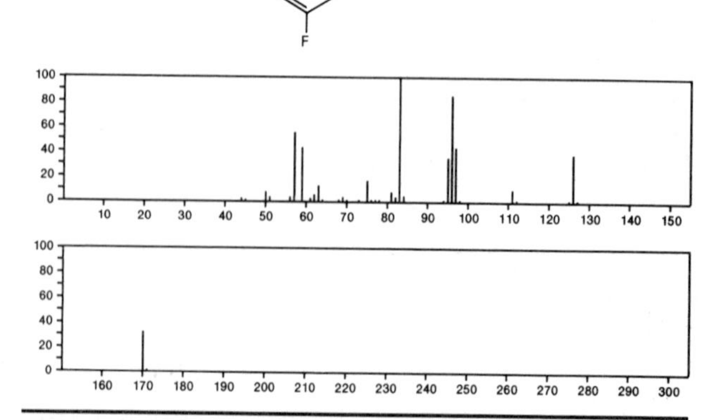

170 $C_8H_{10}O_2S$ 3112-90-1
Benzene, [(methylsulfonyl)methyl]-

$MeSO_2CH_2Ph$

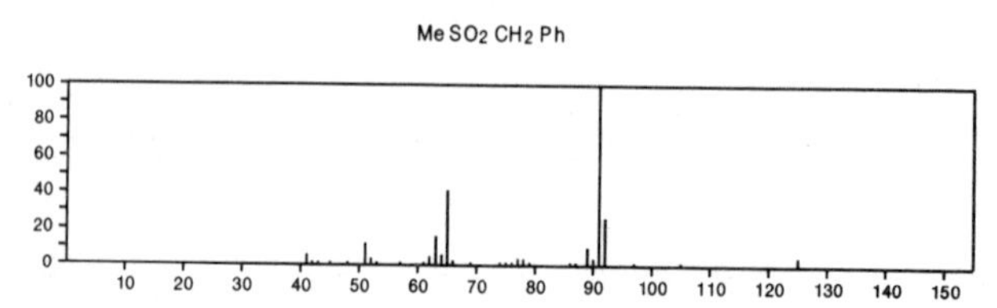

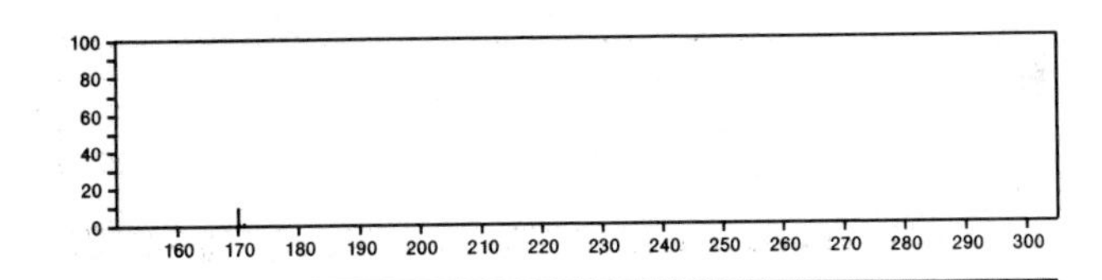

170 $C_8H_{10}O_2S$ 3185–99–7
Benzene, 1–methyl–4–(methylsulfonyl)–

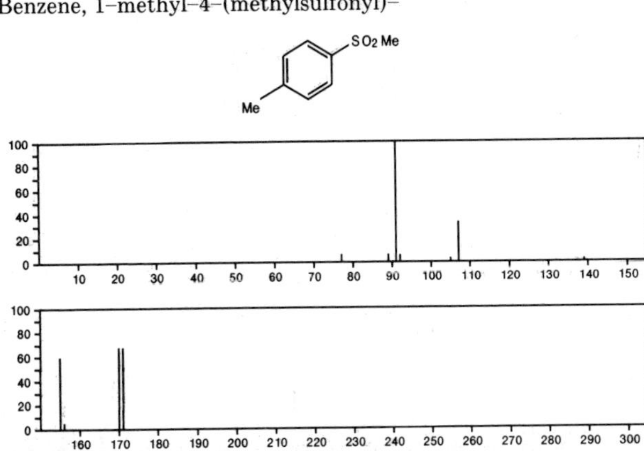

170 $C_8H_{10}O_2S$ 56701–04–3
2–Thiabicyclo[3.1.0]hex–3–ene–6–carboxylic acid, 3–methyl–, methyl ester

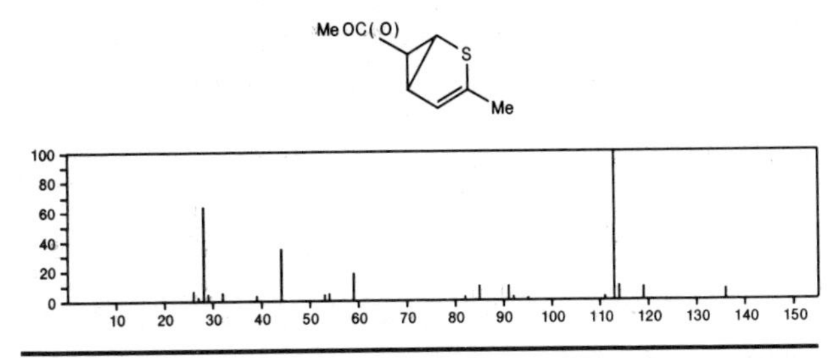

170 $C_8H_{10}O_4$ 635–08–5
1–Cyclohexene–1,2–dicarboxylic acid

170 $C_8H_{10}O_4$ 2305–26–2
4–Cyclohexene–1,2–dicarboxylic acid, *cis*–

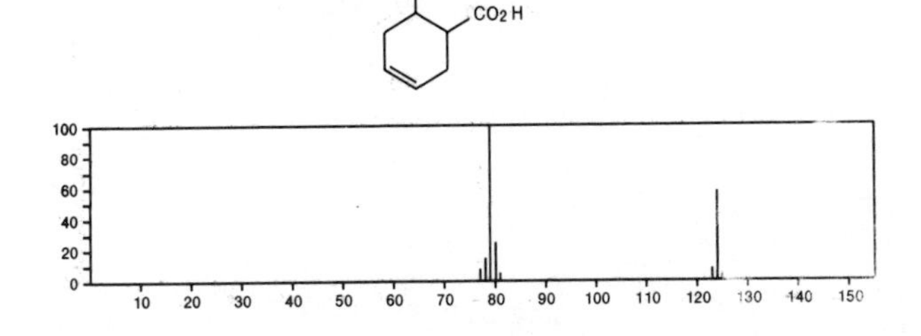

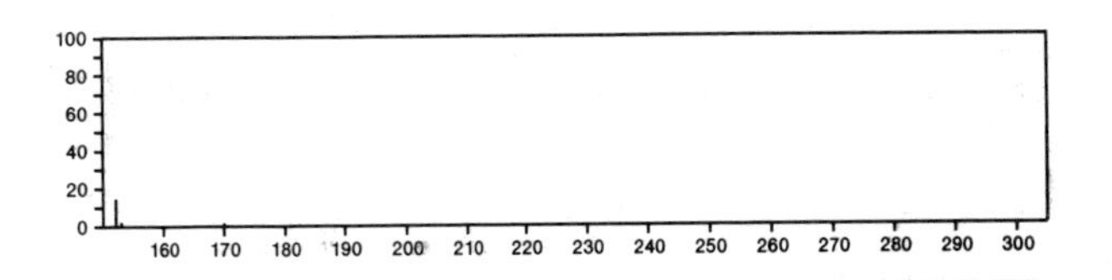

170 $C_8H_{10}O_4$ 23336–82–5
3–Furancarboxylic acid, 2–(methoxymethyl)–, methyl ester

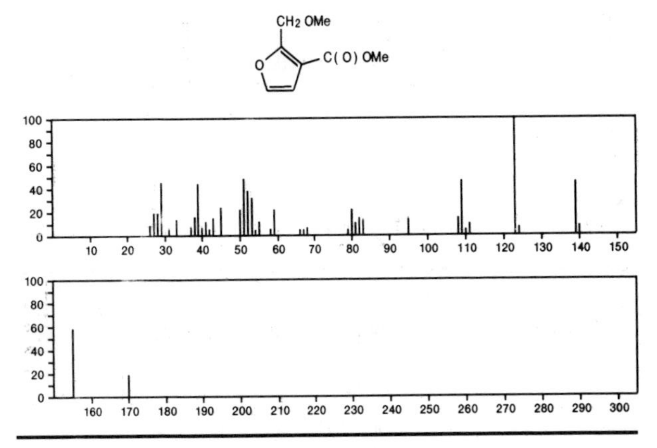

170 $C_8H_{10}O_4$ 28822–73–3
1,2–Benzenediol, 4–(1,2–dihydroxyethyl)–, (±)–

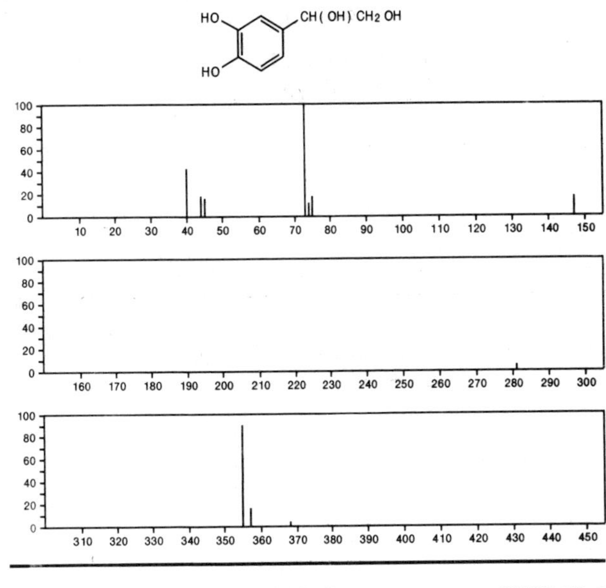

170 $C_8H_{10}O_4$ 38765–78–5
2–Cyclohexene–1,2–dicarboxylic acid

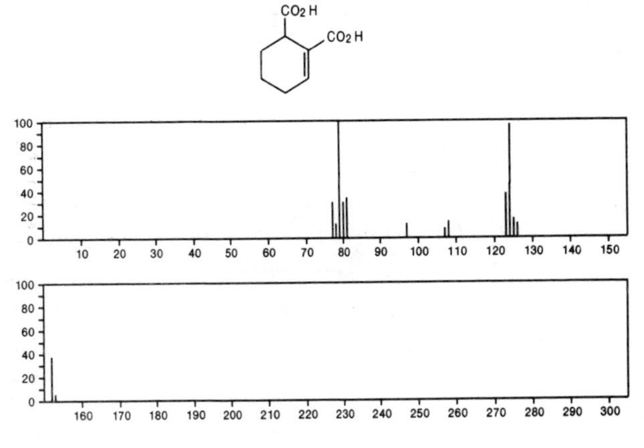

170 $C_8H_{10}O_4$ 52183-77-4
1,5-Cyclohexadiene-1-carboxylic acid, 3,4-dihydroxy-, methyl ester, *trans*-

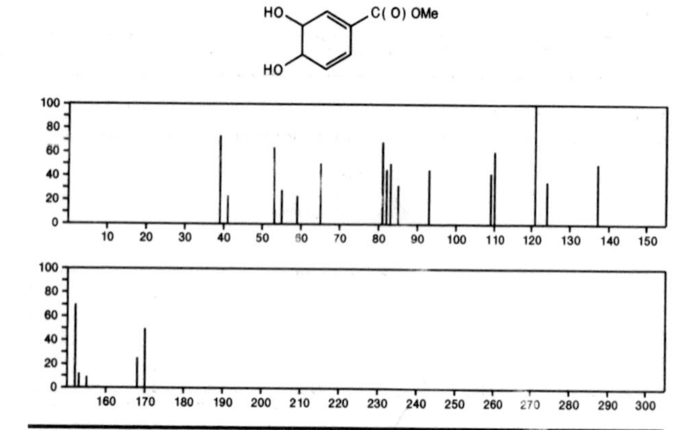

170 $C_8H_{10}O_4$ 55712-75-9
1,3-Cyclohexadiene-1-carboxylic acid, 5-hydroxy-6-methoxy-, *trans*-

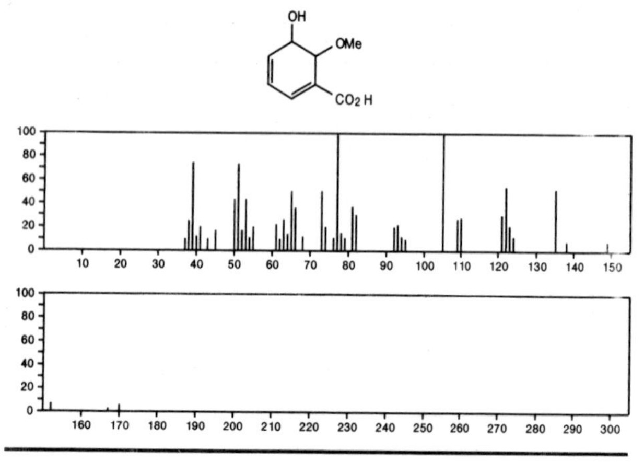

170 $C_8H_{10}S_2$ 15441-54-0
1*H*,3*H*-Thieno[3,4-*c*]thiophene, 4,6-dimethyl-

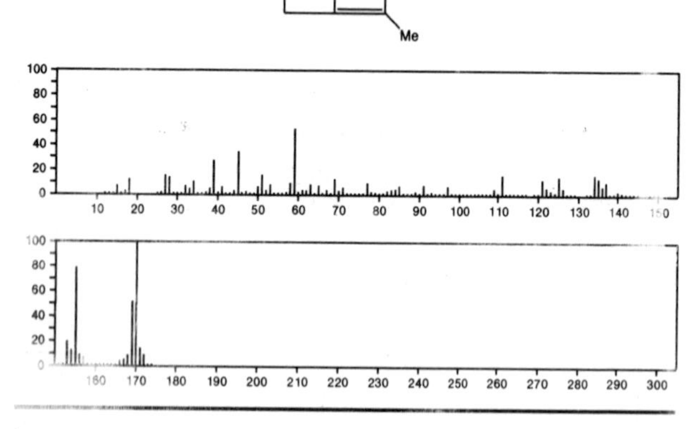

170 $C_8H_{14}N_2O_2$ 14702-42-2
Carbazic acid, 3-cyclohexylidene-, methyl ester

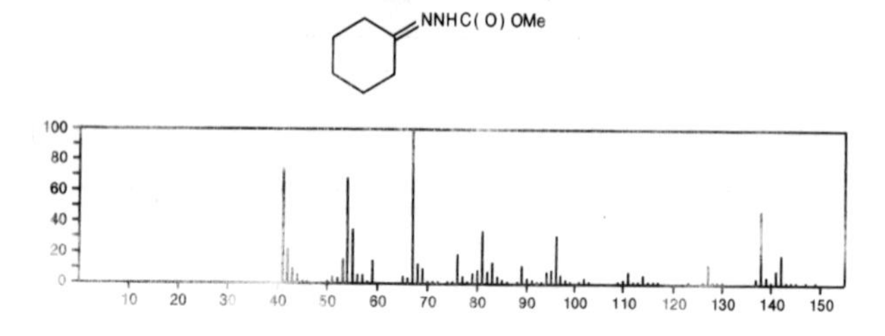

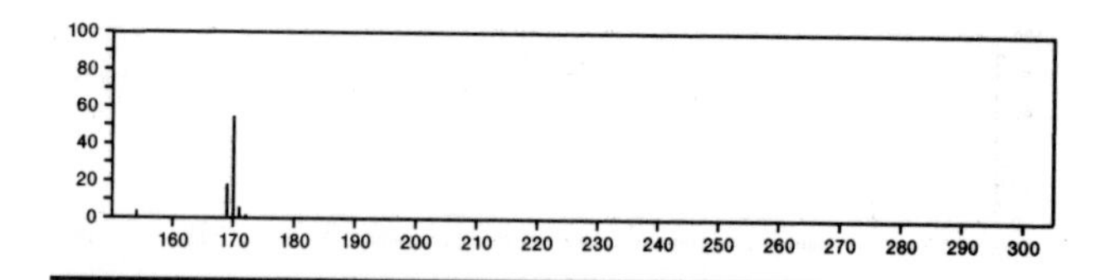

170 $C_8H_{14}N_2O_2$ 26537-49-5
Sydnone, 3-(3,3-dimethylbutyl)-

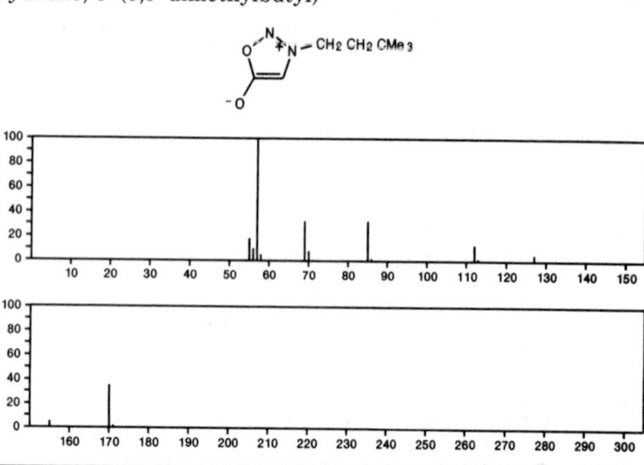

170 $C_8H_{14}N_2O_2$ 27886-67-5
2,4-Imidazolidinedione, 5-methyl-5-(2-methylpropyl)-

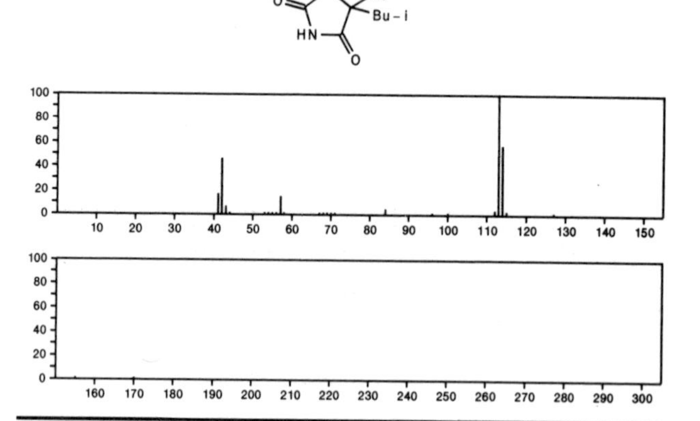

170 $C_8H_{14}N_2O_2$ 29924-68-3
1,4-Diazabicyclo[2.2.2]octane-2-carboxylic acid, methyl ester

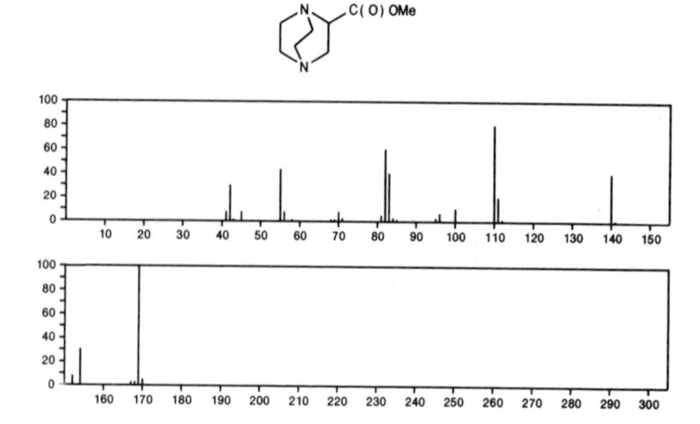

170 $C_8H_{14}N_2O_2$ 49582-44-7

2-Propenoic acid, 3-(1-aziridinyl)-3-(dimethylamino)-, methyl ester

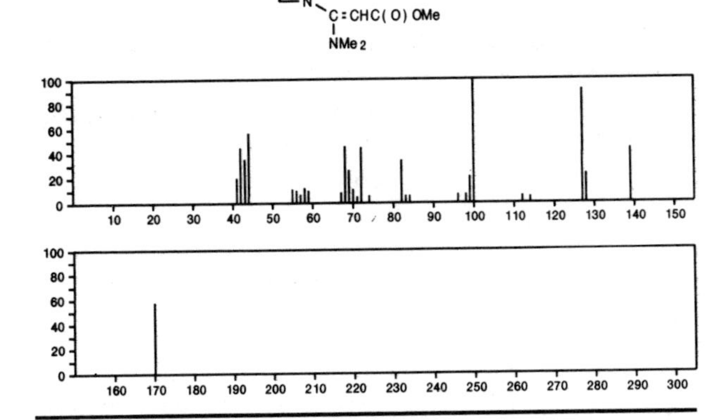

170 $C_8H_{14}N_2O_2$ 54460-94-5

1,2,3-Oxadiazolium, 3-(2,2-dimethylbutyl)-5-hydroxy-, hydroxide, inner salt

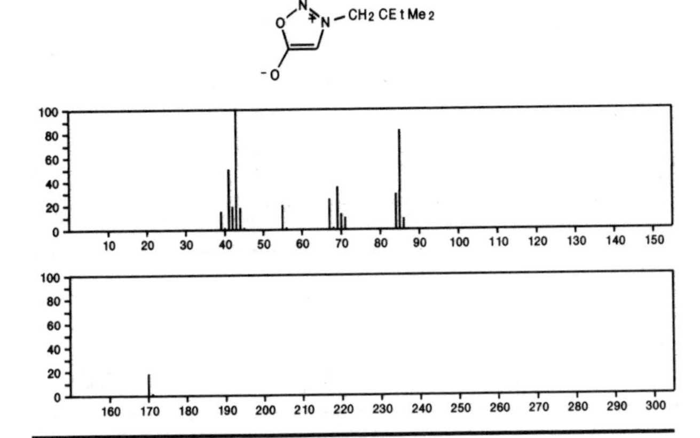

170 $C_8H_{14}N_2O_2$ 55401-89-3

Hydrazinecarboxylic acid, cyclopentylidene-, ethyl ester

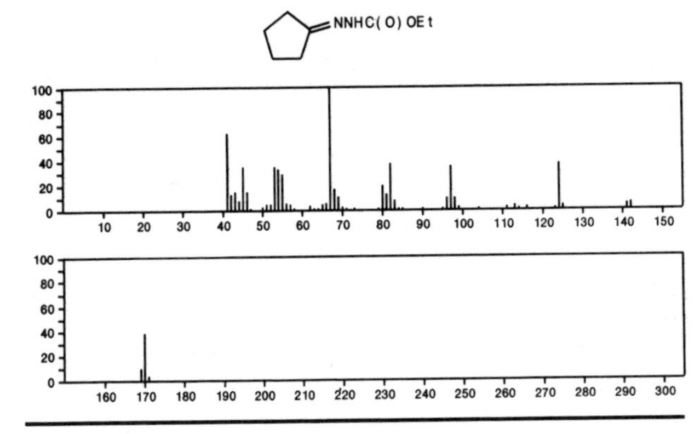

170 $C_8H_{15}BO_3$ 24372-02-9

Lactic acid, 2-methyl-, monoanhydride with 1-butaneboronic acid, cyclic ester

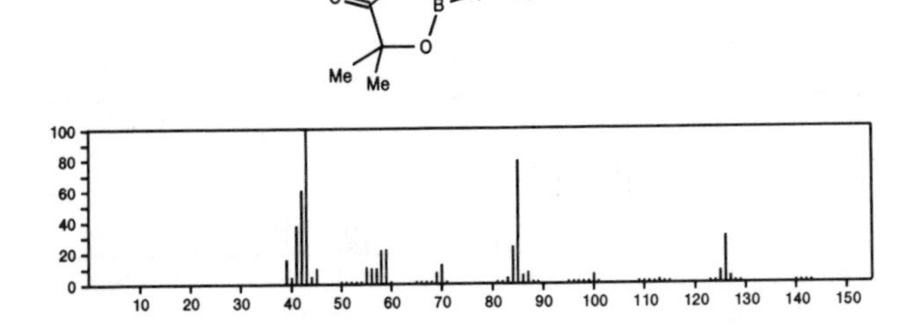

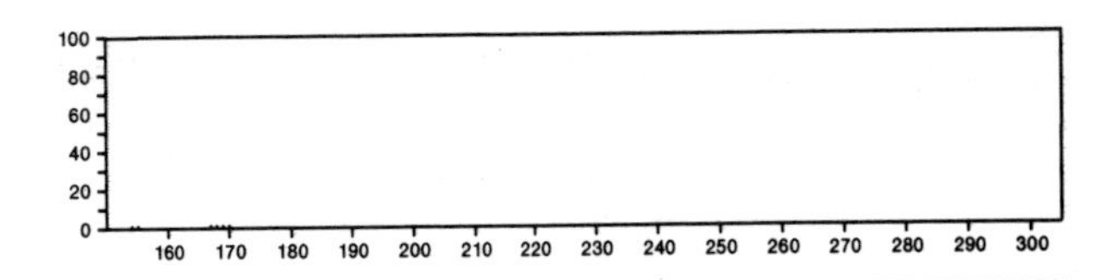

170 $C_8H_{15}BO_3$ 31767-20-1

Butyric acid, 3-hydroxy-, monoanhydride with 1-butaneboronic acid, cyclic ester

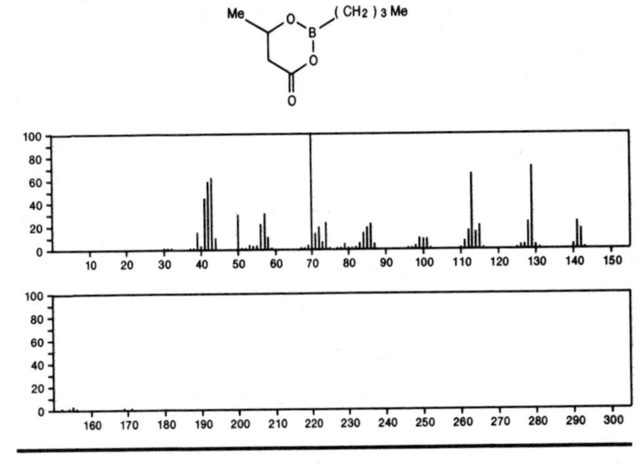

170 $C_8H_{15}BO_3$ 55089-04-8

1,3,2-Dioxaborolane, 2-(cyclohexyloxy)-

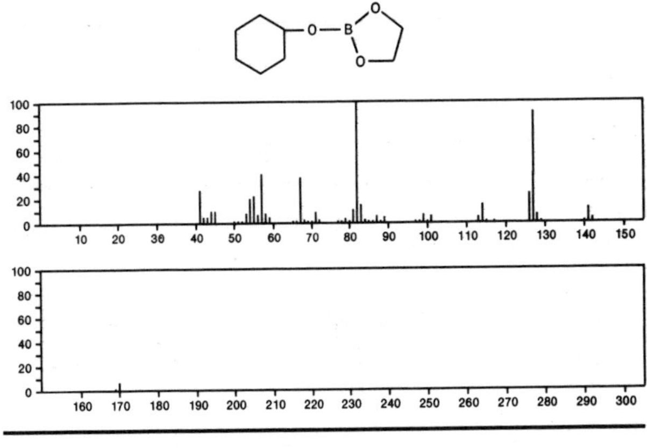

170 $C_9H_6N_4$ 37159-99-2

1,3,6,9b-Tetraazaphenalene

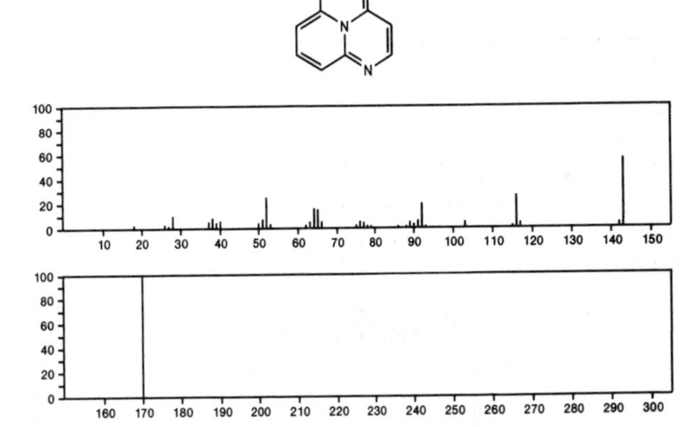

170 $C_9H_{11}ClO$ 21120-79-6
Phenetole, β-chloro-*o*-methyl-

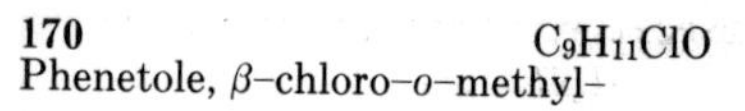

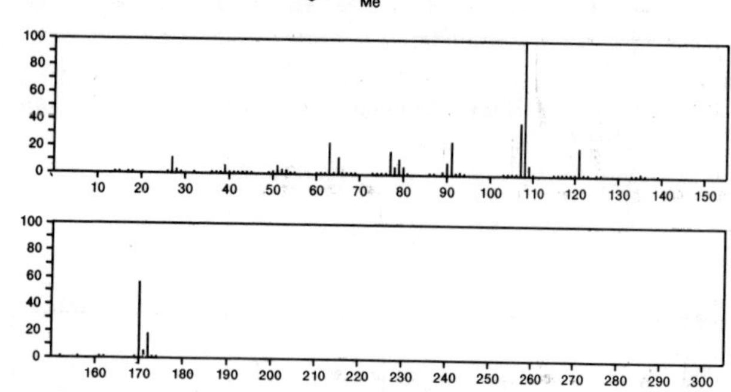

170 $C_9H_{11}ClO$ 35144-25-3
Benzene, 1-(2-chloroethyl)-2-methoxy-

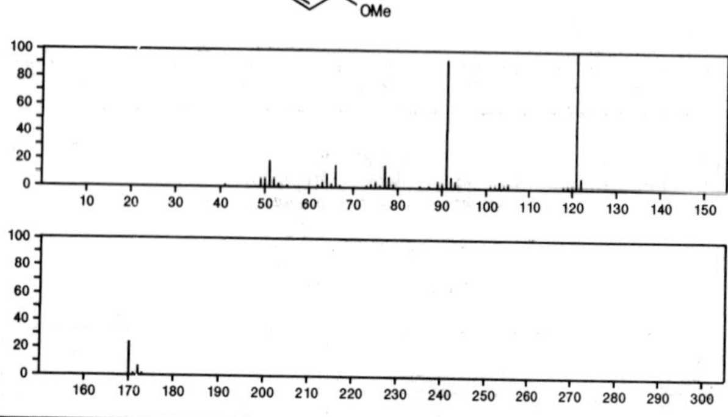

170 $C_9H_{11}ClO$ 54461-05-1
Phenol, 4-chloro-2-(1-methylethyl)-

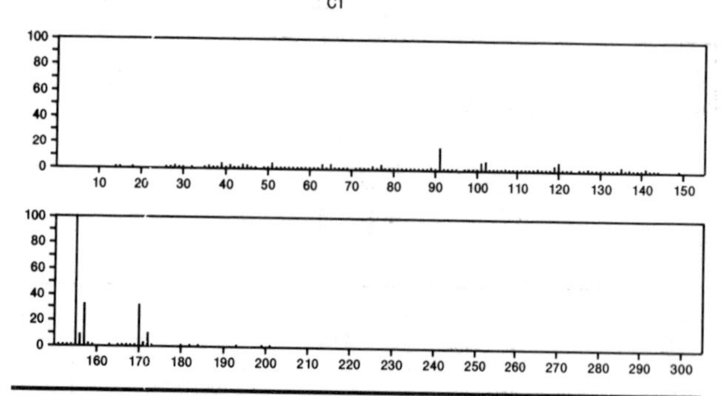

170 $C_9H_{14}O_3$ 610-89-9
4-Pentenoic acid, 2-acetyl-, ethyl ester

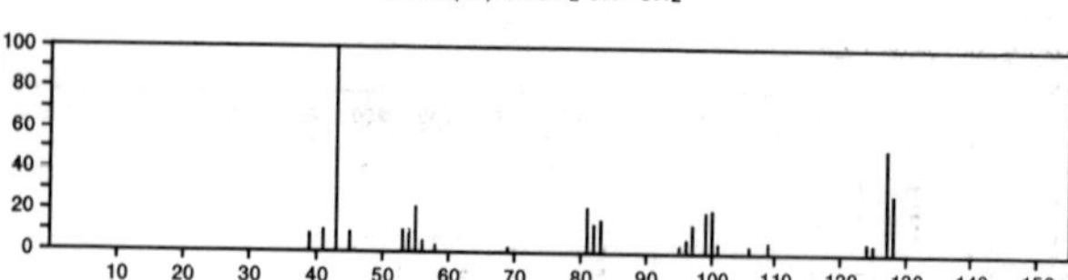

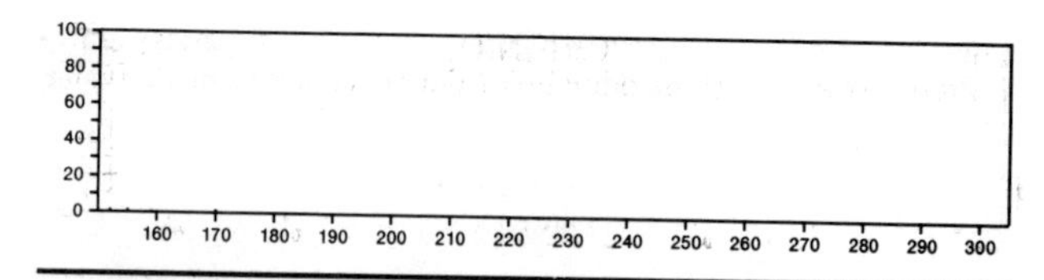

170 $C_9H_{14}O_3$ 10407-36-0
Cyclopentanepropanoic acid, 2-oxo-, methyl ester

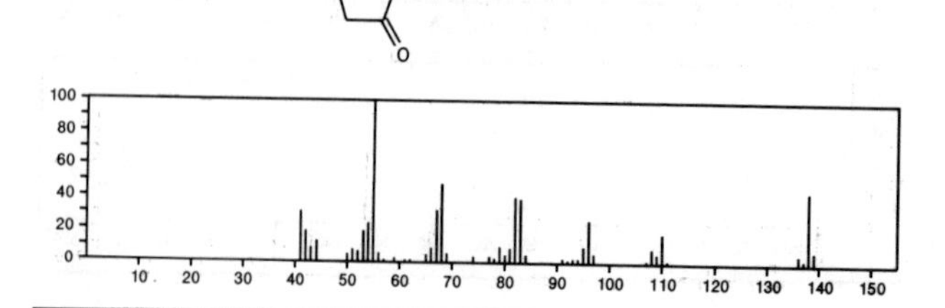

170 $C_9H_{14}O_3$ 18927-21-4
1,2-Butanediol, 1-(2-furyl)-2-methyl-

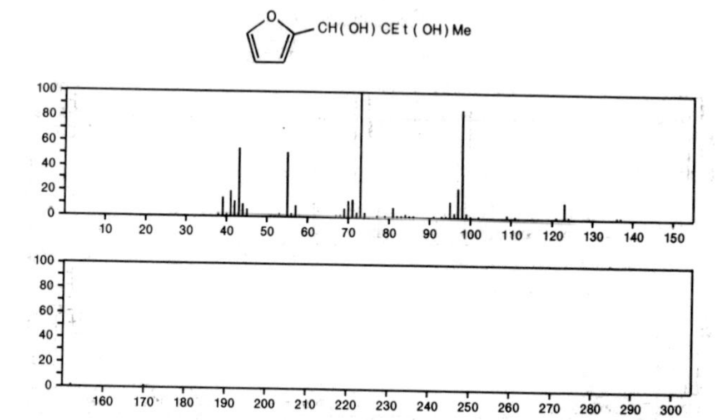

170 $C_9H_{14}O_3$ 20962-71-4
4-Pentenoic acid, 2-acetyl-4-methyl-, methyl ester

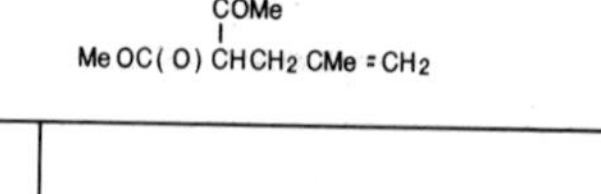

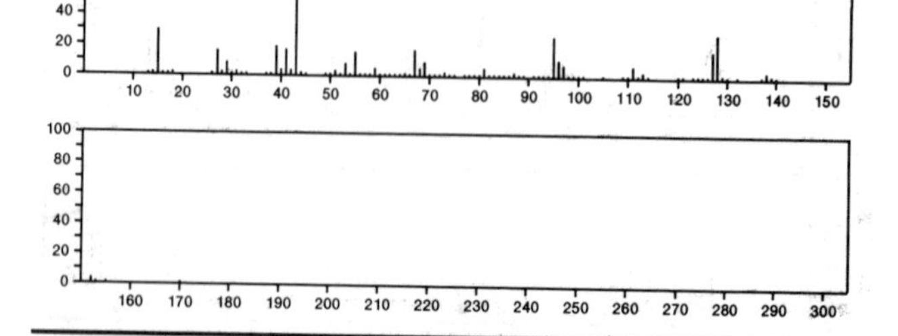

170 $C_9H_{14}O_3$ 24588-61-2
2*H*-Pyran-2-carboxylic acid, 3,6-dihydro-4,5-dimethyl-, methyl ester

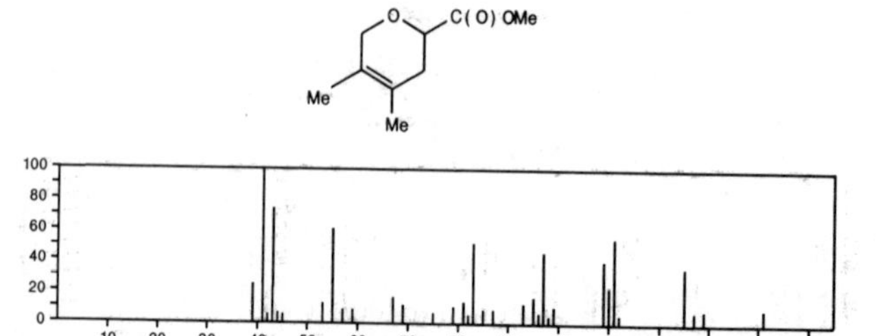

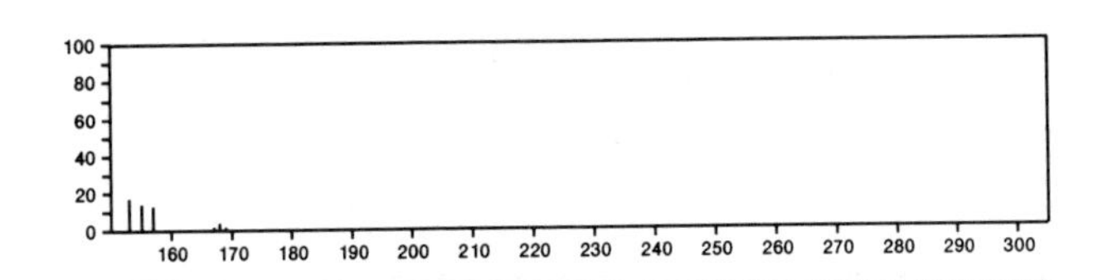

170 $C_9H_{14}O_3$ 38653-27-9
2-Propanol, 1-[(1-methyl-2-propynyl)oxy]-, acetate

HC≡CCHMe OCH2 CHMe OAc

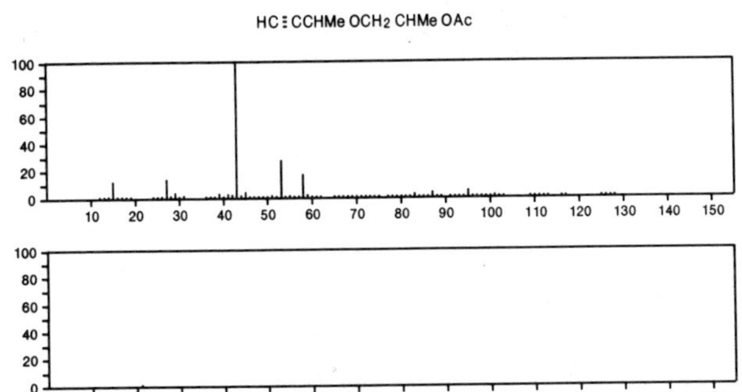

170 $C_9H_{14}O_3$ 38858-64-9
2*H*-Pyran-4-carboxylic acid, 3,4-dihydro-5-methyl-, ethyl ester

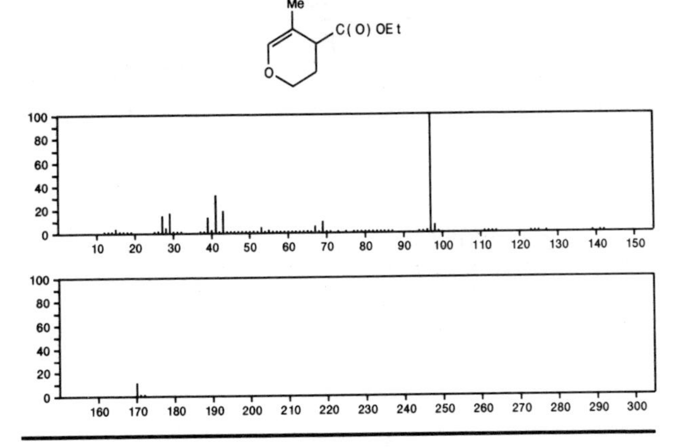

170 $C_9H_{14}O_3$ 38858-66-1
4-Oxepincarboxylic acid, 2,3,6,7-tetrahydro-, ethyl ester

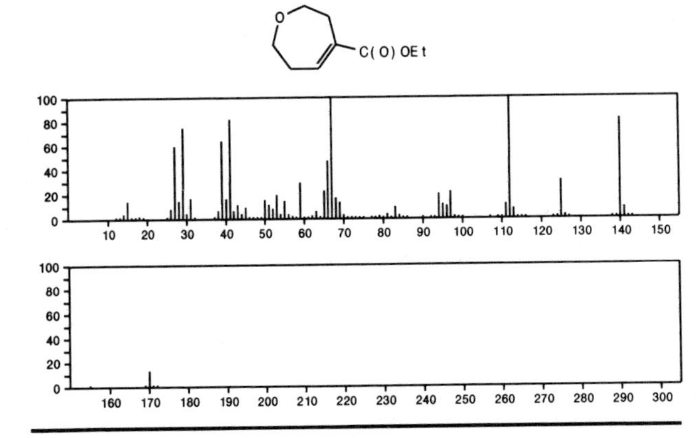

170 $C_9H_{14}O_3$ 55402-04-5
3-Pentyn-2-one, 5,5-diethoxy-

(EtO)2CHC≡CCOMe

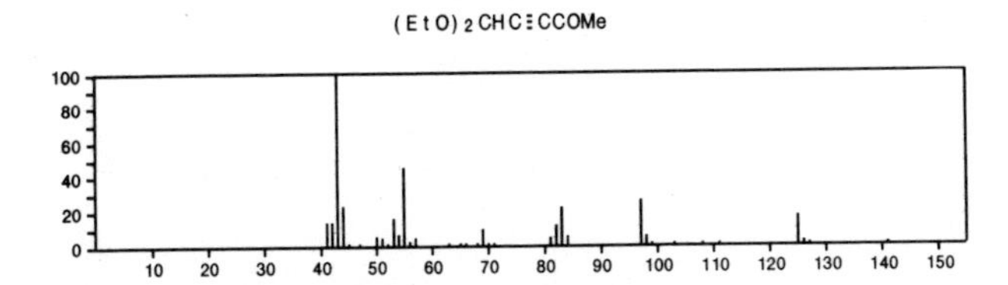

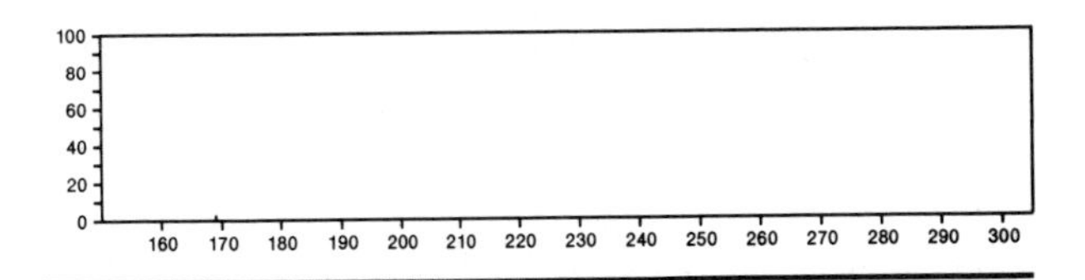

170 $C_9H_{14}O_3$ 56666-76-3
2,5-Furandione, 3-(1,1-dimethylpropyl)dihydro-

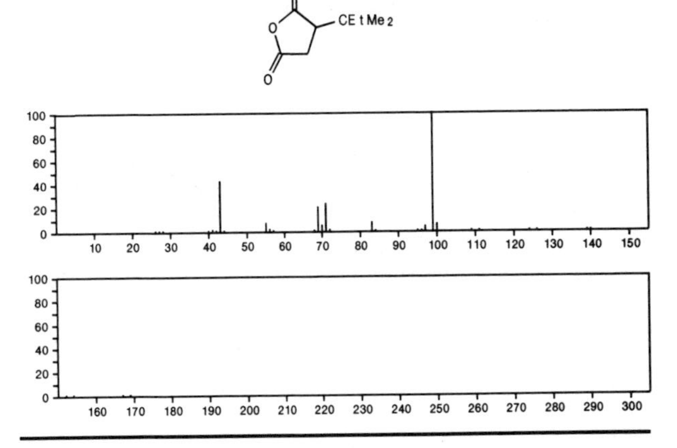

170 $C_9H_{16}NO_2$ 2896-70-0
1-Piperidinyloxy, 2,2,6,6-tetramethyl-4-oxo-

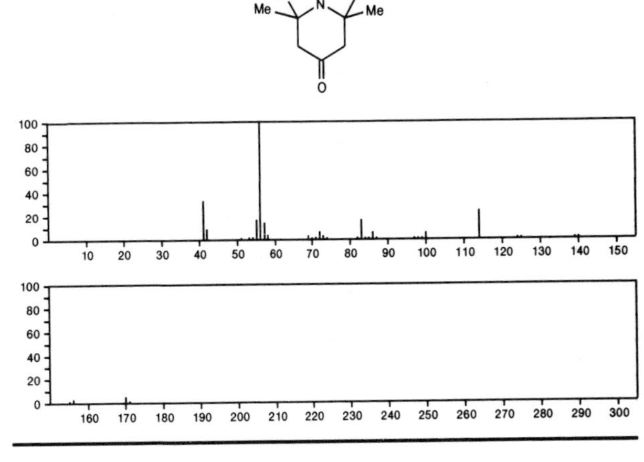

170 $C_9H_{18}N_2O$ 6130-93-4
Piperidine, 2,2,6,6-tetramethyl-1-nitroso-

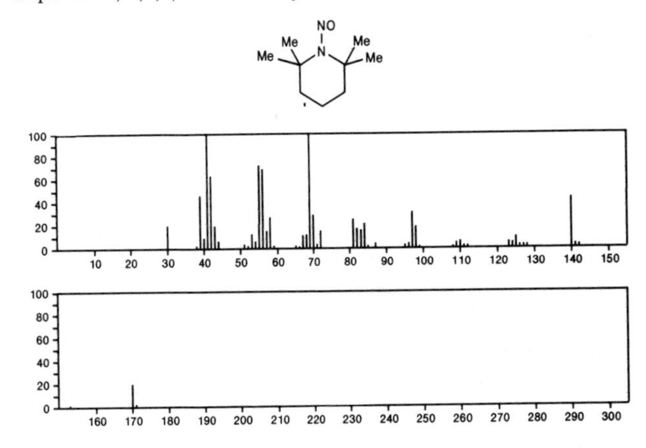

170 $C_9H_{18}N_2O$ 19656-74-7
Diaziridinone, bis(1,1-dimethylethyl)-

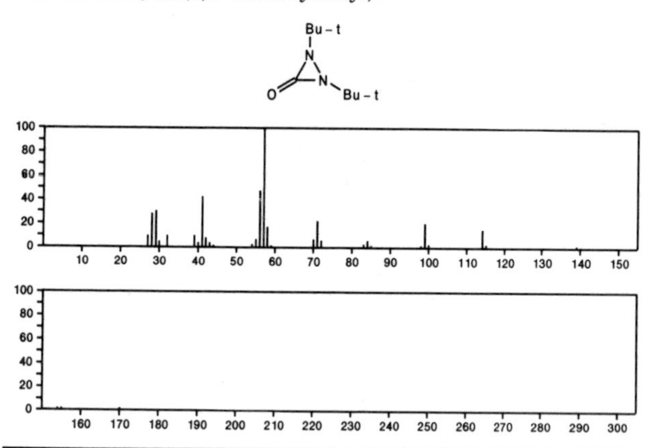

170 $C_9H_{18}N_2O$ 49582-55-0
3-Buten-2-one, 4-(dimethylamino)-4-[(1-methylethyl)amino]-

Me2NC(NHPr-i)=CHCOMe

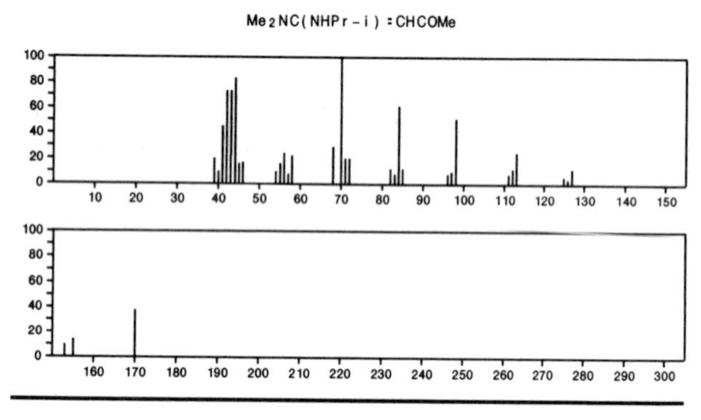

170 $C_9H_{18}N_2O$ 49582-61-8
2-Propenal, 2-(diethylamino)-3-(dimethylamino)-

Me2NCH=C(CHO)NEt2

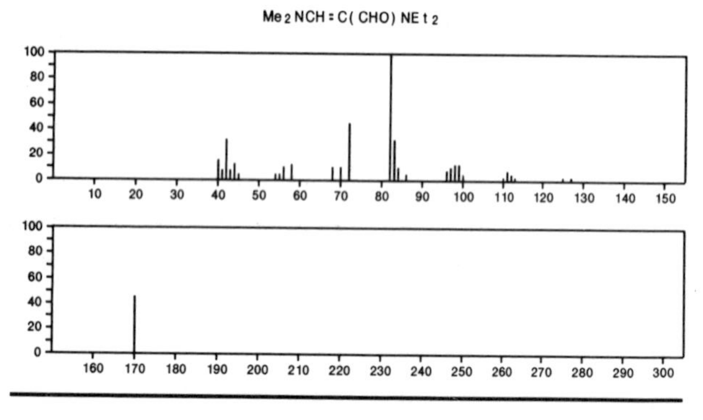

170 $C_9H_{18}N_2O$ 49582-65-2
3-Buten-2-one, 4-(dimethylamino)-3-[(1-methylethyl)amino]-

Me2NCH=C(NHPr-i)COMe

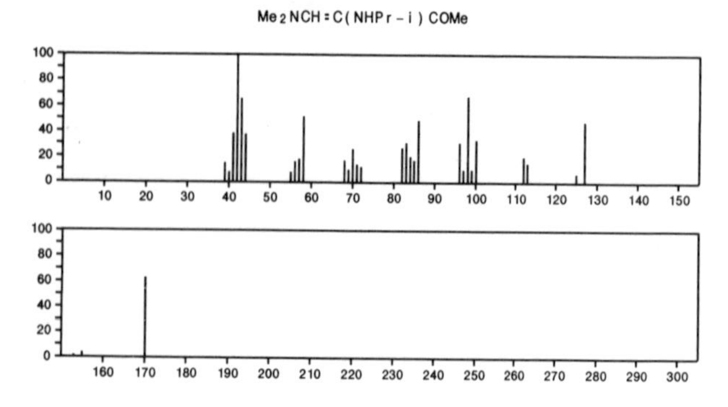

170 $C_9H_{18}OSi$ 6651-36-1
Silane, (1-cyclohexen-1-yloxy)trimethyl-

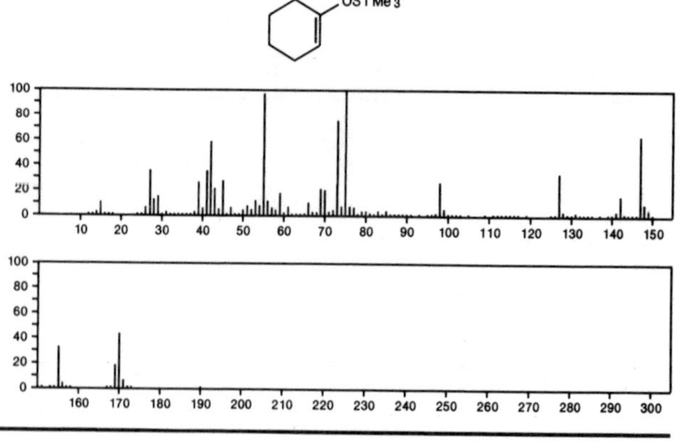

170 $C_9H_{18}OSi$ 54725-71-2
Silane, (2-cyclohexen-1-yloxy)trimethyl-

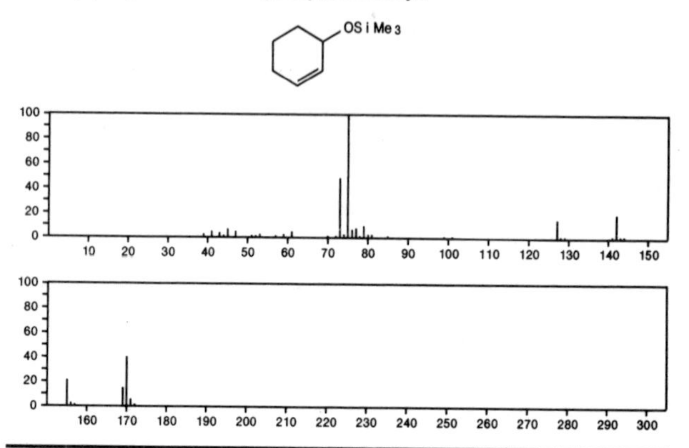

170 $C_{10}H_6N_2O$ 6969-11-5
1-Isoquinolinecarbonitrile, 2-oxide

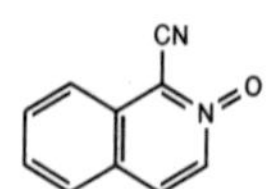

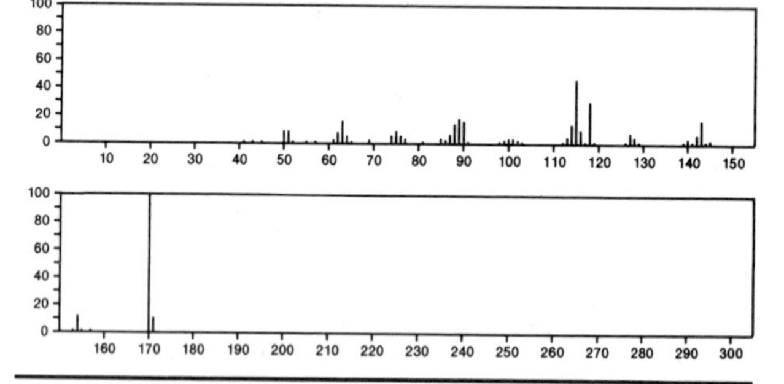

170 $C_{10}H_6N_2O$ 18457-80-2
3,1-Benzoxazepine-2-carbonitrile

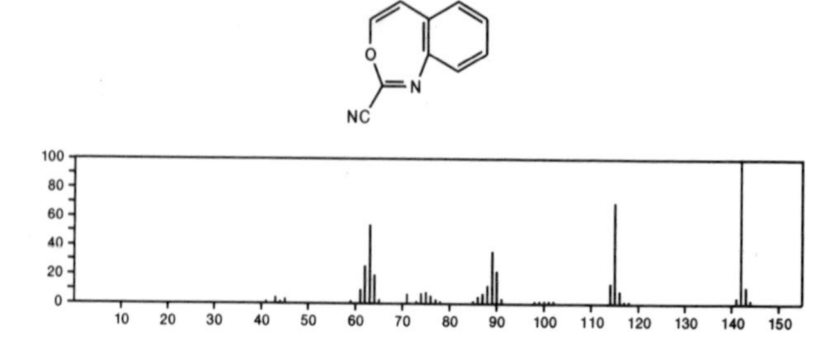

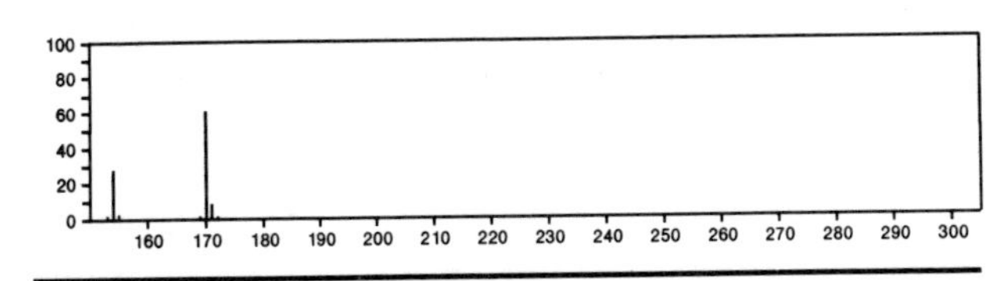

170 $C_{10}H_{18}O_2$ 111-79-5

2-Nonenoic acid, methyl ester

Me (CH2) 5 CH = CHC(O) OMe

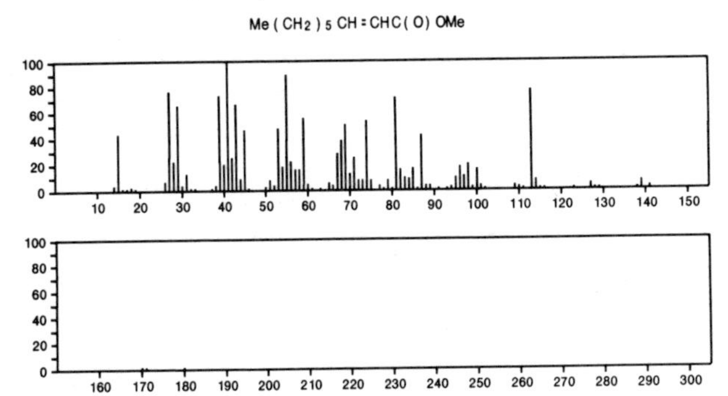

170 $C_{10}H_{18}O_2$ 512-77-6

Cyclopentanecarboxylic acid, 1-methyl-3-(1-methylethyl)-, *cis*-

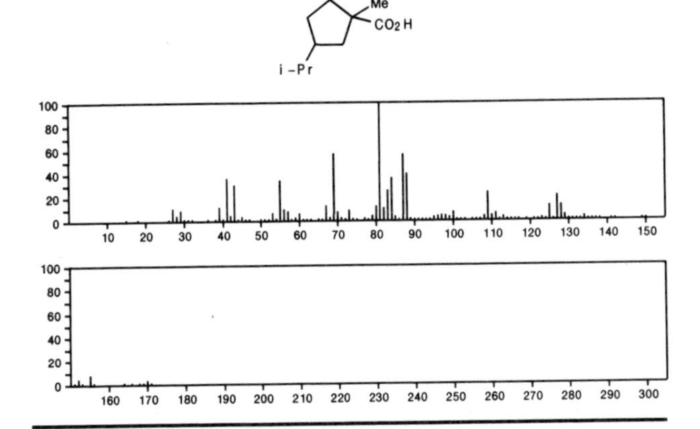

170 $C_{10}H_{18}O_2$ 706-14-9

2(3*H*)-Furanone, 5-hexyldihydro-

170 $C_{10}H_{18}O_2$ 1127-51-1

1,4-Naphthalenediol, decahydro-, (1α,4α,4aα,8aα)-

170 $C_{10}H_{18}O_2$ 1127-52-2

1,4-Naphthalenediol, decahydro-, (1α,4α,4aα,8aβ)-

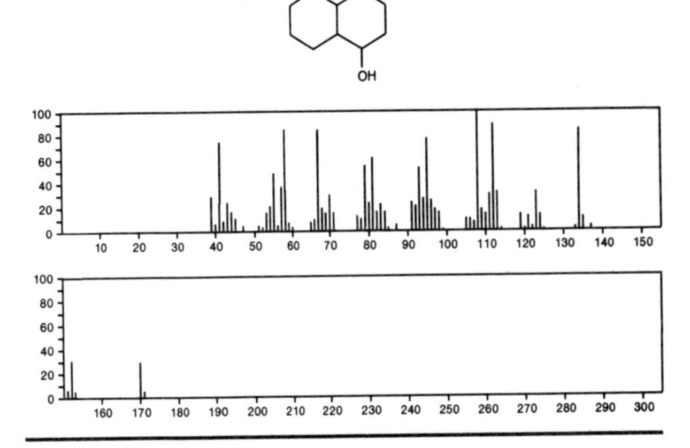

170 $C_{10}H_{18}O_2$ 1127-53-3

1,4-Naphthalenediol, decahydro-, (1α,4β,4aβ,8aα)-

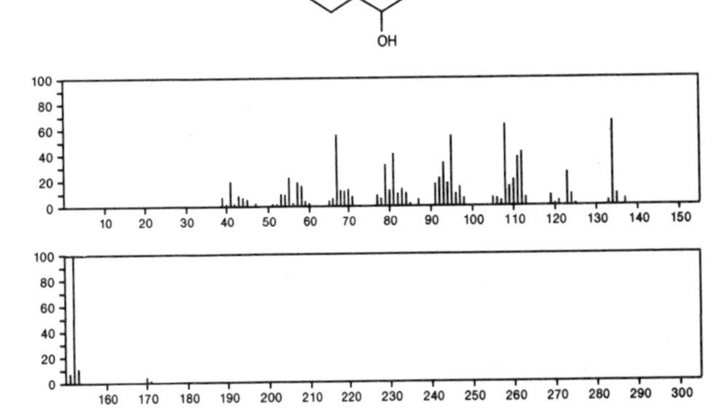

170 $C_{10}H_{18}O_2$ 1127-54-4
1,4-Naphthalenediol, decahydro-, (1α,4β,4aα,8aα)-

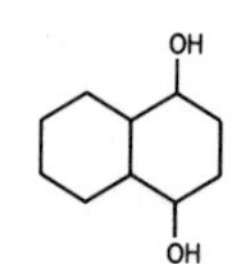

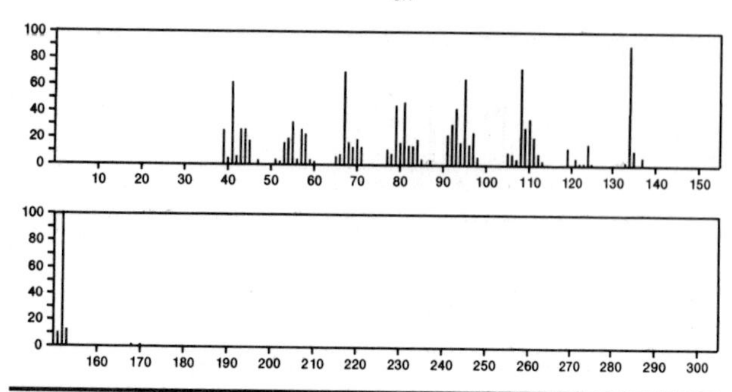

170 $C_{10}H_{18}O_2$ 1127-55-5
1,4-Naphthalenediol, decahydro-, (1α,4β,4aα,8aβ)-

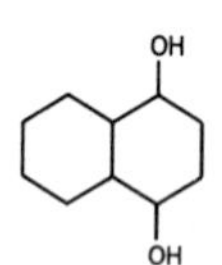

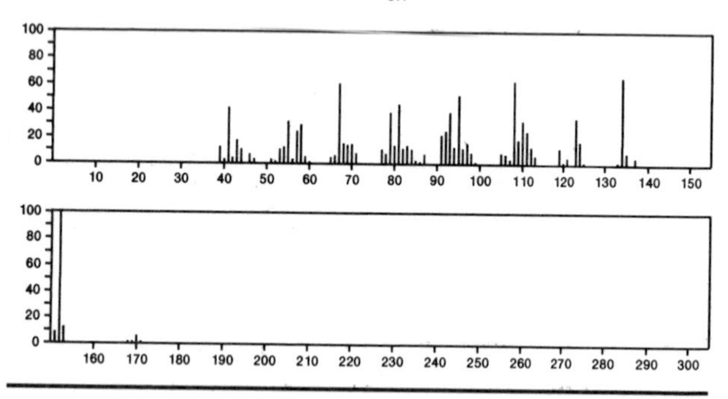

170 $C_{10}H_{18}O_2$ 2955-63-7
3,8-Decanedione

EtCO(CH2)4COEt

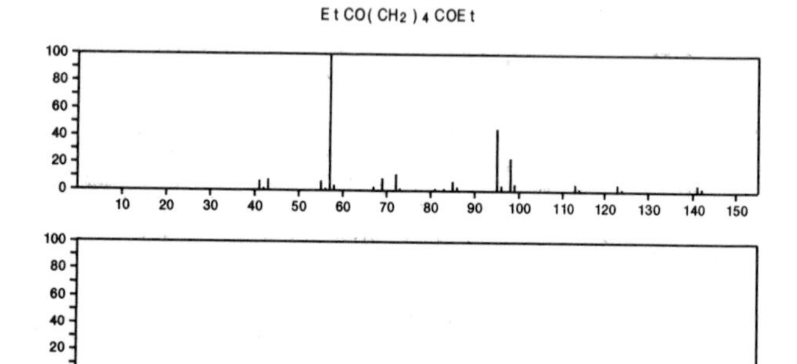

170 $C_{10}H_{18}O_2$ 3618-40-4
Octanoic acid, 2-methylene-, methyl ester

CH2
‖
MeOC(O)C(CH2)5Me

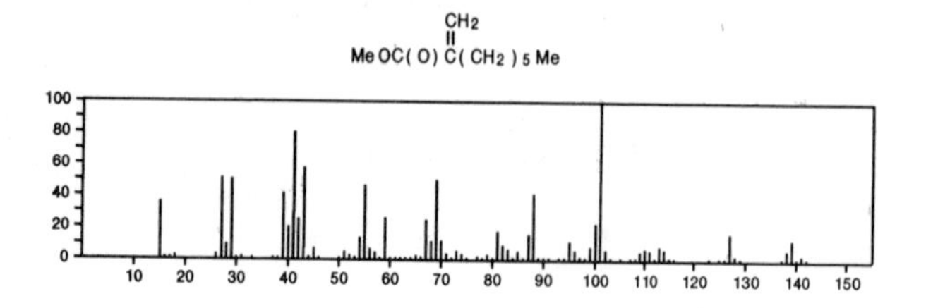

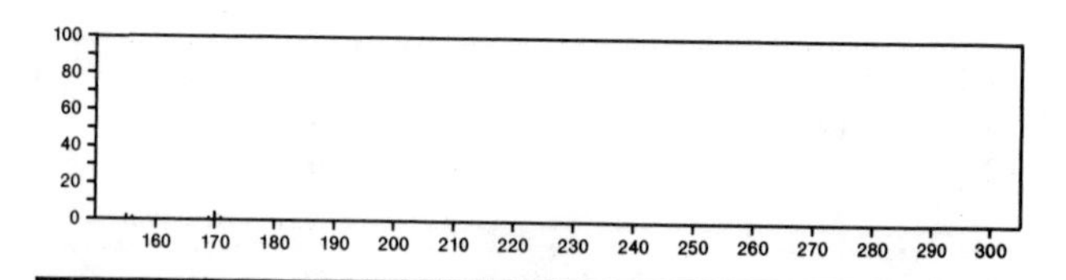

170 $C_{10}H_{18}O_2$ 4388-88-9
3,4-Hexanedione, 2,2,5,5-tetramethyl-

Me3CCOCOCMe3

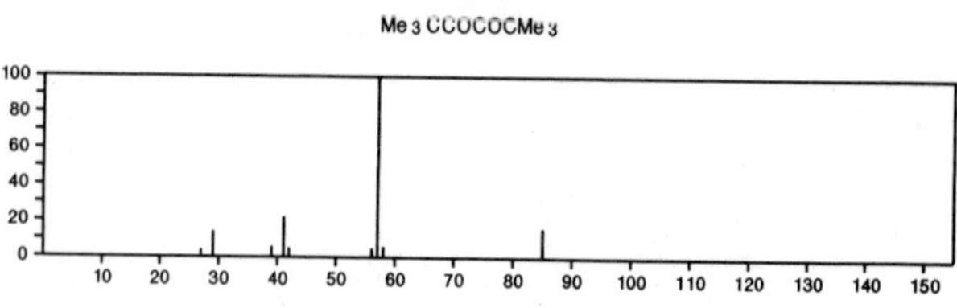

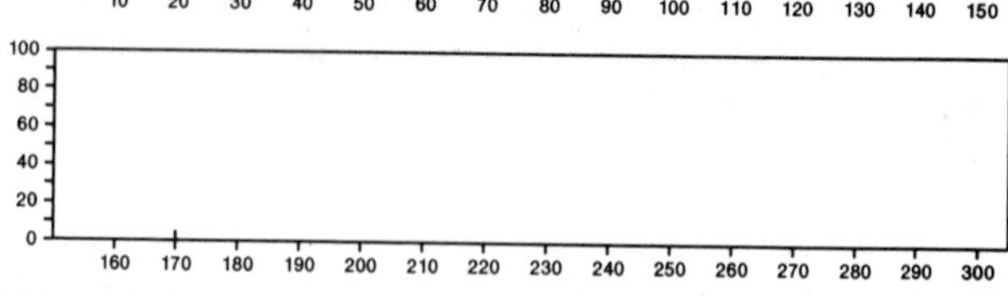

170 $C_{10}H_{18}O_2$ 4441-63-8
Cyclohexanebutanoic acid

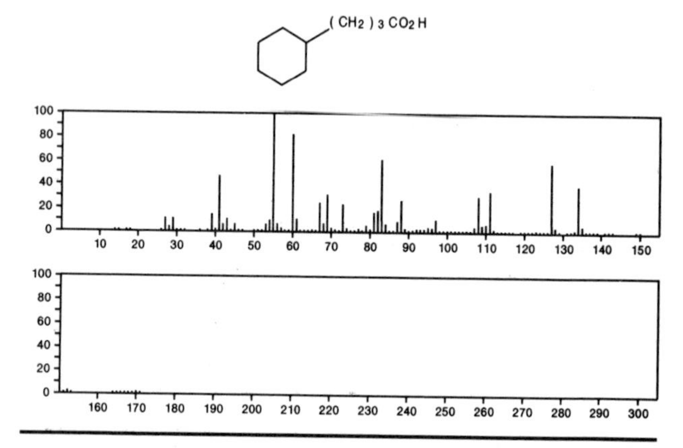

170 $C_{10}H_{18}O_2$ 5989-33-3
2-Furanmethanol, 5-ethenyltetrahydro-α,α,5-trimethyl-, *cis*-

Me2COH O CH=CH2 Me

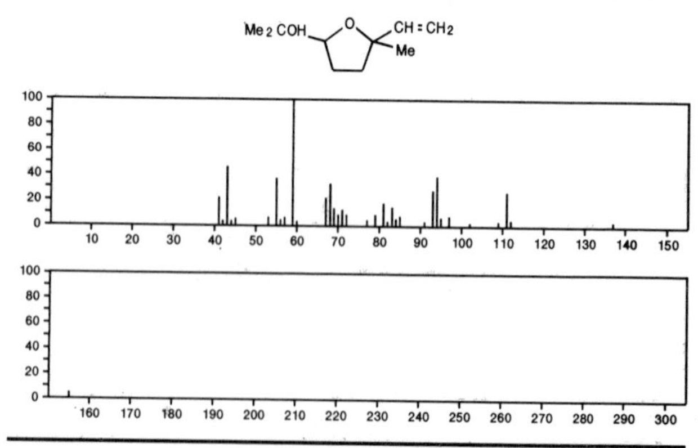

170 $C_{10}H_{18}O_2$ 7560-66-9
3-Cyclohexene-1-carboxaldehyde, 4-methyl-, dimethyl acetal

CH(OMe)2
Me

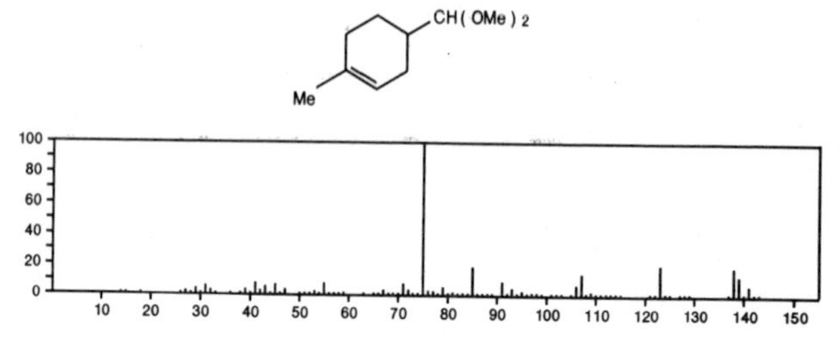

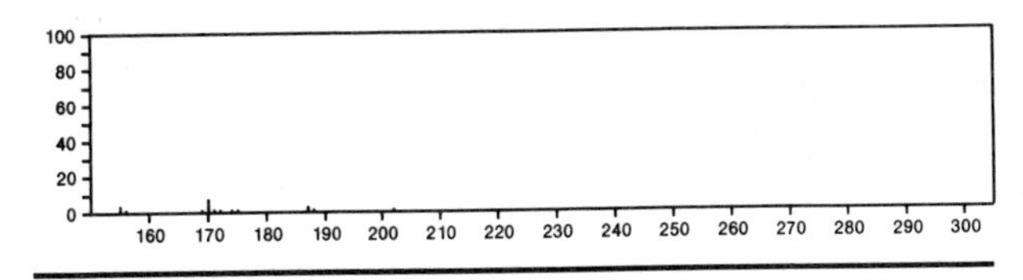

170 $C_{10}H_{18}O_2$ 10225–31–7
2,4–Pentanedione, 3–isopentyl–

COMe
Me2 CHCH2 CH2 CHCOMe

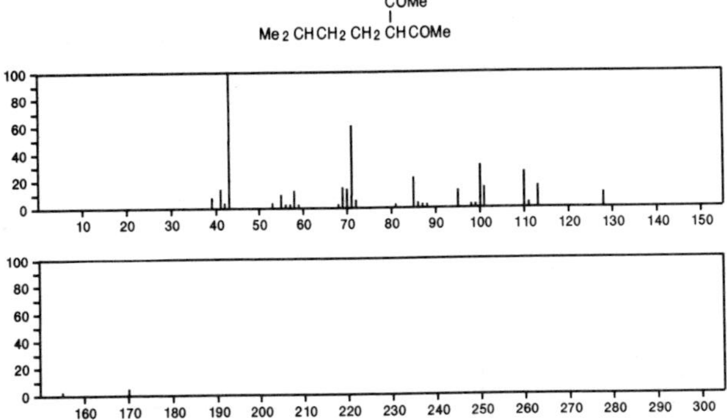

170 $C_{10}H_{18}O_2$ 10317–05–2
2,6–Octadiene–4,5–diol, 3,6–dimethyl–

Me CH = CMe CH(OH) CH(OH) CMe = CHMe

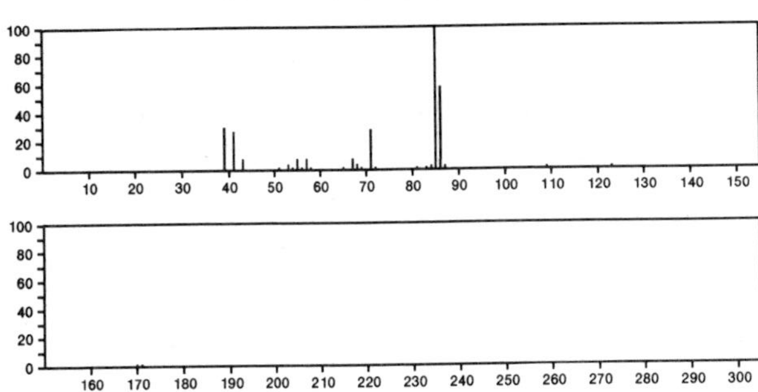

170 $C_{10}H_{18}O_2$ 13429–50–0
2,8–Bornanediol, stereoisomer

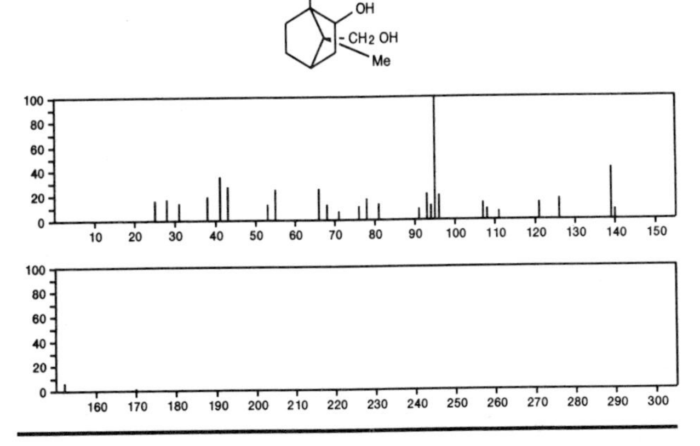

170 $C_{10}H_{18}O_2$ 13481–87–3
3–Nonenoic acid, methyl ester

Me OC(O) CH2 CH = CH(CH2) 4 Me

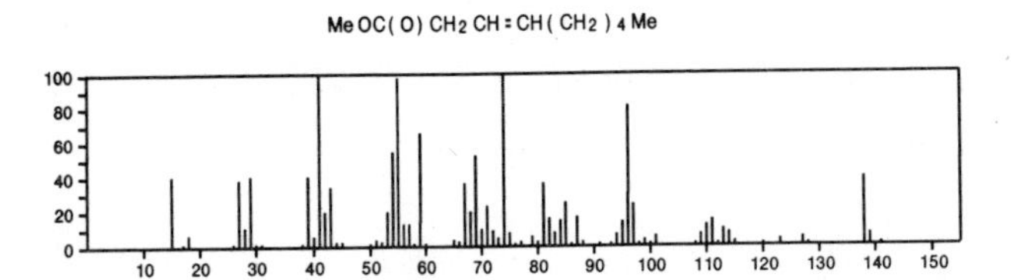

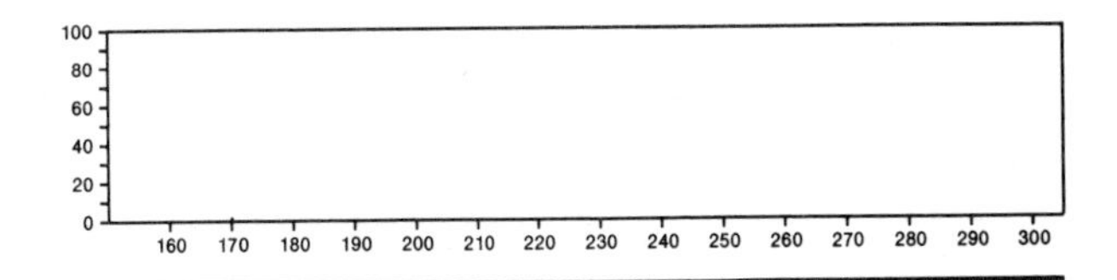

170 $C_{10}H_{18}O_2$ 13837–85–9
2,3–Bornanediol, *endo*–2,*exo*–3–

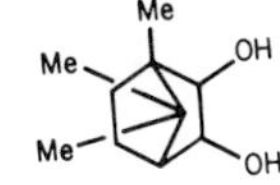

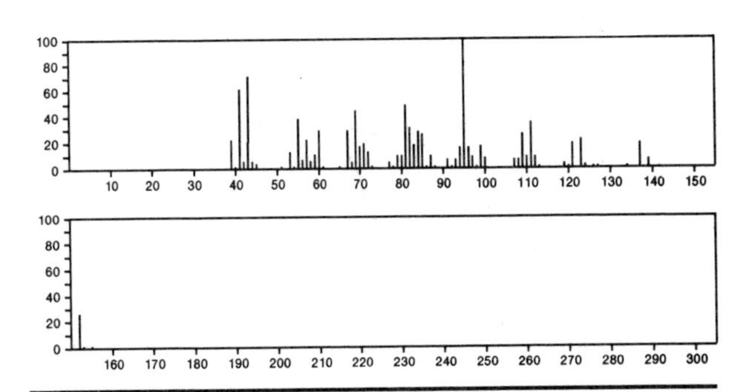

170 $C_{10}H_{18}O_2$ 14049–11–7
2*H*–Pyran–3–ol, 6–ethenyltetrahydro–2,2,6–trimethyl–

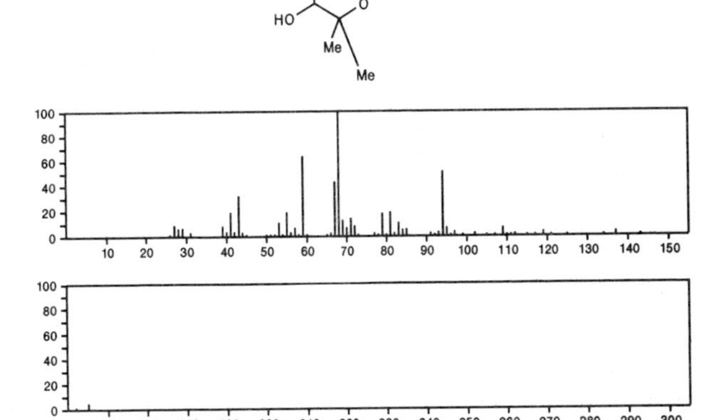

170 $C_{10}H_{18}O_2$ 15287–79–3
Cyclohexanol, 2–methyl–, propionate, *trans*–

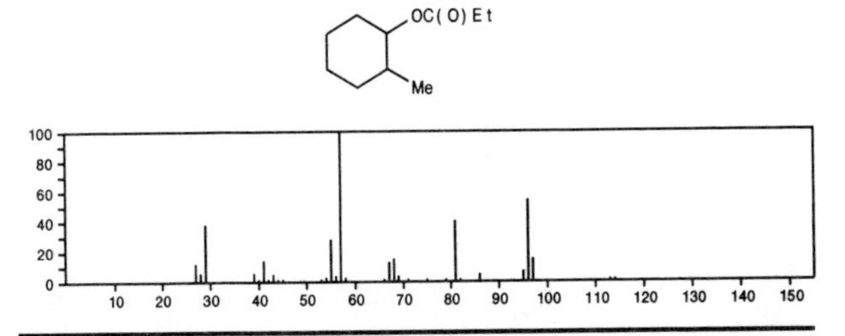

170 $C_{10}H_{18}O_2$ 16812–82–1
2–Pentenoic acid, 3,4,4–trimethyl–, ethyl ester, (*E*)–

E t OC(O) CH = CMe CMe 3

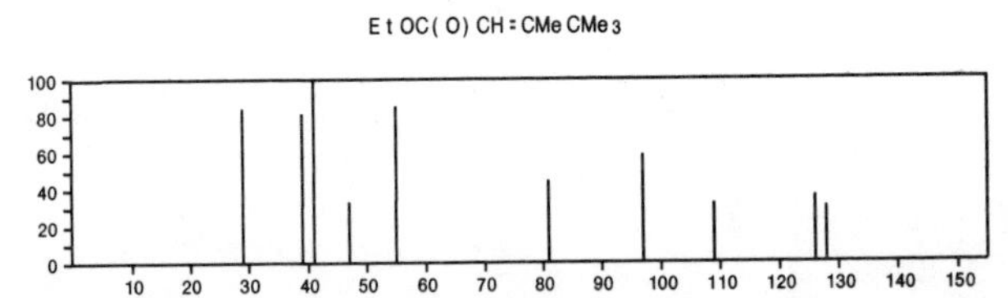

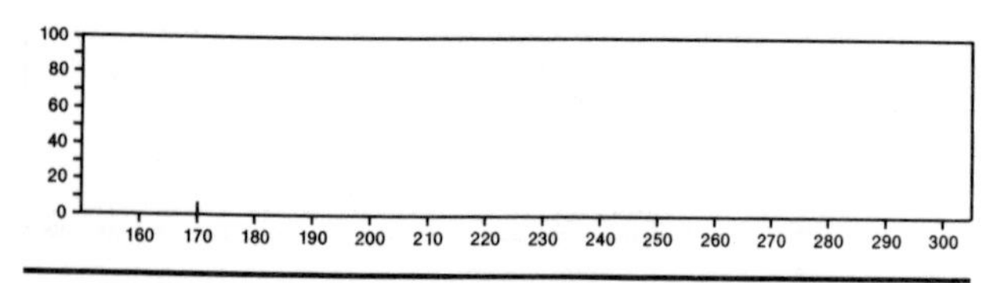

170 $C_{10}H_{18}O_2$ 17429-03-7
Cyclohexanone, 4-methoxy-2,2,6-trimethyl-

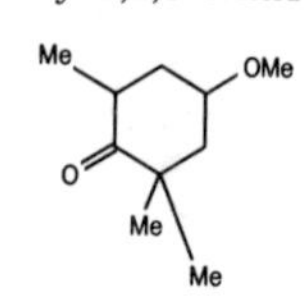

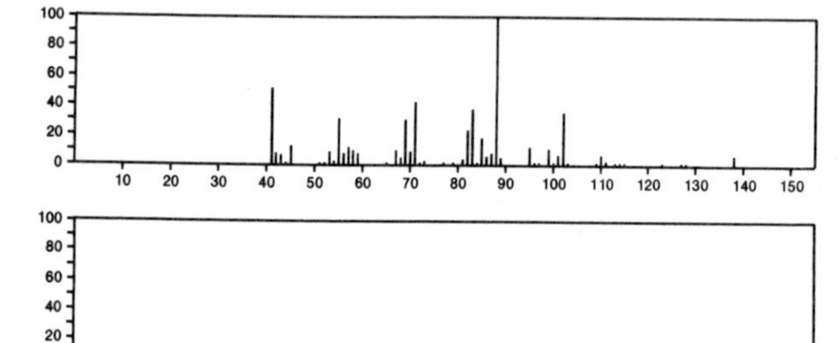

170 $C_{10}H_{18}O_2$ 17429-06-0
Cyclohexanone, 4-hydroxy-3,3,5,5-tetramethyl-

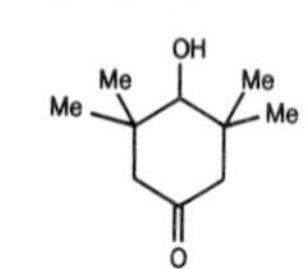

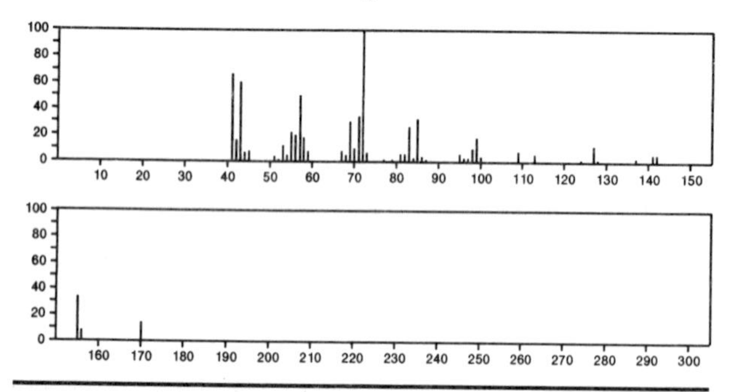

170 $C_{10}H_{18}O_2$ 19089-92-0
2-Butenoic acid, hexyl ester

Me(CH2)5OC(O)CH=CHMe

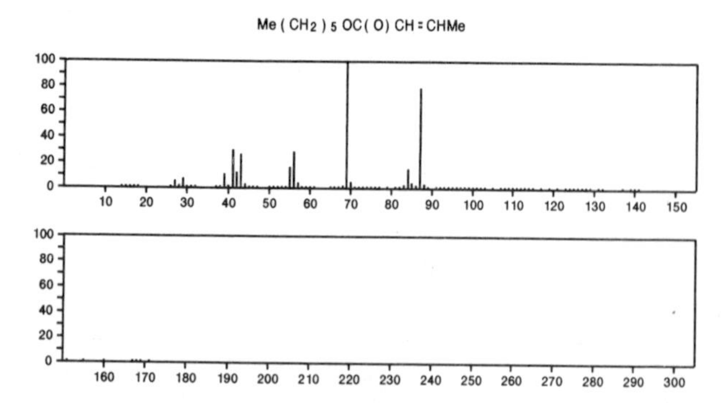

170 $C_{10}H_{18}O_2$ 20731-19-5
4-Nonenoic acid, methyl ester

Me(CH2)3CH=CHCH2CH2C(O)OMe

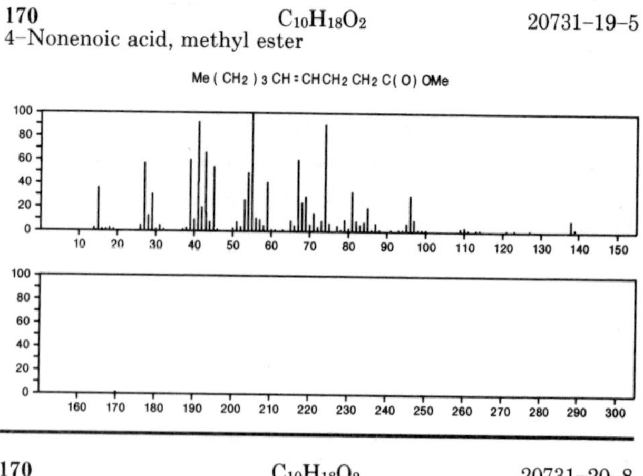

170 $C_{10}H_{18}O_2$ 20731-20-8
5-Nonenoic acid, methyl ester

PrCH=CH(CH2)3C(O)OMe

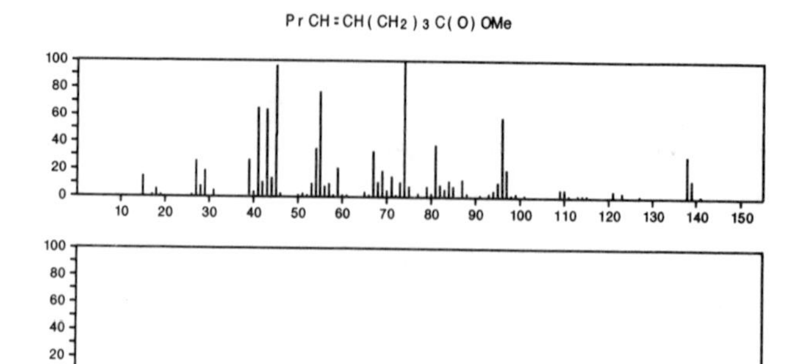

170 $C_{10}H_{18}O_2$ 20731-21-9
6-Nonenoic acid, methyl ester

EtCH=CH(CH2)4C(O)OMe

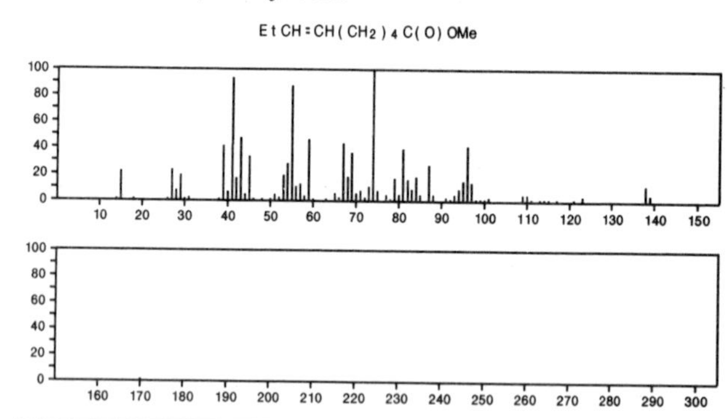

170 $C_{10}H_{18}O_2$ 20731-22-0
7-Nonenoic acid, methyl ester

MeOC(O)(CH2)5CH=CHMe

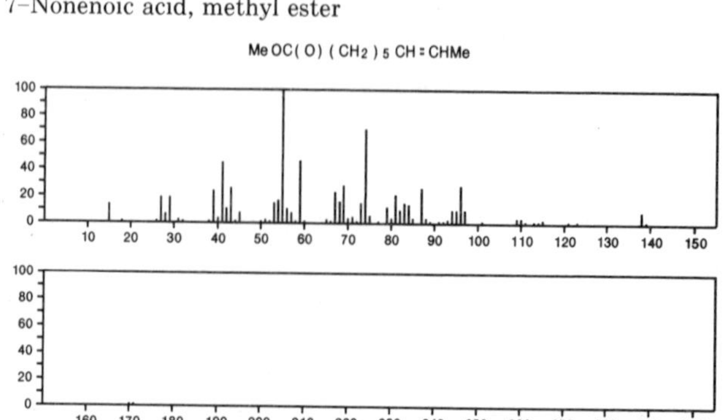

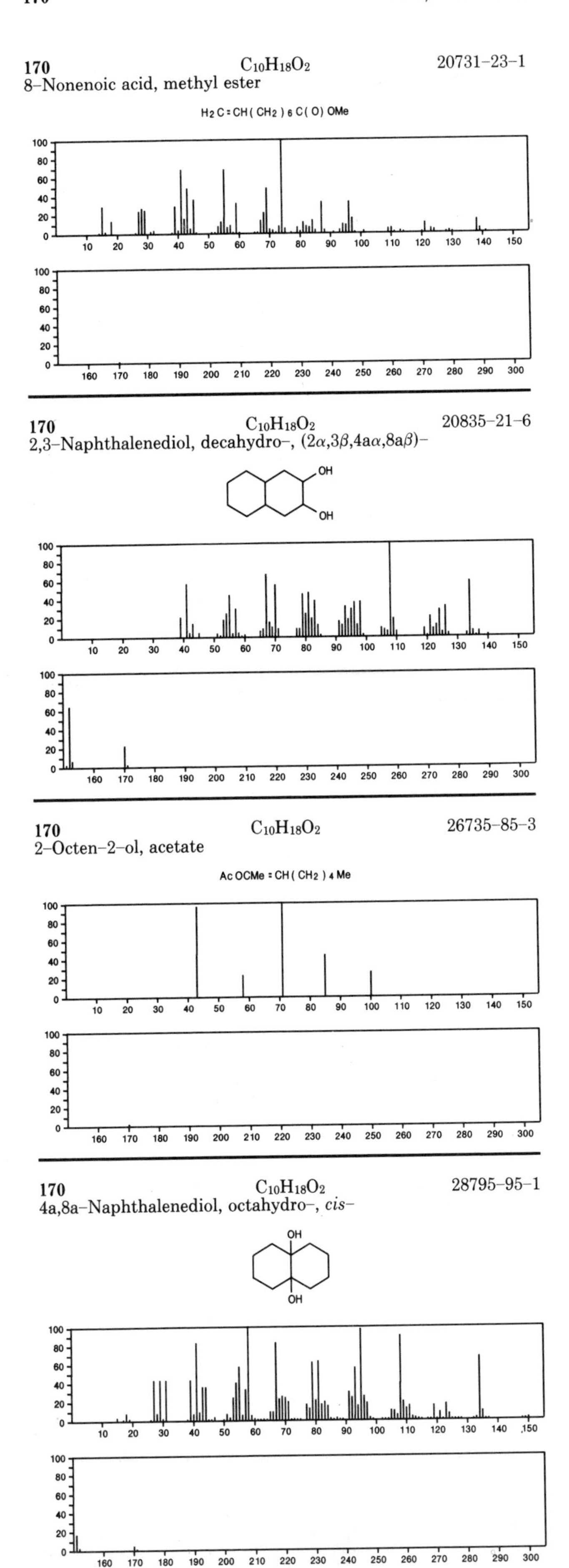
170 $C_{10}H_{18}O_2$ 20731-23-1
8-Nonenoic acid, methyl ester
H2C=CH(CH2)6C(O)OMe
170 $C_{10}H_{18}O_2$ 20835-21-6
2,3-Naphthalenediol, decahydro-, (2α,3β,4aα,8aβ)-
OH
OH
170 $C_{10}H_{18}O_2$ 26735-85-3
2-Octen-2-ol, acetate
AcOCMe=CH(CH2)4Me
170 $C_{10}H_{18}O_2$ 28795-95-1
4a,8a-Naphthalenediol, octahydro-, cis-
OH
OH

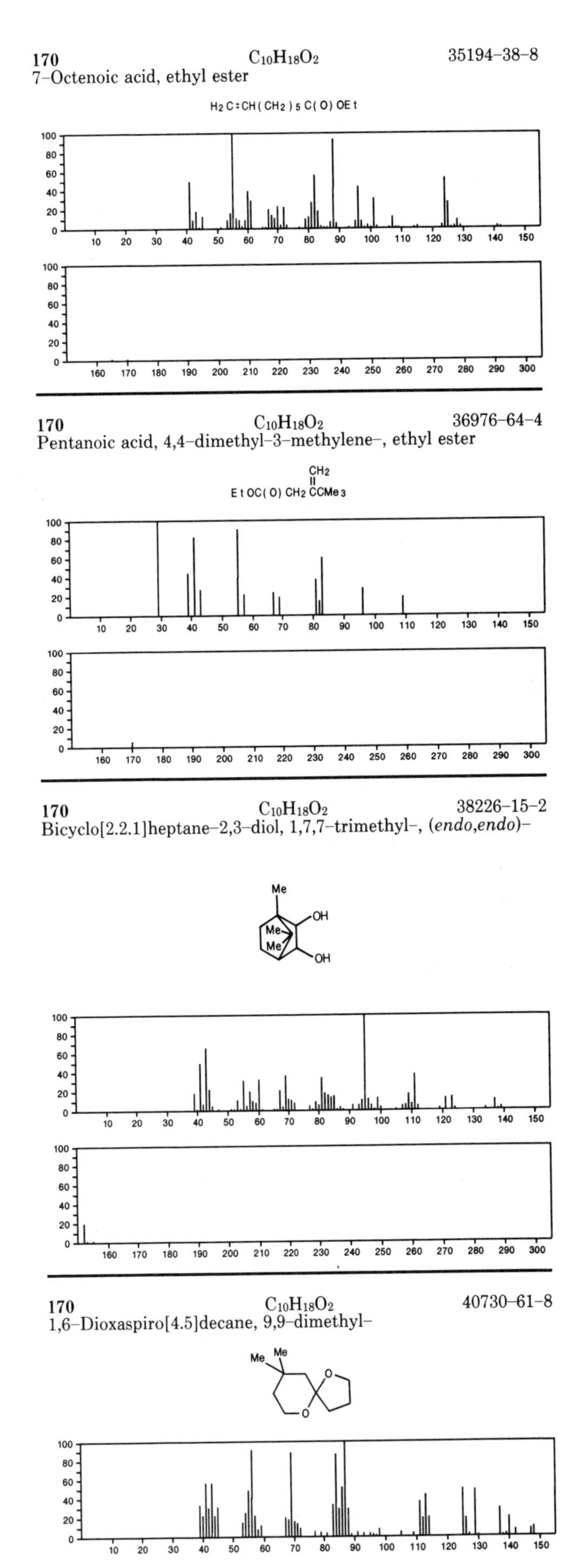
170 $C_{10}H_{18}O_2$ 35194-38-8
7-Octenoic acid, ethyl ester
H2C=CH(CH2)5C(O)OEt
170 $C_{10}H_{18}O_2$ 36976-64-4
Pentanoic acid, 4,4-dimethyl-3-methylene-, ethyl ester
CH2
EtOC(O)CH2CCMe3
170 $C_{10}H_{18}O_2$ 38226-15-2
Bicyclo[2.2.1]heptane-2,3-diol, 1,7,7-trimethyl-, (endo,endo)-
Me
Me
Me
OH
OH
170 $C_{10}H_{18}O_2$ 40730-61-8
1,6-Dioxaspiro[4.5]decane, 9,9-dimethyl-
Me
Me
O
O

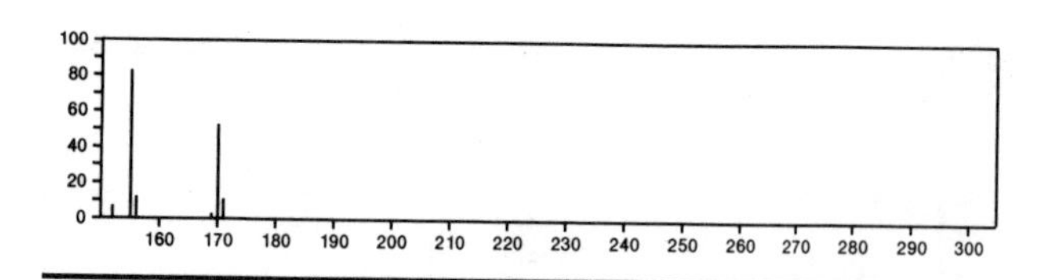

170 $C_{10}H_{18}O_2$ 41654-17-5
6-Nonenoic acid, methyl ester, (*Z*)-

Et CH=CH(CH2)4 C(O) OMe

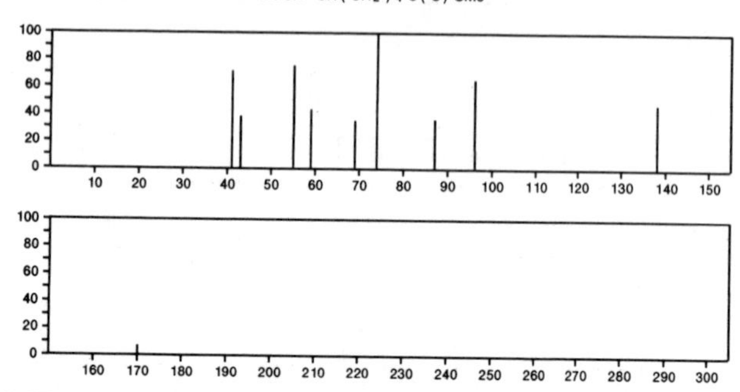

170 $C_{10}H_{18}O_2$ 42177-35-5
2,3-Naphthalenediol, decahydro-, (2α,3β,4aα,8aα)-

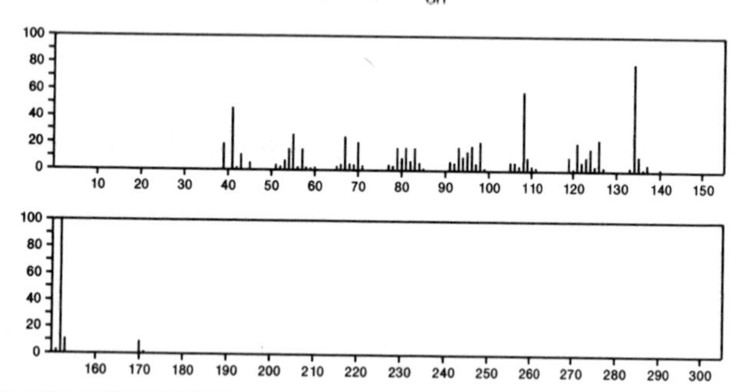

170 $C_{10}H_{18}O_2$ 51149-71-4
2-Propanone, 1-cyclopentyl-3-ethoxy-

CH2 COCH2 OEt

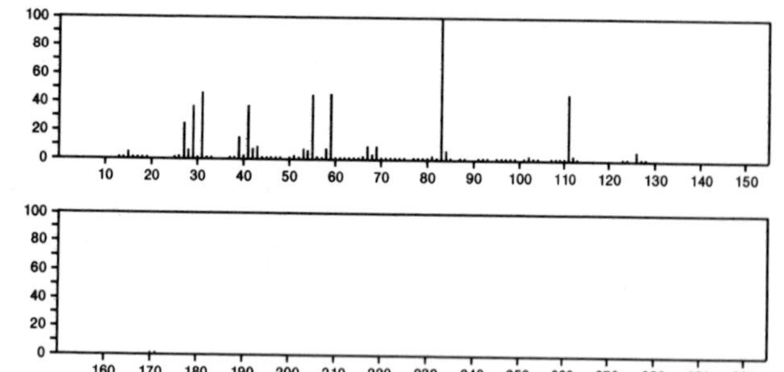

170 $C_{10}H_{18}O_2$ 54382-58-0
4-Isobenzofuranol, octahydro-3a,7a-dimethyl-, (3aα,4β,7aα)-(±)-

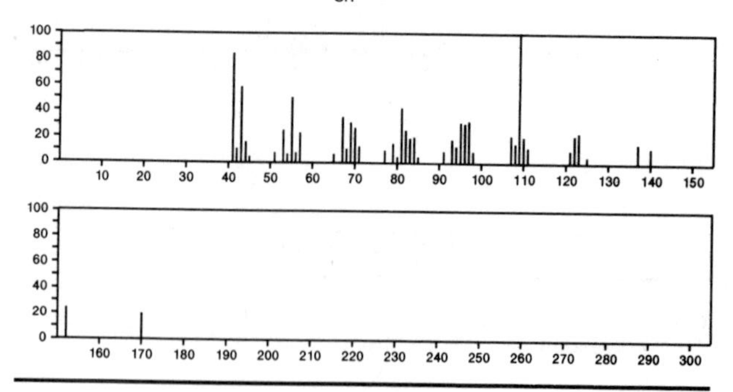

170 $C_{10}H_{18}O_2$ 54410-94-5
Butanoic acid, 3-methyl-, 3-methyl-3-butenyl ester

H2 C=CMe CH2 CH2 OC(O) CH2 CHMe2

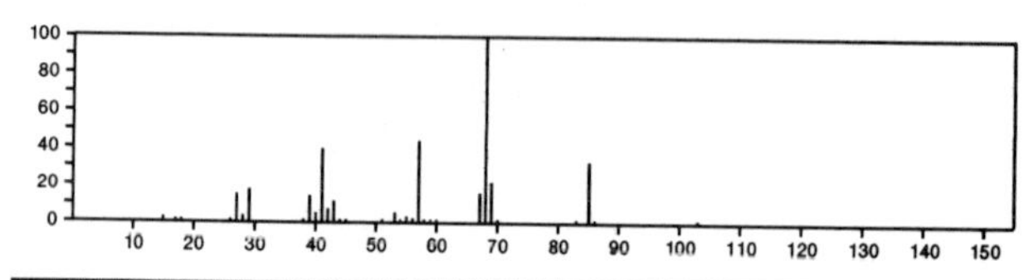

170 $C_{10}H_{18}O_2$ 56614-57-4
Bicyclo[2.2.1]heptane-2,3-diol, 1,7,7-trimethyl-, (*exo,exo*)-

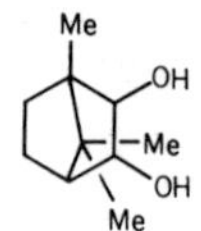

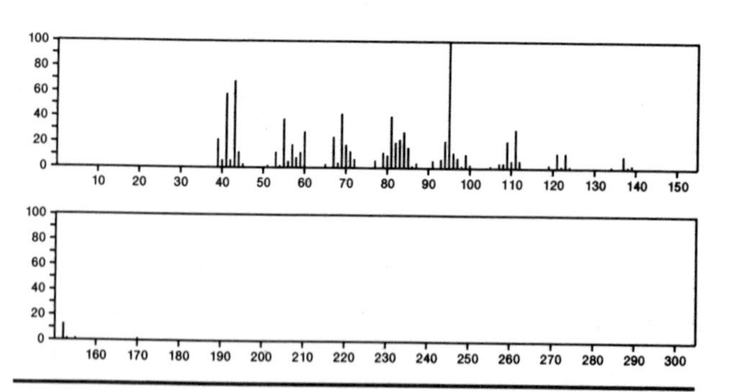

170 $C_{10}H_{18}O_2$ 56614-58-5
Bicyclo[2.2.1]heptane-2,3-diol, 1,7,7-trimethyl-, (2-*exo*,3-*endo*)-

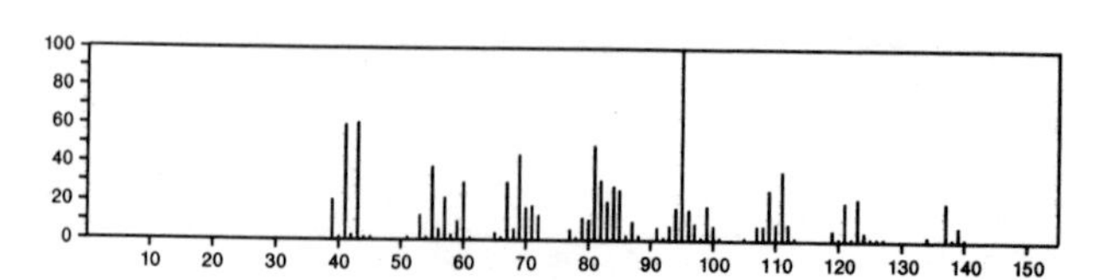

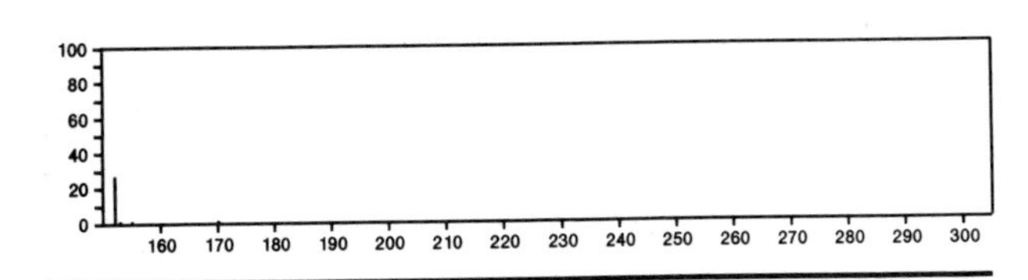

170 $C_{10}H_{18}O_2$ 57289-63-1
4a,8a-Naphthalenediol, octahydro-, *trans*-

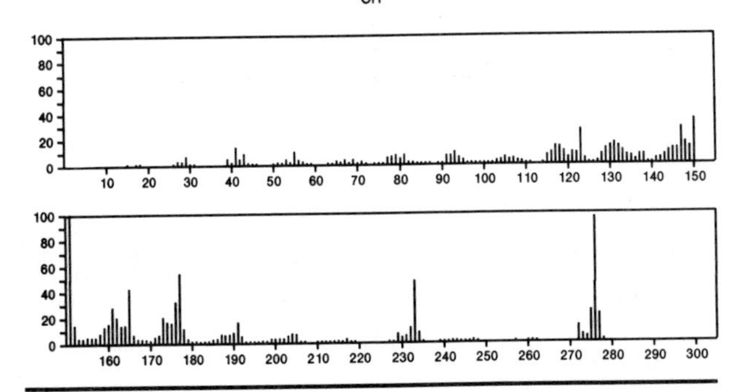

170 $C_{10}H_{18}O_2$ 57397-07-6
2,3-Naphthalenediol, decahydro-

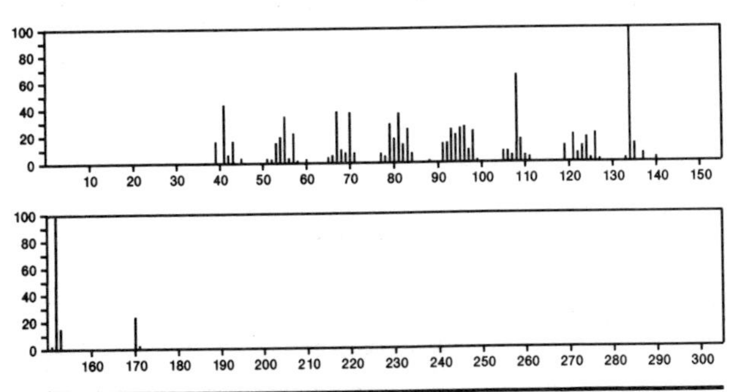

170 $C_{10}H_{18}S$ 1126-65-4
Cyclopentane, 1,1'-thiobis-

170 $C_{10}H_{18}S$ 54461-02-8
Bicyclo[2.2.2]octane, 1-methyl-4-(methylthio)-

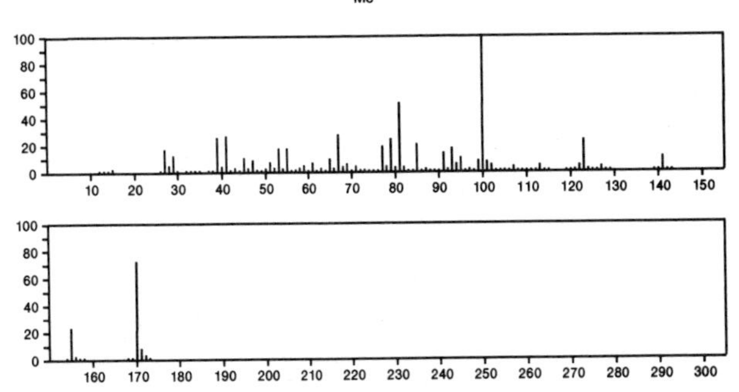

170 $C_{10}H_{22}N_2$ 54410-91-2
Piperazine, 2,3-dimethyl-5-(2-methylpropyl)-

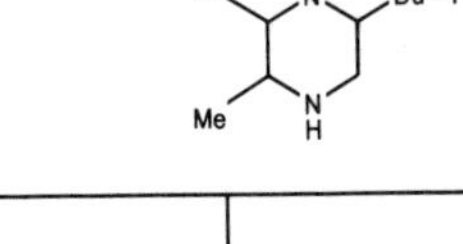

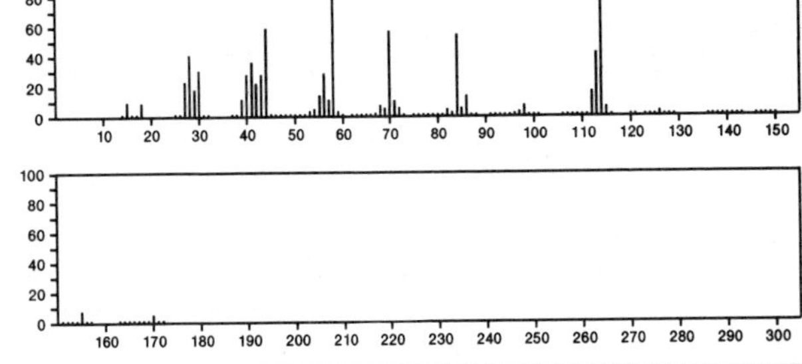

170 $C_{10}H_{22}N_2$ 54410-92-3
Piperazine, 2,5-dimethyl-3-(2-methylpropyl)-

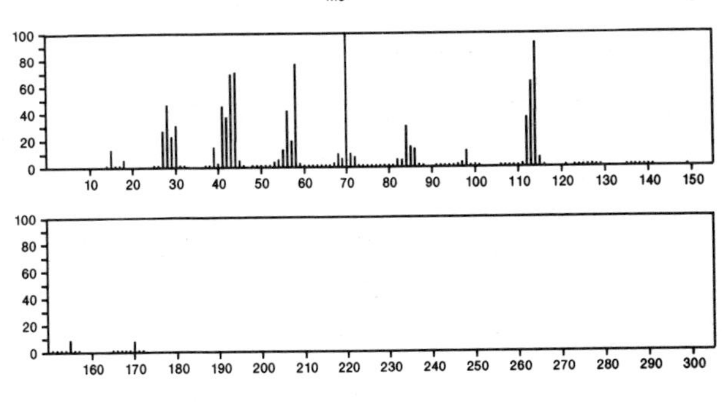

170 $C_{10}H_{22}N_2$ 54410-93-4
Piperazine, 3-butyl-2,5-dimethyl-

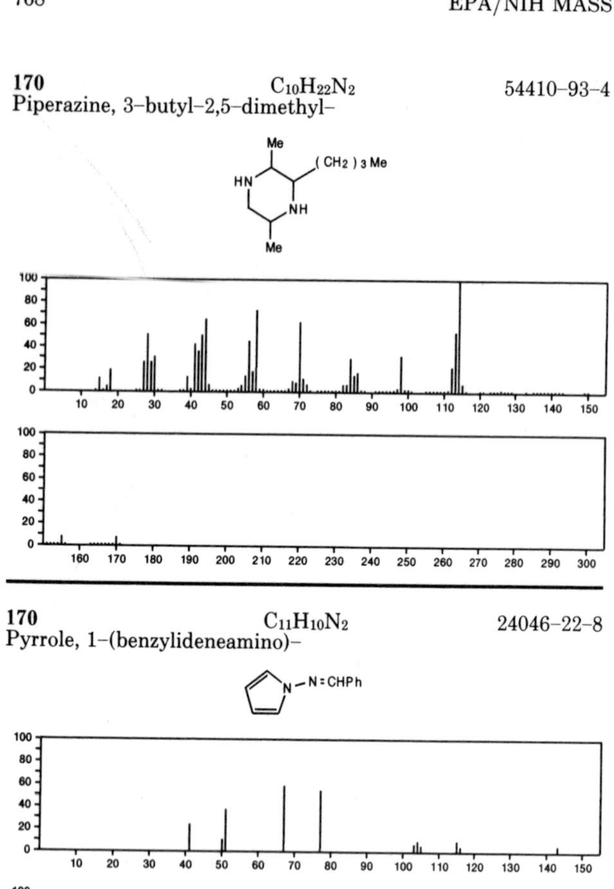

170 $C_{11}H_{10}N_2$ 24046-22-8
Pyrrole, 1-(benzylideneamino)-

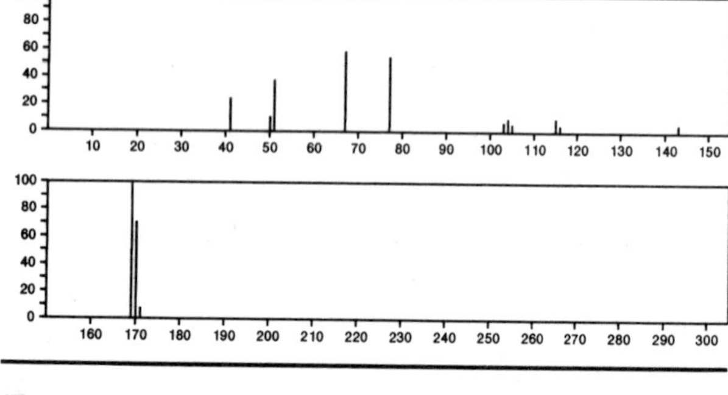

170 $C_{11}H_{10}N_2$ 31378-82-2
Pyridinium, 1-anilino-, hydroxide, inner salt

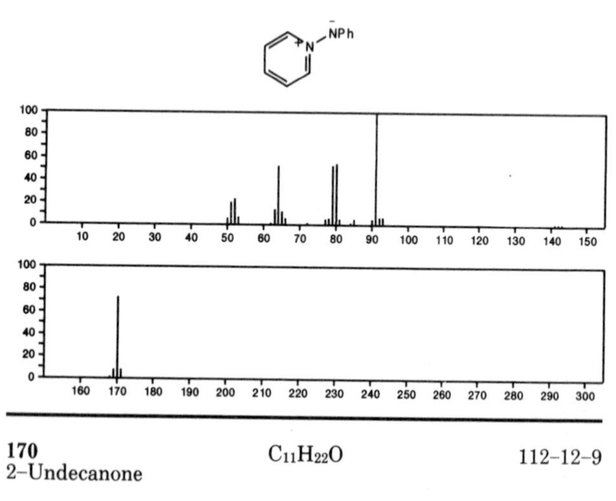

170 $C_{11}H_{22}O$ 112-12-9
2-Undecanone

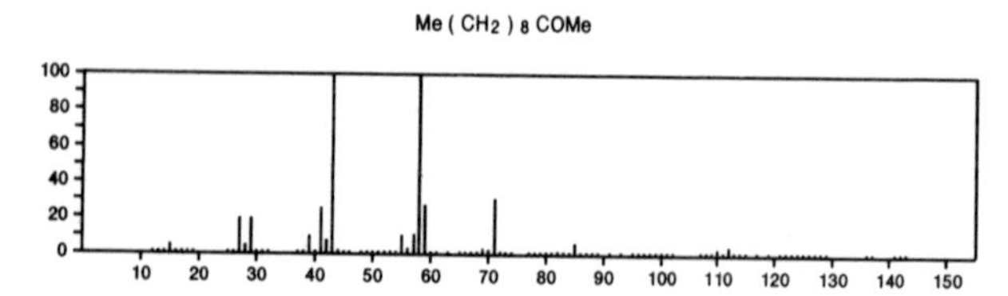

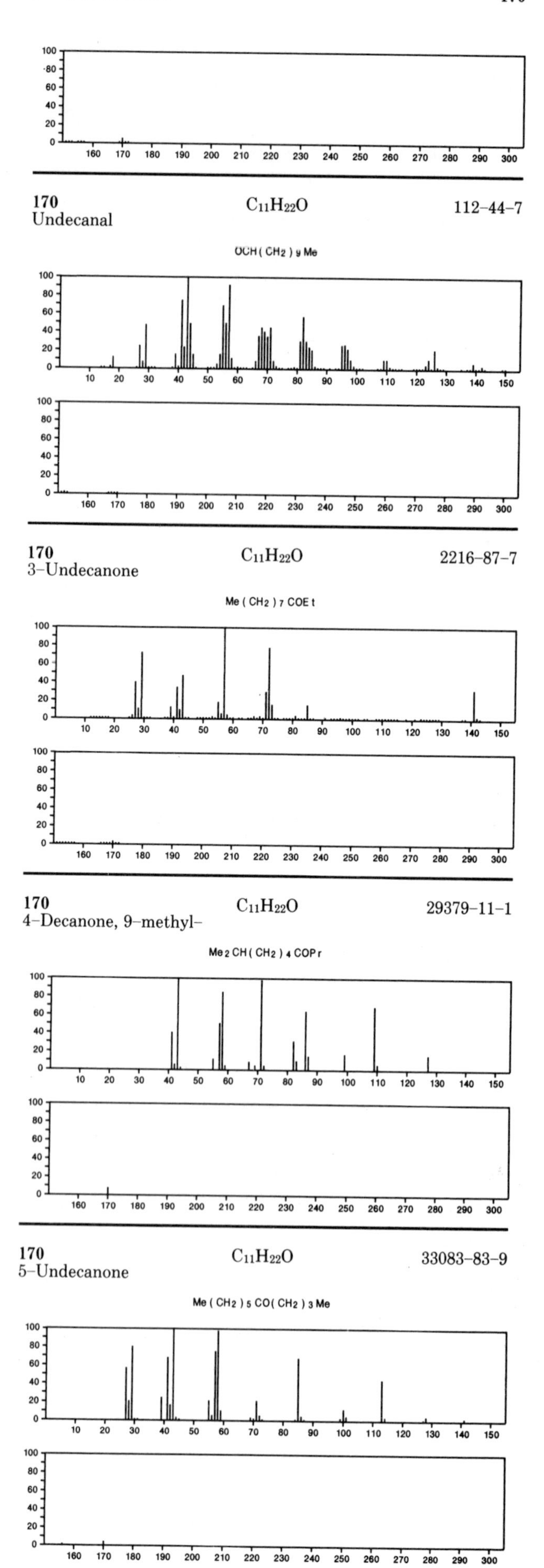

170 $C_{11}H_{22}O$ 112-44-7
Undecanal

170 $C_{11}H_{22}O$ 2216-87-7
3-Undecanone

170 $C_{11}H_{22}O$ 29379-11-1
4-Decanone, 9-methyl-

170 $C_{11}H_{22}O$ 33083-83-9
5-Undecanone

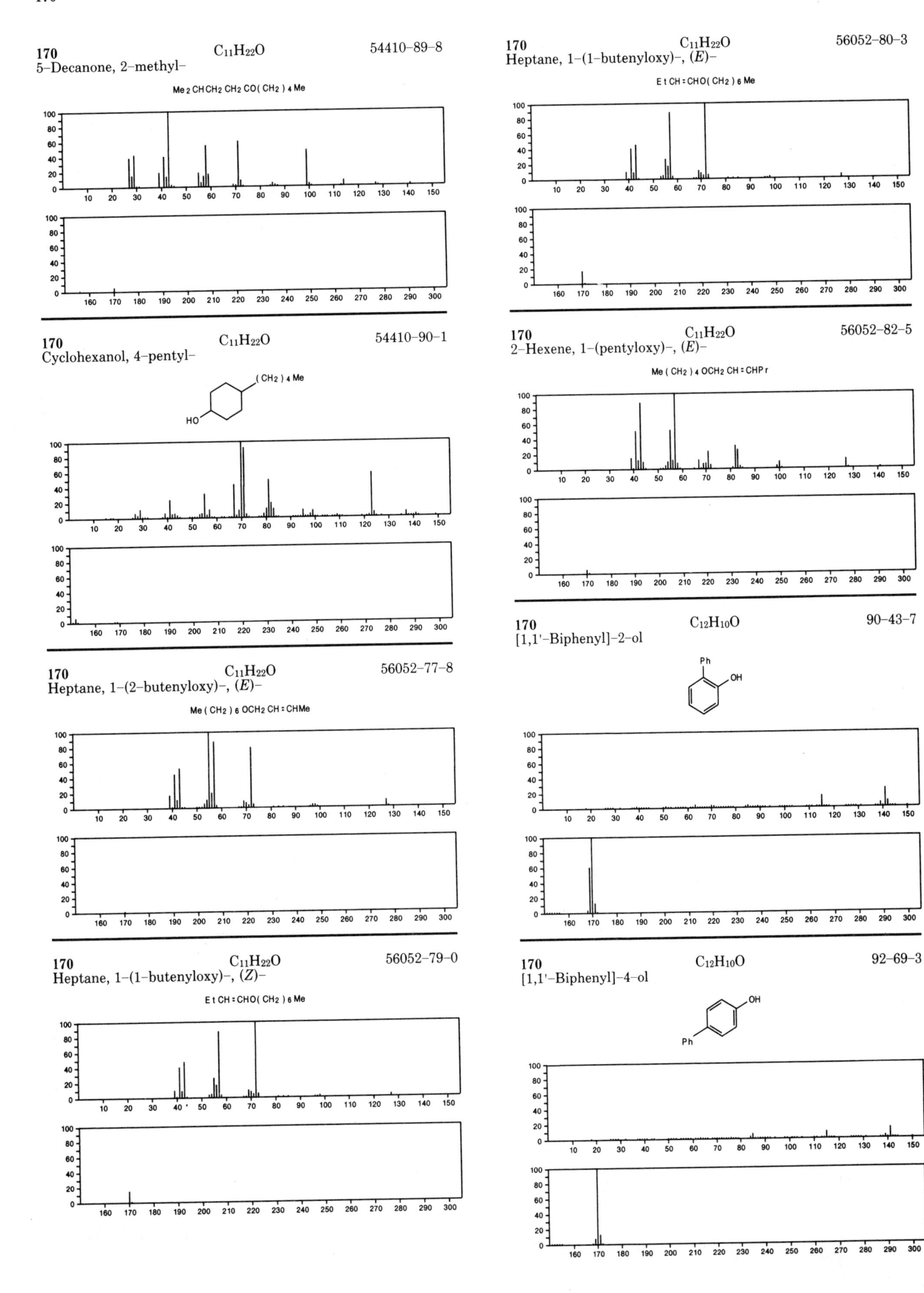
170
C11H22O
54410-89-8
5-Decanone, 2-methyl-
Me2CHCH2CH2CO(CH2)4Me
170
C11H22O
56052-80-3
Heptane, 1-(1-butenyloxy)-, (E)-
EtCH=CHO(CH2)6Me
170
C11H22O
54410-90-1
Cyclohexanol, 4-pentyl-
(CH2)4Me
HO
170
C11H22O
56052-82-5
2-Hexene, 1-(pentyloxy)-, (E)-
Me(CH2)4OCH2CH=CHPr
170
C11H22O
56052-77-8
Heptane, 1-(2-butenyloxy)-, (E)-
Me(CH2)6OCH2CH=CHMe
170
C12H10O
90-43-7
[1,1'-Biphenyl]-2-ol
Ph
OH
170
C11H22O
56052-79-0
Heptane, 1-(1-butenyloxy)-, (Z)-
EtCH=CHO(CH2)6Me
170
C12H10O
92-69-3
[1,1'-Biphenyl]-4-ol
OH
Ph

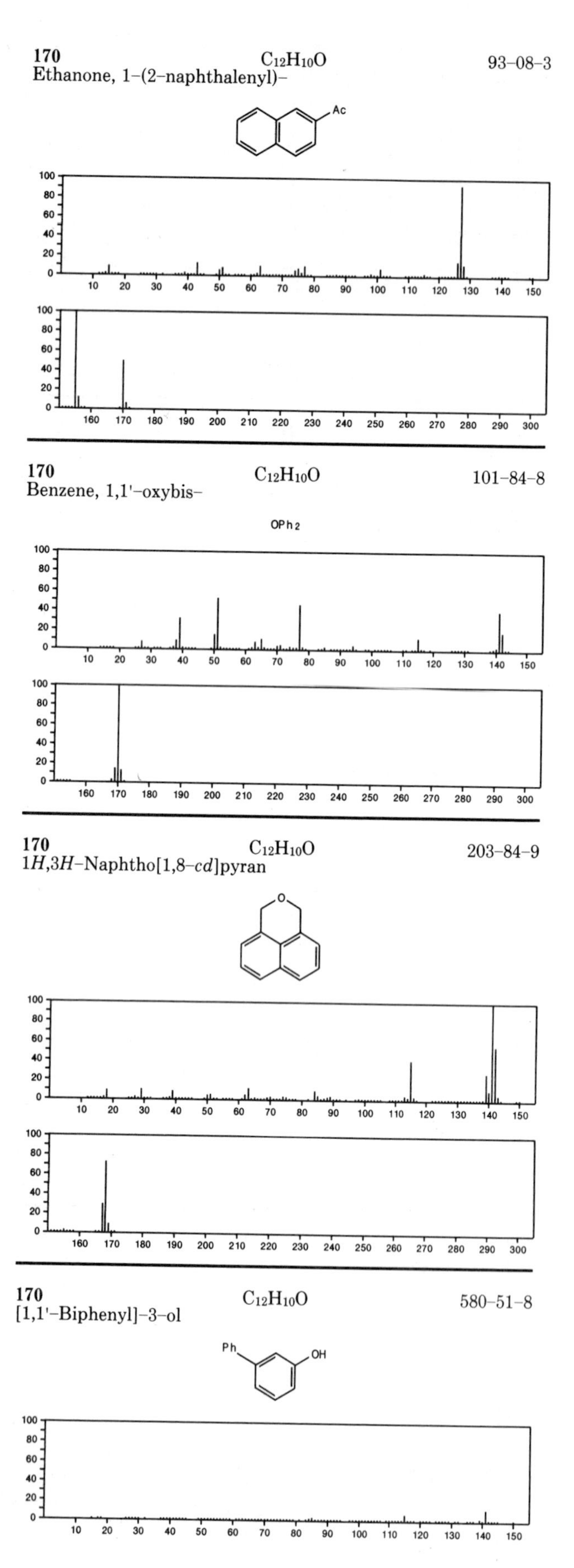
170
C12H10O
93-08-3
Ethanone, 1-(2-naphthalenyl)-
Ac
170
C12H10O
101-84-8
Benzene, 1,1'-oxybis-
OPh2
170
C12H10O
203-84-9
1H,3H-Naphtho[1,8-cd]pyran
O
170
C12H10O
580-51-8
[1,1'-Biphenyl]-3-ol
Ph
OH

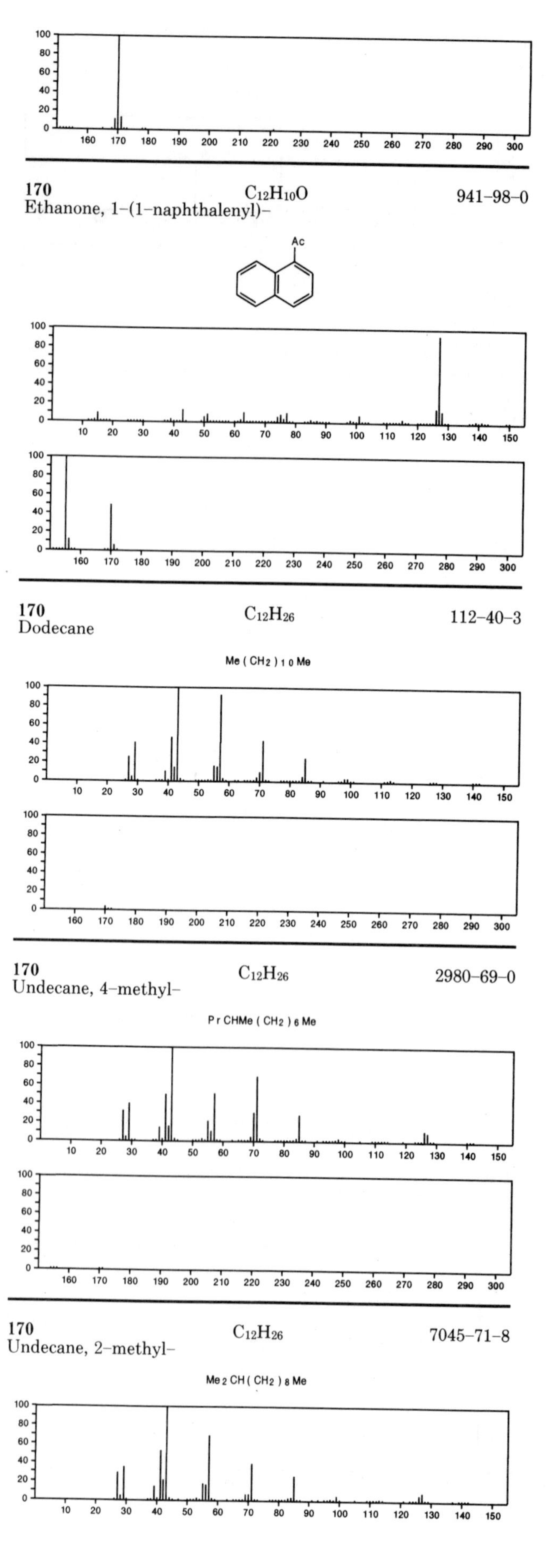
170
C12H10O
941-98-0
Ethanone, 1-(1-naphthalenyl)-
Ac
170
C12H26
112-40-3
Dodecane
Me(CH2)10Me
170
C12H26
2980-69-0
Undecane, 4-methyl-
PrCHMe(CH2)6Me
170
C12H26
7045-71-8
Undecane, 2-methyl-
Me2CH(CH2)8Me

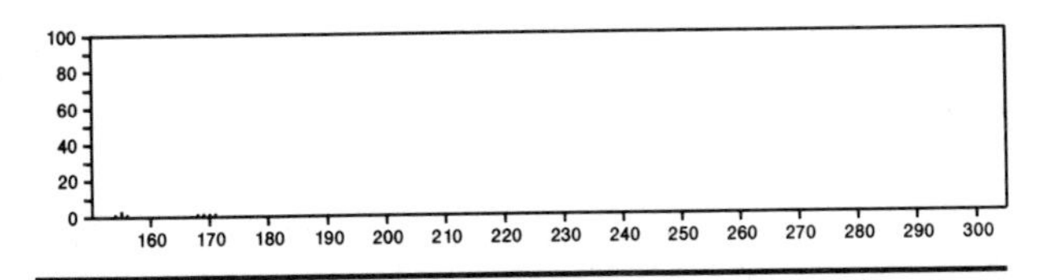

170 $C_{12}H_{26}$ 13475-82-6
Heptane, 2,2,4,6,6-pentamethyl-

$Me_3CCH_2CHMeCH_2CMe_3$

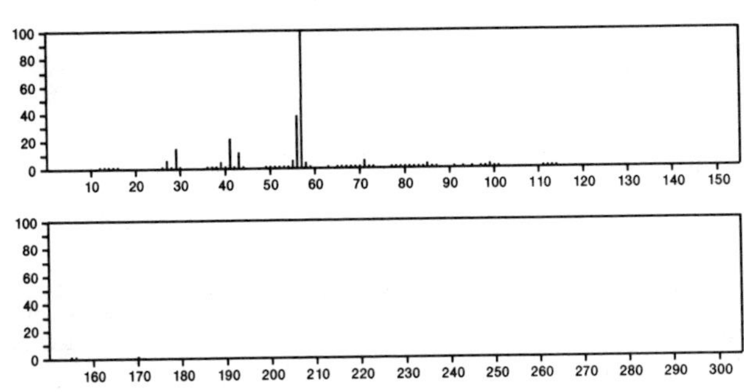

170 $C_{12}H_{26}$ 17312-50-4
Decane, 2,5-dimethyl-

$Me_2CHCH_2CH_2CHMe(CH_2)_4Me$

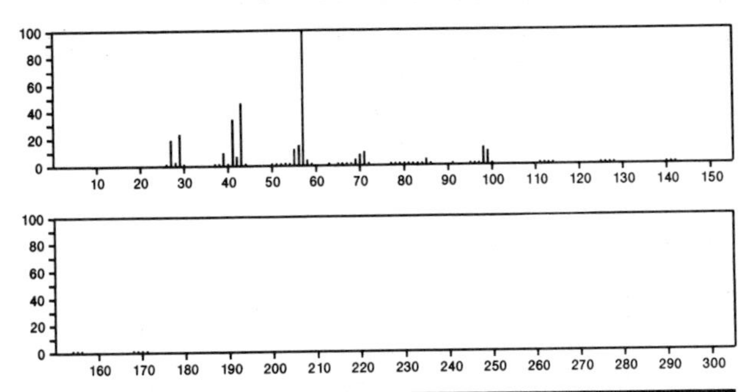

170 $C_{13}H_{14}$ 829-26-5
Naphthalene, 2,3,6-trimethyl-

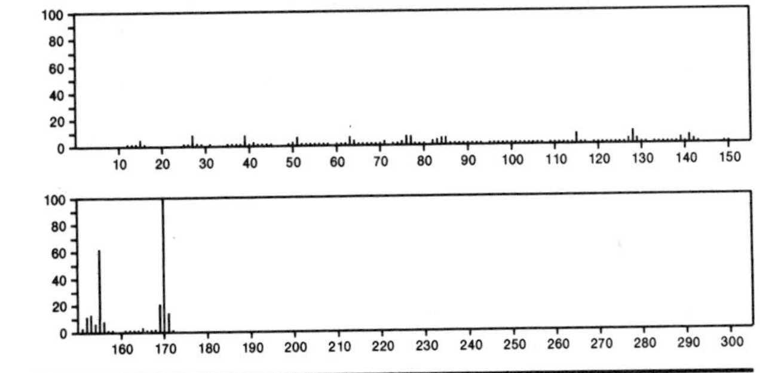

170 $C_{13}H_{14}$ 2027-17-0
Naphthalene, 2-(1-methylethyl)-

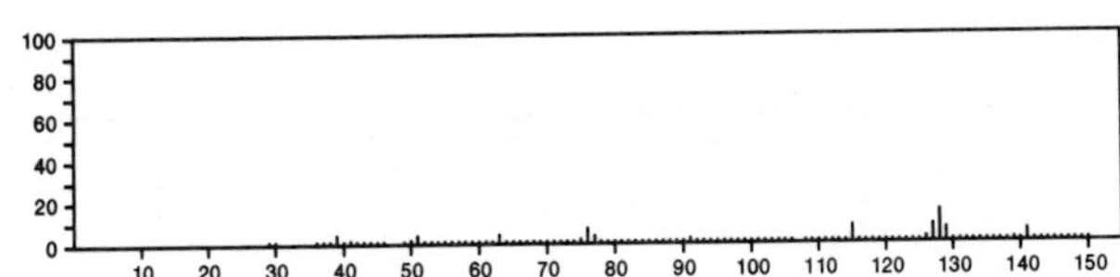

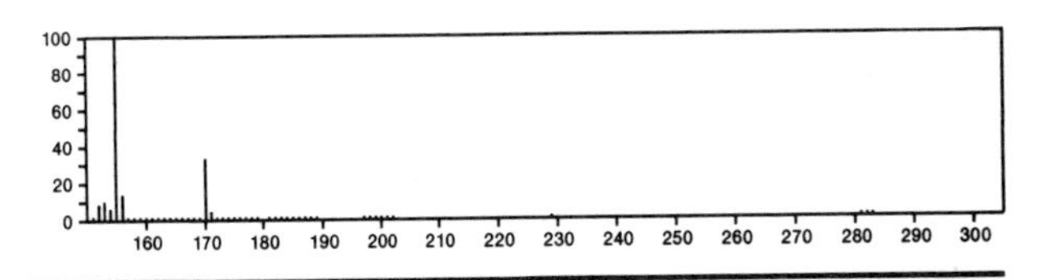

170 $C_{13}H_{14}$ 2245-38-7
Naphthalene, 1,6,7-trimethyl-

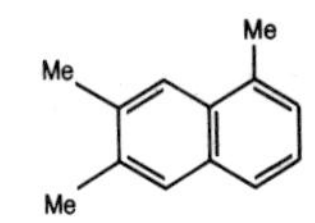

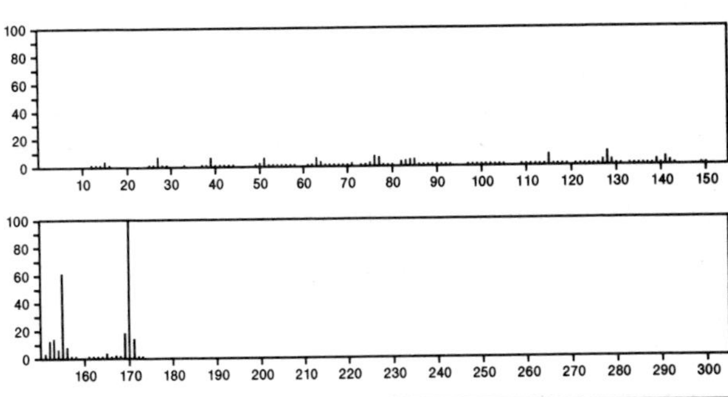

170 $C_{13}H_{14}$ 3031-08-1
Naphthalene, 1,3,6-trimethyl-

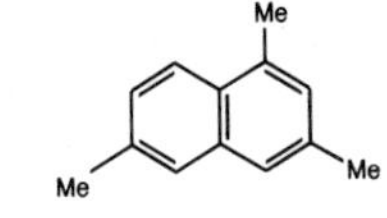

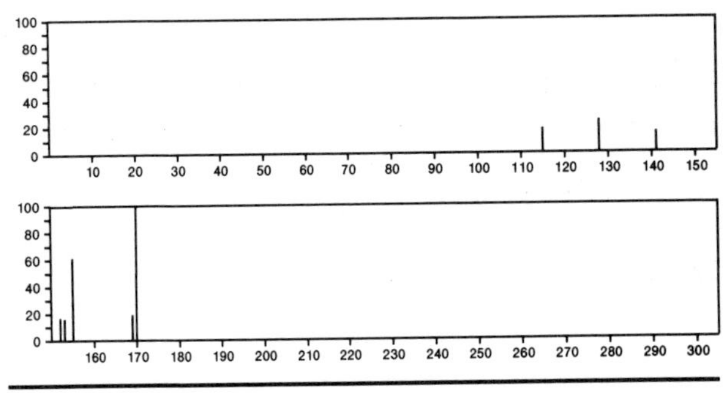

170 $C_{13}H_{14}$ 29253-36-9
Naphthalene, (1-methylethyl)-

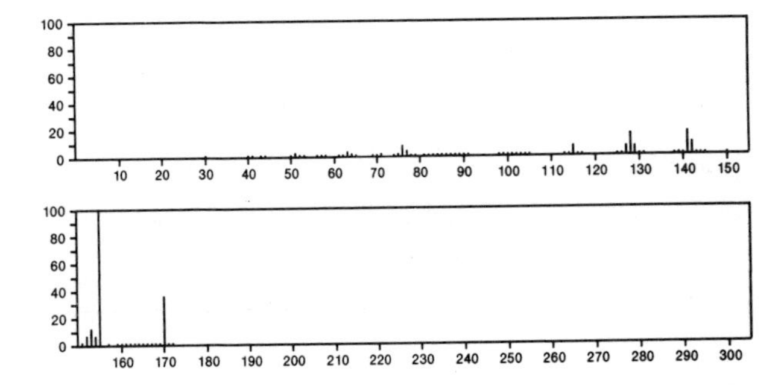

171 $C_5H_2ClN_3S$ 13316–08–0
Thiazolo[5,4–*d*]pyrimidine, 5–chloro–

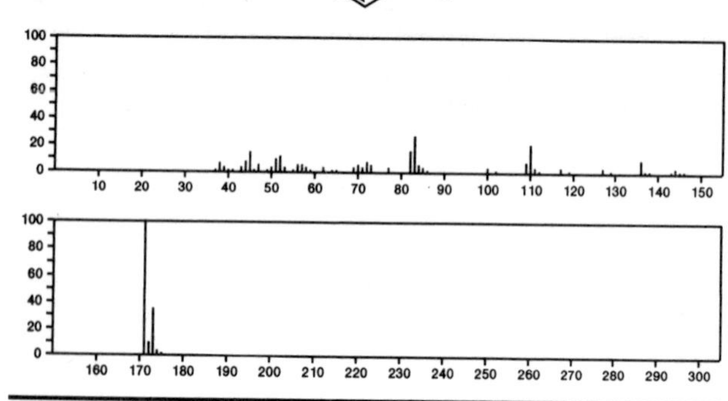

171 $C_5H_2ClN_3S$ 13316–12–6
Thiazolo[5,4–*d*]pyrimidine, 7–chloro–

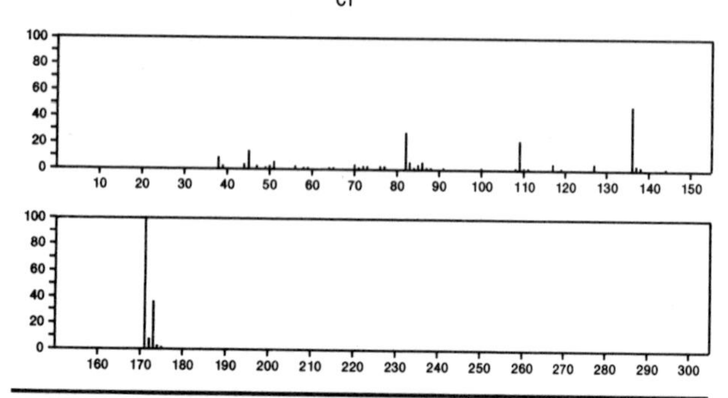

171 $C_6H_5NO_5$ 668–43–9
Sorbic acid, 3,5–dihydroxy–2–nitro–, δ–lactone

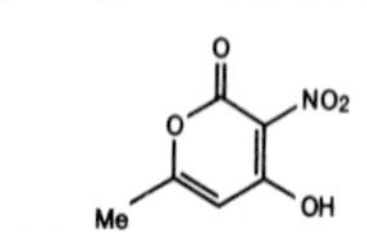

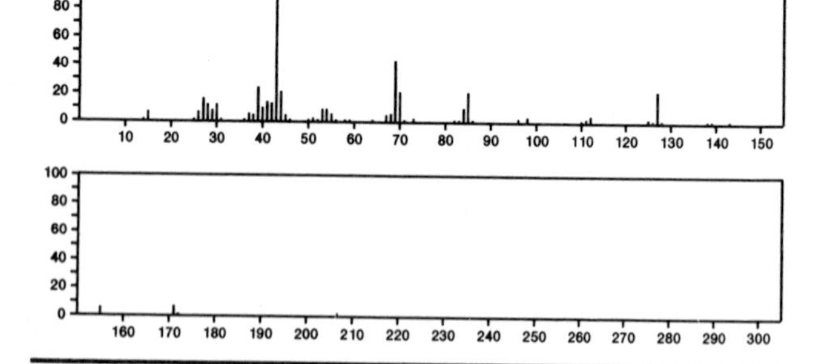

171 C_6H_6BrN 55401–97–3
Pyridine, 2–(bromomethyl)–

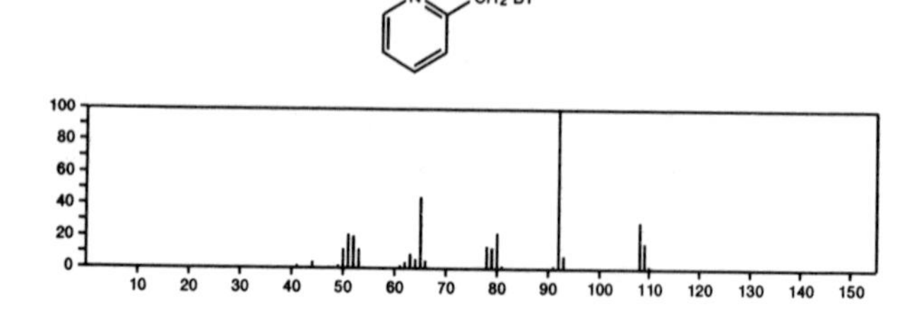

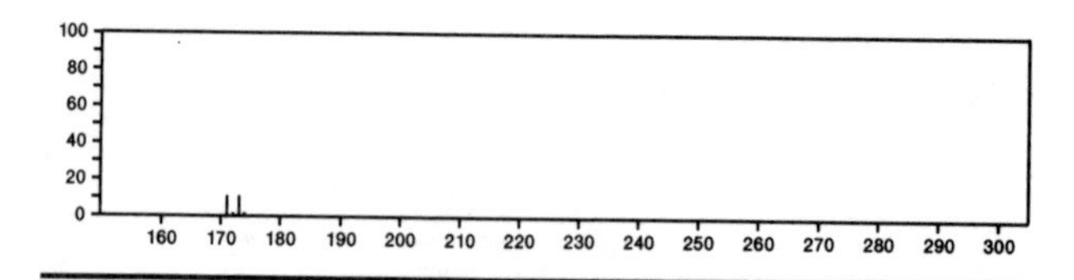

171 $C_6H_9N_3O_3$ 827–16–7
1,3,5–Triazine–2,4,6(1*H*,3*H*,5*H*)–trione, 1,3,5–trimethyl–

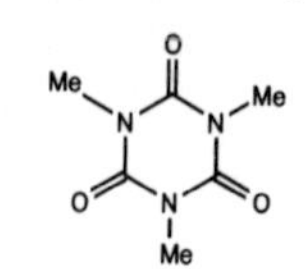

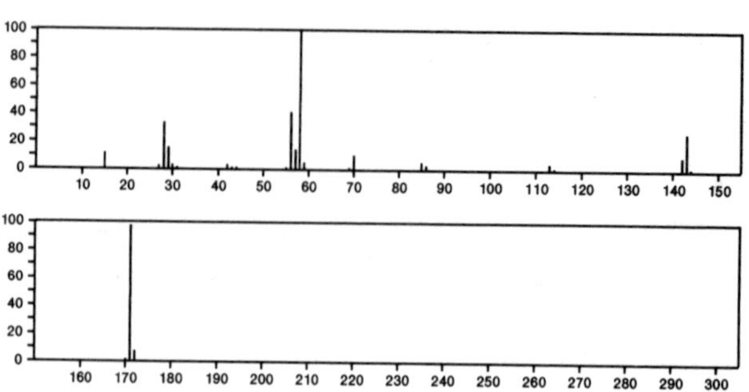

171 $C_6H_9N_3O_3$ 877–89–4
1,3,5–Triazine, 2,4,6–trimethoxy–

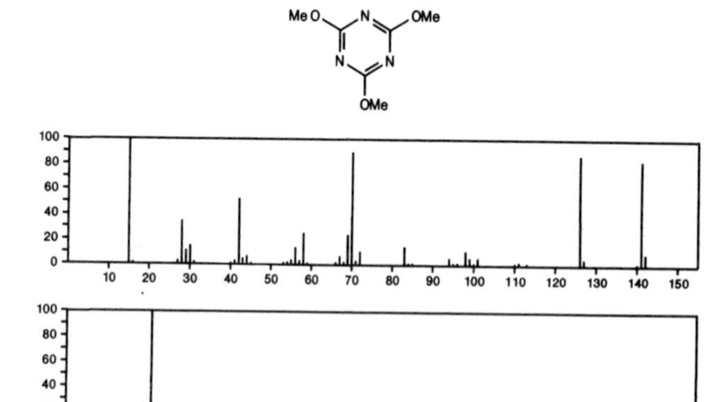

171 $C_6H_9N_3O_3$ 20379–33–3
Ribopyranoside, methyl 2,3–anhydro–4–azido–4–deoxy–, β–L–

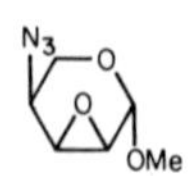

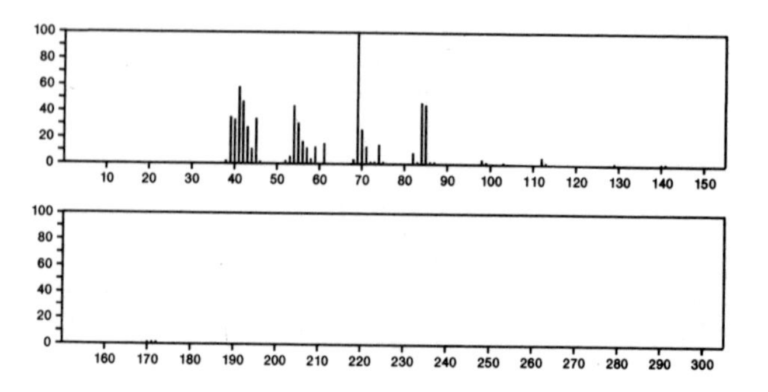

171 $C_7H_3Cl_2N$ 1194-65-6
Benzonitrile, 2,6-dichloro-

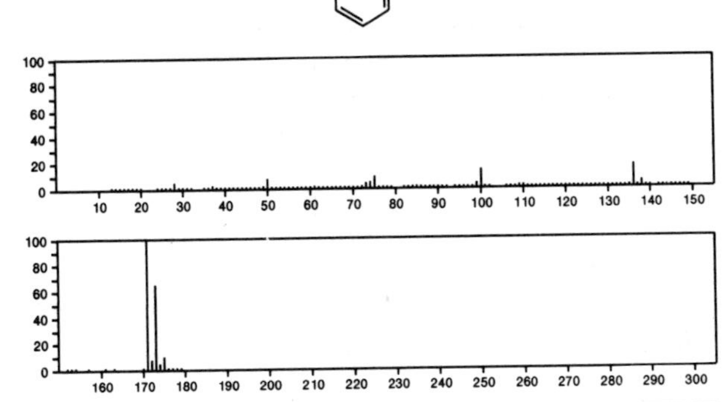

171 $C_7H_6ClNO_2$ 83-42-1
Benzene, 1-chloro-2-methyl-3-nitro-

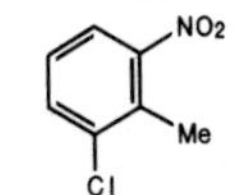

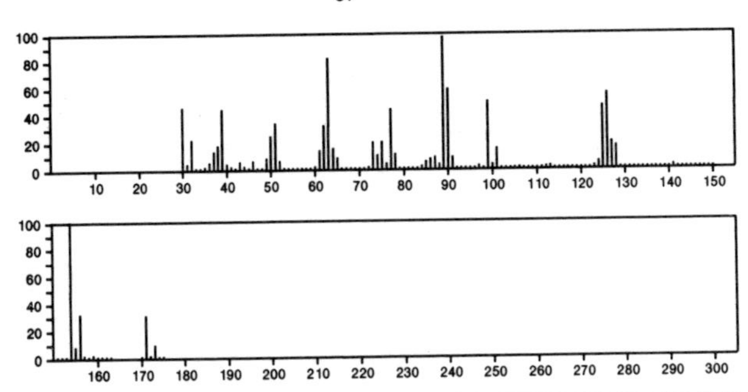

171 C_7H_6ClNS 26074-38-4
Methanethioamide, *N*-(2-chlorophenyl)-

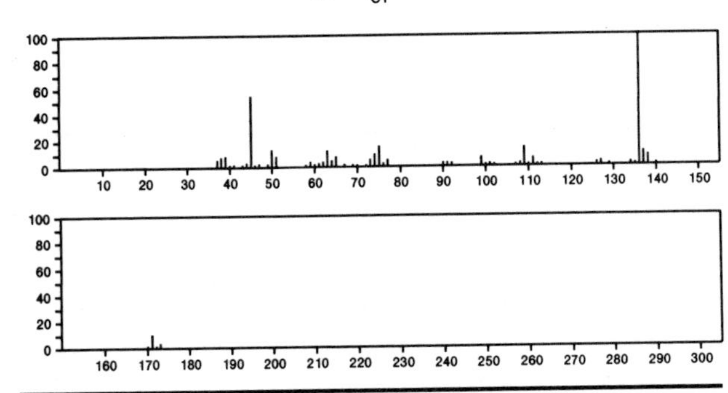

171 $C_7H_9NO_2S$ 70-55-3
Benzenesulfonamide, 4-methyl-

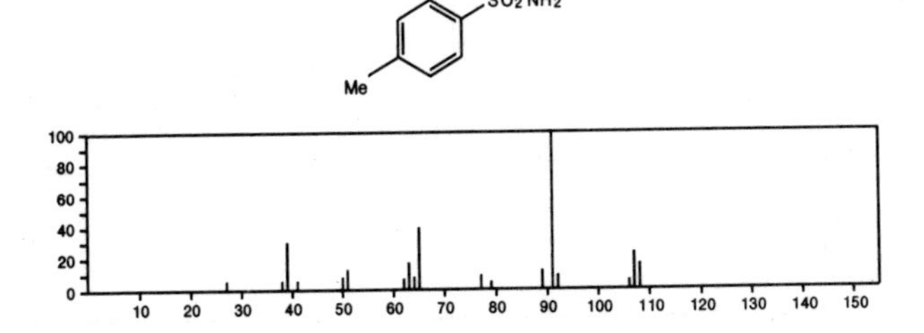

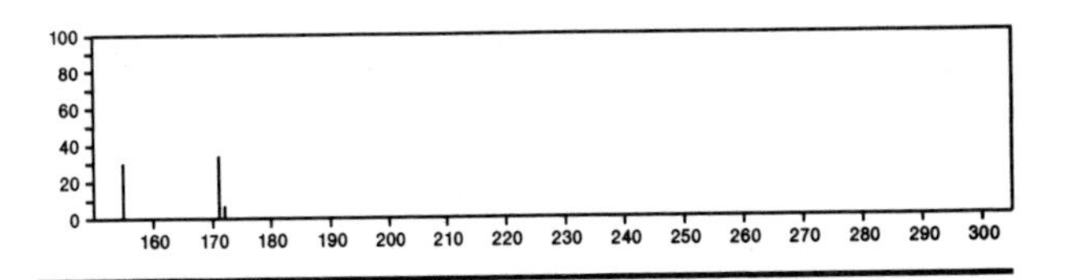

171 $C_7H_9NO_2S$ 88-19-7
Benzenesulfonamide, 2-methyl-

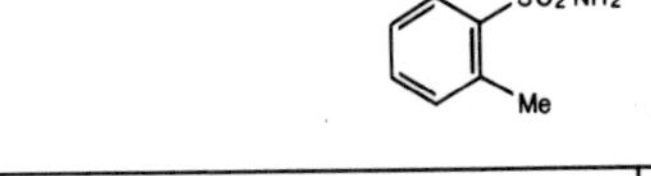

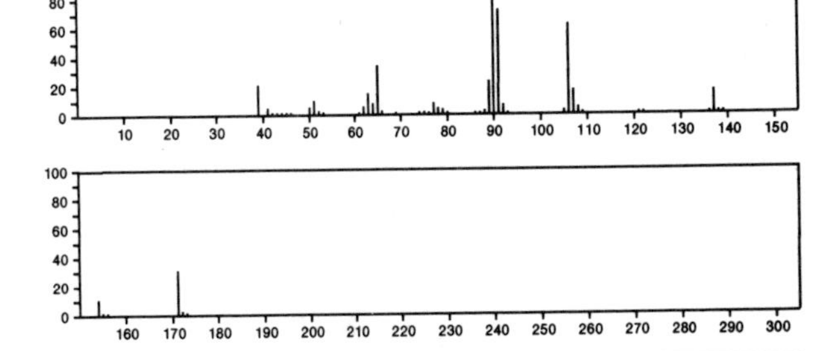

171 $C_7H_9NO_4$ 56805-18-6
Proline, 1-acetyl-5-oxo-

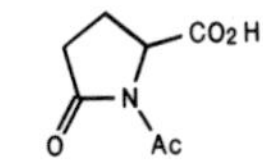

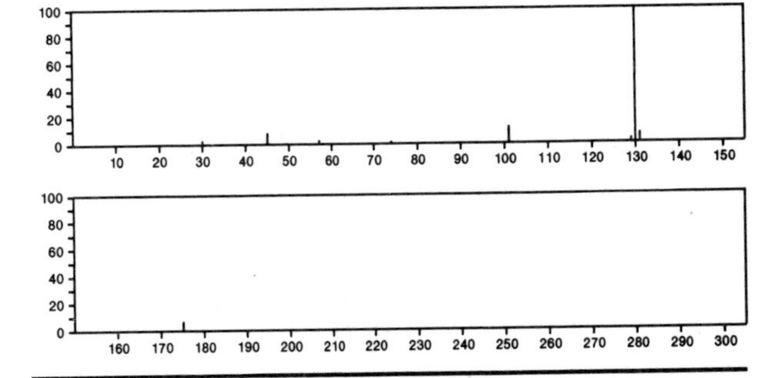

171 $C_7H_{13}N_3S$ 5351-77-9
Hydrazinecarbothioamide, 2-cyclohexylidene-

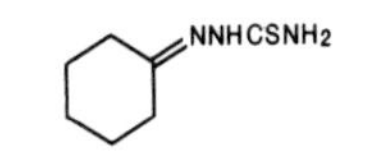

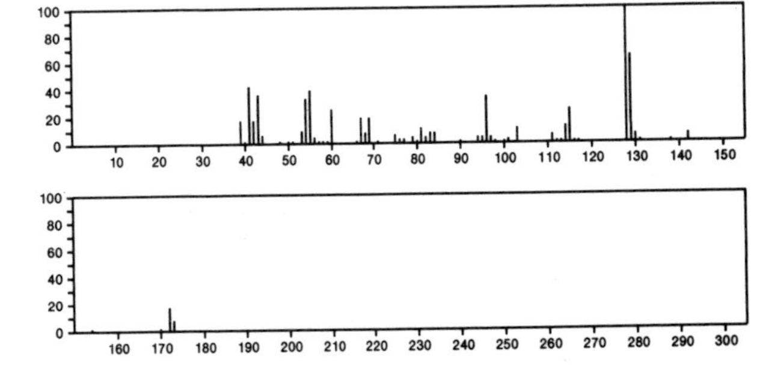

171 $C_8H_{13}NO_3$ 575-63-3
8-Azabicyclo[3.2.1]octan-3-one, 6,7-dihydroxy-8-methyl-, (*exo,exo*)-

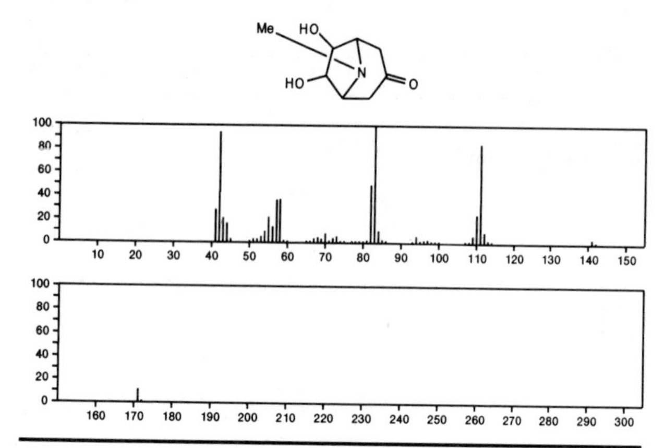

171 $C_8H_{13}NO_3$ 27460-51-1
L-Proline, 1-acetyl-, methyl ester

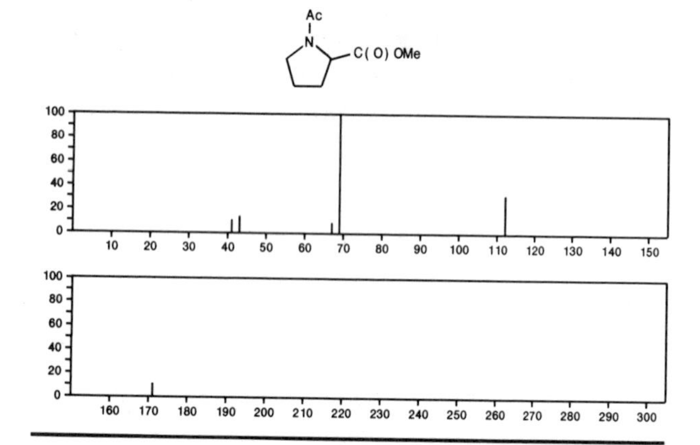

171 $C_8H_{13}NO_3$ 54725-52-9
8-Azabicyclo[3.2.1]octan-3-one, 6,7-dihydroxy-8-methyl-

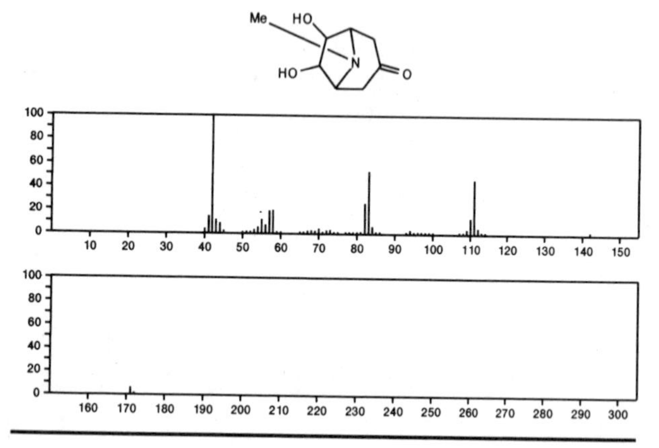

171 $C_8H_{13}NO_3$ 55649-53-1
Glycine, *N*-(2-methyl-1-oxo-2-butenyl)-, methyl ester, (*E*)-

MeCH=CMeCONHCH2C(O)OMe

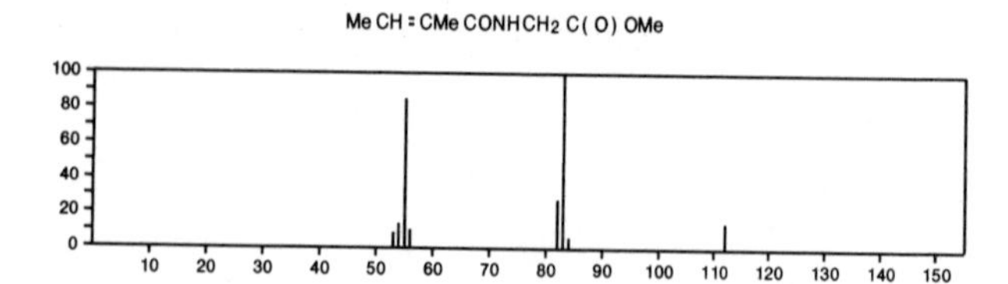

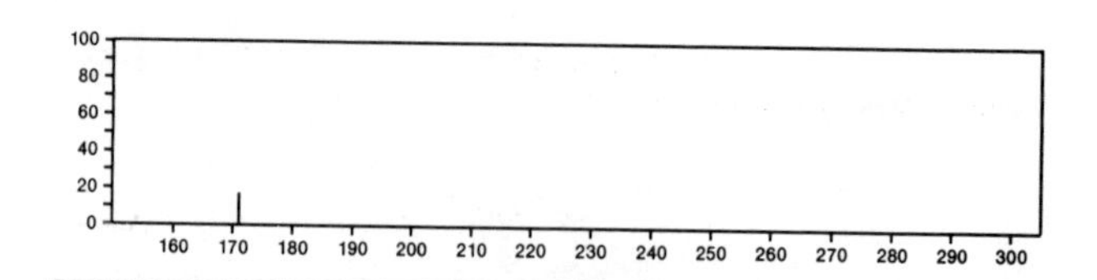

171 $C_8H_{13}NO_3$ 56009-34-8
Glycine, *N*-(3-methyl-1-oxo-2-butenyl)-, methyl ester

MeOC(O)CH2NHCOCH=CMe2

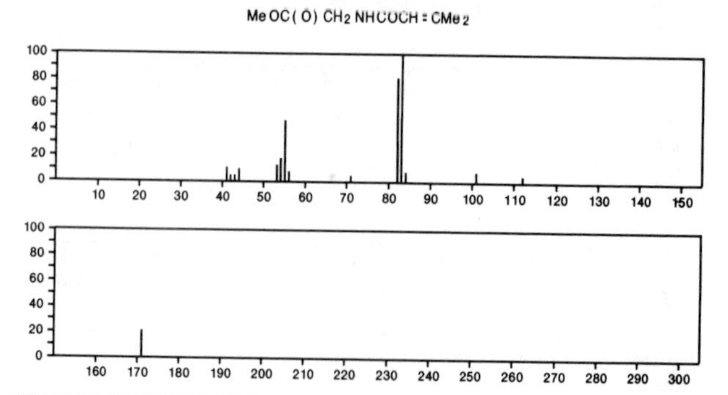

171 $C_8H_{13}NO_3$ 56145-23-4
2-Pyrrolidinecarboxylic acid, 1,2-dimethyl-5-oxo-, methyl ester

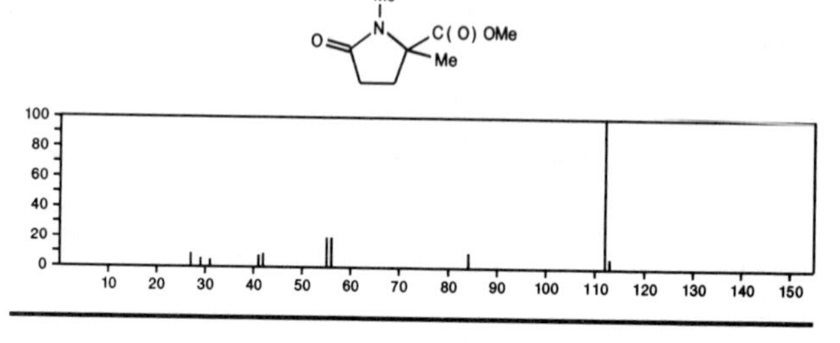

171 $C_8H_{17}NOSi$ 3553-93-3
2-Piperidinone, 1-(trimethylsilyl)-

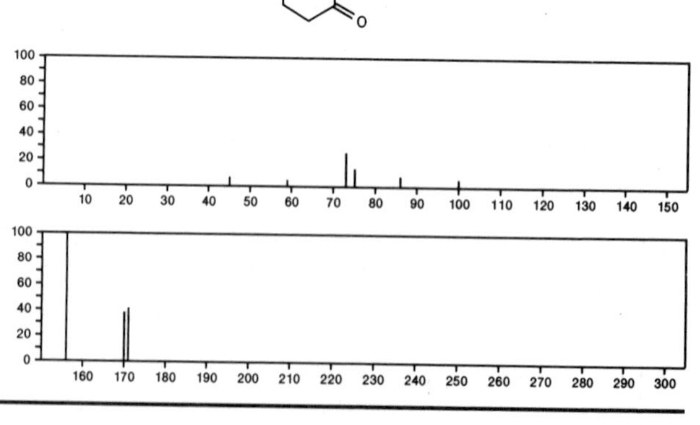

171 $C_8H_{17}N_3O$ 3622-68-2
4-Heptanone, semicarbazone

Pr2C=NNHCONH2

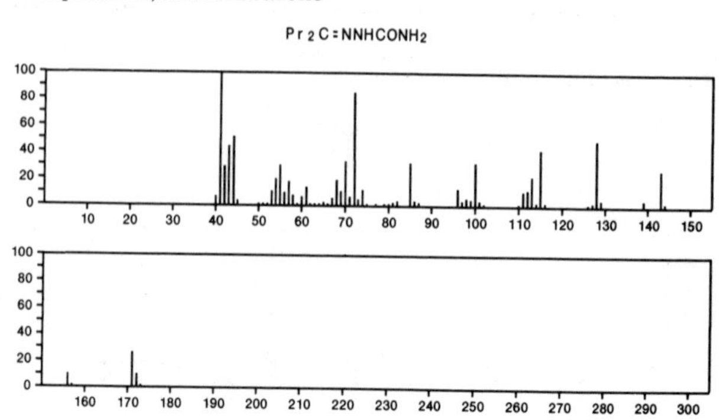

171 $C_8H_{17}N_3O$ 16519-71-4
Valeraldehyde, 2,2-dimethyl-, semicarbazone

$H_2NCONHN{=}CHCPrMe_2$

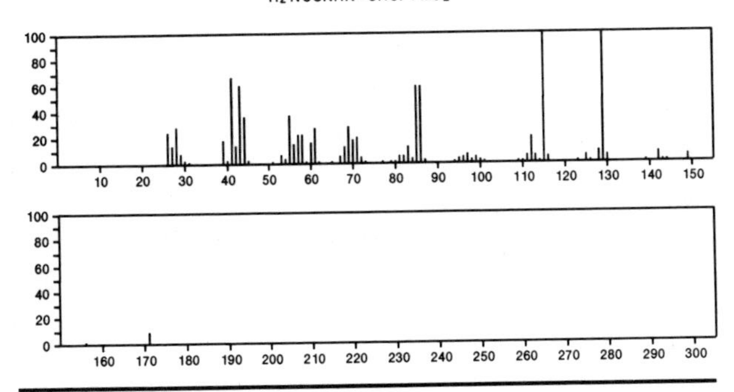

171 $C_9H_5N_3O$ 18457-81-3
2-Quinoxalinecarbonitrile, 1-oxide

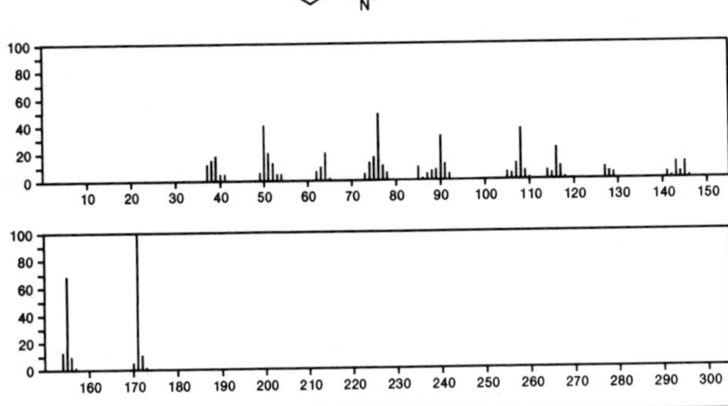

171 $C_9H_{13}N.ClH$ 30684-05-0
Propylamine, 3-phenyl-, hydrochloride

$Ph(CH_2)_3NH_2 \bullet HCl$

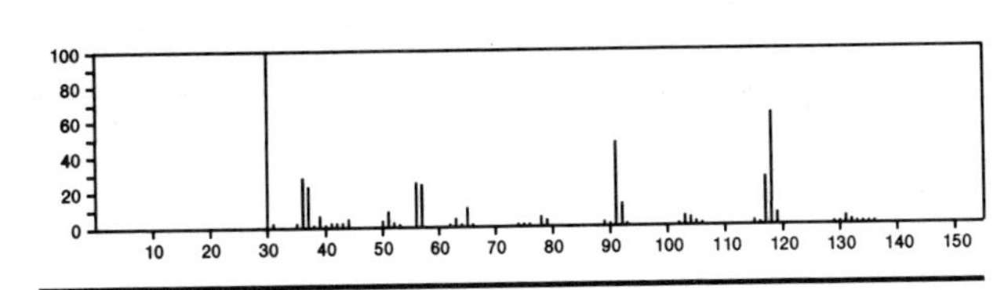

171 $C_9H_{17}NO_2$ 28248-38-6
Cyclooctanecarboxylic acid, 1-amino-

171 $C_9H_{17}NO_2$ 39077-13-9
6-Azabicyclo[3.2.1]octan-8-ol, 5-methoxy-6-methyl-, *syn*-

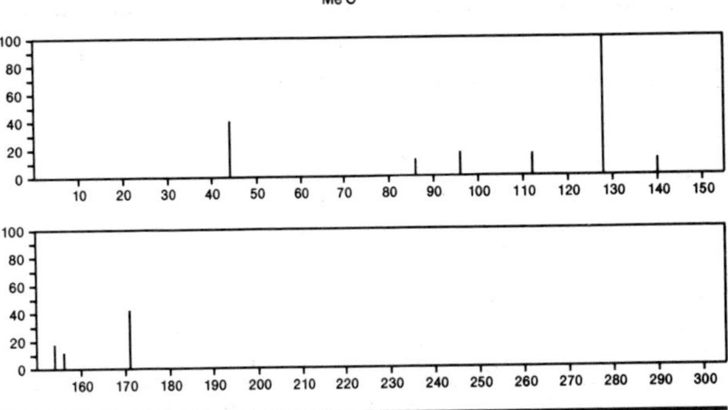

171 $C_9H_{17}NO_2$ 49656-39-5
9-Azabicyclo[4.2.1]nonane-2,5-diol, 9-methyl-, (*endo,endo*)-

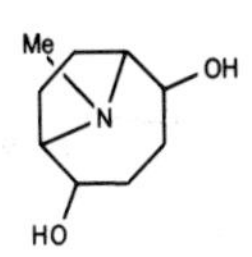

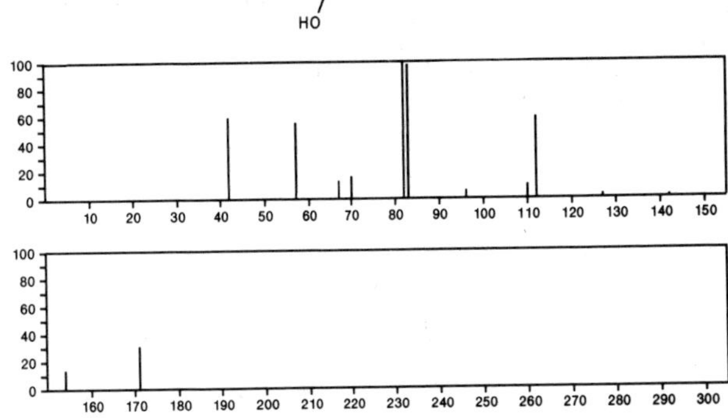

171 $C_9H_{17}NO_2$ 49656-40-8
9-Azabicyclo[3.3.1]nonane-2,6-diol, 9-methyl-, (*endo,endo*)-

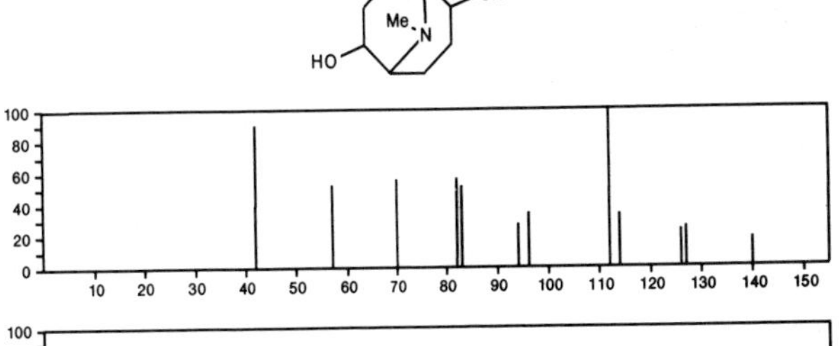

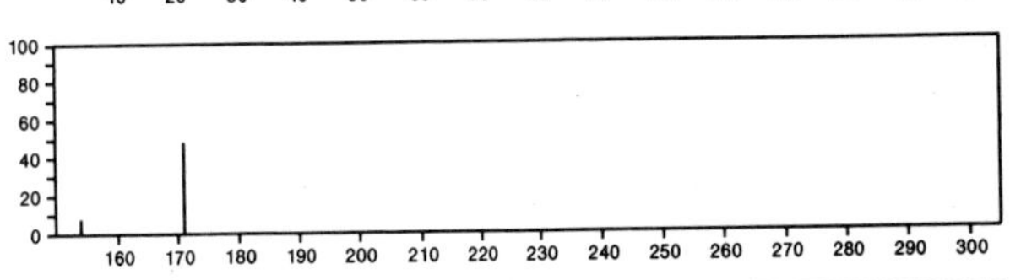

171 $C_9H_{17}NO_2$ 54725-47-2
8-Azabicyclo[3.2.1]octan-3-ol, 6-methoxy-8-methyl-

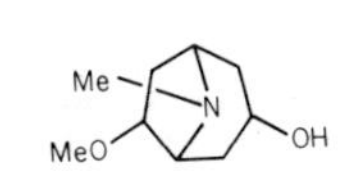

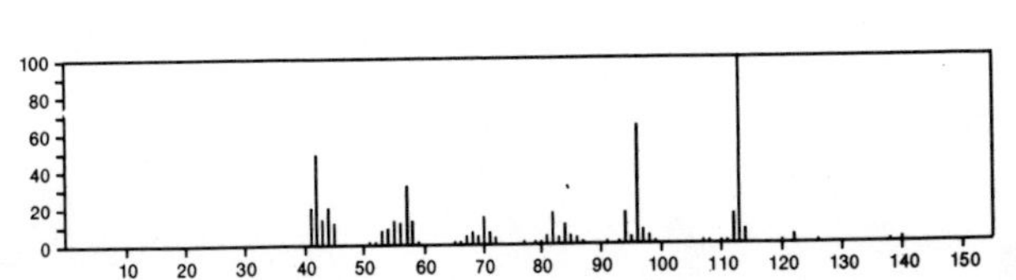

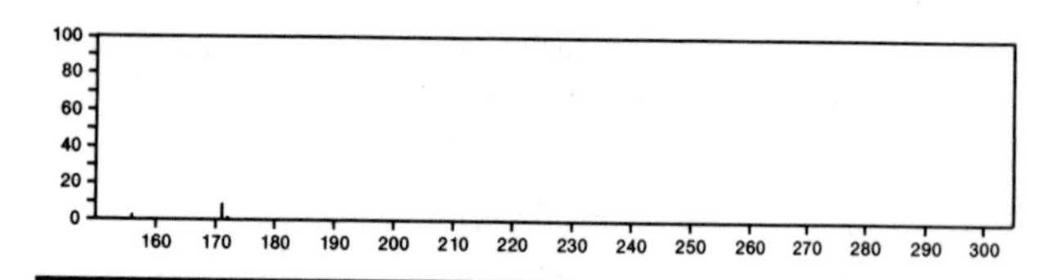

171 $C_9H_{17}NO_2$ 56051-37-7

8-Azabicyclo[3.2.1]octan-3-ol, 6-methoxy-8-methyl-, (3-*endo*,6-*exo*)-

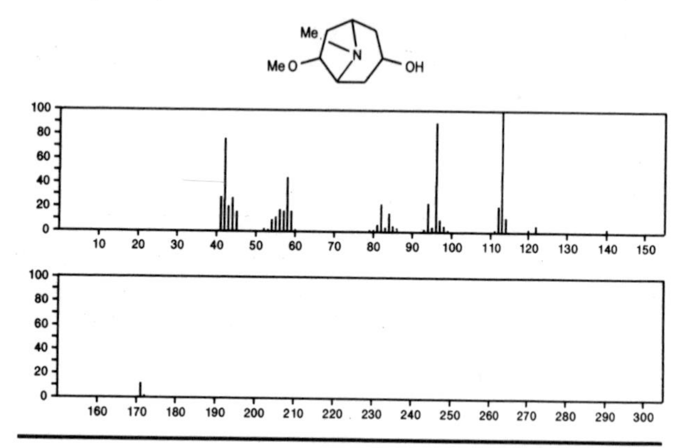

171 $C_9H_{21}N_3$ 7779-27-3

1,3,5-Triazine, 1,3,5-triethylhexahydro-

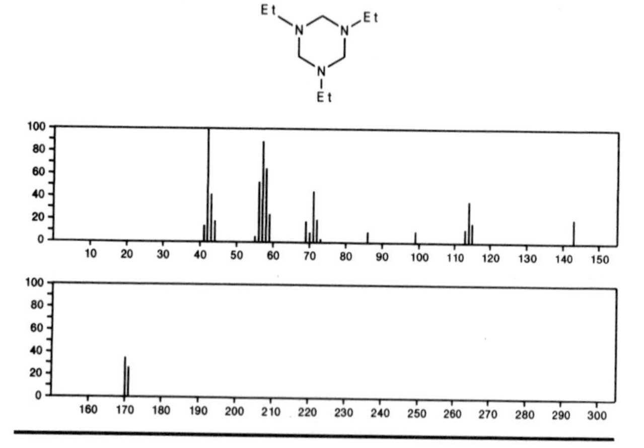

171 $C_{10}H_9N_3$ 3435-23-2

Pyrimidine, 5-amino-4-phenyl-

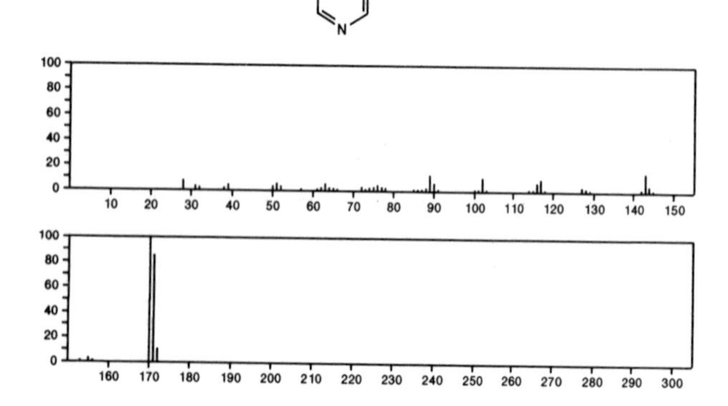

171 $C_{10}H_9N_3$ 13535-13-2

Pyrazinamine, 5-phenyl-

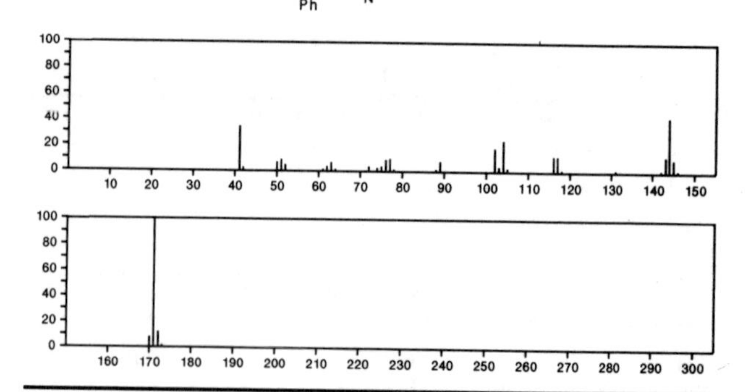

171 $C_{10}H_{21}NO$ 1563-90-2

Acetamide, *N*,*N*-dibutyl-

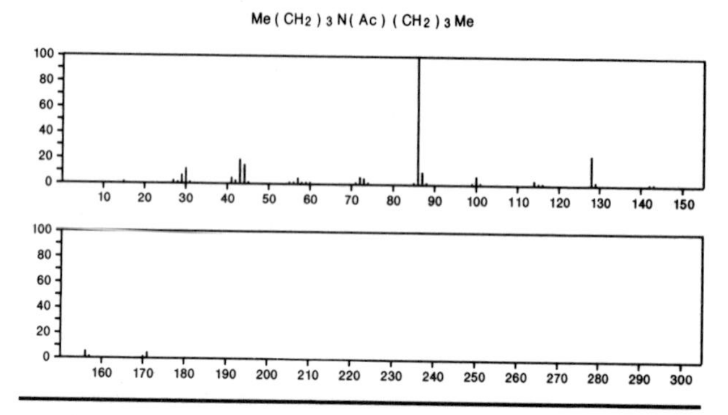

171 $C_{10}H_{21}NO$ 6282-97-9

Hexanamide, *N*,*N*-diethyl-

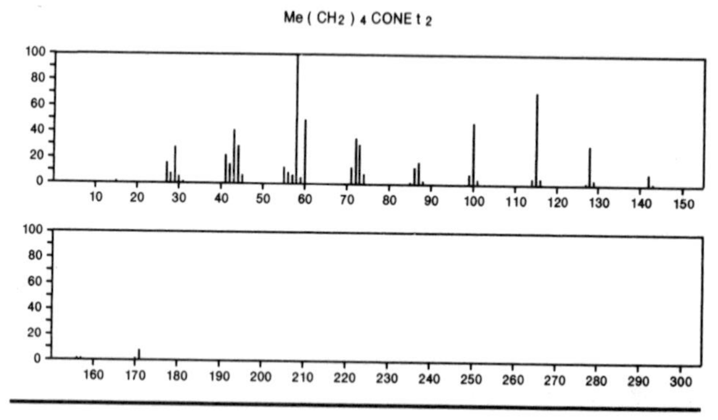

171 $C_{10}H_{21}NO$ 10264-17-2

Butanamide, *N*-hexyl-

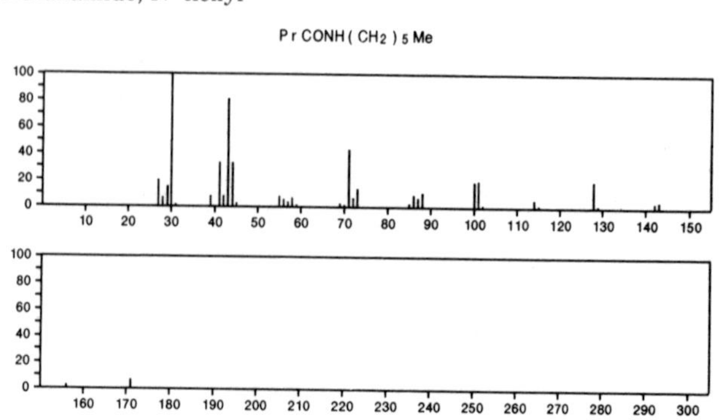

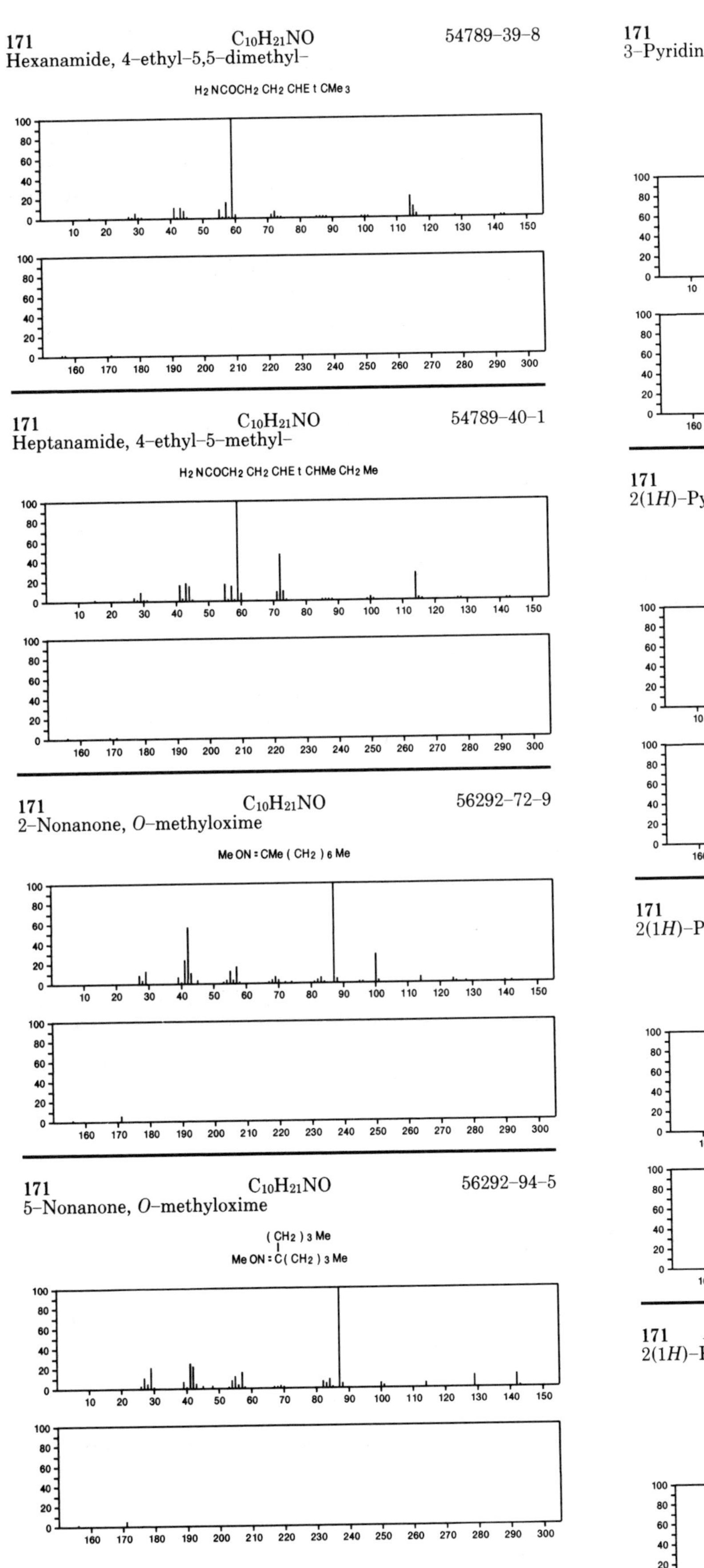
171 C10H21NO 54789-39-8
Hexanamide, 4-ethyl-5,5-dimethyl-
H2NCOCH2CH2CHEtCMe3
171 C10H21NO 54789-40-1
Heptanamide, 4-ethyl-5-methyl-
H2NCOCH2CH2CHEtCHMeCH2Me
171 C10H21NO 56292-72-9
2-Nonanone, O-methyloxime
MeON=CMe(CH2)6Me
171 C10H21NO 56292-94-5
5-Nonanone, O-methyloxime
(CH2)3Me
MeON=C(CH2)3Me

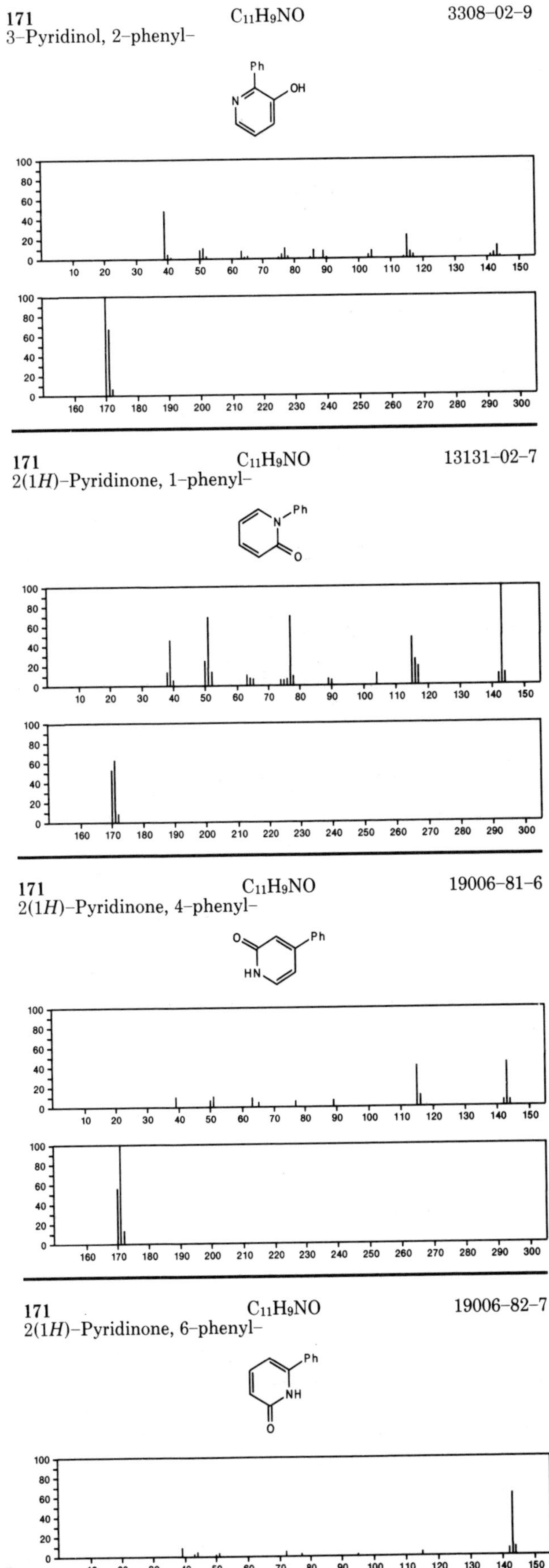
171 C11H9NO 3308-02-9
3-Pyridinol, 2-phenyl-
Ph
N
OH
171 C11H9NO 13131-02-7
2(1H)-Pyridinone, 1-phenyl-
N
Ph
O
171 C11H9NO 19006-81-6
2(1H)-Pyridinone, 4-phenyl-
O
Ph
HN
171 C11H9NO 19006-82-7
2(1H)-Pyridinone, 6-phenyl-
Ph
NH
O

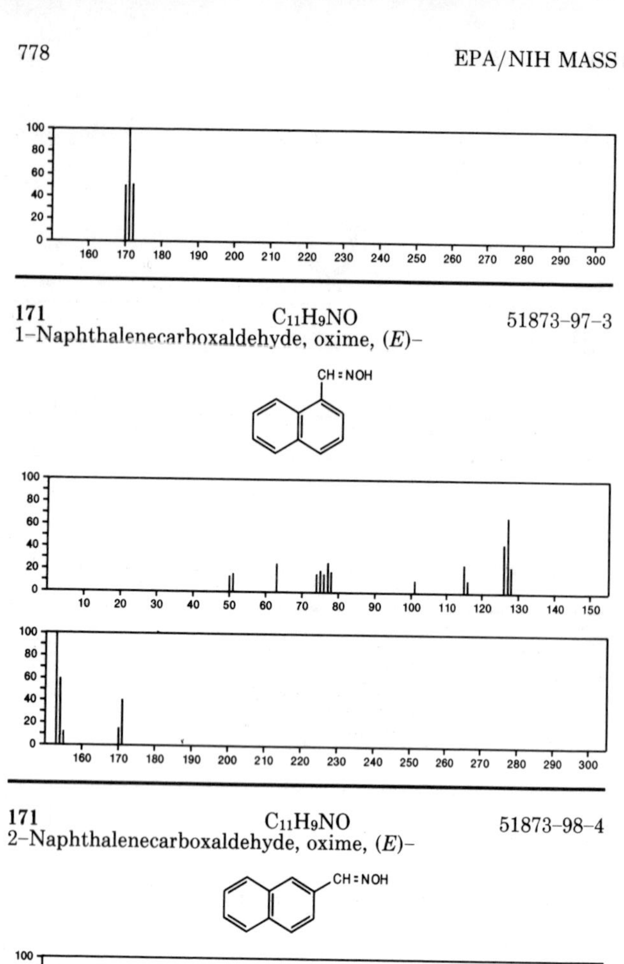

171 $C_{11}H_9NO$ 51873-97-3
1-Naphthalenecarboxaldehyde, oxime, (*E*)-

171 $C_{11}H_9NO$ 51873-98-4
2-Naphthalenecarboxaldehyde, oxime, (*E*)-

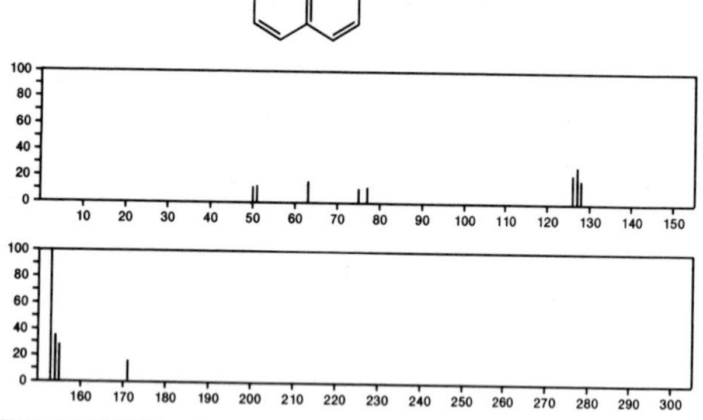

171 $C_{11}H_{25}N$ 7307-55-3
1-Undecanamine

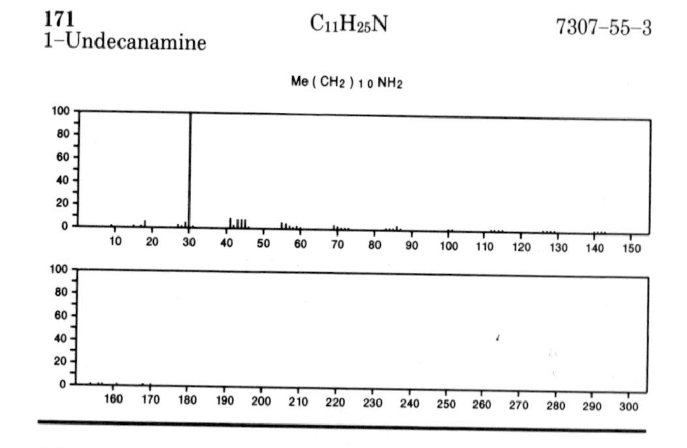

171 $C_{11}H_{25}N$ 41495-45-8
1-Hexanamine, *N*-pentyl-

Me(CH2)5NH(CH2)4Me

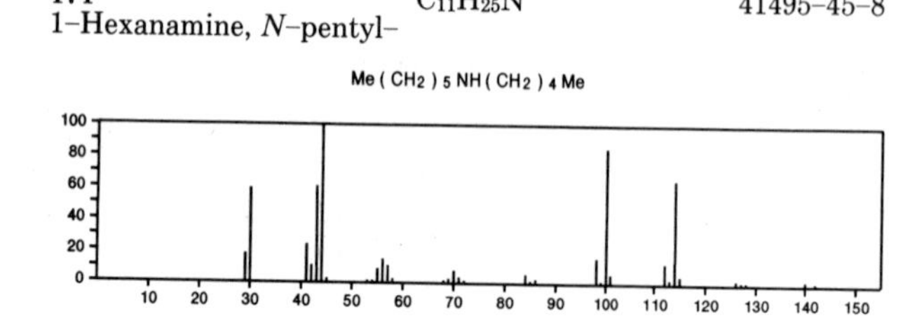

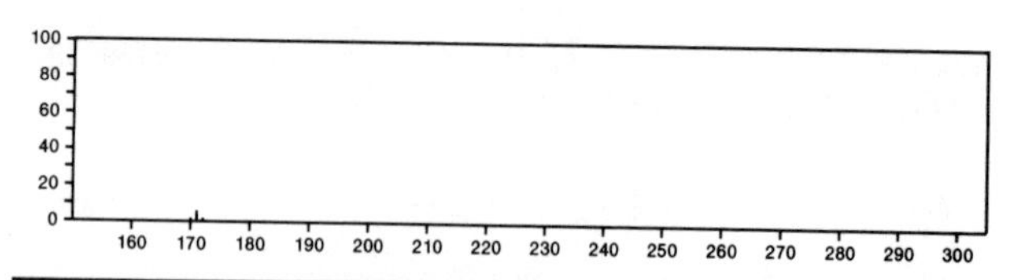

171 $C_{12}H_{13}N$ 942-01-8
1*H*-Carbazole, 2,3,4,9-tetrahydro-

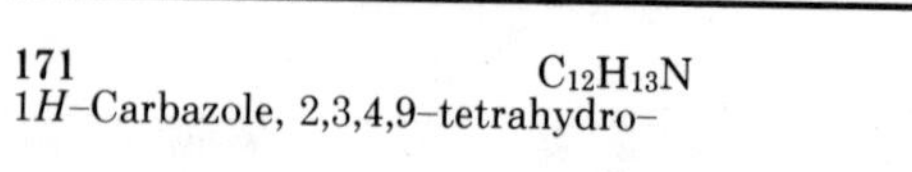

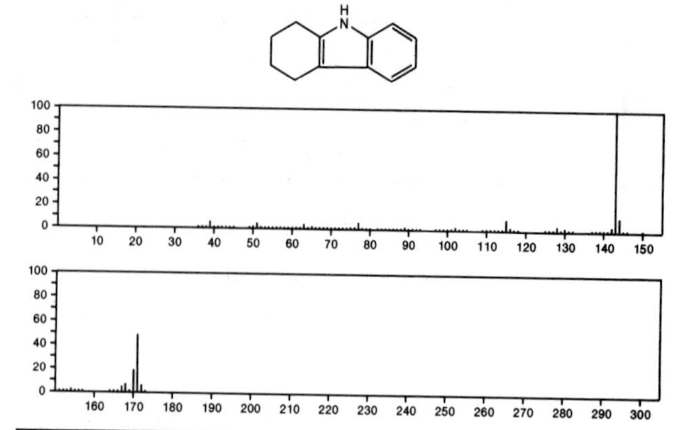

171 $C_{12}H_{13}N$ 1613-32-7
Quinoline, 2-propyl-

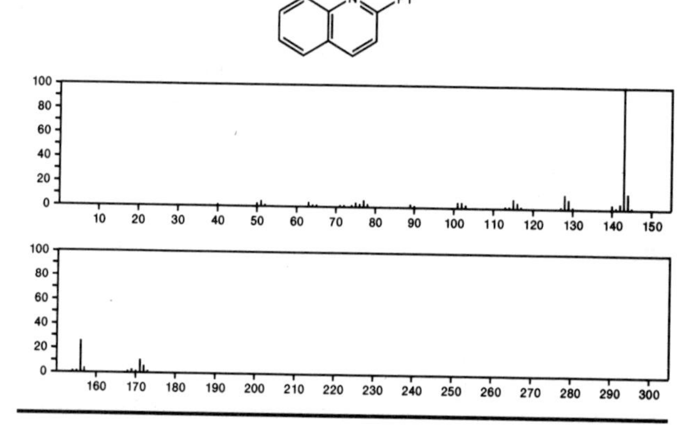

171 $C_{12}H_{13}N$ 2437-72-1
Quinoline, 2,3,4-trimethyl-

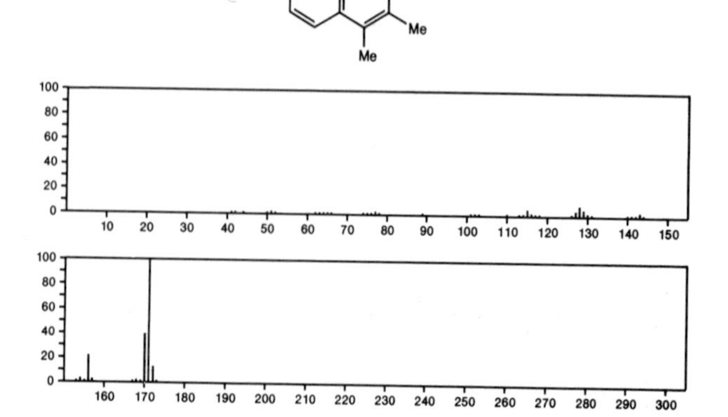

171 $C_{12}H_{13}N$ 7661-37-2
Isoquinoline, 1-propyl-

171 $C_{12}H_{13}N$ 7661-53-2
Quinoline, 8-propyl-

171 $C_{12}H_{13}N$ 7661-58-7
Quinoline, 6-propyl-

171 $C_{12}H_{13}N$ 7661-59-8
Quinoline, 7-propyl-

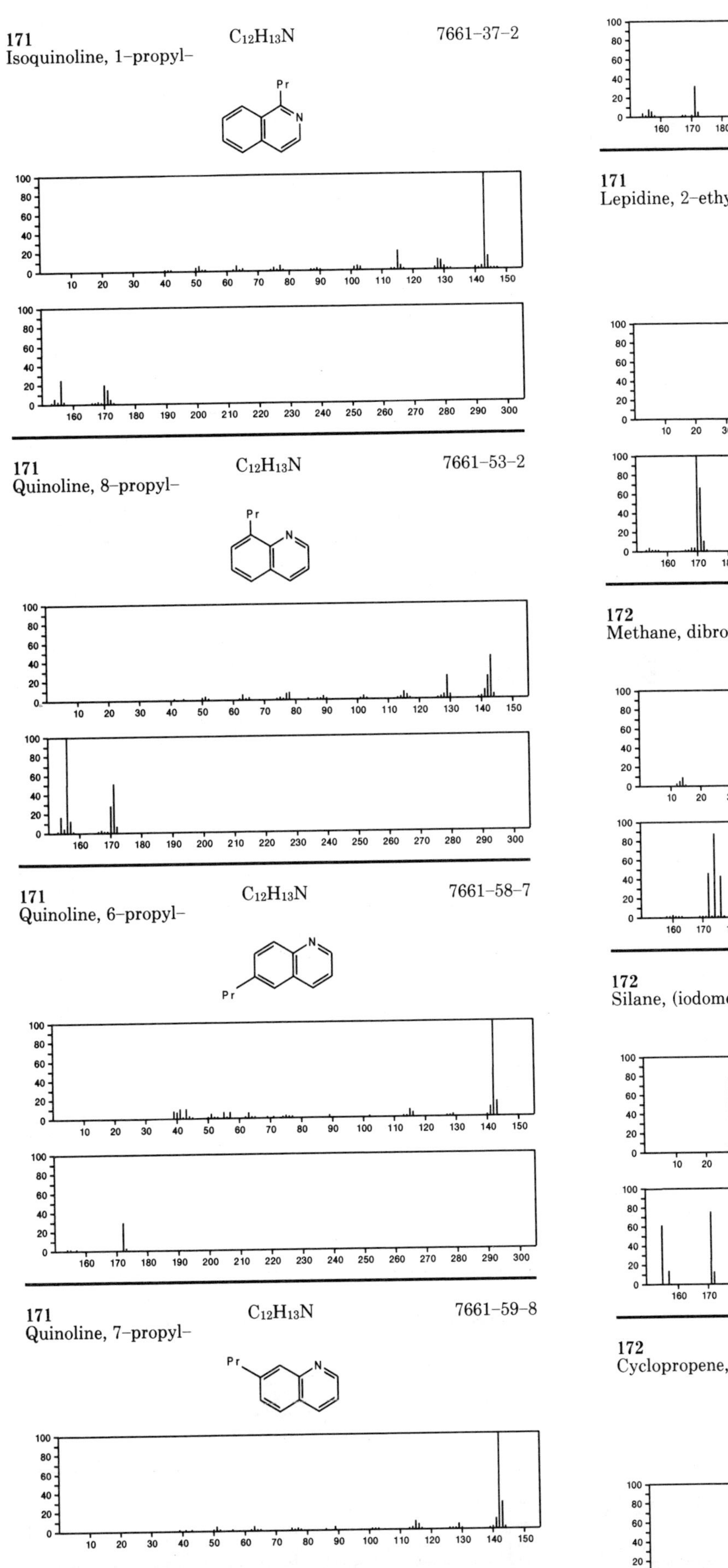

171 $C_{12}H_{13}N$ 33357-44-7
Lepidine, 2-ethyl-

172 CH_2Br_2 74-95-3
Methane, dibromo-

172 CH_5ISi 7570-22-1
Silane, (iodomethyl)-

172 C_3BrF_3 29777-44-4
Cyclopropene, 1-bromo-2,3,3-trifluoro-

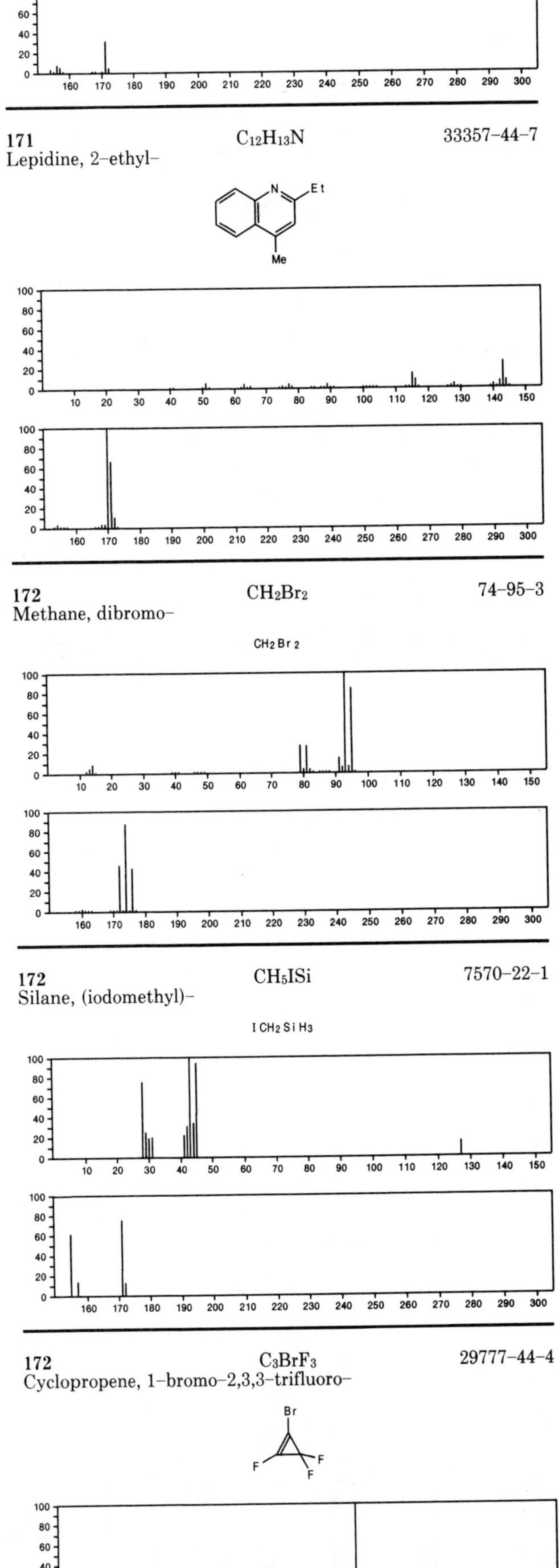

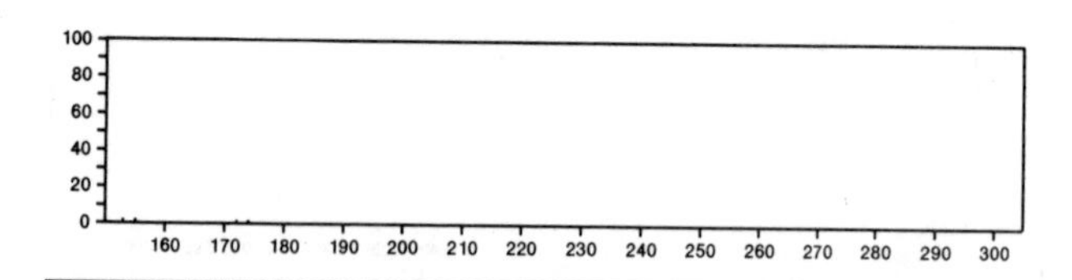

172 C_3H_3BrClF 24071-59-8
Cyclopropane, 1-bromo-1-chloro-2-fluoro-

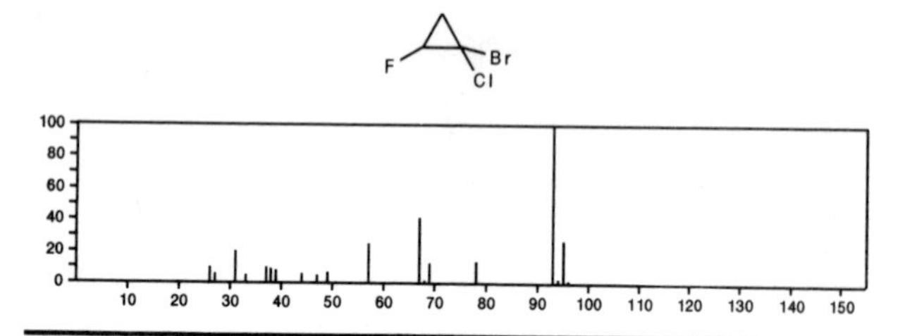

172 $C_3H_9O_2PS_2$ 2953-29-9
Phosphorodithioic acid, *O,O,S*-trimethyl ester

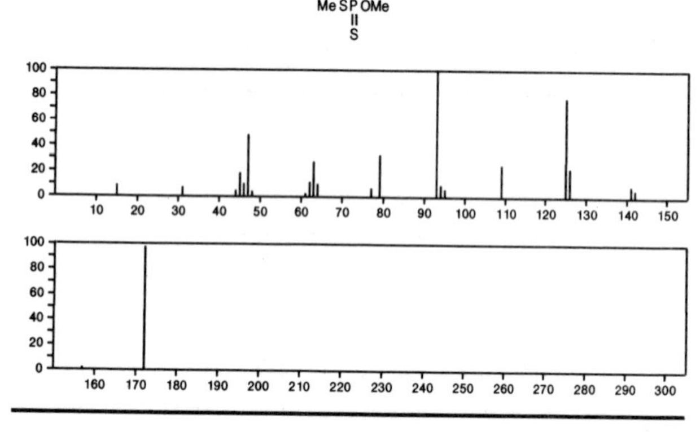

172 $C_3H_9O_2PS_2$ 22608-53-3
Phosphorodithioic acid, *O,S,S*-trimethyl ester

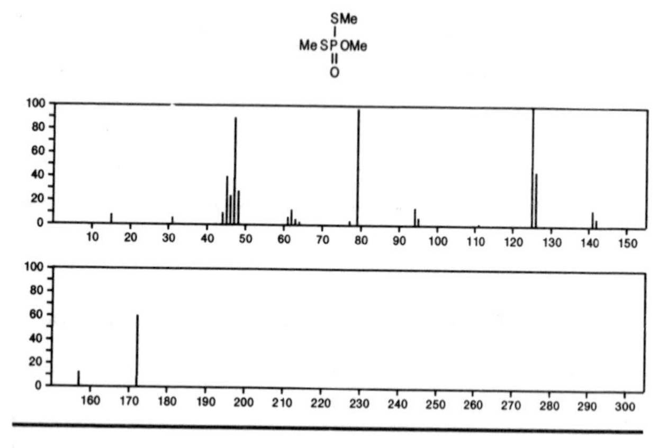

172 C_4HCoO_4 16842-03-8
Cobalt, tetracarbonylhydro-

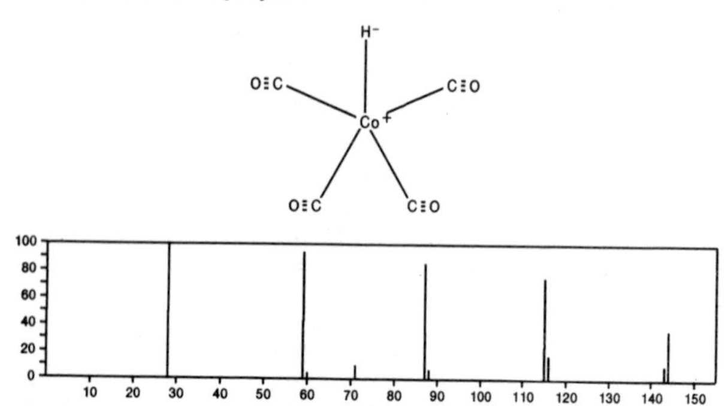

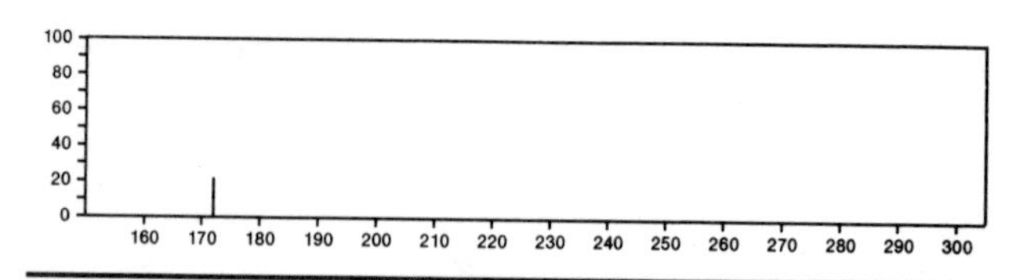

172 $C_4H_{10}ClO_3P$ 814-49-3
Phosphorochloridic acid, diethyl ester

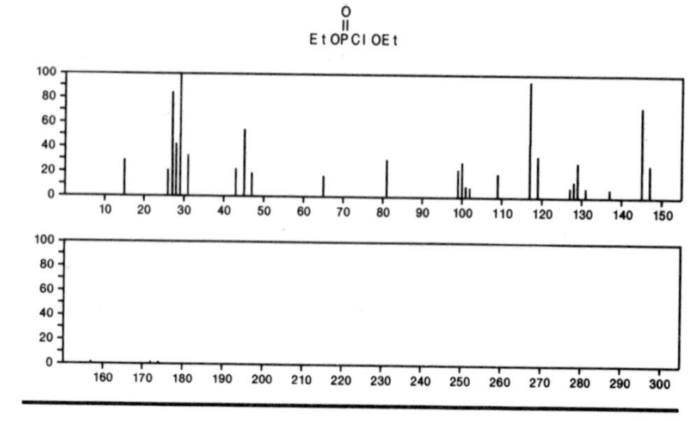

172 $C_5H_7Cl_3$ 2677-33-0
1-Pentene, 1,1,5-trichloro-

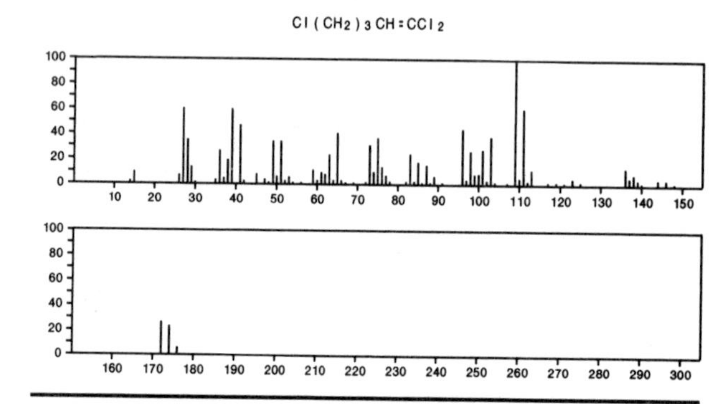

172 $C_5H_{10}Cl_2O_2$ 111-91-1
Ethane, 1,1'-[methylenebis(oxy)]bis[2-chloro-

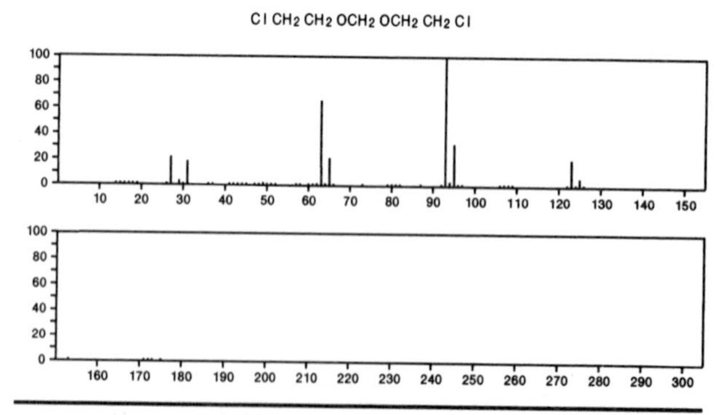

172 $C_6H_4O_6$ 319-89-1
2,5-Cyclohexadiene-1,4-dione, 2,3,5,6-tetrahydroxy-

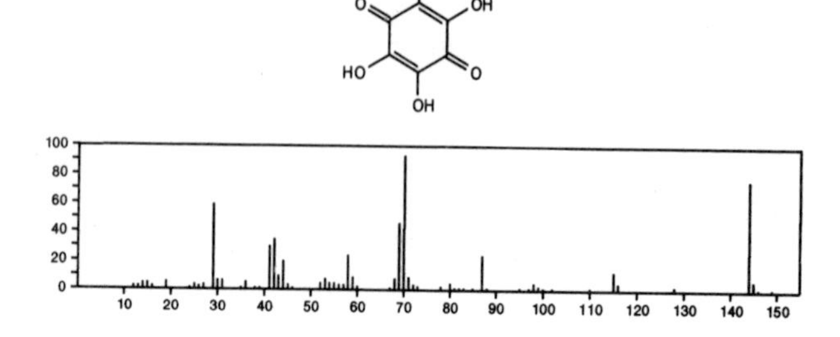

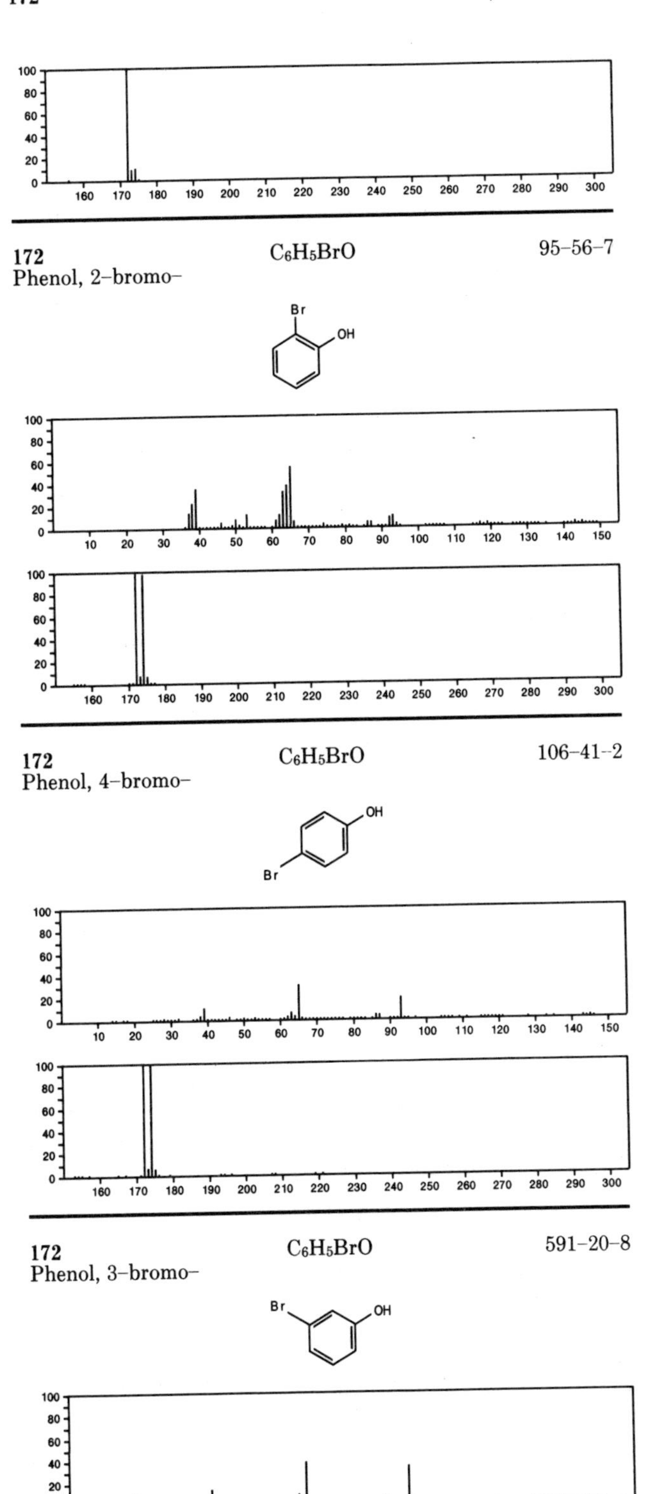

172 C_6H_5BrO 95-56-7
Phenol, 2-bromo-

172 C_6H_5BrO 106-41-2
Phenol, 4-bromo-

172 C_6H_5BrO 591-20-8
Phenol, 3-bromo-

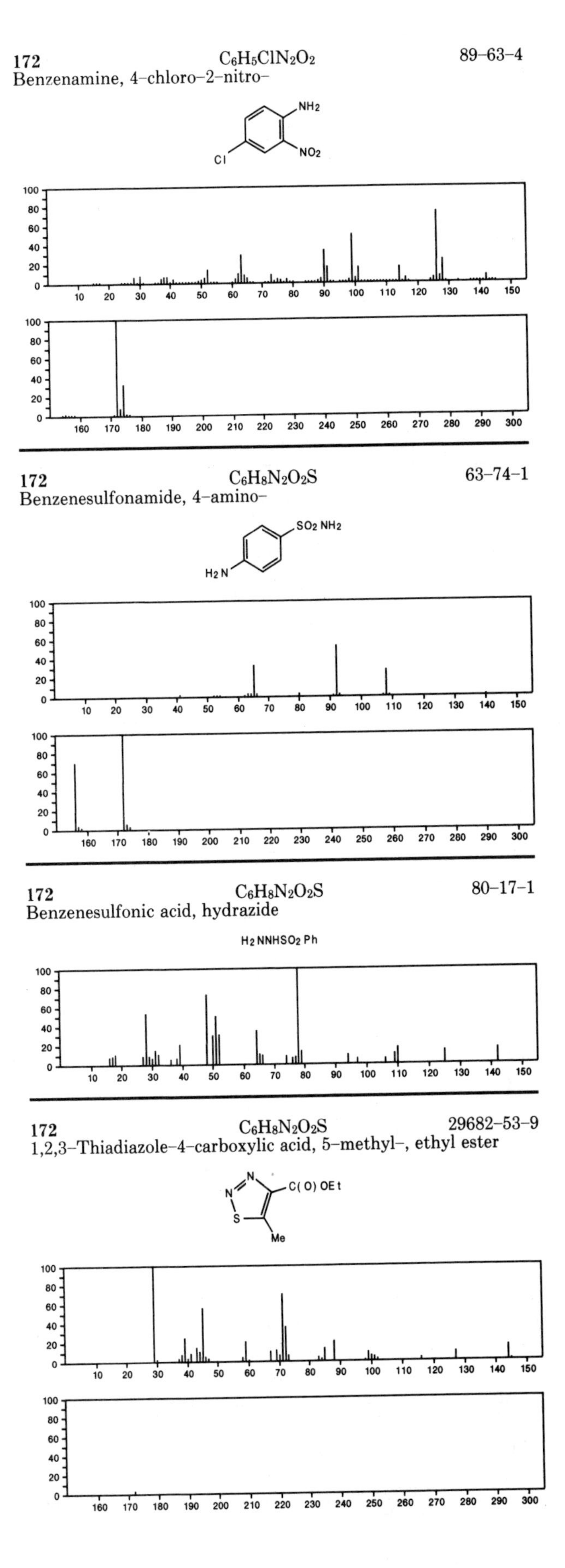

172 $C_6H_5ClN_2O_2$ 89-63-4
Benzenamine, 4-chloro-2-nitro-

172 $C_6H_8N_2O_2S$ 63-74-1
Benzenesulfonamide, 4-amino-

172 $C_6H_8N_2O_2S$ 80-17-1
Benzenesulfonic acid, hydrazide

172 $C_6H_8N_2O_2S$ 29682-53-9
1,2,3-Thiadiazole-4-carboxylic acid, 5-methyl-, ethyl ester

172 $C_6H_{12}N_4O_2$ 55380-34-2
Piperazine, 2,6-dimethyl-1,4-dinitroso-

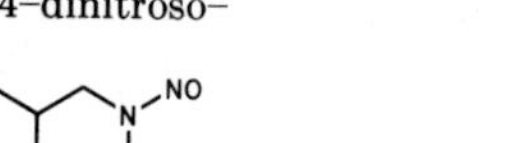

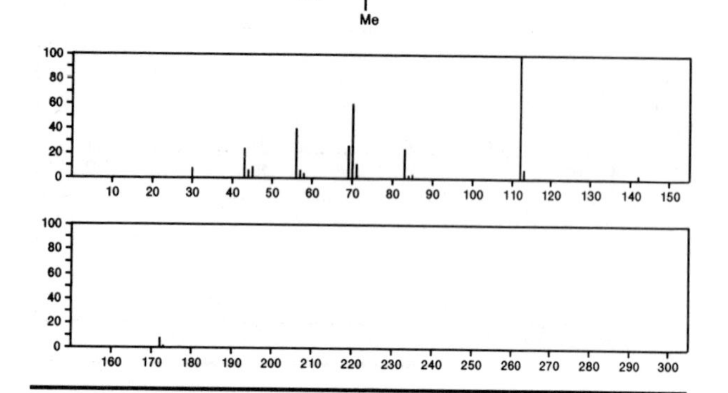

172 $C_6H_{12}N_4O_2$ 55556-88-2
Piperazine, 2,5-dimethyl-1,4-dinitroso-

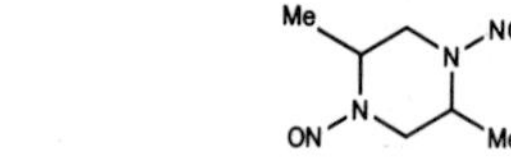

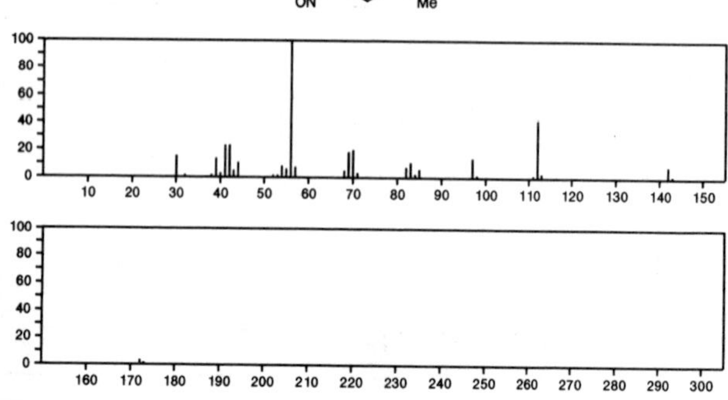

172 $C_6H_{12}N_4O_2$ 55556-89-3
1,5-Diazocine, octahydro-1,5-dinitroso-

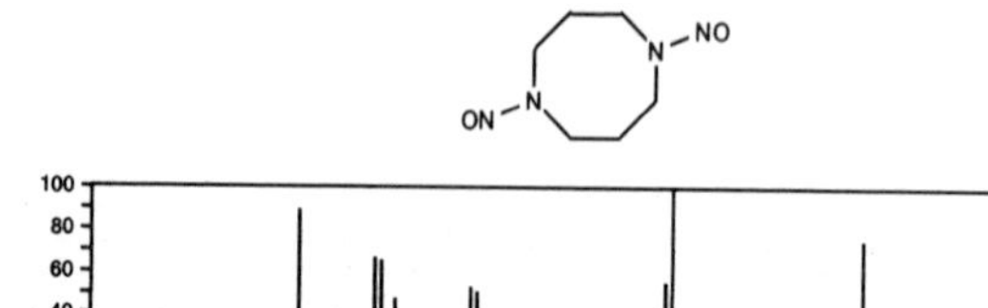

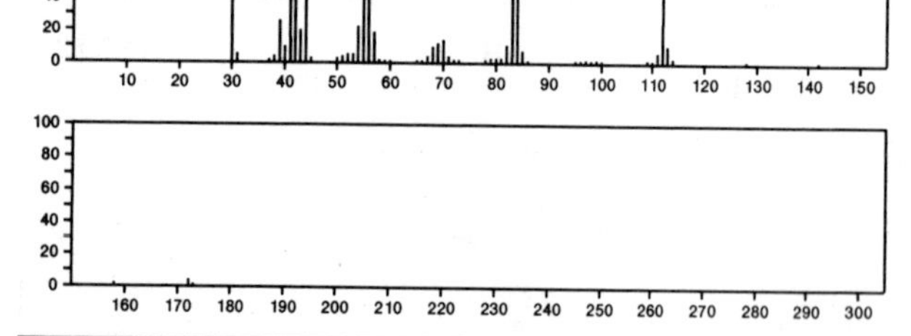

172 C_7H_5ClOS 1005-56-7
Carbonochloridothioic acid, *O*-phenyl ester

Cl C(S) OPh

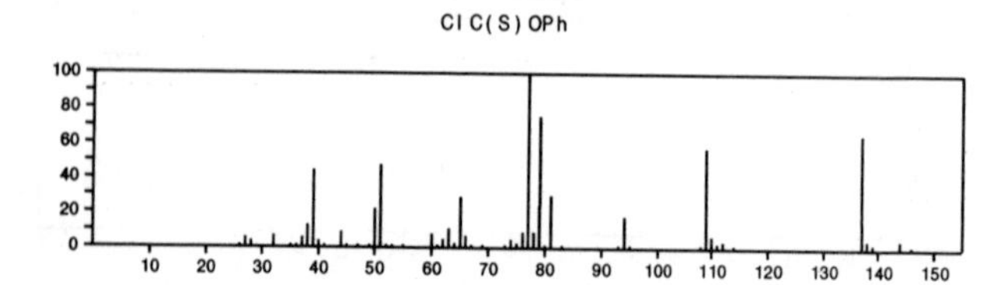

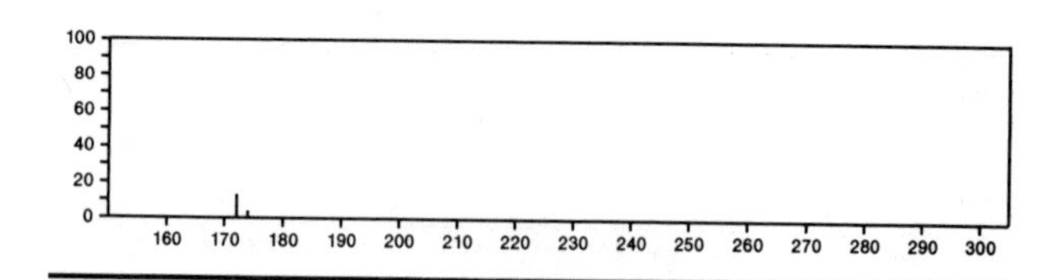

172 C_7H_5ClOS 13464-19-2
Carbonochloridothioic acid, *S*-phenyl ester

Cl C(O) SPh

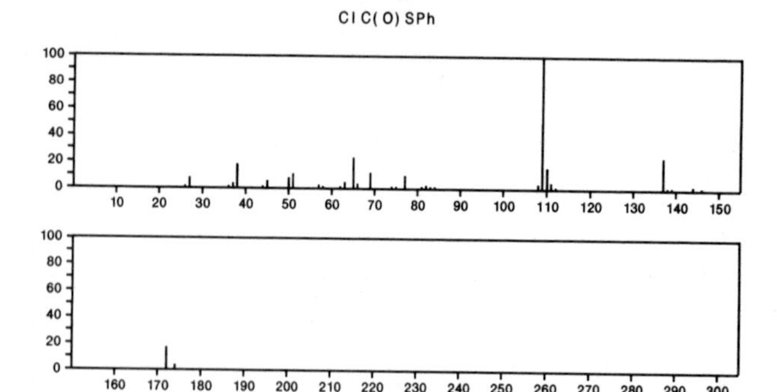

172 $C_7H_8OS_2$ 1005-55-6
2-Propanone, 1-(5-methyl-3*H*-1,2-dithiol-3-ylidene)-

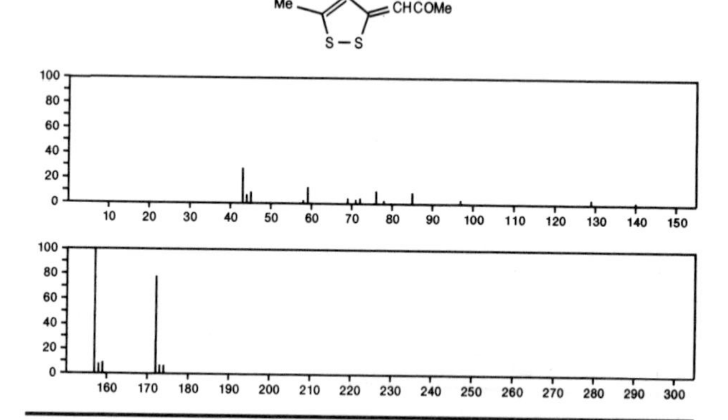

172 $C_7H_8OS_2$ 20849-29-0
2-Furancarbodithioic acid, ethyl ester

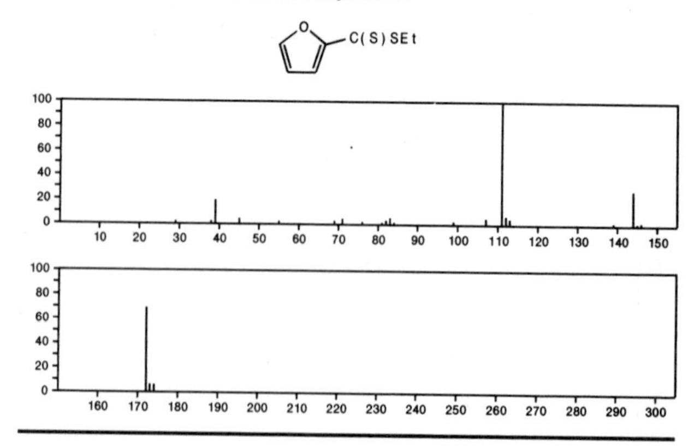

172 $C_7H_8O_3S$ 104-15-4
Benzenesulfonic acid, 4-methyl-

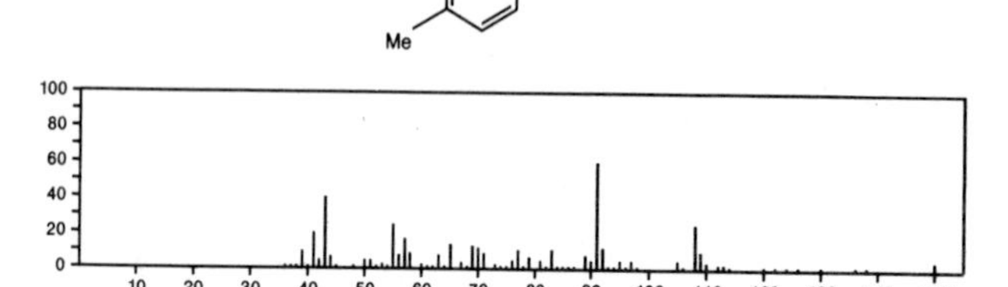

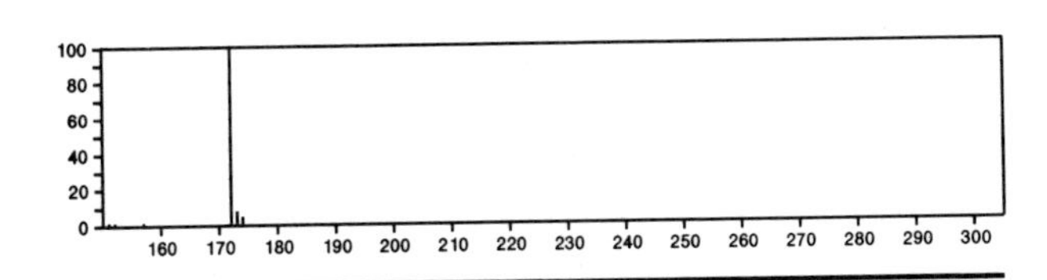

172 $C_7H_8O_3S$ 2158-88-5
2-Thiophenecarboxylic acid, 3-hydroxy-, ethyl ester

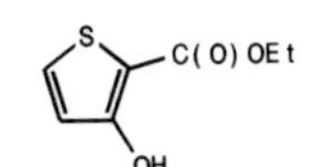

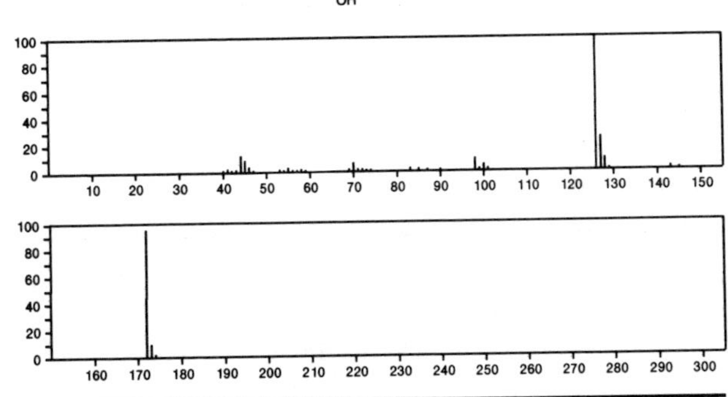

172 $C_7H_8O_3S$ 7210-60-8
2-Thiophenecarboxylic acid, 5-hydroxy-, ethyl ester

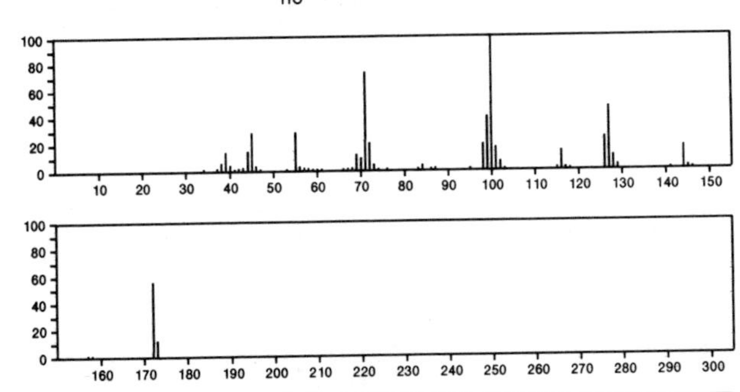

172 $C_7H_8O_3S$ 27489-33-4
Phenol, *o*-(methylsulfonyl)-

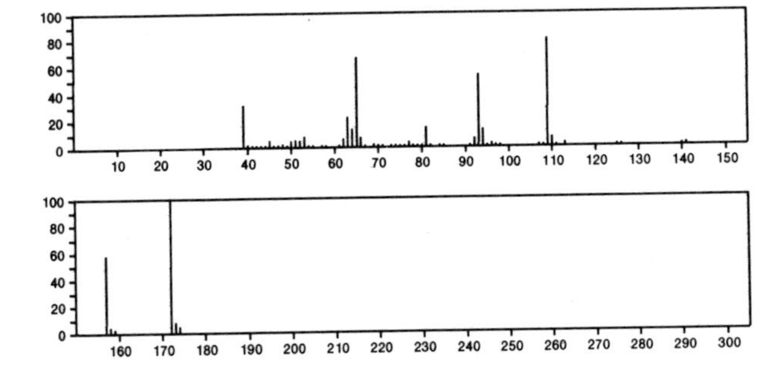

172 $C_7H_9ClN_2O$ 20551-33-1
4(3*H*)-Pyrimidinone, 5-chloro-2-ethyl-6-methyl-

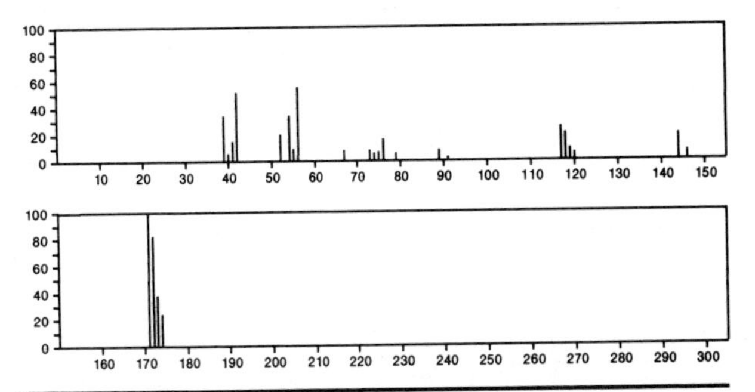

172 $C_7H_9ClN_2O$ 24611-12-9
Pyrimidine, 4-chloro-5-ethoxy-2-methyl-

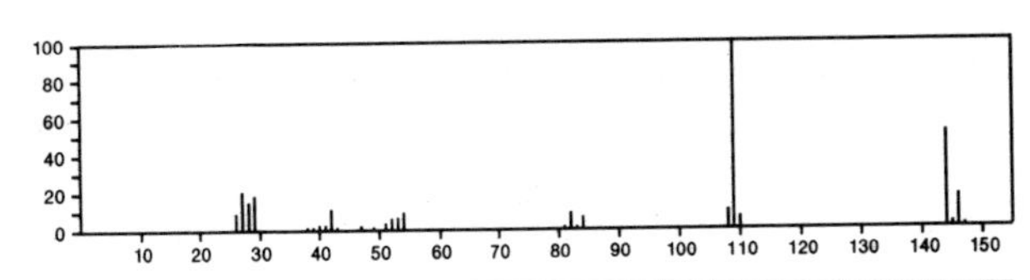

172 $C_7H_{12}N_2OS$ 56805-19-7
4-Imidazolidinone, 5-(2-methylpropyl)-2-thioxo-

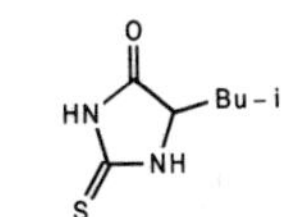

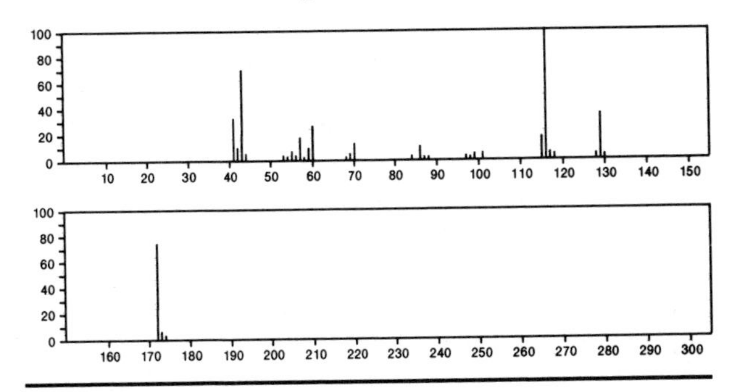

172 $C_7H_{12}N_2OS$ 56830-83-2
4-Imidazolidinone, 5-(1-methylpropyl)-2-thioxo-

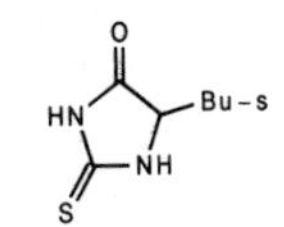

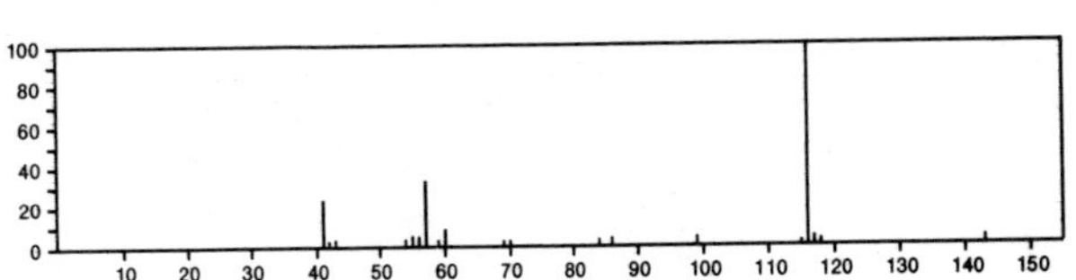

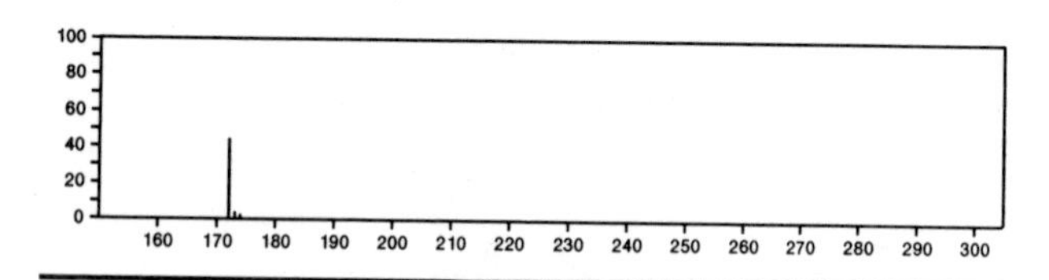

172 $C_7H_{12}N_2O_3$ 5458-06-0
2,4-Imidazolidinedione, 5-(4-hydroxybutyl)-

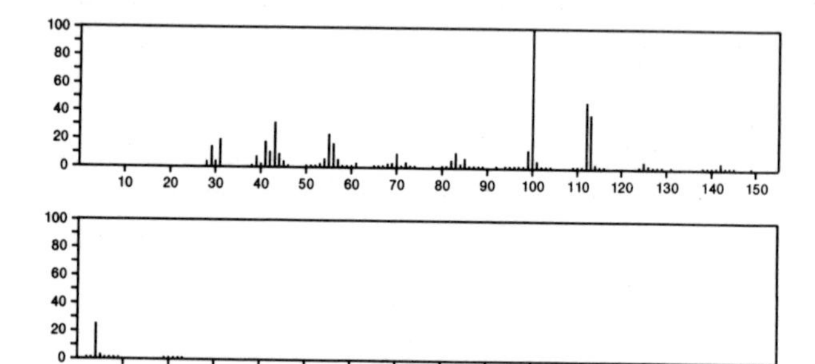

172 $C_7H_{12}N_2O_3$ 29071-93-0
2,4-Imidazolidinedione, 3-(2-hydroxyethyl)-5,5-dimethyl-

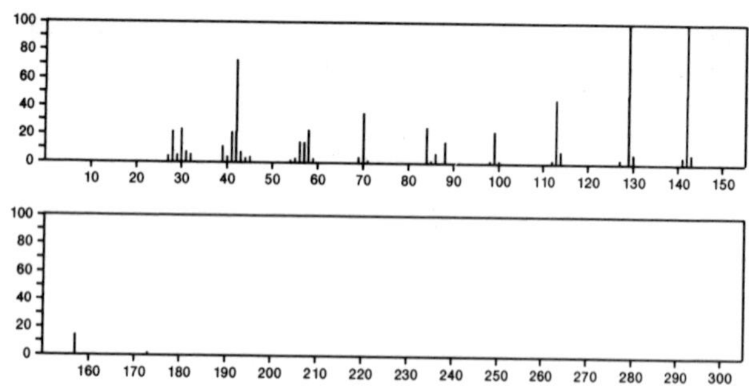

172 $C_7H_{14}Ge$ 4514-07-2
4-Germaspiro[3.4]octane

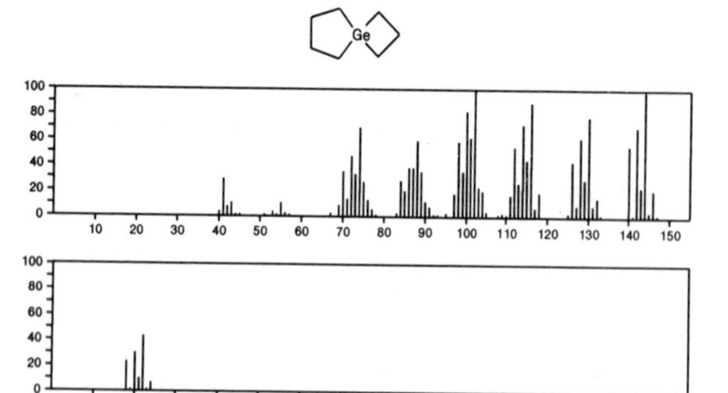

172 $C_8H_6Cl_2$ 1123-84-8
Benzene, 1,4-dichoro-2-ethenyl-

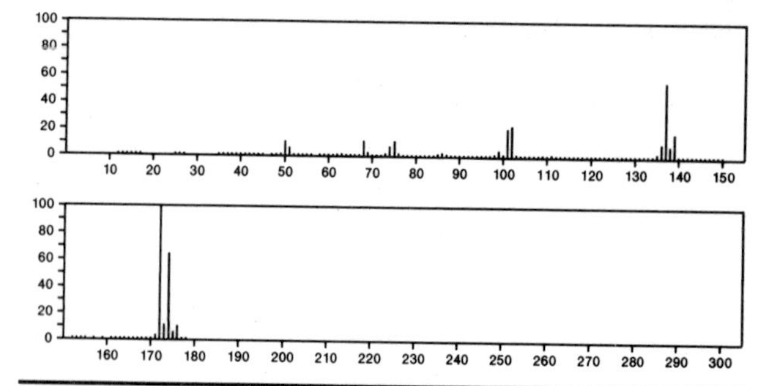

172 $C_8H_9ClO_2$ 1892-43-9
Ethanol, 2-(4-chlorophenoxy)-

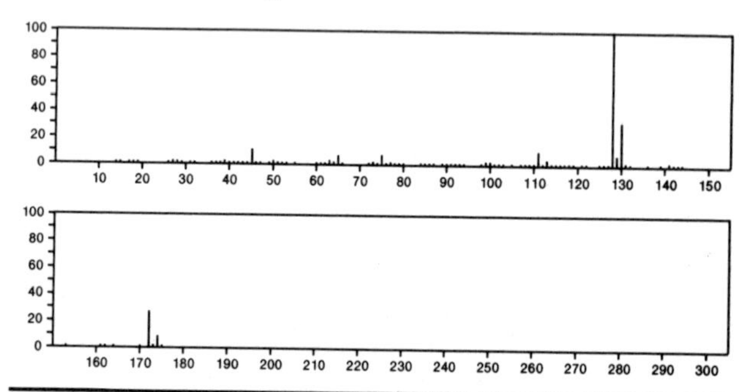

172 $C_8H_{12}O_2S$ 33266-05-6
1-Oxa-4-thiaspiro[4.5]decan-6-one

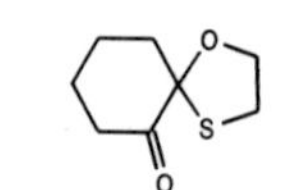

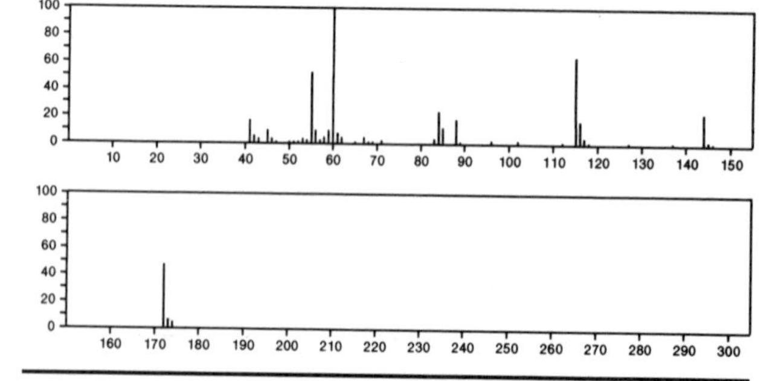

172 $C_8H_{12}O_4$ 610-09-3
1,2-Cyclohexanedicarboxylic acid, *cis*-

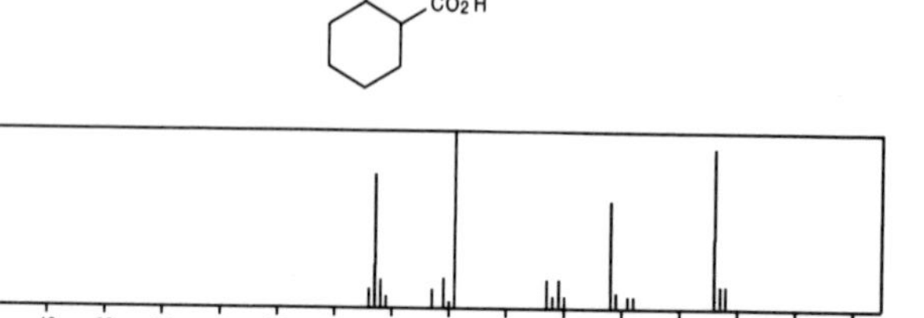

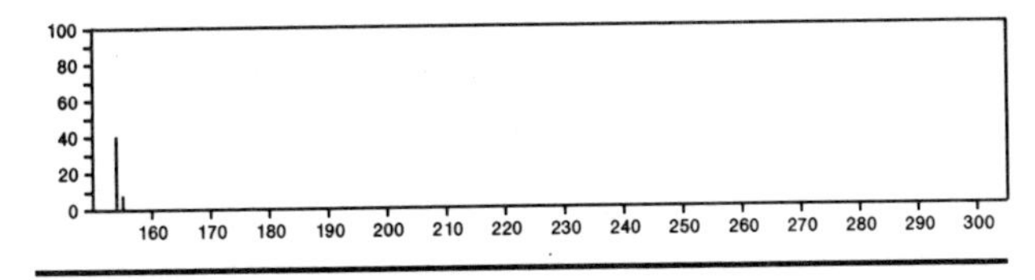

172 $C_8H_{12}O_4$ 623–91–6
2–Butenedioic acid (*E*)–, diethyl ester

EtOC(O)CH=CHC(O)OEt

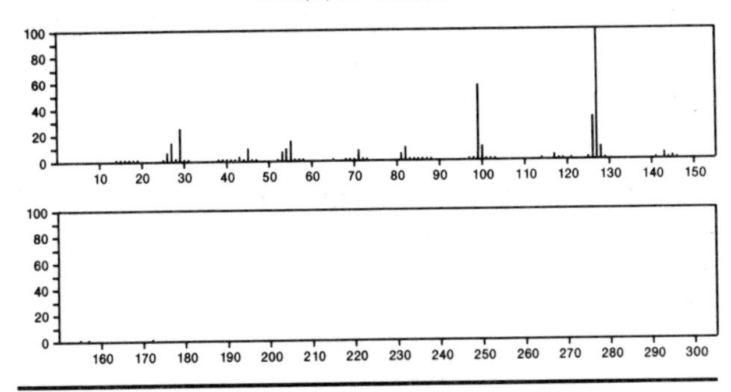

172 $C_8H_{12}O_4$ 925–21–3
2–Butenedioic acid (*Z*)–, monobutyl ester

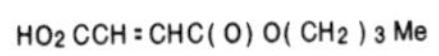

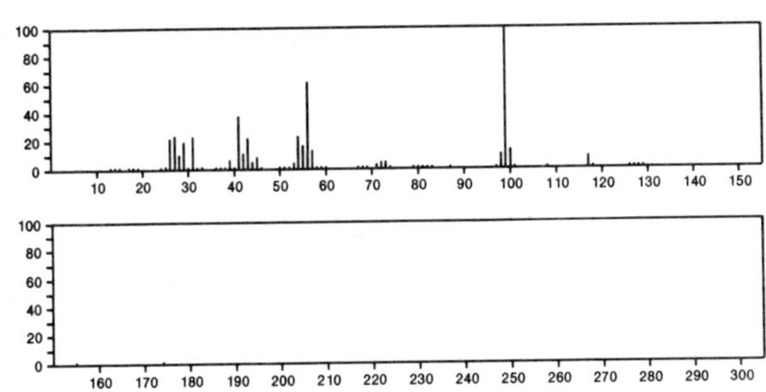

172 $C_8H_{12}O_4$ 2305–32–0
1,2–Cyclohexanedicarboxylic acid, *trans*–

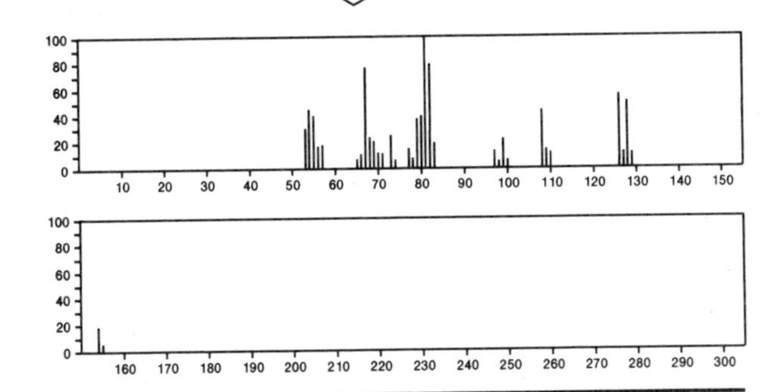

172 $C_8H_{12}O_4$ 2607–03–6
1,2–Cyclobutanedicarboxylic acid, dimethyl ester, *cis*–

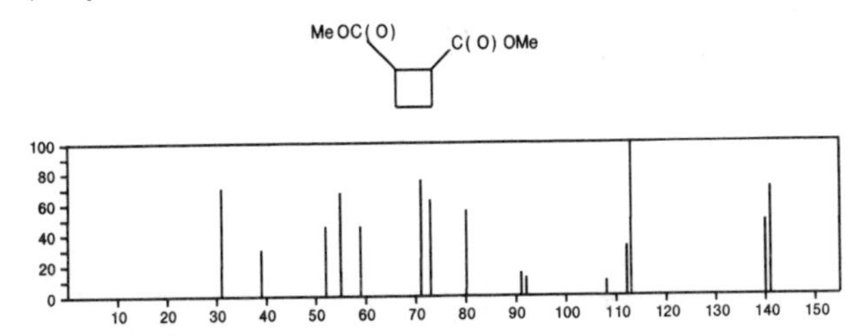

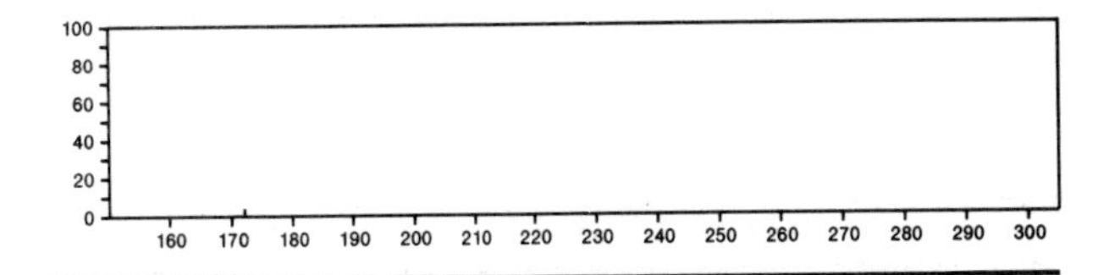

172 $C_8H_{12}O_4$ 4374–57–6
1,4–Dioxane–2,5–dione, 3,6–diethyl–

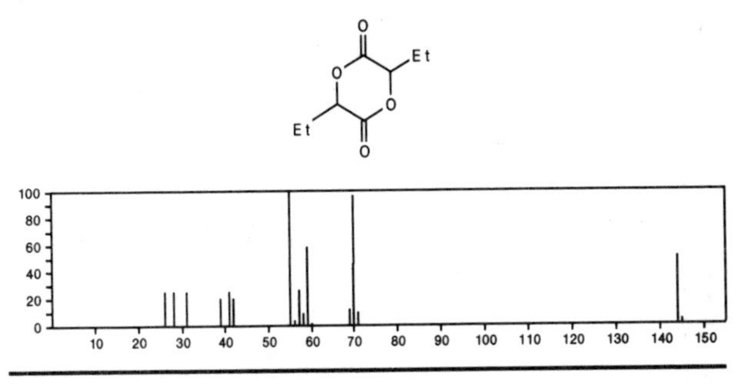

172 $C_8H_{12}O_4$ 4625–13–2
Ribofuranose, 1,5–anhydro–2,3–*O*–isopropylidene–, D–

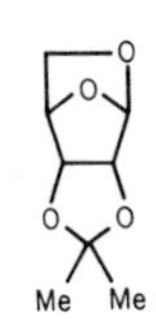

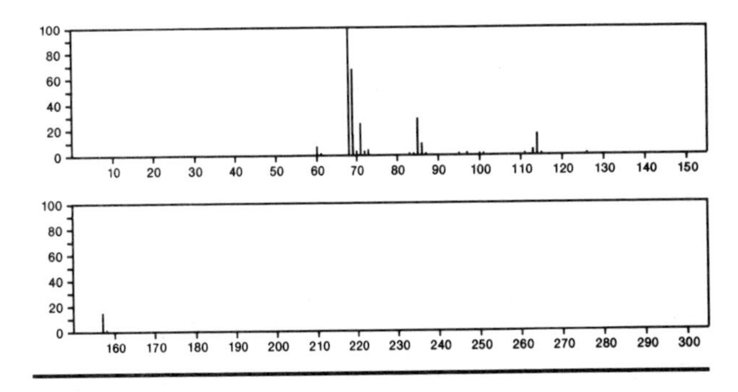

172 $C_8H_{12}O_4$ 6713–72–0
1,4–Dioxane–2,5–dione, 3,3,6,6–tetramethyl–

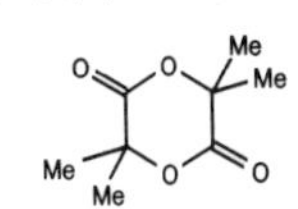

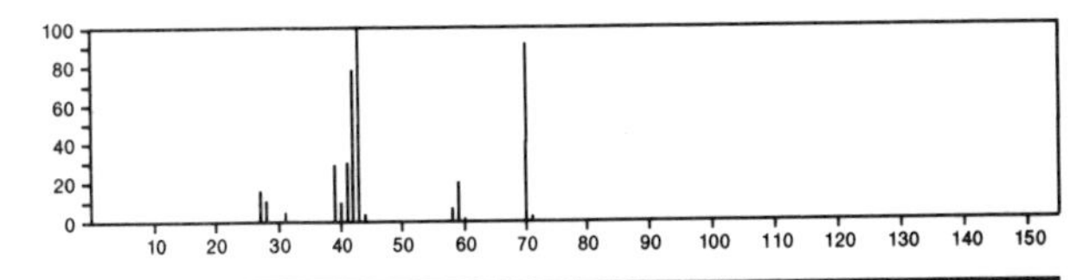

172 $C_8H_{12}O_4$ 7371–67–7
1,2–Cyclobutanedicarboxylic acid, dimethyl ester, *trans*–

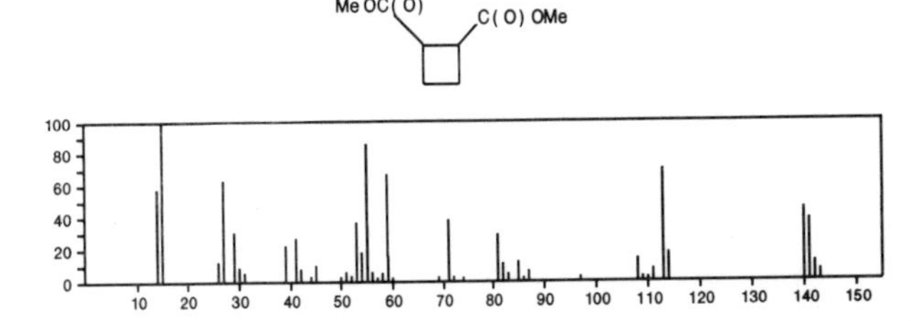

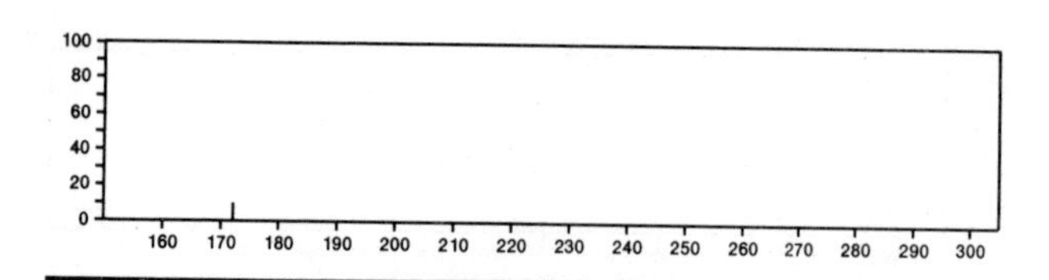

172 $C_8H_{12}O_4$ 10476-95-6
2-Propene-1,1-diol, 2-methyl-, diacetate

OAc
AcOCHCMe = CH2

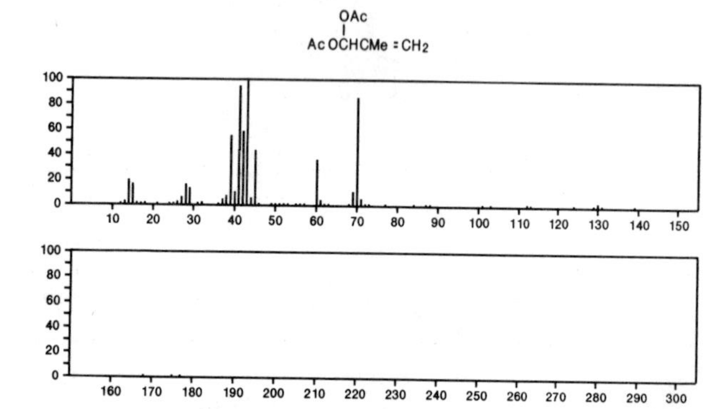

172 $C_8H_{12}O_4$ 22870-43-5
2H-Pyran-2-carboxylic acid, 3,6-dihydro-6-methoxy-, methyl ester

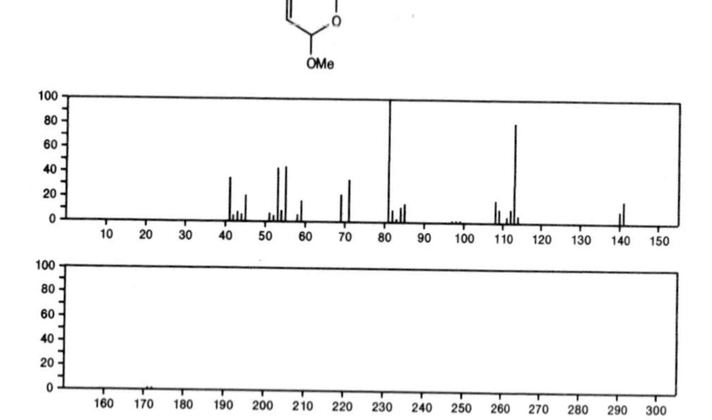

172 $C_8H_{12}O_4$ 25878-56-2
2H-Pyran-3-ol, 3,6-dihydro-6-methoxy-, acetate, (3*S*-*trans*)-

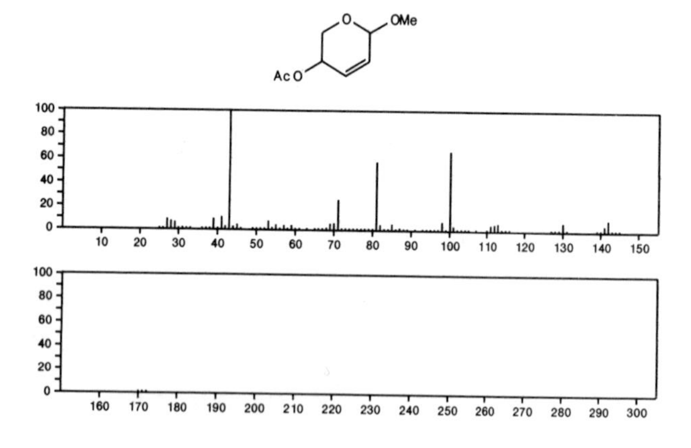

172 $C_8H_{12}O_4$ 26532-19-4
2H-Pyran-3-ol, 3,6-dihydro-6-methoxy-, acetate, (3*R*-*cis*)-

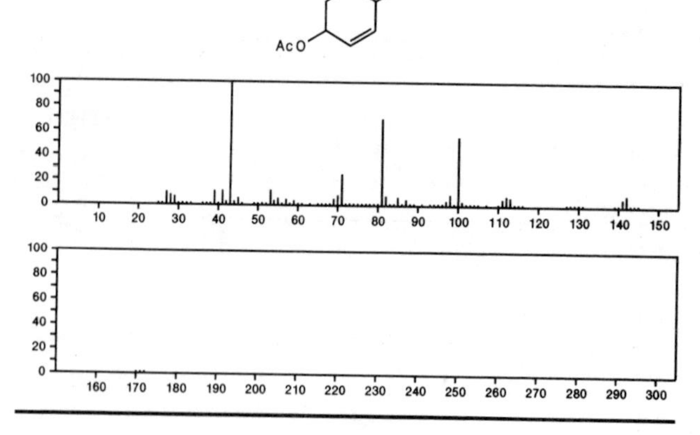

172 $C_8H_{12}O_4$ 40637-56-7
Propanedioic acid, 2-propenyl-, dimethyl ester

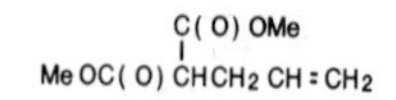

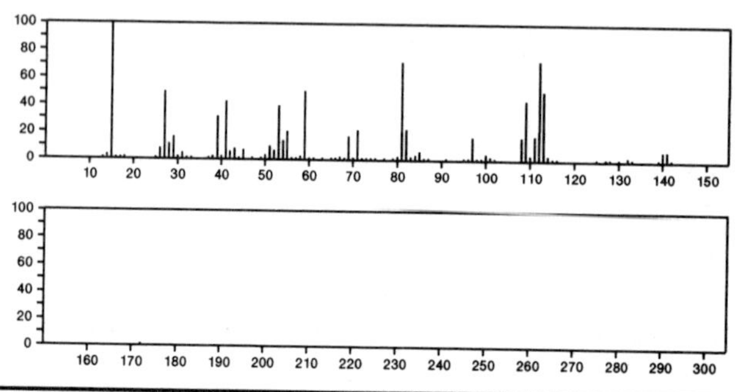

172 $C_8H_{12}O_4$ 54484-67-2
1-Butene-1,4-diol, diacetate

AcOCH = CHCH2 CH2 OAc

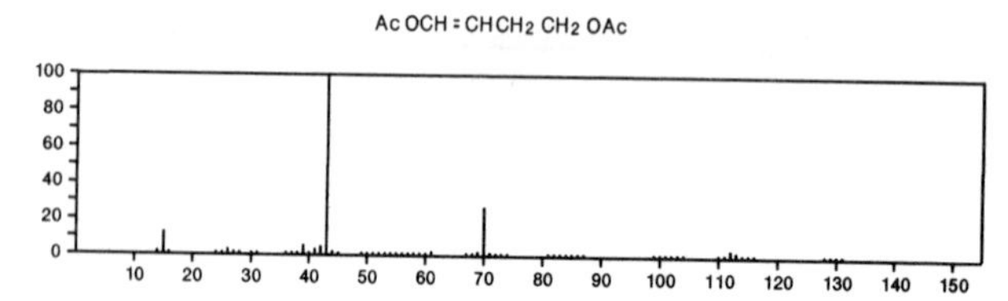

172 $C_8H_{12}S_2$ 3988-71-4
Thiophene, 2-(butylthio)-

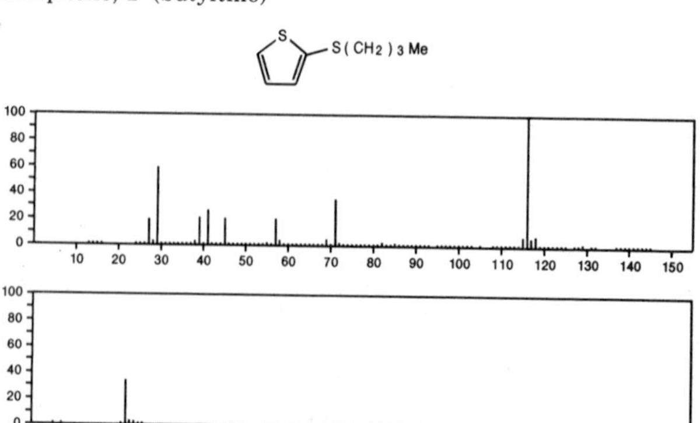

172 $C_8H_{12}S_2$ 40697-97-0
Thiophene, 3-(dihydro-3(2*H*)-thienylidene)tetrahydro-

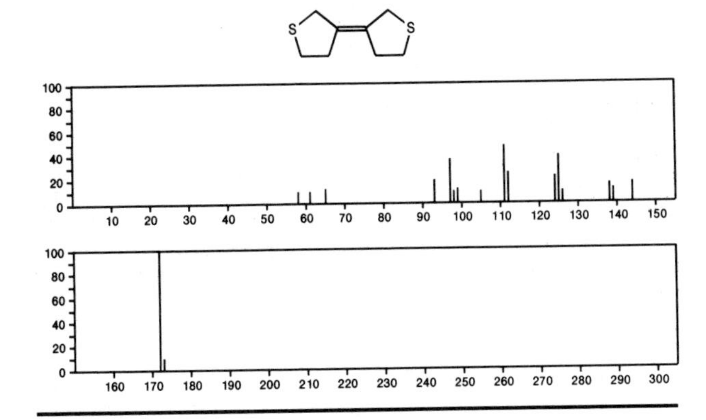

172 $C_8H_{16}N_2O_2$ 14702-38-6
Carbazic acid, 3-pentylidene-, ethyl ester

$Me(CH_2)_3CH{=}NNHC(O)OEt$

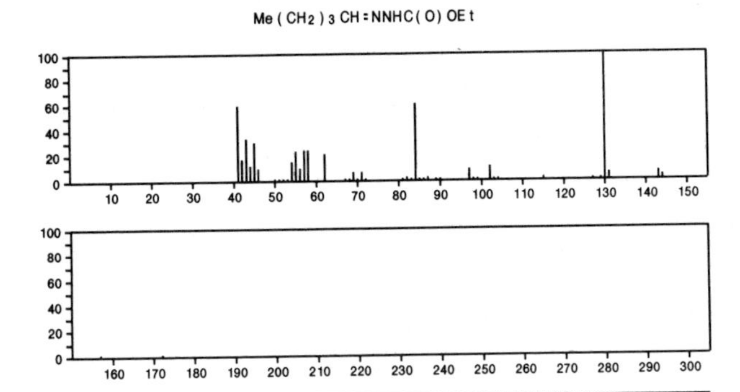

172 $C_8H_{16}N_2O_2$ 26394-95-6
2-Propenoic acid, 3,3-bis(dimethylamino)-, methyl ester

$MeOC(O)CH{=}C(NMe_2)NMe_2$

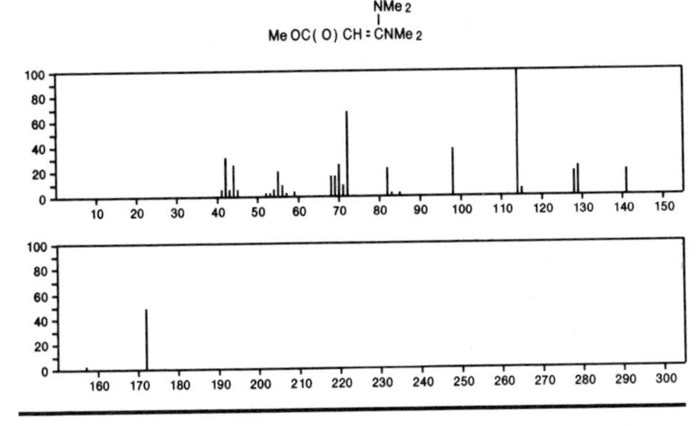

172 $C_8H_{16}O_2Si$ 23523-56-0
4-Pentenoic acid, trimethylsilyl ester

$Me_3SiOC(O)CH_2CH_2CH{=}CH_2$

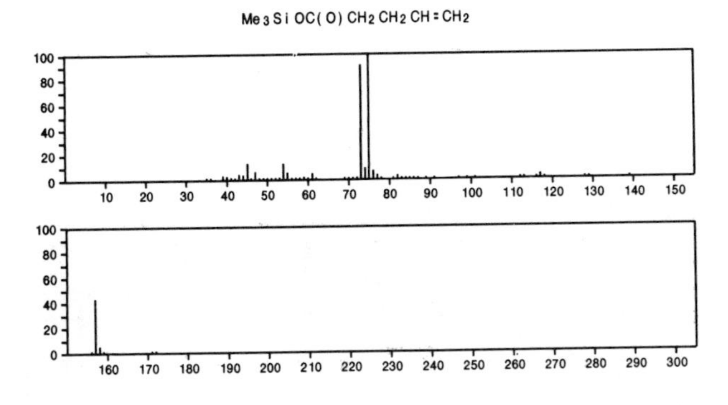

172 $C_8H_{16}O_2Si$ 25436-25-3
Crotonic acid, 3-methyl-, trimethylsilyl ester

$Me_3SiOC(O)CH{=}CMe_2$

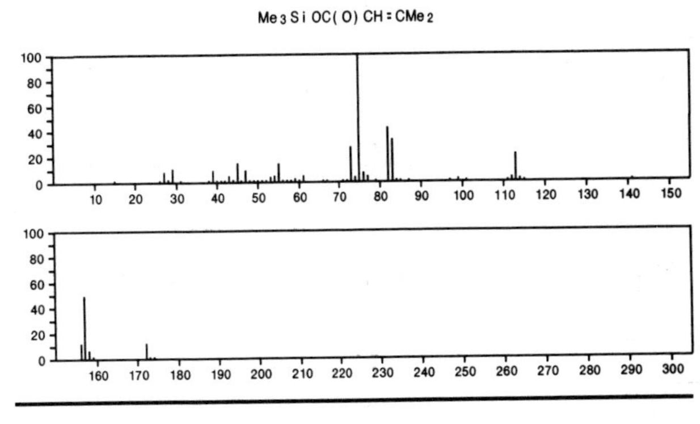

172 $C_8H_{17}BO_3$ 55162-68-0
1,3,2-Dioxaborinane, 2-(pentyloxy)-

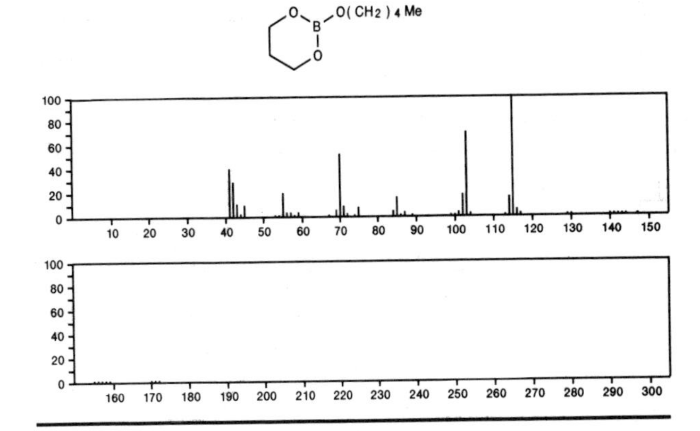

172 $C_8H_{17}BO_3$ 55162-69-1
1,3,2-Dioxaborinane, 2-(1-methylbutoxy)-

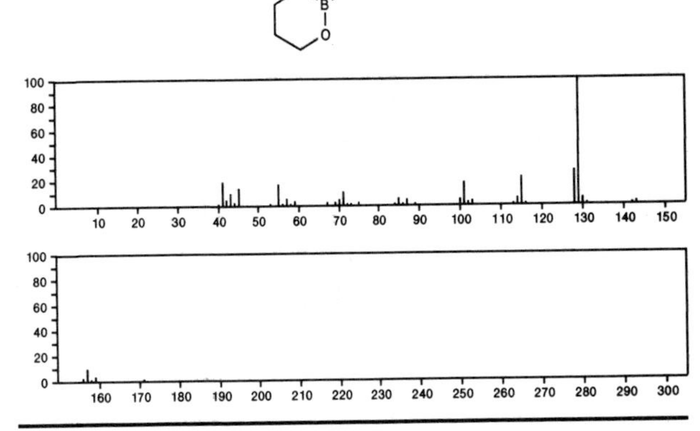

172 $C_9H_4N_2O_2$ 1807-49-4
1*H*-Indene-1,3(2*H*)-dione, 2-diazo-

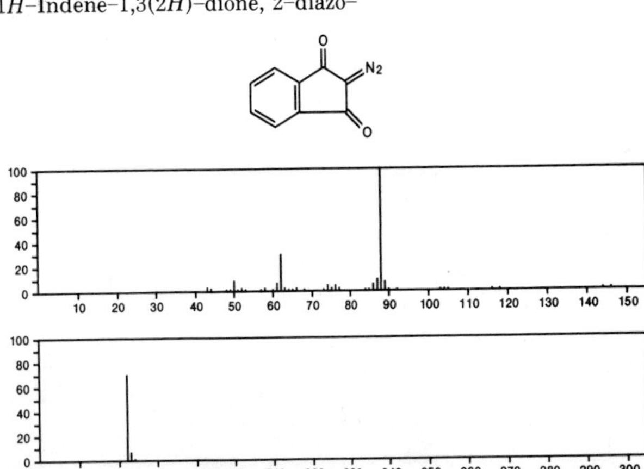

172 $C_9H_{16}OS$ 57156-88-4
1-Oxa-4-thiaspiro[4.4]nonane, 6,9-dimethyl-

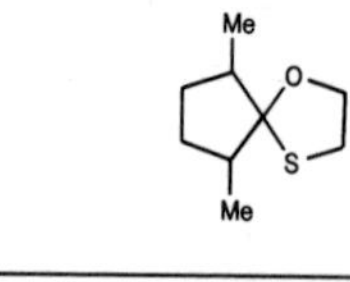

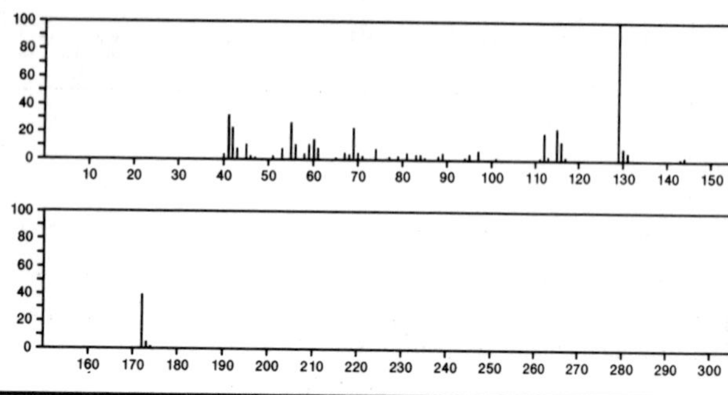

172 $C_9H_{16}O_3$ 16493-42-8
Octanoic acid, 7-oxo-, methyl ester

MeOC(O)$(CH_2)_5$COMe

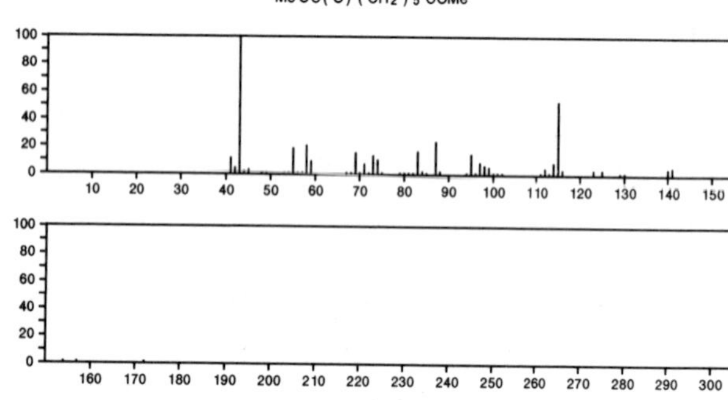

172 $C_9H_{16}O_3$ 24222-05-7
Valeric acid, 2,3-epoxy-3,4-dimethyl-, ethyl ester, *cis*-

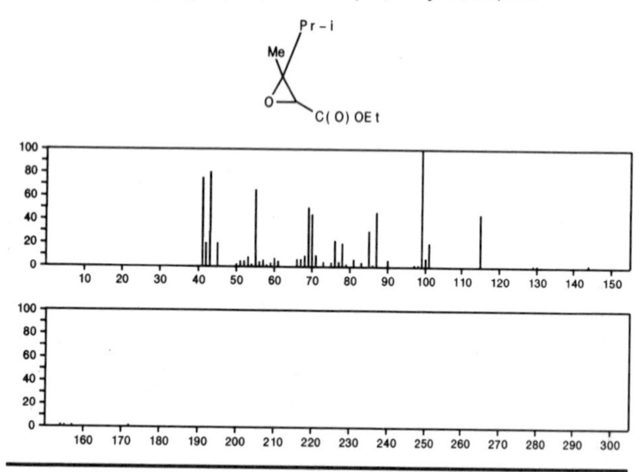

172 $C_9H_{16}O_3$ 30956-41-3
Heptanoic acid, 6-oxo-, ethyl ester

MeCO$(CH_2)_4$C(O)OEt

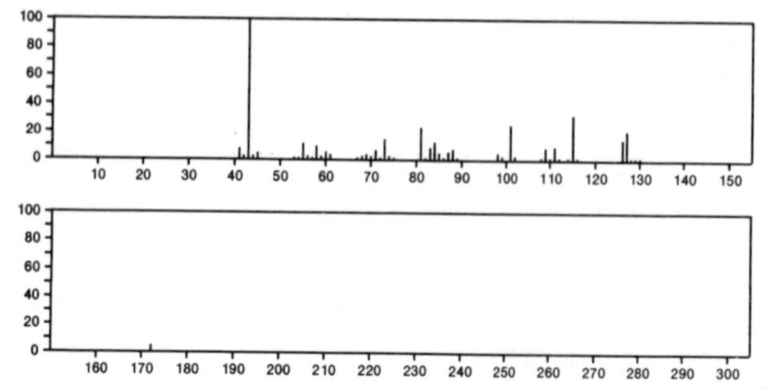

172 $C_9H_{16}O_3$ 51756-09-3
Pentanoic acid, 2-acetyl-4-methyl-, methyl ester

COMe
MeOC(O)$CHCH_2CHMe_2$

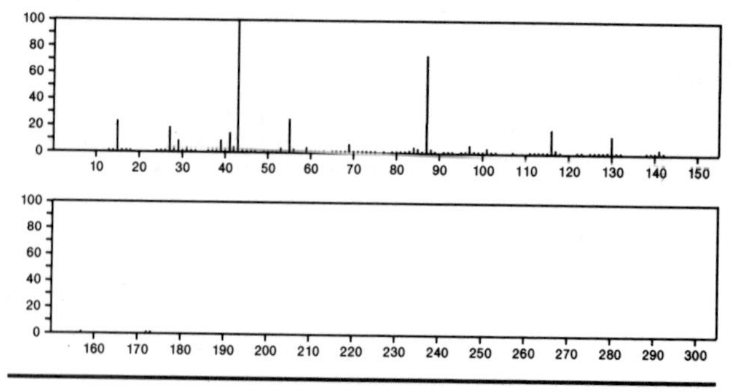

172 $C_9H_{16}O_3$ 55162-84-0
4-Pentenoic acid, 5-ethoxy-, ethyl ester, (*E*)-

EtOC(O)CH_2CH_2CH=CHOEt

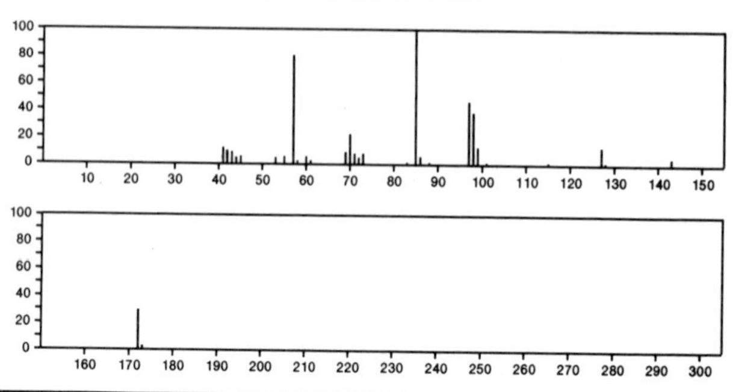

172 $C_9H_{16}O_3$ 56292-99-0
1,4-Dioxaspiro[4.5]decane, 8-methoxy-

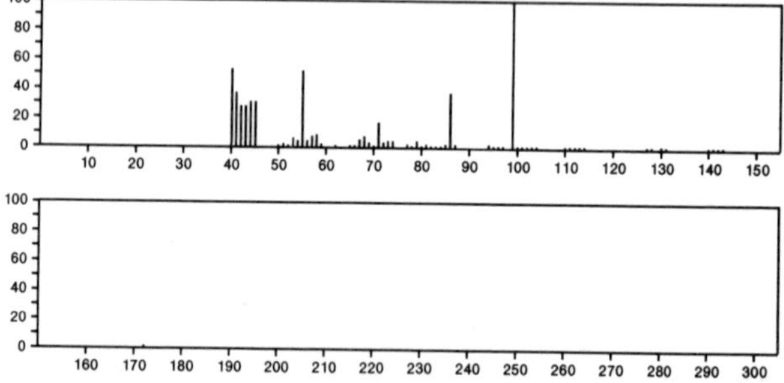

172 $C_9H_{18}NO_2$ 2226-96-2
1-Piperidinyloxy, 4-hydroxy-2,2,6,6-tetramethyl-

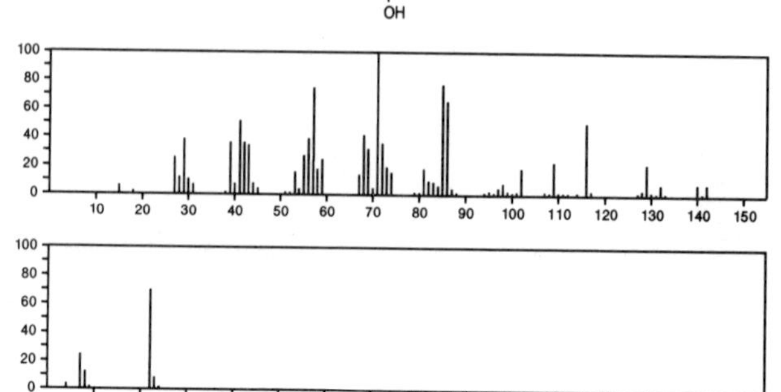

172 $C_9H_{18}NO_2$ 27298–75–5
1–Pyrrolidinyloxy, 3–(hydroxymethyl)–2,2,5,5–tetramethyl–

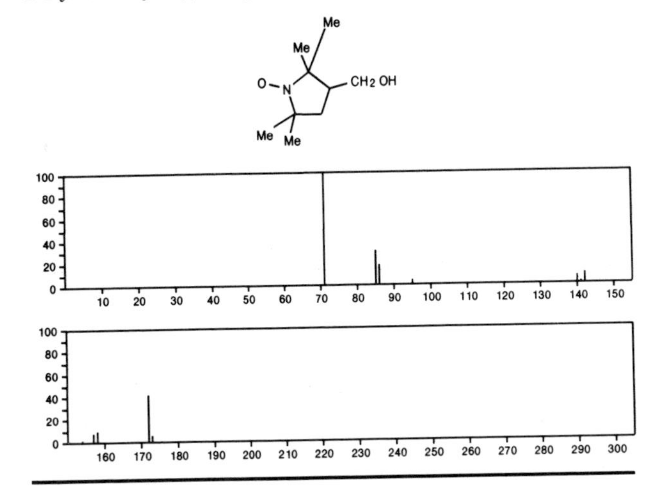

172 $C_9H_{20}N_2O$ 5336–24–3
Urea, *N*,*N'*–bis(1,1–dimethylethyl)–

t-BuNHCONHBu-t

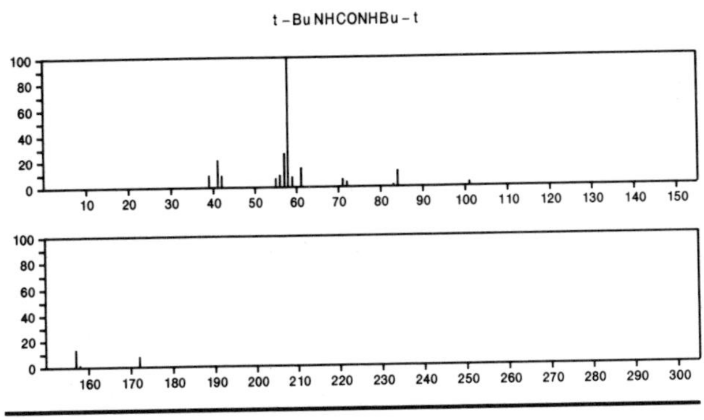

172 $C_9H_{20}N_2O$ 16339–05–2
1–Pentanamine, *N*–butyl–*N*–nitroso–

Me(CH2)4N(NO)(CH2)3Me

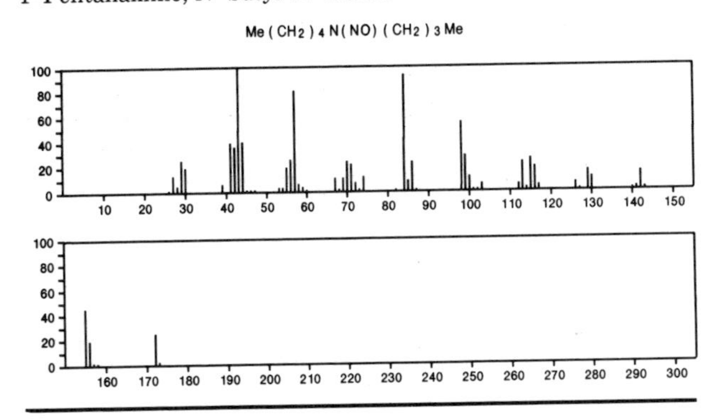

172 $C_9H_{20}N_2O$ 28023–79–2
Pentylamine, *N*–*sec*–butyl–*N*–nitroso–

s-BuN(NO)(CH2)4Me

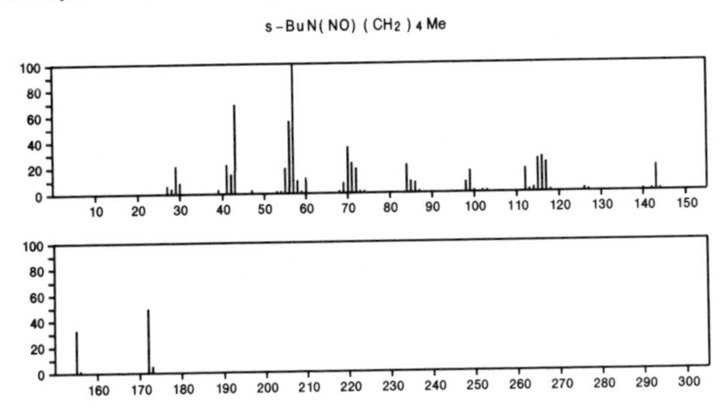

172 $C_9H_{20}N_2O$ 34423–54–6
Octylamine, *N*–methyl–*N*–nitroso–

MeN(NO)(CH2)7Me

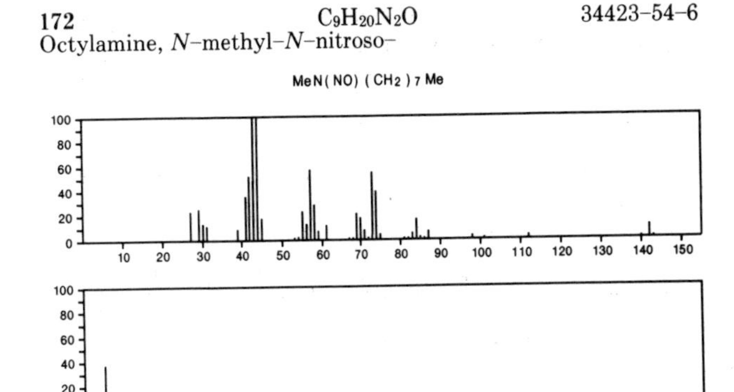

172 $C_9H_{20}OSi$ 13871–89–1
Silane, (cyclohexyloxy)trimethyl–

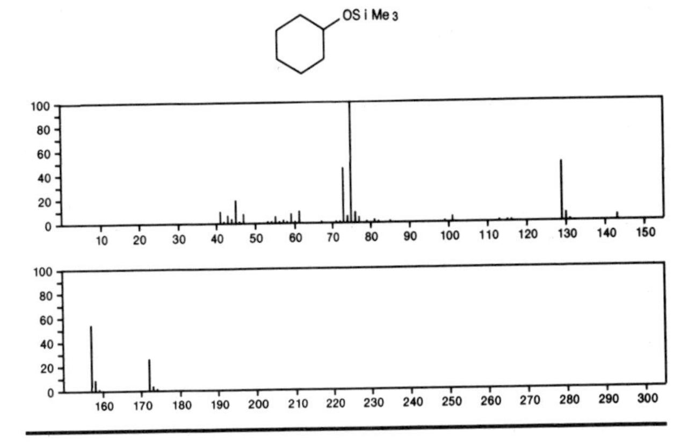

172 $C_9H_{20}OSi$ 20584–41–2
Cyclohexanol, 2–(trimethylsilyl)–, *trans*–

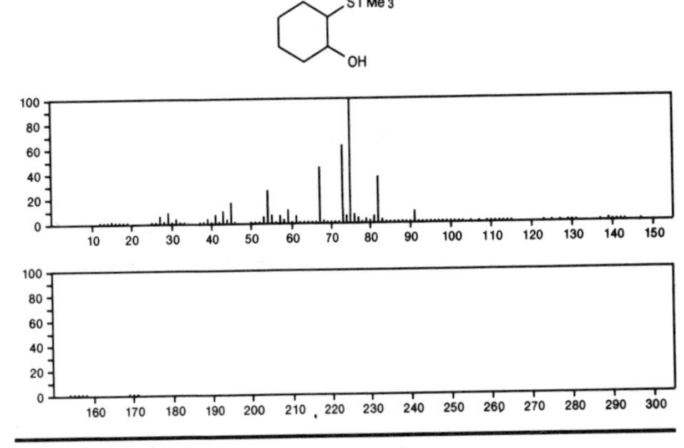

172 $C_9H_{20}OSi$ 20584–43–4
Cyclohexanol, 2–(trimethylsilyl)–, *cis*–

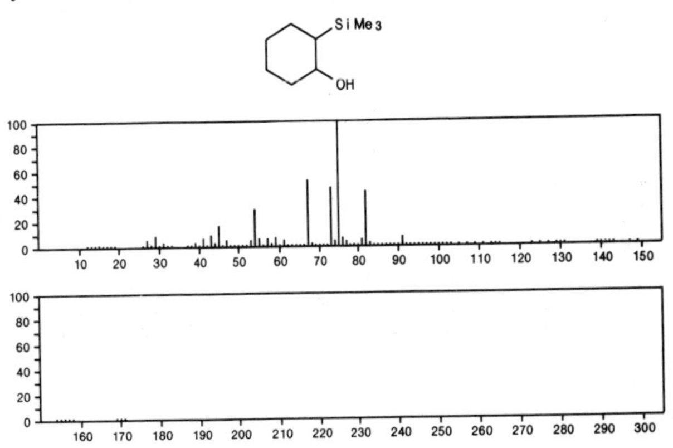

172 $C_{10}H_8N_2O$ 1131-68-6
2-Quinolinecarboxaldehyde, oxime

CH=NOH

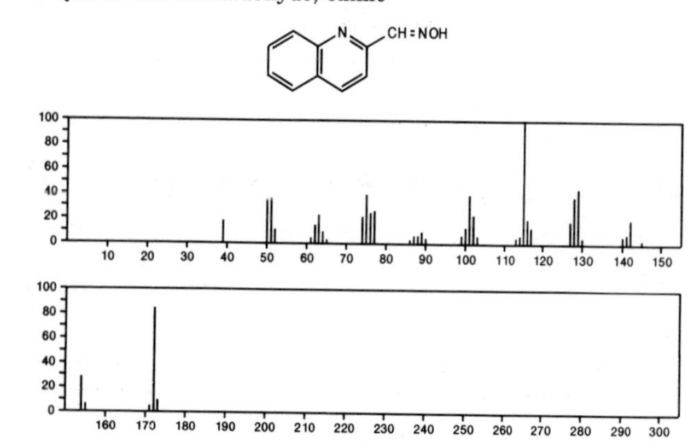

172 $C_{10}H_8N_2O$ 23228-05-9
Formamide, *N*-(β-cyanostyryl)-

NCCPh=CHNHCH=O

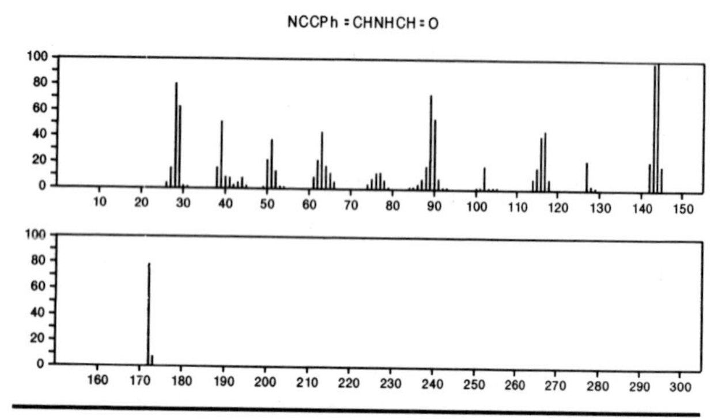

172 $C_{10}H_8N_2O$ 54789-38-7
1*H*-Inden-1-one, 2-diazo-2,3-dihydro-3-methyl-

Me N2 O

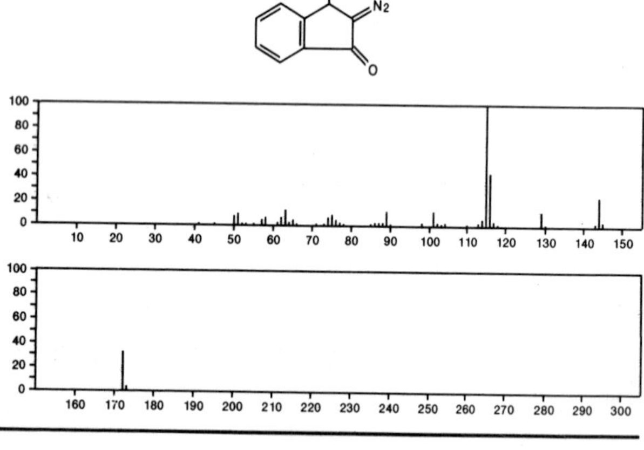

172 $C_{10}H_8N_2O$ 56248-10-3
1*H*-Imidazole-2-carboxaldehyde, 4-phenyl-

H N CH=O N Ph

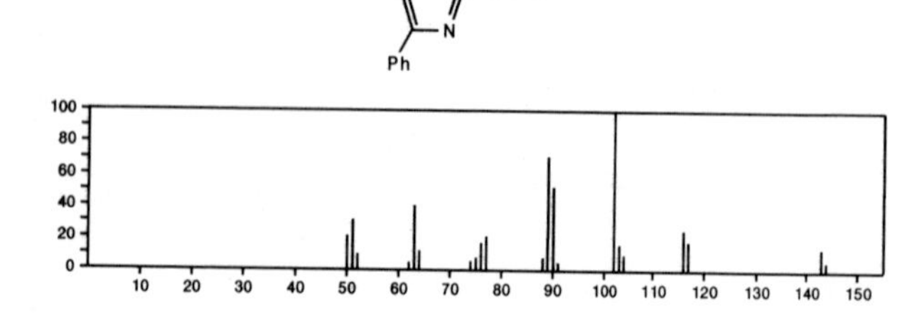

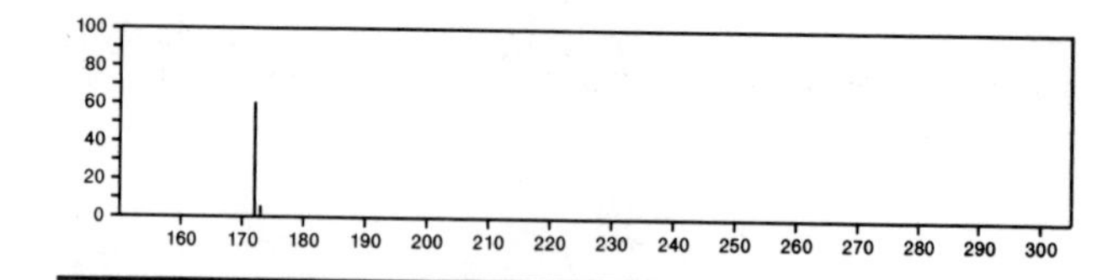

172 $C_{10}H_8N_2O$ 57204-65-6
1*H*-Pyrazole-3-carboxaldehyde, 5-phenyl-

Ph CH=O HN—N

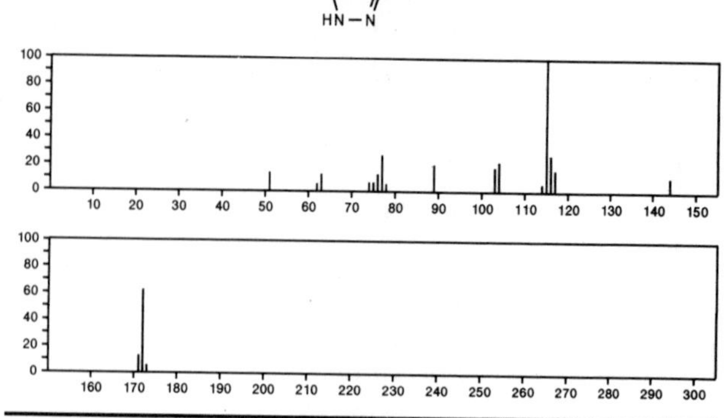

172 $C_{10}H_{17}Cl$ 464-41-5
Bicyclo[2.2.1]heptane, 2-chloro-1,7,7-trimethyl-, *endo*-

Me Cl Me Me

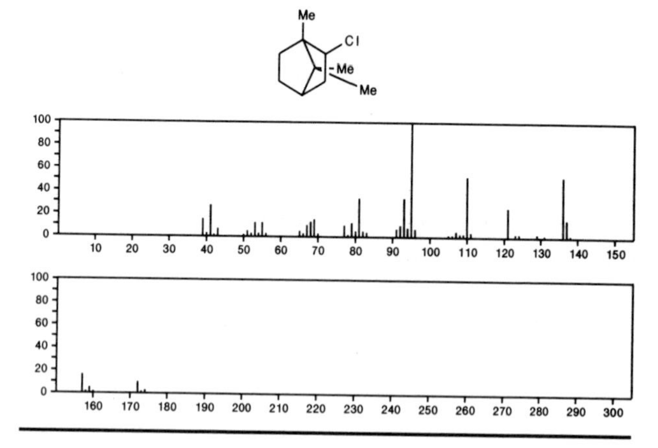

172 $C_{10}H_{17}Cl$ 465-30-5
Bicyclo[2.2.1]heptane, 2-chloro-2,3,3-trimethyl-

Me Me Me Cl

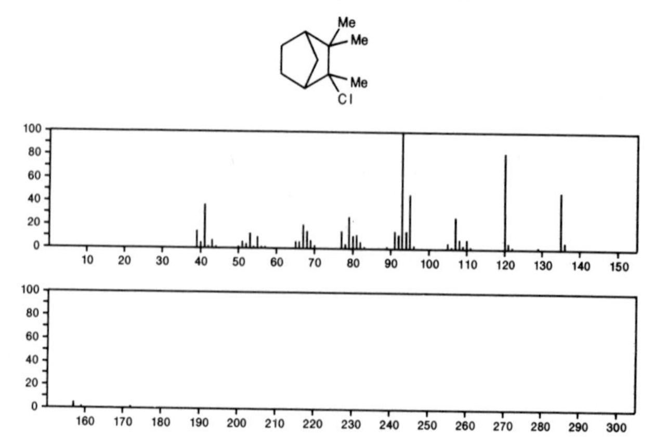

172 $C_{10}H_{17}Cl$ 559–45–5
Bicyclo[2.2.1]heptane, 2–chloro–1,7,7–trimethyl–, *exo*–

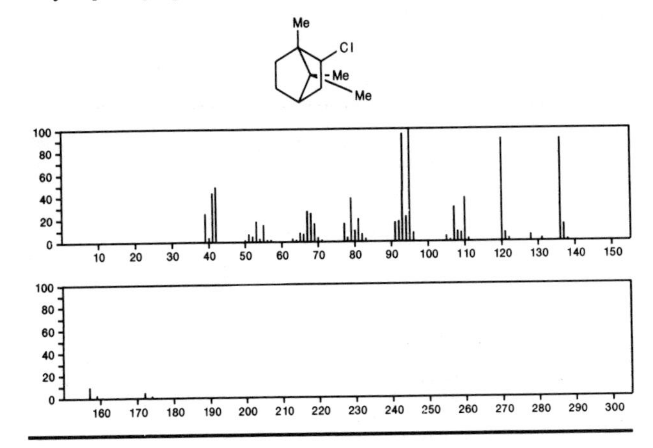

172 $C_{10}H_{17}Cl$ 1126–28–9
Norbornane, 2–chloro–1,5,5–trimethyl–, *exo*–

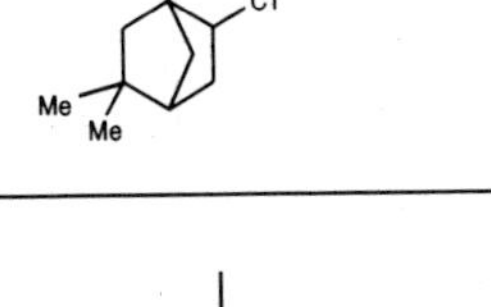

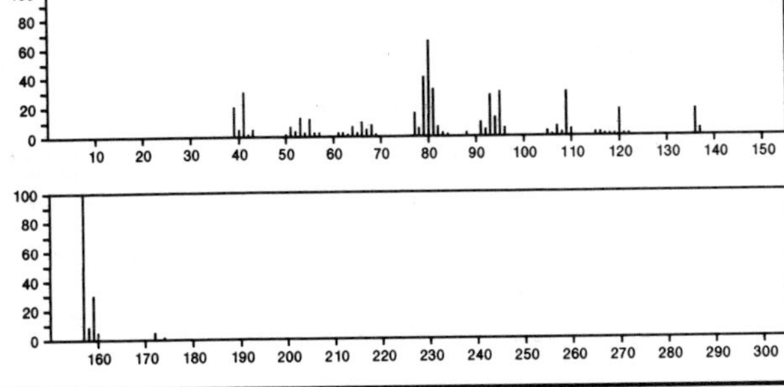

172 $C_{10}H_{17}Cl$ 3372–12–1
Bicyclo[2.2.1]heptane, 2–chloro–1,3,3–trimethyl–, *endo*–

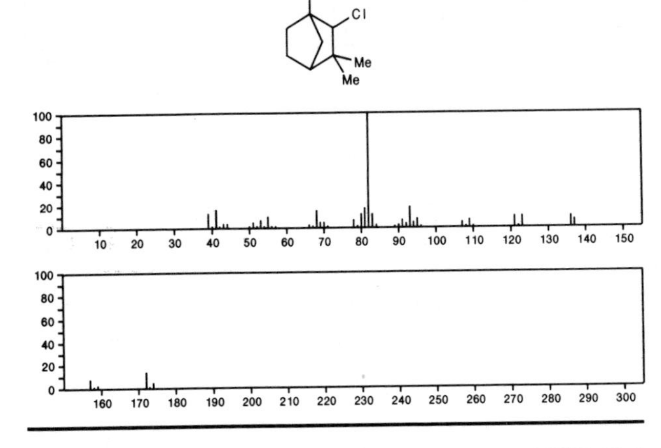

172 $C_{10}H_{17}Cl$ 22768–99–6
Norbornane, 2–chloro–2,5,5–trimethyl–, *exo*–

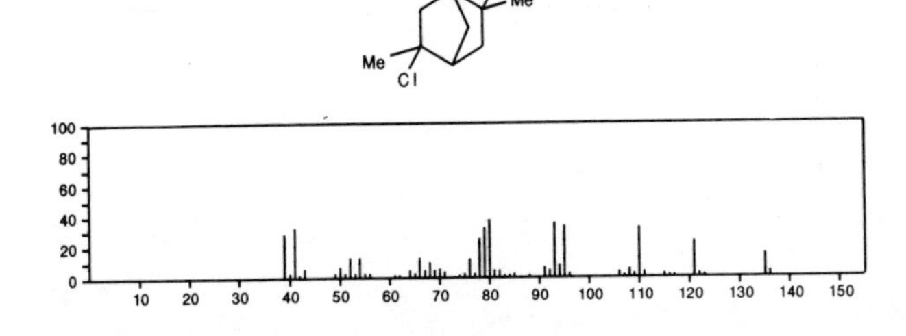

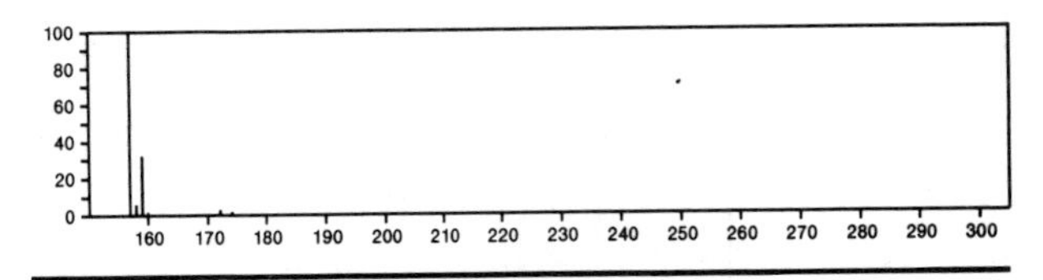

172 $C_{10}H_{17}Cl$ 22852–22–8
Bicyclo[2.2.1]heptane, 2–chloro–2,7,7–trimethyl–, *exo*–

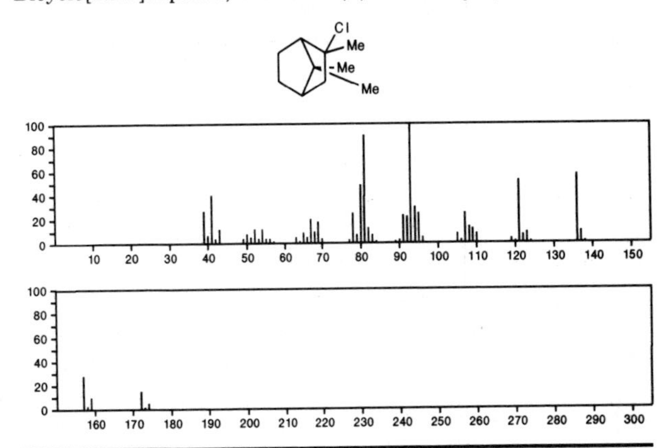

172 $C_{10}H_{20}O_2$ 103–09–3
Acetic acid, 2–ethylhexyl ester

$AcOCH_2CHEt(CH_2)_3Me$

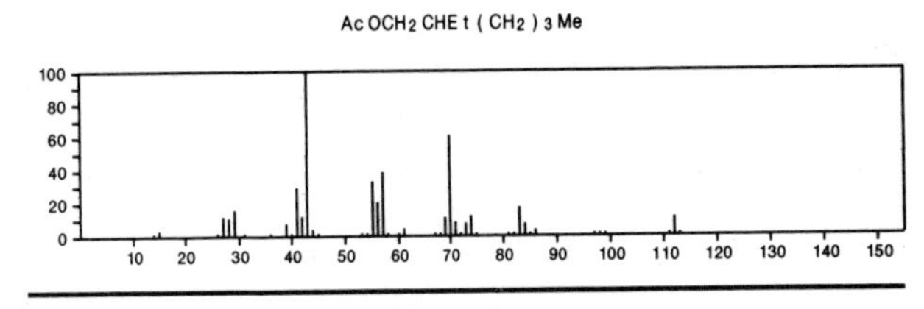

172 $C_{10}H_{20}O_2$ 107–75–5
Octanal, 7–hydroxy–3,7–dimethyl–

$Me_2COH(CH_2)_3CHMeCH_2CHO$

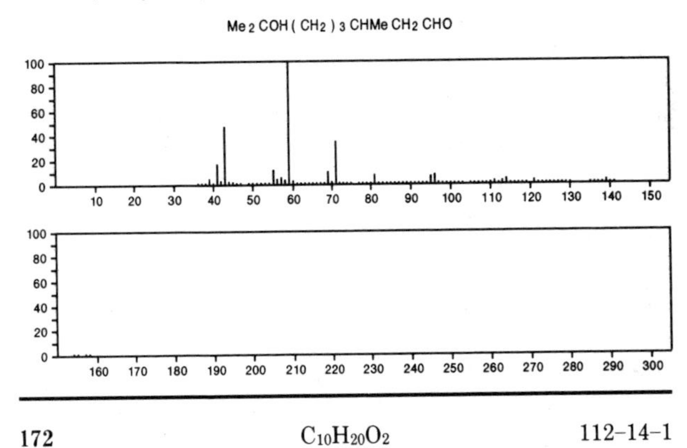

172 $C_{10}H_{20}O_2$ 112–14–1
Acetic acid, octyl ester

$Me(CH_2)_7OAc$

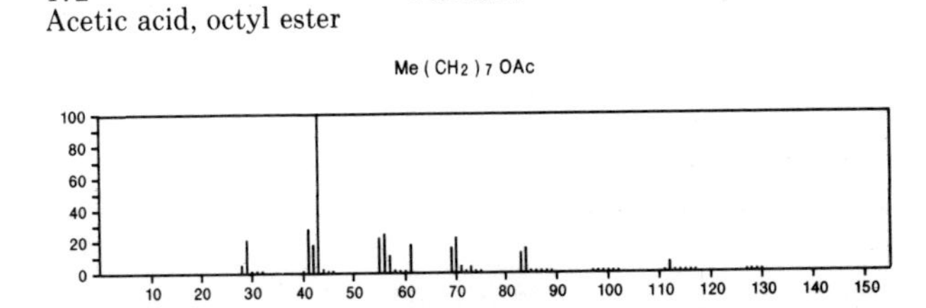

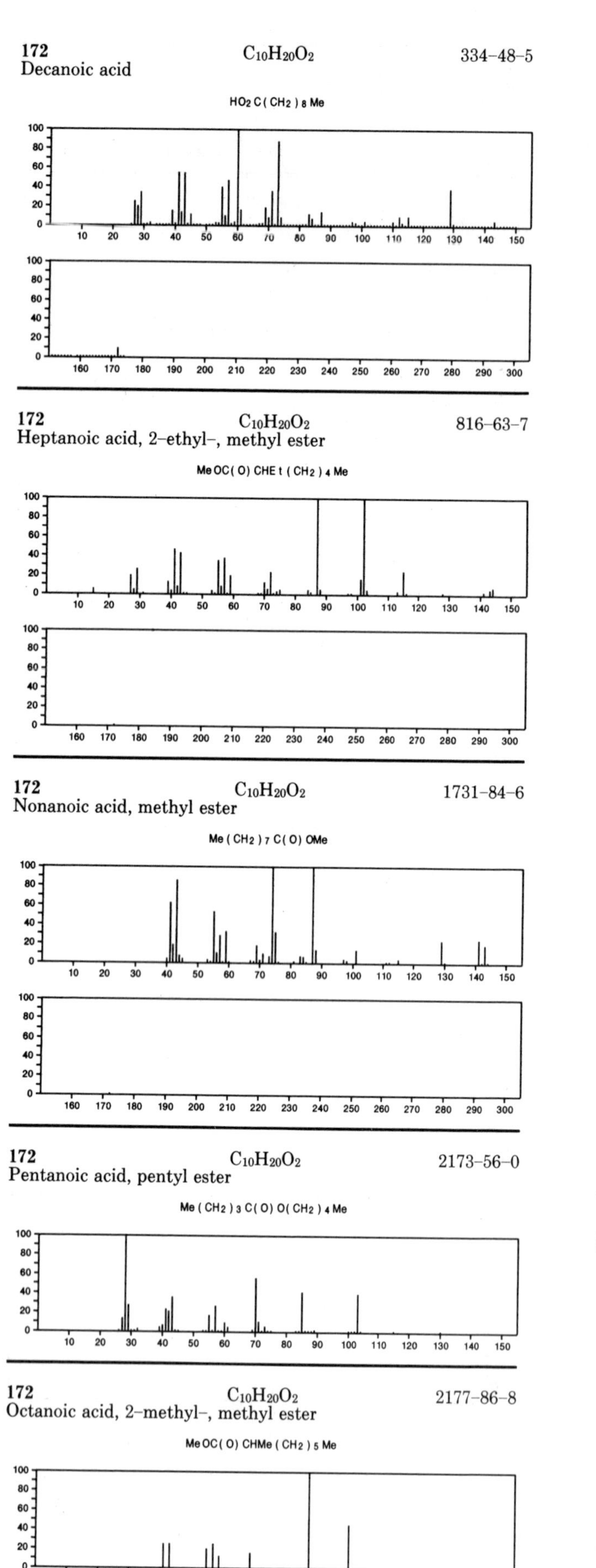
172
Decanoic acid
$C_{10}H_{20}O_2$
334–48–5
HO2 C(CH2) 8 Me
172
Heptanoic acid, 2–ethyl–, methyl ester
$C_{10}H_{20}O_2$
816–63–7
Me OC(O) CHE t (CH2) 4 Me
172
Nonanoic acid, methyl ester
$C_{10}H_{20}O_2$
1731–84–6
Me (CH2) 7 C(O) OMe
172
Pentanoic acid, pentyl ester
$C_{10}H_{20}O_2$
2173–56–0
Me (CH2) 3 C(O) O(CH2) 4 Me
172
Octanoic acid, 2–methyl–, methyl ester
$C_{10}H_{20}O_2$
2177–86–8
Me OC(O) CHMe (CH2) 5 Me

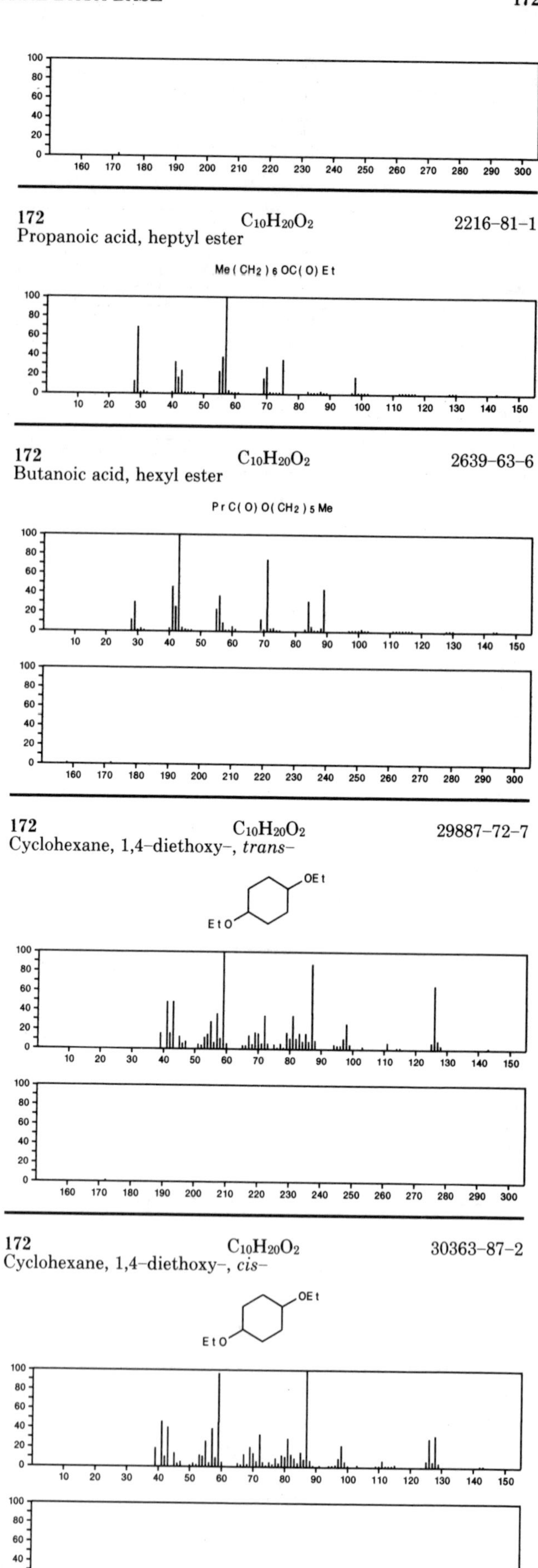
172
Propanoic acid, heptyl ester
$C_{10}H_{20}O_2$
2216–81–1
Me (CH2) 6 OC(O) E t
172
Butanoic acid, hexyl ester
$C_{10}H_{20}O_2$
2639–63–6
P r C(O) O(CH2) 5 Me
172
Cyclohexane, 1,4–diethoxy–, *trans*–
$C_{10}H_{20}O_2$
29887–72–7
OE t
E t O
172
Cyclohexane, 1,4–diethoxy–, *cis*–
$C_{10}H_{20}O_2$
30363–87–2
OE t
E t O

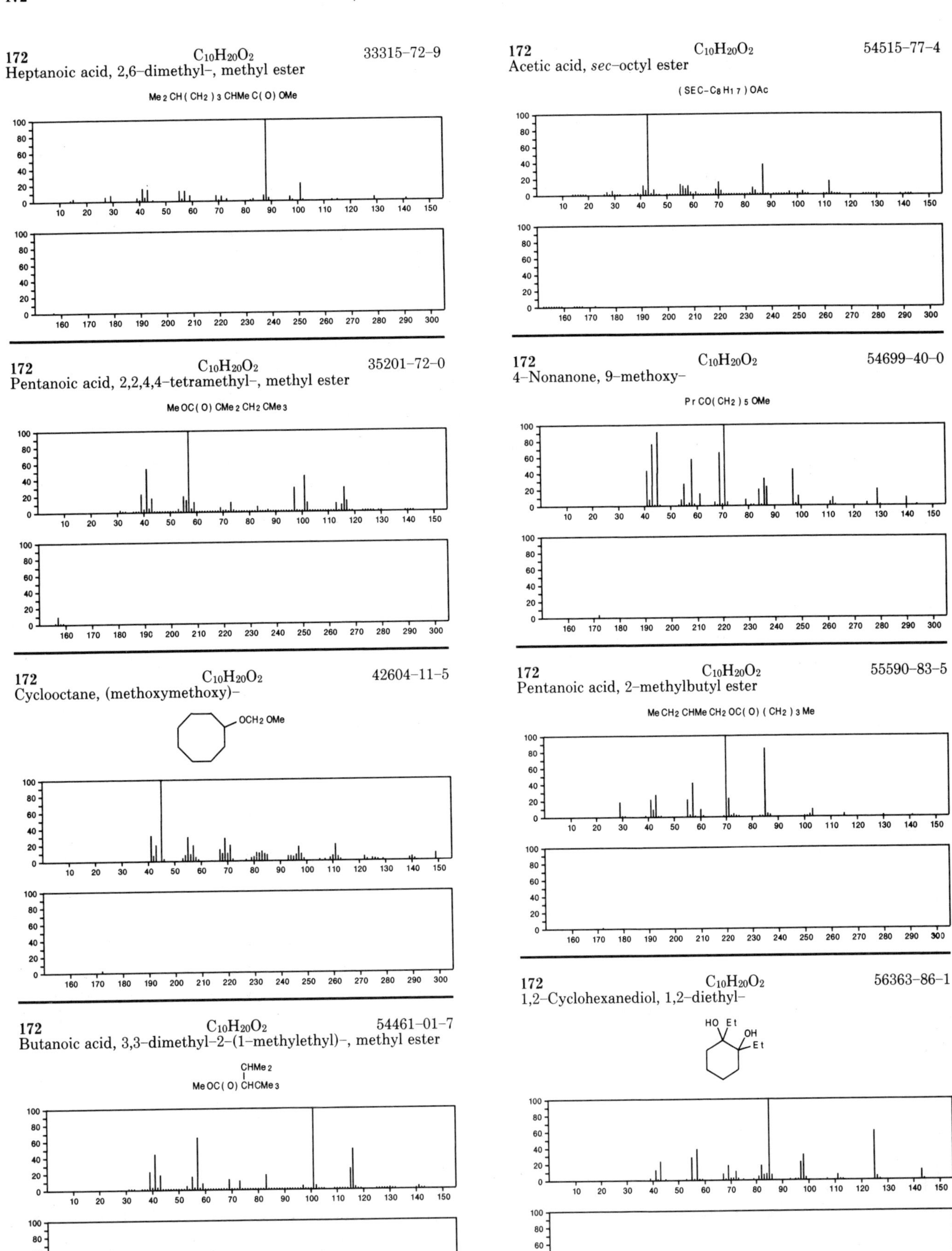
172 $C_{10}H_{20}O_2$ 33315-72-9
Heptanoic acid, 2,6-dimethyl-, methyl ester
Me 2 CH (CH2) 3 CHMe C(O) OMe
172 $C_{10}H_{20}O_2$ 35201-72-0
Pentanoic acid, 2,2,4,4-tetramethyl-, methyl ester
Me OC(O) CMe 2 CH2 CMe 3
172 $C_{10}H_{20}O_2$ 42604-11-5
Cyclooctane, (methoxymethoxy)-
OCH2 OMe
172 $C_{10}H_{20}O_2$ 54461-01-7
Butanoic acid, 3,3-dimethyl-2-(1-methylethyl)-, methyl ester
CHMe 2
Me OC(O) CHCMe 3
172 $C_{10}H_{20}O_2$ 54515-77-4
Acetic acid, *sec*-octyl ester
(SEC-C8 H1 7) OAc
172 $C_{10}H_{20}O_2$ 54699-40-0
4-Nonanone, 9-methoxy-
Pr CO(CH2) 5 OMe
172 $C_{10}H_{20}O_2$ 55590-83-5
Pentanoic acid, 2-methylbutyl ester
Me CH2 CHMe CH2 OC(O) (CH2) 3 Me
172 $C_{10}H_{20}O_2$ 56363-86-1
1,2-Cyclohexanediol, 1,2-diethyl-
HO Et
OH
Et

172 $C_{10}H_{20}O_2$ 57983-17-2
Pentanoic acid, 2-methyl-, 1-methylpropyl ester

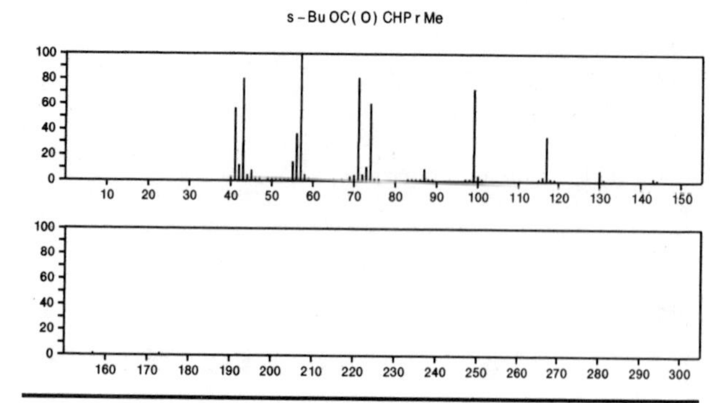

172 $C_{10}H_{20}S$ 7133-20-2
Sulfide, cyclopentyl pentyl

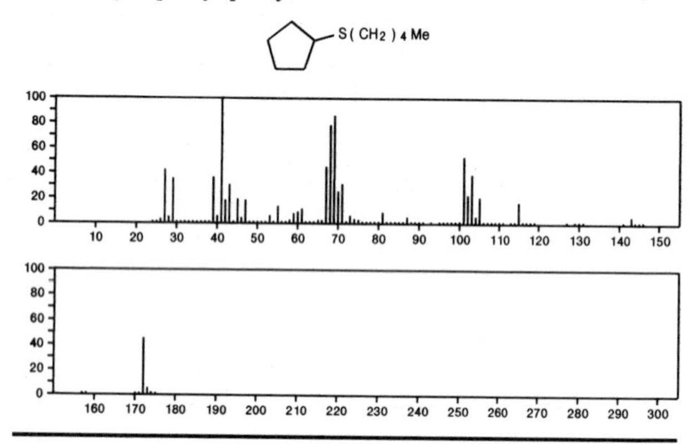

172 $C_{10}H_{20}S$ 7133-22-4
Sulfide, *sec*-butyl cyclohexyl

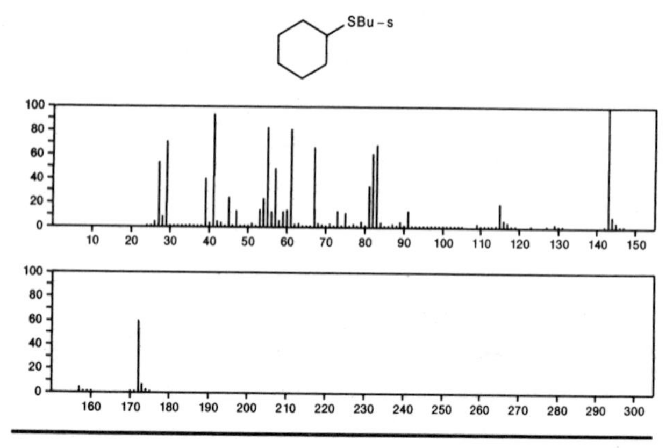

172 $C_{10}H_{20}S$ 7133-23-5
Sulfide, *tert*-butyl cyclohexyl

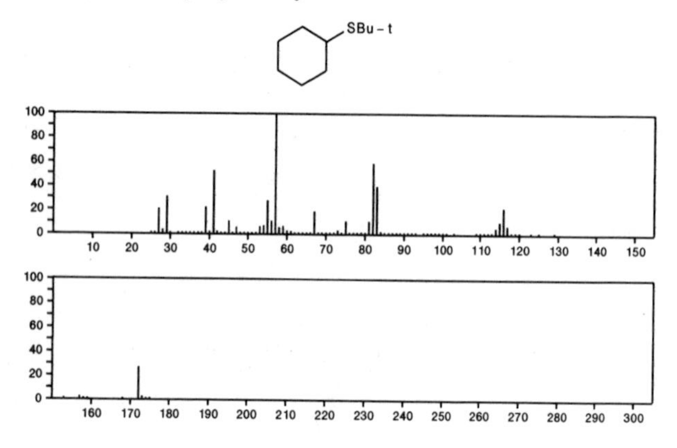

172 $C_{10}H_{20}S$ 7133-40-6
Cyclohexane, (butylthio)-

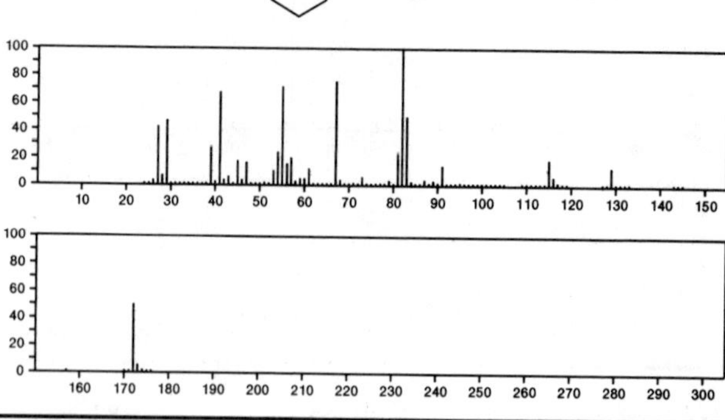

172 $C_{10}H_{20}S$ 42779-08-8
Octane, 1-(ethenylthio)-

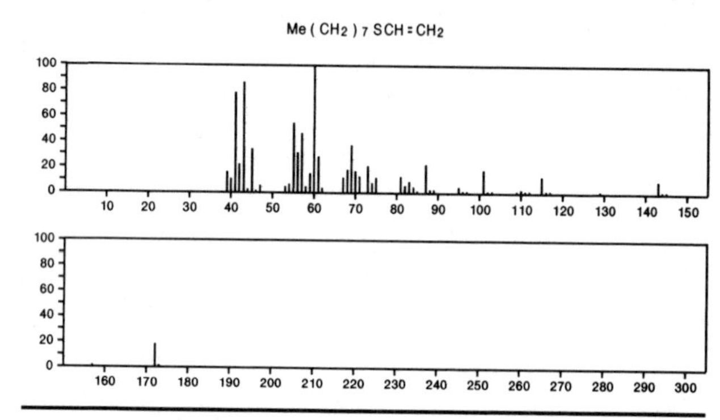

172 $C_{10}H_{24}N_2$ 54966-00-6
1,2-Ethanediamine, *N*,*N'*-dimethyl-*N*,*N'*-bis(1-methylethyl)-

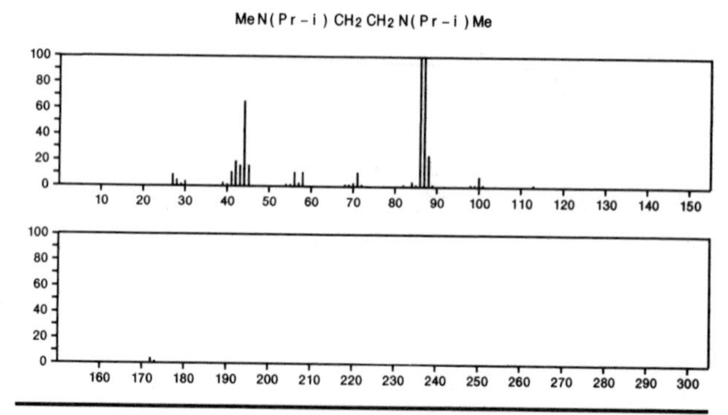

172 $C_{11}H_8O_2$ 58-27-5
1,4-Naphthalenedione, 2-methyl-

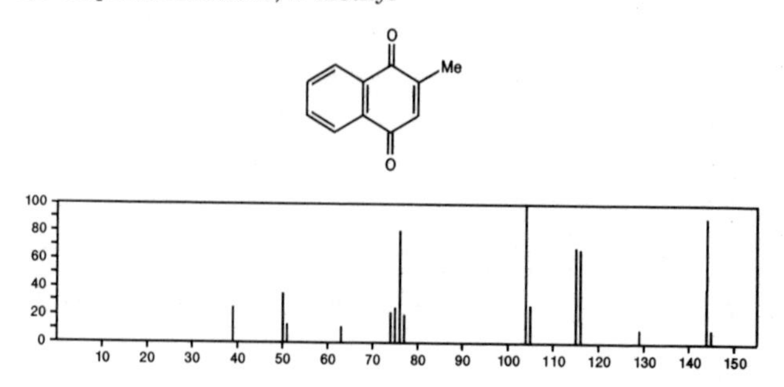

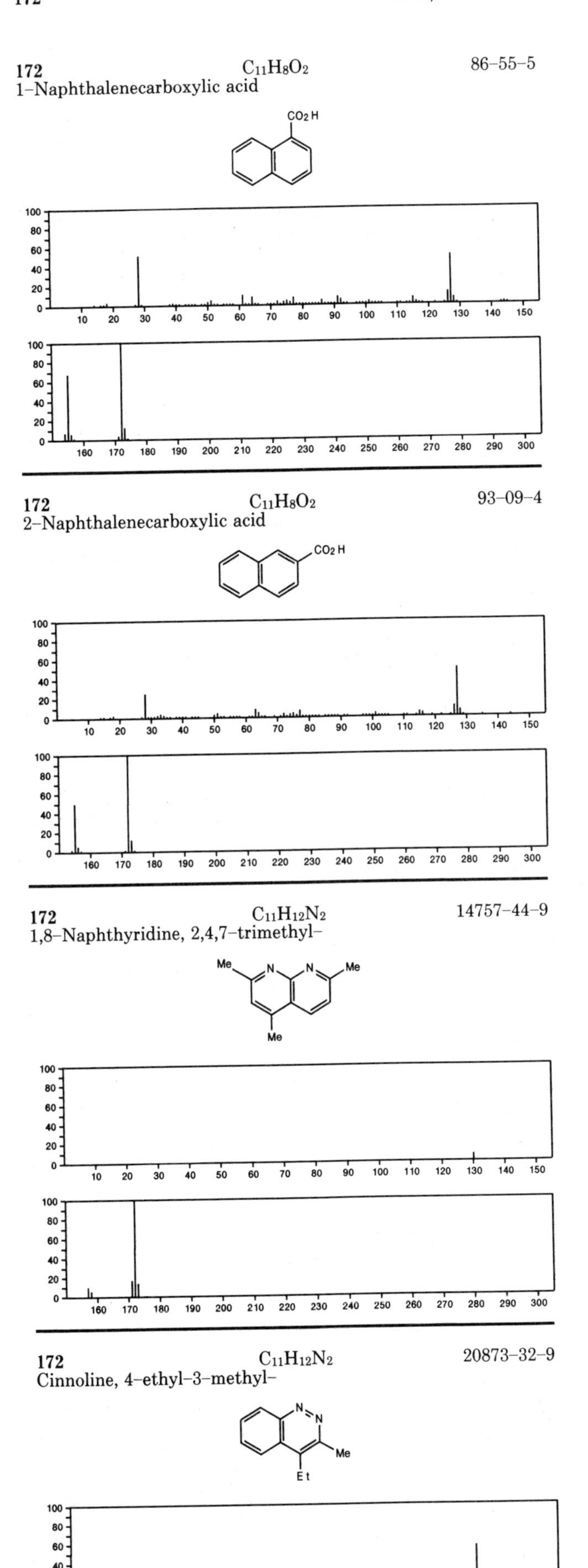
172 $C_{11}H_8O_2$ 86-55-5
1-Naphthalenecarboxylic acid
CO2H
172 $C_{11}H_8O_2$ 93-09-4
2-Naphthalenecarboxylic acid
CO2H
172 $C_{11}H_{12}N_2$ 14757-44-9
1,8-Naphthyridine, 2,4,7-trimethyl-
Me
N
172 $C_{11}H_{12}N_2$ 20873-32-9
Cinnoline, 4-ethyl-3-methyl-
Et

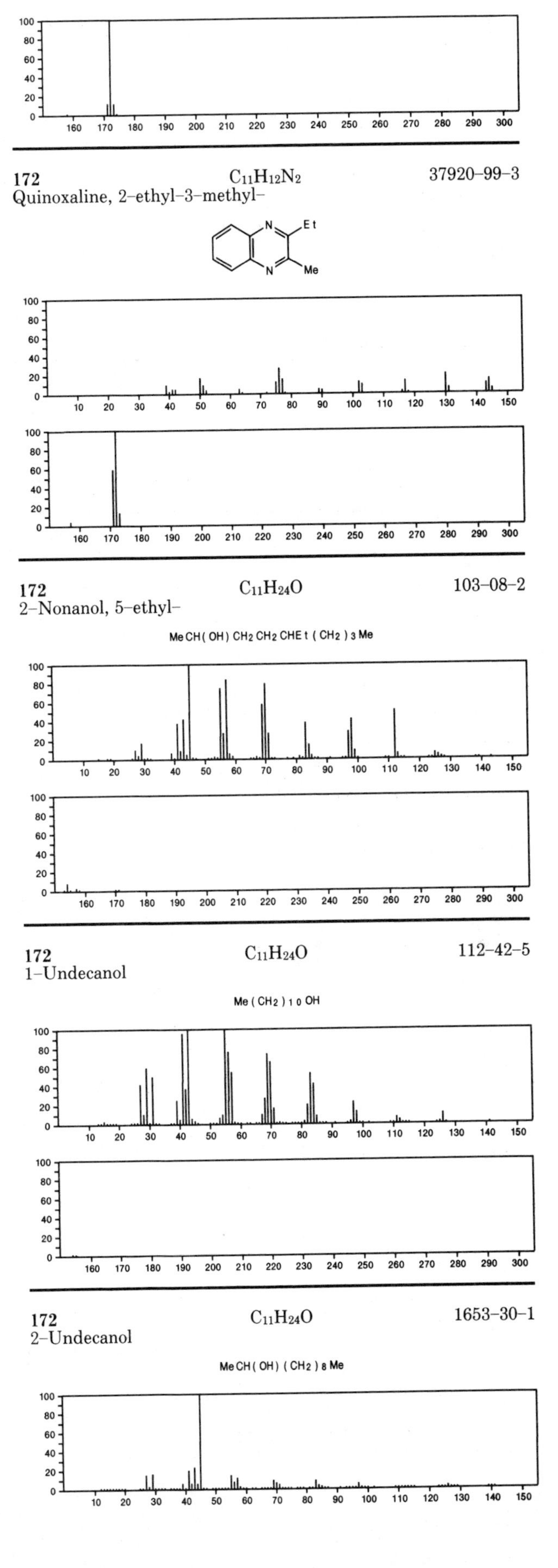
172 $C_{11}H_{12}N_2$ 37920-99-3
Quinoxaline, 2-ethyl-3-methyl-
Et
Me
172 $C_{11}H_{24}O$ 103-08-2
2-Nonanol, 5-ethyl-
MeCH(OH)CH2CH2CHEt(CH2)3Me
172 $C_{11}H_{24}O$ 112-42-5
1-Undecanol
Me(CH2)10OH
172 $C_{11}H_{24}O$ 1653-30-1
2-Undecanol
MeCH(OH)(CH2)8Me

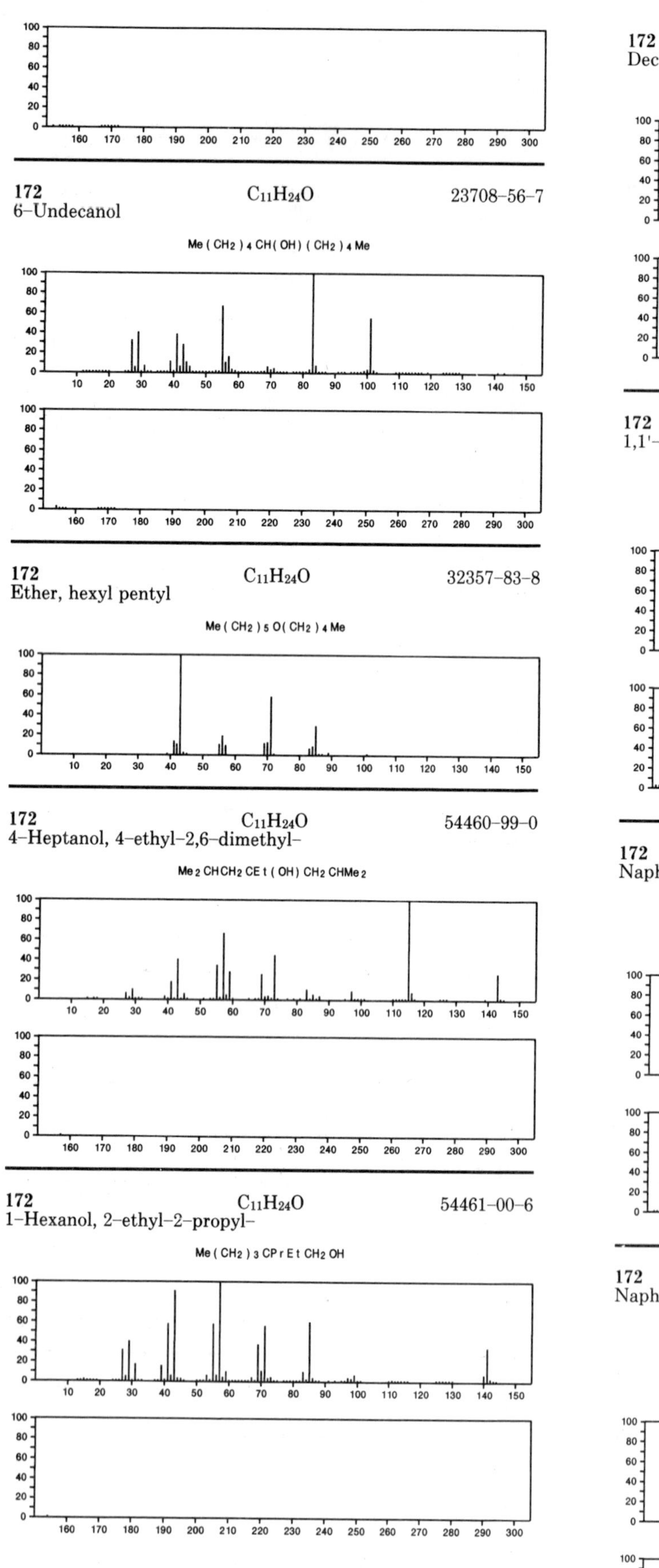

172 $C_{11}H_{24}O$ 23708–56–7
6–Undecanol
Me (CH2) 4 CH(OH) (CH2) 4 Me
172 $C_{11}H_{24}O$ 32357–83–8
Ether, hexyl pentyl
Me (CH2) 5 O(CH2) 4 Me
172 $C_{11}H_{24}O$ 54460–99–0
4–Heptanol, 4–ethyl–2,6–dimethyl–
Me 2 CHCH2 CEt (OH) CH2 CHMe 2
172 $C_{11}H_{24}O$ 54461–00–6
1–Hexanol, 2–ethyl–2–propyl–
Me (CH2) 3 CPr Et CH2 OH

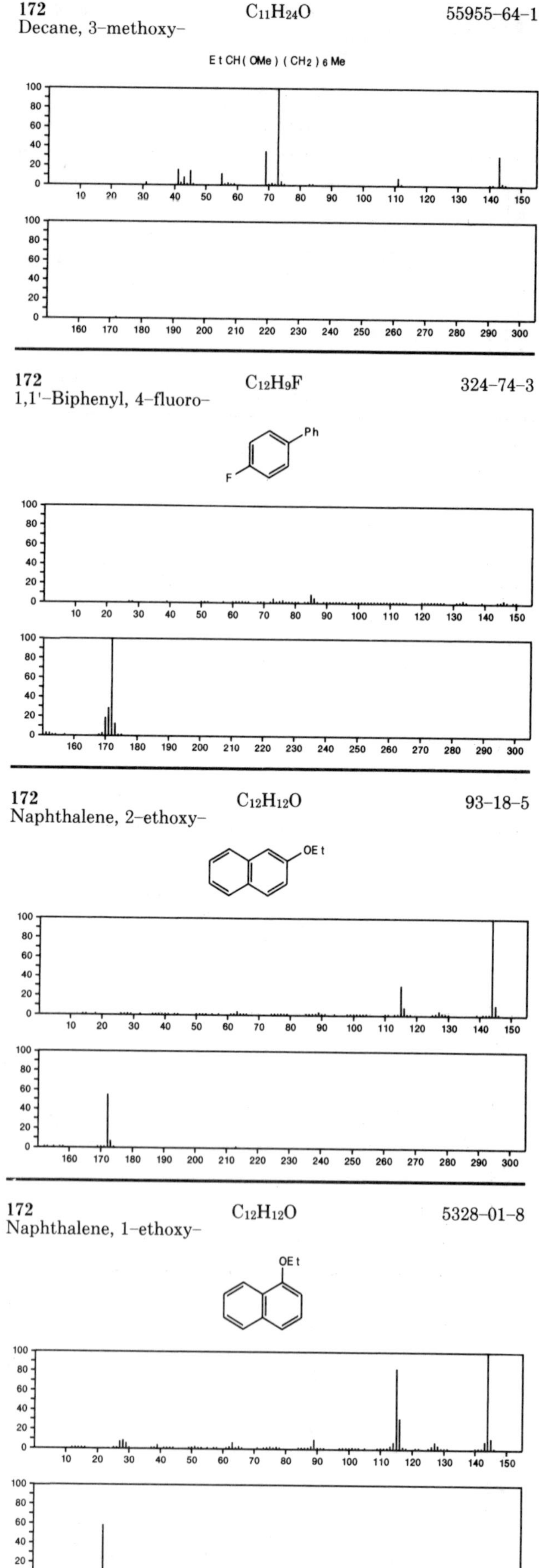

172 $C_{11}H_{24}O$ 55955–64–1
Decane, 3–methoxy–
Et CH(OMe) (CH2) 6 Me
172 $C_{12}H_9F$ 324–74–3
1,1'–Biphenyl, 4–fluoro–
Ph
F
172 $C_{12}H_{12}O$ 93–18–5
Naphthalene, 2–ethoxy–
OEt
172 $C_{12}H_{12}O$ 5328–01–8
Naphthalene, 1–ethoxy–
OEt

172 $C_{12}H_{12}O$ 13153–76–9
1,4–Ethanonaphthalen–2(1*H*)–one, 3,4–dihydro–

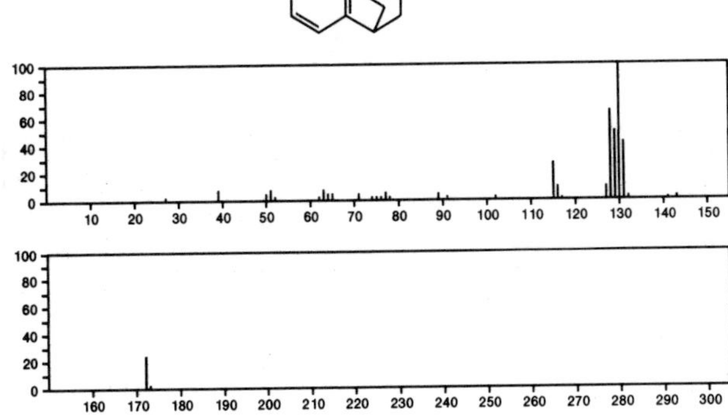

172 $C_{12}H_{12}O$ 23911–58–2
Benzofuran, 3–methyl–2–(1–methylethenyl)–

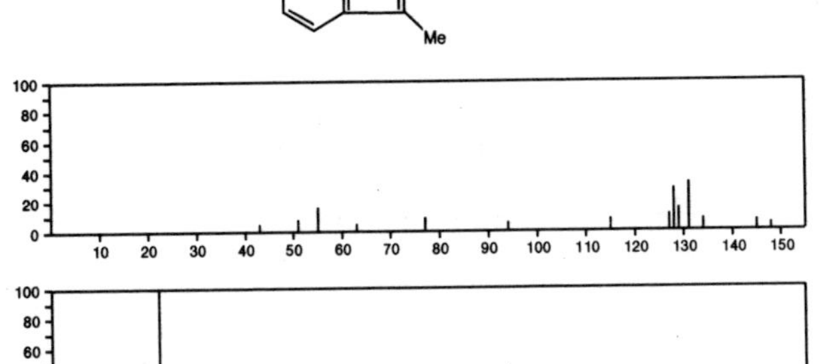

172 $C_{12}H_{12}O$ 31706–76–0
1–Naphthol, 5,7–dimethyl–

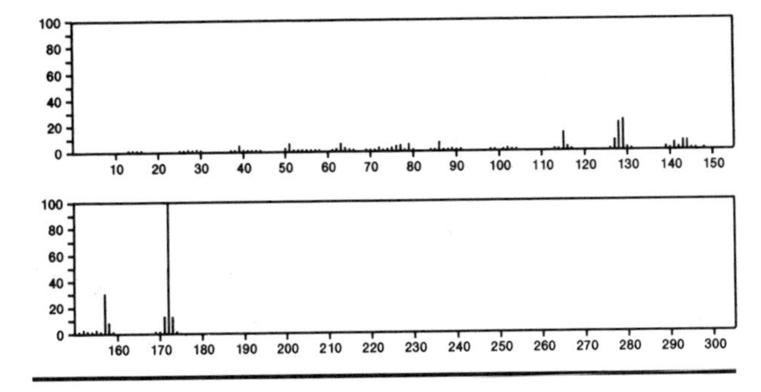

172 $C_{12}H_{12}O$ 31776–14–4
1–Naphthol, 6,7–dimethyl–

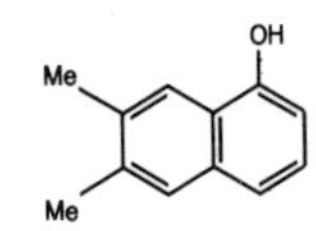

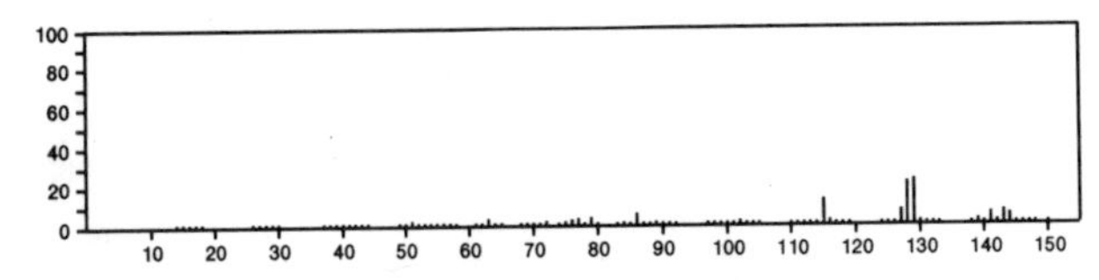

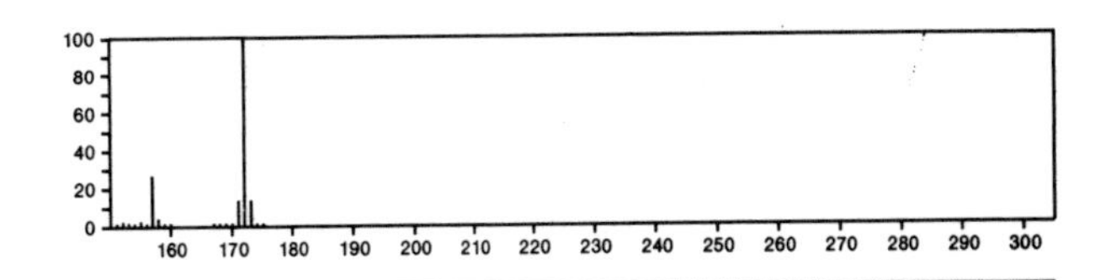

172 $C_{12}H_{12}O$ 54461–07–3
2,7–Ethanonaphth[2,3–*b*]oxirene, 1a,2,7,7a–tetrahydro–, (1aα,2β,7β,7aα)–

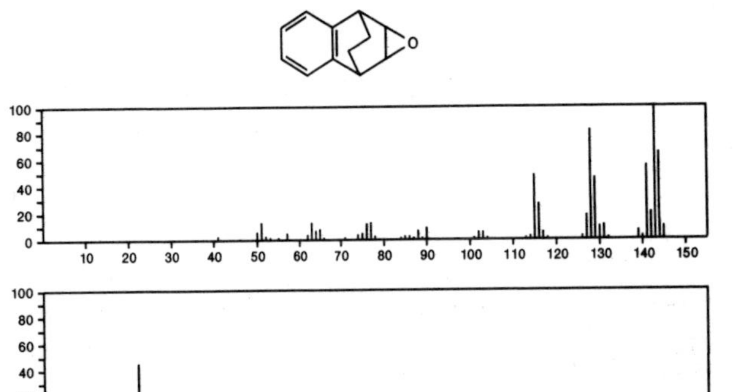

172 $C_{12}H_{12}O$ 54515–76–3
2,7–Ethanonaphth[2,3–*b*]oxirene, 1a,2,7,7a–tetrahydro–, (1aα,2α,7α,7aα)–

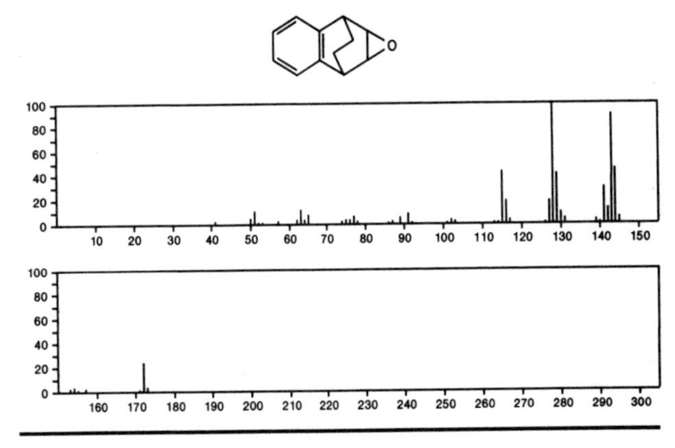

172 $C_{12}H_{12}O$ 56701–48–5
Cyclobut[*c*]indene–1–carboxaldehyde, 1,2,2a,3–tetrahydro–

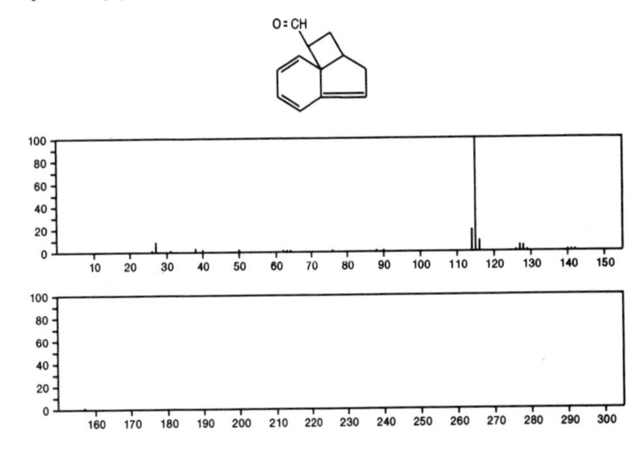

172 $C_{13}H_{16}$ 4506-36-9
Naphthalene, 1,2-dihydro-1,5,8-trimethyl-

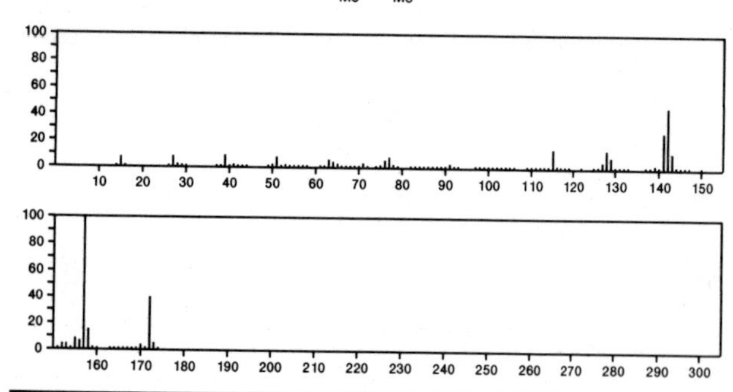

172 $C_{13}H_{16}$ 5732-00-3
Benzene, 2-(1,3-butadienyl)-1,3,5-trimethyl-

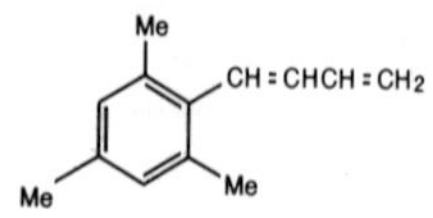

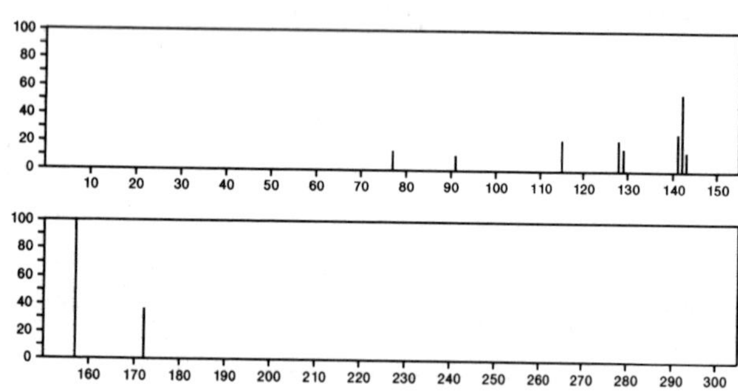

172 $C_{13}H_{16}$ 24578-28-7
Benzene, 1-methyl-4-[(1-methylethylidene)cyclopropyl]-

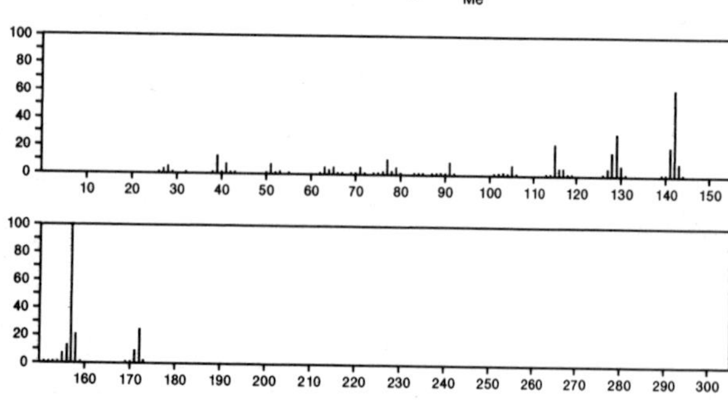

172 $C_{13}H_{16}$ 30316-18-8
Naphthalene, 1,2-dihydro-3,5,8-trimethyl-

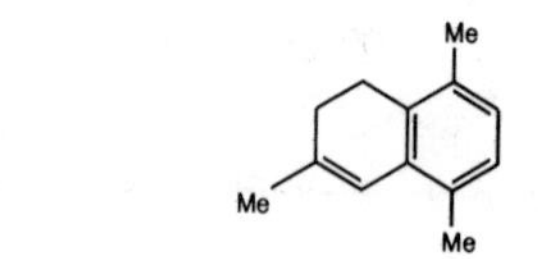

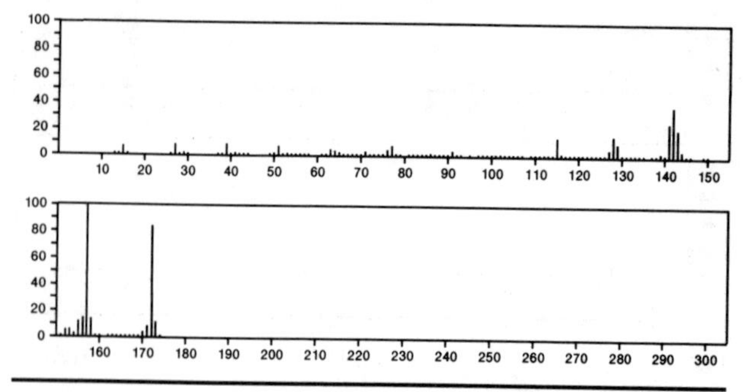

172 $C_{13}H_{16}$ 30316-19-9
Naphthalene, 1,4-dihydro-2,5,8-trimethyl-

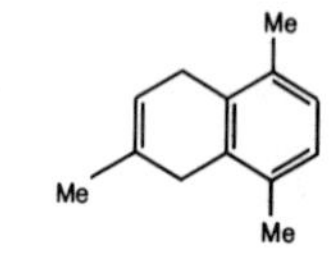

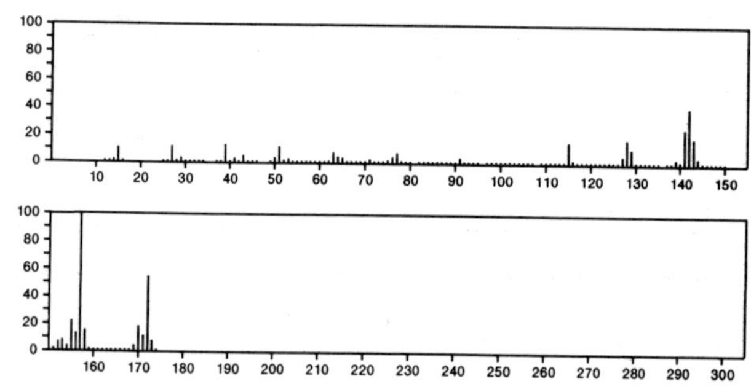

172 $C_{13}H_{16}$ 30316-23-5
Naphthalene, 1,2-dihydro-2,5,8-trimethyl-

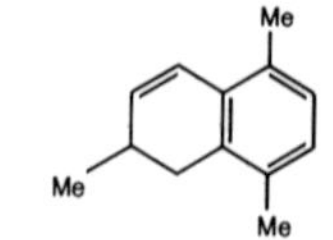

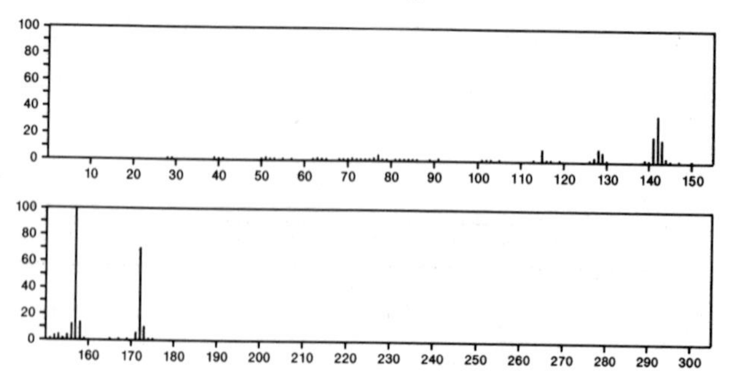

172 $C_{13}H_{16}$ 53156-03-9
Naphthalene, 1,2-dihydro-2,5,7-trimethyl-

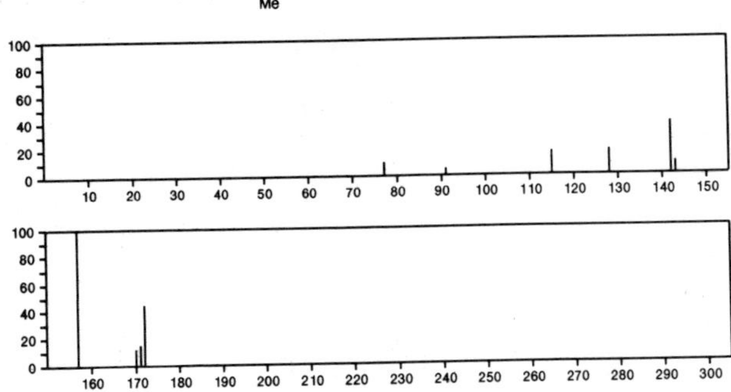

172 $C_{13}H_{16}$ 53156-06-2
Naphthalene, 1,2-dihydro-3,6,8-trimethyl-

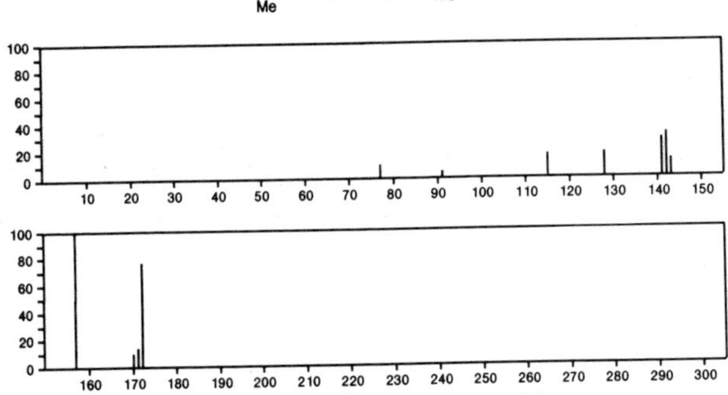

172 $C_{13}H_{16}$ 53156-11-9
Naphthalene, 1,2-dihydro-4,5,7-trimethyl-

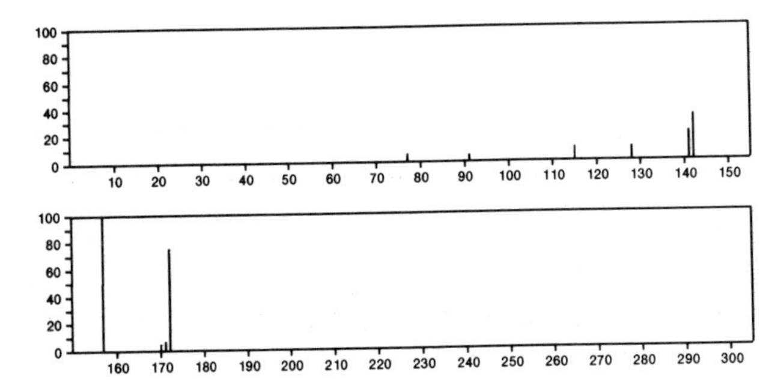

172 $C_{13}H_{16}$ 53156-12-0
Naphthalene, 1,2-dihydro-4,6,8-trimethyl-

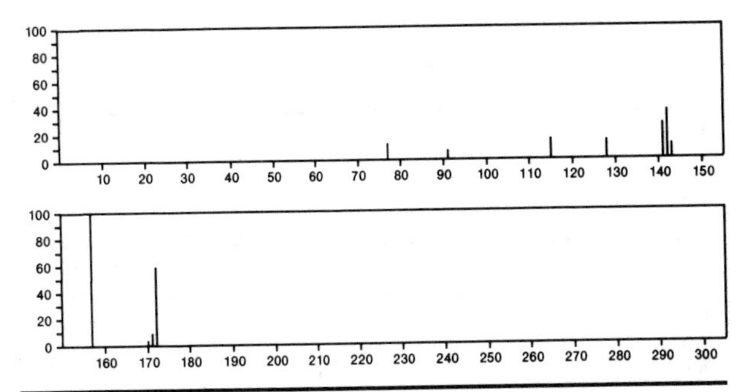

172 $C_{13}H_{16}$ 54725-17-6
Benzene, 2-heptynyl-

$Me(CH_2)_3C{\equiv}CCH_2Ph$

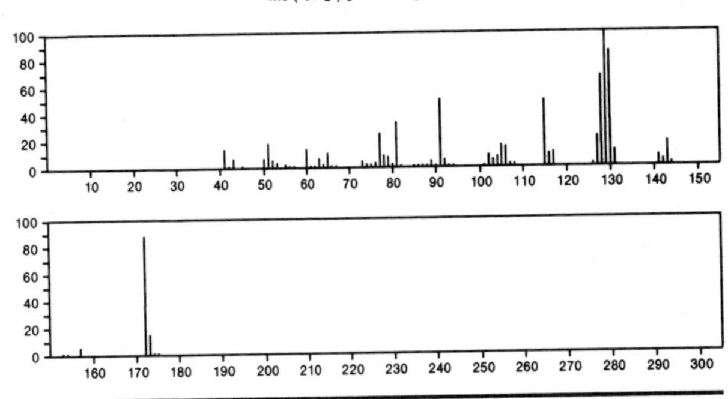

172 $C_{13}H_{16}$ 55682-80-9
Naphthalene, 1,2-dihydro-1,4,6-trimethyl-

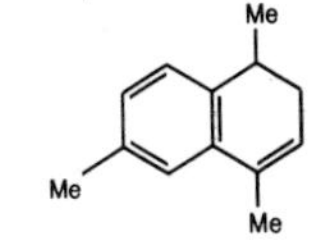

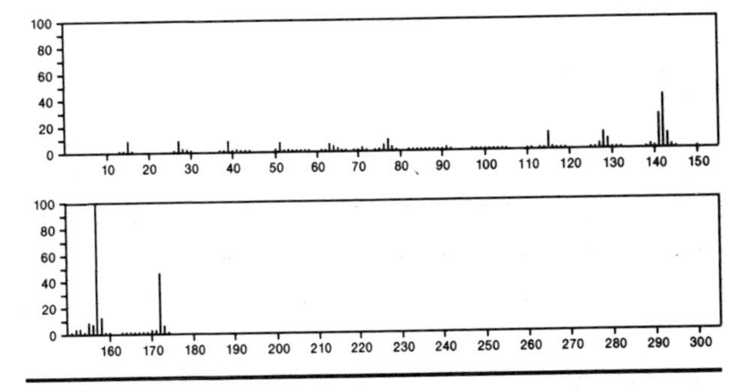

172 $C_{13}H_{16}$ 56293-02-8
Benzene, 6-heptynyl-

$HC{\equiv}C(CH_2)_5Ph$

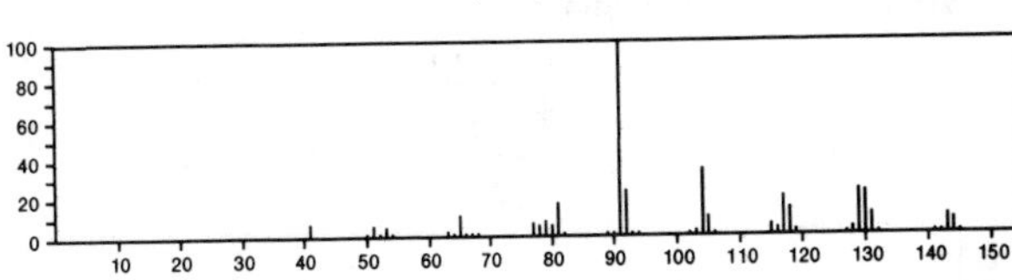

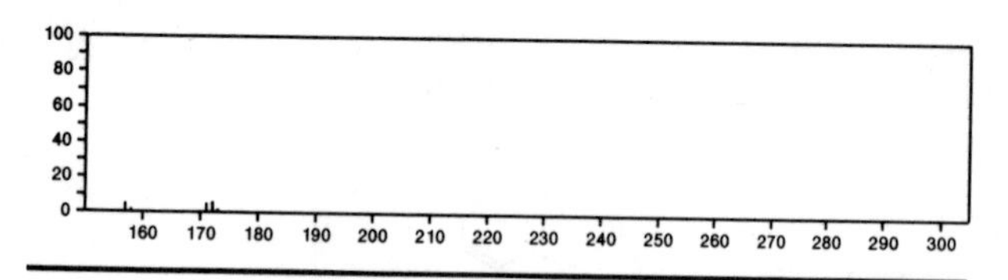

172 $C_{13}H_{16}$ 56293-03-9
Benzene, 4-heptynyl-

Ph(CH_2)$_3$C≡CEt

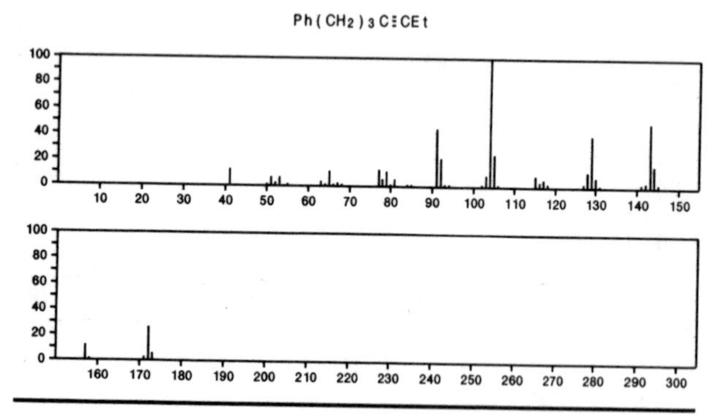

172 $C_{13}H_{16}$ 56293-04-0
Benzene, 3-heptynyl-

PrC≡C$CH_2$$CH_2$Ph

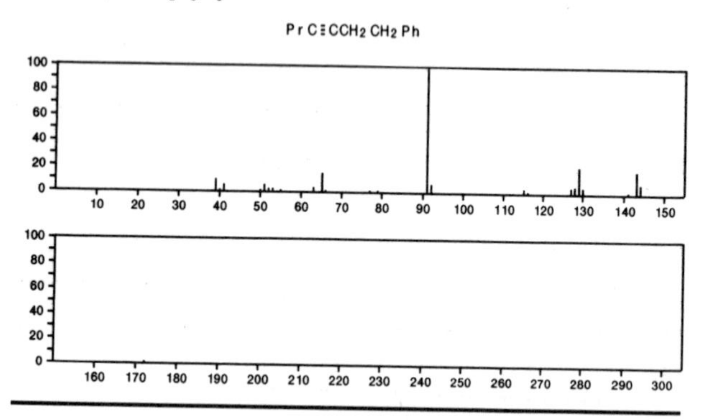

172 $C_{13}H_{16}$ 56701-42-9
Benzene, 1-methyl-3-[(1-methylethylidene)cyclopropyl]-

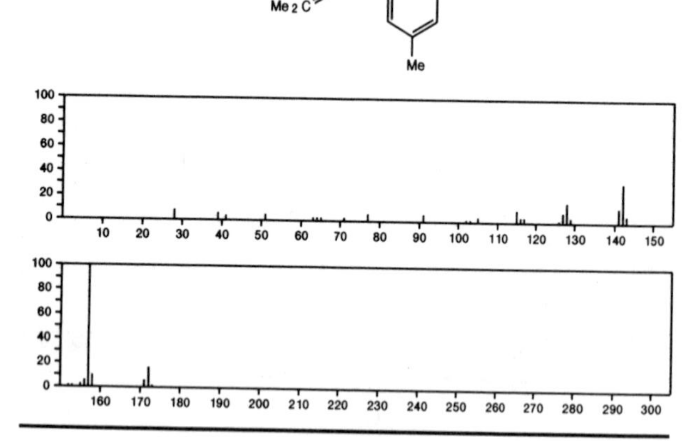

173 $C_4H_4BrN_3$ 7752-82-1
2-Pyrimidinamine, 5-bromo-

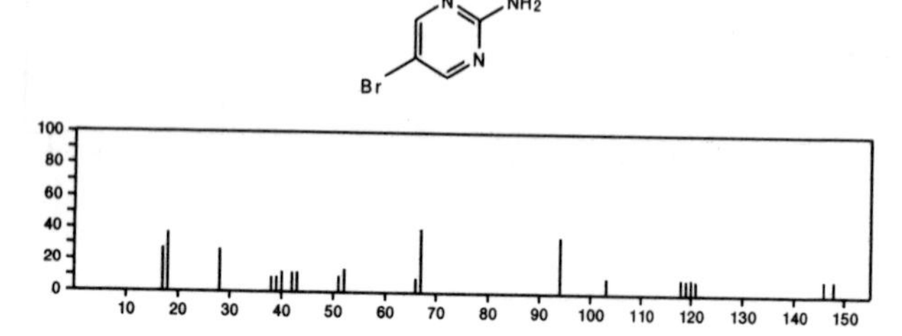

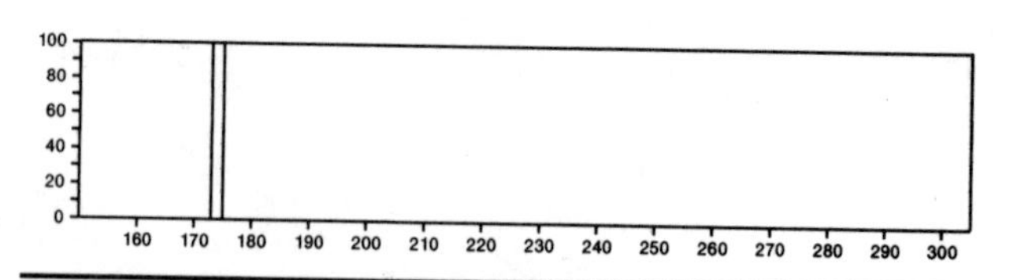

173 $C_5H_4ClN_3O_2$ 31396-27-7
Pyridine, 5-chloro-2-nitramino-

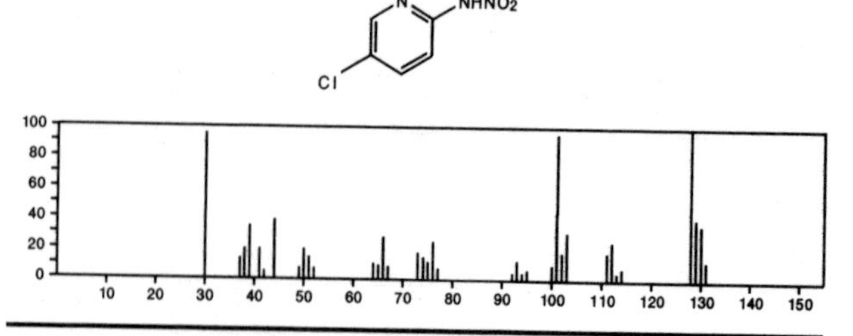

173 $C_5H_8ClN_5$ 1007-28-9
1,3,5-Triazine-2,4-diamine, 6-chloro-*N*-ethyl-

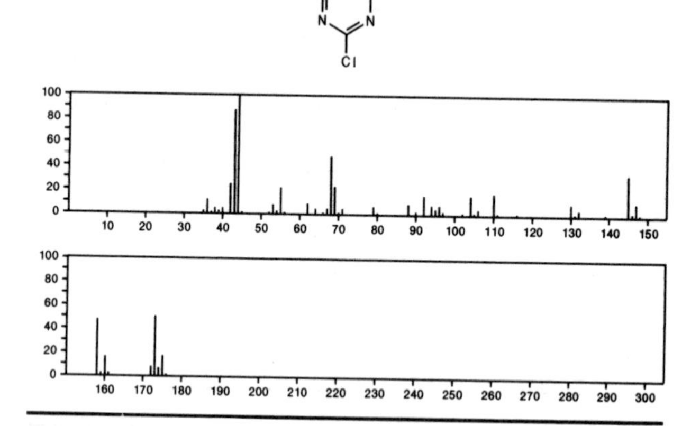

173 C_6H_4ClNOS 13165-68-9
Benzenamine, 4-chloro-*N*-sulfinyl-

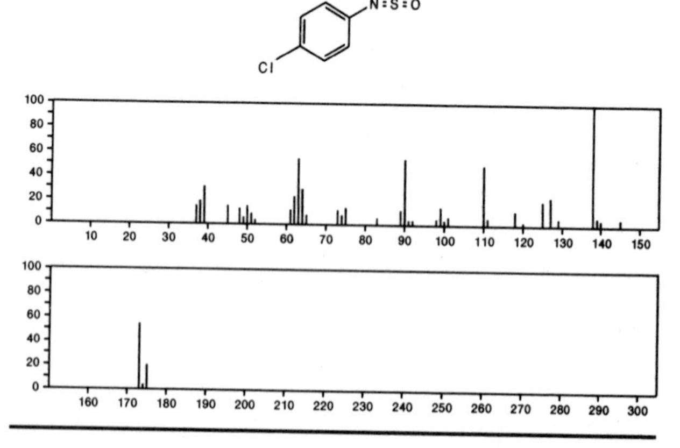

173 C_6H_4ClNOS 15851-82-8
Benzenamine, 3-chloro-*N*-sulfinyl-

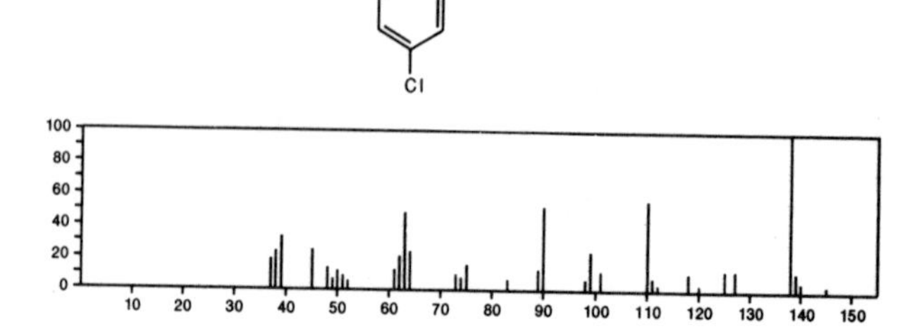

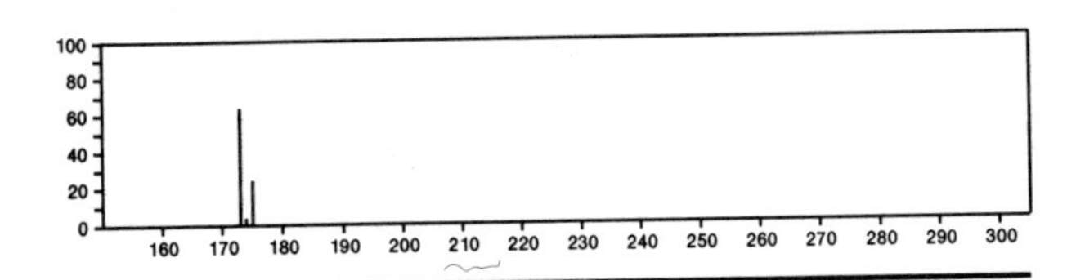

173 $C_6H_7NO_3S$ 56196-66-8

Acetic acid, (4-oxo-2-thiazolidinylidene)-, methyl ester

S, CHC(O)OMe, NH, O

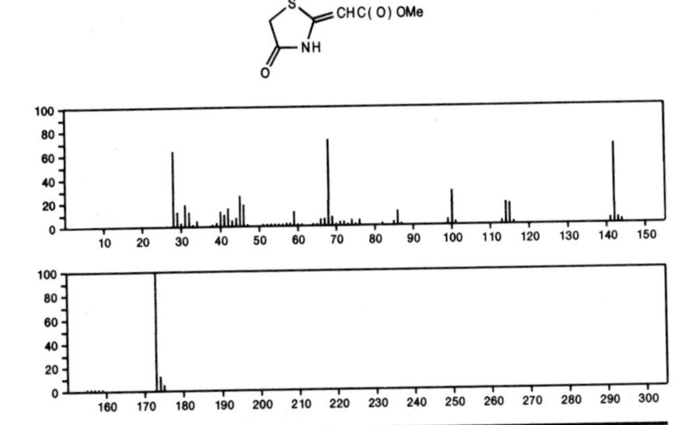

173 $C_6H_8ClN_3O$ 54484-70-7

5-Pyrimidinamine, 2-chloro-4-ethoxy-

Cl, N, OEt, N, NH_2

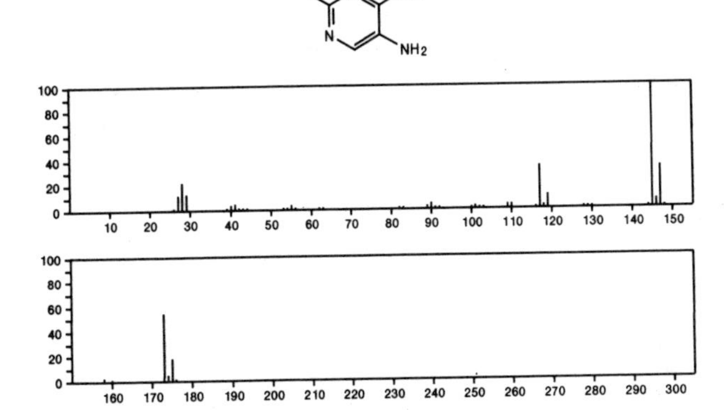

173 $C_7H_{11}NO_2S$ 41783-61-3

2-Propenethioamide, 3-(acetyloxy)-*N*,*N*-dimethyl-, (*Z*)-

AcOCH=CHCSNMe2

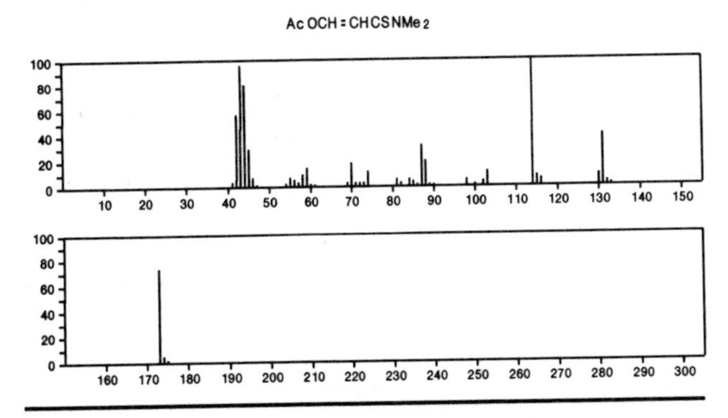

173 $C_7H_{11}NO_2S$ 52118-16-8

2-Propenethioamide, 3-(acetyloxy)-*N*,*N*-dimethyl-, (*E*)-

AcOCH=CHCSNMe2

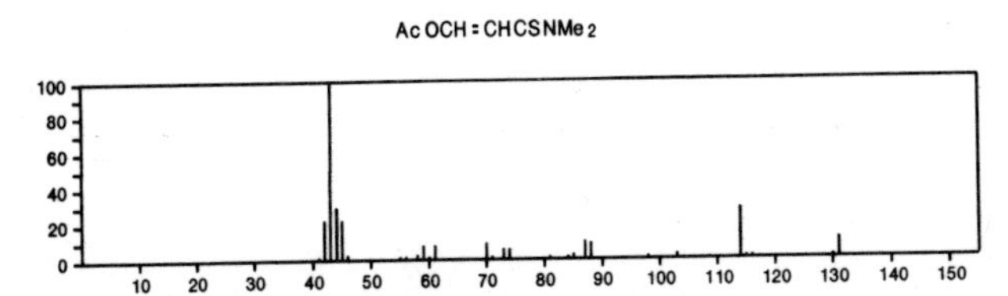

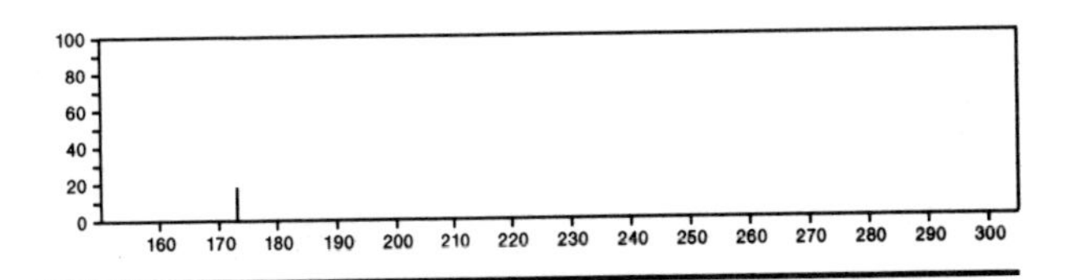

173 $C_7H_{11}NO_4$ 33996-33-7

L-Proline, 1-acetyl-4-hydroxy-, *trans*-

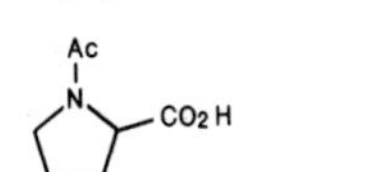

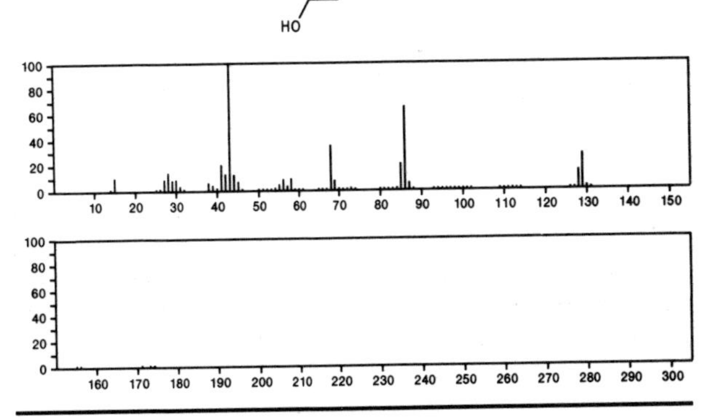

173 $C_7H_{15}N_3O_2$ 50285-70-6

Urea, triethylnitroso-

Et2NCON(NO)Et

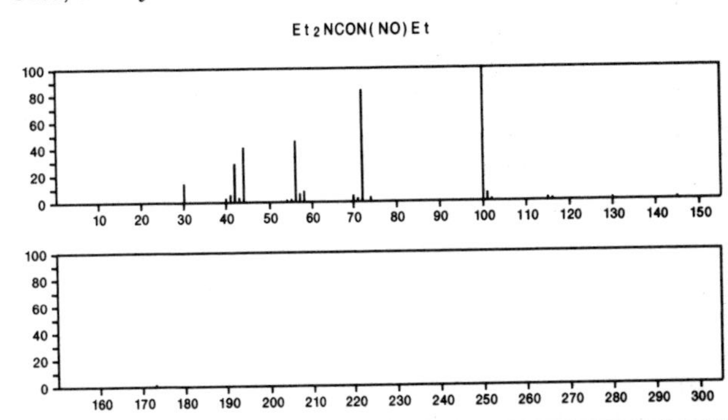

173 $C_8H_{15}NO_3$ 99-15-0

DL-Leucine, *N*-acetyl-

Me2CHCH2CH(CO2H)NHAc

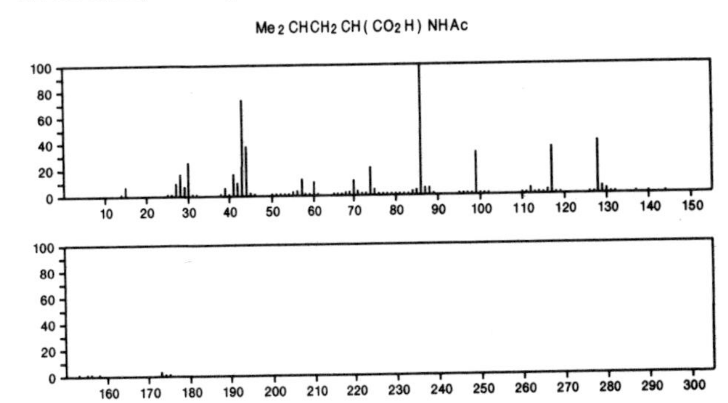

173 $C_8H_{15}NO_3$ 3077-46-1

L-Isoleucine, *N*-acetyl-

MeCH2CHMeCH(CO2H)NHAc

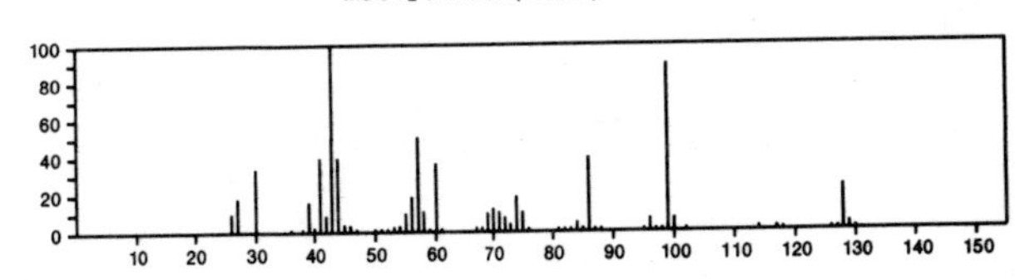

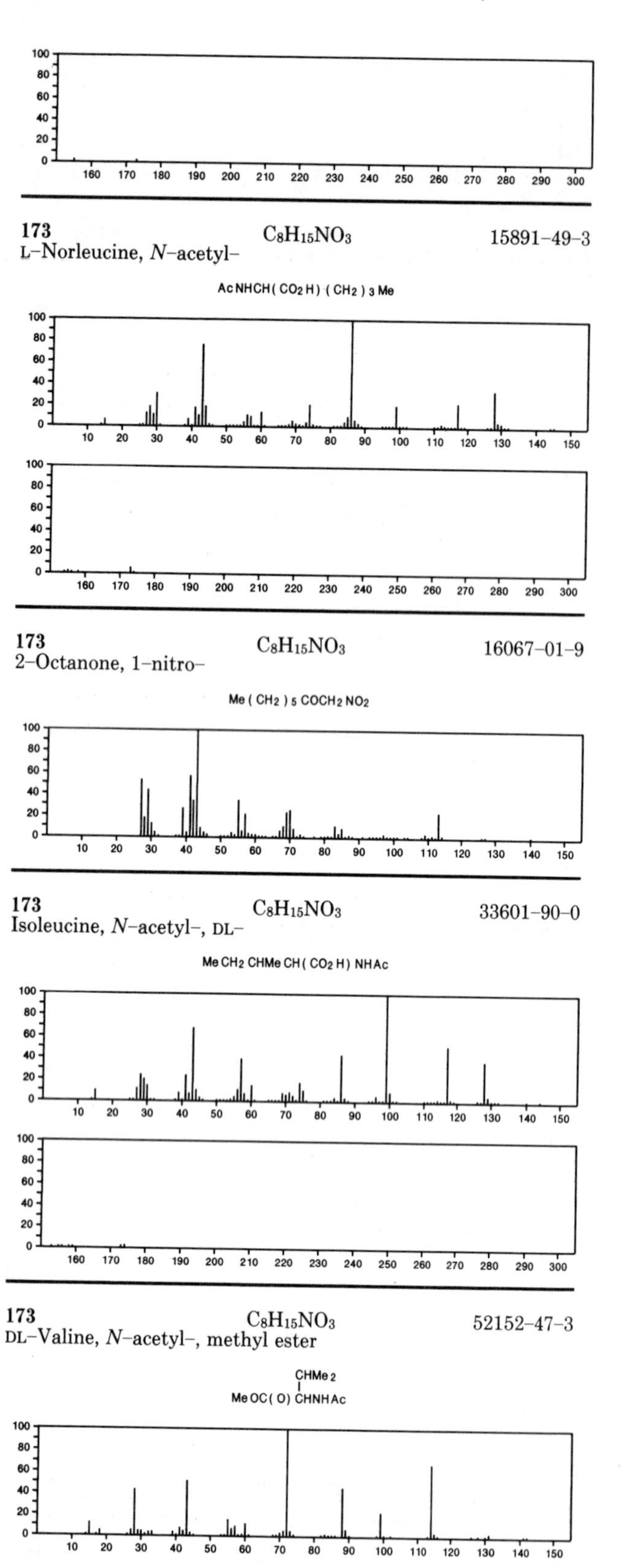

173 $C_8H_{15}NO_3$ 15891-49-3
L-Norleucine, *N*-acetyl-

AcNHCH(CO2H)(CH2)3Me

173 $C_8H_{15}NO_3$ 16067-01-9
2-Octanone, 1-nitro-

Me(CH2)5COCH2NO2

173 $C_8H_{15}NO_3$ 33601-90-0
Isoleucine, *N*-acetyl-, DL-

MeCH2CHMeCH(CO2H)NHAc

173 $C_8H_{15}NO_3$ 52152-47-3
DL-Valine, *N*-acetyl-, methyl ester

CHMe2 | MeOC(O)CHNHAc

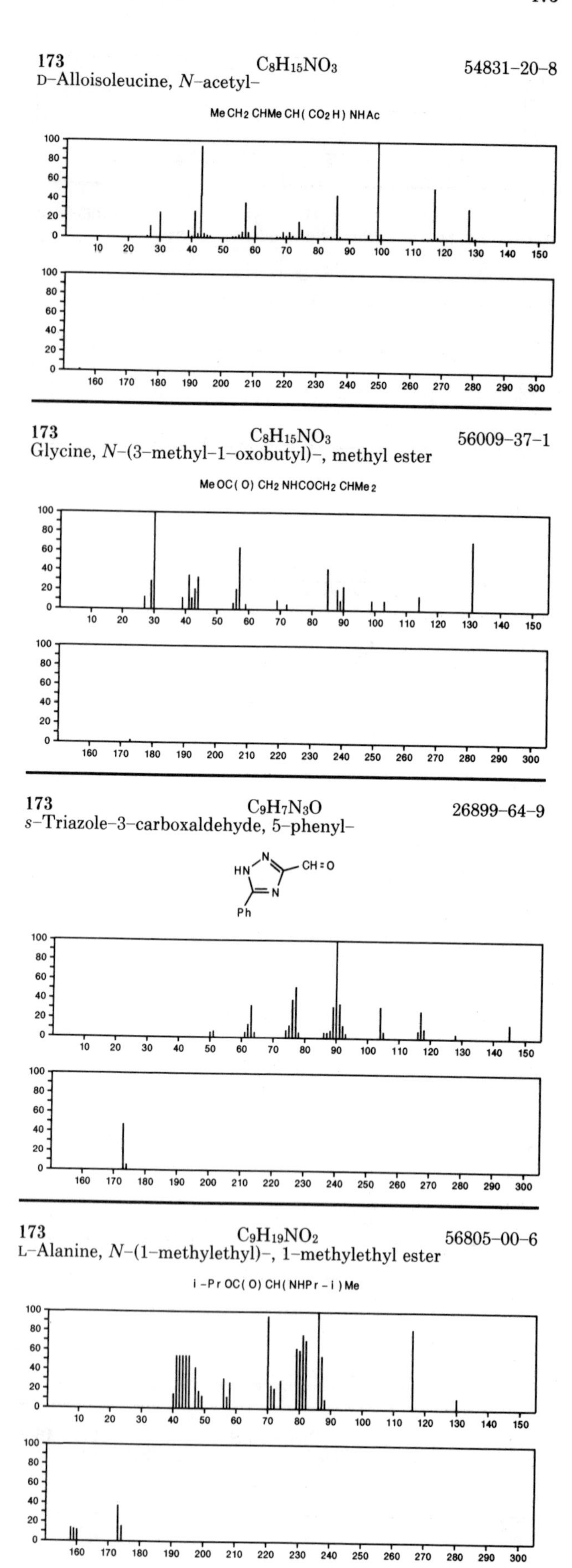

173 $C_8H_{15}NO_3$ 54831-20-8
D-Alloisoleucine, *N*-acetyl-

MeCH2CHMeCH(CO2H)NHAc

173 $C_8H_{15}NO_3$ 56009-37-1
Glycine, *N*-(3-methyl-1-oxobutyl)-, methyl ester

MeOC(O)CH2NHCOCH2CHMe2

173 $C_9H_7N_3O$ 26899-64-9
s-Triazole-3-carboxaldehyde, 5-phenyl-

173 $C_9H_{19}NO_2$ 56805-00-6
L-Alanine, *N*-(1-methylethyl)-, 1-methylethyl ester

i-PrOC(O)CH(NHPr-i)Me

173 $C_{10}H_7NO_2$ 86–57–7
Naphthalene, 1–nitro–

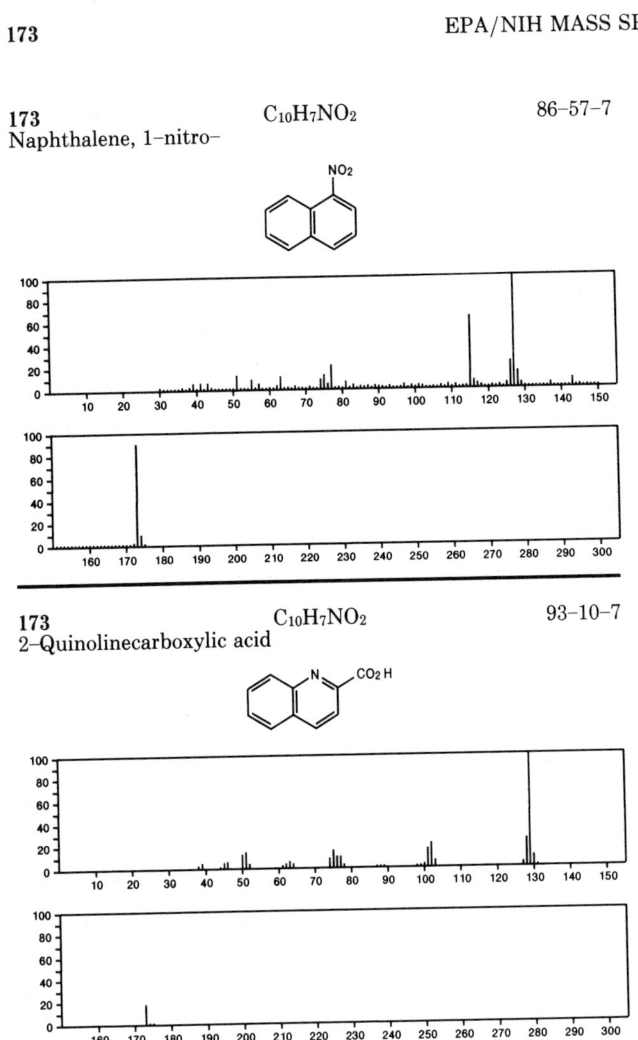

173 $C_{10}H_7NO_2$ 93–10–7
2–Quinolinecarboxylic acid

173 $C_{10}H_7NO_2$ 941–69–5
1*H*–Pyrrole–2,5–dione, 1–phenyl–

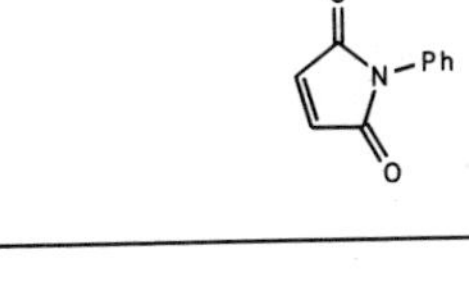

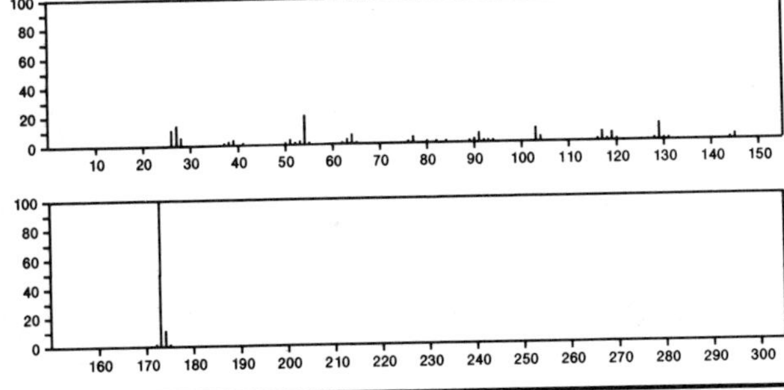

173 $C_{10}H_7NO_2$ 14510–06–6
2–Quinolinecarboxaldehyde, 8–hydroxy–

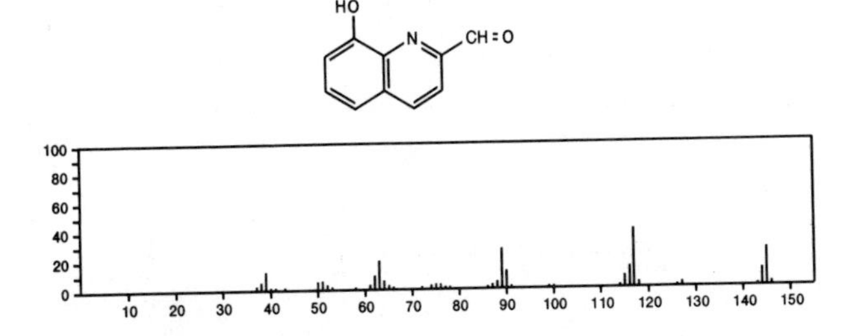

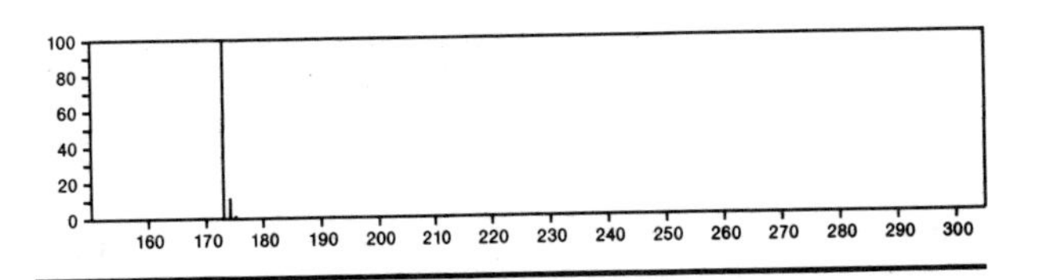

173 $C_{10}H_7NO_2$ 19990–26–2
2(5*H*)–Furanone, 5–(phenylimino)–

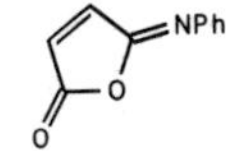

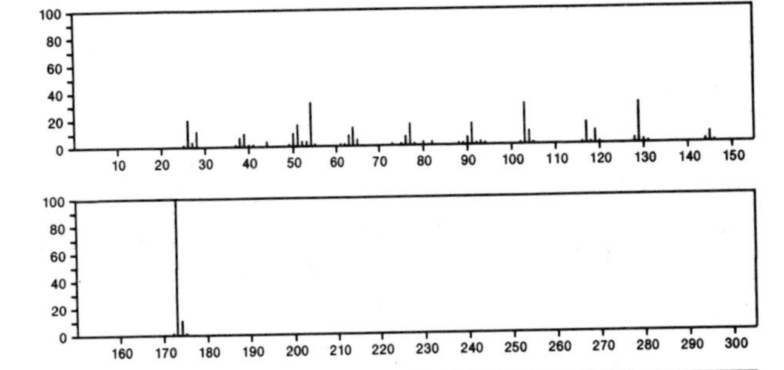

173 $C_{10}H_{23}NO$ 29812–79–1
Hydroxylamine, *O*–decyl–

$H_2NO(CH_2)_9Me$

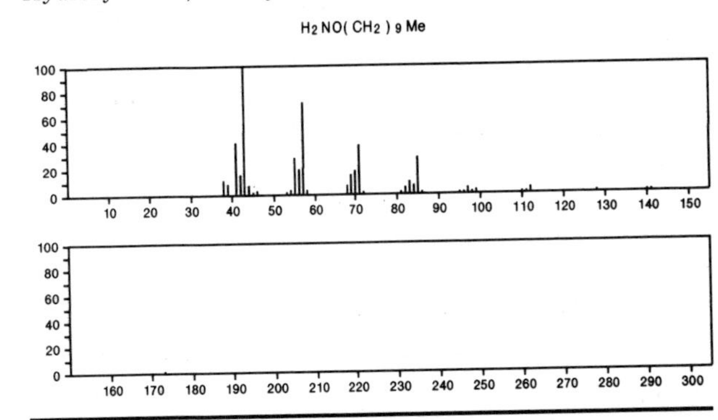

173 $C_{11}H_{11}NO$ 4211–90–9
Isoxazole, 3–ethyl–5–phenyl–

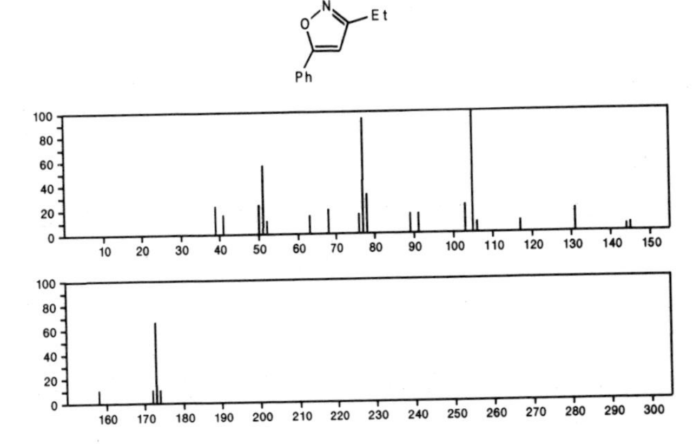

173 $C_{11}H_{11}NO$ 14300–11–9
Quinoline, 2,3–dimethyl–, 1–oxide

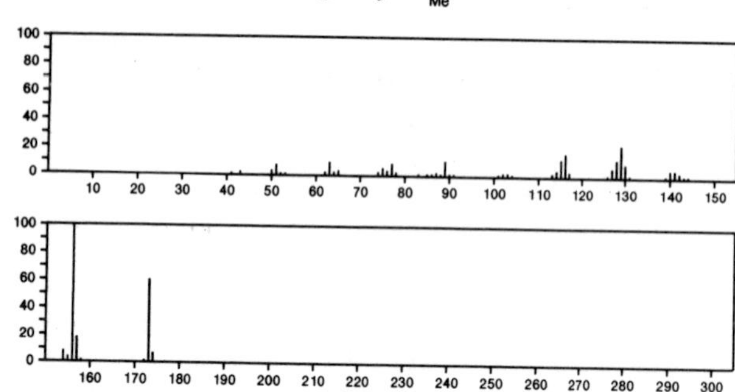

173 $C_{11}H_{11}NO$ 14300–12–0
Quinoline, 2,4–dimethyl–, 1–oxide

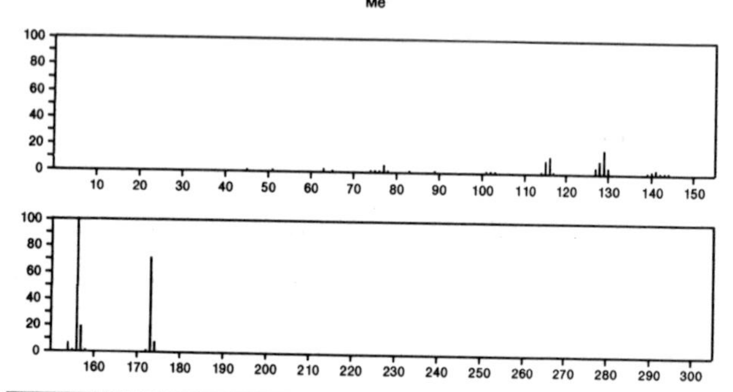

173 $C_{11}H_{11}NO$ 17336–90–2
2(1*H*)–Quinolinone, 3,4–dimethyl–

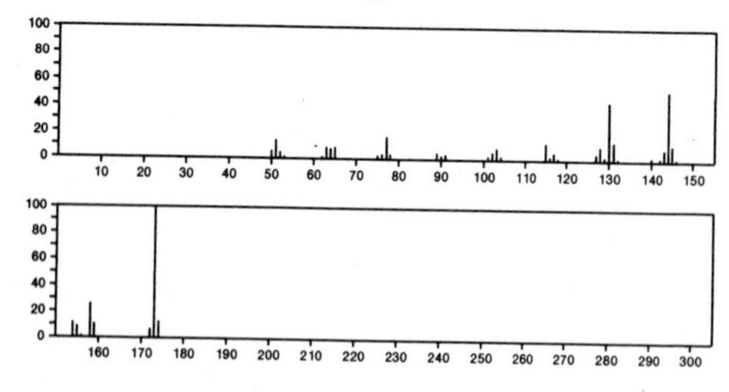

173 $C_{11}H_{11}NO$ 20662–91–3
Oxazole, 5–ethyl–4–phenyl–

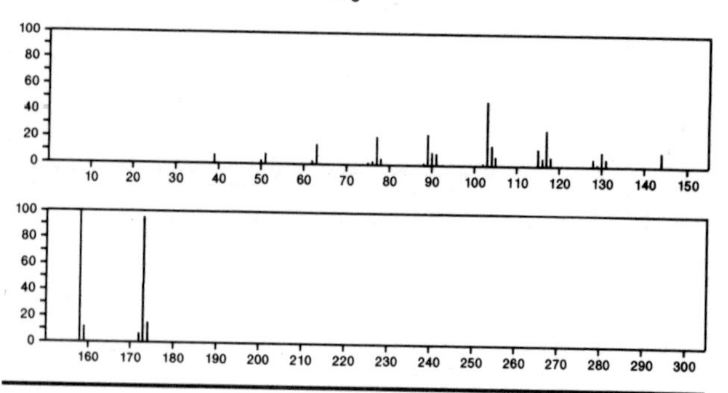

173 $C_{11}H_{11}NO$ 20662–92–4
Oxazole, 2,5–dimethyl–4–phenyl–

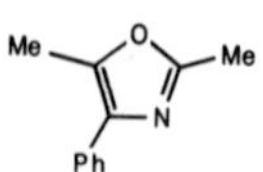

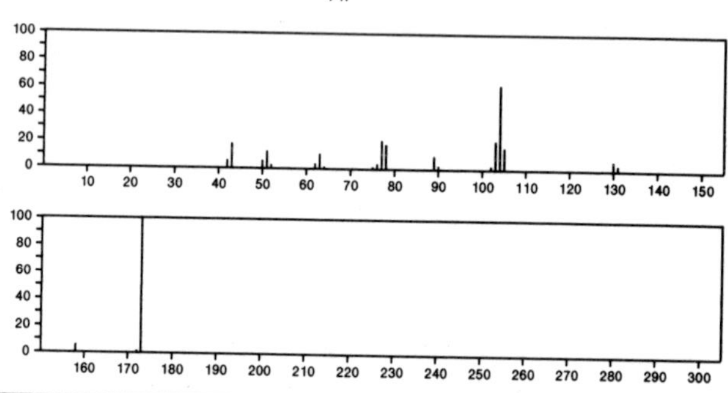

173 $C_{11}H_{11}NO$ 24562–79–6
2*H*–1,4–Ethanoquinolin–3(4*H*)–one

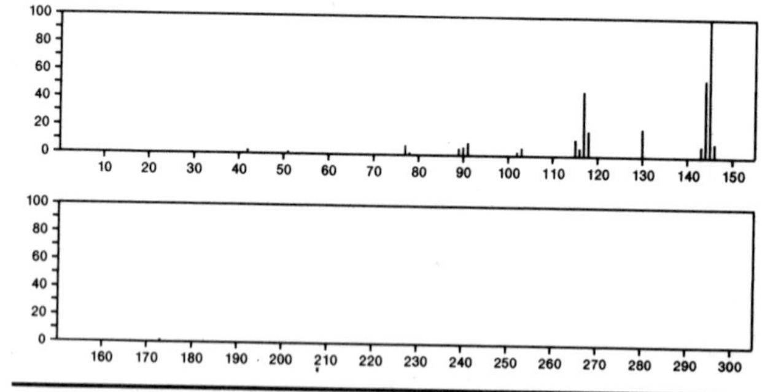

173 $C_{11}H_{11}NO$ 31108–61–9
Ketone, methyl 5–methyl–3–indolizinyl

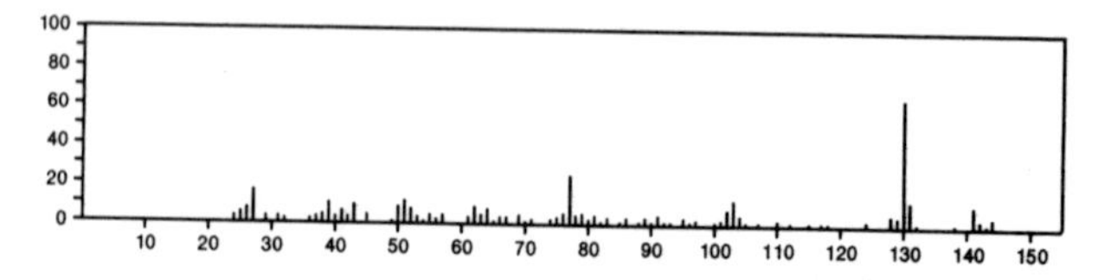

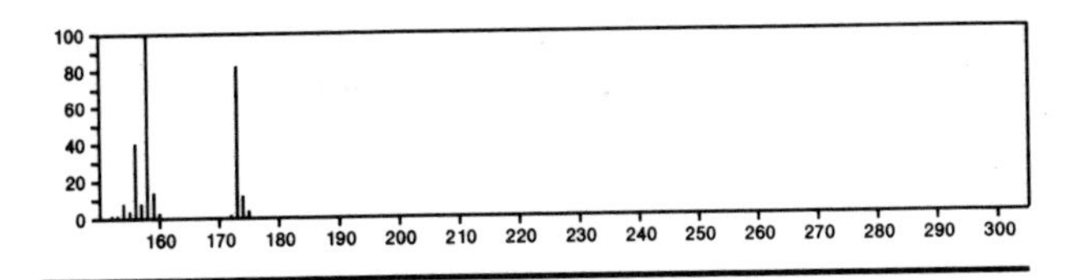

173 $C_{11}H_{11}NO$ 41037-26-7
Quinoline, 6-methoxy-4-methyl-

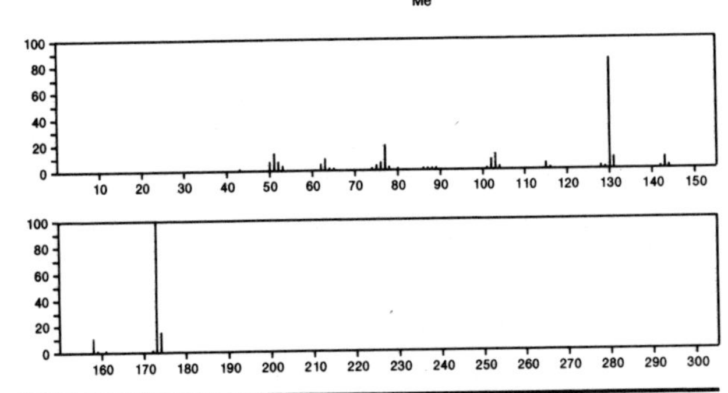

173 $C_{11}H_{11}NO$ 54484-68-3
Hydroxylamine, *O*-(1-naphthalenylmethyl)-

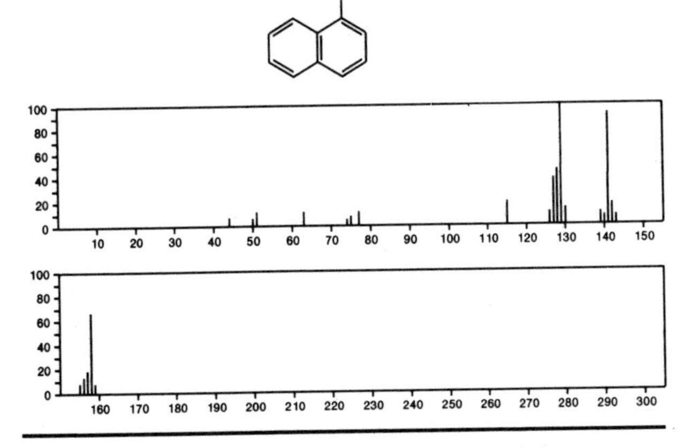

173 $C_{11}H_{11}NO$ 54484-69-4
Hydroxylamine, *O*-(2-naphthalenylmethyl)-

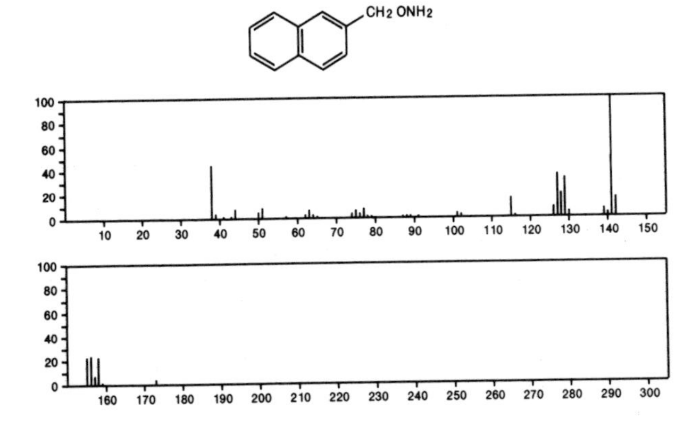

173 $C_{11}H_{11}NO$ 56909-00-3
5-Azatricyclo[7.2.0.0^{1,4}]undeca-2,5,7,10-tetraene, 6-methoxy-

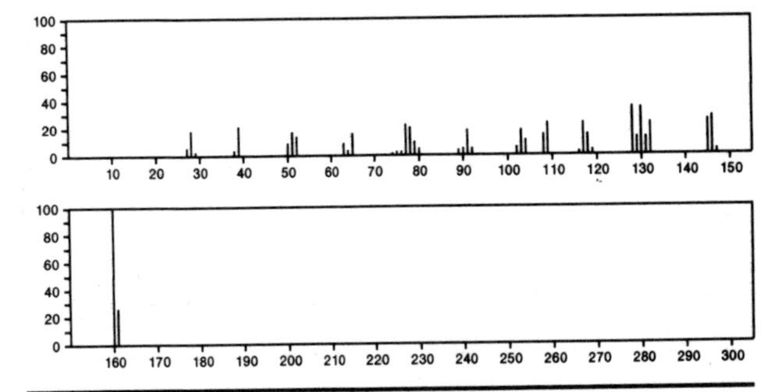

173 $C_{12}H_{15}N$ 147-47-7
Quinoline, 1,2-dihydro-2,2,4-trimethyl-

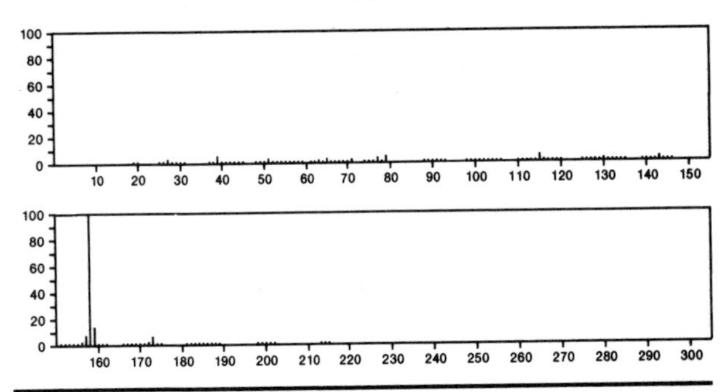

173 $C_{12}H_{15}N$ 1798-39-6
3*H*-Indole, 3-ethyl-2,3-dimethyl-

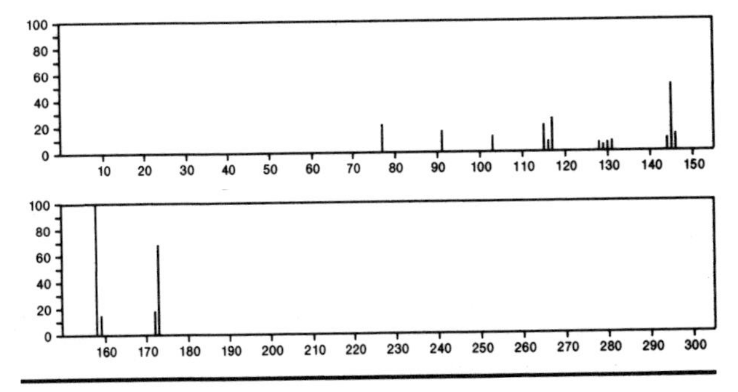

173 $C_{12}H_{15}N$ 18781-53-8
3*H*-Indole, 2-ethyl-3,3-dimethyl-

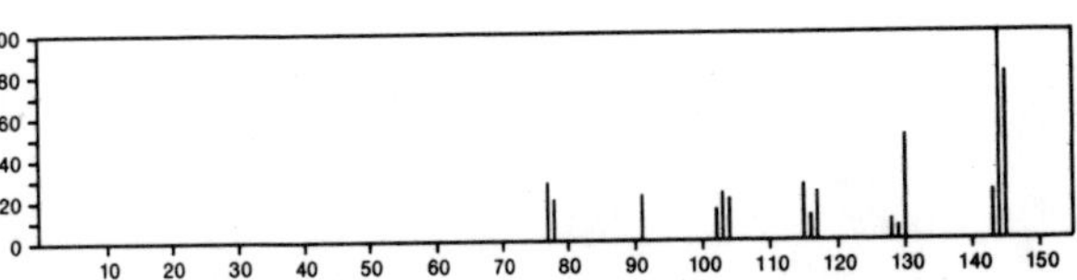

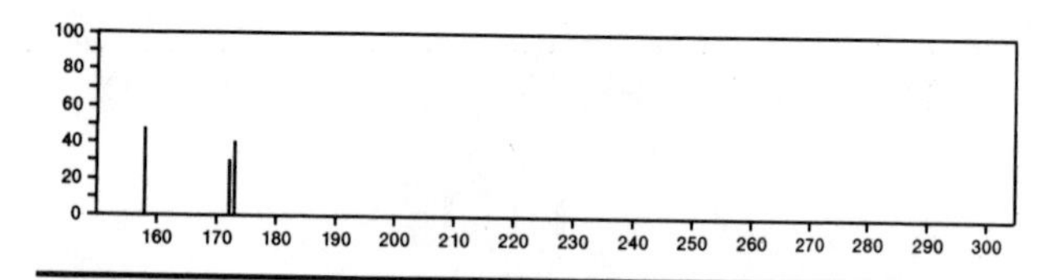

173 $C_{12}H_{15}N$ 23853-53-4
Aziridine, 1-(1,2,3,4-tetrahydro-2-naphthyl)-

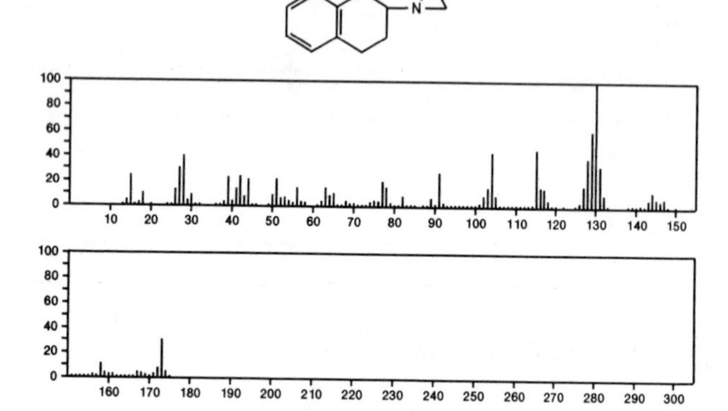

173 $C_{12}H_{15}N$ 31078-98-5
2,6-Xylidine, *N*-methyl-*N*-2-propynyl-

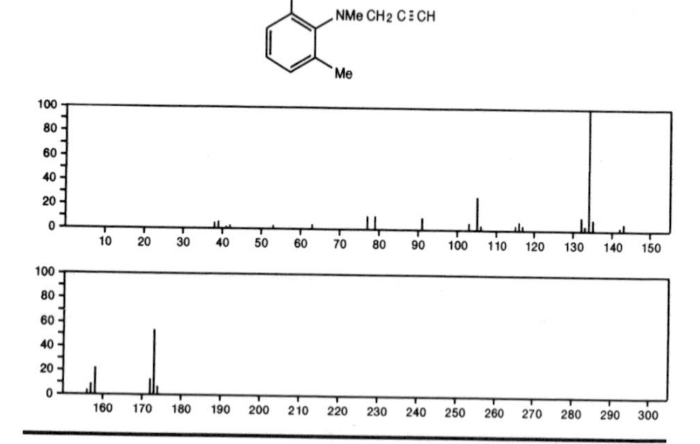

173 $C_{12}H_{15}N$ 40135-93-1
4*H*-Pyrrolo[3,2,1-*ij*]quinoline, 1,2,5,6-tetrahydro-6-methyl-

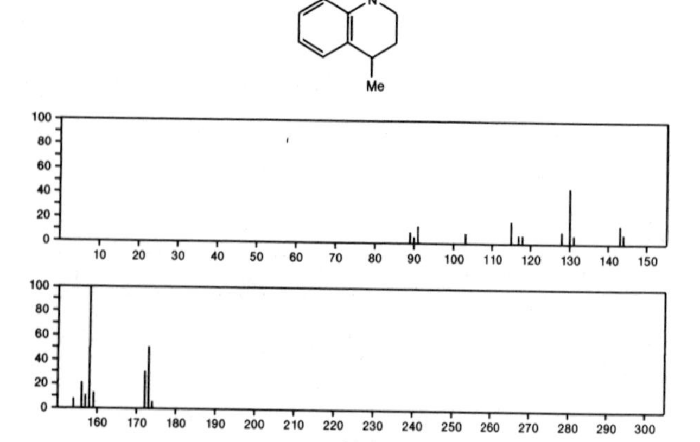

173 $C_{12}H_{15}N$ 40135-99-7
4*H*-Pyrrolo[3,2,1-*ij*]quinoline, 1,2,5,6-tetrahydro-4-methyl-

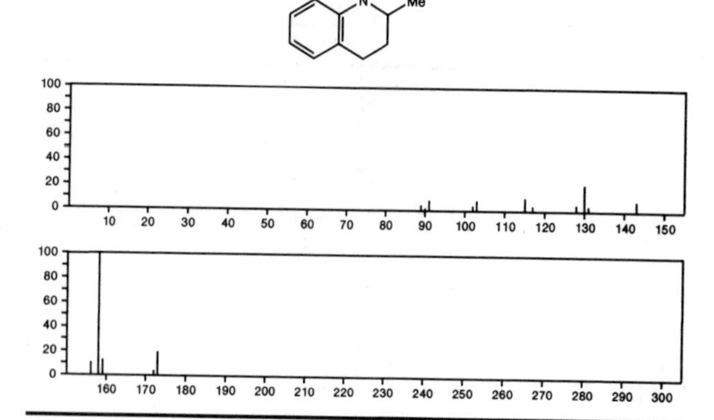

174 $C_3HCl_3O_2$ 2257-35-4
2-Propenoic acid, 2,3,3-trichloro-

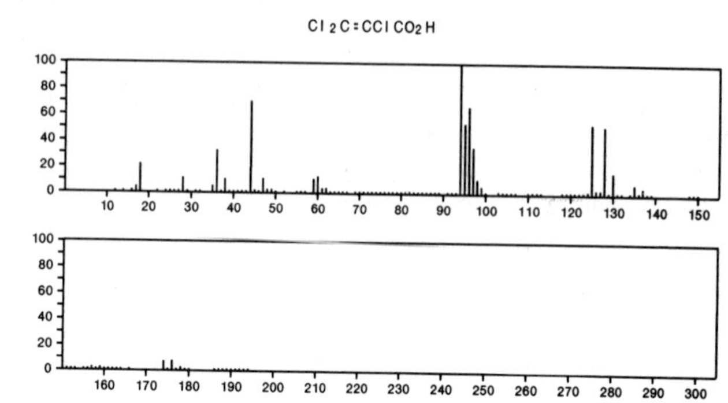

174 $C_3H_9Cl_2N_2P$ 22692-21-3
Phosphorodichloridous hydrazide, trimethyl-

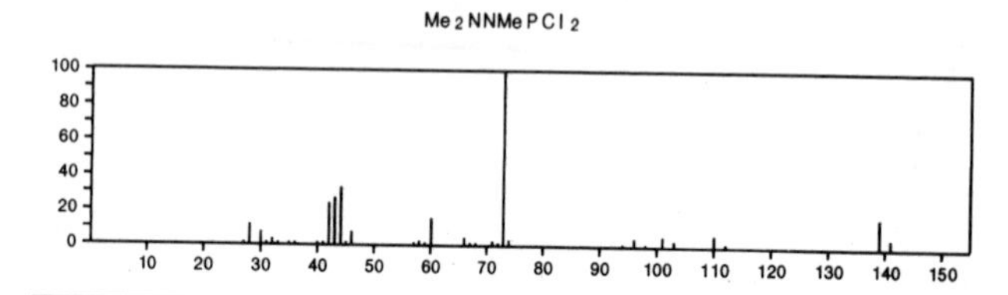

174 $C_4H_3BrN_2O$ 19808-30-1
4(1*H*)-Pyrimidinone, 5-bromo-

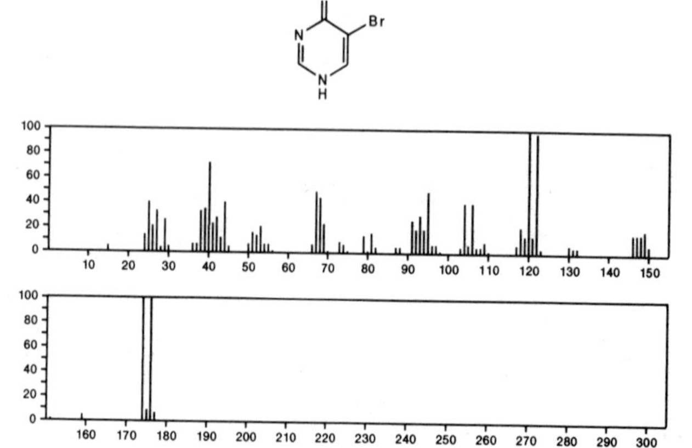

174 $C_4H_5Cl_3O$ 3083–25–8
Oxirane, (2,2,2–trichloroethyl)–

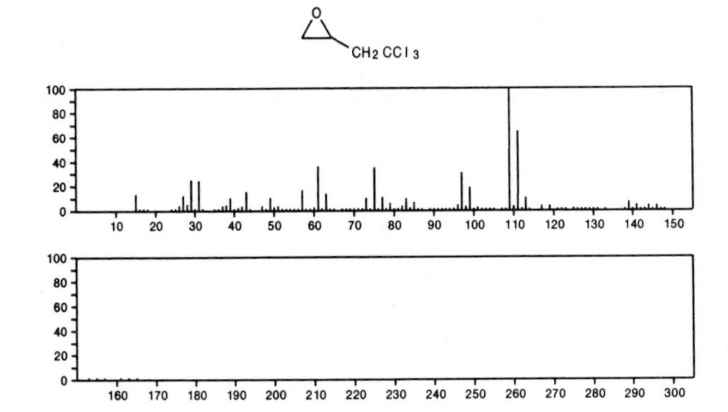

174 $C_4H_5Cl_3O$ 56272–98–1
2–Butanone, 1,1,3(*or* 1,1,4)–trichloro–

Cl2CHCOEt + Cl

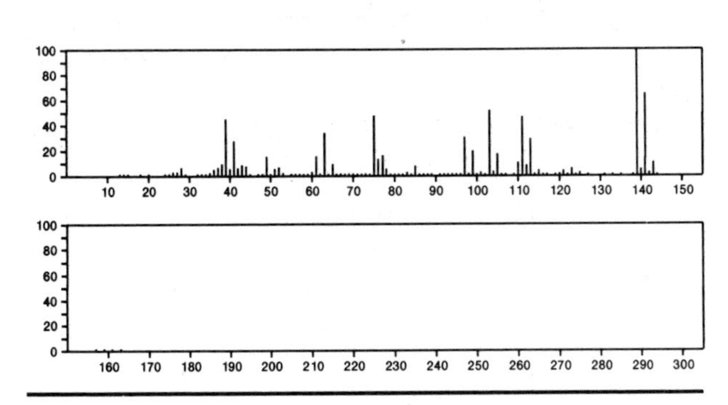

174 $C_5H_3BrO_2$ 1899–24–7
2–Furancarboxaldehyde, 5–bromo–

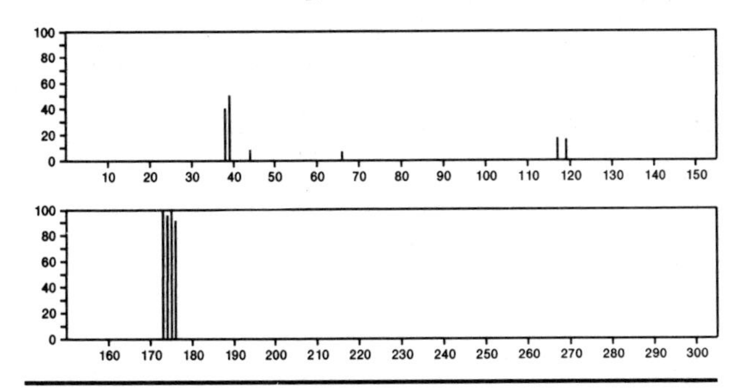

174 $C_5H_6F_4O_2$ 1547–52–0
1,3–Dioxepane, 5,5,6,6–tetrafluoro–

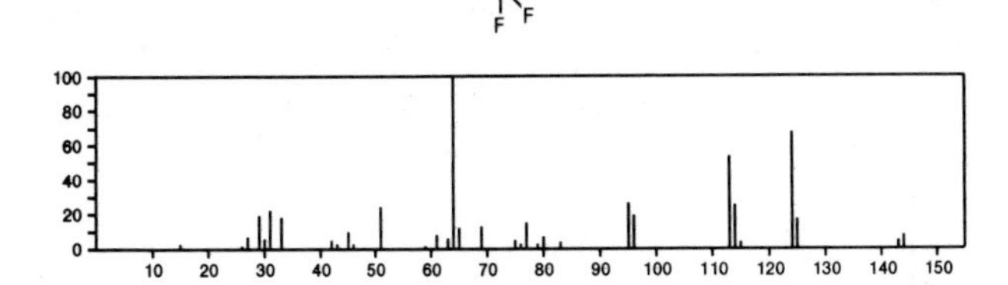

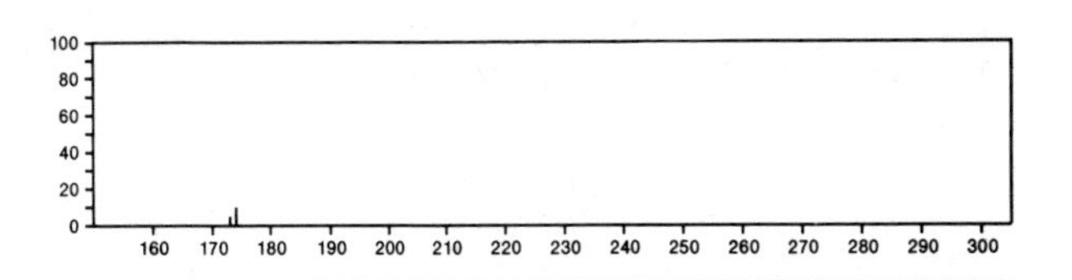

174 C_6H_4BrF 460–00–4
Benzene, 1–bromo–4–fluoro–

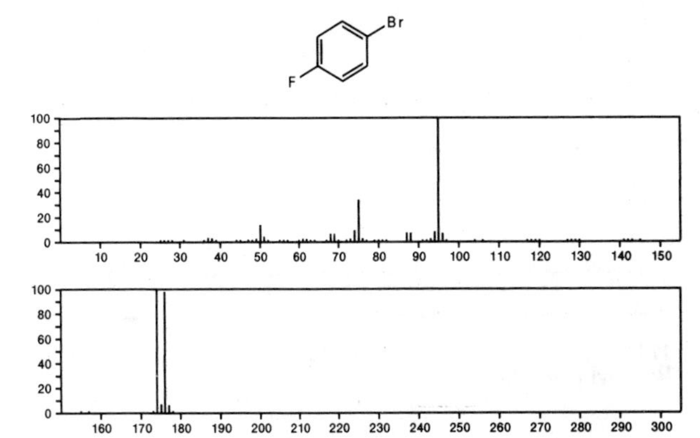

174 C_6H_4BrF 1072–85–1
Benzene, 1–bromo–2–fluoro–

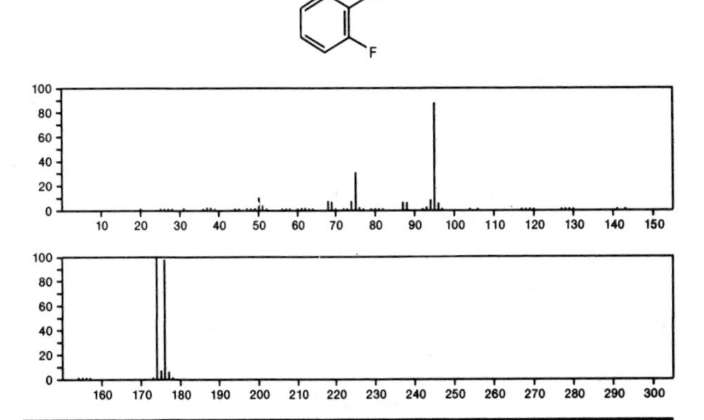

174 C_6H_4BrF 1073–06–9
Benzene, 1–bromo–3–fluoro–

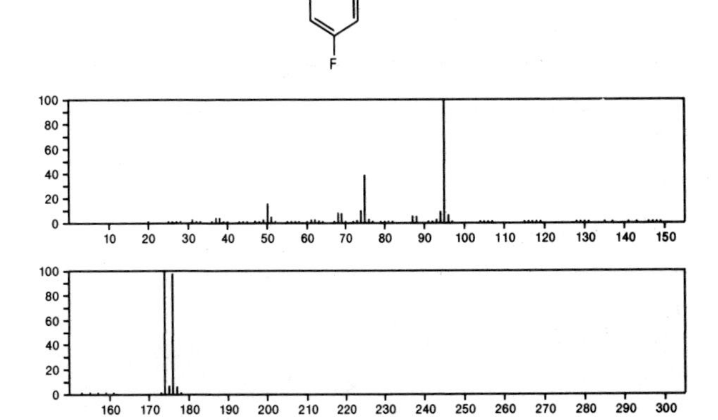

174 $C_6H_{18}Si_3$ 18339-88-3
1,3,5-Trisilacyclohexane, 1,1,3-trimethyl-

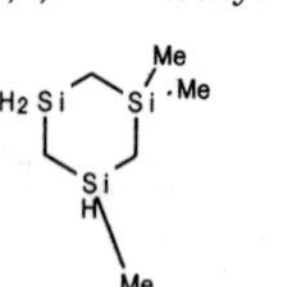

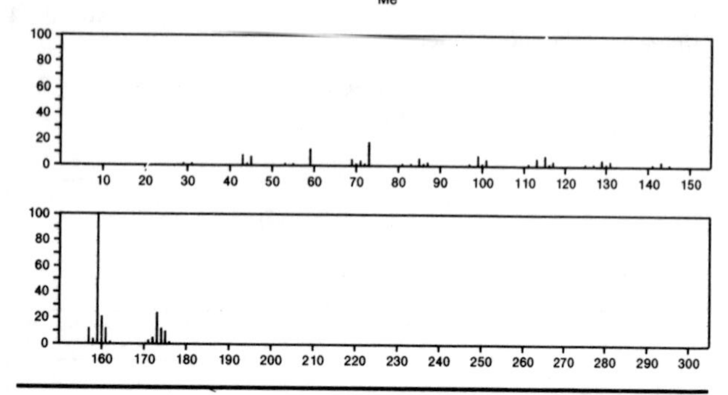

174 $C_7H_4Cl_2O$ 122-01-0
Benzoyl chloride, 4-chloro-

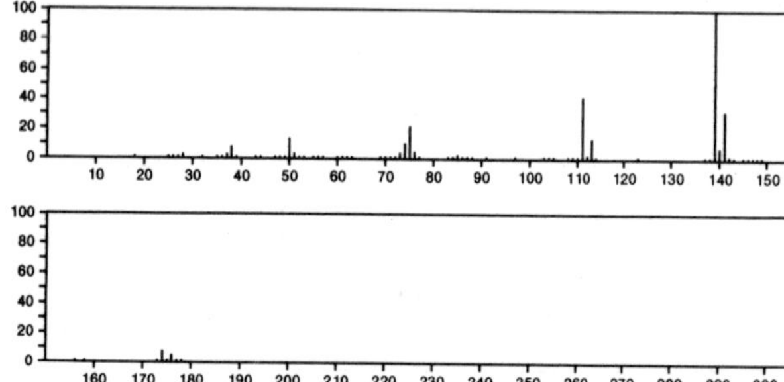

174 $C_7H_4Cl_2O$ 609-65-4
Benzoyl chloride, 2-chloro-

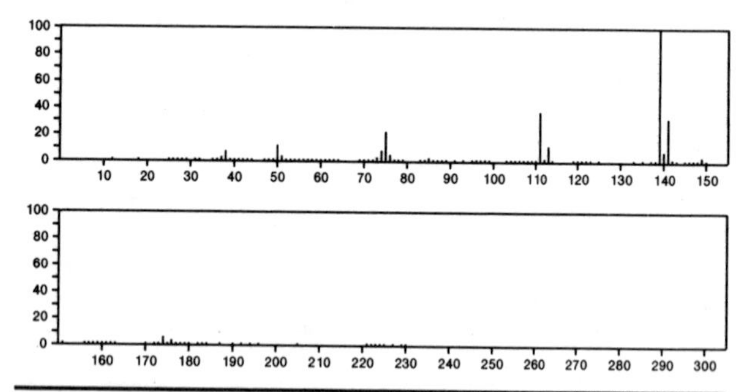

174 $C_7H_4Cl_2O$ 874-42-0
Benzaldehyde, 2,4-dichloro-

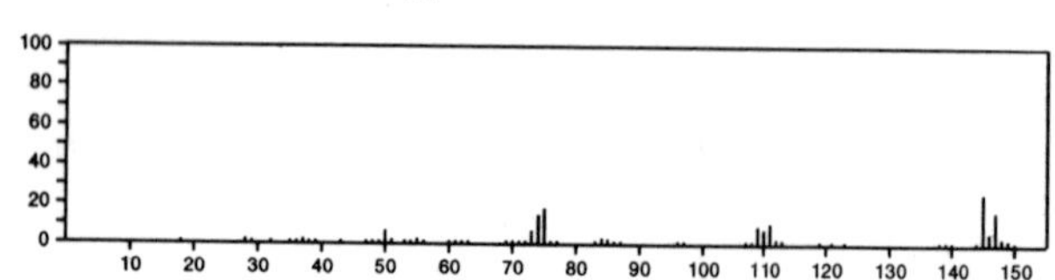

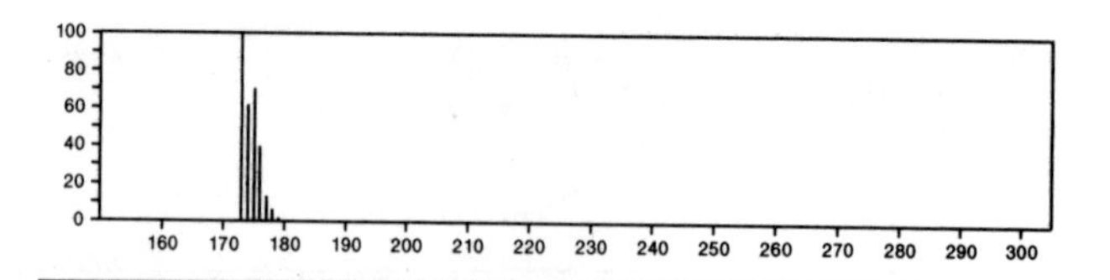

174 $C_7H_4Cl_2O$ 6287-38-3
Benzaldehyde, 3,4-dichloro-

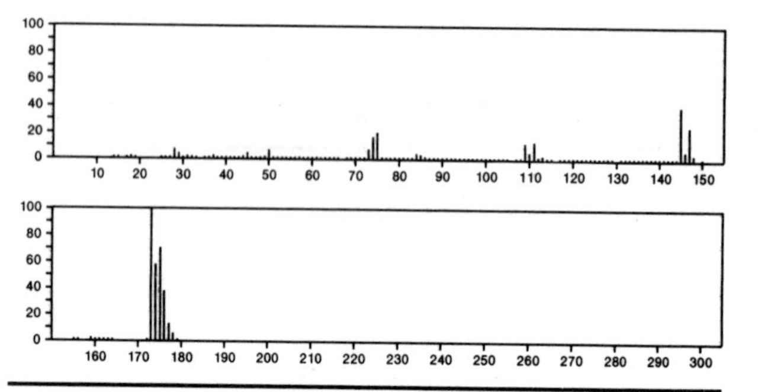

174 $C_7H_7ClO_3$ 32744-80-2
1,4-Benzenediol, 2-chloro-6-(hydroxymethyl)-

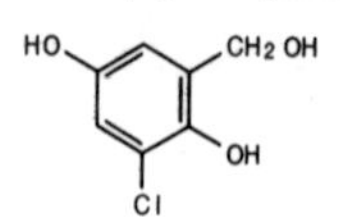

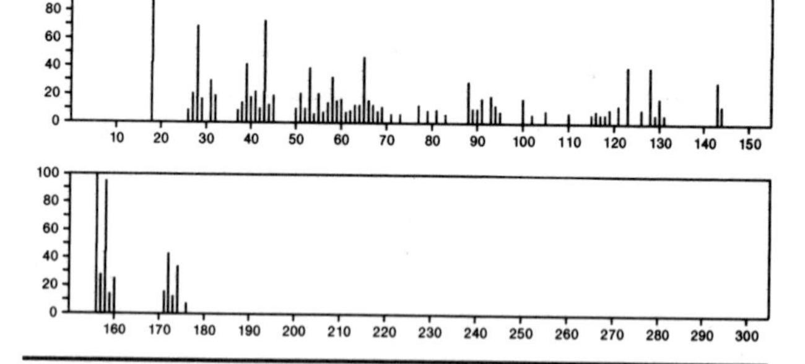

174 $C_7H_{10}N_2O.ClH$ 19501-58-7
Hydrazine, (4-methoxyphenyl)-, monohydrochloride

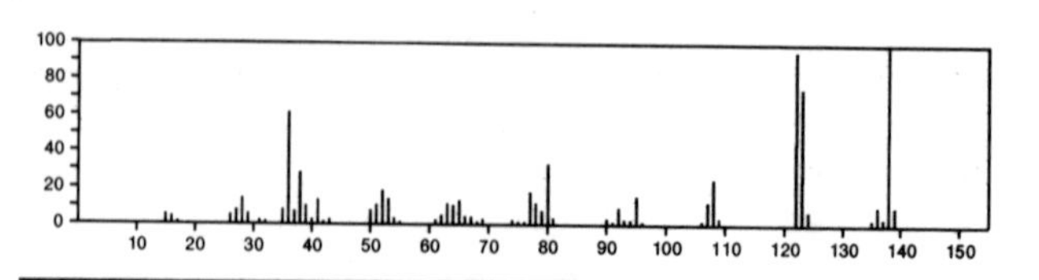

174 $C_7H_{10}O_5$ 13192-04-6
Pentanedioic acid, 2-oxo-, dimethyl ester

MeOC(O)COCH2CH2C(O)OMe

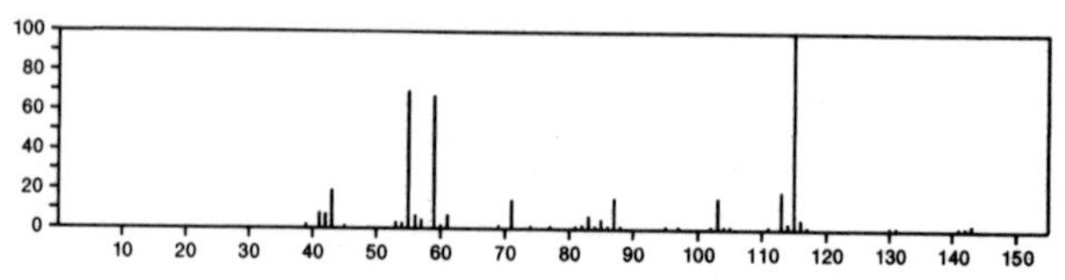

174 $C_7H_{10}O_5$ 26579-97-5
2-Butenedioic acid, 2-methoxy-, dimethyl ester

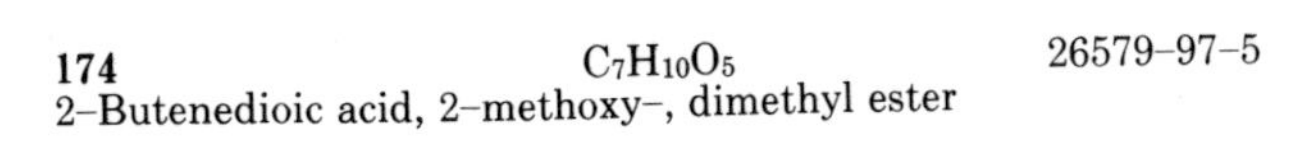

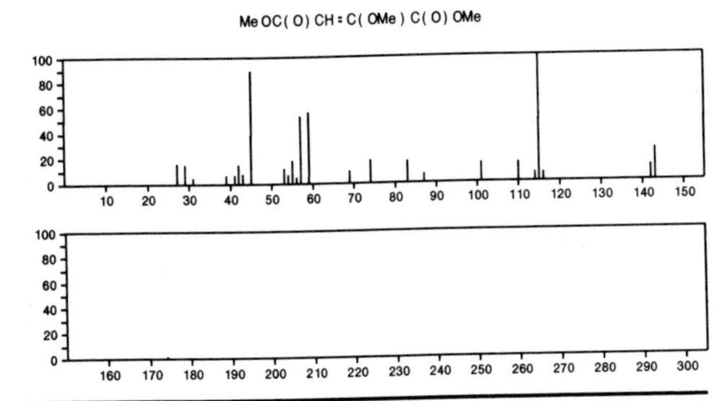

174 $C_7H_{11}Br$ 18317-64-1
Cycloheptene, 1-bromo-

Br

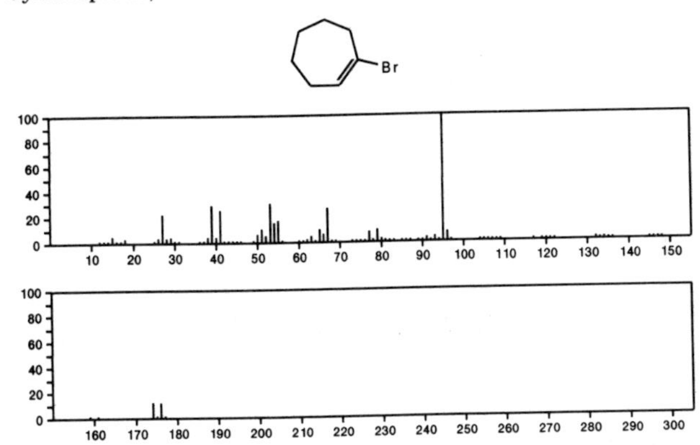

174 $C_7H_{11}Br$ 34825-93-9
Cyclohexene, 3-(bromomethyl)-

CH2 Br

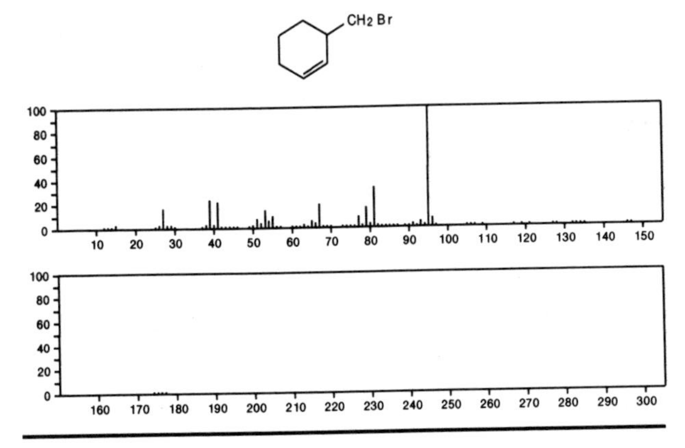

174 $C_7H_{11}Br$ 40648-09-7
Cyclohexene, 1-bromo-6-methyl-

Me
Br

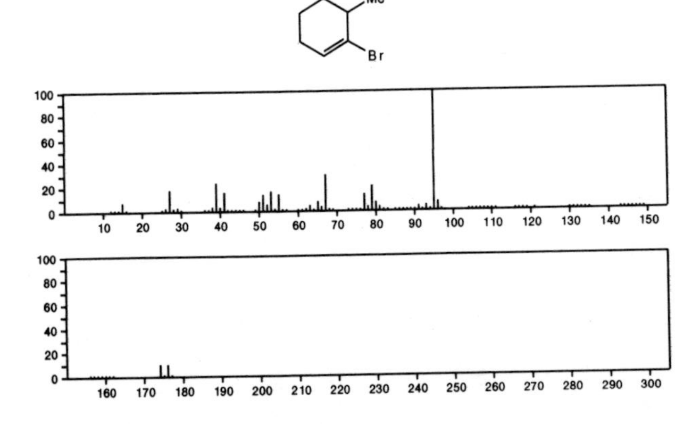

174 $C_7H_{11}Br$ 54484-64-9
Cycloheptene, 5-bromo-

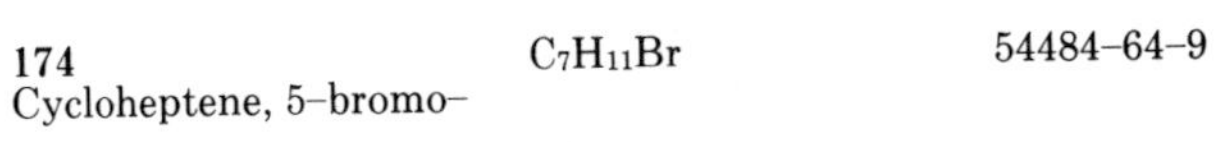

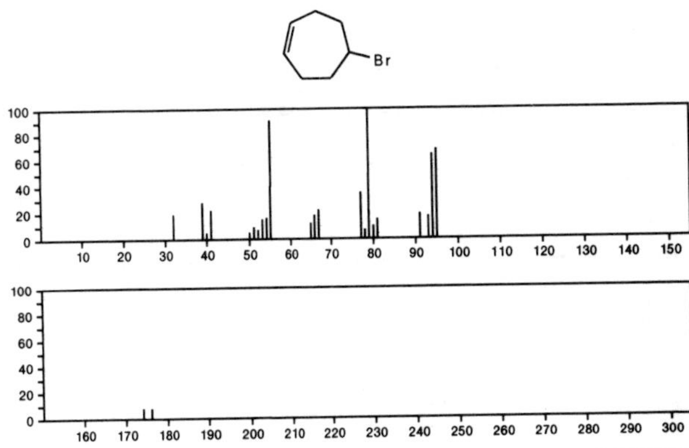

174 $C_7H_{16}Ge$ 4554-78-3
Germacyclopentane, 1-propyl-

GeH Pr

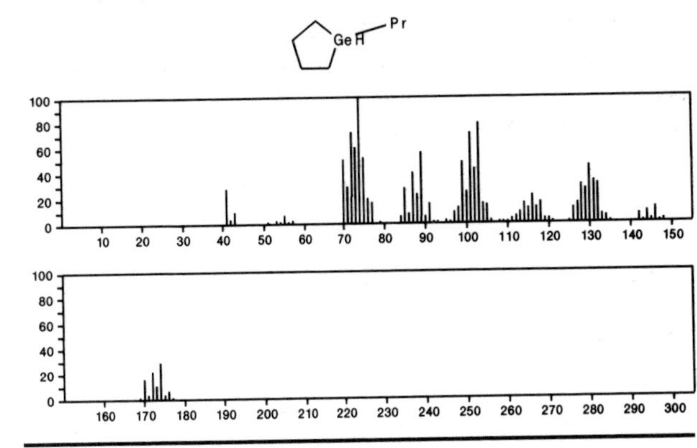

174 $C_8H_6N_4O$ 42786-73-2
1*H*-1,2,4-Triazole-3-carboxaldehyde, 5-(4-pyridinyl)-

O=CH N N-NH N

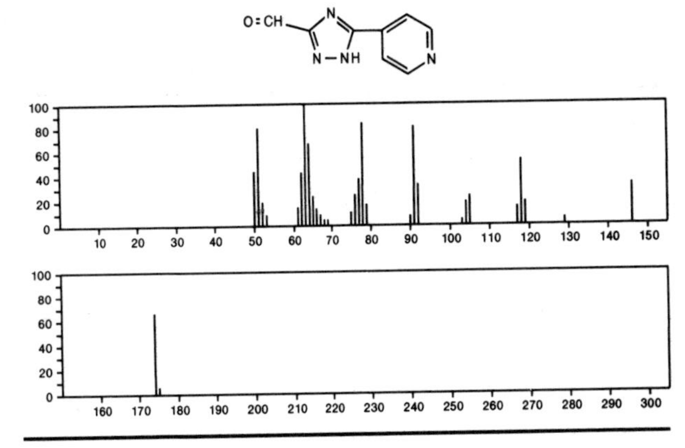

174 $C_8H_8Cl_2$ 626-16-4
Benzene, 1,3-bis(chloromethyl)-

Cl CH2 CH2 Cl

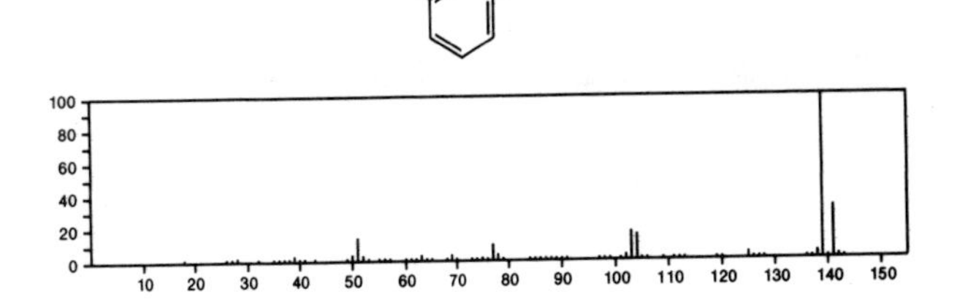

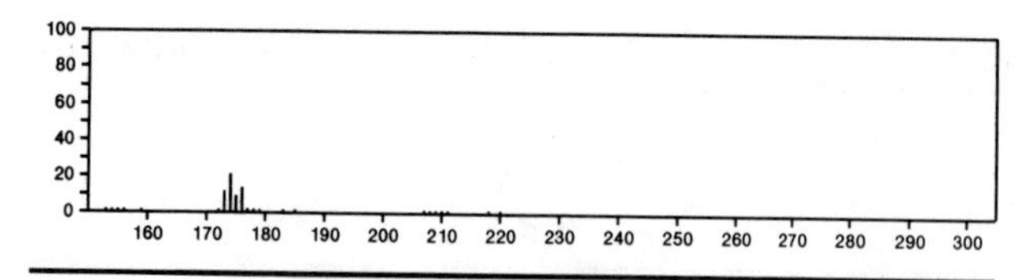

174 $C_8H_8Cl_2$ 1074-11-9
Benzene, (1,2-dichloroethyl)-

ClCH2CHClPh

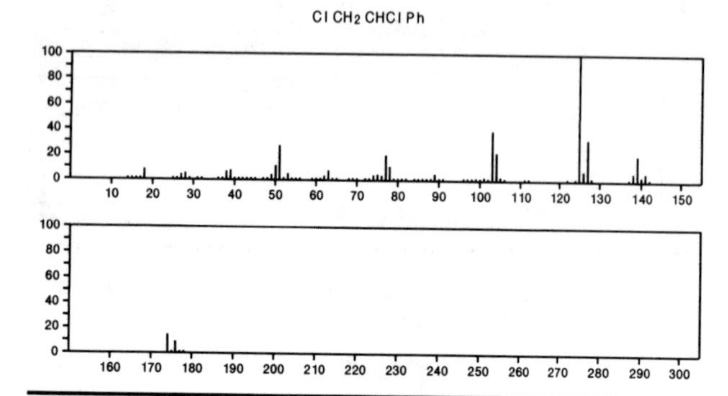

174 $C_8H_8Cl_2$ 1124-05-6
Benzene, 1,4-dichloro-2,5-dimethyl-

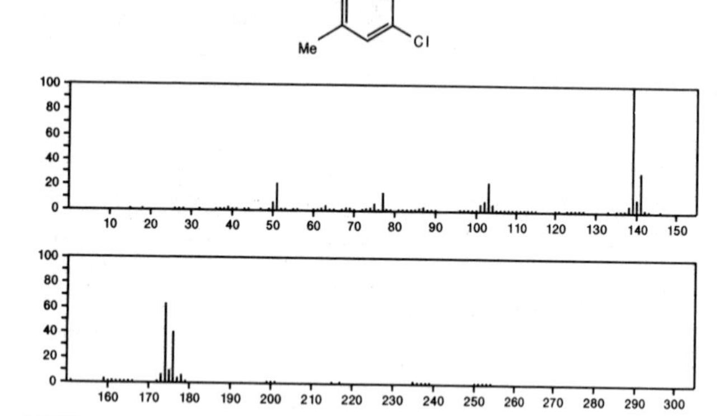

174 $C_8H_8Cl_2$ 2719-42-8
Benzene, 1-(dichloromethyl)-3-methyl-

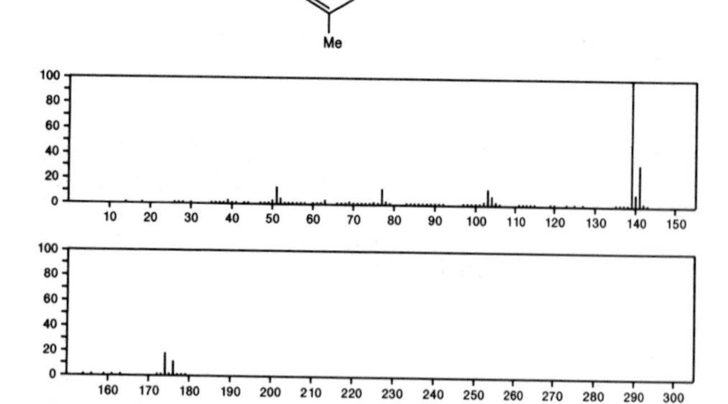

174 $C_8H_8Cl_2$ 6623-59-2
Benzene, 1,2-dichloro-4-ethyl-

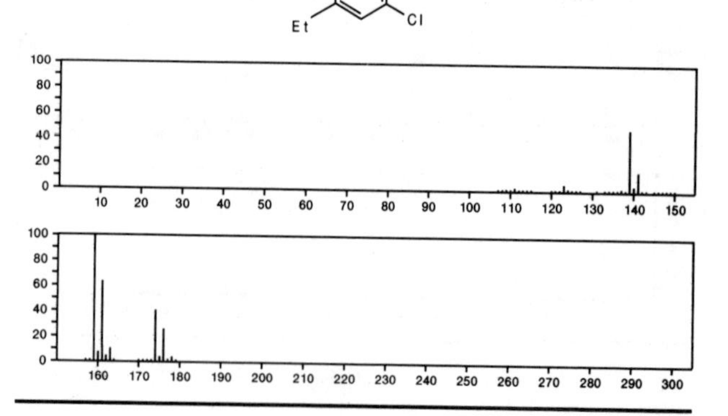

174 $C_8H_8Cl_2$ 20001-64-3
Benzene, 1-chloro-2-(1-chloroethyl)-

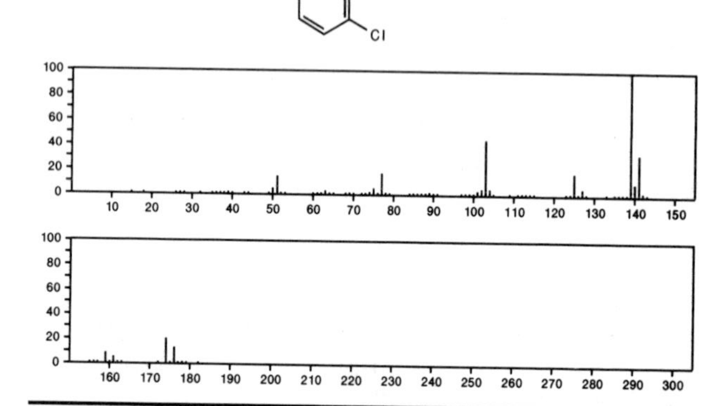

174 $C_8H_8Cl_2$ 23063-36-7
Benzene, 1-(dichloromethyl)-4-methyl-

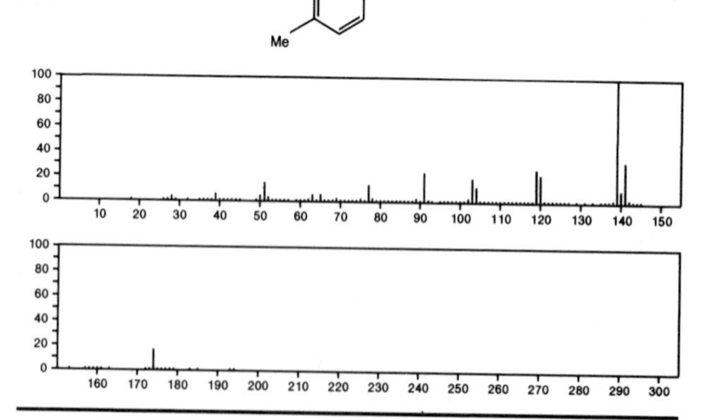

174 $C_8H_8Cl_2$ 33407-02-2
Benzene, 1,3-dichloro-2-ethyl-

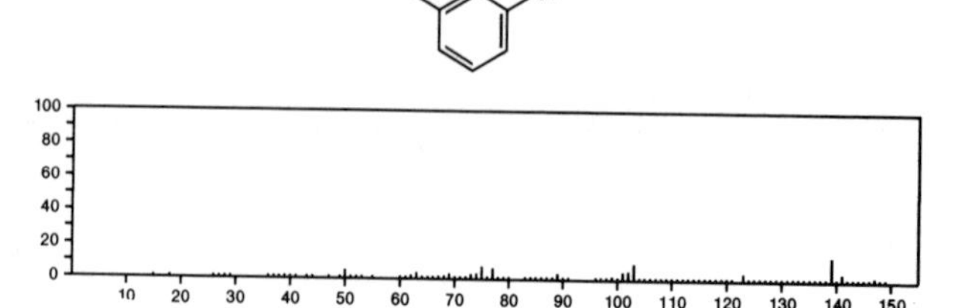

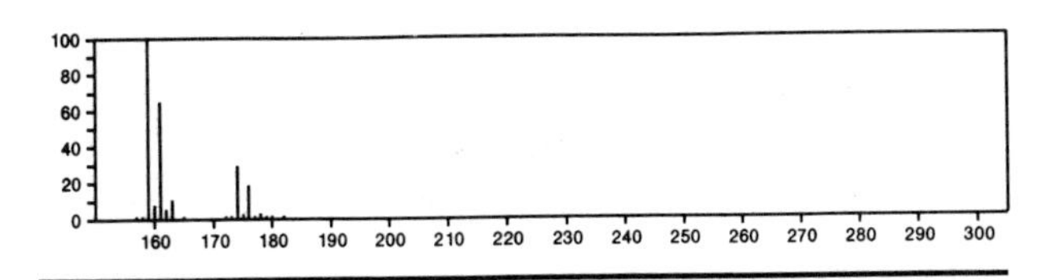

174 $C_8H_8Cl_2$ 54484–55–8
Benzene, chloro(2–chloroethyl)–

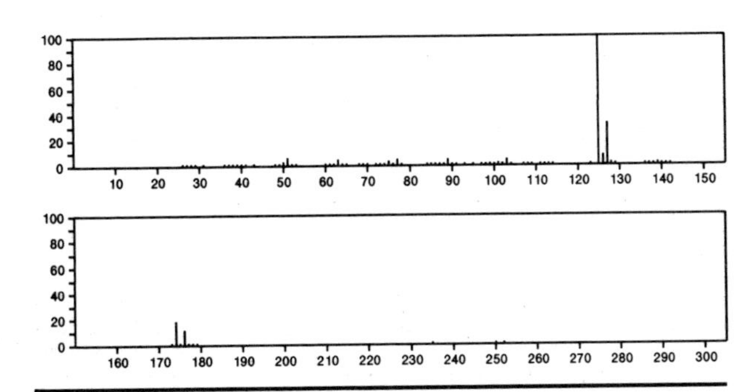

174 $C_8H_8Cl_2$ 54484–61–6
Benzene, 1,2–dichloro–3–ethyl–

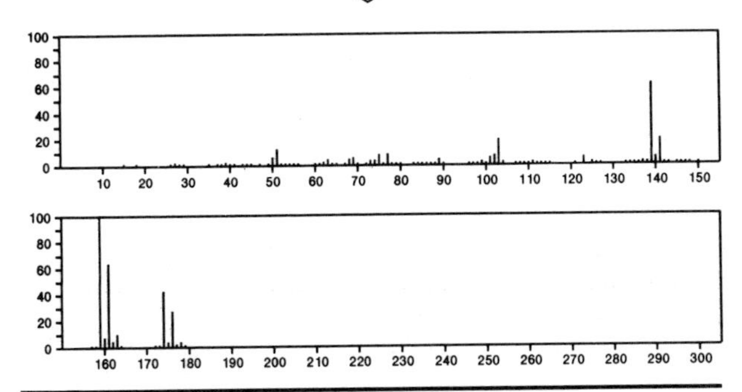

174 $C_8H_8Cl_2$ 54484–62–7
Benzene, 2,4–dichloro–1–ethyl–

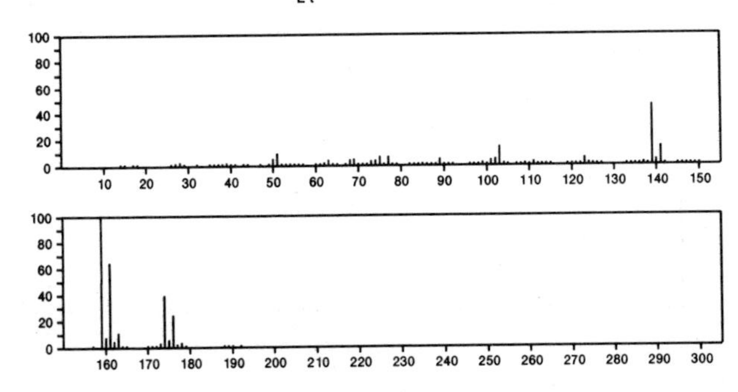

174 $C_8H_8Cl_2$ 54484–63–8
Benzene, 1,4–dichloro–2–ethyl–

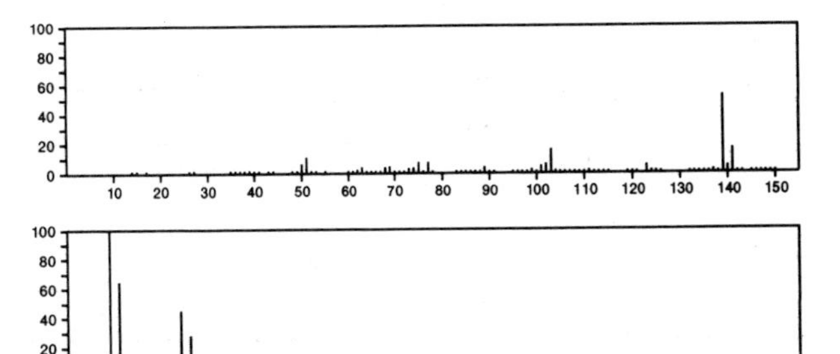

174 $C_8H_{14}O_2S$ 15780–63–9
Butanethioic acid, 3–oxo–, *S*–butyl ester

Me(CH2)3SC(O)CH2COMe

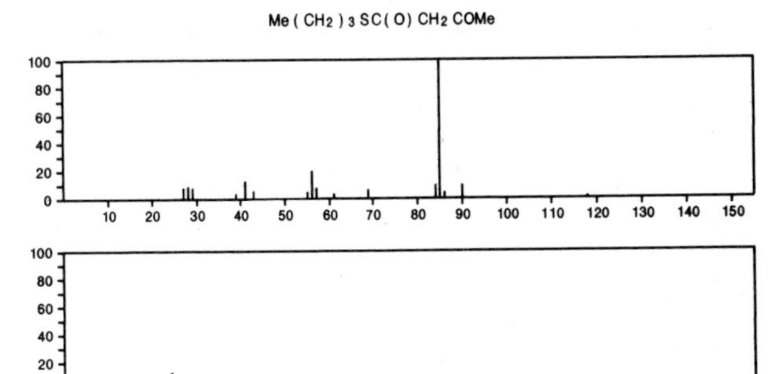

174 $C_8H_{14}O_2S$ 16849–98–2
Acetic acid, mercapto–, cyclohexyl ester

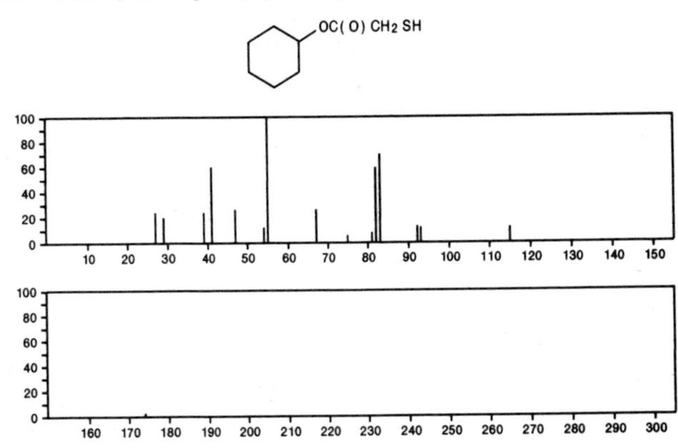

174 $C_8H_{14}O_2S$ 23246–24–4
Propionic acid, 3–(allylthio)–, ethyl ester

EtOC(O)CH2CH2SCH2CH=CH2

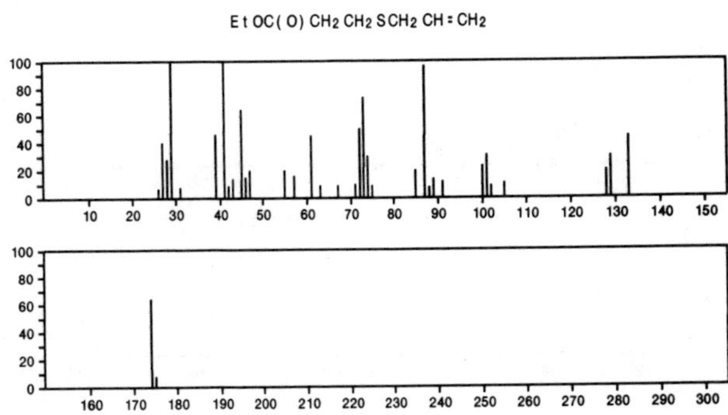

174 $C_8H_{14}O_2S$ 50838-20-5
2-Propenoic acid, 3-[(1,1-dimethylethyl)thio]-, methyl ester

t-BuSCH=CHC(O)OMe

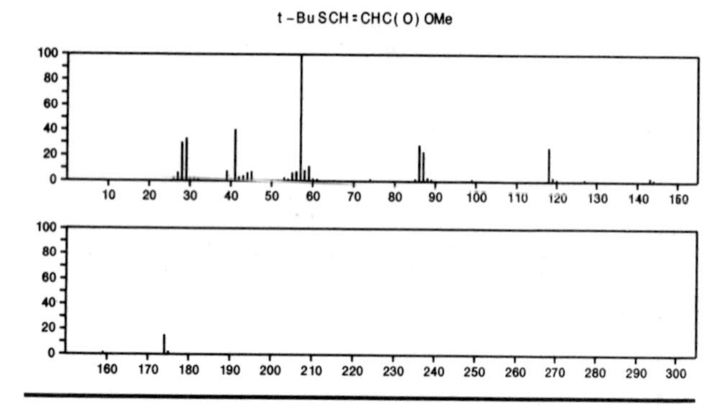

174 $C_8H_{14}O_2S$ 54725-51-8
8-Thiabicyclo[3.2.1]octan-3-ol, 6-methoxy-

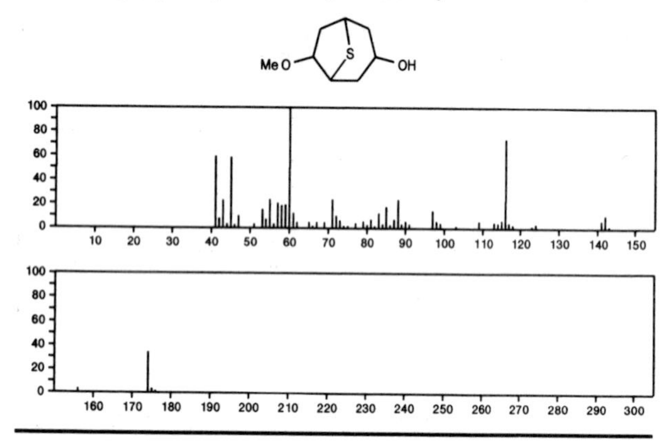

174 $C_8H_{14}O_2S$ 56323-65-0
8-Thiabicyclo[3.2.1]octan-3-ol, 6-methoxy-, (3-*endo*,6-*exo*)-

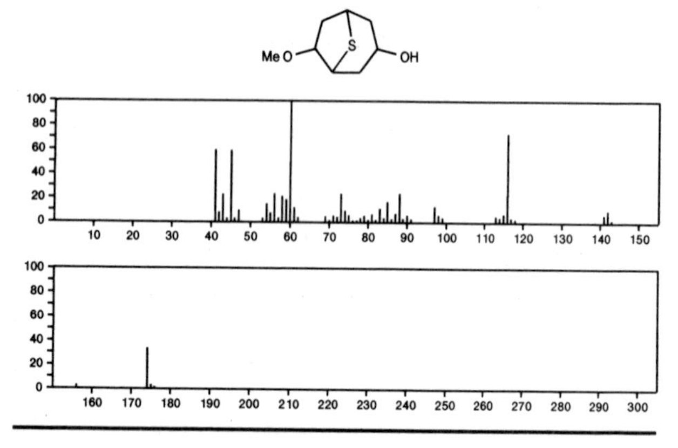

174 $C_8H_{14}O_4$ 123-25-1
Butanedioic acid, diethyl ester

EtOC(O)CH2CH2C(O)OEt

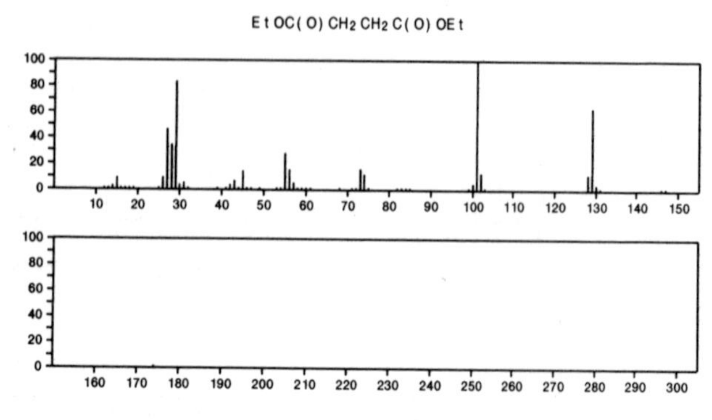

174 $C_8H_{14}O_4$ 123-80-8
1,2-Ethanediol, dipropanoate

EtC(O)OCH2CH2OC(O)Et

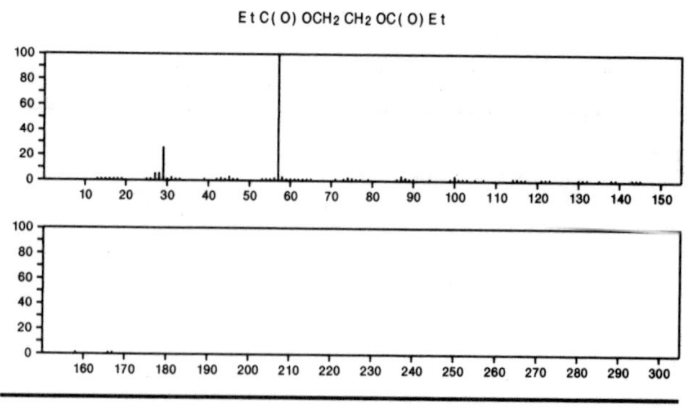

174 $C_8H_{14}O_4$ 505-48-6
Octanedioic acid

HO2C(CH2)6CO2H

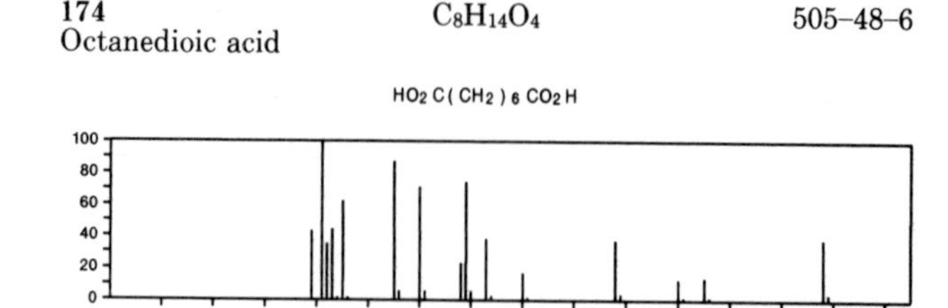

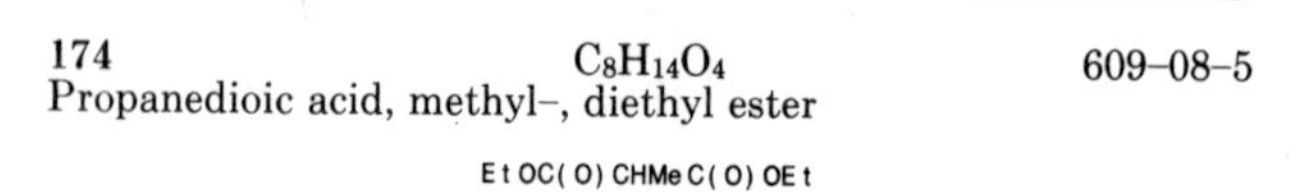

174 $C_8H_{14}O_4$ 609-08-5
Propanedioic acid, methyl-, diethyl ester

EtOC(O)CHMeC(O)OEt

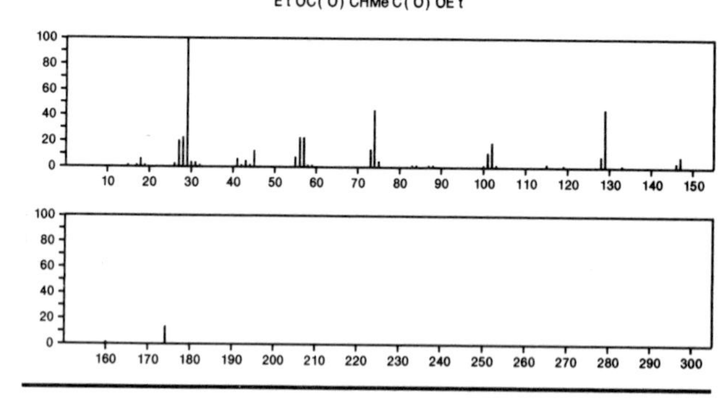

174 $C_8H_{14}O_4$ 627-93-0
Hexanedioic acid, dimethyl ester

MeOC(O)(CH2)4C(O)OMe

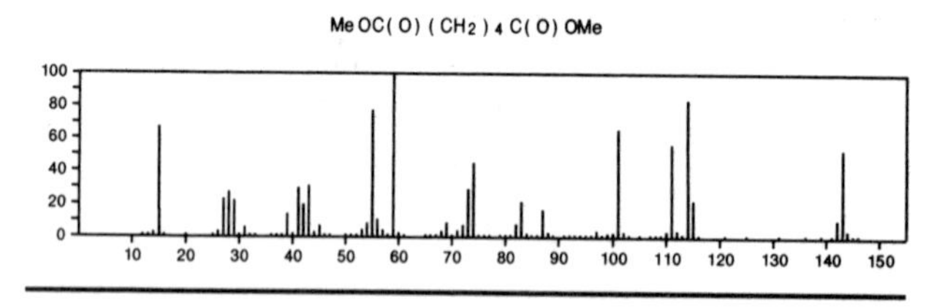

174 $C_8H_{14}O_4$ 628-67-1
1,4-Butanediol, diacetate

AcO(CH2)4OAc

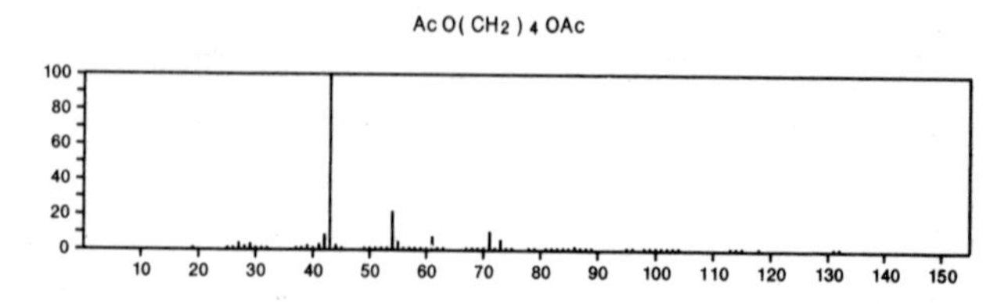

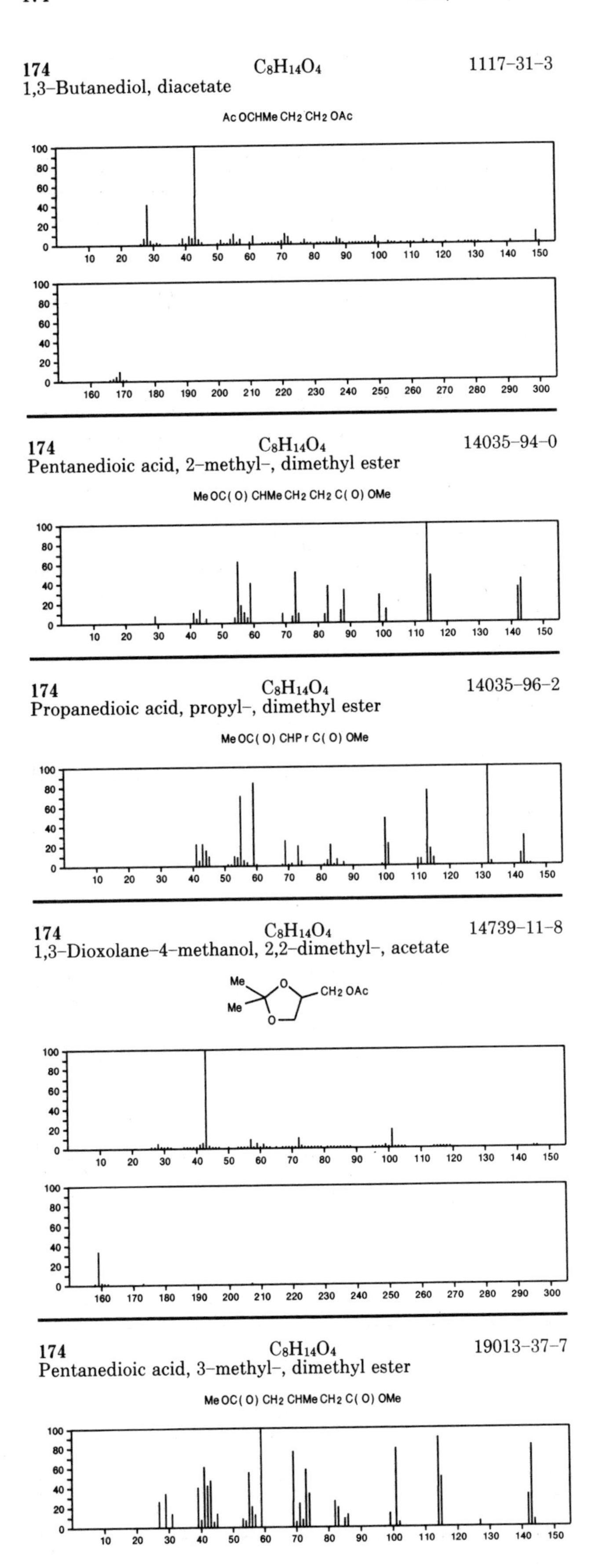

174 $C_8H_{14}O_4$ 1117–31–3
1,3–Butanediol, diacetate

AcOCHMeCH2CH2OAc

174 $C_8H_{14}O_4$ 14035–94–0
Pentanedioic acid, 2–methyl–, dimethyl ester

MeOC(O)CHMeCH2CH2C(O)OMe

174 $C_8H_{14}O_4$ 14035–96–2
Propanedioic acid, propyl–, dimethyl ester

MeOC(O)CHPrC(O)OMe

174 $C_8H_{14}O_4$ 14739–11–8
1,3–Dioxolane–4–methanol, 2,2–dimethyl–, acetate

174 $C_8H_{14}O_4$ 19013–37–7
Pentanedioic acid, 3–methyl–, dimethyl ester

MeOC(O)CH2CHMeCH2C(O)OMe

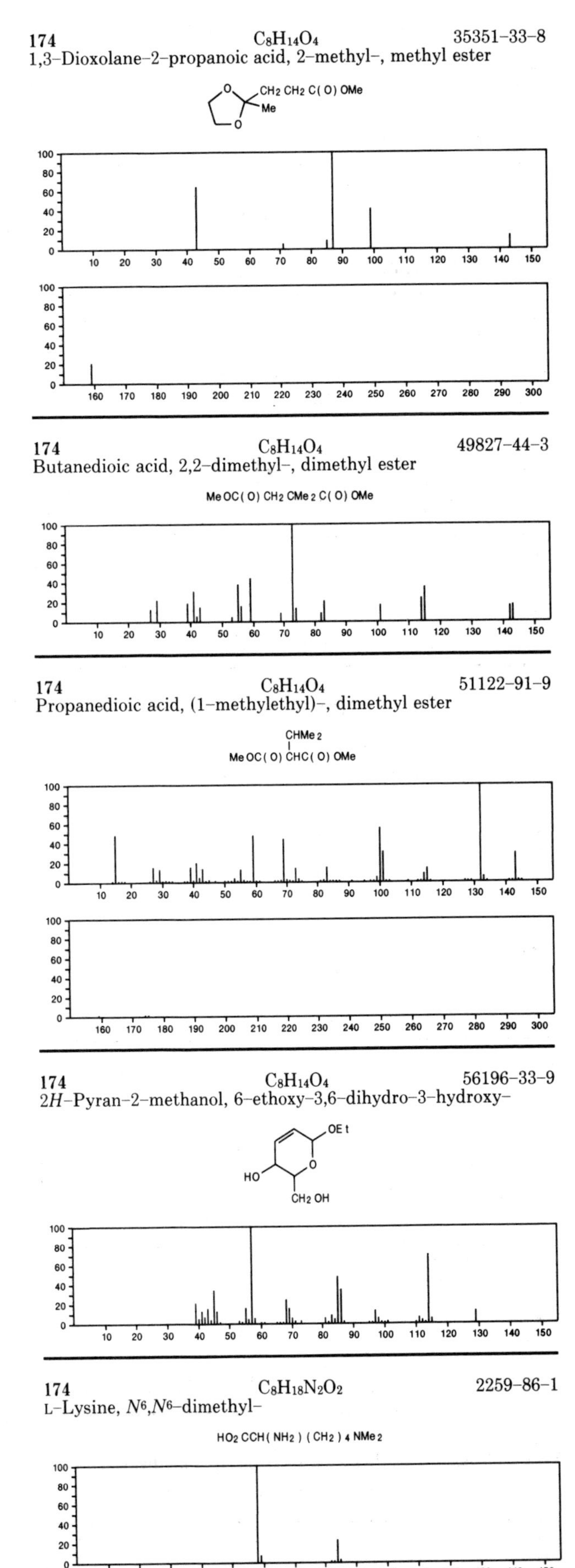

174 $C_8H_{14}O_4$ 35351–33–8
1,3–Dioxolane–2–propanoic acid, 2–methyl–, methyl ester

174 $C_8H_{14}O_4$ 49827–44–3
Butanedioic acid, 2,2–dimethyl–, dimethyl ester

MeOC(O)CH2CMe2C(O)OMe

174 $C_8H_{14}O_4$ 51122–91–9
Propanedioic acid, (1–methylethyl)–, dimethyl ester

174 $C_8H_{14}O_4$ 56196–33–9
2*H*–Pyran–2–methanol, 6–ethoxy–3,6–dihydro–3–hydroxy–

174 $C_8H_{18}N_2O_2$ 2259–86–1
L–Lysine, N^6,N^6–dimethyl–

HO2CCH(NH2)(CH2)4NMe2

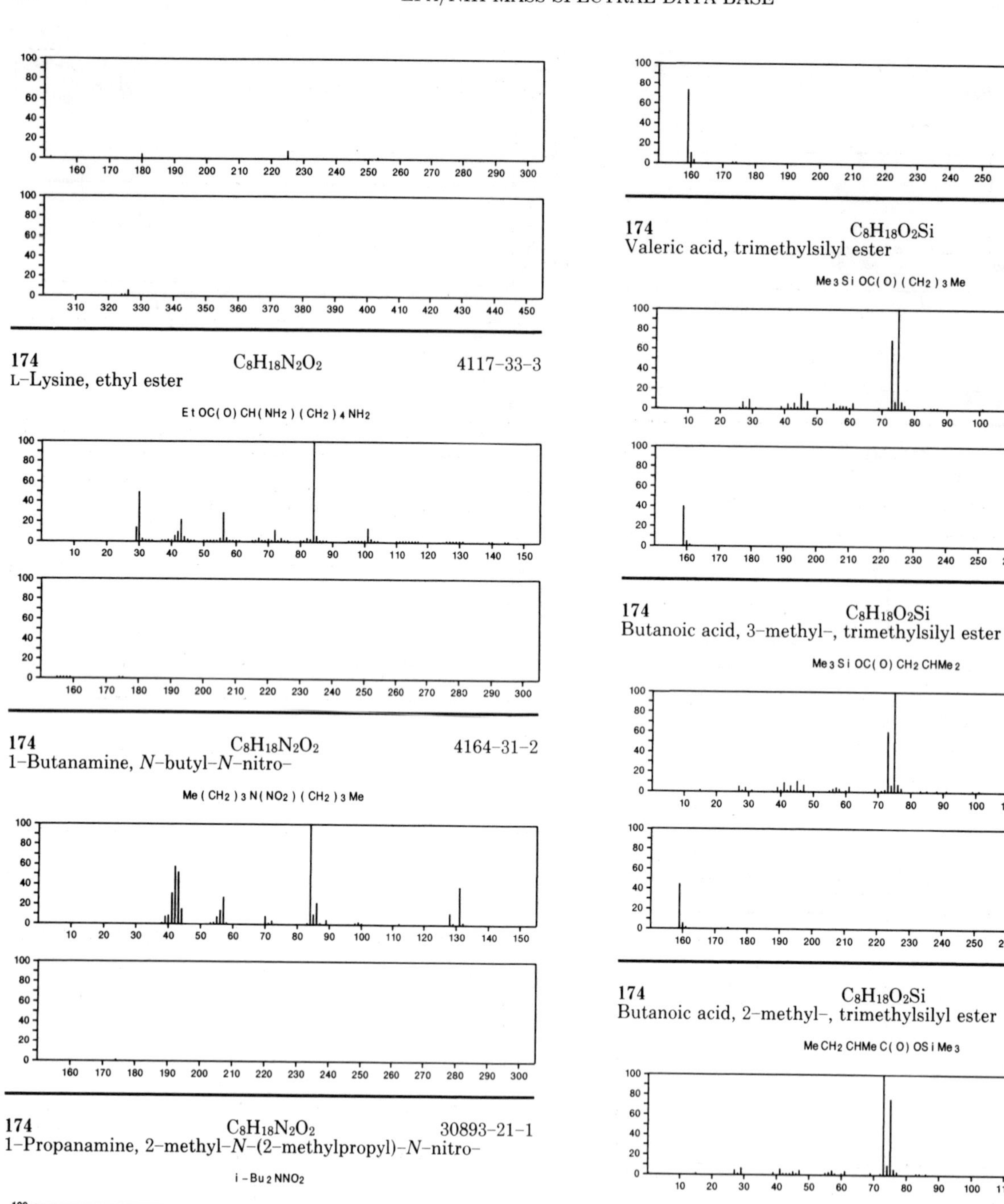

174 $C_8H_{18}N_2O_2$ 4117-33-3
L-Lysine, ethyl ester

EtOC(O)CH(NH2)(CH2)4NH2

174 $C_8H_{18}N_2O_2$ 4164-31-2
1-Butanamine, *N*-butyl-*N*-nitro-

Me(CH2)3N(NO2)(CH2)3Me

174 $C_8H_{18}N_2O_2$ 30893-21-1
1-Propanamine, 2-methyl-*N*-(2-methylpropyl)-*N*-nitro-

i-Bu2NNO2

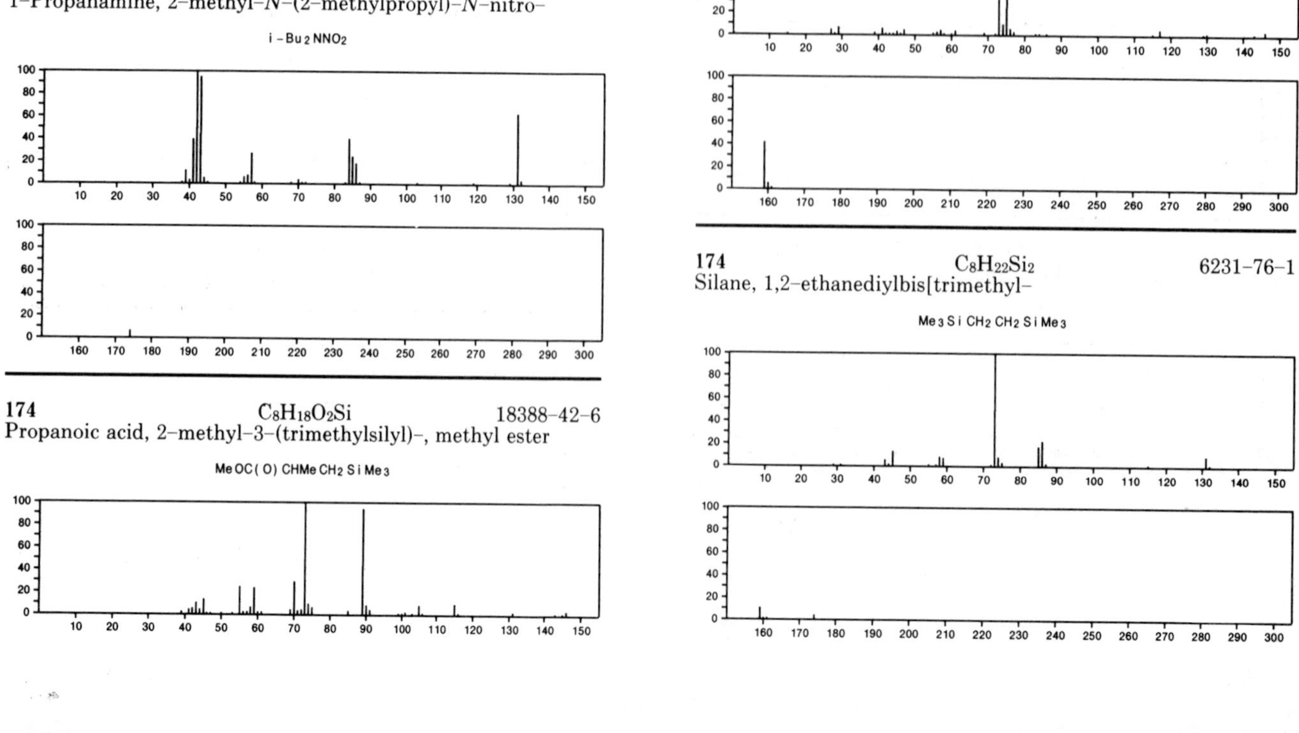

174 $C_8H_{18}O_2Si$ 18388-42-6
Propanoic acid, 2-methyl-3-(trimethylsilyl)-, methyl ester

MeOC(O)CHMeCH2SiMe3

174 $C_8H_{18}O_2Si$ 26429-16-3
Valeric acid, trimethylsilyl ester

Me3SiOC(O)(CH2)3Me

174 $C_8H_{18}O_2Si$ 55557-13-6
Butanoic acid, 3-methyl-, trimethylsilyl ester

Me3SiOC(O)CH2CHMe2

174 $C_8H_{18}O_2Si$ 55557-14-7
Butanoic acid, 2-methyl-, trimethylsilyl ester

MeCH2CHMeC(O)OSiMe3

174 $C_8H_{22}Si_2$ 6231-76-1
Silane, 1,2-ethanediylbis[trimethyl-

Me3SiCH2CH2SiMe3

174 $C_8H_{22}Si_2$ 18406-29-6
Silane, ethylidenebis[trimethyl-

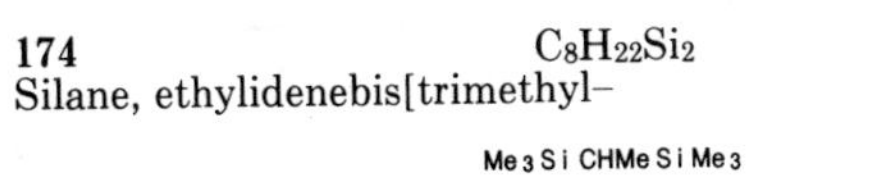

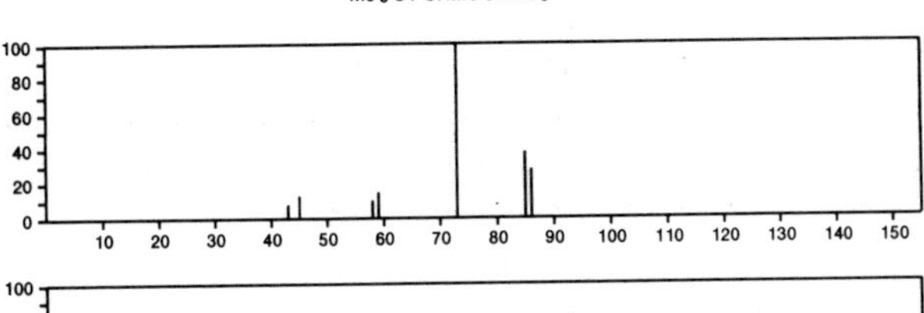

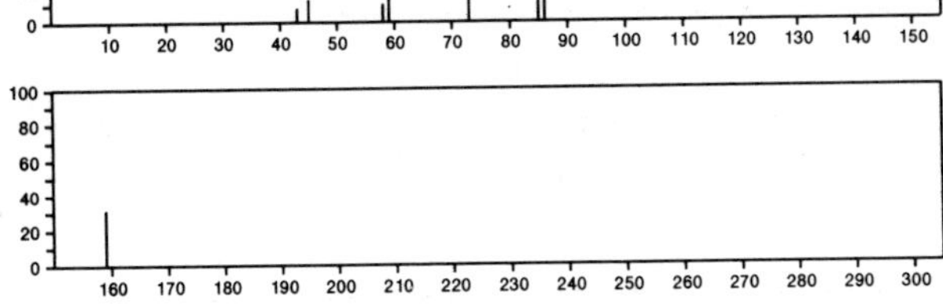

174 $C_9H_6N_2O_2$ 607-34-1
Quinoline, 5-nitro-

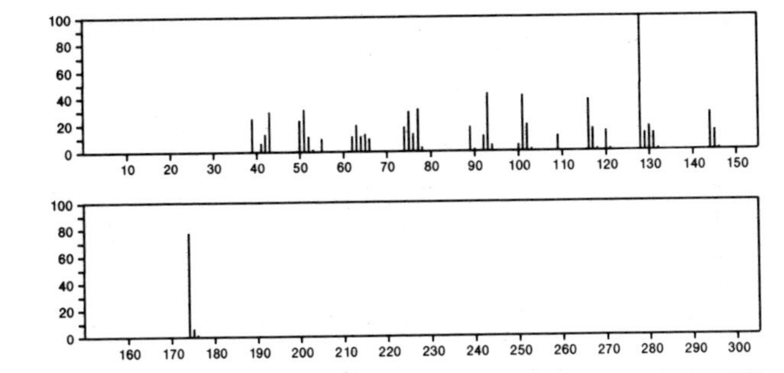

174 $C_9H_6N_2O_2$ 607-35-2
Quinoline, 8-nitro-

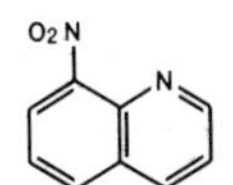

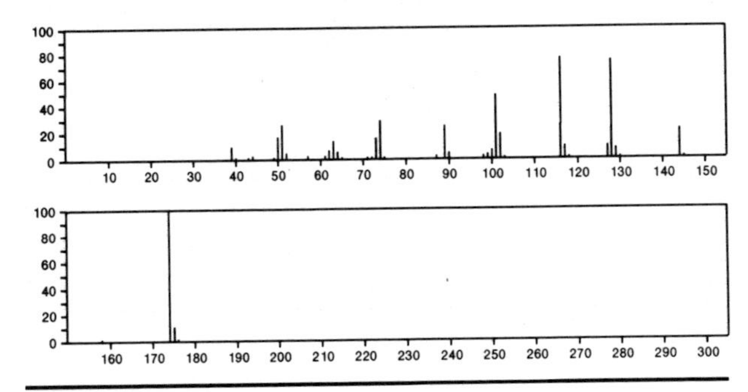

174 $C_9H_6N_2O_2$ 613-51-4
Quinoline, 7-nitro-

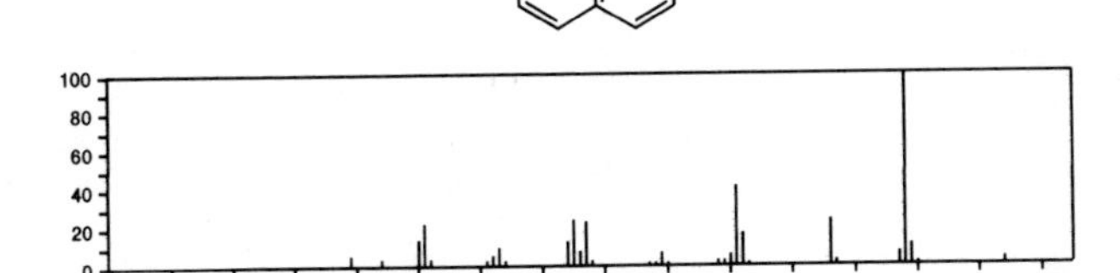

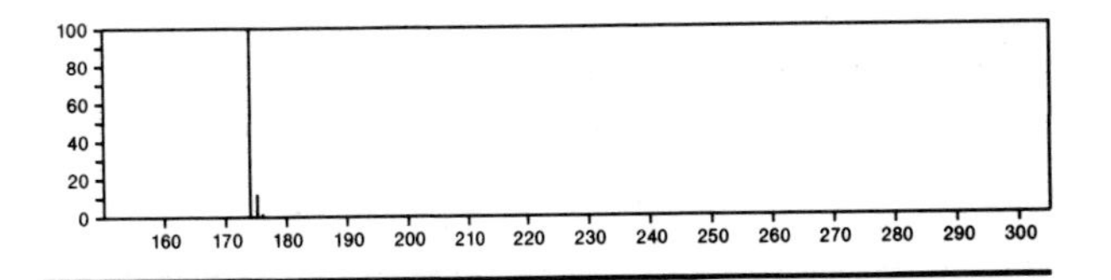

174 $C_9H_6N_2O_2$ 26471-62-5
Benzene, 1,3-diisocyanatomethyl-

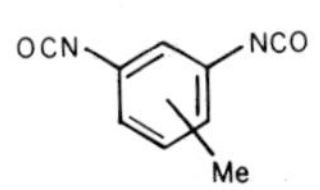

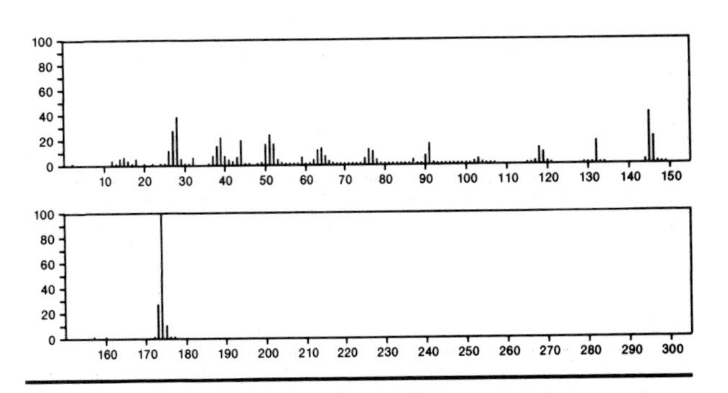

174 $C_9H_6N_2S$ 23706-25-4
Thiocyanic acid, 1*H*-indol-3-yl ester

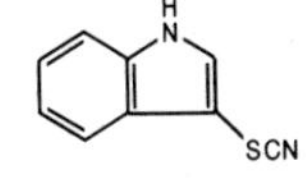

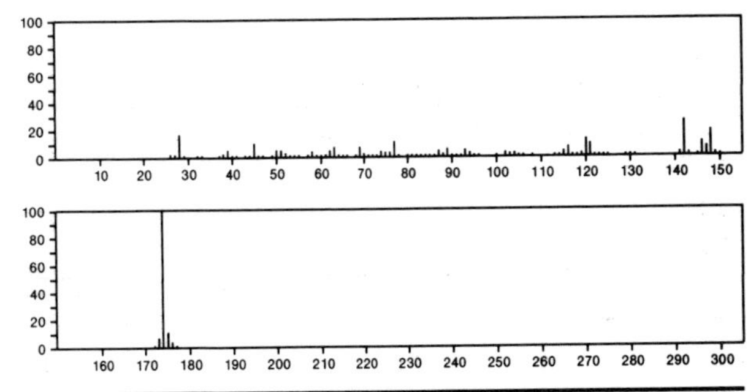

174 $C_9H_{10}N_4$ 14757-63-2
1*H*-1,2,3-Triazole, 1-anilino-4-methyl-

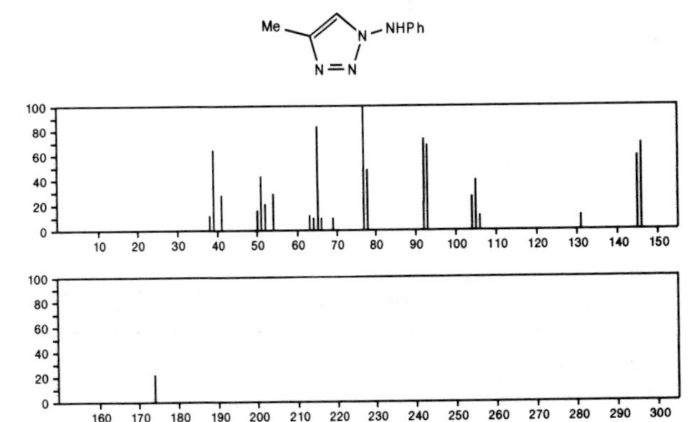

174 $C_9H_{18}OS$ 1927-53-3

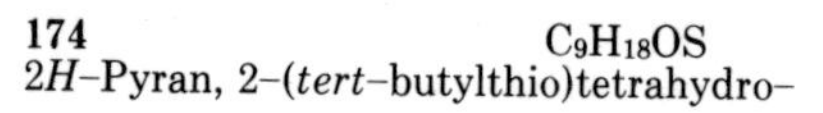
2*H*-Pyran, 2-(*tert*-butylthio)tetrahydro-

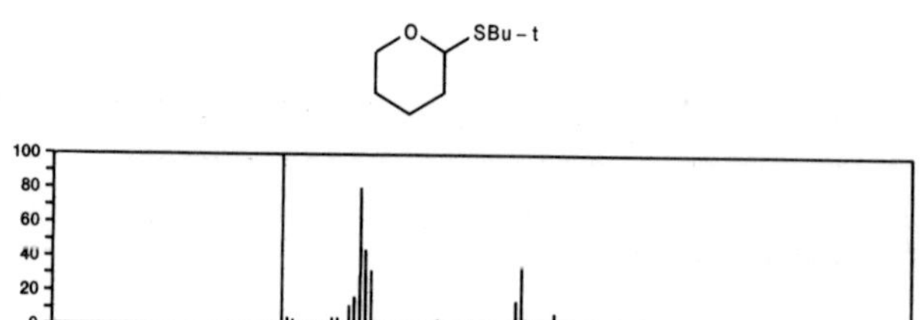

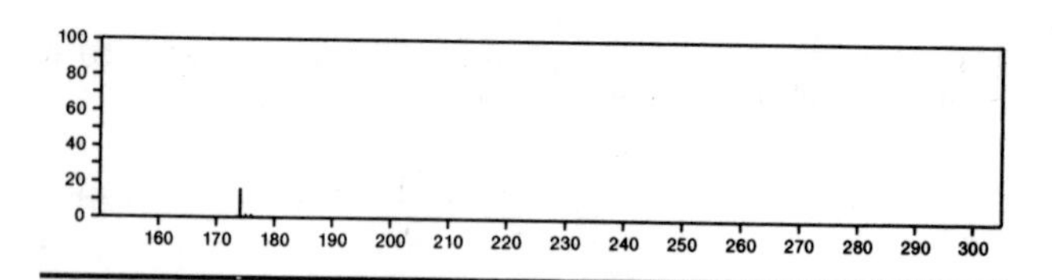

174 $C_9H_{18}OS$ 2432-53-3

Butyric acid, thio-, *S*-pentyl ester

Pr C(O) S(CH2) 4 Me

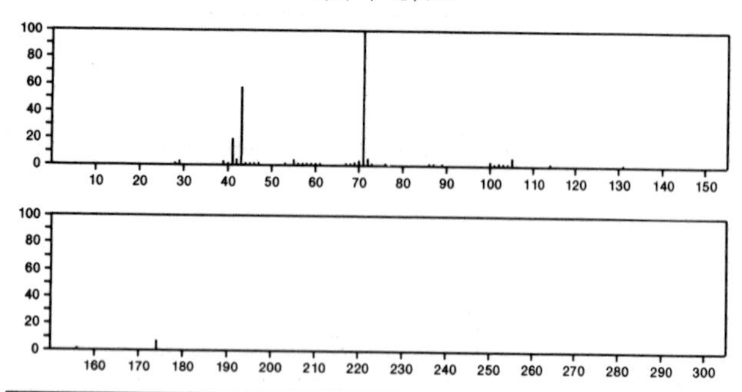

174 $C_9H_{18}OS$ 2432-78-2

Hexanethioic acid, *S*-propyl ester

Me (CH2) 4 C(O) SPr

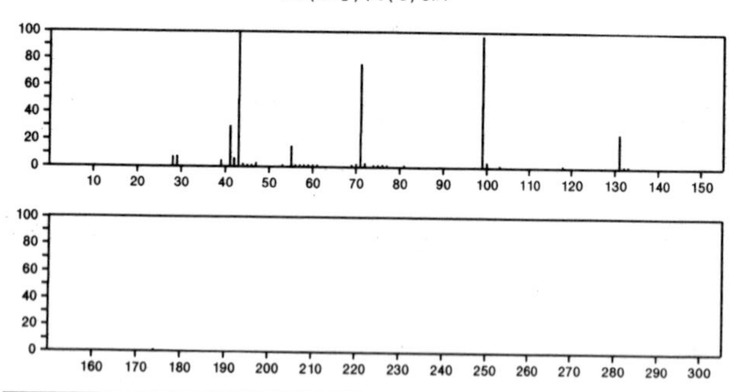

174 $C_9H_{18}OS$ 2432-83-9

Octanethioic acid, *S*-methyl ester

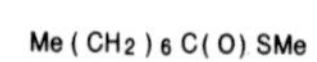

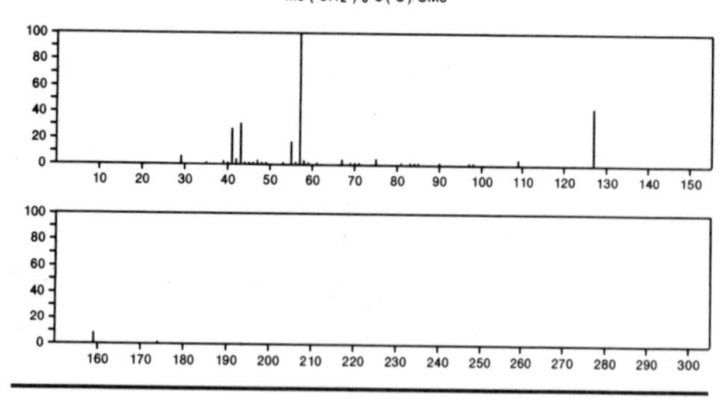

174 $C_9H_{18}OS$ 2432-91-9

Butanethioic acid, 3-methyl-, *S*-(1-methylpropyl) ester

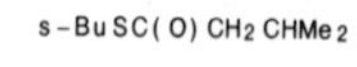

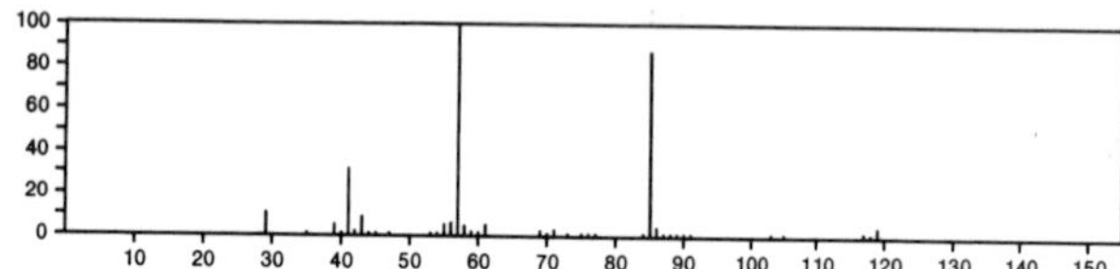

174 $C_9H_{18}OS$ 2450-11-5

Valeric acid, thio-, *S*-*sec*-butyl ester

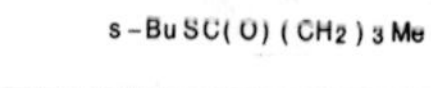

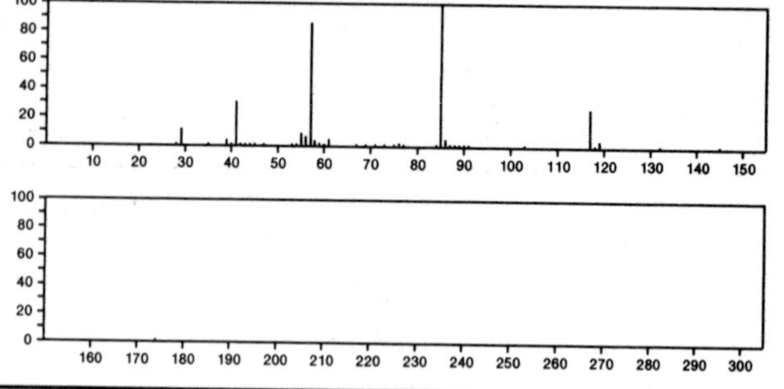

174 $C_9H_{18}OS$ 16315-52-9

2*H*-Pyran, 2-(butylthio)tetrahydro-

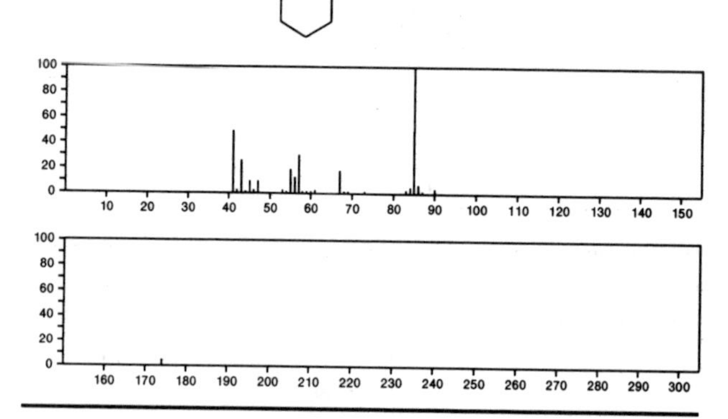

174 $C_9H_{18}OS$ 24699-60-3

1,3-Oxathiane, 2-(1,1-dimethylethyl)-2-methyl-

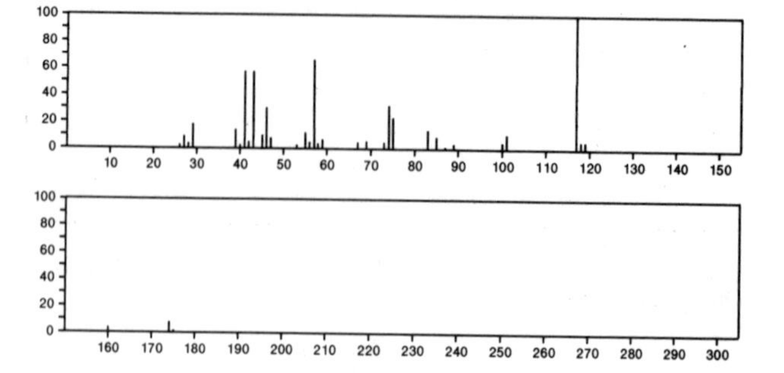

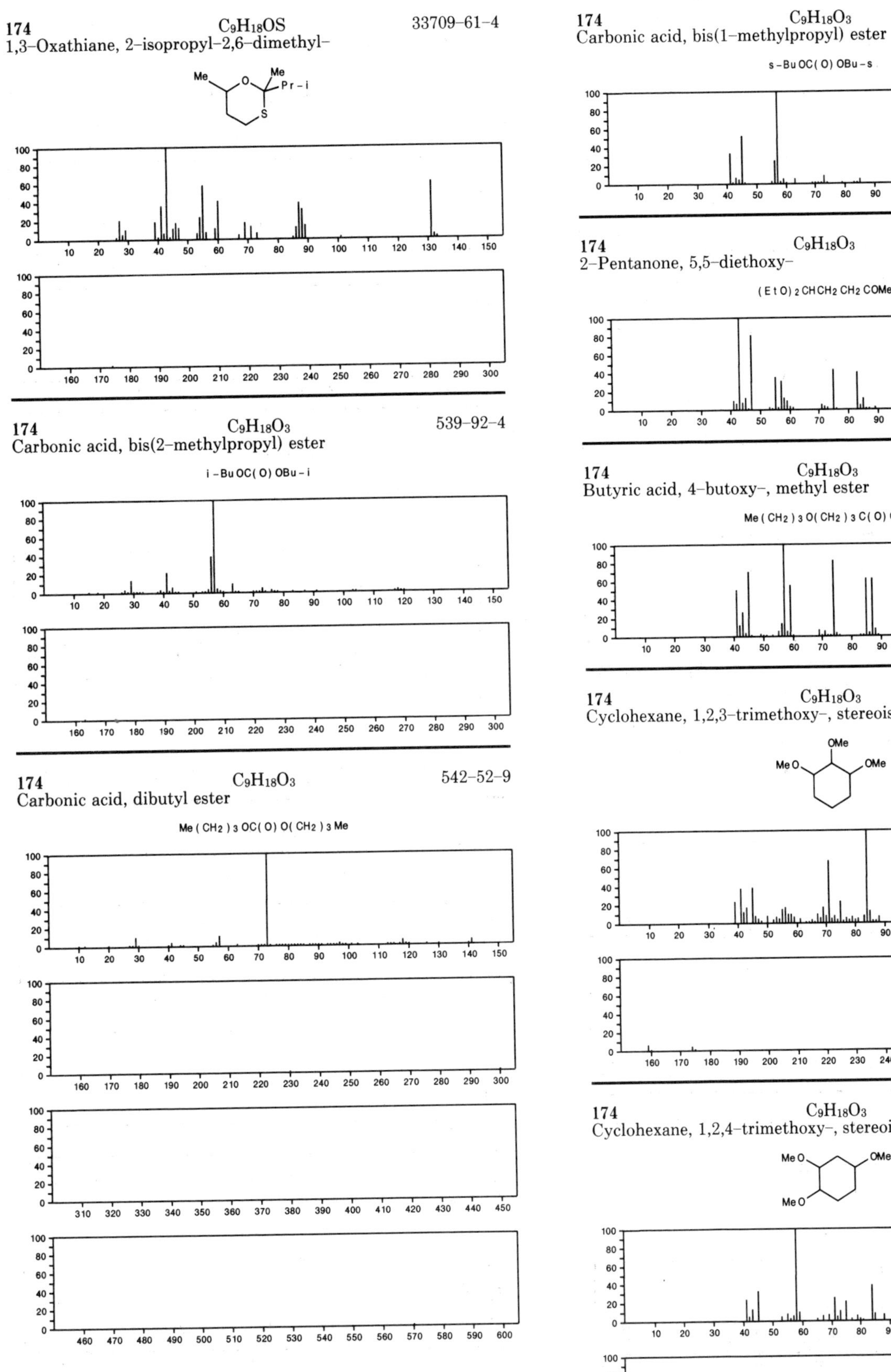

174 $C_9H_{18}OS$ 33709-61-4
1,3-Oxathiane, 2-isopropyl-2,6-dimethyl-

174 $C_9H_{18}O_3$ 539-92-4
Carbonic acid, bis(2-methylpropyl) ester
i-BuOC(O)OBu-i

174 $C_9H_{18}O_3$ 542-52-9
Carbonic acid, dibutyl ester
Me(CH2)3OC(O)O(CH2)3Me

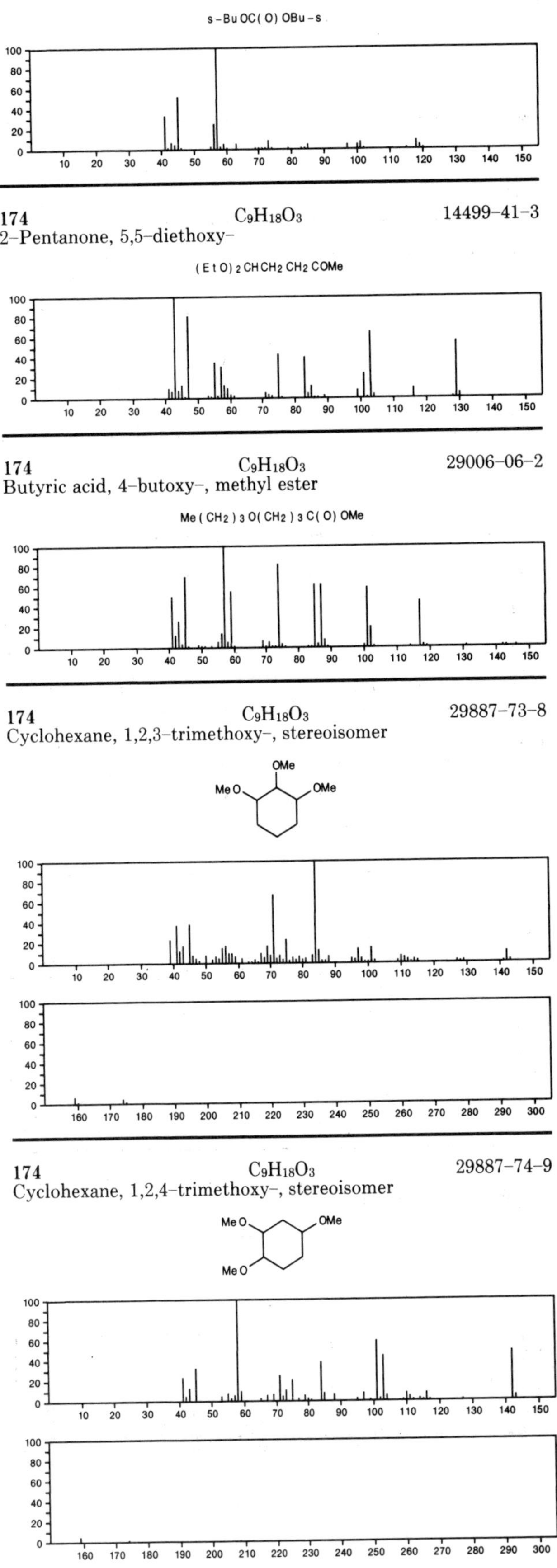

174 $C_9H_{18}O_3$ 623-63-2
Carbonic acid, bis(1-methylpropyl) ester
s-BuOC(O)OBu-s

174 $C_9H_{18}O_3$ 14499-41-3
2-Pentanone, 5,5-diethoxy-
(EtO)2CHCH2CH2COMe

174 $C_9H_{18}O_3$ 29006-06-2
Butyric acid, 4-butoxy-, methyl ester
Me(CH2)3O(CH2)3C(O)OMe

174 $C_9H_{18}O_3$ 29887-73-8
Cyclohexane, 1,2,3-trimethoxy-, stereoisomer

174 $C_9H_{18}O_3$ 29887-74-9
Cyclohexane, 1,2,4-trimethoxy-, stereoisomer

174 $C_9H_{18}O_3$ 29887-75-0
Cyclohexane, 1,3,5-trimethoxy-, $(1\alpha,3\alpha,5\beta)$-

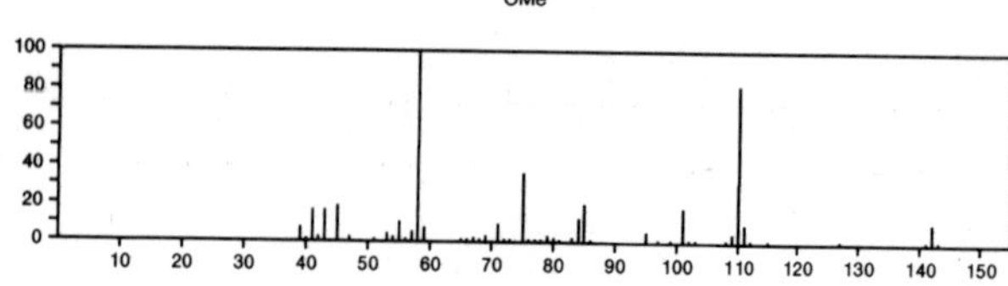

174 $C_9H_{18}O_3$ 30363-70-3
Cyclohexane, 1,2,3-trimethoxy-, stereoisomer

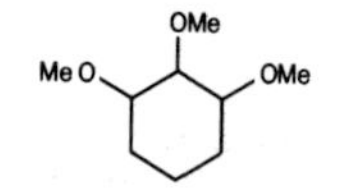

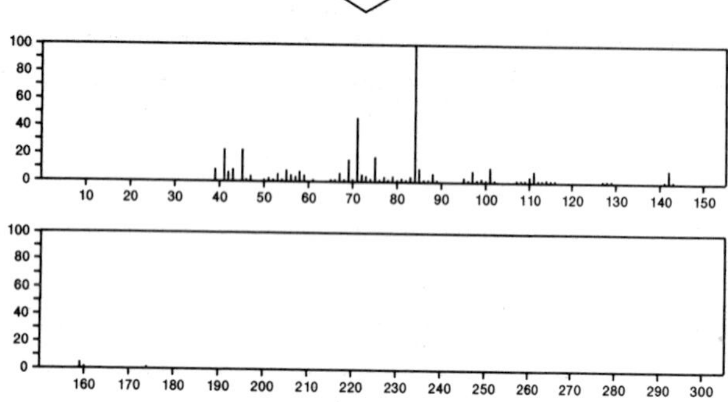

174 $C_9H_{18}O_3$ 30363-72-5
Cyclohexane, 1,2,4-trimethoxy-, stereoisomer

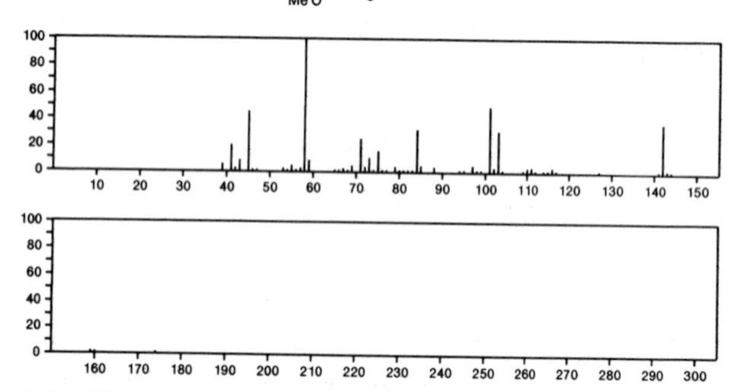

174 $C_9H_{18}O_3$ 30363-89-4
Cyclohexane, 1,3,5-trimethoxy-, $(1\alpha,3\alpha,5\alpha)$-

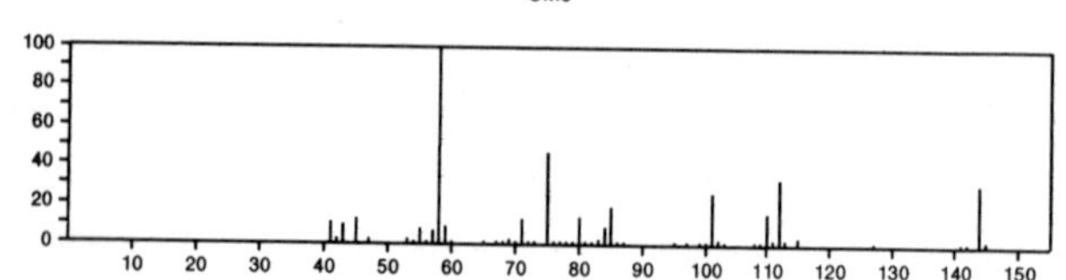

174 $C_9H_{18}O_3$ 30377-28-7
Cyclohexane, 1,2,3-trimethoxy-, stereoisomer

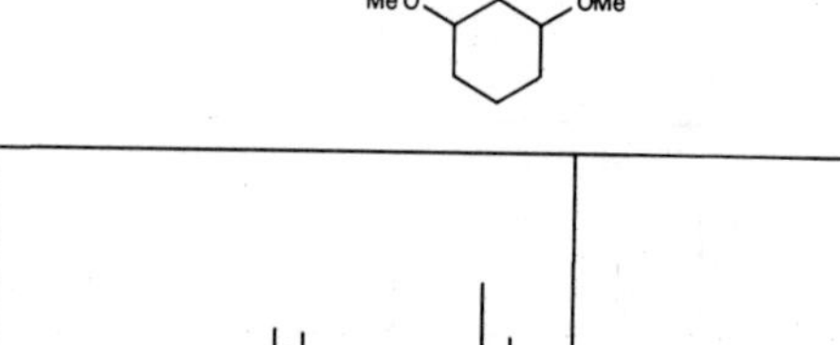

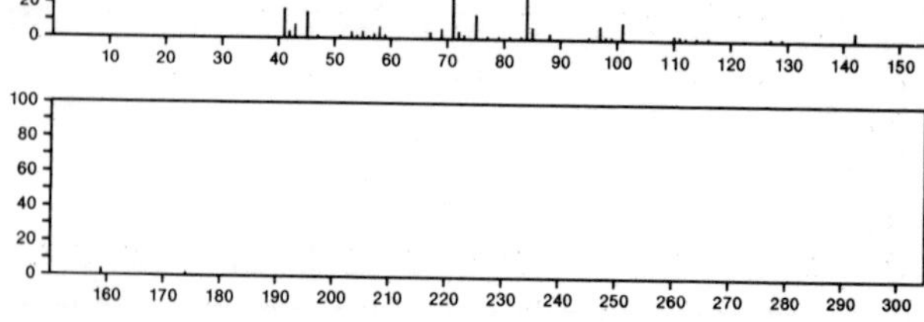

174 $C_9H_{18}O_3$ 36651-23-7
1,3-Dioxolane-2-pentanol, 2-methyl-

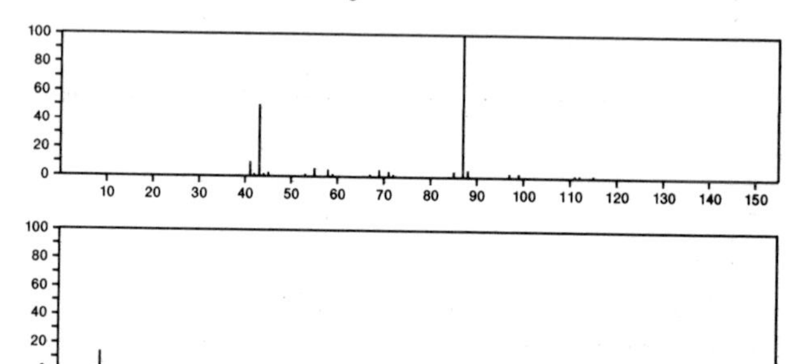

174 $C_9H_{18}O_3$ 55956-25-7
2-Propanol, 1-[1-methyl-2-(2-propenyloxy)ethoxy]-

MeCH(OH)CH2OCHMeCH2OCH2CH=CH2

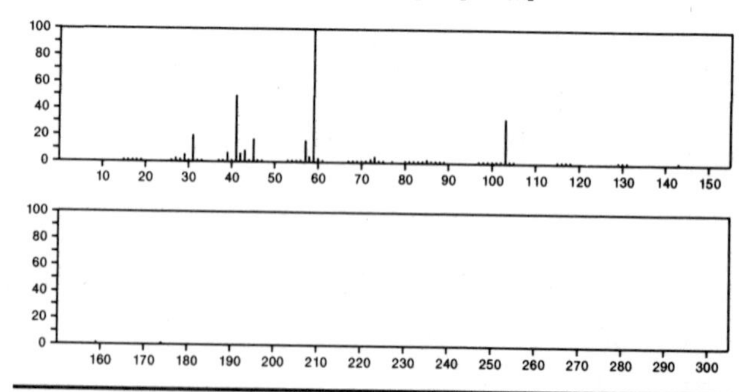

174 $C_9H_{22}OSi$ 17877-22-4
Silane, trimethyl[(2-methylpentyl)oxy]-

Me3SiOCH2CHPrMe

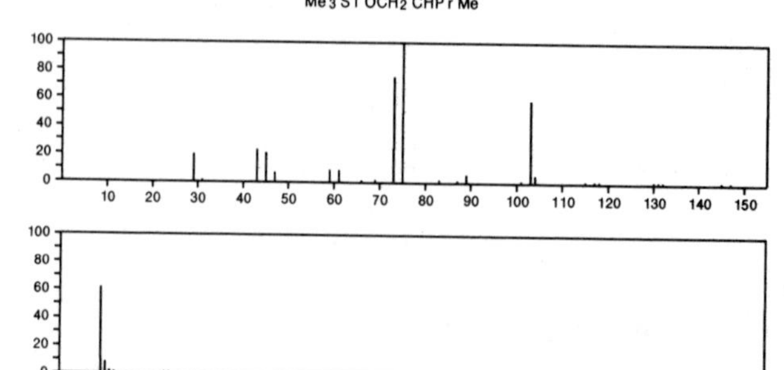

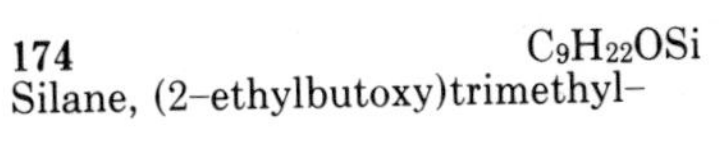

174 $C_9H_{22}OSi$ 17888-61-8
Silane, (2-ethylbutoxy)trimethyl-

$Me_3SiOCH_2CHEt_2$

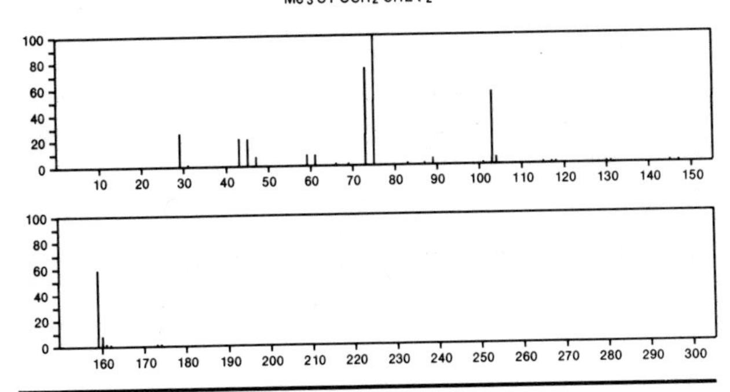

174 $C_9H_{22}OSi$ 17888-62-9
Silane, (hexyloxy)trimethyl-

$Me(CH_2)_5OSiMe_3$

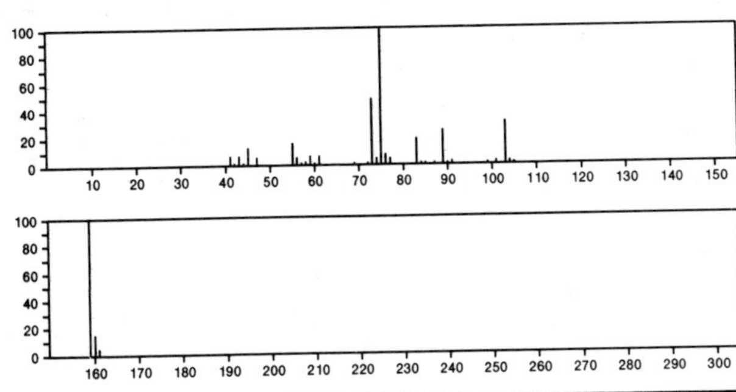

174 $C_9H_{22}OSi$ 17888-63-0
Silane, trimethyl[(1-methylpentyl)oxy]-

$Me_3SiOCHMe(CH_2)_3Me$

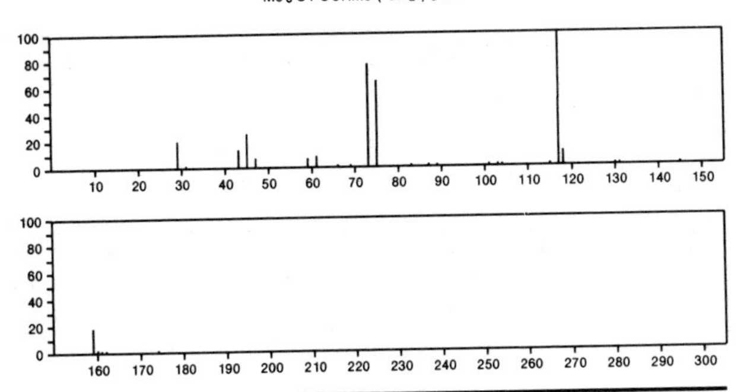

174 $C_9H_{22}OSi$ 17888-64-1
Silane, trimethyl[(3-methylpentyl)oxy]-

$Me_3SiOCH_2CH_2CHMeCH_2Me$

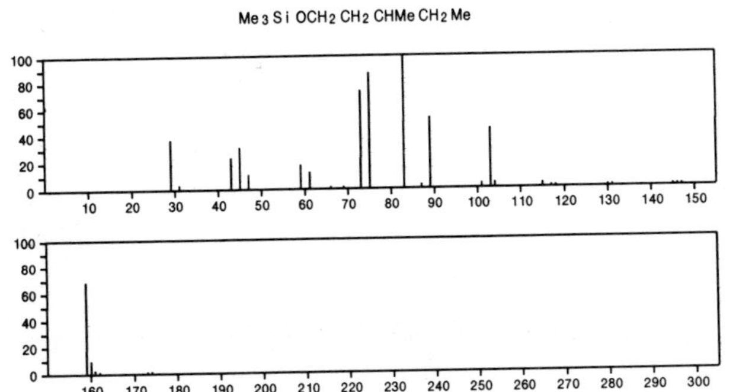

174 $C_{10}H_6O_3$ 83-72-7
1,4-Naphthalenedione, 2-hydroxy-

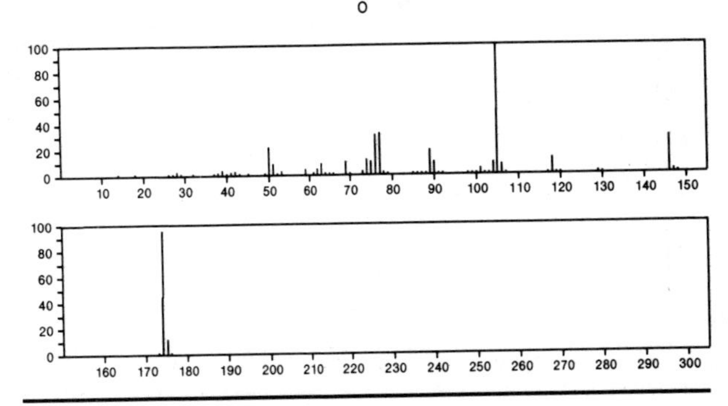

174 $C_{10}H_6O_3$ 481-39-0
1,4-Naphthalenedione, 5-hydroxy-

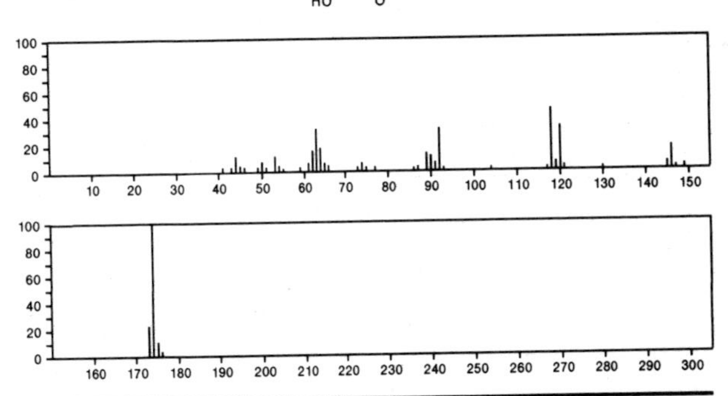

174 $C_{10}H_6O_3$ 607-20-5
1,2-Naphthalenedione, 6-hydroxy-

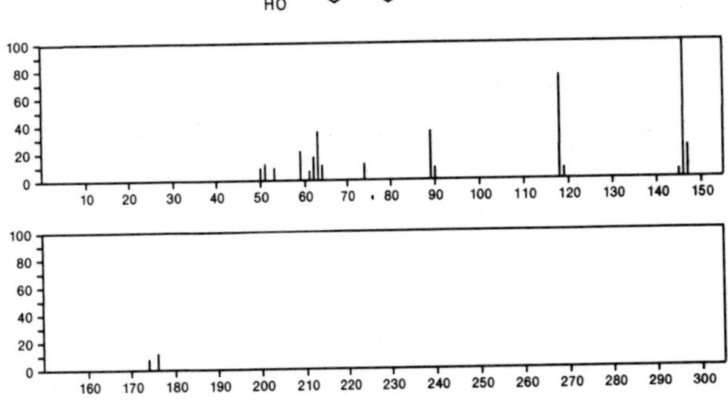

174 $C_{10}H_{10}N_2O$ 89-25-8
3*H*-Pyrazol-3-one, 2,4-dihydro-5-methyl-2-phenyl-

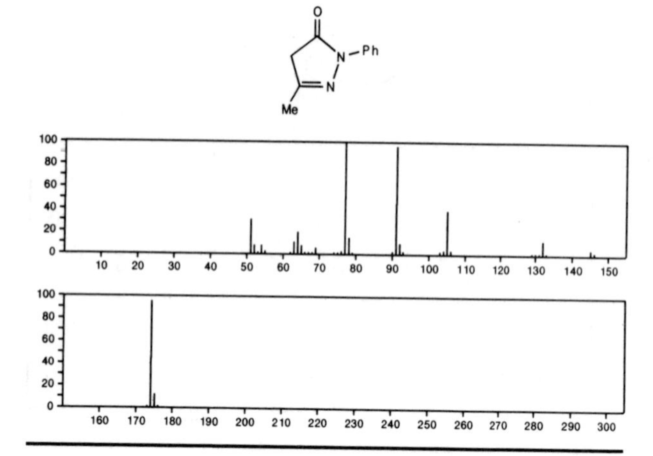

174 $C_{10}H_{10}N_2O$ 879-37-8
1*H*-Indole-3-acetamide

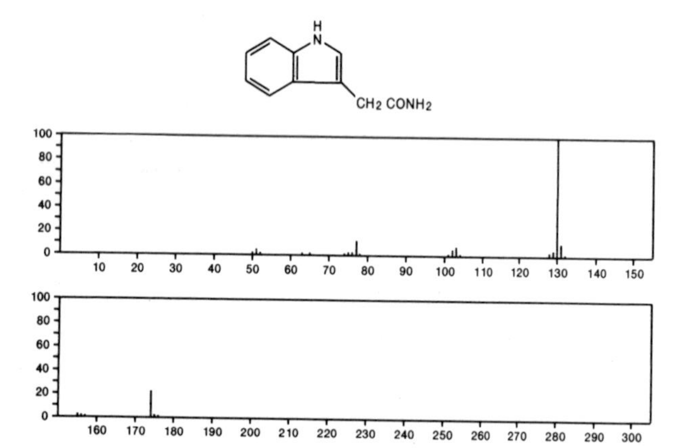

174 $C_{10}H_{10}N_2O$ 6940-11-0
Quinoxaline, 2,3-dimethyl-, 1-oxide

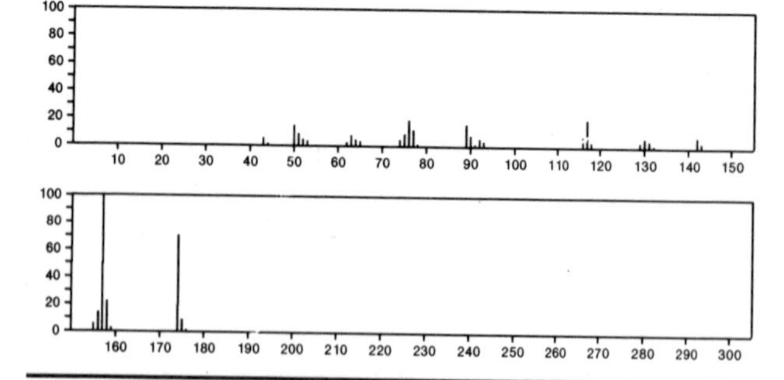

174 $C_{10}H_{10}N_2O$ 16154-82-8
Quinoxaline, 2-ethyl-, 1-oxide

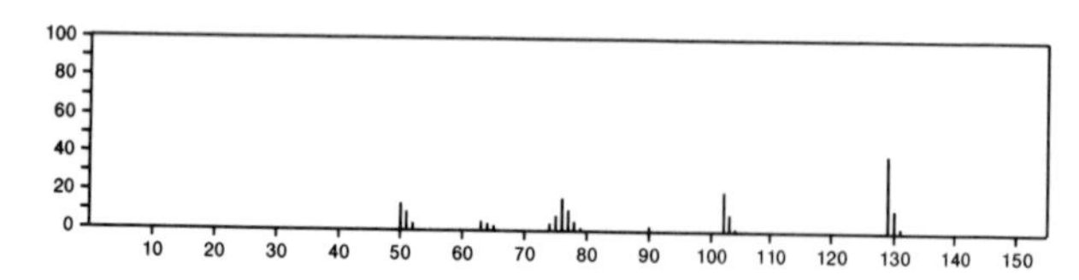

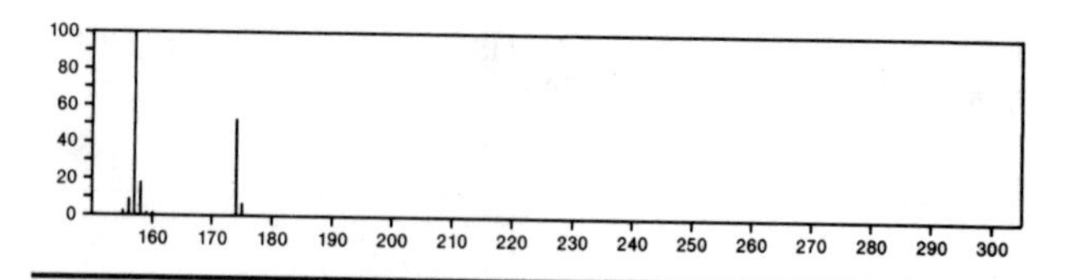

174 $C_{10}H_{10}N_2O$ 16154-83-9
Quinoxaline, 2-ethyl-, 4-oxide

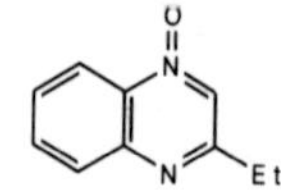

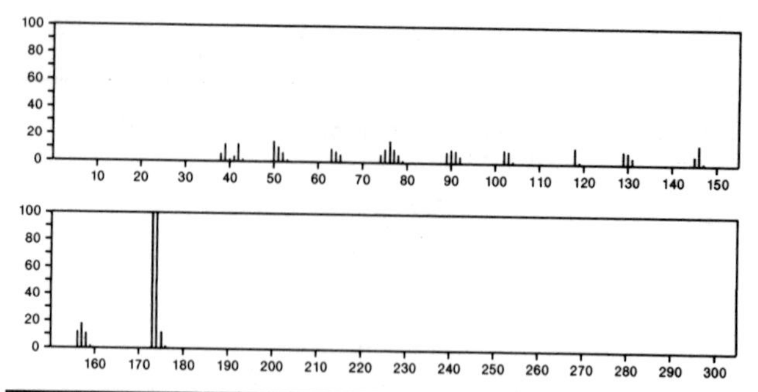

174 $C_{10}H_{10}N_2O$ 17018-81-4
8-Quinolinol, 2-(aminomethyl)-

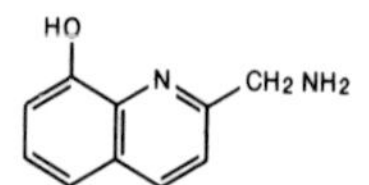

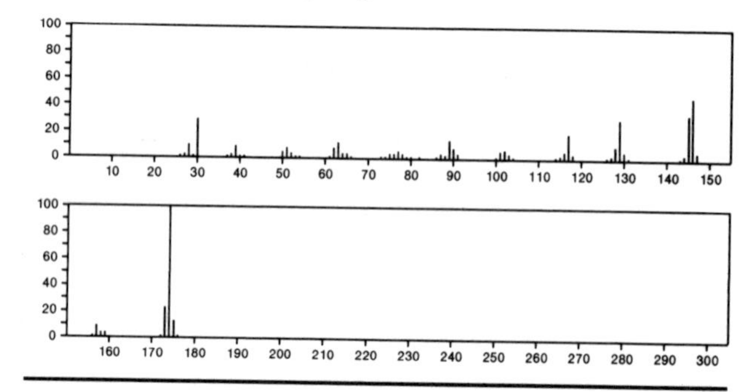

174 $C_{10}H_{10}N_2O$ 17408-29-6
Ethanone, 1-(2-methylpyrazolo[1,5-*a*]pyridin-3-yl)-

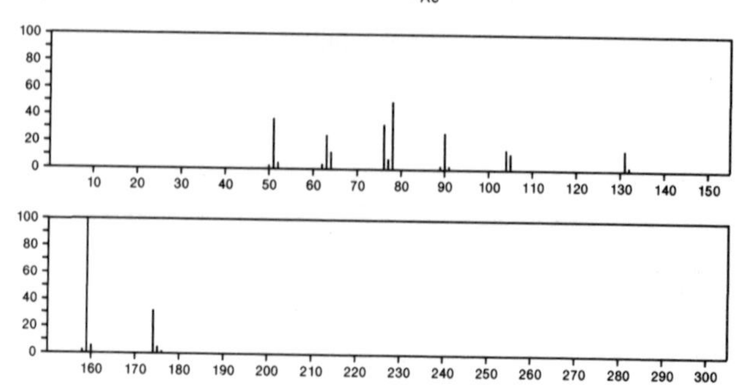

174 $C_{10}H_{10}N_2O$ 22365-23-7
2*H*-Pyrido[1,2-*a*]pyrimidin-2-one, 4,8-dimethyl-

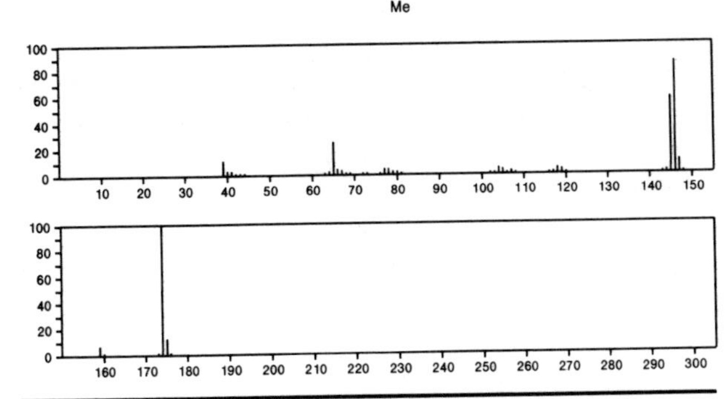

174 $C_{10}H_{10}N_2O$ 22380-20-7
6-Indolizinecarboxamide, 2-methyl-

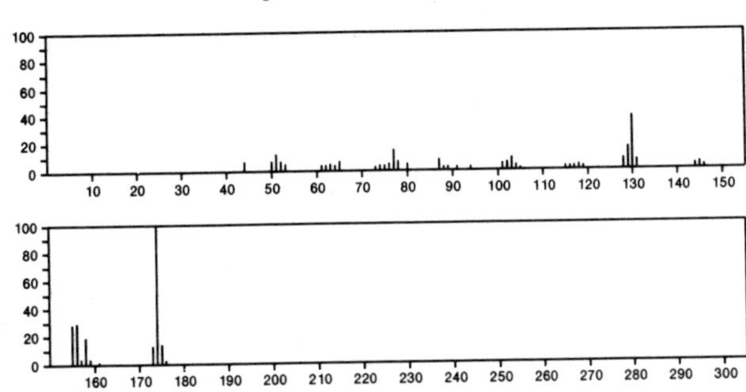

174 $C_{10}H_{10}N_2O$ 28883-91-2
Isoxazole, 5-amino-3-*p*-tolyl-

174 $C_{10}H_{10}N_2O$ 28883-95-6
2*H*-Azirine-2-carboxamide, 3-*p*-tolyl-

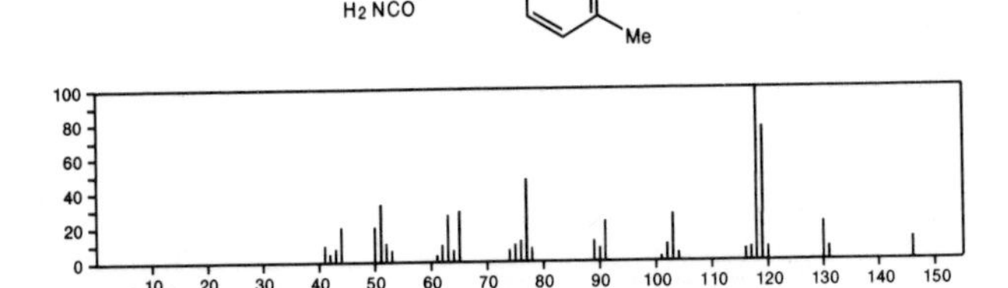

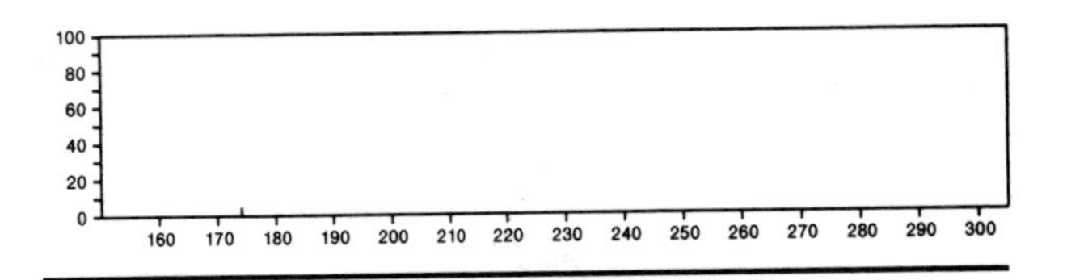

174 $C_{10}H_{10}N_2O$ 30986-10-8
2*H*-1,4-Ethanocinnolin-3(4*H*)-one

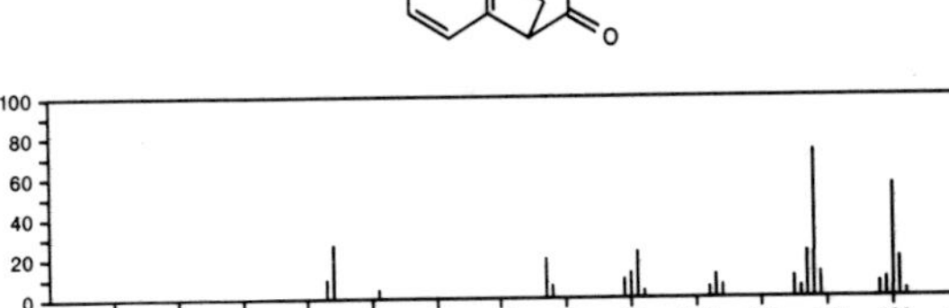

174 $C_{10}H_{10}N_2O$ 37920-74-4
Quinazoline, 4-ethyl-, 3-oxide

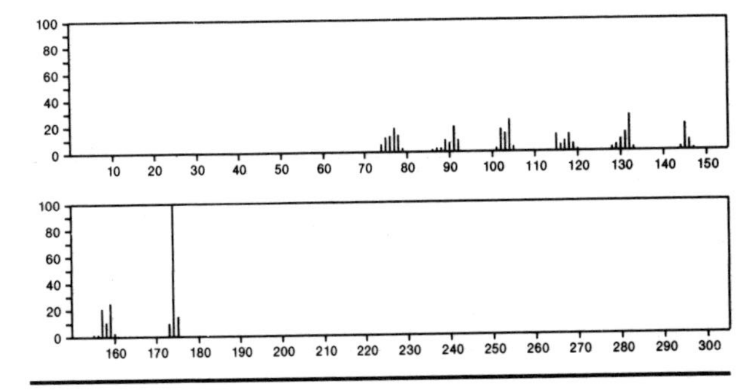

174 $C_{10}H_{10}N_2O$ 37920-75-5
Quinazoline, 4-ethyl-, 1-oxide

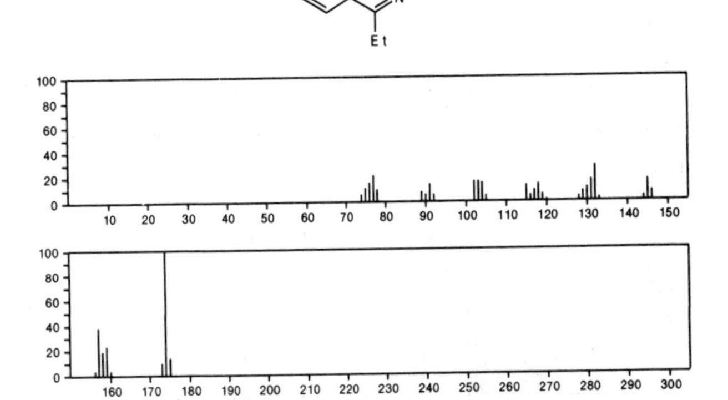

174 $C_{10}H_{10}N_2O$ 41927-50-8
3*H*-Pyrazol-3-one, 2,4-dihydro-2-methyl-5-phenyl-

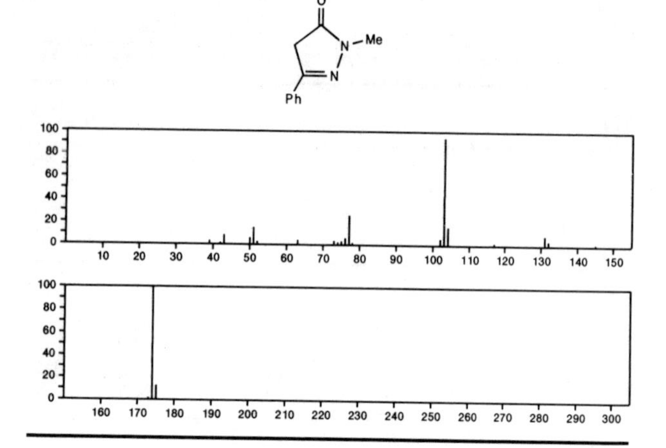

174 $C_{10}H_{15}BN_2$ 31748-14-8
1*H*-1,3,2-Benzodiazaborole, 2-butyl-2,3-dihydro-

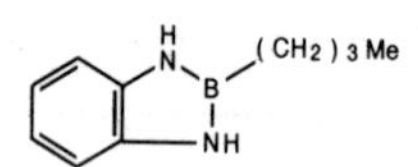

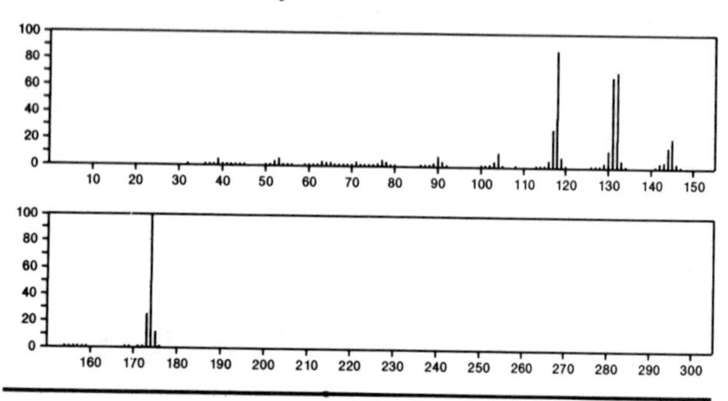

174 $C_{10}H_{22}O_2$ 107-74-4
1,7-Octanediol, 3,7-dimethyl-

HOCH2 CH2 CHMe (CH2) 3 CMe2 OH

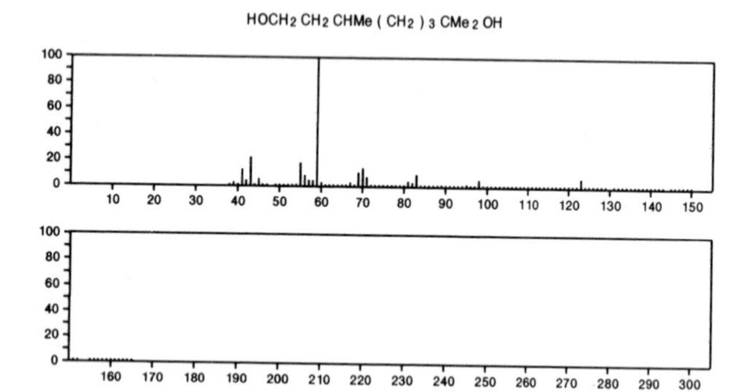

174 $C_{10}H_{22}O_2$ 871-22-7
Butane, 1,1'-[ethylidenebis(oxy)]bis-

Me (CH2) 3 OCHMe O(CH2) 3 Me

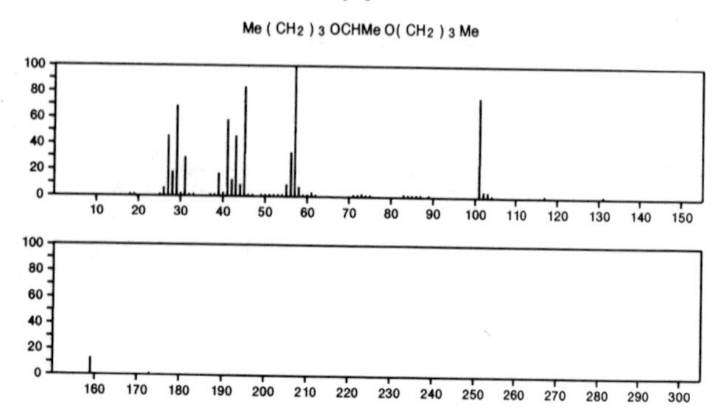

174 $C_{10}H_{22}O_2$ 4541-13-3
1-Butanol, 4-(hexyloxy)-

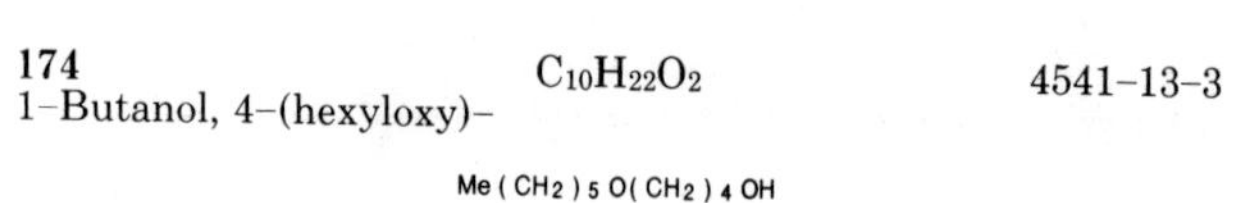

Me (CH2) 5 O(CH2) 4 OH

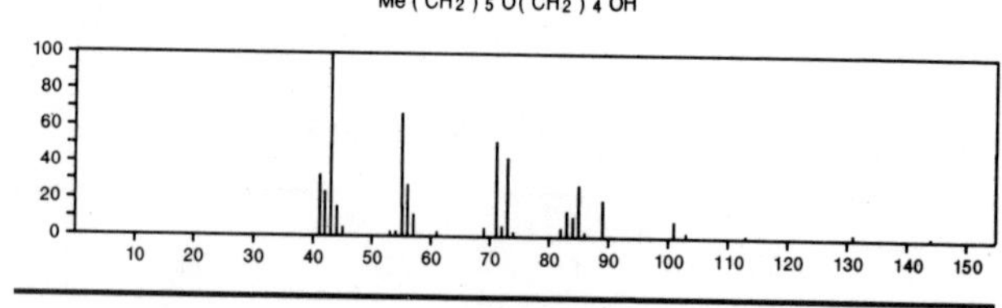

174 $C_{10}H_{22}O_2$ 5314-41-0
Acetaldehyde, di-*sec*-butyl acetal

Me CH(OBu - s) 2

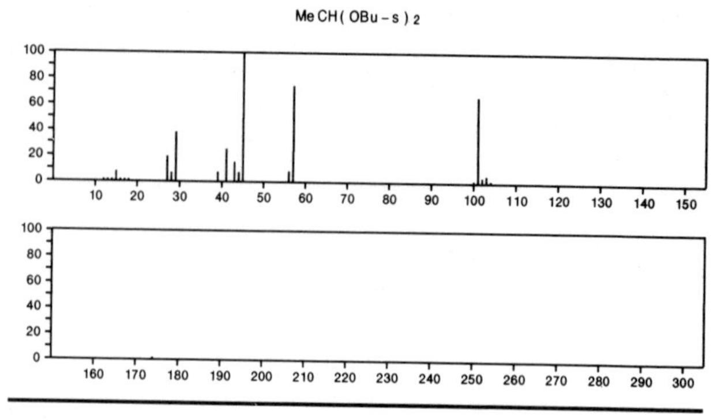

174 $C_{10}H_{22}O_2$ 5669-09-0
Propane, 1,1'-[ethylidenebis(oxy)]bis[2-methyl-

Me CH(OBu - i) 2

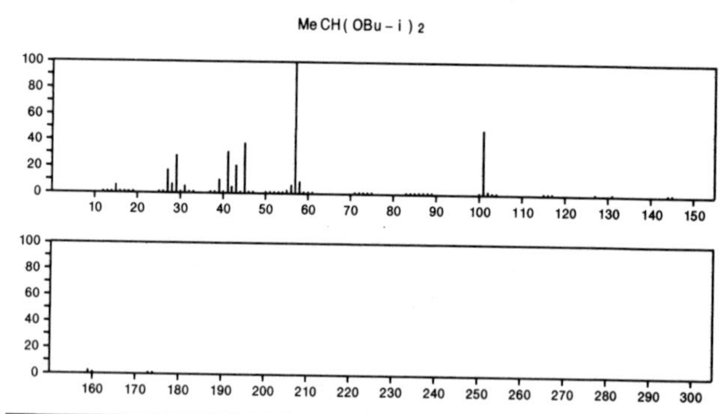

174 $C_{10}H_{22}O_2$ 10020-43-6
Ethanol, 2-(octyloxy)-

HOCH2 CH2 O(CH2) 7 Me

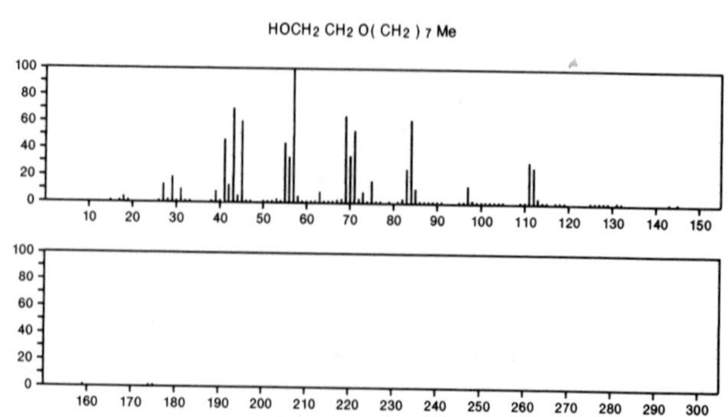

174 $C_{10}H_{22}O_2$ 20637-47-2
Heptane, 4-methoxy-3-(methoxymethyl)-

Pr CH(OMe) CHEt CH2 OMe

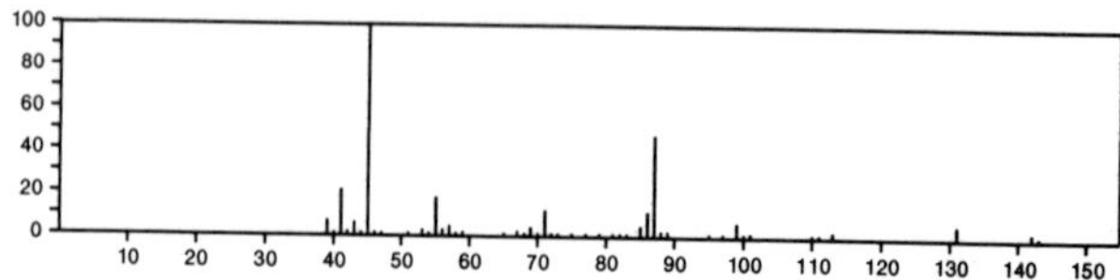

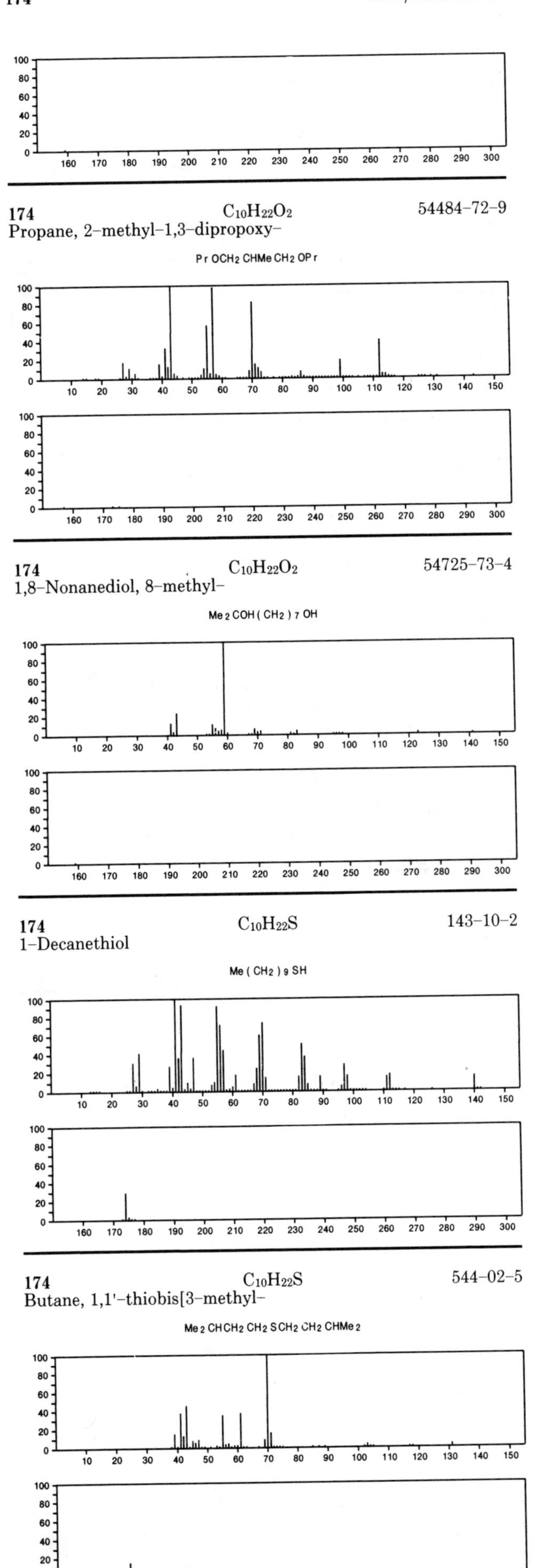
174 $C_{10}H_{22}O_2$ 54484-72-9
Propane, 2-methyl-1,3-dipropoxy-
Pr OCH2 CHMe CH2 OPr
174 $C_{10}H_{22}O_2$ 54725-73-4
1,8-Nonanediol, 8-methyl-
Me2 COH (CH2)7 OH
174 $C_{10}H_{22}S$ 143-10-2
1-Decanethiol
Me (CH2)9 SH
174 $C_{10}H_{22}S$ 544-02-5
Butane, 1,1'-thiobis[3-methyl-
Me2 CHCH2 CH2 SCH2 CH2 CHMe2

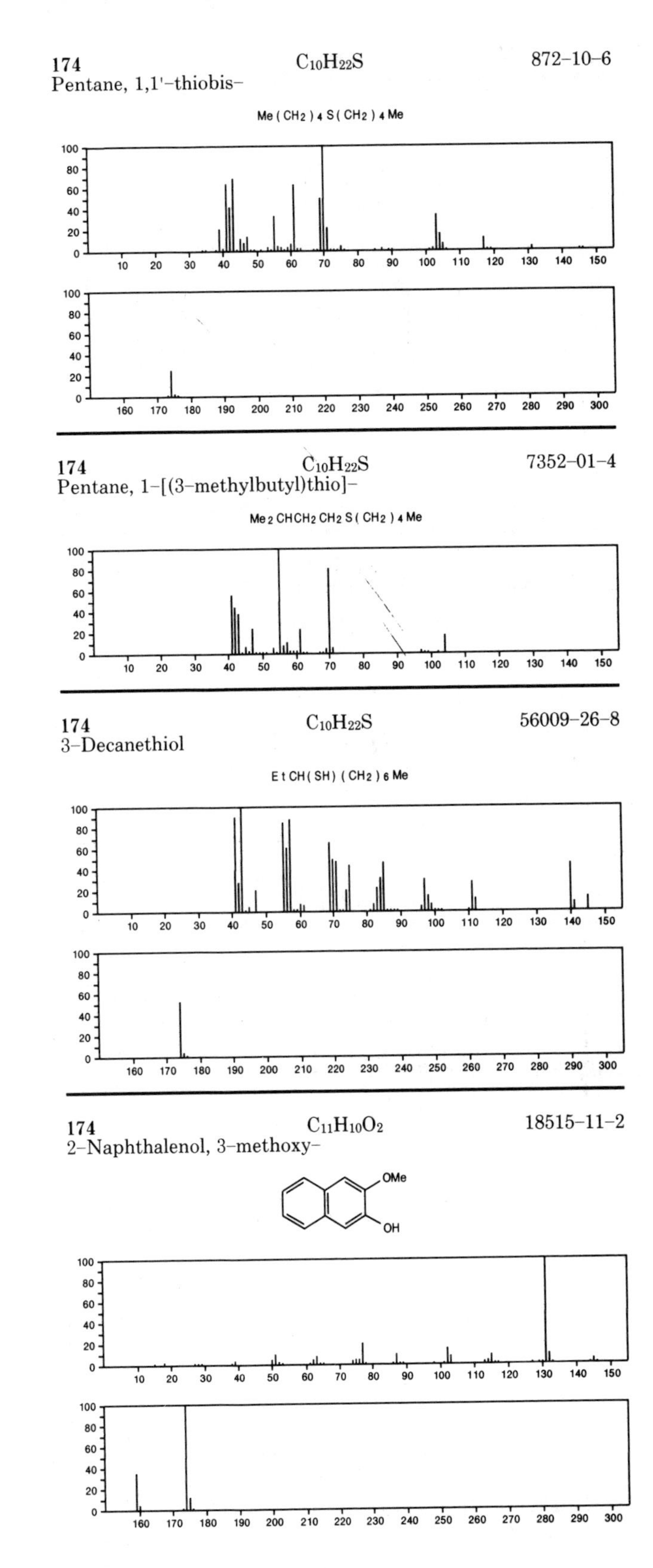
174 $C_{10}H_{22}S$ 872-10-6
Pentane, 1,1'-thiobis-
Me (CH2)4 S (CH2)4 Me
174 $C_{10}H_{22}S$ 7352-01-4
Pentane, 1-[(3-methylbutyl)thio]-
Me2 CHCH2 CH2 S (CH2)4 Me
174 $C_{10}H_{22}S$ 56009-26-8
3-Decanethiol
Et CH (SH) (CH2)6 Me
174 $C_{11}H_{10}O_2$ 18515-11-2
2-Naphthalenol, 3-methoxy-
OMe
OH

174 $C_{11}H_{10}O_2$ 20651-88-1
1*H*-Indene-1,2(3*H*)-dione, 3,3-dimethyl-

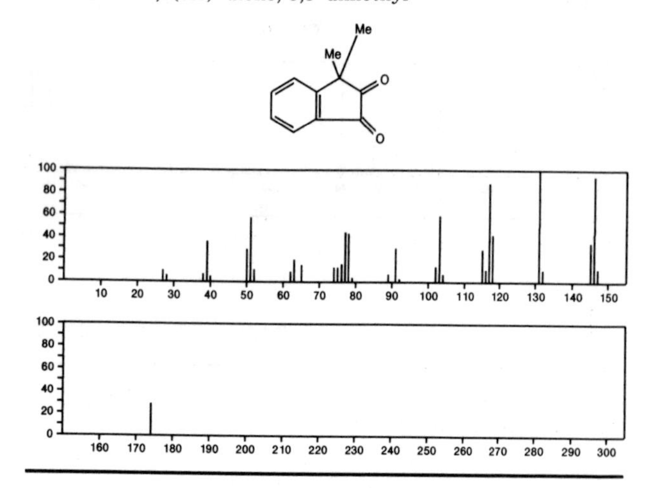

174 $C_{11}H_{10}O_2$ 21053-63-4
2(5*H*)-Furanone, 4-methyl-5-phenyl-

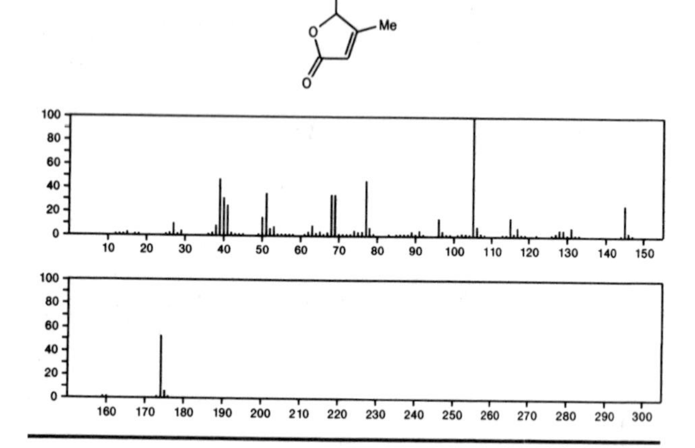

174 $C_{11}H_{10}O_2$ 53774-21-3
2(5*H*)-Furanone, 5-methyl-5-phenyl-

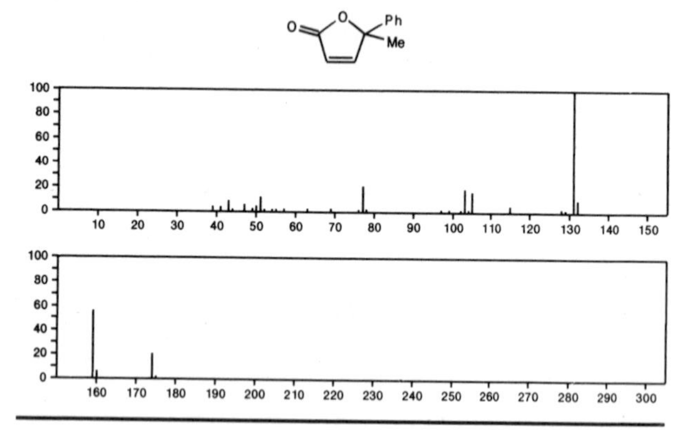

174 $C_{11}H_{10}O_2$ 56909-26-3
2-Oxatricyclo[5.5.0.0^{4,10}]dodeca-5,8,11-trien-3-one

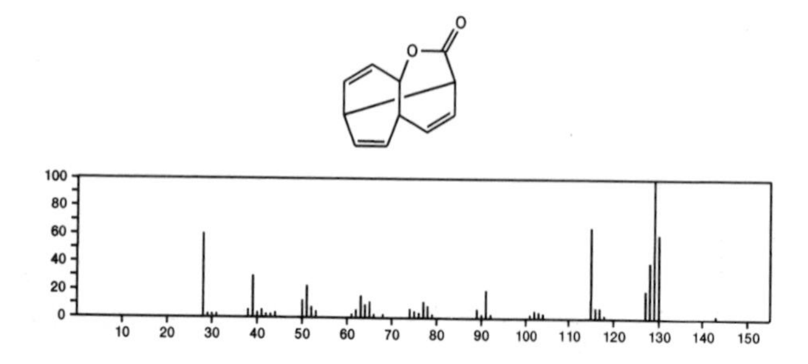

174 $C_{11}H_{10}S$ 13132-15-5
Thiophene, 2-(phenylmethyl)-

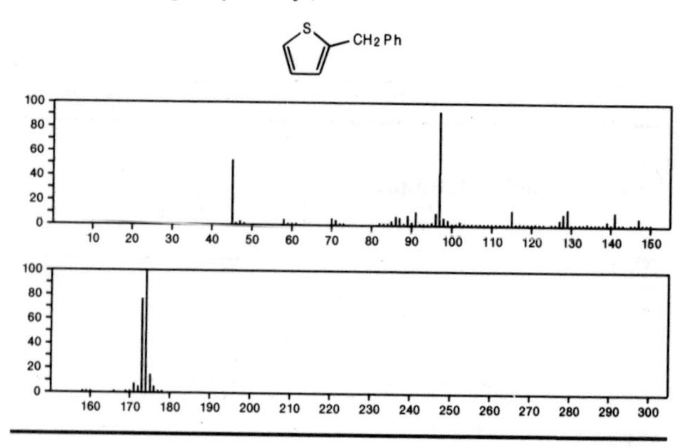

174 $C_{11}H_{14}N_2$ 87-52-5
1*H*-Indole-3-methanamine, *N*,*N*-dimethyl-

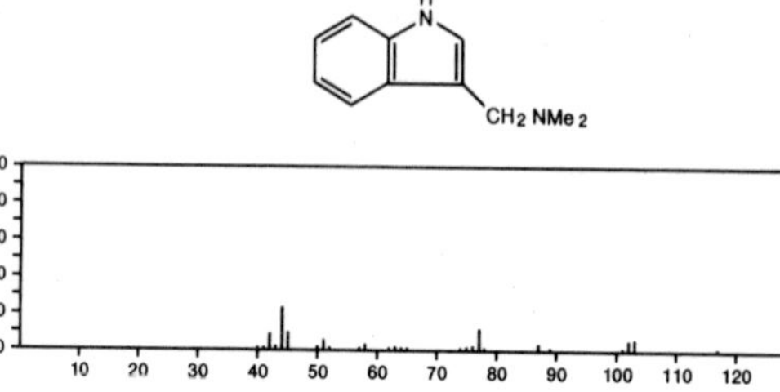

174 $C_{11}H_{14}N_2$ 5851-44-5
1*H*-Benzimidazole, 2-butyl-

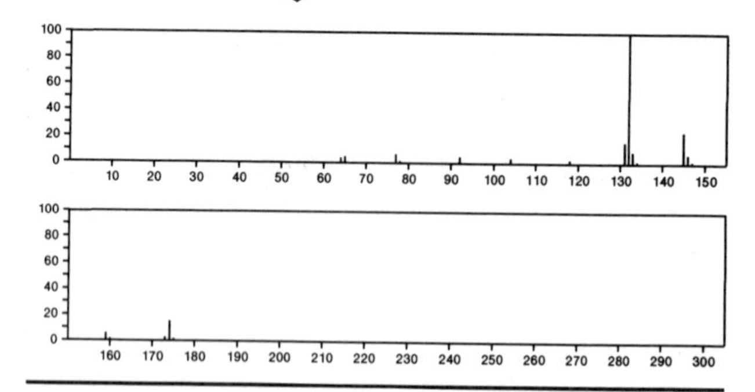

174 $C_{11}H_{14}N_2$ 5851-45-6
1*H*-Benzimidazole, 2-(2-methylpropyl)-

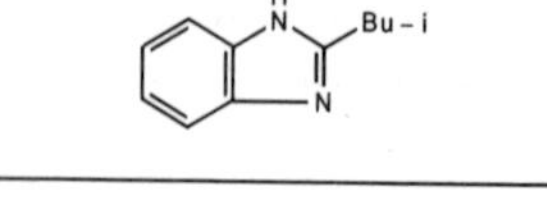

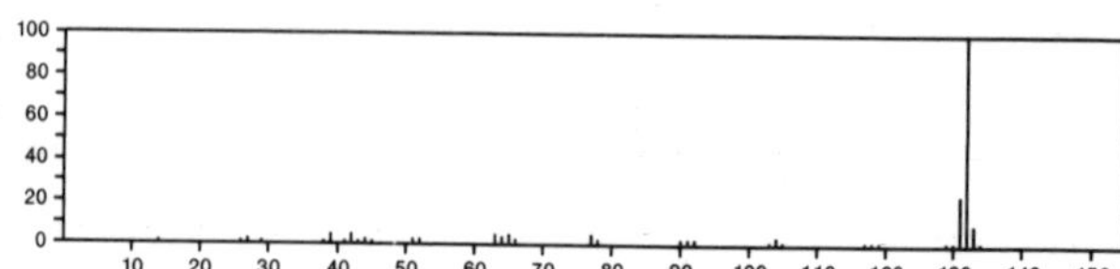

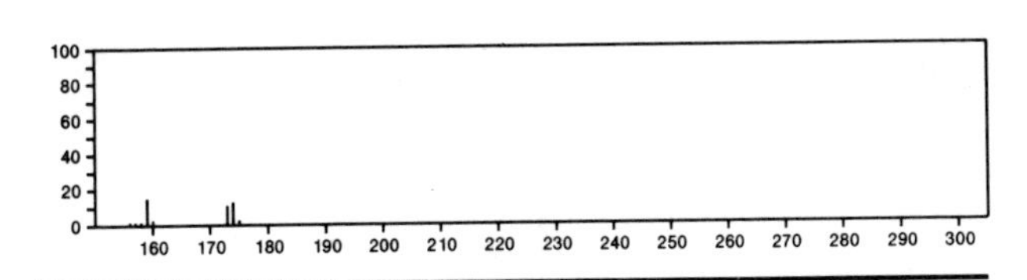

174 $C_{11}H_{14}N_2$ 24425-13-6
1*H*-Benzimidazole, 2-(1,1-dimethylethyl)-

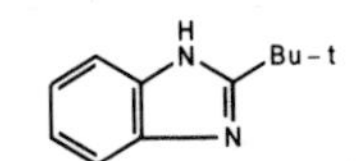

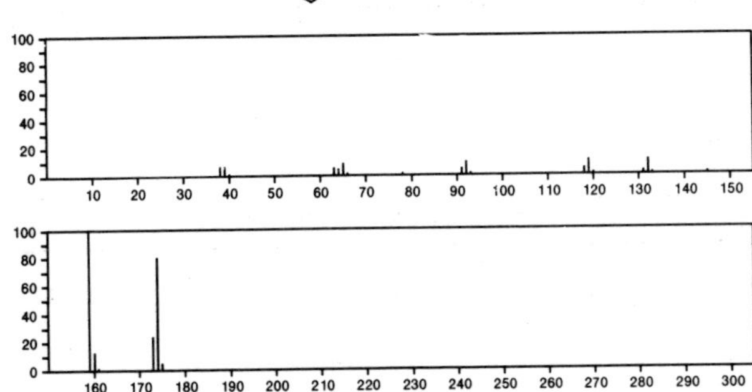

174 $C_{12}H_{14}O$ 3437-89-6
2-Pentanone, 3-(phenylmethylene)-

PhCH=CEt COMe

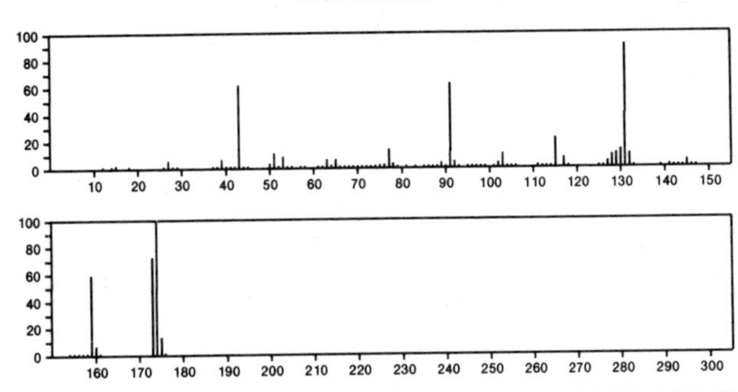

174 $C_{12}H_{14}O$ 5037-63-8
1(2*H*)-Naphthalenone, 3,4-dihydro-5,8-dimethyl-

Me
O Me

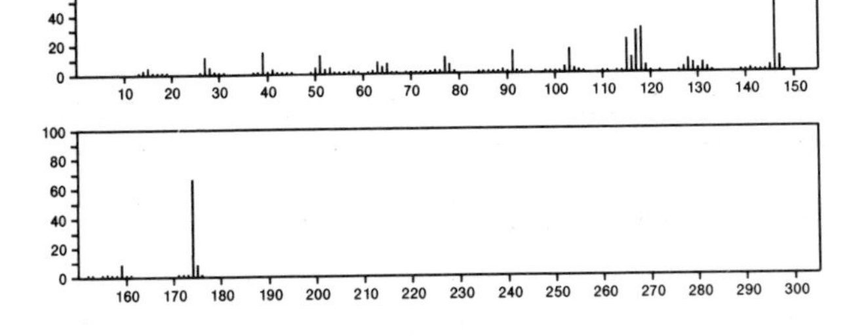

174 $C_{12}H_{14}O$ 13153-77-0
1,4-Ethanonaphthalen-2-ol, 1,2,3,4-tetrahydro-, *exo*-

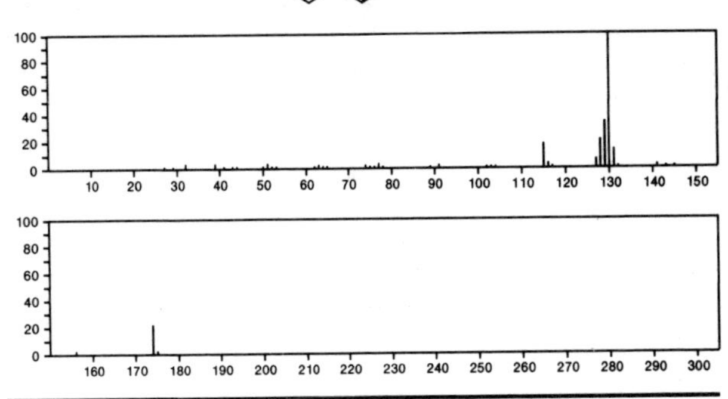

174 $C_{12}H_{14}O$ 13153-78-1
1,4-Ethanonaphthalen-2-ol, 1,2,3,4-tetrahydro-, $(1\alpha,2\beta,4\alpha)$-

OH

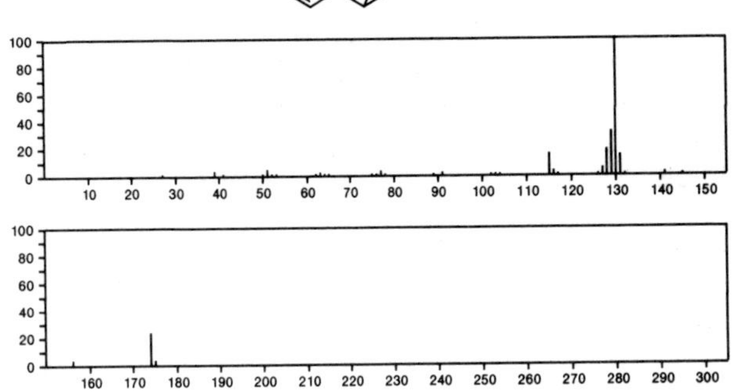

174 $C_{12}H_{14}O$ 13621-25-5
1(2*H*)-Naphthalenone, 3,4-dihydro-5,7-dimethyl-

Me
Me
O

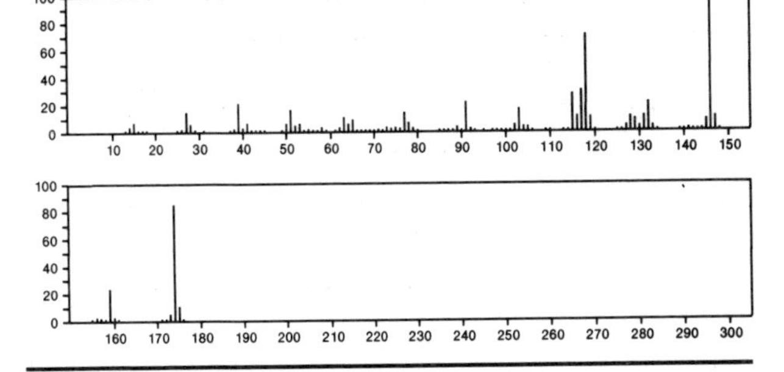

174 $C_{12}H_{14}O$ 15561-15-6
Crotonophenone, 2',5'-dimethyl-

Me
COCH=CHMe
Me

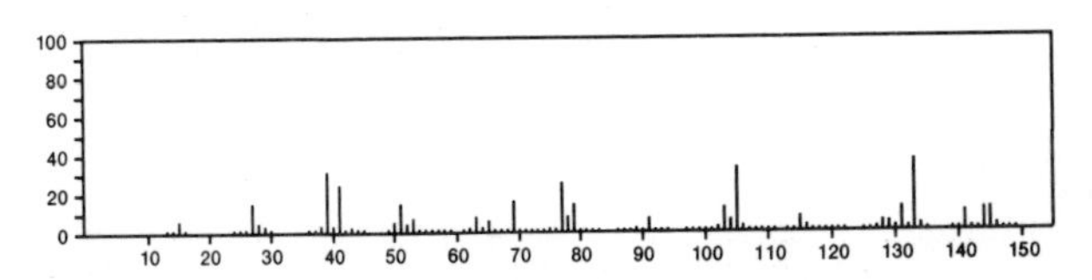

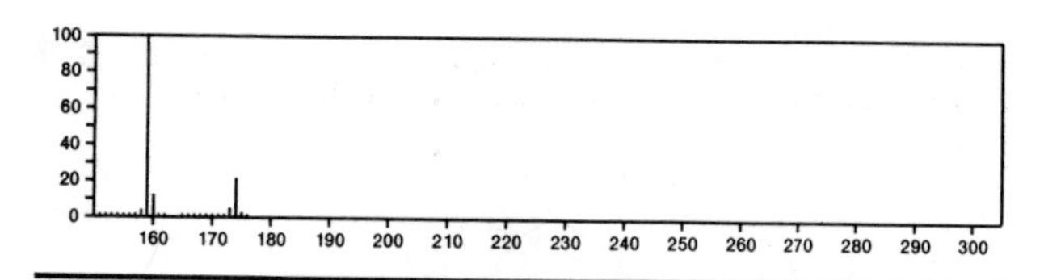

174 $C_{12}H_{14}O$ 16205-96-2
Acrylophenone, 2,2',5'-trimethyl-

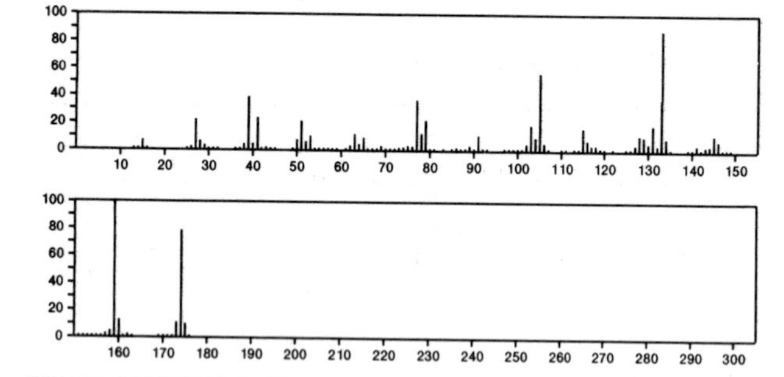

174 $C_{12}H_{14}O$ 19550-57-3
1(2*H*)-Naphthalenone, 3,4-dihydro-6,7-dimethyl-

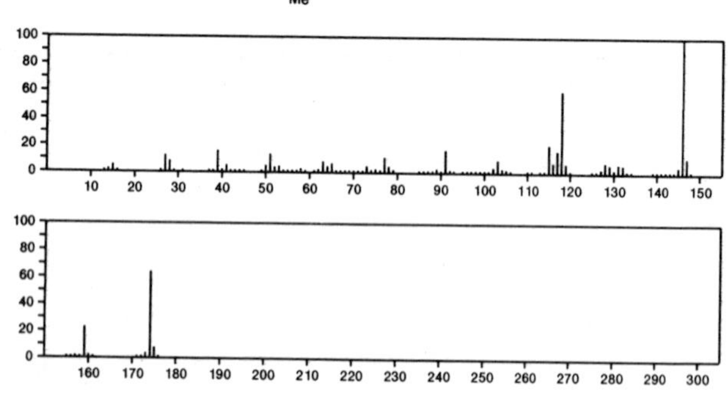

174 $C_{12}H_{14}O$ 30316-30-4
1(2*H*)-Naphthalenone, 3,4-dihydro-6,8-dimethyl-

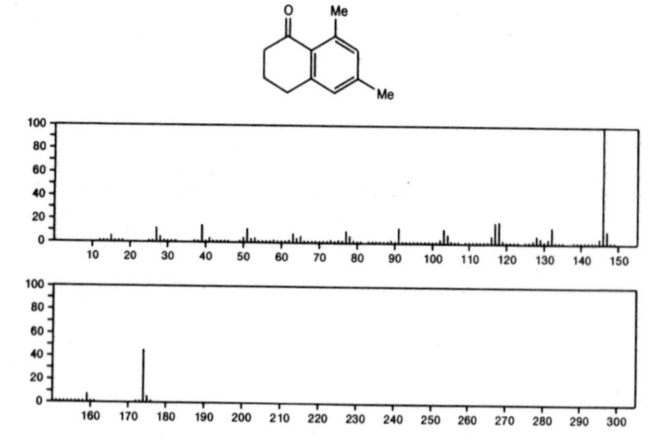

174 $C_{12}H_{14}O$ 32281-65-5
1(2*H*)-Naphthalenone, 3,4-dihydro-5,6-dimethyl-

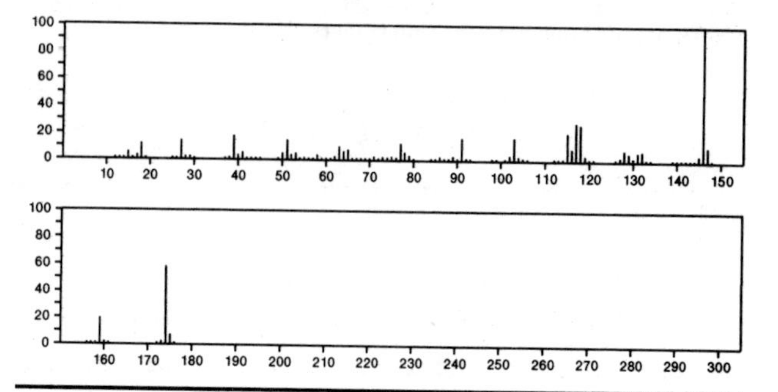

174 $C_{12}H_{14}O$ 33046-41-2
3-Hexen-2-one, 6-phenyl-

$PhCH_2CH_2CH=CHCOMe$

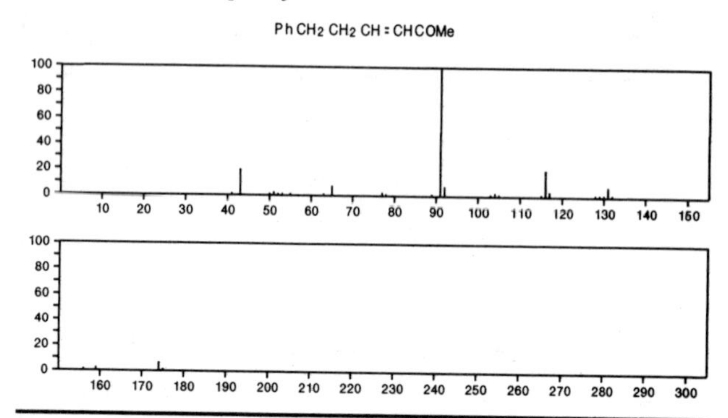

174 $C_{12}H_{14}O$ 35322-84-0
1*H*-Inden-1-one, 2,3-dihydro-3,4,7-trimethyl-

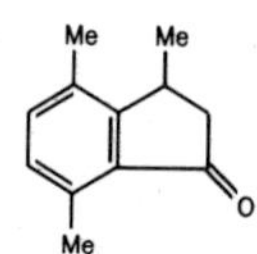

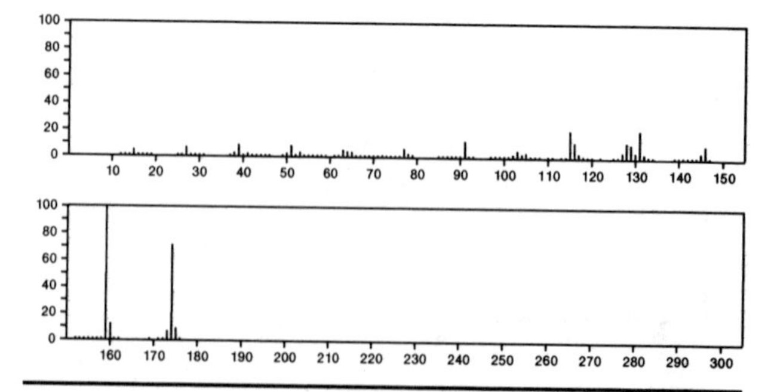

174 $C_{12}H_{14}O$ 36051-81-7
1*H*,3*H*-Naphtho[1,8-*cd*]pyran, 3a,4,5,6-tetrahydro-

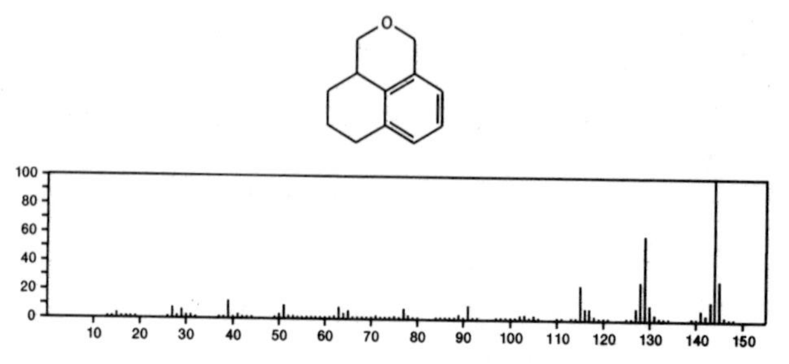

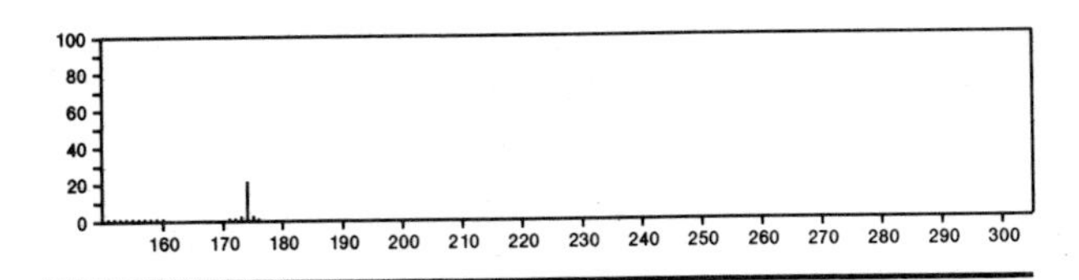

174 $C_{12}H_{14}O$ 39877-86-6
Benzene, 1-(1-cyclopenten-1-yl)-2-methoxy-

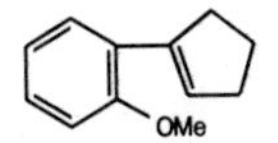

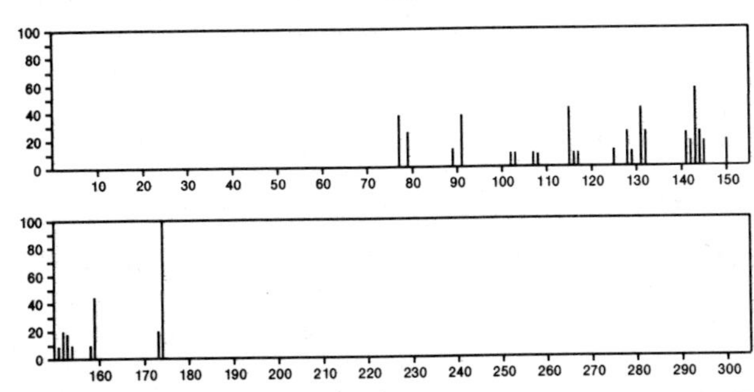

174 $C_{12}H_{14}O$ 51015-31-7
1(2*H*)-Naphthalenone, 5-ethyl-3,4-dihydro-

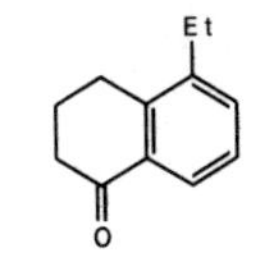

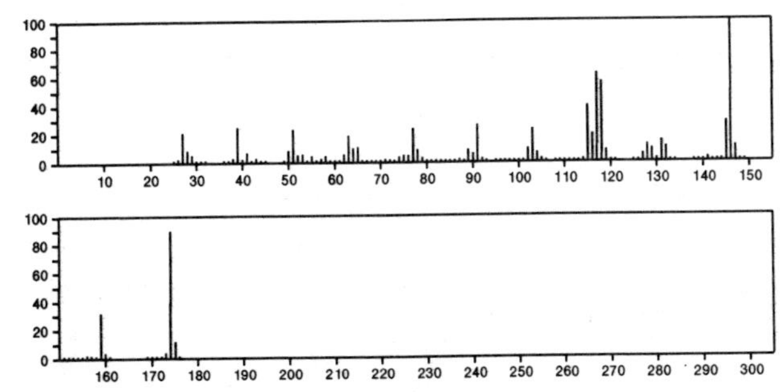

174 $C_{12}H_{14}O$ 51015-33-9
1(2*H*)-Naphthalenone, 8-ethyl-3,4-dihydro-

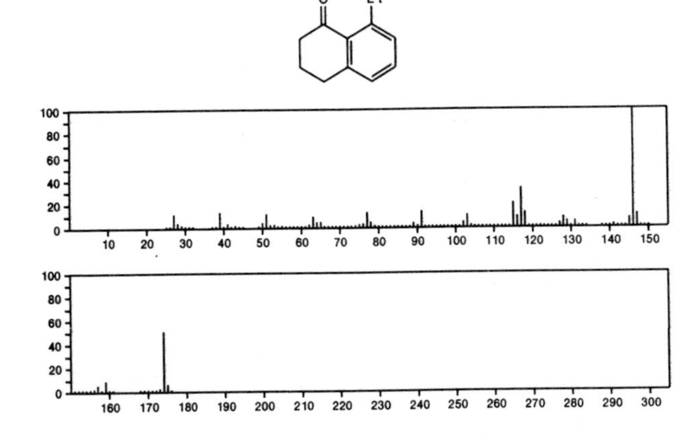

174 $C_{12}H_{14}O$ 54484-71-8
1*H*-Inden-1-one, 2,3-dihydro-3,3,6-trimethyl-

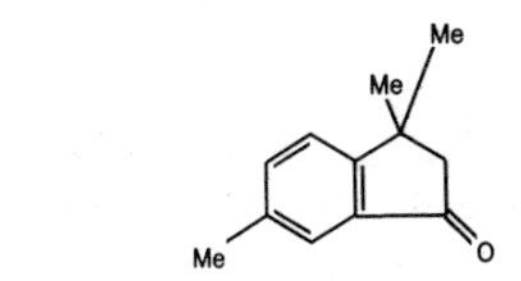

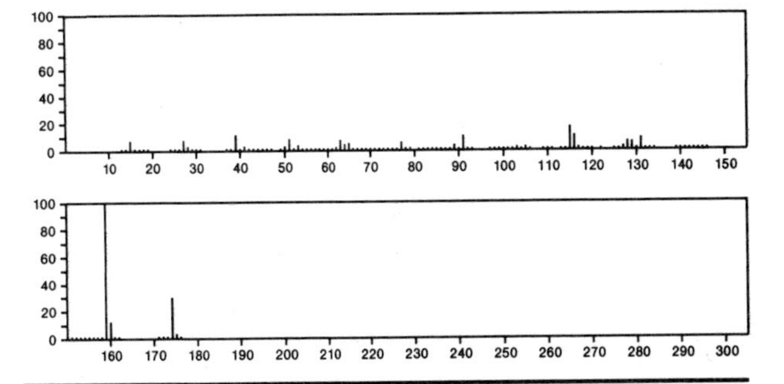

174 $C_{12}H_{14}O$ 55282-87-6
2-Hexenal, 6-phenyl-, (*E*)-

$Ph(CH_2)_3CH{=}CHCHO$

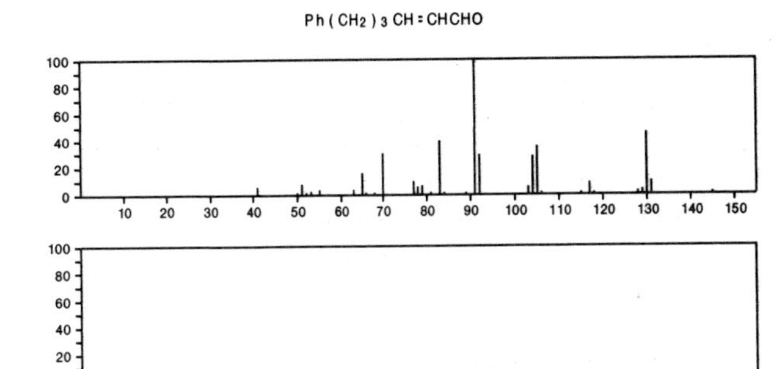

174 $C_{13}H_{18}$ 475-03-6
Naphthalene, 1,2,3,4-tetrahydro-1,1,6-trimethyl-

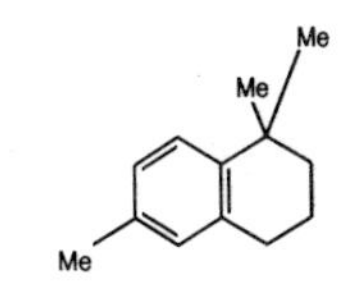

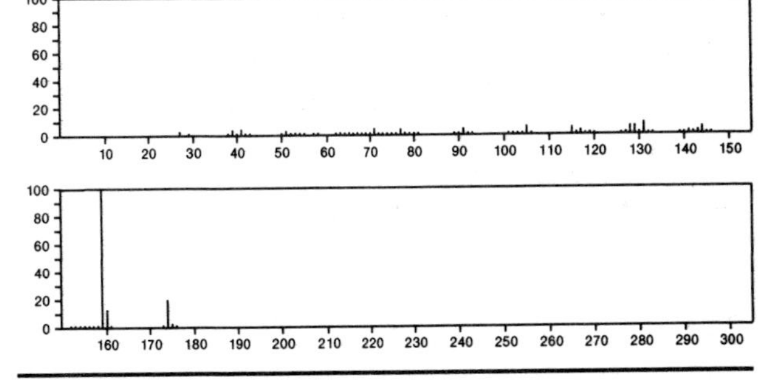

174 $C_{13}H_{18}$ 829-99-2
Benzene, 1-heptenyl-

$PhCH{=}CH(CH_2)_4Me$

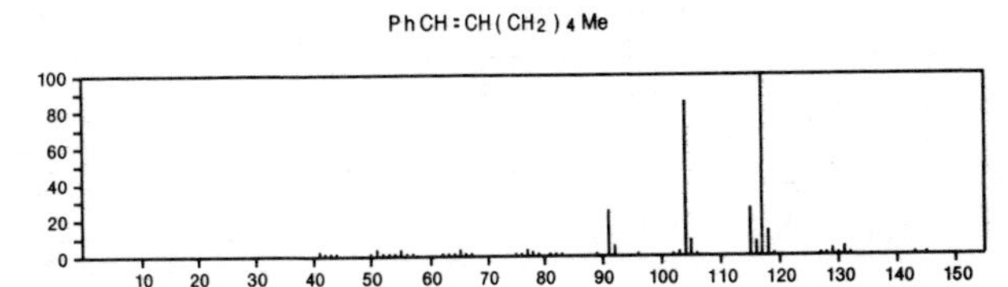

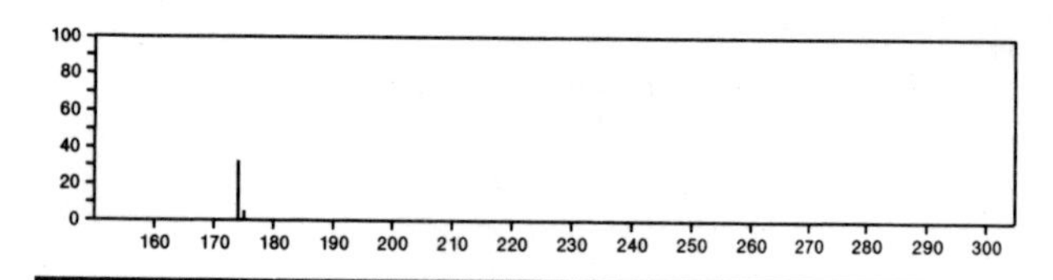

174 $C_{13}H_{18}$ 941–60–6
Indan, 1,1,4,6–tetramethyl–

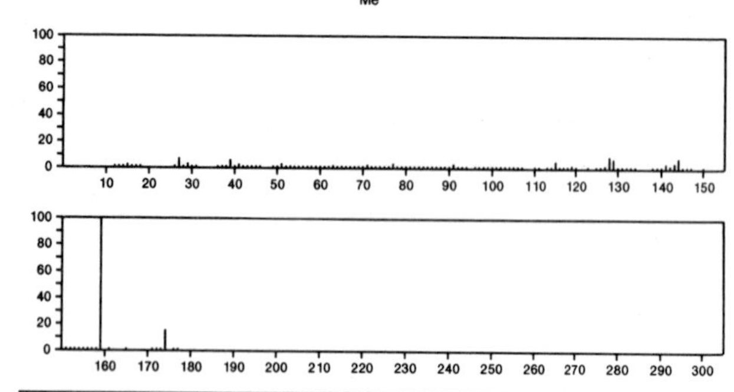

174 $C_{13}H_{18}$ 1078–04–2
1*H*–Indene, 2,3–dihydro–1,1,4,7–tetramethyl–

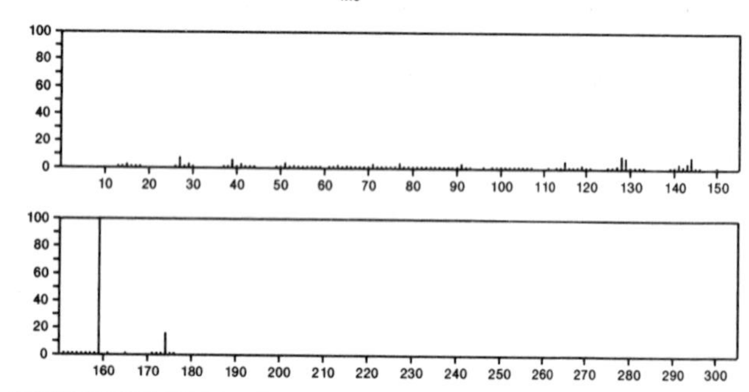

174 $C_{13}H_{18}$ 4410–75–7
Benzene, (cyclohexylmethyl)–

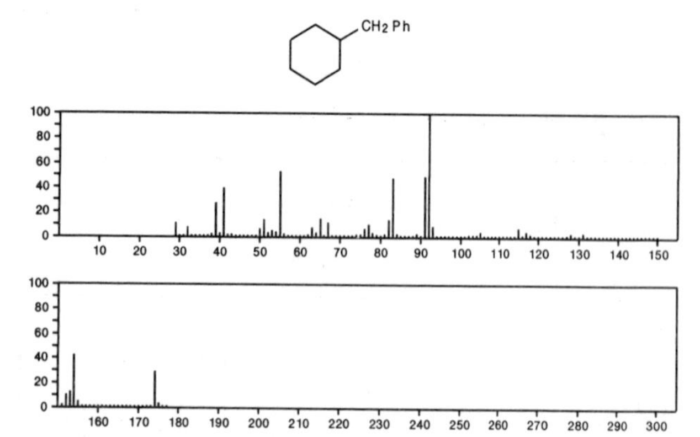

174 $C_{13}H_{18}$ 4575–46–6
Benzene, 1–cyclohexyl–3–methyl–

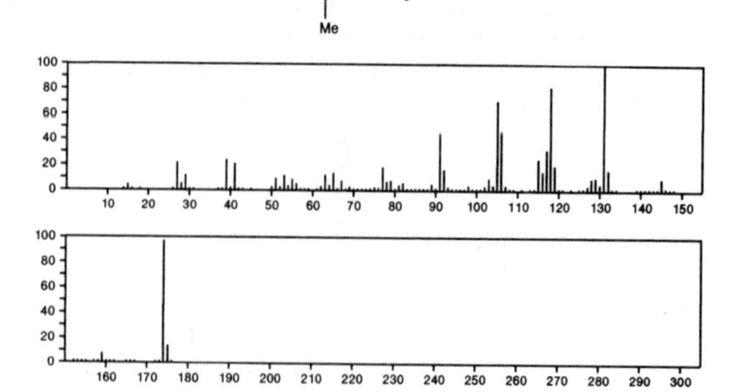

174 $C_{13}H_{18}$ 16204–57–2
1*H*–Indene, 2,3–dihydro–1,1,4,5–tetramethyl–

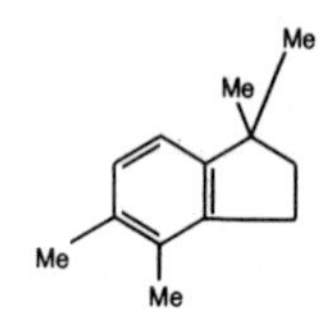

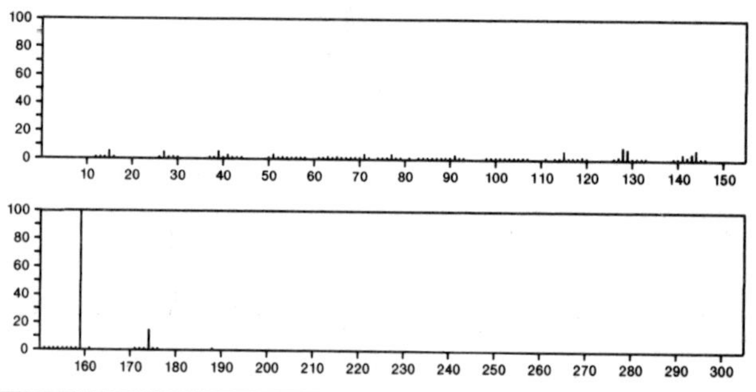

174 $C_{13}H_{18}$ 16204–58–3
Indan, 1,1,6,7–tetramethyl–

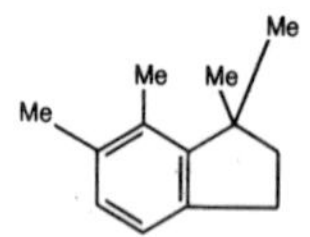

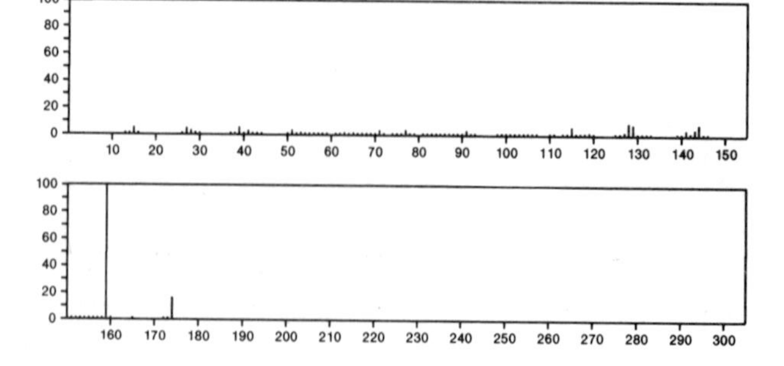

174 $C_{13}H_{18}$ 21693-51-6

Naphthalene, 1,2,3,4-tetrahydro-1,5,8-trimethyl-

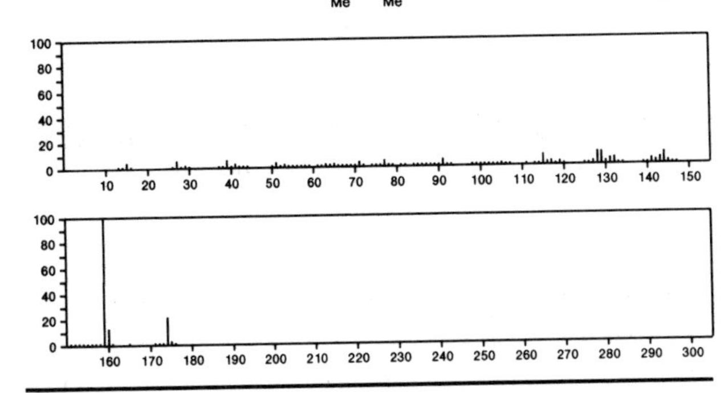

174 $C_{13}H_{18}$ 21693-55-0

Naphthalene, 1,2,3,4-tetrahydro-1,5,7-trimethyl-

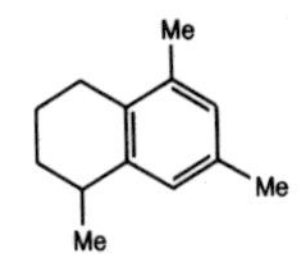

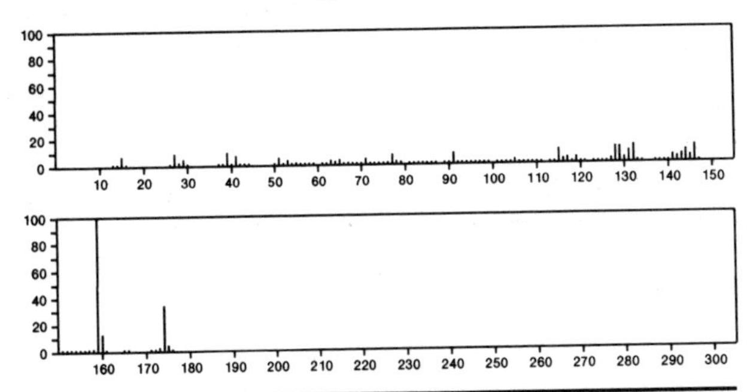

174 $C_{13}H_{18}$ 22824-32-4

Naphthalene, 1,2,3,4-tetrahydro-1,4,6-trimethyl-

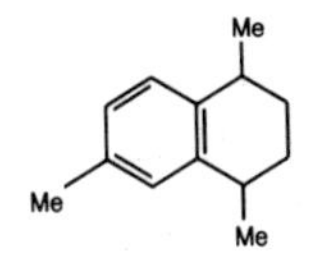

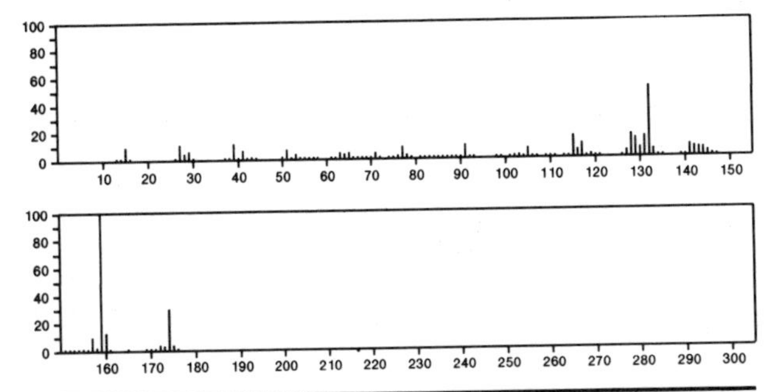

174 $C_{13}H_{18}$ 26447-63-2

Benzene, 2-heptenyl-

Me(CH2)3CH=CHCH2Ph

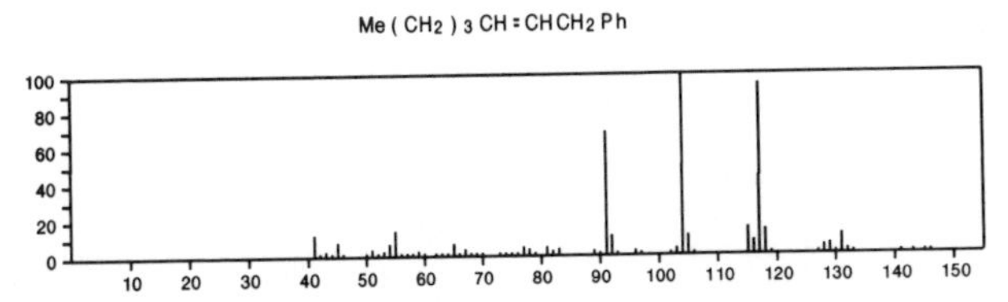

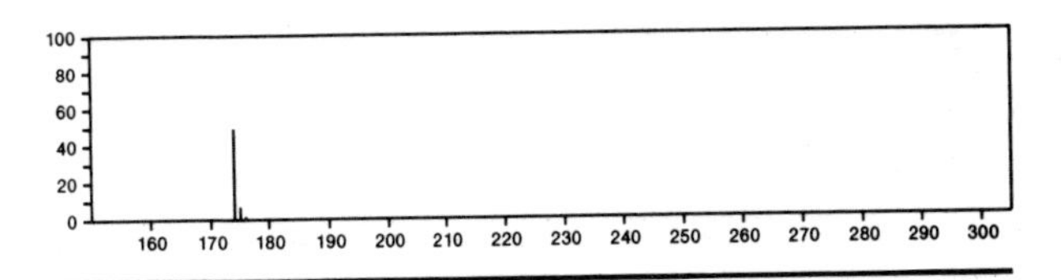

174 $C_{13}H_{18}$ 26447-64-3

Benzene, 3-heptenyl-

PrCH=CHCH2CH2Ph

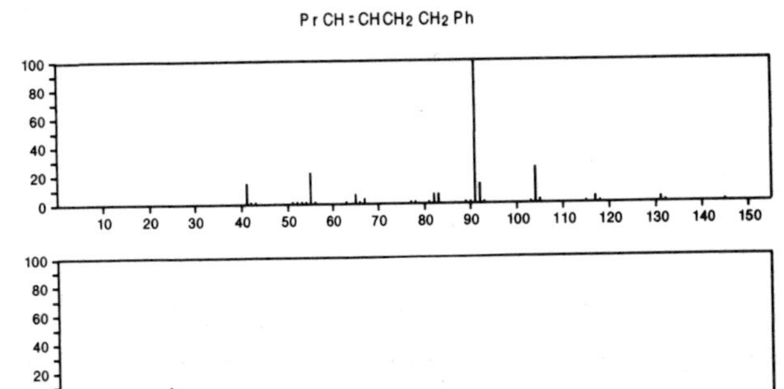

174 $C_{13}H_{18}$ 26447-65-4

3-Heptene, 7-phenyl-

Ph(CH2)3CH=CHEt

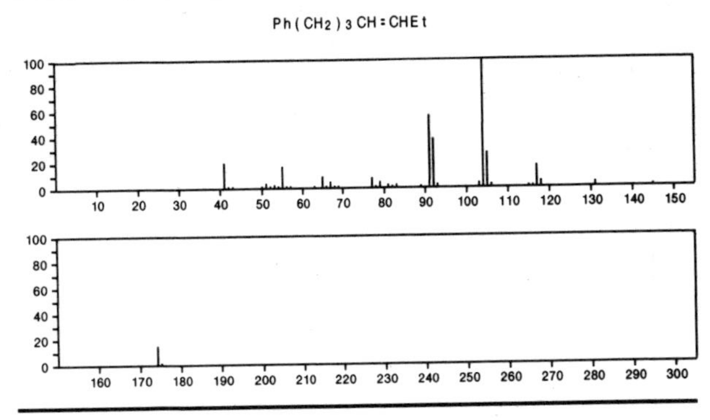

174 $C_{13}H_{18}$ 26447-66-5

Benzene, 5-heptenyl-

Ph(CH2)4CH=CHMe

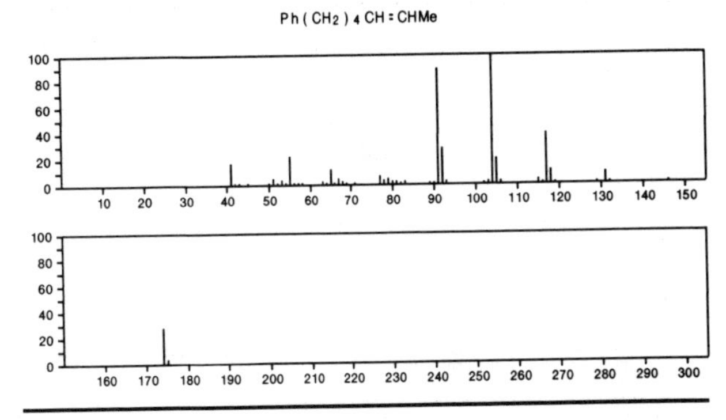

174 $C_{13}H_{18}$ 30316-17-7

Naphthalene, 1,2,3,4-tetrahydro-2,5,8-trimethyl-

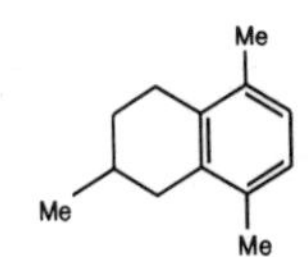

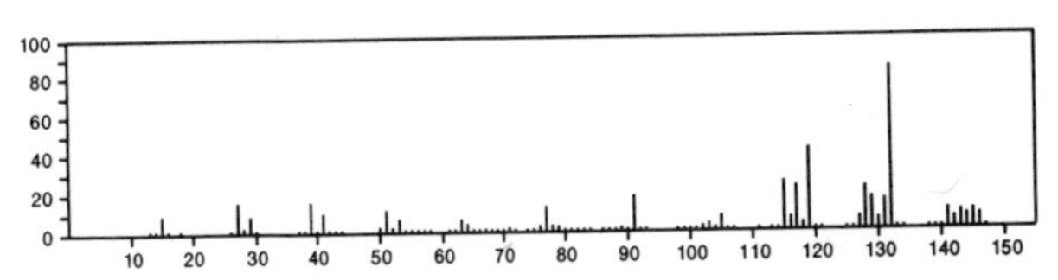

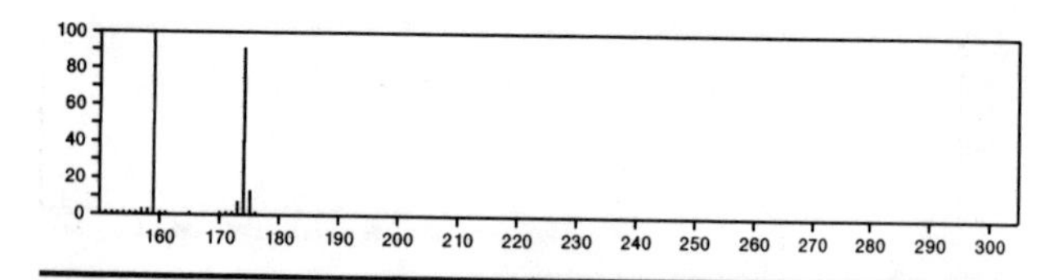

174 $C_{13}H_{18}$ 30316-36-0
Naphthalene, 1,2,3,4-tetrahydro-1,6,8-trimethyl-

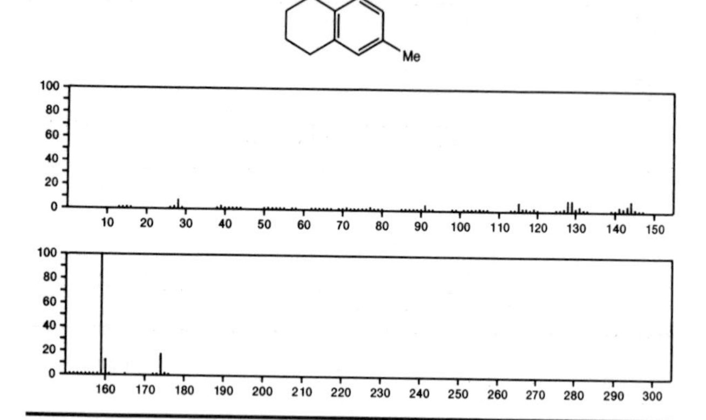

174 $C_{13}H_{18}$ 54725-18-7
Benzene, 2-heptenyl-, (*Z*)-

Me (CH2) 3 CH = CHCH2 Ph

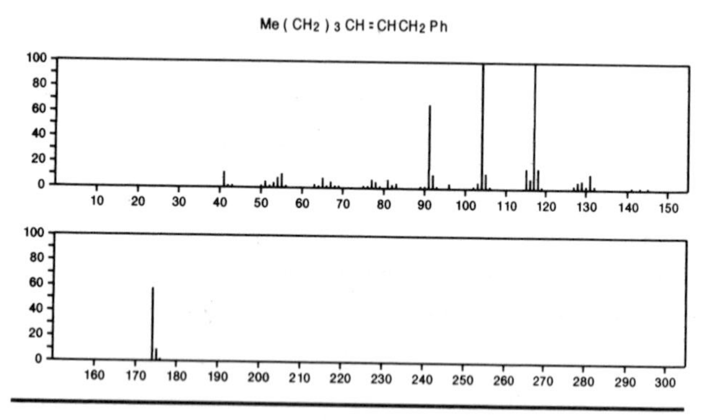

175 $C_5H_9N_3S_2$ 30062-47-6
1,3,4-Thiadiazol-2-amine, 5-[(1-methylethyl)thio]-

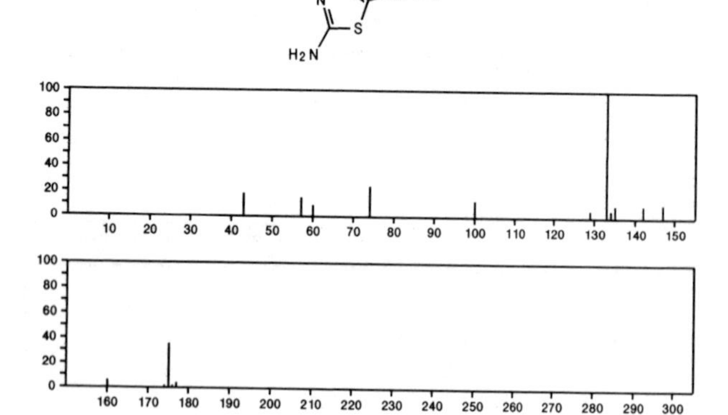

175 $C_5H_9N_3S_2$ 30062-49-8
1,3,4-Thiadiazol-2-amine, 5-(propylthio)-

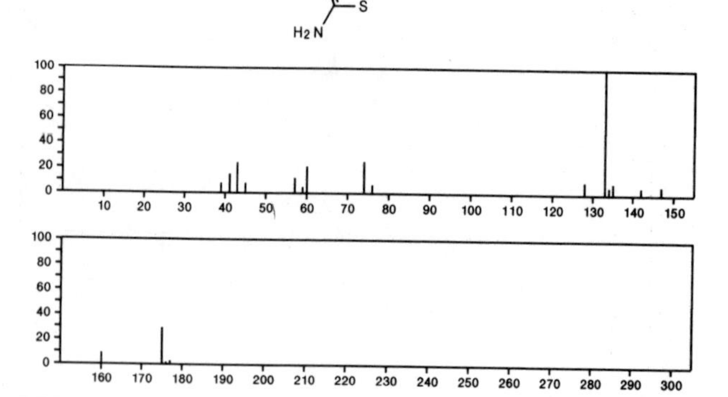

175 $C_6H_9NO_5$ 2545-40-6
DL-Aspartic acid, *N*-acetyl-

HO2 CCH2 CH (CO2 H) NHAc

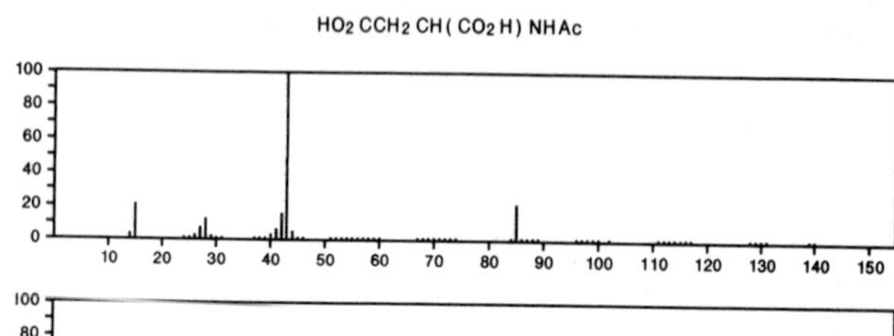

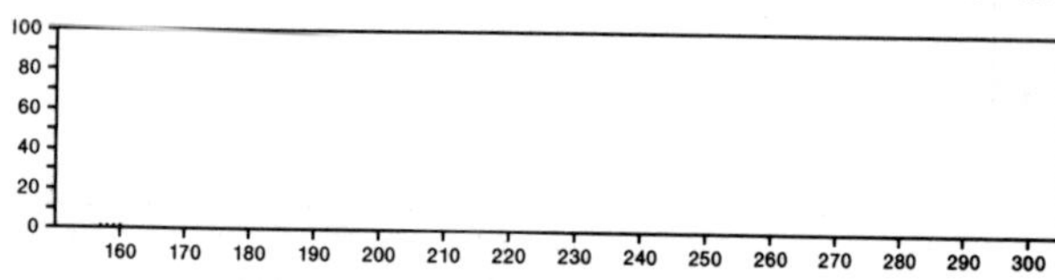

175 $C_6H_9NO_5$ 27160-23-2
Malonic acid, formamido-, dimethyl ester

C(O) OMe
Me OC(O) CHNHCH = O

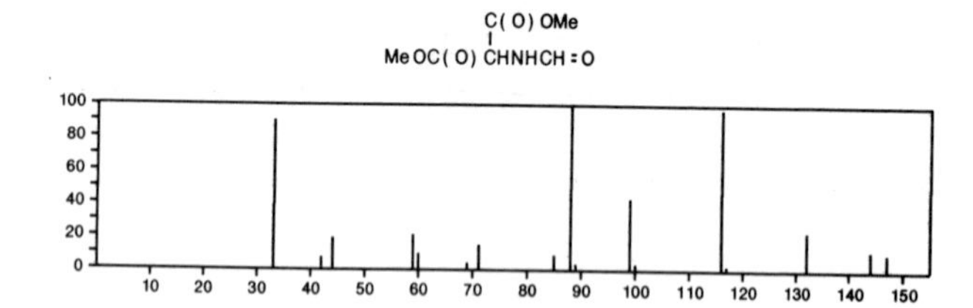

175 $C_6H_9NO_5$ 55590-76-6
Propanedioic acid, (methoxyimino)-, dimethyl ester

C(O) OMe
Me OC(O) C = NOMe

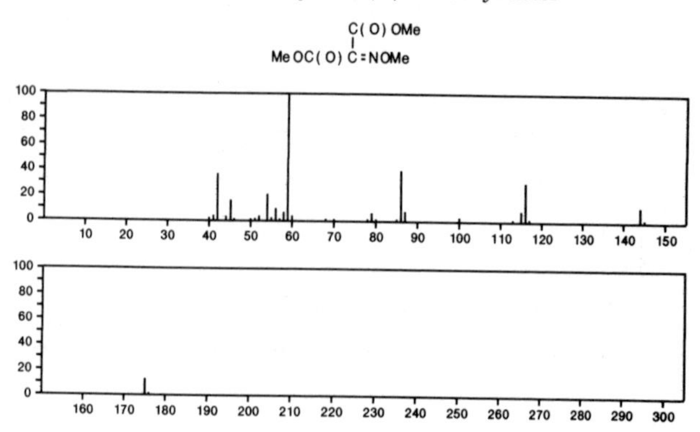

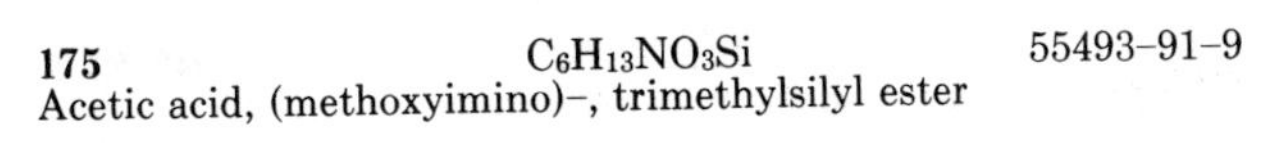

175 $C_6H_{13}NO_3Si$ 55493-91-9

Acetic acid, (methoxyimino)-, trimethylsilyl ester

Me3 Si OC(O) CH=NOMe

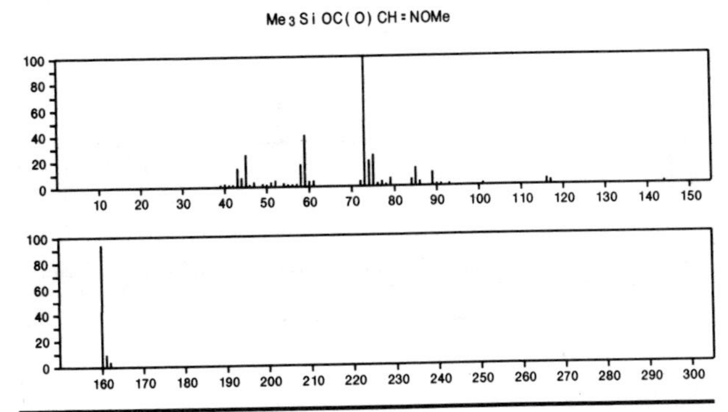

175 $C_6H_{13}NO_3Si$ 55836-37-8

Glycine, *N*-formyl-, trimethylsilyl ester

Me3 Si OC(O) CH2 NHCH=O

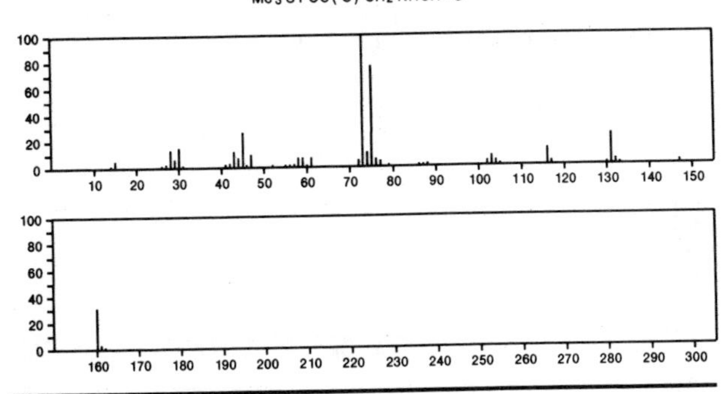

175 $C_7H_{10}ClNO_2$ 54484-57-0

2,4-Pentanedione, 3-(1-aminoethylidene)-1-chloro-

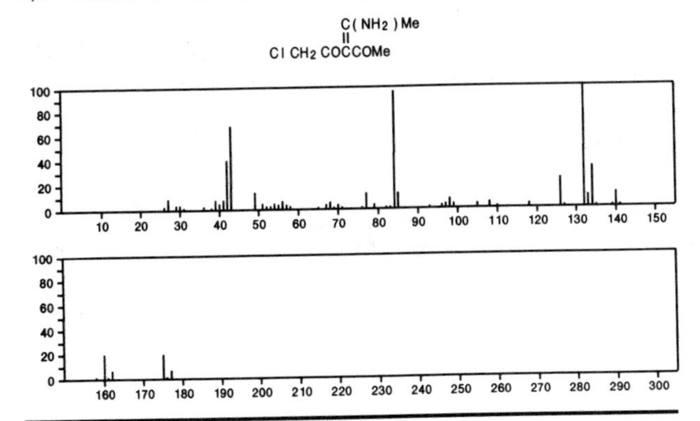

175 $C_7H_{21}NSi_2$ 920-68-3

Silanamine, *N*,1,1,1-tetramethyl-*N*-(trimethylsilyl)-

Me3 Si NMe Si Me3

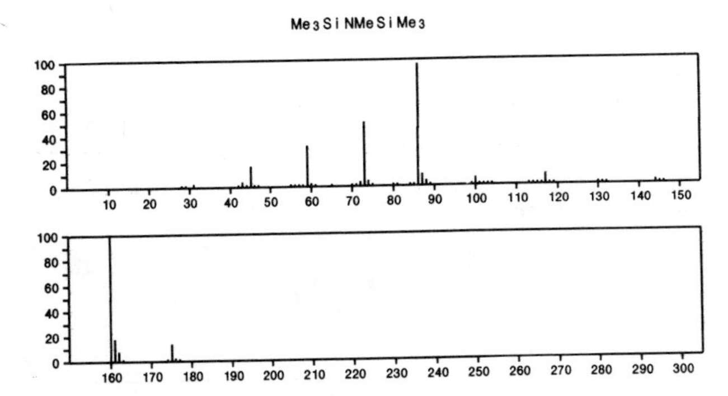

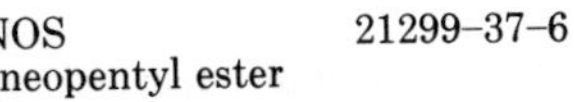

175 $C_8H_{17}NOS$ 21299-37-6

Carbamic acid, dimethylthio-, *O*-neopentyl ester

Me2 NC(S) OCH2 CMe3

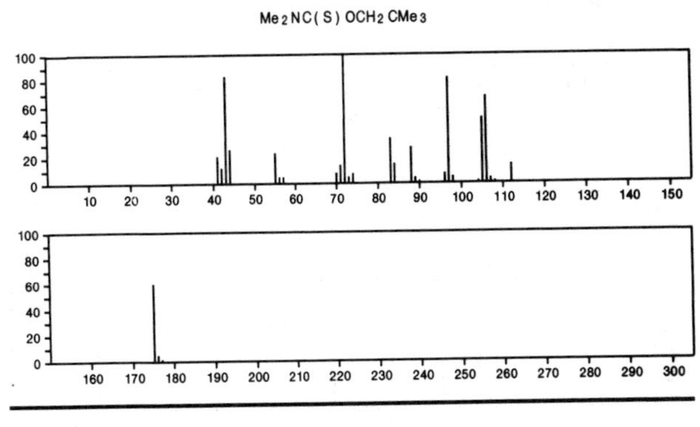

175 $C_8H_{17}NO_3$ 629-39-0

Nitric acid, octyl ester

Me(CH2)7 ONO2

175 $C_9H_9N_3O$ 22378-51-4

Pyrido[3,4-*d*]pyrimidin-4(3*H*)-one, 6,8-dimethyl-

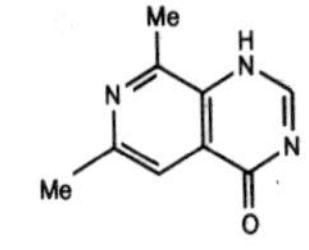

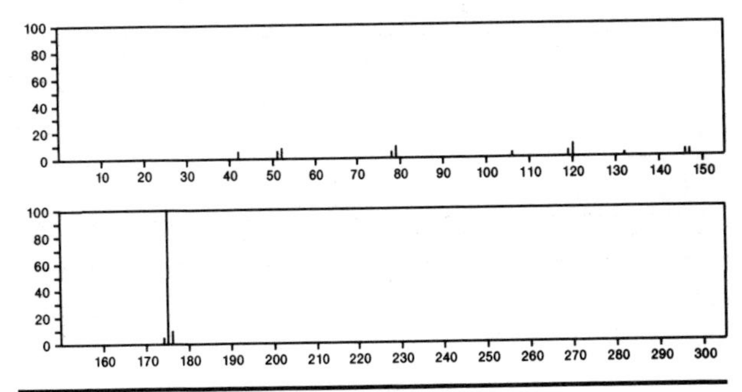

175 $C_9H_9N_3O$ 22863-24-7

3*H*-1,2,4-Triazol-3-one, 2,4-dihydro-5-methyl-2-phenyl-

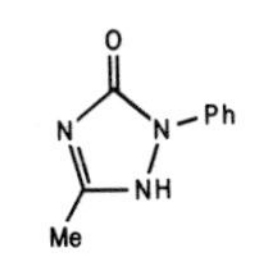

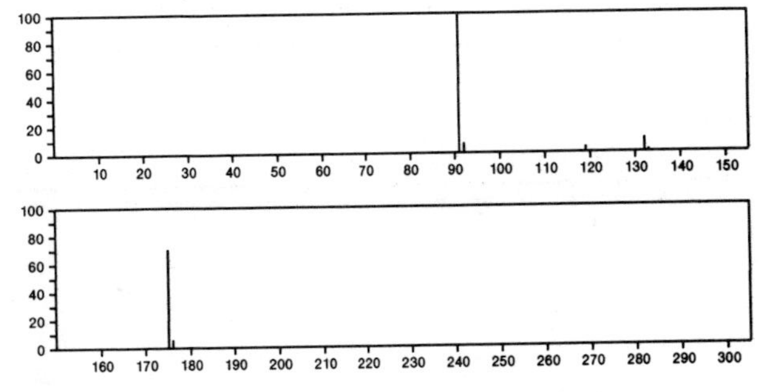

175 $C_9H_9N_3O$ 28718–27–6
1*H*–1,2,4–Triazolium, 3–hydroxy–4–methyl–1–phenyl–, hydroxide, inner salt

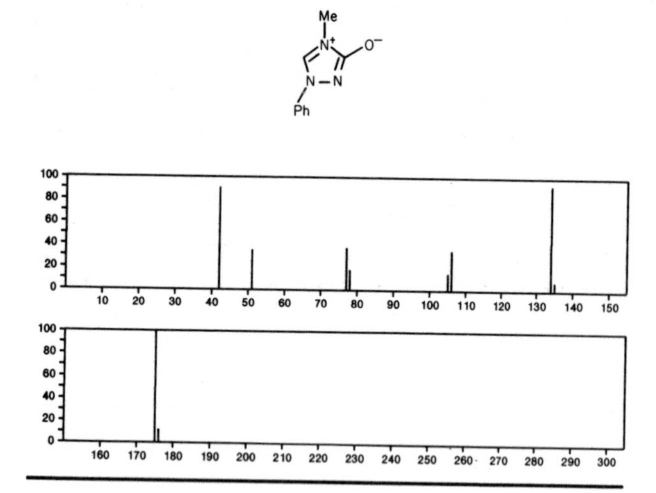

175 $C_9H_9N_3O$ 41536–79–2
3(2*H*)–Isoquinolinone, 1–amino–, oxime

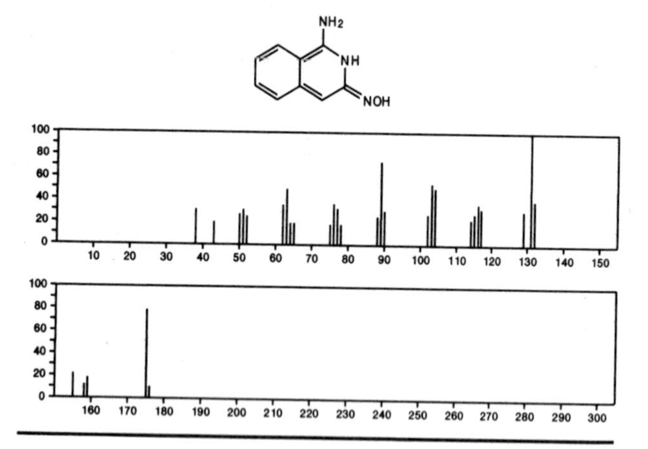

175 $C_{10}H_9NO_2$ 87–51–4
1*H*–Indole–3–acetic acid

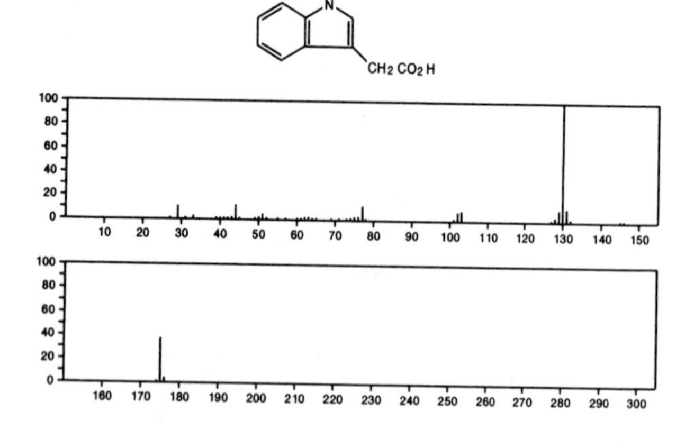

175 $C_{10}H_9NO_2$ 942–24–5
1*H*–Indole–3–carboxylic acid, methyl ester

175 $C_{10}H_9NO_2$ 1892–21–3
1*H*–Indole–3–acetaldehyde, 5–hydroxy–

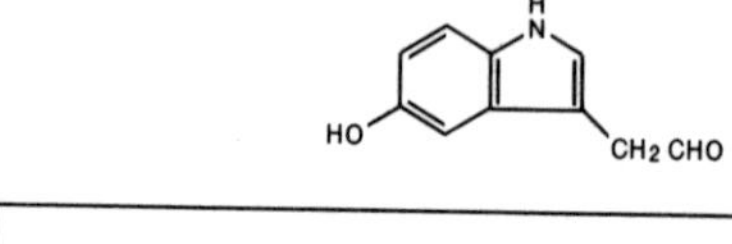

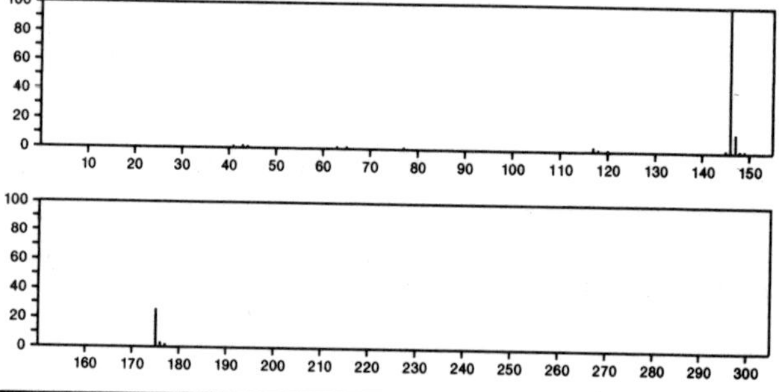

175 $C_{10}H_9NO_2$ 5022–29–7
1*H*–Isoindole–1,3(2*H*)–dione, *N*–ethyl–

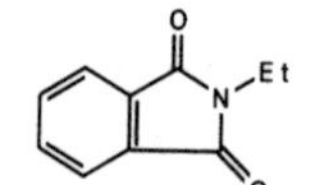

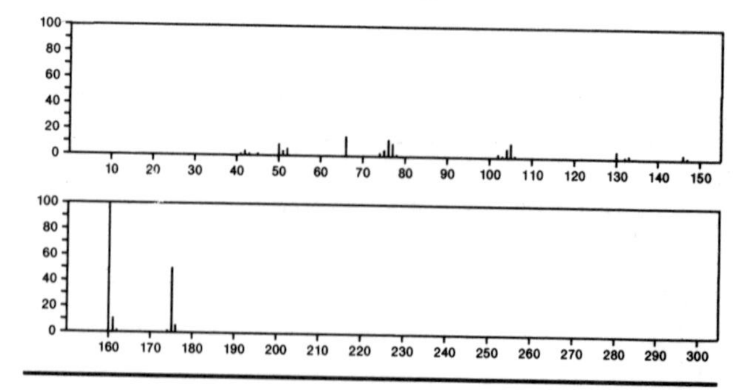

175 $C_{10}H_9NO_2$ 6563–13–9
Quinoline, 6–methoxy–, 1–oxide

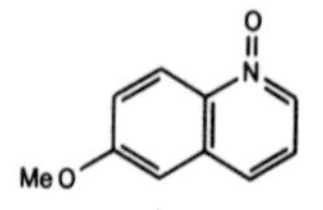

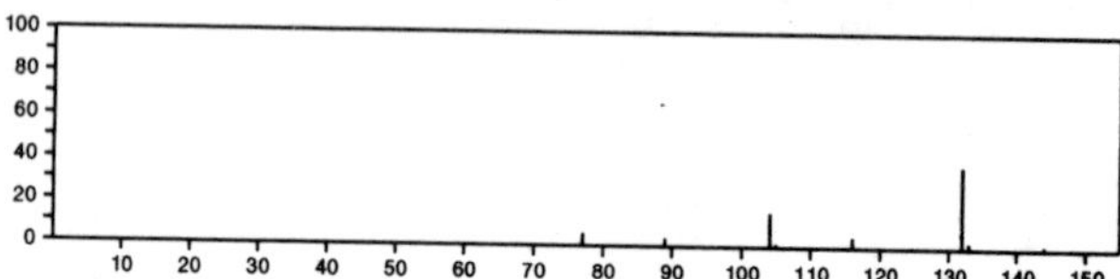

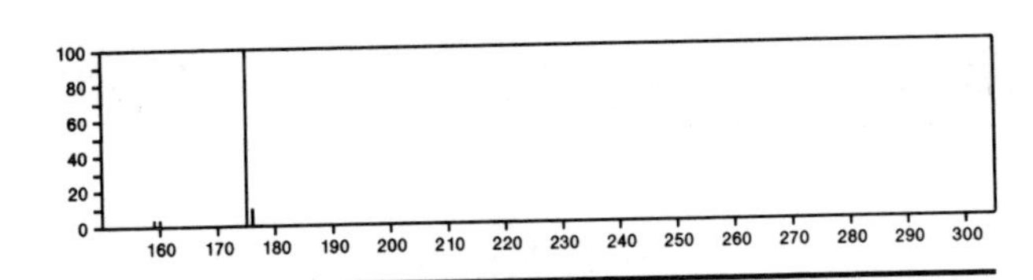

175 $C_{10}H_9NO_2$ 16511–38–9

1*H*–1–Benzazepine–2,5–dione, 3,4–dihydro–

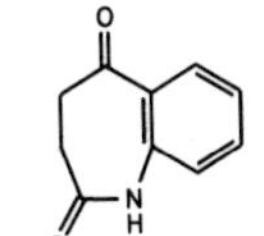

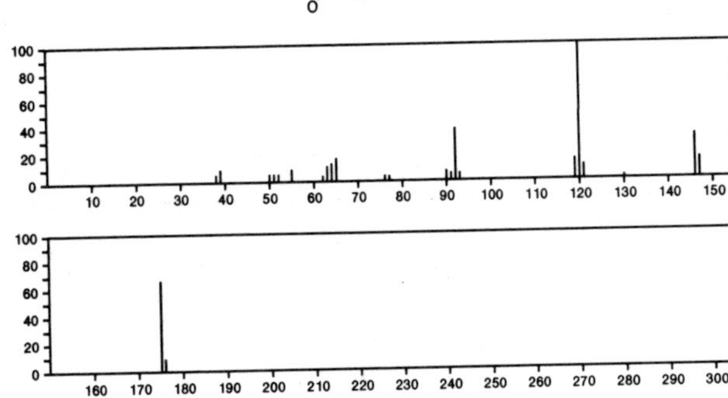

175 $C_{10}H_9NO_2$ 16959–62–9

2–Indolizinecarboxylic acid, methyl ester

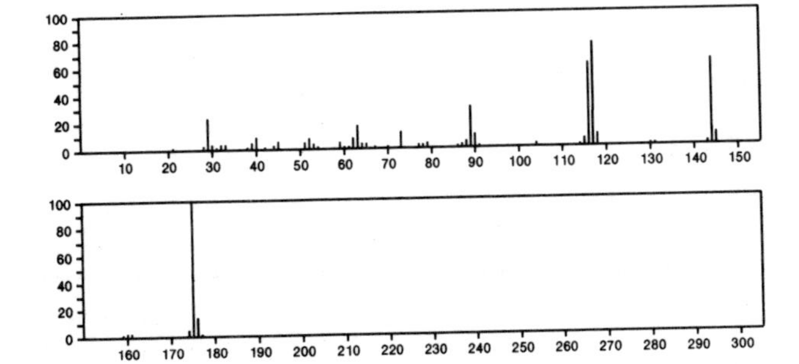

175 $C_{10}H_9NO_2$ 17018–82–5

2–Quinolinemethanol, 8–hydroxy–

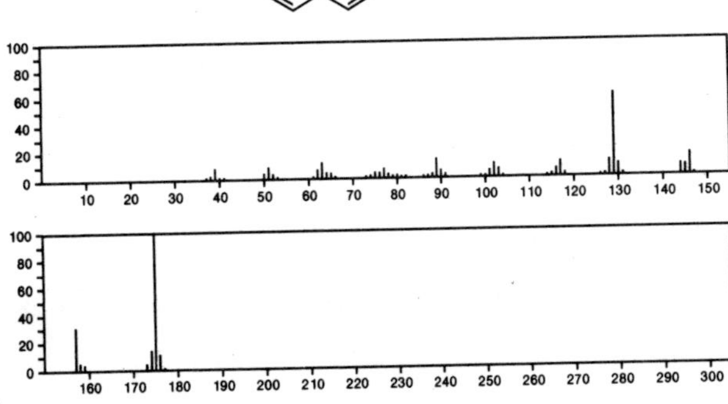

175 $C_{10}H_9NO_2$ 21201–47–8

Carbostyril, 1–hydroxy–4–methyl–

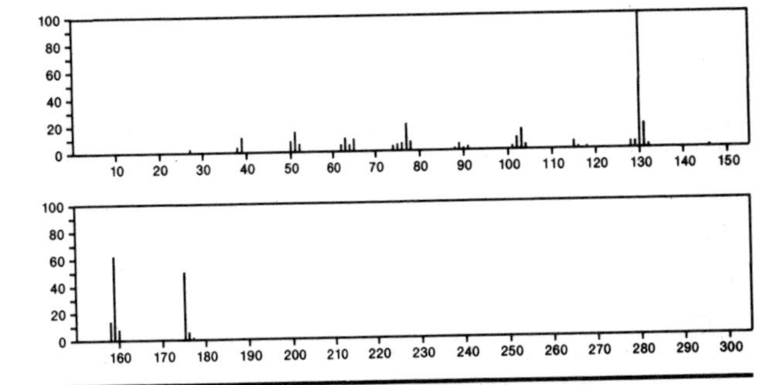

175 $C_{10}H_9NO_2$ 21905–78–2

2*H*–Indol–2–one, 1–acetyl–1,3–dihydro–

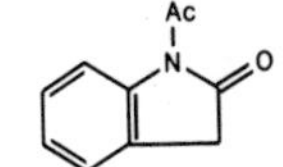

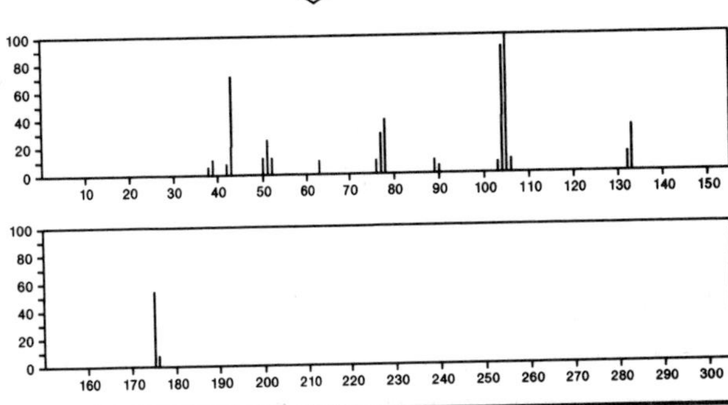

175 $C_{10}H_9NO_2$ 49656–77–1

5(2*H*)–Oxazolone, 4–(phenylmethyl)–

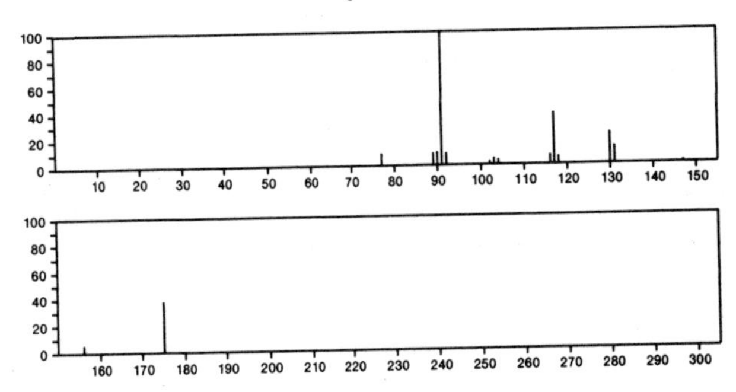

175 $C_{10}H_9NO_2$ 55759-83-6
4(1*H*)-Quinolinone, 3-hydroxy-1-methyl-

175 $C_{10}H_9NS$ 1732-45-2
Isothiazole, 3-methyl-5-phenyl-

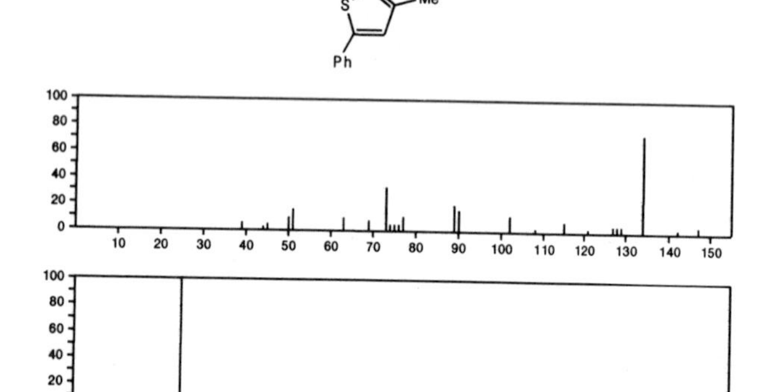

175 $C_{10}H_{13}N_3$ 49629-06-3
6*H*-[1,2,4]Triazolo[1,5-*a*]indole, 4a,5,7,8,8a,9-hexahydro-9-methylene-

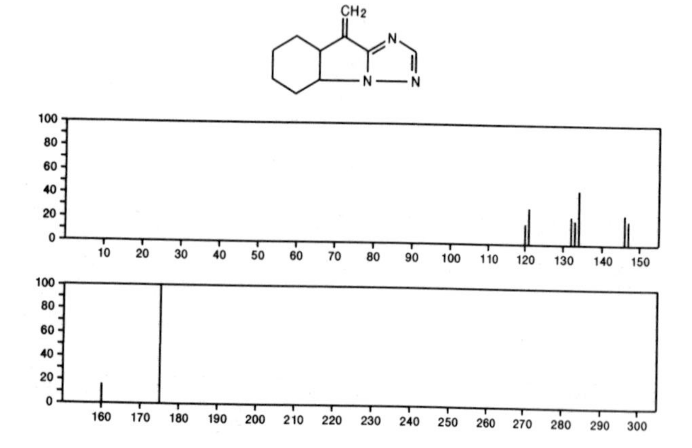

175 $C_{10}H_{13}N_3$ 49629-08-5
5*H*-Pyrrolo[1,2-*b*][1,2,4]triazole, 6,7-dihydro-6-methyl-7-methylene-6-(1-methylethenyl)-

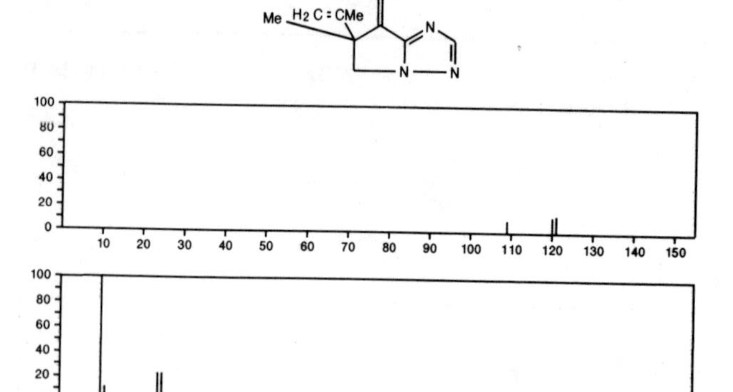

175 $C_{10}H_{13}N_3$ 49629-10-9
5*H*-[1,2,4]Triazolo[1,5-*a*]azepine, 8,9-dihydro-6,7-dimethyl-9-methylene-

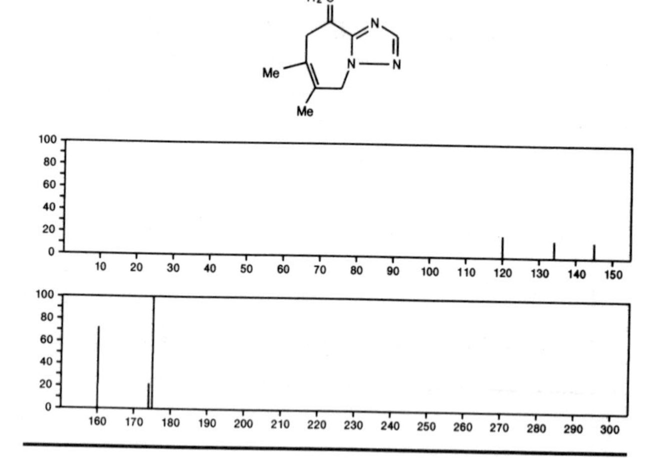

175 $C_{10}H_{13}N_3$ 50873-04-6
6*H*-1,2,4-Triazolo[4,3-*a*]indole, 4a,5,7,8,8a,9-hexahydro-9-methylene-

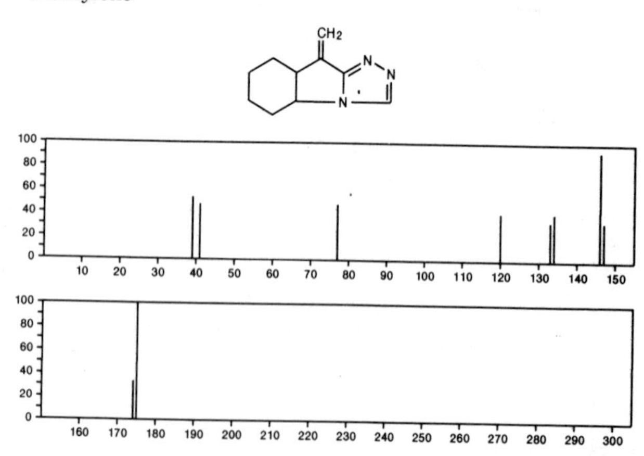

175 $C_{10}H_{14}BNO$ 26535-23-9
1,3,2-Oxazaborolidine, 2,4-dimethyl-5-phenyl-

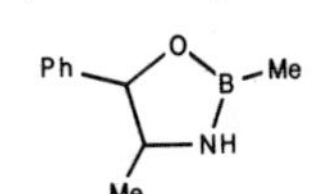

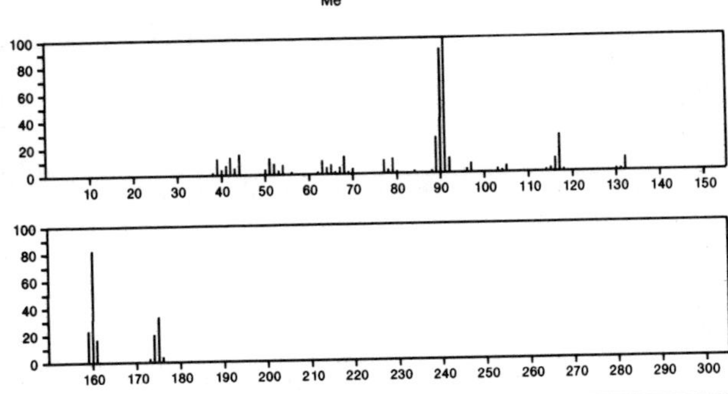

175 $C_{10}H_{14}BNO$ 31748-13-7
1,3,2-Benzoxazaborole, 2-butyl-2,3-dihydro-

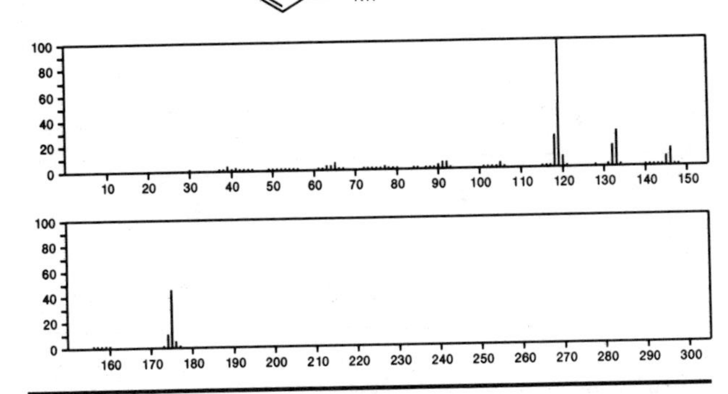

175 $C_{11}H_{13}NO$ 3063-79-4
2-Pyrrolidinone, 1-(4-methylphenyl)-

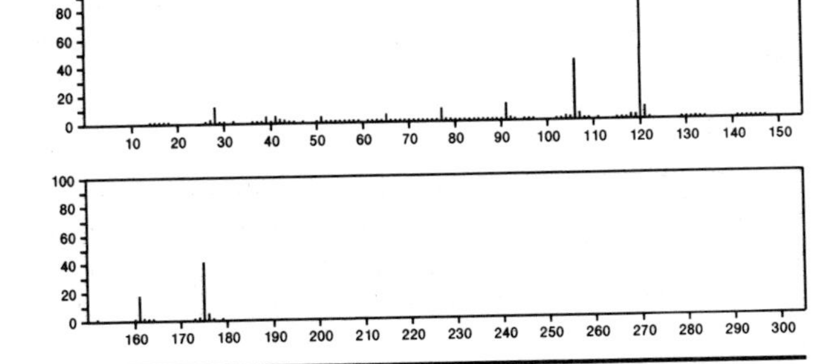

175 $C_{11}H_{13}NO$ 14091-93-1
2-Buten-1-one, 3-(methylamino)-1-phenyl-

MeC(NHMe) :CHCOPh

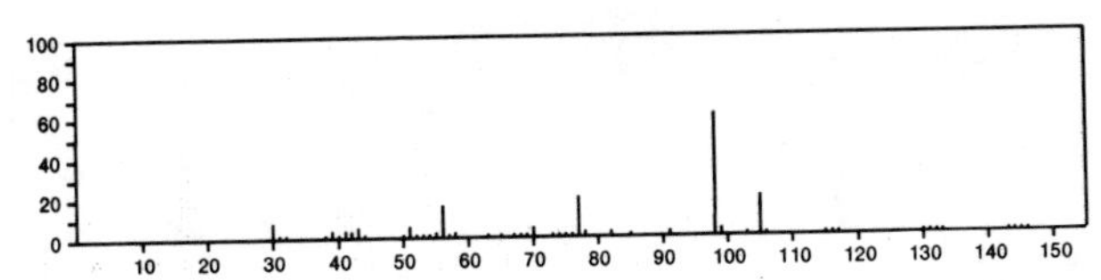

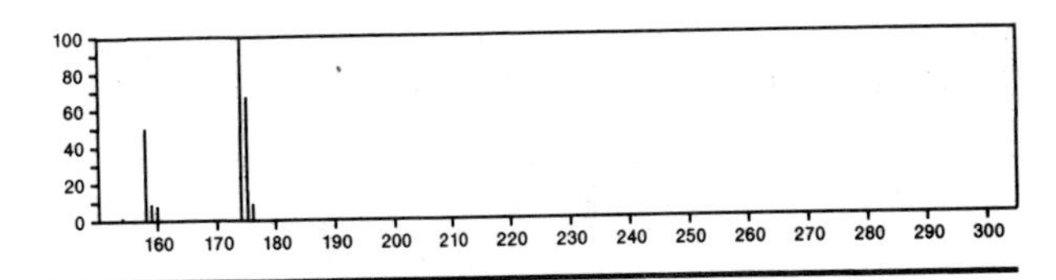

175 $C_{11}H_{13}NO$ 20205-45-2
7-Azabicyclo[4.2.2]deca-2,4,7,9-tetraene, 8-methoxy-10-methyl-

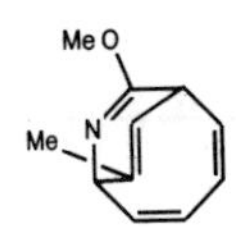

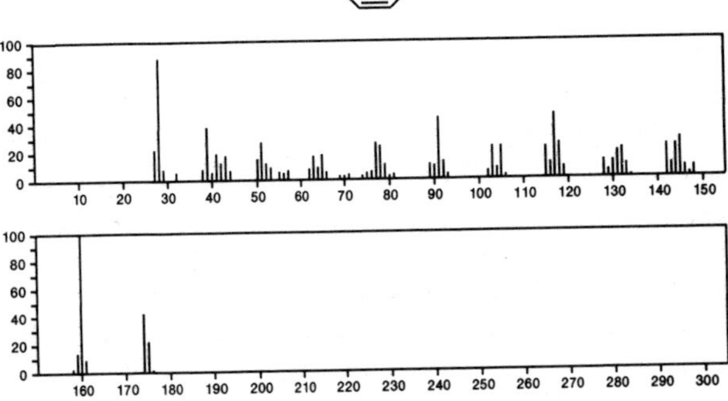

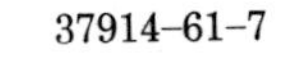

175 $C_{11}H_{13}NO$ 37914-61-7
3*H*-Indole, 3-methoxy-2,3-dimethyl-

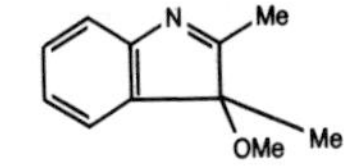

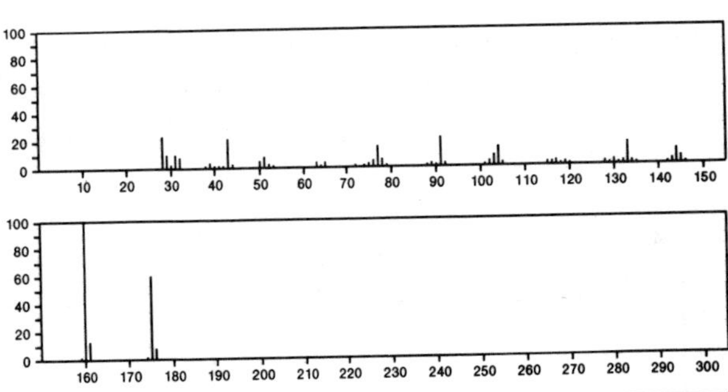

175 $C_{11}H_{13}NO$ 54518-02-4
2*H*-Furo[2,3-*b*]indole, 3,3a,8,8a-tetrahydro-3a-methyl-

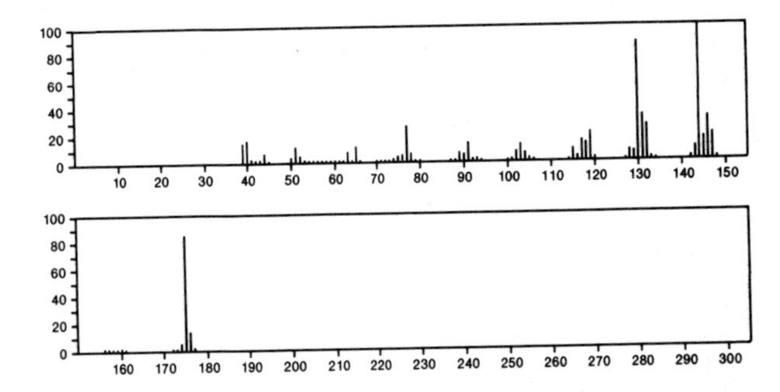

175 $C_{11}H_{13}NO$ 56298-85-2
2*H*-Furo[2,3-*b*]indole, 3,3a,8,8a-tetrahydro-3-methyl-

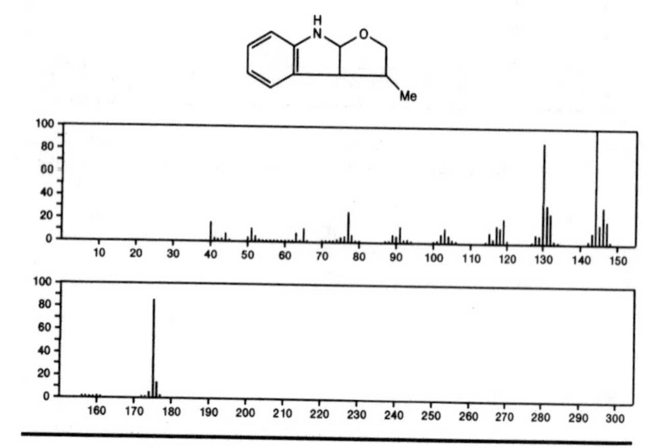

175 $C_{11}H_{18}BN$ 55976-16-4
Boranamine, *N*,*N*-dimethyl-1-phenyl-1-propyl-

$Me_2NBPrPh$

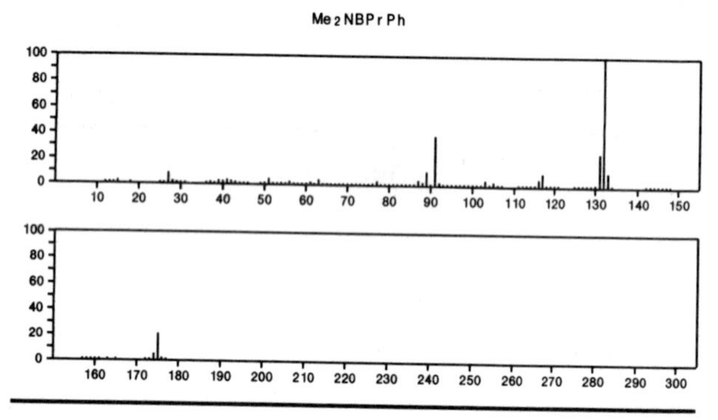

175 $C_{12}H_{17}N$ 4806-81-9
Benzenamine, 2-cyclohexyl-

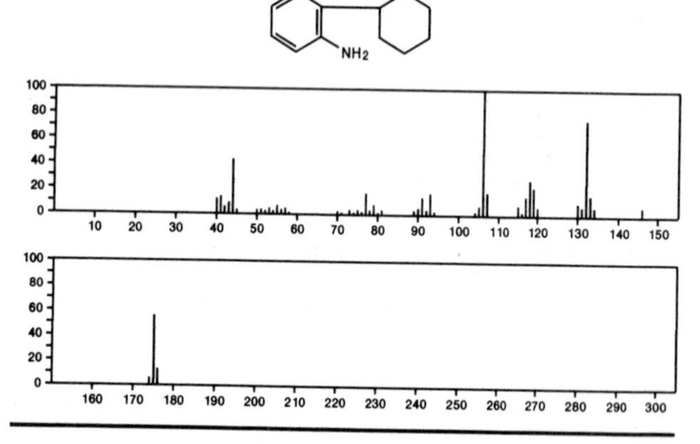

175 $C_{12}H_{17}N$ 6373-50-8
Benzenamine, 4-cyclohexyl-

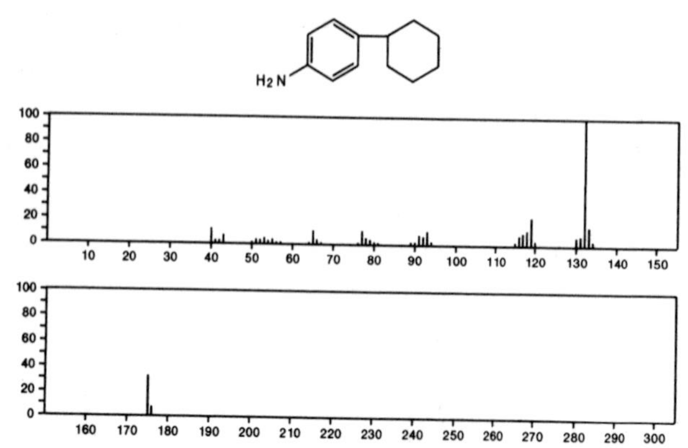

175 $C_{12}H_{17}N$ 18781-59-4
Indoline, 3-ethyl-2,3-dimethyl-

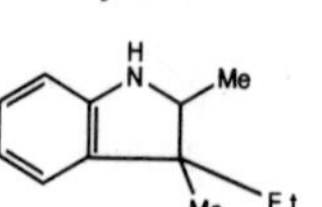

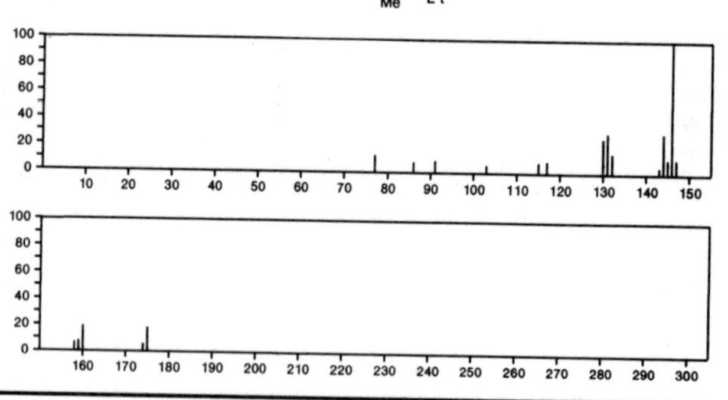

175 $C_{12}H_{17}N$ 22710-00-5
1-Pentanamine, *N*-(phenylmethylene)-

$PhCH{:}N(CH_2)_4Me$

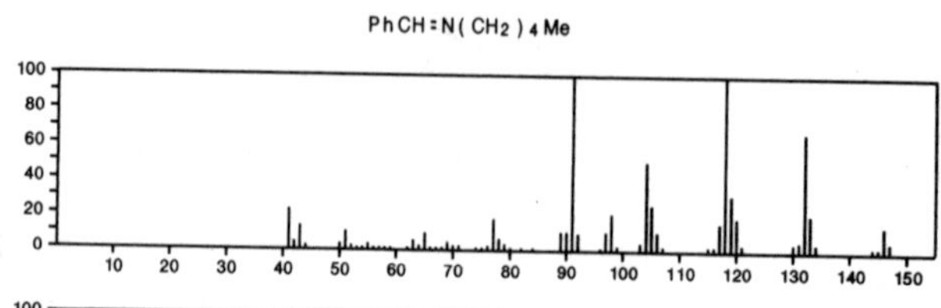

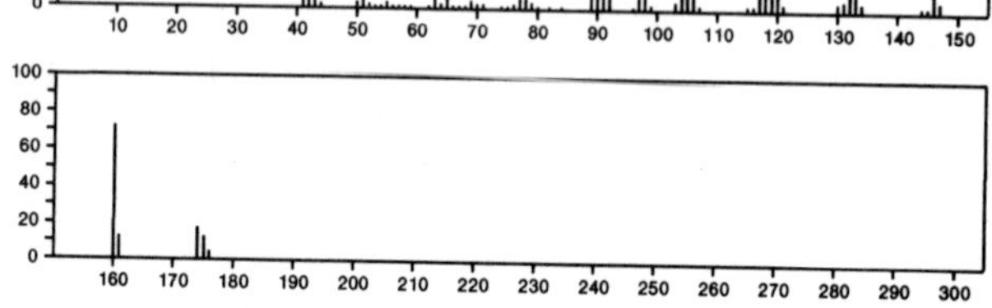

175 $C_{12}H_{17}N$ 41122-65-0
2-Pentanamine, *N*-(phenylmethylene)-

$PrCHMeN{:}CHPh$

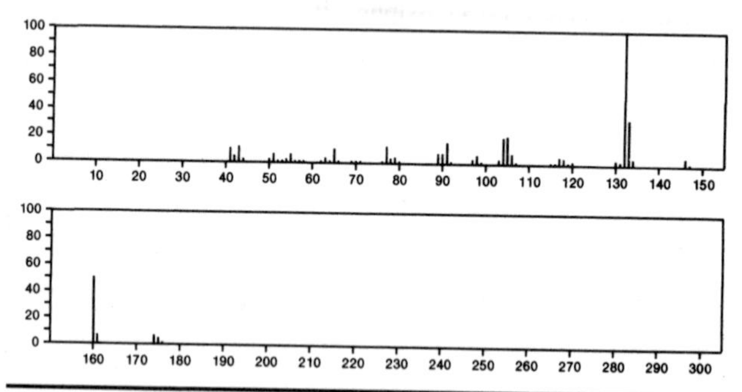

175 $C_{12}H_{17}N$ 54484-56-9
Naphthalenamine, *N*-ethyl-1,2,3,4-tetrahydro-

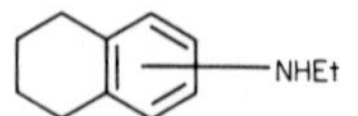

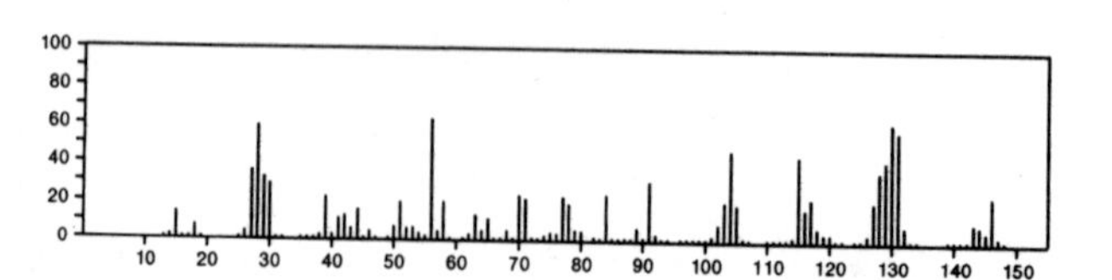

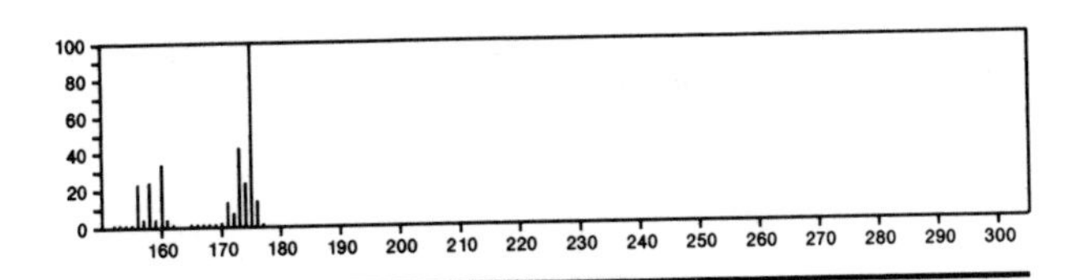

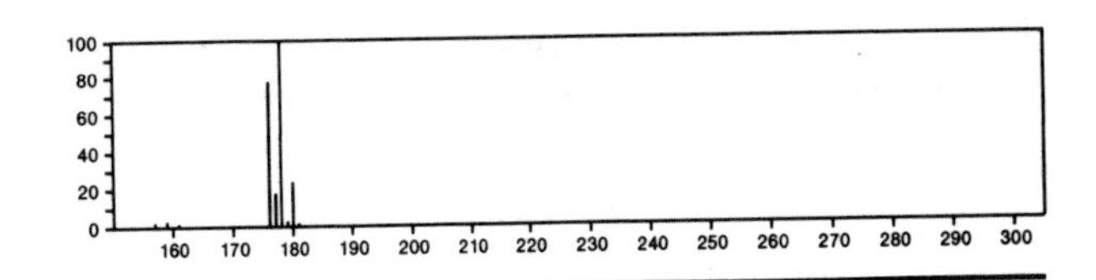

175 $C_{12}H_{17}N$ 54484-65-0
1*H*-Indole, 2-ethyl-2,3-dihydro-3,3-dimethyl-

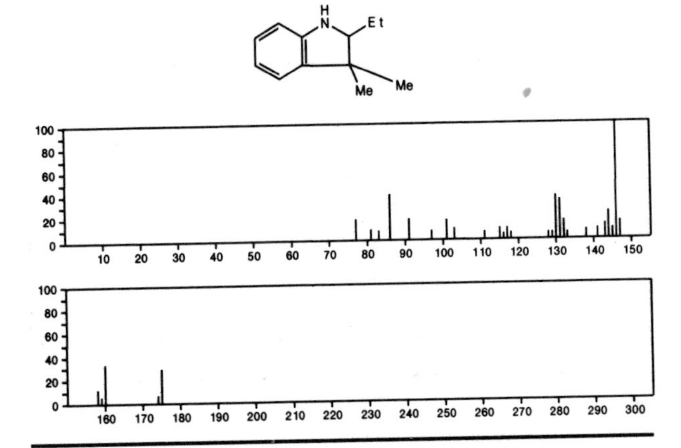

175 $C_{12}H_{17}N$ 55955-58-3
1*H*-Indole, 2,3-dihydro-1-(1-methylpropyl)-

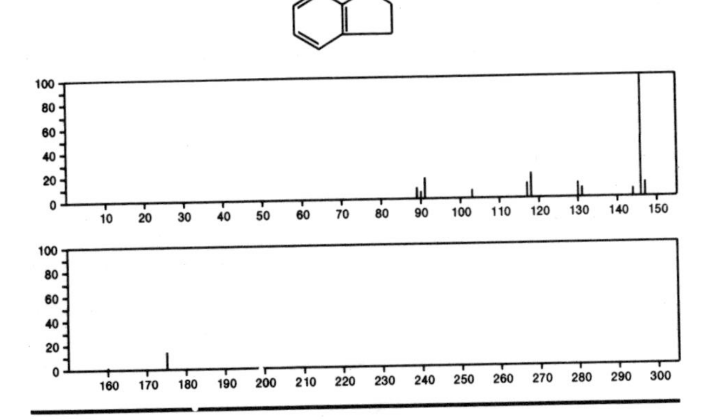

176 C_2BrClF_2 758-24-7
Ethene, 1-bromo-1-chloro-2,2-difluoro-

$F_2C=CBrCl$

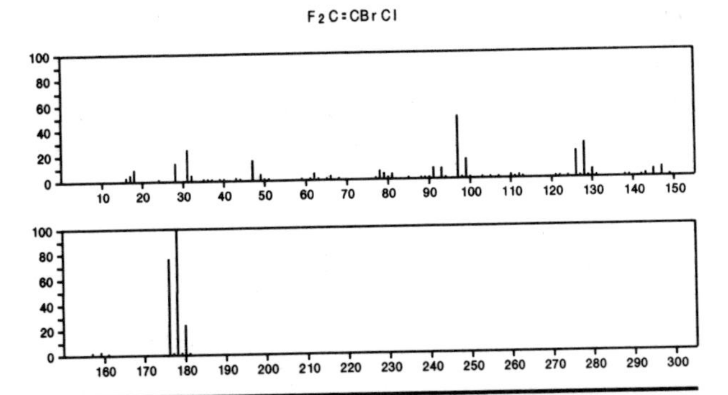

176 C_2BrClF_2 2106-93-6
Ethylene, 1-bromo-2-chloro-1,2-difluoro-

$ClCF=CBrF$

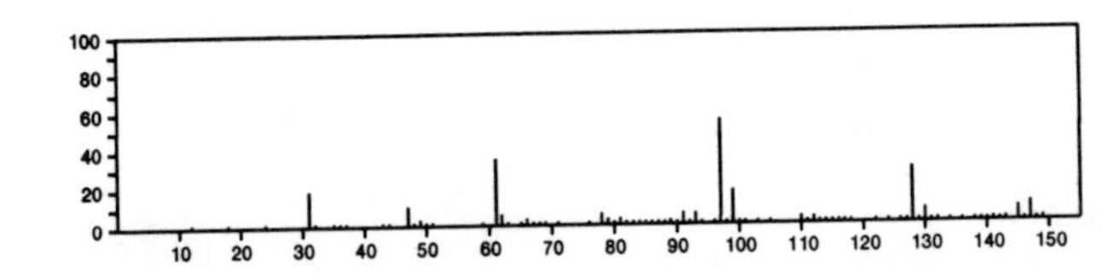

176 $C_2H_3BrCl_2$ 683-53-4
Ethane, 2-bromo-1,1-dichloro-

$BrCH_2CHCl_2$

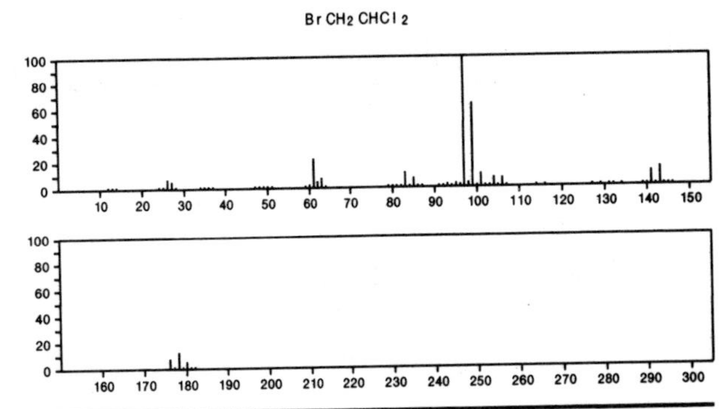

176 $C_3H_3Cl_3O_2$ 598-99-2
Acetic acid, trichloro-, methyl ester

$MeOC(O)CCl_3$

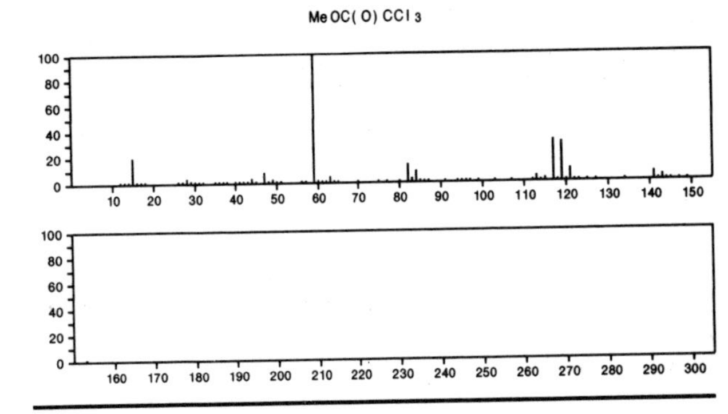

176 $C_3H_3Cl_3O_2$ 3278-46-4
Propanoic acid, 2,2,3-trichloro-

$ClCH_2CCl_2CO_2H$

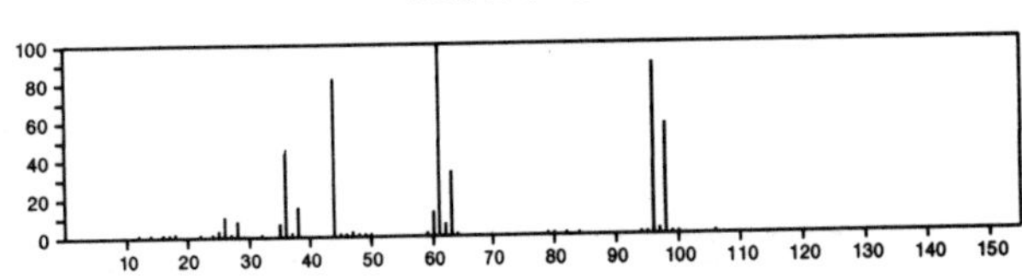

176 $C_4H_8N_4S_2$ 38362-24-2
1,2,4-Thiadiazole, 5-(1-methylhydrazino)-3-(methylthio)-

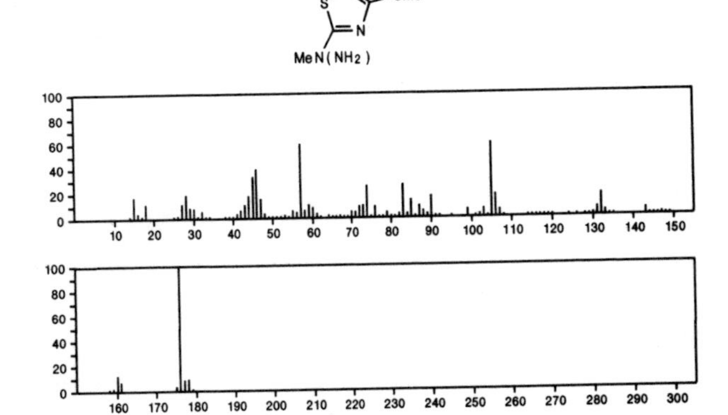

176 $C_5H_5BrO_2$ 14203–24–8
1,3–Cyclopentanedione, 2–bromo–

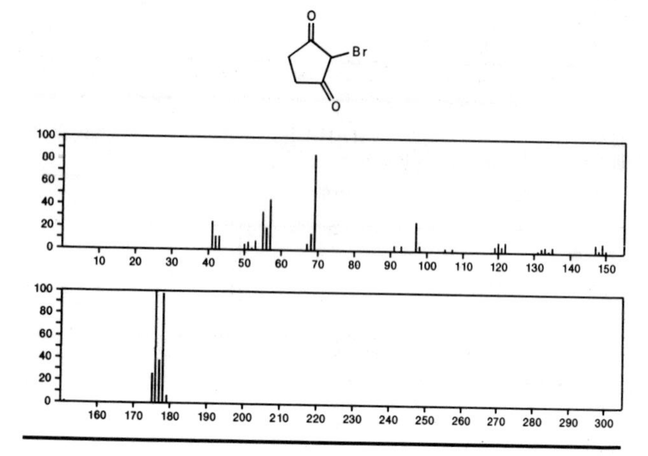

176 $C_6H_5ClO_2S$ 98–09–9
Benzenesulfonyl chloride

Cl SO2 Ph

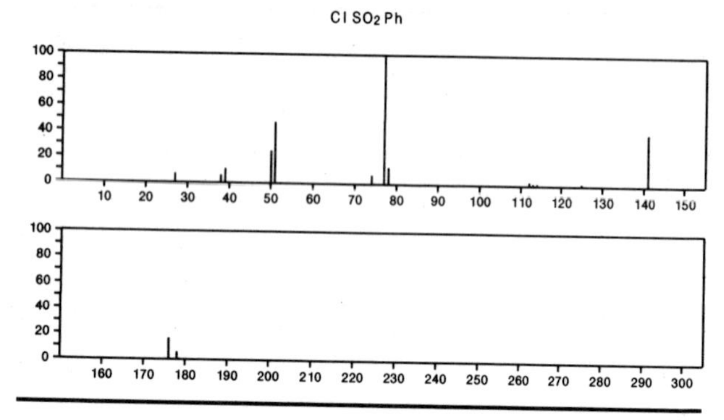

176 $C_6H_8O_6$ 50–81–7
L–Ascorbic acid

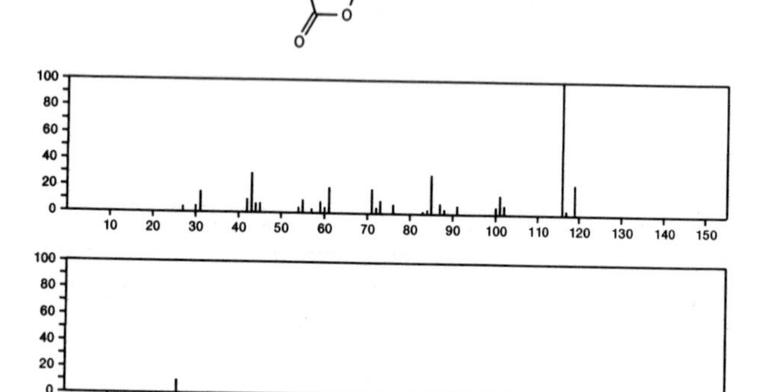

176 $C_6H_8O_6$ 32449–92–6
D–Glucuronic acid, γ–lactone

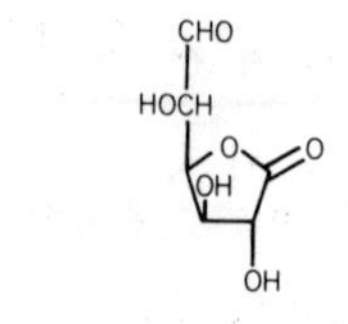

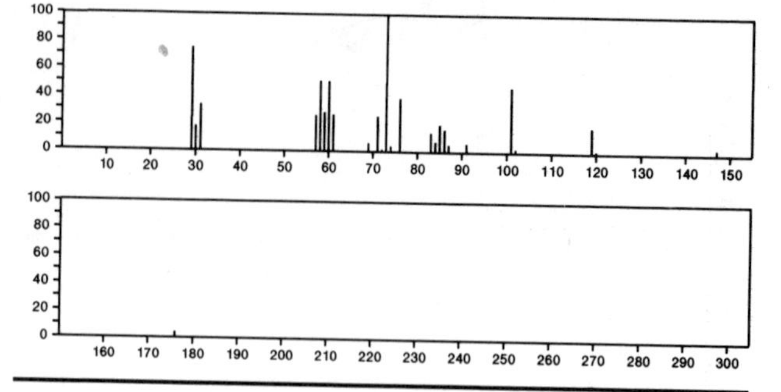

176 $C_6H_{12}N_2S_2$ 16475–50–6
Ethanedithioamide, *N,N'*–diethyl–

EtNHCSCSNHEt

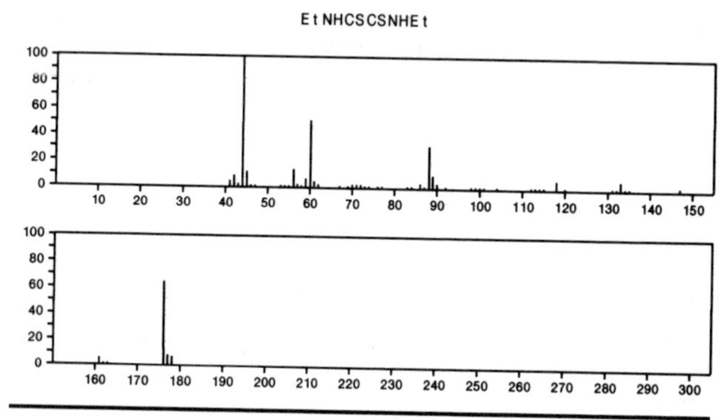

176 $C_7H_6Cl_2O$ 553–82–2
Benzene, 2,4–dichloro–1–methoxy–

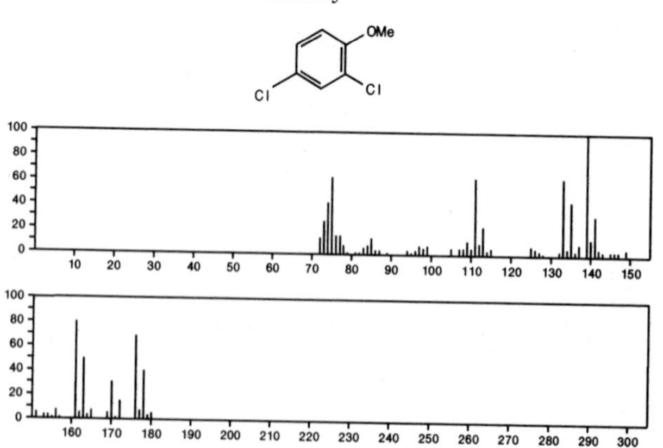

176 $C_7H_6Cl_2O$ 1570-65-6
Phenol, 2,4-dichloro-6-methyl-

Cl Cl HO Me

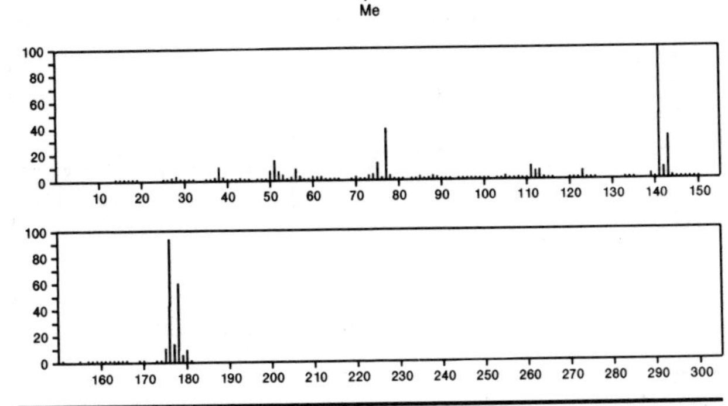

176 $C_7H_6Cl_2O$ 54518-15-9
Benzene, dichloromethoxy-

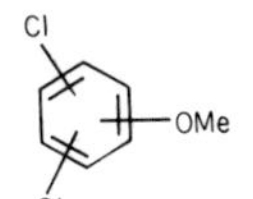

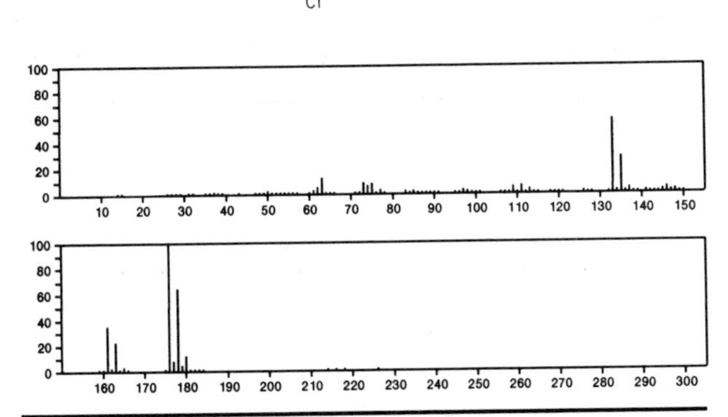

176 $C_7H_{12}OS_2$ 20560-74-1
Acetoacetic acid, 1,3-dithio-, *S*-propyl ester

Pr SC(O) CH2 CSMe

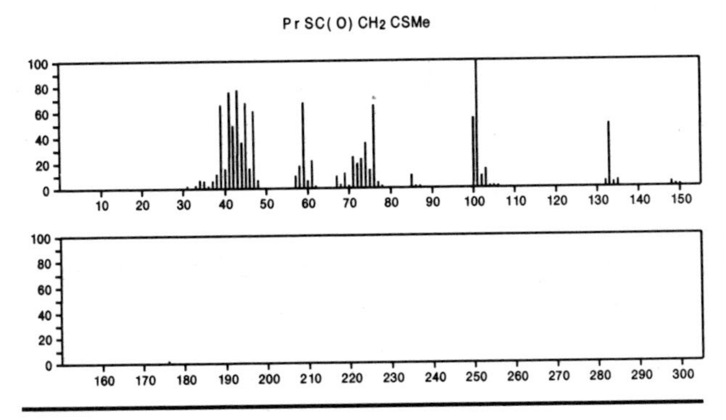

176 $C_7H_{12}O_5$ 102-62-5
1,2,3-Propanetriol, 1,2-diacetate

CH2 OH
AcOCH2 CHOAc

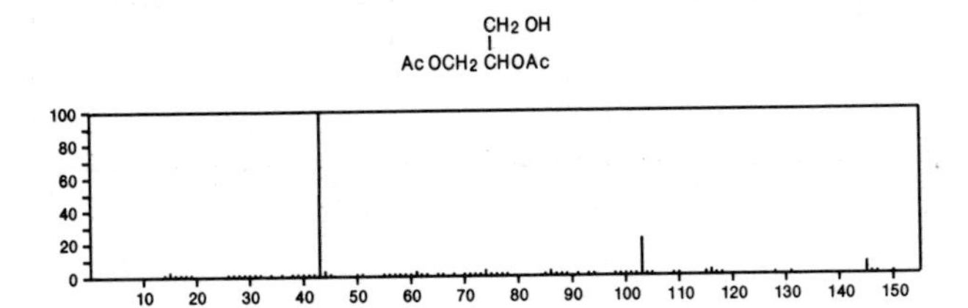

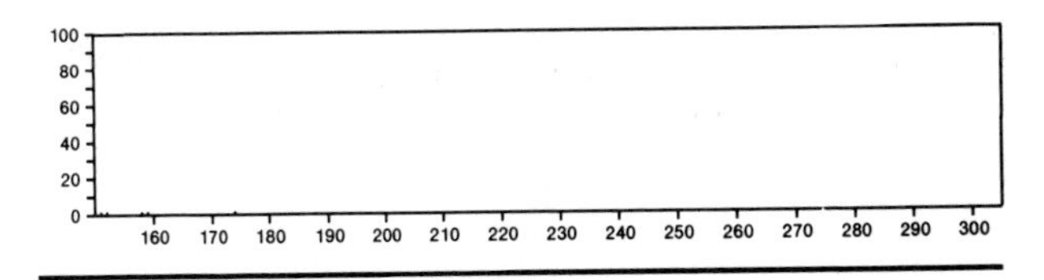

176 $C_7H_{12}O_5$ 3056-46-0
β-D-Glucopyranoside, methyl 3,6-anhydro-

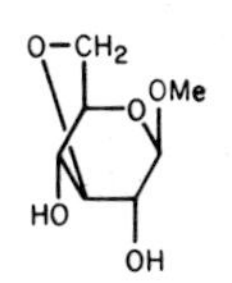

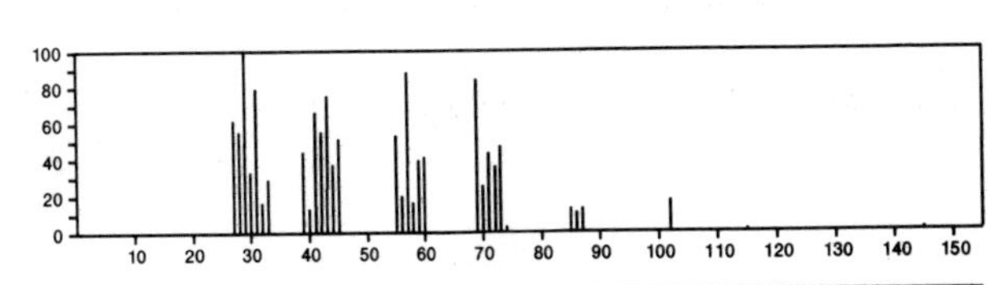

176 $C_7H_{12}O_5$ 4148-97-4
Butanedioic acid, methoxy-, dimethyl ester

MeOC(O) CH2 CH(OMe) C(O) OMe

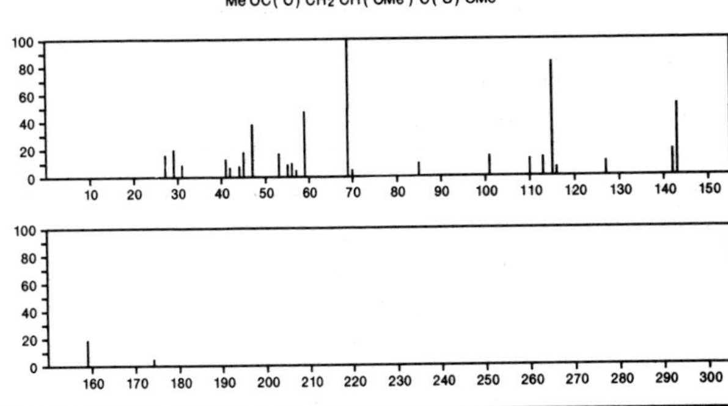

176 $C_7H_{12}O_5$ 5540-31-8
α-D-Galactopyranoside, methyl 3,6-anhydro-

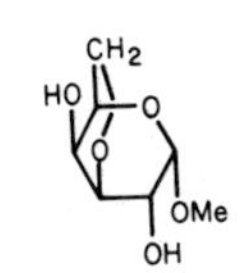

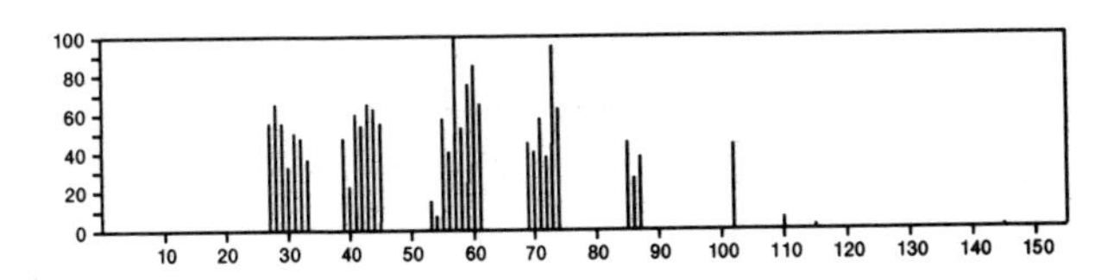

176 $C_7H_{12}O_5$ 13407-60-8
α-D-Glucopyranoside, methyl 3,6-anhydro-

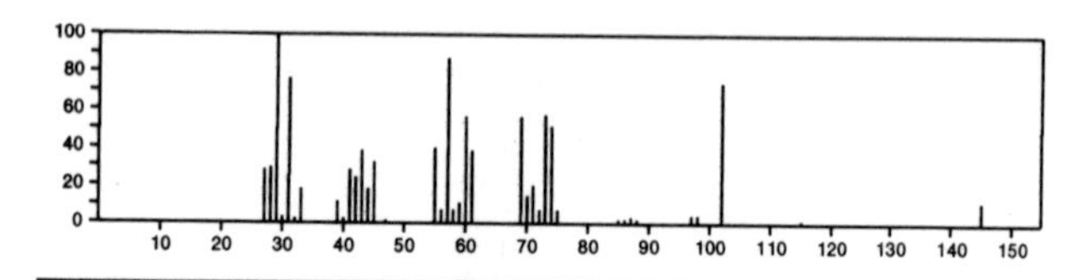

176 $C_7H_{12}O_5$ 15814-56-9
Mannopyranoside, methyl 3,6-anhydro-, α-D-

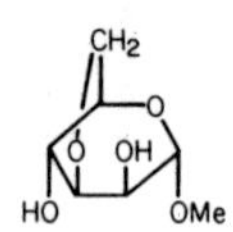

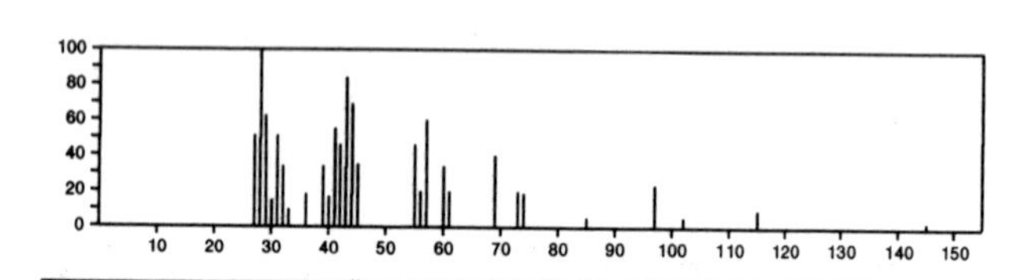

176 $C_7H_{13}Br$ 2404-35-5
Cycloheptane, bromo-

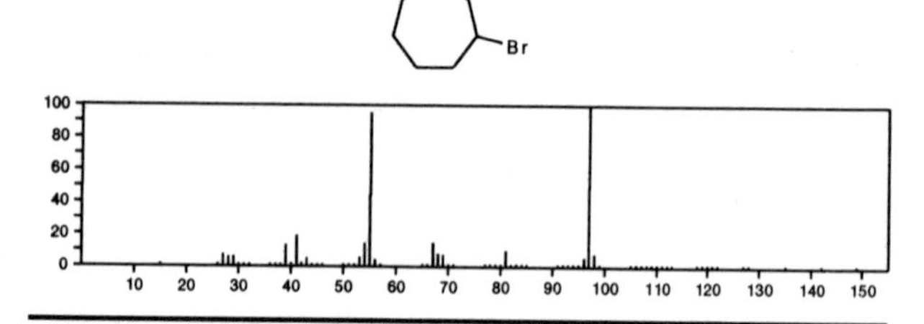

176 $C_7H_{13}Br$ 6294-39-9
Cyclohexane, 1-bromo-2-methyl-

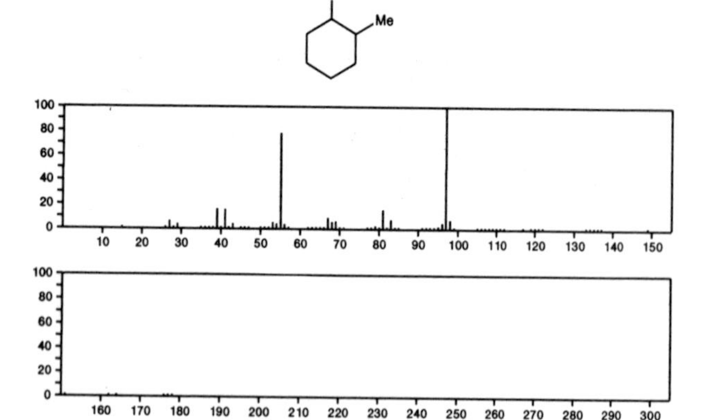

176 $C_7H_{13}Br$ 6294-40-2
Cyclohexane, 1-bromo-4-methyl-

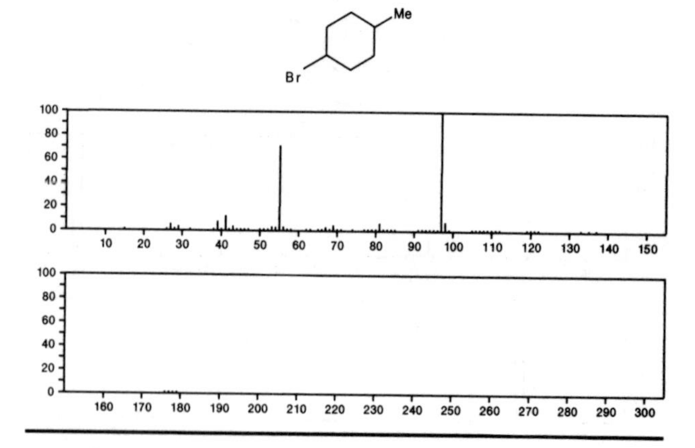

176 $C_7H_{13}Br$ 13905-48-1
Cyclohexane, 1-bromo-3-methyl-

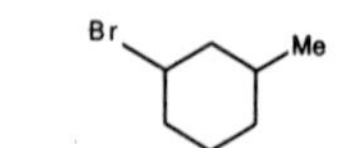

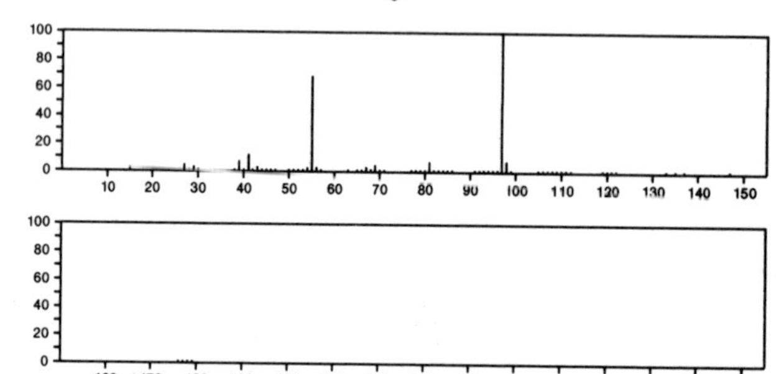

176 $C_7H_{13}Br$ 32816-30-1
Cyclopropane, 1-bromo-2-butyl-, *trans*-

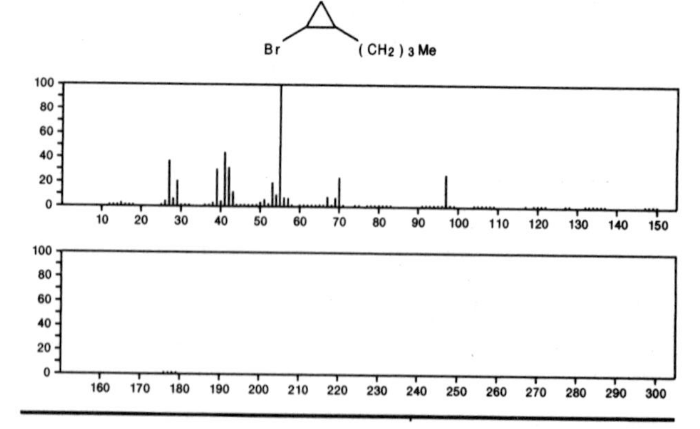

176 $C_7H_{13}Br$ 55682-99-0
Cyclopropane, 1-bromo-2-(1,1-dimethylethyl)-

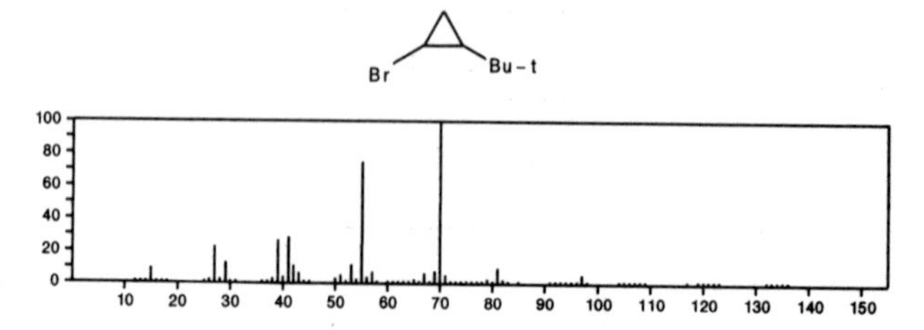

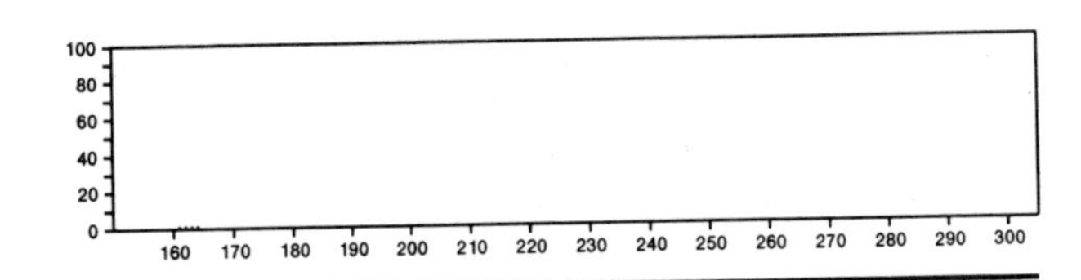

176 $C_7H_{13}Br$ 56312-52-8

2-Pentene, 5-bromo-2,3-dimethyl-

Me2C=CMeCH2CH2Br

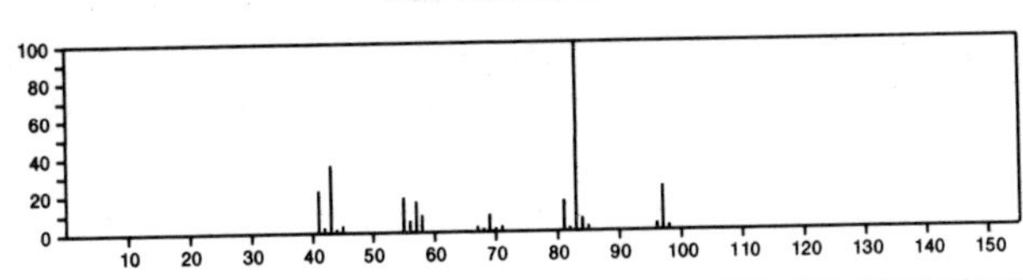

176 $C_7H_{13}O_3P$ 18644-16-1

2,6,7-Trioxa-1-phosphabicyclo[2.2.2]octane, 4-propyl-

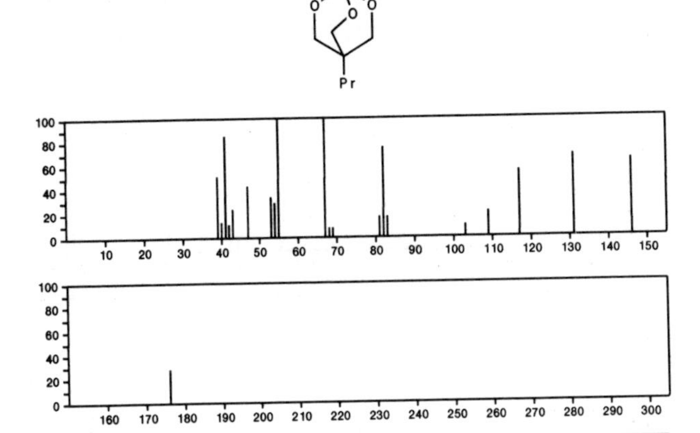

176 $C_8H_4N_2OS$ 54518-09-1

Benzo[*b*]thiophen-2(3*H*)-one, 3-diazo-

176 $C_8H_8N_4O$ 14684-54-9

4(1*H*)-Pteridinone, 6,7-dimethyl-

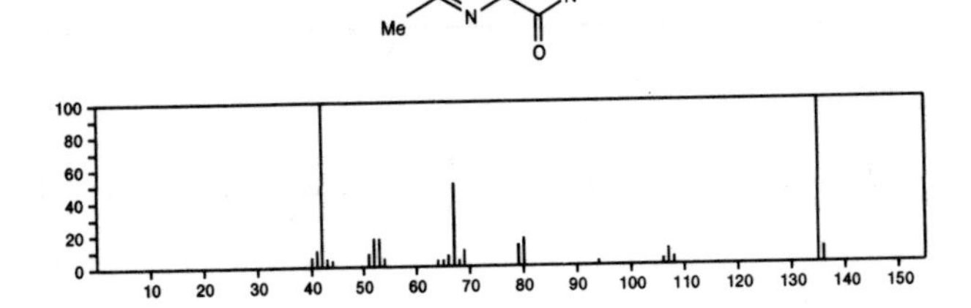

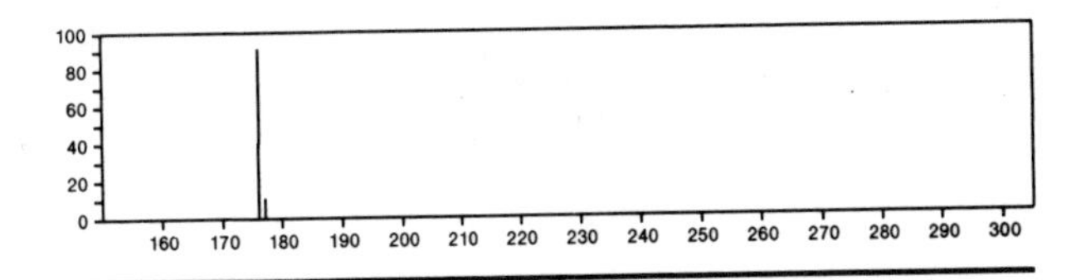

176 $C_8H_8N_4O$ 34244-77-4

4(3*H*)-Pteridinone, 2,6-dimethyl-

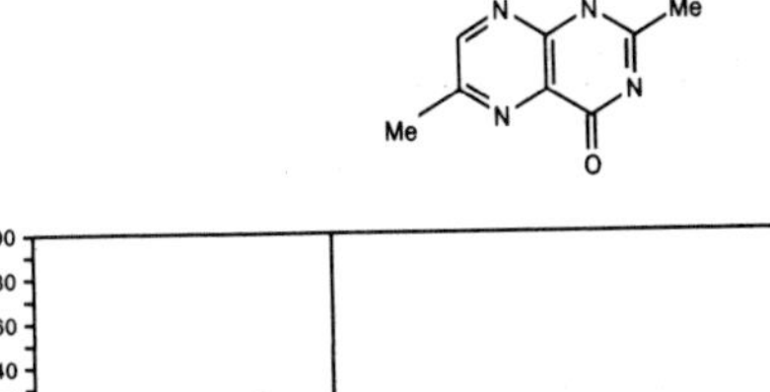

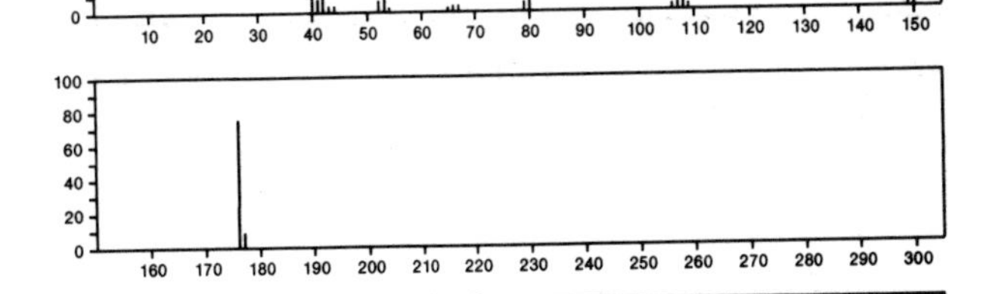

176 $C_8H_8N_4O$ 34244-79-6

4(3*H*)-Pteridinone, 2,7-dimethyl-

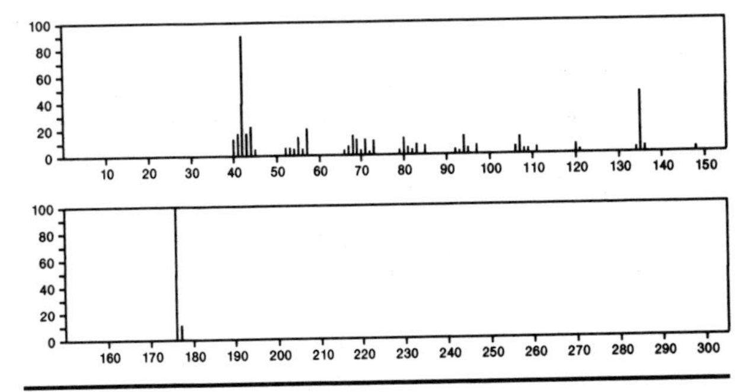

176 $C_8H_{10}Cl_2$ 13547-06-3

Cyclohexene, 1-chloro-4-(1-chloroethenyl)-

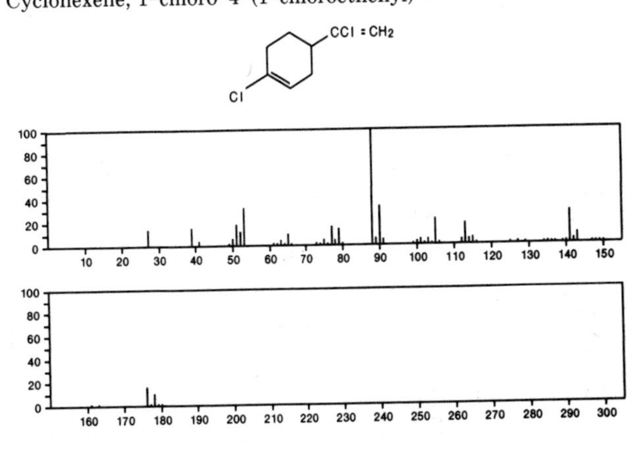

176 $C_8H_{10}Cl_2$ 13547–07–4
Cyclohexene, 1–chloro–5–(1–chloroethenyl)–

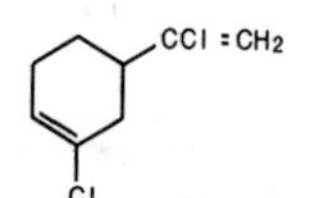

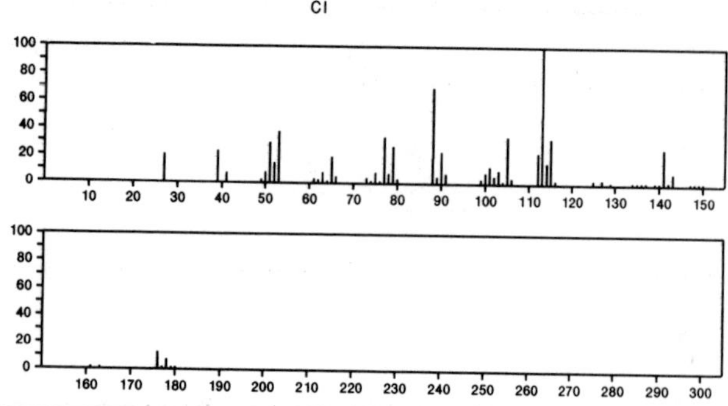

176 $C_8H_{10}Cl_2$ 29480–42–0
1,5–Cyclooctadiene, 1,6–dichloro–

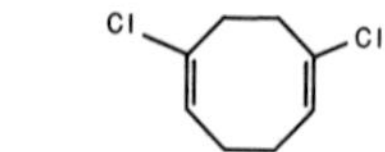

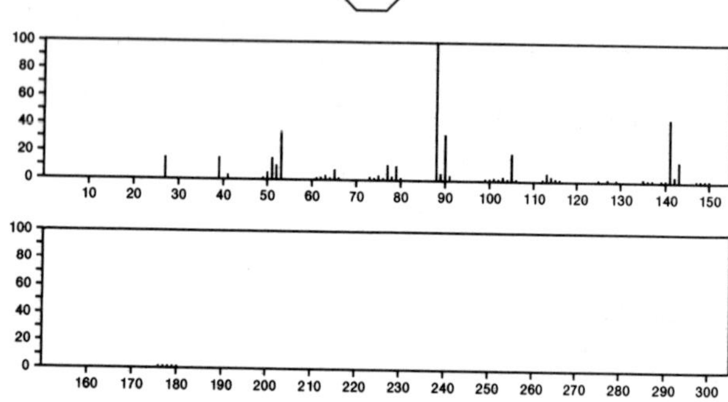

176 $C_8H_{13}ClO_2$ 55724–03–3
1,4–Dioxaspiro[4.5]decane, 8–chloro–

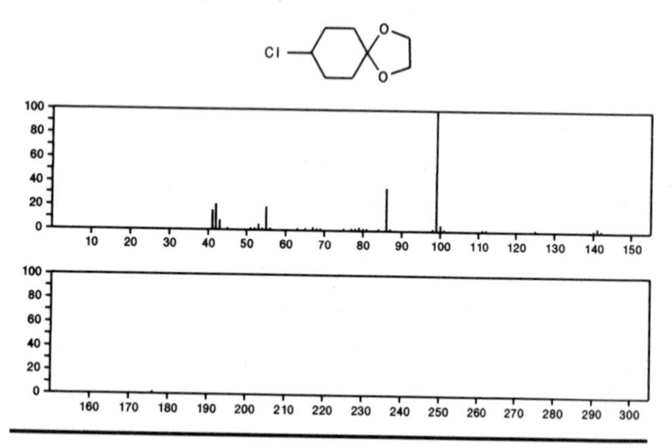

176 $C_8H_{16}O_2S$ 23246–22–2
Propionic acid, 3–(butylthio)–, methyl ester

MeOC(O)CH2CH2S(CH2)3Me

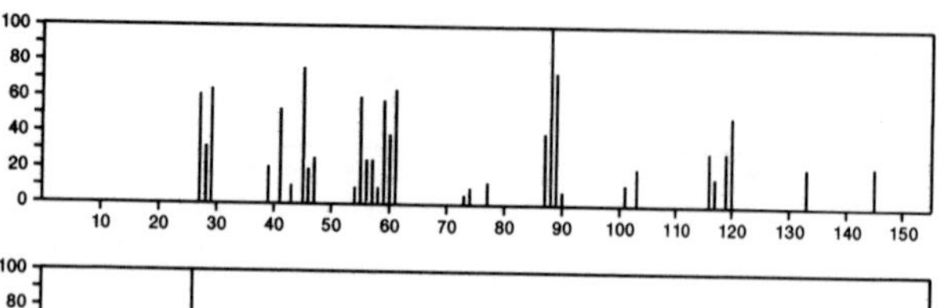

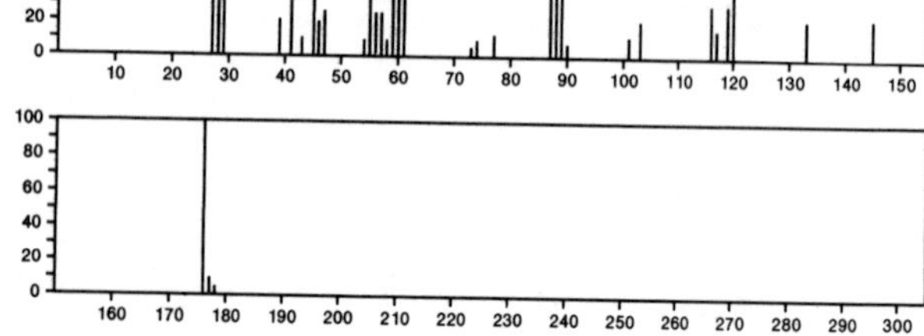

176 $C_8H_{16}O_4$ 112–15–2
Ethanol, 2–(2–ethoxyethoxy)–, acetate

AcOCH2CH2OCH2CH2OEt

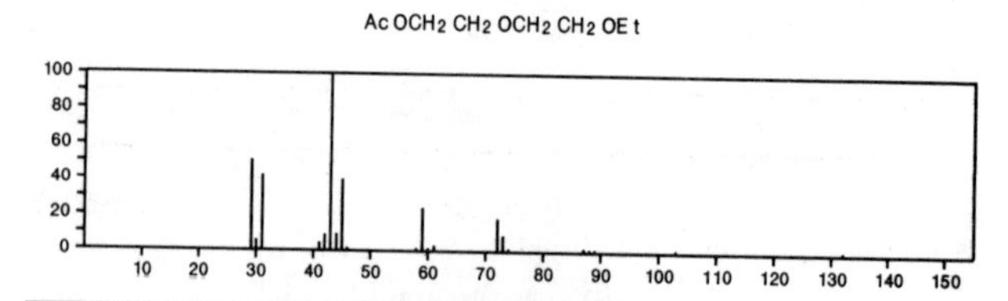

176 $C_8H_{16}O_4$ 25252–24–8
Succinaldehydic acid, 2–methyl–, methyl ester, 4–(dimethyl acetal)

MeOC(O)CHMeCH2CH(OMe)2

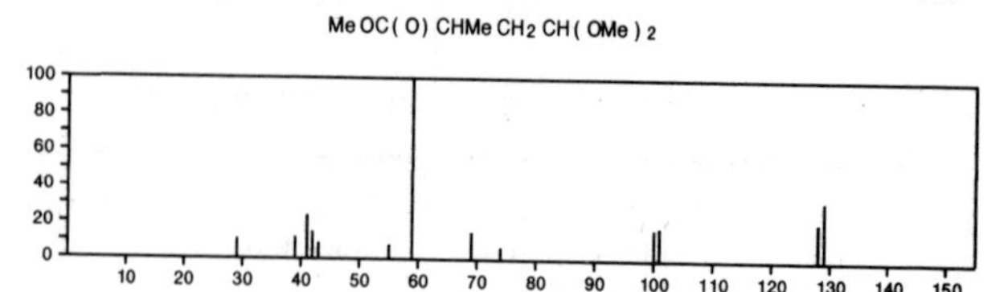

176 $C_8H_{20}O_2Si$ 16654–44–7
Silane, (4–methoxybutoxy)trimethyl–

Me3SiO(CH2)4OMe

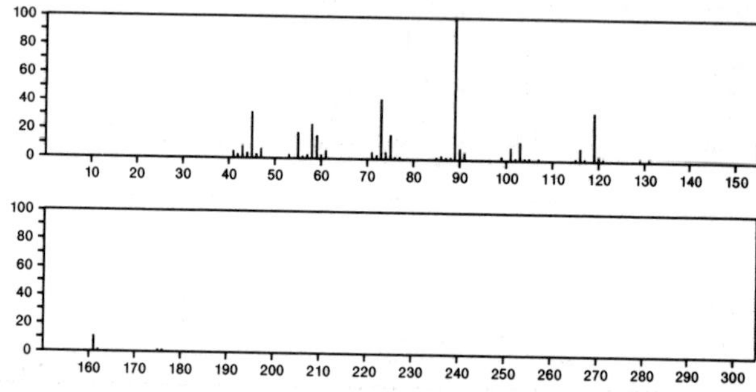

176 $C_8H_{20}O_2Si$ 54550–18–4
Silane, trimethyl[2–(1–methylethoxy)ethoxy]–

i-PrOCH2CH2OSiMe3

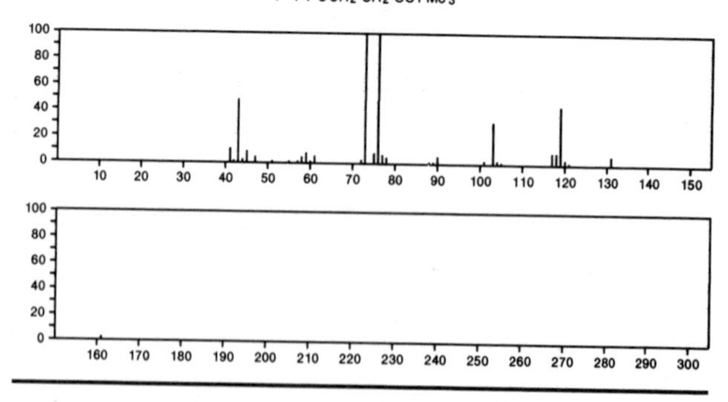

176 $C_9H_8N_2O_2$ 89–24–7
2,4–Imidazolidinedione, 5–phenyl–

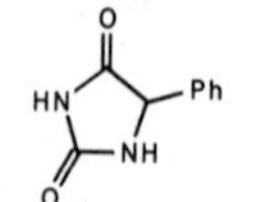

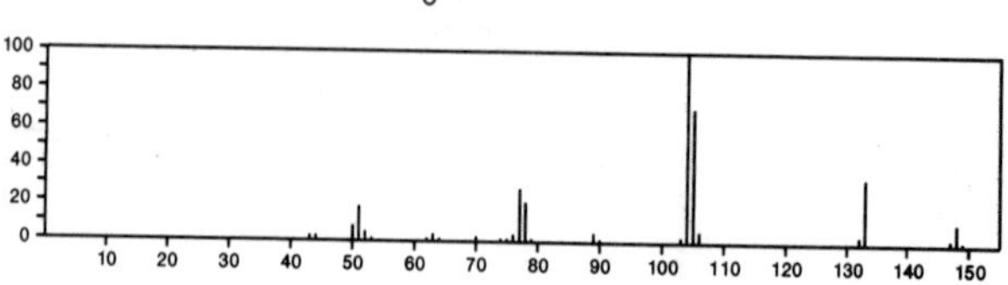

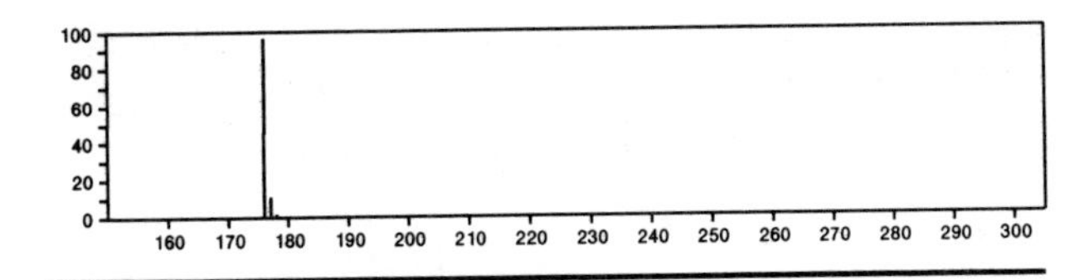

176 $C_9H_8N_2O_2$ 2152-34-3

4(5*H*)-Oxazolone, 2-amino-5-phenyl-

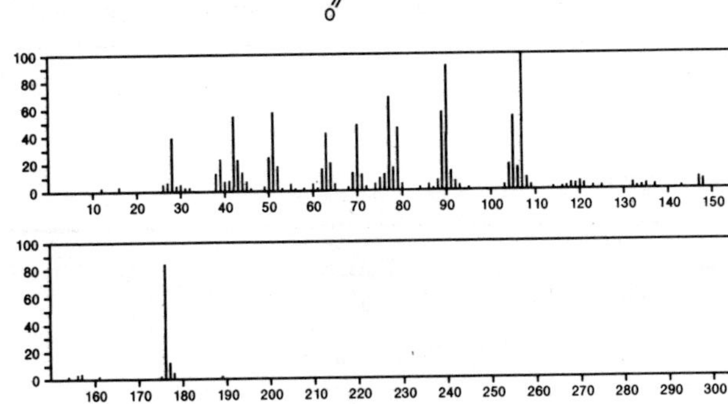

176 $C_9H_8N_2O_2$ 3483-16-7

Sydnone, 4-methyl-3-phenyl-

176 $C_9H_8N_2O_2$ 3483-18-9

Sydnone, 3-(2-methylphenyl)-

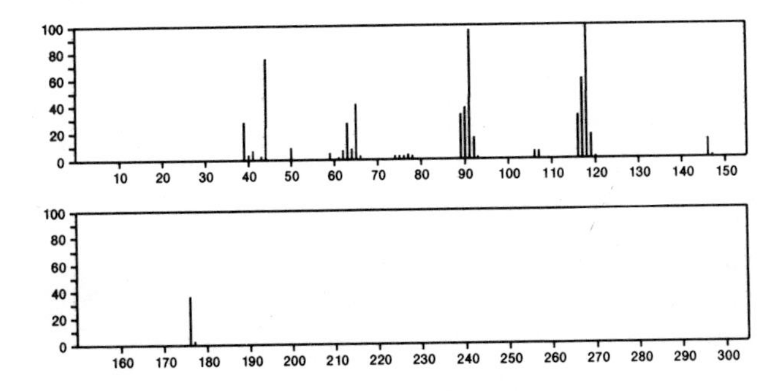

176 $C_9H_8N_2O_2$ 5004-33-1

Cinnoline, 4-methyl-, 1,2-dioxide

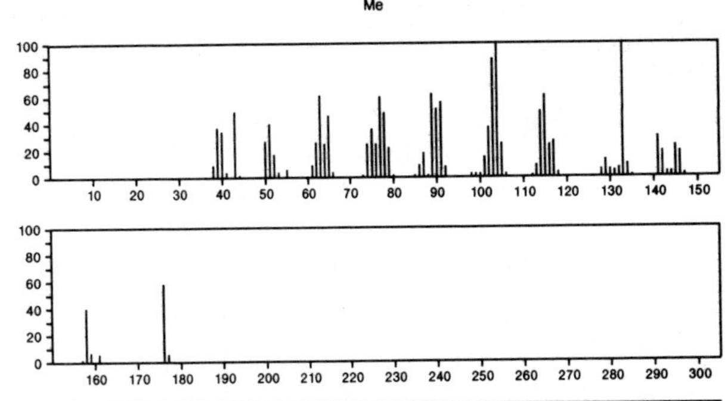

176 $C_9H_8N_2O_2$ 16551-96-5

Carbostyril, 3-amino-1-hydroxy-

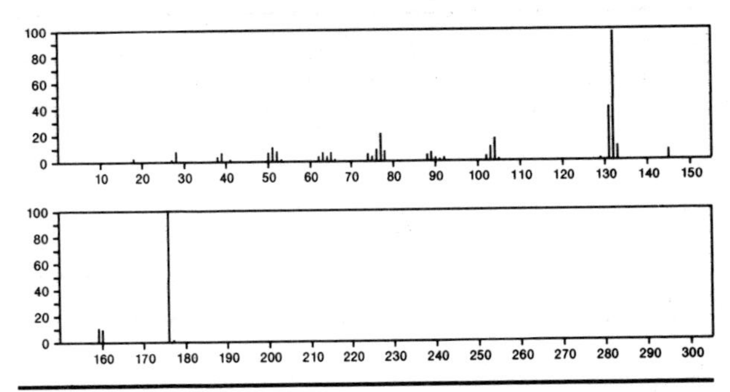

176 $C_9H_8N_2O_2$ 16844-42-1

Sydnone, 3-(phenylmethyl)-

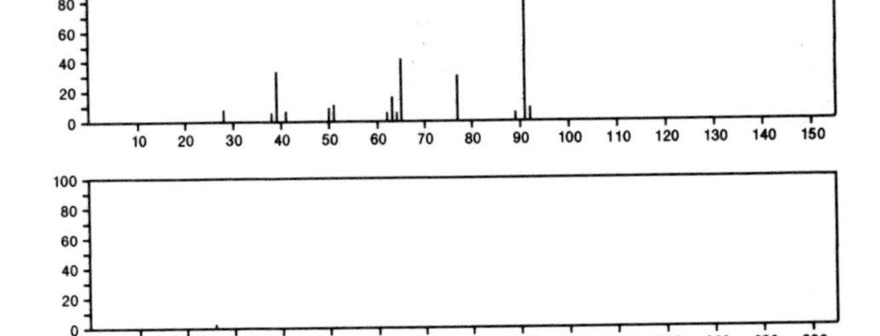

176 $C_9H_8N_2O_2$ 18916-46-6

Quinoxaline, 2-methoxy-, 4-oxide

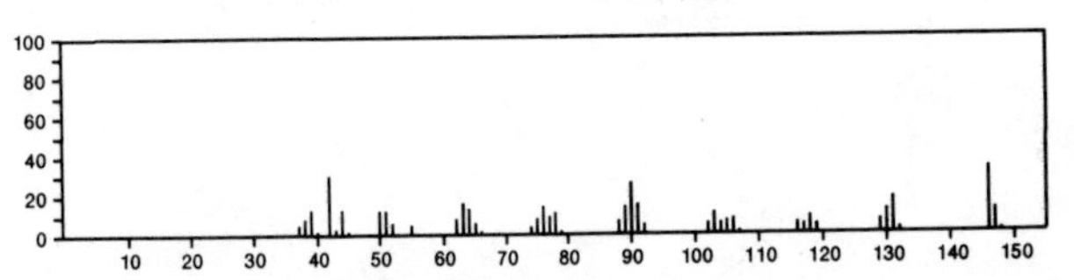

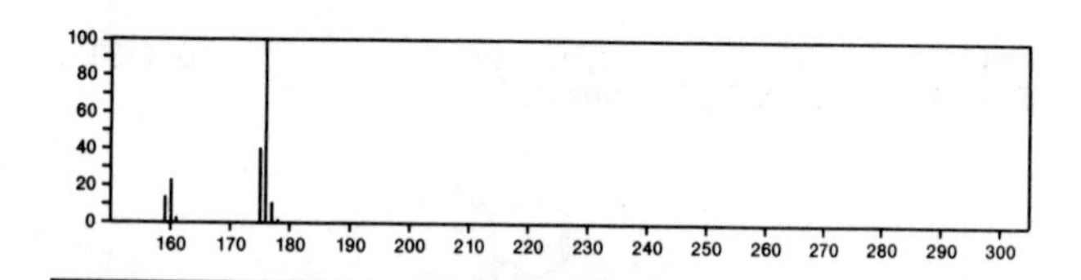

176 $C_9H_8N_2O_2$ 24660-53-5
1,3,4-Oxadiazolium, 5-hydroxy-2-methyl-3-phenyl-, hydroxide, inner salt

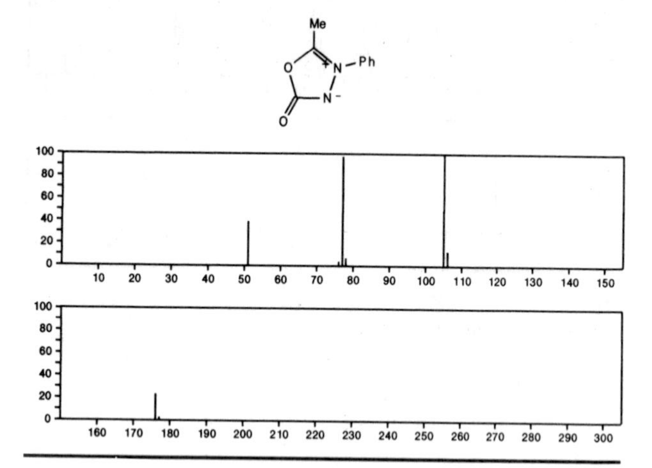

176 $C_9H_8N_2O_2$ 28740-63-8
1,3,4-Oxadiazol-2(3*H*)-one, 5-methyl-3-phenyl-

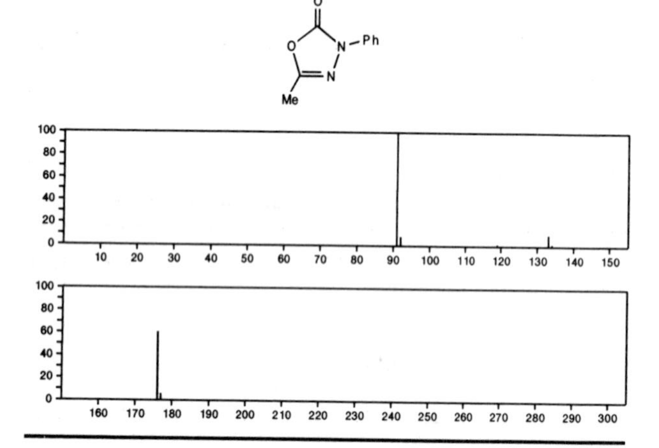

176 $C_9H_8N_2O_2$ 54518-07-9
4-Quinazolinol, 2-methyl-, 3-oxide

176 $C_9H_8N_2S$ 50993-74-3
4*H*-Pyrido[1,2-*a*]pyrimidine-4-thione, 2-methyl-

176 $C_9H_9BO_3$ 54518-06-8
1,3,2-Dioxaborolan-4-one, 2-methyl-5-phenyl-

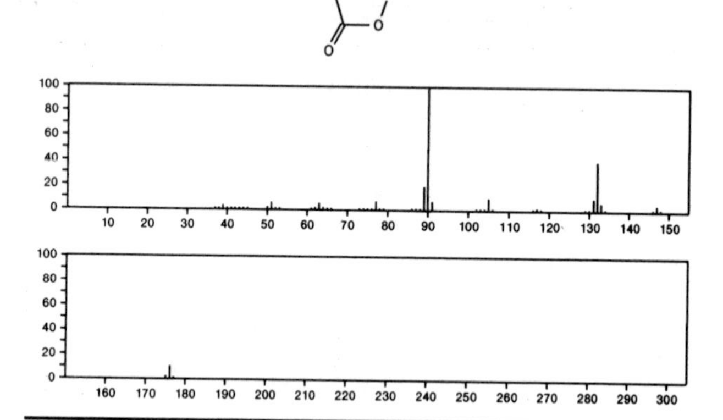

176 $C_9H_{12}N_4$ 19848-81-8
s-Triazolo[4,3-*a*]pyrazine, 3-ethyl-5,8-dimethyl-

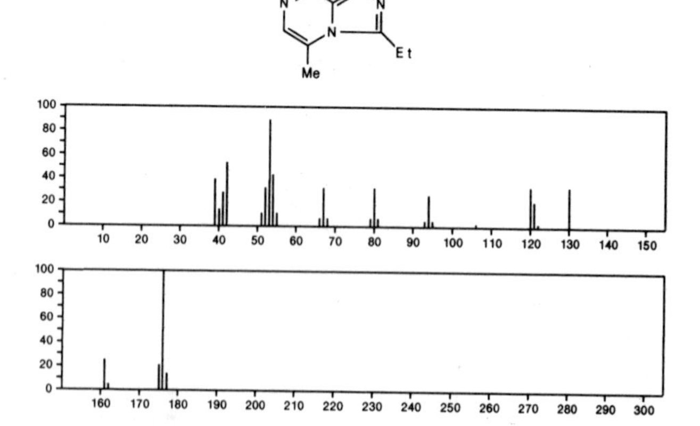

176 $C_9H_{12}N_4$ 54518-05-7
[1,2,4]Triazolo[1,5-*a*]pyrazine, 2-ethyl-5,8-dimethyl-

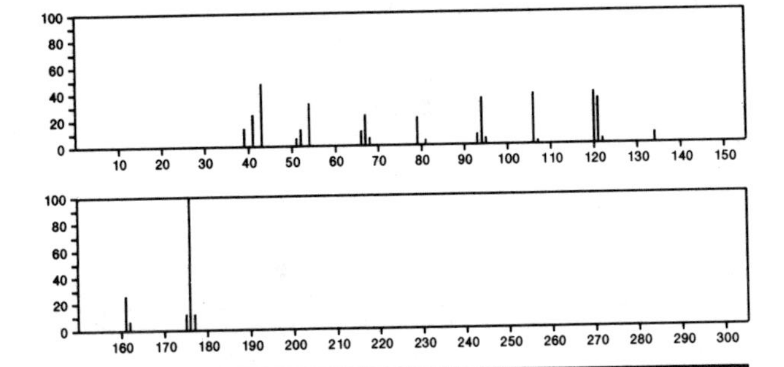

176 $C_9H_{20}O_3$ 115-80-0
Propane, 1,1,1-triethoxy-

$EtC(OEt)_3$

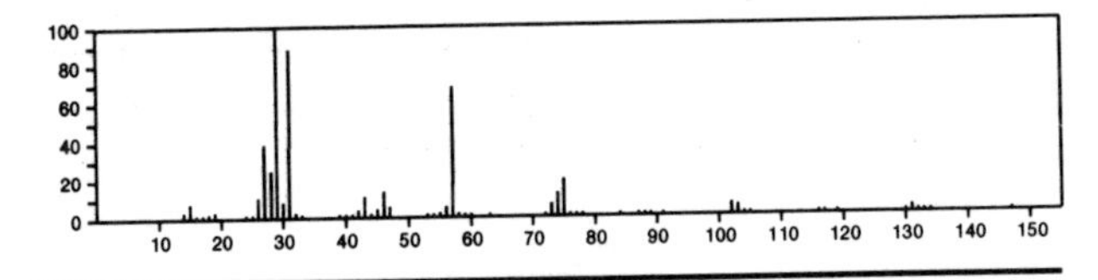

176 $C_9H_{20}O_3$ 20637-29-0
Hexane, 1,2,3-trimethoxy-

$PrCH(OMe)CH(OMe)CH_2OMe$

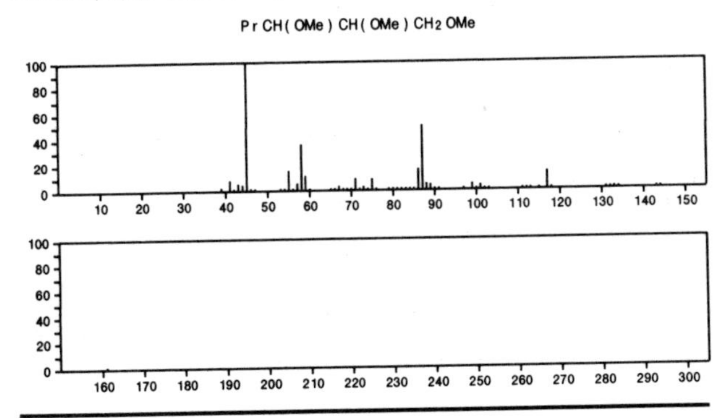

176 $C_9H_{20}O_3$ 54518-03-5
1-Propanol, 3-[3-(1-methylethoxy)propoxy]-

$HO(CH_2)_3O(CH_2)_3OPr\text{-}i$

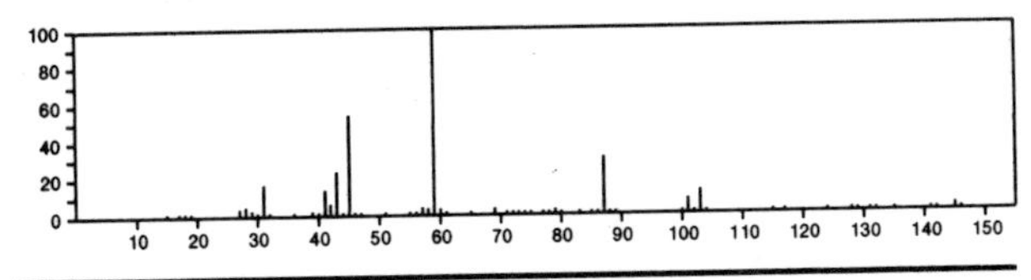

176 $C_9H_{20}O_3$ 54518-04-6
Methanol, dibutoxy-

$Me(CH_2)_3OCH(OH)O(CH_2)_3Me$

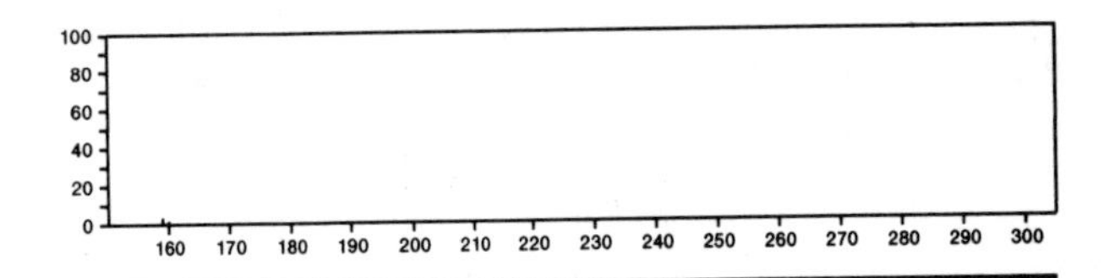

176 $C_{10}H_8O_3$ 90-33-5
2*H*-1-Benzopyran-2-one, 7-hydroxy-4-methyl-

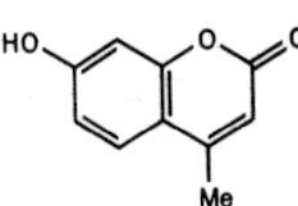

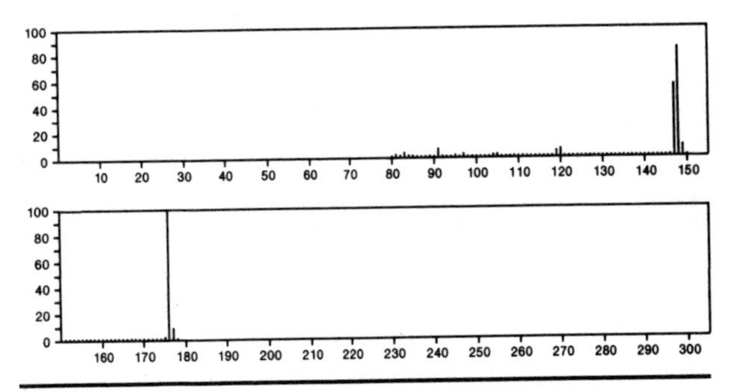

176 $C_{10}H_8O_3$ 5463-50-3
1,3-Isobenzofurandione, 4,7-dimethyl-

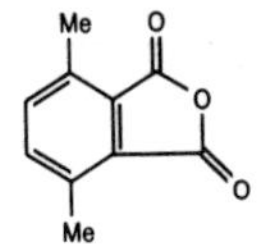

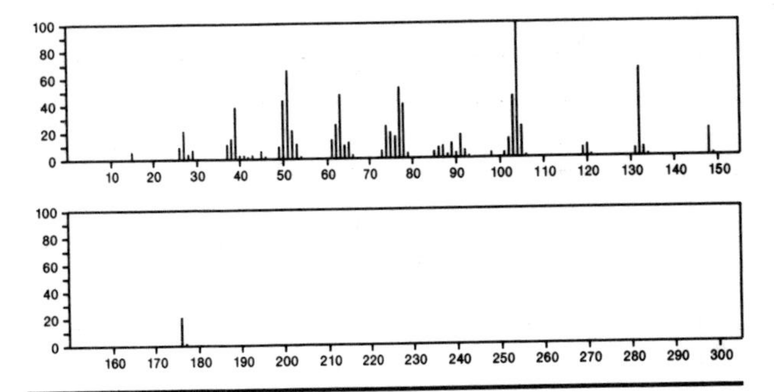

176 $C_{10}H_8O_3$ 22105-12-0
4*H*-1-Benzopyran-4-one, 6-hydroxy-2-methyl-

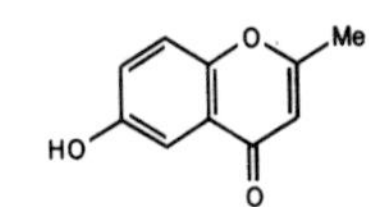

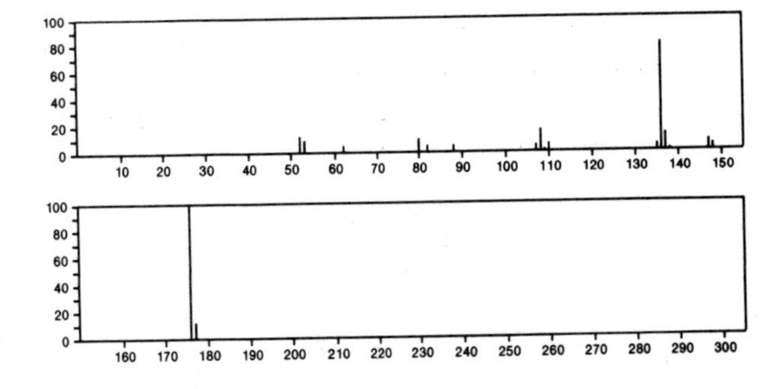

176 $C_{10}H_8O_3$ 33488-56-1
2-Furancarboxaldehyde, 5-(2-furanylmethyl)-

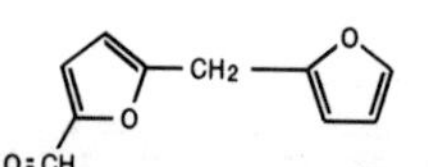

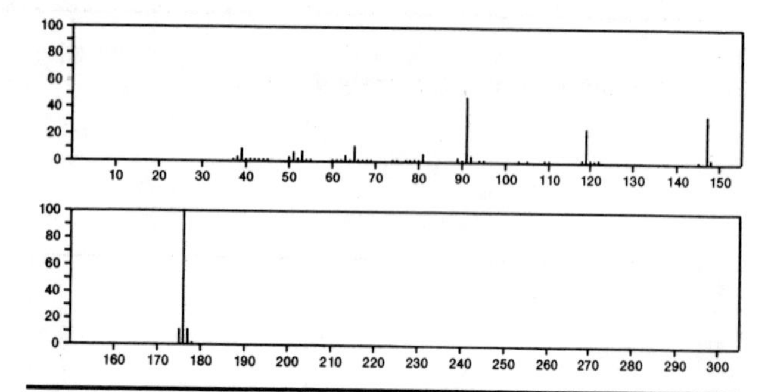

176 $C_{10}H_8O_3$ 40800-89-3
3-Benzofurancarboxaldehyde, 2-methoxy-

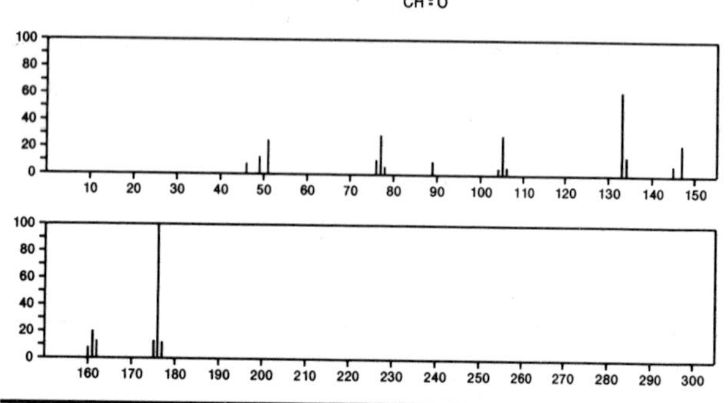

176 $C_{10}H_8O_3$ 40800-90-6
2(3*H*)-Benzofuranone, 3-(methoxymethylene)-

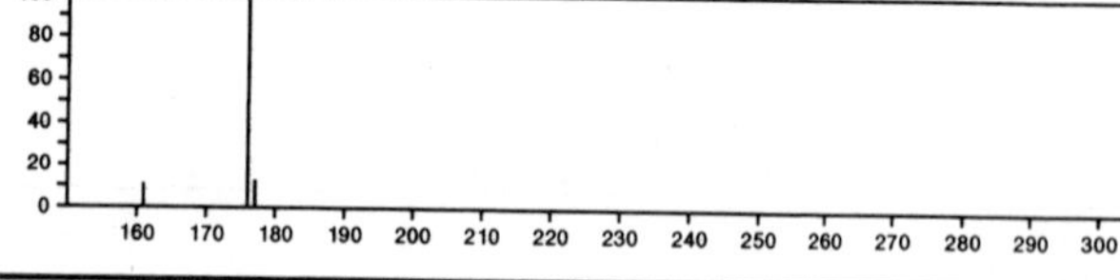

176 $C_{10}H_8O_3$ 50276-98-7
1*H*-2-Benzopyran-5-carboxaldehyde, 3,4-dihydro-1-oxo-

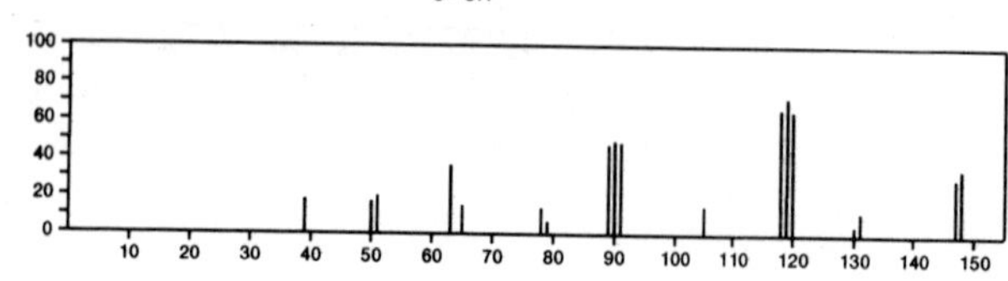

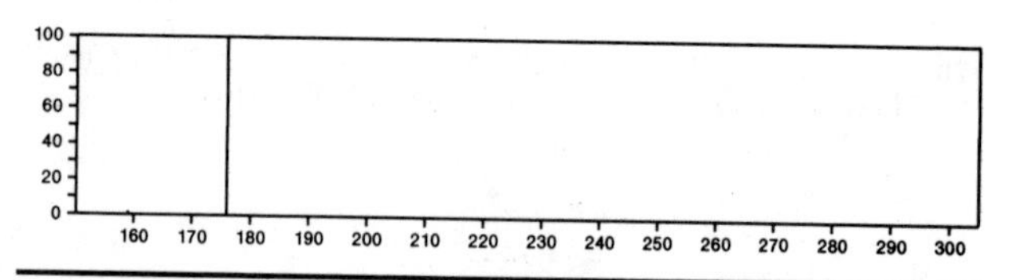

176 $C_{10}H_{12}N_2O$ 50-67-9
1*H*-Indol-5-ol, 3-(2-aminoethyl)-

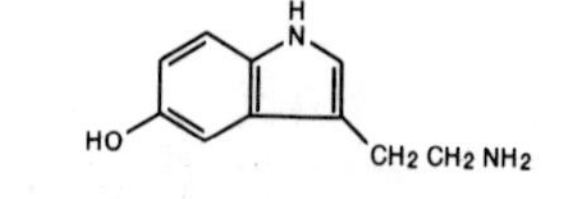

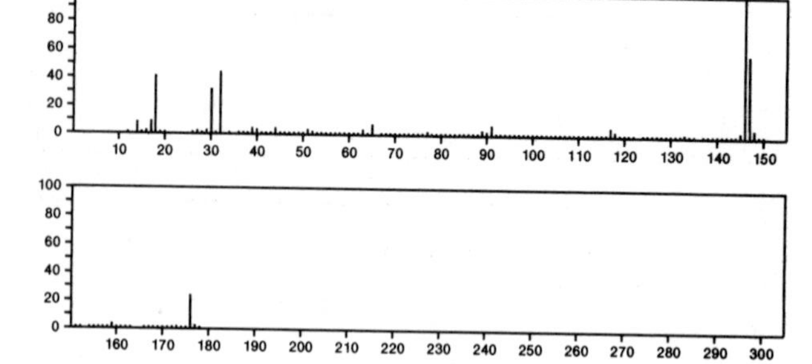

176 $C_{10}H_{12}N_2O$ 486-56-6
2-Pyrrolidinone, 1-methyl-5-(3-pyridinyl)-, (*S*)-

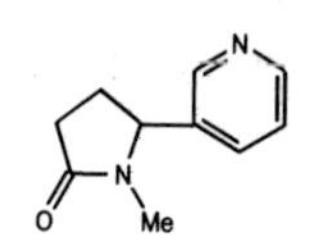

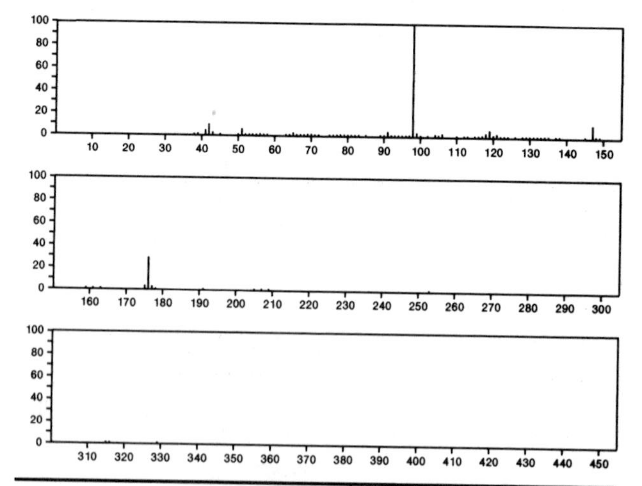

176 $C_{10}H_{12}N_2O$ 829-65-2
1-Aziridinecarboxamide, *N*-(4-methylphenyl)-

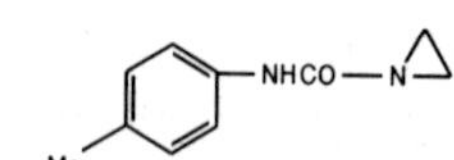

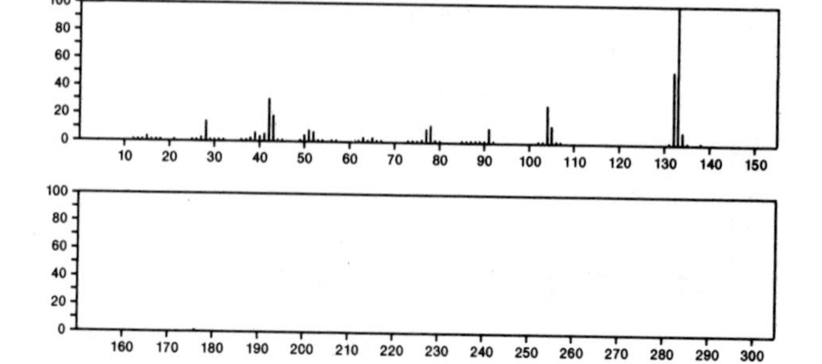

176 $C_{10}H_{12}N_2O$ 6302–84–7
1*H*–Imidazole, 4,5–dihydro–2–(4–methoxyphenyl)–

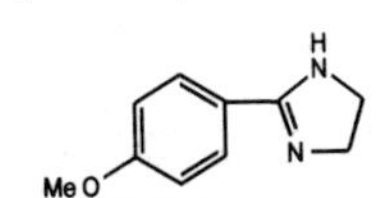

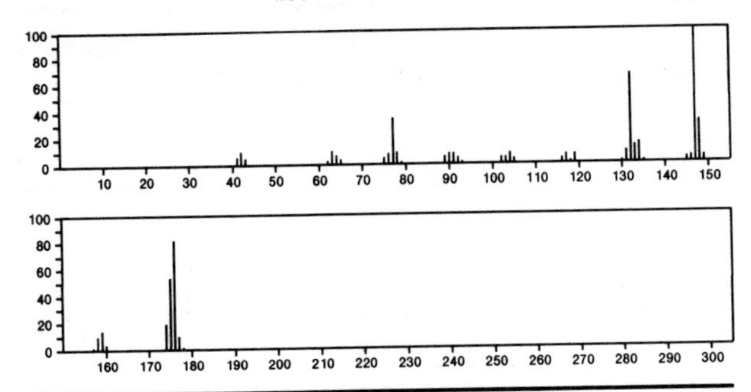

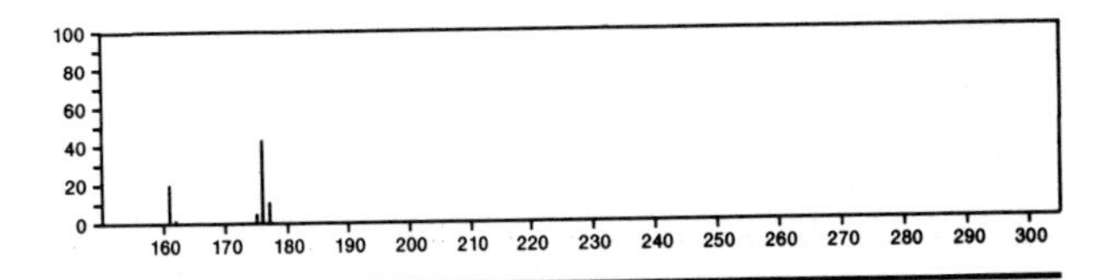

176 $C_{10}H_{12}N_2O$ 51111–02–5
1,5–Benzodiazocin–6(1*H*)–one, 2,3,4,5–tetrahydro–

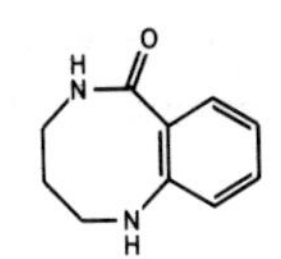

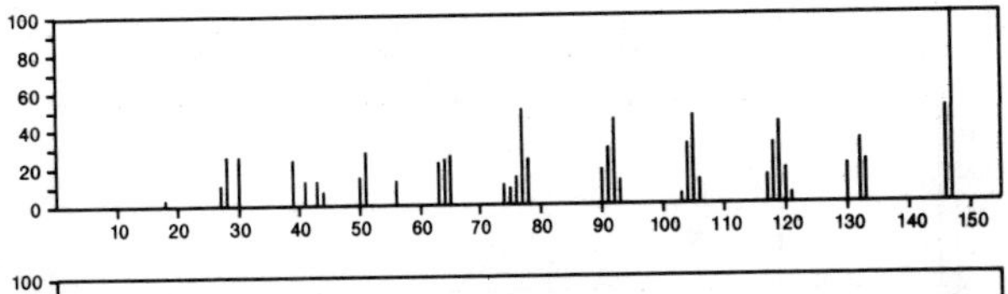

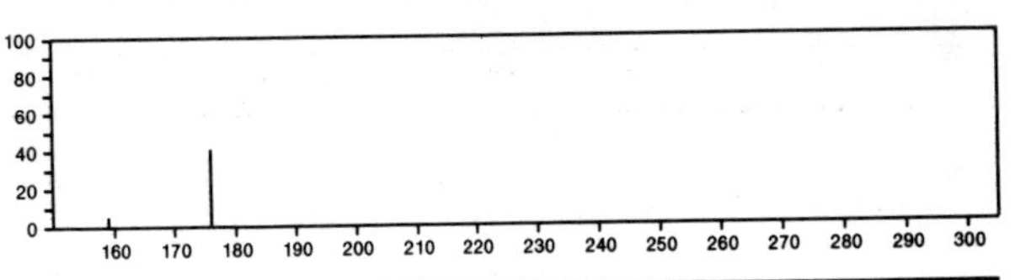

176 $C_{10}H_{12}N_2O$ 16007–54–8
Benzimidazole, 2–isopropyl–, 3–oxide

Pr–i

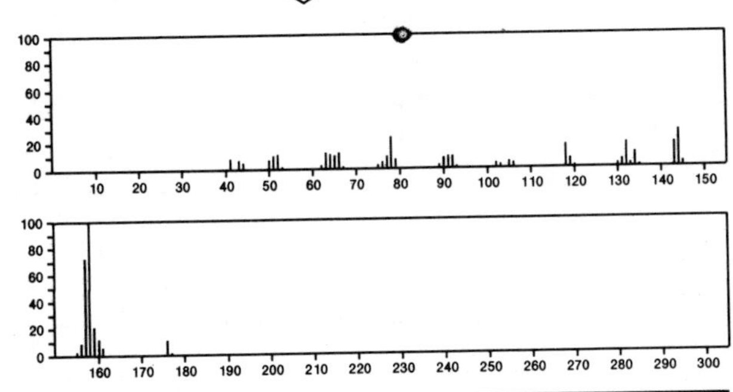

176 $C_{10}H_{12}N_2O$ 28291–80–7
2–Benzoxazolamine, *N*–propyl–

NHPr

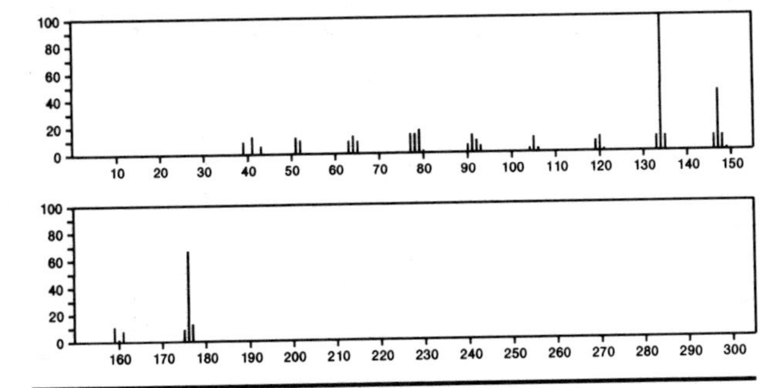

176 $C_{10}H_{12}N_2O$ 28455–42–7
Benzoxazole, 2–(isopropylamino)–

NHPr–i

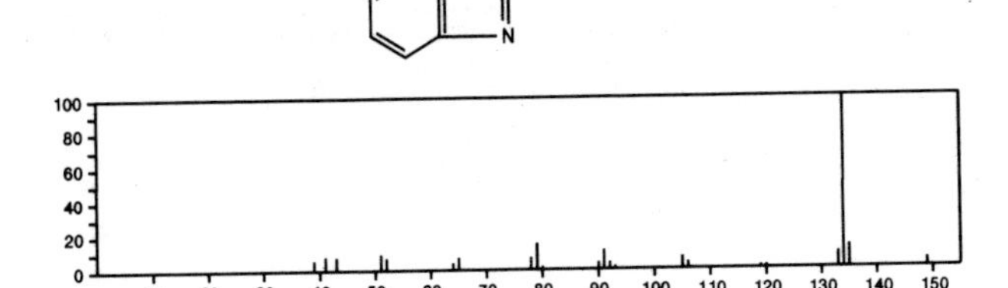

176 $C_{10}H_{21}Cl$ 1002–11–5
Decane, 3–chloro–

$EtCHCl(CH_2)_6Me$

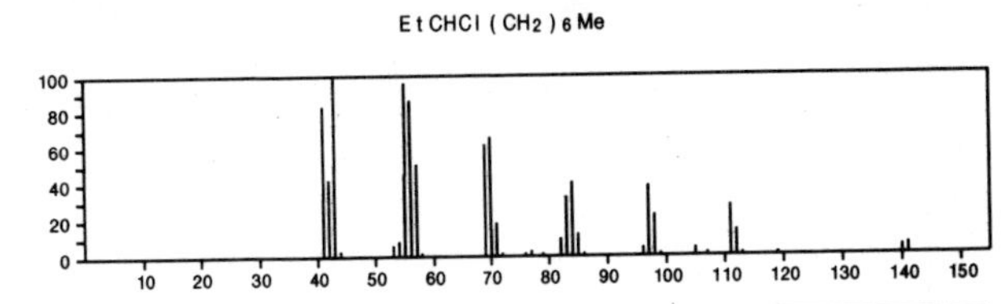

176 $C_{10}H_{21}Cl$ 1002–69–3
Decane, 1–chloro–

$Me(CH_2)_9Cl$

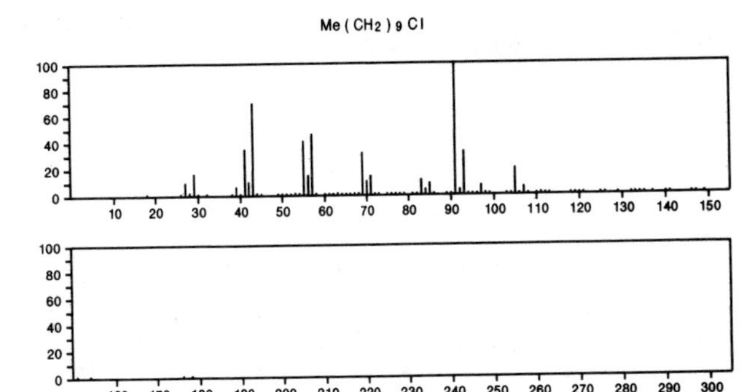

176 $C_{11}H_9Cl$ 86–52–2
Naphthalene, 1–(chloromethyl)–

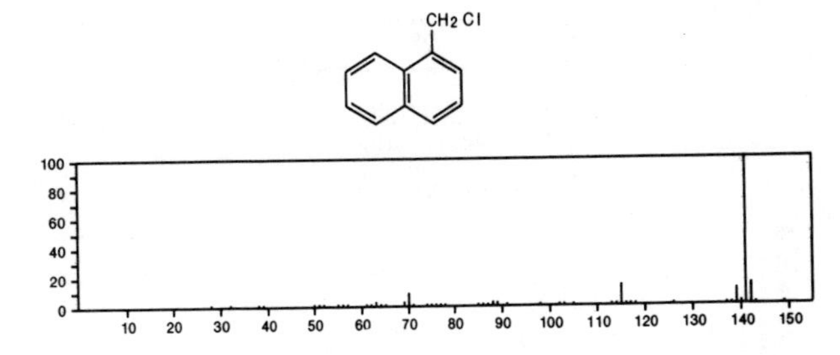

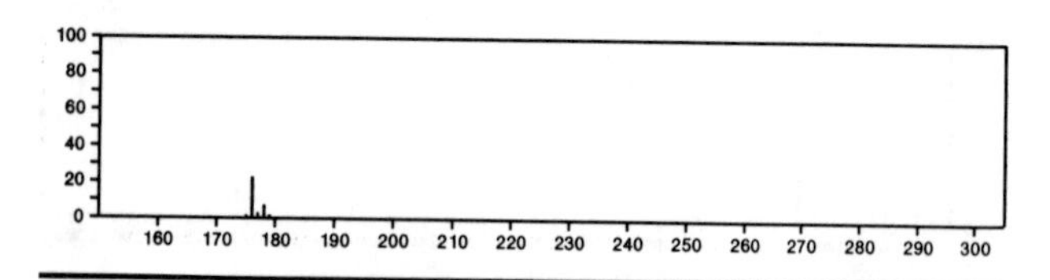

176 $C_{11}H_{12}O_2$ 103-36-6
2-Propenoic acid, 3-phenyl-, ethyl ester

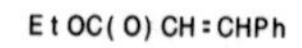

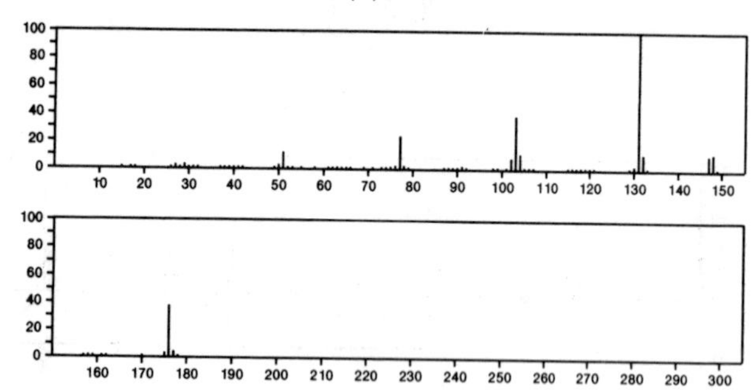

176 $C_{11}H_{12}O_2$ 103-54-8
2-Propen-1-ol, 3-phenyl-, acetate

AcOCH2CH=CHPh

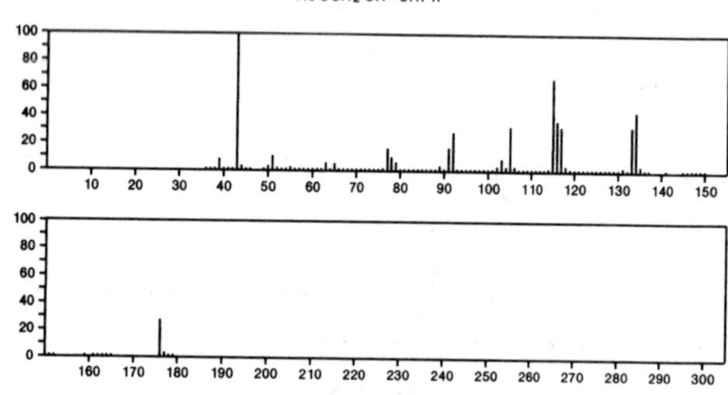

176 $C_{11}H_{12}O_2$ 1078-19-9
1(2*H*)-Naphthalenone, 3,4-dihydro-6-methoxy-

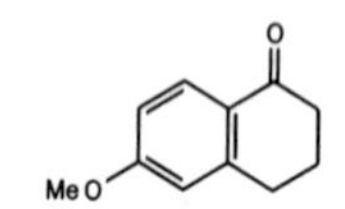

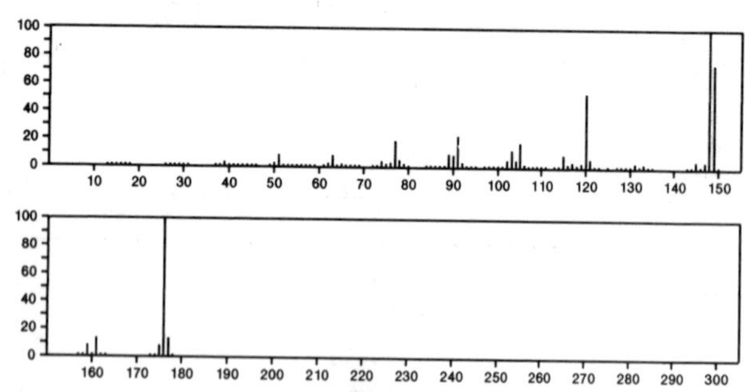

176 $C_{11}H_{12}O_2$ 1797-74-6
Benzeneacetic acid, 2-propenyl ester

PhCH2C(O)OCH2CH=CH2

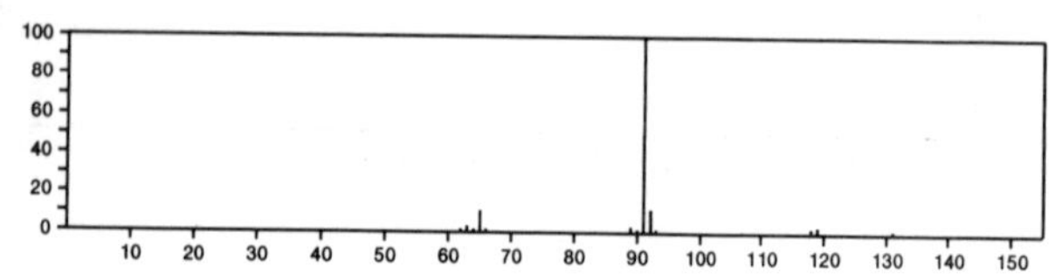

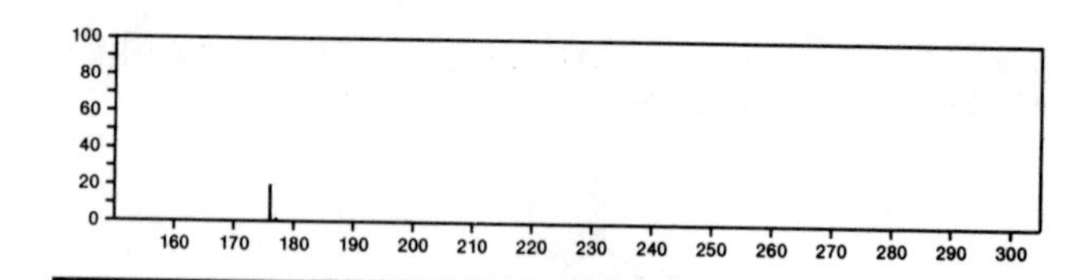

176 $C_{11}H_{12}O_2$ 3070-71-1
Benzenepropanoic acid, α-methylene-, methyl ester

CH2
MeOC(O)CCH2Ph

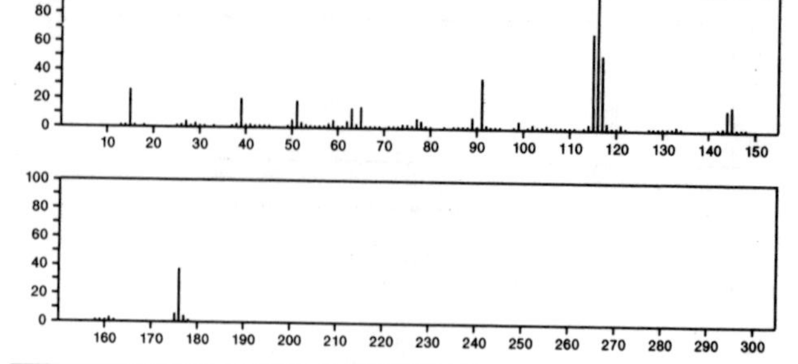

176 $C_{11}H_{12}O_2$ 4125-54-6
Phenol, 2-(2-propenyl)-, acetate

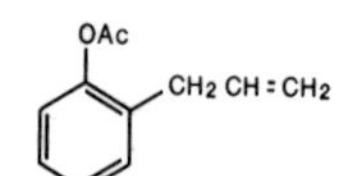

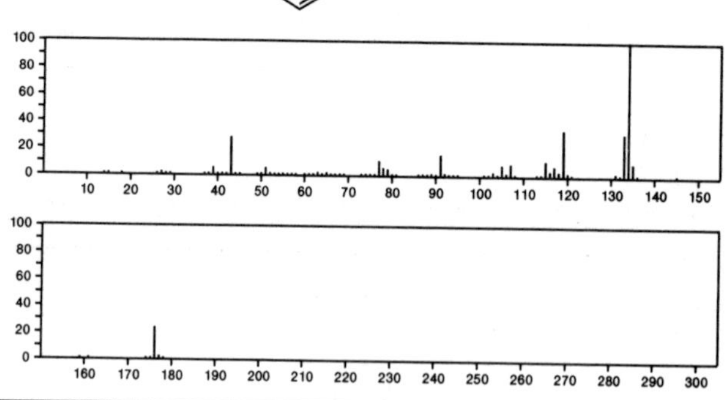

176 $C_{11}H_{12}O_2$ 4242-18-6
1-Naphthalenecarboxylic acid, 5,6,7,8-tetrahydro-

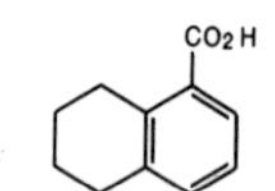

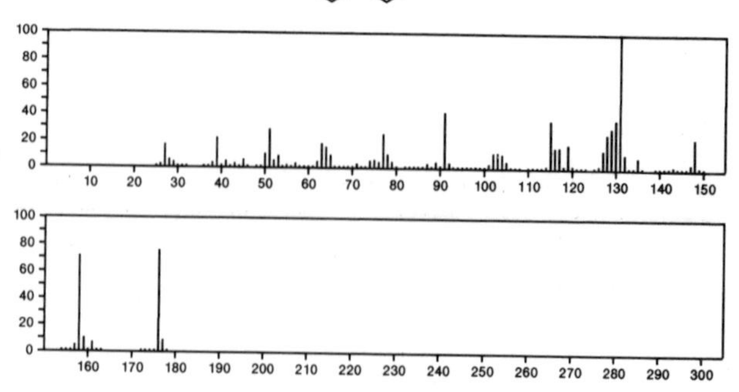

176 $C_{11}H_{12}O_2$ 5631-63-0
2-Buten-1-one, 1-(2-hydroxy-5-methylphenyl)-

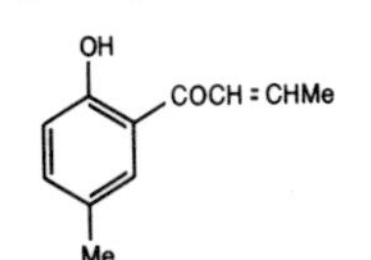

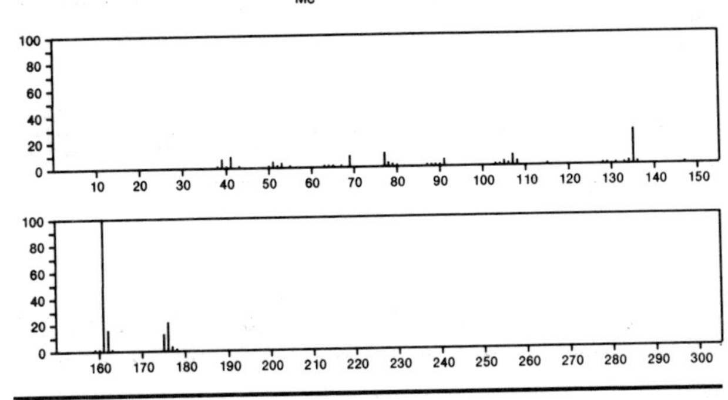

176 $C_{11}H_{12}O_2$ 5910-25-8
2,4-Pentanedione, 3-phenyl-

Me COCHPh COMe

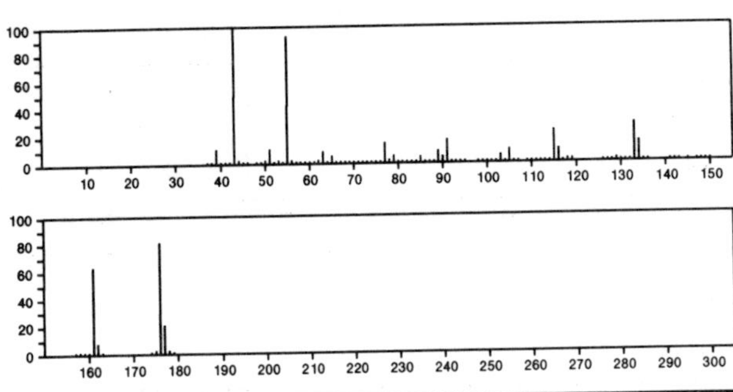

176 $C_{11}H_{12}O_2$ 6668-24-2
1,3-Butanedione, 2-methyl-1-phenyl-

Ph COCHMe COMe

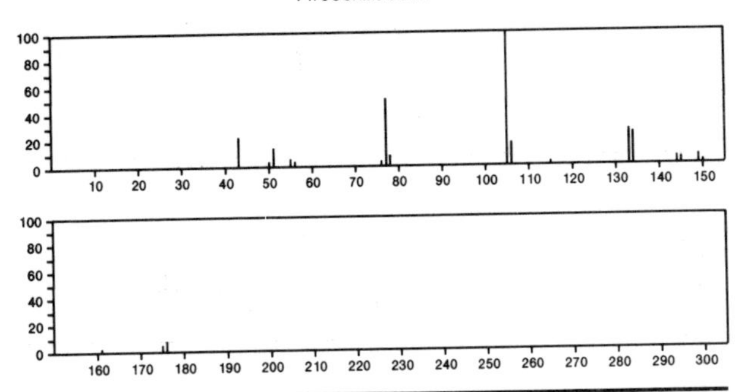

176 $C_{11}H_{12}O_2$ 21303-80-0
2(3*H*)-Furanone, dihydro-5-methyl-5-phenyl-

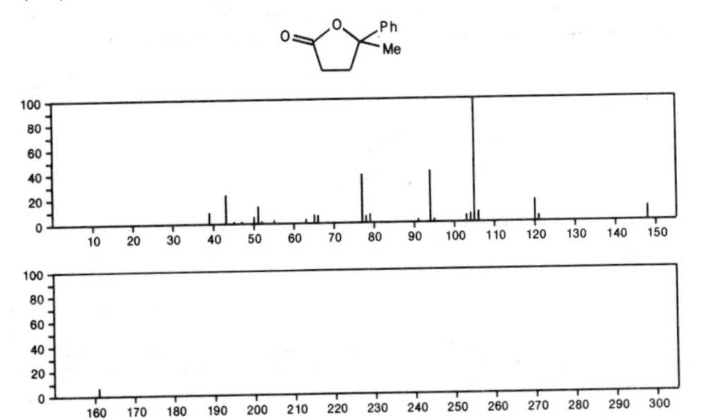

176 $C_{11}H_{12}O_2$ 35355-35-2
Benzofuran, 5-methoxy-6,7-dimethyl-

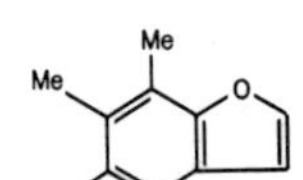

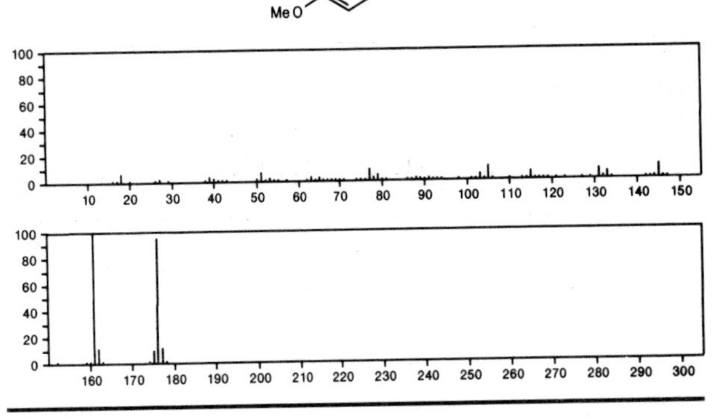

176 $C_{11}H_{12}O_2$ 53774-19-9
3-Pentenoic acid, 4-phenyl-

$HO_2CCH_2CH:CMePh$

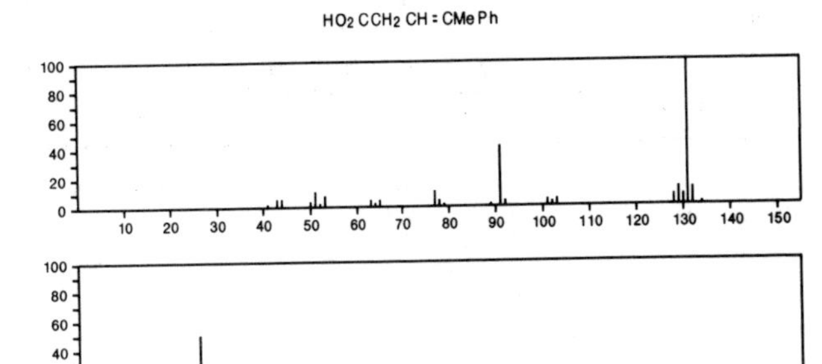

176 $C_{11}H_{12}O_2$ 55320-96-2
2-Pentenoic acid, 5-phenyl-, (*E*)-

$HO_2CCH:CHCH_2CH_2Ph$

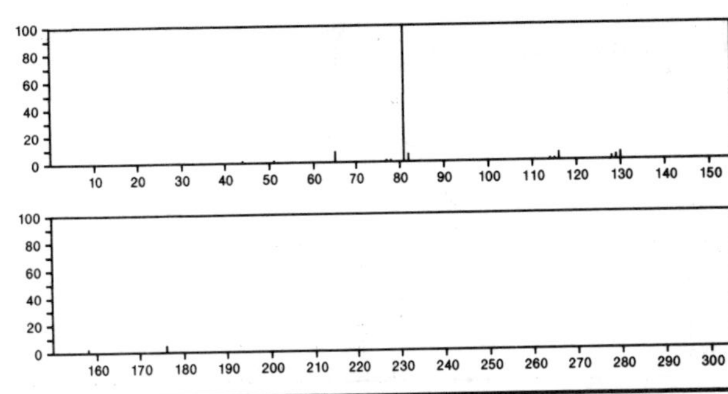

176 $C_{11}H_{12}S$ 16587-32-9
Benzo[*b*]thiophene, 2-propyl-

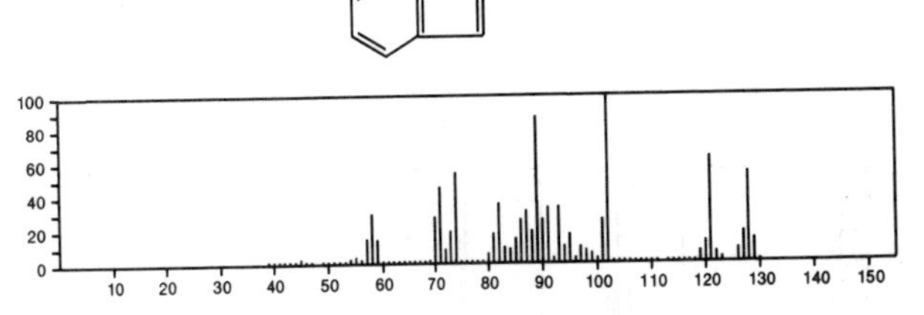

176 $C_{11}H_{12}S$ 16587-43-2
Benzo[*b*]thiophene, 2-ethyl-7-methyl-

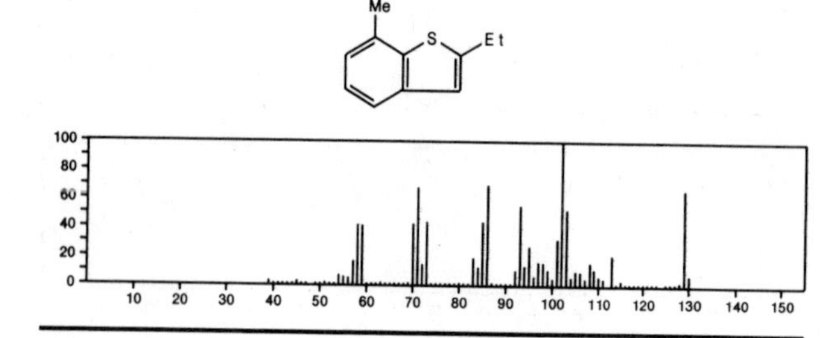

176 $C_{11}H_{12}S$ 16587-44-3
Benzo[*b*]thiophene, 7-ethyl-2-methyl-

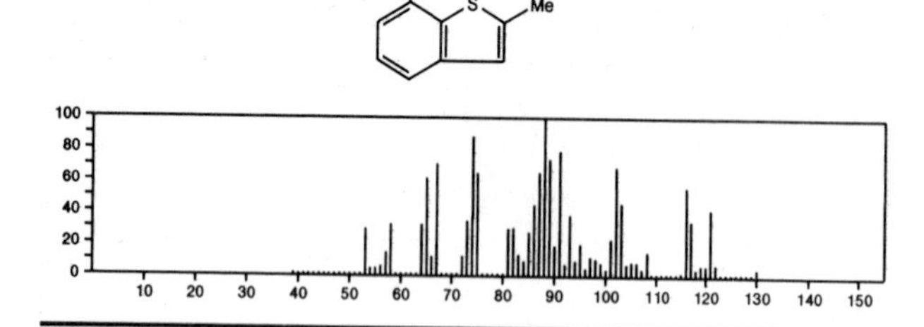

176 $C_{11}H_{12}S$ 16587-51-2
Benzo[*b*]thiophene, 2-ethyl-5-methyl-

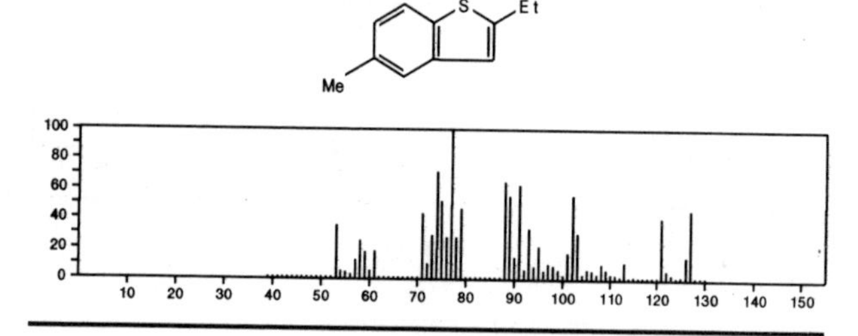

176 $C_{11}H_{12}S$ 16587-65-8
Benzo[*b*]thiophene, 2,5,7-trimethyl-

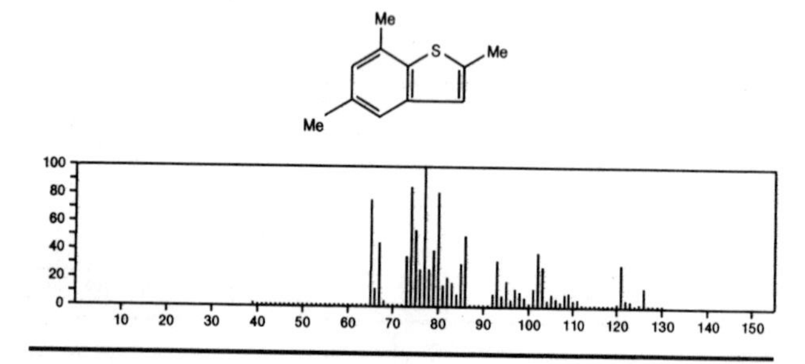

176 $C_{11}H_{16}N_2$ 16738-90-2
1,1-Cyclopropanedicarbonitrile, 2-methyl-2-pentyl-

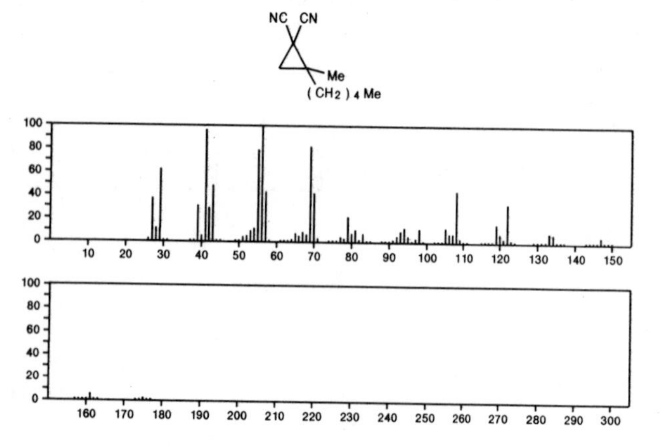

176 $C_{11}H_{16}N_2$ 19730-04-2
Pyridine, 2-(1-methyl-2-pyridinyl)-

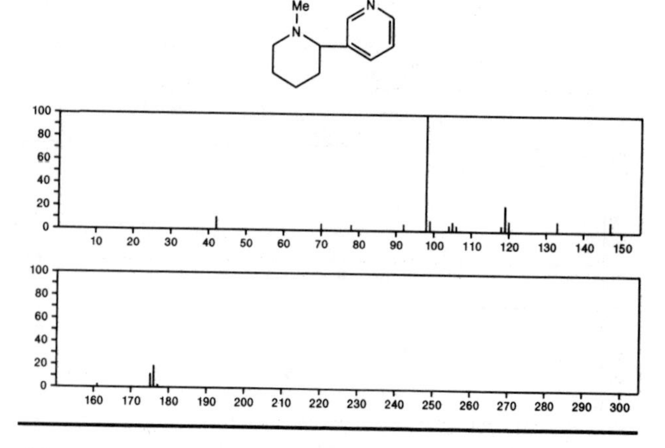

176 $C_{11}H_{16}N_2$ 23229-37-0
Imidazolidine, 1,3-dimethyl-2-phenyl-

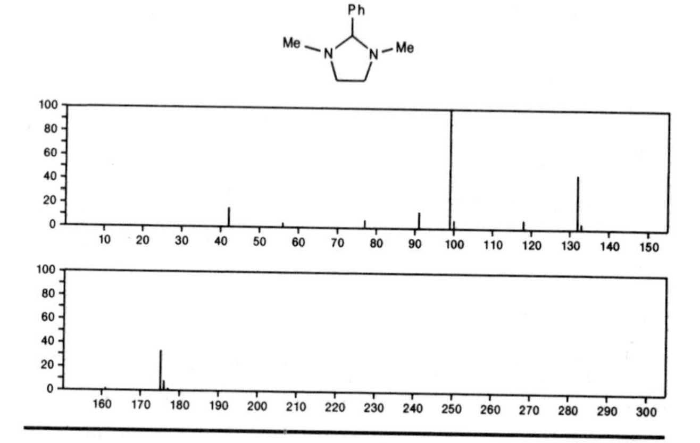

176 $C_{11}H_{16}N_2$ 24380-92-5
Pyridine, 3-(1-methyl-2-piperidinyl)-, (*S*)-

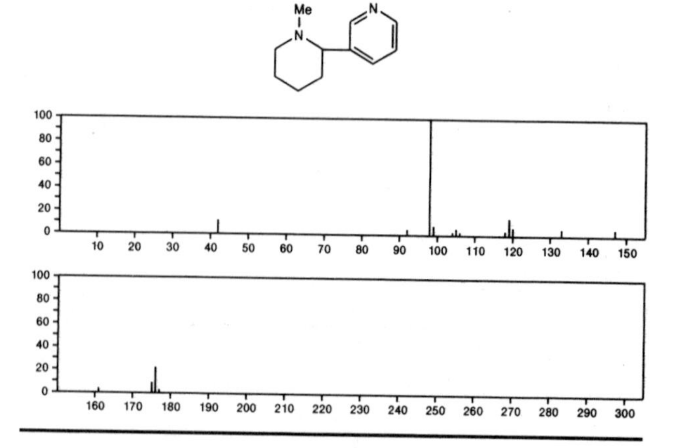

176 $C_{12}H_{16}O$ 119-42-6
Phenol, 2-cyclohexyl-

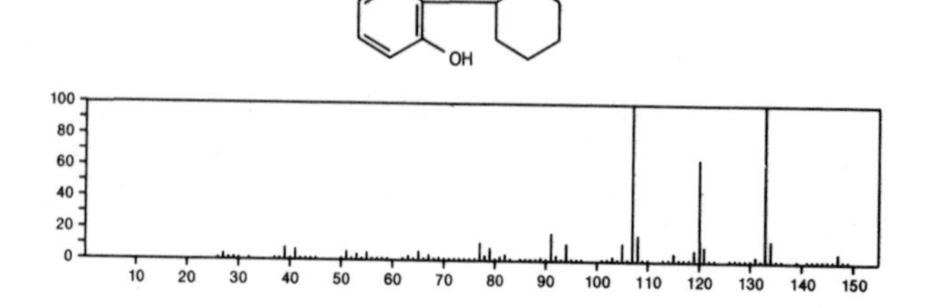

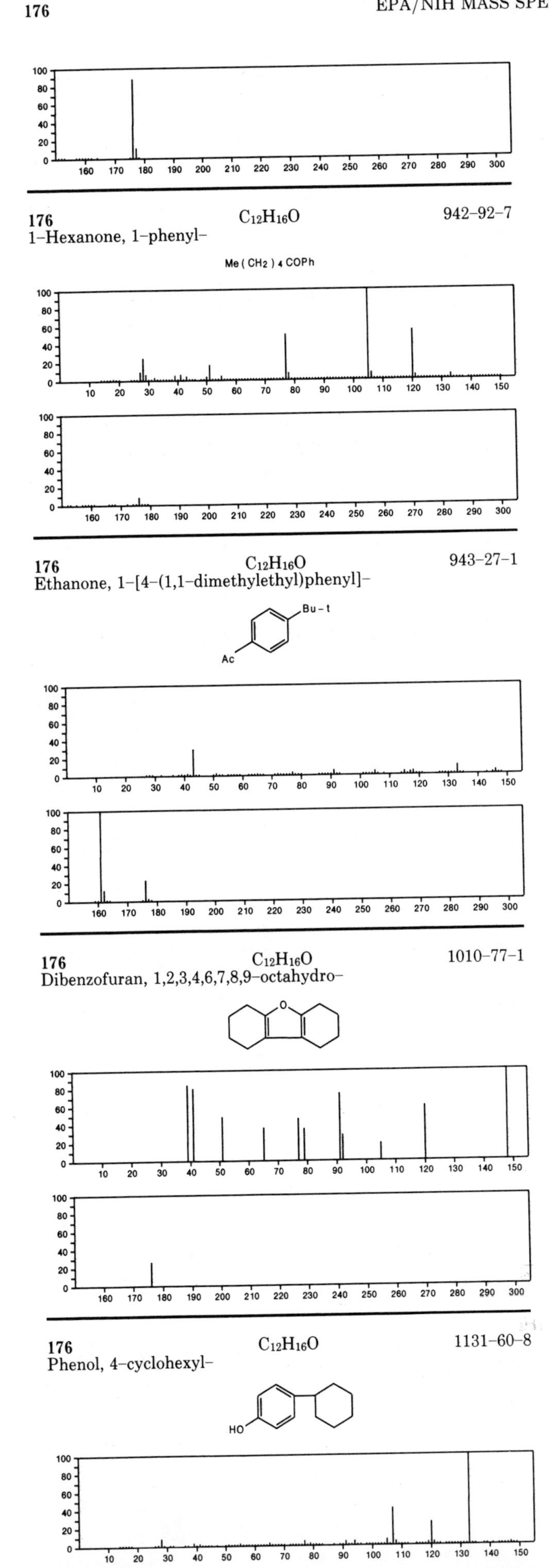

176 $C_{12}H_{16}O$ 942-92-7
1-Hexanone, 1-phenyl-

176 $C_{12}H_{16}O$ 943-27-1
Ethanone, 1-[4-(1,1-dimethylethyl)phenyl]-

176 $C_{12}H_{16}O$ 1010-77-1
Dibenzofuran, 1,2,3,4,6,7,8,9-octahydro-

176 $C_{12}H_{16}O$ 1131-60-8
Phenol, 4-cyclohexyl-

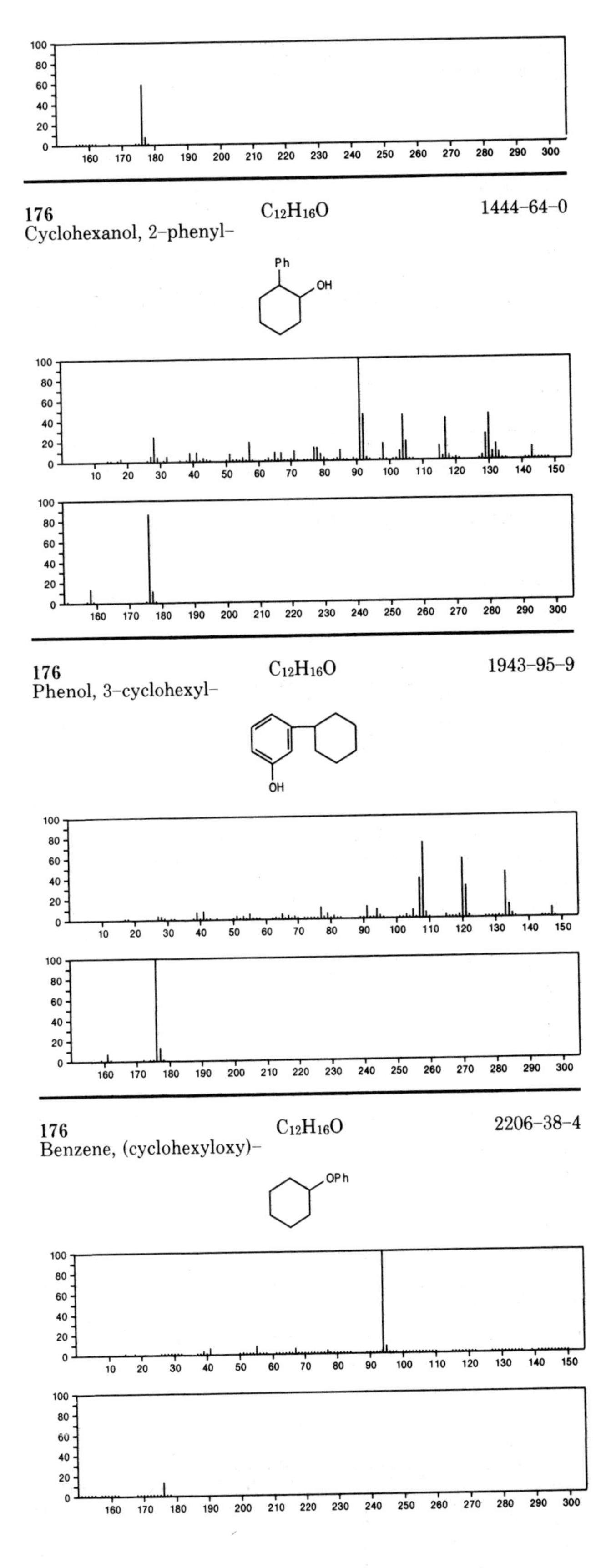

176 $C_{12}H_{16}O$ 1444-64-0
Cyclohexanol, 2-phenyl-

176 $C_{12}H_{16}O$ 1943-95-9
Phenol, 3-cyclohexyl-

176 $C_{12}H_{16}O$ 2206-38-4
Benzene, (cyclohexyloxy)-

176 $C_{12}H_{16}O$ 5437-46-7
Cyclohexanol, 4-phenyl-

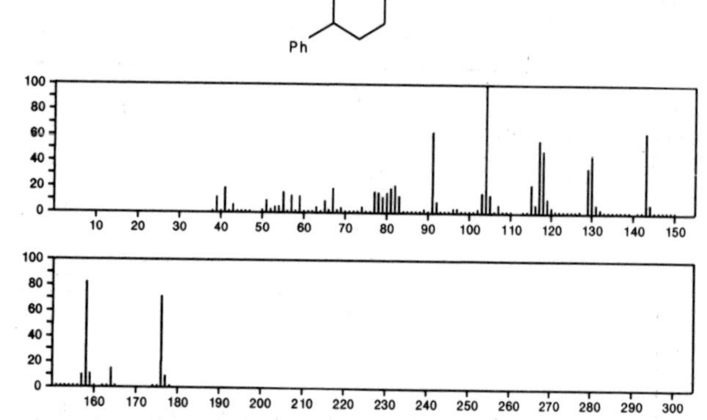

176 $C_{12}H_{16}O$ 7403-42-1
2-Pentanone, 4-methyl-4-phenyl-

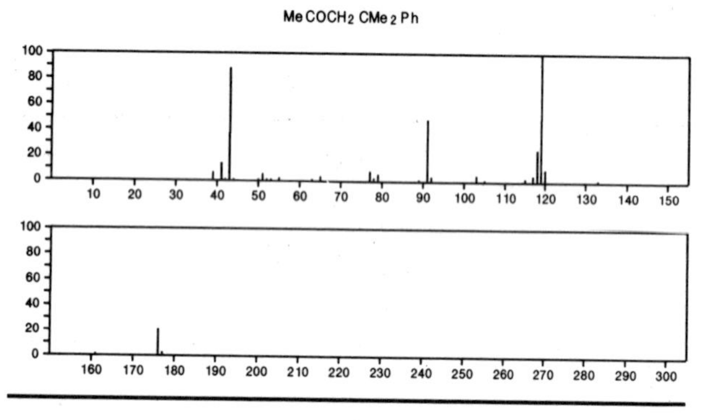

176 $C_{12}H_{16}O$ 14171-89-2
2-Hexanone, 6-phenyl-

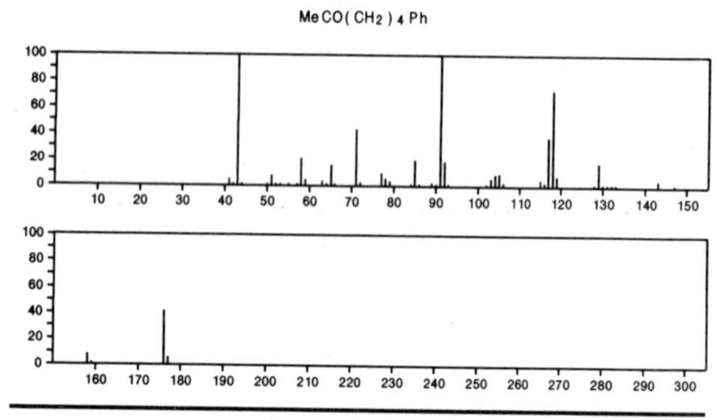

176 $C_{12}H_{16}O$ 16282-15-8
Benzene, (1-ethoxy-2-methyl-1-propenyl)-

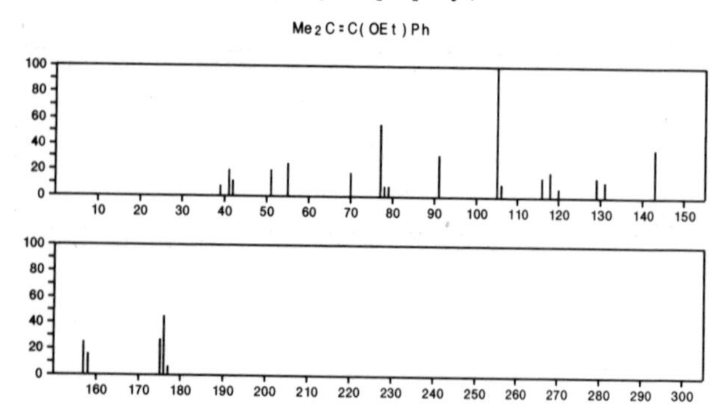

176 $C_{12}H_{16}O$ 28068-45-3
1,4:5,8-Dimethanonaphthalen-9-ol, 1,4,4a,5,6,7,8,8a-octahydro-, stereoisomer

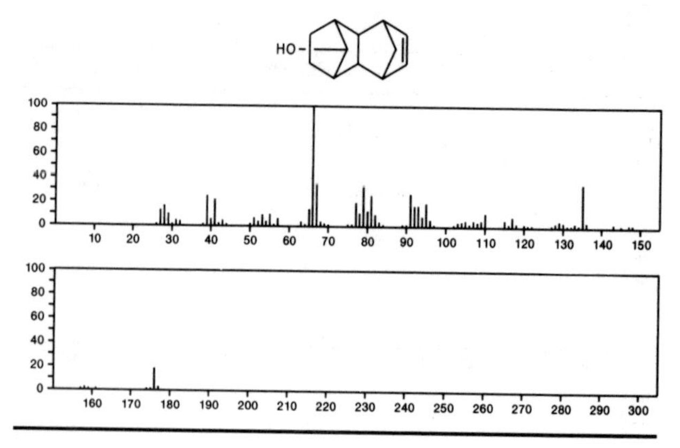

176 $C_{12}H_{16}O$ 29898-25-7
3-Hexanone, 1-phenyl-

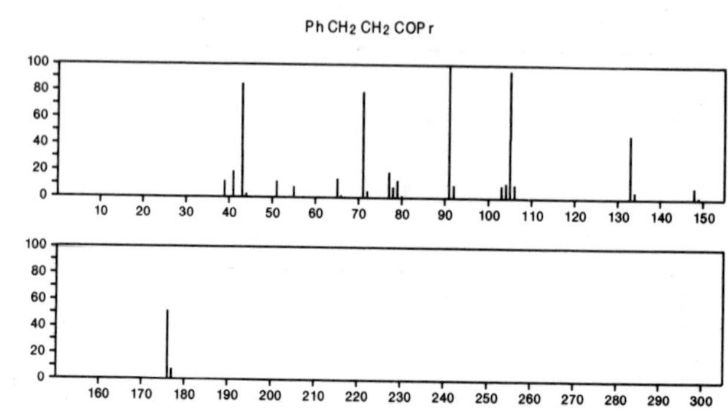

176 $C_{12}H_{16}O$ 36052-28-5
1-Naphthalenemethanol, 1,2,3,4-tetrahydro-8-methyl-

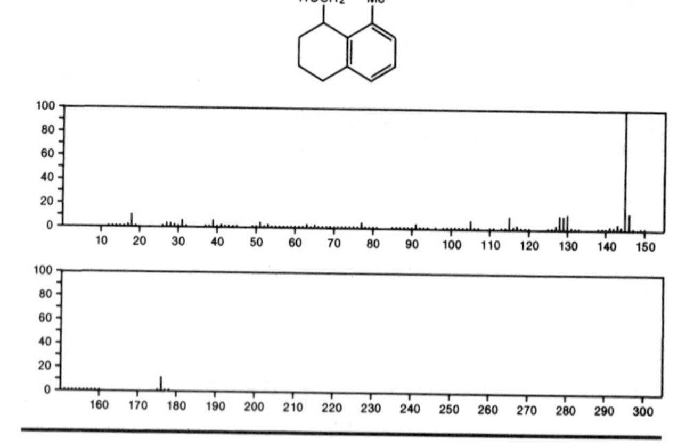

176 $C_{12}H_{16}O$ 38409-40-4
1,4,:5,8-Dimethanonaphthalen-2-ol, 1,2,3,4,4a,5,8,8a-octahydro-, (1α,2α,4α,4aα,5α,8α,8aα)-

OH

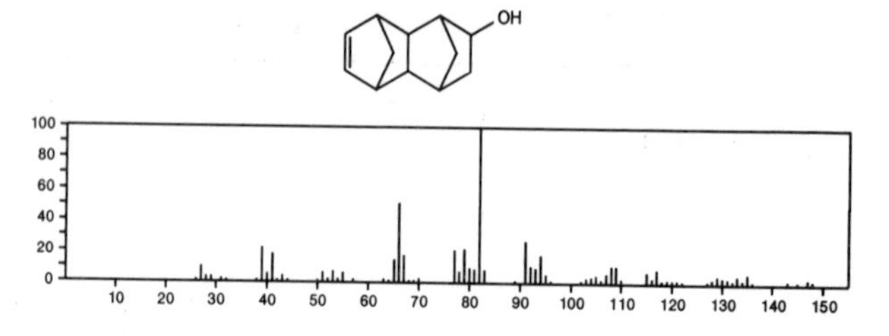

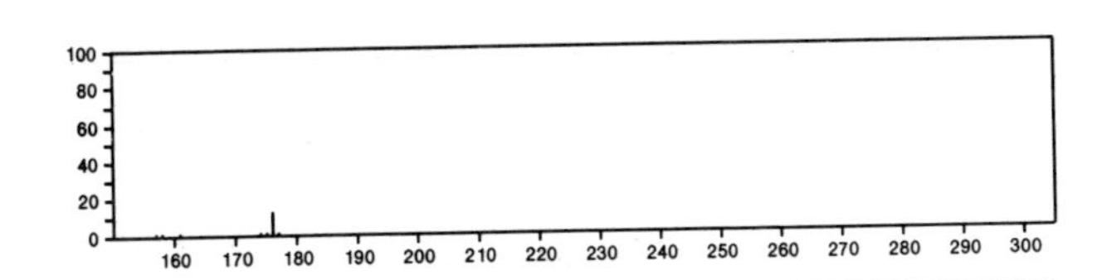

176 $C_{12}H_{16}O$ 50506-60-0
2-Cyclopenten-1-one, 2,3,5-trimethyl-4-methylene-5-(1-methylethenyl)-

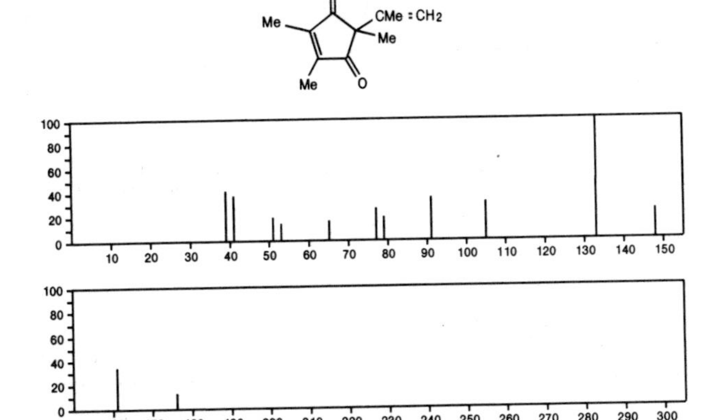

176 $C_{12}H_{16}O$ 54518-01-3
Benzene, (3-propoxy-1-propenyl)-

PrOCH2CH=CHPh

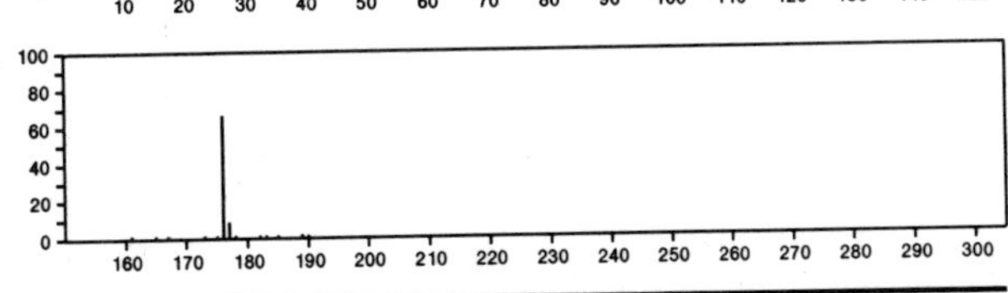

176 $C_{12}H_{16}O$ 55103-77-0
2(1*H*)-Naphthalenone, 4a,5,8,8a-tetrahydro-4,4a-dimethyl-, *trans*-

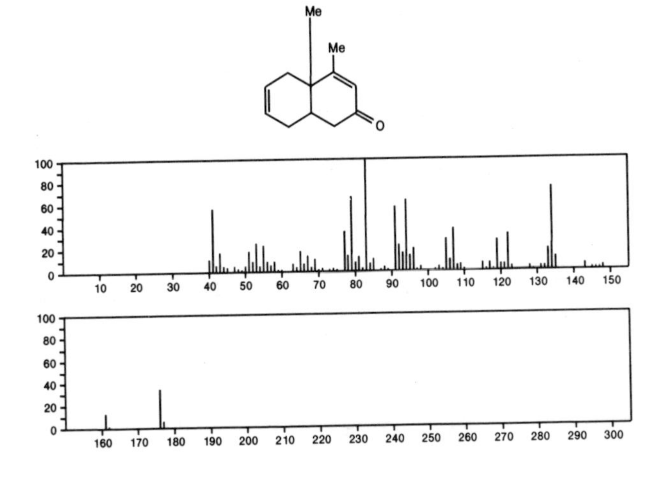

176 $C_{12}H_{16}O$ 55591-09-8
1*H*-Indene-4-methanol, 2,3-dihydro-1,1-dimethyl-

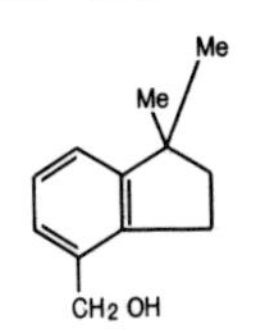

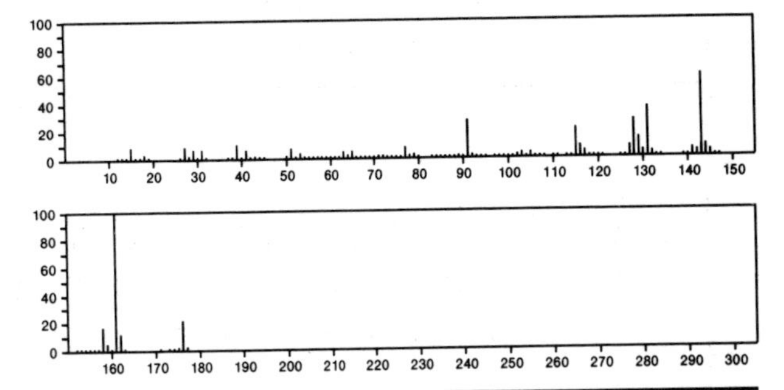

176 $C_{13}H_{20}$ 4170-84-7
Benzene, (1,1-diethylpropyl)-

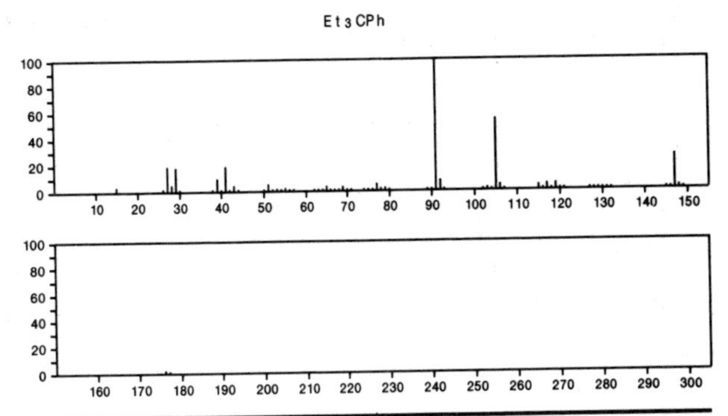

176 $C_{13}H_{20}$ 4468-40-0
Benzene, (1-ethyl-1-methylbutyl)-

CPhEtEtPr

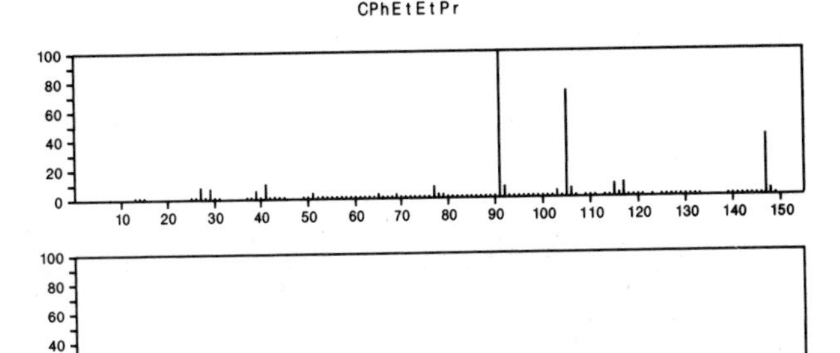

176 $C_{13}H_{20}$ 6630-01-9
Benzene, 1-(1,1-dimethylethyl)-3-ethyl-5-methyl-

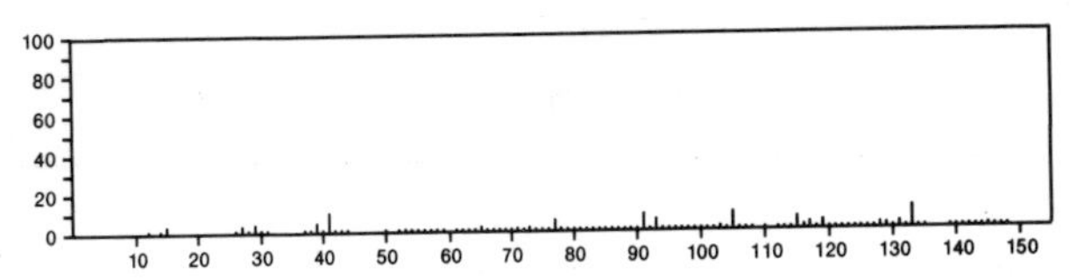

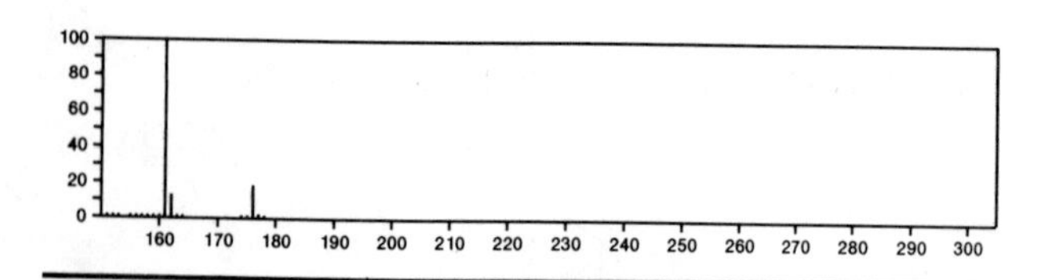

176 $C_{13}H_{20}$ 21777-84-4
Benzene, [2-methyl-1-(1-methylethyl)propyl]-

Me2CHCHPhCHMe2

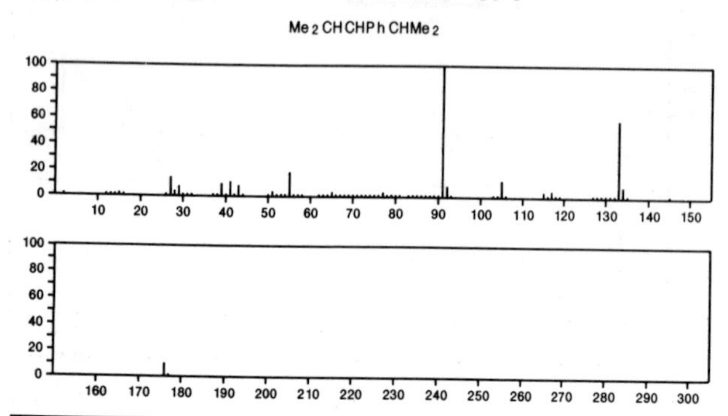

176 $C_{13}H_{20}$ 54518-00-2
Benzene, (2,4-dimethylpentyl)-

PhCH2CHMeCH2CHMe2

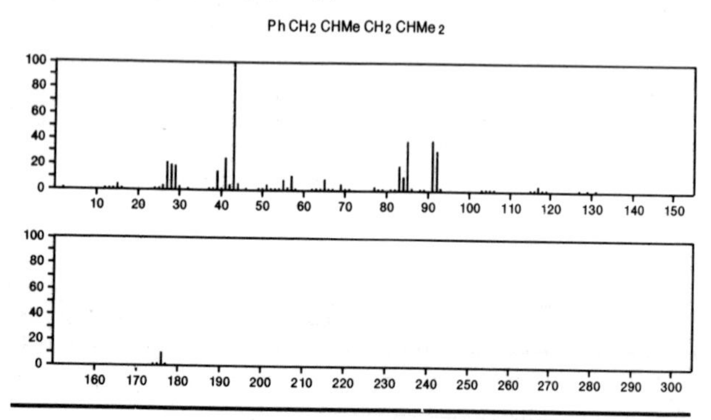

176 $C_{13}H_{20}$ 55638-52-3
3,5-Dodecadiyne, 2-methyl-

Me(CH2)5C≡CC≡CCHMe2

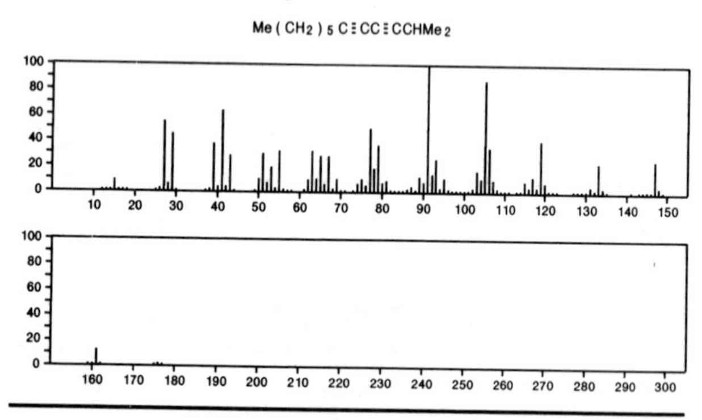

176 $C_{13}H_{20}$ 56248-15-8
Cyclohexene, 6-(1,3-butadienyl)-1,5,5-trimethyl-

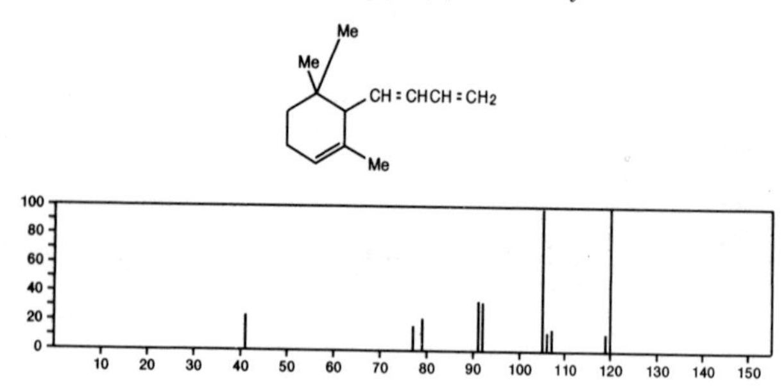

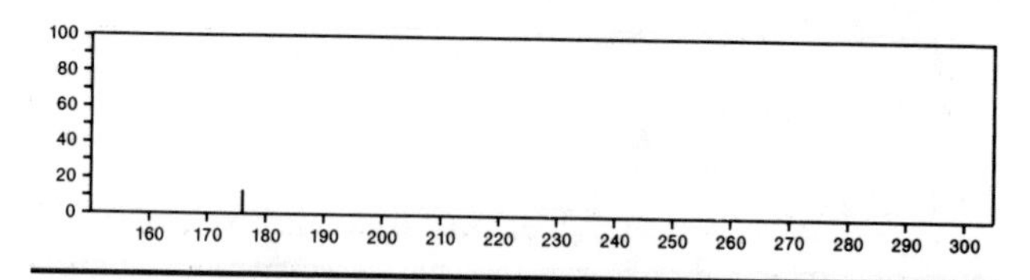

176 $C_{14}H_8$ 5236-46-4
1,3,7,11-Cyclotetradecatetraene-5,9,13-triyne

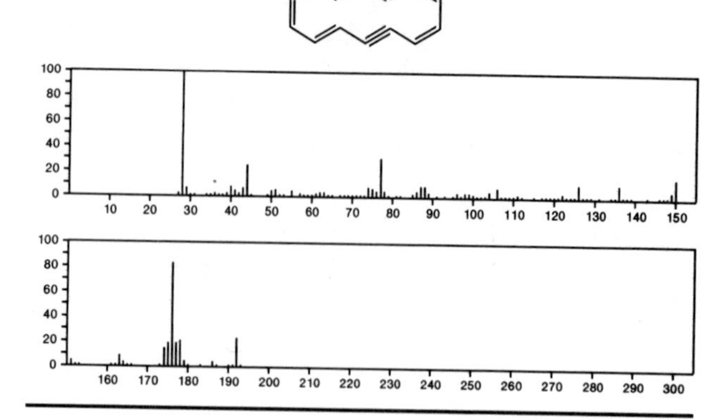

177 C_3IN 2003-32-9
2-Propynenitrile, 3-iodo-

IC≡CCN

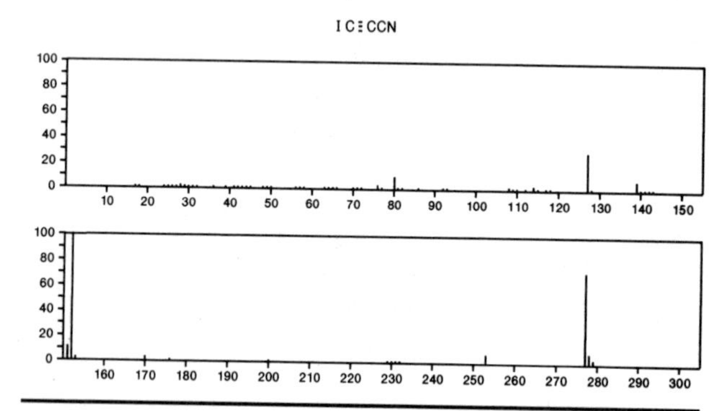

177 C_4H_4BrNS 20493-60-1
Isothiazole, 5-bromo-3-methyl-

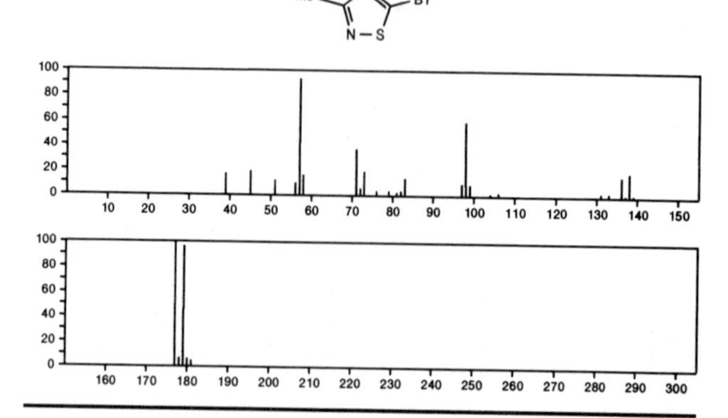

177 $C_5H_7NS_3$ 56248-20-5
Thiazole, 2,5-bis(methylthio)-

MeS SMe

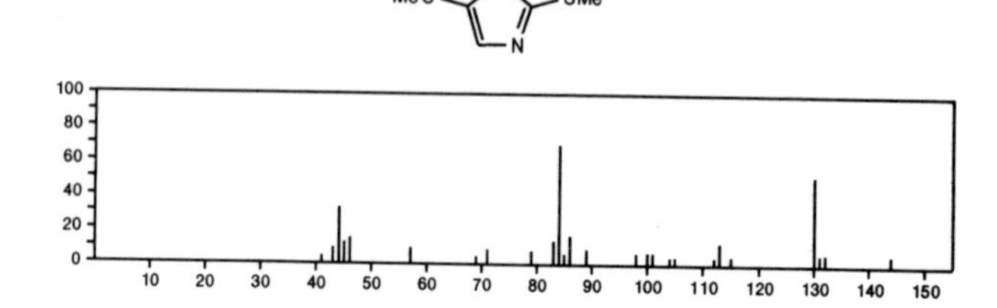

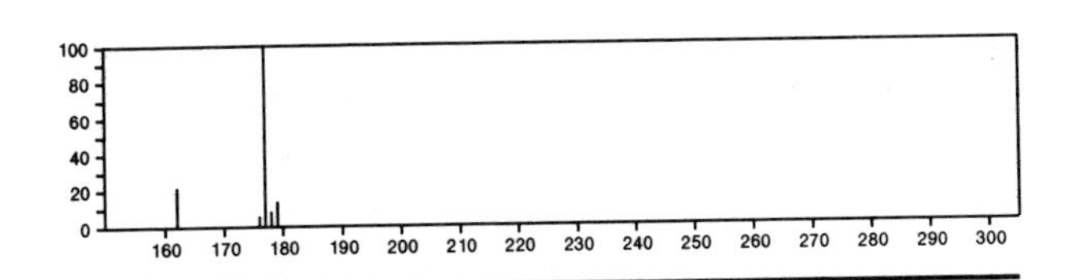

177 $C_6H_6F_3N_3$ 5734-63-4
2-Pyrimidinamine, 4-methyl-6-(trifluoromethyl)-

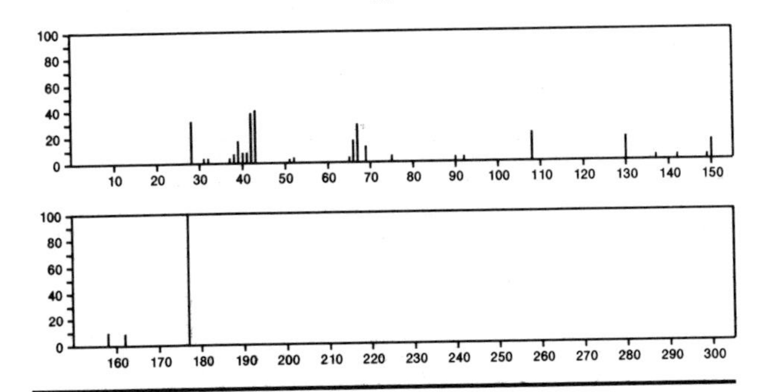

177 $C_6H_6F_3N_3$ 54518-10-4
4-Pyrimidinamine, 2-methyl-6-(trifluoromethyl)-

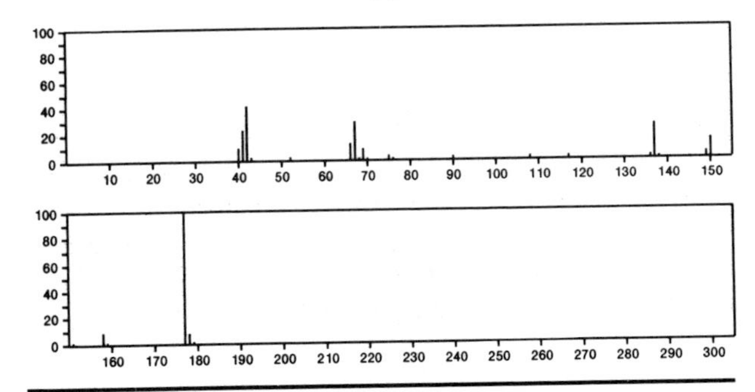

177 $C_6H_{11}NOS_2$ 56909-10-5
2-Oxazolidinethione, 5-[2-(methylthio)ethyl]-

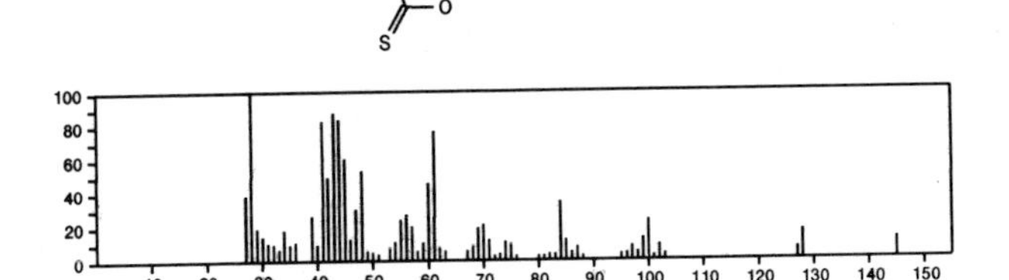

177 $C_7H_7N_5O$ 708-75-8
4(1*H*)-Pteridinone, 2-amino-6-methyl-

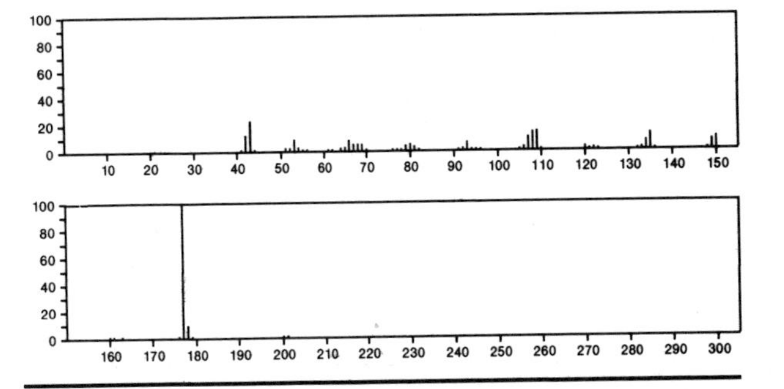

177 $C_7H_7N_5O$ 13040-58-9
4(1*H*)-Pteridinone, 2-amino-7-methyl-

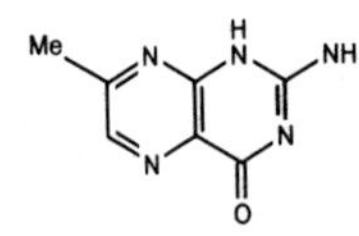

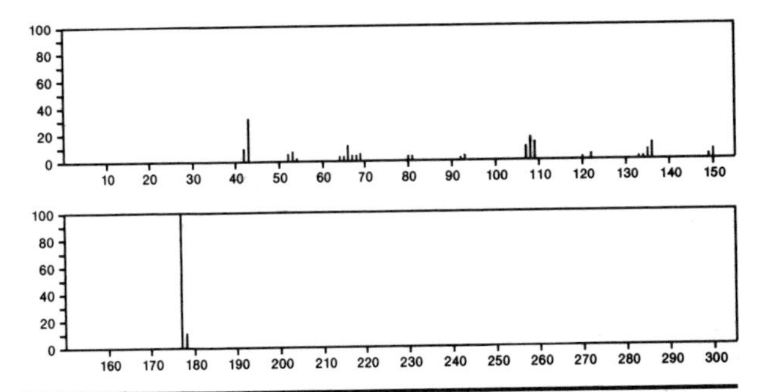

177 $C_7H_{15}NO_2S$ 3082-77-7
L-Methionine, ethyl ester

MeSCH2CH2CH(NH2)C(O)OEt

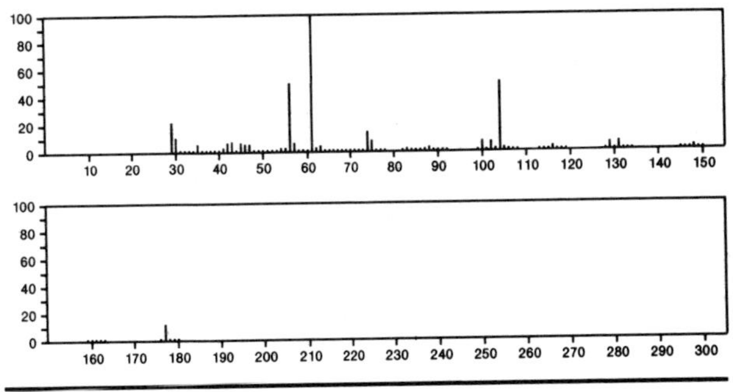

177 $C_7H_{15}NS_2$ 4740-11-8
Carbamodithioic acid, diethyl-, ethyl ester

EtSC(S)NEt2

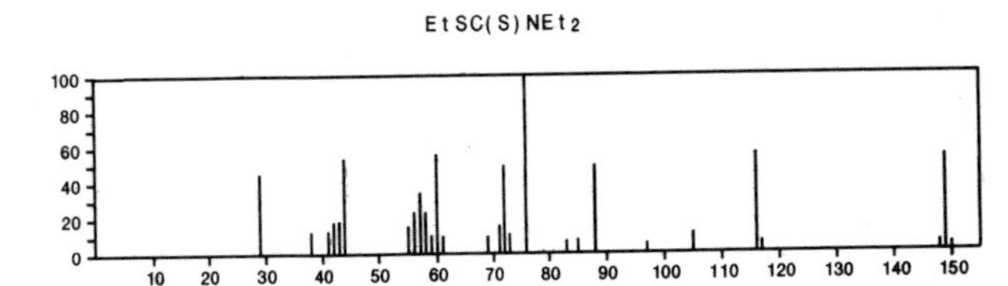

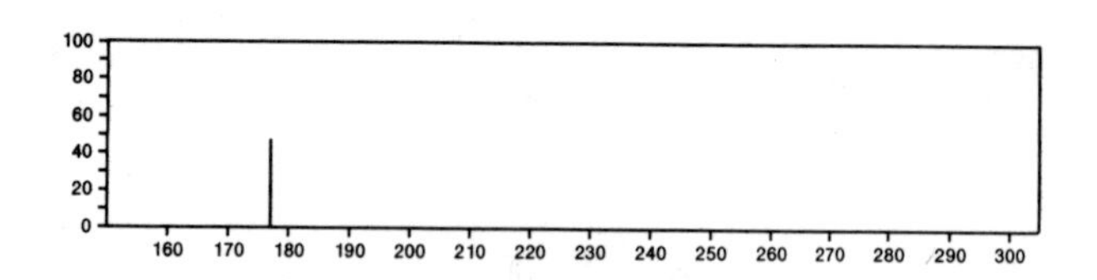

177 $C_8H_7N_3O_2$ 2499-96-9
Pyrido[3,2-*d*]pyrimidine-2,4(1*H*,3*H*)-dione, 6-methyl-

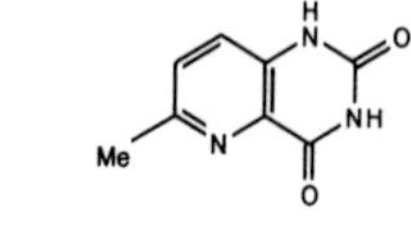

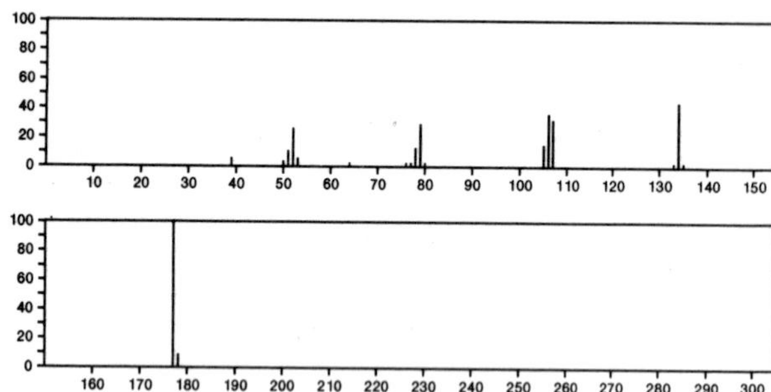

177 $C_8H_7N_3O_2$ 3303-23-9
Pyrido[3,2-*d*]pyrimidin-4(3*H*)-one, 3-hydroxy-2-methyl-

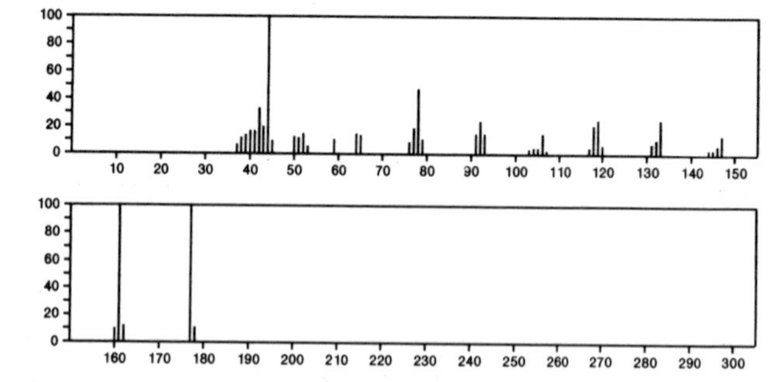

177 $C_8H_7N_3O_2$ 23616-50-4
1*H*-Pyrrolo[2,3-*b*]pyridine, 2-methyl-3-nitro-

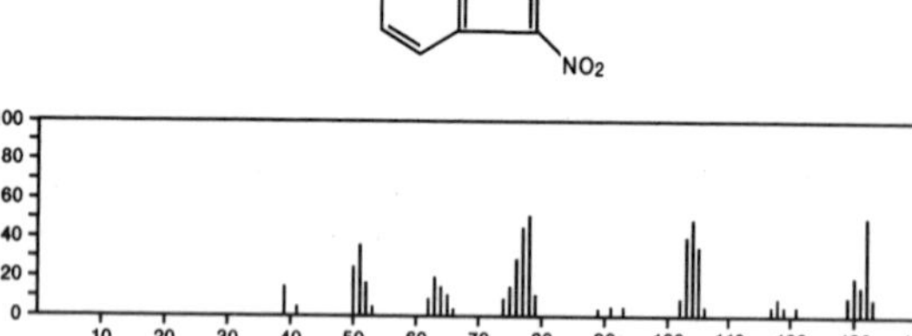

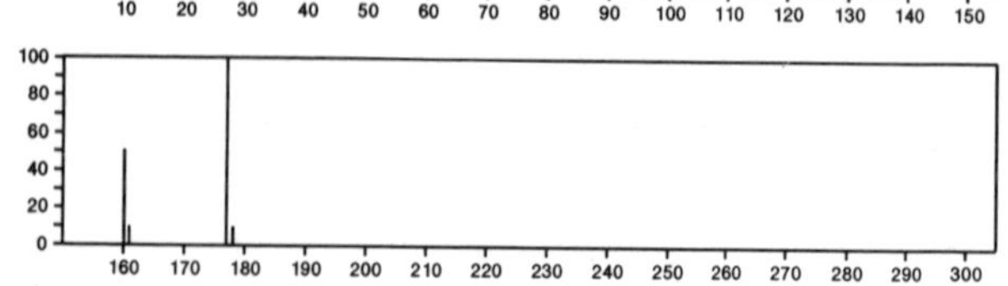

177 $C_8H_7N_3O_2$ 24310-40-5
1,2,3-Benzotriazin-4(3*H*)-one, 3-(hydroxymethyl)-

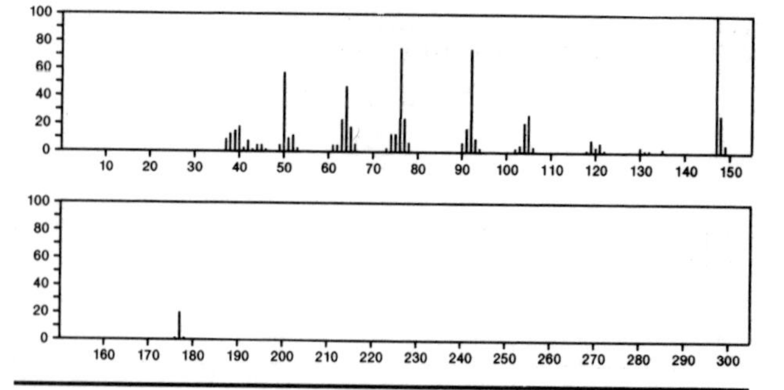

177 $C_8H_7N_3O_2$ 26120-43-4
1*H*-Indazole, 1-methyl-4-nitro-

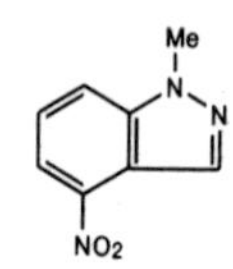

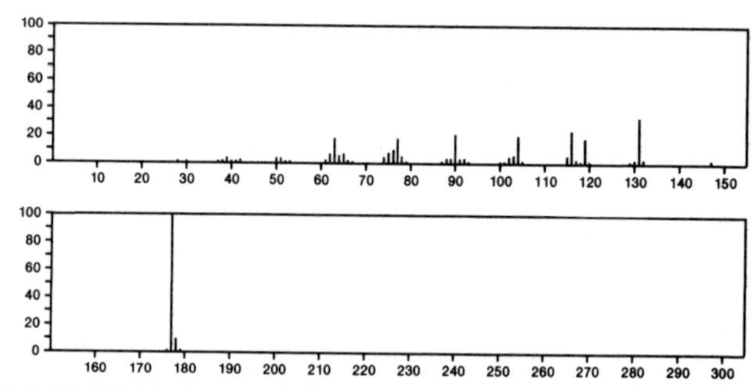

177 $C_8H_7N_3O_2$ 26120-44-5
2*H*-Indazole, 2-methyl-4-nitro-

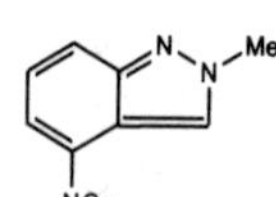

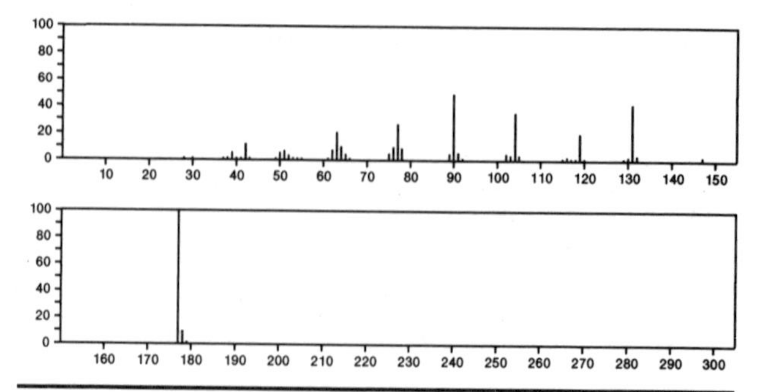

177 $C_8H_7N_3S$ 17467-15-1
1,2,4-Thiadiazol-5-amine, 3-phenyl-

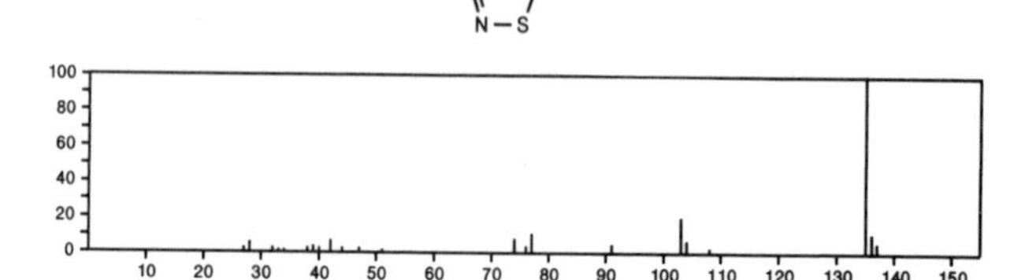

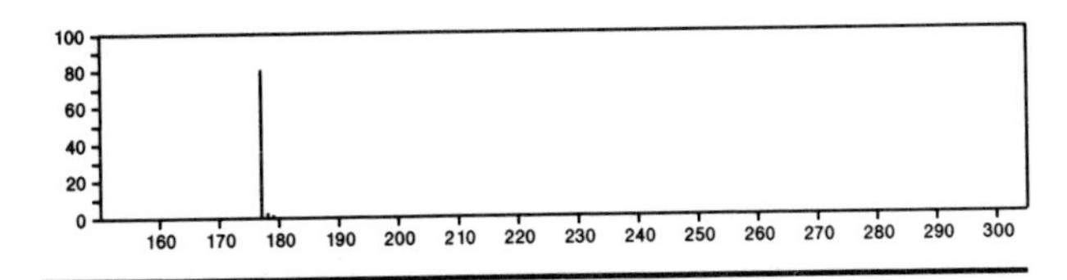

177 $C_8H_7N_3S$ 20970–15–4
Pyrido[2,3–*d*]pyridazine, 5–(methylthio)–

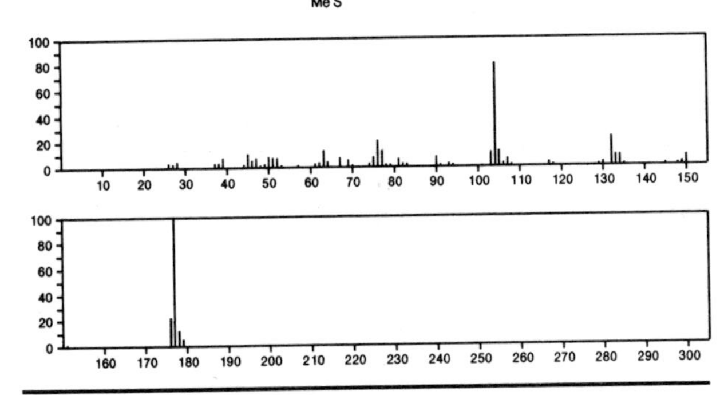

177 C_9H_7NOS 35524–66–4
3*H*–Indol–3–one, 2–(methylthio)–

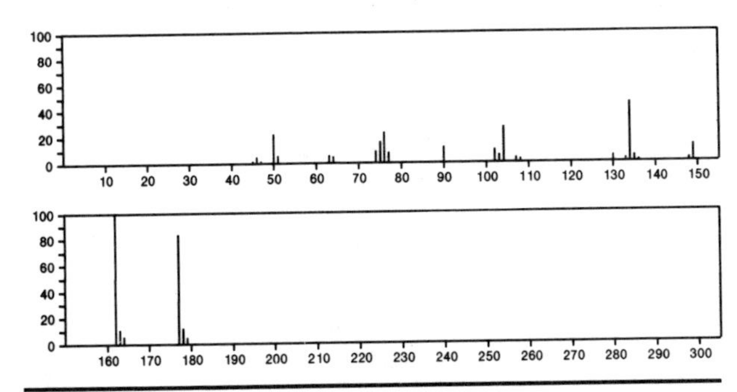

177 $C_9H_7NO_3$ 17175–18–7
Carbonic acid, methyl ester, ester with *p*–hydroxybenzonitrile

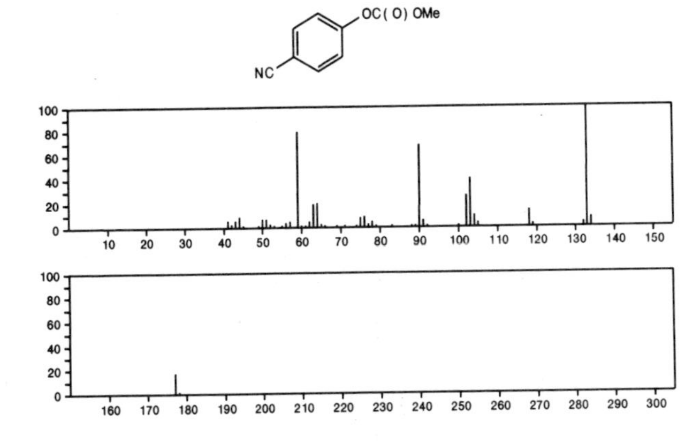

177 $C_9H_7NO_3$ 17175–19–8
Carbonic acid, methyl ester, ester with *m*–hydroxybenzonitrile

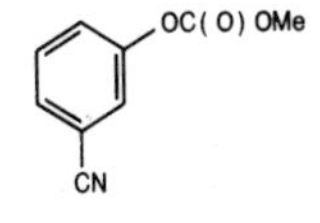

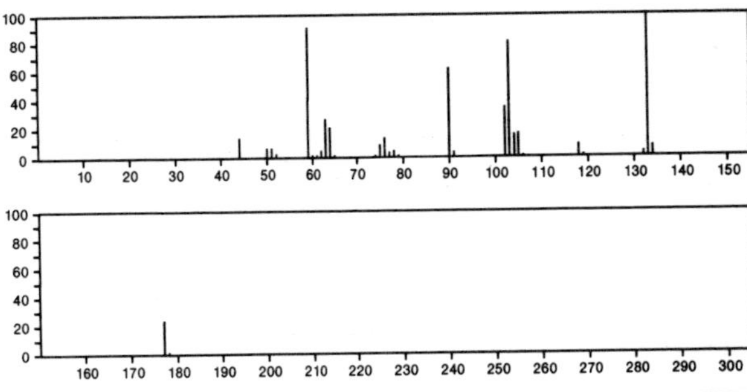

177 $C_9H_7NO_3$ 21201–44–5
Carbostyril, 1,4–dihydroxy–

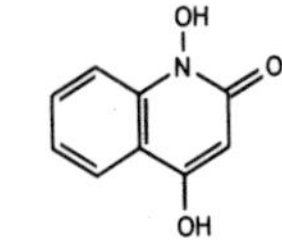

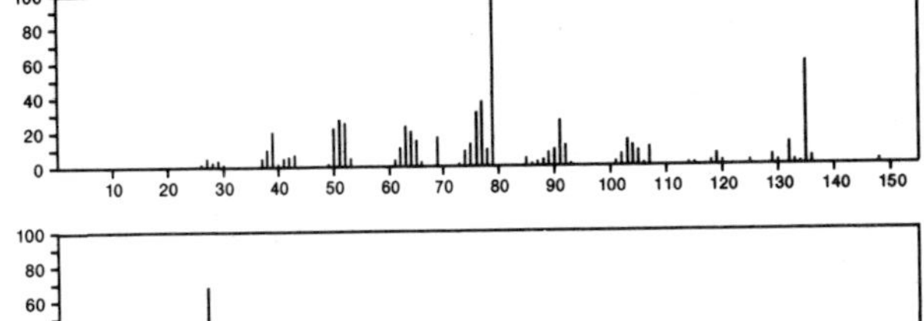

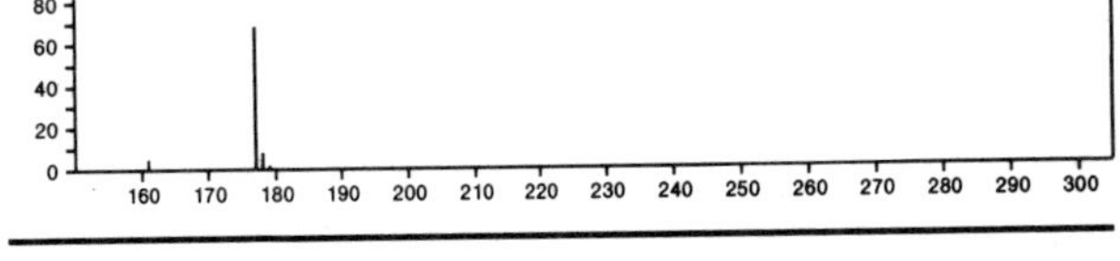

177 $C_9H_{11}N_3O$ 2492–30–0
Hydrazinecarboxamide, 2–(1–phenylethylidene)–

PhCMe:NNHCONH₂

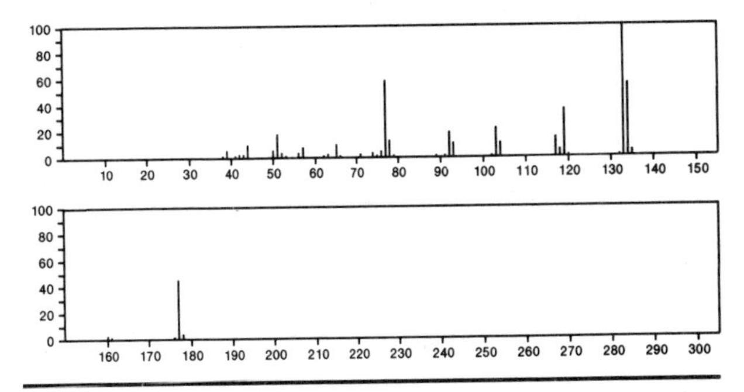

177 $C_{10}H_{11}NO_2$ 154–02–9
1*H*–Indole–3–ethanol, 5–hydroxy–

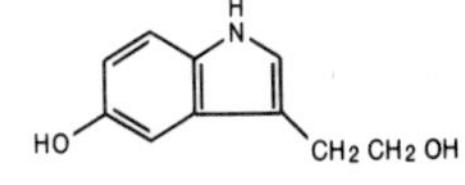

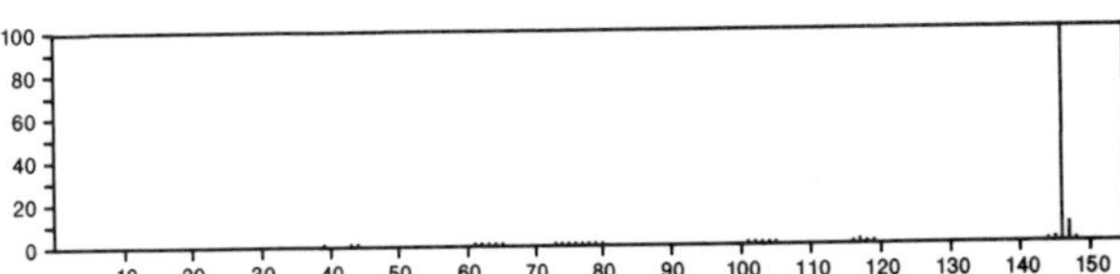

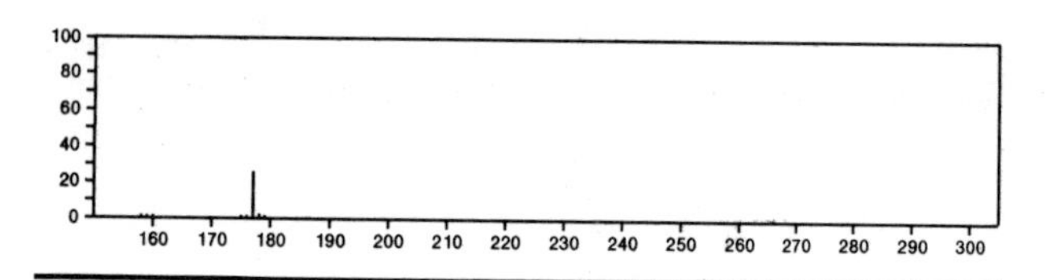

177 $C_{10}H_{11}NO_2$ 1563-87-7
Acetamide, *N*-acetyl-*N*-phenyl-

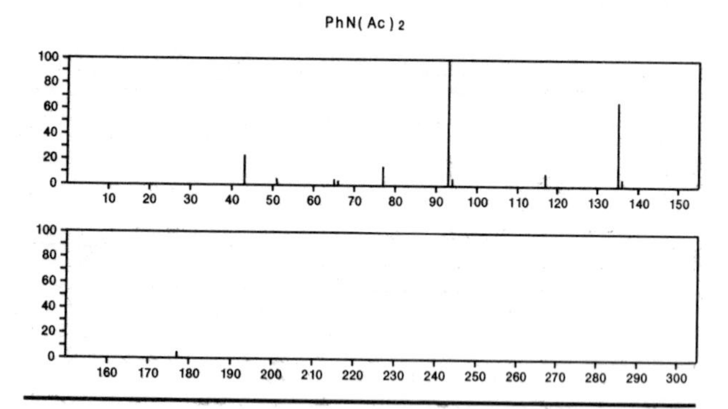

177 $C_{10}H_{11}NO_2$ 5552-45-4
2-Indolinol, 1-acetyl-

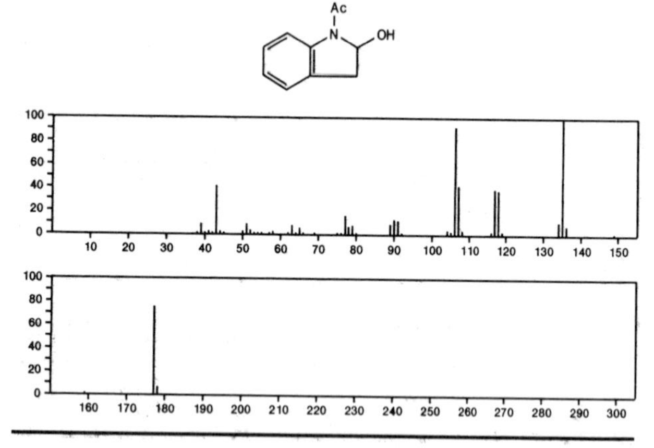

177 $C_{10}H_{11}NO_2$ 13303-69-0
1-Indolinecarboxaldehyde, 2-hydroxy-5-methyl-

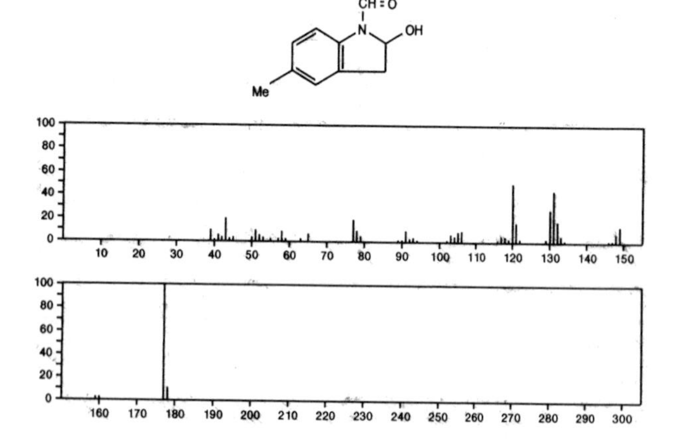

177 $C_{10}H_{11}NO_2$ 19353-86-7
Naphthalene, 1,2,3,4-tetrahydro-6-nitro-

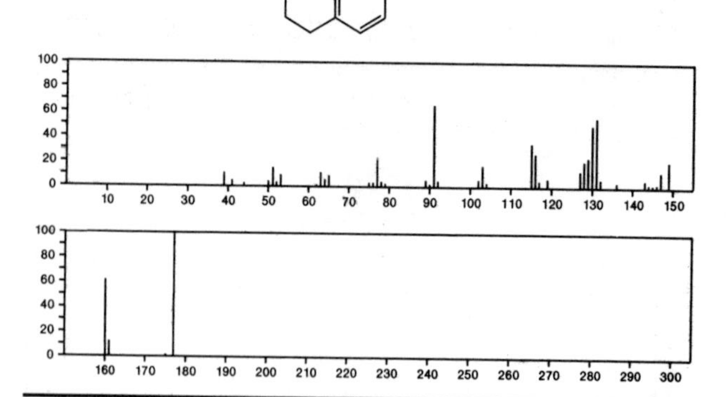

177 $C_{10}H_{11}NO_2$ 19901-85-0
2-Oxazolidinone, 5-methyl-4-phenyl-, *trans*-

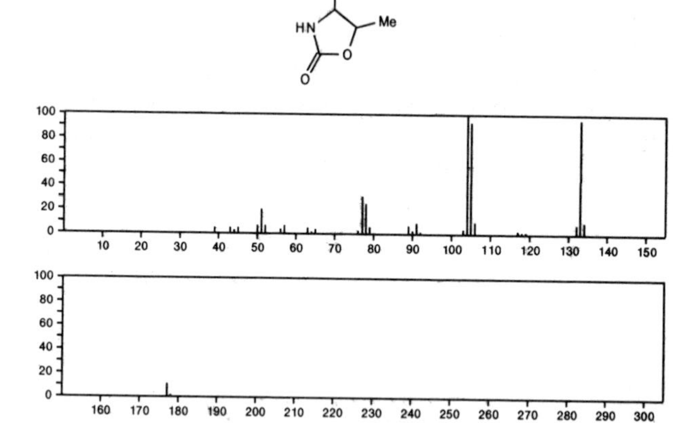

177 $C_{10}H_{11}NO_2$ 19901-86-1
2-Oxazolidinone, 5-methyl-4-phenyl-, *cis*-

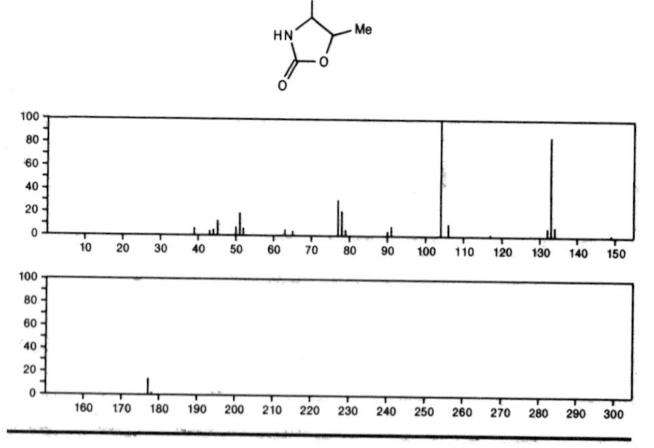

177 $C_{10}H_{11}NO_2$ 22614-65-9
1-Indolinecarboxaldehyde, 2-hydroxy-7-methyl-

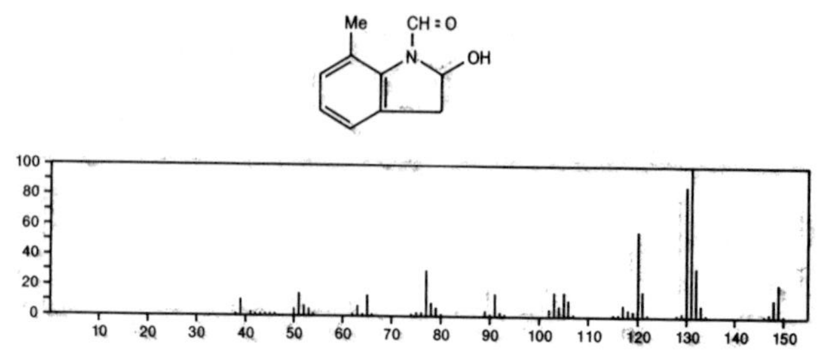

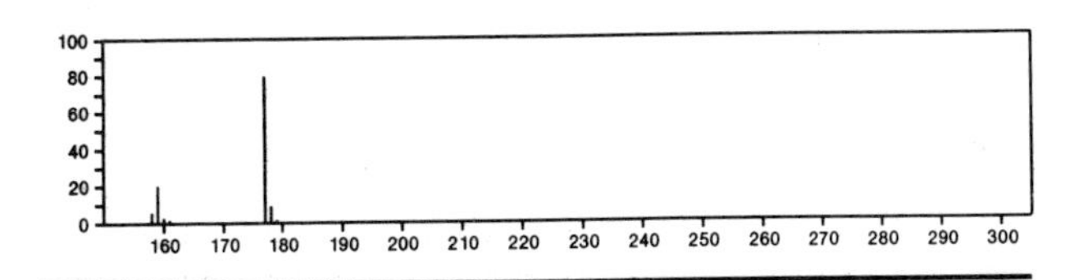

177 $C_{10}H_{11}NO_2$ 28044-22-6
2-Oxazolidinone, 4-methyl-5-phenyl-, *cis*-

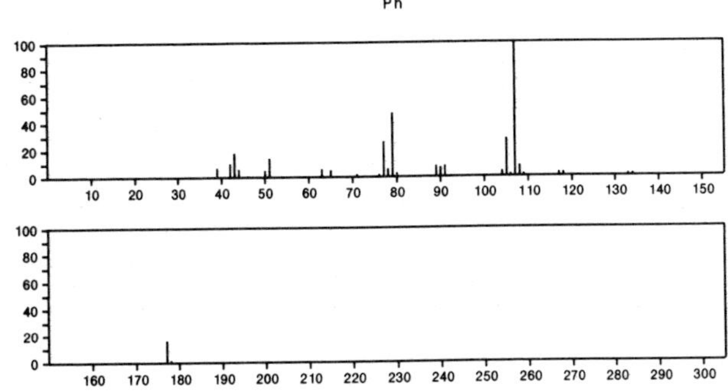

177 $C_{10}H_{11}NO_2$ 29809-14-1
Naphthalene, 1,2,3,4-tetrahydro-5-nitro-

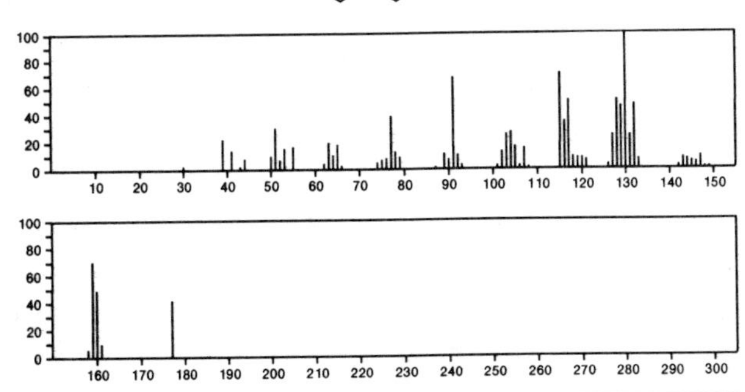

177 $C_{10}H_{11}NO_2$ 50838-16-9
2-Propenoic acid, 3-(di-2-propynylamino)-, methyl ester

$CH_2C\equiv CH$
$MeOC(O)CH=CHNCH_2C\equiv CH$

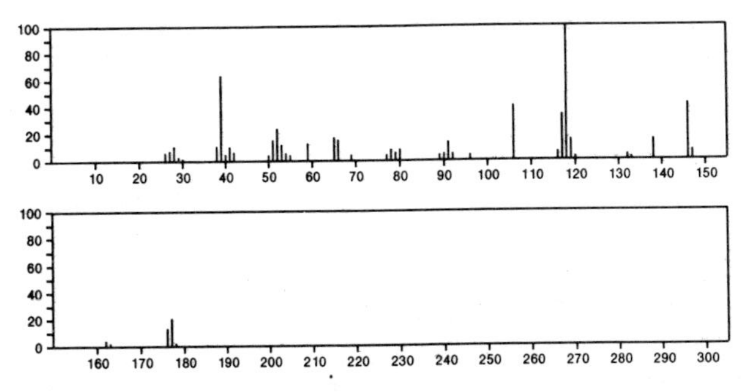

177 $C_{10}H_{11}NO_2$ 51110-93-1
2*H*-1,6-Benzoxazocin-5(6*H*)-one, 3,4-dihydro-

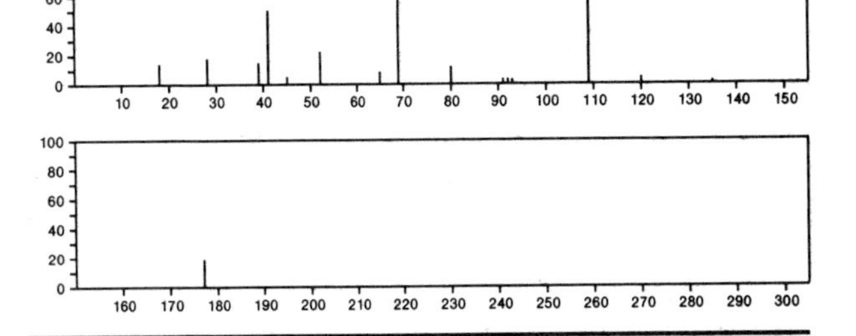

177 $C_{10}H_{11}NO_2$ 51110-99-7
6*H*-1,5-Benzoxazocin-6-one, 2,3,4,5-tetrahydro-

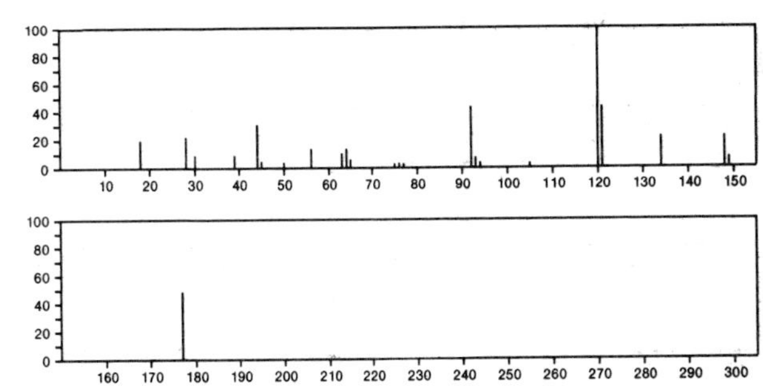

177 $C_{10}H_{11}NS$ 6552-61-0
Isocarbostyril, 3,4-dihydro-2-methylthio-

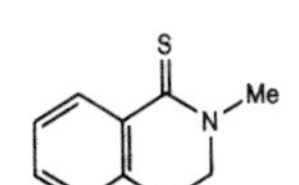

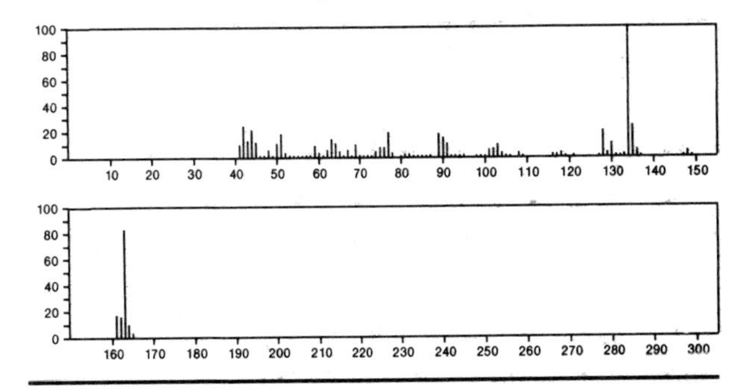

177 $C_{11}H_{15}NO$ 120-21-8
Benzaldehyde, 4-(diethylamino)-

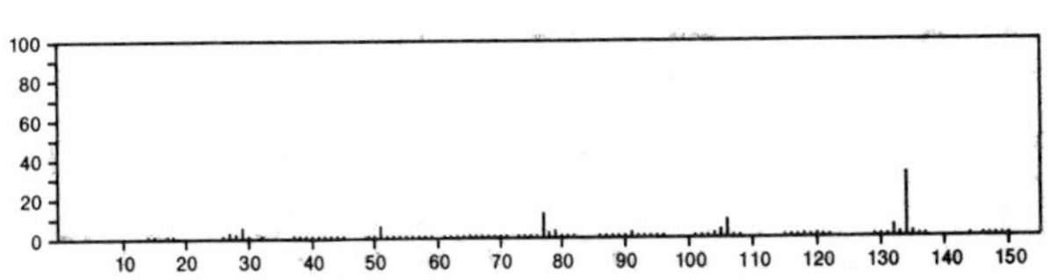

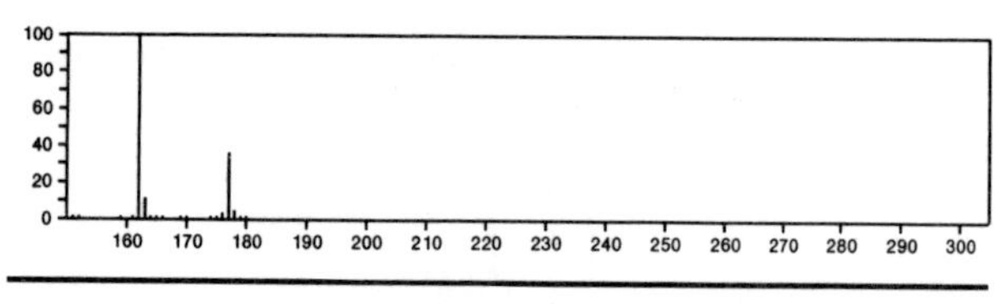

177 $C_{11}H_{15}NO$ 134-49-6
Morpholine, 3-methyl-2-phenyl-

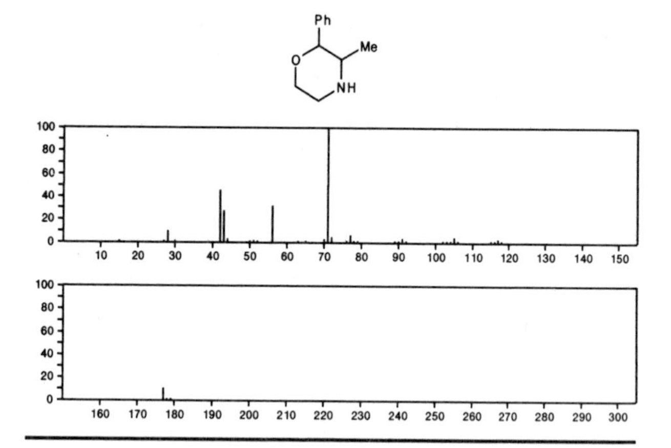

177 $C_{11}H_{15}NO$ 1696-17-9
Benzamide, *N*,*N*-diethyl-

Et₂NCOPh

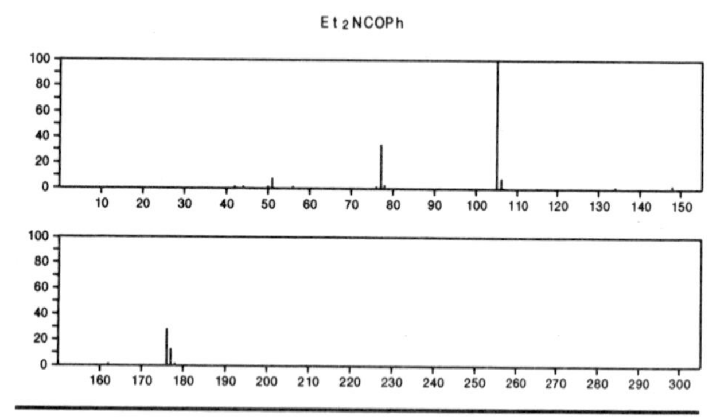

177 $C_{11}H_{15}NO$ 2782-40-3
Benzamide, *N*-butyl-

PhCONH(CH2)3Me

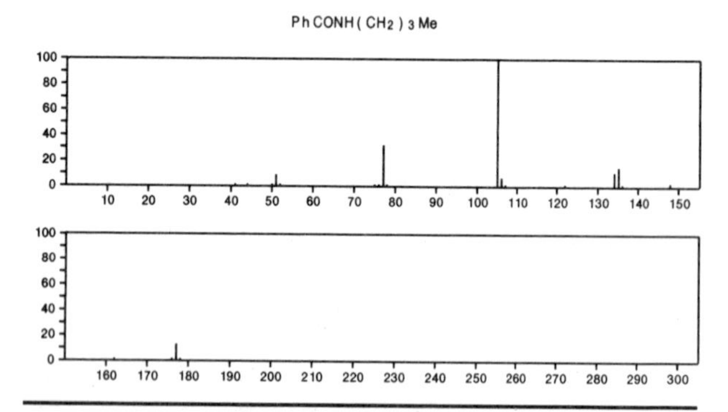

177 $C_{11}H_{15}NO$ 5830-31-9
Benzenepropanamide, *N*,*N*-dimethyl-

PhCH2CH2CONMe2

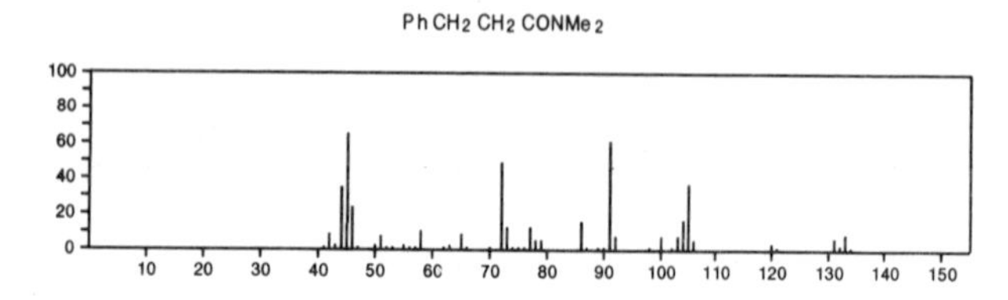

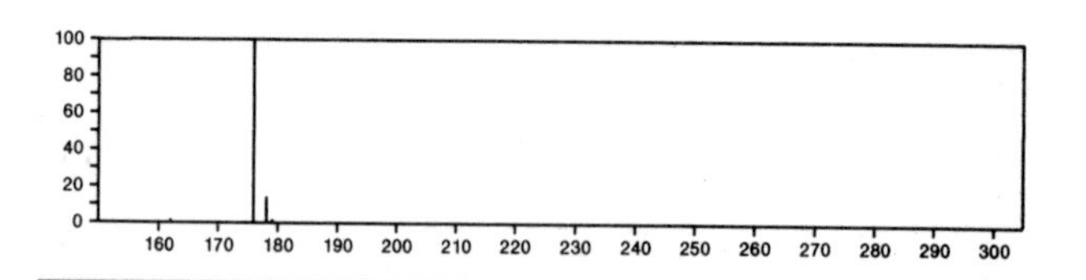

177 $C_{11}H_{15}NO$ 6625-74-7
Propanamide, 2,2-dimethyl-*N*-phenyl-

Me3CCONHPh

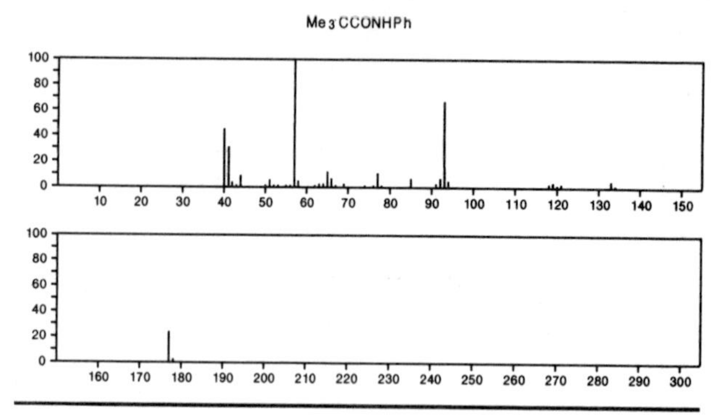

177 $C_{11}H_{15}NO$ 29971-61-7
5-Hexen-3-yn-2-one, 6-(1-piperidinyl)-

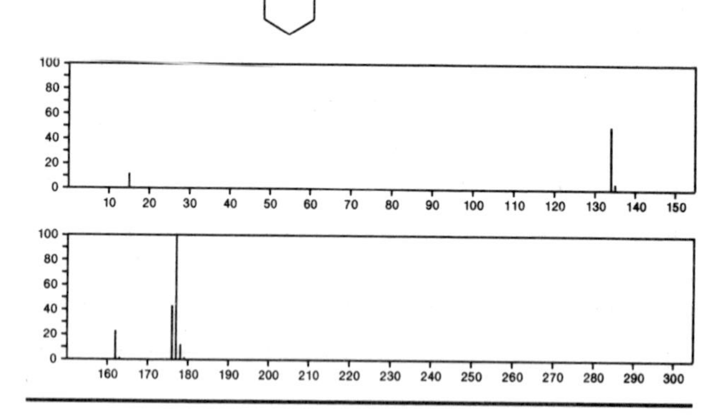

177 $C_{11}H_{15}NO$ 34059-10-4
Acetamide, *N*-(3-phenylpropyl)-

Ph(CH2)3NHAc

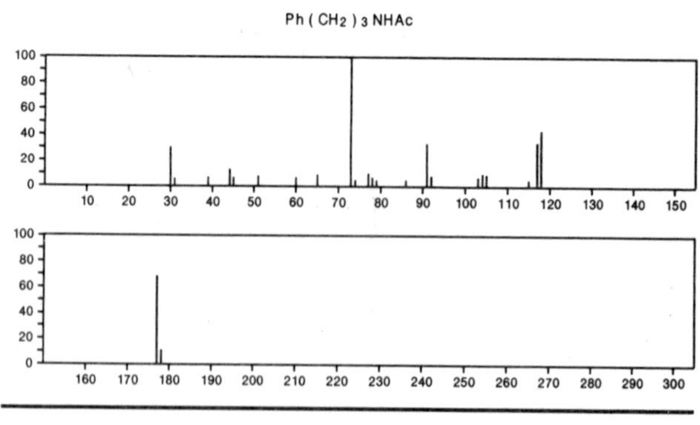

177 $C_{11}H_{15}NO$ 34597-04-1
Acetamide, *N*-ethyl-*N*-(phenylmethyl)-

EtN(Ac)CH2Ph

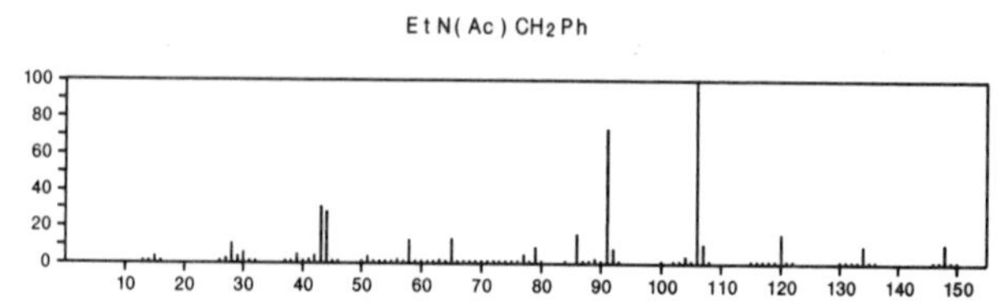

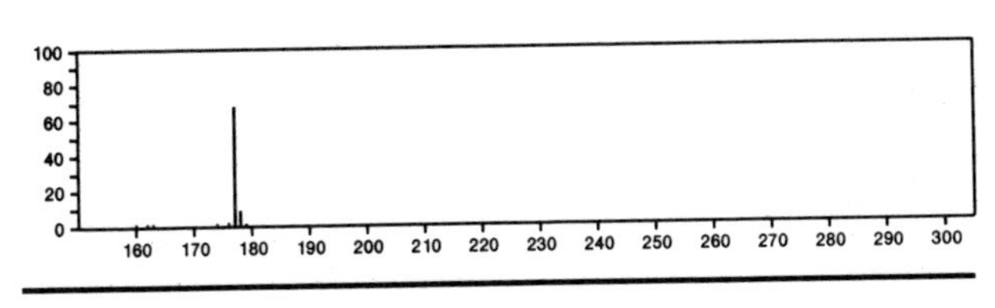

177 $C_{11}H_{15}NO$ 42203-03-2
1-Hexanone, 1-(2-pyridinyl)-

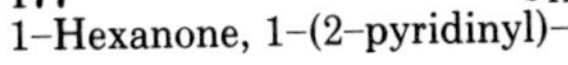

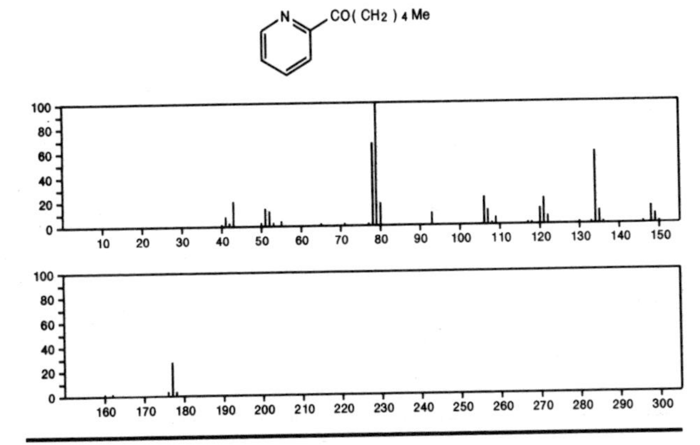

177 $C_{11}H_{15}NO$ 42540-70-5
Benzenamine, *N*-[1-(methoxymethyl)cyclopropyl]-

NHPh
CH2 OMe

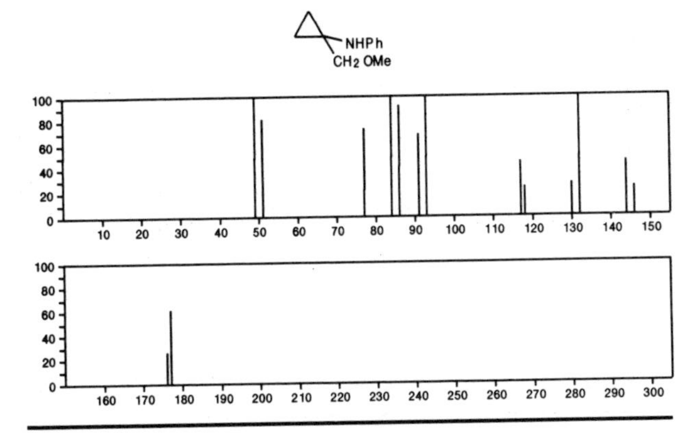

177 $C_{11}H_{15}NO$ 50893-11-3
Acetamide, *N*-methyl-*N*-(2-phenylethyl)-

MeN(Ac)CH2CH2Ph

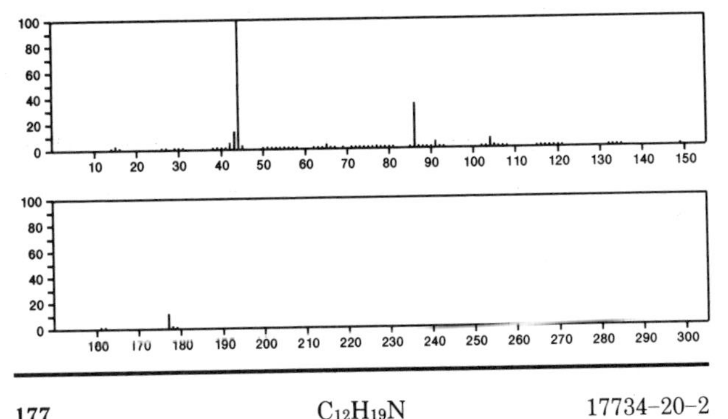

177 $C_{12}H_{19}N$ 17734-20-2
Benzenehexanamine

Ph(CH2)6NH2

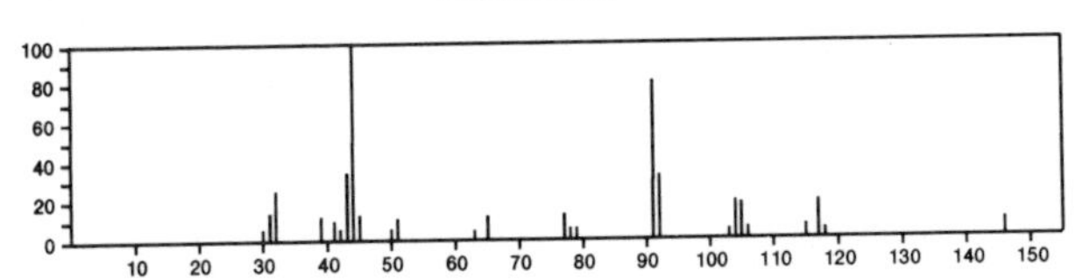

177 $C_{12}H_{19}N$ 49826-47-3
Pyrrolidine, 1-bicyclo[3.2.1]oct-2-en-3-yl-

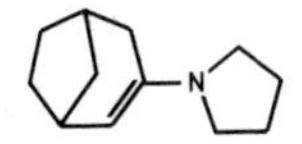

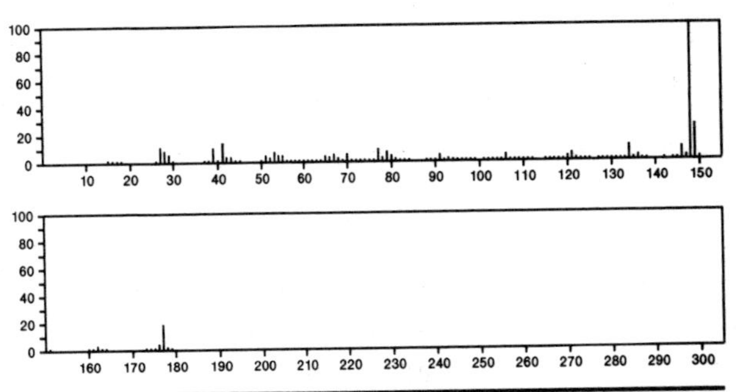

177 $C_{12}H_{19}N$ 53927-61-0
Benzenamine, *N*-(2,2-dimethylpropyl)-*N*-methyl-

PhNMeCH2CMe3

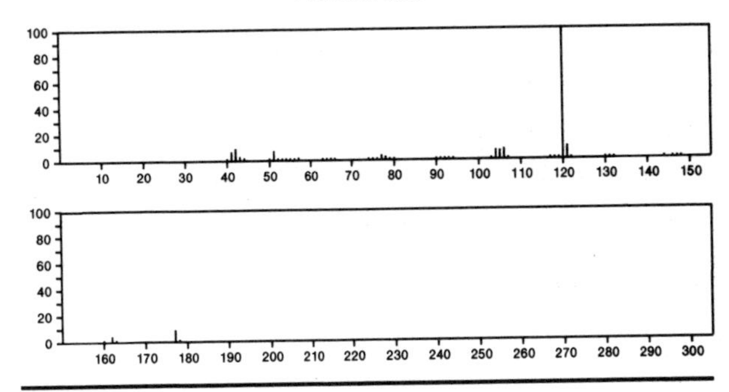

177 $C_{12}H_{19}N$ 54518-13-7
Pyridine, 2-(1-propylbutyl)-

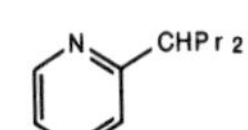

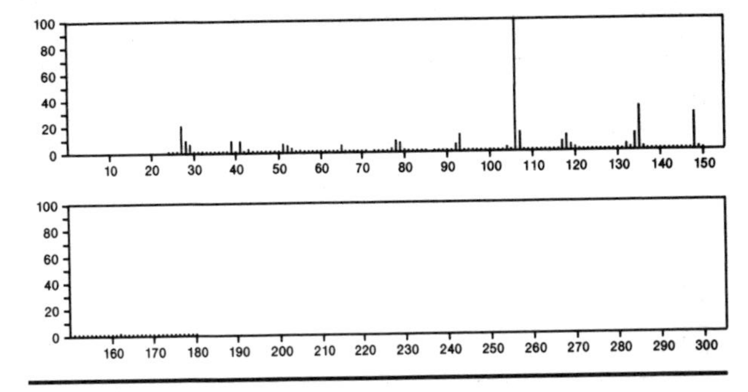

178 $C_3H_2Cl_4$ 20589-85-9
1-Propene, 1,2,3,3-tetrachloro-

ClCH=CClCHCl2

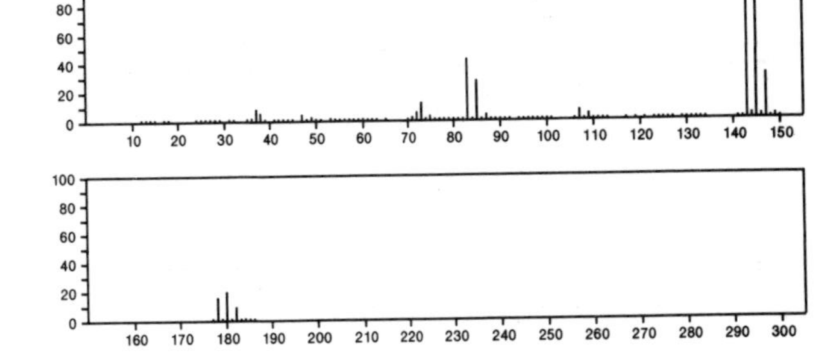

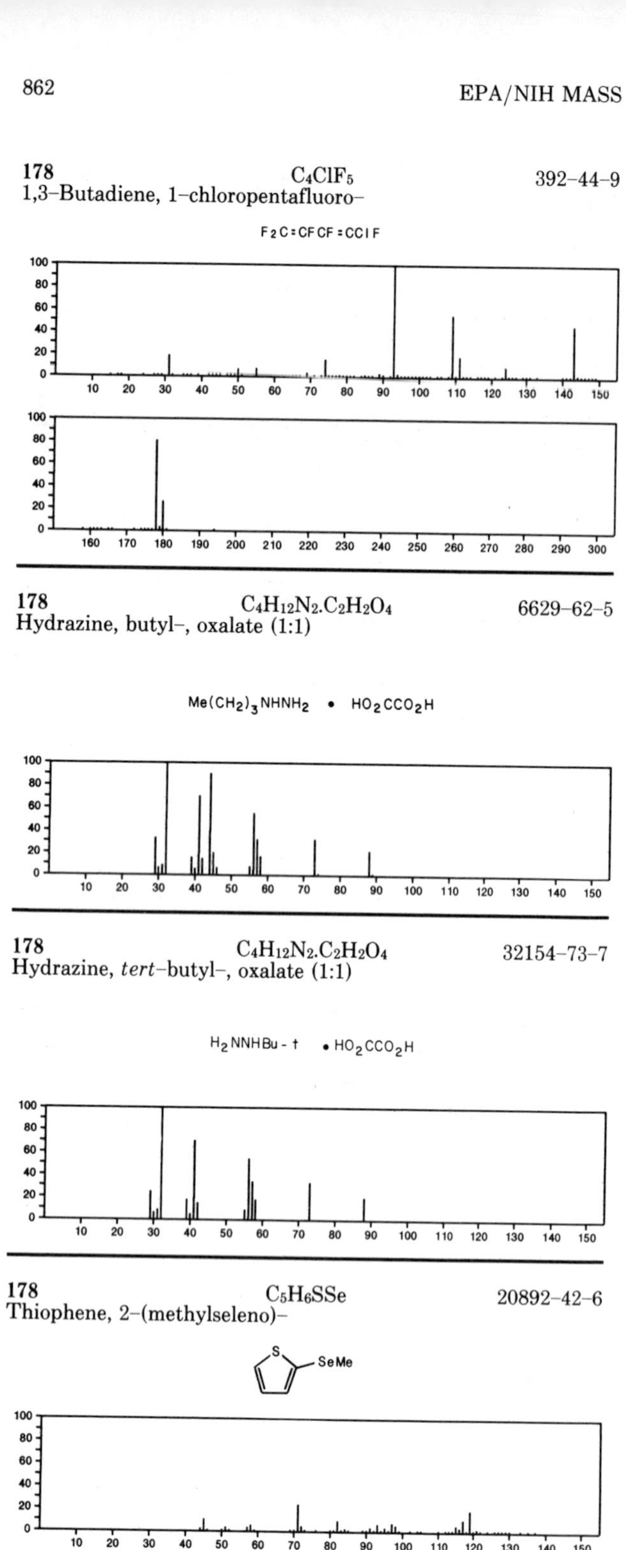

178 C_4ClF_5 392–44–9
1,3–Butadiene, 1–chloropentafluoro–

F2C=CFCF=CClF

178 $C_4H_{12}N_2.C_2H_2O_4$ 6629–62–5
Hydrazine, butyl–, oxalate (1:1)

Me(CH2)3NHNH2 • HO2CCO2H

178 $C_4H_{12}N_2.C_2H_2O_4$ 32154–73–7
Hydrazine, *tert*–butyl–, oxalate (1:1)

H2NNHBu-t • HO2CCO2H

178 C_5H_6SSe 20892–42–6
Thiophene, 2–(methylseleno)–

178 C_5H_6SSe 31053–53–9
Thiophene, 3–(methylseleno)–

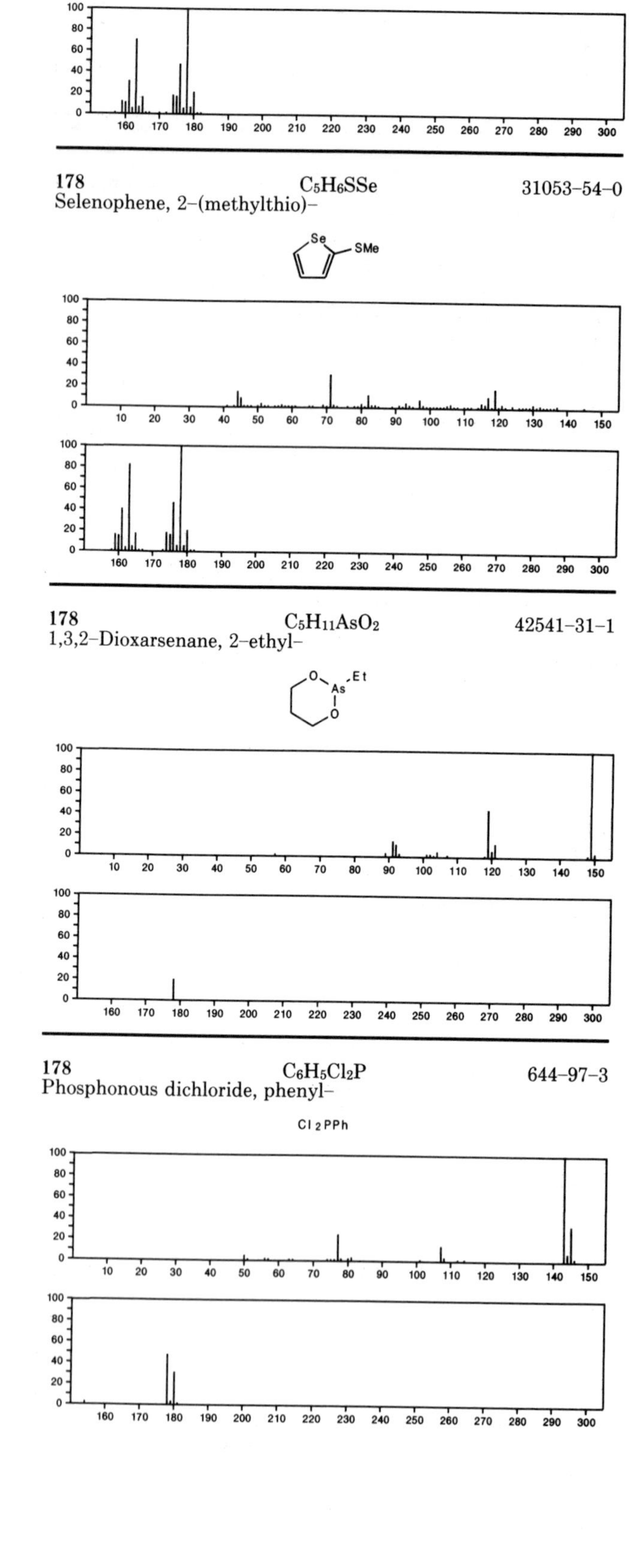

178 C_5H_6SSe 31053–54–0
Selenophene, 2–(methylthio)–

178 $C_5H_{11}AsO_2$ 42541–31–1
1,3,2–Dioxarsenane, 2–ethyl–

178 $C_6H_5Cl_2P$ 644–97–3
Phosphonous dichloride, phenyl–

Cl2PPh

178 $C_6H_5F_3N_2O$ 2557-79-1
4-Pyrimidinol, 6-methyl-2-(trifluoromethyl)-

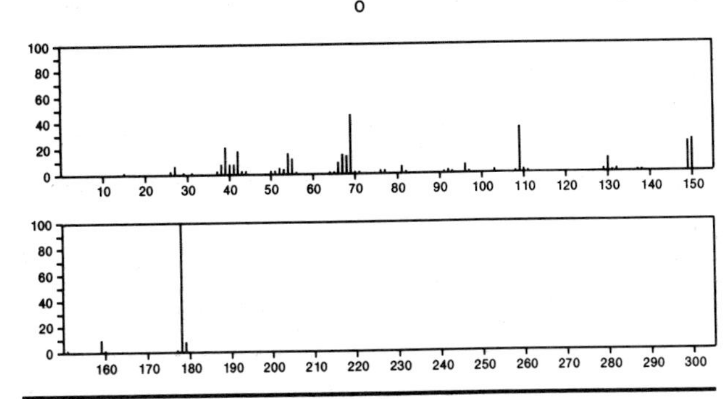

178 $C_6H_5F_3N_2O$ 2836-44-4
4-Pyrimidinol, 2-methyl-6-(trifluoromethyl)-

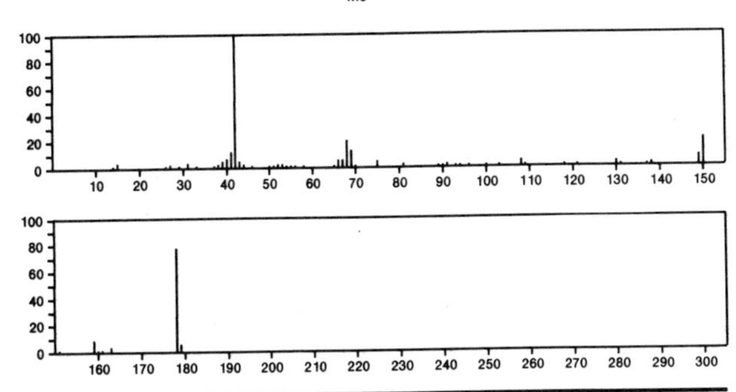

178 $C_6H_7ClN_2.ClH$ 1073-70-7
Hydrazine, (4-chlorophenyl)-, monohydrochloride

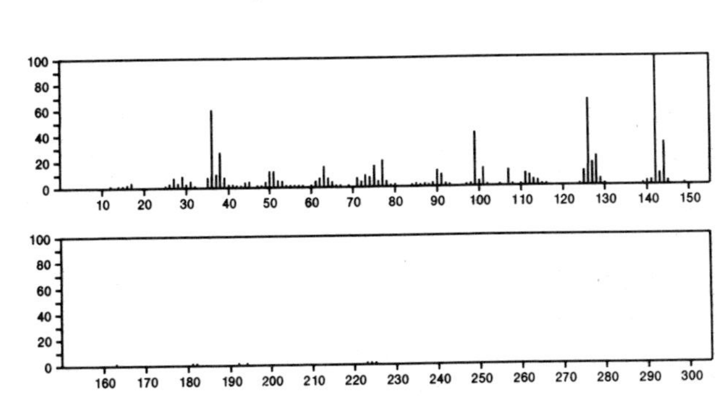

178 $C_6H_7ClN_2.ClH$ 2312-23-4
Hydrazine, (3-chlorophenyl)-, monohydrochloride

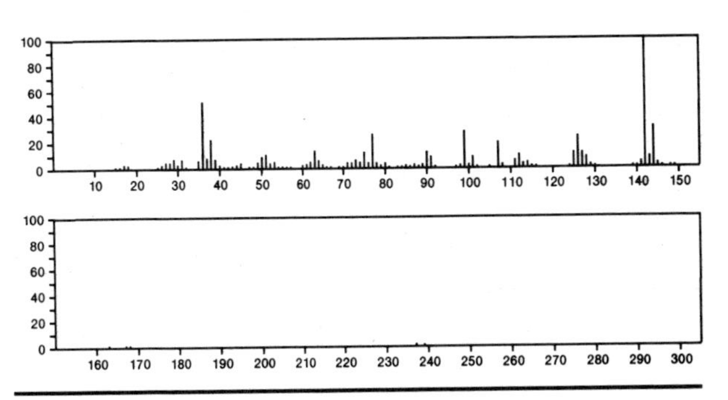

178 $C_6H_{10}O_2S_2$ 13125-44-5
1,2-Dithiolane-3-propanoic acid

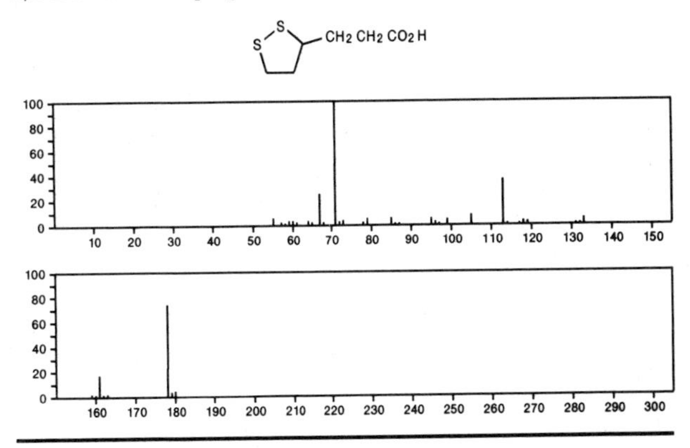

178 $C_6H_{10}O_6$ 608-68-4
Butanedioic acid, 2,3-dihydroxy- [*R*-(*R**,*R**)]-, dimethyl ester

MeOC(O)CH(OH)CH(OH)C(O)OMe

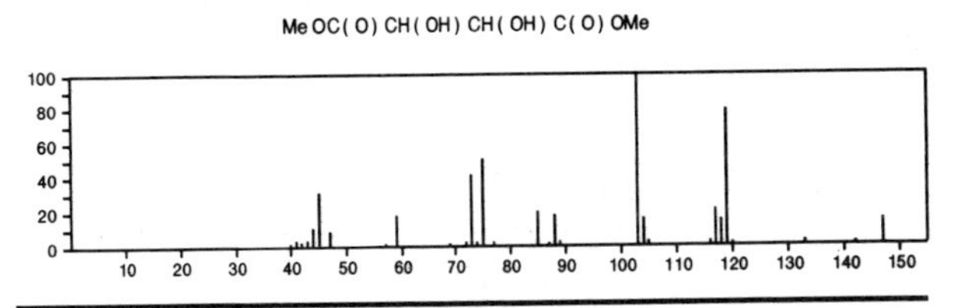

178 $C_6H_{10}S_3$ 2050-87-5
Trisulfide, di-2-propenyl

$H_2C=CHCH_2SSSCH_2CH=CH_2$

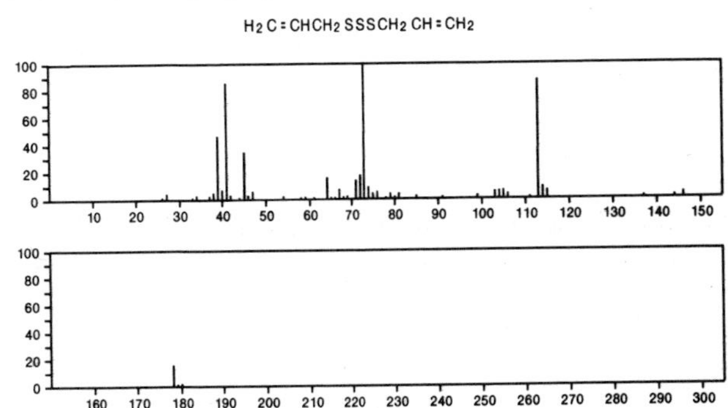

178 $C_6H_{11}BrO$ 2425-33-4
Cyclohexanol, 2-bromo-, *trans*-

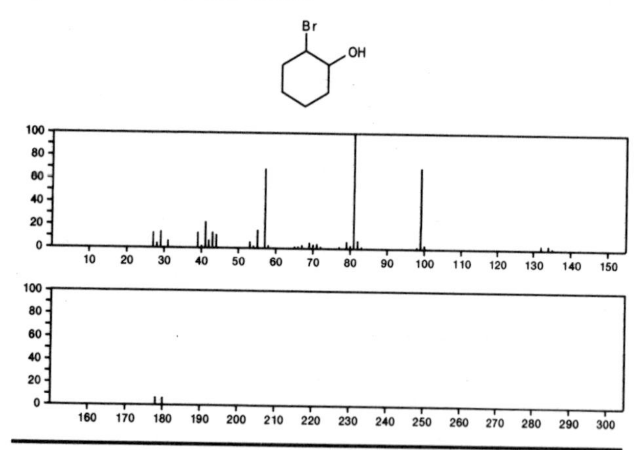

178 $C_6H_{11}BrO$ 10226-29-6
2-Hexanone, 6-bromo-

Br (CH2) 4 COMe

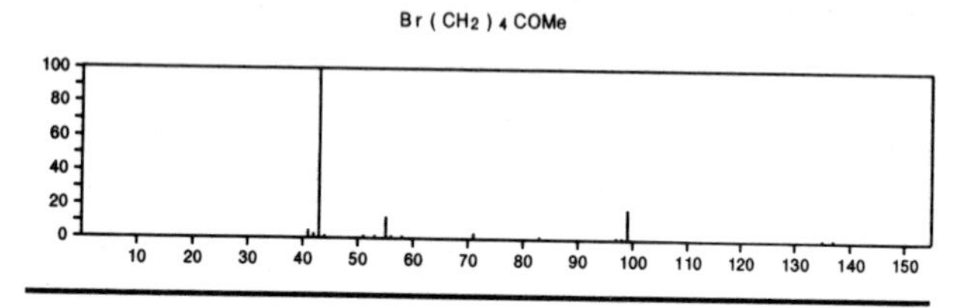

178 $C_6H_{11}BrO$ 16536-57-5
Cyclohexanol, 2-bromo-, *cis*-

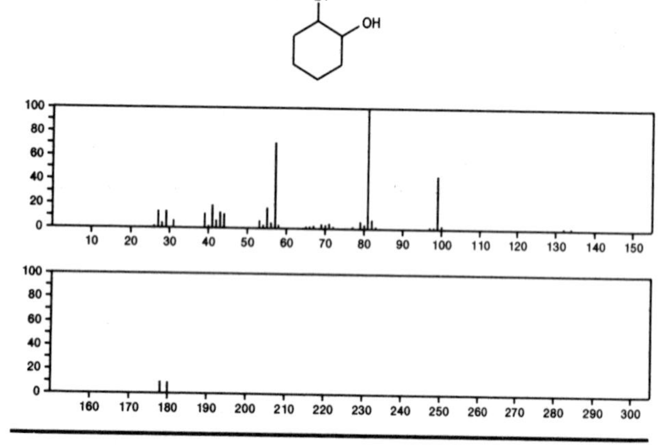

178 $C_6H_{11}BrO$ 51422-76-5
Cyclopentane, 1-bromo-2-methoxy-, *trans*-

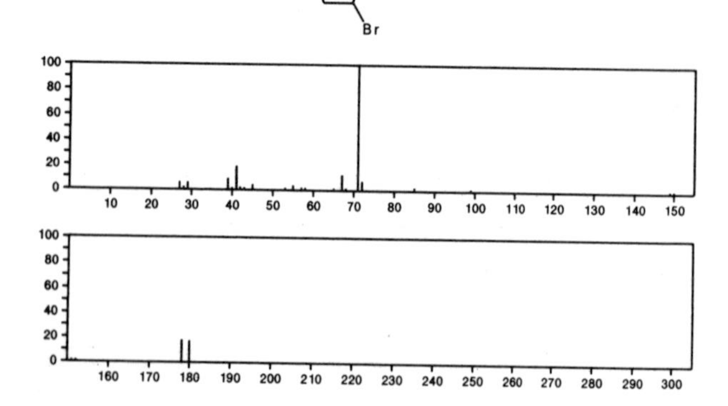

178 $C_6H_{11}BrO$ 51475-11-7
Cyclopentane, 1-bromo-2-methoxy-, *cis*-

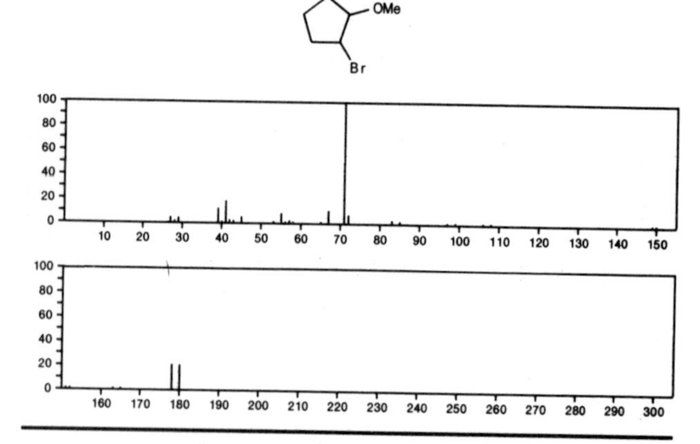

178 $C_7H_6N_4O_2$ 6726-55-2
Pyrazolo[5,1-*c*]-*as*-triazine-3-carboxylic acid, 4-methyl-

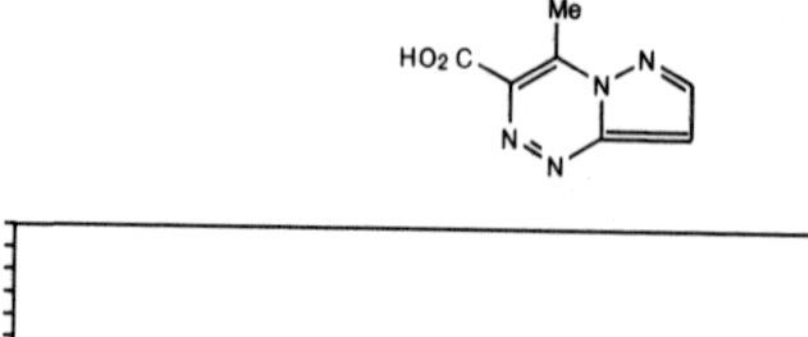

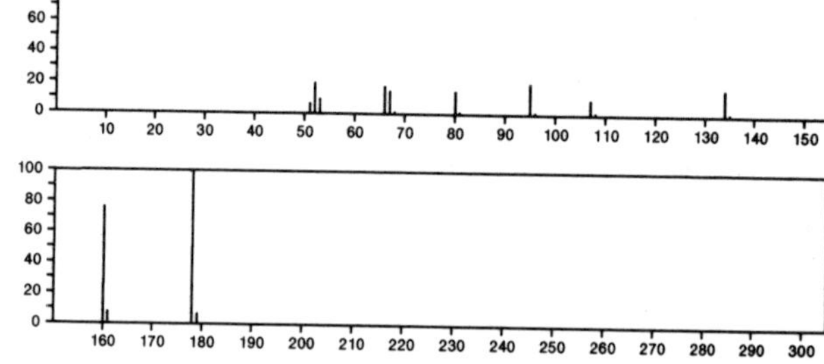

178 $C_7H_6N_4O_2$ 14209-07-5
1*H*-Benzotriazole, 1-methyl-7-nitro-

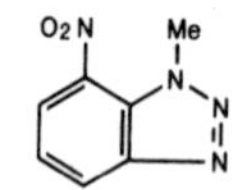

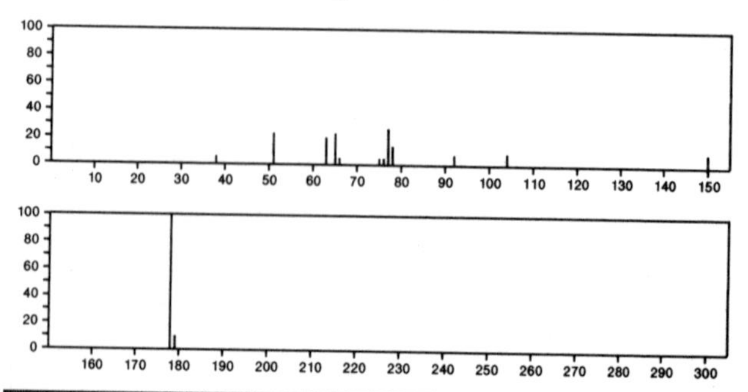

178 $C_7H_6N_4O_2$ 18106-58-6
4(3*H*)-Pteridinone, 3-hydroxy-2-methyl-

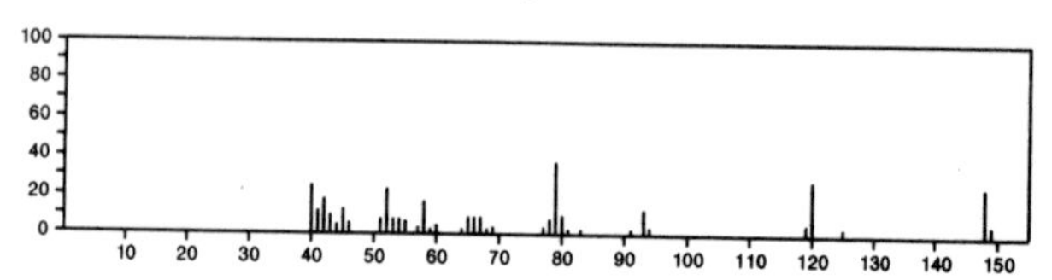

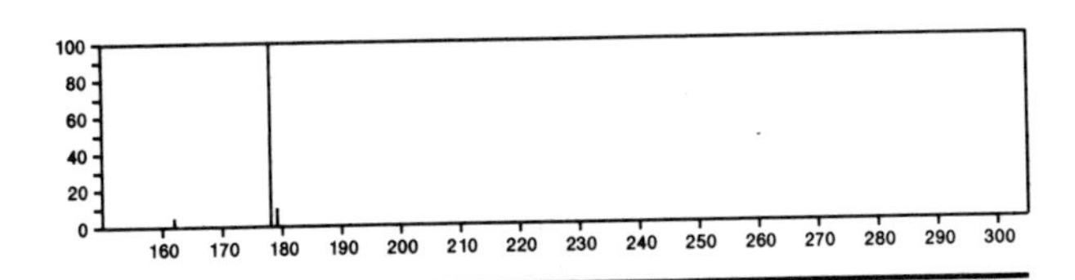

178 $C_7H_6N_4O_2$ 18106-59-7
4(3*H*)-Pteridinone, 3-hydroxy-6-methyl-

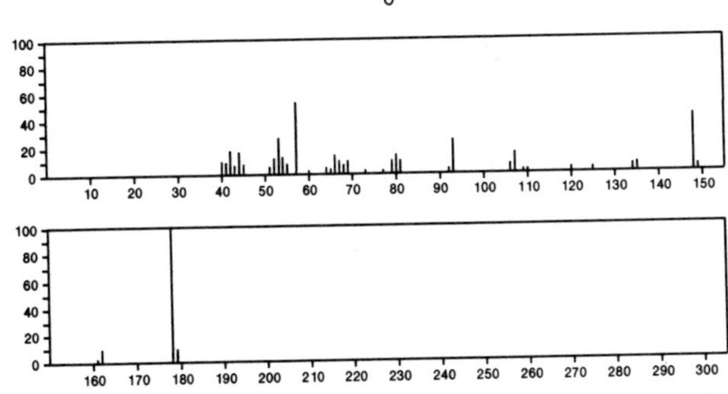

178 $C_7H_6N_4O_2$ 18106-60-0
4(3*H*)-Pteridinone, 3-hydroxy-7-methyl-

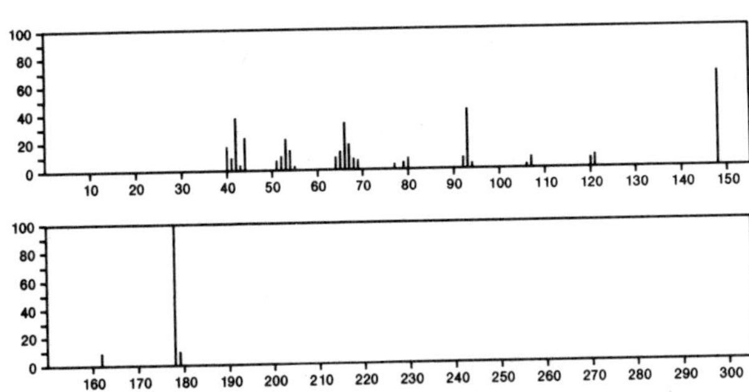

178 $C_7H_6N_4O_2$ 20615-75-2
Benzene, 1-azido-4-methyl-2-nitro-

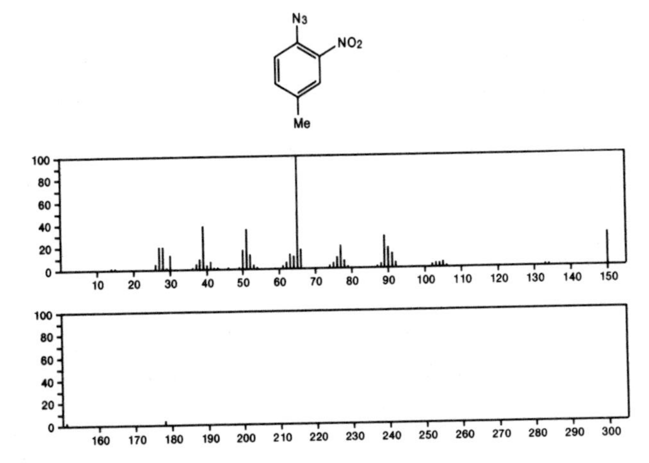

178 $C_7H_6N_4O_2$ 25877-34-3
1*H*-Benzotriazole, 1-methyl-5-nitro-

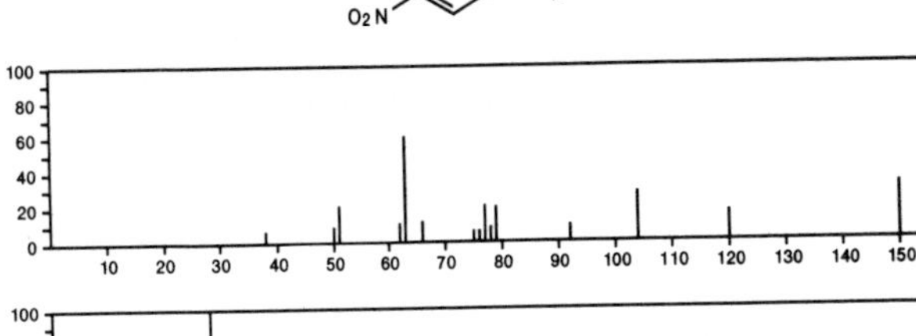

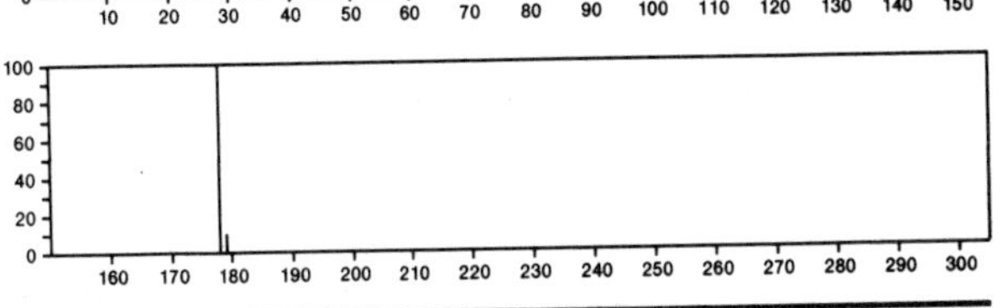

178 $C_7H_6N_4O_2$ 25877-35-4
1*H*-Benzotriazole, 1-methyl-6-nitro-

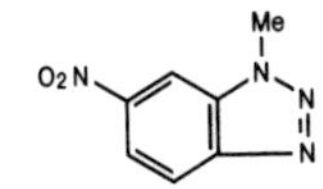

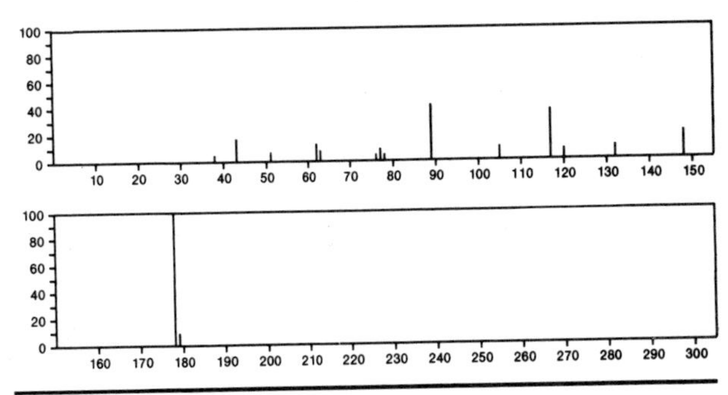

178 $C_7H_6N_4O_2$ 26070-05-3
4(3*H*)-Pteridinone, 3-methoxy-

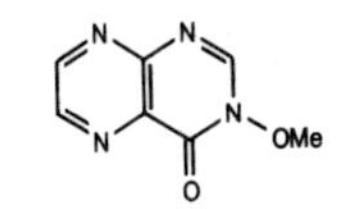

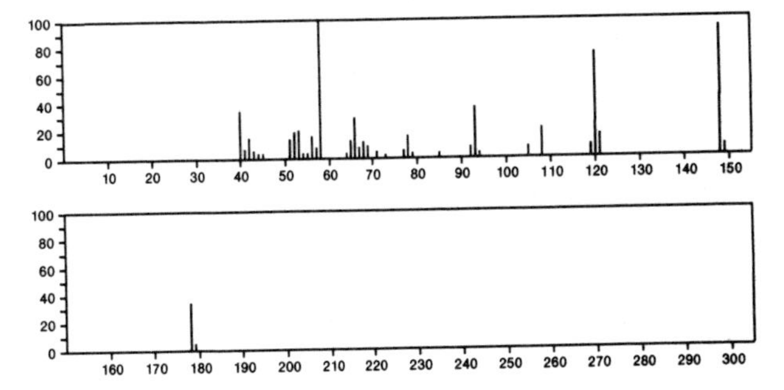

178 $C_7H_6N_4O_2$ 27799-86-6
1*H*-Benzotriazole, 1-methyl-4-nitro-

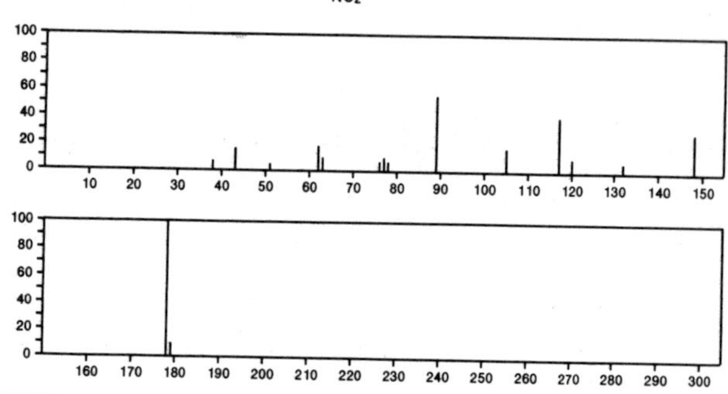

178 $C_7H_6N_4O_2$ 40515-18-2
Benzene, 4-azido-1-methyl-2-nitro-

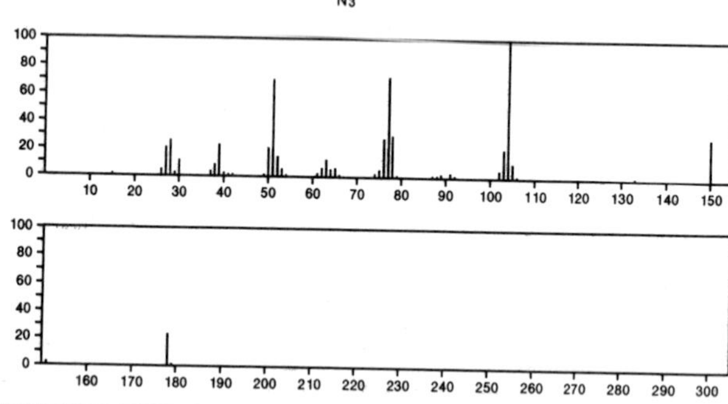

178 $C_7H_6N_4O_2$ 40515-19-3
Benzene, 2-azido-1-methyl-4-nitro-

178 $C_7H_7.BF_4$ 27081-10-3
Cycloheptatrienylium, tetrafluoroborate(1-)

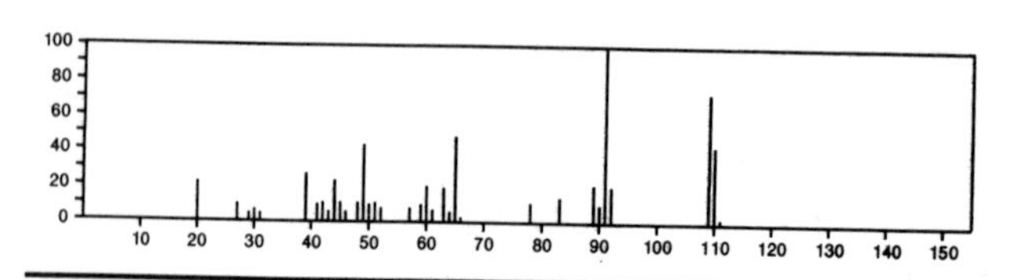

178 $C_7H_{10}N_6$ 50473-86-4
Imidazo[5,1-*f*][1,2,4]triazine-2,7-diamine, 4,5-dimethyl-

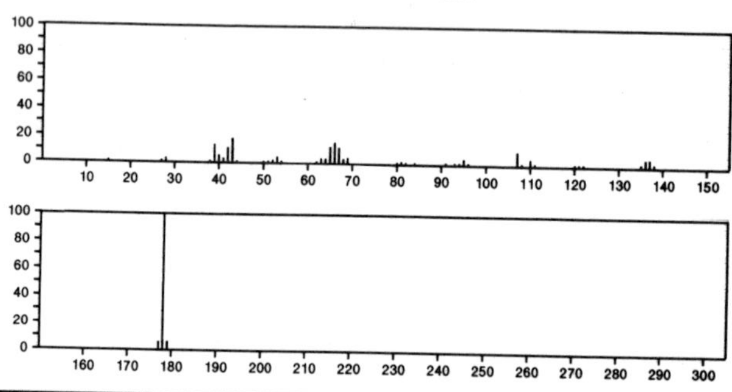

178 $C_7H_{14}OS_2$ 10596-56-2
Carbonic acid, dithio-, *S,S*-dipropyl ester

PrSC(O)SPr

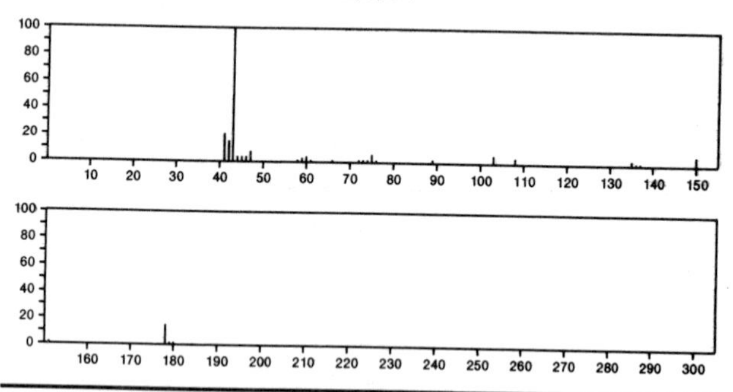

178 $C_7H_{14}OS_2$ 16118-33-5
Carbonic acid, dithio-, *S,S*-diisopropyl ester

i-PrSC(O)SPr-i

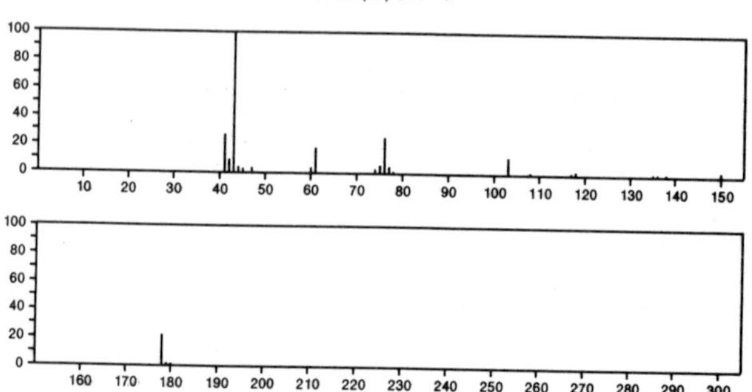

178 $C_7H_{14}OS_2$ 19615-06-6
Carbonodithioic acid, *O,S*-bis(1-methylethyl) ester

i-PrOC(S)SPr-i

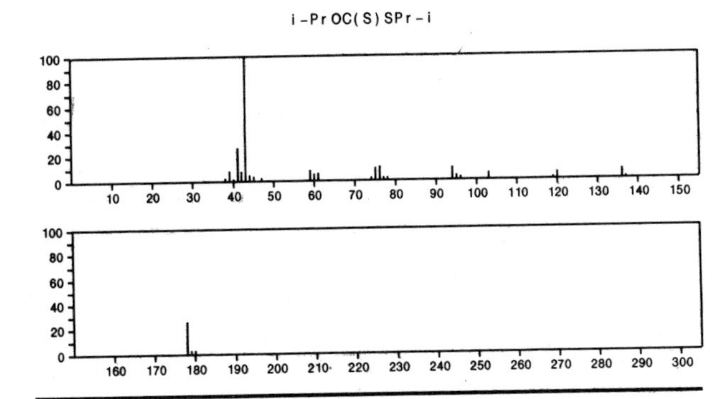

178 $C_7H_{14}O_5$ 626-84-6
Ethanol, 2-methoxy-, carbonate

$MeOCH_2CH_2OC(O)OCH_2CH_2OMe$

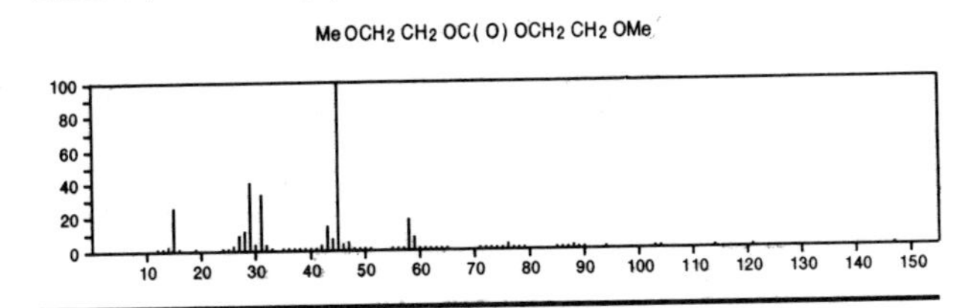

178 $C_7H_{14}O_5$ 14687-15-1
α-L-Galactopyranoside, methyl 6-deoxy-

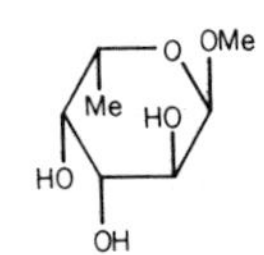

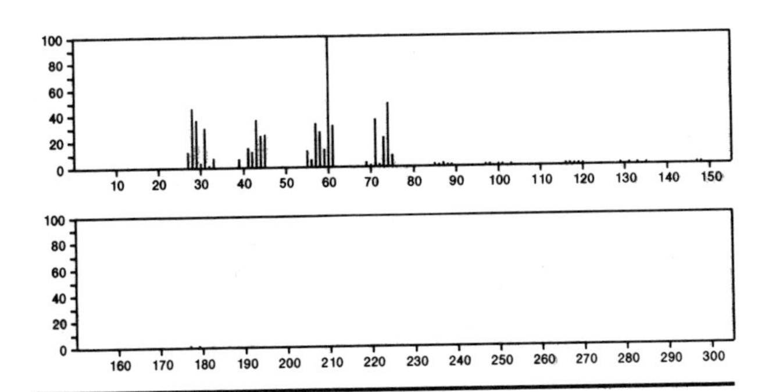

178 $C_7H_{14}O_5$ 18546-09-3
L-Glucose, 6-deoxy-3-*O*-methyl-

MeCH(OH)CH(OH)CH(OMe)CH(OH)CHO

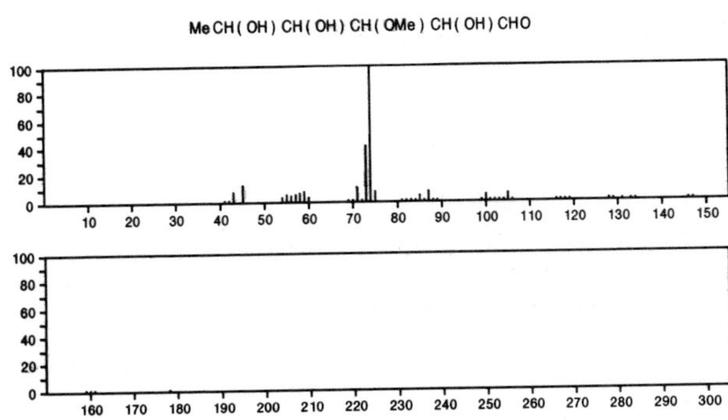

178 $C_7H_{14}O_5$ 32469-86-6
α-D-Xylofuranoside, methyl 2-*O*-methyl-

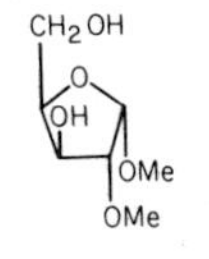

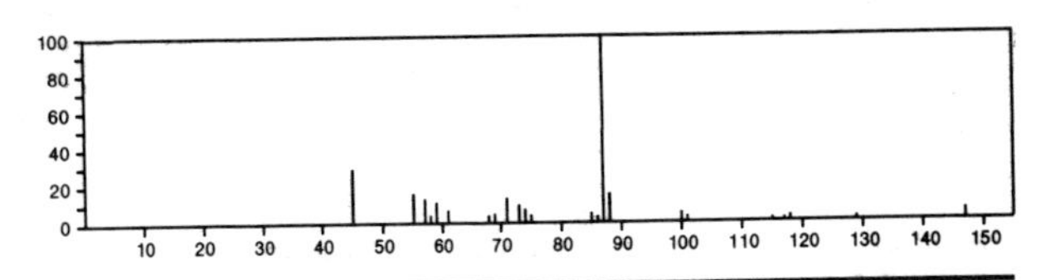

178 $C_7H_{14}O_5$ 34338-86-8
α-D-Xylofuranoside, methyl 3-*O*-methyl-

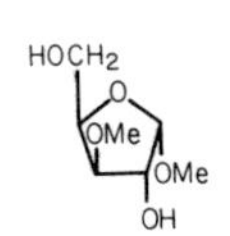

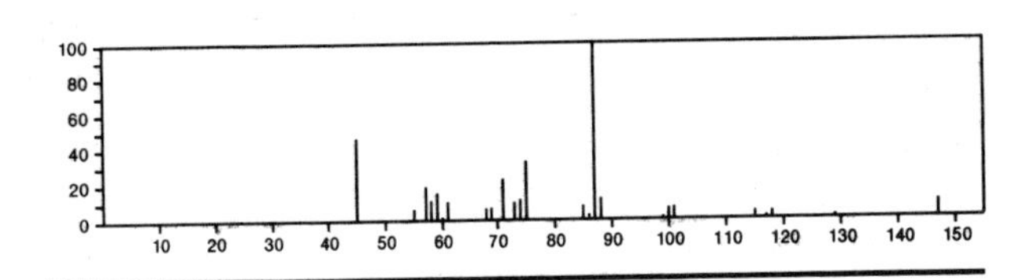

178 $C_7H_{14}O_5$ 35007-57-9
α-D-Xylofuranoside, methyl 5-*O*-methyl-

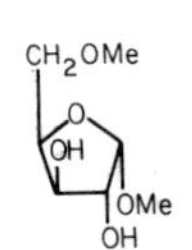

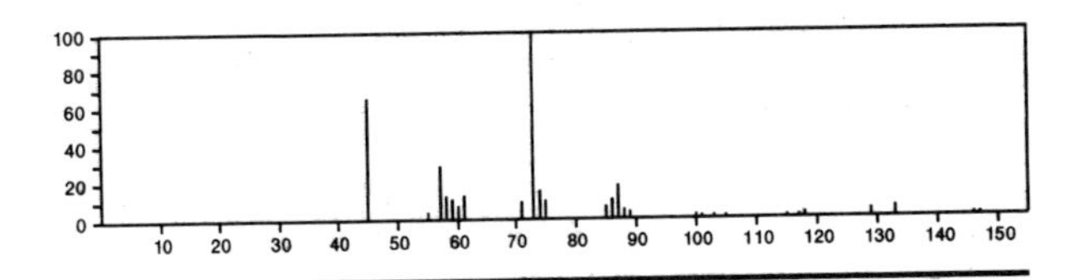

178 $C_7H_{15}Br$ 629-04-9
Heptane, 1-bromo-

$Br(CH_2)_6Me$

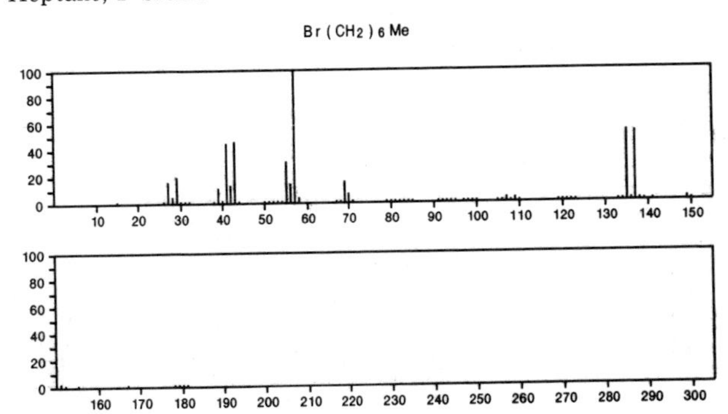

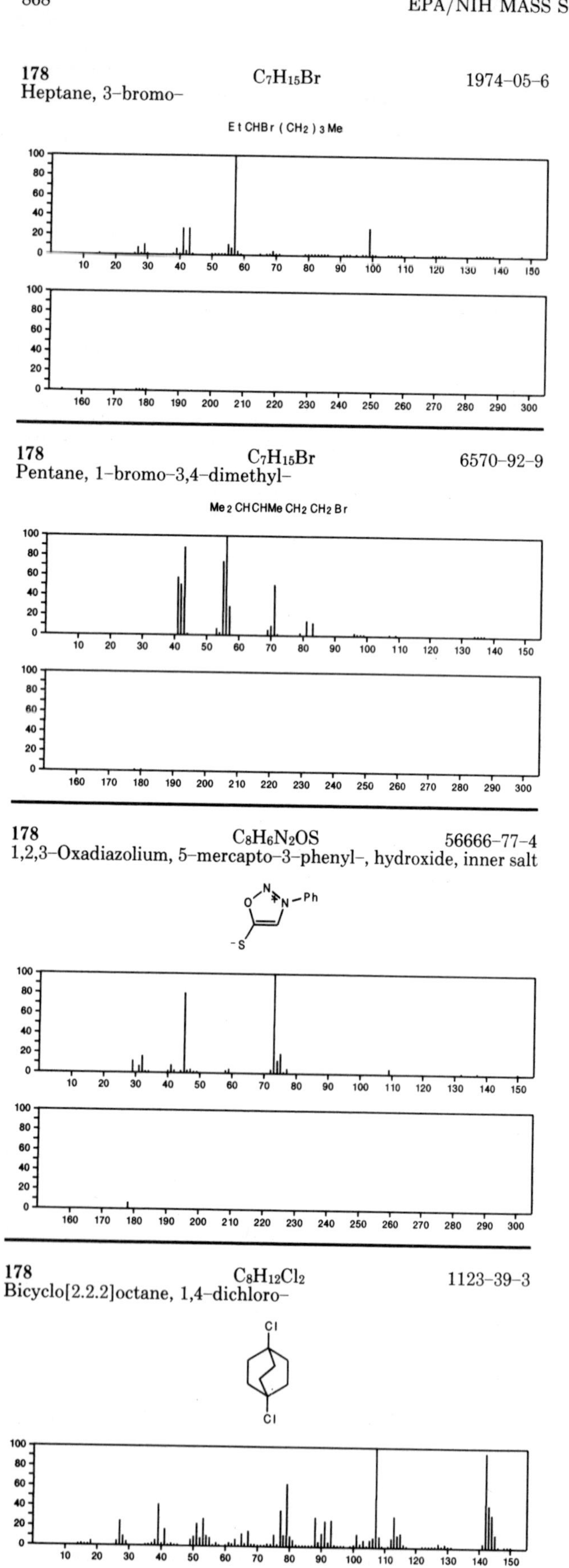
178
Heptane, 3-bromo-
C7H15Br
1974-05-6
Et CHBr (CH2) 3 Me
178
Pentane, 1-bromo-3,4-dimethyl-
C7H15Br
6570-92-9
Me 2 CHCHMe CH2 CH2 Br
178
1,2,3-Oxadiazolium, 5-mercapto-3-phenyl-, hydroxide, inner salt
C8H6N2OS
56666-77-4
178
Bicyclo[2.2.2]octane, 1,4-dichloro-
C8H12Cl2
1123-39-3

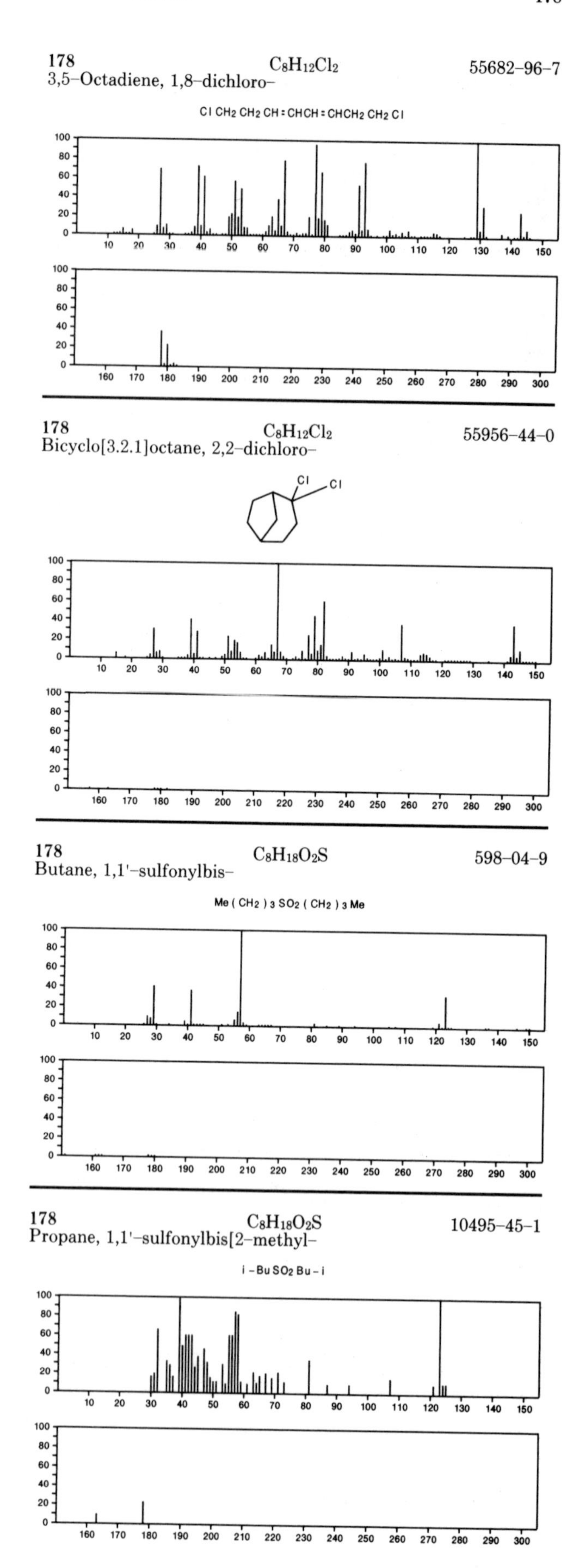
178
3,5-Octadiene, 1,8-dichloro-
C8H12Cl2
55682-96-7
Cl CH2 CH2 CH = CHCH = CHCH2 CH2 Cl
178
Bicyclo[3.2.1]octane, 2,2-dichloro-
C8H12Cl2
55956-44-0
178
Butane, 1,1'-sulfonylbis-
C8H18O2S
598-04-9
Me (CH2) 3 SO2 (CH2) 3 Me
178
Propane, 1,1'-sulfonylbis[2-methyl-
C8H18O2S
10495-45-1
i - Bu SO2 Bu - i

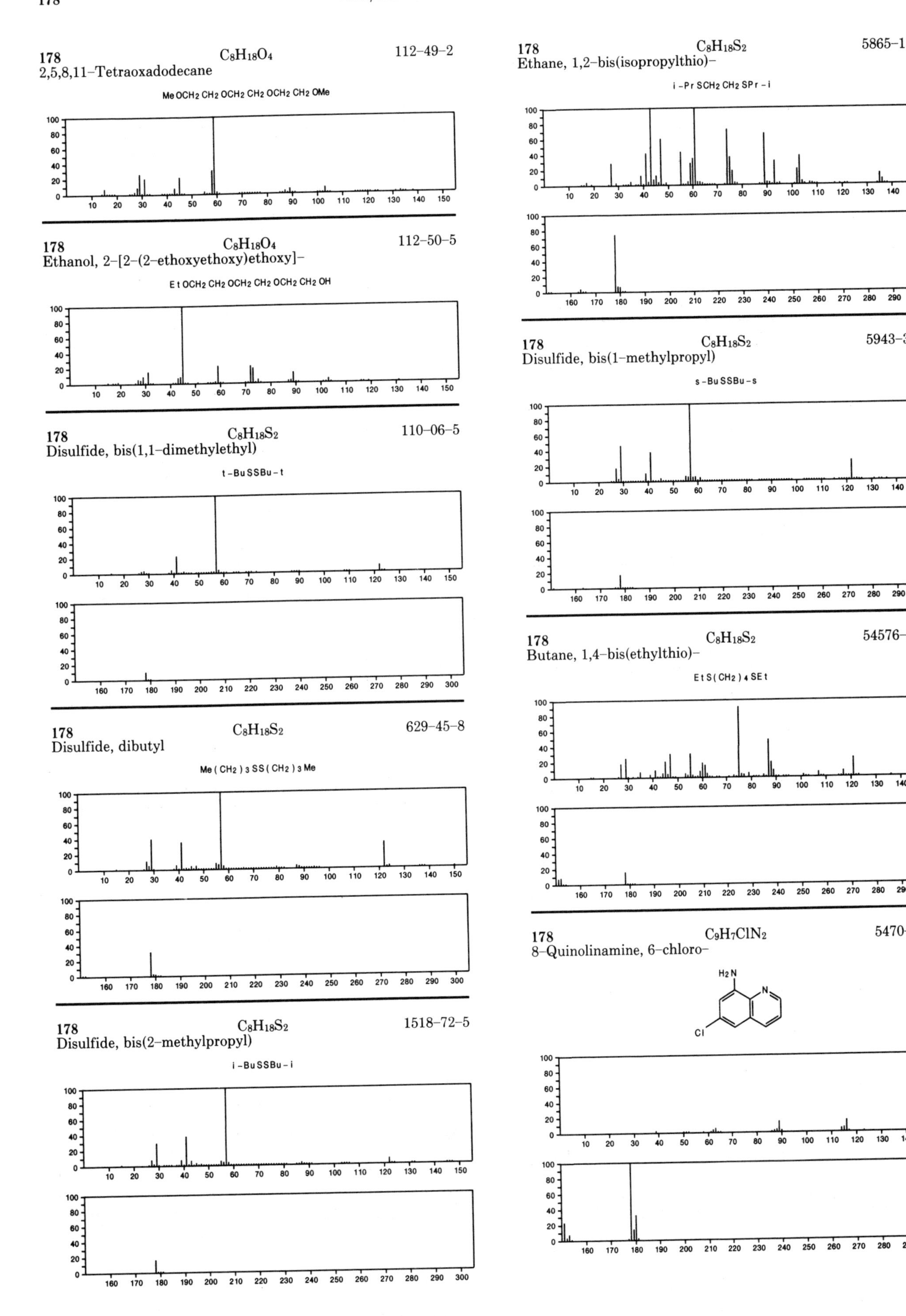

178 $C_8H_{18}O_4$ 112-49-2
2,5,8,11-Tetraoxadodecane
MeOCH2CH2OCH2CH2OCH2CH2OMe

178 $C_8H_{18}O_4$ 112-50-5
Ethanol, 2-[2-(2-ethoxyethoxy)ethoxy]-
EtOCH2CH2OCH2CH2OCH2CH2OH

178 $C_8H_{18}S_2$ 110-06-5
Disulfide, bis(1,1-dimethylethyl)
t-BuSSBu-t

178 $C_8H_{18}S_2$ 629-45-8
Disulfide, dibutyl
Me(CH2)3SS(CH2)3Me

178 $C_8H_{18}S_2$ 1518-72-5
Disulfide, bis(2-methylpropyl)
i-BuSSBu-i

178 $C_8H_{18}S_2$ 5865-15-6
Ethane, 1,2-bis(isopropylthio)-
i-PrSCH2CH2SPr-i

178 $C_8H_{18}S_2$ 5943-30-6
Disulfide, bis(1-methylpropyl)
s-BuSSBu-s

178 $C_8H_{18}S_2$ 54576-32-8
Butane, 1,4-bis(ethylthio)-
EtS(CH2)4SEt

178 $C_9H_7ClN_2$ 5470-75-7
8-Quinolinamine, 6-chloro-

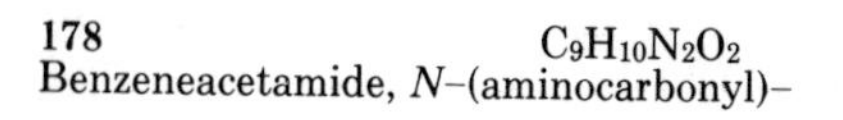

178 $C_9H_{10}N_2O_2$ 63-98-9
Benzeneacetamide, *N*-(aminocarbonyl)-

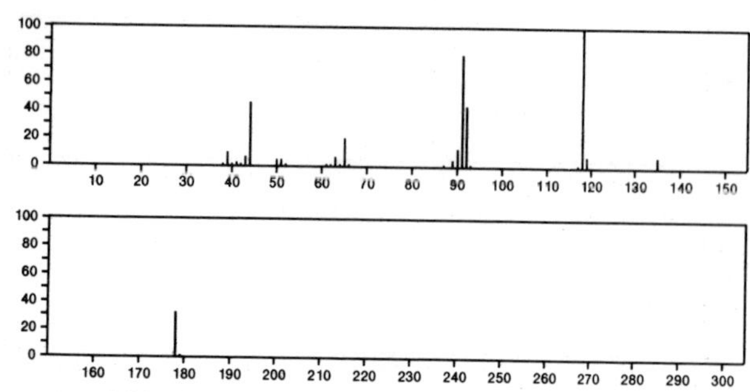

178 $C_9H_{10}N_2O_2$ 16007-57-1
Benzimidazole, 2-ethoxy-, 3-oxide

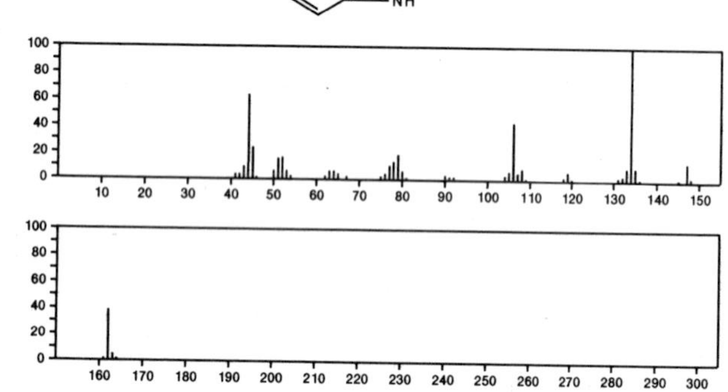

178 $C_9H_{10}N_2O_2$ 37704-51-1
Isoxazole, 5,5'-(1,3-propanediyl)bis-

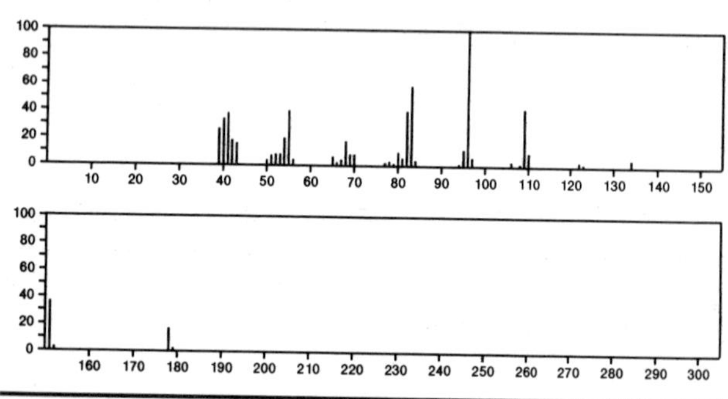

178 $C_9H_{10}N_2O_2$ 51460-33-4
Ethanone, 1,1'-(3-amino-2,4-pyridinediyl)bis-

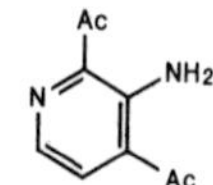

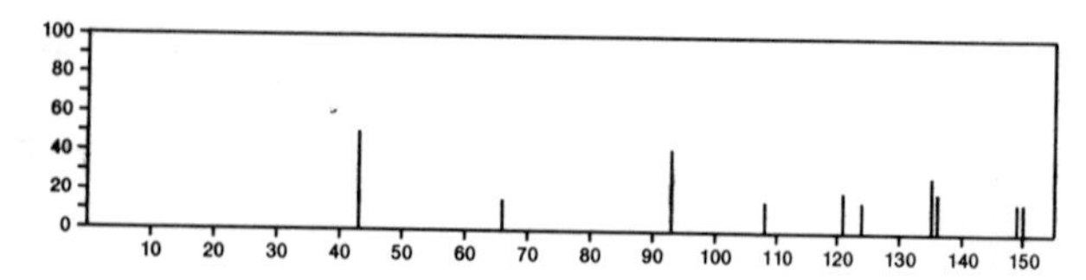

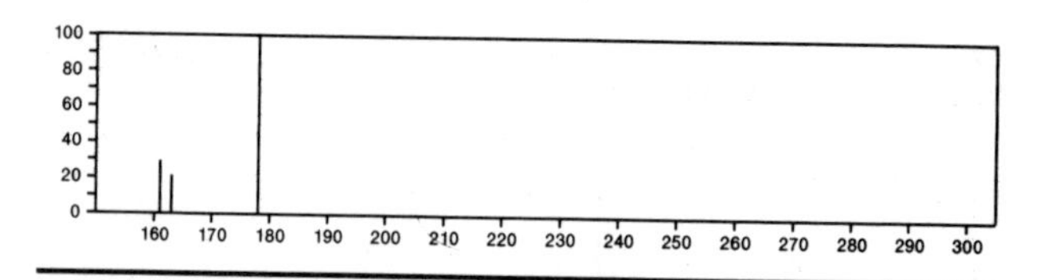

178 $C_9H_{10}N_2S$ 1009-70-7
2-Thiazolamine, 4,5-dihydro-*N*-phenyl-

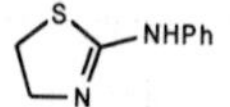

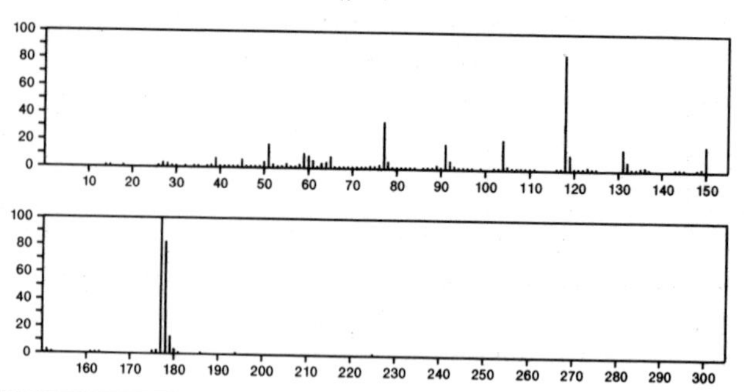

178 $C_9H_{10}N_2S$ 28291-69-2
2-Benzothiazolamine, *N*-ethyl-

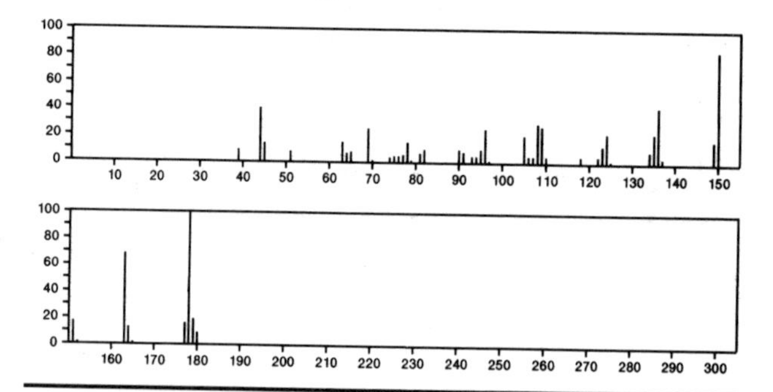

178 $C_9H_{11}Co$ 1271-08-5
Cobalt, π-cyclopentadienyl(1-methylene-π-allyl)-

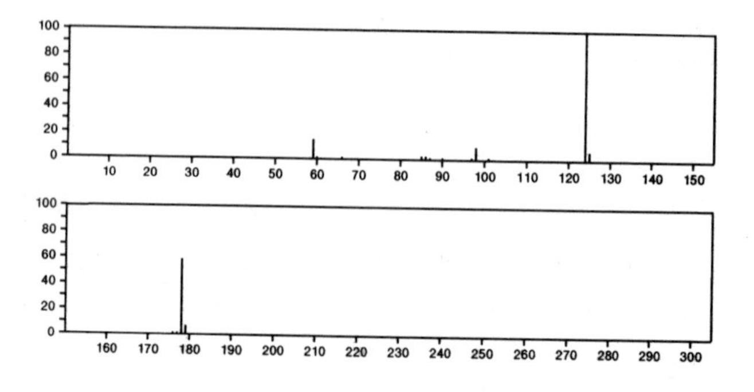

178 $C_9H_{14}Si_2$ 17864-73-2
1,3-Disilaindan, 1,3-dimethyl-

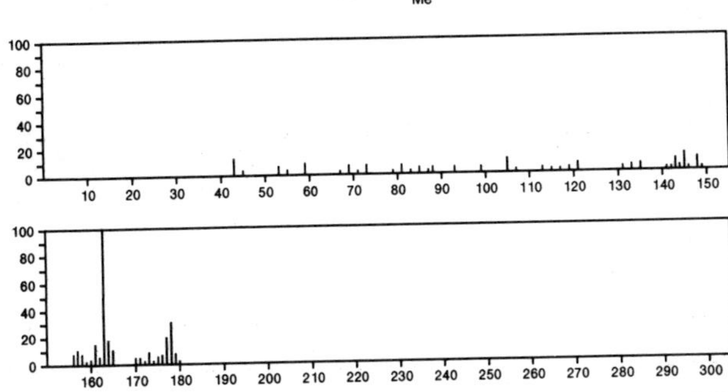

178 $C_{10}H_{10}OS$ 771-17-5
4*H*-1-Benzothiopyran-4-one, 2,3-dihydro-3-methyl-

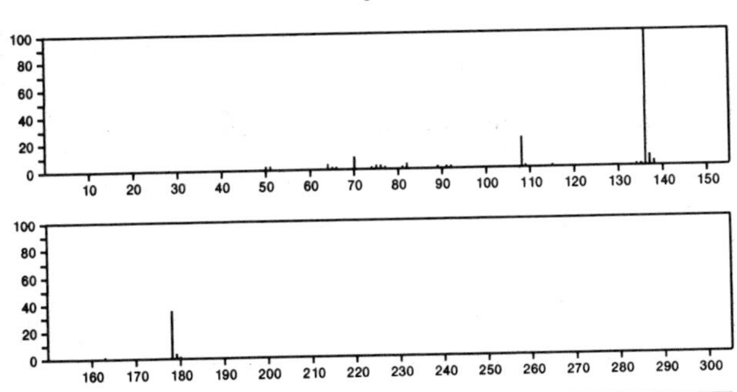

178 $C_{10}H_{10}OS$ 826-86-8
4*H*-1-Benzothiopyran-4-one, 2,3-dihydro-2-methyl-

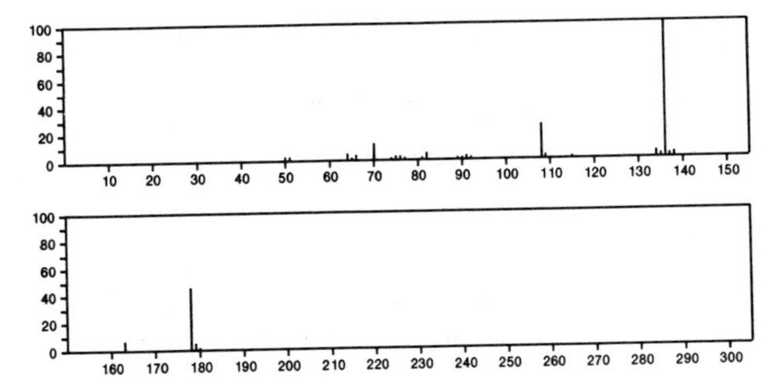

178 $C_{10}H_{10}OS$ 29373-02-2
4*H*-1-Benzothiopyran-4-one, 2,3-dihydro-8-methyl-

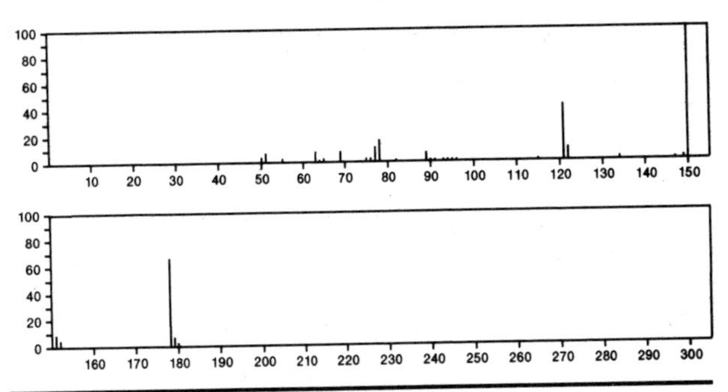

178 $C_{10}H_{10}O_3$ 614-27-7
Benzenepropanoic acid, β-oxo-, methyl ester

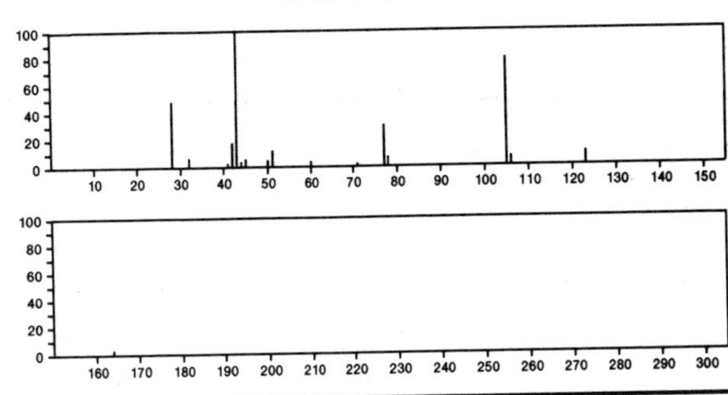

178 $C_{10}H_{10}O_3$ 1603-79-8
Benzeneacetic acid, α-oxo-, ethyl ester

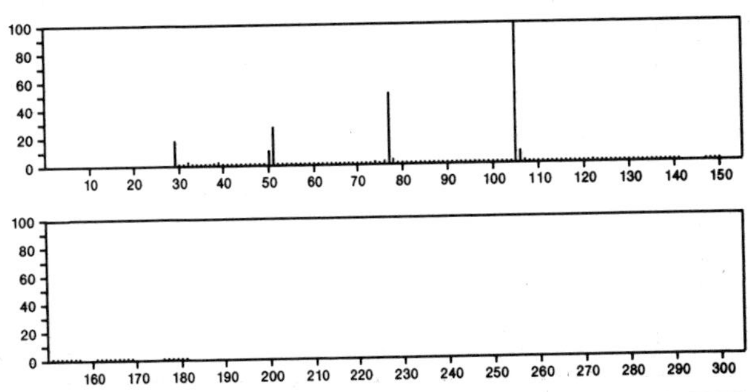

178 $C_{10}H_{10}O_3$ 2051-95-8
Benzenebutanoic acid, γ-oxo-

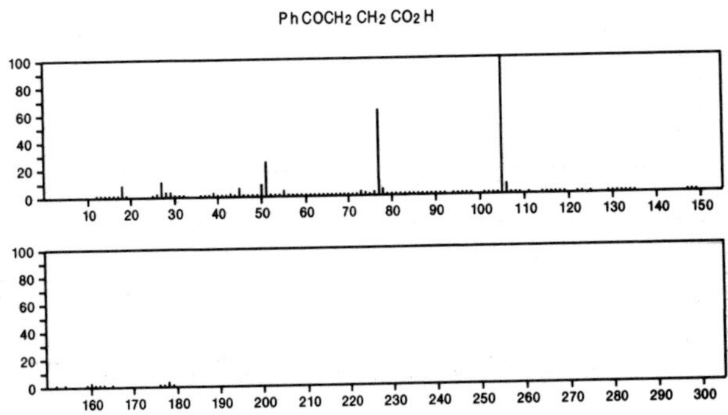

178 $C_{10}H_{10}O_3$ 2243-35-8
Ethanone, 2-(acetyloxy)-1-phenyl-

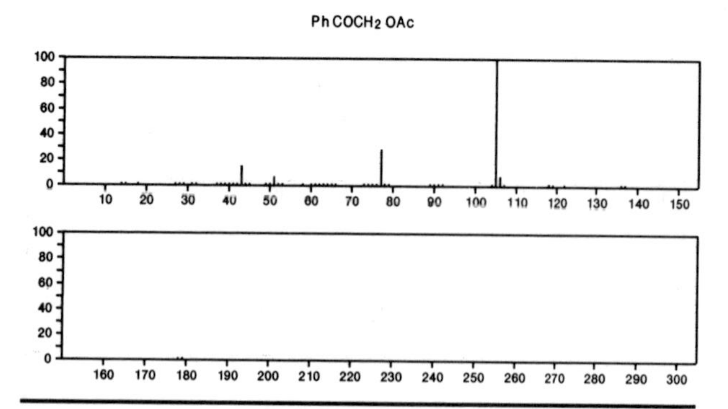

178 $C_{10}H_{10}O_3$ 3943-97-3
2-Propenoic acid, 3-(4-hydroxyphenyl)-, methyl ester

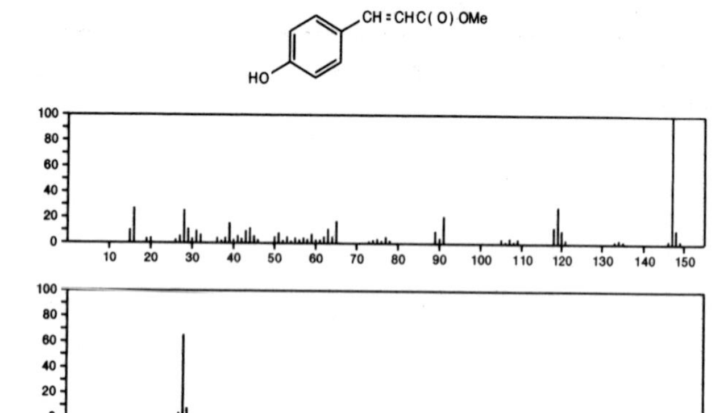

178 $C_{10}H_{10}O_3$ 4437-22-3
Furan, 2,2'-[oxybis(methylene)]bis-

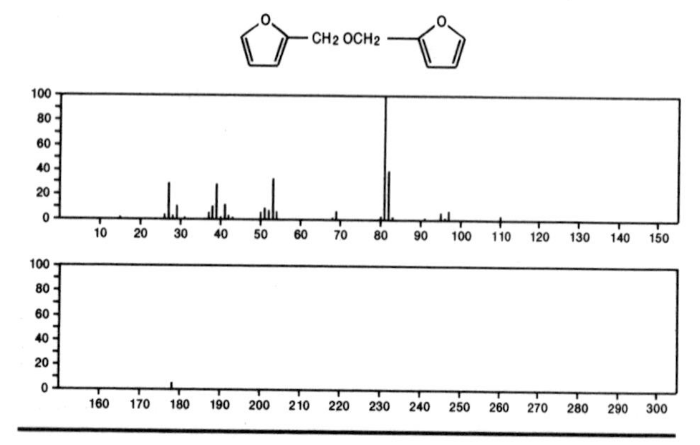

178 $C_{10}H_{10}O_3$ 6362-58-9
Benzenepropanoic acid, α-oxo-, methyl ester

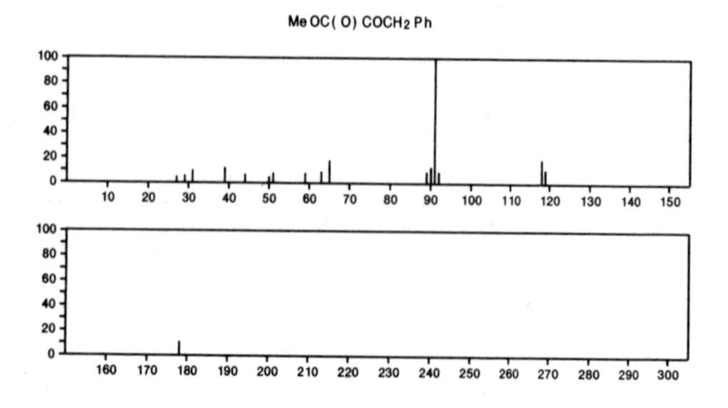

178 $C_{10}H_{10}O_3$ 16824-02-5
1(3*H*)-Isobenzofuranone, 3-ethoxy-

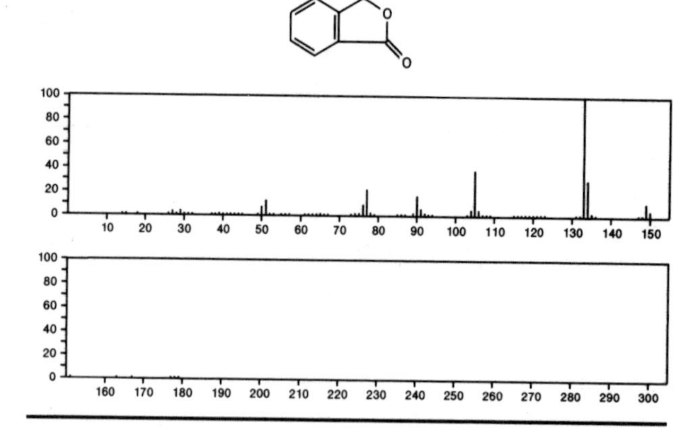

178 $C_{10}H_{10}O_3$ 17397-85-2
1*H*-2-Benzopyran-1-one, 3,4-dihydro-8-hydroxy-3-methyl-

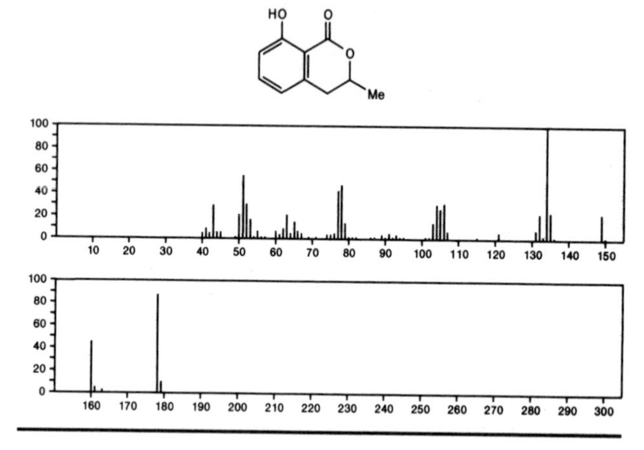

178 $C_{10}H_{10}O_3$ 29953-17-1
2-Furanmethanol, 5-(2-furanylmethyl)-

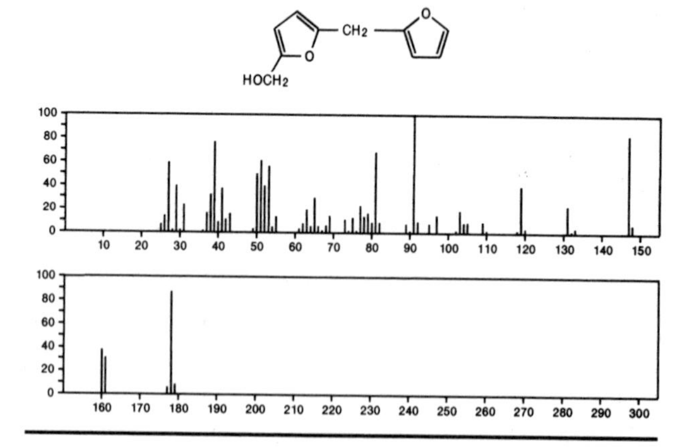

178 $C_{10}H_{10}O_3$ 31969-27-4
2(5*H*)-Furanone, 5-(2-furanylmethyl)-5-methyl-

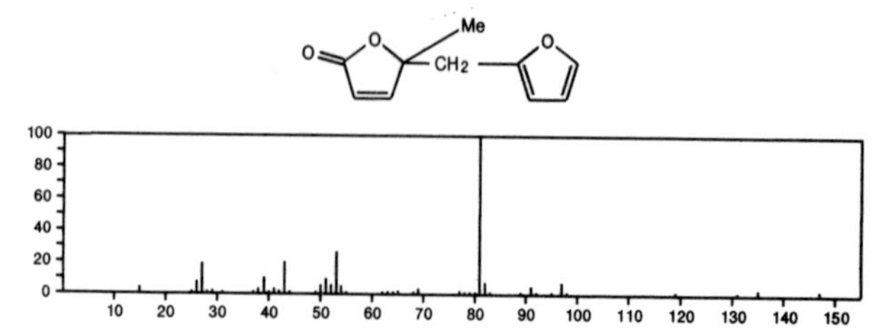

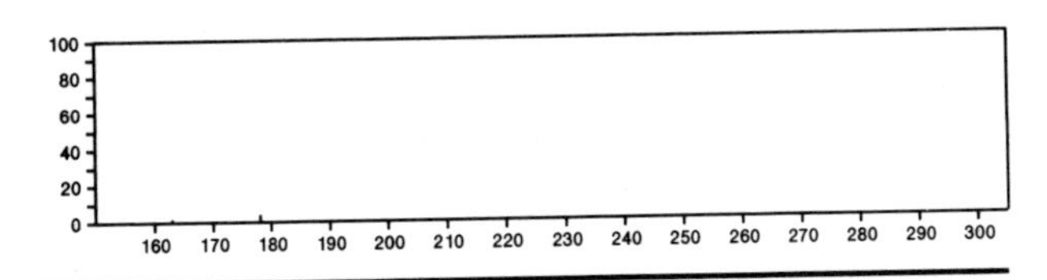

178 $C_{10}H_{10}O_3$ 54549-74-5
Benzaldehyde, 4-[(acetyloxy)methyl]-

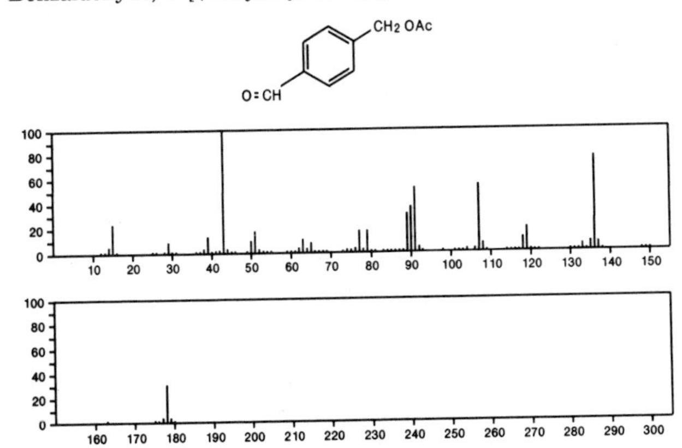

178 $C_{10}H_{10}O_3$ 54549-75-6
1(2*H*)-Naphthalenone, 3,4-dihydro-6,7-dihydroxy-

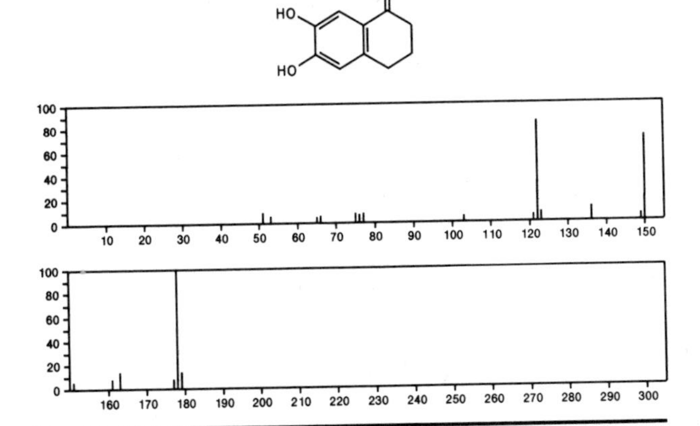

178 $C_{10}H_{14}N_2O$ 59-26-7
3-Pyridinecarboxamide, *N*,*N*-diethyl-

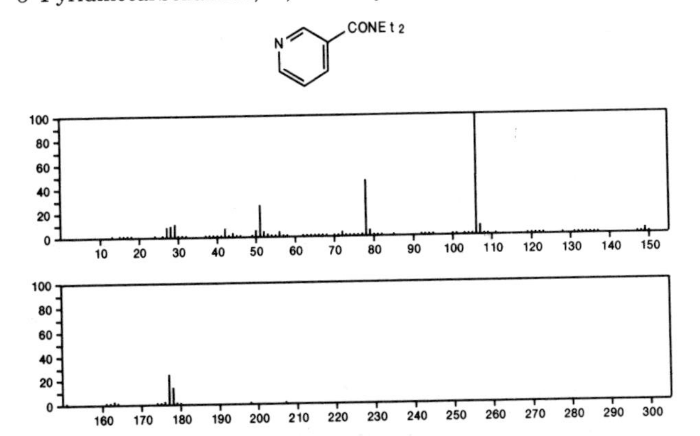

178 $C_{10}H_{14}N_2O$ 491-26-9
Pyridine, 3-(1-methyl-2-pyrrolidinyl)-, *N*-oxide, (2*S*)-

178 $C_{10}H_{14}N_2O$ 1202-42-2
Methanimidamide, *N'*-(3-methoxyphenyl)-*N*,*N*-dimethyl-

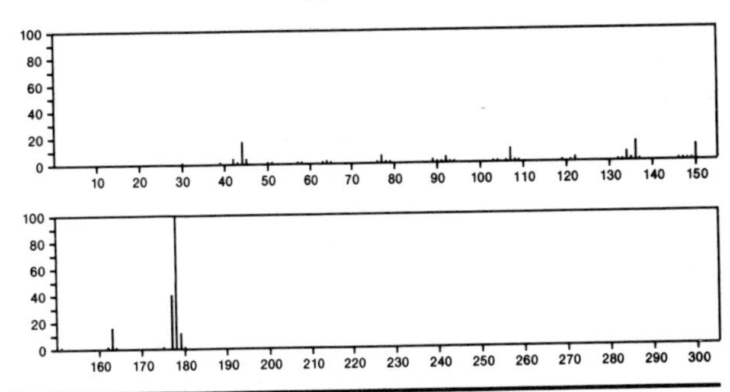

178 $C_{10}H_{14}N_2O$ 1202-62-6
Methanimidamide, *N'*-(4-methoxyphenyl)-*N*,*N*-dimethyl-

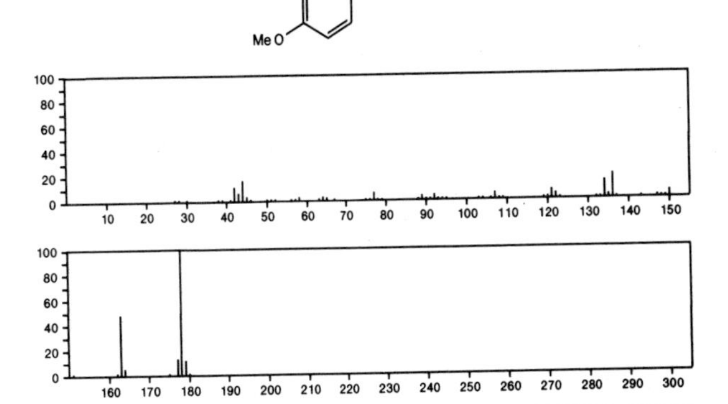

178 $C_{10}H_{14}N_2O$ 2820-55-5
Pyridine, 3-(1-methyl-2-pyrrolidinyl)-, 1-oxide, (*S*)-

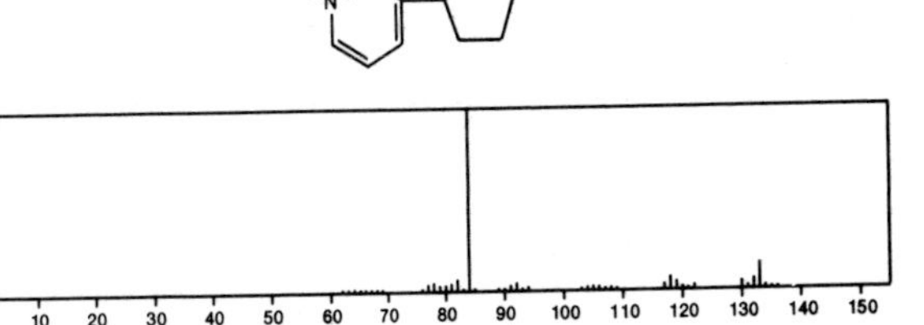

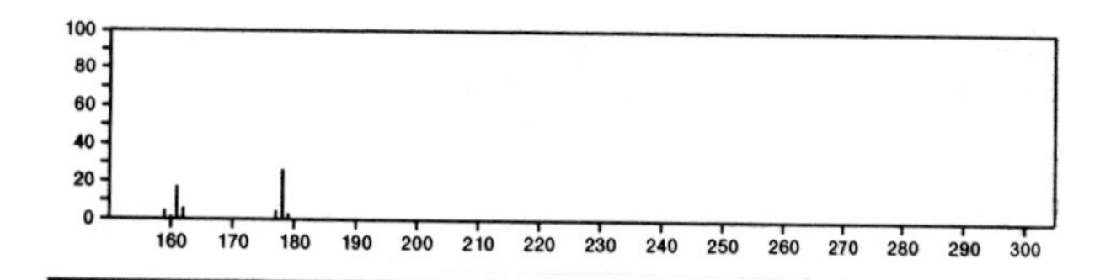

178 $C_{10}H_{14}N_2O$ 13950-22-6
2-Pyrrolidinone, 1-methyl-5-(1-methylpyrrol-2-yl)-

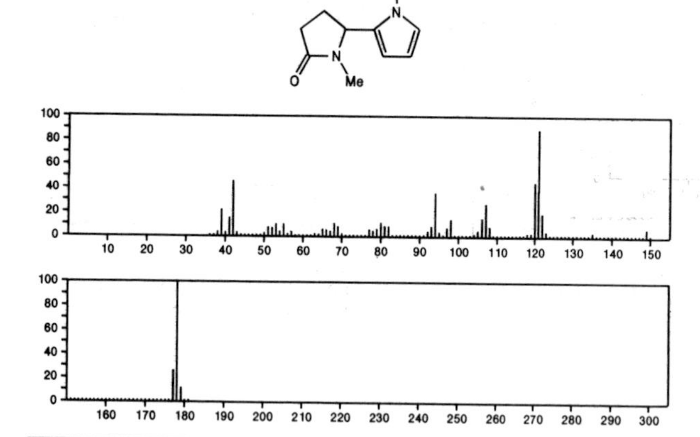

178 $C_{10}H_{14}N_2O$ 13950-23-7
2-Pyrrolidinone, 1-methyl-5-(1-methylpyrrol-3-yl)-

178 $C_{10}H_{14}N_2O$ 15769-88-7
2*H*-1,2-Oxazine, tetrahydro-2-methyl-6-(3-pyridinyl)-, (-)-

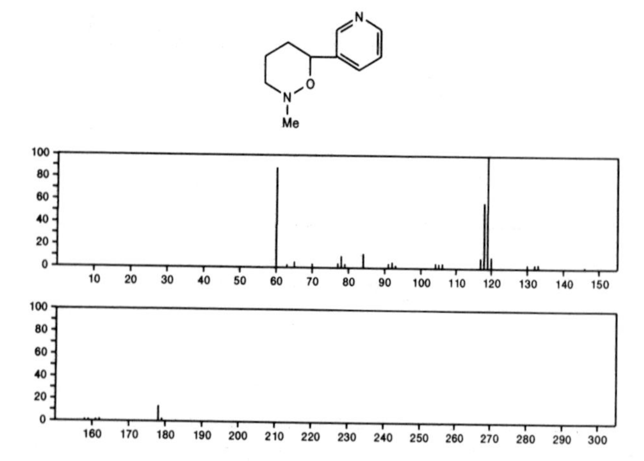

178 $C_{10}H_{14}N_2O$ 19293-74-4
Benzaldehyde, 4-(dimethylamino)-, *O*-methyloxime

178 $C_{10}H_{14}N_2O$ 54966-09-5
1-Piperidinecarboxaldehyde, 2-(1*H*-pyrrol-2-yl)-

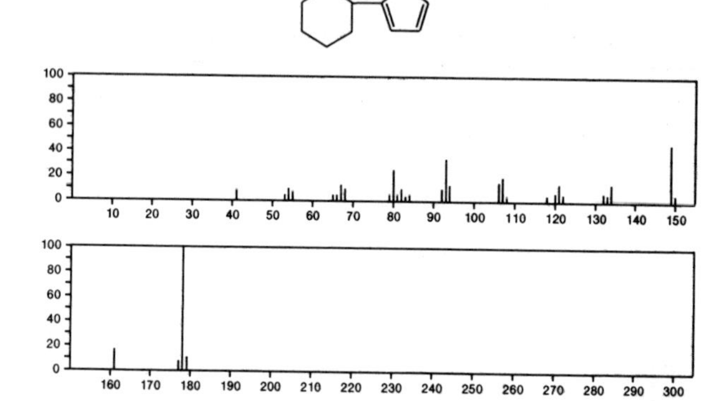

178 $C_{11}H_{14}O_2$ 93-15-2
Benzene, 1,2-dimethoxy-4-(2-propenyl)-

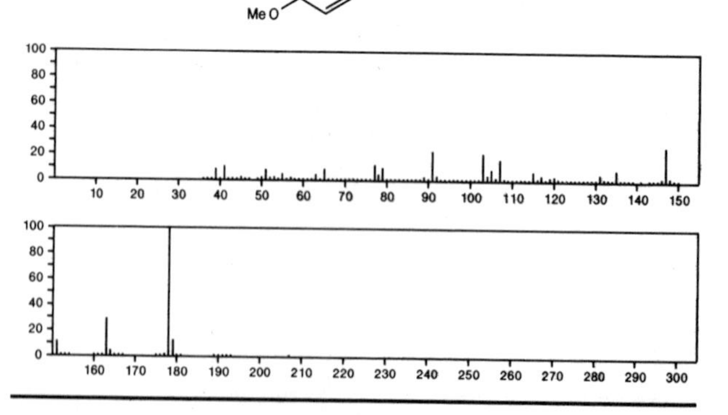

178 $C_{11}H_{14}O_2$ 93-16-3
Benzene, 1,2-dimethoxy-4-(1-propenyl)-

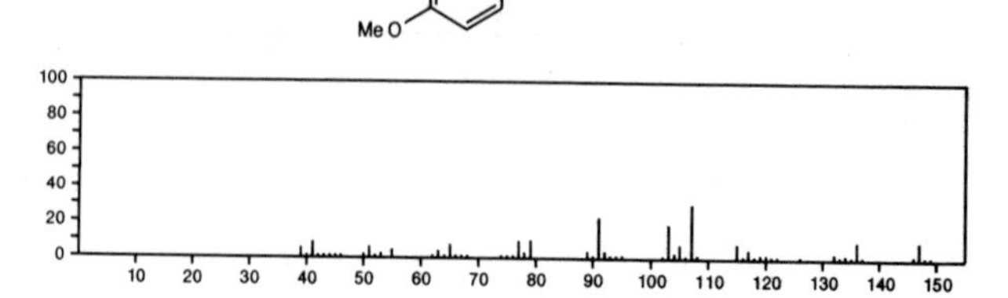

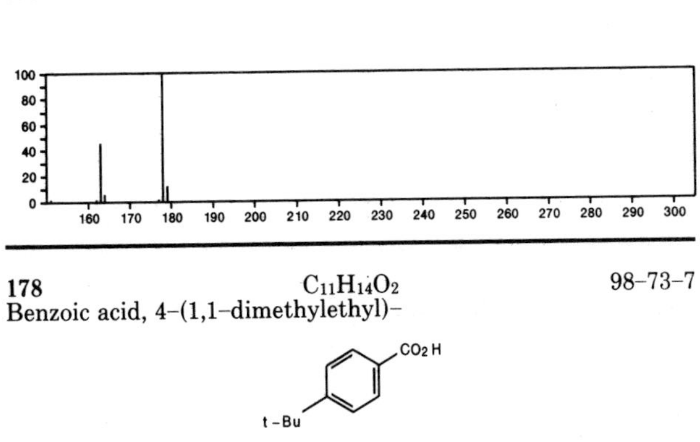

178 $C_{11}H_{14}O_2$ 98-73-7
Benzoic acid, 4-(1,1-dimethylethyl)-

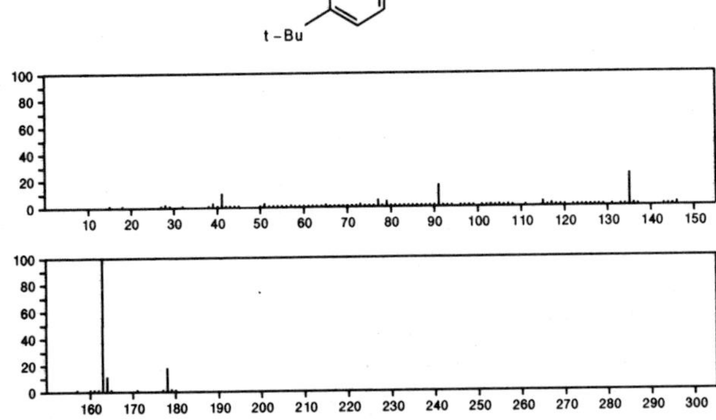

178 $C_{11}H_{14}O_2$ 103-37-7
Butanoic acid, phenylmethyl ester

PhCH2OC(O)Pr

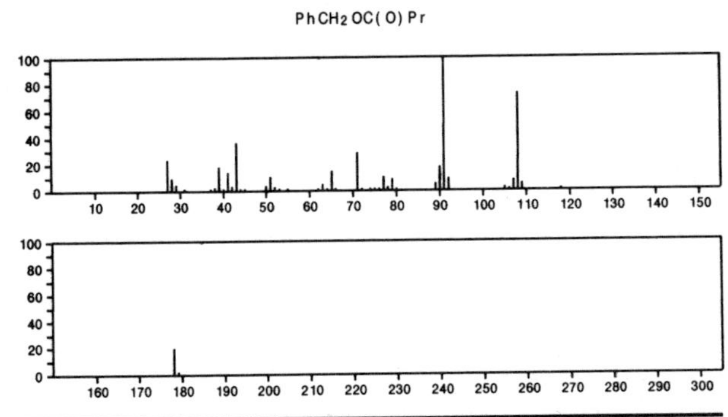

178 $C_{11}H_{14}O_2$ 104-20-1
2-Butanone, 4-(4-methoxyphenyl)-

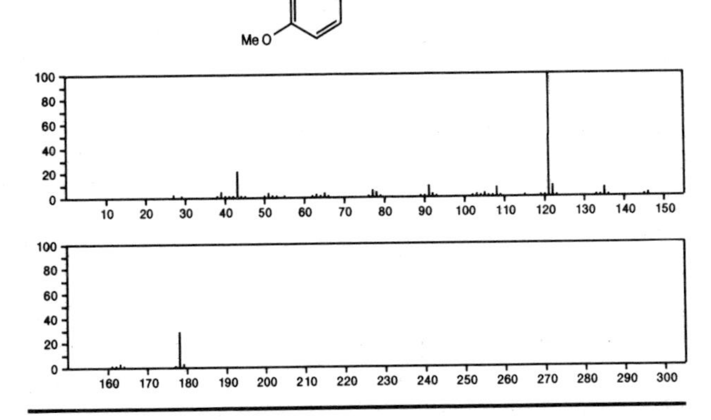

178 $C_{11}H_{14}O_2$ 120-50-3
Benzoic acid, 2-methylpropyl ester

i-BuOC(O)Ph

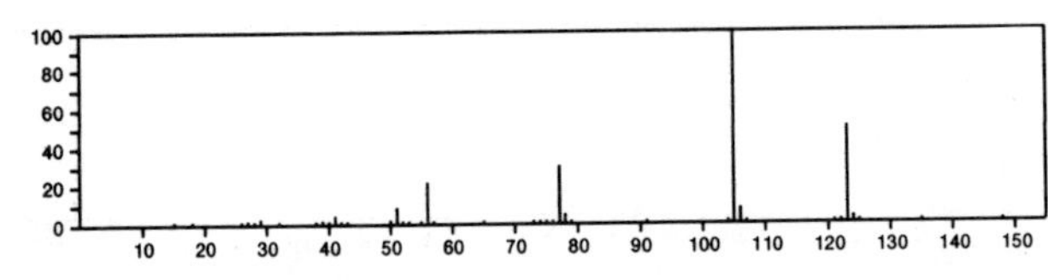

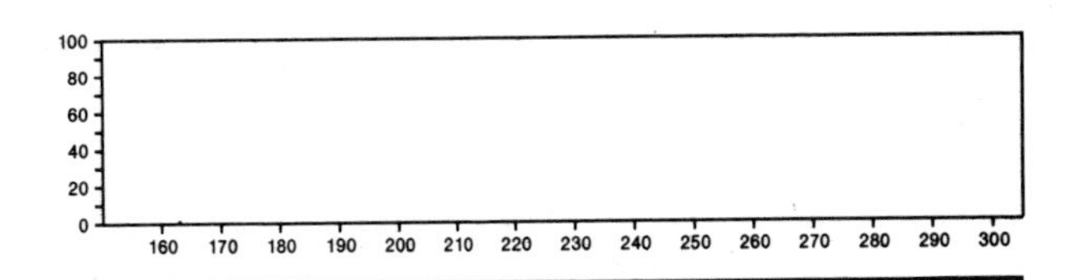

178 $C_{11}H_{14}O_2$ 122-72-5
Benzenepropanol, acetate

AcO(CH2)3Ph

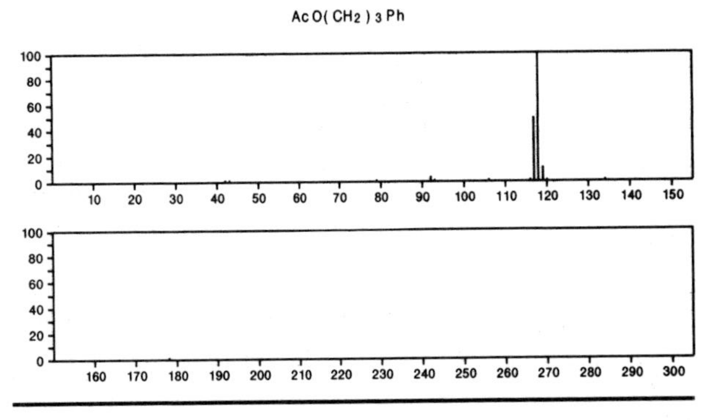

178 $C_{11}H_{14}O_2$ 136-60-7
Benzoic acid, butyl ester

PhC(O)O(CH2)3Me

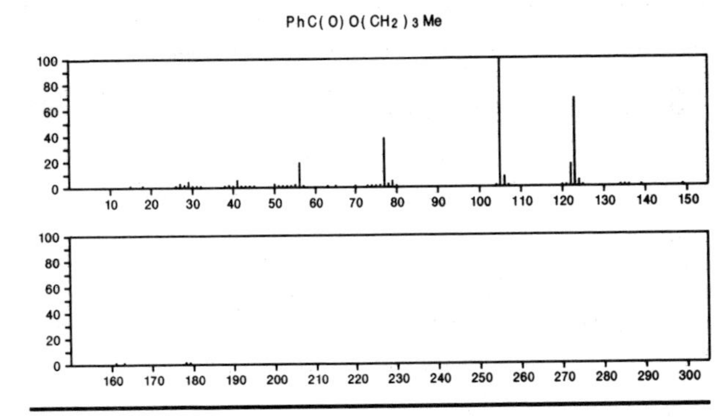

178 $C_{11}H_{14}O_2$ 487-67-2
2-Cyclopenten-1-one, 4-hydroxy-3-methyl-2-(2,4-pentadienyl)-, (Z)-(+)-

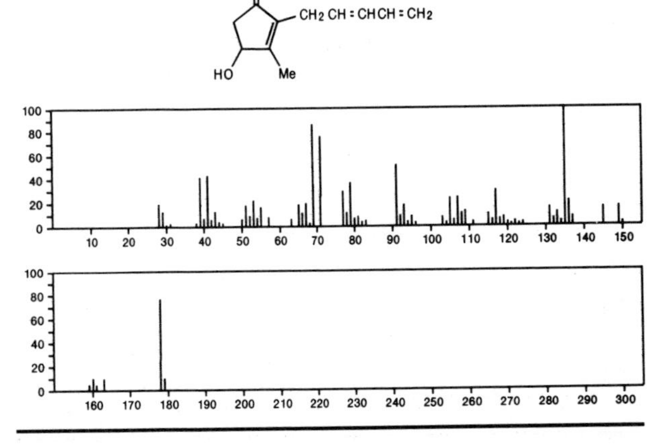

178 $C_{11}H_{14}O_2$ 1010-48-6
Benzenepropanoic acid, β,β-dimethyl-

HO2CCH2CMe2Ph

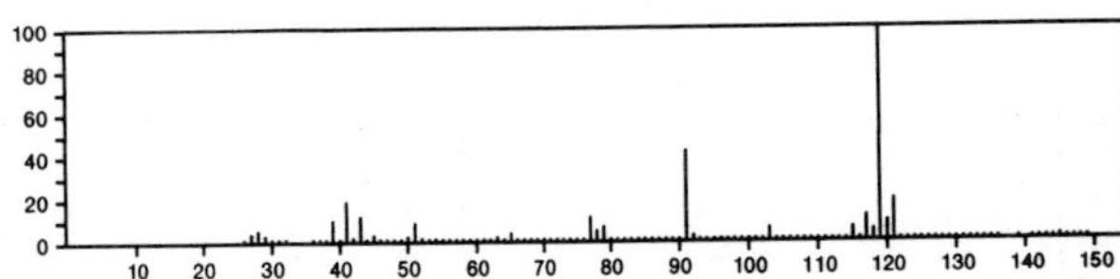

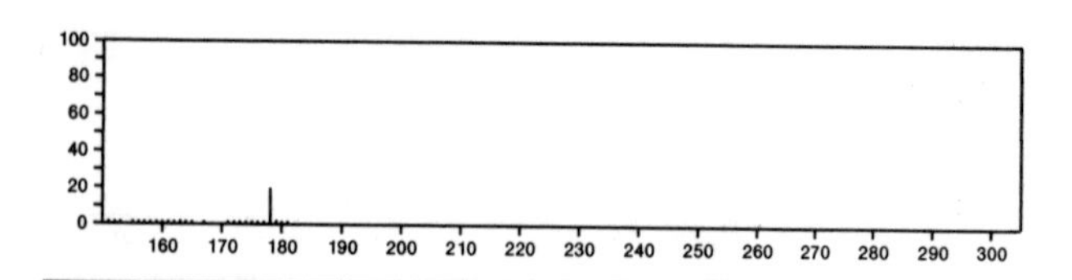

178 $C_{11}H_{14}O_2$ 1521–94–4
1–Phthalanol, 1,3,3–trimethyl–

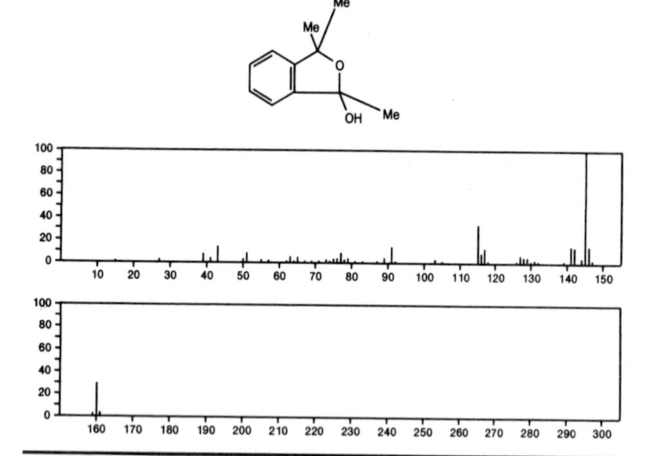

178 $C_{11}H_{14}O_2$ 2046–17–5
Benzenebutanoic acid, methyl ester

Ph (CH_2)$_3$ C(O) OMe

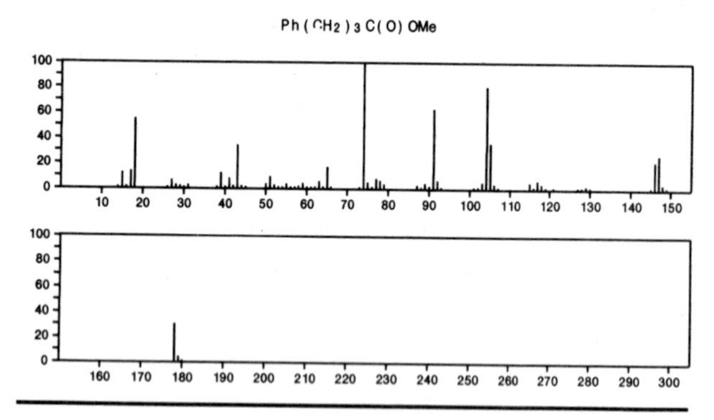

178 $C_{11}H_{14}O_2$ 2294–71–5
Benzeneacetic acid, α–ethyl–, methyl ester

MeOC(O) CHEt Ph

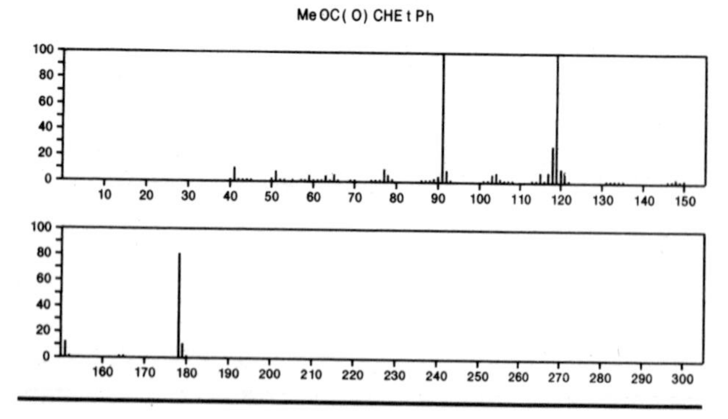

178 $C_{11}H_{14}O_2$ 4362–18–9
1,3–Dioxolane, 2–methyl–2–(phenylmethyl)–

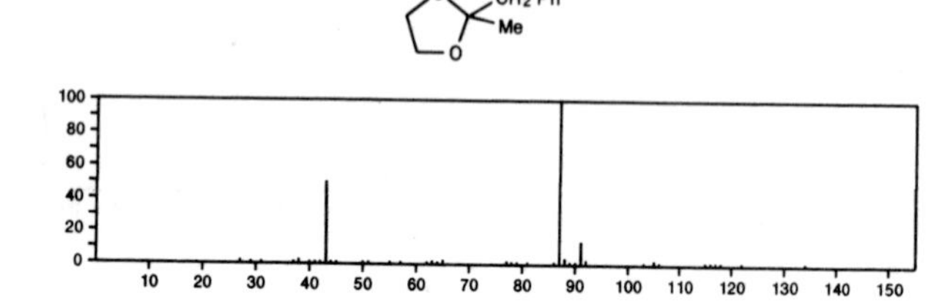

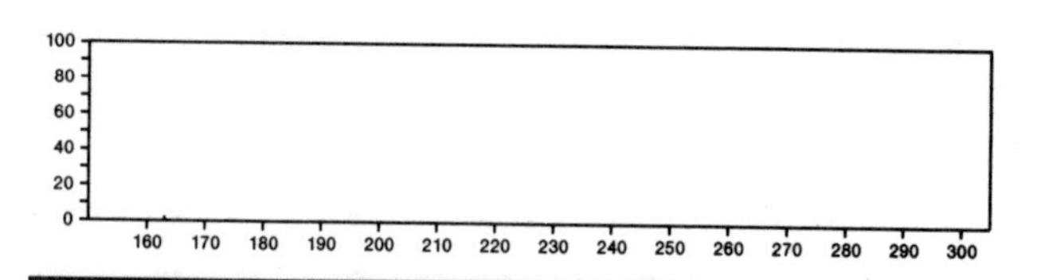

178 $C_{11}H_{14}O_2$ 4920–92–7
Propanoic acid, 2,2–dimethyl–, phenyl ester

$Me_3CC(O)OPh$

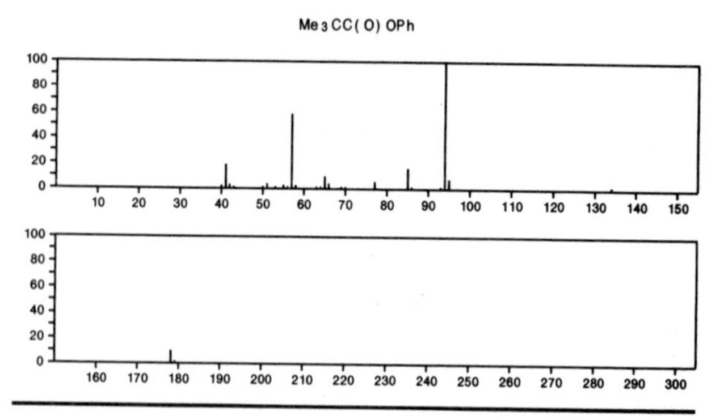

178 $C_{11}H_{14}O_2$ 7124–99–4
2*H*–1,7–Benzodioxonin, 3,4,5,6–tetrahydro–

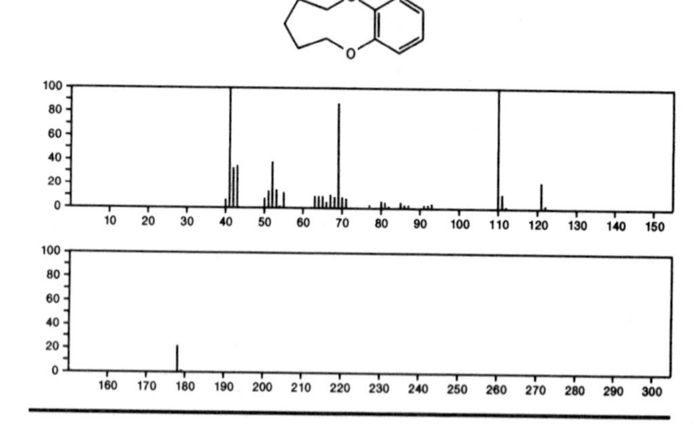

178 $C_{11}H_{14}O_2$ 7315–68–6
Benzenebutanoic acid, β–methyl–

$PhCH_2CHMeCH_2CO_2H$

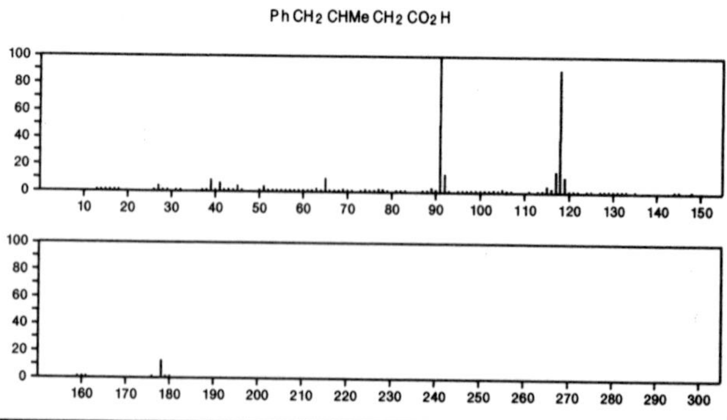

178 $C_{11}H_{14}O_2$ 17235–14–2
Phenol, 2–(3–hydroxy–3–methyl–1–butenyl)–, (*Z*)–

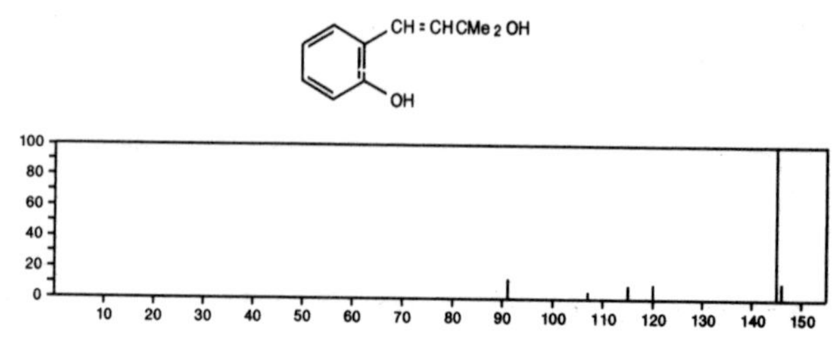

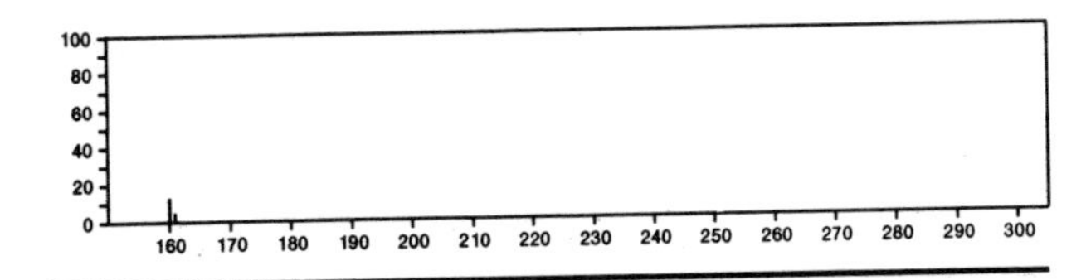

178 $C_{11}H_{14}O_2$ 19731-91-0
Benzenepropanoic acid, α,β-dimethyl-

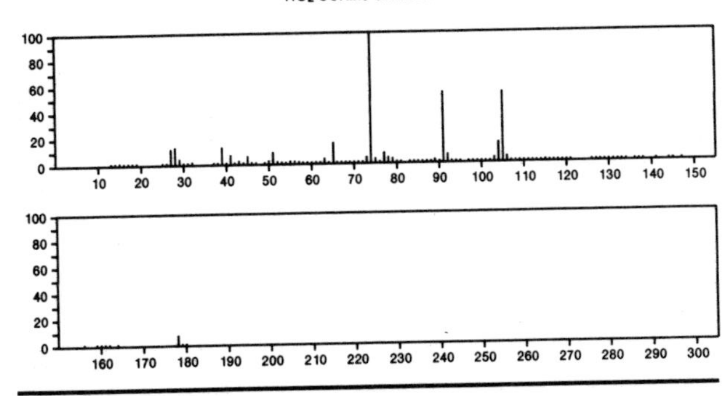

178 $C_{11}H_{14}O_2$ 20115-23-5
Pentanoic acid, phenyl ester

PhOC(O)(CH2)3Me

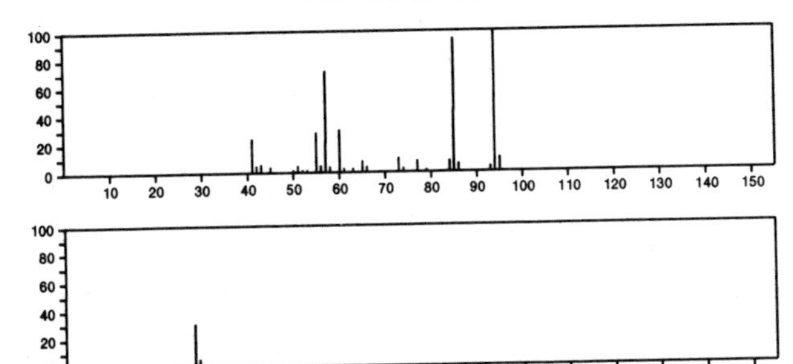

178 $C_{11}H_{14}O_2$ 20185-55-1
Benzoic acid, 4-(1-methylethyl)-, methyl ester

C(O)OMe
i-Pr

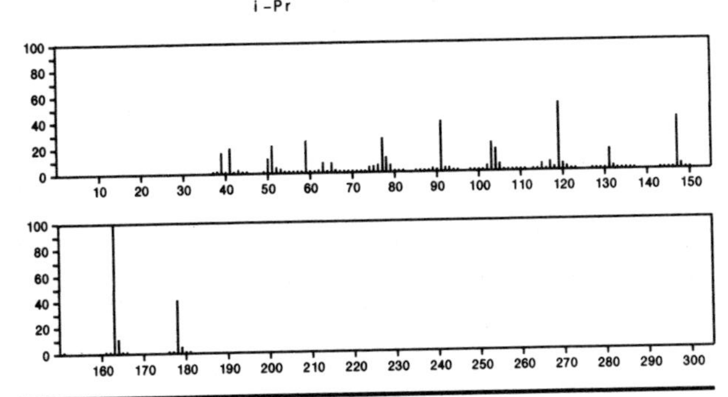

178 $C_{11}H_{14}O_2$ 36646-68-1
Methanone, cyclohexyl-3-furanyl-

CO

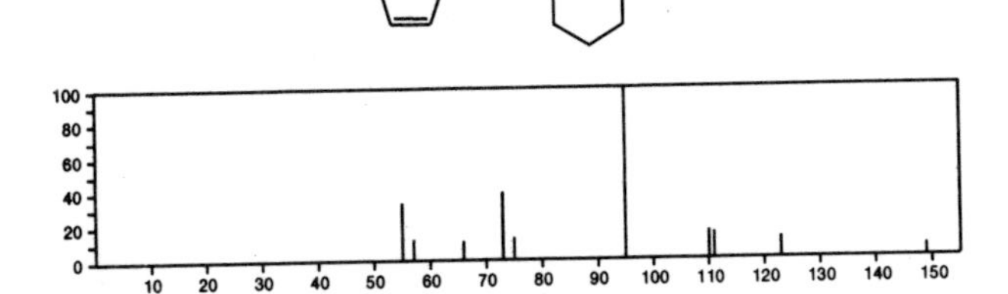

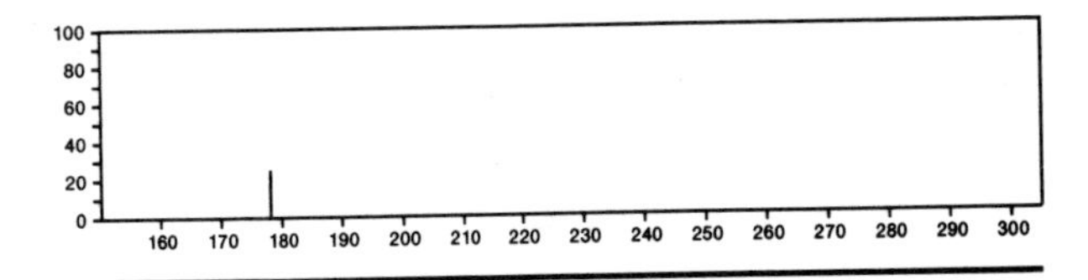

178 $C_{11}H_{14}O_2$ 40924-58-1
2*H*-Pyran, 2-(2,5-hexadiynyloxy)tetrahydro-

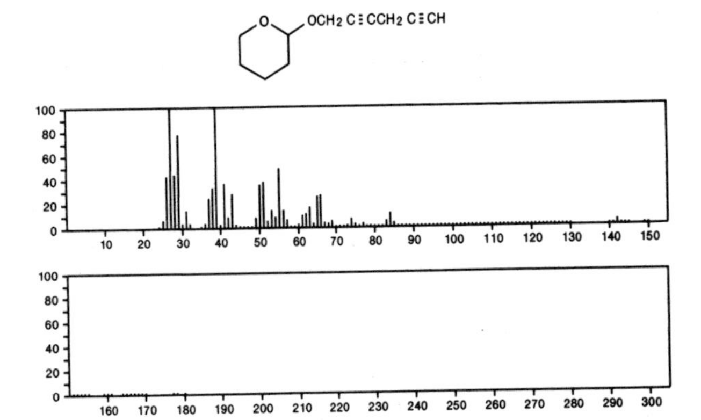

178 $C_{11}H_{14}O_2$ 51086-38-5
1,2-Naphthalenediol, 1,2,3,4-tetrahydro-4-methyl-

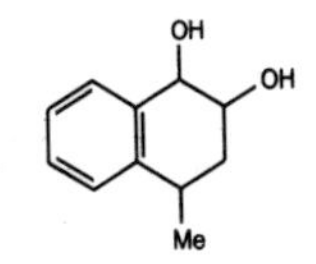

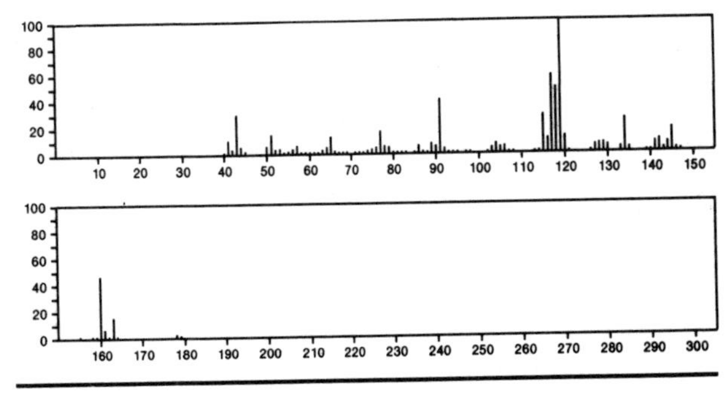

178 $C_{11}H_{14}O_2$ 51149-73-6
2-Propanone, 1-ethoxy-3-phenyl-

EtOCH2COCH2Ph

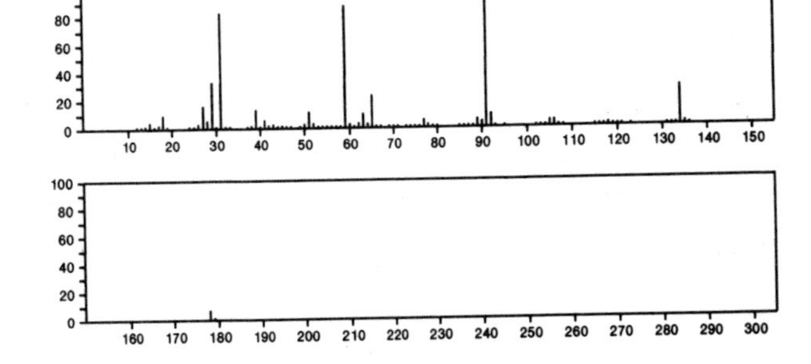

178 $C_{11}H_{14}O_2$ 51664-96-1

Benzoic acid, 2,4,5-trimethyl-, methyl ester

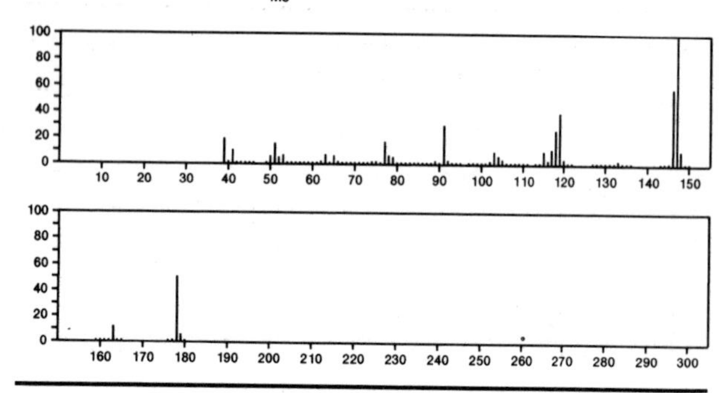

178 $C_{11}H_{14}O_2$ 54549-71-2

Propanoic acid, 4-ethylphenyl ester

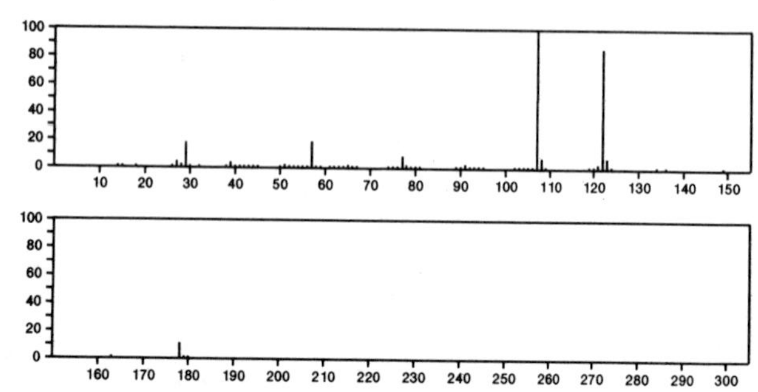

178 $C_{11}H_{14}O_2$ 54549-72-3

Ethanone, 1-[4-(1-hydroxy-1-methylethyl)phenyl]-

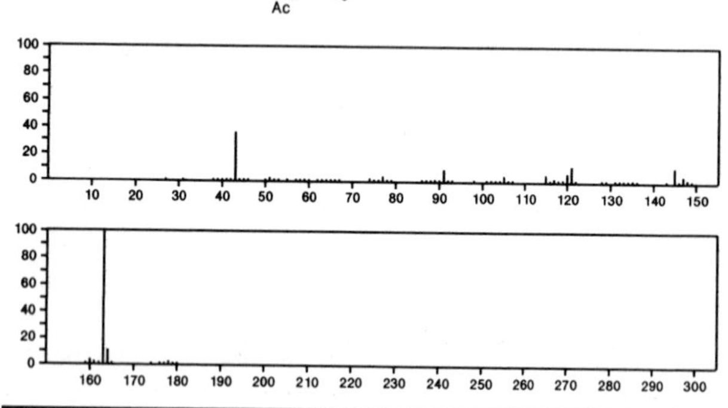

178 $C_{11}H_{14}O_2$ 56588-36-4

1,2-Naphthalenediol, 1,2,3,4-tetrahydro-1-methyl-, *cis*-

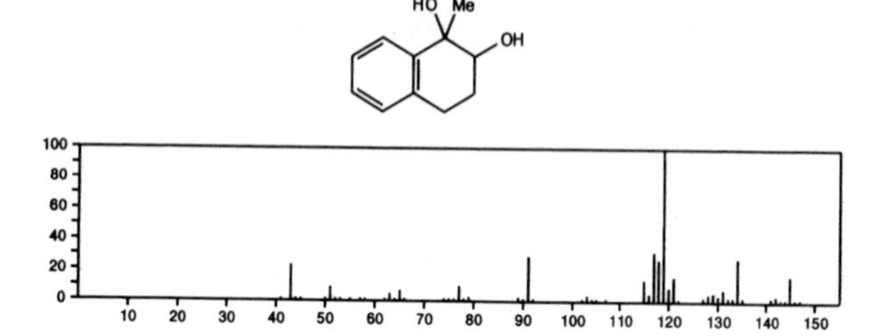

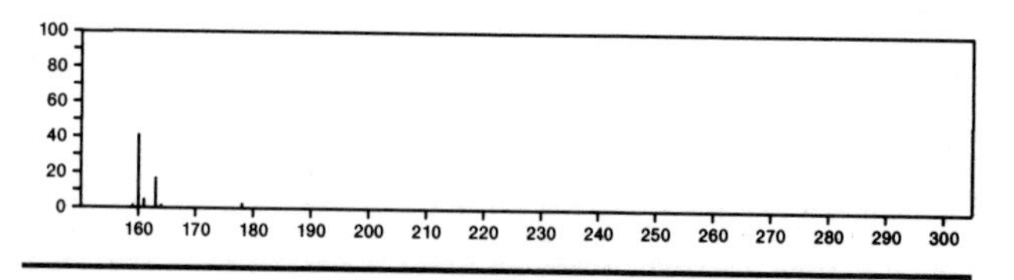

178 $C_{11}H_{14}S$ 10276-04-7

Benzene, [(3-methyl-2-butenyl)thio]-

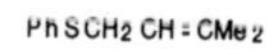

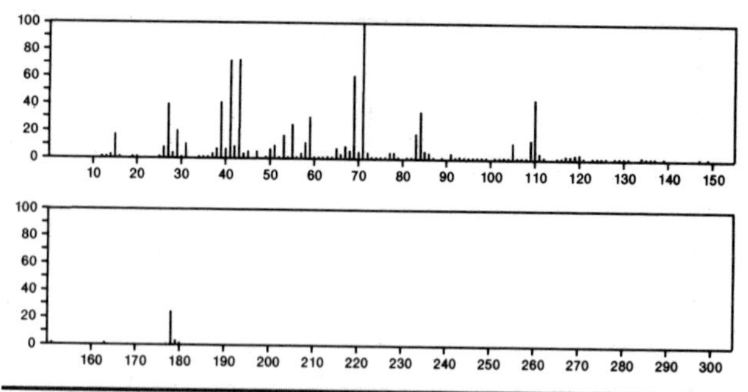

178 $C_{11}H_{14}S$ 54549-73-4

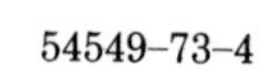

2*H*-1-Benzothiopyran, 2-ethyl-3,4-dihydro-

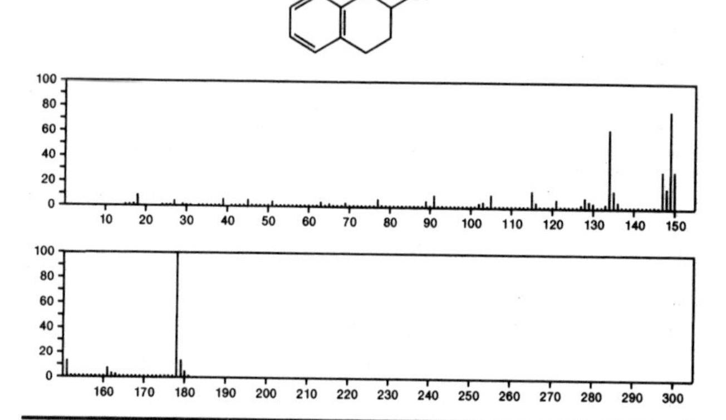

178 $C_{11}H_{18}N_2$ 18433-98-2

Pyrazine, 2,5-dimethyl-3-(3-methylbutyl)-

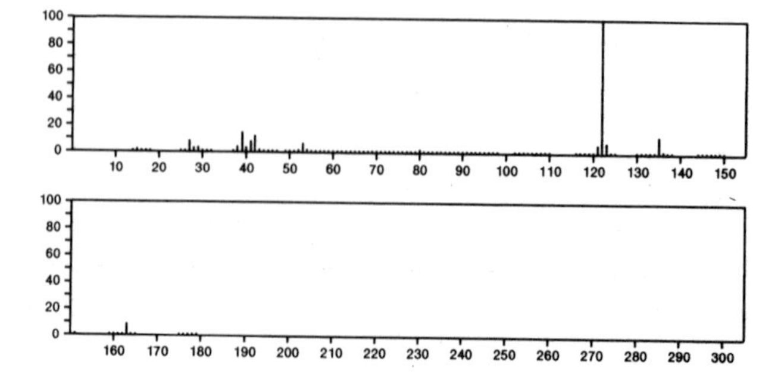

178 $C_{11}H_{18}N_2$ 34176-71-1
1*H*-Cyclodecapyrazole, 4,5,6,7,8,9,10,11-octahydro-

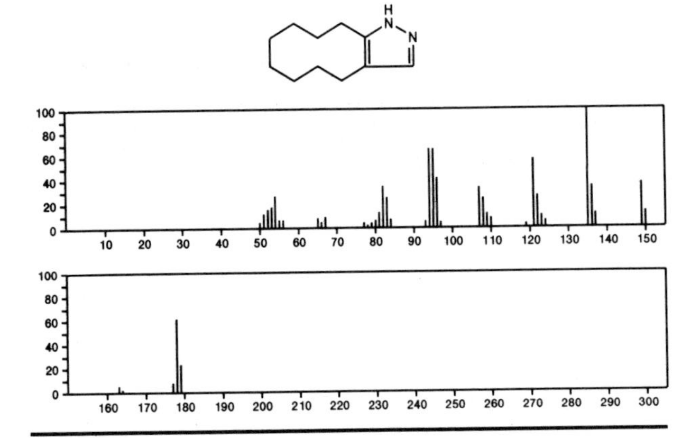

178 $C_{11}H_{18}N_2$ 54518-12-6
Hexanenitrile, 3-(1-pyrrolidinylmethylene)-

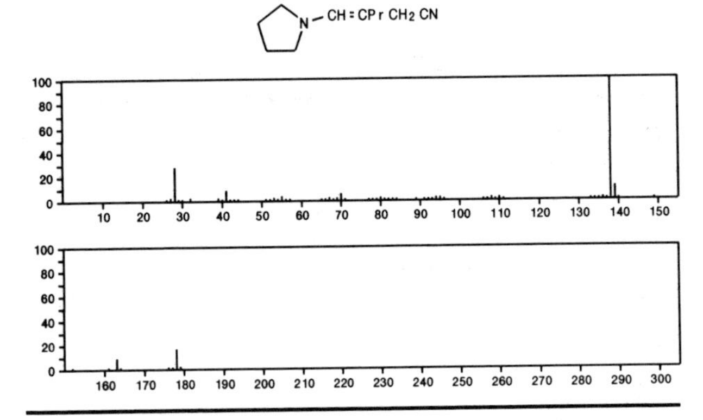

178 $C_{12}H_{18}O$ 1011-12-7
Cyclohexanone, 2-cyclohexylidene-

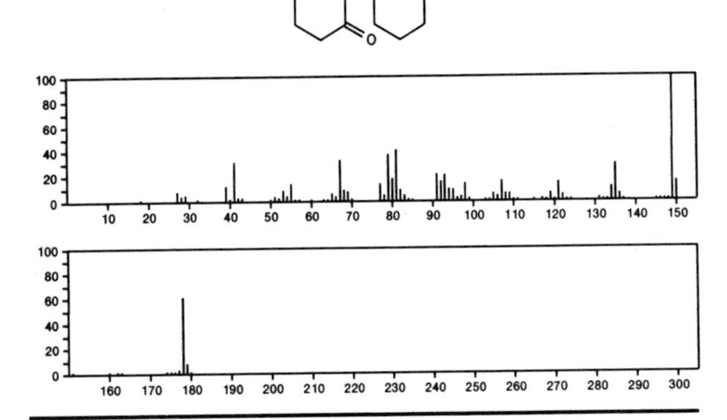

178 $C_{12}H_{18}O$ 1132-66-7
Benzene, (hexyloxy)-

Me(CH2)5OPh

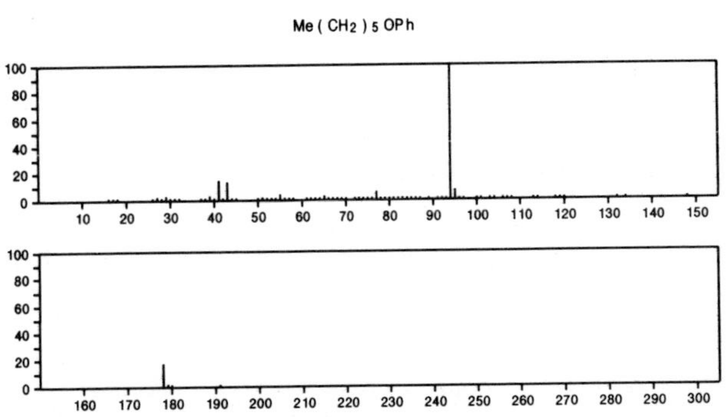

178 $C_{12}H_{18}O$ 2078-54-8
Phenol, 2,6-bis(1-methylethyl)-

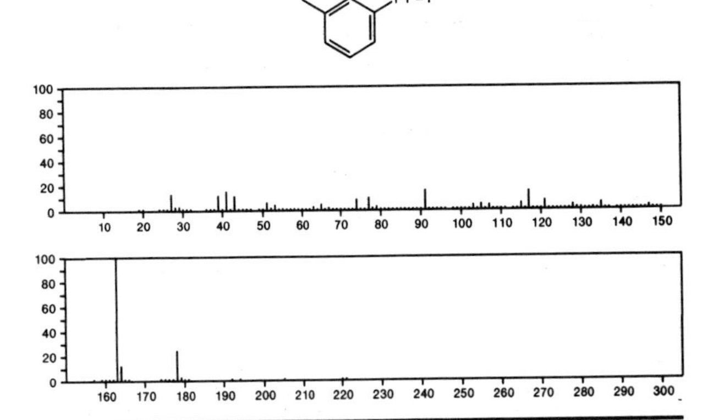

178 $C_{12}H_{18}O$ 2934-05-6
Phenol, 2,4-bis(1-methylethyl)-

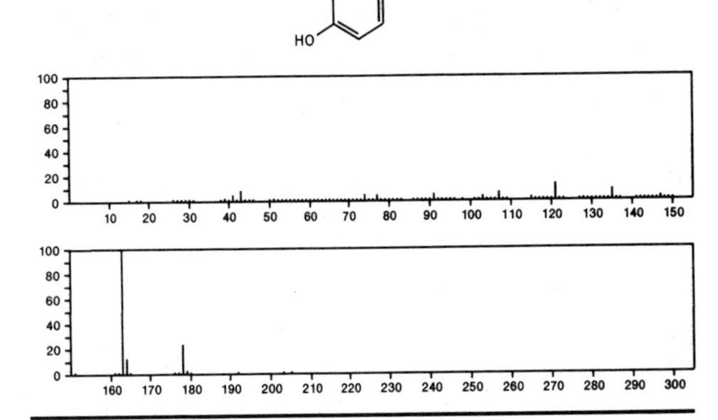

178 $C_{12}H_{18}O$ 13485-66-0
2(1*H*)-Naphthalenone, 4a,5,6,7,8,8a-hexahydro-4a,8a-dimethyl-, *cis*-

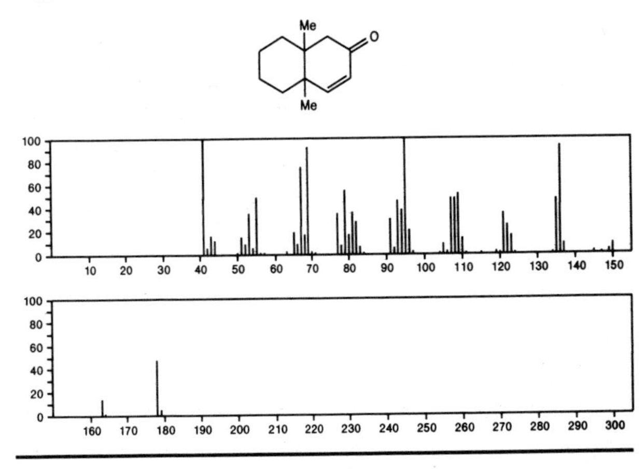

178 $C_{12}H_{18}O$ 14289-73-7
Ether, 3-phenylpropyl propyl

Pr O(CH2)3Ph

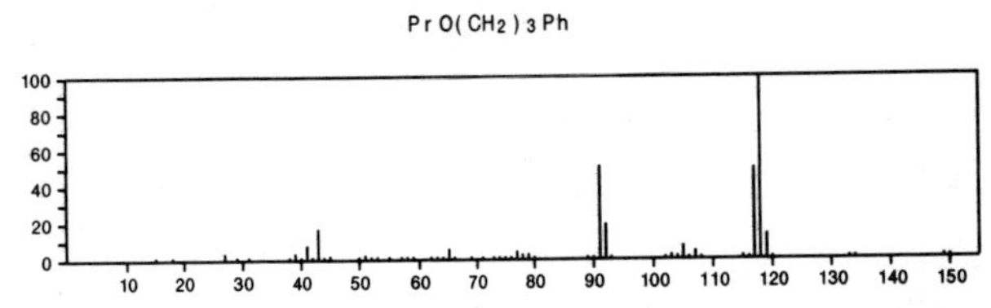

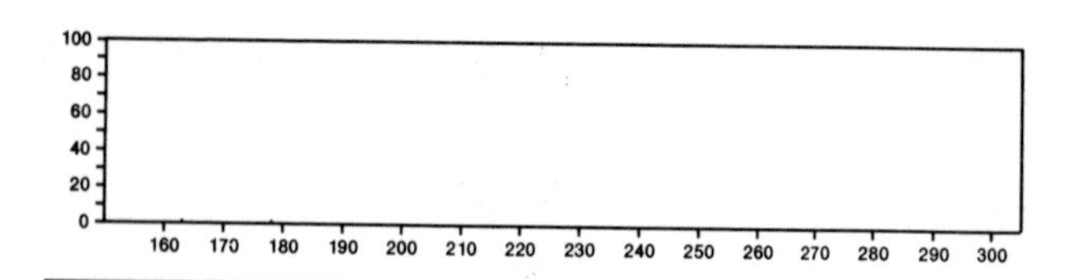

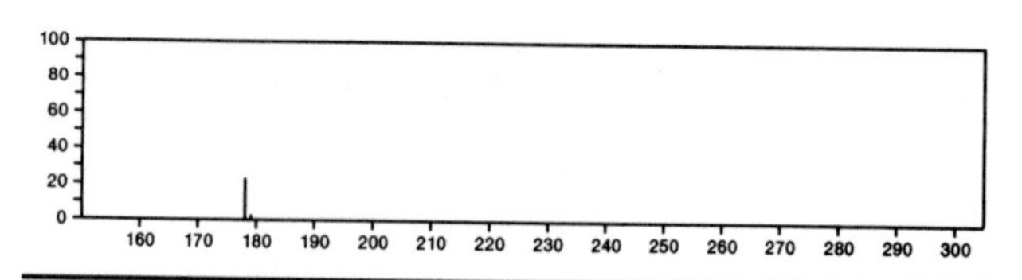

178 $C_{12}H_{18}O$ 17269-94-2
Benzene, 1-(1,1-dimethylethyl)-4-ethoxy-

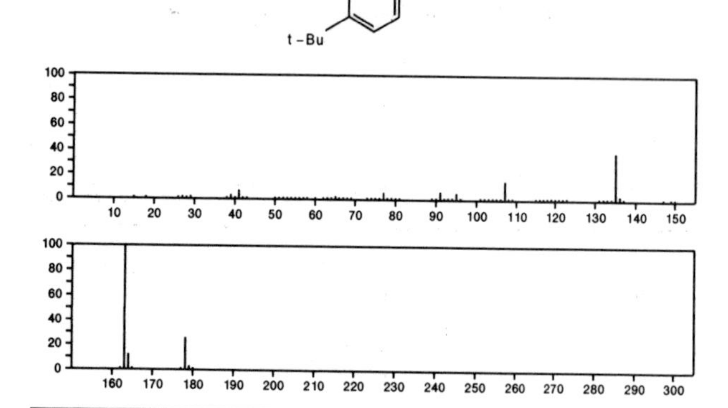

178 $C_{12}H_{18}O$ 17429-26-4
2*H*-Benzocyclohepten-2-one, 1,4a,5,6,7,8,9,9a-octahydro-4a-methyl-, *trans*-

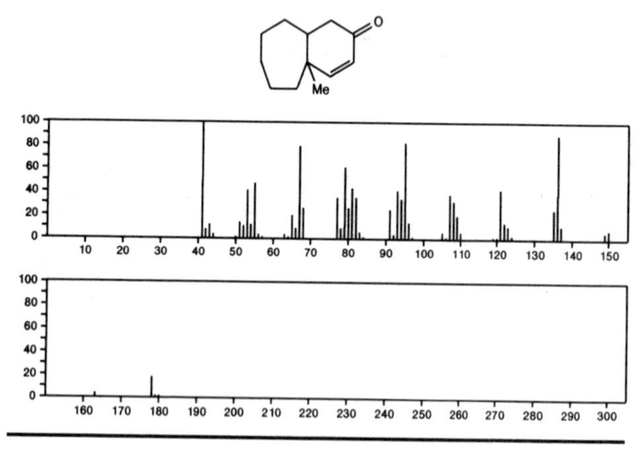

178 $C_{12}H_{18}O$ 18346-78-6
Bicyclo[3.3.1]nonan-2-one, 9-isopropylidene-

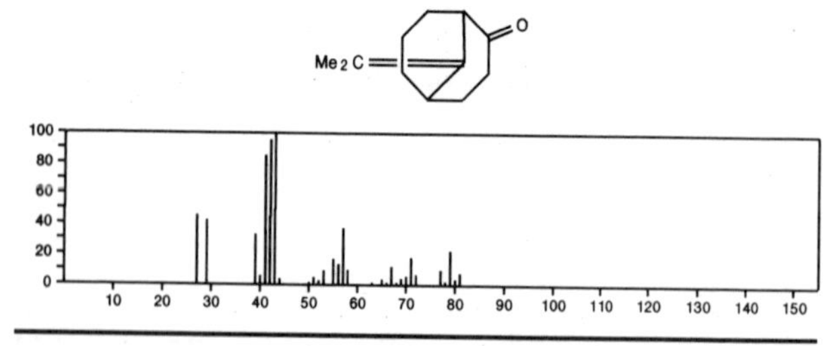

178 $C_{12}H_{18}O$ 20056-56-8
Anisole, *o*-pentyl-

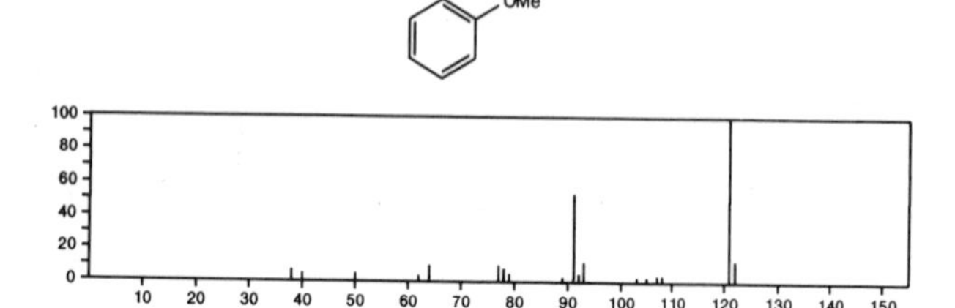

178 $C_{12}H_{18}O$ 20056-57-9
Benzene, 1-methoxy-3-pentyl-

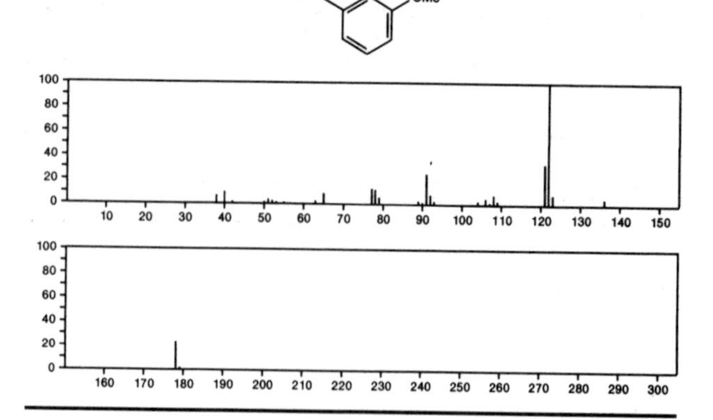

178 $C_{12}H_{18}O$ 20056-58-0
Benzene, 1-methoxy-4-pentyl-

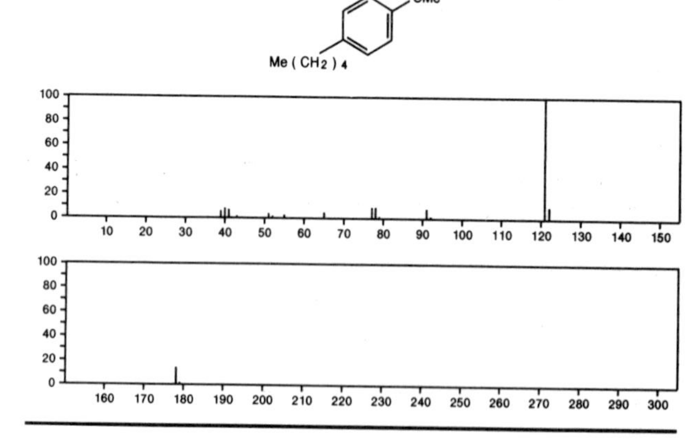

178 $C_{12}H_{18}O$ 24142-77-6
Ether, α,α-dimethylbenzyl propyl

$PhCMe_2(OPr)$

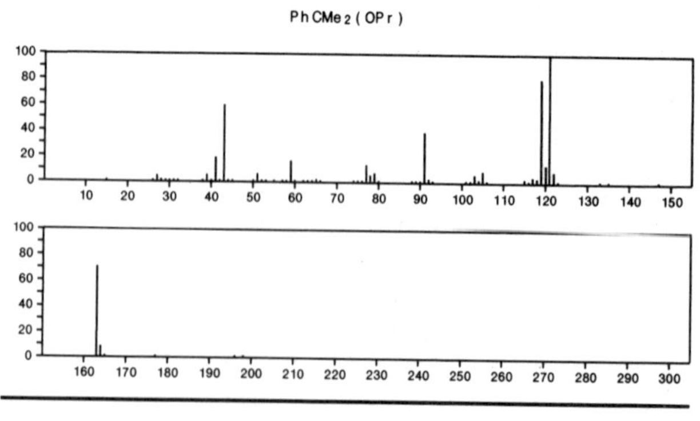

178 $C_{12}H_{18}O$ 24142-78-7
Ether, α,α-dimethylbenzyl isopropyl

$PhCMe_2(OPr\text{-}i)$

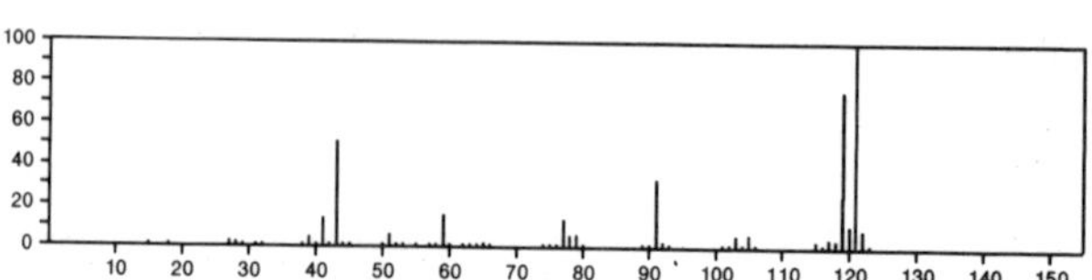

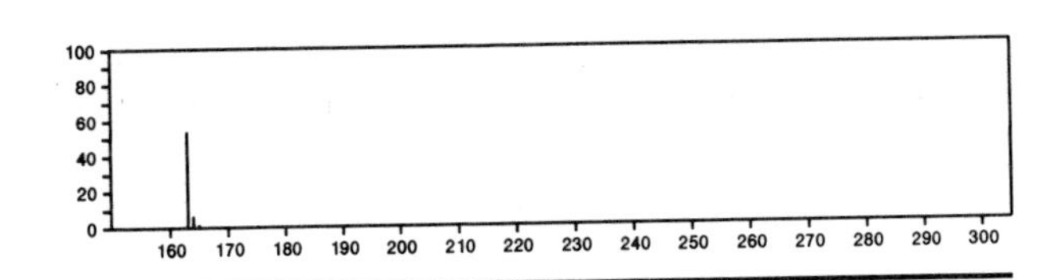

178 $C_{12}H_{18}O$ 29460-68-2
2-Cyclohexene-1-acrolein, 2,6,6-trimethyl-

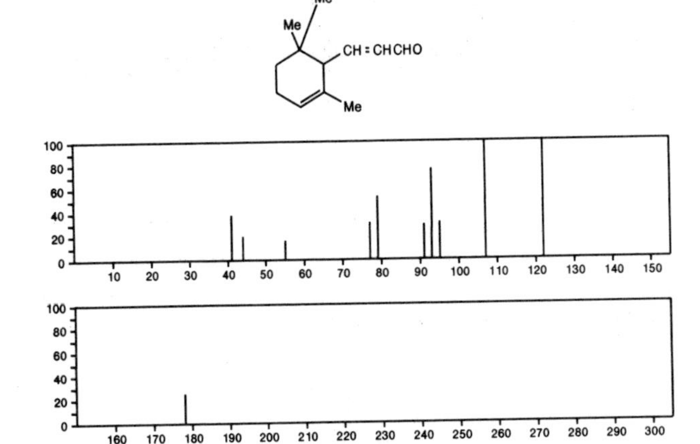

178 $C_{12}H_{18}O$ 34386-42-0
Benzenemethanol, 4-(1,1-dimethylethyl)-α-methyl-

178 $C_{12}H_{18}O$ 54345-68-5
1-Oxaspiro[4.5]deca-3,6-diene, 6,10,10-trimethyl-

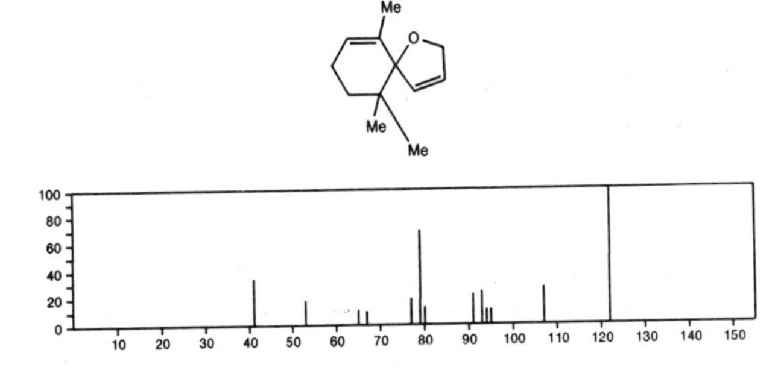

178 $C_{12}H_{18}O$ 54345-69-6
1-Oxaspiro[4.5]dec-3-ene, 6,6-dimethyl-10-methylene-

178 $C_{12}H_{18}O$ 54518-11-5
Benzeneethanol, α-methyl-3-(1-methylethyl)-

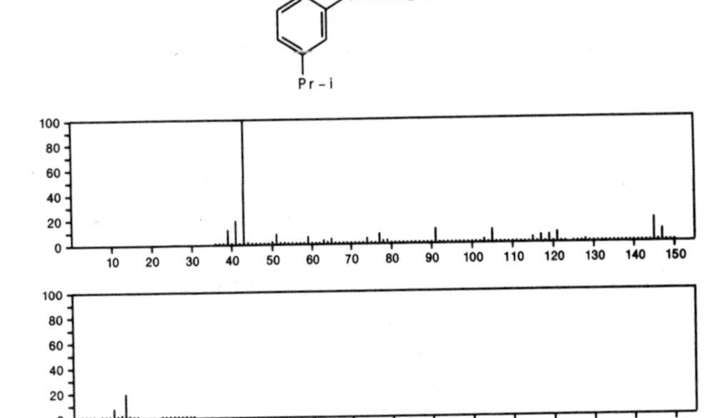

178 $C_{12}H_{18}O$ 54518-14-8
Benzeneethanol, *ar*,*ar*-diethyl-

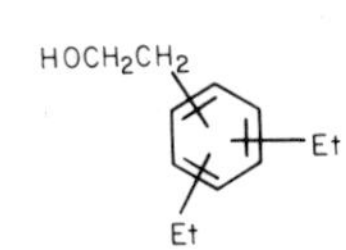

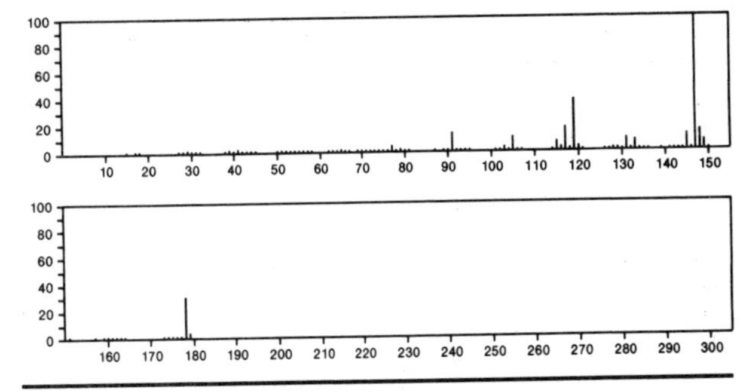

178 $C_{12}H_{18}O$ 55103-71-4
2*H*-Benzocyclohepten-2-one, 3,4,4a,5,6,7,8,9-octahydro-4a-methyl-, (*S*)-

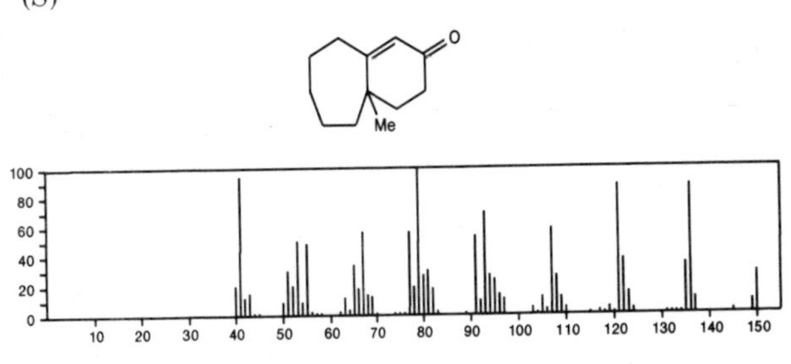

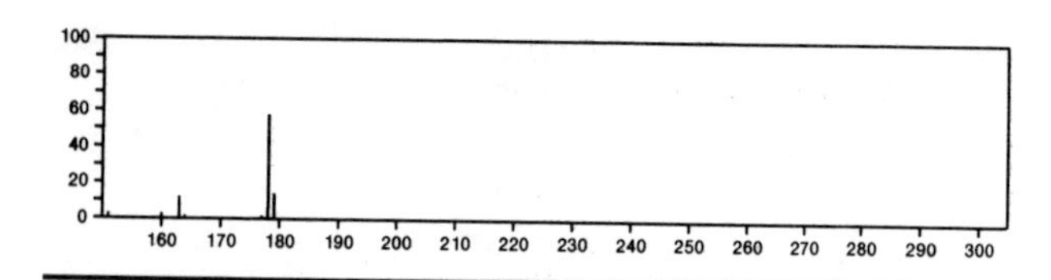

178 $C_{12}H_{18}O$ 56052-33-6
Benzene, 2-butoxy-1,3-dimethyl-

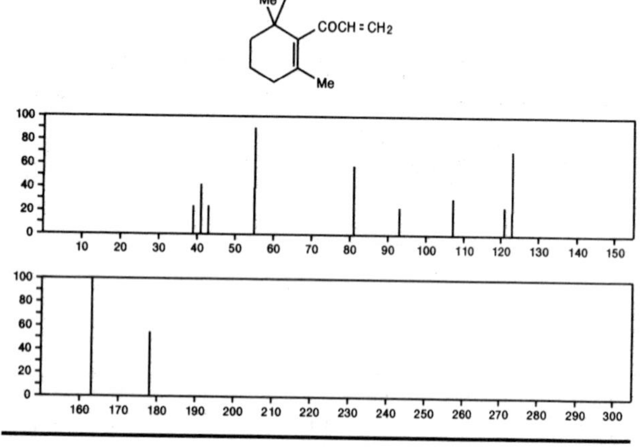

178 $C_{12}H_{18}O$ 56248-16-9
2-Propen-1-one, 1-(2,6,6-trimethyl-1-cyclohexen-1-yl)-

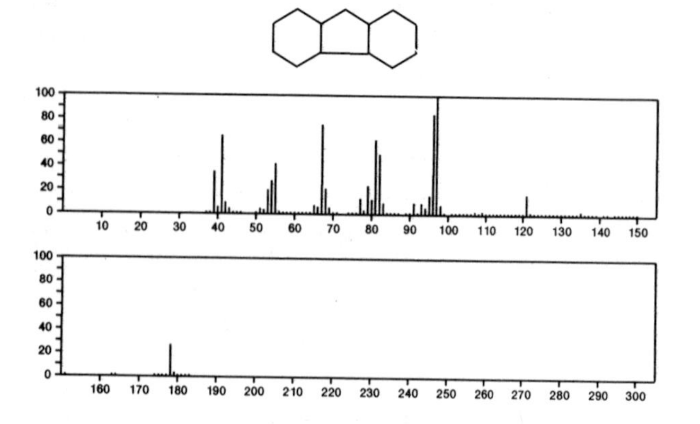

178 $C_{13}H_{22}$ 5744-03-6
1*H*-Fluorene, dodecahydro-

178 $C_{13}H_{22}$ 50746-55-9
Bicyclo[3.1.1]heptane, 2,6,6-trimethyl-3-(2-propenyl)-, (1α,2β,3α,5α)-

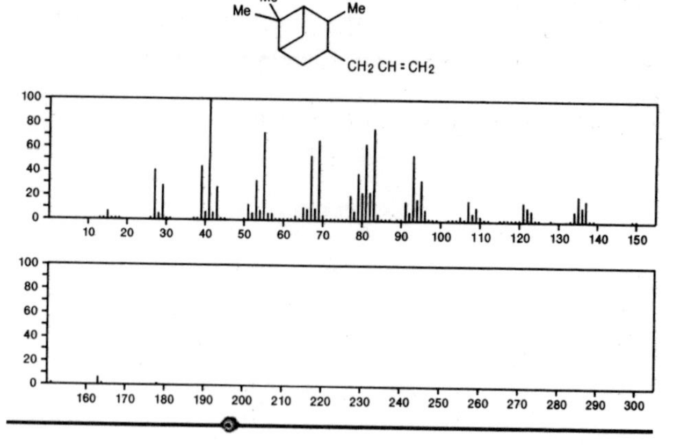

178 $C_{13}H_{22}$ 55103-62-3
Benzocyclooctene, 1,4,4a,5,6,7,8,9,10,10a-decahydro-4a-methyl-, *trans*-

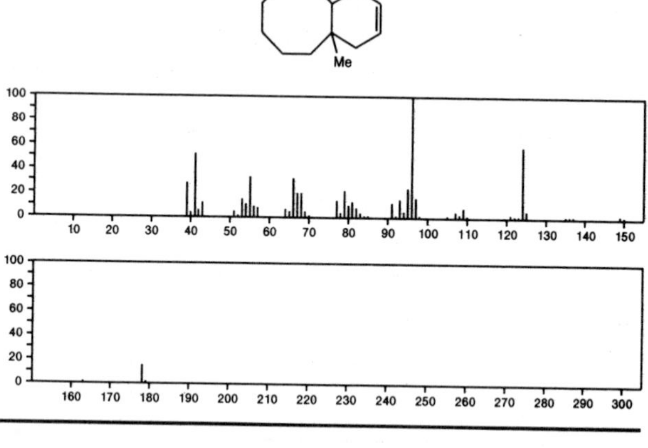

178 $C_{14}H_{10}$ 85-01-8
Phenanthrene

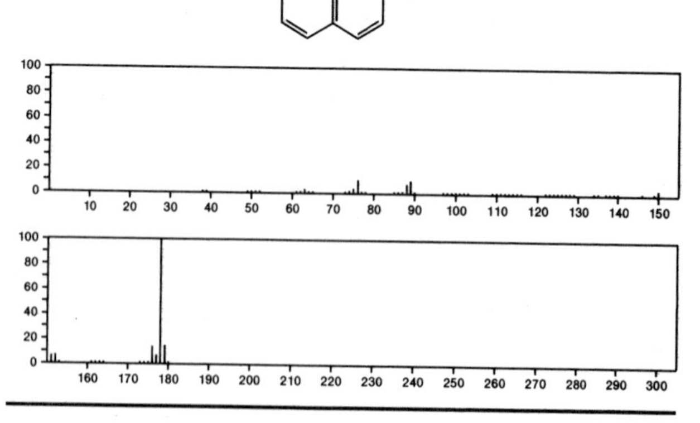

178 $C_{14}H_{10}$ 120-12-7
Anthracene

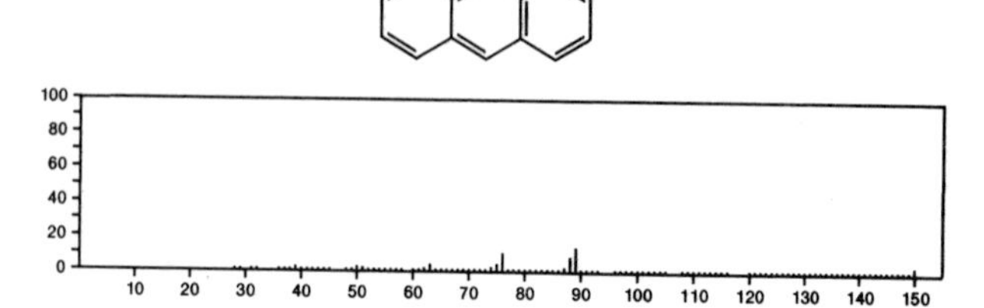

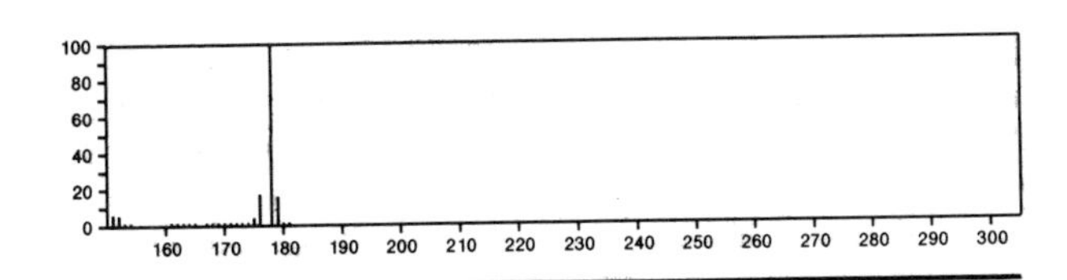

178 $C_{14}H_{10}$ 501–65–5
Benzene, 1,1'–(1,2–ethynediyl)bis–

PhC≡CPh

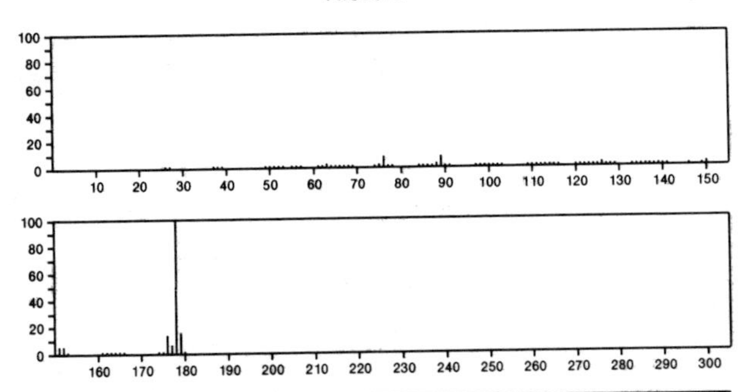

179 $C_4H_6NO_5P$ 3048–73–5
2,6,7–Trioxa–1–phosphabicyclo[2.2.2]octane, 4–nitro–

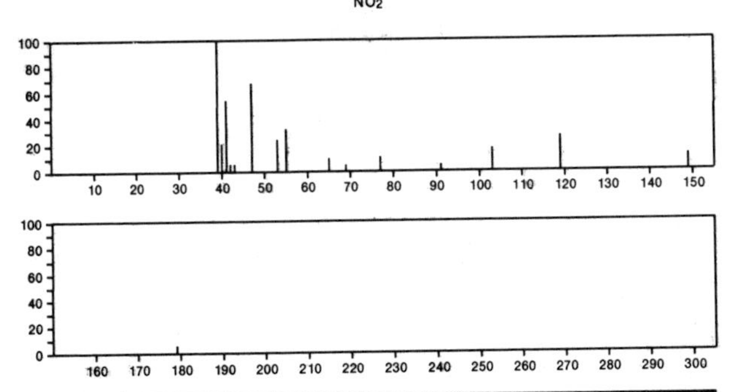

179 $C_6H_5N_5O_2$ 119–44–8
4,6–Pteridinedione, 2–amino–1,5–dihydro–

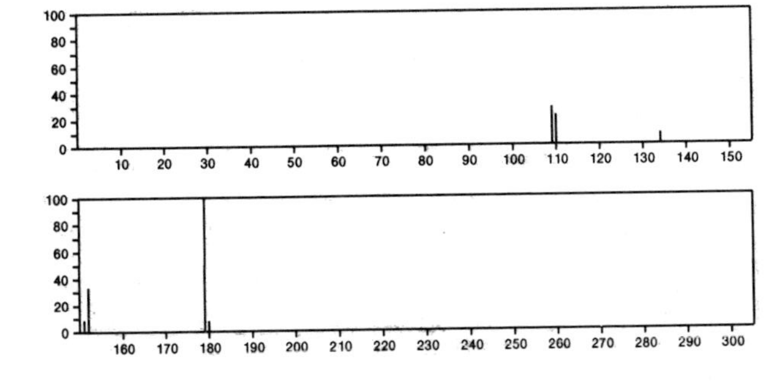

179 $C_6H_5N_5O_2$ 529–69–1
4,7(1*H*,8*H*)–Pteridinedione, 2–amino–

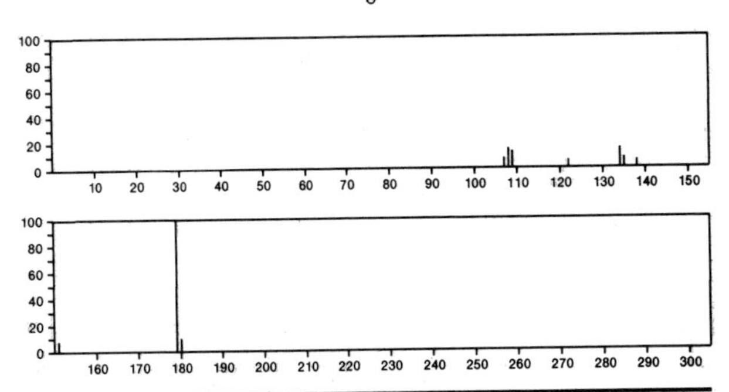

179 $C_6H_{18}N_3OP$ 680–31–9
Phosphoric triamide, hexamethyl–

NMe2
Me2NP=O
Me2N

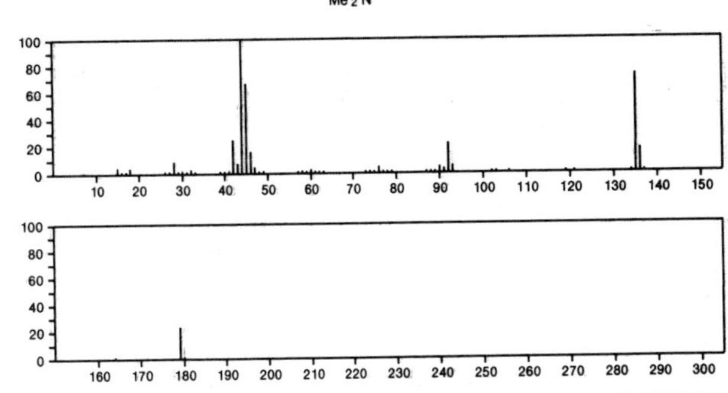

179 $C_7H_5N_3O_3$ 53975–72–7
1,2,4–Triazolo[4,3–*a*]pyridine–8–carboxylic acid, 2,3–dihydro–3–oxo–

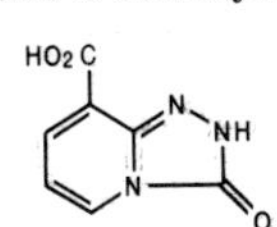

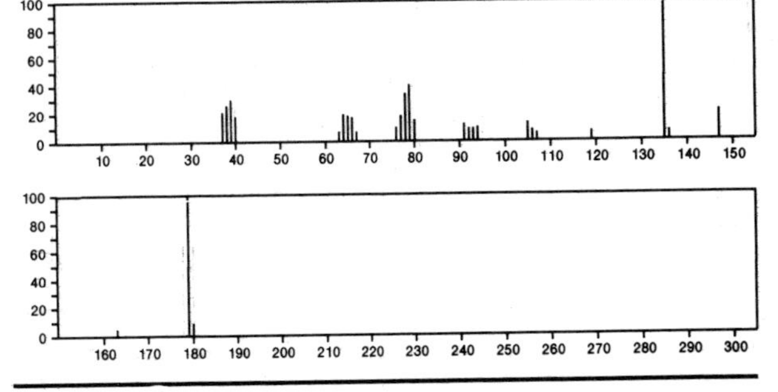

179 $C_7H_9N_5O$ 1445–15–4
6*H*–Purin–6–one, 2–(dimethylamino)–1,7–dihydro–

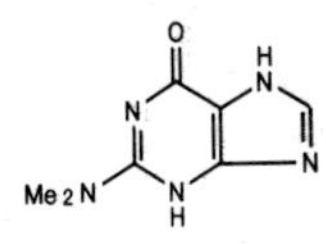

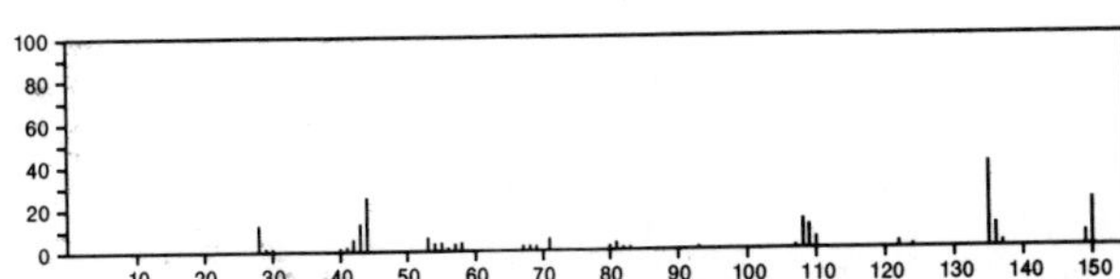

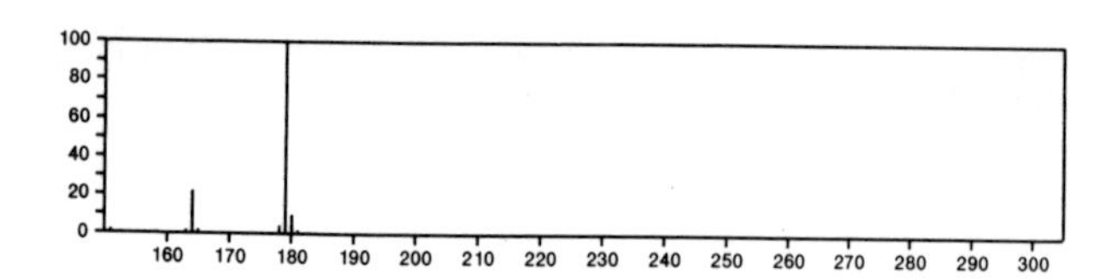

179 $C_7H_9N_5O$ 50704-44-4
1*H*-Purin-2-amine, 6-methoxy-*N*-methyl-

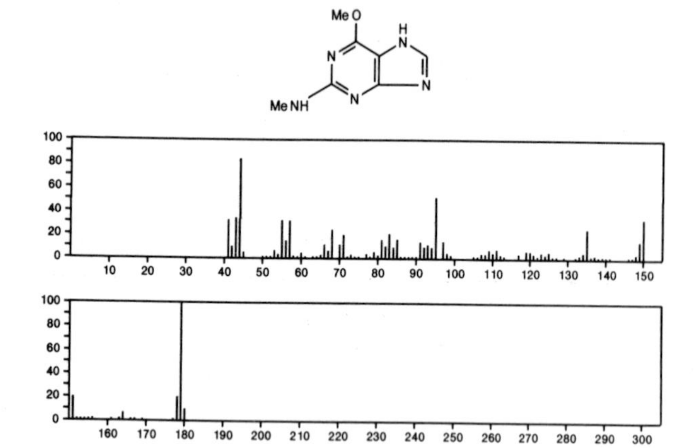

179 $C_8H_5NO_2S$ 40991-34-2
1,2-Benzisothiazole-3-carboxylic acid

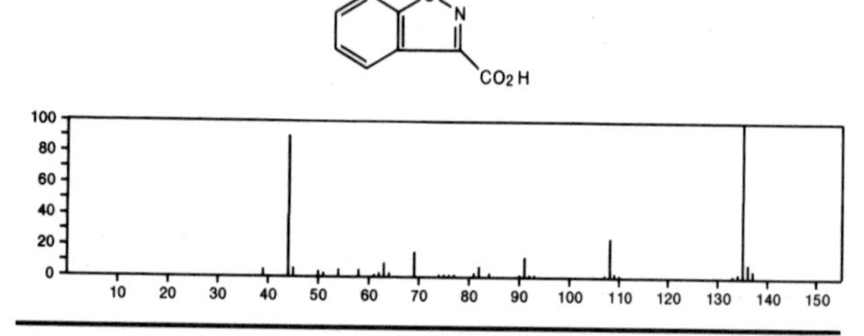

179 $C_8H_9N_3S$ 1627-73-2
Hydrazinecarbothioamide, 2-(phenylmethylene)-

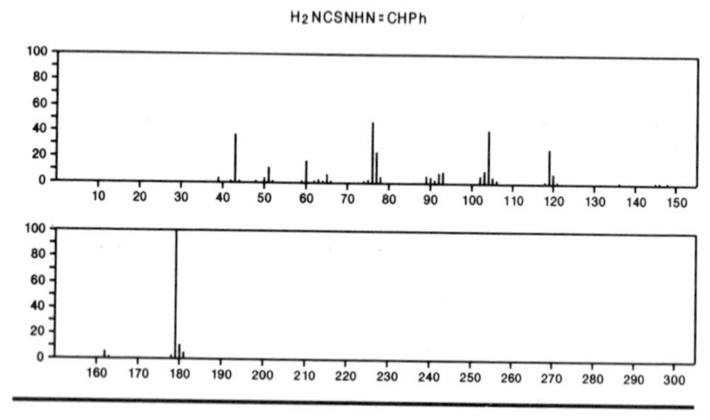

179 C_9HN_5 17638-20-9
2,3,5,6-Pyridinetetracarbonitrile

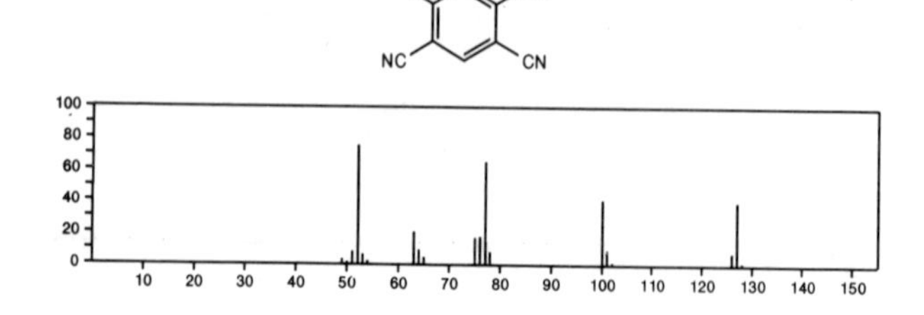

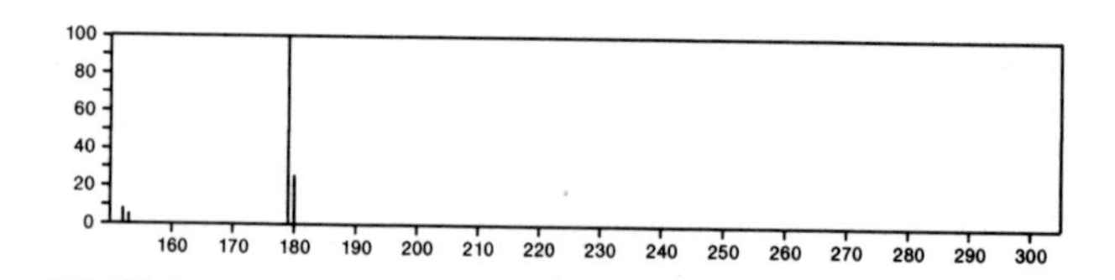

179 C_9H_6ClNO 3356-89-6
Isoxazole, 5-chloro-3-phenyl-

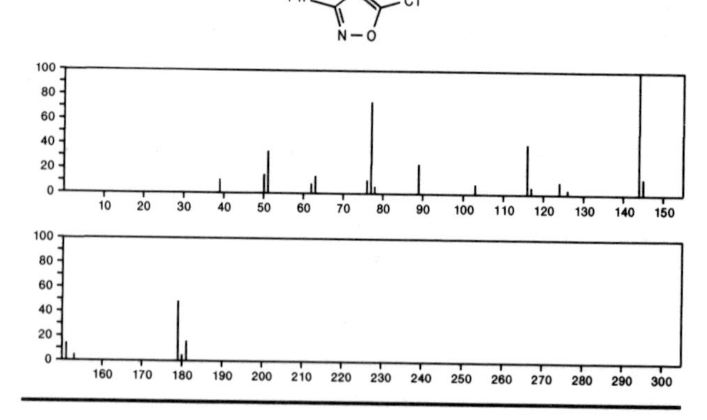

179 C_9H_9NOS 30276-97-2
Thiazolo[3,2-*a*]pyridinium, 8-hydroxy-2,5-dimethyl-, hydroxide, inner salt

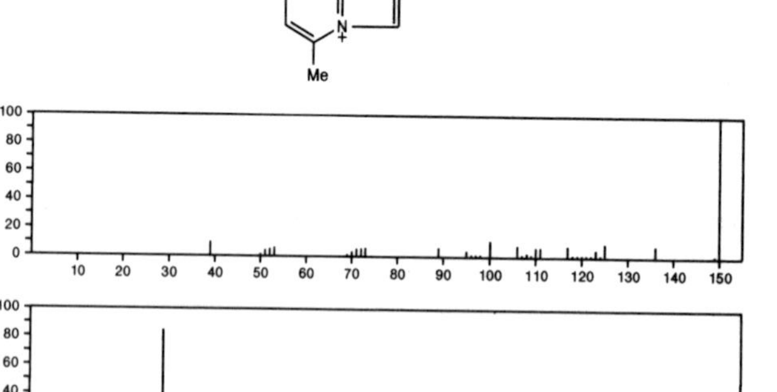

179 C_9H_9NOS 30277-00-0
Thiazolo[3,2-*a*]pyridinium, 8-hydroxy-3,5-dimethyl-, hydroxide, inner salt

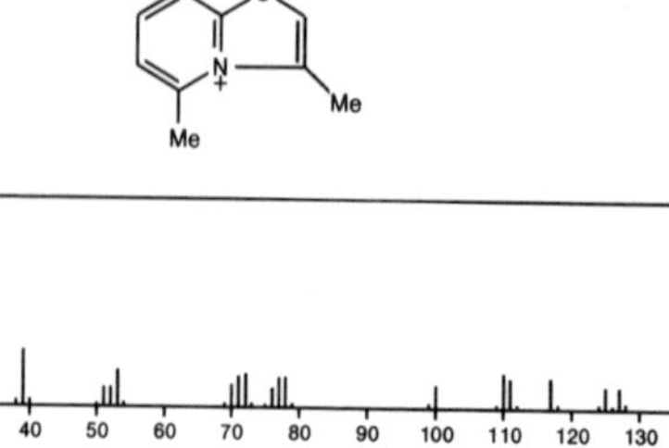

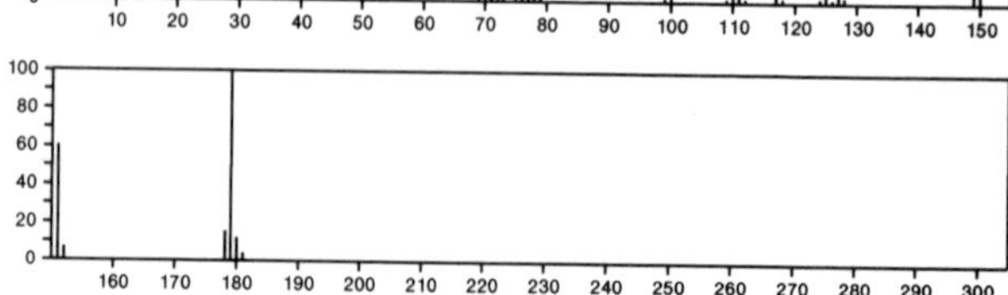

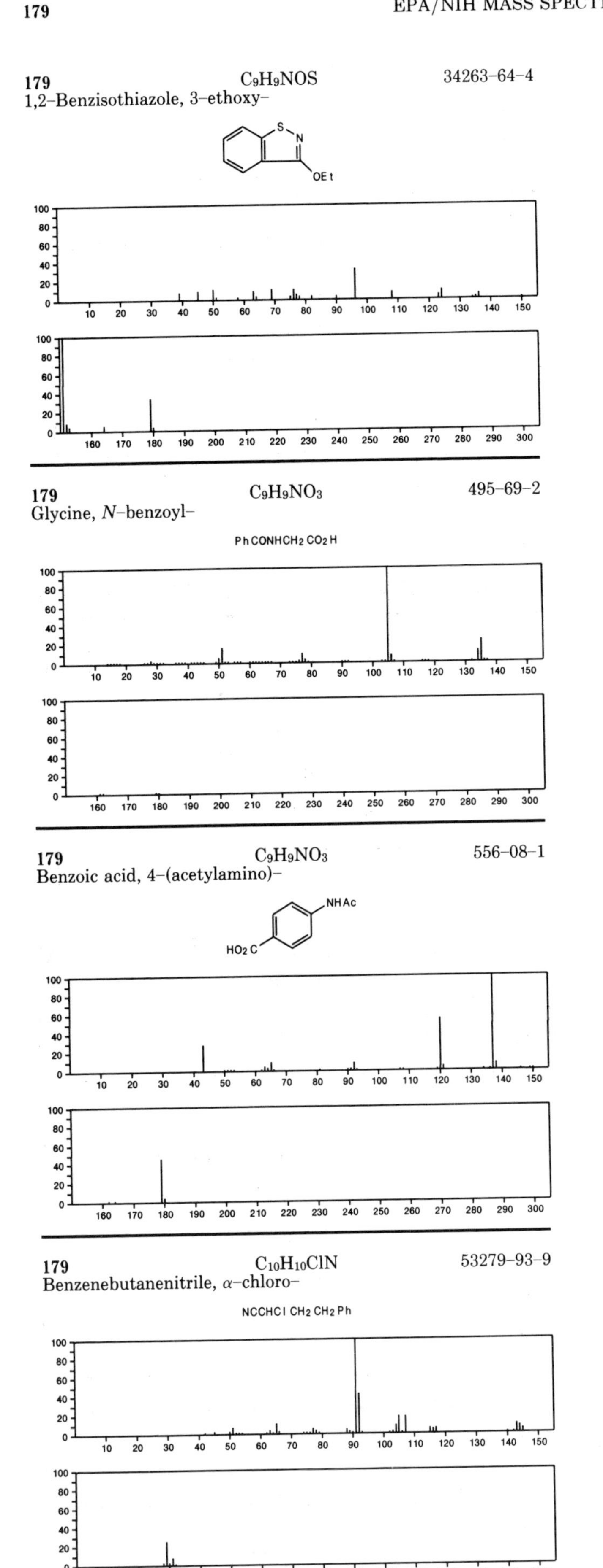
179 C9H9NOS 34263-64-4
1,2-Benzisothiazole, 3-ethoxy-
S
N
OEt
179 C9H9NO3 495-69-2
Glycine, N-benzoyl-
PhCONHCH2CO2H
179 C9H9NO3 556-08-1
Benzoic acid, 4-(acetylamino)-
NHAc
HO2C
179 C10H10ClN 53279-93-9
Benzenebutanenitrile, α-chloro-
NCCHCl CH2 CH2 Ph

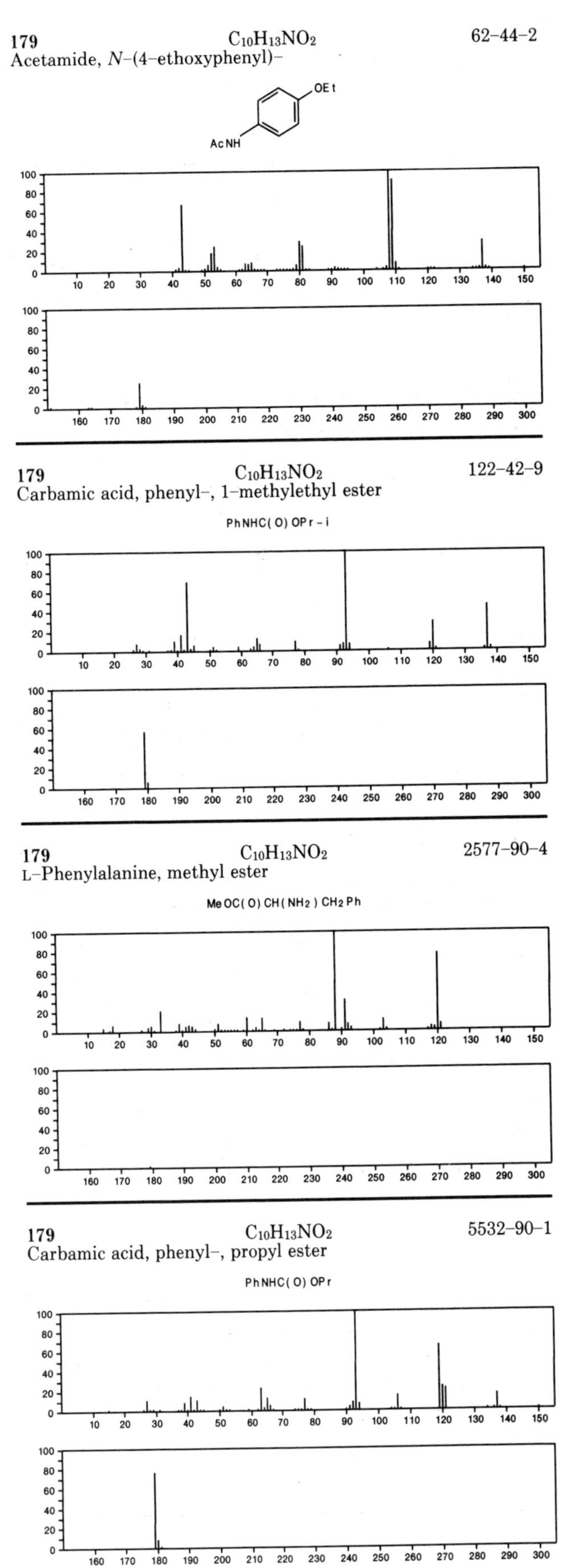
179 C10H13NO2 62-44-2
Acetamide, N-(4-ethoxyphenyl)-
OEt
AcNH
179 C10H13NO2 122-42-9
Carbamic acid, phenyl-, 1-methylethyl ester
PhNHC(O)OPr-i
179 C10H13NO2 2577-90-4
L-Phenylalanine, methyl ester
MeOC(O)CH(NH2)CH2Ph
179 C10H13NO2 5532-90-1
Carbamic acid, phenyl-, propyl ester
PhNHC(O)OPr

179 $C_{10}H_{13}NO_2$ 10072–05–6
Benzoic acid, 2–(dimethylamino)–, methyl ester

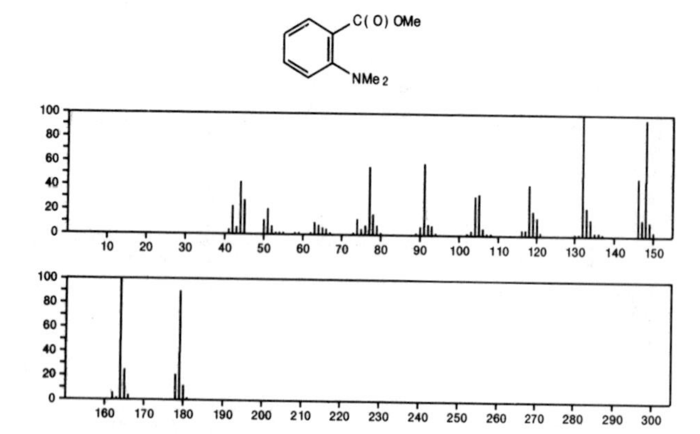

179 $C_{10}H_{13}NO_2$ 10315–42–1
1*H*–Azepine–2,5–dione, 4–isopropyl–7–methyl–

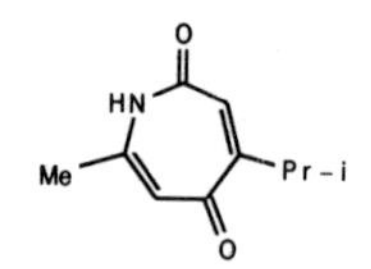

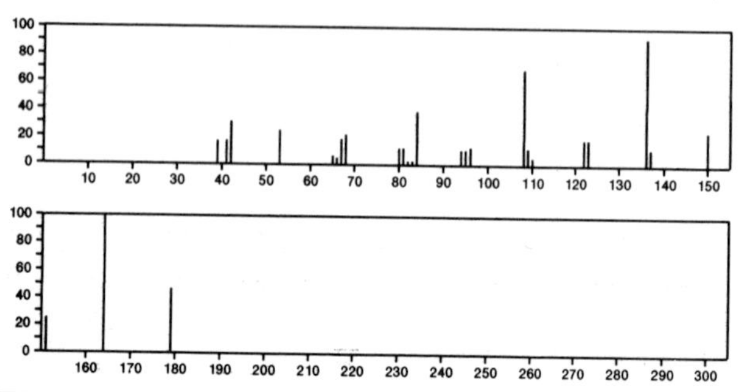

179 $C_{10}H_{13}NO_2$ 18189–02–1
Anthranilic acid, isopropyl ester

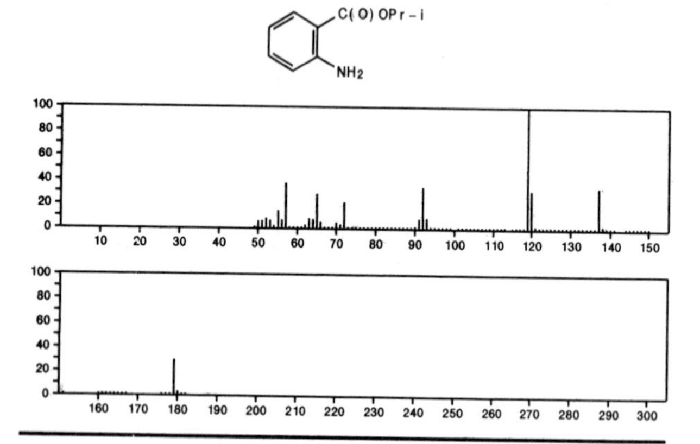

179 $C_{10}H_{13}NO_2$ 20642–93–7
Carbamic acid, (2,6–dimethylphenyl)–, methyl ester

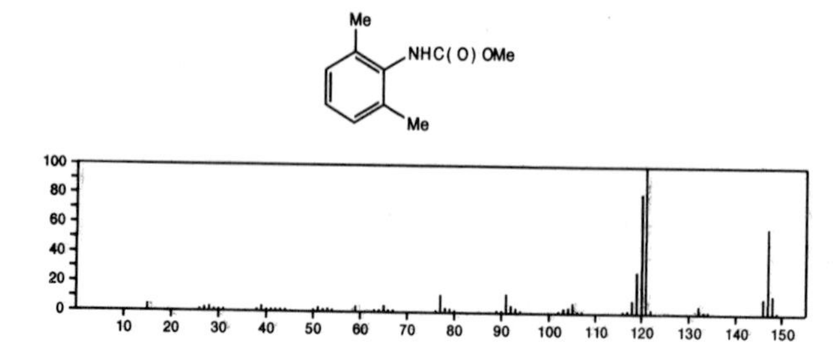

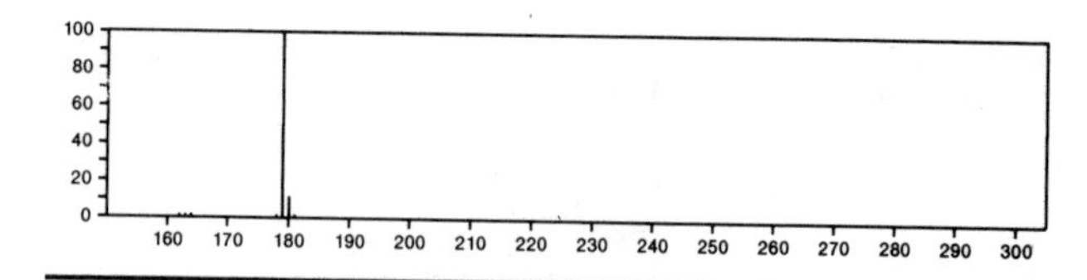

179 $C_{10}H_{13}NO_2$ 20646–44–0
1*H*–Azepine–1–carboxylic acid, 4,5–dimethyl–, methyl ester

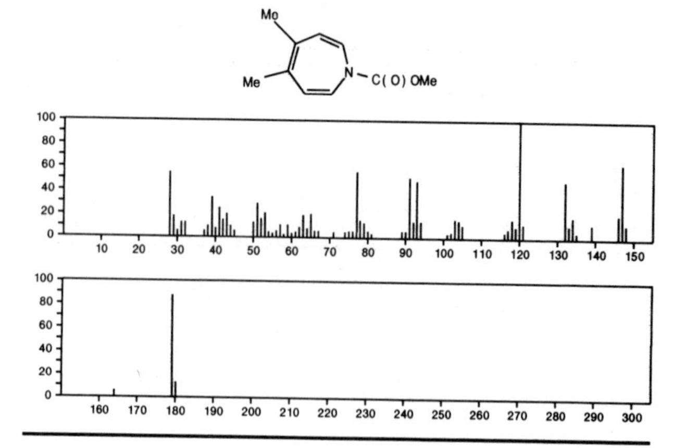

179 $C_{10}H_{13}NO_2$ 21864–63–1
Benzenaminium, 2–carboxy–*N*,*N*,*N*–trimethyl–, hydroxide, inner salt

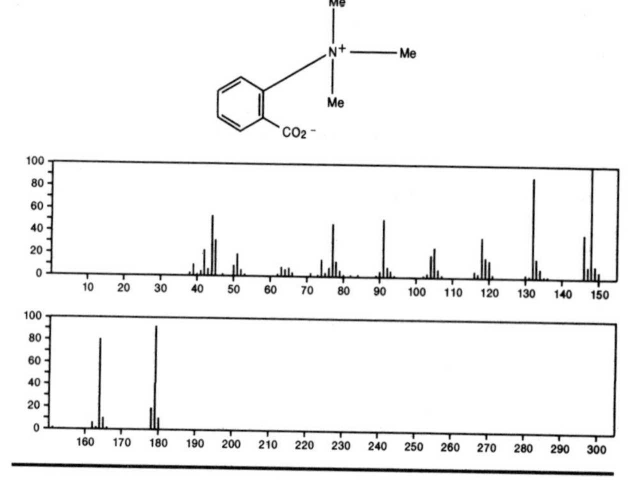

179 $C_{10}H_{13}NO_2$ 27740–96–1
6,7–Isoquinolinediol, 1,2,3,4–tetrahydro–1–methyl–, (*S*)–

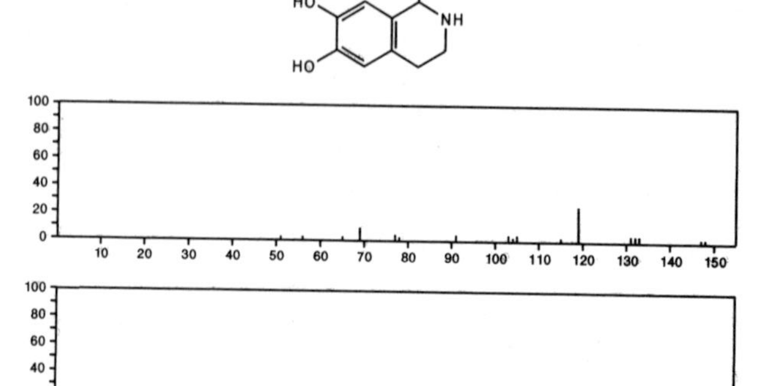

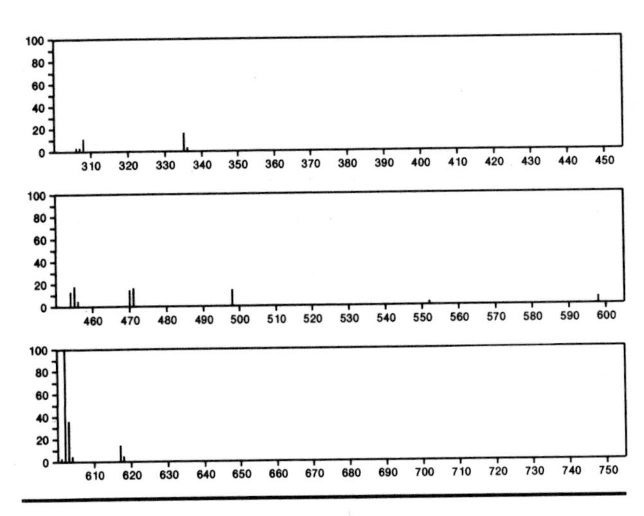

179 $C_{10}H_{13}NO_2$ 28915-98-2

4-Cyclohexene-1,2-dicarboximide, *N*-ethyl-, *cis*-

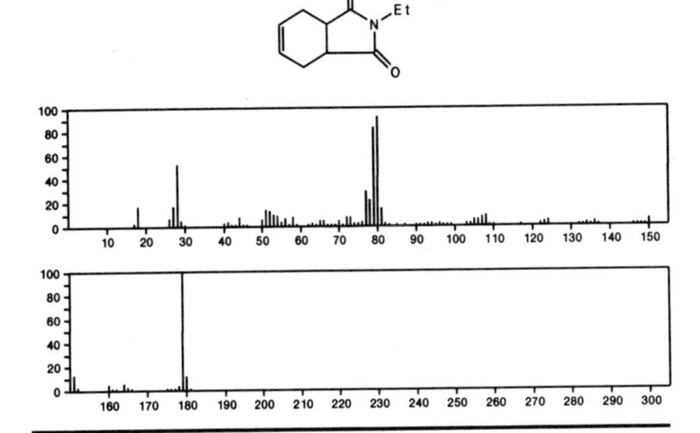

179 $C_{10}H_{13}NO_2$ 33046-28-5

Benzenaminium, 4-carboxy-*N*,*N*,*N*-trimethyl-, hydroxide, inner salt

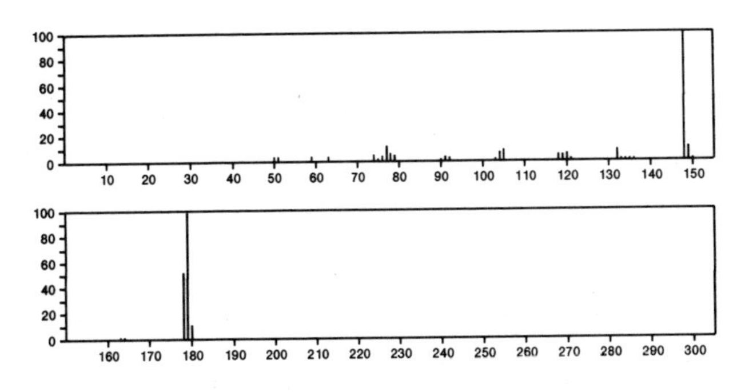

179 $C_{10}H_{13}NO_2$ 33192-03-9

Benzenaminium, 3-carboxy-*N*,*N*,*N*-trimethyl-, hydroxide, inner salt

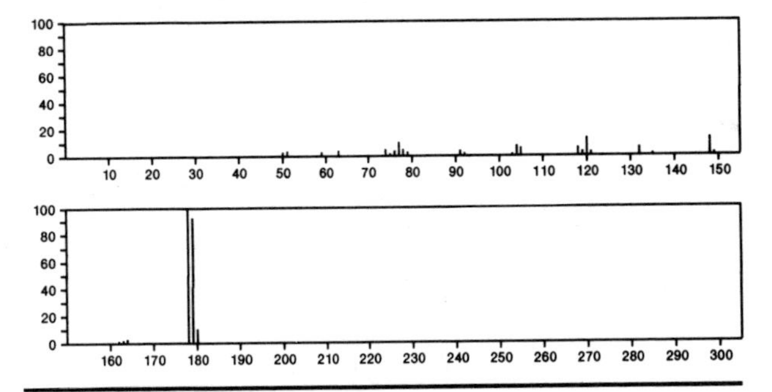

179 $C_{10}H_{13}NO_2$ 35103-34-5

Acetamide, *N*-[(4-methoxyphenyl)methyl]-

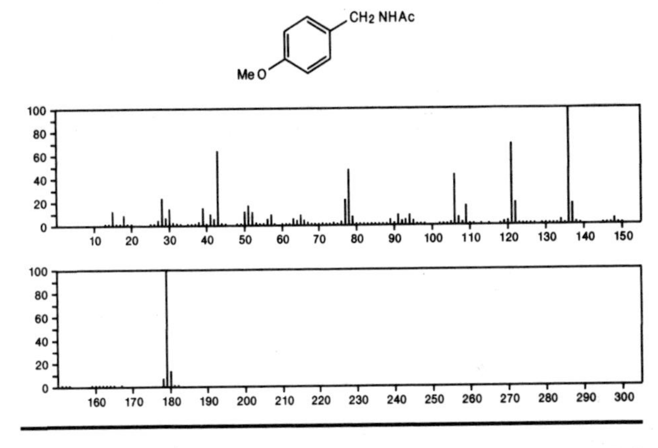

179 $C_{10}H_{13}NO_2$ 43021-97-2

2-Propanone, 1-(2-methoxyphenyl)-, oxime

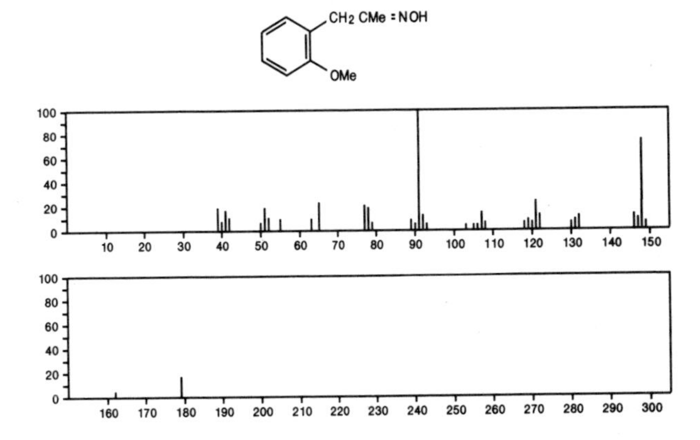

179 $C_{10}H_{13}NO_2$ 52271-41-7
2-Propanone, 1-(4-methoxyphenyl)-, oxime

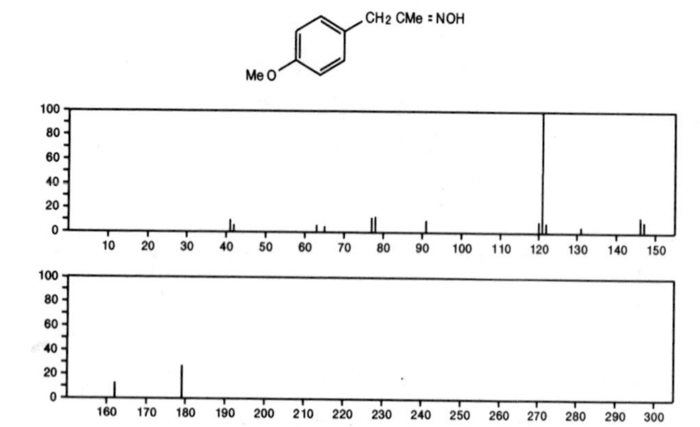

179 $C_{10}H_{13}NO_2$ 56588-19-3
1*H*-Indol-2-ol, 2,3-dihydro-5-methoxy-1-methyl-

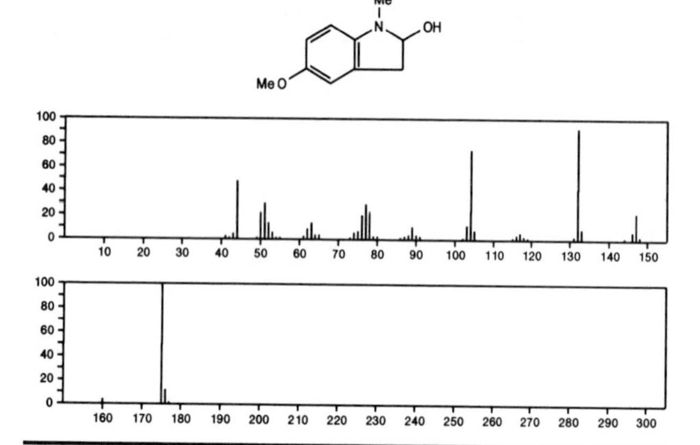

179 $C_{10}H_{13}NO_2$ 56701-07-6
2-Azabicyclo[3.2.0]hepta-3,6-diene-2-carboxylic acid, 1,3-dimethyl-, methyl ester

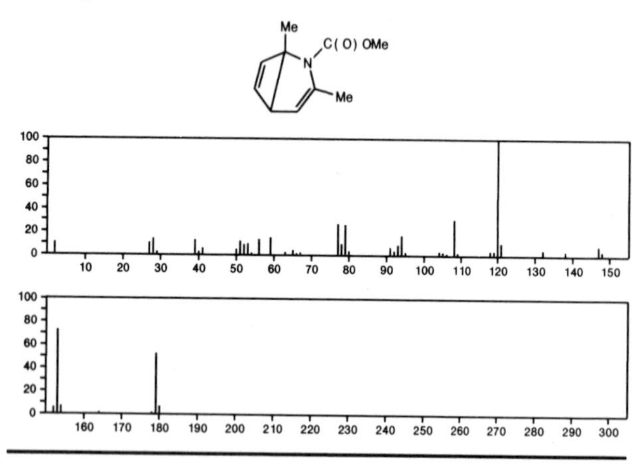

179 $C_{10}H_{13}NO_2$ 56728-08-6
Tricyclo[3.3.1.1^{3,7}]decane-2,6-dione, 4-amino-

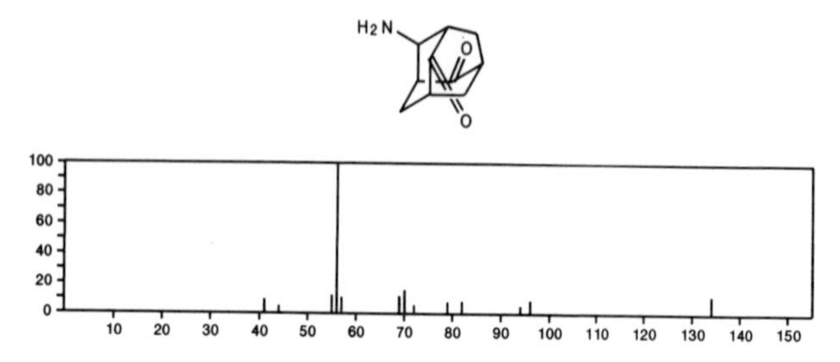

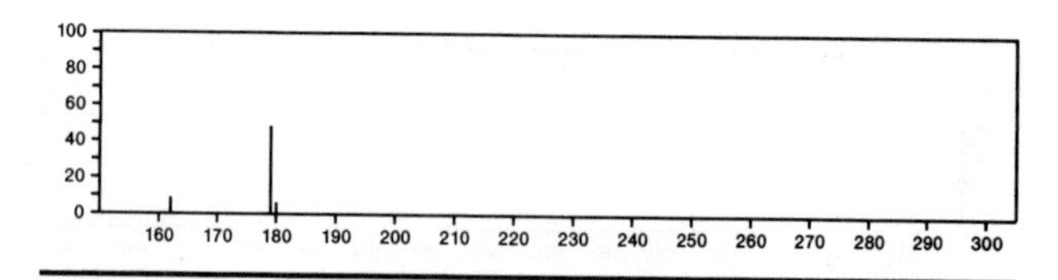

179 $C_{10}H_{17}NSi$ 14856-79-2
Silanamine, 1,1,1-trimethyl-*N*-(phenylmethyl)-

$Me_3SiNHCH_2Ph$

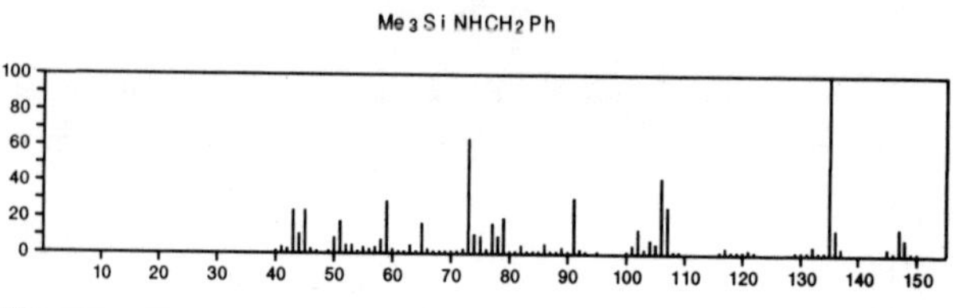

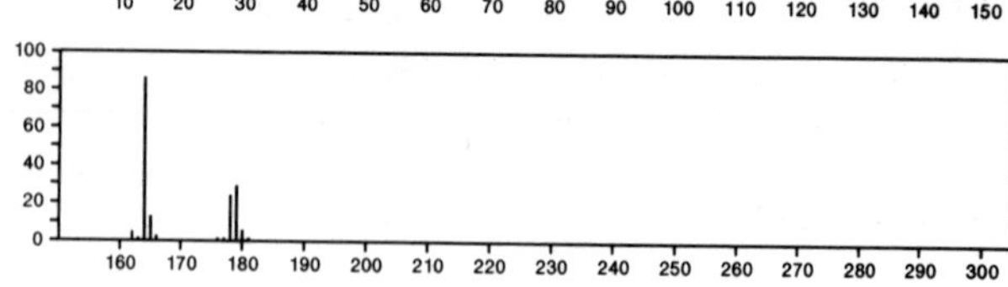

179 $C_{10}H_{17}NSi$ 17890-16-3
Pyridine, 2-[2-(trimethylsilyl)ethyl]-

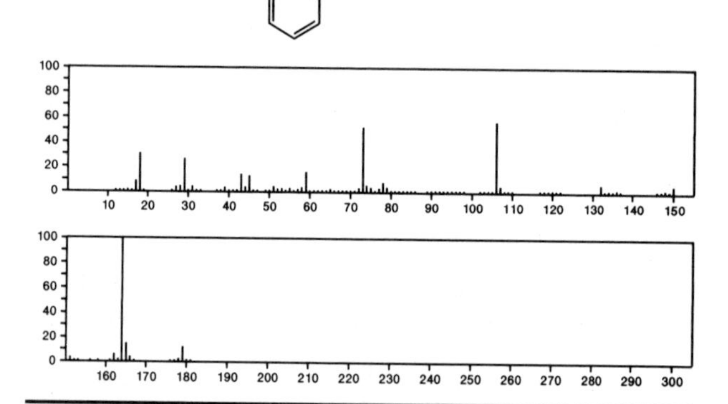

179 $C_{11}H_{17}NO$ 6721-66-0
Benzeneethanamine, β-methoxy-*N*,*N*-dimethyl-

PhCH(OMe)CH_2NMe_2

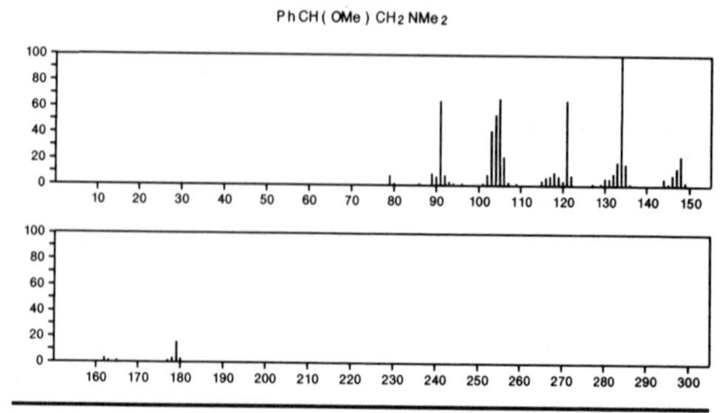

179 $C_{11}H_{17}NO$ 14573-22-9
Ethanamine, 2-(2,6-dimethylphenoxy)-*N*-methyl-

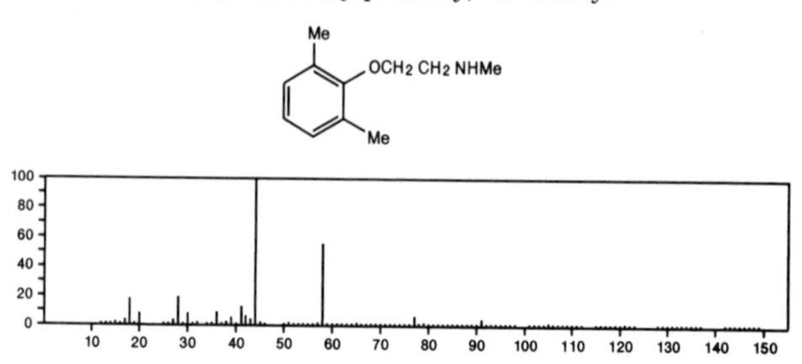

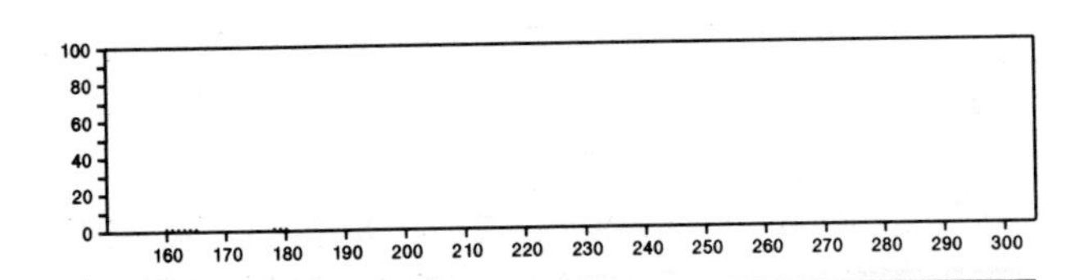

179 $C_{11}H_{17}NO$ 19059-89-3

Phenol, 2-amino-6-(1,1-dimethylethyl)-4-methyl-

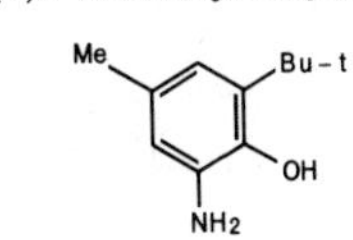

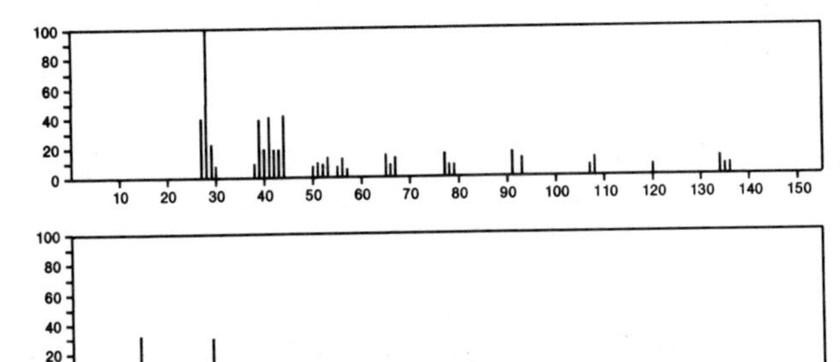

179 $C_{11}H_{17}NO$ 21922-69-0

2-Cyclobuten-1-one, 4,4-dimethyl-3-piperidino-

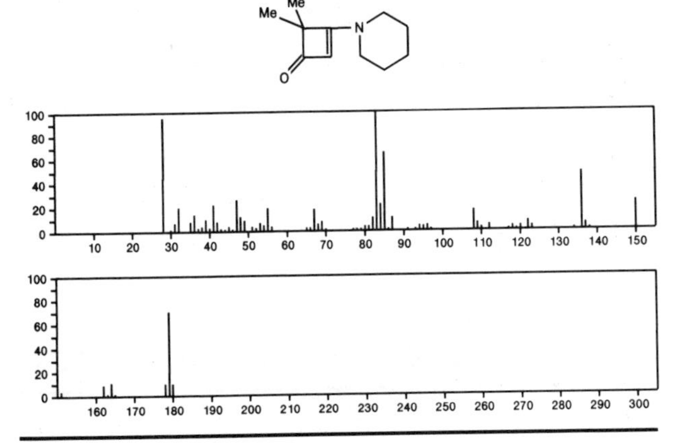

179 $C_{11}H_{17}NO$ 27058-12-4

Ethanamine, *N,N*-dimethyl-2-(phenylmethoxy)-

$Me_2NCH_2CH_2OCH_2Ph$

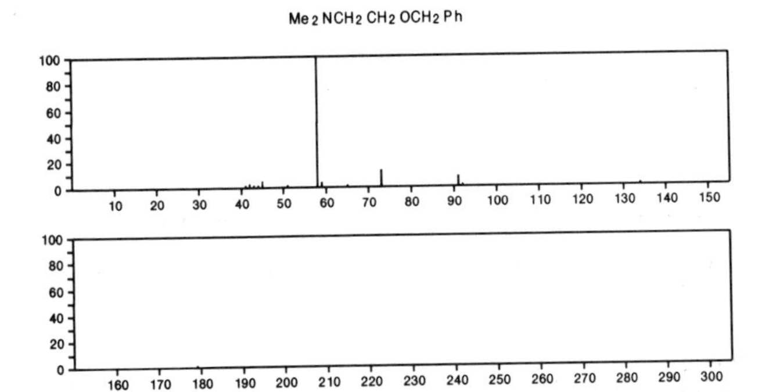

179 $C_{11}H_{17}NO$ 55955-99-2

Phenol, 2-[(dimethylamino)methyl]-4-ethyl-

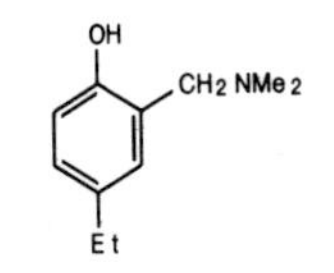

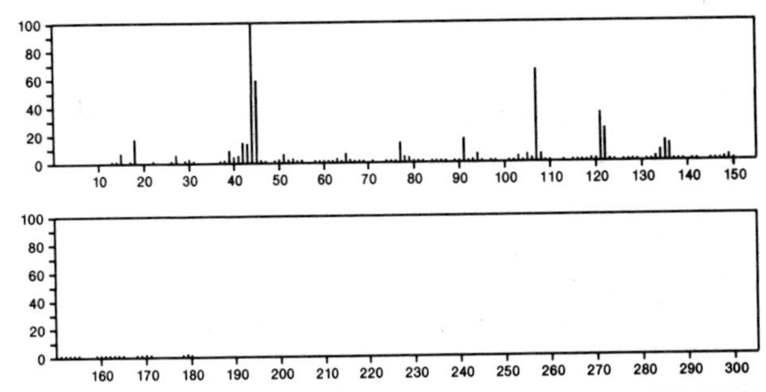

179 $C_{11}H_{17}NO$ 57397-12-3

2-Cyclohexen-1-one, 2-methyl-5-(1-methylethenyl)-, *O*-methyloxime, (+)-

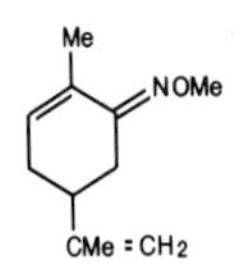

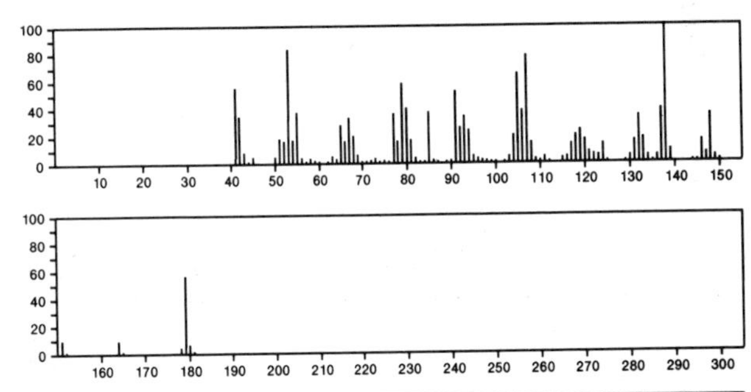

179 $C_{12}H_{21}N$ 23430-63-9

1*H*-Azepine, 1-(1-cyclohexen-1-yl)hexahydro-

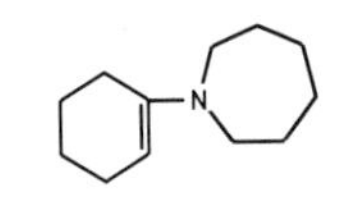

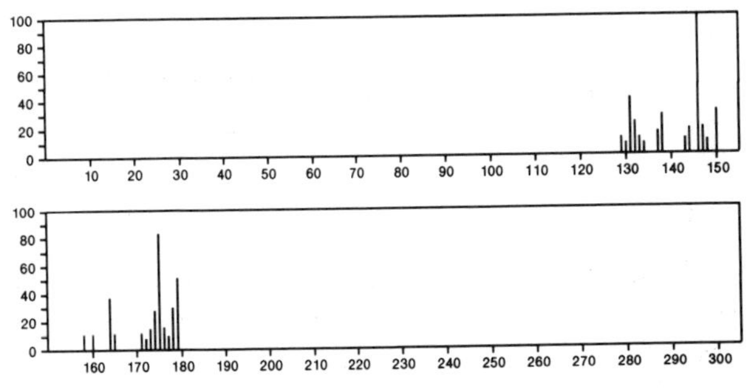

179 $C_{12}H_{21}N$ 26974–21–0
Pyrrolidine, 1–(6–ethyl–1–cyclohexen–1–yl)–

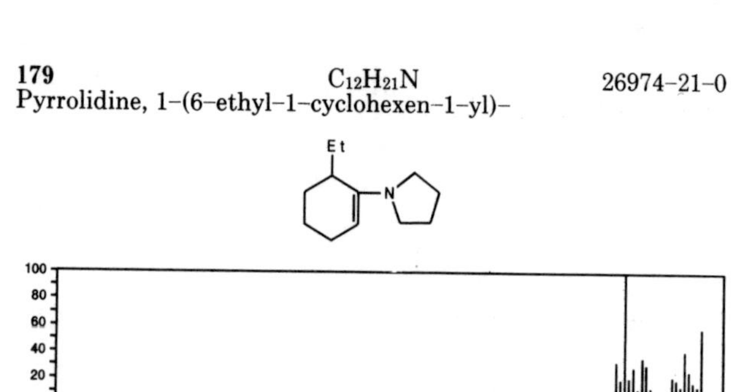
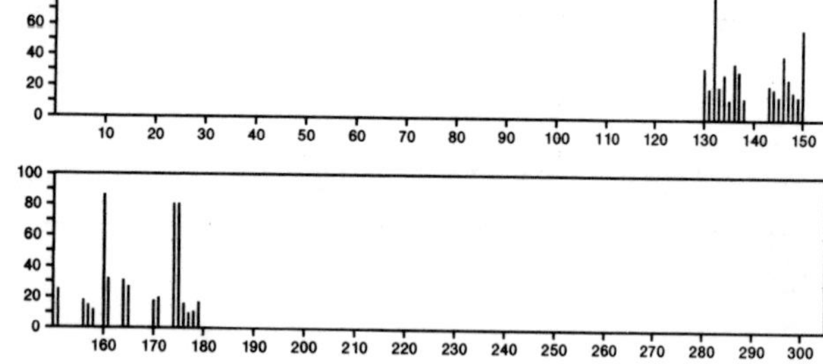

179 $C_{13}H_9N$ 229–87–8
Phenanthridine

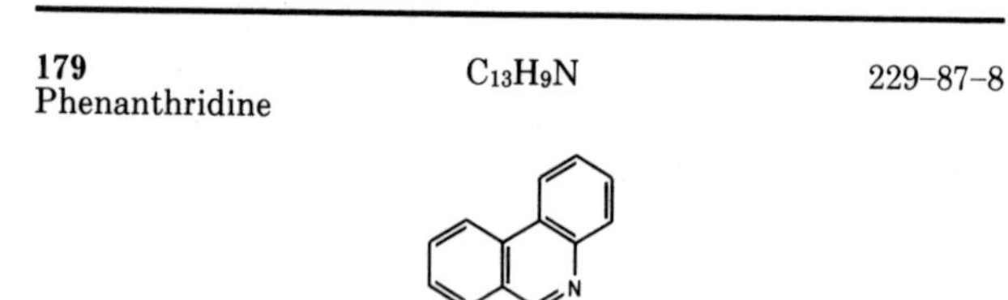
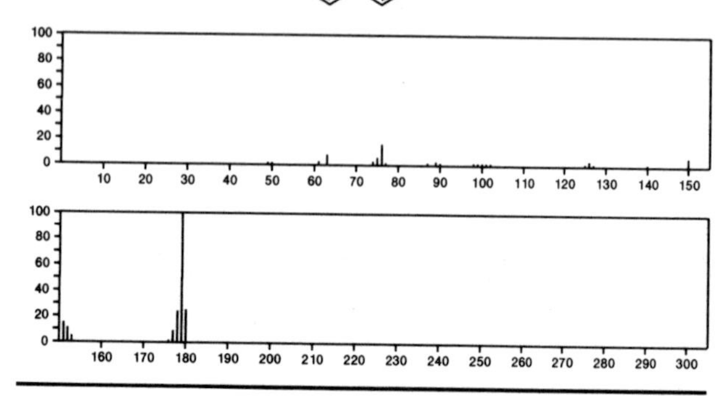

179 $C_{13}H_9N$ 230–27–3
Benzo[*h*]quinoline

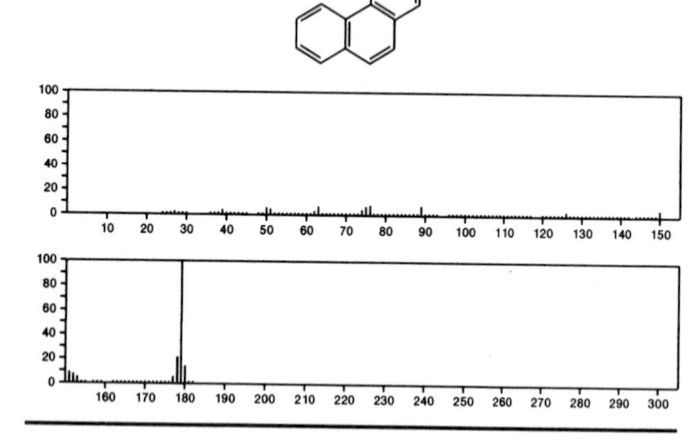

179 $C_{13}H_9N$ 260–94–6
Acridine

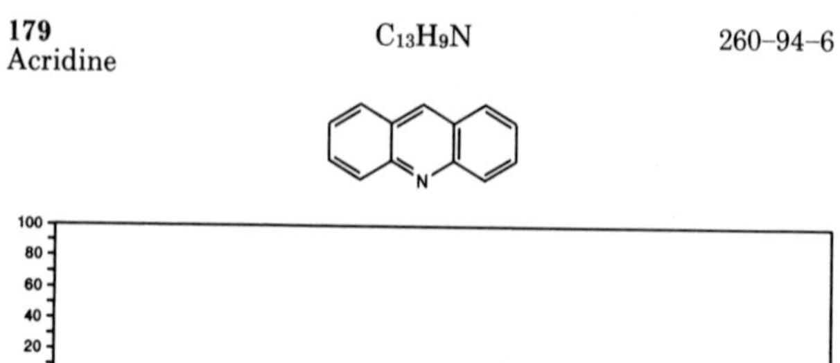
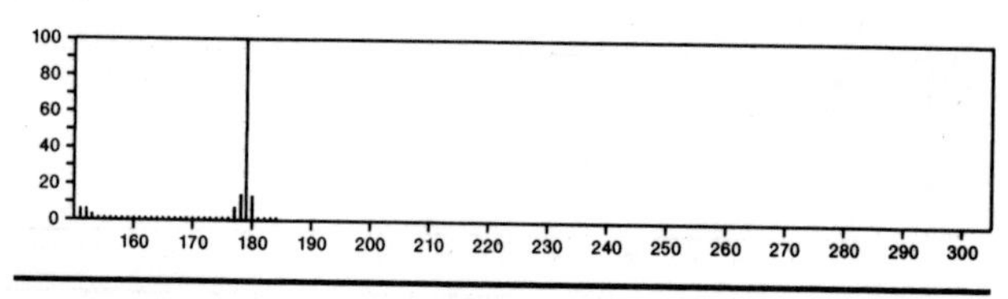

179 $C_{13}H_9N$ 4440–33–9
9*H*–Fluoren–9–imine

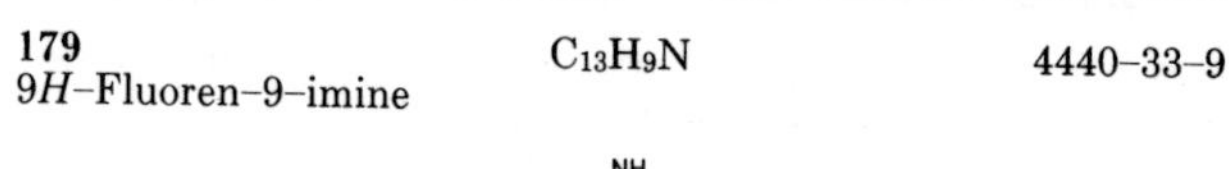
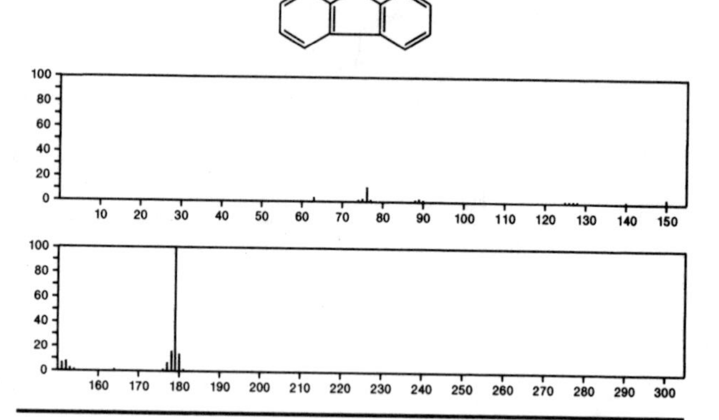

179 $C_{13}H_9N$ 13141–42–9
Pyridine, 2–(phenylethynyl)–

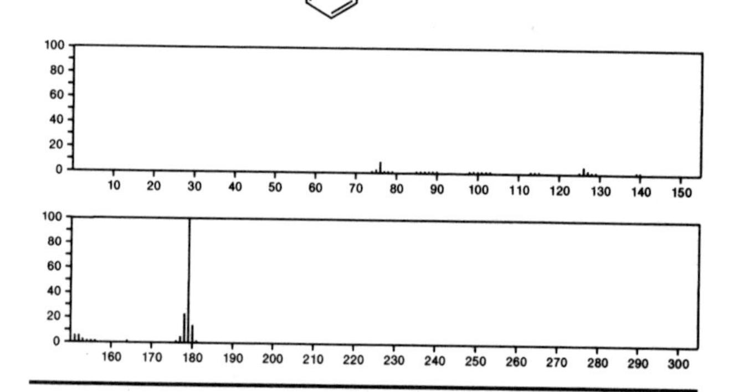

179 $C_{13}H_9N$ 24973–49–7
[1,1'–Biphenyl]–2–carbonitrile

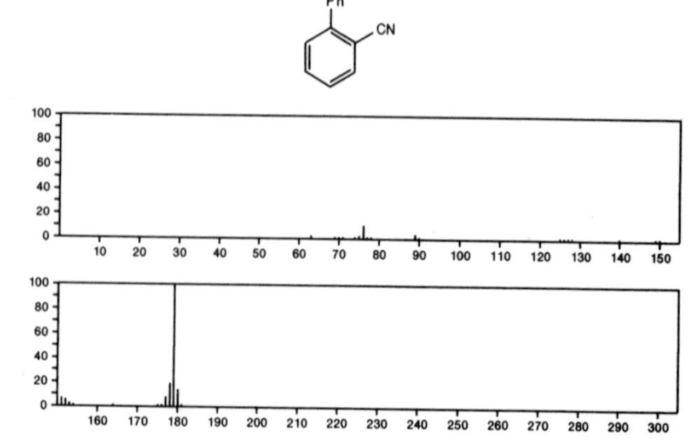

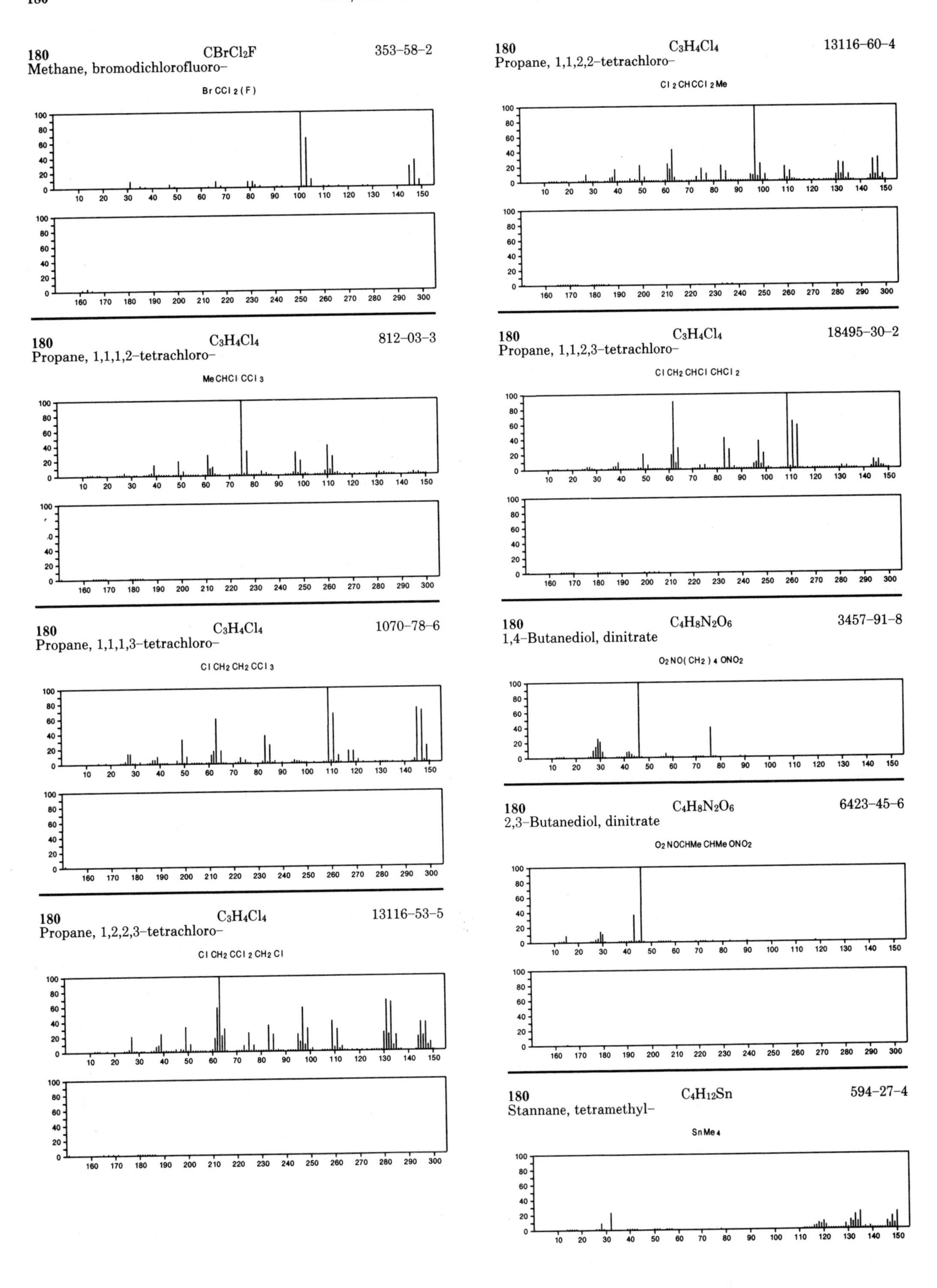
180 CBrCl2F 353-58-2
Methane, bromodichlorofluoro-
Br CCl 2 (F)
180 C3H4Cl4 13116-60-4
Propane, 1,1,2,2-tetrachloro-
Cl 2 CHCCl 2 Me
180 C3H4Cl4 812-03-3
Propane, 1,1,1,2-tetrachloro-
Me CHCl CCl 3
180 C3H4Cl4 18495-30-2
Propane, 1,1,2,3-tetrachloro-
Cl CH2 CHCl CHCl 2
180 C3H4Cl4 1070-78-6
Propane, 1,1,1,3-tetrachloro-
Cl CH2 CH2 CCl 3
180 C4H8N2O6 3457-91-8
1,4-Butanediol, dinitrate
O2 NO(CH2) 4 ONO2
180 C4H8N2O6 6423-45-6
2,3-Butanediol, dinitrate
O2 NOCHMe CHMe ONO2
180 C3H4Cl4 13116-53-5
Propane, 1,2,2,3-tetrachloro-
Cl CH2 CCl 2 CH2 Cl
180 C4H12Sn 594-27-4
Stannane, tetramethyl-
SnMe 4

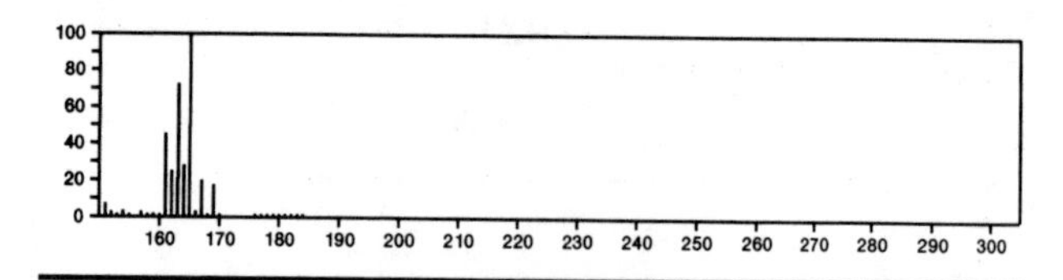

180 $C_5H_3F_3N_2O_2$ 672-45-7
2,4(1*H*,3*H*)-Pyrimidinedione, 6-(trifluoromethyl)-

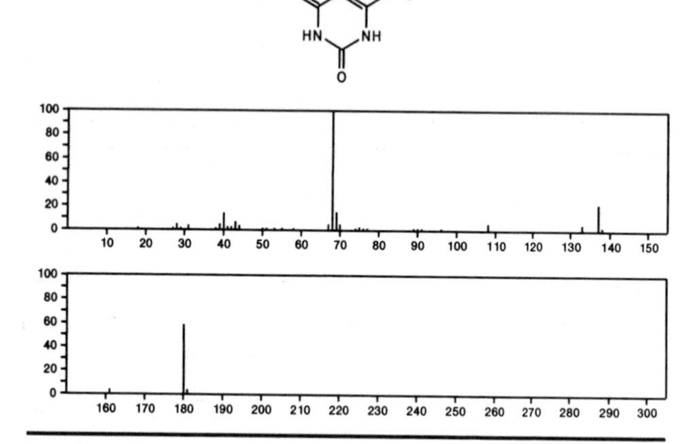

180 $C_5H_9BrO_2$ 535-11-5
Propanoic acid, 2-bromo-, ethyl ester

EtOC(O)CHBrMe

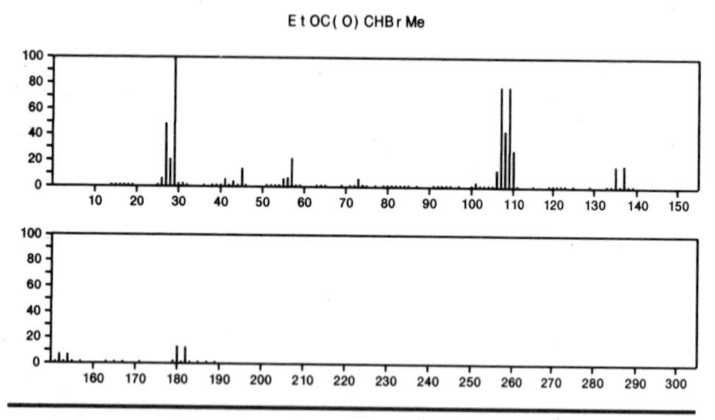

180 $C_5H_9BrO_2$ 539-74-2
Propanoic acid, 3-bromo-, ethyl ester

BrCH2CH2C(O)OEt

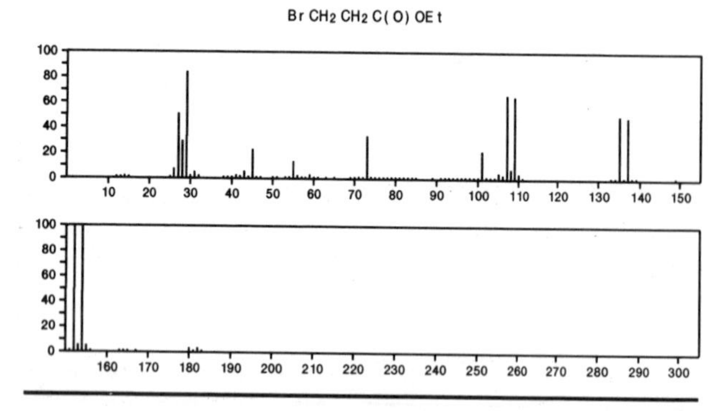

180 $C_5H_9BrO_2$ 4897-84-1
Butanoic acid, 4-bromo-, methyl ester

Br(CH2)3C(O)OMe

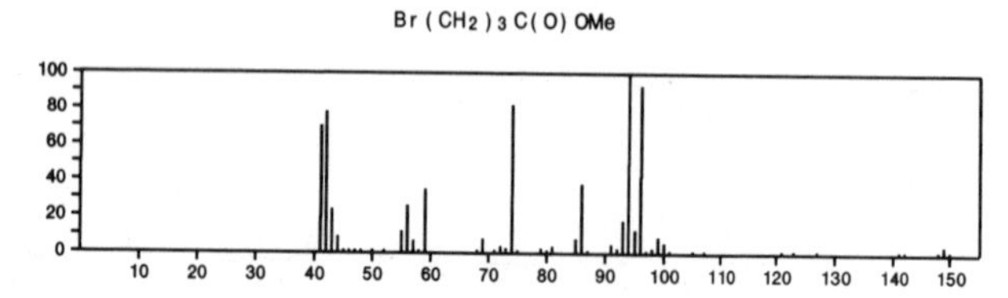

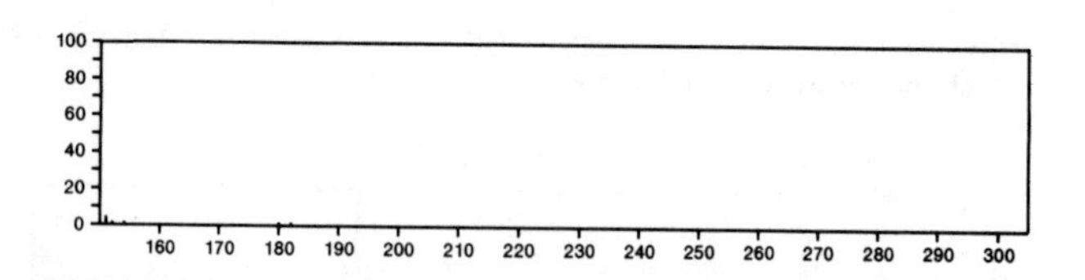

180 $C_5H_9BrO_2$ 10299-39-5
2-Propanol, 1-bromo-, acetate

AcOCHMeCH2Br

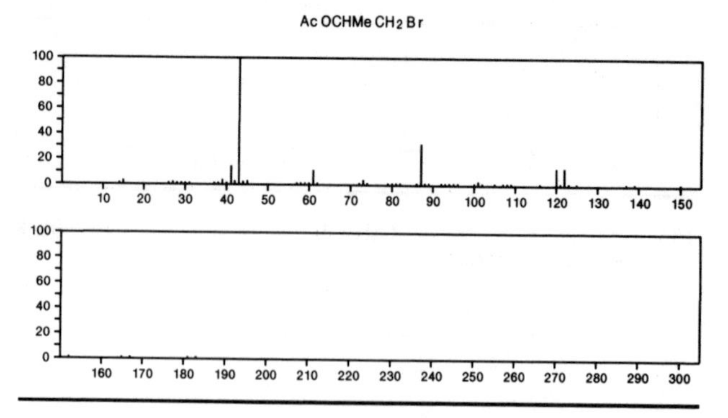

180 $C_5H_9BrO_2$ 35878-05-8
1,3-Dioxane, 5-bromo-2-methyl-, *trans*-

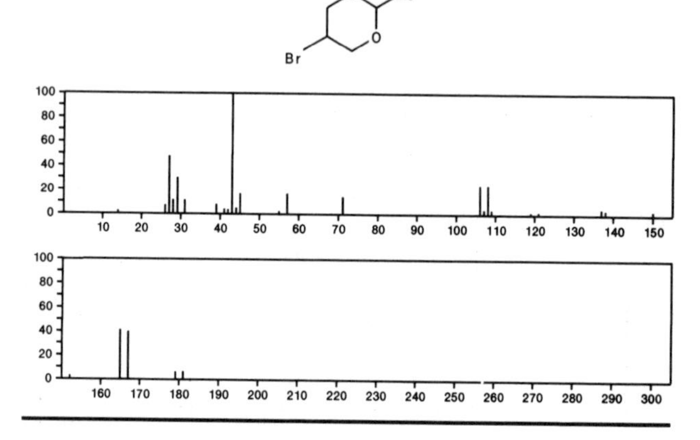

180 $C_5H_{12}N_2Se$ 5943-53-3
Selenourea, tetramethyl-

NMe2
Me2NC=Se

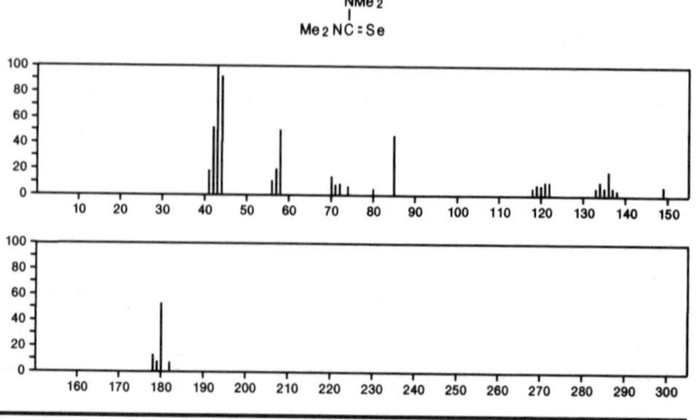

180 $C_5H_{13}AsO_2$ 40515-06-8
Arsonous acid, methyl-, diethyl ester

EtOAsMeOEt

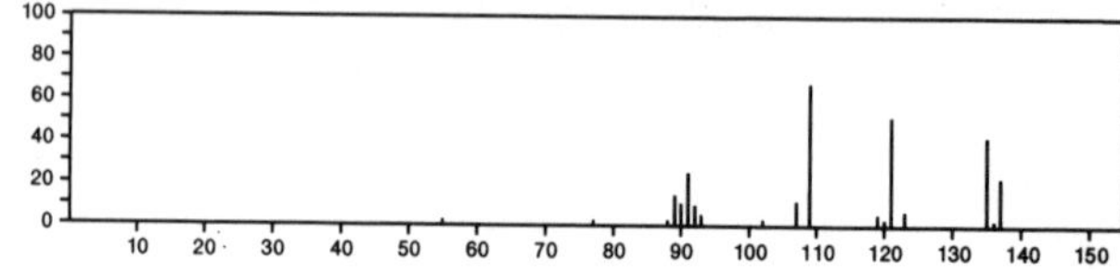

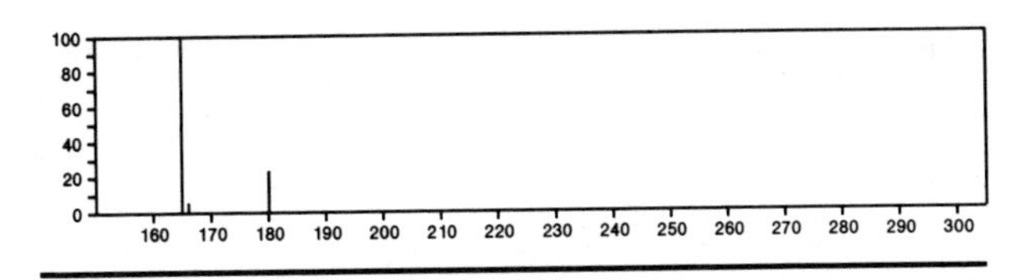

180 $C_6H_3Cl_3$ 87-61-6

Benzene, 1,2,3-trichloro-

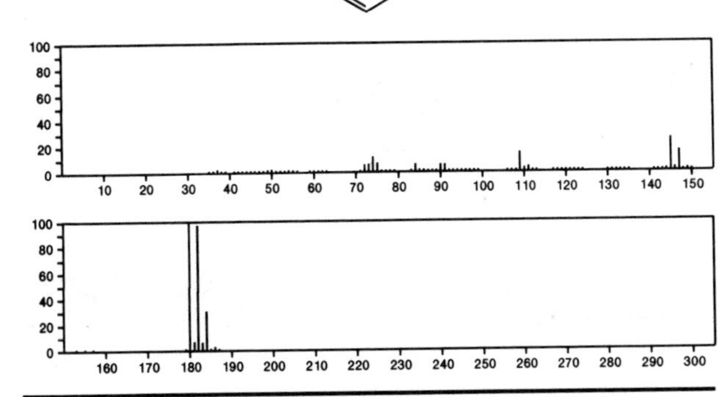

180 $C_6H_3Cl_3$ 108-70-3

Benzene, 1,3,5-trichloro-

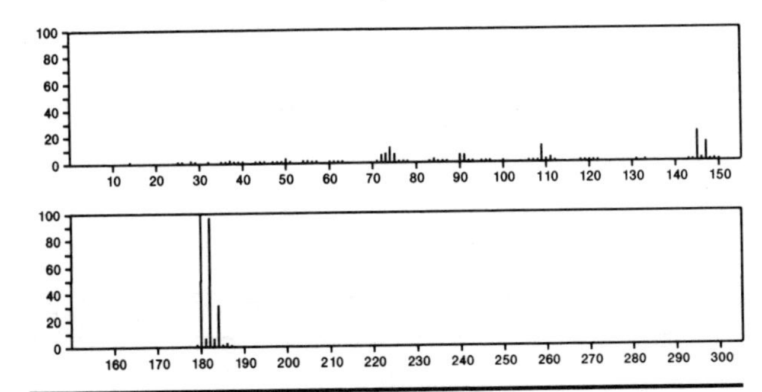

180 $C_6H_3Cl_3$ 120-82-1

Benzene, 1,2,4-trichloro-

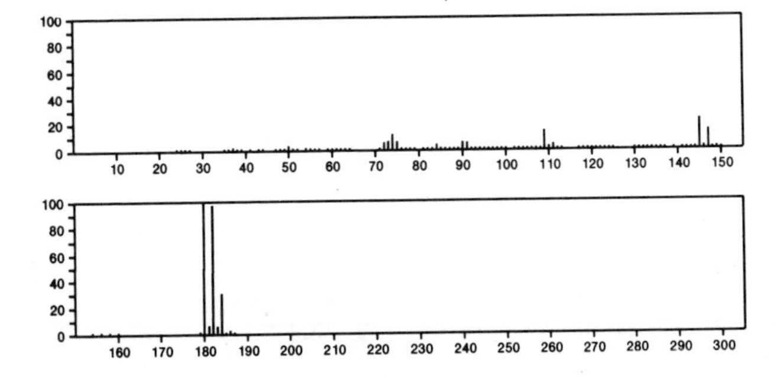

180 $C_6H_4F_4N_2$ 1198-63-6

1,3-Benzenediamine, 2,4,5,6-tetrafluoro-

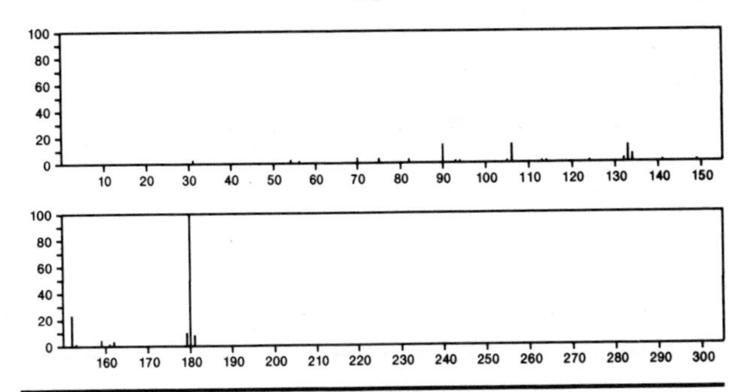

180 $C_6H_4F_4N_2$ 1198-64-7

1,4-Benzenediamine, 2,3,5,6-tetrafluoro-

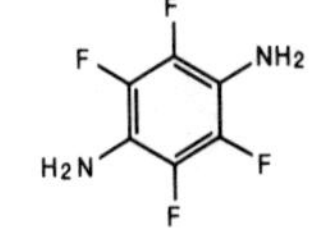

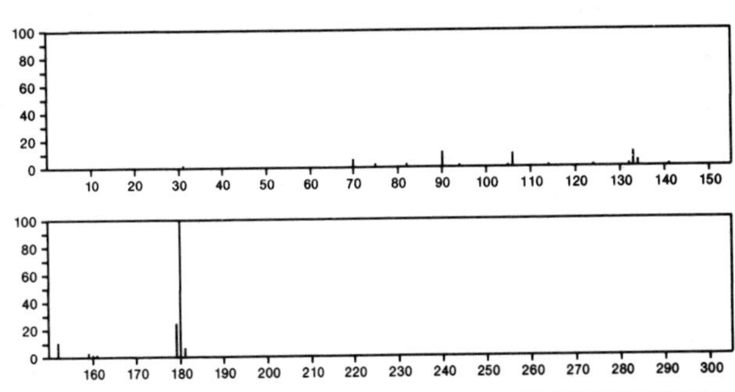

180 $C_6H_4N_4O_3$ 2577-38-0

2,4,7(1*H*,3*H*,8*H*)-Pteridinetrione

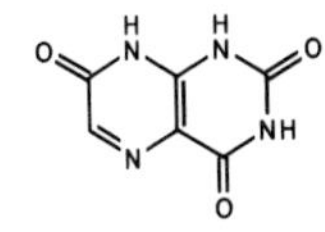

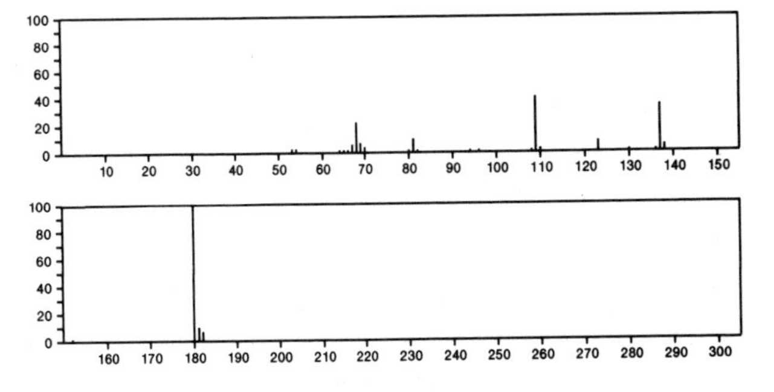

180 $C_6H_4N_4O_3$ 5019-55-6
1*H*-1,2,4-Triazole, 3-(5-nitro-2-furanyl)-

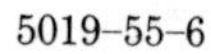

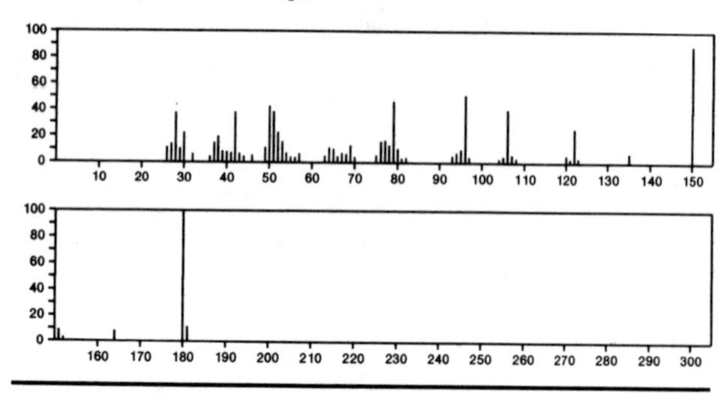

180 $C_6H_6Cl_2O_2$ 56272-99-2
1,2-Cyclopentanedione, dichloromethyl-

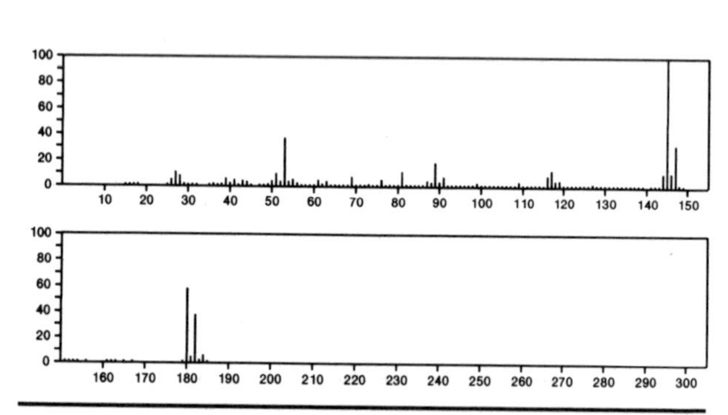

180 $C_6H_{10}BrF$ 17170-96-6
Cyclohexane, 1-bromo-2-fluoro-, *trans*-

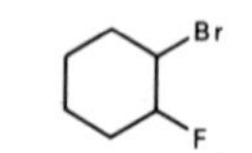

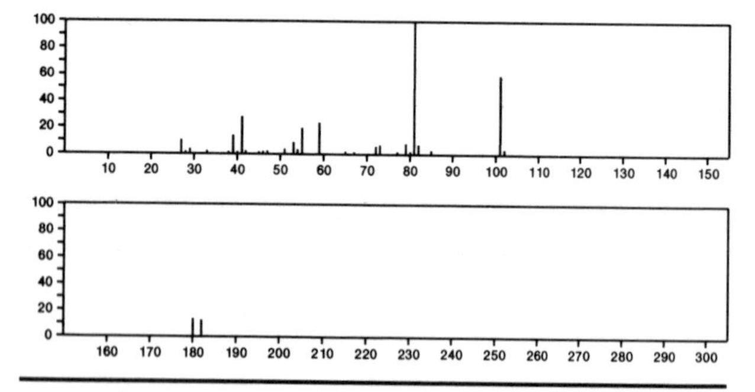

180 $C_6H_{10}BrF$ 51422-74-3
Cyclohexane, 1-bromo-2-fluoro-, *cis*-

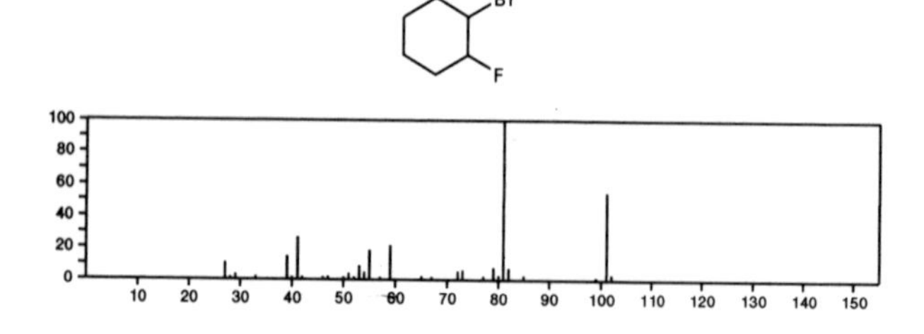

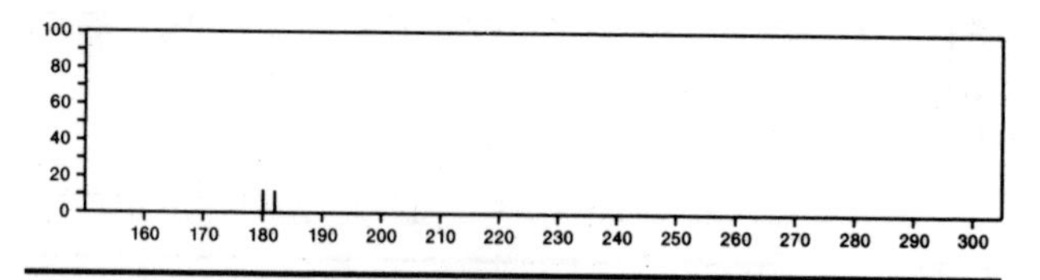

180 $C_6H_{12}O_6$ 50-99-7
D-Glucose

HOCH2 CH(OH) CH(OH) CH(OH) CH(OH) CHO

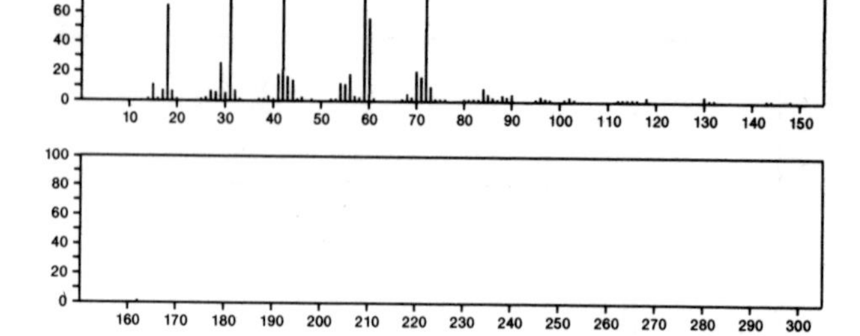

180 $C_6H_{12}O_6$ 87-89-8
myo-Inositol

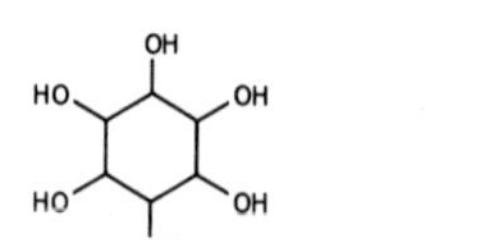

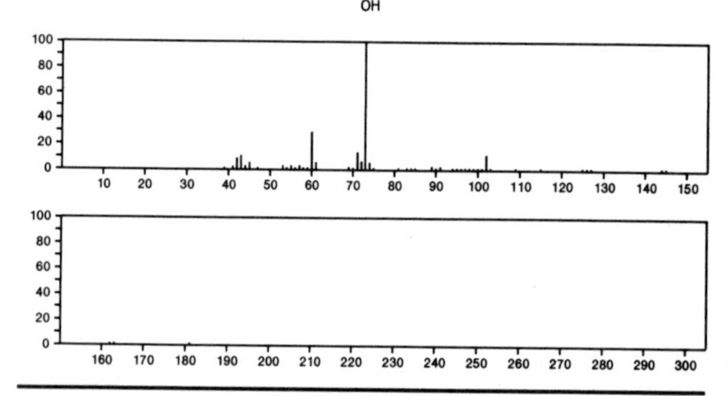

180 $C_6H_{12}O_6$ 488-54-0
neo-Inositol

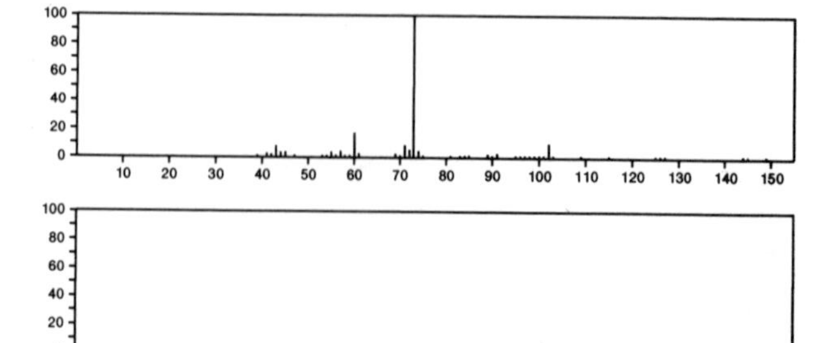

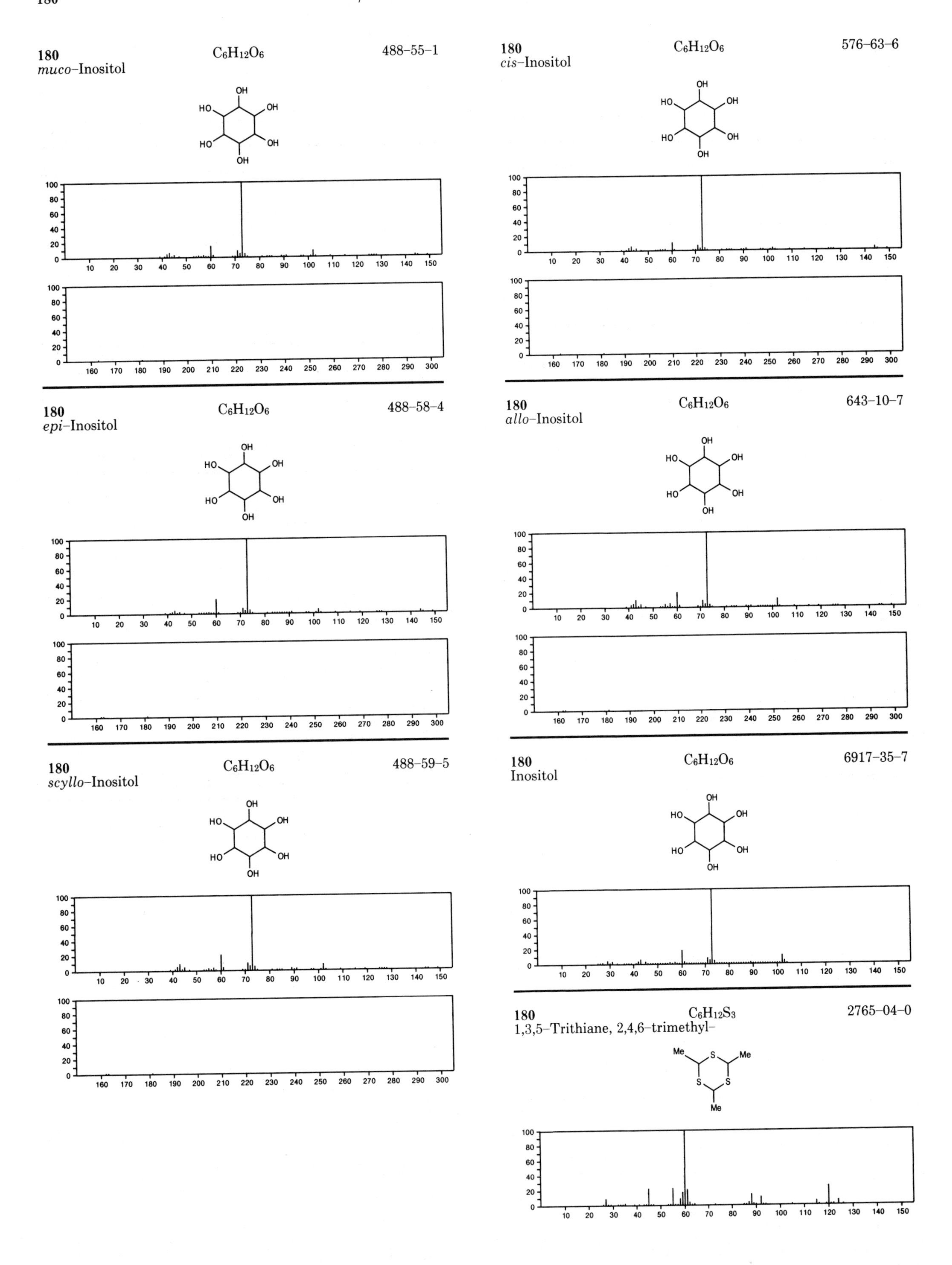
180
$C_6H_{12}O_6$
488-55-1
muco-Inositol
180
$C_6H_{12}O_6$
576-63-6
cis-Inositol
180
$C_6H_{12}O_6$
488-58-4
epi-Inositol
180
$C_6H_{12}O_6$
643-10-7
allo-Inositol
180
$C_6H_{12}O_6$
488-59-5
scyllo-Inositol
180
$C_6H_{12}O_6$
6917-35-7
Inositol
180
$C_6H_{12}S_3$
2765-04-0
1,3,5-Trithiane, 2,4,6-trimethyl-

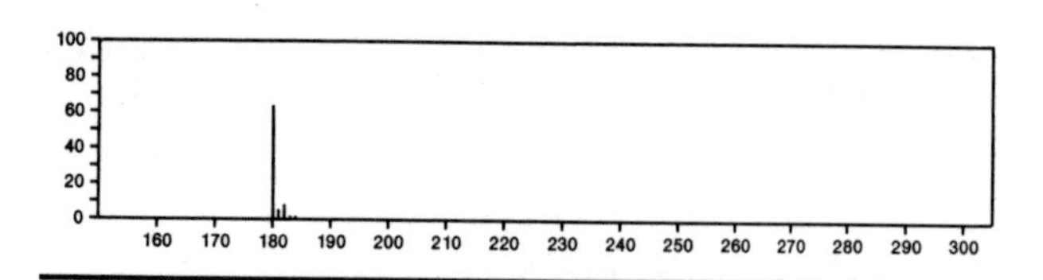

180 $C_6H_{13}BrO$ 14155-86-3
Ether, 5-bromopentyl methyl

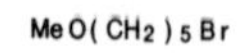

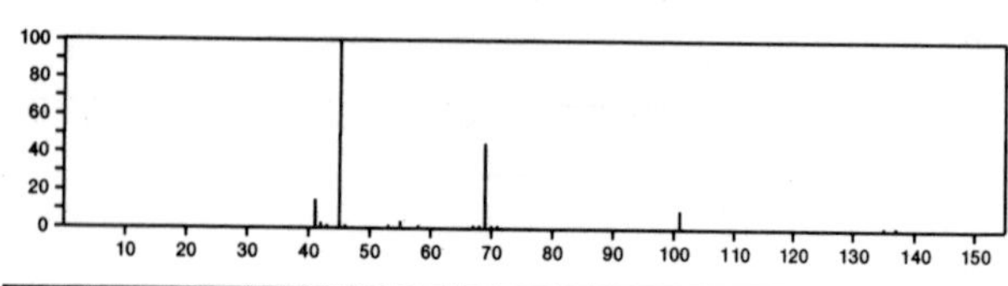

180 $C_6H_{13}O_4P$ 1005-96-5
1,3,2-Dioxaphosphorinane, 2-methoxy-5,5-dimethyl-, 2-oxide

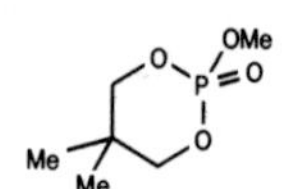

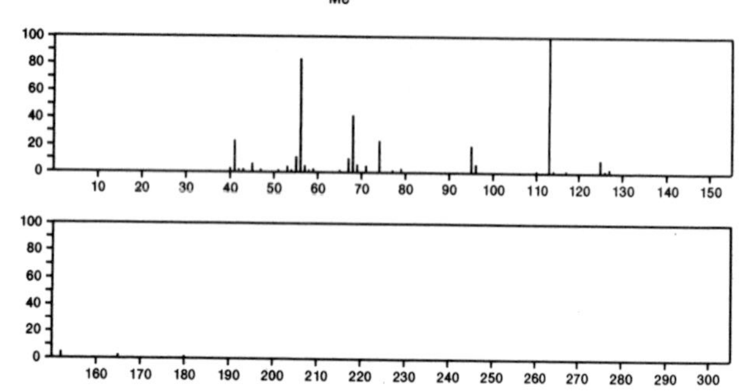

180 $C_6H_{13}O_4P$ 4851-64-3
Phosphoric acid, ethenyl diethyl ester

OEt
H2C=CHOPOEt
O

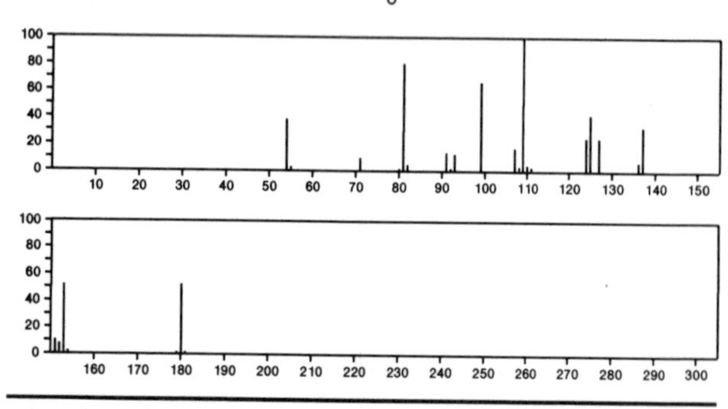

180 $C_6H_{13}O_4P$ 14477-80-6
Phosphoric acid, dimethyl 1-methylpropenyl ester

OMe
MeCH=CMeOPOMe
O

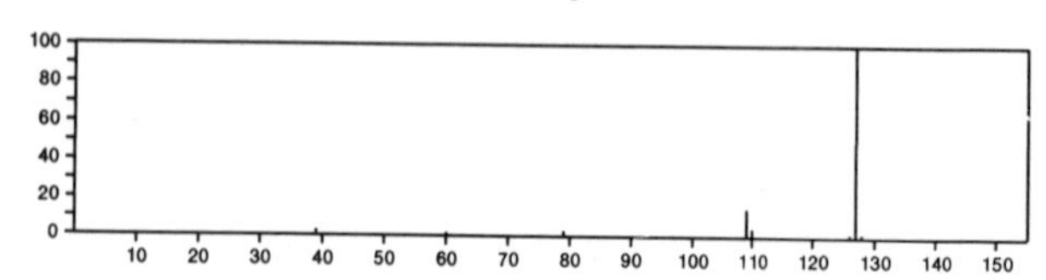

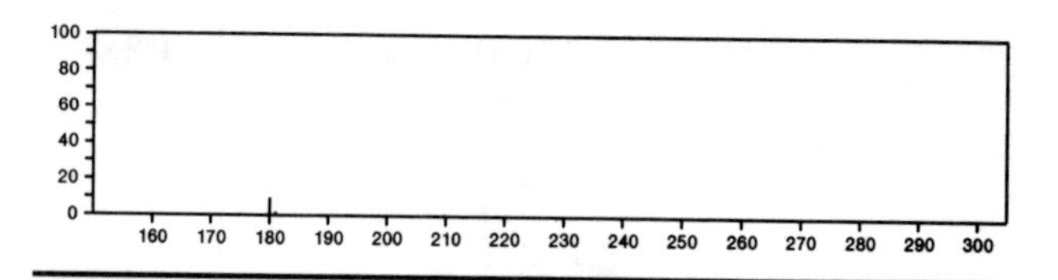

180 $C_7H_4N_2O_2S$ 28783-05-3
Thieno[3,2-*c*]pyridine, 3-nitro-

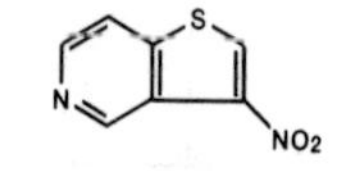

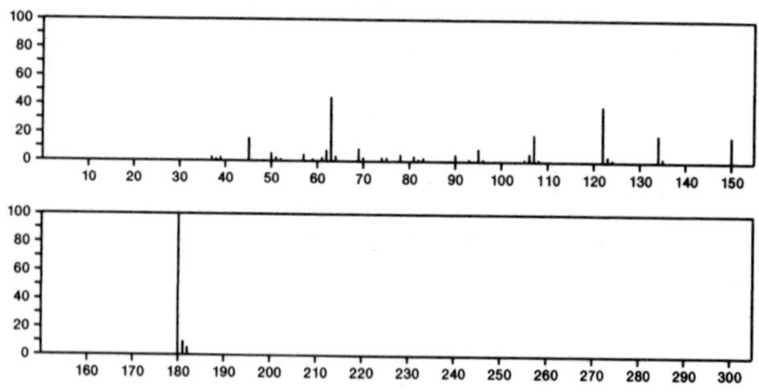

180 $C_7H_4N_2O_2S$ 28783-28-0
Thieno[2,3-*c*]pyridine, 3-nitro-

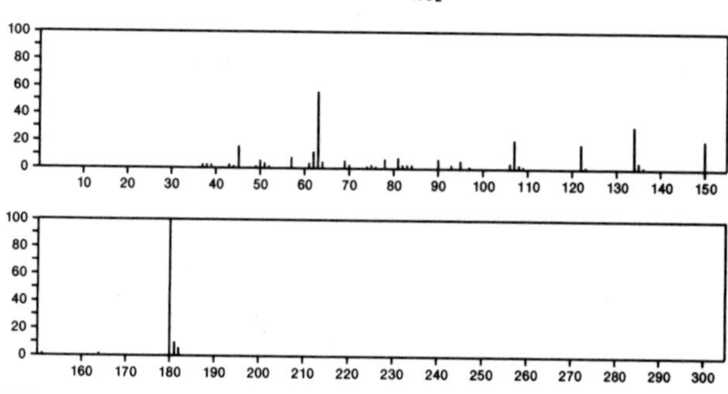

180 $C_7H_5CoO_2$ 12078-25-0
Cobalt, dicarbonyl(η^5-2,4-cyclopentadien-1-yl)-

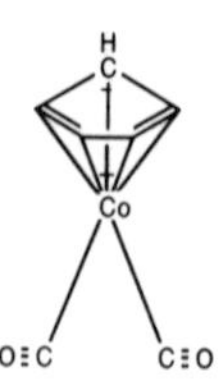

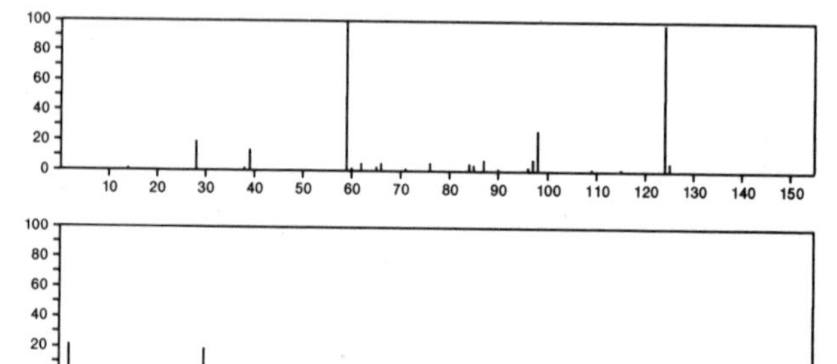

180 $C_7H_7F_3O_2$ 30923-69-4
1,3-Butanedione, 1-cyclopropyl-4,4,4-trifluoro-

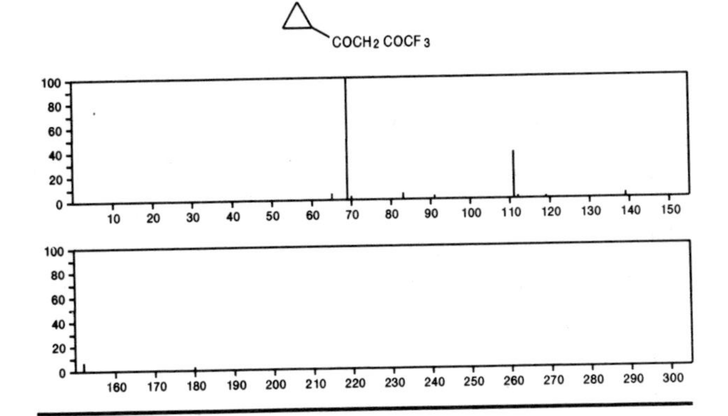

180 $C_7H_8N_4O_2$ 58-55-9
1*H*-Purine-2,6-dione, 3,7-dihydro-1,3-dimethyl-

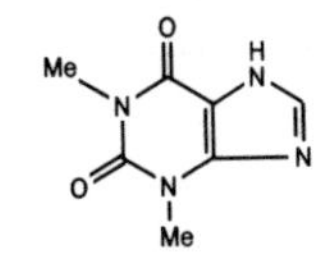

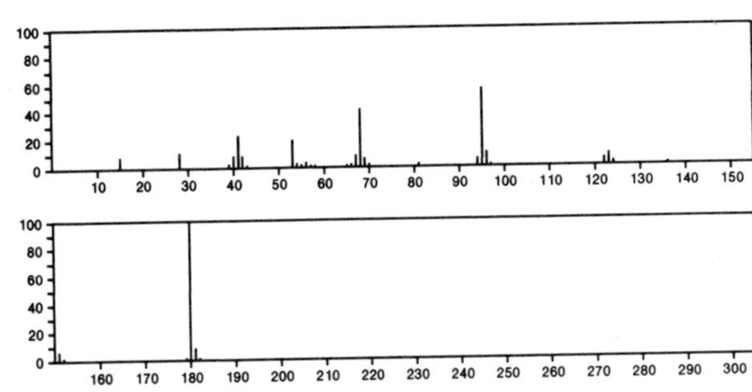

180 $C_7H_8N_4O_2$ 83-67-0
1*H*-Purine-2,6-dione, 3,7-dihydro-3,7-dimethyl-

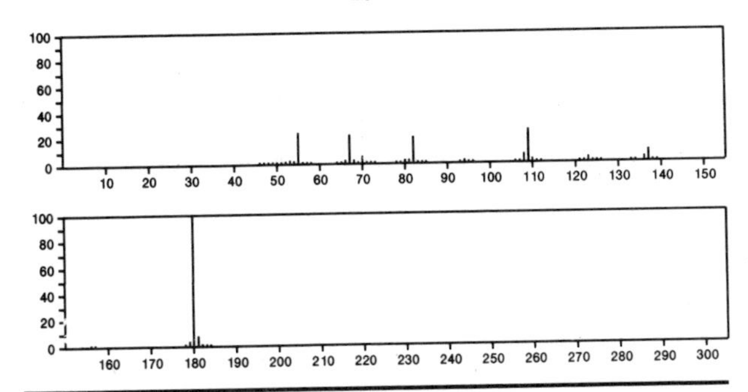

180 $C_7H_8N_4S$ 1008-01-1
7*H*-Purine, 7-methyl-6-(methylthio)-

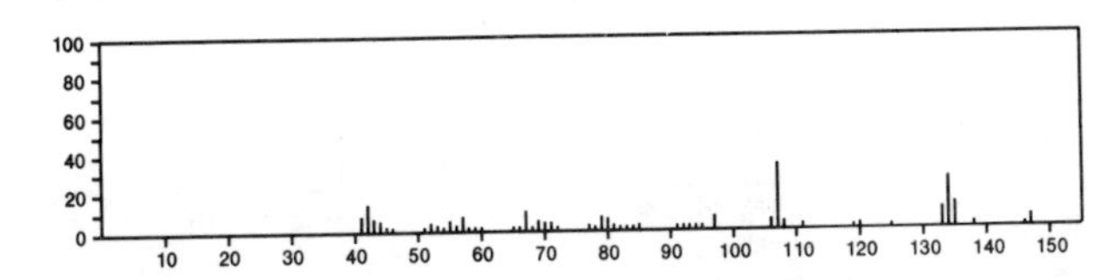

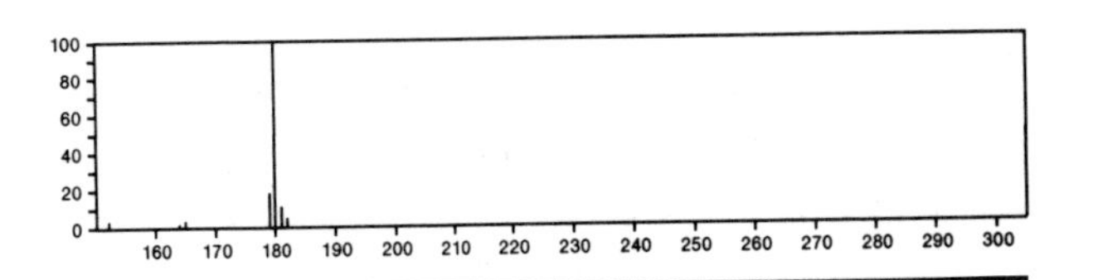

180 $C_7H_8N_4S$ 1008-47-5
1*H*-Purine, 2-methyl-6-(methylthio)-

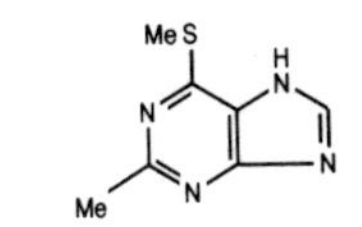

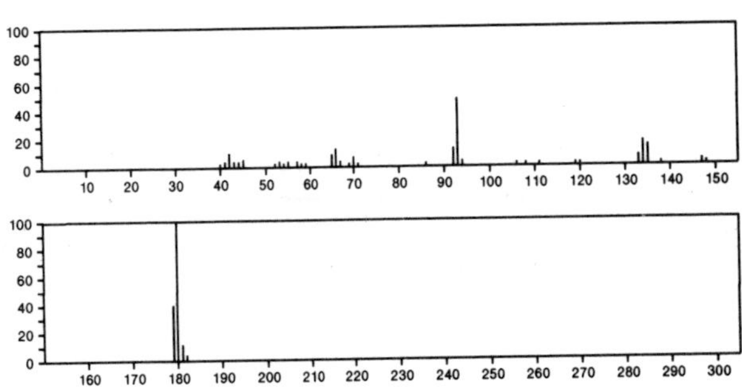

180 $C_7H_8N_4S$ 1008-51-1
1*H*-Purine, 8-methyl-6-(methylthio)-

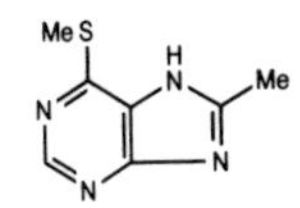

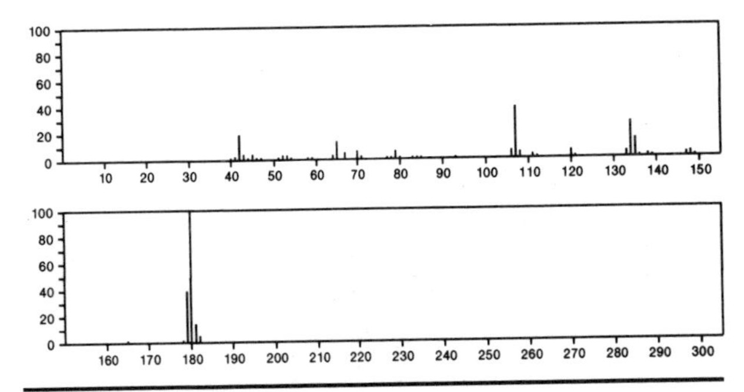

180 $C_7H_8N_4S$ 1127-75-9
9*H*-Purine, 9-methyl-6-(methylthio)-

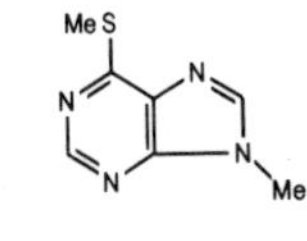

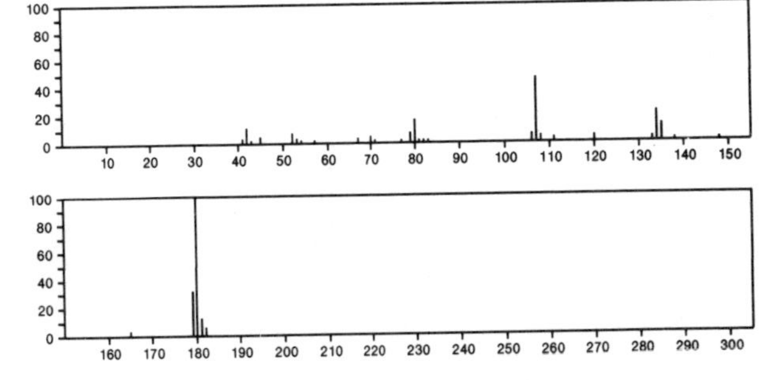

180 $C_7H_8N_4S$ 5752-11-4
7*H*-Purinium, 6-mercapto-7,9-dimethyl-, hydroxide, inner salt

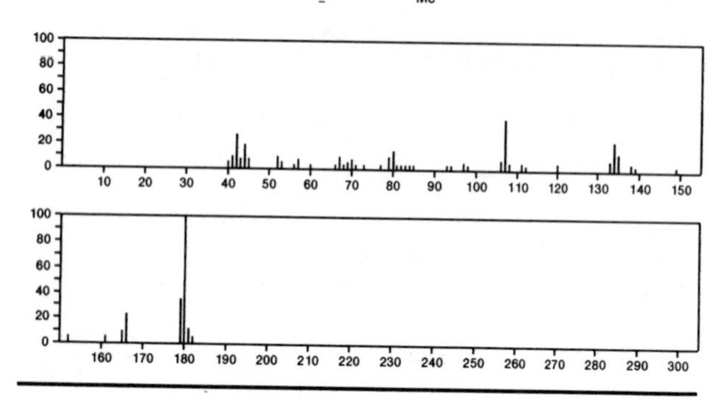

180 $C_7H_8N_4S$ 5759-60-4
6*H*-Purine-6-thione, 3,7-dihydro-3,7-dimethyl-

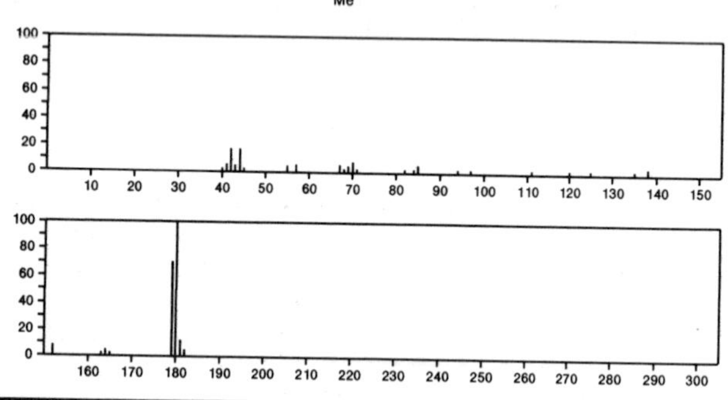

180 $C_7H_8N_4S$ 19835-21-3
Thiazolo[5,4-*d*]pyrimidine, 5-(ethylamino)-

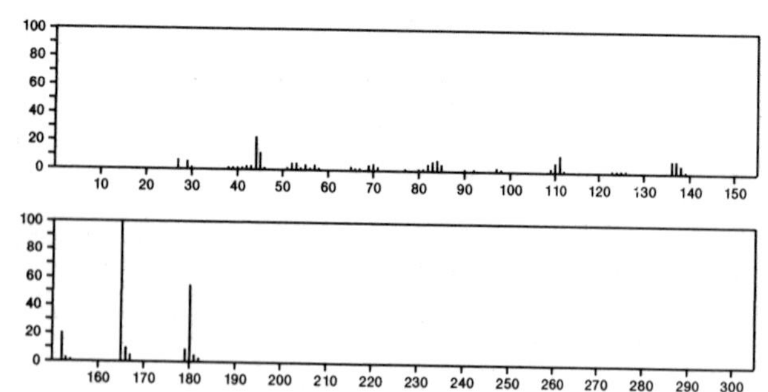

180 $C_7H_8N_4S$ 19854-99-0
s-Triazolo[4,3-*a*]pyrazine-3-thiol, 5,8-dimethyl-

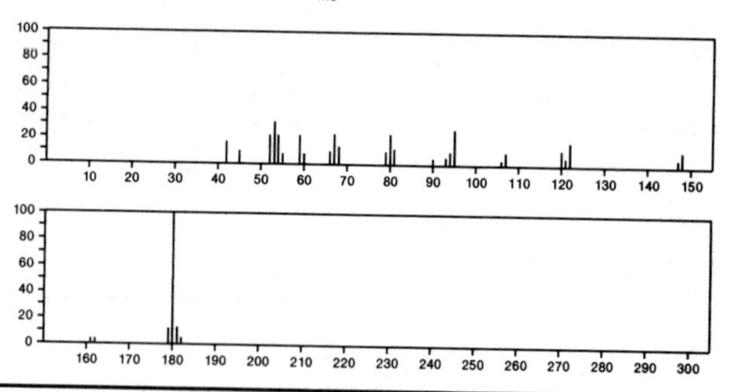

180 C_8H_5Br 932-87-6
Benzene, (bromoethynyl)-

BrC≡CPh

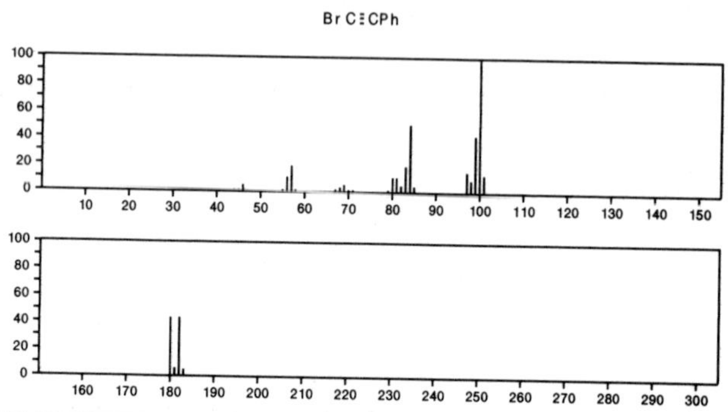

180 $C_8H_5FN_2O_2$ 5352-95-4
Sydnone, 3-(*p*-fluorophenyl)-

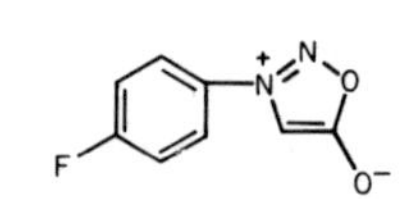

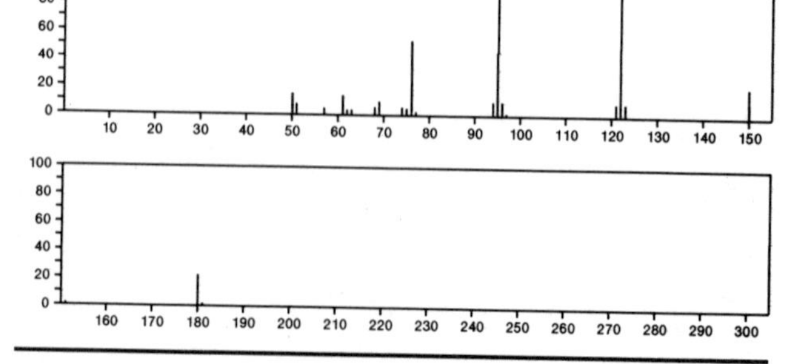

180 $C_8H_8N_2O_3$ 104-04-1
Acetamide, *N*-(4-nitrophenyl)-

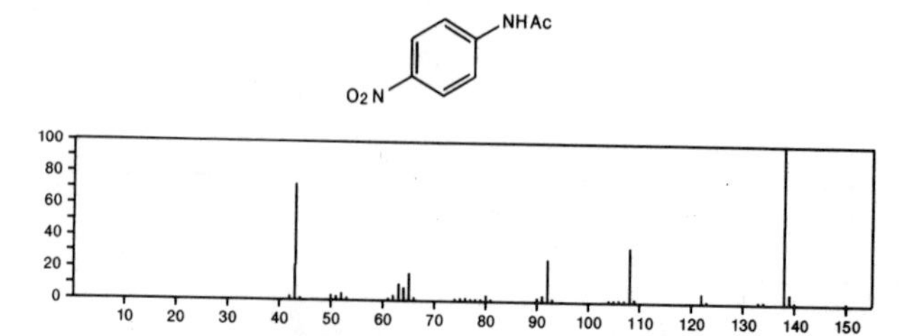

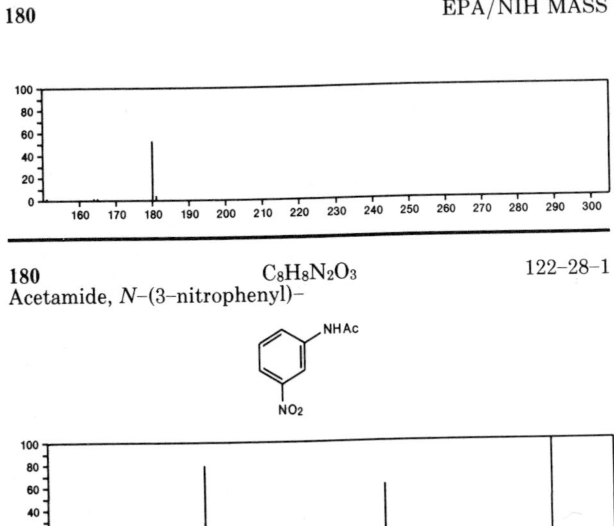

180 $C_8H_8N_2O_3$ 122–28–1
Acetamide, *N*–(3–nitrophenyl)–

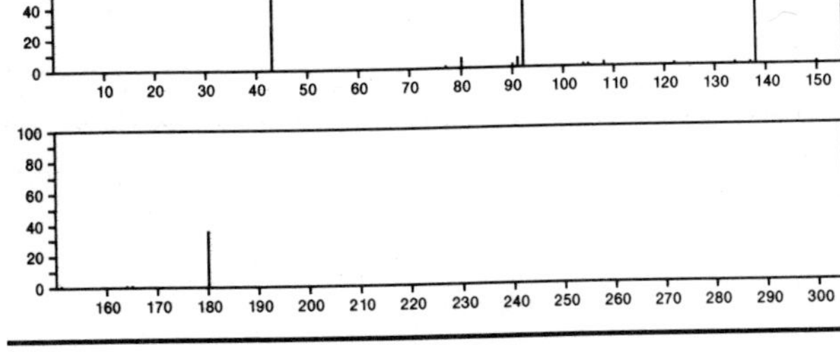

180 $C_8H_8N_2O_3$ 552–32–9
Acetamide, *N*–(2–nitrophenyl)–

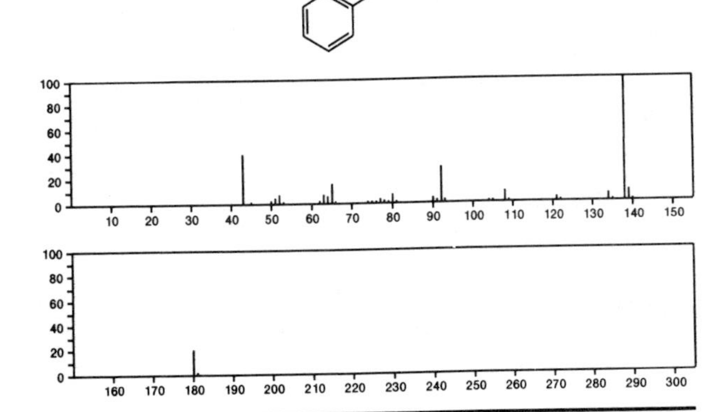

180 $C_8H_8N_2O_3$ 583–08–4
Glycine, *N*–(3–pyridinylcarbonyl)–

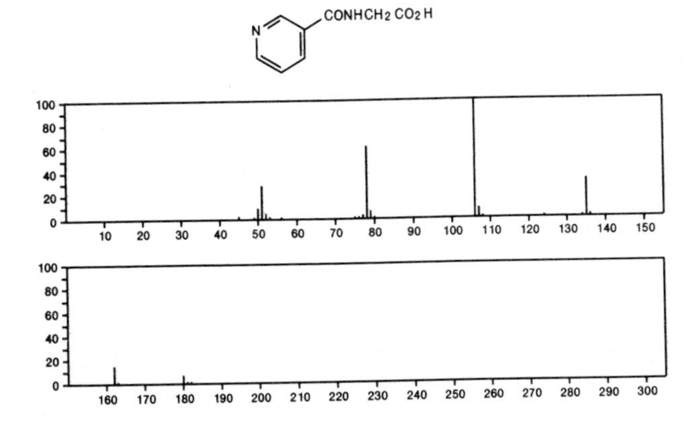

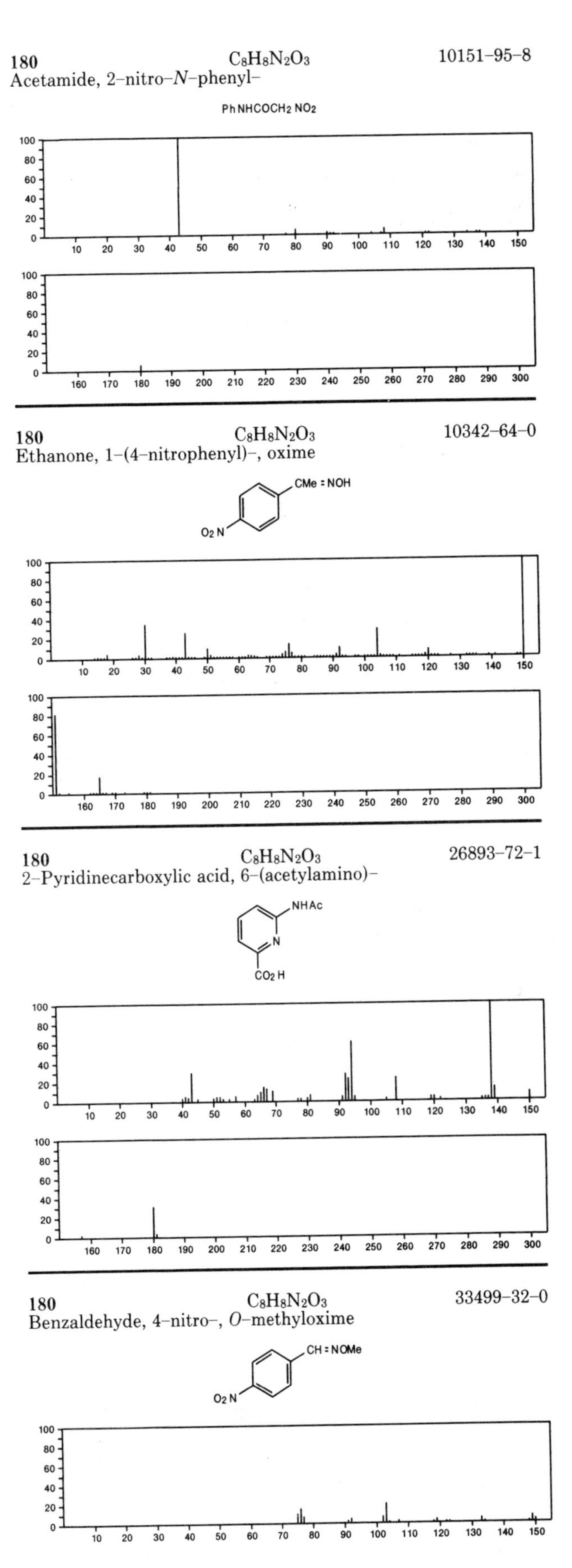

180 $C_8H_8N_2O_3$ 10151–95–8
Acetamide, 2–nitro–*N*–phenyl–

$PhNHCOCH_2NO_2$

180 $C_8H_8N_2O_3$ 10342–64–0
Ethanone, 1–(4–nitrophenyl)–, oxime

180 $C_8H_8N_2O_3$ 26893–72–1
2–Pyridinecarboxylic acid, 6–(acetylamino)–

180 $C_8H_8N_2O_3$ 33499–32–0
Benzaldehyde, 4–nitro–, *O*–methyloxime

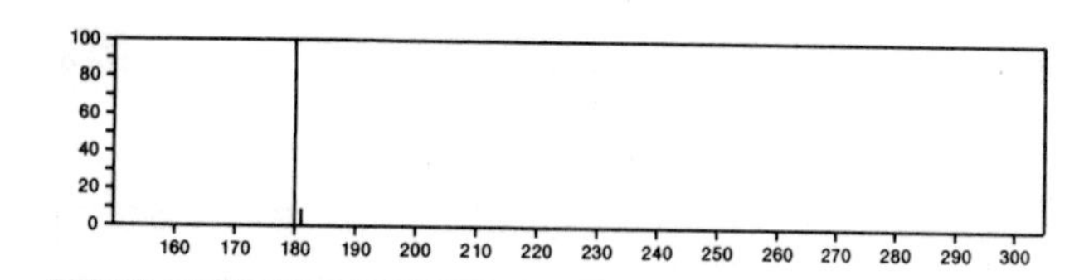

180 $C_8H_8N_2O_3$ 33499-33-1

Benzaldehyde, *m*-nitro-, *O*-methyloxime

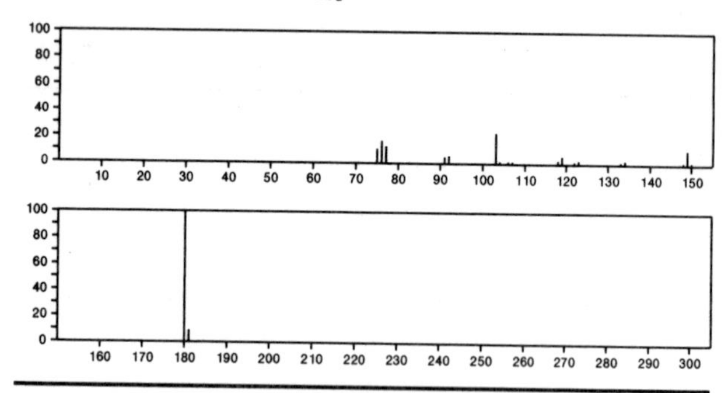

180 $C_8H_{12}N_4O$ 56247-56-4

6*H*-Purine, 5,7-dihydro-6-(1-methylethoxy)-

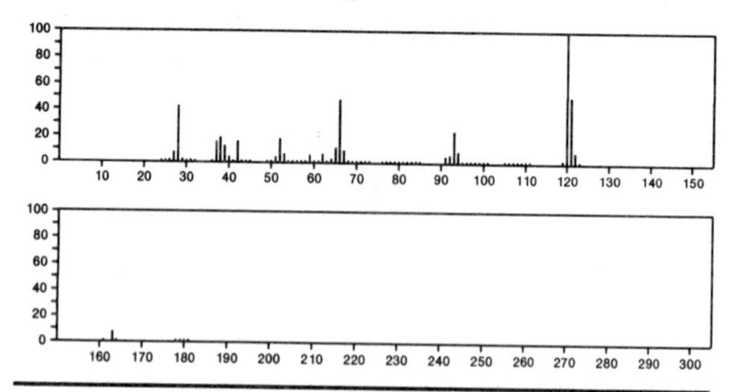

180 $C_8H_{18}ClP$ 13716-10-4

Phosphinous chloride, bis(1,1-dimethylethyl)-

ClP(Bu-t)2

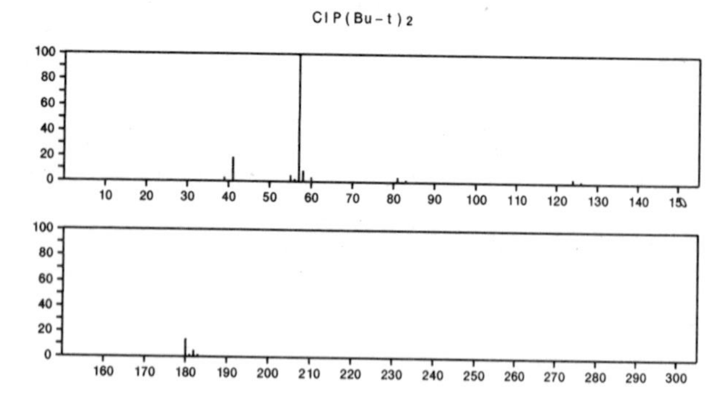

180 $C_9H_8O_2S$ 26524-91-4

4*H*-1-Benzothiopyran-4-one, 2,3-dihydro-, 1-oxide

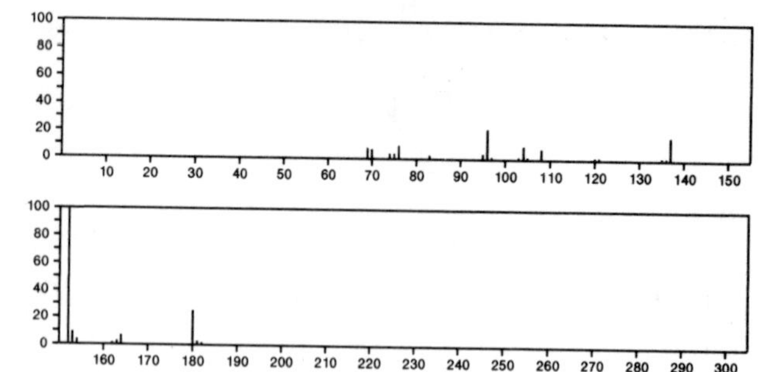

180 $C_9H_8O_2S$ 29399-50-6

1*H*-2-Benzothiopyran-4(3*H*)-one, 2-oxide

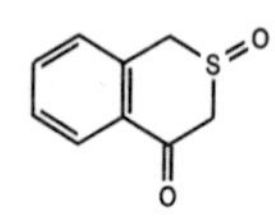

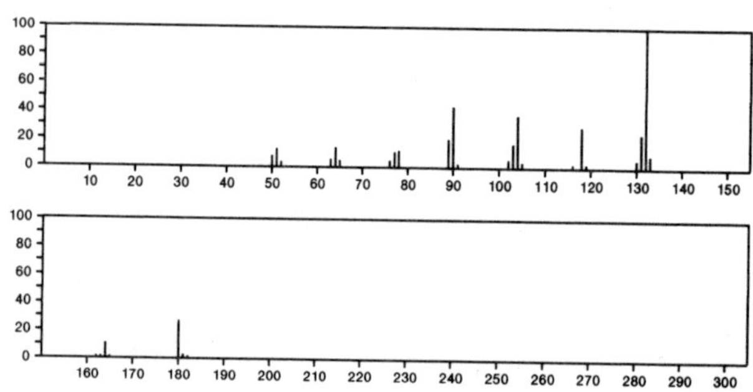

180 $C_9H_8O_4$ 50-78-2

Benzoic acid, 2-(acetyloxy)-

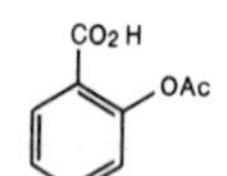

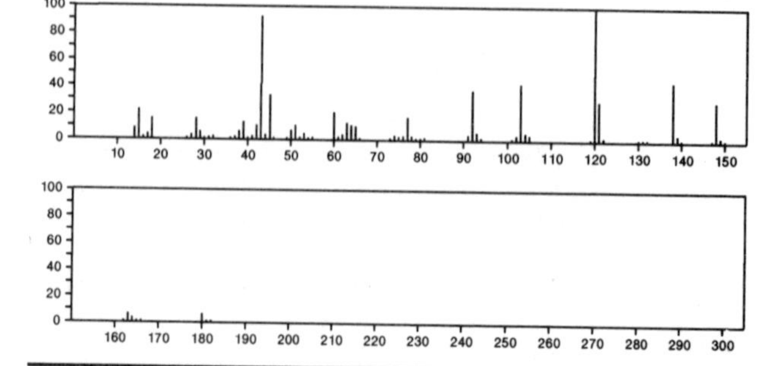

180 $C_9H_8O_4$ 156-39-8

Benzenepropanoic acid, 4-hydroxy-α-oxo-

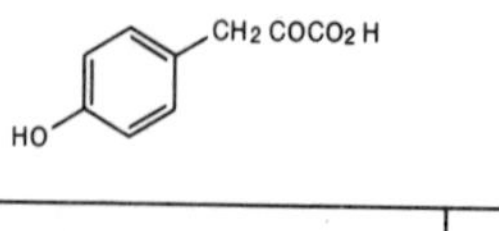

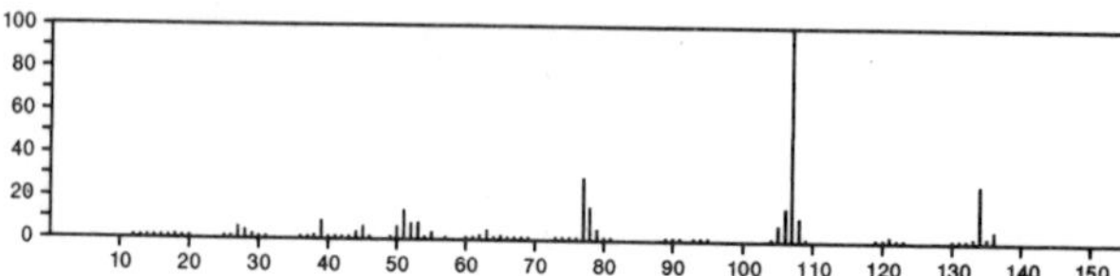

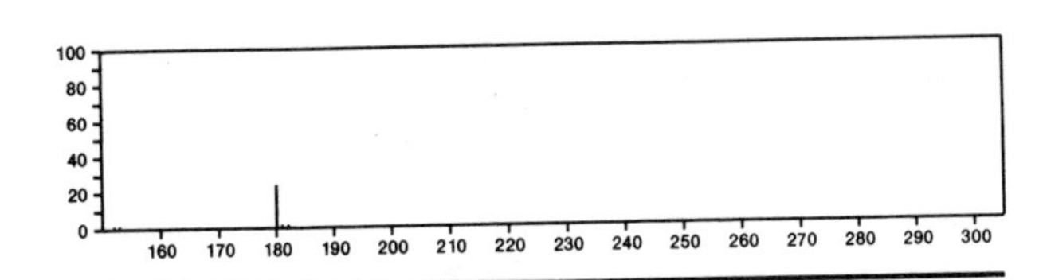

180 $C_9H_8O_4$ 331-39-5

2-Propenoic acid, 3-(3,4-dihydroxyphenyl)-

HO / HO / CH=CHCO2H

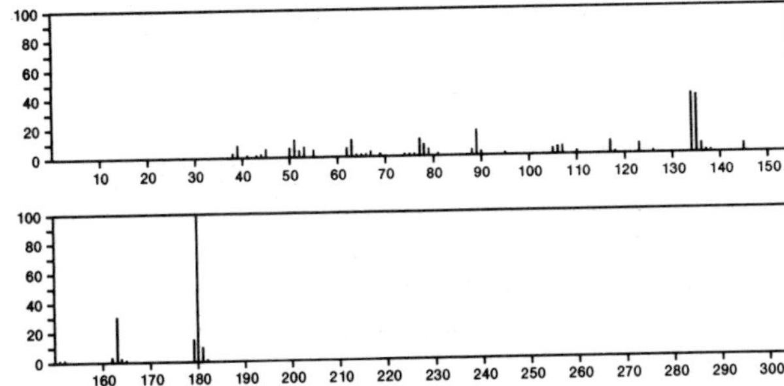

180 $C_9H_8O_4$ 499-49-0

1,3-Benzenedicarboxylic acid, 5-methyl-

Me / CO2H / CO2H

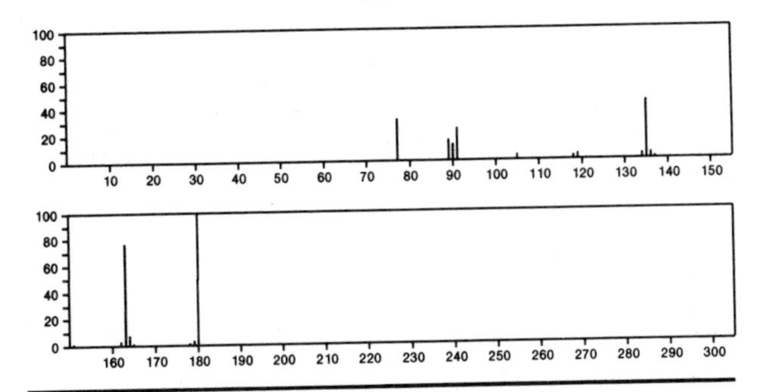

180 $C_9H_8O_4$ 1877-71-0

1,3-Benzenedicarboxylic acid, monomethyl ester

C(O)OMe / CO2H

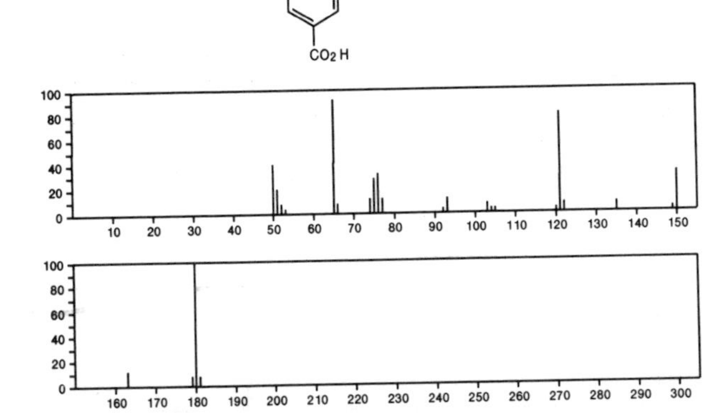

180 $C_9H_8O_4$ 3347-99-7

1,3-Benzenedicarboxylic acid, 4-methyl-

HO2C / CO2H / Me

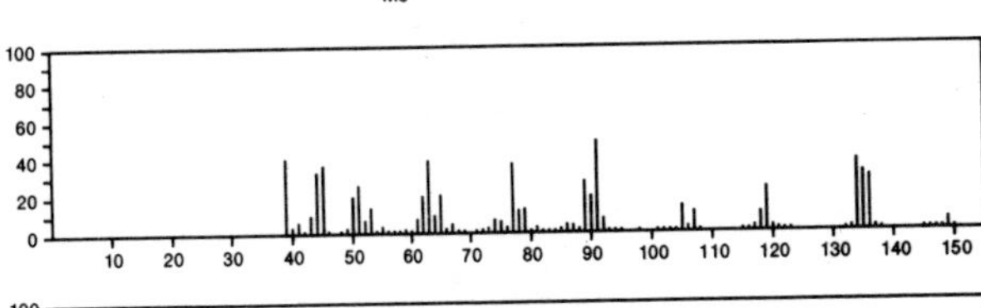

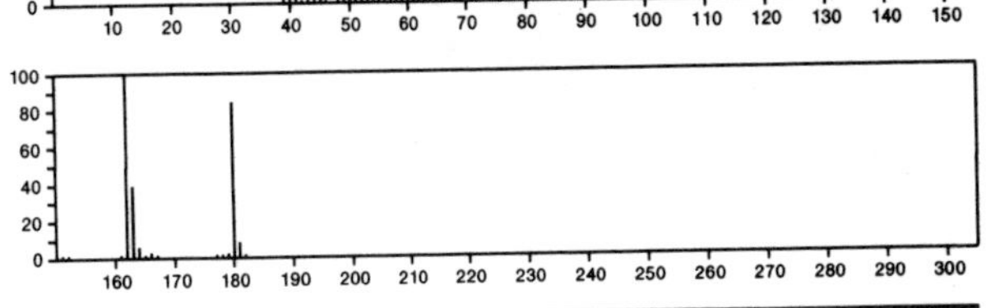

180 $C_9H_8O_4$ 4316-23-8

1,2-Benzenedicarboxylic acid, 4-methyl-

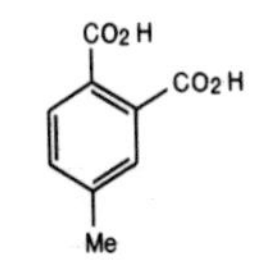

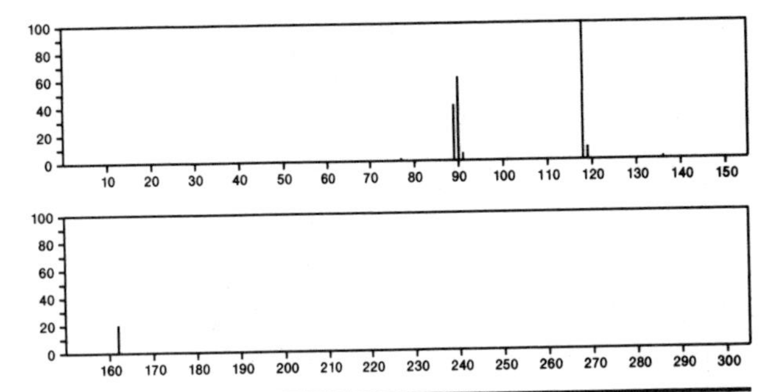

180 $C_9H_8O_4$ 4376-18-5

1,2-Benzenedicarboxylic acid, monomethyl ester

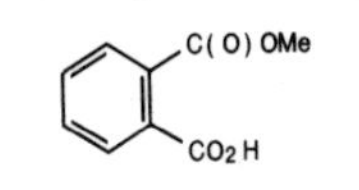

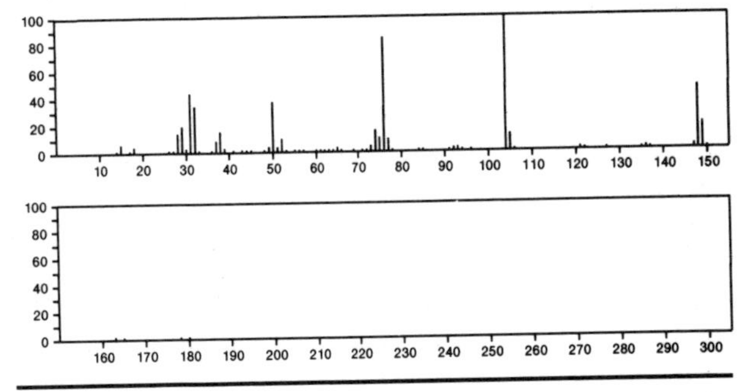

180 $C_9H_8O_4$ 5156-01-4

1,4-Benzenedicarboxylic acid, 2-methyl-

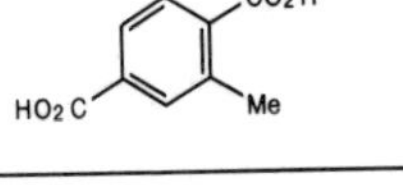

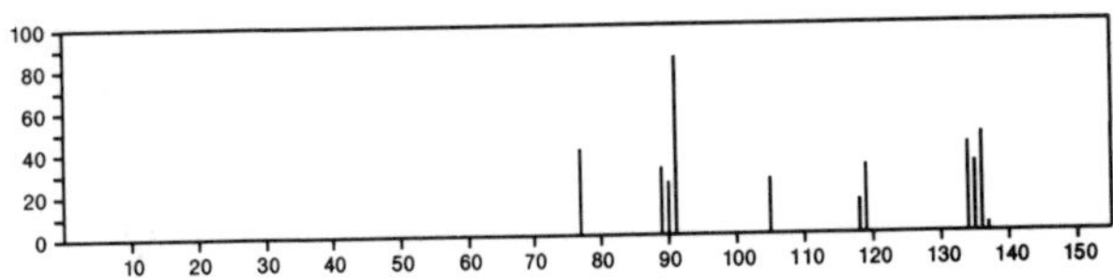

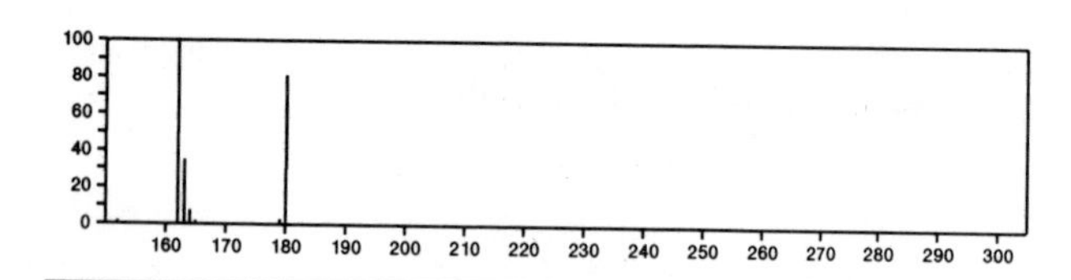

180 $C_9H_8O_4$ 10209-57-1
1,3-Benzenedicarboxaldehyde, 2,4-dihydroxy-6-methyl-

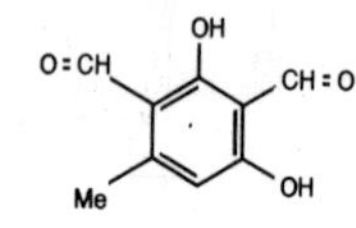

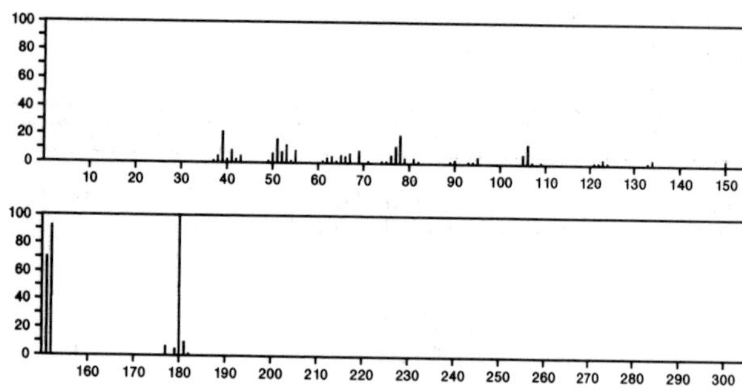

180 $C_9H_8O_4$ 52097-95-7
Spiro[3.3]hepta-1,5-diene-2,6-dicarboxylic acid, (±)-

HO2C—⟨spiro⟩—CO2H

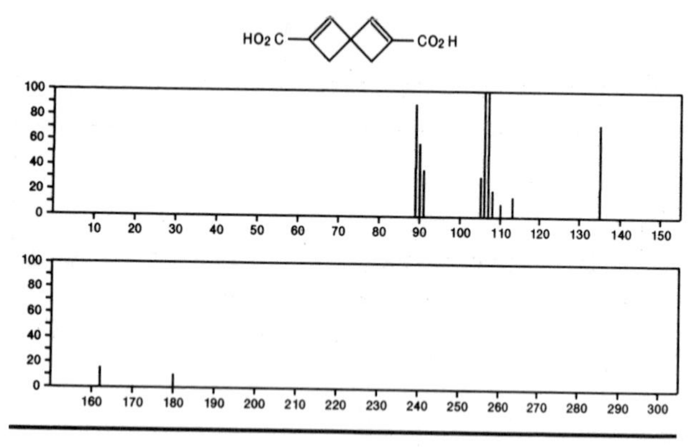

180 $C_9H_{12}N_2O_2$ 150-69-6
Urea, (4-ethoxyphenyl)-

NHCONH2
EtO

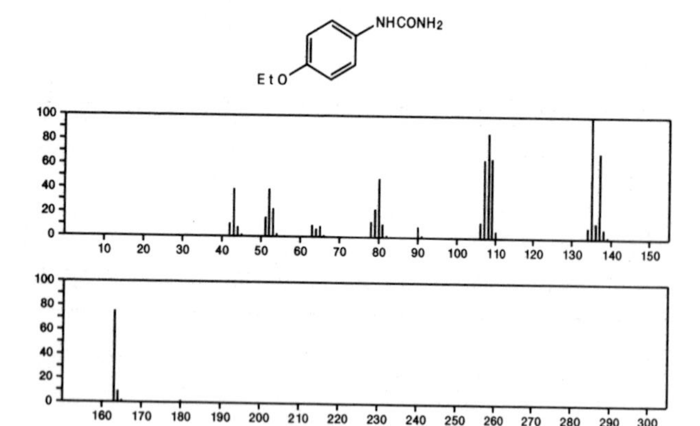

180 $C_9H_{12}N_2O_2$ 1521-60-4
Benzenamine, 2,4,6-trimethyl-3-nitro-

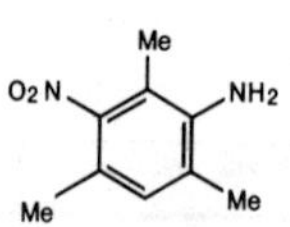

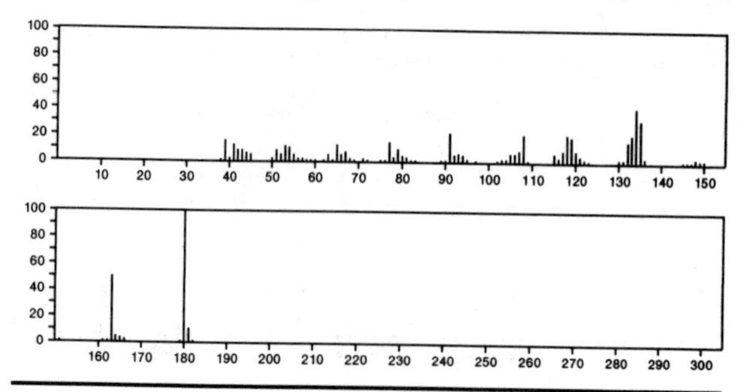

180 $C_9H_{12}N_2S$ 705-62-4
Thiourea, *N,N*-dimethyl-*N'*-phenyl-

Me2NCSNHPh

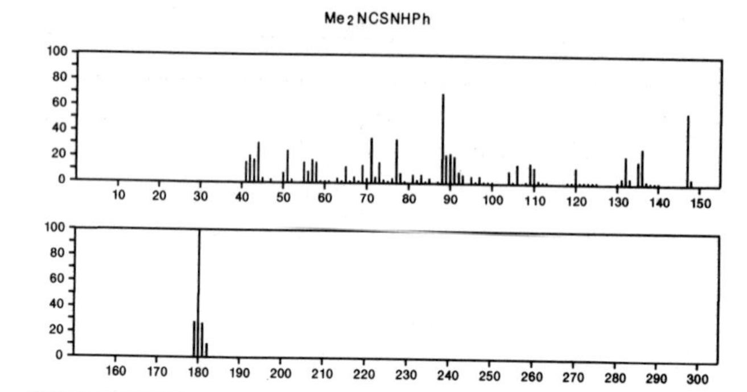

180 $C_9H_{12}N_2S$ 2741-06-2
Thiourea, *N*-ethyl-*N'*-phenyl-

PhNHCSNHEt

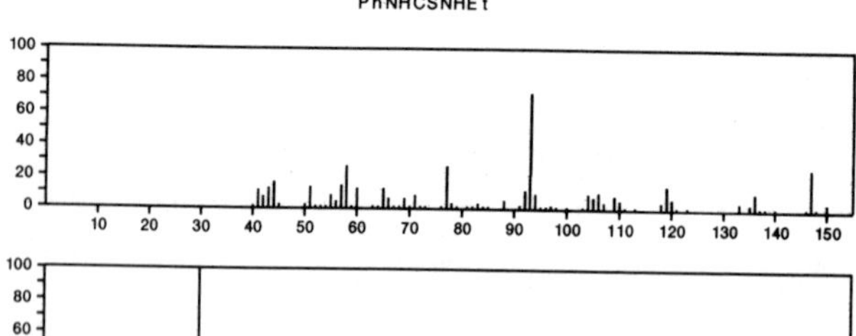

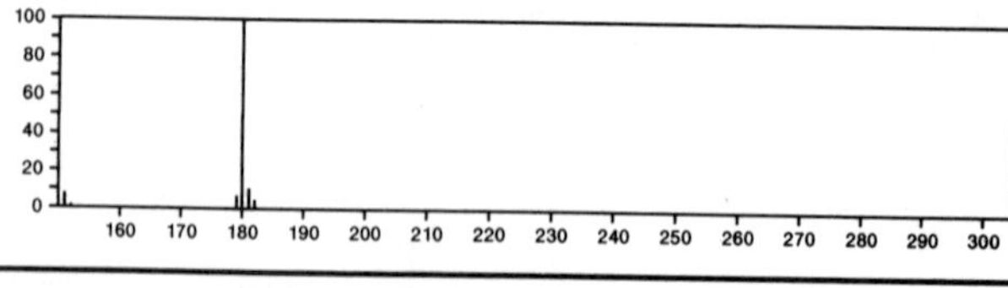

180 $C_{10}H_9ClO$ 3160-40-5
3-Buten-2-one, 4-(4-chlorophenyl)-

CH=CHCOMe
Cl

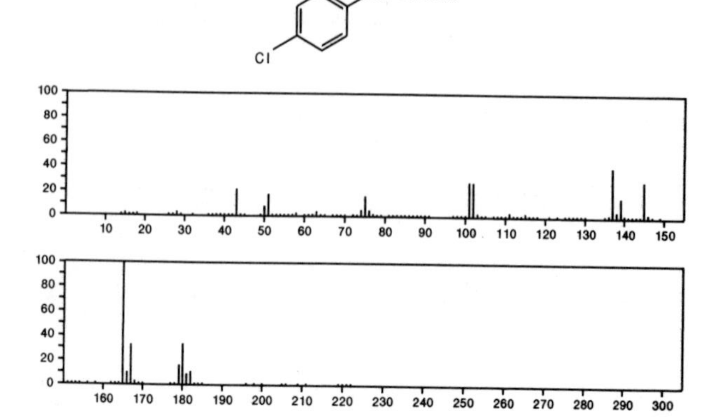

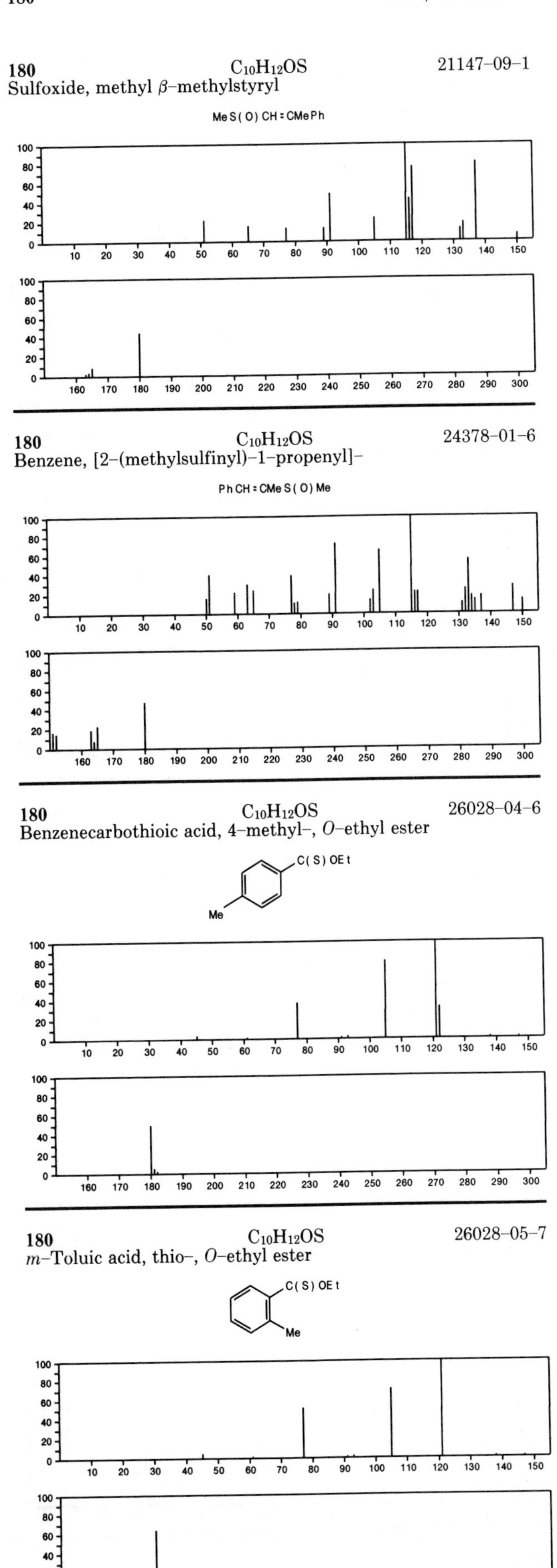

180 $C_{10}H_{12}OS$ 21147-09-1
Sulfoxide, methyl *β*-methylstyryl
MeS(O)CH=CMePh

180 $C_{10}H_{12}OS$ 24378-01-6
Benzene, [2-(methylsulfinyl)-1-propenyl]-
PhCH=CMeS(O)Me

180 $C_{10}H_{12}OS$ 26028-04-6
Benzenecarbothioic acid, 4-methyl-, *O*-ethyl ester
C(S)OEt, Me

180 $C_{10}H_{12}OS$ 26028-05-7
m-Toluic acid, thio-, *O*-ethyl ester
C(S)OEt, Me

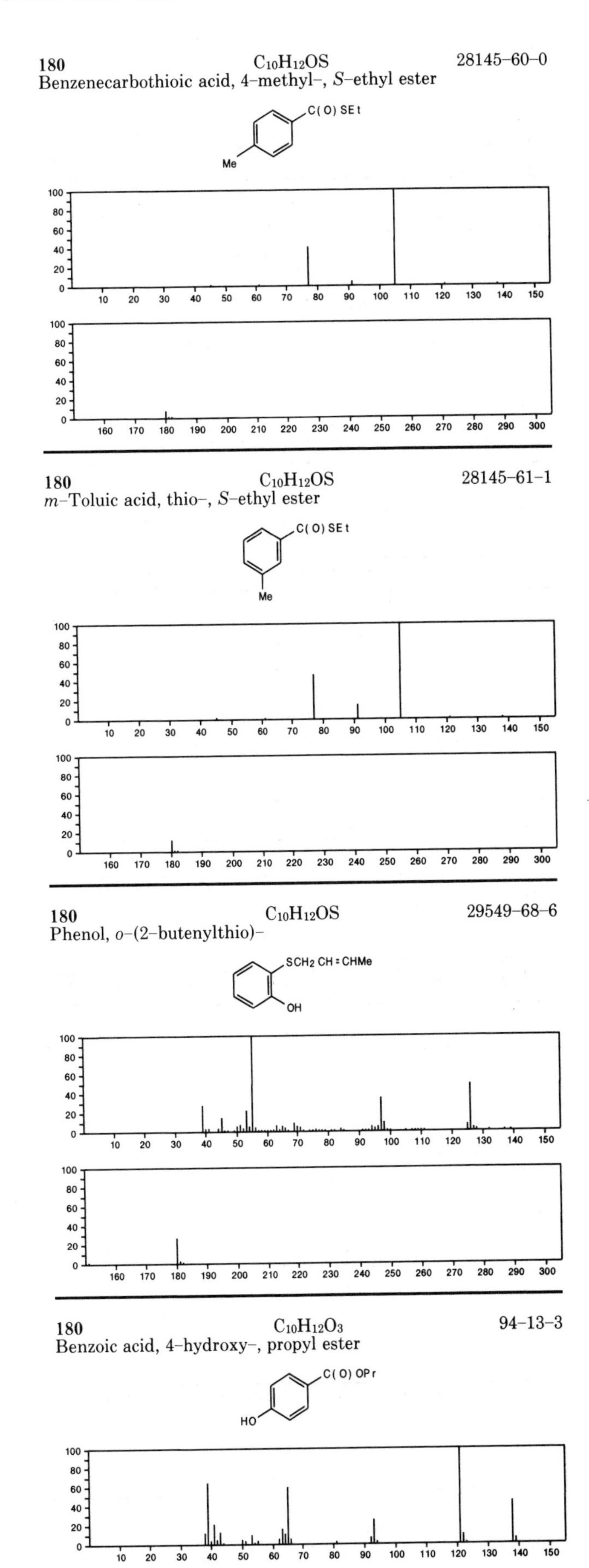

180 $C_{10}H_{12}OS$ 28145-60-0
Benzenecarbothioic acid, 4-methyl-, *S*-ethyl ester
C(O)SEt, Me

180 $C_{10}H_{12}OS$ 28145-61-1
m-Toluic acid, thio-, *S*-ethyl ester
C(O)SEt, Me

180 $C_{10}H_{12}OS$ 29549-68-6
Phenol, *o*-(2-butenylthio)-
SCH2CH=CHMe, OH

180 $C_{10}H_{12}O_3$ 94-13-3
Benzoic acid, 4-hydroxy-, propyl ester
C(O)OPr, HO

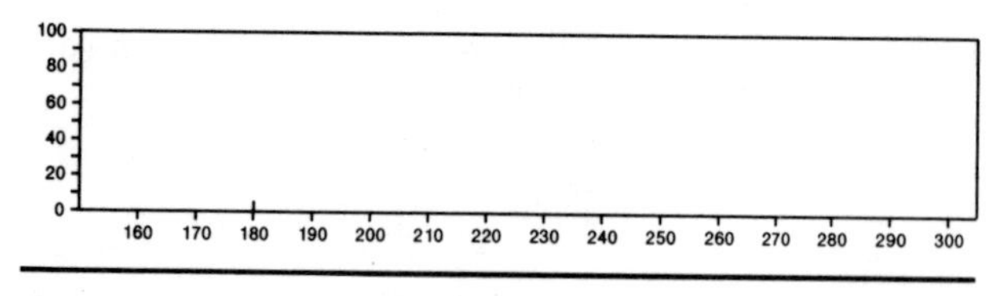

180 $C_{10}H_{12}O_3$ 94-30-4
Benzoic acid, 4-methoxy-, ethyl ester

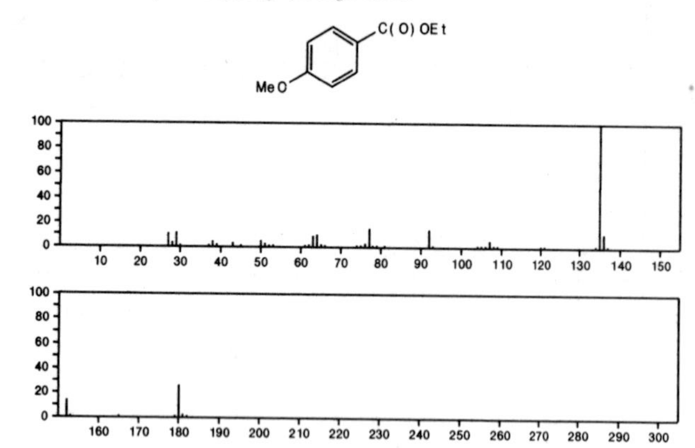

180 $C_{10}H_{12}O_3$ 104-21-2
Benzenemethanol, 4-methoxy-, acetate

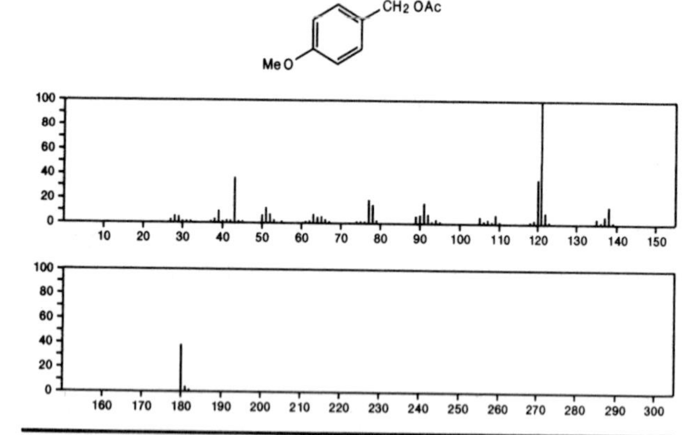

180 $C_{10}H_{12}O_3$ 458-35-5
Phenol, 4-(3-hydroxy-1-propenyl)-2-methoxy-

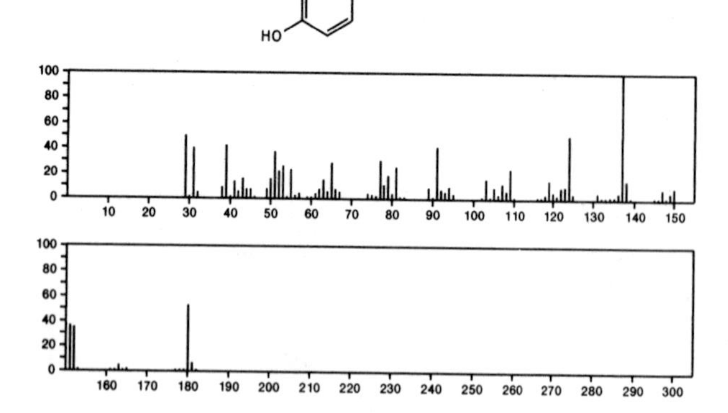

180 $C_{10}H_{12}O_3$ 774-40-3
Benzeneacetic acid, α-hydroxy-, ethyl ester

EtOC(O)CH(OH)Ph

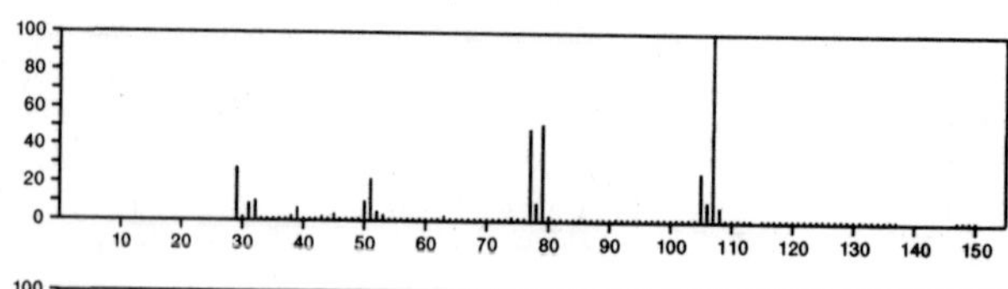

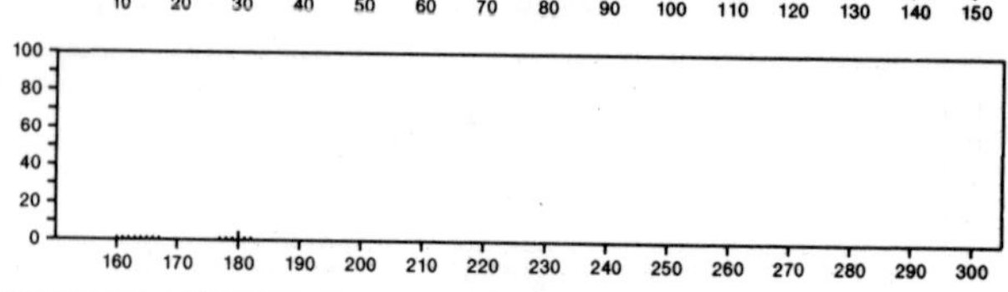

180 $C_{10}H_{12}O_3$ 943-57-7
Carbonic acid, 1-methylethyl phenyl ester

PhOC(O)OPr-i

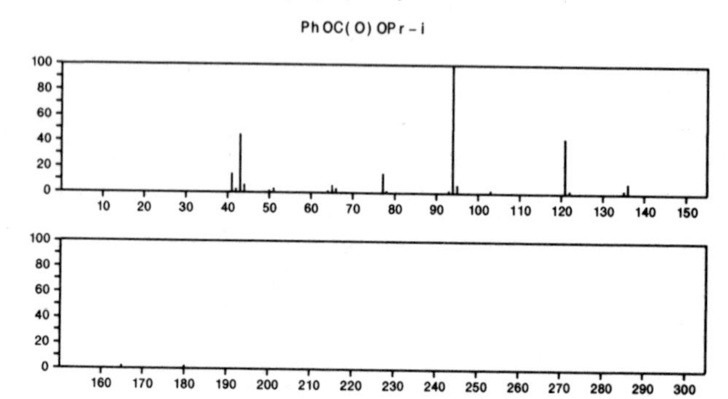

180 $C_{10}H_{12}O_3$ 1131-62-0
Ethanone, 1-(3,4-dimethoxyphenyl)-

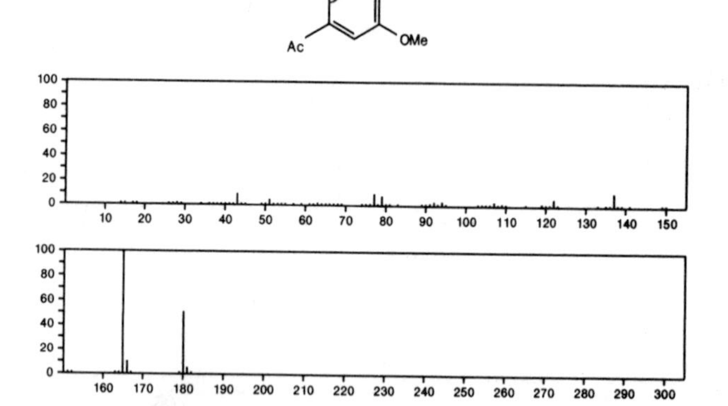

180 $C_{10}H_{12}O_3$ 1331-83-5
Benzenemethanol, *ar*-methoxy-, acetate

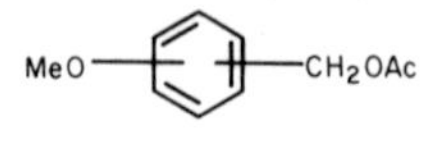

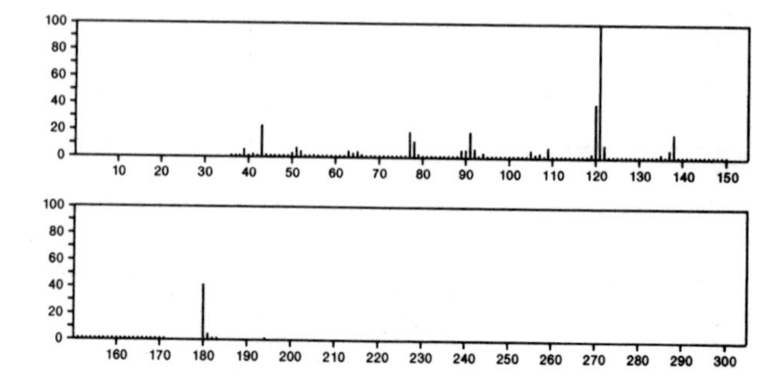

180 $C_{10}H_{12}O_3$ 2403-50-1
1,3-Dioxolane, 2-(4-methoxyphenyl)-

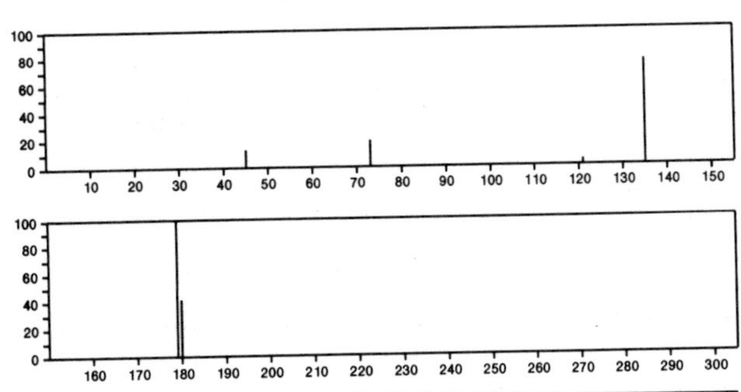

180 $C_{10}H_{12}O_3$ 2503-46-0
2-Propanone, 1-(4-hydroxy-3-methoxyphenyl)-

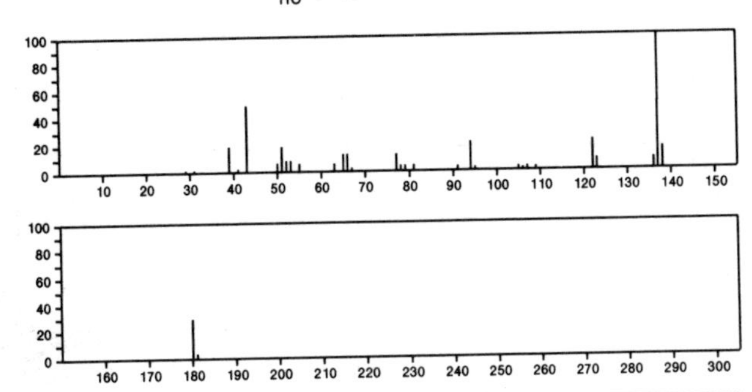

180 $C_{10}H_{12}O_3$ 2651-48-1
1,3-Isobenzofurandione, 3a,4,7,7a-tetrahydro-4,7-dimethyl-

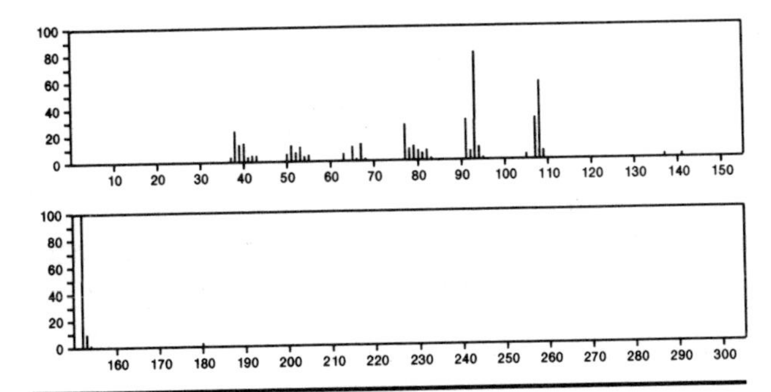

180 $C_{10}H_{12}O_3$ 4223-84-1
Ethanone, 1-(2-hydroxy-5-methoxy-4-methylphenyl)-

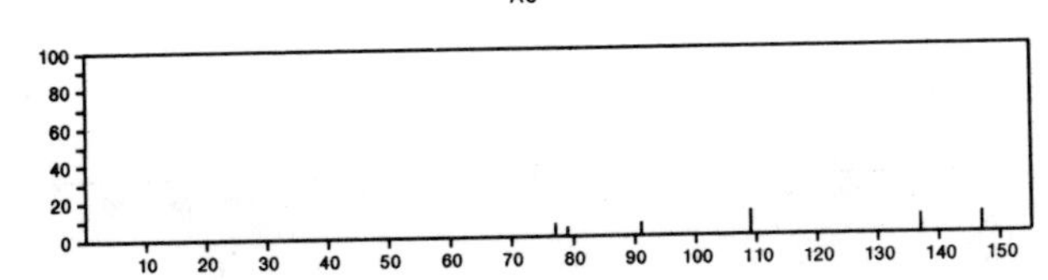

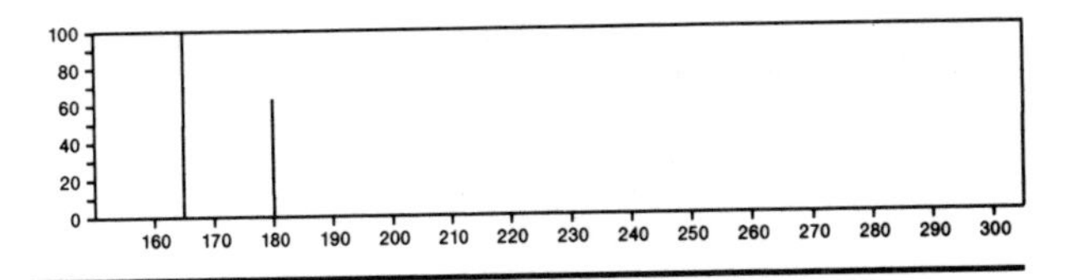

180 $C_{10}H_{12}O_3$ 5438-24-4
1,3-Isobenzofurandione, 3a,4,7,7a-tetrahydro-5,6-dimethyl-

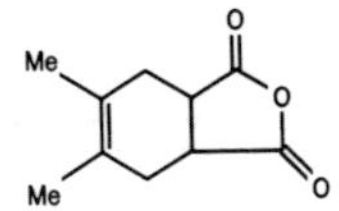

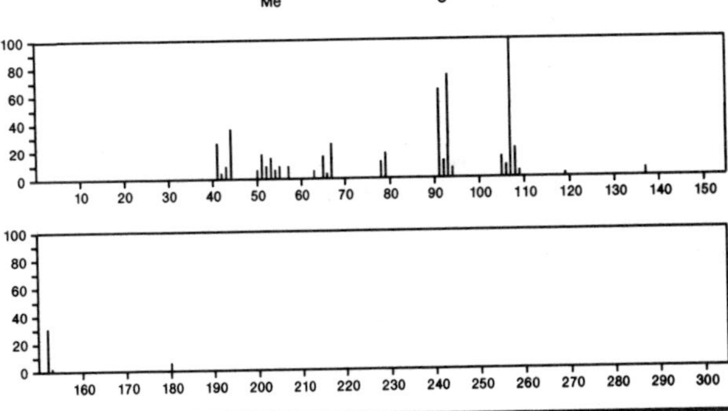

180 $C_{10}H_{12}O_3$ 5597-50-2
Benzenepropanoic acid, 4-hydroxy-, methyl ester

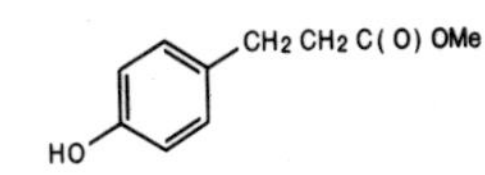

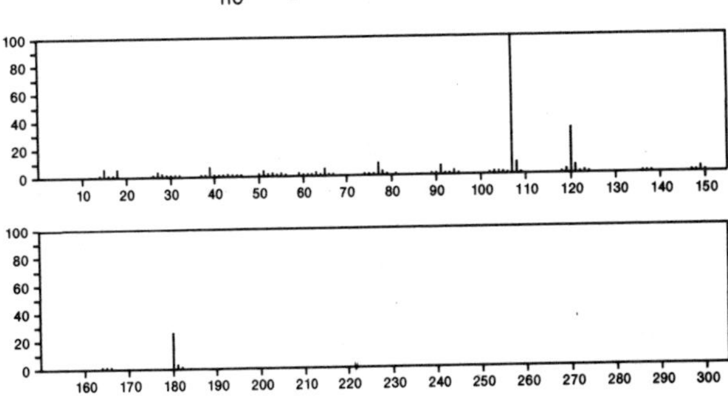

180 $C_{10}H_{12}O_3$ 6303-58-8
Butanoic acid, 4-phenoxy-

$HO_2C(CH_2)_3OPh$

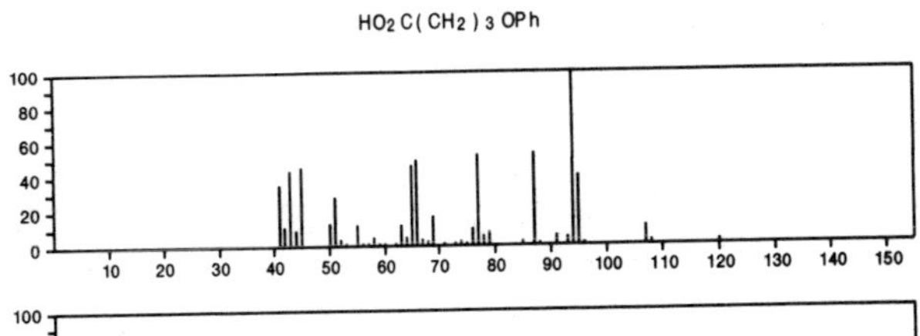

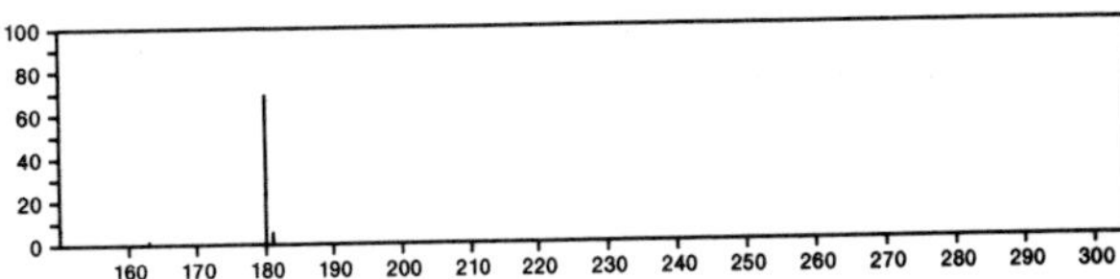

180 $C_{10}H_{12}O_3$ 7149-90-8
Benzaldehyde, 2,4-dimethoxy-6-methyl-

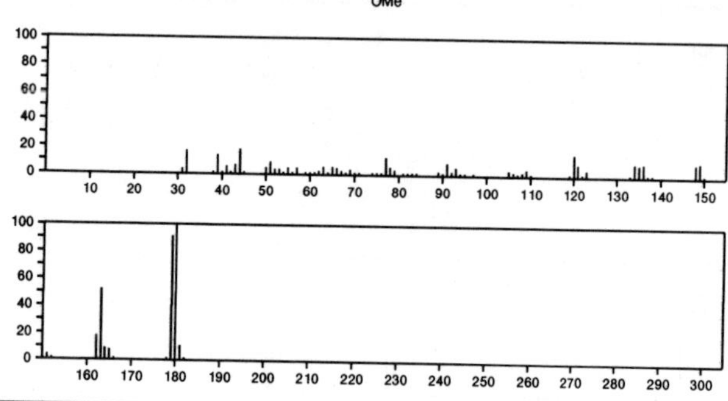

180 $C_{10}H_{12}O_3$ 7335-26-4
Benzoic acid, 2-methoxy-, ethyl ester

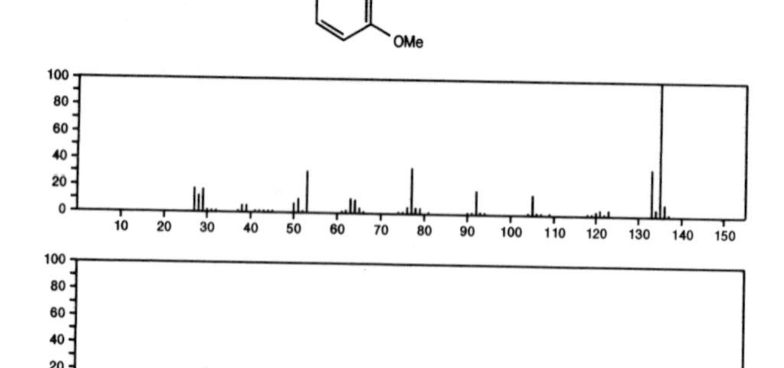

180 $C_{10}H_{12}O_3$ 7497-89-4
Propanoic acid, 3-phenoxy-, methyl ester

PhOCH2 CH2 C(O) OMe

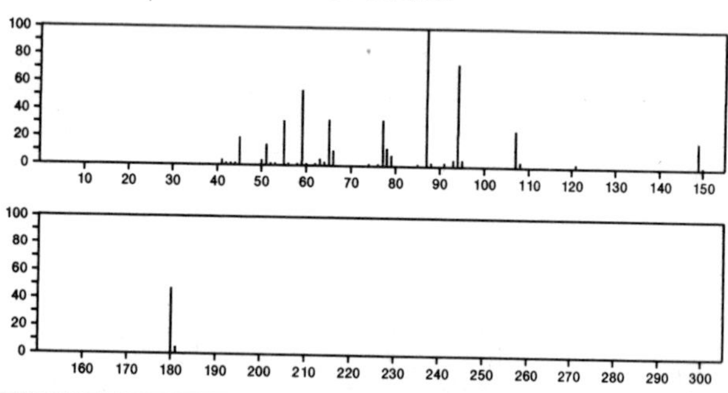

180 $C_{10}H_{12}O_3$ 13183-16-9
Carbonic acid, phenyl propyl ester

PhOC(O) OPr

180 $C_{10}H_{12}O_3$ 13674-16-3
Benzenepropanoic acid, α-hydroxy-, methyl ester

MeOC(O) CH(OH) CH2 Ph

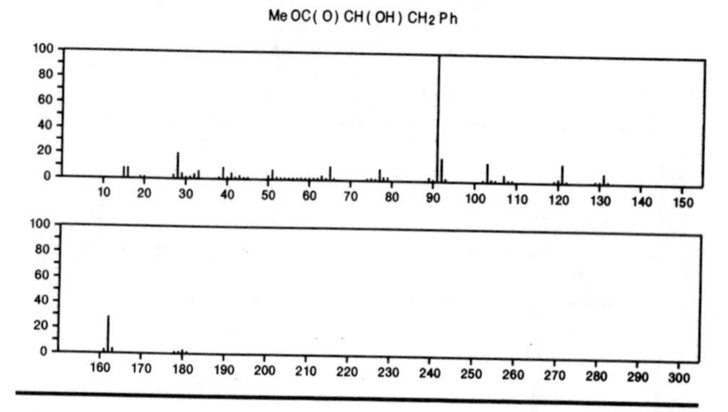

180 $C_{10}H_{12}O_3$ 17138-28-2
Benzeneacetic acid, 4-hydroxy-, ethyl ester

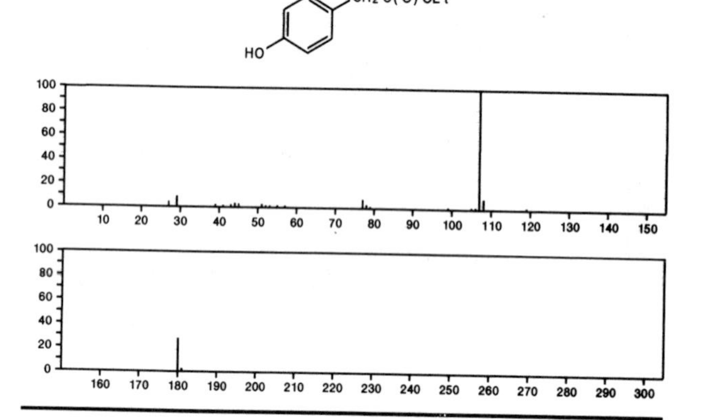

180 $C_{10}H_{12}O_3$ 18927-05-4
Benzeneacetic acid, 3-methoxy-, methyl ester

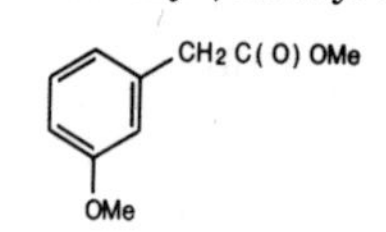

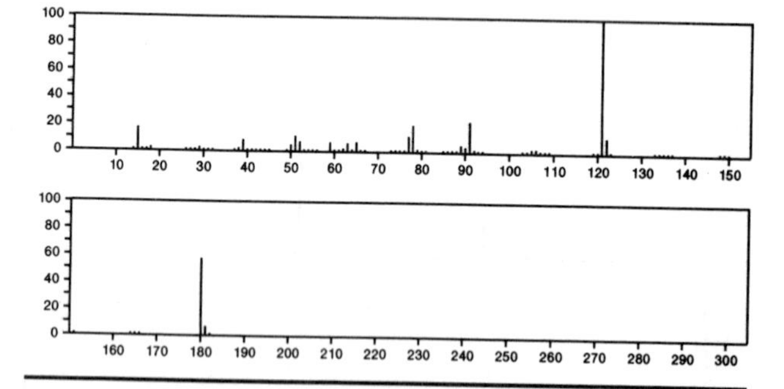

180 $C_{10}H_{12}O_3$ 20731-95-7
Mandelic acid, α-methyl-, methyl ester

MeOC(O) CMe (OH) Ph

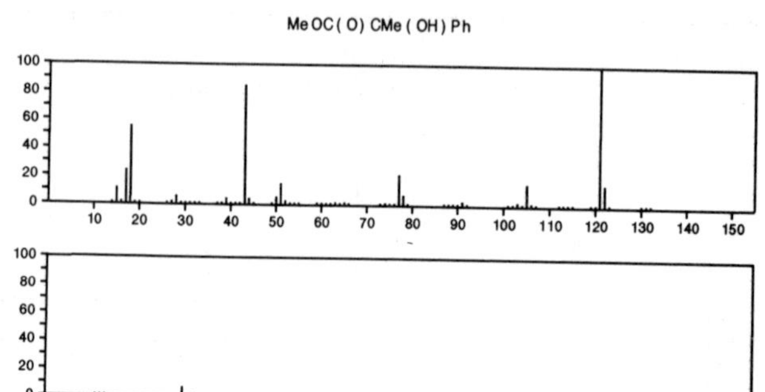

180 $C_{10}H_{12}O_3$ 22446-38-4

Acetic acid, (*m*-hydroxyphenyl)-, ethyl ester

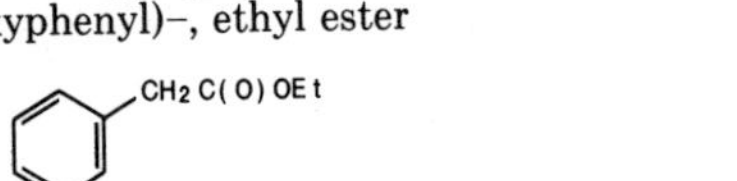

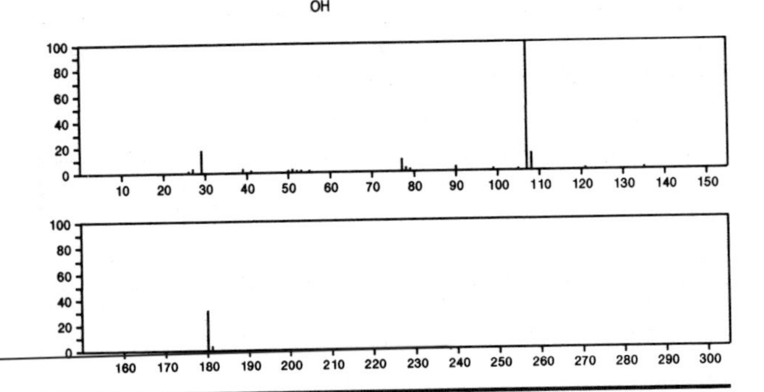

180 $C_{10}H_{12}O_3$ 23786-14-3

Benzeneacetic acid, 4-methoxy-, methyl ester

CH2 C(O) OMe

MeO

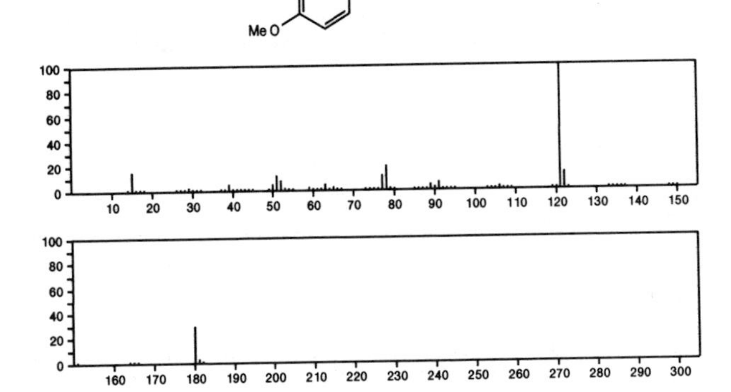

180 $C_{10}H_{12}O_3$ 27798-60-3

Benzeneacetic acid, 2-methoxy-, methyl ester

CH2 C(O) OMe

OMe

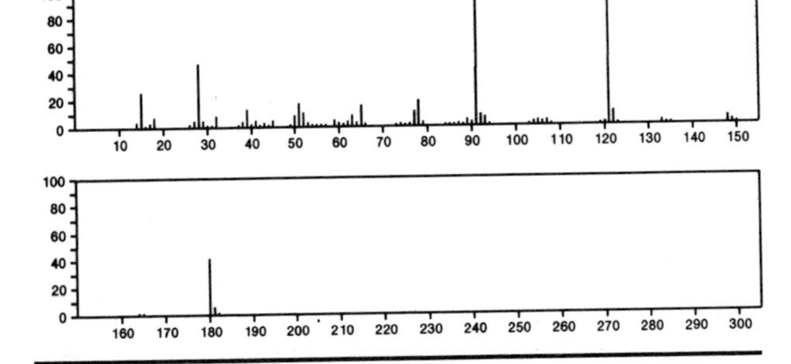

180 $C_{10}H_{12}O_3$ 31268-81-2

Carbonic acid, methyl 3,4-xylyl ester

Me OC(O) OMe

Me

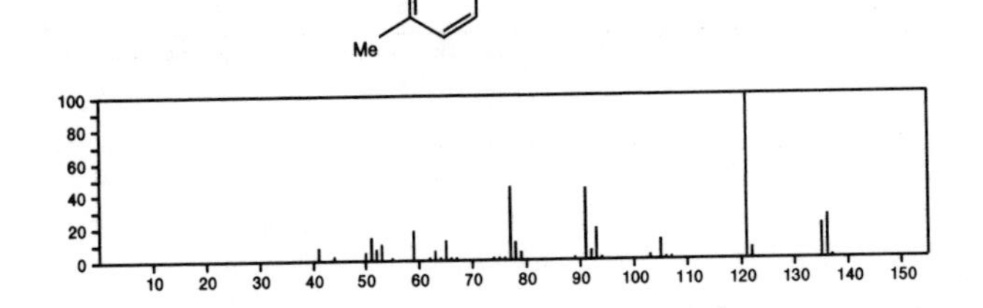

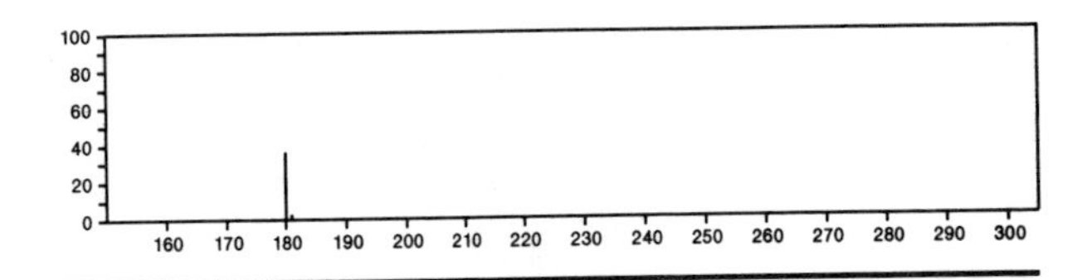

180 $C_{10}H_{12}O_3$ 34883-12-0

Benzaldehyde, 6-hydroxy-4-methoxy-2,3-dimethyl-

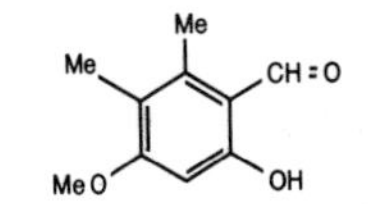

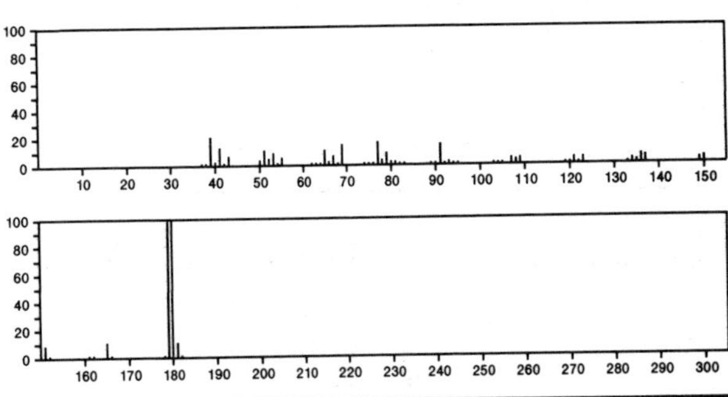

180 $C_{10}H_{12}O_3$ 34883-15-3

Benzaldehyde, 2-hydroxy-4-methoxy-3,6-dimethyl-

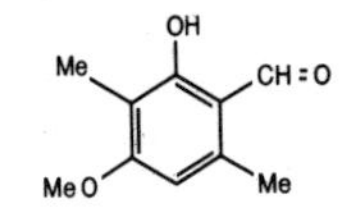

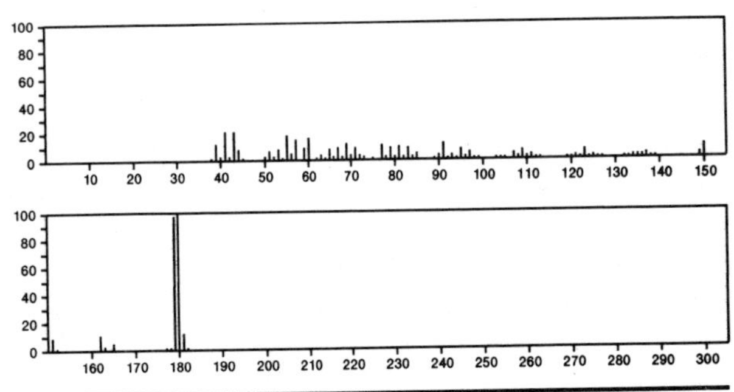

180 $C_{10}H_{12}O_3$ 34949-12-7

Carbonic acid, 2,6-dimethylphenyl methyl ester

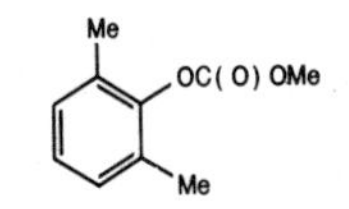

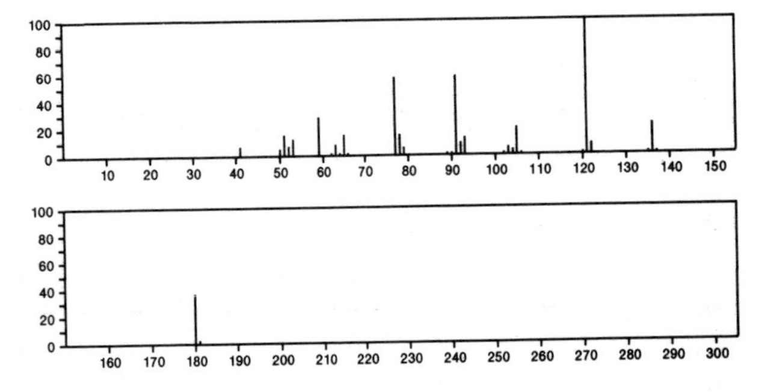

180 $C_{10}H_{12}O_3$ 35942-12-2
4*H*-1-Benzopyran-4-one, 5,6,7,8-tetrahydro-3-hydroxy-2-methyl-

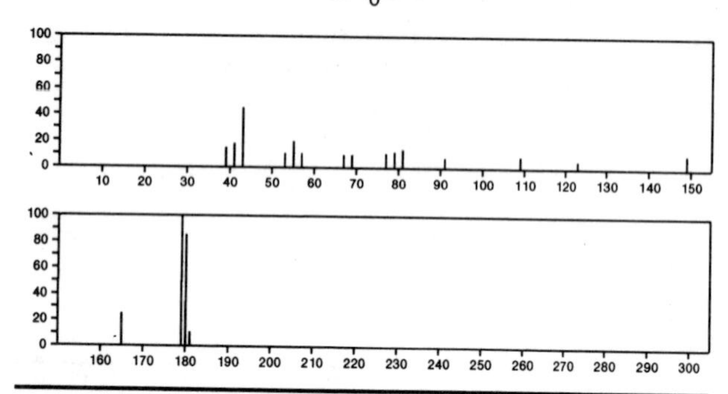

180 $C_{10}H_{12}O_3$ 35973-49-0
1,3-Isobenzofurandione, 3a,4,7,7a-tetrahydro-5,6-dimethyl-, *trans*-

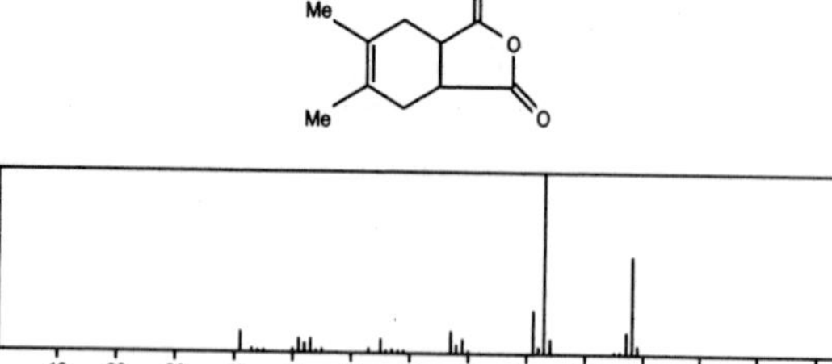

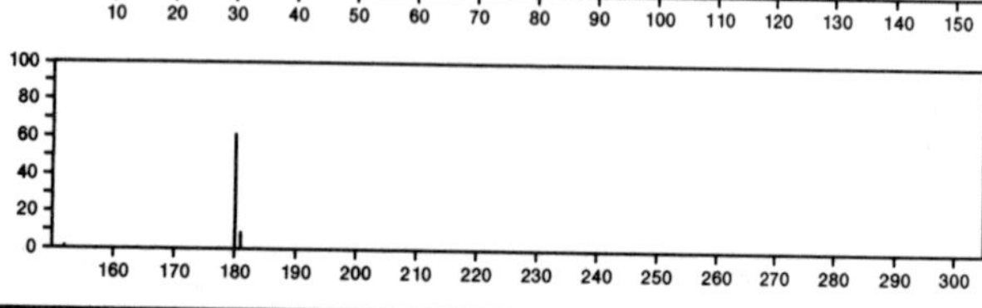

180 $C_{10}H_{12}O_3$ 41873-65-8
Benzeneacetic acid, 2-hydroxy-, ethyl ester

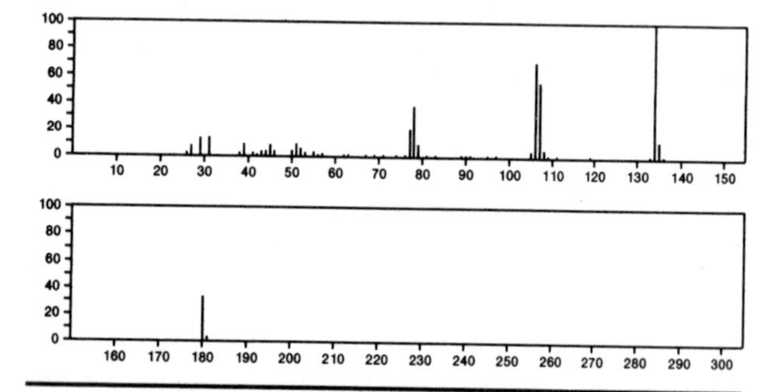

180 $C_{10}H_{12}O_3$ 51835-44-0
Ethanol, 1-methoxy-, benzoate

MeCH(OMe)OC(O)Ph

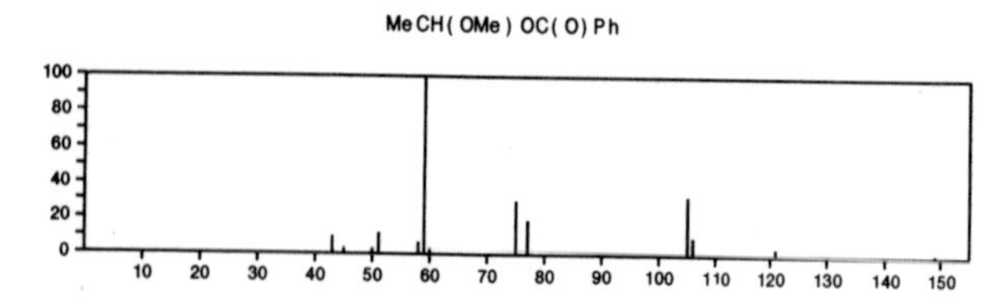

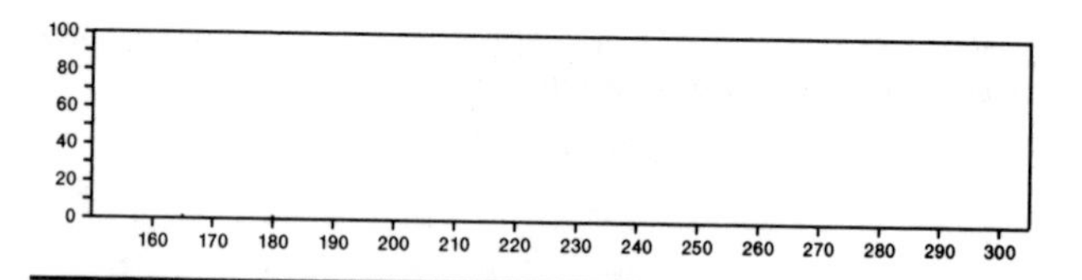

180 $C_{10}H_{12}O_3$ 51835-45-1
Benzenemethanol, α-methoxy-, acetate

AcOCH(OMe)Ph

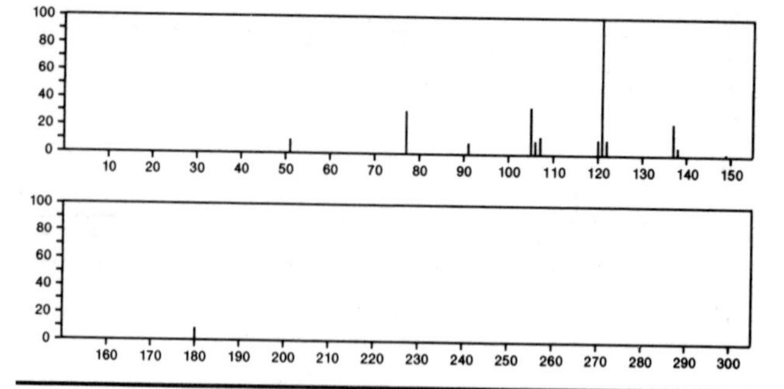

180 $C_{10}H_{12}O_3$ 54576-43-1
1,3-Isobenzofurandione, 4,5,6,7-tetrahydro-4,7-dimethyl-

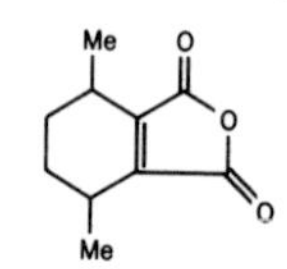

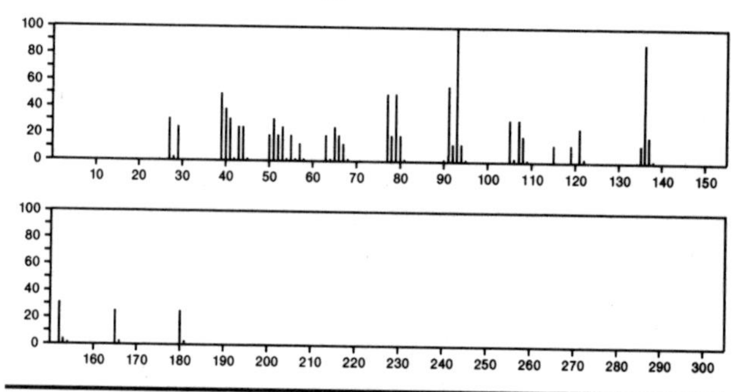

180 $C_{10}H_{12}O_3$ 54644-27-8
1,2,4-Cyclopentanetrione, 3-(2-pentenyl)-

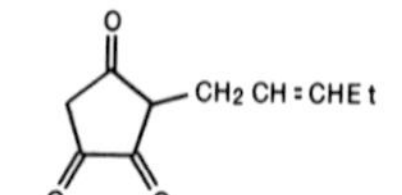

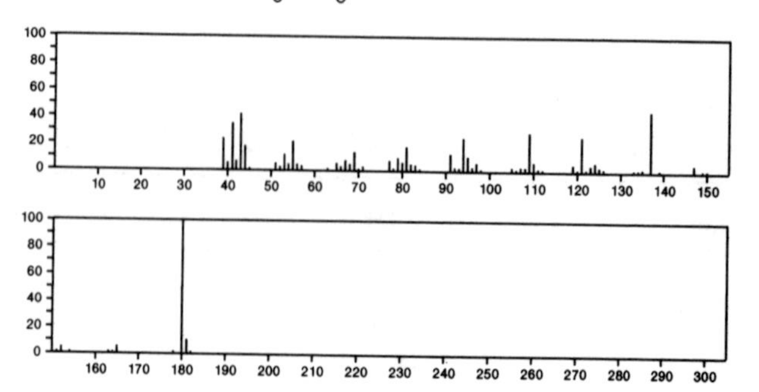

180 $C_{10}H_{12}O_3$ 54644-37-0
Carbonic acid, 2,3-dimethylphenyl methyl ester

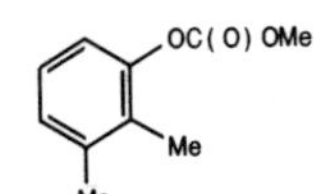

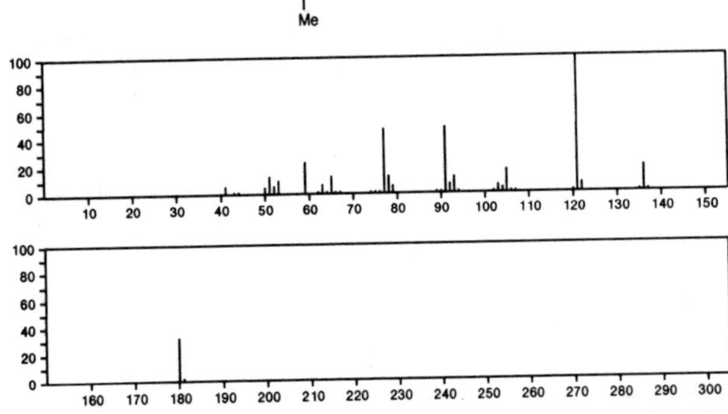

180 $C_{10}H_{12}O_3$ 54966-51-7
2-Butynoic acid, 4-cyclobutyl-4-oxo-, ethyl ester

COC≡CC(O)OEt

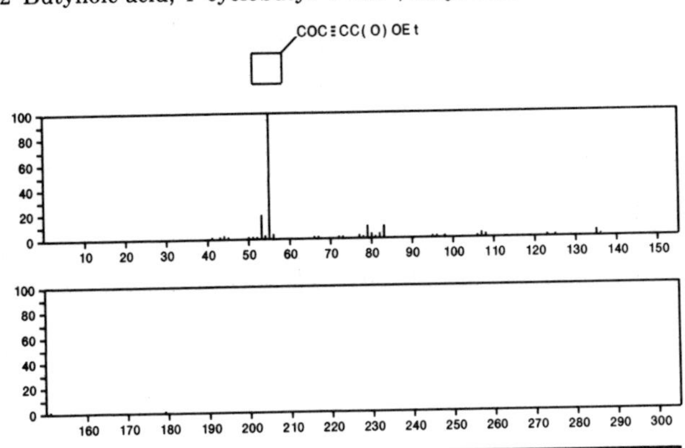

180 $C_{10}H_{12}O_3$ 55836-64-1
Benzoic acid, 2-ethyl-6-hydroxy-, methyl ester

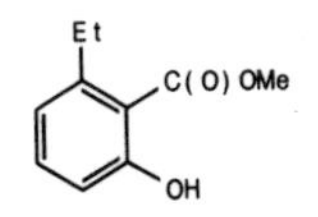

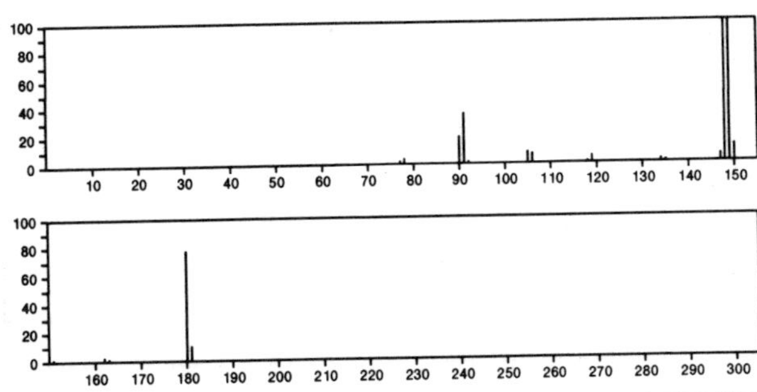

180 $C_{10}H_{12}O_3$ 56143-21-6
Benzeneacetic acid, α-methoxy-, methyl ester, (±)-

MeOC(O)CH(OMe)Ph

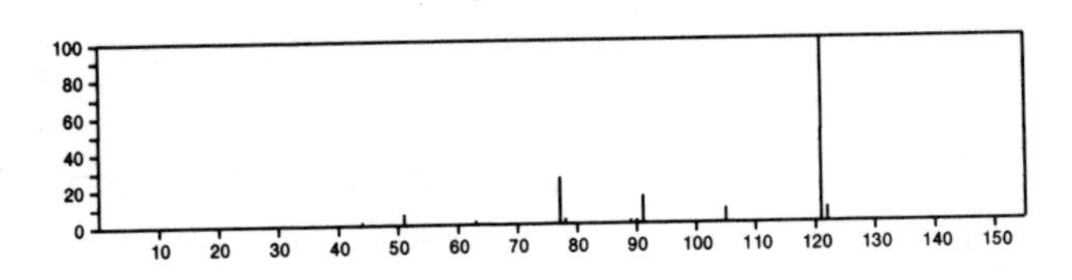

180 $C_{10}H_{12}O_3$ 56781-80-7
Tricyclo[3.3.1.1^{3,7}]decane-2,6-dione, 4-hydroxy-

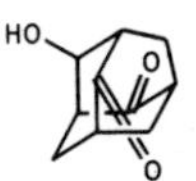

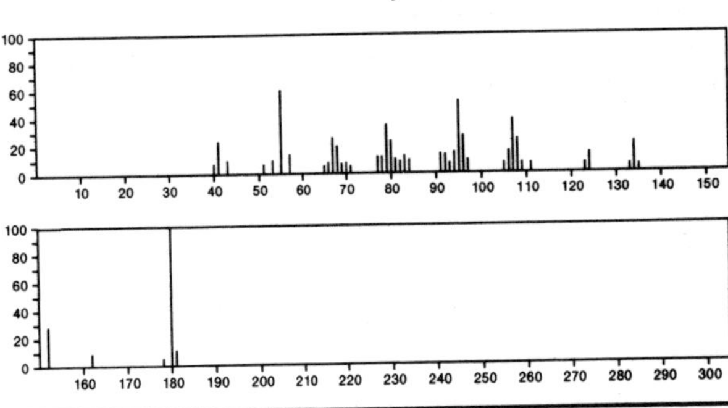

180 $C_{10}H_{16}N_2O$ 52196-11-9
1-Piperidinecarboxaldehyde, 2-(3,4-dihydro-2*H*-pyrrol-5-yl)-

CH=O

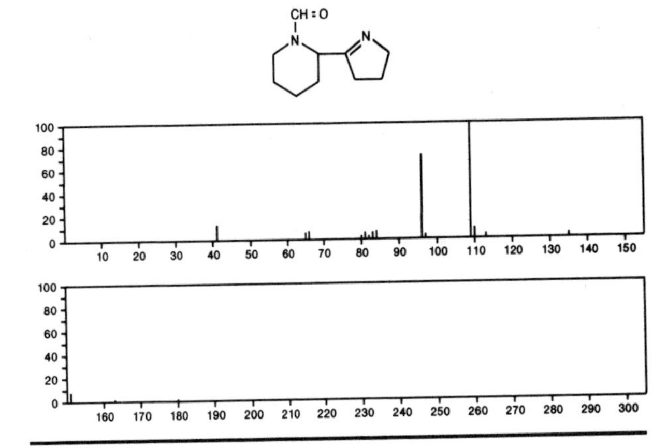

180 $C_{10}H_{16}OSi$ 877-68-9
Silane, (4-methoxyphenyl)trimethyl-

OMe
Me_3Si

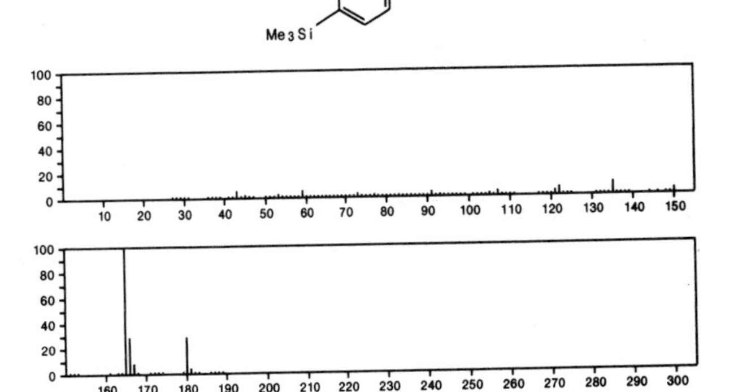

180 $C_{10}H_{16}OSi$ 1009-02-5
Silane, trimethyl(2-methylphenoxy)-

$OSiMe_3$
Me

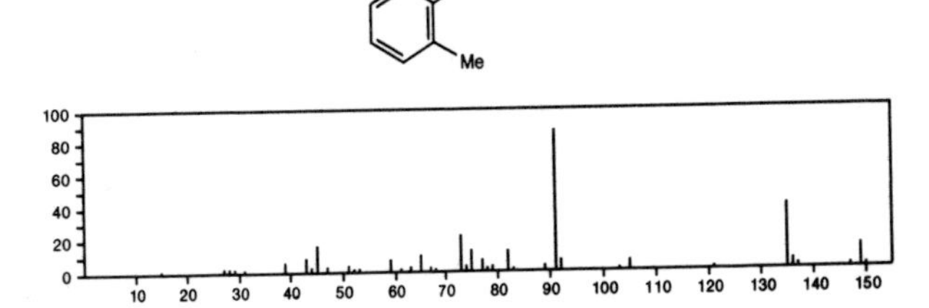

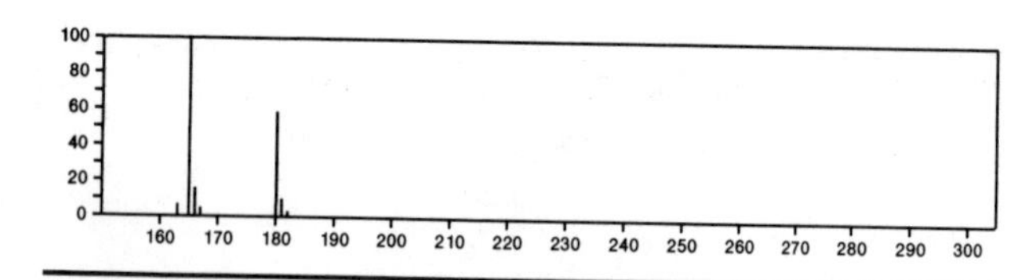

180 $C_{10}H_{16}OSi$ 14642–79–6
Silane, trimethyl(phenylmethoxy)–

Me_3SiOCH_2Ph

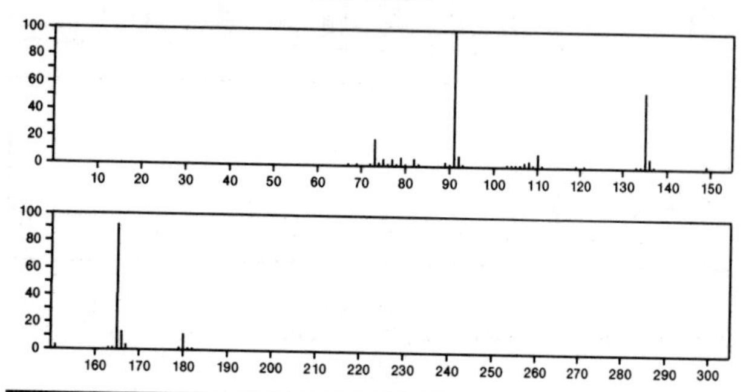

180 $C_{10}H_{16}OSi$ 17902–31–7
Silane, trimethyl(3–methylphenoxy)–

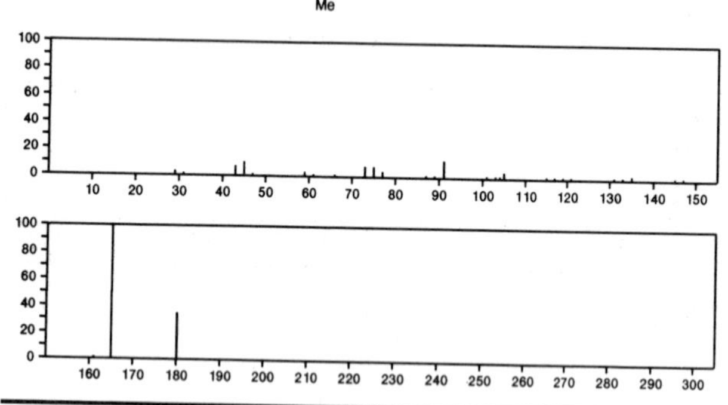

180 $C_{10}H_{16}OSi$ 17902–32–8
Silane, trimethyl(4–methylphenoxy)–

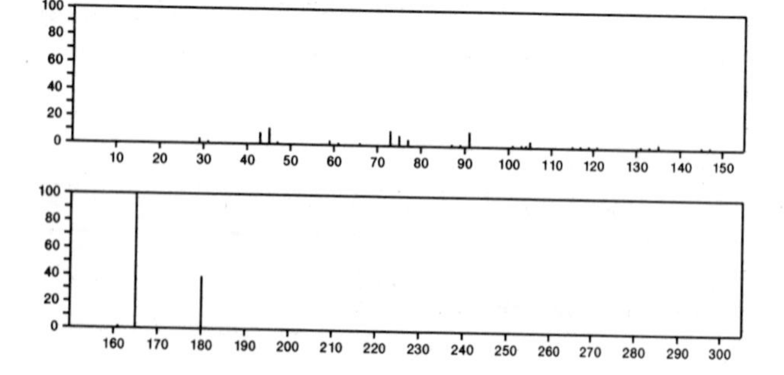

180 $C_{10}H_{17}BO_2$ 26600–82–8
1–Butaneboronic acid, cyclic 1–cyclohexen–1,2–ylene ester

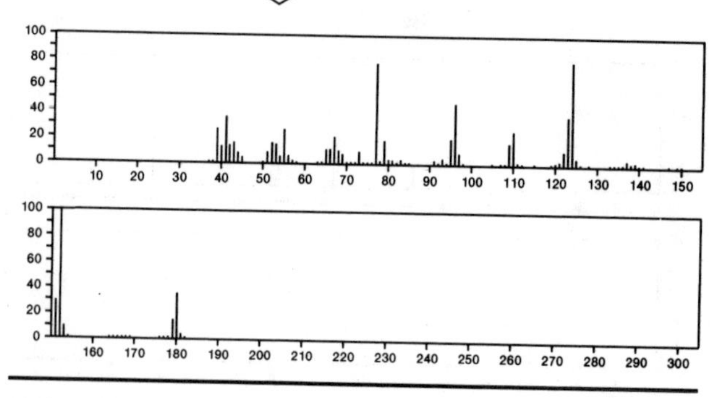

180 $C_{11}H_{16}O_2$ 1202–10–4
4H–Pyran–4–one, 2,6–diethyl–3,5–dimethyl–

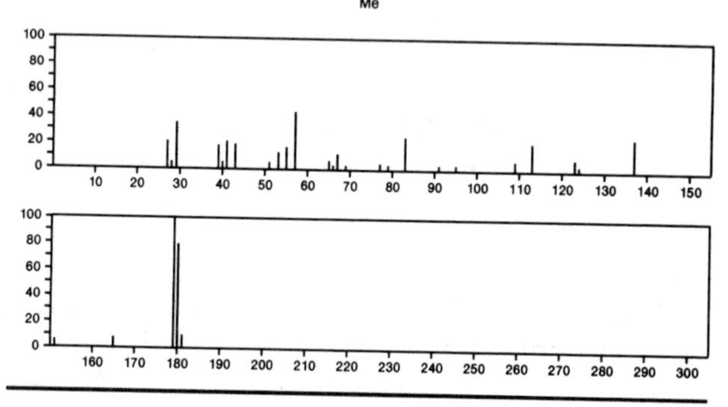

180 $C_{11}H_{16}O_2$ 4541–14–4
1–Butanol, 4–(phenylmethoxy)–

$PhCH_2O(CH_2)_4OH$

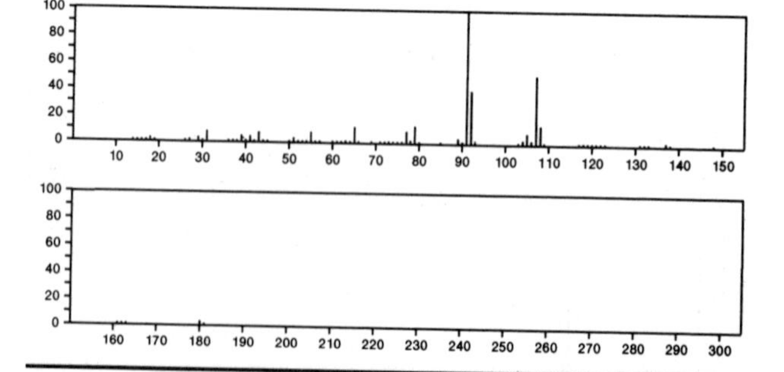

180 $C_{11}H_{16}O_2$ 4799–69–3
2–Butanol, 4–(benzyloxy)–

$MeCH(OH)CH_2CH_2OCH_2Ph$

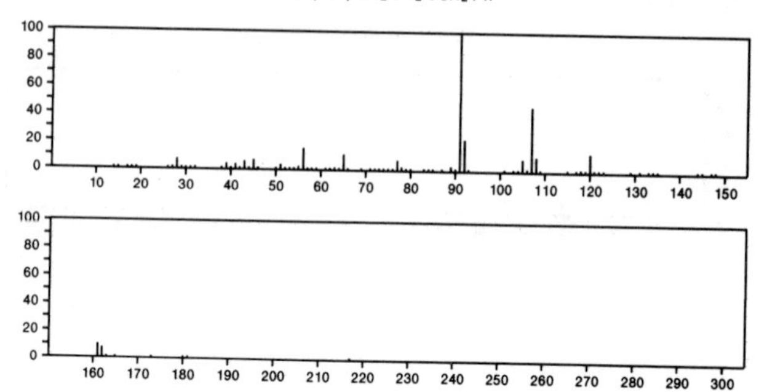

180 $C_{11}H_{16}O_2$ 5107-69-7
Methane, *sec*-butoxyphenoxy-

PhOCH2 OBu-s

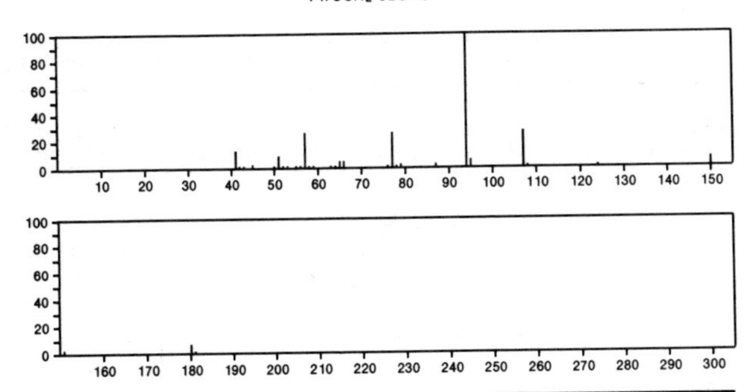

180 $C_{11}H_{16}O_2$ 14869-00-2
2-Butanol, 1-(phenylmethoxy)-

PhCH2 OCH2 CH(OH)Et

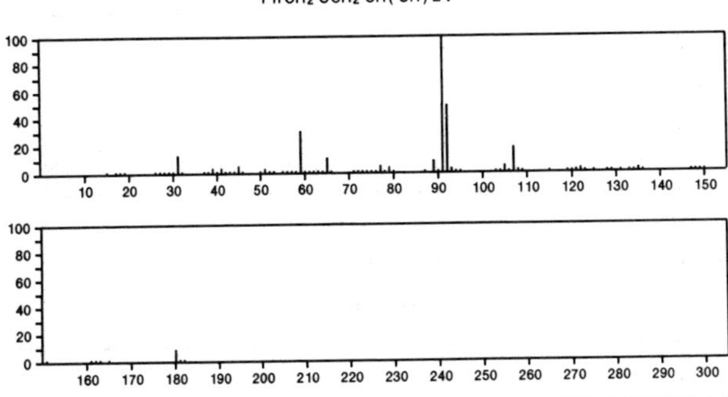

180 $C_{11}H_{16}O_2$ 15356-74-8
2(4*H*)-Benzofuranone, 5,6,7,7a-tetrahydro-4,4,7a-trimethyl-

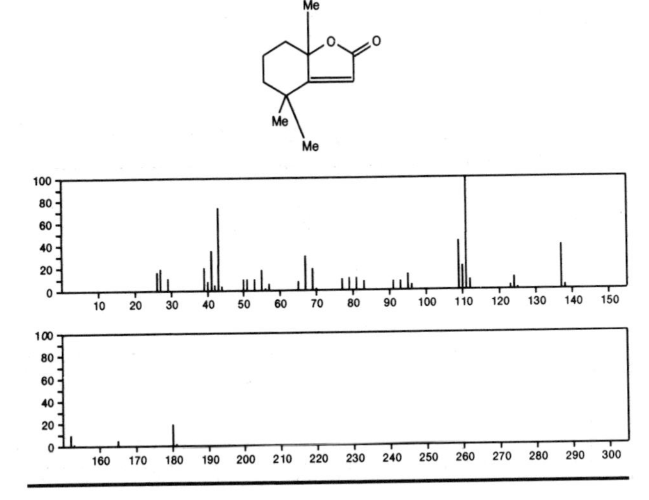

180 $C_{11}H_{16}O_2$ 16778-26-0
2(3*H*)-Benzofuranone, 3a,4,5,6-tetrahydro-3a,6,6-trimethyl-

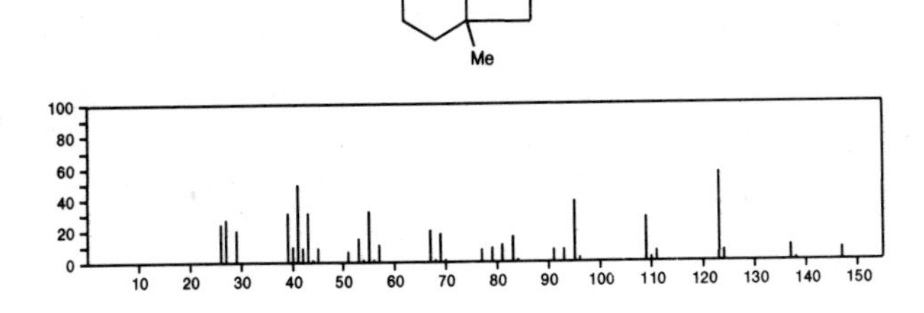

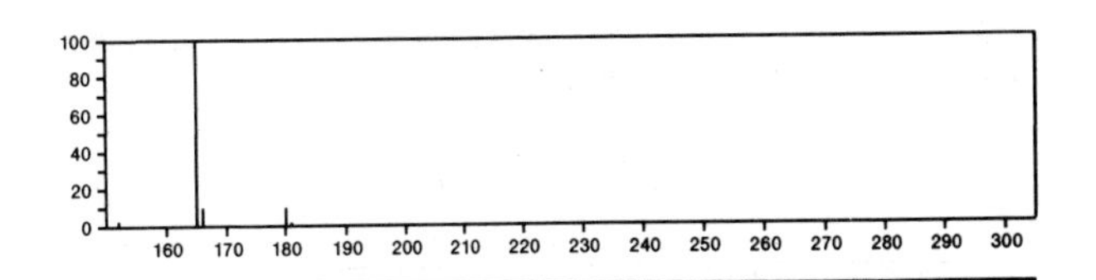

180 $C_{11}H_{16}O_2$ 17092-92-1
2(4*H*)-Benzofuranone, 5,6,7,7a-tetrahydro-4,4,7a-trimethyl-, (*S*)-

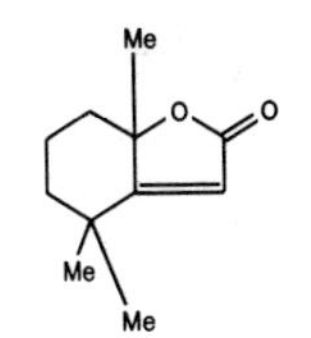

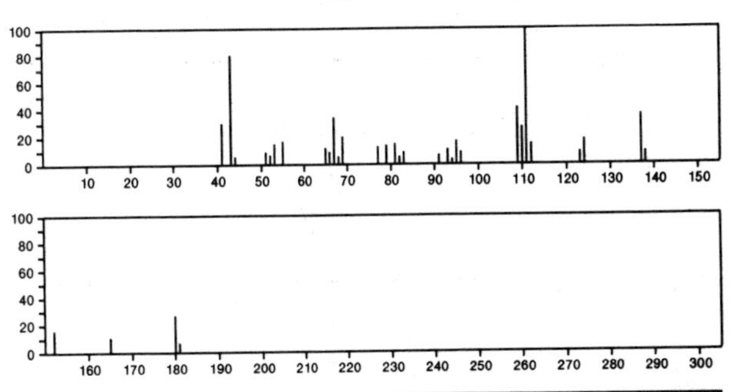

180 $C_{11}H_{16}O_2$ 22054-39-3
2-Cyclopenten-1-one, 4-hydroxy-3-methyl-2-(2-pentenyl)-

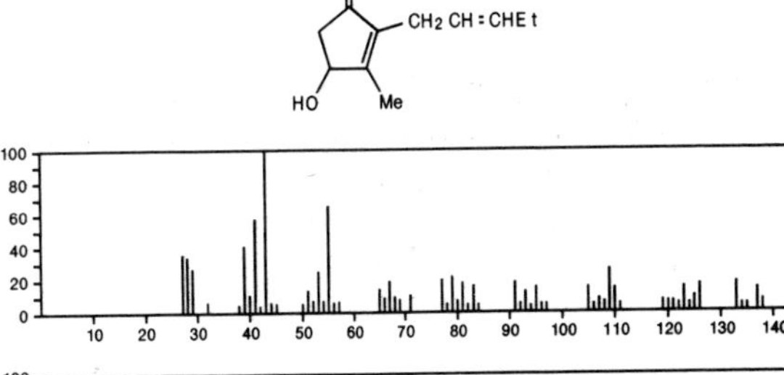

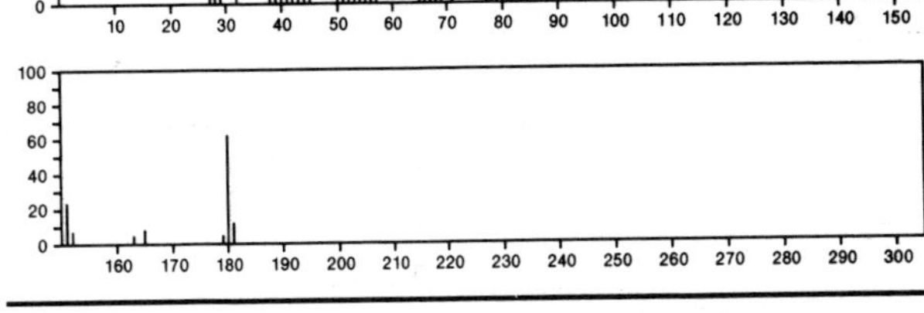

180 $C_{11}H_{16}O_2$ 25013-16-5
Phenol, (1,1-dimethylethyl)-4-methoxy-

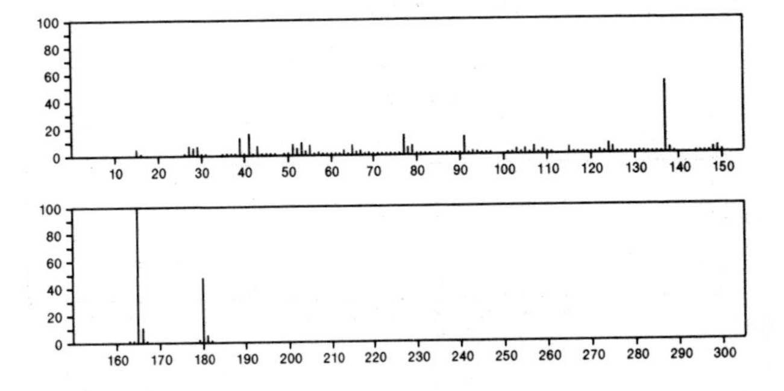

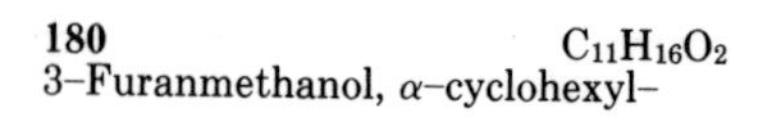

180 $C_{11}H_{16}O_2$ 36646-66-9
3-Furanmethanol, α-cyclohexyl-

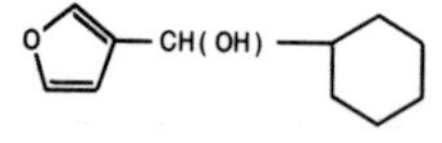

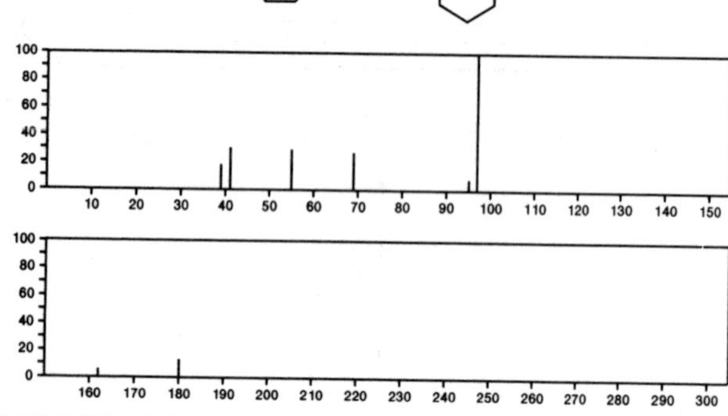

180 $C_{11}H_{16}O_2$ 41398-33-8
Tricyclo[3.3.1.1^{3,7}]decanone, 4-methoxy-, (1α,3β,4α,5α,7β)-

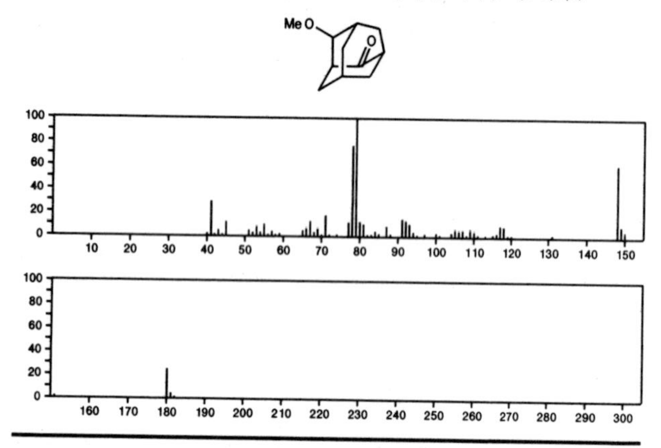

180 $C_{11}H_{16}O_2$ 49833-95-6
1-Propen-2-ol, 3-cyclohexylidene-, acetate

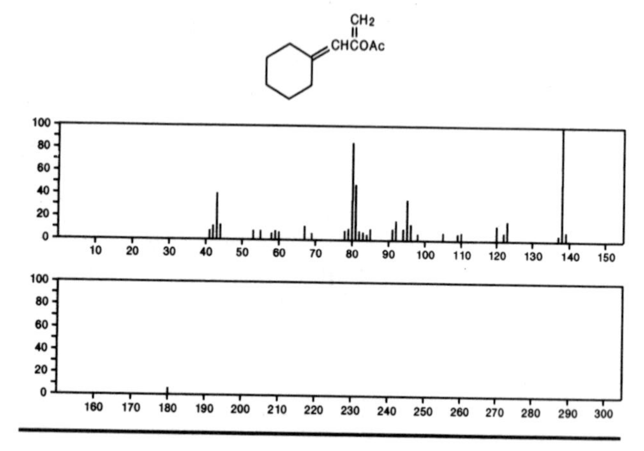

180 $C_{11}H_{16}O_2$ 54576-35-1
Ethanol, 2-[4-(1-methylethyl)phenoxy]-

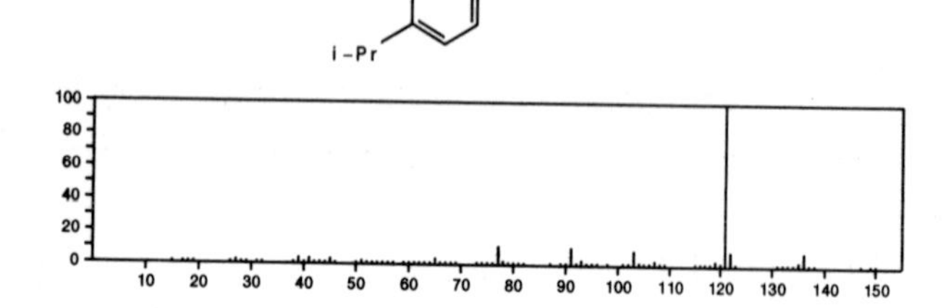

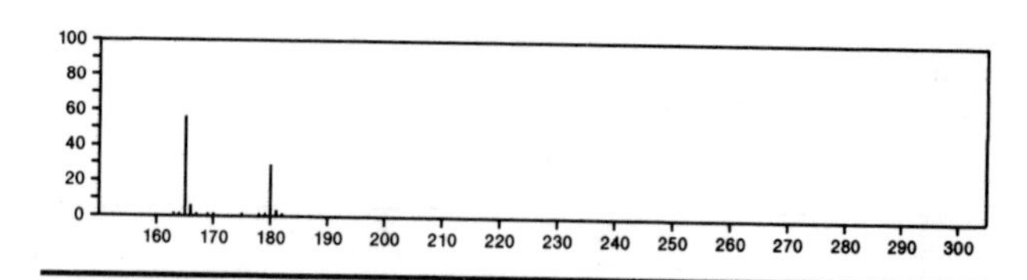

180 $C_{11}H_{16}O_2$ 56781-91-0
Tricyclo[3.3.1.1^{3,7}]decan-2-one, 4-(hydroxymethyl)-, (1α,3β,= 4β,5α,7β)-

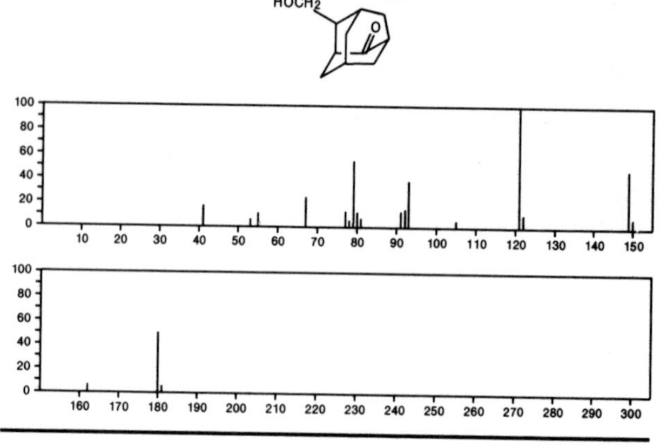

180 $C_{11}H_{16}S$ 7210-80-2
Benzene, [(2,2-dimethylpropyl)thio]-

Me3CCH2SPh

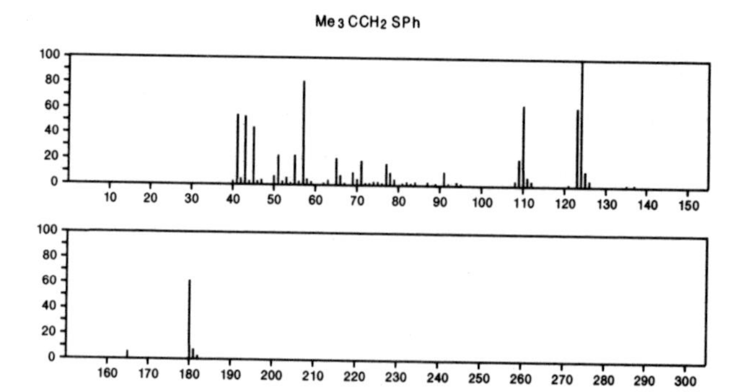

180 $C_{11}H_{16}S$ 7439-10-3
Benzene, 1-[(1,1-dimethylethyl)thio]-4-methyl-

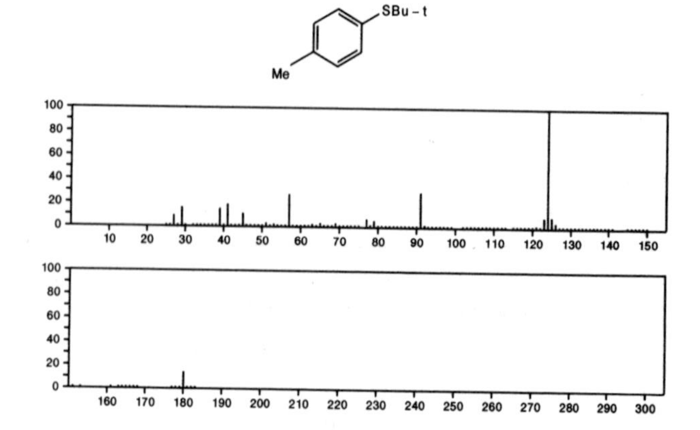

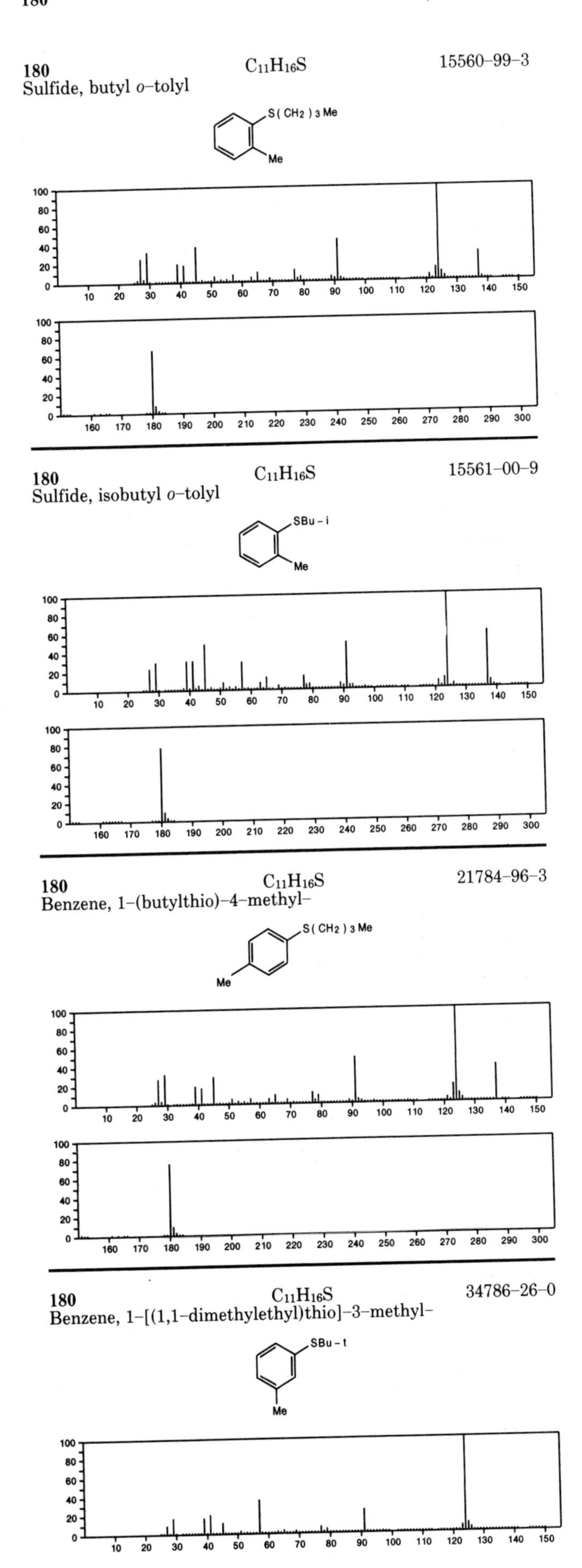

180 $C_{11}H_{16}S$ 15560-99-3
Sulfide, butyl *o*-tolyl

180 $C_{11}H_{16}S$ 15561-00-9
Sulfide, isobutyl *o*-tolyl

180 $C_{11}H_{16}S$ 21784-96-3
Benzene, 1-(butylthio)-4-methyl-

180 $C_{11}H_{16}S$ 34786-26-0
Benzene, 1-[(1,1-dimethylethyl)thio]-3-methyl-

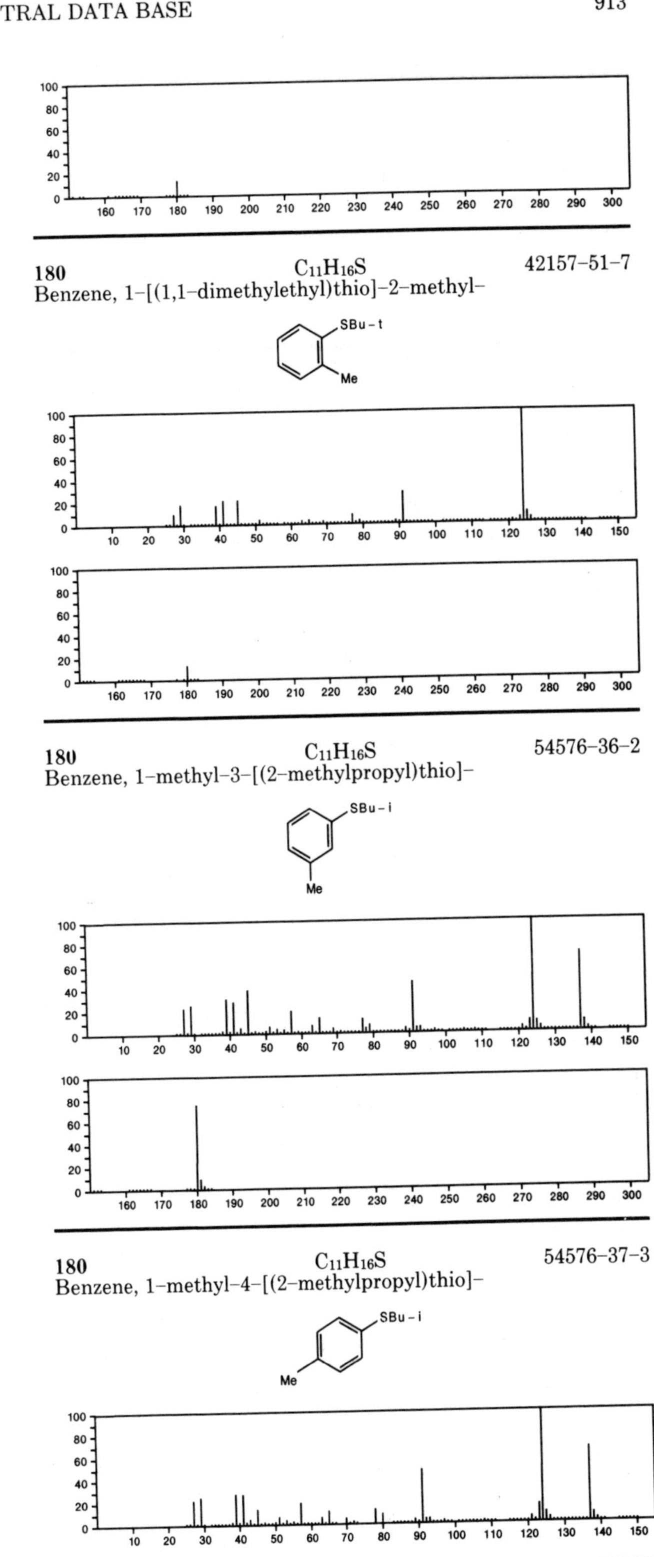

180 $C_{11}H_{16}S$ 42157-51-7
Benzene, 1-[(1,1-dimethylethyl)thio]-2-methyl-

180 $C_{11}H_{16}S$ 54576-36-2
Benzene, 1-methyl-3-[(2-methylpropyl)thio]-

180 $C_{11}H_{16}S$ 54576-37-3
Benzene, 1-methyl-4-[(2-methylpropyl)thio]-

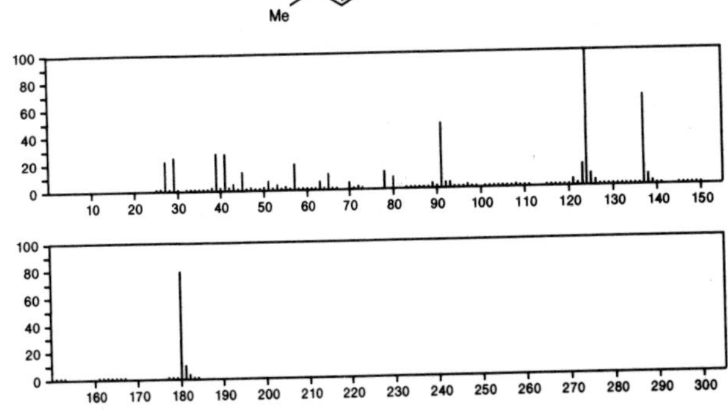

180 $C_{11}H_{16}S$ 54576-38-4
Benzene, 1-methyl-2-[(1-methylpropyl)thio]-

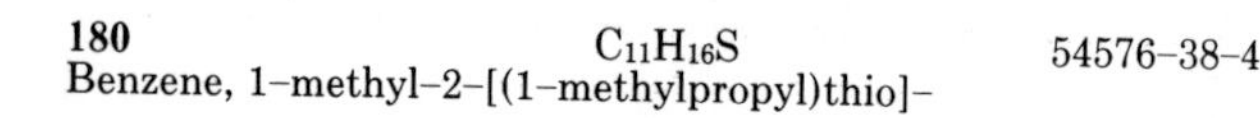

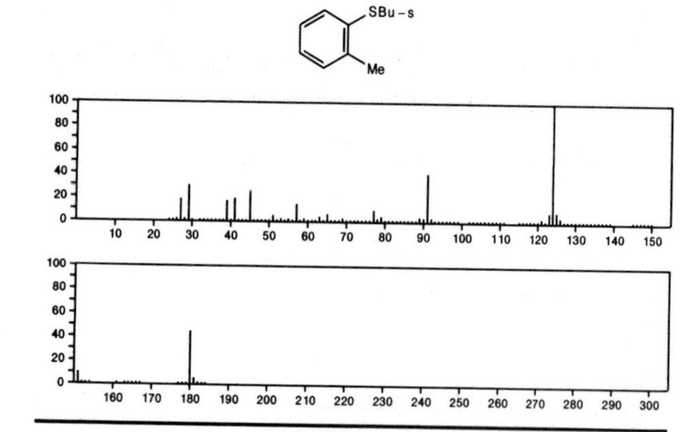

180 $C_{11}H_{16}S$ 54576-39-5
Benzene, 1-methyl-3-[(1-methylpropyl)thio]-

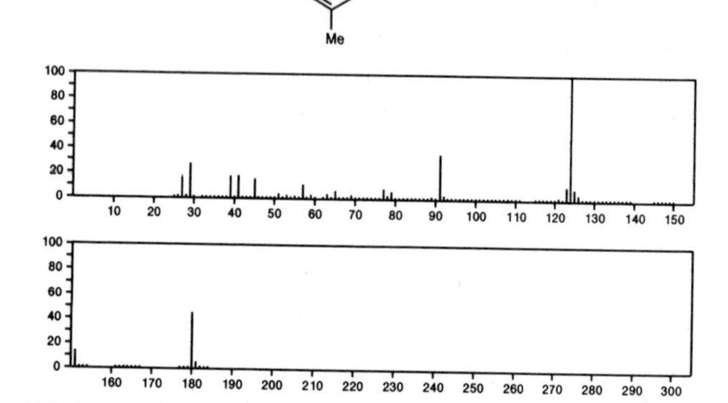

180 $C_{11}H_{16}S$ 54576-40-8
Benzene, 1-methyl-4-[(1-methylpropyl)thio]-

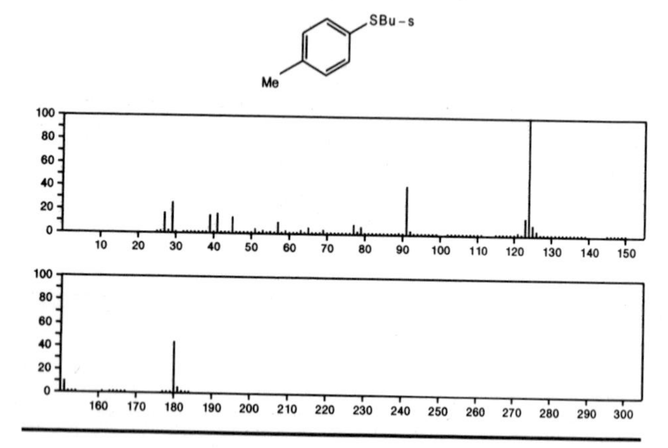

180 $C_{11}H_{16}S$ 54576-41-9
Benzene, 1-(butylthio)-3-methyl-

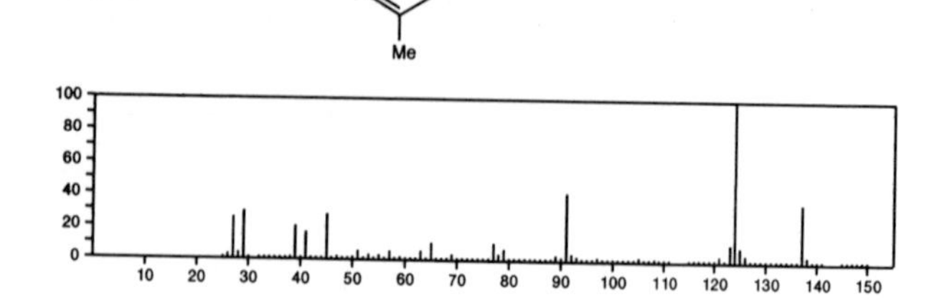

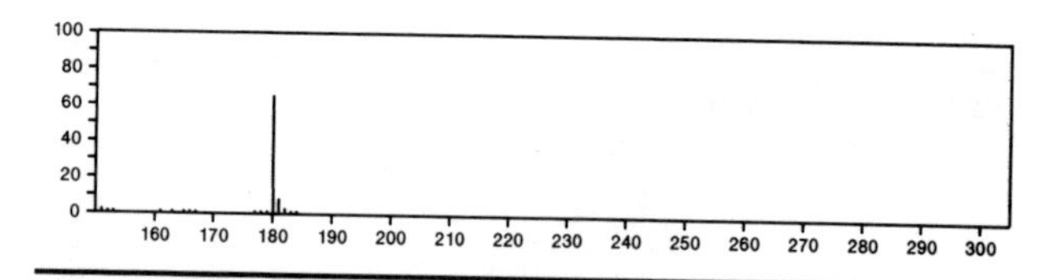

180 $C_{11}H_{16}S$ 54576-42-0
Benzene, [2-[(1-methylethyl)thio]ethyl]-

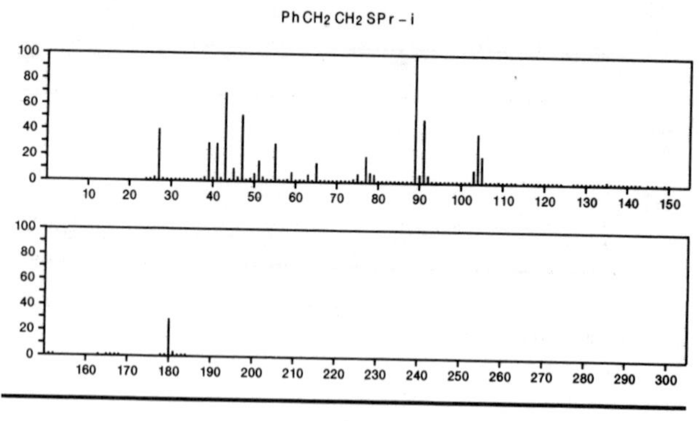

180 $C_{12}H_8N_2$ 230-17-1
Benzo[*c*]cinnoline

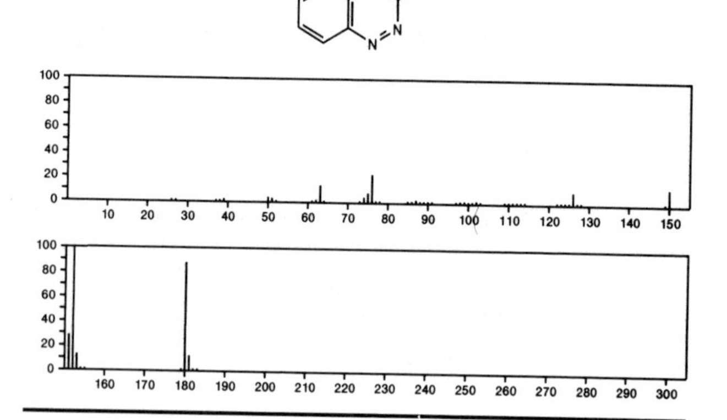

180 $C_{12}H_8N_2$ 50559-45-0
Pyridine, 3,3'-(1,2-ethynediyl)bis-

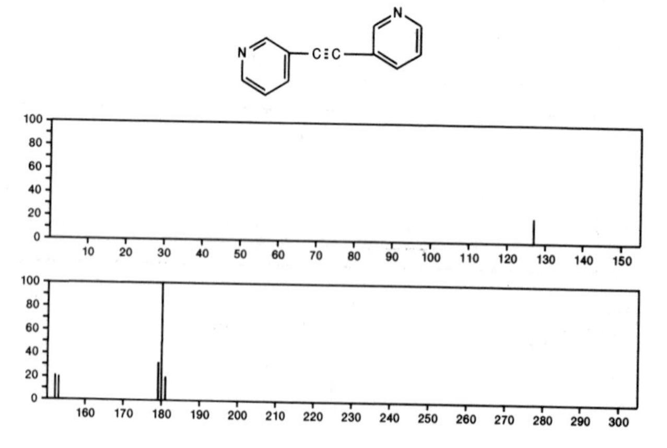

180 $C_{12}H_{20}O$ 90-42-6
[1,1'-Bicyclohexyl]-2-one

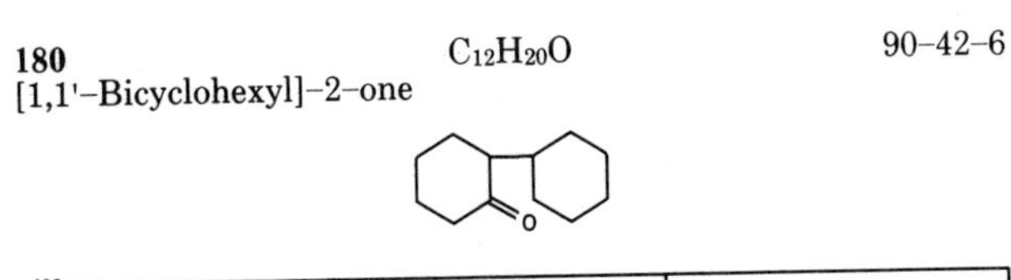

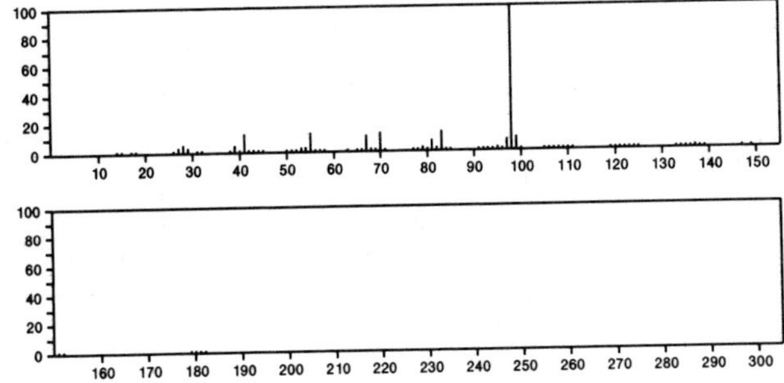

180 $C_{12}H_{20}O$ 92-68-2
[1,1'-Bicyclohexyl]-4-one

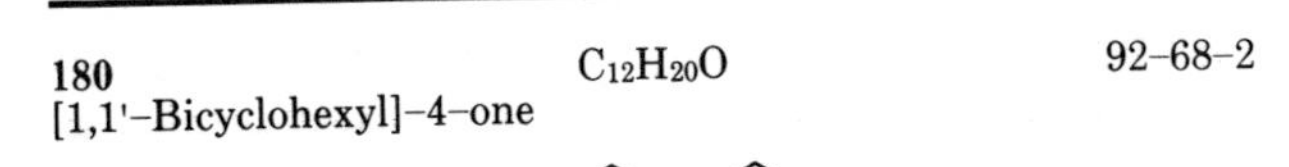

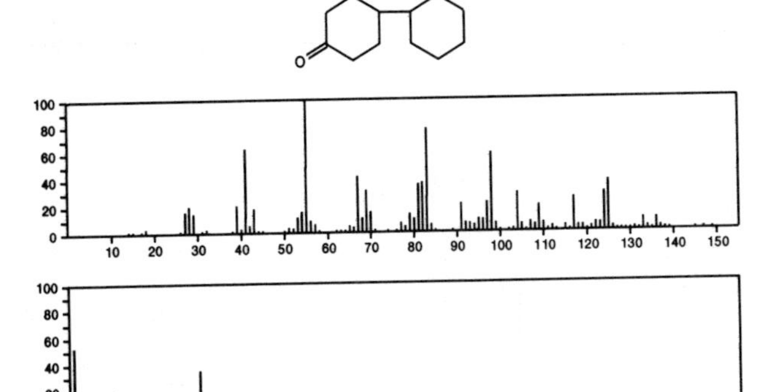

180 $C_{12}H_{20}O$ 941-17-3
2(1*H*)-Naphthalenone, octahydro-8,8a-dimethyl-, (4aα,8β,8aβ)-

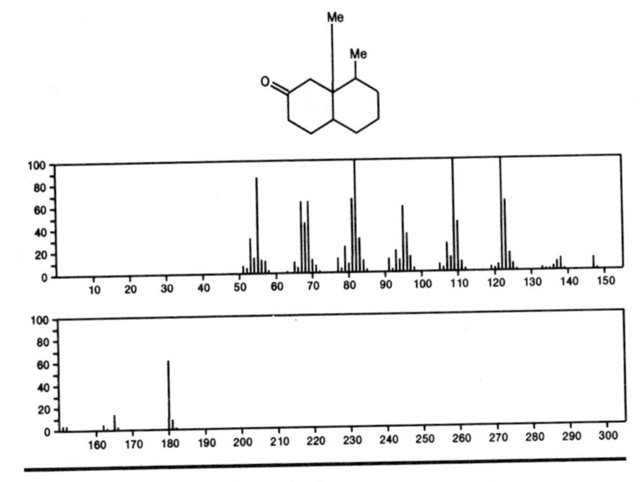

180 $C_{12}H_{20}O$ 941-19-5
2(1*H*)-Naphthalenone, octahydro-4,4a-dimethyl-, (4α,4aα,8aβ)-

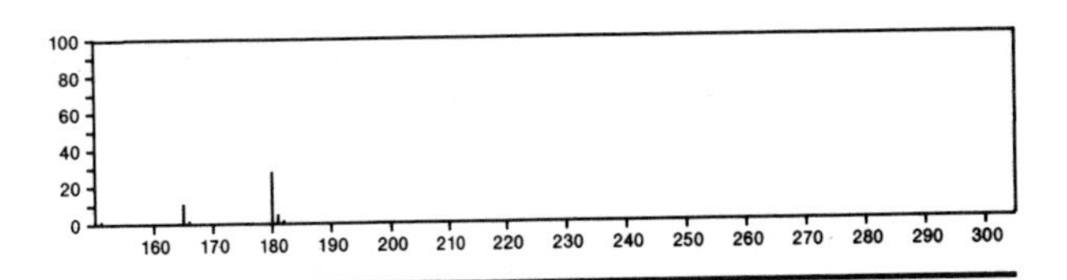

180 $C_{12}H_{20}O$ 941-20-8
1(2*H*)-Naphthalenone, octahydro-3,8a-dimethyl-, (3α,4aβ,8aα)-

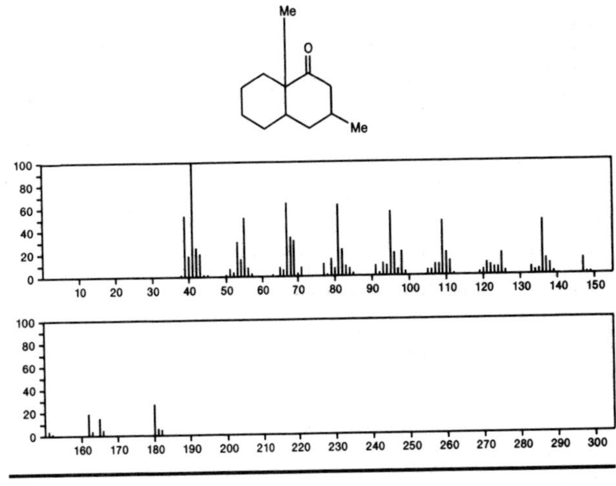

180 $C_{12}H_{20}O$ 4789-40-6
Furan, 2,5-bis(1,1-dimethylethyl)-

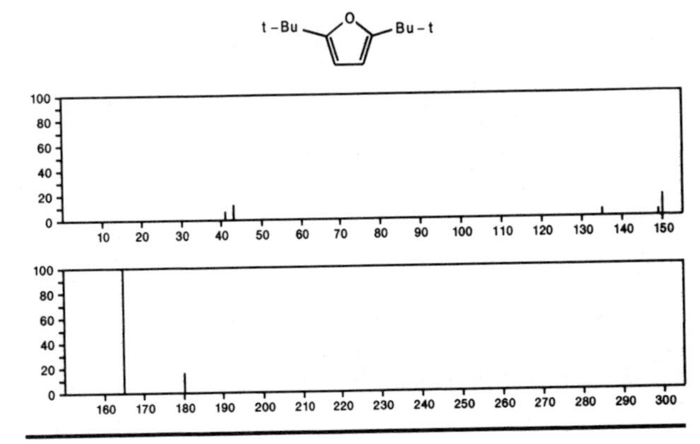

180 $C_{12}H_{20}O$ 4808-01-9
2-Propen-1-ol, 3-(2,6,6-trimethyl-1-cyclohexen-1-yl)-

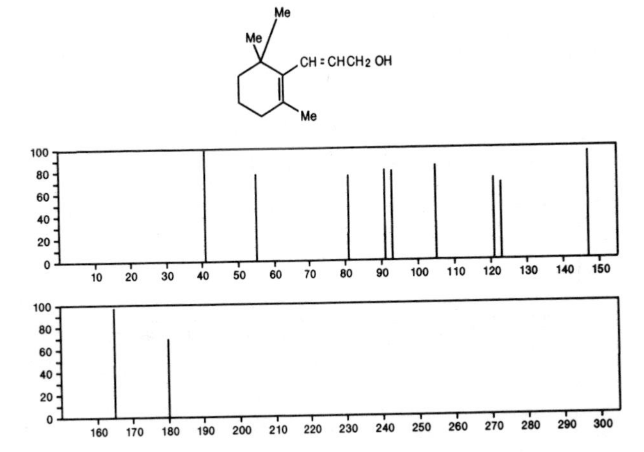

180 $C_{12}H_{20}O$ 16510–55–7
Ethanone, 1–(octahydro–7a–methyl–1*H*–inden–1–yl)–, (1α,3aβ,7aα)–

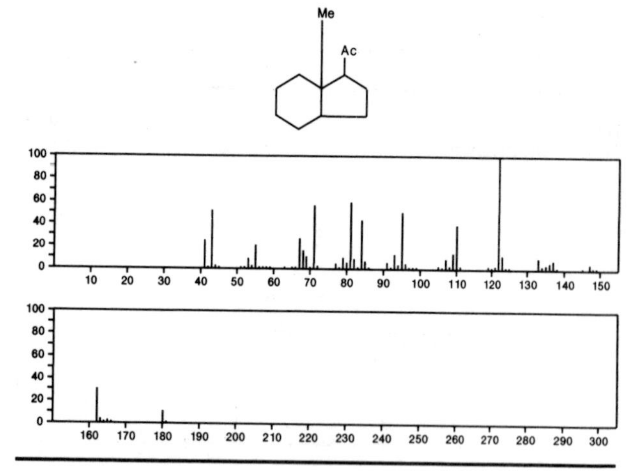

180 $C_{12}H_{20}O$ 16510–56–8
Cyclobut[*c*]inden–2–ol, decahydro–2–methyl–

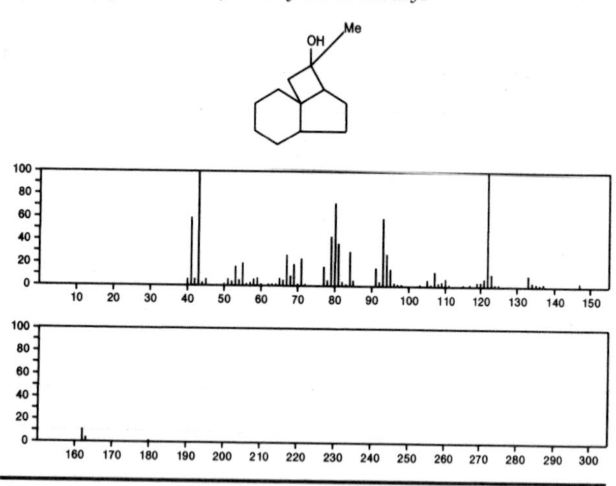

180 $C_{12}H_{20}O$ 17986–87–7
Ketone, 3aβ,4,5,6,7,7a–hexahydro–7aβ–methyl–1β–indanyl methyl

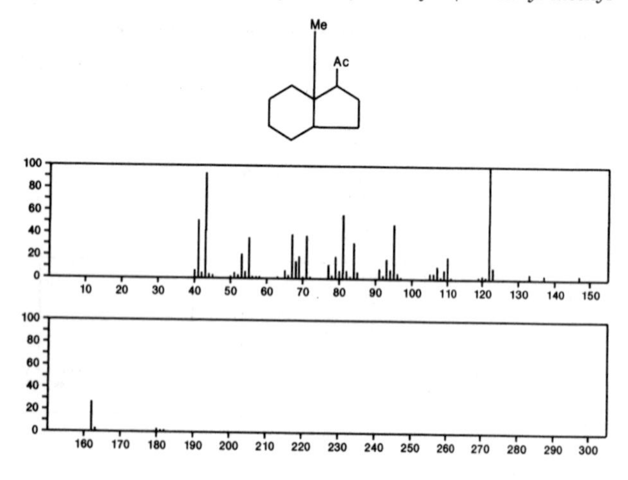

180 $C_{12}H_{20}O$ 17986–96–8
Ethanone, 1–(octahydro–7a–methyl–1*H*–inden–1–yl)–, (1α,3aα,7aβ)–

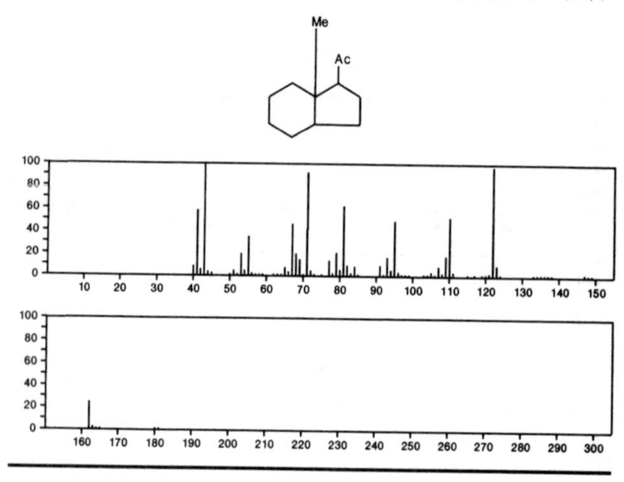

180 $C_{12}H_{20}O$ 17986–97–9
Ketone, 3aβ,4,5,6,7,7a–hexahydro–7aβ–methyl–1α–indanyl methyl

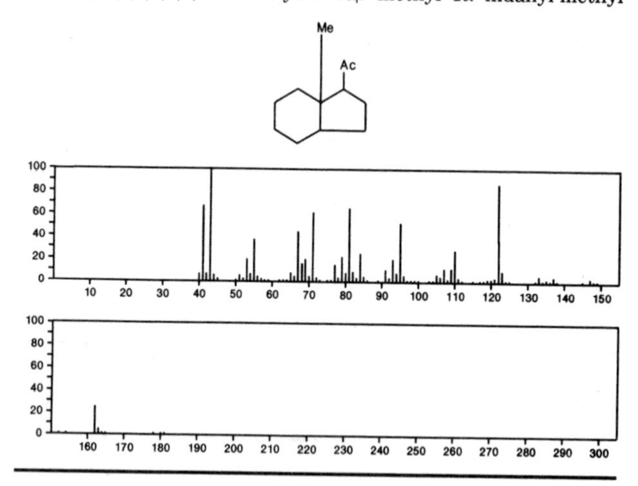

180 $C_{12}H_{20}O$ 19377–97–0
4,5–Octadien–3–one, 2,2,7,7–tetramethyl–

$Me_3CCH=C=CHCOCMe_3$

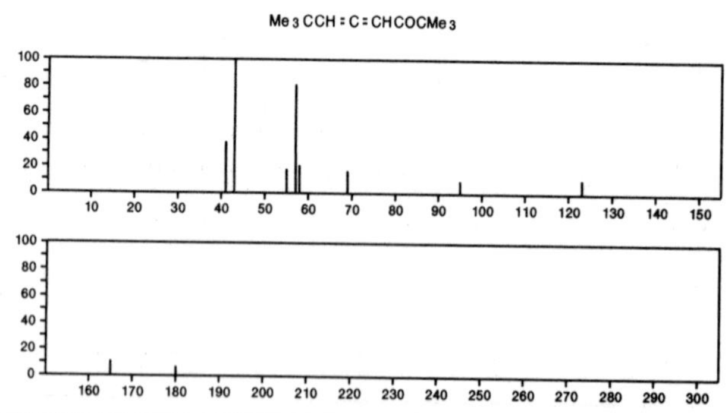

180 $C_{12}H_{20}O$ 19576–21–7
1–Propanone, 1–[2–(1,1–dimethylethyl)–2–cyclopropen–1–yl]–2,2–dimethyl–

t–Bu COCMe3

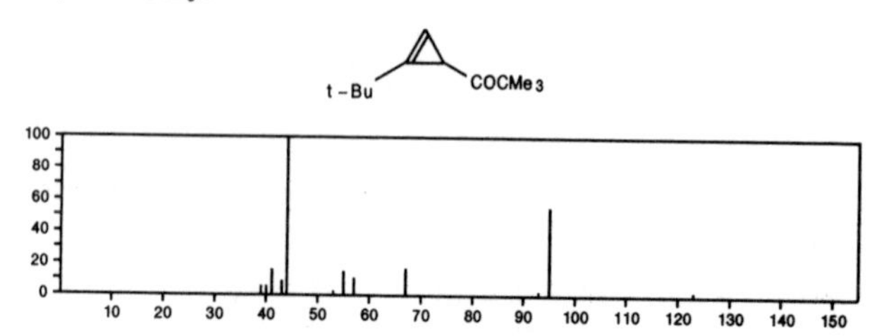

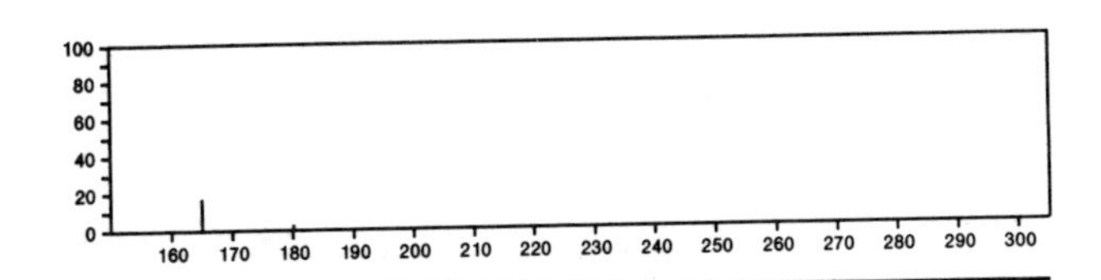

180 $C_{12}H_{20}O$ 21898-92-0
Tricyclo[4.3.1.1^3,8]undecane, 3-methoxy-

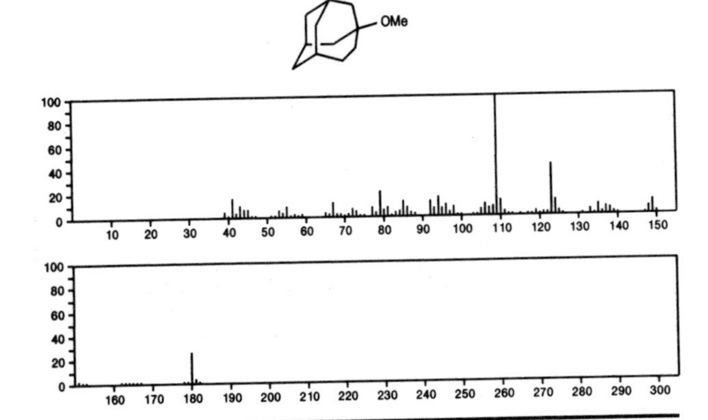

180 $C_{12}H_{20}O$ 21898-95-3
Tricyclo[4.3.1.1^3,8]undecane, 1-methoxy-

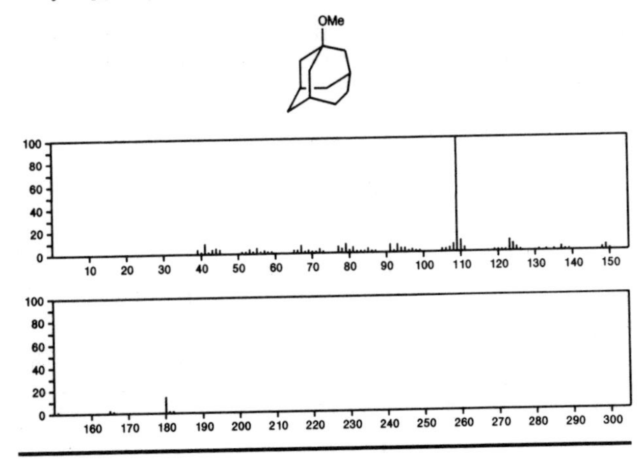

180 $C_{12}H_{20}O$ 22629-28-3
2-Pentanone, 5-(2-methylenecyclohexyl)-, stereoisomer

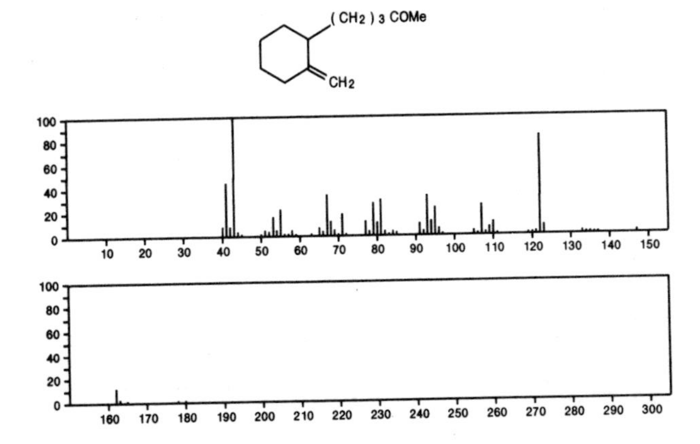

180 $C_{12}H_{20}O$ 22738-31-4
2(1*H*)-Naphthalenone, octahydro-1,4a-dimethyl-, (1α,4aβ,8aα)-

180 $C_{12}H_{20}O$ 28884-89-1
5-Octyn-4-one, 2,2,7,7-tetramethyl-

$Me_3CC{\equiv}CCOCH_2CMe_3$

180 $C_{12}H_{20}O$ 29460-67-1
2-Propen-1-ol, 3-(2,6,6-trimethyl-2-cyclohexen-1-yl)-

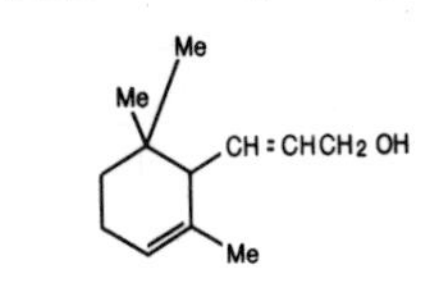

180 $C_{12}H_{20}O$ 37609-41-9
Bicyclo[3.2.1]oct-2-ene, 3-(1,1-dimethylethoxy)-

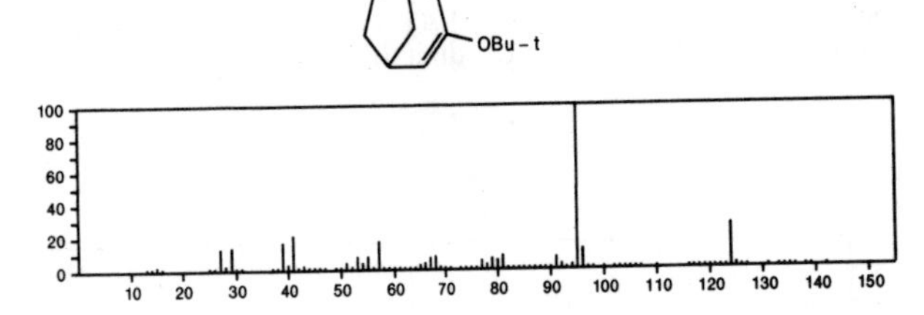

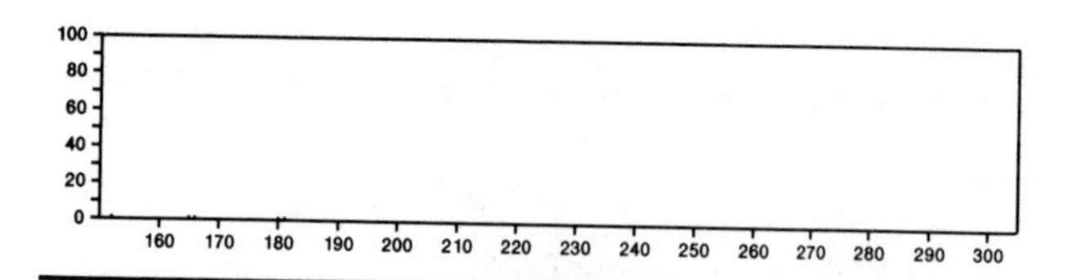

180 $C_{12}H_{20}O$ 37730-45-3
1-Propanone, 1-(1,4-dimethyl-3-cyclohexen-1-yl)-2-methyl-

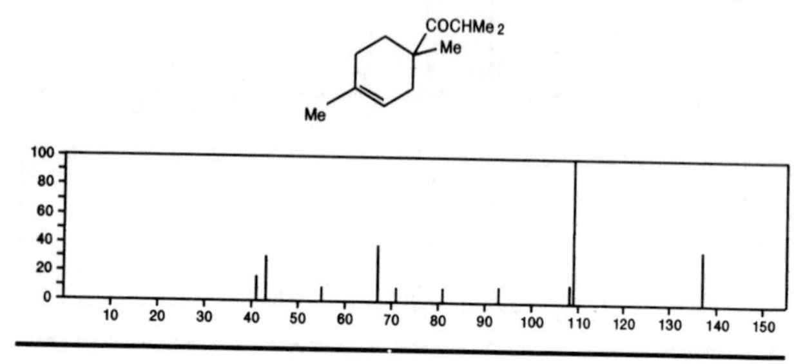

180 $C_{12}H_{20}O$ 38366-85-7
Cyclohexanone, 3-ethylidene-2,2,5,5-tetramethyl-

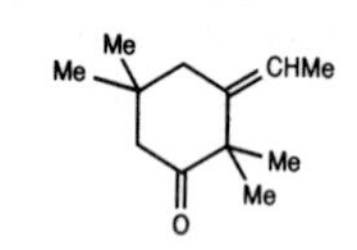

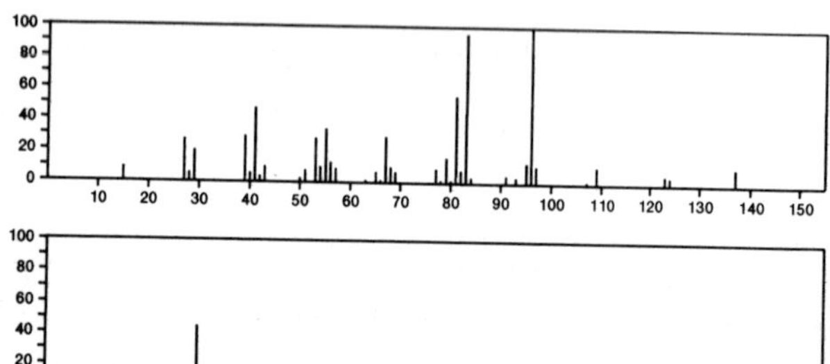

180 $C_{12}H_{20}O$ 38696-32-1
Cyclohexanone, 2,5,5-trimethyl-3-(1-methylethylidene)-

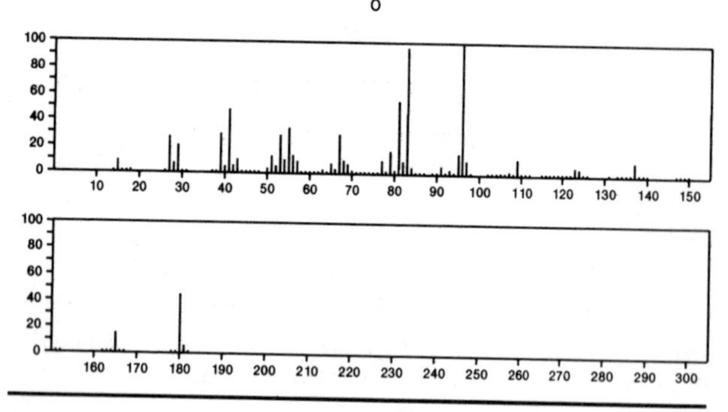

180 $C_{12}H_{20}O$ 49826-51-9
Bicyclo[3.2.1]oct-2-ene, 4-(1,1-dimethylethoxy)-

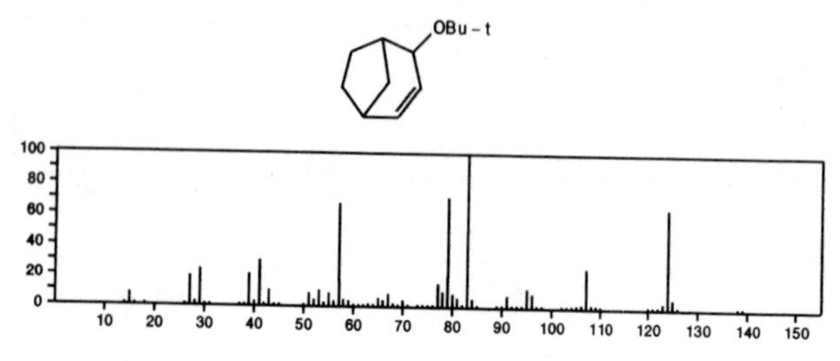

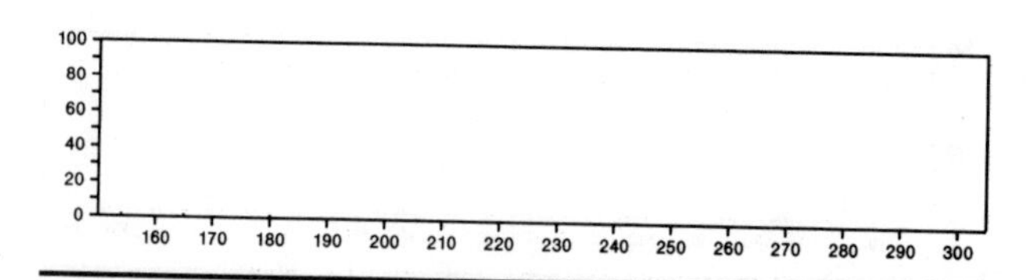

180 $C_{12}H_{20}O$ 51557-64-3
2(1*H*)-Naphthalenone, octahydro-4a,5-dimethyl-, (4aα,5α,8aβ)-

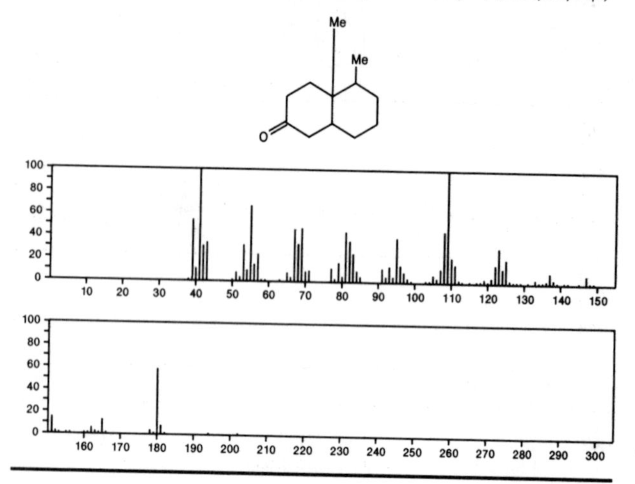

180 $C_{12}H_{20}O$ 51768-87-7
1-Cyclohexene-1-methanol, α-ethenyl-2,6,6-trimethyl-

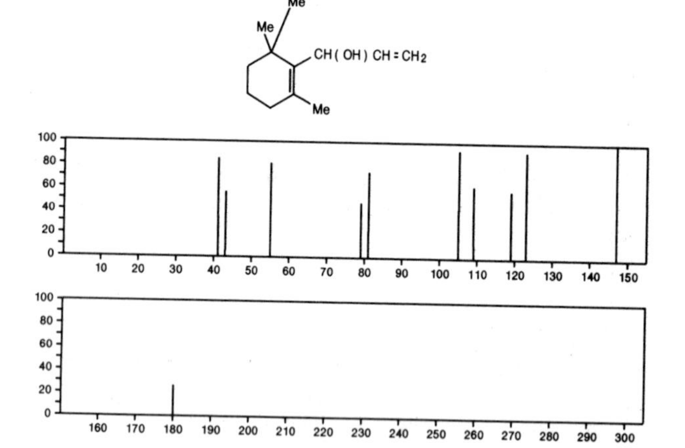

180 $C_{12}H_{20}O$ 54764-59-9
1-Pentanone, 1-bicyclo[4.1.0]hept-7-yl-

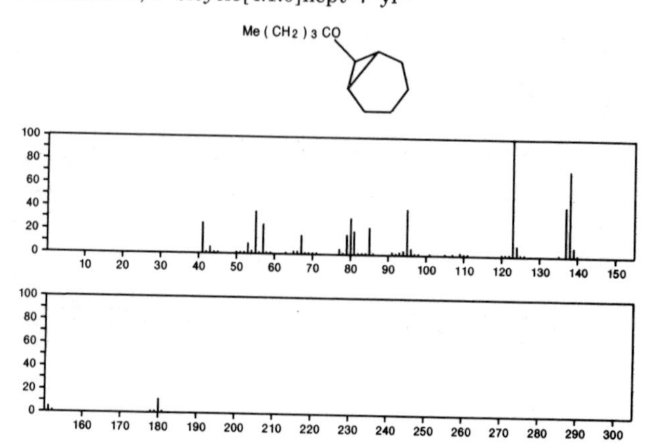

180 $C_{12}H_{20}O$ 55103–64–5
2*H*–Benzocyclohepten–2–one, decahydro–4a–methyl–, *trans*–

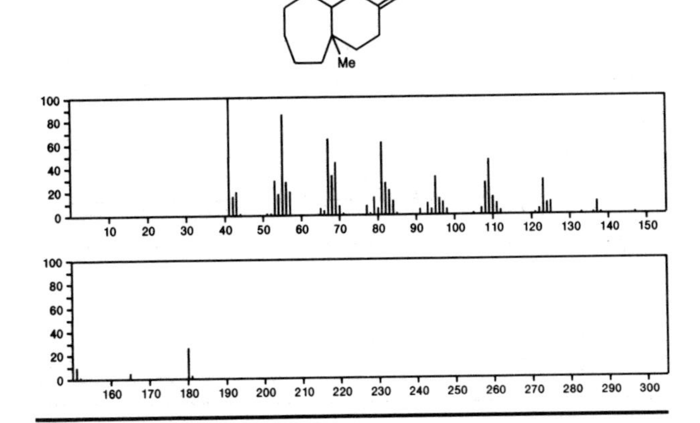

180 $C_{12}H_{20}O$ 55103–65–6
2*H*–Cyclopentacycloocten–2–one, decahydro–3a–methyl–, *trans*–

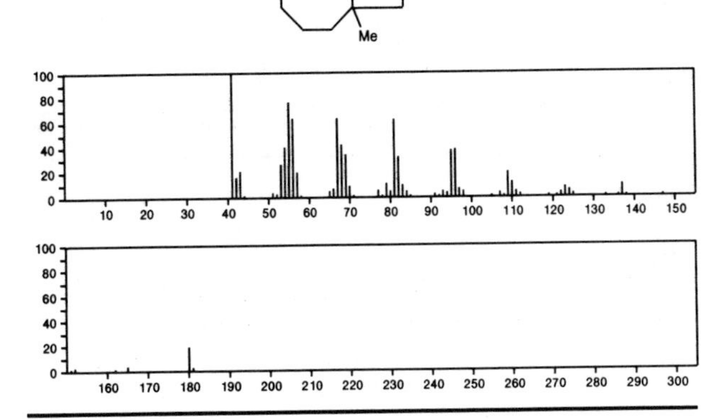

180 $C_{12}H_{20}O$ 55103–67–8
2*H*–Benzocyclohepten–2–one, decahydro–9a–methyl–, *trans*–

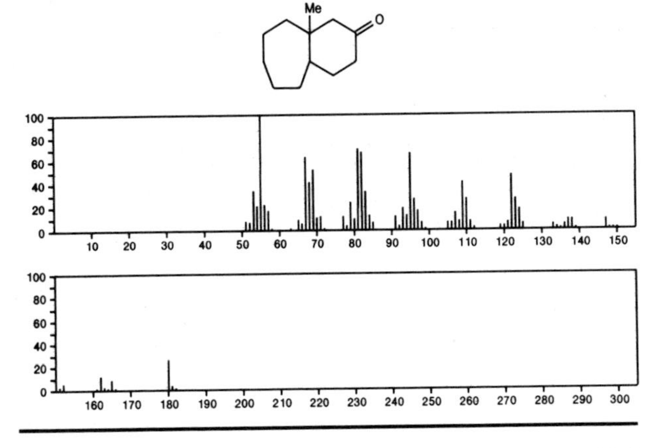

180 $C_{12}H_{20}O$ 55976–08–4
Naphth[1,2–*b*]oxirene, decahydro–1a,7–dimethyl–

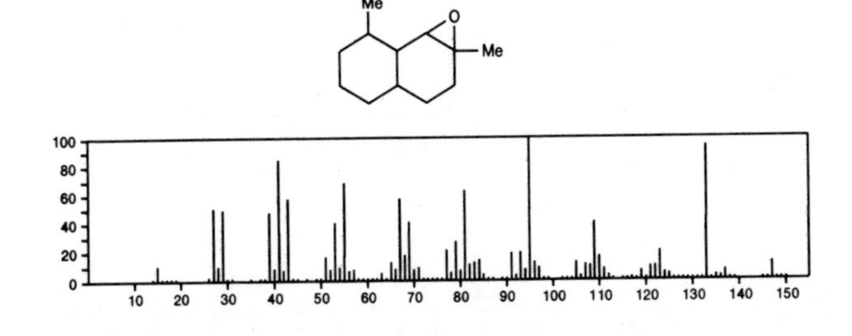

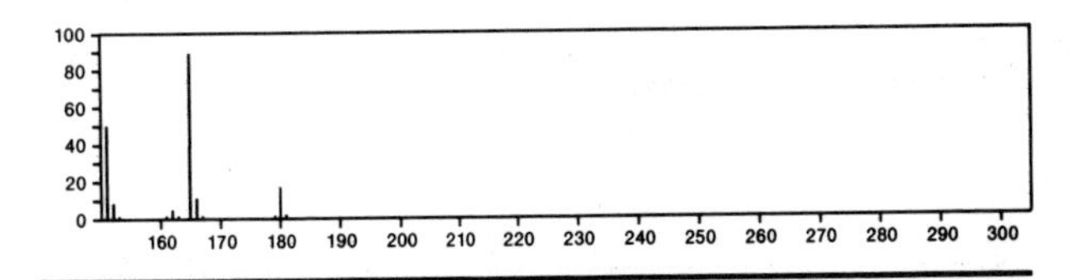

180 $C_{12}H_{20}O$ 56298–45–4
Cyclohexanol, 4–ethenyl–4–methyl–3–(1–methylethenyl)–, (1α,=
3α,4β)–

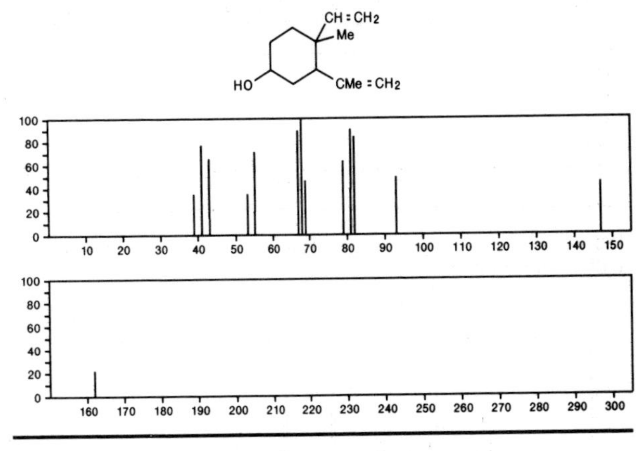

180 $C_{12}H_{20}O$ 56298–46–5
Cyclohexanol, 4–ethenyl–4–methyl–3–(1–methylethenyl)–, (1α,=
3β,4α)–

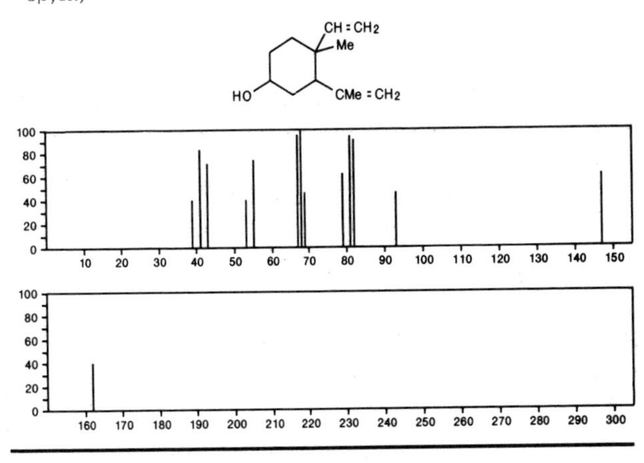

180 $C_{13}H_8O$ 486–25–9
9*H*–Fluoren–9–one

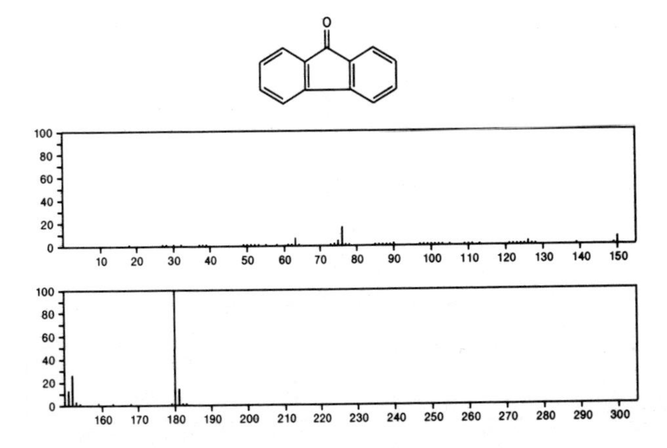

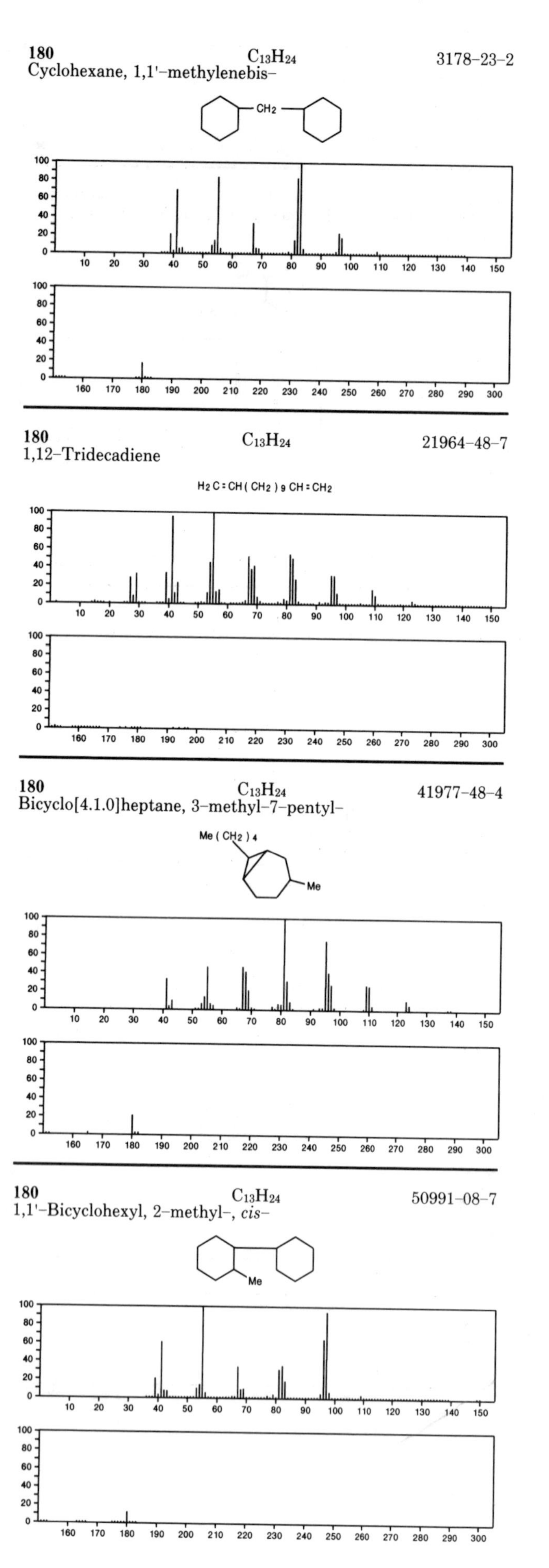

180 $C_{13}H_{24}$ 3178-23-2
Cyclohexane, 1,1'-methylenebis-

180 $C_{13}H_{24}$ 21964-48-7
1,12-Tridecadiene

180 $C_{13}H_{24}$ 41977-48-4
Bicyclo[4.1.0]heptane, 3-methyl-7-pentyl-

180 $C_{13}H_{24}$ 50991-08-7
1,1'-Bicyclohexyl, 2-methyl-, *cis*-

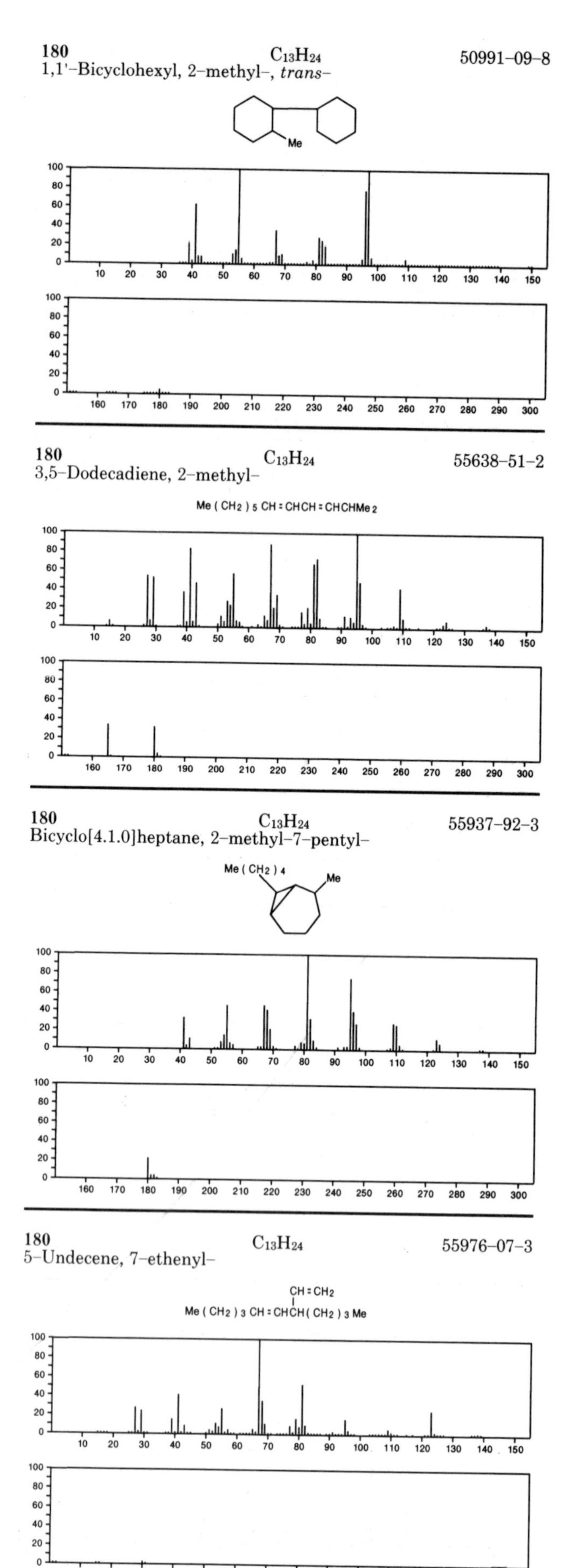

180 $C_{13}H_{24}$ 50991-09-8
1,1'-Bicyclohexyl, 2-methyl-, *trans*-

180 $C_{13}H_{24}$ 55638-51-2
3,5-Dodecadiene, 2-methyl-

180 $C_{13}H_{24}$ 55937-92-3
Bicyclo[4.1.0]heptane, 2-methyl-7-pentyl-

180 $C_{13}H_{24}$ 55976-07-3
5-Undecene, 7-ethenyl-

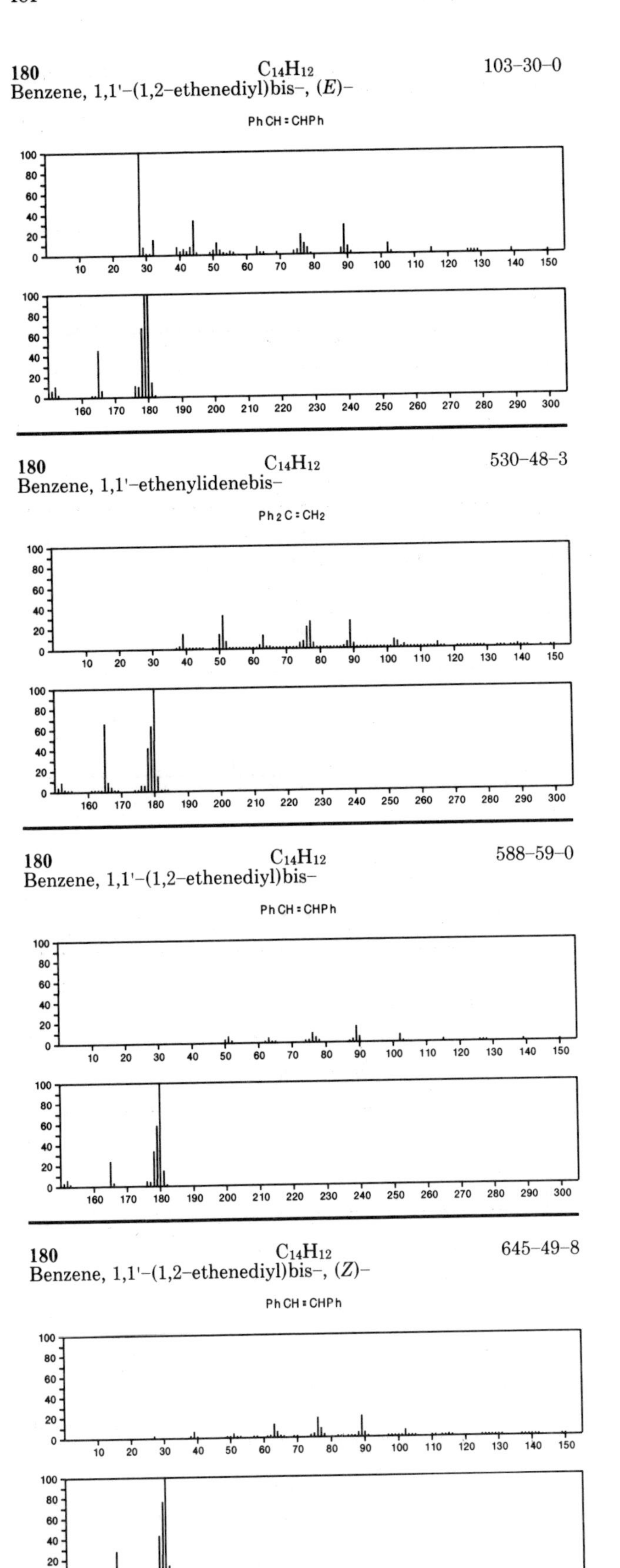
180
C14H12
103–30–0
Benzene, 1,1'–(1,2–ethenediyl)bis–, (E)–
PhCH=CHPh
180
C14H12
530–48–3
Benzene, 1,1'–ethenylidenebis–
Ph2C=CH2
180
C14H12
588–59–0
Benzene, 1,1'–(1,2–ethenediyl)bis–
PhCH=CHPh
180
C14H12
645–49–8
Benzene, 1,1'–(1,2–ethenediyl)bis–, (Z)–
PhCH=CHPh

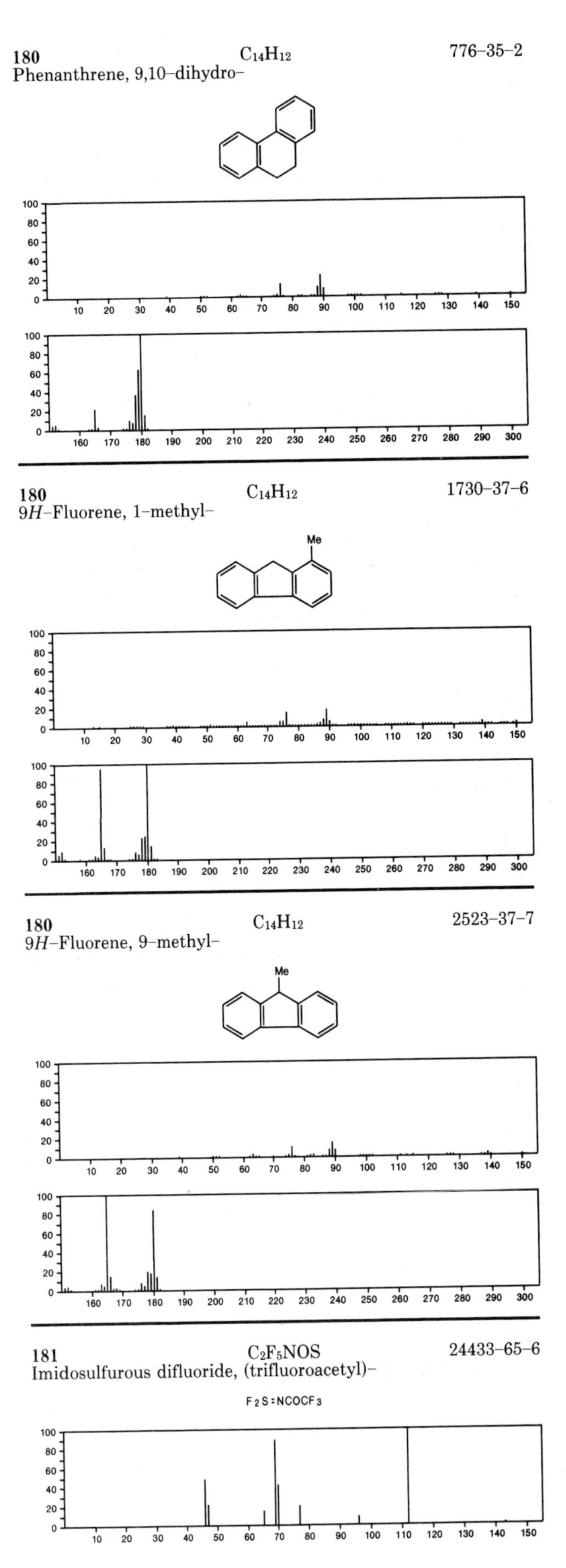
180
C14H12
776–35–2
Phenanthrene, 9,10–dihydro–
180
C14H12
1730–37–6
9H–Fluorene, 1–methyl–
Me
180
C14H12
2523–37–7
9H–Fluorene, 9–methyl–
Me
181
C2F5NOS
24433–65–6
Imidosulfurous difluoride, (trifluoroacetyl)–
F2S=NCOCF3

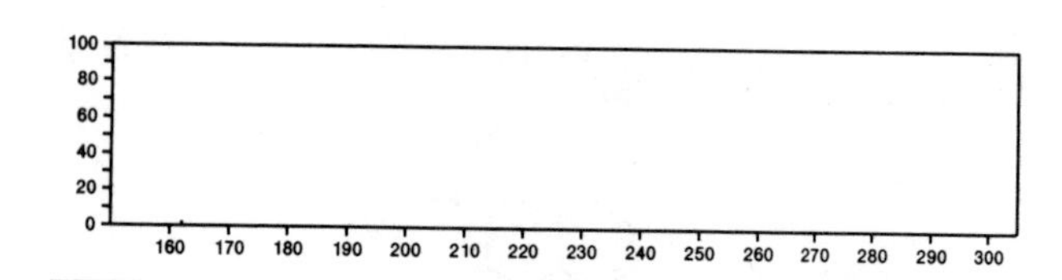

181 $C_6H_{16}NO_3P$ 6415-20-9
Phosphoramidic acid, bis(1-methylethyl) ester

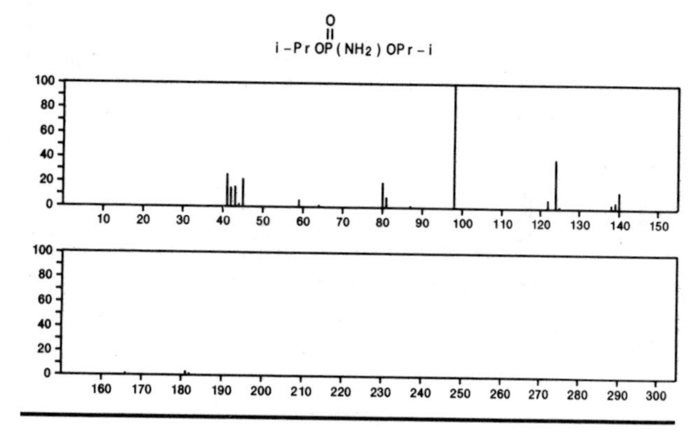

181 $C_6H_{16}NO_3P$ 17123-09-0
Phosphoramidic acid, dipropyl ester

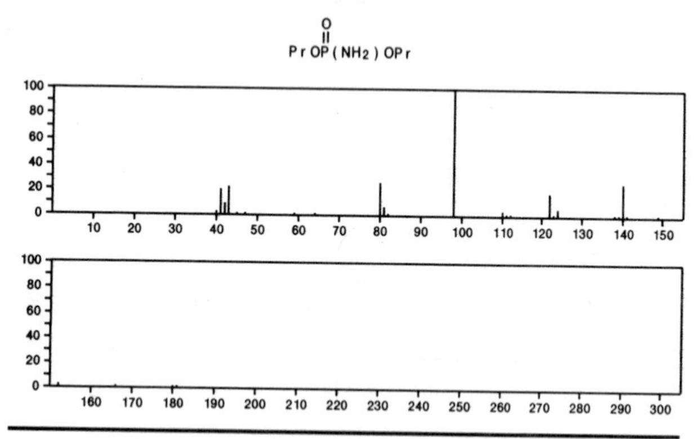

181 $C_7H_7N_3O_3$ 55649-68-8
Benzofurazan, 1,3-dihydro-1-methyl-3-nitro-

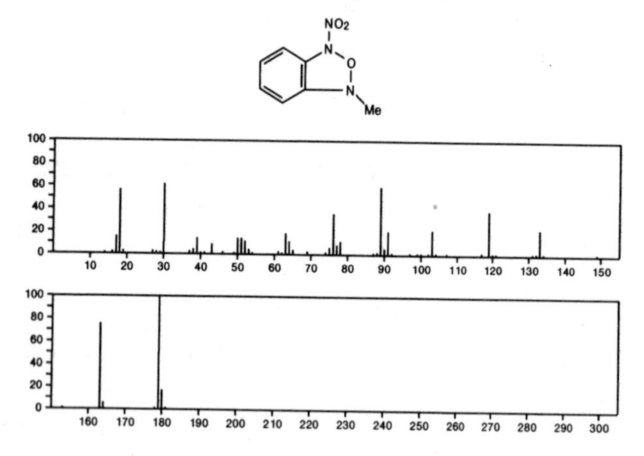

181 $C_8H_7NO_2S$ 21069-05-6
2*H*-1,4-Benzothiazin-3(4*H*)-one, 4-hydroxy-

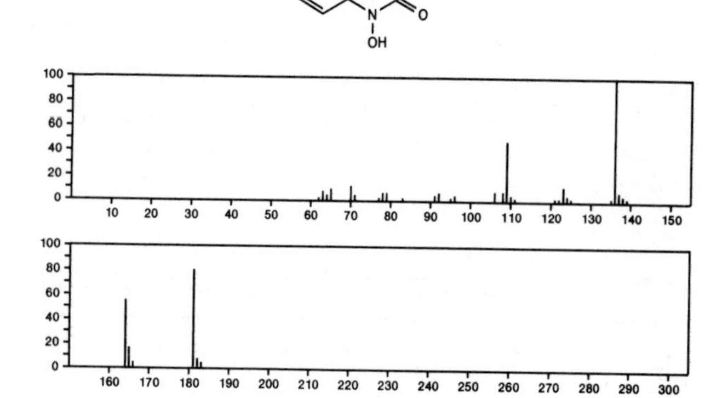

181 $C_8H_7NO_4$ 619-50-1
Benzoic acid, 4-nitro-, methyl ester

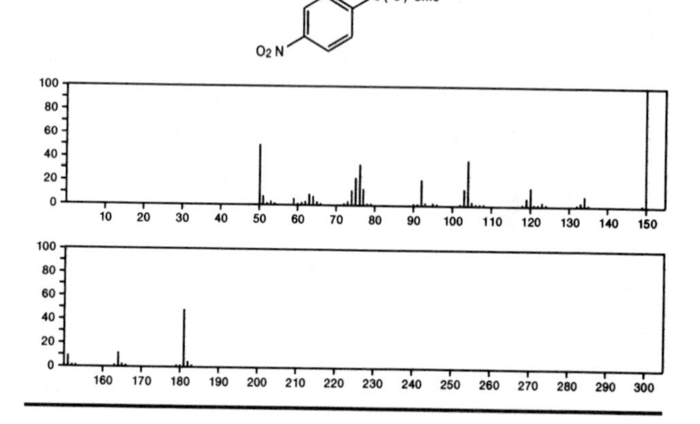

181 $C_8H_7NS_2$ 615-22-5
Benzothiazole, 2-(methylthio)-

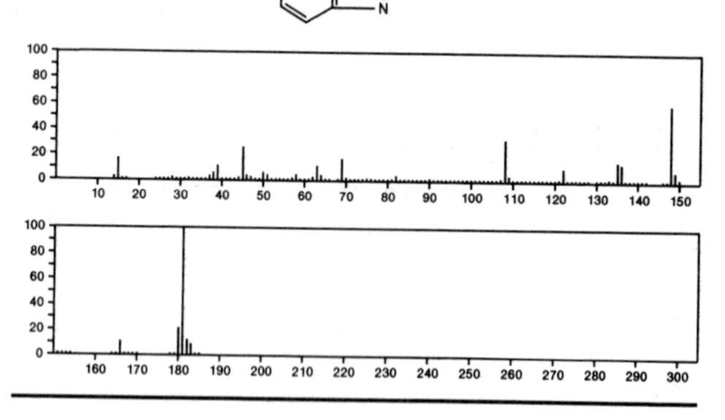

181 $C_8H_{15}N_5$ 1973-07-5
s-Triazine, 2,4-bis(ethylamino)-6-methyl-

EtNH NHEt N N N Me

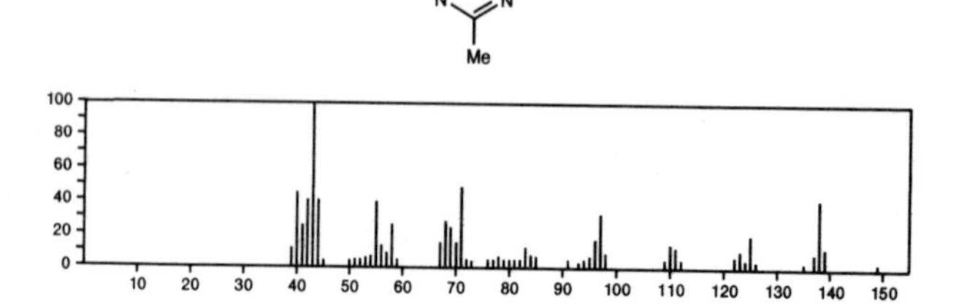

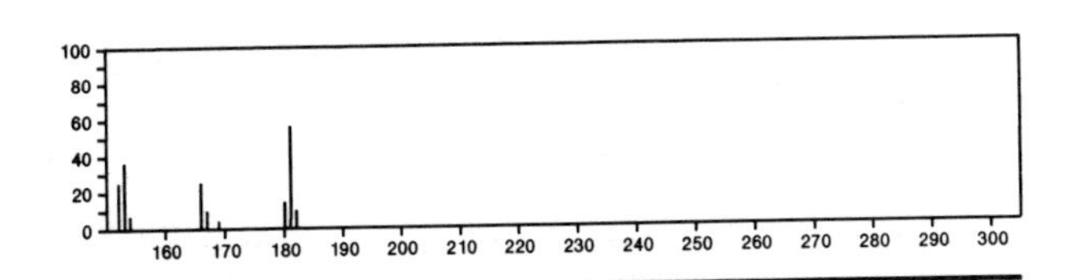

181 $C_9H_{11}NOS$ 7304–68–9
Carbamothioic acid, dimethyl–, *S*–phenyl ester

$Me_2NC(O)SPh$

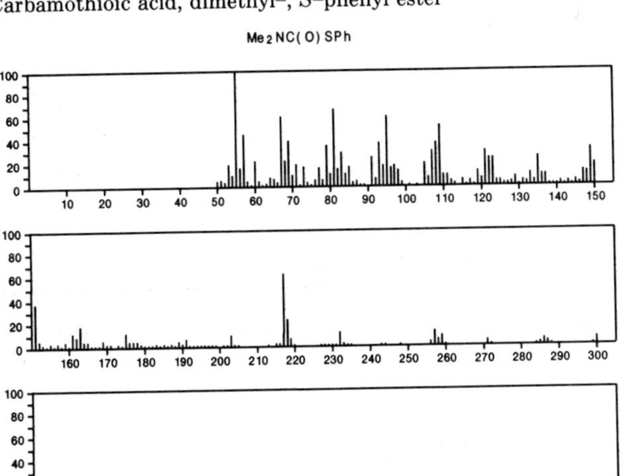

181 $C_9H_{11}NOS$ 23933–08–6
Thiazolo[3,2–*a*]pyridinium, 2,3–dihydro–8–hydroxy–2,5–dimethyl–, hydroxide, inner salt

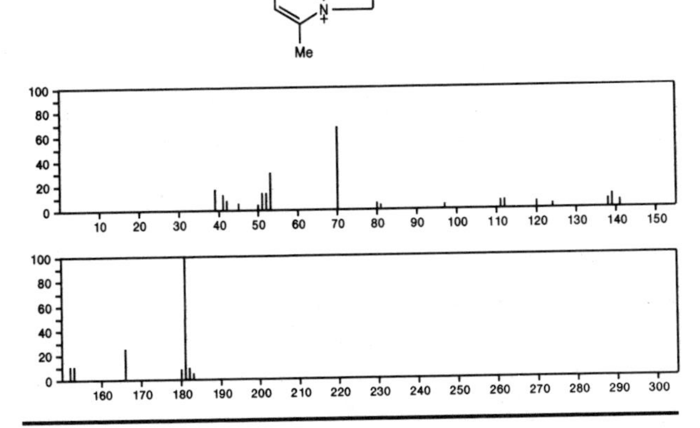

181 $C_9H_{11}NOS$ 55649–96–2
Sulfonium, (benzoylamino)dimethyl–, hydroxide, inner salt

$Me_2\overset{+}{S}$–$\overset{-}{N}COPh$

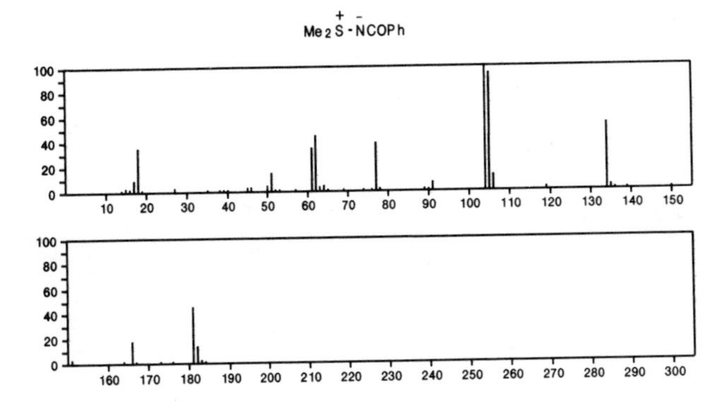

181 $C_9H_{11}NO_3$ 60–18–4
L–Tyrosine

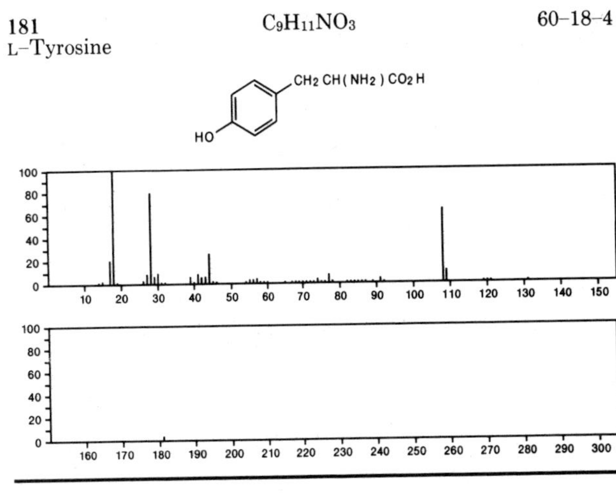

181 $C_9H_{11}NO_3$ 94–35–9
1,2–Ethanediol, 1–phenyl–, 2–carbamate

$PhCH(OH)CH_2OC(O)NH_2$

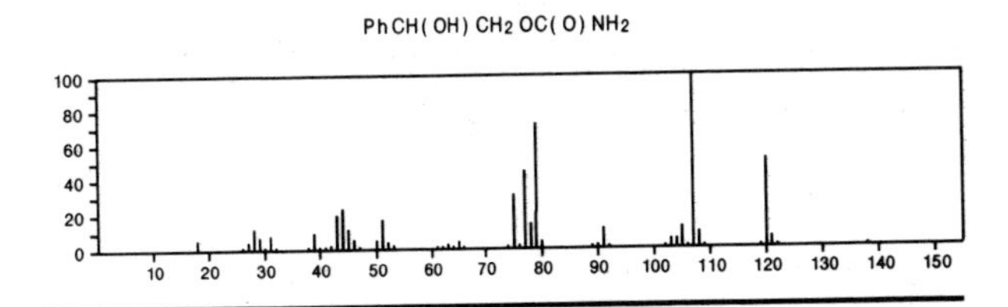

181 $C_9H_{11}NO_3$ 1719–21–7
Phenol, 2,4,6–trimethyl–3–nitro–

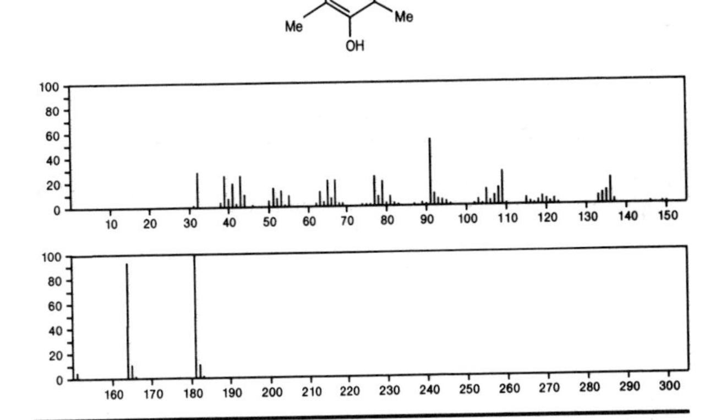

181 $C_9H_{11}NO_3$ 7135–82–2
2(1*H*)–Pyridinone, 4–hydroxy–6–methyl–3–(1–oxopropyl)–

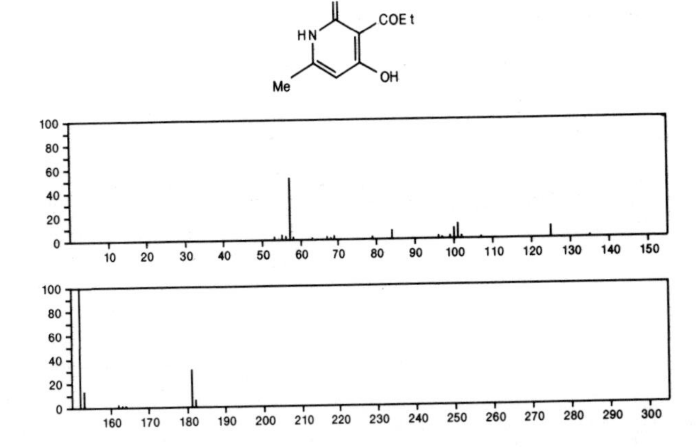

181 $C_9H_{11}NO_3$ 7202-55-3
2(1*H*)-Pyridone, 3-acetyl-4-hydroxy-1,6-dimethyl-

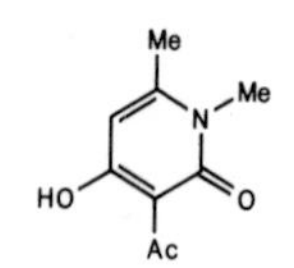

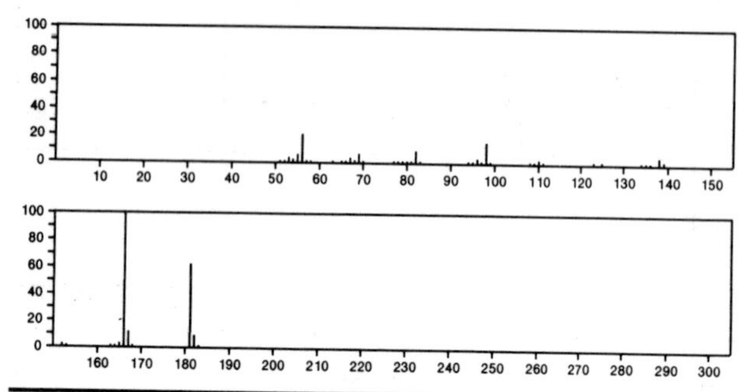

181 $C_9H_{11}NO_3$ 7211-75-8
2(1*H*)-Pyridone, 4-hydroxy-1,6-dimethyl-, acetate (ester)

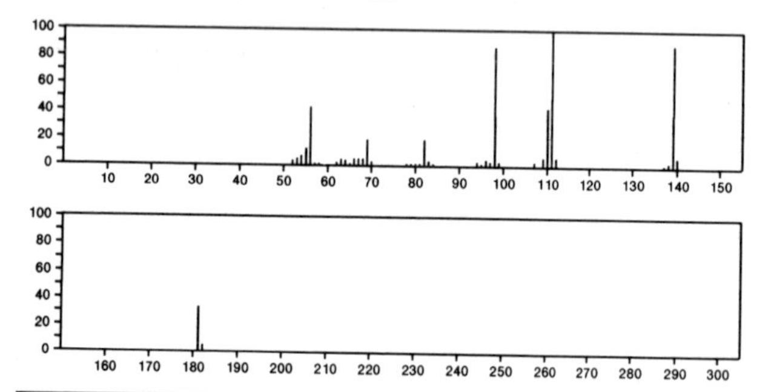

181 $C_9H_{11}NO_3$ 29121-49-1
Acetamide, 2-(4-hydroxy-3-methoxyphenyl)-

MeO CH2 CONH2
HO

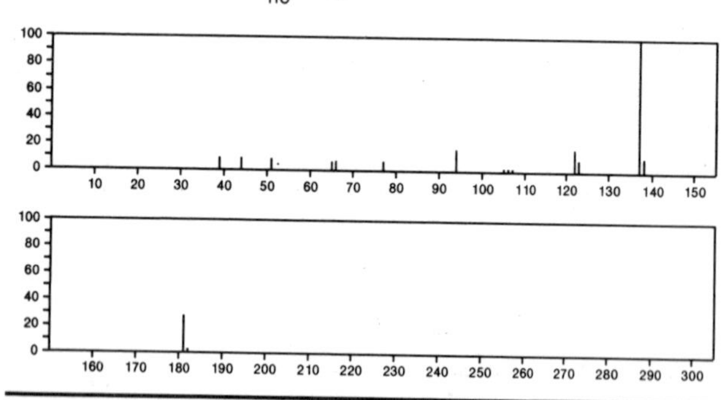

181 $C_9H_{11}NO_3$ 30569-18-7
Pyrrole-2-carboxaldehyde, 5-(hydroxymethyl)-1-methyl-, acetate (ester)

CH2 OAc
O=CH N Me

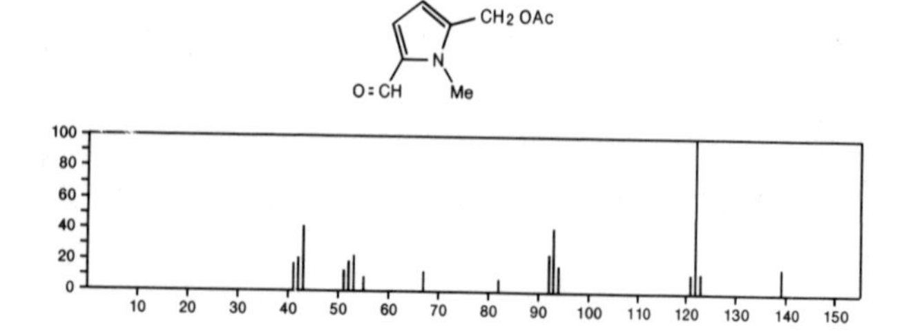

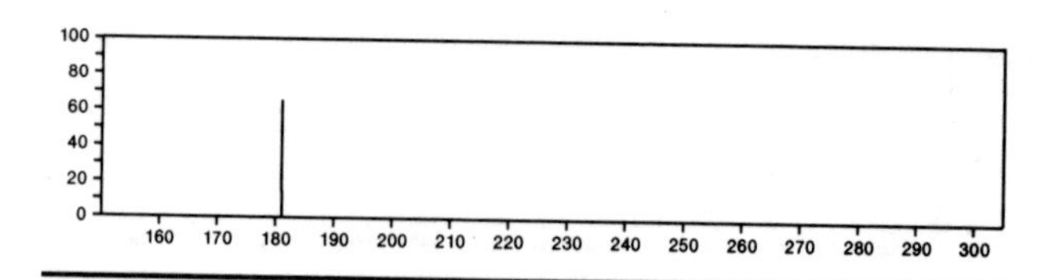

181 $C_9H_{11}NO_3$ 50267-15-7
3,8-Dioxa-11-azatetracyclo[4.4.1.0^{2,4}.0^{7,9}]undecane-11-carbox=
aldehyde, (1α,2β,4β,6α,7β,9β)-

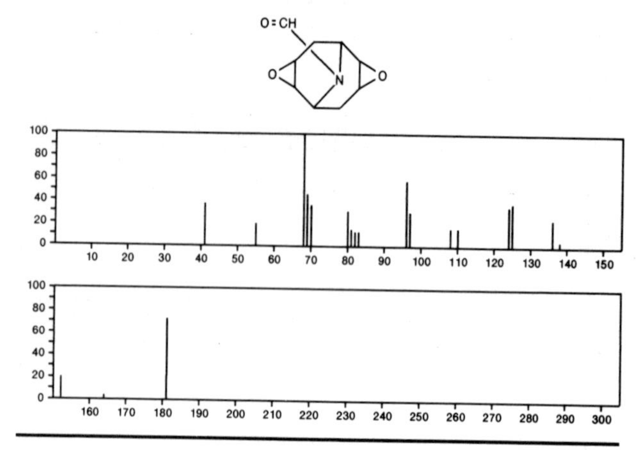

181 $C_9H_{11}NO_3$ 51422-77-6
Carbamic acid, (3-methoxyphenyl)-, methyl ester

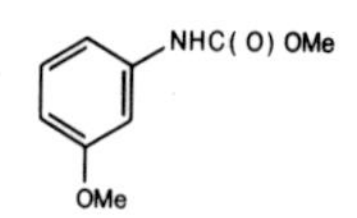

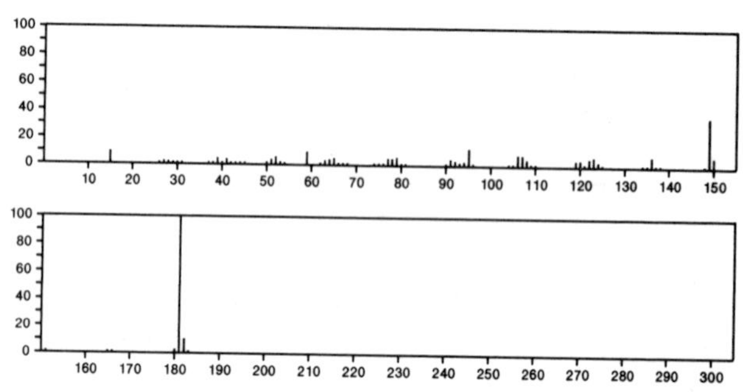

181 $C_9H_{13}BClN$ 55702-64-2
Boranamine, 1-chloro-1-ethyl-*N*-methyl-*N*-phenyl-

EtBCl NMePh

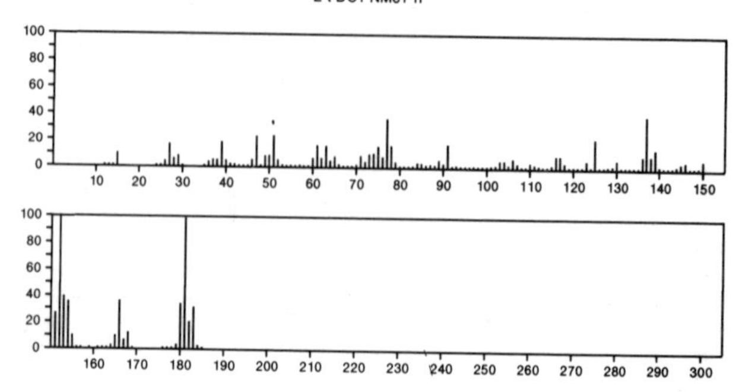

181 $C_{10}H_{12}ClN$ 42540-69-2

Benzenamine, *N*-[1-(chloromethyl)cyclopropyl]-

NHPh
CH_2Cl

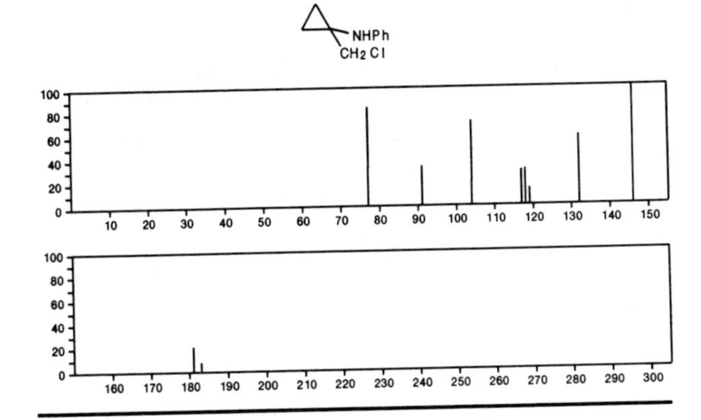

181 $C_{10}H_{15}NO_2$ 120-20-7

Benzeneethanamine, 3,4-dimethoxy-

MeO $CH_2CH_2NH_2$
MeO

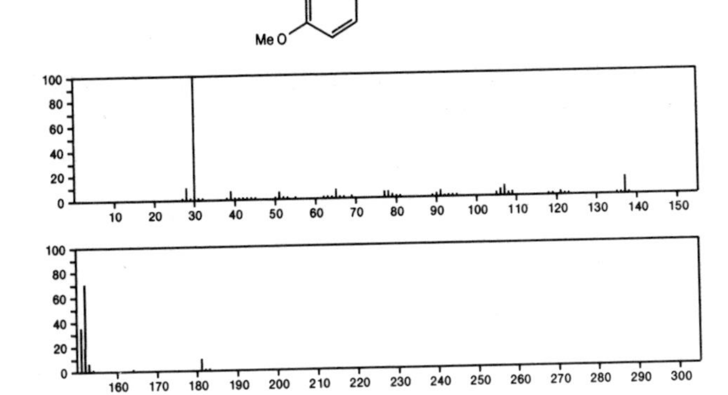

181 $C_{10}H_{15}NO_2$ 358-52-1

Cyclohexanol, 1-(2-propynyl)-, carbamate

$CH_2C{\equiv}CH$
$OC(O)NH_2$

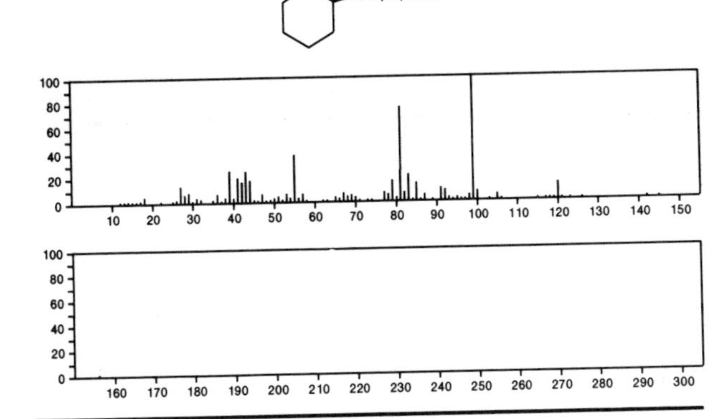

181 $C_{10}H_{15}NO_2$ 3693-69-4

2-Azabicyclo[2.2.2]oct-5-ene-2-carboxylic acid, ethyl ester

N C(O)OEt

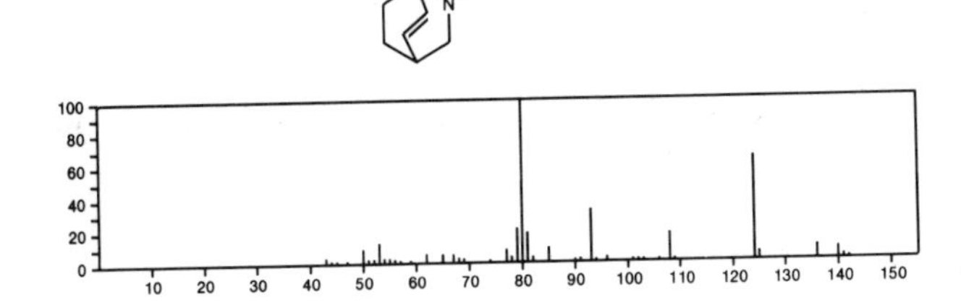

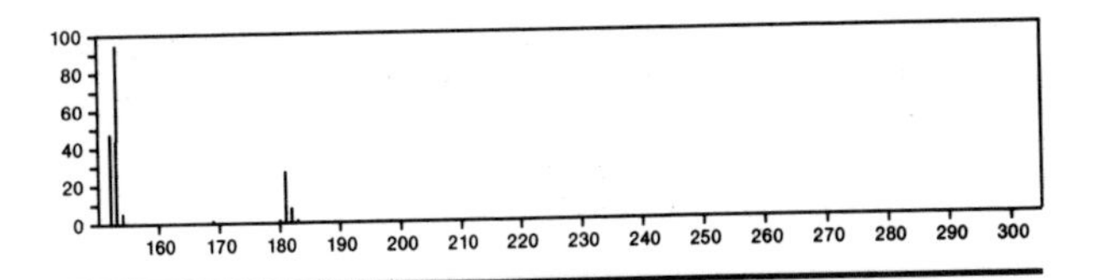

181 $C_{10}H_{15}NO_2$ 6238-32-0

1-Azabicyclo[2.2.2]oct-2-ene-3-carboxylic acid, ethyl ester

N C(O)OEt

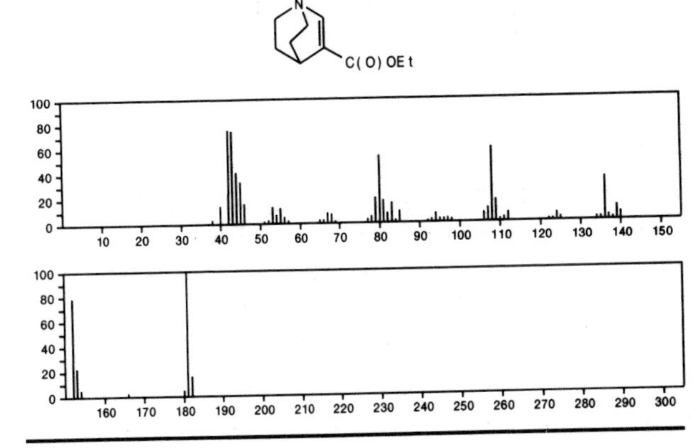

181 $C_{10}H_{15}NO_2$ 6967-70-0

2,4-Pyridinediol, 3-butyl-6-methyl-

O $(CH_2)_3Me$
HN OH
Me

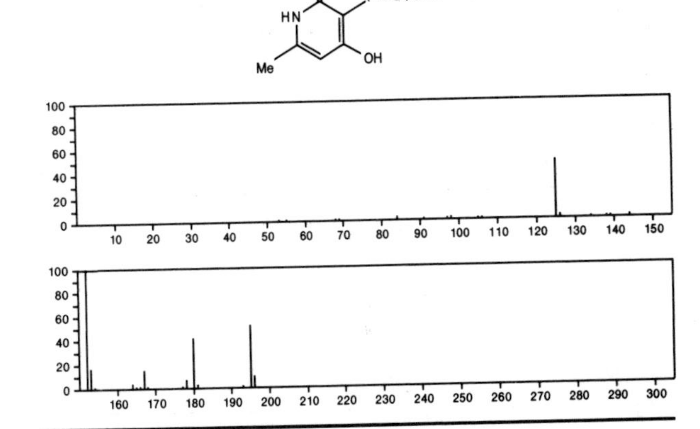

181 $C_{10}H_{15}NO_2$ 23562-77-8

Phenol, 2-[(dimethylamino)methyl]-4-methoxy-

Me_2NCH_2 OMe
HO

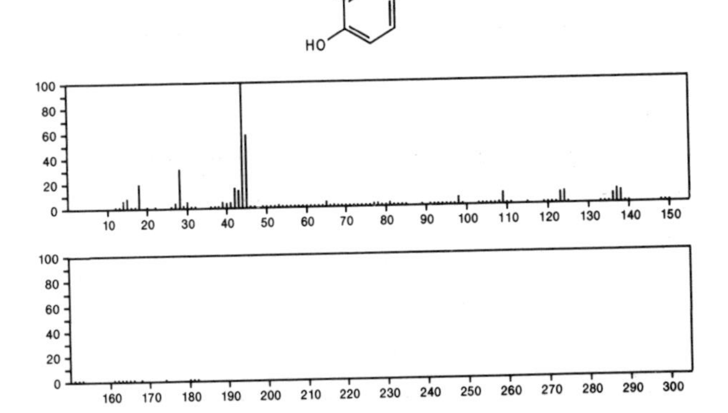

181 $C_{10}H_{15}NO_2$ 54774-92-4
Morpholine, 4-[(3,4-dihydro-2*H*-pyran-2-ylidene)methyl]-

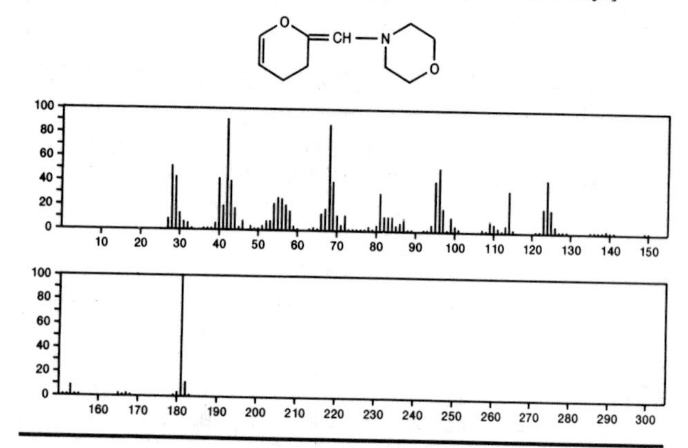

181 $C_{10}H_{15}NS$ 18794-27-9
Pyridine, 4-[(*tert*-butylthio)methyl]-

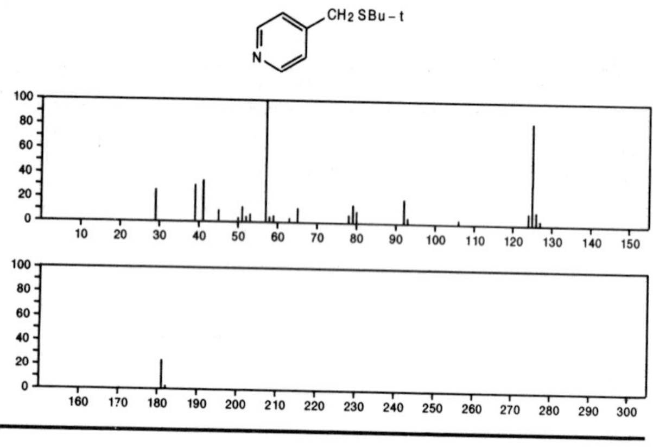

181 $C_{10}H_{15}NS$ 18794-36-0
4-Picoline, 2-(*tert*-butylthio)-

181 $C_{10}H_{15}NS$ 18794-37-1
4-Picoline, 3-(*tert*-butylthio)-

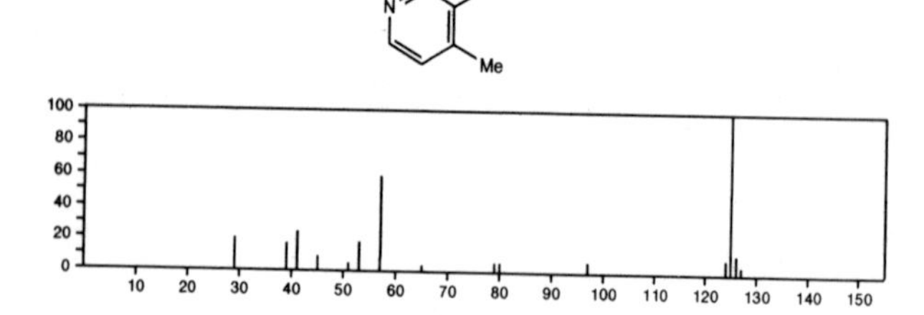

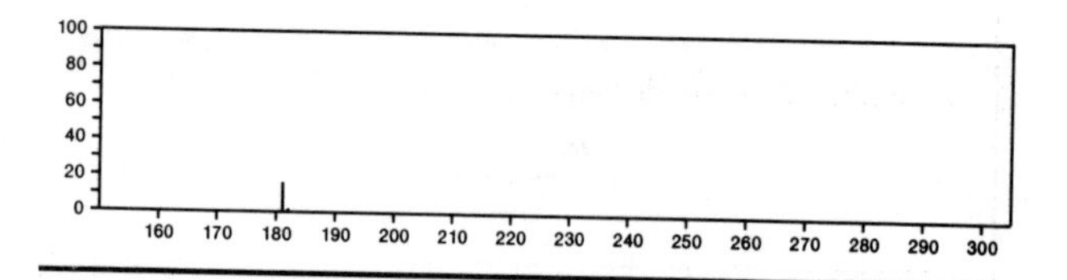

181 $C_{10}H_{15}NS$ 18794-43-9
2-Picoline, 6-(*tert*-butylthio)-

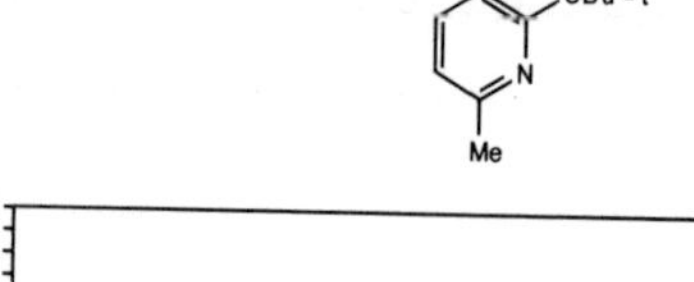

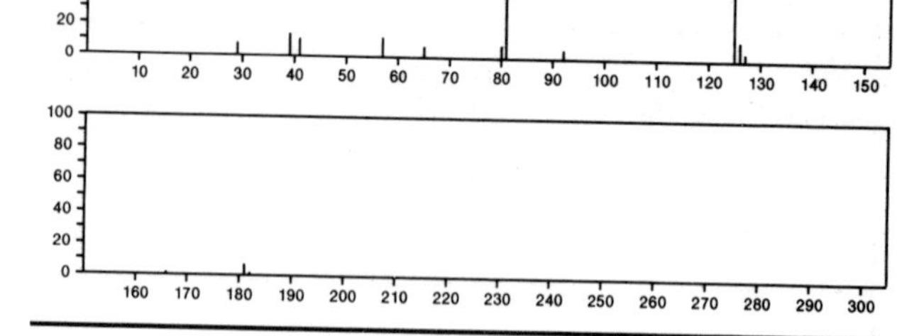

181 $C_{10}H_{15}NS$ 18794-44-0
2-Picoline, 5-(*tert*-butylthio)-

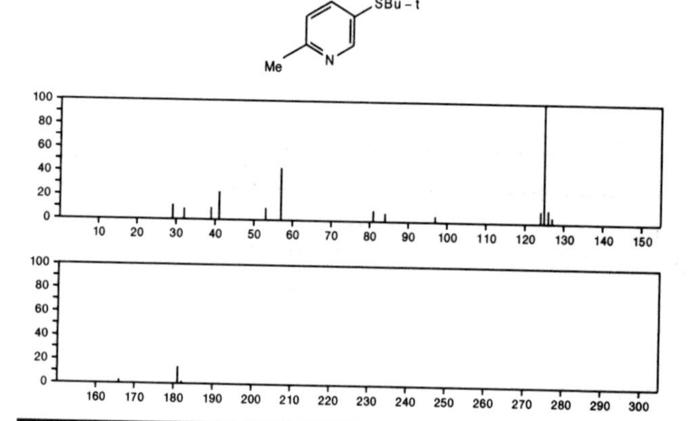

181 $C_{10}H_{15}NS$ 18794-45-1
Pyridine, 2-[(*tert*-butylthio)methyl]-

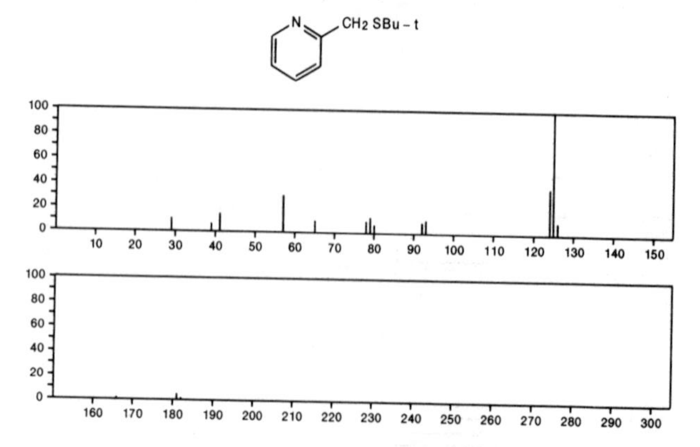

181 $C_{10}H_{15}NS$ 18794-46-2
3-Picoline, 6-(*tert*-butylthio)-

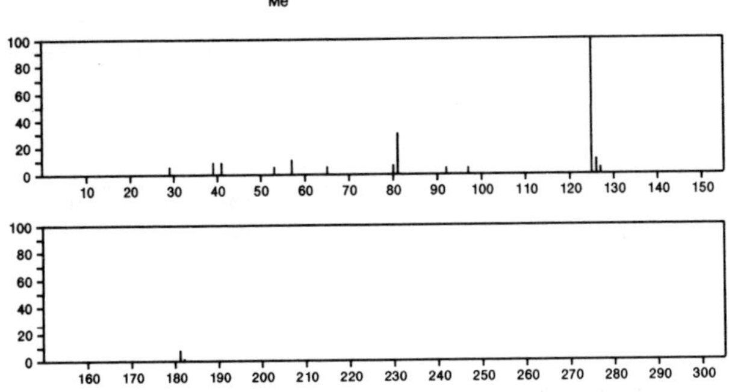

181 $C_{10}H_{15}NS$ 18794-47-3
3-Picoline, 5-(*tert*-butylthio)-

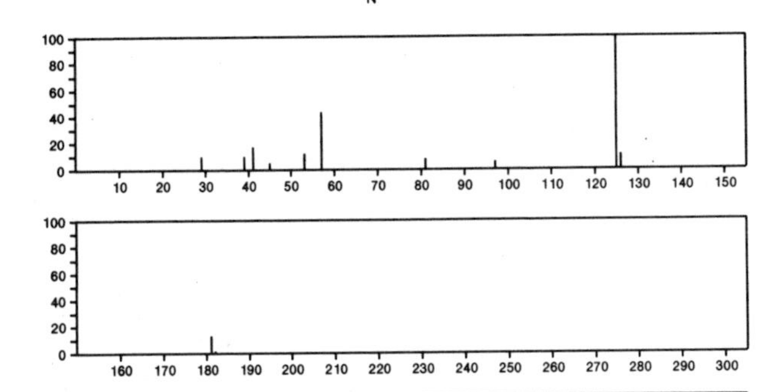

181 $C_{10}H_{15}NS$ 18833-87-9
3-Picoline, 2-(*tert*-butylthio)-

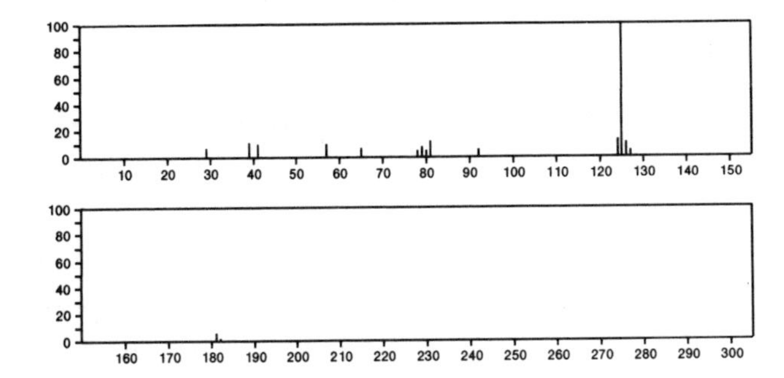

181 $C_{11}H_{19}NO$ 1130-36-5
Acetamide, *N*-(4-methylbicyclo[2.2.2]oct-1-yl)-

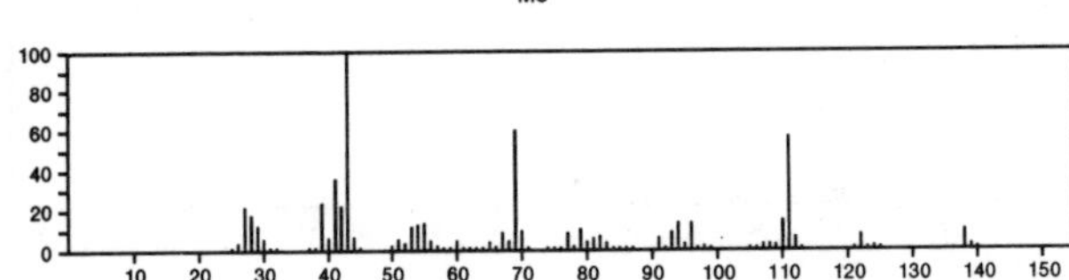

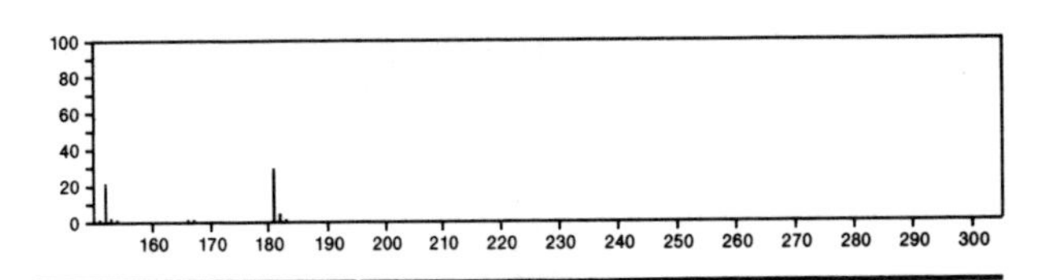

181 $C_{11}H_{19}NO$ 5601-45-6
Morpholine, 4-(4-methyl-1-cyclohexen-1-yl)-

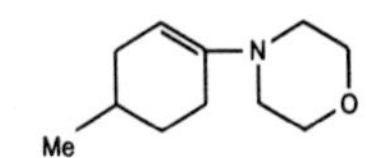

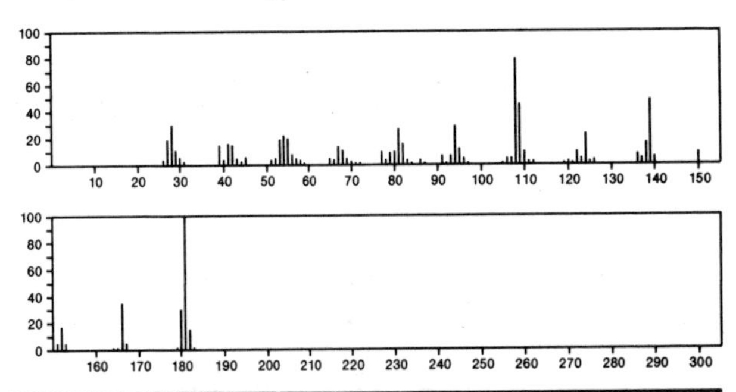

181 $C_{11}H_{19}NO$ 13606-83-2
1-Penten-3-one, 4-methyl-1-piperidino-

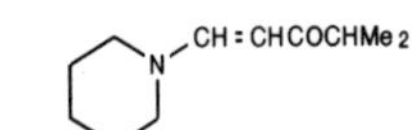

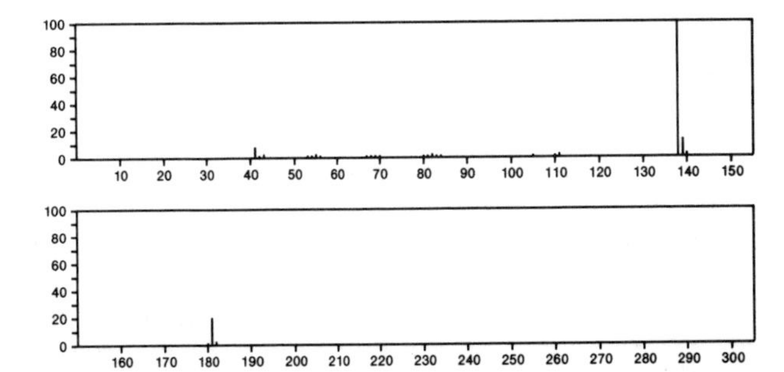

181 $C_{11}H_{19}NO$ 14293-08-4
2-Pyrrolidinone, 5-(cyclohexylmethyl)-

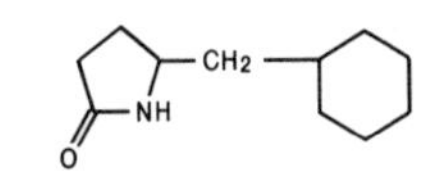

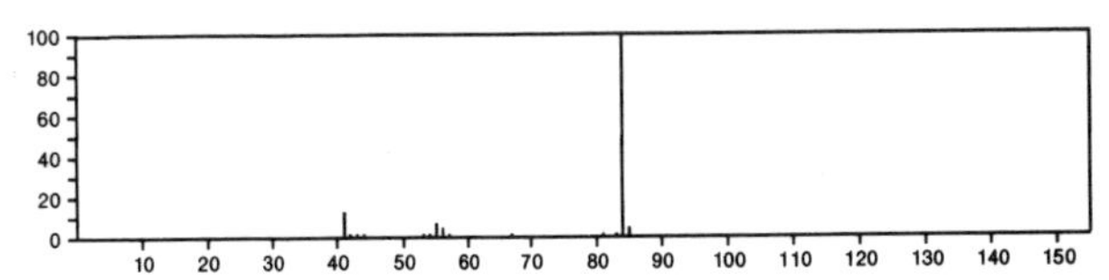

181 $C_{11}H_{19}NO$ 32134-53-5
2-Norbornanone, 1,3,7,7-tetramethyl-, oxime

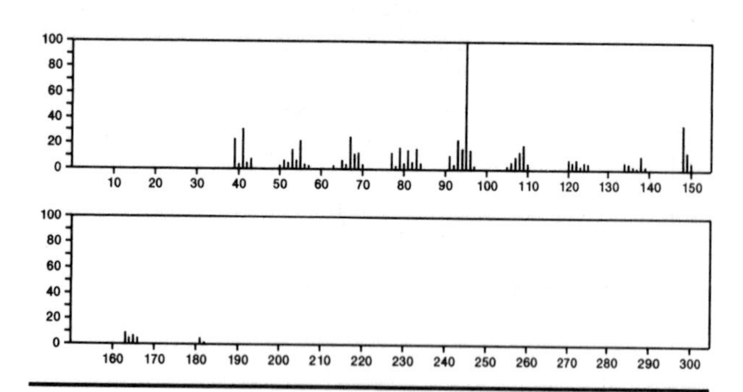

181 $C_{11}H_{19}NO$ 54677-80-4
Ethanone, 1-[3-methyl-2-(1-pyrrolidinyl)cyclobutyl]-

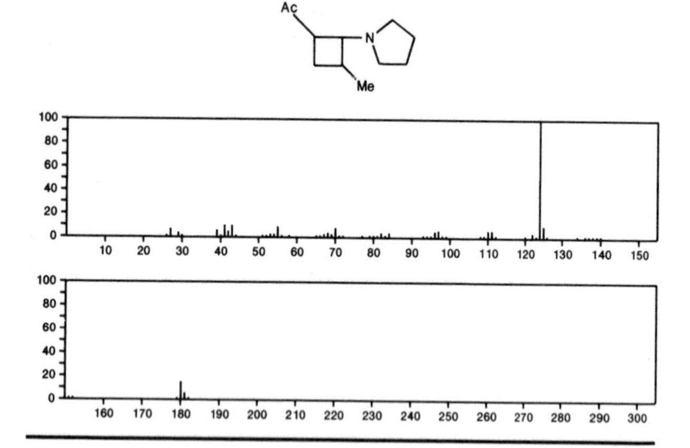

181 $C_{12}H_{23}N$ 101-83-7
Cyclohexanamine, *N*-cyclohexyl-

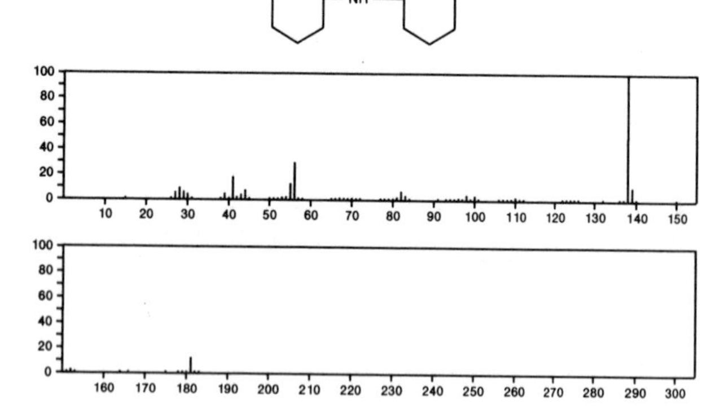

181 $C_{12}H_{23}N$ 3570-07-8
Bicyclo[2.2.1]heptan-2-amine, *N*,*N*,2,3,3-pentamethyl-

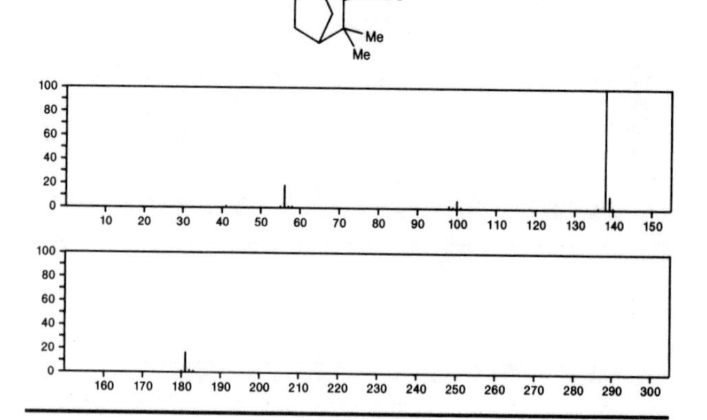

181 $C_{12}H_{23}N$ 3710-93-8
2,6-Octadien-1-amine, *N*,*N*,3,7-tetramethyl-

$Me_2C=CHCH_2CH_2CMe=CHCH_2NMe_2$

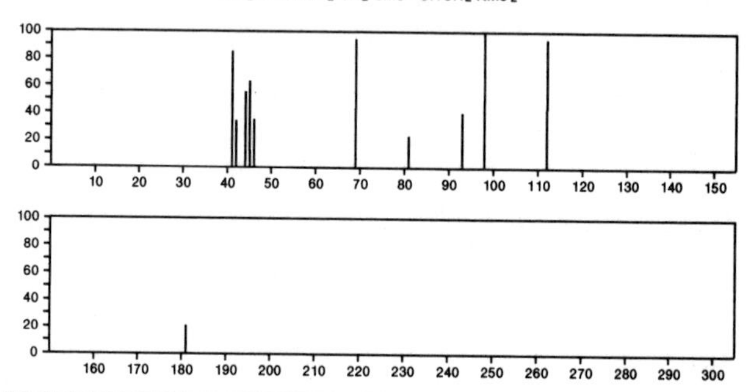

181 $C_{12}H_{23}N$ 14727-50-5
2-Bornanamine, *N*,*N*-dimethyl-, *endo*-

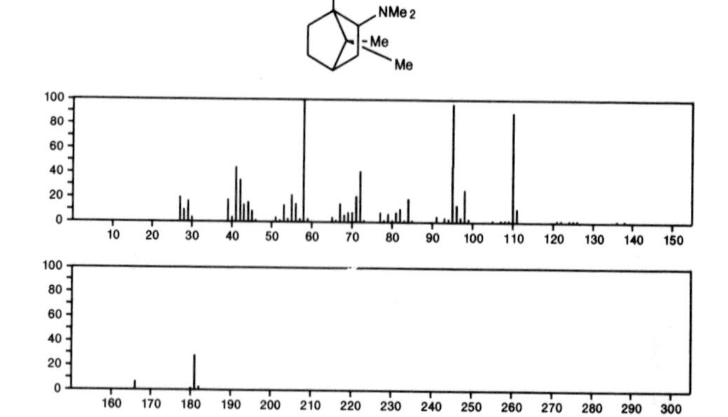

181 $C_{12}H_{23}N$ 17943-83-8
Cyclopentanemethylamine, 2-isopropylidene-*N*,*N*,5-trimethyl-, (1*R*,5*R*)-(−)-

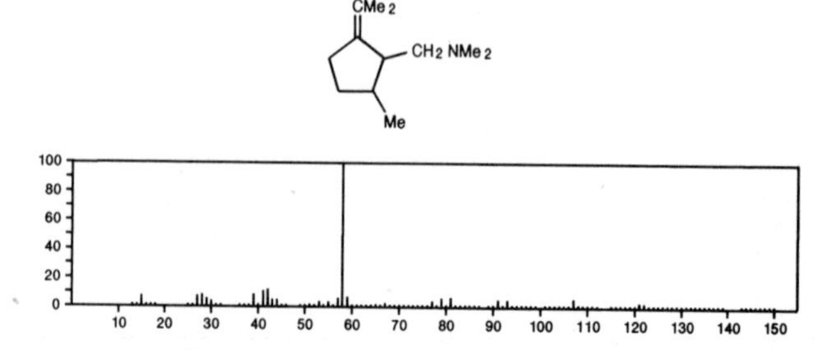

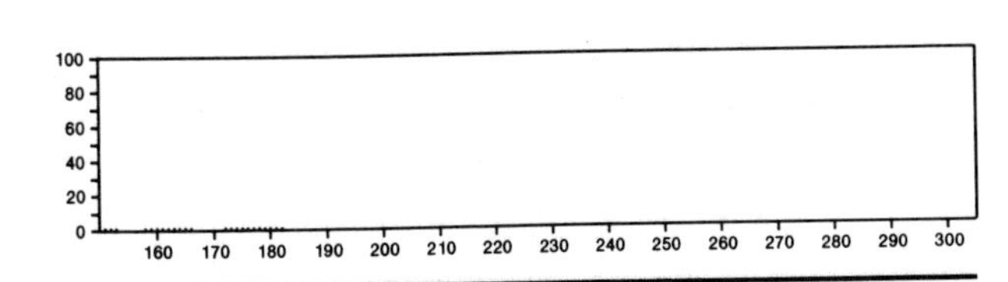

181 $C_{12}H_{23}N$ 17943-85-0

Cyclopentanemethylamine, 2-isopropylidene-*N*,*N*,5-trimethyl-, (1*S*,5*R*)-(+)-

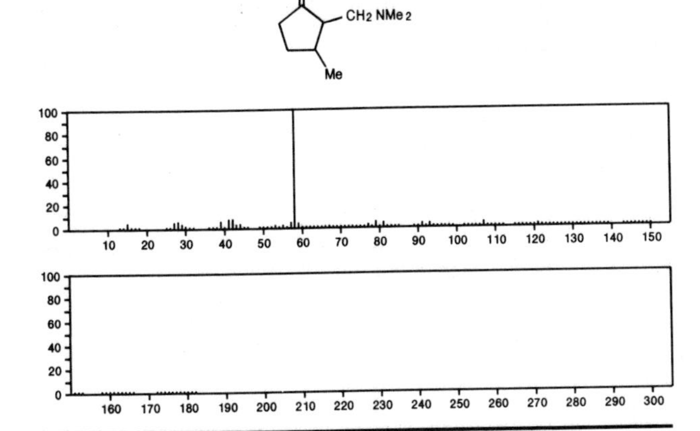

181 $C_{12}H_{23}N$ 35973-45-6

Bicyclo[2.2.1]heptan-2-amine, *N*,*N*,4,7,7-pentamethyl-

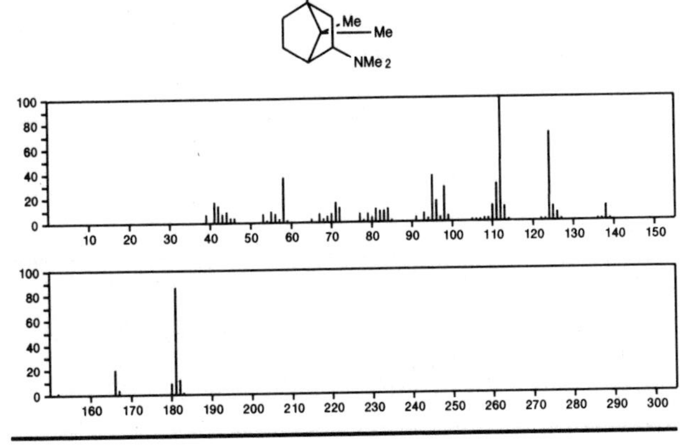

181 $C_{13}H_{11}N$ 153-78-6

9*H*-Fluoren-2-amine

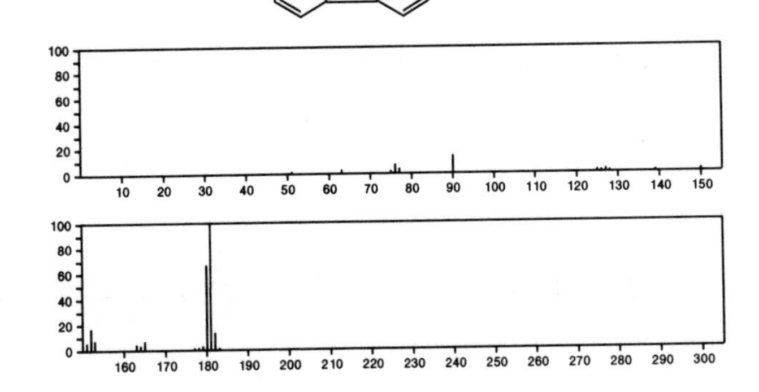

181 $C_{13}H_{11}N$ 538-51-2

Benzenamine, *N*-(phenylmethylene)-

PhN=CHPh

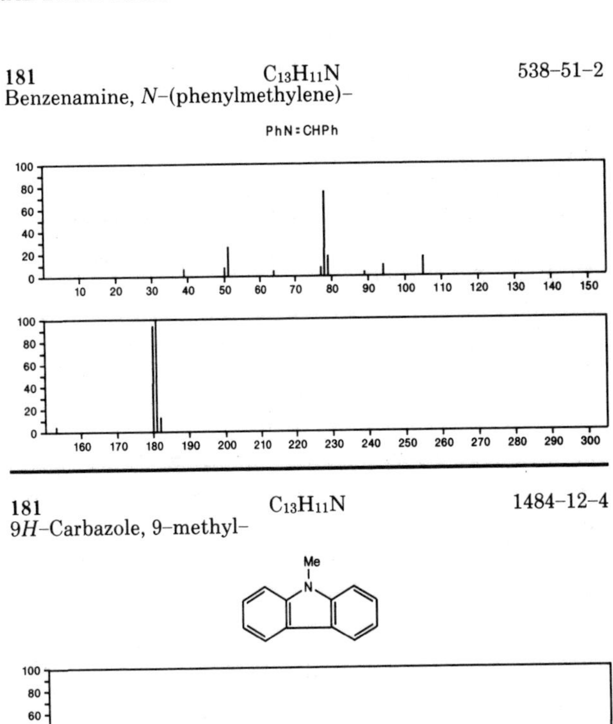

181 $C_{13}H_{11}N$ 1484-12-4

9*H*-Carbazole, 9-methyl-

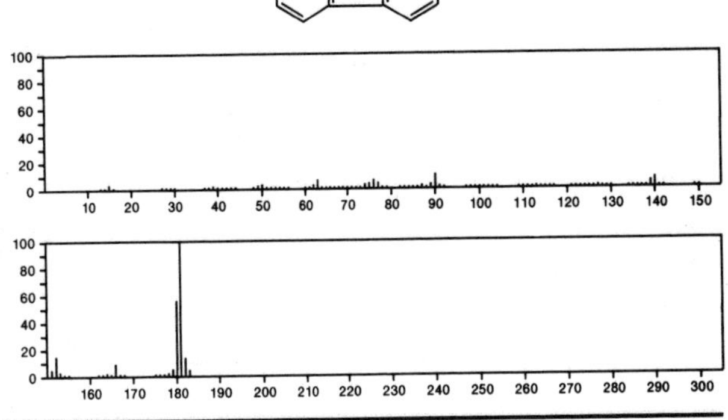

181 $C_{13}H_{11}N$ 3652-91-3

9*H*-Carbazole, 2-methyl-

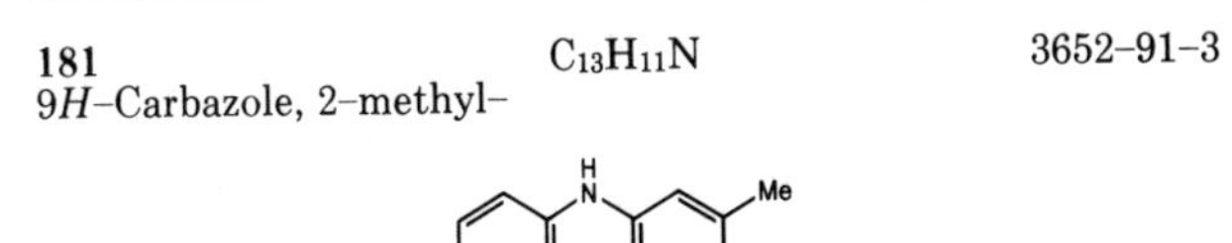

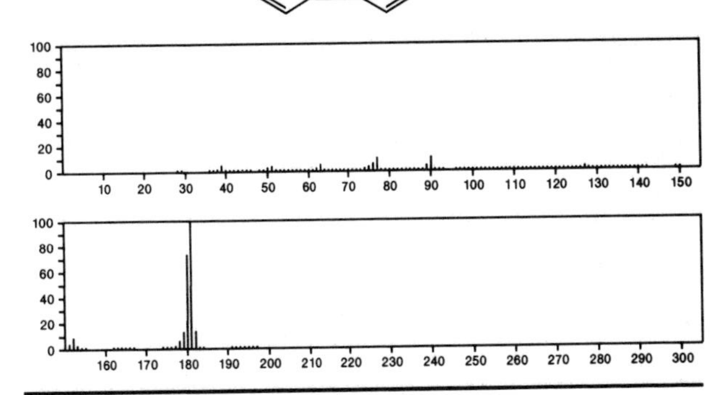

181 $C_{13}H_{11}N$ 27799-79-7

Phenanthridine, 5,6-dihydro-

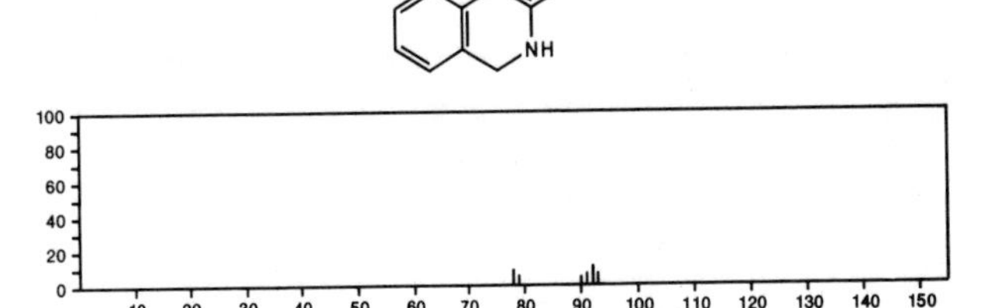

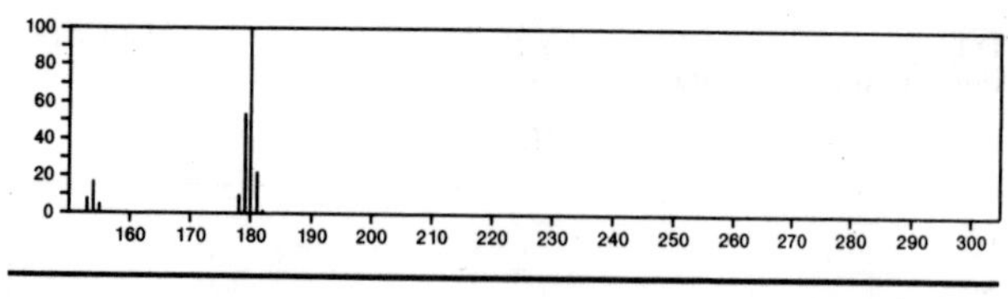

182 $C_2H_2Cl_4O$ 20524-86-1
Methane, oxybis[dichloro-

$Cl_2CHOCHCl_2$

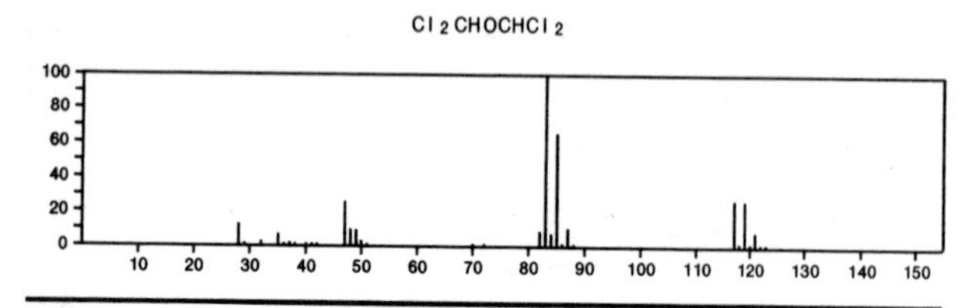

182 $C_2H_{10}Ge_2$ 20549-65-9
Digermane, ethyl-

$EtGeH_2GeH_3$

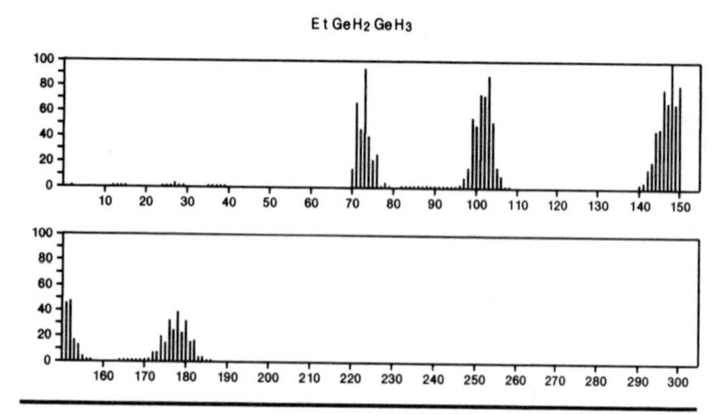

182 $C_3Cl_2F_4$ 431-53-8
Propene, 1,2-dichlorotetrafluoro-

$F_3CCCl=CClF$

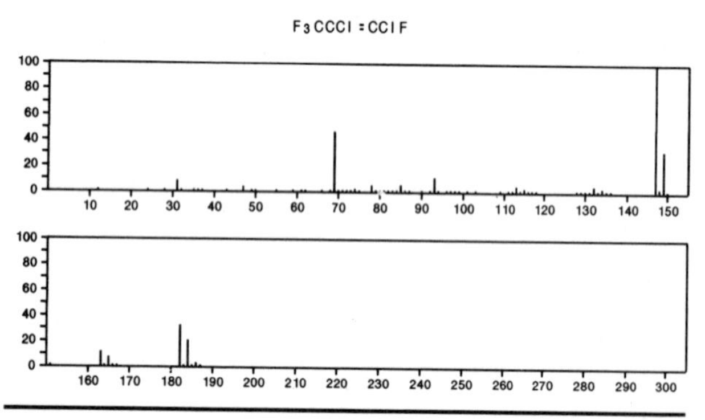

182 C_3F_6S 1490-33-1
2-Propanethione, 1,1,1,3,3,3-hexafluoro-

F_3CCSCF_3

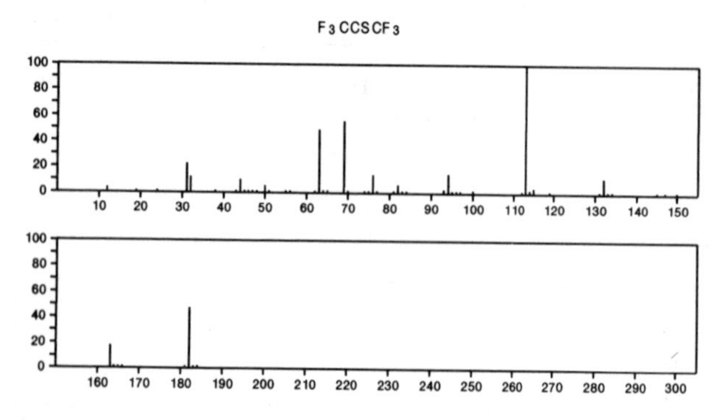

182 C_3F_6S 24345-51-5
Thietane, hexafluoro-

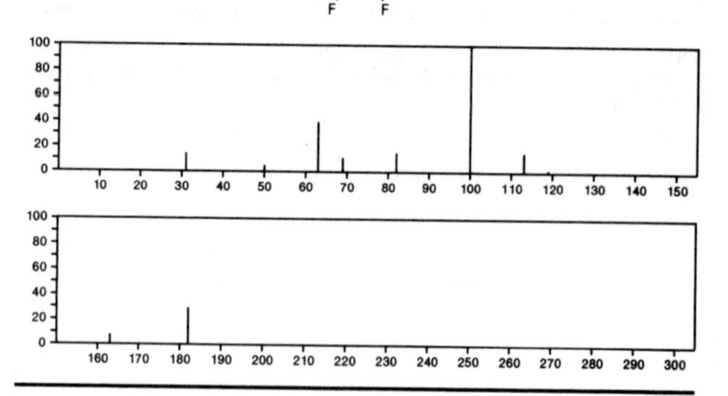

182 $C_3H_3Cl_2F_3O$ 428-92-2
Ethane, 2-chloro-1-(chloromethoxy)-1,1,2-trifluoro-

$FCHClCF_2OCH_2Cl$

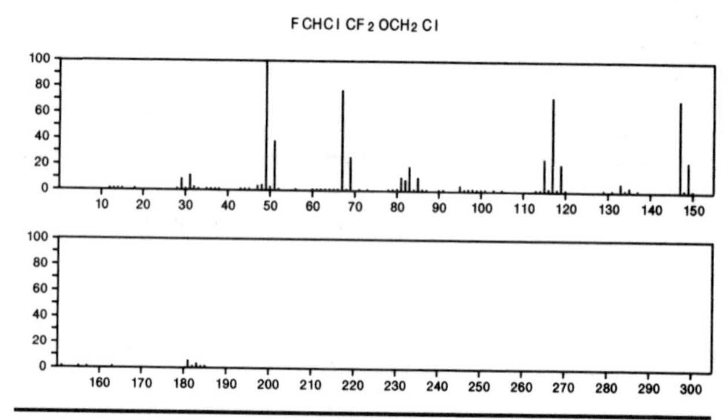

182 C_3H_3IO 54724-99-1
2-Propyn-1-ol, 1-iodo-

$(HO)CHIC{\equiv}CH$

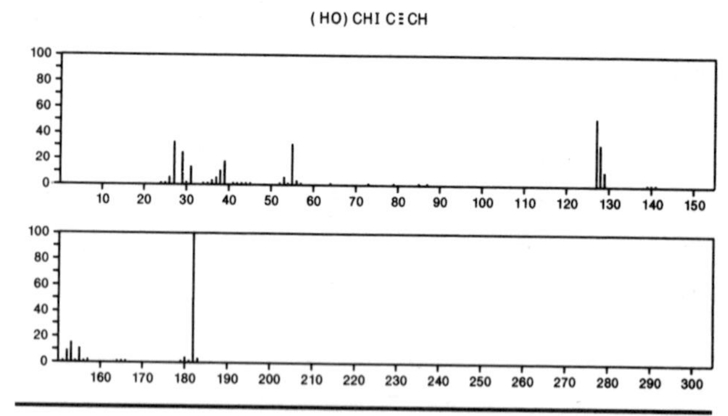

182 $C_3H_6N_2O_7$ 623-87-0
1,2,3-Propanetriol, 1,3-dinitrate

$O_2NOCH_2CH(OH)CH_2ONO_2$

182 $C_4H_{10}N_2O_2S_2$ 6171-08-0
Urea, 1-ethyl-3-(methylsulfonyl)-2-thio-

MeSO2NHCSNHEt

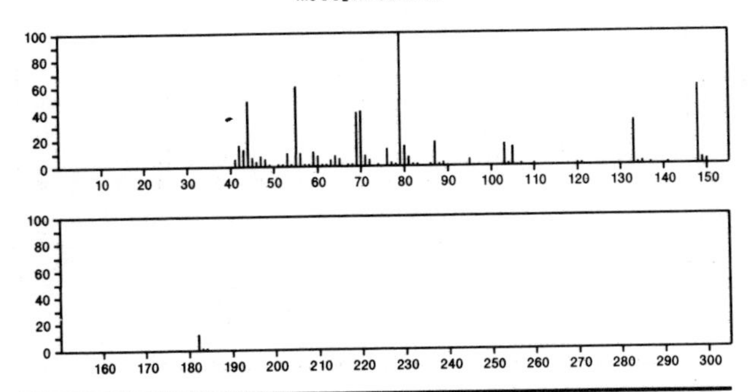

182 C_5H_8BrCl 14376-82-0
Cyclopentane, 1-bromo-2-chloro-, *trans*-

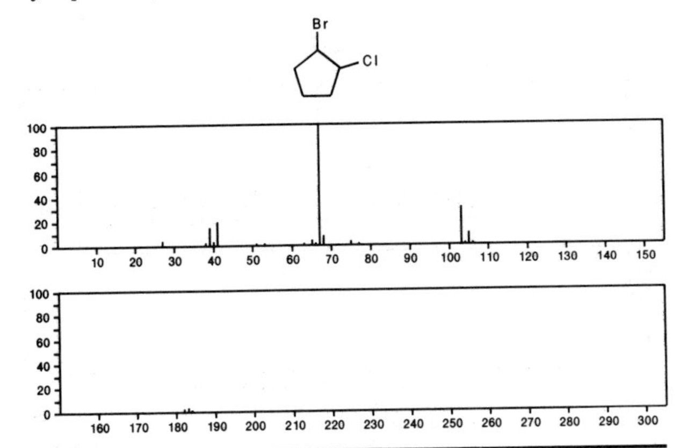

182 C_5H_8BrCl 37722-39-7
Cyclopentane, 1-bromo-2-chloro-, *cis*-

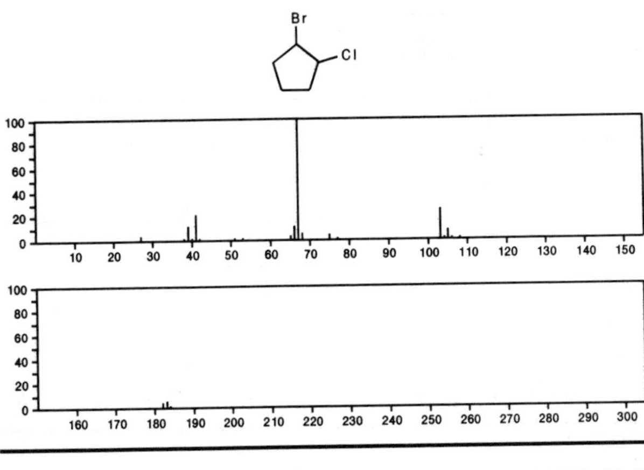

182 C_5H_8BrCl 55683-03-9
1-Pentene, 1-bromo-1-chloro-

ClCBr=CHPr

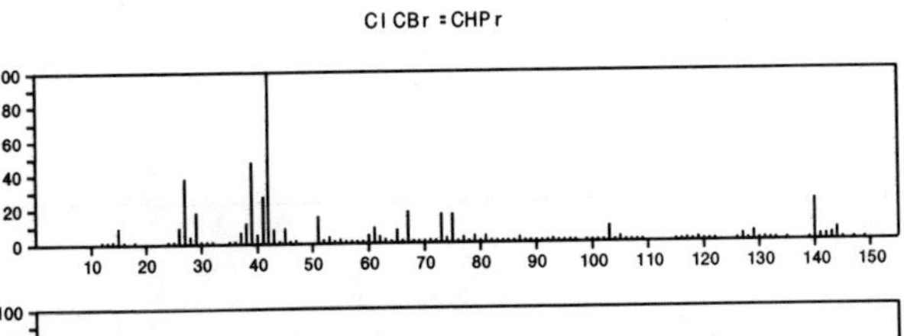

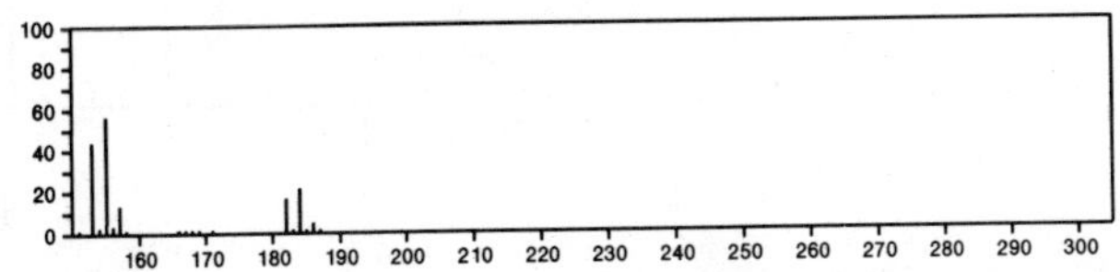

182 $C_6H_2F_4O_2$ 771-63-1
1,4-Benzenediol, 2,3,5,6-tetrafluoro-

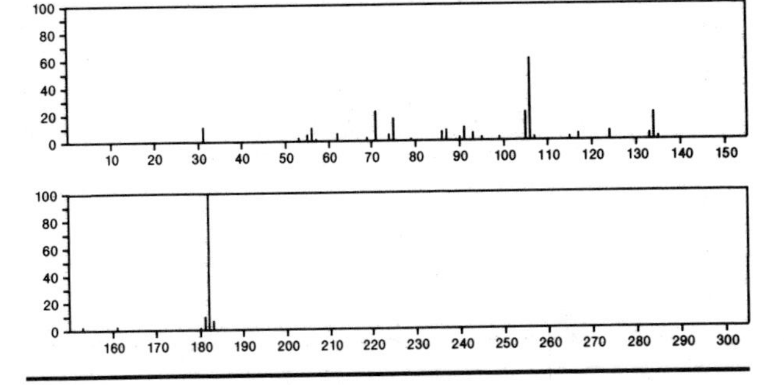

182 $C_6H_2F_4O_2$ 16840-25-8
1,3-Benzenediol, 2,4,5,6-tetrafluoro-

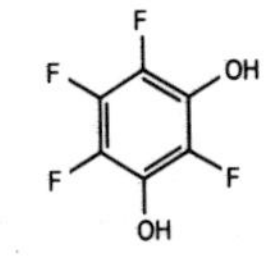

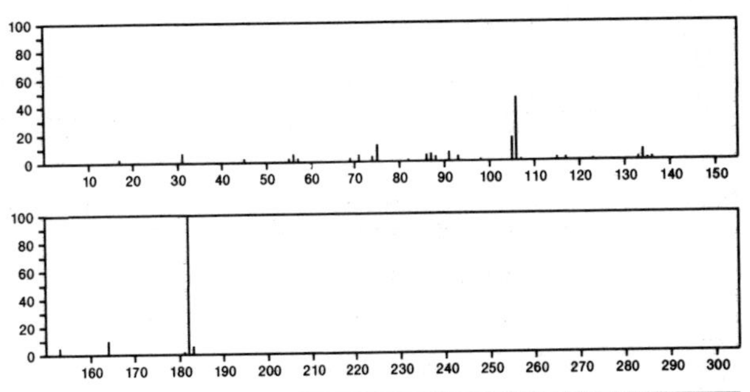

182 $C_6H_6N_4OS$ 28139-02-8
Xanthine, 3-methyl-2-thio-

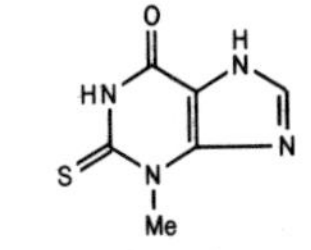

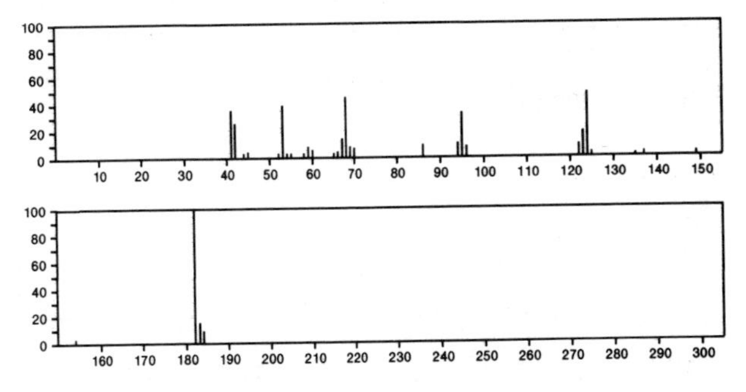

182 $C_6H_6N_4O_3$ 33070-47-2
[1,2,5]Oxadiazolo[3,4-*d*]pyrimidine-5,7(4*H*,6*H*)-dione, 4,6-dimethyl-

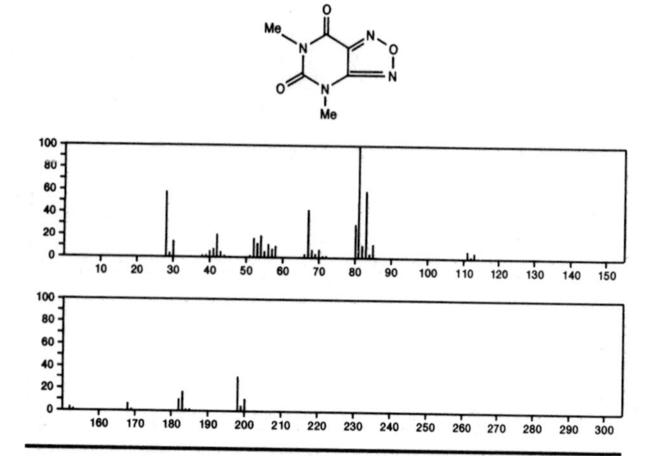

182 $C_6H_{11}ClN_2.ClH$ 40645-62-3
Propanimidamide, *N*-(1-chloro-1-propenyl)-, monohydrochloride

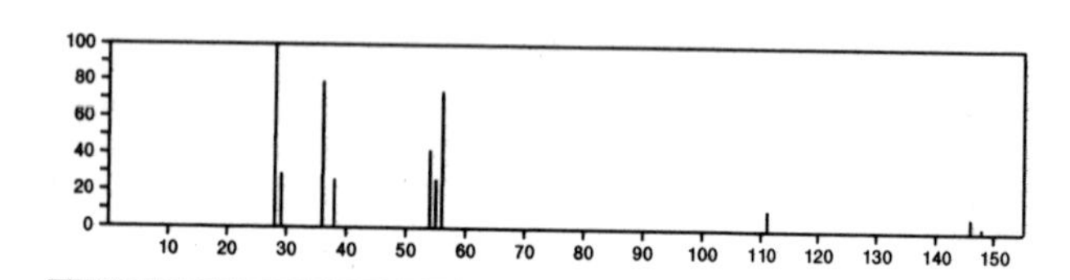

182 $C_6H_{14}O_2S_2$ 5244-34-8
Ethanol, 2,2'-[1,2-ethanediylbis(thio)]bis-

HOCH2 CH2 SCH2 CH2 SCH2 CH2 OH

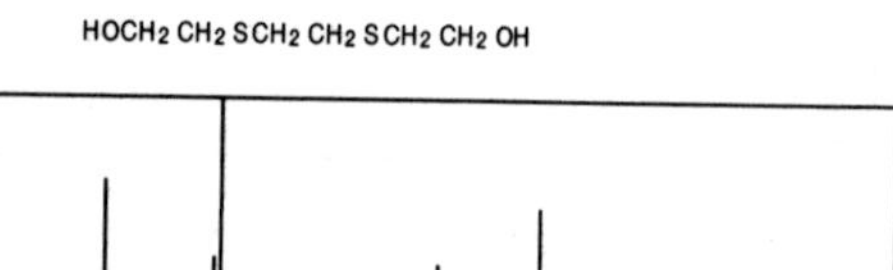

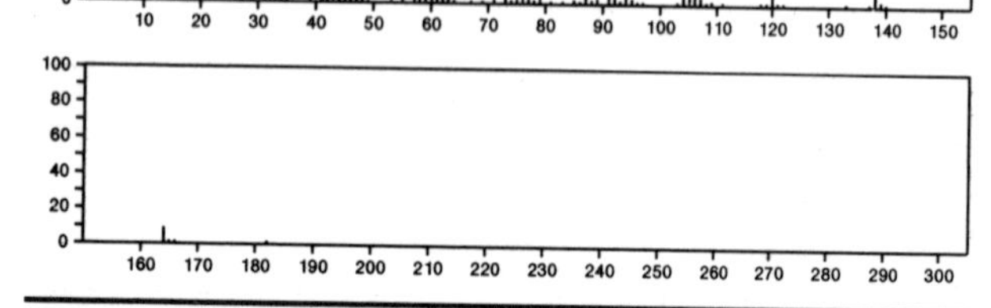

182 $C_6H_{14}S_3$ 6028-61-1
Trisulfide, dipropyl

PrSSSPr

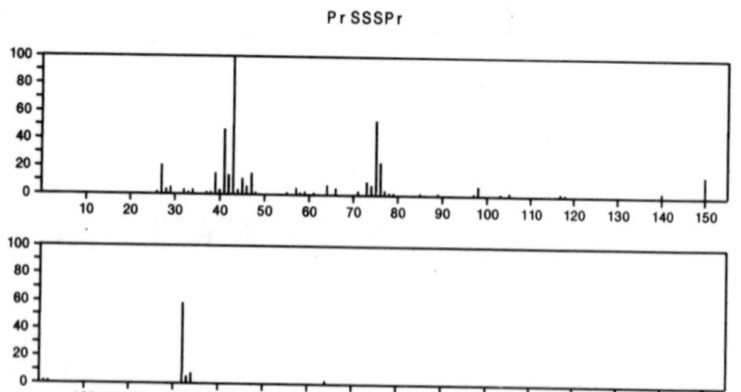

182 $C_6H_{14}S_3$ 54724-98-0
Ethane, 1,1'-[[(methylthio)methylene]bis(thio)]bis-

EtS2CHSMe

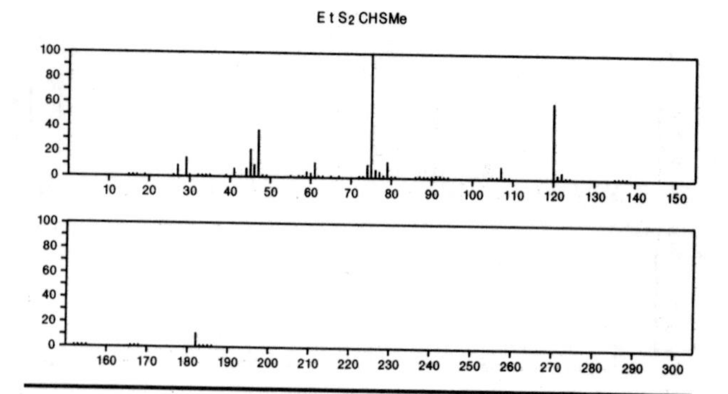

182 $C_6H_{15}O_2PS$ 13088-83-0
Phosphonothioic acid, methyl-, *O*-ethyl *S*-propyl ester

PrSPMeOEt

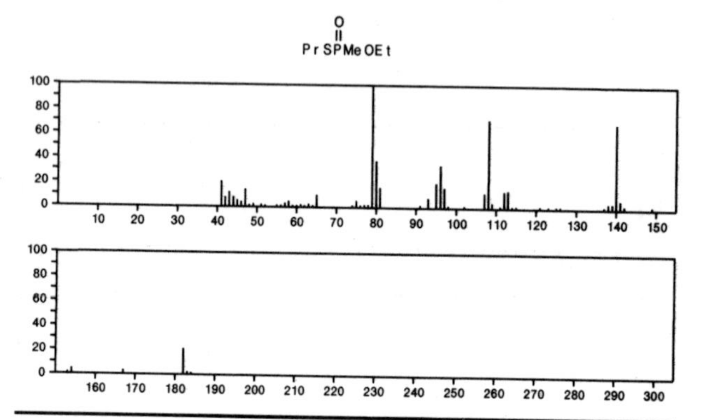

182 $C_7H_3ClN_2O_2$ 6575-07-1
Benzonitrile, 2-chloro-6-nitro-

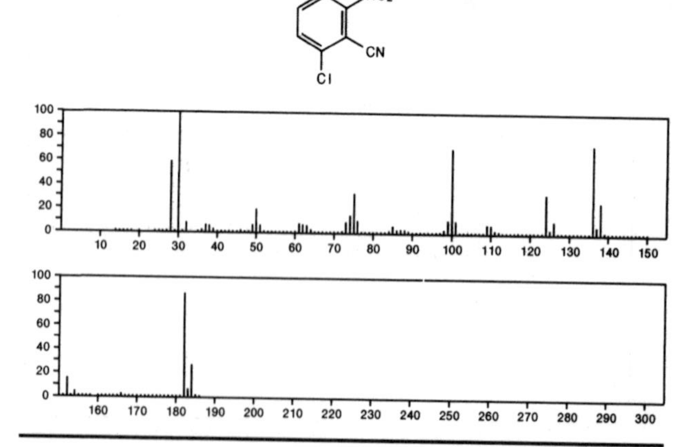

182 $C_7H_3F_5$ 771-56-2
Benzene, pentafluoromethyl-

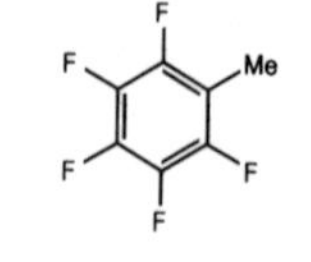

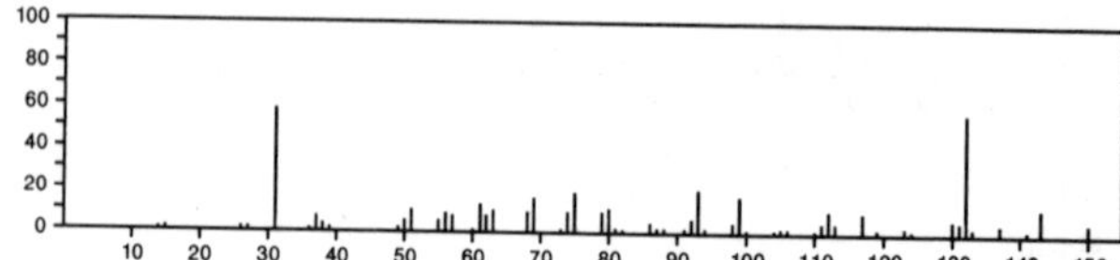

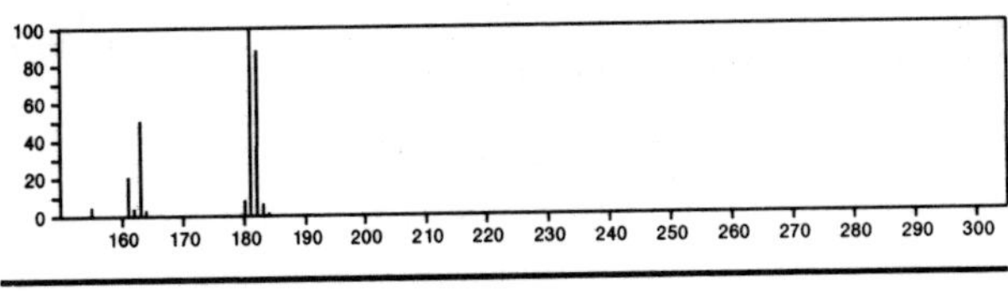

182 $C_7H_6N_2O_4$ 121-14-2
Benzene, 1-methyl-2,4-dinitro-

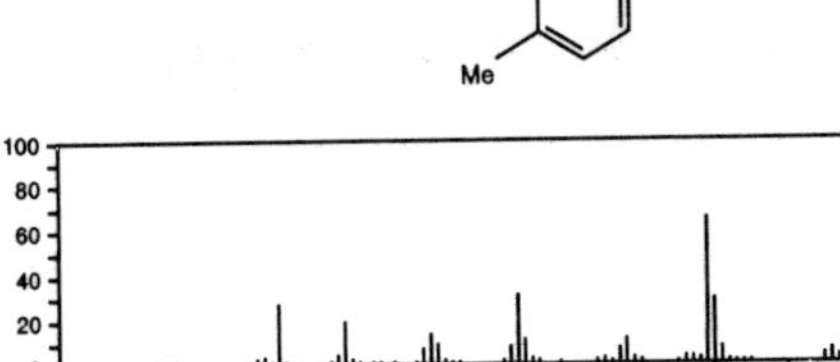

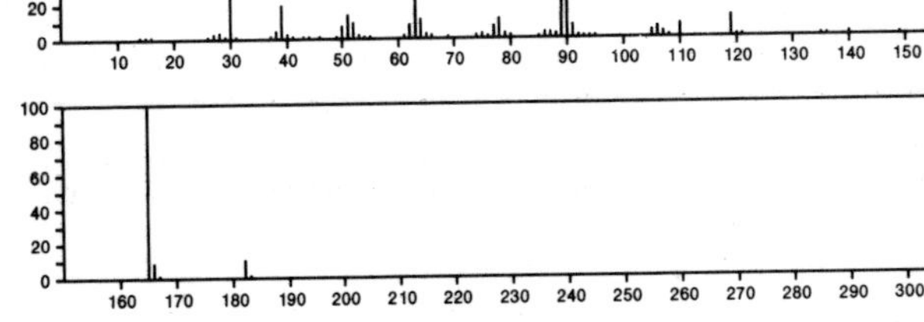

182 $C_7H_6N_2O_4$ 602-01-7
Benzene, 1-methyl-2,3-dinitro-

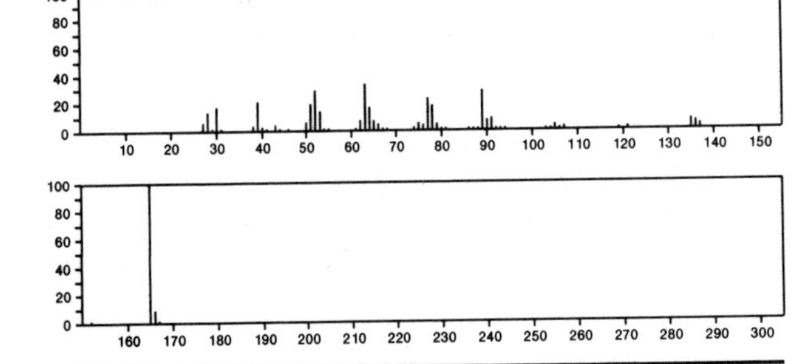

182 $C_7H_6N_2O_4$ 606-20-2
Benzene, 2-methyl-1,3-dinitro-

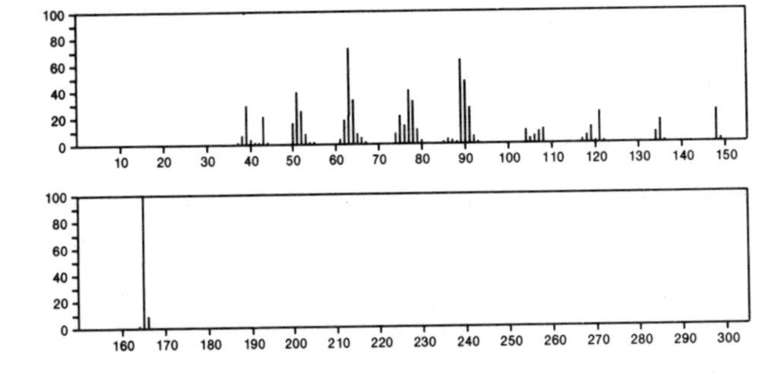

182 $C_7H_6N_2O_4$ 610-39-9
Benzene, 4-methyl-1,2-dinitro-

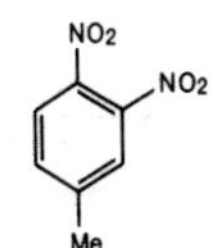

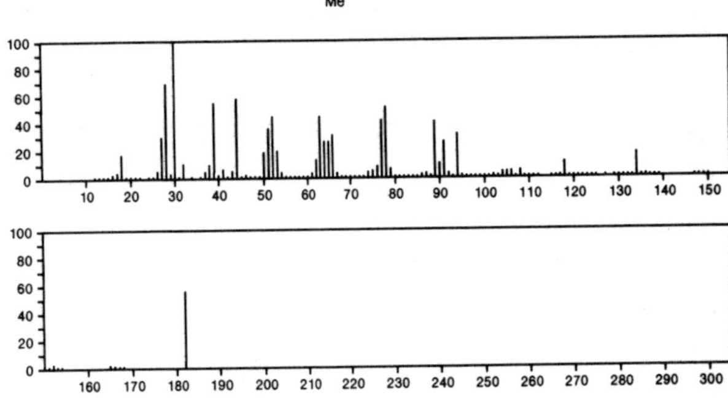

182 $C_7H_6N_2O_4$ 618-85-9
Benzene, 1-methyl-3,5-dinitro-

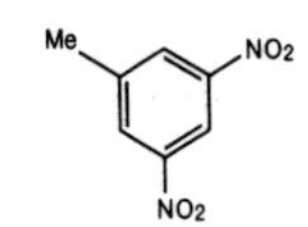

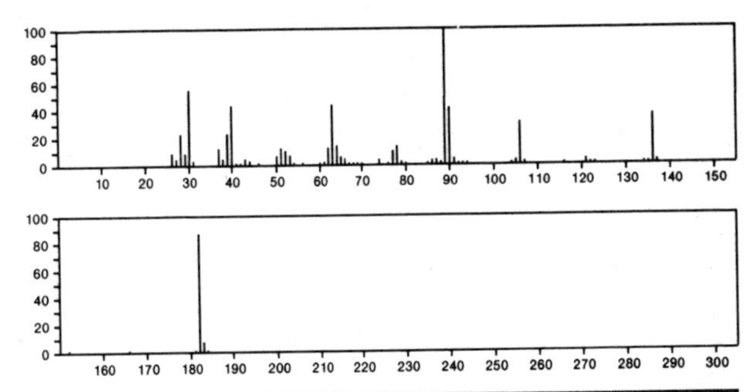

182 $C_7H_6N_2O_4$ 619-15-8
Benzene, 2-methyl-1,4-dinitro-

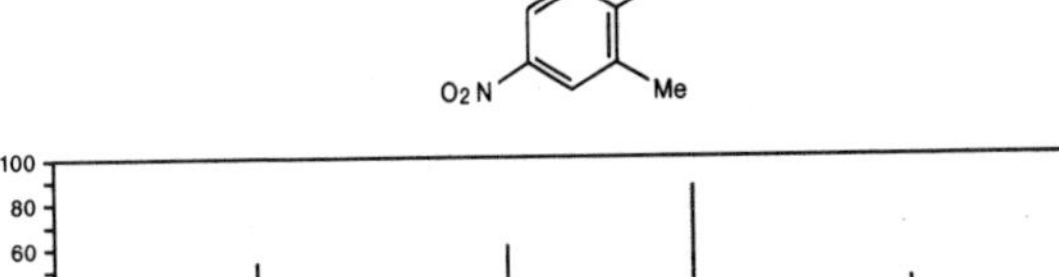

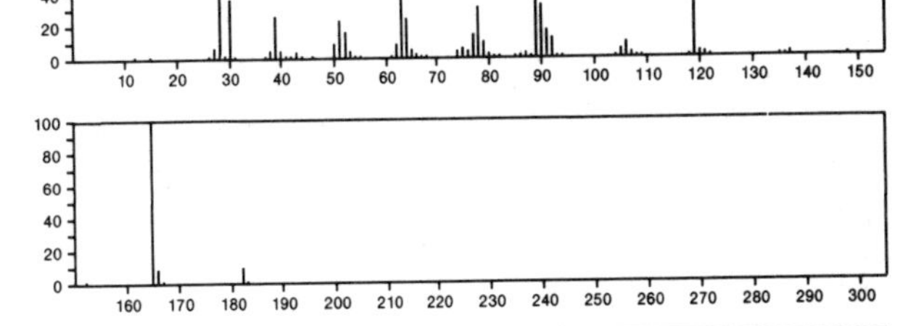

182 $C_7H_9F_3O_2$ 30984-28-2
2,4-Hexanedione, 1,1,1-trifluoro-5-methyl-

$F_3CCOCH_2COCHMe_2$

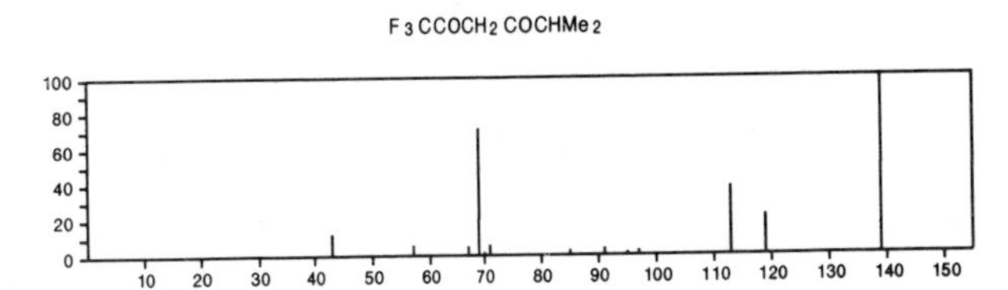

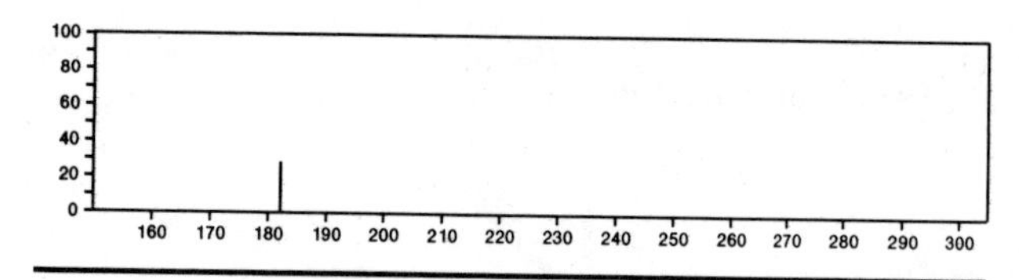

182 $C_7H_9F_3O_2$ 33284-43-4
2,4-Heptanedione, 1,1,1-trifluoro-

F3CCOCH2COPr

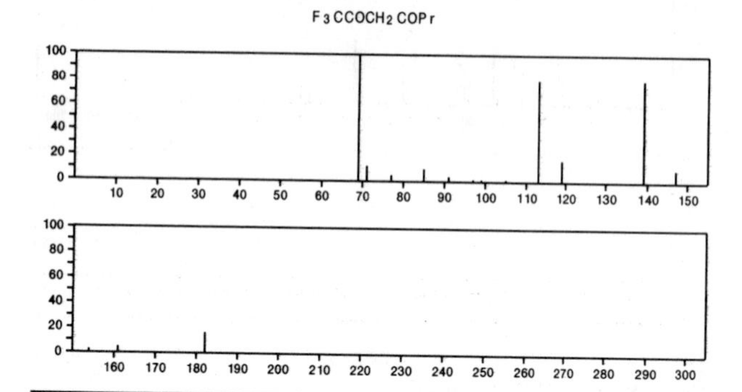

182 $C_7H_{10}N_4S$ 37527-51-8
3*H*-Purine, 6,7-dihydro-3-methyl-6-(methylthio)-

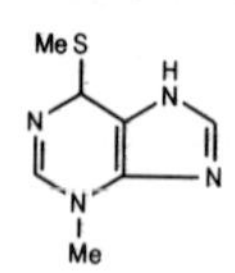

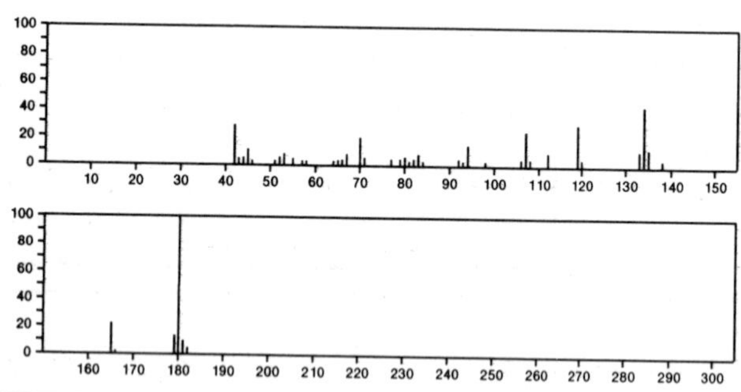

182 $C_7H_{12}Cl_2O$ 40624-07-5
4-Heptanone, 1,7-dichloro-

Cl(CH2)3CO(CH2)3Cl

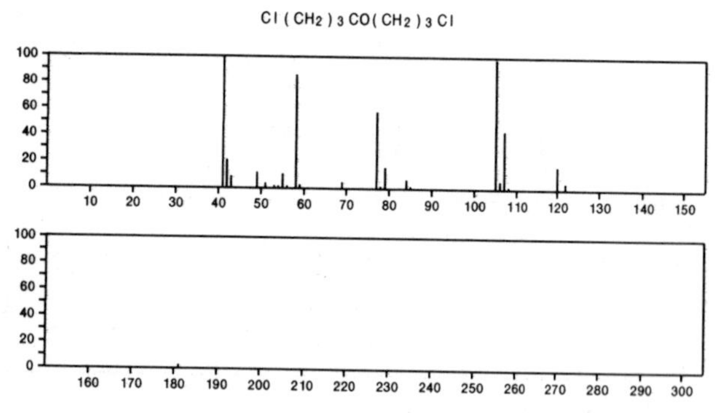

182 $C_7H_{14}N_6$ 5606-16-6
1,3,5-Triazine-2,4,6-triamine, *N,N'*-diethyl-

EtNH NHEt NH2

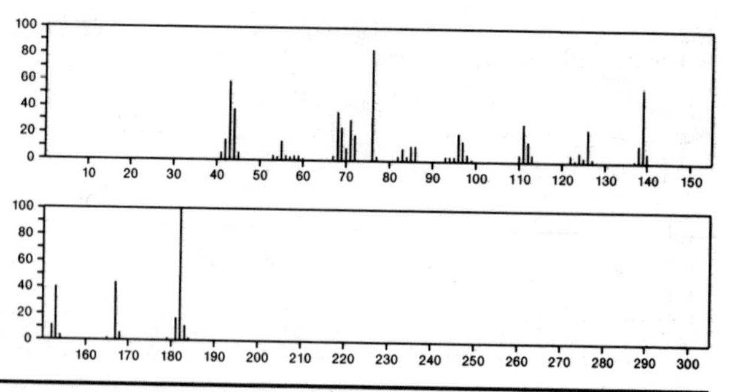

182 $C_8H_6O_5$ 52183-73-0
Bisoxireno[*e,g*]isobenzofuran-3,5-dione, hexahydro-, (1aα,1bβ,= 2aβ,2bβ,5aα,5bα)-

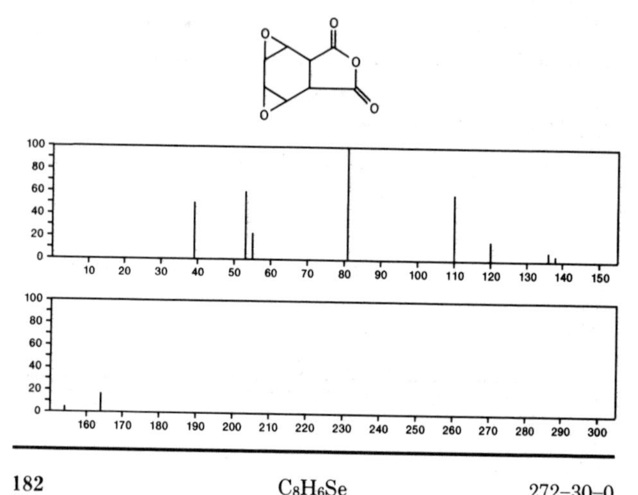

182 C_8H_6Se 272-30-0
Benzo[*b*]selenophene

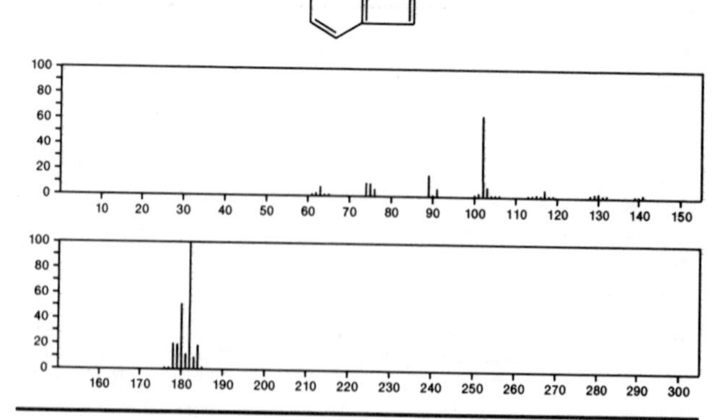

182 C_8H_7Br 103-64-0
Benzene, (2-bromoethenyl)-

BrCH=CHPh

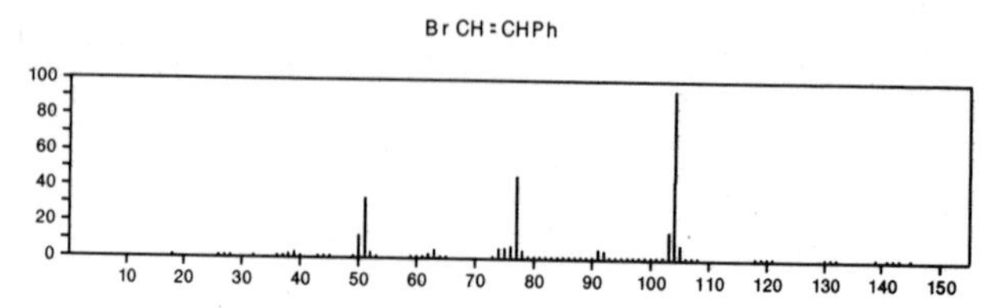

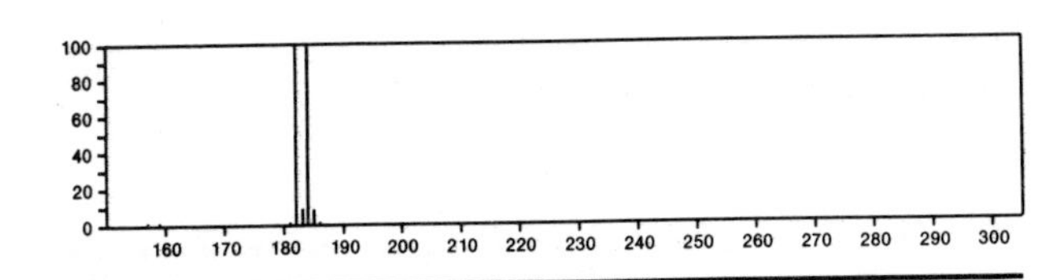

182 $C_8H_{10}N_2OS$ 20184-98-9
Carbazic acid, 3-phenylthio-, *O*-methyl ester

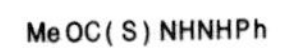

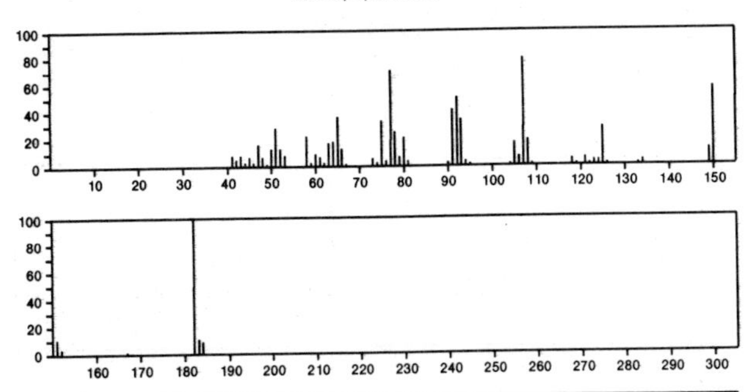

182 $C_9H_{10}O_2S$ 103-46-8
Acetic acid, [(phenylmethyl)thio]-

$PhCH_2SCH_2CO_2H$

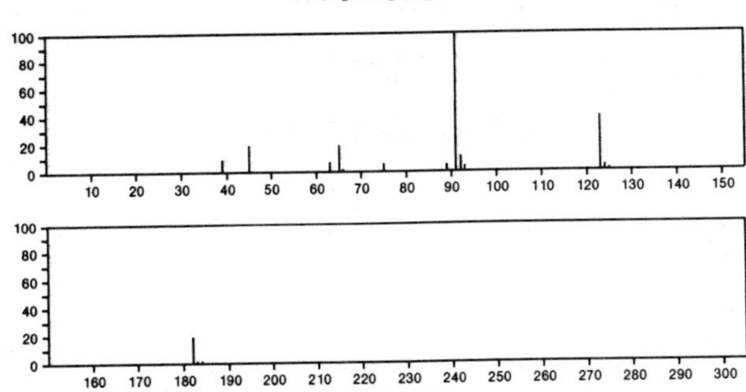

182 $C_9H_{10}O_2S$ 5342-84-7
Benzene, [2-(methylsulfonyl)ethenyl]-

$PhCH{=}CHSO_2Me$

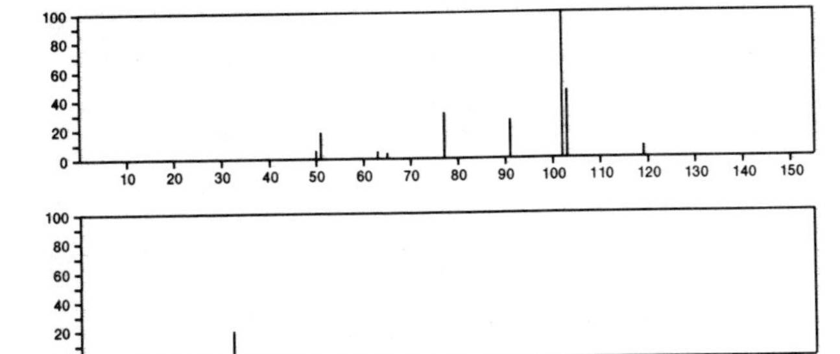

182 $C_9H_{10}O_2S$ 5535-52-4
Benzene, 1-(ethenylsulfonyl)-4-methyl-

$SO_2CH{=}CH_2$
Me

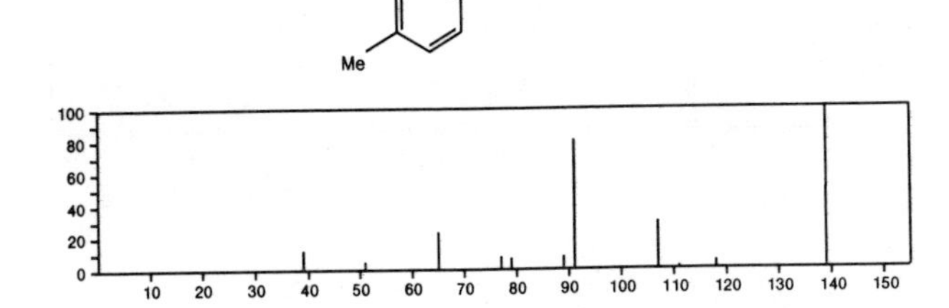

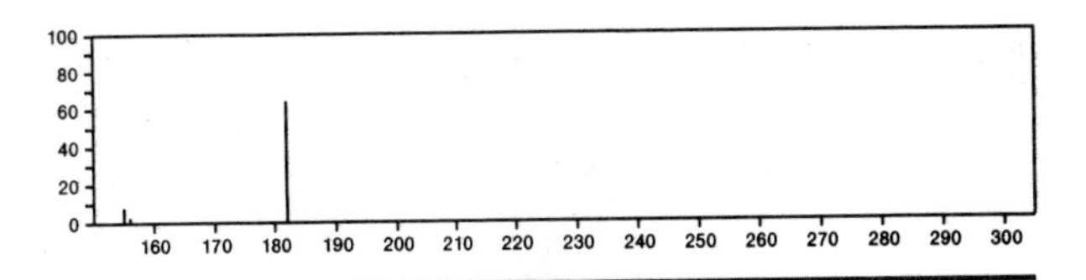

182 $C_9H_{10}O_2S$ 5925-50-8
Benzenecarbothioic acid, 4-methoxy-, *O*-methyl ester

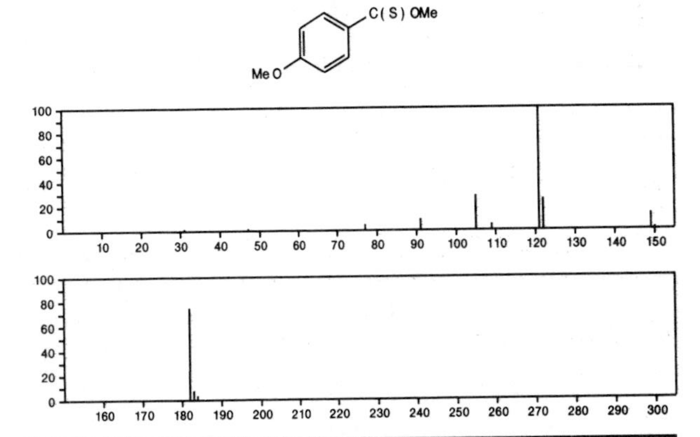

182 $C_9H_{10}O_2S$ 5925-72-4
Benzenecarbothioic acid, 4-methoxy-, *S*-methyl ester

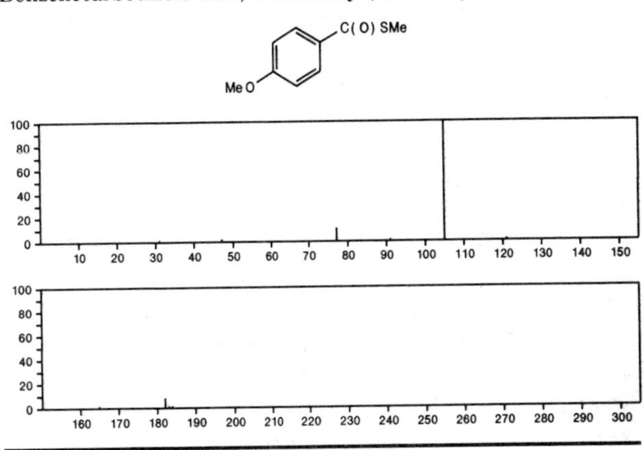

182 $C_9H_{10}O_4$ 93-07-2
Benzoic acid, 3,4-dimethoxy-

OMe
HO_2C
OMe

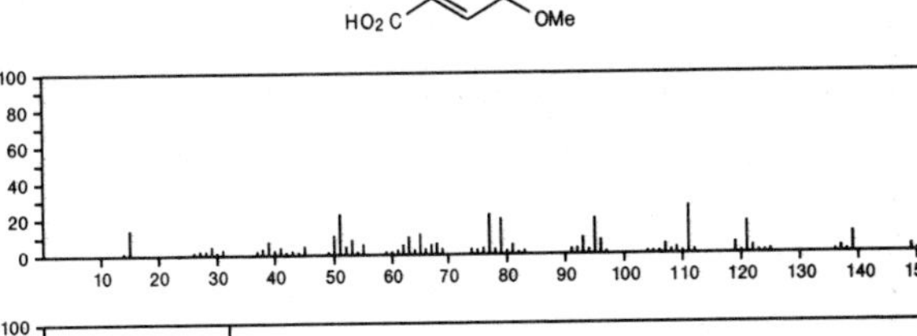

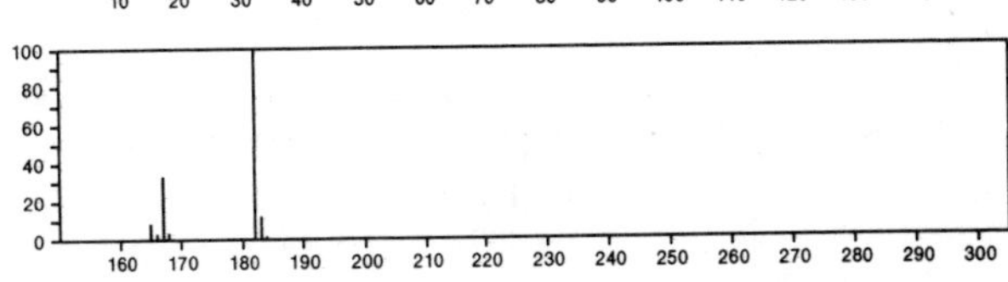

182 $C_9H_{10}O_4$ 134-96-3
Benzaldehyde, 4-hydroxy-3,5-dimethoxy-

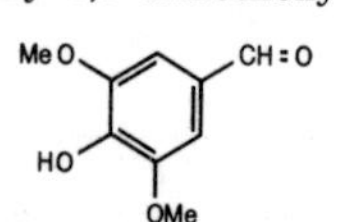

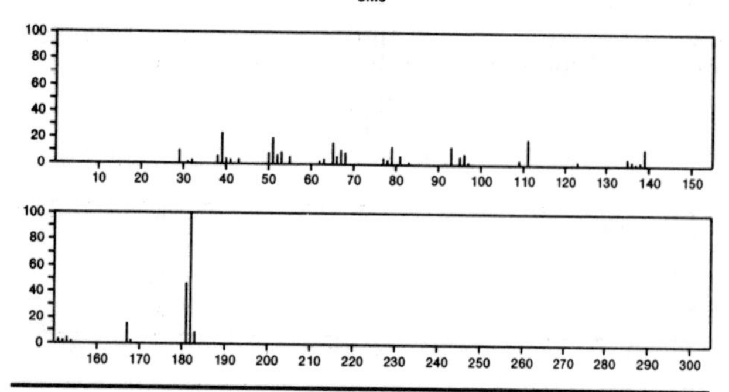

182 $C_9H_{10}O_4$ 306-08-1
Benzeneacetic acid, 4-hydroxy-3-methoxy-

OH
OMe
CH2 CO2 H

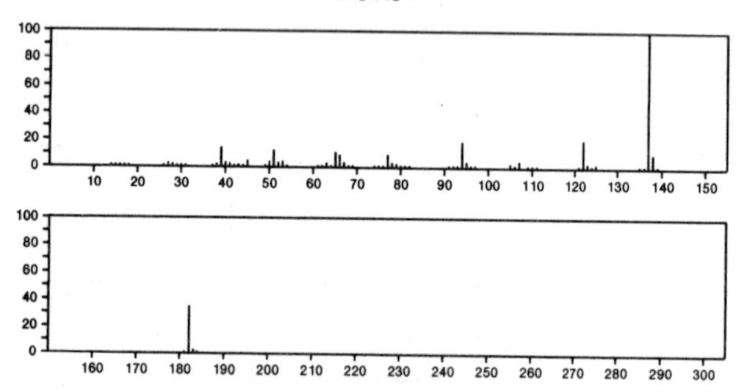

182 $C_9H_{10}O_4$ 306-23-0
Benzenepropanoic acid, α,4-dihydroxy-

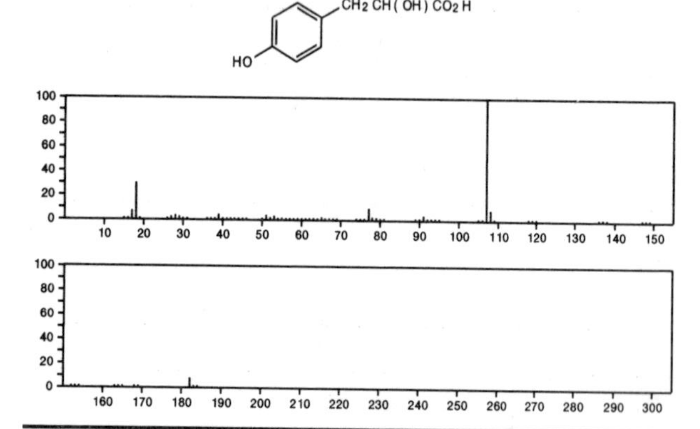

182 $C_9H_{10}O_4$ 570-10-5
Benzoic acid, 2-hydroxy-4-methoxy-6-methyl-

HO
OMe
HO2 C
Me

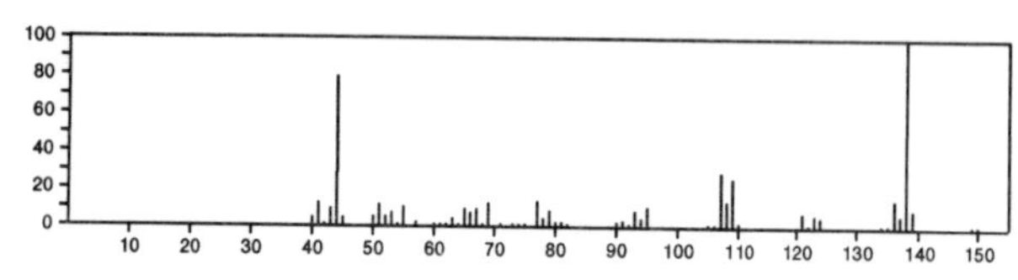

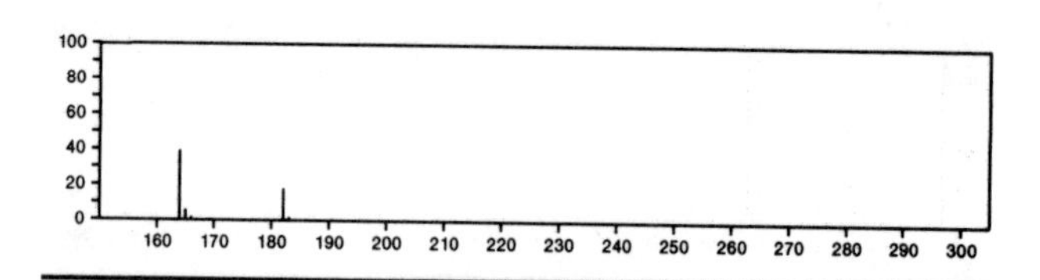

182 $C_9H_{10}O_4$ 605-94-7
2,5-Cyclohexadiene-1,4-dione, 2,3-dimethoxy-5-methyl-

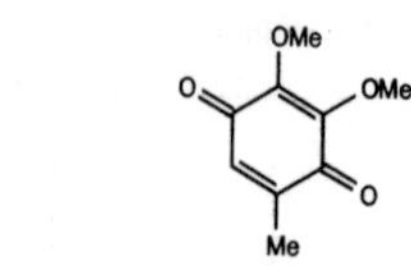

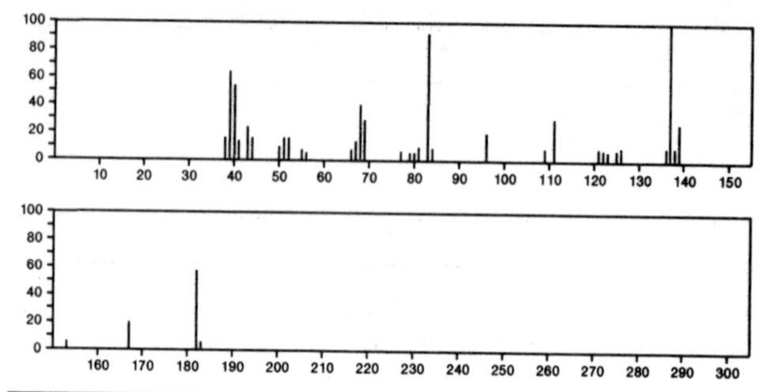

182 $C_9H_{10}O_4$ 1078-61-1
Benzenepropanoic acid, 3,4-dihydroxy-

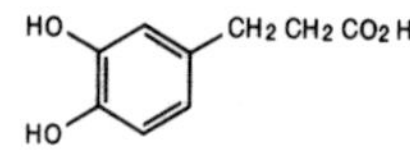

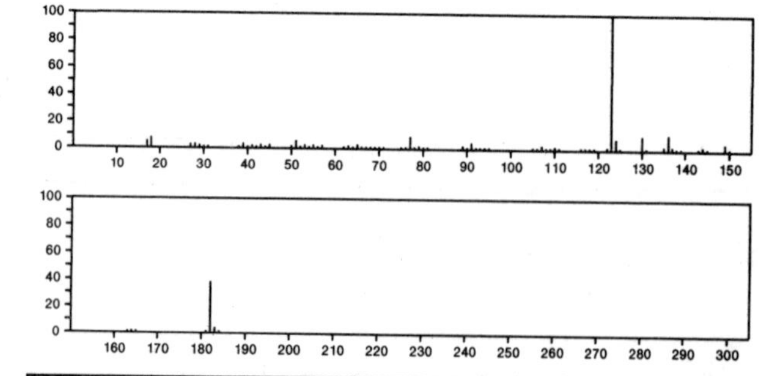

182 $C_9H_{10}O_4$ 3187-58-4
Benzoic acid, 2,4-dihydroxy-6-methyl-, methyl ester

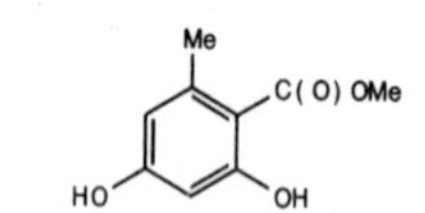

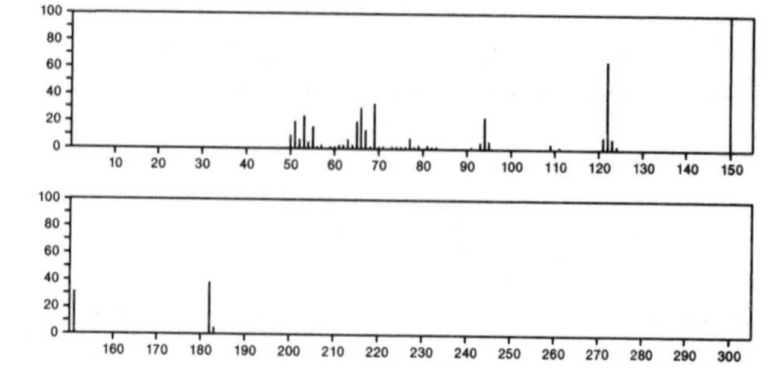

182 $C_9H_{10}O_4$ 3943–74–6
Benzoic acid, 4–hydroxy–3–methoxy–, methyl ester

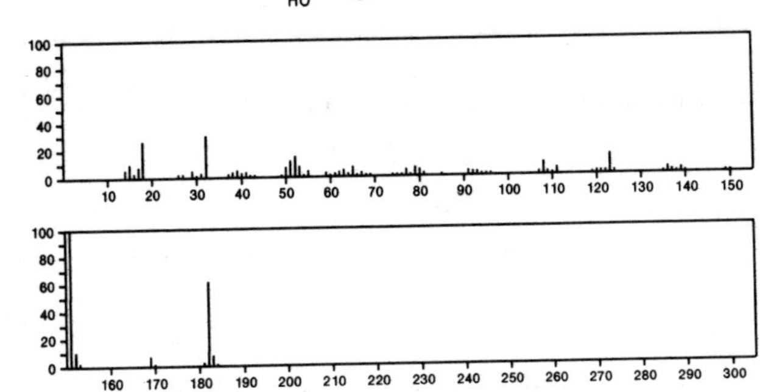

182 $C_9H_{10}O_4$ 7507–89–3
Ethanone, 1–(2,6–dihydroxy–4–methoxyphenyl)–

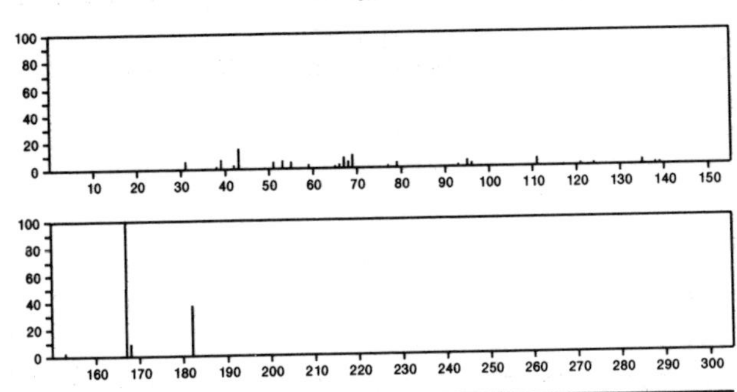

182 $C_9H_{10}O_4$ 10408–29–4
Benzeneacetic acid, α–hydroxy–2–methoxy–

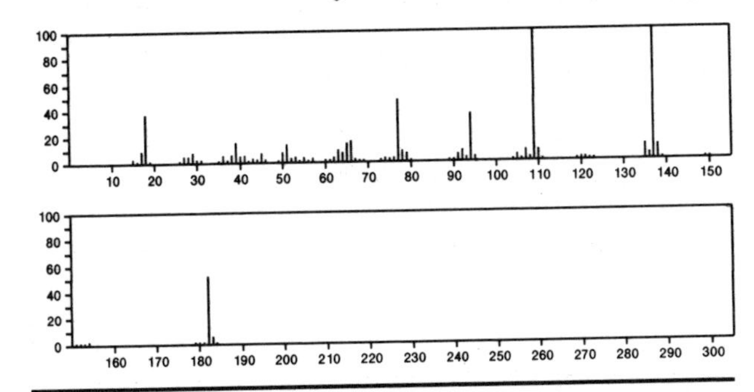

182 $C_9H_{10}O_4$ 10502–44–0
Benzeneacetic acid, α–hydroxy–4–methoxy–

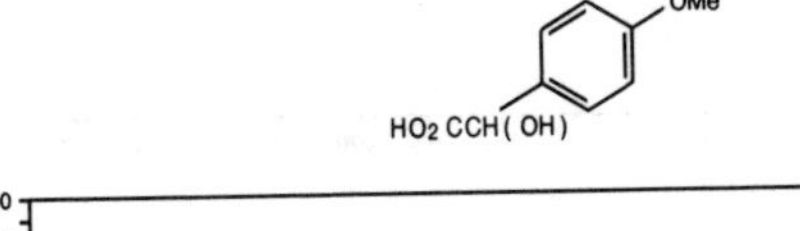

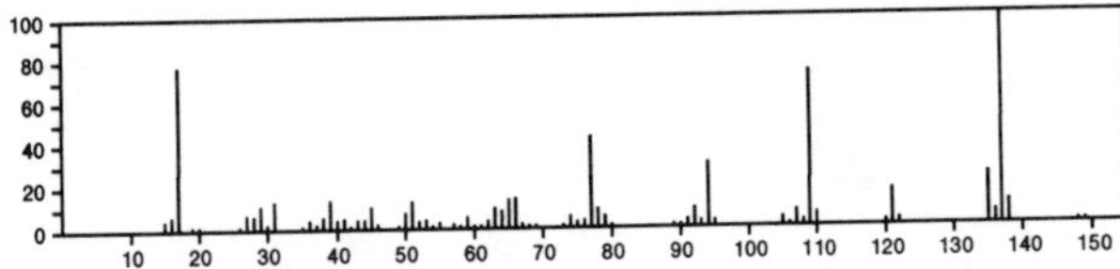

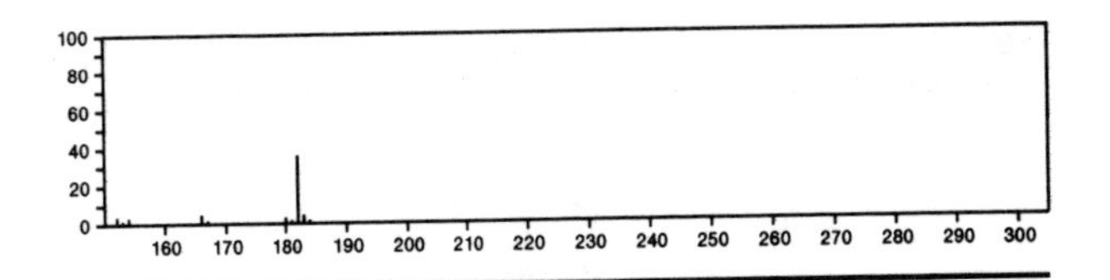

182 $C_9H_{10}O_4$ 21150–12–9
Benzeneacetic acid, α–hydroxy–3–methoxy–

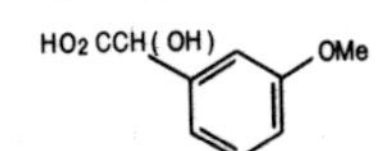

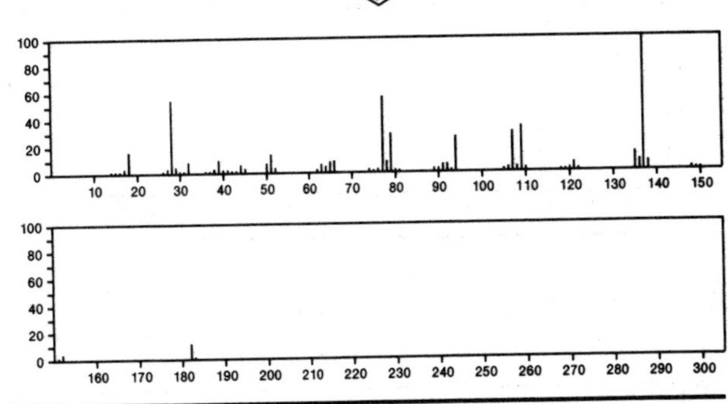

182 $C_9H_{10}O_4$ 22159–41–7
Carbonic acid, 4–methoxyphenyl methyl ester

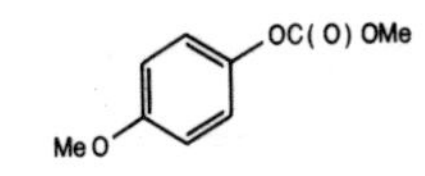

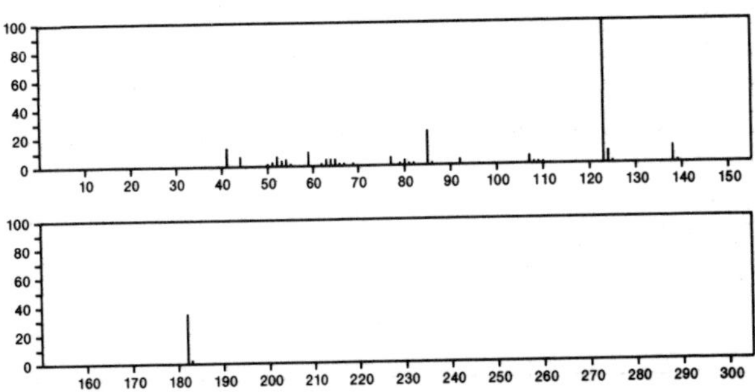

182 $C_9H_{10}O_4$ 54644–49–4
Carbonic acid, 3–methoxyphenyl methyl ester

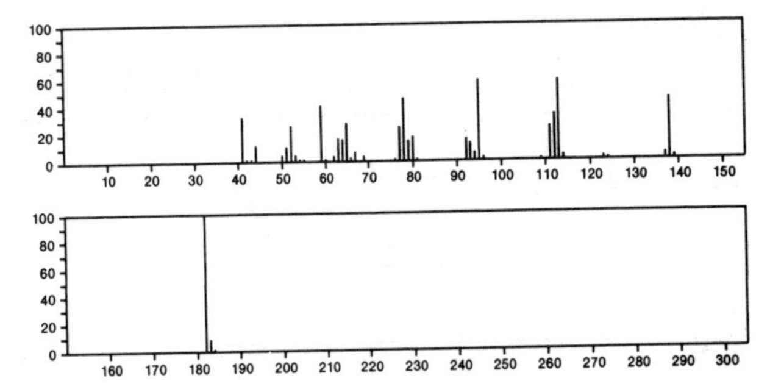

182 $C_9H_{10}O_4$ 57174-14-8
1,2,4-Cyclopentanetrione, 3-methyl-5-(1-oxopropyl)-

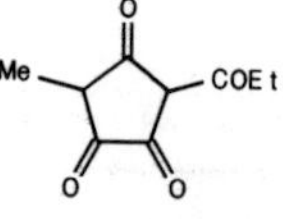

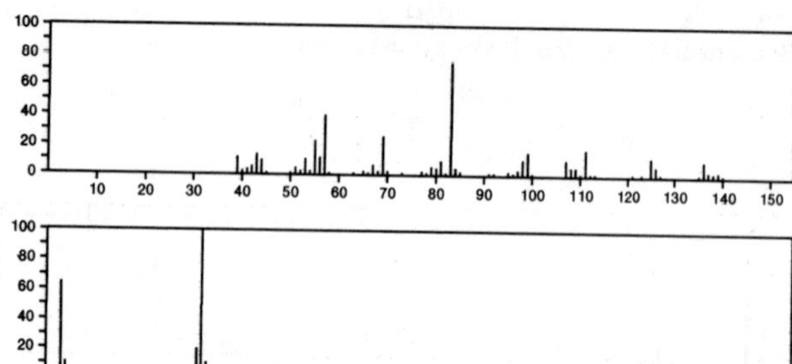

182 $C_9H_{10}S_2$ 936-63-0
Benzenecarbodithioic acid, ethyl ester

EtSC(S)Ph

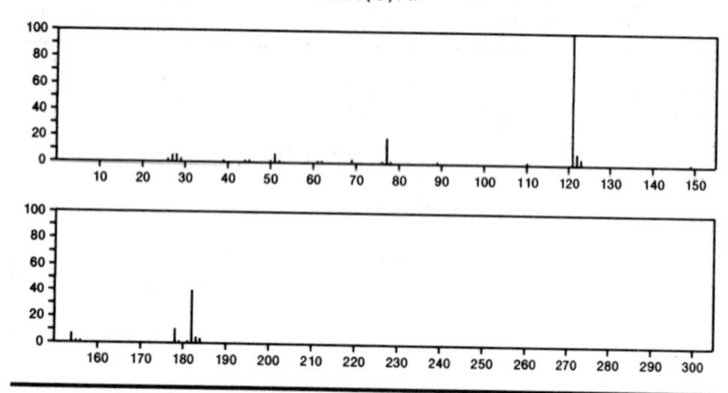

182 $C_9H_{10}S_2$ 5616-55-7
1,3-Dithiolane, 2-phenyl-

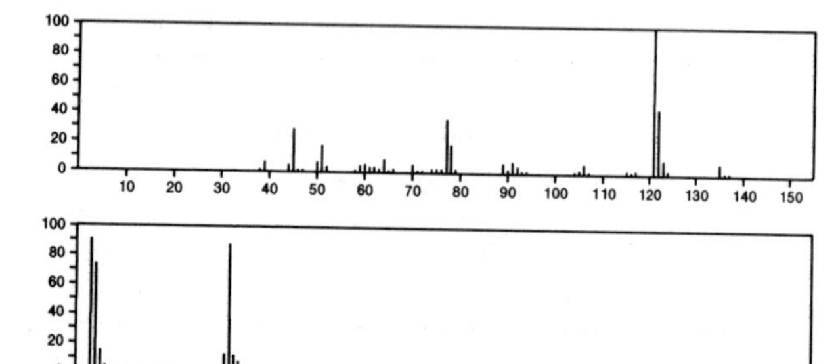

182 $C_9H_{11}ClN_2$ 2103-46-0
Methanimidamide, *N'*-(4-chlorophenyl)-*N*,*N*-dimethyl-

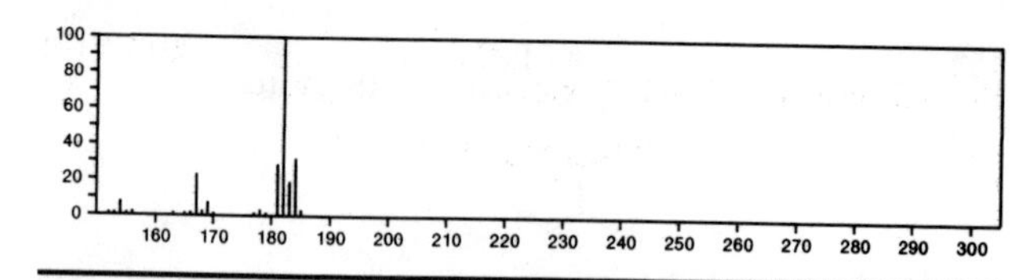

182 $C_9H_{11}ClN_2$ 2103-49-3
Methanimidamide, *N'*-(2-chlorophenyl)-*N*,*N*-dimethyl-

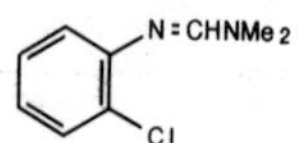

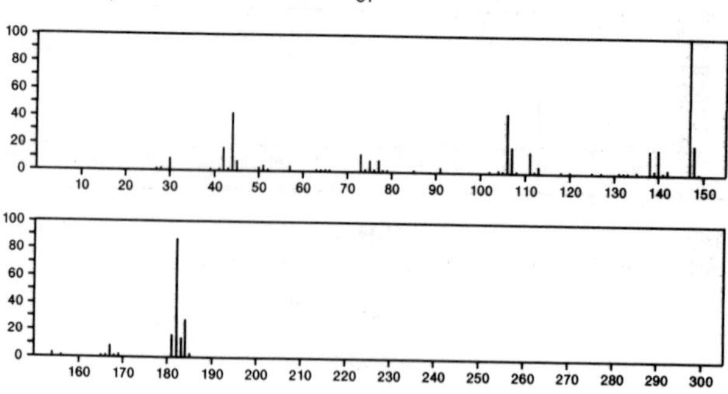

182 $C_9H_{11}ClN_2$ 2103-50-6
Methanimidamide, *N'*-(3-chlorophenyl)-*N*,*N*-dimethyl-

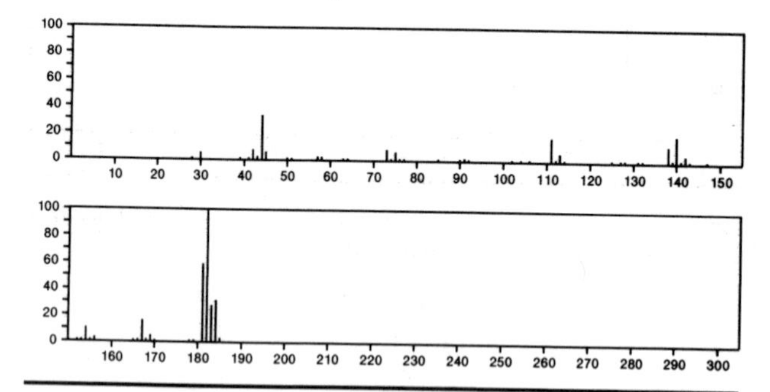

182 $C_9H_{14}N_2O_2$ 707-09-5
1,3-Diazaspiro[4.5]decane-2,4-dione, 3-methyl-

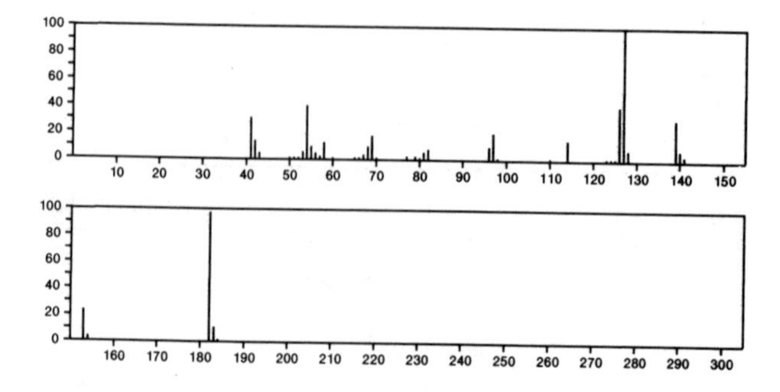

182 $C_9H_{14}N_2O_2$ 878-46-6
1,3-Diazaspiro[4.5]decane-2,4-dione, 1-methyl-

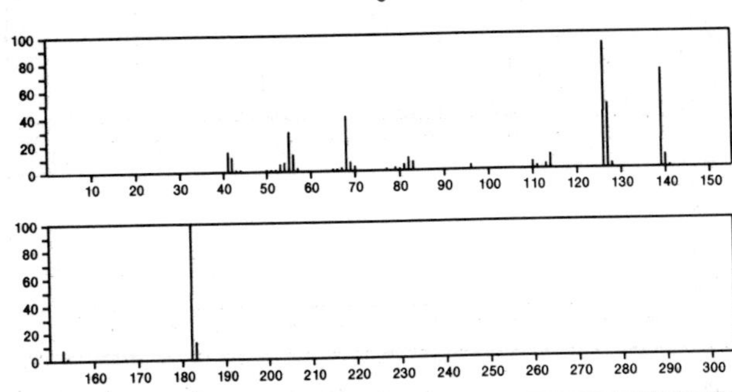

182 $C_9H_{14}N_2O_2$ 1010-89-5
Uracil, 1-butyl-6-methyl-

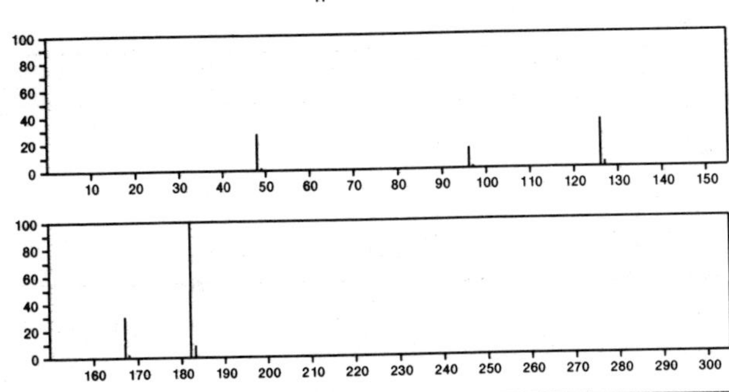

182 $C_9H_{14}N_2O_2$ 1010-90-8
2,4(1*H*,3*H*)-Pyrimidinedione, 3-butyl-6-methyl-

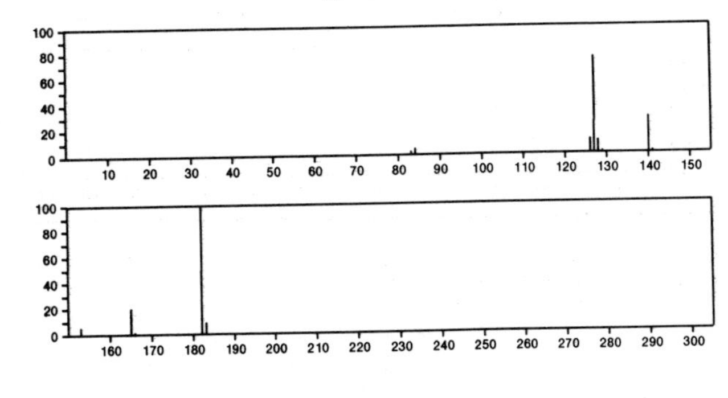

182 $C_{10}H_{11}ClO$ 4981-63-9
1-Butanone, 1-(4-chlorophenyl)-

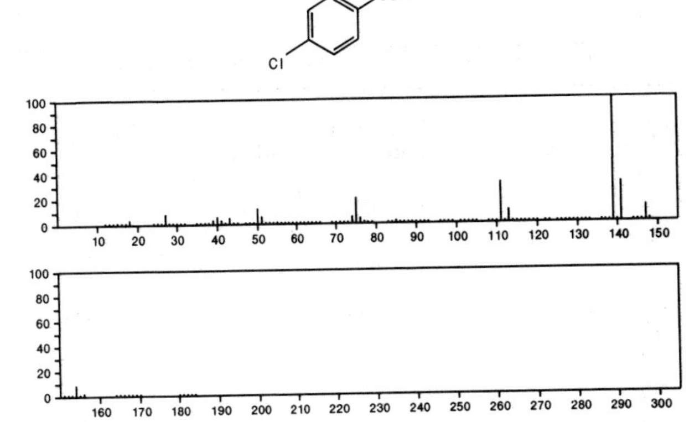

182 $C_{10}H_{11}ClO$ 54644-21-2
Benzene, 1-[(3-chloro-2-propenyl)oxy]-2-methyl-

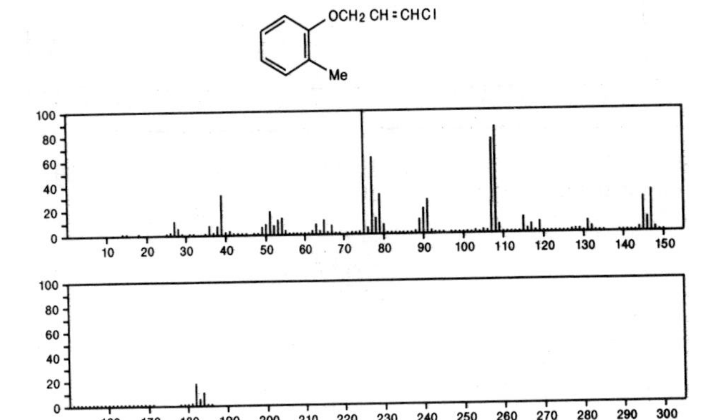

182 $C_{10}H_{11}ClO$ 54644-22-3
Benzene, 1-[(3-chloro-2-propenyl)oxy]-3-methyl-

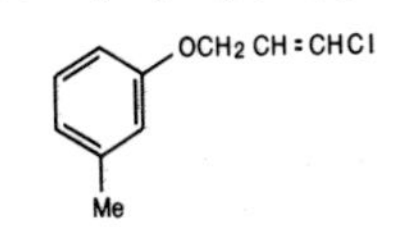

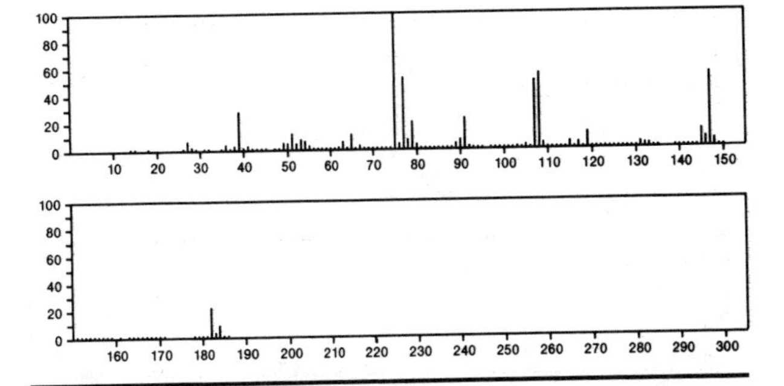

182 $C_{10}H_{11}ClO$ 54644-23-4
Benzene, 1-(3-chloro-2-propenyl)-4-methoxy-

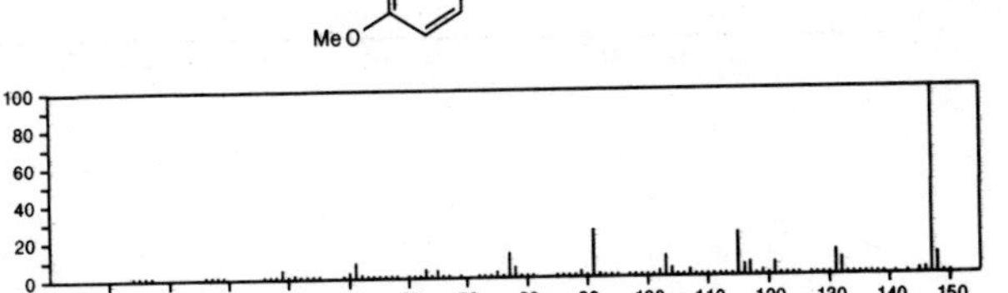

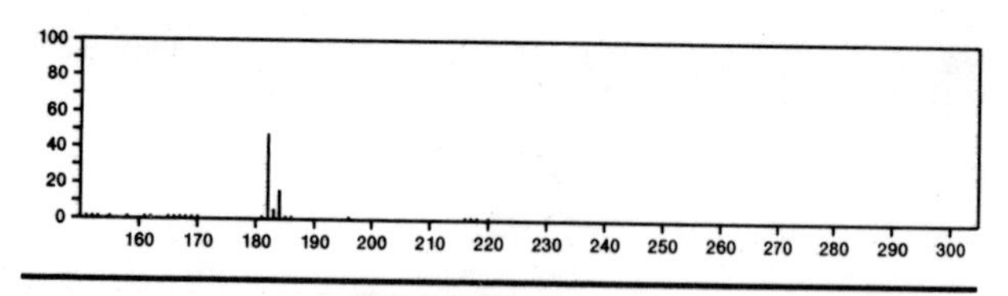

182 $C_{10}H_{11}ClO$ 55649-98-4
1-Propanone, 1-(3-chlorophenyl)-2-methyl-

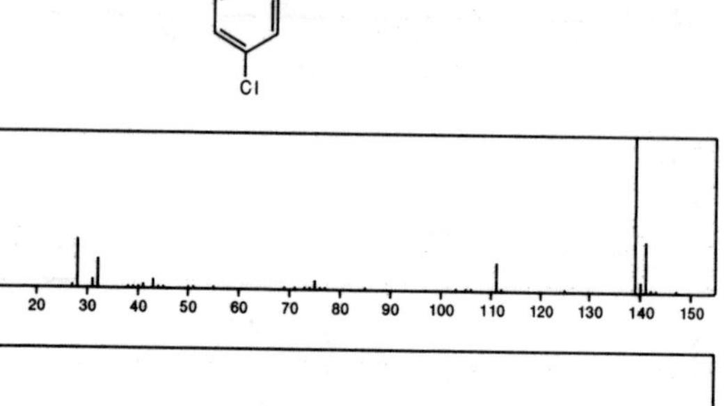

182 $C_{10}H_{11}FO_2$ 19305-99-8
2,6-Adamantanedione, 4-fluoro-

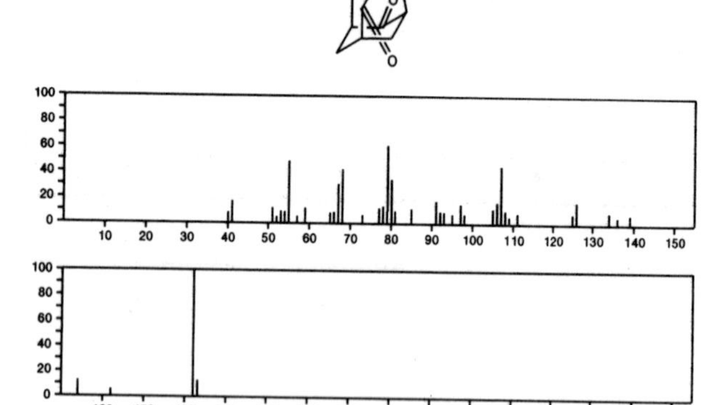

182 $C_{10}H_{14}OS$ 24362-87-6
Phenol, 2-(butylthio)-

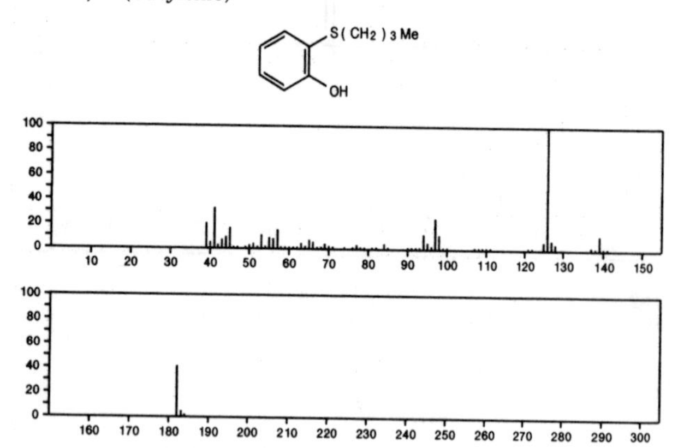

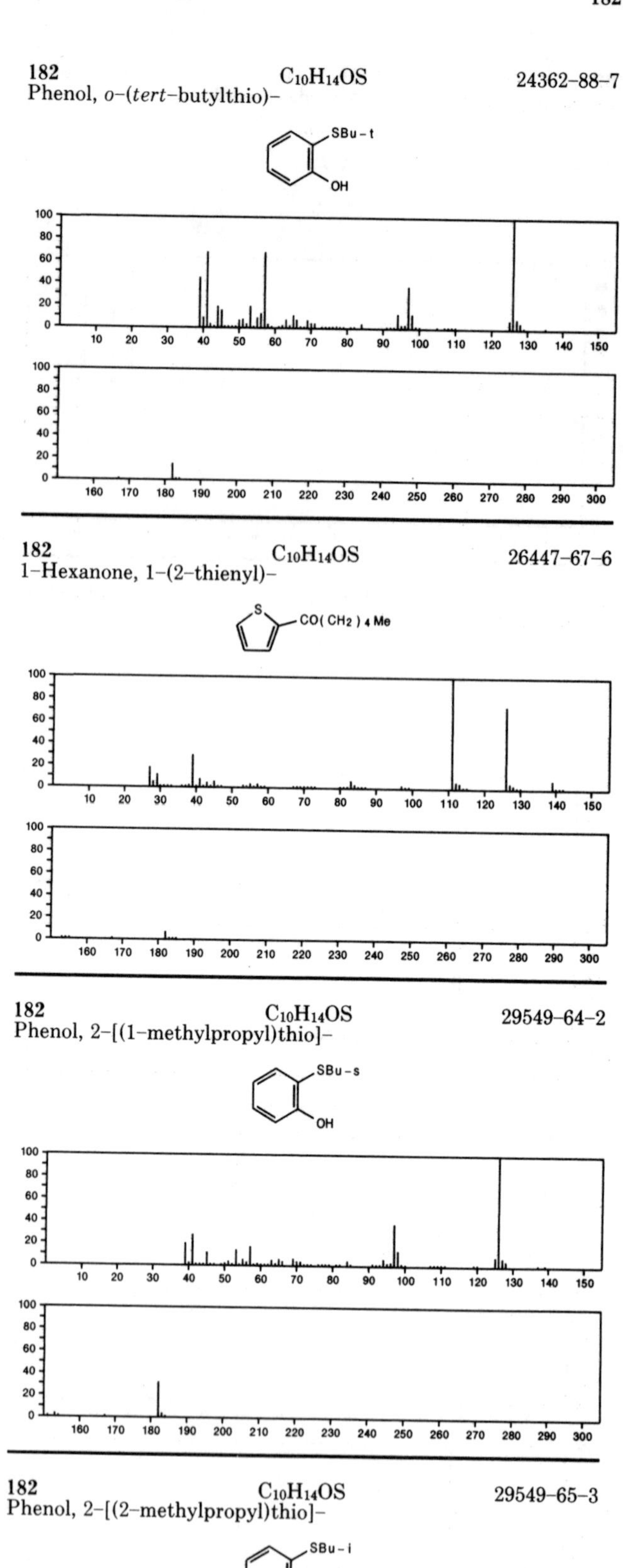

182 $C_{10}H_{14}OS$ 24362-88-7
Phenol, *o*-(*tert*-butylthio)-

182 $C_{10}H_{14}OS$ 26447-67-6
1-Hexanone, 1-(2-thienyl)-

182 $C_{10}H_{14}OS$ 29549-64-2
Phenol, 2-[(1-methylpropyl)thio]-

182 $C_{10}H_{14}OS$ 29549-65-3
Phenol, 2-[(2-methylpropyl)thio]-

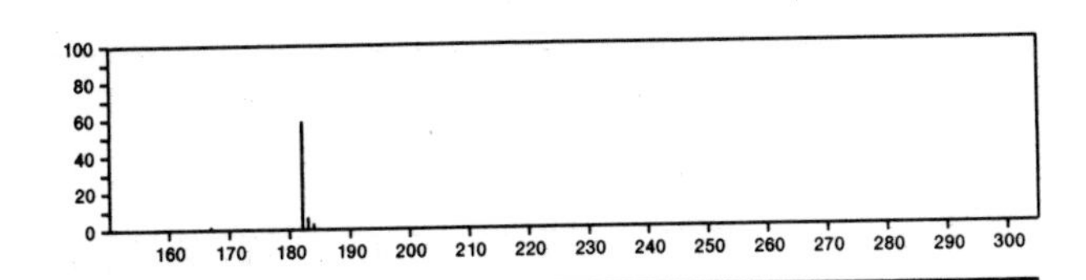

182 $C_{10}H_{14}O_3$ 104-68-7
Ethanol, 2-(2-phenoxyethoxy)-

$HOCH_2CH_2OCH_2CH_2OPh$

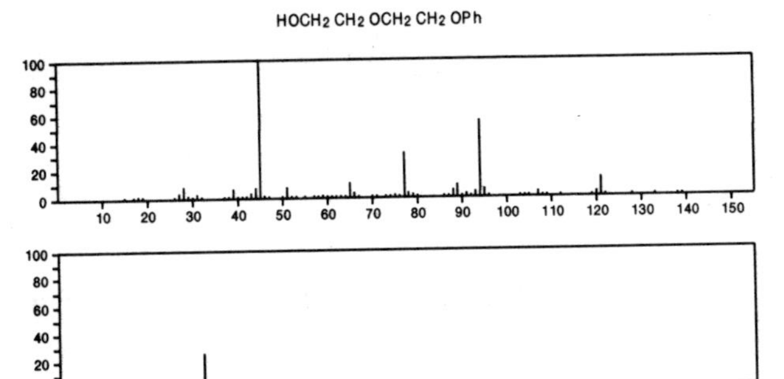

182 $C_{10}H_{14}O_3$ 707-07-3
Benzene, (trimethoxymethyl)-

$PhC(OMe)_3$

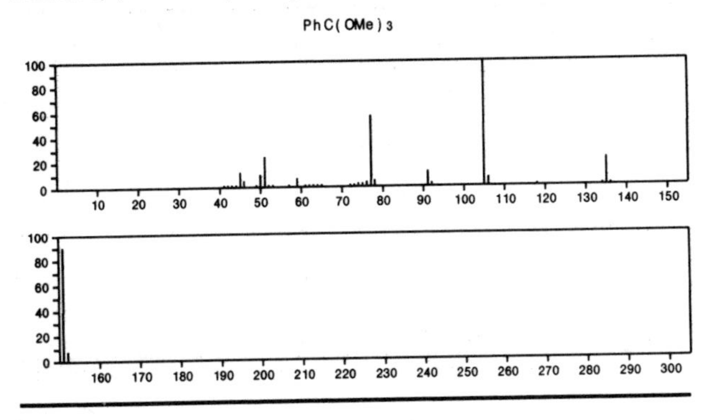

182 $C_{10}H_{14}O_3$ 19757-51-8
3-Butene-1,2-diol, 1-(2-furyl)-2,3-dimethyl-

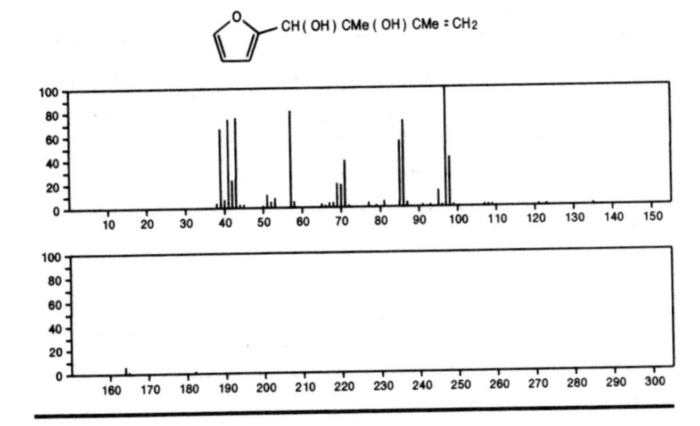

182 $C_{10}H_{14}O_3$ 41654-27-7
2-Cyclohexene-1,4-dione, 2-methoxy-3,5,5-trimethyl-

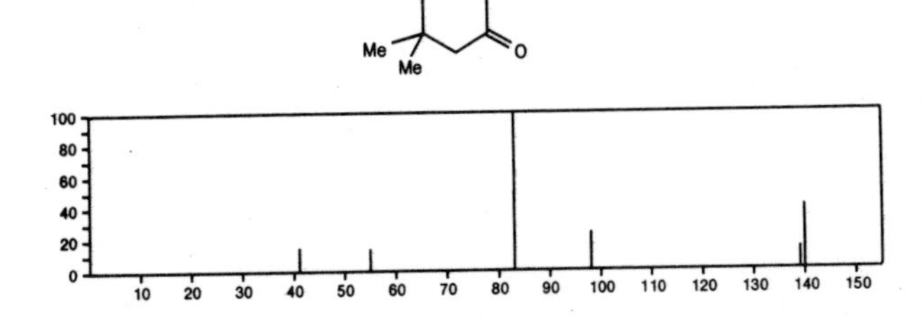

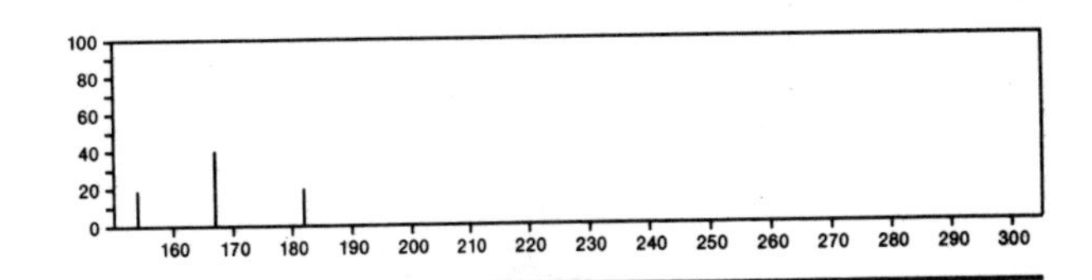

182 $C_{10}H_{14}O_3$ 54346-06-4
1(3*H*)-Isobenzofuranone, 3a,4,5,7a-tetrahydro-4-hydroxy-3a,7a-dimethyl-, (3aα,4β,7aα)-(±)-

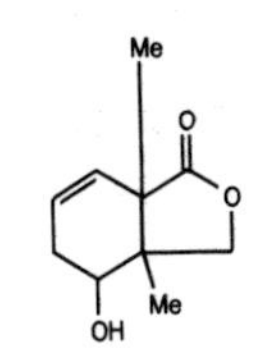

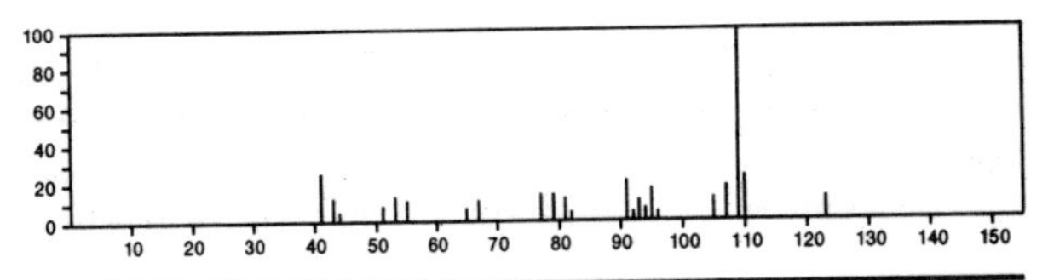

182 $C_{10}H_{14}O_3$ 54644-18-7
1,3-Isobenzofurandione, hexahydro-4,7-dimethyl-

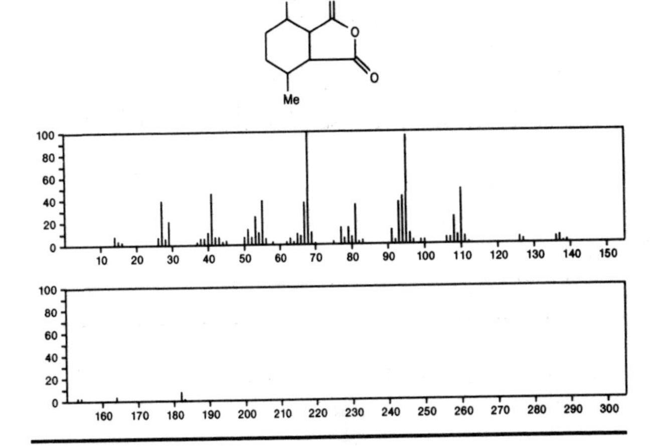

182 $C_{10}H_{14}O_3$ 54644-19-8
1,2,4-Cyclopentanetrione, 3-(1-methylbutyl)-

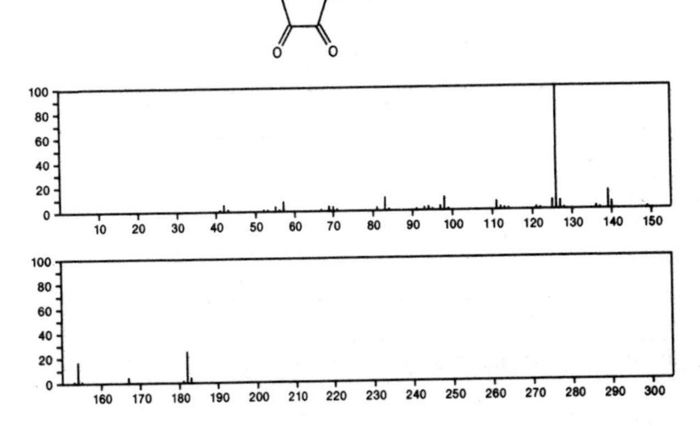

182 $C_{10}H_{14}O_3$ 56324-76-6
Bicyclo[2.2.2]octan-2-one, 4-(acetyloxy)-

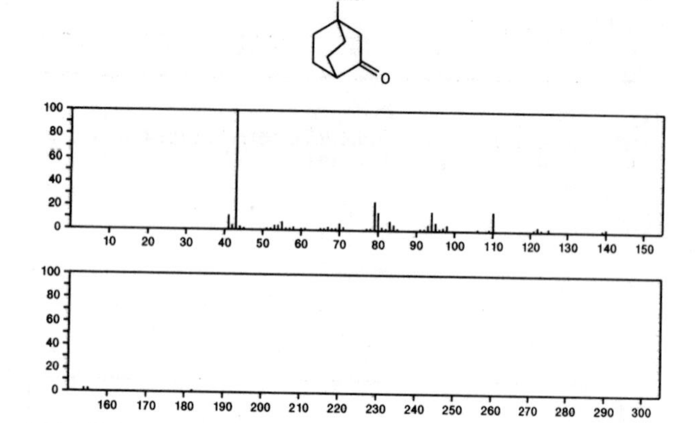

182 $C_{10}H_{14}O_3$ 57156-89-5
1,3-Cyclopentanedione, 4-hydroxy-5-(3-methyl-1-butenyl)-

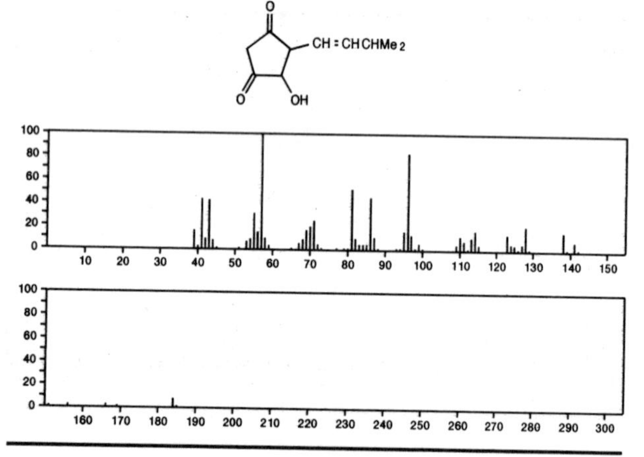

182 $C_{10}H_{18}N_2O$ 4074-30-0
3-Azabicyclo[3.2.1]octane, 1,8,8-trimethyl-3-nitroso-

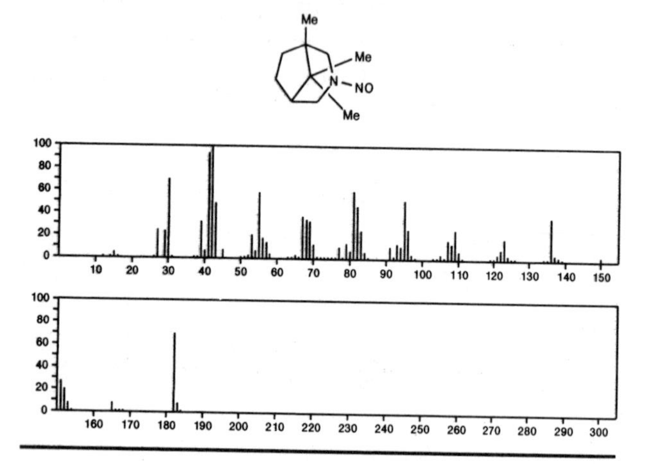

182 $C_{10}H_{18}N_2O$ 28884-14-2
Isoxazole, 5-amino-3-butyl-4-propyl-

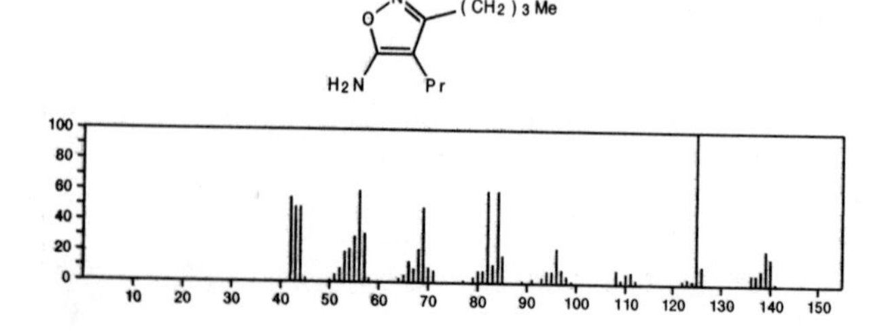

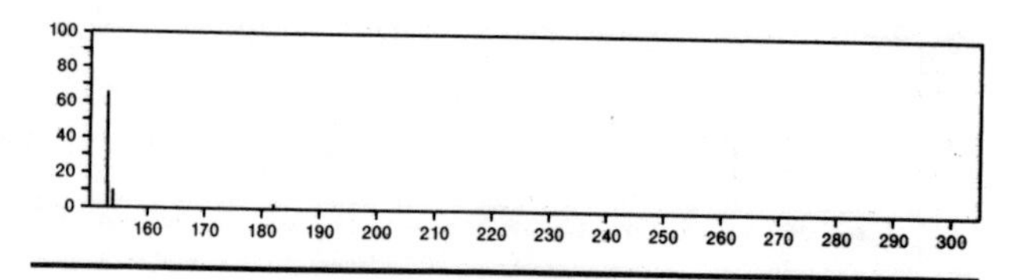

182 $C_{10}H_{18}N_2O$ 29924-75-2
2-Quinuclidinone, 6,6,8,8-tetramethyl-

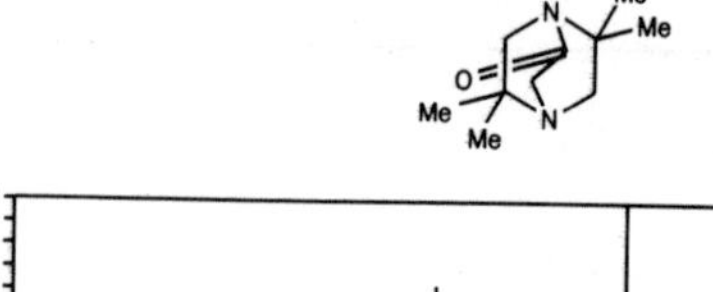

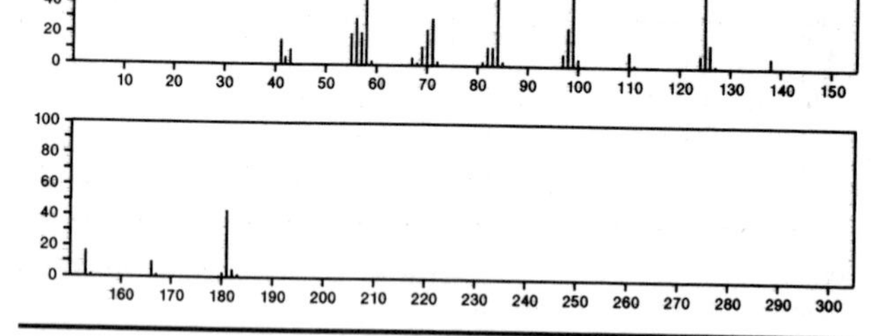

182 $C_{10}H_{18}N_2O$ 49582-39-0
2-Propenal, 3-(dimethylamino)-3-(1-piperidinyl)-

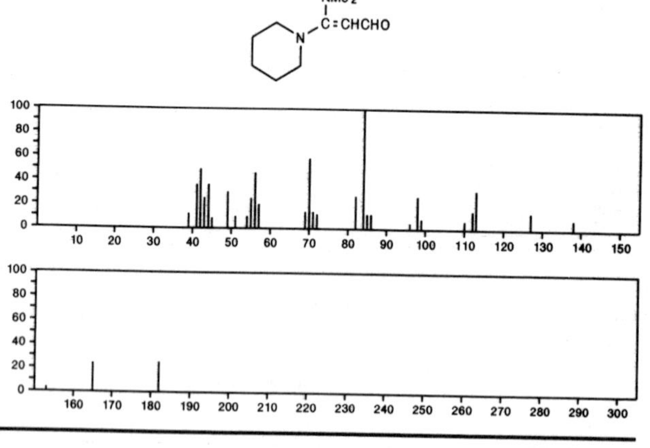

182 $C_{11}H_{15}Cl$ 4830-95-9
Benzene, (3-chloro-3-methylbutyl)-

$Me_2CCl\,CH_2CH_2Ph$

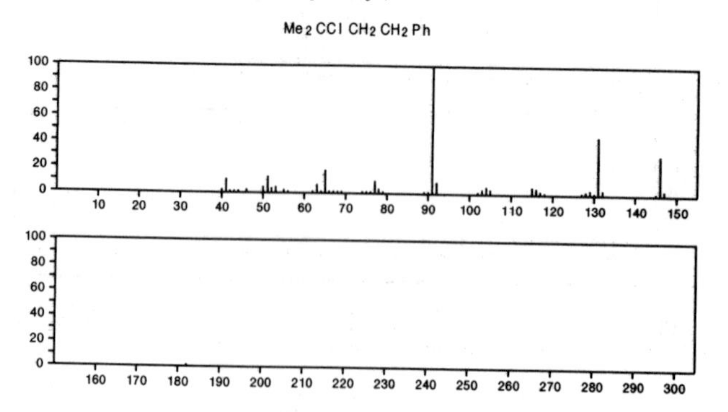

182 $C_{11}H_{15}Cl$ 54657-99-7
Benzene, chloro(1-methylbutyl)-

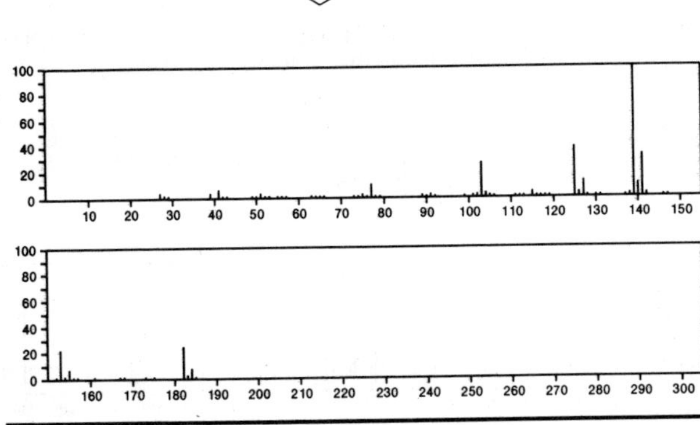

182 $C_{11}H_{15}Cl$ 54658-00-3
Benzene, chlorodiethylmethyl-

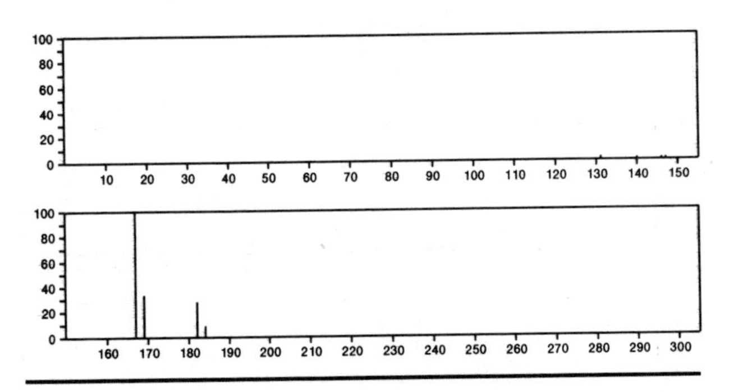

182 $C_{11}H_{18}O_2$ 105-86-2
2,6-Octadien-1-ol, 3,7-dimethyl-, formate, (*E*)-

$O=CHOCH_2CH=CMeCH_2CH_2CH=CMe_2$

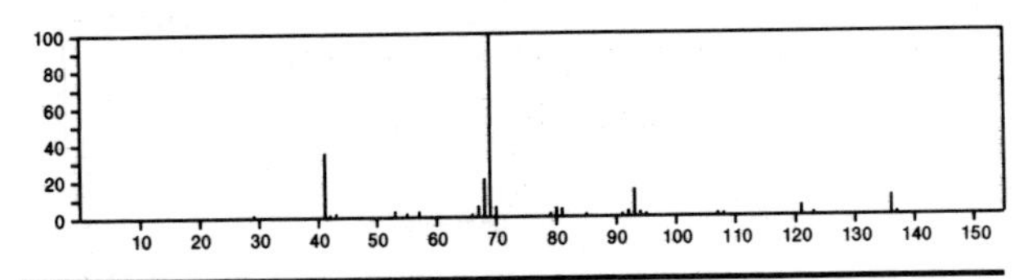

182 $C_{11}H_{18}O_2$ 1189-09-9
2,6-Octadienoic acid, 3,7-dimethyl-, methyl ester, (*E*)-

$MeOC(O)CH=CMeCH_2CH_2CH=CMe_2$

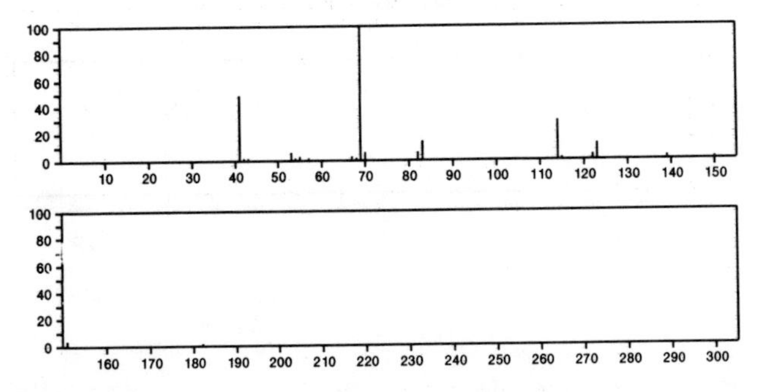

182 $C_{11}H_{18}O_2$ 1862-61-9
2,6-Octadienoic acid, 3,7-dimethyl-, methyl ester, (*Z*)-

$MeOC(O)CH=CMeCH_2CH_2CH=CMe_2$

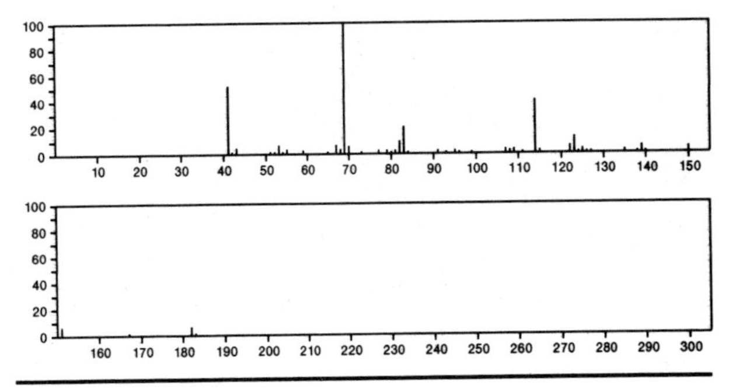

182 $C_{11}H_{18}O_2$ 1919-64-8
1,3-Cyclohexanedione, 5,5-dimethyl-2-propyl-

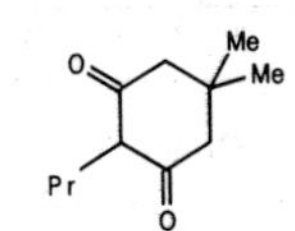

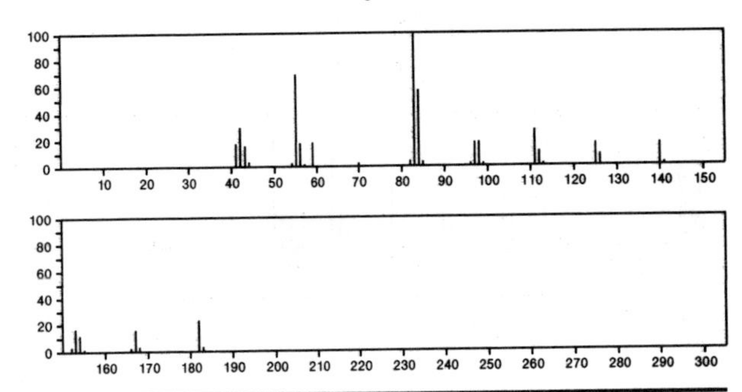

182 $C_{11}H_{18}O_2$ 2349-14-6
2,6-Octadienoic acid, 3,7-dimethyl-, methyl ester

$MeOC(O)CH=CMeCH_2CH_2CH=CMe_2$

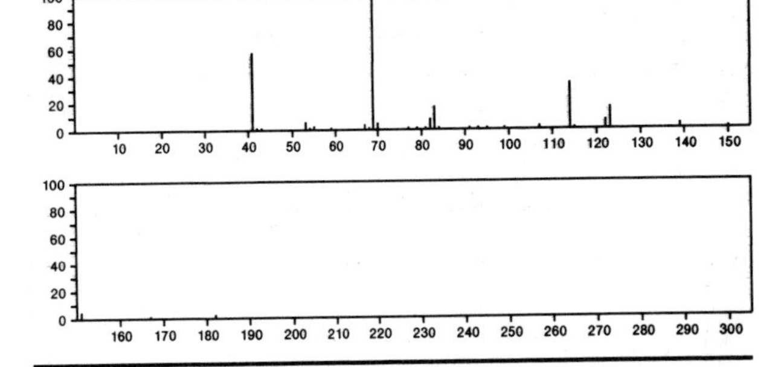

182 $C_{11}H_{18}O_2$ 4707-07-7
2(1*H*)-Naphthalenone, octahydro-8a-hydroxy-4a-methyl-, *cis*-

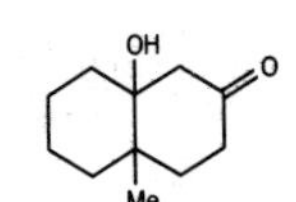

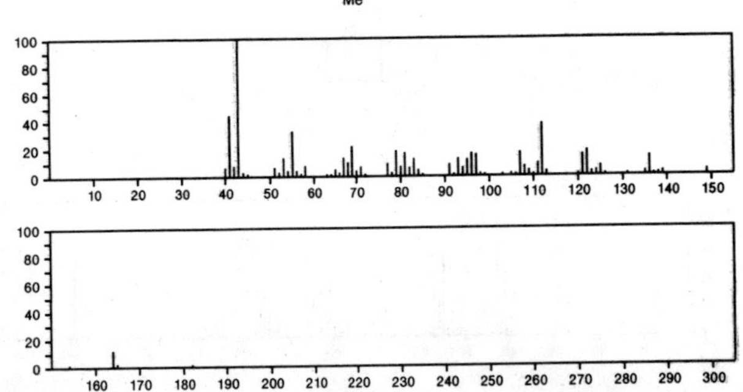

182 $C_{11}H_{18}O_2$ 5460–63–9
Cyclopropanecarboxylic acid, 2,2–dimethyl–3–(2–methyl–1–propenyl)–, methyl ester

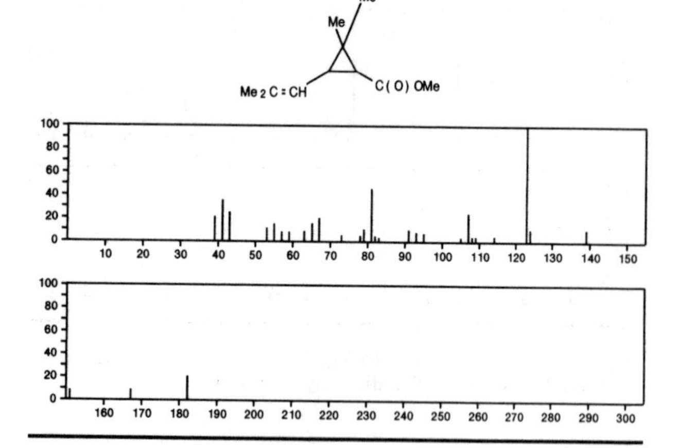

182 $C_{11}H_{18}O_2$ 16642–56–1
1–Cycloheptene–1–acetic acid, α,α–dimethyl–

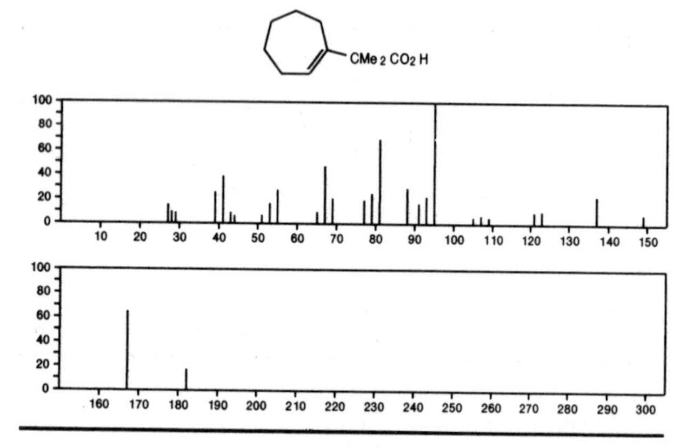

182 $C_{11}H_{18}O_2$ 16750–88–2
3,6–Octadienoic acid, 3,7–dimethyl–, methyl ester, (*Z*)–

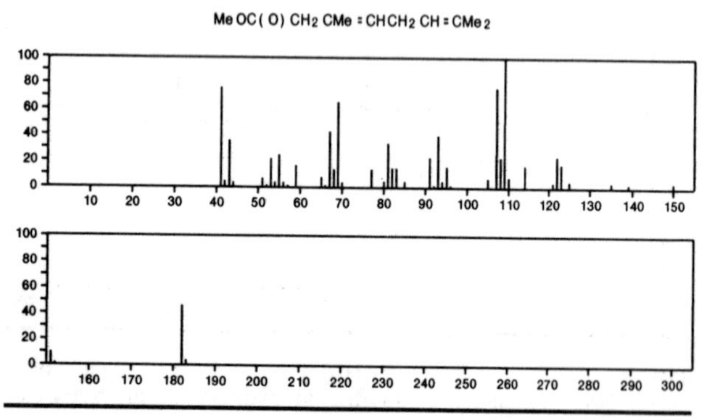

182 $C_{11}H_{18}O_2$ 16778–27–1
2(3*H*)–Benzofuranone, hexahydro–4,4,7a–trimethyl–

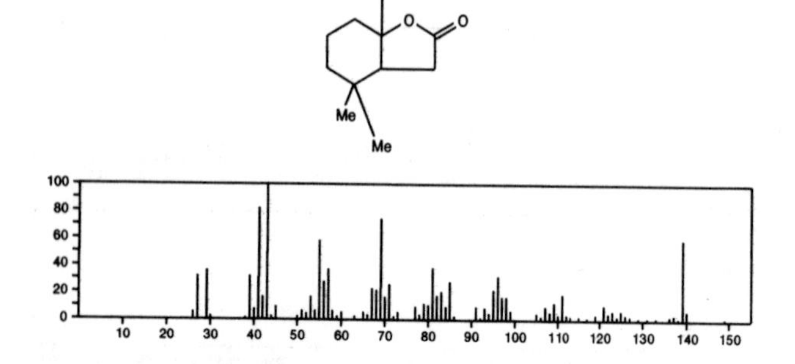

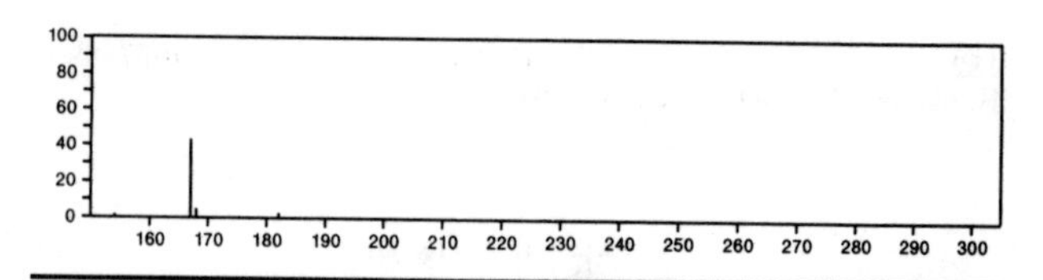

182 $C_{11}H_{18}O_2$ 21727–79–7
1(2*H*)–Naphthalenone, octahydro–4–methoxy–, *trans*–

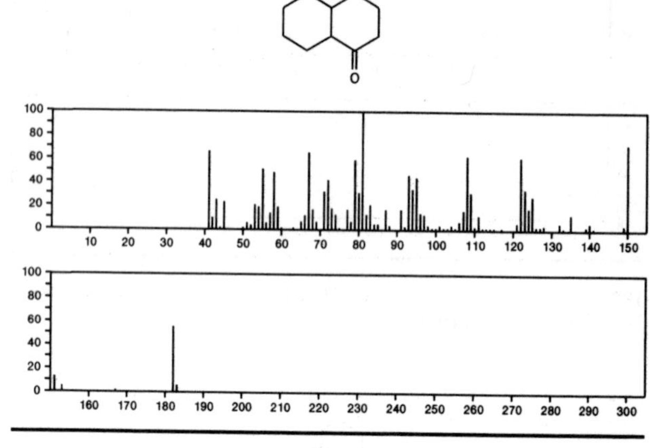

182 $C_{11}H_{18}O_2$ 36334–87–9
1,3–Dioxolane, 2,2–dimethyl–4,5–bis(1–methylethenyl)–

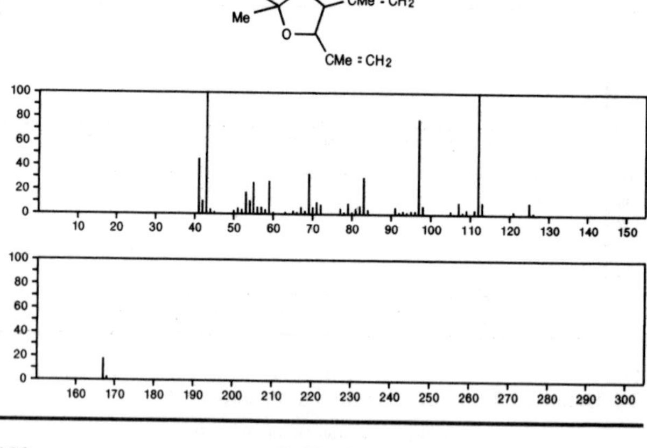

182 $C_{11}H_{18}O_2$ 36334–88–0
1,3–Dioxolane, 2,2–dimethyl–4,5–di–1–propenyl–

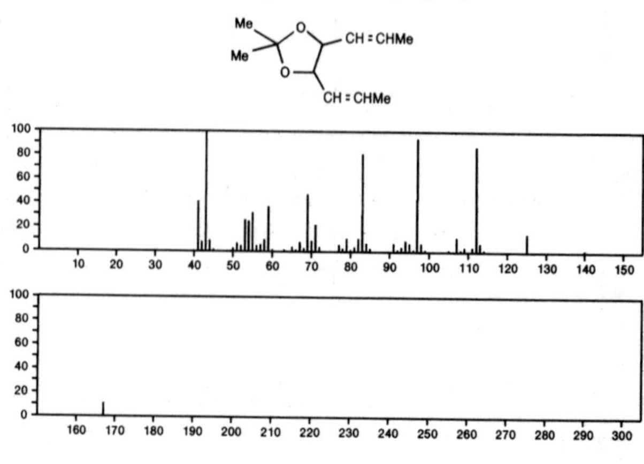

182 $C_{11}H_{18}O_2$ 51149–72–5
2–Propanone, 1–(1–cyclohexen–1–yl)–3–ethoxy–

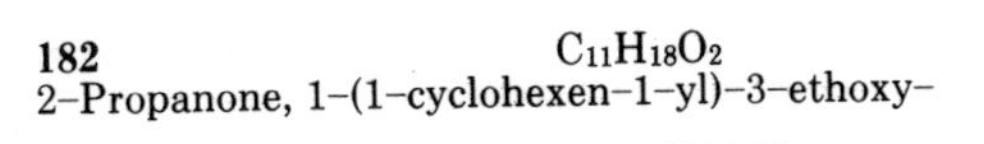

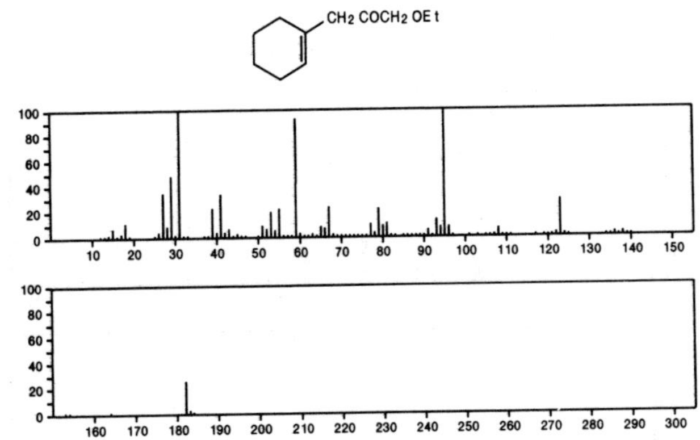

182 $C_{11}H_{18}O_2$ 53690–81–6
2–Cyclohexen–1–one, 6–butyl–3–methoxy–

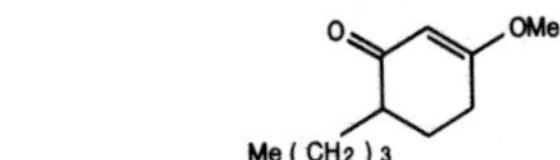

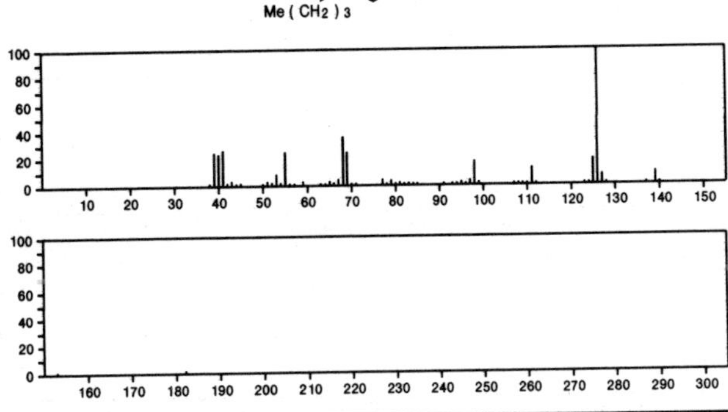

182 $C_{11}H_{18}O_2$ 53690–84–9
2–Cyclohexen–1–one, 4–butyl–3–methoxy–

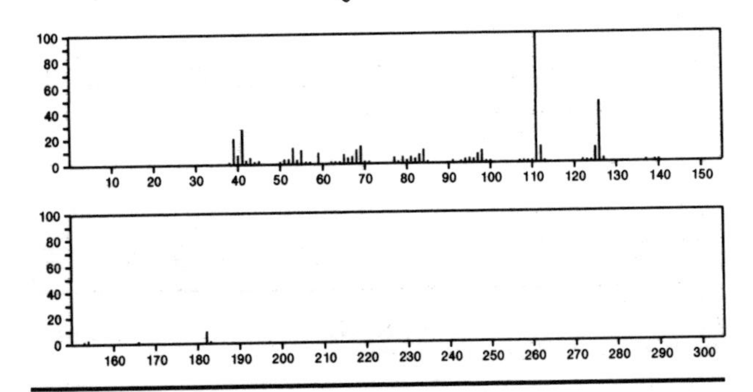

182 $C_{11}H_{18}O_2$ 53690–86–1
2–Cyclohexen–1–one, 2–butyl–3–methoxy–

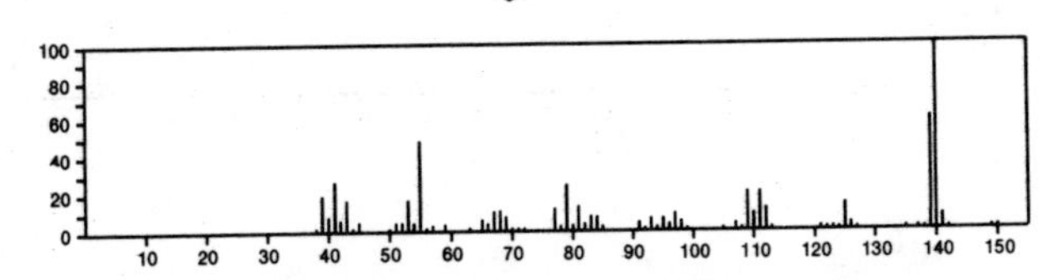

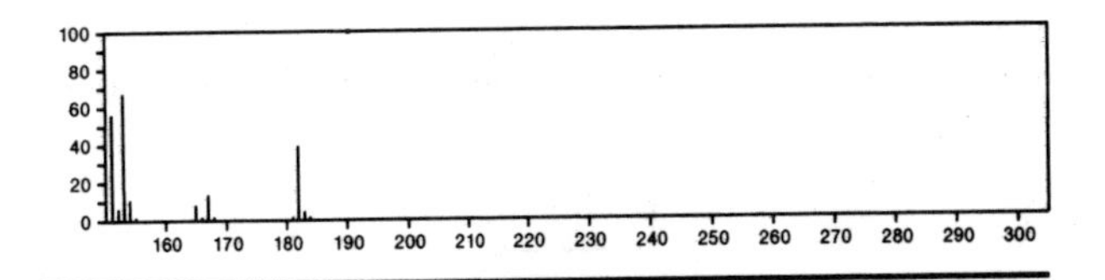

182 $C_{11}H_{18}O_2$ 54345–57–2
2–Cyclohexen–1–ol, 2,4,4–trimethyl–, acetate

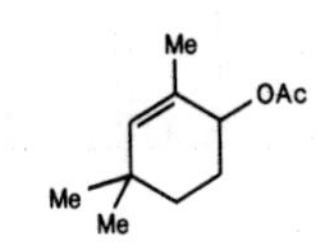

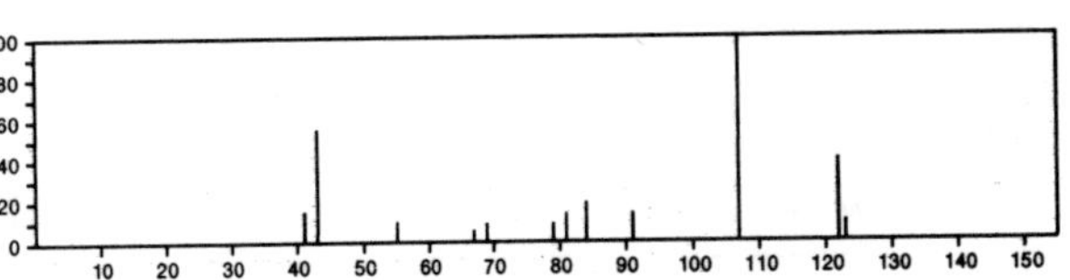

182 $C_{11}H_{18}O_2$ 54345–58–3
2–Cyclohexen–1–ol, 2,6,6–trimethyl–, acetate

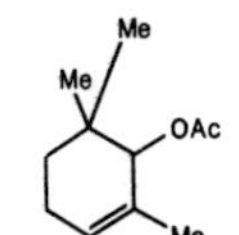

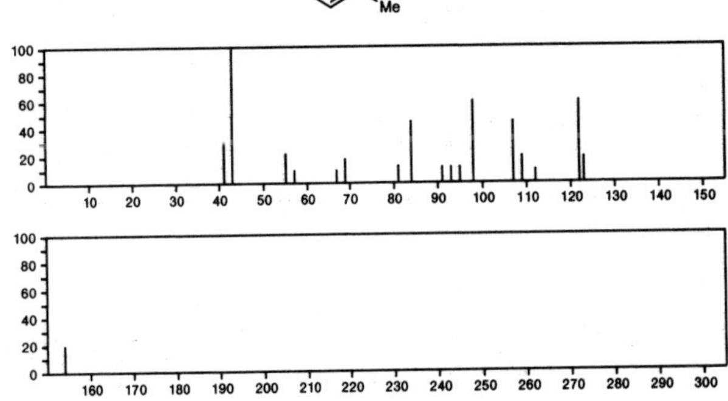

182 $C_{11}H_{18}O_2$ 54644–17–6
2–Propanol, 1–[(1–ethynylcyclohexyl)oxy]–

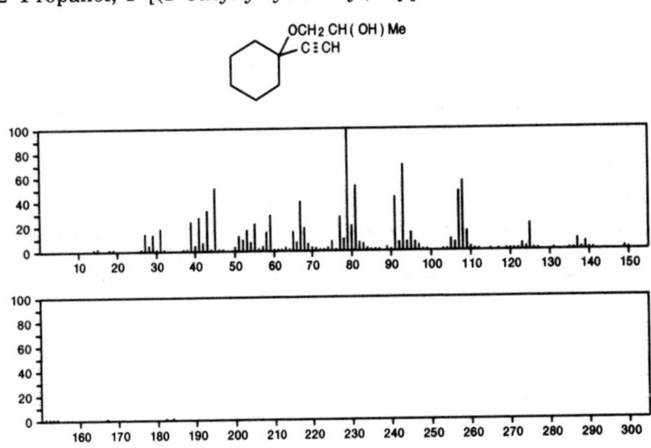

182 $C_{11}H_{18}O_2$ 54644-24-5
1-Pentalenecarboxylic acid, octahydro-3-methyl-, methyl ester

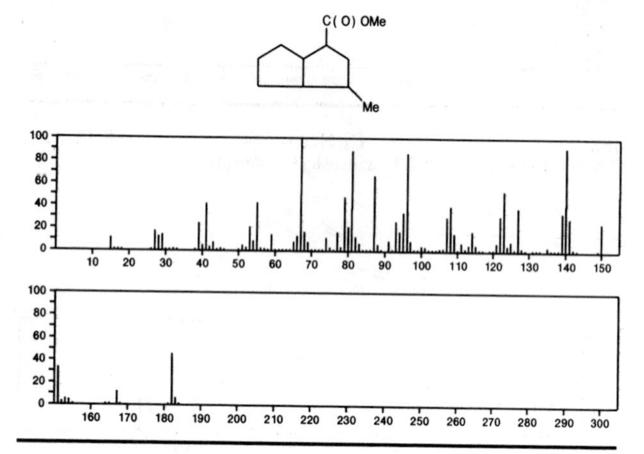

182 $C_{11}H_{18}O_2$ 54644-25-6
Bicyclo[2.2.2]octan-1-ol, 4-methyl-, acetate

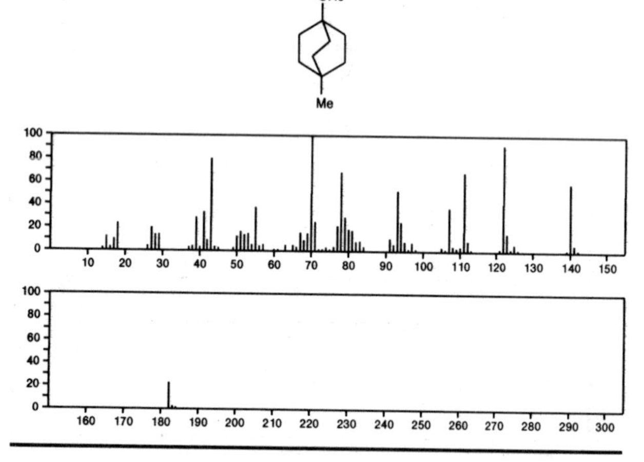

182 $C_{11}H_{18}O_2$ 54764-60-2
Bicyclo[4.1.0]heptane-7-carboxylic acid, 3-methyl-, ethyl ester

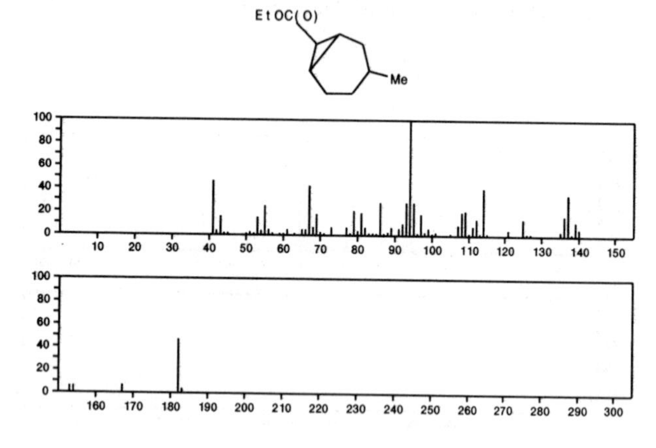

182 $C_{11}H_{18}O_2$ 57157-05-8
1,3-Cyclopentanedione, 4-methyl-5-pentyl-

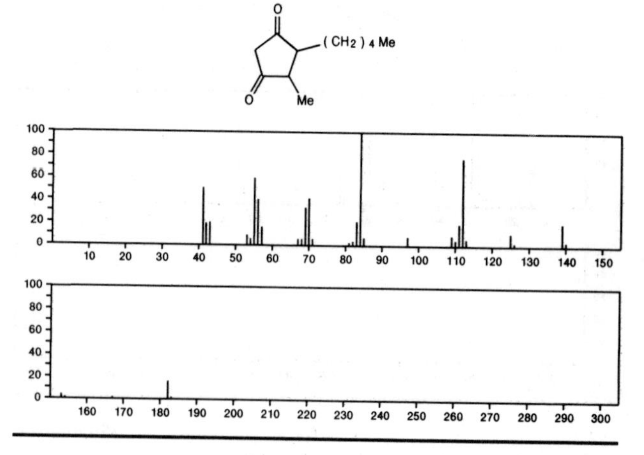

182 $C_{11}H_{22}N_2$ 880-09-1
Piperidine, 1,1'-methylenebis-

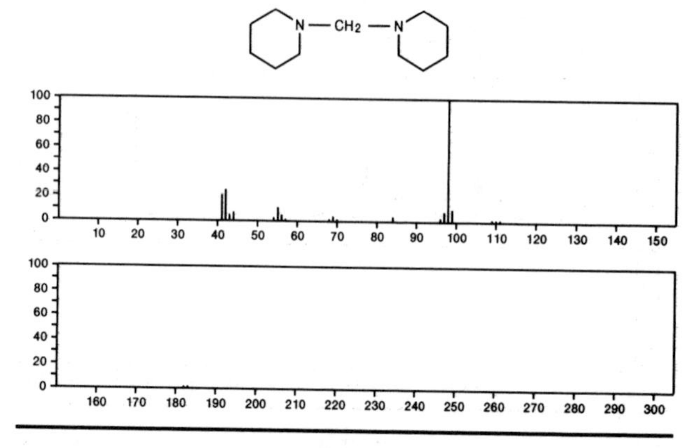

182 $C_{12}H_{10}N_2$ 103-33-3
Diazene, diphenyl-

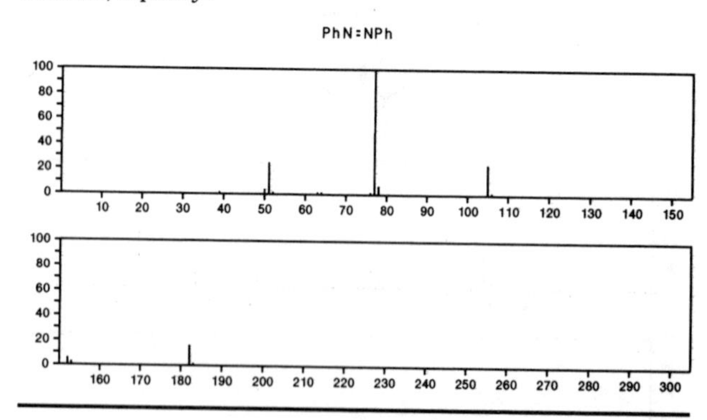

182 $C_{12}H_{10}N_2$ 486-84-0
9*H*-Pyrido[3,4-*b*]indole, 1-methyl-

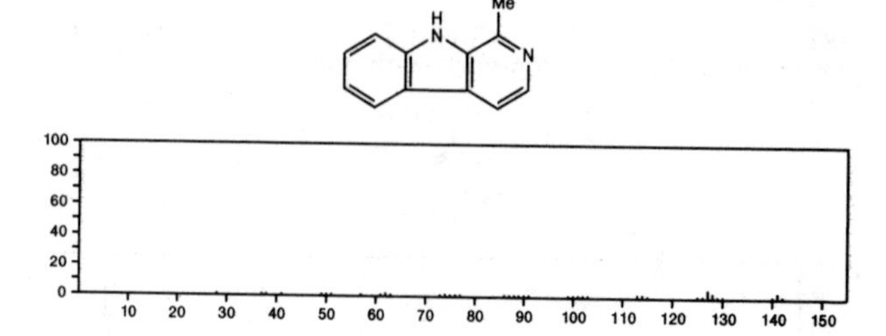

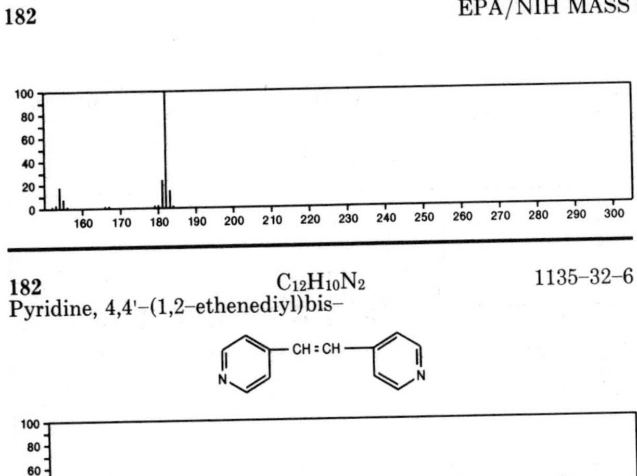

182 $C_{12}H_{10}N_2$ 1135-32-6
Pyridine, 4,4'-(1,2-ethenediyl)bis-

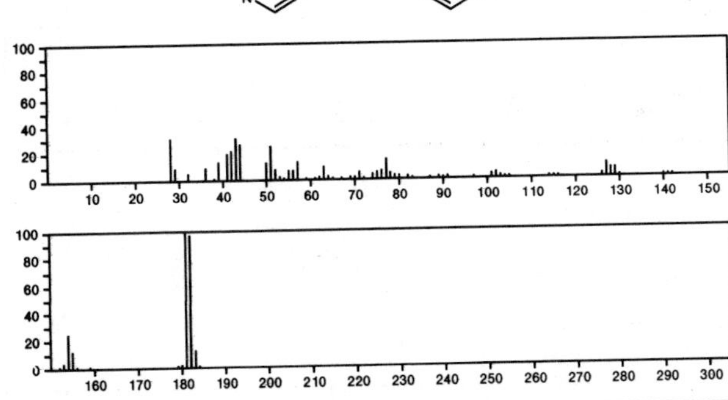

182 $C_{12}H_{10}N_2$ 7032-25-9
Benzenamine, *N*-(2-pyridinylmethylene)-

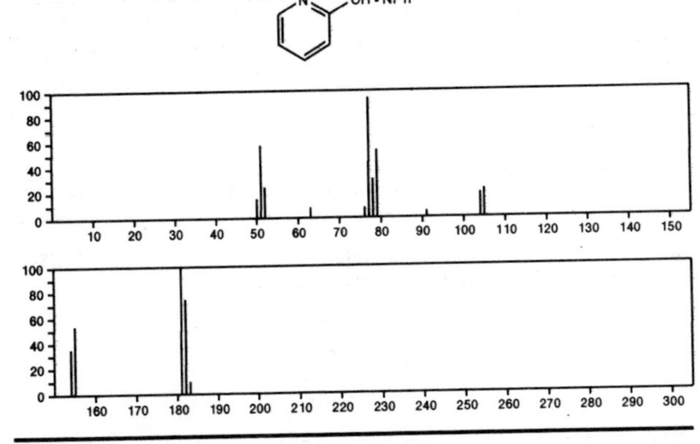

182 $C_{12}H_{10}N_2$ 17082-12-1
Diazene, diphenyl-, (*E*)-

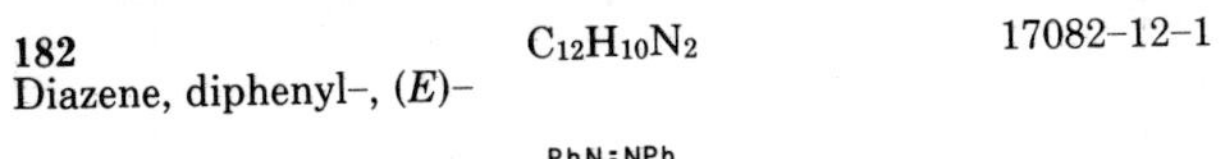

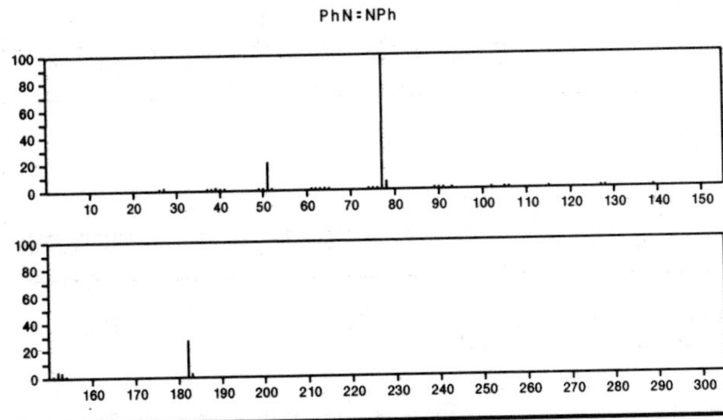

182 $C_{12}H_{10}N_2$ 27768-46-3
Benzenamine, *N*-(4-pyridinylmethylene)-

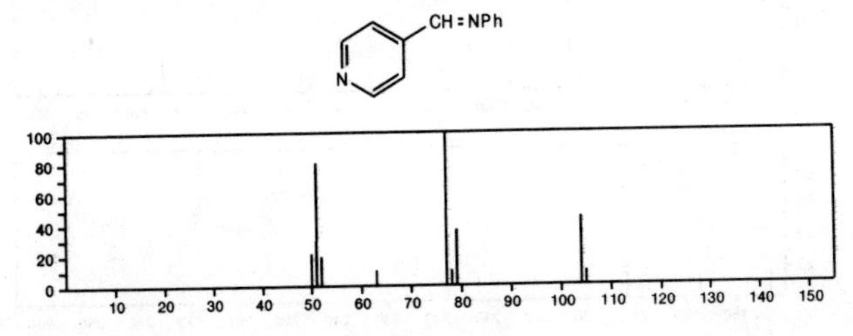

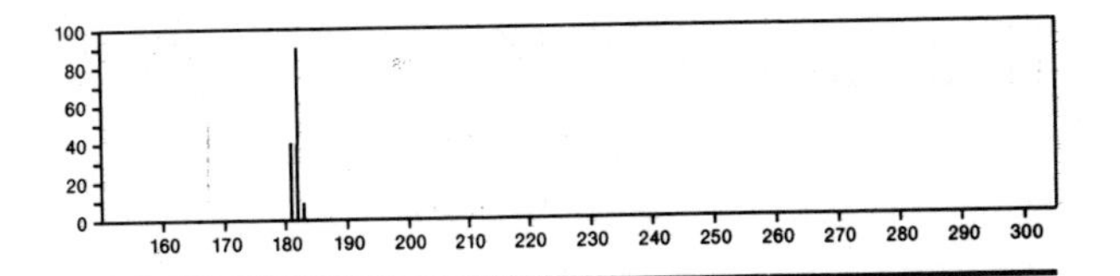

182 $C_{12}H_{10}N_2$ 29722-97-2
Benzenamine, *N*-(3-pyridinylmethylene)-

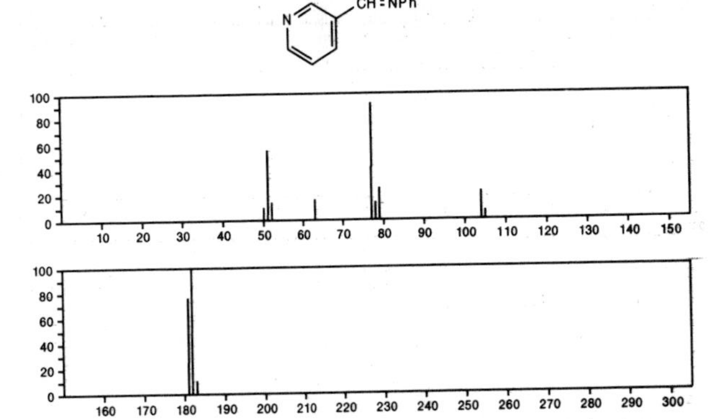

182 $C_{12}H_{22}O$ 286-99-7
13-Oxabicyclo[10.1.0]tridecane

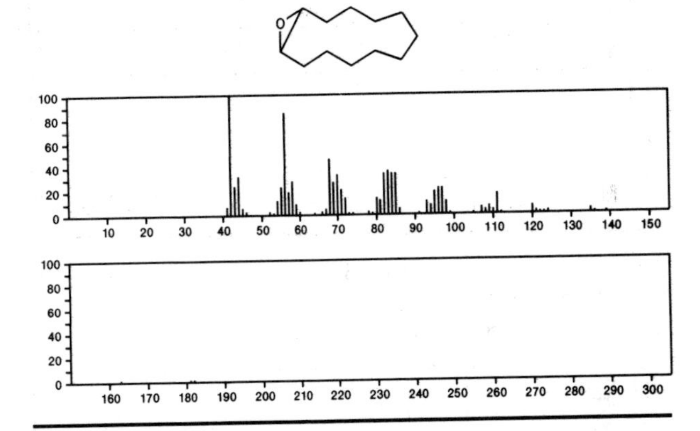

182 $C_{12}H_{22}O$ 830-13-7
Cyclododecanone

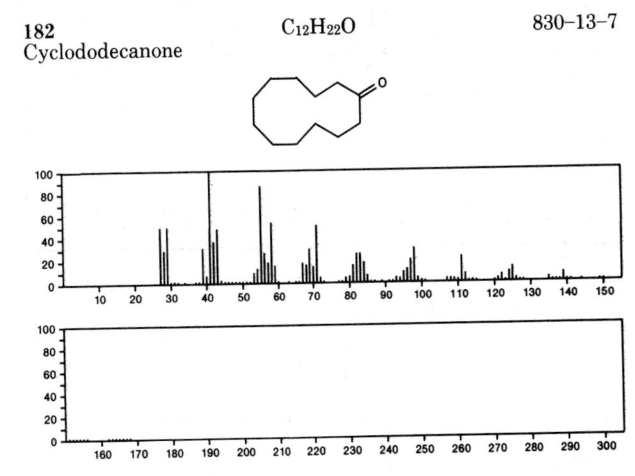

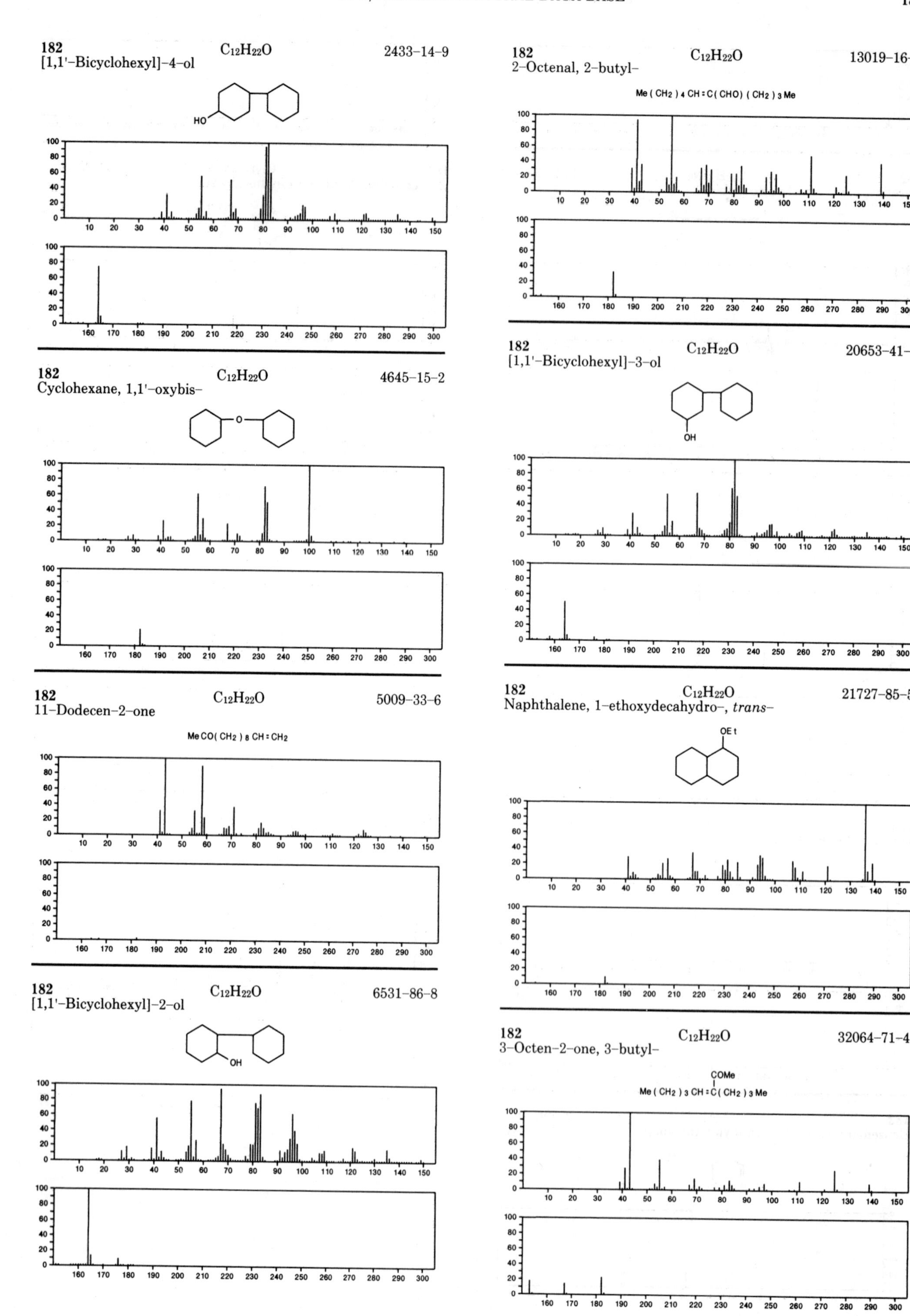

182
[1,1'-Bicyclohexyl]-4-ol
$C_{12}H_{22}O$
2433-14-9
182
Cyclohexane, 1,1'-oxybis-
$C_{12}H_{22}O$
4645-15-2
182
11-Dodecen-2-one
$C_{12}H_{22}O$
5009-33-6
MeCO(CH2)8CH=CH2
182
[1,1'-Bicyclohexyl]-2-ol
$C_{12}H_{22}O$
6531-86-8
182
2-Octenal, 2-butyl-
$C_{12}H_{22}O$
13019-16-4
Me(CH2)4CH=C(CHO)(CH2)3Me
182
[1,1'-Bicyclohexyl]-3-ol
$C_{12}H_{22}O$
20653-41-2
182
Naphthalene, 1-ethoxydecahydro-, *trans*-
$C_{12}H_{22}O$
21727-85-5
182
3-Octen-2-one, 3-butyl-
$C_{12}H_{22}O$
32064-71-4
COMe
Me(CH2)3CH=C(CH2)3Me

182 $C_{12}H_{22}O$ 32064-76-9
7-Dodecen-6-one

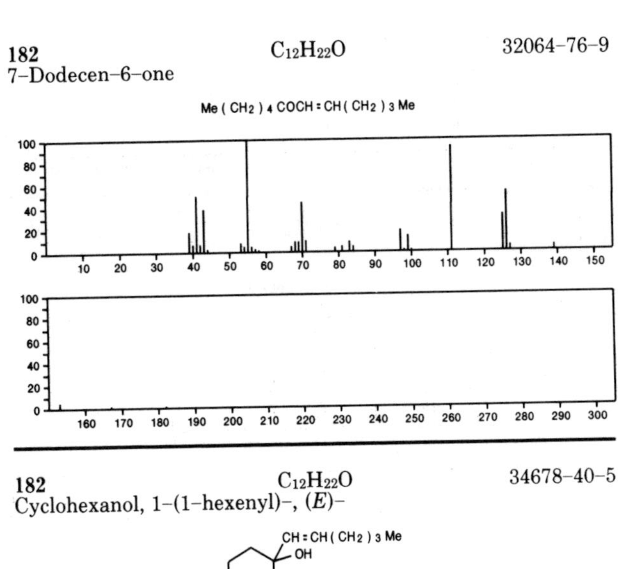

182 $C_{12}H_{22}O$ 34678-40-5
Cyclohexanol, 1-(1-hexenyl)-, (*E*)-

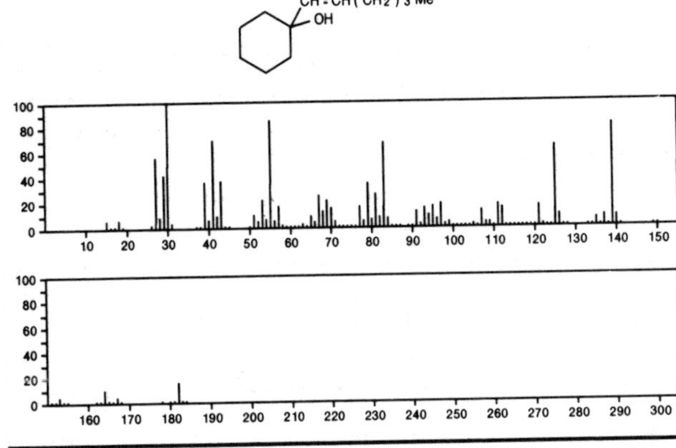

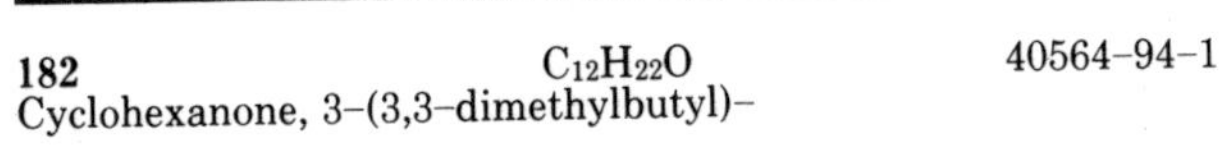

182 $C_{12}H_{22}O$ 40564-94-1
Cyclohexanone, 3-(3,3-dimethylbutyl)-

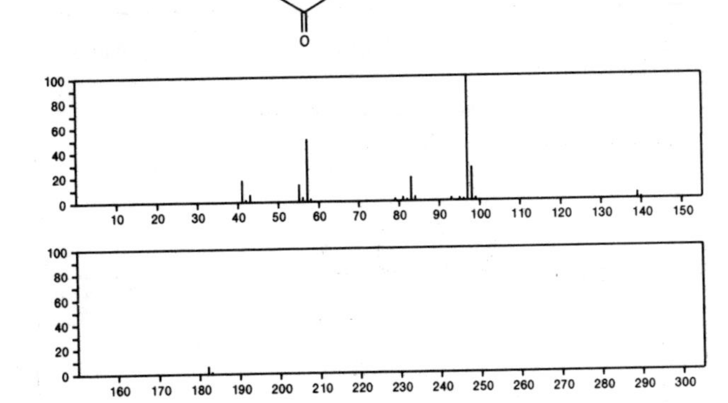

182 $C_{12}H_{22}O$ 55821-16-4
Cyclohexanone, 4-ethyl-4-methyl-3-(1-methylethyl)-, *trans*-

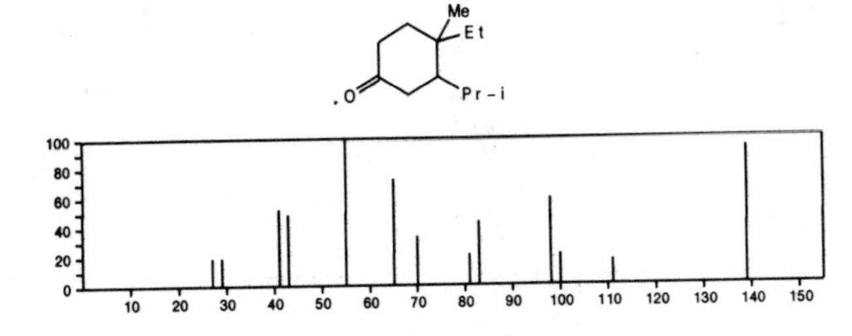

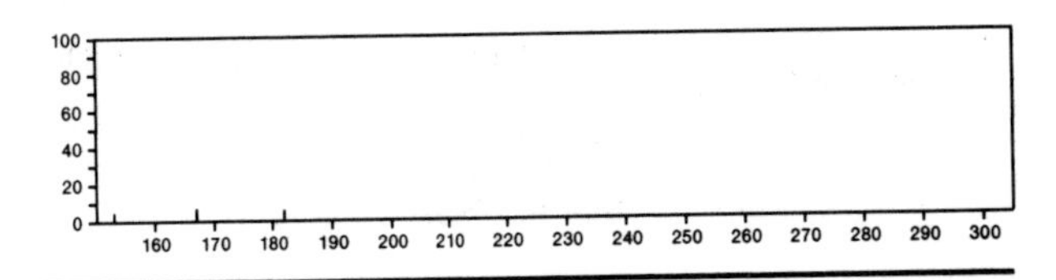

182 $C_{12}H_{22}O$ 56272-08-3
Cyclohexanol, 4-ethyl-4-methyl-3-(1-methylethenyl)-, (1α,3α,4β)-

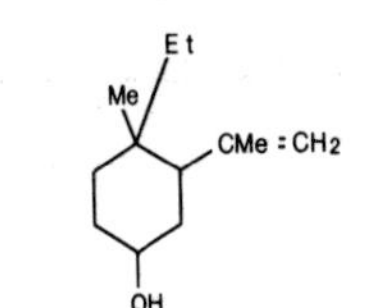

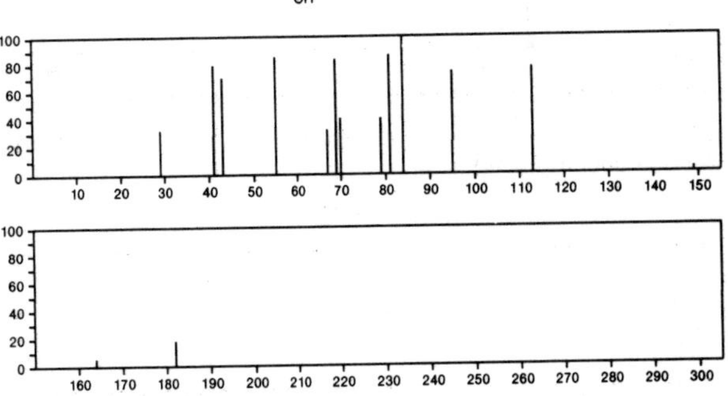

182 $C_{12}H_{22}O$ 56272-09-4
Cyclohexanol, 4-ethyl-4-methyl-3-(1-methylethenyl)-, (1α,3β,4α)-

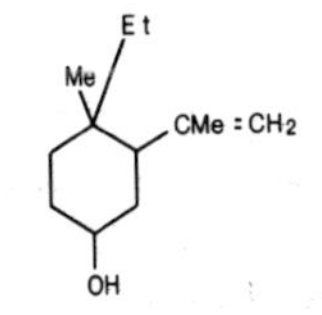

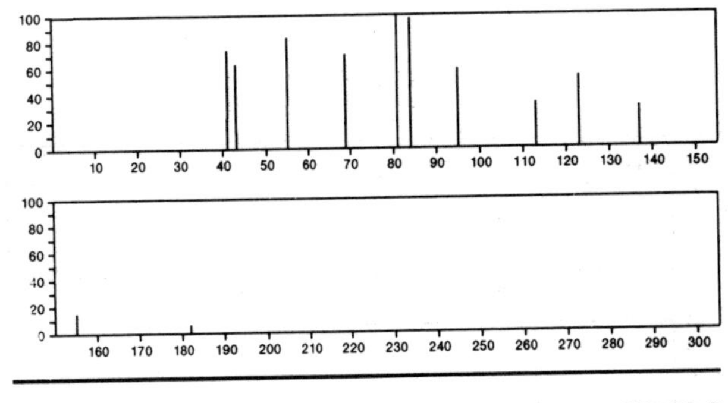

182 $C_{12}H_{27}B$ 122-56-5
Borane, tributyl-

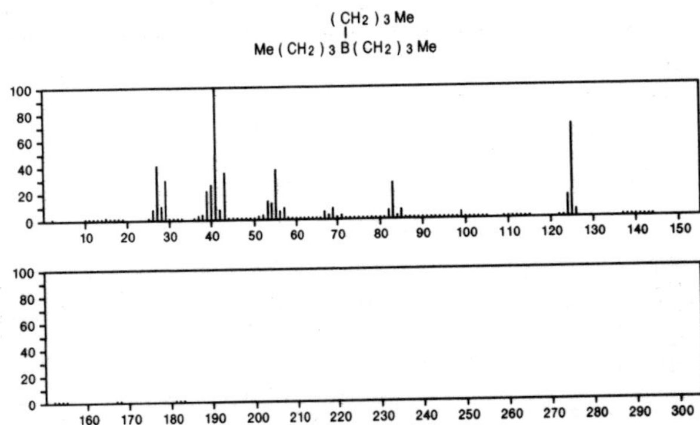

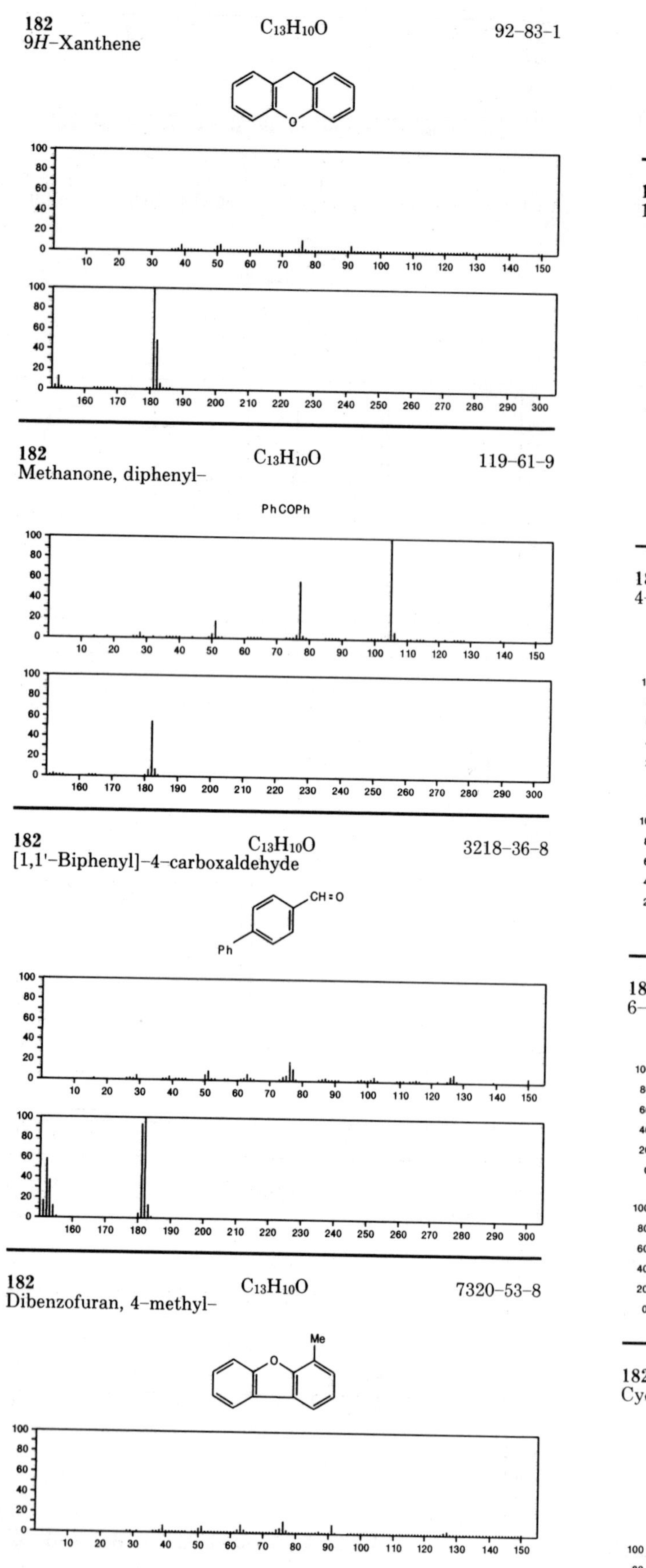

182
9*H*-Xanthene
$C_{13}H_{10}O$
92-83-1

182
Methanone, diphenyl-
$C_{13}H_{10}O$
119-61-9

PhCOPh

182
[1,1'-Biphenyl]-4-carboxaldehyde
$C_{13}H_{10}O$
3218-36-8

182
Dibenzofuran, 4-methyl-
$C_{13}H_{10}O$
7320-53-8

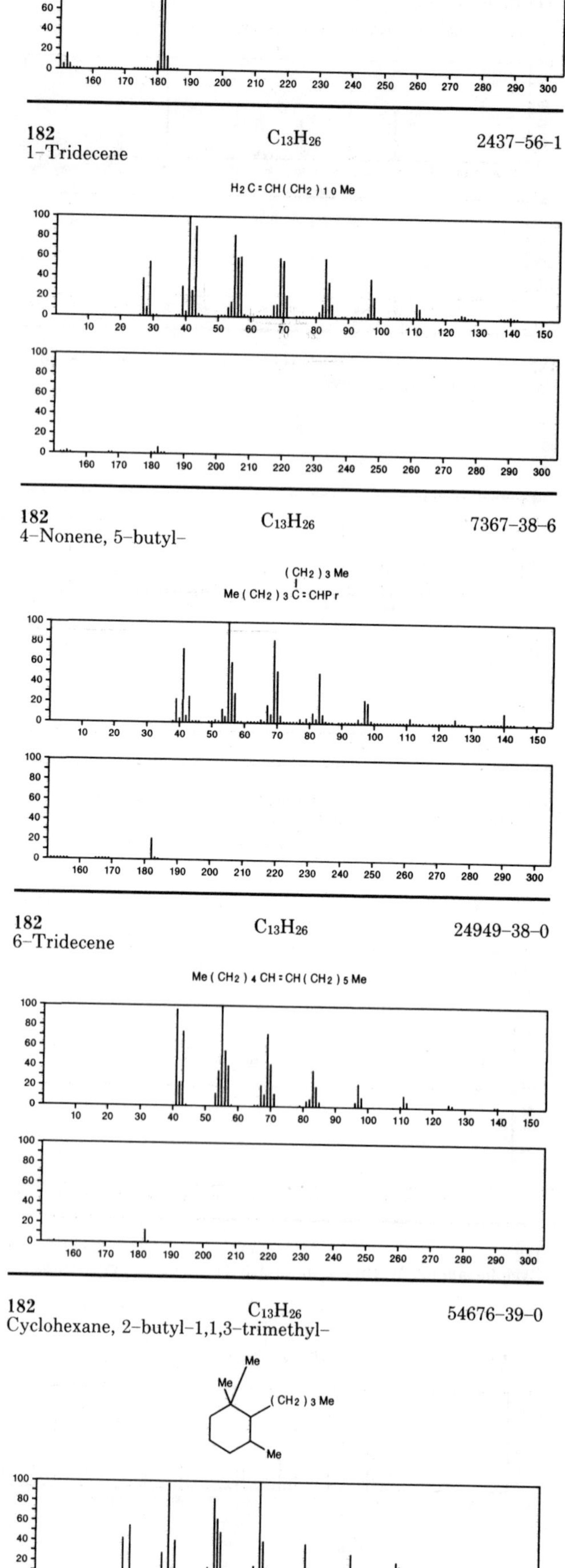

182
1-Tridecene
$C_{13}H_{26}$
2437-56-1

$H_2C=CH(CH_2)_{10}Me$

182
4-Nonene, 5-butyl-
$C_{13}H_{26}$
7367-38-6

$Me(CH_2)_3C(=CHPr)(CH_2)_3Me$

182
6-Tridecene
$C_{13}H_{26}$
24949-38-0

$Me(CH_2)_4CH=CH(CH_2)_5Me$

182
Cyclohexane, 2-butyl-1,1,3-trimethyl-
$C_{13}H_{26}$
54676-39-0

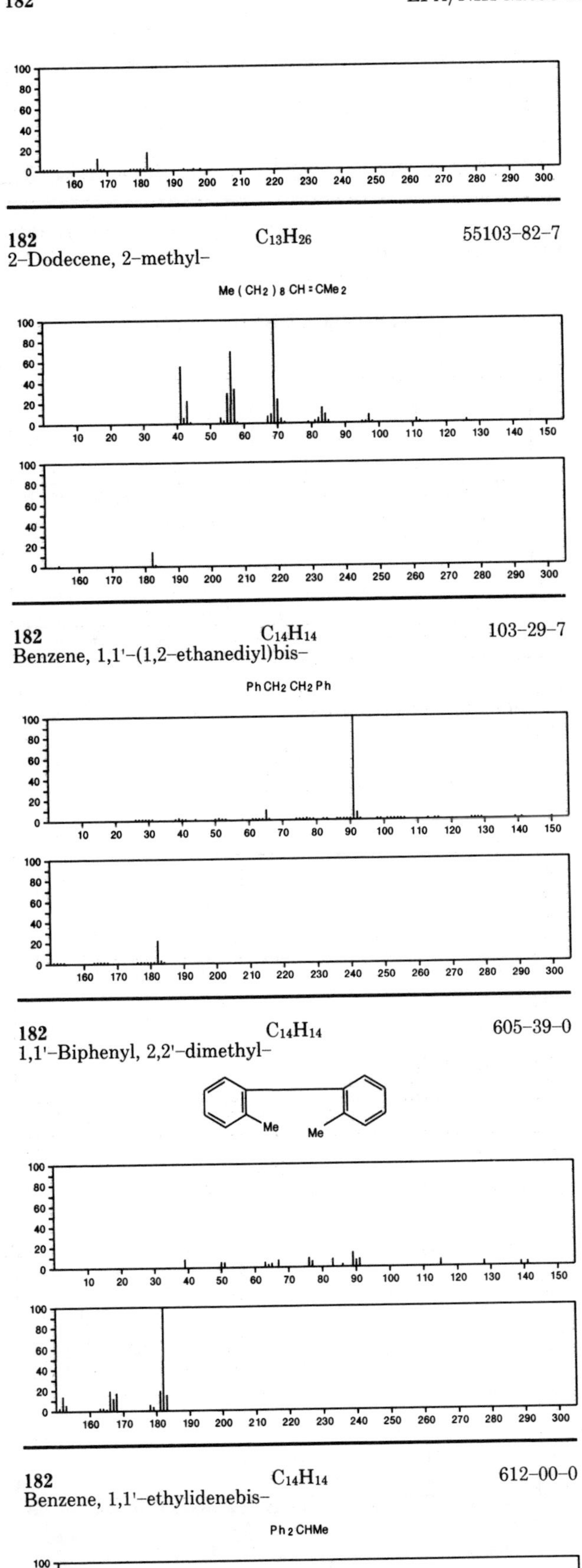

182 $C_{13}H_{26}$ 55103–82–7
2–Dodecene, 2–methyl–

Me(CH2)8CH=CMe2

182 $C_{14}H_{14}$ 103–29–7
Benzene, 1,1'–(1,2–ethanediyl)bis–

PhCH2CH2Ph

182 $C_{14}H_{14}$ 605–39–0
1,1'–Biphenyl, 2,2'–dimethyl–

182 $C_{14}H_{14}$ 612–00–0
Benzene, 1,1'–ethylidenebis–

Ph2CHMe

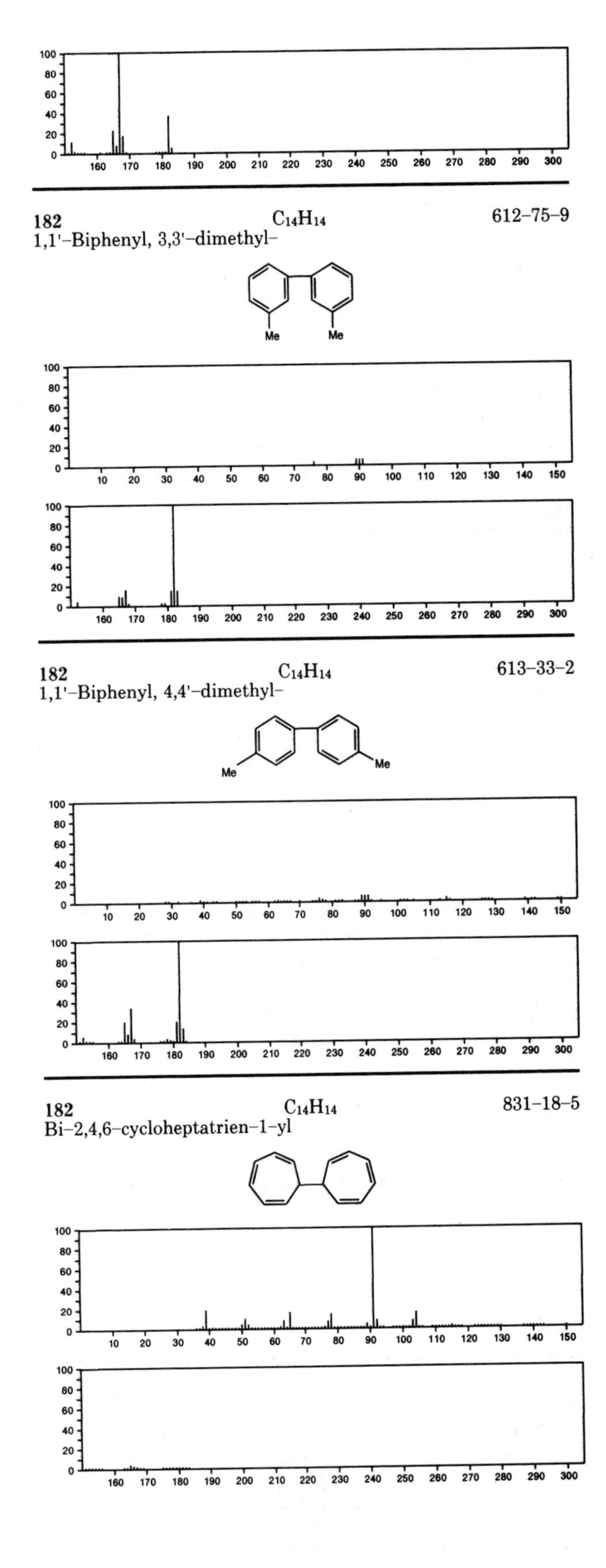

182 $C_{14}H_{14}$ 612–75–9
1,1'–Biphenyl, 3,3'–dimethyl–

182 $C_{14}H_{14}$ 613–33–2
1,1'–Biphenyl, 4,4'–dimethyl–

182 $C_{14}H_{14}$ 831–18–5
Bi–2,4,6–cycloheptatrien–1–yl

182 $C_{14}H_{14}$ 1812-51-7
1,1'-Biphenyl, 2-ethyl-

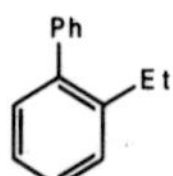

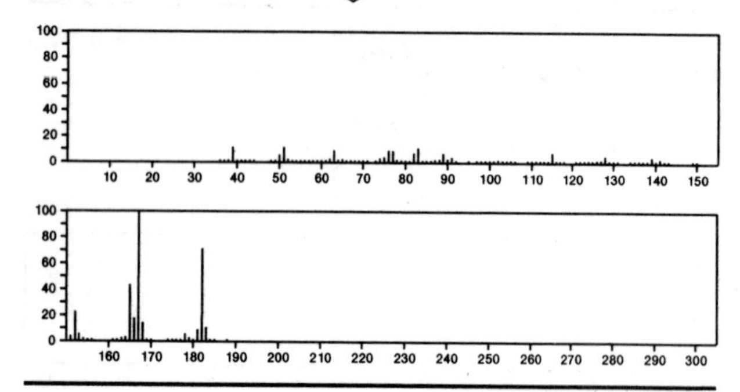

182 $C_{14}H_{14}$ 7383-90-6
1,1'-Biphenyl, 3,4'-dimethyl-

Me Me

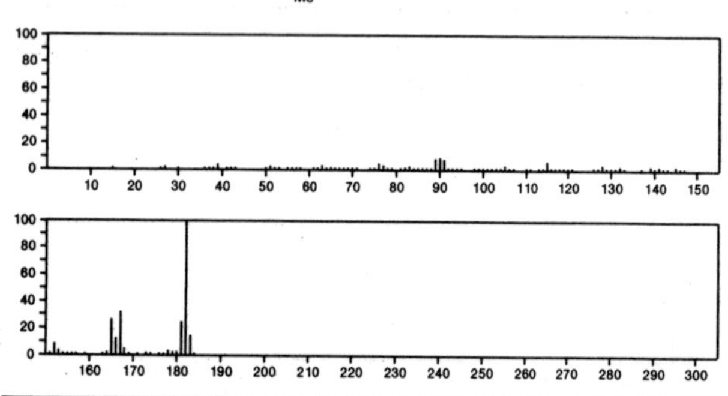

182 $C_{14}H_{14}$ 22245-13-2
Bi-1,3,5-cycloheptatrien-1-yl

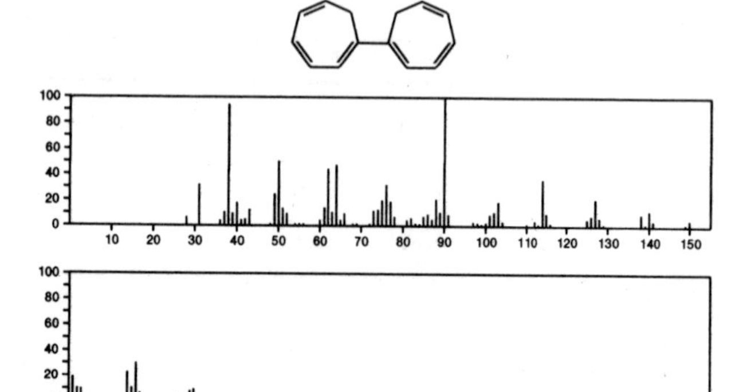

182 $C_{14}H_{14}$ 55836-29-8
Tricyclo[4.4.1.0^{2,5}]undeca-1(10),3,6,8-tetraene, 11-(1-methyl=ethylidene)-

CMe_2

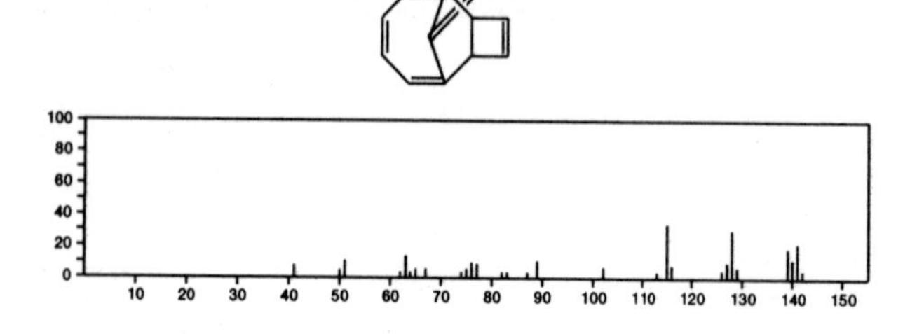

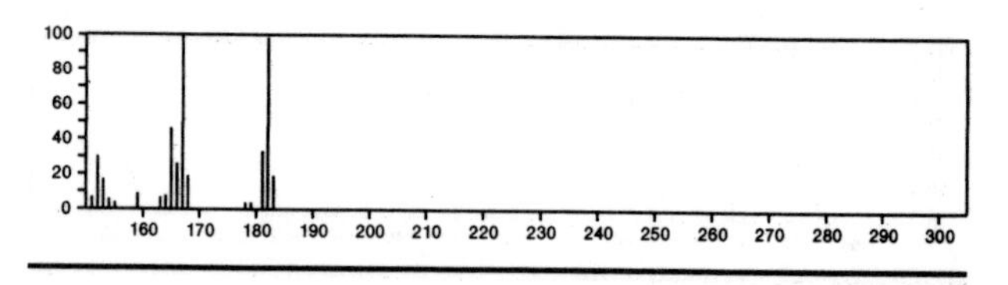

182 Cl_2O_7 10294-48-1
Chlorine oxide (Cl_2O_7)

STRUCTURE UNDEFINED

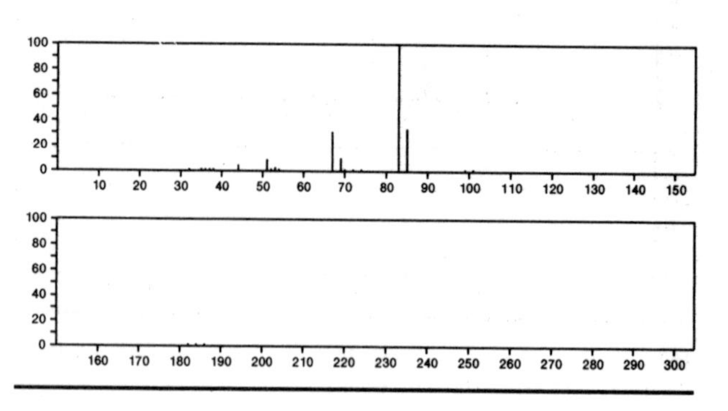

183 $C_3Cl_3N_3$ 108-77-0
1,3,5-Triazine, 2,4,6-trichloro-

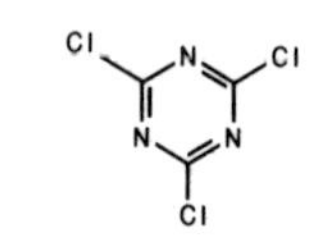

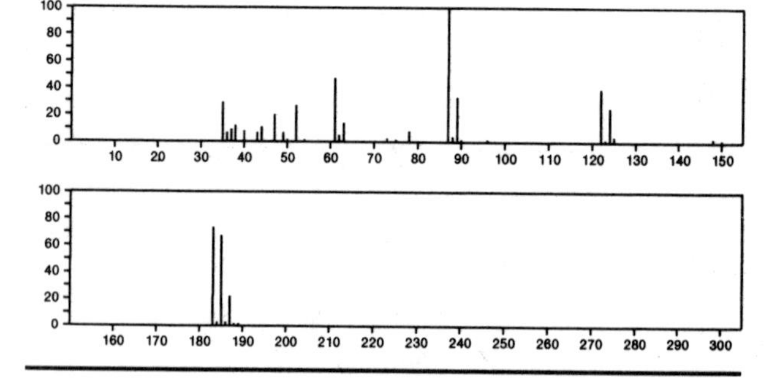

183 C_4H_9NSSe 21347-33-1
Carbamoselenothioic acid, dimethyl-, *S*-methyl ester

$Me_2NC(SMe)=Se$

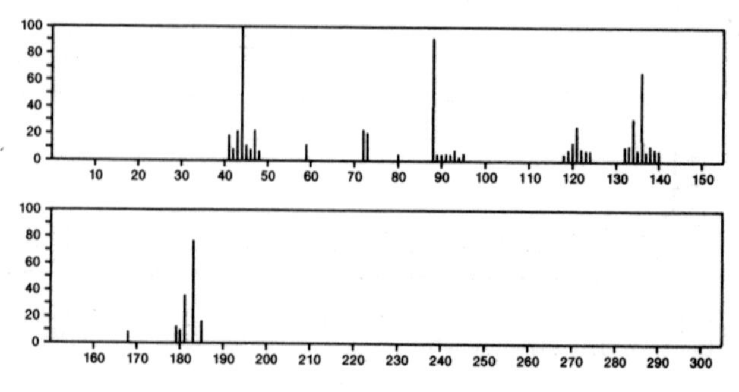

183 $C_6H_2F_5N$ 771-60-8

Benzenamine, 2,3,4,5,6-pentafluoro-

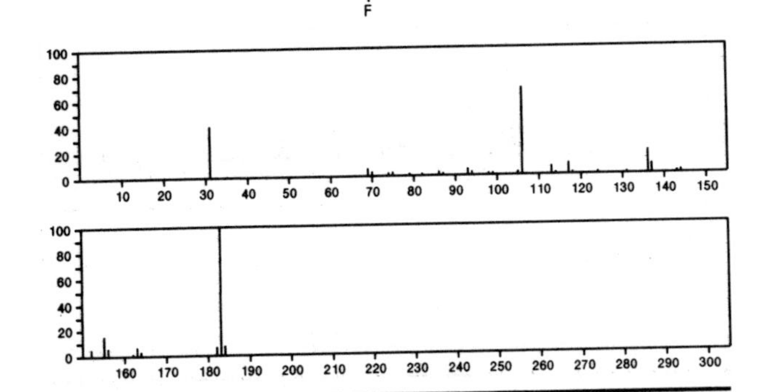

183 $C_6H_5N_3O_2S$ 938-10-3

Benzenesulfonyl azide

N_3SO_2Ph

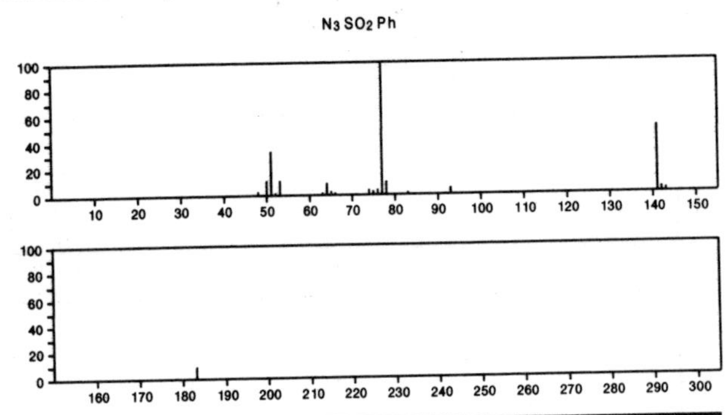

183 $C_6H_5N_3O_4$ 97-02-9

Benzenamine, 2,4-dinitro-

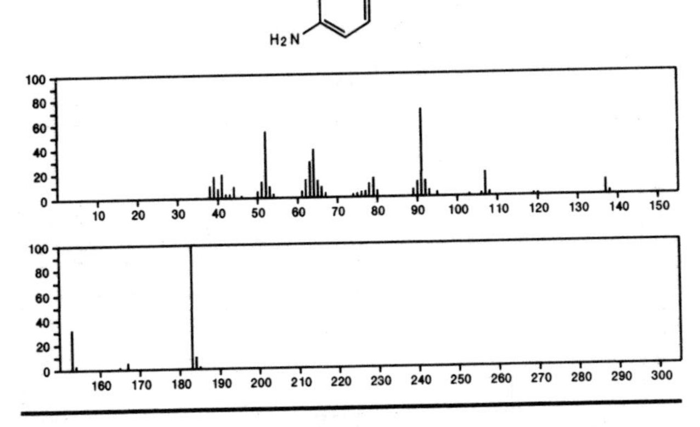

183 $C_6H_5N_3O_4$ 606-22-4

Benzenamine, 2,6-dinitro-

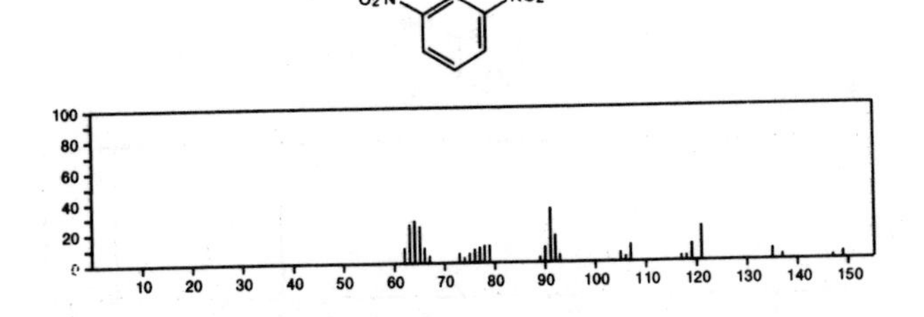

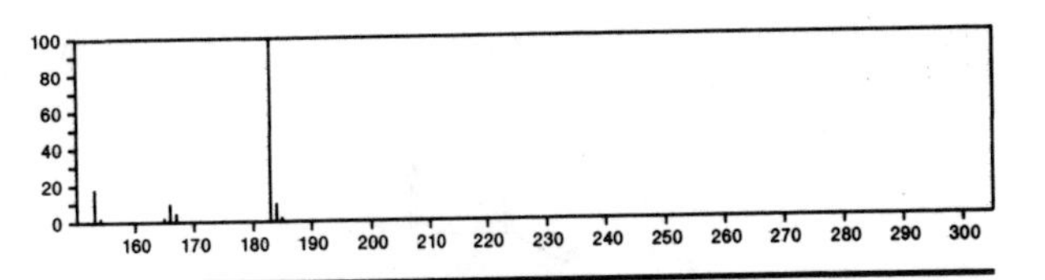

183 $C_6H_5N_3S_2$ 54774-93-5

Thiazolo[5,4-*d*]pyrimidine-7(4*H*)-thione, 5-methyl-

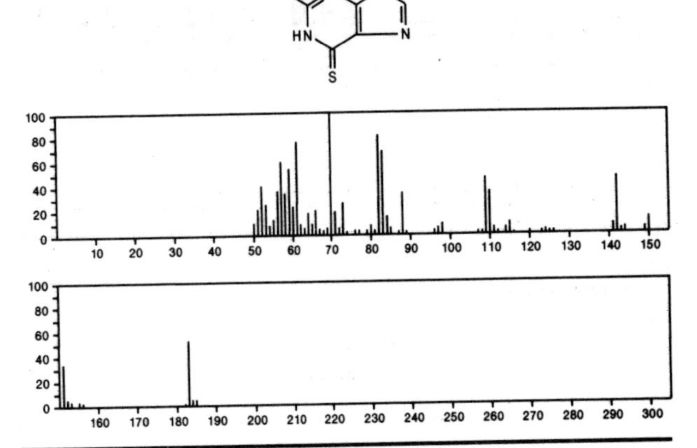

183 $C_7H_5NO_3S$ 81-07-2

1,2-Benzisothiazol-3(2*H*)-one, 1,1-dioxide

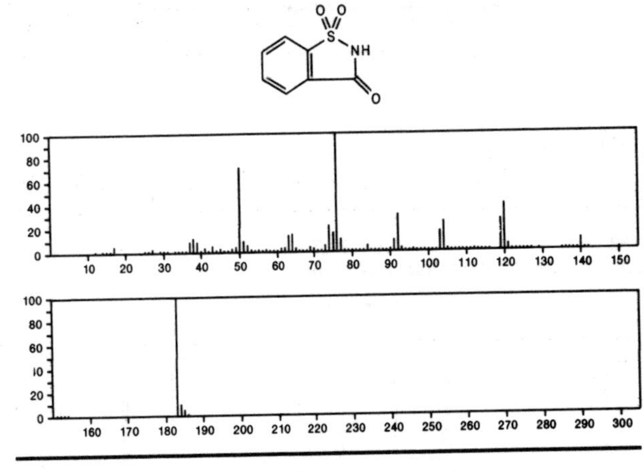

183 $C_7H_5NO_5$ 85-38-1

Benzoic acid, 2-hydroxy-3-nitro-

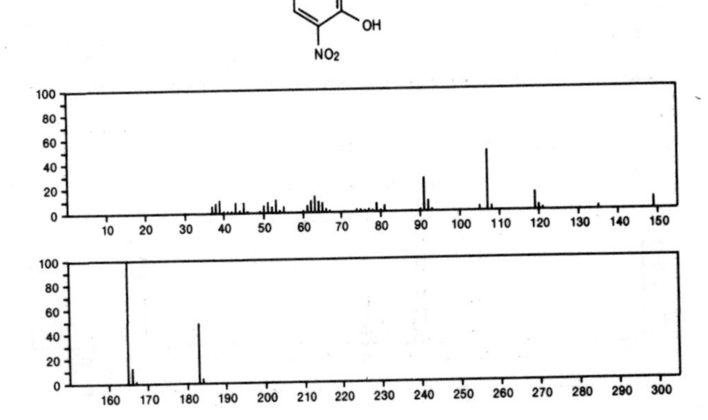

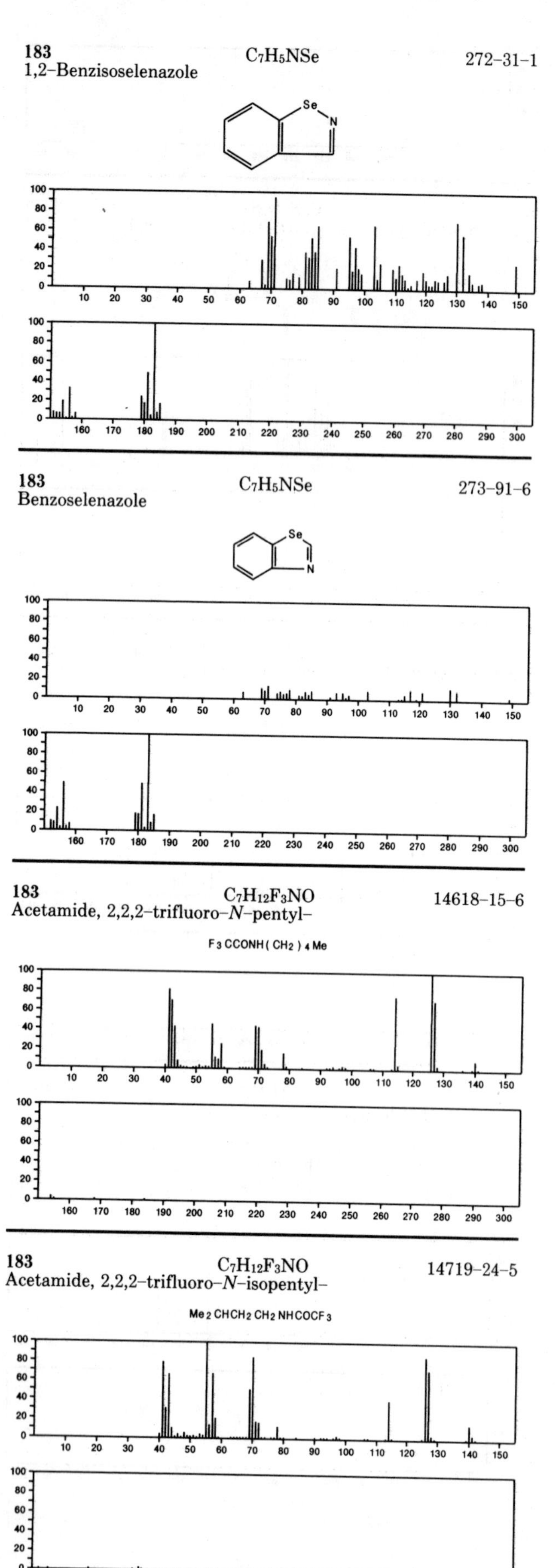

183 C_7H_5NSe 272–31–1
1,2–Benzisoselenazole

183 C_7H_5NSe 273–91–6
Benzoselenazole

183 $C_7H_{12}F_3NO$ 14618–15–6
Acetamide, 2,2,2–trifluoro–*N*–pentyl–

$F_3CCONH(CH_2)_4Me$

183 $C_7H_{12}F_3NO$ 14719–24–5
Acetamide, 2,2,2–trifluoro–*N*–isopentyl–

$Me_2CHCH_2CH_2NHCOCF_3$

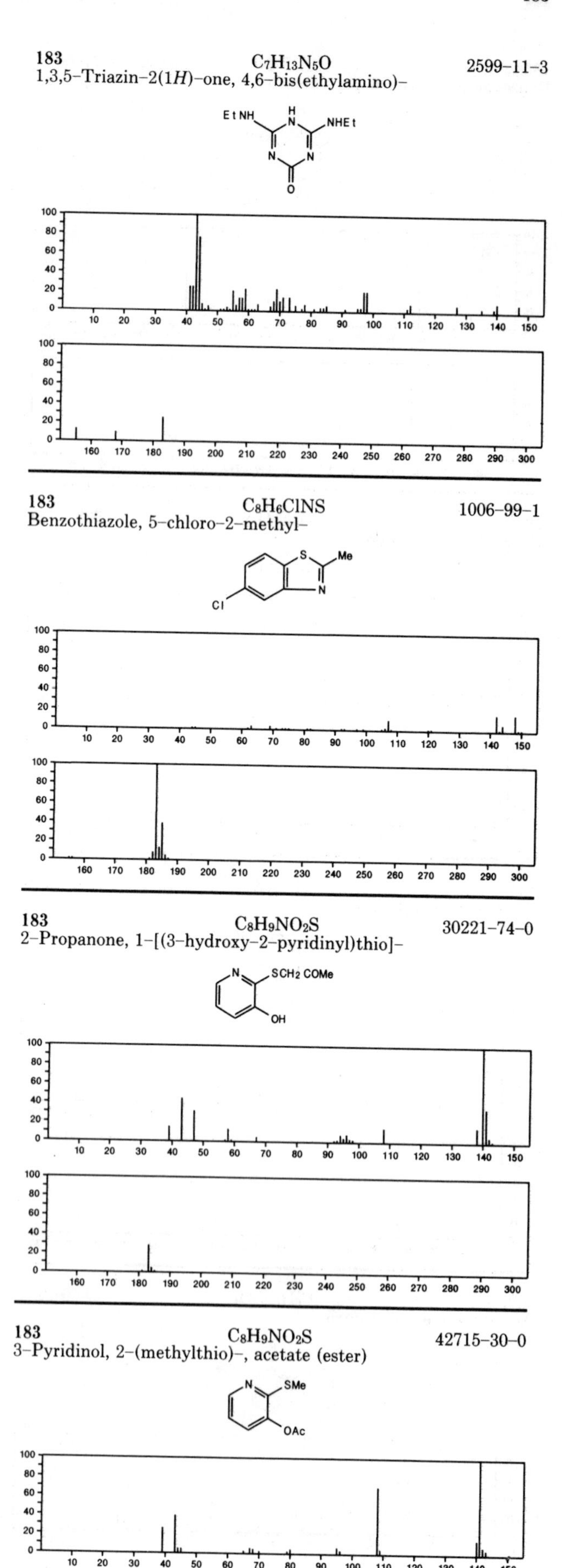

183 $C_7H_{13}N_5O$ 2599–11–3
1,3,5–Triazin–2(1*H*)–one, 4,6–bis(ethylamino)–

183 C_8H_6ClNS 1006–99–1
Benzothiazole, 5–chloro–2–methyl–

183 $C_8H_9NO_2S$ 30221–74–0
2–Propanone, 1–[(3–hydroxy–2–pyridinyl)thio]–

183 $C_8H_9NO_2S$ 42715–30–0
3–Pyridinol, 2–(methylthio)–, acetate (ester)

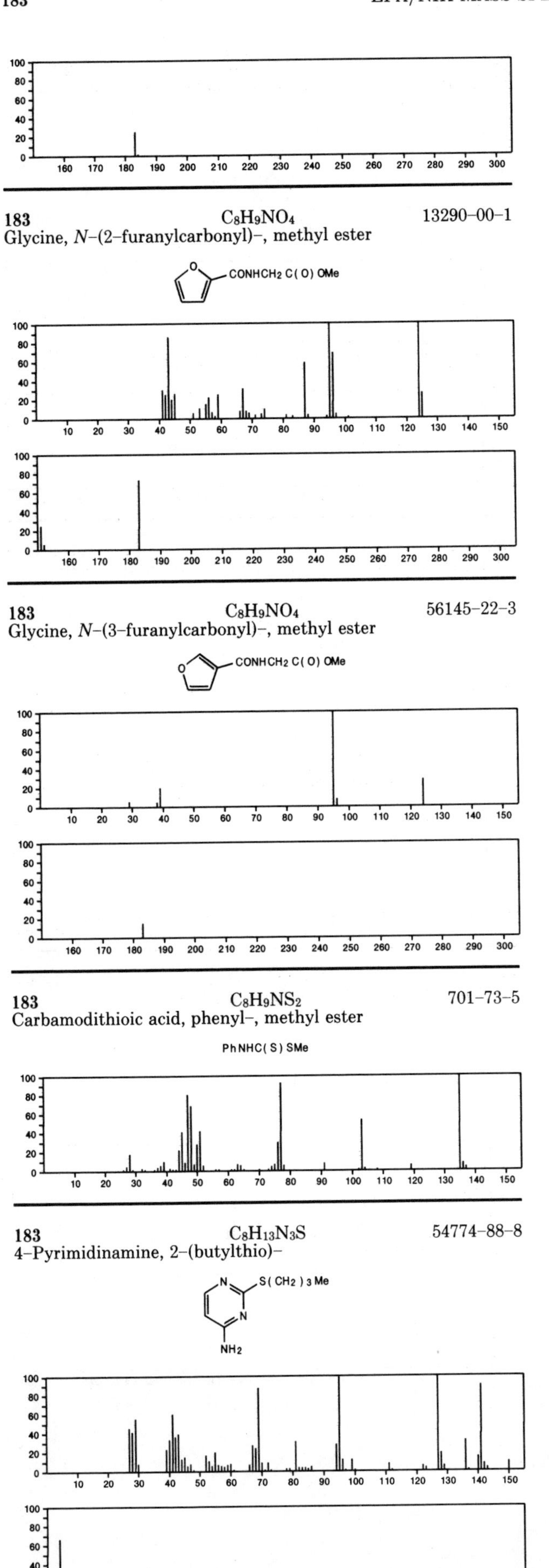

183 $C_8H_9NO_4$ 13290-00-1
Glycine, *N*-(2-furanylcarbonyl)-, methyl ester

183 $C_8H_9NO_4$ 56145-22-3
Glycine, *N*-(3-furanylcarbonyl)-, methyl ester

183 $C_8H_9NS_2$ 701-73-5
Carbamodithioic acid, phenyl-, methyl ester

183 $C_8H_{13}N_3S$ 54774-88-8
4-Pyrimidinamine, 2-(butylthio)-

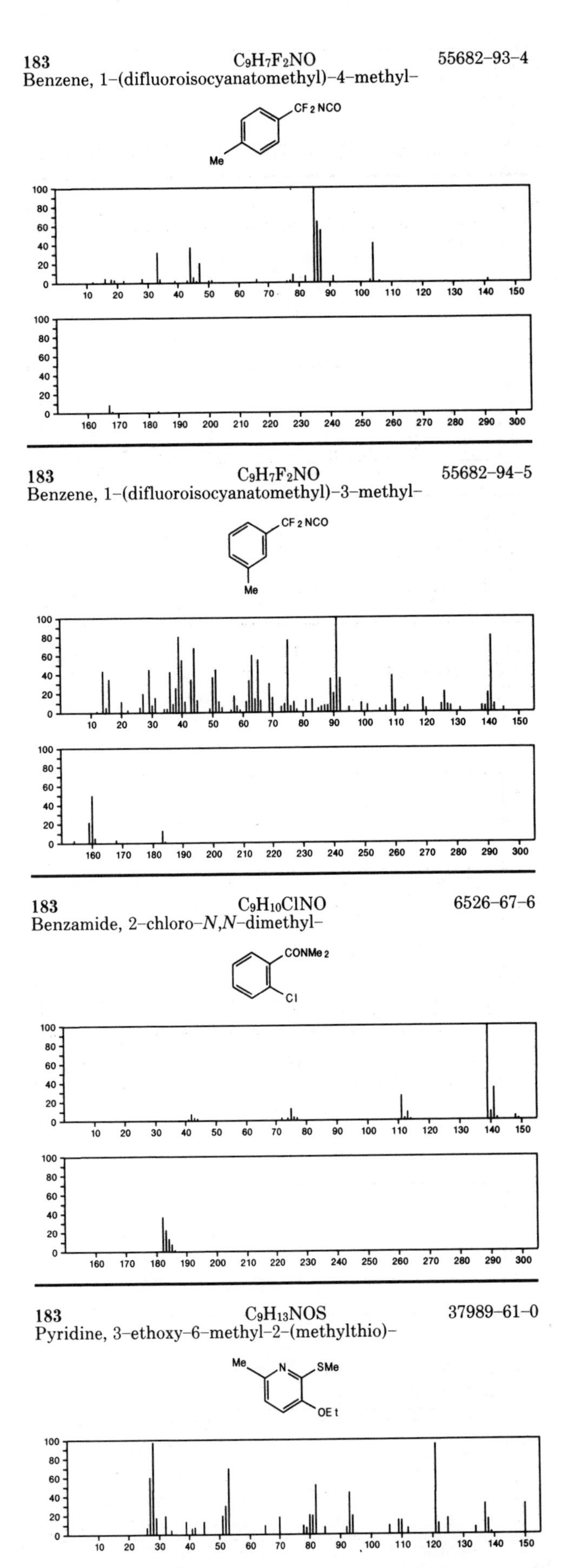

183 $C_9H_7F_2NO$ 55682-93-4
Benzene, 1-(difluoroisocyanatomethyl)-4-methyl-

183 $C_9H_7F_2NO$ 55682-94-5
Benzene, 1-(difluoroisocyanatomethyl)-3-methyl-

183 $C_9H_{10}ClNO$ 6526-67-6
Benzamide, 2-chloro-*N*,*N*-dimethyl-

183 $C_9H_{13}NOS$ 37989-61-0
Pyridine, 3-ethoxy-6-methyl-2-(methylthio)-

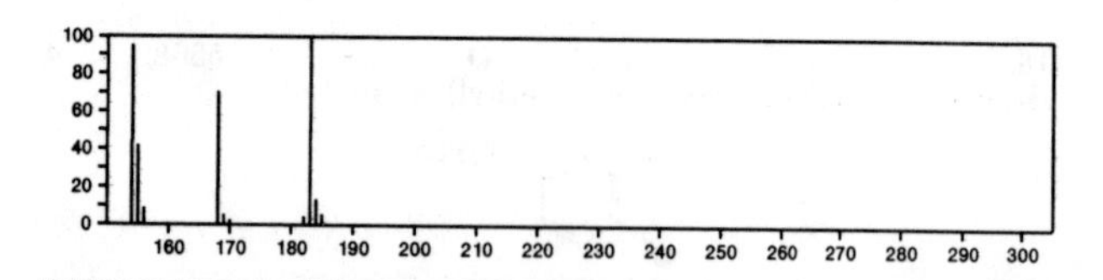

183 $C_9H_{13}NOS$ 39132-48-4
Pyridinium, 2-(ethylthio)-3-hydroxy-1,6-dimethyl-, hydroxide, inner salt

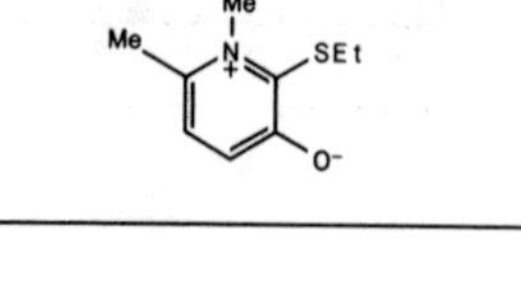

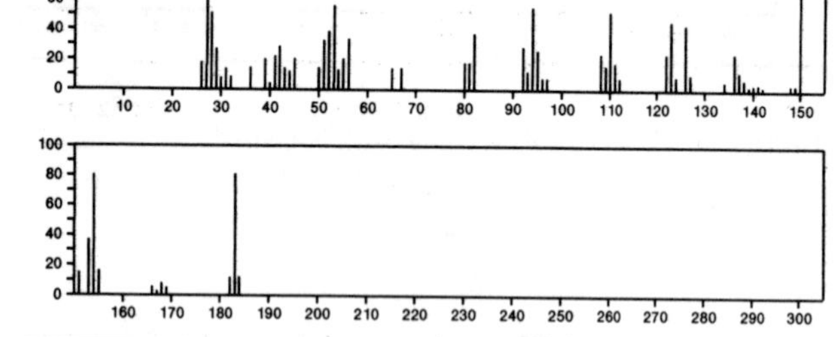

183 $C_9H_{13}NO_3$ 51-43-4
1,2-Benzenediol, 4-[1-hydroxy-2-(methylamino)ethyl]-, (*R*)-

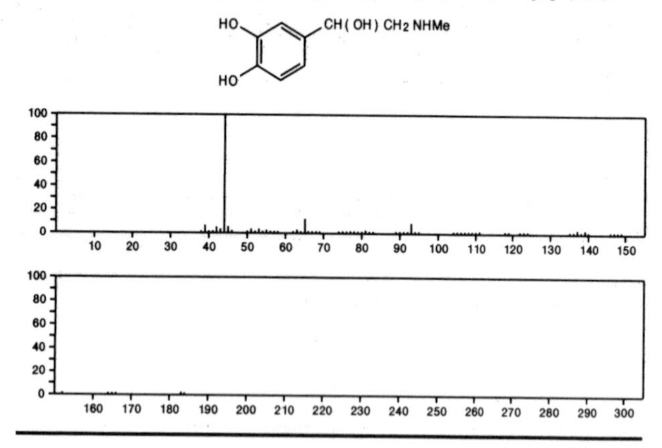

183 $C_9H_{13}NO_3$ 30740-21-7
1-Azabicyclo[2.2.2]octane-2-carboxylic acid, 5-oxo-, methyl ester

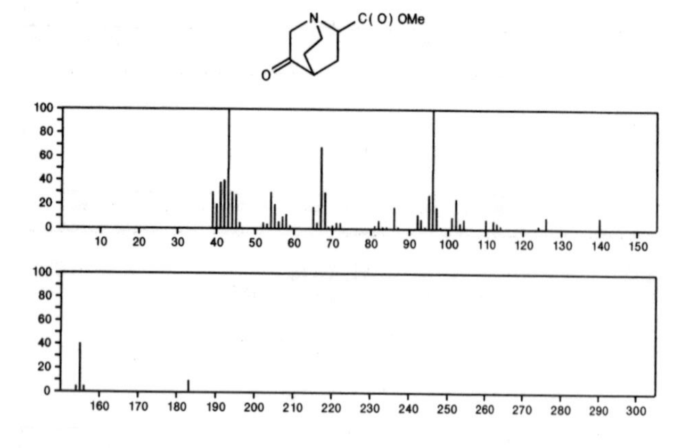

183 $C_9H_{15}N_2O_2$ 3229-73-0
1*H*-Pyrrol-1-yloxy, 3-(aminocarbonyl)-2,5-dihydro-2,2,5,5-tetramethyl-

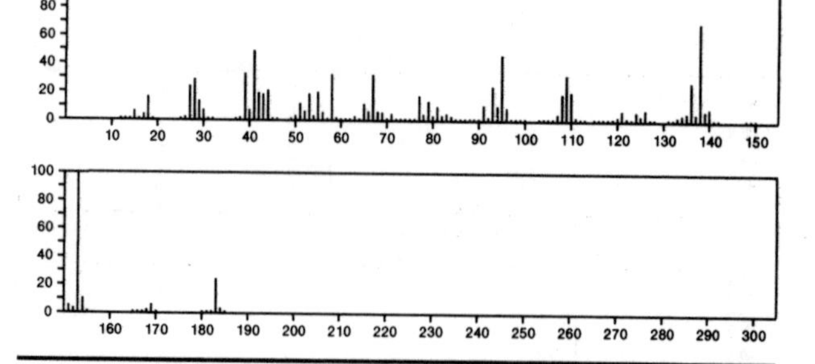

183 $C_9H_{17}N_3O$ 57174-11-5
Hydrazinecarboxamide, 2-(2,6-dimethylcyclohexylidene)-

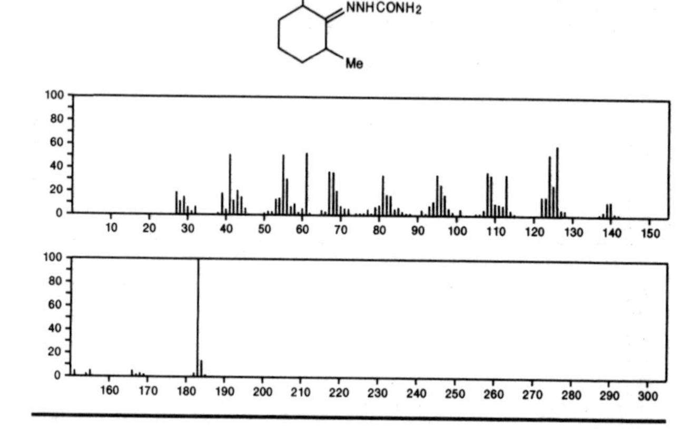

183 $C_{10}H_{17}NO_2$ 125-64-4
2,4-Piperidinedione, 3,3-diethyl-5-methyl-

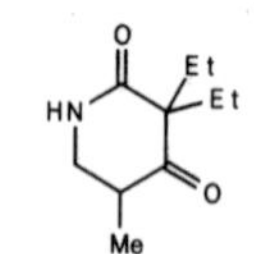

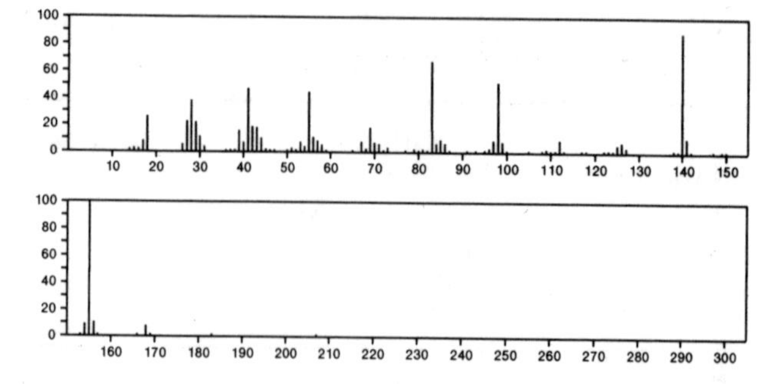

183 $C_{10}H_{17}NO_2$ 22766-68-3
1-Azabicyclo[2.2.2]octane-4-carboxylic acid, ethyl ester

C(O)OEt

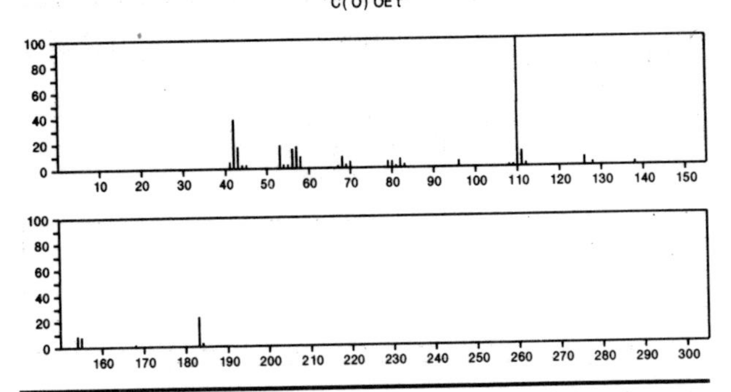

183 $C_{10}H_{17}NO_2$ 53171-59-8
Bicyclo[3.2.0]heptan-3-one, 2-hydroxy-1,4,4-trimethyl-, oxime

Me OH NOH Me Me

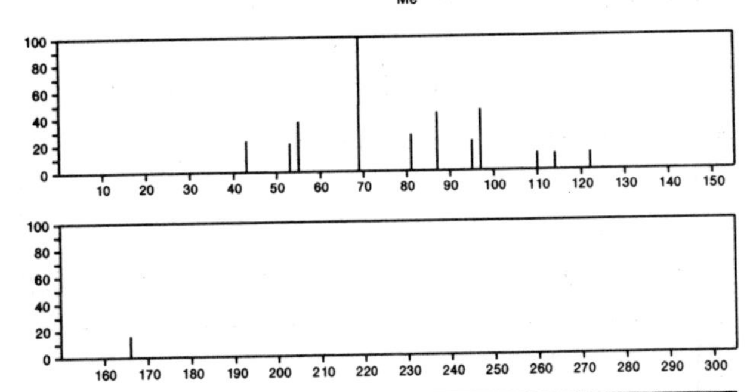

183 $C_{10}H_{17}NS$ 52414-86-5
Thiazole, 2-butyl-5-propyl-

Pr S (CH2)3Me N

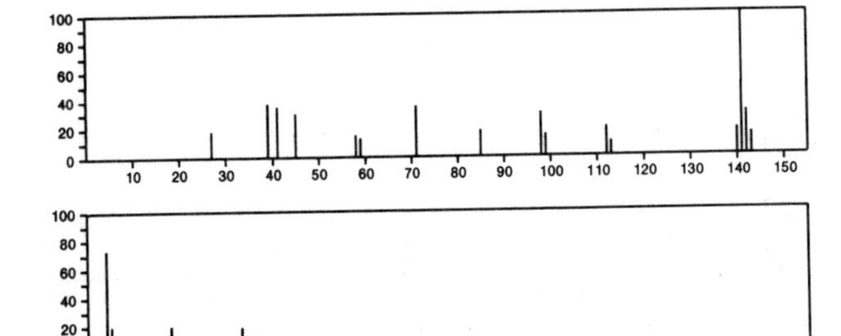

183 $C_{10}H_{17}NS$ 52414-88-7
Thiazole, 2-butyl-4-ethyl-5-methyl-

Me S (CH2)3Me N Et

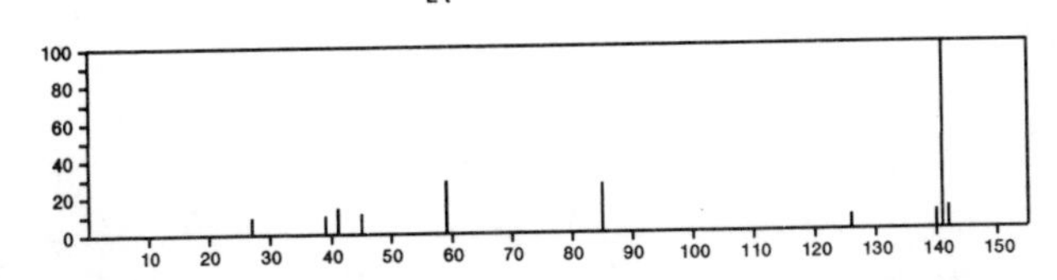

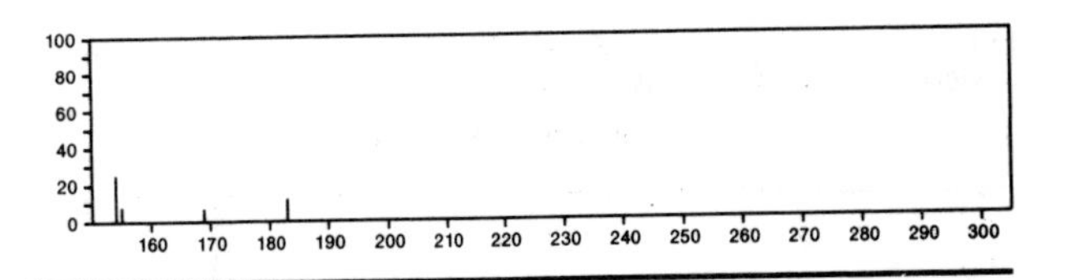

183 $C_{11}H_{21}NO$ 1925-46-8
2-Bornanol, 3-(methylamino)-, *endo,endo*-

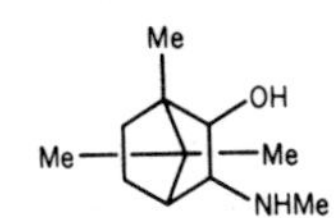

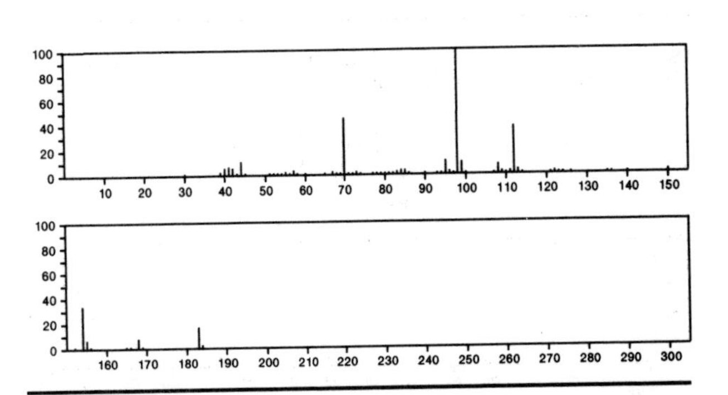

183 $C_{11}H_{21}NO$ 3189-61-5
Cycloundecanone, oxime

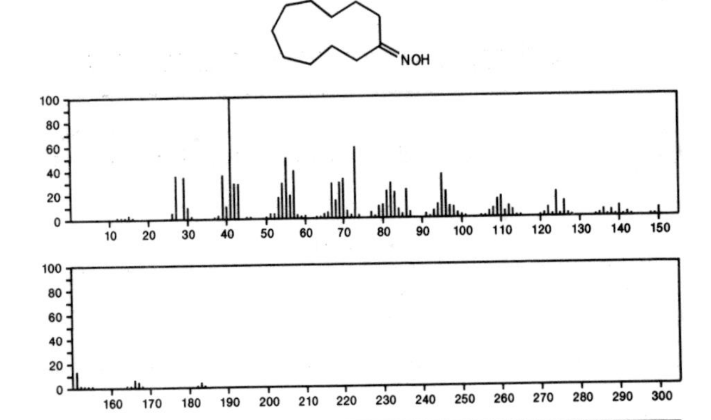

183 $C_{11}H_{21}NO$ 32232-17-0
3-Bornanol, 2-(methylamino)-, *endo,endo*-

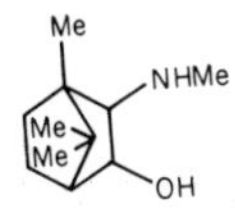

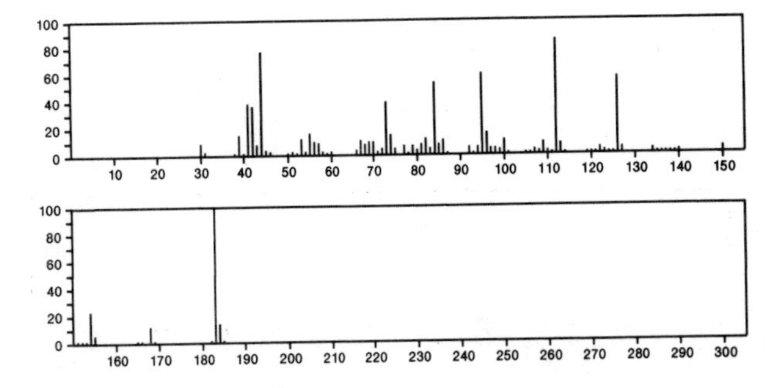

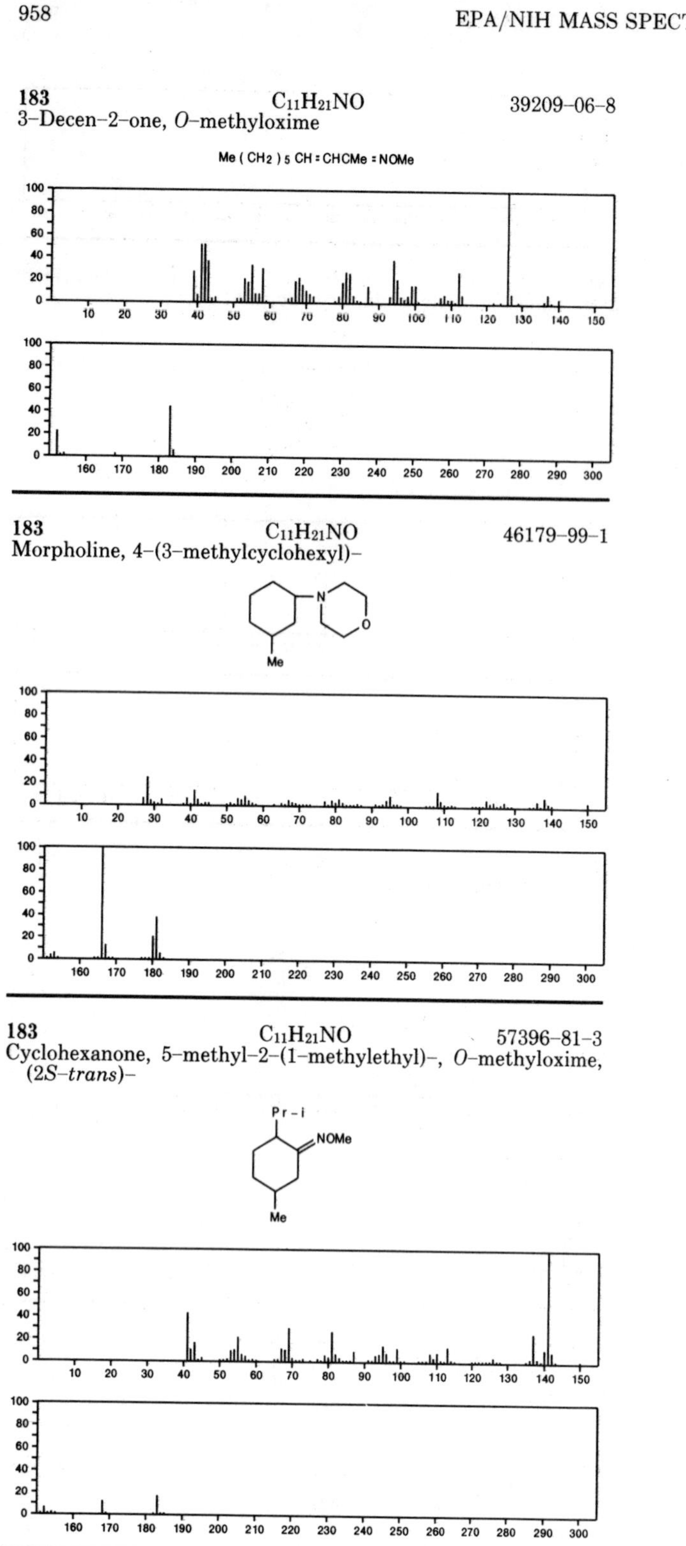

183 $C_{12}H_9NO$ 91-02-1

Methanone, phenyl-2-pyridinyl-

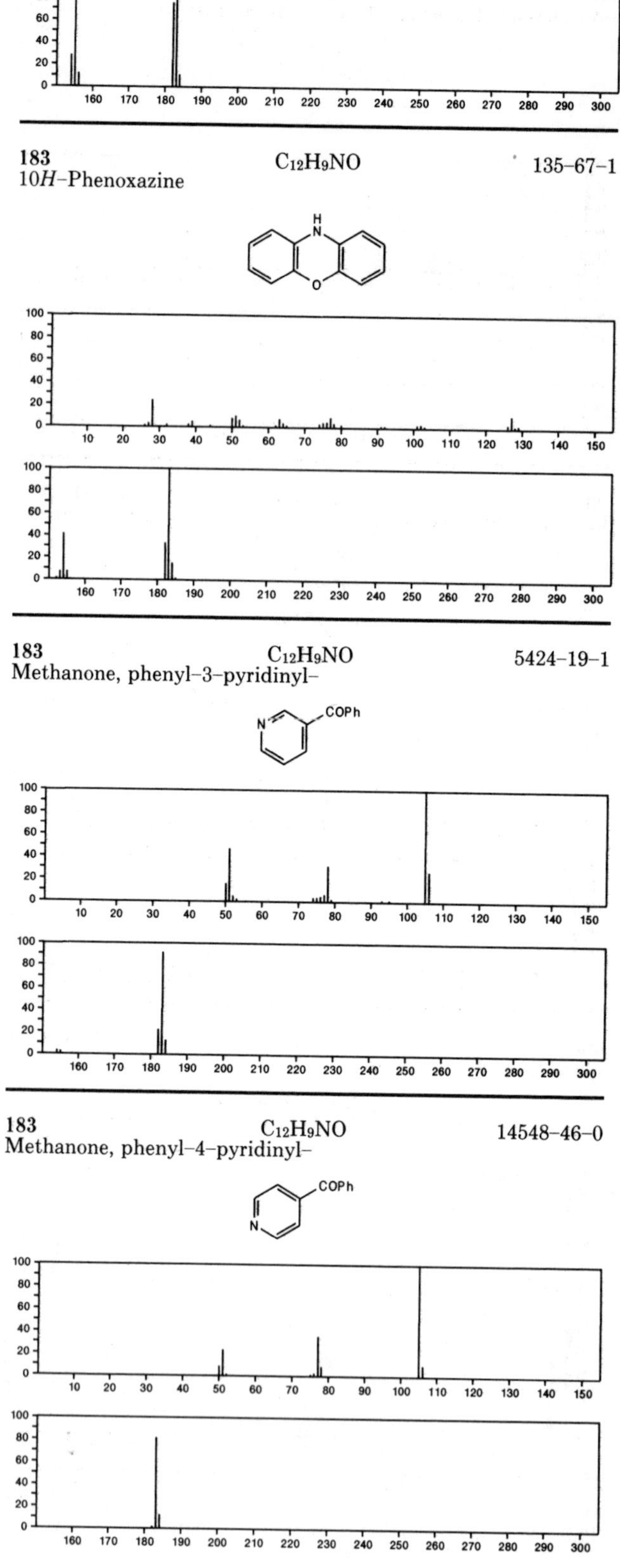

183 $C_{13}H_{13}N$ 552–82–9
Benzenamine, *N*–methyl–*N*–phenyl–

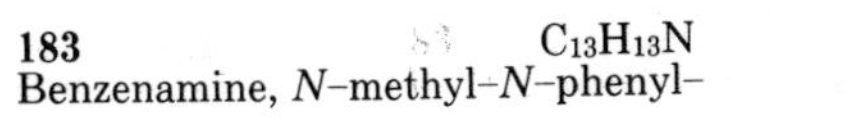

Ph2NMe

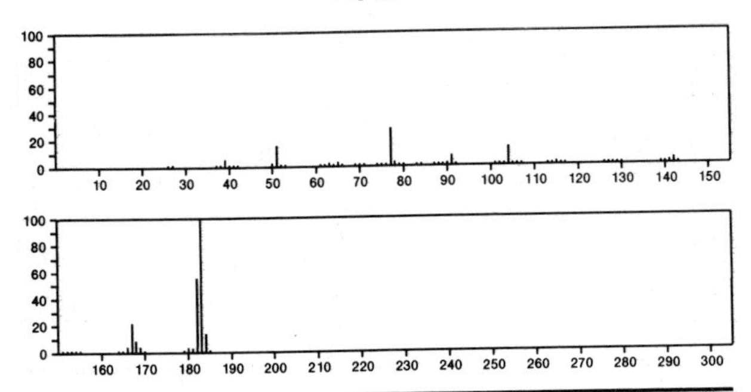

183 $C_{13}H_{13}N$ 14294–33–8
2–Biphenylamine, 3–methyl–

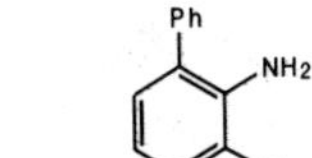

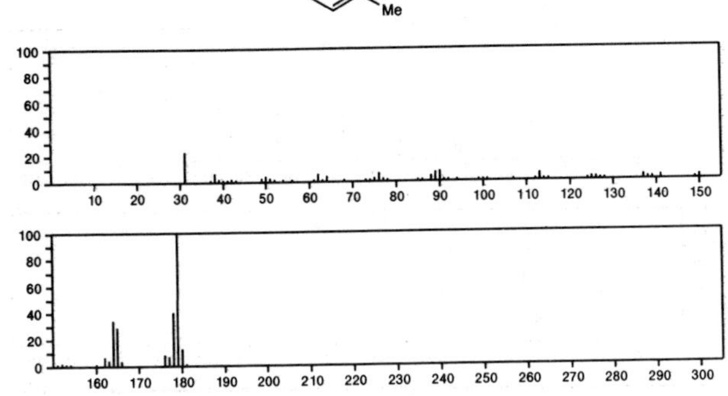

183 $C_{13}H_{13}N$ 27985–90–6
Benzenamine, *ar*–(phenylmethyl)–

PhCH2—(C6H4)—NH2

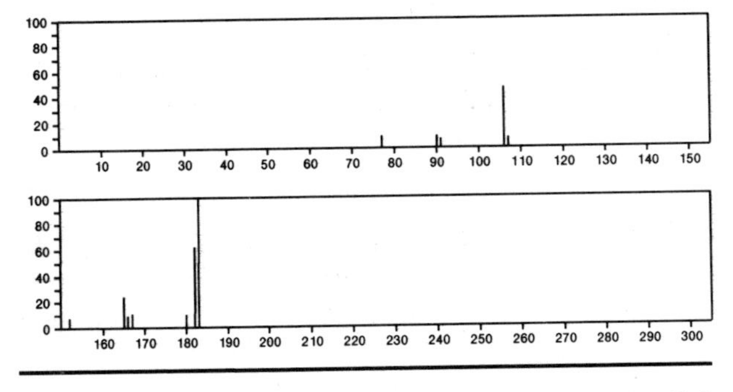

184 $C_2H_2Br_2$ 540–49–8
Ethene, 1,2–dibromo–

BrCH=CHBr

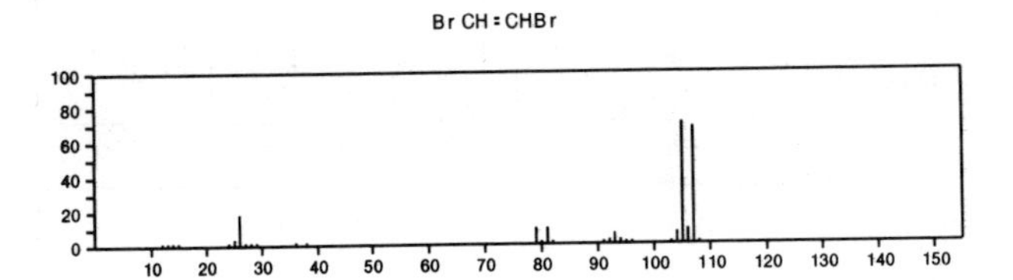

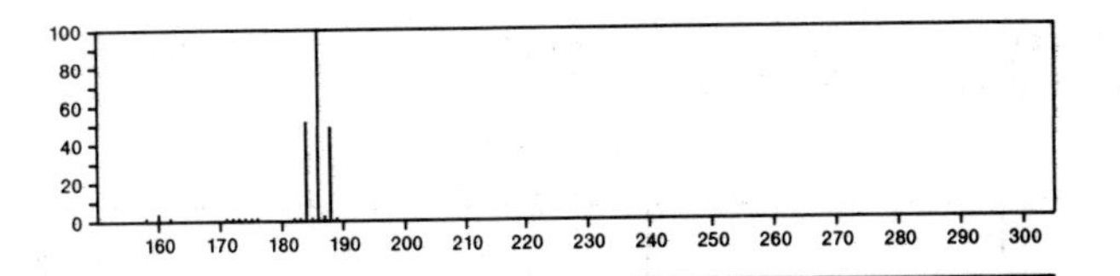

184 $C_3H_2ClF_5O$ 13838–16–9
Ethane, 2–chloro–1–(difluoromethoxy)–1,1,2–trifluoro–

F2CHOCF2CHClF

184 $C_3H_9AsO_4$ 13006–30–9
Arsenic acid (H_3AsO_4), trimethyl ester

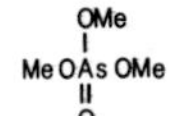

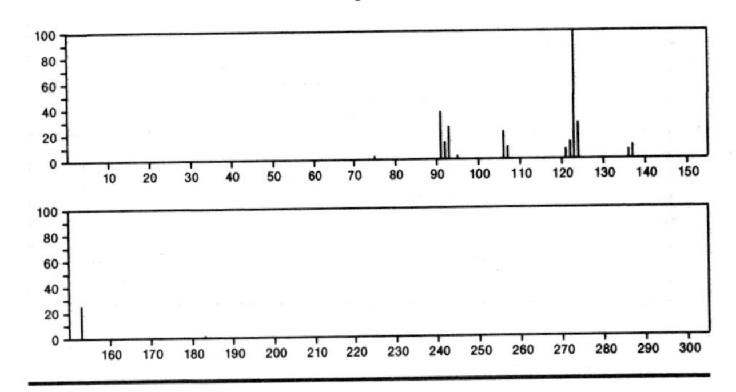

184 C_4H_9I 513–38–2
Propane, 1–iodo–2–methyl–

i–BuI

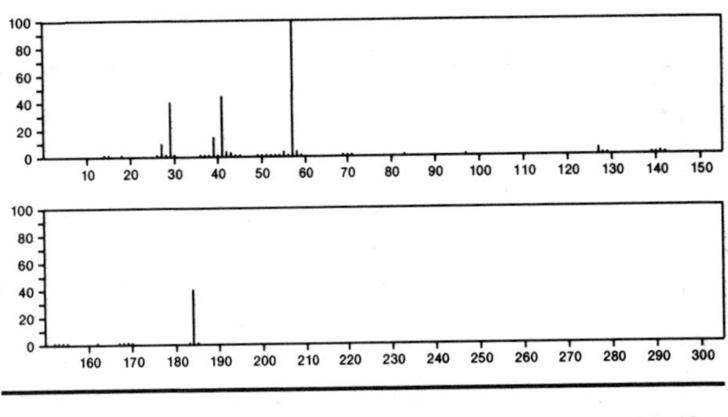

184 C_4H_9I 513–48–4
Butane, 2–iodo–

s–BuI

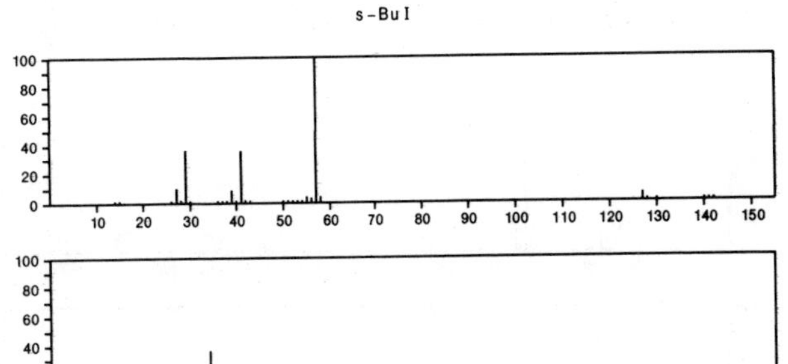

184 C_4H_9I 542–69–8
Butane, 1–iodo–

Me (CH2) 3 I

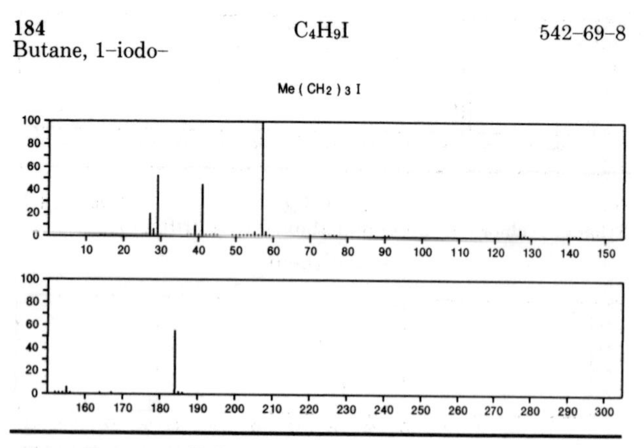

184 C_4H_9I 558–17–8
Propane, 2–iodo–2–methyl–

t – Bu I

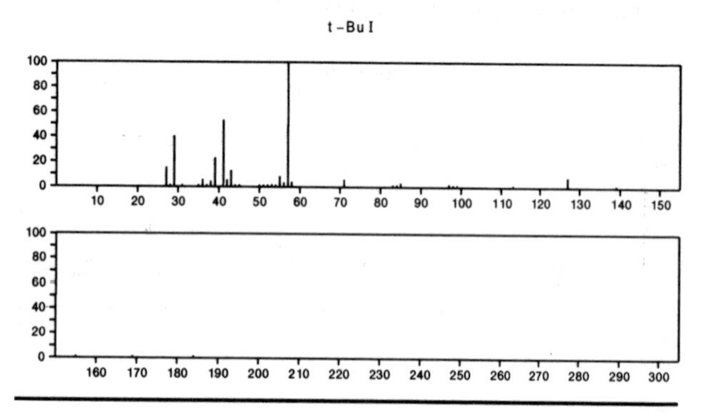

184 C_6HF_5O 771–61–9
Phenol, pentafluoro–

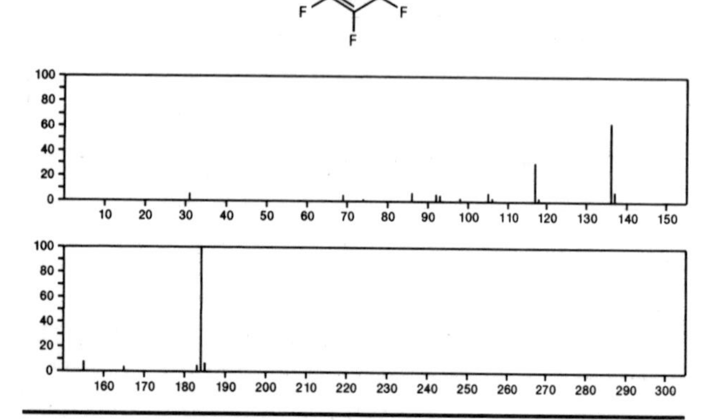

184 $C_6H_{10}Cl_2O_2$ 54587–48–3
Propanoic acid, 2,2–dichloro–, 1–methylethyl ester

i – Pr OC(O) CCl 2 Me

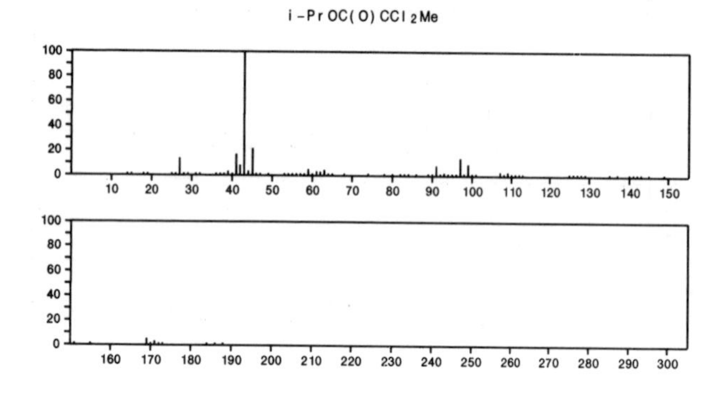

184 $C_6H_{10}Cl_2O_2$ 54774–99–1
Propanoic acid, 2,3–dichloro–, 1–methylethyl ester

Cl CH2 CHCl C(O) OPr – i

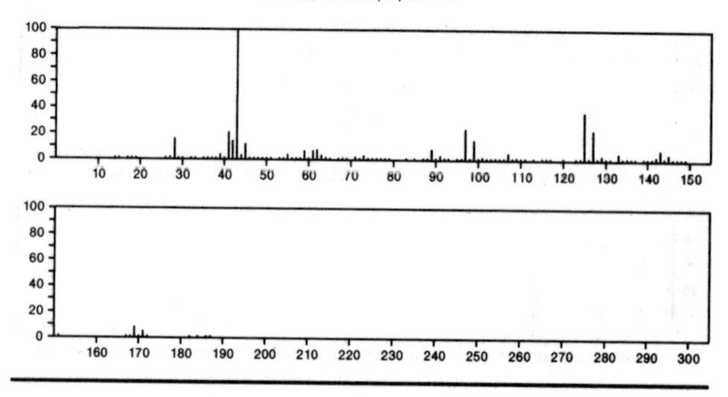

184 $C_7H_8N_2O_2S$ 33853–77–9
1*H*–2,1,3–Benzothiadiazine, 3,4–dihydro–, 2,2–dioxide

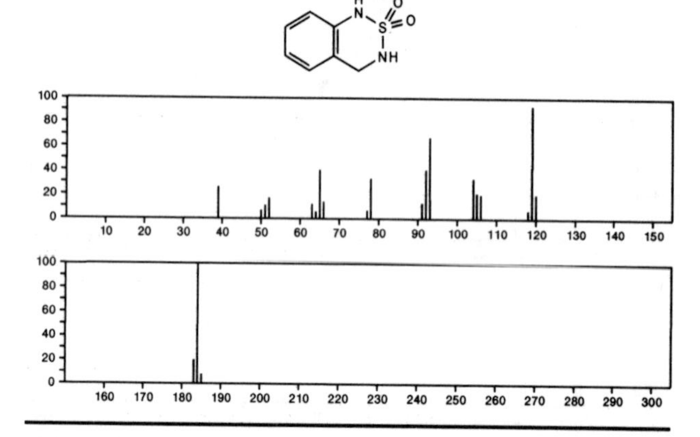

184 $C_7H_{11}F_3O_2$ 16408–83–6
2*H*–Pyran, tetrahydro–2–(2,2,2–trifluoroethoxy)–

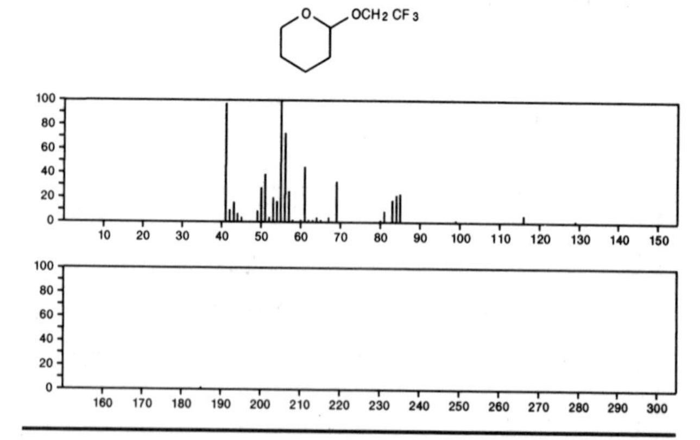

184 $C_8H_8OS_2$ 6047–46–7
Carbonic acid, dithio–, *O*–methyl *S*–phenyl ester

PhSC(S) OMe

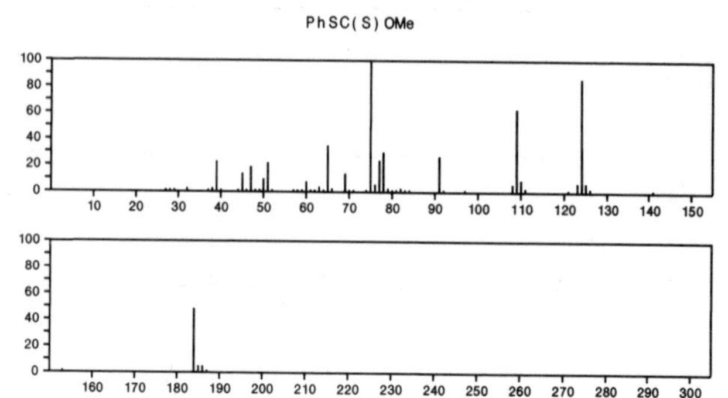

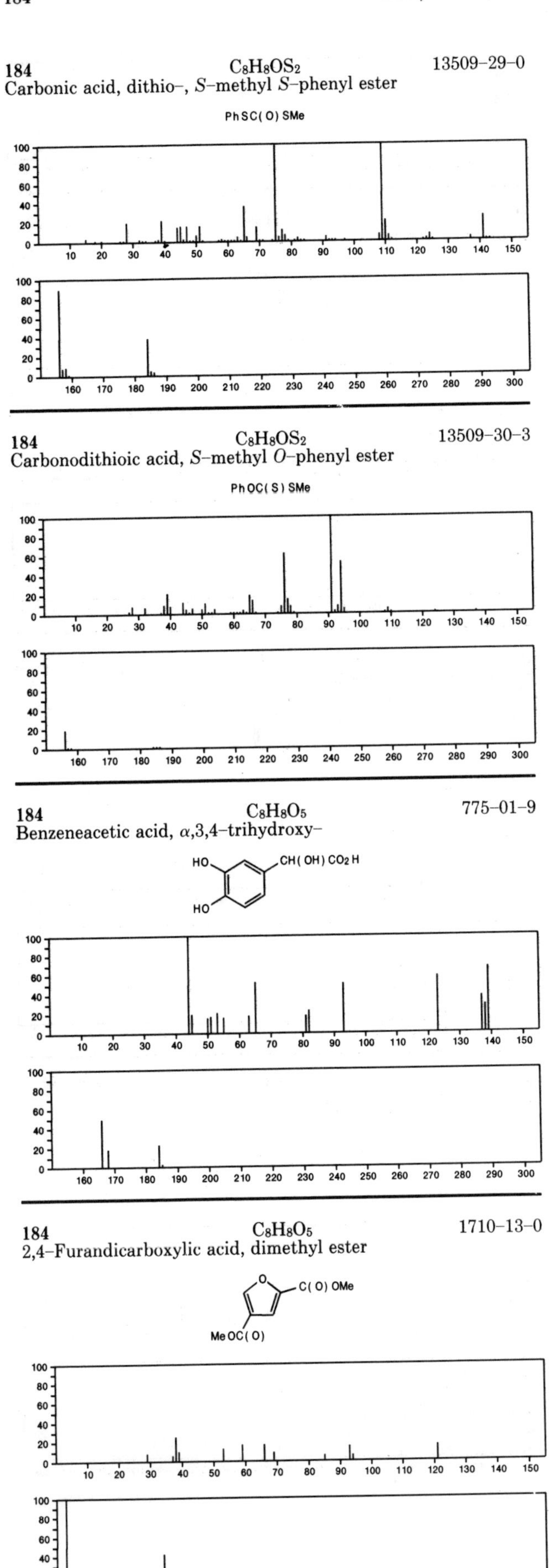
184 C8H8OS2 13509-29-0
Carbonic acid, dithio-, S-methyl S-phenyl ester
PhSC(O)SMe
184 C8H8OS2 13509-30-3
Carbonodithioic acid, S-methyl O-phenyl ester
PhOC(S)SMe
184 C8H8O5 775-01-9
Benzeneacetic acid, α,3,4-trihydroxy-
HO
HO
CH(OH)CO2H
184 C8H8O5 1710-13-0
2,4-Furandicarboxylic acid, dimethyl ester
O
C(O)OMe
MeOC(O)

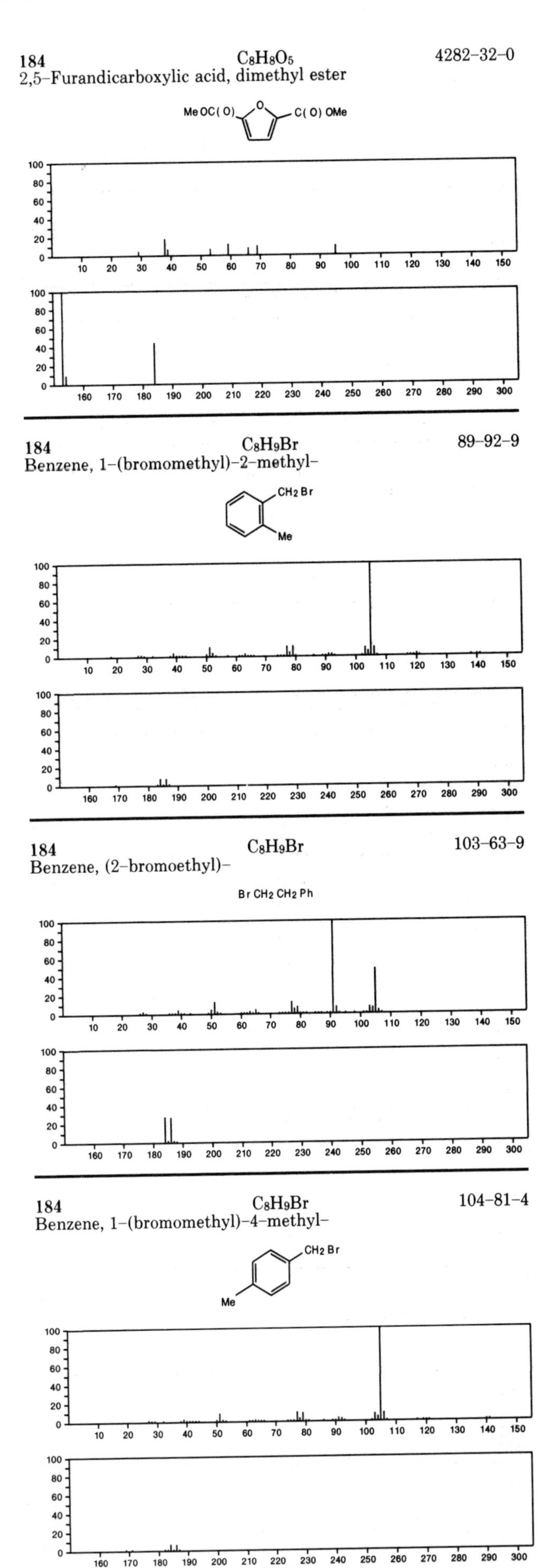
184 C8H8O5 4282-32-0
2,5-Furandicarboxylic acid, dimethyl ester
MeOC(O)
O
C(O)OMe
184 C8H9Br 89-92-9
Benzene, 1-(bromomethyl)-2-methyl-
CH2Br
Me
184 C8H9Br 103-63-9
Benzene, (2-bromoethyl)-
BrCH2CH2Ph
184 C8H9Br 104-81-4
Benzene, 1-(bromomethyl)-4-methyl-
CH2Br
Me

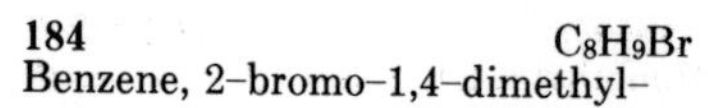

184 C_8H_9Br 553-94-6
Benzene, 2-bromo-1,4-dimethyl-

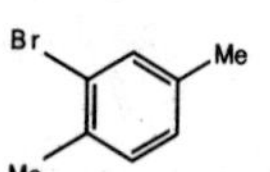

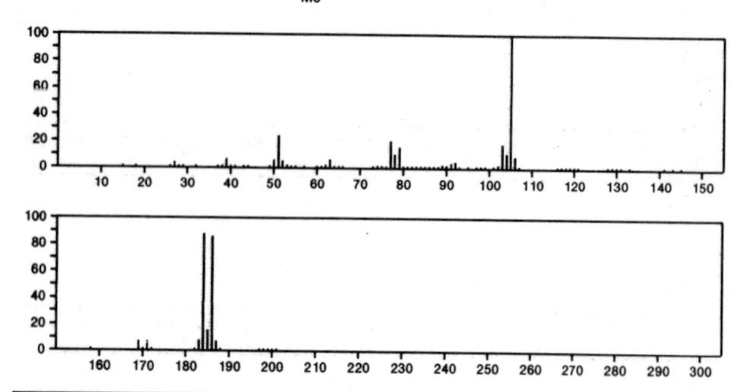

184 C_8H_9Br 583-70-0
Benzene, 1-bromo-2,4-dimethyl-

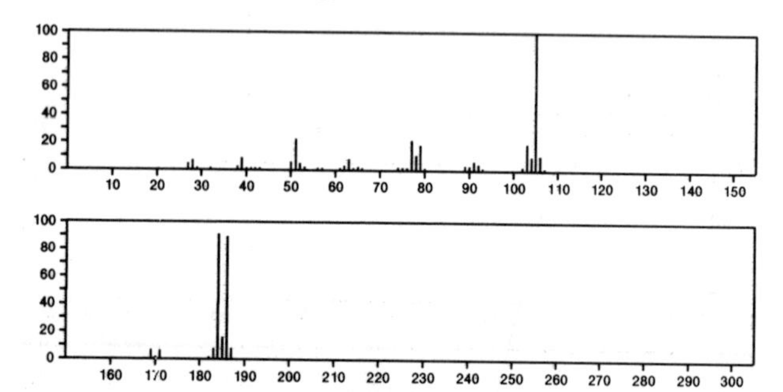

184 C_8H_9Br 583-71-1
Benzene, 4-bromo-1,2-dimethyl-

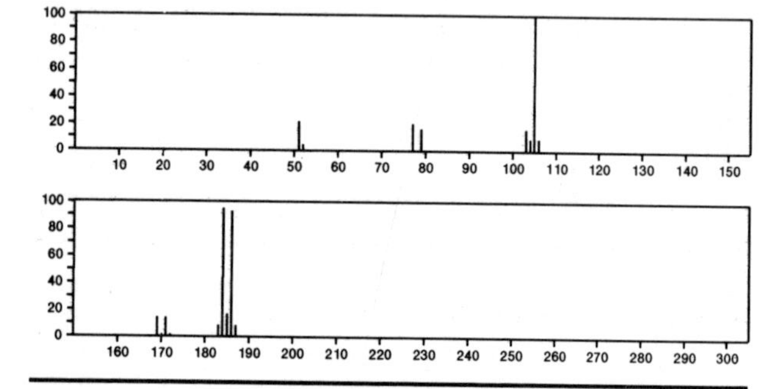

184 C_8H_9Br 585-71-7
Benzene, (1-bromoethyl)-

PhCHBr(Ph)

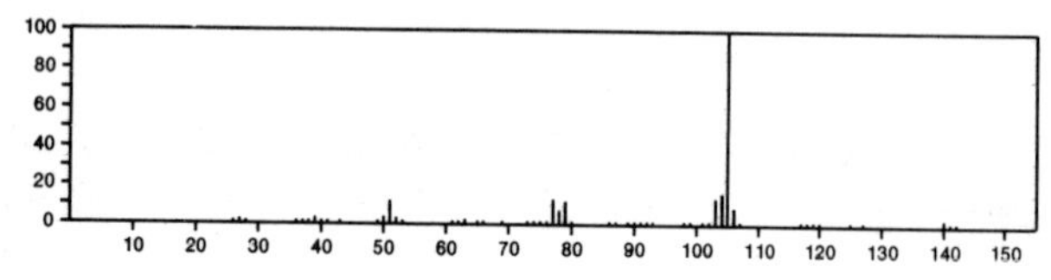

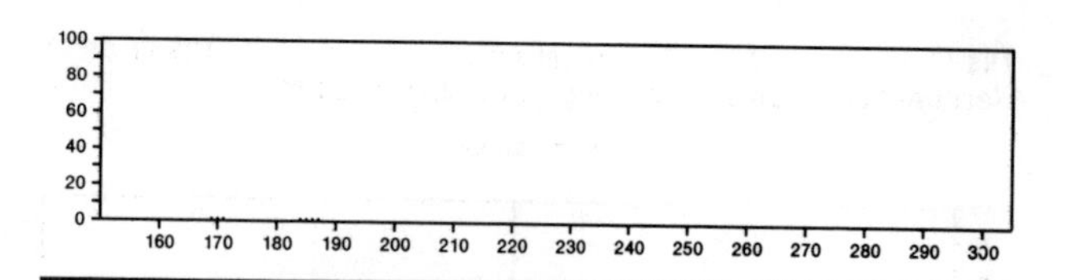

184 C_8H_9Br 620-13-3
Benzene, 1-(bromomethyl)-3-methyl-

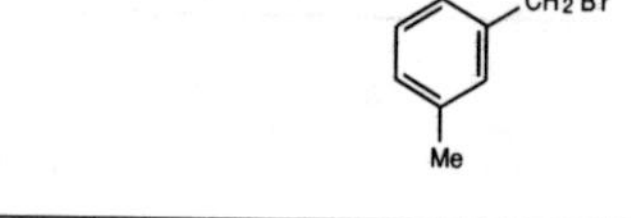

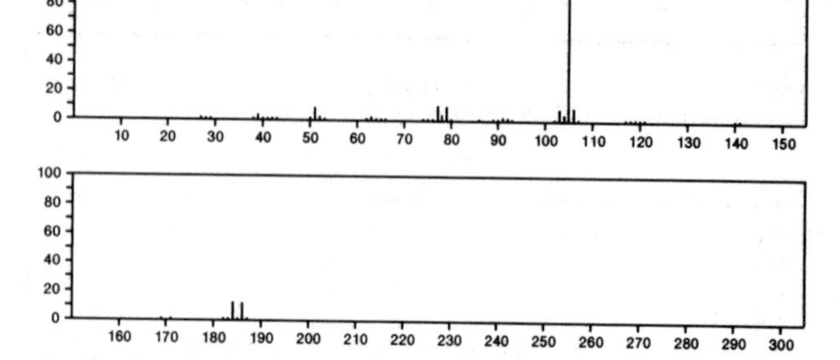

184 C_8H_9Br 1585-07-5
Benzene, 1-bromo-4-ethyl-

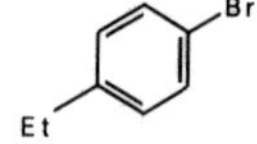

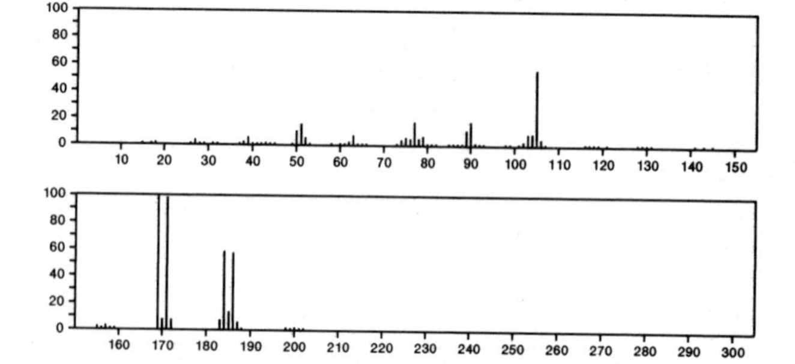

184 C_8H_9Br 1973-22-4
Benzene, 1-bromo-2-ethyl-

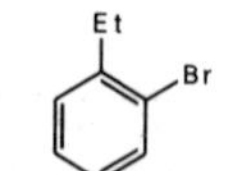

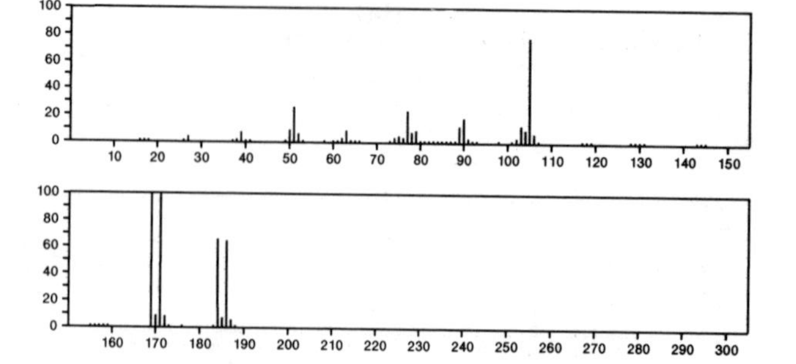

184 C_8H_9Br 28258-59-5

Xylene, bromo-

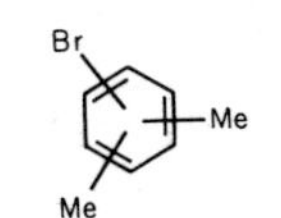

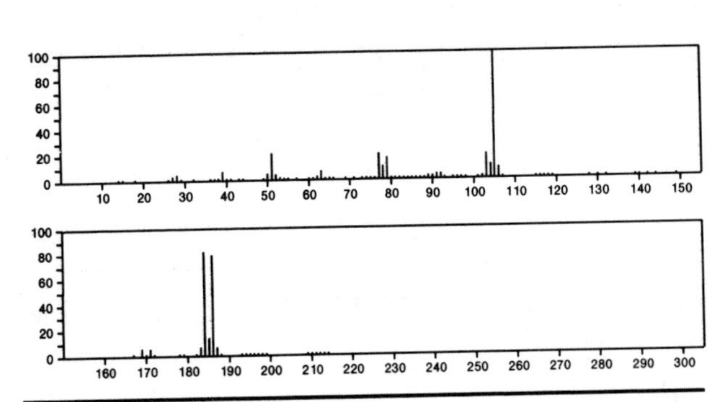

184 C_8H_9Br 31620-80-1

Benzene, bromoethyl-

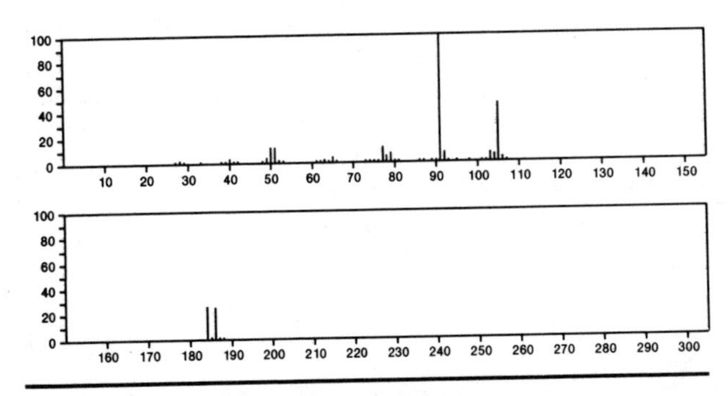

184 $C_8H_9ClN_2O$ 5352-88-5

Urea, *N*-(4-chlorophenyl)-*N'*-methyl-

NHCONHMe

Cl

184 $C_8H_9ClN_2O$ 22517-43-7

Urea, 1-(*p*-chlorophenyl)-1-methyl-

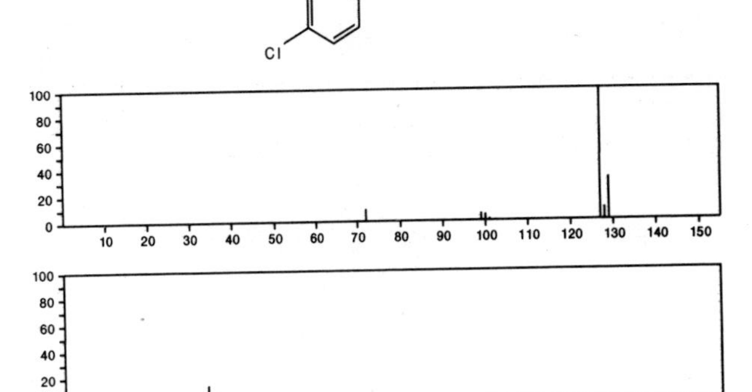

184 $C_8H_{12}N_2OS$ 35231-62-0

Pyrimidine, 5-ethoxy-2-methyl-4-(methylthio)-

Me N SMe N OEt

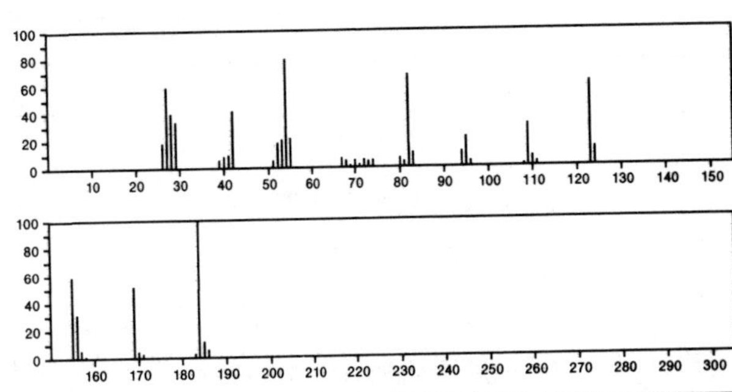

184 $C_8H_{12}N_2OS$ 54774-97-9

4(1*H*)-Pyrimidinone, 2-(butylthio)-

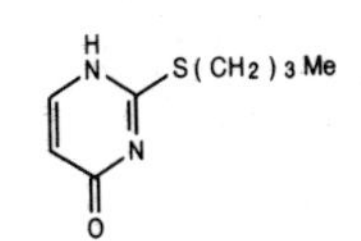

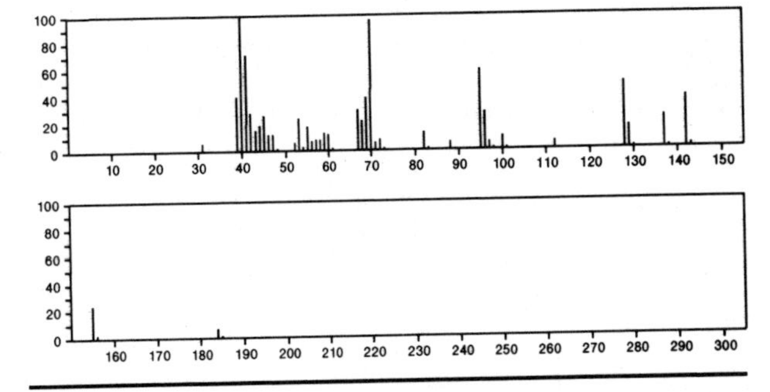

184 $C_8H_{12}N_2OS$ 54774-98-0

4(1*H*)-Pyrimidinone, 5-methyl-2-(propylthio)-

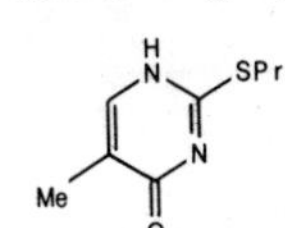

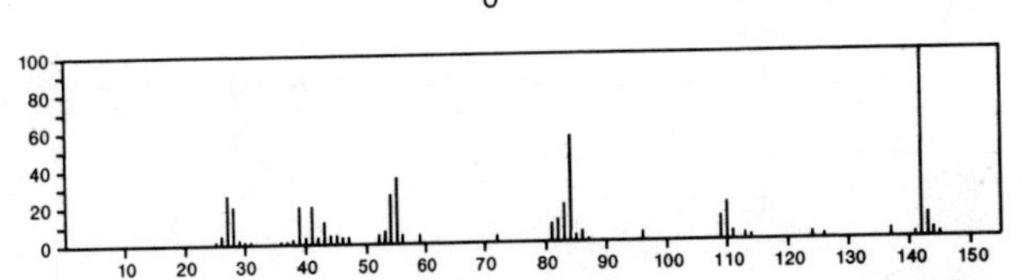

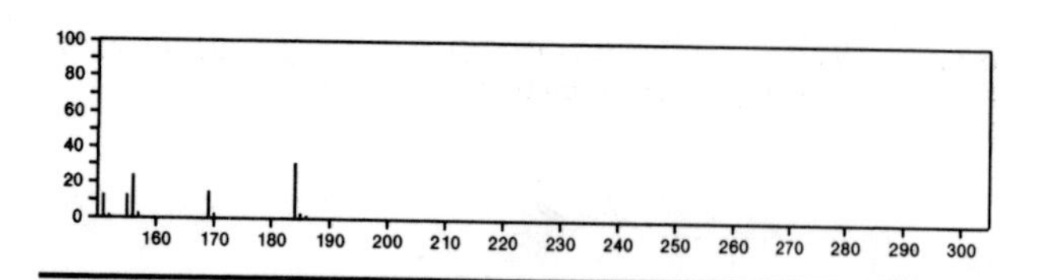

184 $C_8H_{12}N_2O_3$ 57-44-3
2,4,6(1*H*,3*H*,5*H*)-Pyrimidinetrione, 5,5-diethyl-

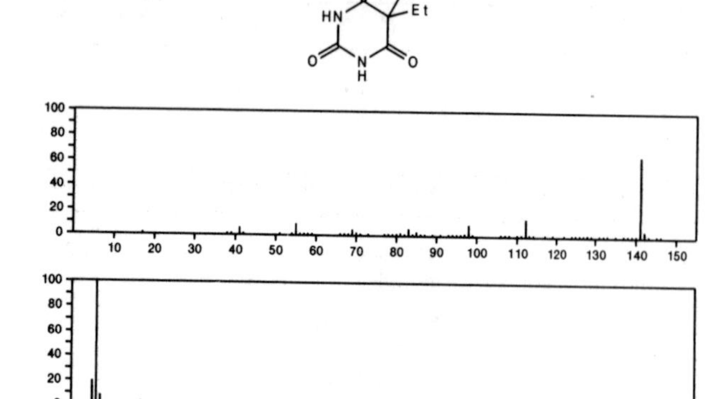

184 $C_8H_{12}N_2O_3$ 1953-33-9
2,4,6(1*H*,3*H*,5*H*)-Pyrimidinetrione, 5-butyl-

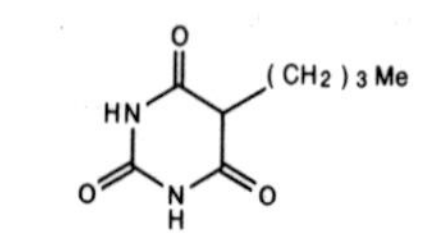

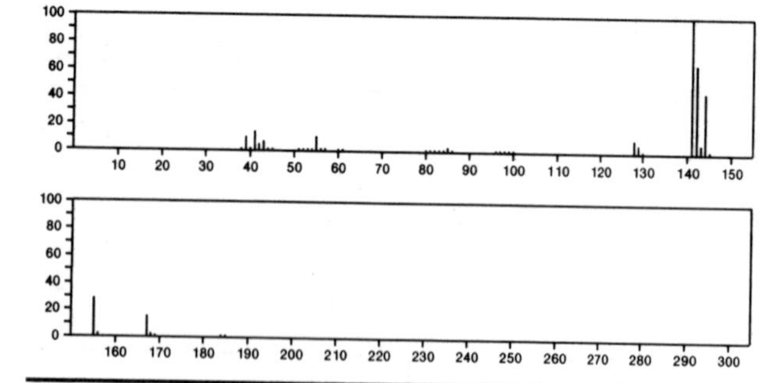

184 $C_8H_{12}N_2O_3$ 7391-61-9
2,4,6(1*H*,3*H*,5*H*)-Pyrimidinetrione, 5-ethyl-1,3-dimethyl-

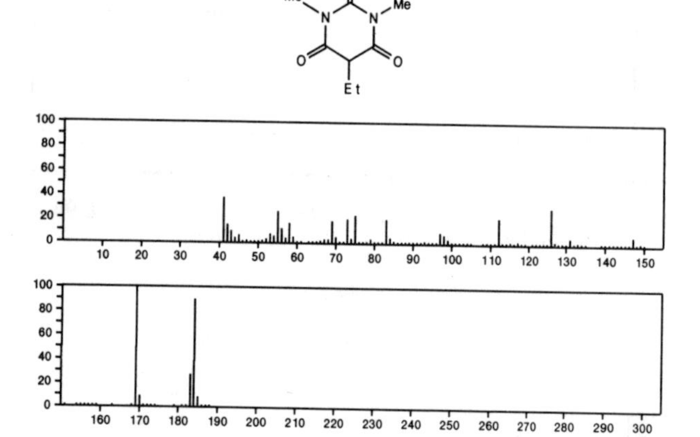

184 $C_8H_{12}O_3Si$ 55887-53-1
2-Furancarboxylic acid, trimethylsilyl ester

C(O)OSiMe3

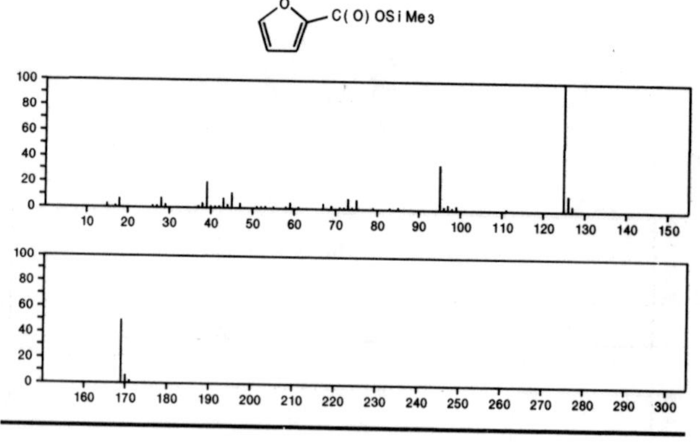

184 $C_8H_{15}F_3O$ 453-43-0
2-Octanol, 1,1,1-trifluoro-

Me(CH2)5CH(OH)CF3

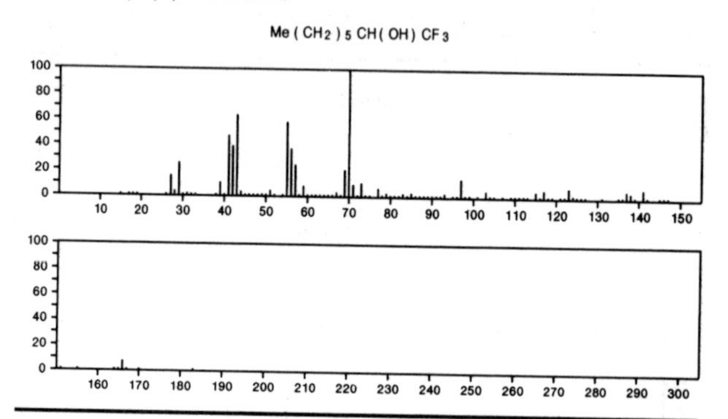

184 $C_9H_9ClO_2$ 140-18-1
Acetic acid, chloro-, phenylmethyl ester

ClCH2C(O)OCH2Ph

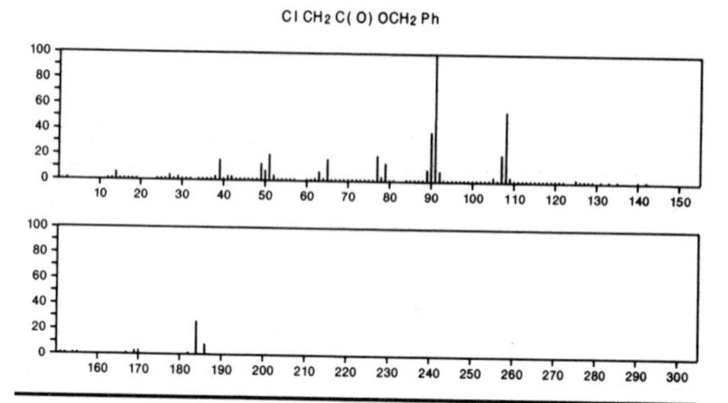

184 $C_9H_9ClO_2$ 1128-76-3
Benzoic acid, 3-chloro-, ethyl ester

C(O)OEt

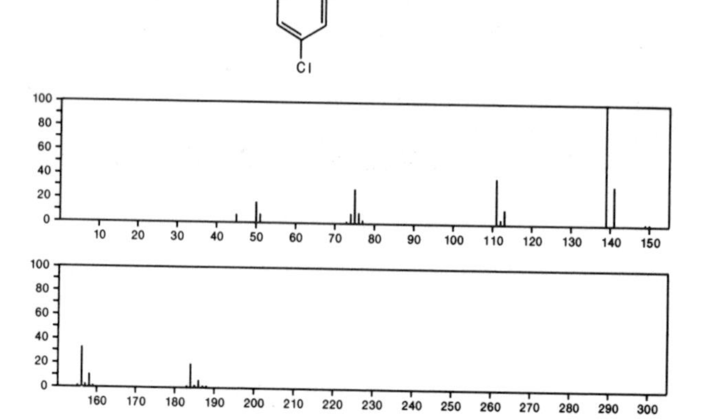

184 $C_9H_9ClO_2$ 2403-54-5
1,3-Dioxolane, 2-(4-chlorophenyl)-

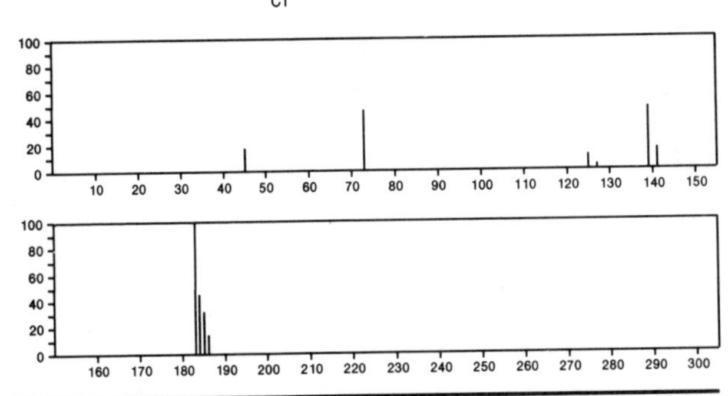

184 $C_9H_9ClO_2$ 6341-98-6
Phenol, 2-chloro-6-methyl-, acetate

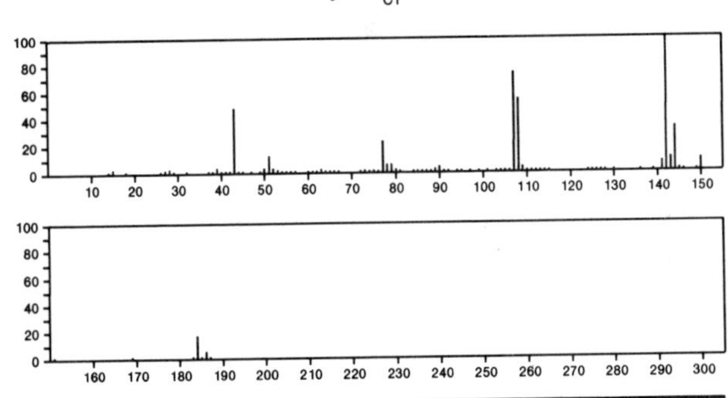

184 $C_9H_9ClO_2$ 34040-64-7
Benzoic acid, 4-(chloromethyl)-, methyl ester

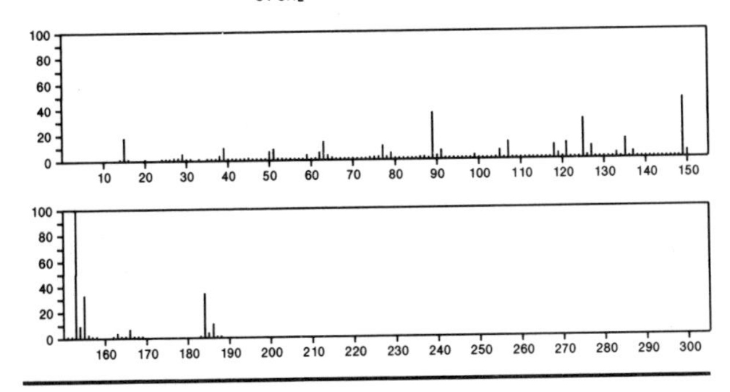

184 $C_9H_9ClO_2$ 37612-52-5
Ethanone, 1-(3-chloro-4-methoxyphenyl)-

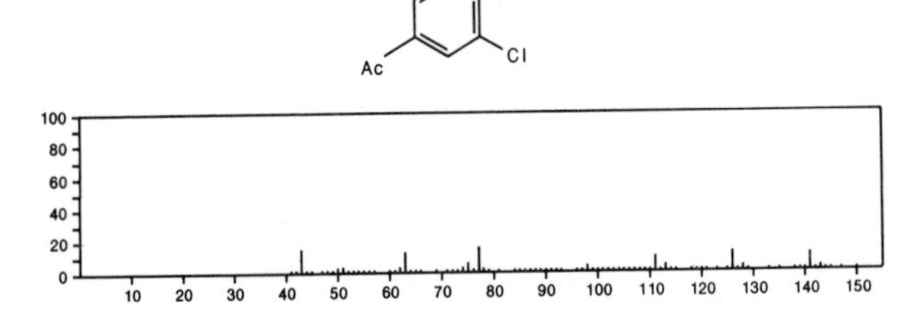

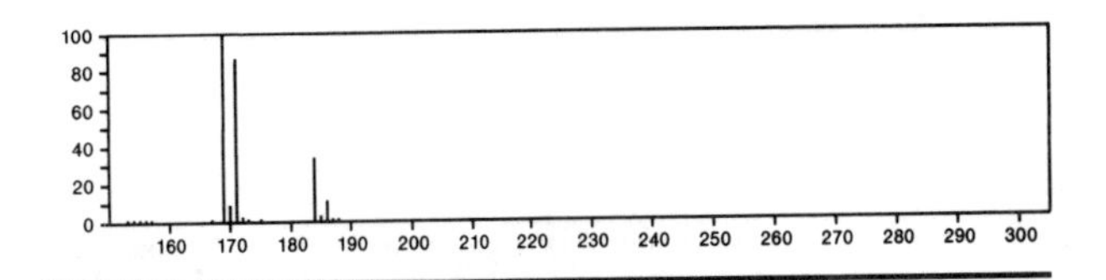

184 $C_9H_{12}O_2S$ 20452-34-0
9-Thiabicyclo[6.2.0]deca-1(8),6-diene, 9,9-dioxide

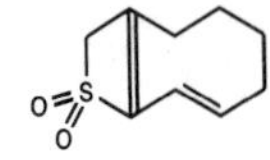

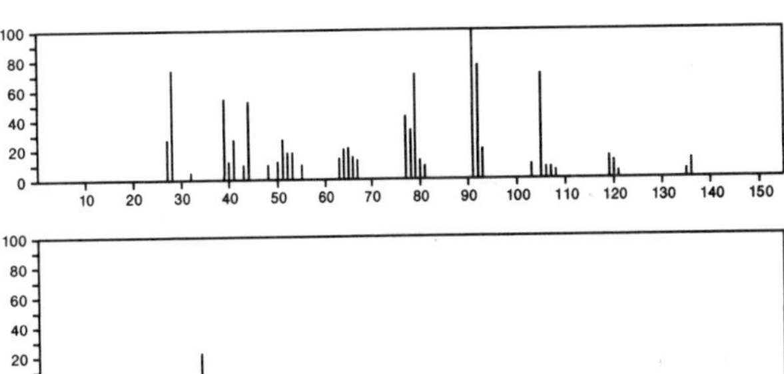

184 $C_9H_{12}O_2S$ 56666-52-5
2-Thiabicyclo[3.1.0]hex-3-ene-6-carboxylic acid, 1,3-dimethyl-, methyl ester

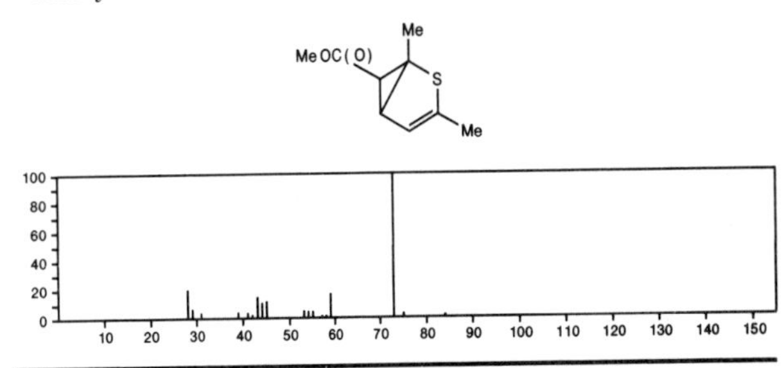

184 $C_9H_{12}O_4$ 35339-97-0
3-Furancarboxylic acid, 2-(methoxymethyl)-5-methyl-, methyl ester

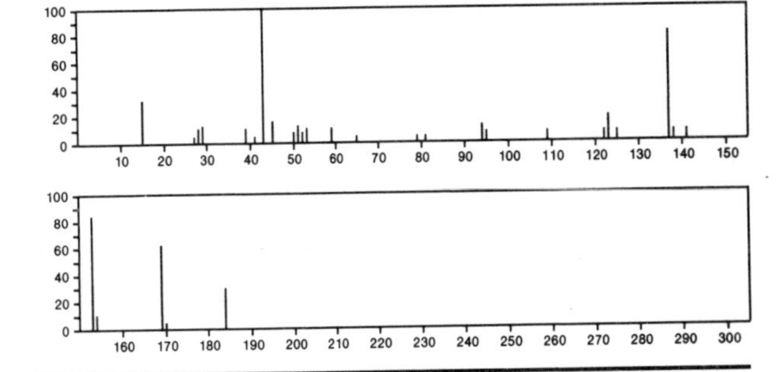

184 $C_9H_{12}O_4$ 35339-99-2
3-Furancarboxylic acid, 2-(ethoxymethyl)-5-methyl-

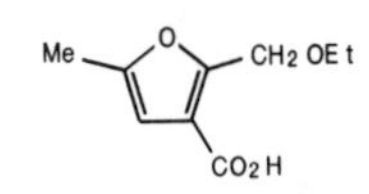

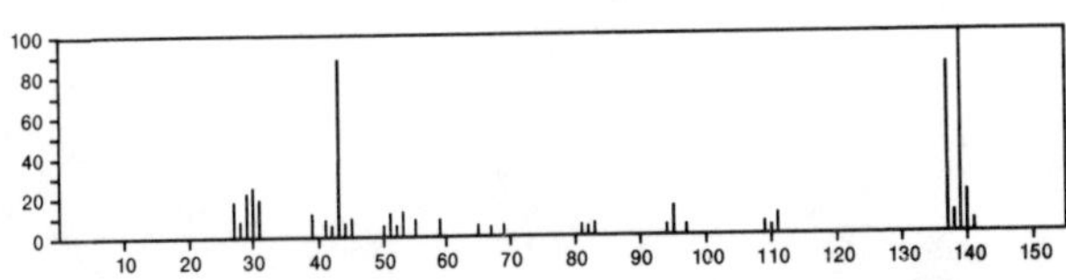

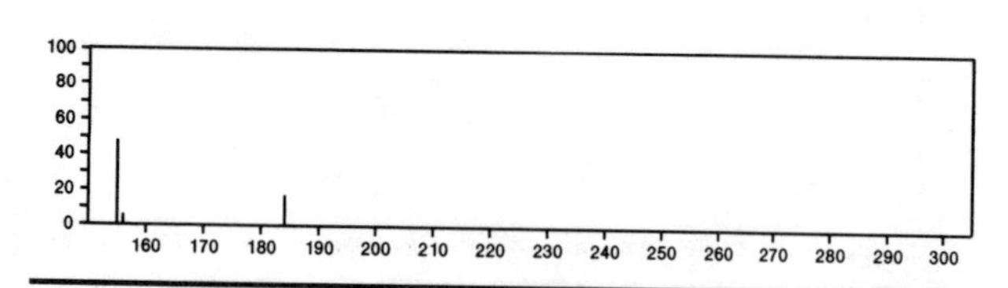

---

184 $C_9H_{12}O_4$ 35340-00-2
3-Furancarboxylic acid, 5-(methoxymethyl)-2-methyl-, methyl ester

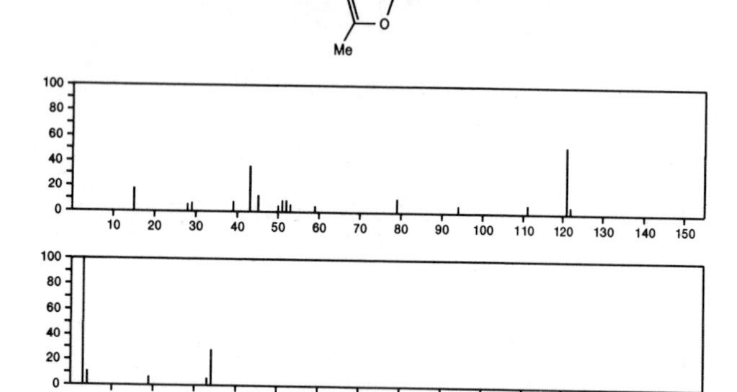

---

184 $C_9H_{12}O_4$ 35340-01-3
3-Furancarboxylic acid, 5-(ethoxymethyl)-2-methyl-

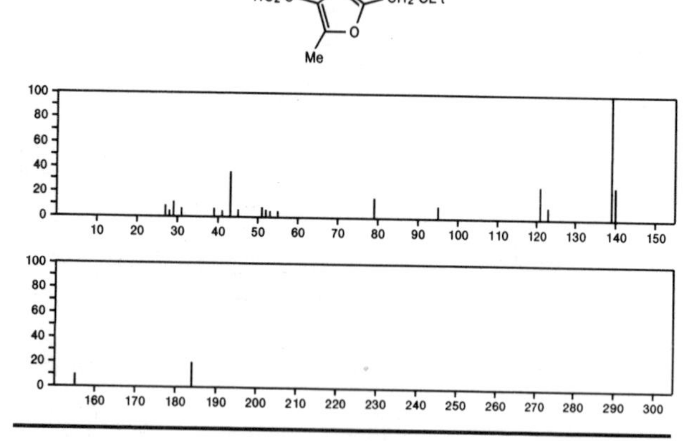

---

184 $C_9H_{12}S_2$ 54774-96-8
Thiophene, 2-(cyclopentylthio)-

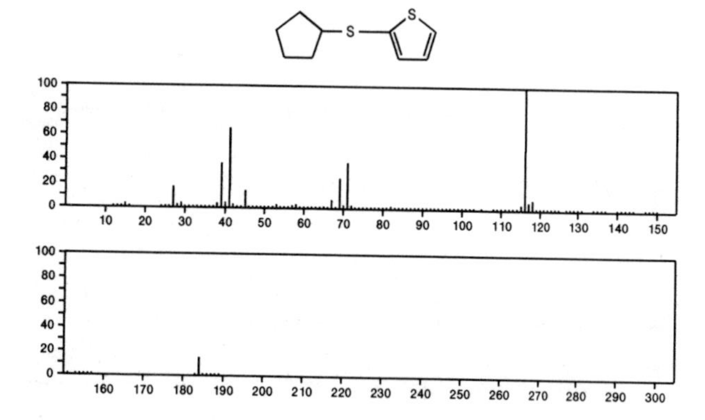

---

184 $C_9H_{16}N_2O_2$ 640-01-7
4-Piperidinone, 2,2,6,6-tetramethyl-1-nitroso-

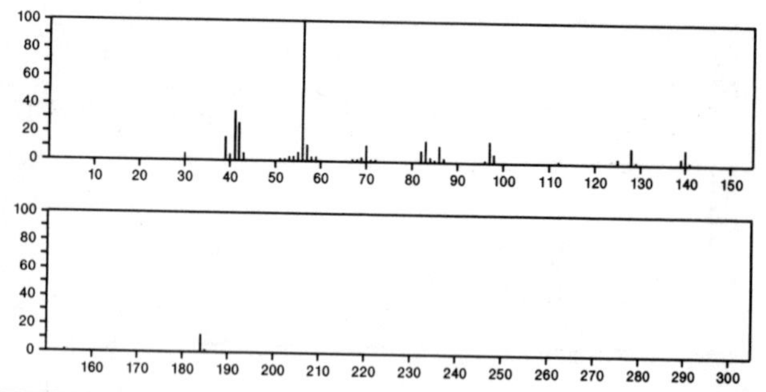

---

184 $C_9H_{17}BO_3$ 55162-66-8
1,3,2-Dioxaborolane, 2-[(2-methylcyclohexyl)oxy]-

---

184 $C_{10}H_8N_4$ 37550-68-8
1,3,6,9b-Tetraazaphenalene, 2-methyl-

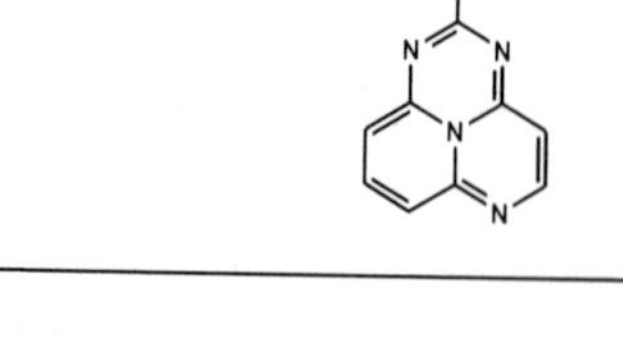

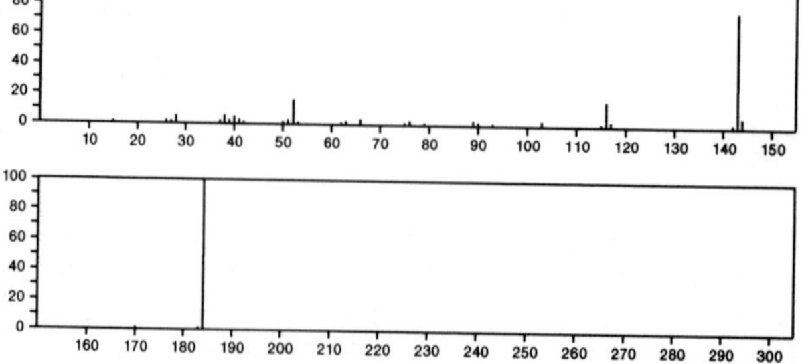

184 $C_{10}H_{13}ClO$ 89-68-9
Phenol, 4-chloro-5-methyl-2-(1-methylethyl)-

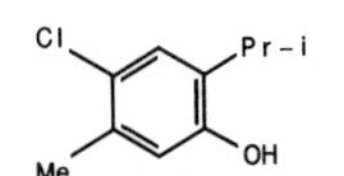

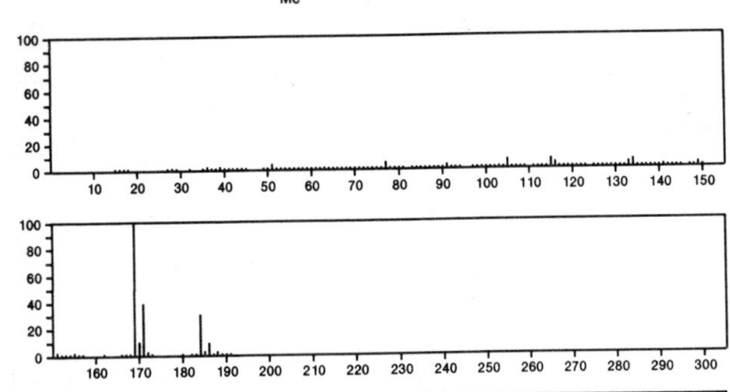

184 $C_{10}H_{13}ClO$ 98-28-2
Phenol, 2-chloro-4-(1,1-dimethylethyl)-

Cl Bu-t HO

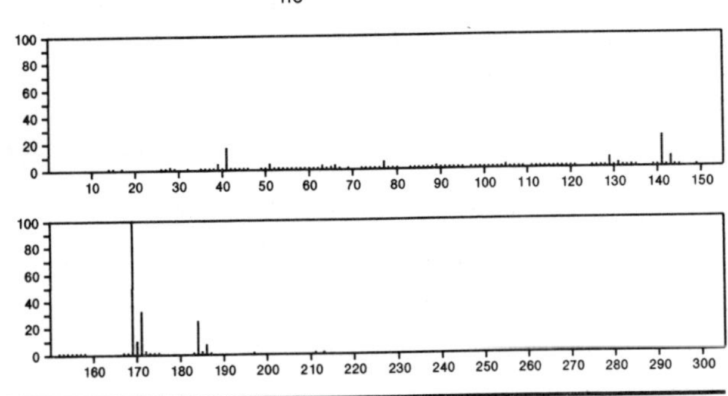

184 $C_{10}H_{13}ClO$ 4237-37-0
Phenol, 2-chloro-6-(1,1-dimethylethyl)-

Bu-t OH Cl

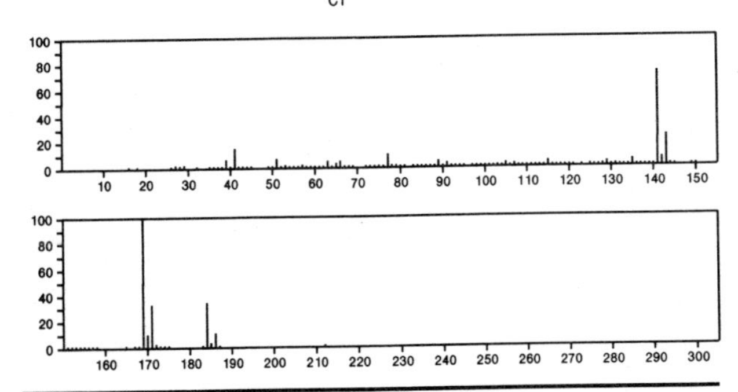

184 $C_{10}H_{13}ClO$ 19301-54-3
Tricyclo[3.3.1.1^{3,7}]decanone, 4-chloro-, (1α,3β,4β,5α,7β)-

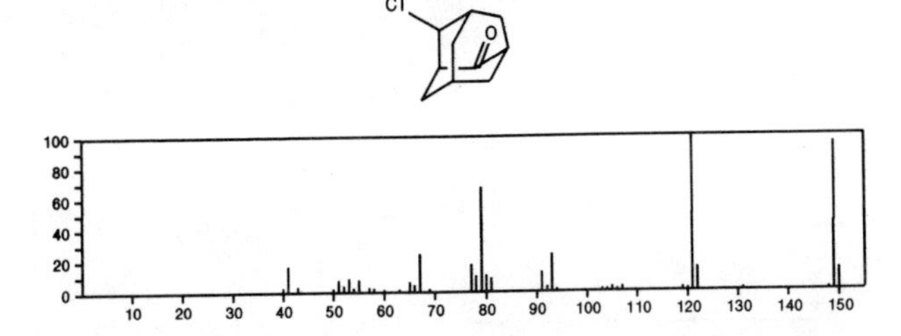

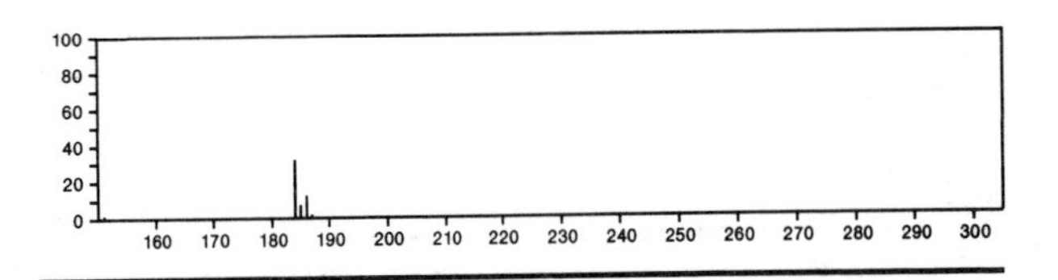

184 $C_{10}H_{13}ClO$ 19543-61-4
Tricyclo[3.3.1.1^{3,7}]decanone, 4-chloro-, [1*S*-(1α,3β,4α,5α,7β)]-

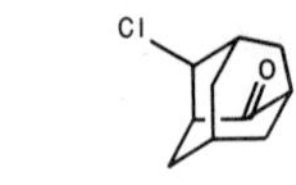

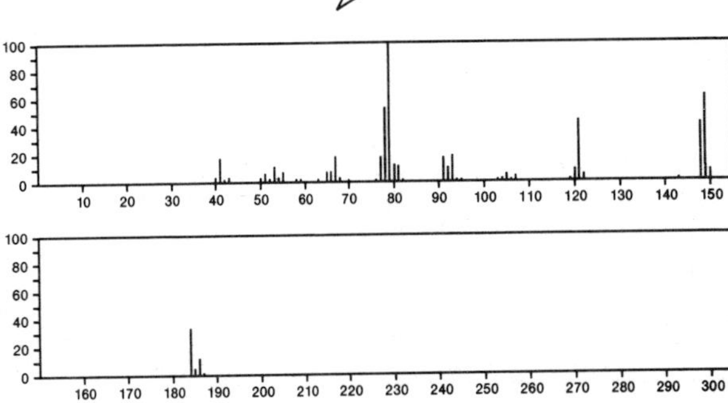

184 $C_{10}H_{13}ClO$ 55955-55-0
Benzene, (1-chloro-3-methoxypropyl)-

PhCHCl CH2 CH2 OMe

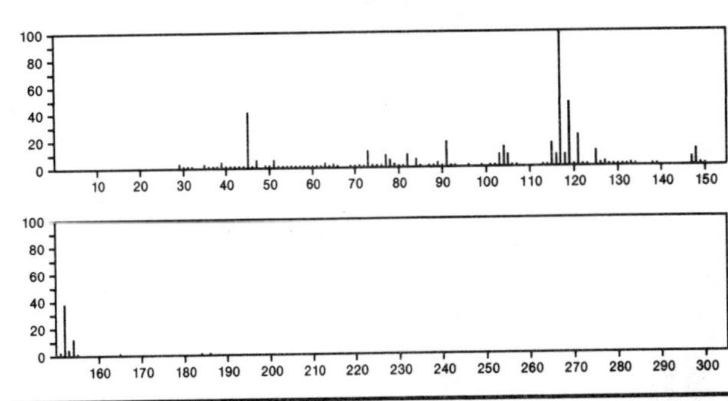

184 $C_{10}H_{16}O_3$ 14128-60-0
Cyclopentanepropionic acid, α-methyl-2-oxo-, methyl ester

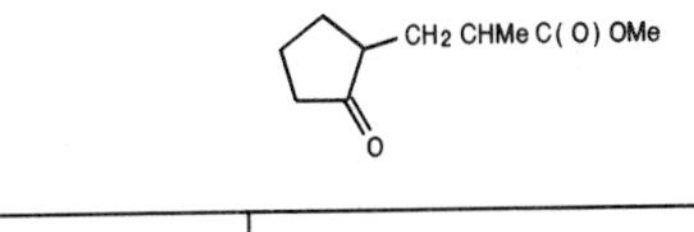

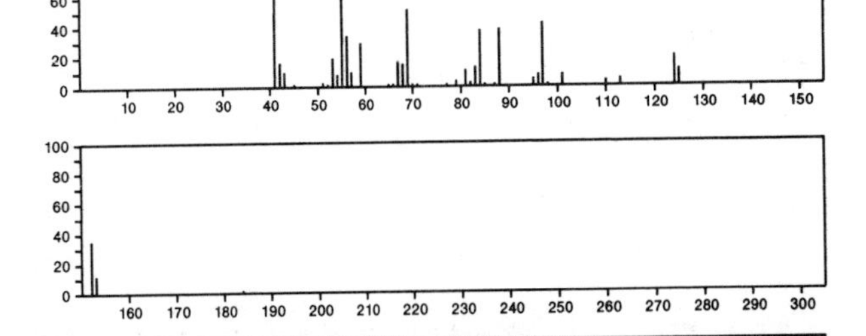

184 $C_{10}H_{16}O_3$ 16754-50-0
1-Propene, 3,3',3''-[methylidynetris(oxy)]tris-

OCH2 CH=CH2
H2C=CHCH2 OCHOCH2 CH=CH2

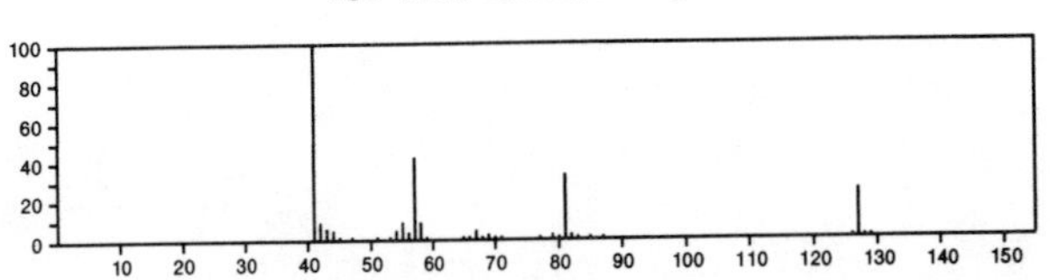

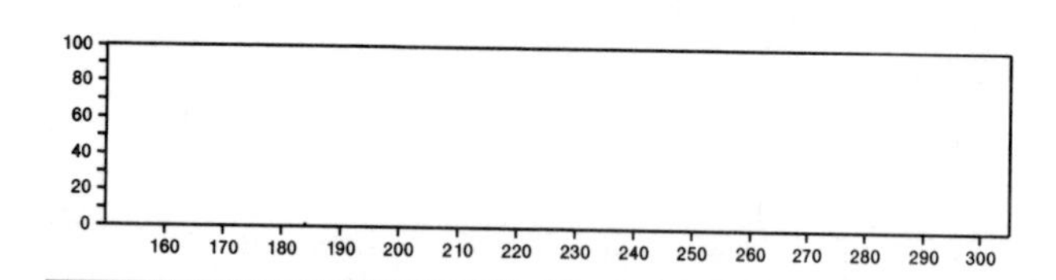

184 $C_{10}H_{16}O_3$ 18927-22-5
1,2-Butanediol, 1-(2-furyl)-2,3-dimethyl-

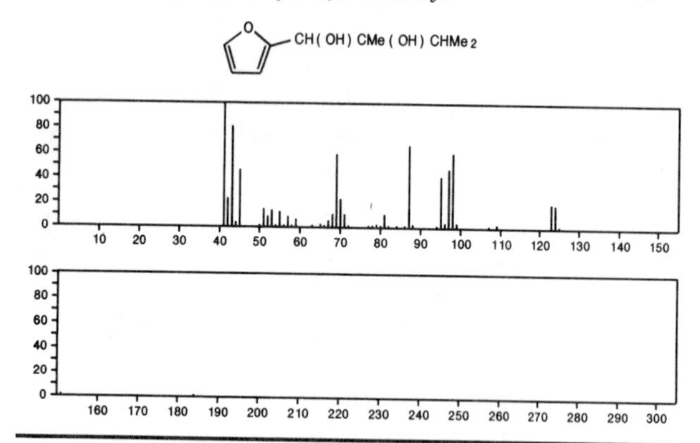

184 $C_{10}H_{16}O_3$ 36903-65-8
1,3-Cyclopentanedione, 4-hydroxy-5-(3-methylbutyl)-

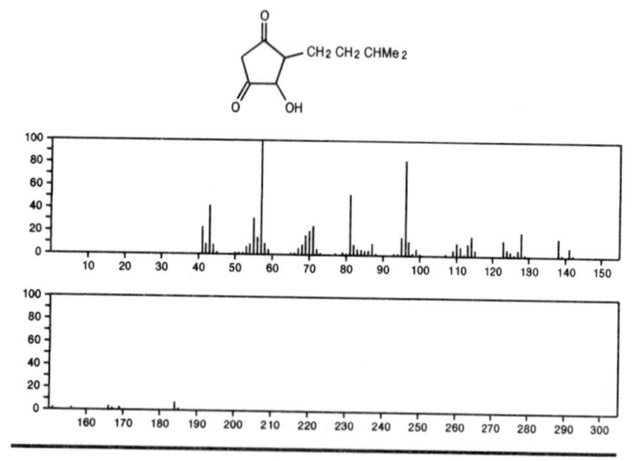

184 $C_{10}H_{16}O_3$ 42031-65-2
4,5-Oxepanedione, 3,3,6,6-tetramethyl-

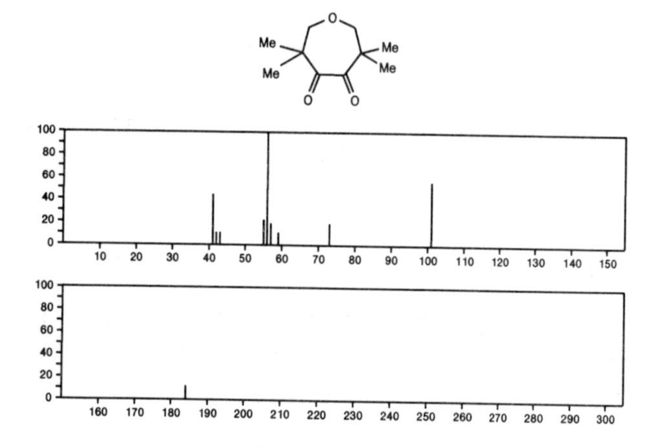

184 $C_{10}H_{16}O_3$ 54774-94-6
Bicyclo[2.2.2]octane-1,4-diol, monoacetate

184 $C_{10}H_{16}O_3$ 54774-95-7
1,3-Cyclopentanedione, 4-hydroxy-2-pentyl-

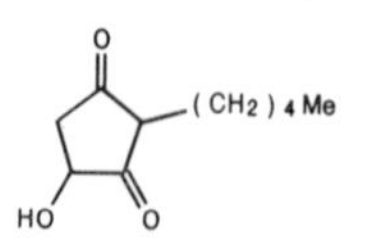

184 $C_{10}H_{20}N_2O$ 23435-03-2
2H-Azepin-2-one, 7-(4-aminobutyl)hexahydro-

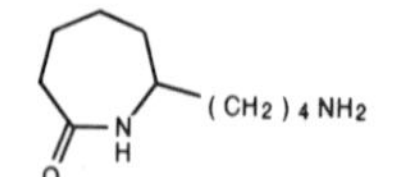

184 $C_{10}H_{20}N_2O$ 49582-49-2
3-Buten-2-one, 4-(diethylamino)-4-(dimethylamino)-

NMe2
MeCOCH=CNEt2

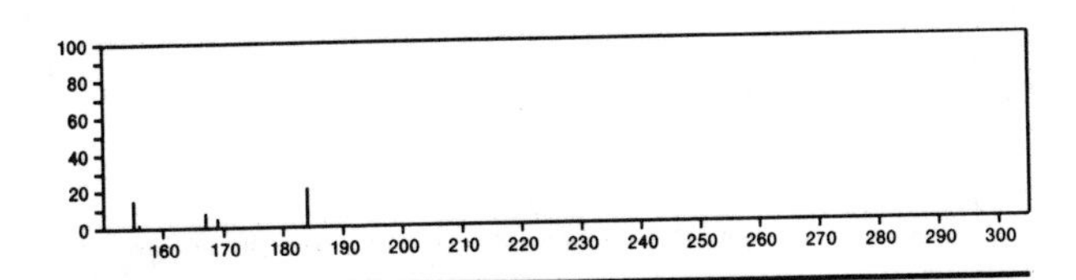

184 $C_{10}H_{20}N_2O$ 55975-99-0
Propanal, 2-(cyclohexylamino)-2-methyl-, oxime

NHCMe2 CH=NOH

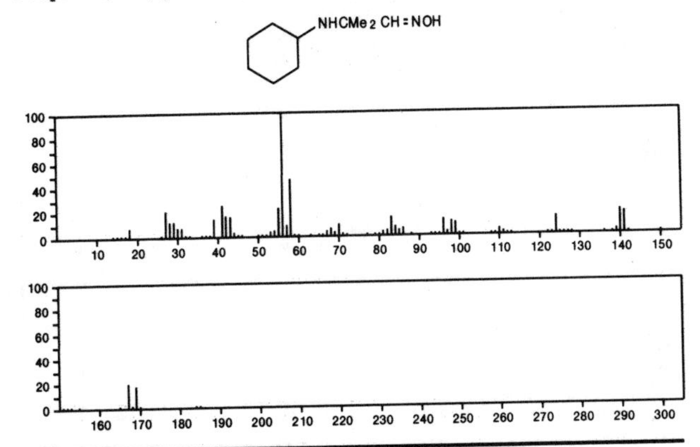

184 $C_{11}H_8N_2O$ 3878-19-1
1*H*-Benzimidazole, 2-(2-furanyl)-

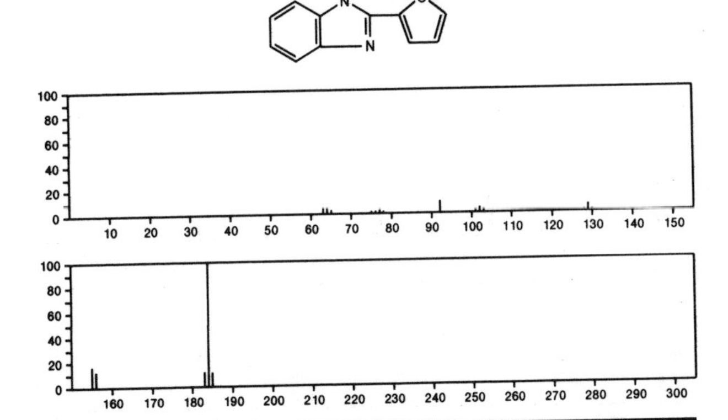

184 $C_{11}H_8N_2O$ 16347-56-1
4(3*H*)-Quinazolinone, 3-(2-propynyl)-

CH2 C≡CH

184 $C_{11}H_8N_2O$ 19062-85-2
3,1-Benzoxazepine-2-carbonitrile, 4-methyl-

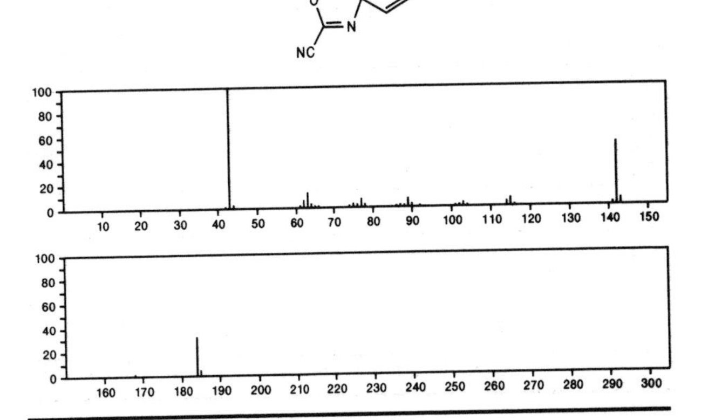

184 $C_{11}H_8N_2O$ 19062-86-3
3,1-Benzoxazepine-2-carbonitrile, 5-methyl-

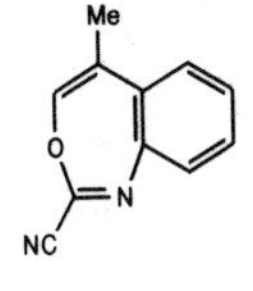

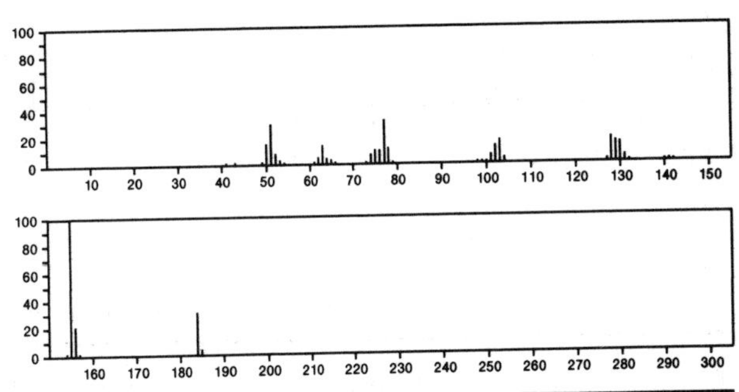

184 $C_{11}H_8N_2O$ 19062-87-4
3,1-Benzoxazepine-2-carbonitrile, 7-methyl-

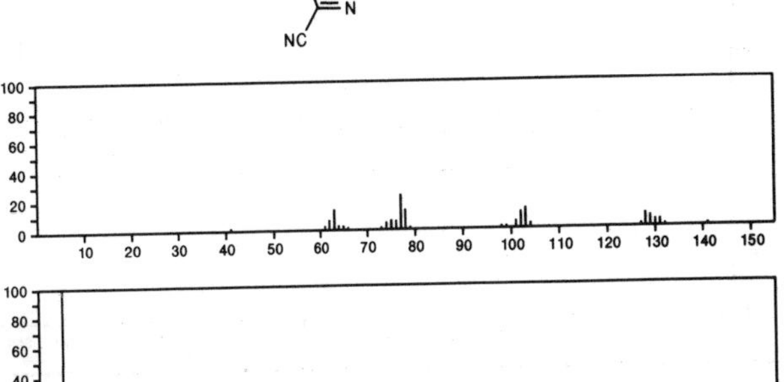

184 $C_{11}H_8N_2O$ 25379-65-1
4(3*H*)-Quinazolinone, 3-propadienyl-

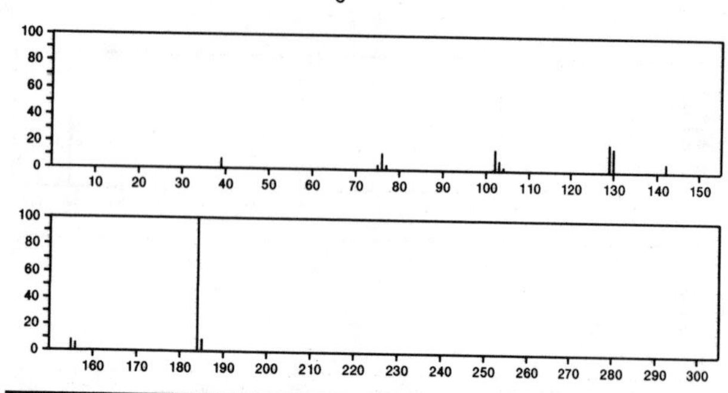

184 $C_{11}H_{17}Cl$ 27011-46-7
Tricyclo[4.3.1.1^{3,8}]undecane, 1-chloro-

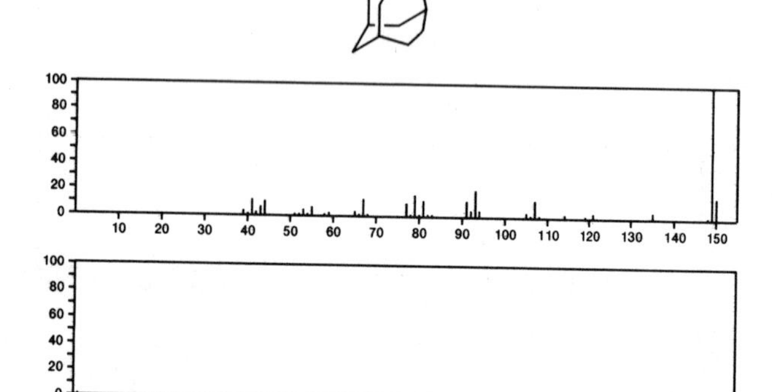

184 $C_{11}H_{17}Cl$ 27011-47-8
Tricyclo[4.3.1.1^{3,8}]undecane, 3-chloro-

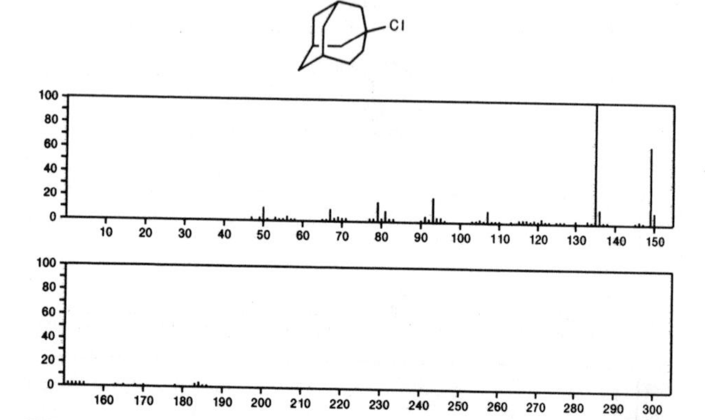

184 $C_{11}H_{20}O_2$ 104-67-6
2(3*H*)-Furanone, 5-heptyldihydro-

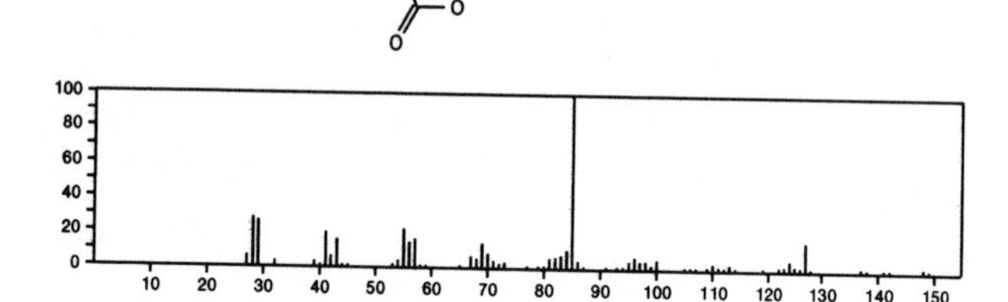

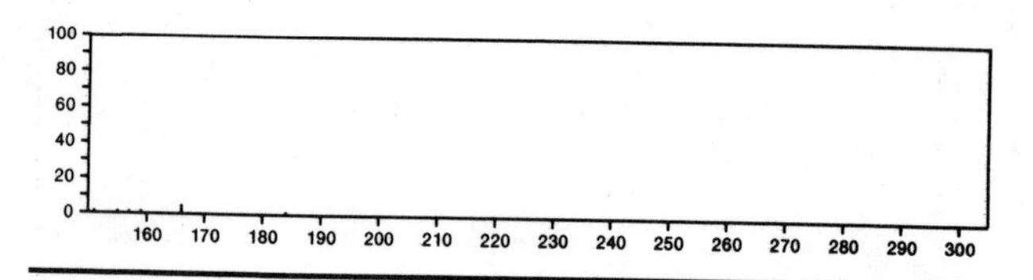

184 $C_{11}H_{20}O_2$ 707-29-9
1,5-Dioxaspiro[5.5]undecane, 3,3-dimethyl-

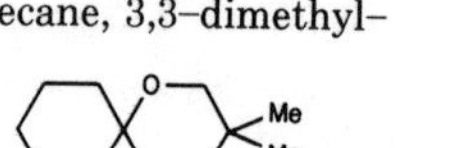

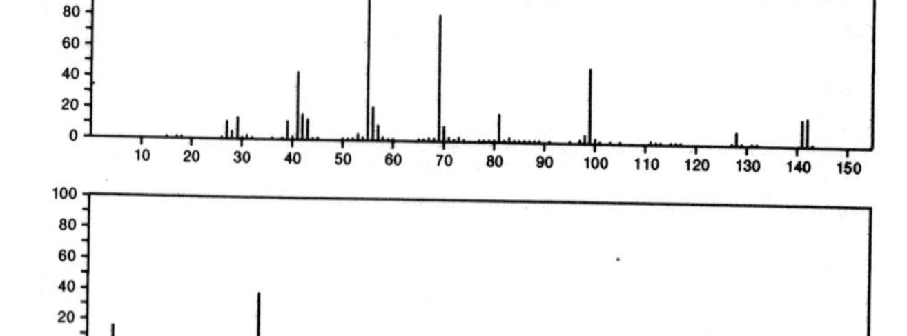

184 $C_{11}H_{20}O_2$ 710-04-3
2*H*-Pyran-2-one, 6-hexyltetrahydro-

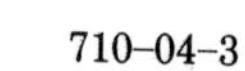

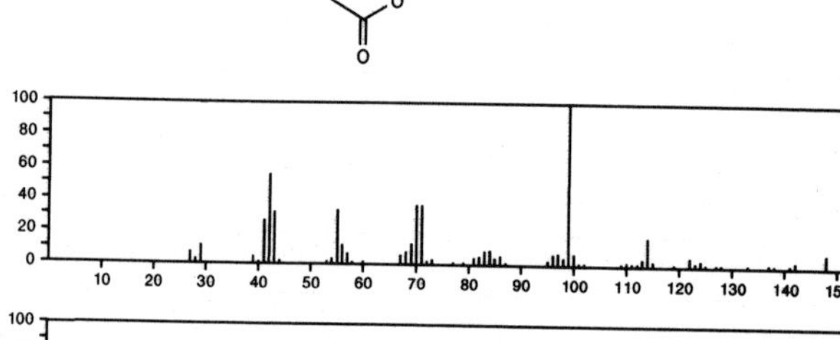

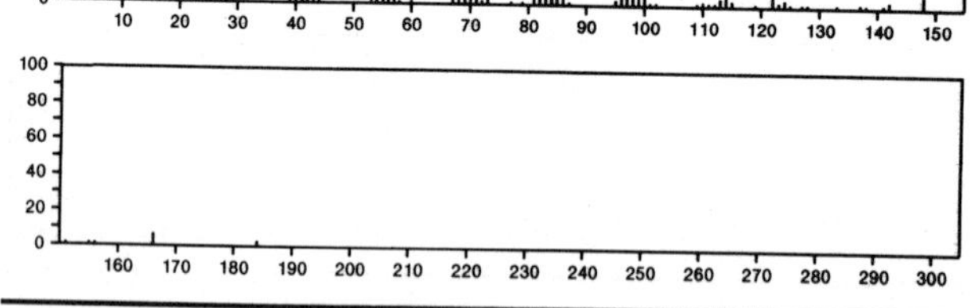

184 $C_{11}H_{20}O_2$ 943-28-2
Cyclohexanecarboxylic acid, 4-(1,1-dimethylethyl)-, *cis*-

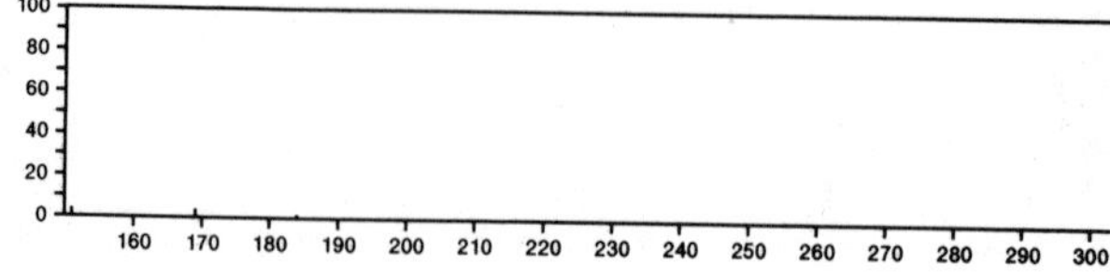

184 $C_{11}H_{20}O_2$ 943–29–3
Cyclohexanecarboxylic acid, 4–(1,1–dimethylethyl)–, *trans*–

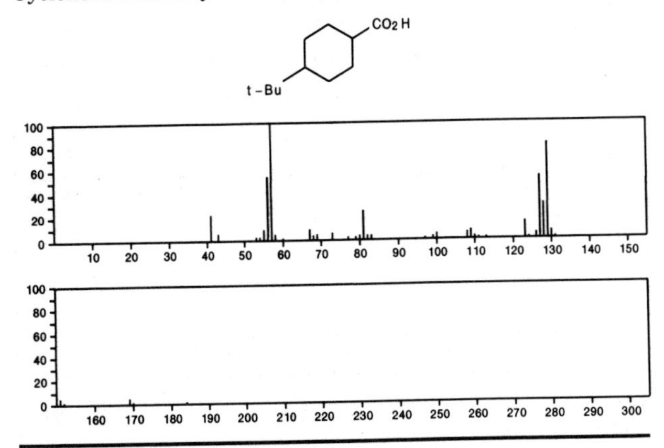

184 $C_{11}H_{20}O_2$ 1191–02–2
4–Decenoic acid, methyl ester

Me OC(O) CH2 CH2 CH = CH(CH2) 4 Me

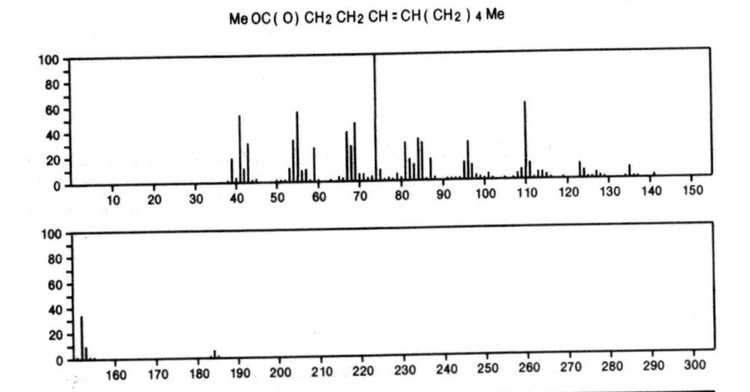

184 $C_{11}H_{20}O_2$ 2482–39–5
2–Decenoic acid, methyl ester

Me (CH2) 6 CH = CHC(O) OMe

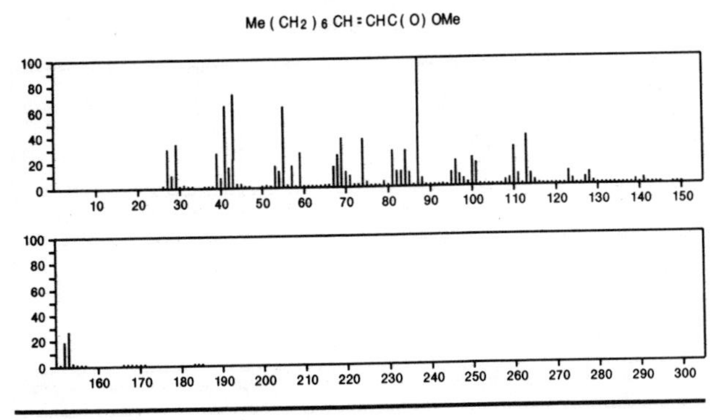

184 $C_{11}H_{20}O_2$ 2499–59–4
2–Propenoic acid, octyl ester

H2 C = CHC(O) O(CH2) 7 Me

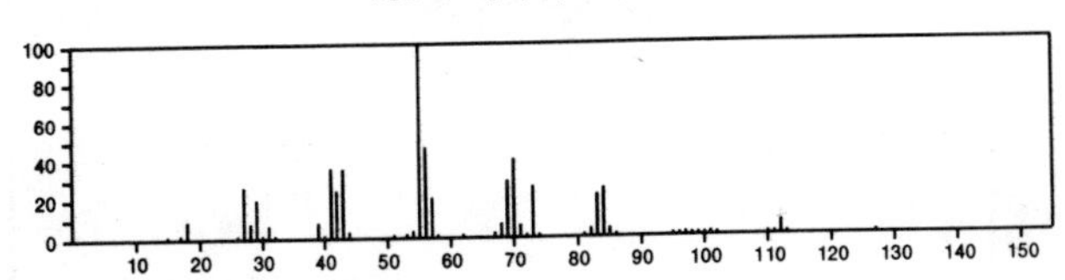

184 $C_{11}H_{20}O_2$ 15287–80–6
Butyric acid, 2–methylcyclohexyl ester, *trans*–

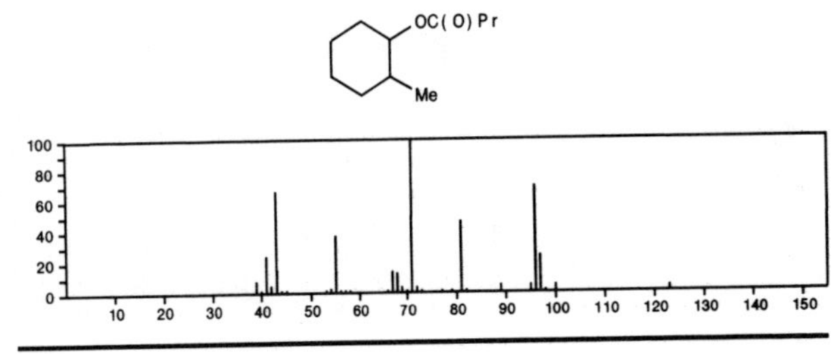

184 $C_{11}H_{20}O_2$ 27334–43–6
Cyclohexanecarboxylic acid, 1–(1,1–dimethylethyl)–

CO2 H
Bu - t

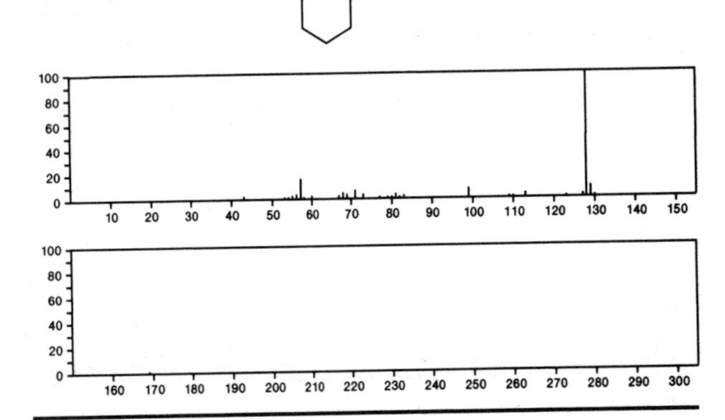

184 $C_{11}H_{20}O_2$ 27392–15–0
Cyclohexanecarboxylic acid, 2–(1,1–dimethylethyl)–, *cis*–

Bu - t
CO2 H

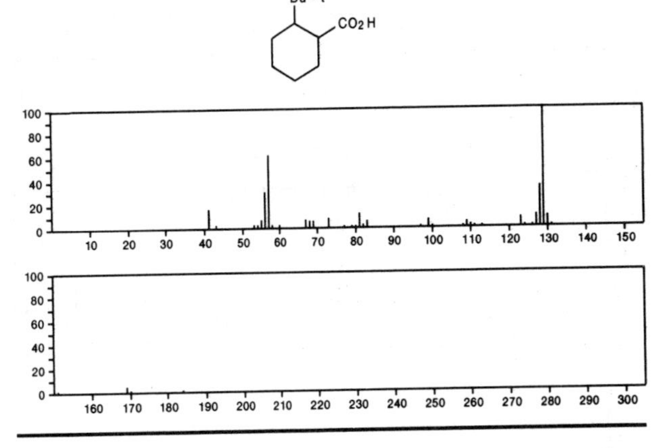

184 $C_{11}H_{20}O_2$ 27392–16–1
Cyclohexanecarboxylic acid, 2–(1,1–dimethylethyl)–, *trans*–

Bu - t
CO2 H

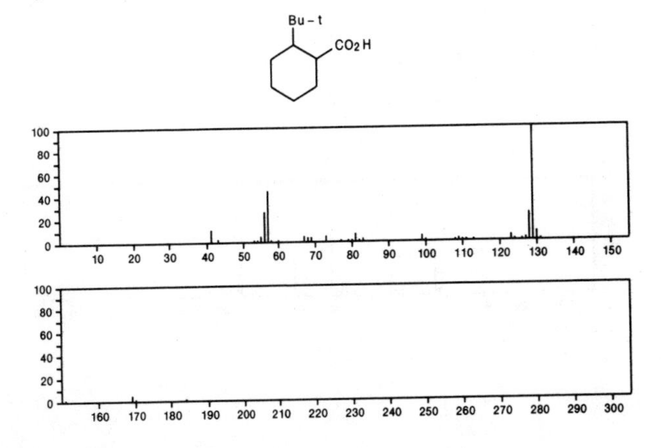

184 $C_{11}H_{20}O_2$ 27392–17–2
Cyclohexanecarboxylic acid, 3–(1,1–dimethylethyl)–, *cis*–

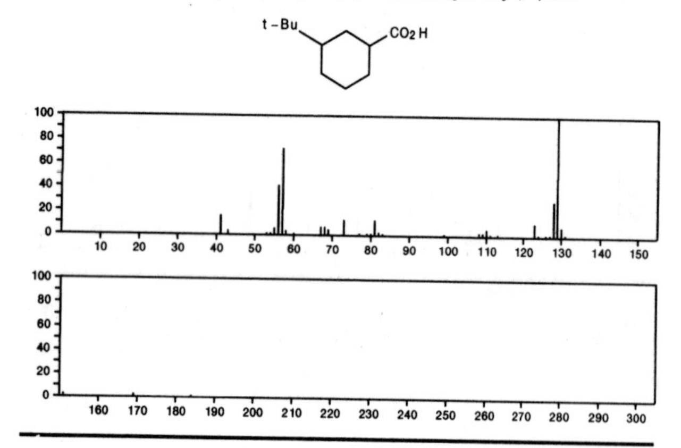

184 $C_{11}H_{20}O_2$ 27392–18–3
Cyclohexanedicarboxylic acid, 3–(1,1–dimethylethyl)–, *trans*–

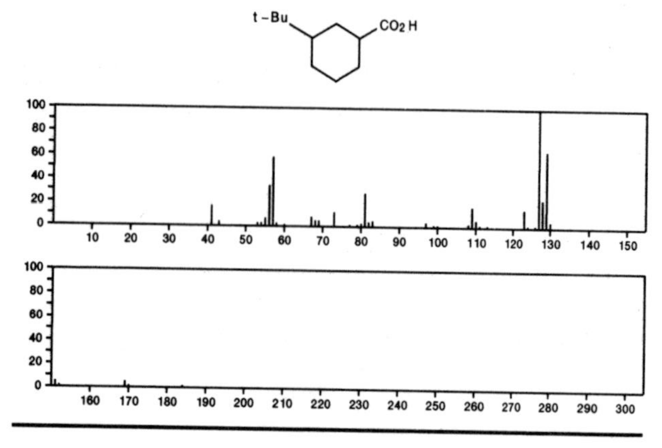

184 $C_{11}H_{20}O_2$ 35194–39–9
8–Nonenoic acid, ethyl ester

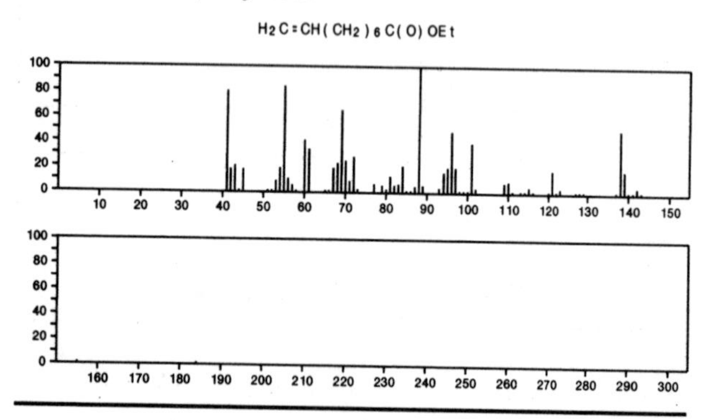

184 $C_{11}H_{20}O_2$ 54714–35–1
Butanoic acid, 2–methylcyclohexyl ester, *cis*–

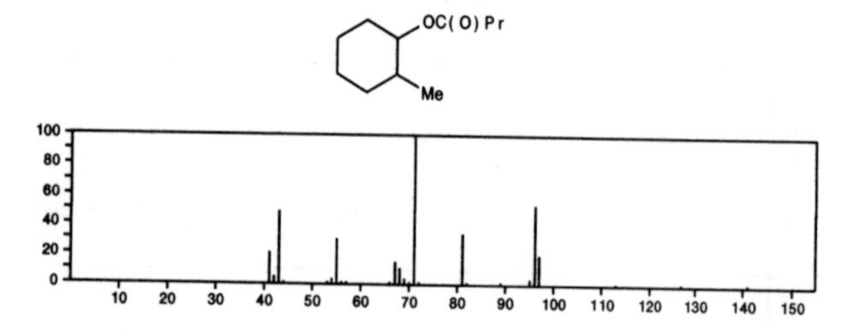

184 $C_{11}H_{20}O_2$ 54774–91–3
2–Propenoic acid, 6–methylheptyl ester

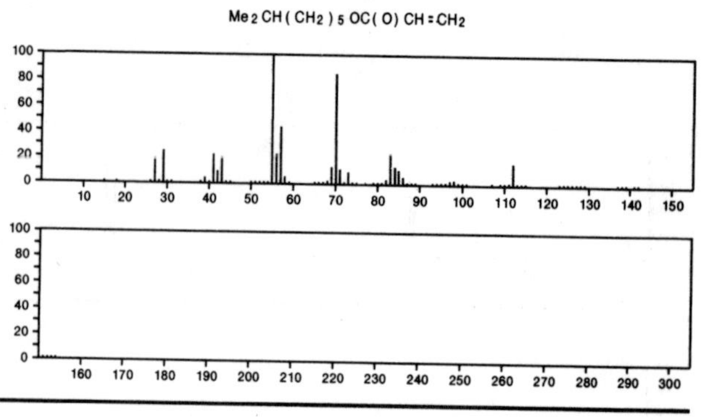

184 $C_{11}H_{20}S$ 7133–21–3
Sulfide, cyclohexyl cyclopentyl

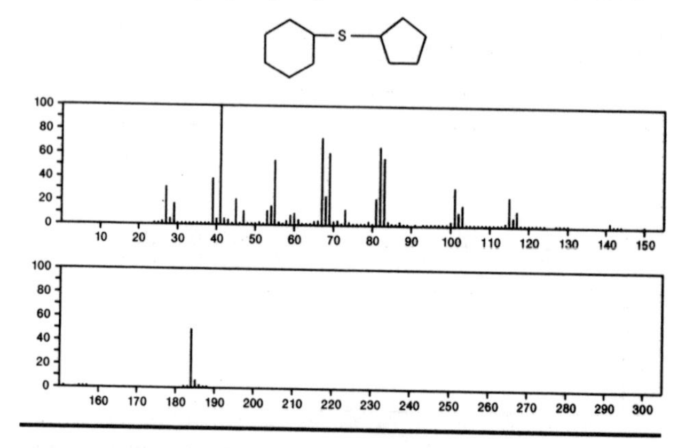

184 $C_{11}H_{24}N_2$ 14090–59–6
5–Nonanone, dimethylhydrazone

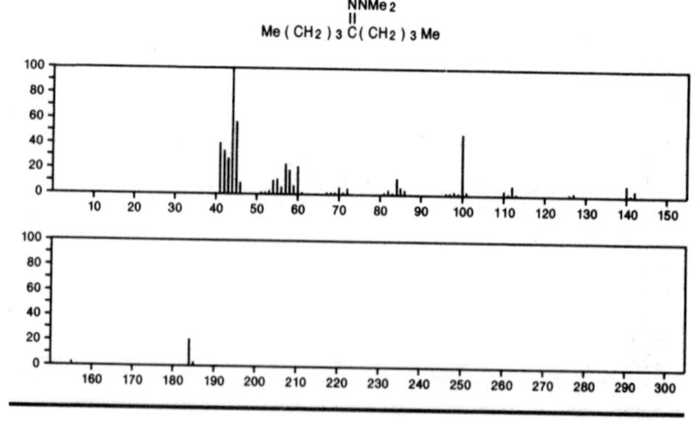

184 $C_{12}H_8O_2$ 262–12–4
Dibenzo[*b*,*e*][1,4]dioxin

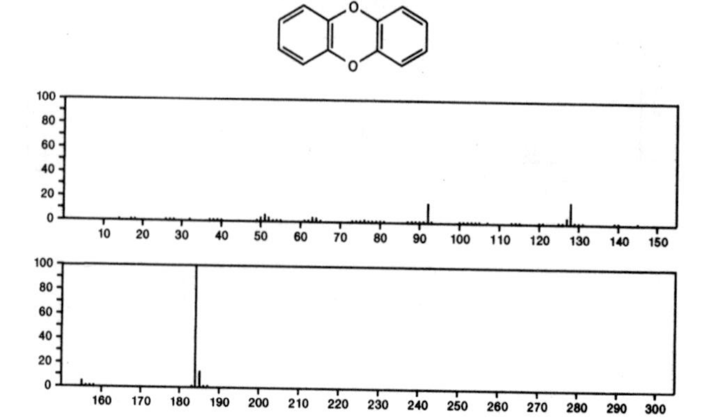

184 $C_{12}H_8O_2$ 363-03-1
2,5-Cyclohexadiene-1,4-dione, 2-phenyl-

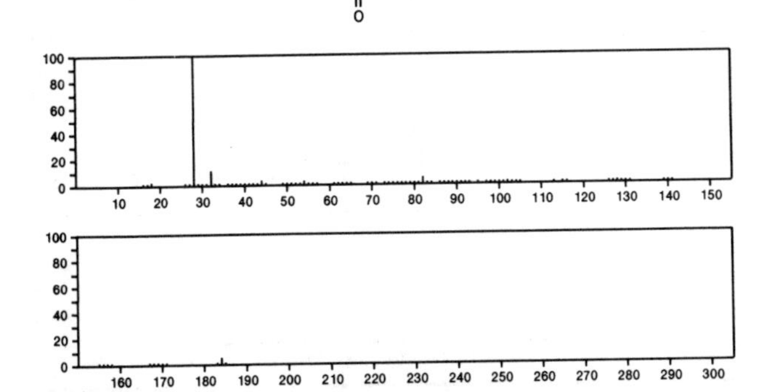

184 $C_{12}H_8O_2$ 494-72-4
2,5-Cyclohexadien-1-one, 4-(4-oxo-2,5-cyclohexadien-1-ylidene)-

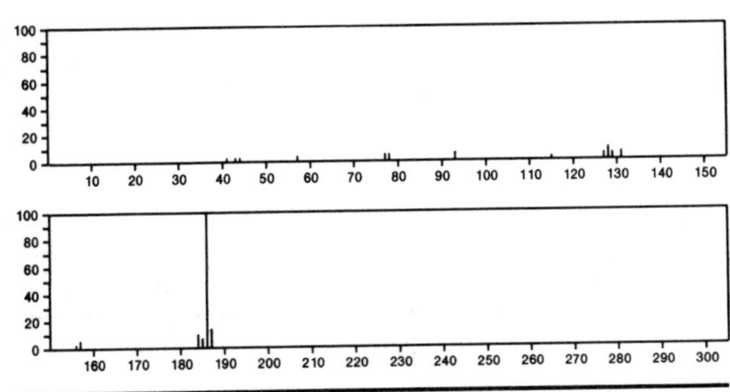

184 $C_{12}H_8O_2$ 518-86-5
1*H*,3*H*-Naphtho[1,8-*cd*]pyran-1-one

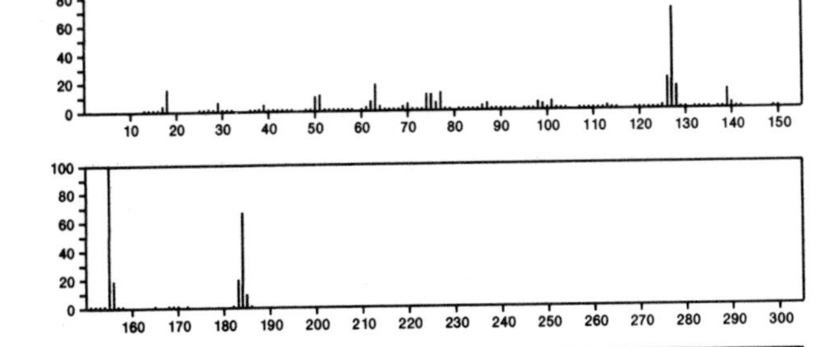

184 $C_{12}H_8S$ 132-65-0
Dibenzothiophene

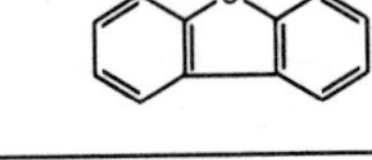

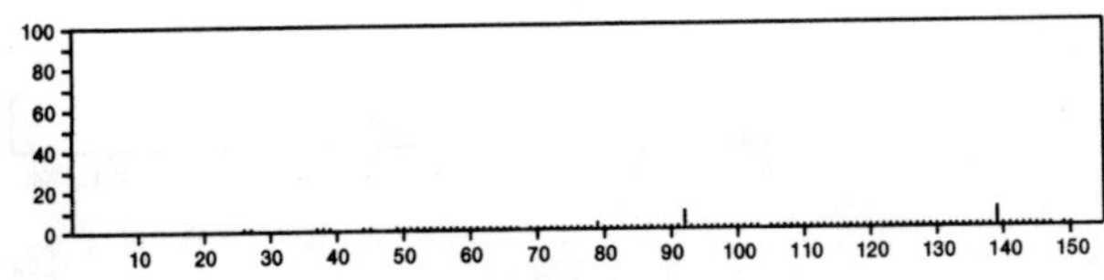

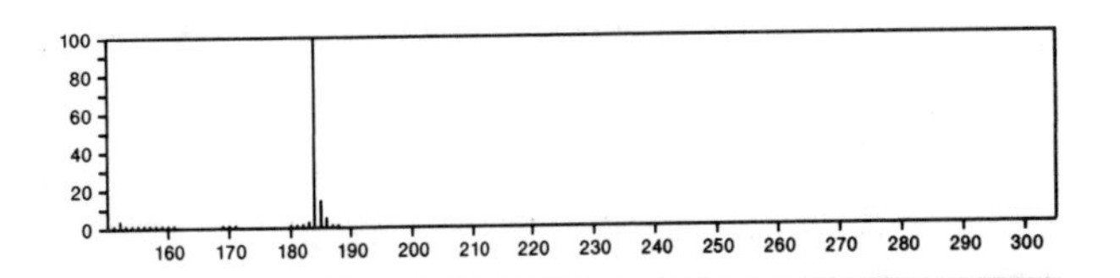

184 $C_{12}H_{12}N_2$ 92-87-5
[1,1'-Biphenyl]-4,4'-diamine

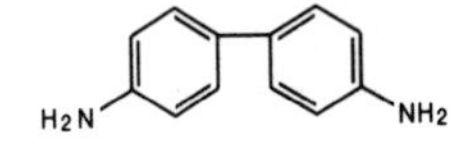

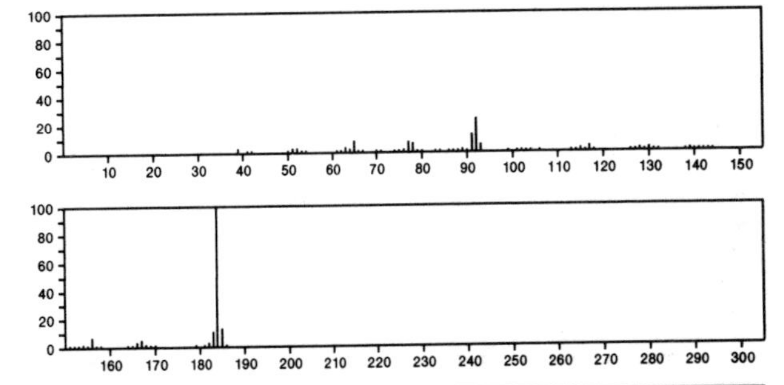

184 $C_{12}H_{12}N_2$ 530-50-7
Hydrazine, 1,1-diphenyl-

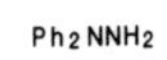

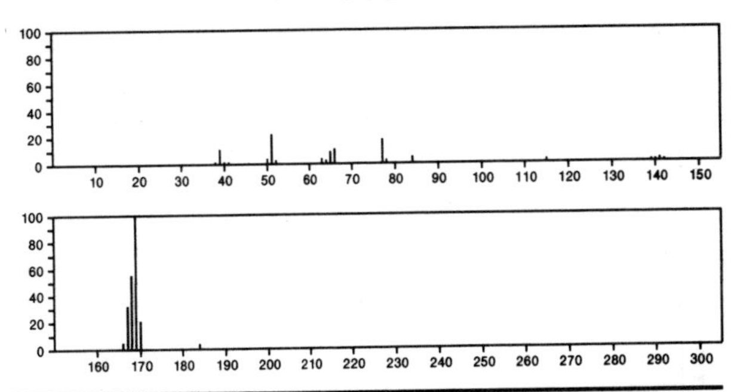

184 $C_{12}H_{12}N_2$ 10183-74-1
4*H*-Pyrrolo[1,2-*b*]pyrazole, 5,6-dihydro-3-phenyl-

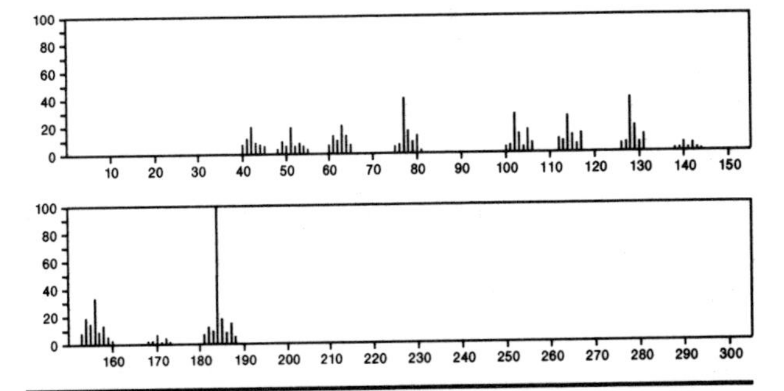

184 $C_{12}H_{12}N_2$ 24046-23-9
Pyrrole, 1-[(α-methylbenzylidene)amino]-

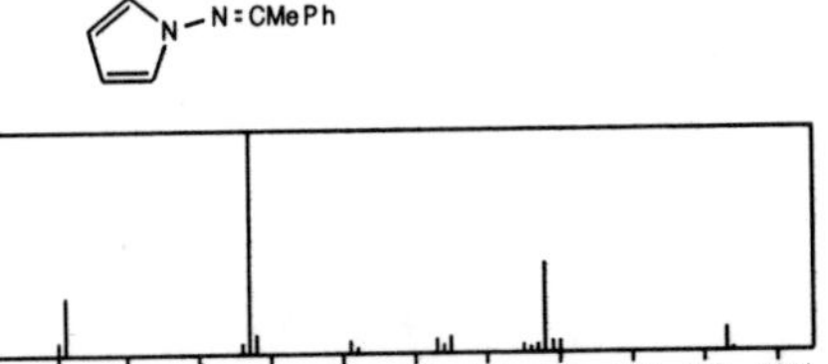

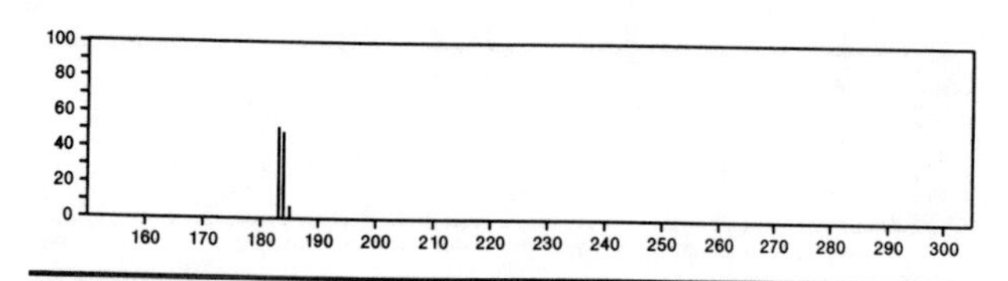

184 $C_{12}H_{12}N_2$ 31378-89-9
Pyridinium, 1-*o*-toluidino-, hydroxide, inner salt

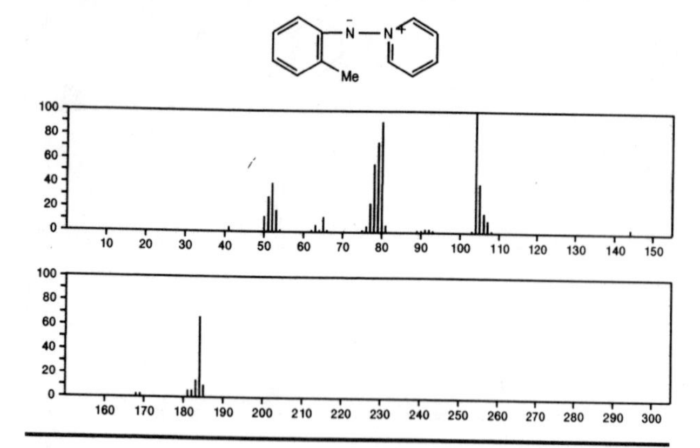

184 $C_{12}H_{12}N_2$ 31378-90-2
Pyridinium, 1-*m*-toluidino-, hydroxide, inner salt

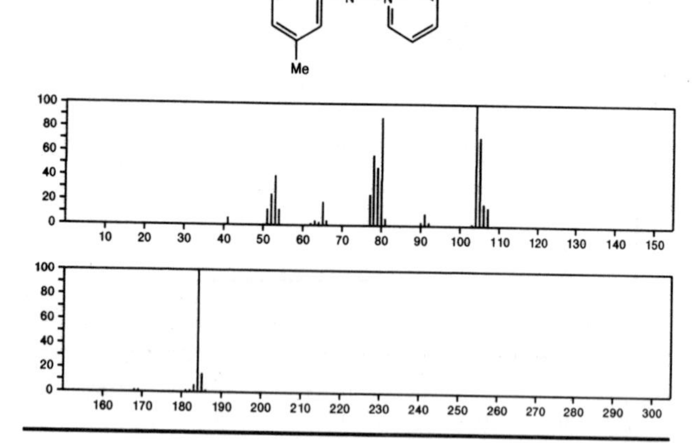

184 $C_{12}H_{12}N_2$ 31378-92-4
Pyridinium, 1-*p*-toluidino-, hydroxide, inner salt

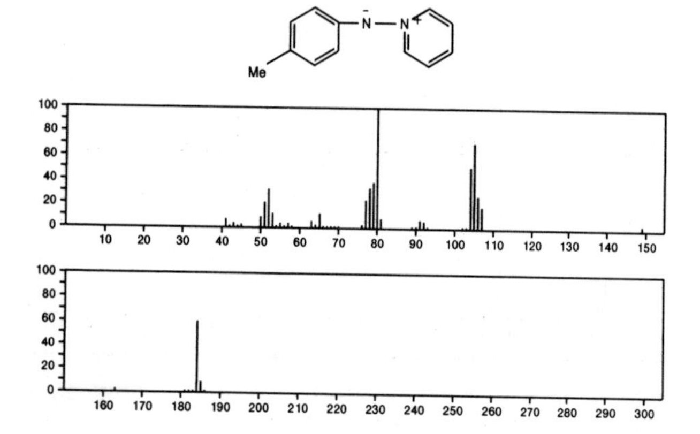

184 $C_{12}H_{12}N_2$ 31382-88-4
3-Picolinium, 1-anilino-, hydroxide, inner salt

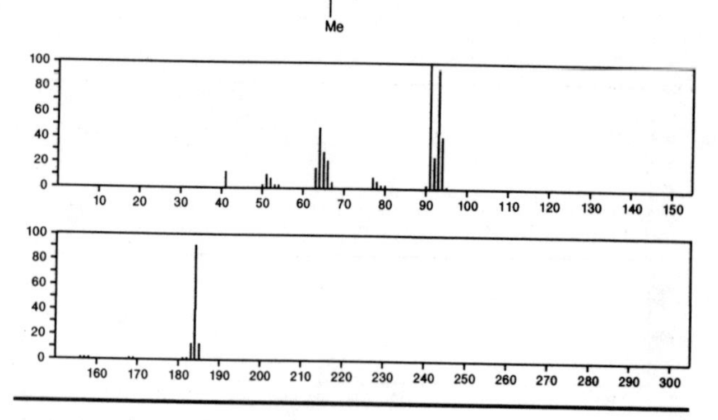

184 $C_{12}H_{24}O$ 110-41-8
Undecanal, 2-methyl-

OCHCHMe (CH2) 8 Me

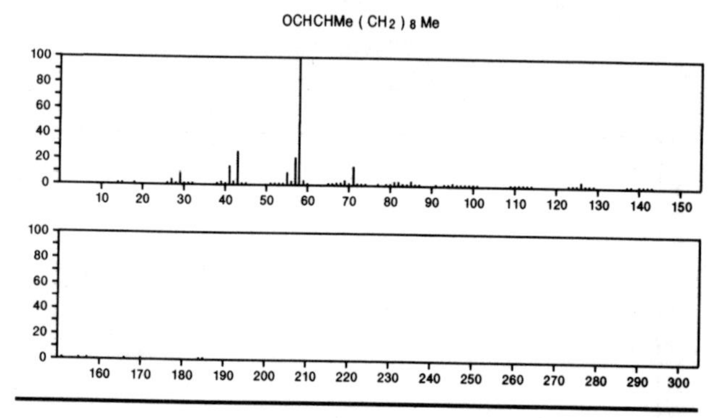

184 $C_{12}H_{24}O$ 112-54-9
Dodecanal

OCH (CH2) 10 Me

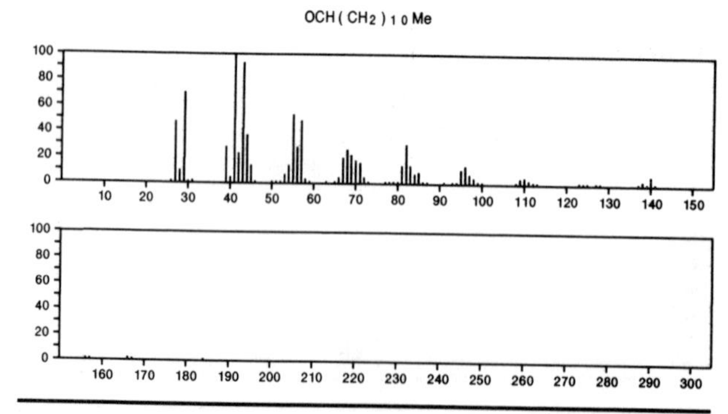

184 $C_{12}H_{24}O$ 765-05-9
Decane, 1-(ethenyloxy)-

Me (CH2) 9 OCH = CH2

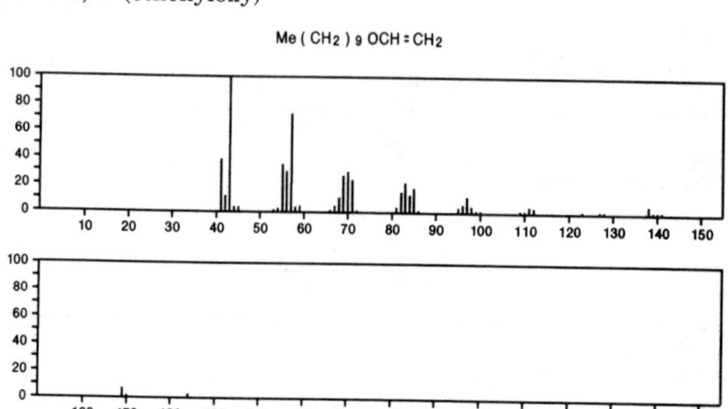

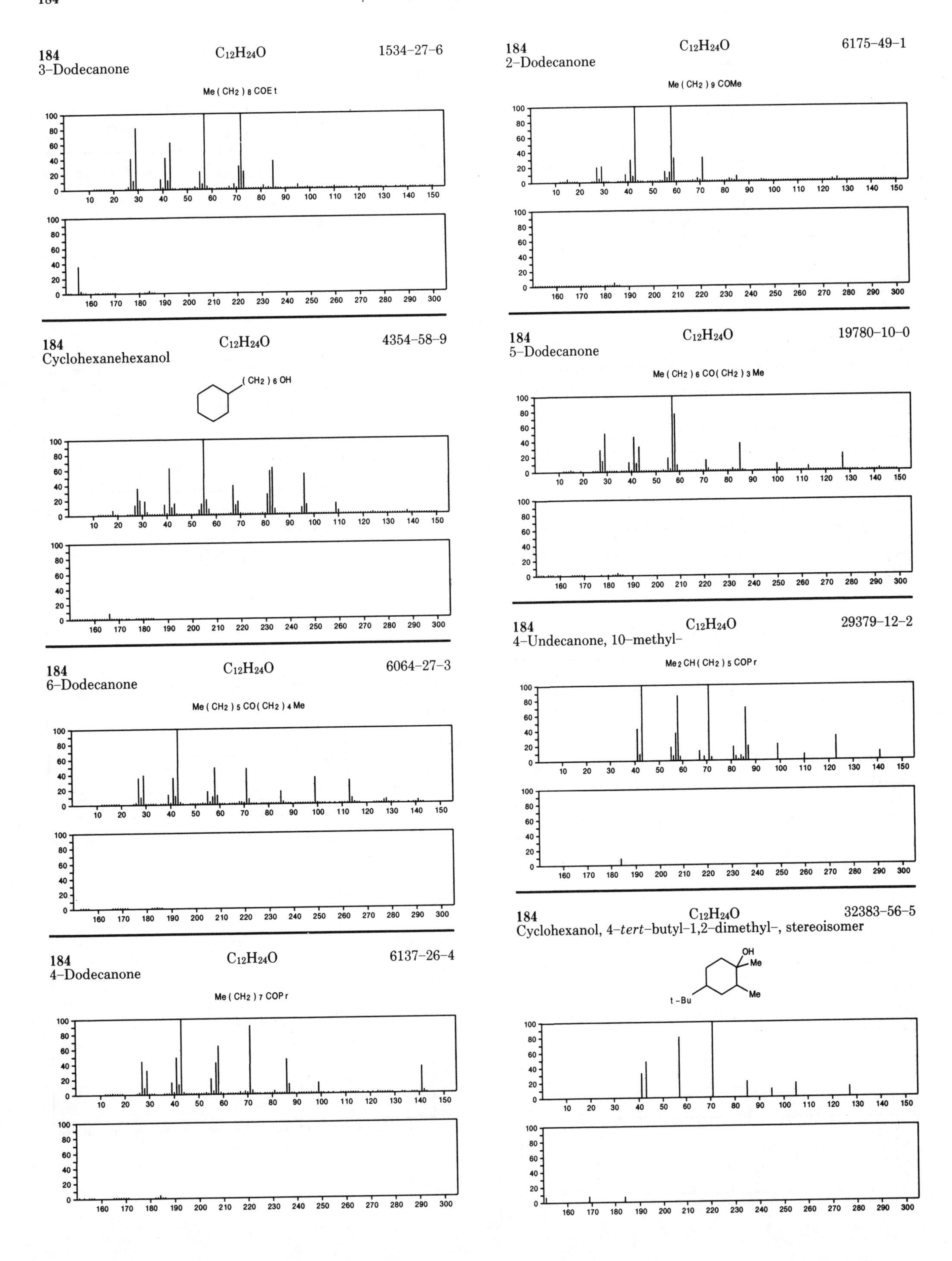
184
C12H24O
1534-27-6
3-Dodecanone
Me (CH2) 8 COEt
184
C12H24O
6175-49-1
2-Dodecanone
Me (CH2) 9 COMe
184
C12H24O
4354-58-9
Cyclohexanehexanol
(CH2) 6 OH
184
C12H24O
19780-10-0
5-Dodecanone
Me (CH2) 6 CO(CH2) 3 Me
184
C12H24O
6064-27-3
6-Dodecanone
Me (CH2) 5 CO(CH2) 4 Me
184
C12H24O
29379-12-2
4-Undecanone, 10-methyl-
Me2 CH(CH2) 5 COPr
184
C12H24O
6137-26-4
4-Dodecanone
Me (CH2) 7 COPr
184
C12H24O
32383-56-5
Cyclohexanol, 4-tert-butyl-1,2-dimethyl-, stereoisomer
OH
Me
Me
t-Bu

184 $C_{12}H_{24}O$ 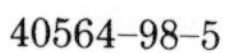 40564-98-5
Cyclohexanol, 3-(3,3-dimethylbutyl)-

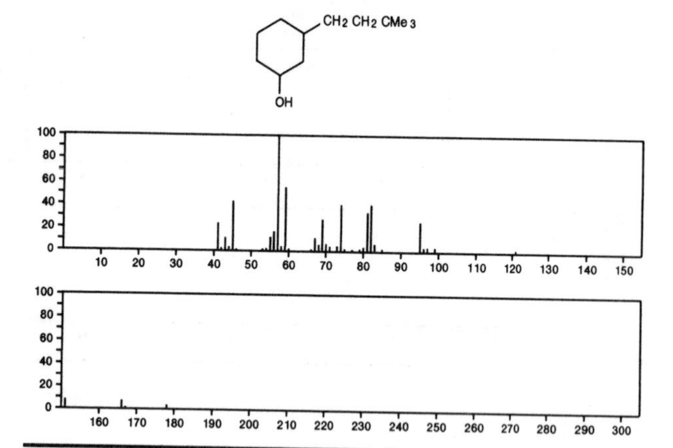

184 $C_{12}H_{24}O$ 50639-02-6
5-Undecanone, 2-methyl-

Me₂CHCH₂CH₂CO(CH₂)₅Me

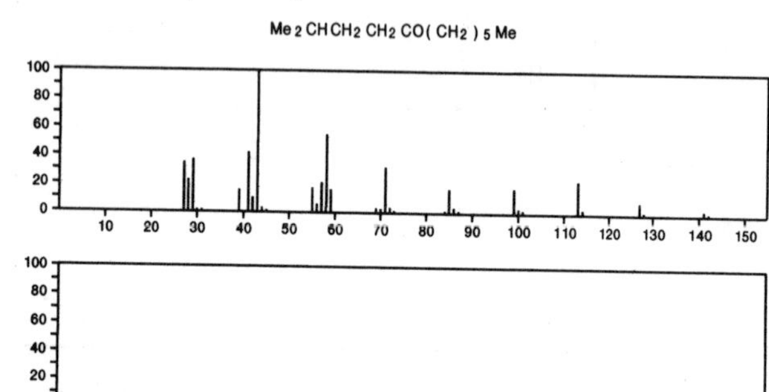

184 $C_{12}H_{24}O$ 55869-52-8
Cyclohexanol, 4-ethyl-4-methyl-3-(1-methylethyl)-, (1α,3α,4β)-

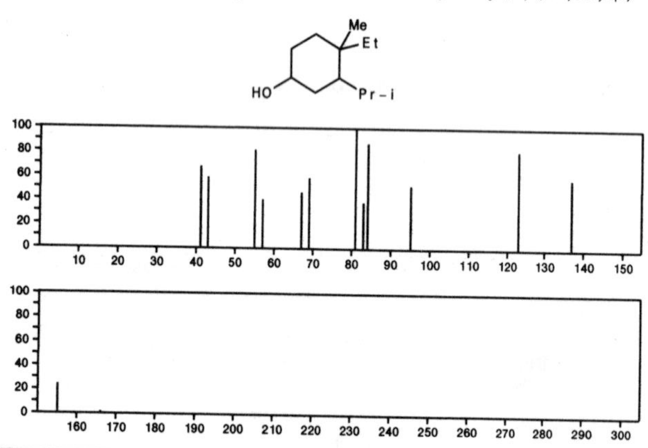

184 $C_{12}H_{24}O$ 55869-53-9
Cyclohexanol, 4-ethyl-4-methyl-3-(1-methylethyl)-, (1α,3β,4α)-

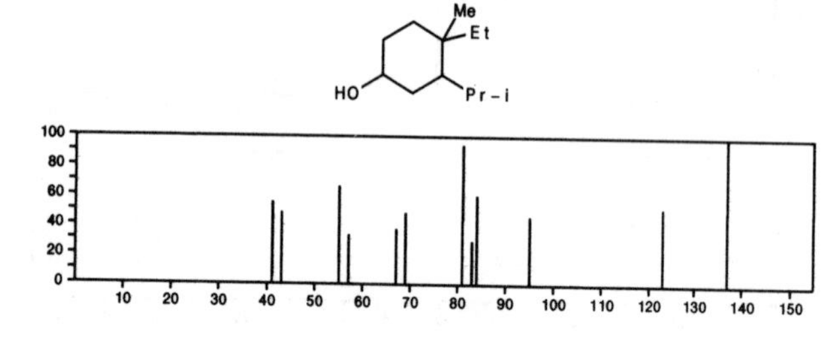

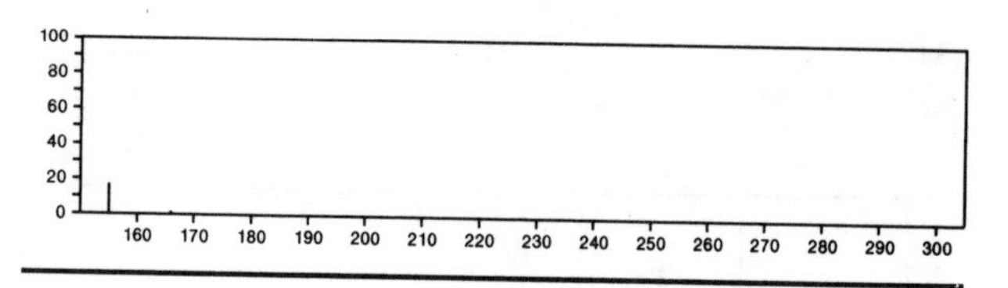

184 $C_{13}H_{12}O$ 86-26-0
1,1'-Biphenyl, 2-methoxy-

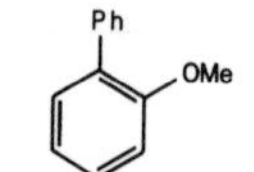

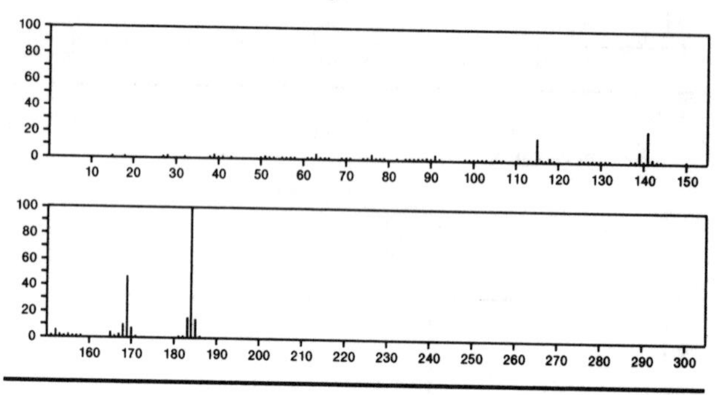

184 $C_{13}H_{12}O$ 101-53-1
Phenol, 4-(phenylmethyl)-

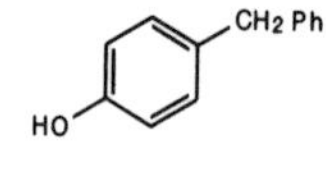

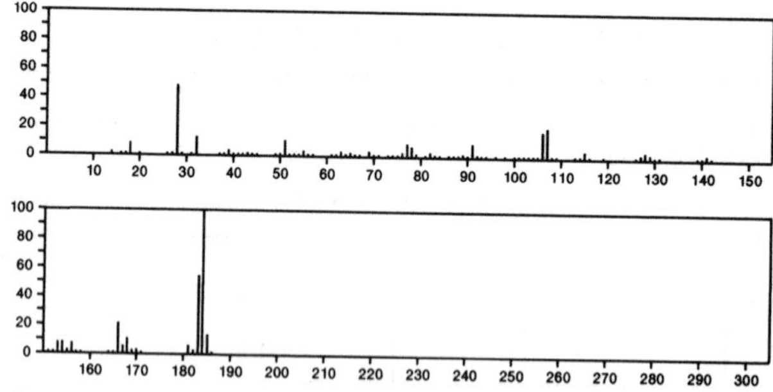

184 $C_{13}H_{12}O$  1706-12-3
Benzene, 1-methyl-4-phenoxy-

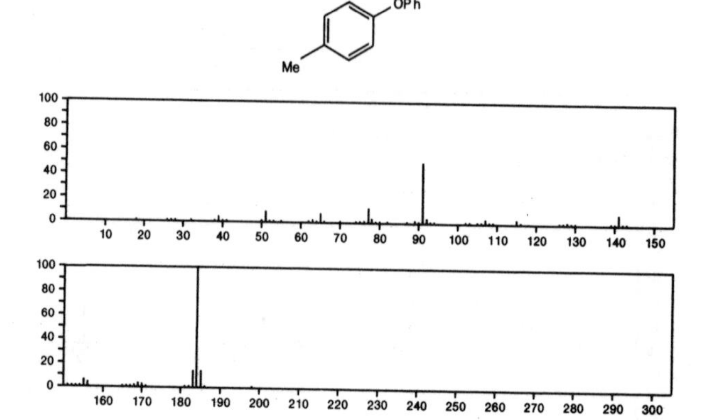

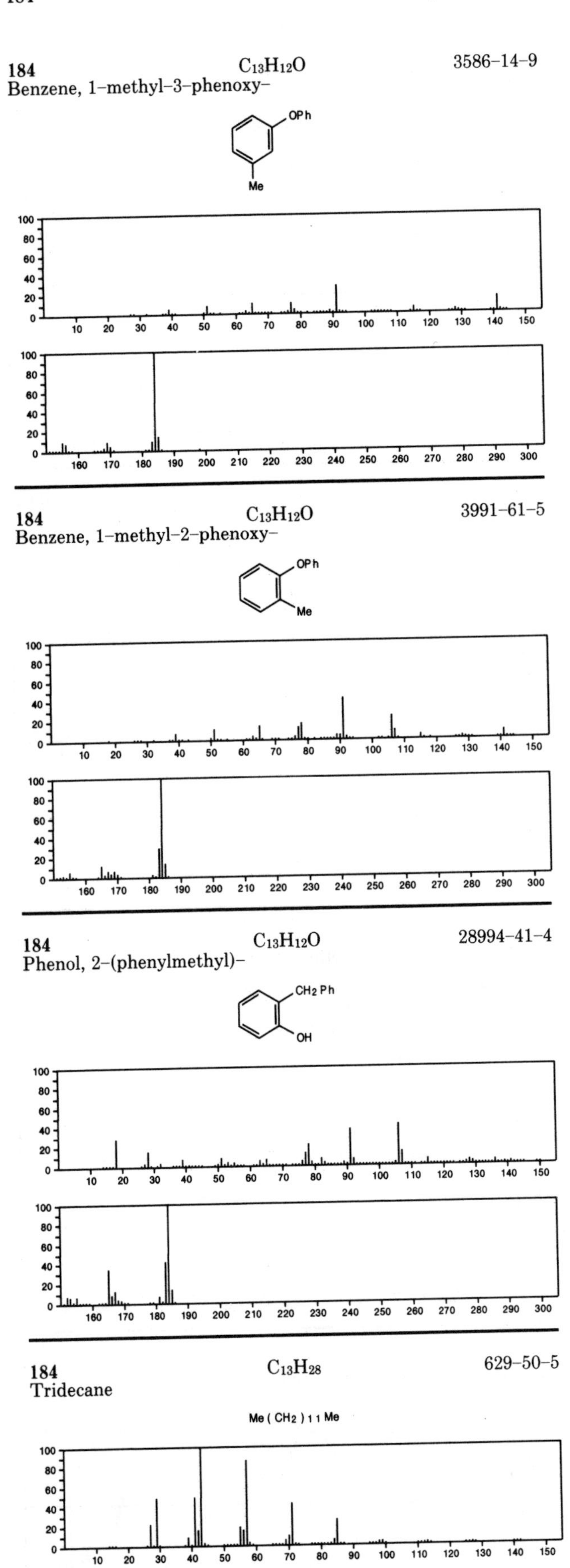

184 $C_{13}H_{12}O$ 3586-14-9
Benzene, 1-methyl-3-phenoxy-

184 $C_{13}H_{12}O$ 3991-61-5
Benzene, 1-methyl-2-phenoxy-

184 $C_{13}H_{12}O$ 28994-41-4
Phenol, 2-(phenylmethyl)-

184 $C_{13}H_{28}$ 629-50-5
Tridecane
Me(CH2)11Me

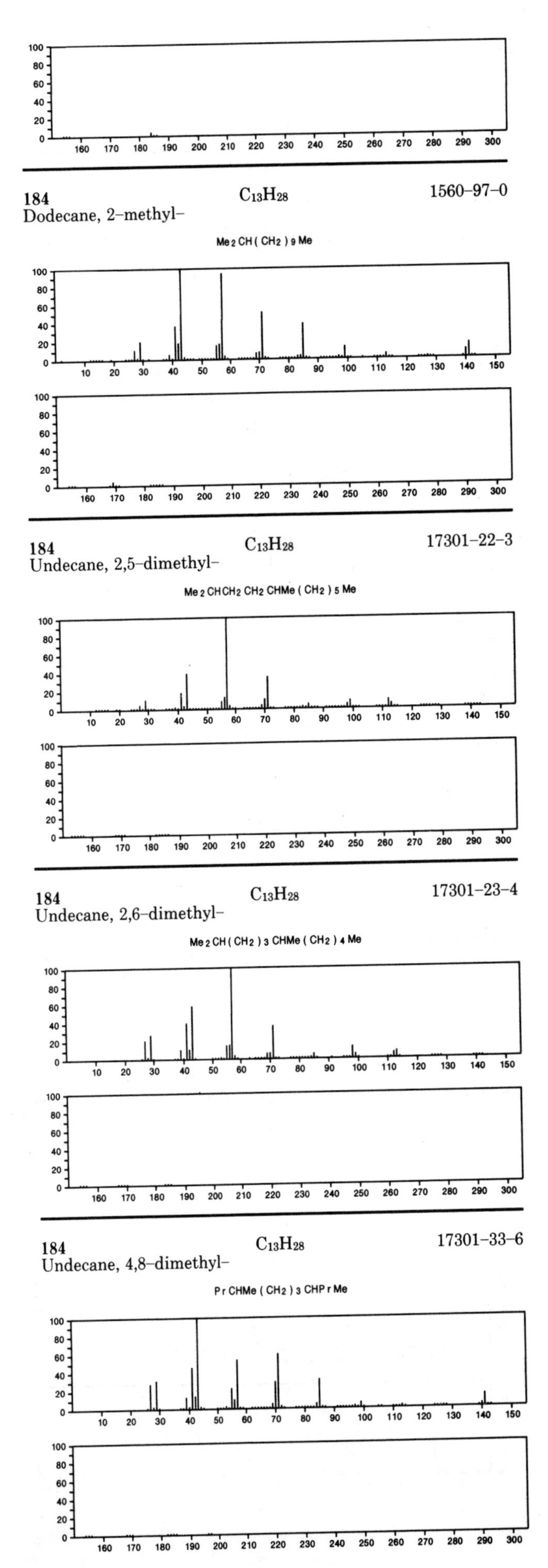

184 $C_{13}H_{28}$ 1560-97-0
Dodecane, 2-methyl-
Me2CH(CH2)9Me

184 $C_{13}H_{28}$ 17301-22-3
Undecane, 2,5-dimethyl-
Me2CHCH2CH2CHMe(CH2)5Me

184 $C_{13}H_{28}$ 17301-23-4
Undecane, 2,6-dimethyl-
Me2CH(CH2)3CHMe(CH2)4Me

184 $C_{13}H_{28}$ 17301-33-6
Undecane, 4,8-dimethyl-
PrCHMe(CH2)3CHPrMe

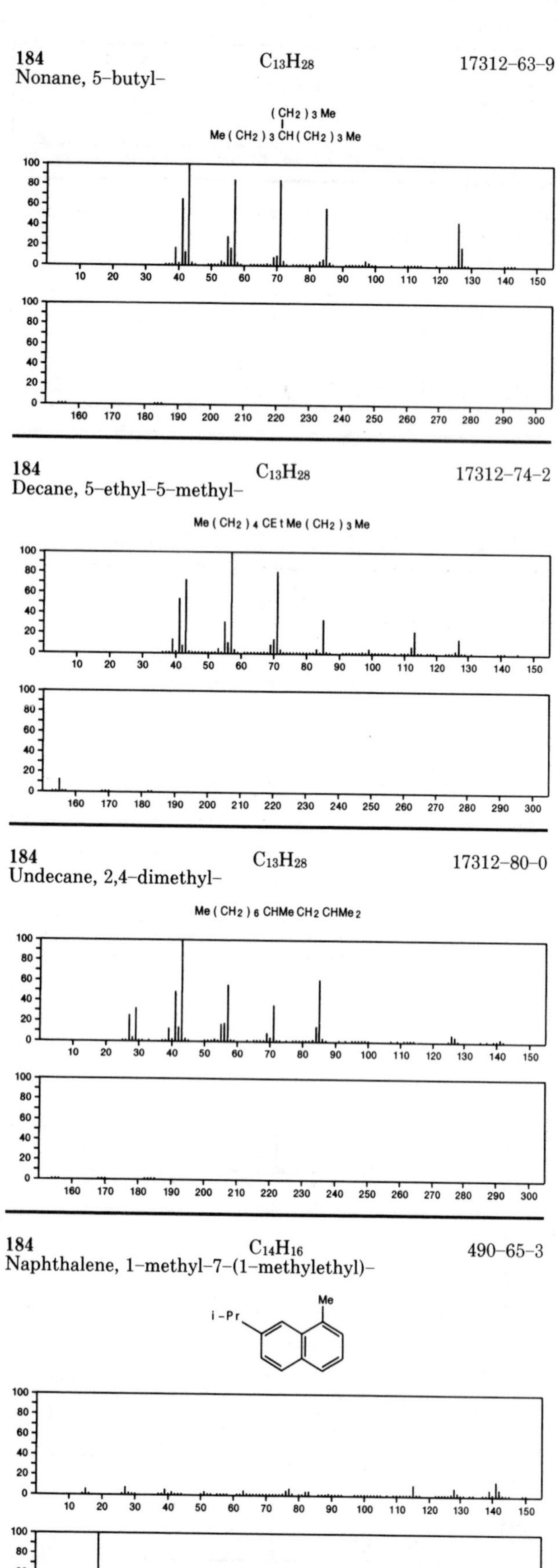

184 Nonane, 5-butyl- $C_{13}H_{28}$ 17312-63-9

184 Decane, 5-ethyl-5-methyl- $C_{13}H_{28}$ 17312-74-2

184 Undecane, 2,4-dimethyl- $C_{13}H_{28}$ 17312-80-0

184 Naphthalene, 1-methyl-7-(1-methylethyl)- $C_{14}H_{16}$ 490-65-3

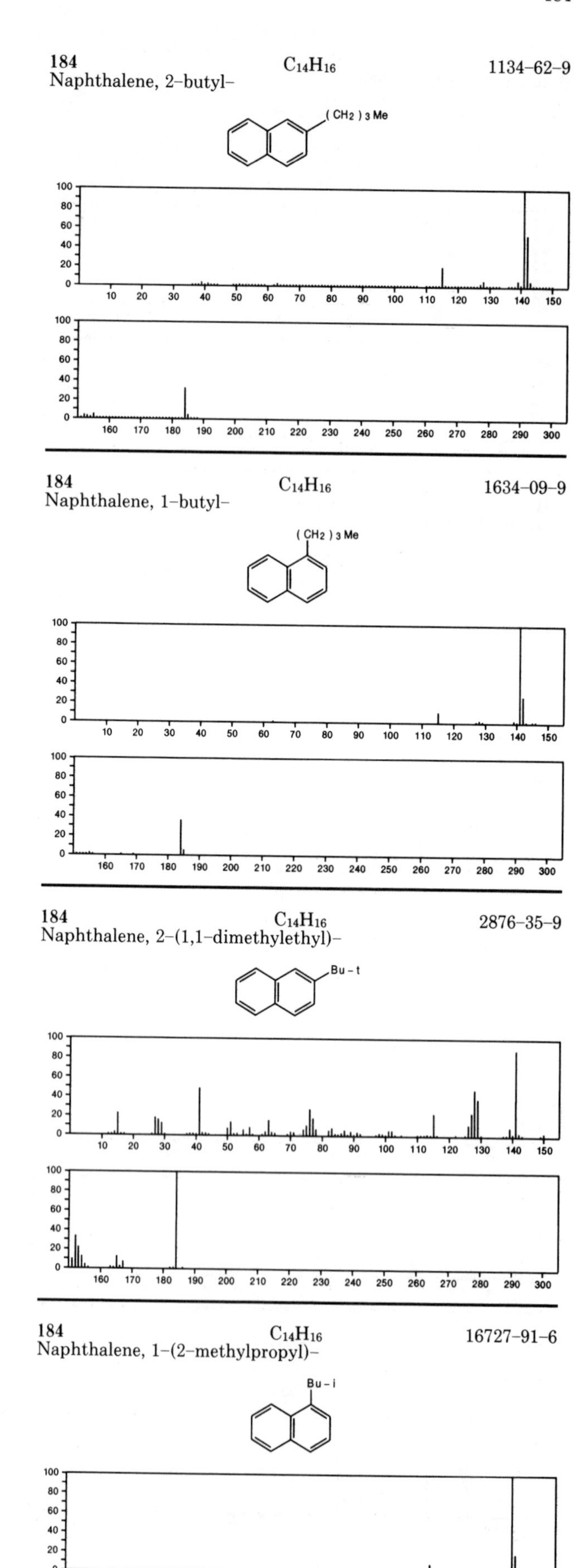

184 Naphthalene, 2-butyl- $C_{14}H_{16}$ 1134-62-9

184 Naphthalene, 1-butyl- $C_{14}H_{16}$ 1634-09-9

184 Naphthalene, 2-(1,1-dimethylethyl)- $C_{14}H_{16}$ 2876-35-9

184 Naphthalene, 1-(2-methylpropyl)- $C_{14}H_{16}$ 16727-91-6

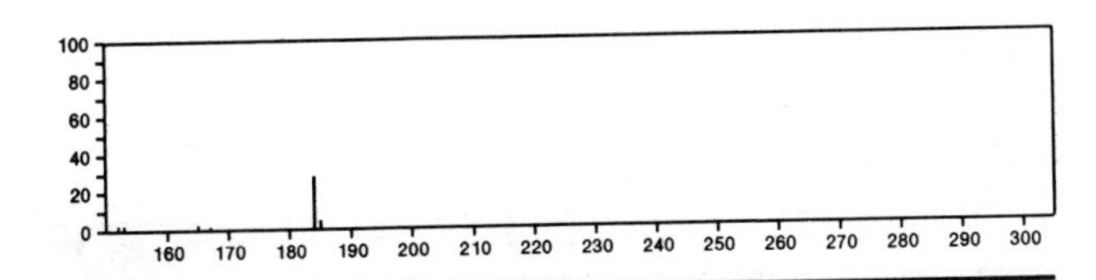

184 $C_{14}H_{16}$ 17085-91-5
Naphthalene, 1-(1,1-dimethylethyl)-

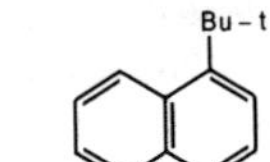

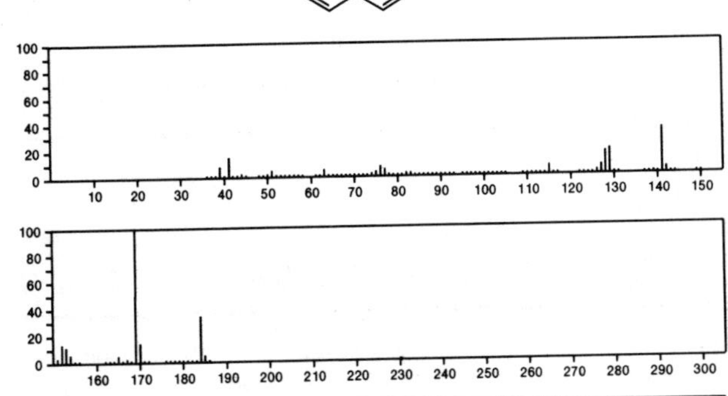

185 $CClN_3O_6$ 1943-16-4
Methane, chlorotrinitro-

Cl C(NO2)3

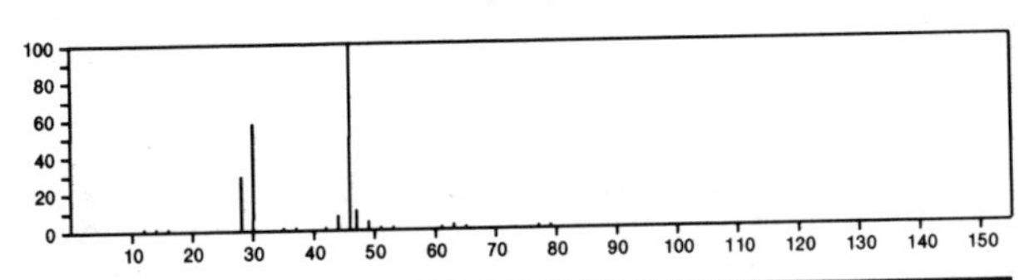

185 C_5ClF_4N 54774-81-1
Pyridine, 2-chloro-3,4,5,6-tetrafluoro-

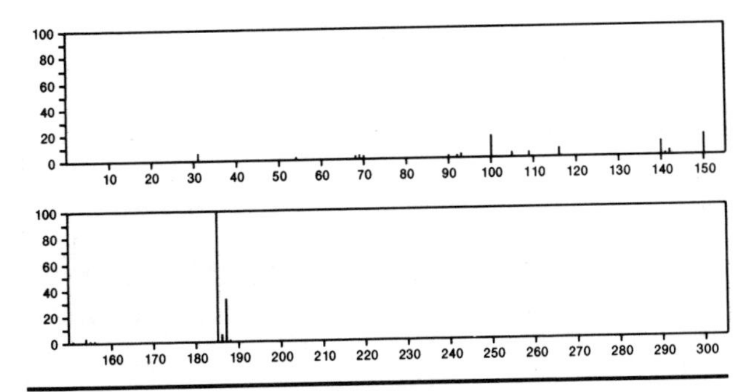

185 $C_5H_{12}NO_2S.Cl$ 29548-68-3
Sulfonium, [(ethoxycarbonyl)amino]dimethyl-, chloride

$Me_2\overset{+}{S}NHC(O)OEt \cdot Cl^-$

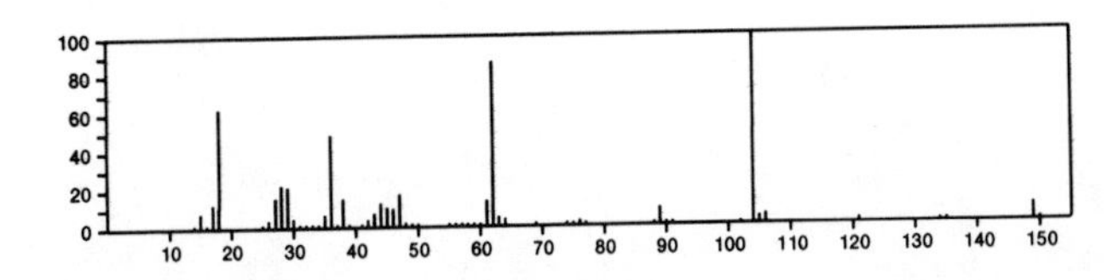

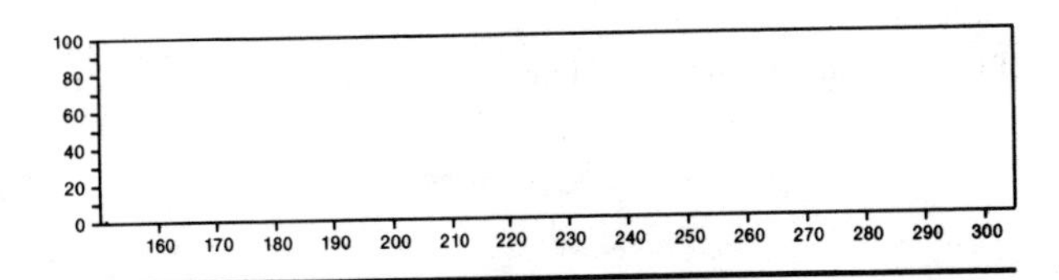

185 $C_6H_4ClN_3S$ 13316-09-1
Thiazolo[5,4-*d*]pyrimidine, 7-chloro-5-methyl-

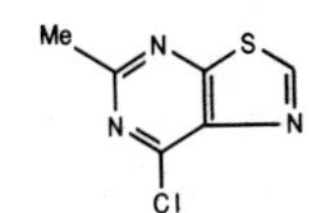

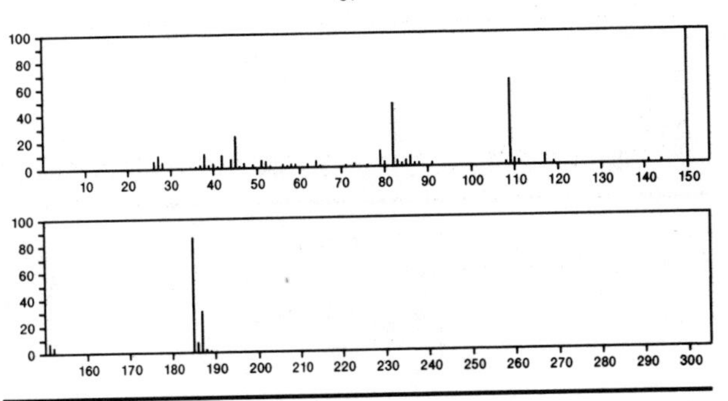

185 $C_7H_7NO_5$ 54774-80-0
2*H*-Pyran-2-one, 4-methoxy-6-methyl-3-nitro-

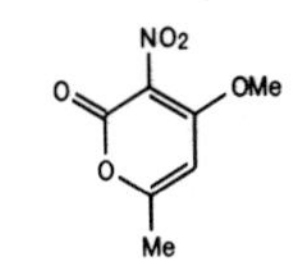

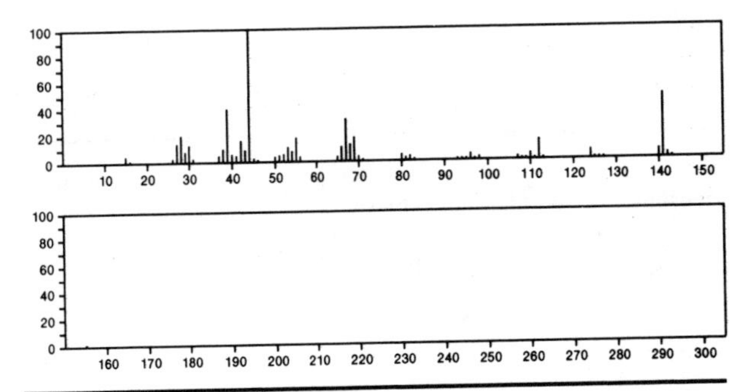

185 $C_8H_8ClNO_2$ 940-36-3
Carbamic acid, (4-chlorophenyl)-, methyl ester

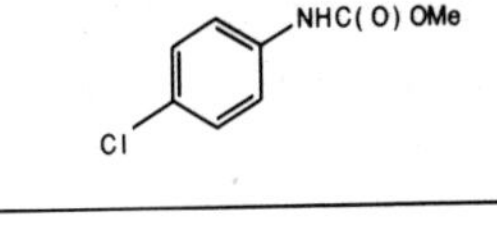

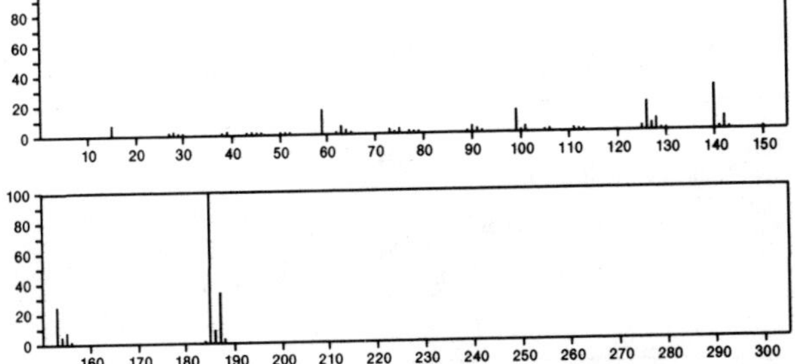

185 $C_8H_8ClNO_2$ 5202-89-1
Benzoic acid, 2-amino-5-chloro-, methyl ester

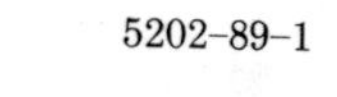

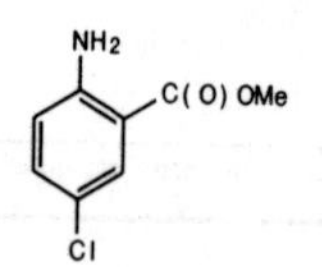

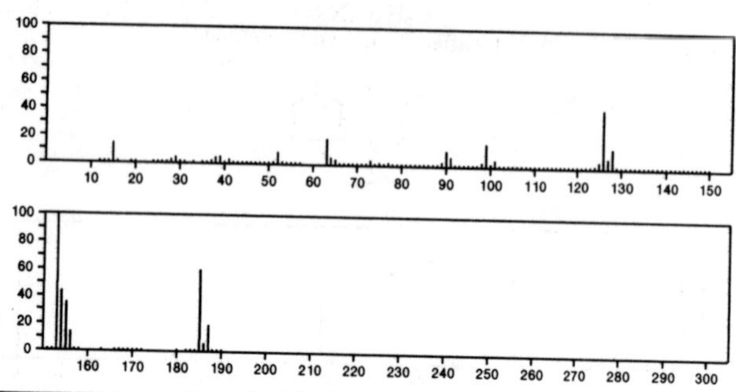

185 $C_8H_8ClNO_2$ 20668-13-7
Carbamic acid, (2-chlorophenyl)-, methyl ester

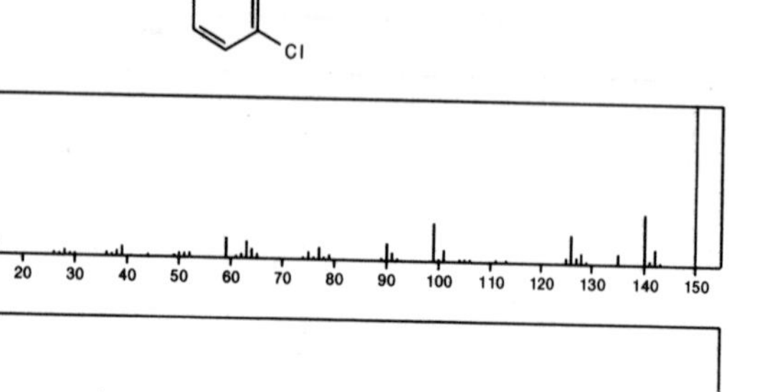

185 $C_8H_{11}NO_2S$ 640-61-9
Benzenesulfonamide, *N*,4-dimethyl-

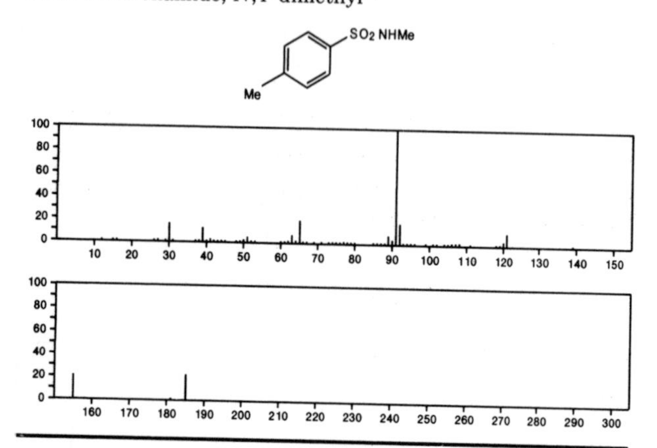

185 $C_8H_{11}NO_2S$ 6326-18-7
3,4-Xylenesulfonamide

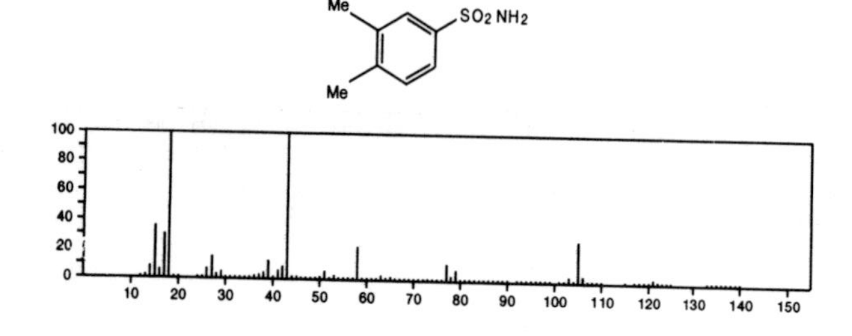

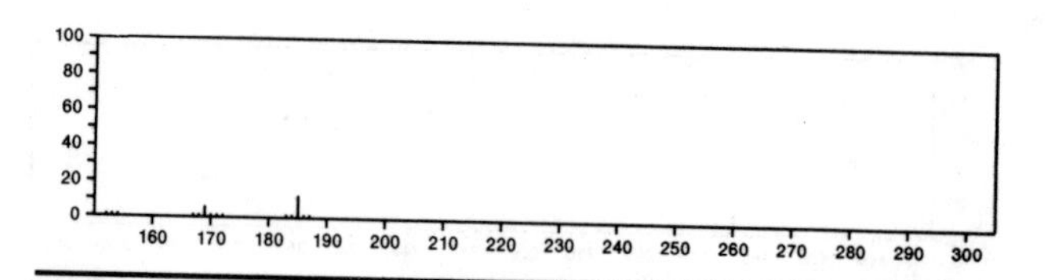

185 $C_8H_{11}NO_2S$ 14417-01-7
Benzenesulfonamide, *N*,*N*-dimethyl-

Me2NSO2Ph

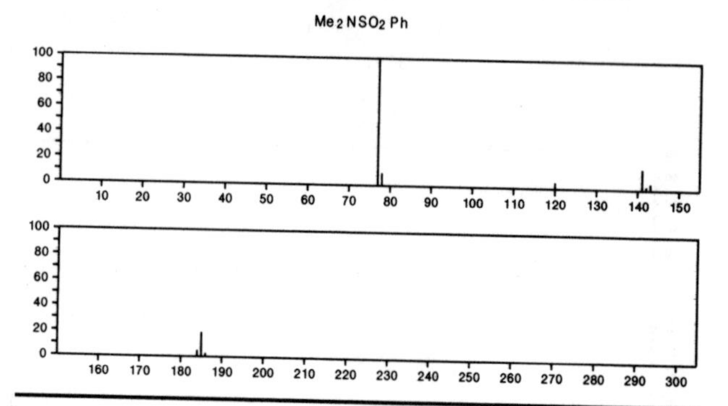

185 $C_8H_{11}NO_2S$ 23003-28-3
3-Pyridinol, 2-[(2-hydroxyethyl)thio]-6-methyl-

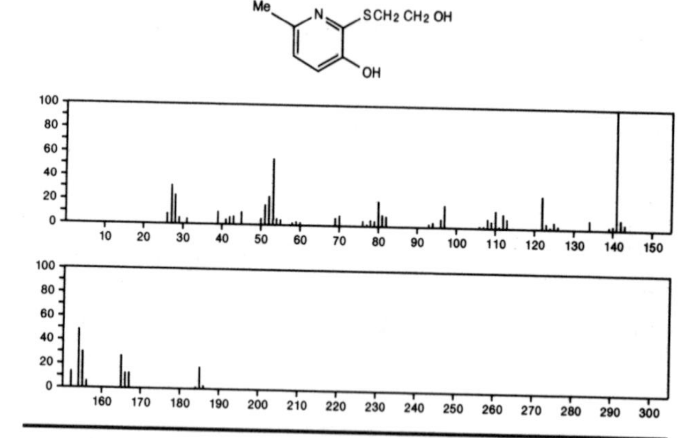

185 $C_8H_{15}N_3S$ 22397-22-4
Cyclohexanone, 4-methyl-, thiosemicarbazone

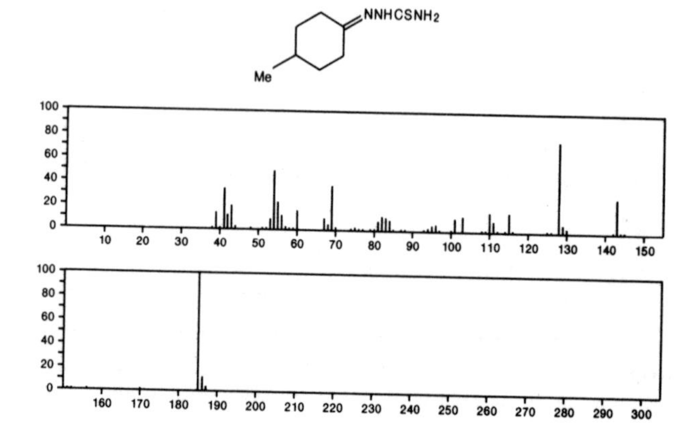

185 $C_8H_{15}N_3S$ 56324-61-9

Hydrazinecarbothioamide, 2-(2-methylcyclohexylidene)-

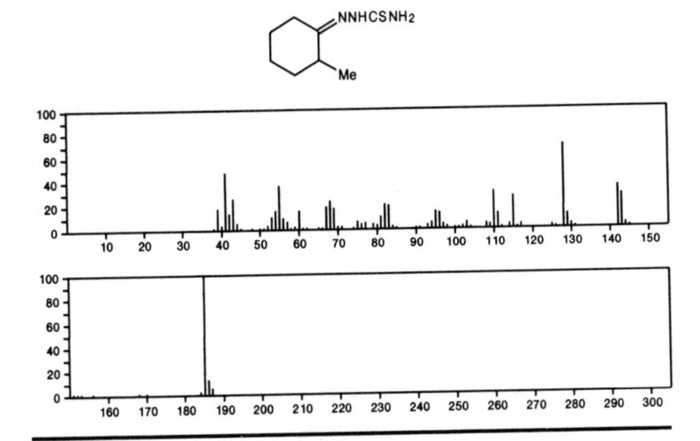

185 $C_8H_{16}BNO_3$ 31970-40-8

1,3,2-Oxazaborolane-4-carboxylic acid, 2-butyl-, methyl ester, L-

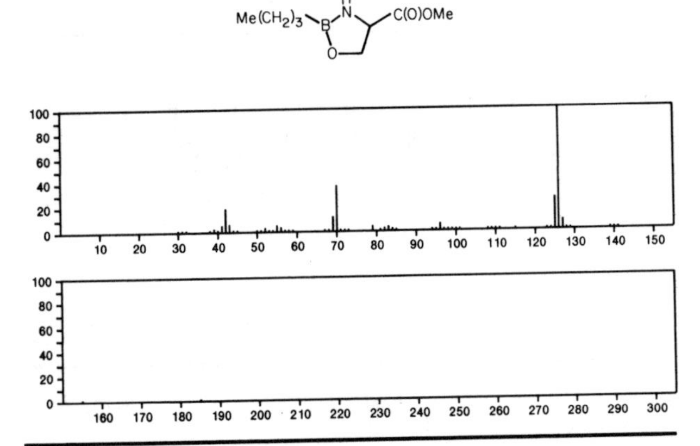

185 $C_9H_{15}NO_3$ 481-37-8

8-Azabicyclo[3.2.1]octane-2-carboxylic acid, 3-hydroxy-8-methyl-, [1*R*-(*exo*,*exo*)]-

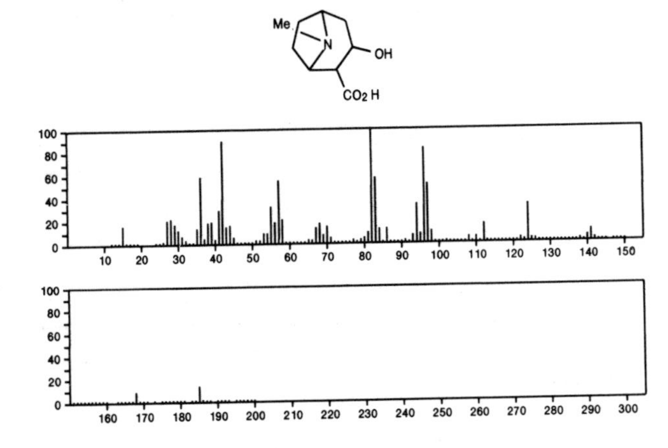

185 $C_9H_{15}NO_3$ 481-38-9

8-Azabicyclo[3.2.1]octane-2-carboxylic acid, 3-hydroxy-8-methyl-, [1*R*-(2-*endo*,3-*exo*)]-

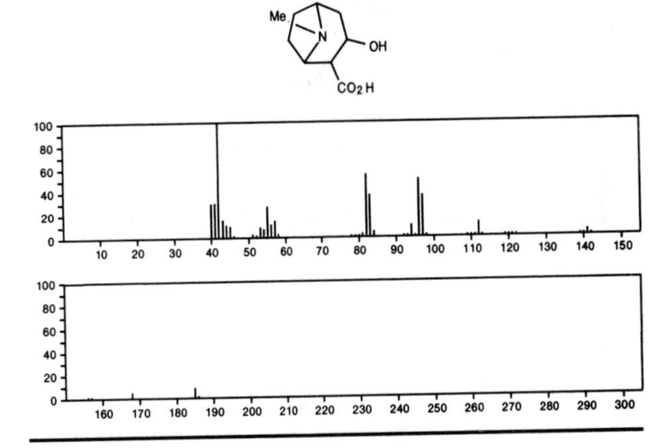

185 $C_9H_{15}NO_3$ 33927-64-9

1-Pyrrolidineacetic acid, 2-methyl-5-oxo-, ethyl ester

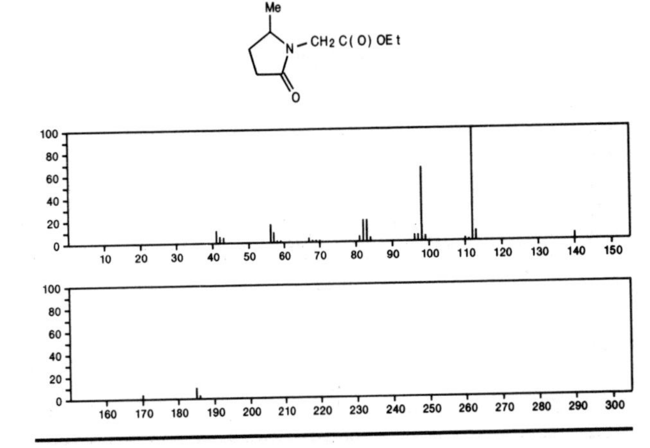

185 $C_9H_{15}NO_3$ 50267-38-4

2-Oxa-6-azatricyclo[3.3.1.1^{3,7}]decane-4,8-diol, 6-methyl-, (1α,3β,4α,5β,7β,8β)-

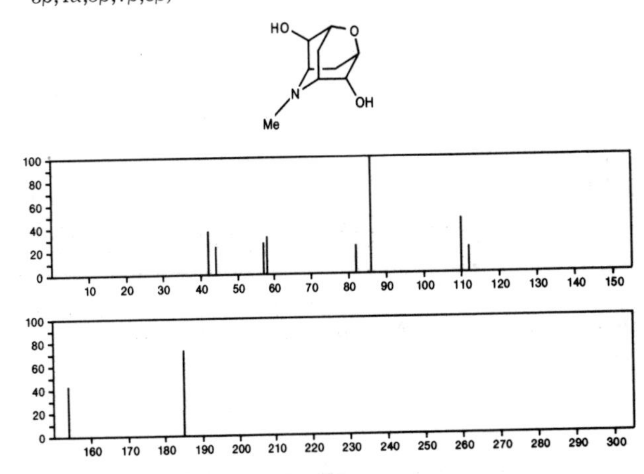

185 $C_9H_{15}NO_3$ 54808-87-6
8-Azabicyclo[3.2.1]octane-2-carboxylic acid, 3-hydroxy-8-methyl-, (*exo,exo*)-

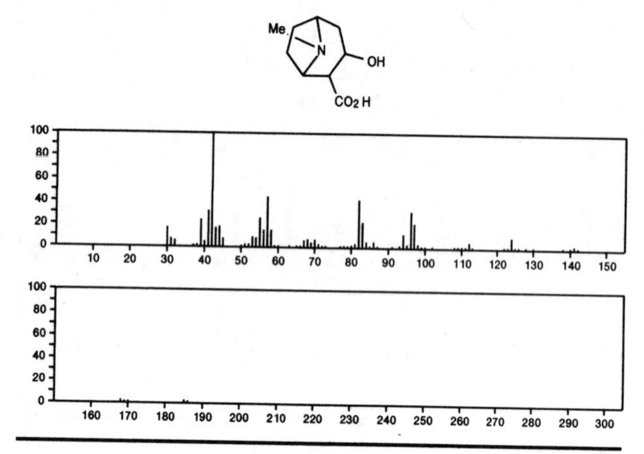

185 $C_9H_{15}NO_3$ 54808-88-7
8-Azabicyclo[3.2.1]octane-2-carboxylic acid, 3-hydroxy-8-methyl-, (2-*endo*,3-*exo*)-

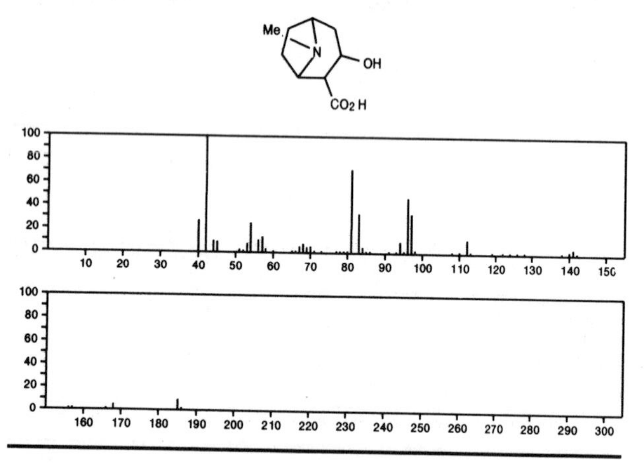

185 $C_{10}H_{15}N.ClH$ 30684-06-1
Butylamine, 4-phenyl-, hydrochloride

Ph(CH2)4NH2 • HCl

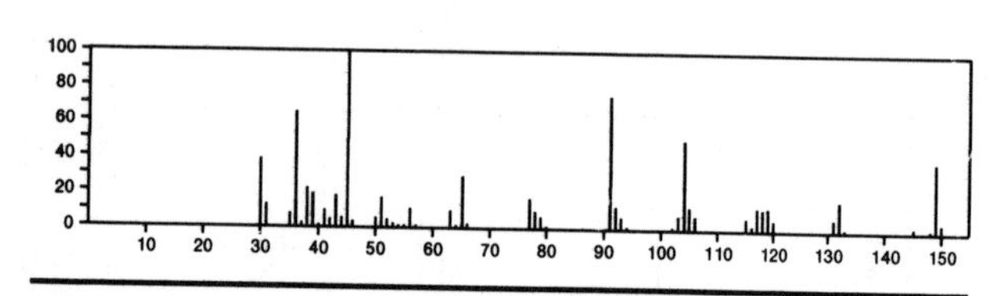

185 $C_{10}H_{15}N.ClH$ 30684-07-2
Benzenepropanamine, *N*-methyl-, hydrochloride

MeNH(CH2)3Ph • HCl

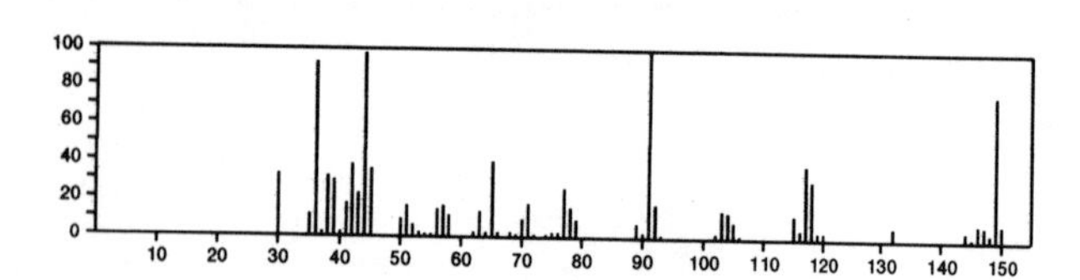

185 $C_{10}H_{19}NO_2$ 3168-91-0
Hexanoic acid, 2-(1-aminoethylidene)-, ethyl ester

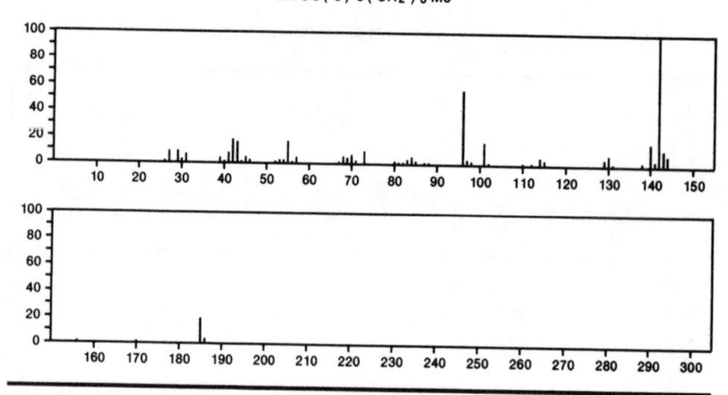

185 $C_{10}H_{19}NO_2$ 23435-00-9
Azacycloundecan-2-one, 7-hydroxy-

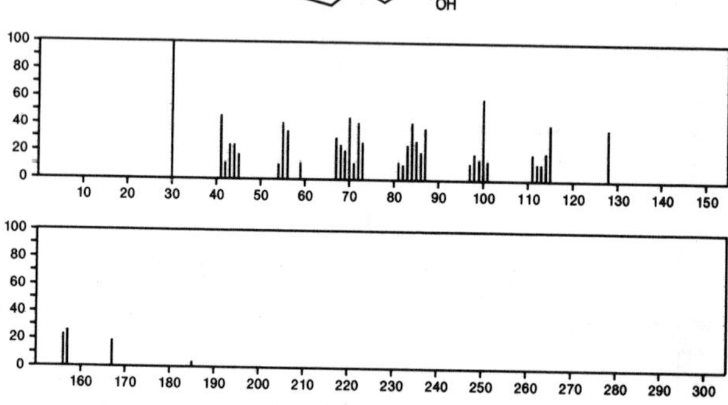

185 $C_{10}H_{19}NO_2$ 54751-92-7
3-Piperidinol, 1-acetyl-6-propyl-, (3*R*-*trans*)-

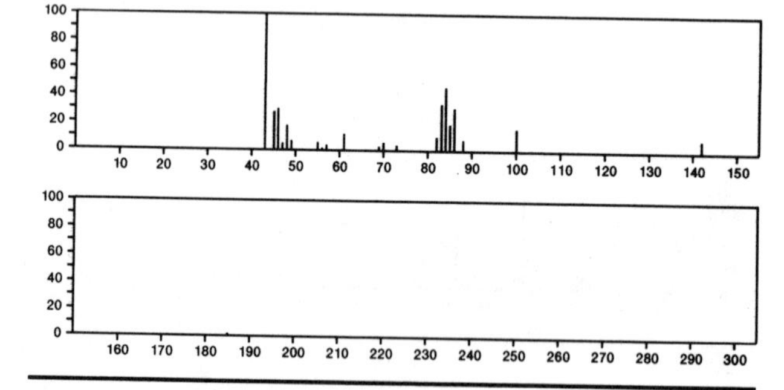

185 $C_{10}H_{19}NO_2$ 55669-84-6
1-Aziridineacetic acid, 2-methyl-3-(1-methylethyl)-, ethyl ester, *trans*-

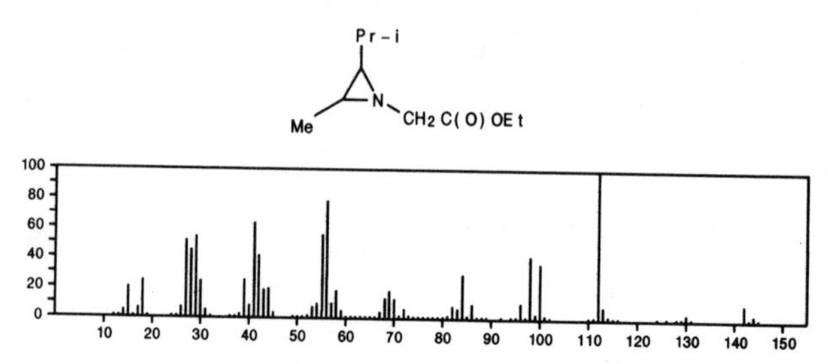

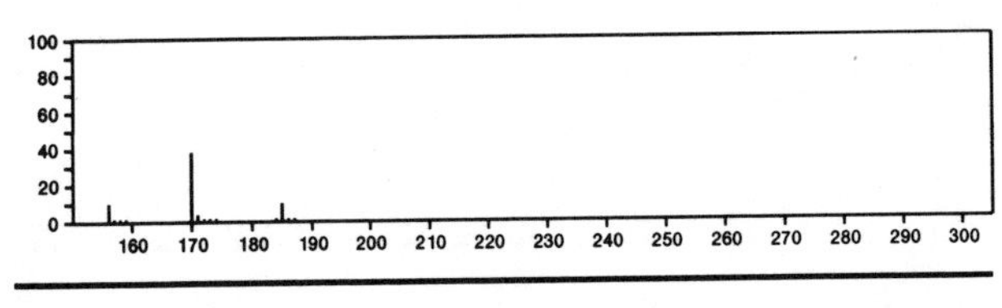

185 $C_{11}H_7NS$ 551–06–4
Naphthalene, 1–isothiocyanato–

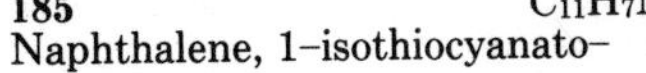

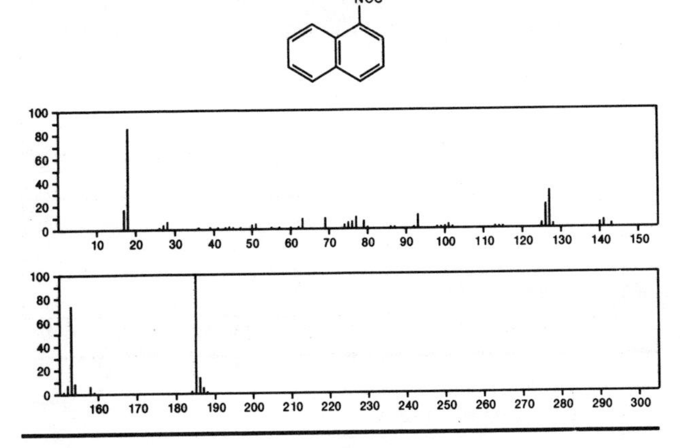

185 $C_{11}H_{11}N_3$ 16728–92–0
Acetonitrile, 2,2'–[(phenylmethyl)imino]bis–

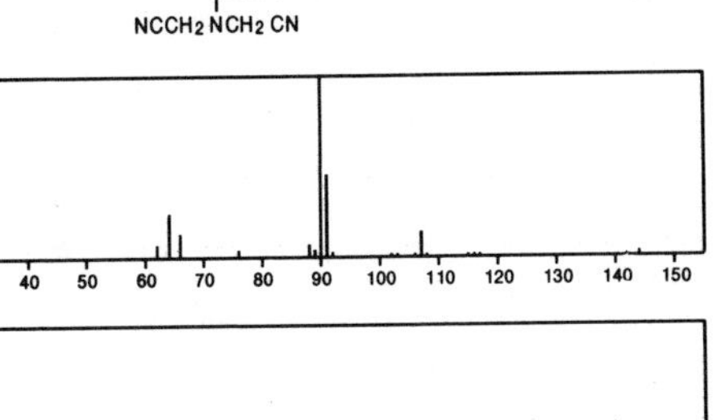

185 $C_{11}H_{23}NO$ 10264–25–2
Valeramide, *N*–hexyl–

Me(CH2)5NHCO(CH2)3Me

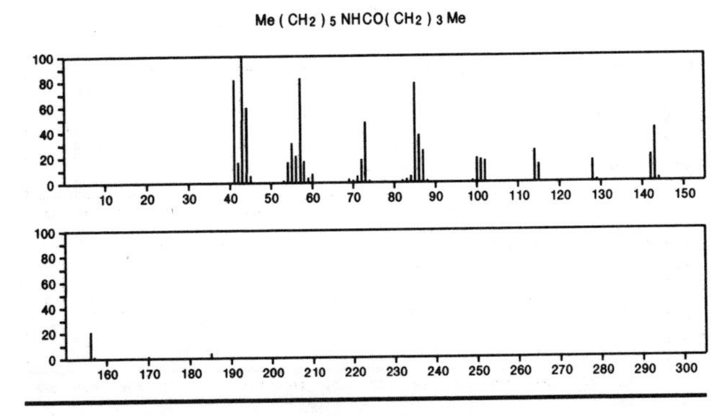

185 $C_{11}H_{23}NO$ 35473–94–0
Decanal, *O*–methyloxime

MeON=CH(CH2)8Me

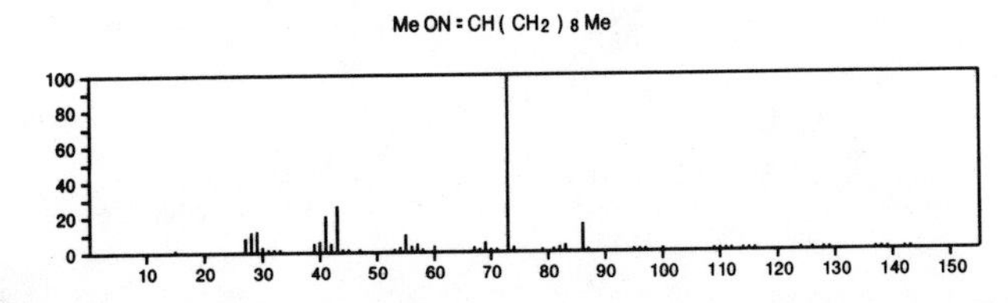

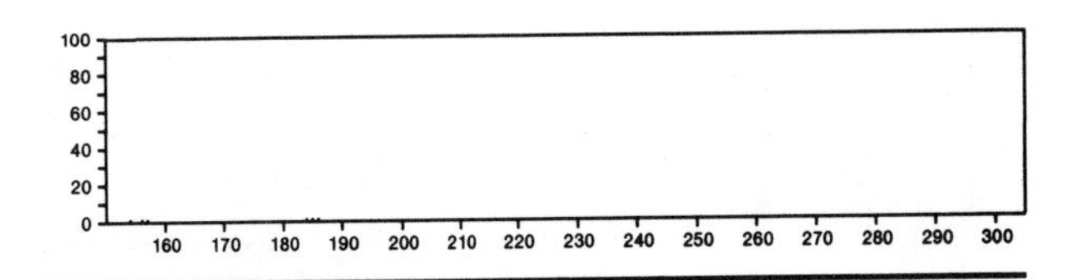

185 $C_{11}H_{23}NO$ 36382–61–3
2–Decanone, *O*–methyloxime

Me(CH2)7CMe=NOMe

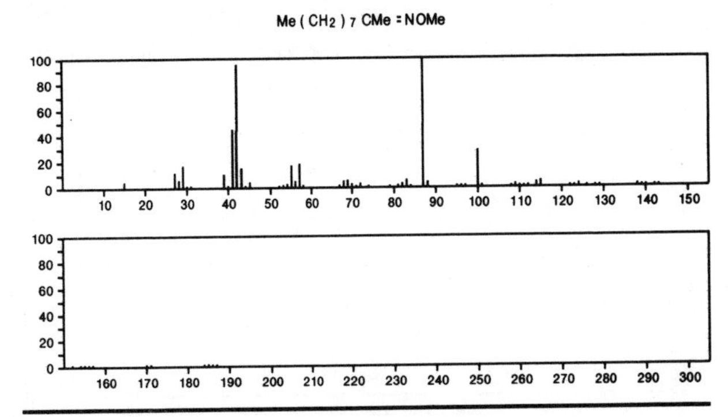

185 $C_{11}H_{23}NO$ 50837–73–5
Propanamide, 2,2–dimethyl–*N*,*N*–bis(1–methylethyl)–

(i-Pr)2NCOCMe3

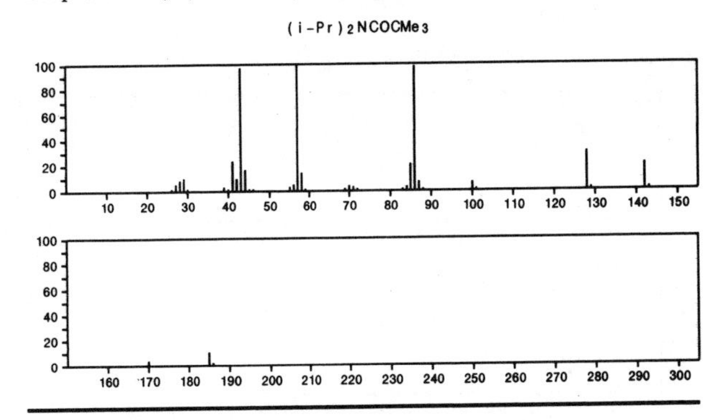

185 $C_{12}H_{11}NO$ 3400–33–7
1–Naphthalenecarboxamide, *N*–methyl–

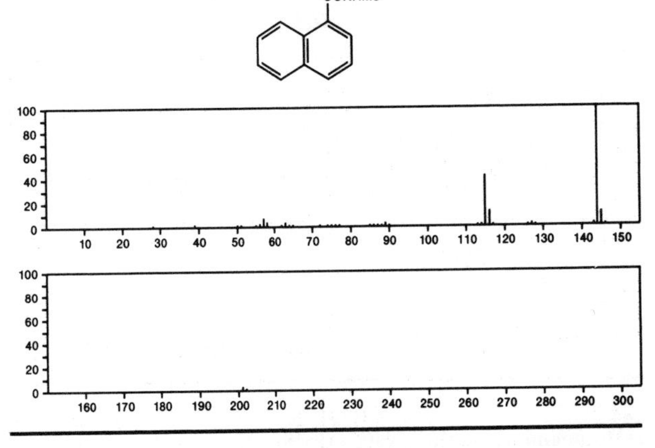

185 $C_{12}H_{11}NO$ 54774–86–6
2(1*H*)–Pyridinone, 1–(2–methylphenyl)–

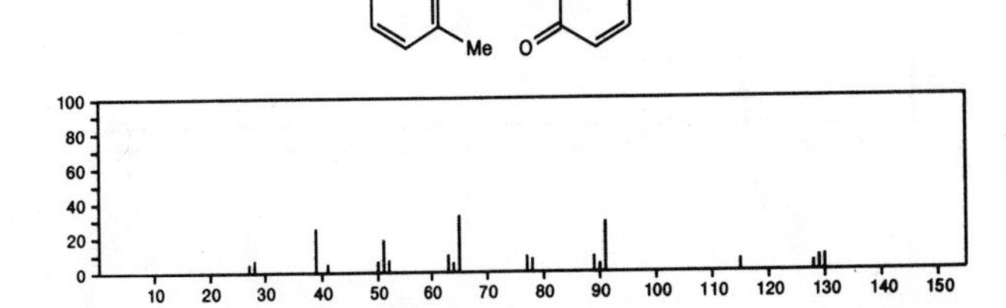

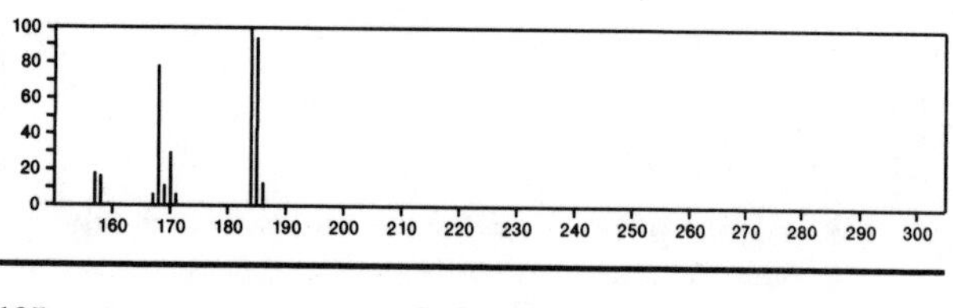

185 $C_{12}H_{27}N$ 102–82–9
1–Butanamine, *N*,*N* dibutyl–

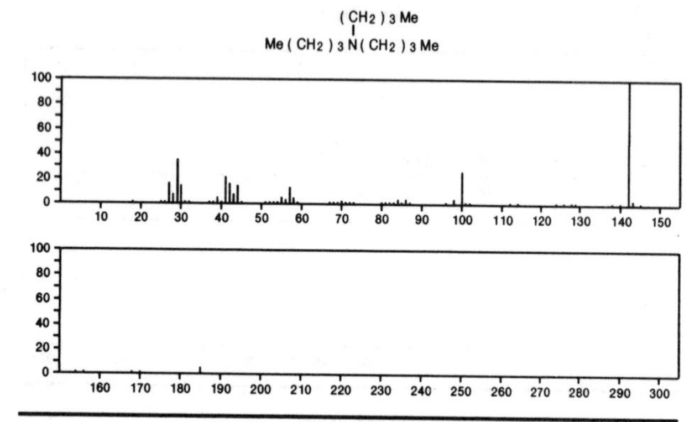

185 $C_{12}H_{27}N$ 124–22–1
1–Dodecanamine

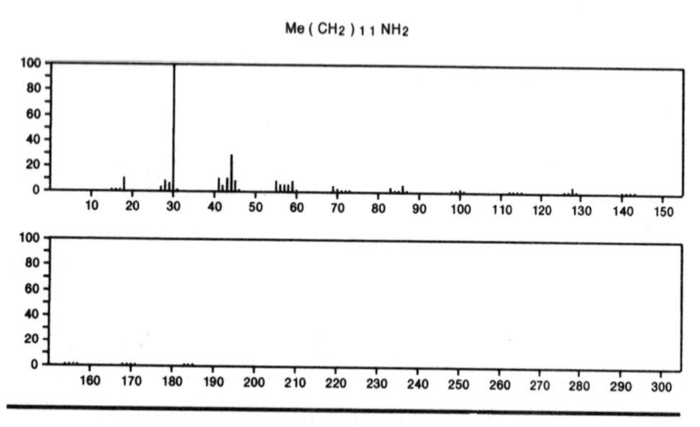

185 $C_{12}H_{27}N$ 143–16–8
1–Hexanamine, *N*–hexyl–

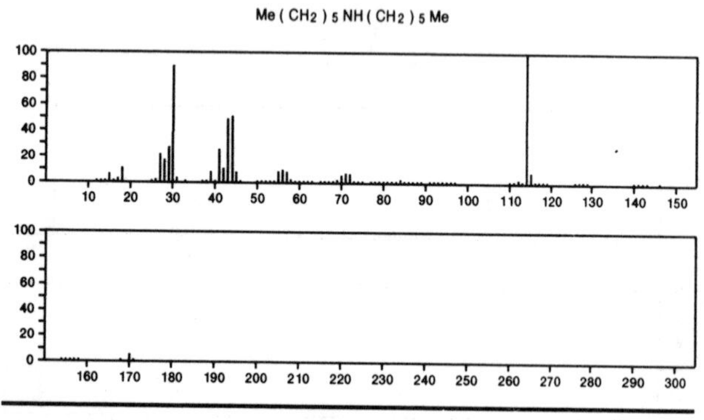

185 $C_{12}H_{27}N$ 54774–85–5
1–Butanamine, 2–ethyl–*N*–(2–ethylbutyl)–

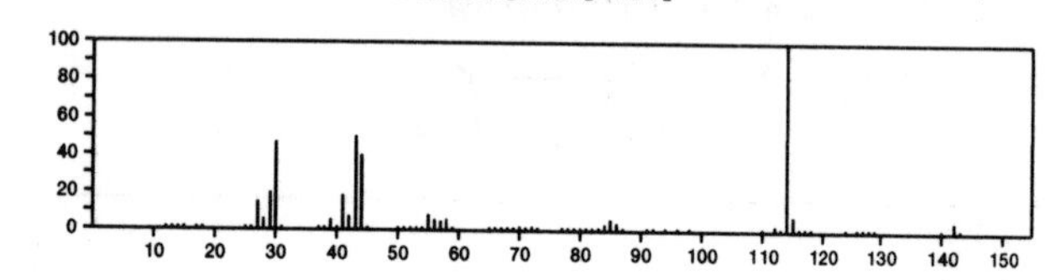

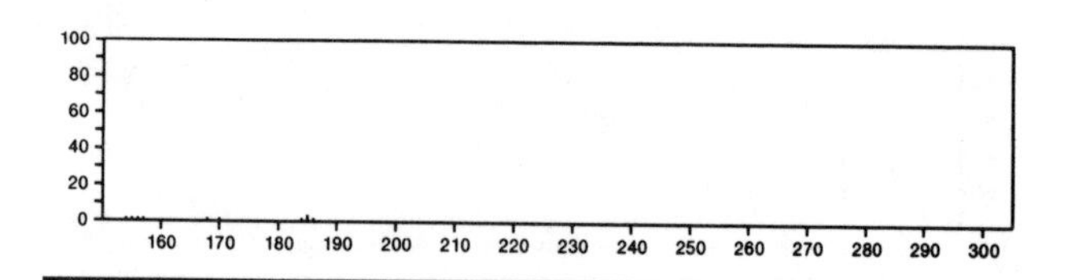

185 $C_{12}H_{27}N$ 54775–00–7
1–Pentanamine, 4–methyl–*N*–(4–methylpentyl)–

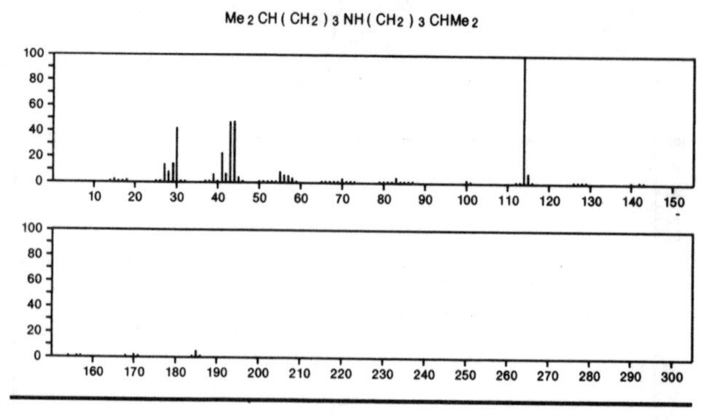

185 $C_{13}H_{15}N$ 93–19–6
Quinoline, 2–(2–methylpropyl)–

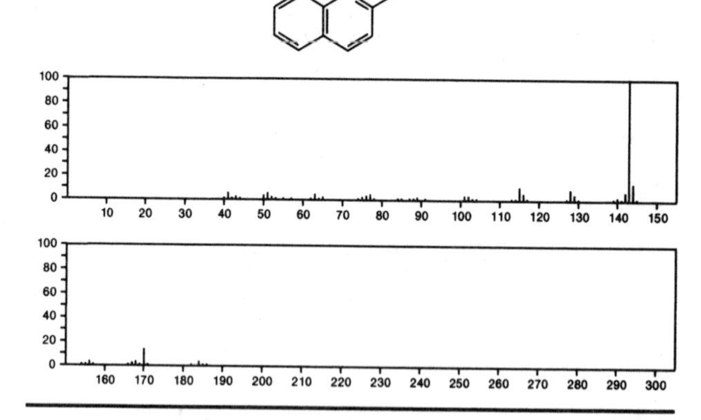

185 $C_{13}H_{15}N$ 7634–74–4
Quinoline, 6–butyl–

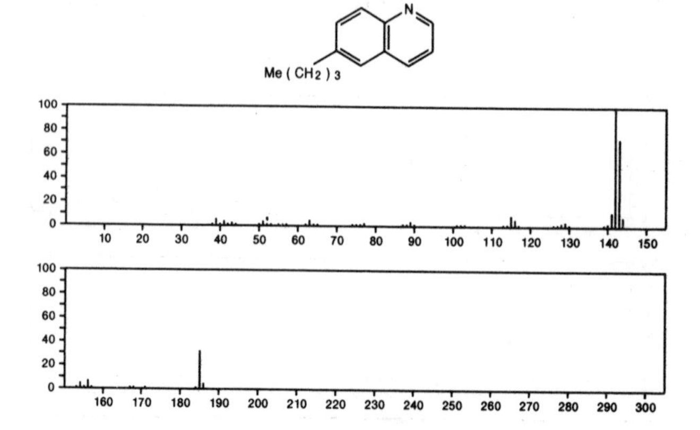

185 $C_{13}H_{15}N$ 7661-38-3
Isoquinoline, 1-butyl-

185 $C_{13}H_{15}N$ 7661-39-4
Quinoline, 2-butyl-

185 $C_{13}H_{15}N$ 7661-40-7
Isoquinoline, 1-isobutyl-

185 $C_{13}H_{15}N$ 7661-42-9
Isoquinoline, 3-butyl-

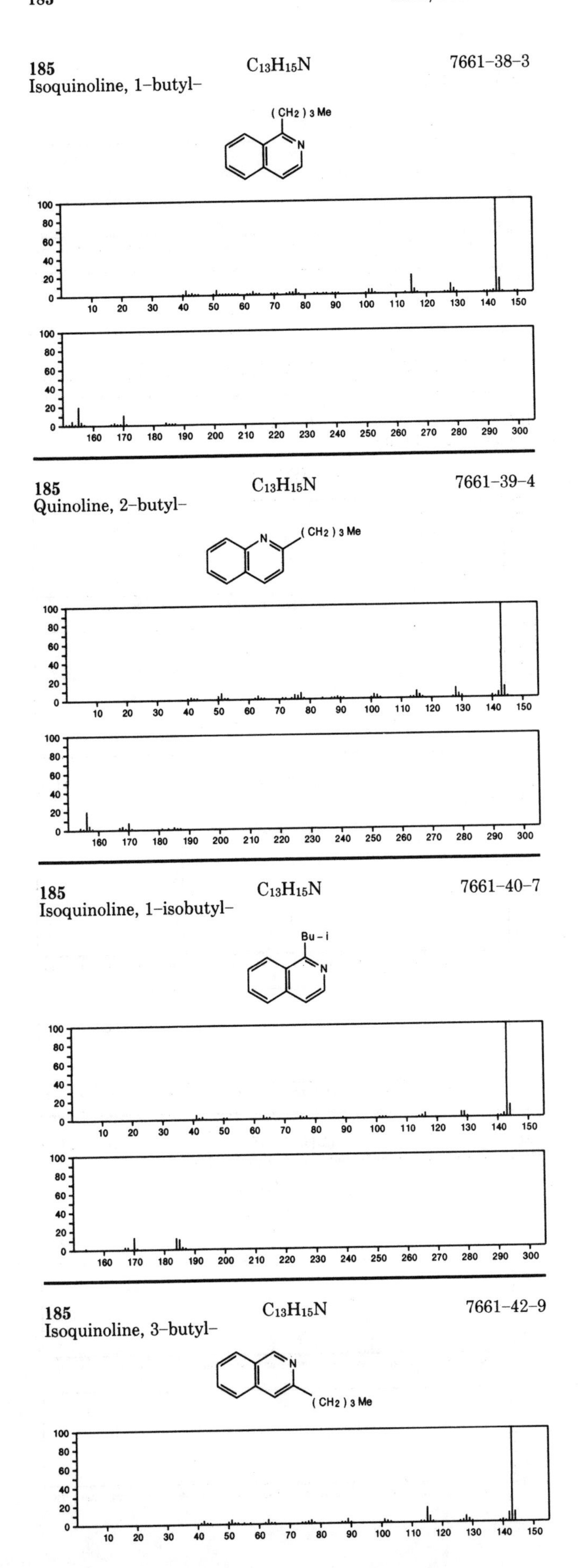

185 $C_{13}H_{15}N$ 7661-51-0
Quinoline, 4-isobutyl-

185 $C_{13}H_{15}N$ 7661-52-1
Quinoline, 7-butyl-

185 $C_{13}H_{15}N$ 18781-72-1
1*H*-Carbazole, 2,3,4,4a-tetrahydro-4a-methyl-

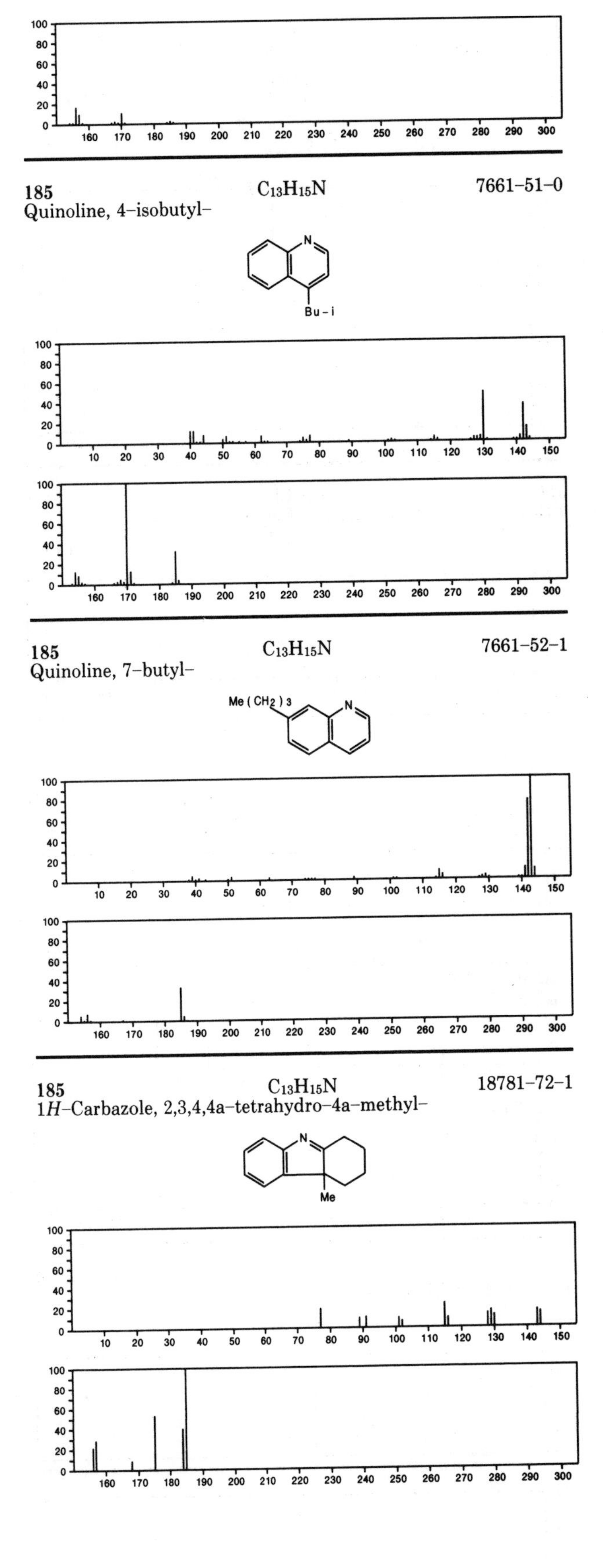

185 $C_{13}H_{15}N$ 23077-27-2
Spiro[cyclopentane-1,3'-[3*H*]indole], 2'-methyl-

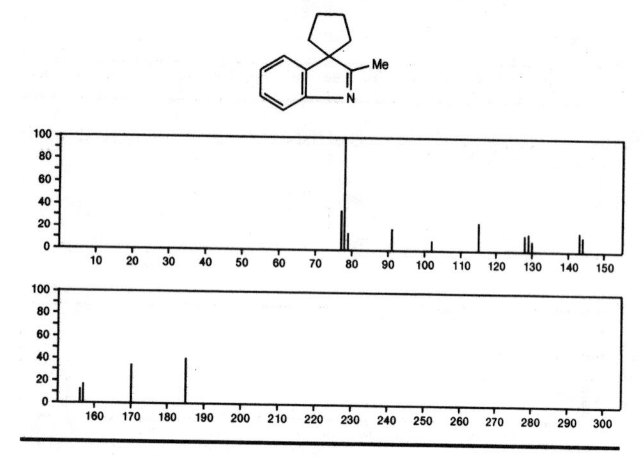

186 $C_2Cl_3F_3$ 76-13-1
Ethane, 1,1,2-trichloro-1,2,2-trifluoro-

$F_2CCl\,CCl_2F$

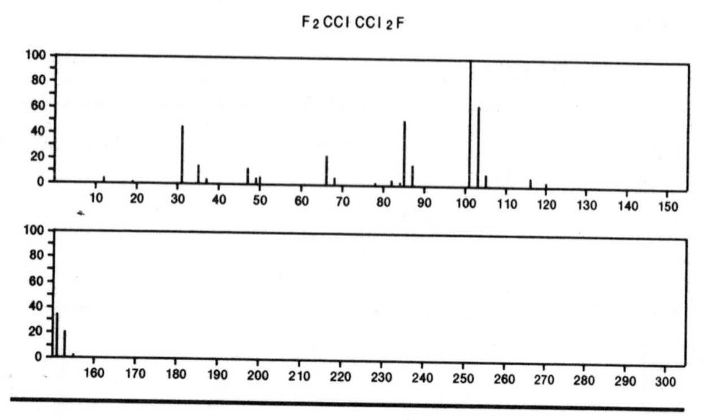

186 $C_2Cl_3F_3$ 354-58-5
Ethane, 1,1,1-trichloro-2,2,2-trifluoro-

F_3CCCl_3

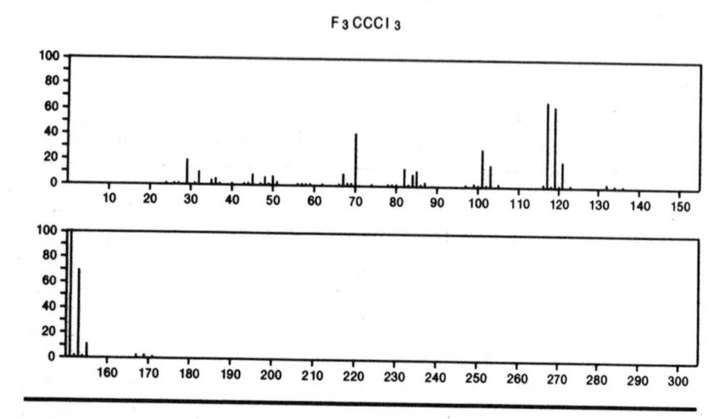

186 $C_2F_6O_3$ 1718-18-9
Trioxide, bis(trifluoromethyl)

$F_3COOOCF_3$

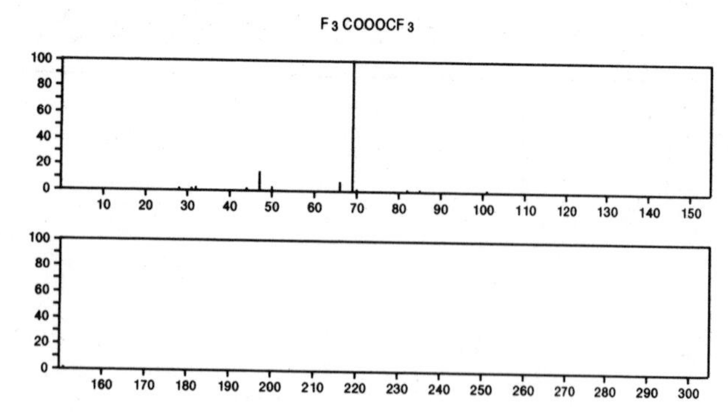

186 $C_2H_3IO_2$ 64-69-7
Acetic acid, iodo-

$I\,CH_2\,CO_2H$

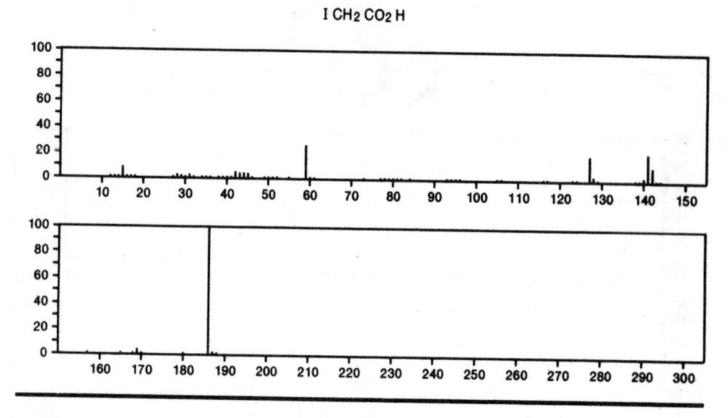

186 $C_2H_4Br_2$ 106-93-4
Ethane, 1,2-dibromo-

$Br\,CH_2\,CH_2\,Br$

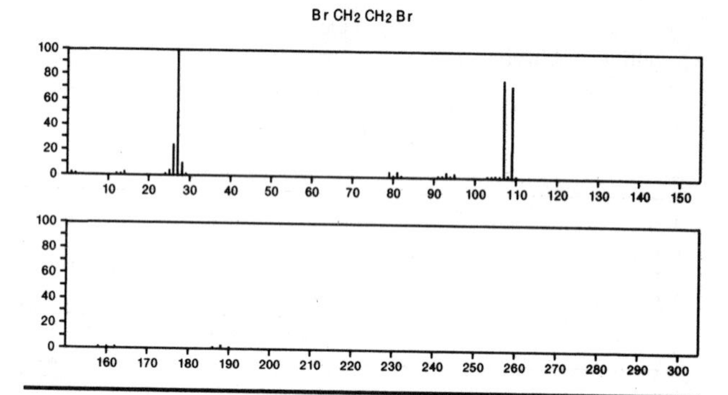

186 $C_2H_4Br_2$ 557-91-5
Ethane, 1,1-dibromo-

Br_2CHMe

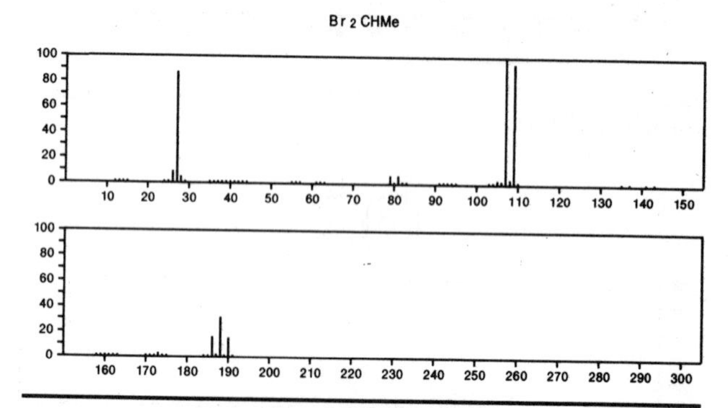

186 $C_4H_8ClO_4P$ 17027-41-7
Phosphoric acid, 2-chloroethenyl dimethyl ester

OMe
Cl CH=CHOP(O)OMe

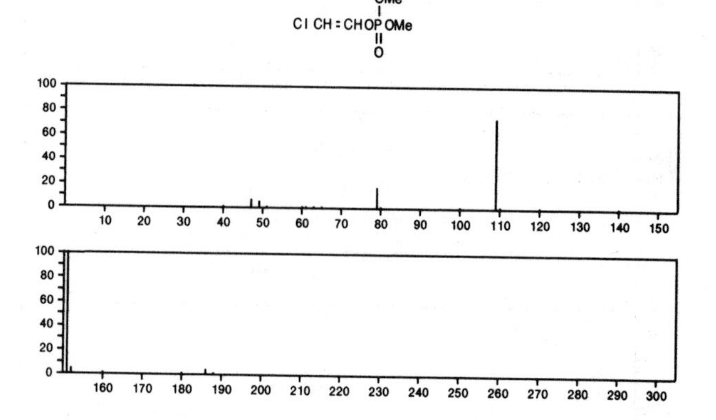

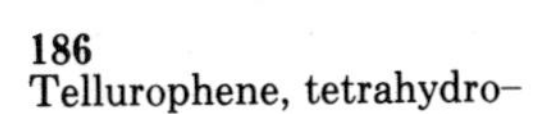

186 C_4H_8Te 3465-99-4
Tellurophene, tetrahydro–

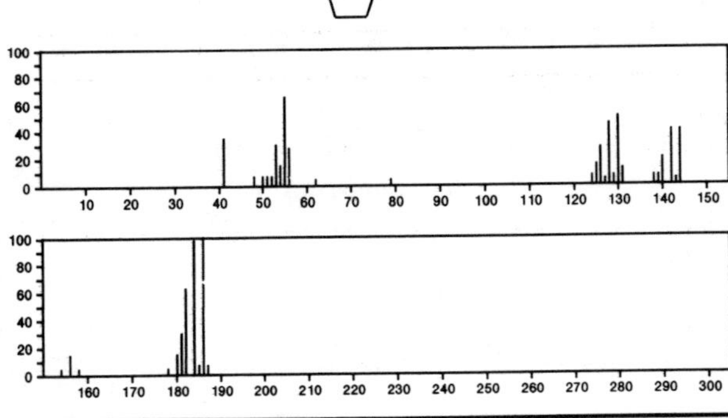

186 $C_4H_{12}ClN_2O_2P$ 22753-44-2
Phosphorodiamidous chloride, *N*,*N'*–dimethoxy–*N*,*N'*–dimethyl–

Me ONMe P Cl NMe OMe

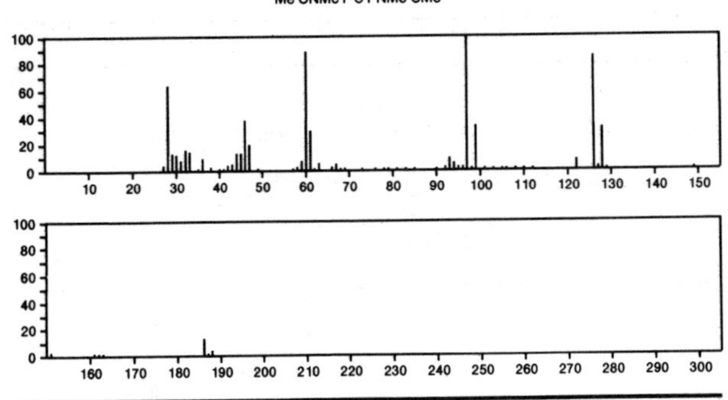

186 $C_4H_{12}ClN_2PS$ 3732-81-8
Phosphorodiamidothioic chloride, tetramethyl–

NMe 2
Me 2 NPCl =S

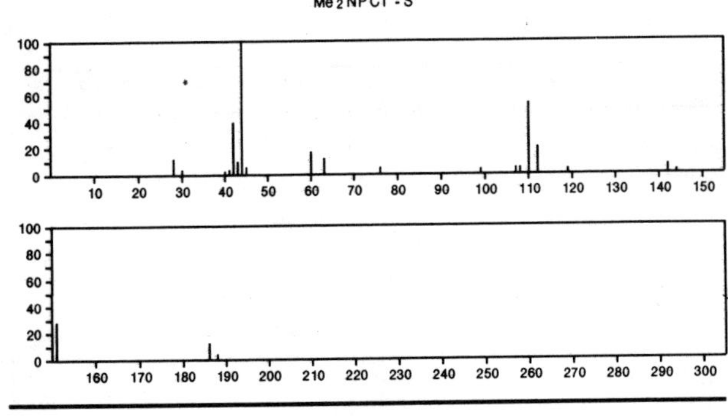

186 $C_5H_{12}ClO_3P$ 3167-63-3
Phosphonic acid, (chloromethyl)–, diethyl ester

OEt
Cl CH2 P OEt
O

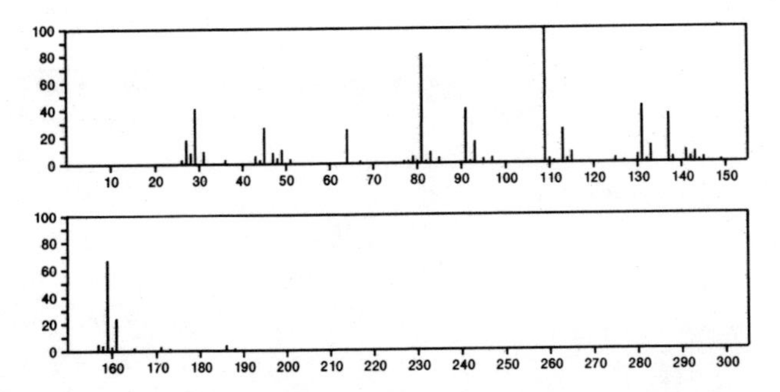

186 C_6F_6 392-56-3
Benzene, hexafluoro–

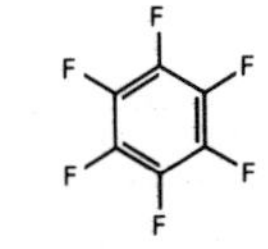

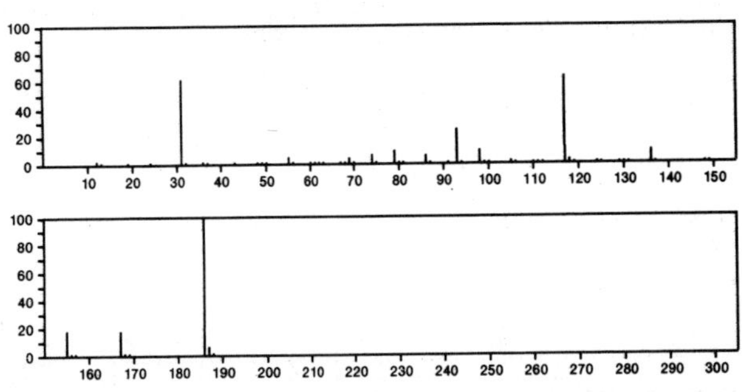

186 $C_6H_3FN_2O_4$ 70-34-8
Benzene, 1–fluoro–2,4–dinitro–

O2N NO2
F

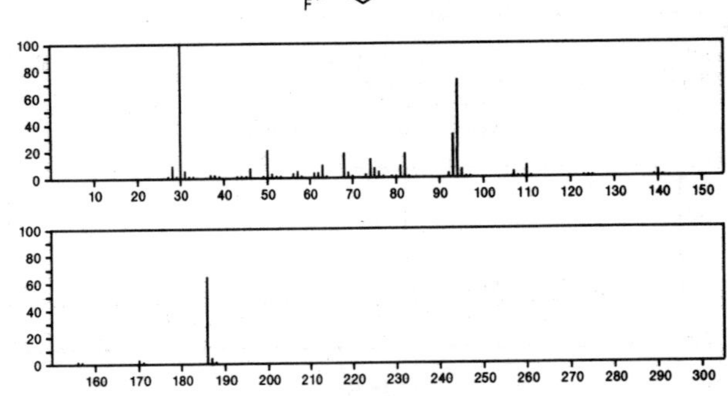

186 $C_6H_6N_2O_3S$ 16238-33-8
4*H*–1,3–Thiazine–6–carboxylic acid, 2–amino–4–oxo–, methyl ester

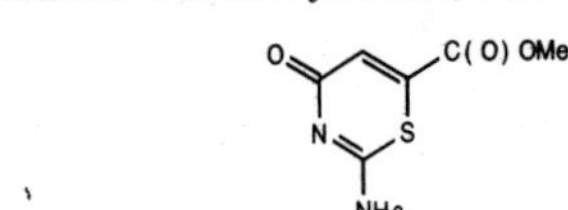

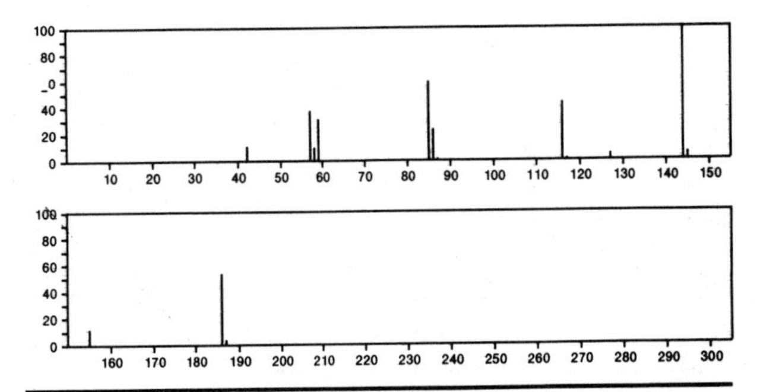

186 $C_6H_{10}N_4OS$ 56247-55-3
1,2,4–Thiadiazole–5–carbohydrazonic acid, 3–methyl–, ethyl ester

Me N C(OEt) =NNH2
N – S

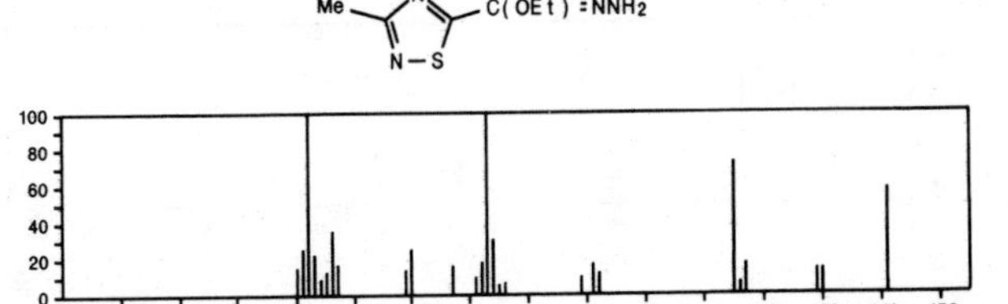

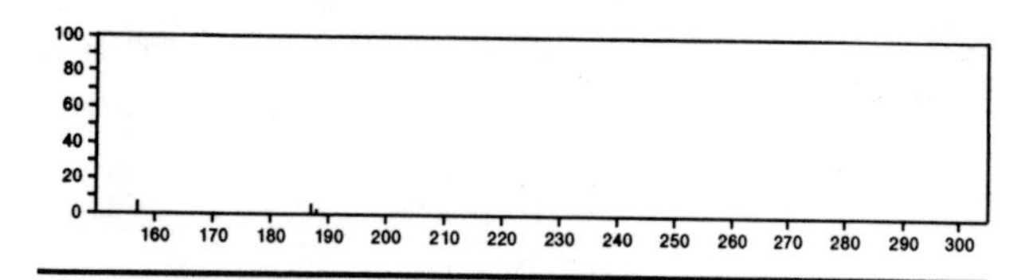

186 $C_6H_{10}N_4O_3$ 35975-26-9
2,6-Piperazinedione, 4-acetyl-, 2,6-dioxime

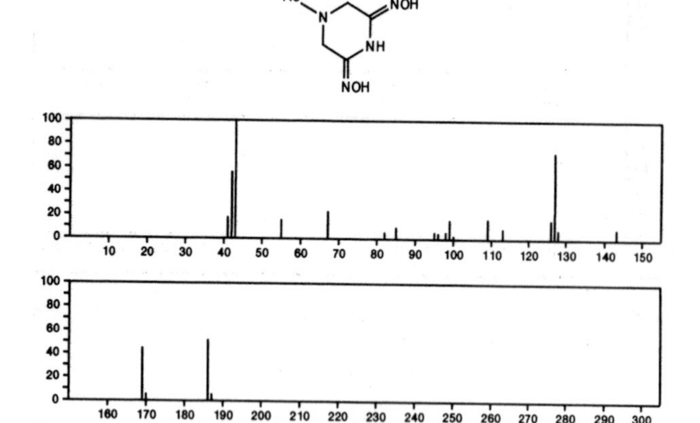

186 $C_6H_{12}Cl_2O_2$ 112-26-5
Ethane, 1,2-bis(2-chloroethoxy)-

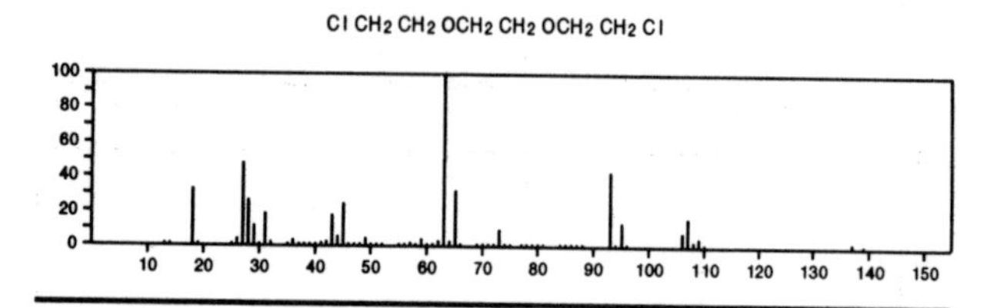

186 C_7H_7BrO 578-57-4
Benzene, 1-bromo-2-methoxy-

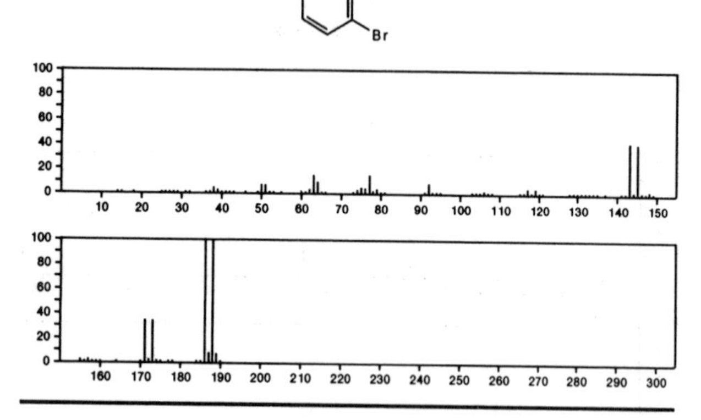

186 $C_7H_{10}N_2O_2S$ 1576-35-8
Benzenesulfonic acid, 4-methyl-, hydrazide

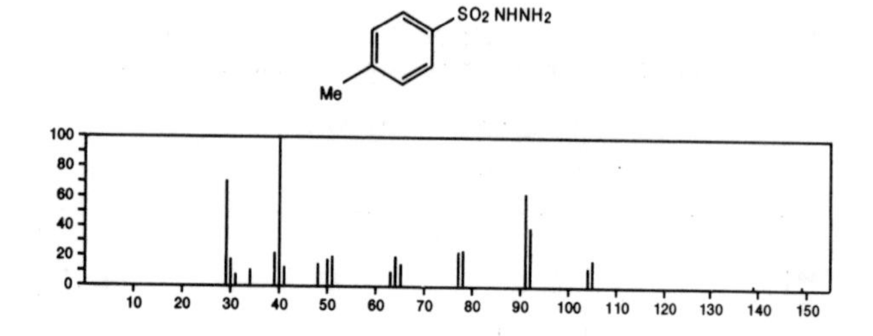

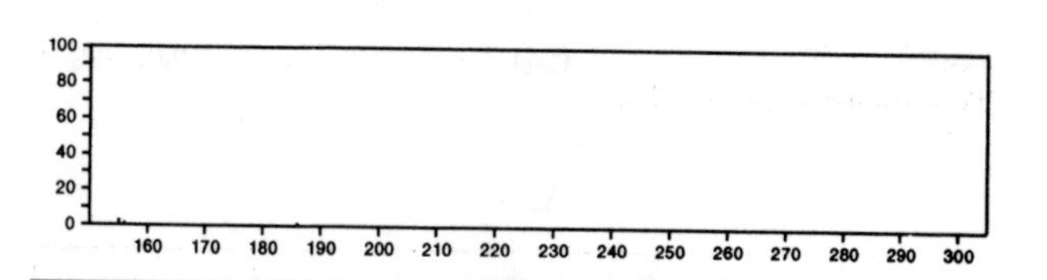

186 $C_7H_{18}N_2Si_2$ 1000-70-0
Silanamine, *N,N'*-methanetetraylbis[1,1,1-trimethyl-

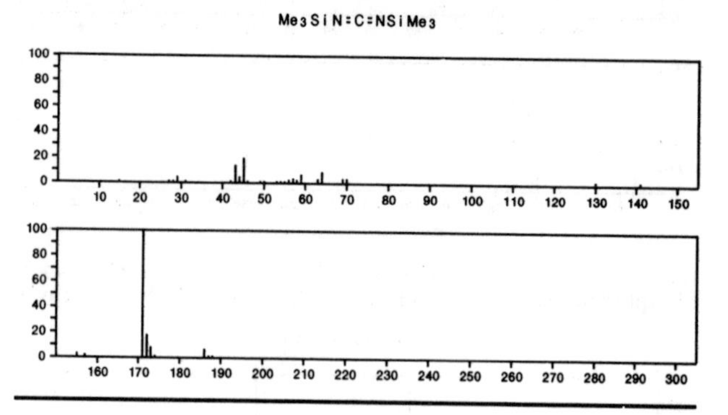

186 C_8H_7ClOS 5925-49-5
Benzenecarbothioic acid, 4-chloro-, *O*-methyl ester

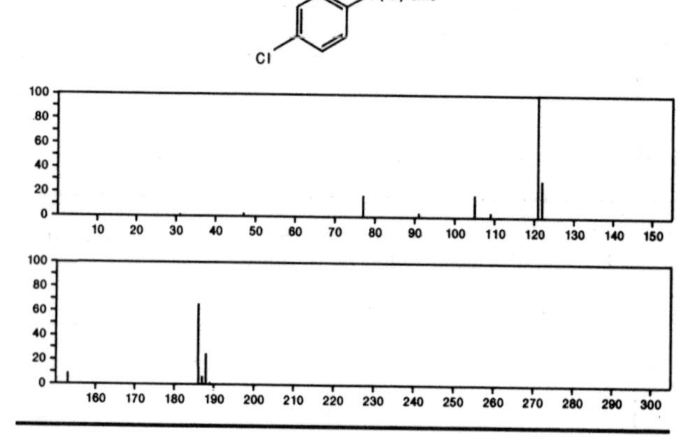

186 C_8H_7ClOS 5925-67-7
Benzenecarbothioic acid, 4-chloro-, *S*-methyl ester

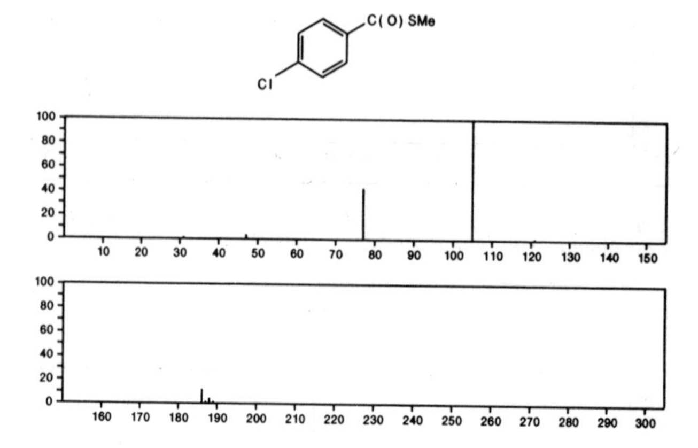

☆ U.S. GOVERNMENT PRINTING OFFICE : 1980 O—309-109

NBS-114A (REV. 11-77)

U.S. DEPT. OF COMM. BIBLIOGRAPHIC DATA SHEET	1. PUBLICATION OR REPORT NO. NSRDS-NBS 63, Vol. 1	2. Gov't Accession No.	3. Recipient's Accession No.

4. TITLE AND SUBTITLE

EPA/NIH Mass Spectral Data Base

Volume 1, Molecular Weights 30-186

5. Publication Date December 1978

6. Performing Organization Code

7. AUTHOR(S) S. R. Heller and G. W. A. Milne

8. Performing Organ. Report No.

9. PERFORMING ORGANIZATION NAME AND ADDRESS

NATIONAL BUREAU OF STANDARDS
DEPARTMENT OF COMMERCE
WASHINGTON, D.C. 20234

10. Project/Task/Work Unit No.

11. Contract/Grant No.

12. Sponsoring Organization Name and Complete Address *(Street, City, State, ZIP)*

Environmental Protection Agency, Washington, DC 20460
and
National Institutes of Health, Bethesda, MD 20014

13. Type of Report & Period Covered N/A

14. Sponsoring Agency Code

15. SUPPLEMENTARY NOTES

Library of Congress Catalog Card Number: 78-606175

16. ABSTRACT *(A 200-word or less factual summary of most significant information. If document includes a significant bibliography or literature survey, mention it here.)*

This publication presents a collection of 25,556 verified mass spectra of individual substances compiled from the EPA/NIH mass spectral file. The spectra are given in bar graph format over the full mass range. Each spectrum is accompanied by a Chemical Abstracts Index substance name, molecular formula, molecular weight, structural formula, and Chemical Abstracts Service Registry Number.

17. KEY WORDS *(six to twelve entries; alphabetical order; capitalize only the first letter of the first key word unless a proper name; separated by semicolons)*

Analytical data; mass spectra; organic substances; verified spectra.

18. AVAILABILITY [X] Unlimited

[ ] For Official Distribution. Do Not Release to NTIS

[X] Order From Sup. of Doc., U.S. Government Printing Office Washington, D.C. 20402, SD Stock No. 003-003-01987-9

[ ] Order From National Technical Information Service (NTIS) Springfield, Virginia 22151

19. SECURITY CLASS (THIS REPORT) UNCLASSIFIED

20. SECURITY CLASS (THIS PAGE) UNCLASSIFIED

21. NO. OF PAGES 1001

22. Price $65.00 per 5 part set; sold in sets only

USCOMM-DC 66035-P78

Announcement of New Publications in
National Standard Reference Data Series

Superintendent of Documents,
Government Printing Office,
Washington, D.C. 20402

Dear Sir:

Please add my name to the announcement list of new publications to be issued in the series: National Standard Reference Data Series—National Bureau of Standards.

Name____

Company____

Address____

City______________State__________Zip Code____________

(Notification Key N-519)